CONCISE POLYMERIC MATERIALS ENCYCLOPEDIA

CONCISE POLYMERIC MATERIALS ENCYCLOPEDIA

Editor-in-Chief

JOSEPH C. SALAMONE

Professor Emeritus
University of Massachusetts, Lowell
and
Vice President, Chemical Research
Global Scientific Affairs, Vision Care
Bausch & Lomb

[redacted]

CRC Press
Boca Raton London New York Washington, D.C.

Library of Congress Cataloging-in-Publication Data

Concise polymeric materials encyclopedia / editor-in-chief, Joseph C.
Salamone.
p. cm.
Includes bibliographical references and index.
ISBN 0-84932-226-X (alk. paper)
1. Plastics--Encyclopedias. 2. Polymers--Encyclopedias.
I. Salamone, Joseph C., 1939—
TP1110.C66 1999
668.9'03--dc21 98-6146
CIP

This book contains information obtained from authentic and highly regarded sources. Reprinted material is quoted with permission, and sources are indicated. A wide variety of references are listed. Reasonable efforts have been made to publish reliable data and information, but the author and the publisher cannot assume responsibility for the validity of all materials or for the consequences of their use.

Direct all inquiries to CRC Press LLC, 2000 Corporate Blvd., N.W., Boca Raton, Florida 33431.

International Standard Book Number 0-84932-226-X
Library of Congress Card Number 98-6146
Printed in the United States of America 1 2 3 4 5 6 7 8 9 0
Printed on acid-free paper

PREFACE

The *Concise Polymeric Materials Encyclopedia* is an abridgment of the *Polymeric Materials Encyclopedia,* a principal reference source in the polymer field that pertains to state-of-the-art research and development on modern polymeric systems. Over 1,100 articles in the original work have been condensed by the editors in order to provide a synopsis of the topics described, with appropriate references.

A vast array of subjects is discussed regarding the synthesis, properties, and applications of polymeric materials, the development of modern catalysts to prepare new and modified polymers, and the modification of existing polymers by chemical and physical processes. Included in this discussion are a large number of articles pertaining to biologically oriented polymers.

The topics covered in this work include the following: additives, adhesives, adsorptive resins, alloys, alternating copolymers, antibacterial polymers, antifoaming agents, antifouling materials, amphiphilic materials, barrier materials, batteries, biodegradable materials, bioerodible materials, biomaterials, biomimetic materials, biopolymers, biosensors, blends, block copolymers, block copolymer micelles, branched polymers, building materials, calixarenes, carbon fibers, carbon monoxide copolymers, catalysts, ceramic precursors, chemical amplification resists, chemical sensors, coatings, colloids, comb-like materials, compatibilizers, composites, conjugated polymers, contact lenses, cosmetic materials, cyclopolymerization, degradation, dendrimers, drag reducing agents, drug delivery systems, elastomers, electroconducting materials, electrooptical materials, electrorheological materials, emulsion polymerization, environmentally degradable polymers, engineering thermoplastics, enzyme-catalyzed polymerization, ferroelectric polymers, fibers, fillers, flame retardants, flocculants, fluorine-containing polymers, foams, food additives, floor finishes, functional materials, foams, food polymers, free radical polymerization, gas permeable polymers, gels, gradient polymers, graft copolymers, group transfer polymerization, gums, hair care polymers, high temperature materials, high modulus materials, high performance polymers, high solids coatings, high strength materials, host-guest materials, hydrogels, hydrophilic materials, hydrophobic materials, hyperbranched polymers, immobilized enzymes, inclusion complexes, initiators, inorganic materials, intelligent materials, interpenetrating polymer networks, inverse-emulsion/suspension polymerization, ion-chelating polymers, ionomers, lacquers, ladder polymers, Langmuir–Blodgett films, liquid crystalline materials, living polymerization, macrocycles, macromonomers, membranes, metathesis polymerization, microbial polymers, microcapsules, microemulsion polymerization, modified natural products, molecular assemblies, monodisperse polymers, monodisperse particles, networks, oligomers, optical fibers, optically active polymers, organic/inorganic materials, organometallic materials, oxidative coupling polymerization, paints, paper, photopolymers, polyampholytes, polyelectrolyte complexes, polyelectrolytes, polymeric compatibilizers, polymeric drugs, polymeric initiators, polymeric surfactants, plastics, printing inks, powder coatings, preceramic polymers, precipitation polymerization, printing inks, recyclable materials, reinforcing agents, release coatings, resins, ring-opening polymerization, rotaxanic materials, rubbers, sealants, scale inhibitors, self-assembled materials, skin care polymers, spiro polymers, star polymers, superabsorbent materials, surface modified materials, suspension polymerization, telechelics, template polymerization, thermosensitive materials, thickeners, topochemical polymerization, ultrathin films, vulcanization, waterborne coatings, water-soluble polymers, wood composites, and zwitterionic polymerization.

The original articles were prepared by internationally recognized polymer scientists. All articles are presented in alphabetical order in the same fashion as the original work. A number of articles may pertain to a specific subject, and cross-reference titles are given, which list all articles in a general subject area. A Subject Index also facilitates searching.

The *Concise Polymeric Materials Encyclopedia* is a comprehensive, easily used, major reference source on modern polymeric materials. The editors believe that it will be a valuable addition to reference collections in the polymer field.

Joseph C. Salamone
Editor-in-Chief

A. Maureen Aller
Chemistry Editor

EDITOR-IN-CHIEF

Joseph C. Salamone received his Ph.D. in Chemistry/Polymer Science from the Polytechnic University (formerly the Polytechnic Institute of Brooklyn) in 1967 under Professor Charles G. Overberger. He was a National Institutes of Health Postdoctoral Fellow at the University of Liverpool, Liverpool, England, from 1966 to 1967. During 1967 to 1970 he was at the University of Michigan, again with Professor Overberger, where he was a Horace H. Rackham Postdoctoral Fellow and Administrative Secretary of the Macromolecular Research Center. In 1970, he joined the University of Massachusetts, Lowell, as an Assistant Professor of Chemistry and helped develop the Polymer Science Program. In 1972 he was appointed Associate Professor, was elected Chairman of the Department of Chemistry in 1975, and named as Professor in 1976. His research interests have included the synthesis of new monomers and polymers and the solution properties of unusual ionic polymers. He is the author of over 160 publications and co-editor of two books, *Catalysis — A Key to Advances in Applied Polymer Science*, ACS Books, 1992, and *Contemporary Topics in Polymer Science (Volume 7), Advances in New Materials*, Plenum Press, 1992.

At the University of Massachusetts, Lowell, he continued in the Administration and was appointed Acting Dean of the College of Pure and Applied Science in 1978. From 1980 to 1984 he was appointed the Dean of the College of Science, Chairman of the Council of Deans, and a member of the President's Cabinet. In 1984 he was also made Distinguished Research Fellow, and in 1989 he received the honorary title of Professor Emeritus. At the Royal Institute of Technology in Stockholm, Sweden, he was a Visiting Professor in 1988 and 1990. Polytechnic University named him a Distinguished Alumnus in 1984, and Kyoto University, Kyoto, Japan, presented him with the Kyoto University Medal in 1984. For the American Biographical Institute, Inc., he has been a Contributing Member of the Honorary Educational Advisory Board.

Dr. Salamone has also been active in professional organizations. From 1974 to 1978 he was the Treasurer of the Division of Polymer Chemistry, Inc., of the American Chemical Society, Vice Chairman in 1980, Chairman-Elect in 1981, and Chairman in 1982. He was a founding member of the Pacific Polymer Federation, where he served as a Council Member from 1987 to 1995, Secretary/Treasurer from 1987 to 1990, and President from 1993 to 1995. Dr. Salamone has served on a number of advisory boards of scientific publications in chemistry and polymer science and he has also organized major meetings in the polymer field. He has served on several award committees of the American Chemical Society, and has received the Outstanding Large Division Award for the ACS Division of Polymer Chemistry, Inc. in 1982 for his tenure as Chairman. The ACS Polymer Division has also presented him with the Distinguished Service Award in 1985 and the 40th Anniversary Special Division Service Award in 1991.

Dr. Salamone has also been active in entrepreneurial activities, primarily in biomedical materials in ophthalmology and in skin care. He was a founder and Director of Polymer Technology Corporation for the development of rigid gas-permeable contact lenses and related solutions, and is a founder, Director, and Vice President of Rochal Industries for the development of oxygen-permeable, non-stinging, skin protection systems, and a founder, Director, and President of Optimers, Inc., for the development of novel hydrogel contact lens materials. He is also active in the development of novel intraocular lenses, anterior chamber refractive lenses, and solutions for related surgical procedures.

Recently, Dr. Salamone joined Bausch & Lomb, where he is Vice President of Chemical Research, Global Scientific Affairs, Vision Care.

ACKNOWLEDGMENTS

The abridgment of the *Polymeric Materials Encyclopedia* to form the *Concise Polymeric Materials Encyclopedia* was a complex undertaking. It involved the cooperation of many individuals within CRC Press LLC. I am particularly grateful to Maureen Aller, who served as Chemistry Editor for this project and assisted in many aspects of the preparation of the *Concise Polymeric Materials Encyclopedia.* Her organizational skills greatly facilitated the condensation and arrangement of the articles presented. In this endeavor, she was assisted by Carole Sweatman, and her services are also greatly appreciated. Of great significance was the work of Andrea Demby, who served as Project Editor on this work, as she had done on the *Polymeric Materials Encyclopedia.* Her activities transposed all the editorialized articles into the final printed format. I am also grateful for additional organizational suggestions provided by Dr. Mark Licker, President of the Electronic Division, and by Karen Feinstein, Marketing Manager.

Joseph C. Salamone
Editor-in-Chief

CONTRIBUTORS

Mohamed S. A. Abdou
Department of Chemistry
Simon Fraser University
Burnaby, British Columbia, Canada

H. Abe
Polymer Chemistry Laboratory
The Institute of Physical and Chemical Research (RIKEN)
Saitama, Japan

Yoshimoto Abe
Department of Industrial Chemistry
Faculty of Science and Technology
Science University of Tokyo
Chiba, Japan

Rym Abidi
EHICS Laboratoire de Chimie Minérale et Analytique
Associé au CNRS
Strasbourg, France

H. R. Acharya
Occidental Chemical Corporation
Grand Island, New York

Ricardo Acosta-Ortiz
Centro de Inventigacion en Quimica Aplicada
Saltillo Coahuila, Mexico

H. Adachi
Central Research Laboratory
Mitsubishi Electric Corporation
Tsukaguchi-Honmachi, Amagasaki, Hyogo, Japan

W. W. Adams
Materials Directorate
Wright Laboratory
Wright Patterson Air Force Base, Ohio

F. A. Adamsky
University of Pittsburgh
Pittsburgh, Pennsylvania

Esbaide Adem
Instituto de Física
UNAM
Mexico D.F., Mexico

Basudam Adhikari
Materials Science Center
Indian Institute of Technology
Kharagpur, India

Mauro Aglietto
Dipartimento di Chimica e Chimica Industriale
University of Pisa
Pisa, Italy

Iqbal Ahmed
Phillips Petroleum Company
Phillips Research Center
Bartlesville, Oklahoma

Kwang-Duk Ahn
Functional Polymer Laboratory
Korea Institute of Science and Technology
Seoul, Korea

Takuzo Aida
Department of Chemistry and Biotechnology
Faculty of Engineering
The University of Tokyo
Tokyo, Japan

Masuo Aizawa
Department of Bioengineering
Faculty of Bioscience and Biotechnology
Tokyo Institute of Technology
Yokohama, Japan

A. Ajji
National Research Council Canada
Industrial Materials Institute
Boucherville, Quebec, Canada

Kazuo Akagi
Institute of Materials Science
University of Tsukuba
Tsukuba, Ibaraki, Japan

Ichimoto Akasaki
Nippon Shokubai Company, Ltd.
Osaka, Japan

Fuminori Akiyama
Institute for Chemical Reaction Science
Tohoku University
Sendai, Japan

Saburo Akiyama
Faculty of Technology
Tokyo University of Agriculture and Technology
Tokyo, Japan

Kazunari Akiyoshi
Department of Synthetic Chemistry and Biological Chemistry
Graduate School of Engineering
Kyoto University
Kyoto, Japan

Joseph A. Akkara
Biotechnology Division
U.S. Army Natick Research Development and Engineering Center
Natick, Massachusetts

Sahar Al-Malaika
Polymer Processing and Performance Group
Department of Chemical Engineering and Applied Chemistry
Aston University
Birmingham, United Kingdom

W. Al-Shahib
Department of Chemistry
Imperial College of Science, Technology and Medicine
London, United Kingdom

Valery Yu Alakhov
Institute Armand-Frappier
Laval, Quebec, Canada

Ann-Christine Albertsson
Department of Polymer Technology
The Royal Institute of Technology (KTH)
Stockholm, Sweden

Martha Albores-Velasco
Facultad de Quimica
UNAM Circuito Interior
Mexico D.F., Mexico

Pierre Aldebert
CEA-Département de Recherche Fondamentale sur la Matière Condensée
SESAM/Laboratoire de Physico-Chimie Moléculaire
Grenoble, France

Spiro D. Alexandratos
Department of Chemistry
University of Tennessee
Knoxville, Tennessee

Paschalis Alexandridis
Department of Chemical Engineering
Massachusetts Institute of Technology
Cambridge, Massachusetts

Larissa Alexandrova
Instituto de Investigaciones en Materiales, UNAM
Coyoacan, Mexico

F. J. Alfan
Facultad de Quimica
UNAM
Mexico D.F., Mexico

K. M. Idriss Ali
Bangladesh Atomic Energy Commission
Dhaka, Bangladesh

Soheir H. Ali
Microbiological Chemistry Department
National Research Center
Dokki, Cairo, Egypt

Norman S. Allen
Chemistry Department
Manchester Metropolitan University
Manchester, United Kingdom

Kazuo Ametani
Tokyo Metropolitan Isotope Research Center
Tokyo, Japan

N. Aminuddin
Department of Textile Engineering, Chemistry, and Science
North Carolina State University
Raleigh, North Carolina

Lea L. Anderson
Akzo Nobel Chemicals Incorporated
Dobbs Ferry, New York

Cristina T. Andrade
Instituto de Macromoleculas
Universidade Federal do Rio de Janeiro
Rio de Janeiro, Brazil

B. A. Kottes Andrews
Southern Regional Research Center
Mid South Area
Agricultural Research Service
U.S. Department of Agriculture
New Orleans, Louisiana

Stephen M. Andrews
Amoco Performance Products Incorporated
Department of Chemical Engineering
University of Delaware
Newark, Delaware

Kirk P. Andriano
APS Research Institute
Redwood City, California

Marianela Andújar
University of Puerto Rico
Mayagüez, Puerto Rico

Luigi Angiolini
Dipartimento di Chimica Industriale e dei Materiali
University of Bologna
Bologna, Italy

Vadim V. Annenkov
Irkutsk State University
Irkutsk, Russia

Markus Antonietti
Max-Planck-Institut für Kolloid- und Grenzflächenforschung
Teltow-Seehof, Germany

A. Antonov
Institute of Synthetic Polymeric Materials
Moscow, Russia

Keigo Aoi
Department of Applied Biological Sciences
Faculty of Agricultural Sciences
Nagoya University
Nagoya, Japan

Toshiki Aoki
Department of Chemistry and Chemical Engineering
Faculty of Engineering
Niigata University
Niigata, Japan

María A. Aponte
University of Puerto Rico
Mayagüez, Puerto Rico

Kenichiro Arai
Faculty of Engineering
Gunma University
Kiryu, Gunma, Japan

Masao Arima
Department of Applied Chemistry
Faculty of Engineering
Kanagawa University
Yokohama, Japan

Valeria Arrighi
Chemistry Department
Heriot-Watt University
London, United Kingdom

N. Arsalani
Institute of Organic Chemistry
Faculty of Chemistry and Pharmacy
University of Tuebingen
Tuebingen, Germany

R. Arshady
Department of Chemistry
Imperial College of Science, Technology, and Medicine
London, United Kingdom

Katsuya Asao
Osaka Prefectural Industrial Technology Research Institute
Osaka, Japan

Zouhair Asfari
EHICS Laboratoire de Chimie Minérale et Analytique
Associé au CNRS
Strasbourg, France

Tamaichi Ashida
Department of Biotechnology
School of Engineering
Nagoya University
Nagoya, Japan

Teresa Dib Zambon Atvars
Instituto de Química
Universidade Estadual de Campinas
Campinas, Sao Paulo, Brazil

Aziza I. Atwa
Department of Microbial Biotechnology
National Research Center
Dokki, Cairo, Egypt

Oddvar Aune
SINTEF
Applied Chemistry
Trondheim, Norway

Kaoru Awazu
Industrial Research Institute of Ishikawa Prefecture
Kanazawa, Ishikawa, Japan

Frederick H. Axtell
Department of Chemistry
Faculty of Science
Mahidol University
Bangkok, Thailand

M. Ayyagari
Biotechnology Division
U.S. Army Natick RD&E Center
Natick, Massachusetts

M. M. Azab
Department of Chemistry
Faculty of Science
Benha University
Benha, Egypt

M. Abdel Azzem
Laboratory of Electrochemistry
Faculty of Science
El-Menoufia University
Shebin, El-Kom, Egypt

Wolfgang Baade
Bayer AG
Dormagen, Germany

Yoshihiro Baba
Laboratory of Chemistry
Department of General Education
Osaka Institute of Technology
Osaka, Japan

David A. Babb
Central Research and Development
Organic Products Research
The Dow Chemical Company
Freeport, Texas

Edward D. Babich
IBM Corporation
T. J. Watson Research Center
IBM Corporation
Yorktown Heights, New York

You Han Bae
Center for Controlled Chemical Delivery
Department of Pharmaceutics and Pharmaceutical Chemistry
University of Utah
Salt Lake City, Utah

Pratap Bahadur
Department of Chemistry
Indiana University-Purdue University at Indianapolis
Indianapolis, Indiana

Donald G. Baird
Department of Chemical Engineering
Virginia Polytechnic Institute and State University
Blacksburg, Virginia

Anjali Bajpai
Department of Chemistry
Government Autonomous Science College
Jabalpur, India

U. D. N. Bajpai
Polymer Research Laboratory
Department of Post Graduate Studies and Research in Chemistry
Rani Durgavati University
Jabalpur, India

Kyle P. Baldwin
Chemistry Department
University of Akron
Akron, Ohio

Stephen T. Balke
Department of Chemical Engineering and Applied Chemistry
University of Toronto
Toronto, Ontario, Canada

Matthias Ballauff
Polymer-Institut
Universität (T.H.) Karlsruhe
Karlsruhe, Germany

Vincenzo Balzani
Dipartimento di Chimica G. Ciamician
Università di Bologna
Bologna, Italy

P. Banerjee
Polymer Science Unit
Indian Association for the Cultivation of Science
Jadavpur, Calcutta, India

Zhenan Bao
Department of Chemistry
University of Chicago
Chicago, Illinois

Eugene S. Barabas
International Specialty Products Corporation
Wayne, New Jersey

Ronilson V. Barbosa
Instituto de Macromoléculas
Universidade Federal do Rio de Janeiro
Rio de Janeiro, Brazil

Edward J. Bartoszek
Norristown, Pennsylvania

Richard A. Bartsch
Department of Chemistry and Biochemistry
Texas Tech University
Lubbock, Texas

Jürgen Barwich
BASF Aktiengesellsuaft
Ludwigshafen/Rhein, Germany

Rafil A. Basheer
Polymer Department
GM Research and Development Center
Warren, Michigan

Zahir Bashir
Courtaulds
Coventry, United Kingdom

Altaf H. Basta
Cellulose and Paper Department
National Research Center
Cairo, Egypt

Jörg Bauer
Fraunhofer Institute of Applied Materials Research
Teltow, Germany

Monika Bauer
Fraunhofer Institute of Applied Materials Research
Teltow, Germany

C. Geraldine Bazuin
Université Laval
Laval, Québec, Canada

George Beall
Science and Technology Group
Corning Incorporated
Corning, New York

Charles L. Beatty
Department of Materials Science and Engineering
University of Florida
Gainesville, Florida

M. A. Becker
National Institute of Sericultural and Entomological Science Oowashi
Tsukuba, Ibaraki, Japan

Robert Becker
Institut für Angewandte Chemie (ACA)
Berlin, Germany

Eric J. Beckman
Department of Chemical and Petroleum Engineering
University of Pittsburgh
Pittsburgh, Pennsylvania

Esen A. Bekturov
Institute of Chemical Sciences
Kazakh Academy of Sciences
Almaty, Republic of Kazakhstan

Mohamed Naceur Belgacem
Matèriaux Polymères
Ecole Française de Papeterie et des Industries Graphiques
Saint Martin D'Hères, France

Gennadii P. Belov
Institute of Chemical Physics
Russian Academy of Sciences
Chernogolovka, Moscow Region, Russia

James N. BeMiller
Whistler Center for Carbohydrate Research
Purdue University
West Lafayette, Indiana

Ettore Benedetti
Biocrystallography Research Center, C.N.R.
Department of Chemistry
University of Naples
Naples, Italy

Milan J. Benes
Institute of Macromolecular Chemistry
Academy of Sciences of The Czech Republic
Prague, Czech Republic

Nicholas Benfaremo
Max-Planck-Institute for Polymer Research
Mainz, Germany

Brian C. Benicewicz
Los Alamos National Laboratory
Los Alamos, New Mexico

Greggory S. Bennett
3M Center
St. Paul, Minnesota

Maria Bercea
Department of Macromolecules
"P. Poni" Institute of Macromolecular Chemistry
Jassy, Romania

Arvid Berge
Department of Industrial Chemistry
University of Trondheim
NTH
Trondheim, Norway

B. Bergenstahl
Institute for Surface Chemistry
Stockholm, Sweden

V. K. Berry
General Electric Plastics
Technology Center
Washington, West Virginia

Roland Beyreuther
Institute of Polymer Research
Dresden, Germany

Suraj N. Bhadani
Department of Chemistry
Ranchi University
Ranchi, India

M. S. Bhatnager
Karanpur, Dehradun, India

S. N. Bhattacharyya
Polymer Science Unit
Indian Association for the Cultivation of Science
Jadavpur, Calcutta, India

Pradip K. Bhowmik
Department of Chemistry
University of Detroit Mercy
Detroit, Michigan

Niyazi Biçak
Department of Chemistry
Instanbul Technical University
Istanbul, Turkey

Blake R. Bichlmeir
Advanced Fibers Systems
DuPont
Wilmington, Delaware

Fabio Bignotti
Dipartimento di Chimica e Fisica per i Materiali
Università degli Studi di Brescia
Brescia, Italy

Tulin Bilgic
Petkim Petrochemicals Holding, Inc.
Research and Development Center
Korfez, Kocaeli, Turkey

L. Billon
Laboratoire de Chimie des Polymères Organiques
Ecole Nationale Supérieure de Chimie et de Physique de Bordeaux
Talence, France

L. A. Bimendina
Institute of Chemical Sciences
Kazakh Academy of Sciences
Almaty, Republic of Kazakhstan

Amrit P. Bindra
Engelhard Corporation
Beachwood, Ohio

Jon Bjørgum
SINTEF
Applied Chemistry
Trondheim, Norway

Kenneth P. Blackmon
Fina Research and Technology Center
Deer Park, Texas

Anne Blayo
Matériaux Polymères
Ecole Francaise de Papeterie et des Industries Graphiques
Saint Martin d'Héres, France

Jonathan P. Blitz
Department of Chemistry
Eastern Illinois University
Charleston, Illinois

Manfred Bochmann
School of Chemical Sciences
University of East Anglia
Norwich, United Kingdom

Leonard E. Bogan, Jr.
Rohm and Haas Company
Spring House, Pennsylvania

Bogdan Georgiev Bogdanov
Central Research Laboratory
Bourgas Technological University
Bourgas, Bulgaria

S. P. Bohan
Bristol Analytical Research
Rohm and Haas Company
Bristol, Pennsylvania

Yuri B. Boiko
School of Physics
The University of Melbourne
Parkville, Victoria, Australia

Raymond E. Bolich, Jr.
The Procter and Gamble Company
Cincinnati, Ohio

Gueorgui Borissov
Institute of Polymers
Bulgarian Academy of Science
Sofia, Bulgaria

Anthony Bosch
Fiberite, Incorporated
Phoenix, Arizona

C. P. Bosnyak
The Dow Chemical Company
Freeport, Texas

Koji B. Bota
Department of Chemistry
High Performance Polymers and Ceramics Center
Clark Atlanta University
Atlanta, Georgia

F. A. Bottino
Istituto Chimico
Facoltà di Ingegneria
Università di Catania
Catania, Italy

J. J. Bou
Departament d'Enginyeria Química
Universitat Politecnica de Catalunya, ETSEIB
Barcelona, Spain

M. Bouquey
Laboratoire de Chimie des Polymères Organiques
Ecole Nationale Supérieure de Chimie et de Physique de Bordeaux
Talence, France

B. Boutevin
Ecole Nationale Supérieure de Chimie de Montpellier
Laboratoire de Chimie Appliqué
Montpellier, France

Hans Bradaczek
Institut für Kristallographie
Freie Universität Berlin
Berlin, Germany

G. Braithwaite
Department of Chemical Engineering and Chemical Technology
Imperial College of Science, Technology and Medicine
London, United Kingdom

Jean-Luc Brédas
Chimie des Matériaux Nouveaux
Centre de Recherche en Electronique et Photonique Moléculaires
Université de Mons-Hainaut
Mons, Belgium

Ludwig Brehmer
Research Group for Thin Organic Films
Institute of Solid State Physics
University of Potsdam
Teltow, Germany

Martin Brehmer
Institute of Organic Chemistry
Universität Mainz
Mainz, Germany

J. Briers
Department of Chemistry
University of Antwerp (UIA)
Wilrijk, Belgium

Linda J. Broadbelt
Department of Chemical Engineering
Northwestern University
Evanston, Illinois

Stephen Brocchini
Department of Chemistry
Rutgers University
New Brunswick, New Jersey

Wilfried M. Brouwer
Akzo Nobel Chemicals Incorporated
Dobbs Ferry, New York

Paul G. Brown
Department of Chemistry
University of New England
Armidale, New South Wales, Australia

G. J. Brownsey
Institute of Food Research
Norwich Laboratory
Norwich Research Park
Colney, Norfolk

Maria Bruma
Institute of Macromolecular Chemistry
Iasi, Romania

Daniel J. Brunelle
General Electric Research and Development Center
Schenectady, New York

Harald Brünig
Institute of Polymer Research
Dresden, Germany

K. J. Bruza
Central Research and Development
Material Science and Development
Dow Chemical Company
Midland, Michigan

S. Bruzaud
Laboratoire de Chimie des Polymères Organiques
Ecole Nationale Supérieure de Chimie et de Physique de Bordeaux
Talence, France

Daniel Bucca
Department of Materials Science and Engineering Polymer Science Program
The Pennsylvania State University
University Park, Pennsylvania

Richard P. Buck
Departments of Chemistry and Medicine
University of North Carolina
Chapel Hill, North Carolina

Warren H. Buck
DuPont Fluoroproducts
DuPont Central Research and Development Experimental Station
Wilmington, Delaware

D. J. Buckley, Jr.
General Electric Research and Development
Schenectady, New York

T. J. Bunning
Science Applications International Corporation
Dayton, Ohio

Peter J. Burchill
Aeronautical and Maritime Research Laboratory
Melbourne, Victoria, Australia

Robert P. Burford
Department of Polymer Science
University of New South Wales
Sydney, New South Wales, Australia

Karen J. L. Burg
Department of Bioengineering
Clemson University
Clemson, South Carolina

A. Burger
Hydrotec AG
Rehan, Germany

Christian Burger
Max-Planck-Institut für Kolloid- und Grenzflächenforschung
Teltow-Seehof, Germany

Guillermina Burillo
Instituto de Ciencias Nucleares UNAM
Civdad Universitarú
Mexico D.F., Mexico

George B. Butler
Department of Chemistry
University of Florida
Gainesville, Florida

Israel Cabasso
Polymer Research Institute
Faculty of Chemistry
College of Environmental Science and Forestry
State University of New York at Syracuse
Syracuse, New York

Kevin M. Cable
Department of Polymer Science
University of Southern Mississippi
Hattiesburg, Mississippi

Jason L. Cain
Department of Chemistry and Center for Materials for Information Technology
The University of Alabama
Tuscaloosa, Alabama

Matthew R. Callstrom
Department of Chemistry
The Ohio State University
Columbus, Ohio

Sebastiano Campagna
Dipartimento di Chimica Inorganica e Struttura Molecolare
Università di Messina
Messina, Italy

F. Candau
Institut Charles Sadron (CRM-EAHP)
CNRS-ULP
Strasbourg, France

G. S. Canessa
Departamento de Polìmeros
Facultad de Ciencias Quìmicas
Universidad de Concepciòn
Concepciòn, Chile

Leonardo Canova
EniChem-Istituto G. Donegani
Novara, Italy

Y. Cao
UNIAX Corporation
Santa Barbara, California

Galo Cárdenas-Trivino
Departmento de Polímeros
Facultad de Ciencias Químicas
Universidad de Concepción,
Concepción, Chile

Judith Cardoso
Universidad Autónoma Metropolitana-Iztapalapa
Mexico D.F., Mexico

Daniele Caretti
Dipartimento di Chimica Industriale e dei Materiali
University of Bologna
Bologna, Italy

Carlo Carlini
Dipartimento di Chimica Industriale e dei Materiali
University of Bologna
Bologna, Italy

Debbie Carpenter
Morton International
Woodstock, Illinois

Charles E. Carraher, Jr.
Department of Chemistry
Florida Atlantic University
Boca Raton, Florida
and
Florida Center for Environmental Studies
Northcorp Center
Palm Beach Gardens, Florida

Shawn M. Carraher
Department of Management
Indiana State University
Terre Haute, Indiana

W. G. Carson
Bristol Analytical Research
Rohm and Haas Company
Bristol, Pennsylvania

Peter Carty
University of Northumbria at Newcastle
Newcastle upon Tyne, United Kingdom

Maria Grazia Cascone
Department of Chemical Engineering
University of Pisa
Pisa, Italy

Walter Caseri
Eidgenössische Materialprüfungs und Forschungsanstalt EMPA
Dübendorf, Switzerland

Mario Casolaro
Department of Chemistry
University of Siena
Siena, Italy

Patrick E. Cassidy
Chemistry Department
Southwest Texas State University
San Marcos, Texas

Gretchen A. Caywood
DynaGen, Inc.
Cambridge, Massachusetts

Peggy Cebe
Department of Physics and Astronomy
Tufts University
Cambridge, Massachusetts

Marcelo Cestari
Materials Engineering
Federal University of São Carlos
São Carlos, São Paulo, Brazil

Won-Ill Cha
Research Center for Biomedical Engineering
Kyoto University
Kyoto, Japan

Antony P. Chacko
Department of Chemistry and Center for Materials for Information Technology
The University of Alabama
Tuscaloosa, Alabama

Grace Y. N. Chan
Department of Chemical Engineering
The University of Melbourne
Parkville, Victoria, Australia

Satish Chandra
Chalmers University of Technology
Goteborg, Sweden

Ji Young Chang
Department of Applied Chemistry
Ajou University
Suwon, Korea

Jin-Hae Chang
Department of Polymer Science and Engineering
Kum-Oh University of Technology
Kumi, Korea

Ken-Yuan Chang
Department of Chemical Engineering
National Tsing Hua University
Hsinchu, Taiwan, Republic of China

Shoou-Jinn Chang
Department of Electrical Engineering
National Cheng Kung University
Tainan, Taiwan, Republic of China

Thomas Ming Swi Chang
Artificial Cells and Organs Research Center
Faculty of Medicine
McGill University
Montreal, Quebec, Canada

Wen-Chi Chang
Materials Science Center
National Tsing Hua University
Tainau, Taiwan, Republic of China

Víctor M. Chapela
Centro de Química
Universidad Autónoma de Puebla
Puebla, Mexico

R. P. Chaplin
School of Chemical Engineering and Industrial Chemistry
University of New South Wales
Sydney, New South Wales, Australia

R. P. Chartoff
The Center for Basic and Applied Polymer Research
The University of Dayton
Dayton, Ohio

Gautam Chatterjee
Polymer Research Institute
Faculty of Chemistry
College of Environmental Science and Forestry
State University of New York at Syracuse
Syracuse, New York

Nadia Chavesi
Materials Engineering
Federal University of São Carlos
São Carlos, São Paulo, Brazil

Hong-Zheng Chen
Department of Polymer Science and Engineering
Zhejiang University
Hangzhou, Republic of China

Jeng-I Chen
University of Massachusetts, Lowell
Lowell, Massachusetts

Jianan Chen
Laboratory of Cellulose and Lignocellulocis Chemistry
Guangzhou Institute of Chemistry
Academia Sinica
Guangzhou, Republic of China

L. Chen
Department of Chemistry
University of Antwerp (UIA)
Wilrijk, Belgium

Leo-Wang Chen
Institute of Materials Engineering
National Taiwan University
Taipei, Taiwan, Republic of China

Meng-Jiu Chen
Department of Chemistry
University of Detroit Mercy
Detroit, Michigan

Show-An Chen
Department of Chemical Engineering
National Tsing-Hua University
Hsincho, Taiwan, Republic of China

Xiaohe Chen
Polymer Science Program
University of Connecticut
Storrs, Connecticut

H. N. Cheng
Research Center
Hercules Incorporated
Wilmington, Delaware

Stephen Z. D. Cheng
Department of Polymer Science
Maurice Morton Institute
The University of Akron
Akron, Ohio

Y. T. Chern
Institute of Chemical Engineering
National Taiwan Institute of Technology
Taipei, Taiwan, Republic of China

H. Michael Cheung
Department of Chemical Engineering
The University of Akron
Akron, Ohio

C. H. Chew
Department of Chemistry
National University of Singapore
Kent Ridge, Republic of Singapore

Wei-Kuo Chin
Department of Chemical Engineering
National Tsing Hua University
Hsinchu, Taiwan, Republic of China

Aurica P. Chiriac
"P'P. Poni" Institute of Macromolecular Chemistry
Jassy, Romania

Mihai V. Chiriac
"P'P. Poni" Institute of Macromolecular Chemistry
Jassy, Romania

Traian V. Chirila
Department of Biomaterials and Polymer Research
Lions Eye Insitute
Nedlands, Western Australia, Australia

K. G. Chittibabu
Center for Advanced Materials
Department of Chemistry and Physics
University of Massachusetts, Lowell
Lowell, Massachusetts

Tae Joon Cho
Department of Applied Chemistry
Ajou University
Suwon, Korea

Won-Jei Cho
Department of Polymer Science and Engineering
Pusan National University
Pusan, Korea

David D. Choe
New Jersey Institute of Technology
Newark, New Jersey

Eui-Won Choe
Robert L. Mitchell Technical Center
Hoechst Celanese Coporation
Summit, New Jersey

J.-O. Choi
Occidental Chemical Corporation
Grand Island, New York

Kil-Yeong Choi
Advanced Polymer Division
Korea Research Institute of Chemical Technology
Yusung, Taejeon, Korea

Kyu Yong Choi, Ph.D.
Department of Chemical Engineering
University of Maryland
College Park, Maryland

Ling-Siu Choi
Chemistry Division
Naval Research Laboratory
Washington, D. C.

Sam-Kwon Choi
Department of Chemistry
Korea Advanced Institute of Science and Technology
Yusong, Daejeon, Korea

Weon-Jung Choi
Polymer Laboratory
Inchon Research Center
Yukong Ltd.
Nam, Inchon, Korea

Julian Chojnowski
Center of Molecular and Macromolecular Studies of the Polish Academy of Sciences
Sienkiewicza, Poland

Michael S. Cholod
Rohm and Haas Company
Bristol, Pennsylvania

Bjorn E. Christensen
Norwegian Biopolymer Laboratory
Department of Biotechnology
University of Trondheim, NTH
Trondheim, Norway

Steven P. Christiano
Dow Corning
Midland, Michigan

Benjamin Chu
Department of Chemistry
State University of New York at Stony Brook
Stony Brook, New York

C. C. Chu
Fiber and Polymer Science Program
Department of Textiles and Apparel
Cornell University
Ithaca, New York

Riichirô Chûjô
Department of Materials Engineering
Teikyo University of Science and Technology
Yamanashi, Japan

Yoshiki Chujo
Division of Polymer Chemistry
Graduate School of Engineering
Kyoto University
Yoshida, Sakyo-ku, Kyoto, Japan

T. C. Chung
Department of Materials Science and Engineering
The Pennsylvania State University
University Park, Pennsylvania

Francesco Ciardelli
Dipartimento di Chimica e Chimica Industriale
University of Pisa
Pisa, Italy

J. F. Ciebien
Department of Chemical Engineering
Massachusetts Institute of Technology
Cambridge, Massachusetts

Sossio Cimmino
Istituto di Ricerca e Tecnologia delle Materie Plastiche
Italian National Council of Research-CNR
Arco Felice (Na), Italy

James H. Clark
Department of Chemistry
University of York
Heslington, York, United Kingdom

Richard J. Clark
Huntsman Corporation
Austin, Texas

Stephen J. Clarson
Department of Materials Science and Engineering
Polymer Research Center
University of Cincinnati
Cincinnati, Ohio

R. T. Clay
Department of Chemical Engineering
Massachusetts Institute of Technology
Cambridge, Massachusetts

Katherine S. Clement
Central Research and Development
Organic Products Research
The Dow Chemical Company
Freeport, Texas

R. A. Clendinning
Amoco Polymers Business Group
Alpharetta, Georgia

Robert E. Cohen
Department of Chemical Engineering
Massachusetts Institute of Technology
Cambridge, Massachusetts

Bradley K. Coltrain
Eastman Kodak Company
Rochester, New York

Anthony H. Conner
Forest Products Laboratory
USDA Forest Service
Madison, Wisconsin

Edwin C. Constable
Institut für Anorganische Chemie der Universität Basel
Basel, Switzerland

Vasile V. Cosofret
Departments of Chemistry and Medicine
University of North Carolina
Chapel Hill, North Carolina

Cornelia Cotzur
"P. Poni" Institute of Macromolecular Chemistry
Jassy, Romania

Ghislaine Coulon
Institut Curie P.S.I.
Paris, France
and
Université des Sciences et Technologies de Lille
L.S.P.E.S.
Villeneuve d' Ascq, France

B. Coutin
Laboratoire de Chimie Macromoléculaire
Université Pierre et Marie Curie
Paris, France

Fernanda M. B. Coutinho
Instituto de Macromoléculeas
Universidade Federal do Rio de Janeiro
Rio de Janeiro, Brazil

J. M. G. Cowie
Department of Chemistry
Heriot-Watt University
Edinburgh, Scotland

O. L. Crees
Sugar Research Institute
Mackay, Australia

J. A. Cuculo
Department of Textile Engineering, Chemistry, and Science
North Carolina State University
Raleigh, North Carolina

Bill M. Culbertson
The Ohio State University
Columbus, Ohio

M. David Curtis
Department of Chemistry
Macromolecular Science and Engineering Center
The University of Michigan
Ann Arbor, Michigan

Marek Cypryk
Center of Molecular and Macromolecular Studies
Polish Academy of Sciences
Sienkiewicza, Poland

Paul D. Dalton
Department of Biomaterials and Polymer Research
Lions Eye Institute
Nedlands, Western Australia, Australia

Daniel E. Damiani
Planta Piloto de Ingeniería Quimica
Universidad Nacional del Sur-CONICET
Bahia Blanca, Argentina

H. Daniell
Department of Botany-Microbiology
Molecular Genetics Program
Auburn University
Auburn, Alabama

Graham D. Darling
Department of Chemistry
McGill University
Montreal, Quebec, Canada

C. K. Das
Materials Science Center
Indian Institute of Technology
Kharagpur, India

Partha S. Das
Materials Science Center
Indian Institute of Technology
Kharagpur, India

Sudhin Datta
Exxon Chemical Company
Baytown, Texas

Martin Dauner
Institut für Textile- und Verfahrenstechnik
Denkendorf Forschungsbereich Blomedizintechnik
Denkendorf, Germany

Geta David
Department of Macromolecules
"Gh. Asachi" Technical University
Jassy, Romania

R. S. Davidson
Chemical Laboratory
The University of Kent at Canterbury
Kent, United Kingdom

Geoffrey R. Davies
Interdisciplinary Research Center in Polymer Science and Technology
University of Leeds
Leeds, United Kingdom

H. Ted Davis
Department of Chemical Engineering and Materials Science
University of Minnesota
Minneapolis, Minnesota

Stephen C. Davis
Polysar Rubber Incorporated
Sarnia, Ontario, Canada

Javier de Abajo
Instituto de Ciencia y Tecnologia de Polimeros, CSIC
Madrid, Spain

Alessandra de Almeida Lucas
Materials Engineering
Federal University of São Carlos
São Paulo, Brazil

Anthony N. de Belder
TdB Consultancy AB
Uppsala, Sweden

José G. de la Campa
Instituto de Ciencia y Tecnologia de Polimeros, CSIC
Madrid, Spain

Marco-Aurelio De Paoli
Instituto de Química
Universidade Estadual de Campinas
Campinas, Sao Paulo, Brazil

Jose Alexandre de Mendonça
Materials Engineering
Federal University of São Carlos
São Paulo, Brazil

N. F. de Rooij
Institute of Microtechnology
University of Neuchâtel
Neuchâtel, Switzerland

Barry D. Dean
Amoco Polymers Business Group
Alpharetta, Georgia

Rudolph D. Deanin
Plastics Engineering Department
University of Massachusetts, Lowell
Lowell, Massachusetts

Gero Decher
C.N.R.S. Institut Charles Sadron (CRM)
Mainz, Germany

Christian Decker
Laboratorie de Photochimie Généale
Université de Haute Alsace
Mulhouse, France

Alain Deffieux
Laboratoire de Chimie des Polymères Organiques
ENSCPB-CNRS
Université Bordeaux-1
Talence, France

Shigeru Deguchi
Department of Synthetic Chemistry and Biological Chemistry
Graduate School of Engineering
Kyoto University
Kyoto, Japan

Genevieve Delmas-Patterson
Chemistry Department
Unitversity of Quebec at Montreal
Montreal, Quebec, Canada

Michel Delmotte
Laboratoire de Microstructures et Mécanique des Matériaux
Center National de la Recherche Scientifique, Research Unit
ENSAM
Paris, France

Robert D. DeMarco
Zeon Chemicals, Inc.
Louisville, Kentucky

Vyacheslav V. Dementiev
Department of Chemistry
University of Texas at El Paso
El Paso, Texas

C. Denger
Max-Planck-Institut für Kohlenforschung
Mulheim an der Ruhr, Germany

James Denness
Department of Chemistry
University of York
Heslington, York, United Kingdom

Gianfranco Denti
Laboratorio di Chimica Inorganica
Istituto di Chimica Agraria
Università di Pisa
Pisa, Italy

Hemant Desai
Defence Research Agency
Fort Halstead, Kent, United Kingdom

Shreeram Deshpande
Department of Chemistry and Biochemistry and Laboratory for Electronic Properties of Materials
The University of Oklahoma
Norman, Oklahoma

K. L. DeVries
University of Utah
Salt Lake City, Utah

R. A. DeVries
Ventures Advanced Materials
The Dow Chemical Company
Midland, Michigan

Emilia Di Pace
Istituto di Ricercae Tecnologia delle Materie Plastiche.CNR.
Arco Felice (Na), Italy

G. Di Pasquale
Istituto Chimico
Facoltà di Ingegneria
Università di Catania
Catania, Italy

Giuseppe Di Silvestro
Dipartimento di Chimica Organica e Industriale
Universit à di Milano
Milano, Italy

Anthony Jay Dias
Butyl Product Development
Exxon Chemical Company
Baytown Polymers Center
Baytown, Texas

Marcos Lopes Dias
Instituto de Macromoléculas
Universidade Federal do Rio de Janeiro
Rio de Janeiro, Brazil

Fernando R. Díaz
Facultad de Química
Pontificia Universidad Católica de Chile
Santiago, Chile

Arthur F. Diaz
IBM Almaden Research Center
San Jose, California

Victor B. Diaz
Syntex S. A.
Buenos Aires, Argentina

Francisco Diaz-Camacho
Instituto de Investigaciones en Materiales
Mexico D.F., Mexico

B. L. Dickinson
Amoco Polymers Business Group
Alpharetta, Georgia

Stefan Disch
Institut für Makromolekulare Chemie der Universität Freiburg
Freiburg, Germany

W. O. S. Doherty
Department of Molecular Sciences
James Cook University
Townsville, Australia

Yoshihara Doi
Polymer Chemistry Laboratory
The Institute of Physical and Chemical Research (RIKEN)
Wako-shi, Saitama, Japan

Takashi Domae
Hyogo Prefectural Institute of Industrial Research
Yukihira-Cho Suma-Ku, Kobe, Japan

Jean-Baptiste Donnet
Ecole Nationale Supérieure de Chimie de Mulhouse, France
Mulhouse, France

My Dotrong
Nonmetallic Materials Division
University of Dayton Research Institute
Dayton, Ohio

William E. Douglas
Unite Mixte CNRS/Rhône-Poulenc/UM II
CNRS UMR 44
Université de Montpellier II
Montpellier, France

Patricia Dreyfuss
Midland, Michigan

X. Drujon
Elf Atochem North America, Inc.
King of Prussia, Pennsylvania

Paul L. Dubin
Indiana University, Purdue
Indianapolis, Indiana

T. R. du Toit
Plastics Technology Department
Cape Technikon
Capetown, South Africa

Philippe Dubois
Center for Education and Research on Macromolecules (CERM)
University of Liège - Institute of Chemistry
Liège, Belgium

Henryk Dudziak
Industrial Chemistry Research Institute
Warsaw University of Technology
Warsaw, Poland

E. Duguet
Laboratoire de Chimie des Polymères Organiques
Ecole Nationale Supérieure de Chimie et de Physique de Bordeaux
Bordeaux, France

Ljerka Duic
Faculty of Chemical Engineering and Technology
University of Zagreb
Zagreb, Croatia

Michel Dumont
Institut d'Optique Thèorique et Appliquèe (Unitè Associèe au CNRS n 14)
Orsay, France

M. M. Dumoulin
National Research Council Canada
Industrial Materials Institute
Boucherville, Quebec, Canada

Manfred Dunky
Krems Chemie AG
Krems, Austria

Manzer J. Durrani
The BF Goodrich Company
Brecksville, Ohio

Petar R. Dvornic
Michigan Molecular Institute
Midland, Michigan

Andrzej Dworak
Polish Academy of Sciences
Institute of Coal Chemistry
Sowlnskiego, Poland

John R. Ebdon
The Polymer Center
School of Physics and Chemistry
Lancaster University
Lancaster, United Kingdom

Peter Eckerle
ERATO Molecular Catalysis Project
Research Development Corporation of Japan
Yachigusa, Yakusa-cho, Toyota, Japan

Yoshiaki Eda
Department of Chemistry Faculty of Science and Engineering
Saga University
Saga, Japan

Toshiro Egashira
Industrial Research Institute of Ishikawa Prefecture
Kanazawa, Ishikawa, Japan

H. Eisazadeh
Intelligent Polymer Research Laboratory
University of Wollongong
Wollongong, New South Wales, Australia

Claus D. Eisenbach
Makromolekulare Chemie II
Bayreuther Institut für Makromolekülforschung
Bayreuth, Germany

Adi Eisenberg
Department of Chemistry
McGill University
Toronto, Ontario, Canada

Sonny A. Ekhorutomwen
Department of Chemistry Polymer Science/Plastics Engineering Option
University of Massachusetts, Lowell
Lowell, Massachusetts

S. H. El-Hamouly
Laboratory of Polymer Chemistry
Faculty of Science
El-Menoufia University
Shebin El-Kom, Egypt

Hoda El-Masry
Microbial Biotechnology Department
National Research Center
Dokki, Cairo, Egypt

F. A. El-Saied
Laboratory of Polymer Chemistry
Faculty of Science
El-Menoufia University
Shebin El-Kom, Egypt

Houssni El -Saied
Cellulose and Paper Department
National Research Center
Dokki, Cairo, Egypt

Magdy M. El-Sayed
Food Technology and Dairy Science
National Research Center
Dokki, Cairo, Egypt

Mohamed A. El-Taraboulsi
Chemical Engineering Department
Faculty of Engineering
Alexandria University
Alexandria, Egypt

Karin Elbl-Weiser
BASF AG
Ludwigshafen, Germany

D. H. Ellington
General Electric Plastics
Technology Center
Washington, West Virginia

Diane L. Ellis
Massachusetts Institute of Technology
Department of Mechanical Engineering
Cambridge, Massachusetts

Mostafa A. Elrahman
Department of Materials Science and Engineering
University of Florida
Gainesville, Florida

Maria Victoria Encinas
Departamento de Quimica
Facultad de Ciencias
Universidad de Santiago de Chile
Santiago, Chile

Kiyoshi Endo
Department of Applied Chemistry
Faculty of Engineering
Osaka City University
Sugimoto, Osaka, Japan

Takeshi Endo
Research Laboratory of Resources Utilization
Tokyo Institute of Technology
Nagatsuta-cho, Midori-ku, Yokohama, Japan

Royce E. Ennis
E. I. DuPont de Nemours and Company, Incorporated
Beaumont Works, Texas

Erol Erbay
Research and Development Center
Petkim Petrochemicals Holding Inc.
Korfez, Kocaeli, Turkey

Candan Erbil
Department of Chemistry
Instanbul Technical University
Istanbul, Turkey

Miklós Erdösy
Departments of Chemistry and Medicine
University of North Carolina
Chapel Hill, North Carolina

Semih Erhan
Albert Einstein Medical Center
Philadelphia, Pennsylvania

Alexandre Espeleta
Instituto de Quimica
Universidade Estadual de Campinas
Campinas, Sao Paulo, Brazil

George Evens
DSM Research
Geleen, The Netherlands

R. C. Evers
Wright Laboratory
Wright-Patterson Air Force Base
Ohio

Bob R. Ezzell
Central Research and Development
Advanced Polymeric Systems Lab
The Dow Chemical Company
Midland, Michigan

Thomas J. Fabish
Alcoa Laboratories
Alcoa Center, Pennsylvania

David R. Fagerburg
Eastman Chemical Company
Kingsport, Tennessee

Darryl R. Fahey
Research and Development
Phillips Petroleum Company
Bartlesville, Oklahoma

Karim Faïd
Département de Chimie
Université de Montréal
Montréal, Québec, Canada

Y. L. Fan
Union Carbide Corporation
Bound Brook, New Jersey

Chin-Kun Fang
Materials Science Center
National Tsing Hua University
Taiwan, China

Tianru Fang
Laboratory of Polymer Physics
Changchun Institute of Applied Chemistry Chinese Academy of Sciences
Changchun, China

George F. Fanta
Plant Polymer Research
National Center For Agricultural Utilization Research USDA
Agricultural Research Service
Peoria, Illinois

Michael F. Farona
Department of Chemistry
The University of North Carolina at Greensboro
Greensboro, North Carolina

Rudolf Faust
Department of Chemistry
University of Massachusetts, Lowell
Lowell, Massachusetts

Dennis G. Fauteux
Arthur D. Little Incorporated
Cambridge, Massachusetts

Dorel Feldman
Center for Building Studies
Concordia University
Montreal, Quebec, Canada

Maria Isabel Felisberti
Instituto de Química
Universidade Estadual de Campinas
Campinas, Sao Paulo, Brazil

Ke Feng
Department of Chemistry and Center for Theoretical Studies of Physical Systems
Clark Atlanta University
Atlanta, Georgia

Zhiliu Feng
Changchun Institute of Applied Chemistry
Chinese Academy of Sciences
Changchun, China

Richard Fengl
Eastman Chemical Company
Kingsport, Tennessee

James Ferguson
University of Strathclyde
Glasgow, Scotland

M. L. Fernandez
Chemical Engineering Department
Imperial College of Science and Technology
London, United Kingdom

J. M. Fernandez-Santin
Universitat Politecnicade Catalunya
Barcelona, Spain

Inés Fernandez de Piérola
Departamento de Fisicoquímica (CTFQ)
Facultad de Ciencias
Universidad a Distancia (UNED)
Madrid, Spain

M. Ferreira
Instituto de Física de São Carlos
Sao Carlos, Brazil

Maria Ines P. Ferreira
Instituto de Macromolèculas
Universida de Federal do Rio de Janeiro
Rio de Janeiro, Brazil

María L. Ferreira
Planta Piloto de Ingenieria Quimica
Universidad Nacional del Sur-CONICET
Bahia Blanca, Argentina

Paolo Ferruti
Dipartimento di Chimica e Fisica per i Materiali via
Brescia, Italy

Wilmer K. Fife
Department of Chemistry
Indiana University-Purdue University at Indianapolis
Indianapolis, Indiana

Garret D. Figuly
DuPont Central Research and Development
Wilmington, Delaware

Gerhard Fink
Max-Planck-Institut für Kohlenforschung
Mulneim an der Ruhr, Germany

Heino Finkelmann
Institut für Makromolekulare Chemie der Universität Freiburg
Freiburg, Germany

L. Fiore
EniChem-Istituto G. Donegani
Novara, Italy

Anthony R. Fiorillo
Morton International
Woodstock, Illinois

Karl Fischer
Makromolekulare Chemie II
Bayreuther Institut für Makromolekülforschung
Bayreuth, Germany

Richard M. Fischer
3M Weathering Resource Center
St. Paul, Minnesota

Adriano Fissi
CNR - Institute of Biophysics
Pisa, Italy

John W. Fitch
Chemistry Department
Southwest Texas State University
San Marcos, Texas

James A. Fitzgerald
DuPont Advanced Fibers Systems
Wilmington, Delaware

Richard J. Flecksteiner
Zeon Chemicals, Inc.
Louisville, Kentucky

Zbigniew Florjanczyk
Faculty of Chemistry
Warsaw University of Technology
Warszawa, Poland

David T. Floyd
Goldschmidt Chemical Corporation
Hopewell, Virginia

Peter D. Folland
Polymer Science Group
School of Chemical Technology
University of South Australia
South Australia, Australia

Michel Fontanille
ENSCPB-Université Bordeaux
Talence, France

Alex Forschirm
Hoechst Celanese Corporation
Summit, New Jersey

Eric Fossum
Department of Chemistry
Carnegie Mellon University
Pittsburgh, Pennsylvania

Daniel A. Foucher
Department of Chemistry
University of Toronto
Toronto, Ontario, Canada

Marye Ann Fox
University of Texas at Austin
Austin, Texas

Alain Fradet
Laboratoire de Synthèse Macromoléculaire - CNRS U.R.A.
Université Pierre et Marie Curie
Paris, France

Jeanne François
Institut Charles Sadron CNRS-ULP
Strasbourg, France

Emile Franta
Institut Charles Sadron/ULP
Strasbourg, France

Paul Frayer
Science and Technology Group
Corning Incorporated
Corning, New York

Jean M. J. Fréchet
Department of Chemistry
Baker Laboratory
Cornell University
Ithaca, New York

Giuliano Freddi
Stazione sperimentale per la Seta
Milano, Italy

R. Freudenberger
Institut für Organische Chemie
Freie Universität Berlin
Berlin, Germany

Harry L. Frisch
Department of Chemistry
State University of New York at Albany
Albany, New York

Esteban P. Fuentes
Syntex S. A.
Buenos Aires, Argentina

Shigetada Fujii
General Education Chemistry
Oita Medical University
Hasama-machi Oita, Japan

Kazuhiro Fujiki
Division of Life and Health Sciences
Joetsu University of Education
Joetsu, Niigata, Japan

Satoru Fujime
Graduate School of Integrated Science
Yokohama City University
Kanazawa, Yokohama, Japan

Kiyohisa Fujimori
Department of Chemistry
University of New England
Armidale, New South Wales, Australia

Masao Fujisawa
Department of Chemistry
Kinki University
Osaka, Japan

Isaburo Fukawa
Asahi Chemical Industry Company
Tokyo, Japan

Akinori Fukuda
Osaka Municipal Technical Research Institute
Osaka, Japan

Mitsuhiro Fukuda
Textile Materials Science Laboratory
Hyogo University of Teacher Education
Yashiro-cho, Hyogo, Japan

W. Stephen Fulton
The Malaysian Rubber Products' Research Association
Tun Abdul Razak Laboratory
Hertford, United Kingdom

Werner Funke
Institute of Technical Chemistry
University of Stuttgart
Mainz, Germany

Gary P. Funkhouser
Department of Chemistry and Biochemistry and Laboratory for Electronic Properties of Materials
The University of Oklahoma
Norman, Oklahoma

Anita Furlani
Department of Chemistry
University "La Sapienza,"
Rome, Italy

Liu Futian
The Research Institute of Polymer Chemistry
Nankai University
Tianjin, Republic of China

Satish K. Gaggar
General Electric Plastics
Technology Center
Washington, West Virginia

Himansu M. Gajiwala
Department of Chemistry
Tuskegee University
Tuskegee, Alabama

James P. Galbo
Ciba Additive
Ardsley, New York

Jean-Claude Galin
Institut Charles Sadron (CNRS-ULP)
Strasbourg, France

Maria C. Gallazzi
Dipartimento di Chimica Industriale e Ingegneria Chimica
Politecnico di Milano
Milano, Italy

Bernard R. Gallot
Laboratoire Des Materiaux Organiques Proprietes Specifiques
Centre National de la Recherche Scientifique
Vernaison, France

Leong Ming Gan
Department of Chemistry
National University of Singapore
Kent Ridge, Republic of Singapore

Alessandro Gandini
Matèriaux Polymères
Ecole Française de Papeterie et des Industries Graphiques
Saint Martin d'Hères, France

P. K. Ganguly
Jute Technological Research Laboratories
Indian Council of Agricultural Research
Calcutta, India

H. Gao
Center for Advanced Materials
Department of Chemistry and Physics
University of Massachusetts, Lowell
Lowell, Massachusetts

Fabio Garbassi
EniChem S.p.A.—Istituto Guido Donegani
Novara, Italy

J. Romero Garcia
Centro de Investigación en Quimica Aplicada
Saltillo, Coahuila, Mexico

Rosangela B. Garcia
Garcia Departamento de Química
Universidade Federal do Rio Grande do Norte
Natal, Brazil

Ligia Gargallo
Depto. Química Física
Facultad de Química
Pontificia Universidad Católica de Chile
Casilla Santiago, Chile

R. I. Gascón
Facultad de Quimica
UNAM Circuito Interior
Ciudad Universitaria
Mexico D.F., Mexico

Russell A. Gaudiana
Polaroid Corporation
Cambridge, Massachusetts

C. Gavach
Centre National de la Recherche Scientifique Laboratoire des Matáriaux et Procédás Membranaires
Montpellier, France

Norman G. Gaylord
Charles A. Dana Research Institute for Scientists Emeriti
Drew University
Madison, New Jersey

Gérard Gebel
CEA-Département de Recherche Fondamentale sur la Matière Condensée
SESAM/Laboratoire de Physico-Chimie Moléculaire
Grenoble, France

Kurt E. Geckeler
Faculty of Chemistry and Pharmacy
Institute of Organic Chemistry
University of Tuebingen
Tuebingen, Germany

P. H. Gedam
Indian Institute of Chemical Technology
Hyderabad, India

Steven C. Gedon
Eastman Chemical Company
Kingsport, Tennessee

Brendan J. Geelan
Uniroyal Chemical Company, Incorporated
Middlebury, Connecticut

Stevin H. Gehrke
Department of Chemical Engineering
University of Cincinnati
Cincinnati, Ohio

Jon F. Geibel
Research and Development
Phillips Petroleum Company
Bartlesville, Oklahoma

H. J. Geise
Department of Chemistry
University of Antwerp (UIA)
Wilrijk, Belgium

Ian R. Gelling
Malaysian Rubber Producers' Research Association
Tyn Abdul Razak Laboratory
Brickendonbury, Hertford, United Kingdom

Graeme A. George
School of Chemistry
Queensland University of Technology
Brisbane, Queensland, Australia

Maurice H. George
Department of Chemistry
Imperial College of Science, Technology and Medicine
London, United Kingdom

Anton L. German
Department of Polymer Chemistry and Technology
Eindhoven University of Technology
Eindhoven, The Netherlands

M. Reza Ghadiri
Departments of Chemistry and Molecular Biology
Scripps Research Institute
La Jolla, California

Premamoy Ghosh
Department of Plastics and Rubber Technology
University of Calcutta
Calcutta, India

Subrata Ghosh
Berger Paints India Ltd.
Howrah, India

W. M. Gibbons
Research Center
Hercules Incorporated
Wilmington, Delaware

Harry W. Gibson
Department of Chemistry
Virginia Polytechnic Institute and State University
Blacksburg, Virginia

M. Helena Gil
Department of Biochemistry
University of Coimbra
Coimbra, Portugal

Richard D. Gilbert
Department of Wood and Paper Science
North Carolina State University
Raleigh, North Carolina

Paolo Giusti
Department of Chemical Engineering
University of Pisa
Pisa, Italy

Wolfgang G. Glasser
Department Wood Science and Forest Products
Virginia Polytechnic Institute and State University
Blacksburg, Virginia

Daniel T. Glatzhofer
Department of Chemistry and Biochemistry and Laboratory for Electronic Properties of Materials
The University of Oklahoma
Norman, Oklahoma

Yves Gnanou
Laboratoire de Chimie des Polymères Organiques NSCPB
Université Bordeaux
Talence, France

Andreas Göldel
Lehrstuhl für Makromolekulare Chemie II and Bayreuther Institut für Makromolekülforschung (BIMF)
Universität Bayreuth
Bayreuth, Germany

Steve Goldschmidt
Ascona Resins
Buenos Aires, Argentina

Ailton de Souza Gomes
Instituto de Macromoléculas
Universidade Federal do Rio de Janeiro
Rio de Janeiro, Brazil

M. Rosa Gómez-Anton
Universidad Nacional de Educación a Distancia
Madrid, Spain

Christian Gondard
PVC Basic Research
Limburgse Vinyl Maatschappij
Tessenderlo, Belgium

J. P. Gong
Division of Biological Sciences
Graduate School of Science
Hokkaido University
Sapporo, Japan

Myoung-Seon Gong
Department of Chemistry
Dankook University
Cheonan, Chungnam, Korea

Kenneth E. Gonsalves
Polymer Science Program
Institute of Materials Science and Department of Chemistry
University of Connecticut
Storrs, Connecticut

Mattheus F. A. Goosen
College of Agriculture
Sultan Qaboos University
Sultanate of Oman

Wolfgang Göpel
Institute of Physical and Theoretical Chemistry and Center of Interface Analysis and Sensors
Tuebingen, Germany

Bernard Gordon III
Department of Materials Science and Engineering Polymer Science Program
The Pennsylvania State University
University Park, Pennsylvania

Sherald H. Gordon
Biopolymer Research Unit
National Center for Agricultural Utilization Research
Agricultural Research Service
U.S. Department of Agriculture
Peoria, Illinois

R. Gosain
Department of Chemistry
Imperial College of Science, Technology and Medicine
London, United Kingdom

Tomoko Goto
Department of Materials Science and Engineering
Nagoya Institute of Technology
Nagoya, Japan

Yaeko Gotoh
The University of Tokyo
Tokyo, Japan

D. C. Gowda
Laboratory of Molecular Biophysics
School of Medicine
University of Alabama at Birmingham
Birmingham, Alabama

Marsha M. Grade
GE Corporate Research and Development Center
Schenectady, New York

Neil B. Graham
Department of Pure and Applied Chemistry
University of Strathclyde
Glasgow, Scotland, United Kingdom

Richard D. Grant
Dow Corning Australia Pty. Ltd.
Blacktown, New South Wales, Australia

Joseph Green
FMC Corporation
Princeton, New Jersey

R. V. Greene
Biopolymer Research Unit
National Center for Agricultural Utilization Research
Agricultural Research Service
U.S. Department of Agriculture
Peoria, Illinois

Rinaldo Gregório Jr.
Materials Engineering
Federal University of São Carlos
São Carlos, São Paulo, Brazil

Andreas Greiner
Philipps-Universität Marburg
Phyikalische Chemie/Polymere
Hans-Meerwein-Strasse
Marburg, Germany

A. M. Griffen
Institute of Food Research
Norwich Laboratory
Colney, Norfolk

G. J. L. Griffin
Ecological Materials Research Institute
EPRON Industries Ltd.
Ketton, Stamford, United Kingdom

James R. Griffith
Naval Research Laboratory
Washington, D. C.

Gabriel Groeninckx
ICI Polyurethanes
Heverlee, Belgium

Werner M. Grootaert
Specialty Fluoropolymers Department
3M Company
St. Paul, Minnesota

Daniel J. Grosse
S. C. Johnson Wax
Racine, Wisconsin

Bruce A. Gruber
Air Products and Chemicals, Incorporated
Allentown, Pennsylvania

H. F. Gruber
Institute of Chemical Technology of Organic Materials
Vienna University of Technology
Vienna, Austria

James V. Gruber
Amerchol Corporation
Edison, New Jersey

Henri J. M. Grünbauer
Urethane Polymers Research and Development
DOW Benelux NV
The Netherlands

C. Guda
Department of Botany-Microbiology
Molecular Genetics Program
Auburn University
Auburn, Alabama

Güngör Gündüz
Kimya Mühendisliği Bölümü
Orta Doğu Teknik and Universitesi
Ankara, Turkey

Takahiro Gunji
Department of Industrial Chemistry
Faculty of Science and Technology
Science University of Tokyo
Noda, Chiba, Japan

X. Andrew Guo
Department of Chemistry
University of Alberta
Edmonton, Alberta, Canada
and
Department of Chemistry
University of Calgary
Calgary, Alberta, Canada

Alka D. Gupta
SETI
NASA Ames Research Center
Moffett Field, California

J. T. Guthrie
Department of Colour Chemistry
University of Leeds
Leeds, United Kingdom

Alain Guyot
CNRS Laboratoire de Chimie et Procédés de Polymerisation
Villeurbanne, France

Chang-Sik Ha
Department of Polymer Science and Engineering
Pusan National University
Pusan, Korea

Osamu Haba
Faculty of Engineering
Hokkaido University
Sapporo, Japan

Nikos Hadjichristidis
Department of Chemistry
University of Athens
Athens, Greece

Stephen F. Hahn
Central Research and Development
The Dow Chemical Company
Midland, Michigan

Adel F. Halasa
Tire Materials and Compound Research
Corporate Research
The Goodyear Tire and Rubber Company
Akron, Ohio

Henry K. Hall, Jr.
C. S. Marvel Laboratories
Department of Chemistry
The University of Arizona
Tucson, Arizona

Gary. R. Hamed
College of Polymer Science and Polymer Engineering
The University of Akron
Akron, Ohio

James G. Hamilton
School of Chemistry
The Queen's University of Belfast
Belfast, Northern Ireland

J. Hammer
Institut für Technische und Makromolekulare Chemie der Universität Hamburg
Hamburg, Germany

Dong Keun Han
Polymer Chemistry Laboratory
Korea Institute of Science and Technology
Seoul, Korea

Haesook Han
Department of Chemistry
University of Detroit Mercy
Detroit, Michigan

Man Jung Han
Department of Applied Chemistry
Ajou University
Suwon, Korea

Michael Hanack
Institut für Organische Chemie
University of Tübingen
Tübingen, Germany

Masanori Hara
Department of Mechanics and Materials Science
Rutgers University
New Brunswick, New Jersey

Valeria Harabagiu
"P. Poni" Institute of Macromolecular Chemistry
Jassy, Romania

Akira Harada
Department of Macromolecular Science
Faculty of Science
Osaka University
Toyonaka, Osaka, Japan

Issifu I. Harruna
Department of Chemistry
High Performance Polymers and Ceramics Center
Clark Atlanta University
Atlanta, Georgia

David C. Harsh
Department of Chemical Engineering
University of Cincinnati
Cincinnati, Ohio

Jeffrey D. Hartgerink
Departments of Chemistry and Molecular Biology
Scripps Research Institute
La Jolla, California

Kiichi Hasegawa
Osaka Municipal Technical Research Institute
Osaka, Japan

Masaki Hasegawa
Department of Materials Science and Technology
Faculty of Engineering
Toin University of Yokohama
Yokohama, Japan

Masatoshi Hasegawa
Department of Chemistry
Faculty of Science
Toho University
Chiba, Japan

Kazuhiko Hashimoto
Department of Applied Chemistry
Faculty of Engineering
Kogakuin University
Nakano-cho, Hachioji, Tokyo, Japan

Masato Hashimoto
Department of Polymer Science and Engineering
Faculty of Textile Science
Kyoto Institute of Technology
Kyoto, Japan

Alfred Hassner
Department of Chemistry
Bar-Ilan University
Ramat-Gan, Israel

Koichi Hatada
Department of Chemistry
Faculty of Engineering Science
Osaka University
Toyonaka, Osaka, Japan

Kenichi Hatanaka
Department of Biomolecular Engineering
Faculty of Bioscience and Biotechnology
Tokyo Institute of Technology
Yokohama, Japan

Yasuo Hatate
Faculty of Engineering
Kagoshima University
Kagoshima, Japan

Galen R. Hatfield
Washington Research Center
W. R. Grace and Company
Columbia, Maryland

Ronald D. Hatfield
USDA Agricultural Research Service
U.S. Dairy Forage Research Center
Madison, Wisconsin

T. Alan Hatton
Department of Chemical Engineering
Massachusetts Institute of Technology
Cambridge, Massachusetts

Isamu Hattori
Production Development Department
Goi Works, Hitachi Chemical Company, Ltd.
Chiba, Japan

Kenjiro Hattori
Department of Industrial Chemistry
Tokyo Institute of Polytechnics
Kanagowa, Japan

Karlheinz Hausmann
DuPont de Nemours International European Technical Center
Geneva, Switzerland

Kathleen O. Havelka
The Lubrizol Corporation
Wickliffe, Ohio

Marvin Havens
Cyrovac Division
W. R. Grace and Company
Duncan, South Carolina

Shuzi Hayase
Materials and Devices Research Laboratories
Research and Development Center
Toshiba Corporation
Saiwai-ku, Kawasaki, Japan

Sadao Hayashi
Faculty of Textile Science and Technology
Shinshu University
Ueda, Japan

Toshio Hayashi
Research Institute for Advanced Science and Technology
Osaka Prefecture University
Sakai City, Osaka, Japan

Heidi Hayen
Makromolekulare Chemie II
Bayreuther Institut für Makromoleküllforschung
Bayreuth, Germany

Byron G. Hays
Engelhard Corporation
Beachwood, Ohio

Baki Hazer
TüBiTAK-Marmara Research Center Department of Chemistry
Kocaeli, Turkey

J. W. S. Hearle
Department of Textiles
UMIST
Manchester, United Kingdom

Alan J. Heeger
UNIAX Corporation
Santa Barbara, California

A. Heinrichs
Max-Planck-Institut fur Kohlenforschung
Mulheim an der Ruhr, Germany

Thomas Heinze
Institute of Organic and Macromolecular Chemistry of the Friedrich-Schiller
University of Jena
Jena, Germany

Walter Heitz
Fb Physikalische Chemie
Polymere und Zentrum für Materialwissenschaften
Philipps-Universität Marburg
Marburg, Germany

Jorge Heller
APS Research Institute
Redwood City, California

C. S. Henkee
Huntsman Corporation
Austin, TX

Edwin R. Hensema
Hercules European Research Center
Barneveld, The Netherlands

Karl-Heinz Hentschel
LS Marketing
Bayer AG
Leverkusen, Germany

James F. Hermann
S. C. Johnson Wax
Racine, Wisconsin

José Hernández-Barajas
Department of Chemical Engineering
Vanderbilt University
Nashville, Tennessee

Walter R. Hertler
Kennett Square, Pennsylvania

Y. Hervaud
Ecole Nationale Supérieure de Chimie de Montpellier
Laboratoire de Chimie Appliquée
Montpellier, France

Michael Hess
Department of Physical Chemistry
Gerhard–Mercator University
Duisburg, Germany

Volker Hessel
Institut für Organische Chemie
Universität Mainz
Mainz, Germany

Mitsuo Higuchi
Kyushu University
Higashi-ku, Fukuoka, Japan

Randal M. Hill
Dow Corning
Midland, Michigan

Hidefumi Hirai
Science University of Tokyo
Tokyo, Japan

M. Hiramitsu
Polymer Chemistry Laboratory
The Institute of Physical and Chemical Research (RIKEN)
Wako-shi, Saitama, Japan

Takashi Hirano
Protein Engineering Laboratory
Molecular Biology Department
National Institute of Bioscience and Human Technology
Ibaraki, Japan

Akira Hirao
Department of Polymer Chemistry
Tokyo Institute of Technology
Tokyo, Japan

Mitsuo Hirata
Department of Industrial Chemistry
Nihon University
Chiba, Japan
Toshimi Hirata
Forestry and Forest Products Research Institute
Ibaraki, Japan
Chuichi Hirayama
Department of Applied Chemistry and Biochemistry
Faculty of Engineering
Kumamoto University
Kumamoto, Japan
Shin Hiwasa
Saitama Institute of Technology
Saitama, Japan
C. C. Ho
Department of Chemistry
University of Malaya
Kuala Lumpur, Malaysia
Rong-Ming Ho
Department of Polymer Science
Maurice Morton Institute
The University of Akron
Akron, Ohio
Hartwig Höcker
Lehrstuhl für Textilchemie und Makromolekulare Chemie der Rheinisch-Westfälischen Technischen Hochschule Aachen
Aachen, Germany
Allan S. Hoffman
Center for Bioengineering
University of Washington
Seattle, Washington
A. Hofland
Institute for Surface Chemistry
Stockholm, Sweden
Steven Holdcroft
Department of Chemistry
Simon Fraser University
Burnaby, British Columbia, Canada
Geoffrey Holden
Holden Polymer Consulting, Incorporated
Prescott, Arizona
Soon Man Hong
Polymer Processing Laboratory
Korea Institute of Science and Technology
Cheongryang, Seoul, Korea
Ye Hong
Lions Eye Institute
Department of Biomaterials and Polymer Research
Nedlands, Western Australia, Australia
Daniel Horák
Institute of Macromolecular Chemistry
Academy of Sciences of The Czech Republic
Prague, Czech Republic
Hans-Heinrich Hörhold
Institut für Organische Chemie und Makromolekulare Chemie
Universität Jena
Jena, Germany
Arturo Horta
Departamento de Fisicoquímica (CTFQ)
Facultad de Ciencias
Universidad a Distancia (UNED)
Madrid, Spain
Satoru Hosoda
Petrochemicals Research Laboratory
Sumitomo Chemical Company Ltd.
Chiba, Japan
D. J. Hourston
Loughborough University of Technology
Leicestershire, United Kingdom
Andrew N. Hrymak
Department of Chemical Engineering
McMaster University
Hamilton, Ontario, Canada
Benjamin S. Hsiao
Central Research and Development Department
Experimental Station
E. I. DuPont de Nemours and Company
Wilmington, Delaware
Bing R. Hsieh
The Wilson Center for Research and Technology
Xerox Corporation
Webster, New York
Kuo-Huang Hsieh
Department of Chemical Engineering
National Taiwan University
Taipei, Taiwan
Ging-Ho Hsiue
Department of Chemical Engineering
National Tsing Hua University
Taiwan, Republic of China
W. L. Hsu
Tire Materials and Compound Research Corporate Research
The Goodyear Tire and Rubber Company
Akron, Ohio
Henry S. W. Hu
Geo-Centers Incorporated
Fort Washington, Maryland
Mu-Yi Hua
Department of Chemical Engineering
National Tsing-Hua University
Taiwan, Republic of China
Baotong Huang
Changchun Institute of Applied Chemistry
Chinese Academy of Sciences
Changchun, Republic of China
Sun-Yi Huang
Cytec Industries
Stamford, Connecticut
Wenhua H. Huang
Department of Chemistry
The Ohio State University
Columbus, Ohio
X. X. Huang
TRI/Princeton
Princeton, New Jersey
Yong Huang
Laboratory of Cellulose and Lignocellulocis Chemistry
Guangzhou Institute of Chemistry
Academia Sinica
Guangzhou, Republic of China
S. M. Hudson
Department of Textile Engineering, Chemistry, and Science
North Carolina State University
Raleigh, North Carolina
Anders Hult
Department of Polymer Technology
Royal Institute of Technology
Stockholm, Sweden
Klaus Hummel
Institute for Chemical Technology of Organic Materials
Technical University Graz
Graz, Austria
J. Steve Humphrey, Jr.
Elf Atochem North America, Inc.
King of Prussia, Pennsylvania
Chi-Cheng Hung
Bristol Analytical Research
Rohm and Haas Company
Bristol, Pennsylvania
Ming-H. Hung
Experimental Station
DuPont Fluoroproducts
DuPont Central Research and Development
Wilmington, Delaware
David Hunkeler
Department of Chemical Engineering
Vanderbilt University
Nashville, Tennessee
Allen D. Hunter
Department of Chemistry
Youngstown State University
Youngstown, Ohio
Dang Mai Huong
Institute of Chemistry
National Center for Natural Science and Technology
Hanoi, Vietnam
Gue-Wuu Hwang
Department of Chemical Engineering
National Tsing-Hua University
Taiwan, Republic of China
Seung Sang Hwang
Polymer Processing Laboratory
Korea Institute of Science and Technology
Cheongryang, Seoul, Korea
Suong-Hyu Hyon
Research Center for Biomedical Engineering
Kyoto University
Kyoto, Japan
Hermis Iatrou
Department of Chemistry
University of Athens
Athens, Greece
Tachio Ichikawa
Industrial Research Institute of Ishikawa Prefecture
Kanazawa, Ishikawa, Japan
Tomoko Ichikawa
Faculty of Home Economics
Otsuma Women's University
Tokyo, Japan
Eiji Ihara
Department of Applied Chemistry
Faculty of Engineering
Hiroshima University
Higashima-Hiroshima, Japan
Hirotaka Ihara
Department of Applied Chemistry and Biochemistry
Faculty of Engineering
Kumamoto University
Kumamoto, Japan
Takao Iijima
Department of Applied Chemistry
Faculty of Engineering
Yokohama National University
Tokiwadai, Hodogaya-ku Yokohama, Japan
Masashi Iino
Institute for Chemical Reaction Science
Tohoku University
Katahira, Aoba-ku, Sendai, Japan
Takashi Iizawa
Department of Chemical Engineering
Faculty of Engineering
Hiroshima University
Higashi-Hiroshima, Japan
Eisaku Iizuka
Department of Functional Polymer Science
Faculty of Textile Science and Technology
Shinshu University
Ueda, Japan
Yoshito Ikada
Research Center for Biomedical Engineering
Kyoto University
Kyoto, Japan

Takao Ikariya
ERATO Molecular Catalysis Project
Research Development Corporation of Japan
Yachigusa, Yakusa-cho, Toyota, Japan

Tokimitsu Ikawa
Department of Chemistry
Faculty of Science
Okayama University of Science
Okayama, Japan

Masahiko Imai
Tokyo Metropolitan Isotope Research Center
Tokyo, Japan

Yoshio Imai
Department of Organic and Polymeric Materials
Tokyo Institute of Technology
Meguro-ku, Tokyo, Japan

S. H. Imam
Biopolymer Research Unit
National Center for Agricultural Utilization Research
Agricultural Research Service
U.S. Department of Agriculture
Peoria, Illinois

Yukio Imanishi
Division of Material Chemistry
Faculty of Engineering
Kyoto University
Kyoto, Japan

Corrie T. Imrie
Department of Chemistry
University of Aberdeen
Old Aberdeen, United Kingdom

Norihiro Inagaki
Faculty of Engineering
Shizuoka University
Johoku, Hamamatsu, Japan

Yoshiaki Inaki
Faculty of Engineering
Osaka University
Suita, Osaka, Japan

Akio Inoue
Forestry and Forest Products Research Institute
Danchi-nai, Japan

Shohei Inoue
Department of Industrial Chemistry
Faculty of Engineering Science
University of Tokyo
Shinjuku-ku, Tokyo, Japan

Takashi Inoue
Department of Organic and Polymeric Materials
Tokyo Institute of Technology
Tokyo, Japan

Toshihide Inoue
Plastics Research Laboratories
Toray Industries, Inc.
Minato-ku, Nagoya, Japan

Silvia Ioan
Department of Macromolecules
"P. Poni" Institute of Macromolecular Chemistry
Jassy, Romania

Jude O. Iroh
Department of Materials Science and Engineering
University of Cincinnati
Cincinnati, Ohio

Fumiaki Ise
Fundamental Research Laboratory of Natural and Synthetic Polymers
Asahi Chemical Industry Company, Ltd.
Hacchownawate, Takatsuki, Osaka, Japan

Norio Ise
Central Laboratory
Rengo Company
Osaka, Japan

Hatsuo Ishida
The NSF Center for Molecular and Microstructure of Composites
Department of Macromolecular Science
Case Western Reserve University
Cleveland, Ohio

Shin-Ichiro Ishida
Department of Chemistry and Chemical Engineering
Faculty of Technology
Kanazawa University
Kodatsuno, Kanazawa, Japan

Shigehisa Ishihara
Wood Research Institute
Kyoto University
Kyoto, Japan

Mitsuo Ishikawa
Department of Applied Chemistry
Faculty of Engineering
Hiroshima University
Higashi-Hiroshima, Japan

Tatsuo Ishikawa
Production Development Department
Goi Works, Hitachi Chemical Company, Ltd.
Goi-Minami-Kaigan, Ichihara, Chiba, Japan

Koji Ishizu
Department of Polymer Science
Tokyo Institute of Technology
Tokyo, Japan

Tomoyuki Itaya
Department of Chemistry
Faculty of Science
Hiroshima University
Higashi-Hiroshima, Japan

Hiroshi Ito
IBM Almaden Research Center
San Jose, California

Koichi Ito
Department of Materials Science
Toyohashi University of Technology
Toyohashi, Japan

Yoshihiro Ito
Division of Material Chemistry
Faculty of Engineering
Kyoto University
Kyoto, Japan

Hiroshi Itoh
Koken Bioscience Institute
Tokyo, Japan

Takashi Itoh
Department of Polymer Science and Engineering
Faculty of Textile Science
Kyoto Institute of Technology
Kyoto, Japan

Yoshihiro Itoh
Faculty of Textile Science and Technology
Shinshu University
Ueda, Japan

Satoshi Itou
Aichi College of Technology
Gamagori, Aichi, Japan

Shinichi Itsuno
Department of Materials Science
Toyohashi University of Technology
Toyohashi, Japan

Carlos C. Iturbe
Grupo Nuevos Materiales
Departamento de Química Física
Universidad del País Vasco
Lejona, Spain

Takeshi Iwasaki
Department of Applied Chemistry
Faculty of Engineering
Kanagawa University
Yokohama, Japan

Hiroo Iwata
Research Center for Biomedical Engineering
Kyoto University
Kyoto, Japan

Satoru Iwata
Department of Molecular Chemistry and Engineering
Faculty of Engineering
Tohoku University
Aoba, Japan

Takashi Iwatsubo
National Institute of Materials and Chemical Research
Ibaraki, Japan

Shouji Iwatsuki
Instrumental Analysis Center
Mie University
Tsu, Japan

Yoshinobu Izumi
Macromolecular Research Laboratory
Faculty of Engineering
Yamagata University
Yonezawa, Japan

Saleh A. Jabarin
Polymer Institute
The University of Toledo
Toledo, Ohio

S. Jahromi
Department of Polymer Technology
Faculty of Chemical Engineering and Materials Science
Delft University of Technology
Delft, The Netherlands

R. L. Jalbert
General Electric Plastics Technology Center
Washington, West Virginia

Bernd Jansen
Institute of Medical Microbiology and Hygiene
University of Cologne
Cologne, Germany

Carlos F. Jasso-Gastinel
Department of Chemical Engineering
University of Guadalajara
Guadalajara, Jalisco, Mexico

Zbigniew Jedlinski
Institute of Polymer Chemistry
Polish Academy of Sciences
Zabrze, Poland

Klaus R. Jenni
Goldschmidt Chemical Corporation
Hopewell, Virginia

Soraya Jericó
Instituto de Quimica
Universidade Estadual de Campinas
Campinas, Sao Paulo, Brazil

Mu Shik Jhon
Department of Chemistry Korea Advanced Institute of Science and Technology
Taejon, Korea

Bingzheng Jiang
Changchun Institute of Applied Chemistry
Chinese Academy of Sciences
Changchun, China

Jung-Il Jin
Department of Chemistry and Advanced Materials
Chemistry Research Center
Korea University
Seoul, Korea

Moon Young Jin
Advanced Polymer Division
Korea Research Institute of Chemical Technology
Yusung, Taejeon, Korea

N. Jing
Laboratory of Molecular Biophysics
School of Medicine
University of Alabama at Birmingham
Birmingham, Alabama

Wu Jinguo
Department of Chemical Engineering
Tianjin Institute of Technology
Tianjin, China

J. F. Joanny
Institut Charles Sadron (CRM-EAHP)
CNRS-ULP
Strasbourg, France

Inés Joekes
Instituto de Química
Universidade Estadual de Campinas
Campinas, Sao Paulo, Brazil

Kristian B. Johansen
Heparin Research Laboratory
Leo Pharmaceutical Products
Ballerup, Denmark

M. Johansson
Department of Polymer Technology
Royal Institute of Technology
Stockholm, Sweden

Vijay T. John
Department of Chemical Engineering
Tulane University
New Orleans, Louisiana

Harry C. Johnson IV
Department of Chemistry
University of North Carolina at Charlotte
Charlotte, North Carolina

Timothy A. Johnson
Departments of Chemistry and Medicine University of North Carolina
Chapel Hill, North Carolina

J. Darrell Jones
Dow Corning Corporation
Midland, Michigan

M. S. Jones
The Polymer Center
School of Physics and Chemistry
Lancaster University
Lancaster, United Kingdom

E. H. Jonsson
Plastics Division Research and Development
Bayer AG
Dermugen, Germany

Alfredo Juan
Planta Piloto de Ingeniería Quimica
Universidad Nacional del Sur-CONICET
Bahia Blanca, Argentina

Henri Jullien
Laboratoire de Microstructures et Mécanique des Matériaux
Center National de la Recherche Scientifique Research Unit
Ecole Nationale Superiour d'Arts etMéticrs
Paris, France

Jae Chang Jung
Department of Chemistry
Korea Advanced Institute of Science and Technology
Taejon, Korea

Jin Chul Jung
Department of Materials Science and Engineering
Pohang University of Science and Technology
Pohang, Korea

K. H. Jungbluth
Department of Trauma and Reconstructive Surgery
University Hospital Hamburg-Eppendorf (UKE)
Hamburg, Germany

B.-J. Jungnickel
Deutsches Kunststoff-Institut (German Plastics Institute)
Darmstadt, Germany

Alberto Juris
Dipartimento di Chimica G. Ciamician
Università di Bologna
Bologna, Italy

H. Justnes
SINTEF
Trondheim, Norway

Alexander V. Kabanov
Department of Pharmaceutical Sciences
College of Pharmacy
University of Nebraska Medical Center
Omaha, Nebraska

Kimiaki Kabuki
Toshiba Battery
Shinagawa-ku, Tokyo, Japan

Mibuko Kaburaki
Department of Chemistry for Materials
Mie University
Mie-Kan, Japan

Jun-Ichi Kadokawa
Department of Materials Science and Engineering
Faculty of Engineering
Yamagata University
Yonezawa, Japan

Kyoji Kaeriyama
Department of Materials Science
Kyoto Institute of Technology
Kyoto, Japan

Isao Kaetsu
Faculty of Science and Technology
Kinki University
Osaka, Japan

Akihiro Kagemoto
Laboratory of Chemistry
Department of General Education
Osaka Institute of Technology
Asahi-ku, Osaka, Japan

Toshikuni Kaino
NTT Opto-electronics Laboratories
Kanagawa, Japan

Kanji Kajiwara
Kyoto Institute of Technology
Koyota, Japan

D. K. Kakati
Department of Chemistry
Imperial College of Science, Technology, and Medicine
London, United Kingdom

Masa-Aki Kakimoto
Department of Organic and Polymeric Materials
Tokyo Institute of Technology
Tokyo, Japan

Toyoji Kakuchi
Graduate School of Environmental Science
Hokkaido University
Sapporo, Japan

Douglas S. Kalika
Department of Chemical and Materials Engineering
University of Kentucky
Lexington, Kentucky

Mikiharu Kamachi
Faculty of Science
Osaka University
Toyonaka, Osaka, Japan

Yashavanth K. Kamath
TRI/Princeton
Princeton, New Jersey

S. Kametekar
Center for Advanced Materials
Department of Chemistry and Physics
University of Massachusetts, Lowell
Lowell, Massachusetts

Atsushi Kameyama
Department of Applied Chemistry
Faculty of Engineering
Kanagawa University
Yokohama, Japan

Leon A. P. Kane-Maguire
Intelligent Polymer Research Laboratory
Department of Chemistry
University of Wollongong
Wollongong, New South Wales, Australia

Kiyotomi Kaneda
Derpartment of Chemical Engineering
Faculty of Engineering Science
Osaka University
Toyonaka, Osaka, Japan

Masao Kaneko
Department of Science
Ibaraki University
Mito, Japan

E. T. Kang
Department of Chemical Engineering
National University of Singapore
Kent Ridge, Singapore

Yoshihisa Kano
Faculty of Technology
Tokyo University of Agriculture and Technology
Tokyo, Japan

R. C. Kanu
Department of Chemical Engineering and Institute of Materials Science
University of Connecticut
Storrs, Connecticut

David Kaplan
Biotechnology Division
U.S. Army Natick Research, Development, and Engineering Center
Natick, Massachusetts

Jozsef Karger-Kocsis
Institut für Verbundwerkstoffe GmbH
Universität Kaiserslautern
Kaiserslautern, Germany

I. Karino
Material and Electronic Devices Laboratory
Mitsubishi Electric Corporation
Amagasaki, Hyogo, Japan

Sigbritt Karlsson
Department of Polymer Technology
The Royal Institute of Technology (KTH)
Stockholm, Sweden

Masato Kasamori
Industrial Research Institute of Ishikawa Prefecture
Ishikawa, Japan

Hideo Kasatani
Asahi Chemical Industry Company, Ltd.
Nobeoka City, Miyazaki, Japan

Gabor Kaszas
Bayer Rubber Incorporated
Sarnia, Ontario, Canada

Ryoichi Katakai
Department of Chemistry
Faculty of Engineering
Gunma University
Kiryu, Gunma, Japan

Fumio Kataoka
Production Engineering Research Laboratory
Hitachi Ltd.
Yokohama, Japan

Erich E. Kathmann
Department of Polymer Science
University of Southern Mississippi
Hattiesburg, Mississippi

Issa A. Katime
Universidad del País Vasco
Bilbao, Spain

Takashi Kato
Institute of Industrial Science
The University of Tokyo
Tokyo, Japan

Sanjeev S. Katti
The Dow Chemical Company
Freeport, Texas

A. Katzer
Department of Trauma and Reconstructive Surgery
University Hospital Hamburg-Eppendorf (UKE)
Hamburg, Germany

Nariyoshi Kawabata
Department of Chemistry and Materials Technology
Faculty of Engineering and Design
Kyoto Institute of Technology
Kyoto, Japan

Makoto Kawagoe
Department of Mechanical Systems Engineering
Faculty of Engineering
Toyama Prefectural University
Toyama, Japan

Haruma Kawaguchi
Department of Applied Chemistry
Keio University
Yokohama, Japan

Seiichi Kawahara
Faculty of Technology
Tokyo University of Agriculture and Technology
Fuchu, Japan

Akihiro Kawai
Department of Applied Chemistry
Faculty of Engineering
Kanagawa University
Yokohama, Japan

Yusuke Kawakami
Graduate School of Materials Science
Japan Advanced Institute of Science and Technology (JAIST)
Tatsunokuchi, Ishikawa, Japan

Shunichi Kawanishi
Osaka Laboratory for Radiation Chemistry
Japan Atomic Energy Research Institute
Neyagawa, Osaka, Japan

Hitoshi Kawasaki
Industrial Technology Center of Okayama Prefecture
Okayama, Japan

Hiroshi Kawazura
Faculty of Pharmaceutical Sciences
Josai University
Sakao, Saitama, Japan

Teddy M. Keller
Materials Chemistry Branch
Naval Research Laboratory
Washington, D. C.

Dewey L. Kerbow
E. I. DuPont de Nemours and Company, Inc.
Parkersburg, West Virginia

Meinoff Kersting
Polymer Research Division
BASF AG
Ludwigshafen, Germany

Robert G. Keske
Amoco Polymers Business Group
Alpharetta, Georgia

Warren D. Ketola
Traffic Control Materials Division
3M Company
St. Paul, Minnesota

Helmut Keul
Lehrstuhl für Textilchemie und Makromolekulare Chemie der Rheinisch-Westfälischen Technischen Hochschule Aachen
Aachen, Germany

Joachim Kötz
Universität Potsdam
Potsdam, Germany

N. Khalturinskij
Institute of Synthetic Polymeric Materials
Moscow, Russia

Mubarak A. Khan
Radiation Chemistry Laboratory
Institute of Nuclear Science and Technology
Bangladesh Atomic Energy Commission
Dhaka, Bangladesh

Milind Khandwe
Polymer Research Laboratory
Department of Post Graduate and Research in Chemistry
Rani Durgavati University
Jabalpur, India

Tapan Kumar Khanra
Materials Science Centre
Indian Institute of Technology
Kharagpur, India

Gregory B. Kharas
Chemistry Department
DePaul University
Chicago, Illinois

Eugene Khor
Department of Chemistry
National University of Singapore
Kent Ridge, Singapore

Viera Khunova
Faculty of Chemical Technology
Department of Plastics and Rubber
Slovak Technical University
Bratislava, Slovak Republic

Junji Kido
Department of Materials Science and Engineering
Yamagata University
Yonezawa, Yamagata, Japan

Hayato Kihara
Petrochemicals Research Laboratory
Sumitomo Chemical Company, Ltd.
Chiba, Japan

Akihiko Kikuchi
Institute of Biomedical Engineering
Tokyo Women's Medical College
Shinjuku, Tokyo, Japan

Lars Kilaas
SINTEF
Applied Chemistry
Trondheim, Norway

Byoung Chul Kim
Division of Polymer Researches
Korea Institute of Science and Technology
Seoul, Korea

Byung-Gu Kim
Department of Polymer Science and Engineering
Dankook University
Cheonan, Chungnam, Korea

Chulhee Kim
Department of Polymer Science and Engineering
Dankook University
Incheon, Korea

Jason Kim
EVAL Company of America
Lisle, Illinois

Joon-Seop Kim
Department of Chemistry
McGill University
Montreal, Quebec, Canada

Junkyung Kim
Macromolecular Science and Engineering Center and Department of Materials Science and Engineering
The University of Michigan
Ann Arbor, Michigan

Ki Ho Kim
Protein Engineering Laboratory
Molecular Biology Department
National Institute of Bioscience and Human Technology
Higashi, Tsukuba, Japan

Kwang Ung Kim
Polymer Processing Laboratory
Korea Institute of Science and Technology
Cheongryang, Seoul, Korea

Oh-Kil Kim
Chemistry Division
Naval Research Laboratory
Washington, D. C.

Sung Wan Kim
Center for Controlled Chemical Delivery
Department of Pharmaceutics and Pharmaceutical Chemistry
University of Utah
Salt Lake City, Utah

Yang-Bae Kim
Korea Chemical Company
Yongin, Kyunggi, Korea

Young H. Kim
DuPont Central Research and Development
Experimental Station
Wilmington, Delaware

Young Ha Kim
Polymer Chemistry Laboratory
Korea Institute of Science and Technology
Seoul, Korea

Derek M. Ole Kiminta
Department of Chemical Engineering and Chemical Technology
Imperial College of Science, Technology, and Medicine
London, United Kingdom

Jun-Ichi Kimura
Propellants and Explosives Laboratory
The 1st Research Center
Technical Research and Development Institute
Japan Defense Agency
Meguro-Ku, Tokyo, Japan

Keiichi Kimura
Chemical Process Engineering
Faculty of Engineering
Osaka University
Suita, Osaka, Japan

Kunio Kimura
Faculty of Environmental Science and Technology
Okayama University Thushima
Okayama, Japan
Mutsumi Kimura
Department of Functional Polymer Science
Faculty of Textile Science and Technology
Shinshu University
Ueda, Japan
Satoshi Kimura
Laboratory of Chemistry
Department of General Education
Osaka Institute of Technology
Asahi-ku, Osaka, Japan
Takayoshi Kimura
Department of Chemistry
Kinki University
Osaka, Japan
Yoshiharu Kimura
Department of Polymer Science and Engineering
Kyoto Institute of Technology
Kyoto, Japan
Roswell Easton King III
Ciba Additives
Ardsley, New York
Takatoshi Kinoshita
Department of Materials Science and Engineering
Nagoya Institute of Technology
Gokiso-cho, Showa-ku, Nagoya, Japan
R. A. Kirchhoff
Central Research and Development
Material Science and Development
Dow Chemical Company
Midland, Michigan
Duygu Kisakurek
Middle East Technical University Department of Chemistry
Ankara, Turkey
V. S. Kishanprasad
Indian Institute of Chemical Technology
Hyderabad, India
Yasuhisa Kishimoto
ERATO Molecular Catalysis Project
Research Development Corporation of Japan
Toyota, Japan
Yasushi Kishimoto
Polyolefins Development Department
Asahi Chemical Industry Company Ltd.
Okayama, Japan
Kaushal Kishore
Inorganic and Physical Chemistry
Indian Institute of Science
IDL-Nitro Nobel Basic Research Institute
Bangalore, India
Shinichi Kitamura
Department of Agricultural Chemistry
Kyoto Prefectural University
Shimogamo, Kyoto, Japan
Hiromi Kitano
Department of Chemical and Biochemical Engineering
Toyama University
Toyama, Japan
Tatsuki Kitayama
Department of Chemistry
Faculty of Engineering Science
Osaka University
Toyonaka, Osaka, Japan
Kenneth J. Klabunde
Department of Chemistry
Kansas State University
Manhattan, Kansas
Joachim E. Klee
De Trey Dentsply
Konstanz, Germany
Michael T. Klein
Amoco Performance Products Incorporated
Department of Chemical Engineering
University of Delaware
Newark, Delaware
Dieter Klemm
Institute of Organic and Macromolecular Chemistry of the Friedrich-Schiller-University of Jena
Jena, Germany
R. Klopsch
Institut für Organische Chemie
Freie Universität Berlin
Berlin, Germany
Ralf Knapp
Polymer Institute
Karlsruhe University
Karlsruhe, Germany
Wolfgang Knoll
Max-Planck-Institut für Polymerforschung and Frontier Research Program
The Institute of Physical and Chemical Research (RIKEN)
Mainz, Germany
Eiichi Kobayashi
Department of Industrial Chemistry
Science University of Tokyo
Noda, Chiba, Japan
Katsumi Kobayashi
Photon Factory
National Laboratory for High Energy Physics
Tsukuba-shi, Ibaraki, Japan
Masamichi Kobayashi
Department of Macromolecular Science
Faculty of Science
Osaka University
Toyonaka, Osaka, Japan
Shigeo Kobayashi
Department of Electrical Engineering
Tokyo University of Agriculture and Technology
Tokyo, Japan
Shiro Kobayashi
Department of Materinals Chemistry
Graduate School of Engineering
Tohoku University
Aoba, Sendi, Japaqn
Yoshinari Kobayashi
Shikoku National Industrial Research Institute
Kagawa-ken, Japan
Toshiyuki Kodaira
Department of Materials Science and Engineering
Faculty of Engineering
Fukui University
Fukui-Shi, Japan
Shinzo Kohjiya
Institute for Chemical Research
Kyoto University
Kyoto, Japan
Joachim Kohn
Department of Chemistry
Rutgers - The State University of New Jersey
New Brunswick, New Jersey
W. Kohnen
Institute of Medical Microbiology and Hygiene
University of Cologne
Cologne, Germany
N. Koizumi
Institute for Chemical Research
Kyoto University
Kyoto, Japan
M. Kojima
Yamazaki Works
Hitachi Chemical Company Ltd.
Hitachi-shi, Japan
Etsuo Kokufuta
Institute of Applied Biochemistry
University of Tsukuba
Tsukuba, Ibaraki, Japan
Joseph V. Koleske
Consolidated Research Incorporated
Charleston, West Virginia
Jeffrey S. Kollodge
3M Company
St. Paul, Minnesota
Kougaku Komamiya
Laboratory of Urban Safety Planning
Tokyo, Japan
Keiichi Komatsu
The Silk Science Research Institute
Tokyo, Japan
Tamikuni Komatsu
Analytical Research Center
Asahi Chemical Industry Company, Ltd.
Fuji, Sizuoka, Japan
Jiro Komiyama
Department of Polymer Chemistry
Tokyo Institute of Technology
Ookayama, Meguro-ku, Tokyo, Japan
Koichi Kondo
Department of Chemistry
Faculty of Science and Technology
Ritsumeikan University
Nojicho, Kusatsu, Shiga, Japan
Shin-Ichi Kondo
Laboratory of Pharmaceutical Physical Chemistry
Gifu Pharmaceutical University
Gifu, Japan
Shuji Kondo
Department of Materials Science and Engineering
Nagoya Institute of Technology
Nagoya, Japan
Yoshiyuki Kondo
Department of Functional Polymer Science
Faculty of Textile Science and Technology
Shinshu University
Ueda, Japan
Kenji Kono
Department of Applied Materials Science
College of Engineering
Osaka Prefecture University
Sakai, Osaka, Japan
Ryusuke Kono
Department of Applied Physics
The National Defense Academy
Yokosuka, Japan
I. C. Konstantakopoulos
Laboratory of Organic Chemical Technology
Department of Chemical Engineering
National Technical University of Athens
Athens, Greece
Vitan Bonev Konsulov
Chemistry Faculty
University "Konstantin Preslavski"
Bulgaria
William J. Koros
Department of Chemical Engineering
The University of Texas at Austin
Austin, Texas
Yuri Korshak
The Mendeleev University of Chemical Technology
Moscow, Russia

Lars Korsnes
DYNAL A/S
Oslo, Norway
Robert Kosfeld
Department of Physical Chemistry
Gerhard–Mercator University
Duisburg, Germany
Joseph Kost
Department of Chemical Engineering
Ben Gurion University
Beer Sheva, Israel
B. A. Kottes Andrews
Southern Regional Research Center
Mid South Area
Agricultural Research Service
U.S. Department of Agriculture
New Orleans, Louisiana
Joachim Kotz
Universität Potsdam
Potsdam, Germany
M. Koudelke-Hep
University of Neuchâtel
Institute of Microtechnology
Neuchâtel, Switzerland
A. L. Kovarskii
Semenov Institute of Chemical Physics
Russian Academy of Sciences
Moscow, Russia
Arturo R. Kraglievich
Syntex S. A.
Buenos Aries, Argentina
Pavel Kratochvíl
Institute of Macromolecular Chemistry
Academy of Sciences of the Czech Republic
Prague, Czech Republic
Peter Kraxner
Institut für Technishche und
Makromolekulare Chemie
Universität Hamburg
Hamburg, Germany
Lowell R. Kreeger
Union Carbide Corporation
Bound Brook, New Jersey
V. N. Krishnamurthy
Vikram Sarabhai Space Center
Tivandrum, India
R. K. Krishnaswamy
Department of Chemical and
Materials Engineering
University of Kentucky
Lexington, Kentucky
Christoph Kröhnke
CIBA-GEIGY Ltd.
Additives Division
Basel, Switzerland
Viktoriya A. Kruglova
Irkutsk State University
Irkutsk, Russia
Duygu Ksakürek
Department of Chemistry
Middle East Technical University
Ankara, Turkey
Masataka Kubo
Instrumental Analysis Center
Mie University
Kamihama-cho, Japan
Shizuo Kubota
Industrial Technology Center of
Wakayama Prefecture
Wakayama-City, Japan
S. E. Kudaibergenov
Institute of Chemical Sciences
Kazakh Academy of Sciences
Almaty, Republic of Kazakhstan
Ju Kumanotani
Institute for Industrial Science
The University of Tokyo
Tokyo, Japan
Devendra Kumar
SETI Institute
NASA Ames Research Center
Moffett Field, California
Jayant Kumar
Center for Advanced Materials
Department of Chemistry and Physics
University of Massachusetts, Lowell
Lowell, Massachusetts
Satish Kumar
School of Textile and Fiber Engineering
Georgia Institute of Technology
Atlanta, Georgia
Robert J. Kumpf
Corporate Polymer Research
Bayer Corporation
Pittsburgh, Pennsylvania
Toyoki Kunitake
Faculty of Engineering
Kyushu University
Fukuoka, Japan
Toshio Kunugi
Department of Applied Chemistry
and Biotechnology
Yamanashi University
Kofu, Japan
Jay F. Künzler
Department of Chemistry and
Polymer Development
Bausch and Lomb Incorporated
Rochester, New York
Masahiko Kuramoto
Polymer Research Laboratory
Idemitsu Petrochemical Company, Ltd.
Chiba, Japan
Noriyuki Kuramoto
Department of Materials Science
and Engineering
Faculty of Engineering
Yamagata University
Yamagata, Japan
Witold Kuran
Department of Chemistry
Warsaw University of Technology
Warsaw, Poland
Julie M. Kure
CSIRO Division of Wool Technology
Belmont, Victoria, Australia
Valery F. Kurenkov
Kazan State Technological University
Kazan, Russia
A. P. Kuriakose
Department of Polymer Science and
Rubber Technology
Cochin University of Science and Technology
Department of Applied Chemistry
Faculty of Engineering
Kanagawa University
Kanagawa-ku, Yokohama, Japan
Minoru Kuriki
Department of Applied Chemistry
Faculty of Engineering
Kanagawa University
Kanagawa, Yokohama, Japan
Yoshimi Kurimura
Department of Science
Ibaraki University
Mito, Japan
Keisuke Kurita
Department of Industrial Chemistry
Faculty of Engineering
Seikei University
Tokyo, Japan
Jenci Kurja
Department of Polymer Chemistry
and Technology
Eindhoven University of Technology
Eindhoven, The Netherlands
S. V. Kurmaz
Institute of Chemical Physics
Russian Academy of Sciences Chernogolovka
Moscow, Russia
Helena Kurowska
Industrial Chemistry Research Institute
Warsaw University of Technology
Warsaw, Poland
Yasuhiko Kurusu
Department of Chemistry
Faculty of Science and Technology
Sophia University
Tokyo, Japan
Shoichi Kutsumizu
Faculty of Engineering
Gifu University
Yanagido, Gifu, Japan
Masayuki Kuzuya
Laboratory of Pharmaceutical
Physical Chemistry
Gifu Pharmaceutical University
Gifu, Japan
T. K. Kwei
Department of Chemistry
Herman F. Mark Polymer Research Institute
Polytechnic University
Brooklyn, New York
F. P. La Mantia
Dipartimento di Ingegneria Chimica dei Processi e dei
Materiali
Viale delle Scienze
Università di Palermo
Palermo, Italy
Ovidio Laguna
Polymer Science and Technology Institute CSIC
Madrid, Spain
Paul M. Lahti
Department of Chemistry
University of Massachusetts
Amherst, Massachusetts
Manuela Lang
Institut für Organische Chemie
University of Tübingen
Tübingen, Germany
Horst Lange
Bayer AG
Krefeld, Germany
Grzegorz Lapienis
Center of Molecular and
Macromolecular Studies
Polish Academy of Sciences
Sienkiewicza, Poland
A. Larnkjaer
Heparin Research Laboratory
Leo Pharmaceutical Products
Ballerup, Denmark
A. Larsson
Insitute for Surface Chemistry
Stockholm, Sweden
Zsuzsa László-Hedvig
Central Research Institute for Chemistry
Hungarian Academy of Sciences
Budapest, Hungary

Aldrick N. K. Lau
Corporate Research and Development
Raychem Corporation
Menlo Park, California

Wayne W. Y. Lau
Department of Chemical Engineering
National University of Singapore
Republic of Singapore

Luigi Lazzeri
Department of Chemical Engineering
University of Pisa
Pisa, Italy

Alain Le Borgne
Laboratoire de Physicochimie des Biopolymères
CNRS-Thiais
Paris, France

J. Le Moigne
Institut de Physique et Chimie des Matériaux de Strasbourg
Strasbourg, France

Jean-François Le Nest
Ecole Française de Papeterie et des Industries Graphiques (IndianaPG)
Saint Martin d'Hères, France

Jean-Pierre Leblanc
National Starch and Chemical Company
Bridgewater, New Jersey

Mario Leclerc
Département de Chimie
Université de Montréal
Montréal, Québec, Canada

Burm-Jong Lee
Department of Chemistry
Inje University
Kimhae, Korea

Dong-Ho Lee
Department of Polymer Science
Engineering College
Kyungpook National University
Taegu, Korea

K. C. Lee
Department of Chemistry
University of Malaya
Kuala Lumpur, Malaysia

Kang I. Lee
Monsanto Company
Springfield, Massachusetts

Sang Mook Lee
Polymer Processing Laboratory
Korea Institute of Science and Technology
Cheongryang, Seoul, Korea

Yu-Der Lee
Department of Chemical Engineering
National Tsing Hua University
Taiwan, Republic of China

Günther Leising
Institut für Festkörperphysik
Technische Universität Graz
Graz, Austria

Subasini Lenka
Department of Chemistry
Ravenshaw College
Orissa, India

Pierre LePoutre
Department of Chemical Engineering
University of Maine
Orono, Maine

Louis M. Leung
Department of Chemistry
Hong Kong Baptist University
Kowloon, Hong Kong

Dongming Li
Baytown Polymers Center
Exxon Chemical Company
Baytown, Texas

Jing-Song Li
Department of Medical Informatics
Faculty of Medicine
Kyoto University
Kyoto, Japan

Kai Li
Corporate Research and Development
3M Canada Incorporated
London, Ontario, Canada

San Xi Li
Istituto di Chimica delle Macromolecole
Milano, Italy

Yingjie Li
Department of Chemistry
Indiana University - Purdue at Indianapolis
Indianapolis, Indiana

Yuliang Li
Changchun Institute of Applied Chemistry
Chinese Academy of Sciences
Changchun, China

Jeng-Li Liang
Department of Chemistry and Center for Materials for Information Technology
The University of Alabama
Tuscaloosa, Alabama

Jun Liao
High Technology Materials Center
Department of Chemistry
Virginia Commonwealth University
Richmond, Virginia

Der-Jang Liaw
Department of Chemical Engineering
National Taiwan Institute of Technology
Taipei, Taiwan, Republic of China

Joseph D. Lichtenhan
Phillips Laboratory
Edwards Air Force Base
California

Rüdiger N. Lichtenthaler
Angewandte Thermodynamik
Physikalisch-Chemisches Institut
Ruprecht-Karls-Universität Heidelberg
Heidelberg, Germany

Raimond Liepins
Materials Science and Technology Division
Los Alamos National Laboratory
Los Alamos, New Mexico

Dmitri Likhatchev
Instituto de Investigaciones en Materiales
UNAM
Coyoacàn, Mexico

Donald T. Liles
Designed Materials Development
Dow Corning Corporation
Midland, Michigan

Kwon-Taek Lim
Pusan National University of Technology
Nam, Pusan, Korea

Zuzana Limpouchová
Department of Physical and Macromolecular Chemistry
Faculty of Science
Charles University in Prague
Prague, Czech Republic

Jiun-Hung Lin
Department of Chemical Engineering
Tatung Institute of Technology
Taipei, Taiwan, Republic of China

Lars-Ake Lindén
Polymer Research Group
Department of Dental Biomaterials Science
Karolinska Institute
Royal Institute of Medicine
Stockholm, Sweden

Tore Lindmo
Department of Physics
University of Trondheim NTH
Trondheim, Norway

Ernö Lindner
Institute for General and Analytical Chemistry
Technical University of Budapest
Budapest, Hungary

David W. Lipp
Cytec Industries
Stamford, Connecticut

E. A. Lissi
Departamento de Quimica
Facultad de Ciencias
Universidad de Santiago de Chile
Casilla, Santiago, Chile

A. A. Litmanovich
Moscow State Automobile and Road Technical University
Moscow, Russia

Morton H. Litt
Department of Macromolecular Science
Case Western Reserve University
Cleveland, Ohio

Guojun Liu
Department of Chemistry
The University of Calgary
Calgary, Alberta, Canada

Jing-Ping Liu
The NSF Center for Molecular and Microstructure of Composites
Department of Macromolecular Science
Case Western Reserve University
Cleveland, Ohio

Jui-Hsiang Liu
Department of Chemical Engineering
National Cheng Kung University
Tainan, Taiwan, Republic of China

Paolo Locatelli
Istituto di Chimica delle Macromolecole
Milano, Italy

M. Löffler
Freie Universität Berlin
Institut für Organische Chemie
Berlin, Germany

A. Löfgren
Royal Institute of Technology
KTH Department of Polymer Technology
Stockholm, Sweden

Robert B. Login
International Specialty Products Corporation
Wayne, New Jersey

Karsten Löhr
Daimler-Benz AG
Research Center Ulm
Ulm, Germany

David J. Lohse
Exxon Research and Engineering Company
East Annandale, New Jersey

Léa Lopes
Instituto de Macromoléculas (IMA/UFRJ)
Rio de Janeiro, Brazil

F. López-Carrasquero
Department d'Enginyeria Química
Universitat Politicnica de Catalunya
Barcelona, Spain

Claude Loucheux
Laboratoire de Chimie Macromoléculaire
Université des Sciences et Technologies de Lille
Villeneuve d'Ascq, France

Douglas A. Loy
Properties of Organic Materials Department
Sandia National Laboratories
Albuquerque, New Mexico

Angel E. Lozano
Instituto de Cienciay
Tecnologia de Polimeros, CSIC
Madrid, Spain

Shaoxiang Lu
Department of Chemistry and The Herman F. Mark Polymer Research Institute
Polytechnic University
Brooklyn, New York

Paul F. Luckham
Department of Chemical Engineering and Chemical Technology
Imperial College of Science, Technology and Medicine
London, United Kingdom

Robert R. Luise
Hytem Consultants, Incorporated
Boothwyn, Pennsylvania

Pierre J. Lutz
Institut Charles Sadron (CNRS/ULP)
Strasbourg, France

Chin-Chi M. Ma
Materials Science Center
National Tsing Hua University
Hsinchu, Taiwan, Republic of China

Jingjing Ma
3M Adhesive Technology Center
St. Paul, Minnesota

John L. MacAdams
Polyolefins Research Center
Quantum Chemical Company
Morris, Illinois

Antonio L. Macanita
Instituto de Tecnología Química e Biológica
Oelvas, Portugal

Alan G. MacDiarmid
Chemistry Department
University of Pennsylvania
Philadelphia, Pennsylvania

W. A. MacDonald
ICI Films
Middlesbrough, Cleveland, United Kingdom

A. Maes
Laboratory of Macromolecular and Physical Organic Chemistry
Catholic University of Lueven
Heverlee (Leuven), Belgium

P. L. Magagnini
Dipartimento di Ingegneria Chimica
Chimica Industriale e Scienza dei Materiali
Università di Pisa
Pisa, Italy

H. Magg
Bayer AG
Leverkusen, Germany

Jun Magoshi
National Institute of Agrobiological Resources
Tsukuba, Ibaraki, Japan

Y. Magoshi
National Institute of Sericultural and Entomological Science Oowashi
Tsukuba, Ibaraki, Japan

C. Maier
General Electric Plastics Technology Center
Washington, West Virginia

Gerhard Maier
Technische Universität München
Lichtenbergstr, Garching, Germany

S. N. Maiti
Center for Polymer Science and Engineering
Indian Institute of Technology
New Delhi, India

Sukumar N. Maiti
Materials Science Center
Indian Institute of Technology
Kharagpur, India

Mir Reza Majidi
Intelligent Polymer Research Laboratory
Department of Chemistry
University of Wollongong
Wollongong, New South Wales, Australia

Syoichi Makimoto
Department of Chemistry
Ritsumeikan University
Kusatsu, Japan

Tatjana Malavašic
EPF, Institute of Technology
Maribor National Institute of Chemistry
University of Maribor
Ljubljana, Slovenia

Debesh Maldas
Department of Wood Science and Technology
Faculty of Agriculture
Kyoto University
Kyoto, Japan

E. Malmström
Department of Polymer Technology
Royal Institute of Technology
Stockholm, Sweden

Sei-Ichi Manabe
Faculty of Human Environmental Science
Fukuoka Women's University
Fukuoka, Japan

B. M. Mandal
Polymer Science Unit
Indian Association for the Cultivation of Science
Jadavpur, Calcutta, India

O. Manero
Instituto de Investigaciones en Materiales
Mexico D.F., Mexico

P. A. Manji
The BF Goodrich Company
Brecksville, Ohio

Ian Manners
Department of Chemistry
University of Toronto
Toronto, Ontario, Canada

A. Mansour
Microanalytical Center
Faculty of Science
Cairo University
Cairo, Egypt

Olfat Y. Mansour
Cellulose and Paper Department
National Research Center
Dokki, Cairo, Egypt

J. Mansouri
Department of Polymer Science
University of New South Wales
Sydney, New South Wales, Australia

Guangzhao Mao
Department of Chemical Engineering and Materials Science
University of Minnesota
Minneapolis, Minnesota

Gary R. Marchand
The Dow Chemical Company
Plaquemine, Louisiana

M. Marcos
Facultad de Química
Pontificia Universidad Católica de Chile
Santiago, Chile

Daniela Mardare
Department of Chemistry
Carnegie Mellon University
Pittsburgh, Pennsylvania

Yitbarek H. Mariam
Department of Chemistry and Center for Theoretical Studies of Physical Systems
Clark Atlanta University
Atlanta, Georgia

James E. Mark
Department of Chemistry
Polymer Research Center
University of Cincinnati
Cincinnati, Ohio

John F. Marko
Center for Studies in Physics and Biology
The Rockefeller University
New York, New York

Maurice J. Marks
Texas Polymer Center
Freeport, Texas

C. E. Marsden
Crosfield Ltd.
Warrington, United Kingdom

Charles W. Martin
Department of Chemistry
Clemson, South Carolina

Oscar E. Martinez
Syntex S. A.
Buenos Aires, Argentina

Sutiyao Marturunkakul
Department of Chemistry
University of Massachusetts, Lowell
Lowell, Massachusetts

K. A. Marx
Center for Advanced Materials
Department of Chemistry and Physics
University of Massachusetts, Lowell
Lowell, Massachusetts

Junzo Masamoto
Polymer Development Laboratory
Asahi Chemical Industry Company, Ltd.
Kurashiki, Japan

Seizo Masuda
Department of Chemical Science and Technology
Faculty of Engineering
University of Tokushim
Tokushima, Japan

Toshio Masuda
Department of Polymer Chemistry
Kyoto University
Kyoto, Japan

D. Mathew
Vikram Sarabhai Space Center
Trivandrum, India

Janis G. Matisons
Polymer Science Group
School of Chemical Technology
University of South Australia
South Australia, Australia

Hiro Matsuda
National Institute of Materials and Chemical Research
Higashi, Tsukuba, Ibaraki, Japan

Kiyohide Matsui
Sagami Chemical Research Center
Kanagawa, Japan
Akihiro Matsumoto
Osaka Municipal Technical Research Institute
Osaka, Japan
Akira Matsumoto
Department of Applied Chemistry
Faculty of Engineering
Kansai University
Suita, Osaka, Japan
Atsushi Matsumoto
Department of Applied Chemistry and Biochemistry
Faculty of Engineering
Kumamoto University
Kumamoto, Japan
Hideo Matsumura
Department of Fixed Prosthodontics
Nagasaki University School of Dentistry
Nagasaki, Japan
Shigeru Matsuo
Central Research Laboratories
Idemitsu Kosan
Chiba, Japan
Luiz H. C. Mattoso
Instituto de Física de São Carlos
São Carlos, Brazil
Krzysztof Matyjaszewski
Department of Chemistry
Carnegie Mellon University
Pittsburgh, Pennsylvania
Kenneth A. Mauritz
Department of Polymer Science
University of Southern Mississippi
Hattiesburg, Mississippi
Ian A. Maxwell
Memtec Ltd.
South Windsor, New South Wales, Australia
Leon Maya
Chemical and Analytical Sciences Division
Oak Ridge National Laboratory
Oak Ridge, Tennessee
Jimmy W. Mays
Department of Chemistry
University of Alabama at Birmingham
Birmingham, Alabama
Francis B. McAndrew
Hoechst Celanese Corporation
Summit, New Jersey
Douglas S. McBain
GenCorp Specialty Polymers Division
Akron, Ohio
Joseph J. McBride Jr.
Panama City, Florida
Mark D. McClain
Department of Chemistry
Macromolecular Science and Engineering Center
The University of Michigan
Ann Arbor, Michigan
Charles L. McCormick
Department of Polymer Science
University of Southern Mississippi
Hattiesburg, Mississippi
Richard L. McCreery
Department of Chemistry
The Ohio State University
Columbus, Ohio
P. T. McGrail
Imperial Chemical Industries
Wilton Materials Research Centre
Middlesbrough, Cleveland, United Kingdom
Frank M. McMillan
Calsec Consultants
Orinda, California
Marion E McNeill
Department of Pure and Applied Chemistry
University of Strathclyde
Glasgow, Scotland, United Kingdom
D. T. McPherson
Laboratory of Molecular Biophysics
School of Medicine
University of Alabama at Birmingham
Birmingham, Alabama
F. J. Medellin-Rodriguez
CIEP-Fac. de C. Quimicas
UASLP, Mexico
San Luis Potosi, Mexico
P. Mehta
Polaroid Corporation
Cambridge, Massachusetts
John J. Meister
Department of Chemistry
University of Detroit Mercy
Detroit, Michigan
Michele Melchiorre
Daimler-Benz AG Research Center
Ulm, Germany
Anastasios P. Melissaris
Materials Engineering Division
Research Institute
University of Dayton
Dayton, Ohio
Charlene M. Mello
Biotechnology Division
U.S. Army Natick Research Development and Engineering Center
Natick, Massachusetts
Wesley Memeger, Jr.
DuPont Central Research and Development
Experimental Station
DuPont Company
Wilmington, Delaware
Eduardo Mendizábal
Departamento de Ingeniería Química
Universidad de Guadalajara
Guadalajara, Mexico
Jose Alexandre de Mendonça
Materials Engineering
Federal University of São Carlos
São Carlos, Sao Paulo, Brazil
Henning Menzel
Institut für Makromolekulare Chemie
Universität Hannover
Hannover, Germany
Craig C. Meverden
Quantum Chemical Company
Cincinnati, Ohio
Khaled Mezghani
Department of Materials Science and Engineering
University of Tennessee
Knoxville, Tennessee
Georg H. Michler
Martin-Luther-Universität Halle-Wittenberg
Institute of Materials Science
Merseburg, Germany
Sanjeev Midha
The Procter and Gamble Company
Cincinnati, Ohio
Michel Milas
Centre de Recherches sur les Macromolácules Végétales (CERMAV/CNRS)
Grenoble, France
K. F. Miller
General Electric Plastics
Technology Center
Washington, West Virginia
Peter Miller
Department of Chemistry
Carnegie Mellon University
Pittsburgh, Pennsylvania
Frank Millich
Department of Chemistry
University of Missouri—Kansas City
Kansas City, Missouri
Noriaki Minamii
Wako Pure Chemical Industries
Osaka, Japan
James L. Minor
USDA Forest Service
Madison, Wisconsin
Norihiko Minoura
National Institute of Materials and Chemical Research
Tsukuba, Japan
A. Mirmohseni
Intelligent Polymer Research Laboratory
Department of Chemistry
University of Wollongong
Wollongong, New South Wales, Australia
Munmaya K. Mishra
Department of Chemistry
Polymer Research
Texaco Research and Development
Beacon, New York
V. Mishra
Department of Chemical Engineering
Center for Polymer Science Materials Research Center
Lehigh University
Bethlehem, Pennsylvania
Michael A. Mitchell
Los Alamos National Laboratory
Los Alamos, New Mexico
Motonori Mitoh
Osaka Prefectural Industrial Technology Research Institute
Higashiosaka, Osaka, Japan
Fukuda Mitsuhiro
Textile Materials Science Laboratory
Hyogo University of Teacher Education
Yashiro-cho, Hyogo, Japan
Kazuta Mitsuishi
Industrial Technology Center of Okayama Prefecture
Okayama, Japan
Yozo Miura
Department of Applied Chemistry
Faculty of Engineering
Osaka City University
Sumiyoshiku, Osaka, Japan
H. Miyagi
Petrochemicals Research Laboratory
Sumitomo Chemical Company, Ltd.
Chiba, Japan
Yasuhiro Miyake
Department of Polymer Science
Faculty of Science
Hokkaido University
Sapporo, Japan
Masatoshi Miyamoto
Department of Polymer Science and Engineering
Kyoto Institute of Technology
Matsugasaki, Sakyo-ku, Kyoto, Japan

Tokuji Miyashita
Department of Molecular Chemistry and Engineering
Tohoku University
Sendai, Japan

Mikiji Miyata
Department of Chemistry
Faculty of Engineering
Gifu University
Gifu, Japan

Tereo Miyata
Koken Bioscience Institute
Tokyo, Japan

Tatsuya Miyatake
ERATO Molecular Catalysis Project
Research Development Corporation of Japan
Toyota, Japan

Hirofumi Miyoshi
Research Institute for Advanced Science and Technology
University of Osaka Prefecture
Sakai, Osaka, Japan

K. Mizoguchi
National Institute of Materials and Chemical Research
Ibaraki, Japan

Zhishen Mo
Laboratory of Polymer Physics
Changchun Institute of Applied Chemistry Chinese Academy of Sciences
Changchun, China

Graeme Moad
CSIRO Division of Chemicals and Polymers
Clayton, Victoria, Australia

Maryam Moaddeb
Department of Chemical Engineering
The University of Texas at Austin
Austin, Texas

P. C. Modnshirk
Department of Industrial Chemistry
NTH University of Trondheim
Trondheim, Norway

Giovanni Moggi
Istituto di Chimica Industriale
Universitè Genova
Ausimont CRS, Bollate
IMAG CNR, Genova
Genova, Italy

V. M. Möhring
Max-Planck-Institut fur Kohlenforschung
Mulheim an der Ruhr, Germany

Andrzej Molek
Industrial Chemistry Research Institute
Warsaw University of Technology
Warsaw, Poland

Edward P. Moore, Jr.
Montell U.S.A., Incorporated
Wilmington, Delaware

Robert B. Moore
Department of Polymer Science
University of Southern Mississippi
Hattiesburg, Mississippi

Ahmad Moradi-Araghi
Phillips Petroleum Company
Phillips Research Center
Bartlesville, Oklahoma

Elena Morales
Universidad Nacional de Educación a Distancia
Madrid, Spain

Simona Morariu
"P. Poni" Institute of Macromolecular Chemistry
Jassy, Romania

Roberto Luiz Moreira
Departamento de Física
University Federal de Minas Gerais
Belo Horizonte MG, Brazil

Atsunori Mori
Graduate School of Materials Science
Japan Advanced Institute of Science and Technology
Tatsunokuchi, Ishikawa, Japan

Kunio Mori
Department of Applied Chemical and Molecular Science
Iwate University
Ueda, Morioka, Japan

Atsushi Morikawa
Department of Materials Science
Faculty of Engineering
Ibaraki University
Hitachi, Ibaraki, Japan

Mitsuhiro Morita
Department of Forest Products
Faculty of Agriculture
Kyushu University
Fukuoka, Japan

Preben C. Mork
Department of Industrial Chemistry, NTH
University of Trondheim
Trondheim, Norway

V. J. Morris
Institute of Food Research
Norwich Laboratory
Colney, Norfolk, United Kingdom

P. H. Mühlenbrock
Max-Planck-Institut fur Kohlenforschung
Mulheim an der Ruhr, Germany

Amit Mukherjee
Berger Paints India Ltd.
Howrah, India

Marcel H. V. Mulder
University of Twente
Enschede, The Netherlands

Rolf Mülhaupt
Freiburger Materialforschungszentrum und Institut für Makromolekulare Chemie der Albert-Ludwigs Universität
Freiburg, Germany

Klaus Müllen
Max-Planck-Institute for Polymer Research
Mainz, Germany

Helmut W. J. Müller
EDM/KH
BASF Aktiengesellschaft
Ludwigshafen, Germany

Petr Munk
Department of Chemistry and Biochemistry and Center for Polymer Research
University of Texas at Austin
Austin, Texas

Sebastián Muñoz-Guerra
Department d'Enginyeria Química
Escola Tècnica
Superior d'Enginuers Industrials de Barcelona
Universitat Politècnica de Catalunya
Barcelona, Spain

Masahide Murata
Tonen Corporate Research and Development Laboratory
Tonen Chemical Corporation
Saitama, Japan

Yukinobu Murata
Osaka Prefectural College of Technology
Osaka-fu, Japan

Yoshio Muroga
Department of Applied Chemistry
Nagoya University
Furo-cho, Chikusa-ku, Nagoya, Japan

K. Shanmugananda Murthy
Inorganic and Physical Chemistry
Indian Institute of Science
IDL-Nitro Nobel Basic Research Institute
Bangalore, India

A. Muscat
Institute of Technology
FAL
Braunschweig, Germany

R. Musch
Bayer AG
Leverkusen, Germany

Mamed I. Mustafaev
Department of Chemistry
Baku State University
Baku, Azerbaijan

Riccardo A. A. Muzzarelli
Faculty of Medicine
University of Ancona
Ancona, Italy

V. A. Myagchenkov
Kazan State Technological University
Kazan, Russia

Ginger G. Myers
Akzo Nobel Chemicals Incorporated
Dobbs Ferry, New York

Terry N. Myers
Buffalo Research and Development Center
Elf Atochem North America Incorporated
Buffalo, New York

Herbert Naarmann
BASF Plastics Research Laboratory
Ludwigshafen, Germany

Akira Nagai
Hitachi Research Laboratory
Hitachi Ltd.
Ibaraki, Japan

Shoji Nagaoka
Department of Applied Chemistry
Faculty of Engineering
Kumamoto University
Kumamoto, Japan

Yukio Nagasaki
Department of Materials Science and Technology
Science University of Tokyo
Noda, Japan

Keisaku Nagasawa
Asahi Chemical Industry Company, Ltd.
Nobeoka City, Miyazaki, Japan

Yu Nagase
Sagami Chemical Research Center
Kanagawa, Japan

Minoru Nagata
Kyoto Prefectural University
Sakyoku, Kyoto, Japan

Z. A. Nagieb
Cellulose and Paper Department
National Research Center
Dokki, Cairo, Egypt

C. P. Reghunadhan Nair
Polymers and Special Chemicals Division
Vikram Sarabhai Space Center
Tivandrum, India

Chitoshi Nakafuku
Faculty of Education
Kochi University
Kochi-shi, Japan

Hiroko Nakagawa
Faculty of Pharmaceutical Sciences
Josai University
Sakao, Saitama, Japan

Osamu Nakagawa
Department of Chemistry
Faculty of Engineering Science
Osaka University
Toyonaka, Osaka, Japan

Seiichi Nakahama
Department of Polymer Chemistry
Tokyo Institute of Technology
Tokyo, Japan

Toshinari Nakajima
School of Human Life and Environmental Science
Ochanomizu University
Tokyo, Japan

Hiromu Nakamichi
University of Shizuoka
Shizuoka, Japan

Yoshiaki Nakamoto
Department of Chemistry and Chemical Engineering
Faculty of Technology
Kanazawa University
Kanazawa, Japan

Shigeo Nakamura
Department of Applied Chemistry
Faculty of Engineering
Kanagawa University
Yokohama, Japan

Naofumi Nakamura
Tezukayama College
Gakuen-Minami Nara, Japan

Tatsuo Nakamura
Department of Artificial Organs
Research Center for Biomedical Engineering
Kyoto University
Kyoto, Japan

Yoshiharu Nakamura
Faculty of Engineering
Fukui University
Fukui, Japan

Hachiro Nakanishi
Institute for Chemical Reaction Science
Tohoku University
Sendai, Japan

Tamaki Nakano
Department of Applied Chemistry
School of Engineering
Nagoya University
Nagoya, Japan

Yasushi Nakao
Technical Research Laboratory
Kansai Paint Company, Ltd.
Kanagawa, Japan

Yoshiro Nakata
Department of Biophysics
Faculty of Engineering
Gunma University
Maebashi, Japan

O. Nalamasu
Lucent Technologies
Bell Laboratories
Murray Hill, New Jersey

Hari Singh Nalwa
Hitachi Research Laboratory
Hitachi Ltd.
Hitachi City, Japan

Atsushi Nanasawa
Styrenic Resins Development Department
Asahi Chemical Industry Company Ltd.
Kawasaki, Japan

Masato Nanasawa
Department of Applied Chemistry and Biotechnology
Yamanashi University
Kofu, Japan

Pramod J. Nandapurkar
The Dow Chemical Company
Freeport, Texas

Tadashi Narita
Department of Environmental Engineering
Saitama Institute of Technology
Okabe, Saitama, Japan

L. V. Natarajan
Science Applications International Corporation
Dayton, Ohio

P. L. Nayak
Laboratory of Polymers and Fibers
Department of Chemistry
Ravenshaw College
Orissa, India

Thomas X. Neenan
AT&T Bell Laboratories
Murray Hill, New Jersey

Charles J. Neff
Department of Chemistry and Biochemistry and Laboratory for Electronic Properties of Materials
The University of Oklahoma
Norman, Oklahoma

Hartmut Nefzger
Makromolekulare Chemie II
Bayreuther Institut für Makromolekülforschung
Bayreuth, Germany

Nobukatsu Nemoto
Sagami Chemical Research Center
Sagamihara, Kanagawa, Japan

K. G. Neoh
Department of Chemical Engineering
National University of Singapore
Kent Ridge, Singapore

H. Phuong Nguyen
Chemistry Department
Unitversity of Quebec at Montreal
Montreal, Quebec, Canada

Mark E. Nichols
Ford Research Laboratory
Dearborn, Michigan

J. B. Nickaf
School of Chemical Engineering and Industrial Chemistry
University of New South Wales
Sydney, New South Wales, Australia

Kenneth A. Nielsen
Union Carbide
South Charleston, West Virginia

David E. Nikles
Department of Chemistry and Center for Materials for Information Technology
The University of Alabama
Tuscaloosa, Alabama

K. N. Ninan
Vikram Sarabhai Space Center
Trivandrum, India

K. A. Niño
Departamento de Microbiologia e Inmunologia
Facultad de Ciencias Biologicas
Universidad Autónoma de Nuevo León
San Nicolás de los Garza, N. L. Mexico

Takehiro Nishikawa
Department of Synthetic Chemistry and Biological Chemistry
Graduate School of Engineering
Kyoto University
Kyoto, Japan

Tadatomi Nishikubo
Department of Applied Chemistry
Faculty of Engineering
Kanagawa University
Yokohama, Japan

Nivedita
Polymer Research Laboratory
Department of Post Graduate Studies and Research in Chemistry
Rani Durgavati University
Jabalpur, India

Marcos Antônio Nobre
Instituto de Quimica
Universidade Estadual de Campinas
Campinas, Sao Paulo, Brazil

Nadia Chaves P. S. Nociti
Materials Engineering
Federal University of São Carlos
São Carlos, São Paulo, Brazil

Ichiro Noda
Department of Applied Chemistry
Nagoya University
Nagoya, Japan

Hiromichi Noguchi
Department of Chemistry and Biotechnology
Faculty of Engineering
The University of Tokyo
Tokyo, Japan

Annegret Noll
Instituto de Quimica
Universidade Estadual de Campinas
Campinas, Sao Paulo, Brazil

Roeland J. M. Nolte
Department of Organic Chemistry
University of Nijmegen, Toernooiveld
Nijmegen, The Netherlands

Ryoji Nomura
Research Laboratory of Resources Utilization
Tokyo Institute of Technology
Yokohama, Japan

Takamasa Nonaka
Department of Applied Chemistry
Faculty of Engineering
Kumamoto University
Kumamoto-shi, Japan

William K. Nonidez
Department of Chemistry
University of Alabama at Birmingham
Birmingham, Alabama

Jaan Noolandi
Xerox Research Center of Canada
Mississauga, Ontario, Canada

Eckhard Nordmeier
Bad Essen/Germany

Takashi Norisuye
Department of Macromolecular Science
Osaka University
Toyonaka, Osaka, Japan

Ryoji Noyori
ERATO Molecular Catalysis Project
Research Development Corporation of Japan
Yakusa-cho, Toyota, Japan

Yasuo Nozawa
School of Pharmaceutical Sciences
University of Shizuoka
Shizuoka, Japan

Oskar Nuyken
Lehrstuhl füautr Makromolekulare Stoffe
Technische Universität München
Garching, Germany

Susan Adams Nye
GE Silicones
Waterford, New York

J. M. O'Reilly
Eastman Kodak Company
Rochester, New York

Maria Obloj-Muzaj
Industrial Chemistry Research Institute
Warsaw University of Technology
Warsaw, Poland

E. Occhiello
EniChem S.p.A.—Istituto Guido Donegani
Novara, Italy

Hiroshi Ochiai
Department of Chemistry
Faculty of Science
Hiroshima University
Higashi-Hirashima, Japan

Takeshi Ogawa
Instituto de Investigaciones en Materiales
Universidad Nacional Autónoma de México
Mexico D.F., Mexico

Craig A. Ogle
Department of Chemistry
University of North Carolina at Charlotte
Charlotte, North Carolina

Yoshiro Ogoma
Department of Functional Polymer Science
Faculty of Textile Science and Technology
Shinshu University
Ueda, Japan

Y. Ohama
Chalmers University of Technology
Sintef Nihon University
Koriyama, Japan

Shinichi Ohashi
Protein Engineering Laboratory
Molecular Biology Department
National Institute of Bioscience and Human Technology
Tsukuba, Ibaraki, Japan

Fujio Ohishi
Faculty of Science
Kanagawa University
Konagawa, Japan

Yasukazu Ohkatsu
Kogakuin University
Tokyo, Japan

Makoto Ohkoshi
Forestry and Forest Products Research Institute
Ibaraki, Japan

Yutaka Ohkoshi
Faculty of Textile Science and Engineering
Shinshu University
Ueda, Japan

Joji Ohshita
Department of Applied Chemistry
Faculty of Engineering
Hiroshima University
Higashi-Hiroshima, Japan

Takayuki Ohta
Research Center
Mitsubishi Kasei Corporation
Kamoshida, Yokohama, Japan

Toshihiko Ohta
Faculty of Human Life Science
Osaka City University
Osaka, Japan

Hajime Ohtani
Department of Applied Chemistry
School of Engineering
Nagoya University
Nagoya, Japan

Noritaka Ohtani
Department of Materials Engineering and Applied Chemistry
Akita University
Akita, Japan

Takashi Ohtsubo
Matsuyama Shinonome Junior College
Ehime, Japan

Eizo Oikawa
Department of Chemistry and Chemical Engineering
Faculty of Engineering
Niigata University
Niigata, Japan

Yoshiyuki Oishi
Department of Applied Chemistry and Molecular Science
Iwate University
Morioka, Iwate, Japan

Yukihiro Oka
Department of Chemistry for Materials
Mie University
Tsu, Mie-Ken, Japan

Masahiko Okada
Department of Applied Biological Sciences
School of Agricultural Sciences
Nagoya University
Chikusa, Nagoya, Japan

Shuji Okada
Institute for Chemical Reaction Science
Tohoku University
Sendai, Japan

Kunihiko Okajima
Fundamental Research Laboratory of Natural and Synthetic Polymers
Asahi Chemical Industry Company, Ltd.
Takatsuki, Osaka, Japan

Masami Okamoto
Toyobo Research Center
Katata, Ohtsu, Shiga, Japan

Miyoshi Okamoto
Toray Industries, Incorporated
Sonoyama, Japan

Yasushi Okamoto
Petrochemicals Research Laboratory
Sumitomo Chemical Company, Ltd.
Sodegaura-shi, Chiba, Japan

Yoshio Okamoto
Department of Applied Chemistry
School of Engineering
Nagoya University
Nagoya, Japan

Yoshiyuki Okamoto
Department of Chemistry
Herman F. Mark Polymer Research Institute
Polytechnic University
Brooklyn, New York

Keizo Okamura
Faculty of Agriculture
Kyoto University
Kyoto, Japan

Shigeru Okita
Plastics Research Laboratories
Toray Industries, Incorporated
Nagoya, Japan

Jun-Ichi Oku
Department of Applied Chemistry
Nagoya Institute of Technology
Nagoya, Japan

Masayoshi Okubo
Chemical Science and Engineering
Faculty of Engineering
Kobe University Rokko, Nada
Kobe, Japan

Tsuneo Okubo
Department of Polymer Chemistry
Faculty of Engineering
Kyoto University
Kyoto, Japan

Tomoko Okuyama
Toyobo Research Center
Katata, Ohtsu, Shiga, Japan

O. N. Oliveira Jr.
Instituto de Física de São Carlos
São Carlos, Brazil

Shinzo Omi
Graduate School of Bio-Applications and Systems Engineering
Tokyo University of Agriculture and Technology
Koganei, Tokyo, Japan

Hideki Omichi
Department of Material Development
Takasaki Radiation Chemistry Research Establishment
Japan Atomic Energy Research Institute
Takasaki, Gunma, Japan

Ahmet M. Önal
Department of Science Education
Middle East Technical University
Ankara, Turkey

Bruce E. Orler
Department of Polymer Science
University of Southern Mississippi
Hattiesburg, Mississippi

T. Orta
Instituto de Ingeniería
UNAM
Mexico D.F., Mexico

Tetsuo Osa
Pharmaceutical Institute
Tohoku University
Sendai, Japan

Yoshihito Osada
Division of Biological Sciences
Graduate School of Science
Hokkaido University
Sapporo, Japan

Kohtaro Osakada
Research Laboratory of Resources Utilization
Tokyo Institute of Technology
Midori-ku, Yokohama, Japan

Gunilla Östberg
Institute for Surface Chemistry
Stockholm, Sweden

Raphael M. Ottenbrite
High Technology Materials Center
Department of Chemistry
Virginia Commonwealth University
Richmond, Virginia

Alexander A. Ovchinnikov
Institute of Chemical Physics
Russian Academy of Sciences
Moscow, Russia

Michael J. Owen
Dow Corning Corporation
Midland, Michigan

Yukio Ozeki
EVAL Company of America
Lisle, Illinois

Samuel D. Pace II
Department of Polymer Science
The University of Southern Mississippi
Hattiesburg, Mississippi

Jerzy Paczkowski
Department of Chemistry and Chemical Engineering
Technical and Agricultural University
Bydgoszcz, Poland

Anne Buyle Padias
Chemistry Department
The University of Arizona
Tucson, Arizona

W. R. Palani Raj
Department of Chemical Engineering
The University of Akron
Akron, Ohio

Sanjay Palsule
Department of Chemistry
Heriot-Watt University
Bhopal, India

R. Pande
Center for Advanced Materials
Department of Chemistry and Physics
University of Massachusetts, Lowell
Lowell, Massachusetts

Keith H. Pannell
Department of Chemistry
University of Texas at El Paso
El Paso, Texas

Constantine D. Papaspyrides
Laboratory of Polymer Technology (Special Chemical Technology)
Department of Chemical Engineering
National Technical University of Athens
Athens, Greece

Peter G. Pape
Dow Corning Corporation
Midland, Michigan

Ivan M. Papisov
Moscow State Automobile and Road Technical University
Moscow, Russia

Dong Ki Park
Polymer Chemistry Laboratory
Korea Institute of Science and Technology
Seouk, Korea

Haesun Park
Purdue University
School of Pharmacy
West Lafayette, Indiana

Ki Hong Park
Department of Organic and Polymeric Materials
Tokyo Institute of Technology
Meguro-ku, Tokyo, Japan

Kinam Park
School of Pharmacy
Purdue University
West Lafayette, Indiana

Tae Suk Park
Polymer Processing Laboratory
Korea Institute of Science and Technology
Cheongryang, Seoul, Korea

Dane K. Parker
The Goodyear Tire and Rubber Company
Akron, Ohio

Hsing-Yeh Parker
Rohm and Haas Company
Bristol, Pennsylvania

T. M. Parker
Bioelastics Research, Ltd.
Birmingham, Alabama

Fabrizio Parodi
Saiag S.p.A.
Novara, Italy

David A. D. Parry
Department of Physics
Massey University
Palmerston North, New Zealand

G. Di Pasquale
Instituto Chimico
Facoltá di Ingogneria
Universitá di Catania
Catania, Italy

A. V. Patel
Institute of Technology
FAL
Braunschweig, Germany

Abhimanyu O. Patil
Corporate Research Laboratory
Exxon Research and Engineering Company
Annandale, New Jersey

D. Paul
GKSS Research Center
Geesthacht, Germany

Attila E. Pavalath
Western Regional Research Center
Agricultural Research Service
U.S. Department of Agriculture
Albany, California

Eli M. Pearce
Department of Chemistry and The Herman F. Mark Polymer Research Institute
Polytechnic University
Brooklyn, New York

P. J. Pearce
Aeronautical and Maritime Research Laboratory
Melbourne, Victoria, Australia

Leighton H. Pebbles, Jr.
Chemistry Division
Naval Research Laboratory
Washington, D. C.

K.-V. Peinemann
GKSS Research Center
Geesthacht, Germany

Piotr Penczek
Industrial Chemistry Research Institute
Warsaw, Poland

Stanislaw Penczek
Center of Molecular and Macromolecular Studies
Polish Academy of Sciences
Lodz, Sienkiewicza, Poland

Jacques Penelle
Laboratoire Cinétique et Macromolécules
Université Catholique de Louvain
Louvain-la-Neuve, Belgium

María Judith Percino
Centro de Química
Universidad Autónoma de Puebla
Puebla, México

Ernesto Pérez
Instituto de Ciencia y Tecnología de Polímeros (CSIC)
Madrid, Spain

M. C. Perry
University of Utah
Salt Lake City, Utah

Robert J. Perry
Imaging Research and Advanced Development
Eastman Kodak Company
Rochester, New York

Ted M. Pettijohn
Witco Corporation
Marshall, Texas

R. L. Pettit
Technology Center
General Electric Plastics
Washington, West Virginia

H. T. Pham
The Dow Chemical Company
Freeport, Texas

Burkart Philipp
Max-Planck-Institute of Colloid and Interface Research Teltow-Seehof
Teltow-Seehof, Germany

Paul J. Phillips
Department of Materials Science and Engineering
University of Tennessee
Knoxville, Tennessee

Pranee Phinyocheep
Department of Chemistry
Faculty of Science
Mahidol University
Bangkok, Thailand

Neelam Phougat
Center for Rural Development and Technology
Indian Institute of Technology
Delhi, India

Joseph W. Pialet
The Lubrizol Corporation
Wickliffe, Ohio

Maurizio Pianca
Istituto di Chimica Industriale
Universitè Genova
Ausimont CRS, Bollate
IMAG CNR
Genova, Italy

A. P. Piedade
Department of Colour Chemistry
University of Leeds
Leeds, United Kingdom

Anthony P. Pierlot
CSIRO Division of Wool Technology
Belmont, Victoria, Australia

Inés Fernandez de Piérola
Departamento de Química Física
Universidad Nacional de Educación a Distancia (UNED)
Madrid, Spain

Osvaldo Pieroni
CNR - Institute of Biophysics and Department of Chemistry and Industrial Chemistry
University of Pisa
Pisa, Italy

Irja Piirma
The University of Akron
Akron, Ohio

Michel Pineri
CEA-Département de Recherche Fondamentale sur la Matière Condensée
SESAM/Laboratoire de Physico-Chimie Moléculaire
Grenoble, France

R. K. Pinschmidt, Jr.
Air Products and Chemicals, Incorporated
Allentown, Pennsylvania

Mariana Pinteală
"P. Poni" Institute of Macromolecular Chemistry
Department of Macromolecules
"Gh. Asachi" Technical University
Jassy, Romania

Maria Pizzoli
Dipartimento di Chimica "G. Ciamician"della Universita di Bologna
Centro di Studio per la Fisica delle Macromolecole del C.N.R.
Bologna, Italy

Heinrich Planck
Institut für Textile- und Verfahrenstechnik (ITVP)
Denkendorf Forschungsbereich Blomedizintechnik
Denkendorf, Germany

Riccardo Po
Enichem S.p.A.
Istituto Guido Donegani
Novara, Italy

Nicolas L. Pocard
Department of Chemistry
The Ohio State University
Columbus, Ohio

Gary W. Poehlein
Georgia Institute of Technology
Atlanta, Georgia

Ranier Polley
Institut für Organische Chemie
University of Tübingen
Tübingen, Germany

A. Pollicino
Istituto Chimico
Facoltà di Ingegneria
Università di Catania
Catania, Italy

Stefan Polowinski
Department of Physical Chemistry of Polymers
Technical University of Lódź
Lodz, Poland

Anatoly D. Pomogailo
Institute of Chemical Physics
USSR Academy of Sciences
Moscow, Russia

S. Amalia Pooley
Departamento de Polìmeros
Facultad de Ciencias Quìmicas
Universidad de Concepciòn
Concepciòn, Chile

Galina Popova
Mendeleyev University of Chemical Technology of Russia
Moscow, Russia

Roger S. Porter
Polymer Science and Engineering Department
University of Massachusetts
Amherst, Massachusetts

Roman Poturalski
Nitrogen (Petrochemical) Works "Wloclawek"
Wloclawek, Poland

John G. Poulakis
Laboratory of Polymer Technology (Special Chemical Technology)
Department of Chemical Engineering
National Technical University of Athens
Zographou, Athens, Greece

Paras N. Prasad
Photonics Research Laboratory
Department of Chemistry
The State University of New York at Buffalo
Buffalo, New York

L. M. Pratt
Fiber and Polymer Science Program
Department of Textiles and Apparel
Cornell University
Ithaca, New York

Jon A. Preece
School of Chemistry
University of Birmingham
Edgbaston, Birmingham, United Kingdom

Jack Preston
Department of Polymer Science
Camille Dreyfus Laboratory
Research Triangle Institute
Research Triangle Park, North Carolina

Dusan C. Prevorsek
AlliedSignal Senior Science Fellow Emeritus
Hood River, Oregon

Gareth J. Price
School of Chemistry
University of Bath
Bath, United Kingdom

W. E. Price
Intelligent Polymer Research Laboratory
Department of Chemistry
University of Wollongong
Wollongong, Australia

Duane B. Priddy
Dow Plastics
The Dow Chemical Company
Midland, Michigan

Karel Procházka
Department of Physical and Macromolecular Chemistry
Faculty of Science
Charles University in Prague
Prague, Czech Republic

Margarita Gonzalez Prolongo
Departamento de Materiales y Producción Aeroespacial
ETSI Aeronauticos
Universidad Politécnica
Madrid, Spain

Arthur Provatas
Polymer Science Group
School of Chemical Technology
University of South Australia
South Australia, Australia

John K. Pudelski
Department of Chemistry
University of Toronto
Toronto, Ontario, Canada

Rudolf Puffr
Institute of Macromolecular Chemistry
Academy of Sciences of the Czech Republic
Prague, Czech Republic

Jorge E. Puig
Departamento de Ingeniería Química
Universidad de Guadalajara
Guadalajara, Mexico

Jordi Puiggalí
Department d'Enginyeria Química
Escola Tècnica
Superior d'Enginuers Industrials de Barcelona
Universitat Politècnica de Catalunya
Barcelona, Spain

Béla Pukánszky
Department of Plastic and Rubber Technology
Technical University of Budapest and Central Research Institute for Chemistry
Hungarian Academy of Sciences
Budapest, Hungary

Judit E. Puskas
Bayer Rubber Incorporated
Sarnia, Ontario, Canada

Thanun M. Pyriadi
Department of Chemistry
College of Science
University of Baghdad
Baghdad, Iraq

Yu Qi
Ottawa-Carleton Chemistry Institute
Department of Chemistry
Carleton University
Ohowa, Ontario, Canada

Xue Qifeng
Department of Chemical Engineering
Tianjin Institute of Technology
Tianjin, China

Pan Jiang Qing
Institute of Chemistry
Academia Sinica
Beijing, China

Baojun Qu
Structure Research Laboratory
University of Science and Technology of China
Hefei, China

Heribert Quante
Max-Planck-Institute for Polymer Research
Mainz, Germany

Candace Jo Quinn
Science and Technology Group
Corning Incorporated
Corning, New York

Francis X. Quinn
Institut Français du Pétrole
Rueil-Malmaison, France

José R. Quintana
Universidad del País Vasco
Bilbao, Spain

Asfia Qureshi
Department of Chemistry
University of Melbourne
Parkville, Victoria, Australia

Franco M. Rabagliati
Departamento de Ciencias Químicas
Facultad de Químicay Biología
Universidad de Santiago de Chile
Santiago, Chile

Deodato Radic
Deptamento Química Física
Facultad de Química
Pontificia Universidad Católica de Chile
Santiago, Chile

Adrzej Rajca
Department of Chemistry
University of Nebraska
Lincoln, Nebraska

Suchada Rajca
Department of Chemistry
University of Nebraska
Lincoln, Nebraska

Slavtcho K. Rakovsky
Institute of Catalysis
Bulgarian Academy of Sciences
Sofia, Bulgaria

M. M. Ramiz
Department of Electronic Engineering
El-Menoufia University
Shebin, El-Kom, Egypt

Ruicheng Ran
Department of Chemistry
Mississippi State University
Mississippi State, Mississippi

Bengt Rånby
Department of Polymer Technology
Royal Institute of Technology
Stockholm, Sweden

Elisabetta Ranucci
Dipartimento di Chimica e Fisica per i Materiali
Università degli Studi di Brescia
Brescia, Italy

J. B. Rappaport
Du Pont Advanced Fibers Systems
Wilmington, Delaware

S. Sh. Rashidova
Institute of Polymer Chemistry and Physics
Toshkent, Uzbekistan

Paul G. Rasmussen
Departments of Chemistry and Macromolecular Science and Engineering
The University of Michigan
Ann Arbor, Michigan

Jatuporn Ratanapaka
Department of Chemistry
Faculty of Science
Mahidol University
Bangkok, Thailand

Buddy D. Ratner
Center for Bioengineering and Department of Chemical Engineering
University of Washington
Seattle, Washington

Iris U. Rau
Kunststofflaboratorium, BASF AG
Ludwigshafen, Germany

James W. Rawlins
Department of Polymer Science
The University of Southern Mississippi
Hattiesburg, Mississippi

Françisco M Raymo
School of Chemistry
University of Birmingham
Edgbaston, Birmingham, United Kingdom

A. Recca
Istituto Chimico
Facoltà di Ingegneria
Università di Catania
Catania, Italy

Ivo Reetz
Hahn-Meitner-Institut Berlin GmbH
Berlin-Wannsee, Germany

Miguel F. Refojo
The Schepens Eye Research Institute
Department of Ophthalmology
Harvard Medical School
Boston, Massachusetts

Matthias Rehahn
Polymer Institute
Karlsruhe University
Karlsruhe, Germany

E. Reichmanis
Lucent Technologies
AT&T Bell Laboratories
Murray Hill, New Jersey

Paul F. Rempp
Institut Charles Sadron (CNRS/ULP)
Strasbourg, France

Luigi Resconi
Montell Polyolefins
G. Natta Research Center
Ferrara, Italy

Paul R. Resnick
DuPont Fluoroproducts
DuPont Central Research and Development
Experimental Station
Wilmington, Delaware

Harry Reynaers
PVC Basic Research
Limburgse Vinyl Maatschappij N.V.
H Hartlaan
Tessenderlo, Belgium

Alexander Ribbe
Makromolekulare Chemie II
Bayreuther Institut für Makromolekülforschung
Bayreuth, Germany

Robert E. Richard
Johnson and Johnson Professional Incorporated
Raynham, Massachusetts

W. Frank Richey
Central Research and Development
Organic Products Research
The Dow Chemical Company
Freeport, Texas

Klaus-Peter Richter
Bayer AG
Leverkusen, Germany

Marguerite Rinaudo
Centre de Recherches sur les Macromolécules Végétales (CERMAV/CNRS)
Grenoble, France

Helmut Ringsdorf
Institut für Organische Chemie
Universität Mainz
Mainz, Germany

Helmut Ritter
Bergische Universität-GH Wuppertal
Organische Chemie und Makromolekulare
Wuppertal, Germany

Bernabe L. Rivas
Departamento de Polìmeros
Facultad de Ciencias Quìmicas
Universidad de Concepciòn
Concepciòn, Chile

Ezio Rizzardo
CSIRO Division of Chemicals and Polymers
Clayton, Victoria, Australia

Douglas R. Robello
Imaging Research and Advanced Development
Eastman Kodak Company
Rochester, New York

George H. Robertson
Western Regional Research Center
Agricultural Research Service
U.S. Department of Agriculture
Albany, California

Richard E. Robertson
Macromolecular Science and Engineering Center and Department of Materials Science and Engineering
The University of Michigan
Ann Arbor, Michigan

Lloyd M. Robeson
Air Products and Chemicals Incorporated
Allentown, Pennsylvania

Marisa C. G. Rocha
Instituto de Macromoléculeas
Universidade Federal do Rio de Janeiro
Rio de Janeiro, Brazil

Cyrille Rochas
Laboratoire de Spectrométrie Physique
Université Joseph Fourier
Saint Martin d'Héres, France

O. A. Rodríguez
Facultad de Quimica
UNAM
Mexico D.F., Mexico

A. Rodríguez-Galán
Departament d'Enginyeria Química
Universitat Politecnica de Catalunya, ETSEIB
Barcelona, Spain

Garcia Jorge Romero
Centro de Investigación en Quimica Aplicada
Coahuila, México

Nelson G. Rondan
Central Research and Development
The Dow Chemical Company
Midland, Michigan

Jacques Roovers
Institute for Environmental Research and Technology
National Research Council of Canada
Ottawa, Ontario, Canada

Jerald C. Rosenfeld
Occidental Chemical Corporation
Grand Island, New York

V. P. Roschupkin
Institute of Chemical Physics
Russian Academy of Sciences
Chernogolovka, Russia

James F. Ross
Polyolefins Research Center
Quantum Chemical Company
Morris, Illinois

Roger M. Rowell
Department of Forestry
Forest Products Laboratory
United States Department of Agriculture
Madison, Wisconsin

I. N. Ruban
Institute of Polymer Chemistry and Physics of Uzbek Academy of Sciences
Tashkent, Uzbekistan

Eli Ruckenstein
Department of Chemical Engineering
State University of New York at Buffalo
Buffalo, New York

S. B. Ruetsch
TRI/Princeton
Princeton, New Jersey

Ian M. Russell
CSIRO Division of Wool Technology
Belmont, Victoria, Australia

Maria Vittoria Russo
Department of Chemistry
University "La Sapienza"
Rome, Italy

Saverio Russo
Istituto di Chimica e Chimica Industriale
Università di Genova
Genova, Italy

Maria Carmela Sacchi
Istituto di Chimica delle Macromolecole CNR
Milan, Italy

Takeo Saegusa
Kansai Research Institute
Kyoto Research Park
Shimogyo-ku, Kyoto, Japan

Yasuo Saegusa
Department of Applied Chemistry
Faculty of Engineering
Kanagawa University
Kanagawa-ku, Yokohama, Japan

Robert Saf
Institute for Chemical Technology of Organic Materials
Technical University Graz
Graz, Austria

Agneza Safranj
Department of Material Development
Takasaki Radiation Chemistry
Research Establishment
Japan Atomic Energy Research Institute
Takasaki, Gunma, Japan

D. J. Sagl
Air Products and Chemicals, Incorporated
Allentown, Pennsylvania

Kazuhiko Saigo
Department of Chemistry and Biotechnology
Graduate School of Engineering
The University of Tokyo
Hongo, Bunkyo-ku, Tokyo, Japan

Enrique Saiz
Departamento de Química Física
Universidad de Alcalá de Henares
Alcalá de Heneres, Spain

Masato Sakaguchi
Ichimura Gakuen College
Uchikubo, Inuyama, Japan

Makoto Sakamoto
Industrial Research Institute of Ishikawa Prefecture
Kanazawa, Ishikawa, Japan

Maria Sakarellos-Daitsiotis
Section of Organic Chemistry and Biochemistry
University of Ioannina
Ioannina, Greece

Isao Sakat
Department of Forest Products
Faculty of Agriculture
Kyushu University
Fukuoka, Japan

Masayo Sakata
Department of Applied Chemistry
Faculty of Engineering
Kumamoto University
Kumamoto, Japan

Catalina Salom
Departamento de Materiales y Producción Aeroespacial
ETSI Aeronauticos
Universidad Politécnica
Madrid, Spain

Lynne A. Samuelson
Biotechnology Division
U.S. Army Natick RD&E Center
Natick, Massachusetts

C. Samyn
Laboratory of Macromolecular and Physical Organic Chemistry
Catholic University of Lueven
Heverlee (Leuven), Belgium

Anand R. Sanadi
Department of Forestry
University of Wisconsin
Forest Products Laboratory, USDA
Madison, Wisconsin

Jose Sanchez
Buffalo Research and Development Center
Elf Atochem North America Incorporated
Buffalo, New York

Fumio Sanda
Research Laboratory of Resources Utilization
Tokyo Institute of Technology
Yokohama, Japan

R. D. Sanderson
Institute for Polymer Science
University of Stellenbosch
Stellenbosch, South Africa

Daniel J. Sandman
Center for Advanced Materials
Department of Chemistry
University of Massachusetts, Lowell
Lowell, Massachusetts

Osamu Sangen
Faculty of Engineering
Himeji Institute of Technology
Shosha, Himeji, Japan

H. Sankarasubramanian
Dr. Bharat Ram Research and Development Centre
SRF Ltd.
Manali, Madras, India

Mashito Sano
π-Electron Materials Project - JRDC
Tsukuba, Ibaraki, Japan

A. Sezai Saraç
Department of Chemistry
Istanbul Technical University
Istanbul, Turkey

Dominique Sarazin
Institut Charles Sadron/ULP
Strasbourg, France

Hasaya Sato
Faculty of Technology
Tokyo University of Agriculture and Technology
Koganei, Tokyo, Japan

Hiroko Sato
Department of Polymer Chemistry
Kyoto University
Kyoto, Japan

Mitsuru Satoh
Department of Polymer Chemistry
Tokyo Institute of Technology
Tokyo, Japan

Toshifumi Satoh
Division of Molecular Chemistry
Graduate School of Engineering
Hokkaido University
Sapporo, Japan

Makoto Satou
Laboratory of Chemistry
Department of General Education
Osaka Institute of Technology
Osaka, Japan

J. A. Sauer
Department of Mechanics and Materials Science
Rutgers University
New Brunswick, New Jersey

Aurelio Savadori
EniChem
Milano, Italy

A. C. Savaoca
Air Products and Chemicals, Inc.
Allentown, Pennsylvania

Hideo Sawada
Department of Chemistry
Nara National College of Technology
Nara, Japan

Samuel P. Sawan
Department of Chemistry Polymer Science/Plastics Engineering Option
University of Massachusetts, Lowell
Lowell, Massachusetts

Mariastella Scandola
Dipartimento di Chimica "G. Ciamician"della Universita di Bologna
Centro di Studio per la Fisica delle Macromolecole del C.N.R.
Bologna, Italy

M. Schappacher
Laboratoire de Chimie des Polymères Organiques
Ecole Nationale Supérieure de Chimie et de Physique de Bordeaux
Talence, France

John Scheirs
Excel Plas Australia
Victoria, New South Wales, Australia

Rolf Scherrenberg
PVC Basic Research
Limburgse Vinyl Maatschappij N.V.
H Hartlaan
Tessenderlo, Belgium

Michael J. Scherrer
Morton International
Woodstock, Illinois

Michael Schimetta
Christian Doppler Laboratorium für Katalytische Polymerisation
Institut für Chemische Technologie
Organischer Stoffe
Graz, Austria

H. Schirmer
Freie Universität Berlin
Institut für Organische Chemie
Berlin, Germany

B. Schlicke
Freie Universität Berlin
Institut für Organische Chemie
Berlin, Germany

Arnulf Dieter Schlüter
Institut für Organische Chemie
Freie Universität Berlin
Berlin, Germany

Ruth Schmid
SINTEF Applied Chemistry
Trondheim, Norway

Claudia Schmidt
Institut für Makromolekulare Chemie der Universität Freiburg
Freiburg, Germany

Ludwig Schmitz
Polymer Institute
Karlsruhe University
Karlsruhe, Germany

Wolfram Schnabel
Hahn-Meitner-Institut Berlin GmbH
Berlin-Wannsee, Germany

F. Howard Schneider
DynaGen, Incorporated
Cambridge, Massachusetts

Hans Adam Schneider
Institut für Makromolekulare Chemie "Hermann-Staudinger-Haus" and Freiburgrer Materialforschungszentrum FMF der Universität
Freiburg, Germany

Gunter Schnurpfeil
Universität Bremen Institute für Organische und Makromolekulare Chemie
Bremen, Germany

Eduardo Schröder
University of Puerto Rico
Mayaguez, Puerto Rico

J. R. Schroeder
General Electric Plastics
Technology Center
Washington, West Virginia

R. H. Schubbe
Max-Planck-Institut für Kohlenforschung
Mülheim an der Ruhr, Germany

Ulrich S. Schubert
Lehrstuhl für Makromolekulare Chemie II and Bayreuther
Institut für Makromolekülforschung (BIMF)
Universität Bayreuth
Bayreuth, Germany

Ulf Schuchardt
Instituto de Quimica
Universidade Estadual de Campinas
Campinas, Sao Paulo, Brazil

Olivier J. A. Schueller
Department of Chemistry
The Ohio State University
Columbus, Ohio

Hans-Rolf Schulten
Department of Trace Analysis
Fachhochschule Fresenius
Wiesbaden, Germany

Burkhard Schulz
Research Group Thin Organic Films
Institute of Solid State Physics
University of Potsdam
Teltow, Germany

S. Schulz
Institute of Organic Chemistry
University of Hamburg
Hamburg, Germany

J. Schuster
Institut für Verbundwerkstoffe GmbH
Universität Kaiserslautern
Kaiserslautern, Germany

Kalyan Sehanobish
The Dow Chemical Company
Freeport, Texas

Takahiro Seki
Research Laboratory of Resources Utilization
Tokyo Institute of Technology
Nagatsuta, Midori-Ku, Yokohama, Japan

Yoshinori Seki
Petrochemicals Research Laboratory
Sumitomo Chemical Company Ltd.
Chiba, Japan

Hikaru Sekiguchi
Laboratoire de Chimie Macromoléculaire
Université Pierre et Marie Curie
Paris, France

Dennis G. Sekutowski
Engelhard Corporation
Iselin, New Jersey

Ibrahim Z. Selim
Department of Physical Chemistry
National Research Center
Cairo, Egypt

Sumanta K. Sen Gupta
Department of Chemistry
Ranchi University
Ranchi, India

Ernest Senogles
Department of Molecular Sciences
James Cook University
Townsville, Queensland, Australia

Yongsok Seo
Polymer Processing Laboratory
Korea Institute of Science and Technology
Cheongryang, Seoul, Korea

Scolastica Serroni
Dipartimento di Chimica Inorganica e Struttura Molecolare
Università di Messina
Messina, Italy

Kishore R. Shah
ConvaTec
Skillman, New Jersey

Waleed S. W. Shalaby
College of Medicine
Medical University of South Carolina
Charleston, South Carolina
and
Poly-Med Incorporated Center for Applied Technology
Pendleton, South Carolina

Robert A. Shanks
Applied Chemistry and Cooperative Research Center for Polymer Blends
Royal Melbourne Institute of Technology
Melbourne, Victoria, Australia

P. J. Shannon
Research Center
Hercules Incorporated
Wilmington, Delaware

Min-Da Shau
Department of Applied Chemistry
Chia Nan Junior College of Pharmacy
Tainan, Taiwan, Republic of China

M. T. Shaw
Department of Chemical Engineering and Institute of Materials Science
University of Connecticut
Storrs, Connecticut

Kenneth J. Shea
Department of Chemistry
University of California, Irvine
Irvine, California

Wenfang Shi
Department of Applied Chemistry
University of Science and Technology of China
Hefei, Anhui, Republic of China

Yoshio Shibasaki
Department of Chemistry
Saitama University
Urawa, Saiatama, Japan

Masaaki Shibata
Department of Applied Chemistry and Biochemistry
Faculty of Engineering
Kumamoto University
Kumamoto, Japan

Mitsuhiro Shibayama
Department of Polymer Science and Engineering
Tyoto Institute of Technology
Kyoto, Japan

Saburo Shimabayashi
Faculty of Pharmaceutical Sciences
University of Tokushima
Tokushima, Tokushima, Japan

E. Shimamura
Polymer Chemistry Laboratory
The Institute of Physical and Chemical Research (RIKEN)
Saitama, Japan

Yasuo Shimano
Department of Chemical and Biological Engineering
Hachinohe National College of Technology
Tamonoki, Hachinohe, Japan

Chiochiro Shimasaki
Department of Chemical and Biochemical Engineering
Toyama University
Toyama, Japan

Harumichi Shimizu
Saitama Institute of Technology
Saitama, Japan

Tetsuo Shimizu
Research and Development Department
Daikin Industries Ltd.
Osaka, Japan

Toshio Shimizu
Department of Information Science
Faculty of Science
Hirosaki University
Hirosaki, Japan

Yasuhiko Shimizu
Department of Artificial Organs
Research Center for Biomedical Engineering
Kyoto University
Shogoin, Kyoto, Japan

Yukio Shimura
Kantogakuin University
Yokohama, Japan

Yoichi Shindo
Department of Chemistry
Faculty of Science
Toho University
Chiba, Japan

Tomoo Shiomi
Department of Material Science and Technology
Nagaoka University of Technology
Nagaoka, Niigata, Japan

Keishiro Shirahama
Department of Chemistry Faculty of Science and Engineering
Saga University
Saga, Japan

Hirofusa Shirai
Department of Functional Polymer Science
Faculty of Textile Science and Technology
Shinshu University
Ueda, Japan

Masamitsu Shirai
Department of Applied Chemistry
College of Engineering
University of Osaka Prefecture
Sakai, Osaka, Japan

Hideki Shirakawa
Institute of Materials Science
University of Tsukuba
Tsukuba, Ibaraki, Japan

Shin-ichiro Shoda
Department of Materials Chemistry
Graduate School of Engineering
Tohoku University
Aoba, Sendai, Japan

F. Shoji
Production Engineering Research Laboratory
Hitachi Ltd.
Yokohama, Japan

Li Shuben
Lanzhou Institute of Chemical Physics
Academia Sinica
Lanzhou, Gansu, Republic of China

Shin Shing Shyu
Department of Chemical Engineering
National Central University
Chung Li, Taiwan, Republic of China

Clara Silvestre
Istituto di Ricerca e Tecnologia delle Materie Plastiche
Italian National Council of Research-CNR
Arco Felice, (Na) Italy

Rosanna Silvestri
Montell Polyolefins
G. Natta Research Center
Ferrara, Italy

Bogdan C. Simionescu
Department of Macromolecules
"Gh. Asachi" Technical University
Jassy, Romania

Cristofor I. Simionescu
Department of Macromolecules
"Gh. Asachi" Technical University
Jassy, Romania

Roberto Simonutti
Dipartimento di Chimica Organica e Industriale
Università di Milano
Milano, Italy

Yu M. Sivergin
Semenov Institute of Chemical Physics
Russian Academy of Sciences
Moscow, Russia

Arne Skjeltrop
Institute for Energy Technology
Kjeller, Norway

Peter V. Smallwood
Zeneca Resins
Cheshire, United Kingdom

Bruce E. Smart
DuPont Fluoroproducts
DuPont Central Research and Development
Experimental Station
Wilmington, Delaware

Johannes Smid
Polymer Research Institute
Faculty of Chemistry
College of Environmental Science and Forestry
State University of New York at Syracuse
Syracuse, New York

Olav Smidsrod
Norwegian Biopolymer Laboratory
Department of Biotechnology
University of Trondheim, NTH
Trondheim, Norway

Diane R. Smith
Institut für Anorganische Chemie der Universität Basel
Basel, Switzerland

P. Smith
UNIAX Corporation
Santa Barbara, California

Paul F. Smith
School of Chemistry
University of Bath
Bath, United Kingdom

Truis Smith-Palmer
Chemistry Department
St. Francis Xavier University
Antigonish, Nova Scotia, Canada

J. Smyllie
Glasgow Caledonian University
Glasgow, Scotland, United Kingdom

Ying-Hung So
The Dow Chemical Company
Midland, Michigan

Bluma G. Soares
Instituto de Macromoléculas
Universidade Federal do Rio de Janeiro
Rio de Janeiro, Brazil

Kazuo Soga
Japan Advanced Institute of Science and Technology
Tatsunokuchi, Ishikawa, Japan

B. H. Sohn
Department of Chemical Engineering
Massachusetts Institute of Technology
Cambridge, Massachusetts

David H. Solomon
Department of Chemistry
University of Melbourne
Parkville, Victoria, Australia

V. S. Soloviev
Image Processing Systems Institute of the Samara Aerospace University
Samara, Russia

A. Sommazzi
Enichem S.p.A.
Instituto G. Donegani
Novara, Italy

David Y. Son
Materials Chemistry Branch
Naval Research Laboratory
Washington, D. C.

Hyun Hoon Song
Department of Macromolecular Science
Han Nam University
Taejon, South Korea

Kenkichi Sonogashira
The Institute of Scientific and Industrial Research
Osaka University
Osaka, Japan

José M. Sosa
Fina Research and Technology Center
Deer Park, Texas

Alain Soum
Laboratoire de Chimie des Polymères Organiques
Ecole Nationale Supérieure de Chimie et de Physique de Bordeaux
Talence, France

Piero Sozzani
Dipartimento di Chimica Organica e Industriale
Universit à di Milano
Milano, Italy

Nicolas Spassky
Laboratoire de Chimie Macromoléculaire
Université Pierre et Marie Curie
Paris, France

Val N. Spector
Department of Electronics of Organic Materials
Russian Academy of Sciences
Moscow, Russia

L. H. Sperling
Materials Research Center
Lehigh University
Bethlehem, Pennsylvania

Stephen H. Spiegelberg
Division of Applied Sciences
Harvard University
Cambridge, Massachusetts

G. M. Spinks
Intelligent Polymer Research Laboratory
University of Wollongong
Wollongong, New South Wales, Australia

Edward P. Squiller
Industrial Chemicals Division
Miles Incorporated
Pittsburgh, Pennsylvania

M. Srinivasan
Dr. Bharat Ram Research and Development Center
SRF Ltd.
Madras, India

Cyrus E. Sroog
Polymer Consultants, Incorporated
Wilmington, Delaware

Edmund Stadler
Makromolekulare Chemie II
Bayreuther Institut für Makromolekülforschung
Bayreuth, Germany

Rayna Stamenova
Institute of Polymers
Bulgarian Academy of Sciences
Sofia, Bulgaria

P. A. Staniland
Imperial Chemical Industries
Wilton Materials Research Centre
Middlesbrough, Cleveland, United Kingdom

W. H. Starnes, Jr.
Applied Science Ph. D. Program
Department of Chemistry
College of William and Mary
Williamsburg, Virginia

Claudia Staudt-Bickel
Angewandte Thermodynamik
Physikalisch-Chemisches Institut
Ruprecht-Karls-Universität Heidelberg
Heidelberg, Germany

P. A. M. Steeman
DSM Research BV
Geleen, The Netherlands

Jaroslav Stehlícek
Institute of Macromolecular Chemistry
Academy of Sciences of the Czech Republic
Prague, Czech Republic

Armin Stein
Institute of Organic and Macromolecular Chemistry of the Friedrich-Schiller-University of Jena
Jena, Germany

Judith Stein
General Electric Corporate Research and Development
Schenectady, New York

Michael Stein
Schering AG
Berlin, Germany

Franz Stelzer
Christian Doppler Laboratorium für Katalytische Polymerisation
Institut für Chemische Technologie Organischer Stoffe
Graz, Austria

Per Stenstad
SINTEF
Applied Chemistry
Trondheim, Norway

Malcolm P. Stevens
Department of Chemistry
University of Hartford
West Hartford, Connecticut

J. Fraser Stoddart
School of Chemistry
University of Birmingham
Edgbaston, Birmingham, United Kingdom

Bjorn T. Stokke
Norwegian Biopolymer Laboratory
Department of Physics and Mathematics
University of Trondheim, NTH
Trondheim, Norway

R. Storbeck
Polymer-Institut
Universität (T.H.) Karlsruhe
Karlsruhe, Germany

Harald D. H. Stöver
Department of Chemistry
McMaster University
Hamilton, Ontario, Canada

Ulrich P. Strauss
Department of Chemistry
Rutgers University
New Brunswick, New Jersey

Frederick H. Strickler
Department of Materials Science and Engineering Polymer Science Program
The Pennsylvania State University
University Park, Pennsylvania

D. J. Strike
University of Neuchâtel
Institute of Microtechnology
Neuchâtel, Switzerland

Wei-Fang A. Su
Westinghouse Science and Technology Center
Pittsburgh, Pennsylvania

Juan A. Subirana
Department d'Enginyeria Química Escola Tècnica Superior d'Enginuers Industrials de Barcelona
Universitat Politècnica de Catalunya
Barcelona, Spain

Ramachandran P. Subrayan
Departments of Chemistry and Macromolecular Science and Engineering
The University of Michigan
Ann Arbor, Michigan

Kazuaki Suehiro
Department of Applied Chemistry
Faculty of Science and Engineering
Saga University
Saga, Japan

Shintaro Sugai
Department of Bioengineering
Faculty of Engineering
Soka University
Hachioji, Tokyo, Japan

Junghun Suh
Department of Chemistry
Seoul National University
Seoul, Korea

Bernt-Åki Sultan
Borealis AB
Stenungsund, Sweden

G. Sun
Department of Chemistry
Auburn University
Auburn, Alabama

Guifeng Sun
Changchun Institute of Applied Chemistry
Chinese Academy of Sciences
Changchun, China

S. T. Sun
Research Center
Hercules Incorporated
Wilmington, Delaware

Yih-Min Sun
Department of Chemical Engineering
National Cheng Kung University
Tainan, Taiwan, Republic of China

Junzo Sunamoto
Department of Synthetic Chemistry and Biological Chemistry
Graduate School of Engineering
Kyoto University
Kyoto, Japan

P. S. Suresh
Dr. Bharat Ram Research and Development Centre
SRF Ltd.
Manali, Madras, India

Ulrich W. Suter
Eidgenössische Technische Hochschule ETH (Institut für Polymere)
Zürich, Switzerland

R. L. Sutherland
Science Applications International Corporation
Dayton, Ohio

Shinichi Suto
Department of Materials Science and Engineering
Yamagata University
Yonezawa, Yamagata, Japan

Mario Suwalsky
Faculty of Chemical Sciences
University of Concepción
Concepción, Chile

Atsuo Suzuki
Department of Biotechnology
School of Engineering
Nagoya University
Nagoya, Japan

Masato Suzuki
Research Laboratory of Resources Utilization
Tokyo Institute of Technology
Midori-ku, Yokohama, Japan

Toshio Suzuki
Sumitomo Bakelite Company, Ltd.
Grand Island, New York

Barbara Szczepaniak
Industrial Chemistry Research Institute
Warsaw, Poland

Marta Szesztay
Central Research Institute for Chemistry
Hungarian Academy of Sciences
Budapest, Hungary

J. O. Tabe
IPTME
Loughborough University of Technology
Loughborough, Leicestershire
United Kingdom

Nanyan-Hwa Tai
Institute of Chemical Engineering
Taiwan, Republic of China

Peter J. T. Tait
UMIST
Manchester, United Kingdom

Sadao Takagi
Department of Chemistry
Kinki University
Osaka, Japan

Toru Takagishi
Department of Applied Materials Science
College of Engineering
Osaka Prefecture University
Sakai, Osaka, Japan

Akio Takahashi
Hitachi Research Laboratory
Hitachi Ltd.
Ibaraki, Japan

Akira Takahashi
Department of Chemistry for Materials
Mie University
Mie-Ken, Japan

Keiko Takahashi
Department of Industrial Chemistry
Faculty of Engineering
Tokyo Institute of Polytechnics
Kanagawa, Japan

Shigetoshi Takahashi
The Institute of Scientific and Industrial Research
Osaka University
Osaka, Japan

Takashi Takahashi
Department of Medical Informatics
Faculty of Medicine
Kyoto University
Kyoto, Japan

Kaoru Takakura
Physics Department
International Christian University
Tokyo, Japan

Teruo Takakura
Functional Products Research and Development Center
Chemicals General Division
ASAHI Glass Company, Ltd.
Yokohama, Japan

Nobuo Takamiya
Advanced Research Center for Science and Engineering
Waseda University
Shinjuku-Ku, Tokyo, Japan

Toshikazu Takata
Department of Applied Chemistry
College of Engineering
Osaka Prefecture University
Osaka, Japan

Hiromu Takeda
Faculty of Human Environmental Science
Fukuoka Women's University
Fukuoka, Japan

Tsutomu Takeichi
School of Materials Science
Toyohashi University of Technology
Toyohashi, Japan

Kiichi Takemoto
Faculty of Engineering
Osaka University
Suita, Osaka, Japan

Noboru Takisawa
Department of Chemistry Faculty of Science and Engineering
Saga University
Saga, Japan

Mostafa A. H. Talukder
Chemistry and Materials Branch
Research and Technology Division
China Lake Naval Air Warfare Center Weapons Division
China Lake, California

Kam Chiu Tam
Advanced Materials Research Center
School of Mechanical and Production Engineering
Nanyang Technological University
Republic of Singapore

Saburo Tamura
Miyagi Polytechnic College
Kurihara, Miyagi, Japan

K. L. Tan
Department Physics
National University of Singapore
Kent Ridge, Republic of Singapore

Akira Tanaka
Department of Materials Science
The University of Shiga Prefecture
Shiga, Japan

Hitoshi Tanaka
Department of Optical Science and Technology
Faculty of Engineering
Tokushima University
Tokushima, Japan

J. Tanaka
Production Engineering Research Laboratory
Hitachi Ltd.
Totsuka-ku Yokohama, Japan

Jun Tanaka
Saitama Institute of Technology
Saitama, Japan

Kazuyoshi Tanaka
Division of Molecular Engineering
Faculty of Engineering
Kyoto University
Sakyoku, Kyoto, Japan

Makoto Tanaka
Department of Applied Chemistry
College of Engineering
University of Osaka Prefecture
Sakai, Osaka, Japan

Yasuyuki Tanaka
Faculty of Technology
Tokyo University of Agriculture and Technology
Koganei, Tokyo, Japan

Hao Tang
Changchun Institute of Applied Chemistry
Chinese Academy of Sciences
Changchun, Republic of China

Takashi Taniguchi
Faculty of Education
Niigata University
Niigata, Japan

Yoshihiro Taniguchi
Department of Chemistry
Ritsumeikan University
Kusatsu, Japan

François René Taravel
CERMAV Université Joseph Fourier
Associé à l'Université Joseph Fourier
Grenoble, France

María Pilar Tarazona
Departamento de Química Física
Universidad de Alcalá de Henares
Madrid, Spain

Kohji Tashiro
Department of Macromolecular Science
Faculty of Science
Osaka University
Toyonaka, Osaka, Japan

Susumu Tate
Toyobo Research Institute
Toyobo Company Ltd.
Katata Ohtsu, Shiga, Japan

Masayoshi Tatemoto
Chemical Division
Daikin Industries Ltd.
Osaka, Japan

Daniel Taton
Laboratoire de Chimie Macromoléculaire
Université Pierre et Marie Curie
Paris, France

Nobuhide Tatsumoto
General Education Chemistry
Oita Medical University
Hasama-machi Oita, Japan

Akio Teramoto
Department of Macromolecular Science
Osaka University
Toyonaka, Osaka, Japan

T. Teramoto
Chemicals Division Advanced Materials and Technology Research Laboratory
Nippon Steel Corporation
Kawasaki, Japan

Monika Terskan-Reinold
Lehrstuhl für Makromolekulare Chemie II and Bayreuther
Institut fü r Makromolekülforschung (BIMF)
Universität Bayreuth
Bayreuth, Germany

Claire A. Tessier
Chemistry Department
University of Akron
Akron, Ohio

Yasuyuki Tezuka
Department of Organic and Polymeric Materials
Tokyo Institute of Technology
Niigata, Japan

Shelby F. Thames
Department of Polymer Science
The University of Southern Mississippi
Hattiesburg, Mississippi

A. G. Theodoropoulos
Laboratory of Organic Chemical Technology
Department of Chemical Engineering
National Technical University of Athens
Athens, Greece

Lothar Thiele
Henkel KGaA
Dusseldorf, Germany

A. Thierry
Institut Charles Sadron, CNRS
Strasbourg, France

Emma Thorn-Csányi
Institut für Technishche und Makromolekulare Chemie
Universität Hamburg
Hamburg, Germany

W. M. H. Thorpe
The Malaysian Rubber Products' Research Association
Tun Abdul Razak Laboratory
Brickendonbury, Hertford, United Kingdom

Jun Tian
Changchun Institute of Applied Chemistry
Chinese Academy of Sciences
Changchun, Republic of China

Anton Ticktin
BASF AG
Ludwigshafen, Germany

Chao-Fong Tien
Air Products and Chemicals, Incorporated
Allentown, Pennsylvania

Y. P. Ting
Department of Chemical Engineering
National University of Singapore
Kent Ridge, Singapore

Matthew Tirrell
Department of Chemical Engineering and Materials Science
University of Minnesota
Minneapolis, Minnesota

C. Tiu
Department of Chemical Engineering
Monash University
Clayton, Victoria, Australia

T. R. du Toit
Plastics Technology Department
Cape Technikon
Capetown, South Africa

Katsuhisa Tokumitsu
Osaka Gas Company, Ltd.
Osaka, Japan

Donald A. Tomalia
Michigan Molecular Institute
Midland, Michigan

A. Tomanek
Wacker-Chemie GmbH
München, Germany

Tahei Tomida
Department of Chemical Science and Technology
Faculty of Engineering
University of Tokushima
Tokushima, Japan

Akira Tominaga
Technical Research Laboratory
Kansai Paint Company, Ltd.
Kanagawa, Japan

Ikuyoshi Tomita
Research Laboratory of Resources Utilization
Tokyo Institute of Technology
Midori-ku, Yokohama, Japan

J. George Tomka
Department of Textile Industries
University of Leeds
Leeds, United Kingdom

Masao Tomoi
Department of Applied Chemistry
Faculty of Engineering
Yokohama National University
Yokohama, Japan

V. Tondigilia
Science Applications International Corporation
Dayton, Ohio

Claudio Toniolo
Department of Organic Chemistry
Biopolymer Research Center, C.N.R.
University of Padova
Padova, Italy

C.O. Too
Intelligent Polymer Research Laboratory Department of Chemistry
University of Wollongong
Wollongong, New South Wales, Australia

Naoki Toshima
The University of Tokyo
Tokyo, Japan

Jean-Marc Toussaint
Chimie des Matériaux Nouveaux
Centre de Recherche en Electronique et Photonique Moléculaires
Université de Mons-Hainaut
Mons, Belgium

Qui Tran-Cong
Department of Polymer Science and Engineering
Kyoto Institute of Technology
Kyoto, Japan

Sukant K. Tripathy
Department of Chemistry
Center of Advanced Materials
University of Massachusetts, Lowell
Lowell, Massachusetts

Incoronata Tritto
Istituto di Chimica delle Macromolecole, CNR
Milano, Italy

Barbara Trzebicka
Polish Academy of Sciences
Institute of Coal Chemistry
Gliwice, Poland

T. Tsanov
Institute of Polymers
Sofia, Bulgaria

Yiannis Tselikas
Department of Chemistry
University of Athens
Athens, Greece

John Tsibouklis
Advanced Polymers and Composites Research Group
University of Portsmouth
Portsmouth, United Kingdom

Christos P. Tsonis
Chemistry Department
King Fahd University of Petroleum and Minerals
Dhahran, Saudi Arabia

Norio Tsubokawa
Department of Chemistry and Chemical Engineering
Faculty of Engineering
Niigata University
Niigata, Japan

Yasuhiko Tsuchitani
Faculty of Dentistry
Osaka University
Suita, Osaka, Japan

Kazuichi Tsuda
Department of Applied Chemistry
Nagoya Institute of Technology
Nagoya, Japan

Tetsuo Tsuda
Department of Polymer Chemistry
Graduate School of Engineering
Kyoto University
Yoshida, Kyoto, Japan

Shin Tsuge
Department of Applied Chemistry
School of Engineering
Nagoya University
Nagoya, Japan

Masuhiro Tsukada
National Institute of Sericultural and Entomological Science
Tsukuba City, Japan

Yasuhisa Tsukahara
Department of Materials Science
Kyoto Institute of Technology
Matsugasaki, Kyoto, Japan

Vladimir V. Tsukruk
College of Engineering and Applied Sciences
Western Michigan University
Kalamazoo, Michigan

K. Tsunashima
Toray Industries Incorporated
Sonoyama, Otsu Shiga-ken, Japan

Masahiro Tsunooka
Department of Applied Chemistry
College of Engineering
University of Osaka Prefecture
Osaka, Japan

Naoto Tsutsumi
Kyoto Institute of Technology
Kyoto, Japan

C. Tsvetanov
Institute of Polymers
Bulgarian Academy of Sciences
Sofia, Bulgaria

F. Tüdős
Department of Chemical Technology
Eötvö Loránd University
Budapest, Hungary

S. Richard Turner
Technische Universität München
Lehrstuhl für Makromoleculare Stoffe
Garching, Germany

J. Van Turnhout
TNO Plastics and Rubber Institute
Delft, The Netherlands

Mituaki Tutiya
Tokyo Metropolitan Isotope Research Center
Tokyo, Japan

Zdenêk Tuzar
Institute of Macromolecular Chemistry Academy of Sciences of the Czech Republic
Prague, Czech Republic

Mitsuru Ueda
Department of Materials Science and Engineering
Faculty of Engineering
Yamagata University
Yonezawa, Yamagata, Japan

Satoshi Ueki
Tonen Corporate Research and Development Laboratory
Tonen Chemical Corporation
Iruma-gun, Saitama, Japan

Hiroshi Ueno
Department of Chemistry and Chemical Engineering
Faculty of Engineering
Niigata University
Niigata, Japan

John Ugelstad
Department of Industrial Chemistry
NTH University of Trondheim
Trondheim, Norway

Jill S. Ullett
The Center for Basic and Applied Polymer Research
The University of Dayton
Dayton, Ohio

Abraham Ulman
Department of Chemistry
Polytechnic Institute of New York
Brooklyn, New York

Tadayuki Uno
Faculty of Pharmaceutical Sciences
University of Tokushima
Tokushima, Japan

Edward J. Urankar
Department of Chemistry
Baker Laboratory
Cornell University
Ithaca, New York

Satoshi Urano
Research Center
Nippon Paint Company, Ltd.
Neyagawa City, Osaka, Japan

Dan W. Urry
Laboratory of Molecular Biophysics
School of Medicine
University of Alabama at Birmingham
Birmingham, Alabama

Toshiyuki Uryu
Institute of Industrial Science
University of Tokyo
Tokyo, Japan

L. A. Utracki
National Research Council Canada
Industrial Materials Institute
Boucherville, Quebec, Canada

Hiroshi Uyama
Department of Molecular Chemistry and Engineering
Faculty of Engineering
Tohoku University
Aoba, Sendai, Japan

N. Valdebenito
Facultad de Química Pontificia Universidad Católica de Chile
Santiago, Chile

M. Valdés
Instituto de Investigaciones en Materiales
Mexico D.F., Mexico

Enrique M. Vallés
Planta Piloto de Ingeniera Quimica
Bahía Blanca, Argentina

M. Van Beylen
Laboratory of Macromolecular and Physical Organic Chemistry
Catholic University of Lueven
Heverlee (Leuven), Belgium

M. F. van Buren
Arthur D. Little Incorporated
Cambridge, Massachusetts

P. van der Wal
University of Neuchâtel
Institute of Microtechnology
Neuchâtel, Switzerland

Dang Van Luyen
Institute of Chemistry
National Center for Natural Science and Technology
Hanoi, Vietnam

A. J. van Reenen
Institute for Polymer Science
University of Stellenbosch
Matieland, South Africa

J. Van Turnhout
TNO Plastics and Rubber Institute
Delft, The Netherlands

József Varga
Department of Plastics and Rubber Technology
Technical University of Budapest
Budapest, Hungary

Padma Vasudevan
Center for Rural Development and Technology
Indian Institute of Technology
Delhi, India

A. M. Vekselman
Department of Chemistry
McGill University
Montreal, Quebec, Canada

Martha Albores Velasco
Facultad de Quimica
UNAM
Circuito Interior
Ciudad Universitaria
Mexico D.F., Mexico

S. Venkatachalam
Vikram Sarabhai Space Center
Trivandrum, India

Margherita Venturi
Dipartimento di Chimica G. Ciamician
Università di Bologna
Bologna, Italy

Ricardo Vera-Graziano
Instituto de Investigaciones en Materiales UNAM
Coyoacan, D.F., Mexico

I. Vermeesch
ICI Polyurethanes
Heverlee, Belgium

Michel Vert
CRBA-URA CNRS 1465
Faculty of Pharmacy
University of Montpellier 1
Montpellier, France

Jacques Vicens
EHICS Laboratoire de Chimie Minérale et Analytique
Associé au CNRS
Strasbourg, France

David E. Vietti
Morton International
Woodstock, Illinois

Veena Vijayanathan
Vikram Sarabhai Space Center
Tivandrum, India

S. Vilasagar
General Electric Plastics
Technology Center
Washington, West Virginia

Manuel Villacampa
Universidad del País Vasco
Bilbao, Spain

Marcelo A. Villar
Planta Piloto de Ingeniería Química
UNS-CONICET
Bahía Blanca, Argentina

Roland Vogel
Institute of Polymer Research
Dresden, Germany

I. I. Vointseva
Institute of Organo-Element Compounds
Moscow, Russia

Brigitte I. Voit
Technische Universität München
Lehrstuhl für Makromoleculare Stoffe
Garching, Germany

Nicola Volpi
Department of Animal Biology
University of Modena
Modena, Italy

Walter Von Hellens
Bayer Rubber Incorporated
Sarnia, Ontario, Canada

K. D. Vorlop
Institute of Technology
FAL
Braunschweig, Germany

N. L. Voropayeva
Institute of Polymer Chemistry and Physics
Tashkent, Uzbekistan

Menas S. Vratsanos
Air Products and Chemicals, Incorporated
Allentown, Pennsylvania

Moriyasu Wada
Toshiba Battery
Tokyo, Japan

Takehiko Wada
Faculty of Engineering
Osaka University
Suita, Osaka, Japan

Tohru Wada
Kuraray Company, Ltd.
Kurashiki, Okayama, Japan

Wolfgang Wagenknecht
Max-Planck-Institute of Colloid and Interface Research Teltow-Seehof
Teltow-Seehof, Germany

David Wails
Department of Chemistry
University of York
Heslington, York, United Kingdom

Phillip A. Waitkus
Plastics Engineering Company
Sheboygan, Wisconsin

Masahide Wakakura
Kanagawa Industrial Research Institute
Ebinasi, Japan

Masa-aki Wakita
Kurita Central Laboratories
Kurita Water Industries Ltd.
Morinosato, Atsugi City, Kanagawa, Japan

Wojciech Walach
Polish Academy of Sciences
Institute of Coal Chemistry
Gliwice, Poland

Gordon G. Wallace
Intelligent Polymer Research Laboratory
Department of Chemistry
University of Wollongong
Wollongong, New South Wales, Australia

Chun-Shan Wang
Department of Chemical Engineering
National Cheng Kung University
Tainan, Taiwan, Republic of China

Chyi-Shan Wang
University of Dayton Research Institute
Nonmetallic Materials Division
Dayton, Ohio

Der-Wun Wang
Department of Chemical Engineering
National Taiwan Institute of Technology
Taipei, Taiwan, Republic of China

Fei Wang
EIC Laboratories, Incorporated
Norwood, Massachusetts

H.-C. Wang
Baytown Polymers Center
Exxon Chemical Company
Baytown, Texas

Lili Wang
Department of Chemistry and Chemical Engineering
Xiamen University
Xiamen, Republic of China

Mang Wang
Department of Polymer Science and Engineering
Zhejiang University
Hangzhou, Republic of China

Shanger Wang
Laboratory of Polymer Physics
Changchun Institute of Applied Chemistry Chinese Academy of Sciences
Changchun, China

Tong Kuan Wang
Ecole Nationale Supérieure de Chimie de Mulhouse
ICSI
Mulhouse, France

Xiao Li Wang
Department of Chemistry
University of North Carolina at Charlotte
Charlotte, North Carolina

Zhi Yuan Wang
Ottawa-Carleton Chemistry Institute
Department of Chemistry
Carleton University
Ottawa, Ontario, Canada

Kazuhiro Watanabe
Chemicals Division
Advanced Materials and Technology Research Laboratory
Nippon Steel Corporation
Kawasaki, Japan

Shinji Watanabe
Department of Materials Science
Faculty of Engineering
Kitami Institute of Technology
Hokkaido, Japan

Mineo Watase
Applied Biological Chemistry
Faculty of Agriculture
Shizuoha University
Shizuoha, Japan

Diana M. Watkins
Department of Chemistry
University of Texas at Austin
Austin, Texas

Russell I. Webb
Department of Chemistry and Center for Materials for Information Technology
The University of Alabama
Tuscaloosa, Alabama

Stephen E. Webber
Department of Chemistry and Biochemistry and Center for Polymer Research
University of Texas at Austin
Austin, Texas

Stephen G. Weber
Department of Chemistry
Chevron Science Center
University of Pittsburgh
Pittsburgh, Pennsylvania

Rongbao Wei
Department of Chemical Engineering
Tianjin Institute of Technology
Tianjin, China

H.-D. Weigmann
TRI/Princeton
Princeton, New Jersey

Werner Weitschies
Schering AG
Berlin, Germany

Jianye Wen
Department of Chemical Engineering and Polymer Materials and Interfaces Laboratory
Virginia Polytechnic Institute and State University
Blacksburg, Virginia

J. V. Wening
Department of Trauma and Reconstructive Surgery
University Hospital Hamburg-Eppendorf (UKE)
Hamburg, Germany

Peter J. West
School of Chemistry
University of Bath
Bath, United Kingdom

E. Westerweele
UNIAX Corporation
Santa Barbara, California

J. R. White
Department of Mechanical Materials and Manufacturing Engineering
University of Newcastle upon Tyne
New Castle upon Tyne, United Kingdom

Michael L. White
GE Corporate Research and Development
Schenectady, New York

Douglas A. Wicks
Bayer Corporation
Pittsburgh, Pennsylvania

Władysław Wieczorek
Faculty of Chemistry
Warsaw University of Technology
Warsaw, Poland

Chi Wi
Department of Chemistry
The Chinese University of Hong Kong
Shatin, N. T., Hong Kong

J. S. Wiggins
Plastics New Product Development
Bayer Corporation
Pittsburgh, Pennsylvania

W. M. K. P. Wijekoon
Photonics Research Laboratory
Department of Chemistry
The State University of New York at Buffalo
Buffalo, New York

Hauke Wilcken
Department of Trace Analysis
Fachhochschule Fresenius
Wiesbaden, Germany

R. H. Wildi
General Electric Plastics Technology Center
Washington, West Virginia

M. A. Wilding
Department of Textiles
UMIST
Manchester, United Kingdom

Garth L. Wilkes
Department of Chemical Engineering and Polymer Materials and Interfaces Laboratory
Virginia Polytechnic Institute and State University
Blacksburg, Virginia

Roberto J. J. Williams
Institute of Materials Science and Technology
University of Mar del Plata and National Research Council
Mar del Plata, Argentina

A. V. Wilson
Department of Textile Engineering, Chemistry, and Science
North Carolina State University
Raleigh, North Carolina

David J. Wilson
Research and Development Laboratory
EniChem Elastomers Ltd.
Grangemouth, Stirlingshire, Scotland, United Kingdom

Eugene Wilusz
U.S. Army Natick Research Development and Engineering Center Natick
Natick, Massachusetts

Gerhard Winter
Bayer AG
Krefeld, Germany

Elizabeth T. Wise
Department of Chemistry
Chevron Science Center
University of Pittsburgh
Pittsburgh, Pennsylvania

Dieter Wöhrle
Universität Bremen
Institut für Organische und Makromolekulare Chemie
Bremen, Germany

Peter Wolf
BASF AG
Ludwigshafen, Germany

Dominic W. S. Wong
Western Regional Research Center
Agricultural Research Service
U.S. Department of Agriculture
Albany, California

Fei Wong
EIC Laboratories, Incorporated
Norwood, Massachusetts

John Woods
Loctite Corporation
Rocky Hill, Connecticut

S. D. Worley
Department of Chemistry
Auburn University
Auburn, Alabama

Gerhard Wötting
Bayer AG
Krefeld, Germany

Chi Wu
Department of Chemistry
The Chinese University of Hong Kong
Shatin, N. T., Hong Kong

Guangwei Wu
Department of Chemistry
State University of New York at Stony Brook
Stony Brook, New York

Yu Xianda
Lanzhou Institute of Chemical Physics
Academia Sinica
Beijing, China

Rui Xie
Changchun Institute of Applied Chemistry
Chinese Academy of Sciences
Changchun, China

Gu Xu
McMaster University
Hamilton, Ontario, Canada

J. Xu
Laboratory of Molecular Biophysics
School of Medicine
University of Alabama at Birmingham
Birmingham, Alabama

Yongpeng Xue
Department of Chemistry
State University of New York at Albany
Albany, New York

Liang Ya
Department of Chemical Engineering
Tianjin Institute of Technology
Tianjin, China

Motoshi Yabuta
Technical Research Laboratory
Kansai Paint Company, Ltd.
Hiratsuka, Kanagawa, Japan

Yusuf Yăgci
Department of Chemistry
Istanbul Technical University
Istanbul, Turkey

Tokio Yamabe
Division of Molecular Engineering
Faculty of Engineering
Kyoto University
Sakyoku, Kyoto, Japan

Bunichiro Yamada
Material Chemistry Laboratory
Faculty of Engineering
Osaka City University
Osaka, Japan

Koh-ichi Yamada
Faculty of Pharmaceutical Sciences
Josai University
Sakao, Saitama, Japan

Masayuki Yamagami
Research Institute for Advanced Science and Technology
University of Osaka Prefecture
Osaka, Japan

Tada-Aki Yamagishi
Department of Chemistry and Chemical Engineering
Faculty of Technology
Kanazawa University
Kanazawa, Japan

Chiharu Yamaguchi
Osaka Gas Company, Ltd.
Osaka, Japan

Isao Yamaguchi
Research Laboratory of Resources Utilization
Tokyo Institute of Technology
Midori-ku, Yokohama, Japan

Kazuo Yamaguchi
Department of Materials Science
Faculty of Science
Kanagawa University
Hiratsuka, Kanagawa, Japan

Wataru Yamahara
Saitama Institute of Technology
Saitama, Japan

Hiroyuki Yamamoto
Institute of High Polymer Research
Faculty of Textile Science and Technology
Shinshu University
Ueda, Japan

S. Yamamoto
Central Research Laboratory
Mitsubishi Electric Corporation
Amagasaki, Hyogo, Japan

Takakazu Yamamoto
Research Laboratory of Resources Utilization
Tokyo Institute of Technology
Midori-ku, Yokohama, Japan

Tohei Yamamoto
Faculty of Engineering
Himeji Institute of Technology
Shosha, Himeji, Japan

Yasuhiko Yamamoto
Department of Chemistry
University of Tsukuba
Tsukuba, Japan

Chihiro Yamane
Fundamental Research Laboratory of Natural and Synthetic Polymers
Asahi Chemical Industry Company, Ltd.
Takatsuki, Osaka, Japan

Hideki Yamane
Department of Polymer Engineering and Science
Kyoto Institute of Technology
Kyoto, Japan

Takeshi Yamanobe
Department of Chemistry
Gunma University
Kiryu Gunma, Japan

Kiwamu Yamaoka
Faculty of Science
Hiroshima University
Higashi-Hiroshima, Japan

Tetsuji Yamaoka
Department of Polymer Science and Engineering
Kyoto Institute of Technology
Kyoto, Japan

A. Yamasaki
National Institute of Materials and Chemical Research
Ibaraki, Japan

Toshihiro Yamase
Research Laboratory of Resources Utilization
Tokyo Institute of Technology
Tokyo, Japan

Iwao Yamashita
Hyogo Prefectural Institute of Industrial Research
Yukihira-Cho
Suma-Ku, Japan

Natsuki Yamashita
Department of Applied Chemistry
Kinki University
Osaka, Japan

Keiji Yamashita
Department of Applied Chemistry
Nagoya Institute of Technology
Nagoya, Japan

Yuhiko Yamashita
Faculty of Environmental Science and Technology
Okayama University Thushima
Okayama, Japan

Yuya Yamashita
Professor Emeritus of Nagoya University
Nagoya, Japan

Kiyoshi Yamauchi
Department of Bioapplied Chemistry
Faculty of Engineering
Osaka City University
Osaka, Japan

Kazuo Yamura
Faculty of Textile Science and Technology
Shinshu University
Ueda-city, Nagano-Prefecture, Japan

Deyue Yan
Department of Applied Chemistry
Shanghai Jiao Tong University
Shanghai, Republic of China

Rui-Fang Yan
Institute of Chemistry
Academia Sinica
Beijing, China

Chin-Ping Yang
Department of Chemical Engineering
Tatung Institute of Technology
Taipei, Taiwan, Republic of China

Shi-Lin Yang
Department of Polymer Science and Engineering
Zhejiang University
Hangzhou, Republic of China

Y. Yang
UNIAX Corporation
Santa Barbara, California

I. V. Yannas
Department of Mechanical Engineering
Massachusetts Institute of Technology
Cambridge, Massachusetts

S. Yano
Faculty of Engineering
Gifu University
Gifu, Japan

A. A. E. Yaseen
Food Technology Department
National Research Center
Dokki, Cairo, Egypt

Hajime Yasuda
Department of Applied Chemistry
Faculty of Engineering
Hiroshima University
Higashi-Hiroshima, Japan

Sachio Yasufuku
Department of Electrical Engineering
Tokyo Denki University
Chiyoda-ku, Tokyo, Japan
Mitsuo Yasui
Hyogo Prefectural Institute of Industrial Research
Kobe, Japan
Shizukuni Yata
Battery Business Promotion
Kanebo Ltd.
Osaka, Japan
E. B. Yeap
Department of Geology
University of Malaya
Kuala Lumpur, Malaysia
A. F. Yee
Department of Materials Science and Engineering
University of Michigan
Ann Arbor, Michigan
Philip E. Yeske
Bayer Corporation
Pittsburgh, Pennsylvania
Mi Hie Yi
Advanced Polymer Division
Korea Research Institute of Chemical Technology
Yusung, Taejeon, Korea
Kazuaki Yokota
Faculty of Engineering
Hokkaido University
Sapporo, Japan
Kenji Yokota
Department of Materials Science and Engineering
Nagoya Institute of Technology
Showa-ku, Nagoya, Japan
Rikio Yokota
Institute of Space and Astronautical Science
Kanagawa, Japan
Toshio Yokota
Department of Polymer Science and Engineering
Kyoto Institute of Technology
Kyoto, Japan
Masayuki Yokoyama
Institute of Biomedical Engineering
Tokyo Women's Medical College
Tokyo, Japan
Haruyuki Yoneda
Asahi Chemical Industry Company
Shiga-Prefecture, Japan
Masaru Yoneyama
Department of Biological and Chemical Engineering
Faculty of Engineering
Gunma University
Kiryu, Gunma, Japan
Dorie J. Yontz
Department of Polymer Science
University of Southern Mississippi
Hattiesburg, Mississippi
Eda Yoshiaki
Saga University
Saga, Japan
Masaru Yoshida
Department of Material Development
Takasaki Radiation Chemistry Research Establishment
Japan Atomic Energy Research Institute
Gunma, Japan
Toshio Yoshihara
Department of Chemistry and Chemical Engineering
Faculty of Engineering
Niigata University
Niigata, Japan
Norio Yoshino
Science University of Tokyo
Tokyo, Japan
Toshinori Yoshioka
Department of Macromolecular Science
Faculty of Science
Osaka University
Osaka, Japan
Hidekazu Yoshizawa
Department of Applied Chemistry and Chemical Engineering
Faculty of Engineering
Kagoshima University
Korimoto, Kagoshima, Japan
Tai-Horng Young
Center for Biomedical Engineering
College of Medicine
National Taiwan University
Taipei, Taiwan, Republic of China
Wiley J. Youngs
Chemistry Department
University of Akron
Akron, Ohio
U. S. Yousef
Laboratory of Electrochemistry
Faculty of Science
El-Menoufia University
Shebin, El-Kom, Egypt
Guangqian Yu
Changchun Institute of Applied Chemistry
Chinese Academy of Sciences
Changchun, China
Long Yu
Applied Chemistry and Cooperative Research Center for Polymer Blends
Royal Melbourne Institute of Technology
Melbourne, Queensland, Australia
Luping Yu
Department of Chemistry
University of Chicago
Chicago, Illinois
Xian Da Yu
Lanzhou Institute of Chemical Physics
Academia Sinica
Lanzhou, Gansu, Republic of China
L. Y. Yuan
National Central University
Department of Chemical Engineering
Taiwan, Republic of China
Nobuhiko Yui
Japan Advanced Institute of Science and Technology
Ishikawa, Japan
Hayder A. Zahalka
Polysar Rubber Corporation
Sarnia, Ontario, Canada
Craig Zaluski
McMaster University
Hamilton, Ontario, Canada
Z. Zamorsky
Technical University Brno
Faculty of Technology Zlín
Zlin, Czech Republic
R. Zana
Institut C. Sadron
Strasbourg, France
Robert Zand
The Macromolecular Research Center
The University of Michigan
Ann Arbor, Michigan
Peter Zarras
Commander
Naval Air Warfare Center Weapons Division
China Lake, California
Rudolf Zentel
Institute of Organic Chemistry
Universität Mainz
Mainz, Germany
G. Zerbi
Dipartimento di Chimica Industrialee
Ingegneria Chimica
Politecnico di Milano
Milano, Italy
B. Žerjal
University of Maribor, EPF
Institute of Technology, Maribor
National Institute of Chemistry
Maribor, Slovanic
C. Zhang
UNIAX Corporation
Santa Barbara, California
Hongfang Zhang
Laboratory of Polymer Physics
Changchun Institute of Applied Chemistry Chinese Academy of Sciences
Changchun, China
Xichen Zhang
Department of Chemical Engineering
Queen's University
Kingston, Ontario, Canada
H. Zhao
Department of Chemistry
University of Wollongong
Wollongong, New South Wales, Australia
Ji Zheng
Division of Material Chemistry
Faculty of Engineering
Kyoto University
Kyoto, Japan
Jian Zhou
Department of Chemistry
University of Alabama at Birmingham
Birmingham, Alabama
Rongnong Zhou
Institute of Organic Chemistry
University of Tuebingen
Tuebingen, Germany
Jan Zielinski
Nitrogen (Petrochemical) Works "Wloclawek"
Wloclawek, Poland
J. U. Zilles
Institut für Technische und Makromolekulare Chemie der Universität Hamburg
Hamburg, Germany
Hennie F. Zirkzee
Department of Polymer Chemistry and Technology
Eindhoven University of Technology
Eindhoven, The Netherlands
Rita A. Zoppi
Instituto de Química
Universidade Estadual de Campinas
Campinas, Sao Paulo, Brazil
P. Zugenmaier
Institute of Physical Chemistry
Technical University Clausthal
Clausthal-Zellerfeld, Germany

A

AAS

See: Acrylic-Styrene-Acrylonitrile

ABS

See: Acrylonitrile-Styrene-Butadiene

Absorbable Polymers

See: Bioabsorbable Polymers (Tissue Engineering)

Accelerators

See: Rubber Additives, Multifunctional
Unsaturated Polyester Resins (Overview)
Vulcanization (Overview)

ACENAPHTHYLENE, POLYMERIZATION

Shouji Iwatsuki and Masataka Kubo
Instrumental Analysis Center
Mie University

HOMOPOLYMERIZATION OF ACENAPHTHYLENE

In general, 1,2-distributed ethylenes are said to be less reactive than 1,1-distributed and monosubstituted ethylenes due to steric hindrance between the 1,2-substituents in the transition state of propagation in their polymerization.[1] Only a few 1,2-disubstituted ethylenes are homopolymerizable (e.g., maleimides, vinylene carbonate, fumarates, and acenaphthylene).[2-5] Many studies on the polymerization of acenaphthylene have been published (e.g., Dziewonski and Flowers and Miller).[6,7]

COPOLYMERIZATION OF ACENAPHTHYLENE

Ueberreiter and Krull carried out the copolymerization of acenaphthylene (M_1) with styrene (M_2) in bulk at 90°C to obtain monomer reactivity ratios (MRRs) or $r_1 = 0.33$ and $r_2 = 3.81$.[8] Saotome and Imoto found in the copolymerization of acenaphthylene with styrene in benzene using boron trifluoride etherate as an initiator that the rate of the copolymerization was much lower than that of either homopolymerization.[9] Noma et al. carried out radical copolymerizations of acenapthylene (M_1) with methacrylonitrile (MN), acrylonitrile (AN), methyl methacrylate (MMA), and methyl acrylate (MA) in benzene at 60°C. They obtained MRRs of $r_1 = 2.38$ and $r_2 = 0.15$ for an acenapthylene–MN system, $r_1 = 2.56$ and $r_2 = 0.02$ for an acenapthylene–AN system, $r_1 = 2.25$ and $r_2 = 0.44$ for an acenaphthylene–MMA system, and $r_1 = 4.00$ and $r_2 = 0.10$ for an acenapthylene–MA system. These MRRs were used to calculate Alfrey–Price Q-e values of acenaphthylene: Q = 0.70 and e = 2.00.[10]

THERMAL PROPERTY OF POLYACENAPHTHYLENE

Jones reported that polyacenaphthylene prepared in thermal polymerization was not subject to softening below 250°C.[11] Barb measured a softening temperature of polyacenaphthylene to be above 205°C.[12,13] Ballesteros et al. found that in the differential scanning calorimeter (DSC) measurement of glass transition temperature for polyacenaphthylene and copolymers of acenaphthylene with methyl methacrylate, styrene, maleic anhydride, diethyl maleate, *N*-vinylpyrrolidone, or *trans*-stilbene the glass transition temperatures of the copolymers are much lower than that of polyacenaphthylene.[14] Scaffhauser et al. also observed that the glass transition temperature of acenaphthylene-styrene copolymers increases with an increase of acenaphthylene.[15] Dunham et al. pointed out that the copolymerization of acenaphthylene with styrene is effective in increasing the softening temperature of polystyrene.[16] Markevich found in the thermal decomposition of poly(acenaphthylene) that the decomposition starts to take place at 335–345°C and that acenaphthene, acenaphthylene, and the conjugated polymer were obtained as degradation products.[17]

EQUILIBRIUM POLYMERIZATION OF ACENAPHTHYLENE[18]

In the radical polymerization of acenaphthylene in toluene using AIBN is an initiator, the time-conversion plots for monomer concentration of 0.066 mol/L at 50, 60, and 70°C indicated there is no induction period. The radical polymerization of acenaphthylene follows a square-root dependence of polymerization rate on initiator concentration similar to that of a conventional radical vinyl polymerization.

Acenaphthylene is polymerizable as a highly conjugative, electron-donating monomer, but it is subject to equilibrium polymerization, having a fairly high equilibrium monomer concentration at a conventional temperature of radical polymerization. Moreover, in the case of a radical polymerization influenced by depolymerization, a reaction other than polymerization and depolymerization takes place to give low molecular products, indicating an incomplete type of equilibrium polymerization behavior. However, in the case of a monomer concentration that is much higher than an equilibrium monomer concentrations, polymerization exclusively takes place and other side reactions do not occur. Acenaphthylene is high crystalline with a melting point of 92°C, and its polymerization usually has to be carried out in states of solution. Therefore, careful consideration has to be given to the monomer concentration for the intended polymerization.

REFERENCES

1. Odian, G. *Principles of Polymerization,* 2nd ed., John Wiley & Sons: New York, 1981; p. 266.
2. Tawney, P. O.; Synder, R. H.; Conger, R. P.; Steteler, K. A.; Williams, A. R. *J. Org. Chem.* **1961**, *26*, 15.
3. Newman, M. S.; Addor, R. J. *J. Am. Soc.* **1953**, *75*, 1263.
4. Otsu, T.; Ito, O.; Toyada, N.; Mori, S. *Makromol. Chem., Rapid Commun.* **1981**, *2*, 725.
5. Otsu, T.; Yasuhara, T.; Shiraishi, K.; Mori, S. *Polym. Bull.* **1984**, *12*, 449.
6. Dziewonski, K. *Ber.* **1903**, *36*, 962.

7. Flowers, R. G.; Miller, H. F. *J. Chem. Soc.* **1947**, *69*, 1388.
8. Ueberreiter, K.; Krull, W. Z. *Phys. Chem. (Frankfort)* **1957**, *12*, 303.
9. Saotome, K.; Imoto, M. *Kobunshi Kagaku* **1958**, *15*, 368.
10. Noma, K.; Niwa, M.; Norisada, H. *Doshisha Daigaku Rikogaku Kenkyu Hokoku* **1970**, *10*, 349.
11. Jones, J. I. *J. Appl. Chem.* **1951**, *1*, 568.
12. Barb, W. G. *J. Polym. Sci.* **1959**, *37*, 515.
13. Kaufman, M.; Williams, A. F. *J. Appl. Chem.* (London) **1951**, *1*, 489.
14. Ballesteros, J.; Howard, G. J.; Teasdale, L. *J. Makromol. Sci. Chem.* **1977**, *A11*, 39.
15. Schaffhauser, R. J.; Shen, M. G.; Tobolsky, A. V. *J. Appl. Polym. Sci.* **1964**, *8*, 2825.
16. Dunham, K. R.; Vandenberghe, J.; Faber, J. W. H.; Flower, W. F. *J. Appl. Polym. Sci.* **1963**, *7*, 143.
17. Markevich, I. N. *Dokl. Akad. SSSR* **1970**, *191*, 362.
18. Iwatsuki, S.; Kubo, M.; Iwayama, H. *Macromolecules* **1993**, *26*, 7309.

ACETAL RESINS

Alex Forschirm and Francis B. McAndrew
Hoechst Celanese Corporation

With the introduction in 1960 of the first acetal polymer, DuPont's (Wilmington, DE) Delrin™ acetal homopolymer, acetal resins have grown in number, commercial importance, and stature such that they have become a truly significant component of the family of crystalline engineering thermoplastic resins.[1,2] Following closely on the heels of Delrin's introduction, Celanese Corporation (now Hoechst Celanese Corporation, Bridgewater, NJ) introduced an acetal copolymer (Celcon™) in late 1961.[3-5]

The introduction of these two acetal products was followed by a number of other acetal resins, both homo- and copolyacetals. Although each type of acetal resin has certain property and/or processing advantages, copolymers hold the major portion of the total acetal market for these engineering resins.

As a group acetal resins offer a high level of mechanical properties, excellent long-term dimensional stability, and, because they are highly crystalline, unusually good chemical resistance. Polyacetals have other desirable attributes, particularly compared with metals, such as low specific gravity, low electrical and thermal conductivities (a "warm-to-the-touch" feel), easy colorability, desirable acoustical properties, and they are easily processed into finished articles via all of the common plasticprocessing techniques.

SYNTHESIS OF ACETAL POLYMERS

The synthesis involves combining trioxane, the cyclic trimer of formaldehyde, with a few percent by weight of a cyclic ether such as ethylene oxide or 1,3-dioxolane in a ring-opening copolymerization, catalyzed by any of a large number of strong Lewis acids.[2,3]

To stabilize the above molecule, which has two unstable hemiacetal end groups, the polymer was heated under controlled conditions to remove the unstable end segments thermally or subjected to basic hydrolysis to remove them chemically. The final stabilized polymer had hydroxyethyl end groups as depicted in the following formula (**Equation 1**):

$$HO\text{–}[\text{–}CH_2\text{–}CH_2\text{–}O\text{–}(\text{–}CH_2\text{–}O\text{–})_m\text{–}]_n\text{–}CH_2\text{–}CH_2\text{–}OH \qquad 1$$

together with a small concentration of oxyethylene units randomly dispersed along the backbone of the polymer chain.

PROPERTIES

Acetal engineering thermoplastic resins have a more than 30-year record of providing predictable performance at high loading in adverse conditions in many applications. They are highly crystalline and thus are resistant to many solvents. They have a sharp crystalline melting point, giving relatively low viscosity melts which are easy to process by injection molding, which is the primary fabrication method for these resins.

Of considerable commercial importance is that acetal resins are among the lowest costing true engineering thermoplastics on a cents-per-cubic-inch basis.

In many applications, acetal resins have replaced metals because of their unique balance of properties, which include lower cost on a per-part basis, lighter weight, simpler processing, ability to get desirable combinations of properties, and generally improved benefit/cost ratios.

Stability

Both homopolymer and copolymer are stable at their recommended processing and use temperatures, but the performance of both can be enhanced by the incorporation of suitable additives. Further stability is achieved by incorporation of stabilizer additives.

Because acids are deleterious to acetal stability, various basis substances such as amines, amides, and basic salts such as calcium carboxylates may be added to the formulation as calcium carboxylates may be added to the formulation to counteract the acids.

Mechanical Properties

Acetal homopolymer has a higher level of short-term mechanical properties compared with the copolymer because of its higher level of crystallinity.

Chemical Properties

Both homopolymer and copolymer are relatively impervious to neutral inorganic and organic chemicals, including aliphatic and aromatic hydrocarbons, due to their highly crystalline nature. However, under certain conditions, such as high concentration and high temperature, some chemicals such as acids and oxidizing agents can cause acetal resins to degrade.

Flammability

Both homopolymer and copolymer are rated HB (horizontal burning test) by UL94 laboratory testing (Underwriters Laboratories).

APPLICATIONS

Because of their range of properties, acetal resins are used in a wide variety of applications.[6-10] Domestic consumption of polyacetal resins in industrial, plumbing/irrigation, automotive, and consumer applications accounts for approximately 77% of the acetal resin sold. The remaining 23% is divided among appliance/tools, electrical/electronic; hardware and other, including extruded products; medical; and miscellaneous applications.

EMERGING TRENDS

Acetal producers continue to develop more energy- and cost-efficient production processes, with a heavy emphasis on minimization of waste, improved safety, and protection of the environment. Acetal suppliers continue to improve the quality and performance of their products. Although this may involve a certain amount of polymer backbone chemistry, the principal route is through the use of additives. One of the limitations of such technology is that the high crystallinity of the resins limits how much additive can be incorporated because the crystalline volume is impermeable.

The emphasis is now on providing specialty resins for new application areas and working with customers on joint projects in which the customers have early and continuing input to product development.

REFERENCES

1. Persak, K. J.; Fleming, R. A. *High Performance Polymers: Their Origin and Development;* Seymour, R. B.; Kirshenbaum, G. S., Eds.; Elsevier: New York, 1986; pp 105–114.
2. *Polyaldehydes*, Vogl, O., Ed.; Marcel Dekker: New York, 1967; Chapter 2, pp 9–30.
3. Dolce, T. J.; McAndrew, F. B. In *High Performance Polymers: Their Origin and Development;* Seymour R. B.; Kirshenbaum, G. S., Eds.; Elsevier: New York, 1986; pp 115–124.
4. Serle, A. G. In *Engineering Thermoplastics, Properties and Applications;* Margolis, J. M., Ed.; Marcel Dekker: New York, 1985; pp 151–175.
5. *Polyaldehydes*, Vogl, O., Ed.; Marcel Dekker: New York, 1967; pp 31–42.
6. *DuPont Bulletin: Delrin P-Product and Properties Guide,* #233207A, 1993.
7. Hoechst Celanese Advanced Materials Group, *Celcon Acetal Copolymer, Properties,* CE-1A, 1992.
8. *DuPont Bulletin: Delrin II Properties Comparator,* #E-95274, 1987.
9. Sheehy, J.; Jakobi, R.; Yoshida, Y. *Chemical Economics Handbook*; SRI International, 1992.
10. *DuPont Bulletin: Delrin Product and Properties Guide,* #189168E.

ACETAL RESINS (Homopolymers Copolymers, and Block Copolymers)

Junzo Masamoto
Polymer Development Laboratory
Asahi Chemical Industry

Acetal resin (sometimes called polyoxymethylenes, polyacetals, or aldehyde resins) is a term used to describe high molecular weight polymers and copolymers of formaldehyde. First commercialized as a homopolymer in 1960 by Du Pont, acetal resins are engineering thermoplastics that have found broad use in traditional metal applications. The molecular structure of these resins is a repeating carbon-oxygen linkage (**Structure 1**).

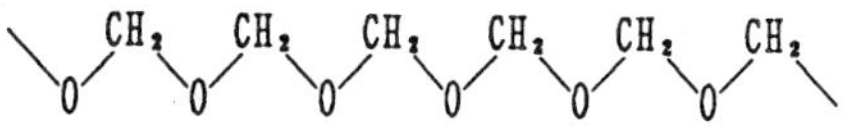

1

The generic name, acetal resin, has been accepted to describe these polymers because the alternating oxymethylene structure, ($-OCH_2-)_n$, gives them a chemistry similar to that for simple acetals. Acetal homopolymer refers to a resin consisting solely of this carbon–oxygen backbone. An acetal copolymer resin has the oxymethylene structure occasionally interrupted by a comonomer unit, such as an oxyethylene linkage.

In the early 1950s, Du Pont discovered that tough, solid polymers of formaldehyde were readily prepared from high-purity formaldehyde using ionic initiators. These resins were stabilized by replacing the hydroxyl radicals on the polymer chain ends with ester groups. The resulting polymer having excellent tensile, impact, and compression strengths and good abrasion and wear resistance appeared capable of filling many applications then reserved for metals.[1-3]

Shortly thereafter, Celanese (presently Hoechst Celanese) researchers developed an acetal resin based on the copolymerization of trioxane and cyclic ethers, such as ethylene oxide.[4]

Until 1971, Du Pont, Celanese, and Celanese joint ventures were the sole producers of acetal resins. At present, Asahi Chemical, BASF, Mitsubishi Gas Chemical and Ube Industries have commercialized acetal resins. Only Asahi Chemical has the acetal homo-, co-, and block copolymer.[5,6]

PRODUCTION OF POLYACETAL RESINS

Commercially, polyacetal resins are produced by polymerizing formaldehyde or trioxane. In general, polyacetal resins are said to be promising for the future because their starting raw material is methanol, an inexpensive base material of C_1 chemistry.

Acetal Homopolymer

At present, Du Pont and Asahi Chemical are the only producers of the acetal homopolymer, which is obtained from polymerizing formaldehyde. The general processing steps can be outlined as monomer purification, polymerization, end-capping, and finishing.

There are numerous reports about the polymerization of formaldehyde.[6]

Acetal Copolymer from Trioxane

Commercially, trioxane is copolymerized with ethylene oxide or cyclic formal such as dioxolane or 1,4-butanediol formal using boron trifluoxide to initiate the production of acetal copolymer.

Recently, the precise initiation mechanism for the copolymerization of trioxane and ethylene oxide was discovered.[14]

The initiation mechanism during bulk copolymerization of trioxane and ethylene oxide using boron trifluoride dibutyl ether as the initiator was studied. The new intermediates or novel compounds, 1,3,5,7-tetraoxacyclononane (TOCN) and 1,3,5,7,10-pentaoxacyclododecane (POCD), disclosed the precise initiation mechanism.[14]

Acetal Block Copolymer

Acetal block copolymer is obtained by polymerizing formaldehyde in the presence of the functional polymer, $RO(X)_mH$, which has an active hydrogen atom.[6]

PROPERTIES AND APPLICATIONS

Since their introduction in 1960, acetal resins have been modified in many ways to fit the needs of various applications.

The addition of elastomeric polymers have created toughened grades of polyacetal which retain a high degree of impact resistance even at very low temperatures.[19]

Among the acetal homopolymers, a new product line has been introduced, beginning with basic grades, and was developed for use at a wider range of processing temperatures. Furthermore, two new products, polyacetal film and super drawn acetal fiber, were developed by Asahi Chemical, and they are now under market research.[6]

Stabilizer

For these new products, Asahi Chemical and Du Pont developed new stabilizers, Du Pont has already proposed copolyamide of nylon 6-66-610 as a formaldehyde scavenger.

Asahi Chemical researchers discovered that nylon 3 was an excellent stabilizer for polyoxymethylene.[20]

Characteristics of polyacetal resins compared with other plastics are as follows:

- Excellent impact strength, especially under repeated impact
- High resistance to fatigue
- Low friction and wear
- Resistance to organic solvents and oils as well as to cold and hot water with excellent resistance to alkalis (it is, however, affected by some acids)

Applications

There are many applications of polyacetals for replacing metals as an engineering plastic. High molecular weight polyacetals can be used in extruded rods and sheets, roller conveyor parts, ski parts, pipe fittings, rollers for building materials, and exterior automotive components. Medium molecular weight polyacetals (general purpose grade) can be used in gears, cams, hose joints, clips, switches, mechanical parts of audio and video equipment, interior and exterior automotive components (e.g., window regulator handles, seat hooks, outer door handles), cassette hubs, and bearings. High flow polyacetals can be used in lighters, aerosol bottles, fasteners, VTR reels, fasteners, aerosol valve parts, other multicavity molded parts, and cassette hubs.

REFERENCES

1. Schweitzer, C. E.; MacDonald, R. N.; Punderson, J. O. *J. Appl. Polym. Sci.* **1959**, *1*, 158.
2. Koch, T. A.; Lindvig, P. E. *J. Appl. Polym. Sci.* **1959**, *1*, 164.
3. Linton, W. H.; Goodman, H. H. *J. Appl. Polym. Sci.* **1959**, *1*, 179.
4. Walling, C.; Brown, F.; Bartz, K. U.S. Patent 3 027 352, 1962.
5. Masamoto, J.; Iwaisako, T.; Matsuzaki, K. *J. Macromol. Sci., Pure Appl. Chem.* **1992**, *A29*, 441.
6. Masamoto, J. *Progr. Polym. Sci.* **1993**, *18*, 1.
7. Masamoto, J.; Iwaisako, T.; Chohono, M., et al. *J. Appl. Polym. Sci.* **1993,** *50*, 1289.
8. Morishita, H.; Masamoto, J.; Hata, T. U.S. Patent 4 962 235, 1990.
9. Masamoto, J.; Matsuzaki, K.; Morishita, H. *J. Appl. Polym. Sci.* **1993,** *50*, 1307.
10. Masamoto, J. Japanese Patent 3-128910, 1991.
11. Masamoto, J.; Matsuzaki, K. *Polym. Plast. Technol. Eng.* **1942**, *33*, 233.
12. Matsuzaki, K.; Masamoto, J. Japanese Patent 89-11025, 1989.
13. Hamanaka, K., Iwaisako, T.; Masamoto, J., et al. U.S. Patent 4 332 644, 1982.
14. Masamoto, J.; Iwaisako, T.; Yoshida, K., et al. *Makromol. Chem. Macromol. Sym.* **1991**, *42143*, 409.
15. Weissermel, K.; Fisher, E.; Gutweiler, K. *Kunstoffe*, **1964**, *54*, 410.
16. Weissermel, K.; Fisher, E.; Gutweiler, K., et al. *Agnew Chem. Int. Ed.* (English), **1967**, *6*, 526.
17. Collins, G. L.; Greene, R. K.; Beradinelle, F. M., et al. *J. Polym. Sci., Polym. Chem. Ed.* **1981**, *19*, 1597.
18. Tsunemi, K.; Suzuki, T.; Yamamoto, Y. Japanese Patent 87-31727, 1987; U.S. Patent 4 504 636, 1985; U.K. Patent 2 132 213, 1984.
19. Flexmann, E. A.; Huang, D. D.; Snyder, H. L. *ACS Polym. Prepr.* **1988**, *29*(2), 189.
20. Yamamoto, F.; Misumi, T. U.S. Patent 4 855 365, 1989; U.S. Patent 5 015 707, 1991.

Acetals

See: Acetal Resins
Acetal Resins (Homopolymers, Copolymers, and Block Copolymers)
Bicyclic Acetals, Ring-Opening Polymerization
Engineering Thermoplastics (Survey of Industrial Polymers)
Poly(1,3-dioxolane)
Polyoxymethylene Fiber, Superdrawn

ACETAN

G. J. Brownsey, A. M. Griffin, and V. J. Morris
Institute of Food Research
Norwich Laboratory

Polysaccharide hydrocolloids obtained from land plants and seaweeds are used successfully in foods, pharmaceuticals, and numerous industrial applications. Their usefulness stems from the ability of the biopolymer to structure the rheological properties of the solvent (usually water) phase.

Bacteria in general can produce a vast range of neutral and charged polysaccharides. They provide a means of converting waste or excess agricultural material into new polysaccharides that supplement those extractable directly from plants. Bacterial polysaccharides offer the advantages of regular chemical structures, reproducible chemical and physical properties, and a constant source of supply. Two useful industrial products recently identified in this fashion include gellan gum and xanthan gum. Gellan gum is being marketed as a broad-spectrum gelling agent, and xanthan gum has cornered a unique market as a thickening and suspending agent.[1,2]

Xanthan gum, perhaps the most studied of the bacterial polysaccharides, is one member of a family of substituted cellulosics having a cellulose backbone solubilized with a trisaccharide side chain substituted on alternate glucose residues.[3] With xanthan it has been possible to alter the fermentation conditions and isolate mutant bacteria that synthesize particular members of the xanthan family.[4-6]

Genetic engineering has been used to isolate a series of mutants synthesizing the whole family of xanthan variants.[7]

ACETAN AND ACETAN VARIANTS

Acetobacter are acetic acid-producing bacteria, which are often identified as the contaminant that turns wine sour. Certain strains of *Acetobacter xylinum* provide a primary source of bacterial cellulose.[8] Other polysaccharides can be produced, and their concentration can be boosted at the expense of cellulose production merely by swirling or shaking the culture broth.[9]

Acetan is an anionic heteropolysaccharide secreted by *Acetobacter xylinum*.[10,11] The biochemical pathway for the biosynthesis of acetan has been elucidated but, as yet, little is known about the genetics of this process. Couso and co-workers postulated that the repeat unit of the polysaccharide was assembled as a heptasaccharide linked to a lipid carrier.[12]

Variants of the acetan structure can be prepared by a variety of methods. Deacetylated acetan can be prepared by treatment in the cold by using 0.1 *M* NaOH for 24 h at a pH of 12.5. Chemical methods can also be used to truncate the pentasaccharide side chain.[15,16]

Random mutagenesis and screening can be used to select for mutants that secrete acetan variant structures. Several such mutants have been prepared by chemical (*N*-methyl-*N′*-nitro-*N*-nitrosoguanidine) mutagenesis.[14]

Rhamnose-negative mutants were isolated as candidates for producing acetan variant structures. Only one such mutant (CR1/4) has been analyzed in great detail. The chemical structure of CR1/4 has been determined and shown to be a partially acetylated acetan variant with a side chain terminating in glucuronic acid.[13,14] Deacetylated CR1/4 can be prepared by alkali treatment.

FUNCTIONALITY OF ACETAN AND CR1/4

All three polysaccharides—xanthan, acetan, and CR1/4—share a common cellulosic backbone substituted on alternate glucose residues with different side chains. The nature of the side-chain backbone linkage is common, and the carbohydrate sequence for the first two sugars in the side chains is identical. The noncarbohydrate decoration of the structure is different. All three polymers have the same acetylation site on the (1 → 2) mannose residue. However, whereas xanthan may contain pyruvate or acetate on the terminal mannose residue, both acetan and CR1/4 share an acetylation site on the polymer backbone.[13]

Comparative studies of the xanthan, xanthan tetramer, and xanthan trimer have shown that the (1 → 2) mannose residue is sufficient to alter the conformation of the cellulosic backbone.[17] Thus, it would be expected that acetan and CR1/4 should adopt a xanthan-like fivefold helical structure. Chiroptical and differential scanning calorimetry studies of aqueous solutions of acetan and CR1/4 show clear evidence for an order-disorder transition.[18-20] The ordered conformations are favored by low temperatures and high ionic strength.[18,20] Deacetylation of acetan has been found to stabilize the ordered structure.[20] By analogy, with similar studies on xanthan these order-disorder transitions have been attributed to a helix-coil transition.[18,20,21] Physicochemical studies of both acetan and CR1/4 suggest that they exist as stiff coils in solution.[19,20] This has been confirmed for acetan by electron microscopy and atomic force microscopy (AFM).[22,23] AFM studies directly revealed the helical structure of acetan.[23]

Solutions of acetan and CR1/4 in the helical conformation show high viscosities and marked shear thinning behavior.[16,18-20] At higher polymer concentrations, and particularly at higher ionic strengths, there is increasing evidence for intermolecular association and weak gellike behavior.[19,20] For acetan, this associative behavior is reduced upon deacetylation.[20] The viscosity of solutions of acetan will be determined by the stiffness of the helical structure and the molecular weight.

By partial acid hydrolysis of acetan it is possible to prepare a polymer approximating CR1/4.

Mixtures of xanthan with certain galactomannans or glucomannans result in the formation of thermoreversible gels, under conditions for which the individual polymers do not gel.[24] Such synergistic interactions have been shown to result from specific intermolecular binding between the two polysaccharides.[25-27] Mixtures of acetan with carob or glucomannan (konjac mannan) show no detectable interaction unless acetan is deacetylated. Gelation of deacetylated acetan with carob or konjac mannan has been confirmed at higher polymer concentrations.[20] Because deacetylation stabilizes the acetan helix, these studies emphasize the potential steric effects of acetyl substituents on the polymer backbone and therefore on functionality.

FUTURE PROSPECTS

Genetic engineering of microbial polysaccharide structures resulting in the deletion of sugars or noncarbohydrate substituents can be used to produce families of structurally related polysaccharides. The families based on acetan will encompass most of the xanthan family of polysaccharides. The similarity of the xanthan and acetan structures, and their biosynthetic pathways, offers an opportunity to test the feasibility of swapping genes between microorganisms for the *de novo* design of new polysaccharide structures.

REFERENCES

1. Sanderson, G. R. *Food Gels*, Elsevier Applied Science: London, 1990; Chapter 6.
2. Morris, V. J. *Food Biotechnology*; Elsevier Applied Science: London, 1987; Chapter 5.
3. Jansson, P.-E.; Kenne, L.; Lindberg, B. *Carbohydr. Res.* **1975**, *45*, 275.
4. Thorne, L.; Tansey, L.; Pollock, T. J. *J. Bacter,* **1987**, *169*, 3563.

5. Barrere, G. C.; Barber, C. E.; Daniels, M. J. *Int. J. Biol. Macromol.* **1986**, *8*, 372.
6. Harding, N. E.; Cleary, J. M.; Cabanas, D. K.; Rosen, I. G.; Kang, K. S. *J. Bacter.* **1987**, *169*, 2854.
7. Hassler, R. A.; Doherty, D. H. *Biotechnol. Progr.* **1990**, *6*, 182.
8. Ross, P.; Mayer, R.; Benziman, M. *Microbiol. Rev.* **1991**, *55*, 35.
9. Schramm, M.; Hestrin, S. *J. Gen. Microbiol.* **1954**, *11*, 123.
10. Couso, R. O.; Ielpi, L.; Dankert, M. A. *J. Gen. Micorbiol.* **1987**, *133*, 2123.
11. Jansson, P.-E.; Lindberg, J.; Wilmalasiri, K.; Danker, M. A. *Carbohydr. Res.* **1993**, *245*, 303.
12. Couso, R. O.; Ielpi, L.; Garcia, R. C.; Dankert, M. A. *Eur. J. Biochem.* **1982**, *123*, 617.
13. Colquhoun, I. J.; Defernez, M.; Morris, V. J. *Carbohydr. Res.* **1995**, *269*, 319.
14. McCormick, C. A.; Harris, J. E.; Gunning, A. P.; Morris, V. J. *J. Appl. Bact.* **1993**, *74*, 196.
15. Christensen, B. E.; Smidsrod, O.; Stokke, B. T. *Carbohydr. Polym.* **1994**, *25*, 25.
16. Morris, V. J. *Biotechnology and Bioactive Polymers*; Plenum: New York, 1994; pp 9–16.
17. Millane, R. P. *Frontiers in Carbohydrate Research-2*; Elsevier Applied Science: London, 1992; pp 168–190.
18. Morris, V. J.; Brownsey, G. J.; Cairns, P.; Chilvers, G. R.; Miles, M. J. *Int. J. Biol. Macromol.* **1989**, *11*, 326.
19. Ridout, M. J.; Brownsey, G. J.; Morris, V. J.; Cairns, P. *Int. J. Biol. Macromol.* **1994**, *16*, 324.
20. Ojinnaka, C.; Morris, E. R.; Morris, V. J.; Brownsey, G. J. *Gums and Stabilisers for the Food Industry 7*; IRL: Oxford, England, 1994; pp 15–26.
21. Morris, E. R.; Rees, D. A.; Young, G.; Walkinshaw, M. D.; Darke, D. *J. Mol. Biol.* **1977**, *110*, 1.
22. Stokke, B. T.; Elgsaeter, A.; Skjak-Braek, G.; Smidsrod, O. *Carbohydr. Res.* **1987**, *160*, 13.
23. Kirby, A. R.; Gunning, A. P.; Morris, V. J.; Ridout, M. J. *Biophys. J.* **1995**, *68*, 360.
24. Dea, I. C. M.; Morris, E. R.; Rees, D. A.; Welsh, E. J.; Barnes, H. A.; Price, J. *Carbohydr. Res.* **1977**, *57*, 249.
25. Brownsey, G. J.; Cairns, P.; Miles, M. J.; Morris, V. J. *Carbohydr. Res.* **1988**, *176*, 329.
26. Cairns, P.; Miles, M. J.; Morris, V. J. *Nature*, **1986**, *322*, 89.
27. Cairns, P.; Miles, M. J.; Morris, V. J.; Brownsey, G. J. *Carbohydr. Res.* **1987**, *160*, 411.

Acetylene Polymers

See: *Acetylene-Terminated Monomers (Catalyzed Cure)*
Acetylenic Polymers
Cobalt Carbonyl Complexes (With Main Chain Acetylenic Groups)
Inorganic/Organic Hybrid Polymers (High Temperature, Oxidatively Stable)
Personal Care Application Polymers (Acetylene-Derived)
Solid-State Polymerization (Crosslinked Di-p-Ethynyl Substituted Polymers)
Star-Branched and Crosslinked Polymers (Cyclotriphosphazene Cores)

ACETYLENE-TERMINATED MONOMERS (Catalyzed Cure)

William E. Douglas
Unite Mixte CNRS/Rhône-Poulenc/UM II
CNRS UMR 44
Université de Montpellier II

High-performance resins formed from acetylene-terminated monomers (ATMs) are currently being developed to replace epoxies for use under hot, wet conditions.[1-3] ATMs fall into two main categories: those terminated by ethynylarylene (e.g., **Structure 1**) and those terminated by arylenepropargyl ether groups (e.g., **Structure 2**). Both types of ATM are usually cured thermally in the absence of catalysts.

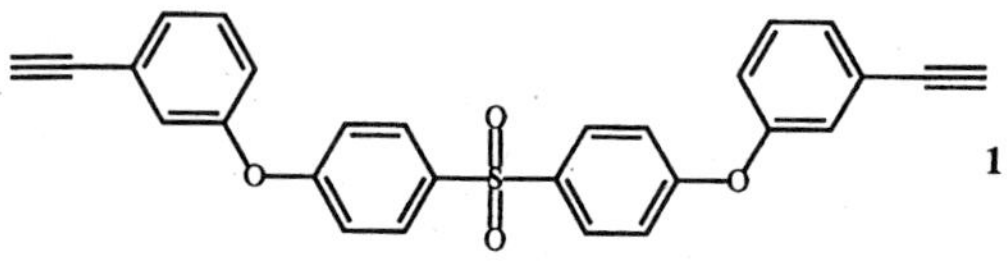
1

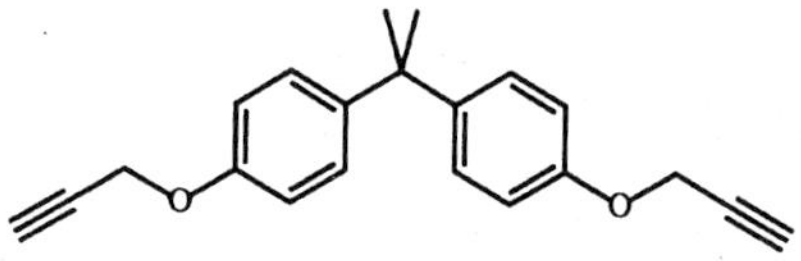
2

Many transition-metal complexes are well known to catalyze acetylene reactions such as cyclotrimerization, cyclotetramerization, and linear polyene formation,[5] and their use in the cure of ATMs might therefore be expected to affect the structure and properties of the resulting resins.

PREPARATION AND PROPERTIES

Arylenepropargyl Ether-Terminated Monomer 2

Structure 2 and the corresponding **Structure 3** were prepared following the previously described procedure.[6]

^{1}H and ^{13}C NMR analysis of the products resulting from a partial cure of Structure 2 at 220°C, and a complete cure of Structure 3, showed that crosslinking involves prior sigmatropic rearrangement of the arylenepropargyl ether groups to mono-2*H*-1-benzopyran (**Structure 4**) and di-2*H*-1-benzopyran (**Structure 5**) structures (**Scheme I**), which subsequently polymerize. The various intermediate 2*H*-1-benzopyran-containing species were isolated and described by Douglas and Overend.[4]

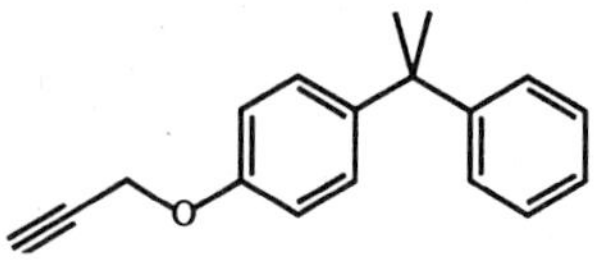
3

The resin glass transition temperature (T_g) values, as determined by thermochemical analysis (TMA), show that the structure of the uncatalyzed resin must be different from the structures obtained by using catalysts, because the T_g for the uncatalyzed resin (258°C) is considerably higher than those for the catalyzed resins (118–170°C). The specific network formed

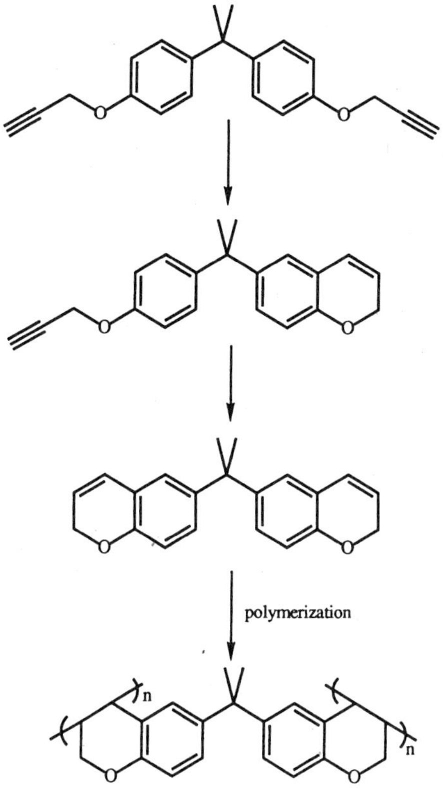

2

4

5

SCHEME I. Uncatalyzed cure of Structure 2.

via the 2*H*-1-benzopyran route is thus particularly rigid and has high thermal stability. This is supported by IR and NMR studies into the differing chemical structures formed in the presence of the various catalysts.[8]

APPLICATIONS

Few reports of the catalyzed cure of ATMs have appeared. $(PPh_3)_2NiCl_2$ has been found to lower the cure temperature of ethynylarylene-terminated monomers sufficiently (<177°C) for the resins to be used as adhesives for aluminum alloys.[9,10] Two of the catalysts that had originally been reported by Douglas and Overend[7] as being active have been used in the cure of arylenepropargyl ether ATMs: $(\eta\text{-Cp})_2Co$, which was found to give rise to cyclotrimerization,[11] and bis(triphenylphosphine) palladium dichloride.[12]

It is to be expected that the fundamental differences in chemical structure arising from the use of different catalysts will affect the physical properties of the cured resins. In particular, resins cured in the presence of cyclotrimerization catalysts such as $(\eta\text{-Cp})Co(CO)_2$ should contain highly stable benzene crosslink sites. However, overall thermal stability of the resin may depend on the structure of the monomer itself rather than on that of the crosslinks.

REFERENCES

1. Sergeev, V. A.; Chernomordik, Y. A.; Kurapov, A. A. *Uspekhi Khimii* **1984**, *53*, 518. [Russ. Chem. Rev. **1984**, *53*, 307].
2. Hergenrother, P. M. In *Encyclopedia of Polymer Science and Engineering,* 2nd ed.; Mark, H. F. et al., Eds.; John Wiley & Sons: New York, 1985; Vol. 1, pp 61–86.
3. Lee, C. Y.-C. In *Developments in Reinforced Plastics;* Pritchard, G., Ed.; Elsevier: Barking G. B., 1986; pp 121–150.
4. Douglas, W. E.; Overend, A. S. *Eur. Polym. J.* **1991**, *27*, 1279.
5. Collman, J. P.; Hegedus, L. S.; Norton, J. R.; Finke, R. G. *Principles and Applications of Organotransition Metal Chemistry;* University Science Books; Mill Valley, CA, 1987.
6. Picklesimer, L. G. U.S. Patent 4 226 800, 1981; Chem. Abstr. **1981**, *94*, 31283q.
7. Douglas, W. E.; Overend, A. J. *Organomet. Chem.* **1986**, *308*, C14.
8. Douglas, W. E.; Overend, A. S. *J. Mater. Chem.* **1993**, *3*, 1019.
9. Picklesimer, L. G.; Lucarelli, M. A.; Jones, W. B.; Helminiak, T. E.; Kang, C. C. *Polym. Prepr.* **1981**, *22*, 97.
10. Lucarelli, M. A.; Jones, W. B.; Picklesimer, L. G.; Helminiak, T. E. In ACS Symposium Series 195; American Chemical Society: Washington, DC, 1982, p 237.
11. Sergeyev, V. A.; Shitikov, V. K.; Kurapov, A. S.; Antonova-Antipova, I. P. *Vys. Soyed.* **1989**, *A31*, 1188. [Polym. Sci. USSR, **1989**, *31*, 1300].
12. Pigneri, A. P.; Bauer, R. S. U.S. Patent Appl. 386 083, 1989; *Chem. Abstr.* **1991**, *115*, 30129j.

ACETYLENIC POLYMERS

Toshio Masuda
Department of Polymer Chemistry
Kyoto University

In general, acetylenic polymers can be obtained by the polymerization of acetylenes using transition-metal catalysts (**Equation 1**). The most popular way of preparing polyacetylene is the Shirakawa method, which uses the $Ti(O\text{-}n\text{-}Bu)_4\text{-}Et_3Al$ catalyst and provides polyacetylene film directly. In recent years, various polymers have been obtained from substituted acetylenes by using group 5–8 transition-metal catalysts.

$$RC{\equiv}CR' \xrightarrow[\text{(R, R': H or substituent)}]{\text{catalyst}} \left[\underset{\substack{| \\ R}}{C} = \underset{\substack{| \\ R'}}{C} \right]_n \quad 1$$

Polyacetylene is one of the conjugated polymers that possess the simplest structures, and there have been many studies on the electrical conductivity of doped polyacetylenes. The extent of conjugation in substituted polyacetylenes depends on the kind and number of substituent in the repeat unit. These polymers are stiffer than vinyl polymers, which gives rise to unique properties such as high gas permeability.

SYNTHESIS

Monomers

Acetylene

Polyacetylene membrane can be directly obtained by the Shirakawa method, which features high catalyst concentrations.[1] Typical polymerization conditions are $[Ti(O\text{-}n\text{-}Bu)_4] = 0.25\ M$, $[Et_3Al] = 1.0\ M$, with toluene, a temperature of −78°C, and the pressure of acetylene at 500–600 mmHg.

Syntheses of polyacetylene via other routes are also known.

Monosubstituted Acetylenes

Typical examples of the polymerization of monosubstituted acetylenes are shown in **Table 1**.[2-20] Note that the polymer molecular weights strongly depend on both monomer and catalyst.

TABLE 1 Examples of the Polymerization of Monosubsituted Acetylenes

Monomer	Catalyst	*Mw* or [η], 10^3 dL/g	Ref.
Hydrocarbon acetylenes			
HC≡C-*n*-Bu	$WCl_2(OC_6H_4\text{-}o,o\text{-}Me_2)_4$	170 (M_n)	2
HC≡C-*s*-Bu	$Fe(acac)_3-i\text{-}Bu_3Al$	1.0 ([η])	3
HC≡C-*t*-Bu	$MoCl_5$	33 (M_n)	4
HC≡CPh	WCl_6-Ph_4Sn	15 (M_n)	5
HC≡CPh	$W(CO)_6-CCl_4-h\nu$	80 (M_n)	6
HC≡CPh	$(NBD{\cdot}RhC1)_2$	350	7
$HC{\equiv}CC_6H_2\text{-}o,o\text{-}Me_2\text{-}p\text{-}t\text{-}Bu$	$W(CO)_6-CCl_4-h\nu$	2600	8
Heteroatom-containing acetylenes			
$HC{\equiv}CCH(SiMe_3)\text{-}n\text{-}C_5H_{11}$	$MoCl_5-Et_3SiH$	4500	9
$HC{\equiv}CC_6H_4\text{-}o\text{-}SiMe_3$	$W(CO)_6-CCl_4-h\nu$	3400	10
$HC{\equiv}CC_6H_4\text{-}o\text{-}GeMe_3$	WCl_6	690	11
$HC{\equiv}CCH_2OH$	$PdCl_2$	Insoluble	12
$HC{\equiv}CCO_2\text{-}n\text{-}Bu$	$(NBD{\cdot}RhC1)_2$	84	13
HC≡C-α-thiophene	$WCl_6-n\text{-}Bu_4Sn$	20 (M_n)	14
$HC{\equiv}CCH_2NH_2$	$Mo(OEt)_5-EtAlCl_2$	Insoluble	15
HC≡CCN	$Ti(OBu)_4-Et_3Al$	0.11 ([η])	16
$HC{\equiv}C(CF_2)_5CF_3$	WCl_6-Ph_4Sn	0.047 ([η])	17
$HC{\equiv}CC_6H_4\text{-}o\text{-}CF_3$	$W(CO)_6-CCl_4-h\nu$	1600	18
$HC{\equiv}CC_6F_5$	WCl_6-Ph_4Sn	0.61 ([η])	19
$HC{\equiv}CCH_2Cl$	$MoCl_5$	—	20

Typical catalysts for the polymerization of phenylacetylene include W, Rh, and Fe catalysts. W catalysts produce an auburn, soluble, trans-rich polymer.[5] Polymerization by Rh catalysts forms a yellow, soluble, cis-transoidal polymer.[7,21] In contrast, the polyphenylacetylene obtained with a $Fe(acac)_3$–Et_3Al catalyst is crimson, insoluble in any solvent, and cis-cisoidal.[22]

An interesting point in the polymerization of phenylacetylenes by W and Mo catalysts is that derivatives with bulky ortho-substituents (e.g., $SiMe_3$ and CF_3) yield polymers whose molecular weights reach about 1 million. This ortho-substituent effect is remarkable because the molecular weight of polyphenylacetylene with the same catalysts is about 2 orders of magnitude smaller. This demonstrates that W and Mo catalysts are especially effective in sterically crowded monosubstituted acetylenes.

Disubstituted Acetylenes

Because disubstituted acetylenes in general exhibit more steric hindrance than do monosubstituted ones, the catalysts for the former are virtually restricted to group 5 and 6 transition-metal catalysts. Disubstituted monomers with less steric hindrance tend to polymerize with Mo and W catalysts, whereas sterically crowded counterparts polymerize only with Nb and Ta catalysts.

One of the most reactive and interesting heteroatom-containing disubstituted acetylenes is 1-(trimethylsilyl)-1-propyne. It quantitatively polymerizes with Nb and Ta catalysts.[23] The weight-average molecular weight of the polymer obtained with $TaCl_5$–Ph_3Bi reaches 4 million.[24]

α,ω-Diacetylenes

There is a possibility that the cyclopolymerization of α,ω-diacetylenes affords soluble, long conjugated polymers with ring-closed structures. In fact, Choi et al. have reported many examples of the synthesis of such polymers from dipropargyl compounds.[25-33]

Control of Polymerization

Stereospecific Polymerization

The cis-transoidal and trans-transoidal structures have been found for polyacetylene. The geometric structure of polyphenylacetylene varies depending on the catalyst: cis-cisoidal with Fe catalysts, cis-transoidal with Rh, cis-rich with Mo, and trans-rich with W.[7,22,34]

Living Polymerization

A mixture of $MoOCl_4$, *n*-Bu_4Sn, and EtOH at a 1:1:1 mole ratio produces polymers with narrow MWDs from 1-chloro-1-octyne, various phenylacetylenes with bulky ortho-substituents, and *tert*-butylacetylene. In these polymerizations, the polymer molecular weight increases with the monomer conversion, which verifies that these are living polymerizations.

PROPERTIES AND FUNCTIONS

Properties

General Features

Most of the substituted polyacetylenes are soluble, due to the interaction between solvents and substituents. Furthermore, they are stable in air at room temperature. Although polyacetylene is a semiconductor and paramagnetic, its derivatives do not necessarily show such properties. This is attributable to twisted conformations of the main chain due to the presence of side groups.

Color

Although polyacetylene strongly absorbs visible light, its thick film looks like a metal, which is due to reflection of visible light by free electrons.[35,36] The color of substituted polyacetylenes varies widely from dark purple to colorlessness.

Solubility

Polyacetylene is insoluble, but most substituted polyacetylenes are soluble in common organic solvents such as toluene and chloroform. Exceptions are the polymers from symmetrically disubstituted acetylenes (e.g., 4-octyne and diphenylacetylene), which are usually insoluble in any solvent.

Thermal Stability

Aromatic polymers are more stable than aliphatic ones, and disubstituted acetylene polymers are more stable than their monosubstituted counterparts.

Mechanical Properties

As-polymerized *cis*-polyacetylene has a small Young's modulus (E = 200 MPa) and a fairly large elongation at break (γ_B = 200%).[37] However, an oriented *trans*-polyacetylene has very large E (100 GPa) and tensile strength (σ_B = 0.9 GPa) values.[38] Among substituted polyacetylenes, aromatic ones are generally hard and brittle (E = 2000–1000 MPa, γ_B <10%), whereas aliphatic ones, especially those with long alkyl groups, are soft and ductile (E <500 MPa; γ_B, 50–400%).[39,40]

Functions

Electrical Conductivity

The σ of the doped Shirakawa polyacetylene is in the range 10^2–10^3 S·cm^{-1}. It is claimed that Naarmann's stretched film shows a σ value that is higher than 10^4 S·cm^{-1}.[41] The polyacetylenes with ring structures show σ values of 10^{-12}–10^{-6} S·cm^{-1}, which increase to 10^{-1}–10^{-4} S·cm^{-1} when the polymers are doped with I_2 or AsF_5.

Photoelectronic Functions

The third-order nonlinear optical susceptibilities of polyacetylenes are as follows: polyacetylene, 5 × 10^{-10} esu; poly(HC≡CPh), 3 × 10^{-12} esu; poly(HC≡CC_6H_4-*o*-Si-Me_3), 12 × 10^{-12} esu; and poly(HC≡CC_6H_4-*o*-GeMe_3), 26 × 10^{-12} esu.[42-45]

Gas Permeability

In general, substituted polyacetylenes are very permeable to gases because of their stiff main-chain and bulky side groups.[46-55] Two types of the most gas-permeable polymers are (i) poly[1-(trimethylsilyl)-1-propyne] and its analogues and (ii) polydiphenylacetylenes with round-shaped para- or meta-substituents such as $SiMe_3$, $GeMe_3$, and *t*-Bu groups.

Separation of Liquid Mixtures

Ethanol concentration of dilute aqueous solutions by using a membrane is under intensive research. It has been revealed that poly(MeC≡C$SiMe_3$) and poly(dimethylsiloxane) are among the rare polymer membranes that permeate ethanol preferentially.[56,57]

REFERENCES

1. Ito, T.; Shirakawa, H.; Ikeda, S. *J. Polym. Sci., Polym. Chem. Ed.* **1974**, *12*, 11.
2. Nakayama, Y.; Mashima, K.; Nakamura, A. *J. Chem. Soc., Chem. Commun.* **1992**, 1496.
3. Ciardelli, T.; Lanzillo, S.; Pieroni, O. *Macromolecules* **1974**, *7*, 175.
4. Masuda, T.; Okano, Y.; Kuwane, Y.; Higashimura, T. *Polym. J.* **1980**, *12*, 907.
5. Masuda, T.; Thieu, K.-Q.; Sasaki, N.; Higashimura, T. *Macromolecules* **1976**, *9*, 661.
6. Masuda, T.; Yamamoto, K.; Higashimura, T. *Polymer* **1982**, *23*, 1663.
7. Tabata, M.; Yang, W., Yokota, K. *J. Polym. Sci., Part A, Polym. Chem.* **1994**, *32*, 1113.
8. Yoshida, T.; Abe, Y., Masuda, T.; Higashimura, T. *Polym. Preprs. Jpn* **1988**, *37*(2), 144.
9. Masuda, T.; Tajima, H.; Yoshimura, T.; Higashimura, T. *Macromolecules* **1987**, *20*, 1467.
10. Masuda, T.; Hamano, T.; Tsuchihara, K.; Higashimura, T. *Macromolecules* **1990**, *23*, 1374.
11. Mizumoto, T.; Masuda, T.; Higashimura, T. *J. Polym. Sci., Part A, Polym. Chem.* **1993**, *31*, 2555.
12. Akopyan, I. A.; Grigoryan, S. G.; Zhamkochyan, G. A.; Matsoyan, S. G. *Vysokomol. Soed.* **1975**, *A17*, 2517.
13. Tabata, M.; Inaba, Y.; Yokota, K.; Nozaki, Y. *J. Macromol. Sci., Pure Appl. Chem.* **1994**, *A31*, 465.
14. Gal, Y-S.; Aho, H-N.; Choi, S-K. *J. Polym Sci., Part A, Polym. Chem.* **1986**, *24*, 2021.
15. Gal, Y-S.; Jung, B.; Lee, W-C.; Choi, S-K. *J. Polym. Sci., Part A, Polym. Chem.* **1992**, *30*, 2657.
16. Carlini, C.; Chien, J. C. W. *J. Polym. Sci., Polym. Chem. Ed.* **1984**, *22*, 2749.
17. Tsuchihara, K.; Masuda, T.; Higashimura, T. *Polym. Bull.* **1988**, *20*, 343.
18. Masuda, T.; Hamano, T.; Higashimura, T.; Ueda, T.; Muramatsu, H. *Macromolecules* **1988**, *21*, 281.
19. Yoshimura, T.; Masuda, T.; Higashimura, T.; Okuhara, K.; Ueda, T. *Macromolecules* **1991**, *24*, 6053.
20. Kunzler, J.; Percec, V. *J. Polym. Sci., Part A, Polym. Chem.* **1990**, *28*, 1043.
21. Furlani, A.; Napoletano, C.; Russo, M. V.; Camus, A.; Marsich, N. *J. Polym. Sci., Part A, Polym. Chem.* **1989**, *27*, 75.
22. Kern, R. J. *J. Polym. Sci., Part A-1* **1969**, *7*, 621.
23. Masuda, T.; Isobe, E.; Higashimura, T. *Macromolecules* **1985**, *18*, 841.
24. Masuda, T.; Isobe, E.; Hamano, T.; Higashimura, T. *Macromolecules* **1986**, *19*, 2448.
25. Gibson, H. W.; Bailey, F. C.; Epstein, A. J.; Rommelmann, H.; Kapla, S.; Harbour, J.; Yang, X-Q.; Tanner, D. B.; Pochan, J. M. *J. Am. Chem. Soc.* **1983**, *105*, 4417.
26. Jang, M-S.; Kwon, S-K.; Choi, S-K. *Macromolecules* **1990**, *23*, 4135.
27. Kim, Y-H.; Gal, Y-S.; Kim, U-Y.; Choi, S-K. *Macromolecules* **1988**, *21*, 1991.
28. Cho, O-K.; Kum, Y-H.; Choi, K-Y.; Choi, S-K. *Macromolecules* **1990**, *23*, 12.
29. Ryoo, M-S.; Lee, W-C.; Choi, S-K. *Macromolecules* **1990**, *23*, 3029.
30. Han, S-H.; Kim, U-Y.; Kang, Y-S.; Choi, S-K. *Macromolecules* **1991**, *24*, 974.
31. Park, J-W.; Lee, J-H.; Cho, H-N.; Choi, S-K. *Macromolecules* **1993**, *26*, 1191.
32. Jin, S-H.; Choi, S-J.; Ahn, W.; Cho, H-N.; Choi, S-K. *Macromolecules* **1993**, *26*, 1487.
33. Kim, S-H.; Choi, S-J.; Park, J-W.; Cho, H-N.; Choi, S-K. *Macromolecules* **1994**, *27*, 2339.
34. Masuda, T.; Higashimura, T. *Adv. Polym. Sci.* **1986**, *81*, 121.
35. Shirakawa, H.; Ito, T.; Ikeda, S. *Polym. J.* **1973**, *4*, 460.
36. Fincher, C. R., Jr.; Ozaki, M.; Tanaka, M.; Peebles, D. L.; Lauchlau, L.; Heeger, A. J.; MacDiarmid, A. G. *Phys. Rev. B* **1979**, *20*, 1589.
37. Druy, M. A.; Tsang, C-H.; Brown, N.; Heeger, A. J.; MacDiarmid, A. G. *J. Polym. Sci., Polym. Phys. Ed.* **1980**, *18*, 429.
38. Akagi, K.; Suezaki, M.; Shirakawa, H.; Kyotani, H.; Shimomura, M.; Tanabe, Y. *Synth. Met.* **1989**, *28*, D1.
39. Masuda, T.; Tang, B-Z.; Tanaka, A.; Higashimura, T. *Macromolecules* **1986**, *19*, 1459.
40. Seki, H.; Masuda, T.; Tang, B-Z.; Tanaka, A.; Higashimura, T. *Polymer* **1994**, *35*, 3456.
41. Naarmann, H. *Synth. Met.* **1989**, *17*, 223.
42. Sinclair, M.; Moses, D.; Heeger, A. J.; Vilhelmsson, K.; Vakl, B., Salour, M. *Solid State Commun.* **1987**, *61*, 221.

43. Neher, D.; Wolf, A.; Bubeck, C.; Wegner, G. *Chem. Phys. Lett.* **1989**, *163*, 116.
44. Neher, D.; Wolf, A.; Leclerc, M.; Kaltbeitzel, A.; Bubeck, C.; Wegner, G. *Synth. Met.* **1990**, *37*, 249.
45. Wada, T.; Masuda, T.; Sasabe, H. *Mol. Cryst. Liq. Cryst.* **1994**, *247*, 139.
46. Masuda, T.; Iguchi, Y.; Tang, B-Z.; Higashimura, T. *Polymer* **1988**, *29*, 2041.
47. Masuda, T.; Isobe, E., Higashimura, T.; Takada, K. *J. Am. Chem. Soc.* **1983**, *105*, 7473.
48. Takada, K.; Matsuya, H.; Masuda, T.; Higashimura, T. *J. Appl. Polym. Sci.* **1985**, *30*, 1605.
49. Ichiraku, Y.; Stern, S. A.; Nakagawa, T. *J. Membr. Sci.* **1987**, *34*, 5.
50. Witchey-Lakshmanan, L. C.; Hopfenberg, H. B.; Chern, R. T. *J. Membr. Sci.* **1990**, *48*, 321.
51. Plate, N. A.; Bokarev, A. K.; Kalieuzhnyi, N. E.; Litvinova, E. G.; Khotimskii, V. S.; Volkov, V. V.; Yampolskii, Y. P. *J. Membr. Sci.* **1991**, *60*, 13.
52. Ito, H.; Tsuchihara, K.; Masuda, T.; Higashimura, T. *Polym. Preprts., Jpn.* **1992**, *41*(2), 251.
53. Isobe, E.; Masuda, T.; Higashimura, T. Yamamoto, A. *J. Polym. Sci., Polym. Chem. Ed.* **1986**, *24*, 1839.
54. Hayakawa, Y.; Nishida, M.; Aoki, T.; Muramatsu, H. *J. Polym. Sci., Part A, Polym. Chem.* **1992**, *30*, 873.
55. Aoki, T.; Nakahara, H.; Hayakawa, Y.; Kokai, M.; Oikawa, E. *J. Polym. Sci., Part A, Polym. Chem.* **1994**, *32*, 849.
56. Masuda, T.; Takatsuka, M.; Tang, B-Z.; Higashimura, T. *J. Membr. Sci.* **1990**, *49*, 69.
57. Aminabhavi, T. M.; Khinnavar, R. S.; Harogoppad, S. B.; Aithal, U. S.; Nguyen, Q. T.; Hansen, K. C. *J. Macromol. Sci. Rev.* **1994**, *C34*, 139.

ACROLEIN (Polymerization, Utilization)

Natsuki Yamashita
Department of Applied Chemistry
Kinki University

Since its discovery in the mid-1840s, acrolein (AL, acrylaldehyde, propenal) has been known to polymerize very readily in the presence of anionic, cationic, or free-radical agents.

AL is very reactive; reactions can be carried out at the aldehyde group as well as at the C=C double bond. AL can be polymerized by a 1,2-addition at the vinyl group or at the carbonyl group, or it may be polymerized as a conjugated diene by 1,4-addition across the carbonyl and α,β-unsaturation. Three different repeat units are possible: 1,2-addition, 3,4-addition, and 1,4-addition. The polymers formed by spontaneous or thermal polymerization are color-to-yellow powders or a horny mass of undefined structure. They are insoluble, infusible, and of no practical interest.

NEW INITIATION SYSTEMS FOR AL

Since 1968, in order to obtain soluble poly-AL, we have investigated some new initiation systems such as a pyridine-water mixture and imidazole (IM) derivatives. Copolymers of AL with methyl vinyl ketone (MVK), acrylamide, and methyl, methacrylate and the grafting of AL (or MVK) onto IM-containing polymers have also been described.

Polymerization Induced by a Pyridine-Water System

AL polymerizes readily in the presence of pyridine (Py) and water at and below room temperature. The polymerization of AL in the presence of Py and water has been carried out in tetrahydrofuran (THF) below room temperature.[2] The polymer, obtained as a white powder, was found to be composed of polymer units through both vinyl (1,2-addition) and aldehyde (3,4-addition) polymerization. The degree of polymerization (dp) of the polymer was determined to be in the range of 10–35, a low molecular weight. The contents of 1,2-addition and 3,4-addition in the polymer and the dp of the polymer were dependent on the polymerization temperature.

Polymerization Induced by an Imidazole Catalyst

The polymerization of AL in the presence of Im or benzimidazole gives a soluble poly-AL containing the Im ring at the end of the polymer.[3,4] The polymerization of AL was carried out in THF below room temperature. The polymer obtained, white or pale yellow powder, was composed of vinyl polymer with one Im group and an aldehyde side chain, of which 70–80 mol% of the aldehyde revealed a bridge structure. The number-average molecular weight (M) of these poly-ALs was in the range of 317–691.

To investigate the polymerization mechanism (**Scheme I**), we conducted kinetic studies of the polymerization of AL and MVK in the presence of Im in THF at 0°C. R_p is expressed by the equations $R_p = k[Im][AL]^2$ and $R_p = k[Im][MVK]$.[1]

I

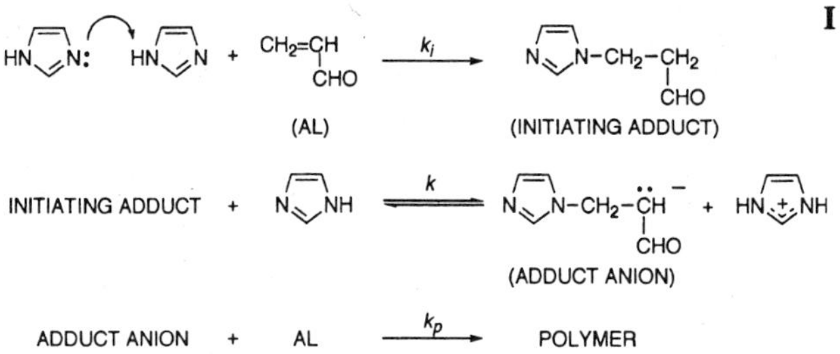

Additive Effects of the Polymerization System Induced by Im Catalyst

The addition of water, phenol, and DMSO accelerated the polymerization reaction.

The ability to polymerize is also increased by the presence of polyacrylamide (poly-AAm).[6-8] The R_p in the presence of poly-AAm increased markedly in the Im and Py-water systems.

INITIATING ADDUCT (IM-MONOMER ADDUCT)

Im-AL and Im-MVK Adducts

In the polymerization of AL or MVK, an initiation adduct is formed in the initial polymerization step. NMR observations support the addition of the NH proton of Im to the vinyl group of monomer.

UTILIZATION OF POLY-AL ANION (COPOLYMERIZATION)

In the presence of an Im catalyst, copolymerizations of AL (M_1) with MVK (M_2) were studied in THF at 0°C.[9,10] The decrease of AL concentration in the monomer feed decreased the copolymerization rate. The monomer reactivity ratio (r_1 = 2.02 and r_2 = 0.06) was determined from Fineman-Ross plots.

Copolymerizations of AL (M_1) with AAm (M_2) were also carried out in THF at 0°C.[9] Both monomers homopolymerized slowly in these polymerization conditions, although the copolymerization rate was very high. The structure of the copolymer is shown in **Structure 1.**

1

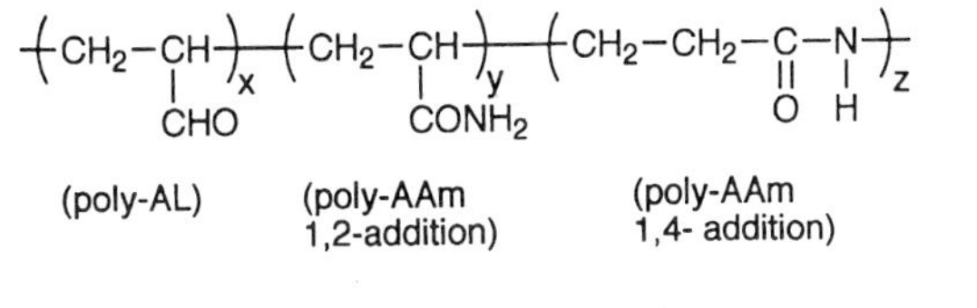

It is well known that AAm polymerizes with intermolecular hydrogen transfer which might explain the polymerization mechanism in the copolymerization of AL with AAm.[9,10]

UTILIZATION OF THE IM INITIATION SYSTEM (GRAFT POLYMERIZATION AND POLYMER EFFECT)

Graft Polymerization of AL onto Poly(4-Vinylimidazole) and Its Polymer Effect

We investigated graft polymerizability[5,12-15] of AL onto several Im-containing polymers such as 4-vinylimidazole (VIm) homopolymer and its copolymers with 4-vinyl-pyridine (VPy), 1-vinyl-2-pyrrolidone (VPr), and styrene (St).

These results clearly suggest that the AL polymer grafted to the Im groups in the parent polymer.

Graft Polymerization of AL onto Several Im-Containing Polymers

We conducted kinetic studies on the graft polymerizabilities of AL onto Im-containing polymers such as the homopolymer of VIm and several copolymers of VIm–4-vinylPy (VPy), VIm–VPr, VIm–St. VIm–acrylic acid (AA), VIm–vinyl acetate (VAc), VIm–vinyl alcohol (VA), and VIm–AAm in a water–ethanol mixture at 0°C.[14]

UTILIZATION OF THE CARBONYL-AMIDE INTERACTION (MATRIX POLYMERIZATION)

Matrix Polymerizability of AL in the Presence of Poly-AAm or Poly(1-Vinyl-2-Pyrrolidone)

The polymerizability of AL induced by the Im catalyst was investigated in the presence of poly-AAm and poly-VPr in the water–ethanol mixed solvent at 0°C.[16]

A poly-AL with the high molecular weight and low content of a free-aldehyde group was formed by this system.

Matrix Polymerization of AAm in the Presence of Homo- and Copolymer of MVK

Matrix polymerization of AAm is observed by the MVK polymer as the model of AL. AAm is also polymerized without catalysts in the presence of several copolymers of MVK, which was synthesized by use of a radical initiator.[11-17]

UTILIZATION OF NETWORK POLYMERS CONTAINING AL

It is well known that the aldehyde group in several polyaldehydes such as dialdehyde starch (DAS) and poly-AL indicates an adsorptive activity for urea and ammonia.[18-20] This adsorption effect has been studied as a candidate for removal of uremic waste metabolites.[21,22]

It emerges that several network polymers containing AL show a tendency to adsorb urea and ammonia. The appropriate contents of AL in the network polymer were ~ 30–50 mol% in charged monomers. The results indicate that it may be possible to remove uremic waste metabolites, but confirmation awaits future research. Some adsorptive activity for phenol was also indicated in this study. These results indicate the possibility of the removal of trace amounts of phenol from industrial waste-water.

REFERENCES

1. *Vinyl Polymerization*; Ham, G. E., Ed; Marcel Dekker: New York 1967; Vol. 1, Chapter 7.
2. Yamashita, N.; Sumitomo, H.; Maeshima, T. *Kogyo Kagaku Zassh (The Chemical Soc. of Japan)* **1968**, *71*, 1723.
3. Yamashita, N.; Morita, S.; Maeshima, T. *J. Macromol. Sci. Chem.* **1978**, *A12*(9), 1261.
4. Yamashita, N.; Morita, S.; Yoneyama, H.; Maeshima, T. *J. Polym. Sci., Polym. Lett. Ed.* **1983**, *21*, 13.
5. Horiba, M.; Yamashita, N.; Maeshima, T. *J. Macromol. Sci. Chem.* **1986**, *A23*(9), 1117.
6. Yamashita, N.; Morita, S.; Nishino, M.; Maeshima, T. *J. Polym. Chem. Sci., Polym. Chem. Ed.* **1983**, *21*, 239.
7. Yamashita, N.; Tanaka, H.; Deguchi, S.; Maeshima, T.; Wei, L. S. *J. Macromol. Sci. Chem.* **1987**, *A24*(9), 1121.
8. Ozu, E.; Yamashita, N.; Inoue, H.; Maeshima, T.; Baiane, I. C.; Wei, L. S. *J. Macromol. Sci. Chem.* **1989**, *A26*(11), 1525.
9. Yamashita, N.; Ikezawa, K.; Yamamoto, Y.; Kinugasa, H.; Maeshima, T. *J. Macromol. Sci. Chem.* **1984**, *A21*(3), 291.
10. Morita, S.; Ikezawa, K.; Inoue, H.; Yamashita, N.; Maeshima, T. *J. Macromol. Sci. Chem.* **1982**, *17*(9), 1495.
11. Ozu, E.; Yamashita, N.; Kinoshita, T.; Maeshima, T. *Chemistry Express (The Kinki Chemical Soc. of Japan)* **1989**, *4*, 133.
12. Morita, S.; Yamashita, N.; Maeshima, T. *J. Polym. Sci., Polym. Chem. Ed.* **1980**, *18*, 1599.
13. Yamashita, N.; Morita, S.; Kanzaki, K.; Maeshima, T. *J. Polym. Sci., Polym. Chem. Ed.* **1983**, *21*, 191.
14. Morita, S.; Murakami, S.; Yamashita, N.; Maeshima, T. *Nippon Kagaku Kaishi (The Chemical Soc. of Japan)* **1980**, *1980*(9), 1410.
15. Yamashita, N.; Yamada, S.; Maeshima, T. *Nippon Kagaku Kaishi (The Chemical Soc. of Japan)* **1985**, *1985*(4), 771.
16. Yamashita, N.; Omori, N.; Koumi, Y.; Maeshima, T. *Nippon Kagaku Kaishi (The Chemical Soc. of Japan)* **1985**, *1985*(4), 776.
17. Tsuneka, T.; Ishifune, M.; Yamashita, N.; Kashimura, S.; Wei, L. S. *Macromol. Sci. Pure Appl. Chem.* **1994**, *A31*(9), 1169.
18. Obayashi, T.; Yamashita, N.; Yuasa, H.; Maeshima, T. *J. Polym. Sci., Polym. Lett. Ed.* **1985**, *23*, 5930.
19. Giordano, C.; *Adv. Nephoro. Neeker Hosp.* **1972**, *2*, 251.
20. Sparks, R. E.; Mason, N. S.; Meier, P. M. *Trans. Am. Soc. Artif. Intern. Organs.* **1972**, *18*, 458.
21. Nakabayashi, N. *Zinko-Zoki (Artif. Organs)* **1975**, *4*, 253.
22. Matsumoto, H. *Kagaku-Sosetsu (Chemical Rev., The Chemical Soc. of Japan)* **1978**, *21*, 35.

ACRYLAMIDE (Polymerization and Applications)

V. F. Kurenkov* and V. A. Myagchenkov
Kazan State Technological University

Acrylamide polymers are versatile synthetic water-soluble polymers. They are used worldwide in the various fields of engineering and technology as flocculants for separation and clarification of liquid-solid phases, thickening and binding agents, and for film formation and lubrication. Acrylamide polymers began to be manufactured in the 1950s, and for the past 30 years their production has been intensively developing. The growth of production of acrylamide polymers has not met the demand; therefore, the development of new methods and the perfection of existing ones have been necessary. To develop and improve methods, knowledge of the kinetic peculiarities and mechanism of polymerization and copolymerization, the properties of the medium, the influence of acrylamide as a monomer, and several other factors are required. The principal production method for acrylamide polymers is free-radical polymerization in solutions, emulsions, or suspensions.[1,2]

POLYMERIZATION

Homogeneous Polymerization in Solutions

This includes polymerization processes that occur in solvents carrying both polymers and monomers. For acrylamide and acrylamide polymers, the list of such solvents is limited: it includes water, formamide, acetic and formic acids, dimethyl sulfoxide, and some water-organic mixtures. The kinetics of the polymerization of acrylamide and the properties of the polymers formed depend on the nature of the solvent, pH, complexing agents, surfactants, chain-transfer agents, and temperature.[3] In estimating monomer reactivity the influence of hydrogen bonding, dipolar interaction, complex formation, the conformation of growing macroradicals, and various side reactions would be taken into account.

Homogeneous polymerization of acrylamide is usually performed in aqueous solutions. The principal factors that make this technique popular are a high rate of polymer formation and the possibility of obtaining a polymer with a high molecular weight (MW).

Acrylamide has a high value of the ratio constants $k_p/k_t^{1/2}$ = 3.2–4.4 $(1\ mol^{-1}sec^{-1})^{1/2}$ at 30–60°C which, along with small values of the constant of chain transfer to monomer and water, stimulates acrylamide polymer formation in aqueous solutions at rates and MWs unachievable in the case of polymerization in organic solvents.[4]

Various methods of polymerization of acrylamide in aqueous solution were reviewed.[4-6]

Precipitation Polymerization

This type of polymerization of acrylamide is conducted in either organic solvents (e.g., acetone acetonitrile, dioxane, ethanol, tert-butanol, and tetrahydrofuran) or aqueous organic media, which serve as solvents for monomers and as precipitates for polymers. The commonly used initiators are persulfates, perborates, benzoyl peroxide, and AIBN. In precipitation polymerization the medium never gets very viscous and the polymer is relatively easy to isolate and dry.

Suspension Polymerization

The initial system is obtained by dispersion (in the form of droplets with diameters of 0.1–5.0 mm) of an aqueous monomer solution in an organic liquid by mechanical stirring in the presence of stabilizers. The polymerization is initiated by water-soluble initiators: UV and γ-radiation. The process occurs in droplets of an aqueous monomer solution acting as microreactors, and its kinetics resemble in some respects solution polymerization, although it is still affected by the stabilizers.

The polymerization rate and MW in suspensions are lower than those observed in aqueous solutions. This can be attributed to the effect of organic solvents. Depending on the concentration and nature of the stabilizer and on the stirring conditions, the polymers are obtained in the form of powder or granules.

Emulsion Polymerization

For polymerization in inverse emulsions, the aqueous solution of acrylamide is dispersed (to provide the particle size of 1–10 μm) in a continuous hydrophobic organic phase in the presence of a water-in-oil type of emulsifier (e.g., sorbitan monooleate, stearates, polyalkyl derivatives of higher monoatomic alcohols and monocarbon acids, fatty acids, and polyatomic alcohols). The process is initiated by an oil- or water-soluble initiator. An important advantage of the polymerization in inverse emulsions is the possibility of using concentrated monomer solutions in conditions of facilitated heat removal and polymerization in low-viscosity media. Moreover, the polymerization may occur with a high rate to yield a high molecular, water-soluble polymeric product. The reaction results in the formation of a colloidal emulsion of hydrophilic polymeric particles dispersed in a continuous organic phase. This latex is characterized by a broad distribution of the particle size and retains stability over several hours or days. It can either be used directly as the final product or be subjected to azeotropic distillation, solvent removal, and drying.

The rate of polymerization in inverse emulsions is, under comparable conditions, higher, but MW is lower than that obtained in aqueous solutions. The use of an oil-soluble initiator usually results in a greater MW compared with the water-soluble initiators.

Microemulsion Polymerization

Polymerization of acrylamide is carried out in microemulsions stabilized by surfactants. In this way it is possible to obtain thermodynamically stable and optically transparent microlatex (with a particle size of 0.005–0.01 μm) and uniform particle-size distribution containing up to 25% high molecular fraction (MW is 10^6 to 10^7) exhibiting good rheological properties.

Copolymerization

Acrylamide easily enters reaction radical copolymerization with vinyl monomers. Anionic copolymers are received by radical copolymerization of acrylamide with acrylic, methacrylic, maleic, fumaric, and styrenesulfonic acids and its salts and by hydrolysis of polyacrylamide (PAA). Cationic copolymers are received by radical copolymerization of acrylamide with *N*-dialkylaminoalkyl acrylates and methacrylates and 1,2-dimethyl-5-vinylpyridinum

*Author to whom correspondence should be addressed.

sulfate or by postreactions of PAA (i.e., the Mannich reaction and Hofman degradation). The methods of providing copolymerization are the same as the polymerization of acrylamide.

Graft and Block Copolymerization

Acrylamide can be grafted onto numerous substrates for the purposes of increasing hydrophilicity, altering crystallinity, reducing susceptibility to degradation, and providing a reactive site. Acrylamide is grafted by using free-radical sources, UV light, or X-rays.

Block copolymers of acrylamide may be obtained by mechanical or UV degradation of the polymer in the presence of acrylamide or by incorporation of a reactive end group in the base polymer.

PROPERTIES

Physical Properties

PAA is a linear polymer. It is white; odorless; exhibits very low toxicity; is soluble in water, water-salt mixtures, and formamide and ethylene glycol; partially dissolves in acetic and formic acids and glycerin; swells in DMSO (dimethyl sulfoxide), DMFA (dimethylformamide), and propylene glycol; and is insoluble in acetone, dioxane, alcohols, hexane, and heptane. The main properties of aqueous solutions of PAA and copolymers of acrylamide have been reviewed.[7,8]

Chemical Properties

PAA solutions undergo the general reactions of the aliphatic amide group.[9] The most important reactions of PAA are hydrolysis, Hofman degradation, and Mannich reaction.

Hydrolysis of PAA can be carried out in an acidic or basic medium. Acidic hydrolysis of the amide group at pH 4.5 is a very slow reaction. Strong acidic conditions lead to a progressive insolubilization of the reaction product because of formation of cyclic imide structures.

Hydrolysis of PAA under basic conditions is rapid and can be used to introduce acrylate groups into macromolecules.

A Hofman degradation of PAA caused by the use of a very slight excess of sodium hypochlorite and a large excess of sodium hydroxide at 0°C to –15°C for ~ 15 h makes polyvinylamine (95 mol% amine units) obtainable.

The Mannich reaction of acrylamide polymers with formaldehyde and dimethylamine may be used for obtaining polymers that contain *N*-methylol groups.

The *N*-methylol derivatives of PAA may be made cationic by heating with amines or anionic by heating with aqueous bisulfite solution.

PAA may be crosslinked by imide formation at temperatures >100°C. PAA starts to decompose at 220°C.

APPLICATION

Acrylamide polymers are used in many industrial applications.[4,9,10] Anionic copolymers of acrylamide are used for enhanced oil recovery as a multipurpose additive. They improve the rheological properties of the fluids in question, positively affect the size of suspended particles, and act as filters, accelerating the process of well preparation for operation. Another important function is that of soil-formation agents, which impart additional strength to well walls.

Acrylamide polymers are used to improve the printing qualities of paper by more fully retaining the filler (usually kaoline) in the paper pulp which improves the structure of the paper surface layer and the tensile properties.

One of the most promising applications of acrylamide polymers is their use in the reduction of hydraulic resistance in turbulent flows. This effect has extensive practical use for increasing the range covered by water jets from fire hose lances in fire fighting and for raising the speed of travel of ships and submarines. They are also used in high-speed turbulent pumping of water suspensions and emulsions through pipelines.

Nonthrombogenic granulated gels based on acrylamide polymers are used in medicine for filling postsurgical voids, as a material in contact lenses, as a coating for water-soluble pharmacological preparations (microencapsulating), and in the manufacture of high-quality products such as tampons and diapers.

Acrylamide polymers are used as a protective layer on granules of mineral fertilizers. This allows fungicides and herbicides to be applied onto the surface of the granules.

The main use for acrylamide polymers is as flocculants in mining, papermaking, metallurgy, light, food, coal, and oil industries.

CONCLUSION

Taking into account the complicating factors of polymerization and copolymerization of acrylamide will help in optimizing the technology of obtaining acrylamide polymers to meet the requirements of industry.

REFERENCES

1. MacWilliams, D. C. In *Functional Monomers*; Marcel Dekker: New York, 1973; Vol. 1, p 1.
2. Bikales, N. M. In *Water Soluble Polymers*; Polymers Science and Technology: New York, 1973; Vol. 2, p 213.
3. Kurenkov, V. F.; Myagchenkov, V. A. *Eur Polym. J.* **1980**, *16*, 1229.
4. *Polyacrylamide*; Kurenkov, V. F., Ed.; Khimiya, Moscow, **1992**.
5. Gromov, V. F.; Teleshov, E. N. *Plastmassy* **1984**, *10*, 9.
6. Kurenkov, V. F.; Abramova, L. I. *Polym., Plast. Technol. Eng.* **1992**, *31*, 659.
7. Kulicke, W-M.; Kniewske, R.; Klein, J. *Porgr. Polym. Sci.* **1982**, *8*, 373.
8. Myagchenkov, V. A.; Kurenkov, V. F. *Acta Polymerica.* **1991**, *42*, 475.
9. Morris, J. D.; Penzenstadler, R. J. In *Kirk-Othmer Encyclopedia of Chemical Technology*; John Wiley & Sons: New York, 1978; Vol. 1, p 312.
10. Myagchenkov, V. A.; Kurenkov, V. F. *Polym., Plast. Technol. Eng.* **1991**, *30*, 109.

ACRYLAMIDE, INVERSE EMULSION POLYMERIZATION (In Supercritical Carbon Dioxide)

F. A. Adamsky and E. J. Beckman
University of Pittsburgh

The study of polyacrylamide is significant because of its commercial importance in the mining, water treatment, and paper industries.[1-3] The acrylamide monomer is also unique in its ability to polymerize to very high molecular weights, with a high degree of linearity. The hydrogen-bonding properties of the resulting polymer make it useful in the industries mentioned above by allowing the polymer to act as a coagulating or flocculating agent for suspended solids in aqueous systems. The molecular weight of a polyacrylamide largely determines its performance in these systems, with increasing molecular weight almost invariably improving performance. Polymerizing acrylamide to very high molecular weights (>1 million) is facilitated by using the inverse emulsion process. A secondary benefit of the inverse emulsion is reflected in the ease of use of the product. The inverse emulsion has a much lower viscosity than the corresponding solution of equal strength and greater stability than a similar dispersion.

The kinetics of inverse emulsion polymerizations are not as well understood as standard emulsion kinetics.[4] A review of current literature reveals studies that describe a variety of potential mechanisms, the bulk of which can be divided into three main areas. In the first area, inverse emulsion kinetics follow those of conventional emulsions.

The second classification has been termed an inversemicrosuspension polymerization mechanism by Hunkeler et al., the principal investigators in this area.[5]

In the third area, Vanderhoff et al. utilized an aromatic continuous phase, oil-soluble initiators, and moderate to low single surfactant concentrations (<6 weight % of continuous phase).[6] The resulting kinetics had elements of solution, precipitation, and emulsion polymerization. There are several other areas in which inverse emulsions have been explored for the polymerization of acrylamide, such as microemulsions, but these reactions produced predominately lower molecular weight (<1 million) polymers, which are not of primary interest in the present work.[7]

In commercial inverse emulsion processes, the major shortcoming involves the properties of the continuous phase. Use of light-weight hydrocarbons and aromatics requires accounting for volatile organic compounds (VOCs) as mandated by the U.S. Environmental Protection Agency.[8] Because the hydrocarbon has no use in the final polymer application, a second surfactant emulsifies the oil phase so that the polymer can be diluted in water for use. Phase separation, gel formation, and compaction occur during storage. Although these difficulties are reason for concern, they do not prevent the use of millions of pounds of inverse emulsion polymer annually.[9] There would be a sizable economic incentive, however, to be able to deliver all of the benefits of an inverse emulsion polymer without the oil phase in the final product.

Consequently, there is a need for improvement of the process through either the replacement of the hydrocarbon phase or its elimination after polymerization. The requirements of hydrophobicity, initiator solubility, chemical neutrality, and low viscosity also limit the choices for the continuous phase. Supercritical fluids offer an intriguing option, because they exhibit liquidlike densities and gaslike viscosities under the same conditions.[10] It has been shown that supercritical and near-critical alkanes can be used in the inverse emulsion polymerization of acrylamide to high molecular weight by using ethoxylated alcohol surfactants at 65°C.[11,12] Although alkanes are technologically suitable, they exhibit both VOC and flammability drawbacks. Supercritical carbon dioxide meets all of the requirements for use as the continuous phase because it is environmentally friendly, not having a VOC classification, inexpensive, nonflammable, and of relatively low toxicity. Supercritical carbon dioxide also has a relatively low critical temperature and pressure (T_c = 31.1°C, P_c = 7.38 MPa). The pressure-solubility relationship of the emulsion polymer in supercritical carbon dioxide is an additional advantage because the polymer can be phase separated from the emulsion and recovered if desired. All of these properties are required for or enhance the usefulness of supercritical carbon dioxide as a replacement for hydrocarbons in the continuous phase.

The principal challenge of working with supercritical carbon dioxide is the relatively high pressures (30–50 MPa) required to approximate hydrocarbon densities. Previous work by Consani and Smith reveals a second problem — that conventional alkyl surfactants are poorly soluble in carbon dioxide.[13] We have determined that solute molecules containing fluoroethers or dimethylsiloxanes are highly soluble in carbon dioxide. Hence, novel carbon-dioxide-soluble amphiphiles have been developed by substituting the alkyl chain with perfluoroethers and siloxanes.[14,15]

The objective of this work is to determine the effects of surfactant concentration and structure on the inverse emulsion polymerization of acrylamide in which supercritical carbon dioxide replaces hydrocarbon as the continuous phase. Previous work has shown the efficacy of such a process.[16] Custom synthesized surfactants are used to create the inverse emulsion. These surfactants use a perfluoro(propylene oxide) tail (molecular weight of 2500), in place of the alkanes to give the required carbon dioxide solubility in combination with a poly(ethylene glycol) (molecular weight of 200–1500) head. Standard thermal free-radical initiators, such as azo bis(isobutyrnitrile) (AIBN), which are soluble and active in carbon dioxide, are used.[17] Determination of reaction rate is made by measuring pressure and temperature changes during the reaction process.[4,18] Characterization of the product involves the determination of molecular weight by viscometric methods.

CONCLUSIONS

Inverse emulsion polymerizations of acrylamide have been conducted, with reaction rates and resulting polymer molecular weights determined, for a series of novel surfactants in supercritical CO_2. The surfactants showed significant solubility in supercritical CO_2. Increases in both the molar surfactant concentration and hydrophile chain length increased the average reaction rate: The molecular weights of the polyacrylamides produced are lowered by increasing the surfactant concentrations; this effect may be the result of autoacceleration effects, particle size, and chain transfer (to which acrylamide is prone).

Previous work by Hunkeler et al. suggests the ability to change the mechanism of polymerization (e.g., from emulsion to suspension) by changing the identity of the continuous phase (hence changing continuous phase solvent power).[19] It is therefore conceivable that simply changing reaction pressure could change the mechanism of polymerization in the present system, because of the strong pressure/solvent power relationship of CO_2.

REFERENCES

1. Richardson, P. F.; Connely, L. J. *Reagents in Mineral Technology*; Marcel Decker: New York, 1988.
2. *Betz Handbook of Industrial Water Conditioning*; Betz Laboratories: Tevois, PA, 1980.
3. Casey, J. P. *Pulp and Paper Handbook*; John Wiley & Sons: New York, 1981; Vol. 3.
4. Poehlein, G. W. *Encyclo. Pol. Sci. & Eng.* John Wiley & Sons: New York, 1985.
5. Baade, W.; Hunkeler, D.; Hammielec, A. E. *J. Applied Poly. Sci.* **1989**, *38*, 185.
6. Vanderhoff, J. W.; DiStefano, F. V.; El-Asser, M. S.; O'Leary, R.; Shaffer, O. M.; Visioli, D. L. *J. Disp. Sci. & Tech.* **1984**, *5*(3 and 4), 323.
7. Candeau, F.; Leong, Y. S. *J. Poly. Sci., Poly. Chem.* **1985**, *23*, 193.
8. *Chemical Engineering Progress* **1993**, *89*(7), 20–41.
9. Bradley, R.; Jackel, M.; Yoshida, Y. *Specialty Chemical Review for Paper Chemicals,* 1995, SRI International, San Antonio, TX; pp 141–145.
10. McHugh, M. A.; Krukonis, V. J. *Supercritical Fluid Extraction*; Butterworths: 1986.
11. Beckman, E. J.; Smith, R. D. *J. Phys. Chem.* **1990**, *94*, 345.
12. Beckman, E. J.; Smith, R. D. *J. Supercrit. Fluids* **1990**, *3*, 205.
13. Konsani, K. A.; Smith, R. D. *J. Supercrit. Fluids* **1990**, *3*, 51.
14. Hoefling, T. A.; Enick, R. M.; Beckman, E. J. *J. Phys. Chem.* **1991**, *95*, 7127.
15. Hoefling, T. A.; Beitle, R. R.; Enick, R. M.; Beckman, E. J. *Fluid Phase Equilib.* **1993**, *83*, 203.
16. Adamsky, F. A.; Beckman, E. J. *Macromolecules* **1994**, *27*, 312.
17. Guan, Z.; Combes, J. R.; Menceloglu, Y. A.; DeSimone, J. M. *Macromolecules* **1993**, *23*, 2663.
18. Thomas, W. M.; Wei Wang, *D. Encylco. Pol. Sci. & Eng.*; John Wiley & Sons: New York; 1, 169.
19. Hunkeler, D.; Hamielec, A. E.; Baade, W. *Polymer* **1989**, *30*, 127.

Acrylamide Polymers

See: Acrylamide (Polymerization and Applications)
Acrylamide, Inverse Emulsion Polymerization (In Supercritical Carbon Dioxide)
Water-Soluble Polymers (Oil Recovery Applications)

Acrylate Polymerization

See: Peroxide Initiators (Overview)
Rare Earth Polymerizatrion Initiators

ACRYLIC ELASTOMER CURING

C. K. Das
Materials Science Centre
Indian Institute of Technology

Acrylic elastomers are designated as ACM by the American Society for Testing and Materials. Among specialty elastomers, ACM ranks above epichlorohydrin, chlorosulfonated polyethylene, polychloroprene, and nitrile rubbers but below silicones and fluorocarbon elastomers.[1] ACMs cost significantly less than silicones and fluorocarbons and are very well known for their resistance to sulfur-bearing oils and hypoid additives containing oils. Their uniqueness results primarily from the saturated backbone structure that makes them inherently resistant to UV radiation, ozone, and heat. A tight cure is often essential to impart good properties in the vulcanizates of ACMs.

The first polyacrylic elastomers were homopolymers of ethyl acrylate or methyl acrylate. These polymers had certain disadvantages concerning curing. Poly(ethyl acrylate) has a saturated backbone but reactive side groups as ester and active α-hydrogen.[2,3] It does not respond to peroxide curing because a bulky ester group protects the α-hydrogen.[4] Hence, in the course of development of a cure site, a monomer is introduced during the polymerization process. The first cure site monomers for polyacrylic rubber were butadiene and isoprene.[5] These polymers, with unsaturated side chains, can be cured with common sulfur vulcanization ingredients and with organic peroxide. However, these polymers suffer from the disadvantage of being cured while used in sulfur-bearing oils, which causes them to stiffen.

The most used commercially available ACMs have chlorine cure sites. These ACMs include the copolymer of ethyl acetate and chloroethyl vinyl either.[6] A big improvement in the reactivity of the chlorine in polyacrylic elastomers came with the introduction of vinyl chloroacetate as a cure site monomer in which a carboxyl group in the acetate activates like an allylic group.[7] Although chlorine-containing polymers dominate the commercial field, other cure site monomers containing polymers need special mention.

Other cure site monomers include glycidal ether, methylol compounds, and imidoesters. The comonomer chosen should give a compromise between cure rate and the degree of polymerization. Better polymerization reactivity is obtained from glycidyl methacrylate or acrylate.[8]

All the commercial ACMs contain a small amount (1–5%) of a reactive cure site monomer, usually a chlorine type, the reactivity of which governs the cure behavior of the polymer. The conventional chloroethyl vinyl ether ACM grades were mostly crosslinked with thiourea and polyamine cure systems. For vinyl chloroacetate grades, reactive cure systems (e.g., metal soaps and sulfur) are used.[9] Crosslinking with metal soaps or sulfur can be inhibited by acids such as stearic acid, or the reaction can be accelerated by bases such as MgO. Although glycidyl-containing ACM grades are sometimes crosslinked with metal soaps or sulfur, the preferred cure system for these grades is ammonium benzoate, or ammonium adipate, if a low compression set is required.[10]

Despite significant advancement in cure technology, the current acrylic rubbers need a relatively longer postcuring cycle to achieve optimum properties. Although metal oxides are commonly used in curing acrylic rubber, their exact role in the curing mechanism still remains to be explored. The type of metal oxides used changes depending on the type of accelerator used. Here we discuss the curing mechanism of polyacrylic rubber by the use of accelerators which as thiourea, blocked diamine, and ammonium benzoate in conjunction with group IIB and IVB metal oxides alone or in combination.

ETU CURING OF ACM

Thioureas are special products used in the vulcanization of polychloroprene, epichlorohydrin, and ethylene propylene diene monomer (EPDM) rubbers. The oldest accelerator in this class is diphenyl thiourea. Other known materials in this group are especially active ethylene thiourea (ETU), diethyl thiourea, and dibutyl thiourea.

MERCAPTOBENZOTHIAZOLE AND SULFENAMIDE CURING OF POLYACRYLIC RUBBER

The basic compound among the mercapto accelerators is 2-mercaptobenzothiazole (MBT). Dibenzothiazyl disulfide (MBTS) and zinc mercaptobenzothiazole are derived from MBT. MBT and MBTS are used broadly in many types of rubber and may be called all-purpose accelerators. The vulcanization begins rapidly with MBT, whereas MBTS gives a slightly delayed start. The benzothiazole sulfenamide acceleration are synthesized from MBT. The effects of various metal oxides on the curing of polyacrylic rubber by thiazoles and sulfenamides are shown to elucidate the curing mechanism.

In the absence of PbO, MBT is not effective in the curing of polyacrylic rubber. However, in the presence of PbO, MBT is able to cure polyacrylic rubber, although to varying extents depending on the MBT/PbO mole ratios. Both state and rate of cure increase with the increase of MBT content at a fixed level of PbO. The state and rate of cure attain a maximum and then decrease with a further increase of MBT. The hardness, modulus, and tensile strength are higher for the MBT system than for MBTS system, and they are further increased considerably with the addition of MDA. The elongation at break value of the MBT system is lower than that for the MBTS system, and it lowers more with the incorporation of MDA.

BLOCKED DIAMINE CURING OF POLYACRYLIC RUBBER

Polyfunctional amines are important crosslinking agents for some special elastomers such as chlorosulfonated polyethylene, epichlorohydrin, and polymers containing carboxyl groups. Nevertheless, the technically most important vulcanizing agents for these rubbers are metal oxides. Polyfunctional amines, particularly blocked diamines such as hexamethylene diamine carbamate and *N,N′*-dicinnamylidene hexamethylene diamine (Diak-3) are the most important crosslinking agents for polyacrylic and fluoro rubbers.

State and rate of cure increase with the increase of blocked diamine content at a fixed level of PbO (4 phr).

Metal oxides facilitate the crosslinking by adsorbing hydrogen chloride.

AMMONIUM BENZOATE CURING OF POLYACRYLIC RUBBER

Ammonium benzoate is an effective curing agent for ACMs. Its activity is enhanced in the presence of metal oxides (e.g., PbO and CdO). PbO and CdO are more active than PbO_2 in conjunction with ammonium benzoate. Metal oxides facilitate crosslinking by absorbing benzoic acid and ammonium chloride from the system.

CONCLUSION

In the curing of polyacrylic rubber by ETU, the role of metal oxide is to release active ethylene urea and metal sulfide and to absorb the hydrogen chloride produced. Thus, the metal oxide initiates and facilitates the crosslinking reaction through the formation of metal sulfide and metal chloride, respectively.

The role of PbO in the curing of polyacrylic rubber by thiazole or sulfenamide is to produce active 2-hydroxybenzothiazole and PbS and to withdraw chloride from the system. Thus, PbO initiates and facilitates the crosslinking reaction through the formation of PbS and $PbCl_2$, respectively. In the metal oxide/thiazole and metal oxide/sulfenamide cure systems in ether type of crosslink is formed at the C-Cl cure site of the polyacrylic rubber.

Metal oxides increase the efficiency of Diak-3. PbO is more active than CdO and PbO_2. The efficiency of ammonium benzoate in the curing of polyacrylic rubber is increased by the addition of metal oxides. Metal oxides help to form the C-C type of crosslinks by absorbing benzoic acid and ammonium chloride from the system.

REFERENCES

1. DeMarco, R. D. *Rubb. Chem. Technol.* **1979**, *54*, 173.
2. Semengen, S. T.; Wakelin, J. H. *Rubb. Age* **1952**, *71*, 57.
3. Schultz, A. R.; Bovey, F. A. *J. Polym. Sci.* **1956**, *22*, 485.
4. Hansen, R. H.; Martin, W. M. *Trans. Inst. Rubb. Ind.* **1963**, *39*, 301.
5. Fisher, C. H.; Mast, W. C.; Rehberg, C. E.; Smith, L. T. *Ind. Eng. Chem.* **1944**, *36*, 1032.
6. Mast, W. C.; Fisher, C. H. *Ind. Eng. Chem.* **1948**, *40*, 107.
7. Kaizerman, S. U.S. Patent 3 201 373, 1965.
8. Sims, J. A. *J. Appl. Polym. Sci.* **1961**, *5*, 58.
9. Holley, H. W.; Mihal, F. F.; Starer, I. *Rubb. Age* **1965**, *96*, 565.
10. Thiokol Chemical Corporation U.S. Patent 3 317 491, 1964.

ACRYLIC RUBBER

Robert D. DeMarco and Richard J. Fleckstein er
Zeon Chemicals, Incorporated

Acrylic elastomers are classified as high-temperature, oil-resistant, specialty rubbers. The American Society for Testing Materials (ASTM) approved designation for acrylic elastomers is ACM for acrylic monomer. Almost 80% of acrylic rubbers sold go into automotive components. They are generally formulated with reinforcing, curing, and other modifying agents. These thermoset compositions are processed into parts designed for applications requiring performance in the temperature range of –40°C to 175°C.

The earliest development of acrylic rubber can be traced to work in the 1940s at the U.S. Department of Agriculture's Eastern Region Research Laboratory.[1,2] The first commercial product was introduced in 1947.[3] Many excellent review articles have been written that describe the numerous developments since that time.[4-9]

POLYMER MANUFACTURING TECHNIQUES

Acrylates can be polymerized in several ways, including bulk, solution, suspension, and emulsion polymerization. In industry, suspension and emulsion polymerization are the most common methods used because they have aqueous phases with the ability to aid in the removal of the heat produced during the course of the polymerization.

The rubber is recovered from the latex by coagulation, usually by using a di- or trivalent metal salt to form a crumb. The crumb is washed and dried and then packaged either in a slab or particulate form.

POLYMER COMPOSITION

The first polyacrylate elastomers made were homopolymers of EA. These polymers were saturated and hard to cure, relying on curatives such as sodium metasilicate pentahydrate.[10] The use of this curative caused processing problems and gave poor dispersion.[11] Because of the difficulty of the vulcanization of the EA homopolymers, small amounts of crosslinking or cure-site monomers were added to the polymer. The vulcanization reactions could then take place at these reactive sites. This process is similar to that of the copolymerization of small amounts of isoprene with isobutylene to make butyl rubber.[12] Therefore, two different types of monomers are used in the production of polyacrylate elastomers: backbone monomers and cure-site monomers.

Backbone Monomers

Backbone monomers account for 95–99% of the weight of a normal polyacrylate elastomer. The major backbone monomers are ethyl acrylate, *n*-butyl acrylate, and 2-methoxyethyl acrylate (MEA). These monomers are responsible for the physical properties of the elastomer, especially the oil, low temperature, and heat resistances. As we mentioned previously, the first polyacrylate elastomers were based on EA. These elastomers gave excellent oil and heat resistance. However, the T_g of poly(ethyl acrylate) is only –24°C.[12] In many applications lower temperatures are required for the rubber, therefore, other backbone monomers are added to lower the T_g of the copolymer.

The design of the composition of polyacrylate elastomers is a compromise between low-temperature performance, high-temperature resistance, and low-volume swell in oil.

Cure-Site Monomers

Cure-site monomers are copolymerized with backbone monomers to produce a usable polyacrylate elastomer. Butadiene and isoprene were the first cure-site monomers to be incorporated into acrylic rubbers. This was to take advantage of the common sulfur cure systems that had been developed for the commodity rubbers.[13] Other monomers containing unsaturated cure sites such as dicyclopentadiene, tetrahydrobenzyl acrylate, and ethylidenenorbornene were also used to prepare polyacrylate elastomers.[14-16] The use of these polymers was limited because they would continue to cure in the presence of sulfur-containing oils.

The largest volume polyacrylate elastomers produced today contain chlorine cure sites. The first commercial chlorine-containing polyacrylate was a copolymer of EA and chloroethyl vinyl ether.[17]

The second generation of chlorine-containing cure-site monomers came with the introduction of vinyl chloroacetate.[18] Vinyl chloroacetate is used in a large volume of the polyacrylate elastomers produced today.

The other major type of cure-site monomer is the epoxide-containing monomers. The two most widely used are allyl glycidyl ether and glycidyl methacrylate.[19] The use of combinations of allyl glycidyl ether and glycidyl methacrylate gives better physical properties than the use of either monomer alone.[20]

A novel approach to the cure of polyacrylate elastomers was the development of dual cure sites along the polymer backbone. The two types that have achieved commercial importance are the chlorine/carboxyl and epoxide/carboxyl.[21,22] The emphasis for the development of the dual cure-site polymers was to eliminate the need for postcuring of the rubber article. The dual cure-site polymers are able to give good physical properties, especially compression set resistance, without the use of a postcure. In some applications a postcure may still be necessary, however.

PERFORMANCE

Polyacrylics are considered one of the specialty elastomer family. These are rubbery materials that have some unique physical characteristics that can solve particularly difficult sealing, containment, dampening, or environmental protection problems. They generally offer lower cost ($s/unit volume) compared with the other excellent oil-resistant types.

CHARACTERISTIC TEMPERATURE RESISTANCE

The mechanisms of thermal and oxidative degradation of acrylic gums have been studied in detail.[23]

Low temperature resistance is most dependent on the specific grade of polyacrylic gum rubber.

CHARACTERISTIC FLUID RESISTANCE

Polyacrylics are known to have outstanding resistance to hot oils, including those containing enhanced performance additive packages. These additives often contain sulfur, zinc, and other chemicals that can act as curatives and/or prodegradants for many rubbers. Polyacrylics also perform very well in petroleum-based engine-lubricating oils and automatic transmission fluids.

However, they have only fair resistance to gasoline, water, and most aromatic hydrocarbons. They would do poorly and would never be recommended for direct contact with strong acids, bases, alcohols, glycols, or steam.

WEATHERING AND OZONE RESISTANCE

This is an area where polyacrylics excel. Without the addition of any protective agent typical formulations easily pass standard tests (e.g., 168 h exposure to 100 pphm ozone at 49°C and 6 months roof exposure under 20% stretch test conditions).

APPLICATIONS

Automotive seals, gaskets, and hoses represent the largest market area for polyacrylics. They find more limited use in adhesives and as binders for propellants and flexible magnets. In the area of seals/gaskets, polyacrylics are used in transmissions, valve stems, crankshafts, pinions, oil pans, rocker covers,

O-rings, cork binders, grommets and packings, and sponges. They are also used in oil cooler hoses, transmission oil cooler hoses, spark plug boots, dust boots, belting, rolls, vibration dampeners, and fabric coating.

REFERENCES

1. Mast, W. C.; Rehberg, C. E.; Dietz, T. J.; Fisher, C. H. *Ind. Eng. Chem.* **1944**, *36*, 1022.
2. Mast, W. C.; Fisher, C. H. U.S. Patent 2 509 513, 1950.
3. "Polyacrylic Ester-Experimental Product"; B.F. Goodrich Chemical Division: 1947; Service Bulletin 47-SD3: Cleveland.
4. Fisher, C. H.; Whitby, G. S.; Beavers, E. M. In *Synthetic Rubber*; Whitby, G. S. Ed.; John Wiley & Sons: New York, 1954; p 900.
5. Vial, T. M. *Rubber Chem. Technol.* **1971**, *44*, 334.
6. DeMarco, R. D. *Rubber Chem. Technol.* **1979**, *52*(1), 173.
7. Starmer, P. H.; Wolf, F. W. *Encyclopedia of Polymer Science and Engineering,* 2nd ed.; John Wiley & Sons: New York, 1985; p 306.
8. Starmer, P. H. *Prog. in Rubber Plastics Technol.* **1987**, *3*(1), 1.
9. DeMarco, R. D. *European Rubber Journal* **1989**, *17*(3), 25.
10. Owen, H. *Rubber Age* **1950**, *66*, 544.
11. Starmer, P. *Progress in Rubber and Plastics Technology* **1987**, *3*(1), 1.
12. Thomas, R. *Rubber Chem. Technol.* **1969**, *42*, G40.
13. Fisher, C.; Mast, W.; Rehberg, C.; Smith, L. *Ind. Eng. Chem.* **1944**, *36*, 1032.
14. Thiokol Chemical Co. U.S. Patent 3 402 158, 1968.
15. Tellier, P.; Grimand, E. U.S. Patent 3 497 591, 1970.
16. Fukumori, T.; Okuya, E.; Enyo, H. Sulfur Curable Acrylic Rubber, Rubber Division Meeting, Toronto, May 11, 1983.
17. Mast, W.; Fisher, C. *Ind. Eng. Chem.* **1949**, *40*, 107.
18. Kaizerman, S. U.S. Patent 3 201 373, 1965.
19. Sims, J. *J. Appl. Polym. Sci.* **1961**, *5*, 58.
20. Uchiyama Kogyo Kaisha Ltd. Japanese Patent 60 120 708, 1985.
21. Morris, R. U.S. Patent 3 875 092, 1975.
22. Giannetti, E.; Mazzocchi, R.; Fiore, L.; Crespi, E. *Rubber Chem. Tech.* **1983**, *56*, 21.
23. Grassie, N.; Speakman, J. G. *J. Polym Sci.* **1971**, *A-1.9*, 919.

Acrylic-Styrene-Acrylonitrile

See: Acrylonitrile-Acrylic Elastomer-Styrene Terpolymer

Acrylics

See: Acrylic Elastomer Curing
Acrylic Rubber
Fibers, Synthetic (Unspun Natural Pulp-Like)
Floor Finishes
Fluoropolymers, Acrylates and Epoxies (With Low Dielectric Constants)
Photocrosslinking (Overview)
Polycarbonate/Acrylic Blends (Phase Morphology)
Poly(Methyl Methacrylate)
Powder Coatings (Overview)
Resins (Applications in Coatings Industry)
Textile Fibers (Structure and Properties)
Vinylidene Fluoride-Based Thermoplastics (Blends with Other Polymers)
Water-Borne Coatings (Urethane/Acrylic Hybrid Polymers)

ACRYLONITRILE-ACRYLIC ELASTOMER-STYRENE TERPOLYMER

Tatsuo Ishikawa and Isamu Hattori
Production Development Department
Goi Works
Hitachi Chemical Company, Ltd.

Acrylonitrile-acrylic elastomer-styrene terpolymer (AAS) (CAS Registry Number: 26299-47-8) is an elastomer-modified thermoplastic that has excellent strength, weather resistance, and shape-retaining characteristics. This material is sometimes called acrylic-styrene-acrylonitrile. In forming AAS, acrylic elastomer disperses as particles in an acrylonitrile-styrene copolymer (SAN) (CAS Registry Number: 9003-54-7) matrix. Because acrylic elastomer is a saturated elastomer, it is resistant to sunlight and oxygen: therefore, it has not only impact strength but also weather resistance. SAN provides rigidity, shape-retaining characteristics, and moldability. Grafting acrylonitrile and styrene onto acrylic elastomer enables sufficient compatibility between the elastomer phase and the matrix phase and provides stable morphology during processing. Because of the many variations of AAS properties, resulting from changing the ratio of the three monomers, the molecular weight, the degree of grafting, and morphology, AAS can be tailormade to meet the specific requirements of a customer. This flexibility can be further enhanced by substituting α-methylstyrene or *N*-phenylmalemide for styrene to get a higher heat deflection temperature and by alloying with such polymers as polycarbonate (PC) (CAS Registry Number: 25971-63-5), poly(vinyl chloride) (PVC) (CAS Registry Number: 9002-86-2), and poly(butylene telephtalate) (PBT) (CAS Registry Number: 26062-94-2).[1-3]

PREPARATIONS

AAS is usually manufactured by a graft polymerization of acrylonitrile and styrene onto acrylic elastomer by using an emulsion polymerization method. The AAS emulsion is then coagulated, filtrated, and dried. Next, it is mixed with stabilizers, pigments, and lubricants and, finally, the mixture is extruded to produce the finished product.

PROPERTIES

Weather Resistance

AAS has excellent weather resistance. Its properties and color retention are highly superior to that of ABS. This is because acrylic elastomer is resistant to sunlight and oxygen.

Thermal Stability

AAS provides excellent stability against thermal aging.

Chemical Resistance

AAS resists saturated hydrocarbons, oils, water, aqueous salt solutions, dilute acids, and alkalis, but it is attacked by concentrated inorganic acids, aromatic and chlorinated hydrocarbons, esters, and ketones.

AAS provides better resistance to environmental stress cracking than ABS and polycarbonate.

Electrical Properties

AAS has excellent electrical properties. Typical values are volume resistivity (1–2 × 10^{16} Ω-cm), arc resistance (85–100 s), and dielectric constant (3.0–4.5, 60 Hz–1 MHz).

APPLICATIONS

Because of its excellent weather resistance, mechanical properties, and moldability, AAS can be used without painting in many fields. In automobiles, AAS is used for door mirror housings, lamp housings, front grills, rear spoilers, and bumper covers. In building/construction, it is used for rain conduits, balustrades, shutters, sidings, bathtubs, parts of window profile, hot-water drainpipes and fittings, mailbox, and flower boxes. In appliances, it is used for parts of air conditioners, lamps, outlets, distributors, and electronic antennas. Other areas of use are sign boards, surfboards, lawnmower housings, garden furniture, swimming pool pump and filter housings, and boats.

REFERENCES

1. Mitulla, K.; Swoboda, J.; Echte, A.; Frank, H.; Hambrecht, J.; Schwabb, J.; Siebel, P. U.S. Patent 4 605 699, 1986.
2. Lindner, C.; Braese, H.; Ott, K. U.S. Patent 4 376 843, 1983.
3. Lindner, C.; Binsack, R.; Rempel, D.; Ott, K. U.S. Patent 4 417 026, 1983.

ACRYLONITRILE-BUTADIENE-STYRENE

V. K. Berry, D. H. Ellington, S. K. Gaggar,* R. L. Jalbert, C. Maier, K. F. Miller, R. L. Pettit, J. R. Schroeder, S. Vilasagar, and R. H. Wildi
General Electric Plastics
Technology Center

D. J. Buckley, Jr.
General Electric Research and Development

The most striking feature of the family of polymers known as Acrylonitrile–Butadiene–Styrene (ABS) is the rich diversity of its compositions, microstructures, methods of preparation, markets, and manufacturers. The availability of the three key building blocks of ABS, coupled with its broad utility, has attracted many producers to the market. There are more than 35 suppliers worldwide. With so many capable suppliers in all geographic regions, ABS is undoubtedly the most competitive of all engineering thermoplastics.

ABS is found in an extensive range of applications because of its excellent balance of mechanical properties, processing latitude, recyclability, and economics. Applications include automotive interior components, major appliances, house and personal care products, toys, sports equipment, computer and business equipment, medical devices, communication devices, and building and construction. The specific performance requirements of these different markets are achieved by manipulation of monomer composition, microstructure, morphology, or additives. This extensive material differentiation is matched by a broad range of process options for producing ABS.

*Author to whom correspondence should be addressed.

All the applications illustrate the use of ABS as a thermoplastic, but ABS also finds utility as any alloy or modifier of other thermoplastics. Although various alloys with ABS have been commercialized, the system enjoying the most success is the blend of polycarbonate and ABS, where ABS provides not only impact modification, but also improved processability and reduced cost. ABS resins are also well-known modifiers of poly(vinyl chloride) (PVC), where it is used to enhance heat and impact resistance.

ABS materials and markets are intensely competitive. Competition comes not only from ABS manufacturers, but also from traditionally lower performance polymers such as polypropylene and high-impact polystyrene that have been upgraded through technological advances. This intensely competitive environment assures that ABS will continue to evolve technically to bring improved cost and value to the markets it serves.

PREPARATION AND CHARACTERIZATION OF ABS

Preparation of ABS

Most commercially important ABS products consist of blends of a grafted terpolymer, acrylonitrile–butadiene–styrene, dispersed in a glassy matrix of styrene–acrylonitrile (SAN) copolymer. The preparation of ABS refers to the polymerization of two individual components, the graft terpolymer and SAN. The graft terpolymer consists of a polybutadiene (PBD) rubber core and grafted SAN shell.

Analysis and Characterization of ABS

Characterization of ABS materials is similar to that of other polymers. Two restricting aspects in analysis of ABS are the insolubility of the rubber phase and the inhomogeneity of the SAN and grafted-rubber-phases. Although many techniques can use a two-phase or insoluble sample, other measurements, such as those made with chromatographic systems, often require the isolation of phases prior to analysis. Two important analytical techniques are microscopy and infrared spectroscopy.

Microscopy

The key factors determining the properties of ABS are the size and distribution of the rubber particles, which together characterize the morphology of an ABS polymer. Characterization of morphology is usually carried out by electron microscopy techniques.[1]

Spectroscopy

Compositional control by Fourier transform infrared spectroscopy (FTIR) is the most often used tool for analysis of ABS polymers, with applications in many areas, including quality assurance and on-line and near-line process control.

Thermal and Rheological Analysis

The whole polymer viscosity of ABS is measured typically by capillary viscometry or parallel plate rheology. In both cases, the viscosity versus shear rate data is dependent on several parameters, including rubber particle size and crosslink density, M_n and M_w of SAN, additive type, and additive concentration.

Whole polymer differential scanning calorimetry (DSC) analysis of ABS shows two distinct T_g's, one for the rigid SAN phases at about 108°C and one for the polybutadiene phase at about –80°C.

Compounding Methods

Compounding plays a variety of important roles in the manufacturing of ABS. Its most basic function is to change the physical state of the plastic so that a subsequent processor can economically form the material into a product. When the operation includes the addition of other materials, the compounder must obtain a sufficiently homogeneous mixture to impart uniform visual and physical properties in the final product.

Compounding Characteristics of ABS

ABS has unique characteristics that must be taken into account in a compounding operation: the first is temperature effects. While molten ABS is highly shear thinning, at temperatures just above the glass transition point of the SAN, the polymer has an extremely high viscosity.

Compared to other polymers, ABS displays strong shear thinning, a trait that has both advantages and disadvantages. On the plus side, shear thinning minimizes degradation in molds with tight clearances since materials of higher viscosity are exposed to higher levels of heat from viscous dissipation.

The rubber phase is prone to degradation and cross-linking in abusive conditions such as exposure to high temperatures for extended periods, oxygen exposure, and very high shear fields. Rubber degradation will result in discoloration and, under extreme circumstances, loss of impact properties.

The hygroscopic nature of ABS means that the compounds must be adequately vented to produce pellets that are not foamed. However, the high solubility of water in ABS at elevated temperatures and pressures means that the liquid can be used to advantage as a temporary lubricant, stripping agent, and cooling medium. Water is sometimes also used to improve color by mitigating yellowing.[2]

PROPERTIES

Physical and Mechanical Properties

A wide range of properties is achievable for this class of polymeric materials, one of the most versatile of all engineering thermoplastics. The latitude possible with ABS is a consequence of its structure, a blend of grafted polybutadiene rubber and SAN matrix.

Effect of Rubber Content, Particle Size, and Particle Size Distribution

The mechanical properties of ABS are controlled by the compositional variables of its three phases: the SAN matrix, the polybutadiene (PBD) rubber particles, and the grafted SAN at the interface between the matrix and the particles. The most important rubber phase variables are the rubber content (more precisely, the volume fraction occupied by the rubber-containing particles), the rubber particle size, and the particle size distribution.[3]

Increasing the rubber volume fraction decreases the modulus, yield strength, and harness monotonically,[4] but increases impact strength to chemical stress cracking.[5]

Particle size and particle size distribution affect both the mechanical and physical characteristics of ABS. The mean particle size controls the impact behavior. The mean particle size also affects melt viscosity and appearance parameters such as gloss, opacity, and colorability. An associated parameter, the interparticle spacing (IPS), is a function of both the rubber content and the particle size. Rubber toughening is ineffective if this spacing exceeds a critical value related to the inherent ductility of the matrix.[6,7]

The particle size distribution affects the efficiency of impact modification and the transmission of light through unpigmented material. Because the maximum impact energy in a given SAN matrix occurs at a definite particle size, distributions containing many particles with nonoptimal sizes are less efficient for impact toughening.

Influence of Molecular Weight and AN Content

Grafted rubber particles promote localized craze and shear deformations in the SAN matrix phase throughout the volume of material subjected to external stresses.[3] So, the mechanical behavior of the matrix is critical in imparting toughness to ABS or any rubber-toughened polymer.[8] The most important matrix variables affecting overall material performance are the molecular weight and acrylonitrile (AN) content of the SAN. For lower M_n, no significant crazing occurs and rubber addition does not toughen the matrix. Most commercial ABS contains SAN with $M_n > 25{,}000$ and $M_w > 70{,}000$ g/mol; injection molding grades of ABS generally use lower molecular weights than extrusion grades.

A second matrix variable affecting the behavior of ABS is the AN content of the SAN, which in most commercial materials ranges between 25 and 35% by weight. The toughness of ABS increases with increasing AN content in the matrix SAN within the commercial range of AN levels.[9,10]

Specialty ABS Grades

ABS resins are very versatile materials and can be formulated in a wide variety of grades, blends, and alloys. ABS, especially when emulsion polymerized, can be produced with rubber contents between 12 and 85%, and styrene-to-acrylonitrile ratios between 90:10 and 60:40. Specialty grades demonstrate the versatility of ABS.

PRINCIPAL MARKETS

Because of excellent balance of mechanical properties and desirable flow and surface appearance, ABS and ABS blends have numerous uses. Some important applications are listed below:

- Automotive/transportation: interior trim, instrument panels, wheel covers, grills, headlight bezels
- Appliances/consumer electronics: refrigerator liners, hand-held appliances, vacuum cleaner housings, TV cabinets and components, telephones and related items, radio and speaker housings
- Business machines: monitors, computer housings, keyboards, laptop computers
- Miscellaneous: luggage and cases, furniture, protective headgear, packaging, toys, irrigation

- Building and construction: DWV pipe and fittings, faucets, window lintels and frames, tubs and shower surrounds

ACKNOWLEDGMENTS

The authors would like to gratefully acknowledge the support and assistance of the following individuals: Kay Wilhelm, Jo Ellen Butcher, Mark Reeder, Wally Bennett, Marian Smith, Louise Macuirles, and Nora Balcer. We would also like to thank GE Plastics for supporting publication of this article.

REFERENCES

1. Sawyer, L. C.; Grubb, D. T. *Polymer Microscopy*; Chapman and Hall: London, U.K., 1987.
2. Sun, Y.-C. U.S. Patent 4 387 179, 1983.
3. Bucknall, C. B. *Toughened Plastics*; Applied Science: London, U.K., 1977.
4. *Rubber Toughened Engineering Plastics*; Collyer, A. A. Ed.; Chapman and Hall: London, U.K., 1994.
5. Gaggar, S. K. *Am. Chem. Soc. Prepr., Org. Coatings and Appl. Polym. Sci. Div.* **1982**, *47*, 292.
6. Wu, S. H. *Polymer* **1985**, *26*, 1855.
7. Wu, S. H. *J. Appl. Polym. Sci.* **1988**, *35*, 549.
8. Wu, S. H. *Polym. Mater. Sci. Eng.* **1990**, *63*, 220.
9. Kim, H.; Keskkula, H.; Paul, D. R. *Polym. Mater. Sci. Eng.* **1990**, *63*, 210.
10. Kim, H.; Keskkula, H.; Paul, D. R. *Polym.* **1990**, *31*, 869–876.

Acrylonitrile Polymers

See: Acrylonitrile-Acrylic Elastomer-Styrene Terpolymers
Acrylonitrile-Butadiene-Styrene
Blends (Thermoplastic Polyurethane-San)
Bulk Polymerization
Polyacrylonitrile Gel Fibers
Precipitation Polymerization (Overview)
Pyroelectricity

Acrylonitrile-Styrene-Butadiene

See: Acrylonitrile-Butadiene-Styrene
Bulk Polymerization

ACRYLOYL-L-PROLINE ALKYL ESTER HYDROGELS

Masaru Yoshida, Agneza Safranj, and Hideki Omichi
Department of Material Development
Takasaki Radiation Chemistry Research Establishment
Japan Atomic Energy Research Institute

Ryoichi Katakai
Department of Chemistry
Faculty of Engineering Gunma University

Hydrogels are crosslinked macromolecular networks that swell in water and biological fluids. The polymer network can change its volume in response to a change in the environment such as temperature, solvent composition, electric or magnetic field, and exposure to light. This change may occur discontinuously at a specific stimulus level (phase transition), or gradually over a range of stimulus values. The temperature-induced phase transition is of great interest because of its relevance to physiological systems.

Our interest in thermoreversible gels stems from their intended application in controlled drug delivery devices, bioseparation processes, and artificial organs. The characteristics we seek are biocompatibility, elasticity, nonbiodegradability, and ability to control the release of various bioactive materials over an extended period of time. We have found that polymers with certain α-amino acids in their side chain are good candidates for our purposes. Hydrogels based on polymers with L-alanine, glycine, or proline in the side chain exhibit reversible swelling-deswelling behavior when cycled in solution between certain temperatures.[1-5] These gels have a potential use as pulsatile drug delivery devices.[5]

In this article, we summarize our recent work on the synthesis, properties, and applications of hydrogels based on poly(acryloyl-L-proline alkyl esters), [poly(A-ProOR)].

PREPARATION

We synthesize acryloyl-L-proline alkyl esters (A-ProOR) by a condensation reaction, in tetrahydrofuran, of L-proline alkyl ester hydrochlorides (HCl·H-ProOR) and acrylic acid. Dicyclohexylcarbodimide is the condensing agent.[6] The monomers can also be synthesized by a coupling reaction of HCl·H-ProOR and acryloylchloride in chloroform.[7] From A-ProOR monomers, linear polymers and crosslinked hydrogels can be prepared by the method of radiation polymerization and crosslinking. This method has several advantages over conventional methods: it is a simple and additive-free process at all temperatures; the degree of crosslinking can be easily controlled by irradiation conditions; and in some cases, it is also possible to combine into one step the synthesis and sterilization of the product.[8] This method is especially useful for the synthesis of hydrogels for biomedical applications in which the residual chemical initiators, usually toxic materials, may contaminate the product.

When an alcoholic solution of the A-ProOR monomer or the pure monomer is irradiated in nitrogen atmosphere, a linear polymer is obtained (**Figure 1**).[6,7] However, gels can be synthesized by irradiation of the alcoholic monomer solutions in the presence of a small amount of multifunctional monomer to act as a crosslinker. Recently, we found that gels could be obtained even without a crosslinker, when pure monomers were irradiated up to high doses or when aqueous monomer solutions were irradiated.

PROPERTIES

The most important property hydrogels based on polymers with amino acid side chains is the inverse volume phase transition in aqueous solutions. At low temperatures strong hydrogen bonding between hydrophilic groups and the surrounding water will cause the formation of a highly organized water layer around the polymer chains.[9] The formation of this structured water contributes to the enthalpy of mixing, which outweighs the unfavorable free energy related to the exposure of hydrophobic proline groups

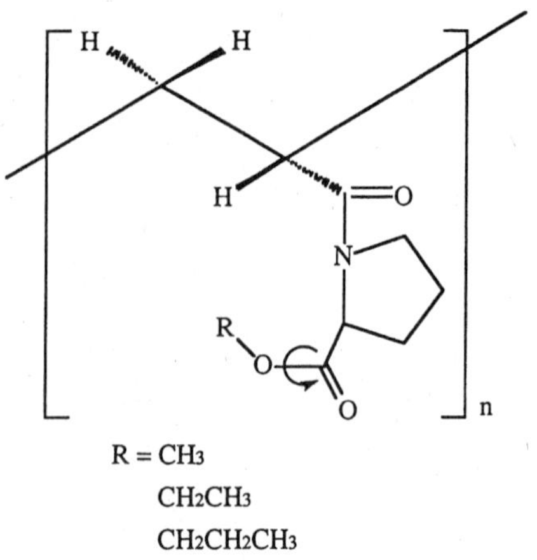

FIGURE 1. Structural formula of A-ProOR polymers.

of the side chain to water.[10-13] This enables good solubility of the polymer. We measured the radius of gyration for polymers with molecular weights of ~ 1 million and found it to be ~ 40 nm at 10°C, with an apparent volume of 270,000 cm^3.[7] This shows the extended coil conformation of the molecules. With increasing temperature, the hydrogen bonding weakens, thus reducing the structuring of water around the hydrophobic groups. With this water released, the inter- and intramolecular interactions between the neighboring proline groups increase. Above the lower critical solution temperature (LCST), which is ~ 14°C for A-ProOMe, the hydrophobic interactions become dominant, leading to an entropy-driven collapse of polymer chains to a compact globule conformation (e.g., at 50°C, the radius of gyration becomes ~ 8 nm, and the apparent volume is 2100 cm^3) and aggregation of the molecules. The aggregation can be seen from the increase of the apparent molecular weight of the polymer at temperatures above the phase transition.

In a crosslinked hydrogel below the volume phase transition temperature the polymer chains, although held together at the crosslink points, still keep the extended coil conformation, and the gel is swollen. Above the phase transition temperature, as the polymer chains fold and aggregate, the gel shrinks. The swelling ability of the gels—its magnitude, kinetics, and continuous or discontinuous nature—depends on the synthesis conditions. The gels synthesized from pure monomer, for example, do not show volume phase transition. The swelling behavior of gels synthesized from aqueous solution, however, is dependent on the monomer concentration, irradiation temperature, and absorbed dose.

The swelling behavior of the gels could also be influenced by copolymerization with monomers of different hydrophobicity or charge.

The volume phase transition temperature of the gels can also be influenced by adding to the aqueous solution salts or surfactants.[9,15-20]

APPLICATIONS

The major field of proposed applications for A-ProOR gels is the field of drug delivery systems. The gels based on polymers with α-amino acids in side chains are especially suited for implantable drug delivery devices because they are biocompatible and could be custom made to the desired elasticity, hydrophilicity, and swelling kinetics in response to changes in temperature.[14] In this way, the drug could be released according to the condition of the disease.

Another proposed application of A-ProOR gels is for flow regulation and recognition–separation purposes. This approach is also relevant to drug delivery devices.

REFERENCES

1. Yoshida, M.; Suzuki, Y.; Tamada, M.; Kumakura, M.; Katakai, R. *Eur. Polym J.* **1991**, *27*, 493.
2. Yoshida, M.; Yang, J-S.; Kumakura, M.; Hagiwara, M.; Katakai, R. *Eur. Polym. J.* **1991**, *27*, 997.
3. Yoshida, M.; Tamada, M.; Kumakura, M.; Katakai, R. *Radiat. Phys. Chem.* **1991**, *38*, 7.
4. Yoshida, M.; Sakurai, Y.; Tamada, M.; Kumakura, M.; Hagiwara, M.; Katakai, R. *Radiat. Phys. Chem.* **1992**, *39*, 469.
5. Ding, Z-L.; Yoshida, M.; Asano, M.; Ma, Z-T.; Omichi, H.; Katakai, R. *Radiat. Phys. Chem.* **1994**, *44*, 263.
6. Yoshida, M.; Omichi, H.; Katakai, R. *Eur. Polym. J.* **1992**, *28*, 1141.
7. Yoshida, M.; Omichi, H.; Kubota, H.; Katakai, R. *J. Intel. Mater. Sys. Struc.* **1993**, *4*, 223.
8. Rosiak, J. M.; Olejniczak, J. *Radiat. Phys. Chem.* **1993**, *42*, 903.
9. Schild, H. G.; Tirrell, D. A. *J. Phys. Chem.* **1990**, *94*, 4352.
10. Otake, K.; Inomata, H.; Konno, M.; Saito, S. *Macromolecules* **1990**, *23*, 283.
11. Bae, Y. H.; Okano, T.; Kim, S. W. *J. Polym. Sci., Polym. Phys. Ed.*, **1990**, *28*, 923.
12. Ringsdorf, H.; Venzmer, J.; Winnik, F. M. *Macromolecules* **1991**, *24*, 1678.
13. Feil, H.; Bae, Y. H.; Feijen, J.; Kim, S. W. *Macromolecules* **1993**, *26*, 2496.
14. Yoshida, M.; Asano, M.; Kumakura, M.; Katakai, R.; Mashimo, T.; Yuasa, H.; Yamanaka, H. *Drug Design. Del.* **1991**, *7*, 159.
15. Inomata, H.; Goto, S.; Saito, S. *Langmuir* **1992**, *8*, 687.
16. Park, T. G.; Hoffman, A. S. *Macromolecules* **1993**, *26*, 5045.
17. Wada, N.; Kajima, Y.; Yagi, Y.; Inomata, H.; Saito, S. *Langmuir* **1993**, *9*, 46.
18. Kokufuta, E.; Zhang, Y-Q.; Tanaka, T.; Mamada, A. *Macromolecules* **1993**, *26*, 1053.
19. Schild, H. G.; Tirrell, D. A. *Langmuir* **1991**, *7*, 665.
20. Winnik, F. M.; Ringsdorf, H.; Venzmer, J. *Langmuir* **1991**, *7*, 905.

Activated Monomer Polymerization

See: Block Copolymers (By Changing Polymerization Mechanism)
Telechelic Polyoxetanes

Acyclic Diene Metathesis

See: p-Phenylene Vinylene Oligomers, Homo- and Copolymers (Metathesis Preparation)
Ring-Opening Metathesis Polymerization (Formation of Cyclic Butadiene Oligomers)

Additives

See: Additives (Agents for Sustaining Properties)
Additives (Property and Processing Modifiers)

Additives (Types and Applications)
Antifoaming Agents
Antioxidants
Antioxidants and Stabilizers
Carbon Black
Carbon Black (Graft Copolymers)
Colorants (Overview)
Ethylene-Vinyl Acetate Copolymer (Chemical Modification, Compatibilizer in Blends)
Fillers (Glass Bead Reinforcement)
Fillers and Reinforcing Agents
Fillers, Polypropylene
Fillers, Surface Modification
Fire Retardants, Antimony-Containing
Flame Retardant Finishing
Flame Retardants (Overview)
Flame Retarding Polymers
Flammability (Char Formation)
Foaming Agents
Hindered Amine Light Stabilizers, Monomeric
Interphases, Ternary Polymer Composites
Polyethylene (Stabilization and Compounding)
Powder Coatings (Overview)
Rigid Polyurethane Foams
Rubber Additives, Multifunctional
Silicone Rubber Latex
Silicone Sealants
Siloxane Elastomers
Surface Modification (Overview)
Thermoplastic Composites
Unsaturated Polyester Resins (Overview)
Unsaturated Polyester Resins (Polyester Composites)
Unsaturated Polyester Resins (Toughening with Elastomers)
Unsaturated Polyester Resins (Toughened with Liquid Rubber)
Vegetal Biomass (2. Oligomers and their Polymerization)

ADDITIVES (Agents for Sustaining Properties)

Yasukazu Ohkatsu
Kogakuin University

Plastics are used in many fields because they can be processed easily, are relatively light, and possess excellent chemical and physical properties. However, they also have undesirable properties from a combination of the material itself and processing. Therefore, it is important to hide or reduce the defects of plastics while sustaining or enhancing the merits as much as possible. To attain that aim, many kinds of polymer additives have been developed.

This article focuses only on antioxidants.

FUNCTION-SUSTAINING AGENTS OF POLYMERS

Properties of Polymeric Materials

The degradation of polymeric materials is complicated but generally is induced by heat, light, mechanical shear, and ozone, which essentially initiates autoxidation.[1]

The degradation of polymer materials by autoxidation depends on the polymer and starts wherever bonds have low dissociation energy. Polymers with photosensitive functional groups such as carbonyls are likely to degrade when exposed to light.

Polymers degrade mainly through use, although some degradation occurs in preparing and processing stages. The primary route for degradation is autoxidation. Additives that prevent degradation can be classified into two groups: inhibitors of initiation steps such as photostabilizers, heat stabilizers, metal deactivators, and peroxide decomposers, and inhibitors of chain-propagating steps such as radical scavengers.

Phenolic Antioxidants

Phenolic antioxidants are radical scavengers combined with aromatic amine antioxidants. The latter antioxidant is used almost entirely in rubber because it colors the materials during use. A phenolic antioxidant, however, is used in many fields and is an indispensable component for stabilizing plastics. A phenolic antioxidant is effective for scavenging oxygen-containing radicals such as RO· and RO_2·.

Characteristics

A phenolic antioxidant shows synergistic effects when it is used with one or more additives. A phenol, for example, can contribute to heat stabilization with a sulfur- or phosphorus-containing compound. However, phenols color polymer materials such as fabrics, which have frequent contact with NO_x gases.[2]

Peroxide Decomposers

Sulfur-Containing Antioxidants

A sulfur-containing antioxidant is known as an effective hydroperoxide decomposer and is added to various polymers with a phenol because it is ineffective when used alone. A sulfur-containing antioxidant decomposes hydroperoxides ionically to inhibit initiation reactions, which would otherwise be induced by homolytic cleavage. Of all such antioxidants only a thiopropionic acid derivative is very effective, and this may be because of the action mechanism of a sulfur compound.

A sulfur compound does not provide good stabilization at a higher temperature because of the formation of an intermediate, which forms during decomposition of hydroperoxides and works as a pro-oxidant.[3]

A sulfur compound acts on hindered amine light stabilizers with antagonism, resulting in the decreased weather resistance of plastics.

However, it exhibits remarkable synergism with a phenolic antioxidant.

Phosphorus-Containing Antioxidants

A phosphorus-containing antioxidant functions similarly to one containing sulfur. Thus, it also is used advantageously as a processing stabilizer.

Phosphorus compounds are used to stabilize plastics by ionic decomposition of hydroperoxides, the rate of which depends on the molecular structure of a phosphorus compound.

A phosphorus compound behaves like a sulfur compound in terms of the hydroperoxide decomposition, but the former compound shows remarkably higher effects, especially in processing at a temperature above 200°C.[4] A phosphorus compound is also characterized by no formation of any pro-oxidant like a sulfur compound after the hydroperoxide decomposition.

Although the synergism with a phenol is not so remarkable as that of a sulfur compound, a phosphorus compound often can inhibit coloration in plastics stabilized by a phenol.

UV Absorbers

There are many kinds of photostabilizers that protect plastics from degradation by light, especially harmful UV light. A UV absorber is added to various plastics to keep the bulk and surface of plastic materials protected from light. It can absorb light with a wavelength of about 290 to 310 nm by transferring a hydrogen atom and then releasing the energy as heat.

Hindered Amine Light Stabilizers

HALS were developed as light stabilizers as their name indicates. Because they exhibit many functions, the stabilizers are used in plastics, fabrics, and paints and are called multifunctional additives. HALS have been thought of as photostabilizers, but they have little or no ability to stabilize materials against UV light. HALS can scavenge free radicals (R·), decompose hydroperoxides, and trap metal ions.

HALS inhibit surface deterioration of plastics better than UV absorbers and are used in polyolefins, polystyrene, and polyurethanes. However, they cannot be used in base-sensitive plastics such as polycarbonates, polyethylene therephthalates, and epoxy resins.

CONCLUSIONS

It is important to evaluate additives for the purpose intended. A degrading polymer does not always result in limited use. The evaluation of polymer degradation is carried out by observing chemical and physical changes of a material's properties. Degradation occurs during processing or use with the corresponding estimation procedures established.

REFERENCES

1. Ohkatsu, Y. *Theory and Practice of Autoxidation* (in Japanese); Kagakukogyosha: Tokyo, 1987.
2. Smelts, K. C. *Textile Chemist and Colorist* **1983**, *15*(4), 17.
3. Scott, G. *Develop. Poly. Stab.* **1981**, *4*, 16.
4. Haruna, T. Asahi Denka Kogyo K. K. (Urawa Lab.), personal communication.

ADDITIVES (Property and Processing Modifiers)

Malcolm P. Stevens
University of Hartford

Polymers are rarely used in their pure form in commercial applications. They may be too brittle or too high melting for processing, for example, or they may break down under the influence of oxygen or sunlight. Additives render polymers suitable for commercial use, or they expand a polymer's range of applications. Polyethylene and poly(methyl methacrylate) are among the few polymers that may, in certain applications, be used without additives. In contrast poly(vinyl chloride), a material virtually useless by itself, is one of the major commodity plastics of the judicious use of additives.

Additives encompass a variety of organic and inorganic compounds and a broad spectrum of applications. **Table 1** lists by function the additive types now in widespread commercial use. Depending on the application, most commercial polymers contain a combination of additives.

Some additives, including fillers and reinforcing agents, are immiscible with the polymer; others, such as antioxidants and ultraviolet stabilizers, are completely soluble. Surface-property

TABLE 1. Polymer Additives

Type	Function
Mechanical property modifiers	
Fillers	Increase strength, reduce cost
Impact modifiers	Improve impact strength
Nucleating agents	Promote crystallinity
Plasticizers	Increase flexibility
Reinforcing fibers	Increase strength annd stiffness
Surface property modifiers	
Antifogging agents	Prevent moisture from obscuring film clarity
Antistatic agents	Prevent static charge build-up
Coupling agents	Improve bonding to filler or reinforcing fiber
Release agents	Prevent sticking
Chemical property modifiers	
Antioxidants	Prevent oxidative degradation
Biocides	Prevent microbial attack and mildew
Flame retardants	Reduce flammability
Ultraviolet stabilizers	Prevent degradation by sunlight
Aesthetic property modifiers	
Biocides	Prevent development of odor
Coloring agents	Impart color
Nucleating agents	Improve light transmission
Odorants	Add fragrance, mask objectionable odors
Optical brighteners	Impart a more vivid appearance
Processing modifiers	
Blowing agents	Manufacture foams
Crosslinking agents	Promote crosslinking (curing)
Defoaming agents	Reduce foaming, remove trapped air
Emulsifiers	Stabilize polymer emulsions
Heat stabilizers	Prevent thermal degradation
Low profile additives	Prevent shrinkage and warpage
Plasticizers	Reduce melt viscosity
Release agents	Prevent sticking to processing machinery
Thickening agents	Increase viscosity

modifiers have limited compatibility with the polymer and migrate to the surface where their influence is needed. They are normally used in low concentrations (< 2% by weight) to avoid forming undesirable films on the polymer surface. In some instances, surface-property modifiers are applied externally to the finished product.

This is a brief overview of the types of compounds that comprise this very broad area of polymer chemistry. More detailed information on individual additives may be found elsewhere in this encyclopedia or in other publications covering the field.[1-10]

MECHANICAL PROPERTY MODIFIERS

Fillers

Fillers (also called extenders) comprise the most widely used group of additives. They are used primarily in plastics for improving strength or reducing cost but are also used in coatings and paper to enhance hiding power or gloss.[11] Metallic fillers are used to increase conductivity in plastics or carpeting and to facilitate heat transfer or metal plating. Sometimes fillers are used to reduce shrinkage during polymer crosslinking.

Inorganic fillers are used mainly with thermoplastic polymers; calcium carbonate is the most important in terms of consumption.

Organic fillers include wood flour, cellulose, and ground corn cobs or nut shells.

Impact Modifiers

Impact modifiers are, in general, flexible polymers that are blended with rigid polymers to reduce brittleness and improve impact resistance.

Nucleating Agents

Nucleating agents, blended uniformly into amorphous polymers, provide a surface area to initiate crystallization. Polymer strength is optimal when the crystallites are small and uniformly distributed. Nucleating agents also enhance light transmission in translucent plastics.

Plasticizers

Plasticizers are added to lower a polymer's glass transition temperature and to reduce the melt viscosity during processing. Both external and internal plasticizers are used.

Reinforcing Fibers

Three types of fiber are currently used in polymer composites: glass, carbon (or graphite), and aramid (aromatic polyamide). The fibers are available in different forms, including continuous filament (roving), chopped filament, mat, woven cloth, and pulp.[12]

Major commercial outlets for both aramid- and carbon-fiber composites are in aerospace and military applications and in sporting goods. Aramid fiber is also used for reinforcing automobile tires. Glass fiber is lowest in cost and is the most widely used.

Antifogging Agents

"Fogging" is the condensation of minute water droplets on the inside surface of film packaging. Antifogging agents prevent fogging by "wetting" the film surface and causing the condensate to form a continuous film. The most widely used antifogging agents are nonionic surfactants, such as ethoxylated fatty acid esters.

Antistatic Agents

Polymer surface charge attracts dirt, can cause film to stick together and fabrics to cling, and may cause static interference in electrical equipment. Antistatic agents dissipate surface charge.[13]

Two types of antistatic agents are used commercially: those that form a hygroscopic surface to attract atmospheric moisture and thereby increase surface conduction and those that provide a conducting mechanism through the polymer.

Coupling Agents

Coupling agents are used with reinforcing fibers or fillers to improve adhesion to the polymer. The two most common types of coupling agents are silanes and titanates.[14]

Release Agents

Release agents serve two purposes: to prevent polymer from sticking to processing machinery and to prevent film or sheet from sticking to itself or to other surfaces. They are normally categorized according to their function: mold-release agents prevent sticking in compression or injection molding; slip agents prevent sticking in extrusion and calendaring; and antiblocking agents (also called flatting agents) are used to prevent plastic film or sheet from sticking together. Similar compounds are used in each case.

Calcium and zinc salts of fatty acids, especially stearic acid, are the most widely applied release agents.

CHEMICAL PROPERTY MODIFIERS

Antioxidants

Oxidative degradation may be initiated by free radicals formed under the influence of shear stress or high temperatures during processing or by exposure to ultraviolet or ionizing radiation. The radicals subsequently react with oxygen to form unstable hydroperoxides which, in turn, initiate further radical abstraction and disproportionation reactions.[15] Polymers containing tertiary and allylic hydrogens, such as polypropylene and polybutadiene, respectively, are the most susceptible to oxidative degradation, and such polymers represent the major markets for antioxidants.[16]

Antioxidants are of two types: primary, which react with the radicals as they form, and secondary, which reduce hydroperoxides. Usually a combination of primary and secondary types are used.

Primary antioxidants are either hindered phenols or secondary aromatic amines.

Secondary antioxidants include divalent sulfur and trivalent phosphorus compounds. These compounds reduce hydroperoxides.

Biocides

The main function of biocides is to prevent attack by microorganisms that cause mildew and objectionable odor. Unlike natural polymers, most synthetic polymers are resistant to microbial attack, but they are often rendered susceptible by plasticizers and other additives.

Flame Retardants

Polymer flammability can be reduced significantly by incorporating halogen- or phosphorus-containing monomers or by using flame-retardant additives.[17,18]

Additives containing chlorine and bromine form hydrogen halides on combustion, which reduce flammability by inhibiting the free-radical combustion reactions.

Phosphorus-containing additives, which are widely used in cellulosic textile fibers, reduce flammability by promoting formation of a carbonaceous char on the polymer surface that inhibits the release of combustible gases and shields the surface from the heat flux.

Ultraviolet Stabilizers

Photooxidation and photodegradation are common causes of failure in polymers used in exterior settings. Several types of stabilizers are used: ultraviolet absorbers shield the polymer in much the same way sunscreen lotions protect the body, by absorbing radiation and dissipating the energy via vibrational transitions; radical scavengers intercept free radicals formed during photochemical reactions; and quenchers reduce photoexcited states of polymers to their ground states.

AESTHETIC PROPERTY MODIFIERS

Two kinds of aesthetic property modifiers, biocides and nucleating agents, were described earlier.

Coloring Agents

Pigments are inorganic or organic coloring agents that are insoluble in polymers. Dyes, which are usually organic, are soluble. Both types are commonly blended into the polymer as color concentrates.

A variety of inorganic pigments are available. They are used in pure form or mixed for color variations.

Organic dyes and pigments for polymers encompass most of the common complex dyestuff categories: phthalocyanine, xanthene, azo, azine, anthraquinone, indigo, quinoline, and quinacridine, as well as specialty fluorescent and phosphorescent dyes.[19,20]

Odorants

Odorants are fragrances (e.g., alcohols, esters, and terpenes) added to plastics either to mask odors (e.g., in garbage cans or garbage disposal units) or to add fragrance to plastics for aesthetic appeal.

Optical Brighteners

Most of the more than 1000 commercially available brighteners are used in the detergent, paper, and textile industries. A relatively small amount, ~ 4% of the total production, is used in plastics and synthetic fibers.[21,22] By absorbing ultraviolet radiation and emitting a blue-violet fluorescence, brighteners make the polymer surfaces appear brighter or cleaner.

PROCESSING MODIFIERS

Blowing Agents

Blowing agents are used in the manufacture of foamed plastics. Two types, physical and chemical, are used.[23,24] Physical blowing agents are either gases dissolved in the polymer under pressure, or low-boiling liquids that are converted to their gaseous state by external heating or by a reaction exotherm during processing, or both. Chemical blowing agents are compounds that decompose upon heating with the evolution of gaseous products. The finished foam results from the trapping of gas bubbles in the viscous polymer matrix by cooling or gelation.

Nitrogen is the most commonly used compressed-gas blowing agent, but air and carbon dioxide are also used.

The most common types of chemical blowing agents are compounds containing nitrogen–nitrogen bonds that evolve nitrogen gas upon heating. About 90% of the market is held by 1,1-azobisformamide (decomposition range: 195–216°C), also known as azodicarbonamide or ADC.

Crosslinking Agents

Vulcanization of rubber by using sulfur is probably the oldest example of polymer crosslinking.[25] Although sulfur is still widely used with natural rubber, numerous crosslinking agents have been developed for crosslinking (curing) synthetic polymers, the most important being peroxides.[26]

Defoaming Agents

Trapped foams may create weak spots in fabricated polymers, or foaming may cause processing difficulties. To reduce surface tension and thereby break up the foams, a variety of additives are used, including polysiloxane oils (often in combination with polyethers), anionic and nonionic surfactants, and acetylenic glycols.[27]

Emulsifiers

Emulsifying agents are used whenever a polymer is synthesized in emulsion form or when it is formulated as a latex coating or adhesive.

Heat Stabilizers

Dehydrochlorination of poly(vinyl chloride) during high-temperature fabrication results in discoloration of the polymer. Heat stabilizers are combinations of compounds that prevent decomposition by reacting with the HCl as it is formed and by rendering inactive the resultant allylic chloride sites.

Low-Profile Additives

Shrinkage of a thermosetting polymer during crosslinking may cause cracking or warping of the finished product or, in the case of composites, a weakened interface between the polymer and the reinforcing fiber. Polymeric low-profile additives are used to prevent shrinkage, most commonly with unsaturated polyesters.

Thickening Agents

Thickening agents prevent sagging of thermosetting polymers during molding by increasing the polymer's viscosity. As such, they are also referred to as antisag agents. Thickening agents also impart thixotropic properties to coatings and adhesives.

REFERENCES

1. *Plastics Additives Handbook* 3rd ed.; Gächter, R.; Müller, H. Eds.; Hanser/Oxford: New York, 1990.
2. *Thermoplastics Additives Handbook*; Lutz, J. T., Ed.; Marcel Dekker: New York, 1988.
3. Flick, E. W. *Plastics Additives: An Industrial Guide*; Noyes: Park Ridge, NJ, 1986.
4. Štepek, J.; Daoust, H. *Additives for Plastics*; Springer-Verlag: New York, 1983.
5. Seymour, R. B. *Additives for Plastics*; Academic: New York, 1978.
6. Mascia, L. *The Role of Additives in Plastics*; Halstead: New York, 1975.
7. *Additives for Polymers*; Shelton, J. Ed.; Elsevier: New York.
8. Seymour, R. B. In *Encyclopedia of Polymer Science and Engineering*, 2nd ed; Mark, H. F. et al., Eds.; Wiley-Interscience: New York, 1985.
9. Greek, B. F. *Chem. Eng. News* **1988**, *66*(23), 35.
10. *Modern Plastics Encyclopedia: 1984–85*; Agranoff, J., Ed.; McGraw-Hill: New York, 1984; pp 106–180.
11. *Handbook of Fillers and Reinforcements for Plastics*; Katz, H. S.; Milewski, J. J., Eds., Reinhold: New York, 1978.
12. Seymour, R. B.; Deanin, P. D. *History of Polymeric Composites*; VNU; Utrecht, Netherlands, 1987.
13. Johnson, K. *Antistatic Agents: Technology and Applications*; Noyes: Park Ridge, NJ, 1972.
14. Plueddenmann, E. P. *Silane Coupling Agents*; Plenum: New York, 1982.
15. *Autoxidation and Antioxidants*; Lundberg, W. O. Ed.; Wiley-Interscience: New York, 1961.
16. Dexter, M. In *Encyclopedia of Polymer Science and Engineering*, 2nd ed; Mark, H. F. et al., Eds.; Wiley-Interscience: New York, 1985; Vol. 2, pp 86–90.
17. Lewin, M.; Atlas, S. M.; Pearce, E. M. *Flame Retardant Polymeric Materials*; Plenum: New York, 1975.
18. Troitzsch, H. J. In *Plastics Additives Handbook*, 3rd ed.; Gächter, R. and Müller, H. Eds.; Hanser/Oxford: New York, 1990; Chapter 12.
19. McKlaren, K. *The Color Science of Dyes and Pigments*; Hilger: Bristol, U.K., 1983.
20. Billmeyer, F. W., Saltzman, M. *Principles of Color Technology*; Wiley-Interscience: New York, 1981.
21. Martini, T. In *Thermoplastic Polymer Additives*; Lutz, J. T., Ed.; Marcel Dekker: New York, 1989; pp 315–344.
22. Berger, K. In *Plastics Additives Handbook*, 3rd ed.; Gächter, R. and Müller, H. Eds.; Hanser/Oxford: New York, 1990; Chapter 14.
23. Frisch, K. C.; Saunders, J. H. *Plastic Foams*; Marcel Dekker: New York, 1972–73.
24. Bikerman, S. J. *Foams*; Springer-Verlag: New York, 1973.
25. *Vulcanization of Elastomers*; Alliger, G.; Sjothun, I. J., Eds.; Van Nostrand: New York, 1964.
26. *The Chemistry of Peroxides*; Patai, S., Ed.; Wiley-Interscience: New York, 1983.
27. Owen, M. J. In *Encyclopedia of Polymer Science and Engineering*, 2nd ed.; Mark, H. F. et al., Eds.; Wiley-Interscience: New York, 1988; Vol. 12, pp 286–288.

ADDITIVES (Types and Applications)

S. N. Maiti
Centre for Polymer Science and Engineering
Indian Institute of Technology

Plastics have the widest range of mechanical, physical, and other useful properties combined with easy fabricability and economical costs.[1,2] However, plastics are seldom used alone and, to suit end-use requirements such as colorability and fire resistance, certain additives are almost always mixed with them.[2-8]

CLASSIFICATION

Although additives may be solid, rubbery, liquid, or gaseous, the following nine classes of materials, the most frequently used, will be discussed here.

STABILIZERS

The properties of polymers deteriorate under the combined effects of high and ambient temperatures, atmospheric radiation, oxygen, ozone, water, microorganisms, and other atmospheric agents. The overall effects on the polymer are loss of strength, hardening and embrittlement, color formation and/or reduction of optical clarity, changes in chemical activity, and a decrease in electrical insulation properties.

Antioxidants

The degradation of polymers by heat, oxygen, ozone, mechanical shearing, UV exposure, metal ions, and other agents normally occurs by a free-radical mechanism.[2,3,9] A host of radical species are produced by initiation and propagation mechanisms in which the hydroperoxide radical is by far the most reactive, decomposing to give rise to chain scission. Radical species recombine to form extended-chain or crosslinked polymers in the termination step.

Effective antioxidants interrupt the sequence of chain reactions and are of two types depending on their mode of action: preventive antioxidants that prevent the formation of radicals R· and ROO· and chain-breaking antioxidants that directly intervene in the propagation cycle reaction with R· and ROO· radicals and introduce new termination reactions.

Preventive antioxidants prevent the formation of free radicals and are classed as: peroxide decomposers, metal deactivators, and UV light protectors.

Chain-breaking antioxidants are nonstaining hindered phenols and aromatic amines (phr 0.02–1).[2,3,6] Used in excess oxidation may be facilitated by these antioxidants.

Stabilizers for Halogenated Polymers

Polymers of vinyl and vinylidene chloride undergo rapid and autocatalytic dehydrochlorination, generating conjugated unsaturation, which in turn undergoes an oxidation reaction. Stabilizers should arrest HCl evolution by absorbing and neutralizing it, prevent oxidation and other free-radical processes, displace active and labile tertiary and allylic Cl atoms with more stable substituents, and disrupt the conjugation of the polymer.[2,3,10,11]

Stabilizer Synergism, Autosynergism, and Antagonism

Some stabilizer combinations are more powerful than the sum of the individual members; in some the effect of each component is doubled. This phenomenon is known as synergism.

Some antioxidants functioning via more than one mechanism are autosynergistic. An antioxidant that is synergistic with one member may be antagonistic to another antioxidant.

FILLERS

Fillers are usually solid additives mixed with plastics to improve general properties, to introduce specific characteristics, or to reduce the cost of the compound.[2-4,12] In the process, the physical properties of the polymers are modified to a varied extent.

Fillers may be organic or inorganic, and each group consists of fibrous and nonfibrous types.[2,4] Nonfibrous fillers (or particulates) may be an inert (or loading) type or a reinforcing type. Reinforcing fillers increase properties such as tensile strength and modulus through enhanced interaction with the polymer.

Fibrous fillers are primarily used for thermosets to improve impact behavior and rigidity.[4,12] Inorganic fibrous fillers are used in thermoplastics and in thermosets to enhance heat resistance and impact strength.

Coupling Agents

A stress-bearing interphase necessary for a high-quality composite is achievable through enhanced interfacial adhesion. This adhesion is accomplished by surface treatment of the inclusion with a suitable coupling agent, which also promotes physical and/or chemical interaction with the polymer.[2,3,13]

PLASTICIZERS

Plasticizers are added to polymers primarily to achieve flexibility, although many other polymer properties are also modified.[2,3,7] In general, with an increase in plasticizer content there are corresponding reductions in modulus, tensile strength, hardness, density, softening temperatures, brittle temperatures, and volume resistivity. Breaking elongation, toughness, softness, dielectric constant, and power factor show increases.

They function through a solvent action opposing and reducing aggregation of the polymer molecules. They are a type of nonvolatile solvent with a molecular weight of at least 300 and a solubility parameter close to the polymer.

LUBRICANTS AND FLOW PROMOTERS

Lubricants decrease the frictional forces between two rubbing surfaces. Depending on the type of rubbing surfaces and lubricating functions required, the lubricants are classified mainly into external and internal types.

There are two categories of external lubricants. Solid layer lubricants reduce the friction of moldings and finished parts rubbed against surface with an identical or different composition.[2]

Boundary lubricants are required when extraordinarily high loads are encountered at the interface (e.g., during polymer processing). The lubricant forms a stationary thin film linked with the metal of the equipment.

Used in excess, boundary lubricants can cause detrimental side effects, including haze and greasiness, and difficulties in printing, heat sealing, gelation, and fusion of the compound.

Internal lubricants bring about plasticization at high temperatures, increasing the fusion rate of polymer molecules that rub against one another, which enhances the melt flow.

COLORANTS

Colorants are added to plastics to make them attractive in appearance.[6] They produce color by selectively absorbing, transmitting, reflecting, and scattering specific areas of light energy from wave bands that constitute while light.[3,14]

Two broad classes of colorants are dyes and pigments.[14] Dyes are synthetic organics that have good transparency, high tinctorial strength, low specific gravity, and good solubility in most solvents and plastics. Pigments are intensely colored chemicals of both organic and inorganic origin and are insoluble in common solvents and plastics.

FLAME RETARDERS

Plastic materials are required for fire retardance in buildings, furniture, and fittings, and are used in transportation and other industrial and domestic applications.[2]

Two broad classes of materials, additive and reactive types, have evolved. In volume terms additive types supercede the reactives and include antimony oxide halogen combinations, organohalogen compounds, organophosphorous and halophosphorous compounds, boron compounds, and alumina trihydrate.[2,5]

BLOWING AGENTS

These additives (also called foaming agents) are used to produce cellular plastics by filling varying proportions of gas-filled cells in the plastics.[2,3,15] The cells may or may not be intercommunicating, resulting in "open-cell" or "closed cell" structures, respectively. The foamed plastics exhibit decreased density and thermal and electrical conductivity than do the corresponding noncellular plastics. They also possess mechanical and acoustical energy dissipation characteristics.

Blowing agents may be physical or chemical types. Physical blowing agents undergo a phase change to form cells (e.g., a liquid may be volatilized, or a gas diffused in a polymer matrix under high pressure may be desorbed by decompression and/or heating at a high temperature). Chemical blowing agents either decompose thermally or react chemically with active groups in a polymer to produce gas. Blowing agents should satisfy three criteria.[2] They should generate gas within a narrow, clearly defined temperature range in a controlled and reproducible manner.

CROSSLINKING AGENTS

Crosslinking is used to make plastics insoluble and infusible with enhanced general properties by joining the polymers together to form a three-dimensional network structure.[2,16] There are five classes of crosslinking brought about by crosslinking agents: bridging agents, crosslinking initiators, catalytic crosslinking agents, active-site generators, and reversible crosslinking agents.

PHOTODEGRADANTS

Plastics are used extensively in packaging applications, which creates environmental pollution.[2] If polymers are biodegradable they can mix easily with soil after they are discarded.[17] These additives stabilize the polymer during processing and degrade them upon UV exposure.

Photodegradation has to be carefully controlled so that when it is stored indoors the packaging material will not degrade.

Photodegradable plastics compositions are 5–15% more expensive than normal polymers, although recent reports claim the contrary. Their use is predicted to increase globally to attenuate the litter problem.[18]

REFERENCES

1. Nielsen, L. E. *Mechanical Properties of Polymers and Composites*; Marcel Dekker: New York, 1974; Vols. 1 and 2.
2. Brydson, J. A. *Plastics Materials*, 5th ed.; Butterworths: London, 1989.
3. Mascia, L. *The Role of Additives in Plastics*; Arnold: London, 1974.
4. *Plasticizers, Stabilizers and Fillers*; Ritchie, P. D., Ed.; Iliffe: London, 1972.
5. Gann, R. G.; Dipert, R. A.; Drews, M. J. In *Encyclopedia of Polymer Science and Engineering*; Mark, H. F. et al., Eds.; Wiley-Interscience: New York, 1987; Vol. 7, pp 154–210.
6. *Degradation and Stabilization of Polyolefins*; Allen, N. S., Ed.; Applied Science: London, 1983; Chapter 8.
7. Matthews, G. *Vinyl and Allied Polymers*; Iliffe: London, 1972.
8. Sohma, J.; Sakaguchi, M. In *Degradation and Stabilization of Polymers*; Geuskins, G., Ed., John Wiley & Sons: New York, 1975.
9. Nicholas, P. P.; Luxeder, A. M.; Brooks, L. A.; Hammes, P. A. In *Kirk-Olthmer Encyclopedia of Chemical Technology*, 3rd ed.; John Wiley & Sons: New York, 1982; Vol. 3, pp 128–148.
10. Wypych, J. *Polyvinyl Chloride Stabilization*; Elsevier: Amsterdam, 1986.
11. *Degradation and Stabilization of PVC*; Owen, E. D., Ed.; Elsevier Applied Science: London, 1984.
12. Stoy, W. S.; Washabangh, F. J. In *Encyclopedia of Polymer Science and Engineering*; Mark, H. F. et al., Eds.; Wiley-Interscience: New York 1987; Vol. 7, 53.
13. Mallick, P. K. *Fibre-Reinforced Plastics*; Marcel Dekker: New York, 1988.
14. Ahmed, M. *Coloring of Plastics—Theory and Practice*; Van Nostrand Reinhold: New York, 1979.
15. Heck, R. L. III; Peascoe, W. J. In *Encyclopedia of Polymer Science and Engineering*; Mark, H. F. et al., Eds.; Wiley-Interscience: New York, 1987; Vol. 2, pp 434–436.
16. Odian, G. *Principles of Polymerization*; John Wiley & Sons: New York, 1981; pp 112–140.
17. Henman, T. J. In *Degradation and Stabilization of Polyolefins*; Allen, N. S., Ed., Applied Science: London, 1983; Chapter 2.
18. Omichi, H. In *Degradation and Stabilization of Polyolefins*; Allen, N. S. Ed., Applied Science: London, 1983; Chapter 4, pp 204–206.

Adhesion Promoters

See: Silane Coupling Agents (Adhesion Promoters)
Silicone Sealants

Adhesive Tack

See: Tack, Elastomers

Adhesives

See: Cyanoacrylates
Dental Adhesives (for Alloys and Ceramics)
Epoxidized Natural Rubber
Epoxies, Rubber-Modified
Epoxy Acrylate-Based Resins
Epoxy Resins (Overview)
Ethylenimine-Modified Polymers
Fluoro-Acrylate Adhesive Blends (Surface Segregation Behavior)
Fluoropolymers (Surface Modification by Eximer-Laser Irradiation)
Furan Resins
Marine Adhesive Proteins, Synthetic
Melamine Resins (Overview)
Poly(2-cyanoacrylates)
Poly(imide siloxane)s
Polyimides (Containing Ether Linkages; Adhesive Properties)
Polyimides (Introduction and Overview)
Polysulfides (Prepared from Sulfur Dioxide)
Polysulfides (Use as Modifiers in Epoxy Systems)
Polyterpene Resins
Polyurethanes (Overview)
Reactive Liquid Rubbers (Epoxy Toughening Agents)
Silicone Release Coatings
Styrene-Butadiene-Styrene Triblock Copolymer
Urea-Formaldehyde Adhesive Resins
Urea-Formaldehyde Glue Resins
Vinyl Chloride Copolymers
Vinyl Ether Polymers

ADSORPTIVE RESINS

Takamasa Nonaka
Department of Applied Chemistry
Faculty of Engineering
Kumamoto University

A method for selective adsorption of certain metal ions from their mixture solution has been developed to separate similar ions, remove harmful metal ions, and recover valuable metal ions. Macroporous chelating resins are very useful for this purpose. The adsorptive resins are essentially insoluble in water and have complexing agents as functional groups. The functional groups fixed to the resin matrix have electron donor atoms such as O, N, or S in ligand groups to make metal chelate. Some electron-donor atoms have a higher complexing ability for various metal ions than other atoms.[1] It is possible to make chelating resins that have selective adsorption ability for a certain metal ion by fixing the desired ligand groups into the resins. The selective adsorption behavior of these resins for metal ions is based on the different stability of metal complexes on the resins at appropriate pH values. However, the selective adsorption ability for metal ions is not only affected by the chelating groups attached to the resin matrix but also by the physical structure of the resins.

For practical use, the resin should have, in addition to the demanded selectivity, sufficient mechanical and chemical stability and be easy to handle. Therefore, the resins that have been practically used are not many.[2]

There are two types of polymer beads: homogeneous-gel types and macroporous-type (or macroreticular) resins. The gel-type resins are swollen in water and unswollen in air. The pore formed by swelling in water is called a micropore, in which metal complexes are formed in the gel-type resins. The size of the micropore depends on the hydrophilicity-hydrophobicity balance of the polymer matrix, the hydrophilicity of functional groups, and the degree of crosslinking of the resins. Experiment data suggest that the formation of metal complexes is affected by the size of micropores in the resins. The gel-type resins have large micropores in water but have low mechanical strength when they are wet. However, the microporous chelating resins with sufficient pore sizes for the formation of metal complexes in the resins have a high adsorption rate and adsorption capacity for metal ions. These resins also have high mechanical strength because they are crosslinked with an appropriate amount of crosslinking agent and the resins have almost the same pore size in both water and air. These pores are called macropores. The macroporous chelating resins with various pore sizes can be prepared by varying the amount of crosslinking agent and diluent.

SYNTHESIS OF CHELATING RESINS

The chelating resins have mainly been prepared by following two methods: incorporation of functional groups into insoluble polymer matrix and polymerization or copolymerization of monomers bearing functional groups. The resins made by the second method have several disadvantages such as low complexing capacity, low mechanical and chemical strength, and difficulty in making resins with uniform dimensions.

Most of the chelating resins used for practical purposes are made by the first method.

ADSORPTION PROPERTIES OF CHELATING RESINS

Selective Adsorption

The most important characteristic of chelating resins is selective adsorption of certain metal ions. The selective behavior of the resins is based on the different stabilities of the metal complexes on the resins at appropriate pH values. However, the determination of a chelate-forming constant (stability constant) of the resins by use of a chelate-forming theory for chelating agents of low molecular weight is very difficult because the chelate formation or the resins is not only affected by pH, but also by the inhomogeneous distribution of functional groups in the resin, the kind of matrix, the degree of crosslinking, and the specific surface area.

The adsorption ability of the resins for metal ions is not only affected by the chemical structure but also by the physical structure of the resins. Ion-exchange capacity of the ion-exchange resins usually corresponds to the amount of ion-exchange groups introduced into the resins. In the chelating resins, all the functional groups do not always take part in the formation of the chelate. This is attributed to the difficulty of diffusion of metal ions and the necessity of certain space for chelate formation in the resins. Therefore, the adsorption capacity of the gel-type resin for metal ions does not always correspond to the ion-exchange capacity of the resins. However, the adsorption of metal ions by the macroporous chelating resins increases as the functional groups introduced are increased because the macropores are favorable for chelate formation even in resins with a high degree of crosslinking.

APPLICATIONS

The chelating resins are widely used for the removal or recovery of harmful or valuable metal ions in various wastes from industry processes and for the separation of metal ions in analytical chemistry. The chelating resins are usually used in a column method. Applications for the resins are listed in **Table 1**.[2] Some of the resins are used in the form of metal chelate resins to adsorb other solutes. Metal chelate resins are used for the removal of fluoride ion, arsenic ion, or phosphoric acid in waste and for the separation of proteins and optical resolution of amino acids. In some cases, the resins are used for adsorbent for mercury gas in air or hydrogen gas. Iminodiacetic acid resin is also used for adsorbent for nitrogen oxide.

REFERENCES

1. Sakaguchi, T.; Ueno, K. *Kinzoku Kireto [1]*; Nankodo: Kyoto, 1965, pp 22, 23.
2. *Koubunshi Shinsozai Binran*; Koubunshi Gakkai, Ed.; Maruzen: Tokyo, 1989; pp 214, 218.

Aerogels

See: Bridged Polysilsesquioxanes (Highly Porous Hybrid Organic-Inorganic Materials)

Aerospace Materials

See: Anoxic Polymer Materials

Agar

See: Agarose
Gums (Overview)

Agarose

See: Agarose (Overview)
Gelling Agents (Agarose and Carrageenan)
Gums (Overview)

TABLE 1. Application of Chelating Resins

Purpose	Example of practical use	Available resins[a]
Determination of trace amount of metal ions	Determination of metal ions in industrial waste, river water, or seawater	A, C
Purification of reagents	Removal of mercury from sulfuric acid	E, F, G
	Removal of Ta from KNO_3	B
Polution prevention	Removal of mercury or heavy metal ions from various wastes	A, B, D E, F, G
	Removal or recovery of U and F from radioisotope wastes	A, B
Removal of metal ions from manufacturing process	Removal of Ca or Mg from saline in manufacturing process of NaOH	A, B
	Removal of Fe and Cu in Ni plate bath	A
	Removal of heavy metal in amino acid solution	A, B, D
Recovery of metal ion	Recovery of Ga from aluminic acid solution in Bayer's process	C
	Recovery of U from seawater	C
	Recovery of Ga, In, and Ge from Zn smelt wastes	A
	Recovery of Cu, Ni, Zn, or Cd from various metal plate wastes	A
Separation	Separation of Pd, Cu, and Ni	D
	Separation and purification of protein	
	Optical resolution of amino acids	

[a]A, iminodiacetic acid type or iminopropionic acid type; B, aminophosphoric acid type; C, amidoxime type; D, polyamine or cryptand type; E, thiol type; F, dithiocarbamic acid type; G, dithizone or thiourea type.

AGAROSE (Overview)

Cristina T. Andrade
Instituto de Macromoléculas
Universidade Federal do Rio de Janeiro

Rosangela B. Garcia
Departamento de Química
Universidade Federal do Rio Grande do Norte

Agarose [9012–36–6] is a gelling polysaccharide extracted from certain marine algae of the class *Rhodophyceae*. It is usually obtained from the species *Gelidium*, *Gracilaria*, and *Pterocladia*.[1] Its chemical structure consists of alternating (1 → 3)-β-D-galactopyranosyl and (1 → 4)-3,6-anhydro-α-L-galactopyranosyl repeat units.[2] The backbone is occasionally substituted with neutral and charged groups such as sulfate esters, methyl ethers, and pyruvates.[3-5] More appropriately, the term "agar" refers to the family of polysaccharides existent in the algae and agarose to the component of this family that has the greatest gelling tendency.[4,5]

The properties and the mechanism of formation of agarose gels have been studied extensively because this polysaccharide has a great commercial importance in foods products and pharmaceuticals, and also because its hydrogel is the simplest one of biological origin, which makes it an ideal model for study. Agarose can form firm gels at concentrations as low as 1% when hot aqueous solutions are cooled to room temperature.

In general, the gels from aqueous solutions of biopolymers are set by conformational changes of the macromolecules, which are followed by an association process.[7-12] In agarose, gelation is followed by a sharp change in optical rotation during the cooling of the aqueous solution, which has been attributed to a coil-helix conformational transition.[13,14] This transition shows a marked thermal hysteresis between melting and setting that can be justified by the extensive helix-helix aggregation. According to these experimental results, some mechanisms have been proposed to explain gelation.

PREPARATION

Several proceedings of isolation and purification for agar polymers have been used.[16-23] For each algae species, depending on the country of origin as well as on the desired degree of purity for the finished product, a specific method should be used.

PROPERTIES

Carbohydrate polymers present important functional properties either in the solid state or in solution because of their structural characteristics and ability to interact with each other or with other molecules.[24] The irregularities in the agarose chains produced by the presence of two types of sugars, which are combined through $1 \rightarrow 4$ and $1 \rightarrow 3$ linkages, corroborate to the reduction of the intermolecular fit.[6] so that agarose molecules have intermediate solubility in water. Agarose swells in cold water and dissolves in boiling water.

Phase-Behavior and Gelation

When hot aqueous solutions of agarose are cooled to room temperature, the result can be turbid solutions or turbid gels, depending on the concentration. Turbidity has been observed in solutions of concentrations as low as 0.05%. A decay of the transmitted light intensity has been observed during the cooling which can be attributed to interhelical packing effects.[15] Increasing the concentration of agarose in the solution, the turbidity increases, too, and well-defined and concentration-dependent cloud points are observed.[25]

Young's Modulus Dependence on Concentration

Macroscopic gelation is observed above a critical concentration[12,26,27] that depends on the molecular weight of the polysaccharide.[27]

Effect of Solvents

The formation of agarose gels is disturbed by the presence of some compounds, which can break hydrogen bonds between OH groups in the polysaccharide or can increase the solubility of the polymer. Some examples of those substances are urea,[17] guanidine hydrochloride,[17] formamide,[17] dimethyl sulfoxide,[22] and sodium thiocyanate.[23,28] The decreasing ability of gelation has been attributed to the lack of helical stability.

APPLICATIONS

Agarose performs useful and practical functions in a wide spectrum of products, mainly to increase viscosity, give solution stability, suspendability, and emulsifying action, and produce gelation and compatibility among several ingredients.[29]

In the food industry, agarose is used predominantly for its stabilizing and gelling characteristics. In pharmaceutical products, it is used in emulsions of many types, as an ingredient of slow-release capsules, as a carrier of topical medicaments, and others. Agarose is extensively used in microbiology as culture media and in several laboratory techniques as gel filtration, affinity chromatography,[30] and hydrophobic chromatography.[31]

REFERENCES

1. Selby, H. H.; Wynne, W. H. *Industrial Gums—Polysaccharides and Their Derivatives*, 2nd ed.; Whistler, R. L.; BeMiller, J. N. Eds.; Academic: New York, 1973; Chapter 3.
2. Araki, C. *Bull. Chem. Soc. Jpn.* **1956**, *29*, 543.
3. Duckworth, M.; Yaphe, W. *Carbohydr. Res.* **1971**, *16*, 189.
4. Lahaye, M.; Yaphe, W. *Carbohydr. Polym.* **1988**, *8*, 285.
5. Rochas, C.; Lahaye, M. *Carbohydr. Polym.* **1989**, *10*, 289.
6. Whistler, R. L. *Advances in Chemistry Series* **1973**, *117*, 242.
7. Jones, R. A.; Staples, E. J.; Penman, A. *J. Chem. Soc.* **1973**, 1608.
8. Bryce, T. A.; McKinnosn, A. A.; Morris, E. R.; Rees, D. A.; Thom, D. *Faraday Discussions of the Chemical Society* **1974**, *57*, 221.
9. Vento, G.; Palma, M. U.; Indovina, P. *J. Chem. Phys.* **1979**, *70*(6), 2848.
10. Indovina, P. L.; Tettamanti, E.; Micciancio-Grammarinaro, M. S.; Palma, M. U. *J. Chem. Phys.* **1979**, *70*(6), 2841.
11. Ross-Murphy, S. B. *Food Hydrocolloids* **1987**, *1*(5/6), 485.
12. Clark, A. H.; Ross-Murphy, S. B. *Adv. Polym. Sci.* **1987**, *83*, 56.
13. Dea, I. C. M.; McKinnon, A. A.; Rees, D. A. *J. Mol. Biol.* **1972**, *68*, 153.
14. Arnott, S.; Fulmer, A.; Scott, W. E.; Dea, I. C. M.; Moorhouse, R.; Rees, D. A. *J. Mol. Biol.* **1974**, *90*, 269.
15. Pines, E.; Prins, W. *Macromolecules* **1973**, *6*, 888.
16. Watase, M.; Nishinari, K. *J. Texture Studies* **1981**, *12*, 427.
17. Watase, M.; Nishinari, K. *Food Hydrocolloids* **1986**, *1*, 25.
18. Watase, M.; Nishinari, K. *Polymer J.* **1986**, *18*, 1017.
19. Watase, M.; Nishinari, K. In *Gums and Stabilisers for the Food Industry 3*, G. O. Phillips, D. J. Wedlock, and P. A. Williams, Eds., Elsevier: London, 1986; p 535.
20. Lahaye, M.; Rochas, C.; Yaphe, W. *Can J. Bot.* **1986**, *64*, 579.
21. Watase, M.; Nishinari, K. *Makromol. Chem.* **1987**, *188*, 1177.
22. Watase, M.; Nishinari, K. *Polymer J.* **1988**, *20*, 1125.
23. Watase, M.; Nishinari, K. *Carbohydr. Polym.* **1989**, *11*, 55.
24. Yalpani, M. *Polysaccharides—Syntheses, Modifications and Structure/Property Relations*; Elsevier: New York, 1988.
25. Andrade, C. T.; Garcia, R. B.; Abritta, T. *Polym. Bull.* **1991**, *27*, 297.
26. Clark, A. H.; Ross-Murphy, S. B. *Brit. Polym J.* **1985**, *17*, 164.
27. Clark, A. H.; Ross-Murphy, S. B.; Nishinari, K.; Watase, M. In Physical Networks—Polymers and Gels, W. Burchard and S. B. Ross-Murphy, Eds.; Elsevier: London, 1990; Chapter 17.
28. Rochas, C.; Lahaye, M. *Carbohydr. Polym.* **1989**, *10*, 289.
29. Davidson, R. L. In *Handbook of Water-Soluble Gums and Resins*, McGraw-Hill: New York, 1980; Chapter 7.
30. Cuatrecasas, P.; Willchek, M.; Anfinsen, C. B. *Proc. Nat. Acad. Sci. U.S.A.* **1968**, *61*, 636.
31. Yon, R. J. *Biochem. J.* **1972**, *126*, 765.

Agricultural Polymers

See: Biologically Active Agricultural Polymers (Mechanism of Action on Plants)

AGRO-FIBER/THERMOPLASTIC BLENDS AND ALLOYS

Anand R. Sanadi
Department of Forestry
University of Wisconsin

Roger M. Rowell
Forest Products Laboratory
United States Department of Agriculture

Blends and alloys have revolutionized the plastics industry because they offer materials with properties never before available and can be tailored for specific end uses. The agro-fiber composite industry has the opportunity to follow this trend and greatly expand the market for its products by offering new materials based on blends and alloys with other resources.

Our recent research has focused on blends and alloys of agro-fiber with thermoplastics—specifically, composites made using kenaf (*Hibiscus cannabinus*, a fast-growing annual plant harvested for its bast fibers) with polypropylene and a compatibilizer. We have also begun research on thermoplasticizing the kenaf fiber matrix and reacting it with a modified polypropylene to develop thermoplastic alloys reinforced with cellulose and polymers.

AGRO-FIBER/THERMOPLASTIC BLENDS

Recent research on lignocellulosic fiber derived from certain annual crop plants suggests that this fiber has good potential for use as a reinforcing filler in thermoplastics; a brief preliminary account was published earlier.[6] Fibers like kenaf were found to have significant property advantages over typical wood-based fillers and fibers such as wood flour, wood fiber, and recycled newspaper.[1-5]

The primary advantages of annual-growth lignocellulosic fiber as a thermoplastics additive are low density, low cost, lack of abrasiveness, high filling levels, low energy consumption, renewability, wide distribution, biodegradability, and the benefits it can bring to rural, agriculture-based economics. The two main disadvantages are its high absorbency[6] and the low processing temperatures permissible.

Moisture absorbance and corresponding dimensional changes can be largely prevented if the fibers are thoroughly encapsulated in the plastic and there is good adhesion between the fiber and the polypropylene (PP) matrix. If necessary, absorbency can be significantly reduced by acetylating the hydroxyl groups in the fiber,[7] although this somewhat increases costs. The disadvantage of absorbency can also be minimized by selecting applications where it is not a major drawback.

The processing temperature of composites is restricted because lignocellosic fibers degrade at higher temperatures. This limits the plastics that can be used to those with low melting temperatures. In general, we have not found thermal deterioration problems when processing temperatures are kept below about 200°C except for short periods.

Kenaf bast fibers with filaments longer than 1 m are common. Filament and individual fiber properties can vary depending on the source, age, separating techniques used, and the history of the fiber.

MOLECULAR INTERACTIONS AT THE FIBER SURFACE

The dispersion and adhesion between polar lignocellulosic fibers and the nonpolar PP matrix are critical factors in determining the properties of the resulting composite. Several different types of functionalized additives have been used to improve the dispersion and interaction[4,8] between cellulose-based fibers and polyolefins.

Maleic anhydride grafted polypropylene (MAPP) has been reported to function efficiently for lignocellulosic-PP systems. Earlier results suggest that the amount of MA grafted and the molecular weight are both important in determining the efficiency of the additive.[9,10] The MA not only provides polar interactions but can covalently link to the hydroxyl groups on the lignocellulosic fiber.

EFFECT OF MAPP ON COMPOSITE PROPERTIES

A small amount of the MAPP (0.5% by weight) improves the composites' flexural and tensile strength, tensile energy absorption, failure strain, and unnotched Izod impact strength. The anhydride groups in the MAPP can covalently bond to the hydroxyl groups of the fiber surface. Any MA that has been converted to the acid form can interact with the fiber surface through acid-base interactions. The improved interaction and adhesion between the fibers and the matrix leads to better matrix-to-fiber stress transfer.

AGRO-FIBER MATRIX THERMOPLASTICIZATION

Over the years there have been many research projects studying ways to thermoform lignocellulosics. Most have concentrated on film formation and thermoplastic composites. They took the approach of chemically modifying the cellulose, lignin, and hemicelluloses—to decrystallize or otherwise modify the cellulose and to thermoplasticize the lignin and hemicellulose matrix—in order to mold the entire lignocellulosic resource into films or thermoplastic composites.[11-16]

Agro-based fibers are composites made up of a rigid polymer (cellulose) in a thermoplastic matrix (lignin and the hemicelluloses). Our research program centers on modifying the matrix of agro-fibers using succinic anhydride.[17] If a non-decrystallizing reaction condition is used, it is possible to chemically modify the lignin and hemicellulose and not the cellulose. This selective reactivity has been shown to occur if uncatalyzed anhydrides are reacted with wood fiber.[18]

Our goal is to modify only the matrix of agro-based fiber, allowing thermoplastic flow but keeping the cellulose backbone as a reinforcing filler. This type of composite should have less of the heat-induced deformation (creep) that restricts structural uses of thermoplastic-based composites. Initial results from reacting succinic anhydride with wood or kenaf fiber show that it is possible to thermoplasticize the matrix of the fiber and form a composite bonded together by thermoplastic flow.

AGRO-FIBER THERMOPLASTIC ALLOYS

Research to develop agro-fiber thermoplastic alloys is based on first thermoplasticizing the fiber matrix, then grafting the modified fiber with a reactive thermoplastic. In this type of composite, the thermoplastic is bonded onto the lignocellulosic so there is only one continued phase in the molecule.

CONCLUSIONS

Combining agro-fibers with thermoplastics provides a strategy for producing advanced composites that take advantage of the properties of both types of resources. It allows the scientist to design materials based on end-use requirements within a framework of cost, availability, recyclability, energy use, and environmental considerations. These new composites make it possible to explore new applications and new markets in the packaging, furniture, housing, and automotive sectors.

REFERENCES

1. Woodhams, R.; Thomas, T. G.; Rodgers, D. K. "Wood Fibers as Reinforcing Fillers for Polyolefins," *Polym. Eng. Sci.* **1984**, *24*, 1166.
2. Klason, C.; Kubat, J. *Cellulose in Polymer Composites, Composite Systems from Natural and Synthetic Polymers*; Salmen, L. et al., Eds.; Elsevier Science: Amsterdam, 1986.
3. Myers, G.; Clemons, E. C. M.; Balatinecz, J. J.; Woodhams, R. T.; "Effects of Composition and Polypropylene Melt Flow on Polypropylene-Waste Newspaper Composites." *Proceed. Annual Technical Conference*, Society of Plastics Industry, **1992**, *602*.

4. Kokta, B. V.; Raj, R. G.; Daneault, C. "Use of Wood Flour as Filler in Polypropylene; Studies in Mechanical Properties." *Polym.-Plast. Technol. Eng.* **1989**, *28*, 247.
5. Sanadi, A. R.; Young, R. A.; Clemons, C.; Rowell, R. M.; "Recycled Newspaper Fibers as Reinforcing Fillers in Thermoplastics: Analysis of Tensile and Impact Properties in Polypropylene." *J. Rein. Plast. Compos.* **1994**, *13*, 54.
6. Sanadi, A. R.; Caulfield, D. F.; Rowell, R. M. "Reinforcing Polypropylene with Natural Fibers." *Plast. Eng.* **1994**, *April*, 27.
7. Rowell, R. M.; Tillman, A. M.; Simonson, R. "Simplified Procedure for the Acetylation of Hardwood and Softwood Flakes for Flakeboard Production." *J. Wood Chem. Tech.* **1986**, *6*, 427.
8. Dalvag, H.; Klason, C.; Stromvall, H. E. "The Efficiency of Cellulosic Fillers in Common Thermoplastics, Part II, Filling with Processing Aids and Coupling Agents," *Intern. J. Polymeric Mater.* **1985**, *11*, 9.
9. Felix, J. M.; Gatenholm, P.; Schreiber, H. P. "Controlled Interactions in Cellulose-Polymer Composites. I: Effect on Mechanical Properties." *Polym. Compos.* **1993**, *14*, 449.
10. Sanadi, A. R.; Rowell, R. M.; Young, R. A. "Interphase Modification in Lignocellulosic Fiber-Thermoplastic Composites." *Engineering for Sustainable Development: AICHE Summer National Meeting* 1993, paper 24f.
11. Hon, D. N-S.; Ou, N-H. "Thermoplasticization of Wood. I. Benzylation of Wood." *J. Applied Polymer Science: Part A: Polymer Chemistry* **1989**, *27*, 2457.
12. Hon, D. N-S.; Xing. L. M. Thermoplasticization of wood, In *Viscoelasticity of Biomaterials*, Glasser, W. G., Ed.; Am. Chem. Soc.: Washington, DC, 1992; p 118.
13. Matsuda, H.; Ueda, M. "Preparation and Utilization of Esterified Woods Bearing Carboxyl Groups IV. Plasticization of Esterified Woods." *Mokuzai Gakkaishi*, **1995**, *31*(3), 215.
14. Matsuda, H. "Preparation and Utilization of Esterified Woods Bearing Carboxyl Groups. " *Wood Sci. Technol.* **1987**, *21*, 75.
15. Shiraishi, N. "Wood plasticization," In *Wood and Cellulose Chemistry*; Hon, N. S.; Shiraishi, N., Eds.; Marcel Dekker: New York, 1991; pp 861–906.
16. Ohkoshi, M.; Hayashi, N.; Ishihara, M. "Bonding of Wood by Thermoplasticizing the Surfaces. III. Mechanism of Thermoplasticization of Wood by Allylation." *Mokuzai Gakkaishi* **1992**, *38*(9), 854.
17. Rowell, R. M.; Clemons, C. M. Chemical modification of wood fiber for thermoplasticity, compatibilization with plastics, and dimensional stability, In *Proceedings, International Particleboard/Composite Materials Symposium*, Maloney, T. M., Ed.; Pullman, WA, 1992; p 251.
18. Rowell, R. M.; Caulfield, D. F.; Sanadi, A.; O'Dell, J.; Rials, T. G. Thermoplasticization of kenaf and compatibilization with other materials, In *Proceedings, Sixth Annual International Kenaf Conference*, New Orleans, LA, March 1994.

Alcaligenes Biopolymer

See: Water-Soluble Polymers (Oil Recovery Applications)

Aldehyde Resins

See: Acetal Resins
Acetal Resins (Homopolymers and Copolymers)

ALGINATE FIBERS (High Performance Papers)

Yoshinari Kobayashi
Shikoku National Industrial Research Institute

CHEMICAL STRUCTURE AND CHARACTERISTICS OF ALGINATES

Alginatic acid is an acidic linear polysaccharide composed of a cell wall and an intercellular cementing matrix of brown algae (the content of which is ~20–40%).

Until recently, the main applications of the polymer have been, in comparison with those of cellulose, limited to such narrow uses as a glue for textile printing and a viscous stabilizer, thickener, and emulsifier for food and cosmetics, with aid of the properties of high molecular weight, high viscosity, hydrophilicity, and water solubility.[1]

NEW APPLICATIONS OF ALGINATES FOR PAPERMAKING

The development of a new alginate industry as one of the large marine-biomass-based industries has required research and development of applications for the textile and paper industries. These applications correspond to those of cellulosic fibers, a representative polysaccharide fiber from land plants.

Making paper from alginate was one of the targets of the high-performance paper fields. It is impossible to extract from brown algae alginate pulp (or fibers) directly from the cells and intercellular matrix. This ability differentiates paper made of alginate and that made of land plants, because land plants require cooking to produce fibrous cellulose, or pulping. Paper made from alginate is produced via a spinning process to obtain fibers analogous to dissolved cellulose (i.e., viscose), because this polysaccharide is a linear polymer without branched chains.

Sporadic reports of producing paper by using short-cut alginate fibers instead of continuous alginate filaments were found in patents. Kobayashi et al. found that the wet-fiber web had interfiber self-bonding properties when slender fibers with suitable fiber length were homogeneously dispersed in water without entanglement and that alginate paper could be made with good formation that was very similar to that of cellulosic paper.[3]

SPINNING ALGINATE DOPE

Research of solution rheology of Na alginate revealed that it showed non-Newtonian flow and was viscoelastic.[4] When the Na alginate dope was spun into hydrophilic solvents of lower molecular weight (e.g., EtOH, isopropanol, and acetone as coagulants), water-soluble fibers were produced that had a tenacity between 0.24 and 0.97 g/d, depending on the solvents used.[5] These water-soluble fibers are under application to be used as a nonwoven, slowly dissolvable hemostatic for covering wounds.

CHARACTERIZATION OF ALGINATE-FIBER PAPER

Natural cellulosic fibers such as wood pulps have strong interfiber self-bondability, but regenerated cellulosic fibers such as rayon almost lose this property because of transformation of the crystal structure, which hinders the formation of interfiber hydrogen bonding. Alginate has the self-bonding property even after it is regenerated.[3] In paper making, therefore, alginate fibers can be treated in the same manner as wood pulp, except that they have a higher moisture content and the tendency to shrink a lot during drying.

NEW APPLICATIONS OF ALGINATE-FIBER PAPERS

Alginate Fiber Paper as Speaker Cone Materials

Alginate fiber paper was chosen as a speaker cone (or diaphragm) material and was commercialized under the trade name Bio-Cone™ by Pioneer (Japan). The alginate-fiber paper did not emit a friction sound, just as natural cellulosic paper, and furthermore fulfilled the essential requirements for the speaker cone: low paper density for high speaker efficiency and sensitivity, high modulus of elasticity for broader regenerative bandwidth, and moderate internal loss for damping resonance.[2,6] This paper also has other merits: applicability of conventional papermaking equipment for one paper, a unique timber, and a higher possibility of better timber by entrapping foreign fine powders within alginate fibers.

Entrapping Enzymes in Alginate Fibers and Papers

Various sorts of enzymes were entrapped in Ca alginate fibers and papers.[7] Compared with corresponding enzymes immobilized in beads with ~2–3 mm diameter, fiber and paper forms had higher entrapping yields than did the corresponding beads. Entrapping medications in the form of alginate-fiber paper is believed to open the way to developing new drug delivery systems.

FUTURE PROSPECTS

For further development of marine biomasses in industrial fields, applications of alginate-fiber papers and paper products are expected to be one of the promising frontier fields. In the area of high-performance paper, paper resources are not limited to cellulosic fibers from land plants. Building a paper-making industry through the use of marine biomasses is a key target for the development of marine resources.

REFERENCES

1. McNeely, W. H.; Pettit, D. *Industrial Gums—Polysaccharides and Their Derivatives,* Academic: 1973; pp 49–81.
2. Kobayashi, Y. *J. Appl. Polymer Sci., Appl. Polymer Symposium* **1991**, *47*, 521.
3. Kobayashi, Y.; Matsuo, R.; Kawakatsu, H. *J. Appl. Polymer Sci.* **1986**, *31*, 1735.
4. Hashimoto, K.; Imai, T. *Polymer J.* **1990**, *22*(4), 331.
5. Fukuoka, S.; Obika, H.; Kamishima, H.; Kobayashi, Y.; Tenma, K. *Sen-i Gakkaishi* **1992**, *48*(1), 42 (in Japanese).
6. Kobayashi, Y.; Kawahara, F.; Kogyo Zairyo (*Industrial Materials*), **1987**, *35*(3), 19 (in Japanese).
7. Kobayashi, Y.; Matsuo, R.; Ohya, T.; Yokoi, N. *Biotech Bioeng.* **1987**, *30*, 451.

Alginates

See: Alginate Fibers (High Performance Papers)
Gums (Overview)

Alkali-Soluble Polymers

See: Floor Finishes

ALKYD EMULSIONS

B. Bergenståhl, A. Hofland, G. Östberg, and A. Larsson
Institute for Surface Chemistry

The use of waterborne alkyds as binders in coatings is not new; their history goes back some 30 years. Mostly water-soluble alkyds with high acid numbers have been used because they provide the water solubility of the resins after neutralization with amines. Such alkyds are used mostly in industrial coatings. Emulsified, water-insoluble alkyds have so far been used most in decorative and protective coatings, although their use is still limited. A major factor limiting the use of alkyd emulsions is their low rate of oxidative drying along with some remaining problems with colloidal stability. Several factors may contribute to the reduced drying properties of alkyd emulsions, such as the interaction between the dryer and other paint components or the dryer's solubility properties. A waterborne paint is a complex system that contains, besides alkyd droplets in water pigments, dispersants (to stabilize the pigments), surfactants (to stabilize the alkyd droplets), thickeners, and several other additives.

The alkyd emulsion is mostly used as a co-binder, especially in exterior primers and for hiding stains. Usually with these applications, alkyds in white spirit are, however, still the binder of choice. Alkyd resins used in decorative coatings can be environmentally friendly as either high solids or waterborne emulsions. With current technology, it is unlikely that alkyd emulsions can match the performance of solvent-based, siliconized alkyds and other extreme, outdoor, durable systems. Because of their conventional permeability and additional layer of thickness, high solids seem more suitable.

At lower end of the market, high-solids systems can be applied to almost every segment, the only limitation being their price. For the lowest segments (primers and stains) their application is limited. In these segments, acrylic dispersions have acquired a good reputation because of their performance and price. In the middle section, ranging from outdoor durable alkyds to the better-performing stains, alkyd emulsions can be positioned. Their versatility and excellent environmental score (no volatile organic components are necessary) make them exceptionally suitable.

In many outdoor applications alkyd emulsions have been compared with acrylate dispersions. Typical advantages of acrylate dispersions are their low particle size, quick drying, and outdoor durability.

HYDROLYTIC STABILITY

Alkyd resins are polyesters and, consequently, susceptible to hydrolysis reactions in water. The hydrolysis of the ester groups can be catalyzed by either acidic or alkaline conditions and usually shows a minimum rate at pH 7. Because of hydrolysis, the molecular weight of the alkyd decreases and the acid number increases. Also, the produced carboxylic-acid groups decrease the emulsion's pH, retarding the hydrolysis if the original pH was above 7 and accelerating it if the emulsion was initially acidic. The reduced molecular weight affects the film properties of the emulsion-based coating (most noticeably the film hardness), while the changes in acid number and pH may affect stability and drying properties.

Apart from pH and temperature, the hydrolytic resistance of an alkyd emulsion depends strongly on the alkyd's chemical composition.

EMULSIFICATION

Mechanisms

Alkyds can be emulsified in two different ways: by direct emulsification of the alkyd in water using some type of high-shear-rate mixer or homogeneizer (often a dissolver) or by phase inversion emulsification. For the latter, water is gradually added to the alkyd and stirred with an anchor stirrer, producing a water-in-alkyd emulsion. By adding more water (or by lowering the temperature), an inversion of the emulsion is provoked, giving the desired alkyd-in-water emulsion. Phase inversion emulsification is usually the method of choice for short and medium oil length alkyds of high viscosities that cannot be handled in homogenizers.

Emulsifiers

The most commonly used emulsifiers in alkyd emulsions are of nonionic and anionic character. The choice of emulsifier is critical for emulsion stability and for the performance of the final paint formulation.

For alkyd emulsions, anionic emulsifiers are more effective than nonionic emulsifiers for producing small droplets with good storage and shear stability at low concentrations.[1,2]

MAKING PAINT FROM ALKYD EMULSIONS

A waterborne paint contains a binder, dispersed in water with surfactants, and pigments, also used with suitable dispersants. In addition, the paint will contain other surface-active additives, such as antiskin, antifoam, and antiwrinkling agents. To retain the paint's colloidal stability, the additives must stay on the surface. In particular, the migration of surfactants from the binder particle surfaces to the pigments is a major cause of "pigment shock", for example, the flocculation of paints while mixing the components.

To ensure that no problems arise with the alkyd emulsion during the paint preparation, it is wise to be acquainted with two ways to introduce instability. High temperatures can destroy the stabilizing effect of nonionic emulsifiers. Adding electrolytes impairs the electrostatic stabilization of anionic emulsifiers, and so does a low pH (below pH 3). Small amounts of salts like the dryers do not present a problem. Most commercial alkyd emulsions are stabilized in a mixed fashion with nonionic and anionic emulsifiers. As long as either the temperature or salt criteria are met, there is no problem.

Applying high shear to an emulsion induces stabilization. Consequently, grinding in the emulsion should be avoided.

Finally, while freezing an emulsion, impurities in the water are concentrated. Because of the pure nature ice crystals, impurities are driven out. Consequently, cosolvent instabilization or charge instabilization can occur (freeze-thaw stability), unless this is prevented by cosolvents that cause freezing-point depression.

DRYING PROPERTIES

The drying properties of alkyd emulsion paints are one of their major drawbacks. Compared to white spirit-based formulations, emulsions generally have a longer drying time. The drying properties also deteriorate rapidly when an alkyd emulsion paint is stored.

REFERENCES

1. Östberg, G.; Bergenståhl, B.; Huldén, M. *J. Coatings Tech.* **1994**, *66*, 37.
2. Östberg, G.; Bergenståhl, B.; Huldén, M. *Colloids and Surfaces* **1995**, *94*, 161.

Alkyds

See: Alkyd Emulsions
Resins (Applications in Coatings Industry)

ALLENE POLYMERIZATION

Takeshi Endo and Ikuyoshi Tomita
Research Laboratory of Resources Utilization
Tokyo Institute of Technology

Allene derivatives have cumulated double bonds that can be regarded as the isomers of propargyl (i.e., acetylene) derivatives, and they have been used in organic synthesis. However, these compounds have been investigated little for polymerization possibly because of the complexity of the reactions and the inavailability of these monomers commercially. If the selective polymerization of either part of the cumulated double bonds (i.e., 1,2- or 2,3-polymerization) were possible, polymers having reactive exomethylene substituents directly attached to the polymer backbone or those having internal double bonds in the main chain might be obtained, and the obtained polymers would be served as novel reactive polymers.

To obtain polymers from allene derivatives, radical, cationic, and coordination polymerizations were examined.

PREPARATION AND PROPERTIES

Radical Polymerization

Allene derivatives with appropriate substituents such as allenyl ethers (**1**), phenylallene (**2**), *N*-allenylpyrrolidone (**3**), and *N*-allenyloxazolidone (**4**) can be used to synthesize soluble polymers by radical polymerization.[1-5] Unlike vinyl ethers, the radical polymerization of allenyl ethers produces polymers bearing double bonds directly attached to the main chain (**Scheme I**).[6] Allene derivatives can also undergo radical copolymerization with vinyl monomers such as styrene, methyl methacrylate, and acrylonitrile.

The radical polymerization of allene derivatives can be used to synthesize functional polymers. For instance, the radical polymerization of allenyl glycidyl ether (**5**) produced a polymer bearing the epoxy group and the exomethylenes.[7]

Because allene derivatives are normally obtained by the isomerization reaction of the corresponding propargyl precursors, the radical polymerizations of propargyl derivatives were carried

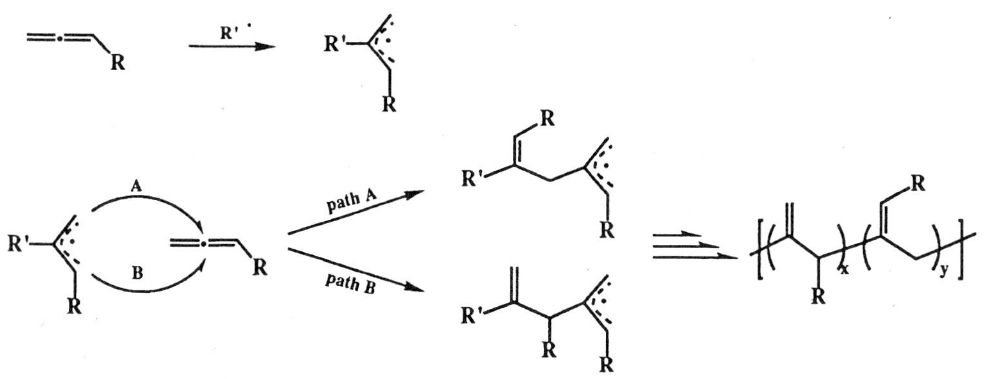

I

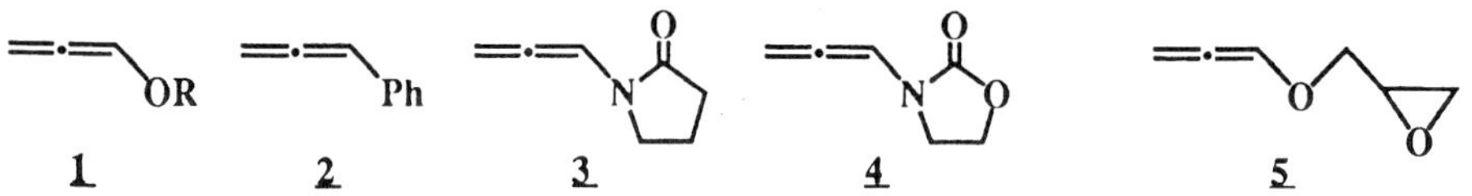

1 2 3 4 5

out in the presence of the isomerization catalyst to obtain polymers with structures identical to those from allene derivatives.[9-11]

Cationic Polymerization

Canionic polymerizations of alkoxyallenes (**1**) as well as phenyl-substituted allenes (including **2**) with Lewis or protic acids produced corresponding polymers with structures similar to those obtained by the radical method.[12-15] Cationic polymerization seems to have more severe limitations than the radical methods to obtain soluble polymers from allene derivatives.

Allylnickel Catalyst

The coordination polymerizations of some allene derivatives have reportedly produced polymers with desired structures. In unsubstituted and alkyl-substituted allenes, polymerization by Ni(0) as well as $[(\eta^3\text{-allyl})NiBr]_2$ catalysts produced highly crystalline polymers as colorless solids that dissolved in organic solvents.[16,17]

It is well-known that π-allylmetals such as π-allyl-palladium and π-allynickel complexes are important intermediates in organic synthesis that can provide selective reaction systems. By successively inserting one part of double bonds of allene derivatives to the carbon-metal bond to form π-allylmetal intermediates, selective polymerization methods may be developed.

APPLICATIONS

Various polymerization methods of allene derivatives have been developed to obtain polymers containing double bonds directly attached to the main chain. Such polymers can be regarded as a reactive polymer homologue of those from vinyl monomers. Living coordination polymerization especially may produce well-defined polymers from allenes with various substituents.

Because allenes can be obtained by the isomerization of the corresponding acetylenes, monomers may be supplied abundantly. For the applications of polymers from allene derivatives as novel reactive polymers, cationic crosslinking and grafting reactions have been reported.[7,8,18] Further study of the reactivity of polyallenes is needed to determine whether they can be used as reactive polymers.

REFERENCES

1. Yokozawa, T.; Tanaka, M.; Endo, T. *Chem. Lett.* **1987**, 1831.
2. Yokozawa, T.; Ito, N.; Endo, T. *Chem. Lett.* **1988**, 1955.
3. Ito, N.; Yokozawa, T.; Endo, T. *Polym. Prepr. Jpn.* **1988**, *37*, 346.
4. Leland, J.; Boucher, J.; Anderson, K. *J. Polym. Sci., Polym. Chem. Ed.* **1977**, *15*, 2785.
5. Tachibana, Y.; Takata, T.; Endo, T. *Polym. Prepr. Jpn.* **1992**, *41*, 219.
6. Matsumoto, A.; Nakana, T.; Oiwa, M. *Makromol. Chem., Rapid Commun.* **1983**, *4*, 277.
7. Mizuya, J.; Yokozawa, T.; Endo, T. *J. Polym. Sci., Polym. Chem. Ed.* **1988**, *26*, 3119.
8. Aggour, Y. A.; Tomita, I.; Endo, T. *J. Polym. Sci., Part A: Polym. Chem. Ed.* **1994**, *32*, 1991.
9. Tomita, I.; Yamamura, I.; Endo, T. *J. Polym. Sci., Part A: Polym. Chem.* in press.
10. Yamamura, I.; Tomita, I.; Endo, T. *The 65th Annual Meeting of the Chemical Society of Japan, Prepr. II*; **1993**, 266.
11. Tomita, I.; Yamamura, I.; Endo, T. *Polym. Prepr. Jpn.* **1993**, *42*, 1856.
12. Okuyama, T.; Izawa, K.; Fueno, T. *J. Am. Chem. Soc.* **1973**, *95*, 6749.
13. Hoff, S.; Brandsma, L.; Arenes, J. F. *Recl. Trav. Chem. Pays-Bas.* **1968**, *87*, 1179.
14. Takahashi, T.; Yokozawa, T.; Endo, T. *J. Polym. Sci., Part A: Polym. Chem.* in press.
15. Takahashi, T.; Yokozawa, T.; Endo, T. *Makromol. Chem.* **1992**, *193*, 1493.
16. Otsuka, S.; Mori, K.; Imaizumi, F. *J. Am. Chem. Soc.* **1965**, *87*, 3017.
17. Otsuka, S.; Mori.; Imaizumi, F. *Eur. Polym. J.* **1967**, *3*, 73.
18. Tomita, I.; Igarashi, T.; Endo, T. *J. Jpn. Soc. Color Material* **1994**, *67*, 754.

Alloys

See: Agro-Fiber/Thermoplastic Blends and Alloys
Alloys and Blends
Dental Adhesives (for Alloys and Ceramics)
Interpenetrating Polymer Networks, Rubber-Based

Liquid Crystalline Polymers (Polyesters)
Maleic Anhydride, Grafting
Poly(butylene terephthalate)/Polycarbonate Blends
Poly(phenylene sulfide) (Elastomer Toughened)
Polypropylene (Commercial)
Solid Polymer Electrolytes
Water-Borne Coatings (Urethane/Acrylic Hybrid Polymers)

ALLOYS AND BLENDS

L. A. Utracki, A. Ajji, and M. M. Dumoulin
National Research Council Canada
Industrial Materials Institute

Polymers are the fastest growing structural materials. About one-third current plastics production is used to produce composites and another third for polymer blends.

Blending offers numerous economic benefits. It is a necessity for several engineering resins, either to improve processability (e.g., for PPE) or impact strength (e.g., semicrystalline resins). Blending makes it possible to generate, rapidly and economically, a desired set of properties: mechanical, chemical, barrier, and others. It also offers better processability of difficult-to-form, high-performance polymers by reducing the viscosity and the processing temperature. Furthermore, in view of environmental concerns, blending technology is used to enhance polymer waste recycling.[1,2]

FUNDAMENTALS

Thermodynamics, Miscibility, and Compatibilization

Thermodynamics are essential for understanding the behavior of polymer blends. The key question to ask is whether the blend is miscible or not.

Polymer-polymer miscibility depends on the level of specific interchain interactions and intramolecular repulsions. Specific interactions may result in $\Delta H_n < 0$, thus miscibility. Several types of specific interactions, such as hydrogen bonding or ionic and dipole interactions, have been identified. The other source of miscibility is the presence of strong internal stresses in the macromolecules. Interactions with another polymer may reduce these internal stresses and be energetically favorable. This idea has been particularly useful for explaining the window of miscibility in homopolymer-copolymer blends.[3,4] The concept has been extended to polymer blends, considering that one or more polymers can be treated as a copolymer (e.g., PS as a copolymer of $-CH_2-$ and $-C(\phi)H-$ structural units).

The control of morphology and its stability requires a well-thought strategy, which may be either to add a compatibilizer or initiate a chemical reaction between the principal blend components during the compounding or processing steps. The most frequently used method of compatibilization has been to add a block or graft copolymer. The copolymer often comprises segments that interact with the main polymers and lead to segmental miscibility. The compatibilizer must be designed to migrate to the interface and modify the interfacial characteristics of the blend.

Morphology

The morphology of polymer blends depends on thermodynamics, rheology, interfacial properties of components, composition, as well as compounding and processing conditions. Because of the strong property–structure relationship, knowledge of the morphology is paramount. In immiscible blends, the performance depends on the interface as well as on the size and shape of the dispersed phase.

The lattice theory of the interface predicts that there is a reciprocity between the interfacial tension coefficient and the interphase thickness, and it offers good guidance for developing compatibilization strategies. A semi-empirical dependence of the interfacial tension coefficient on compatibilizer concentration was derived assuming analogy of a blend compatibilization to titration of an emulsion with a surfactant.[5]

Rheology

For miscible as well as for homologous blends, a positive deviation from the log-additivity rule, PDB, is usually predicted.[6-8] However, there is mounting evidence that PDB is not a rule for these blends. Depending on the system and method of preparation, polymer blends can show either a positive deviation, a negative deviation, or additivity. Miscibility in polymeric systems requires strong specific interactions, which in turn affects the free volume, thus the rheological behavior.

With immiscible blends, the rheological properties depend on a complex manner on composition and properties of the constituents. The best models used to describe their behavior are emulsions. Like blends, the emulsions comprise one liquid dispersed in another, with morphology stabilized by surfactant.

Performance

The generating optimum morphology is critical for ensuring the expected performance. Commercial blends' morphology is far from equilibrium; it is imposed by compounding, forming, and cooling. Thus, the properties can be quite different from those expected from equilibrium thermodynamics. Although considerable work is still needed to fully understand the morphology–performance relationships, some general rules can be listed:

- Brittle polymer is toughened by dispersing up to 10 vol% of a ductile resin in the form of drops.
- Phase co-continuity provides the best balance of properties (e.g., maximum ductility, high rigidity, and large deformability or elongation). Also here, the properties depend on the thickness of the interpenetrating strands, thus the interface.
- For blends used as barrier resins to gases, vapors, and liquids permeation, the lamellar structure is desired.

This section aims at discussing the main types of morphology observed in polymer blends and the relationship between structure and performance.

Toughness is probably the most examined property for polymer blends.

Crazing seems to be the principal toughening mechanism in brittle resins, with the shear yielding more important in ductile resins.[9] Crazes and shear bands are initiated at rubber particles.

The two mechanisms can be synergistic, as shear bands act as obstacles to the crazes and keep them smaller.

To obtain maximum impact strength from a matrix-dispersed phase blend, the latter phase must be in the form of uniformly distributed spheres.

Phase co-continuity provides the best balance of properties (e.g., maximum ductility, high rigidity, and large deformability).

TECHNOLOGY

Blends have been developed for many different reasons: improving a specific property (e.g., impact strength), engendering a set of properties, extending engineering resin performance, processability, and recycling. For the resin manufacturer, blending provides a means to improve and broaden the resin's performance, thereby enhancing sales. By contrast, compounders or processors aim at generating a set of performance parameters for the required application. Although extending engineering resin performance constitutes the largest part of the high-performance blends production, the most difficult and interesting task is the development of blends with a full set of desired properties. A systematic procedure is needed to achieve this goal. The procedure starts with the selection of blend components possessing at least one desired property for the blend. This task can be facilitated with a tabulated list of resins and their principal contribution in the blend (see **Table 1**).

Half a century of experience indicates that adding one polymer to another can induce a well-predicted, specific response. Thus, the strategy for modifying a polymer is to select a polymeric modifier known to cause the desired effect: an elastomer to improve impact strength, an inflammable polymer to induce flame retardancy, or a stiff polymer to improve modulus.

In the second step the best modifier is selected from the list of possible ones. The selection is based on the principle of compensation of properties.

In the third and final stage, the method of manufacturing must be selected. Because a blend's performance depends not only on the ingredients' properties and composition but also on morphology, the principal goal of this stage is to generate and maintain the desired morphology by selecting the method of compatibilization, compounding, and forming.

The economics must be considered at each development step.

Mixing and Compounding

Several technologies have been used to manufacture polymer blends: chemical modification or resins such as grafting and blending that yielded HIPS, ABS, or reactor-modified PP, melt blending of polymer mixtures, usually with a copolymer compatibilizer, and reactive processing that combines advantages of melt blending and reactor modification.

Clearly, the morphology and its development inside the processing equipment depend not only on the system's thermal and rheological characteristics but also on the process. The morphology is strongly affected by the phenomena occurring during all three processing stages: melting, mixing, and forming.

COMMERCIAL BLENDS

About one-third of polymer production is used in blends. There are hundreds of trade-named polymeric blends. However, these mainly engineering polymer alloys constitute only part of the manufactured blends because others are rarely identified as such. In most countries the national norms specify the limiting concentration as below which the resin "is modified" and above which it "is blended." Frequently, the limit is set as high as 20 wt%. Thus, the majority of "modified" resins are blends. Furthermore, most polyolefins have properties upgraded and adjusted to consumer needs through blending. Because different types of polyolefins are immiscible, these materials must be considered blends.

Currently, the major efforts of the industry are directed toward developing of blends with high-performance resins and alloys with up to six polymers. These materials were most frequently obtained in reactive processing or by controlling and stabilizing the morphology. The use of blending technology for polymer recycling is also gaining ground.

Because the principal factor for the polymer blend industry is the interrelationship of processing, morphology, and performance, there is a major effort to develop computational methods that can predict and control the morphology evolution during compounding and processing.

Properties

Not surprisingly, properties of polymer blends depend on those of the ingredients. Although this is a good baseline to start with, in reality the additivity is rarely observed.[6,7] There are several mixing rules used in the field. However, it is impossible

TABLE 1. How to Modify Polymer Properties by Blending

Property	Matrix	Modifier
Impact strength	PVC, PP, PC, PA, PPE, PEST	ABS, SBS, SAN, SMA, MBA, HIPS, EPR, PE
HDT, stiffness	PC, PA	PEST, PEI, PPE
	ABS, SAN	PC, PSO, ...
Flame retardancy	ABS, PMMA	PVC, CPE
	PA, PC	Arom.-PA, PSO, siloxanes, phosphazenes
Chem. resistance	PC, PA, PPE	PEST, siloxanes, polyphosphates
Barrier properties	PO	PA, EVAl, PVDC
Processability	PPE	Styrenics (HIPS, SBS, ABS, ...)
	PET, PA, PC, ...	LCP, TPU, PO, PBR, MBS, EVOH
	PVC	CPE, acrylics
	PO	PTFE, siloxanes

a priori to predict which one will be observed; only general statements can be made.

SUMMARY

The polymer blends industry already consumes over 30 wt% of all polymers, and with a stable growth rate of 9% per year, the role of blends can only increase.

There is rapid growth in polymer science (e.g., thermodynamics, rheology, and performance) and technology (e.g., process modeling, control, and automation for improved generation and stabilization of morphology). Present efforts are directed at developing multifunctional, multiphase blends, especially those with high-performance, specialty resins, for processing, aging, and recycling. There is an acute need to understand interphase and morphology generation for blends with different characteristics. Future technology development will be based on automation, process control, and intelligent design of stable blend morphology.

REFERENCES

1. *Encyclopaedic Dictionary of Commercial Polymer Blends*; Utracki, L. A., Ed.; ChemTech: Toronto, 1994.
2. Utracki, L. A. In *Interpenetrating Polymer Networks*, Advances in Chemistry 239; Klempner, D., Sperling, L. H., and Utracki, L. A., Eds.; American Chemical Society: Washington, DC, 1994.
3. Barlow, J. W.; Paul, D. R. *Polym. Eng. Sci.* **1987**, *27*, 1482.
4. Ellis, T. S. *Polym. Eng. Sci.* **1990**, *30*, 998.
5. Utracki, L. A.; Shi, Z.-H. *Polym. Eng. Sci.* **1992**, *32*, 1824.
6. Utracki, L. A. *Polymer Alloys and Blends*; Hanser Verlag: Munich, 1989.
7. Utracki, L. A. In *Multiphase Polymers: Blends and Ionomers*, Utracki, L. A.; Weiss, R. A., Eds.; ACS Symposium Series 395; American Chemical Society: Washington, DC, 1989.
8. Utracki, L. A. In *Rheological Fundamentals of Polymer Processing*; Covas, J. A., Ed.; Kluver Academic: Dordrecht, 1995.
9. Bucknall, C. B. *Toughened Plastics*, Applied Science: London, 1977.

ALLYL DIGLYCOL CARBONATE

David H. Solomon and Asfia Qureshi
Department of Chemistry
University of Melbourne

Diethylene glycol bis(allyl carbonate) or 4,6,9,12,14-pentaoxia-1,16-heptadecadiene-5,13-dione, is a diallyl monomer known commercially as CR39® (Columbia Resin 39 or Allymer 39, PPG Industries; **Structure 1**). It is the most important diallyl carbonate known to date. The objectives of better scratch resistance and optical properties than those of methyl methacrylate polymers were accomplished in clear case sheets after immediate difficulties.[1-25]

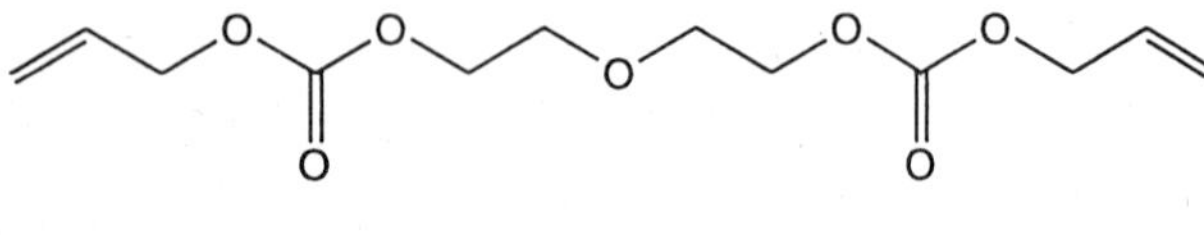

1

PREPARATION

Various methods are available for preparing diallyl diglycol carbonate monomer. The reaction of diethylene glycol bis(chloroformate), prepared from diethylene glycol and phosgene, with allyl alcohol in the presence of an alkali such as sodium hydroxide, gives the required monomer.[28]

However, Yonemori and co-workers[31] recently discovered that this method for producing commercial CR39 monomer may also form oligomers with excess ethoxy ethyl carbonate units.

Commercial CR39 monomer was found to consist of about 80% "real" monomer, 15–16% oligomers, and the polymer.

Another method of preparing the diallyl carbonate ester which bypasses this problem uses allyl chloroformate and diethylene glycol in the presence of base with cooling. In our laboratories as elsewhere, pyridine is the base of choice as the pyridine nitrogen's one pair of electrons can be donated to the formed acid (HCl), thereby precluding any side reactions from taking place. Distilling the crude material affords "pure" CR39.

Recently, we developed an alternative approach for synthesizing CR39. This approach came about as we required deuterated CR39 with the deuterium labeling at the allylic position.[13] Rather than using phosgene to prepare allyl chloroformate, we chose instead to prepare alyloxy *N*-carbonyl imidazole from deuterated allyl alcohol and *N,N'*-carbonyl dimidazole. This method altogether bypasses the use of phosgene, and even di- and tri-phosgene, which are slightly safer alternatives. The allyloxy *N*-carbonyl imidazole was then reacted with diethylene glycol, yielding deuterated CR39 monomer.

PROPERTIES

The CR39 monomer was moderately toxic by ingestion (LD_{50} in rats = 349 mg/kg) and is known to cause skin irritations in some people.[27] **Table 1** lists numerous properties of the CR39 monomer.[26]

TABLE 1. Properties of Commercial CR39 Monomer

Property	Value
Appearance	Clear, colorless liquid
Color, APHA	10
Odor	None to slight
Specific gravity	1.15_4^{20}
Refractive index, n_4^{20}	1.452
Boiling point at 266 Pa[a], °C	166
Melting point, supercooled, °C	−4 to 0
Viscosity at 25 °C, mm^2/s (= cSt)	15
Flash point—Seta closed cup, °C	173
Flash point—Cleveland open cup, °C	186
Water content, slightly hygroscopic, %	0.1

[a]To convert Pa to mmHg, multiply by 0.0075.

APPLICATIONS

The earliest recorded mention of allyl diglycol carbonate (CR39) in the literature was in 1940 when Strain and Muskat put forth a patent covering their initial discoveries.[28] In 1948 Eastman Kodak used CR39 and butyl-α-methacrylate as a thermosetting lens cement and formed a soft gelatinous material to cement multicomponent lenses.[29]

In the following years, the potential seen in CR39 for controlling deposit formation of motor fuels was exploited when it was discovered that adding 0.1% volume of CR39 decreased deposit formation in automotive intake systems and on valves in Diesel engines and minimized the octane requirement increase.[2] It was postulated that its addition to jet fuels would decrease formation on heat exchangers.

In 1958, de Gooreynd and Saxty[3] showed that vinyl or allyl polymers using CR39 have increased resistance to abrasion and impact and thus, CR39 was used as a thermosetting, resin-forming constituent.[3] The year 1959 saw the development of partially polymerized, malleable, plastic materials. Finally in 1960, Combined Optical Industries and John Johnson put forth a British patent covering the casting of synthetic resins for ophthalmic lenses.[7]

In the optical industry, CR39 is used as a casting resin to produce lenses, rods, tubes, and flat sheets of high quality.[24] Cast polymerization is one of the conventional production processes for the CR39 polymer. The lengthy cycle of peroxide-radical-initiated polymerization, necessary to avoid the formation of defects such as strain, stress, foam, and surface shrinkage, is an important drawback of the thermal process.[30] In the past, benzoyl peroxide was used to initiate polymerization of CR39 in a casting process. Important factors in the choice of free-radical initiators include cure temperature, half-life considerations, residual color, and cure efficiency. The popular choice in industry today is a peroxydicarbonate because of its low color and decomposition temperature.

A recent application of CR39 polymer is in solid-state detectors of nuclear particles, including α-particles, fast neutrons, cosmic rays, and ions of elements with atomic number 10 and above.[25]

There are many varied applications in building, consumer products, and industrial safety, office, laboratory, transportation, and navigation equipment. The following list briefly summarizes the current uses of CR39:

- transparent enclosures—airplane canopies, marine canopies, and crane enclosures;
- safety equipment—ophthalmic lenses, safety lenses, face shields, welding hood cover plates, industrial gas mask lenses and screens, and saw guards in steel mills;
- instrumentation—dial and gauge covers, fluid-level gauge tubes, lens filters for cathode ray tubes, liquid-crystal displays, and light-emitting diodes;
- navigation equipment—transparent plotting boards, edge-lighted instruments, map and chart covers, and polar roses;
- laboratory—glove boxes, animal cages, pharmaceutical trays and plaques, biological test chambers, and X-ray equipment and accessories; and
- others—typewriters, copying machines, computer tape reels, business machine observation sheets or panels, viewing sheets for scales, chemical and laundry equipment, vending machine and parking meter viewing sheets, watch crystals, projection slides, photographic filters, and mirrors.

REFERENCES

1. Jeppensen, M. A. *J. Polymer Sci.* **1956**, *19*, 331.
2. Heisler, R. Y.; Newman, S. R.; Alpert, N. U.S. Patent 2 844 448, 1958; *Chem. Abstr.* **1959**, *53*, 2598a.
3. de Gooreynd, P. M. J. K.; Saxty, L. J. U.K. Patent 796 867, 1958; *Chem. Abstr.* **1959**, *53*, 2685e.
4. de Gooreynd, P. M. J. K.; Saxty, L. J.; Ross, M. U.K. Patent 818 471, 1959; *Chem. Abstr.* **1960**, *54*, 11565i.
5. de Gooreynd, P. M. J. K. German Patent 1 016 446, 1957; *Chem. Abstr.* **1960**, *54*, 13727c.
6. de Gooreynd, P. M. J. K.; Saxty, L. J. U.S. 2 910 456, 1959; *Chem. Abstr.* **1960**, *54*, 26024g.
7. Combined Optical Industries Ltd., Johnson, J. U.K. Patent 847 797, 1960; *Chem. Abstr.* **1961**, *55*, 7928c.
8. Kamenskii, V.; Grigor'ev, A. P. *Trudy Moskov. Khim.-Tekhnol. Inst. im. D. I. Mendeleeva* **1959**, *29*, 50; *Chem. Abstr.* **1961**, *55*, 8928c.
9. Hungerford, A. O.; Mullane, P. J. German Patent 1 062 003, 1959; *Chem. Abstr.* **1961**, *55*, 9946f.
10. Sarofeen, G. M. J. U.S. Patent 2 964 501, 1960; *Chem. Abstr.* **1961**, *55*, 10961a.
11. Dille, K. J.; Newman, S. R.; Heisler, R. Y. et al. U.S. Patent 3 001 941, 1955; *Chem. Abstr.* **1962**, *56*, 6259i.
12. Hungerford, A. O.; Mullane, P. J. U.S. Patent 3 038 210, 1962; *Chem. Abstr.* **1962**, *57*, 12745d.
13. Philipson, J. U.S. Patent 3 092 526, 1962; *Chem. Abstr.* **1963**, *59*, 6192a.
14. Shirakawa, T. Japanese Patent 27 866, 1964; *Chem. Abstr.* **1965**, *63*, 756a.
15. Ecole, J. *Offic. Plastiques Caoutchouc* **1965**, *12*(125), 349; *Chem. Abstr.* **1965**, *63*, 4393c.
16. Greshes, M. *Plastics Technol.* **1965**, *11*(7) 38; *Chem. Abstr.* **1965**, *63*, 7169h.
17. Gudzinowicz, B. J. *Anal. Chem.* **1965**, *37*(8), 1051.
18. Windsor, M. W. U.S. Patent 3 214 382, 1965; *Chem. Abstr.* **1966**, *64*, 3797a.
19. Sheld, C. A. U.S. Patent 3 216 958, 1965; *Chem. Abstr.* **1966**, *64*, 3797h.
20. Grandperret, R. U.S. Patent 3 222 432, 1965; *Chem. Abstr.* **1966**, *64*, 8433a.
21. Wallis, B. L. *Encycl. Polymer Sci. Technol.* **1964**, *3*, 1.
22. Dial, W. R. U.S. Patent 3 256 113, 1966; *Chem. Abstr.* **1966**, *65*, 10741h.
23. Bugnon, M. J. M. *Offic. Plastiques Caoutchouc* **1964**, *11*, 181; *Chem. Abstr.* **1966**, *65*, 15601a.
24. Schildknecht, C. E. *Allyl Compounds and Their Polymers, High Polymers*; Wiley-Interscience: New York, 1973; 28.
25. Dowbenko, R. *Encyclopedia of Chemical Technology*, 4th ed.; Kroschwitz, J. I.; Wiley-Interscience: New York, 1992; Vol. 2.
26. *Nouryset 200: The Key to Top Qualtiy Organic Glass*; Akzo Chemie Handout.
27. *Concise Encyclopedia of Polymer Science and Engineering*; Kroschwitz, J. I.; Wiley-Interscience: New York, 1990.
28. Muskat, I. E.; Strain, F. U.S. Patent 2 592 058.
29. Parsons, W. F.; Dann, J. R. U.S. Patent 2 445 535–6, *Chem. Abstr.* **1948**, *42*, 8528f.
30. Dial, W. R.; Bissinger, W. E.; DeWitt, B. J.; et al. *Ind. Eng. Chem.* **1955**, *47*(12), 2447.
31. Yonemori, S.; Masui, A.; Noshiro, M. *J. Appl. Polym. Sci., Appl. Polym. Symp.* **1991**, *48*, 523.

Alternating Copolymers

See: *Alternating Ethylene-Vinyl Alcohol Copolymer*
Alternating Radical Copolymerization
Captodative Compounds, Polymerization
Charge-Transfer Polymerization
Comb-Like Polymers, Widely Spaced
Cyclic Carbonates (Ring-Opening Polymerization)
Cyclic Imino Ethers (Ring-Opening Polymerization)
Cyclic Imino Ethers (Ring-Opening Polymerization of 1,3-Oxazo Monomers)
Functional Polymaleimides, N-Protected
Group-Transfer Alternating Copolymerization
Immortal Polymerization
Maleic Anhydride Alternating Copolymers (Configurations of Cyclic Units)
Melaimide Copolymers (N-Substituted)
Olefin-Carbon Monoxide Copolymers (Overview)
Olefin-Carbon Monoxide Copolymers, Alternating
Palladium-Catalyzed Synthesis (Monomers and Polymers)
Periodic Polymers
Poly(N-acylethylenimine)s
Polys[(disilanylene) oligophenylene]s
Poly(propylene carbonate)
Polysulfones
Ring-Opening-Closing Polymerization
Styrene-Maleic Anhydride Copolymer
Sulfur Dioxide Copolymerization
Tetrafluoroethylene Copolymers (Overview)
Trisubstituted Ethylene Copolymers
Vinyl Polyperoxides (Oxygen Copolymerization)
Zwitterionic Polymerization (Overview)
Zwitterionic Polymerization (Aziridine, Oxazoline Derivatives with Anhydrides)

ALTERNATING ETHYLENE-VINYL ALCOHOL COPOLYMER

Kenji Yokota
Department of Materials Science and Engineering
Nagoya Institute of Technology

Because the monomer pairs that produce alternating copolymers are limited and the synthesis of periodic copolymers with sequences of three or more monomer units is hardly possible at present, we studied the synthesis of periodic copolymers by polymer reactions and the properties of their products.

In this chapter, a synthesis of alternating ethylene–vinyl alcohol and ethylene–vinyl acetate copolymers by polymer reactions of poly(1-trimethylsilyloxy-1,3-butadiene) and the copolymer properties are described.[3] These alternating copolymers have never before been obtained by directly copolymerizing related monomers.

POLYMERIZATION OF 1-TRIMETHYLSILYLOXY-1,3-BUTADIENE

1-Trimethylsilyloxy-1,3-butadiene is considered both a diene monomer and an unsaturated ether monomer. Its radical copolymerization experiment with styrene or methyl methacrylate gave the Alfrey-Price's parameters Q = 1.6~1.9 and e = −1.4~−1.6.[6] As these Q and e values indicate, this monomer did not polymerize alone via a radical mechanism to a high molecular weight polymer but instead polymerized via a cationic mechanism. We found that this monomer can also polymerize via an aldol type group transfer mechanism to produce high molecular weight polymers, essentially in 1,4-addition fashion.[1,2,4-9]

The group transfer polymerization of this monomer is a living-type polymerization in nature, and the polymer's molecular weight is essentially controlled by the monomer-to-initiator ratio in feed.

We found that adding small amounts of diethyl ether to the polymerization mixture made the zinc bromide catalyst soluble and the polymerization mixture homogeneous. The homogeneous polymerization occurred at lower temperatures than the heterogeneous polymerization and suppressed the side reactions that often led to polymers with a low molecular weight. Under conditions used for homogeneous polymerization, polymers of high and uniform molecular weight were available.

Three important facts should be pointed out here. First, only the monomer of *E*-configuration could polymerize; the *Z*-monomer could not polymerize. Second, when the reaction vessel contained no water and an aldehyde initiator was absent, a Lewis acid alone could not polymerize this monomer. Third, the group transfer polymerization accompanied a 3,4-addition propagation.

SYNTHESIS

We used the polymer reactions of poly(1-trimethyl-silyloxy-1,3-butadiene) to synthesize alternating ethylenevinyl alcohol and ethylene-vinyl acetate copolymers (**Equation 1**).

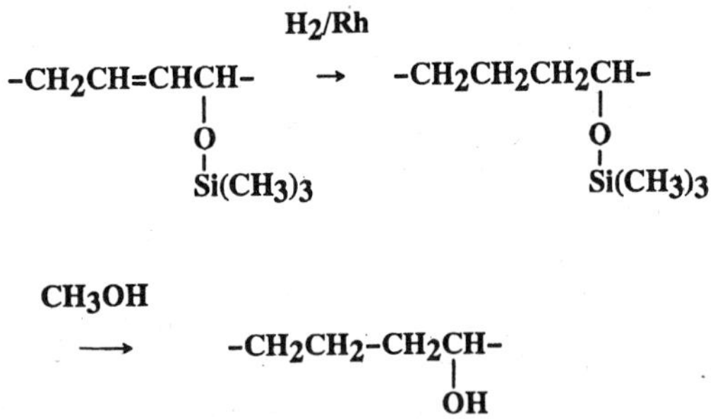

1

Hydrogenation of poly(1-trimethylsilyloxy-1,3-butadiene) was carried out by medium pressure hydrogen and a soluble Wilkinson's rhodium catalyst.

The hydrogenated product was then desilylated by stirring in methanol to give the alternating ethylene–vinyl alcohol copolymer.

A mixture of alternating ethylene–vinyl alcohol copolymer and an excess of acetic anhydride in pyridine was stirred. Then, methanol was added and cooled with ice to decompose the excess acetic anhydride. The reaction mixture was concentrated, and the acetylated polymer product was recovered by reprecipitating with ether.

PROPERTIES

Alternating ethylene–vinyl alcohol copolymer was a white, tough, crystalline material. A sample as precipitated from its solution showed in differential scanning calorimetry (DSC) studies that the polymer melting temperature was 140°C and the glass transition temperature 46°C. When a melt sample was quenched by liquid nitrogen, the sample became amorphous. It was soluble in dimethyl sulfoxide or hot methanol and insoluble in diethyl ether, chloroform, or water. The melt sample showed a diffraction peak at 2 θ = 21°C to the nickel-filtered copper K_α X-ray. The crystalline-to-amorphous area ratio indicating the extent of crystallization was 42%.

Alternating ethylene–vinyl acetate copolymer was a pale yellow, rubbery material and showed glass transition temperature at –40°C. The copolymer was soluble in benzene, chloroform, and tetrahydrofuran and insoluble in methanol.

REFERENCES

1. Hirabayashi, T.; Itoh, T.; Yokota, K. *Polym. J.* **1988**, *20*, 1041.
2. Hirabayashi, T.; Kawasaki, T.; Yokota, K. *Polym. J.* **1990**, *22*, 287.
3. Mori, Y.; Sumi, H.; Hirabayashi, T.; et al. *Macromolecules* **1994**, *27*, 1051.
4. Sumi, H.; Hirabayashi, T.; Inai, Y.; et al. *Polym. J.* **1992**, *24*, 669.
5. Sumi, H.; Suzuki, A.; Hirabayashi, T.; et al. *Polym. J.* **1994**, *26*, 705.
6. Sumi, H.; Noto, T.; Hirabayashi, T.; et al. *Polym. J.* **1994**, *26*, 1080.
7. Sumi, H.; Haraguchi, K.; Hirabayashi, T.; et al. *Polym. J.* **1994**, *26*, 1262.
8. Sumi, H.; Haraguchi, K.; Hirabayashi, T.; et al. *Polym. J.* **1995**, *27*, 26.
9. Sumi, H.; Ishikawa, K.; Hirabayashi, T.; et al. *Polym. J.* **1995**, *27*, 34.

ALTERNATING RADICAL COPOLYMERIZATION

Hidefumi Hirai
Science University of Tokyo

Yaeko Gotoh
The University of Tokyo

Radical polymerizations of two monomers A and D produce comonomer-sequence (-AA-, -AD-, and -DD-) distribution in the resulting copolymers, according to the Lewis-Mayo equation. An alternating copolymer $(\text{-AD-})_n$ is termed poly(A-*alt*-D). The individual alternating copolymer, however, usually deviates from the ideal structure of poly(A-*alt*-D). In this paper, the term "alternating A-D copolymer" is used to designate the copolymer mainly composed of alternating comonomer sequence.

Alternating radical copolymerizations are classified into two categories: (a) copolymerization in the absence of metal halide, and (b) copolymerization in the presence of metal halide. The former case is expressed by the phrase "alternating radical copolymerization"; the latter by "alternating copolymerization using metal halide".

The phenomenon of alternating radical copolymerization has been studied on various combinations of electron acceptor and donor monomers. The acceptor monomers are strong electron acceptors. The alternating copolymerization proceeds commonly in the presence of a free radical initiator, γ, or ultraviolet (UV) irradiation. The combination of strong electron acceptor and donor monomers produces an alternating copolymer by spontaneous initiation. The processes of radical initiated alternating copolymerization was reviewed by Cowie.[2]

In systems of alternating copolymerization using metal halides, most of the monomers are conventional monomers and the donor monomers are olefins. When alkyl aluminum halides and zinc chloride are used as metal halide, the alternating copolymers are formed by spontaneous initiation. In general, the alternating copolymerization using metal halide is promoted by using a free radical initiator, UV, or γ irradiation. Alternating radical copolymerization using metal halide was reviewed by Furukawa,[6] Hirai,[7] and Bamford.[8]

PREPARATION

Alternating Radical Copolymerization

Several examples of preparation of alternating copolymers in the absence of metal halide are maleic anhydride—styrene, maleic anhydride—divinyl ether, chlorotrifluoroethylene—propylene, and benzoquinone—germylene.

Mechanisms for Alternating Radical Copolymerization

To mechanisms have been proposed for alternating radical copolymerization: (a) enhanced cross-propagation, and (b) polymerization of a binary complex composed of both monomers.

The cross-propagation is enhanced by stabilization of the transition state of the cross-propagation step. The stabilization may be caused either by electrostatic interation[9] or by electron transfer[10] between the growing radical and the incoming comonomer.

The binary complexes between electron acceptor and donor monomers are frequently formed. The homopolymerization of a binary complex produces a strictly alternating copolymer. The reactivity of the binary complex is enhanced by the charge transfer interaction between electron acceptor and donor monomers.[1,11,12] When the concentration of the binary complex and/or the enhancement of reactivity are not sufficient, the participation of donor monomer in the polymerization causes a deviation from the structure of the strictly alternating copolymer.

Alternating Radical Copolymerization by Using Metal Halide

Some examples of preparation of alternating radical copolymerization by using metal halide are acrylonitrile–propylene, acrylonitrile–butadiene, and methyl methacrylate–styrene.

Terpolymerization in Alternating Copolymerization

When an acceptor monomer (A) and two kinds of donor monomers (D and D′) are polymerized in the absence or presence of metal halide, an alternating copolymer is obtained in which the ratio of (the content of A) (that of [D + D′]) is kept as 1 regardless of the monomer feed ratio. The terpolymer consists of -AD- and -AD′- sequences randomly distributed in the polymer chain.

When two kinds of acceptor monomer (A and A′) and a donor monomer (D) are used, the resulting copolymer is an alternating copolymer with an (A + A′)/D ratio of 1.

Alternation-Regulating Activities of Metal Halide

In alternating copolymerization the amounts of metal halides required for comonomer-sequence regulation depend on the species of metal halides and the mole ratios of metal-to-acceptor monomers are usually larger than 0.2.

Mechanisms for Alternating Copolymerization Using Metal Halide

Three mechanisms have been proposed for alternating copolymerization using metal halide: (a) enhanced cross propagation; (b) polymerization of a ternary complex composed of both monomers and metal halide; and (c) radical complex mechanism.

The cross-propagation should be greatly enhanced by the complex formation between acceptor monomer and metal halide.[4,5] The copolymer with a high degree of alternation may not be produced by the cross-propagation mechanism.

The radical complex mechanism[13] postulates the stabilization of the radical end of the acceptor monomer by complexing with metal halide and capping with donor monomer. The resulting radical complex forms a double complex with the incoming acceptor-metal halide complex.

The ternary complex mechanism[3,14,15] has been supported by determination of the ternary complex[16] and by the kinetic studies on copolymerization.[17,18]

PROPERTIES

It is important for commercial application of alternating copolymers to compare the properties of an alternating copolymer with those of equimolar random copolymers. Physical properties of alternating copolymers are reviewed by McEwen and Johnson.[19]

APPLICATIONS

Applications of the alternating copolymers of MAn are excellently and extensively reviewed by Trivedi and Culbertson.[20] The common applications of these copolymers are coatings, dispersant, paper modifier, fabric sizings, adhesives, and enzyme and medicinal carrier. Alternating copolymers containing fluorine are useful in preparation of specialty elastomers and coatings. Alternating copolymers of acrylic monomer with electron-donating monomer copolymers have aroused interest as a possible alternative to the random copolymers, mainly in the field of elastomers. Alternating copolymers of MAn with specific monomer have found application in medicine.

REFERENCES

1. Kokubo, T.; Iwatsuki, S.; Yamashita, Y. *Macromolecules* **1968**, *1*, 482.
2. Cowie, J. M. G. *Alternating Copolymers*; Plenum: New York and London, 1985; Chapter 2.
3. Ikegami, T.; Hirai, H. *J. Polym. Sci., A-1*, **1970**, *8*, 195.
4. Golubev, V. B.; Zubov, V. P.; Georgiev, G. S.; Stoyachenko, I. L.; Kabanov, V. A. *J. Polym. Sci. Polym. Chem. Ed.* **1973**, *11*, 2463.
5. Yabumoto, S.; Ishii, KI.; Arita, K. *J. Polym. Sci. Polym. Chem. Ed.* **1969**, *7*, 1577.
6. Furukawa, J. *J. Macromol. Sci. Chem.* **1975**, *A9*, 867.
7. Hirai, H. *J. Polym. Sci. Macromol. Rev.* **1976**, *11*, 47.
8. Bamford, C. H. *Alternating Copolymers*; Plenum: New York and London, 1985; Chapter 3.
9. Price, C. C. *J. Polym. Sci.* **1946**, *1*, 83.
10. Walling, C.; Briggs, E. R.; Wokfstirn, K. B.; Mayo, F. R. *J. Am. Chem. Soc.* **1948**, *70*, 1537.
11. Barb, W. G. *Proc. R. Soc. (London)*, **A 212 1952**, *66*, 167.
12. Booth, D.; Dainton, F. S.; Ivin, K. J. *Tran. Faraday Soc.* **1959**, *55*, 1293.
13. Hirooka, M. *J. Polym. Sci.* **1972**, *B10*, 171.
14. Furukawa, J.; Kobayashi, E.; Iseda, Y. *Polym. J. (Tokyo)* **1970**, *1*, 155.
15. Gaylord, N. G.; Matyska, B. *J. Macromol. Sci. Chem.* **1970**, *A4*, 1507.
16. Hirai, H.; Komiyama, M. *J. Polym. Sci. Polym. Chem. Ed.* **1974**, *12*, 2701.
17. Hirai, H.; Komiyama, M. *J. Polym. Sci. Polym. Chem. Ed.* **1975**, *13*, 2419.
18. Furukawa, J.; Arai, Y.; Kobayashi, E. *J. Polym. Sci. Polym. Chem. Ed.* **1976**, *14*, 2243.
19. McEwen, I. J.; Johnson, A. F. *Alternating Copolymers*; Plenum: New York and London, 1985; Chapter 6.
20. Trivedi, B. C.; Culbertson, B. M. *Maleic Anhydride*; Plenum: New York, 1982; Chapter 10.

ALUMINA FIBERS (from Poly[(acyloxy)aloxane]s)

Hideki Yamane and Yoshiharu Kimura
Department of Polymer Engineering and Science
Kyoto Institute of Technology

Ceramics fibers have been recently attracting interest because of their superior tensile and thermal properties. From the viewpoint of polymer science and engineering, the use of polymeric preceramic materials seems to be most interesting for preparing ceramic fibers. The performance of these ceramic fibers obtained by pyrolysis is determined in part by the properties of their precursors which must contain the same elements as the product ceramics. Therefore, one of the key steps in the method should be preparation of precursor polymers which are both processable and calcine into the desired ceramics.

Alumina fibers have also been produced by the precursor method. The precursors utilized are polymeric ion species formed from hydroxoaluminum chloride and polyaloxane by hydrolysis of organoaluminum compounds. Since both polymers are built up from an Al–O backbone, they presumably have a strong tendency to calcine into alumina. However, the yield of alumina fibers from the polymers is rather low because of their low processability during spinning. Recently, the synthesis of a novel series of poly[(acyloxy)aloxane]s has been developed and the possibility of developing alumina fiber was demonstrated with these preceramic polymers.[1-3]

PREPARATION

Poly[(acyloxy)aloxane]s are readily prepared by a one-pot, sequential reaction of organoaluminum compounds with water and carboxylic acids (**Scheme I**). Although, both trialkyl- and trialkoxyaluminums are useful as the starting materials, triethyl aluminum (TEA) is more easily utilized because the reaction proceeds of low temperatures and the conversion is quantitative. It was already shown that utilization of n-alkyl carboxylic acid,

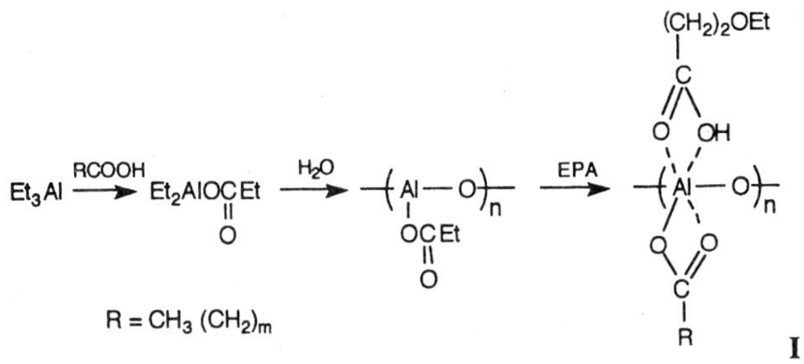

I

which is a bidentate ligand, produced a polymer with high molecular weight, whereas 3-ethoxy-propanoic acid (EPA), a tridentate ligand, gave an excellent spinnability of the solution and melt into fiber. Poly[(acyloxy)aloxane]s obtained utilizing both two ligands do not only have a high molecular weight but also show an excellent fiber making property. These are either solution or melt processable depending on the length of alkyl chain in the n-alkanoic acid utilized.

PYROLYSIS OF POLY[(ACYLOXY)ALOXANE]S

Filaments were pyrolyzed in nitrogen atmosphere up to 700–800°C at a heating rate of ca. 10°C/min. In this pyrolysis condition, no crack in the fiber due to the sudden removal of the organic residue was observed, although these fibers were still contaminated by the residual carbon. This contamination by the residual carbon greatly affect the mechanical properties of the fibers even sintered up to higher temperature. Elimination of the residual carbon was performed by switching the atmosphere from nitrogen to air at this stage, and continuing the heating up to the temperatures desired.

Both tensile strength and modulus gradually increase. These deterioration in mechanical properties at higher temperature are considered to be associated with the sudden phase transition ($\gamma \rightarrow \alpha$) and grain growth of α-alumina.

EFFECT OF SINTERING AGENTS

Magnesia is known as a good sintering agent for alumina ceramics,[4] which acts as a nucleating agent and prevents the grain growth of the alumina crystal at the interface between crystal grains forming a stable mesophase called spinel.

As a conclusion, the poly[(acyloxy)aloxane]s synthesized are either melt or solution processable into fine filaments and those can be easily converted into alumina fibers with excellent mechanical properties. The deterioration of the fibers in mechanical properties due to the grain growth of α-alumina can be suppressed by the addition of M_g component in these preceramic polymers.

REFERENCES

1. Kimura, Y.; Sugaya, S.; Ichimura, T.; Taniguchi, I. *Macromolecules* **1987**, *20*, 2329.
2. Kimura, Y.; Furukawa, M.; Yamane, H.; Kitao, T. *Macromolecules* **1989**, *22*, 79.
3. Morita, H.; Yamane, H.; Kimura, Y.; Kitao, T. *J. Appl. Polym. Sci.* **1990**, *40*, 753.
4. Coble, R. L.; Burke, L. E. *Progress in Ceramic Science*, Bueke, J. E., Ed.; Macmillan: New York, 1963; Vol. 3.

Aluminoxanes

See: Metallocene Catalysts
Polystyrene, Stereoregular
Polystyrene, Syndiotactic
Styrene, Stereospecific Polymerization

Aluminum-Containing Polymers

See: Alumina Fibers (from Poly[(acyloxy)aloxane]s)
Inorganic Nanostructured Materials
Preceramic Polymers

Amine-Quinone Polymers

See: Quinone-Amine Polyurethanes
Quinone-Diamine Polymers

AMINIMIDES—MONOMERS AND POLYMERS

Bill M. Culbertson
The Ohio State University

The trialkylamine acylimide (ylid) functionality (-$CONNR_3$), first discovered in 1959, provides a unique route to producing functionalized monomers and polymers. Trialkylamine acylimides, most commonly called aminimides, are a family of relatively exotic chemical specialties developed by Ashland Chemical in the early 1970s. Materials with this functionality have many potential uses, depending upon the availability of monomers and polymers.

Aminimides belong to the class of compounds known as ylids, since the functionality consists of a carbanion attached to a heteroelement bearing a positive charge. The synthesis and the chemical and physical properties of several different types of aminimide ylids have been the subject of reviews.[1-7] Aminimide functionalized monomers and polymers have also received attention.[3-5,8,11]

The nomenclature of the trialkylamine acylimides, which is both varied and complex, has been adequately described.[3] The aminimide nomenclature is generally used because this type of ylid functionality (I) is isolectric with an amine oxide[3] (II). Generalized aminimide structures (I and II) provide some insight into the substances' chemical versatility. Absorption bands in the IR at 1555-1600 cm^{-1} support the resonance structures shown (**Scheme I**).

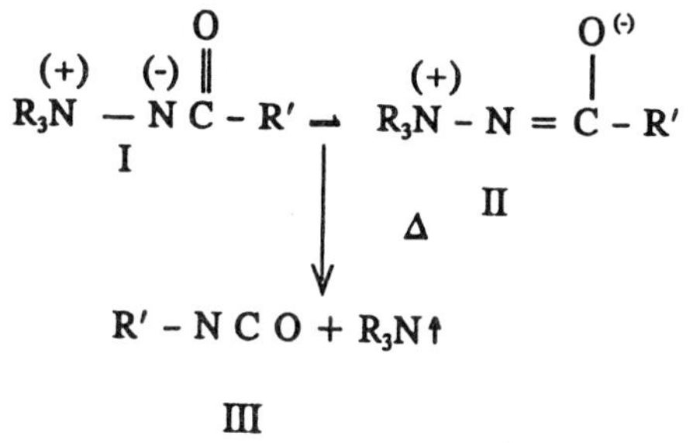

I

As shown above, the dipolar ion (ammonium acylimine) residue, when heated, suffers a carbon–nitrogen migration reaction (rearrangement), making aminimide's isocyanate precursors. Thus monomers and polymers with aminimide residues provide a clever method for chemists to prepare a wide variety of monomeric and polymeric isocyanates (III). The surface active aminimide compounds also function as amine precursors, providing an easily handled source of tertiary amine.

PREPARATION OF MONOMERS

Many different types of monofunctional, difunctional, and multifunctional aminimides have been prepared at Ashland Chemical and other locations.

Several procedures for preparing monomers and polymers with the aminimide residue are known.[3,7,12]

PROPERTIES OF MONOMERS

Aminimides are generally highly soluble in water and a number of polar organic solvents. Aqueous solutions of aminimides are neutral (pH 6.9–7.2), staying unchanged at room temperature or at 50°C for 24 hours, and exhibit low conductivity.[3,7]

Aminimides are acid-accepting compounds. This important feature is very useful in nonaqueous titration procedures to determine percent aminimide functionality for both monomeric and polymeric materials.[10,14,15]

Aminimides exhibit good shelf stability when properly stored. However, when heated at or well above their melting points, they undergo nitrogen–nitrogen bond cleavage, rearrangement, or elimination reactions.

Aminimides are not a good source of aromatic isocyanates or vinyl isocyanates due to the tertiary amine-catalyzed trimerization reaction.[16-18]

POLYMERIZATION

Aminimide monomers with acrylic and methacrylic residues readily homo- and copolymerize, using nonoxidative, free-radical initiators.[3,8,13] Peroxides, such as hydrogen and benzoyl peroxide, are not efficient initiators, since monomer-initiator reactions destroy both the monomer and the initiator much faster than radical polymerization occurs.

In strongly acidic solutions, the aminimide monomer or polymer exists as a hydrazinium salt.[7] The hydrazinium monomer exhibits homo- and copolymerization characteristics significantly different from those of the patent aminimide monomer. Peroxides work well as polymerization initiators for the hydrazinium monomers. In general, the hydrazinium monomer polymerizes more readily to higher molecular weight than does the corresponding parent aminimide.[15]

The great worth of aminimide monomers in preparing reactive and/or functional polymers is additionally illustrated by the synthesis of poly(n-butyl) acrylate-coisopropenyl isocyanate).[19] Both the isocyanate copolymer and its *N,N*-ethylene ureido derivative, obtained from reaction of the isocyanate groups with ethylenimine, are very useful as wool shrink-proofing agents.

Polymeric materials such as aminimide functionalized polystyrenes may also be used to crosslink epoxy resins. In these systems, the aminimide functions as a latent hardener/promoter to crosslink the epoxy resin, giving adhesives and coatings with high tensile and high impact strength. The compound 1,1-dimethyl-1-(2-hydroxy-3-allyloxypropyl)-amine β-lactimide is particularly useful as a latent epoxy resin hardener.

Variations of the above scheme employ epoxides, glycols, and other active hydrogen materials to crosslink polymers with pendent aminimide residues. Also, the use of difunctional aminimides to crosslink polymers with pendent active hydrogens is useful for producing thermosetting materials.[20]

Several monomers are useful for preparing water-soluble polyurethanes.

Ethylenimine (aziridinyl) substituted aminimide monomers are useful for modifying pendent carboxylic acid functional groups on polymer chains.

POLYMER PROPERTIES

Aminimide polymer films are hard, brittle materials. When heated at or above 125°C, the films become insoluble in water, alcohol, and other common organic solvents. Infrared spectra taken during heating show that the aminimide carbonyl band at 1560 cm^{-1} vanishes and a new absorption band appears at 2260 cm^{-1}, indicating isocyanate formation.

Heating the films isothermally at various temperatures and following the rate of disappearance of the 1560 cm^{-1} band shows that the carbon–nitrogen migration reaction is a temperature-dependent, first-order reaction.[21]

The temperature at which the nitrogen–nitrogen bond cleavage reaction occurs is somewhat lowered for polymers substituted with the 1,1-dimethyl-1-(2,3-diacetoxypropyl)amine moieties.[22] No suitable catalysts have been discovered to significantly lower the thermolysis temperature of aminimides.

The trialkylamine acylimides and the hydrazinium salt form of the aminimide monomers have great influence on the solubility characteristics of polymers.[3,15,23] In general, the hydrophilicity of copolymers is directly proportional to the percent of dipolar aminimide or hydrazinium salt residues on the backbone of the polymer molecule. All aminimide polymers are water-soluble. Copolymers may also be water-soluble, depending on aminimide or hydrazinium salt content.

The dipolar ylid residue has a great effect on the adhesion of copolymers to various surface.[9,23] Variations of the adhesion character are readily achieved by changing the type of alkyl substituent on the aminimide functionality.[36,37]

Copolymers with aminimide residues also suffer nitrogen–nitrogen bond-breaking reactions and elimination of tertiary amine when heated at 125°C. Thermolysis reactions of the copolymers may be achieved either neat or in refluxing solvents.[9,15,24,25]

APPLICATIONS

Uses for aminimide monomers and polymers take advantage of their dipolar ion structure, water-solubility, or their nature as isocyanate precursors and latent amine sources. In all areas of possible uses for aminimides—adhesives, coatings, films, elastomers, and so on—these materials continue to offer opportunities to design new products if monomers and aminimide functionalized polymers become available. Since the last reviews of possible aminimide applications,[26] several patents have been issued that give additional insight on potential uses of aminimide monomers to prepare highly modified perfluoralkyl acrylate coatings,[27-30] photographic films,[31] photopolymerizable compositions,[32] thermosetting cationic electro phoretic coating systems,[33,39] 2-acrylamido-2-methylpropane sulfonic acid-containing copolymers with surfactant-suspending and electrical conductivity properties,[34] modified toners for electrostatic imaging,[35] and sizing for glass-fiber composites.[38]

REFERENCES

1. Timpe, H. J. *Z. Chem.* **1972**, *12*(7), 250.
2. Sasaki, T. *Kagaku Kogyo* **1972**, *23*, 504.
3. McKillip, W. J.; Sedor, E. A.; Culbertson, B. M.; Wawzonek, S. *Chem. Rev.* *73*, **1973**, 255.
4. Smith, P. A. S. *Annu. Rep. Inorg. Gen. Sym.* **1973**, *1*, 201.
5. McKillip, W. J. *Adv. Urethane Sci. Technol.* **1974**, *3*, 81.
6. Kameyama, E. *Yukagaku* **1978**, *27*, 197.
7. Wawzonek, S. *Ind. Eng. Prod. Res. Dev.* **1980**, *19*(3), 338.
8. Culbertson, B. M. *Polym. News* **1978**, *5*, 104.
9. Slagel, R. C. *J. Org. Chem.* **1968**, *83*, 1374.
10. Inubuse, S.; Endo, T.; Tazuke, S. *Fine Chemicals J.* **1985**, *12*, 14–24.
11. Shevchenko, V. V.; Klimenko, N. S. *Kompoz. Polim. Mater.* **1992**, *53*, 3–13.
12. U.S. Patent 3 565 1868 (Feb. 23, 1971) E. A. Sedor and R. C. Slagel (to Ashland Oil, Inc.); U.S. Patent 3 485 806 (Dec. 23, 1969), A. E. Bloomquist, E. A. Sedor, and R. C. Slagel (to Ashland Oil, Inc.).
13. Mehta, A. C.; Rickter, D. O.; Kolesinski, H. S.; Taylor, L. D. *J. Polym. Sci.* **1983**, *21*(4), 1159; U.S. Patent 4 105 694 (Aug. 8, 1978), A. C. Mehta, D. O. Rickter, L. D. Taylor (to Polaroid Corp.).
14. Culbertson, B. M.; Sedor, E. A.; Dietz, S.; Freis, R. E. *J. Polym. Sci.* **1968**, *A-1 6*, 2197.
15. Culbertson, B. M.; Randen, N. A. *J. Appl. Polym. Sci.* **1971**, *15*, 2609.
16. Smith, R. F.; Briggs, P. C. *Chem. Comm.* **1965**, *120*.
17. McKillip, W. J.; Clemens, L. M.; Haugland, R. *Can. J. Chem.* **1967**, *45*, 2613; U.S. Patent 3 706 797 (Dec. 19, 1972), W. J. McKillip. L. M. Clemens (to Ashland Oil, Inc.).
18. Slagel, R. C.; Bloomquist, A. E. *Can. J. Chem.* **1967**, *45*, 2625–2628.
19. U.S. Patent 3 640 676 (Feb. 8, 1972), W. J. McKillip, B. M. Culbertson, C. N. Impola (to Ashland Oil, Inc.).
20. U.S. Patent 3 989 752 (Feb. 11, 1976), W. D. Emmons (to Rohm & Haas Co.).
21. McKillip, W. J.; Culbertson, B. M.; Gynn, G. M.; Menardi, P. F. *Prod. Res. Develop.* **1974**, *13*, 197.
22. Culbertson, B. M.; Langer, H. J. *J. Appl. Polym. Symp.* **1975**, *26*, 399.
23. Culbertson, B. M.; Sedor, E. A.; Slagel, R. C. *Macromolecules* **1968**, *1*, 254.
24. Culbertson, B. M.; Freis, R. E.; Grote, D. *J. Polym. Sci.* **1971**, *A-1 9*, 3453; U.S. Patent 3 728 387 (April 17, 1973), R. E. Freis and B. M. Culbertson (to Ashland Oil, Inc.).
25. Langer, H. J.; Culbertson, B. M. "Copolymers, Polyblends, and Composites," *Adv. Chem. Ser.* **1975**, *142*, 129.
26. Temin, S. C. *Rev. Macromol. Chem. Phys.* **1982-1983**, *C22*(1), 158.
27. Jpn. Patent 79 128 991 (Oct. 5, 1979), M. Hisasue, T. Hayashi, H. Matsuo (to Asahi Glass Co., Ltd.).
28. Jpn. Patent 79 128 992 (Oct. 5, 1979), T. Hayashi, H. Matsuo (to Asahi Glass Co., Ltd.).
29. Jpn. Patent 79 131 579 (Oct. 12, 1979), T. Hayashi, H. Matsuo (to Asahi Glass Co., Ltd.).
30. Jpn. Patent 79 139 641 (Oct. 30, 1979), T. Hayashi, H. Matsuo (to Asahi Glass Co., Ltd.).
31. U.S. Patent 4 022 623 (May 10, 1977), M. J. Fitzgerald, H. S. Kilensinski, L. D. Taylor (to Polaroid Corp.).
32. U.S. Patent 3 898 087 (Aug. 5, 1975), G. W. Brutchen, G. O. Fanger (to Ball Corp.).
33. Jpn. Patent 80 21 459 (Feb. 15, 1980), O. Nakachi, A. Osawa (to Nippon Oil and Fats Co., Ltd.).
34. U.S. Patent 4 140 680 (Feb. 20, 1979), C. I. Sullivan (to Polaroid Corp.).
35. Jpn. Patent 9 088 741A (Nov. 11, 1982) to Konishiroku Photo KK.
36. Ger. Patent 2 327 452 (Dec. 6, 1973), S. F. Spencer, L. E. Winslow, A. R. Zigman (to Minnesota Mining and Mgf. Co.).
37. U.S. Patent 3 691 140 (Sept. 12, 1972) and U.S. Patent 4 049 483 (Sept. 10, 1977), S. P. Spencer (to Minnesota Mining and Mfg. Co.).
38. U.S. Patent 3 946 131 (Mar. 23, 1976) to Owens-Corning Fiber Co.
39. U.S. Patent 4 046 658 (Sept. 6, 1977) to General Motors Corp.

Amino Resins

See: *Melamine Resins (Overview)*
Powder Coatings (Overview)
Resins (Applications in Coatings Industry)
Urea-Formaldehyde Adhesive Resins
Urea-Formaldehyde Glue Resins

Aminoethylation

See: *Ethylenimine-Modified Polymers*

AMINOSILOXANES

J. G. Matisons and A. Provatas
Polymer Science Group
School of Chemical Technology
University of South Australia

Aminofunctional siloxanes are a relatively new class of silicon polymers and can be represented by these formulas (**Equations 1** and **2**):

$$Me_3SiO(Me_2SiO)_x(MeRSiO)_ySiMe_3 \qquad 1$$

$$RMe_2SiO(Me_2SiO)_xSiMe_2R \qquad 2$$

The R group is generally $(CH_2)_nNH_2$, and is either pendant along the chain or attached to the end of the siloxane backbone. These polymers couple the unique properties of polydimethylsiloxanes with the intrinsic reactivity of amino groups, giving the resultant polymers many commercial applications.

PREPARATION OF AMINOSILOXANES

Hydrosilylation

The hydrosilylation reaction, first reported in 1945, has become the major route for synthesizing many siloxane products in use today, including the aminosiloxanes.[1] The hydrosilylation reaction involves adding Si–H functional compounds to unsaturated organic compounds in the presence of a transition-metal catalyst or UV light.

The most common catalyst is hexachloroplatinic acid,[2] other transition-metal catalysts used include rhodium[3] and palladium complexes[4] as well as divinyltetramethyldisiloxane platinum and its derivatives.[5]

Well-characterized prepolymers containing Si–H groups are now readily available, allowing the production of well-defined functional siloxane products in high yields. Disadvantages to hydrosilylation include the high cost of the metal catalyst and the difficulty of recovering it, though the latter problem has to some extent been overcome by the use of recyclable catalysts on solid supports.[6]

Synthesis of Aminosiloxanes Via the Equilibration Polymerization Process

The equilibration reaction involves heating a mixture of a cyclic tetramer like octamethyltetracyclosiloxane with a cyclic siloxane monomer containing amino groups and hexamethyldisiloxane (as an end-capper) together in the presence of a basic catalyst.

The basic catalyst must be picked to minimize its interaction with the reactive amino groups on the siloxane; in most cases it is either potassium silanolate or tetramethylammonium silanolate.[8] More recently, crown-ether initiators[42] and alkali-metal hydrides[9] have been used.

A large amount of work on the mechanism and kinetics of the equilibration reactions has been accomplished with several papers reviewing the mechanism in detail.[8,10]

Amino-Hydroxy Functional Siloxanes

The synthesis of bifunctional siloxanes containing both amino and hydroxyl groups can be achieved by reacting an amine with an epoxysiloxane.[7,11,12] Bifunctional siloxanes show properties of adhesion to inorganic substrates[12] better than those of similar nonionic aminosiloxanes.[13] The starting epoxysiloxane reactant is prepared by hydrosilylation of the appropriate Si–H functional siloxane with allyl glycidyl ether. The epoxysiloxane is then reacted with amines, resulting in the amine's attack on the epoxy group to form α-amino-β-hydroxy functional siloxanes.

PROPERTIES OF AMINO-FUNCTIONAL SILOXANES

The presence of a pendant amino groups on the siloxane[3] allows strong physical interactions to occur between the siloxanes and surfaces with hydroxyl groups. Hydrogen bonding between surface hydroxyl groups and/or adsorbed moisture with the amino groups of the functionalized siloxanes drives the strong adhesion forces linking siloxane polymers to inorganic oxide surfaces. Amino-siloxanes adhere quite readily and generate improved surface properties on glass[13] and titania.[15]

USES OF AMINOSILOXANES

Aminosiloxanes have a variety of applications, due in part to the reactivity of the hydrophilic amino group, as distinct from the relative inertness of the hydrophobic siloxane backbone. The surface properties of aminosiloxanes have been little studied, despite their potential in applications such as lubrication.

Cosmetics

Functionalized siloxanes are widely used in cosmetic applications due to their unique properties; in particular, they exhibit a lowering of surface tension in cosmetic formulations, providing better wetting characteristics.[14] Not surprisingly, aminosiloxanes are commonly used in hair conditioners, where they impart a smooth, silky feel and easy hair-combing characteristics.[16,17]

Aminosiloxanes can also be applied as one part moisture-cured hair conditioners,[18] where the product sets when it is applied to the hair, giving a soft, conditioned feeling, yet can subsequently be washed off easily under running water. Aminosiloxanes, particularly the quaternary aminosiloxanes, have also been used in hair dye products, giving improved depth and color fastness.[19]

Textile Applications

Textile manufacturers have long since recognized the use of aminosiloxanes to impart water repellency to garments while still allowing them to "breathe" and feel soft.[20] In fact, amino-functional silicone softeners dominate the textile market.[21] Typically, to dramatically alter the texture of wool garments and prevent shrinkage, only a small amount of aminosiloxane (~1%) need be added.

Polyester fibers requiring water repellency have been treated similarly.[22,23] Pendant diamino siloxane is used to give the polyester fibers wool-like suppleness and good mechanical spinnability. These fibers are commonly used as fillings for continental quilts, giving the quilt added elasticity and enhanced air entrapment.

Recent softeners use aminosiloxane microemulsions, which are transparent, water-clear formulations with average particle sizes lower than 40 nm and which give an excellent finish.[24] The only disadvantage of these softeners appears to be a slight yellowing caused by prolonged exposure to high temperatures during production. This yellowing, however, can be controlled by acylating the amino group.[24] or by reacting with lactones[25] such as butyrolactone, to give hydroxyl amido functional polymers. Further research by Habereder[24] has uncovered new aminosilicone softeners—the cyclohexyl-amino-functional siloxanes—which show no yellowing, good hand, and give improved soil-release properties.

AMPHOTERIC SURFACTANTS

Amphoteric betaines retain their surfactant properties in water types[26] ranging from hard to soft water and high to low pH. This allows betaine siloxane surfactants to be employed in cleaning formulations used under a wide variety of conditions.

Block Copolymers

Siloxanes with amino end groups have been used to form a variety of block copolymers (including polyamides,[27] polyurethanes,[28] polyimides,[29] and polycarbonates,[30,31]) that have increased flexibility, solubility, thermal stability, and even biocidal properties.[32] These copolymers are of enormous value to industry, with uses ranging from adhesives to medicinal implants.

The importance of incorporating siloxane segments into copolymers has become increasingly important in the manufacture of high-performance industrial copolymeric materials.

Aminosiloxanes can be copolymerized with thermoplastics like polyurethanes to form block copolymers that are particularly useful in contact lens applications as well as *in vivo* applications like medical implants and catheters.[33]

Siloxane block copolymers find uses as flexible spacers in liquid crystal technology, where a rigid liquid crystal backbone benefits from the freedom gained by the addition of siloxane units.

These polymers reduce the transition temperatures[34] and the response times of liquid crystals to applied thermal, optical, or electrical fields.[35] Furthermore, the siloxane blocks impart excellent elastomeric properties to the copolymer backbone and improve material processing.

Other Applications

Aminosiloxanes are also used as mold-release agents,[36,37] adhesive polymide films for use in the manufacture of flexible printed circuit boards,[38] epoxy-resin potting compositions for electronic parts,[39] primers for rubber-based automotive components,[40] and water-repellent coatings for automobile window glass.[41]

CONCLUSION

Aminosiloxanes have become important chemicals for the synthesis of a wide range of polymers. Generally these siloxanes are used either to confer silicon characteristics to a host copolymeric system or to produce aminosiloxanes with new properties not attainable from conventional polydimethylsiloxanes. Currently, amino-functional siloxanes are used in a large variety of commercial applications, particularly in the textile and cosmetic industries. Their use in the formation of siloxane block copolymers is one of the more active research areas in polymer chemistry today.

REFERENCES

1. Ojima, I. *The Chemistry of Organic Silicon Compounds*; Patai, S., Rappoport, Z., Eds.; Wiley: New York, 1989; Chapter XXV.
2. Speier, J. L.; Webster, J. A.; Barnes, G. H. *J. Am. Chem. Soc.* **1957**, *79*, 974.
3. Chung, P. H.; Crivello, J. V.; Fan, M. *J. Polym. Sci.* **1993**, *31*, 1741.
4. Hara, M.; Ohno, K.; Tsuiji, J. *J. Am. Chem. Soc., D. Trans.* **1971**, *247*.
5. Karstedt, B. D. U.S. Patent 3 775 452, 1973.
6. Ejike, E. N.; Parish, R. V. *J. Organomet. Chem.* **1987**, *321*, 135.
7. Ryang, H. S. U.S. Patent 4 892 918, 1990.
8. Kendrick, T. C.; Parbhoo, B.; White, J. W. *The Silicon-Heteroatom Bond*; Patai, S., Rappoport, Z., Eds., Wiley, England, 1991; Chapter III.
9. Stauffer Chemicals GB Patent 1 439 037, 1976.
10. McGrath, J. E.; Sormani, P. M.; Elsbernd, C. S.; Kilic, S. *Makromol. Chem.* Macromol. Symp. **1986**, *6*, 67.
11. Pleudemann, E. P. U.S. Patent 2 946 701, 1960.
12. Matisons, J. G.; Provatas, A. *Macromolecules* **1994**, *27*, 3397.
13. Bennett, D. R.; Matisons, J. G.; Netting, A. K. O.; Smart, R. St. C.; Swincer, A. G. *Polym. Int.* **1992**, *27*, 147.
14. Owen, M. J. *Ind. Eng. Chem. Prod. Res. Dev.* **1980**, *19*, 97.
15. Ashmead, B. V.; Bowrey, M.; Burrill, P. M.; Kendrick, T. C.; Owen, M. J. *J. Oil Colour Chem. Assoc.* **1971**, *54*, 403.
16. Chandra, G.; Kohl, G. S.; Tassoff, J. A. U.S. Patent 4 559 227, 1985.
17. Cornwall, S. M.; Homan, G. R. U.S. Patent 4 586 518, 1986.
18. Homan, G. R.; Smithhart, T. J. Eur. Patent 113 992, 1983.
19. Fridd, R.; Taylor, R. M. GB Patent 2 186 889, 1987.
20. Cook, J. R. *J. Text. Inst.* **1984**, *75*, 191.
21. Joyner, M. M. *Text. Chem. Color.* **1986**, *18*, 34.
22. Nomura, K.; Tsurumi, H. Jap. Patent 2 045 789, 1987; *Chem. Abstr.* 106(26), 215473z.
23. Makino, S.; Morgia, H. Jap. Patent 3 294 522, 1991; *Chem. Abstr.* 116(20), 196181r.
24. Habereder, P.; Lautenschlager, H. *VTT Symp.* 133 (Text. Comps. '92), **1992**, 277.
25. Cray, S. E.; Yianni, P.; McVie, J. Eur Patent 342 834, 1989.
26. Myers, D. *Surfactant Science and Technology,* VCH: NY, 1980; Chapter II.
27. Kajiyama, M.; Nishikata, Y.; Kakimoto, M.; Imai, Y. *Polym. J.* **1986**, *18*, 735.
28. Ikaga, S. Jap. Patent 2 302 420, 1990.
29. Arnold, C. A.; Summers, J. D.; McGrath, J. E. *Polym. Eng. Sci.* **1989**, *29*, 1413.
30. Riffle, J. S.; Banthia, A. K.; Webster, D. C.; McGrath, J. E. *Org. Coat. Chem.* **1980**, *42*, 122.
31. Tang, S. H.; Meinecke, E. A.; Riffle, J. S.; McGrath, J. E. *Rubber Chem. Technol.* **1984**, *57*, 184.
32. Sauvet, G.; Helary, G.; Hazziza-Laskan, J. FR Patent 2 686 610, 1993; *Chem. Abstr.* 120, 32449r.
33. Su, K. C.; Robertson, R. J. U.S. Patent 4 740 533, 1988.
34. Aguilera, C.; Bartulin, J.; Hisgen, B.; Ringsdorf, H. *Makromol. Chem.* **1983**, *184*, 253.
35. Simon, R.; Coles, H. J. *Polymer* **1986**, *27*, 811.
36. Fujikawa, T.; Kimura, T. Jap. Patent 2 117 922, 1990; *Chem. Abstr.* 113(10), 79286f.
37. Piskotti, C. U.S. Patent 4 633 002, 1986.
38. Kunimune, K.; Soeda, Y.; Itami, S.; Kikuta, K. Eur. Patent 538 075, 1993.
39. Kagawa, H.; Kyotani, Y. Jap. Patent 4 031 425, 1992; *Chem. Abstr.* 117, 71253e.
40. Traver, F. J.; Simoneav, E. T. U.S. Patent 4 618 689, 1986.
41. Uchida, T. Jap. Patent 5 301 742, 1993; *Chem. Abstr.* 94, 148443.
42. Zavin, B. G.; Zhdanov, A. A.; Blokhina, O. G.; Scibiorek, M.; Chojinowski, J. *Otkrytika, Izobret.* **1990**, *26*, 92; *Chem. Abstr.* 114, 17430w.

AMPHIPHILIC GAS SENSORS (Based on Polymer-Coated Piezoelectric Crystals)

Yoshiaki Eda, Noboru Takisawa, and Keishiro Shirahama
Department of Chemistry
Faculty of Science and Engineering
Saga University

A number of drugs are essentially amphiphilic because they must dissolve in aqueous body fluid and permeate through lipophilic biomembranes to reach their receptors. Inhalation anesthetics, in particular, are thought to be recepted nonsite specifically to synaptic membranes judging from the good correlation between anesthetic potency and the oil–gas partition coefficient.[1] Middle-chain alcohols and inhalation anesthetics are regarded as volatile amphiphiles for the following reasons: alcohols consist of slender hydrophobic chain and hydrophilic –OH groups; inhalation anesthetics have bulky hydrophobic groups consisting of halogenated carbons and large dipole moments; and alcohols and anesthetics show limited solubilities in water and oil. We studied amphiphile behavior in solution by using amphiphilic ion-selective electrodes based on modified poly(vinyl chloride) films.[2,3] Unfortunately, these electrodes cannot detect nonionic species, such as alcohols and inhalation anesthetics, which we were also interested in. We paid attention to a gravimetric sensor with a piezoelectric quartz oscillator. Because piezoelectric sensors can detect ionic and nonionic substances, these sensors enable studies of volatile amphiphile behavior in solution through the liquid–vapor equilibria.

The piezoelectric sensor is called the Quartz-Crystal Microbalance (QCM) because of its high sensitivity. Since King demonstrated that a coated quartz oscillator functions as a highly sensitive sorption sensor, many researchers have studied piezoelectric gas sensors coated with thin polymer films.[4] In this study, we designed piezoelectric sensors for the amphiphilic gases by coating with a hydrophobic polymer. As an application for colloid science, partitioning of an alcohol to a surfactant micelle was measured by the sensors.

MATERIALS AND METHODS

A hydrophobic polymer, ELVALOY HP441® (Du Pont) was used as the sensing films on piezoelectric crystals. ELVALOY HP441 should have almost the same solubility parameter (δ) as poly(vinyl chloride) (PVC, $\delta = 9.4$ cal$^{1/2}$ mL$^{-1/2}$), because it was designed originally as a plasticizer for PVC.

1-Butanol (C4OH), 1-pentanol (C5OH), 1-hexanol (C6OH), and 1-heptanol (C7OH) were used as received from Wako Pure Chemical. The inhalation anesthetics used were halothane (2-bromo-2-chloro-1,1,1-trifluoroethane), enflurane (2-chloro-1,1,2-trifluoroethyl difluoromethyl ether), and isoflurane (1-chloro-2,2,2-trifluoroethyl difluoromethyl ether). Water was purified by deionization and double distillation, once from alkaline $KMnO_4$ solution. A surfactant sodium dodecyl sulfate (SDS) was obtained from Pierce and used as received.

CONCLUSIONS

The quartz oscillators coated with ELVALOY HP441 functioned as useful sensors responsive to amphiphilic gases such as alcohols and inhalation anesthetics. The solubilization of 1-pentanol to an SDS micelle could be measured using these sensors. Studies of how volatile anesthetics and their analogs interact with micelles and other colloidal systems (models of biomembranes and receptors) are in progress using these sensors in our laboratory. These analyses are expected to produce useful information about the molecular mechanism of anesthetic action.

REFERENCES

1. Eger, E. I.; Lundgren, C.; Miller, S. L.; et al. *Anesthesiology* **1969**, *30*, 129.
2. Shirahama, K.; Kameyama, K.; Takagi, T. *J. Phys. Chem.* **1992**, *92*, 6817.
3. Takisawa, N.; Shirahama, K.; Tanaka, I. *Colloid Polym. Sci.* **1993**, *271*, 499.
4. King, W. H., Jr. *Anal. Chem.* **1964**, *36*, 1735.

Amphiphilic Oligomers

See: Fluoroacrylic Acid Oligomers

Amphiphilic Polymers

See: Amphiphilic Polymers (Binding Properties for Small Molecules)
Amphiphilic Polymers (Fluorescence)
Block Copolymer Micelles (Microcontainers for Drug Targeting)
Block Copolymer Micelles (Overview)
Block Copolymer Micelles (1. Structure and Properties)
Block Copolymer Micelles (2. Fluorometric Studies and Computer Modelling)
Block Copolymers (Micellization in Solution)
Cellulosic Ethers, Cationic
Colloidal Dispersions, Nanogels
Comb-Like Amphiphilic Polymers (Supramolecular Structure)
Comb-Like Polymers
Globular Polybases, Microemulsion-Like
Hydrophobic Polyelectrolytes (Conformational Transitions)
Hydrophobized Polysaccharides (Versatile Functions)
Langmuir-Blodgett Film Polymerization
Molecular Assemblies (Polymerization in Aqueous Solution)
Poly(ethylene oxide) Macromonomers (Micellar Polymerization)
Poly(ethylene oxide) Side-Chain Acrylates (Synthesis and Polymerization)
Polysoaps
Surface Active Monomers
Surfactants, Polymeric (Overview)

AMPHIPHILIC POLYMERS (Binding Properties for Small Molecules)

Toru Takagishi
Department of Applied Materials Science
University of Osaka Prefecture

The binding of small organic compounds and ions by synthetic polymers have been studied extensively not only because of their pharmaceutical significance, but also as a model for biomacromolecular interactions with substrates. For this purpose, a successful approach has been developed to increase the binding ability for small substrates. The strategy is to attach pendant groups (ionic groups, aromatic or aliphatic hydrophobic residues, macrocyclic residues, or chelate moieties) to the polymer chains to enhance electrostatic interaction, hydrophobic interaction, or coordinate bond.[1-9] Amphiphilic polymers, which involve hydrophilic and hydrophobic groups in their structure, are particularly interesting when macromolecules and small molecules interact and play a significant role in enhancing binding affinity for small substrates.

The most representative amphiphilic polymer is an amphiphilic polyelectrolyte, known as a polysoap, which behaves like micelles of surfactants and solubilizes small co-solutes. The amphiphilic polyelectrolyte is of interest as it relates to the structure and function of proteins, in particular globular ones. The most characteristic property of amphiphilic polyelectrolytes is their ability to form intramolecular aggregates and hydrophobic microdomains, that is, microscopic phase separation occurs.

No water-soluble, synthetic macromolecule binds small molecules with an avidity comparable to that of serum albumin.[1,10] Polyvinylpyrrolidone, an amphiphilic polymer, binds far better than any of the other synthetic macromolecules, except modified synthetic ones, and displays a strong binding for dissolved, small molecules.[10,11] This macromolecule is particularly interesting because it parallels serum albumin in many respects. The nature of the interaction and the thermodynamic parameters of binding are similar, making polyvinylpyrrolidone an important material physiologically.

In general, synthetic macromolecules do not bind small molecules with an avidity comparable to that of serum albumin probably because they have swollen, extended, unfolded conformations in aqueous solution, whereas serum albumin has a compact structure. In 1969, Klotz successfully enhanced the binding of small molecules by using synthetic polymers.[1,2] He attached apolar pendant groups to side chains of a highly branched polymer, thereby increasing the local concentration of residues. From this research, it was found that polymers with both hydrophilic and hydrophobic residues, that is, amphiphilic polymers, bound small substrates strongly. Thus, this research showed a guiding principle for preparing polymers with enhanced binding.

Methyl orange is a good reference anion and is used as a "binding probe" because apolar and ionic groups make up its chemical structure and provide a suitable balance of hydrophobicity and hydrophilicity.

PREPARATION AND PROPERTIES

Polycations

The binding of methyl, ethyl, and propyl orange by amphiphilic polycations involving various apolar pendant groups such as methyl, ethyl, benzyl, or dodecylbenzyl has been examined.[3] All these polymers were obtained from Nitto Boseki Company and were either commercially available or prepared in the company's Research Laboratory for Chemical Fibers.

In all cases the enthalpy of binding was an exothermic quantity, and the entropy was a substantial positive number. Evidently, the binding process was accompanied by an entropy gain and an exothermic enthalpy change. The shorter the alkyl chain of the small molecules or polymers, the more negative was the enthalpy change, and hence the smaller the entropy change. The favorable enthalpy change in the binding was primarily because of such electrostatic interactions. Hydrophobic interaction can be a positive entropy source.

Copolymers

The amphiphilic copolymers of 2-hydroxyethyl methacrylate (HEMA) and N-vinyl-2-pyrrolidone (VPy) with various compositions were prepared.[12] The copolymers obtained were examined for their ability to bind a homologous series of methyl orange derivatives (methyl orange, ethyl orange, propyl orange, and butyl orange). The contribution of the entropy term to the free-energy change tends to increase, and the absolute magnitude of the enthalpy change decreases with increasing hydrophobicity of the small molecule and the copolymer used. All these observations are explicable considering the hydrophobic interaction involved in these bindings.

Water-soluble polymers with an extended conformation do not bind small molecules as strongly as compact macromolecules.[1,10,13]

APPLICATIONS

Amphiphilic polymers with a strong binding affinity for small substrates should be useful for new materials such as strong adsorbents in industry and pharmacology.

REFERENCES

1. Klotz, I. M.; Sloniewsky, A. R. *Biochem. Biophys. Res. Commun.* **1968**, *31*, 421.
2. Klotz, I. M.; Royer, G. P.; Sloniewsky, A. R. *Biochemistry* **1969**, *8*, 4752.
3. Takagishi, T.; Nakata, Y.; Kuroki, N. *J. Polym. Sci. Polym. Chem. Ed.* **1974**, *12*, 807.
4. Takagishi, T.; Nakata, Y.; Kuroki, N. *J. Polym. Sci. Polym. Chem. Ed.* **1975**, *13*, 2411.
5. Takagishi, T.; Naoi, Y.; Kuroki, N. *J. Polym. Sci. Polym. Chem. Ed.* **1979**, *17*, 1865.
6. Takagishi, T.; Naoi, Y.; Kuroki, N. *J. Polym. Sci. Polym. Chem. Ed.* **1979**, *17*, 1953.
7. Kobayashi, K.; Sumitomo, H. *Macromolecules* **1980**, *13*, 234.
8. Wong, L.; Smid, J. *J. Am. Chem. Soc.* **1977**, *99*, 5637.
9. Sinta, R.; Smid, J. *Macromolecules* **1980**, *13*, 339.
10. Klotz, I. M.; Shikama, K. *Arch. Biochem. Biophys.* **1968**, *123*, 551.
11. Takagishi, T.; Kuroki, N. *J. Polym. Sci. Polym. Chem. Ed.* **1973**, *11*, 1889.
12. Kozuka, H.; Takagishi, T.; Hamano, H.; et al. *J. Polym. Sci. Polym. Chem. Ed.* **1985**, *23*, 1243.
13. Klotz, I. M.; Harris, J. U. *Biochemistry* **1971**, *10*, 923.

AMPHIPHILIC POLYMERS (Fluorescence)

Yoshihiro Itoh
Faculty of Textile Science and Technology
Shinshua University

In this article, we are concerned with the structure and properties, in particular fluorescent ones, of water-soluble, synthetic polymers in aqueous solution. These polymers are often called hydrophobically modified polymers.[1] After the pioneering work on "polysoap",[2] various types of amphiphilic polymers have been prepared with potential applications in diverse fields. For convenience, these polymers have been grouped into amphiphilic polyelectrolytes (polyanions and polycations), nonionic polymers, and others. For amphiphilic molecules, water is a solvent for the hydrophilic portions and a precipitant for the hydrophobic portions, the latter self-aggregating to minimize their exposure to water and forming supramolecular assemblies such as micelles and bilayers.

The situation is the same for amphiphilic polymers. When these polymers are dissolved in water, hydrophobic interactions, as well as ionic and hydrogen bondings, occurred intra- and/or inter-molecularly, forming specific structures.

PREPARATION

Amphiphilic polymers are usually prepared by polymerizing hydrophobic monomers with hydrophilic monomers or by modifying preformed polymers. These methods produce random, block, alternating, or graft copolymers.

PROPERTIES

In an amphiphilic polyelectrolyte, if the attractive hydrophobic interaction between the hydrophobic groups prevails over the electrostatic repulsion between the charged groups, the hydrophobic groups can self-aggregate intramolecularly to form a micellelike structure in aqueous solution. This compact conformation is reflected in a sharp decrease in solution viscosity. Such intramolecular micellation largely depends on chemical structure and composition. Many properties behave differently around a critical point.

In contrast, amphiphilic nonionic polymers such as hydrophobically modified cellulose, polyacrylamide, and PEG derivatives undergo interpolymer association primarily because of hydrophobic interaction.[3]

Solution Properties

At high polymer concentrations (typically higher than 0.1 g/dl), numerous experimental techniques, similar to those used in studies of surfactant micelles, are used to elucidate the formation, structure, and properties of polymer assemblies.[4]

Fluorescence Spectroscopy

With low concentrations (typically less than 0.1 g/dl) of polymers, these techniques obtain little or no information about the polymer assemblies. Under such conditions, fluorescence spectroscopy, a small amount of fluorescent dye is incorporated physically or covalently into amphiphilic polymers. This "covalent" method is more suitable for investigating the microenvironment in dilute polymer solutions because one can locate the dye molecule in the assemblies.

Photophysical analysis for several types of random, block, and alternating copolymers containing polycyclic aromatic groups in aqueous solution indicates that strong interaction between the chromophores in the hydrophobic microdomains leads to remarkably enhanced excimer formation or self-quenching, primarily because of efficient energy migration among the densely packed chromophores. Such favorable migration also facilitates energy transfer, photoinduced electron transfer, and other photochemical reactions in these polymer assemblies. In this context, these polymers serve as models for artificial photosynthetic systems and are called "photon-harvesting" or "antenna" polymers.[5,6]

APPLICATIONS

Studies of amphiphilic polymers generally have been directed toward improved performance of synthetic water-soluble polymers to be used as additives for oil-recovery fluids, waste-water flocculation, stabilizers for latex particles, and associative thickeners.[1] Such applications are based primarily upon the bulk solution properties of these polymers, including solubility and viscosity. Chemically modified biopolymers are potentially important in medical, pharmaceutical, and agricultural applications such as immobilized enzymes, drug encapsulation, and biodegradable materials.[7]

Photoinduced electron transfer from aromatic chromophores covalently attached to amphiphilic polyelectrolytes to water-soluble electron acceptors or donors is facilitated by energy migration over a cluster of the chromophores and by hydrophobic interaction between the polymer and the substrates. In addition, efficient charge separation has been achieved by electrostatic repulsion between the charged products and the charged surface of the polymer or by "compartmentalization" of chromophores by hydrophobic clusters.[8] Energy transfer from poly(styrenesulfonate-*co*-2-vinylnaphthalene) to soluble molecules in the hydrophobic microdomains induces various photosensitized chemical reactions such as unimolecular photolysis of 2-undecanone, bimolecular oxidation of polynuclear aromatic compounds or styrene, and dechlorination of polychlorinated biphenyls.[9] This system referred to as a "photozyme" is expected to provide an environmentally safe way to remove pollutants in aqueous solution, although there is still room for further improvement.

REFERENCES

1. McCormick, C. L.; Bock, J.; Schulz, D. N. *Encycl. Polym. Sci. Eng.* **1989**, *17*, 730.
2. Strauss, U. P.; Gershfield, N. L. *J. Phys. Chem.* **1954**, *58*, 747.
3. Winnik, F. M. *Macromolecules* **1989**, *22*, 734.
4. Fendler, J. H.; Fendler, E. J. *Catalysis in Micellar and Macromolecular Systems* Academic: New York, 1975.
5. Guillet, J. E. *Polymer Photophysics and Photochemistry* Cambridge University: Cambridge, U.K. 1985.
6. Webber, S. E. *Chem. Rev.* **1990**, *90*, 1469.
7. Peppas, N. A.; Langer, R. S., Ed. *Adv. Polym. Sci.* **1993**, 107.
8. Morishima, Y. *Adv. Polym. Sci.* **1992**, *104*, 51.
9. Nowakowska, M.; Sustar, E.; Guillet, J. E. *J. Am. Chem. Soc.* **1991**, *113*, 253.

AMPHOTERIC LATEX

Haruma Kawaguchi
Department of Applied Chemistry
Keio University

Amphoteric latex is defined as the dispersion in which the sign and magnitude of particle's zeta potential change with pH. In addition to the inversion of the potential zeta sign, the pH change of latices affects dispersion stability. Namely, many amphoteric latices exhibit reversible dispersion–aggregation–dispersion upon changing pH through IEP. The least viscosity, the lowest osmotic pressure, the highest foaming, and so on are also generally observed at the IEP of latices.

PREPARATION

Amphoteric latices can be prepared directly by emulsion copolymerization, soap-free emulsion copolymerization, or dispersion copolymerization using cationic and anionic monomers with a matrix-composing monomer, or indirectly by modifying existing non-amphoteric latices. Styrene is often used as the matrix-forming monomer except for the preparation of hydrogel microspheres. The ratio of ionic groups on the particle surfaces, and consequently the electrophoretic property of particles, depend on the ratio of ionic monomers in the polymerization recipe.[1,2] In some cases, the ratio of ionic groups on the particle surface does not coincide with the ratio in whole particles because their ionic groups are unevenly distributed.

Amphoteric hydrogel particles can be obtained when hydrophilic monomers are used as a matrix-composing material with a crosslinking reagent.

When it is difficult to prepare amphoteric latices from direct copolymerization of anionic and cationic monomers, an alternative is to convert existing non-amphoteric latices to amphoteic ones.

CHARACTERIZATION

The determining ions of amphoteric particles are generally proton or hydroxyl ions. Therefore, each latex has a specific pH at which the particles become electrically neutral. Potentiometric titration indicates the pH at which the net charge of particles is zero. The pH is referred to as the point of zero charge (POZ). Electrophoresis, that is, measurement of electrophoretic mobility, is another technique for determining the electrically neutral condition of particles. But the pH at which the mobility of particles is zero, the isoelectric point (IEP) does not necessarily coincide with the POZ. These points do not coincide because potentiometric titration determines the net charge at the actual surface, whereas electrophoresis measures the net charge inside the surface of shear. Therefore, the kind and amount of ions existing in the area between the actual surface and the surface of shear determine the discrepancy between the POZ and IEP.[3]

IEP is more useful than POZ for discussing the dispersion stability of particles. Namely, at IEP, the zeta potential of particles is zero and the potential energy of repulsion between particles is minimal. Thus, IEP indicates optimum conditions for flocculation. Flocculation is measured simply by turbidimetry, the condition at which the latex loses its stability.[4] Conductometric titration is used often to determine the charge of latex particles.

APPLICATIONS

Amphoteric latices are used as reliable models for naturally occurring amphoteric materials, such as amphoteric oxides and proteins, because of their excellent uniformity, stability, and reproducibility. In industry, they are used as binders and fillers for paper and various composites because their binding mode and strength are easily controlled with pH. They are expected to have many uses in the growing nano-technology, and chemo-electromechanical devices seem to be one of their promising applications.

REFERENCES

1. Harding, I. H.; Healy, T. W. *J. Colloid Interface Sci.* **1985**, *107*, 382.
2. Homola, A.; James, R. O. *J. Colloid Interface Sci.* **1977**, *59*, 123.
3. Bijsterbosch, B. H.; Lyklema, J. *Adv. Colloid Interface Sci.* **1978**, *9*, 147.
4. Healy, T. W.; Homola, A.; James, R. O. *Faraday Disc. Chem. Soc.* **1978**, *65*, 156.

Amylopectin

See: Starch Polymers, Natural and Synthetic

Amylose

See: Inclusion Complexes (Overview)
Starch Polymers, Natural and Synthetic

ANIONIC INITIATORS (Lithium Aggregate Effects, Anionic Polymerization)

Craig A. Ogle, Harry C. Johnson IV, and Xiao Li Wang
Department of Chemistry
University of North Carolina at Charlotte

Frederick H. Strickler, Daniel Bucca, and Bernard Gordon III
Department of Materials Science and Engineering
Pennsylvania State University

Organolithium reagents are an important class of anionic inititators. They provide an important route to the synthesis of monodispersed polymers and in preparing *cis*-1,4-polyisoprene in hydrocarbon solvent.[1]

The study of organolithium reagents is complicated by their propensity toward self-association. These reagents form bridged structures in solid, solution, and even in the gas phase causing sometimes dramatic reactivity differences between isomers. The tendency for self-aggregation is so great that virtually all organolithiums are known to be in a dynamic self-aggregated equilibrium state in both polar or non-polar solvents with a few exceptions.[2-4] Several factors influence the degree to which these compounds self-aggregate, namely: the alkyl group, the solvent, temperature, and concentration.

The aggregation of organolithium reagents is important in understanding their reactivity. Since they can react in aggregates and yield lithium products, formation of mixed/product aggregates is possible. The formation of cross-associated species is another complication in the study of organolithium reagents.

With an eye toward understanding the influence aggregation plays in anionic polymerization, we report our findings on the polymerization of styrene and 3-methylstyrene with *n*-, *sec*-, and *tert*-BuLi in THF, with and without lithium alkoxides, by ^{1}H and ^{7}Li rapid injection NMR spectroscopy.[5] RINMR experiments provide a method to simultaneously observe all reagents during the reaction, including the initiator and its aggregated state. RINMR is essentially a stop-flow NMR experiment; although inherently slower and less sensitive than stop-flow IR and UV-visible techniques, it offers the advantage of more structural information and quantification of all reagents with distinct NMR signals.[6]

CONCLUSIONS

Organolithium reagents form aggregated structures with themselves and can form mixed aggregates with lithium alkoxides. These mixed aggregates cause dramatic changes in the reactivity of organolithium reagents toward the polymerization of styrene. The reaction of styrene with various BuLi aggregates was studied in THF at −80°C. The initiation rate is dependent upon the type of butylithium used as the initiator. The initiation of styrene in THF follows the order *sec*-BuLi > *tert*-BuLi > *n*-BuLi. In the initiation of styrene by *n*-BuLi, the *n*-BuLi reacts as an aggregate, while *sec*- and *tert*-BuLi react from their monomeric forms. Computationally it was found that the order for initiation of styrene by the solvated isomeric butylithium would be $(n\text{-BuLi})_1 > (sec\text{-BuLi})_1 > (tert\text{-BuLi})_1 > (n\text{-BuLi})_2$.

Lithium alkoxides had a dramatic effect on the initiation of styrene. The mixed aggregates formed with *n*-BuOLi increased the initiation rate of styrene over *n*-BuLi alone. However, *tert*-BuOLi decreased the initiation rate of *n*-BuLi toward styrene. This effect was accentuated as the concentration of the alkoxide was increased. In fact, a *tert*-BuOLi/*n*-BuLi ratio of five prevented initiation at –80°C. The electronic effects due to the alkoxy groups should be the same for all of the alkoxides. That there are differences in reactivity shows that the steric bulk of the alkoxide greatly affects reactivity.

An induction period is not observed with *sec*- and *tert*-BuLi reacting smoothly with styrene. With *n*-BuLi there is a substantial induction period. The induction period also appears with *n*-BuLi and lithium alkoxides. This effect is probably related to the aggregation state of organolithiums. The first step involves coordination of the styrene to the butyllithium aggregate. Steric bulk has a major effect on this complexation. The bulky *tert*-butoxy groups greatly increase the induction period, while *n*-BuOLi decreases the length relative to *n*-BuLi alone. The less sterically hindered *sec*- and *tert*-BuLi have no induction period since they are monomeric.

Further evidence for reaction of the organolithium through aggregates is obtained by the presence of the mixed aggregate between styrene and *n*-BuLi. This aggregate only appears upon reaction of *n*-BuLi with styrene. A mixture of *n*-BuLi and polystyryllithium failed to produce a mixed aggregate indicating it is only formed during the reaction.

REFERENCES

1. Stavely, F. W. *Ind. and Eng. Chem.* **1956**, *48*, 778.
2. Bauer, W.; Winchester, W. R.; Schleyer, P. V. R. *Organometallics* **1987**, *6*, 2371.
3. Stowell, J. C. *Carbanions in Organic Synthesis*, John Wiley & Sons: New York, 1979.
4. Heinzer, J.; Orth, J. F. M.; Seebach, D. *Helv. Chim. Acta.* **1985**, *68*, 1848.
5. Ogle, C. A.; Wang, X. L.; Johnson, H. C.; Strickler, F. H.; Bucca, D.; Gordon III, B. G. *Macromolecules* **1995**, *28*, 5184.
6. McGarrity, J. F.; Prodolliet, J.; Smyth, T. *Org. Magn. Reson.* **1983**, *17*, 59.

Anionic Polymerization

See: Anionic Initiators (Lithium Aggregate Effects, Anionic Polymerization)
Block Copolymers (By Changing Polymerization Mechanism)
Branched Polymers
Cyclic Siloxanes (Ring-Opening Polymerization)
Cyclopolymerization (Overview)
Living Anionic Polymerization
Living Anionic Polymerization (Protected Functional Monomers)
Molecular Weight Distribution (Ionic Polymerization)
Poly(alkylene phosphate)s
Ring-Opening Polymerization, Anionic (With Expansion in Volume)
Silacryclobutanes and Related Compounds (Ring-Opening Polymerization)

Annealing

See: Injection Moldings (Residual Stresses)

ANOXIC POLYMER MATERIALS

Sanjay Palsule
Department of Chemistry
Heriot-Watt University

AEROSPACE POLYMER MATERIALS

The choice of materials for space structures is governed by their ability to form an aerodynamically stable structure, to retain functional properties in an exo-space environment, and to perform successfully throughout all phases of their designed lifetime.[1] In general, materials for space structures are required to have low weight, high stiffness, low coefficient of thermal expansion, dimensional stability, and resistance to the effects of the space environment. Carbon/epoxy composites and polyimides are the state-of-the-art aerospace materials.

The international manned space station will be established in the low Earth orbit (LEO) (200–500 Km). The 30-year long life of the space station in the low Earth orbit environment governs the choice of the materials for the space station. Atomic oxygen forms a major component of the low Earth orbit environment. Atomic oxygen reacts only with the surface of the exposed material, and the thermo-oxidative effects of atomic oxygen lead to polymer surface recession and degradation. Polymer materials resistant to the thermo-oxidative effects of the low Earth orbit atomic oxygen have been termed as anoxic polymer materials.[1]

MECHANISM OF ATOMIC OXYGEN-INDUCED POLYMER DEGRADATION

Low Earth orbit missions have shown that the interaction between spacecraft materials and ambient atomic oxygen (ATOX) can produce significant changes in surface properties of many materials. These changes are directly related to the atomic oxygen fluence or total integrated flux incident on the material surface. The ATOX fluence incident on the surface is a function of orbital altitudes, atmosphere density, phase of solar cycle, the spacecraft velocity, the mission duration, and the orientation of the surface relative to the flux.

The degradation of polymer materials by ATOX is an oxidation reaction involving breaking of bonds within the polymer backbone. The degradation reaction proceeds via the formation of an intermediate hydroperoxide, which then decomposes into low molecular weight fragments that are volatile under space vacuum conditions. Any proposed mechanism for ATOX-induced polymer degradation is still highly speculative.

ATOMIC OXYGEN-INDUCED POLYIMIDE DEGRADATION

Polyimides (e.g., Du Pont's Kapton based on pyromellitic dianhydride-oxydianiline, PMDA-ODA) have been used for the electrical insulation layer of the spacecraft solar power generators. Kapton polyimide has comparatively weak bonds in its backbone structure. During ATOX exposure, CO and CO_2 gases were detected by mass spectroscopy, which indicates that probably Kapton degrades under ATOX owing to an oxidation reaction involving the breaking of bonds within the macromolecular backbone. The mechanism of Kapton degradation is highly speculative. Investigations are underway to understand mechanism of ATOX-induced Kapton degradation.

ANOXIC POLYMER MATERIALS

Polymer materials resistant to the low Earth orbit atomic oxygen have been termed "anoxic" polymer materials.[1] Polysiloxane, although not a conventional aerospace polymer, is known to resist the thermo-oxidative attack of atomic oxygen.[2] Efforts are under way to develop anoxic materials based on polysiloxanes.

POLYSILOXANE-BASED ANOXIC POLYMER MATERIALS: APPLICATIONS AND LIMITATIONS

Although polysiloxanes are anoxic, thermodynamic and spectroscopic studies indicate that in polydimethylsiloxanes, the methyl groups rotate freely around the Si-O bonds leading to a large molar volume, low cohesive energy density, and low intramolecular forces, which in turn impart flexibility to the chain backbone. Such a molecular structure lowers the glass transition temperature of the polymer to well below 0°C and limits its mechanical properties.[3]

Owing to the low intramolecular forces, polysiloxanes gain beneficial properties such as low surface energy, low solubility parameter, and low dielectric constant. Additionally, polydimethylsiloxanes have extremely low surface tensions. In the case of siloxane-containing polymer materials, the lower surface energy siloxane component migrates toward the surface to minimize interfacial and/or surface energy. This equilibrium segregation occurs at very low concentrations—also termed as the critical concentration—of siloxanes (ca. 1%) and efficiently leads to siloxane-enriched surface in the case of siloxane-containing materials without any significant change in the bulk properties.[3]

Organosiloxane copolymers, inter-penetrating networks based on organosiloxanes, and blends of organosiloxane copolymers with high performance polymers do not demonstrate thermomechanical properties required for LEO space applications.

ANOXIC POLYSILOXANE COATINGS: APPLICATIONS AND LIMITATIONS

A solar power generator is the best power source for low Earth orbit spacecrafts. The substrate/insulation layer of the power system is based on a polyimide such as Du Pont's Kapton. As Kapton-polyimide degrades under ATOX, the only way to protect Kapton-polyimide from ATOX is by means of an anoxic polysiloxane coating.

Kapton polyimide coated with polysiloxane has been exposed to ATOX.[4-6] During the initial stages of exposure ATOX reacted with siloxane forming a silica (SiO_2) layer that protected the material from further ATOX attack. The coatings have limitations for long duration space applications. Micrometeoroid and/or space debris impact leads to pin holes/erosion of the coating and creates a site for ATOX-induced degradation of underlying Kapton-polyimide. Moreover, thermal cycling of the spacecraft leads to cracks in the coatings.

As an extension of the principle of a coating, a special laminate has been developed for the Hubble Space Telescope's Solar Arrays.[7] In essence, the silver mesh used for power transfer was surrounded by a symmetrical sandwich of silicone adhesive, Kapton, and glass fiber laminate with silicone coating that provided ATOX protection.

ANOXIC MOLECULAR COMPOSITE

To take the advantage of the atomic oxygen resistance of polysiloxanes and to improve their thermomechanical properties, Palsule developed a novel concept of anoxic siloxane molecular composite.[8,9] Anoxic molecular composite is defined as an isotropic miscible blend of an organosiloxane copolymer and a rigid macromolecule in which the molecular level thermal and mechanical reinforcement of the organosiloxane copolymer by the rigid macromolecule is achieved along with the siloxane-enriched surface that provides stability to the material in harsh atomic oxygen dominant low Earth orbit environment.[9]

Palsule developed two anoxic molecular composites of an organosiloxane copolymer—polysiloxane-etherimide (PSI)-reinforced molecularly by polyamide-imide (PAI) and poly-p-phenylene-oxydiphenylenetere-phthalamide (PPOT).[9,10]

ACKNOWLEDGMENTS

Support of the European Space Agency CESA-ESTEC-Materials and Processes Division is gratefully acknowledged.

REFERENCES

1. Palsule, S. *Aerospace Polymers and Composites,* 1st ed., Praxis: Chichester, UK, due 1997.
2. Leger, L. J.; Visentine, J. T.; Santos Mason, B. *SAMPE Quart* **1987**, *18*, 48.
3. Yilgor, I.; McGrath, J. E. *Advances in Polymer Science* **1988**, *86*, 1.
4. Arnold, C. A.; Chen, D. H.; Chen, Y. P.; Waldbauer, R. O.; Rogers, M. E.; McGrath, J. E. *High Performance Polymers* **1990**, *2*, 83.
5. Yilgor, I. *Proc. ACS Symp on High Performance Polymers for Harsh Environments*, April 1987, 249.
6. Rutledge, S. K.; Mihileie, J. A. *Surf. & Coat. Technol.* **1989**, *39/40*, 607.
7. Gerlach, L. *ESA Journal* **1990**, *14*, 149.
8. Palsule, S. *ESA Journal* **1993**, *17*, 133.
9. Palsule, S. European Patent Application (To ESA), EP-0632100A1, 1995.
10. Palsule, S. Ph.D. Thesis, Heriot-Watt University, Edinburgh, UK, 1994.

Antibacterial Polymers

See: Antibacterial Resins
Marine-Antifouling Coating Materials
Supported Polythioethers and Polythiacrownethers

ANTIBACTERIAL RESINS

P. L. Nayaka and S. Lenka
Laboratory of Polymers and Fibers
Department of Chemistry
Ravenshaw College

Phenols have been well reviewed for their antimicrobial activities and found to be effective eradicants. A survey of the literature reveals that resins from aromatic compounds having hydroxy, amino, and carboxyl groups are considered potential substrates for bacteriocides and fungicides. We describe the preparation of some resin copolymers from mono-di-hydroxy and amino-substituted acetophenones by reacting them with numerous hydroxy and amino aromatic substrates and formaldehyde in the presence of an acid and base catalyst. The bacteriocidal and fungicidal properties of the resins are also described.

PREPARATION

We used *o*-, *m*-, and *p*-substituted hydroxy acetophenones (OHAP, MHAP, PHAP); 2,4-, 2,5-dihydroxy acetophenone (RAP, QAP); 2-hydroxy-1-acetonaphthalene (BNAP); and *m*- and *p*-amino acetophenone (MAAP, PAAP) to prepare the resin polymers and reacted them with formaldehyde (F); *o*-, *m*-, and *p*-hydroxy phenols (C,R,H); *o*-, *m*-cresol (OC, MC); *o*-, *p*-hydroxy benzoic acids (OHBA, PHBA); *o*-, *m*-toluic acids (OTA, MTA); *o*-, *p*-aminobenzoic acid (OABA, PABA); *p*-nitrobenzoic acid (PNBA); *o*-, *p*-chloroanilines (OCA, PCA); *o*-, *p*-chlorobenzoic acid (OCBA, PCBA); salicylic acid (SA) and its acetyl derivative (ASA); and phenolphthalein (Ph) in the presence of acid and base catalysts with standard procedures.

MICROBICIDAL ANALYSIS

Pathogens

The bacterium (human pathogens) used in the assay were primarily pyrogenic except *Bacillus subtilis* (a nonpathogenic, laboratory contaminant). Both coagulse (+ve) and (–ve) strains of *Staphylococcus, aureus* and *albus*, were used. The former is a pyrogenic bacteria specifically effective in infections, abscesses, and wound sepses; the latter has only low-grade pathogenicity. The other species, *Klebsiella, Escherichia coli, Pseudomonas pyocyanea*, and *Proteus*, are pyrogenic and responsible for many urinary tract infections. *Streptococcus viridans* is known to cause acute endocarditis, and *Salmonella typhosa para B* causes typhoid.

Plant pathogens such as *Penicillium expansum Botrydepladia theobromae, Nigrospora* species, *Tricothesum* species, and *Rhizopus nigricans* were used to screen the resins for their effectiveness.

PROPERTIES AND APPLICATIONS

The resins studied are not quite effective bacteriocides, but they are perfect fungistats. Depending upon their charge distribution, molecular weight, conformation, and tacticity, the polymers may interact with cell membrane components or become disproportionate through chemical reactions by endocytosis.[1-3] Overall, most of the resins strikingly inhibited spore germination of *Culvularia lunata* and were effective fungistats.

REFERENCES

1. Tirrell, D.; Boyd, P. M. *Makromol. Chem. Rapid Commun.* **1981**, *2*, 193.
2. Marwaha, L. K.; Tirrell, D. *ACS Symposium Series 186* American Chemical Society: Washington, DC, 1982, 163.
3. Drobnik, J.; Rypacek, F. *Adv. Polym. Sci.* **1983**, *57*, 1.

ANTIBODY-POLYMER (Porphyrin) COMPLEXES

Akira Harada
Department of Macromolecular Science
Osaka University

Recently, with the advent of genetic engineering and hybridama technology, it has become possible to generate chemically homogeneous, monoclonal antibodies.[1] Antibodies are unique in their ability to recognize diverse substrates. More recently, the diversity of the immune system has been used to generate catalysts, called catalytic antibodies, that have specific binding properties.[2] Most catalytic antibodies have been produced by immunizing mice with "transition state analog" compounds. In these instances, product inhibition was indispensable. Another approach has been to use a cofactor, such as enzymes, as an active site.[3] One of the most important cofactors is a family of porphyrins, which function as oxygen carriers and cause redox reactions and photo-induced electron transfer. I consider monoclonal antibodies tailor-made hosts for artificial guest polymers. This chapter describes the preparation, characterization, and properties of monoclonal antibodies against meso-tetrakis(4-carboxyphenyl)porphine (TCPP) and carboxyphenyltriphenylporphine (CATPP) and its metal complexes.

PREPARATION OF MONOCLONAL ANTIBODIES

TCPP and CATPP (**Figure 1**) were covalently attached to the carrier proteins, keyhole limpet hemocyanin (KLH) and bovine serum albumin (BSA), using water-soluble carbodiimide, 1-(3-dimethylamino)propyl-3-ethylcarbodimide, or carbonyldiimidazol. The conjugates were purified by column chromatography on Sephadex G-50®.

SUMMARY

Monoclonal antibodies for TCPP and CATPP were prepared by immunizing Balb/c mice with TCPP and CATPP keyhole

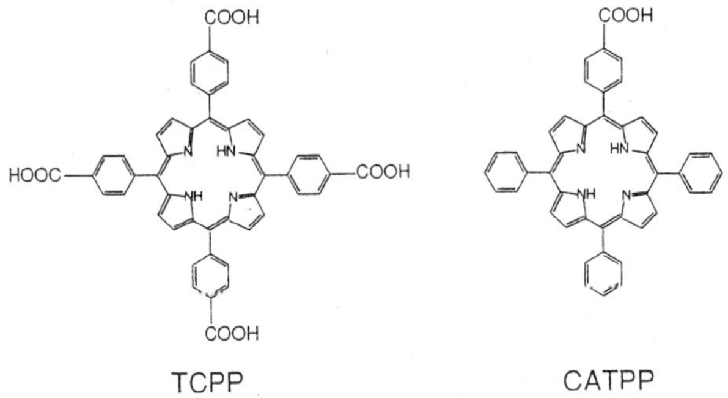

FIGURE 1. TCPP and CATPP structures.

limpet hemocyanin conjugates and fusing the spleen cells with myeloma cells using poly(ethylene glycol). The antibodies bound TCPP and CATPP strongly with dissociation constants of 10^{-6} to 10^{-8} *M*. One of the antibodies shifted the Soret and Q bands of TCPP to a longer wavelength and caused large, induced cotton effects. Quantitative analyses of TCPP binding to the antibody by ICD showed that not only 1:1, but higher orders of binding, such as 2:1, took place with the antibodies exceeding the amount of TCPP.

REFERENCES

1. Campbell, A. M. *Monoclonal Antibody Technology*; Elsevier: Amsterdam, 1984.
2. Schultz, P. G.; Lerner, R. A.; Benkovic, S. J. *Chem. Eng. News* **1990**, *68*, 26.
3. Iverson, B. L.; Lerner, R. A. *Science* **1989**, *243*, 1184.

ANTICANCER POLYMERIC PRODRUGS, TARGETABLE

Ki Ho Kim, Takashi Hirano, and Shinichi Ohashi*
Protein Engineering Laboratory
Molecular Biology Department
National Institute of Bioscience and Human Technology

Many attempts have been made to improve cancer chemotherapy over the last decade, to decrease undesirable side effects, and to increase the targetibility of drugs to the cancer cell. Although low molecular weight anticancer agents exhibit high cytotoxicities against cancer cells, they also display their activities against normal cells, which is attributable to their non-specificities for cancer cells. They are also rapidly excluded by renal excretion due to their low molecular weight and are not protected from enzymic attack during delivery to target cells.

One of the most promising methods to overcome the drawbacks of low molecular weight anticancer agents and subsequently to attain increased therapeutic effect is to incorporate drugs into a drug carrier. Many kinds of drug carriers, such as soluble synthetic or natural polymers,[1-4] liposomes,[5-7] microspheres,[8,9] and nanospheres,[10,11] have been employed to increase drug concentration in target cells by altering the body distribution of drugs (targeting of drug) and by sustaining a therapeutic concentration in the body for an extended time, owing to their physiochemical properties (controlled release of drug). Though liposomes and microparticles have been used to target and sustain drug release, their utility is restricted because they have difficulty gaining access to most tumors and because of their phagocytic capture by the reticuloendothelial system.[12,13] In contrast, soluble polymers used for anticancer polymeric prodrugs seem to offer greater potential because they can traverse compartmental barriers in the body[14] and, therefore, gain access to a greater number of cell types; in most cases they are not subjected to rapid clearance by the reticuloendothelial cells.[15,16]

Over the last two decades, many water-soluble polymers, such as human serum albumin (HSA),[3] dextran,[17] lectins,[1] poly(ethylene glycol) (PEG),[18,19] styrene-maleic anhydride copolymer (SMA),[20-22] N-hydroxypropylmethacrylamide copolymer (HPMA),[23,24] and poly(divinyl ether-co-maleic anhydride) (DIVEMA)[24,25] have been used to prepare anticancer polymeric prodrugs. Such drugs have often shown good solubility in water, increased half-life in the body, and high antitumor effects.

As mentioned above, the aim of using polymers as a drug carrier is to increase antitumor activity by sustained drug release. But antitumor activity of anticancer polymeric prodrugs also depends on the distribution of active drug, which is affected not only by the physiochemical properties of the polymer and the biological properties of some components conjugated to the polymer chain, such as targeting moiety and spacer, but also by anatomical characteristics of target issue.

In this review, therefore, we have subdivided anticancer polymeric prodrugs into two categories according to drug transport mechanism in the body.

Passive targeting drug delivery systems based on the anatomical characteristics of tumor tissue for macromolecules.[26-28]

Active targeting drug delivery systems based on specific interactions between receptors on cell surface and targeting moieties conjugated to the polymer backbone.

PASSIVE TARGETING DRUG DELIVERY SYSTEMS

Many natural or synthetic polymers have been used as polymer backbones of passive targeting drug delivery systems to improve sustained drug release in the body and to increase drug concentration in tumor cells. To be used as a polymer backbone, polymers generally should be biodegradable, biocompatible, non-toxic, and non-immunogenic in the body. To avoid accumulation of polymers composed of non-degradable C–C bonds in the main chain, polymers with appropriate molecular weights can be used to permit glomerular filtration in the kidney, or those containing biodegradable side chains on the polymer backbone can be employed to result in low molecular weights of the polymers after hydrolysis.[14]

Tumor issues have several characteristics different from normal tissues, such as enhanced permeability of neovasculature to macromolecules, hypervasculature, and little recovery or macromolecules through either blood vessels or lympatic vessels.[26-28] The injected macromolecules, therefore, can permeate tumor tissues more easily than normal tissues and accumulate for a long time to exhibit the desired phamacological activities.

Water-soluble polymers are internalized through the cell membrane by pinocytosis,[35,81] and transferred via endosomes[36,37] to the lysosomal compartment of the cell. Therefore, anticancer agents should be covalently conjugated to polymers in which the bond between polymer and drug is resistant to enzymic attack while traveling to tumor tissues, but susceptible to enzyme-catalyzed hydrolysis inside the lyosomes (lysosomotropic agent). The rate of drug release can be controlled by the detailed structure of the bond between the drug and the polymer.[38]

On the other hand, some other polymer-drug conjugates have been designed to be degradable only by chemical hydrolysis to exhibit antitumor activity in the body.[39] Antitumor activities of the conjugate mainly depend on the rate of drug release from

*Author to whom correspondence should be addressed.

the conjugate and the extent of cellular interaction of the conjugate.

Although restricting drug distribution and/or action to a specific site, i.e., neoplastic target tissue, is still an unresolved problem, therapeutic advantages of the polymeric prodrugs by passive targeting even in the absence of specific targeting moieties have often been found. In this section, we describe antitumor activities of several kinds of polymeric prodrugs and their clinical trials according to the kind of polymers used.

DEXTRAN

Dextran is a synthetic polymer of linear chains of α-D-glucose molecules; 95% of the chains consist of 1,6-α-linked linear glucose units whereas the side chains consist of 1,3-α-linked moieties. Dextrans are colloidal, hydrophilic, and water-soluble substances, inert in biological systems, and do not affect cell viability. These properties have enabled dextrans to be used for many years as blood expanders to maintain or replace lost blood volume. Owing to the retarded clearance of dextrans from the body, they have also been developed as drug carriers to improve therapeutic efficacy.[17]

HPMA [POLY(N-(HYDROXYPROPYL)-METHACRYLAMIDE)]

Homopolymers of N-(2-hydroxypropyl)methacrylamide (HPMA) were originally synthesized to serve as blood plasma expanders,[40,41] but more recently HPMA copolymers have undergone considerable development as polymer backbones of anticancer polymeric prodrugs.

PEG [POLY(ETHYLENE GLYCOL)]

PEG is water-soluble polymer that exhibits protein resistance, low toxicity, and non-immunogenicity. Thus, it has been extensively studied for use as a biocompatible polymer in the field of biotechnical and biomedical application.[45-50,82]

SMA [POLY(STYRENE-*co*-MALEIC ANHYDRIDE)]

Styrene-maleic anhydride copolymer (SMA) has been used to synthesize tailor-made protein drugs for tumor targeting.[21] Conjugation of protein drugs with SMA copolymer is expected to increase the half-life of the drugs and decrease their toxicity. In the synthesis of polymer-enzyme conjugates, it is important to synthesize such conjugates with their enzymatic activity retained, because the decrease in enzymatic activity is in general one of the most serious problems in the conjugation of enzymes with polymers.

DIVEMA [POLY(DIVINYL ETHER-*co*-MALEIC ANHYDRIDE)] AND RELATED POLYMERS

DIVEMA, an alternating copolymer of divinyl ether and maleic anhydride, has been extensively studied as an antitumor or antiviral agent itself,[51] and has also been used as a polymer backbone for preparing targetable anticancer polymeric prodrugs.[52-55] Because the main chain of DIVEMA is connected together by C–C bonds, it cannot be cleaved under physiological conditions. But most of DIVEMAs used as polymer backbones of anticancer polymeric prodrug have molecular weights low enough to permit body clearance through glomerular filtration in the kidney. Synergy might be expected in the treatment of tumor disease owing to the antitumor activities of both polymer itself and the drug conjugated to the polymer backbone.

ACTIVE TARGETING DRUG DELIVERY SYSTEM

Although we can obtain a certain degree of antitumor activity by using passive targeting drug delivery systems, the ultimate aim of delivering active drug only to the target site is not satisfied. Because the antitumor activity of polymeric prodrugs by passive targeting is mainly dominated by physiocochemical properties of the polymer-drug conjugate and anatomical characteristics of tumor tissues for macromolecules, it is hard to control the delivery of drugs to the target site by these means. Therefore, it is necessary to design more precise drug delivery systems that can deliver drugs mainly to cancer cells. To meet this requirement, various investigators have attached certain kinds of molecules, i.e., a targeting moiety that can recognize receptors[56-58] on the existing cell-surface membrane, to the polymer backbone. Antibodies,[54-66] sugars,[14,15,23,43,44] glicolipids,[29,30] and other ligands[31-34] have been used as targeting moieties for specific interactions with target cells. Among them, antibodies and sugars have thus far been most commonly used. In this section, therefore, we will deal with targetable polymeric prodrugs bearing antibody or a sugar molecule as the targeting moiety.

DEXTRAN-DRUG CONJUGATES

To increase the targetability of dextran-drug conjugates to target issue, many attempts have been made to use monoclonal antibodies as a targeting moiety, especially owing to the development of the hybridoma technique,[67] which permits the production of large amounts of monoclonal antibodies.

One of the monoclonal antibodies used as a targeting moiety for conjugation with dextran is anti-α-fetoprotein antibody. α-Fetoprotein (AFP) is a major serum protein synthesized during fetal life. The serum concentration of AFP decreases rapidly during fetal development to a residual level in adulthood, but rises again in individuals with hepatomas or germ-cell tumors.[68-70]

HPMA[POLY(N-HYDROXYPROPYL)-METHACRYLAMIDE] COPOLYMER

The HPMA copolymer has been extensively studied by Kopecek et al. for use as the polymer backbone of targetable polymeric prodrugs.[71] Systems employing it are composed of several components, such as polymer backbone, spacer, active drug, and targeting moiety. Because the repeating units of HPMA are bound together by C–C bonds, this backbone cannot be cleaved into fragments small enough to be excreted via glomerular filtration of kidney. Therefore, polymers must be designed to have low molecular weights or to contain biodegradable side chains on the polymer backbone to result in low molecular weights of polymers after hydrolysis.[14] In order to meet these conditions, various investigators have synthesized soluble crosslinked HPMA copolymers containing biodegradable oligopeptide side chains and biodegradable oligopeptide crosslinks.[42,72,73] As oligopeptides, di-, tri-, or tetrapeptide chains have been used, and their biodegradability has been investigated under various conditions.[74-80] By detailed studies on the relationship between the structure of oligopeptide sequences in HPMA copolymers and their ease of biodegradation, polymeric prodrugs that are stable in blood plasma and serum[78] but release drugs by intralysosomal enzymes[75,76] could be synthesized.

REFERENCES

1. Molteni, L. *Lecints as Drug Carriers*, In *Drug Carriers in Biology and Medicine*; Gregoriadis, G., Eds., Academic: London, 1979; p 43.
2. Arnon, R. *Antibodies and Dextrans as an Anti-Tumor Drug Carrier in Targeting of Drugs* Gregoriadis et al., Eds.; Plenum: New York, 1982; p 31.
3. Trouet, A.; Masquelier, M.; Baurain, R.; Compeneere, D. *Proc. Natl. Acad. Sci., USA* **1982**, *79*, 626.
4. Kopecek, J.; Rejmanova, P.; Duncan, R.; Lloyd, J. B. *Ann. N.Y. Acad. Sci* **1985**, *446*, 93.
5. Gregoriadis, G.; Florence, A. T. *Drugs, Liposomes in Drug Delivery: Clinical Diagnostic and Opthalmic Potential* **1993**, *45*, 15.
6. Kim, S. *Drugs*, **1993**, *46*, 618.
7. Fielding, R. M. *Clin. Pharmacokinet., Liposomal Drug Delivery; Advantages and Limitations from a Clinical, Pharmacokinetic and Therapeutic Perspective* **1991**, *21*, 155.
8. Davis, S. S.; Illum, L.; McVie, J. G.; Tomlinson, E., Eds.; *Microspheres as Drug Therapy. Pharmaceutical, Immunological and Medical Aspect*, Elsevier: Amserdam, 1984.
9. Arshady, R. *J. Controll. Rel.* **1990**, *14*, 111.
10. Wasser, P. G.; Muller, U.; Kreuter, J.; Berger, S.; Munz, K.; Kaiser, E.; Pfluger, B. *Int. J. Pharma.* **1987**, *39*, 213.
11. Kreuter, J. *J. Controll. Rel.* **1991**, *16*, 169.
12. Kato, A. *Encapsulated Drugs in Targeted Cancer Therapy.* In *Controlled Drug Delivery* Bruck, S. D. Ed., CRC: Boca Raton, FL, 1983; Vol II, p 189.
13. Poste, G. *Biol. Cell, Liposome Targeting in vivo: Problems and Opportunities* **1983**, *47*, 19.
14. Cartlidge, S. A.; Duncan, R.; Lloyd, J. B.; Kopeckova-Rejmanova, P.; Kopecek, J. *J. Controll. Rel.* **1987**, *4*, 265.
15. Duncan, R.; Seymour, L. W.; Scarlett, L.; Lloyd, J. B.; Rejmanova, P.; Kopecek, J. *Biochem. Biophys. Acta* **1986**, *880*, 62.
16. Seymour, L. W.; Duncan, R.; Strohalm, J.; Kopecek, J. *J. Biochem. Mater. Res.* **1987**, *21*, 1341.
17. Molteni, L. *Dextrans as Drug Carriers*, In *Drug Carriers in Biology and Medicine;* Gregoriadis, G. Eds.; Academic: London, 1979; p 107.
18. Ashihara, Y.; Kono, T.; Yamazaki, S.; Inada, Y. *Biochem. Biophys. Res. Commun.* **1978**, *83*, 385.
19. Park, Y. K.; Abuchowski, A.; Davis, S.; Davis F. *Anticancer Res.* **1981**, *1*, 373.
20. Maeda, H.; Takeshita, J.; Kanamaru, R. *Int. J. Protein Peptide Res.* **1979**, *14*, 81.
21. Maeda, H.; Matsumoto, T.; Konno, T.; Iwai, K.; Ueda, M. *J. Protein Chem.* **1984**, *3*, 181.
22. Maeda, H.; Ueda, M.; Morinaga, T.; Matsumoto, T. *J. Med. Chem.* **1985**, *28*, 455.
23. Duncan, R.; Kopecek, J. *Adv. Polym. Sci.* **1984**, *57*, 51.
24. Hirano, T.; Ohashi, S.; Morimoto, S.; Tsuda, K.; Kobayashi, T.; Tsukagoshi, S. *Makromol. Chem.* **1986**, *187*, 2815.
25. Zunino, F.; Pratesi, G.; Pezzoni, G. *Cancer Treat. Rep.* **1987**, *71*, 367.
26. Matsumura, Y.; Maeda, H. *Cancer Res.* **1986**, *46*, 6387.
27. Matsumura, Y.; Maeda, H. *Cancer Chemother.* **1989**, *6*, 3323.
28. Folkmann, J.; Klagsbrun, M. *Science* **1987**, *235*, 442.
29. Ghosh, P.; Bachhawat, B. K. *Arch. Biochem. Biophys.* **1981**, *206*, 454.
30. Ghosh, P.; Das, P. K.; Bachhawat, B. K. *Arch. Biochem. Biophys.* **1982**, *213*, 266.
31. Czop, J. K.; Fearon, D. T., Austen, K. F. *Proc. Natl. Acad. Sci. USA* **1978**, *75*, 3831.
32. Kipnis, T. L.; David, J. R.; Alper, C. A.; Sher, A.; Da Silva, W. D. *Proc. Natl. Acad. USA* **1981**, *78*, 602.
33. Shichijo, S.; Alving, C. R. *Biochim. Biophys. Acta* **1985**, *820*, 289.
34. Livingston, P. O.; Natoli, E. J.; Calves, M. J.; Stockert, E.; Oettgen, H. F.; Old, L. J. *Proc. Natl. Acad. Sci. USA* **1987**, *84*, 2911.
35. Pratten, M. K.; Duncan, R.; Lloyd, J. B. *Coated Vesicles*; Ockleford, C. D.; Whyte, A., Eds.; Cambridge University: 1980; p 179.
36. Helenius, A.; Mellman, I.; Wall, D.; Hubbard, A. *TIBS* 1983, 245–250.
37. Pastan, I.; Willingham, M. C. *TIBS* **1983**, 250–254.
38. Kopecek, J.; Rejmanova, P. *Enzymatically Degradable Bonds in Synthetic Polymers.* In *Controlled Drug Delivery*; Bruck, S. D., Ed., CRC: Boca Raton, FL, 1983; p 81.
39. Matsumoto, S.; Arase, Y.; Takakura, Y.; Hashida, M.; Sezaki, H. *Chem. Pharm. Bull.* **1985**, *33*, 2941.
40. Kopecek, J.; Sprincl, L.; Lim, D. *J. Biomed. Mater. Res.* **1973**, *7*, 179.
41. Sprincl, L.; Exner, J.; Sterba, O.; Kopecek, J. *J. Biomed. Mater. Res.* **1976**, *10*, 953.
42. Rejmanova, P.; Obereigner, B.; Kopecek, J. *Makromol. Chem.* **1981**, *182*, 1899.
43. Duncan, R.; Kopecekova, P.; Strohalm, J.; Hume, I. C.; Lloyd, J. B.; Kopecek, J. *Br. J. Cancer* **1988**, *57*, 147.
44. Duncan, R.; Hume, I. C.; Kopecekova, P.; Ulbrich, K.; Strohalm, J.; Kopecek, J. *J. Controll. Rel.* **1988**, *10*, 51.
45. Kondo, A.; Kishmura, M.; Katoh, S.; Sada, E. *Biotech. Bicengin* **1989**, *34*, 532.
46. Suzuki, T.; Ikeda, K.; Tomono, T. *J. Biomater. Sci. Polymer Ed.* **1989**, *1*, 75.
47. Zalipsky, S.; Gilon, C.; Zilkha, A. *Eur. Polym. J.* **1983**, *19*, 1177.
48. Topchieva, I. N. *Russ. Chem. Rev.* **1980**, *49*, 260.
49. Abuchewski, A.; van Es, Ta; Palczuk, N. C.; Davis, F. F. *J. Biol. Chem.* **1977**, *252*, 3578.
50. Harris, J. M. *Poly(ethylene glycol) Chemistry; Biotechnical and Biomedical Applications* Plenum: New York, 1992.
51. Regelson, W.; Shnider, B. I.; Colsky, J.; Holland, J. F.; Johnston, C. L. Jr.; Dennis, L. H. *Immune Modulation and Control of Neoplasia by Adjuvant Therapy*; Chirigos, M. A., Ed.; Raven: New York, 1978; p 469.
52. Ueda, H.; Hirano, T.; Todoroki, T.; Ohashi, S.; Tsukagoshi, S.; Iwasaki, Y. *Drug Delivery Systems* **1991**, *6*, 51.
53. Yamamoto, H.; Miki, T.; Oda, T.; Hirano, T.; Sera, Y.; Akagi, M.; Maeda, H. *Eur. J. Cancer* **1990**, *26*, 253.
54. Han, M. J.; Cho, T. J.; Park, S. J.; Sohn, Y. S.; Lee, C. O.; Choi, S. U. *J. Bioact. Compat. Polym.* **1992**, *7*, 358.
55. Han, M. J.; Park, S. J.; Cho, T. J.; Chang, J. Y.; Sohn, Y. S.; Lee, C. O.; Choi, S. U. *J. Bioact. Compat. Polym.* **1994**, *9*, 142.
56. Kawasaki, T.; Ashwell, G. *J. Biol. Chem.* **1977**, *252*, 6536.
57. Monsigny, M.; Roche, A.-C.; Midoux, P. *Biol. Cell* **1984**, *51*, 187.
58. Wileman, T.; Boshaus, R.; Stahl, P. *J. Biol. Chem.* **1985**, *260*, 7387.
59. Tsukada, Y.; Ohkawa, K.; Hibi, N. *Cancer Res.* **1987**, *47*, 4293.
60. Yu, D.-S.; Chu, T. M.; Yeh, M.-Y.; Chang, S.-Y.; Ma, C.-P.; Han, S.-H. *J. Urology* **1988**, *140*, 415.

61. Aboud-Pirak, E.; Hurwitz, E.; Bellot, F.; Schlessinger, J.; Sela, M. *Proc. Natl. Acad. Sci. USA* **1989**, *86*, 3778.
62. Noguchi, A.; Takahashi, T.; Yamaguchi, T.; Kitamura, K.; Takakura, Y.; Hashida, M.; Sezaki, H. *Jpn. J. Cancer Res.* **1991**, *82*, 219.
63. Noguchi, A.; Takahashi, T.; Yamaguchi, T.; Kitamura, K.; Takakura, Y.; Hashida, M.; Sezaki, H. *Bioconjug. Chem.* **1992**, *3*, 132.
64. Kotanagi, H.; Takahashi, T.; Masuko, T.; Hashimoto, Y.; Koyama, K. *Tohoku J. Exp. Med.* **1986**, *148*, 353.
65. Yeh, M.-Y.; Roffler, S. R.; Yu, M.-H. *Int. J. Cancer* **1992**, *51*, 274.
66. Hata, Y.; Takada, N.; Sasaki, F.; Abe, T.; Hamada, H.; Takahashi, H.; Uchino, J.; Tsukada, Y. *J. Pediatric Surgery* **1992**, *27*, 724.
67. Kohler, G.; Milstein, C. *Nature* **1975**, *256*, 495.
68. Abelev, G. I. *Transplantn Rev.* **1974**, *20*, 1.
69. Rouslahti, E.; Pihko, H.; Seppala, M. *Transplantn Rev.* **1974**, *20*, 38.
70. Sell, S.; Becker, F.; Leffert, H.; Waterbe, H. *Cancer Res.* **1976**, *36*, 4239.
71. Kopecek, J.; Duncan, R. *J. Controll. Rel.* **1990**, *11*, 279.
72. Kopecek, J.; Rejmanova, P.; Chytry, V. *Makromol Chem.* **1981**, *182*, 799.
73. Ulbrich, K.; Strohalm, J.; Kopecek, J. *Makromol Chem.* **1982**, *182*, 1917.
74. Duncan, R.; Lloyd, J. B.; Kopecek, J. *Biochem. Biophys. Res. Comm.* **1980**, *94*, 28.
75. Duncan, R.; Cable, H. C.; Lloyd, J. B.; Rejmanova, P.; Kopecek, J. *Bioscience Reps.* **1982**, *1041*, 2.
76. Duncan, R.; Cable, H. C.; Lloyd, J. B.; Rejmanova, P.; Kopecek, J. *Makromol Chem.* **1983**, *184*, 1997.
77. Rejmanova, P.; Pohl, J.; Baudy, M.; Kostka, V.; Kopecek, J. *Makromol Chem.* **1983**, *184*, 2009.
78. Rejmanova, P.; Kopecek, J.; Duncan, R.; Lloyd, J. B.; *Biomaterials* **1985**, *6*, 45.
79. Subr. V.; Kopecek, J.; Pohl, J.; Baudys, M.; Kostka, V. *J. Controll. Rel.* **1988**, *8*, 133.
80. Subr. V.; Duncan, R.; Kopecek, J. *J. Biomat. Sci. Polym. Ed.* **1990**, *1*, 261.
81. Duncan, R. Selective Endocytosis, Sustained and Controlled Drug Delivery; Robinson, J. R.; Lee, V-H. Eds.; Marcel Dekker: New York, 1986; p 581.
82. Abuchowski, A.; McCoy, J. R.; Palczuk, N. C.; Van Es. T.; Davis, F. F. *J. Biol. Chem.* **1977**, *252*, 3582.

Anticorrosive Polymers

See: Quinone-Amine Polyurethanes
Quinone-Diamine Polymers

Antidegradants

See: Rubber Additives, Multifunctional

ANTIFOAMING AGENTS

Randal M. Hill and Steven P. Christano
Dow Corning

Chemical additives designed to reduce foam problems are called antifoam agents, foam inhibitors, foam suppressants, foam control agents, deaerators, or air release agents.[1] This article describes the use of polymeric substances to control foam and discusses several important applications, including foam control in polymer processing.

FOAM STABILITY

Stable foams are always associated with the presence of some surface-active agent (surfactant). Transient foams can also occur in distillation processes near miscibility boundaries and in viscous liquids. The stability of foam, like that of an emulsion or clay dispersion, depends on the adsorbed film at the particle surface.

Investigation of foaming and foam control problems usually focuses on the thin film (bubble wall) rather than on the bubble as a particle. The stability of this thin film, and therefore of the foam, depends on surface elasticity and viscosity, electrostatic and steric interactions, and disjoining pressure effects.[1,3] The drainage rate of liquid from the foam (caused by the density difference between the liquid and air) also influences foam stability; thinner films are more fragile. Bulk viscosity slows drainage and stabilizes foam. Evaporation of the liquid or diffusion of gas in and out of the bubbles also decreases foam stability.

Foam is often divided into two types: kugelschaum, a dispersion of small, spherical bubbles separated by thick films, and polyhederschaum, closely packed polyhedral bubbles separated by thin films.

ANTIFOAMING AND DEFOAMING

An agent that inhibits initial foam formation is called an antifoam. With antifoam any foam which does form may be very stable in the absence of agitation. An antifoam acts on kugelschaum foam. An agent that ruptures already formed foam is called a defoamer. The action of a defoamer is easily observed in the rapid decay of foam when agitation is stopped. Defoamers act on polyhederschaum foam.

A defoamer should exhibit rapid knockdown of a foam, but persistence of action is more characteristic of antifoaming. In practice, the same type of chemical additives are used for both defoaming and antifoaming. We will use foam control to refer generically to all types and degrees of foam control exhibited by such process aids. We will use foam rupture and film rupture to refer to the mechanistic aspects of foam control.

ROAM RUPTURE MECHANISMS

There are four basic concepts that help us understand the ability of antifoam agents to rupture foam: entering, bridging, dewetting, and rupture.

STATISTICS OF PARTICLE SIZE VS. NUMBER

Often, the performance of an antifoam depends on its particle size as dispersed into the foaming medium.[2] It is not so often realized that changing the particle size also changes the number of particles and therefore affects the statistics of the foam rupture process.[2,4]

TIME DECAY OF ANTIFOAM EFFICACY

The efficacy of antifoams decreases with time, and this is believed to occur because of the particles gradually breaking down into smaller and smaller particles which eventually become too small to be effective.[2] Increasing the viscosity of the oil, incorporating fumed silicas which build the viscosity of the mixture, and crosslinking the silicone oil are all documented ways of improving the resistance of the foam control agent to this process, therefore improving its persistence.[5-8] Incorporation of resins and isocyanate condensation have similar effects.[9]

SOLUBLE FOAM CONTROL AGENTS

There is a whole class of soluble foam control agents, mostly block copolymers of polyoxyethylene (EO) and polyoxypropylene (PO), which function by phase separating above their cloud point to form insoluble droplets of a hydrophobic liquid.[10] Examples include certain of the well-known Pluronic® surfactants.[11] Soaps and fatty acids are also used as soluble foam control agents in aqueous systems. On contact with hard water, soap forms insoluble hydrophobic particles, which defoam quite well as long as there is sufficient other surfactant present in the system to keep them dispersed. Some soluble polymeric surfactants are believed to destabilize foam by entering the surface as a soluble surfaceactive species and altering the surface elasticity and viscosity required for stable foam. Short-chain alcohols are also sometimes used as soluble foam control agents, primarily for their defoaming ability.[1]

APPLICATIONS OF POLYMERIC ANTIFOAMS

Formulation Trends

Two general trends emerge from the patent literature of the last decade. Much of the new development taking place in foam control technology involves silicone foam control agents.

The first trend is a continuing process of expanding the silicone chemistry used in foam control agents. This trend includes a wide range of approaches that fall into two general categories: modification of silicone rheological properties to provide for enhanced foam control longevity and modification of silicone physical and chemical properties by chemically grafting functional groups onto the siloxane backbone.

The second general trend, and the area which has seen the greatest change in recent years, is in the development of delivery vehicles for foam control agents. At this point, an antifoam compound of general utility is often customized. Thus, an antifoam may be selected and then prepared through dilution or emulsification to be used when most suitable by a particular customer.

Examples of Applications

Many foam control agents are used primarily as processing aids in a wide variety of manufacturing processes. Foam control is used in aqueous applications such as polymer processing, gas and oil separation, diesel fuel, pulp and paper manufacture, powdered and liquid detergents, textile manufacture, paint and ink manufacture, phosphoric acid manufacture, metal working, lubricants, sugar processing, food manufacture (frying), waste water treatment, fermentation (pharmaceuticals, food, and industrial chemicals), agrochemicals, leather manufacture, chemical processing, and distillation. As this list proves, the applications include a wide variety of aqueous and nonaqueous environments.

SUMMARY

A new consensus has emerged concerning the dominant mechanism of aqueous foam rupture that involves entering, bridging, and dewetting. These concepts were elucidated by introducing particle size statistics.

In reviewing the technologies of foam control we focused on silicone polymers, which combine their unique surface activity with opportunity for an almost endless variety of chemical and physical property modifications.

Technologies for granulated forms, emulsions, and self emulsifiable antifoam blends are well-documented and described in the patient literature. In the future, antifoam development will continue to be complex and diversified through a wide range of markets and applications. Complexity in the surfactant foaming conditions and the dynamics of air and water mixing have very practical impact. Thus, one of the key problems the formulator must address is the need for laboratory evaluation methods that meaningfully predict the performance of different foam control agents under actual use conditions. Market complexity will always exist, in that different industries and different end use customers require distinctly different solutions to their foam control problems. Thus, a wide array of foam control agents will continue to be used in the market including silicone-based polymers and many different organic compounds to address the many foaming conditions and cost benefit requirements.

REFERENCES

1. Byron, K. J. *Crit. Rep. Appl. Chem.* **1990**, *30*, 163.
2. Garrett, P. R. *Defoaming: Theory and Industrial Applications*; Marcel Dekker: New York, 1993; p 1.
3. Lucassen, J. In *Anionic Surfactants—Physical Chemistry of Surfactant Action*; Lucassen-Reynders, E. H. Ed.; Marcel Dekker: New York, 1981: p 217.
4. Frye, G. C.; Berg, J. C. *J. Coll. Interface Sci.* **1989**, *127*, 222.
5. Aizawa, K.; Sewa, S.; Hideki, N. U.S. Patent 4 749 740, 1988.
6. John, V. B.; Sawicki, G. C.; Pope, R.; et al. European Patent 0 217 501, 1986.
7. Aizawa, K.; Sewa, S.; Nakahara, H. U.S. Patent 4 639 489, 1987.
8. Hill, R. M.; Starch, M. S.; Gaul, M. S. U.S. Patent 5 262 088, 1993.
9. Pirson, E.; Schmidlkofer, J. BE Patent 0 866 716, 1978.
10. Nakagawa, T. In *Nonionic Surfactants*; Schick, M. J. Eds.; Marcel Dekker: New York, 1966; p 572.
11. Ash, M.; Ash, I. *The Condensed Encyclopedia of Surfactants*; Chemical: New York, 1989; p 353.

Antifogging Agents

See: Additives (Property and Processing Modifiers)

Antifouling Polymers

See: Marine-Antifouling Coating Materials

ANTIINFECTIVE BIOMATERIALS

W. Kohnen and B. Jansen
Institute of Medical Microbiology and Hygiene
University of Cologne

Today polymers are essential in many fields of medicine. The benefits of these materials (biomaterials) are beyond doubt. However, aside from complications like mechanical irritation and thrombosis, infection of the materials (foreign-body infection) is the main problem associated with their use.[1-3]

The main causative organisms of these so-called foreign-body infections are coagulase-negative staphylococci.[4] In most cases, antimicrobial therapy alone cannot cure the infection, and removal of the biomaterial becomes necessary.[5] Along with severe consequences for the individual patient, foreign-body infections lead to prolongation of the hospital stay and thus to an increase in therapy costs. Therefore it is of outstanding importance to prevent this kind of infection.

HOW TO REDUCE FOREIGN-BODY INFECTIONS

Bacterial adhesion is the first step in the pathogenesis of foreign-body infections. After settling to the biomaterial, many bacterial strains (e.g., coagulase-negative staphylococci) are able to produce extracellular substances ("slime") that protect them against host defense mechanisms and antibiotics.[6-9]

Prevention of bacterial adherence or at least significant lowering of the number of viable adhering bacteria seems to be the most effective way to avoid foreign-body infections. There are two principal approaches to preventing foreign-body infections by influencing the interaction between biomaterial and bacterium:

1. Development of polymers (polymer surfaces) with antiadhesive properties
2. Development of polymers (polymer surfaces) with antimicrobial properties

For effective development of antiinfective medical devices, detailed knowledge about the interactions between the surface of a biomaterial and bacterial adherence is a prerequisite. Therefore, many scientists have tried to describe bacterial adherence in dependence of surface properties of a biomaterial and to establish physico-chemical models for the adhesion process.

POLYMERS WITH ANTIMICROBIAL PROPERTIES

Another strategy to prevent foreign-body infections than the development of "antiadhesive" surfaces by polymer modification is loading polymers with antimicrobial substances. As active agents antibiotics, metal salts, and disinfectants or detergents have been used. The aim is to achieve polymeric drug delivery systems which gradually release the antimicrobial agent and therefore are capable of preventing either primary adhesion of microorganisms or—more likely—colonization of an implant or medical device.

Well known and of clinically documented value is the system poly(methyl methacrylate)–Gentamicin (Septopal®) for the treatment of chronic osteomyelitis and soft tissue infection.[10,11] Antibiotics have also been incorporated in bone cement,[12,13] vascular prosthesis,[14-16] ventricular shunts,[17] as well as in prosthetic heart valves.[18]

Antimicrobials other than "classic" antibiotics have also been investigated as potential candidates to generate antimicrobial activity in medical devices, e.g., chlorhexidine[19] and IRGASAN®.[20]

Because of their lack of cross-resistance to "classic" antibiotics and their broad antimicrobial activity against bacteria, fungi, viruses, and parasites, metals or metal salts have been studied. Silver especially has raised the interest of many investigators, perhaps because of its low toxicity and good biocompatibility.

Other investigators have tried to produce polymers with "intrinsic" antimicrobial activity by covalently bonding antiinfectives to polymers.[21-27]

In our group we have also developed polymer systems with antimicrobial activity. In earlier studies, antibiotics like flucloxacillin, clindamycin, and vancomycin have been incorporated into polyurethanes.[28] It seems that in the near future, more antimicrobial polymer systems will be brought onto the market and into clinical use.

CREATING POLYMERS SUITABLE FOR THE INVESTIGATION OF BACTERIAL ADHESION AND THE DEVELOPMENT OF ANTIMICROBIAL BIOMATERIALS: THE GLOW DISCHARGE TECHNIQUE

In order to investigate correlations between bacterial adhesion and biomaterial characteristics and whether the described physico-chemical models are suitable for a description of adhesion, polymers with different surface properties are needed. An elegant and versatile method to achieve polymers with varying surface properties is surface modification by the glow discharge technique, where the surface of a synthetic material is exposed to a glow discharge under reduced pressure.[29-32] With this method, the surface of most polymers can be modified in a great variety of ways without changing bulk properties like mechanical stability and elasticity. Additionally, any medical device can be equipped with functional groups able to bind antibiotics.

Depending on the nature of the gas that is used in glow discharge treatment, the polymer surface may be modified in different ways. Modifications include etching, cross-linking, functionalization, plasma polymerization, and plasma-induced grafting.

APPLICATIONS

Modified polymers generated by the glow discharge technique are very helpful to study the interaction between bacteria and biomaterials. By using the glow discharge technique, with the knowledge of the correlation between bacterial adhesion and the surface properties of a biomaterial, it is possible to produce medical devices with reduced adherence.

Another possibility for creating antiinfective devices by glow discharge treatment results from the different function of serum and tissue proteins in bacterial adherence. After insertion of implantation of a medical device into the body, the material becomes coated by host factors, e.g., serum and tissue proteins.[33] Whereas for some proteins (e.g., fibrinogen, fibronectin, and vitronectin) adherence-promoting properties are described, it is well known that albumin reduces bacterial adherence.[34-36] Generation of surfaces that selectively bind albumin and thus reduce adherence of bacteria is one of our main interests and is currently under investigation.

REFERENCES

1. Hirshman, H. P.; Schurman, D. J. *Current Clinical Topics in Infectious Diseases*, Remington, J. S.; Swartz, M. S., Eds., McGraw-Hill: New York, 1982, p 206.
2. Maki, D. G. *Current Clinical Topics in Infectious Diseases*, Remington, J. S.; Swartz, M. S., Eds., McGraw-Hill: New York, 1982, p 309.

3. Dankert, J.: Hogt, A. H.; Feijen, J. *CRC Critical Rev. Biocompatibility*, **1986**, *2*, 219.
4. Jansen, B.; Peters, G.; Pulverer, G. *J. Biomater. Appl.* **1988**, 2(4), 520.
5. Schoen, F. J. *ASAIO Trans.* **1987**, *33*, 8.
6. Peters, G.; Locci, R.; Pulverer, G. *J. Infect. Dis.* **1982**, *146*, 579.
7. Christensen, G. D.; Simpson, W. A.; Bisno, A. L.; Beachey, E. H. *Infect. Immun.* **1982**, *37*, 318.
8. Franson, T. R.; Sheth, N. D.; Rose, H. D.; Sohnle, P. G. *J. Clin. Microbiol.* **1984**, *20*, 500.
9. Marrie, T. J.; Costerton, J. W. *J. Clin. Microbiol.* **1984**, *19*, 687.
10. Marcinko, D. E. *J. Foot Surgery* **1985**, *24*, 116.
11. Aderhold, L. *Dtsch. Z. Mund Kiefer Gesichtschir.* **1985**, *9*, 94.
12. Buchholtz, H. W.; Engelbrecht, H. *Der Chirurg.* **1970**, *11*, 511.
13. Welch, A. G. *J. Biomed. Mater. Res.* **1978**, *12*, 679.
14. Moore, W. A.; Charpil, M.; Seiffert, G.; Koewn, K. *Arch. Surg.* **1981**, *116*, 1403.
15. McDougal, E. G.; Burnham, S. J.; Johnson, G. *Vasc. Surgery* **1986**, *4*, 5.
16. Greco, R. S.; Harvey, R. A. *Ann. Surg.* **1982**, *195*, 168.
17. Bayston, R.; Grove, N.; Siegel, J.; Lawellin, D. *J. Neurology Neurosurgery Psychiatry* **1989**, *52*, 605.
18. Olanoff, L. S.; Anderson, J. M.; Jones, R. D. *Trans. Am-Soc. Artif. Intern. Org.* **1979**, 25, 334.
19. Nakano, H.; Seko, S.; Sumii, T.; Nihira, H.; Fujii, M.; Okada, K.; Kitano, T.; Kodama, M. *Hinyokoika Kiyo* **1986**, *32*, 567.
20. Kingston, D.; Seal, D.; Hill, I. D. *J. Hyg. Camb.* **1986**, *96*, 185.
21. Solowskij, M. V.; Ulbrich, K.; Kopecek, J. *Biomaterials* **1983**, *4*, 44.
22. Ghedini, N.; Scapini, G.; Ferruti, P.; Rigidi, P.; Matteuzi, D. *Il Farmaco Ed. Sc.* **1985**, *40*, 102.
23. Geckler, K.; Bayer, E. *Naturwissenschaften* **1985**, *72*, 88.
24. Messinger, P.; Schimpke-Meier, A. *Arch. Pharm.* **1988**, *32*, 89.
25. Simionescu, C. I.; Dimitriu, S. *Makrom. Chem. Suppl.* **1985**, *9*, 179.
26. Demitriu, S.; Popa, M. I. *Colloid Polym. Sci.* **1989**, *267*, 595.
27. Reddy, B. S. R. *Int. J. Biochem. Biophys.* **1989**, *26*, 80.
28. Jansen, B.; Schareina, S.; Treitz, U.; Peters, G.; Schumacher-Perdreau, F.; Pulverer, G. *Progress in Biomedical Polymers* Plenum: New York, 1990; p. 347.
29. Gombotz, W. R.; Hoffman, A. S. *CRC Critical Reviews Biocompatibility* **1986**, *4*, 1.
30. Hoffman, A. S. *Advances in Polymer Science 57*, Dusek, K., Ed., Springer Verlag: Berlin, 1984; p 141.
31. Yasuda, H.; Gazicki, M. *Biomaterials*, **1982**, *31*, 68.
32. Yasuda, H. *Thin Film Processes*, Vossen, J. L.; Kern, W., Eds., Academic: New York, 1978; p 316.
33. Bantjes, A. *Brit. Polym. J.* **1978**, *10*, 267.
34. Vaudaux, P.; Yasuda, H.; Velazco, M. I.; Huggler, E.; Ratti, I.; Waldvogel, F. A.; Lew, D. P. *J. Biomater. Appl.* **1990**, *5*, 134.
35. Vaudaux, P.; Lerch, P.; Velazco, M.; Nydegger, U. E.; Waldvogel, F. A. *Adv. in Biomaterials* **1986**, *6*, 355.
36. Jansen, B.; Ellinghorst, G. *J. Biomed. Mater. Res.* **1984**, *18*, 655.

Antimicrobial Polymers

See: Antiinfective Biomaterials

Antioxidants

See: Additives (Agents for Sustaining Properties)
Additives (Property and Processing Modifiers)
Antioxidants and Stabilizers
Antioxidants (Overview)
Photo-oxidation, Polyolefins
Polyethylene (Stabilization and Compounding)
Rubber Additives, Multifunctional

ANTIOXIDANTS (OVERVIEW)

R. E. King III
Ciba Additives

Everyone is familiar with the oxidation of iron; the oxidation reaction product is the reddish, dusty material called rust. With polymers, the oxidation reaction product is not as easily identified or recognized. Polymers have more subtle changes such as indirect changes in color, loss of initial gloss or clarity, crazing, chalking, or brittleness. With iron, rust spreads and penetrates deeper into the body of the material. Eventually, the iron becomes brittle, small pieces and flakes start to break off, and the article begins to lose its original strength and physical properties. A less obvious process occurs with polymers; nevertheless, the end result is still an overall loss of physical properties.

Slowing down this process of oxidation is what many scientists strive to do. Trace amounts of transition metals and carbon are used to turn iron into stainless steel. For polymers, trace amounts of antioxidants are used. These antioxidants do not transform polymers into "stainless" polymers; however, they do significantly inhibit the oxidation process, thereby increasing the useful lifetime of the polymer. Antioxidants have been well reviewed.[1-6]

Since the invention of plastics, there has been a need for antioxidants. Plastics have become high quality materials, primarily through polymer structure but also through the use of antioxidants to preserve that structure.

POLYMER AUTOXIDATION

Often, free radicals are generated at the onset of high temperatures and high shear associated with melt compounding the polymer. Still, the melt compounding of the polymer is virtually unavoidable if the polymer powder, granules, or beads are to be transformed into a useful article.

Because it is generally accepted that oxidation is the key process by which a polymer loses its original properties, it is important to discuss the chemistry of oxidation.[1-6] Actually, the most appropriate term is autoxidation as the process is automatic once polymers are exposed to oxygen. Autoxidation then feeds upon itself because of the byproducts of the process giving it a catalytic aspect. In addition, impurities in the polymer also tend to accelerate the process.

Changes in molecular weight and molecular weight distribution occur. This alteration of the polymer is the mechanism by which the original properties of the polymer are significantly transformed. These types of changes in molecular weight and molecular weight distribution not only affect polymer processing characteristics but also significantly affect physical properties. Ultimately, without interrupting the free-radical chemistry, the polymer is oxidized to the point where discoloration, brittleness, crazing, cracking, and chalking are eventually observed.

POLYMER STABILIZATION

To prevent this undesirable chain of events, various chemistries can be used to interrupt the free-radical process of oxidation. One class of chemistries is based on chain breaking or primary antioxidants. Another class is based on preventive or secondary antioxidants, which decompose hydroperoxides before they are transformed into free radicals.

CHAIN-BREAKING OR PRIMARY ANTIOXIDANTS

This general class of antioxidants is capable of interrupting free-radical process by donating labile hydrogen atoms that neutralize or quench the free radical. These hydrogen-donating antioxidants, slow down oxidation by effectively competing with the polymer for free radicals, thereby abbreviating the chain length of the propagation reactions.

Phenolics

The chemistry of phenolic antioxidants has been extensively reviewed and consequently, will not be discussed in detail here other than to describe how they can be used in practice.[7-12] The point of incorporating them into the polymer is to stop the chemistry associated with free radicals by first donating hydrogen atoms, not a proton or a hydride, to interrupt the autoxidation process. Phenolics typically react with oxygen-centered free radicals, and consequently, can interrupt the autoxidation cycle. In general, phenolic antioxidants quench free radicals by donating hydrogen atoms.

Aromatic Amines

Certain classes of antioxidants, such as aromatic amines, are effective at scavenging free radicals as are the phenolic antioxidants. Aromatic amines are even more powerful in easily oxidized polymers like unsaturated elastomers. Although potent, aromatic amines tend to stain or discolor. Aromatic amines are typically used in elastomeric systems that are pigmented or loaded with carbon black to mask staining.

Hindered Amines

This class of chemistry is typically associated with light stabilizers because they are extremely effective at protecting certain classes of polymers from the damaging effects of ultraviolet radiation. However, hindered amines also belong to the family of chain-breaking antioxidants because they are also capable of scavenging free radicals. The most common hindered amine is based on 2,2,6,6-tertra-methylpiperidine.

Metal Deactivators

Polymers that come into contact with metals that have low oxidation potentials, such as copper, are susceptible to oxidation from the metal catalyzed decomposition of hydroproxides. One way to avoid these types of free radical activators is by using metal deactivators.[13] Metal deactivators are designed to contain hydrazide or amine functional groups, which can complex to the metal.

PREVENTIVE OR SECONDARY ANTIOXIDANTS

One of the most damaging species in the autoxidation process is the hydroperoxide, ROOH. Under elevated temperatures, hydroperoxides decompose via a homolytic cleavage to yield two free radicals. This step demonstrates the catalytic nature of autoxidation. The destruction of the hydroperoxides, which continually build up in the polymer, is essential in protecting the polymer. Most commercially available peroxide decomposers are based on trivalent phosphorus compounds and divalent sulfur compounds.

SYNERGIST MIXTURES OF ANTIOXIDANTS

When used alone, neither trivalent phosphorus compounds nor divalent sulfur compounds are capable of providing complete melt processing and thermal stability of the polymer. However, combined with a phenolic antioxidant, the results are better than the sum of the performance levels of each additive type. This is known as synergism.

ANTAGONISTIC MIXTURES OF ANTIOXIDANTS

Mixtures of antioxidants can work together synergistically; they can also work against each other. Chemistries that interfere with each other may not necessarily be obvious until the evidence is presented. For example, a phenolic antioxidant combined with a divalent sulfur compound for thermal stability and a hindered amine for light stability ensures long-term thermal and good light stability.

ANCILLARY PROPERTIES

In reality, there is more to antioxidants than providing stability to the polymer by quenching free radicals and decomposing hydroperoxides. Other key issues besides rates of reactivity and efficiency include performance parameters such as volatility, compatibility, color stability, physical form, propensity to form transformation products with taste or odor, regulatory issues associated with products used in food packaging, and in-polymer performance versus cost.

CONCLUSION

Antioxidants are used to preserve the properties of polymers that were designed to meet certain specifications. Antioxidants are comprised of various chemistries that perform by different mechanisms. Understanding the chemistry combines with the judicious selection of the appropriate antioxidants for a product's use is the key to success.

REFERENCES

1. Dexter, M. *Encyclopedia of Chemical Technology*, 4th ed.; Wiley-Interscience: New York, 1992; Vol. 3, p 424.
2. Gugumus, F. *Plastics Additives Handbook*, 3rd ed; Oxford University: New York, 1990; Chapter 1.
3. Al-Malaika, S. *Comprehensive Polymer Science*, 1st ed.; Pergammon: New York, 1988; Vol. 6, p. 539.
4. Klemchuk, P. P. *Ullmann's Encyclopedia of Industrial Chemistry*, 5th ed.; VCH: Deerfield Beach, FL, 1985: p 91.
5. Shelton, J. R. *Polymer Stabilization*; Wiley-Interscience: New York, 1972.
6. Scott, G. *Atmosphere Oxidation and Antioxidants*; Elsevier: Amersterdam, 1965; p 109.
7. Altwicker, E. R. *Chem. Rev.*, **1967**, *67*, 475.
8. Pospíšil, J. *Pure Appl. Chem.* **1973**, *36*, 207.

9. Pospíšil, J. *Polym. Degrad. Stab.* **1988**, *20*, 181.
10. Ibid. **1993**, *40*, 217.
11. Ibid. **1993**, *39*, 103.
12. Scott, G. *Atmospheric Oxidation and Antioxidants*; 1993; p 431; Elsevier: Amsterdam, The Netherlands.
13. Muller, H. *Plastics Additives Handbook*, 3rd ed; Oxford University: New York, 1990; Chapter 2.

ANTIOXIDANTS AND STABILIZERS

S. Al-Malaika
Polymer Processing and Performance Group
Department of Chemical Engineering and Applied Chemistry
Aston University

Oxidation of polymers, for example, plastics, rubbers, and fibers, is invariably manifested by loss of physical properties and ultimate failure of polymer artifacts. Polymer stabilization deals primarily with the inhibition of oxidative processes throughout the lifetime of the polymeric material. The underlying mechanisms of polymer oxidation (and degradation) are well known, and the selection and design of stabilizers (and antioxidants) for polymers are made in the light of current understanding of their mode of action.

Environmental and physical factors, for example, high temperatures, stress, UV light, ozone, and high-energy radiation, exert detrimental effects on polymer performance. These effects, however, can be mitigated by the incorporation of low levels of stabilizers during the fabrication process; the more demanding the application, the greater the need or more efficient stabilizers to achieve economic optimum properties of polymer products. Antioxidants and stabilizers, therefore, occupy a key position in the market of compounding ingredients for polymers, in particular, commodity polymers, such as polypropylene, polyethylene, and poly(vinyl chloride).

The terms "stabilizers" and "antioxidants" are generally used in the plastics industry to describe chemical agents that inhibit degradative effects of oxygen, light, heat, and high temperatures "antidegradants," "antifatigue agents," and "antiozonants," are widely employed by rubber technologists to refer to similar chemical agents, as well as to inhibitors, that combat the effects of stress and ozone.

"Antioxidant" is used herein to comprehensively describe all chemical agents that act to inhibit oxidation of a polymer matrix arising from the adverse effects of mechanical, thermal, photochemical, and environmental factors during the manufacture of the polymeric material and throughout the service life of the end-use product.

EFFECTS OF PROCESSING AND ENVIRONMENTAL FACTORS ON OXIDATIVE DETERIORATION OF POLYMERS

Thermooxidative degradation of polymers can occur at all stages of their life cycle (polymerization, storage, fabrication, weathering), but its effect is most pronounced during conversion processes of the polymer to finished products. Oxidative degradation of polymer articles in the outdoor environment is often exacerbated by combined factors in the environment, such as sunlight, rain, ozone, temperature, humidity, atmospheric pollutants, and microorganisms.

Polymer oxidation is best described by a cyclical free-radical chain reaction.[1-3] This autooxidation process normally starts slowly but autoaccelerates, leading in most cases to catastrophic failure of the polymer product. It is generally accepted that this radical chain reaction, which involves both alkyl and alkylperoxyl radicals as propagating species, is similarly involved in both thermal- and photooxidation of polymers, albeit at faster rates of initiation for the latter.

Hydroperoxides are the main initiators in both thermal- and photooxidation. The initiating species, hydroperoxides and their decomposition products, for example, are responsible for the changes in molecular structure and overall molar mass of the polymer that are manifested in practice by the loss of mechanical properties (e.g., impact, flexural, tensile strengths, and elongation) and by changes in the physical properties of the polymer surface (e.g., loss of gloss, reduced transparency, cracking, chalking and yellowing).

The extent of oxidative degradation of the macromolecular chain during melt processing and in service depends ultimately on the nature and structure of the base polymer.

INHIBITION OF OXIDATIVE DEGRADATION: CLASSIFICATION OF ANTIOXIDANTS AND THE BASIS OF THEIR MECHANISMS OF ACTION

The terms "antioxidants" and "stabilizers" cover a number of chemical classes of compounds that can interfere with the oxidative cycles to inhibit or retard oxidative degradation of polymers. Two major classes have been identified according to the way they interrupt the overall oxidation process: chain-breaking and preventive antioxidants.

Quinones and stable free radicals that can act as alkyl radical trapping agents are good examples of CB-A antioxidants.

Preventive antioxidants (sometimes referred to as secondary antioxidants), interrupt the second oxidative cycle by preventing or inhibiting the generation of free radicals.[4] The most important preventive mechanism is the nonradical hydroperoxide decomposition. Phosphite esters and sulfur-containing compounds are the most important classes of peroxide decomposers. Sterically hindered aryl phosphites have an additional chain-breaking activity: they react with peroxyl and alkoxyl radicals during their function as antioxidants.[15]

Sulfur compounds, e.g., thiopropionate esters and metal dithiolates, decompose hydroperoxides catalytically whereby one antioxidant molecule destroys several hydroperoxides through the intermediacy of sulfur acids.[6,7]

Metal deactivators primarily act by retarding metal-catalyzed oxidation of polymers; they are, therefore, important under conditions in which polymers are in contact with metals. Metal deactivators are normally polyfunctional metal-chelating compounds that can chelate with metals and decrease their catalytic activity.[18]

UV absorbers act by absorbing UV light, hence retarding the photolysis of hydroperoxides. Typical examples are based on 2-hydroxybenzophenones and 2-hydroxybenztriazoles; both are relatively stable to light between 300 and 360 nm and have high molar absorptions in this region.

PROCESSING ANTIOXIDANTS

Stabilization against mechano-oxidative degradation during high-temperature processing is essential in order to stabilize the polymer melt and to minimize the formation of adverse molecular impurities and defects that may contribute to early mechanical failure of finished articles during service. The choice of antioxidants for melt stabilization varies depending on the level of oxidizability of the base polymer, the extrusion temperature, and the performance target of the end-use application.

The effectiveness of melt-processing antioxidants is normally measured by their ability to minimize changes in the melt flow index (MFI) of the polymer that occur in their absence. Chain-breaking antioxidants are generally used to stabilize the melt in most hydrocarbon polymers. Hindered phenols are very effective processing antioxidants for polyolefins. Aromatic amines, however, have limited use because they give rise to highly colored conjugated quinonoid structures during their antioxidant function.

THERMOOXIDATIVE ANTIOXIDANTS

Stabilizers with high molar masses have lower volatility and are potentially more effective than those with lower molar masses containing the same antioxidant function for thermooxidative stabilization in service. Furthermore, peroxide decomposers, such as sulfur-containing compounds enhance the performance of high molar-mass phenols under high-temperature conditions in service.

PHOTO ANTIOXIDANTS

The pressure of light-absorbing impurities, trace levels of metals, and adventitious species arising from commercial production processes (e.g., polymer manufacture, processing, and fabrication) renders many commercial polymers (e.g., PE and PP) susceptible to outdoor weathering. Outdoor performance of polymers can, however, be markedly improved by a suitable choice of stabilizers. In practical applications, the end-use performance is governed by both physical parameters and chemical factors. Other factors that can affect the ultimate photostability of polymers include sample thickness, polymer crystallinity, and the presence of other additives (e.g., pigments and fillers). The largest market for light stabilizers is in polyolefins; for example, photostabilization of commercial PP is essential for outdoor and indoor end-use applications because of its extensive sensitivity to UV light.

An effective photoantioxidant must satisfy not only the basic chemical and physical requirements mentioned above but must also be stable to UV light and withstand continuous periods of exposure to it without being destroyed or effectively transformed into sensitizing products. Many metal thiolates are very effective UV stabilizers because of their much UV stability.

ANTIOXIDANTS FOR TIME-CONTROLLED STABILIZATION

Certain applications, for example agricultural mulching films, require precise time-controlled stabilization initially, followed by rapid degradation and embrittlement of the films at the precise time of cropping; inadequate time control would give rise to premature failure of the film and damage to the crops. Metal thiolates such as dithiocarbamates and dithiophosphates, where the metal is iron, exhibit this dual behavior.

The photoantioxidant activity of metal complexes of dithiocarbamic and dithiophosphoric acids depends on the nature of the metal ion and their stability to UV light.[17] Understanding the minor differences in the details of the mechanism of action of different metal dithiolates, particularly iron and nickel dithiocarbamates, has led to the design of very effective antioxidant-photoactivator systems and the enabling technology of time-controlled stabilization.[9] An ideal antioxidant-photoactivator system should exhibit an initial controlled photoinduction period followed by rapid photodegradation.

REACTIVE AND BIOLOGICAL ANTIOXIDANTS

The effectiveness of antioxidants depends not only on their intrinsic activity but also on their physical retention in the polymer. Migration of antioxidants into the surrounding environment, for example, leaching out into food constituents in contact with oils and fats, leads not only to premature failure of the polymer article but also to problems associated with health hazards and toxicological effects. A number of approaches are available to reduce the loss of antioxidants: increasing the molar mass of antioxidants, copolymerizing antioxidant functions in polymers (during polymer synthesis), and *in situ* grafting of reactive antioxidants onto polymer backbones.

The biological antioxidant, vitamin E, may be considered as a safe migrant and should have no problem in clearance for use in food packaging. It is essentially a hindered phenol that acts as an effective chain-breaking donor antioxidant where it donates a hydrogen to ROO· to yield a very stable tocopheroxyl radical. α-Tocopherol, (the active form of vitamin E), is a very effective melt stabilizer in polyolefins that offers high protection to the polymer at very low concentration.[10]

SYNERGISM AND ANTAGONISM

A cooperative interaction between two or more antioxidants (or antioxidant functions) that leads to an overall antioxidant effect greater than the sum of the individual effects of each antioxidant is referred to as synergism. Synergism can arise from the combined action of

- two chemically similar antioxidants, for example, two hindered phenols (homosynergism);
- two different antioxidant functions that are present in the same molecule; or
- the cooperative effects between mechanistically different classes of antioxidants, such as the combined effects of chain breaking antioxidants and peroxide decomposers (heterosynergism).[11]
- Antisynergistic effects, on the other hand, arise when antioxidants show antagonistic effects and give rise to a reduced net effect when compared to the sum of their individual effects.[11]

REFERENCES

1. Bolland, J. L. *Quart. Rev.* **1949**, *3*, 1.
2. Bateman, L. *Quart. Rev.* **11954**, *8*, 147.
3. Al-Malaika, S.; Scott, G. In *Degradation and Stabilization of Polyolefins*; Allen, N. S., Ed., App. Sci. London, 1983; Chapter 7.
4. Al-Malaika, S. In *Atmospheric Oxidation and Antioxidants*; Scott, G., Ed.; Elsevier Science: Amsterdam, 1993; Vol. 1, Chapter 5.

5. Schwetlick, K. In *Mechanisms of Polymer Degradation and Stabilization*; Scott, G., Ed.; Elsevier Science: New York, 1990; Chapter 2.
6. Al-Malaika, S. In *Mechanisms of Polymer Degradation and Stabilization*; Scott, G., Ed.; Elsevier Science: New York, 1990; Chapter 3.
7. Al-Malaika, S.; Chakraborty, K. B.; Scott, G. In *Developments in Polymer Stabilization*; Scott, G., Ed.; App. Sci.: London, 1983; Vol. 6, Chapter 3.
8. Muller, H. In *Plastics Additives Handbook*; Gachter, R.; Muller, H., Eds.; Hanser: Munich, 1987, Chapter 2.
9. Gilead, D.; Scott, G. In *Developments in Polymer Stabilization*; Scott, G., Ed.; App. Sci.: London, 1982; Vol. 5, Chapter 4.
10. Al-Malaika, S.; Issenhuth, S. In *Advances in Chemistry Series*; Clough, R. L. et al., Eds.; American Chemical Society: Washington, DC, in press.
11. Scott, G. In *Atmospheric Oxidation and Antioxidants*; Scott, G., Ed.; Elsevier Science: Amsterdam, 1993; Vol. 2, Chapter 9.

Antiozonants

See: Diene Rubbers, Conventional (Ozone Degradation and Stabilization)

ANTIPYRINE-ACRYLONITRILE COPOLYMERS (Thermal Characteristics)

S. H. El-Hamouly* and F. A. El-Saied
Laboratory of Polymer Chemistry
El-Menoufia University

M. M. Ramiz
Department of Electronic Engineering
El-Menoufia University

*Author to whom correspondence should be addressed.

Although technologically important, polyacrylonitrile (PAN) has had problems in three areas: processing, dyeing, and degrading. We resolved these problems by introducing comonomers into the PAN chains.

SYNTHESIS

N-antipyryl acrylamide monomer (NAA) [2,3-dimethyl-1-phenyl-5-oxo-3-pyrazolin-4-yl acryloyl amine] and antipyrine acryloyl thiourea monomer (APAT) [2,3-dimethyl-1-phenyl-5-oxo-3-pyrazolin-4-yl acryloyl thiourea] were synthesized as shown in **Schemes I** and **II**.[1,2]

COPOLYMERIZATION

Copolymers of each of the NAA and APAT monomers with acrylontrile (AN) were obtained through solution polymerization in dimethylformamide. The overall conversions were limited to less than 10% in every case. The copolymers dissolved in the reaction medium and recovered by precipitations (NAA in water and APAT in methanol). They were redissolved, reprecipitated, washed, dried, and weighed. The obtained copolymers were a yellowish powder soluble in dimethyl sulfoxide and dimethylformamide.

Thermal Analysis

The DTA curve of PNAA shows exotherms with a maximum at 300, 390, 440, and 530°C, which may represent the homolytic scission and complete decomposition. The proposed homolytic scission of PNAA was suggested from the obtained fragmentation patterns of mass spectrum of the NAA monomer.

The DTA curves of the NAA-AN copolymers show exotherms at 290, 360, 400, and 560–600°C, reflecting many processes

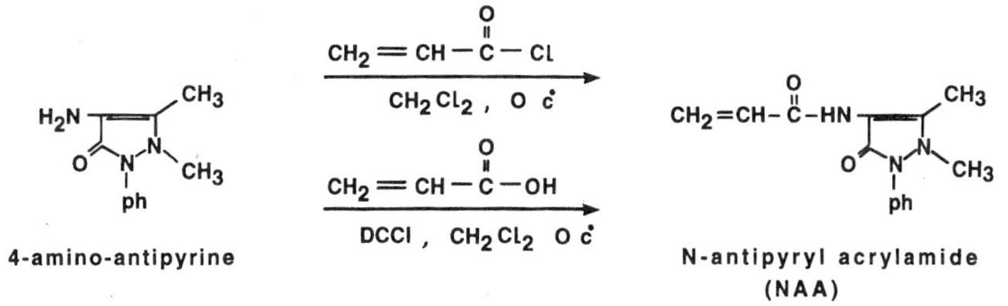

I

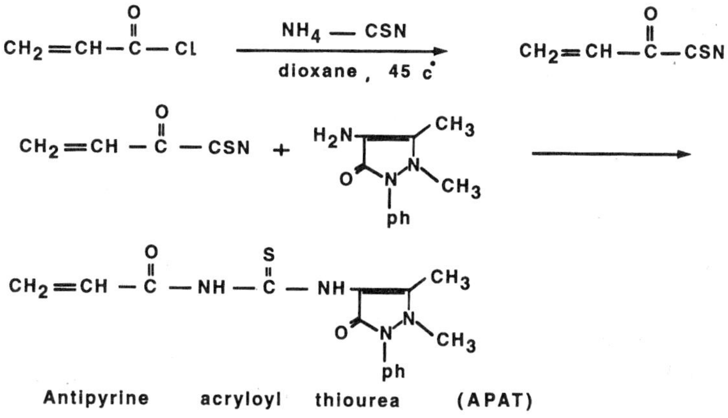

II

including scission, reaction cycle crosslinking, and decomposition at the higher temperature exotherm.

The DTA curve of PAPAT shows exotherms with maxima at 165, 175, 260, 335, and 530°C.

The DTA curves of the APAT-AN copolymers in air show exotherms at 175, 215, 245, 305, and 660°C, reflecting many processes, including scission, cyclization, crosslinking, and decomposition at the higher temperature exotherm.

REFERENCES

1. El-Hamouly, S. H.; El-Kafrawi, S. A.; Messiha, N. N. *Eur. Polym. J.* **1992**, *28*(11), 1405.
2. El-Hamouly, S. H.; El-Saied, F. A.; Ramiz, M. M. M. *Thermochim. Acta* **1989**, *153*, 237.

Antipyrine Polymers

See: Antipyrine-Acrylonitrile Copolymers (Thermal Characteristics)
Antipyryl Acrylamide (Electrochemical Polymerization)
Antipyryl Copolymers
Antipyryl Terpolymers

ANTIPYRYL ACRYLAMIDE (Electrochemical Polymerization)

S. H. El-Hamouly*
Laboratory of Polymer Chemistry
Faculty of Science
El-Menoufia University

M. Abdel Azzem and U. S. Yousef
Laboratory of Electrochemistry
Faculty of Science
El-Menoufia University

4-Aminoantipyrine and its derivatives are being used as drugs (antipyretic, analgesic,[1] and antischistosomiasis[2-4]), as well as in biomedical and biological analytical reagents.[5,6] Although there is an abundance of data on the chemistry of electropolymerization, relatively little is known about the electroinitiated polymerization of vinyl heterocyclic monomers.

This investigation focuses on the electroinitiated polymerization of a new synthetic monomer, N-antipyryl acrylamide (NAA), in the hope of discovering interesting new material properties.

PREPARATION OF NAA MONOMER

NAA monomer has been prepared either by the reaction of 4-aminoantipyrine with acryloyl chloride in dry methylene chloride or with acrylic acid in the presence of *N-N* dicyclohexylcarbodimide (DCCI) in dry methylene chloride.

Electropolymerization Procedure

The electrochemical polymerization was carried out in single and two-compartment reaction cells. Ni electrodes were used as cathodes and anodes.

*Author to whom correspondence should be addressed.

Properties of PNAA

The polymers obtained were yellowish brown powders. They were partially insoluble in most organic solvents, suggesting branched structures.

FACTORS AFFECTING THE YIELD OF POLYMERIZATION

Preliminary experiments showed that a flow of electrolytic current was necessary to initiate the polymerization. Addition of certain electrolytes to the medium not only increased the conductance of the medium but also initiated the polymerization reaction.

The effect of NAA concentration on the polymerization process is characterized by an initial fast rate that is followed by a slower one and then levels off. This was noticed regardless of NAA concentration and reaction time. The rate of polymerization (Rp) is proportional to the $[NAA]^{1.4}$.

The effect of the magnitude of the electrolytic current on polymer yield over a fixed period of electrolysis was studied. The polymer yield increased with increasing current up to 20 mA; the nickel dissolution also increased.

The increase of temperature in the range of 40–70°C enhances the yield of the electropolymer.

The experimental results indicate the enhancing effect of stirring on the polymer yield.

REFERENCES

1. Akiro, N.; Seizabura, K. *Nippon Yakurigaku Zasshi Japan* **1981**, *77*(1), 9–31.
2. Ksenga, J. N.; Wilson, H. R.; Robins, R. K. *J. Med. Chem.* **1981**, *24*, 610.
3. Faham, H. A.; Galil, F. M.; Ibraheim, Y. R.; El Nagdi, M. H. *J. Heterocyclic Chem.* **1983**, *20*, 667.
4. Kandeel, E. M.; Baghos, V. B. Mohareb, L. S.; El-Nagdi, M. H. *Arch. Pharm. (Weinheim)* **1983**, *376*, 713.
5. Kabayashi, K.; Sakugoshi, T.; Kimura, M.; Haito, K.; Matasuka, A. *Chem. Pharm. Bull.* **1980**, *28*(10), 2960.
6. Konishiroku Photo Industries Co. Ltd., Appl., 1982, 85115, 162.

ANTIPYRYL COPOLYMERS

S. H. El-Hamouly**
Laboratory of Polymer Chemistry
Faculty of Science
El-Menoufia University

M. M. Azab
Department of Chemistry
Faculty of Science
Benha University

O. A. Mansour
Microanalytical Center
Faculty of Science
Cairo University

Interest in multifunctional synthetic polymers or copolymers is steadily increasing because of their possible use as macromolecular catalysts,[1-4] macromolecular drugs (antiheparin),[5-8]

**Author to whom correspondence should be addressed.

or antimetastatic agents.[9] 4-Aminoantipyrine and its derivatives are being used as antipyretics and analgesics.[10] Unlike that of salicylates, the antipyretic effect of antipyrine derivatives does not affect the central nervous system. Antipyrine derivatives may also have considerable antischistosomiasis activity.[11-13] and herbicidal and growth-regulating activity for wheat,[14] and use as biomedical and biological analytical reagents.[15,16]

The aim of the present work is to correlate the effect of the structure of two newly prepared monomers, N-antipyryl-acrylamide (NAA) and antipyrine acryloyl thiourea (APAT), on the copolymerization parameters and the sequence distributions of the copolymers obtained.

COPOLYMERIZATION PROCEDURE

NAA or APAT copolymers with each of methyl methacrylate (MMA), acrylonitrile (AN), and vinyl acetate (VA), and NAA copolymers with each alkyl acrylate (MA, EA, BuA) and styrene (ST), were obtained by solution polymerization in DMF (2 mol%) under N_2 at 65°C in the presence of 1 mol% AIBN. The overall conversions were limited to less than 10% in every case.

PROPERTIES OF THE COPOLYMERS

The copolymer samples of NAA–MMA, NAA–MA, NAA–EA, NAA–BuA, NAA–VA, and APAT–VA are obtained as white powder soluble in most organic solvents, whereas NAA–AN, NAA–ST, and APAT–AN systems are obtained as yellowish solids soluble in DMF and DMSO.

APPLICATIONS[18-20]

The results show that the four copolymer systems of NAA–AE[MA–EA- and BuA] and NAA–ST have azeotropic composition at f_1 = 0.61, 0.69, 0.73, and 0.73, calculated from the Kelen and Tudos results.[17]

The Q and e values for NAA and APAT were calculated using the Alfrey-Price equation (1963).[21] They were found to be Q = 0.51 and e = –0.43 for NAA, and Q = 0.75 and e = 1.38 for APAT. These values are in good agreement with those reported in the literature for both the amide and thiocyanate derivatives.[22]

REFERENCES

1. Overberger, C. G.; Sitarmaiah, R.; St. Pierre, T.; Yaroslwsky, S. *J. Am. Chem. Soc.* **1965**, *87*, 3270.
2. Overberger, C. G.; St. Pierre, T.; Yaroslwsky, C.; Yaroslwsky, S. *J. Am. Chem. Soc.* **1966**, *88*, 1184.
3. Overberger, C. G.; Salamone, J. C.; Yaroslwsky, S. *J. Am. Chem. Soc.* **1967**, *89*, 6181.
4. Overberger, C. G.; Maki, H. *Macromolecules* **1970**, *3*, 220.
5. Marchisio, M. A.; Sbertoli, C.; Farina, G.; Ferruti, P. *Eur. J. Pharmac.* **1970**, *12*, 236.
6. Marchisio, M. A.; Longo, T.; Ferruti, P.; Dunusso, F. *Eur. Surg. Res.* **1971**, *3*, 249.
7. Marchisio, M. A.; Ferruti, P.; Longo, T. *Eur. Surg. Res.* **1972**, *4*, 312.
8. Marchisio, M. A.; Longo, T.; Ferruti, P. *Experienta* **1973**, *29*, 93.
9. Ferruti, P.; Danusso, P.; Franchi, G.; Plantarntti, N.; Garattini, S. *J. Med. Chem.* **1974**, *16*, 496.
10. Akiro, N.; Seizabura, K. *Nippon Yakurigaku Zasshi Japan* **1981**, *77*(1), 9.
11. Senga, K.; Novinson, J.; Wilson, H. R.; Robins, R. K. *J. Med. Chem.* **1981**, *24*, 610.
12. El-Faham, H. A.; Galil, F. M. E.; Ibraheim, Y. R.; El-Nagdi, M. H. *J. Heterocyclic Chem.* **1983**, *20*, 667.
13. Kandeel, E. M.; Baghos, V. B., Mohareb, L. S.; El-Nagdi, M. H. *Arch. Pharm. (Weinheim)* **1983**, *376*, 713.
14. Vasilev, G.; Vasilev, N. *Doki. Bolg. Akad. Nauk.* **1980**, *33*(7), 969.
15. Kabayashi, K.; Sakugoshi, T.; Kimura, M.; Haito, K.; Matssuka, A. M. *Chem. Pharm. Bull.* **1980**, *28*(10), 2960.
16. Konishiroku Photo Industries Co. Ltd. *Appl.* 85115, 192, 1982.
17. Kelen, T.; Tudos, F. *J. Macromolec. Sci. Chem.* **1975**, *9*, 1.
18. El-Hamouly, S. H.; El-Kafrawi, S. A.; Messiha, N. N. *Eur. Polym. J.* **1992**, *28*(11), 1405.
19. El-Hamouly, S. H.; Mansour, O. A. *J. Polym. Sci.* **1993**, *A.31*, 1335.
20. El-Hamouly, S. H.; Azab, M. M.; Mansour, O. A. *Polymer* **1994**, *35*(15), 3329–3331.
21. Alfrey, Jr. T.; Price, C. *J. Polym. Sci.* **1963**, *A1*, 1137.
22. Brandrup, J.; Immergut, E. H. *Polymer Handbook,* 2nd ed., John Wiley & Sons: New York, 1975, p 400.

ANTIPYRYL TERPOLYMERS

S. H. El-Hamouly
Laboratory of Polymer Chemistry
EL-Menoufia University

M. M. Azab
Department of Chemistry
Benha University

Our objective was to determine unitary, binary, and ternary azeotrope for the *N*-antipyrylacrylamide monomer (NAA) and antipyrine acryloyl thiourea monomer (APAT) terpolymers.

TERPOLYMERIZATION PROCEDURE

Ternary copolymerization of NAA and APAT was carried out with methyl methacrylate or vinyl acetate and acrylonitrile at 65°C under N_2 using 2 mol% as initiator in 2 mol/L dimethylformamide. Ternary copolymerization of NAA with different alkyl acrylate (MA-EA-BuA) or styrene were carried out similarly.

APPLICATIONS

The relationship between composition of the feed and the composition of the resulting polymer is represented in the form of triangular plots. Slocombe proposed the representation of experimental points by arrows; the heads of arrows indicate the initial terpolymer composition and the tails the composition of the feed.[1]

Terpolymer systems NAA-MA-AN, NAA-EA-AN, NAA-BuA-AN, NAA-ST-AN, and APAT-MMA-AN produced compositions represented by arrows pointing towards a well-defined point corresponding to the ternay azeotropic compositions.

The azeotropic compositions of the five terpolymer systems were polymerized for a wide range of conversions. The results indicated that the terpolymer composition of all systems was constant over a wide range of conversion proving that the monomer reactivity ratio values are reliable.

Selective feed compositions corresponding to unitary and binary azeotropy for each system were polymerized to low conversion. The relationship between the experimental and predicted compositions indicate that for each system the experimental values agree with those obtained from theoretical calculations.

REFERENCE

1. Slocombe, R. J. *J. Polym. Sci.* **1957**, *26*, 9.

Antistatic Agents

See: Additives (Property and Processing Modifiers)
Antistatic Agents, Polymeric

Antistatic Agents, Polymeric

Takayuki Ohta
Research Center
Mitsubishi Kasei Corporation

The inherent insulating properties of plastics have introduced various problems with electrostatic charge accumulation. In the plastics industry, many antistatic agents have been used to promote static discharge from plastics to reduce fabric cling, eliminate spark discharge, and prevent accumulation of dust and dirt.[1]

Recently, a new method for dissipating static using alloys and coatings has been introduced. It was originally developed to modify fibers for which mostly ionically conductive copolyethers were used as antistatic agents. These agents are grouped as follows: poly- and copolyethers such as high molecular weight polyoxyethylene, ethylene oxide-prophylene oxide copolymers, and ethylene oxide-epichloro-hydrin copolymers; polyoxyalkylene(meth)acrylate copolymers prepared by copolymerizing the monomers with such comonomers as styrene and methyl methacrylate; poly(amide-ethers) and poly(amide-esters) prepared by copolymerizing polyethylene glycol dicarboxylic acids and diamines with lactams or the diacid-diamine salts; block and graft copolymers containing such ionic groups as quaternary ammonium and sulfonate salts synthesized by reacting monomers containing ionic groups with a polymeric peroxy, azo compound, or macromonomer, ionene polymers; and others.[2-17]

Reported here is a novel antistatic agent based on a graft copolymer having both segments compatible with the matrix polymers and ionic segments. The agent has excellent antistatic properties and an anomalous formation of ion-conducting polymer networks by polymer films.

PREPARATION

The graft copolymer, PMAC-g-PMV, was synthesized (as shown in **Scheme I**).

PROPERTIES

The graft copolymer PMAC-g-PMV was amphiphilic, possessing both aliphatic polyester segments, which are compatible

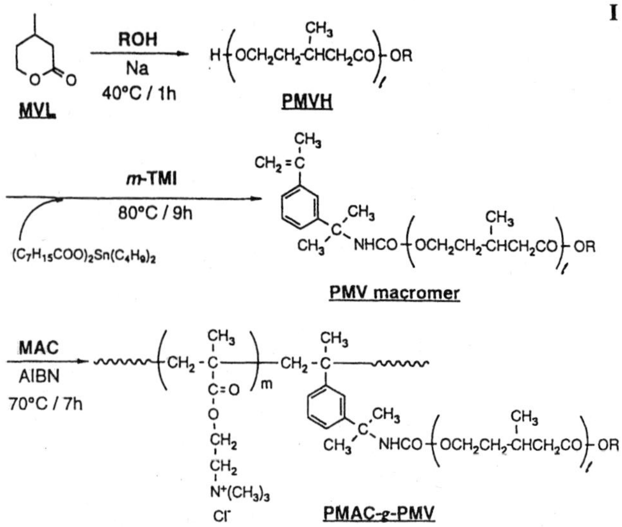

I

with matrix polymers, and ionic segments. Polymer blends from PMAC-g-PMV have transparent coatings with excellent antistatic properties. Therefore, PMAC-g-PMV can be used for antistatic modification of various polymers. However, the antistatic properties and transparent of the polymer coatings consisting of PMAC-g-PMV were affected by various factors such as the blend ratio, the graft segment's molecular weight, and the matrix polymer's properties.

APPLICATIONS

The antistatic coatings consisting of PMAC-g-PMV and various matrix polymers have many unusual properties: low surface resistance of approximately 10^8 ohm/sq, volume resistance as low as 10^9 ohm·cm, stable and superior antistatic properties that withstand changes in humidity, resistance to water, permanent antistatic properties, colorlessness and transparency, no stickiness or bleeding, and appropriate matrix polymers for substrates. Therefore, these antistatic coatings can prevent electrostatic discharge in plastic moldings such as films, sheets, plates, and forms; packages and containers for transporting and storing electronic parts; industrial uses like rollers, conveyor belts, and hoses; and office equipment such as copy machines and printers.

REFERENCES

1. Reck, R. A. *Encyclopedia of Polymer Science and Engineering,* 2nd ed.; John Wiley & Sons: New York, 1985; Vol. 2, pp 99–115.
2. Harding, J. A.; Richards, B. D. *ANTEC '90 Symposium Proceedings*; **1990**; 409.
3. Lee, B.-L. *ANTEC '90 Symposium Proceedings*; 1990; p 412.
4. Lee, B.-L. *J. Appl. Polym. Sci.* **1983**, *47*, 587.
5. Mass; T. R.; Fahey, T. E. *ANTEC '91 Symposium Proceedings*; 1991; p 551.
6. U.S. Patent 4 719 263, 1988.
7. Kokai Tokkyo Koho, Japanese Patent 01126357, 1989.
8. European Patent 124347, 1984.
9. Kokai Tokkyo Koho, Japanese Patent 63095251, 1988.
10. Kokai Tokkyo Koho, Japanese Patent 58164635, 1983.

11. Kokai Tokkyo Koho, Japanese Patent 57131253, 1982.
12. Ohmura, H.; Dohya, H.; Oshibe, Y. et al. *Kobunshi Ronbunshu* **1988**, *45*(11), 857.
13. U.S. Patent 5 082 730, 1992.
14. Ohta, T.; Sano, S.; Goto, J. et al. *ACS Polym. Prepr.* **1993**, *34*(2), 590.
15. Kokai Tokkyo Koho, Japanese Patent 03255139, 1991.
16. Kokai Tokkyo Koho, Japanese Patent 57123255, 1982.
17. German Patent 296974, 1991.

ANTITUMOR ACTIVE POLYMERS (DIVEMA-Like)

Won-Jei Cho and Chang-Sik Ha
Department of Polymer Science and Engineering
Pusan National University

The high molecular weight antitumor compounmds (HMAC) are attracting much interest from the standpoint of polymer synthesis as well as drug development, because of their high specificity of action, low toxic side effects, and longer duration of drug action when compared with low molecular weight antitumor compounds (LMAC). HMAC may be combined with large molecules in the body and be slowly released into biological fluids, giving effective concentrations over long periods. Unlike LMAC, HMAC are reported to be engulfed by cells and thereby reach reactive sites by an endocytosis process.[1]

Many polyanions exhibit antitumor activities. Among the polyanionic polymers, the 1:2 regularly alternating copolymer (DIVEMA) of divinyl ether (DVE) and maleic anhydride (MAH) first reported by Butler has been extensively studied for its structure and antitumor activity.[2-4] The structural feature of the hydrolyzed form of DIVEMA contains a carboxylic acid group as a hydrophilic part and sugar moieties such as a pyran or furan ring as a hydrophobic part. DIVEMA performs antitumor, antiviral, antibacterial, and antifungal activities and has a interferon-inducing ability. However, it also has toxicity that can cause enlargement of the liver and spleen.[5] Many attempts have been made to obtain DIVEMA-like HMAC to reduce side effects and to enhance antitumor activity.[6-9] We synthesized and characterized several antitumor-active homopolymers and copolymers of *exo*-3,6-epoxy-1,2,3,6-tetrahydrophthalic anhydride (ETA), *exo*-3,6-epoxy-1,2,3,6-tetrahydrophthalic glycinyl maleimide (ETGI), and *N*-glycinyl maleimide (GMI) by photopolymerizations or thermal radical copolymerizations.[10-14] Several terpolymers were also synthesized by radical copolymerizations of ETA with vinylacetate (VAc), methacrylic acid (MA), acrylamide (AAm), and vinylpyrrolidone (VP). These polymers exhibited good antitumor activity.

PREPARATION

The important factors for molecular design in the preparation of the new HMAC with excellent antitumor activity and very low toxicity include the proper balance of water and fat solubility to achieve good distribution in the body, proper target moiety, stereo-arrangement of atoms, and electronic characteristics to be able to react specifically with the reactive site, and proper molecular weight to be capable of elimination by the kidneys.

DIVEMA and Poly(DVE-*co*-CTA)[15]

One of the most extensively studied synthetic polyanions having antitumor activity is DIVEMA, prepared from the 1:2 mole ratio of DVE and MAH through the formation of a charge transfer complex by conventional copolymerization. The propagation occurs by means of a cyclopolymerization process.

Poly(DVE-*co*-CTA)[7]

Poly(DVE-*co*-CTA) containing 1:1 composition of DVE and citraconic anhydride (CTA) was prepared by a manner similar to that used for the synthesis of DIVEMA.

Antitumor Activity of HMAC

When LMAC are injected and/or administrated into the body, some of the drug dose reaches the disease target and treats it, but most does not. Instead, it interacts with the other parts of the body, resulting in adverse side effects.

The delivery of HMAC to specific sites within the human body for the treatment of localized diseases, with a minimum of adverse side effects, is one of the greatest challenges in chemotherapy. Target specificity in HMAC would considerably increase dose effectiveness, allowing the dosage and frequency of administration of a given HMAC to be reduced. Most LMAC lack target specificity, a limitation that is particularly evident in cancer chemotherapy. Target specificity is required for a drug-carrier system, which is able to recognize the desired target site.

The bioactivities of the synthesized HMAC can be evaluated by *in vitro* or *in vivo* tests.[17,18] *In vitro* tests commonly involve cell-culture techniques, and the primary purpose of these tests is to identify the toxic behavior of a material. *In vivo* tests obviously use implantation procedures and provide a more realistic appraisal of the behavior of the material. *In vivo* tests also permit the examination of both the local tissue response and the systemic response. HMAC to be tested are injected according to a definite schedule for several days. The survival time is checked, or the tumor removed from killed animals is weighed.

DIVEMA has received the greatest attention with respect to its antitumor activity.[19,20] DIVEMA inhibited tumors induced by the Friend leukemia virus, the Maloney sarcoma virus, the Rauscher leukemia virus, mamary carcinoma, Lewis lung carcinoma, Madision lung carcinoma, and fibrosarcomas.[21-29] It has antibacterial, antifungal, and interferon-inducing activities.[6,30-32]

Antitumor activity of DIVEMA is better than that of copolymers of DVE and substituted maleimides against L-1210 lymphoid leukemia, but the effect of substitutes on antitumor activity is not significant in *N*-alkyl or *N*-aminoacid substituted maleimides. Antitumor activity of polymers containing ETA against sarcoma 180 is better than that of 5-fluorouracil, which has been used in chemotherapy for human cancer.[10]

REFERENCES

1. Hodnett, E. M. *Polym. News* **1983**, *8*, 323.
2. Butler, G. B. *J. Polym. Sci.* **1960**, *48*, 279.
3. Butler, G. B. U.S. Patent 3 320 216, 1967.
4. Butler, G. B. *J. Macromol. Sci. Chem.* **1971**, *A5*(1), 219.
5. Ottenbrite, R. M.; Goodell, E.; Munson, A. *Polymer* **1977**, *18*, 461.
6. Butler, G. B.; Badgett, J. T.; Sharabash, M. *J. Macromol. Sci., Chem.* **1970**, *A4*, 51.

7. Scharpe, Jr., A. J. Ph.D Thesis, University of Florida, 1970.
8. Butler, G. B.; Zampini, A. *J. Macromol. Sci., Chem.* **1977**, *A11*, 491.
9. Breslow, D. S.; Edwards, E. I.; Newberg, N. R. *Nature* **1973**, *346*, 160.
10. Ha, C. S.; Cho, W. J. *Proceedings of the 6th Pusan-Kyushu Joint Symposium on High Polymers*; Miyazaki, Japan, June 4, 1993; p 51.
11. Jeong, J. G.; Lee, N. J.; Choi, W. M.; Ha, C. S.; Cho, W. J. *Proceeding of the 3rd Pacific Polymer Conf.* **1993**; p 649.
12. Lee, N. J.; Jeong, C. S.; Ha, C. S.; Cho, W. J. *Proc. 2nd IUPAC Int'l. Symp. On Biolog. Chem.*; Fukuoka, Japan, June 6-10, **1993**; p 216.
13. Lee, D. Y.; Jeung, J. G.; Lee, N. J.; Ha, C. S.; Cho, W. J. Polymer Preprints, Japan (English Edition); **1994**, *43*(5–11), E797.
14. Gam, G. T.; Lee, D. Y.; Jeong, J. G.; Lee, N. J.; Ha, C. S.; Cho, W. J. *Proceeding of the 35th IUPAC International Symposium on Macromolecules*; Akron, Ohio, U.S.A., July 11–16, 1994; p 507.
15. Butler, G. B. *JMS-REV, Macromol. Chem. Phys.* **1982–83**, *C22*(1), 89.
16. Butler, G. B.; Fujimori, K. *J. Macromol. Sci., Chem.* **1972**, *A6*, 1533.
17. Autian, J. *Artif. Organs* **1977**, *1*, 531.
18. Turner, J. E.; Lawrence, W. H.; Autian, J. *J. Biomed. Mater. Res.* **1973**, *7*, 39.
19. Breslow, D. S. *Pure Appl. Chem.* **1976**, *46*, 103.
20. Breinig, M. C.; Morahan, P. S. In *Interferon and Interferon Inducers;* Stringfellow, Marcel Dekker: New York, 1980; pp 239–261.
21. Chirigos, M. A.; Turner, W.; Pearson, J.; Griffin, W. *Int. J. Cancer* **1969**, *4*, 267.
22. Regelson, W. *Adv. Exp. Med. Biol.* **1967**, *1*, 315.
23. Chirigos, M. A. *Comparative Leukemia Research 1969, Bibl. Haemat*; Dutcher, M. Ed., Karger; Basel, Switzerland, **1970**, p. 278.
24. DeClerg, E.; Merigan, T. C. *J. Gen. Virol.* **1969**, *5*, 359.
25. Hirsch, M. S.; Black, P. H.; Wood, M. L.; Monaco, A. P. *J. Immunol.* **1972**, *108*, 1312.
26. Sandberg, J.; Goldin, A. *Cancer Chemother. Rep.* **1971** *55*, 233.
27. Morahan, P. S.; Munson, J. A.; Baird, L. G.; Kaplan, A. M.; Regelson, W. *Cancer Res.* **1974**, *34*, 506.
28. Schulz, R. M.; Papamatheakis, J. D.; Luetzeler, J.; Ruiz, P.; Chirigos, M. A. *Cancer Res.* **1977**, *37*, 358.
29. Morahan, P. S.; Kaplan, A. M. *Int. J. Cancer* **1976**, *17*, 82.
30. Pindak, F. F. *Infect. Immun.* **1970**, *1*, 271.
31. Munson, A.; Regelson, W.; Wooles, W. R. *J. Reticuloendothel. Soc.* **1969**, *6*, 623.
32. Merigan, T. C.; Finkelstein, M. S. *Virology* **1968**, *35*, 363.

Antitumor Polymers

See: *Anticancer Polymeric Prodrugs, Targetable*
Antitumor Active Polymers (DIVEMA-Like)
Antitumoral and Antiviral Polyoxometalates (Inorganic Discrete Polymers of Metal Oxide)
Biologically Active Polyanions
Nucleic Acid Interactions (with cis-platin)
Polymeric Drugs

ANTITUMORAL AND ANTIVIRAL POLYOXOMETALATES (Inorganic Discrete Polymers of Metal Oxide)

Toshihiro Yamase
Research Laboratory of Resources Utilization
Tokyo Institute of Technology

Polyoxometalates (PM), the discrete polymeric anions of early transition metal oxide, form a large class of inorganic compounds with great molecular diversity and significant potential applications in analytical chemistry, material science, catalysis, and medicine. The polyoxometalates can usually be isolated as ammonium, metallic, or organometallic salts.

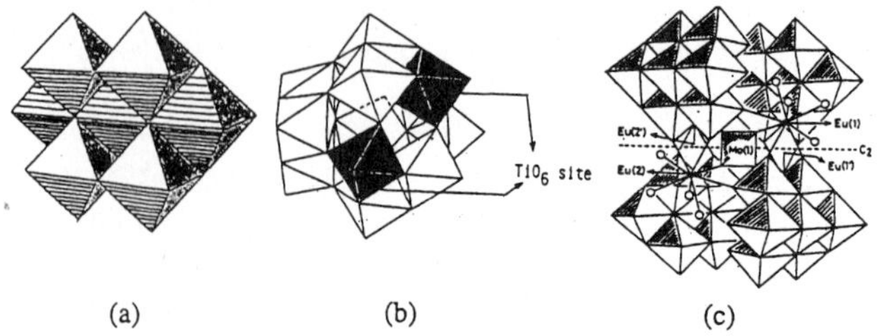

FIGURE 1. Anion structures of (a) PM-8, (b) PM-19, and (c) PM-104.

Our application of polyoxometalate chemistry to medical fields originates from our finding of *in vivo* activity of V-shaped heptamolybdate, $[Mo_7O_{24}]^{6-}$ against solid tumors.[1] Recent studies of antiviral activities have focused on finding polyoxometalates that are effective against HIV and have improved clinical safety profiles. $K_7[PTi_2W_{10}O_{40}].7H_2O$ (PM-19), $[NH_4]_{12}H_2[Eu_4(MoO_4)(H_2O)_{16}(Mo_7O_{24})_4].13H_2O$ (PM-104), Cs^+ and $[NMe_4]^+$ salts of $[(RSi)_2O]SiW_{11}O_{39}^{4-}$ (R=CH_2CH_2COMe, $(CH_2)_3CN$, and $CH{=}CH_2$), and $K_6[BVW_{11}O_{40}].xH_2O$ (E3925) were selected as promising compounds for treating HIV.[2-5]

This review concentrates on antitumoral and antiviral activities of polyoxometalates that our group found. The discussion of antitumors addresses a potent activity of polyoxomolybdates, especially, $[Pr^iNH_3]_6[Mo_7O_{24}].4H_2O$ (PM-8), against animal transplanted tumors and human cancer xenografts.[1] The antitumor activity is explained by the treated redox cycles of $[Mo_7O_{24}]^{6-}$ in tumor cells, which prevent the electron transfer in the mitochondria from generating ATP. The discussion of antiviral chemotherapy includes *in vitro* and *in vivo* antiherpes activities of polyoxotungstates, especially PM-19 and PM-104, and also evaluates their effectiveness as anti-HIV agents. The anion structures of three families of polyoxometalates discussed here are given in polyhedra notations in **Figure 1**: $[Mo_7O_{24}]^{6-}$ (V-shaped heptamolybdate), $[PTi_2W_{10}O_{40}]^{7-}$ (a representative of the common Keggin class of polyoxo-metalates), and $[Eu_4(MoO_4)(H_2O)_{16}(Mo_7O_{24})_4]^{14-}$.[6-10]

MATERIALS AND METHODS

The polyoxometalates were prepared and purified as described in the literature: $K_5[BW_{12}O_{40}].15H_2O$ (PM-1), PM-8, $[NH_3Pr^i]_6[H_{x(=1-2)}Mo_7O_{24}]$ (PM-17), PM-19, $Na_5[IMo_6O_{24}].3H_2O$ (PM-32), $K_{13}[Eu(SiW_{11}O_{39})_2]$. $30H_2O$ (PM-48), and PM-104.[1,8,10-14]

Antitumor Activity

Meth-A and MM-46

Test compounds were administered intraperitoneally (i.p.) daily from day 1 to 9 after the subcutaneous (s.c.) or i.p. implantation of tumor cells on day 0.[1] Tumor weight inhibition was estimated by measuring the length (l) and width (w) of each tumor with a venier caliper (mm) and using a formula of $lw^2/2$. An increase in life span (ILS) was calculated using ILS (in%) =

100(*t*-*c*)/*c*, where *t* and *c* were mean survival times for polyoxometalate-treated and control groups, respectively.

RESULTS AND DISCUSSION

Antitumor Activity

The i.p. administration of PM-8 significantly inhibited Meth-A sarcoma and MM-46 adenocarcinoma.[1] PM-8 administrations of 50 mg/kg provided 32% ILS for s.c. implants and 111% ILS for i.p. implants. Values of ILS for PM-8 were higher than 5-fluorouracil (5-FU) and 1-(4-amino-2-methylpyrimidin-5-yl)methyl-3-(2-chloroethyl)-3-nitrosourea (ACNU), which are clinically approved drugs effective against breast, gastrointestinal, and intracranial tumors.

Antiviral Activities

DNA Virus Inhibition

Various nucleotide analogs such as acyclovir (ACV), ganciclovir, and carbocyclic derivatives exhibit activity against herpes simplex virus type 1 (HSV-1) and type 2 (HSV-2).[16] The anti-HSV activity of these compounds largely depends on phosphorylation by the virus-encoded thymidine kinase (TK). Virus-drug resistance arises from the mutation in the viral TK gene.[17-19] The inhibitory effects of PM compounds on HSV-induced cytopathogenicity are exemplified for PM-1, PM-19, PM-48, and PM-104 where Keggin structural polyoxotungstates (PM-1 and PM-19) and their derivative (PM-48) and a polyoxomolybdate PM-104 exhibit significant antiherpes activities.[3,15]

RNA Virus, HIV Inhibition

PM compounds were also assayed for their anti-HIV-1 activities. The Keggin and dimerized deficient Keggin structures proved potent inhibitors of HIV-1. The most efficient PM compounds were PM-19 and PM-104.[3] These compounds were efficient only when added simultaneously with HIV-1 or 1 minute after infection. This result implied a membrane-related mechanism of the antiviral pathway that interfered with the early steps of HIV infection.

Keggin structured PM compounds were also active against Coxsackie type B5, EMC-M, and influenza B Lee, when $\Delta\log_{10}TCKD_{50} \geq 1$ was considered necessary to demonstrate the activity. PM 8, which was inactive against the DNA viruses tested, was active against certain RNA viruses such as Coxsackie and influenza but not EMC.

Although the antiviral mechanism of PM compounds remains unknown, they continue to be an important candidate for non-nucleoside antiviral agents.

ACKNOWLEDGMENT

I thank Yoshiko Seto and Haruhisa Fujita at the Pharmaceutical Institute, School of Medicine, Keio University; Yoshio Inouye at the Institute of Pharmaceutical Sciences, School of Medicine, Hiroshima University; and Shigeki Nishiya of Terumo Corporation for their help with the biological assays and for their comments.

REFERENCES

1. Yamase, T.; Fujita, H.; Fukushima, K. *Inorg. Chim. Acta* **1988**, *151*, L15.
2. Take, Y.; Tokutake, Y.; Inouye, Y. et al. *Antiviral Res.* **1991**, *15*, 113.
3. Inouye, Y.; Tokutake, Y.; Yoshida, T. et al. *Antiviral Res.* **1993**, *20*, 317.
4. Weeks, M. S.; Hill, C. L.; Schinazi, R. F. *J. Med. Chem.* **1992**, *35*, 1216.
5. Blasecki, J. W. *Topics in Molecular Organization and Engineering 10: Polyoxometalates: From Platonic Solids to Antiretroviral Activity*; Pope, M. T.; Müller, A., Eds.; Kluwer Academic: Dordrecht, The Netherlands, 1994; pp 373.
6. Evans, H. T., Jr.; Gatehouse, B.; Levrett, P. *J. Chem. Soc. Dalton Trans.* **1975**, 505.
7. Ohashi, Y.; Yanagi, K.; Sasada, Y. et al. *Bull. Chem. Soc. Jpn.* **1982**, *55*, 254.
8. Domaille, P. J.; Knoth, W. H. *Inorg. Chem.* **1983**, *22*, 818.
9. Ozeki, T.; Yamase, T. *Acta Crystallogr.* **1991**, *C47*, 693.
10. Naruke, H.; Ozeki, T.; Yamase, T. *Acta Crystallogr.* **1991**, *C47*, 489.
11. Yamase, T.; Ikawa, T. *Bull. Chem. Soc. Jpn.* **1977**, *50*, 746.
12. Yamase, T.; Sasaki, R.; Ikawa, T. *J. Chem. Soc. Dalton Trans.* **1981**, 628.
13. Filowitz, M.; Ho. R. K. C.; Klemperer, W. C. et al. *Inorg. Chem.* **1979**, *18*, 93.
14. Ballardini, R.; Chiorboli, E.; Balzani, V. *Inorg. Chim. Acta* **1984**, *95*, 323.
15. Fukuma, M.; Seto, Y.; Yamase, T. *Antiviral Res.* **1991**, *16*, 327.
16. De Clercq, E. *Curr. Opinion Infect. Dis.* **1991**, *4*, 795.
17. Coen, M. D.; Schaffer, P. A. *Proc. Natl. Acad. Sci. U.S.A.* **1980**, *77*, 2265.
18. Field, H. J.; Darby, G.; Wildy, P. *J. Gen. Virol.* **1980**, *49*, 115.
19. Schnipper, L. E.; Crumpacker, C. S. *Proc. Natl. Acad. Sci. U.S.A.* **1980**, *77*, 2270.

Antiviral Polymers

See: Antitumoral and Antiviral Polyoxometalates
(Inorganic Discrete Polymers of Metal Oxide)
Biologically Active Polyanions

Arabinan

See: Hemicelluloses

Arabinogalactans

See: Hemicelluloses

ARAMID FIBERS AND NONWOVENS

B. R. Bichlmeir and J. B. Rappaport
DuPont Advanced Fibers Systems

Since about 1940, nylon and other synthetic fibers have been the engine for one of the most massive material substitutions of our time: the replacement of cellulosic fibers by noncellulosic fibers also shows that another substitution, which began in the 1960s, appears to be underway: nonwovens are replacing standard woven and knit fabrics. Strong functional capabilities, coupled with attractive economics, have allowed nonwovens to be adopted in many woven and knit fabric applications, as well as in traditional paper, paperboard, and film end uses.

The success of nylon and subsequent synthetic textile fibers provided a powerful vision and direction for research efforts aimed at organic "super fibers" with two special properties: the heat resistance of asbestos and the stiffness of glass. It was envisioned that fibers with these properties could fill many market needs. Early experimental work indicated that the route to achieve this vision lay with aromatic polyamides, or aramids. These materials are similar to nylon but contain mostly aromatic carbon rings.

Aramids are classified in two types: those with flexible molecules and those with unusually rigid molecules. The flexible type can be formed into fibers without much difficulty because of its solubility in conventional organic solvents. This property led directly to the development of poly(*meta*-phenylene isophthalamide), CAS RN 24938-60-1 or MPDI, by DuPont in 1957, which was commercialized in the early 1960s as Nomex® aramid.

Aramids with rigid molecules, however, neither dissolve nor melt, precluding traditional fiber spinning. This problem finally yielded to Stephanie Kwolek, a research scientist at DuPont. In 1965, Kwolek discovered a rigid rod system that could be polymerized and solubilized under special conditions to yield a spinnable polymer. Unfortunately, this system was too expensive to be commercially viable. The task then became one of finding ingredients that could provide the best possible properties at the lowest cost while maintaining processability.

Continuing efforts eventually led DuPont to commercialize its current Kevlar® aramid composition, poly(*para*-phenylene-terephthalamide), CAS RN 2125-61-1 or PPTA. Ingredients for both *meta*-aramid and *para*-aramid fibers are derived from petroleum or natural gas feedstocks.

para-ARAMIDS

Product Forms

para-Aramids are available from fiber manufacturers in three basic forms: continuous multifilament yarn; cut and crimped fiber called staple which is usually processed into yarns and textiles; and short fibrillated fibers called pulp, which is specially engineered to have high surface area. *para*-Aramids pulp has found many applications as an asbestos replacement material. *para*-Aramid materials are also available in the marketplace as woven, knit, and nonwoven fabrics; as spun yearns, prepregs, and tapes; and in thermoplastic pellets of nylon and other resins for injection molding of wear-resistant parts.

Commercial *para*-aramid fibers have a high tensile strength that is 5–6 times that of steel on an equal weight basis. They also have high stiffness at low weight and temperature stability to greater than 400°C, depending on environmental conditions and mechanical requirements. In addition, *para*-aramids do not support combustion, do not corrode, resist attack by many chemicals, and are electrically nonconductive.

The *para*-aramid yarn family now stretches from its original, relatively high tenacity (strength), high modulus (stiffness) products to very high tenacity and very high modulus products made available through copolymer and other technologies. Yarn engineering has also led to the development of properties such as high elongation and high toughness. Today, *para*-aramids offer more of a continuum of properties, from which products can be selected and/or tailored for many applications.

meta-ARAMIDS

Product Forms

meta-Aramid fiber products are currently commercially available and from DuPont as Nomex®, from Teijin as Teijinconex®. From these fiber products, *meta*-aramids are unavailable in five basic forms: continuous, multifilament yarn; cut and crimped fiber called staple; paper; pressboard, a thick rigid three-dimensional form of paper; and spunlaced, nonwoven sheets.

meta-Aramid materials share many of the same characteristics of *para*-aramids: excellent thermal stability, inherent flame resistance, and good chemical and corrosion resistance. However, *meta*-aramid fibers, because of their chemical bonding arrangement, are softer and more textile-like than are *para*-aramid fibers. Therefore, *meta*-aramid fibers are favored for use in apparel. Mechanical toughness and excellent dielectric characteristics are additional *meta*-aramid properties that have especially high value in paper applications.

Key *meta*-aramid fiber applications include woven fabrics for protective apparel; needled felts for hot gas filtration; webs used in business machines, such as copiers; while *meta*-aramid paper applications include honeycomb structures and electrical insulation.

SPUNLACED ARAMIDS

Nonwoven spunlacing technology was pioneered by DuPont more than 30 years ago. Spunlacing involves the formation of a fiber web in which the fibers are entangled via water jet interlacing. A wide spectrum of fabrics can be made by blending different fibers, fiber deniers, and cut lengths, along with various entangling arrangements, all without the use of binders.

A broad range of 100% and blend spunlaced fabrics has been made from *meta*- and *para*-aramid fibers. Spunlaced aramid nonwovens have excellent uniformity, cover, and porosity, as well as all the thermal performance advantages of aramids.

Protective clothing applications for aramid blends include substrates for moisture barriers, quilted thermal insulation liners, and limited wear garments. Fire-blocking regulations for aircraft seating and contract furnishings led to the development of fire-blocking *para*-aramid fabrics. These spunlaced fabrics are used to cover flammable interior cushioning materials, thereby forming a barrier to inhibit flame spread.

The excellent uniformity and high-temperature stability of the nonwoven *meta*-aramids have led to their use in high-temperature filtration, as a support substrate for cartridge filters, and other specialized applications. Their good electrical insulation and mechanical properties, coupled with resin saturability, have also led to their use in electrical transformer insulation systems.

Other applications include automotive engine hood liners where nonwovens are used as fabric facing over formed fiberglass thermal insulation to enhance appearance and durability. Spunlaced aramids work well in resin composite structures to provide surface smoothness and crack resistance and have been used as an interface bonding agent to improve adhesion. Another application is as an asbestos replacement for fiber-bonded, asphalt-coated culvert pipe.

CONCLUSION

Although aramid fibers have been in existence for more than 30 years and face continuing functional competition, their diverse applications appear to position them well for broad economic expansion. The outlook for these materials is generally optimistic, with growth rates of 5–10%/year expected worldwide.

Aramid fiber development has been driven primarily by fiber products, who maintain high levels of interest and continue to invest in aramide materials. These fiber producers are working with downstream manufacturers and end users to introduce new, higher value products and systems. Although considerable growth is expected in traditional markets, the industry continues to develop new applications. For growth to continue, fundamental properties will have to be altered. Materials and applications engineering will have to continue. Should these developments continue, they will fuel the existing trends of fiber and composite substitution for metals and nonwoven substitution for wovens for the foreseeable future.

ARAMID FIBERS, HIGH STRENGTH

James A. Fitzgerald
DuPont Advanced Fibers System

For thousands of years, fibers or fibrous products have been used to increase the strength, stiffness, or performance of structural materials. From using straw to reinforce the clay in bricks to using graphite fibers in thermoset plastics for the Concorde and space shuttles, fibers have added tremendous versatility to materials available to designers.

Until 1965, modern, reinforcing industrial fibers included rayon (cellulose), nylon, polyester, steel wire, fiberglass, and to a limited extent carbon fibers. In 1971, the first high-strength aromatic polyamide (aramid) fiber was introduced commercially. This introduction indicated a new focus on high-performance fiber reinforcement.

In synthetic organic fibers, there are several parameters that influence strength and stiffness, the most important of which are polymer chain length (molecular weight) and polymer chain extension and its orientation relative to the fiber axis.

Polymer chain orientation is usually achieved by stretching (drawing) the fiber after it has been formed. The fiber is heated almost to its softening or melting point and stretched almost to its breaking point and, while still under tension, cooled down.

A significant advance was made in attainable tenacity and modulus levels from 1965–1970, when DuPont workers prepared solutions of rigid rod, all para aromatic polyamides such as poly(*p*-benzamide) (PBA) and poly(*p*-phenylene-terephthalamide) (PPTA), to spin them into fibers.[1,3] This discovery ushered in the next generation of high-tenacity, high-modulus reinforcing fibers and led to the commercialization of DuPont's Kevlar®, Akzo's Twaron, and Teijin's Technora® aramid fibers.

LIQUID CRYSTALLINE POLYMER SOLUTIONS

para-Aramid polymers, unlike most other synthetic polymers, are rigid molecules and are rodlike in solution. When these polymers are dissolved above a critical concentration, the rods spontaneously self-assemble into microscopic domains. Within the domains, the molecules align perfectly with each other, although the collections of domains themselves are randomly oriented. If the solution is sheared, the domains line up in the direction of the shear with overall molecular orientation parallel to the direction of solution flow.

The solvent used in PPTA spinning is 100% H_2SO_4. The critical concentration for PPTA to form liquid-crystalline solutions is about 10% polymer in solution. The concentration usually used for spinning is 19–20%. Because the solution is solid at room temperature and melts at 70°C, spinning temperature is normally about 80°C.

FIBER STRUCTURE

Fibers from gap-spun liquid-crystalline solutions of PPTA are highly crystalline, about 85% for the spun yarn and up to 95% for the heat-treated, high-modulus varieties. This crystallinity is facilitated by the all *para* structure of the molecules and the existence of liquid-crystalline structure before filament formation.[4] Molecular orientation is also high, with angles from 12 to 20 degrees for spun fibers and less than 12 degrees for heat-treated fibers. Because the polymer molecules are closely aligned with the fiber axis, they also hydrogen bond to the adjacent polymer chains and sheets that are radially oriented.

CHEMICAL AND THERMAL PROPERTIES

The all aromatic structure imparts significantly improved chemical and thermal stability to polyamides. The all *meta* oriented aramid, poly(*m* phenylene isophthalamide) (MPDI) is not a high-tenacity fiber in the same sense as PPTA.

SUMMARY

The introduction of high-strength aramid fibers in the 1970s significantly enhanced the potential for reinforcing materials at a time when the cost of energy was increasing very rapidly with a simultaneous increase in the value of lighter weight material substitutions. As a result, substituting fiber-reinforced plastics for metal allowed for enormous savings in weight and other benefits in hundreds of applications. New polymerization and spinning technologies have also led to several new fibers based on nonaramid polymers, many of which have higher tensile properties than aramids.[2,3] As these fibers become commercially available, the assault on heavy construction materials will become even more intense.

REFERENCES

1. Preston, J. *In Kirk-Othmer Encyclopedia of Chemical Technology*, 3rd. ed.; John Wiley & Sons: New York, 1978; Vol. 3, *Aramid Fibers*.
2. Yang, H. H. *High Strength Aromatic Fibers*; John Wiley & Sons: New York, 1989.
3. Kwolek, S. L.; Morgan, P. W.; Schaefgen, J. R. In *Encyclopedia of Polymer Science and Engineering*, 2nd. ed.; John Wiley & Sons: New York, 1987; Vol. 9, *Liquid Crystalline Polymers*.
4. Yang, H. H. *Kevlar® Aramid Fiber*; John Wiley & Sons: Chichester, England, 1993.

ARAMID FILM

Keisaku Nagasawa and Hideo Kasatani
Asahi Chemical Industry Company Limited

Aromatic polyamides (also called aramids) are noted for their high mechanical properties and thermal resistance. Commercial production of aramid fibers started before aramid films. Among these aramids, poly(*p*-phenylene-terephthalamide) (PPTA) has a rigid and symmetrical molecular structure. Therefore, PPTA has a highly ordered crystalline structure, and its products are expected to have excellent mechanical properties. PPTA fibers, Kevlar and Twaron, possess outstanding tensile strength and modulus.

High-performance polymer films are required for highly technological electronic, electric, and packaging applications. The mechanical and thermal properties of films are needed in electronics to produce smaller equipment with materials performing the same functions. Poly(ethylene terephthalate) (PET) films are widely used for electrical and electronic applications, but the melting point is too low for applications that involve high temperatures.

Among several heat-resistant polymer films, aramid films possess the highest thermal properties, solvent resistance, and dimensional stability.

Aramids have been candidates for high-performance films and PPTA especially is expected to provide a film with excellent mechanical and thermal properties. Unfortunately, poor solubility and solution characteristics disrupt PPTA commercial production. Asahi Chemical Industry, Co. Ltd. recently developed a new process to produce an isotropic film, Aramica®, from PPTA anisotropic solution.[1,2]

LIQUID-CRYSTALLINE STATE

PPTA is insoluble in most organic solvents and soluble only in strong acids such as concentrated sulfuric, chlorosulfuric, or fluorosulfuric acids. Concentrated solutions of PPTA show a liquid-crystalline state. In a liquid-crystalline state, polymers have a characteristic flow, and the aggregation of molecules differs from that of the isotropic state. In a liquid-crystalline state, polymers are oriented easily along the flow's direction by extrusion of polymer melts or dopes. The flow-induced molecular arrangement has a big advantage for fiber formation but a serious disadvantage for attaining biaxial orientation during the film formation process.

FORMATION PROCESS

When the sulfuric acid solution was 10–15% of the polymer concentration, the solution was extruded through a slit die onto a flat plate. This dope was optically anistropic at the setting temperature. The apparent solution viscosity was lower than that of the optically isotropic dope at the same polymer concentrations. The cast dope passed through a conditioning chamber where the temperature and humidity were adjusted. In the chamber, the optically anisotropic-to-isotropic-phase transition was generated by varying the dope's temperature or water content. The isotropic dope obtained was immersed into the coagulation bath, washed, and then dried. Drying was carried out while holding the film's width and length constant. Uniaxial or biaxial stretching can be performed when the film is wet, if necessary. Thermal annealing remarkably improved the film properties. This process can be designed as a continuous part of an industrial scale process.

The obtained film had isotropically high tensile properties as shown by similar stress–strain curves in shear and transverse directions.

PROPERTIES

Aramica is a trade name of the PPTA film produced commercially by the method reported. It is light yellow and perfectly transparent. Transmission of visible rays is better than most heat-resistant films including polyimides.

Mechanical

Aramica has excellent mechanical properties. Tensile strength and modulus are superior to those of any other plastic films now available, and resistance to tear is considerable. Therefore, thinner films can substitute for other plastic films. Moreover, Aramica can reinforce composite materials.

Thermal

Aramica has no melting temperature, and no serious decline in formation or properties occurs after brief exposure to temperatures of 400°C. In addition to its mechanical properties at ambient temperatures, Aramica features excellent mechanical properties at elevated temperatures. Furthermore, metal or ceramic layers can be deposited on Aramica by evaporation or the sputtering method at high temperatures.

Chemical

Aramica is insoluble in all organic solvents and in most inorganic chemicals. It dissolves in only high concentrations of strong acids. Aramica's most remarkable feature is its resistance to alkaline solution, in which polyimide films degrade quickly.

Aramica loses weight from thermal decomposition in temperatures above 400°C, below that, only water vapor is evaporated. This feature is advantageous for the metal evaporation or sputtering process because volatile organic substances such as residues from organic solvents or oligomers evaporate from most other organic films and can interfere with the adhesion of both layers. Permeability of oxygen and nitrogen gas is very small, so Aramica is available for cover and barrier films.

Electrical

Besides its general electrical properties for insulating films. Aramica shows excellent tracking resistance. Tracking is the permanent conductive path generated by the film's continual carbonization when the current leaks because of surface contamination.

APPLICATIONS

Electrical and Electronic

Aramica is used in flexible printed circuits; cover films; masking, pressure sensitive, film carrier, and electrical insulation tapes; and other electrical or electronic applications.

Magnetic Recording Media

Among several data recording systems, data storage tapes are expected to be important recording media for computer back-up systems. In recording tapes, poly(ethylene terephthalate) has been used extensively for base films. But increased tape capacity will be indispensable for future information systems and will require a thinner base film that can endure high temperatures. Aramica is expected to be an optimum base film for data storage tapes.

Composing Materials

Novel composite materials reinforced by Aramica film are currently used in fishing rods and golf club shafts. Epoxy resin-coated Aramica tapes ("prepreg tapes") are used to create the composite materials by laminating and bonding the tapes with a fiber-reinforced resin layer.

Acoustic Materials

Aramica has excellent characteristics for acoustic materials, that is, a high sound velocity from a high tensile modulus and dissipation. Press-formed acoustic diaphragms and voice coil bobbins are used for high-performance audio speakers.

REFERENCES

1. Fujita, T.; Fujiwara, T.; Sata, E. et al. *Polym. Eng. Sci.* **1989**, *29*, 1237.
2. Kasatani, H.; Fujita, T.; Fujiwara, T. et al. *Sen'i Gakkaishi* **1992**, *48*, 1.

Aramids

See: Aramid Fibers and Nonwovens
Aramid Fibers, High Strength
Aramid Film
Aramids (Processable Precursors)
Aramids and Aromatic Polyimides, Soluble (Bulky Substituents)
Carbonylation Polymerization (Palladium-Catalyzed)
Condensation Polymers (Synthesis from Silyated Diamines)
Cyclic Oligomers of Engineering Thermoplastics
Fibers, Synthetic (Unspun Natural Pulp-Like)
Macrocyclic Aramids
Polyisophthalamides (Thermal Stability Enhancement by Pendant Groups)
Polyisophthalamides (Water-Uptake Enhancement)
Poly[N,N′-(sulfo-p-phenylene) terephthalamide]
Textile Fibers (Structure and Properties)

ARAMIDS (Processable Precursors)

Issifu I. Harruna and Kofi B. Bota
Department of Chemistry
High Performance Polymers and Ceramics Center
Clark Atlanta University

Aramids are high-temperature polymers. The principal properties required for high-temperature polymers include retaining structural integrity and dimensional, chemical, and oxidative stability at elevated temperatures, and tractability under processing conditions. In aramids, an inverse relationship exists between thermal and solvent resistance and processability. Although meritorious, these approaches are also self-defeating in that they compromise the desired high solvent resistance and thermal and thermooxidative stability of aramids. Techniques for synthesizing processable precursors or prepolymers to aramids, which could be fabricated into films or fibers, should alleviate these problems.

SYNTHESIS, PROPERTIES, AND APPLICATIONS

Solution and melt processable precursors to aramids can be subsequently converted to aramids with better thermal and solution stability through thermal, photochemical, or chemical means using the following techniques: rearrangement reactions, reducing or eliminating aromatic ring pendant groups, dealkylating *N*-alkyl substituents, ring-opening of macrocyclics, dehydrohalogenation or dehydroacetoxylation followed by retro Diels–Alder reaction, crosslinking via additional reactions, doping followed by polymerization accelerators, and decomplexation of transition metal π-complexes.[1-36]

ACKNOWLEDGMENT

This work was supported in part by the U.S. Army Research Office, under Grant No. DAAL03-90-G-0190 and by the Office of Naval Research under Grant No. N00014-91-3-1643.

REFERENCES

1. Naruchi, K.; Tanaka, S.; Takemori, T. et al. *Makromol. Chem., Rapid Commun.* **1986**, *7*, 607.
2. Shigeyoshi, H. CA 97322, *107*, 1987.
3. De Abajo, J.; Guijarro, E.; Serna, F. J. et al. *J. Polym. Sci., Polym. Chem. Ed.* **1986**, *24*, 483.
4. De La Campa, J. G.; Guijarro, E.; Serna, F. J. *Eur. Polym. J.* **1985**, *21*, 1013.
5. Memeger, W. U.S. Patent 4 178 419, 1979.
6. Memeger, W. U.S. Patent 4 226 949, 1980.
7. Memeger, W.; Lazar, J.; Overnall, D. et al. *Macromolecules* **1993**, *26*, 3476.
8. Guggenheim, T. L.; McCormick, S. J.; Guiles, J. W. et al. *Polym. Prepr., Am. Chem. Soc. Div. Polym. Chem.* **1989**, *30*(2), 138.
9. Guggenheim, T. L.; McCormick, S. J.; Kelly, J. J. et al. *Polym. Prepr., Am. Chem. Soc. Div. Polym. Chem.* **1989**, *30*(2), 579.
10. Harruna, I. I.; Bota, K. B.; McLamore, S. D. *Polymer* **1993**, *34*(15), 3328.
11. Harruna, I. I.; Bota, K. B.; Akinseye, T. et al. *Polym. Prepr., Am. Chem. Soc. Div. Polym. Chem.* **1993**, *34*(1), 453.
12. Grin, D. L.; Conticello, V. P.; Grubbs, R. H. *Polym. Mater. Sci. Eng.* **1992**, *67*, 87.
13. Mikroyannidis, J. A. *J. Polym. Sci., Polym. Chem. Ed.* **1992**, *30*, 2371.
14. Tsuyoshi, K.; Tsutsumi, N.; Okada, H. *Polymer* **1992**, *33*(23), 4990.
15. Yu, L.; Han, W.; Bao, Z. *Macromolecules* **1992**, *25*, 5609.
16. Miller, W. T.; Ringsdorf, H. *Macromolecules* **1990**, *23*, 2825.
17. Sankaran, V.; Lin, S.-C.; Marvel, C. S. *J. Polym. Sci., Polym. Chem. Ed.* **1980**, *18*, 495.

18. Sankaran, V.; Marvel, C. S. *J. Polym. Sci., Polym. Chem. Ed.* **1980**, *18*, 1835.
19. Sankaran, V.; Marvel, C. S. *J. Polym. Sci., Polym. Chem. Ed.* **1980**, *18*, 1821.
20. Sankaran, V. Chemical Abstracts 218740v, 96, 1982.
21. Marvel, C. S. Chemical Abstracts 200735n, 96, 1982.
22. Takeichi, N.; Kobayashi, A.; Takayama, Y. *J. Polym. Sci., Polym. Chem. Ed.* **1992**, *30*, 2645.
23. Hosogane, T.; Takiyama, E. Chemical Abstracts 129843v, 116, 1992.
24. Hosogane, T.; Takiyama, E. Chemical Abstracts 130434u, 116, 1992.
25. Hosogane, T.; Hiroshi, N.; Takiyama, E. Chemical Abstracts 151636x, 117, 1992.
26. Hosogane, T.; Takiyama, E. Chemical Abstracts 93802u, 115, 1991.
27. Hosogane, T.; Takiyama, E. Chemical Abstracts 30140f, 115, 1991.
28. Spilman, G.; Jones, M.-C.; Mavkoski, L. J. et al. *Abstracts of Papers*, 35th International Union of Pure and Applied Chemistry International Symposium on Macromolecules, Akron, OH: University of Akron: Akron, OH, 1994; 652.
29. Swedo, R.; Marvel, C. S. *J. Polym. Sci., Polym. Chem. Ed.* **1978**, *16*, 2711.
30. Diakoumakos, C. D.; Mikroyannidis, J. A. *Polymer* **1993**, *34*(10), 2227.
31. Korshak, V. V.; Kutepov, D. F.; Pankrotov, V. A. et al. *Vyskomol. Soedin* **1974**, *16*(3), 156.
32. Korshak, V. V.; Vinogradova, S. V.; Pankrotov, V. A. et al. Chemical Abstracts 217321t, 97, 1982.
33. Park, J. J.; Rhim, M. S.; Park, S. B.; et al. Chemical Abstracts 8664w, 112, 1990.
34. Park, J. J.; Rhim, M. S.; Park, S. B. Chemical Abstracts 155774q, 111, 1989.
35. Dembek, A. A.; Burch, R. R.; Feining, A. E. *Polym. Prepr., Am. Chem. Soc. Div. Polym. Chem.* **1993**, *34*(1), 171.
36. Davis, R.; Kane-Magurie, L. A. P. In *Comprehensive Organometallic Chemistry*; Wilkinson, G. et al., Eds.; Permagon: Oxford, 1982; Vol. 3, Chapter 26.2.

ARAMIDS AND AROMATIC POLYIMIDES, SOLUBLE (Bulky Substituents)

Yoshio Imai and Ki Hong Park
Department of Organic and Polymeric Materials
Tokyo Institute of Technology

Aromatic polyamides (aramids) and aromatic polyimides have been accepted as high-performance materials exhibiting high-thermal stability and excellent mechanical properties.[1,2] Hence, they are of great technological importance. However, their technological applications are limited because of infusibiliy at high glass transition temperatures (T_g) or high melting temperatures (T_m) and insolubility in organic solvents.

Numerous efforts have been made over the past three decades to produce soluble or processable thermoplastic aramids and aromatic polyimides. These approaches involved angular linkages or flexible spacers, kinked or asymmetrical structure, bulky substituents such as phenyl group into polymer backbone, and the disruption of a repeating unit through copolymerization.[3,4] The *N*-alkylation or *N*-phenylation of aramids was also promising in order to weaken hydrogen bonding between polyamide main chains.[5-7]

Although all the methods improved solubility in organic solvents, some approaches drastically reduced the high decomposition temperatures and T_g values that are important thermal properties of aramids and polyimides. One successful approach, with a good balance of properties, was to introduce bulky substituents. These bulky groups reduced polymer cohesive energy and allowed large free volume in solution. Therefore, the amorphous aramids and polyimides containing bulky pendant groups exhibited good solubility while maintaining high T_g values. These pendant groups contained phenyl, naphthyl, fluorene, benzoyl, and trifluoromethyl units. The phenyl group especially worked well because of its efficient bulkiness and good thermal stability.

Our strategy for producing soluble aromatic polymers while simultaneously maintaining their high T_g and high thermal stability was to introduce bulky phenyl pendants. In addition, introducing crank and twisted non-coplanar structures, such as biphenyl-2,2′-diyl and 1,1′-binaphthyl-2,2′-diyl units, was also successful for preparing polymers with a good balance of properties.[8-24]

In conclusion, we confirmed that bulky phenyl substituents, as well as introducing crank and twisted non-coplanar structure into the polymer backbone, produced soluble aramids and polyimides with high glass transition and decomposition temperatures. These soluble aramids and polyimides are promising, new materials with many useful applications.

REFERENCES

1. Cassidy, P. E. *Thermal Stable Polymers*; Marcel Dekker: New York, 1980, pp 67–140.
2. Yang, H. H. *Aromatic High Strength Fibers*; Wiley-Interscience: New York, 1989, pp 66–289 and 673–795.
3. Arnold, C. A.; Summers, J. D.; Chen, Y. D. et al. *Polymr* **1989**, *30*, 986.
4. Shiang, W. R.; Woo, E. P. *J. Polym. Sci., Part A: Polym. Chem.* **1993**, *31*, 2081.
5. Greenwood, T. D.; Kahley, R. A.; Wolfe, J. F. *J. Polym. Sci., Polym. Chem. Ed.* **1980**, *18*, 1047.
6. Burch, R. R.; Manring, L. E. *Macromolecules* **1991**, *24*, 1731.
7. Kakimoto, M.; Oishi, Y.; Imai, Y. *Makromol. Chem. Rapid Commun.* **1985**, *6*, 557.
8. Imai, Y.; Maldar, N. N.; Kakimoto, M. *J. Polym. Sci., Polym. Chem. Ed.* **1984**, *22*, 2189.
9. Imai, Y.; Maldar, N. N.; Kakimoto, M. *J. Polym. Sci., Polym. Chem. Ed.* **1985**, *23*, 1797.
10. Jeong, H. J.; Kobayashi, K.; Kakimoto, M. *Polym. J.* **1994**, *26*, 99, 373.
11. Park, K. H.; Kakimoto, M. M.; Imai, Y. *J. Polym. Sci., Part A: Polym. Chem.* **1995**, *33*, 1031.
12. Jeong, H. J.; Oishi, Y.; Kakimoto, M. et al. *J. Polym. Sci., Part A: Polym. Chem.* **1990**, *28*, 3293.
13. Jeong, H. J.; Oishi, Y.; Kakimoto, M. et al. *J. Polym. Sci., Part A: Polym. Chem.* **1991**, *28*, 39.
14. Jeong, H. J.; Kakimoto, M.; Imai, Y. *J. Polym. Sci., Part A: Polym. Chem.* **1991**, *29*, 767, 1691.
15. Oishi, Y.; Zhang, X.-M.; Kakimoto, M. et al. *Polym. Prepr. Jpn.* **1990**, *39*, 799.

16. Oishi, Y.; Zhang, X.-M.; Kakimoto, M. et al. Presented at the *59th* Japanese Chemical Society, *Spring Annual Meeting,* Japan; 1990, 1008,
17. Xie, M.; Oishi, Y.; Kakimoto, M. et al. *Polym. Prepr. Jpn.* **1988**, *37*, 2966.
18. Xie, M.; Oishi, Y.; Kakimoto, M. et al. *J. Polym. Sci., Part A: Polym. Chem.* **1991**, *29*, 55.
19. Oishi, Y.; Takado, H.; Yoneyama, M. et al. *J. Polym. Sci., Part A: Polym. Chem.* **1990**, *28*, 1763.
20. Oishi, Y.; Ishida, M.; Yoneyama, M. et al. *J. Polym. Sci., Part A: Polym. Chem.* **1992**, *30*, 1027.
21. Yamashita, M.; Kakimoto, M.; Imai, Y. *J. Polym. Sci., Part A: Polym. Chem.* **1993**, *31*, 1513.
22. Liou, G.-S.; Maruyama, M.; Kakimoto, M. et al. *J. Polym. Sci., Part A: Polym. Chem.* **1993**, *31*, 2499, 3273.
23. Liou, G.-S.; Kakimoto, M; Imai, Y. *J. Polym. Sci., Part A: Polym. Chem.* **1993**, *31*, 3265.
24. Liou, G.-S.; Maruyama, M.; Kakimoto, M. et al. *Polym. Prepr. Jpn.* **1994**, *43*, 1102.

Arborols

See: Highly Branched Polymers

AROMATIC HYDROCARBON-BASED POLYMERS

Christos P. Tsonis
Chemistry Department
King Fahd University of Petroleum and Minerals

One method for improving polymer properties in high temperatures is to incorporate *p*-phenylene rings into the polymer chain. These aromatic rings hinder rotation of the chemical bonds along the chain to increase the polymer's stiffness, crystallinity and, hence, thermal stability. Candidates that may meet these requirements are polyphenylenes, poly(*p*-xylylene)s, poly(phenlylene vinylene)s, and polybenzyls.

SYNTHESIS

Polyphenylenes

Direct Oxidative Coupling

This method is the most common for synthesizing polyphenylenes. It involves the polycondensation of benzene or arene rings through direct oxidative coupling of aromatic C-C bonds in the presence of a metal–copper chloride catalytic system.[1,2] If toulene or chlorobenzene is substituted for benzene, the resulting polyphenylene has 1,2-linkages. The polymer structure consists of 1,2- and 1,4-phenylene units, and the average chain length contains 10–15 phenylene residues. This method's drawback is the difficulty removing $CuCl_2$ from the polymer product.

Grignard Method

The Grignard reaction technique is also used to synthesize polyphenylenes.[3,4] For example, 1,4-dibromobenzene reacts with an equivalent amount of magesium in tetrahydrofuran, followed by treatment with an organometallic compound such as $Ni(acac)_2$ or $NiCl_2(bipy)$ to produce polyphenylene. The molecular weight of the resulting polymer is about 800 because the product precipitates out of solution as it forms and does not allow long-chain growth to occur.

Dehydrogenation of Cyclohexadiene

A relatively inexpensive method for preparing polyphenylene is the anionic polymerization of 1,3-cyclohexadiene by *n*-butyl lithium.[5,6] The unsaturated polymeric olefin is treated with bromine, followed by dehydrobromination with molten sodium or potassium hydroxides to give the desired polyphenylene product.

Poly(p-xylylene)s

Industrially, poly (*p*-xylylene) is prepared using Gorham's method.[8] This technique requires that di-*p*-xylylene is heated at 650°C in a vacuum to form *p*-xylylene. The *p*-xylylene intermediate is then cooled to 30°C by contact with a cold surface, and polymerization occurs through a diradical process.

Poly(phenylene vinylene)

The synthesis of poly(*p*-phenylene vinylene) has been investigated many times because this polymeric material has tremendous potential in electronic applications.[9,10] A successful way to achieve high-molecular-weight poly(*p*-phenylene vinylene) and its copolymers has been base-initiated polymerization of bis-sulfonium salt from *p*-xylene, followed by thermal elimination of dialkyl sulfide and hydrogen halide.[11-25]

This technique forms high-molecular-weight polymers because the formed poly(*p*-xylylene sulfornium salt) is soluble in the polymerization reaction and allows the polymer chain to grow. Homopolymers and copolymers are prepared using this method and are classified as liquid-crystalline, main-chain polyhydrocarbons.[15,19,20,26] Laser techniques have also been used successfully to induce the direct transformation of some poly(phenylene vinylene) precursor polyelectrolytes into conjugated poly(phenylene vinylene).[9,10,25]

Polybenzyls

Unsubstituted Benzyl Halides

The polycondensation of benzyl halides in the presence of a Lewis acid catalyst leads to a family of polymers known as polybenzyls.[27] Benzyl chlorides and bromides are generally used because they are inexpensive. Low and high-temperature catalytic systems are used to polymerize benzyl chloride and its derivatives.[28-39]

Regardless of polymerization temperature and catalyst used, the polybenzyl synthesized from the self-polymerization of benzyl halides are amorphous and highly branched with low molecular weight, and a low melting point.[27,40] These properties develop because the polymerization proceeds using an electrophilic polyalkylation mechanism.[40]

APPLICATIONS

Polyphenylene is used as a coating material in the packaging industry to protect integrated circuits from breakage, humidity, and corrosion.[41] Although polyphenylene has no conductivity in its pure state, it becomes a good electrical conductor when *n*- and *p*-doped methods are used. Polyxylylene has excellent mechanical and electrical insulating behavior between –200 and

+200°C.[7,42] Consequently, it has been found in many applications in the coating industry. In particular, polyxylylene is used extensively in electronics for coating and encapsulating various types of circuitry to protect them from their surroundings. There has been considerable interest in poly(phenylene vinylene) materials because of their electrical, thermal, and photo characteristics and because they can be doped with either oxidizing or reducing agents to produce films with good electrical conductivity.[25,43,44] Polybenzyls are still at experimental stages without any commercial applications.

REFERENCES

1. Kovacic, P.; Jones, M. B. *Chem. Rev.* **1987**, *87*, 357.
2. Kovacic, P.; Wu, C. *J. Polym. Sci.* **1960**, *47*, 448.
3. Yamamoto, T.; Yamamoto, A. *Chem. Lett.* **1977**, 353.
4. Yamamoto, T.; Hayashi, Y.; Yamamoto, A. *Bull. Chem. Soc. Jpn* **1978**, *51*, 2091.
5. Cassidy, P. E.; Marvel, C. S. *Macromol. Synth.* **1972**, *4*, 7.
6. Cassidy, P. E. *Thermally Stable Polymers*; Marcel Dekker: New York, 1980; Chapter 2.
7. Gorham, W. F.; Niegish, W. D. *Encyclopedia of Polymer Science Technology*; 1971, Vol. 15, p 98.
8. Gorham, W. F. *J. Polym. Sci.* **1966**, *A-14*, 3027.
9. Taguchi, S.; Tanaka, T. European Patent 261991, 1988.
10. Yoshimo, K.; Kuwabara, T.; Iwasa, T. et al. *Jpn. J. Appl. Phys.* **1990**, *29*, L1514.
11. Wessling, R. A.; Zimmerman, R. G. U.S. Patent 3 401 152, 1968.
12. Wessling, R. A.; Zimmerman, R. G. U.S. Patent 3 706 677, 1972.
13. Wessling, R. A. *J. Polym. Sci., Polym. Symp.* **1986**, *72*, 55.
14. Gagnon, D. R.; Capistan, J. D.; Karasz, F. E. et al. *Polym. Bull.* **1984**, *12*, 293.
15. Antoun, S.; Gagnon, D. R.; Karasz, F. E. et al. *Polym. Bull.* **1986**, *15*, 181.
16. Granier, T.; Thomas, E. L.; Gagnon, D. R.; Karasz, F. E. et al. *J. Polym. Sci., Polym. Phys. Ed.* **1986**, *24*, 2793.
17. Murase, I.; Ohnishi, T.; Noguchi, T. U.S. Patent 4 528 118, 1985.
18. Murase, I.; Ohnishi, T.; Hiroka, M. *Polym. Commun.* **1984**, *25*, 327.
19. Lenz, R. W.; Han, C.-C.; Stenger-Smith, J. et al. *J. Polym. Sci., Polym. Chem. Ed.* **1988**, *26*, 3241.
20. Memeger, W., Jr. *Macromolecules* **1989**, *22*, 1577.
21. Singh, B. P.; Prasad, P. N.; Karasz, F. E. *Polymer* **1988**, *29*, 1940.
22. Liang, W. P.; Lenz, R. W.; Karasz, F. E. *J. Polym. Sci., Polym. Chem. Ed.* **1990**, *28*, 2867.
23. Garay, R. O.; Lenz, R. W. *J. Polym. Sci., Polym. Chem. Ed.* **1992**, *30*, 977.
24. Denton, F. R.; Sarker, A.; Lahti, P. M. et al. *J. Polym. Sci., Polym. Chem. Ed.* **1992**, *30*, 2233.
25. Torres-Filho, A.; Lenz, R. W. *J. Polym. Sci., Polym. Phys. Ed.* **1993**, *31*, 959.
26. Karasz, F. E.; Capistran, J. D.; Gagnon, D. R. et al. *Mol. Cryst. Liq. Cryst.* **1985**, *118*, 327.
27. Tsonis, C. P. *Comprehensive Polymer Science* Allen, G.; Bevington, J. C., Eds.; Pergamon: Oxford, England, 1989; Vol. 5, Chapter 26, p 455.
28. Kenny, J. P.; Isaacson, R. B. *J. Macromol. Chem.* **1946**, *1*, 541.
29. Montando, G.; Bottino, F.; Caccamese, S. et al. *J. Polym. Sci.* **1970**, A-1, *8*, 2453.
30. Montando, G.; Finocchiaro, P.; Caccamese, S. et al. *J. Polym. Sci.* A-1, **1970**, *8*, 2475.
31. Grassie, N.; Meldrum, I. G. *Eur. Polym. J.* **1971**, *7*, 629.
32. Kuo, J.; Lenz, R. W. *J. Polym. Sci., Polym. Chem. Ed.* **1976**, *14*, 2749.
33. Sharpless, T. W.; Stevens, M. J. *J. Polym. Sci., Polym. Chem.* **1993**, *31*, 1075.
34. Spanier, E. J.; Caropreso, F. E. *J. Polym. Sci.* **1969**, A-1, *7*, 2679.
35. Hiro, M.; Arasta, K. *Chem. Lett.* **1979**, *9*, 1141.
36. Tsonis, C. P.; Hassan, M. U. *Polymer* **1983**, *24*, 707.
37. Tsonis, C. P. *Poly. Bull. (Berlin)* **1983**, *9*, 349.
38. Tsonis, C. P. *J. Mol. Catal.* **1985**, *33*, 61.
39. Baumberger, T. R.; Woolsen, N. F. *J. Polym. Sci., Polym. Chem. Ed.* **1992**, *30*, 1717.
40. Tsonis, C. P. *J. Mol. Catal.* **1990**, *57*, 313.
41. Cowie, J. M. G. *Polymer: Chemistry and Physics of Modern Materials*, 42nd ed.; Blackie and Son Ltd: London, 1991, p 418.
42. Rodriguez, F. *Principles of Polymer Systems* 3rd ed.; Hemisphere: New York, 1989, p 354.
43. Taguchi, S.; Tanaka, T. European Patent 2 619 91, 1988.
44. Yoshimo, K.; Kuwabara, T.; Iwasa, T. et al. *J. Appl. Phys. (Japan)* **1990**, *29*, L1514.

Artificial Antigens

See: Polyelectrolyte Complexes (In Immunology)

Artificial Blood

See: Blood Compatible Polymers
Blood Substitutes
Red Blood Cell Substitutes, Artificial (for Transfusion)

Artificial Cells

See: Microcapsules (From Two Dimension Networks)
Microencapsulation (Artificial Cells)

Artificial Cornea

See: Cornea, Artificial (Hydrophilic Polymeric Sponges)

Artificial Glycoproteins

See: Glycoproteins (Synthesis by NCA Method)

Artificial Heart

See: Blood Compatible Polymers

Artificial Heparinoids

See: Heparinoids, Artificial (Biological Activities)

Artificial Ligaments

See: Ligament Replacement, Artificial
Ligament Replacement Polymers (Commercial Products)
Ligament Replacement Polymers (Technological and Design Aspects)

Artificial Lung

See: Lung, Artificial

Artificial Metalloenzymes

See: Metalloenzymes, Artificial (Polyethylenimine-Based)

Artificial Nerves

See: Regeneration Templates, Artificial Skin and Nerves

Artificial Organs

See: Hybrid Artificial Organs

Artificial Polysaccharides

See: Polysaccharides, Synthetic (Hypoglycemic Activity)
Polysaccharides, Synthetic (Preparation and Biological Activities)

Artificial Skin

See: Regeneration Templates, Artificial Skin and Nerves

Artificial Vitreous Body

See: Vitreous Body, Artificial

Associated Polymers

See: Ionomers (Solution Behavior)
Polyampholytes (Overview)
Polyampholytes (Properties in Aqueous Solution)
Poly [N,N′-(Sulfo-p-Phenylene)terephthalamide]
Polyzwitterions (Overview)
Zwitterionic Polymers (Dilute Solution and Rheological Properties)

Asymmetric Polymerization

See: Asymmetric Polymerization (Overview)
Cyclopolymerization (Cyclization Efficiency and Configuration Control)
Cyclopolymerization, Enantioselective Cationic
Optically Active Polymers (Overview)
Topochemical Photopolymerization (Types of Topochemical Behavior)

ASYMMETRIC POLYMERIZATION (Overview)

Tamaki Nakano and Yoshio Okamoto
Department of Applied Chemistry
School of Engineering
Nagoya University

There are two ways to synthesize optically active polymers: one is to polymerize monomers having optically active moiety (e.g., vinyl monomers with an optically active side group) and the other is to introduce chirality to the polymer structure using asymmetric polymerization. Here, asymmetric polymerization is described.

INDUCING CONFIGURATIONAL CHIRALITY

It is difficult to obtain an optically active polymer from a simple vinyl monomer such as styrene and methyl methacrylate; for instance, an isotactic or syndiotactic vinyl polymer cannot be chiral because a mirror plane is present. Even when some chiral induction occurs through polymerization, an isotactic or syndiotactic vinyl polymer only shows optical activity from the asymmetric center near the chain ends. Chiral configuration of a vinyl polymer is possible for hexad or longer stereosequences.[1] In addition, the extent of chiral induction has not been determined in most cases for the polymers having configurational chirality.

Polydienes and Poly(cyclic olefin)s

The first successful example of configurational chiral induction was the polymerization of 1,3-diene monomers using optically active organolithium initiators and the polymerization of 1,3-pentadiene using titanium, aluminum, and vanadium complexes.[2]

Benzofuran gave an optically active polymer by polymerization with optically active aluminum complexes.[2] The polymer did not have mirror planes regardless of the tacticity.

Polymethacrylates

An optically active dimethacrylate monomer cyclopolymerizes with a radical initiator.[3] The poly(methyl methacrylate) (PMMA) derived from the cyclopolymer was rich in heterotacticity and showed optical activity and CD absorption. The chiroptical properties could be based on some chiral configuration of the polymer chain.

Polystyrene

In radical copolymerization of an optically active distyryl monomer with styrene, the monomer cyclizes to make chiral racemo units based on the chiral side group in the copolymer chain.[4]

Polyolefin

1,5-Hexadiene has been polymerized with an optically active zirconium catalyst.[5] This polymerization proceeds via a cyclization mechanism to afford a polymer with 68% trans structure. The polymer showed molecular rotation, $[\Phi]_{405}$ −49°.

INDUCING CHIRAL CONFORMATION

Polymethacrylates and Related Polymers

Poly(methyl methacrylate) has a random conformation in solution. However, one can get a rigid polymethacrylate with helical conformation when a monomer with a bulky ester group is used for polymerization. Steric repulsion between the bulky side groups suppresses the internal rotation around the bonds in the main chain. Therefore, asymmetric synthesis of single-handed helix is possible using chiral initiators (helix-sense-selective polymerization) in anionic polymerization (**Scheme I**).

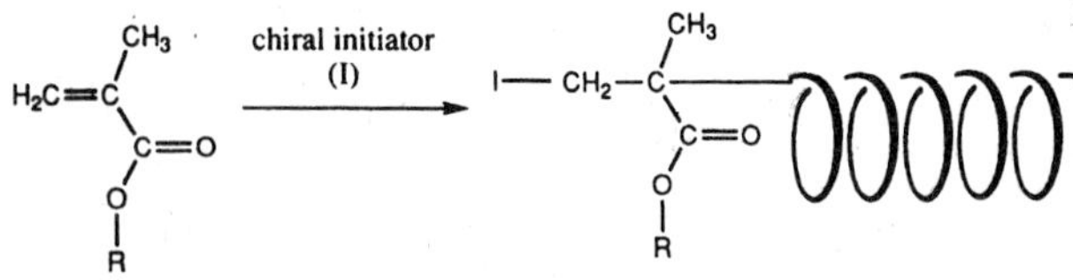

I

Polychloral

Another example of helix-sense-selective polymerization was for chloral (trichloroacetaldehyde).[6]

Polyisocyanates

Polyisocyanates with an achiral side group have a rigid conformation consisting of racemic mixture of right- and left-handed dynamic helices. It is possible to obtain optically active polyisocyanates by introducing an optically active side group in the polymer chain or an optically active end group to the chain terminal.

Polyisocyanides

A 4/1 helical polymer was obtained by polymerizing a bulky isocyanide (R-NC) with Ni(II) catalysts. In contrast to the other polymers described here, in polyisocyanides all the carbon atoms of the main chain had a substituent, and therefore, the polymer structure was sterically crowded.

Cyclopolymerization of a 1,2-phenylene diisocyanide compound with palladium and nickel complexes also gives a cone-handed helical, optically active cyclopolymer.[7]

Chiral Crosslinked Gel

Asymmetric polymerization has also been achieved for the synthesis of crosslinked polymers. This was done using a template polymerization technique as reviewed by Okamoto and Nakano.[1]

REFERENCES

1. Okamoto, Y.; Nakano, T. *Chem. Rev.* **1994**, *94*, 329.
2. Pino, P. *Adv. Polym. Sci.* **1965**, *4*, 393.
3. Kakuchi, T.; Kawai, H.; Katoh, S. et al. *Macromolecules* **1992**, *25*, 5545.
4. Wulff, G.; Dhal, P. K. *Angew. Chem.* **1989**, *28*, 196.
5. Coates, G. W.; Waymouth, R. M. *J. Am. Chem. Soc.* **1991**, *113*, 6207.
6. Jaycox, G. D.; Vogl, O. *Polym. J.* **1991**, *23*, 1213.
7. Ito, Y.; Ihara, E.; Murakami, M. *Angew. Chem.* **1992**, *31*, 1509.

Asymmetric Reactions

See: Asymmetric Synthesis (Using Chiral Polymers From (+)-Camphor)
Supported Chiral Catalysts

ASYMMETRIC SYNTHESIS (Using Chiral Polymers From (+)-Camphor)

Jui-Hsiang Liu
Department of Chemical Engineering
National Cheng Kung University

Synthesizing chiral compounds from achiral reagents always yields the racemic modification. This is simply one aspect of the general rule: optically inactive reactants yield optically inactive products. Optically active compounds, however, can be induced in the presence of chiral catalyst. Homogeneous low molecular weight models often have the advantages of giving high stereoselectivity. Also, every molecule of the catalyst is available to the reactants. In most cases, however, the asymmetric source is expensive and preparing asymmetric catalysts is so troublesome that it becomes desirable to use the catalyst repeatedly. This leads to the preparation of polymeric chiral catalysts.[1]

During our investigations of the synthesis of both monomers and polymers containing the (+)-bornane group, we found that polymers with pendant chiral bornyl groups are effective for asymmetric induction reactions.[2-6] In connection with the studies on the catalytic functionality of the polymers having bornyl groups, we also studied the synthesis and polymerization of new chiral monomers, cis,endo-3-dimethylamino-2-bornyl methacrylate (DABM) and *N,N*-dimethyl[cis,endo-2-(2-vinyloxyethoxy)-3-bornyl]amine (DVEBA).

For this article, the optically active DABM and DVEBA were synthesized, homopolymerized, and copolymerized with methyl methacrylate (MMA) or styrene (St). Effects of temperature, solvents, and reaction time on the specific rotation of the chiral polymers and/or copolymers are discussed. We investigated application of the chiral polymers on the asymmetric induction of chiral alcohols from the addition of the n-butyllithium to aldehydes, and studied the effects of temperature and solvent on the asymmetric induction.

Results and Discussion

Chiral monomers cis, endo-3-dimethylamino-2-bornyl methacrylate (DABM) and *N,N*-dimethyl[cis,endo-2-(2-vinyloxyethoxy)-3-bornyl]-amine (DVEBA) were synthesized from (+)-camphor, as shown in **Scheme I** and **Scheme II.**

The free radical polymerization of the chiral monomers cis,endo-3-dimethylamino-2-bornyl methacrylate (**4**) and/or *N,N*-dimethyl-[cis,endo-2-(2-vinyloxyethoxy)-3-bornyl]-amine (**6**) were carried out in various solvents for a certain polymerization time.

To study the steric hindrance on the chiral polymers, the copolymerization of both chiral monomers with the achiral monomers of methylmethacrylate (MMA) and Styrene (St) were carried out. The results are summarized in Table 1.

CONCLUSION

Chiral monomers cis,endo-3-dimethylamino-2-bornyl methacrylate (DABM) and *N,N*-dimethyl[cis,endo-2-(2-vinyloxyethoxy)-3-bornyl]amine (DVEBA) were synthesized from (+)-camphor. We conducted the homopolymerization of both DABM and DVEBA, and the copolymerization of both chiral monomers with achiral methyl methacrylate (MMA) and styrene (St), with 2,2-azo-bisisobutyronitrile (AIBN) in various organic solvents. The conversion of the polymerization is affected by reaction temperature, solvent, and reaction time. We found that the chiral polymers so-synthesized in this investigation are effective for the enantioface differentiated. The enantiomeric excess was found to increase with the reaction temperature. The chiral polymer can be reused without decreasing enantiomeric excess.

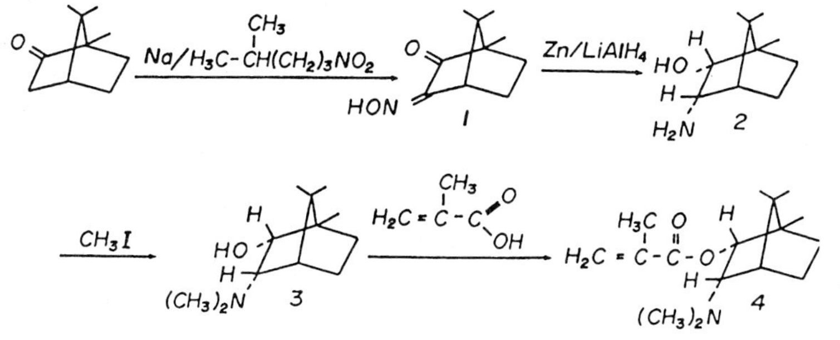

I

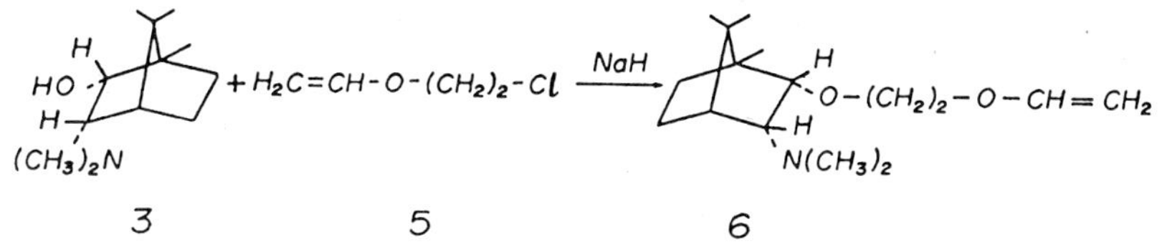

II

TABLE 1. Copolymerization of DABM and DVEBA with Various Conditions[a]

Entry	Feed monomer (mol-%)	Comonomer	Conversion[b] in %	Content of chiral monomer[c]	$[\alpha]_D^{20}$ (cg/ml)[d]	M.W.[e] ($\times 10^{-4}$)
1	DABM(100)	—	46.2	100	+25.3(2.57)	6.8
2	(45)	MMA	79.1	19	+12.8(3.71)	9.2
3	(35)	St	58.7	12	+10.2(3.12)	7.4
4	DEVBA(100)	—	49.5	100	+18.3(2.16)	8.6
5	(40)	MMA	76.2	22	+ 8.7(2.42)	12.3
6	(20)	St	62.1	14	+ 5.3(3.27)	9.8

[a]At 60 °C in benzene, with 1 wt-% AIBN for 6 h.
[b](Weight of polymer/total weight of monomer) × 100%.
[c]Evaluation by elemental analysis, in %.
[d]Concentration, in benzene.
[e]Molecular weight, evaluated by GPC.

REFERENCES

1. Aglietto, M. *Pure Appl. Chem.* **1988,** *60,* 415.
2. Liu, J. H.; Lin, R.; Kuo, J. H. *J. Polym. Sci., Polym. Chem.* **1987,** *25,* 2521.
3. Liu, J. H. *Angew. Makromol. Chem.* **1988,** *164,* 35.
4. Liu, J. H.; Wang, J. H.; Kuo, J. C. *Makromol. Chem.* **1989,** *190,* 2269.
5. Liu, J. H.; Kuo, J. C. *J. Appl. Polym. Sci.* **1990,** *40,* 2207.
6. Liu, J. H.; Kuo, J. C.; Chin, J. *I. Ch. E.* **1992,** *23*(2), 111.

Atatic Polypropylene

See: Metallocene Catalysts
Polypropylene (Commercial)
Polypropylene, Atatic (High Molecular Weight)

Atatic Polystyrene

See: Thermoreversible Gels (Isotactic, Syndiotactic, and Atactic Polystyrene)

Autohesion

See: Tack, Elastomers

AZIDATION POLYMER

Alfred Hassner
Department of Chemistry
Bar-Ilan University

This chapter describes an azidation polymer ($P\text{-}NR_3^+N_3^{-1}$) that contains azide ions bound to a quaternary ammonium ion exchange resin and can be used conveniently in displacement reactions with alkyl halides for the synthesis of organic azides.

A polymeric reagent to introduce the azide (N_3) function into organic molecules represents an ideal case of taking advantage of the unique properties of polymers in the synthesis of sensitive materials. The azidation polymer shows all the desirable features of such a reagent, namely, it is reactive at ambient temperature; it can be used in excess without wasting material; it can be removed by filtration; it can be readily regenerated; and when monitored for the disappearance of the starting alkyl halide, an essentially pure organic product that requires little or no purification remains at the end of the reaction and after filtration of the polymer. Thus, the polymer reagent is of great value because it provides a method for the preparation of certain organic azides in relatively pure form that avoids heating these heat sensitive products.

The polymeric resin in the azidation polymer is readily prepared from a quaternary ammonium resin such as Amberlyte IR-400 by exchange with azide ions and proper washing.

The polymer is also available from Aldrich as an azide exchange resin containing ~3 meq of azide ions per gram of resin.

The polymer is used in displacements of alkyl halides or alkyl sulfonates by azide ions at room temperature in a solvent of choice. Because the polymer beads are insoluble, a suitable solvent in which the starting alkyl handle and the alkyl azide product are soluble should be used. Polar solvents are more desirable because the displacement reactions occur more readily in these solvents than in nonpolar solvents.

Caution is advised when methylene cloride is used as a solvent for the displacement reacton and the reaction time is more than a few hours.

Iodides are displaced faster than the corresponding bromides, which react faster than tosylates. α-Heloketones undergo displacement readily. Acyl chlorides react smoothly with the polymer, but alkyl chlorides react very slowly.[1] Secondary halides react slower than primary ones, and tertiary halides do not undergo displacements.[1] The more dilute the slurry, the slower the reaction.

REFERENCES

1. Hassner, A.; Stern, N. *Angew. Chem. Inter. Ed.* **1986**, *25*, 478.

Aziridine-Based Polymers

See: *Ethylenimine-Modified Polymers*
Polyethylenimine (Overview)
Zwitterionic Polymerization (Aziridine, Oxazoline Derivatives with Anhydrides)

Azo Initiators

See: *Azo Initiators (Function and Application)*
Block Copolymers (From Macroinitiators)

AZO INITIATORS (Function and Application)

Noriaki Minamii
Wako Pure Chemical Industries

Bunichiro Yamada
Material Chemistry Labortory
Osaka City University

Azo or peroxide compounds, which are used as radical initiators, involve chemical bonds with relatively low dissociation energy. The azo compounds can be represented by a generalized formula, R–N=N–R′, in which R may be the same as R′ to constitute a symmetrical molecular structure. An aliphatic azo compound homolytically decomposes thermally or photochemically, releasing nitrogen, and produces the corresponding alkyl radicals.[1,2] The generation of a stable nitrogen molecule lowers the activation energy of the decomposition and increases the frequency factor. The primary radicals, R· and ·R′, may initiate the polymerization of a vinyl monomer.

Table 1 (Structures 1-14) shows a range of major commercially available azo initiators with different decomposition temperatures. Most commercial azontrile initiators are the symmetrical type. Azonitriles, including 2,2′-azobisisobutyronitrile (AIBN), which is one of the most common initiators, have the following advantages compared with peroxides.[3]

- The decomposition follows first-order kinetics regardless of solvent, and induced decomposition is ruled out.
- Chain transfer to initiator is not important.
- The decomposition in aqueous media is not influenced by pH, metal ion, or additives.
- The primary radical is less reactive in hydrogen abstraction and is unlikely to cause branching, crosslinking, and graft polymerization.
- Azonitriles are stable against friction and impact.

The selection of azo initiators is made on the basis of decomposition activity, solubility in the polymerization system, or the type of functional group. If the half-life time of an initiator is too short at the polymerization temperature, polymerization cannot be completed because of exhaustion of the initiator.

APPLICATIONS

Commercial azo initiators, listed in Table 1, have been used in polymerizations of various vinyl monomers, including bulk, solution, suspension, and emulsion polymerization. The decomposition activity of azonitriles (Structures 1–5 in Table 1) is suitable to polymerization over a wide range of temperatures, and these initiators are used alone or in combination to be matched with a desired polymerization rate. AIBN (Structure 3 in Table 1) and 2,2′-azobis(2-methylbutyronitrile) (Structure 4 in Table 1) are similar in the rate of decomposition, but the former is inferior in solubility in solvents and high in toxicity of its decomposition product, tetramethylsuccinonitrile. The latter could be used instead of the former depending on polymerization system and field of application.

Dimethyl 2,2′-azoisbutyrate (Structure 8 in Table 1), which does not have a nitrile group, is low in toxicity of its decomposition product, and excellent in solubility in solvent. This azoester has been used in fields where safety is particularly important (e.g., resins in food processing, medicated poultice preparations, toners, and hybrid paints).[4]

Azoamide, azonitrile bearing a carboxyl group, azoamidine, and azoimidazoline are nonionic, anionic, and cationic water-soluble initiators, respectively, and are used in the polymerization of water-soluble monomers to synthesize water-soluble polymers and polymer microspheres.[3,5,6]

Azoalkane has a high decomposition temperature, and hence it can be introduced into the polymerization system by distillation.

TABLE 1. Commercial Azo Initiators

Category	Chemical name	Structure	Melting point (°C)	10-h half-life time Temp (°C)	10-h half-life time Solvent	Solubility[a] Toluene	Solubility[a] Methanol	Solubility[a] Water
Azonitrile	2,2'-Azobis (4-methoxy-2,4-dimethylvaleronitrile) (Structure (**1**)	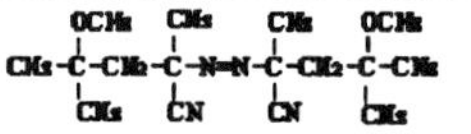	52–88[b]) (Dec.)	30	Toluene	2.5	2.0	0.1[c]
	2,2'-Azobis(2,4-dimethyl-valelonitrile) (Structure **2**)	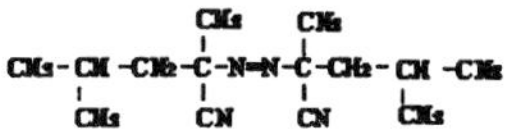	46–62[d] (Dec.)	51	Toluene	72	14	<0.01
	2,2'-Azobis(2-methyl-propionitrile) (Structure **3**) (2,2'-Azobisisobutyro-nitrile)	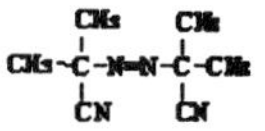	100–103 (Dec.)	65	Toluene	7	7.5	0.04[c]
	2,2'-Azobis(2-methylbutyro-nitrile) (Structure **4**)	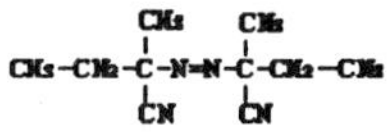	49–50 (Dec.)	67	Toluene	> 100	62	< 0.1[c]
	1,1'-Azobis(cyclohexane-1-carbonitrile) (Structure **5**)	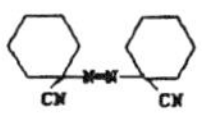	113–115 (Dec.)	88	Toluene	22	3.2	< 0.01
	4,4'-Azobis(4-cyanovaleic-acid) (Structure **6**) (4,4'-Azobis(4-cyanopentan-oic acid)	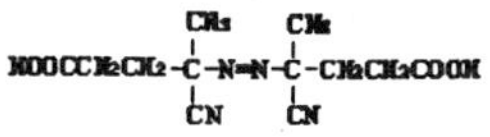	120–123 (Dec.)	69	Water	< 0.1	8.5	< 0.1
	2,2'-Azobis[2-(hydroxymethyl)-propionitrile] (Structure **7**)	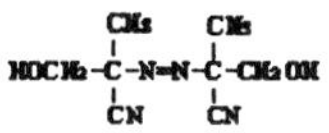	114–117[e]	77	Methyl-cellosolve	< 0.01	1.6	0.22
Azoester	Dimethyl 2,2'azobis(2-methyl-propionate (Structure **8**) (Dimethyl 2,2'-azobisisobuty-rate)	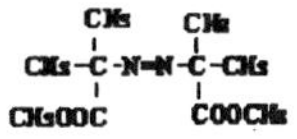	22–28	66	Toluene	80	86	0.3
Azoamide	2,2'-Azobis[2-methyl-*N*-(2-hydroxyethyl)propionamide] (Structure **9**)	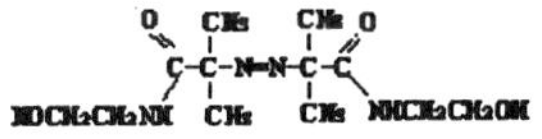	140–145 (Dec.)	86	Water	< 0.01	4.6	2.4
Azoalkane	2,2'-Azobis(2,4,4-trimethyl-pentane (Structure **10**)	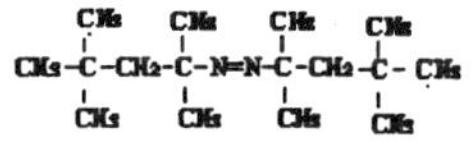	23–24	110	Diphenyl ether	> 10	< 1	< 0.01
	2,2'-Azobis(2-methylpropane) (Structure **11**)	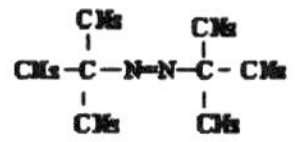	109–110 (bp)	160	Gas phase	> 10	< 1	< 0.01
Azoamidine	2,2'-Azobis(2-methylpropion-amidine) dihydrochloride (Structure **12**)	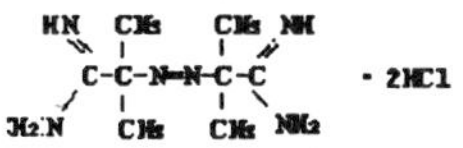	163–170 (Dec.)	56	water	< 0.1	2.1	23
	2,2'Azobis [*N*-(2-hydroxyethyl)-2-methylpropionamidine] dihydrochloride (Structure **13**)	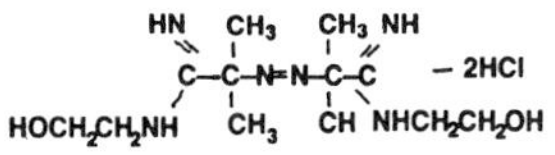	123–127 (Dec.)	58	Water	< 1	2	> 10
Azoimida-zoline	2,2'-Azobis[2-(2-imidazolin-2-yl) propane] dihydrochlo-ride (Structure **14**)	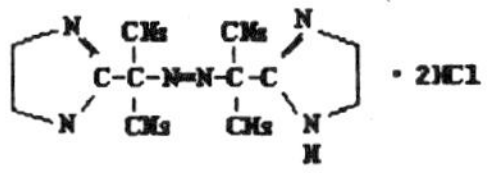	189–193 (Dec.)	44	Water	< 0.01	1.7	35

[a]g/100 g of solvent at 20 °C.
[b]Mixture of racemic (mp: 56–59.5 °C) and meso (mp: 98–100 °C) isomers.
[c]At 25 °C.
[d]Mixture of racemic (mp: 56–58 °C) and meso (mp: 76–78 °C) isomers.
[e]Mixture of racemic (mp: 109.5 °C) and meso (mp: 129.5 °C) isomers.
Source: Reference 12.

It is therefore used as an initiator to make plastic optical fibers mainly composed of MMA.

The azo initiator functioning as a polymerizable monomer can be successfully used in the synthesis of block and graft polymers.[7-10]

Various macroazoinitiators (MAIs) with azo linkage in the polymer backbone have been used to synthesize block copolymers.[10,11]

Azo initiators generate carbon-centered radicals, which are generally less reactive than oxygen-centered radicals in hydrogen abstraction. Decomposition of an azo initiator under the coexistence of a polymer and a monomer could not produce a high yield of a graft copolymer, whereas the oxygen-centered radical abstracts hydrogen from the polymer chain. Thus, azo initiators are known to induce mild graft polymerization on appropriate materials or to assist such reactions.

REFERENCES

1. Engel, P. S. *Chem. Rev.* **1990**, *80*, 99.
2. Leffler, J. E. *An Introduction to Free Radicals*; John Wiley & Sons: New York, 1993; Chapter 4.
3. Minamii, N.; Yamada, B.; Otsu, T. *Mem. Fac. Eng.* **1987**, *28*, 101.
4. Otsuka, S.; Ito, Y. Jpn. Tokkyo Koho JP 034 523, Jan. 23, 1991.
5. Goodwin, J. W.; Ottewill, R. H.; Pelton, R. *Colloid Polym. Sci.* **1979**, *257*, 61.
6. Breitenbach, J. W.; Kuchner, K.; Fretze, H.; Tarnowiecki, H. *Br. Polym. J.* **1970**, *2*, 13.
7. Yamada, B.; Otsu, T. *Makromol. Chem.* **1989**, *190*, 915.
8. Otsu, T.; Yamada, B.; Minamii, N. Jpn. Tokkyo Koho JP 0 556 361, Aug. 19, 1993.
9. Ota, T.; Miki, Y. Jpn. Kokai Tokkyo Koho JP 01 141 905, June 2, 1989.
10. Nuyken, O.; Weidner, R. *Adv. Polym. Sci.* **1986**, *73–74*, 145.
11. Ueda, A.; Nagai, S. *Nippon Settyaku Gakkaishi* **1990**, *26*, 112.
12. "Azo Polymerization Initiators"; Wako Pure Chemical Industries: Osaka, Japan, 1993.

B

BACTERIAL POLYESTERS (Containing Unusual Substituent Groups)

María A. Aponte, Marianela Andújar, and
Eduardo Schröder
Mayagüez Campus
University of Puerto Rico

Many prokaryotic microorganisms synthesize an optically active polyester, poly(3-hydroxybutyrate) (PHB), and accumulate it as a reserve material when grown under limitation of a key nutrient. For many bacteria, PHB functions either as a carbon reserve or as a sink for excess-reducing equivalents. In addition, some bacteria accumulate various poly(3-hydroxyalkanoates) (PHA) by incorporating monomeric units other than 3-hydroxybutyrate.[1]

In 1981, Imperial Chemical Industries developed a controlled fermentation process in which PHA copolymers could be produced. A copolymer of 3-hydroxybutyrate (HB) and 3-hydroxyvalerate (HV) has been produced commercially under the trade name Biopol®.[1] These microbial polymers are biodegradable and biocompatible thermoplastics. Interestingly, their physical properties can be varied by changing the copolymers' compositions.

When alkanes and alkanoic acids are fed to *Pseudomonas oleovorans*, it can produce PHA copolymers containing 3-hydroxyalkanoate units with various alkyl pendant groups ranging in size from methyl to nonyl residues.[4] Fuller and Lenz reported that PHA with branched, unsaturated, or phenyl groups can be obtained from *P. oleovorans* if the bacteria are forced to grow with organic substrates containing such chemical substituents.[2,3,5]

Because novel bacterial polyesters containing aromatic ring substituents have been isolated in the past by Lenz and coworkers, the purpose of this study was to investigate if new polymers of this kind could be obtained with an aliphatic ring as a substituent.[5] This study would offer additional information about the enzymatic specificity of the polymerase system. Varying the distance of an aliphatic ring (cyclohexyl or cyclopentyl) from the carboxylic end of the carbon source fed to *P. oleovorans* was of special interest to see if the bulky aliphatic group could be incorporated into the polymer as well as planar phenyl group, and if there is a constraint on the proximity of the substituent it was restricted to the active end of the molecule. The physical properties of the polymer were also of interest.

PREPARATION

P. oleovorans was grown using five unusual carbon sources: 2-cyclohexylacetic acid (CHA), 3-cyclohexylpropionic acid (CHP), 4-cyclohexylbutyric acid (CHB), 5-cyclohexylvaleric acid (CHV), and 3-cyclopentylpropionic acid (CPP) either pure or in varying proportions with nonanoic acid as a co-feed.

PROPERTIES

This research suggested that new polymers containing cycloalkyl units were isolated when *P. oleovorans* was grown with either pure or 80% CHV and 90% or 50% CHB, using nonanoic acid as a co-feed. This result contrasted with the case that used shorter chain acids as the unusual carbon sources. In these acids, the cyclohexyl ring was progressively closer to the carboxyl group. This proximity apparently affected the proper use of these acids as carbon sources and provided information about the specificity that the active site of the polymerase system in this microorganism exhibited.

REFERENCES

1. Doi, Y. *Microbial Polyesters*; VCH: New York, 1990.
2. Fritzsche, K.; Lenz, R. W.; Fuller, R. C. *Int. J. Biol. Macromol.* **1990**, *12*, 92.
3. Fritzsche, K.; Lenz, R. W.; Fuller, R. C. *Int. J. Biol. Macromol.* **1990**, *12*, 85.
4. Gross, R. A.; DeMello, C.; Lenz, R. W. et al. *Macromolecules* **1989**, *22*, 1106.
5. Kim, Y. B.; Lenz, R. W.; Fuller, R. C. *Macromolecules* **1991**, *24*, 5256.

Bacteriocidal Polymers

See: Antibacterial Resins

Bakelite

See: Calixarenes (as Supramolecular Cyclic Oligomers)

Balata

See: Eucommia Ulmoide Gum

Barrier Polymers

See: Gas Barrier Polymers

Bast Fibers

See: Agro-Fiber/Thermoplastic Blends and Alloys
Jute

Batteries

See: Batteries (Materials for High Performance)
Polyacenic Semiconductor Materials (Applications to Rechargeable Batteries)
Solid Polymer Electrolytes (Overview)
Solid Polymer Electrolytes (Preparation, Characterization, and Properties)
Solid Polymer Electrolytes (Polyether Blends and Composites)

BATTERIES (Materials for High Performance)

Moriyasu Wada and Kimiaki Kabuki
Toshiba Battery

Batteries can convert direct current energy from electrochemical or physical energy. This ability is based on an important fundamental technology. The polymeric materials in batteries are also an important component in higher performance involving the battery's durability, reliability, and usability.

At present, about 1.5 billion batteries are produced annually worldwide. About 95% of them are primary (nonrechargeable) batteries, but secondary batteries, such as nickel-cadmium (Ni-Cd), nickel-metal hydride (Ni-MH), and lithium-ion (Li-ion), have created a large market by sales networks, and the market for these types of batteries is expected to increase in the future.

Here we discuss the polymer materials we investigated and which chemicals are used in batteries to achieve the best results for high-performance equipment.

THE FUNCTION OF POLYMERIC MATERIALS IN BATTERIES

Batteries are manufactured in cylindrical, prismatic, button, sheet, and coin shapes suitable for various instruments. Chemical batteries, however, operate on a similar basic chemical reaction.

Polymeric materials are used in the following structures: for the inner battery, separators, gaskets, electrodes, insulating parts, and safety devices; for the outer battery, insulated disks, insulating tubes, insulating sheets, and plastic housings; and for materials for battery production, binders and carbons.

Separators

The most important function of a separator is to decrease self-discharge and increase current and voltage efficiency. Toward that purpose, functional separators made of nonwoven fabrics or minute porous polymeric films must satisfy the following criteria: allow only ions in electrolytes to pass freely through the separators; prevent any other positive or negative materials from spreading and mixing; and have a smaller contact angle with each of the electrolytes because of excellent wettability properties.

Separators in most batteries are made of nonwoven fabrics that have diameters of about 10 μm, such as one or a combination of polyolefins, which includes polypropylene (PP), high-density polyethylene (HDPE), and cured polyethylene (PE); polyamides (PA), polytetrafluoroethylene (PTFE); poly(vinylidene fluoride) (PVDF); and poly(vinyl chloride) (PVC).

Gaskets, Vents, and Plastic Washers

Gaskets, vents, and plastic washers in batteries are generally used as electrical insulators and packing for the prevention of electrolyte leakage. Gaskets aid in the electrical insulation between positive and negative metal parts and simultaneously prevent electrolytes from leaking. Usually, gaskets are mechanically held in place with the outer metal electrode parts. Therefore, materials for gaskets must satisfy many criteria of stress conditioning, mechanical strength, and chemical structure when in contact with electrolytes. They must show no evidence of stress breaking, solvent cracking, creeping, solubility, swelling, or fatigue. In the past, suitable polymeric materials were the thermoset rubbers such as acrylonitrile butadiene rubber (NB R), chloroprene (CR), and isobutylene-isoproprene rubber (IIR), but recently thermoplastics such as PA, PP, PE, HDPE, partial crosslinked PE, polyfluoro materials (PFM), and ethylene propylene diene rubber (EPDR) have been used because they have fewer additives and are, therefore, purer for higher battery performance.

Vents are one of the safety mechanisms for batteries. When the inner pressure of the battery rises above the prescribed value, the vents automatically decrease the pressure to avoid the dangers of battery explosion and ignition. A plastic washer made of polymeric materials in manganese batteries also acts as a vent.

Insulating Materials

Many insulating materials such as shrink labels, shrink tubes, and insulating spacers are used in the batteries. PVC is unrivaled for them by the reason of good chemical resistance, nonflammability, and economical cost. However, PVC should be replaced with another material to alleviate the environmental problem of each protection.

The replacement materials utilized in battery industry are mainly PET and crosslinked PE.

Because the market for insulating materials is very large, chemical manufacturers are competing with each other to develop new products. Soon, a new multilayer film based on nonflammable PA is expected.

As a part of highly functional alkaline-manganese batteries, a polymeric tube jacket or a thin shrink label is used instead of a metal outer tube to improve battery mass-productivity and economization.

Electrodes and Electrolytes

Polymeric materials that generally have higher electrical resistance have not been widely used in electrodes and electrolytes except in polymeric batteries. Some polymeric materials, however, greatly improve performance.

Carbon and carbon fiber are often used for electrodes. For higher performance of secondary batteries, glassy carbon is used for electrodes.

Positive Temperature Coefficient (PTC) Parts

PTC parts of applied batteries are well-known for their structure, which consists of a uniform mixture of high-electroconductive carbon powder in polyolefins partly crosslinked by electron beams of radiant rays.

APPLICATIONS OF BATTERIES

Applications in Advanced Portable Electronic Products

Applications of secondary batteries, such as Ni-MH and Li-ion, to portable electronic products are increasing remarkably. To extend the use of portable products and keep their portability, secondary batteries are indispensable because they provide a larger capacity than do conventional Ni-Cd batteries and lower the consumption of the products.

Polymer Batteries

Polymer batteries use electroconductive polymers as active materials for electrodes and electrolytes. Recharging takes place by the reaction of the polymer materials and their doping ions. Therefore, the mechanism of a polymer battery reaction is fundamentally different from that of electrochemical batteries. The polymer battery has some of the same characteristics as electrochemical batteries such as long life, lightness, minimization, no electrolyte leakage, good preservation, long-term use, and clean elements without heavy metals for environmental protection.

The conductive polymeric materials for polymer batteries have a fundamentally conjugated double bond through the chemical structure causing π electrons to increase and form an ion complex with a donor (e.g., alkaline metal and alkyl ammonium ion) or with an acceptor (e.g., halogen, Lewis acid, proton acid, transition metal, halide, tetracyanoquinodimethane (TCNQ), and tetracyanoethane (TCNE). Because the ion complex reaction is reversible by electrochemical doping and is accompanied by a change in chemical potential, application to electrodes of secondary batteries has been proposed.[1] Coin-type secondary polymer batteries use polyaniline or polyacene for electrodes have been commercialized, but real use is just beginning.[2-9]

CONCLUSION

In response to the advance in cordless electronic products, increased battery performance is needed to rapidly advance technical progress. To this end, the contribution of polymeric materials to batteries for consumer electronic products is a remarkable technical innovation and is widely affecting the development of many fields, including industry, automobile, and power storage source.

REFERENCES

1. Nigrey, P. J.; MacLanes, C. Jr.; Navns, D. P.; MacDiarmid, A. G. *J. Electrochem. Soc.* **1981**, *128*, 1651.
2. Nakajima, T.; Kawagoe, T. *Synth. Metals* **1989**, *28*, 629.
3. Osaka, T.; Naoi, K.; Ogano, S. *J. Electrochem. Soc.* **1988**, *135*, 1071.
4. Osaka, T.; Ogano, S.; Naoi, K.; Oyama, N. *J. Electrochem. Soc.* **1989**, *136*, 306.
5. Osaka, T.; Nakajima, T.; Naoi, K.; Owens, B. B. *J. Electrochem. Soc.* **1990**, *137*, 2139.
6. Naoi, K.; Menda, M.; Ooike, H.; Oyma, N. *J. Electroanal. Chem.* **1991**, *318*, 395.
7. Yata, S.; Hato, Y.; Sakurai, K.; Osaka, T.; Tanaka, K.; Yamabe, T. *Synth. Metals* **1987**, *18*, 645.
8. Yata, S.; Sakurai, K.; Osaka, T.; Inoue, Y.; Yamaguchi, K.; Tanaka, K.; Yabe, T. *Synth. Metals* **1990**, *38*, 185.
9. Yata, S.; Kinoshita, H.; Komori, M.; Ando, N.; Kashiwamura, T.; Harada, T.; Tanaka, K.; Yambe, T. *Synth. Metals* **1993**, *55–57*, 388.

Bead Polymerization

See: Suspension Polymerization

BENZOCYCLOBUTENE-BASED MONOMERS (Diels-Alder Ring-Opening Polymerization)

Stephen F. Hahn and Nelson G. Rondan
Central Research and Development
The Dow Chemical Company

The hydrocarbon benzocyclobutene (BCB, monomer 1 **Structure 1**) bears appreciable ring strain due to the geometric constraints imposed by the fused four-membered ring.[1-4] When heated, this moiety undergoes an electrocyclic rearrangement to provide the labile intermediate *o*-quinodimethane (**Structure 2**).

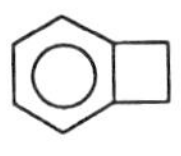

1

2

The propensity of this species to dimerize via a Diels–Alder reaction[5-8] or to act as a diene in the Diels–Alder cycloaddition reaction in the presence of an added dienophile has been well documented.[9-12]

The high stereospecificities observed in the cycloaddition reaction suggest that the ring opening proceeds in a conrotatory fashion. Furthermore, it supports the concerted nature of the Diels–Alder [4+2]-cycloaddition. In the absence of a dienophile or trapping agent, *o*-quinodimethane can either dimerize or revert back to BCB.[5]

The thermally induced rearrangement of a wide array of substituted benzocyclobutenes has been utilized to construct novel polymeric systems.[13-22] Benzocyclobutene functional monomers are unique in that the polymerization is simply thermally activated, with the result that these systems can polymerize without the release of volatile components. Benzocyclobutene functional monomers have been prepared without incorporation of functional groups with large dipole moments; as a result, low dielectric constant thermoset resins designed for use in electronic applications have often been targeted for synthesis using these monomers.[23-28]

POLYMERIZATION STRATEGIES

Approaches to the polymerization of benzocyclobutene-based monomers fall into two distinct categories. The simplest polymerization scheme is that in which ring opened *o*-quinodimethanes react with other *o*-quinodimethanes.[32-35] In the second general category, a Diels–Alder dienophile is included which is intended to react with the *o*-quinodimethane. A variety of dienophiles including maleimides,[14,18,36-42] stilbenes,[29,31,43-45] and alkynes[16,30,46] have been used. This category can be further differentiated between those monomers in which the benzocyclobutene and dienophile reside on the same monomer.

POLYMERIZATION PROCESS

Diels–Alder polymerization of benzocyclobutene functional monomers with alkenes involves at least two discrete steps: opening of the four-membered ring, followed by reaction of the intermediate *o*-quinodimethane with a dienophile to provide a tetrahydronaphthalene repeat unit. The electrocyclic ring-opening reaction is of central importance to the Diels–Alder polymerization of benzocyclobutene functional monomers.

Diels–Alder dimerization could also potentially compete with the *o*-quinodimethane/*trans*-alkene reaction. The dimerization of *o*-quinodimethane can be divided into two pathways. The formation of the spirodimer occurs via a concerted [4+2] Diels–Alder cycloaddition, while the formation of dibenzocyclooctadiene seems to occur in steps involving a diradical intermediate. In addition, the dimerization process was found to be highly exothermic.

PHYSICAL PROPERTIES OF BENZOCYCLOBUTENE DIELS–ALDER POLYMERS

Polymers prepared from monomers with 3 or more functional groups provide crosslinked polymers; no glass-transition temperature is observed for these materials before thermal composition. A–B monomers do polymerize to give materials with glass-transition temperatures below the decomposition range. A–B monomers in which the dienophile is a maleimide do not require a multifunctional crosslinking agent to polymerize to useful molecular weights.

DISUBSTITUTED BENZOCYCLOBUTENES

It has recently been discovered that benzocyclobutene functional molecules that are substituted at both the 2 and 5 positions on the arene ring, such as bisamide react quite differently than benzocyclobutenes monosubstituted at the 3 position.[47-50] In previous studies of substituted benzocyclobutenes, only minor differences in the temperature required for the ring-opening process have been observed for benzocyclobutenes with substituents on the arene ring, while substituents on the cyclobutene ring are known to lower the reaction temperature considerably.[51]

REFERENCES

1. Finkelstein, H. *Chem. Ber.* **1910**, *43*, 1528.
2. Cava, M. R.; Napier, D. R.; *J. Am. Chem. Soc.* **1956**, *78*, 500.
3. Klundt, I. L. *Chem. Rev.* **1970**, *70*, 471.
4. Thummel, R. P. *Acc. Chem. Res.* **1980**, *13*, 70.
5. Errede, L. A. *J. Am. Chem. Soc.* **1961**, *83*, 949.
6. Stephan, D.; Gorgues, A.; Le Coq, A. *Tet. Lett.* **1984**, *25*, 5649.
7. Trahanovsky, W. S.; Macias, J. R. *J. Am. Chem. Soc.* **1986**, *108*, 6820.
8. Martin, N.; Seone, C.; Hanack, M. *Org. Propr. Int.* **1991**, *23*, 237.
9. Jensen, F. R.; Coleman, W. E. *J. Am. Chem. Soc.* **1958**, *80*, 6149.
10. Cava, M. P.; Deana, A. A. *J. Am. Chem. Soc.* **1959**, *81*, 4266.
11. Huisgen, R.; Seidl, H. *Tet. Lett.* **1964**, *46*, 3381.
12. Charlton, J. L.; Alauddin, M. M. *Tetrahedron*, **1987**, *43*, 2873.
13. Iwatsuki, S. *Advances in Polymer Science*, Springer-Verlag, **1984**, *85*, 93.
14. Kirchhoff, R. A. U.S. Patent 4 540 763, 1985.
15. Kirchhoff, R. A.; Baker, C. E.; Gilpin, J. A.; Hanhn, S. F.; Schrock, A. K. *Proc. of the 18th International SAMPE Conference* 1986, 478.
16. Tan, L. S.; Arnold, F. E. *J. Polym. Sci., Polym. Chem. Ed.* **1987**, *25*, 3159.
17. Tan, L. S.; Arnold, F. E. *J. Polym. Sci., Polym. Chem. Ed.* **1988**, *26*, 1819.
18. Tan, L. S.; Arnold, F. E. *J. Polym. Sci., Polym. Chem. Ed.* **1988**, *26*, 3103.
19. Kirchhoff, R. A.; Carriere, C. J.; Bruza, K. J.; Rondan, N. G.; Sammler, R. L. *J. Macromol. Sci.-Chem.* **1991**, *A28* (11 and 12), 1079.
20. Kirchhoff, R. A.; Bruza, K. J. *Prog. Polym. Sci.* **1993**, *18*, 85.
21. Kirchhoff, R. A.; Bruza, K. J. *CHEMTECH*, **1993**, *23*, 22.
22. Corley, L. S. U.S. Patent 4 927 907, 1990.
23. Hahn, S. F.; Townsend, P. H.; Burdeaux, D. C.; Gilpin, J. A. *Polymeric Materials for Electronics and Interconnection*; Lupinski, J. H.; Moore, R. S. Eds.; ACS Symposium Series 407, Am. Chem. Soc'y: District of Columbia, 1989.
24. Johnson, R. W.; Phillips, T. L.; Jaeger, R. C.; Hahn, S. F.; Burdeaux, D. C. *IEEE Trans. Comp. Hybrid Manu. Technol.* **1989**, *12*, 185.
25. Johnson, R. W.; Phillips, T. L.; Weidner, W. K.; Hahn, S. F.; Burdeaux, D. C.; Townsend, P. H. *IEEE Trans. Comp. Hybrid Manu. Technol.* **1990**, *13*, 347.
26. Townsend, P. H.; Hahn, S. F.; Burdeaux, D. C.; Thomsen, M.; McGee, R. L.; Gilpin, J. A.; Carr, J. N.; Johnson, R. W.; Phillips, T. L. *Proc. 3rd Int. SAMPE Electron. Conf.* 1989; p 67.
27. Townsend, P. H.; Burdeaux, D. C.; Hahn, S. F.; Thomsen, M.; McGee, R. L.; Carr, J. N.; Johnson, R. W.; Weidner, K. *Proc. Int. Soc. Hybrid Microelectron* 1989; p 447.
28. Burdeaux, D. C.; Townsend, P. H.; Carr, J. N.; Garrou, P. *J. Electron. Mater.* **1990**, *19*, 1357.
29. Kirchhoff, R. A.; Schrock, A. K.; Hahn, S. F. U.S. Patent 4 724 260, 1988.
30. Kirchhoff, R. A.; Hahn, S. F. U.S. Patent 4 687 823, 1987.
31. Schrock, A. K. U.S. Patent 4812588, 1989.
32. Marks, M. *J. Am. Chem. Soc., Polym. Mat. Sci. Eng. Prepr.* **1992**, *66*, 362.
33. Marks, M. *J. Am. Chem. Soc., Polym. Mat. Sci. Eng. Prepr.* **1992**, *66*, 365.
34. Marks, M. J.; Sekinger, J. K. *Macromolecules* **1994**, *27*, 4106.
35. Marks, M. J.; Erskine, J. S.; McCrery, D. A. *Macromolecules* **1994**, *27*, 4114.
36. Tan, L.-S.; Soloski, E. J.; Arnold, F. E. *Polym. Prep. (ACS)* **1986**, *27*(1), 453.
37. Bartmann, M. U.S. Patent 4 719 283, 1988.
38. Corley, L. S. U.S. Patent 4 973 636, 1990.
39. Corley, L. S. U.S. Patent 5 032 451, 1991.
40. Sue, H.-J.; Garcia-Meitin, E. I.; Yang, P. C.; Bishop, M. T. *J. Mat. Sci. Lett.* **1993**, *12*, 1463.
41. Bruza, K. J.; Bell, K. A.; Bishop, M. T.; Woo, E. P. *Polym. Prepr. (ACS)* **1994**, *35*(1), 373.
42. Hahn, S. F.; Kirchhoff, R. A. U.S. Patent 4 730 030, 1988.
43. Kirchhoff, R. A.; Schrock, A. K.; Hahn, S. F. U.S. Patent 4 783 514, 1988.
44. Hahn, S. F.; Martin, S. J.; McKelvy, M. L. *Macromolecules* **1992**, *25*, 1539.
45. Hahn, S. F.; Martin, S. J.; McKelvy, M. L.; Patrick, D. W. *Macromolecules* **1993**, *26*, 3870.
46. Tan, L.-S.; Arnold, F. E. *Polym. Prepr. (ACS)* **1985**, *26*(2), 176.
47. Walker, K. A.; Markoski, L. J.; Moore, J. S. *Synthesis* 1992, p 1265.
48. Walker, K. A.; Markoski, L. J.; Moore, J. S. *Macromolecules* **1993**, *26*, 3713.

49. Markoski, L. J.; Walker, K. A.; Deeter, G. A.; Spilman, G. E.; Martin, D. C.; Moore, J. S. *Chem. Mater.* **1993**, *5*, 248.

50. Deeter, G. A.; Venkataraman, D.; Kampf, J. W.; Moore, J. S. *Macromolecules* **1994**, *27*, 2647.

51. Kametani, T.; Kajwara, M.; Takahashi, T.; Fukumoto, K. *Tetrahedron* **1975**, *31*, 949.

BENZOCYCLOBUTENE HOMOPOLYMERIZATION (Chemistry and Applications)

M. J. Marks
Texas Polymer Center

R. A. Kirchhoff
Central Research and Development

R. A. DeVries
Ventures Advanced Materials
The Dow Chemical Co.

Benzocyclobutene (Structure **1**, BCB) is a latently reactive functional group that has only recently been exploited in the field of polymer chemistry. Although the use of BCBs as latent dienes (Structure **2**, **Equation 1**) in Diels–Alder cycloaddition reactions for the synthesis of organic compounds has long been known,[1-3] not until pioneering research at the Dow Chemical Company was this moiety used to prepare polymeric materials by additional reactions with itself and with other functional groups.[4-6] Its thermally induced reactivity and otherwise relative chemical inertness, the absence of reaction by-products, and the mostly hydrocarbon nature of its addition reaction products have combined to make available a variety of BCB-based polymers that are of interest in a wide range of applications.

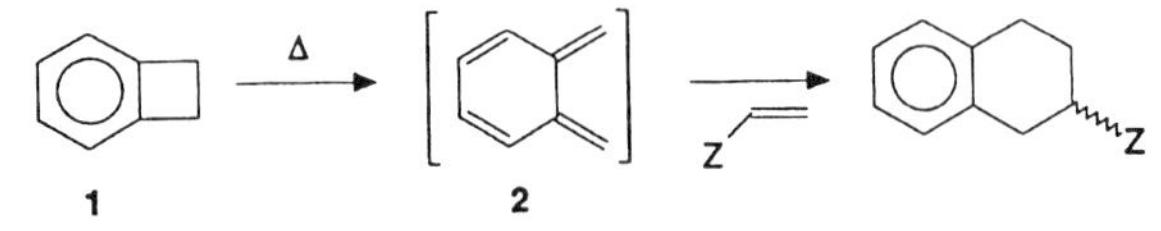

(1)

BCB-SUBSTITUTED MONOMERS AND POLYMERS

Mono- and Bis-BCB Monomers

Mono-BCBs lacking olefinic unsaturation have been only of minor interest in polymer synthesis because when heated they produce only low-molecular-weight oligomers.

The major use of BCBs in polymer synthesis has been in the area of reactive oligomers. The first and still one of the largest groups of BCB monomers are those in which two or more BCB moieties are connected by some linking group X. Upon heating, these monomers undergo an exothermic polymerization to give a complex, three-dimensional network polymer. In the most technologically useful manifestations of this reaction, the monomer has a melting point significantly below the onset temperature of polymerization. This is an important consideration in processing, where one wishes to have a fairly large temperature difference between melting and polymerization.

The polymers derived from bis-BCBs are typical thermoset resins, similar in some ways to those derived from epoxies and polycyanates. What distinguishes them is their generally higher use temperatures.

BCB Functionalized Polymers

A wide range of BCB functionalized polymers having the BCB moiety bonded at various chain positions have been reported.[8-32]

ADVANCEMENT OF *BIS*-BCB MONOMERS

The processing BCB monomers has generally been carried out as a melt polymerization by techniques common to a wide variety of other thermoset resins. Most of the poly-BCB monomers reported thus far in the literature have been *bis*-BCBs and, with a few exceptions, are solids at room temperature.[4-6] For most applications, therefore, the first step in processing the monomer into polymer is the melting of the monomer.

With some BCBs, the solution viscosity of the monomer can be too low to make useful coatings and thin films, or its molecular weight can be so low that it is volatile under polymerization conditions. In these cases, B-staging the monomer to low molecular weight oligomers provides a practical method to make non-volatile, relatively high-viscosity materials suitable for coatings.

PRODUCTS AND MECHANISM OF THERMAL BCB HOMOPOLYMERIZATION

Understanding of the products formed by thermally induced BCB homopolymerization and the reaction mechanisms involved is still developing. The first stages of this chemistry are reasonably well established. BCBs ring open at temperatures above about 200°C to *o*-quinodimethane (Structure 2) (*o*-xylylene) (Equation 1).[1-3] This intermediate, which has been generated from other precursors, usually forms one of two products, depending on the reaction temperature. At 25°C and below, Structure 2 dimerizes by a Diels–Alder cycloaddition reaction to form "spirodimer" (Structure **24**),[7,33,34] while at 25–200°C, dibenzocyclooctadiene (Structure 3) and higher molecular weight oligomers are formed (**Scheme I**). Spirodimer Structure 24 itself homopolymerizes slowly at room temperature to poly(*o*-xylylene) (Structure **25**), by a postulated free-radical mechanism, and copolymerizes with assorted vinyl compound in the presence of free-radical initiators.[35]

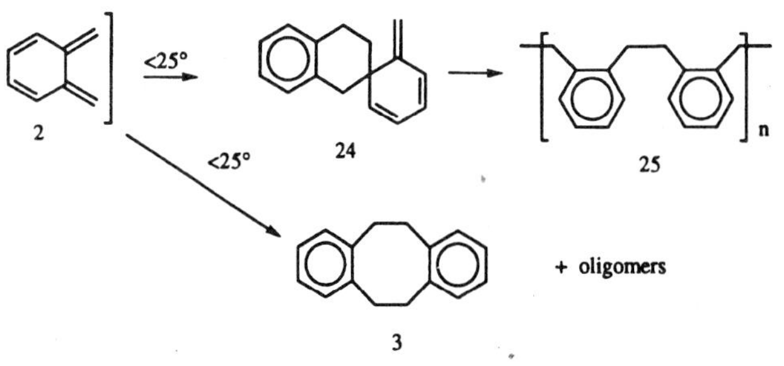

SCHEME I. Reactions of *o*-quinodimethane (2). *Source:* References 7, 33, and 34.

ACID-CATALYZED BCB HOMOPOLYMERIZATION

Most BCB polymerization reactions have involved the thermally induced oligomerization of BCB-substituted monomers or polymers, BCB is, however, an acid-sensitive material, and under the appropriate conditions it can be cationically polymerized.

FUTURE TRENDS

The chemistry and applications of BCB homopolymerization are just beginning to blossom, and both extensions of the concept described here and altogether new discoveries can be expected. Indeed, the recent commercialization of the first BCB-based compounds, Dow Chemical's Cyclotene™ resins,[36] should stimulate a wide range of fundamental and developmental activities. The development of new BCB-based monomers for use in high-value applications can be foreseen, and BCB-functionalized polymers will probably provide a host of new polymeric materials, including crosslinked, branched, block, and graft (co)polymers. More detailed analyses of the products formed upon BCB homopolymerization and the effects of chemical structure and reaction conditions on the range and types of products formed are under way. The results will lead to a better understanding of the BCB homopolymerization mechanism(s), and thereby (perhaps) to the ability to control the products as desired. Most of these studies will pertain to thermally induced BCB homopolymerizations, but new methods and polymers made using acid-catalyzed routes can be expected as well.

REFERENCES

1. Oppolzer, W. *Synthesis* **1978**, 793.
2. Charton, J.; Alauddin, M. M. *Tetrahedron* **1987**, *43*, 2873.
3. McCollough, J. J. *Acc. Chem. Res.* **1980**, *13*, 270.
4. Kirchhoff, R. A.; Carriere, C. J.; Bruza, K. J.; Rondan, N. G.; Sammler, R. L. *J. Macromol. Sci.-Chem.* **1991**, *A28*(11 and 12), 1079.
5. Kirchhoff, R. A.; Bruza, K. J. *Chemtech* **1993**, Sept. 22.
6. Kirchhoff, R. A.; Bruza, K. J. *Prog. Polym. Sci.* **1993**, *18*, 85.
7. Errede, L. A. *J. Am. Chem. Soc.* **1961**, *83*, 949.
8. Tan, L. S.; Arnold, F. E. *J. Polym. Sci. Part A: Polym. Chem.* **1988**, *26*, 1819.
9. Tan, L. S.; Arnold, F. E. *Polym. Preprints (Amer. Chem. Soc., Div. Polym. Chem.)* **1985**, *26*, 176.
10. Denny, L. R.; Goldfarb, I. J.; Farr, M. P. *Polym. Mater. Sci. Eng.* **1987**, *56*, 656.
11. Cuah, H. H.; Tan, L. S.; Arnold, F. E. *Polym. Eng. Sci.* **1989**, *28*, 107.
12. Upshaw, T. A.; Stille, J. K.; Droske, J. P. *Macromolecules* **1991**, *24*, 2143.
13. Wong, P. K. U.S. Patent 4 708 994, 1987.
14. Wong, P. K. U.S. Patent 4 994 548, 2-19-91.
15. Marks, M. J.; Sekinger, J. K. *Macromolecules* **1994**, *27*, 4106.
16. Marks, M. J.; Schrock, A. K.; Newman, T. H. U.S. Patent 5 171 824, 1992.
17. Marks, M. J.; Schrock, A. K.; Newman, T. H. U.S. Patent 5 198 527, 1993.
18. Wong, P. K. U.S. Patent 4 708 990, 1987.
19. Delassus, S. L.; Howell, B. A.; Cummings, C. J.; Dais, V. A.; Nelson, R. M.; Priddy, D. B. *Macromolecules* **1994**, *27*(6), 1307.
20. Walker, K. A.; Markoski, L. J.; Moore, J. S. *Macromolecules* **1993**, *26*, 3713
21. Spilman, G. E.; Markoski, L. J.; Walker, K. A.; Deeter, G. A.; Martin, D. C.; Moore, J. S. *Polym. Mat. Sci. Eng.* **1993**, *68*, 139.
22. Markoski, L. J.; Walker, K. A.; Deeter, G. A.; Spilman, G. E.; Martin, D. C.; Moore, J. S. *Chem. Mater.* **1993**, *5*, 248.
23. Ricket, C.; Neuenschwander, P.; Suter, U. W. *Macromol. Chem. Phys.* **1994**, *195*, 511.
24. Wong, P. K. U.S. Patent 4 698 394, 1987.
25. Wong, P. K. U.S. Patent 4 798 990, 1987.
26. Wong, P. K. U.S. Patent 4 722 974, 1988.
27. Wong, P. K. U.S. Patent 4 687 815, 1987.
28. Pabon, R. A. Jr.; DeVries, R. A. U.S. Patent 5 310 809, 1994.
29. Campbell, R. E.; DeVries, R. A. U.S. Patent 5 077 367, 1991.
30. Wong, P. K. U.S. Patent 4 667 004, 1987.
31. Wong, P. K. U.S. Patent 4 667 005, 1987.
32. Wong, P. K. U.S. Patent 4 622 375, 1986.
33. Ito, Y.; Nakatsuka, M.; Saegusa, T. *J. Am. Chem. Soc.* **1982**, *104*, 7609.
34. Ito, Y.; Nakatsuka, M.; Saegusa, T. *J. Am. Chem. Soc.* **1980**, *102*, 863.
35. Errede, L. A. *J. Polym. Sci.* **1961**, *49*, 253.
36. *Mod. Plast. Int.* **1992**, *22*(10), 30.

BENZOCYCLOBUTENES (Crosslinking and Related Reactions)

Michael F. Farona
Department of Chemistry
The University of North Carolina at Greensboro

Benzocyclobutene (BCB) and its derivatives are thermally activated at around 200°C, where the strained ring undergoes an electrocyclic ring opening to *o*-xylylene, which is highly reactive. In the presence of a suitable dienophile, the *o*-xylylene participates in a Diels–Alder reaction, or in the absence of such a species, reacts with itself to form an eight-membered ring.

Functionalized BCBs can be incorporated into polymers in several ways to give polymers or oligomers containing the BCB functionality. Depending on the location and number of BCB groups incorporated, these species can be used in grafting or crosslinking reactions.

BCB monomers may be prepared by straightforward organic reactions or by catalytic methods. Organic synthesis of these molecules usually requires several steps, resulting in overall low yields.[1]

PREPARATION AND PROPERTIES

A paper by Deeken and Farona reported the preparation and characterization of a graft polymer, in which polystyrene was grafted onto a hydrocarbon copolymer by means of a BCB group.[2]

A second paper by Deeken and Farona reported the preparation of 4-(benzocyclobutenyl)methanol and the application of this molecule to bicap a polyurethane composed of 1,9-nonane diol and toluene-2,4-diisocyanate.[3] The bicapped polyurethane was then used to crosslink styrene-butadiene rubber (SBR).

Another example of a two-headed BCB molecule used for crosslinking was reported by Fishback and Farona.[4] In this case, bis(benzocyclobutenyl)ethane, prepared from 1,5-hexadiyne using $NbCl_5$ as the catalyst, was used to crosslink SBR.

A third method of crosslinking with BCB is to incorporate a BCB-containing monomer into a polymer with reactive sites. To this end, several ethynyl-BCB derivatives were prepared and copolymerized with alkynes of similar structures. The final polyalkynes are conjugated structures, and curing was induced thermally, resulting in Diels–Alder reactions of the BCB groups with double bonds in the polyconjugated backbone.

Fishback et al. synthesized allyl-BCB by a cross trimerization of 1,5-hexadiyne with 5-chloro-1-pentyne, followed by a dehydrohalogenation to give the product.[5] This compound was copolymerized with 1-hexene and a diene,7-methyl-1,6-octadiene (MOD) or 5-methyl-1,4-hexadiene (MHD), by the Ziegler-Natta catalyst $TiCl_3/Al(C_2H_5)_3$. Another copolymer of 1-hexene (97%) and MHD (3%) was prepared and cured with sulfur accelerators to compare certain mechanical properties.

The purpose of the research was to investigate an alternative for sulfur curing in a biomedical material. The final material would be considered for use as an artificial spinal disk; therefore, the mechanical properties of interest were ultimate elongation, tensile strength, and modulus.

APPLICATIONS

The majority of research carried out to date on BCBs in polymers has focused on crosslinking and chain extension in high performance composites.[6-9] In these systems, functionalized BCBs are typically used to end-cap low-molecular-weight polyimide prepolymers, which are then converted to high-molecular-weight networks thermally. A typical example of this type of material is bis(BCB) and bis(maleimide) mixtures; the carbon–carbon crosslinks provide high strength and chemical and heat stability. These properties are of interest to industries that use structural materials, for example, the aerospace industry. Other examples are composites of diketone, bis(BCB) and graphite fiber and BCBs as thermoset matrix hosts for rigid-rod molecular composites.[10-13] Polymers containing BCBs have also been investigated as adhesives, coatings, and as dielectric materials.[14-16,22]

An application of great potential is using BCBs in biocompatible materials. Hexsyn (trade name, Goodyear), a copolymer of 1-hexene and 5-methyl-1,4-hexadiene (97 and 3 molar ratio, respectively), is biocompatible. Vulcanized Hexsyn is used to construct the blood-pumping diaphragm in the Cleveland Clinic's left-ventricular assist device and is also used as the flexural component in artificial finger joints, artificial spinal discs and hip protheses.[17-20]

Unfortunately, the sulfur vulcanization process employs two accelerators, 2-mercaptobenzothiazole and tetramethylthiuram disulfide, both of which are suspected carcinogens.[21] Hexsyn crosslinked with BCB was subjected to cytotoxicity tests and found to be nontoxic.[5] The study showed that the traditional accelerated sulfur cure vulcanization systems can be replaced with nontoxic BCB materials in some polymers without a great sacrifice in physical or mechanical properties.

REFERENCES

1. Kirchhoff, R. A. U.S. Patent 4 540 763, 1985.
2. Deeken, J. S.; Farona, M. F. *J. Polym. Sci., Part A: Polym. Chem.* **1993**, *31*, 2863.
3. Deeken, J. S.; Farona, M. F. *Polym. Bull.* **1992**, *29*, 295.
4. Fishback, T. L.; Farona, M. F. *J. Polym. Sci., Part A: Polym. Chem.* **1993**, *31*, 2747.
5. Fishback, T. L.; McMillin, C. R.; Farona, M. F. *Bio-Med. Mat. Eng.* **1992**, *2*, 83.
6. Corely, L. S. U.S. Patent 4 972 907, 1990.
7. Wang, P. C. U.S. Patent 4 968 810, 1990.
8. Tan, L.; Soloski, E. J.; Arnold, F. E. *ACS Symp. Ser.* **1988**, *367*, 24.
9. Mueller, W. H. European Patent 88119370.0, 1989.
10. Lee, W.; Laman, S.; Houle, S. et al. *36th Int. SAMPE Symp.* San Diego, CA, 1991, 1207.
11. Laman, S.; Yalvac, S.; McGee, R. *36th Int. SAMPE Symp.* San Diego, CA, 1991, 469.
12. Chuah, H. H.; Tan, L.; Arnold, F. E. *Polym. Eng. Sci.* **1989**, *29*, 107.
13. Kirchhoff, R.; Bruza, K.; Harris, R. et al. *36th Int. SAMPE Sym.* San Diego, CA, 1991, 457.
14. Hahn, S. F.; Krell, D. J. U.S. Patent 4 831 172, 1989.
15. Kimura, T.; Japanese Patent 01 313 520, 1988.
16. Hahn, S. F.; Townsend, P. H.; Burdeaux, D. C. et al. *ACS Symp. Ser.* **1989**, *407*, 199.
17. Hillegas, D. V.; Kiraly, R. J. *In Polyolefin Blood Pump Components, Synthetic Biomedical Polymers*; Szycher, M., Ed.; Technomic: CT, 1980, 59.
18. Lee, H.; Quach, H.; Berry, D.; Stitch, W. Polymeric Materials and Artificial Organs, ACS Symposium Series 1984, *256*, 99.
19. Steffe, A. D. *The Development and Use of an Artificial Disc: Case Presentation, Camp Back Issues* **1989**, *2:1*, 4.
20. Cook, S. D.; Anderson, R. C.; French, H. G. et al. *Biomater. Med. Devices Artific. Org.* **1989**, *2:1*, 4.
21. Dieter, M. P. *Toxicology and Carcinogenesis Studies of 2-Mercaptobenzothiazole in Rats and Mice,* Pub. No. 88-2588, National Toxicology Program, Technical Report No. 332, U.S. Dept. of Health and Human Services, Government Printing Office: Washington, DC, 1988.
22. Fong, S. O.; Keister, F. Z.; Peters, J. W. *Int. SAMPE Tech. Conf.* **1990**, *22*, 602.

BENZOCYCLOBUTENES (Diels–Alder Polymerizations)

R. A. Kirchhoff and K. J. Bruza
Central Research and Development
Material Science and Development
Dow Chemical Company

The Diels–Alder chemistry of benzocyclobutenes is well established in the literature and was among the first benzocyclobutene reactions studied. In 1958, Jensen and Coleman reported that the room temperature reaction of 1,2-diphenylbenzocyclobutene and maleic anhydride yielded 3,4-diphenyltetralin-2,3-dicarboxylic acid anhydride.[1]

The rate-limiting step in the Diels–Alder reaction of benzocyclobutenes with olefins is the opening of the four-membered ring to yield the highly reactive diene, *o*-quinodimethane, as the key intermediate. The ring-opening reaction is seen as a first-order process with an activation energy of 39 kcal-mole. The reaction is also exothermic to the extent of 25 kcal/mole.[2,3] The half lives of benzocyclobutene have been determined at various temperatures. For example, the half life is 84 minutes at 200°C, which is commonly referred to as the ring-opening temperature of benzocyclobutene.[4,5] Also, the half lives of benzocyclobutenes at various temperatures strongly depend upon the substituents on the four-membered ring.

The Diels–Alder reaction of benzocyclobutene has been shown to proceed through an *o*-quinodimethane intermediate. When a suitable unsaturated site for reaction is absent, *o*-quinodimethane can revert to benzocyclobutene, dimerize to 1,2,5,6-dibenzocyclooctadiene, or oligomerize into a complex mixture of products. The reaction's course depends strongly upon reaction conditions.

Numerous reports of benzocyclobutenes used in polymer synthesis have appeared in recent years.[6,7] Two major classes of polymer-forming reactions employing this reaction has been investigated. One class uses AB-type monomers, which has benzocyclobutene and the double bond contained in the same molecule. The other class involves bisbenzocyclobutenes (BCB) reacting with bisolefins or bisalkynes and is the subject of this review.

Kirchhoff et al. reported the earliest examples of copolymerizing a bisbenzocyclobutene and a bisdieneophile.[8,9] Their studies primarily focused on DSC and thermogravimetric analysis (TGA) experiments of bismaleimide (BMI) and bisbenzocyclobutene (BCB) mixtures.

A particularly novel copolymerization of a bismaleimide with a bisbenzocyclobutene was reported in a patient by Bartmann. Bartmann used the hydrocarbon benzocyclobutene as the bisbenzocyclobutene monomer, which was copolymerized with a series of bismaleimides to provide soluble copolymers (**Figure 1**). The polymerizations were all carried out in boiling solvents such as sulfolane or diphenyl ether. In addition, the patent states that polymerization in the presence of a free-radical inhibitor suppressed the homopolymerization of the bismaleimide.

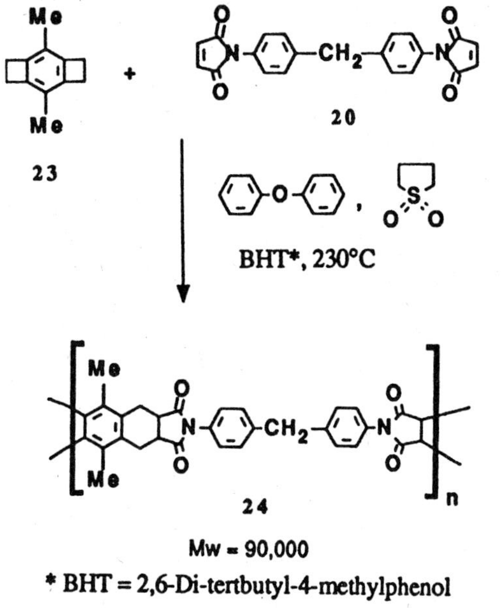

FIGURE 1. Copolymerization of benzodicylobutene 23 with bismaleimide 20.

Other patents elaborated further on the effects of free-radical inhibitors, which greatly enhanced the properties of bismaleimide and bisbenzocyclobutene copolymers.[10-12]

The Diels–Alder copolymerization of a bisbenzocyclobutene with diacetylenes was also reported.[9] Using benzocyclobutenes as latent dienes in the Diels–Alder reaction has provided a new way of applying this well-established reaction in polymer synthesis. The Diels–Alder polymerization of bisbenzocyclobutenes with bismaleimides proceeds well and has been the subject of several patents and papers. The properties of the polyimides prepared in this manner depend strongly upon the reactants' molar ratios and the presence of free-radical inhibitors during polymerization. Enhanced fracture toughness was obtained using equimolar amounts of reactants and by polymerizing in the presence of phenothiazine. Diacetylenes have also been copolymerized with bisbenzocyclobutenes, but the data here is not as extensive as for bismaleimides. Finally, the Diels–Alder polymerization of bisbenzocyclobutenes with aryl cyanates and allylcyanurates have also been studied to a lesser degree.

REFERENCES

1. Jensen, F. R.; Coleman, W. E. *J. Am. Chem. Soc.* **1958**, *89*, 6149.
2. Kirchhoff, R. A.; Bruza, K. J.; Carriere, C. J. et al. *J. Macromol. Sci., Chem.* **1991**, *A28*(11,12), 1079.
3. Bruza, K. J.; Bonk, P. J.; Harris, R. F. et al. In *Proceedings of the 36th International SAMPE Symposium*; Sampe: Covina, 1991; p 457.
4. Wong, P. K. U.S. Patent 4 623 375, 1986.
5. Roth, W. R.; Bierman, M.; Dekker, H. et al. *Chem. Ber.* **1978**, *111*, 3892.
6. Kirchhoff, R. A.; Bruza, K. J. *Prog. Polym. Sco.* **1993**, *18*, 85.
7. Arnold, F. E.; Tan. L.-S. In *31st International SAMPE Symposium*; Sampe: Covina, 1986; p 968.
8. Hahn, S. F.; Kirchhoff, R. A. U.S. Patent 4 730 030, 1988.
9. Tan, L.-S.; Arnold, F. E. *Polym. Prepr. Am. Chem. Soc. Div. Polym. Chem.* **1986**, *27*(1), 453.
10. Corley, L. S. U.S. Patent 4 927 907, 1990.
11. Corley, L. S. U.S. Patent 4 973 636, 1990.
12. Corley, L. S. U.S. Patent 5 032 451, 1991.

BENZOXAZINE MONOMERS AND POLYMERS (New Phenolic Resins by Ring-Opening Polymerization)

Jingping Liu and Hatsuo Ishida
The NSF Center for Molecular and Microstructure of Composites
Department of Macromolecular Science
Case Western Reserve University

Oxazine-derived phenolics offer an alternative to traditional phenolics. They are synthesized by ring-opening polymerization of the aromatic oxazines. This family of phenolics provides tremendous freedom in molecular design. Furthermore, they produce no polymerization by-products and no strong catalysts are required.[9-13]

Aromatic oxazines were first synthesized in 1944 by Holly and Cope through Mannich reactions from phenols, formaldehyde, and amines.[4] From the 1950s to 1960s, Burke and co-workers synthesized many benzoxazines and naphthoxazines.[5,18,19,24-29] Additionally, benzoxazines were also found as the intermediates in the synthesis of novolac resins.[12] The heat-cured polymers derived from benzoxazines were first achieved by Schreiber.[3,13,32] Later, Reiss and co-workers investigated the polymerization of monofunctional benzoxazines with and without phenol as an initiator, resulting in linear polymers under 4000 molecular weight.[11]

Recently, polyfunctional benzoxazines were synthesized in our laboratory.[12,14,15,17,20] Phenolic resins cured from such benzoxazines proved to be crosslinked polymers and exhibited good mechanical properties.[14,17,31]

Benzoxazine is a single benzene ring fused to another six-membered heterocycle containing one oxygen atom and a single nitrogen atom. There are a number of possible isomeric benzoxazines depending on the relative positions of the two heteroatoms and the degrees of oxidation of this oxazine ring system. 3,4-Dihydro-2H-1,3-benzoxazine is one kind of hydrogenated derivative of benzoxazine.[1] When the benzene ring is replaced by naphthalene, the corresponding oxazine becomes naphthoxazine.

As a specific kind of hydrogenated benzoxazine, 3-substituted-3,4-dihydro-2H-1,3-benzoxazine was studied mainly as an alternative product of the Mannich reaction in the middle of the century.[5,6,16] Few applications of this compound as a polymer had been reported until it was identified as an intermediate of the amine-catalyzed phenolic resin.[7] Recently, this type of benzoxazine was found as a new type of precursor of phenolic resin.[8-13] Some of polybenzoxazine materials have excellent physical and mechanical properties as well as processability.[14,15,17]

SYNTHESIS METHODS OF 3,4-DIHYDRO-2H-1,3-BENZOXAZINES

Most of 3-substituted-3,4-dihydro-2H-1,3-benzoxazines resulted from the reaction of p-substituted phenols with formaldehyde and a primary amine in a molar ratio of 1:2:1, respectively.[5,6] This reaction may be considered a variant of the Mannich reaction. The reaction is best carried out by first condensing the primary amine with formaldehyde to form the *N,N*-dimethylolamine (see Structure **1**, **Scheme I**), which is then allowed to react with the phenol. Alternatively, when a *p*-substituted phenol, formaldehyde, and a primary amine were allowed to react in a molar ratio of 1:1:1, *p*-aminomethylphenols, that is, Mannich's base, were formed (Structure **2**, Scheme I).[2] These compounds condensed with formaldehyde in the presence of base to yield the 3,4-dihydro-2H-1,3,-benzoxazine (Structure **3**, Scheme I).[3]

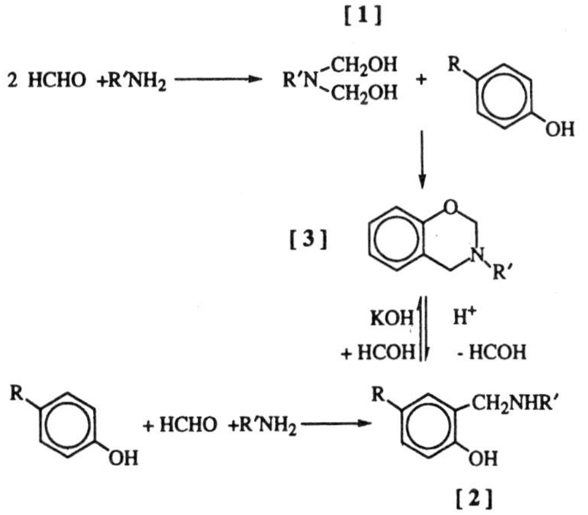

I

R: CH_3, $C(CH_3)_3$, C_6H_5, $NHCOCH_3$, Br, Cl, OH
$C(CH_3)C_6H_5OH$, $(OC_6H_5OH$, C_6H_5OH
R′: CH_3, $C(CH_3)_3$, C_6H_{12}, $C_6H_5CH_2$,
$CoH_5(CH_3)CH$, C_6H_5, $HOCCH_2)_2$
$CH_2 = CHCH_2$

Most of these benzoxazines were also synthesized in the absence of solvent by reactions between paraformaldehyde, phenols, and amines at elevated temperatures.[20]

REACTIVITY OF 3,4-DIHYDRO-2H-1,3-BENZOXAZINE

Stability of Benzoxazine Ring

Reaction of benzoxazine with aqueous hydrochloric acid resulted in the elimination of formaldehyde and formation of isomeric hydrochloride salts in high yield. The hydrochloride salt was treated with aqueous sodium bicarbonate and formed free base. The benzoxazine was quite stable toward hot aqueous alkali.[1,5,16] For some benzoxazines, the ring can be opened even in the presence of phenol.[18] In an acidic condition, naphthoxazines yielded formaldehyde; the other product was a red resinous substance. Hydrochlorides of naphthoxazines from aromatic amines were unstable, whereas the related naphthoxazines from aliphatic amines formed highly stable hydrochloride salts.[29] The stability of the benzoxazine in alcohol was found to depend on the particular phenol and primary amine used in the synthesis. Substituents on the benzoxazine ring have a certain effect on the ring stability. Structures bearing carbonyl groups are generally more sensitive to base-catalyzed hydrolysis. In the presence of compounds with active hydrogen, such as phenol, naphthol, indoles, carbazole, imides, and aliphatic nitro compounds, the ring opening reaction of benzoxazine occurred to form the Mannich bridge structure (Structure **26**, **Scheme II**).

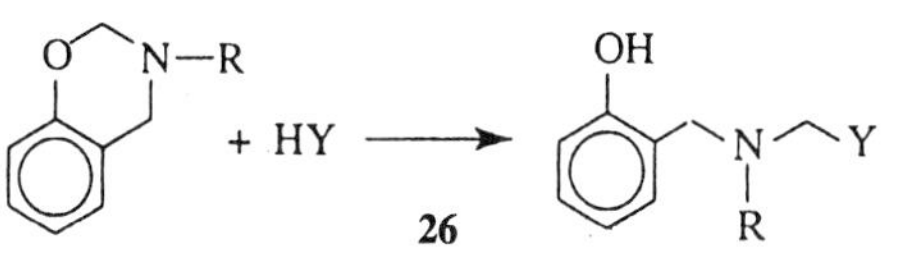

II

This new type of aminoalkylation always happens at the ortho position. A possible explanation for this may be related to the free ortho position, which puts the latter in a favorable position for reaction as a result of intermolecular hydrogen bonding in the initial step. Benzoxazines derived from methylamine were found to be much more reactive than those from benzylamine, whereas those from cyclohexylamine were of intermediate reactivity; only starting materials were isolated when the analogous benzoxazine derived from cyclohexylamine was employed.

Polymerization

Benzoxazine monomers can be cured thermally.[3,12,14,15,17,31] The self-dissociation of the benzoxazine ring at elevated temperatures offered the initiating species.[11] Benzoxazine would polymerize more easily in the presence of initiators or a catalyst, compounds with active hydrogen (HY), or those characterized by the presence of a highly nucleophilic carbon or nitrogen atom.[9,11,30] The reaction aptitude depends on the structure of benzoxazine, the structure of the initiator, and the reaction condition. Polymerization of benzoxazines with a free ortho position (C-8) can occur by following the ring-opening reaction. The corresponding polymer was called polybenzoxazine (see Structures **27** and **28**, **Scheme III**).

Reiss et al. studied the reactions of monofunctional benzoxazines with and without a phenol initiator in the bulk by NMR and gel permeation chromatography (GPC).[11] Only oligomers with an average molecular weight ~ 1000 were obtained in the

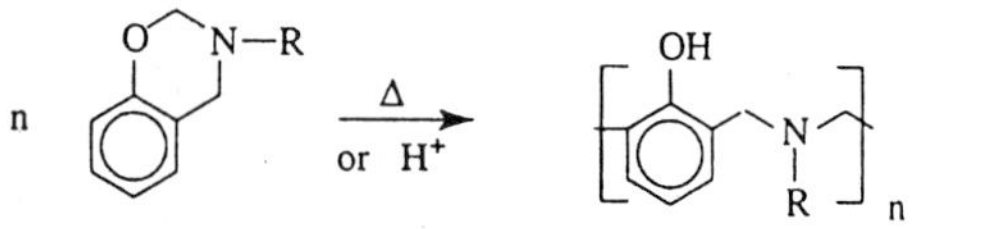

27

III

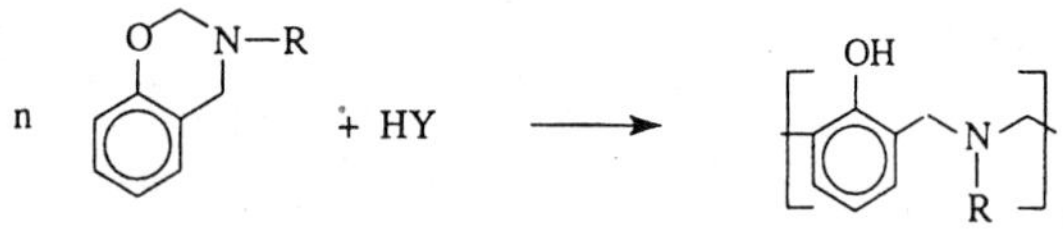

28

absence of initiator because the thermodissociation of the monomer competed with the chain-propagation reaction. The compositions consisting of both polyfunctional benzoxazines and reactive polyamines cured more rapidly than difunctional benzoxazine alone.[10]

Ning and Ishida polymerized several kinds of difunctional benzoxazine precursors thermally by using dimers and higher oligomers as initiators; these were the by-products during the synthesis of benzoxazine. These dimers and oligomers contained phenolic structures with free ortho positions rendering the precursor to be self-initiating toward polymerization and crosslinking reactions.[12,31]

Shen and Ishida successfully cured pure difunctional benzoxazine and nathphoxazine monomers into a crosslinked matrix of carbon composites in an autoclave.[17]

Ishida and Rodriguez surveyed potential catalyst systems ranging from strong acid to weak carboxylic acids to phenol for benzoxazine polymerization. They found that the weak carboxylic acids, such as sebacic acid, produced the best results in terms of final properties and reaction rate.[15] Polybenzoxazine can also be prepared with an aminomethylation reaction between the difunctional benzoxazine and compounds that contain more than two active hydrogen (HY–Y′H), such as polyamines.[9]

PHYSICAL AND MECHANICAL PROPERTIES OF POLYBENZOXAZINE AND THE COMPOSITES

Depending on the structure of benzoxazine monomers, initiators, and the curing condition, a wide range of cured properties can be obtained. The typical strength values of polybenzoxazine were apparently superior to conventional phenolic materials. The glass-transition temperatures were comparable with some of the widely used high-performance epoxy materials. Some carbon/polybenzoxazine composites possessed excellent mechanical properties and were heat resistant, which were close to the properties of polyimide composites with better processability.[14,15,17,31]

The glass-transition temperatures of polybenzoxazine were much higher than the curing temperature because polymerization didn't cease after vitrification.[15,31] The shrinkage of polybenzoxazine was near zero and its water uptake was very low.[14]

REFERENCES

1. Sainsbury, M. In *Comprehensive Heterocyclic Chemistry*; Katritzky, A. R.; Rees, C. W., Eds.; Pergamon: New York, 1984; p 995.
2. Bernadi, L.; Codon, S.; Suchowsky, G. K.; Pegrassi, L. South African Patent 6 804 925, 1969; *Chem. Abstr.* **1969**, *71*, 112948.
3. Schreiber, H. German Patent 2 255 504, 1973.
4. Holly, F. W.; Cope *J. Am. Chem. Soc.* **1944**, *66*, 1875.
5. Burke, W. J. *J. Am. Chem. Soc.* **1949**, *71*, 609.
6. Tramontini, M. *Synthesis* **1973**, 703.
7. Kopf, P. W.; Wagner, E. R. *J. Polym. Sci., Polym. Chem. Ed.* **1973**, *11*, 939.
8. Grabarnik, L. G.; Leonova, M. B.; Korotin, M. M.; Vorosova, T. G.; Osetrova, E. N. *Izy. Alad. Nauk. Kaz. SSR, Ser. Khim* **1990**, *3*, 69; *Chem. Abst.* **1990**, *113*, 133439y.
9. Higgnbottom, U.S. Patent 4 501 864, 1985.
10. Higgnbottom, U.S. Patent 4 557 979, 1985.
11. Reiss, G.; Schwob, J. M.; Guth, G.; Roche, M.; Lande, B. In *Advances in Polymer Synthesis*; Culbertson, B. M.; McGrath, J. E., Eds.; Plenum: New York, 1985.
12. Ning, X.; Ishida, H. *J. Polym. Sci., Poly. Chem. Ed.* **1994**, *32*, 1121.
13. Schreiber, H.; Saur, W. European Patent EP 356 379, 1990; *Chem. Abst.* **1990**, *113*, 41934c.
14. Ishida, H.; Allen, D. *ACS Polym. Mat. Sci. Eng.* **1995**, *73*, 496.
15. Ishida, H.; Roduriguez, Y. *Polymer* **1995**, *36*, 3151.
16. Elderfield, R. C.; Todd, W. H.; Gerber, S. In *Heterocyclic Compounds*; Elderfield, R. K., Ed.; Wiley: New York, 1957; Vol. 6, Chapter 12.
17. Shen, S. B.; Ishida, H. to be published.
18. Burke, W. J.; Kolbezen, M. J.; Stephens, C. W. *J. Am. Chem. Soc.* **1952**, *74*, 3601.
19. Burke, W. J.; Nurdock, K. C. *J. Am. Chem. Soc.* **1954**, *76*, 1677.
20. Liu, J.; Ishida, H. to be published.
21. Hicks, W. L. *J. Chem. Soc.* **1910**, *97*, 1032.
22. Uno, H.; Kurokawa, M. *Chem. Pharm. Bull.* **1978**, *26*, 549.
23. Deck; Dains, *J. Am. Chem. Soc.* **1933**, *55*, 4986.
24. Burke, W. J.; Smith, R. P.; Weatherbee, C. *J. Am. Chem. Soc.* **1952**, *74*, 602.
25. Burke, W. J.; Mortenson Glennie, E. L.; Weatherbee, C. *J. Organ. Chem.* **1964**, *29*, 909.
26. Burke, W. J.; Hammer, C. R.; Weatherbee, C. *J. Organ. Chem.* **1961**, *26*, 4403.
27. Burke, W. J.; Nasutuvicus, W. A.; Weatherbee, C. *J. Organ. Chem.* **1964**, *29*, 407.
28. Burke, W. J.; Weatherbee, C. *J. Am. Chem. Soc.* **1950**, *72*, 4691.
29. Burke, W. J.; Reynolds, R. J. *J. Am. Chem. Soc.* **1954**, *76*, 1291.
30. Burke, W. J.; Bishop, J. L.; Mortenson Glennie, E. L.; Bauer, W. N., Jr., *J. Org. Chem.* **1965**, *30*, 3423.
31. Ning, X.; Ishida, H. *J. Polym. Sci., Polym. Phys. Ed.* **1994**, *32*, 92.
32. Schreiber, U.S. Patent 3 274 5, 1988.

BICYCLIC ACETALS, RING-OPENING POLYMERIZATION

Masahiko Okada
Department of Applied Biological Sciences
School of Agricultural Sciences
Nagoya University

Ring-opening polymerization of bicyclic acetals is of particular importance in that it is closely related to chemical synthesis of polysaccharides. This synthesis has been achieved by three different techniques: stepwise condensation, polycondensation, and ring-opening polymerization. Among these, ring-opening polymerization of anhydro sugar derivatives is the most effective method for synthesizing stereoregular polysaccharides of high molecular weights.[1-4] The skeletons of these anhydro sugar derivatives are bicyclic acetals.

The stereoregularity of synthetic polysaccharides derived from anhydro sugar derivatives by ring-opening polymerization depends primarily on the bicyclic acetal skeleton, although in some cases substituents and initiators dramatically alter the steric course of the polymerization. Therefore, ring-opening polymerization of bicyclic acetals can be regarded as model reactions for polysaccharide synthesis from anhydro sugar derivatives, and it is of great significance to elucidate the reactivities and steric control of this process in relation to their structures.

Most bicyclic acetals are more strained than typical monocyclic acetals such as 1,3-dioxolane and 1,3-dioxepane, and they possess much higher ring-opening polymerizability. Synthesis of bicyclic acetals, especially highly strained ones, it is often difficult, but once they are formed, their ring-opening polymerizations with cationic initiators readily take place to give high molecular weight polyacetals in high yields. The backbones of polyacetals derived from bicyclic acetals, unlike those derived from monocyclic acetals, contain cyclic ether rings that may impart desirable mechanical and thermal properties. Reactivities and polymerization mechanisms of bicyclic acetals have been reviewed by several authors.[3,5-8]

2,6- AND 2,7-DIOXABICYCLO[2.2.1]HEPTANES

2,6- and 2,7-Dioxabicyclo[2,2.1]heptane (**Structure 1**) is one of the most reactive bicyclic acetals as judged by the relative rate of acid-catalyzed solvolysis.[9] It is polymerized at –78°C with phosphorus pentafluoride, boron trifluoride, and silicon tetrafluoride to give a mixture of soluble polymer and gel.

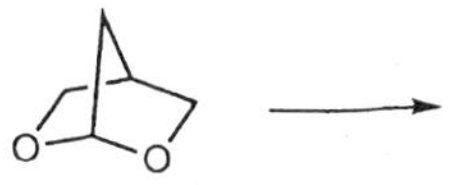

1

2,7-Dioxabicyclo[2.2.1]heptane (**Structure 2**) also is highly reactive. However, its reactivity in acid-catalyzed solvolysis is one order lower than that of Structure **I**.[10] The bicyclic acetal (Structure **3**) readily polymerizes with various cationic initiators such as phosphorus pentafluoride, boron trifluoride, silicon tetrafluoride, fluorosulfonic acid, methanesulfonic acid, and methyl trifluoromethanesulfonate.

The polymers obtained at –78°C are exclusively composed of tetrahydrofuran rings (**Structure 3**), whereas the polymers obtained at higher temperatures contain significant amounts of tetrahydropyran rings (**Structure 4**), although the tetrahydrofuran rings still prevail.

2,6-DIOXABICYCLO[2.2.2]OCTANE

2,6-Dioxabicyclo[2.2.2]octane (**Structure 5**), readily undergoes cationic polymerization to give polyacetal (**Structure 6**) containing tetrahydropyran rings.[9]

6,8-DIOXABICYCLO[3.2.1]OCTANE

Among bicyclic acetals, the polymerization of 6,8-dioxabicyclo[3.2.1] octane (**Structure 7**) has been most extensively studied.[11-14] It is readily polymerized with various Lewis acid initiators. Polyacetal (**Structure 8**) has been obtained in dichloromethane at or below –78°C with boron trifluoride etherate as initiator.[12]

6,8-DIOXABICYCLO[3.2.1]OCT-3-ENE AND 6,8-DIOXABICYCLO[3.2.1]OCT-2-ENE

Two unsaturated bicyclic acetals, 6,8-dioxabicyclo[3.2.1]oct-3-ene (**Structure 9**) and 6,8-dioxabicyclo[3.2.1]oct-2-ene (**Structure 11**), prepared from acrolein dimer, undergo cationic polymerization at –78°C in dilute dichloromethane solution to yield polymers that are completely (**Structure 10**) or nearly completely (**Structure 12**) soluble.

Polymerization at higher temperatures or in concentrated solutions leads to crosslinked insoluble polymers because of the concurrent side reactions involving the unsaturated bonds.

Because the polyacetals in Structures 10 and 12 have a reactive dihydropyran ring in their repeating units, they are converted to various functional polymers by chemical modifications. Of particular interest is the chemical synthesis of polysaccharides and their derivatives. The epoxidation of the olefinic bond with *m*-chloroperbenzoic acid or hydrogen peroxide/benzonitrile, followed by alkaline hydrolysis of the resulting epoxide, gives the structural unit in which two hydroxyl groups are *trans*-diequatorially or *trans*-diaxially oriented.[15] On the other hand, the oxidation of the olefinic bond with osmium tetroxide/hydrogen peroxide leads to the structural unit in which two hydroxyl groups are *cis*-oriented.[16]

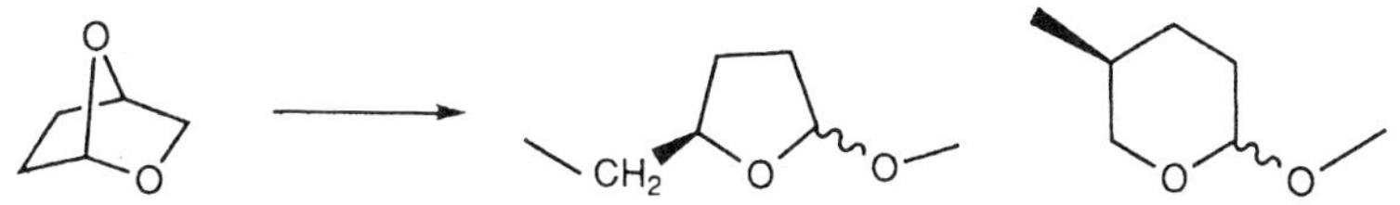

2 3 4

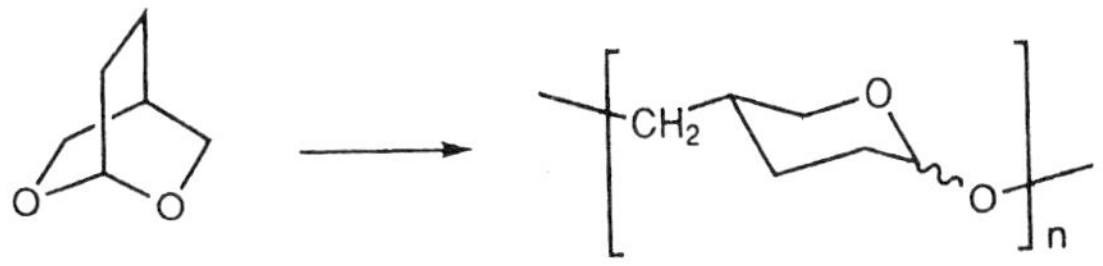

5 6

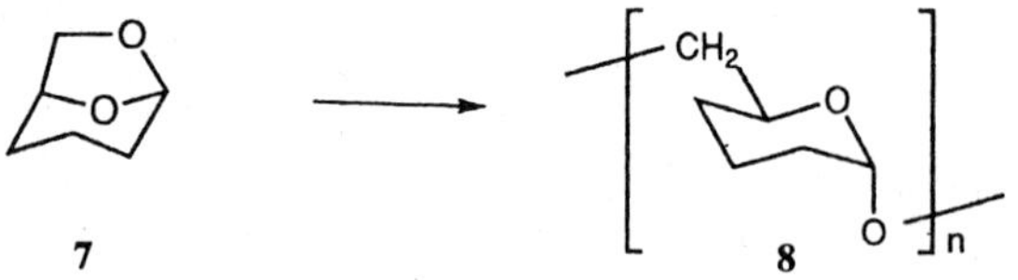

7 8

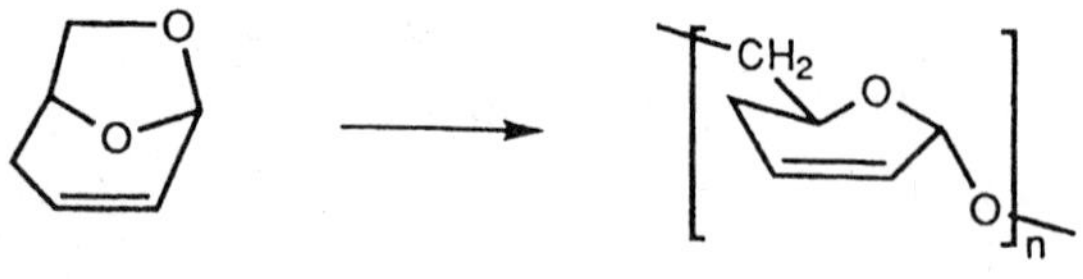

9 10

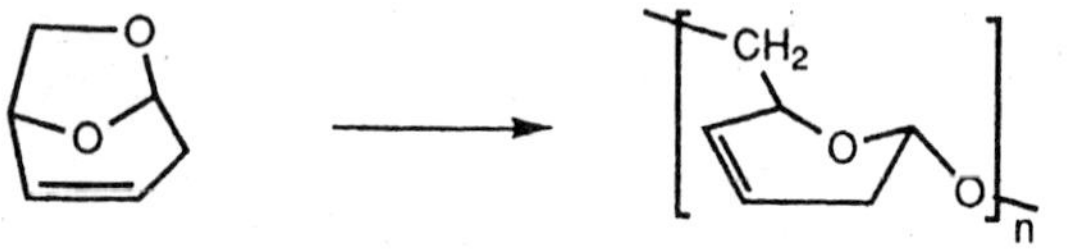

11 12

trans- AND *cis*-7,9-DIOXABICYCLO[4.3.0]NONANES

Substituted 1,3-dioxolanes are thermodynamically unfavorable for homopolymerization. Thus, attempted polymerization of *cis*-4,5-dimethyl-1,3-dioxolane and *trans*-4,5-dimethyl-1,3-dioxolane in bulk at low temperature gave only low molecular weight oligomers in very low yields.[17,18] *cis*-7,9-Dioxabicyclo[4.3.0]nonane corresponding to *cis*-4,5-disubstituted 1,3-dioxolane, gave only cyclic dimer in the attempted polymerization with phosphorus pentafluoride as initiator at temperatures from –25–0°C.[19] In contrast, *trans*-7,9-dioxabicyclo[4.3.0]nonane rapidly polymerized in bulk or in solution in the presence of phosphorus pentafluoride and triethyloxonium hexachloroantimonate to yield linear polyacetal, with number average molecular weights of several thousands, along with a small amount of a cyclic dimer.

The difference in the polymerization behavior of these *cis*- and *trans*-isomers can be unequivocally ascribed to the greater ring strain in the *trans*-isomer, that is, a more favorable free energy change for the polymerization of the *trans*-isomer.

REFERENCES

1. Schuerch, C. *Adv. Carbohydr. Chem. Biochem.* **1981**, *39*, 157.
2. Schuerch, C. *Encyclopedia of Polymer Science and Engineering*, 2nd ed.; 1985; Vol. 13, 162.
3. Sumitomo, H.; Okada, M. *Ring-Opening Polymerization*; Ivin, K.; Saegusa, T., Eds.; Elsevier-Applied Science: London, 1984; Vol. 1, Chapter 5.
4. Uryu, T. *Models of Biopolymers by Ring-Opening Polymerization*; Penczek, S. Ed., CRC: Boca Raton, 1989; Chapter 2.
5. Sumitomo, H.; Okada, M. *Adv. Polym. Sci.* **1978**, *28*, 47.
6. Yokoyama, Y.; Hall, H. K., Jr. *Adv. Polym. Sci.* **1982**, 42, 107.
7. Okada, M. *Prog. Polym. Sci.* **1991**, *16*, 1027.
8. Okada, M. *Adv. Polym. Sci.* **1992**, *102*, 1.
9. Hall, H. K., Jr.; Carr, L. J.; Kellman, R.; DeBlauwe, F. *J. Am. Chem. Soc.* **1974**, *96*, 7265.
10. Hall, H. K., Jr.; DeBlauwe, F. *J. Am. Chem. Soc.* **1975**, *97*, 655.
11. Kops, J. *J. Polym. Sci. A-1* **1972**, *10*, 1275.
12. Sumitomo, H.; Okada, M.; Hibino, Y. *J. Polym. Sci., Polym. Lett.* **1972**, *10*, 871.
13. Hall, H. K., Jr.; Steuck, M. J. *J. Polym. Sci., Polym. Chem.* **1973**, *11*, 1035.
14. Okada, M.; Sumitomo, H.; Komada, H. *Macromolecules* **1979**, *12*, 395.
15. Okada, M.; Sumitomo, H.; Komada, H. *Makromol. Chem.* **1978**, *179*, 949.
16. Komada, H.; Okada, M.; Sumitomo, H. *Makromol. Chem.* **1978**, *179*, 2859.
17. Okada, M.; Mita, K.; Sumitomo, H. *Makromol. Chem.* **1975**, *176*, 859.
18. Okada, M.; Mita, K.; Sumitomo, H. *Makromol. Chem.* **1976**, *177*, 2055.
19. Kops, J.; Spanggaard, H. *Makromol. Chem.* **1975**, *176*, 299.

BICYCLIC AND SPIROCYCLIC MONOMERS, OXYGEN-CONTAINING (Ring-Opening Polymerization)

Toshikazu Takata and Takeshi Endo
Department of Applied Chemistry
College of Engineering
Osaka Prefecture University
and
Research Laboratory of Resources Utilization
Tokyo Institute of Technology

Polymerization chemistry of oxygen-containing bicyclic and spirocyclic monomers has been mainly developed as that of monomers and materials showing volume expansion on expand polymerization and curing.[1] Monomers always polymerize with considerable volume shrinkage, but bicyclic and spirocyclic orthoesters polymerize without volume shrinkage or with slight volume expansion.[1-4] These orthoesters are spiroorthoesters (SOE), bicycloorthoesters (BOE), and spiroorthocarbonates (SOC) and have the following features:

- monomers reported before 1990 are only polymerized cationically;
- polymerization usually accompanies volume expansion;
- molecular weight for the polymers formed is low;
- monomers undergo tandem double-ring-opening polymerization through isomerization, and no volume shrinkage takes place;
- the polymerization mechanism is complicated because it involves isomerization, but some mechanistic features have been identified; and
- the monomers produce polymers formed via single-ring-opening polymerization under mild conditions during which volume shrinks.

CATIONIC POLYMERIZATION

Spiroorthoesters

Double-Ring-Opening Polymerization

Although SOE is easily synthesized from lactone and epoxide, SOEs obtained by this method are structurally limited by a five-membered 1,3-dioxolane skeleton.[5] Ring-opening polymerization of SOE with cationic initiators such as BF_3OEt_2

reportedly affords poly(ether ester).[1,4,7] Because epoxide and five-membered lactone do not copolymerize, SOE polymerization can be regarded as a completely alternating copolymerization or epoxide and lactone through an SOE intermediate. Or, SOE can be considered a hybrid of 2,2-disubstituted 1,3-dioxolane and 2,2-disubstituted tetrahydrofuran, neither of which can homopolymerize.[6]

In the polymerization mechanism, an oxonium or a stabilized carbenium ion, presumably formed by initial single-ring-opening with acid catalysis, are attacked by another SOE monomer.[1-4]

An initial attack on R^+ on only acetal oxygen but also ether oxygen can occur, eventually yielding a poly(ether ester) shown in Structure **2** of Scheme I.[8] The presence of 10–20% of head-to-head polymer indicates that the two mechanisms (**Schemes I** and **II**) occur simultaneously during SOE polymerization.[3,6,8,9]

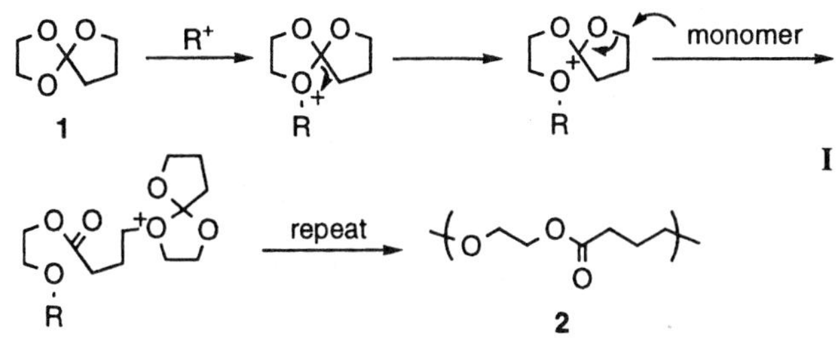

I

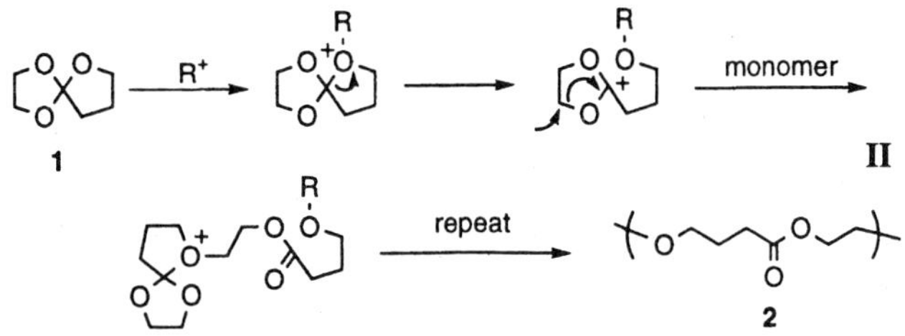

II

Single-Ring-Opening Polymerization

Single-ring-opening polymerization of SOEs was confirmed with *cationic catalysts* such as $SnCl_4$ at low temperatures.[10-14] For the single-ring-opening polymerization, the SOEs contained seven-membered ether rings because SOEs with five- and six-membered ether rings were not susceptible to the cationic catalyst.

Bicycloorthoesters

Double-Ring-Opening Polymerization

Bicycloorthoesters (BOE) are readily prepared by reacting a triol with an orthoester, a carboxylic acid via transesterification, or from acid chloride *via* oxetane.[1] BOEs polymerized readily with a Lewis acid such as BF_3OEt_2 and gave polyethers containing pendant ester groups.[1-4,15-18]

Single-Ring-Opening Polymerization

Cationic polymerization of 2,6,7-trioxabicyclo[2.2.1]-heptanes at low temperatures gave poly five- or six-membered (monocyclic orthoester)s depending on the substituents.[19-22] In the cationic polymerization of tricyclic BOEs, three possible monocyclic orthoester structures for the main chain units were produced.[20] Polymers obtained from the single-ring-opening polymerization were kinetically controlled products and those obtained from the double-ring-opening were thermodynamically controlled products.

Spiroorthocarbonates (SOC)

Double-Ring-Opening Polymerization

Of the three orthoesters, SOC polymerization have been studied the most extensively. Various substituted six-membered SOCs were synthesized and polymerized.[23-28] Synthesis of symmetrical SOCs was performed by the transesterification of tetraalkyl orthocarbonate with 2 moles of diols. In cationic polymerization of six-membered SOC, the corresponding poly(ether carbonate) was obtained selectively by a double-ring-opening isomerization polymerization. Methyl substitution on the SOC ring caused the polymerization rate to increase greatly.

The polymerization mechanism of SOC suggested that a monocyclic trialkoxycarbenium ion intermediate was involved. It gave poly(ether carbonate) (Structure **30**) through a nucleophilic attack of an SOC monomer on the ring methylene carbon adjacent to the oxygen (**Scheme X** and **III**), similar to the SOE and BOE mechanisms.

Effect of Ring Size

SOCs with other than six-membered rings do not polymerize without eliminating small molecules. Ring size-dependent polymerization behaviors of five- to seven-membered SOCs have been reported.[23,25,29,30]

Aromatic Spiroorthocarbonates

By synthesizing materials with high T_g by polymerization of SOCs, cationic polymerizations of totally aromatic mono- and bifunctional SOCs were examined.[31] Although some crosslinked polymers were obtained from the bifunctional SOCs, polymerization efficiency and polymer yield were very low even under drastic conditions.[31] Aromatic SOCs were stable when exposed to heat and acid, a feature also demonstrated by aromatic poly(spiroorthocarbonate)s, which had an SOC function in the main chain.[32,33]

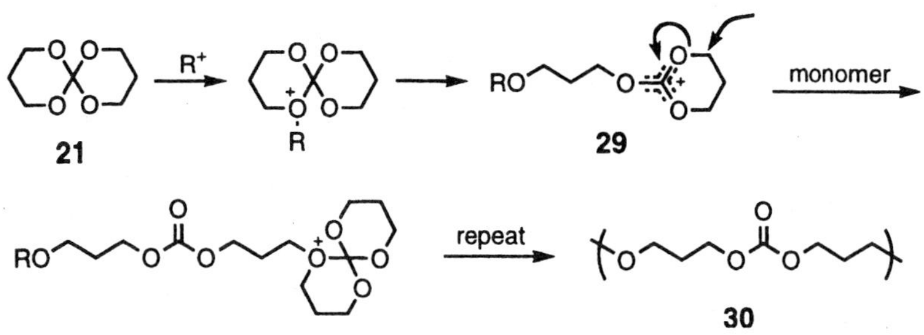

III

Copolymerization with Epoxy Compounds

A few reports described the possibility of copolymerizing some epoxy resins and SOCs by mixing and curing them with BF_3 complexes.[34,35] Takata et al. established that the copolymerization proceeded in a random mode.[36]

ANIONIC POLYMERIZATION

Homopolymerization

SOEs, BOEs, and SOCs are all alkali-resistant, so an electrophilic position was needed to prepare anionically polymerizable bicyclic and spirocyclic monomers. Lactone structure was an excellent candidate, although chain transfer reactions such as fast backbiting were encountered. Double-ring-opening polymerization of a bicyclic lactone was reported to proceed by either cationic and anionic initiators, although no detailed data were given. However, it could not be traced.

Copolymerization with Epoxide

Although bicyclic bislactones (BBLs) and spirocyclic bislactones (SBLs) did not homopolymerize under various anionic conditions, both BBLs and SBLs underwent complete alternating copolymerization with epoxy compounds, including epoxy resins.[37-40] The copolymerizations proceeded via an anionic tandem double-ring-opening process of BBL or SBL along with an opening of the epoxy ring. This polymerization proceeded via a living-like mechanism and could be applied to synthesizing block copolymers consisting of two alternating copolymer units.

Radical Polymerization

Several attempts have been made but, currently, no oxygen-containing bicyclic or spirocyclic monomer undergoes complete double-ring-opening polymerization under radical conditions.[41-51]

CONCLUSION

Cationic polymerizations of oxygen-containing bicyclic and spirocyclic monomers (SOE, BOE, and SOC) proceeded via single-ring- or double-ring-opening processes with isomerization depending on polymerization conditions and structure. Although the polymerizations of these monomers have been studied to develop expanding monomers, their detailed polymerization mechanisms are still unclear. Meanwhile, further development of monomers that can polymerize via a double-ring-opening isomerization process under radical and anionic conditions is needed. Perfect suppression of vinyl polymerization, which competes with the ring-opening polymerization, is expected because the radically polymerizable monomers are quite versatile. In addition to clarifying polymerization mechanisms, researchers must design monomers capable of undergoing double-ring-opening polymerization.

REFERENCES

1. Takata, T.; Endo, T. *Expanding Monomers - Synthesis, Characterization, and Applications*; Sadhir, R. K.; Luck, R. M., Eds.; CRC: Boca Raton, FL, 1992, Chapter 3, p 63.
2. Bailey, W. J.; Sun, R. L.; Katuski, H. et al. *Ring-Opening Polymerization*; Saegusa, T.; Goethals, E., Ed.; ACS Symposium Series 59; American Society: Washington, DC 1977; p 38.
3. Bailey, W. J.; Endo, T. *J. Polym. Sci., Polym. Symp.* **1978**, *64*, 17.
4. Takata, T.; Endo, T. *Prog. Polym. Sci.* **1993**, *18*, 839.
5. Bodenbenner, K. *Ann.* **1955**, *625*, 183.
6. Matyjaszewski, K. *J. Polym. Sci., Polym. Chem. Ed.* **1984**, *22*, 29.
7. Endo, T. *Polym. Prepr. Jpn.* **1984**, *33*, 39.
8. Bailey, W. J.; Iwama, H.; Tsushima, R. *J. Polym. Sci., Polym. Symp.* **1976**, *56*, 117.
9. Bailey, W. J. *Macromol. Sci. Chem.* **1975**, *A9*(5), 849.
10. Chikaoka, S.; Takata, T.; Endo, T. *Macromolecules* **1991**, *24*, 6557.
11. Chikaoka, S.; Takata, T.; Endo, T. *Macromolecules* **1991**, *24*, 331.
12. Chikaoka, S.; Takata, T.; Endo, T. *Macromolecules* **1991**, *24*, 6563.
13. Chikaoka, S.; Takata, T.; Endo, T. *Macromolecules* **1992**, *25*, 625.
14. Chikaoka, S.; Takata, T.; Endo, T. *Macromolecules* **1994**, *27*, 2380.
15. Endo, T.; Saigo, K.; Bailey, W. J. *J. Polym. Sci., Polym. Lett. Ed.* **1980**, *18*, 457.
16. Saigo, K.; Bailey, W. J.; Endo, T. et al. *J. Polym. Sci., Polym. Chem. Ed.* **1983**, *21*, 1453.
17. Uno, H.; Endo, T.; Okawara, M. *J. Polym. Sci., Polym. Chem. Ed.* **1985**, *23*, 63.
18. Endo, T.; Uno, H. *J. Polym. Sci., Polym. Lett. Ed.* **1985**, *23*, 359.
19. Yokoyama, Y.; Padias, A. B.; DeBlauwe, F. et al. *Macromolecules* **1980**, *13*, 252.
20. Yokoyama, Y.; Hall, H. K., Jr. *J. Polym. Sci., Polym. Chem. Ed.* **1980**, *18*, 3133.
21. Yokoyama, Y.; Padias, A. B.; Bratoeff, E. A. et al. *Macromolecules* **1982**, *15*, 11.
22. Hall, H. K., Jr.; Yokoyama, Y. *Polym. Bull.* **1980**, *2*, 281.
23. Saka, S.; Fujinami, T.; Sakurai, S. *J. Polym. Sci., Polym. Lett. Ed.* **1973**, *11*, 631.
24. Endo, T.; Arita, H. *Makromol. Chem. Rapid Commun.* **1985**, *6*, 137.
25. Fujinami, T.; Tsuji, H.; Sakai, S. *Polym. J.* **1977**, *9*, 552.
26. Endo, T.; Katsuki, H. *Makromol. Chem.* **1976**, *177*, 3231.
27. Endo, T.; Bailey, W. J. *Makromol. Chem.* **1975**, *176*, 2897.
28. Endo, T.; Sato, H.; Takata, T. *Macromolecules* **1987**, *20*, 1416.
29. Argia, T.; Takata, T.; Endo, T. *Macromolecules* **1992**, *25*, 3829.
30. Takata, T.; Endo, T. *Macromolecules* **1988**, *21*, 900.
31. Bailey, W. J.; Amone, M. J. *Polym. Prepr. Am. Chem. Soc. Div. Polym. Chem.* **1987**, *28*(1), 45.
32. Komatsu, S.; Takata, T.; Endo, T. *Macromolecules* **1991**, *24*, 2132.
33. Komatsu, S.; Takata, T.; Endo, T. *Polym. Prepr. Jpn.* **1991**, *40*, 302.
34. Takata, T.; Amachi, K.; Kitazawa, K. et al. *Macromolecules* **1989**, *22*, 3188.
35. Bailey, W. J.; Amone, M. J.; Issari, B. et al. *Polym. Prepr. Am. Chem. Soc. Div. Poly. Mat.* **1986**, *54*, 23.
36. Piggott, M. R.; Phshnov, T. *Polym. Mater. Sci. Eng.* **1986**, *54*, 18: *Chem. Abstr. 104*, 169310c.
37. Takata, T.; Tadokoro, A.; Endo, T. *Macromolecules* **1992**, *26*, 2782.
38. Tadokoro, A.; Takata, T.; Endo, T. *Macromolecules* **1993**, *26*, 4400.
39. Takata, T.; Tadokoro, A.; Chung, K. et al. *Macromolecules* in press.
40. Chung, K.; Takata, T.; Endo, T. *Macromolecules* in press.
41. Pan, C.-Y.; Wang, Y.; Bailey, W. J. *J. Polym. Sci., Polym. Chem. Ed.* **1988**, *26*, 2737.
42. Endo, T.; Bailey, W. J. *J. Polym. Sci., Polym. Lett. Ed.* **1975**, *13*, 193.
43. Endo, T.; Bailey, W. J. *J. Polym. Sci., Polym. Chem. Ed.* **1975**, *13*, 2525.

44. Sugiyama, J.; Yokozawa, T.; Endo, T. *Polym. Prepr. Jpn.* **1989**, *38*, 276.
45. Tagoshi, H.; Endo, T. *J. Polym. Sci., Polym. Chem. Ed.* **1989**, *27*, 1415.
46. Sugiyama, J.; Yokozawa, T.; Endo, T. *J. Polym. Sci., Polym. Chem. Ed.* **1990**, *28*, 3529.
47. Stansbury, J. W. *J. Dent. Res.* **1992**, *71*, 1408.
48. Sanda, F.; Takata, T.; Endo, T. *Macromolecules* **1993**, *26*, 729.
49. Sanda, F.; Takata, T.; Endo, T. *Macromolecules* **1993**, *26*, 737.
50. Sanda, F.; Takata, T.; Endo, T. *J. Polym. Sci., PART A: J. Polym. Chem.* **1994**, *32*, 323.
51. Sanda, F.; Takata, T.; Endo, T. *Macromolecules* **1995**, *28*, 1346.

BICYCLIC ETHERS, RING-OPENING POLYMERIZATION

Masahiko Okada
Department of Applied Biological Sciences
School of Agricultural Sciences
Nagoya University

Ring-opening polymerization of bicyclic ethers provides a convenient and efficient route to synthesizing polyethers containing cyclic structures in their backbones. There are, however, some drawbacks. Syntheses of highly strained bicyclic monomers are often difficult. Conversely, less strained bicyclic compounds, although readily prepared, are sometimes reluctant to polymerize or even do not homopolymerize for thermodynamic reasons. In other words, a ring-chain equilibrium leans far toward the cyclic monomer side.

Since some of the polyethers derived from bicyclic ethers can also be prepared by polycondensations of appropriate bifunctional compounds, one may say that polycondensation is preferable to ring-opening polymerization. However, ring-opening polymerization of strained bicyclic monomers clearly has advantages over polycondensation and polyaddition in that it generally produces polymers of high molecular weights under very mild conditions.[1,2]

The subject here is ring-opening polymerization of some bicyclic ethers to polyethers containing cycloalkane structures in their backbones or polymers. The cycloalkane moieties in the backbones can give proper rigidity to the polyether chains and improve the mechanical strengths and thermal stabilities. Reactivities of some bicyclic ethers and their polymerization mechanisms have been described in detail in original and review articles.[3-5]

7-OXABICYCLO[4.1.0]HEPTANE (CYCLOHEXENE OXIDE)

7-Oxabicyclo[4.1.0]heptene (Structure **1**), which is commonly called cyclohexene oxide, is usually classified under the epoxide category. However, this monomer gives a polyether (Structure **2**) that contains cyclohexane rings in the main chain, as other bicyclic ethers described below do, and hence the ring-opening polymerization of Structure 1 is included here. Since this monomer is an α,β-disubstituted epoxide, its reactivity in ring-opening polymerization is similar to other α,β-disubstituted epoxides—that is, it polymerizes both cationically and anionically (**Scheme I**).

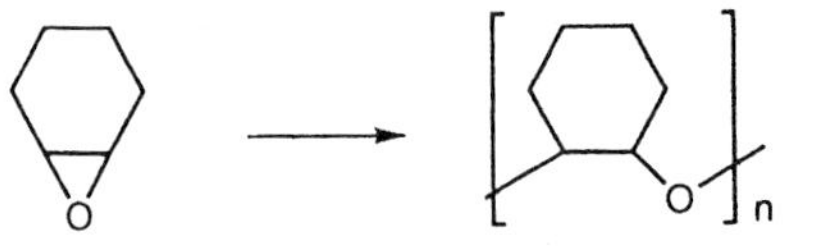

1 2 I

Stereospecific polymerization of **1** was first reported by Vandenberg.[6]

1,2-Cyclohexene oxide (**1**) exhibits polymerization behavior with several types of organometallic compounds that is quite different from that of simple epoxides such as propylene oxide.

Carbon dioxide copolymerizes with epoxides to give, in some cases, the alternating copolymers (aliphatic polycarbonates) under mild conditions. The copolymer from carbon dioxide and **1** by the $ZnEt_2$-H_2O system gives a *trans* (*threo*) diol upon alkaline hydrolysis, indicating the inversion of configuration at the carbon atom of the epoxide ring where the ring is cleft.[7]

Phthalic anhydride and **1** also gives an alternating copolymer in the presence of the (TPP)AlCl-$EtPh_3P^+Br^-$ system.[8]

Copolymerization of **1** and sulfur dioxide takes place in the presence of a Lewis acid or Lewis base catalyst. With pyridine as catalyst, the copolymerization gives poly(1,2-cyclohexylene sufite).[9]

7-OXABICYCLO[2.2.1]HEPTANE

Ring-opening polymerization of 7-oxabicyclo[2.2.1]heptane (Structure **3**) and its various alkyl-substituted derivatives have been investigated.[10-14] Polymerization of **3** gives a polyether (Structure **4**) containing cyclohexane rings with a 1,4-*trans* configuration. This is a powdery polymer with a melting point of 450°C (**Scheme II**).

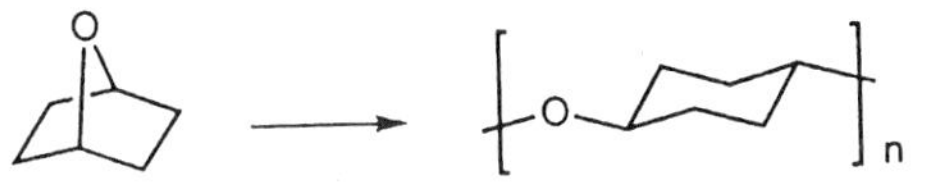

3 4 II

2-OXABICYCLO[2.2.2]OCTANE

2-Oxabicyclo[2.2.2]octane polymerizes with antimony pentachloride or tin tetrachloride in the presence of a small amount of epichlorohydrin as the cocatalyst.[15] The polymer is sticky solids that are mostly soluble in dichloromethane.

3-OXABICYCLO[3.2.2]NONANE

Phosphorus pentafluoride in combination with epichlorohydrin as a promoter initiates the polymerization of 3-oxabicyclo[3.2.2]nonane (**Structure 5**) in dichloromethane to give amorphous and sticky polymers (**Structure 6**) with number average molecular weights of several thousands. These polymers consist of repeating units in which the methylene ether groups on positions 1 and 4 of the cyclohexane ring are *cis* to each other.[16]

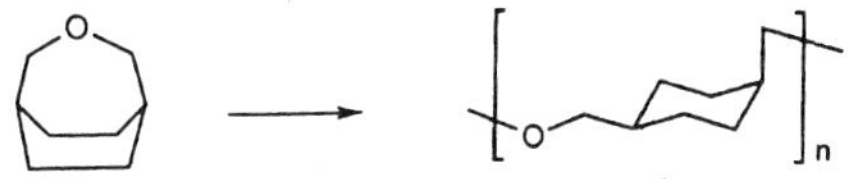

5 6

7- AND 8-OXABICYCLO[4.3.0]NONANES

Polymerization of *trans*-7-oxabicyclo[4.3.0]nonane (**Structure 7**) was first reported by Kops et al.[17] Phosphorus pentafluoride was used as the initiator for the polymerization and products with weight average molecular weights over 1×10^5 were obtained. These polymers are elastomeric solids at room temperature (Tg~0°C). NMR analysis revealed that the chain propagation occurred by predominant cleavage of the O–CH_2 bond, leading to a polyether (**Structure 8**) containing *trans*-1,2-linked cyclohexane rings.

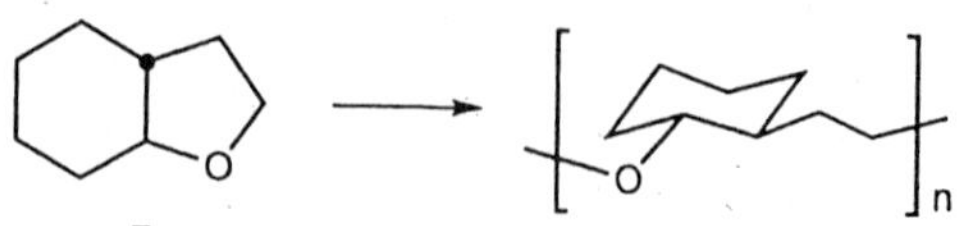

7 8

cis-7-Oxabicyclo[4.3.0]nonane is less strained than the *trans*-counterpart (Structure 7) and it is not homopolymerizable.

REFERENCES

1. Okada, M. *Prog. Polym. Sci.* **1991**, *16*, 1027.
2. Okada, M. *Adv. Polym. Sci.* **1992**, *102*, 1.
3. Hall, H. K. Jr. *J. Am. Chem. Soc.* **1958**, *80*, 6412.
4. Hall, H. K. Jr. *J. Am. Chem. Soc.* **1960**, *82*, 1209.
5. Yokoyama, Y.; Hall, H. K. Jr. *Adv. Polym. Sci.* **1982**, *42*, 107.
6. Vandenberg, E. J. *J. Polym. Sci.* **1969**, *Part A-1, 7*, 525.
7. Inoue, S.; Matsumoto, K.; Yoshida, Y. *Makromol. Chem.* **1980**, *181*, 2287.
8. Aida, Y.; Sanuki, K.; Inoue, S. *Macromolecules* **1985**, *18*, 1049.
9. Soga, K.; Kiyohara, K.; Hattori, I.; Ikeda, S. *Makromol. Chem.* **1980**, *181*, 2151.
10. Wittbecker, E. L.; Hall, H. K. Jr.; Campbell, T. W. *J. Am. Chem. Soc.* **1960**, *82*, 1218.
11. Saegusa, T.; Motoi, M.; Suda, H. *Macromolecules* **1976**, *9*, 231 and 526.
12. Baccaredda, M.; Giusti, P.; Cerrai, F.; Dimaina, M. *J. Polym. Sci., Part C* **1979**, *31*, 157.
13. Chapoy, L. L.; Matsuo, K.; Stockmayer, W. M. *Macromolecules* **1985**, *18*, 188.
14. Kops, J.; Spanggaard, H. *Polym. Bull.* **1986**, *16*, 507.
15. Saegusa, T.; Hodaka, T.; Fujii, H. *Polym. J.* **1971**, *2*, 670.
16. Andruzzi, F.; Ceccarelli, G.; Paci, M. *Polymer* **1980**, *21*, 1180.
17. Kops, J.; Larsen, E.; Spangaard, H. *J. Polym. Sci. Polym. Symp.* **1976**, *56*, 91.

BICYCLIC LACTAMS, RING-OPENING POLYMERIZATION

Kazuhiko Hashimoto
Department of Applied Chemistry
Faculty of Engineering
Kogakuin University

The pioneering and systematic investigation on the polymerizability of bicyclic lactams was reported by H. K. Hall, Jr.,[1,2] and was cited in review.[3] The driving force in the polymerization of bicyclic lactams is the release of strain, which is composed of the ring strain, the repulsion between hydrogen atoms on the neighboring carbon atoms, and the inhibition to the coplanarity of the amide bond.

A bicyclic lactam, 6-azabicyclo[3.2.1]octan-7-one (Structure **1**, **Scheme I**), can be polymerized easily in bulk, using sodium hydride as a catalyst at 210°C for only 10 min, to give a polyamide (**2**) in high yield: [η],0.41 (in sulfuric acid); m.p., >400°C.[1,2]

The very ready polymerization of Structure **1** is brought about mainly by the requirement of coplanarity for nitrogen-carbonyl interaction.[1]

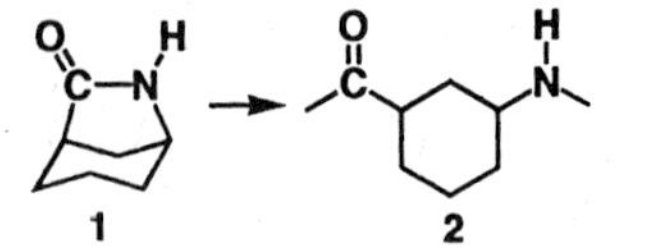

1 2 I

However, another bicyclo[3.2.1]octane-type lactam, 2-azabicyclo[3.2.1]octan-3-one has no polymerizability, because of the lack of ring strain in the seven- and six-membered rings containing the amide bond.[2]

The anionic polymerization of a bicyclo[2.2.2]octane-type lactam, 2-azabicyclo[2.2.2]octan-3-one (**Structure 3, Scheme II**), using sodium hydride and *N*-acetylcaprolactam as a catalyst and an activator at 260°C for 1.5 h gives a polyamide (**Structure 4, Scheme II**) in 21% yield. Lactam 3 has two boat forms in the skeleton, which are slightly twisted. During the polymerization, the strain in the system is relieved on conversion to the polymeric chair form.[1,2]

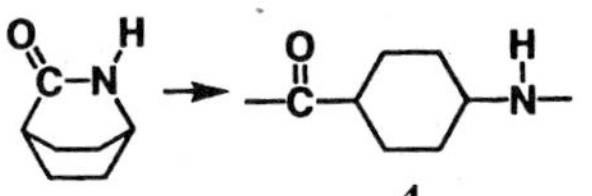

3 4 II

Ring-opening polymerization of an unsaturated bicyclic lactam, 2-azabicyclo[2.2.1]hept-5-en-3-one proceeds with the use of metathesis catalysts such as tungsten hexachloride-triethylaluminum (1:4) in chlorobenzene at 60°C for 12 hr. This yields a polymer consisting of alternating pyrrolidone rings and C=C moieties.[4]

Bicyclic lactams containing oxygen atoms in the skeletons show various and frequently conspicuous reactivities.[5,6] Ring-opening polymerization of a bicyclic oxalactam, 8-oxa-6-azabicyclo[3.2.1]octan-7-one (Structure **5**, **Scheme III**), has been extensively investigated.

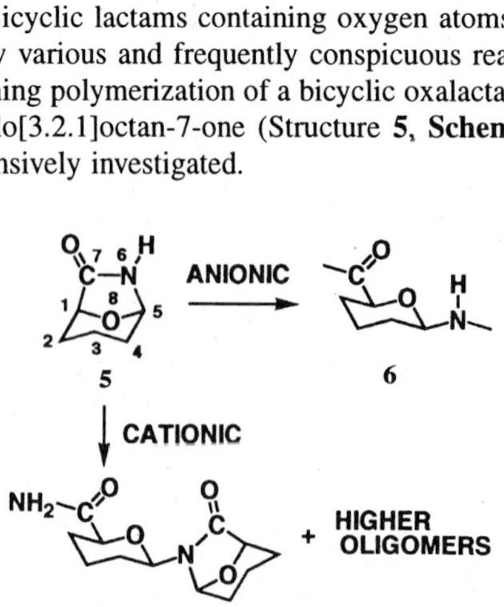

III

The high molecular weight polyamide (Structure **6**, **Scheme III**) is obtained in high yield by the anionic solution polymerization of Structure **5** at 25°C, using a trace amount of an activator (0.01–0.05 mol%)[8] [η],2.1 (in *m*-cresol at 25°C).[7] The product is soluble not only in *m*-cresol, 2,2,2-trifluoroethanol, but also in dimethyl sulfoxide and a mixture of chloroform and methanol, at room temperature. Thermal transitions appear at about 130°C, 260–285°C, and 315°C, due to glass transition, fusion, and decomposition, respectively. However, the bulk polymerization of structure **5** at 100°C gives a crosslinked polymer insoluble even in *m*-cresol.[7]

The optically active high molecular weight polyamide was also prepared by the anionic polymerization of an enantiomer, (+)-(1*R*,5*S*)-8-oxa-6-azabicyclo[3.2.1]octan-7-one, under a similar condition.[8]

Bridgehead substitution of a bicyclo[3.2.1]octane-type lactam reduces the ring-opening ability. The lactam ring in 1-methyl-8-oxa-6-azabicyclo[3.2.1]octan-7-one does not open with any anionic catalysts.[9]

An optically active bicyclic lactam synthesized from D-glucuronic acid, (1*S*,4*S*,5*R*)-4-(benzyloxy)-8-oxa-6-azabicyclo[3.2.1] oct-2-en-7-one is easily polymerized anionically at different temperatures (ranging from –60° to 25°C) in various solvents, to give acetone-insoluble polymers.[10]

Skeletons of bicyclic lactams have important effects upon their ring-opening polymerizability and properties of the resulting polyamides. Therefore, the ring-opening polymerization of bicyclic lactams can be useful for molecular design of novel polyamide materials.

REFERENCES

1. Hall, H. K. Jr. *J. Am. Chem. Soc.* **1958**, *80*, 6412.
2. Hall, H. K. Jr. *J. Am. Chem. Soc.* **1960**, *82*, 1209.
3. Sekiguchi, H. *Ring-Opening Polymerization*; Ivin, K. J.; Saegusa, T., Eds.; Elsevier: London and New York, 1984; Vol. 2, p 809.
4. Cho, H.-N.; Choi, K.-Y.; Choi, S.-K. *J. Polym. Sci., Polym. Chem. Ed.* **1985**, *23*, 1469.
5. Okada, M. *Prog. Polym. Sci.* **1991**, *16*, 1027.
6. Okada, M. *Prog. Polym. Sci.* **1991**, *102*, 1.
7. Sumitomo, H.; Hashimoto, K. *Macromolecules* **1977**, *10*, 1327.
8. Hashimoto, K.; Sumitomo, H. *Macromolecules* **1980**, *13*, 786.
9. Hashimoto, K.; Mori, K.; Okada, M., unpublished data.
10. Hashimoto, K. et al. *Macromolecules* **1992**, *25*, 2592.

BICYCLIC LACTONES, RING-OPENING POLYMERIZATION

Masahiko Okada
Department of Applied Biological Sciences
School of Agricultural Sciences
Nagoya University

Bicyclic lactones undergo ring-opening polymerization cationically and anionically go give polyesters having ring structures in their backbones.[1-3] The ring structures often improve the thermal and mechanical properties of the polyesters. In addition, oxacycles such as tetrahydrofuran and tetrahydropyran rings enhance the polarity of the polymers, and as a consequence, they are expected to impart adequate hydrophilicity to the polymers. This part deals with the synthesis and properties of two types of compounds—cyclic oligoesters and linear polyesters containing structures in the backbones—produced by ring-opening polymerization of several bicyclic lactones.

MACROCYCLIC OLIGOESTERS CONTAINING TETRAHYDROPYRAN RINGS

Preparation

Cyclic oligomers of various ring sizes are often formed in ring-opening polymerization of cyclic monomers.[4-6] For example, the anionic polymerization of ε-caprolactone by potassium tert-butoxide in tetrahydrofuran at first gives a polyester of high molecular weight, but is followed by slower depolymerization that yields a mixture of cyclic oligoesters of various ring sizes.[7]

Scheme I illustrates the unusual cationic polymerization behavior of a bicyclic lactone, 6,8-dioxabicyclo[3.2.1]octan-7-one (Structure **1**), which is prepared from sodium 3,4-dihydro-2*H*-pyran-2-carboxylate simply by acidification followed by distillation under reduced pressure.[8] Polymerization of **1** at 0°C in dichloromethane with boron trifluoride etherate as initiator gives a polymer containing both *trans*- and *cis*-linked five-membered 1,3-dioxolan-4-one rings (Structure **2**) in the backbone. The unusual formation of polymer **2** has been interpreted by a mechanism involving isomerization of a once-formed polyester backbone.[9]

Polymerization of **1** at –60°C produces, as expected, polyester **6**, with number-average molecular weights up to 3×10^4 in the initial stage of the polymerization. However, the polyester is gradually converted to cyclic oligomers, predominantly 20-membered cyclic tetramer **4** and 25-membered cyclic pentamer **5**, after a prolonged reaction.

In the polymerization of **1** at –40°C, only cyclic oligomers are produced, except in the initial stage, where higher oligomers and a linear polymer are formed. The sizes of the cyclic oligomers depend on the solvents employed for the polymerization.[10-11]

Properties

Since the epoch-making discovery of crown ethers by Pedersen in 1967,[12] a number of artificial macrocyclic ligands with specific ion and substrate selectivities have been designed, and their binding behavior has been investigated in detail.[13-15] The macrocyclic oligoesters described above consist of alternating tetrahydropyran and ester moieties, and they bear a structural resemblance to nonactin, which is a 32-membered, naturally occurring cyclic tetraester having a tetrahydrofuran ring in its repeating unit. Nonactin has been shown to play important roles in the regulation of metabolic behavior and is thought to selectively transport potassium ions through cell membranes.[16]

POLYESTERS CONTAINING CYCLOALKANE OR CYCLIC ETHER STRUCTURES IN THEIR BACKBONES

Hall and his co-workers synthesized several 2-oxabicyclo[2.1.1]hexan-3-ones from the corresponding 3-chlorocyclobutanecarboxylic acids.[17] These bicyclic lactones undergo

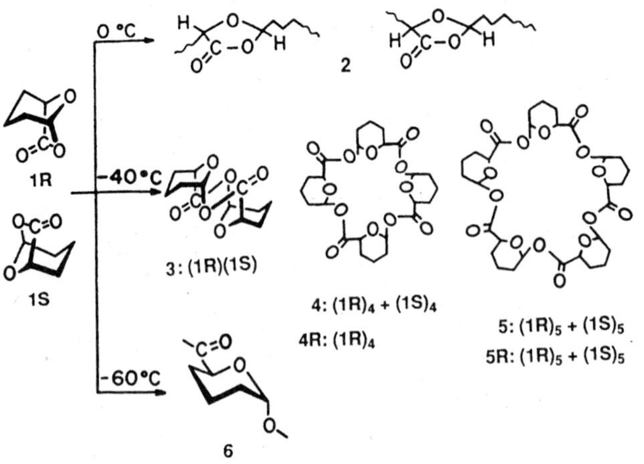

I

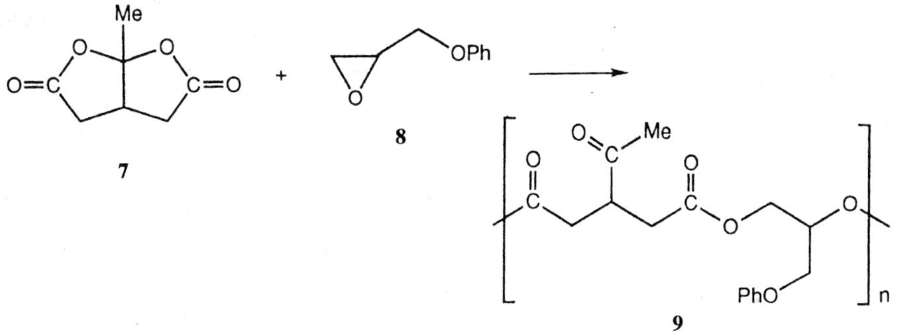

II

polymerization with a variety of anionic and cationic initiaors such as sodium–potassium alloy, sodium methoxide, and boron fluoride etherate, polyesters with inherent viscosities up to 3.43 dL/g and a melting point of 245°C have been obtained. They are soluble in chloroform and trifluoroacetic acid, and can be melt-pressed or cast into films or spun into fibers.

In marked contrast to 6,8-dioxabicyclo[3.2.1]octan-7-one (Structure **1**), described in the foregoing section, 6-oxabicyclo[3.2.1]octan-7-one polymerizes sluggishly to provide a low molecular weight polymer.[18]

2,6-Dioxabicyclo[2.2.2]octan-3-one is synthesized from acrolein and dimethyl malonate as starting materials via five step reactions.[19] The bicyclic lactone is readily polymerized with boron trifluoride etherate at or below –60°C to give a high molecular weight polyester in a high yield. It is also polymerized with anionic initiators such as lithium benzophenone ketyl, although higher temperature and longer reaction time are required.

A bis(γ-butyrolactone), 1-methyl-2,8-dioxabicyclo[3.3.0]octan-3,7-dione (Structure 7), does not homopolymerize with anionic initiators such as potassium *tert*-butoxide and sodium methoxide. However, it copolymerizes with phenyl glycidyl ether (Structure 8) to yield an alternating copolymer in the presence of potassium *tert*-butoxide at 120°C.[20] The 1:1 copolymer composition is not altered when the monomer feed ratio is varied from 20:80 to 80:20 (**Scheme II**).

CAS REGISTRY NO.

1, 1R, 3, 4R, 5, 5R.

REFERENCES

1. Sumitomo, H.; Okada, M. *Adv. Polym. Sci.* **1978**, *28*, 47.
2. Okada, M. *Prog. Sci.* **1991**, *16*, 1027.
3. Okada, M. *Adv. Polym. Sci.* **1992**, *102*, 1.
4. Goethals, E. J. *Adv. Polym. Sci.* **1977**, *23*, 103.
5. Penczek, S.; Kubisa, P.; Matyjaszewski, K. *Adv. Polym. Sci.* **1985**, *68/69*, 1.
6. Penczek, S.; Slomkowski, S. *Comprehensive Polymer Science*; Eastmond, G. C. et al., Eds.; Pergammon, Oxford, 1989; Vol. 3, p 725.
7. Ito, K.; Hashizuka, Y.; Yamashita, Y. *Macromolecules* **1977**, *10*, 821.
8. Okada, M.; Sumitomo, H. *Makromol. Chem. Suppl.* **1985**, *14*, 29.
9. Okada, M.; Sumitomo, H.; Atsumi, M. *Macromolecules* **1984**, *17*, 1840.
10. Tajima, I.; Okada, M.; Sumitomo, H. *Macromolecules* **1981**, *14*, 1180.
11. Okada, M.; Sumitomo, H.; Atsumi, M. *J. Am. Chem. Soc.* **1984**, *106*, 2101.
12. Pedersen, C. J. *J. Am. Chem. Soc.* **1967**, *89*, 7017.
13. Izatt, R. M.; Christensen, J. J. *Synthetic Multidentate Macrocyclic Compounds*; Academic: New York, 1978.
14. Izatt, R. M.; Christensen, J. J. *Synthesis of Macrocycles, the Design of Selective Complexing Agents*; John Wiley & Sons: New York, 1987.
15. Hiraoka, M. *Crown Compounds, Their Characteristics and Applications*; Elsevier: Amsterdam, 1982.
16. Haynes, D. H.; Pressman, B. C. *J. Membr. Biol.* **1974**, *18*, 1.
17. Hall, H. K. Jr.; Banchard, E. P. Jr.; Martin, E. L. *Macromolecules* **1971**, *4*, 142.
18. Hall, H. K. Jr. *J. Am. Chem. Soc.* **1958**, *80*, 6412.
19. Okada, M.; Sumitomo, H.; Atsumi, M.; Hall, H. K. Jr.; Ortega, R. B. *Macromolecules* **1986**, *19*, 503.
20. Takata, T.; Tadokoro, A.; Endo, T. *Macromolecules* **1992**, *25*, 2782.

Bicyclic Monomers

See: Bicyclic Acetals, Ring-Opening Polymerization
Bicyclic and Spirocyclic Monomers, Oxygen-Containing (Ring-Opening Polymerization)
Bicyclic Ethers, Ring-Opening Polymerization
Bicyclic Lactams, Ring-Opening Polymerization
Bicyclic Lactones, Ring-Opening Polymerization
Bicyclobutanes, Ring-Opening Polymerization

BICYCLOBUTANES, RING-OPENING POLYMERIZATION

Anne Buyle Padias and H. K. Hall, Jr.
Chemistry Department
The University of Arizona

Incorporation of a cyclic repeat unit in a polymer backbone has often been shown to give desirable properties to a polymer; these may include high glass-transition temperature and increased crystallinity, and therefore, improved mechanical properties. Bicyclobutanes polymerize by the breaking of the central 1,3-bond; as a result, their polymerization produces a polymer backbone consisting of 1,3-linked cyclobutane units (**Structure 1**).

X = CN, COOMe, ...

1

Bicyclobutanes are highly strained monomers, which makes them difficult to synthesize; however, this same high ring strain usually makes them easy to polymerize. The central bond acts in many respects more like a π-bond than like a carbon–carbon single bond, as shown in studies of nucleophilic addition to this bond.[1-5] The parent [1.1.0]bicyclobutane and donor-substituted bicyclobutanes are very sensitive to protonic conditions and cannot be isolated. Electron-withdrawing substituents such as cyano or ester groups stabilize the central 1,3-bond and make the synthesis of these bicyclobutanes possible.

The polymerization of bicyclobutanes was reviewed in the mid 1970s by Hall and Ykman[6] and in 1984 by Hall and Snow.[7]

POLYMERIZATIONS

Based on known thermochemical data, the enthalpy of polymerization of bicyclobutanes can be estimated at approximately −35 kcal/mol.[6] This extremely high enthalpy value is due to the ring strain in this small ring system, and explains the high polymerizability of this class of monomers.

Both free-radical and anionic polymerization mechanisms have been observed for bicyclobutane derivatives. The presence of least one electron-withdrawing substituent at the bridgeheads is a necessary condition for polymerization for both the free radical and the anionic mechanisms.

As reviewed by Hall[6,7] the following monomers led to high molecular weight in free-radical conditions: 1-cyanobicyclobutane, 2,2-dimethyl-1-cyanobicyclobutane, 1,3-dicyanobicyclobutane, methyl and *t*-butyl bicyclobutane-1-carboxylate and their 3-methyl derivatives, dimethyl bicyclobutane-1,3-dicarboxylate, 1-acetylbicyclobutane, and bicyclobutane-1-carboxylic acid and -1-carboxamide. All these monomers led to high yields of soluble polymers in standard free-radical conditions. It should be noted that the methyl 3-methylbicyclobutane-1-carboxylate polymerized, but not 1-cyano-3-methylbicyclobutane.

Most bicyclobutane monomers readily copolymerize with vinyl monomers under free-radical conditions.

Anionic homopolymerizations have usually been performed using 15% BuLi in THF at 0°C, and the following monomers polymerized well: 1-cyanobicyclobutane and its 3-methyl derivative, 1,3-dicyanobicyclobutane,[9] and methyl bicyclobutane-1-carboxylate.[8] The anionic polymerization of 1-cyano-3-methylbicyclobutane proceeded in high yield, in contrast to the free-radical polymerization. Bicyclobutane-1-carboxamide underwent anionic polymerization by proton transfer, leading to a polyamide.[8]

The geometry of the cyclobutane ring permits either 1,3-*cis* or 1,3-*trans*-enchainments during polymerization.

CONCLUSIONS

Many bicyclobutane polymers show very promising physical properties. The cyano-substituted bicyclobutanes lead to polymers, with excellent thermal and mechanical properties, superior to polyacrylonitrile. The improved thermal stability is due to the structure of the polymer, namely the increased distance between the consecutive nitrile groups. PCNB also has excellent piezoelectric properties.

The ester-substituted bicyclobutane polymers have been shown to have excellent optical properties. They are optically clear and have a high refractive index, which would make them suitable for such applications as optical fibers. However, their relatively low glass-transition temperatures do not allow their use in high-temperature environments.

The limiting factor in the commercial use of these polymers is the cost of monomers. These monomers require rather sophisticated synthesis methods, such as the high-pressure cycloaddition of allene to acrylonitrile, and the base-catalyzed ring closure to form the 1,3-central bond, which has to be performed in scrupulously anhydrous conditions. Until a more practical and lower-cost route has been developed, these polymers will only be useful for very high-tech applications.

REFERENCES

1. Blanchard, E. P. Jr.; Cairncross, A. *J. Am. Chem. Soc.* **1966**, *88*, 487.
2. Blanchard, E. P. Jr.; Cairncross, A. *J. Am. Chem. Soc.* **1966**, *88*, 496.
3. Hoz, S.; Aurbach, D. *J. Am. Chem. Soc.* **1983**, *105*, 7685.
4. Hoz, S.; Aurbach, D. *J. Org. Chem.* **1984**, *49*, 3285.
5. Hoz, S.; Aurbach, D. *J. Org. Chem.* **1984**, *49*, 4144.
6. Hall, H. K. Jr.; Ykman, P. J. *Polym. Sci., Macrom. Rev.* **1976**, *11*, 1.
7. Hall, H. K. Jr.; Snow, L. G. *Ring-Opening Polymerization*: Ivin, K. J.; Saegusa, T., Eds.; Elsevier: London, 1984; p 83.
8. Hall, H. K. Jr.; Smith C. D.; Blanchard, E. P. Jr.; Cherkofsky, S. C.; Sieja, S. B. *J. Am. Chem. Soc.* **1971**, *93*, 121.
9. Hall, H. K. Jr.; Fischer, W. *Helvet. Chim. Acta* **1977**, *60*, 1897.

Bilayers

See: Molecular Assemblies (Polymerization in Aqueous Solution)
Phospholipid Bilayers
Ultrathin Films (Self-Assembled)

Binders

See: Alkyd Emulsions
Furan Resins
Paper Coatings
Poly(ester amide) and Poly(ester imide) Resins (Binders for Air-Drying Protective Coatings)
Quinone-Amine Polyurethanes
Thermodegradable Polymers (Azo Groups in Main Chain)
Thermosensitive Paper (Thermosensitive Dyes, Thermochromic Compounds)

Binding Isotherms

See: Polyelectrolyte-Surfactant Complexes (Binding Isotherms)

BIOABSORBABLE POLYMERS (Tissue Engineering)

K. J. L. Burg
Department of Bioengineering
Clemson University

S. W. Shalaby
Poly-Med Incorporated
Center for Applied Technology

The field of tissue engineering his grown substantially over the last two years, especially the use of biodegradable polymers as scaffolds or templates. Cells are seeded on an absorbable polymeric or organic matrix, the system is implanted *in vivo*, and the matrix is gradually resorbed as the time develops.

PROPERTIES

Absorbable polymers that find use in scaffolding include *polyglycolide acid*[3-7] 90/10 poly(glycolide-co-l-lactide) or "polyglactin-910,[3,8,9] polyanhydrides and polyorthoesters,[8] and *poly-L-lactide acid.*[7,10-16]

Organic scaffolding materials include collagen gel[17] and hyaluronate gel. Faster-absorbing materials such as *polyglycolide acid* lead to a relatively unstable tissue because long-term tissue development has not occurred. Various coatings can be applied to the scaffold surface to promote *cell adhesion*, and cell survival including poly-L-lysine[18] and cell matrix proteins.[19]

APPLICATIONS

Cartilage Reconstruction

Cartilage degeneration has been a major focus of research because cartilage has minimal ability for self-repair and is therefore a common subject of reconstructive surgery. Attempts to repair cartilage defects have been made using direct chondrocyte transplantation as well as *naturally derived or synthetic seeded* scaffolds. In direct cartilage transplantation, excised healthy cartilage is used to repair the damaged area. To minimize the need for donor tissue and meet a particular shape requirement, scaffolding or direct chondrocyte transplantation is used. The applications include articular cartilage repair in the knee, reconstruction of nasoseptal cartilage,[2] and cosmetic and reconstructive surgery[1] where a three-dimensional custom-shaped implant is required.

The cell distribution on the matrix depends on the porosity or fiber spacing, the surface characteristics of the scaffold material and the viscosity of the cell suspension.

Liver Reconstruction

Cell transplantation would be ideal for liver replacement because living donors could be used, which would increase the donor pool. Collagenous scaffolds for hepatocytes have been found to lack sufficient mechanical integrity and induce an immune response. Various absorbable foams[15] and water discs[8] have also been tested as possible scaffolds. Advanced absorbable foams[20] and gels[21] have been described as suitable matrices for tissue regeneration.

Pancreas Reconstruction

Islets are clusters of cells in the pancreas whose damage may lead to diabetes or hypoglycemia. Direct islet allotransplantation has been examined in some detail.[22]

Osteoblasts

Bone regeneration by osteoblast transplantation may be useful in the repair of skeletal defects. Osteoblasts appear to attach and proliferate on a variety of synthetic absorbable polymers.[7]

Skin Reconstruction

Skin replacement research has expanded, largely in response to the needs of burn victims. An effective skin graft is quite difficult to manufacture because of the composite nature of the dermis and epidermis and the difficulty of revascularizing the system quickly enough to support the new tissue growth. Collagen-GAG (col-GAG) grafts seeded with fibroblasts and keratinocytes has been utilized,[3,23] however, they are susceptible to enzymatic degradation, possibly causing the breakdown of newly formed collagen. This problem led to the concept of seeding fibroblasts onto synthetic, hydrolytically degrading surfaces.[3]

Vascular Grafts

Ongoing research in this particular area involves implanting a biodegradable scaffold and allowing the tissue to regenerate *in vivo* as the material breaks down.[9-12,24]

CONCLUSIONS

Although progress has been made in the area of tissue engineering, the attainment of major future milestones will depend on the availability of novel forms of absorbable material and rigorous techniques for handling cells and implantable constructs.

REFERENCES

1. Kim, W. S.; Vacanti, J. P.; Cima, L.; Mooney, D.; Upton, J.; Puelacher, W. C.; Vacanti, C. A. *Plastic Reconstr. Surg.* **1994**, *94*(2), 233.
2. Puelacher, W. C.; Mooney, D.; Langer, R.; Upton, J.; Vacanti, J. P.; Vacanti, C. A. *Biomaterials* **1994**, *15*(10), 774.
3. Cooper, M. L.; Hansbrough, J. F.; Spielvogel, R. L.; Cohen, R.; Bartel, R. L.; Naughton, G. *Biomaterials* **1991**, *12*, 243.
4. Atala, A.; Vacanti, J. P.; Peters, C. A.; Mandell, J.; Retik, A. B.; Freeman, M. R. *J. Urology* **1992**, *148*, 658.
5. Freed, L. E.; Vunjak-Novakovic, G.; Langer, R. *J. Cellular Biochem.* **1993**, *51*, 257.
6. Freed, L. E.; Vunjak-Novakovic, G.; Biron, R. J.; Eagles, D. B.; Lesnoy, D. C.; Barlow, S. K.; Langer, R. *Bio/Technology* **1994**, *12*, 689.
7. Ishaug, S. L.; Yaszemski, M. J.; Bizios, R.; Mikos, A. G. *J. Biomed. Mater. Res.* **1994**, *28*, 1445.
8. Vacanti, J. P.; Morse, M. A.; Saltzman, W. M.; Domb, A. J.; Perez-Atayde, A.; Langer, R. *J. Pedia. Surg.* **1988**, *23*(1), 3.
9. Vacanti, J. P.; Shepard, J.; Retik, A. B. J. *Urology* **1993**, *150*, 608. This makes the *Surgery* article by Lei (currently numbered 12) NUMBER 13 and so forth.
10. Lei, B. v. d. B.; Bartels, H. L.; Nieuwenhuis, P.; Wildevuur, C. R. H. *Surgery* **1985**, *98*(5), 955.
11. Lei, B. v. d. B.; Wildevuur, C. R. H.; Nieuwenhuis, P.; Blaauw, E. H.; Dijk, F.; Hulstaert, C. E.; Molenaar, I. *Cell & Tissue Res.* **1985**, *242*, 569.
12. Lei, B. v. d. B.; Wildevuur, C. R. H.; Dijk, F.; Blaauw, E. H.; Molenaar, I.; Nieuwenhuis, P. *J. Thorac. & Cardiovasc. Surg.* **1987**, *93*(5), 695.
13. Beumer, G. J.; van Blitterswijk, C. A.; Bakker, D.; Ponec, M. *Clin. Mater.* **1993**, *14*, 21.
14. Beumer, G. J.; van Blitterswijk, C. A.; Ponec, M. *Biomaterials* **1994**, *15*(7), 551.
15. Wald, H. L.; Sarakinos, G.; Lyman, M. D.; Mikos, A. G.; Vacanti, J. P.; Langer, R. *Biomaterials* **1993**, *14*(4), 270.
16. Wake, M. C.; Patrick, C. W., Jr.; Mikos, A. G. *Cell Transplant.* **1994**, *3*(4), 339.
17. Wakitani, S.; Kimura, T.; Hirooka, A.; Ochi, T.; Yoneda, M.; Yasui, N.; Owaki H.; Ono, K. *J. Bone & Joint Surg.* **1989**, *71-B*(1), 74.
18. Sittinger, M.; Bujia, J.; Minuth, W. W.; Hammer, C.; Burmester, G. R. *Biomaterials* **1994**, *15*(6), 451.
19. Gilbert, J. C.; Takada, T.; Stein, J. E.; Langer, R.; Vacanti, J. P. *Transplantation* **1993**, *56*, 423.
20. Shalaby, S. W.; Roweton, S. L. U.S. wo Patent 05, 083, 1995.
21. Shalaby, S. W.; U.S. Patent Application 08/421, 222, 1995.
22. Kretschmer, G. J.; Sutherland, D. E. R.; Matas, A. J.; Cain, T. L.; Najarian, J. S. *Diabetologia* **1977**, *13*, 495.
23. Bell, E., Sher, S.; Hull, B.; Merrill, C.; Rosen, S.; Chamson, A.; Asselineau, D.; Dubertret, L.; Coulomb, B.; Lapiere, C.; Nusgens, B.; Neveux, Y. *J. Investig. Dermatol.* **1983**, *81* 2s.
24. Niu, S.; Kurumatani, H.; Satoh, S.; Kanda, K.; Oka, T.; Watanabe, K. *Am. Soc. Artific. Intern. Organs* **1993**, *39*(3), M750.

Bioadhesives

See: Water-Swellable Polymers (Carbomer Resins)

BIOARTIFICIAL MATERIALS

Paolo Giusti,* Luigi Lazzeri, and Maria Grazia Cascone
Department of Chemical Engineering
University of Pisa

Polymers and polymeric composite materials of natural and synthetic origin, constitute the broadest class of biomaterials, which have been defined as those materials intended to interface with biological systems to evaluate, treat, augment, or replace any tissue, organ, or function of the body. The main medical areas where polymers and composites have found a wide use are artificial organs, the cardiovascular system, orthopaedics, dentistry, ophthalmology, drugs, and drug delivery systems.[1]

In order to overcome the poor biological performance of synthetic polymers, and to enhance the mechanical characteristics of biopolymers, a new class of specifically designed biomaterials, called "Bioartificial Polymeric Materials," has been introduced.[2-4] On one hand, the physico-chemical and mechanical characteristics of these bioartificial materials, based on blends, composites, or interpenetrating polymer networks of both synthetic and biological polymers, would hopefully be enhanced owing to the presence of the synthetic component. On the other hand, the improvement of the biological characteristics relies on the idea of smoothing away the interactions of biomaterials with the living system by the use of a material inside which changes at a molecular level, due to synthetic-biological polymer interactions, have been already accomplished before getting in touch with the living tissue.

Biopolymers such as fibrin (FBN), collagen (CLG), or hyaluronic acid (HA) were used in preparing bioartificial materials in which the synthetic component was chosen among commercially available polymers such as polyurethanes (PU), poly(vinyl alcohol) (PVA) (9002-89-5), poly(acrylic acid) (PAA) (9003-04-7), poly(vinylsulfonic acid) (PVSA) (25053-27-4), poly(styrenesulfonic acid) (PSSA) (25704-18-1), and ethylene-vinyl alcohol copolymers (EVAL, Clarene®).

PREPARATION AND PROPERTIES

Fibrin-Based Bioartificial Polymeric Materials

Fibrin-based materials represent the first class of bioartificial materials that were studied. Fibrin by iself has been used as biomaterial for a number of applications.[5] Reputed to be involved in hemostasis and wound healing, fibrin has been applied as a hemostatic agent. As a biodegradable material, fibrin has been used for temporary tissue replacement and as an absorbable implant material in a number of surgical situations.

Blends of FBN and PU were used to produce highly porous small-diameter vascular prostheses, by way of a combined phase-inversion and spraying process.

Collagen-based Bioartificial Polymer Materials

Collagen is one of the most important structural protein in the vertebrate body and is the principal extracellular protein in the connective tissues. Because of its important role in the living systems, collagen has been extensively studied as a polymer for

*Author to whom correspondence should be addressed.

use in manufactured materials and has attracted great attention for its potential applications in the biomedical field including dialysis membranes, heart valves, wound dressing, and artificial skin.[6,7]

In preparing bioartificial polymeric materials containing soluble collagen (SC) as biological component, water soluble polymers such as PVA, PAA, PVSA, and PSSA have been used as synthetic components. Blends between SC and the above synthetic polymers have been prepared by mixing their aqueous solutions. These blends have been used to prepare materials in form of films or hydrogels.[8-12]

SC/PVA Films

PVA has shown a stabilizing effect on SC as a result of weak hydrogen bond interactions.

The results of the biodegradation analysis performed by using the collagenase test show that the presence of PVA causes a decrease of the digested collagen in the uncrosslinked films. This could be due to a steric hindrance screening produced by the PVA molecules interacting with the protein. As one would expect, the crosslinked samples are much more resistant to the enzymatic attack.

SC/Polyelectrolyte Blends

In the attempt to enhance the compatibility between collagen and the synthetic component, blends of SC and PAA, PVSA or PSSA have been investigated.[13] It is known from thermodynamics that the miscibility of a polymeric system can be improved by allowing some kind of interaction to take place between the components. The most common interactions that have been identified and found to enhance miscibility are: hydrogen bonding, ionic and dipolar interactions, π-electron interactions, and charge-transfer complexes. Since hydrogen bonds, such as those present in SC/PVA systems, are rather weak interactions, more promising could be the SC/PAA, SC/PVSA, and SC/PSSA systems owing to stronger acid-base like interactions and ionic interactions due to the polyelectrolyte nature of the polymers.

Hyaluronic Acid-based Bioartificial Polymeric Materials

Hyaluronic acid is a polymer belonging to the polysaccharide class. As with the proteins that so far have been considered, the interest related to biomedical applications of hyaluronic acid has grown rapidly.[14]

In preparing bioartificial polymeric materials containing HA as biological component, PVA and ethylene vinyl-alcohol copolymers (EVAL L6, EVAL P10, and EVAL R20 with low-, moderate, and high-ethylene content respectively) have been used as synthetic components, with a HA weight content ranging from 0 to 100%, have been prepared by mixing their aqueous solutions. These blends have been used to prepare materials in form of films or hydrogels.[15-17]

APPLICATIONS

Bioartificial polymeric materials can be used for a variety of biomedical applications including vascular prostheses, dyalisis membranes, wound dressing, and drug delivery systems.

Bioartificial polymer materials could be used in implantable devices to release biologically active substances; such additives could be growth factors, antibiotics, or other therapeutic agents, used in a controlled manner. Blends of SC and HA with PVA and PAA have been used to prepare hydrogels and sponges. These materials have been used as drug delivery systems for growth hormone (GH).[18,19] Systemic treatment with GH has been shown to cause an increase in bone formation and direct stimulation of chondrocytes. The development of controllable, long-term, effective release systems for the delivery of growth hormone and other growth factors may improve wound healing and tissue repair in a variety of biomedical applications.

HA/PAA sponges have been also evaluated as wound dressing.[20] The results of *in vitro* degradation tests indicate that the amount of HA digested by the enzyme hyaluronidase is dependent from the composition of the blends but is in general very small. *In vitro* experiments have shown that these materials are able to inhibit bacterial growth. The antibacterial effect, exerted mainly by PAA, and the stimulatory effect, exerted mainly by PAA, and the stimulatory effect on the tissue growth and repair, exerted by HA, make these sponges very promising materials in wound dressing applications.

REFERENCES

1. Giusti, P. *Biomaterials and Clinical Applications, Adv. in Biomaterials*; Elsevier: Amsterdam, NL, 1987; Chapter I.
2. Giusti, P.; Lazzeri, L.; Lelli, L. *Trends in Polymer Science* **1993**, *1*, 261.
3. Giusti, P.; Lazzeri, L.; Barbani, N.; Lelli, L.; De Petris, S.; Cascone, M. G. *Makromol. Chem. Macromol. Symp.* **1994**, *78*, 277.
4. Giusti, P.; Lazzeri, L.; De Petris, S.; Palla, M.; Cascone, M. G. *Biomaterials* **1994**, *15*, 1229.
5. Kerény, G. *Macromolecular Biomaterials*: CRC: Boca Raton, FL, 1984; Chapter IV.
6. Parkany, M. *Macromolecular Biomaterials*; CRC: Boca Raton, FL, 1984; Chapter V.
7. Pentapharm, A. G. U.S. Patent 4 066 083, 1978.
8. Barbani, N.; Giusti, P.; Lazzeri, L.; Polacco, G.; Pizzirani, G. *J. Biomater. Sci. Polymer Edn.* **1995**, *7*, 461.
9. Barbani, N.; Cascone, M. G.; Giusti, P.; Lazzeri, L.; Polacco, G.; Pizzirani, G. *J. Biomater. Sci. Polymer Edn.* **1995**, in press.
10. Barbani, N.; Lazzeri, L.; Lelli, L.; Bonaretti, A.; Seggiani, M.; Narducci, P.; Pizzirani, G.; Giusti, P. *J. Matr. Sci. Mater. Med.* **1994**, *5*, 882.
11. Cascone, M. G.; Lazzeri, L.; Barbani, N.; Polacco, G.; Pollicino, A.; Giusti, P. *J. Matr. Sci. Mater. Med.* **1995**, (in press).
12. Giusti, P.; Lazzeri, L.; Barbani, N.; Narducci, P.; Bonaretti, A.; Palla, M.; Lelli, L. *J. Matr. Sci. Mater. Med.* **1995**, *4*, 538.
13. Lazzeri, L.; Barbani, N.; Cascone, M. G.; Polacco, G.; Palla, M.; Giusti, P. *Proc. 12th European Conference on Biomaterials*; Porto: Portugal, September 10-13, 1995; p 59.
14. Swann, D. A.; Kuo, J. W. *Biomaterials, Novel Materials From Biological Sources:* Stockton: New York, 1991; Chapter VI.
15. Seggiani, M.; Lazzeri, L.; Giusti, P. *J. Mater. Sci. Mater. Med.* **1994**, *5*, 877.
16. Lazzeri, L.; Barbani, N.; Cascone M. G.; Lupinacci, D., Giusti, P.; Laus, M. *J. Mater. Sci. Mater. Med.* **1994**, *5*, 862.
17. Sbarbati Del Guerra, R.; Cascone, M. G.; Barbani, N.; Lazzeri, L. *J. Sci. Mater. Med.* **1994**, *5*, 613.
18. Cascone, M. G.; Sim, B.; Downes, S. *Biomaterials* 1995, *16*, 569.
19. Cascone, M. G.; Di Silvio L.; Sim, B.; Downes, S. *J. Mater. Sci. Mater. Med.* **1994**, *5*, 1970.

20. Lazzeri, L.; Barbani, N.; Lelli, L.; Bonaretti, A.; Centonze, P.; Giusti, P. *Proc. 10th European Conference on Biomaterials*; Davos: Switzerland, September 8-11, 1993; p 78.

Biocatalysts

See: *Enzymatic Polymerization*
Enzyme-Catalyzed Oxidative Polymerization (Aromatic Compounds)
Enzyme-Catalyzed Polymerization
Enzyme-Degradation Hydrogels
Enzyme Immobilization (Polyion Complexes)
Enzyme Immobilization (Porous Polymer Beads)
Enzyme Mimics (Cyclodextrins)
Enzyme Modification (Use in Organic Media)
Enzyme-Poly(N-isopropylacrylamide) Conjugates (Thermoreversibly soluble/Insoluble)
Functional Immobilized Biocatalysts (Stimulus-Sensitive Gels)
Globular Polybases, Microemulsion-like
Immobilized Enzymes, Biosensors
Metal Complex Catalysts, Polymeric
Metalloenzymes, Artificial (Polyethylenimine-Based)
Microgels, Macrogels (by Photocrosslinking)
Reverse Micelles (Microcontainers for Functional Polymers)
Synzymes

Biocidal Polymers

See: *Additives (Property and Processing Modifiers)*
Biocidal Polystyrene Hydantoins
Interpolymers

BIOCIDAL POLYSTYRENE HYDANTOINS

S. D. Worley* and G. Sun
Department of Chemistry
Auburn University

Research for developing new water-soluble biocidal materials has been performed at our laboratories since 1980. All of the materials studied can be classified as cyclic, organic, *N*-halamine compounds, primarily in the classes oxazolidinones and imidazolidinones.[1-6] Although these new compounds have proved biocidal and stable in water for extensive periods, they are soluble over 1000 mg l^{-1}, which subjects them to considerable toxicity evaluation before they can be approved for commercial use. Thus, the emphasis on biocide development here has shifted to preparing and investigating insoluble biocidal polymers. The new polymers described here were prepared from inexpensive commercial polystyrene or poly(α-methylstyrene) and contained substituted *N*-chlorohydantoin functional groups. The monomeric *N*-halohydantoins have been used for many years as water disinfectants, particularly for swimming pools and hot tubs, and a polystyrene hydantoin should obtain regulatory approval easily. A preliminary account describing the preparation and some of the properties for one of the polymers, poly[-1,3-dichloro-5-methyl-5-(4′-vinylphenyl)hydantoin] (I-Me), has appeared.[7]

*Author to whom correspondence should be addressed.

PREPARATION

Complete details for all polymer preparations are presented elsewhere.[8]

APPLICATIONS

The applications for the *N*-halamine polymers are numerous. Certainly, a primary use would be in filter units for potable water, hot tubs, swimming pools, and gas streams, all of which are supported by existing data. A second application would be in biocidal fibers and elastomers for fabrics, bandages, and condoms. Insoluble biocidal polymers could be used as preservatives in paints, waxes, and oils as long as those materials did not contain reducing agents as additives. If the biocidal *N*-halamine polymers could be anchored to material surfaces, a means of preventing biofilm formation would be at hand. Work in all of these areas is progressing in our laboratories at this time.

ACKNOWLEDGMENT

The authors wish to thank W. B. Wheatley for the biocidal testing results discussed here. We also acknowledge partial support of this work by the Department of the Interior, U.S. Geological Survey, through the Alabama Water Resources Research Institute at Auburn University. The contents of this publication do not necessarily reflect the views and policies of the Department of the Interior, nor does the mention of trade names or commercial products constitute their endorsement by the U.S. government.

REFERENCES

1. Worley, S. D.; Williams, D. E. *CRC Crit. Rev. Environ. Contr.* **1988**, *18*, 133.
2. Worley, S. D.; Burkett, H. D. *Wat. Res. Bull.* **1984**, *20*, 365.
3. Barnela, S. B.; Worley, S. D.; Williams, D. E. *J. Pharm. Sci.* **1987**, *76*, 245.
4. Tsao, T. C.; Williams, D. E.; Worley, S. D. *Biotechnol. Prog.* **1991**, *7*, 60.
5. Williams, D. E.; Worley, S. D.; Barnela, S. B. et al. *Appl. Environ. Microbiol.* **1987**, *53*, 2082.
6. Worley, S. D.; Williams, D. E.; Barnela, S. B. *Wat. Res.* **1987**, *21*, 983.
7. Sun, G.; Wheatley, W. B.; Worley, S. D. *Ind. Eng. Chem. Res.* **1994**, *33*, 168.
8. Sun, G. Ph.D. Thesis., Auburn University, 1994.

BIODEGRADABLE COPOLYMERS [Compromising Poly(L-Lactic Acid)]

Tetsuji Yamaoka and Yoshiharu Kimura
Department of Polymer Science and Engineering
Kyoto Institute of Technology

Poly(L-lactic acid) (PLLA) is one of the most widely used biodegradable polymers, with a variety of biomedical and pharmaceutical applications. PLLA has excellent mechanical properties and can be slowly broken down into nontoxic metabolites by bioorganisms. However, the range of its applications has been somewhat limited because of difficulty in controlling the hydrolysis rate,[1] poor hydrophilicity,[2] and high rigidity and crystallinity.[3,4]

To solve these problems, some copolymers of PLLA have been developed.

Here, two kinds of copolymer systems comprising PLLA are dealt with. One is the A–B–A block copolymers comprising PLLA (A) and polyether (B); examples are poly(propylene glycol) (PPG) and poly(ethylene glycol) (PEG). The other is a random copolymer of L-lactic acid and another hydroxy acid metabolite, malic acid. The pendant carboxyl groups incorporated in the latter can work as the sites of water absorption, as the catalyst for hydrolysis, and as functional groups for immobilizing bioactive agents. These derivatives have recently been developed for well-controlled polymerization and new monomer design. Their potential applications as functional biodegradable polymers cover not only the biomedical field but also commodity plastics.

PREPARATION

PLLA-PPG-PLLA Copolymers

PLLA-PPG-PLLA copolymers (**Scheme I**) were prepared using mono-dispersed PPGs whose number average molecular weight (M_n) are 2000 and 4000, with a molecular weight distribution (M_w/M_n) around 1.1.

I

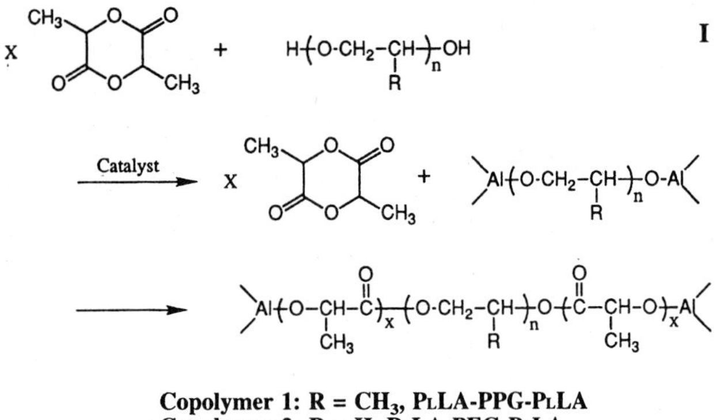

Copolymer 1: R = CH_3, PLLA-PPG-PLLA
Copolymer 2: R = H, PLLA-PEG-PLLA

SCHEME I. Copolymerization of l-lactide and PPG or PEG.

PLLA-PEG-PLLA Copolymers

PLLA-PEG-PLLA copolymers **2** (Scheme I) were prepared by bulk polymerization of L-lactide and PEGs with number average molecular weights (M_n) of 200, 1000, and 2000, using triethylaluminum (TEA) as catalyst.

Copolymer with Pendant Carboxyl Groups

Copolymers with pendant carboxyl groups as side chains were prepared by copolymerization of L-lactide and 3-(*S*)-[(benzyloxycarbonyl)-methyl]-1,4-dioxane-2,5-dione (BMD), a novel cyclic diester monomer consisting of both glycolate and benzyl-α-L-malate units. The copolymerization was carried out in bulk, with stannous 2-ethylhexanoate as catalyst.

APPLICATIONS

In Vitro and *In Vivo* Degradation of PLLA-PPG-PLLA Copolymer Fiber

Melt-spinning of copolymer **1** was carried out using a ram extruder having an orifice of 0.5 mm in diameter. The polymer melt extruded at 150°C was drawn and wound on a winder to obtain a filament about 50–80 μm in diameter. The filament was then stretched to various draw ratios at 80–90°C and was heat-treated at the same temperature for 10 min. Both the spinnability and drawability were excellent for all the copolymers except the one whose molecular weight was less than 20,000.

The tensile strength and modulus decreased with increasing PPG. This implied that the incorporation of PPG segments in PLLA is significantly effective for increasing the flexibility of the fiber.

The drawn fiber was coiled to a small loop and fixed with an absorbable suture of silk. It was then sterilized with ethylene oxide gas and implanted in the subdermal tissue of male wister rats.

After implantation for 1 week, the fiber was found to be enfolded by the tissues, but without adhesion of any cell and tissue. Histological observation of fiber cross-sections revealed a slight infiltration of inflammatory cells. The decrease in tensile strength was faster for the fiber comprising a higher content of PPG. This tendency was parallel to the results of the *in vitro* degradations using a phosphate buffer. We concluded that the biodegradability of PLLA can be much improved by the introduction of a small amount of PPG segments and that the degradation rate can possibly be regulated by the PPG content.

Drug Release from a PLLA-PEG-PLLA Copolymer Film

Films of PLLA-PEG-PLLA which contained the hydrophilic anti-cancer drug mitomycin C (MMC), were prepared by casting a chloroform solution of the copolymers and MMC.

In the case of PLLA homopolymer, a rapid release of MMC (about 20%) was observed in the early stage, perhaps because of a heterogeneity of the film. Thereafter, no more drug was released, although MMC remained in the matrix.

In contrast, from the copolymer films, more than 70% of MMC was released at the initial stage, which is called a burst effect. The release percentage did not reach at 100% because MMC was denatured or degraded in PBS with time.

On the other hand, blend matrices of these polymers exhibit excellent release profiles of MMC for 4 days. Moreover, the release rate of MMC could be controlled by the blending ratio, and the higher PEG content gives a higher release rate with no burst effect. The reason for the successfully controlled release seems to be the miscibility of the copolymer and so highly hydrophilic a drug as MMC, based on the hydrophilicity of polymer comprising the hydrophilic PEG segment.

Cell Attachment of Copolymers with Pendent Carboxyl Groups

As an example of biofunctionalization of a biodegradable polymer, the bioactive peptide RGDS was immobilized on the surface of the copolymer to increase its affinity with cells or tissues. The results show that cell adhesion can be greatly enhanced by immobilization of RGDS.

REFERENCES

1. Miller, R. A.; Brady, J. M.; Cutright, D. E. *J. Biomed. Mater. Res.* **1977**, *11*, 711.
2. Chu, C. C. *Biomaterials* Winter, G. D.; Gibbons, D. F.; Plenk, H. Jr., Eds.; Wiley: New York, 1982; p 781.
3. Eling, B.; Gogloewski, S.; Pennings, A. J. *Polymer* **1982**, *23*, 1587.
4. Zwiers, R. J. M.; Gogloewski, S.; Pennings, A. J. *Polymer* **1983**, *24*, 167.

BIODEGRADABLE OPTICALLY ACTIVE POLYAMIDES (Stereoregular Polyamides from L-Tartaric Acid)

J. J. Bou, A. Rodríguez-Galán, and S. Muñoz-Guerra*
Department d'Enginyeria Química
Universitat Politècnica de Catalunya
ETSEIB

Interest in the development of biodegradable polymers is increasing rapidly owing to both the expanding demand for them as ecologically sound disposable materials and their potential use in biomedical and pharmaceutical applications.

Modification of the chemical structure of nylons with the purpose of enhancing their susceptibility to biodegradation without excessively diminishing their genuine physical properties is a current issue of research.

Polyamides based on L-tartaric acid were first reported by Minoura.[1] Since then, a number of papers on the polycondensation of differently *O*-protected and unprotected dimethyl-L-tartrates with 1,6-hexamethylenediamine have been published by Ogata.[2-4] More recent works on polytartaramides mainly focus on the use of these polymers as releasing pesticide carriers[5] and explore their hydrolytic and microbial degradability.[6]

Polycondensation of tartaric acid derivatives is by no means straightforward. Adequate protection of the lateral hydroxyl groups together with efficient activation of the condensing functions are required in order to avoid severe polymerization conditions leading to epimerization and crosslinking side reactions. Otherwise, noncrystalline low molecular weight products are usually obtained.

*Author to whom correspondence should be addressed.

The polycondensation method developed some years ago by Katsarava[7] for synthesizing polyamides from sensitive monomers is based on activating of the diacid as a diactive ester and the diamine as a bis(trimethylsilyl) derivative. By this method, high polymerization conversions can be achieved in solution at room temperature. Combining of this procedure with adequate protection of the hydroxyl groups is the strategy we followed to obtain linear polyamides from L-tartaric acid without distorting the initial configuration of this compound.[8] The three types of polyamides based on L-tartaric acid that we investigated are shown in **Table 1**.

PROPERTIES

These polyamides are white solids that occasionally display a slightly yellowish color. They all are highly crystalline polymers that yield transparent, consistent films with a spherulitic texture when cast in chloroform or formic acid. They may be also stretched in well-oriented fibers either from solution or from the melt. Some chemical, thermal, and optical properties of these polymers have been systematically studied and found to correlate well with the molecular structure.

Solubility

The presence of the tartaric moiety largely modifies the solubility behavior typically found in conventional nylons. The effect is more pronounced at both extremes of each series, that is, when the concentration of such moieties in the polymer chain is either high or low. Thus, certain lower members may be soluble in solvents unusual for polyamides as they are water or ethanol. On the other hand, solubility in chloroform steadily increases with the length of the polymethylene segment to the point that P*n*DMLT with $n > 4$ as well as P9- and P12MLT become readily soluble at room temperature.

Thermal Behavior

The differential scanning calorimetry (DSC) analysis of these polyamides reveals that the first heating runs all display well-defined endotherms corresponding to melting transitions. Multiple peaks are common, a phenomenon often found in polyamides and interpreted as due to the melting of crystallites of different sizes. Melting points decrease with the concentration of amido groups in the polymer chain and, in the case of the P*n*DMLT series, they follow the zigzag curve typical of polycondensation polymers containing flexible polymethylene segments.

TABLE 1. Polyamides Based on L-tartaric Acid

Polyamide	Abbreviation	Monomers	*n*	Reference
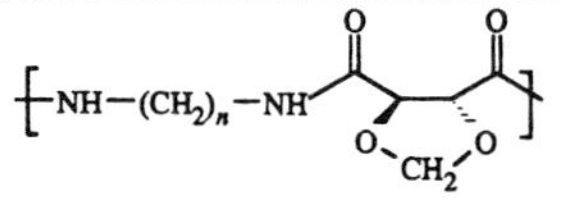	P*n*MLT	IX*n* + IVa	9, 12	9
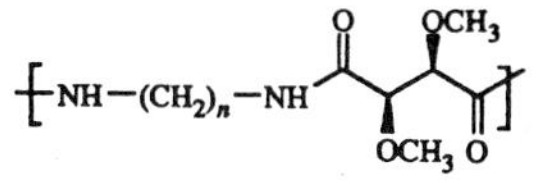	P*n*DMLT	IX*n* + IVb	2–9, 12	10
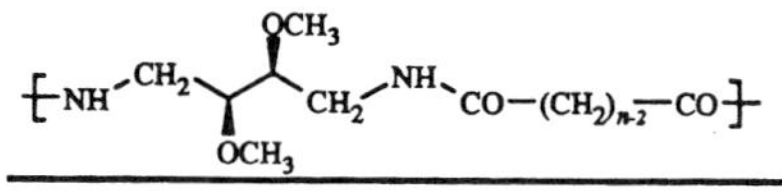	PDMBA*n*	VII + VIII*n*	4,6,8,10,12	11

Optical Activity

As logically expected from the asymmetric constitution of these polyamides, in solution all of them display optical activity that steadily increases with the concentration of stereocenters in the polymer chain.

Hydrolytic Degradation

The hydrolytic degradation of polyamides P*n*DMLT has been investigated on discs immersed in buffered aqueous solutions, as a function of time, pH, and temperature.[12] The loss of mass taking place at degradation was measured and variations in the molecular weight were followed by viscosimetry and gel permeation chromatography analysis.

ACKNOWLEDGMENT

Financial support of this project from the CICYT (Grant numbers MAT90-0779-CO2-02 and MAT93-555-CO2-02) is gratefully acknowledged.

REFERENCES

1. Minoura, Y.; Urayama, S.; Noda, Y. *J. Polym. Sci., Polym. Chem. Ed.* **1967**, *3*, 2441.
2. Ogata, N.; Hosoda, Y. *Polym. Lett.* **1974**, *12*, 355.
3. Ogata, N.; Hosoda, Y. *J. Polym. Sci., Polym. Chem. Ed.* **1975**, *13*, 1793.
4. Ogata, N.; Sanui, K.; Kuwahara, M.; Nakamura, H. *J. Polym. Sci., Polym. Chem. Ed.* **1980**, *18*, 939.
5. Akelah, A.; Kenawy, E. R.; Sherrington, D. C. *Eur. Polym. J.* **1992**, *28*, 453.
6. Aikawa, T.; Fuchino, H.; Sanui, K.; Kusuru, K.; Higashihara, T.; Sato, A. *Polym. Prep. Japan (English Edition)* **1992**, *39*, E962.
7. Katsarava, R. M.; Kharadze, D. P.; Avalishvili, L. M.; Zaalishvili, M. M. *Makromol. Chem. Rapid Commun.* **1984**, *5*, 585.
8. Bou, J. J.; Rodríguez-Galán, A.; Iribarren, I.; Muñoz-Guerra, S. *Biogradable Polymers and Plastics*; Royal Soc. Chemistry: Cambridge, UK, 1992.
9. Rodríguez-Galán, A.; Bou, J. J.; Muñoz-Guerra, S. *J. Polym. Sci., Polym. Chem. Ed.* **1992**, *30*, 713.
10. Bou, J. J.; Rodríguez-Galán, A.; Muñoz-Guerra, S. *Macromolecules* **1993**, *26*, 5664.
11. Bou, J. J.; Iribarren, I.; Muñoz-Guerra, S. *Macromolecules* **1994**, *27*, 5263.
12. Ruíz-Donaire, P.; Carné, R.; Muñoz-Guerra, S.; Rodríguez-Galán, A. *Proceedings* 6th Int. Mediterranean Congress, Barcelona, 1993; Vol. II. p 749.

BIODEGRADABLE POLYDEPSIPEPTIDES (Synthesis, Characterization, and Degradation)

C. Samyn, M. Van Beylen, and A. Maes
Laboratory of Macromolecular and Physical Organic Chemistry
Catholic University of Lueven

The requirements in the selection of any biomedical material, besides biodegradability and absorbability, are that it should have suitable mechanical properties and lack toxic degradation products.

Two important classes currently being investigated for a wide variety of surgical and pharmaceutical applications are poly(α-hydroxy acid)s and poly(α-amino acids)s.[1,2]

Copolymers of (α-hydroxy acid)s and (α-amino acids)s, which are called depsipeptides, could be a valuable addition to the existing list of synthetic biodegradable polymers. Since these polymers contain both ester and amide groups, their biodegradation behavior will be different from the homopolymers.

These linear, alternating polydepsipeptides were initially produced by stepwise active ester-peptide coupling reactions.[3-13]

Ring-opening polymerization of 2,5-morpholinedione derivatives (cyclic dimers of α-hydroxy and α-amino acid) is a second and more attractive way to produce these poly(ester amide)s.

In this section we report on the synthesis of copolymers of L,L-lactide and 2,5-morpholinedione derivatives, namely 3-methyl-, 3,6-dimethyl-, and 3-isobutyl-6-methyl, L-lactide acid was also synthesized for comparison studies. The copolymers were characterized by ^{1}H NMR or elementary analysis for small concentrations of depsipeptide (< 5%), differential scanning calorimetry (DSC) (glass transition temperatures, melting temperatures, and heat of fusion). Degradation measurements were done *in vitro* with phosphate buffer at 37°C.

PREPARATION

Results and Properties

The copolymerizations of L,L-lactide with 2,5-morpholinedione derivatives 2a–c were performed in bulk using stannous 2-ethyl hexanoate as initiator, mole ratio M/I = 10,000.

The scheme is presented as follows (**Scheme I**):

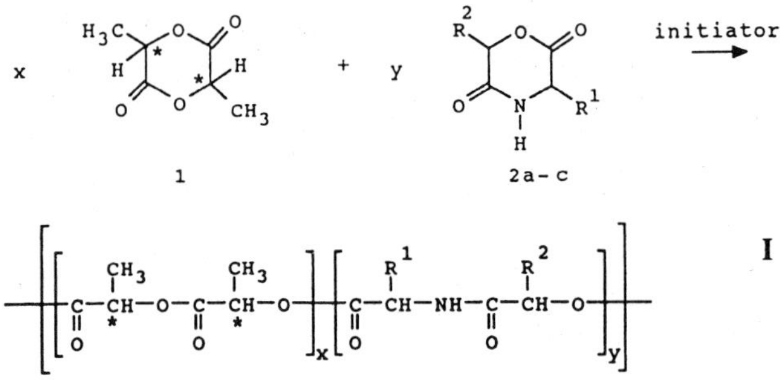

I

a: $R^1 = CH_3, R^2 = H$
b: $R^1 = CH_3, R^2 = CH_3$
c: $R^1 = CH_2CH(CH_3)_2, R^2 = CH_3$

During the copolymerization process, the ring-opening of L,L-lactide proceeds with retention of configuration, resulting in the formation of optically active copolymers.

DEGRADATION STUDIES

Polymers with a low degree of crystallinity degrade faster than polymers with a high degree of crystallinity. We also found an increase in crystallinity during degradation; this was confirmed by X-ray measurements (wide angle X-ray diffraction). It means that the depsipeptide sequence breaks down first, which results in an increase of the poly(L-lactide acid) content in the remaining copolymer, and as a consequence higher values of ΔH.

Because of the formation of more highly crystalline copolymers, the glass-transition temperature increases, and an increase in T_m is also observed.

ACKNOWLEDGMENT

The authors are indebted to the "Instituut voor Wetenshappelijk Onderzoek in Nijverheid en Landbouw" (I.W.O.N.L.) for a fellowship (A. Maes) and to the University of Leuven for financial support.

REFERENCES

1. Daniels, A. U.; Chang, M. K. O.; Andriano, K. P.; Heller, J. *J. Appl. Biomater.* **1990**, *1*, 57.
2. Vert, M.; Li, S. M.; Splenhauer, G.; Guerin, P. *J. Mater. Sci. Med.* **1992**, *3*, 432.
3. Stewart, F. H. C. *Aust. J. Chem.* **1969**, *22*, 1291.
4. Ridge, B.; Rydon, H. N.; Snell, C. R. *J. Chem. Soc., Perkin Trans.* **1972**, *1*, 2041.
5. Nissen, D.; Gilon, C.; Goodman, M. *Makromol. Chem. Suppl.* **1975**, *1*, 23.
6. Mathias, L. J.; Fuller, W. D.; Nissen, D.; Goodman, M. *Macromolecules* **1978**, *11*, 534.
7. Ingwall, R. T.; Gilon, C.; Becktel, W. J.; Goodman, M. *Macromolecules* **1978**, *11*, 540.
8. Goodman, M. *J. Polym. Sci., Polym. Symp.* **1978**, *62*, 173.
9. Katakai, R.; Goodman, M. *Macromolecules* **1982**, *15*, 25.
10. Katakai, R. *J. Chem. Soc. Perkin Trans.* **1987**, *1*, 2249.
11. Yoshida, M.; Asano, M.; Kumakura, M.; Katakai, R.; Mashimo, T.; Yuasu, H.; Imai, K.; Yamanaka, H. *J. Biomed. Mater. Res.* **1990**, *24*, 1173.
12. Yoshida, M.; Asano, M.; Kumakura, M.; Katakai, R.; Mashimo, T.; Yuasu, H.; Imai, K.; Yamanaka, H. *Colloid Polym. Sci.* **1990**, *268*, 726.
13. Yoshida, M.; Asano, M.; Kumakura, M.; Katakai, R.; Mashimo, T.; Yuasu, H.; Yamanaka, H. *Eur. Polym. J.* **1991**, *27*, 325.

BIODEGRADABLE POLYESTERS (Chemically Induced Surface Degradation)

L. M. Pratt and C. C. Chu
Department of Textiles and Apparel
Cornell University

The synthetic biodegradable biomedical polymers that are currently commercially available are linear aliphatic polyesters such as poly(glycolic acid) (Dexon), glycolide-lactide random copolymer (Vicryl), trimethylene carbonate block copolymer (Maxon), and poly-p-dioxanone (PDS). The mechanism of degradation of these polymers is simple hydrolytic chain scission. Several intrinsic and extrinsic factors have been found to affect this hydrolytic degradation, such as pH,[1,2] enzymes,[3-6] γ-irradiation,[3,5,7-9] electrolytes,[10] cell medium,[11] annealing treatment,[12] plasma surface treatment,[13] external stress,[14,15] and polymer morphology.[16,17]

These intrinsic and extrinsic factors in the biodegradation of synthetic absorbable fibers and polymers have recently been reviewed.[18-20] Due to the inherent structure–property relationship, the loss of strength always occurs well before the loss of mass during hydrolysis. Since this class of biodegradable polymers can stimulate regeneration of arterial tissues through hydrolytic degradation, Chu postulated that the degradation products and their rate of release into the surrounding tissues must bear a relationship to the experimentally observed tissue-regeneration capability. This hypothesis suggests that an ideal biodegradable polymer for arterial tissue regeneration should be able to timely release enough degradation products to activate the macrophages to produce the required growth factors for accelerated wound healing, without premature strength loss of the material during the early stages of hydrolytic degradation.

Recently, we were able to selectively alter the rate of degradation on the surface of linear aliphatic polymers, without adversely affecting their bulk properties, by using functional-group transformation. We used dimethyltitanocene to convert the surface ester linkages to vinyl ether functionalities, which are rapidly hydrolyzed to alcohol and carbonyl functional groups, resulting in chain cleavages.[21] This surface chemical functional group conversion was based on the method of Petasis.[22] Here we will describe the chemical, physical, thermal, surface morphological, and mechanical properties of poly(glycolic acid) (PGA) affecting the surface metathesis reaction and the effect of the transformation on the PGA's hydrolytic degradation properties.

PREPARATION

The dimethyltitanocene reagent was prepared from dichlorotitanocene and methylmagnesium bromide, using the method described by Petasis.[22] PGA chips were injection-molded using a Mini-Max injection molder. The surface was chemically modified by placing the PGA disks in a solution of dimethyltitanocene in tetrahydrofuran.

The bulk properties examined in determining the merits of the surface chemical functional group conversion included intrinsic viscosity, average molecular weight, level of crystallinity, melting point, and compression strength.

PROPERTIES

For biomedical applications, one of the two most important properties of the material is its mechanical strength. Partially hydrolyzed PGA disks showed no statistically significant difference between treated and untreated disks at the 95% confidence level. We concluded that the chemical surface modification had no significant effect on either the initial strength of the material or its strength-loss profile under hydrolytic conditions.

Using a relatively innovative technique, we were able to determine the depth of the PGA layer that was chemically modified by dimethyltitanocene reagents. The light-intensity profiles of a fluorescent dye as a function of depth from the surface of the PGA disks were examined after various durations of surface chemical reaction.

It appears that longer duration of either chemical treatment of hydrolysis, in general, increased the maximum depth of dye penetration (MDDP) in both treated and untreated PGA disks, with some variation for the treated samples at longer durations of both titanocene reaction and hydrolysis.

Longer chemical treatment times, however, generated more lower molecular weight species on the surface. These were more easily removed by hydrolysis to expose the underlying core material, and hence a reduction in MDDP was expected and was observed at a shorter hydrolysis time.

In addition, acidity in the local environment, caused by the hydrolytic scissions of main-chain ester linkages of the low molecular weight species, may also accelerate hydrolytic degradation of the underlying material. This local acid-catalyzed hydrolysis would result in a higher dye penetration.

The surface morphology of the various treatment times of titanocene-treated and untreated (but unhydrolyzed PGA disks) showed the untreated disk exhibited surface roughness that resembled casting flash on a cast metal surface.

In addition to cracking, the treated surface layers showed distinctly different morphologies from the control surfaces. The treated samples had a flaky appearance that seemed to result from the polymer chain cleavage shown by the intrinsic viscosity and infrared spectra of the treated films.

Overall, the surface morphology observations support the conclusion that the titanocene treatment caused degradation of the outermost surface layer.

In conclusion, surface degradation of poly(glycolic acid) disks has been achieved using dimethyl titanocene as a reagent for functional group conversion, resulting in the formation of lower molecular weight species on the titanocene-treated surface. A profile of this degradation has been seen using fluorescent dye confocal microscopy and by observing the changes in surface morphology during the initial stages of hydrolysis. Treatment times of 4 to 6 hours maximized the length of time of retention of the partially degraded surface layer. Infrared spectroscopic analysis and intrinsic viscosity measurements on thin films showed that the surface degradation was a result of polymer chain cleavage, presumably via the enol ether intermediate.

Since no change in the compressive bursting strength, bulk intrinsic viscosity, or thermal properties were observed upon treatment of the samples, we conclude that the bulk properties of this material were not adversely affected by this surface treatment. Analysis of the mechanical and thermal properties of the bulk material during hydrolysis showed no significant difference in the rate of degradation of the bulk material between the chemically treated and untreated disks.

APPLICATIONS

Biodegradable polyesters like polyglycolic acid have been commercially used as wound-closure biomaterials. These polymers have been under extensive study for other surgical applications, including use in surgical meshes for hernia and body repair, vascular grafts for injured or diseased blood vessels, components of artificial skin, guidance tubes for peripheral nerve regeneration, bone plates for fixing fractured bones, tendon/ligament prostheses, tissue engineering or regeneration, and drug control and release devices. However, the rate of tissue regeneration on biodegradable polyesters is far slower than the rate of strength loss of the biodegradable device. It is hoped that the chemical modification described in this chapter will be able to promote tissue regeneration without compromising the strength-loss profile of the underlying biodegradable scaffold.

REFERENCES

1. Chu, C. C. *J. Biomed. Mater. Res.* **1981,** *15,* 795.
2. Chu, C. C. *J. Biomed. Mater. Res.* **1982,** *16,* 117.
3. Chu, C. C.; Williams, D. F. *J. Biomed. Mater. Res.* **1983,** *17,* 1029.
4. Williams, D. F.; Mort, E. *Bioeng.* **1977,** *1,* 231.
5. Williams, D. F.; Chu, C. C. *Appl. Polym. Sci.* **1984,** *29,* 1865.
6. Williams, D. F. *ASTM Spec. Tech. Publ.* **1979,** *684,* 61.
7. Chu, C. C.; Campbell, N. D. *J. Biomed. Mater. Res.* **1982,** *16,* 417.
8. Chu, C. C.; Louie, M. *J. Appl. Polym. Sci.* **1985,** *30,* 3133.
9. Campbell, N. D.; Chu, C. C. *27th Inter. Symp. Macromol.,* Abstracts of communications, Vol. II, 1348–52, Strasbourg, France, July 6–9, 1981.
10. Pratt, L. M.; Chu, A.; Kim, J.; Hsu, A.; Chu, C. C. *J. Polym. Sci. Chem. Ed.* **1993,** *31,* 1759.
11. Chu, C. C.; Hsu, A.; Appel, M.; Beth, M. *4th World Biomaterials Congress,* April 27–May 1, 1992, Berlin, Germany.
12. Chu, C. C.; Browning, A. *J. Biomed. Mater. Res.* **1988,** *22,* 699.
13. Loh, I. H.; Chu, C. C.; Lin, H. L. *J. Appl. Biomater.* **1992,** *3*(2), 131.
14. Chu, C. C. In *Surgical Research Recent Development*; Hall, C. W., Ed.; Pergamon: San Antonio, TX, 1985.
15. Miller, N. D.; Williams, D. F. *Biomaterials* **1984,** *5*(6), 365.
16. Chu, C. C.; Kizil, Z. *3rd International ITV Conference on Biomaterials—Medical Textiles,* Stuttgart, W. Germany, June 14–16, 1989.
17. Ginde, R. M.; Gupta, R. K. *J. Appl. Polym. Sci.* **1987,** *33,* 2411.
18. Chu, C. C. In *Handbook of Biomaterials and Applications*; Wise, D. L., et al., Eds.; Marcel Dekker: New York, in press.
19. Chu, C. C. In *Biomedical Applications of Synthetic Biodegradable Polymers*; Hollinger, J. Ed.; CRC: Boca Raton, FL, in press.
20. Chu, C. C. In *Biomedical Engineering Handbook*; Bronzino, J. D., Ed.; CRC: Boca Raton, FL, (in press).
21. Pratt, L. M.; Chu, C. C. *J. Polym. Sci. Chem. Ed.* **1994,** *32,* 949.
22. Petasis, N. A.; Bzowej, E. I. *J. Am. Chem. Soc.* **1990,** *112,* 6392.

BIODEGRADABLE POLYESTERS (Theoretical Modeling of Degradation)

L. M. Pratt and C. C. Chu
Fiber and Polymer Science
Department of Textiles and Apparel
Cornell University

Poly(glycolic acid) (PGA), glycolide–lactide copolymers, and longer-chain homologues find wide use in the field of absorbable biomaterials.[1,2] All of these materials degrade in the human body as a result of hydrolytic cleavage of the ester linkages. Depending on the composition ratio, glycolide–lactide copolymers degrade at a different rate from PGA and show some advantages in properties related to processing, such as lower melting points and greater solubility in some organic solvents.

The most successful commercial application of absorbable biomaterials is in wound closure. Although several types of absorbable suture materials are commercially available, none retain adequate tensile strength long enough, without the sacrificing of handling properties and knot security, to be used in clinical situations where wound healing is delayed for prolonged periods of time, as in cases involving infection, cancer, or AIDS. There is a need for new absorbable materials that will retain their strength for substantially longer periods under physiological conditions, which are slightly alkaline.

The degradation rates of linear aliphatic absorbable polymer fibers are determined by the chemical structure of the material, fiber morphology, and environmental factors such as medium, temperature, and pH. Fibers consist of alternating highly ordered crystalline and amorphous regions; the polymer chains are either partially oriented by drawing or completely random.[3,4] The relatively open amorphous regions of PGA fibers consist of short tie-chain segments of up to roughly 30Å in length, corresponding to six repeat units.[5] It is in these regions that the initial hydrolytic degradation takes place by random chain scissioning, resulting in the loss of tensile strength.

Previous experimental work has shown substantial substituent effects on hydrolytic rate constants and activation energy parameters from substituents in both the acyl and alkyl ends of a variety of esters.[6]

This section describes our recently published studies in molecular modeling the effect of different substituents on the hydrolysis of substituted polyglycolic acid by using small molecular analogues, substituted (2-methoxy) methyl acetate, as a model.[7,8] It is believed that this computational chemical approach to examining the effect of substituents on the hydrolytic degradation of biodegradable polymers provides an intelligent way to screen the feasibility of experimental designs of new or modified biodegradable polymers for specific medical purposes.

PROPERTIES

In conclusion, the rate of ester hydrolysis has been found to be greatly affected by both alkyl and halogen substituents, due primarily to either steric hindrance or charge delocalization. In the steric effect, alkyl substituents in the glycolic esters cause an increase in activation enthalpies, and a corresponding decrease in reaction rate, with up to about three carbons, while alkyl substituents bulkier than isopropyl make the rate-determining elimination step easier.

APPLICATIONS

Poly(glycolic acid) and its lactide derivatives and copolymers have been commercially used as wound-closure biomaterials and surgical meshes for hernia and body-wall repair. These materials are currently used for experimental vascular grafts, orthopedic implants like fixation devices for fractured bones, guidance tubes for nerve regeneration, and drug control/release devices.

REFERENCES

1. Chu, C. C. In *High-Tech Fibrous Material*; Vigo, T. R.; Turbak, A. F., Eds.; Am. Chem. Soc. Symposium Ser. 457; Am. Chem. Soc.: Washington, 1991; Chapter 12, pp 167–213.
2. Barrows, T. H. In *High Performance Biomaterials*; Szycher, M., Ed.; Technomic: Lancaster, 1991; pp 243–257.
3. Peterlin, A. *Text. Res. J.* **1972**, *42*, 20.
4. Prevorsek, D. C.; Harget, P. J.; Sharma, R. K.; Reimschuessel, A. C. *J. Macromol. Sci. Phys.* **1973**, *B8*(1-2), 127.
5. Murthy, N. S.; Reimschuessel, A. C.; Kramer, V. *J. Appl. Polym. Sci.* **1990**, *40*, 249.
6. Sun, S.; Connors, K. *J. Pharm. Sci.* **1969**, *58*(9), 1150.
7. Pratt, L. M.; Chu, C. C. *J. Computational Chemistry* **1993**, *14*(7), 809.
8. Pratt, L. M.; Chu, C. C. *J. Molecular Structure* **1994**, *304*, 213.

BIODEGRADABLE POLYESTER, CYCLIC ETHERS

Masahiko Okada
Department of Biological Sciences
School of Agricultural Sciences
Nagoya University

Commercially available, biodegradable synthetic polymers (including bioabsorbable polymers) are mostly limited to aliphatic polyesters such as poly(glycolic acid), poly(lactic acid), poly(ε-caprolactone), poly(ethylene succinate)s, etc. In order to meet diverse requirements for practical application, it is desirable to develop new biodegradable polymers with controllable degradability as well as specific physical and chemical properties that would not be fulfilled by aliphatic polyesters.

Naturally occurring polysaccharides such as cellulose, starch, and chitin are composed of pyranose rings whose skeletons are six-membered tetrahydropyran rings. Nucleic acids contain furanose rings whose skeletons are five-membered tetrahydrofuran rings. This suggests that biodegradable polyesters of a new type could be created by incorporating cyclic ether units into the polymer backbones. These cyclic ether structures impart polarity to the polymers and increase hydrophilicity. Increasing the hydrophilicity of a polymer generally improves its biodegradability.[1]

This part deals with the synthesis and degradation of polyesters with cyclic ether moieties in their backbones. Only a very limited number of such polyesters have been investigated so far, but some of them have been found biodegradable. As to the synthesis and properties of the related polyesters, see also Bicyclic Lactones, Ring-Opening Polymerization.

PREPARATIONS OF POLYESTERS

Polyesters with Tetrahydropyran Rings

A series of polyesters (Structures **2** and **7-10**, **Scheme I**) having tetrahydropyran rings in their backbones have been synthesized by cationic ring-opening polymerization of bicyclic lactones, 2,6-dioxabicyclo[2.2.2]octan-3-one[2] (Structure **1**, Scheme I) and their 4-alkoxycarbonyl derivatives[3,4] (Structures **3–6**). A structural isomer of Structure **1**, 6,8-dioxabicyclo[3.2.1]octan-7-one (Structure **11**), also gives a polyester **12** with tetrahydropyran rings in the main chain.[7,8] (see Scheme I).

These monomers are readily polymerized with boron trifluoride etherate at temperatures below –60°C to give polyesters of an "acetal–ester" type with alternating tetrahydropyran rings and ester moieties.[2,3]

All these polyesters begin to decompose at 140–160°C. Transparent films of these polyesters can be formed by casting their chloroform solutions.

Polyesters with Fused Tetrahydrofuran Rings

Recently, much attention has been directed toward biomass resources as starting materials for polymer synthesis, in expectation that "environmentally friendly" biodegradable polymers could be derived therefrom.

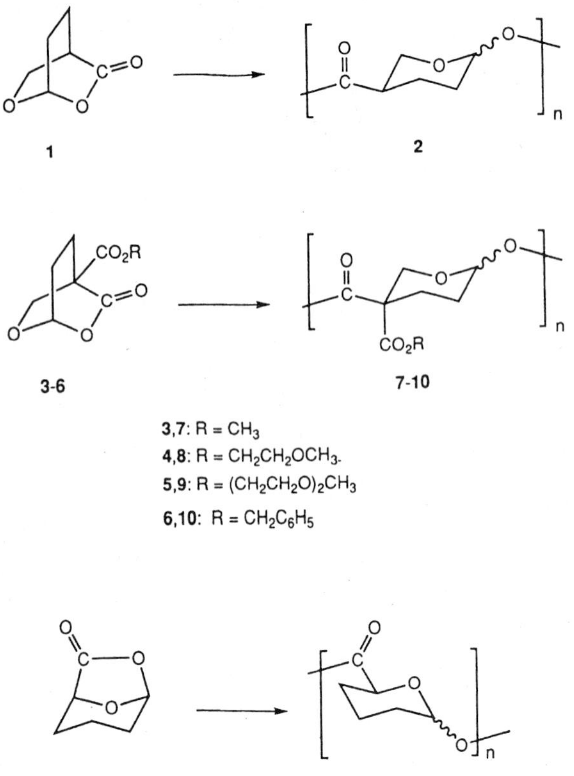

I

3,7: R = CH_3
4,8: R = $CH_2CH_2OCH_3$.
5,9: R = $(CH_2CH_2O)_2CH_3$
6,10: R = $CH_2C_6H_5$

1,4:3,6-Dianhydro-D-glucitol and its stereoisomer, 1,4:3,6-dianhydro-D-mannitol possess only two hydroxyl groups as reactive functional groups. In the first of the two, one of the hydroxyl groups is located in the *exo* position and the other in the *endo* position, whereas in the second, both of the hydroxyl groups are located in the *endo* position. They have been used as difunctional monomers for poly-condensation and polyaddition.[7-11] The rigid fused tetrahydrofuran rings are expected to impart desirable thermal and mechanical properties to the resulting polymers.

DEGRADABILITIES OF POLYESTERS

Hydrolytic Degradability in Neutral Phosphate Buffer

Since polyesters **2,7–10**, and **12** are of an "acetal–ester" type, they may undergo hydrolysis under relatively mild conditions. In fact, some of these polyesters are spontaneously hydrolyzed even in a neutral phosphate buffer solution (pH 7.5) at ambient temperature.[6]

Polyester **12**, composed of 2,6-linked tetrahydropyran rings, is much more rapidly hydrolyzed than polyester **2**, composed of 2,5-linked tetrahydropyran rings, although both polyesters are of a similar "acetal–ester" type. The difference in the spontaneous hydrolysis rate seems to arise from the higher hydrophilicity of **12**.[6]

In marked contrast, polyester **14** and (**Scheme II**, of an "ether–ester" type that is prepared by polycondensation of the corresponding hydroxytetrahydropyran carboxylic acid **13**),[12] is hardly hydrolyzed in a phosphate buffer solution of pH 7.5 at 27°C.

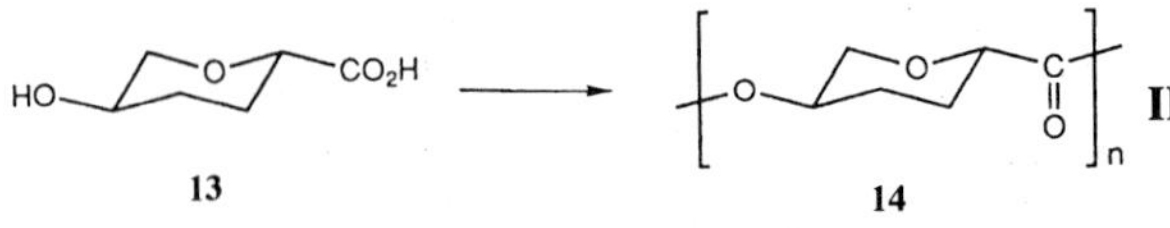

Polyesters containing fused tetrahydrofuran rings in the main chains, are reluctantly hydrolyzed in a phosphate buffer solution (pH 7.5) at ambient temperature. However, these polyesters are relatively easily hydrolyzed at 50°C.

These data clearly demonstrate that the spontaneous hydrolyzability of polyesters under neutral conditions varies significantly depending on the linkage modes of the backbones ("acetal–ester" type and "ether–ester" type), the linkage modes of the cyclic ether moieties (2,5-linked and 2,6-linked, and *cis*-linked and *trans*-linked), the structures of the oxacycles, the pendant groups, and the copolymer compositions.

Biodegradability of Polyesters

Degradation in Soil

A thin film (thickness 50 μm) of polyester **2** that had been buried in soil for one to three months had disintegrated into such small pieces that its original shape was completely lost.[13] Degradation of the film in soil was faster than the heterogeneous hydrolysis in the phosphate buffer solution of pH 7.5. The acidity of the soil used for the experiment was pH 5.2–5.9, and therefore the weakly acidic nature of the soil may be responsible for the accelerated degradation of the film of **2**. However, degradation rate of a film of polyester **7** (with pendant methoxycarbonyl groups) in the same soil was similar to that in the neutral phosphate buffer solution. In addition, copolyesters having pendant benzyoxycarbonyl groups or carboxyl groups degraded more slowly in the soil than in the phosphate buffer solution. Taking these findings into consideration, biodegradation may contribute, at least to some extent, to the soil-burial degradation of polyester **2**.

REFERENCES

1. Marck, K. W.; Widevuur, C. R.; Sederel, W. L.; Bentjes, A.; Feijen, J. *Biomed. Mat. Res.* **1977**, *11*, 405.
2. Okada, M.; Sumitomo, H.; Atsumi, M.; Hall, H. K.; Ortega, R. B. *Macromolecules* **1986**, *19*, 503.
3. Okada, M.; Sumitomo, H.; Atsumi, M.; Hall, H. K. Jr. *Macromolecules* **1987**, *20*, 1199.
4. Atsumi, M.; Okada, M. *Polym. J.* **1992**, *24*, 1109.
5. Okada, M.; Aoi, K.; Ito, S.; Atsumi, M. *Koubunshi Ronbunshu* **1993**, *30*, 703.
6. Okada, M.; Ito, S.; Aoi, K.; Atsumi, M. *J. Appl. Polym. Sci.* **1994**, *51*, 1035.
7. Thiem, J.; Luders, H. *Starch/Starke* **1984**, *36*, 170.
8. Thiem, J.; Luders, H. *Polym. Bull.* **1984**, *11*, 365.
9. Dirlikov, S. K. *Agricultural and Synthetic Polymers*; Glass, J. E. and Swift, G. Eds.; *ACS Symposium Series*, **1990**, *433*, 176.
10. Dirlikov, S. K. *Emerging Technologies for Materials and Chemicals from Biomass*, Rowell, R. M. et al., Eds., *ACS Symposium Series*, **1992**, *476*, 231.
11. Storbeck, R.; Rehahn, M.; Ballauf, M. *Makromol. Chem.* **1993**, *194*, 53.

12. Atsumi, M.; Okada, M.; Sumitomo, H.; Yamada, S. *Makromol. Chem.* **1991**, *192*, 1715.
13. Okada, M.; Ito, S.; Aoi, K.; Atsumi, M. *J. Appl. Polym. Sci.* **1994**, *51*, 1045.

Biodegradable Polymers

See: *Bacterial Polyesters (Containing Unusual Substituent Groups)*
Bioabsorbable Polymers (Tissue Engineering)
Biodegradable Copolymers [Comprising Poly(L-lactic acid)]
Biodegradable Optically Active Polyamides (Stereoregular Polyamides from L-Tartaric Acid)
Biodegradable Polydepsipeptides (Synthesis, Characterization, and Degradation)
Biodegradable Polymers (Orthopedic Applications)
Biodegradable Polyesters (Chemically-Induced Surface Degradation)
Biodegradable Polyesters (Theoretical Modeling of Degradation)
Biodegradable Polyesters, Cyclic Ethers
Bioerodible Polymers A. Controlled Drug Release B. Fracture Fixation
Biomedically Degradable Polymers
Bone Cements, Bioresorbable
Controlled Drug Delivery Systems
Environmentally Degradable Polymers
Enzyme-Degradable Hydrogels
Gluten Plastic, Biodegradable
Lignin Graft Copolymers
Microbial Polyhydroxyalkanoates (Poly[(R)-3-hydroxybutyrate-co-3-hydroxypropionate])
Microencapsulation (Artificial Cells)
Polyamides, Glycine-Containing
Poly(α-amino acids) (Biodegradation, Medical Applications)
Polyesters (Derived from Renewable Resources)
Poly(ethylene oxide) (Applications in Drug Delivery)
Poly(ethylene terephthalate) Copolymers
Poly(β-hydroxyalkanoates) (Microbial Synthesis and Chemical Modifications)
Polyisopeptides
Pseudo-Poly(amino acid)s (Overview)
Ring-Opening Polymerization, Free Radical
Starch-Based Plastics (Measurement of Biodegradability)
Starch Biodegradation (in Starch-Plastic Blends)
Starch-Polymer Composites
Stereo-Block Copolymers (of L- and D-Lactides)
Suture Materials (Current Research and Development)
Wound Dressing Materials
Xanthan, Depolymerization
Xanthan Gum (Overview)

BIODEGRADABLE POLYMERS (Orthopedic Applications)

Xichen Zhang
Department of Chemical Engineering
Queen's University

Mattheus F. A. Goosen*
College of Agriculture
Sultan Qaboos University
Sultanate of Oman

Biodegradable polymers are used in a variety of applications in the medical field primarily because secondary surgery is not required for implant removal.[1-10] In the case of orthopedic applications, key areas include tissue reconstruction and controlled drug release.[1-10] Here, we concentrate on orthopedic applications, with a brief introductory discussion on the preparation and properties of biodegradable polymers.

PREPARATION AND PROPERTIES

Many polymeric materials are biodegradable. However, only a few have been developed as candidates for medical applications, primarily because of biocompatibility requirements and a long certification procedure. These polymers include polyesters [e.g., polylactide (PLA), polygycolide (PGA) (26202-08-4), Poly(ε-caprolactone) (25038-54-4)], poly(*p*-dioxanone) (PDS) (29223-92-5), polycarbonate, polyanhydrides, poly(ortho ester)s, polyphosphazenes, and poly(amino acid)s and natural polymers such as collagen (9059-25-0) and chitosan (9012-76-4).[1] Biodegradation in polymers usually occurs through cleavage of a hydrolyzable linkage. A number of labile bonds can be used to make biodegradable polymers. Among those polymers, PGA, PLA, and their copolymers (PGLA, 26780-50-7) are the most widely documented because of their proven biocompatibility and good material properties.

APPLICATIONS

Tissue Reconstruction

Many applications exist in tissue reconstruction, including surgical sutures, internal bone fracture fixation devices (such as plates, screws, and pins), scaffolds in tissue regeneration, and local hemostatics.

Surgical Sutures

Synthetic bioabsorbable sutures have been used since the 1960s. Several sutures are commercially available: Dexon (PGA), Vicryl (PGLA with a small amount of lactide) and Maxon (glycolide/trimethylene carbonate copolymer).[9] Manufacturing of high-strength PDS and PLLA fiber has been reported.[11-16] The mechanical strength of PGA and PGLA sutures is lost in about a month. PDS and PGA/TMC take longer to degrade, and PLLA is the slowest to degrade. Bioabsorbable sutures have been widely used in soft-tissue closure and in fixations of tendons, ligaments, joints, and bone fractures.[17-19]

*Author to whom correspondence should be addressed.

Internal Bone Fracture Fixation Devices

Compared with traditional metal-fixation devices, biodegradable devices have two advantages: absorbability and gradual transfer of stress to healing bones. Current devices include screws, plates, pins, and rods. The most important polymers used in fracture fixation are poly(L-lactide) (PLLA) PGA, and PDS, mainly because of their good mechanical strength (compared with other biodegradable polymers), proven biocompatibility, controllable degradability, and processability.[7-10]

Supportive Scaffolds in Tissue Regeneration

Biodegradable polymers have been used as scaffolds for repairing tissues such as cartilage, bone, ligaments, and tendons.[20-27]

Another promising application of biodegradable lactide and glycolide polymers, as a way to solve the problem of organ shortage for transplantation, is to help transplanted cells to grow into a tissue organ such as bone or cartilage.[28,29]

An important issue that has to be considered is how to make a well-defined three-dimensional biodegradable polymeric matrix.[28,29] This matrix should satisfy requirements such as cell adhesion, sufficient pores for cell growth and nutrient transport, adequate mechanical strength and geometric shape to support and guide cell growth, and absorption over a desired time period. Controlled release of tissue-inducing factors, growth factors, and angiogenesis stimulators from the matrix may also be helpful in tissue formation.

Local Hemostatics and Bioadhesives

Glycolic acid and lactic acid oligomers, in the form of a paste, have been formulated as local hemostats in bone surgery.[30,31]

A variety of biodegradable adhesive polymers may be used as bone cements, covers, sealants, and hemostats. These materials include fibrin glues, cyanoacrylates, isocyanate-terminated DL-lactide-ε-caprolactone-poly(ethylene glycol) prepolymers, poly(proplene glycol fumarate), polyphosphorates, and polyphosphates.[32-35]

Controlled Drug Release

Besides delivery of growth factors, a major application of biodegradable polymers is in local antibiotic administration for treatment/prevention of bone infection.[36] Bone infection in orthopedic bone surgery can be life threatening and is very hard to treat. A variety of preventive or protective techniques have been used. A prominent method is the incorporation of antibiotics into methyl methacrylate bone cement, first reported in 1970.[37] Although generally thought of as an effective method, controversy persists about the best use of different antibiotics, cements, and their combinations.[38] Most antibiotic delivery devices for local treatment consist of antibiotic impregnated poly(methyl methacrylate) (PMMA, 9011-14-7).[39,40] Although effective in the short term, PMMA is a nonabsorbable material. It has been shown to actually support bacterial growth after the antibiotic has been leached out. The polymer, therefore, must eventually be removed. Thus, a biodegradable antibiotic delivery system, which can release antibiotics at a sustained rate for 4–6 weeks, has the advantage of requiring no secondary surgery.

Several biodegradable controlled antibiotic release devices have been developed.[41-53]

Biodegradable polymers have great potential in the fields of tissue engineering and orthopedic surgery in which they can provide, for example, temporary fixation of prosthetic joints, as well as allowing for the localized release of antibiotics. To be able to make the best use of biodegradable polymers, however, it is necessary to have a good grasp of the material's preparation and properties. This will ensure that the final device will have optimum physiochemical, mechanical, and biological properties.

REFERENCES

1. Chason, M.; Langer, R. *Biodegradable Polymers as Drug Release Systems*; Marcel Dekker: New York, 1990.
2. Jalil, R. U. *Drug Dev. Ind. Pharm.* **1990**, *16*, 2353.
3. Baker, R. *Controlled Release of Biologically Active Agents*; John Wiley & Sons: New York, 1987.
4. Schindler, A.; Jeffcoat, R.; Kimmel, G. L.; Pitt, C. G.; Wall, M. E.; Zweidinger, R. *Contemp. Top. Polym. Sci.* **1987**, *2*, 251.
5. Pitt, C. G. *Int. J. Pharm.* **1990**, *59*, 173.
6. Privalova, L. G.; Zaikov, G. E. *Polym. Plast. Technol. Eng.* **1990**, *29*, 455.
7. Bostman, O. M. *J. Bone Jt. Surg.* **1991**, *73A*, 148.
8. Hofmann, G. O. *Clin. Mater.* **1992**, *10*, 75.
9. Vainionpaa, S.; Rokkanen, P.; Tormala, P. *Prog. Polym. Sci.* **1989**, *14*, 679.
10. Vert, M.; Christel, P.; Chabot, F.; Leray, J. In *Macromolecular Biomaterials*; Hasting, G. W.; Ducheyen, Eds.; CRC: Boca Raton, 1984; Chapter 6.
11. Doddi, N.; Versfelt, C. C.; Wasserman, D. U.S. Patent 4052988, 1977.
12. Postema, A. R.; Luiten, A. H.; Pennings, A. J. *Polymer* **1982**, *23*, 1587.
13. Eling, B.; Gogolewski, S.; Pennings, A. J. *Polymer* **1982**, *23*, 1587.
14. Gogolewski, S.; Pennings, A. J. *J. Appl. Polym. Sci.* **1983**, *28*, 1045.
15. Leenslag, J. W.; Gogolewski, S.; Pennings, A. J. *J. Appl. Polym. Sci.* **1984**, *29*, 2829.
16. Leenslag, J. W.; Pennings, A. J. *Polymer* **1987**, *28*, 1695.
17. Caband, H. E.; Feagin, J. A.; Rodkey, W. G. *Am. J. Sports Med.* **1982**, *10*, 259.
18. Park, J. P.; Grana, W. A.; Chitwood, J. S. *Clin. Orthop.* **1985**, *196*, 175.
19. Vainionpaa, S. Academic Dissertation, University of Helsinki, Helsinki, Finland, March 20, 1987.
20. von Schroeder, H. P.; Kwan, M.; Amiel, D.; Coutts, R. D. *J. Biomed. Mater. Res.* **1991**, *25*, 329.
21. Anselme, K.; Flautre, B.; Hardouin, P.; Chanavaz, M.; Ustariz, C.; Vert, M. *Biomater.* **1993**, *14*, 44.
22. Winet, H.; Hollinger, I. O. *J. Biomed. Mater. Res.* **1993**, *27*, 667.
23. Coombes, A. G. A.; J. D. K. *Biomater*, **1993**, *13*, 297.
24. Kennedy, J. U.S. Appl. 618, 652, Nov. 1990.
25. Laitinen, O.; Tormala, P.; Taurio, R.; Skutnabb, K.; Saarelainen, K.; Iivonen, T.; Vainionpaa, S. *Biomater.* **1992**, *13*, 1012.
26. Davis, P. A.; Huang, S. J.; Ambrosio, L.; Ronca, D.; Nicolais, L. *J. Mater. Sci.; Mater. Med.* **1992**, *3*, 359.
27. Kato, Y. P.; Dunn, M. G.; Zawadsky, J. P.; Tria, A. J.; Silver, F. H. *J. Bone Jt. Surg.* **1991**, *73A*, 561.
28. Langer, R.; Vancanti, J. P. *Science* **1993**, *260*, 920.

29. Cima, L. G.; Langer, R. *Chem. Eng. Prog.* **1993**, *89*, 46.
30. Fues, G. F. Ger. Offen De 3825211, 1990.
31. Henderson, A. M. Can. CA 1260488, 1989.
32. Jaffe, H.; Wade, C. W. R.; Hegyeli, A. F.; Rice, R. M.; Hodge, J. *J. Biomed. Mater. Res.* **1986**, *20*, 213.
33. Kobayashi, H.; Hyon, S. H.; Ikada, Y. *J. Biomed. Mater. Res.* **1991**, *25*, 1481.
34. Domb, A. J. U.S. Appl. 142471, 1988.
35. Richards, M.; Dahiyat, B. I.; Arm, D. M.; Brown, P. R.; Leong, K. W. *J. Biomed. Mater. Res.* **1991**, *25*, 1151.
36. Lucas, P. A.; Laurencin, C.; Syftestad, G. T.; Domb, A.; Goldberg, V. A.; Caplan, A. I.; Langer, R. *J. Biomed. Mater. Res.* **1990**, *24*, 901.
37. Buchholz, H. W.; Engelbrecht, *Chirurgi* **1970**, *41*, 511.
38. Setterstrom, J. A.; Tice, T. R.; Meyers, W. E.; Wincent, I. W.; Battistone, G. C. *Polym. Mater. Eng.* **1985**, *53*, 620.
39. Wahlig, H.; Dingeldein, E.; Buchholz, H.; Buchholz, M.; Bachman, F. *J. Bone Jt. Surg.* **1978**, *60B*, 270.
40. Vecsei, V.; Barquet, A. *Clin. Orthop. Rel. Res.* **1981**, *159*, 201.
41. Lewis, D. H.; Dappert, T. O.; Meyers, W. E.; Pritchett, G.; Suling, W. J.; *Proc. Int. Symp. Contr. Rel. Bioact. Mater.* **1980**, *7*, 129.
42. Tice, T. R.; Rowe, C. E.; Setterstrom, J. A. *Int. Symp. Contr. Rel. Bioact. Mater.* **1984**, *11*, 6.
43. Setterstrom, J. A.; Tice, T. R.; Meyers, W. E.; Wincent, J. W.; Battistone, G. C. *Polym. Mater. Eng.* **1985**, *53*, 620.
44. Baker, R. W.; Krisko, E. A.; Kochinke, F.; Grassi, M.; Armitage, G.; Robertson, P. *Int. Symp. Contr. Rel. Bioact. Mater.* **1988**, *15*, 238.
45. Ikada, Y.; Hyon, S. H.; Jamshidi, K.; Higaski, S.; Yamamuro, T.; Katutani, Y.; Kitsugi, T. *J. Contr. Rel.* **1985**, *2*, 179.
46. Wei, G.; Kotoura, Y.; Oka, M.; Yamamuro, T.; Walda, R.; Hyon, S. H.; Ikada, Y. *J. Bone Jt. Surg.* **1991**, *73B*, 246.
47. Laurencin, C. T.; Gerhart, T.; Witschger, P.; Satcher, R.; Domb, A.; Rosenberg, A. E.; Hanff, P.; Edsberg, L.; Hayes, W.; Langer, R. *J. Orthop Res.* **1993**, *11*, 256.
48. Zhang, X.; Wyss, U. P.; Pichora, D.; Goosen, M. F. A. *J. Pharm. Pharmcol.* **1994**, *46*, 1.
49. Zhang, X.; Wyss, U. P.; Pichora, D.; Goosen, M. F. A. *J. Contr. Rel.* in press.
50. Firsov, A. A.; Nazaro, A. D.; Fomina, I. P. *Drug Dev. Ind. Pharm.* **1987**, *13*, 1651.
51. Sampath, S. S.; Garvin, K.; Robinson, D. H. *Int. J. Pharmace.* **1992**, *78*, 165.
52. Yu, D.; Wong, J.; Matsuda, Y.; Fox, J. F.; Higuchi, W. I.; Otsuka, M. *J. Pharm. Sci.* **1992**, *81*, 529.
53. Yamamura, K.; Iwata, H.; Yotsuyanagi, T. *J. Biomed. Mater. Res.* **1992**, *26*, 1053.

BIOERODIBLE POLYMERS (In Controlled Drug Release)

Jorge Heller
APS Research Institute

The development of bioerodible drug-delivery implants is assuming an ever-increasing importance in research on controlled drug-delivery. A major driving force in this development is the need to deliver therapeutic agents directly to the circulatory system, which is important with drugs that undergo significant inactivation by the liver. Another advantage of bioerodible drug-delivery implants is that small, well-tolerated implants can be left in place for very long periods, making delivery regimes lasting one or more years possible.

However, perhaps the major interest in developing bioerodible drug delivery devices arises in connection with protein delivery.[1] Proteins are not active orally and have very short half-lives. In the absence of a suitable controlled systemic delivery device, they must be administered by daily injection—clearly an undesirable therapy.

In this article we will only consider delivery systems where a drug has been homogeneously dispersed in a polymer matrix, and we will divide these into hydrophilic systems and hydrophobic systems. The latter class will be further subdivided into predominantly diffusion-controlled systems and predominantly polymer-hydrolysis-controlled systems.

HYDROPHILIC SYSTEMS

Bioerodible hydrogels are of considerable interest in delivering macromolecules that can be physically entangled in the hydrogel structure and, because of their large size, can diffuse out only slowly, if at all. The subject of bioerodible hydrogels has recently been exhaustively reviewed.[2]

Bioerodible hydrogels can be constructed with water-labile bonds in either the polymer backbone or the crosslink segment.

Water-Labile Bonds in Polymer Backbone

A number of systems involving crosslinked polysaccharides have been described, principally for release of proteins.[3] In these systems, the protein is physically entangled in the hydrogel and is released, principally by diffusion, in proportion to its molecular weight.

A more desirable hydrogel delivery system would be one with a crosslink density high enough so that diffusional release is prevented and rate of release can be controlled by chemical hydrolysis of the hydrogel. Such a system has been synthesized by crosslinking a water-soluble polyester prepared from fumaric acid and poly(ethylene glycol) with *N*-vinylpyrrolidone (VP).[4]

Water-Labile Bonds in Crosslink Segment

One interesting application of such hydrogels is the development of colon-specific delivery systems. There are two major situations where delivery to the colon is important. The first is treatment of inflammatory bowel disease, where in order to reduce systemic toxicity, anti-inflammatory agents need to be delivered directly. The second is enhancement of oral bioavailability of peptides and proteins.

One such hydrogel is shown in **Scheme I**.[5,6] It is prepared by reacting the linear precursor with a crosslinker, in this case N,N′-(ω-aminocaproyl)-4′-diaminoazobenzene, which reacts with the activated *p*-nitrophenyl ester to form crosslinks containing azo linkages. Because the hydrogel contains free carboxylic acid groups, it swells very little in the acidic environment of the stomach, which protects incorporated agents from digestion by enzymes. In the higher pH of the intestines, swelling begins, and when the hydrogel reaches the colon, swelling has increased to a point where azoreductases can cleave the azo bonds, allowing degradation of the hydrogel and release of its contents.

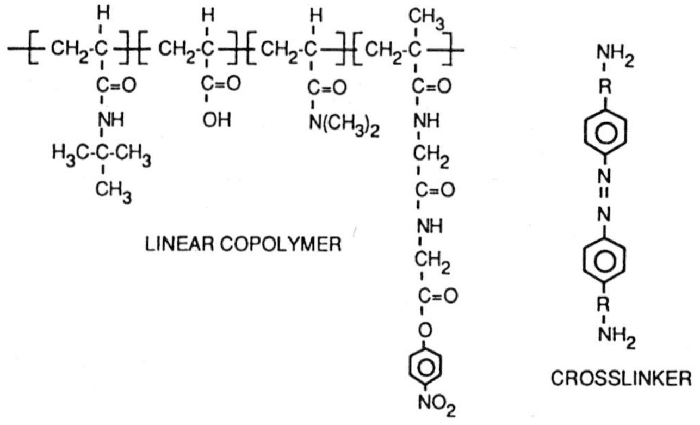

I

HYDROPHOBIC POLYMERS

The bulk of activity concerning the use of bioerodible polymers for drug delivery centers on hydrophobic polymers.

Drug Release Predominantly Controlled by Diffusion

In this type of system, the drug is homogeneously dispersed in a biodegradable polymer matrix and is mainly released by simple Fickian diffusion. If the rate of polymer hydrolysis is very slow, release can occur entirely by diffusion. On the other hand, if hydrolysis is relatively rapid, initial release is diffusion-controlled, but as the process continues, polymer hydrolysis becomes an important factor in determining the rate of release.[7] The most extensively investigated bioerodible polymer system in this class consists of poly(lactide-co-glycolide) copolymers[8] prepared from the respective glycolides or lactides by heating with an acidic catalyst such as $SnCl_4$.

Because of their nontoxic nature and long history of safe use, poly(lactide-*co*-glycolide) copolymers are under intense investigation as bioerodible drug-delivery systems. Their use in drug delivery has been recently reviewed.[8]

Drug Release Predominantly Controlled by Polymer Hydrolysis

In contrast to the poly(lactide-co-glycolide) copolymers or poly(glycolic acid), which were developed for use as bioerodible sutures and were then later adapted for drug release, two bioerodible polymer systems were specifically designed for drug-delivery applications. These are poly(ortho ester)s and polyanhydrides.

Poly(ortho ester)s

As of this writing, three major families of poly(ortho ester)s have been prepared. Their preparation and applications have been comprehensively reviewed.[9]

Polyanhydrides

The use of polyanhydrides as bioerodible matrices for the controlled release of therapeutic agents was first reported in 1983.[10] Although anhydride linkages are highly susceptible to hydrolysis, crystalline polymers are very stable because water is unable to penetrate the crystalline regions. However, when crystallinity is disrupted by copolymerization of an aliphatic and aromatic diacid, erosion rates that vary from days to projected years can be achieved.

Polyanhydrides based on aromatic diacids become brittle and eventually fragment after exposure to water, causing water-soluble drugs to be released more rapidly than by polymer erosion. For this reason, a new class of polyanhydrides was prepared from fatty acid dimers derived from naturally occurring oleic and sebacic acids.[11]

REFERENCES

1. Heller, J. *Adv. Drug Deliv. Reviews* **1993**, *10*, 163.
2. Park, K.; Shalaby, W. S. W.; Park, H. *Biodegradable Hydrogels for Drug Delivery*; Technomic: Lancaster, PA, 1993.
3. Edman, P.; Ekman, B.; Sjoholm, I. *J. Pharm. Sci.* **1980**, *69*, 328.
4. Heller, J.; Helwing, R. F.; Baker, R. W.; Tuttle, M. E. *Biomaterials* **1983**, *4*, 262.
5. Kopecek, J.; Kopeckova, P.; Bronsted, H.; Rathi, R.; Rihova, B.; Yeh, P.-Y.; Ikesue, K. *J. Controlled Release* **1992**, *19*, 121.
6. Kopeckova, P.; Rathi, R.; Takada, S.; Rihova, B.; Berenson, M. M.; Kopecek, J.; *J. Controlled Release* **1994**, *28*, 211.
7. Heller, J.; Baker, R. W. *Controlled Release of Bioactive Materials*; Academic: New York, NY, 1980; Chapter 1.
8. Chasin, M.; Langer, R. *Biodegradable Polymers as Drug Delivery Systems*; Marcel Dekker: New York, NY, 1990; Chapter 1.
9. Heller, J. *Adv. in Polymer Sci.* **1993**, *107*, 41.
10. Rosen, H. G.; Chang, J.; Wnek, G. E.; Linhardt, G. E.; Langer, R. *Biomaterials* **1985**, *19*, 941.
11. Tabata, Y.; Langer, R. *Pharm. Res.* **1993**, *10*, 391.

BIOERODIBLE POLYMERS (In Fracture Fixation)

Kirk P. Andriano and Jorge Heller
APS Research Institute

The prospect of replacing metal fracture-fixation devices with bioadsorbable polymer composite devices that have an appropriate combination of initial strength, stiffness, and biocompatibility is of great interest because eventual device absorption has two important advantages. First, as absorption reduces the device's cross-section or the materials elastic modulus, the load is transferred gradually to the healing bone.

Second, because the device will be completely absorbed, a second surgical procedure is not necessary.

Currently, only a few materials have been intensively investigated for use in bioabsorbable fracture-fixation devices. These are poly(lactic acid), poly(glycolic acid), polydioxanone, and poly(ortho ester)s. Comprehensive reviews of the use of biodegradable polymers in surgical applications have been published.[1-4]

A review covering bioerodible fracture-fixation devices, emphasizing polymer characterization and fabrication procedures, has been published.[5]

CONCLUSIONS

The development of bioerodible polymers for fracture fixation is a challenging problem receiving increased attention. Ideally, such materials would have the initial biocompatibility, strength, stiffness, and ductility of stainless steel; retain these properties for several weeks or months; and then undergo benign and complete biodegradation and absorption or excretion. The completely bioerodible polymers that have been described to date do not meet all these requirements, especially for stiffness and ductility. In spite of this, successful clinical fracture-fixation has been reported with poly(lactic acid), poly(glycolic acid), their copolymers, and polydioxanone. The key to this success has been to pick the clinical applications carefully and to design the method of fixation to suit available polymer properties.

The development of tyrosine-based polycarbonates and poly(ortho ester) composite materials were based on rational design processes, rather than using random screening or trial and error to improve biocompatibility performance. These polymers represent new classes of materials that may find applications as "second generation" orthopedic fracture-fixation devices in about 10 to 15 years.

REFERENCES

1. Daniels, A. U.; Chang, M. K. O.; Andriano, K. P.; Heller, J. *J. Appl. Biomater.* **1990**, *1*, 57.
2. Engelberg, I.; Kohn, J. *Biomaterials* **1991**, *12*, 292.
3. Pulapura, S.; Kohn, J. *Biomater. Applications* **1992**, *6*, 216.
4. Vainionpaa, S.; Rokkanen, P.; Tormala, P. *Prog. Polym. Sci.* **1989**, *14*, 679.
5. Hastings, G. W.; Ducheyne, P. *Macromolecular Biomaterials*; CRC: Boca Raton, FL, 1984; pp 119–142.

BIOLOGICALLY ACTIVE AGRICULTURAL POLYMERS (Mechanism of Action on Plants)

S. Sh. Rashidova, N. L. Voropayeva, and I. N. Ruban
Institute of Polymer Chemistry and Physics

Polymeric materials intended for use in the development of agricultural production have filled the markets of the majority of countries. Here, we define countries and monopolists in the production of goods such as films, any type of covers based on polymers, polymeric sand anchors and soil anchors, structure formers for some types of soil polymeric matrixes that prolong fertilizer action, and remedies for plant development.

Problems concerning the increase of the chemical load on ecological systems come into play. Polymeric materials degrade to produce toxic products, and there is a possibility that these products, upon interaction with other chemical agents in the ecosystem, will produce different complexes, which do not degrade. These problems occur most often with water nonsoluble polymeric substances, which most probably can be solved by biotechnological methods. Our discussion will center on water-soluble polymeric systems that are used in agricultural production as biologically active substance carriers capable of regulation plant growth and development and stability against diseases. These systems are easily utilized, unlike water nonsoluble ones, under the influence of different environment factors, including temperature,[3] illumination and enzimer activity of microorganisms.

Physiologically active substances that are bonded with natural and synthetic water-soluble polymers are released slowly as they degrade due to the nature of the bonds. The degradation of these systems leads to the delivery of the acting substances to the needed place at the necessary time, although these mechanisms are altered by many constantly changing environmental factors. The synthesis should be obtained by using polymeric systems that have physical and chemical parameters that would satisfy the utilization conditions. In addition, it is necessary to understand the influence of macromolecular structures on the physiological processes of living systems, for example, seed sprouting and plant development.

THE PROBLEM

Looking through the literature it is clear that all synthetic and natural water-soluble polymers are used in seed keeping agriculture and plant breeding to protect against phytopathogens and pests as well as to regulate seed growth especially at the last stage of growth of agricultural crops. Poly(vinyl alcohol) and compositions based on it have been used widely in seed keeping.[1-9]

In addition to poly(vinyl alcohol), water-soluble cellulose ethers are used in seed breeding.[10-13] The advantages of water-soluble cellulose ethers are solubility within a range of temperatures, ability to combine with salt solutions of low concentration, absence of foaming, the ability to form films, and comparative inexpensiveness and nontoxicity (e.g., sodium salts of carboxymethylcellulose).[10,12-15] However, several shortcomings limit wide use of cellulose ethers in seed growing. For example, oxyethylcellulose is toxic to warm-blooded organisms and films adhere at relative humidities of 50% and greater. These disadvantages make the technology of presowing seed treatment complex.

Recently, polymeric binding based on polyethers has been used for presowing seed preparation.[16-21] Polyethers are soluble in water and nontoxic. They can also form complexes with alkaline and alkaline earth metals, which solves problems of soil improvement.[21] Polyethers degrade in several ways (e.g., thermal, chemical, biological, and under the influence of different wave lengths of radiation), which makes them desirable for use in the ground and on plants.

Some researchers have studied the use of polyacrylamide and of polyacrylamide mixed with oxyethylcellulose for presowing preparation of seeds.[22-30] But in spite of the advantages of this polymer (e.g., nontoxic, large production volume, inexpensive, and good water solubility), its wide use is limited by chemical, thermal, and biological stability. There are prospects for the use of polyacrylamide in producing water-swollen polymer systems in which moisture absorption, and desorption and vapor permeability depend on amide group conversion.[24]

There have been attempts to use polyacrylamide or its copolymers as film-forming substances for covering seeds.[31-33] In spite of good water solubility and convenience for use in seed breeding, the wide use of polyacrylamide and its copolymers is restrained because of the high cost of monomers and toxicity.[34,35] Low water absorption (up to 5%) is another disadvantage in the use of these polymers.

The use of several polymers and copolymers based on *N*-vinyllactams or on their mixtures with other water-soluble polymers as materials for seed covers has been recommended.[34-46]

Thin polymer seed covers attach active substances to the seed surface and prolong their action; advantages and disadvantages of these covers have been analyzed. However, water-soluble polymers, which provide a thick cover to maximize seed size, to achieve quickness and precise sowing, and to heighten the ability of the seed material to adapt to environmental conditions, are also used for coatings.

The coating systems include fillers such as peat, ceramsit, kaolin, lignin, and other binding substances.[47-58] They also include micro- and macroelements for seed feeding, complexes of organic-mineral fertilizers, insecticides, and phytohormones and their analogues. Liquid polymers, pectins or polyacrylamide, carboxymethylcellulose, and combinations of carbon acid polymers, including hydroxyl groups, are used as polymer-film-forming substances.[53-57]

Of main interest are the water-soluble polymeric bindings. Such systems greatly extend the area in which coated seeds can be used, including the northern as well as the southern regions of agriculture.

The important part in the creation of polymer systems for plant breeding, together with the production of mixtures of water-soluble polymers, fertilizers, a means of seed protection, and growth and development regulators, is the synthesis of polymers with their own biological activity. In all cases, the biologically active substances are isolated either as a result of hydrolytic or fermentative breaking up of the labile bond. The choice of synthesis method mainly depends on the chemical nature of the functional group of the biologically active substance.

We have examined the concepts of creating and applying water-soluble polymer systems for seed treatment in plant breeding. In agriculture production there are two ways to create polymeric systems that are biologically active. These are devising polymeric compositions based on water-soluble polymers and synthesizing polymers that are biologically active. To extend the scope of these polymers, it is possible to include in the polymeric system some components that conduct different types of activity (e.g., fungicides, bacteriocides, insectoacarycides, and plant growth and development regulators). The control of seed sprouting processes in different zones of agriculture, including those with extreme conditions, depends on the nature of the cover. Moreover, it is possible to create polymeric systems that increase the ability of the seed to adapt to environmental factors by using methods of synthesis and modification.

CONCLUSION

As international practice shows, ecologically safe technologies in agricultural production are associated with the use of water-soluble polymer carriers of physiologically active material. Many polymer matrixes are used for plant protection. In this case, most matrixes are safe for the environment. However, producing polymer matrixes with planned physicochemical and mechanical properties capable of biodegradation and utilization in a definite time interval has not occurred. Producing systems that allow for availability and adaptability to manufacture and to take into consideration influence on living systems is proposed.

We are convinced that in creating plant protection by chemical means, in particular polymer seed coatings, polymers must have properties of adhesion, complex formation, and sorption. They must be considered only in combination.

We have defined the set of important characteristics connected with the use of matrixes in plant protection during seed growing. Sorption characteristics of polymer coatings allow for the regulation of water entering the seed, the binding of protoplast cells to the biopolymer surface, the direction and speed of life processes and, ultimately, germination.

However, physicochemical properties of the new polymer seed coatings that satisfy the requirements of seed growing do not yet conform with manufacturing requirements. It is most reasonable to use thin-layer, water-soluble polyfunctional polymeric coatings that have planned physicochemical properties and a controlled release of acting materials. Moreover, polymer systems must be utilized in soil within an extraordinarily short time by all possible means (e.g., biodegradation and failure under the action of fertilizers). This is why finding polymers that biodegrade is important. Biodegradation must be effective for a wide range of microorganisms, soil differences, and cultivating regions of agricultural zones. Nitrogen-containing polymers are the most probable models capable for biodegradation.

Polymer molecules must, evidently, be simple, and their molecular masses must not exceed 50,000–70,000.

Polymer decomposition products must not be toxic to warm-blooded organisms, microfauna, and flora. It is desirable that the products of biodegradation can be metabolized by living systems. Polymer synthesis must allow for the mechanical properties of polymer systems. It is possible to produce polymer systems that are physiologically active and ecologically safe, and are capable of resisting phytopathogens and of regulating water flow into the seed.

REFERENCES

1. Fr. Demand 2 473 254, 1981.
2. ChSSR Patent 231 294, 1984.
3. USSR Patent 680 709, 1979.
4. GDR Patent 208 526, 1984.
5. U.S. Patent N 3703 404, 1972.
6. Fr. Demand 211 972, 1972.
7. Br. Patent 1 313 234, 1973.
8. PPR Demand 249 262.

9. Nikolayev, A. F. *Chemistry* **1964**, 784.
10. Rosselhozizdat, **1985**, 32.
11. Nikolskaya, G. V. *M. Chemistry* **1987**, 48.
12. Nikolayev, A. F.; Ohrymenko, G. I. *L. Chemistry* **1979**, 145.
13. Fr. Demand 2 430 713, 1980.
14. Rogovic, I. A. *M. Chemistry* **1972**, 520.
15. USSR 952 897, 1982.
16. Fr. Demand 2 556 173, 1985.
17. Fr. Demand 2 556 172, 1985.
18. EPV Patent 0 148 522, 1986.
19. USSR 1 178 342, 1985.
20. Patent 242 545, 1987.
21. Dument, O. N.; Kazansky, K. S.; Miroshnicov, A. M. *M. Chemistry* **1976**, 376.
22. USSR 912 093, 1982.
23. USSR 952 897, 1982.
24. Gakparova, R.; Kasynova, A. V. *Technology of light production and every day service.* 1974, Alma-Ata. 1, 136.
25. Janyagina, N. A.; Makin, Y. I.; Neskirova, Y. N. *Agrochemistry* **1992**, *7*, 145.
26. Fr. Demand 2 277 529, 1975.
27. U.S. Patent 4 249 343, 1981.
28. Fr. Demand 2 468 861, 1981.
29. Fr. Patent 2 555 400, 1985.
30. U.S. Claim 570 611, 1984.
31. USSR Author License 537 639, 1976.
32. Br. Claim 2 095 115, 1982.
33. HPR Claim 32 703T, 1984.
34. Sheftel, V. O. *L. Chemistry* **1981**, 232.
35. Izmerov, N. J.; Anotsky, I. V.; Sidorov, K. K. *M. Medicine* **1977**, 240.
36. Rashidova, S. Sh.; Ruban, I. N. *Tashkent. FAS* **1987**, 107.
37. Ovcharov, K. E.; Shtilman, M. I. *Successes of Chemistry* **1974**, *28*(7), 1282.
38. USSR Author License 352 486, 1976.
39. Br. Claim 134 523, 1973.
40. Fr. Claim 2 156 531, 1973.
41. U.S. Claim 5 480 288, 1976.
42. U.S. Patent 4 367 609, 1983.
43. EVP Patent 0 067 479, 1982.
44. U.S. Patent 4 452 008, 1984.
45. USSR Author license 8 843 806, 1981.
46. USSR Author license 1 355 148, 1987.
47. Rashidova, S. Sh.; Voropayeva, N. L.; Kalantarova, T. D.; Tager, A. A. *VMC* **1993**, (2B) 253–255.
48. USSR Author License 1 250 188, 1986.
49. PRB Claim 39 471, 1986.
50. GDR Claim 217 085, 1985.
51. GDR Claim 257 221, 1984.
52. Ovenarov, K. E.; Shtilman, M. I. *Successes of Chemistry* **1974**, *28*(7), 1282.
53. EPV Patent 145 086, 1985.
54. USSR Author License 372 157, 1973.
55. USSR Author License 685 260, 1979.
56. USSR Author License 1 015 837, 1983.
57. FRG Claim 3 439 932, 1985.
58. Miki, Y.; Kocu, C., 1977.

BIOLOGICALLY ACTIVE POLYANIONS

Raphael M. Ottenbrite and Jun Liao
High Technology Materials Center
Department of Chemistry
Virginia Commonwealth University

Polyanionic polymers are polyelectrolytes with negative charges located along the polymer chain. They are involved in a wide range of biological activity and have been investigated for use in the areas of oncology, virology, and immunology.[1-5] The biological effects of the polyanions include antitumor and antiviral activities as well as immunological activity. The immune response of the host organism to the polyanions plays an essential role in the activities observed.

ANTITUMOR ACTIVITY

The antitumor activity of the polyanions is based on the modulation of the immune system by the polyanions. Among the polyanions with antitumor activity, two have been extensively studied: pyran copolymer and a low molecular weight copolymer of ethylene and maleic anhydride, Carbetimer (carboxyimamides). Pyran copolymer, also known as DIVEMA (divinyl ether - maleic anhydride copolymer), was first reported by Butler who was followed by Breslow.[7-8] The antitumor activity appears to be mediated via a number of mechanisms, such as the induction of interferon (IFN) production, macrophage activation, stimulation of antibody-dependent cellular cytotoxicity, and via the cell function activation combined with the induction of IFN.[9-12]

The specific chemical structures of the polyanionic polymers have been found to affect biological activity. The lipophilicity, chain rigidity, carboxylic acid strength, charge density, and charge distribution were found to have significant effects on the activity.[2,6,14,15]

The immunomodulation effects of the polyanions are related to their structural properties. A series of polyanions that differ in molecular weight, lipophilicity, chain rigidity, and surface charge were evaluated for their ability to induce specific states of macrophage activation.[2,17] Liposome encapsulation of polyanions enhances the efficiency of macrophage activation of the specific polyanions.[16]

ANTIVIRAL ACTIVITY

Naturally occurring and synthetic polyanions have been investigated for antiviral activity in a number of systems.[18-22]

The mechanism of antiviral action of the polyanions is not known. Some proposed modes of antiviral activity include: direct inactivation of the virus; inhibition of virus replication; interferon induction; stimulation of phagocytosis and inflammation; specific immunoenhancement of humoral or cell mediated immune responses against the virus; and enhancement of macrophage antiviral functions.[13,23-33]

SUMMARY

Polyanions exhibit a broad spectrum of biological activity. Several clinical studies have been carried out, however, not one polyanion has gone a phase II trial. Their importance lies in the fact that polyanions do exhibit significant antitumor activity as well as acceptable toxicity, but they are not sufficiently potent to deliver outstanding clinical results. Their key function will probably be in the elucidation of various biological mechanisms.

REFERENCES

1. Donaruma, L. G.; Vogl, O. *Polymeric Drugs*; Academic: New York, 1978.
2. Ottenbrite, R. M.; Kuus, K.; Kaplan, A. M. In *Polymers in Medicine*; Chiellini, E.; Giusti, P. Eds.; Plenum: New York, 1983; p 3.
3. Kuus, K.; Ottenbrite, R. M.; Kaplan, A. M. *J. Biol. Reponse Mod.* **1985**, 46.
4. Fenichel, R. L.; Chirigos, M. A. *Immune Modulation Agents and Their Mechanisms*; Marcel Dekker: New York, 1984.
5. Turowski, R. C.; Triozzi, P. L. *Cancer Invest.* **1994**, *12*, 620.
6. Regelson, W.; Kuhar, S.; Tuniz, M.; Fields, J. E.; Johnson, J. J.; Glusenkamp, E. W. *Nature (London)* **1960**, *186*, 778.
7. Butler, G. B. *J. Polym. Sci.* **1960**, *48*, 279.
8. Breslow, D. S. *Pure Appl. Chem.* **1976**, *46*, 103.
9. Merigan, T. C. *Nature (London)* **1967**, *214*, 416.
10. Dean, J. H.; Padarathsingh, M. L.; Keys, L. *Cancer Treat. Rep.* **1978**, *62*, 1807.
11. Tagliabue, A.; Mantovani, A.; Polentarutti, N. et al. *J. Natl. Cancer Inst.* **1977**, *59*, 1019.
12. Papamatheakis, J. D.; Schultz, R. M.; Chirigos, M. A.; Massicot, J. G. *Cancer Treat Rep.* **1978**, *62*, 1845.
13. Morahan, P. S.; Munson, J. A.; Baird, L. B.; Kaplan, A. M.; Regelson, W. *Cancer Res.* **1974**, *34*, 506.
14. Fields, J. E.; Ascular, S.; Johnson, J. H.; Johnson, P. K. *J. Med. Chem.* **1982**, *25*, 1060.
15. Kaplan, A. M.; Kuns, K.; Ottenbrite, R. M. *Ann. NY Acad. Sci.* **1985**, *446*, 169.
16. Sato, T.; Kojima, K.; Ihda, T.; Sunamoto, J.; Ottenbrite, R. M. *J. Bioact. Compat. Polym.* **1986**, *1*, 448.
17. Kuus, K.; Ottenbrite, R. M.; Kaplan, A. M. *J. Biol. Response Mod.* **1985**, *4*, 46.
18. Brown, J. W.; Firshein, W. *Bacteriol. Rev.* **1967**, *31*, 83.
19. Regelson, W. *Adv. Can. Res.* **1968**, *11*, 223.
20. Regelson, W. *Adv. Chemotherp.* **1968**, *3*, 303.
21. Breinig, M. C.; Munson, A. T.; Morahan, P. S. In *Anionic Polymeric Drugs*; Donaruma, L. G. et al. Eds.; Wiley, New York, 1980; p 211.
22. Gianinazzi, S.; Kassanis, B. *J. Gen. Virol.* **1974**, *23*, 1.
23. Merigan, T. C.; Finkelstein, M. S. *Virology* **1968**, *35*, 363.
24. McCord, R. S.; Breinig, M. K.; Morahan, P. S. *Antimicrob. Agents Chemother.* **1976**, *10*, 28.
25. Papas, T. S.; Pry, T. W.; Chirigos, M. A. *Proc. Natl. Acad. Sci. U.S.A.* **1974**, *71*, 367.
26. Merigan, T. C.; Regelson, W. *N. Engl. J. Med.* **1967**, *277*, 1283.
27. Regelson, W.; Morahan, P. S.; Kaplan, A. In *Polyelectrolytes and their Applications*; Rembaum, A.; Selegny, E., Eds.; Reidel: Holland, 1975; Vol. 2.
28. Breslow, D. S.; Edwards, E.; Newburg, N. *Nature (London)* **1973**, *246*, 160.
29. Declerq, E.; Desomer, P. *Infec. Immunol.* **1973**, *8*, 669.
30. Hirsch, M. S.; Black, P. H.; Wood, M. L.; Monaco, A. P. *J. Immunol.* **1973**, *111*, 91.
31. Carrano, R. A.; Iuliucci, J. D.; Luce, J. K.; Page, J. A.; Imondi, A. R. In *Immune Modulation Agents and their Mechanisms*; Fenichel, R. L.; Chirigos, M. A. Eds.; Marcel Dekker: New York, 1984, p 247.
32. Billiau, A.; Muyembe, J. J.; Desomer, P. *Nature (London)* **1971**, *232*, 183.
33. Morahan, P. S.; Schuller, G. B.; Snodgrass, M. J.; Kaplan, A. M. *J. Infect. Dis.* **1976**, *133*, A249.

Biomaterials

See: *Acetan*
Acryloyl-L-Proline Alkyl Ester Hydrogels
Agarose (Overview)
Alginate Fibers (High Performance Papers)
Alternating Radical Copolymerization
Antibacterial Resins
Antibody-Polymer(Porphyrin) Complexes
Anticancer Polymeric Prodrugs, Targetable
Antiinfective Biomaterials
Antitumoral and Antiviral Polyoxometalates (Inorganic Discrete Polymers of Metal Oxide)
Bioabsorbable Polymers (Tissue Engineering)
Bioartificial Materials
Biodegradable Copolymers [Comprising Poly(L-lactic acid)]
Biodegradable Polydepsipeptides (Synthesis, Characterization and Degradation)
Biodegradable Polymers (Orthopedic Applications)
Biodegradable Polyesters (Chemically-Induced Surface Degradation)
Biodegradable Polyesters (Theoretical Modeling of Degradation)
Bioerodible Polymers A. Controlled Drug Release B. Fracture Fixation
Biologically Active Agricultural Polymers (Mechanism of Action on Plants)
Biomedically Degradable Polymers
Biopolymers (Electrochemical Behavior)
Block Copolymer Micelles (Microcontainers for Drug Targeting)
Blood Compatible Polymers
Blood Substitutes
Bone Cements, Bioresorbable
Catalyses (by Polyelectrolytes and Colloidal Particles)
Catalysts, Polymeric (Higher Activity than Monomeric Analogs)
Catalysts, Polymeric (Pressure Effects)
Cell Adhesion-Promoting Biomaterials
Cell Entrapment [Poly(carbamoyl sulfonate) Hydrogels]
Cell Growth-Promoting Biomaterials
Cell-Separation Materials
Chitin and Chitosan, Graft Copolymers
Chitin and Derivatives
Chitosan (Preparation, Structure, and Properties)
Chondroitin Sulfate (Overview)
Collagen
Contact Lenses, Gas Permeable
Contact Lenses, Hydrogels
Cornea, Artificial (Hydrophilic Polymeric Sponges)
Cyanoacrylates
Cyclic Imino Ethers (Ring-Opening Polymerization)

Dental Polymers (Overview)
Dental Polymers (Glass-Ionomer Cements)
Dental Polymers (Hydrogels)
Dental Sealants
Dextran
Enzyme-Degradable Hydrogels
Enzyme Immobilization (Polyion Complexes)
Enzyme Immobilization (Porous Polymer Beads)
Enzyme Mimics (Cyclodextrins)
Enzyme Modification (Use in Organic Media)
Fluorine-Containing Poly(amino acids) (Conformation and Gas Permeability)
Fluorosilicones (Monomer Synthesis and Applications)
Fracture Immobilization Polymers
Functional Immobilized Biocatalysts (Stimulus-Sensitive Gels)
Gelatin (Molecular Weight Distributions)
Gels (Overview)
Globular Polybases, Microemulsion-like
Glucose Responsive Polymers (Medical Applications)
Guar (Graft Copolymerization)
Gums (Overview)
Hair and Skin Care Biomaterials (Cosmetology and Dermatology)
Hair and Skin Cosmetics
Heparin (Overview)
Heparinoids, Artificial (Biological Activities)
Hyaluronic Acid (Macroionic, Rheological, and Lubricity Properties)
Hybrid Artificial Organs
Hydrogel Biomaterials (HEMA-Based)
Hydrogels, Microphase Separated Blends (Biomaterials and Drug Delivery System)
Hydrophilic Polymers (for Friction Reduction)
Hydrophobized Polysaccharides (Versatile Functions)
Hydroxyapatite-Biopolymer Interactions
Immobilized Enzymes, Biosensors
Immobilized Microbial Cells
Intelligent Polymers (in Medicine and Biotechnology)
Ion-Chelating Polymers (Medical Application)
Ion-Selective Electrodes and Biosensors
Ion-Selective Microchemical Sensors (Cardiology)
Ionene Polymers
N-Isopropylacrylamide Copolymers (Drug Delivery)
Ligament Replacement, Artificial
Ligament Replacement Polymers (Commercial Products)
Ligament Replacement Polymers (Technological and Design Aspects)
Lung, Artificial
Membranes (Overview)
Metal-Containing Monomers (Polymerization)
Microbial Polyhydroxyalkanoates (Poly[(R)-3-hydroxybutyrate-co-3-hydroxypropionate])
Microcapsules (From Two Dimensional Networks)
Microcapsules, Molecular Complexes (Dissolution of Drugs by Cyclodextrin)
Microencapsulated Particles
Microencapsulation (Artificial Cells)
Microencapsulation (Drug Release)
Monodisperse Polymer Particles (Nonmagnetic and Magnetic)
Nonthrombogenic Materials
Nonthrombogenic Polymers (Multiblock Copolymers of Polyether and Polyamide)
Nucleic Acid Analogs
Nucleic Acid Interactions (with cis-platin)
Polyamides, Glycine-Containing
Poly(α-amino acid) Spherical Particles
Poly(α-amino acids) (Biodegradation, Medical Applications)
Poly(aryl ether ketone)s (Overview)
Poly(2-cyanoacrylates)
Polyelectrolyte Complexes (Overview)
Polyelectrolyte Complexes (in Immunology)
Poly(ether sulfone)s (Properties, Applications, and Trends)
Polyethylene (Surface Functionalization)
Poly(ethylene oxide) (Applications in Drug Delivery)
Polyisopeptides
Poly(N-isopropylacrylamide) Hydrogels
Polymeric Drugs
Polysaccharides, Synthetic (Hypoglycemic Activity)
Polysaccharides, Synthetic (Preparation and Biological Activities)
Polyurethane Ionomers
Protein-Based Polymeric Materials (Synthesis and Properties)
Pseudo-Poly(amino acid)s (Overview)
Radiation Induced Polymerization (Biomedical Applications)
Reactive Oligomers (Overview)
Red Blood Cell Substitutes, Artificial (for Transfusion)
Regeneration Templates, Artificial Skin and Nerves
Signal Transduction Composites (Biomaterials with Electroactive Polymers)
Silk Fibroin (Soft Tissue Compatible Polymer)
Smart Hydrogels
Starch Graft Copolymers
Starch-Polymer Composites
Stimuli-Responsive Microcapsules
Stimuli-Responsive Polymers (Thermodynamics in Drug Delivery Technology)
Surface-Modified Polymers
Suture Materials (Current Research and Development)
Vitreous Body, Artificial
Water-Swellable Polymers (Carbomer Resins)
Wound Dressing Materials

BIOMEDICALLY DEGRADABLE POLYMERS

A.-C. Albertsson and A. Löfgren
Department of Polymer Technology, KTH
Royal Institute of Technology

Degradable polymers are well suited for use as temporary aids in wound healing, tissue replacement, and for use as a drug delivery matrix. Because the material is gradually absorbed by the body, surgical removal of the implant is not needed. Depending on the site and purpose of the implantation, the material must possess different properties, from hard, stiff materials for replacing bone to soft, flexible materials for replacing soft tissues. Other criteria include variations in hydrophilicity to match drugs that are to be incorporated.

Biomaterials can be of natural or synthetic origin. To differentiate between natural and synthetic biomaterials, the latter class is often called biomedical materials. Biomedical polymers can be classified as either stable or absorbable. The stable biomedical polymers are made to permanently replace human tissue without losing their properties over time, whereas the absorbable polymers are designed to gradually lose their properties, dissolve or metabolize, and leave the body by natural pathways.

By far the most common class of synthetic absorbable polymers used today is the poly(α-hydroxy acid)s. These aliphatic polyesters undergo simple hydrolysis *in vivo* and form natural metabolites as degradation products. M. Vert[4] has proposed that these materials be called "bioresorbable" since they degrade and are further resorbed *in vivo.*

Synthesis of poly(α-hydroxy acid)s involves polycondensation, but generally ring-opening polymerization of the cyclic condensation product of two α-hydroxy acid molecules (e.g., lactide or glycolide) is preferred. Many different applications have been proposed, although only a few have reached commercial practice so far.[1] Examples of commercial products are bone fracture fixation devices, drug-delivery devices, and suture filaments.

It was recognized early on that microorganisms had the ability to produce polymers.[5,6] In recent years, microorganisms have been used to produce aliphatic polyesters commercially.[7] Microorganisms produce, for example, poly(hydroxy butyrate)/(poly(hydroxy valerate) (PHB/PHV) copolymers of perfectly isotactic nature and of high molecular weights.[8]

RING-OPENING POLYMERIZATION OF LACTONES

Carothers and colleagues were the first to systematically explore the ring-opening polymerization of various lactones.[20] Due to the increasing interest in the production of degradable materials, many laboratories have since then, and particularly in recent years, been involved in this research area.

Investigations reported in the literature have been made using all major polymerization mechanisms to develop the ROP method and make available as many different polymeric structures as possible. The different initiation mechanisms can be divided into anionic, coordinative, cationic, radical, zwitterionic, and active hydrogen processes. The highest yields and molecular weights have, however, been obtained mainly by the anionic and coordinative ROP.

Coordinative ROP of Lactones

During the past 10 years, our research group has synthesized a variety of different degradable polymers including polyesters, polyanhydrides,[12] polycarbonates,[13] poly(ether esters)s,[11] and copolymers thereof.[10] From the very start we realized how important it is that the properties of degradable polymers be modifiable. Early attempts to synthesize degradable elastomeric materials in the form of block copolymers were based on condensation polymerization,[14] but later studies revealed the many advantages of ring-opening polymerization.[11] Poly(β-propiolactone) (PPL [25037-58-5]) was investigated and shown to be formed in high yield and high molecular weights.[15] To make the perfectly alternating copolymer of β-PL and ethylene glycol, a new synthetic route was developed to produce the monomer 1,5-dioxepan-2-one (DXO), a cyclic seven-membered ether-lactone ring.[11] Although the homopolymer (PDXO, [121425-66-9]) of this ether lactone is an amorphous material with a glass-transition at –36 to –39°C, it has proven to yield elastomeric, degradable materials when copolymerized with various lactones.[16,17]

DEGRADABLE COPOLYMERS FROM 1,5-DIOXEPAN-2-ONE

Properties

Poly(L- or D,L-lactic acid) [26811-96-1] and [31587-11-8] are indeed some of the most common and well-studied degradable polymers on the market today. The high strength and brittleness that characterize these poly(α-hydroxy acid)s can be substantially modified by copolymerization with, for example, 1,5-dioxepan-2-one. The large difference in glass-transition temperatures (T_g) of poly(DXO) (–36 to –39°C) and polylactides (+55 to +58°C) results in copolymers with a wide range of intermediate properties.

Applications

The use of synthetic degradable polymers and copolymers in biomedical devices has been extensively explored during the past decades. The first commercially successful application was the synthetic absorbable surgical suture.[18] The poly(glycolic acid) used was soon followed by other poly(α-hydroxy acid)s, and copolymers were developed to modify properties like degree of crystallinity and hydrolysis rate.[3,4] Other applications include orthopedic pins, screws, and plates,[19,20] artificial blood vessels,[21] burn dressings,[22,23] nerve guides,[24,25] artificial tendons,[26] and in the regeneration of periodontal tissue.[27] New, more startling ideas also began to emerge, such as "hybrid artificial organs," which may lead to revolutionary advances in transplantation surgery.[28]

Another area of major importance is the controlled release of drugs. The benefits of releasing a predetermined amount of drug at a controllable rate are obvious, as are the needs for various kinds of materials to match different drugs and to provide specific release profiles. The drug can be incorporated into a degradable polymer matrix or be attached via a cleavable bond to a polymer chain.[29] Future prospects include "targeted" drug delivery, where the device can be directed by various means to the pathological site of interest.[30]

As mentioned in the introduction, degradable synthetic polymers began to attain a more widespread use because of increasing environmental concern. Great interest has been shown by the agricultural and packaging industries, and although price has been a limiting factor, large-scale production[31] can bring costs down and open the way for new, exciting applications.

REFERENCES

1. Vert, M. *J. Mater. Sci.: Mater. in Med.* **1992**, *3*, 432.
2. Albertsson, A.-C.; Lundmark, S. *Brit. Polym. J.* **1990**, *23*, 205.
3. Wasserman, D.; Versfelt, C. C. U.S. Patent 3 889 297, 1974.
4. Casey, D. J.; Roby, M. S. Eur. Patent 0098394, 1983.
5. Ellar, D.; Lundgren, D. G.; Okamura, K.; Marchessault, R. H. *J. Mol. Biol.* **1968**, *35*, 489.
6. Doi, Y.; Tamaki, A.; Kunioka, M.; Soga, K. *J. Chem. Soc., Chem. Commun.* **1987**, 1635.
7. Holmes, P. A.; Wright, L. F.; Collins, S. H. (ICI), Eur. Pat. App. 0052459, 1981; Eur. Pat. App. 0069497, 1983.
8. Holmes, P. A. *Phys. Technol.* **1985**, *16*, 32.
9. Johns, D. B.; Lenz, R. W.; Luecke, A. In *Ring-Opening Polymerization*; Ivin, K. J.; Saegusa, T., Eds.; 1984; Vol. 1, 464.
10. Löfgren, A.; Albertsson, A.-C.; Dubois, P.; Jéröme, R.; Teyssié, P. *Macromolecules* **1994a**, *27*, 5556.
11. Mathisen, T.; Masus, K.; Albertsson, A.-C. *Macromolecules* **1989**, *22*, 3842.
12. Albertsson, A.-C.; Ljungquist, O. *J. Macromol. Sci. Chem.* **1986**, *A23*, 393.
13. Albertsson, A.-C.; Sjöling, M. *J. Macromol. Sci. Chem.* **1992**, *A29*, 43.
14. Albertsson, A.-C.; Ljungquist, O. *J. Macromol. Sci. Chem.* **1986**, *A23*, 411.
15. Mathisen, T.; Albertsson, A.-C. *J. Appl. Polym. Sci.* **1990**, *39*, 591.
16. Albertsson, A.-C.; Löfgren, A. *Makromol. Chem., Macromol. Symp.* **1992**, *53*, 221.
17. Löfgren, A.; Renstad, R.; Albertsson, A.-C. *J. Appl. Polym. Sci.* in press.
18. Schmitt and Polistina, 1967.
19. Vert, M.; Chabot, F.; Leray, J.; Christel, P. *Makromol. Chem. Suppl.* **1981**, *5*, 30.
20. Vasenius, J.; Vainionpää, S.; Vihtonen, K.; Mero, M.; Makelä, P.; Törmälä, P.; Rokkanen, P. *J. Biomed. Mater. Res.* **1990**, *24*, 1615.
21. Hinrichs, W. L. J.; Zweep, H.-P.; Satoh, S.; Feijen, J.; Wildevuur, R. H. *Biomaterials* **1994**, *15*, 83.
22. Beumer, G. J.; Blitterswijk, C. A.; Ponec, M. *J. Biomed. Mater. Res.* **1994**, *28*, 545.
23. Gatti, A. M.; Pinchiorri, P.; Monari, E. *J. Mater. Sci., Mater. Med.* **1994**, *5*, 190.
24. den Dunnen, W. F. A.; Schakenraad, J. M.; Zondervan, G. J.; Pennings, A. J.; van der Lei, B.; Robinson, P. H. *J. Mater. Sci., Mater. Med.* **1993**, *4*, 521.
25. Perego, G.; Cella, G. D.; Aldini, N. N.; Fini, M.; Giardino, R. *Biomaterials* **1994**, *15*, 189.
26. Davis, P. A.; Huang, S. J.; Ambrosio, L.; Ronca, D.; Nicolais, L. *J. Mater. Sci., Mater. Med.* **1991**, *3*, 359.
27. Lundgren et al., 1994.
28. Barrera, D.; Langer, R. S.; Lansbury, P. T.; Vacanti, J. P. WO patent 94/09760, May 1994.
29. Langer, R. *Chem. Eng. Commun.* **1980**, *6*, 1.
30. Ottenbrite, R. M. *Encyclopedia of Polymer Science and Engineering, Suppl. Vol.*, 2nd ed.; Mark, H. F. et al., Eds.; Wiley: New York, 1989; p 164.
31. Gruber, P. R.; Hall, E. S.; Kolstad, J. J.; Iwen, M. L.; Benson, R. D.; Borchardt, R. L. WO patent 93/15127, August 1993.

Biopolymers

See: *Acetan*
Agarose (Overview)
Antibody-Polymer (Porphyrin) Complexes
Anticancer Polymeric Prodrugs, Targetable
Bioabsorbable Polymers (Tissue Engineering)
Bioartificial Materials
Biopolymers (Overview)
Biopolymers (Electrochemical Behavior)
Biospinning (Silk Fiber Formation, Multiple Spinning Mechanisms)
Cell Adhesion-Promoting Biomaterials
Cell Entrapment [Poly(carbamoyl sulfonate) Hydrogels]
Cell Growth-Promoting Biomaterials
Cell Separation Materials
Cellulose (Overview)
Cellulose (Pyrolysis)
Cellulose (Regioselectively Substituted Esters and Ethers)
Cellulose Acetate Nitrate
Cellulose Derivative-Protein Complex (in Ice Cream)
Cellulose Derivative-Protein Complexes (Whey and Corn Steep Liquor, Enrichment of Macaroni)
Cellulose Derivatives (Reclamation of Proteins from Cheese Whey)
Cellulosic Liquid Crystals
Chitin (Metal Ion Chelation and Enzyme Immobilization)
Chitin and Chitosan, Graft Copolymers
Chitin and Derivatives
Chitin Chemistry
Chitin Synthesis
Chitinases
Chitosan (Preparation, Structure, and Properties)
Chondroitin Sulfate (Overview)
Collagen
Connectin (Titin) Large Filamentous Protein)
Dextran
DNA (Bending and Twisting Elasticity)
DNA (Intercalation with Drugs)
DNA Degradation (Atomic Target Method using Synchrotron Radiation)
Enzyme-Degradable Hydrogels
Enzyme Immobilization (Polyion Complexes)
Enzyme Immobilization (Porous Polymer Beads)
Enzyme Mimics (Cyclodextrins)
Enzyme Modification (Use in Organic Media)
Gelling Agents (Agarose and Carrageenan)
Globular Proteins (Structures and Functions)
Glycoproteins (Synthesis by NCA Method)
Glycosaminoglycans (Overview)

Guar (Graft Copolymerization)
Gums (Overview)
Hemicelluloses
Hemoproteins
Heparin (Overview)
Hyaluronic Acid (Macroionic, Rheological Lubricity Properties)
Hydrophobized Polysaccharides (Versatile Functions)
Hydroxyapatite-Biopolymer Interactions
Hydroxyethyl Cellulose (Infrared Spectra and Dielectric Properties)
Hydroxyethyl Cellulose (Preparation and Use in Protein Precipitation)
Hydroxyethyl Cellulose (Variables Affecting Hydroxyethylation Reaction)
Jute
Keratins
Lignin Graft Copolymers
Liquid Crystalline Polymers (Polypeptides)
Marine Adhesive Proteins, Synthetic
Melanin (Structure and Chemico-Physical Properties)
Microbial Polyhydroxalkanoates (Poly[(R)-3-hydroxybutyrate-co-3-hydroxypropionate])
Molecular Recognition (Peptide-Based Systems)
Molecular Shape Recognition
Monodisperse Polymer Particles (Nonmagnetic and Magnetic)
Monolayer and Langmuir-Blodgett Films (of Protein)
Nonthrombogenic Materials
Nucleic Acid Analogs
Nucleic Acid Interactions (with cis-platin)
Nucleohistone Complexes (Protein-Nucleic Acid Interactions)
Paracrystals
Peptide-Based Nanotubes (New Class of Functional Biomaterials)
Phospholipid Bilayers
Phospholipid-Immobilized Surfaces
Photochromic Poly(α-aminoacid)s
Poly(n-alkyl L-glutamate) (Structure and Molecular Mobility in Solid Liquid Crystalline State)
Polyamides, Glycine-Containing
Poly(α-amino acid) Spherical Particles
Polyelectrolyte Complexes (Overview)
Polyelectrolyte Complexes (In Immunology)
Polyelectrolyte Complexes (Targeting of Nucleic Acids)
Polyesters (Derived from Renewable Resources)
Poly(β-hydroxyalkanoates) (Microbial Synthesis and Chemical Modifications)
Polyisopeptides
Polymeric Drugs
Polynucleotide Analogues
Polypeptide-Polypeptide Complexes
Polypeptide Restricted Backbone Conformation α,α-Symmetrically Disubstituted Glycines)
Polysaccharide-Polysaccharide Interactions
Polysaccharides, Synthetic (Hypoglycemic Activity)
Polysaccharides, Synthetic (Preparation and Biological Activities)
Protein-Based Polymeric Materials (Synthesis and Properties)
Pseudo-Poly(amino acid)s (Overview)
Self-Assembled Polymers (at Interfaces)
Signal Transduction Composites (Biomaterials with Electroactive Polymers)
Silk (Its Formation, Structure, Character and Utilization)
Silk (Physico-Chemical Properties)
Silk Fibers (Chemical Modification)
Spider Silk (Production of Polypeptide Polymers)
Starch-Based Plastics (Measurement of Biodegradability)
Starch Biodegradation (in Starch-Plastic Blends)
Starch Graft Copolymers
Starch-Polymer Composites
Starch Polymers, Natural and Synthetic
Stereo-Block Copolymers (of L- and D-Lactides)
Superabsorptive Polymers (from Natural Polysaccharides and Polypeptides)
Vegetal Biomass (1. Monomers and their Polymerization)
Vegetal Biomass (2. Oligomers and their Polymerization)
Vegetal Biomass (3. Polymers, Derivatives, and Composites)
Water-Soluble Polymers (Oil Recovery Applications)
Water-Soluble Polymers (Rheological Properties)
Welan Microbial Polysaccharide
Wood (Chemical Modification)
Wood, Chemically Modified (Thermoplaticization and Solubilization)
Wood Graft Copolymers
Wood-Plastic Composites
Wood-Polymer Composites
Wool (Overview)
Wool Keratin
Xanthan, Depolymerization
Xanthan Gum (Overview)
Xanthan Gum (Conformation and Solution Properties)

BIOPOLYMERS (Overview)

Tadayuki Uno and Saburo Shimabayashi
Faculty of Pharmaceutical Sciences
University of Tokushima

"Biopolymer" is the generalized name for polymers that are contained in, and constitute, living organisms; proteins, nucleic acids, and polysaccharides are the predominant members of the class. The biopolymers consist of repetitive structural units such as amino acids (proteins), nucleotides (nucleic acids), and sugars (polysaccharides), which are linked by dehydration. The major difference between biopolymers and synthetic polymers is that the former are isotactic polymers, exclusively, with well-defined molecular structures, and generally are polyelectrolytes. The stereochemistry in the biopolymers is very important for understanding their biological functions. In this section, the general properties of proteins and nucleic acids are briefly reviewed.

PROTEINS

Proteins are usually found in cytosol and in the nucleus of a cell, and are concerned in most biological reactions. The proteins consist of twenty L-α-amino acids that are linked linearly with a chemical bond (peptide bond, –CO–NH–). The unique structure and properties of the proteins are brought about by the amino acid side-chains. Information on the amino acid sequence is encoded for in DNA, which is transcribed and translated through mRNA. It has been estimated that more than 100,000 kinds of proteins are contained in the human body. Many proteins in eukaryotes are processed after translation (digestion, phosphorylation, and/or acylation), which is closely related to the unique function of the particular protein.

Properties and Applications

The amino acids are classified into four groups by the nature of the side chains. Hydrophobic amino acids (Ala, Val, Leu, Ile, Phe, Met, Pro); charged amino acids (Asp, Glu, Lys, Arg); polar amino acids (Ser, Thr, Cys, Asn, Gln, Tyr, Trp, His); and Gly. Proteins can bind with acidic and basic compounds at the charged residues. The net charge is cancelled at a specific pH, which is called the isoelectric point. The occurrence of amino acids is not homogeneous along the sequence. The hydrophobic anion acids are usually located in the interior of the protein, while hydrophilic ones are on the exterior.

Primary Structure

The architecture of proteins refers to four levels of structure. The primary structure defines the sequence of amino acids and location of disulfide bonds. The primary structure describes the covalent connection of the amino acids in the protein.

Secondary Structure

The secondary structure refers to the steric relationship of amino acids in the linear sequence, and the steric structure of a protein is mainly determined by the secondary structure. The α helix and the β pleated sheet, which are regular and periodic structure units, are examples of the secondary structure. Proline is seldom found in the α-helical region because of its unique five-membered ring structure, while asparagine frequently occurs at the amino terminal region of the α-helix.[1]

Tertiary Structure

The tertiary structure refers to the three-dimensional conformation of the polypeptide backbone and amino acid side chains in the protein molecule. Proteins with the same primary structure should fold into a unique tertiary structure, while those with the equivalent tertiary structure does not necessarily mean the same amino acid sequence. The site-directed mutants are usually isomorphous to each other. The tertiary structure has been conventionally determined by X-ray diffraction studies of protein single crystals, although recent developments in multi-dimensional NMR spectroscopy enable to elucidate the solution structure.

X-ray diffraction — The tertiary structure of the protein was first determined by an X-ray analysis,[2-4] and the methodology is well established now.[5] The X-ray crystallography is applicable to large molecules such as viruses[6-8] and membrane proteins with many subunits.[9]

NMR — The conformation of a water-soluble protein with molecular weight less than 20,000 could now be determined by multi-dimensional NMR spectroscopy.[10,11] The NMR spectroscopy reveals the solution structure of the protein and the crystallization is not required.

Quaternary Structure

In the proteins with more than one polypeptide chain, the way in which chains are packed together is called quaternary structure. Therefore, the quaternary structure refers to the organization of subunits from the structural point of view. The dissociation of the subunits is facilitated by the change of protein concentration, pH, ionic strength, and by the subtraction or addition of prosthetic groups or denaturants such as urea and guanidine hydrochloride. The association and dissociation of subunits would regulate the metabolism and signal transduction under physiological conditions.

NUCLEIC ACIDS

Nucleic acid is a linear polynucleotide that is composed of phosphate, sugar, and a base. The nucleic acid is classified structurally and functionally into RNA (ribonucleic acid) and DNA (deoxyribonucleic acid), the former contains ribose while the latter contains deoxyribose as the sugar. They are characterized by the four bases. RNA contains adenine, guanine, and cytosine and uracil, while thymine is contained in DNA instead of uracil. The 5′-hydroxyl group of the sugar is linked together with the 3-hydroxyl of the neighboring sugar by a phosphodiester bond.

Chemical Synthesis

The polynucleotides which consist of up to 100 bases are now chemically synthesized by a commercial DNA synthesizer. Solid-phase synthesis by phosphoroamidite method[12,13] is frequently used.

Properties and Applications

Structure

DNA is a linear polymer and the 5′- and 3′-hydroxyl groups of a nucleoside are covalently linked with the neighboring 3′- and 5′-hydroxide by phosphodiester bonds, respectively. The content of adenine base equals to that of thymine, while the same relation also holds between guanine and cytosine. Therefore, the respective DNA is characterized by the GC (or AT) content. The higher organisms contain trace amounts of 5-methylcytosine, which is assumed to control the DNA expression. The stereochemical structure of the polynucleotides, as well as proteins, could be determined by X-ray and NMR studies.

REFERENCES

1. Richardson, J. S.; Richardson, D. C. *Science* **1988**, *240*, 1648.
2. Kendrew, J. C.; Bodo, G.; Dintzis, H. M.; Parrish, H.; Wyckoff, H.; Phillips, D. C. *Nature* **1958**, *181*, 662.
3. Kendrew, J. C.; Dickerson, R. E.; Strandberg, B. E.; Hart, R. J.; Davies, D. R.; Phillips, D. C.; Shore, V. C. *Nature* **1960**, *185*, 422.

4. Perutz, M. F.; Rossmann, M. G.; Cullis, A. F.; Muirhead, G.; Will, G.; North, A. T. *Nature* **1960**, *185*, 416.
5. Stout, G. H.; Jensen, L. H., *X-ray Structure Determination: A Practical Guide*, 2nd ed.; Wiley: New York, 1989.
6. Giranda, V. L.; Chapman, M. S.; Rossmann, M. G.; *Proteins* **1990**, *7*, 227.
7. Salunke, D. M.; Caspar, D. L. D.; Garcea, R. L. *Cell* **1986**, *46*, 895.
8. Harrison, S. C.; Olson, A. J.; Schutt, C. E.; Winckler, F. K.; Bricogne, G. *Nature* **1978**, *276*, 368.
9. Deisenhofer, J.; Epp, O.; Miki, K.; Huber, R.; Michel, H. *Nature* **1985**, *318*, 618.
10. Wuthrich, K. *NMR of Proteins and Nucleic Acids*; Wiley: New York, 1986.
11. Bax, A. *Annu. Rev. Biochem.* **1989**, *58*, 223.
12. Matteuci, M. D.; Caruthers, M. H. *Tetrahedron Lett.* **1980**, *21*, 3243.
13. Caruthers, M. H. *Science* **1985**, *230*, 281.

BIOPOLYMERS (Electrochemical Behavior)

Shigetada Fujii and Nobuhide Tatsumoto
General Education Chemistry
Oita Medical University

Kiwamu Yamaoka
Faculty of Science
Hiroshima University

Because biopolymers have such great importance in basic and applied fields of polymer science, many electrochemical studies of biopolymers have been reported. The numbers of these studies is constantly growing as the measuring apparatus and methodology advance. In particular, many studies have been directed toward deoxyribonucleic acid (DNA) and ribonucleic acid (RNA), heme proteins such as cytochromes, several enzymes, and peptide hormones, for elucidation of catalytic actions, adsorptions and conformational changes at the electrode surface, and electron-transfer reactions, as will be described below.

ADSORPTION BEHAVIOR AND INTERACTION OF BIOPOLYMERS WITH OTHER COMPOUNDS

Adsorption of DNA

Using differential capacity measurements, Miller[3] examined the steric orientation of RNA and DNA on the mercury electrode surface. He deduced that the native DNA preserves its double-helical structure when absorbed on a negatively charged surface, whereas the DNA molecules unfold on the positively charged surface.

Ever since, many studies concerning the adsorption of DNA in the solution-electrode interface have been carried out in conjunction with DNA structures on the electrode. Paleček[1,4,5] has reviewed the studies for the adsorption of DNA in detail.

Adsorption of Poly(amino acid)s and Proteins

Scheller et al.[6] performed AC polarographic measurements for poly(L-lysine) (PLL). They found that the adsorption of PLL on the electrode surface is closely related to the helix-coil transition of the molecule.

Generally, globular proteins are adsorbed strongly on the electrode surface. It has been mentioned that, especially for some enzymes, the enzymatic activity may be influenced by adsorption-induced structural changes on the electrode surface.[7]

DNA

Calendi et al.[8] studied the interaction of DNA with daunomycin, one of the antibiotics, by DC polarographic measurement together with sedimentation, viscosity, and optical measurements.

The reduction current of daunomycin in the polarogram was lessened remarkably by the addition of DNA to the solution, indicates the complex formation of daunomycin with DNA molecules. Some studies with biologically active substances other than daunomycin have used anthracyclines[9] and distamycin.[10]

DNA also interacts with metal ions such as Zn^{2+}, Cd^{2+}, Cu^{2+}, Fe^{2+}, Co^{2+}, Ni^{2+}, Os^{2+}, Mn^{3+}, and also with metal chelates.[10-16] Rodriguez and Bard et al. reported a series of studies on the interaction of DNA with metal–chelate complexes.

Polyamino Acids and Proteins; Interactions with Metal Ions

Interactions between polyamino acids and metal ions have been reported for various physicochemical methods such as spectroscopy, circular dichroism, electron spin resonance, and light scattering. For the electrochemical methods, interactions between proteins and metal ions have been reported.[17-21]

REDOX REACTIONS OF THE PROSTHETIC GROUPS IN PROTEINS

There are some reports that direct electron transfer between the electrode and proteins does not occur.[22-25] However, Cecil et al.[26] and Pavlovic et al.[27] have reported that electrochemical reduction can occur in the disulfide group containing proteins.

Proteins

For the electrochemical behavior of globular proteins, Dryhurst et al.[7] reviewed the literature in detail.

Metal Ions in Heme Protein

Kadish et al.[28] examined the redox reactions of heme and heme proteins in aqueous and in alcoholic aqueous solutions in detail, using the methods of polarography, controlled potential coulometry, and cyclic voltammetry.

Flavin and Its Derivatives

Flavoproteins act as the catalyzers for several kinds of redox reactions. These proteins can be found in animals and plants. The prosthetic groups that can be found in the flavoproteins are riboflavin (RF), flavin mononucleotide (FMN), and flavin adenine dinucleotide (FAD).

Free RF, FMN, and FAD are adsorbed strongly at the solution-electrode interface and reveal reversible one-electron redox processes.[29]

Protonated Prosthetic Groups

Protonated Bases

Among nucleic acid—the bases that play a very important role in ribonucleic acid—the protonated species of adenine and cytosine can be reduced electrochemically.

Electrochemistry of DNA

Numerous studies on the electrochemistry of DNA have been reported. Paleček[1,30,31] and Dryhurst[2] reviewed these results in great detail.

RELATIONS BETWEEN ADSORPTION AND REDOX REACTIONS OF BIOPOLYMERS ON ELECTRODE

Some redox reactions of biopolymers are closely related to the adsorption behavior on the electrode surface. In the cyclic voltammetry for high potential iron—sulfur proteins,[32] there has been a report that an adsorption process is included in the reduction of the proteins.

APPLICATIONS

Modified Electrodes

Modified electrodes have opened some new functions. The surface of the usual solid (or liquid) electrode is modified with functional substances.

Enzyme Electrodes

Ever since Clark et al.[33] and Updike et al.[34] introduced enzymes to the electrode surface, this field has developed remarkably.[36-37] In the early stage of the biosensor research, the surface of these materials was simply covered with a unimolecular adsorption layer of enzymes. More recently, the enzyme molecules have been fixed to the electrode by means of covalent bonds, or covered with polymer membranes incorporating enzyme molecules.

Immunosensors

An immunosensor is based on the covalent immobilization of antigen or antibody molecules on a clean electrode surface. With such an electrode, the electric polarization or electron-transfer reaction between antibody and antigen molecules may be induced by binding; thus a new electric double layer may be formed on the electrode surface.

Electropolymerization

Poly(amino acid)s are commercially produced by the N-carboxy anhydride (NCA) method.[38] In this method, NCAs are usually polymerized with base catalysts of primary or tertiary amines. The yield and quality of poly(amino acid)s are greatly affected by impurities in NCAs and the kind of base catalysts.[39] By using an electrogenerated base (EGB) catalyst,[40] Komori et al. obtained polyvaline from HCA more efficiently (98% yield).[41,42]

With this method, Komori et al.[42] also obtained poly(L-leucine), poly(amino acid)-urethane copolymers, and electroconductive poly(amino acid)-polypyrole complex films.

REFERENCES

1. Paleček, E. *Prog. Nucleic Acid Res. Mol. Biol.* **1969**, *9*, 31.
2. Dryhurst, G. *Electrochemistry of Biological Molecules* Academic: New York, 1977; p 6.
3. Miller, I. R. *J. Mol. Biol.* **1961**, *3*, 229.
4. Paleček, E. *Electrochem. Bioenerg.* **1986**, *15*, 275.
5. Paleček, E. *Electrochem. Bioenerg.* **1988**, *20*, 179.
6. Scheller, F.; Jänchen, M.; Etzold, G.; Will, H. *Bioelectrochem. Bioenerg.* **1974**, *1*, 478.
7. Dryhurst, G.; Kadish, K. M.; Scheller, F.; Renneberg, R. *Biological Electrochemistry* Academic: New York, 1982; Vol. I, p 426.
8. Calendi, E.; Macro, A. D.; Reggiani, R.; Scarpinato, B.; Valentini, L. *Biochim. Biophys. Acta* **1965**, *103*, 25.
9. Molinier-Jumel, C.; Molfoy, B.; Reynaud, J. A.; Aubel-Sadron, G. *Biochem. Biophys. Res. Commun.* **1978**, *84*, 441.
10. Rodriguez, M.; Bard, A. J. *Anal. Chem.* **1990**, *62*, 2658.
11. Miller, J. R.; Bach, D. *Biopolymers,* **1966**, *4*, 705.
12. Carter, M. T.; Bard, A. J. *J. Am. Chem. Soc.* **1987**, *109*, 7528.
13. Carter, M. T.; Rodriguez, M.; Bard, A. J. *J. Am. Chem. Soc.* **1989**, *111*, 8901.
14. Rodriguez, M.; Kodadek, T.; Torres, M.; Bard, A. J. *J. Bioconjugate Chem.* **1990**, *1*, 123.
15. Carter, M. T.; Bard, A. J. *J. Bioconjugate Chem.* **1990**, *1*, 257.
16. Swiatek, J.; Pawlowski, T.; Kozlowski, H. *Stud. Biophys.* **1988**, *122*, 195.
17. Tanford, C. *J. Am. Chem. Soc.* **1952**, *74*, 211.
18. Saroff, H. A.; Mark, H. J. *J. Am. Chem. Soc.* **1953**, *75*, 1420.
19. Rao, M. S. N.; Lal, H. *J. Am. Chem. Soc.* **1958**, *80*, 3222.
20. Malik, W. U.; Muzaffaruddin, M. *J. Electroanal. Chem.* **1963**, *6*, 214.
21. Malik, W. U.; Jindal, M. R. *J. Electroanal. Chem.* **1968**, *19*, 436.
22. Brown, G. L. *Arch. Biochem. Biophys.* **1954**, *49*, 303.
23. Gygax, H. R.; Jordan, J. *Discuss. Faraday Soc.* **1968**, *45*, 227.
24. Ostowski, W.; Krawczyk, A. *Acta Chem. Scand.* **1963**, *17*, Suppl. 241.
25. Duke, P. R.; Kust, R. N.; King, L. A. *J. Electrochem. Soc.* **1969**, *116*, 32.
26. Cecil, R.; Weitzman, P. D. *J. Biochem. J.* **1964**, *93*, 1.
27. Pavlovic, O.; Miller, I. R. *Experientia, Suppl.* **1971**, *18*, 513.
28. Kadish, K. M.; Jordan, J. *J. Electrochem. Soc.* **1978**, *125*, 1250.
29. Janik, B.; Elving, P. L. *Chem. Rev.* **1968**, *68*, 295.
30. Paleček, E. In *Methods in Enzymology* Grossman, L.; Moldave, K., Eds.; Academic: New York, 1971; Vol. 21, Part D, p 3.
31. Paleček, E. *Bioelectrochem. Bioenerg.* **1981**, *8*, 469.
32. Feinberg, B.; Lau, Y. K. *Bioelectrochem. Bioenerg.* **1980**, *7*, 187.
33. Clark, L. C.; Lyons, C. Jr. *Ann. NY Acad. Sci.* **1962**, *102*, 29.
34. Updike, S. J.; Hicks, G. P. *Nature* **1967**, *214*, 986.
35. Ikariyama, Y.; Yamaguchi, S.; Aizawa, M.; Yukiashi, T.; Ushioda, H. *Bull Chem. Soc. Jpn.* **1988**, *61*, 3525.
36. Ikariyama, Y.; Yamaguchi, S.; Aizawa, M.; Yukiashi, T.; Ushioda, H. *J. Electrochem. Soc.* **1989**, *136*, 702.
37. Ikeda, T. *Denki Kagaku* **1992**, *60*, 872.
38. Katakai, R.; Iizuka, Y. *J. Org. Chem.* **1985**, *50*, 715.
39. Fujimoto, Y. *Poly-Amino-San* Fujimoto, Y. Ed.: Kodansha: Tokyo, 1974, Chap 3.
40. Baizer, M. M. *Tetrahedron* **1984**, *40*, 935.
41. Komori, T.; Nonaka, T.; Fuchigami, T. *Chem. Lett.* **1986**, 11.
42. Komori, T.; Nonaka, T.; Fuchigami, T.; Zhang, K. *Bull. Chem. Soc. Jpn.* **1987**, *60*, 3315.

Biosensors

See: Immobilized Enzymes, Biosensors
Immobilized Microbial Cells
Ion-Selective Electrodes and Biosensors
Polyethylene (Surface Functionalization)
Signal Transduction Composites (Biomaterials with Electroactive Polymers)

BIOSPINNING (Silk Fiber Formation, Multiple Spinning Mechanisms)

J. Magoshi
National Institute of Agrobiological Resources

Y. Magoshi and M. A. Becker
National Institute of Sericultural and Entomological Science Oowashi

S. Nakamura
Department of Applied Chemistry
Faculty of Engineering
Kanagawa University

Silk fiber is a fine, lustrous filament produced by the silkworm, *Bombyx mori*, and other insect and arachnid species, including moths, wasps, bees, butterflies, and spiders. Silk has fascinating fiber characteristics including its pearl-like gloss, light velvety touch, fashionableness, and comfort to the wearer. Silkworms construct the cocoons to protect themselves during metamorphosis. Farmers cultivate the silkworms by feeding them mulberry leaves until they are ready to spin cocoons. The fiber spun is a long double monofilament (1000–1600 m).

Silk fiber contains two proteins, fibroin and sericin. The two fibroin filaments are covered with a layer of sericin. Three silk fibroin conformations have been found by X-ray diffraction and infrared spectroscopy, random coil, α-form (silk fibroin I).[15] and β-form (silk fibroin II).[1,2,13,14] The α-form is a crankshaft pleated structure and the β-form is an antiparallel-chain pleated-sheet structure. More than 80% of the amino acid composition of fibroin consists of glycine, alanine, and serine.[3]

The mechanism of fiber formation from the liquid silk in the silk gland of the silkworm was first clarified by Hiratsuka[4] and Foa.[5] They found that the shear and elongation caused by the way the silkworm draws its head are part of the silk formation process. Since their studies, many investigators have studied the spinning mechanism,[6-8] but the process of spinning silk filaments is still not completely understood.

The liquid silk of the silkworm is a highly viscous aqueous solution of two separate proteins, fibroin surrounded by sericin. The mechanism of silk fiber formation is the action of shear stress and elongational stress acting on the silk fibroin, causing the liquid silk to crystallize.

The silkworm forms fibers from the liquid crystal in the nematic state, which is then transformed from a gel state to a sol state during spinning. Our work indicates that the thread spun by the silkworm is the product of a liquid-crystal spinning process.[9]

The larvae of domestic and wild silkworms control the molecular orientation of the cocoon fibers using several sophisticated spinning techniques.[10]

RESULTS

Fiber Formation from Liquid Silk Fibroin

For liquid silk of anterior division consisting of about 23.6% fibroin and about 7% sericin coating, coagulation and crystallization are induced by the physical action of drawing, which exerts shear stress and elongation on the fibroin during cocoon spinning—that is, by the mechanical denaturation of the silk fibroin.

Drawing of Liquid Silk Fibroin

We have measured the stress–strain behavior of liquid silk at various extension rates, and determined the critical extension rate (q_c) above which silk fiber is produced. We have investigated the structure of liquid silk—before and after extension—by X-ray diffraction, optical birefringence, and Raman spectroscopy.[11]

Liquid silk was obtained from the posterior part of the middle division of silk glands from full-grown *Bombyx mori* larvae (one day before spinning). The sericin in the liquid silk was removed by washing the silk gland thoroughly with deionized water.

Stress–strain curves show that undiluted liquid silk (23% fibroin), taken from the silk gland of *Bombyx mori* one day before spinning, behaves as a non-Newtonian fluid at extension rates of 10–1000 mm/min. As the extension rate is increased, the stress required to stretch the liquid silk increases and the yield point becomes more prominent. At lower rates, 10–75 mm/min, no conformational change in the orientation of silk fibroin molecules can be seen using X-ray diffraction and birefringence.

Structure and Properties of Drawn Liquid Silk

Birefringence of drawn liquid silk is nearly zero, and is independent of extension rate up to 400 mm/min. The birefringence starts to increase suddenly at an extension rate of 450 mm/min and reaches a constant value above 500 mm/min.

The laser Raman spectrum of liquid silk shows that it is mostly in the random-coil conformation, with a small amount of the α-form conformation also present.[12] The amide I peak at 1660 cm^{-1} and the amide II peak at 1260 cm^{-1} correspond to the random-coil conformation. In the skeletal stretching region, sharp peaks at 1109 and 947 cm^{-1} are due to the α-conformation. In the spectrum of drawn liquid silk, peaks due to amide I at 1664 cm^{-1} and amide III at 1233 cm^{-1} indicate that the conformation is transmuted into the β-form by drawing.[12]

High-Speed Spinning

We concluded that until the extension rate becomes greater than 450 mm/min no detectable conformational change occurs. At rates greater than 500 mm/min, the conformation of fibroin in liquid silk is transformed from the random-soil to the β-form, and crystallization to well-oriented fibrils is induced by elongation. Therefore, fibrous silk fibroin is produced by drawing liquid silk at rates higher than 550 mm/min at 20°C.

Liquid Crystal Spinning

We have found that silk fibroin molecules flowing from the anterior division of the silk gland of *Bombyx mori* assume a liquid-crystal order.[9]

The nematic liquid crystalline phase can be produced in the liquid silk coming out from the anterior division of the silk gland. The liquid crystals of silk fibroin are dispersed in water after up to one hour. The gel state of liquid crystalline silk dissolves easily in water.[12]

Zone Drawing and Self-Drawing

A silkworm moves its head in a figure-eight pattern during cocoon spinning. Immediately after the liquid silk comes out of the spinneret, it is fixed at the outside walls with the sericin layer. With this point fixed, liquid silk in the sol state is clamped by the silk press of the nozzle and is drawn out by the motion of the silkworm's head. Once the cocoon filament is fixed at the outside walls and the spinneret is moved, the fibroin is physically pulled out of the spinneret by the strength of the silkworm itself. If the movements of a silkworm are restricted or its spinneret comes in contact with an obstacle during cocoon spinning, the silkworm does not produce a filament. This is referred to as zone drawing.

In the first state of spinning, the silk fibroin in solution is anchored to a substrate. The silk substrate is not oriented under polarized light, so the silkworm is traction spinning.

Summary (Multiple Spinning of Silkworm)

Silkworm larvae control the molecular orientation of their cocoon fibroin with several sophisticated spinning techniques.[10]

Silkworms perform molecular-orientation control very accurately using methods involving numerous sophisticated spinning techniques, such as gel spinning, liquid-crystal spinning, a high-speed spinning, self-exerted (traction) spinning, zone drawing, porous spinning, ion spinning, dry spinning, crimp spinning, and low-energy spinning. These cannot yet be duplicated by advanced environmentally friendly spinning technologies.

REFERENCES

1. Shimizu, M. *Protein Chemistry*, Kyoritsu Shuppan: Tokyo, 1975; Vol. 5, 317.
2. Kratky, O.; Kuriyama, S. Z. *Physik. Chem.* **1931**, *11*, 363.
3. Kirimura, J. *Protein Chemistry*, Kyoritsu Shuppan: Tokyo, 1957; Vol. 5, 339.
4. Hiratsuka, E. *Bull. Sericult. Experiment Station* **1916**, *1*, 181.
5. Foa, C. *Kolloid Z.* **1912**, *10*, 72.
6. Oka, S. *J. Phys. Soc. Japan* **1964**, *19*, 1381.
7. Ogiwara, S. *In Structure of Silk*; Ito, T., Ed.; Chikuma-kai: Ueda, Japan, 1957; p 20.
8. Iizuka, E. *Biorheology* **1965**, *3*, 1.
9. Magoshi, J.; Magoshi, Y.; Nakamura, S. *Repts. Prog. Polym. Phys. Japan* **1973**, *23*, 747.
10. Magoshi, J. *New Technology and Production* **1992**, *11*, 41.
11. Magoshi, J.; Magoshi, Y.; Nakamura, S. *Polymer Com.* **1985**, *26*, 309.
12. Ito, K.; Magoshi, J.; Magoshi, Y.; Nakamura, S. *Proceedings of Sixth International Conference on Raman Spectroscopy* 1978, 2, 290.
13. Marsh, R. E.; Corey, R. B.; Paukig, L. *Biochem. Biophy. Acta* **1955**, *16*, 1.
14. Takahumhi, Y. ASC. Symp. Ser. 1994; 544, 169.
15. Lotz, B. Bioch., **1976**, *61*, 205.

BISALLENES, POLYADDITIONS

Takeshi Endo and Ikuyoshi Tomita
Research Laboratory of Resources Utilization
Tokyo Institute of Technology

Allene derivatives, which have cumulated double bonds, may be candidates for constructing materials with double bonds in the main or side chain.

PREPARATION AND PROPERTIES

Dithiols

The radical chain additions of aromatic thiols to alkocyallenes proceed via 2,3-addition selectively.[1,2] Like the radical polymerization of allene monomers, the reaction may proceed with the thiyl radicals initially attacking the center carbon atom of allene moieties, and the resulting allyl radicals may abstract the hydrogen radical from the thiols. Using this reaction, polymers with vinyl sulfide moieties in the main chain were obtained from bisallenes and dithiols (**Scheme I**).[3-5] Although polymers prepared by radical addition polymerization of dithiols with terminal diynes have poor solubility,[6] those obtained from bisallenes dissolved in organic solvents probably because of the methyl substituents in the side chain.

Diols

Allenes bearing electron-donating substituents, such as allenyl ethers, are susceptible to electrophilic attack. Like vinyl ethers, 1,2-addition of alcohols toward alkoxyallenes took place by using an acid catalyst such as *p*-toluenesulfonic acid. Similar to the polymers prepared by reacting bisallenes with dithiols, the polymers easily degraded to acrolein and the corresponding diols under the acidic conditions.[7]

Coupling with Three Components

The palladium-catalyzed three component coupling reactions of allenes, aryl halides, and nucleophiles reportedly have produced products with high yields for these three units.[8] Polymerizing bisallenes, aryl dihalides, and nucleophiles in the

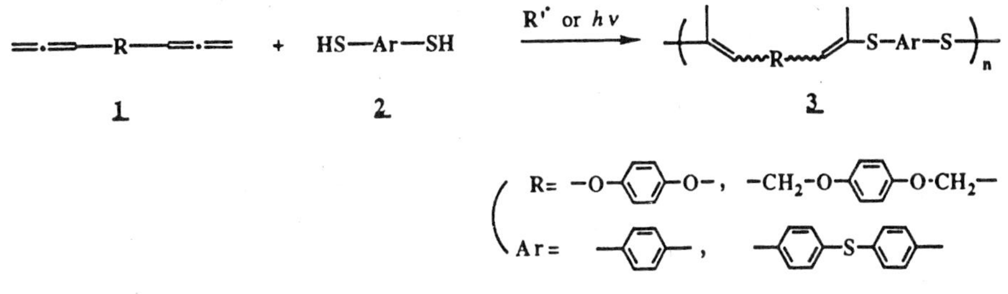

I

presence of a palladium catalyst produced linear polymers in high yields.[9]

By changing nucleophiles, polymers with various functional groups such as amines, ketones, and esters in the side chain were obtained. The geometric ratio of the double bonds in the polymer depended on the nucleophiles.

APPLICATIONS

Although applications for these polymers have not been developed in detail, they may have potential. Polymers from bisallenes and dithiols showed interesting degradation and reaction abilities originating from vinyl sulfide moieties and may show electron-conductive properties when bisallenes and dithiols are designed appropriately. Polymers containing acetal units by reacting bis(alkoxyallene)s with diols can degrade depending on the high reactivity of the acetal linkage. The polymers' structures obtained from coupling polymerization may be designed easily by three different monomers, which may simplify synthesis designs for functional materials.

REFERENCES

1. Yokozawa, T.; Tanala, M.; Endo, T. *Chem. Lett.* **1987**, 1831.
2. Pasto, D. J.; Hermine, G. L. *J. Org. Chem.* **1990**, *55*, 685.
3. Yokozawa, T.; Sato, E.; Endo, T. *Chem. Lett.* **1991**, 823.
4. Sato, E.; Yokozawa, T.; Endo, T. *Macromolecules* **1993**, *26*, 5185.
5. Sato, E.; Yokozawa, T.; Endo, T. *Macromolecules* **1993**, *26*, 5187.
6. Kobayashi, E.; Ohashi, T.; Furukawa, J. *Makromol. Chem.* **1987**, *187*, 2525.
7. Sato, E.; Yokozawa, T.; Endo, T. *Chem. Lett.* **1993**, 1113.
8. Shimizu, I.; Tuji, J. *Chem. Lett.* **1984**, 233.
9. Miyaki, N.; Tomita, I.; Endo, T. *Polym. Prepr. Jpn.* **1992**, *42*, 1952.

Bismaleimide Resins

See: *Bismaleimide Resins (Modification with Engineering Plastics)*
Maleimides, Fluorine-Containing

BISMALEIMIDE RESINS (Modification with Engineering Plastics)

Takao Iijima
Department of Applied Chemistry
Faculty of Engineering
Yokohama National University

Polyimide resins are one of the most important thermosetting polymers and have outstanding mechanical properties and thermal stability. They have wide use as adhesives, coatings, composite matrices, films, fibers, membranes, and so on. The main difficulties with polyimide resins are that they are brittle and difficult to process. Bismaleimide resins are addition-type polyimide resins and have received attention because of good processability and nonvolatility. Bismaleimide resins have recently come to be considered as candidate matrix resins for primary aerospace structural materials, but toughness of the neat resins must be improved for use in advanced composites.

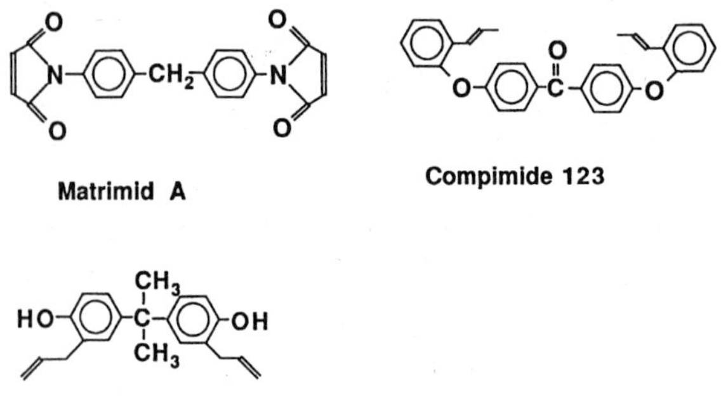

Matrimid A

Compimide 123

Matrimid B

FIGURE 1. Structure of bismaleimide compositions.

A major component of bismaleimide resins is 4,4′-bismaleimidodiphenyl methane (BM). BMI is a crystalline solid with a melting temperature of 155–156°C. When using BMI alone, BMI polymerizes radically over the melting point and the resulting networks are very brittle. Two-component bismaleimide systems (Matrimid and Compimide) have recently been developed as good processable and tough resins (**Figure 1**).[1]

(Engineering thermoplastics are interesting materials as modifiers for improving the toughness of bismaleimide resins from the viewpoint of the maintenance of mechanical and thermal properties for the matrix resins. Recently, various types of ductile thermoplastics have been used as alternatives to reactive rubbers for improving the toughness of bismaleimide resins. The advance in modification of bismaleimide resins has been reviewed.[2,3]

PREPARATION AND PROPERTIES

Commercial polysulfone (PSF) (Udel 1700™, UCC), poly(ether imide) (PEI) (Ultem 1000™, GE), and polyhydantoin (PH) (PH 10™, Bayer AG) were used as modifiers in the modification of the Compimide 796™ (Technochemie Co.)/4,4′-bis(o-propenylphenoxy)benzophenone (Compimide 123™, Technochemie Co.) (65/35 wt. ratio) system.[4] PSF was less effective than the other thermoplastics. Furthermore, the Compimide/PEI modification systems were examined in detail.[5]

Various kinds of thermoplastics were attempted as modifiers for the BMI/diallyl compounds system, where the diallyl compounds included DBA (Matrimid B™), bis-(allylphenyl)ether (Compimide 121™, Technochemie Co.), and Compimide 123™,[6] PEI (Ultem 1000™) and poly(ether sulfone) (PES) (Victrex 3600G™, ICI) were used as effective modifiers.

Three structurally different poly(ether ketone)s (PEKs) were used in the modification of the Compimide 796™/Compimide 123™ system.[7] The effects of PEK structure, molecular weight, and concentration on the fracture behavior were examined. Bisphenol A type PEK (Mw 53,800, Tg 152°C) was most effective; PeKs having bulky moieties or side groups were less effective.

Two kinds of fluorene-type PEKs having functionalities, allyl and maleimide end groups, were prepared and used as modifiers for the Compimide 796™/Compimide 123™ system.[8] The modified resin had a two-phase morphology and its interfacial adhesion was improved, but, unexpectedly, toughness of the reactive-PEK modified resin could not be improved compared with the modification results with the nonreactive PEK.

Amorphous functionalized engineering thermoplastics, poly(arylene ether sulfone) (PSFF), poly(arylene ether phosphine oxide) (PEPO), and 6F polyimide were prepared and used as modifiers for toughening the Matrimid 5292 A/B™ system.[9] The effects of PSF molecular weight, concentration, and functionality (terminal maleimide and amine groups) on the fracture toughness of the modified resins were examined. Nonreactive phthalimide-terminated PSF was less effective than reactive PSF.

N-Phenylmaleimide-styrene copolymer (PMS) was an effective modifier for the Matrimid 5292 A and B™ system.[10] Toughening could be achieved based on a co-continuous phase structure.

There are a few studies on modification of bismaleimide resins compared to epoxies. Modification of bismaleimide resins was often carried out on the basis of the information on toughening of epoxies. Toughening in the modification of bismaleimide resins by engineering thermoplastics could be attained based on co-continuous or phase-inverted structures in some cases.

REFERENCES

1. Stenzenberger, H. D. *Brit. Polym. J.* **1988**, *20*, 383.
2. Ishii, K. *Netsukoukasei Jushi* **1993**, *14*, 167.
3. Pascal, T.; Sillion, B. *Advances in Interpenetrating Polymer Networks*; Technomic: Lancaster, PA, 1994; p 141.
4. Stenzenberger, H. D.; Romer, W.; Herzog, M.; Konig, P. *Int. SAMPE Symp.* **1988**, *33*, 1546.
5. Rakutt, D.; Fitzer, E.; Stenzenberger, H. D. *High Perform. Polym.* **1990**, *2*, 133.
6. Lin, C.-R.; Liu, W.-L.; Hu, J.-T. *Int. SAMPE Symp.* **1989**, *34*, 1803.
7. Stenzenberger, H. D.; Romer, W.; Hergenrother, P. M.; Jensen, B. J. *Int. SAMPE Symp.* **1989**, *34*, 2054.
8. Stenzenberger, H. D.; Romer, W.; Hergenrother, P. M.; Jensen, B. J.; Breiigam, W. *Int. SAMPE Symp.* **1990**, *35*, 2175.
9. Wilkinson, S. P.; Ward, T. C.; McGrath, J. E. *Polymer* **1993**, *34*, 870.
10. Iijima, T.; Hirano, M.; Fukuda, W.; Tomoi, M. *Eur. Polym. J.* **1993**, *29*, 1399.

Bismuth-Containing Polymers

See: Radiopaque Polymers

BISPHENOL-A-POLYCARBONATE/ POLYESTER BLENDS

Jeffrey S. Kollodge
3M Company

The commercial significance of blends of bisphenol-A-polycarbonate (PC) and polyester has been the impetus for a tremendous amount of research over the past two decades. Blends composed of PC with poly(butylene terephthalate) (PBT) and poly(ethylene terephthalate) (PET) have been extensively studied relative to phase behavior and mechanical properties. Other aliphatic, aliphatic–aromatic, and aromatic polyesters have also been examined.

In general, the addition of the solvent-resistant polyesters to the tough PC produces blends with unique impact and stress-cracking resistance. Inclusion of impact modifiers in the blend results in enhanced low-temperature properties, a critical factor in commercial applications. Research on blends of PC and thermotropic liquid crystalline polymer (TLCP) is being conducted to produce "molecular composites." The LCPs improve tensile strength and modulus while lowering viscosity.

The ability of PC/polyester blends to undergo transreaction has spawned further research. Transreaction leads to the formation of block and random copolymers from the initial binary blend, altering the phase behavior and subsequently mechanical properties. Research has focused on the identification, quantitative measurement, and inhibition of these reactions.

PROPERTIES

PC/PBT and PC/PET Blends

Blend phase behavior and morphology are key parameters relative to properties; as a result, much effort has been expended to determine the phase and morphological characteristics of these two blends. Since the polyesters are semicrystalline, the blends have the potential for three phases: crystalline polyester, amorphous PC, and amorphous polyester. A fourth phase, a PC crystalline region, has been reported,[2] but its formation is caused by solvent-induced crystallization during casting. It has not been reported in melt-blended samples. There is no doubt that the polyesters can crystallize in these blends under the appropriate thermal conditions.[4-9] However, results pertaining to the phase behavior of the amorphous regions are conflicting.

Wahrmund et al.[4] show evidence of partial miscibility of the amorphous regions in melt-blended PC/PBT, indicated by shifting of component T_gs in differential thermal analysis (DTA) and (DMTA) measurements. Hanrahan et al.[1] in an effort to minimize exposure to high temperatures, prepared PC/PBT blends by solution casting from HFIP. Differential scanning calorimetry (DSC) measurements conducted on cast samples and samples that had been quenched from the melt show little change in the T_gs compared to pure component values, implying nearly complete immiscibility.

The studies on PC/PET blends appear equally conflicting. From thermal analysis, complete miscibility in melt blends containing over 60% PET is reported by Nassar et al.[5] and Murff et al.,[6] partial miscibility is observed in compositions containing less PET. Linder et al.[12] examined a 25/75 PC/PET blend cast from a 25/75 HFIP/dichloromethane solution. Their carbon–carbon nuclear spin diffusion study indicated that the blend was homogeneously mixed at distances of 4.5–6 Å.

Overall, phase behavior discrepancies can result from a variety of factors: preparation effects (melt blending vs. solution casting), solvent effects, analysis techniques, the extent to which equilibrium-phase behavior is reached (thermal history), component molecular weight, and interchange reaction. These factors are undoubtedly responsible for a majority of the discrepancies described above. Although often conflicting, the bulk of the data indicates that the amorphous regions in both PC/PBT and PC/PET blends are partially miscible over a wide composition range.

The morphology of PC/polyester blends has been investigated by a variety of techniques, including phase-contrast microscopy (SEM), and transmission electron microscopy (TEM). TEM combined with selective staining has revealed the most detailed information about the morphological structure. The morphology of PC/PBT blends as a function of composition has been examined by Delimoy et al.[11] Samples are stained with RuO_4 to enhance the phase contrast in the TEM study. At a PC/PBT ratio of 80/20, a continuous PC phase with included PBT particles is found. At 60/40, a co-continuous morphology is observed. At blend compositions of 40/60 and 20/80, a continuous PBT phase with included PC particles occurs. Excessive staining of the sample reveals amorphous PC incorporated into the interlamellar regions of PBT spherulites. Hobbs et al.[10] also report that PC is located in the interlamellar regions of PBT spherulites. Both authors indicate this as evidence for partial miscibility in the melt, with PC being forced into the interlamellar regions during PBT crystallization.

SEM studies on PC/PET blends[13] show included PET in a continuous PC phase at a composition ratio of 80/20. At ratios of 30/70 and 10/90, the phase behavior is inverted, with included PC particles in a continuous PET matrix. At a 60/40 composition, it is difficult to conclude from the SEM whether a discrete PET phase or a co-continuous morphology is present.

Due to their critical importance in commercial applications, the feature and impact behavior of PBT/PC blends has been extensively studied. Within these contexts, impact-modified PBT/ PC blends have also received attention. Dekkers et al.[14] use tensile dilatometry and notched impact tests to examine the failure mechanisms in pure PBT, PBT/IM, and PC/PBT/IM blends. They correlate the mode of failure in impact tests to the deformation mechanisms identified by tensile dilatometry.

Overall, major increases in volume strain are associated with internal cavitation of the impact modifiers, not crazing around the particles.[14,15] Internal cavitation is undesirable in terms of optimizing toughness, since it leads to brittle or semi-brittle failure.

PC/Polyarylate Blends

Polyarylate (PAr) is an aromatic thermoplastic synthesized from bisphenol-A and terephthalic and isophthalic acid. The terephthalate–isophthalate ratio is typically on the order of 50/50, which yields an amorphous polyester. Due to the lack of crystallinity, PAr is soluble in chloroform, so blends with PC are routinely prepared from codissolution techniques using this solvent. Mondragon et al.[3] report finding miscible blends via DSC over a composition range varying from 10/90 to 90/10 for both solution-cast and melt-blended samples. However, blending and DSC annealing procedures are conducted at temperatures (300°C) well in excess of those required for transreaction.

Later studies by Kimura and Porter[16] indicate this blend to have two phases, a nearly pure PC phase and a PAr-rich phase. Upon annealing at temperatures of 250°C, transreaction is indicated by the formation of a third T_g, and after sufficient annealing time, a single-phase blend is produced.

Both the modulus and yield stress of these blends have maximums that are above predictions based on additivity. Golovoy et al.[17] found that the blend mechanical properties depend on preparation technique.

PC/Poly(ε-caprolactone) Blends

Thermal analysis results on PC/(ε-caprolactone) (PCL) blends reveals single-phase behavior over the entire composition range.[18,19] Results on both solution-case samples from methylene chloride and melt-blended samples are identical.

The melting behavior of these blends is somewhat different from previously discussed PC/polyester blends. Due to the miscible nature of the blend, PCL has a plasticizing effect on PC, enabling it to crystallize from the melt.[18,20] PC percent crystallinity exhibits a maximum at a blend composition ratio of 50/50. PCL crystallinity decreases with increasing PC content in compositions containing more than 30% PC.[18]

PC/Poly(2-ethyl-2-methylpropylene terephthalate) Blends

Kollodge and Porter[21,22] studied the phase behavior of PC/poly(2-ethyl-2-methylpropylene terephthalate) (PEM/PT) blends as a function of component molecular weight and end-group type. In the absence of interchange reactions, the researchers found the blend to follow the general trends of Flory–Huggins theory—improved miscibility with decreasing molecular weight. Some blends containing low molecular weight components were miscible.

Kollodge and Porter[22] also showed that the phase behavior of the blend could be altered by end-group type. The effect was more significant at low molecular weights where the concentration of end-groups is higher.

PC/Thermotropic Liquid Crystalline Polymer Blends

The goal of blending PC with TLCPs is to form molecular composites with enhanced mechanical and rheological properties. The high strength and modulus of TLCPs combined with their shear-thinning behavior due to orientation during flow make them ideal reinforcing agents. The TLCPs are usually copolymers of various aromatic-ester mesogens. The copolymer ratio is selected to prevent the formation of the infusible crystals that the pure homopolymers themselves would have. This allows the polyesters to be processed at conventional temperatures by conventional means, including extrusion, fiber/film drawing, injection molding, and compression molding. A TLCP based on 4-hydroxy benzoic acid (HBA) and PET at a 60/40 mol ratio (HBA/PET) and TLCP composed of HBA and 6-hydroxy-2-naphthoic acid (HNA) at a mol ratio of ~73/27 (HBA/HNA) have been studied extensively in blends with PC.

SEMs of fracture surfaces is one of the most common techniques used to examine the morphology of these blends. In blends containing low concentrations of HBA/PET, a two-phase morphology is observed.[23-25] However, the morphological texture (whether or not a fibrillar structure of the TLCP is obtained) and subsequent mechanical properties depends on both the process temperature and the draw ratio.

TRANSREACTION IN PC/POLYESTER BLENDS

Qualitatively, it is known that transreaction in PC/PBT and PC/PET blends can produce miscible blends.[26,27] Quantitatively, there is little information relating the extent of transreaction to the phase behavior of PC/polyester blends. Interchange reaction leads to an interesting problem in regard to phase-behavior identification. In order to identify the equilibrium phase behavior, the temperatures of the blends must be elevated above the component T_gs and T_ms. However, at these temperatures exchange reactions can occur, altering phase behavior. Without specific examination for the presence or absence of transreaction, phase-behavior results may not represent those of the pure blend, but may already be modified due to reactions.

COMMERCIAL PC/POLYESTER BLENDS AND APPLICATIONS

PC/polyester blends have been commercialized by a variety of companies. Blends composed of PC with PBT and PET are the most predominant. Various grades are reported, including glass- and fiber-reinforced constructions. Blends containing impact modifiers are available. Blends of PC with PCIT are also being marketed. Other companies offer blends containing a certain percentage of recycled material.

Applications take advantage of the blends' good thermal properties, toughness, and chemical resistance. Areas where PC/polyester blends have been employed include the lawn and garden, telecommunications, electrical, commercial food service, sporting goods, and automotive industries.

REFERENCES

1. Hanrahan, B. D.; Angeli, S. R.; Runt, J. *Polym. Bull.* **1985**, *14*, 399.
2. Hobbs, S. Y.; Groshans, V. L.; Dekkers, M. E. J.; Shultz, A. R. *Polym. Bull.* **1987**, *17*, 335.
3. Mondragon, I.; Cortazar, M.; Guzmán *Makromol. Chem.* **1983**, *184*, 1741.
4. Wahrmund, D. C.; Paul, D. R.; Barlow, J. W. *J. Appl. Polym. Sci.* **1978**, *22*, 2155.
5. Nassar, T. R.; Paul, D. R.; Barlow, J. W. *J. Appl. Polym. Sci.* **1979**, *23*, 85.
6. Murff, S. R.; Barlow, J. W.; Paul, D. R. *J. Appl. Polym. Sci.* **1984**, *29*, 3231.
7. Birley, A. W.; Chen, X. Y. *Br. Polym. J.* **1984**, *16*, 77.
8. Birley, A. W.; Chen, X. Y. *Br. Polym. J.* **1985**, *17*(3), 297.
9. Halder, R. S.; Joshi, M.; Misra, A. *J. Appl. Polym. Sci.* **1990**, *39*, 1251.
10. Hobbs, S. Y.; Dekkers, M. E. J.; Watkins, V. H. *J. Mater. Sci.* **1988**, *23*, 1219.
11. Delimoy, D.; Bailly, C.; Devaux, J.; Legras, R. *Polym. Eng. Sci.* **1988**, *28*(2), 104.
12. Linder, M.; Henrichs, P. M.; Hewitt, J. M.; Massa, D. J. *J. Chem. Phys.* **1985**, *82*, 1585.
13. Kim, W. N.; Burns, C. M. *J. Polym. Sci., Polym. Phys.* **1990**, *28*, 1409.
14. Dekkers, M. E. J.; Hobbs, S. Y.; V. H. *J. Mater. Sci.* **1988**, *23*, 1225.
15. Bertilsson, H.; Franzén, B.; Kubát, J. *Plast. Rubber Process Appl.* **1989**, *11*, 167.
16. Kimura, M.; Porter, R. S. *Anal. Calorim* **1984**, *5*, 25.
17. Golovoy, A.; Cheung, M. F.; Van Oene, H. *Polym. Eng. Sci.* **1987**, *27*(20), 1642.
18. Cruz, C. A.; Paul, D. R.; Barlow, J.W. *J. Appl. Polym. Sci.* **1979**, *23*, 589.
19. Jonza, J. M.; Porter, R. S. *Macromolecules* **1986**, *19*, 1946.
20. Varnell, D. F.; Runt, J. P.; Coleman, M. M. *Macromolecules* **1981**, *14*, 1350.
21. Kollodge, J. S.; Porter, R. S. *Macromolecules* **1995**, *28*, 4089.
22. Kollodge, J. S.; Porter, R. S. *Macromolecules* **1995**, *28*, 4097.
23. Acierno, D.; Amendola, E.; Carfagna, C.; Nicolais, L.; Nobile, R. *Mol. Cryst. Liq. Cryst.* **1987**, *153*(Pt. A), 553.
24. Blizard, K. G.; Baird, D. G. *Polym. Eng. Sci.* **1987**, *27*(9), 653.
25. Jung, S. H.; Kim, S. C. *Polymer J.* **1988**, *20*(1), 73.
26. Huang, Z. H.; Wang, L. H. *Macromol. Chem. Rapid Commun.* **1986**, *7*, 255.
27. Golovoy, A.; Cheung, M.-F.; Carduner, K. R.; Rokosz, M. J. *Polym. Eng. Sci.* **1989**, *29*(18), 1226.

Blends

See: *Acrylonitrile-Butadiene-Styrene*
Agro-Fiber/Thermoplastic Blends and Alloys
Alloys and Blends
Bioartificial Materials
Bisphenol-A-Polycarbonate/Polyester Blends
Blends (Amorphous Rubbery Mixtures)
Blends (Interchain Crosslinking)
Blends (Neutron Scattering)
Blends (Silanol Functional Groups)
Blends (Thermoplastic Polyurethane-SAN)
Block Copolymer-Like Structures (End-Group Ionic Interactions in Blends)
Block Copolymers (Ordered Bicontinuous Double-Diamond Morphology)
Compatibilizers (for Blends)
Compatibilizers, Polymeric
Conductive Elastomeric Blends
Crystallizable Polymer Blends
Epoxidized Natural Rubber
Ethylene-Vinyl Acetate Copolymer (Chemical Modification, Compatibilizer in Blends)
Fluorinated High Performance Polymers (Imide and Hexafluoroisopropylidene Groups)
Glass-Polymer Melt Blends
Glass Transition (Theoretical Aspects)
Hydrogels, Microphase Separated Blends (Biomaterials and Drug Delivery System)
Hydrogen-Bonded Blends
Interpenetrating Polymer Networks (Overview)
Interpenetrating Polymer Networks, Rubber-Based
Ionomers (Compatibilization Blends)
Ionomers (Mechanical Properties)
Liquid Crystalline Polymer-Based Blends (Thermotropic)
Maleic Anhydride, Grafting
Methacrylic Resins (Improved Heat Resistance)
Molecular Composites (High Modulus Polyimide/Polyimide Films)
Pervaporation Membranes (Materials)
Polyamides, Lactam-Based
Poly(butylene terephthalate)/Polycarbonate Blends
Polycarbonate/Acrylic Blends (Phase Morphology)
Polycarbonates (Overview)
Poly(ester imide) Compatibilizer (in Thermotropic Liquid Crystalline Polymer Blends)
Polyimide Blends
Polyimides (Introduction and Overview)
Polymer-Polymer Interactions (Fluorescence Studies)
Polyolefin Graft Copolymers (Prepared by Borane Approach)

Polyolefins Overview
Polypropylene (Crystallization Kinetics)
Polypropylene Blends and Composites
Poly(styrene-co-N-alkylmaleimide)s (by Reactive Extrusion: Miscibility in Blends)
Reactive Processing, Thermoplastics
Recycling, Plastics
Solid Polymer Electrolytes (Polyether Blends and Composites)
Solubility Parameters (A Synthetic Approach)
Starch-Based Plastics (Measurement of Biodegradability)
Starch Biodegradation (in Starch-Plastic Blends)
Surfactants, Polymeric (Compatibilizers for Blends)
Thermoplastic Elastomers (Overview)
Unsaturated Polyester Resins (Toughened with Liquid Rubber)
Vegetal Biomass (3. Polymers, Derivatives and Composites)
Vinylidene Fluoride-Based Thermoplastics (Overview and Commercial Aspects)
Vinylidene Fluoride-Based Thermoplastics (Blends with Other Polymers)
Water-Borne Coatings (Urethane/Acrylic Hybrid Polymers)

BLENDS (Amorphous Rubbery Mixtures)

Saburo Akiyama and Seiichi Kawahara
Faculty of Technology
Tokyo University of Agriculture and Technology

In the past few decades, many polymers have been found to be miscible with dissimilar polymers, although immiscibility of a pair of high molecular weight polymers was predicted, except for the pair possessing polar substituents that interact favorably with one another.[1-5] In this regard, we have described the following methods for achieving polymer miscibility:

- effect of favorable interactions between polar polymer pairs;
- effect of a random copolymer on polymer miscibility;
- utilization of a compatibilizer (block or graft copolymers);
- formation of an interpenetrating polymer network (IPN); and
- effect of high pressure a miscible blend.[6,7]

The effect of a random copolymer on polymer miscibility is interesting because the blends containing a random copolymer exhibit miscibility phenomena, such as the miscibility window, the miscibility valley, and the miscibility door.[7-11] This is because the Flory–Huggins interaction parameter, χ_{12}, depends on the copolymer composition.[12-14] However, in many blends, liquid–liquid phase-transition behavior did not reflect only the random copolymer effect, because of strong specific interactions (i.e., hydrogen bonding, charge transfer, and complex formations). To investigate only the random copolymer effect, we have chosen a pair of diene rubbers without strong specific interactions.[7] Here we discuss our study on the miscibility of the diene rubber blend with a soft interaction caused by van der Waals dispersion forces.

MISCIBILITY AND PHASE BEHAVIOR OF RUBBER BLENDS

Although rubber is well known to be immiscible with dissimilar rubbers, a few miscible rubber blends have been reported. Examples of miscible rubber–rubber blends include styrene–butadiene rubbers (SBR) with different styrene levels, acrylonitrile-butadiene rubbers (NBR) with different acrylonitrile contents, butadiene rubbers of different 1,2-unit contents, SBR and BR, natural rubber (NR) and BR (1,2-unit: 71%), and IR and BR (1,2-unit; 92%).[15-20] In SBR/SBR, NBR/NBR, SBR/BR, and BR/BR blends, the butadiene unit is comonomer contained, whereas NR and IR have nothing in common with BR. Therefore, it was not obvious how NR and IR were miscible with BR.

To clarify the factors contributing to the enhancement of miscibility, it was necessary to find the lower critical solution temperature (LCST) and the upper critical solution temperature (UCST) within a range between the glass transition and degradation temperatures. In the lower molecular weight rubber blends of SBR/BR and IR/BR, UCST and LCST phase behavior was found, respectively.[21,22] However, the phase behavior has not been observed for the high molecular weight rubber–rubber blends. In particular, high molecular weight IR and NR have been reported to be immiscible with *cis*-1,4-BR.

POLYBUTADIENE

The microstructure of polybutadiene, such as isomeric units of *cis*-1,4, *trans*-1,4, and 1,2, has been widely known to change with the addition of a polar solvent such as diethylene glycol dimethyl ether and tetrahydrofuran when alkyllithium is used as an initiator of anion polymerization. BR with 1,2-unit content has been provided as a random copolymer[i.e., poly(vinyl ethylene-*co*-1,4-butadiene) (PVB)].[7,23]

PVB/IR BLEND

High molecular weight BR with a lower 1,2-unit content was reported to be immiscible with IR, whereas it was miscible with a higher 1,2-unit content.[19,20] The blend of BR containing a 1,2-unit (PVB) and IR was regarded as the ideal system.

Volume Contraction on Mixing

Excess volume on mixing is one of the important parameters that provides information on how molecular mixing is achieved. For instance, the excess volume is negative when there are strong intermolecular attractive forces, positive when there is a balance of strong attractive and partial repulsion forces, and zero when there is a random copolymer effect on the miscibility. In this regard, LCST phase behavior of the PVB (1,2-unit content: 32.3%)/IR blend cannot be classified by the above three categories because only random copolymer effect was taken into account. Thus, it was necessary to obtain experimentally the excess volume on mixing. The result suggests that a geometrically stable structure formed between PVB (1,2-unit content: 32.3%) and IR gives rise to the volume contraction.

If there is a geometrically stable structure between blend components, conformational variation during mixing should not reflect conformations of the components, because distortions in the polymer chain may take place with the conformational change.

Role of Free Volume in the Miscibility

LCST phase behavior is expected to occur when a free-volume term overcomes an energy interaction term in the Flory equation of state theory: the interaction parameter tending toward a positive value. In this case, the energy interaction term is significantly negative. However, in the PVB/IR blend, the energy interaction term was estimated to be a small positive value because of van der Waals dispersion forces.

CONCLUSION

The high molecular weight PVB/IR blend was found to be miscible when a volume contraction on mixing occurred. In particular, this condition was achieved as the 1,2-unit content increased. The PVB/IR blend exhibited LCST phase behavior and the miscibility door. The appearance of the LCST was presumed to be due to the balance of volume contraction and free-volume expansion on mixing. The miscibility of high molecular weight rubber–rubber blends has been a noteworthy addition to our interest in the miscibility of multicomponent polymers.

REFERENCES

1. Olabishi, O.; Robeson, M.; Shaw, M. T. *Polymer-Polymer Miscibility*; Academic: New York, 1979.
2. Akiyama, S.; Inoue, T.; Nishi, T. *Polymer Blends — Compatibility and Interface* (in Japanese); CMC: Tokyo, Japan, 1979.
3. *Polymer Blends*: Paul, D. R.; Newman, S., Eds.; Academic: New York, 1978; Vols. 1 and 2.
4. Utracki, L. A. *Polymer Alloys and Blends*; C. Hanser: Munich, Germany, 1990.
5. Flory, P. J. *Principles of Polymer Chemistry*; Cornell University: 1953.
6. Akiyama, S. *Kaigaikoubunshikenkyu* **1985**, *21*, 186.
7. Kawahara, S.; Akiyama, S.; Ueda, A. *Polym J.* **1989**, *21*, 221.
8. Alexandrovich, P. R.; Karasz, F. E.; MacKnight, W. J. *Polymer* **1977**, *18*, 1022.
9. Vukovic, R.; Karasz, F. E.; MacKnight, W. J. *Polymer* **1983**, *24*, 529.
10. Akiyama, S.; Ishikawa, K.; Fujiishi, H. *Polymer* **1991**, *32*, 1673.
11. Kawahara, S.; Akiyama, S. *Macromolecules* **1993**, *26*, 2428.
12. Kambour, R. P.; Bendler, J. T.; Bopp, R. C. *Macromolecules* **1983**, *16*, 753.
13. Brinke, G.; Karasz, F. E.; MacKnight, W. J. *Macromolecules* **1983**, *16*, 1827.
14. Paul, D. R.; Barlow, J. W. *Polymer* **1984**, *25*, 487.
15. Livingston, D. I.; Rongone, R. L. Presented at the International Rubber Conference, Brighton, England, 1967; paper 22.
16. Bartenev, G. M.; Kongarov, G. S. *Rubber Chem. Technol.* **1983**, *36*, 668.
17. Watanabe, T.; Fujiwara, Y.; Sumi, Y.; Nishi, T. *Rep. Prog. Polym. Phys. Jpn.* **1982**, *25*, 289.
18. Marsh, P. A.; Voet, A.; Price, L. O.; Mullens, T. J. *Rubber Chem. Technol.* **1968**, *41*, 344.
19. Yoshioka, A.; Komuro, K.; Ueda, A.; Watanabe, H.; Akita, S.; Masuda, T.; Nakajima, A. *Pure Appl. Chem.* **1986**, *58*, 1697.
20. Roland, C. M. *Macromolecules* **1987**, *20*, 2557.
21. Ougizawa, T.; Inoue, T.; Kammer, H. W. *Macromolecules* **1985**, *18*, 2089.
22. Trask, C. A.; Roland, C. M. *Polym. Commun. 29*, 332.
23. Kawahara, S.; Akiyama, S. *Polym. J.* **1990**, *22*, 361.

Blends (Interchain Crosslinking)

C. K. Das
Materials Science Centre
Indian Institute of Technology

Technological problems with polymer blends are presently occurring because the blend partners have incompatible viscosities, thermodynamics, and cure rates. A good composition can be achieved if differences in the polymers' surface energies are small enough to permit tiny microdomains to form and if there is sufficient adhesion between polymer phases. Cure rate incompatibility can be overcome by changing the vulcanization chemistry. Designers of polymer blends need to ensure good stress transfer between components of a system that can only guarantee efficient use of a component's desirable physical properties. A blend that lacks symptoms of phase segregation and that balances a combination of properties is considered compatible. Partial compatibility implies that above a certain level either the major or minor viscous component is a dispersed phase and when two phases are present, exhibits a transition.

The new polyblends or polymer alloys resulted from introducing crosslinking agents, which permitted otherwise incompatible polymers to blend, yielding unique properties. Many of these blends exhibited strong intermolecular association, which provided favorable energy for the mixing process. Energetic interactions are extremely helpful when blending polymers with high molecular weights. Using functional groups to induce strong interactions, such as covalent and ionic bonds, between dissimilar chains is one of the most important and effective ways to enhance macromolecular miscibility.

Interchain crosslinking is merely a chemical crosslinking between two dissimilar polymers without the use of crosslinking agents. Polymers with active functional groups during heating and molding can be crosslinked.[1,2] This type of crosslinking is called self-crosslinking. A recent discovery was hydrogen bonding between lactam carbonyl and acidic groups (such as carboxylic acid and sulfonic acid) to create reinforced domain structures.[13]

Here, we will discuss the following blend systems: nitrile rubber (NBR) and chlorosulfonated polyethylene rubber (hypalon); nitrile rubber and polyacrylic rubber (ACM); nitrile rubber and epichlorohydrin rubber (EPH); nitrile rubber and polysulfide rubber (thiokol); carboxylated nitrile rubber (XNBR) and fluoroelastomers (viton); and carboxylated nitrile rubber and ethylene acrylic rubber (VAMAC). We will discuss the effects of blending sequence on the blend properties with special attention paid to interchain crosslinking. Two types of blending techniques are considered here. One method is preblending virgin polymers with curatives and fillers added to the preblended stock. Another method is the masterbatch technique, which loads virgin polymers separately with curatives and fillers and then mixed together.

DISCUSSION

NBR and Hypalon

In masterbatch blends, large dosages of hypalon increased the rate and state of cure up to 60% before decreasing. Tensile strength, however, improved as the NBR:hypalon ratio decreased. Differential solvent swelling in ethyl acetate (where NBR phase was soluble) and the corresponding Kraus plot confirmed the lack of adhesion between phases and the absence of swelling restrictions. The carbon tetrachloride extracted sample revealed the easy extraction of the hypalon phase. Surprisingly, the blends in the middle range had poor oil aging properties probably because of a non-uniform phase distribution of either partner.

We observed that heating the preblends reduced the rate and state of cure. This observation was confirmed by the loss of functional groups during interchain crosslinking, which were usually responsible for polymer curing.

NBR and ACM

Masterbatch blends showed a decrease in the cure state as ACM replaced NBR by 80%. The rate of cure, however, showed an inflexion at 40% of ACM in the blend. The preblended blend exhibited a lowering trend in the state of cure as ACM increased and also an inflexion in the cure rate at 30% of ACM. The rate and state of cure were higher for preblended blends, and the oil swelling (in ASTM-3) and solvent swelling (in toluene) were less for preblended stock than for masterbatch stock. Differential swelling in acetone suggested the swelling restriction, which resulted in phase adhesion for the preblended blends as the negative slope in the Kraus plot indicated. However, a positive slope in the Kraus plot for masterbatch stock suggested there was a lack of phase adhesion, and easy extraction of ACM phase. A subsequent SEM study confirmed these findings.[4]

In the filled vulcanizate, the masterbatch technique produced rubber with a higher band than that of the preblended blend. However, better reinforcement and a higher state of cure were observed when carbon black was added to the preblended stock instead of blending the black masterbatches. Migration of carbon black from NBR to ACM phase was suggested by the filled system. Interchain crosslinking occurred for preblending, followed by black addition, which was hampered when the masterbatch technique was used.

NBR and Thiokol

Maximum state of cure for the preblends was observed at 80:20 for NBR/thiokol blend and decreased steadily for up to 100% of thiokol rubber. Rate of cure followed the same trend except for a slow increase from 50% of thiokol rubber for unheated preblends. Adding carbon black did not affect behavior.

Modulus values agreed with the state of cure by having a maximum at 80:20 for NBR/thiokol blend. Tensile strength increased 20% and then decreased with additional thiokol rubber in NBR. Differential solvent swelling was performed in toluene for all preblends. The swelling coefficients were below the additive average line, suggesting the swelling restriction. The Kraus plot showed a negative slope which also suggested phase adhesion.

To assess the crosslinked blends' technical properties we added curatives. The state and rate of cure for the preheated stock were higher than those for corresponding preblends, particularly at the higher NBR region. Differential swelling in toluene at ambient temperatures also indicated the swelling restriction for the NBR-rich blend. The Kraus plot's negative slope illustrated the phase adhesion between two polymers with interchain crosslinking in the preheated blends. Preheating the blends was always associated with increased modulus and decreased tensile strength.

NBR and EPH

For the preblends, the state of cure increased as EPH was added, attained a maximum at 70%, and then decreased. However, the cure rate had an inflexion point at 50:50 blend ratio with a decreasing trend at both ends. State of cure was higher for the preblending technique than for the masterbatch technique and followed the same trend. Tensile strength decreased for the blends as EPH was added to NBR, attained a minimum at a 50:50 ratio, and then increased at a faster rate with additional EPH. This trend was similar for both techniques, but preblending was associated with higher tensile strength. Modulus was reduced for masterbatch blending. Differential solvent swelling followed by Kraus plotting suggested the lack of phase adhesion for masterbatch blends.

To study the heating effects in the blend, we heated preblends of virgin polymers at 170°C for 40 minutes before adding curatives. Even without curatives, the state of cure increased with EPH, attained a maximum at 50:50 ratio before decreasing. This result clearly suggested that there was interchain crosslinking between the two phases. This phase adhesion was also confirmed by the Kraus plot's negative slope and by restricted extraction of one phase after extracting solvent and using SEM.

XNBR and Viton

For this blend, we considered metal oxide [$Ca(OH)_2$/MgO] curature systems using viton E 60 C and XNBR. In masterbatch blends, the state of cure and tensile strength gradually decreased as viton replaced XNBR and attained a minimum at a 50:50 ratio. Whereas in preblended blends, the state of cure, hardness, and tensile strength gradually increased as viton replaced XNBR.

XNBR and Vamac

When preblending techniques were used, the blend's cure state decreased as XNBR replaced vomac. The hardness, modulus, and tensile strength decreased, and the elongation at break increased gradually as XNBR was added to the blend. The percentage of volume swell in oil improved as XNBR increased. Studies of solvent swelling were carried out to observe phase adhesion between the blend's two elastomers.

CONCLUSION

The idea of varying blending sequence arises from the need for different properties with the same compounding formulation. The migration of ingredients between the phases, depending on polar groups and the main chain, has made this technology acceptable to industry. Furthermore, heat can link phases to produce a third entity with some improved properties in the blend system. This interchain linking depends on blending sequence. Preblending offers more intimate contact between polymers than the masterbatch technique. However, additional experiments are needed before recommending blend system for a particular application.

REFERENCES

1. Dudgeon, C. D. U.S. Patent 4 275 190, 1989.
2. Alex, R.; De, S. K. *Kauts. Chuk. Gummi Kunsts* **1991**, *44*, 333.
3. Shah, K. R.; U.S. Patent 4 400 820, 1981.
4. Tripathy, A. R.; Das, C. K. *J. Polym. Eng.* **1994**, *13*, 49.

BLENDS (Neutron Scattering)

V. Arrighi
Chemistry Department
Heriot-Watt University

M. L. Fernandez
Chemical Engineering Department
Imperial College of Science and Technology

Blending of two or more polymer species is a well-established method aimed at the design of new materials with specifically tailored properties. Some limited compatibility among the constituents is often the only requirement for a blend to be of practical interest. However, the detailed study and design of polymer blends involves a knowledge of the behavior of miscibility and the thermodynamics of mixing.[1,2] Numerous parameters need to be investigated. Such parameters include the composition, temperature, and molecular weight dependence of the Gibbs free energy of mixing; the influence of intermolecular interactions; the kinetics of the phase separation process; the morphology of the two-phase system formed; and the domain sizes.

In these areas neutron scattering (NS) can contribute to the understanding of the blend behavior in both the one- and two-phase regions of the phase diagram. The spatial range available with NS varies between a few Angstroms and thousands, and it is suitable for studying the local structure, polymer chain conformations, and blend morphology.

Moreover, neutrons make it possible to investigate the molecular properties of each of the constituents in the presence of a second component. The influence of a polymer additive on the local structure, conformations, and dynamics of a polymer species can be investigated by carefully employing the deuterium labeling method.

POLYMER BLEND INTERFACES[3,4-6]

Over the past decade, scientific interest in polymer–polymer interfaces has grown at a remarkably fast pace. From an academic point of view, the study of the interface in polymer blends is invaluable as a test for general thermodynamic and interdiffusion theories. For technological applications, a knowledge of the interfacial shape and size is crucial because most polymer blend properties rely on good interpenetration between the phases.

Various techniques have been used to analyze the interfacial concentration profile. Some of them, such as transmission electron microscopy (TEM) or infra-red densitometry, require the sample to be quenched after annealing and a section taken for analysis. Another group of methods, the so-called ion-beam techniques are based on the bombardment of the bilayers with ions; they give information on the "depth profile" of a particular atom and therefore on the interfacial shape and size. While these techniques have proved to be very useful in polymer–polymer interfacial studies, their resolution of the order of tens of nanometers is not enough to study the detailed shape and size of the interface (in some cases the entire interface is below the resolution of these techniques).

Over the past ten years, neutron specular reflection (NSR) has emerged as one of the most powerful techniques for the study of interfaces, not only on polymer blends but in a myriad of systems—for semiconductors to biological membranes or surface magnetism. NSR presents several advantages: it is nondestructive; it can be used to study "buried" interfaces (since the penetration depth of neurons is relatively large); it can be used in conjunction with selective deuteration to highlight specific layers in a specimen; and, most importantly, its resolution is astonishing, of the order of a fraction of a nanometer.

The basis of the technique is very simple. When a neutron beam impinges on an interface, part of it will be reflected and part will be refracted. The propagation of neutrons can be treated in the same way as the propagation of electromagnetic radiation, and a neutron refractive index n, can be defined.

The reflectivity can be calculated exactly for any given interfacial profile, but unfortunately the opposite process is not possible to date. The extraction of interfacial profiles from reflectivity data is carried out by model fitting. An interfacial profile is proposed, and the reflectivity from such a profile is calculated and compared with the experimental data. If the agreement is not satisfactory, the interfacial profile is modified and the process repeated. The problem with this approach is that while a particular interfacial profile might fit the reflectivity data, it is not certain that the profile is unique or indeed the "real" profile. Another method to analyze reflectivity data is the "kinematic approximation," but its applicability is restricted. To date it has only been used in a handful of cases. The latest development in data analysis is the Bayesian method. It attempts to include the prior information available on the system in the fitting procedure. In any case, given the possible ambiguity of the data analysis, it is always advisable to carry out complementary measurements by using, for example, ion-beam techniques. Although their resolution might not be enough to resolve fine details in the interfacial shape, they can help rule out some interfacial profiles and therefore reduce the ambiguity.

DYNAMICS

Neutron scattering can be employed to study molecular motion that occurs in the frequency range from 10^7 to 10^{14} Hz, including vibrational motion, rotation of side groups, and main chain motion above the glass transition temperature. Although the various neutron scattering techniques, inelastic neutron scattering, quasielastic neutron scattering and neuron spin echo have proved to be successful in the study of the dynamics of pure polymers there is a lack of similar studies in blends.[7] To date only the effect of blending on local motions below T_g has been investigated in some detail although the use of the labeling method and the suitable frequency range could give useful information on the dynamics of each of the blend components.[7]

ACKNOWLEDGMENTS

The authors wish to thank Professor J. S. Higgins for her comments and suggestions.

REFERENCES

1. Utracki, L. A. *Polymer Alloys and Blends. Thermodynamics and Rheology*; Hauser: Munich, 1989.
2. Olabisi, O.; Robison, L. M.; Shaw, N. T. *Polymer-Polymer Miscibility*; Academic: New York, 1979.
3. Higgins, J. S.; Benoit, H. C. *Beoit Polymers and Neutron Scattering*; Oxford University: Oxford, 1993.
4. Lekner, J. *Theory of Reflection*, Martinus Nijhoff: Dordrecht, 1987.
5. Born, M.; Wolf, E. *Principles of Optics*. 6th ed.; Pergamon: Oxford, 1980.
6. Russell, T. P. *Mat. Sci. Rep.* **1990**, *5*, 171.
7. Floudas, G.; Higgins, J. S. **1992**, *33*, 4121.

BLENDS (Silanol Functional Groups)

Eli M. Pearce, T. K. Kwei, and Shaoxiang Lu
Department of Chemistry
and
The Herman F. Mark Polymer Research Institute
Polytechnic University

Recently, we carried out research on the synthesis of silanol polymers and a study of polymer blends involving silanol functional groups.[1] A new convenient route for the synthesis of silanol-containing polymers has been developed through a polymer modification reaction by direct oxidation of precursor polymers containing a Si–H function with dioxirane. Dioxiranes constitute a new class of organic peroxides that possess great potential as oxidants with a variety of applications in synthetic organic chemistry.[2,3] The oxyfunctionalization of the silane polymer with dioxirane results in a rapid, quantitative, and selective conversion of the Si–H function to Si–OH.[4] Moreover, the silanol polymers obtained *in situ* show no tendency for self-condensation to siloxane and the stability of the polymers scan be tailored by replacement of the alkyl groups with more steric bulky and electronegative silicon substituents.[5] Miscibility studies demonstrated that the function was a strong hydrogen-bond donor.[6] Molecular-level penetrated semi-interpenetrating polymer networks were prepared in selected hydrogen-bonded polymer blends through complexation of component polymers and subsequent self-condensation of the silanols.[7]

SYNTHESIS OF SILANOL POLYMERS BY SELECTIVE OXIDATION OF PRECURSOR POLYMERS CONTAINING THE SI–H FUNCTION

The conventional methods for oganosilanol synthesis can be accomplished by the hydrolysis of the corresponding silanes with various silicon functional groups in the presence of an acid or a base. This route, however, presents some practical difficulty in being applied to the synthesis of silanol polymers, which demands not only high conversion for polymer modification reaction but also the resistance of the silanol to self- or catalytic condensation during the preparation.

A new convenient way to prepare silanol polymers through a polymer modification reaction is by direct oxidation of the corresponding precursor polymers containing the Si–H moiety.[4] Recently we have synthesized 4-vinyl-phenyldimethylsilanol polymer (PVPDMS) and its styrene copolymers (ST-VPDMS) by selective oxidation of the corresponding precursor silane polymer and copolymers via a reaction with dimethyldioxirane solution in acetone (**Scheme I**).

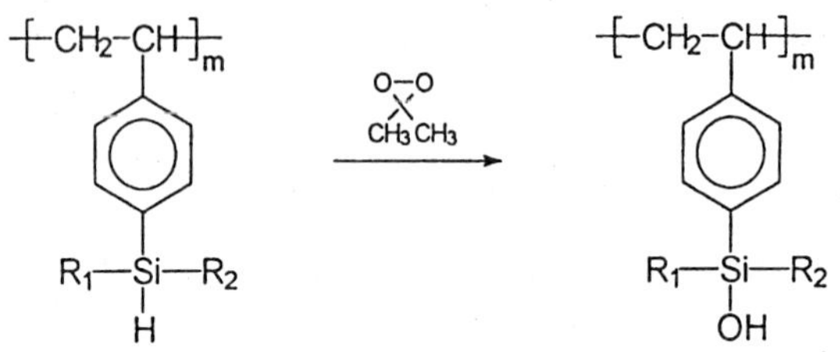

1. $R_1 = CH_3$, $R_2 = CH_3$
2. $R_1 = CH_3$, R_2 = phenyl
3. R_1 = phenyl, R_2 = phenyl

SCHEME I. Selective oxidation of precursor polymers containing Si–H function.

MISCIBLE BLENDS INVOLVING SILANOL-CONTAINING POLYMERS

In the studies of the silanol-containing polymer blends, copolymers consisting of varying amounts of VPDMS monomer units were prepared. The blends of the ST-VPDMS copolymers with poly(*n*-butyl methacrylate) (PBMA) and poly(*N*-vinylpyrrolidone) (PVPr) were prepared by solution cast films.[1,8] Miscibility of the blends was investigated by glass transition temperature measurement and FTIR spectroscopy.

CONCLUSION

We have described a new convenient synthetic route for the conversion of the Si–H function to Si–OH by the selective oxidation of the Si–H bond by dimethyldioxirane. The oxyfunctionalization of the silane proceeds rapidly and quantitatively, and it can be practically applied to the synthesis of a wide variety of novel silanol-containing polymers from precursor silicon polymers containing the Si–H function. Control over the properties of these silanol-containing polymers is possible through the variations of the composition and the positions of the Si–H function in the precursor silicone polymers.

Studies of polymer blends involving silanol-containing polymers showed that the silanol function was a strong hydrogen-bond donor in promoting polymer–polymer miscibility. However, silanols are also strongly self-associated through silanol hydrogen bonds, which compete with the hetero-associated hydrogen bonds and induce phase separation.

Control over the miscibility behaviors of the silanol-containing polymer blends were achieved through the variations of the strength of the hetero-associated hydrogen bond with the use of polymer bearing a strong hydrogen-bond acceptor group. Molecular-level penetrated semi-IPNs for the ST-VPDMS/PVPr were realized by complexation of the component polymers and subsequent condensation of the self-associated silanols during drying when the copolymers contained more than 60 mol% VPDMS. In this approach, the silanol function not only acts as a strong hydrogen donor to enhance polymer–polymer miscibility but also as a chemical crosslinker to form siloxane networks. The formation of siloxane networks by the condensation of the silanols proceeds either spontaneously or by heating without the presence of any external crosslinkers or catalysts.

REFERENCES

1. Lu, S.; Pearce, E. M.; Kwei, T. K. *J. Macromol. Sci. Pure Appl. Chem.* **1994**, *A31*, 1535.
2. Adam, W.; Curci, R.; Edwards, J. O. *Acc. Chem. Res.* **1989**, *22*, 205.
3. Murry, R. W. *Chem. Rev.* **1989**, *89*, 1187.
4. Lu, S.; Pearce, E. M.; Kwei, T. K. *Macromolecules* **1993**, *26*, 3514.
5. Lu, S.; Pearce, E. M.; Kwei, T. K. *J. Polm. Sci., Polym. Chem. Ed.* **1994**, *32*, 2597.
6. Lu, S.; Pearce, E. M.; Kwei, T. K. *J. Polm. Sci., Polym. Chem. Ed.* **1994**, *32*, 2607.
7. Lu, S.; Pearce, E. M.; Kwei, T. K. *Polymer* **1995**, *36*, 2435.
8. Lu, S.; Pearce, E. M.; Kwei, T. K. *J. Polym. Eng. Sci.*, **1995**, *35*, 1113.

BLENDS (Thermoplastic Polyurethane-SAN)

B. Žerjal and T. Malavašič
University of Maribor
EPF
Institute of Technology
National Institute of Chemistry

The impact resistance of thermoplastic polymers is improved by blending them with elastomers. Some performances of elastomers (i.e., the temperature of deformation under flexural load) can be improved by blending the elastomers with thermoplastics. Blends of high molar mass polymers are usually immiscible because of the unfavorable entropic conditions, but specific interactions in the system can result in a negative enthalpy of mixing, thus enhancing the miscibility.[1,2] In copolymer blends, the blend interaction parameter X is very complex and is composed of the intermolecular and intramolecular interaction parameters between individual monomers. The crystallinity of one of the components influences the miscibility, as well.

In blends of poly(styrene-*co*-acrylonitrile) (SAN) with thermoplastic poly(ester urethane) (TPU), all these factors are present. Furthermore, TPU is a block copolymer consisting of soft polyester and hard polyurethane segments that are limited miscible. Their segregation in the soft and in the hard domains is an additional characteristic of this system.[3]

In TPU/SAN blends, the properties vary depending on the type of polyurethane (polyether, polyester, ratio of the soft to the hard segments) and on percentage of the acrylonitrile in SAN.[4-8] In addition, numerous techniques used for determining the (im)miscibility may be the reason for discrepant results. Mechanical, thermal, and spectroscopic studies[5] of melt blended TPU with SAN (20 wt% acrylonitrile) seem to confirm the immiscibility within the investigated composition range, even though good mechanical properties are observed over the wide composition range.

In this work, the blends of thermoplastic poly(ester urethane) with two types of SAN (34 and 24 wt% acrylonitrile) were studied. Miscibility and properties were determined by measurement of mechanical properties, by wide angle X-ray diffraction (WAXD), by scanning electron microscopy (SEM), and by thermal properties.

The samples were prepared using two techniques: extrusion and molding. The influence of the mode of the sample preparation on the properties was also studied. Part of this work has been previously published.[9,10]

PREPARATION OF SAMPLES

A thermoplastic poly(ester urethane) (TPU, Elastollan C 90A, BASF) and two different poly(styrene-co-acrylonitrile) copolymers (SAN24 with 24 wt% acrylonitrile, Luran 368 R and SAN34 with 34 wt% acrylonitrile, Luran 388 S, BASF) were used.

MECHANICAL PROPERTIES

Tensile data were obtained by an Instron tensile tester that used standard methods.

The mechanical properties of the starting copolymers and blends are given in **TABLE 1**.

TABLE 1. Mechanical Properties of TPU/SAN Blends

Composition [g/g]	Molded			Extruded		
	σ_B [MPa]	ε_B [%]	E [MPa]	σ_B [MPa]	ε_B [%]	E [MPa]
TPU/SAN34						
100/0	19.1	435	24	18.0	680	23
90/10	15.7	370	37			
75/25	14.3	326	58	16.0	280	69
60/40	16.3	69	208			
50/50	20.6	62	456	33.0	80	400
40/60	26.0	8	699			
25/75	32.0	3.5	967	44.0	16	1175
10/90	32.9	2.3	1262			
0/100	34.5	2.0	1494	83.0	4.3	2300
TPU/SAN24						
100/0	19.1	435	24	18.0	680	23
90/10	17.1	273	31			
75/25	18.5	137	64	19.4	370	86
60/40	18.8	45	279			
50/50	20.6	20	504	32.3	130	367
40/60	26.8	9.5	727			
25/75	29.4	3.1	812	39.5	28	1090
10/90	30.0	2.7	1062			
0/100	30.5	1.8	1286	69.4	3.8	2134

σ_B = Tensile strength at break.
ε = Elongation at break.
E = Modulus.

X-RAY DIFFRACTION

Wide-angle X-ray diffraction experiments were performed by using a Phillips diffractometer with monochromatized CuK_a radiation. Intensity was scanned in the diffraction angle range of 2Θ = 4–60°. The TPU diffractogram shows two diffuse maxima with Bragg spacings d ≅ 2.23 Å (40.4° 2Θ), which corresponds to the intrachain distances, and d ≅ 4.35 Å (20.4° 2Θ), which corresponds to the interchain character. Weak, discrete maxima on the TPU WAXD amorphous halo correspond to the reflections of crystalline hard segment microdomains.[11,12]

The weight contents of crystalline domains (i.e., degrees of crystallinity) X_{cr} are calculated as the integral of the degree of crystallinity for polyurethane.[13]

The small crystallinity of TPU refers to a small amount of ordered-crystalline, hard-segment, microdomain separated from the amorphous matrix in TPU, but the sample does not give us information about the total amount of the hard segments. The relatively narrow large diffuse maximum (~ 4.35 Å) not only indicates an amorphous matrix (probably hard-soft segment mixture)[14] and a crystalline hard segment domain, but also a tendency for ordering the soft segments.

The degree of crystallinity remains constant by the dilution of TPU with 25 wt% SAN34 and SAN24 (75/25 TPU/SAN).

SCANNING ELECTRON MICROSCOPY

The morphology of the TPU/SAN blends was investigated by using a Jeol JSM scanning electron microscope. The samples, in the form of thin plates, were immersed in liquid nitrogen for 10 min, and then broken. The fractured surfaces were etched with methyl ethyl ketone and afterward metallized with an Au–Pd alloy. SEM measurements were performed on broken and broken-and-etched samples.

SEM micrographs of the cryofractured and etched surfaces of the TPU/SAN blends show a continuous and a disperse phase for all blends. The density and average dimensions of the cavities increase with the increasing amount of SAN in both series of TPU/SAN blends. For the same blends composition, the dimension of the cavities is larger in the TPU/SAN34 than in the TPU/SAN24 blends.

The regular distribution of SAN in TPU proves that the selected technological conditions in preparing the TPU/SAN blends are suitable to obtain a fine dispersion of the two polymers. The preparation of the samples (molding, extrusion) influences the morphology of the blends to some extent.

THERMAL PROPERTIES

Differential scanning calorimetry (DSC-7, Perkin Elmer) was used to determine the glass transition temperatures (T_g) of the TPU soft segments and of SAN in pure copolymers, and in the blends. The miscibility of the blends was estimated on the basis of T_g shifts in blends. T_gs of TPU in TPU/SAN blends were found to increase linearly with increasing weight fraction of SAN, while T_gs of SAN in TPU/SAN blends are found to decrease with increasing weight fraction of TPU.

In the DSC curves, the melting peaks of the TPU hard segments at around 200°C are seen as well, and beside these small peaks at higher temperature, which could be connected with the degradation of TPU.

REFERENCES

1. Vanneste, M.; Groeninckx, G. **1994**, *35*, 162.
2. Xu, M.; MacKnight, W. J.; Chen, C. H. Y.; Thomas, E. L. **1983**, *24*, 1327.
3. Dieterich, D.; Hespe, H. *Perceptions on the Physical Chemistry of the Structure of Polyurethanes*; In *Polyurethane Handbook*; Oertl, G., Ed.; Hanser: Munich, 1985; Chapter 2.5.
4. Paul, D. R.; Newman, S., Eds. *Polymer Blends*; Academic: New York, 1978; Vol. 1, p 35.
5. Iskandar, M.; Tran, C.; McGrath, J. E. *Polym. Prepr. Am. Chem. Soc. Div. Polym. Chem.* **1983**, *24*, 126.
6. Bonk, H. W.; Drzal, R.; Georgacopoulos, C.; Shah, T. M. *Thermoplastic Polyurethanes as Modifiers for Plastic and Elastomers. Antec '85* Washington, DC, 1985; 1300.
7. Ratzsch, M.; Handel, G.; Pompe, G.; Meyer, E. *Macromol. J. Science* **1990**, *A27*, 1631.
8. Ratzsch, M.; Pionteck, J. *Reactive Mixing of Polymers*. Presented at Conference on Reactive Production of Polymeric Goods. Conference Proceedings, Zagreb, September 1990.
9. Žerjal, B.; Jelčič, Z; Šmit, I.; Malavašič, T. *Intern.* **1992**, *2*, 123.
10. Žerjal, B.; Musil, V.; Šmit, I.; Jelčič, Z.; Malavašič, T. *J. Appl. Polym. Sci.* **1993**, *50*, 719.
11. Blackwell, J.; Rose, M. *J. Polym. Lett. Ed.* **1979**, *17*, 447.
12. van Bogart, J. W. C.; Gibson, P. E.; Cooper, S. L. *Polym. Sci., Polym. Phys. Ed.* **1983**, *21*, 65.
13. Kilian, H. G.; Jenckel, E. *Koll.-Zeit Zeit. Polym.* **1959**, *165*, 25.
14. Chau, K. W.; Geil, P. H. *Polymer* **1985**, *26*, 490.

BLOCK COPOLYMER-LIKE STRUCTURES (End-Group Ionic Interactions in Blends)

Akira Hirao and Seiichi Nakahama
Department of Polymer Chemistry
Tokyo Institute of Technology

Polymers, which have complementary binding sites on each repeating unit, such as acid–base, hydrogen, bonding, electron donor–acceptor, and coordination bond, associate with each other to form the stoichiometric polymer complex.[1] Such interactions between polymer chains have been employed by many researchers to enhance compatibility of the otherwise immiscible polymer pair.[2]

However, when the binding sites are linked to the definite position of the polymer chain, for example, telechelic and block polymers, characteristic regular structures may be formed through the interaction between the binding sites in the polymer blends. A combination of telechelic polymer and random copolymer might induce a graft copolymer-like structure through end-to-pendant interaction, resulting in regular structure.

The formation of a block copolymer-like structure might also be expected by the combination of the two telechelic polymers with different interactionable end groups.

We have also demonstrated the formation of multiblock copolymer-like structures in the polymer blends of interesting combinations of α,ω-diaminopolystyrenes and α,ω-dicarboxypoly(ethylene oxide)s, which are amorphous and crystalline polymers, respectively. Of greater importance, we have successfully observed for the first time by electron microscopy that these block copolymers are completely and orderly microphase-separated at the molecular level.[3,4]

The block copolymer of styrene and ethylene oxide was previously prepared by anionic sequential polymerization and observed to induce various types of morphology depending not only on the fractional compositions of the segments but also on the relative rate of crystallization of the poly(ethylene oxide) (PEO) segment in the ordering process.[5] Accordingly, the ionic interaction of end groups of the two telechelic polymers, their polymer chain lengths, and crystallization of the PEO part may proceed competitively to produce interesting ordered microphase-separated structures of the resulting block copolymers formed in this blend system.

PROPERTIES

In summary, by blending α,ω-diaminated polystyrene and dicarboxylated PEO with narrow molecular weight distributions we have successfully demonstrated the formation of multiblock copolymer-like structures through ionic interaction of their end groups. We observed 3-D ordered structures arising from microphase separation. Particularly, direct evidences for such structures were provided for the first time by both TEM and SEM measurements.

We have recently demonstrated the formation of a graft copolymer-like structure by blending α,ω-dicarboxylated PEO with the block copolymer of 3-(2-propenyl)aniline and styrene by means of anionic living polymerization technique.[6] In the blends, microphase-separated lamellar and cylindrical structures have also been directly observed by TEM and SEM.

APPLICATION

Both multiblock and graftlike copolymers are successfully prepared by blending well-regulated polymers with ionic interactionable functional groups at specific points of their chain lengths. Their morphological behaviors are very similar to those real block and graft copolymers in which each of polymer segments is bonded by covalent bonds. Therefore, they can be used in some cases instead of block and graft copolymers. The easy preparation of these polymers by blending only is an attractive advantage from a synthetic viewpoint. Furthermore, novel-type copolymers with more complicated structures difficult to prepare by other methods can potentially be prepared by changing the structures of the starting polymers.

REFERENCES

1. Tsuchida, E.; Abe, K. *Adv. Polym. Sci., Interactions between Macromolecules in Solution and Intermacromolecular Complexes* **1982**, *45*, 1.
2. Eisenberg, A.; Smith, P.; Zhou, Z.-L. *Poly. Eng. Sci., Compatibilization of the Polystyrene/Poly(ethyl acrylate) and Polystyrene/Polyisoprene Systems through Ionic Interaction* **1982**, *22*, 1117.
3. Iwasaki, K.; Hirao, A.; Nakahama, S. *Macromolecules, Morphology of Blends of α,ω-Diaminopolystyrene* with *α,ω-Dicarboxypoly(ethylene oxide)* **1993**, *26*, 2126.
4. Nakahama, S.; Hirao, A. In *New Functionality Materials, Volume C, Synthetic Process and Control of Functionality Materials, Synthesis of Telechelic and Block Polymers with Functional Groups and Formation of Ordered Structures in the Blends of the Polymers.* Tsuruta, T.; Doyama, M.; Seno, M. Eds. Elsevier: 1993; p 307.
5. Hirata, E.; Ijitsu, T.; Soen, T.; Hashimoto, T.; Kawai, H. *Polymer, Domain Structure and Crystalline Morphology of AB and ABA type Block Copolymers of Ethylene Oxide and Isoprene Cast from Solutions* **1975**, *16*, 249.
6. Iwasaki, K.; Tokiwa, T.; Hirao, A.; Nakahama, S. *New Polymeric Mater., Morphology of Blends of Poly(3-(2-propenyl)aniline-b-styrene-b-3-(2-propenyl)aniline) with α,ω-Dicarboxypoly(ethylene oxide)* **1993**, *4*, 53.

Block Copolymer Micelles

See: *Block Copolymer Micelles (Overview)*
Block Copolymer Micelles (For Drug Targeting)
Block Copolymer Micelles (1. Structure and Properties)
Block Copolymer Micelles (2. Fluorometric Studies and Computer Modeling)
Block Copolymers (Micellization in Solution)

BLOCK COPOLYMER MICELLES (Overview)

Paschalis Alexandridis and T. Alan Hatton
Department of Chemical Engineering
Massachusetts Institute of Technology

Copolymers are synthesized by the polymerization of more than one type of monomer. The result of such synthesis is called a block copolymer if the individual monomers occur as blocks of various lengths in the copolymer molecule. The different types of blocks within the copolymer are usually incompatible with one another and, as a consequence, block copolymers tend to self-assemble in melts and solutions. Block copolymers behave like typical amphiphiles when dissolved in a selective solvent (i.e., a good solvent for one block but a precipitant for the other) and associate reversibly to form miscelles with a core composed of the insoluble block and a corona of solvated soluble blocks. The micelles are generally spherical with narrow size distribution but may change in shape and size distribution under certain conditions. The block copolymers exhibit phenomenon of adsorption at interfaces and are capable of stabilizing colloidal dispersions. The colloidal aspects of block copolymers in solutions have been reviewed by Price, Riess et al., Tuzar and Kratochvil, and Alexandridis and Hatton.[1-6]

ASSOCIATION OF BLOCK COPOLYMERS IN THE FORM OF MICELLES

The critical micellization concentration (CMC), the amphiphile concentration at which micelles (i.e., thermodynamically stable polymolecular aggregates) start forming, is a parameter of great fundamental value.[7,8] The micellization of amphiphilic block copolymers is inherently more complex than that of conventional, low molecular weight surfactants. The composition polydispersity could be appreciable even for a copolymer with a narrow molecular weight distribution, and accordingly, to sharp CMC or critical micellization temperature (CMT), the copolymer solution temperature at which micelles form, has been observed for block copolymers. In practice, a certain CMC range with some notable uncertainty is usually detected; a large difference often occurs between the CMC values determined by different methods because the sensitivity of the techniques to the quantity of molecularly dissolved copolymers (unimers) present in solution may vary.[9]

Light scattering has been the most widely used technique in the study of block copolymer micelles.[1,4,11,15] A critical concentration, similar to CMC, has been found for a polystyrene-*block*-polybutadiene-*block*-polystyrene copolymer in dioxane–ethanol mixtures by using light scattering.[11] For copolymers that are heterogeneous in chemical composition and for which there are significant differences in refractive index between the polymer blocks, light-scattering data provide only an apparent particle weight. Numerous static-light-scattering data (micelle molar masses and radii of gyration) have been reported for micelles of block copolymers, such as polystyrene-*block*-poly(methyl methacrylate), polystyrene-*block*-polybutadiene-*block*-polystyrene, polystyrene-*block*-polyisoprene, and polystyrene-*block*-poly(ethylene oxide) in various solvents.[10,16-20] Information on the segment distribution of a particular polymer component in a micelle can be obtained by rendering the other block "invisible" by using an isorefractive solvents.

Surface tension has been used to determine CMC in most of the studies on aqueous block copolymer solutions.[6] The surface tension of the block copolymer solution below CMC decreases linearly with increasing polymer concentration and attains an almost constant value when the concentration exceeds CMC. Surface tension versus concentration plots, which have been well-known for low molecular weight surfactants, have been obtained for aqueous solutions of polystyrene-poly(ethylene oxide) block copolymers, ionene-tetrahydrofuran block copolymers with ionizable groups, and star-shaped block copolymers.[21-24]

Small-angle X-ray scattering (SAXS) may be used to determine the molecular weight, size, and structure of micelles. Small-angle neutron scattering (SANS) can be more powerful than X-ray scattering in investigating micelle structure because of the possibility of producing large contrasts by deuteration.[1] Neutron-scattering studies have been done on micelles of styrene-ethylene oxide graft copolymer in water.[26]

Fluorescence spectroscopy techniques have been developed and optimized over the past 15 years for the study of colloidal solutions, and many fluorescent molecules are now available for probing structural information in such systems.[27-30]

Dilute solutions of block copolymers in selective solvents form practically uniform polymolecular micelles having a spherical shape, a core/corona structure, and a high average volume fraction of polymer segments in a micelle (i.e., ~10 times higher than a polymer coil). The micelle aggregation number (typically 10–100) in a particular solution decreases with an increasing number of blocks and increases with impairing quality of the solvent towards the core-forming block. Micelles with small association numbers or even so-called unimolecular micelles have been reported for multiblock copolymers (see references in Reference 5).

THERMODYNAMICS OF MICELLE FORMATION

It is well established that block copolymers of the A-B diblock or A-B-A triblock type form micelles in selective solvents that are thermodynamically good solvents for one block and precipitants for the other. In general, micellization of block copolymers, as in the case of conventional surfactants, obeys the closed association model, which assumes an equilibrium between molecularly dispersed copolymer (unimer) and multimolecular aggregates (micelles).[2,4] There are two main approaches to the thermodynamic analysis of the micellization process: the phase separation model, in which the micelles are considered to form a separate phase at the CMC, and the mass-action model that considers micelles and unassociated unimers to be in an association–dissociation equilibrium.[31]

MODELING OF BLOCK COPOLYMER MICELLIZATION AND MICELLE STRUCTURE

Several research groups have striven to derive relations between structural parameters of a micelle (i.e., CMC, association number, and core and corona dimensions) and basic characteristics of the constituent block copolymer (i.e., molar mass and composition) by using various modeling approaches.[32-42]

SURFACE ACTIVITY

The presence of two incompatible blocks make block copolymers interfacially active. The adsorption characteristics and surface activity of various water-soluble block copolymers have been studied by surface tension and determination of pressure-area isotherms in the case of adsorbed copolymer monolayers. The monolayer formation of poly(vinyl alcohol)-*block*-polystyrene at the air–water interface was examined using pressure-area isotherms. The observed orientations of the molecule was such that the poly(vinyl alcohol) chains were dissolved in water and supported at the interface by a compact, monomolecular particle of polystyrene on the surface.[43] The surface activity of PEO-PPO block copolymers has been studied in detail and reviewed.[6,44,45] The values for surface area per PEO-PPO-PEO copolymer molecule at the air–water interface are generally small, indicating that there is considerable folding of the polymers at the air–water interface and/or desorption of PEO segments in the water phase.[44,46,47]

APPLICATIONS

Water-soluble triblock copolymers of PEO and PPO are an important class of surfactants and are used widely in industrial applications such as detergency, dispersion stabilization, foaming, emulsification, lubrication, and formulation of cosmetics and inks.[48-53] They are also used in more specialized applications such as pharmaceuticals (e.g., drug solubilization and controlled release and burn wound covering), bioprocessing (e.g., protecting microorganisms against mechanical damage), and separations (e.g., solubilization of organics in aqueous solutions).[54-65]

In interesting property of aqueous micellar systems is their ability to enhance the solubility of water of otherwise water-soluble hydrophobic compounds. This occurs because the core of the micelle provides a hydrophobic microenvironment, suitable for solubilizing such molecules. The phenomenon of solubilization forms the basis for many practical applications of amphiphiles. The enhancement in the solubility of lyophobic solutes in solvents afforded by amphiphilic copolymer micelles has shown promise in many industrial and biomedical applications.[63] Other very important applications of block copolymers are steric stabilization and dispersion polymerization.[25]

REFERENCES

1. Price, C. *Developments in Block Copolymers-1*; Goodman, I., Ed.; Applied Science: London, 1982; Chapter 2.
2. Price, C. *Pure & Appl. Chem.* **1983**, *55*, 1563.

3. Riess, G.; Hurtrez, G.; Bahadur, P.; *Encyclopedia of Polymer Science and Engineering*: John Wiley & Sons: New York, 1985; Vol. 2, 324.
4. Tuzar, Z.; Kratochvil, P. *Adv. Colloid Interface Sci.* **1976**, *6*, 201.
5. Tuzar, Z.; Kratochvil, P. *Surface and Colloid Science*; Matijevic, E., Ed.; Plenum: New York 1993; Vol. 15, Chapter 1.
6. Alexandridis, P.; Hatton, T. A. *Colloids Surfaces A: Physico-chem. Eng. Aspects* **1995**, *96*, 1.
7. Hunter, R. J. *Foundations of Colloid Science*; Oxford University: New York, 1987; Vol. 1.
8. Goddard, E. D.; Hoeve, C. A. J.; Benson, G. C. *J. Phys. Chem.* **1957**, *61*, 593.
9. Zhou, Z.; Chu, B. *J. Colloid Interface Sci.* **1988**, *126*, 171.
10. Utiyama, H.; Taakenaka, K.; Mizumori, M.; Fukuda, M.; Tsunashima, Y.; Kurata, M. *Macromolecules* **1974**, *7*, 515.
11. Tuzar, Z.; Kratochvil, P. *Macromol. Chem.* **1972**, *160*, 301.
12. Krause, S.; Reismiller, P. A. *J. Polym. Sci., Polym. Phys. Ed.* **1975**, *13*, 663.
13. Kotaka, T.; Tanaka, T.; Hattori, M.; Inagaaki, H. *Macromolecules* **1978**, *11*, 138.
14. Selb, J.; Gallot, Y. *Makromol. Chem.* **1981**, *181*, 2605.
15. Selb, J.; Gallot, Y. *Makromol. Chem.* **1981**, *182*, 1491.
16. Kotaka, T.; Tanaka, T.; Inagaki, H. *Polym. J.* **1972**, *3*, 327.
17. Tanaka, T.; Kotaka, T.; Inagaki, H. *Polym. J.* **1972**, *3*, 338.
18. Enyiegbulam, M.; Hourston, D. J. *Polymer* **1978**, *19*, 727.
19. Price, C.; McAdam, J. D. G.; Lally, T. P.; Wood, D. *Polymer* **1974**, *15*, 228.
20. Franta, E. *J. Chim. Phys.* **1996**, *63*, 595.
21. Nakamura, K.; Endo, R.; Takeda, M. *J. Polym. Sci., Polym. Phys. Ed.* **1976**, *14*, 1287.
22. Riess, G.; Rogez, D. *Polym. Prepr.* **1982**, *23*, 19.
23. Taakahashi, A.; Kawaguchi, M.; Kato, T.; Kuno, M.; Matsumoto. S. *J. Macromol. Sci. Phys.* **1980**, *17*, 747.
24. Huynh, B.-G.; Jerome, R.; Teyssie, P. *J. Polym. Sci., Polym Phys. Ed.* **1980**, *18*, 2391.
25. Price, C.; Hudd, A. L.; Booth, C.; Wright, B. *Polymer, 23* **1982**, *23*, 650.
26. Candau, F.; Guenet, J. M.; Boutillier, J.; Picot, C. *Polymer* **1979**, *20*, 1227.
27. Ananthapadmanabhan, K. P.; Goddard, E. D.; Turro, N. J.; Kuo, P. L. *Langmuir* **1985**, *1*, 352.
28. Thomas, J. K. *The Chemistry of Excitation at Interfaces*; ACS Monograph 181; American Chemical Society: Washington DC, 1984; Chapter 5.
29. *Surfactant Solutions: New Methods of Investigation*; Zana, R., Ed.; Marcel Dekker: New York, 1986.
30. Winnik, M. A.; Pekcan, O.; Croucher, M. D. *Scientific Methods for the Study of Polymer Colloids and their Applications*; Candau, F.; Ottewill, R. H. Eds.; Kluwer Academic Publishers: Dortrecht, 1990; 225–245.
31. Yang, Y.-W.; Deng, N.-J.; Yu, G.-E.; Zhou, Z.-K.; Attwood, D.; Booth, C. *Langmuir* **1995**, *11*, 4703.
32. deGennes, P. G. Solid State Physics Liebert, J. Ed.; Academic: New York, 1977; Suppl. 14, 1–18.
33. Leibler, L.; Orland, H.; Wheeler, J. C. *J. Chem. Phys.* **1983**, *79*, 3550.
34. Noolandi, J.; Hong, K. M. *Macromolecules* **1983**, *16*, 1443.
35. Whitmore, M. D.; Noolandi, J. *Macromolecules* **1985**, *18*, 657.
36. Zhulina, E. B.; Birshtein, T. M. *Vysokomol. Soedin.* **1985**, *27*, 511.
37. Halperin, A. *Macromolecules* **1987**, *20*, 2943.
38. Daoud, M.; Cotton, J. P. *J. Physique* **1982**, *43*, 531.
39. Birshtein, T. M.; Zhulina, E. B. *Polymer* **1984**, *25*, 1453.
40. Witten, T. A.; Pincus, P. A. *Macromolecules* **1986**, *19*, 2509.
41. Roe, R.-J. *Macromolecules* **1986**, *19*, 728, 156.
42. Nagarajan, R.; Ganesh, K. *J. Phys. Chem.* **1989**, *90*, 5843.
43. Ikada, Y.; Iwata, H.; Nagaoka, S.; Horii, F.; Hatada, M. *J. Macromol. Sci. Phys.* **1980**, *17*, 191.
44. Prasad, K. N.; Luong, T. T.; Florence, A. T.; Paris, J.; Vaution, C.; Seiller, M.; Puisieux, F. *J. Colloid Interface Sci.* **1979**, *69*, 225.
45. Alexandridis, P.; Athanassiou, V.; Fukuda, S.; Hatton, T. A. *Langmuir* **1994**, *10*, 2604.
46. Aston, M. S.; Herrington, T. M.; Tadros, T. F. *Colloids Surfaces* **1990**, *51*, 115.
47. Santos Magalhaes, N. S.; Benita, S.; Baszkin, A. *Colloids Surfaces* **1991**, *52*, 195.
48. Schmolka, I. R. *J. Am. Oil Chem. Soc.* **1977**, *54*, 110.
49. Bahadur, P.; Riess, G. *Tenside Surf. Det.* **1991**, *28*, 173.
50. Schmolka, I. R. *Cosmetics & Toiletries 1980, 95*, 77.
51. Schmolka, I. R. *Cosmetics & Toiletries 1984, 99*, 69.
52. Winnik, F. M.; Breton, M. P.; Riske, W. U.S. Patent 5 139 574, 1992.
53. Winnik, F. M.; Davidson, A. R.; Lin, J. W.-P.; Croucher, M. D. U.S. Patent 5 145 518, 1992.
54. Lin, S.-Y.; Kawashima, Y. *Pharm. Acta Helv.* **1985**, *60*, 339.
55. Yokoyama, M. *Crit. Rev. Therapeutic Drug Carrier Systems* **1992**, *9*, 213.
56. Guzman, M.; Garcia, F. F.; Molpeceres, J.; Aberturas, M. R. *Int. J. Pharm.* **1992**, *80*, 119.
57. Kabanov, A. V.; Batrakova, E. V.; Melik-Nubarov, N. S.; Fedoseev, N. A.; Dorodnich, T. Y.; Alakhov, V. Y.; Chekhonin, V. P.; Nazalova, I. R.; Kabanov, V. A. *J. Controlled Release* **1992**, *22*, 141.
58. Henry, R. L.; Schmolka, I. R. *Crit. Rev. Biocompatibility* **1989**, *5*, 207.
59. Murhammer, D. W.; Goochee, C. F. *Biotechnol. Prog.* **1990**, *6*, 142.
60. Zhang, Z.; Al-Rubeai, M.; Thomas, C. R. *Enzyme Microb. Technol.* **1992**, *14*, 980.
61. Nagarajan, R.; Barry, M.; Ruckenstein, E. *Langmuir* **1986**, *2*, 210.
62. Hurter, P. N.; Hatton, T. A. *Langmuir* **1992**, *8*, 1291.
63. Hurter, P. N.; Alexandridis, P.; Hatton, T. A. *Sobulization in Surfactant Aggregates* Christian, S. D.; Scamehorn, J. F.; Eds.; Marcel Dekker: New York, 1995; Chapter 6.
64. Colett, J. H.; Tobin, E. A. *J. Pharm. Pharmacol.* **1979**, *31*, 174.
65. Gadele, F.; Koros, W. J.; Schechter, R. S. *Macromolecules* **1995**, *28*, 4883.

BLOCK COPOLYMER MICELLES (For Drug Targeting)

Masayuki Yokoyama
Institute of Biomedical Engineering
Tokyo Women's Medical College

Polymeric micelles were introduced to the field of drug targeting in 1984.[1] Their use in this field is thus recent compared with other types of drug carrier systems, such as natural polymers (e.g., antibodies), water-soluble synthetic polymers, liposomes, and microspheres.[2,3]

PROPERTIES

High Structural Stability

It is known that polymeric micelles possess high structural stability created by an entanglement of polymer chains in the inner core.[4,5] This stability consists of two aspects: static and dynamic.[6-8] Static stability is characterized by an equilibrium constant between single polymer chains and micelle structures, or more conveniently by the critical micellar concentration (CMC). Generally, polymeric micelles show low CMC values (usually in μg/mL order or below); these values are much smaller than the typical CMC values of micelles that form from low molecular weight surfactants.

The other aspect, dynamic stability, may be important—particularly for *in vivo* delivery under physiological conditions where many factors (such as cells, proteins, and lipids) can interact with and destroy micellar structures. Slow dissociation rates of polymeric micelles allow the micelles to maintain these structures for periods long enough to achieve delivery to targets.

Additionally, polymeric micelles are formed only in one defined structure, although in some cases there are possible contributions of secondary associations of the micelles with slightly larger diameters.[9,10]

Phase Separation

Each part of polymeric micelles—the inner core and the outer shell—is made by phase separation between two constituent polymer segments. This advantage is distinct compared to conventional (nonmicelle-forming) polymeric carrier systems.

The third advantage obtained from "phase separation" is a "separated functionality." The inner core is employed for drug loading and release (if release is required). The outer shell is responsible for interfacial interactions with the biocomponents, such as cells and proteins.

Size

Polymeric micelles are in a suitable size range to maintain long-term circulation in the bloodstream by escaping the renal excretion and avoiding the nonspecific capture. This size range is also suitable to achieve direct extravazation from blood vessels to target organs or tissues.[11]

VARIOUS WAYS FOR DRUG INCORPORATION

Drugs can be incorporated into the inner cores by chemical conjugation or by physical entrapment.[10,12] Various interactions can be used as a driving force for micelle formation and for drug incorporation by physical entrapment. The many ways to incorporated drugs expands the application of polymeric micelles as drug carriers for many kinds of drugs. Particularly, the application is not limited to low molecular weight drugs; it can also include the delivery of high molecular weight bioactive compounds, such as nucleic acids and peptides.

REFERENCES

1. Bader, H.; Ringsdorf, H. *Die Angew. Makro. Chem.* **1984**, *123/124*, 457.
2. Yokoyama, M. *Critical Reviews in Therapeutic Drug Carrier Systems* **1992**, *9*, 213.
3. Yokoyama, M. *Advance in Polymeric Systems for Drug Delivery*, Gordon and Breach Science: Yverdon, Switzerland, 1994, Chapter 2.
4. Wilhelm, M.; Zhao, C.-L.; Wang, Y.; Xu, R.; Winnik, M.; Mura, J.L.; Riess, G.; Croucher, M. D. *Macromolecules* **1991**, *24*, 1033.
5. Kwon, G. S.; Naito, M.; Yokoyama, M.; Okano, T.; Sakurai, Y.; Kataoka, K. *Langmuir* **1993**, *9*, 895.
6. Calderara, F.; Hruska, Z.; Hurtrez, G.; Lerch, J.-P., Nugay, T.; Riess, G. *Macromolecules* **1994**, *27*, 1210.
7. Tuzar, Z.; Kratochvil, P. *Adv. Colloid. Interfacial. Sci.* **1976**, *6*, 20.
8. Wang, Y.; Kaush, C. M.; Chun, M.; Quirk, R. P.; Mattice, W. L. *Macromolecules* **1995**, *28*, 904.
9. Xu, R.; Winnik, M. A. *Macromolecules* **1991**, *24*, 87.
10. Yokoyama, M.; Okano, T.; Sakurai, Y.; Kataoka, K. *J. Contr. Rel.* **1994**, *32*, 269.
11. Tomlinson, E. *Site-Specific Drug Delivery*; John Wiley & Sons: Chichester, UK, 1994; Chapter 1.
12. Kwon, G.; Naito, M.; Yokoyama, M.; Okano, T.; Sakurai, Y.; Kataoka, K. *Pharm. Res.* **1995**, *12*, 192.

BLOCK COPOLYMER MICELLES (Microcontainers for Drug Targeting)

Alexander V. Kabanov
Department of Pharmaceutical Sciences
College of Pharmacy
University of Nebraska Medical Center

Valery Yu. Alakhov
Supratele Pharma, Incorporated

The idea of drug targeting, formulated by Paul Ehrlich about a century ago, is currently of great importance for medical and pharmaceutical sciences.[1] Many drugs are toxic and produce side effects in nontarget cells and systems in humans.[2]

One promising approach that uses block copolymer micelles as a basic element of a delivery system (microcontainer) has recently experienced rapid development.[3] The micellar microcontainers represent complexes of a few to several hundred block copolymer molecules. These complexes spontaneously assemble in aqueous solutions as a result of the cooperative interaction of the hydrophobic chain blocks of the copolymer. These hydrophobic blocks form a dense core, which is surrounded by a shell of hydrophilic chains (e.g., polyoxyethylene). Compared to the micelles formed by low molecular weight surfactants block copolymer micelles are usually characterized by higher stability preventing their destruction during dissolution in the body fluids. These micelles are small enough to penetrate into tissue capillaries and enter cells. Drug molecules are either noncovalently incorporated into a hydrophobic core of the micelle, or covalently linked to it. The vector molecules (e.g., antibodies or ligands capable of interaction with specific cells) are conjugated to the micelles aiding specific delivery of the micelle-incorporated drug to the target.

The first work on block copolymer micelles as drug carriers was reported in the early 1980s by the laboratory of H. Ringsdorf in Mainz.[4-6] Kabanov and co-workers proposed pluronic micelles as microcontainers for drug targeting and were the first to demonstrate *in vivo* the efficacy of this approach.[9] In 1989 they reported that by incorporating a drug into a micellar microcontainer one can selectively target a drug to the brain by crossing the blood-brain barrier (BBB).[9,10]

DRUGS SOLUBILIZED IN BLOCK COPOLYMER MICELLES

Preparation and Properties of Micellar Drugs

The poly(oxyethylene-*b*-oxypropylene-*b*-oxyethylene) triblock copolymers, usually referred to as pluronic copolymers (or poloxamers), have been used for the design of drug-delivery systems.[3] In aqueous solutions these block copolymers form micelles, with a core comprised of oxypropylene chain blocks surrounded by a hydrophilic shell of polyoxyethylene,[11] The size of these micelles ranges from 10 to 100 nm. They are characterized by a fairly low critical micelle concentration (CMC) and can solubilize substantial amounts of low molecular compounds.[12-14] Intensive studies have been performed on pluronic gels and emulsions as components of "artificial blood" formulations[15,16] drug release systems,[17] immunoadjuvant,[18,20] anti-tumor,[20,21] and anti-inflammatory agents.[22] These studies also require administration of pluronic copolymers in humans,[23] and demonstrate their fairly low toxicity. The credit for utilizing pluronic micelles for solubilization of hydrophobic drugs is given to Lin and Kawashima.[7,8,23] The concept of utilization of pluronic micelles as microcontainers for drug targeting, however, was first proposed by Kabanov et al.[9] The essence of this approach is as follows. Drug molecules spontaneously incorporate into micelles during solubilization of the drug in pluoronic micellar solutions. They then become masked from the external media, specifically from serum proteins that usually bind the drug and decrease its therapeutic efficacy. To selectively target these micellar microcontainers to specific cells in the organism, a "vector" molecule is covalently attached to a block copolymer.

Contrary to Ringsdorf's scheme, this approach implies that the drug is noncovalently incorporated into the micelles. The stability of such a drug delivery system is determined by the pluronic CMC and the partitioning coefficient of the drug.[14]

Micelle Interactions with Cells

Studies on the interaction of pluronic micelles with mammalian cells have been recently reported.[24,25] These studies revealed that the pluronic micelles, containing a fluorophore are taken up into a cell via an endocytosis mechanism.[25] The uptake is dramatically increased if a protein molecule capable of binding with a cell receptor is covalently linked to a micellar microcontainer.

Drug Targeting *In Vivo*

The tissue distribution of fluorescein solubilized in pluronic micelles was studied using animal models.[10]

CONCLUSION

The decade of the studies on block copolymer micelles as drug carriers has revealed the great potential of this approach and its importance for the targeted drug delivery. Studies in this area are expected to intensify in the near future, and new and important results will occur.

REFERENCES

1. Raso, V. *Semin. Cancer Biol.* **1990**, *1*, 227.
2. *The Pharmacological Basis of Therapeutics*; Goodman, A.; Gilman, A.; Rall, T. W.; Murad, F., Eds. Macmillan: New York, Toronto, London, 1985.
3. Kabanov, A. V.; Alakhov, V. Yu.; Chekhonin, V. P., In *Sov. Sci. Rev. D. Physicochem. Biol.*, Skulachev, V. P., Ed.; Harwood: Glasgow, 1992; Vol. II, Part 2, p 1.
4. Hoerpel, G. Dissertation, Mainz, West Germany, 1983.
5. Bader, H., Ringsdorf, H.; Shmidt, B. *Die Angewandte Makromolekulare Chemie* **1984**, *123/124*, 457.
6. Pratten, M. K.; Lioyd, J. B.; Horpel, G.; Ringsdorf, H. *Makromol. Chem.* **1985**, *186*, 725.
7. Lin, S. Y.; Kawashima, Y. *Pharm. Acta. Helv.* **1985**, *60* 339.
8. Lin, S. Y.; Kawashima, Y. *Pharm. Acta. Helv.* **1985**, *60* 345.
9. Kabanov, A. V.; Chekhonin, V. P.; Alakov, V. Yu.; Batrakova, E. V.; Lebedev, A. S.; Melik-Nubarov, N. S.; Arzhakov, S. A.; Levashov, A. V.; Morozov, G. V.; Severin, E. S.; Kabanov, V. A. *FEBS Lett.* **1989**, *258*, 343.
10. Kabanov, A. V.; Batrakova, E. V.; Melik-Nubarov, N. S.; Fedoseev, N. A.; Dorodnich, T. Yu.; Alakhov, V. Yu.; Chekhonin, V. P.; Nazarova, I. R.; Kabanov, V. A. *J. Contr. Release* **1992**, *22*, 141.
11. Schmolka, I. R. *J. Am. Oil Chem. Soc.* **1977**, *54*, 110.
12. Zhou, Z.; Chu, B. *J. Colloid Interface Sci.* **1988**, *126*, 171.
13. Alexandris, P.; Holzwarth, J. F.; Hanton, T. A. *Macromolecules* **1994**, *27*, 2414.
14. Kabanov, A. V.; Nazarova, I. R.; Astafieva, I. V.; Batrakova, E. V.; Alakhov, V. Yu.; Yaroslavov, A. A.; Kabanov, V. A. *Macromolecules* **1995**, in press.
15. Chubb, C.; Draper, R. *Proc. Soc. Exp. Biol. Med.* **1987**, *184*, 489.
16. Nogata, Y.; Mondon, C. E.; Cooper, A. D. *Metabolism* **1990**, *39*, 682.
17. Morikawa, K.; Okada, F.; Hosokawa, M.; Kobayashi, H. *Cancer Res.* **1987**, *47*, 37.
18. Lowe, K. C.; Armstrong, F. H. *Adv. Exp. Med. Biol.* **1990**, *277*, 267.
19. Allison, A. C.; Byars, N. E. *Mol. Immunol.* **1991**, *28*, 279.
20. Topchieva, I. N.; Erokhin, V. N.; Osipova, S. V.; Khrutskaya, M. M.; Kupriyanova, T. A.; Bukovskaya, S. N. *Biomed. Sci.* **1991**, *2*, 38.
21. Chaplin, D. J.; Horsman, M. R.; Aoki, D. S. *Br. J. Cancer* **1991**, *63*, 109.
22. Chen-Chow, P. C.; Frank, S. G. *Int. J. Pharm.* **1981**, *8*, 89.
23. Kodama, T.; Arima, K.; Sakamoto, H.; Suga, M.; Yachi, A. *Gan To Kagaku Ryoho* **1988**, *15*, 2544.
24. Slepnev, V. I.; Kuznetsova, L. E.; Gubin, A. N.; Batrakova, E. V.; Alakhov, V. Yu; Kabanov, A. V. *Biochem. Internat.* **1992**, *26*, 587.
25. Kabanov, A. V.; Slepnev, V. I.; Kuznetsova, L. E.; Batrakova, E. V.; Alakhov, V. Yu.; Melik-Nubarov, N. S.; Sveshnikov, P. G.; Kabanov, V. A. *Biochem. Internat.* **1992**, *26*, 1035.

BLOCK COPOLYMER MICELLES (1. Structure and Properties)

Zdeněk Tuzar and Pavel Kratochvíl
Institute of Macromolecular Chemistry
Academy of Sciences of the Czech Republic

Petr Munk
Department of Chemistry and Biochemistry and Center for Polymer Research
University of Texas at Austin

Block copolymers, when dissolved in a selective solvent that is thermodynamically good for one block and poor for the other, undergo a self-association in which spherical micelles of a nearly uniform size are formed with insoluble blocks in the core and soluble blocks in the shell. A micelle typically consists of tens to a few hundreds of copolymer molecules. Reviews of experimental data have been done by Riess et al. (in 1985) and Tuzar and Kratochvíl (in 1994).[1,2]

Only in recent years has proper attention been paid to micelles of hydrophobic/hydrophilic block copolymers in aqueous media. Researchers primarily have studied two types of these copolymers: di- and triblock copolymers of ethylene oxide and propylene oxide, known as commercial products Poloxamers® and Pluronics®, and laboratory-prepared block copolymers of styrene and ethylene oxide or methacrylic acid. These systems combine the thermodynamic and interfacial properties of strongly hydrophobic/strongly hydrophilic low-molar mass substances (like soaps and surfactants) with the stability and structures that are based on long polymeric chains.[3-5]

MICELLAR STRUCTURE AND PROPERTIES

Micellar Molar Mass and Structure

Like soaps and surfactants in water, block copolymers in selective solvents form micelles via the so-called closed association, characterized by a certain critical micelle concentration (CMC), below which only molecularly dissolved copolymer (usually called unimer in this context) is present in solution and above which multimolecular spherical micelles are in a dynamic equilibrium with the unimer. In comparison with soaps and surfactants, CMC in block copolymer micellar systems is very small, in most cases even imperceptible.

Micellar Equilibrium

The micellar equilibrium state is given by a minimum of the Gibbs energy of the system. It has been found that polymer micelle formation in organic selective solvents is an enthalpy-driven and entropy-controlled process. Surfactant micelle formation is, on the other hand, an entropy-driven process, due to the effect of hydrophobic interactions and the subsequent change of water structure.[6]

Unimer-micelle equilibrium represents a dynamic process, in which copolymer molecules migrate at a given rate between micellar and unimer states (and also between micelles themselves).

Kinetics of Micelle Formation and Dissociation

The kinetics of micelle formation and dissociation has been studied by the stopped-flow technique with light scattering detection on di- and triblock copolymers AB and ABA, A being polystyrene and B hydrogenated polybutadiene.[7]

Polyelectrolyte Micelles

Our laboratories have conducted a systematic study of polyelectrolyte micelles: polystyrene-*block*-poly(methacrylic acid) and polystyrene-*block*-poly(polymethacrylic acid)-*block*-polystyrene copolymer micelles in various aqueous solvents have been studied by static and dynamic light scattering, sedimentation velocity, fluorescence, and electrophoresis.[8-11] Certain specific features in behavior of aqueous micellar solutions of the copolymer mentioned differ from those of other block copolymer systems.

APPLICATIONS

While soaps and low-molar mass surfactants in aqueous solutions have found—due to their surface and colloidal properties (micelle formation included)—wide practical application, block copolymer solutions have not yet been used on such a large scale. Two classes of block copolymers are produced and employed commercially. First are copolymers of ethylene oxide and propylene oxide (commercial products Pluronics®, and Poloxamers®), colloidal properties of which (such as surfactancy, wettability, micelle formation) depend on their molar mass and composition.[12,13] Because of their very low toxicity, they are used in pharmaceutical and cosmetics industries.[14,15]

Second are copolymers of a (hydrogenated) diene and styrene (produced by the Shell Co. as Kratons® and Shellvises®), which are used as viscosity buffers for motor oils. At higher temperatures they are in the unimer form and as such they enhance low oil viscosity; at lower temperatures they are in the micellar form and virtually do not affect the oil viscosity.

Two phenomena have been lately studied, the practical exploitation of which is obvious: solubilization of low-molar-mass substances into micellar cores and their release.

Solubilization

In a comprehensive study dealing with the solubilization by block copolymer micelles, poly(ethylene oxide)-poly(propylene oxide) and poly(*N*-vinylpyrrolidone)-polystyrene were employed as solubilizers and a series of aliphatic and aromatic hydrocarbons as solubilizates.[16]

Release

Polymer micelles have lately been considered as a potential vehicle for controlled release of organic substances into aqueous media.[17]

Obviously, phenomena of the uptake and release of organic substances scarcely soluble in water, from and into aqueous environment, may find practical applications in pharmacology, agriculture, and ecology.

REFERENCES

1. Riess, G.; Hurtrez, G.; Bahadur, P. *Encyclopedia of Polymer Science and Engineering*, 2nd ed.; Mark, H. F. et al., Eds.; John Wiley & Sons: New York, 1985; Vol. 2, pp 324–434.
2. Tuzar, Z.; Kratochvíl, P. *Encyclopedia of Advanced Materials*; Bloor, D., Ed.; Pergamon: Oxford, 1994; pp 2047–2052.
3. Zhou, Z.; Chu, B. *Macromolecules* **1988**, *21*, 2548.
4. Wilhelm. M. et al. *Macromolecules* **1991**, *24*, 1033.
5. Tuzar, Z. et al. *Collect. Czech. Chem. Commun.* **1993**, *58*, 2362.
6. Tuzar, Z.; Kratochvíl, P. *Surface Colloid Sci.* Matijevic, E., Ed. **1993**, *15*, 1.
7. Bednár, B. et al. *Makromol. Chem. Rapid Commun.* **1988**, *9*, 785.
8. Tuzar, Z. et al. *Polym. Preprints* **1991**, *32*(1), 525.
9. Munk, P. et al. *Makromol. Chem. Macromol. Symp.*, **1992**, *58*, 195.
10. Cao, T. et al. *Macromolecules* **1991**, *24*, 6300.
11. Tuzar, Z. et al. *Polym. Preprints* **1993**, *34*(1), 1038.
12. Piirma, I. In *Surfactant Science Series*; Dekker: New York, 1992; Vol. 42, "Polymeric Surfactants."
13. Hurter, P. N.; Hatton, T. A. *Langmuir* **1992**, *8*, 1291.
14. Schmolka, I. R. *Perfume Cosmet.* **1967**, *82*, 25.
15. Chen-chow, P.-C.; Frank, S. G. *Int. J. Pharm.* **1981**, *8*, 89.
16. Nagarajan, R.; Barry, M.; Ruckenstein, E. *Langmuir* **1986**, *2*, 210.
17. Grief, R. et al. *Science* **1994**, *263*, 1600.

BLOCK COPOLYMER MICELLES (2. Fluorimetric Studies and Computer Modeling)

Karel Procházka and Zuzana Limpouchová
Department of Physical and Macromolecular Chemistry
Faculty of Science
Charles University in Prague

Stephen E. Webber
Department of Chemistry and Biochemistry and Center for Polymer Research
University of Texas at Austin

FLUORIMETRIC STUDIES OF MICELLIZING COPOLYMERS

Fluorimetric Studies with Dispersed Polarity-Sensitive Probes

Critical micelle concentrations (CMC) for high molar mass copolymers are usually very low and cannot be accurately determined by classical experimental techniques of polymer chemistry, such as light scattering.[1] Fluorimetry is one of the few experimental methods that can be used to determine the CMC in systems of amphiphilic copolymers in polar or aqueous media.

Fluorescence from polarity-sensitive probes, such as phenylnaphthylamine (PNA), 8-anilinonaphthalene-1-sulfonic acid (ANS), and 6-(4-toluidino)naphthalene-2-sulfonic acid (TNS), is almost completely quenched in aqueous media and its quantum yield is virtually zero, whereas in nonpolar media an intense emission is observed. These probes are soluble in water, but when micelles (or any other hydrophobic domains) are formed, they are strongly solubilized into nonpolar structures. Studies with probes such as PNA, ANS, and TNS have enabled the association of water-soluble polymers and copolymers to be characterized.[2,3] The finite solubility of those probes in water does not affect the data because they do not contribute appreciably to the fluorescence.

Certain cationic probes, such as octadecyl rhodamin B (ORB), are fluorescent in monomeric form in nopolar media and form nonfluorescent dimers in water. This particular probe was used for studying micellization of pluoronics, that is, poly(ethylene oxide)-*block*-poly(propylene oxide) copolymers.[4]

Fluorimetric Studies with Specifically Tagged Copolymers

Tailor-made copolymers with fluorophores specifically attached at the ends of one or both blocks or at the block junctions offer rich possibilities to study various parts of micellar structures. Procházka et al. have studied micellization of a copolymer sample of polystyrene-*block*-poly(methacrylic acid) or *t*-butyl methacrylate which was tagged by several vinyl-2-napththalene units at the end of the polystyrene block.[5] Formation of micelles may be monitored by measurements of the excimer kinetics and fluorescence lifetimes.

Time-Resolved Fluorescence Anisotropy Measurements

Procházka et al. have studied micellization of anthracene-tagged copolymer polystyrene-*block*-poly(ethylene-*co*-propylene) with a M_w of ~10^5 g·mol^{-1} and a w_{PS} of ~0.4, in solvent mixtures of 1,4-dioxane-heptane by a combination of light-scattering and polarized fluorimetry.[6,7] The sample was randomly tagged by methylene-9-anthryl groups on the polystyrene block (on average less than one tag per polymer chain). 1,4-Dioxane is a good solvent for polystyrene and a nonsolvent for the aliphatic block, whereas heptane is a good solvent for the aliphatic part and a nonsolvent for polystyrene. In solvents rich in 1,4-dioxane, micelles with poly(ethylene-*co*-propylene) cores are formed, and in solvents rich in heptane, micelles with polystyrene cores are formed. Mixtures with comparable ratios of both components are good solvents for the whole copolymer chain, and the sample dissolves to form individual coils.

In heptane-rich solvents, fluorophores are trapped in rigid and compact polystyrene cores and are almost immobilized. In good solvents (40–60 vol% of 1,4-dioxane) the fluorophores in free coils are exposed to solvent molecules and are free to rotate around single covalent bonds. A complete depolarization of fluorescence occurs before the depletion of the excited state. The rotational depolarization correlation time is fast, and the residual anistropy is zero. In 1,4-dioxane-rich solvents, fluorophores are located in micellar shells. The polystyrene blocks in shells are partially radially stretched, which creates a locally anisotropic microenvironment for individual pendant fluorophores. The average segment density in the shells is considerably higher than that in isolated copolymer coils in good solvents, and the rotation of fluorophores is strongly hindered. Depolarization is slower compared with good solvents and does not proceed in all dimensions. Significant values of residual and steady-state anisotropies are observed.

COMPUTER SIMULATIONS OF CHAIN CONFORMATIONS

With the advent of powerful supercomputers, Monte Carlo and molecular dynamics computer simulations have become promising tools for studying polymer chain conformations in solution and in supermolecular structures. However, a rigorous Monte Carlo study of realistic systems of micellizing block copolymers exceeds the capabilities of even the most up-to-date supercomputer.

ACKNOWLEDGMENT

SEW acknowledges the financial support of the U.S. Army Research Office (grant DAAAL03-90—G-0147) and the NSF Polymers Program (grant DMR-9308307). I would also like to express my gratitude to the financial support to the Grant Agency of the Czech Republic (for the Grant No. GACR 203/94/973.

REFERENCES

1. Tuzar, Z.; Kratochvíl, P. *Surface and Colloid Science*; Matijevic, E., Ed.; Plenum: New York, 1993; Vol. 15, p 1.
2. Akiyoshi, D.; Deguchi, S.; Moriguchi, N.; Yamaguchi, S.; Sunamoto, J. *Macromolecules* **1993**, *26*, 3602.
3. Ikemi, M.; Odagiri, N.; Tanaka, S.; Shinohara, I.; Chiba, A. *Macromolecules* **1981**, *14*, 34.
4. Nakashima, K.; Anzai, T.; Fujimoto, Y. *Langmuir* **1994**, *10*, 658.
5. Procházka, K.; Kiserow, D.; Ramireddy, C.; Tuzar, Z.; Munk, P.; Webber, S. E. *Macromolecules* **1992**, *25*, 454.
6. Procházka, K.; Vajda, Š.; Fidler, V.; Bednář, B.; Mukhtar, E.; Almgren, M.; Holmes, S. *J. Mol. Struct.* **1990**, *219*, 337.
7. Procházka, K.; Methage, B.; Almgren, M.; Mukhtar, E.; Svoboda, P.; Tnená, J. T.; Bednár, B. *Polymer* **1993**, *34*, 103.

BLOCK COPOLYMER THIN FILMS

Ghislaine Coulon
Institut Curie P.S.I. and
Université des Sciences et Technologies de Lille L.S.P.E.S.

Block copolymers are increasingly being used as surfactants and thin-film adhesives in biomedical and microelectronics applications.[1] Thus, a fundamental understanding of the behavior of diblock copolymers at surfaces and interfaces is essential. The behavior of diblock copolymers AB is quite well known.[2] Because of the incompatibility between blocks A and B, diblock copolymers experience a microphase separation for temperatures T between the glass transition temperature T_g of the blocks and the order-disorder transition temperature (T_{ODT}). Depending on the relative molecular weights of the blocks in the copolymer, various equilibrium morphologies are encountered: spherical, cylindrical, bicontinuous double-diamond, and lamellar.[3-6] The lamellar structure corresponds to the case of symmetric diblock copolymers and the period L equals the thickness of a -ABBA- or -BAAB- bilayer.

As mentioned, diblock copolymers exhibit interesting surface activity. Recently, numerous studies have focused on the behaviors of diblock copolymers near surfaces and more especially on the influence of the external surfaces on the lamellar ordering in symmetric diblock copolymer thin films.[7-27]

In thin films where the thickness is a few times the characteristic period L in the bulk, there is a strong orientation of the lamellae parallel to the external surfaces, since minimization of the surface energies dictates which block is located at both external surfaces.[7-24] Moreover, the morphology of the free surface of such ordered thin films depends on their initial thickness in the disordered state. The presence of islands or holes on the free surface has been revealed for the very first time by interference microscopy.[8]

The present article shows how the structure of diblock copolymer thin films can be investigated and easily revealed by optical microscopy and by atomic force microscopy.

PREPARATION

Three different symmetric diblock copolymers have been investigated. The first one is a poly(styrene-*b*-methyl-methacrylate), P(S-*b*-MMA). The second one is a poly(styrene-*b*-butyl-methacrylate), (P(S-*b*-BMA). The last one is a P(S-*b*-d-MMA) where PMMA is perdeuterated.

Thin films are prepared by spin-coating a toluene solution of the copolymer onto a clean silicon wafer.

The observation of the free surface of the copolymer film is performed either by optical microscopy or by atomic force microscopy.

PROPERTIES

Lamellar Structure as Evidenced by Interference Microscopy

The spin-coating process gives the perimeter of the film thickness variations that can be large (up to 6000 Å for films *ca* 1000 Å in thickness). For example, an optical micrograph of the perimeter of an as-cast P(S-*b*-MMA) sample (M_w = 91,500) is obtained by using the optical microscope under reflection conditions to get interference colors from the white light source. The source lamp iris is closed down to ensure a parallel beam of light. The interference colors are diffuse, proof that the thickness varies continuously. On the contrary, after annealing at 170°C for 24 hours and quenching, well-defined contours of interference colors are present, proof that the thickness varies in a stepwise manner.[18] Only the existence of a lamellar stacking parallel to a substrate can explain the presence of the terraces on the free surface. Minimization of the interfacial energies dictates which block is present at the two interfaces (air/polymer and substrate/polymer).[7-15]

By using the Newton's scale, the observed successive interference colors can be indexed, assuming the steps have the same height H.

The ordering of symmetric diblock copolymers depends on temperature and film thickness.[23] Thin films undergo a transition from a partially to a fully ordered state at a temperature that is thickness-dependent. Neutron reflectivity is a powerful tool to examine this kind of behavior, especially to obtain quantitative density profiles; the observation of the terraces at the perimeter of the film by interference microscopy is a very simple way of looking at the film's thickness effect on the ordering of the copolymer.

Surface Morphology of Symmetric Diblock Copolymer Thin Films

The free surface of as-cast thin films is flat and smooth.[24,27] The root-mean square (rms) roughness R, measured by AFM, is about 5–10 Å. As the copolymer film is annealed, the morphology of the free surface evolves with the annealing time; first, the rms roughness increases and then, when the lamellar ordering is established, islands or holes of height L are seen on the free surface. Indeed, the morphology of the free surface of the ordered film depends on its initial thickness, e_{ac}, in the as-cast state. Previous studies show that for very thin films (few lamellae) the lamellar order starts at the Si substrate and then propagates through the entire film thickness; thus, the formation of islands or holes reflects the lamellar ordering.[15] Once they are formed, the islands of height L occupy a fraction x of the film surface; this amount is directly related to the initial thickness, e_{ac}, because of the conservation of the total amount of copolymer.

In summary, islands or holes form on the free surface of symmetric diblock copolymer thin films because of the quantization of the thickness in the ordered state. This phenomenon has been observed for copolymers other than the ones presented here. The kinetics of formation of islands at the free surface is parallel to the kinetics of the inner lamellar ordering process; it depends both on the annealing temperature and the initial thickness.

Time-Evolution of the Film Free Surface as Shown by *In Situ* Interference Microscopy

Thin films of diblock copolymers are often used to enhance the adhesion between incompatible polymers.[28,29] Since the adhesion depends strongly on the quality of the surfaces, the presence of such defects as islands or holes can be very unfavorable. One way to avoid their formation is to prepare as-cast films with initial thickness, e_{ac}. It is an open question whether the islands (or holes), of height L and, created on the free surface, correspond to an equilibrium state or evolve with the annealing time.

A close examination of the digitized images obtained during the kinetics of growth (up to 22,000 min) of all the different systems of islands or holes leads to the following qualitative remarks:

- The centers of mass of islands or holes do not move with time. Thus, there is no diffusive motion of these islands or holes in the plane of the free surface.
- Three processes are observed: individual growth, coalescence of neighboring islands (or holes), and disappearance of the smallest islands (or holes) to the benefit of the neighboring largest ones.
- A line free energy exists: in the case of "dilute" systems, islands and holes exhibit a perfect round shape. In addition, after coalescence, the resulting unique (or hole) tends to go back to a circular shape as the annealing time progresses. An interesting point: When the copolymer film is not fully ordered, the line free energy decreases strongly and the contours of the islands (or holes) become tortuous.

The patterns observed on the free surface are reminiscent of those observed in systems that exhibit first-order phase transition with conserved order parameter. For such systems, the growth of clusters obey power laws versus time in the asymptotic time-range.[30]

SUMMARY

The time evolution of the free surface of symmetric diblock copolymer films results from the competition between three parameters: the interfacial energies (air/polymer and substrate/polymer), the line tension of the defects, and the elastic free energy of the copolymer chains. During the formation stage of the defects (islands or holes [i.e., during the lamellar ordering]) the interfacial energies are prevailing.

For intermediate annealing times at a given temperature, the system tends to decrease its total free energy by reducing the line tension of the defects. It proceeds mainly by coalescences and dissolutions of the defects. In the case of islands, their surface coverage is constant. There is no noticeable permeation of molecules; only the shear between lamellae can lead to the coalescences and the dissolutions of islands. In the case of holes, the surface coverage is not constant; it varies slightly with the annealing time and it denotes some permeation of the copolymer molecules. However, the coalescence and the dissolution mechanisms are much more active than the permeation mechanism, and the kinetics of growth of holes results mainly from the shear between lamellae.

For long annealing times and higher temperatures, the kinetics of growth in the ultimate stage is mainly governed by the permeation mechanism. The final morphology of the free surface will depend on the balance between the line tension of the defects and the elastic free energy of the molecules.

It is important to notice that, while the equilibrium state of the inner morphology of copolymer thin films is quite easy to define and to reach within reasonable annealing times, the morphology of their free surface evolves very slowly with time; one has to be careful not to conclude hastily that the equilibrium state is achieved. Nevertheless, it is possible to get rid of the surface defects, such as islands or holes, by heating the copolymer thin films at sufficiently high temperatures.

Finally, it is clear that the observation of the hole free surface of copolymer thin films is very useful to predict the structure and the behavior of the whole film. Interference microscopy and atomic force microscopy are fruitful tools to perform such a study; they are complementary techniques that allow the direct observation of the surface over a large-scale range.

REFERENCES

1. Meier, D. J., Ed.; *Block Copolymers: Science and Technology*; MMI Hardwood Academic: Midland, 1983.
2. Goodman, I. Ed.; *Developments in Block Copolymers - 1*, Applied Science: London, 1982.
3. Leibler, L. *Macromolecules* **1980**, *13*, 1602.
4. Helfand, E. *Macromolecules* **1975**, *8*, 1975.
5. Thomas, E. L.; Alward, D. B.; Kinnig, D. J.; Martin, D. C.; Handlin, D. L.; Fetters, L. J. *Macromolecules* **1986**, *19*, 2197.
6. Hasegawa, H.; Tanaka, K.; Yamasaki, K.; Hashimoto, T. *Macromolecules* **1987**, *20*, 1651.
7. Bradford, E.; Vanzo, E. *J. Polym. Sci.* **1968**, *Part A-1, 6*, 1661.
8. Rastogi, A. K.; St. Pierre, L. E. *J. Coll. Int. Sci.* **1969**, *31*, 168.
9. Thomas, H. R.; O'Malley, J. J. *Macromolecules* **1979**, *12*, 323.
10. Hasegawa, H.; Hashimoto, T. *Macromolecules* **1985**, *18*, 589.
11. Frederickson, G. H. *Macromolecules* **1987**, *20*, 589.
12. Henkee, C. S.; Thomas, E. L.; Fetters, L. J. *J. Mat. Sci.* **1988**, *23*, 1685.
13. Green, P. F.; Christensen, T. M.; Russell, T. P.; Jerome, R. *Macromolecules* **1989**, *22*, 2189.
14. Coulon, G.; Russell, T. P.; Deline, V. R.; Green, P. F. *Macromolecules* **1989**, *22*, 2581.
15. Russell, T. P.; Coulon, G.; Deline, V. R.; Miller, D. C. *Macromolecules* **1989**, *22*, 4600.
16. Anastasiadis, S. H.; Russell, T. P.; Satija, S. K.; Majkrzak, C. F. *Phys. Rev. Lett.* **1989**, *62*, 1852.
17. Brown, H. R. *Macromolecules* **1989**, *22*, 2859.
18. Coulon, G.; Ausseré, D.; Russell, T. P.; *J. Physique France* **1990**, *51*, 777.
19. Anastasiadis, S. H.; Russell, T. P.; Satija, S. K.; Majkrzak, C. F. *J. Chem. Phys.* **1990**, *92*, 5677.
20. Coulon, G.; Collin, B.; Chatenay, D.; Ausseré, D.; Russell, T. P. *J. Physique France* **1990**, *51*, 2801.
21. Ausseré, D.; Chatenay, D.; Coulon, G.; Collin, B. *J. Physique France* **1990**, *51*, 2571.
22. Green, P. F.; Christensen, T. M.; Russell, T. P. *Macromolecules* **1991**, *24*, 252.
23. Menelle, A.; Russell, S. H.; Anastasiadis, S. H.; Satija, S. K.; Majkrzak, C. F. *Phys. Rev. Lett.* **1992**, *68*, 67.
24. Collin, B.; Chatenay, D.; Coulon, G.; Ausseré, D.; Gallot, Y. *Macromolecules* **1992**, *25*, 1621.
25. Maaloum, M.; Ausseré, D.; Chatenay, D.; Coulon, G.; Gallot, Y. *Phys. Rev. Lett.* **1992**, *68*, 1575.
26. Coulon, G.; Collin, B.; Chatenay, D.; Gallot, Y. *J. Phys. 2 France* **1993**, *3*, 697.
27. Coulon, G.; Daillant, J.; Collin, B.; Benattar, J.-J.; Gallot, Y. *Macromolecules* **1993**, *26*, 1582.
28. Fayt, R.; Jêrome, R.; Teyssié, Ph. *Polym. Eng. Sci.* **1987**, 27, 328.
29. Cohen, R. E.; Ramos, A. R. *Macromolecules* **1979**, *12*, 89.
30. Gunton, J. D.; San Miguel, M.; Sahni, P. S. *Phase Transition and Critical Phenomena*; Domb, L.; Lebovitz, J. L., Eds.: Academic Press: London, 1983; Vol. 3.

Block Copolymers

See: *Aminosiloxanes*
Biodegradable Copolymers [Comprising Poly(L-lactic acid)]
Block Copolymer-Like Structures (End-Group Ionic Interactions in Blends)
Block Copolymer Micelles (Overview)
Block Copolymer Micelles (For Drug Targeting)
Block Copolymer Micelles (Microcontainers for Drug Targeting)
Block Copolymer Micelles (1. Structure and Properties)
Block Copolymer Micelles (2. Fluorometric Studies and Computer Modeling)
Block Copolymer Thin Films
Block Copolymers (Anisotropic Conductivity)
Block Copolymers (By Changing Polymerization Mechanism)
Block Copolymers (By Step-Growth Polymerization)
Block Copolymers (From Macroinitiators)
Block Copolymers (Micellization in Solution)
Block Copolymers (Ordered Bicontinuous Double-Diamond Morphology)
Compatibilizers (for Blends)
Compatibilizers, Polymeric (Recycling of Multilayer Structures)
Crosslinked Polymer Brushes
Cyclic Carbonates (Ring-Opening Polymerization)
Cyclic Imino Ethers (Ring-Opening Polymerization)
Cyclic Imino Ethers (Ring-Opening Polymerization of 1,3-Oxazo Monomers)
Free Radical Iniferters
Free Radical Initiators, Bifunctional
Ion-Chelating Polymers (Medical Applications)
Living Anionic Polymerization
Living Carbocationic Polymerization, Olefins
Living Coordination Polymerization
Living Radical Polymerization
Living Radical Polymerization, Iodine Transfer
Macrocyclic Block Copolymers
Macromonomeric Initiators (Macroinimers)
Metal Nanoclusters (within Block Copolymer Domains)
Microspheres, Core-Shell Type (from Block Copolymers)
Molecular Composites (Liquid Crystalline Polyamide and Amorphous Polyimide)
Nonthrombogenic Polymers (Multiblock Copolymers of Polyether and Polyamide)
2-Oxazolines, Polymerization
Poly(N-acylethylenimine)s
Poly(tert-butystyrene)
Poly(ether amide) Triblock and Star Block Copolymers
Poly(ether-block-amide) Copolymers (Characterization of Structure and Polymorphism)
Poly(ethylene oxide) (Applications in Drug Delivery)
Poly(imide siloxane)s
Polyimides (from N-Silylated Aromatic Diamines)
Polyphosphazene Block Copolymers
Polysilanes (Ring-Opening Polymerization of Cyclotetrasilanes)
Polyurethane Elastomers, Segmented (Non-Hydrogen Bonding Systems)
Reactive Oligomers (Overview)
Ring-Opening Coordination Polymerization (by Soluble Multinuclear Aldoxides)
Siloxane-Containing Polymers
Star-Block Copolymers
Star-Shaped Polymers
Stereoregular Polymethacrylates
Styrene-Butadiene-Styrene Elastomer, Hyrogenated
Styrene-Butadiene-Styrene (General Overview)
Styrene-Butadiene-Styrene Triblock Copolymer
Supramolecular Formation and Structure (PEO-PPO-PEO Block Copolymers)
Surfactants, Polymeric (Overview)
Surfactants, Polymeric (Colloid Stabilization)
Surfactants, Polymeric (Compatibilizers for Blends)
Tailor-Made Polymers
Thermoplastic Elastomers (Overview)
Ultrasonic Degradation (Preparation of Block Copolymers)
Vinyl Polyperoxides (Oxygen Copolymerization)
Vinylidene Fluoride-Based Thermoplastics (Homopolymerization and Copolymerization)

BLOCK COPOLYMERS (Anisotropic Conductivity)

Koji Ishizu
Department of Polymer Science
Tokyo Institute of Technology

Many of the available conductive polymers possess some undesirable characteristics, such as environmental instability, poor processability, or poor physical properties.[1-4] Attempts have been made to improve the physical properties of these conductive polymers by using block copolymers. In these attempts one of the blocks consists of a conducting sequence.[5,6] Generally, such block copolymers form a microphase-separated structure in the solid state, owing to incompatible block chains.

It recently became clear that poly[styrene (S)-*b*-2-vinylpyridine(2VP)] diblock copolymers (about 50 wt% polystyrene (PS) blocks and relatively narrow molecular weight distribution) formed horizontally oriented lamellar microdomains by interacting with the air-copolymer and the substrate-copolymer.[6] Subsequently, semiconducting materials were obtained when a block copolymer film was exposed by the vapor of alkyl dihalide. The quaternized P2VP domains of these films exhibited the nature of ionic conduction.[7] Colloidal silver could be introduced into these quaternized P2VP phases by reduction of silver iodide. The anisotropic conductivity on these materials was studied in detail.

PROPERTIES

The sample film behaved as a two-dimensional conductor (conductivity parallel to the film plane $\sigma//$ and conductivity perpendicular to film plane $\sigma\perp$) because of the structure of horizontally oriented lamellae.

It seems that the values of $\sigma//$ reflect electron conduction among colloidal silver. However, the observed $\sigma//$ value is extremely small compared with that of silver metal. It was found that silver particles distribute at intervals of a few nm from each other in the quaternized P2VP layer. Such an arrangement of silver particles leads to a decrease of the mean free path of electrons.

REFERENCES

1. Chance, R. R.; Schacklette, L. W.; Miller, G. G.; Ivory, D. M.; Sowa, J. M.; Elsenbaumer, R. L.; Baughman, R. H. *Chem. Commun.* **1980**, 348.
2. Chien, J. C. W. *Polyacetylene Chemistry, Physics and Material Science*; Academic: New York, 1984.
3. Wnek, G. E.; Chien, J. C. W.; Karasz, F. E.; Lilya, C. P. *Polymer* **1979**, *20*, 1441.
4. Gibson, H. W.; Bailey, F. C.; Epstein, A. J.; Rommelmann, H.; Pochaw, J. M. *Chem. Commun.* **1980**, 426.
5. Lee, K. I.; Jopson, H. *Polym. Bull.* **1983**, *10*, 105.
6. Ikeno, S.; Yokoyama, M.; Mikawa, H. *Polym. J.* **1978**, *10*, 123.
7. Ishizu, K.; Yamada, Y.; Fukutomi, T. *Polymer* **1990**, *31*, 2047.
8. Ishizu, K.; Yamada, Y.; Saito, R.; Kanbara, T.; Yamamoto, T. *Polymer* **1992**, *33*, 1816.

BLOCK COPOLYMERS (By Changing Polymerization Mechanism)

Yusuf Yağci and Munmaya K. Mishra
Department of Chemistry
Istanbul Technical University of Polymer Research
Texaco Research and Development

The synthesis of novel polymeric materials that have a combination of physical properties is very attractive. The field has been described in a recent book edited by Mishra.[1] It is well established that a desired combination of physical properties could be achieved by designing tailor-made block and graft copolymers. Block copolymers are conveniently prepared by living ionic polymerization, in which monomers of a desired block segment are added sequentially.[2] There are various drawbacks that often retard the practical application of living systems to prepare block copolymers. These drawbacks relate to limitation of the method to certain monomers and exclude monomers that polymerize by other mechanisms. Since the pioneering work of Richards and co-workers on the preparation of block copolymers by changing polymerization mechanisms, the range of possible monomer combinations in block copolymers have expanded and many examples have been reported.[3-7] The following transformation reactions have been successfully investigated (**Scheme I**).

ANIONIC TO RADICAL TRANSFORMATION

This approach is the most widely applied process of the transformation technology.

This method is exemplified in the work of Eastmond et al.[8] They had prepared bromine terminated polystyrene (PSt) by living anionic polymerization and then used this polymer as a macroinitiator with Mn_2CO_{10} for the radical polymerization of monomers.

CATIONIC TO RADICAL TRANSFORMATION

General synthetic approach in this particular transformation involves introduction of a common radical initiator functional group, such as peroxide or azo groups, into the center of the macromolecular chain, which is suitable for the subsequent radical generation. Recently the polymerization of THF was achieved by using functional azoinitiators.[9-11] Spectroscopic and thermolysis studies showed that polymers obtained via this initiation method contain one azo linkage per chain.[12] The central azo group present in PTHF is then decomposed in the presence of a comonomer to give block copolymers. Using this type of transformation, it has been possible to obtain new block copolymers composed of crystalline and liquid crystalline blocks.

The other technique to incorporate azoinitiators uses living cationic polymerization of vinyl ethers. The main advantage of these systems is the absence of side reactions, such as chain transfer dominating vinyl ether polymerization observed with classical initiators.

Hizal et al.[13] have achieved an interesting variation of cation to radical transformation by using photochemistry of pyridinium salts in radical generating process.

RADICAL TO CATIONIC TRANSFORMATION

So far, all applied radical to ionic transformations are based on functionalization of polymers (suitable to generate cationic species) by using initiators, redox, or transfer agents. This theme is typically accomplished with photoactive polymers obtained by using azo-benzoin initiators.[14-16]

Irradiation of benzoin terminated polymers in conjunction with pyridinium salt in the presence of cyclohexene oxide (CHO) as a cationically polymerizable monomer leads to the formation of block copolymers by the electron transfer process.[17-19]

RADICAL TO ANIONIC TRANSFORMATION

Most of the reported studies of this transformation involve initiation of anionic polymerization of the amino acid N-carboxy anhydrides by amino telechelics prepared by various free radical initiation systems.[20-22] Several grafting studies using polymeric alcholates to initiate anionic polymerization of ethylene oxides were also reported.[23-26]

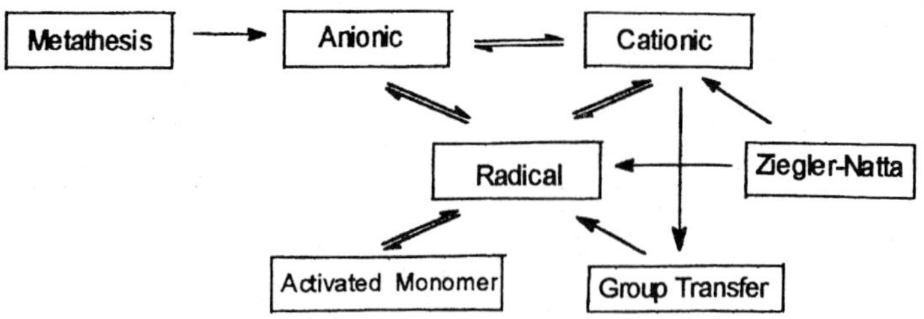

I

TRANSFORMATIONS INVOLVING ANIONIC AND CATIONIC POLYMERIZATION

Anion to cation transformation or reverse transformation reactions were successfully employed to prepare block copolymers. The particular advantage of these transformations is that both anionic and cationic blocks can be prepared under living polymerization conditions.

ANIONIC TO METATHESIS TRANSFORMATION

The only reported example of this transformation is the work of Amass and Gregory,[27] who prepared block copolymers of styrene and cyclopentene.

TRANSFORMATIONS INVOLVING ZIEGLER-NATTA POLYMERIZATION

Active site transformation from anion to Ziegler-Natta and from Ziegler-Natta to radical and cation have been reported. Anion to Ziegler-Natta transformation involves alkylation of aluminum halide with living anionic polymers.[28,29] The resulting polymeric aluminum compounds were successfully employed in Ziegler-Natta polymerization in a manner similar to low-molar mass alkyl aluminum compounds. Ziegler-Natta to radical and cation transformations were achieved by incorporation of peroxy and terminal halide groups, respectively.[30,31]

TRANSFORMATIONS INVOLVING GROUP TRANSFER POLYMERIZATION

Telechelic poly(MMA) having bromine end groups was prepared by Eastmond and Grigon[32] with the aid of group transfer polymerization. A metal carbonyl photoinitiating system in conjunction with the halogen containing polymer provided the formation of polymeric radicals for the desired transformation. The other approach of this theme involves switching active centers to group transfer polymerization.

APPLICATION OF TRANSFORMATION REACTIONS FOR THE SYNTHESIS OF LIQUID CRYSTALLINE AND CRYSTALLINE BLOCK COPOLYMERS

Although a variety of graft, random, and alternating copolymers of liquid crystalline (LC) monomers have been prepared, block copolymer systems have gained interest only recently. Several new classes of block copolymers have been prepared by adapting cation to radical transformation reactions.[33-40]

REFERENCES

1. Mishra, M. K. *"Macromolecular Design: Concept and Practice"*; Polymer Frontiers International: Hopewell Jct., New York, 1994.
2. Van Beylen, M.; Scwarz, M. *Ionic Polymerization and Living Systems*; Chapman and Hall: New York, 1993.
3. Burgess, F.; Cunliffe, A. V.; Richards, D. C.; Sherrington, D. C. *J. Polym. Sci., Polym. Lett. Ed.* **1979**, *14*, 471.
4. Richards, D. H. *Brit. Polym. J.* **1980**, *12*, 89.
5. Schue, F. *Comprehensive Polymer Science*; Pergamon: Oxford, 1991; Vol. 7, Chapter X.
6. Steward, M. J. *New Methods of Polymer Synthesis*: Blackie & Sons: New York, 1991; Chapter IV.
7. Yagci, Y.; Mishra, M. K. *Macromolecular Design: Concept and Practice*, Polymer Frontiers International: Hopewell Jct., New York, 1994; Chapter X.
8. Eastmond, G. C.; Parr, K. J.; Woo, J. *Polymer* **1988**, *29*, 950.
9. Yagci, Y. *Polym. Commun.* **1985**, *26*, 25.
10. Serhatli, I. E.; Hizal, G.; Yagci, Y. *Turk. J. Chem.* **1992**, *16*, 91.
11. Akar, A.; Aydogan, A. C.; Talinli, N.; Yagci, Y. *Polym Bull.* **1986**, *15*, 293.
12. Hizal, G.; Yagci, Y. *Polymer* **1989**, *30*, 722.
13. Hizal, G.; Yagci, Y.; Schnabel, W. *Polymer* **1994**, *35*, 4443.
14. Onen, A.; Yagci, Y. *J. Macromol. Sci. Chem.* **1990**, *A27*, 755.
15. Onen, A.; Yagci, Y. Angew. *Makromol. Chem.* **1990**, *181*, 191.
16. Yagci, Y.; Onen, A. *J. Macromol. Sci. Chem.* **1991**, *A28*, 129.
17. Cai, G.; Yan. D. *Makromol. Chem.* **1987**, *188*, 1005.
18. Ishikawa, S.; Ishizu, K.; Fukutomi, T. *Polym. Bull.* **1986**, *16*, 223.
19. Uno, H.; Endo, T. *Chem. Lett.* **1986**, 1869.
20. Vlasov, G. P.; Rudkovskaya, G. D.; Ovsyannikova, L. A. *Makromol. Chem.* **1982**, *183*, 2635.
21. Tanaka, M.; Mori, A.; Imanishi, Y.; Bamford, C. H. *Int. J. Biol. Macromol.* **1985**, *7*, 173.
22. Higushi, S.; Mozawa, T.; Maeada, M.; Inoue, S. *Macromolecules* **1986**, *19*, 2263.
23. Mulvaney, J. E.; Ottaviani, R. O. *J. Polym. Sci., Polym. Chem. Ed.* **1982**, *20*, 1941.
24. Rempp, P.; Lutz, P.; Masson, P.; Chaumont, P.; Franta, E. *Makromol. Chem. Suppl.* **1985**, *13*, 47.
25. Ito, K.; Hashimura, K.; Itsuno, S.; Yamada, E. *Macromolecules* **1991**, *24*, 3977.
26. Jannasch, P.; Wesslen, B. *J. Polym. Sci., Polym. Chem.* **1993**, *V31*, 1519.
27. Amass, A. J.; Gregory, D. *Brit. Polym. J.* **1987**, *19*, 263.
28. Aldissi, M. *J. Chem. Soc. Chem. Commun.* **1984**, *20*, 1347. *Synthetic Metals* **1986**, *13*, 87.
29. Aldissi, M.; Bishop, A. K. *Polymer* **1985**, *26*, 622.

30. Agnuri, E.; Favier, G.; Laputte, R.; Philardea, T.; Rideau, J. *"Symposium on Block and Graft Copolymerization,"* Preprints, GFP; Mulhouse; **1972,** 55.
31. Doi, Y.; Watanabe, Y.; Ueki, S.; Soga, K. *Makromol. Chem. Rapid Commun.* **1983,** *4,* 533.
32. Eastmond, G. C.; Grigor, *Makromol. Chem., Rapid Commun.* **1986,** *7,* 375.
33. Galli, G.; Chiellini, E.; Yagci, Y.; Serhatli, I. E.; Laus, M.; Bignozzi, M. C.; Angeloni, A. S. *Makromol. Chem. Rapid Commun.* **1993,** *14,* 185.
34. Chiellini, E.; Galli, G.; Serhatli, I. E.; Yagci, Y.; Laus, M.; Angelloni, A. S. *Ferroelectrics* **1993,** *148,* 311.
35. Serhatli, I. E.; Galli, G.; Chiellini, E.; Yagci, Y. *Polym. Bull.* **1994,** *234,* 539.
36. Reeb, R.; Vinchon, Y.; Riess, G.; Catula, J. M.; Brossas, J. *J. Bull. Soc. Chim. France* **1975,** *11-12,* 2717.
37. Catula, J. M.; Riess, G.; Brossas, J. *Makromol. Chem.* **1979,** *178,* 1249.
38. Nicolova-Nankova, Z.; Palacin, F.; Raviola, F.; Kiess, G. *Eur. Polym. J.* **1975,** *11,* 301.
39. Riess, G.; Reeb, R. *Polym. Prep.* **1980,** *21,* 55.
40. Yagci, Y.; Menceloglu, Y. Z.; Baysal, B.; Gungor, A. *Polym. Bull.* **1989,** *21,* 259.

BLOCK COPOLYMERS (By Step-Growth Polymerization)

Alain Fradet
Université Pierre et Marie Curie

Unlike block polymers synthesized by chain polymerization, triblock copolymers,[1] block copolymers synthesized by step polymerization are $(AB)_n$ multiblock copolymers.[2-9]

Most syntheses involve the step-growth polymerization of difunctional oligomers or prepolymers, which can be regarded as the elementary units of the block copolymer. Because it is difficult to control the functionality of high-molecular weight compounds, the block molecular weight is usually much lower in step-polymerization block copolymers ($\overline{Mn}$ = 500 to 5000) than in those obtained by chain polymerization. On the other hand, step polymerization is a versatile method. It allows the synthesis of copolymers having a wide range of blocks and properties, using a wide range of chemical reactions in simple and economical conditions, such as bulk reactions or reactive processing.

SYNTHETIC METHODS AND STRUCTURAL REGULARITY

Starting from reactive difunctional oligomers, there are three possibilities for synthesizing block copolymers by polycondensation.

Method **a** is a reaction between oligomers bearing mutually reactive end groups (**Equation 1**).

Method **b** is a reaction between one oligomer and the monomers of the other block (**Equation 2**).

Method **c** consists of oligomer coupling reactions (**Equation 3**).

There are some differences in the structural regularity of polymers obtained by these methods.

Polydispersity

If the starting oligomers are monodisperse, the classical relations or step polymerization apply, leading to a final polydispersity of 2. If one or more of the starting oligomers are polydisperse, they can be regarded as mixtures of monomers with various molecular weights.

Block Length Distribution

The reaction between two oligomers, A and B (method **a**), yields an $(AB)_n$ poly(poly A-*block*-poly B) alternating block copolymer. The polydispersity of each block is that of the corresponding starting oligomer.

In method **b**, oligomer A is a comonomer in the step polymerization that builds both block B and the final copolymer. This copolymer is alternating because blocks A are connected through blocks B. Methods **a** and **b** should therefore give similar copolymers.

In the coupling reaction of functional oligomers (method **c**), somewhat different copolymers are obtained. The final copolymer is a statistical copolymer with a random distribution of block A, block B, and coupling agent moieties along the chains. Constitutional regularity is therefore lower than for copolymers obtained by methods A or B, especially when coupling agent moieties are very different from A or B constitutional units.

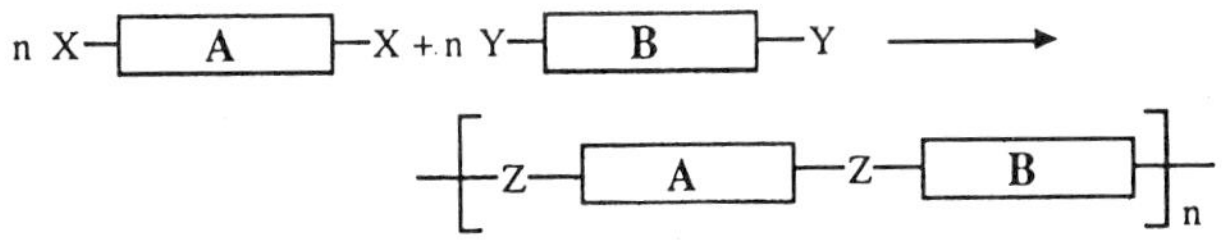

1

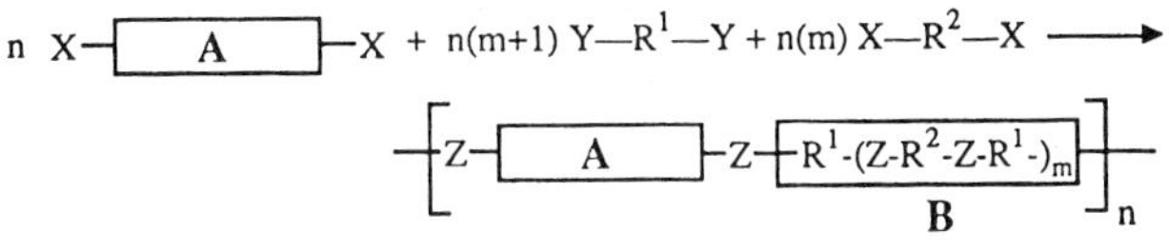

2

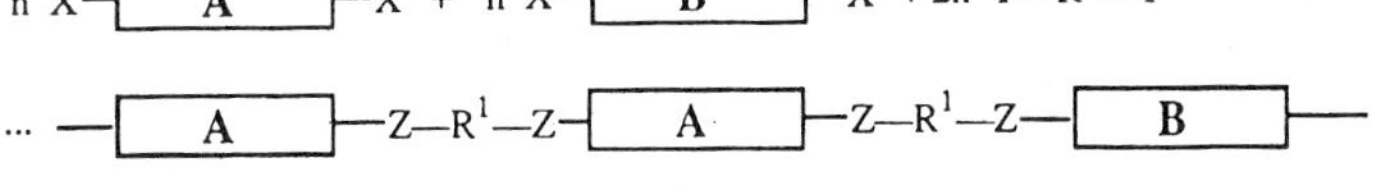

3

Two other methods allow the preparation of block copolymers by step-growth polymerization. Method **d** involves interchange or rearrangement of polymer chains, and method **e** is relative to the direct formation of block copolymers from mixture of monomers. These two methods are limited to a small number of polymers and monomers, and have consequently not been extensively used nor studied. The achievable structural regularity is strongly dependent on the system under consideration and on experimental conditions.

SYNTHESIS

Any classical technique of step polymerization (i.e., bulk, solution, and heterogeneous reactions) can be applied to the synthesis of block copolymers.

The most important step-polymerization reactions that can be carried out in bulk are polyesterification and polyamidation. At the high temperatures necessary (180–250°C), and especially in the presence of catalysts, these reactions are accompanied by interchange reactions between amide or ester groups in the chains.[10-12] Therefore, if two polyesters or two polyamides are reacted, random copolymers will be produced. The degree of randomization will depend on reaction conditions and catalyst. Moreover, at high temperature polymer chains may undergo side or degradation reactions. Bulk reactions are therefore restricted to cases where side reactions do not take place and where at least one block is resistant to interchange reactions.

Solution step polymerization is much more versatile, because one can use oligomers with very high melting points or oligomers sensitive to high-temperature interchange or side reactions. On the other hand, the use of a solvent increases the cost of the process and strongly limits the possible commercial applications. The same holds for interfacial processes.

Reaction Between Oligomers Bearing Mutually Reactive End Groups (Method a)

Because oligomers can be purified, fractionated, and characterized before being reacted, this method should lead to the best results in terms of molecular weight and structural regularity. However, the syntheses are often carried out as one-pot two-step reactions, one of the oligomers being synthesized just before the addition of the other one.

Reaction Between One Oligomer and the Monomers of the Other Block (Method b)

Reaction of an oligomer as a step-polymerization comonomer is probably the most versatile method for the synthesis of block copolymers. The polymerization can be carried out by conventional processes, with only minor modifications, and any type of step-growth polymerization may be used. A number of compounds have found commercial applications, such as polyether/polyester (Hytrel®, DuPont) or polyether/polyurethane (Lycra®, DuPont; Desmopan®, Bayer) block copolymers. Polyesterification, polyamidation, and polymerizations involving diisocyanates are the most important reactions.

Block Copolymers by Oligomer-Coupling Reactions (Method c)

The early uses of oligomer-coupling reactions were related to the synthesis of block copolymers that would otherwise undergo side or rearrangement reactions at high temperature. A lot of work has been done on low-temperature polyester/polyester coupling by diisocyanates.[3] This approach has recently been extended to high-temperature melt reactions. With a proper choice of the coupling agent, the advantages are a shorter reaction time, the absence of byproducts such as water or HCl, and an easier block ratio adjustment because stoichiometry between blocks is no longer necessary.

Synthesis of Block Copolymers by Rearrangement, Interchange, or Post-Polycondensation Reactions (Method d)

Oligomer–Polymer or Polymer–Polymer Reactions

The process involves a high-temperature bulk reaction between oligomer reactive end groups and polymer constitutional units, or a reaction between the constitutional units of two polymers. A typical example is the hydroxy/ester reaction between OH-terminated polyoxyethylene and a high-MW poly(ethylene terephthalate). In the first step, a copolymer with a lower MW is obtained. In the second step, reactions involving hydroxyethyl end groups lead to a MW increase with elimination of enthanediol and to the expected block copolymer.[13]

Direct Synthesis of Block Copolymers from Monomers (Method e)

The direct synthesis of block copolymers via nonequilibrium solution step polymerization is possible under some conditions of comonomer reactivities and rate of introduction.[14,15] Other references are relative to methods with sequential addition[16] or multistep unbalanced addition of comonomers.[17] Method e is actually intermediate between methods **b** and **c**. The control of block length is *a priori* difficult.

Block-like distributions were observed when thermotropic polymers were prepared from melt reactions of comonomer mixtures.[18] A similar phenomenon, more related to chemical modification than to step polymerization, has been reported in the cyclization of poly(amic acid) to polyimides.[19]

REFERENCES

1. Quirk, R. P. et al. In *Comprehensive Polymer Science* Allen, G.; Bevington, J. C. Eds. Pergamon: Oxford; 1989; Vol. 7, Chapter 1.
2. Allport, D. C.; Mojaher, A. A. In *Block Copolymers* Allport, D. C.; Janes, W. H., Eds.; Applied Science: London, 1973; 208.
3. Noshay, A.; McGarth, J. E. In *Block Copolymers, Overview and Critical Survey* Academic: New York, 1977; Chapter 7.
4. Valetskii, P. M.; Storozhuk, I. P. *Usp. Khim.* **1979**, *48*, 75.
5. Goodman, I. In *Comprehensive Polymer Science* Ed. Allen, G.; Bevington, J. C. Pergamon: Oxford; 1989; Vol. 6, Chapter 11.
6. Nguyen, H. A.; Maréchal, E. *J. Macromol. Sci., Rev. Macromol. Chem. Phys.* **1988**, *C28*, 187.
7. Gibson, P. E. et al. In *Developments in Block Copolymers* Goodman, I. Ed. Applied Science: London, 1982; Chapter 6.
8. Riess, G. et al. *Block Copolymers in Encyclopedia of Polymer Science and Engineering* John Wiley & Sons: New York, 1985; Vo. 2, p 324.
9. Bosch, P.; Mateo, J. L. *Rev. Plast. Mod.* **1993**, *65*, 361.
10. Gaymans, R. J.; Sikkema, D. J. In "*Comprehensive Polymer Science*" Allen, G.; Bevington, J. C., Eds. Pergamon: Oxford, 1989; Vol. 5, Chapter 21.
11. Pilati, F. In "*Comprehensive Polymer Science*" Allen, G.; Bevington, J. C., Eds. Pergamon: Oxford, 1989; Vol. 5, Chapter 17.
12. Fradet, A.; Maréchal, E. *Eur. Polym.* **1978**, *14*, 755.

13. Konoo, Y. et al. (Kaneguchi Chemical Industry, Co.), *Jpn. Kobai Tokkyo Koho* JP 86-313140, **1986**, *Chem. Abstr.* **1988**, *109*, 231760.
14. Vasnev, V. A.; Kuchanov, S. I. *Usp. Khim.* **1973**, *42*, 2194.
15. Korshak, V. V. et al. *J. Polym. Sci. Polym. Chem. Ed.* **1973**, *11*, 2209.
16. Curnuk, P. A.; Jones, M. E. B. *Br. Polym. J.* **1973**, *5*, 21.
17. Jho, J. Y.; Yee, A. F. *Macromolecules* **1991**, *24*, 1590.
18. Sato, M. et al. *Makromol. Chem.*, **1991**, *192*, 1139.
19. Vygodskii, Ya. S. et al. *Acta Polym.* **1982**, *33*, 131.

BLOCK COPOLYMERS (From Macroinitiators)

Munmaya K. Mishra
Polymers Research
Texaco Research and Development

Yusuf Yağci
Department of Chemistry
Istanbul Technical University

Designing block copolymers via the macroinitiator approach can be a fascinating and challenging pathway. "Macroinitiator" is an abbreviation of "*macro*molecular *initiator*" and is based on the functional feature of the macromolecule. The selected functional groups of the macromolecule can be activated chemically, thermally, mechanochemically, photochemically, and so forth to yield active centers that initiate polymerization of other monomers to produce block and graft copolymers. The field of macroinitiators was reviewed by several authors in a recent book.[1]

Macroinitiators are precursors for the synthesis of block and graft copolymers depending on the site of initiation, that is, at the chain end or in the side chain, respectively.

BLOCK COPOLYMERS (FROM MACROINITIATORS) VIA RADICAL POLYMERIZATION

The preparation of block copolymers by radical techniques has the advantages of a wide choice of monomer combinations and a low sensitivity toward impurities. However, various transfer as well as termination reactions may lead to homopolymerization, which generally proceeds in parallel with block copolymerization. Block copolymers resulting from free-radical macroinitiators have an overall structure that is dependent on the mode of termination, that is, disproportionation and/or combination.

Despite these problems, the synthesis of block copolymers via radical techniques is one of the promising methods. Block copolymers can be prepared from various types of macroinitiators and can be classified[2] by the mode of radical generation, that is, azo macroinitiators, peroxy macroinitiators, redox macroinitiators, macrophotoinitiators, mechanochemical macroinitiators, coordination-type macroinitiators, and so on.

Block Copolymers from Macroazoinitiators

The field of polymeric azoinitiators was reviewed in detail[3,4] in a recent book. The use of polymeric azoinitiators is probably the best known and the oldest method for producing block copolymers.

Numerous examples of block copolymers from macroazoinitiators have been reported in the literature. Most common is the combination of a polyester or poly(ethylene glycol) prepolymers with styrene,[13-17] methyl methacrylate (MMA),[6,10,11] vinyl chloride,[12] or vinylpyrrolidone.[11]

Block Copolymers from Macroperoxyinitiators

A recent review[3] covers the details of macroperoxyinitiators.

Block Copolymers by Redox Macroinitiators

Polymerizations of vinyl monomers initiated by the reaction between an oxidizing and a reducing agent are called redox polymerizations.[13-18] When a polymeric substrate is used as a redox system in conjunction with an oxidant (i.e., peroxides, persulfates, peroxydiphosphates, the salts of transition metals) for the polymerization of vinyl monomers to form block or graft copolymers, the radical generated on the polymer substrate acts as a macroinitiator.

Block Copolymers from Macrophotoinitiators

Macrophotoinitiators may offer various advantages over the low-molecular weight photoinitiators, including greater reactivity, low volatility, and less migration. Macrophotoinitiators are used as either photocrosslinking agents or precursors for block and graft copolymers. Compared with most of the thermally induced macroinitiators, radiation-sensitive macroinitiators have the advantage of being applicable at low temperatures, which may reduce side reactions or, in some cases, homopolymerization. Moreover, because of the selective absorptivity of certain groups (chromophores), it is possible to produce reactive sites at definite positions in the macromolecule.

Block Copolymers by "Mechanochemical Macroinitiators"

"Mechanochemical macroinitiators" are formed from the polymers that produce macroradicals when subjected to mechanical treatment. In principle, most of the synthetic and natural polymers can act as initiators under mechanochemical influence to yield, block and graft copolymers. Mechanochemical scission may occur at any position of the polymer. It is frequently difficult to produce pure block copolymers. The end products consist of a mixture of block, graft, and homopolymers.

We recently reviewed this field.[2] For further information, refer to the specialized books.[19,20]

BLOCK COPOLYMERS (FROM MACROINITIATORS) VIA CATIONIC POLYMERIZATION

The understanding and controlling of elementary steps of carbocationic polymerization has helped to develop methods to design unique and useful block or graft copolymers using the macroinitiator concept. The design of block/graft copolymers using the macroinitiator concepts may be divided into two categories: by controlled initiation, and by sequential living polymerization.

BLOCK COPOLYMERS (FROM MACROINITIATORS) VIA ANIONIC POLYMERIZATION

Obviously, sequential monomer addition, in which monomers are added in the order of increasing electroaffinity, is the most widely used method to prepare block copolymers via anionic living polymerization. The reader's attention is directed to previous reviews devoted solely to anionic polymerization.[21-24]

BLOCK COPOLYMERS (FROM MACROINITIATORS) VIA GROUP TRANSFER POLYMERIZATION

Group transfer polymerization (GTP) involves generation of keten acetal at the chain end at each monomer unit addition.[25-27] Preparation of block copolymers by GTP is restricted to methacrylate-type monomers. Block copolymers can be prepared by using monomers with appropriate reactivity ratios in a batch process at sequential monomer addition technique. A wide range of block copolymers with AB- or ABA-type structures, using mono and difunctional initiators, respectively, were prepared and reviewed elsewhere.[28-31] In principle, any polymer, independent of polymerization mode, possessing terminal or side-chain silyl keten acetal groups would act as macroinitiators for GTP.

REFERENCES

1. *Macromolecular Design: Concept and Practice*; Mishra, M. K., Ed.; Polymer Frontiers International: Hopewell Junction, NY, 1994.
2. Yagci, Y.; In *Macromolecular Design: Concept and Practice*; Mishra, M. K., Ed.; Polymer Frontiers International: Hopewell Junction, NY, 1994; Chapters 6, 10.
3. Ueda, A.; Nagai, S. In *Macromolecular Design: Concept and Practice*; Mishra, M. K., Ed.; Polymer Frontiers International: Hopewell Junction, NY, 1994; Chapter 7.
4. Nuyken, O.; Voit, B. In *Macromolecular Design: Concept and Practice*; Mishra, M. K., Ed.; Polymer Frontiers International: Hopewell Junction, NY, 1994; Chapter 8.
5. Furukawa, J.; Takamori, S.; Yamashita, S. *Angew Makromol. Chem.* **1967**, *1*, 92.
6. Ueda, A. Nagai, S. *J. Polym. Sci., Polym. Chem.* **1986**, *24*, 405; **1987**, *25*, 3495.
7. Oppenheimer, C.; Heitz, W. *Angew. Makromol. Chem.* **1981**, *98*, 167.
8. Walz, R.; Heitz, W. *J. Polym. Sci., Polym. Chem.* **1978**, *16*, 1897.
9. Hazer, B. *Makromol. Chem.* **1992**, *193*, 1081.
10. Matsukawa, K.; Ueda, A.; Inoue, H. *J. Polym. Sci., Polym. Chem.* **1990**, *28*, 2107.
11. Heitz, W.; Stahl, H.-G.; Dicke, R. Ger. Offen., 3 005 889, Sept. 3, 1981.
12. Laverty, J.; Gardlund, Z. *Polym. Prep.*, American Chemical Society, **1974**, *15*, 306.
13. Mishra, M. K. *J. Macromol. Sci.-Rev. Macromol. Chem.* **1980**, *C19*(2), 193.
14. Mishra, M. K. *J. Macromol. Sci.-Rev. Macromol. Chem.* **1982-83**, *C22*(3), 471.
15. Misra, G. S.; Bajpai, U. D. N. *Prog. Polym. Sci.* **1982**, *8*, 61.
16. Nayak, P. L.; Lenka, S. *J. Macromol. Sci.-Rev. Macromol. Chem.* **1980**, *C19*(1), 83.
17. Samal, R. K.; Sahoo, P. K.; Samantaray, H. S. *J. Macromol. Sci.-Rev. Macromol. Chem.* **1986**, *C26*(1), 81.
18. Mukherjee, A. K.; Goel, H. R. *J. Macromol. Sci.-Rev. Macromol. Chem.* **1985**, *C25*(1), 99.
19. Baramboim, N. K. In *Mechanochemistry of Polymers*, Rubber and Plastic Res Association of Great Britain: Maclaren, 1964.
20. Castle, A.; Porter, R. S. *Polymer Stress Reactions*; Academy: New York, 1978.
21. Rempp, P.; Franta, E.; Herz, J. E. *Adv. Polym. Sci.* **1988**, *86*, 145.
22. Penczek, S.; Kubisa, P.; Matyjaszewski, K. *Adv. Polym. Sci.* **1980**, *37*, 1.
23. Doi, Y.; Keii, T. *Adv. Polym. Sci.* **1986**, *73/74*, 201.
24. Morton, M. *Anionic Polymerization: Principles and Practice*; Academic: New York, 1983.
25. Webster, O. W.; Hertler, W. R.; Sogah, D. Y.; Farnham, W. B.; Rajan Babu, T. V. *J. Am. Chem. Soc.* **1983**, *105*, 5706.
26. Dicker, I. B.; Cohen, G. M.; Farnham, W. B.; Hertler, W. R.; Laganis, E. D.; Sogah, D. Y. *Macromolecules* **1990**, *23*, 4034.
27. Hertler, W. R.; Sogah, D. Y.; Webster, O. W.; Trost, B. M. *Macromolecules* **1984**, *17*, 1415.
28. Webster, O. W. In *Encyclopedia of Polymer Science and Engineering*, Wiley-Interscience: New York, 1987; Vol. 7, p 580.
29. Webster, O. W. *Makromol. Chem. Macromol. Symp.* **1990**, *33*, 133.
30. Eastmond, G. C.; Webster, O. W. In *New Methods of Polymer Synthesis*: Blackie: Glasgow, U.K., 1991; p. 22.
31. Brittain, W. J. *J. Rubber Chem. Tech.* **1992**, *64*, 580.

BLOCK COPOLYMERS (Micellization in Solution)

José R. Quintana, Manuel Villacampa, and Issa Katime
Universidad del País Vasco

When a block copolymer is dissolved in a liquid that is a thermodynamically good solvent for one block and a precipitant for the other, the copolymer chains may associate reversibly to form stable aggregates. These kinds of liquids are called selective solvents. The copolymer aggregates are called micelles because they resemble in all essential aspects the well-known micellar aggregates formed by low-molecular mass surfactants. The micelles formed are constituted by a relatively compact core of insoluble blocks surrounded by a higher swollen shell of soluble blocks. In most cases, block copolymer micelles have a spherical shape and a narrow distribution of both mass and size. Similar to conventional surfactants, micellization of block copolymers obeys the closed association model.

Several reviews of experimental studies of block copolymers in solution have been published in recent years.[1-3]

MICELLAR SIZE DISTRIBUTION

Size exclusion chromatography (SEC) is the technique most used to study the size distribution of micellar solutions.[4-7]

For copolymer solutions in which the equilibrium micelle—molecule overwhelmingly favors micelle formation, the SEC curves show a single, narrow, and symmetrical peak that corresponds to monodisperse copolymer micelles.

The essentially uniform character of copolymer micelles has also been verified by dynamic light scattering (DLS). When the critical micelle concentration is very low compared to the concentration of the copolymer solution (i.e., the micelles are the predominant species in the solution), the autocorrelation function can truly be represented by a single exponential decay function.[8]

Although in most cases copolymer micelles are monodisperse, copolymer solutions containing more than one kind of multimolecular particles have been reported.

MICELLIZATION EQUILIBRIUM

The essentially uniform character of micelles and the constant value of the micelle molar mass at different copolymer concentrations justify the applicability of the closed association model for the block copolymer association.[9]

The existence of a single-stage equilibrium implies that the micellar solutions show in general a bimodal size distribution, since both copolymer molecules and monodisperse micelles coexist in solution.

In the case of closed association there is a concentration at which some given properties of the system show a sharp change. This concentration, called critical micelle concentration (CMC), is the one at which micelles can be experimentally detected by the technique used in the study. Therefore CMC will depend on the experimental technique used.

DYNAMICS OF THE MICELLIZATION EQUILIBRIUM

The equilibrium micelle-molecule represents a dynamic process, in which the copolymer molecules migrate at a given rate between unimer and micelle states. The dynamics of this equilibrium have been analyzed by mixing two kinds of micelles with different molar masses and sizes and monitoring the appearance of mixed micelles.

KINETICS OF MICELLIZATION AND DISSOCIATION

Bednar et al.[10] have studied the rates of association and dissociation of diblock and triblock copolymer micelles by light-scattering stopped-flow methods. A micelle formation was obtained by mixing a copolymer solution in a good solvent with a precipitant, so that the solvent mixture became a selective solvent. Micelle dissociation was carried out by mixing a micellar solution with a good solvent of both copolymer blocks, so that the solvent mixture became a good solvent for both copolymer blocks.

MICELLAR STRUCTURAL PARAMETERS

Static light scattering has been the most widely used technique in the study of block copolymer micelles. When the critical micelle concentration is very low compared to the concentrations of the dilute solutions used in the measurements, the influence of the equilibrium micelle-molecule on the experimental results can be neglected. In such a case, the determination of the Rayleight ratio over a range of concentrations and scattering angles provides information about the molar mass, size, and shape of micelles.

The internal structure of micelles can be studied by small-angle X-ray scattering (SAXS). The core radius can be determined from the position of the side maximum on the SAXS scattering curve. Once one knows the radius of gyration of the micelle, the radius of the micelle core, and the scattering contrasts of both micelle core and shell with respect to the solvent, the micelle geometrical radius and the thickness of the shell can be calculated.

The copolymer micelles show spherical shape in general. However, there exist some copolymer micellar solutions in which worm-like micelles have been detected.[11,12] The worm-like micelles had a wide distribution of contour lengths, but possessed a narrow size distribution in the radial direction.

REFERENCES

1. Tuzar, Z.; Kratochvíl, P. *Adv. Colloid Interface Sci.* **1976**, *6*, 201.
2. Brown, R. A. et al. *Comprehensive Polymer Science*; Pergamon: Oxford, 1989; Vol. 2, Chapter 6.
3. Quintana, J. R.; Villacampa, M.; Katime, I. *Rev. Iberoamericana de Polimeros* **1992**, *1*, 5.
4. Booth, C.; Naylor, T. D.; Price, C. *J. Chem. Soc. Faraday Trans. 1* **1978**, *74*, 2352.
5. Spacek, P.; Kubin, M. *J. Appl. Polym. Sci.* **1985**, *30*, 143.
6. Wang, Q.; Price, C.; Booth, C. *J. Chem. Soc. Faraday Trans.* **1992**, *88*, 1437.
7. Katime, I.; Villacampa, M.; Quintana, J. R. *Macromol. Symp.* **1994**, *84*, 255.
8. Yeung, A. S.; Frank, C. W. *Polymer* **1990**, *31*, 2089.
9. Elias, H.-G. *Light Scattering from Polymer Solutions*; Academic: London, 1972; Chapter 9.
10. Bednar, B. et al. *Makromol. Chem. Rapid Comm.* **1988**, *9*, 785.
11. Canham, P. et al. *J. Chem. Soc. Faraday Trans. 1* **1980**, *76*, 1857.
12. Price, C. et al. *Polymer Comm.* **1986**, *27*, 196.

BLOCK COPOLYMERS (Ordered Bicontinuous Double-Diamond Morphology)

Rui Xie,* Guifeng Sun, and Bingzheng Jiang
Changchun Institute of Applied Chemistry
Chinese Academy of Sciences

Block copolymers, for example, AB and ABA, consist of two or three large segments chemically bonded together. Because of the immiscibility between different segments, such materials may separate into distinct microphase, creating microdomains of components A and B. The ordered bicontinuous double-diamond (OBDD) morphology is a new microdomain structure observed in block copolymers,[1-3] their blends with homoploymers,[4-6] and two-block copolymer blends.[3]

The OBDD morphology was first reported in 1976 by Aggarwal[7] in a polystyrene–polyisoprene (SI) star block copolymer; it was referred to as the "wagon-wheel" image. In the OBDD structure, one phase (e.g., PI) resides in two interwoven but distinct "tetrapod" networks, each exhibiting diamond cubic symmetry. The other phase (e.g., PS) resides in the continuous matrix between the two diamond channels. Combining the results of small-angle X-ray scattering with those of the systematic tilt series of the electron micrographs, Thomas et al.[1] found that the tetrapods are located in a double-diamond lattice of cubic lattice symmetry. Thus, they named the wagon-wheel bicontinuous microdomain structure the ordered bicontinuous double-diamond, on the basis of its symmetry.

Great efforts have been focused on generating the OBDD structure in an easy way. The structure exists in narrow composition regimes, for example, within the composition ranges of 28–33 vol%. PS and 62–66 vol% PS in linear SI diblock copolymers. Such narrow composition requires precise control of the copolymer synthesis, which has hindered the study of the microstructure. Furthermore, the Thomas[1] and Hasegawa[3] groups both made the OBDD structure in star of diblock copolymers of

*Author to whom all correspondence should be addressed.

narrow molecular weight distribution; this stringent constraint hampers the potential uses of the structure in industry. Xie et al.[4] successfully obtained the OBDD morphology in the blends of triblock copolymer polystyrene–polybutadiene–polystyrene (SBS) and poly(vinyl methyl ether) (PVME) over a relatively wide composition range (30–38 vol% of PS and PVME). As the two polymers are commercial products, of broad molecular weight distribution, it seems that the system possesses advantages compared with those studied by other groups.

PREPARATION

Until now, the OBDD morphology were all observed in solid films prepared by slowly casting from solutions in a nonpreferential solvent for the components in the block copolymers. A selective solvent for one component has been found to have effect on the microdomain morphology.[3] Such nonequilibrium effect can be overcome by annealing the specimen at a temperature (e.g., 120°C for the SI star block copolymer)[1] for a period of about 7 d under high vacuum.

The dependence of the OBDD structure on the molecular weight of block copolymers and homopolymers is a prime consideration in the production of the structure.

The composition dependence of the OBDD morphology is a critical factor in producing the morphology.

PROPERTIES

Symmetry

The basic unit of the OBDD structure is a tetrahedral arrangement of short rods, referred to as a "tetrapod." Such units are interconnected on a cubic lattice that has the symmetry of the *Pn3m* space group. The resultant structure consists of two translationally displaced, mutually interwoven but unconnected 3-D networks in which the minor component rods are embedded in the major component matrix. Each of these separated networks exhibits the symmetry of a diamond cubic lattice, as schematically shown in **Figure 1**. Therefore, as the structure projects in different directions, it will produce various 2-D images that possess different symmetries.

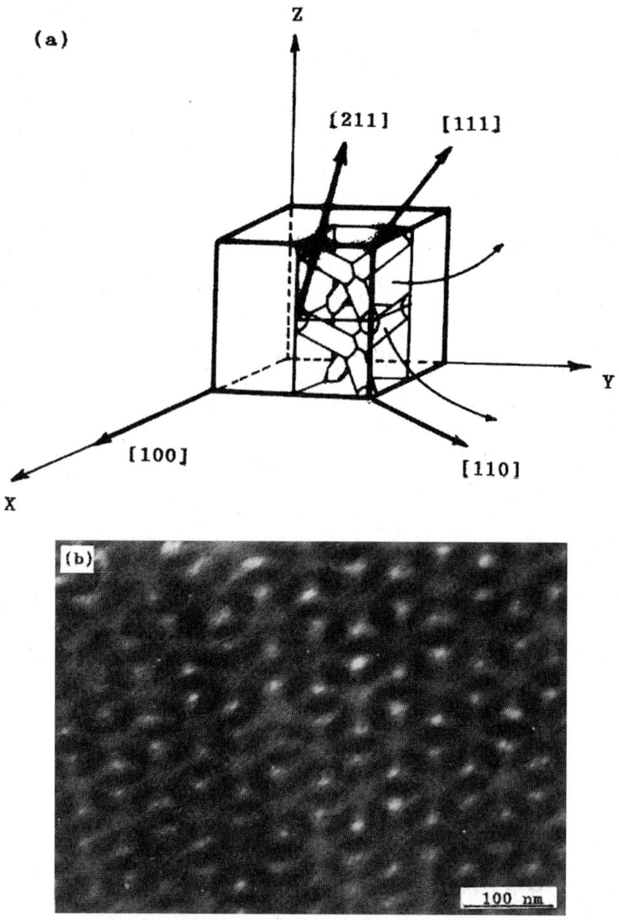

FIGURE 1. (a) Schematic of the OBDD structure, showing the arrangement of the basic tetrahedral units. These interconnect to form two interwoven but independent diamond networks. (b) Electron micrograph of the OBDD structure in a SBS/PVME blend containing 15 wt % of PVME. Bicontinuous structure is visually proved; 3- and 6-fold rotational symmetry are evident.

APPLICATIONS

As an ordered microcomposite, there are many potential industrial uses of the OBDD structure. However, exploration of this aspect has not been reported. Applications may arise from the fact that both components in OBDD morphology are continuous, for example, separation devices, membranes, and self-assembled composites. An A-conducting/B-insulating diblock copolymer with OBDD morphology would provide a good electrical transport with a relatively minor amount of conducting block. Most important, by means of changing the properties of the channels and the matrix (e.g., alternating the two components chemically), one could produce high-selectivity membranes, which is of great importance in the life sciences and a host of other fields.

In summary, the OBDD structure is intriguing and complicated, and the potential industrial uses of materials with such structure are very exciting. That is why investigations are active. In the near future we may find new properties and applications of the structure, but those are beyond the scope of the present paper.

ACKNOWLEDGMENT

The financial support of the "National Basic Research Project-Macromolecular Condensed State" is greatly appreciated.

REFERENCES

1. Thomas, E. L.; Alward, D. B.; Kinning, D. J.; Martin, D. C.; Handlin, D. L. Jr.; Fetters, L. J. *Macromolecules* **1986**, *19*, 2197.
2. Kinning, D. J.; Alward, D. B., Fetters, L. J.; Handlin, D. L. Jr. *Macromolecules* **1986**, *19*, 1288.
3. Hasegawa, H.; Tanaka, H.; Yamasaki, K.; Hashimoto, T. *Macromolecules* **1987**, *20*, 1651.
4. Xie, R.; Yang, B. X.; Jiang, B. Z. *Macromolecules* **1993**, *26*, 7097.
5. Winey, K. I.; Thomas E. L.; Fetters, L. J. *Macromolecules* **1992**, *25*, 422.
6. Spontak, R. J.; Smith, S. D.; Ashraf, A. *Macromolecules* **1993a**, *26*, 956.
7. Aggarwal, S. L. *Polymer* **1976**, *17*, 938.

Blocked Polyurethane Copolymers

See: Polyurethanes, Blocked Copolymer (Reactive Modifiers for Epoxy Resins)

Blood

See: Blood Compatible Polymers
Blood Substitutes
Cell Separation Materials
Red Blood Cell Substitutes, Artificial (for Transfusion)

BLOOD COMPATIBLE POLYMERS

Young Ha Kim, Ki Dong Park, and Dong Keun Han
Polymer Chemistry Laboratory
Korea Institute of Science and Technology

Polymer materials are extending their applications to biomedical uses. In addition to packaging for medical devices and instruments, polymers are widely used as medical devices and instruments, artificial organs and implants, and in drug formulations and delivery systems. Such materials are called biomaterials and are categorized as metals, ceramics, polymers, or composites.

Practical polymer applications include artificial kidneys, membrane oxygenators, and blood vessels. When polymers are used in medical devices contacting blood, coagulation is a serious problem. The blood compatibility of polymers differs depending on their chemical composition and morphology. Despite an enormous amount of research, blood-compatible polymers are still a long way off. That is because the mechanism of thrombus formation is too complicated to find correlations between blood compatibility and material structure.

BLOOD–MATERIAL INTERACTION

Rationale: Antithrombogenicity Vs. Pseudoneointima

There have been many approaches to preventing or decreasing activation of the thrombogenic pathways by tailoring the physicochemical properties of the polymer either to minimize thrombus formation or to selectively adsorb a passivating albumin layer. These approaches include smooth surfaces, hydrophilic surfaces, microdomained structures, inert surfaces, and incorporation of negative ions. In addition, pharmacologically active agents may be incorporated into the polymer to yield a slow-release system or immobilized onto a surface. From these researches we have made great progress in producing chronically nonthrombogenic cardiovascular devices. However, it appears to be impossible to produce long-lasting, completely nonthrombogenic material. Therefore, the other approaches consist of using rough surfaces, such as velours, flocked, or integrally textured surfaces, to encourage the formation of a living biological lining derived from blood itself.

BLOOD COMPATIBILITY OF POLYMERS

Medical Polymers

Blood compatibility is one of the most important properties of biomedical polymers. In general, blood-compatible polymers should possess two characteristics: they should not induce thrombus formation, immune response, inflammatory reaction, or infection; and they must be nontoxic, noncarcinogenic, and nonmutagenic.[1] Despite the several *in vitro* and *ex vivo/in vivo* methods reported, there is no standard method for evaluating blood compatibility.

Many polymers from natural rubber and cellulose to synthetic elastomers, polyurethanes (PUs), and hydrogel have been used in biomedical applications ranging from disposable syringes to materials for artificial organs.[2] Polymers currently being used are categorized according to their characteristics and end-use applications as follows: synthetic nondegradable polymers that were used in most long-term implantable devices and disposables, bioadsorbable or soluble polymers that were used as temporary scaffolding and barrier and drug delivery matrices, and experimental polymers.[3]

Cardiovascular Application of Polymers

Blood compatibility is of primary importance for those polymers that interact with blood. The polymers used in cardiovascular systems such as blood-contacting devices include a wide variety of commercial materials. There are three kinds of cardiovascular applications:[4] replacements that are permanently implanted in the circulatory system (e.g., artificial hearts, heart valves, and vascular grafts); devices that are inserted into a blood vessel for varying periods of time (e.g., catheters, sensors, fibroscopes, and other imaging agents); and extracorporeal devices that remove and return blood from the body (e.g., blood oxygenators, hemodialysis units [so-called artificial kidneys], cardiopulmonary bypass, and liver perfusion systems).

DESIGN OF BLOOD-COMPATIBLE POLYMERS

Current approaches to blood-compatible polymers can be divided into three categories: new polymer synthesis, surface modification of existing polymers, and biological approaches.

Tailoring New Polymers

One approach to minimizing thrombus formation is to synthesize nonthrombogenic polymers. These nonthrombogenic polymers prevent activation of the thrombogenic pathway by tailoring the polymer surface to minimize blood interaction.[5]

For many years, microdomain-structured polymers have received much attention for their improved blood compatibility. Several research groups have synthesized and evaluated polymers with microheterogeneous surfaces. Okano et al.[6,7,9] and Shimada et al.[8] reported that block copolymers having hydrophilic–hydrophobic microdomain structures show antithrombogenicity both *in vivo* and *ex vivo/in vivo*.

Other polymer systems with microdomain structure include segmented polyurethanes (SPUs), which are widely used for medical applications because of their excellent physical and mechanical properties and relatively good blood compatibility. SPUs are segmented heterophase elastomers that exhibit a variety of physical and chemical properties depending on their synthetic conditions.

The surface modification of polymeric materials has been recently reviewed by Ikada.[10]

Surface Modification of Polymers

Another approach of considerable interest is to modify surfaces of existing polymers without changing bulk properties. The modifications are based on factors such as surface free energy, protein adsorption, platelet adhesion, and other blood coagulation factors. Modification techniques have been designed to achieve increased hydrophilicity, chemical modification, and attachment of pharmacologically active agents. Surface modification techniques may be divided into two categories: physicochemical and biological.

Negatively charged polymer surfaces have proved effective in thrombus formation. This is because cellular materials such as platelets are also negatively charged and therefore electrostatically repelled from the negatively charged surfaces.

In our laboratory, novel surface modification of sulfonated polyurethane has been developed to improve blood compatibility, biostability, and anticalcification.

Hypotheses conceived with this research state that hydrophilic PEO chains are expected to reduce protein adsorption and platelet adhesion because of the unique behaviors of PEO and because the pendent negatively charged sulfonate group provides anticoagulant activity, resulting in prevention of fibrin net formation. This "negative cilia surface" should curtail surface-induced thrombus formation because of the synergistic effect of hydrophilic PEO and the pendant negative charge of sulfonate groups. Sulfonated PEO-grafted PUs (PU-PEO-SO_3) have been prepared by a direct surface grafting of PU medical devices or by a solution reaction that can be applied as a coating material. The proposed hypotheses have been proven in *in vitro*, *ex vivo*, and *in vivo* studies.[11-14] In addition, researchers demonstrated reduction in surface crack formation and calcium deposition *in vivo*.[15] The results (improved blood compatibility, biostability, and anticalcification) attest to the usefulness of this negative cilia model for the design of blood-contacting medical devices.

Another approach is to design polymers of selectively increased surface affinity toward albumin. Albumin has been shown to adsorb to surfaces and passivate or neutralize surfaces to further protein adsorption (albumin hypothesis).

The physical coating techniques use relatively nonthrombogenic polymers over existing blood-contacting devices or substrates without changing bulk properties.[16,17]

Biological Approach

Biological surface modification techniques involve preadsorption or protein,[18,19] cell seeding,[20,21] and release or immobilization of antithrombotic agents to create biologically inert surfaces.

In summary, a variety of surface modification techniques are being used to improve the blood compatibility of polymeric surfaces. However, many current blood-compatible polymers and devices need improvement, and there are many unfilled needs for new uses of polymeric biomaterials. Understanding the relationship between blood and material would promote better design of new polymers and use of existing polymers for surface modification, and finally the development of improved blood-compatible polymers.

REFERENCES

1. von Recum, A. F. *Handbook of Biomaterials Evaluation: Scientific, Technical, and Clinical Testing of Implant Materials*; Macmillan, New York, 1986.
2. Szycher, M. *High-Performance Biomaterials*; Technomic: Lancaster, PA, 1991.
3. Barenberg, S. A. *J. Biomed. Mater. Res.* **1988**, *22*, 1267.
4. Helmus, M. N.; Hubbell, J. A. *Cardiovasc. Pathol.* **1993**, 2(3), 53S.
5. Cooper, S. L. et al. *The Physics and Chemistry of Protein-Surface Interaction*; Marcel-Dekker: New York, 1981.
6. Okano, T. et al. *Polym.* **1978**, *10*, 233.
7. Okano, T. et al. *Biomed Mater. Res.* **1981**, *15*, 393.
8. Shimada, M. et al. *Polym.* **1983**, *15*, 649.
9. Okano, T. et al. *Biomed Mater. Res.* **1986**, *20*, 919.
10. Ikada, K. *Biomaterials* **1994**, *15*, 725.
11. Han, D. K. et al. *J. Biomed Mater. Res.* **1991**, *25*, 561.
12. Han, D. K. et al. *J. Biomater. Sci. Polym. Ed.* **1993**, *4*, 401.
13. Han, D. K. et al. *ASAIO J.* **1993**, *39*, 537.
14. Han, D. K. et al. *Biomaterials* **1995**, *16*, 467.
15. Han, D. K. et al. *J. Biomed Mater. Res.* **1993**, *27*, 1063.
16. Okano, T. et al. *Artificial Heart II*; Springer Verlag: Tokyo, 1988.
17. Okano, T. et al. *Proc. Artif. Org.* **1987**, 863.
18. Kim, S. W. et al. *Trans. ASAIO* **1974**, *20*, 449.
19. Plate, N. A.; Matrosovich, M. M. *Akad. Nauk (USSR)* **1976**, *220*, 496.
20. Burkel, W. E. et al. *Ann. N. Y. Acad. Sci.* **1986**, *516*, 131.
21. Stanley, J. C. et al. *Acta. Chir. Scand. Suppl.* **1985**, *529*, 17.

BLOOD SUBSTITUTES

Eckhard Nordmeier
Department of Physical Chemistry
University of Osnabrück

In most instances, replacement of blood serves two important purposes. First, it serves to maintain the volume of the circulation. Second, like blood itself, the blood substitute needs to be able to carry oxygen from the lungs to the cells and tissues of the body, so as to maintain the functions of the various organs. Other desirable characteristics of a blood substitute include the possibility of sterilization and administration without need of blood typing.

The main approaches to red blood cell substitutes are stroma-free hemoglobin (SF-Hb) solutions, synthetic organic materials such as perfluorocarbons, water-soluble covalent Hb-polymer conjugates, and artificial oxygen carriers composed of porphinato–iron complexes.

PROPERTIES AND APPLICATION

Stroma-Free Hemoglobin

SF-Hb solutions embody many attributes of an ideal blood substitute.[1] The advantages compared with blood are

- Blood typing is not needed. Hb is not responsible for blood typing. Blood types are determined by certain antigens present in the red blood cells and certain natural antibodies present in the blood serum.
- Hb can be stored much more easily than blood. It does not deteriorate as quickly as blood. Stocks of blood have

to be discarded after a short period of time. Hb can be easily and purely isolated from blood. It can be freeze dried and stored for long periods.

- SF-Hb solutions possess the colloidal osmotic pressure, oxygen transport, and exchange properties of blood.

However, there are also some disadvantages:

- SF-Hb is rapidly excreted from the kidneys as urine.[2] Massive transfusions of SF-Hb solutions must therefore be used.
- There is another problem: 2,3-diphosphoglycerate (2,3-DPG) is present in red blood cells. Its binding to Hb results in high P_{50} and low oxygen affinity. In this form Hb readily releases oxygen to the tissues.

Perfluorocarbons

An entirely different approach for blood substitutes is achieved through the use of perfluorochemicals, such as perfluorotributylamine, perfluorodecalin, and perfluorocyclicether FX-80. They can be prepared in stable emulsions that have a high capacity for delivering oxygen. Although these compounds appear to be free from short-term side effects other than incidences of lung lesions and thrombocytopenia, their long-term toxicity is unknown. Also, the compounds are difficult to synthesize and purify.

Hemoglobin–Polymer Conjugates

Unfortunately, Hb has a short *in vivo* half-life in plasma because of rapid urinary excretion, insufficient intravascular persistence, and significant transcapillary leakage. To overcome these difficulties, Hb is coupled to polymeric materials.

Hemoglobin Encapsulation

Various methods have been proposed for encapsulation of Hb in liposomes. They are reverse-phase evaporation,[3] hand-shaken method,[4] detergent dialysis,[5] filter extrusion,[6] cell disruption bomb,[7] and French-press extrusion.[8]

A major problem is that artificial blood cells stay in the circulation for only a short time after intravenous injection. Attempts to solve this problem include changing cell diameters, membrane compositions, and surface charges. The most recent promising result comes from the use of lipid membranes to form cells of submicron size.[9] They survive in the circulation for relatively long periods of time.

Soluble Polymer-Bound Hemoglobin Conjugates

Many compounds between Hb and polymeric carriers have been synthesized. These include carriers such as hydroxyethylstarch,[10] inulin,[11] polyvinylpyrrolidone,[12] and poly(ethylene glycol).[13] Of most interest are soluble extraneous carriers in the form of dextran–Hb complexes. Dextrans are nonantigenic and therefore are widely used in the preparation of plasma expanders.[14]

Recently, we have examined dextran–Hb complexes.[15] The results obtained agree quite well with those reached by other groups.

In conclusion, the therapeutic requirements for a blood substitute are that it should be capable of carrying oxygen, reducing viscosity of blood through microcirculation, and maintaining osmotic pressure. Hb satisfies the first two requirements, but not the third. Coupling of Hb to dextran or dextran derivates produces complexes with prolonged *in vivo* half-life that satisfy all three basic requirements. In addition, dextran-Hb conjugates can be stored easily, are nontoxic, and do not need cross-matching prior to administration. More extensive investigations concerning the proper molecular structure of the starting dextran derivate and the hematological properties of the complex are necessary. However, it seems that dextran–Hb conjugates can be erythrocyte substitutes.

OXYGEN CARRIER COMPOSED OF PORPHINATO–IRON COMPLEXES

Much research has been directed to mimicking the function of Hb by synthesizing modified porphinato–iron derivatives.[16,17] These artificial oxygen carriers have been successful in organic solvents and solid states, but under physiological conditions they have been irreversibly oxidized.

In summary, reversible oxygen binding of a porphinato–iron complex can be achieved in aqueous media by combining porphinato–iron with water-soluble polymers, but until now reversibility has not been realized at 37°C with sufficient lifetime under physiological conditions. In addition, many of the polymer-bound porphinato–iron complexes are composed of non-biocompatible compounds.

Liposome-Embedded Porphinato–Irons

Liposome-embedded/heme is clearly a suitable model to mimic the mechanism of oxygen uptake and release by red blood cells. Unfortunately, it cannot be used as a blood substitute in living animals. Tests conducted in dogs and rodents show that the liposome-embedded/heme transported oxygen for several hours, but after that the animals died (*Chem. Eng. News*, Jan. 14, 42, 1985). As far as we know, this problem has not yet been solved. However, there are other applications for these oxygen carriers. For instance, they can isolate pure oxygen from air.[18] Other applications include catalysts for reactions of molecular oxygen *in vitro* and *in vivo* and reagents for drug metabolism and anticancer agents.

GENERAL CONCLUSION

No ideal blood substitute exists as yet. Hb has a too-short *in vivo* life in plasma, insufficient intravascular persistence, and significant transcapillary leakage. Perfluorochemical solvents can be used as oxygen carriers, but because of the limitation of a linear dependence of dissolved oxygen on oxygen pressure, elevated oxygen tension is required for perfluorochemicals to support life. The polylipid liposome/lipid heme complexes are ideal artificial oxygen carriers, but substitution tests with animals show that they are not bioacceptable. The animals died, presumably because of a disturbance in the blood coagulation system. We believe the most important blood substitutes are dextran–Hb derivatives. They have well-defined oxygen-binding characteristics that are close to Hb when dextran is suitably derivatized. Dextran–Hb complexes are uniformly nonimmunogenic in hemologous species, and their overall antigenic properties in heterologous species are unchanged. They do not present any hemagglutination phenomena or toxicity when applied to animals. Dextran–Hb complexes are available for potential use as blood substitutes, but the optimal size and molecular architecture of the complex has yet to be found.

REFERENCES

1. Rabiner, S. F. et al. *J. Exp. Med.* **1967**, *126*, 1127.
2. Savitsky, J. P. et al. *Clin. Pharm. Ther.* **1978**, *23*, 73.
3. Hunt, C. A. et al. *Science* **1985**, *230*, 1165.
4. Szebeni, J. et al. *Biochemistry* **1985**, *24*, 2827.
5. Jaroni, H. W. Ph.D. Dissertation, Tübingen, 1984.
6. Gaber, B. P. et al. *Febs. Lett.* **1983**, *153*, 285.
7. Djordjevich, L. et al. *Crit. Case Med.* **1987**, *15*, 318.
8. Brandl, M. et al. *Drug Dev. Ind. Pharm.* **1989**, *15*, 655.
9. Gaber, B. P.; Farmer, M. C. *Prog. Clin. Biol. Res.* **1984**, *165*, 179.
10. Baldwin, J. E. et al. *Tetrahedron* **1981**, *37*, 1723.
11. Iwasaki, K. et al. *Biochem. Biophys. Res. Comm.* **1983**, *113*, 513.
12. Schmidt, K.; Klin, L. *Wochenschrift* **1979**, *57*, 1169.
13. Leonard, M. et al. *Tetrahedron* **1984**, *40*, 1581.
14. Grönwall, A. et al. *Proc. 11th Congr. Int. Soc. Blood Transfus.* **1968**, *29*, 874.
15. Nordmeier, E.; Rad, D. *J. Polym. Sci., Part B; Polym. Phys.* **1994**, *32*, 63.
16. Collman, J. P. *Acc. Chem. Res.* **1977**, *10*, 265.
17. Traylor, T. G.; Traylor, P. S. *Rev. Biophys. Bioeng.* **1982**, *11*, 105.
18. Tsuchida, E.; Nishide, H. *Makromol. Chem. Rapid Commun.* **1982**, *3*, 693.

Blowing Agents

See: Additives (Property and Processing Modifiers)
Additives (Types and Applications)
Foaming Agents
Polyethylene Foams
Rigid Polyurethane Foams

BONE CEMENTS, BIORESORBABLE

Gretchen A. Caywood* and F. Howard Schneider
DynaGen. Incorporated

The function of a bone cement is to distribute the stress at the bone/cement interface over a wider area. The cement acts as a mechanical adhesive, interlocking (usually) adjacent porous surfaces, such as a prosthesis to bone or adjacent fractures prosthesis material and bone. The cement acts as a mechanical adhesive, interlocking adjacent (usually) porous surface, such as a prosthesis to bone or adjacent fractured bone elements. The key feature of a cement that allows this interlocking is its application as a viscous liquid or dough, allowing for flow of the material into the porous surfaces, followed by completion of cure *in vivo* to form a solid material.

Using a biocompatible, bioresorbable cement system provides the opportunity for bone ingrowth into the repaired area as the cement resorbs, thereby allowing further strengthening of the repaired area over time. Resorbability also eliminates concerns of long-term material fatigue and cracking seen with permanent acrylic materials. Additionally, the various bioresorbable polymers under investigation are less brittle, reflecting the mechanical strength characteristics of the biodegradable polymer structure. Curing of the resorbable systems is often much less exothermic since cure mechanisms are often ionic (not involving exothermic radical polymerization) or the content of the curing component is less.

*Author to whom correspondence should be addressed.

RESORBABLE CEMENT SYSTEMS: PREPARATION AND PROPERTIES

A variety of bioceramics, organics, and composites of the two, as well as homologous and autogenous bone, have been used as bioresorbable bone cements, particularly in fracture repair.

Bone Allograft

Homologous bone tissue is employed in surgical procedures because of its biomechanical and biological properties. Bone can be processed to yield dense cortical bone or porous cancellous bone or a composite of both. Grafton™ Allogenic Bone Matrix (Osteotech, Incorporated; Shrewsbury, NJ) is a demineralized allogenic bone matrix in a glycerine gel consistency. It was introduced to oral/maxillofacial and periodontal markets in late 1991, and to orthopedic markets in 1992.

Inorganics/Ceramics

Hydroxyapatite, or HA ($Ca_{10}(PO_4)(OH)_2$), is a natural choice since it is the major component of bone. However, its brittleness renders it unsuitable for major load-bearing implants.[1]

Sintered porous or dense HA is basically inert with respect to resorption, while corraline porous HA (CHA) slowly resorbs at the rate of 2% per year.[2] CHA is made by the conversion of the calcium carbonate structure of sea coral into pure HA. The most significant features of this material are its totally interconnected pore structure and the high degree of uniformity in pore diameter.[3] These unique features may be responsible for its ability to support bone ingrowth, in contrast to sintered or dense HA.[2] Systems that are partially resorbable can be designed by combining HA with resorbable ceramics such as tricalcium phosphate (TCP), calcium biphosphate (CBP), or calcium sulfate, leaving a nonresorbed inert portion behind that serves as a lattice for osteogenesis.[4,5]

One fast-setting, calcium phosphate-based resorbable cement under investigation is SuperBone Fracture Grout (Norian Corporation). The mixture of hilgenstockite, calcite, and orthophosphoric acid crystals forms a wet paste suitable for injection when mixed with dilute sodium phosphate solution.

Fatigue fracture may be an area of concern with this material, in view of the brittleness of ceramic systems. This may suggest limitations to its use in load-bearing applications.

Calcium sulfate, both as the hemihydrate and dihydrate, has been used for many years to fill bone defects, with the first reported use of the hemihydrate in 1892.[6] Although the material alone is not osteogenic, regeneration of bone is accelerated when it is in contact with the periosteum or bone.

Organic/Ceramic Composites

The next advance in resorbable bone cements was designed to mimic the viscoelastic properties of bone by incorporating a polymer into the ceramic system. Polymers of the lactic-glycolic family have been used for this purpose, as have other polyesters such as polycaprolactone and polyorthoesters.

The first step extensively investigated was the incorporation of natural biopolymers such as purified fibrillar collagen (PFC) into a cement system. Various collagen types have been combined with HA, TCP, and tetra CP.[2-9]

One concern with these collagen-containing materials is the possibility of hypersensitivity reactions of bovine collagen. This concern is eliminated with the use of synthetic polymers. Several cement systems that use synthetic polymers are under investigation, most of these being lactic and/or glycolic-based systems.[10-12]

In recent years, systems have been developed that use Krebs cycle acid-based polyesters.[13] In our laboratories a fast-curing, bioresorbable, biocompatible composite bone cement based on poly(propylene fumarate) (PPF) has been developed.[14] Our OrthoDyn™ bone cement used PPF of $M_w > 20,000$ g/mole to take advantage of near maximum polymer mechanical strength.[15]

OrthoDyn™ bone cement is a two-component powder-liquid system. The powder consists of one-fourth to one-third (various formulations) PPF in $CaSO_4{\cdot}2H_2O$. The liquid component commonly used in our laboratories is *N*-vinylpyrrolidinone (NVP), although one may use a variety of liquid monomers for property variation. The resulting system may best be described as a water-swellable crosslinked organic matrix embedded in $CaSO_4{\cdot}2H_2O$.

Compressive and torsional stiffness of OrthoDyn™ and PMMA were similar. However, the torque versus rotation curves demonstrated the brittleness of PMMA and the improved toughness of OrthoDyn™. These results indicate that OrthoDyn™ is strong enough to withstand normal physiological loads safely.[16]

In summary, while there are a variety of organic ceramic composite systems under investigation, with a few commercially available, only the OrthoDyn™ bioresorbable bone cement system combines the versatility of moldability with the strength of a fast-curing polymer-containing system. Since the fast cure is only mildly exothermic, curing is not achieved at the expense of increased tissue necrosis in the repair area.

APPLICATIONS OF BIORESORBABLE BONE CEMENT SYSTEMS

Bone cements are used in a variety of applications including facial reconstruction, periodontal applications, fracture repair, bone defect filling, and total joint replacement. The more load-bearing the applications, the higher the strength demands on the cement—which determines the material chosen.

Bioresorbable ceramic bone cements have generally been limited to low-load bone filler and extender applications. The development of OrthoDyn™ fast-curing bioresorbable systems represents a major step toward the availability of a bioresorbable system with the strength and moldability for such an application.

ACKNOWLEDGMENT

The authors wish to thank Dr. Fred Quimby and Dr. Alan Nixon of the Center for Research, Animal Resources, Cornell University for the *in vivo* testing, and Dr. Richard Coutts of the Malcolm and Dorothy Coutts Institute for Joint Reconstruction and Research, San Diego, for the study using OrthoDyn™ for intramedullary rod fixation. We also thank Dr. Gregory Kharas (DePaul University, Department of Chemistry) for material strength measurements and for valuable discussion.

REFERENCES

1. Bonfield, W. *Joint Replacement: State of the Art*; Mosby Year Book: St. Louis, MO, 1990; Chapter 2.
2. Shors, E. C. et al. *17th Ann. S. F. B.* **1991**, *14*, 305.
3. Shimazaki, K.; Mooney, K. *11th Ann. S. F. B.* **1985**, *8*, 159.
4. Ricci, J. L. et al. *12th Ann. S. F. B.* **1986**, *9*, 56.
5. Moore, D. C.; Chapman, M. W.; Manske, D. J. *11th Ann. S. F. B.* **1985**, *8*, 160.
6. Peltier, L. F. *Clin. Orthopaed.* **1961**, *21*, 1–29.
7. Cox, H. M. et al. *11th Ann. S. F. B.* **1985**, *8*, 169.
8. Oonishi, H. et al. *17th Ann. S. F. B.* **1991**, *14*, 306.
9. Sugihara, F. et al. *17th Ann. S. F. B.* **1991**, *14*, 188.
10. Verheyen, C. C. P. M. et al. *17th Ann. S. F. B.* **1991**, *14*, 284.
11. Conjeevaram, S. et al. *12th Ann. S. F. B.* **1986**, *9*, 163.
12. Alexander, J. et al. *11th Ann. S. F. B.* **1985**, *8*, 215.
13. Wise, D. L. et al., In *Biopolymeric Controlled in Release Systems*; CRC: Boca Raton, FL, 1984; Vol. II. Chapter 2.
14. Sanderson, J. E. U.S. Patent 4 722 948, 1988.
15. Caywood, G. A. et al. "Synthesis and Biomedical Applications of Polyesters Based on a Krebs Cycle Acid," presented at Advances in Treatment of Osteoporosis and Bone Repair, Boston, MA–June 1993.
16. Coutts, R. D. et al. *40th Ann. O. R. S.* **1994**, *19*(2), 536.

Borate Ion Complex Gels

See: Poly(vinyl alcohol)-Ion Complex Gels

Boron-Containing Polymers

See: Inorganic Nanostructured Materials
Organoboron Main Chain Polymers
Polyborosiloxanes
Polyolefin Graft Copolymers (Prepared by Borane Approach)
Preceramic Polymers

Boron Nitride

See: Inorganic Nanostructured Materials

BOVINE SERUM ALBUMIN (Adsorption Studies by Surface Plasmon Resonance)

Toshio Shimizu
Department of Information Science
Faculty of Science
Hirosaki University

Ellipsometry and the surface plasmon resonance (SPR) method are suitable techniques to measure simultaneously thickness and refractive index of adsorption layers on metal surfaces.[1] Adsorptions of synthetic polymers, organic monolayers assemblies, and proteins have been widely investigated with these techniques.[2-8]

A surface plasmon is a surface charge density wave at a metal boundary that propagates along the interface. The electromagnetic field connected with the motion of the charges peaks at the boundary and decays exponentially on both sides.[9,10]

Researchers have shown that growing an adsorbed layer on the metal surface produces a shift of the resonance in the reflectivity and a broadening of the resonance curve.[11,12]

In the present study, we have used SPR to investigate the shift of resonance angle due to the adsorption of BSA onto a silver film surface from a salt-free aqueous solution, and to determine the thickness of the adsorbed layer.

PROPERTIES

Proteins are known to adsorb on metal surfaces from the solution phase through electrostatic interaction and/or hydrophobic interaction, and this is the case for BSA as well.

The BSA molecule has been understood as a prolate ellipsoid with the dimension of 2.2 nm × 2.9 nm × 13.5 nm. Considering that the size of the BSA molecule along the line of apsides is 13.5 nm, this result indicates that the adsorbed layer is a monolayer and BSA molecules are adsorbed on the silver surface vertically.

It should be mentioned here that we haven't investigated the effects of factors such as added salt concentration and solution pH in this work, since the silver film surface is easily damaged by salts and pH buffer agents. Gold films would be suitable for these purposes.

ACKNOWLEDGMENT

We are grateful to Professor Y. Kanda of Tokyo University for giving us a chance to start this investigation by using the SPR technique. We also thank Dr. M. Kawaguchi of Mie University for his continued discussions. This research was partly supported by a Grant-in Aid for Developmental Scientific Research (62850163) of the Ministry of Education, Science, and Culture of Japan.

REFERENCES

1. Sadana, A. *Chem. Rev.* **1992**, *92*, 1799.
2. Pockrand, I. et al. *Surf. Sci.* **1977**, *74*, 237.
3. Wähling, G. Z. *Naturforsch* **1978**, *33a*, 536.
4. Nylander, G.; Liedberg, B.; Lind, T. *Sensors and Actuators* **1982–83**, *3*, 79.
5. Leidberg, B.; Nylander, C.; Lundstrom, I. *Sensors and Actuators* **1983**, *4*, 299.
6. Loulergue, J. C.; Levy, Y.; Allain, C. *Micromolecules* **1985**, *18*, 306.
7. Fischer, B. et al. *Langmuir* **1993**, *9*, 136.
8. Spinke, J. et al. *Langmuir* **1993**, *9*, 1821.
9. Raether, H. *Physics of Thin Films*; Academic: New York, 1977; Vol. 9, p 145.
10. Girlando, A. et al. *J. Chem. Phys.* **1980**, *72*, 5187.
11. Abelès, F.; Lopez-Rios, T. *Proc. Taormina Conf. Structure of Matter; Polaritons*; Pergamon: New York, 1974, p 241.
12. Lopez-Rios, T.; Vuye, G. *Surf. Sci.* **1979**, *81*, 529.

BRANCHED POLYMERS

Jacques Roovers
Institute for Environmental Research and Technology
National Research Council of Canada

Branched polymers are high-molecular weight homopolymers with more than two chain ends. Within each molecule there are one or more atoms or small multifunctional units that are directly linked to more than two long chains. These are branch points. If a branched polymer has only one branch point, it is a star polymer. Branched polymers with two or more branches are combs when a backbone and the branches can be distinguished. If no chain hierarchy can be identified, the branched polymer has a treelike structure and branches carry branches. Such polymers are usually randomly branched. Recently, dentritic or cascade polymers have been synthesized that combine a central branch point with a regular repetitive branching pattern.

Branched polymers are distinguished from crosslinked polymers (or networks) by their finite size and solubility.

Here we review new results on branched polymers. For information about the basic aspects of branched polymers, the reader is referred to previous reviews.[1-4]

SOLUTION PROPERTIES

At constant molecular weight, the average segment–segment distance is smaller in a branched polymer than in the linear polymer. As a consequence, the segment density in the polymer coil is increased, the branched polymer is more spherical, and the reduction of the radius of gyration is defined by **Equation 1**.

$$g = \langle R_G^2 \rangle_{br} / \langle R_G^2 \rangle_{in} \qquad 1$$

There are many theoretical studies on branched polymers. The static properties describe the radius of gyration, the intramolecular scattering function, and the second virial coefficient, A_2.

The scaling method for branched polymers has been introduced by Daoud and Cotton[5] and expanded upon by others.[6,7] It does not calculate $\langle R_G^2 \rangle$ of a star polymer explicitly, but derives the power law dependence of the radial segment distribution and the radius of gyration on M, f, and solvent quality.

The application of normalization group methods to branched polymers is relatively new.[8] As shown by Douglas and Freed,[9] it allows for the evaluation of dynamic properties such as $[\eta]$, (g_η), D_o, and (g_h), as well as static properties. The results of the normalization group theory have been compared with experimental results.[10]

Since the original Monte Carlo simulations of stars, these techniques have been applied by many groups to branched polymers—mostly to stars, but also to combs.[11-13] Molecular dynamics (MD) simulations require larger computer capabilities, but they allow for the study with up to 50 arms in good solvent and θ condition.[14]

Because size exclusion chromatography (SEC) separates polymers according to their hydrodynamic volume $[\eta]\cdot M$, it is used extensively to detect and quantify the presence of long-chain branching. Two detectors are required: one is a mass detector, and the other measures either [η] or M as a function of elution volume.

Much progress has recently been made in the analysis of the very broad distributions of randomly branched polymers. This has allowed the study of the growth of randomly branched polymers up to the gel point.

BULK PROPERTIES

The physical properties of branched polymers that are dominated by their chemistry or chemical interactions are rarely very different from those of the linear polymer. Density and refractive index are two examples. The same is true for the glass-transition temperature and melting temperature, which may be slightly increased or decreased depending on whether the "chemical effects" of branch points or chain ends dominate.

Dynamic properties, which reflect large-scale chain motions, are greatly affected by branching, however. It was observed in 1965 that the molecular weight dependence of the melt viscosity of three- and four-arm star polymers is very different from that of linear polymers.[15] Low-molecular weight stars with $M_{arm} \approx M_e$, the entanglement molecular weight, have $(\eta_o)_{star} \approx g(\eta_o)_{lin}$. When each arm participates in more than a few entanglements ($M_{arm}/M_e > 3$ or 4), the melt viscosity increases exponentially

$$(\eta_o)_{star} \approx (M_{arm}/M_e)^{1/2} \exp\left(\alpha \frac{M_{arm}}{M_e}\right) \qquad 2$$

in contrast with the $M^{3.4}$ dependence observed for linear polymers (**Equation 2**). The molecular process underlying Equation 2 is thought to be arm retraction from the free-chain end inward to the branch point.[16] Direct experimental evidence from the optorheological measurements with locally deuterated star polymers supports the longer, wider relaxation times of star polymers, at least for intermediate relaxation times.[17] Relaxation times and melt viscosities of star-branched polymers rapidly decrease when they are imbedded in a linear polymer matrix.[18]

Although Equation 2 suggests that the viscoelastic properties of star polymers are dependent only on M_{arm} and not on the number of arms, it has been observed experimentally that the melt viscosity and longest relaxation time increase when $f > 15$.[19,20] Similar viscosity enhancement has been found in melts of end-associating polymers.[21]

REFERENCES

1. Flory, P. J. *Principles of Polymer Chemistry*; Cornell University: Ithaca, NY, 1953.
2. Small, P. A. *Adv. Polym. Sci.* **1978**, *18*, 1.
3. Scholte, T. G. *Developments in Polymer Characterization* **1983**, *4*, 1.
4. Roovers, J. *Encyclopedia of Polymer Science and Engineering* **1983**, 2, 478.
5. Daoud, M.; Cotton, J. P. *J. Phys. (Les Ulis, Fr)* **1982**, *43*, 531.
6. Birshtein, T. M.; Zhulina, E. B. *Polymer* **1984**, *25*, 1453.
7. Ohno, K.; Binder, K. *J. Chem. Phys.* **1991**, *95*, 5444.
8. Miyake, A.; Freed, K. F. *Macromolecules* **1983**, *16*, 1228; **1984**, *17*, 678.
9. Douglas, J. F.; Freed, K. F. *Macromolecules* **1984**, *17*, 1854.
10. Douglas, J. F.; Roovers, J.; Freed, K. F. *Macromolecules* **1990**, *23*, 4168.
11. Mazur, J. McCrackin, F. *Macromolecules* **1977**, *10*, 327, **1981**, *14*, 1214.
12. Freire, E. J. J.; Rey, A.; Garcia de la Torre, J. *Macromolecules* **1986**, *19*, 457.
13. Lipson, J. E. G. *Macromolecules* **1991**, *24*, 1327, **1993**, *26*, 203.
14. Grest, G. S. *Macromolecules* **1993**, *27*, 3493.
15. Kraus, G.; Gruver, J. T. *J. Polym. Sci.* **1965**, *A3*, 105.
16. de Gennes, P.-G. *J. Phys.* **1975**, *36*, 1199.
17. Lantman, C. W.; Tassin, J. F.; Monnerie, L.; Fetters, L. J.; Helfand, E.; Pearson, D. S. *Macromolecules* **1989**, *22*, 1184.
18. Brochart-Wyart, F.; Ajdani, A.; Leibler, L.; Rubinstein, M.; Viovy, J. L. *Macromolecules* **1994**, *27*, 803.
19. Roovers, J. *Macromolecules* **1991**, *24*, 5895.
20. Roovers, J. *J. Non-Cryst. Solids* **1991**, *131–133*, 793.
21. Fetters, L. J.; Graessley, W. W.; Hadjichristidis, N.; Kiss, A. D.; Pearson, D. S.; Younghouse, L. B. *Macromolecules* **1988**, *21*, 1644.

Branched Structures

See: *Branched Polymers*
Hyperbranched Aliphatic Polyesters
Hyperbranched Polyesters
Macromonomers (Preparation, Polymerizability, and Applications)
Polyalkylenimines
Polystyrene Manufacture (Using Bulk Free Radical Polymerization)

BRIDGED POLYSILSESQUIOXANES (Highly Porous Hybrid Organic–Inorganic Materials)

Douglas A. Loy
Properties of Organic Materials Department
Sandia National Laboratories

Kenneth J. Shea
Department of Chemistry
University of California, Irvine

Sol-gel polymerization of tetraalkoxysilanes is a mild and convenient method for synthesizing silica gels having a wide range of properties that depend upon the reaction and processing conditions used.[1] The range of possible physical and chemical properties can be extended enormously by introducing organic substituents into the silica network polymer. One of the simplest methods for preparing such *hybrid organic-inorganic* materials is to replace one of the alkoxide groups on a tetraalkoxysilane with an organic substituent.[2-5] The result is a trialkoxysilane monomer with the fourth coordination site occupied by an organic group that is attached through a hydrolytically stable silicon–carbon bond. With sol-gel polymerization of the trialkoxysilane a *silsesquioxane* material is obtained with three siloxane bonds to each silicon. It was discovered that trialkoxysilanes with sterically bulky organic substituents often form soluble oligosilsequioxanes rather than network polymer (**Scheme I**).[6]

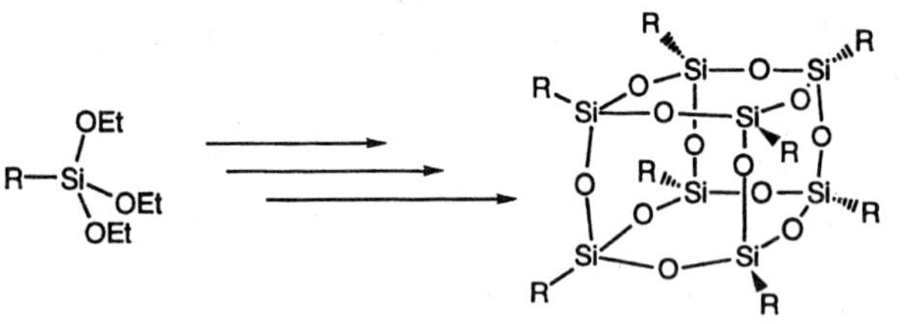

$[RSiO_{1.5}]_n$

1

SCHEME I. Hydrolysis and condensation of triethoxysilanes to form polyhedral oligosilsesquioxanes.

In fact, with the exception of monomers with small substituents (R = H, CH_3, etc.), most trialkoxysilanes oligomerize under typical sol-gel conditions to form polyhedral oligosilsesquioxanes **1**, or soluble randomly linked arrays of polycyclic cages.[7]

BRIDGED POLYSILSESQUIOXANES

Highly condensed hybrid organic–inorganic materials can be prepared from molecular building blocks that contain a variable organic fragment attached to two or more trialkoxysilyl groups. Sol-gel polymerization of such polytrialkoxysilyl monomers leads to a novel class of network materials; bridged polysilsesquioxanes (**Scheme II**).[8-18]

Sol-Gel Polymerization

Bis(trialkoxysilyl) monomers are hydrolyzed and condensed under relatively mild conditions that are typical for sol-gel polymerizations. Monomers (0.4 M) are dissolved in ethanol or tetrahydrofuran and the polymerizations are initiated with the addition of aqueous acid, base, or fluoride catalyst. Alkoxide groups on the silicons are hydrolyzed to silanols that condense with each other or with ethoxysilanes to give rise to siloxane bonds. As hydrolysis and condensation progress, branched polysilsesquioxanes grow in size, causing the solutions to become more viscous. Prior to gelation, the sol containing the growing polysilsesquioxanes can be cast as thin films or drawn into fibers. If undisturbed, the polymerization of most bridged monomers affords gels within a few hours. After gelation the solvent can be removed. Leaving a three-dimensional network that resembles, to some degree, the original polymeric structure that formed in the pre-gel solution.

One of the more striking differences between bridged polysilsesquioxanes and sol-gel polymers derived from triethoxysilanes and tetraethoxysilane (TEOS) is the low concentrations at which the former produces gels.

Xerogels are gels that have been air dried. However, the most important defining characteristic of a xerogel is that it undergoes considerable shrinkage during the drying process. Silica gels prepared by sol-gel polymerization of TEOS and TMOS under shrinkage on the order of 50–70%.[19] Bridged polysilsesquioxane xerogels undergo shrinkage between 90–95%. Much of the shrinkage in bridged polysilsesquioxanes is due to the loss of six equivalents of ethanol for each equivalent of monomer and the low monomer concentrations used in the preparation of these materials. As a result of the large volume loss and stresses incurred during evaporation of solvent, cracking of monolithic polysilsesquioxane xerogels is difficult to avoid. However, monoliths can be prepared by slow air drying of gels prepared with a drying-control chemical agent (DCCA).[20] For most applications, however, monoliths would not be necessary. Xerogels of bridged polysilsesquioxanes can be prepared quickly as powders by crushing and rinsing the wet gel with water before drying under vacuum at 100°C. Xerogels prepared in this fashion have similar porosities to those prepared by the more time-consuming air-drying process.

Aerogels are gels that have been dried in such a fashion as to avoid much of the shrinkage caused by the drying stress. As a result, aerogels are closer in volume and structure to the original wet gel. Historically, aerogels have been made by drying the wet get (alcogel) at a temperature and pressure above the supercritical point of the alcohol in the gel. In addition, supercritical carbon dioxide extraction has also proven to be an effective alternative method for preparing aerogels. Carbon dioxide extractions involve first replacing the original solvent in the gel with supercritical carbon dioxide, then slowly venting the carbon dioxide to afford a dry aerogel. Carbon dioxide extractions are increasingly popular because of the safety concerns of working at high pressure with supercritical alcohols.

ARYLENE-BRIDGED POLYSILSESQUIOXANES

A family of arylene-bridged polysilsesquioxane xerogels were the first representatives of these materials to be prepared.[8,11,12,16] The objective of this research was to determine if a rigid molecular spacer would sustain the porosity in the xerogel and if the length of the spacer would have any effect on the pore size distribution in sol-gel processed materials. A variety of arylene-bridged monomers were prepared and polymerized to afford gels that were subsequently processed to both xerogels and aerogels. Most important, a series of arylene-bridged xerogels with 1,4-phenylene-(**2**), 4,4′-biphenylene-(**3**) and 4,4″-terphenylene-(**4**) bridging groups was prepared in order to examine the effect of rigid molecular spacers on the amorphous network polymers structure.

Molecular Structure

Bridged polysilsesquioxane xerogels are brittle, glassy materials that do not swell or dissolve in either water or organic solvents. Powder X-ray diffraction confirms that the materials are amorphous. Their chemical composition has been determined by a number of spectroscopic techniques including solid-state NMR spectroscopy.

Porosity

Bridged polysilsesquioxanes are porous materials.

One of the most distinguishing features of arylene-bridged polysilsesquioxanes is the discovery that these materials are porous with extremely high surface areas compared to non-bridged polysilsesquioxanes.

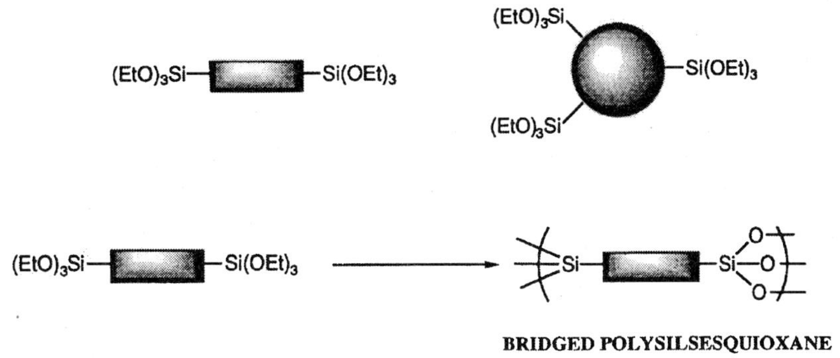

SCHEME II. Schematic of monomeric building blocks containing two or more triethoxysilyl groups as precursors to bridged polysilsesquioxanes.

ALKYLENE-BRIDGED POLYSILSESQUIOXANES

When a flexible organic bridging group is used instead of a rigid arylene, a whole new dimension of control over porosity in amorphous materials is revealed. Alkylene-bridged polysilsesquioxanes are composed of rigid silica-like regions interconnected by pliant straight-chain hydrocarbon segments ranging from 2 to 14 carbons in length.[14,17,18]

The most pronounced effect of a flexible hydrocarbon group in the bridged polysilsesquioxane xerogels is to cause the collapse of porosity with long alkylene groups.

NANO-CLUSTER SYNTHESIS IN BRIDGED POLYSILSESQUIOXANES

Porous bridged-polysilsesquioxanes have been used as a confinement matrix for the growth of semiconductor particles and transition metal clusters.

An internal doping procedure has been developed to prepare porous dried xerogels containing nano-sized transition metal clusters. The technique, involves copolymerizing triethoxysilylphenylchromium-(0)tricarbonyl with 1,4-bis(triethoxysilyl)benzene or other bridged monomers. Following sol-gel processing and drying to the xerogel, the metal tricarbonyl is decomposed with heat (125°C) under vacuum to afford nanoclusters of chromium metal dispersed throughout the gel.[21]

OPTICAL APPLICATIONS

Introduction of dyes into transparent glasses has been an important area of applied sol-gel research.[22] Monolithic gels containing organic dyes have been prepared for wave-guide,[23,24] nonlinear optical (NLO),[25-28] and laser applications.[29-31]

Bridged polysilsesquioxanes allow the dye functionality to be built into the organic bridging group, assuring no phase separation, aggregate, or loss by leaching. A number of arylene-bridging polysilsesquioxanes have been prepared with arylene groups that have promising ultraviolet absorption and fluorescence characteristics.

CONCLUSIONS

Bridged polysilsesquioxanes are a new family of hybrid organic–inorganic materials prepared by sol-gel chemistry. Trialkoxysilyl groups bridged by organic spacers permit the formation of network polymers capable of forming gels with a wide variety of organic functionalities, including arylene-, alkylene-, alkenylene, and alkynylene- groups. The properties of these materials can be readily manipulated by changing the reaction or processing conditions to afford high-surface area aerogels and xerogels or nonporous polymers. The capability for molecular engineering based on selection of the organic spacer distinguishes bridged polysilsesquioxanes from other sol-gel processed materials.

ACKNOWLEDGMENTS

The authors gratefully thank the National Science Foundation (Division of Materials Research) and the Air Force Office of Scientific Research for financially supporting the research at UC Irvine and the Laboratory Directed Research and Development program for supporting the research at Sandia National Labs. This research was supported by the United States Department of Energy under Contract No. DE-AC04-94AL85000.

REFERENCES

1. Brinker, C. J.; Scherer, G. W. *Sol-Gel Science: The Physics and Chemistry of Sol-Gel Processing*; Academic: San Diego, 1990.
2. Chujo, Y.; Saegusa, T. *Nyu Seramikkusu* **1990**, *3*, 43.
3. Novak, B. M. *Adv. Mater.* **1933**, *5*, 422.
4. Schmidt, H.; *J. Non-Cryst. Solids* **1993**, *1*, 13.
5. Wilkes, G. L.; Huang, H. H.; Glaser, R. H. *Adv. Chem. Ser.* **1990**, *224*, 207.
6. Voronkov, M. G.; Lavrent'yev, V. I. *Top. Curr. Chem.* **1982**, *102*, 199.
7. Frye, C. L. *J. Am. Chem. Soc.* **1971**, *93*, 4599.
8. Shea, K. J.; Loy, D. A.; Webster, O. W. *Chem. Mater.* **1989**, *1*, 572.
9. Shea, K. J.; Loy, D. A.; Webster, O. W. *Polym. Mater. Sci. Eng.* **1990**, *63*, 281.
10. Shea, K. J.; Webster, O.; Loy, D. A. *Mater. Res. Soc. Symp. Proc.* **1990**, *180*, 975.
11. Shea, K. J.; Loy, D. A.; Webster, O. J. *Am. Chem. Soc.* **1992**, *114*, 6700.
12. Corriu, R. J. P. et al. *Chem. Mater.* **1993**, *6*, 640.
13. Loy, D. A. et al. *Polym. Prepr.* **1993**, *34*, 244.
14. Oviatt, H. W. Jr.; Shea, K. J.; Small, J. H. *Chem. Mater.* **1993**, *5*, 943.
15. Loy, D. A.; Shea, K. J.; Russick, E. M. *Mater. Res. Soc. Symp. Proc.* **1992**, *271*, 699.

16. Loy, D. A.; Shea, K. J.; Russick, E. M. *Mater. Res. Soc. Symp. Proc.* **1992**, *271*, 699.
17. Loy, D. A.; Jamison, G. M.; et al. *J. Non-Cryst Solids* **1995**, *186*, 44.
18. Small, J. H.; Shea, K. J.; Loy, D. A. *Non-Cryst. Solids* **1993**, *160*, 234.
19. Novak, B. M.; Ellsworth, M. W. *Mater. Sci. Eng.* **1993**, *A162*, 257.
20. Orcel, G.; Hench, L. *Non-Cryst. Solids* **1986**, *79*, 177.
21. Choi, K. M.; Shea, K. *J. Am. Chem. Soc.* **1994**, *116*, 9052.
22. Dunn, B. et al. *Mater. Res. Soc. Symp. Proc.* **1988**, *121*, 331.
23. MacChesney, J. B. *Proc. Spie Int. Soc. Opt. Eng.* **1989**, *988*, 131.
24. Sanchez, C. *Soc. Esp. Ceram. Vidrio* **1992**, *31*, 5.
25. Mackenzie, J. D. *J. Ceram. Soc. Jpn.* **1993**, *101*, 1.
26. Prasad, P. N. *Proc. Int. Symp. Electrets, 7th* **1991**.
27. Prasad, P. N. *Mater. Res. Soc. Symp. Proc.* **1992**, *255*, 247.
28. Schmidt, H. *Top. Issues Glass* **1933**, *1*, 13.
29. Dunn, B. et al. *Proc. Spie Int. Soc. Opt. Eng.* **1990**, *1328*, 174.
30. Hench, L. L. et al. *Proc. Spie Int. Soc. Opt. Eng.* **1990**, *1328*, 230.
31. Canva, M.; Georges, P.; et al. *Applied Optics* **1995**, *34*, 428.

Bromobutyl Rubber

See: Butyl and Halobutyl Rubbers
Isobutylene Copolymers (Commercial)

Brushes, Polymeric

See: Crosslinked Polymer Brushes

Bulk Polymerization

See: Bulk Polymerization
Poly(methyl methacrylate)
Polystyrene Manufacture (Using Bulk Free Radical Polymerization)

BULK POLYMERIZATION

Shinzo Omi
Graduate School of Bio-Applications and Systems Engineering
Tokyo University of Agriculture and Technology

Bulk polymerization is regarded as the ultimate commercial scale process. It requires practically no solvents, and uses only small amounts of initiators and chain transfer agents. Mass production of polystyrene (PS) has been carried out with the thermal initiation kinetics, the mechanism of which has been thoroughly discussed.[1] Viscosity build-up during the progress of polymerization, however, by 10^4-fold or more, results in poor mixing and substantially reduces rates of mass transfer and heat removal. In polymethyl methacrylate (PMMA) or even PS polymerization, a runaway reaction can occur due to the extensive gel effect (the Trommsdorff effect), an autoacceleration of polymerization accompanied with the accumulation of heat. Solid design of the reactor, mixing, cooling, and controlling devices is required to prevent these problems. High plant cost is compensated for by high process efficiency and production rate.

CLASSIFICATION OF CONTINUOUS BULK PROCESSES

Gas-Phase Process

Low-density polyethylene (LDPE) processes at very high pressure were developed in the 1940s by two firms: the British ICI and German BASF.[2] Essentially a low conversion process, under supercritical conditions it requires no solvent, and an unconverted monomer is recovered and directly mixed with the fresh feed without any purification. A small amount of oxygen or peroxide is used as an initiator.

Linear low-density polyethylene (LLDPE), high density polyethylene (HDPE), polypropylene (PP), and PP rubber (block copolymer) are gradually replacing high-pressure (LDPE) as the most common products of gas-phase polymerization.

Liquid-Phase Bulk Process of PP

The critical point of propylene is 365 K and 4.61 MPa, and bulk-slurry processes were developed at first by Dart Company using a Ziegler-type catalyst, and later by Phillips Petroleum Company using its own Cr_2O_3 catalyst.[3] Polymers growing from active sites of the suspended catalysts are not soluble in the liquefied monomer phase, resulting in a viscous slurry process.

Polycondensation Process

Recent development of new nylon products, in particular softer and more flexible nylon 11 and nylon 12 have created a molten-bulk process due to the high melting point of monomers.

Earlier production systems of polyesters, poly(ethylene telephthalate) (PET), and poly(butylene telephthalate) (PBT) consisted of two stages: trans-esterification of dimethyl telephthalate (DMT) with diols, and later polycondensation. Now that high-purity telephthalic acid is available, the direct esterification process is replacing the DMT process. The PET process was easily transformed to the continuous process, unlike the difficult transition of the PBT process; rapid polymerization to manufacture PET bottles achieved with a thin-film evaporator (Luwa Company) is a typical example.

Polycarbonate (PC) has been produced by the phosgene process using toxic materials such as phosgene and methylene chloride. The trans-esterification process using bisphenol-A and diphenyl carbonate is environmentally acceptable, and will become an important alternative. It is a molten-bulk process employing base catalysts such as KOH.

Polystyrene and Related Processes

Productions of polystyrene (PS), rubber-grafted PS (HIPS), copolymers of styrene (AS or SAN), and rubber-grafted AS (ABS) are at least in part produced by the bulk polymerization or radical mechanism. Although the processes of suspension polymerization (PS, HIPS, and ABS) and emulsion (AS and ABS) are also popular, the bulk process has been regarded as ultimate because of its efficiency and pollution-free character.

PS BULK PROCESS

Bulk polymerization processes are no doubt most efficient in producing commodity polymers, and as PET bottle production shows, further efforts will lead to the ultimate process in which polymerization and subsequent processing for final products converge in a single flow.

REFERENCES

1. Kirchner, K.; Buchholz, K. *Angew. Makromol. Chem.* **1970**, *13*, 127.
2. Kaufman, M. *The First Century of Plastics*; Plastics Institute: London, United Kingdom, 1963.
3. "Collection of Polymer Flowsheets," *Hydrocarb. Process* **1979**, *November*, 210–238.

Butadiene-Nitrile Rubber

See: Diene Rubbers, Conventional (Ozone Degradation and Stabilization)

BUTADIENE POLYMERIZATION (Polymer-Supported Lanthanide Catalysts)

Guangqian Yu and Yuliang Li
Changchun Institute of Applied Chemistry
Chinese Academy of Sciences

The interest in lanthanide complexes was largely initiated by the discovery of the Ziegler-Natta type catalysts.[1,2] It is necessary to continue to investigate and develop an understanding of the fundamental mechanisms involved in the polymerization reactions and new lanthanide catalysts that exhibit excellent activity. Polymer-supported lanthanide catalysts have been developed in order to better utilize their potential catalytic activity.[3-23] The polymer-supported catalyst is a new kind of lanthanide coordination catalyst. It offers several advantages, including significant improvement in catalytic activity compared to corresponding low molecular weight catalysts, acceptable thermal and air stability, and ready separation from reaction products and easy recovery of the catalysts. It is also worthy of note that the polymer-supported catalysis may be used as a tool for a new approach to heterogeneous catalysts. Polymer-supported catalysis is particularly pertinent to the environment of the catalytic sites and of the interactions between these sites: factors that influence the activities of heterogeneous catalysts can be analyzed separately and progressively. This approach is now considered complementary to traditional notions of the properties of the lanthanide catalysts. Polymer-supported transition metal catalysts have been widely used in many reactions,[24,25] nevertheless, there have been fewer attempts at diene polymerization with polymer-supported catalysts,[26] especially, with lanthanide catalysts,[27] except research done in our laboratory.

CATALYTIC PROPERTIES

Catalyst Activity

Improved polymerization activity probably represents the single most important advantage of polymer-supported catalysts.

The reason for such a high activity in polymer-supported rare earth catalysts may be due to the bonding of the polymer supports with rare earth metals in the form of coordinative or ionic bond. The coordinative or ionic bond makes it possible to anchor and uniformly distribute rare earth ions on the polymer chains. In the aged solution of the catalysts especially, the swollen polymer-supported rare earth complexes are easy to contract with aluminum alkyl to form the active species. Most of the highly active catalysts based on rare earth elements are heterogeneous and so are low molecular weight earth catalysts. In the heterogeneous catalysts, only the rare earth ions on the surface of the catalysts can be activated and a portion of rare earth is embedded. Embedding of the metals for the polymer-supported rare earth catalysts can be overcome, in principle, for the reason mentioned above.

Some Factors Affecting Catalytic Activity

It is well known that the efficiency of the catalyst is influenced by the nature, valence state of the transition metal, the type of ligands attached to the transition metal, and the catalyst morphology, among other factors. Varying the ligands attached to the metal can change the distribution of the electron density of the metal (i.e., the bond polarity) in the active species so that the catalyst activity is affected. The situation will become more complicated when polymer replaces low molecular weight compound as the ligand.

The Catalytic Activity of Various Individual Rare Earth Elements

The activities of various individual lanthanide elements are quite different from one another. A peak in activity is observed for neodymium. Europium, samarium, and the heavy elements exhibit very low or no catalytic activities. The different activities of various lanthanide elements reveal the difference in complexing ability due to the electronic structure or the number of 4f orbital electrons in each kind of element in the series. The *cis*-1,4 content of polybutadiene produced by different lanthanide elements is the same, more than 98%. The same stereoregularity of polybutadiene prepared with different lanthanide elements reflects the similarity in chemical nature of 4f electrons in all the lanthanides.

REFERENCES

1. Shen, Z. Q. et al. *Sci. Sin.* **1964**, *13*, 1339.
2. Shen, Z. Q.; Gong, C. Y.; Ouyang, J. *Gaofenzi Tongxun (Polymer Communication)* **1965**, *7*, 193.
3. Yu, G. Q. and Li, Y. L. *Cuihua Xuebao* **1988**, *9*, 190.
4. Yu, G. Q. et al. *Macromolecules* **1993**, *26*, 6702.
5. Li, Y. L.; Ouyang, J. *J. Macromol. Sci. Chem.* **1987**, *A24*(3,4), 227.
6. Li, Y. L.; Ouyang, J. *Acta Polymerica Sinica* **1988**, *1*, 39.
7. Li, Y. L.; Liu, G. D.; Yu, G. Q. *J. Macromol. Sci. Chem.* **1989**, *A26*(2,3), 405.
8. Li, Y. L.; Yu, G. Q. *J. Macromol. Sci. Chem.* **1990**, *A27*(9–11), 1335.
9. Yu, G. Q.; Li, Y. L. *Inorganic and Metal-Containing Polymeric Materials*; Plenum: New York, NY, 1990; Chapter III.
10. Yu, G. Q.; Li, Y. L.; Liu, C. M. *Chinese J. of Polym. Sci.* **1992**, *10*(1), 49.
11. Yu, G. Q.; Li, Y. L.; Liu, C. M. *J. of Rare Earth* **1993**, *10*(2), 5.
12. Yin, Z. R. et al. *Yingyong Huaxue* **1993**, *10*(2), 5.
13. Li, Y. L.; Yu, G. Q.; Yao, J. *Acta Polymerica Sinica* **1992**, *5*, 572.
14. Li, Y. L.; Yu, G. Q.; Ouyang, J. *Rare Earth* **1989**, *3*, 53.
15. Li, X. L. et al. *Chinese J. of Polym. Sci.* in press.
16. Li, X. L. et al. *Yingyong Huaxue* **1993**, *10*(6), 92.

17. Yu, G. Q. et al., submitted for publication in *J. Polym. Sci., Polym. Chem. Ed.*
18. Li, X. L. et al. *Chem. J. Chin. Univ.* in press.
19. Li, Y. L. et al. *Functional Polym.* **1993**, *6*(2), 139.
20. Zhu, Y. K.; Li, Y. L.; Yu, G. Q.; *China Synthetic Rubber Industry* **1993**, *16*(6), 139.
21. Qu, Y. H. et al. *Chem. J. Chin. Univ.* in press.
22. Li, Y. L. et al. *China Synthetic Rubber Ind.* **1994**, *17*(3), 146.
23. Zhu, Y. K.; Li, Y. L.; Yu, G. Q. *China Synthetic Rubber Indust.* **1993**, *16*(6), 340.
24. Bailey, D. C.; Langer, S. H. *Chem. Rev.* **1981**, *81*, 109.
25. Chauvin, Y.; Commereuc, D.; Dawans, F. *Prog. Polym. Sci.* **1977**, *5*, 45.
26. Dawans, F.; Morel, D. *J. Mol. Catal.* **1978**, *3*, 403.
27. Bergbreiter, D. E.; Chen, L. B.; Chandran, R. *Macromolecules* **1985**, *18*, 1055.

BUTADIENE POLYMERIZATION, NEODYMIUM-CATALYZED

R. P. Burford, J. B. Nickaf, and R. P. Chaplin
School of Chemical Engineering and Industrial Chemistry
The University of New South Wales

Coordinative compounds of lanthanides (Ln), especially those of neodymium, which are the most active, have been widely used in the stereoregular homo- and copolymerization of butadiene and other dienes. Products of high steric purity are obtained.[1-4] Researchers have examined the effects of varying catalytic components and changing polymerization conditions on catalyst activity, kinetics, and polymer microstructure; a detailed review of research up to 1982 exists.[5] Subsequent studies have focused on the nature of the catalyst components and catalyst preparation,[6,7] polymerization medium,[8,9] active center determination and structure,[10] mechanism of stereochemical control,[11,12] and the mechanism of molecular weight control.[13]

Two types of catalyst systems are in wide use. The first are modified binary systems, which consist of a neodymium halide (usually chloride) complexed with an electron donor compound (usually containing an O or N donor atom) and an aluminum alkyl cocatalyst. The second and most widely used group of catalysts are the ternary systems. These consist of a neodymium compound (usually a carboxylate or naphthenate, for reasons of solubility) with two aluminum alkyls. At least one of the aluminum alkyls must possess halogen atom(s), necessary for both catalytic activity and sterospecificity and a chain transfer agent.[14] The efficiency of these catalysts (as with most Ziegler–Natta catalysts) is low, typically with ≤7% of the total neodymium potential sites used. The ternary systems have efficiencies that are slightly better than the modified binary systems.

KINETICS AND MOLECULAR WEIGHT DEVELOPMENT

Polymerizations can be conducted in temperatures from −70°C to 130°C.[14,15] Rate increases with temperature but average molecular weight decreases.[16] Sylvester and Wielder have suggested that temperature can be used to control molecular weight, with normal very high molecular weights being moderated, by using higher temperatures.[17]

"Pergasol," a high boiling solvent of mixed hydrocarbon fractions is a suitable polymerization diluent, as it allows efficient heat transfer without changing polymer characteristics.[16]

The rate of polymerization is first order with respect to both monomer and neodymium concentration. Molecular weight decreases as the concentration of neodymium increases, due to an increase in active concentration centers.[16-19] However, Oehme et al. have reported that both catalyst activity and polybutadiene molecular weight decrease with increases in neodymium concentration.[20] This would result in lower activities and consequently high polymer molecular weights.

Polymerization of butadiene with neodymium catalysts is highly dependent on the type and concentration of AIR_3. Branched alkyl groups provide more efficient catalysts.[6] Usually an AIR_3/Nd molar ratio of between 20 and 30 leads to optimal molecular weights.[10] As AIR_3 concentration is increased, catalyst activity increases, but molecular weight decreases. These results are attributed to chain transfer to the AIR_3, such as triisobutyl aluminium (TIBA).[10,14,16,18]

Aluminoxanes can be used as effective alkylating agents. Polymerization activity is high and the MWD of the polymers with these cocatalysts is generally narrower than with other aluminum alkyls.

MICROSTRUCTURE

One interesting feature of Nd catalysts is that polymer microstructure is consistently end predominantly *cis*-1,4 (up to 99%) and the vinyl (1,2) content is always low, ≤1%. The *cis* content decreases with increase in transition metal concentration, as with most Ziegler-Natta catalysts.[20,21] Several contradictions remain concerning microstructure dependence on temperature.

MECHANISM OF POLYMERIZATION AND STEREOCHEMICAL CONTROL

Coordination polymerization of butadiene with neodymium catalysts is now widely thought to proceed via an anionic mechanism, with the growing end being in the form of a π^3 betenyl unit.[6,10,14] Polymer growth then occurs by the repetitive insertion of monomer into the Nd–C bond of the growing end. The resultant microstructure of the polymer depends on the manner in which the incoming monomer units coordinates and inserts into the Nd–C bond of the butenyl unit. The cis bidentate coordination leading to an "anti" butenyl unit gives rise to *cis*-1,4 polymers, whereas monodenate coordination leading to "syn" butenyl groups gives rise to *trans*-1,4 polymers.[22-24]

APPLICATIONS

High *cis*-1,4 polybutadiene is employed mainly in the tire industry, where it is blended with natural rubber and/or with SBR and applied in either sidewalls, threads, or rims of tires. Neodymium-catalyzed polybutadiene exhibits superior processing and physical properties, suitable for tire production.[25] Other secondary applications include garden hoses and conveyor belts.

REFERENCES

1. Yun, O.; Fusong, W.; Zhiquan, S. *Proc. China–U.S. Symp. Polym. Chem. Phys.* **1979**, Oct. 5–10.
2. Jihua, Y. et al. *Sci. Sin.* **1980**, *23*, 734.
3. Yang, J. H. et al. *Macromolecules* **1982**, *15*, 230.
4. Zhiquan, S. et al. *J. Polym. Sci., Polym. Chem.* **1980**, *18*, 3345.
5. Yu, B. et al. *DOKI, Akad. Nauk. S.S.S.R.* **1982**, *265*(6), 1431.
6. Cabassi, F. et al. *Transition Metal Catalysed Polymerization*; Quirk, R. P., Ed.; Cambridge University: Cambridge, United Kingdom, 1988.
7. Jenkins, D. K. *Polymer* **1985**, *26*, 147.
8. Ricci, G.; Boffa, G.; Porri, L. *Macromol. Chem. Rapid Commun.* **1986**, *7*, 355.
9. Ricci, G. et al. *Polym. Comm.* **1987**, *28*, 83.
10. Jun, O.; Fusong, W.; Baotong, H. In *Transition Metal Catalysed Polymerizations*; Quirk, R. P. Ed., Harwood Academic: New York, NY, 1983; Part A, p 293.
11. Sabirov, Z. M.; Minchenekova, N. K.; Monakov, Y. B. *Inorg. Chim. Acta* **1989**, *16*, 99.
12. Sabirov, Z. M. et al. *J. Polym. Sci., Part A, Polym. Chem.* **1993**, *31*, 2419.
13. Kozlov, V. G. et al. *J. Polym. Sci., Part A, Polym. Chem.* **1994**, *32*, 1237.
14. Bruzzone, M.; Giordini, S. Presented at the American Chemical Society Rubber Division Meeting, Detroit, MI, October 8–11, 1991; paper No. 15.
15. Li, X.; Sun, Y.; Jin, Y. *Acta Chim. Sin.* **1986**, *44*, 1163.
16. Nickaf, J. B. Ph.D. Thesis, University of New South Wales, Sydney, Australia, 1994.
17. Sylvester, G.; Wieder, W.; Mark, J.; Lal, J. Eds.; Elastomers and Rubba Elasticity; American Chemical Society Symposium Series 193; American Chemical Society: Washington, DC, 1982; p 57.
18. Hsieh, H. L.; Yeh, H. C. *Rubber Chem. Technol.* **1985**, *58*, 117.
19. Hsu, C. C.; Ng, L. *AIChE J.* **1976**, *22*, 66.
20. Oehme, A.; Gebauer, U.; Gehrke, K. *J. Mol. Catal.* **1993**, *82*, 83.
21. Honig, J. A. J.; Burford, R. P.; Chaplin, R. P. *J. Polymer Sci., Polymer Chem. Ed.* **1993**, *21*, 2559.
22. Destri, S.; Gallazzi, M. C.; Giarrusso, A.; Pori, L. *Makromol. Chem. Rapid Commun.* **1980**, *1*, 293.
23. Taube, R.; Gehrke, J. P.; Schmidt, U. *Macromol. Chemie–Macromol. Syrup* **1986**, *3*, 389.
24. Porri, L.; Giarrusso, A. *Comprehensive Polymer Science,* Volume 4, **1989**, *4*, 53–108.
25. Stollfub, D. I. Use of Polybutadiene in Tyre Manufacture with Special Reference to New BUNA Grades, Bayer AG, KA Group Tech. Report Ed. 5.87, 1990.

BUTYL AND HALOBUTYL RUBBERS

Gabor Kaszas* and Judit E. Puskas
Bayer Rubber Incorporated

Wolfgang Baade
Bayer AG

Butyl rubber (CAS No. 9010-85-9) is a copolymer of isobutylene and a small amount of isoprene in the 1–3 mol% range.

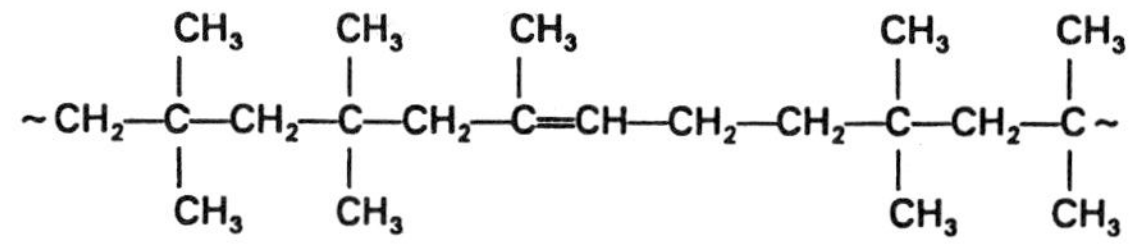

FIGURE 1. Diagram showing the chemical structure of butyl rubber.

The isoprene introduces unsaturated sites into the polymer chain, which allows for vulcanization of the rubber. The chemical structure of the polymer is shown in **Figure 1**. Cure versatility, cure rate, cure compatibility, adhesion to other elastomers, and flex life, as well as heat stability of the butyl rubber, are improved by bromination or chlorination of the IP enchainments.

Butyl rubber and halobutyl rubbers, (bromobutyl (CAS No. 68441-14-5), and chlorobutyl (CAS No. 68081-82-3)) have a unique combination of properties: they have the lowest permeability to air or moisture of known rubbers, as well as high vibrational damping; excellent resistance to acids, bases, and other chemicals, good ozone and weather resistance; good electrical properties; and resistance toward animal and vegetable oils and fats. Butyl rubber is also biocompatible.[1]

Major producers of butyl and halobutyl rubbers are Exxon and Bayer Rubber Inc. Butyl rubber is also produced in the former Soviet Union.

BUTYL RUBBER

Preparation

Butyl rubber is mostly made by a carbocationic slurry copolymerization of isobutylene and isoprene, initiated by $AlCl_3$ and activated by protogenic compounds at –95°C in methyl chloride diluent.[2] The solution process uses *i*-pentane as a solvent and alkyl aluminum chloride as a catalyst activated by water or hydrogen sulfide, at –90 to –60°C.

Whereas the isobutylene units show an exclusively head-to-tail configuration, more than 90% of the isoprene incorporates in trans-1,4-enchainment with ~ 10% of another structure, possibly 1,2-enchainment.[3-5] This other structure has recently been identified as a substituted isoprenyl structure, i.e., a branching point.[23] The fact that no 1,2-isoprene incorporation is found in butyl rubbers was explained by the existence of steric hindrance from the methyl side groups.[6]

HALOBUTYL RUBBERS

Just after World War II, B.F. Goodrich introduced the tubeless tire. In this tire, the air was retained by an inner liner that was part of the tire rather than by a separate inner tube. It was found that butyl rubber could not be used in inner liners because compounds containing butyl rubber would not adhere to the carcass of the tire. The inner liners were therefore made from styrene-butadiene rubber (SBR) by using butyl reclaim to improve the air retention. In the early 1950s, R. T. Morissey at B.F. Goodrich tried to improve the adhesion of butyl rubber by halogenating it with elemental bromine and chlorine, and with bromine-containing compounds.[7] This worked, and B.F. Goodrich started to produce brominated butyl rubber, HYCAR

*Author to whom correspondence should be addressed.

2202, by using a batch process in which *N*-bromo succinimide was blended with butyl rubber on a mill and the mixture heated in an oven.

During the same time period Michelin had introduced the radial passenger tire. By the 1960s, Michelin tires contained a bromobutyl inner liner. Michelin produced the bromobutyl rubber themselves by using a process similar, in principle, to that of B.F. Goodrich. The introduction of tubeless tires in the mid-1950s led to a decrease of the 8–10% per year growth rate of butyl production as innertubes for passenger cars began to disappear.

To commercialize halobutyl, the most important problem to overcome was to minimize polymer degradation during halogenation. The degree of chemical degradation decreases in the order F > Cl > Br. This is probably the reason why *N*-bromosuccinimide was used as the brominating agent instead of liquid bromine in the B.F. Goodrich process.[8]

In about 1965 Bayer Rubber Inc. made a major evaluation of the future of the butyl business. This evaluation predicted that, as tubeless tires became more prevalent and if more tires were radial, over time the growth of butyl rubber would be very slow and would eventually become negative, whereas the growth of halobutyl would be much larger. In 1967 Bayer Rubber Inc. obtained a license from B. F. Goodrich to produce brominated butyl rubber. Between 1968 and 1970, Bayer Rubber Inc. developed a pilot plant process to produce brominated butyl rubber by using elemental bromine and carried out laboratory studies to overcome problems with polymer stability and compound scorch.

PROPERTIES

Butyl Rubber

Butyl rubber is an amorphous material that crystallizes only under stress. The polymer chains are dominantly linear, but some degree of branching is present, which increases with the isoprene enchainment, that is the unsaturation level of the polymer.

Butyl rubber shows good solubility in nonpolar organic liquids, especially in hydrocarbon solvents (e.g., isopentane, hexane, cyclohexane, or toluene). Normal butyl elastomers are gel-free. Insoluble residues, observed sometimes, are antiagglomerants that are added during the manufacturing process. A good summary of the physical properties of butyl rubber was provided by Wood.[9] Butyl rubber has a low thermal expansion coefficient compared with other hydrocarbon elastomers.[10]

Butyl rubber can be used over a very wide temperature range, even down to its lowest glass transition temperature (T_g) value of –72°C to –66°C. The most characteristic properties of butyl rubber are high impermeability to gases and moisture; good chemical resistance, especially against mineral acids, caustics, and alcohols; high weathering and ozone resistance because of the very small amount of unsaturation present in the chain; low resilience and therefore high damping properties; high hysteresis; and good flex cracking resistance over a wide temperature range.[11]

In all applications butyl rubber is crosslinked into a network. This is done by curing, for which the unsaturation of the incorporated isoprene units is used. Regular butyl rubber can be cured with three systems: sulfur, phenol-formaldehyde resins, and quinone dioximes.[12,13] Most common is sulfur curing, for which thiuram disulfides or dithiocarbamates are used to achieve a good network. For special heat-resistance networks, the quinone dioxime system is used. Resin cures are used for making tire-curing bladders. A good overview of efficient curing systems and their use is given by Edwards.[14]

Halobutyl Rubbers

Commercial bromobutyl rubbers typically contains ~ 1.3 mol% of brominated isoprene structures, within which the *exo*-allylic form is dominant. About 0.4 mol% of the isoprene units remain unbrominated. Chlorobutyl rubber contains, on average, 1.9 mol% chlorinated isoprene units, of which ~ 1.4 mol% is in the *exo*-methylene form. On average, bromobutyl rubber contains ~ 1.4 mol% bromine.

The physical properties of halobutyl rubber are similar to those of butyl rubber, with the exception of slightly higher densities (butyl rubber, 0.917 g/cm^3 and brominated butyl rubber, 0.93 g/cm^3). Halobutyls are more reactive than are butyl rubbers because of the presence of allylic halide groups. In general, halobutyl rubber has the following advantages over regular butyl rubber: greater cure versatility; faster cure rates with lower curative levels; good cure compatibility with other rubbers; good adhesion to other elastomers, RFL dipped textiles, and brass-coated steel; and better heat resistance.

Adhesion quality and cure rate of bromobutyl are higher than those of chlorobutyl as a result of the lower stability and therefore higher reactivity of the brominated structures. Whereas bromobutyl can be cured with sulfur alone, curing chlorobutyl requires the addition of zinc oxide (ZnO). ZnO is the most often used curing agent for halobutyl rubbers. It is believed to lead to the formation of intermolecular C–C bonds by an ionic mechanism.

Other curing agents for halobutyl rubber are amines, dienophiles, and peroxides. A comparison of the various curing agents and their efficiencies is given by Walker, Edwards, and Ho.[15-17] Because of a tendency of dehydrohalogenation and chain scission, halobutyl rubbers are usually stabilized with stearates and epoxydized soyabean oil.

APPLICATIONS

Butyl rubber is used primarily in inner tube and halobutyl in the inner liner of tires. About 85% of butyl and halobutyl rubbers are used in tire-related applications. Examples of non-tire-related applications are technical rubber goods and pharmaceutical products. In the past decades the butyl market has shown a decrease and the halobutyl market a moderate growth.

Butyl Rubber

The main use of butyl rubber is for inner tubes of cars, buses, normal trucks and motors, and bicycles. In these applications low permeability to air and good heat resistance are the advantages of using butyl rubber.

In nontire applications, vulcanized butyl rubber is ideal for tire-curing bags and bladders, conveyor belts, wire and cable, and high-temperature service hoses because of its good heat resitance.[20] The high vibration damping property of butyl rubber is an advantage in isolator applications, such as rail pads, dock fenders, suspension bumpers, and automotive body mounts. Whereas good weather and ozone resistance are advantages in reservoir liners, approval for contact with food and adhesion behavior are beneficial for use in packaging films and adhesives. Butyl rubber is also used in pharmaceutical closures because of its impermeability and good chemical resistance. This field and the area of medical implants are growing markets for butyl rubber.[21] In addition, butyl rubber is used in a specially produced grade of chewing gum base.

Halobutyl Rubbers and Specialties

The main use of halobutyl rubber is in tire inner liners of tubeless tires and in heat-resistant tubes for heavy-duty truck tires. In these applications, low permeability to air, good convulcanization compatibility, and excellent adhesion to textile or steel carcasses are beneficial qualities of halobutyl rubber. Nontire applications of halobutyl rubber include:

- pharmaceutical stoppers where low curative levels, which can not be achieved with regular butyl rubber, are needed;
- water seals and membranes;
- gaskets, because of the lower compression set;
- tank linings because of the superior cure rate and good bonding capability;
- vibration insulators;
- protective clothing, because of high impermeability, good adhesion, and fast curing rate; and
- sports ball bladders

Divinylbenzene terpolymers are used in preformed tapes for the car and construction industries, sealants, caulks, pipe wrapping tapes, and condenser caps. Blends of high concentrations of butyl and halobutyl rubber with polyolefins are showing growing importance as thermoplastic elastomers.[22] Star-branched butyl rubber is applied where low die swell, reduced shrinkage, and high green strength are needed.[18] Brominated poly(isobutylene-*co-p*-methylstyrene) is used in tire and nontire applications where special heat and weathering resistance and enhanced cure reactivity are needed.[19]

REFERENCES

1. Kennedy, J. P. *Cationic Polymerization of Olefins: A Critical Inventory*; Wiley-Interscience: New York, 1975.
2. Thomas, R. M. *Rubber Chem. Technol.* **1969**, *42*, G90.
3. Che, H. Y.; Field, J. E. *J. Polym. Sci., Part B* **1957**, *5*, 501.
4. Chu, C. Y.; Vukov, R. *Macromolecules* **1985**, *18*, 1423.
5. Cheng, D. M.; Gardner, I. J.; Wang, H. C.; Frederick, C. B.; Dekmezian, A. H. *Rubber Chem. Technol.* **1990**, *63*, 265.
6. Kaszas, G.; Puskas, J. E.; Kennedy, J. P. *Macromolecules* **1992**, *25*, 1775.
7. Morrissey, R. T. *Rubber World* **1958**, *138*, 725.
8. Crawford, R. A.; Morrissey, R. T. U.S. Patent 2 631 984, 1953.
9. Wood, L. A. In *Polymer Handbook*, 3rd ed.; Brandrup, J.; Immergut, E. H., Eds.; John Wiley & Sons; New York, 1989; Vol. 5, 9.
10. Wood, L. A. *Rubber Chem. Technol.* **1976**, *49*, 199.
11. Yin, J. P.; Pariser, R. *J. Appl. Polym. Sci.* **1964**, *8*, 2427.
12. Fusco, J. V.; House, P. *Rubber Technology*, 3rd ed.; Morton, M., Ed.; Van Norstrand Reinhold Company, Inc. New York, 1987; p 294.
13. Hofmann, W. *Vulcanization and Vulcanizing Agents:* MacLaren: 1967.
14. Edwards, D. C. *Elastomerics* **1990**, *122*(3), 19.
15. Walker, J.; Hopkins, W.; Jones, R. H. Presented at the *International Rubber Conference,* Kyoto, Japan, 1985, IGA10.
16. Edwards, D. C. *Rubber Chem. Technol.* **1987**, *60*, 62.
17. Ho, K.; Steevensz, R. *Rubber Chem. Technol.* **1989**, *62*, 42.
18. Duvdevani, I.; Gursky, L.; Gardner, I. J. Presented at the American Chemical Society Rubber Division Meeting, Mexico City, Mexico, 1989; paper 22.
19. Parente, A. E.; Fusco, J. V. Presented at the 35th Annual General Meeting of the IISRP, Phoenix, AZ, 1994, TS.23.
20. Harmsworth, N. *Butyl and Halobutyl Compounding Guide for Nontire Applications*; Bayer AG, Antwerp, Belgium, 1992.
21. Newman, S. Presented at the American Chemical Society, Rubber Division Meeting, Las Vegas, NV, 1990, 13.
22. Kresge, E. *Rubber Chem. Technol.* **1991**, *64*, 469.
23. White, J. L.; Shaffer, T. D.; Ruff, C. J.; Cross, J. P. *Macromolecules* **1995**, *28*, 3290.

Butyl Rubber

See: *Butyl and Halobutyl Rubbers*
Butyl Rubber (for Chemical Protective Clothing)
Isobutylene Copolymers (Commercial)
Supported Catalysts (Lewis Acid and Ziegler-Natta)

BUTYL RUBBER (for Chemical Protective Clothing)

Eugene Wilusz
U.S. Army Natick Research Development and Engineering Center

Chemical protective clothing is widely used today throughout industry, in medicine, in agriculture, and in the chemicals in solid, liquid, aerosol, and vapor forms and must provide protection against a wide range of industrial chemicals, pesticides, and toxic wastes. In the military, protection must be provided against such hazards as chemical warfare agents, petroleum products, and rocket fuels.

To provide necessary protection, clothing is designed in the form of ensembles that consist of several articles of clothing and a respirator. Clothing articles include items such as jackets, trousers, coveralls, gloves, boots, and hoods. In military applications where material breathability and comfort are important, air-permeable textiles are used for jackets and trousers. In these garments activated charcoal is used to adsorb hazardous chemicals. Heat stress is a major problem associated with chemical protective clothing. Breathable garments can be worn for a longer period of time and allow longer periods of physical activity compared with those made from air-impermeable barrier materials. Ensembles made from impermeable elastomeric or thermoplastic materials are used to provide protection in situations where contamination with gross liquid chemicals is a serious threat. Impermeable materials are also the materials of choice for gloves and overboots and are used with both breathable and impermeable ensembles.

For many years butyl rubber (CAS No. 9010-85-9), poly(isobutylene-*co*-isoprene), has been used in many applications of chemical protective clothing, including suits, gloves, and overboots. Additional polymers have become available recently. These include star-branched butyl and a brominated copolymer of isobutylene and *p*-methylstyrene.[1,2]

PREPARATION AND PROPERTIES

The many excellent properties of butyl rubber are responsible for the widespread uses of the material. Among these properties are chemical resistance; low gas permeability; resistance to abrasion, ozone, weathering, and heat; and nonstaining and nondiscoloring characteristics. Low gas permeability, chemical resistance, and other excellent properties have made butyl rubber the elastomer of choice for chemical protective clothing applications.

Other barrier materials are also used in chemical protective clothing applications. These materials include elastomers such as polychloroprene, poly(acrylonitrile-*co*-butadiene), and others. Thermoplastic films and fluorocarbon polymers are also used.

Recent work with butyl rubber has involved detailed characterization of barrier properties, attempts at improving certain properties of the material, and evaluation of the newly available butyl polymers. Toluene diffusion in butyl rubber and barrier properties as a function of temperature and biaxial tensile strain have been recently reported.[3-5] Self-extinguishing formulations of butyl rubber have been developed using a combination of additives. The properties of these materials and prototype gloves have been reported.[6]

APPLICATIONS

Butyl rubber is widely used in chemical protective clothing. Applications include gloves, footwear covers, overboots, hoods, coveralls, and helmet covers. In clothing items such as footwear covers, hoods, and coveralls the rubber is generally supported with a fabric such as nylon. The base fabric may be solution coated with a thin layer of butyl rubber. A thicker layer of butyl rubber may then be calendared onto the material to arrive at the product, which is then ready for curing.

Specific items of protective clothing made from butyl rubber and used by the military include glove set, chemical protective; tactile CB glove; gloves, toxicological agents protective (TAP) and a toxicological agents protective outfit consisting of overalls, hood, and footwear covers. Butyl-coated nylon is used in the TAP items. A chemical protective helmet cover is also available. Similar protective clothing items are commercially available.

The self-contained toxic environment protective outfit, a completely encapsulated ensemble, is currently under development. In summary, items of protective clothing made from butyl rubber or butyl-coated fabric have been used for many years to provide excellent protection from exposure to hazardous chemicals.

REFERENCES

1. Gursky, L.; Fusco, J. V.; Duvdevani, I. Presented at the American Chemical Society Rubber Division Meeting, Detroit, MI, October 17–20, 1989. Abstract in *Rubber Chem. Technol.* **1990**, *63*, 309.
2. Wang, H.-C.; Powers, K. W. Presented at the American Chemical Society Rubber Division Meeting, Toronto, Canada, May 21–24, 1991. Abstract in *Rubber Chem. Technol.* **1991**, *64*, 680.
3. Schneider, N. S.; Moseman, J. A.; Sung, N.-H. *J. Polym. Sci., Polym. Phys. Ed.* **1994**, *32*, 491.
4. Hassler, K.; Wilusz, E. Presented at the ASTM International Symposium on Performance of Protective Clothing, San Francisco, January 1994.
5. Lee, B. L.; Yang, T. W.; Hassler, K.; Wilusz, E. Presented at the ASTM International Symposium on Performance of Protective Clothing, San Francisco, January 1994.
6. Gulliani, D.; Wilusz, E.; Galezewski, A. *J. Fire Sci.* **1994**, *12*, 246.

C

Calixarenes

See: *Calixarenes (as Supramolecular Cyclic Oligomers)*
Phenolics (Linear and Cyclic Oligomers)

CALIXARENES (As Supramolecular Cyclic Oligomers)

Jacques Vicens,* Rym Abidi, and Zouhair Asfari
EHICS
Laboratoire de Chimie Minérale et Analytique Associé au CNRS

The calixarenes are macrocycles made up of phenolic units meta-linked by methylene bridges (**Figure 1a**) and possessing basket-shaped cavities.[1,2] They have been named calixarenes by Gutsche because of the resemblance of the four-membered ring to a chalice (in Greek, calix) (**Figure 1b**).[1] The suffix arene indicates the presence of aryl rings in the molecular framework.[1] For specifying the size of the macrocycle, one intercalates between calix and arene a number between brackets that represents the number of phenolic units constituting the calixarene. In this article the calixarenes are designated in the following manner: The calixarenes substituted on *para*-position of the hydroxyl groups in the phenolic rings are denoted as *p*-substituent calix[*n*]arenes.

In 1970, Gutsche showed that direct base-catalyzed condensation of *p-tert*-butylphenol with formaldehyde led to a mixture of cyclic oligomers containing methylene bridges.[3]

More recently, the selective preparation of *p-tert*-butylcalix[*n*]arenes, with $n = 4$, 6, and 8 in high yields, has been described.[4–6] In the majority of cases, the base-catalyzed condensation of *p*-alkylphenols lead to the formation of calixarenes with an even number of phenolic units.

CONFORMATIONAL STUDIES

The spatial representation from molecular models suggests that calixarenes can exist in several conformations. With an increase in the number of aromatic rings, the macrocycle becomes more flexible. Conformational analysis of calixarenes is usually carried out by NMR and by crystallographic methods. These two techniques indicate that at a lower temperature and/or in the solid state, the calix[4]- and calix[5]arenes adopt the "cone conformation."[7] Calix[6]arenes adopt the "pinched cone conformation," and calix[7]arenes show no symmetry.[7] The calix[8]arenes are highly symmetrical, the molecule being almost flat and adopting the "pleated loop conformation."[7] In spite of the different ring sizes, the interconversion energies are in the same range of values (~ 12–14 kcal/mole).[7] This narrow range of values is explained by circular intramolecular hydrogen bonding between the hydroxyl groups. This array of hydrogen bonds directs the interconversion process.[7]

*Author to whom correspondence should be sent.

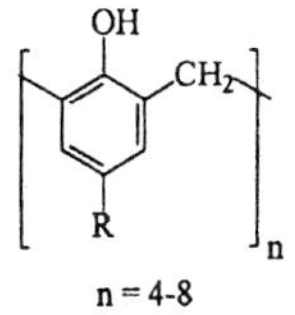

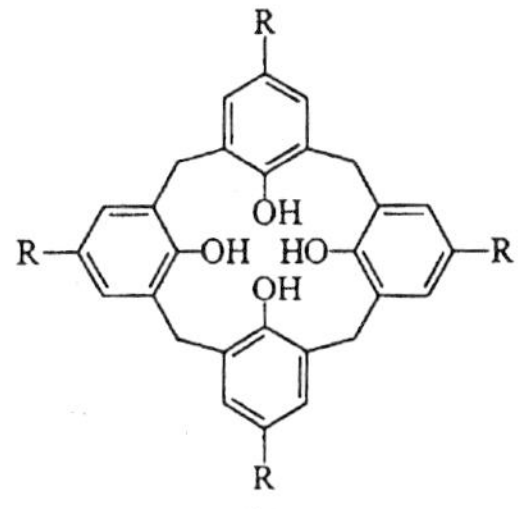

FIGURE 1. (a) General formula of calixarenes. (b) Cyclic tetramer.

NEUTRAL MOLECULAR RECOGNITION

The presence of aromatic rings in a cyclic arrangement gives the calixarenes a similar structure as that of cyclodextrins. For example, the tetramer in cone conformation presents a hydrophobic cavity able to include aromatic molecules such as benzene or toluene.[8]

ION RECOGNITION

The calixarene isolated during the base-catalyzed condensations can be chemically modified at three reacting sites: the hydroxyl functions, the aromatic rings, and the methylene bridges.[1,2] The introduction on the OH groups of ionophoric groups able to coordinate metal cations in association to the calixarene conformational structures leads to the elaboration of receptors or ligands that selectively complex cations.[9] the *O*-alkylation of *p-tert*-butylcalix[4]arene with ethyl bromoacetate produced a receptor selective for sodium cation; due to the cone conformation of calixarene the carbonyl groups converge toward metal cation.[9] This convergence is shown for the related tetraamide *p-tert*-butylcalix[4]arene derivative by X-ray diffraction of KSCN complex in the solid state.[10] And, more generally, reactions with α-halogeno-carbonyl compounds have afforded a series of calix[4]- and calix[6]arenes, bearing ester, carbonyl, amide, thioamide, and carboxylic groups that complex cations with proper selectivities.[11] The introduction of a bridge onto the 1,3-O-positions (the bridge links two opposite phenolic oxygen atoms) produced macropolycyclic molecules such as calixcrowns, calixspherands, calixcryptand, and Schiff-base calixarenes offering some unique properties of complexation.[12–15]

Taking into account the high selectivities in cation recognition observed during complexation and transport experiments, the tetraestercalix[4]arenes were first applied in analytical chemistry in preparing selective electrodes and chemical sensors for alkaline and alkaline earth cations.[16,17]

Recently, anion receptors have been produced. The 1,3-*p*-*tert*-butylcalix[4]arene-bis-4-amidopyridine complexes the halogenides as their 4-amidopyridinium complexes PF_6^{2-}.[18] The introduction of sulfonamide groups on the para position of aromatic rings of *O*-tetraglycolic derivatives leads to a receptor in cone conformation that can complex $H_2PO_4^-$, HSO_4^-, Cl^-, NO_3^- and ClO_4^-.[19] The tetraedric HSO_4^- ion is complexed by a three-dimensional cavity better than are the spherical Cl^- ion and the planar NO_3^- ion.[19]

SYNTHESIS AND METALLOCALIXARENES

The OH proton lability allowed the synthesis of organometallic compounds in which the metal is directly coordinated to the calixarene oxygen atoms. The reactions of metal amides $[Ti(NMe_2)_4]$, $[Fe\{N(SiMe_3)_2\}_3]$, and $[Co(\{N(SiMe_3)_2\}_2)_2]$, with *p*-*tert*-butylcalix[4]arene resulted in the isolation and structural characterization of the first three ς-bonded transition-metal derivatives of calixarenes.[20]

WATER-SOLUBLE CALIXARENES

p-Sulfonate calixarenes soluble in water were obtained by reacting calixarenes with H_2SO_4 at 70°C.[21] They formed complexes in water with organic molecules.[21] A biomimetic application of this ability to maintain organic molecules in water is the acid-catalyzed hydrolysis of 1,4-dihydronicotinamides.[21] From a mechanism point of view the presence of proton-donor groups in calixarene stabilizes the transition state by charge transfer.[21] In addition, *p*-sulfonate calixarenes have been claimed to be superuranophiles.

PHOTOPHYSICAL PROPERTIES

Because of their luminescence property, the complexation of lanthanides has been studied by using the preorganization properties of calixarenes.

MOLECULAR SELF-ASSEMBLY

p-Octadecylcalixarenes, prepared from phenol substituted in para position by a linear aliphatic group containing 18 carbons, formed Langmuir-Blodgett monolayers at the air–water interface. The introduction of ester groups onto OH functions of *p*-*tert*-octylcalixarenes leads to the formation of monolayers sensitive to the cation identity in the water phase. By introducing a photoreactive group in calixarenic monolayers, polymerized films have been prepared that are permeable to gas and that are able to be separated.

The introduction of aliphatic chains on the para positions with rigidification of the calixarenes allowed the growth of liquid crystals.

SELF-ASSEMBLY REACTIONS

We have seen that calixarenes show inclusion properties and molecular self-organization by the formation of noncovalent bonds in three-dimensional complementarity with the formation of weak interactions involving energies similar to that of enzyme–substrate systems. Probably because of their high symmetry, the calixarenes are able to achieve self-assembly reactions in which several covalent bonds are formed at the same time. For example, double-calix-double-crown-ether has been obtained by a one-step reaction of *p*-*tert*-butylcalix[4]arene with the pentaethylene glycol ditosylate.[22] A very recent example of the use of calixarene to produce highly symmetrical and large molecules is the synthesis of "holand."[23]

CONCLUSION

Chemical modifications of the primitive calixarenes allowed for tailor-made derivatives with given properties. The introduction of chosen functions led to molecular receptors designed for the complexation of ionic and neutral species. The subsequent recognition properties allow the development of biomimetic models, supramolecular catalysts, high selective receptors for separation of organic mixtures, solvent purification, and carriers able to separate metallic species. their complexation–recognition processes are now investigated from a theoretical point of view.[24] Modified calixarenes have been used as constituents of specific electrodes. They are also photonic compounds used as fluorescent probes, and recently they have been used in transistor systems. Other modifications provided chemists with self-assembling molecules that can be organized in a supramolecular architecture. Some of them can be organized in monolayers that polymerize to give porous materials for gas separation. Some others can organize in columnar mesophases to yield liquid crystals. Finally, the calixarenes showed antituberculous effects.[25] The present developments of calixarenes make them synthetic organic molecules, which are used in the different fields of chemistry (e.g., organic chemistry, supramolecular chemistry, analytical chemistry, physicochemistry, and industrial chemistry), in physics (e.g., molecular materials, photophysics, and nonlinear optics), and in biology (e.g., enzyme models, physiological properties, and clinical uses).

REFERENCES

1. Gutsche, C. D. *Calixarenes, Monographs in Supramolecular Chemistry*; Stoddart, J. F., Ed.; The Royal Society of Chemistry; 1989; Vol. 1.
2. Vicens, J.; Böhmer, V. *Calixarenes: A Versatile Class of Macrocyclic Compounds: Topics in Inclusion Science*; Kluwer Academic: Dordrecht, The Netherlands, 1991; Vol. 3.
3. Gutsche, C. D.; Muthukrishan, R. J. *J. Org. Chem.* **1978**, *43*, 4905.
4. Gutsche, C. D.; Iqbal, M. *Org. Syn.* **1989**, *68*, 234.
5. Gutsche, C. D.; Dhawan, B.; Leonis, M.; Stewart, D. *Org. Syn.* **1989**, *68*, 238.
6. Munch, J. H.; Gutsche, C. D. *Org. Syn.* **1989**, *68*, 243.
7. Gutsche, C. D. *Calixarenes: A Versatile Class of Macrocyclic Compounds: Topics in Inclusion Science*; Kluwer Academic: Dordrecht, The Netherlands, 1991; Vol. 3, pp 3–37; Perrin, M. *Calixarenes: A Versatile Class of Macrocyclic Compounds: Topics in Inclusion Science*; Kluwer Academic: Dordrecht, The Netherlands, 1991; Vol. 3, pp 65–85.
8. Andreetti, G. D.; Ugozzoli, F. *Calixarenes: A Versatile Class of Macrocyclic Compounds: Topics in Inclusion Science*; Kluwer Academic: Dordrecht, The Netherlands, 1991; Vol. 3, pp 87–123.
9. Ungaro, R.; Pochini, A. *Calixarenes: A Versatile Class of Macrocyclic Compounds: Topics in Inclusion Science*; Kluwer Academic: Dordrecht, The Netherlands, 1991; Vol. 3, pp 127–147.
10. Arduini, A.; Ghidini, E.; Pochini, A.; Ungaro, R.; Andreetti, G. D.; Calestani, G.; Ugozzoli, F. *J. Incl. Phenom* **1988**, *6*, 119.
11. Schwing, M. J.; Mc Kervey, M. A. *Calixarenes: A Versatile Class of Macrocyclic Compounds: Topics in Inclusion Science*; Kluwer Academic: Dordrecht, The Netherlands, 1991; Vol. 3; Arnaud-Neu, F.;

Collins, E. M.; Deasy, M.; Ferguson, G.; Harris, S.; Kaitner, B.; Lough, A. J.; Mc Kervey, M. A.; Marques, E.; Ruhl, B. L.; Schwing, M-J.; Seward, E. M. *J. Am. Chem. Soc.* **1989**, *111*, 8691.

12. Alfieri, C.; Dradi, E.; Pochini, A.; Ungaro, R.; Andreetti, G. D. *J. Chem. Soc. Chem. Commun.* **1983**, 1075.
13. Djistra, P.; Brunnink, J. A.; Bugge, K. E.; Reinhoudt, D. N.; Harkema, S.; Ungaro, R.; Ugozzoli, F.; Ghidini, E. *J. Am. Chem. Soc.* **1989**, *111*, 7567.
14. Beer, P. D.; Martin, J. P.; Drew, M. G. B. *Tetrahedron* **1992**, *48*, 9917.
15. Seangprasertkij, R.; Asfari, Z.; Arnaud, F.; Weiss, J.; Vicens, J. *J. Incl. Phenom.* **1992**, *14*, 141.
16. Diamond, D.; Svelha, G.; Seward, E. M.; Mc Kervey, M. A. *Anal. Chim. Acta* **1988**, *204*, *223*; Telting-Diaz, M.; Diamond, D.; Smyth, M. R. *Anal. Chem. Acta*, **1991**, *251*, 149; Sakaki, T.; Harada, T.; Deng, G.; Kawabata, H.; Kawahara, Y.; Shinkai, S. *J. Incl. Phenom.* **1992**, *14*, 285; Perez-Jimenez, C., Harris, S. J., Diamond, D. *J. Chem. Soc. Chem. Commun.* **1993**, 480; Grigg, R.; Holmes, J. M.; Jones, S. K.; Norbert, W. D. J. A. *J. Chem. Soc. Commun.* **1994**, 185.
17. Foster, R. J.; Cadogan, A.; Telting-Diaz, M.; Diamond, D.; Harris, S. J.; McKervey, M. A. *Sensors and Actuators B* **1991**, *4*, 325.
18. Beer, P. D.; Dickson, C. A. P.; Fletcher, N.; Goulden, A. J.; Grieve, A.; Hodacova, J.; Wear, T. *J. Chem. Soc. Chem. Commun.* **1993**, 828.
19. Morzherin, Y.; Rudkevicht, D. M.; Verboom, W.; Reinhoudt, D. N. *J. Org. Chem.* **1993**, *58*, 7602.
20. Olmstead, M. M.; Sigel, G.; Hope, H.; Xu, X.; Power, P. *J. Am. Chem. Soc.* **1985**, *107*, 8087.
21. Shinkai, S.; *Calixarenes: A Versatile Class of Macrocyclic Compounds: Topics in Inclusion Science*; Kluwer Academic: Dordrecht, The Netherlands, 1991; Vol. 3, pp 173–198.
22. Asfari, Z.; Abidi, R.; Arnaud, F.; Vicens, J. *J. Incl. Phenom.* **1991**, *13*, 163.
23. Timmerman; Verboom, W.; van Veggel, F. C. J. M.; van Hoor, W. P.; Reinhoudt, D. N. *Angew. Chem. Inter. Ed. Engl.* **1994**, *33*, 1292.
24. Miyamoto, S.; Kollman, P. A. *J. Am. Chem. Soc.* **1992**, *115*, 3668.
25. Jain, M. K.; Jhagirdar, D. V. *Biochemistry* **1985**, *227*, 789.

Calmodulin

See: Monolayer and Langmuir–Blodgett Films (of Protein)

Capillary Electrophoresis Materials

See: Host–Guest Chromatography (with Cyclodextrin Derivatives)

CAPTODATIVE COMPOUNDS, POLYMERIZATION

Hitoshi Tanaka
Department of Optical Science and Technology
Faculty of Engineering
Tokushima University

Captodative (cd) substitution is the geminal substitution with both electron-withdrawing (captive) and -donating (dative) groups on the same atom (a carbon atom in most cases). In general, olefinic monomers have been classified into two categories, electron donors and acceptors, according to the density of the electric charge. The monomer and the related compounds with such cd substituents therefore belong to a new class of compounds, and are expected to display a special reactivity in free-radical polymerization due to the possibility of obtaining a stabilized cd radical.[1,2] An anomalous property due to such a synergistic polarization effect is also expected in the cd-substituted polymers.

At present, the cd concept seems to be widely accepted both theoretically and experimentally, and it is believed that such synergistic electronic effects of polar substituents bring about the thermodynamic stability of cd radicals.[3] This is shown abstractly in **Figure 1**, where radicals are stabilized by simultaneous substitution, with both a donor and an acceptor group whereas carbocations are stabilized by electron donors and carbanions by electron acceptors.

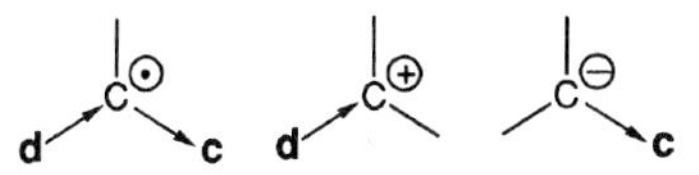

FIGURE 1. Stabilization of carbon radical, carbocation, and carbanion by electron-donating and electron-accepting groups.

PREPARATION AND PROPERTIES

Polymerization of cd Olefins

Compared with general organic chemistry, the cd concept has not been widely applied in polymer chemistry, probably because of the poor radical polymerizability of the cd monomers and because the systematic study in this field started only in about 1984.[4]

Principally, however, persistency of the cd radical is influenced by the ability of the substituents to delocalize the unpaired electron; the combination of electron-donating and -accepting groups often plays an important role in the stability and reactivity of the radical. Thus, some of the cd olefins combining appropriate donor and acceptor substituents are capable of undergoing a propagation step leading to homo-oligomerization or homopolymerization.

Captodative substitution sometimes activates an inert olefin to induce radical homopolymerization. In contrast to α-methylstyrene, α-cyanostyrenes, especially *o*-methoxy-α-cyanostyrene, do undergo radical homopolymerization.[5] Such polar effect is also seen in fully conjugated vicinal push–pull olefins.[6]

An important feature of the polymerizable cd olefins is that these olefins often undergo polymerization spontaneously without the presence of an initiator. Another important feature of the polymerizable cd olefins is that radical polymerization of these olefins is influenced by the solvents used in spite of the radical mechanism. The relative reactivity of cd olefins toward a polystyrl radical increases with increasing polarity of the solvents used.[7] This is probably due to the enhanced stabilization of the cd-substituted radical intermediate by the solvation.[8]

Captodative olefins generally constitute efficient radicophiles for the preparative trapping of various transient radicals and for a spin-trapping reagent for electron spin resonance (ESR).[9]

Polymerization of cd Methanes

Oxidative polyrecombination has provided polymers with a sterically hindered 1,1-disubstituted polymethylene structure, in contrast to a less-hindered polyethylene structure obtained by the usual vinyl polymerization, and has attracted much attention in connection with the syntheses of functional aromatic polymers such as electric conducting polydiarylphenol and polyaniline. The oxidative polymerization proceeds through the repeated coupling reactions of each radical produced by the oxidation.

Polymerization by cd Initiators

Coupling products of cd-substituted carbon radicals dissociate at moderate temperatures (50–130°C) to initiate the radical polymerization of vinyl monomers. In contrast to aryl ethanes such as tetraphenylethanes, aliphatic cd dimers cannot induce the inferter-type polymerization, but they can initiate even the nonconjugated vinyl monomers such as vinyl acetate.

Depolymerization of cd Polymer

The polymers obtained from the cd monomers (i.e., the cd polymers) generally undergo a thermal degradation easily. The polymethylene obtained by the photo-oxidative polymerization has a sterically hindered structure and a cd-substituted carbon sequence. Hence, polymethylene degrades or cyclizes even at relatively low temperatures (60–100°C), and it can initiate radical polymerization of vinyl monomers to give a block copolymer.[10]

APPLICATIONS

The cd compounds constitute a class of compounds that serve the unique polymerization and the polymer with excellent properties. The stimulating behaviors of cd molecules are applicable to the activation of an inert monomer in the radical polymerization and the preparation of sterically hindred polymer. the cd polymers obtained sometimes easily perform a polymer reaction to give the functional polymers such as spin polymer. It is also expected that the cd polymers show a high polarity and a strong dipole–dipole intermolecular stacking that gives orientated material. High refraction and SHG properties observed in poly(MAA) may be due to these characteristics of cd polymers. Therefore, cd polymers provide functional materials that have excellent optical, mechanical, and electronic properties.

REFERENCES

1. Tanaka, H. *Trends in Polym. Sci.* **1993**, *1*, 3618.
2. Penelle, J.; Padias, A. B.; Hall, H. K.; Tanaka, H. *Adv. Polym. Sci.* **1992**, *102*, 73.
3. Viehe, H. G.; Merenyi, R.; Stella, L.; Janousek, Z. *Angew. Chem., Int. Ed. Engl.* **1979**, *18*, 917.
4. Tanaka, H.; Miyake, H.; Ota, T. *J. Macromol. Sci., Chem.* **1984**, *A21*, 1523.
5. Tanaka, H.; Teraoka, Y.; Yoshida, S. *Polym. Inter.* **1995**, *38*, 199.
6. Tanaka, H.; Sakaguchi, M.; Kikukawa, Y.; Sato, T.; Ota, T. *Macromol. Chem. Phys.* **1994**, *195*, 2083.
7. Tanaka, H.; Kikukawa, Y.; Kameshima, T.; Sato, T.; Ota, T. *Makromol. Chem., Rapid Commun.* **1991**, *12*, 535.
8. Katritzky, A. R.; Zerner, M. C.; Karelson, M. M. *J. Am. Chem. Soc.* **1986**, *108*, 7213.
9. Viehe, H. G.; Janousek, Z.; Merenyi, R.; Stella, L. *Acc. Chem. Res.* **1985**, *18*, 148.
10. Tanaka, H.; Tsunemine, N.; Sato, T.; Ota, T. *Polym. Int.*. **1993**, *32*, 83.

CAPTODATIVE OLEFINS

Jacques Penelle
Laboratoire Cinétique et Macromolécules
Département de Chimie
Université Catholique de Louvain

Anne B. Padias and Henry K. Hall, Jr.
C.S. Marvel Laboratories
Department of Chemistry
The University of Arizona

By definition, captodative olefins are olefins geminally substituted by both an electron-donating (donor) and an electron-withdrawing (acceptor) substituent (by contrast, 1,2-disubstitution by the donor and acceptor groups leads to a push-pull substitution). This definition allows a large range of structural possibilities and, in practice, several thousands of organic molecules can be considered captodative olefins.

As this review is concerned with polymerizable olefins, only captodative olefins of the methylene type—that is, $H_2C=C(d)(c)$ with d as donor and c as receptor substituents—will be discussed here. Indeed, only these molecules have been described as polymerizable. In this respect, captodative olefins behave exactly like other olefins for which examples of acceptable polymerizability when tri- or tetra-substituted are very uncommon. Penelle et al. and Tanaka have published two review papers on captodative olefins polymerization.[1,2]

CAPTODATIVE RADICALS STABILIZATION

The main difference between captodative and classical olefins is that the former generate carbon-centered radicals simultaneously substituted by an acceptor and a donor substituent when subjected to free-radical polymerization. It is now generally accepted that the generated captodative radicals benefit from a special synergistic stabilization. In the literature, this stabilization has had many names: push-pull, mero-, and captodative stabilization. Only the last term is used now.

It is important to realize that captodative stabilization is only an extra stabilization energy—that is a stabilization energy that is added to the simple sum of the stabilizations conferred by the two polar substituents. In other words, captodative radicals are appreciably more stable than expected on the basis of the simple addition of the individual effects of each substituent.

An important conclusion that can be drawn from the above discussion is that *not all captodative radicals are highly stabilized.* The final stabilization will depend on the extra stabilization of course, but also on the magnitude of the stabilization of the individual substituents. If the individual stabilization energies are low, the final overall stabilization will also be low. This explains why an olefin-like methyl methacrylate (MMA) is not generally considered as a captodative olefin. MMA has the structural requirements (d = methyl, c = ester), but the stabilization energy provided by a methyl is so low that no extra stabilization can be observed for the radical.

Only captodative olefins with strong donor and acceptor substituents will be considered here. A large variety of combinations is possible for the donor and the acceptor substituents. In practice, however, the discussion will be restricted to only four types of olefins: α-substituted acrylic acids, acrylates, acrylamides, and acrylonitriles.

PREPARATION AND PROPERTIES OF THE POLYMERS

Polymerization of Captodative Olefins

In theory, both anionic and cationic polymerizations should be possible. In practice, examples of ionic polymerizations are very rare[4] and sometimes give polymers with a different structure.[5] It is difficult to say if this lack of examples results from an intrinsically low polymerizability under ionic conditions or (as yet) from a disinterest on the part of investigators in the field. Whatever the answer, most studies published to date on the polymerizability of captodative olefins deal with free-radical polymerizations.

As long as the steric hindrance due to the α-substituent is not substantial, captodative olefins give medium- to high-molecular weight polymers.

Very high molecular weights (M_n up to 10^6) have been obtained for acrylates α-substituted by some –NH(C=O)R substituents.[57] These high molecular weights were attributed to the captodative nature of the olefins, but this conclusion is questionable.

Polymerizations occurring without added initiator have been observed for some captodative acrylates.[37,38] These authors concluded that a strong tendency to spontaneous polymerization is one of the main characteristics of captodative olefins. However, this assertion would have to be confirmed by a careful reexamination of the purity of the concerned olefins.

Disproportionation of low molecular weight captodative radicals is a very uncommon reaction. Most captodative radicals prefer to recombine.[3] As a consequence, it is reasonable to consider that the termination process during the polymerization of captodative olefins occurs mainly by recombination.

Properties of the Polymers

Polymers derived from captodative olefins have been synthesized for various purposes and investigation of the properties has been very dependent of the context of the research involved.

An exception to this rule is thermal degradation. Most studies to date have shown that polymers derived from captodative olefins degrade at lower temperatures than their α-unsubstituted analogs. This probably results from the lower bond dissociation energies of a C–C bond in a (c)(d)C–C(c)(d) system (c and d represent the electron acceptor and donor substituent, respectively). Such head-to-head defects are produced when termination of the free-radical polymerization occurs by a recombination.

APPLICATIONS

The main interest of captodative olefins resides in their polyfunctional nature. By using captodative monomers, one can generate structure-dependent polymer properties. Applications that have been considered up to now include: metallic and non-metallic ion complexation (including metal recovery),[6–27,35,36] photophysics,[28,29,46,53–55,59] polymer complexation,[53–55,59] polynucleotide analogs,[30–34,47,58] polymeric monolayer synthesis,[41–44] nonlinear optics and high refractive index polymer,[37–40] oil recovery agent,[49] adhesive,[48,56] water-thickening agent,[48] electrostatographic toner,[50] gelatin substitute in photoemulsion,[10,51,52] dosimetry and spectrometry of fast neutrons[10,51,52] and hair treatment.[37–40]

REFERENCES

1. Penelle, J.; Padias, A. B.; Hall, H. K. Jr., Tanaka, H. *Adv. Polym. Sci.* **1992**, *102*, 73.
2. Tanaka, H. *Trends Polym. Sci.* **1993**, *1*, 361.
3. Viehe, H. G.; Merényi, R.; Stella, L.; Janousek, Z. *Angew. Chem. Int. Ed. Engl.* **1979**, *18*, 917.
4. Mathias, L. J.; Hermes, A. E. U.S. Patent 4 613 658; *Chem. Abstr.* **1987**, *106*, 33628W.
5. Mathias, L. J.; Hermes, R. E. *Polym. Prepr., Am. Chem. Soc. Div. Polym. Chem.* **1986**, *27*(1), 202.
6. Ivanov, S. S.; Koton, M. M. *Vysokomol. Soedin.* **1961**, *3*, 248; *Chem. Abstr.* **1961**, 55, 26516a.
7. Adams, R.; Johnson, J. L.; Englund, B. *J. Am. Chem. Soc.* **1950**, *72*, 5080.
8. Ivanov, S. S.; Gavryuchenkova, L. P.; Koton, M. M. *Vysokomol. Soedin. Ser B* **1966**, *8*, 470; *Chem. Abstr.* **1966**, *65*, 810h.
9. Ivanov, S. S.; Gavryuchenkova, L. P.; Koton, M. M. *Vysokomol. Soedin., Ser A* **1967**, *9*, 103; *Chem. Abstr.* **1967**, *67*, 11710r.
10. Babkin, V. V.; Ivanov, S. S. *Tr. Leiningr. Inst. Kinoinzhenerov* **1963**, *25; Chem. Abstr.* **1964**, *61*, 9087c.
11. Sopova, O. I.; Ivanov, S. S.; Gavryuchenkova, L. P. *U.S.S.R., 227*, 086; *Chem. Abstr.* **1969**, *70*, 53017n.
12. Boldyrev. A. G.; Gavryuchenkova, L. P.; Kuvshinskii, E. V. *Sin., Strukt. Svoistva Polim.* **1970**, *149*; *Chem. Abstr.* **1972**, *76*, 94010x.
13. Gavryuchenkova, L. P.; Jvanov, S. S.; Karavaeva, V. A.; Koton, M. M.; Sokolova, E. A. *Tr. Metrol. Inst. SSSR* **1968**, *93*, 119; *Chem. Abstr.* **1969**, *70*, 110012h.
14. Gavryuchenkova, L. P.; Ivanov, S. S.; Koton, M. M. *Polym. Sci. USSR (Engl. Transl.)* **1972**, *A14*, 175; *Chem. Abstr.* **1976**, *76*, 127586v.
15. Boldyrev., A. G.; Gavryuchenkova, L. P.; Kuvshinskii, E. V. *Zh. Obshch. Khim.* **1974**, *44*, 914; *Chem. Abstr.* **1974**, *81*, 92162c.
16. Tolmachev, V. N.; Yushko, V. K. *Polym. Sci. USSR* (*Engl. Transl.*) **1975**, *A17*, 1990; *Chem. Abstr.* **1975**, *83*, 193780z.
17. Yushko, V. K.; Gnidenko, V. I.; Lugovaya, Z. A. *Sint. Fiz.-Khim. Polim.* **1975**, *16*, 129; *Chem. Abstr.* **1976**, *84*, 136242x.
18. Asquith, R. S.; Gardner, K. L.; Yeung, K. W. *J. Polym. Sci., Polym. Chem.* **1978**, *16*, 3275.
19. Brasswell, D.; Nelson, G.; St. Pierre, T.; Reams, R.; Lewis, E. A. *Proc. IUPAC, IUPAC 28th Macromol. Symp.* **1982**, 354.
20. Nelson, G.; Chang, C.; St. Pierre, T. *Macromol. Synth.* **1985**, *9*, 89.
21. Lee, C. K. *Pollimo* **1987**, *11*, 42; *Chem. Abstr.* **1987**, *106*, 214427a.
22. Park, I. H.; Lee, C. K.; Choi, J. H.; Jim, J. J. *Taehan Hwahakhoe Chi* **1982**, *26*, 235; *Chem. Abstr.* **1982**, *97*, 198603t.
23. Tomida, T.; Ikawa, T.; Masuda, S. *J. Chem. Eng. Jpn.* **1993**, 26, 575.
24. Masuda, S.; Kobayashi, T.; Tomida, T.; Inoue, T.; Tanaka, M.; Asahi, Y. *Polymer* **1993**, *34*, 4562.
25. Ivanov, S. S. *Vysokomol. Soedin.* **1963**, *5*, 1140; *Chem. Abstr.* **1963**, *59*, 11671e.

26. Ivanov, S. S.; Gavryuchenkova, L. P.; Koton, M. M. *Vysokomol. Soedin.* **1965**, *7*, 1693; *Chem. Abstr.* **1966**, *64*, 3699b.
27. Ivanov, S. S.; Gavryuchenkova, L. P.; Koton, M. M. *Vysokomol. Soedin., Ser. B.* **1967**, *9*, 406; *Chem. Abstr.* **1967**, *6*, 76362c.
28. Arora, K. S.; Overberger, C. G. *J. Polym. Sci., Polym. Lett.* **1983**, *21*, 189.
29. Arora, K. S.; Overberger, C. G. *J. Polym. Sci., Polym. Chem.* **1985**, *23*, 3007.
30. Brandt, K. A.; Overberger, C. G. *J. Polym. Sci., Polym. Chem.* **1985**, *23*, 1981.
31. Overberger, C. G.; Brandt, K. A.; Kikyotani, S.; Ludwick, A. G. *Polym. Bull.* **1981**, *5*, 481.
32. Overberger, C. G.; Brandt, K. A.; Kikyotani, S.; Ludwick, A. G.; Morishima, Y. *Proc. IUPAC, IUPAC 29th Macromol. Symp.* **1982**, 376.
33. Brandt, K. A.; Overberger, C. G. *Nouv. J. Chim.* **1982**, *6*, 673.
34. Fricke, K. W.; Zand, R. *Polym. Prepr., Am. Chem. Soc. Div. Polym. Chem.* **1987**, *28*(2), 300.
35. Yushko, V. K.; Churilo, A. A. *Vestn, Khar'k. Univ.* **1988**, *319*, 81; *Chem. Abstr.* **1991**, *114*, 186618h.
36. Chung, P. S.; Lee, C. K. *Pollimo* **1991**, *15*, 749; *Chem. Abstr.* **1992**, *116*, 256125n.
37. Tanaka, H.; Kameshima, T.; Sasai, K.; Sato, T.; Ota, T. *Makromol. Chem.* **1991**, *192*, 427.
38. Tanaka, H.; Yikukawa, Y.; Kameshina, T.; Sato, T.; Ota, T. *Makromol. Chem., Rapid Commun.* **1991**, *12*, 535.
39. Tanaka, H.; Hall, H. K. Jr. *Macromol. Chem. Phys.* **1994**, *195*, 2073.
40. Tanaka, H. *J. Mater. Sci. Lett.* **1994**, *13*, 545.
41. Mathias, L. J.; Hermes, R. E. *Macromolecules* **1988**, *21*, 11.
42. Hermes, R. E.; Mathias, L. J.; Virden, J. W. Jr. *Macromolecules* **1987**, *20*, 901.
43. Hermes, R. E.; Mathias, L. J.; Virden, J. W. Jr. *Polym. Prepr., Am. Chem. Soc. Div. Polym. Chem.* **1987**, *28*(2), 443.
44. Mathias, L. J. *Polym. Prepr., Am. Chem. Soc. Div. Poly. Chem.* **1988**, *29*(1), 48.
45. Mathias, L. J. *Poly. Commun.* **1988**, *29*, 352.
46. Arora, K. S.; Overberger, C. G. *J. Polym. Sci., Polym. Chem.* **1984**, *22*, 1587.
47. Overberger, C. G. *J. Macromol. Sci.-Chem.* **1984**, *A21*, 1607.
48. Mullins, M. J.; Brondsema, P. J. U.S. Patent 4 734 453; *Chem. Abstr.* **1988**, *109*, 129814s.
49. Platt, J. L., Jr. U.S. Patent 4 282 343, 1981; *Chem. Abstr.* **1981**, *95*, 187892u.
50. Canon, K. K. *Jpn. Kokai Tokkyo Koho 81*, 69, 643; *Chem. Abstr.* **1981**, *95*, 229286g.
51. Bradna, F. *Photogr. Korresp.* **1966**, *102*, 175; *Chem. Abstr.* **1967**, *65*, 51429s.
52. Bradna, F. Czech. Patent 159 389; *Chem. Abstr.* **1976**, *85*, 26635r.
53. Sivadasan, K.; Somasundaran, P. *J. Polym. Sci.: A.: Polym. Chem.* **1991**, *29*, 911.
54. Turro, N. J.; Arora, K. S. *Polymer* **1986**, *27*, 783.
55. Arora, K. S.; Turro, N. J. *J. Polym. Sci., Polym. Phys.* **1987**, *25*, 243.
56. Wegemund, B.; Galinke, J. (Henkel und Cie) Ger. Offen. 1 916 648; *Chem. Abstr.* **1971**, *74*, 112802x.
57. Mathias, L. J.; Hermes, R. E. *Macromolecules* **1986**, *19*, 1536.
58. Zand, R.; Jiang, Z. Z.; Overberger, C. G. *Polym. Prepr., Am. Chem. Soc. Div. Polym. Chem.* **1985**, *26*(2), 216.
59. Chandar, P.; Somasundaran, P.; Turro, N. J. *Macromolecules* **1988**, *21*, 950.

CARBENOID POLYCONDENSATION

Yasuyuki Tezuka
Department of Organic and Polymeric Materials
Tokyo Institute of Technology

Carbenoid polycondensation, a new polymerization reaction system in which an exceptionally selective carbenoid coupling reaction constitutes an elementary step in the polymerization process, can offer a new methodology for macromolecular synthesis and can extend the choice of monomer reagent in polymerization reactions.

Carbenoids are recognized as key reactive intermediates in a variety of transition-metal-catalyzed polymerization reactions, including the ring-opening metathesis polymerization of cyclic alkenes and the metathesis polycondensation of acyclic dienes.[1,2] There have been, however, few examples of efficient polymerization systems involving a carbenoid coupling reaction as an elementary polymerization process.[3]

NEW CARBENOID COUPLING REACTION WITH HIGH SELECTIVITY

A novel reaction system has been recently developed in which a carbenoid coupling reaction takes place in an almost quantitative yield.[4] Thus, a stoichiometric reaction between copper metal and a number of organic halides, in particular *gem*-dichlorides, in dimethyl sulfoxide (DMSO) was found to generate copper carbenoid intermediates under extremely mild conditions. With the formation of a copper complex, $CuCl_2(DMSO)_2$, as a common reaction product, olefinic compounds were obtained from cichloracetic acid esters and dichlorodiphenylmethane via the coupling reaction of carbenoid intermediates generated by α,α-dichloro elimination of the starting *gem*-dichlorides.

By making use of this new carbenoid chemistry, a novel polymerization system has been developed by using dichloroacetic acid diesters as monomers, in which a successive carbenoid coupling reaction takes place with high selectivity, leading to the formation of macromolecular products. Thus, this system is termed carbenoid polycondensation.[5,6]

HOMOPOLYMERIZATION

A series of dichloracetic acid diesters was readily prepared by the reaction of the precursor diols, such as alkane diols and diethylene glycohol, with dichloroacetyl chloride in the presence of pyridine and were purified to the polymerization grade by a dry-column chromatography technique on silica gel. The carbenoid polycondensation of these "monomers" was carried out with copper metal in DMSO at 25–50°C, resulting in a gradual dissolution of copper metal into the reaction medium to form $CuCl_2\,(DMSO)_2$. The resinous or powdery product was isolated after the treatment with dry-column chromatography on silica gel. The subsequent spectroscopic and chromatographic analysis (NMR, IR, and gel permeation chromatography) provided the structure of the product to be poly(alkylene fumarate/maleate), which was formed through the coupling reaction of the carbalkoxy carbenoid intermediates generated by the α,α-dichloro elimination from the starting dichloracetates. The degree of polymerization of the product increased

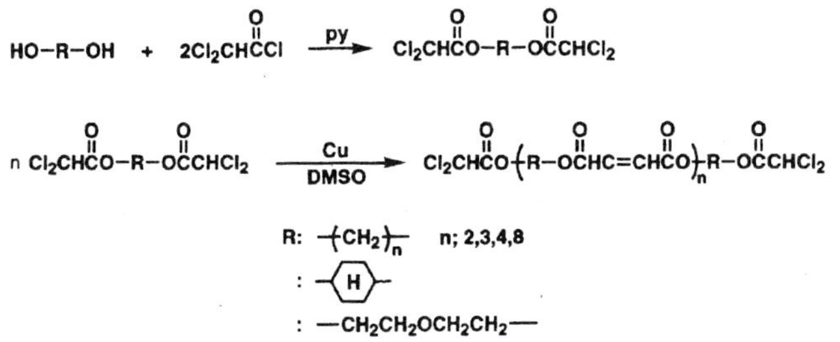

$$HO-R-OH + 2Cl_2CHCOCl \xrightarrow{py} Cl_2CHCO-R-OCCHCl_2$$

$$n\ Cl_2CHCO-R-OCCHCl_2 \xrightarrow[DMSO]{Cu} Cl_2CHCO\{R-OCHC=CHCO\}_n R-OCCHCl_2 \quad \mathbf{I}$$

$$R: \ +(CH_2)_n + \quad n;\ 2,3,4,8$$
$$: \ -\langle H \rangle-$$
$$: \ -CH_2CH_2OCH_2CH_2-$$

either by the increase of the feed amount of copper against the monomer, by the extension of the reaction time, or by the elevation of the reaction temperature.

COPOLYCONDENSATION

Copolycondensation of two different types of cichloroacetic acid diesters (monomers) was carried out in two procedures: by the simultaneous addition of a mixture of the different types of monomers to start the reaction or by the consecutive addition of the different types of monomers in the course of the reaction. The former process produces random copolyesters and the latter block copolyesters.

CARBENOID POLYCONDENSATION OF TELECHELIC POLYETHERS

The carbenoid polycondensation process was applied to a macromolecular coupling reaction of telechelic polyethers having dichloracetic acid ester end groups.[7] Thus, a series of poly(ethylene oxide) and polytetrahydrofuran having dichloracetic acid ester end groups were prepared and were subjected to the carbenoid polycondensation process. The exceptionally selective carbenoid coupling process could produce segmented polyethers containing reactive fumarate/maleate groups within the main chain at predetermined intervals.

The segmented polyethers having reactive fumaric/maleic ester groups were subsequently utilized as a macromolecular crosslinking (curing) reagent. Gel products were obtained by adding them during the polymerization reaction of styrene. This is relevant to the preparation of commercial unsaturated polyester resin.

REFERENCES

1. Grubbs, R. H.; Tumas, W. *Science* **1989**, *243*, 907.
2. Patton, J. T.; Boncella, J. M.; Wagener, K. B. *Macromolecules* **1992**, *25*, 3862.
3. DeKoninck, L.; Smets, G. *J. Polym. Sci., Part-A* **1969**, *7*, 3313.
4. Tezuka, Y.; Hashimoto, A.; Ushizaka, K.; Imai, K. *J. Org. Chem.* **1990**, *55*, 329.
5. Tezuka, Y.; Ushizaka, K.; Nakayama, Y.; Imai, K. *J. Chem. Soc., Chem. Commun.* **1991**, 153.
6. Tezuka, Y.; Nakayama, Y.; Ushizaka, K.; Imai, K. *Macromolecules* **1993**, *26*, 921.
7. Tezuka, Y.; Hoshino, C. *J. Polym. Sci., Part-A, Polym. Chem.* **1994**, *32*, 897.

Carbomer Resins

See: Water-Swellable Polymers (Carbomer Resins)

Carbon Black

See: Carbon Black (Overview)
Carbon Black (Graft Copolymers)
Electrically Conducting Composites
Microencapsulated Particles

CARBON BLACK (Overview)

Jean-Baptiste Donnet and Ton Kuan Wang
Ecole Nationale Supérieure de Chimie de Mulhouse and ICSI

The term "carbon black" refers to a group of powdered solid materials manufactured industrially by the incomplete combustion or the thermal decomposition of aromatic hydrocarbons and in which carbon is the essential component element.

World carbon black production has reached nearly 7 million tons. It is used mainly (about 90%) in the rubber industry as a rubber reinforcing agent and filler. Other applications include plastics as additives for different functions, xerographic toners, coatings, and ink pigment.

SYNTHESIS AND FORMATION MECHANISMS, FULLERENES

Different raw materials, such as natural gas, acetylene, oils from coal or petroleum containing aromatic hydrocarbons, can be used to manufacture carbon black. Manufacturing methods can be classed according to type of chemical reaction into thermal-oxidation decomposition processes (including furnace, lamp, and channel black) and thermal processes. The difference between the two categories is the presence or absence of oxygen in the carbon black formation.[1] The furnace process is by far the most important for manufacturing the rubber grade carbon blacks.

Recently, a new carbon molecule family called "fullerenes" (C_{60}) was discovered and extensively studied.[3] It has been proposed that the same precursor "corannulene" is responsible for both carbon black and fullerene formation.[4]

PROPERTIES

Physical Properties

Morphology

Carbon black is made up of spherical elementary particles. The commercial rubber-grade blacks have a mean particle diameter ranging from 17 nm for N110 to 107 nm for N762.

In most cases the carbon black particles coalesce to form aggregates that constitute true morphological units. *Aggregate size and shape* differ. Typically, the average aggregate diameter for commercial rubber grade blacks varies between 70–480 nm for N110–N990.[5] The aggregates have a more or less open structure, called high- or low-structure.

The *porosity* of carbon black is still a problem not completely resolved because of its complexity, such as open and closed pores, pore shape, size, and distribution.

Surface area is an important parameter for carbon black.

The *density* of carbon black is defined in different ways: True density, determined by crystallographic methods such as X-ray diffraction, ranges from 2.04 for untreated black to 2.25 for graphitized ones.

Intrinsically carbon black is a semiconductor material and its *electrical properties* in the dry state depend on graphitization degree, compression pressure, surface cleanness, porosity, and particle geometry. In the polymer matrix, there is a threshold loading (establishment of particle percolation) to conducting electricity and the conductivity increases with the charge loading until a limiting value is reached.[7] Some industrial carbon blacks of conducting grade have been developed recently for applications.[7]

The *magnetic properties* are twofold: they can be diamagnetic due to the graphitic layer and paramagnetic due to the unpaired electrons.

The crystallite size, especially the L_a value, is an important factor influencing the *thermal conductivity.* In general, the industrial grade carbon blacks have little difference between them.

Carbon black is one of the darkest materials ever found. Particle size is the main parameter determining its *optical properties*: light absorption, diffusion, and coloring strength.

According to the application domains, *classification* of carbon blacks falls roughly into rubber- and nonrubber-grade categories. They can also be divided into furnace, lamp, channel, and acetylene blacks, according to their production processes. Each producer has its own *nomenclature* system; however, the American Society for Testing Materials (ASTM) has proposed a four-digit system for carbon black classification (D1765): the first digit is a letter indicating the relative curing rate, N (normal) or S (slow); the next digit (1–9) is a measure of particle size range; and the final two digits are designated by the supplier.

Microstructure

The crystallites inside the carbon black elementary particles have a graphite-like *bulk structure* that has been studied extensively by X-ray diffraction.

The recently developed scanning probe microscopy (STM, AFM) is an excellent method to characterize carbon black *surface microstructure*. Carbon blacks are particularly suitable for STM measurement because of their electrical conductivity.[8]

Chemical Properties

Bulk Chemical Composition

Apart from the predominant carbon, oxygen and hydrogen are always present in a small proportion. The impurities in water for quenching and pelletizing are mainly responsible for ash content (<0.5%) and eventually some sulfur. Sulfur comes mainly from the feedstocks.

Surface Groups

The elemental composition of most carbon black surfaces is very different from the bulk content and the presence of some elements, especially oxygen, is much richer on the surface. The oxygen atoms form different surface groups that may play an important role in the carbon black properties.

Chemical Modifications

Heat treatment has two relevant effects on carbon black chemical structure: increasing crystallinity and decomposing surface active groups. *Plasma treatment* is an efficient method to modify carbon black surface. It can destroy partly the surface crystalline zones to increase the particle rugosity and create new function groups in the presence of other reactants.[10] *Chemical oxidation* has been done with different oxidizing agents such as air, ozone, air plus nitric oxide, and nitric acid.[11]

Surface grafting is another method to supply the black surface with different chemical functions. Molecules can be grafted in several ways, for example, by free radicals to form macromolecules or condensation reaction to give esters.[12] *Other surface reactions* can introduce different surface groups: halogens,[13,14] nitrogen-[13] or sulfur-containing[9] groups. *Molecular intercalation* in carbon black has not drawn much attention in research, as is the case for graphite. However, a graphitelike crystallite structure is present, and research may be rewarding.

Physico-Chemical Properties of Carbon Black Surface

These properties have a primary importance because carbon blacks are used basically as charges in the polymer matrix. The material performance can be greatly influenced by the polymer–carbon black surface interactions. Several techniques have been developed for this study: immersion calorimetry[6] and especially inverse gas chromatography (IGC) which, by measuring the adsorption behaviors of different small probe molecules, can give rich information about these interactions.[15]

Superficial Site Energy Distribution

The superficial site energy distribution is a pseudointrinsic property of the surface which can be deduced from IGC measurement. The energetic adsorption sites of physical and chemical nature as well as their occurrence frequency are easily detected.

Reinforcing Effect in Rubbers

When the carbon black is incorporated into a rubber matrix, the strength-related properties of the rubber can be largely improved, depending on the nature of the black. Naturally the influence of carbon blacks on rubber properties can be directly tested on rubber compounds either at a fixed loading or at a loading that gives optimum reinforcement. Heat treatment of carbon blacks has a significant effect, reducing their reinforcing capacities while several other chemical modifications (such as oxidation, reduction, grafting) have little influence.[16]

APPLICATIONS

Rubber Formulation

Carbon black is the most important additive in a rubber formulation. Its loading can vary between 30–70 parts per hundred parts of rubber (phr). Tire and mechanical goods consume about 65% and 25% of total black production, respectively. Furnace blacks are mostly used for rubber reinforcement. They can be roughly classed into reinforcing and semireinforcing grades by their particle size (<35 or >45 nm). The antiabrasion property of carbon black increases with the decrease in particle size.

Other Applications

Only about 10% of carbon black products are used in nonrubber applications, these consist of: plastics (36%), printing inks (30%), coating (9%), paper (4%), and miscellaneous applications. The blacks for these applications are produced primarily by processes other than the furnace method.

In plastics, carbon black can have several functions: electrical conductance, UV shielding, opacity, pigment, and mechanical performance improvement. These properties have been used in wire and cable manufacture for electromagnetic interference and electrostatic shielding, in plastics protection of UV degradation or to provide color.[17] Each application needs special carbon black properties.

With the development of xerography carbon black continues to be the most important pigment in printing or duplicating toner applications.[18] The black loading in toner particle formulation is 5–10%. High blackness, good dispersion, and electric properties are necessary in this case. Use of carbon black in inks, paints, and coatings have similar requirements as in the xerographic toners, but with specific rheological property in liquid media.

REFERENCES

1. Kühner, G.; Voll, M. In *Carbon Black*, 2nd ed.; Marcel Dekker: New York, 1993; Chapter 1.
2. Bansal, R. C.; Donnet, J. B. . In *Carbon Black*, 2nd ed.; Marcel Dekker: New York, 1993; Chapters 2 and 4.
3. Kroto, H. W.; Allaf, A. W.; Balm, S. P. *Chem. Rev.* **1991**, *91*, 1213–1235.
4. Donnet, J. B.; Custodéro, E. *Bull. Soc. Chim. de France* **1984**, *131*, 1115.
5. Hess, W. M.; Herd, C. R. . In *Carbon Black*, 2nd ed.; Marcel Dekker: New York, 1993; Chapter 3.
6. Stoeckli, F.; Huguenin, D.; Laederach, A. *Proceedings of the Second International Conference on Carbon Black*; Mulhouse: 1993; p 43.
7. Probst, N. In *Carbon Black*, 2nd ed.; Marcel Dekker: New York, 1993; Chapter 8.
8. Donnet, J. B.; Custodéro, E. *Carbon* **1992**, *30*, 813.
9. Chang, C. H. *Carbon* **1981**, *19*, 175.
10. Wang, W. D. et al. *Proceedings of the Second International Conference on Carbon Black*; Mulhouse: 1993; p 207.
11. Donnet, J. B.; Voet, A. In *Carbon Black*, 1st ed.; Marcel Dekker: New York, 1976; p 147.
12. Donnet, J. B.; Vidal, A.; Papirer, E. *Chemistry and Physics of Carbon*; Walker, P., Thrower, P., Eds.; 1975; Vol. 12, pp 171–207.
13. de Bruin, W. J.; van der Plas, T. *Rev. Gen. Caout.* **1964**, *41*, 543.
14. Rivin, D.; Aron, J. *Abstracts of Papers*, Seventh Conference on Carbon, Cleveland, OH, 1965.
15. Donnet, J. B.; Lansinger, C. M. *Kautschuk Gummi Kunstoffe* **1992**, *45*, 459–486.
16. Donnet, J. B.; Eckhardt, C.; Voet, A. *Rev. Gen. Caout. Plast.* **1967**, *44*, 627–633.
17. Funt, J. M.; Sifleet, W. L.; Tomme, M. In *Carbon Black*, 2nd ed.; Marcel Dekker: New York, 1993; Chapter 12.
18. Julien, P. C. In *Carbon Black*, 2nd ed.; Marcel Dekker: New York, 1993; Chapter 13.

CARBON BLACK (Graft Copolymers)

Norio Tsubokawa
Department of Chemistry and Chemical Engineering
Faculty of Engineering
Niigata University

Carbon black is heat resistant, chemical proof, weather proof, lightweight, electroconductive, a deep black pigment, and a low thermal expansion. It is widely used industrially as a filler of polymer-composite. Carbon blacks are also excellent reinforcing agents for rubbers.

By surface grafting polymers onto carbon black, the dispersibility in solvents and compatibility in polymer matrices are markedly improved.[1–4] IN addition, grafting onto carbon black enables us to give various functions to carbon black such as photosensitivity, bioactivity, crosslinking ability, and an amphiphilic property.[3,4] Therefore, the surface grafting of various polymers onto carbon black has been widely investigated.

PREPARATION

Reactivity of Carbon Black

Carbon black is composed of ultrafine particles of 10–400 nm diameter that aggregates: this aggregate is the so-called "structure."[5] Carbon black is 90–99% carbon, 0.1–10% oxygen, 0.2–1.0% hydrogen, and a small amount of sulfur and ash.[6,7]

Carbon black has oxygen-containing groups, such as carboxyl, phenolic hydroxyl, and quinonic oxygen groups on the surface.[6–9] It is well known that these functional groups undergo normal organic reactions, such as methylation, esterification, neutralization, and nucleophilic substitution.[9–11]

Grafting onto Carbon Black

It is well known that during the radical polymerization of vinyl monomers in the presence of carbon black by using an initiator, marked retardation or inhibition of the polymerization is observed.[2,4] In addition, a part of the polymer formed is grafted onto the carbon black surface depending on the capture of growing polymer radicals by carbon black, whereas the percentage of grafting is very small.

Grafting from Carbon Black

Radical Grafting from Carbon Black Surface

For the design of the graft polymerization from carbon black, the introduction of various functional groups onto the surface has been investigated.

We have reported the preparation of carbon black having peroxyester groups by the reaction of surface acyl chloride groups with *tert*-butyl hydroperoxide.[12]

Recently, the radical graft polymerization of vinyl monomers, initiated by a system consisting of ceric ion and alcoholic hydroxyl groups on the surface was reported.[13]

Anionic Grafting from Carbon Black Surface

We have reported that alkali metal carboxylate groups introduced onto carbon black surface initiate the anionic ring-opening polymerization of β-propiolactone (PL) to give the polyPL-grafted carbon black.[14] In addition, we succeeded in the anionic ring-opening alternating copolymerization of epoxides with cyclic acid anhydrides initiated by alkali metal carboxylate groups on carbon black surface.[15,16]

Cationic Grafting from Carbon Black Surface

Carboxyl groups on the surface of carbon black have an ability to initiate the cationic polymerization of vinyl monomers, whereas the percentage of grafting is less than 15% and the initiating activity is limited to a part of vinyl monomer.[17,18]

Grafting Reaction of Carbon Black with Polymers

By reacting polymers with surface functional groups on carbon black, various polymers can be grafted onto the surface. This grafting was achieved by either the reaction of reactive carbon black, (i.e., carbon black having reactive functional groups) with commercially available polymers having hydroxyl or amino groups or surface carboxyl and/or phenolic hydroxyl groups on carbon black with functional polymers having terminal reactive groups.

Reaction of Reactive Polymers with Carbon Black

Papirer and co-workers have reported the grafting of polymers by the reaction of carbon black having methyl ester groups with a living polymer anion.[19]

Carbon blacks are known to act as a strong radical scavenger.[20,21]

APPLICATIONS

Electrical behavior of carbon black–polymer composition was widely investigated by many researchers because it can be used in electric shielding, heating, and the prevention of static electricity accumulation.[22,23] Because polymer-grafted carbon blacks disperse uniformly in organic solvents and polymer matrices, it is possible to make conductive materials with uniform conductivity over the entire surface. In addition, polymer-grafted carbon black–polymer composition, which was crosslinked with a variety of crosslinking agents, shows a large positive temperature coefficient (i.e., at near the glass-transition transition temperature or melting point of the matrix polymer). This is considered to be a result of widening the gap between carbon black particles and/or decreasing the carbon content by thermal expansion of the polymer matrix.

REFERENCES

1. Laible, R.; Hamann, K. *Adv. Colloid Interface Sci.* **1980**, *13*, 65.
2. Ohkita, K. *Grafting of Carbon Black*; Rubber Digest: Tokyo, 1983.
3. Tsubokawa, N. *Shikizai Kyokaishi* **1992**, *65*, 72.
4. Tsubokawa, N. *Prog. Polym. Sci.* **1992**, *17*, 417.
5. Switzer, C. W.; Goodrich, W. C. *Rubber Age* **1944**, *55*, 469.
6. Studebaker, M. L. *Rubber Chem. Technol.* **1957**, *30*, 1400.
7. Donnet, J. B.; Voet, A. *Carbon Black*; Marcel Dekker: New York, 1976.
8. Boehm, H. P. *Adv. Catalysis* **1966**, *16*, 198.
9. Donnet, J. B. *Carbon* **1982**, *20*, 266.
10. Fujiki, K.; Magara, K.; Tsubokawa, N.; Sone, Y. *Nippon Gomu Kyokaishi* **1991**, *64*, 378.
11. Tsubokawa, N.; Muramatsu, M.; Fujiki, K. *Shikizai Kyokaishi* **1993**, *66*, 405.
12. Tsubokawa, N.; Fujiki, K.; Sone, Y. *Polym. J.* **1988**, *20*, 213.
13. Tsubokawa, N.; Fujiki, K.; Sone, Y. *J. Macromol. Sci., Chem.* **1988**, *A25*, 1159.
14. Tsubokawa, N.; Fujnaki, A.; Hada, Y.; Sone, Y. *J. Polym., Sci., Polym. Chem. Ed.* **1982**, *20*, 3297.
15. Tsubokawa, N.; Tamada, A.; Sone, Y. *Polym. Bull.* **1983**, *10*, 63.
16. Tsubokawa, N.; Yamada, A.; Sone, Y. *Polym. J.* **1984**, *16*, 333.
17. Tsubokawa, N. *J. Polym. Sci., Polym. Lett. Ed.* **1980**, *18*, 461.
18. Tsubokawa, N.; Takeda, N.; Iwasa, T. *Polym. J.* **1981**, *13*, 1093.
19. Papirer, E.; Tao, N.; Donnet, J. B. *J. Polym. Sci., Polym. Lett. Ed.* **1971**, *9*, 195.
20. Hey, D.; Williams, G. *Discuss Faraday Soc.* **1953**, *14*, 216.
21. Ley, M.; Szwarz, M. *J. Chem. Phys.* **1954**, *22*, 1621.
22. Ohkita, K.; Fukushima, K. *Japan Plastics* **1969**; *3*(3), 6 **1969**, *3*(4), 25.
23. Miyauchi, S.; Togashi, T. *J. Appl. Polym. Sci.* **1985**, *30*, 2743.

CARBON/CARBON COMPOSITES (With Surface Treated Carbon Fibers)

Chen-Chi M. Ma, Nyan-Hwa Tai, Wen-Chi Chang, and Chin-Kun Fang
Institute of Chemical Engineering
Nyan-Hwa Tai Materials Science Center
National Tsing Hua University

Carbon–carbon (C/C) composites exhibit superior mechanical properties at elevated temperatures. They consist of a carbonaceous matrix reinforced with carbon or graphite fibers, having uni-, bi- or multidirectional alignment. One can obtain the desired mechanical properties of the composites by controlling the microstructure of the matrix, the type of fiber used, and the interaction between fiber and matrix.[1–5]

Preparation of carbon–carbon composites involves complicated multistage processes.[6,7] Interactions between the carbon fiber and matrix are important factors that control the mechanical properties of carbon–carbon composites. A good adhesion between carbon fiber and matrix is a precondition for the stress transfer from matrix to fiber reinforcements. The oxide groups on the surface of carbon fibers can enhance the chemical bonding between the carbon fiber and some polymer matrices such as epoxy and phenolic resin.[8–10] The improvement in adhesion at the fiber–matrix interface is not only a result of functional groups but also due to the increase of surface roughness created during oxidation treatment of the fibers.[9–12]

Polymer matrix composites can have very high fiber strength when good bonding exists between the fiber and matrix. However, carbon–carbon composites have brittle matrices and special treatment is needed to ensure interfacial bonding.[13–17] Strong bonding between the fiber and matrix can cause cracking when the matrix shrinks during carbonization processes. Transverse cracking in the matrix may propagate through the fiber and results in weak mechanical properties in these composites. On the other hand, a weak interfacial bonding will lead to a crack deflection at the fiber–matrix interface and the fiber pull-out will increase fracture toughness of carbon–carbon composites.

However, weak interfacial bonding may obstruct the stress transfer from matrix to fibers. If the interfacial bonding is in the right range, a crack will deflect at the interface and the desired strength will be achieved.

In this research, we used surface-treated carbon fibers with controlling tensile strength to study the effect of fiber surface functional groups on the mechanical properties and fracture behavior of carbon–carbon composites carbonized at 1000°C and graphitized at 2200°C.

CONCLUSIONS

When carbon fiber is treated with nitric acid, a lower tensile strength with broader distribution is observed with long treatment times. The tensile strength of carbon fiber depends on the flaws on the surface and the structure. The surface morphology changes significantly after acid treatment, hence, it broadens the distribution of fiber strength. Furthermore, the oxide groups on the surface of the carbon fiber increase with the increase in treatment time. Because of the formation of residual tensile stress during the carbonization process and good interfacial bonding, C–C composites made with treated fibers show lower flexural strength than those made with untreated fibers. After graphitization, the C–C composites made with treated fibers show higher flexural strength than those made with untreated fibers due to the formation of a stress-graphitized layer in the matrix around the fiber–matrix interfacial region.

APPLICATIONS

Carbon–carbon composites are widely used in aircraft braking systems (e.g., in the Concorde SST, Airbus A320, Boeing 747-400 and military aircraft). Replacing conventional brakes with carbon–carbon brakes saves about 450 kg on a commercial aircraft. Carbon/carbon disc brakes reduce the stopping distance and have a long wear life. The use of carbon–carbon composites brakes has been exploited or postulated for a number of vehicles such as Formula 1 racing cars, high-speed trains, and other land vehicles.[6] Use of carbon–carbon composites on the nose cone and leading edge of the Space Shuttle is one of the most successful applications of the material.[18] Carbon/carbon composites possess the best mechanical properties at elevated temperature that other materials cannot withstand. Recently, carbon–carbon composites have been successfully tested during the course of the time–temperature–loading cycles representative of the National Aerospace Plane (NASP) airframe.[19] The materials utilized in MASP must be strong, of low density, and able to operate at elevated temperatures. Carbon–carbon composites may be the best choice for the airframe for the future hypersonic aerospace plane. These composites also are being used in rocket nozzles, hot gas dusts, high-temperature fasteners, hot press dies, molds, and biomedical devices.[20]

ACKNOWLEDGMENTS

This research was financially supported by the National Science Council, Taiwan, Republic of China under contract No. NSC81-0405-E-007-521. The authors would like to thank Ms. En Huert Liou in the Materials Science Center at National Tsing Hua University for her assistance in SEM manipulation.

REFERENCES

1. Fitzer, E.; Hüttner, W. *J. Phys. D: Appl. Phys.* **1981**, *14*, 347.
2. Fitzer, E.; Hüttner, W. *Carbon* **1987**, *25*, 163.
3. Fitzer, E. *Pure & Appl. Chem.* **1988**, *60*, 287.
4. Buckley, J. D. *Ceramic Bull.* **1988**, *67*, 364.
5. Hüttinger, W. *Adv. Mater.* **1990**, *2*, 349.
6. Savage, G. *Carbon–Carbon Composites* 1st ed.; Chapman & Hall: London, 1993.
7. Thomas, R. *Essentials of Carbon–Carbon Composites*, 1st ed.; Royal Society of Chemistry; Thomas Graham House: Science Park, United Kingdom, 1993.
8. Fitzer, E.; Weiss, R. *Carbon* **1987**, *25*, 455.
9. Bahl, O. P.; Mathur, R. B.; Dhani, T. L. *Polym. Eng. Sci.* **1984**, *24*.
10. Manocha, L. M. *J. Mater. Sci.* **1982**, *17*, 3039.
11. Neffe, S. *Carbon* **1987**, *25*, 761.
12. Blackketter, D. M. et al. *Polymer Composites* **1993**, *14*, 430.
13. Manocha, L. M. *Composites* **1988**, *19*, 311.
14. Manocha, L. M.; Bahl, O. P.; Singh, Y. K. *Carbon* **1989**, *27*, 381.
15. Manocha, L. M. *Carbon* **1988**, *26*, 333.
16. Fitzer, E.; Geigl, K. H.; Hüttner, W. *Carbon* **1980**, *18*, 265.
17. Kowbel, W.; Shan, C. H. *Carbon* **1990**, *28*, 287.
18. Lewis, C. F. *Mater. Eng.* **1989**, January, 27.
19. Ronald, T. M. F. *Adv. Mater. & Proc.* **1993**, *9*, 24.
20. Klein, A. J. *Adv. Mater. & Proc.* **1986**, *11*, 24.

CARBON COMPOSITES

Shigehisa Ishihara
Wood Research Institute
Kyoto University

CHARCOAL AS RAW MATERIAL OR HIGH-PERFORMANCE WOOD COMPOSITES AND THEIR APPLICATIONS

During the past 30 years, there has been rapid growth in the use of carbon fibers and their related composites, in particular, in the ever-widening use of these materials in the construction industry. Carbon fibers and their composites currently play a leading role in the development of a whole range of new building materials. Carbon materials, in general, which neither melt nor soften under a red heat, are highly resistant to heat and fire and merely glow slowly with mild oxidation at high temperatures, their thermal conversion being due to conductivity rather than combustion.

Charcoal is the chief product of the pyrolysis of wood. It is nor pure carbon, but a complex combination of primary carbon plus a secondary deposit of tar. The amount of secondary material with the carbon decreases rapidly as carbonization temperature increases.

High-performance or high-function properties of charcoals carbonized under various temperatures have been identified and their carbonization temperature, related chemical and physical changes of wood and charcoal, functional changes of charcoal, high performance properties of charcoal as raw material of advanced or high-function carbon composites, and their uses and applications have also been studied.[1,2]

CARBON COMPOSITES FROM WOOD CHARCOAL AS AN ELECTROMAGNETIC SHIELD AND FIRE-RESISTANT MATERIAL[1,3]

In order to develop new advanced materials from wood-industry residue and wastes, forestry wastes, unused wood species, and other biomass wastes, carbon composites were produced from charcoal of wood material. The raw material of carbon composites was made from charcoal powder coated with B-stage phenol-formaldehyde resin. The powders of Mousouchiku bamboo charcoal, cellulose charcoal, graphites, and a charcoal of phenol-formaldehyde resin were also used as raw materials in the manufacture of the carbon composites.

The electromagnetic shielding efficiency of carbon composite boards made from charcoals was found to increase with a rise in the carbonization temperature of the raw wooden materials.

The ignitability and flammability of the composite boards (measured by oxygen index) decreased with a rise in carbonization temperature of wooden raw material.

Fire tolerance of the composite boards was improved by the time-delay in the burn-through of the composite board, due to the resistance, high thermal conductivity, and slow oxidation of charcoals carbonized by high-temperature carbonization.

The resistance of the high-temperature carbonized charcoal or graphitized carbon composite boards to cutting by an oxygen-acetylene torch was greater than that of the stainless steel plate. These carbon composites should soon be available as advanced material for a wide range of new materials in the fields of building, automotive, electrical and electronic, aerospace, engineering, and various other applications.

CARBON-OVERLAID PARTICLEBOARD AS AN ELECTROMAGNETIC SHIELD AND FIRE RESISTANT MATERIAL

The particleboard industry has been making efforts in the past decade to develop a production process to reduce density and swelling of board. It has also intensified its efforts to develop treatments for fire resistance and a high-performance property of particleboards as used in some walls, ceilings, floors, and cores of the fire door.

Surface application has been proved[4–6] very effective in imparting high fire tolerance and high-performance function to the boards.

The electromagnetic shielding and the heat- and flame-shielding performance of graphite-overlaid particleboards has also been developed. The use of graphite as a carbon material costs appreciably less than charcoal carbonized at high temperature.

The shielding efficiency of 30-mm-thick and 0.60-g/cm^3-dense wood particleboard overlaid with a 10% shield of graphite phenol-formaldehyde resinsphere (GPS) was over 40 dB against electric fields, which satisfied a standard for shielding materials for partitions and enclosures of office buildings, and computer aids.

A preliminary study of the electromagnetic shielding efficiency of both 30-mm-thick conventional particleboard and GPS boards with a density of 0.6 and 1.2 g/cm^3, respectively, in accordance with ASTM ES 7-83 of a dual chamber method showed that the electromagnetic shielding efficiency of the GPS boards was over 80 dB (300 MH2), whereas that of particleboards was less than 5 dB. Therefore, shielding effectiveness of overlaid boards against electric field was mainly attributed to GPS (carbon material) overlays.

The fire tolerance of GPS-overlaid particleboards was improved with respect to the time delay both in ignition and in temperature rise at the unexposed surface when compared with controls (conventional particleboards) at the same board thickness and density, due to low combustibility of the GPS overlay.

Non-combustible commercial boards such as gypsum board and autoclaved asbestos calcium silicate board performed similarly to wood-based panels, but their fire tolerance times were rather lessened when compared at the same surface density. On the other hand, GPS overlays prolonged this endurance time 10 to 35%.[7,8]

NEW CHARCOAL-BASED COMPOSITES FOR ENVIRONMENTAL CLEAN-UP

Charcoal and activated charcoal are suitable materials for various kinds of adsorption, both from gas phase and liquid phase. Active charcoal is charcoal that has been subjected to a high-temperature heat treatment in the presence of air or steam, in some cases after a special chemical treatment to increase its adsorbancy.

For a long time, attempts to purify drainage, sewage, and small streams using charcoal have been made and discussed.[9–11] Recently, to purify water and to clarify various aqueous solutions or polluted air, new composites of charcoal and metal compounds are being developed. For example, high-temperature carbonized charcoals were "impregnated" with various kinds of metal alcoholates or metal hydroxides to produce composites by means of a heat treatment. Methods of removal of NOx from polluted air (e.g., gas from automobiles and conversion of NOx or SOx into harmless gases by the composites made from charcoal and metal compounds) are being developed.[12] Carbonized material adsorbents for the removal of mercury from aqueous solutions and vapors are also being developed.[13]

REFERENCES

1. Ishihara, S. Proceedings, All Division Conference on "Forest Products," Iufro, Nancy, France, **1992**, *1*, 125–127.
2. Ishihara, S. *J. Wood Industry and Wood Machinery*; **1994**, *12*(139), 33–37.
3. Ishihara, S.; Ide, I. et al. *J. Soc. Material Science Japan*; **1993**, *43*(473), 147–152.
4. Ishihara, S.; Kawai, S.; Yoshida, S.; Rakamatsu, A. Proc. of S5.0304 Sessions, IUFRO Div. 5, Forest Products Conference, 18 May 1988, Sao Paulo, Brazil, pp 1–9.
5. Ishihara, S.; Kawai, S. *J. Japan Wood Res. Soc.* **1989**, *35*(3), 234–242.
6. Kawai, S.; Ishihara, S.; Yoshida, S.; Takamatsu, A. *J. Soc. Material Science* **1989**, *38*(430), 758–764.
7. Kawai, S.; Ishihara, S.; Ide, I.; Yoshida, Y.; Nakaji, M. Proc. of the 1990 International Timber Engineering Conference, Tokyo, Japan, October 23–25, 1990; Vol. 1, pp 74–79.
8. Ide, I.; Ishihara, S.; Kawai, S.; Yoshida, Y.; Nakajii, M.; Takamatsu, A. *J. Japan Wood Res. Soc.* **1992**, *38*(8), 777–785.

9. Yatagai, M. *Report for investigation into multi-uses of wood carbonization product*; The Forestry Agency, Tsukuba Norim Kenkyu Danchi-Nai, Ibaraki, 1993; pp 19–43.
10. Kato, F. *Water Pollution Research* **1988**, *11*(1), 31–33.
11. Dept. of Forestry and Fisheries, Kumamoto Prefecture, "Manual for Charcoal Uses," Kumamoto Prefecture 1991; 4–34.
12. Ishihara, S.; Furutsuka, T. To be presented, 1996 ACS Spring National Meeting on "Production and Use of Carbon-Based Materials for Environmental Cleanup," New Orleans, Louisiana, March 24–29, 1996.
13. Ishihara, S.; Pulido, L. L.; Kajimoto, T. To be presented, 1996 ACS Spring National Meeting on "Production and Use of Carbon-Based Materials for Environmental Cleanup," New Orleans, Louisiana, March 24–29, 1996.

Carbon Dioxide Copolymers

See: Diyne Cycloaddition Copolymerization (Transition Metal Catalyzed)
Ladder Polymers (Cycloaddition Copolymerization of Cyclic Diynes)
Poly(propylene carbonate)

Carbon Fibers

See: Carbon/Carbon Composites (with Surface Treated Carbon Fibers)
Carbon Fibers (Overview)
Carbon Fibers (Plasma Surface Treatments)
Electrically Conducting Composites
Polyacrylonitrile Gel Fibers
Textile Fibers (Structure and Properties)

CARBON FIBERS (Overview)

Leighton H. Peebles, Jr.
Chemistry Division
Naval Research Laboratory

Carbon fibers are commercially available with a variety of mechanical properties: low strength accompanied by low modulus (0.72 GPa and 40 GPa, respectively), high strength and intermediate modulus (Hercules Magnamite IM7, 5.3 GPa, 303 GPa, respectively, $\epsilon_b = 1.8\%$), and high modulus and intermediate strength (Amoco Thornel P-120, 827 GPa, 2.37 GPa, $\epsilon_b = 0.29\%$). The low-strength, low-modulus fibers are used in filtration, static dissipation, EMI shielding, thermal insulation, as an asbestos substitute in insulation and roofing materials, and as activated gas adsorbers with high selectivity.[1,2] The low strength, low-modulus types are inexpensive, compared with the high-strength, high-modulus fibers, and are used in aircraft brakes and clutch materials because of their ability to dissipate heat. High-strength, high-modulus-type fibers are used primarily as reinforcements in composite structures because stiffer, lightweight materials result. Their original use was in aerospace structures where weight-saving is of premium concern. Commercial applications have led to the development of many types of sports equipment, where again light-weight and stiff composites have advantages.

A brief mention of forming rayon-based carbon fabric for use in carbon–carbon composites may be found in Peebles.[4] Books on carbon fibers are listed in the references.[4–12]

FIBER PRECURSORS AND CARBON FIBER FORMATION

The two major precursors for carbon fibers are based either on copolymers of acrylonitrile or on the mesophase form of either petroleum pitch or coal tar pitch.

Carbonization of both PAN- and MP-stabilized fibers can be accomplished rapidly, in the order of minutes, at temperatures of about 1500°C in a nitrogen atmosphere. The rate of carbonization must be controlled to prevent the exiting gases from forming defects in the fiber. Pores are created in the fiber owing to the formation and perfection of the layered structure of carbon.

CARBON FIBER STRUCTURE

The graphitic structure does exist in some high-temperature heat-treated carbon fibers, but the bulk of the material is in the turbostratic form, where the distance between layer planes is about 0.34 nm and the layers are not planar.[13]

The basic structural unit of carbon fibers consists of a stack of turbostratic layers.[14]

Guigon et al. also studied nine samples of commercial, high-modulus PAN-based carbon fibers. In this case, the irregular collection of basic structural units has become more perfect in orientation and alignment. A skin and core effect is indicated with thicker sections of aligned material near the surface and r_t is larger (less convoluted folding of layers). The change in perfection of layers from the high-strength is consistent with all carbon fibers and carbons in general. In this model, the layers are smoothly curved and parallel to the fiber axis.

The fine structure of four MP-based carbon fibers has been studied by Guigon and Oberlin.[15] They report three different fine structures: a graphitic phase, a microporous turbostratic phase, and a phase similar to high-modulus PAN-based fibers. The three phases are distributed at random, either parallel to the fiber axis or perpendicular to it. The pores in this material are polyhedral, as in other high temperature heat-treated carbons.

TENSILE PROPERTY VARIATIONS

There is a great variation in the tensile strengths of filaments, both along and between given yarns within a spool of carbon fibers, as well as variation between spools and lot numbers.

GAUGE LENGTH DISTRIBUTION

The tensile strength of the fiber also depends strongly on the gauge length. Asloun et al. have examined the Weibull probability of failure for a set of fibers at various gauge lengths.[16]

COMPRESSIVE STRENGTH

The compressive strength of carbon fibers is about half that of the tensile strength.[27] A major reason for improving the compressive strength and reducing the variation in compressive variability is that composites are subject to flexing, which with low compressive properties requires relatively thicker composites with accompanying increases in weight.

SURFACE MODIFICATION AND CHARACTERIZATION

A number of researchers have reviewed the modification of carbon fibers for improving composite properties.[3,6,7,17–20] Donnet and Guilpain, among others, state that commercial carbon fibers are treated by anodic oxidation to improve composite properties.[21] Fibers can also be treated for enhanced fiber–matrix adhesion by low-power plasma treatments.[22] The concentration of surface groups on carbon fibers is usually determined by X-ray photoelectron spectroscopy (XPS, also known as electron spectroscopy for chemical analysis, ESCA) and attempts are made to correlate these results with the interfacial shear strength (IFSS, determined from single filaments embedded in a matrix) or the interlaminar shear strength (ILSS, determined from composite properties). Harvey et al. show that there is no correlation between oxygen concentration determined by XPS and ILSS for different fibers subjected to different treatments.[23]

Because of the lack of correlation between surface oxygen content and ILSS for various surface treatments, the retention of shear strength following removal of surface oxygen by heat treatment.[23,26] and the microrough surfaces evident from scanning tunneling microscopy following surface treatment,[13,24] it is highly probable that microroughness plays a more important role in adhesion than previously considered and that chemical reactions with the fiber surface play a smaller role.[25]

Carbon fibers are also important reinforcement materials for metal matrix composites. Unfortunately, most molten metals of interest do not wet carbon. Even short-term contact of carbon with molten metals will lead to the formation of carbides that can embrittle the composite.

REFERENCES

1. Liu, G. Z.; Edie, D. D.; Fain, C. C. *Extended Abstracts*, 20th Biennial Conference on Carbon, Santa Barbara, CA; American Carbon Society: St. Mary's, PA, 1991; p 264.
2. Suzuki, M. *Carbon* **1994**, *32*, 577.
3. Riggs, D. M.; Shuford, R. J.; Lewis, R. W. In *Handbook of Composites*; Lubin, G., Ed.; Van Nostrand Reinhold: New York, 1982; p 196.
4. Peebles, L. H. Jr. *Carbon Fibers: Formation, Structure, and Properties*; CRC: Boca Raton, FL, 1995.
5. *Carbon and Graphite Fibers: Manufacture and Applications*; Sittig, M., Ed.; Noyes Data: Park Ridge, NJ, 1980.
6. Delmonte, J. *Technology of Carbon and Graphite Fiber Composites*: Van Nostrand Reinhold: New York, NY, 1981.
7. Donnet, J. B.; Bansal, R. C. *Carbon Fibers*: Marcel Dekker: New York, NY, 1984.
8. *Carbon Fibres and Their Composites*; Fitzer, E., Ed.; Springer–Verlag: New York, NY, 1985.
9. *Handbook of Composites*; Watt, W., Perov, B. V., Eds.; Elsevier: New York, NY, 1985; Vol. 1.
10. Dresselhaus, M. S. et al. *Graphite Fibers and Filaments*; Springer–Verlag: New York, NY, 1988.
11. Figueiredo, J. L. et al. *Carbon Fibers, Filaments, and Composites*; Kluwer Academic: The Netherlands, 1990.
12. Donnet, J. B.; Bansal, R. C. *Carbon Fibers*, 2nd ed.; Marcel Dekker: New York, NY, 1990.
13. Hoffman, W. P. et al. *J. Mater. Res.* **1991**, *6*, 1685.
14. Guigon, M.; Oberlin, A.; Desarmot, G. *Fiber Sci. Technol.* **1984**, *20*, 177.
15. Guigon, M.; Oberlin, A. *Composite Sci. Technol.* **1986**, *25*, 231.
16. Asloun, El. M. et al. *J. Mater. Sci.* **1989**, *24*, 3504.
17. Ehrburger, P.; Donnet, F. B. In *Handbook of Composites*; Watt, W., Perov, B. V., Eds.; Elsevier: New York, NY, 1985; Vol. 1, p 577.
18. Riggs, J. P. In *Encyclopedia of Polymer Science and Engineering*; John Wiley & Sons: New York, NY, 1985; Vol. 2, p 640.
19. Wright, W. W. *Composite Polym.* **1990**, *3*, 231.
20. Wright, W. W. *Composite Polym.* **1990**, *3*, 360.
21. Donnet, J. B.; Guilpan, G. *Carbon* **1989**, *27*, 749.
22. Liston, E. M. *J. Adhesion* **1989**, *30*, 199.
23. Harvey, J.; Kozlowski, C.; Sherwood, P. M. A. *J. Mater. Sci.* **1987**, *22*, 1585.
24. Drzal, L. T. et al. *Compos. Structures* **1994**, *27*, 65.
25. Peebles, L. H. Jr. *J. Adhesion* **1995**, *54*, 1.
26. Drzal, L. T.; Rich, M. J.; Lloyd, P. F. *J. Adhesion* **1982**, *16*, 1.
27. Peebles, L. J. Jr. *Int. Mater. Rev.* **1994**, *39* 75.

CARBON FIBERS (Plasma Surface Treatments)

Shin Shing Shyu and L. Y. Yuan
Department of Chemical Engineering
National Central University

Two major processes are used in carbon fiber plasma surface treatments: polymer-forming plasma and nonpolymer-forming plasma. Polymer-forming plasma, organic vapors are polymerized in a plasma reactor in complicated routines. When the plasma ions impinge on the surface atoms, efficient energy transfer creates high-density radicals that increase free energy and wettability of the surface. Pin-hole-free, amorphous, crosslinked structures, plus better thermal stability and higher density are typical features of these polymers. Plasma-polymerized polymers can be oily low molecular weight films and powders—their appearance dependent upon power input, reactor pressure, the gas source, and its flow rates. Generally, the activated polymer surface around the fiber surface enhances adhesion in the resin matrix. Because of these the pin-hole-free characters, insolubility, and thinness, the plasma-polymerized films have found wide application as surface treatments in the past 20 years, in solid-state, micromechanical, and electronic industries.[1]

Nonpolymer-forming plasma surface treatments rely on several mechanisms. Plasma ions from inert gases, such as argon and helium, impinge on the substrate materials to activate them. Ions of chemically reactive gases, such as oxygen, ammonia, and fluorine, cause chemical reactions with the superficial molecules. Chemically nonreactive gases such as water vapor and nitrogen bring about reactions because these gases become more reactive in gas plasma. In addition to their reactivity, plasma ions also scratch the fiber surface, so an ablative cleaning effect is usually observed. Ablation is the most prominent effect when inert gas, oxygen, or CF_4 is used as plasma source.

PLASMA SURFACE TREATMENTS ON CARBON FIBERS

During the past couple of decades, interest has grown in improving and controlling the bonding between fiber and matrix in carbon fiber–synthetic resin composites. One of the main objectives of a successful fiber surface treatment is to alter the fiber surface chemistry while retaining the mechanical properties of the fibers themselves. Most plasma treatments used to date, however, remove a substantial amount of material and cause pits to form in the fiber surface. This allows a greater degree of control over composite properties such as interlaminar shear and compressive strength. When comparing plasma and other chemical treatments of carbon fibers, one finds that plasma treatments increase the surface concentration of functional groups and hence may cause a greater number of chemical bonds to be formed between fiber and resin.

SURFACE TREATMENT PROPERTIES

rf Gas Plasma Treatments

Air Plasma

Polyacrylonitrile- (PAN-) based carbon fibers treated by rf air plasma alter the surface energies remarkably.[4,5] The surface energy of dispersive (γ^d_s) and (γ^p_s) components of carbon fibers are determined by wetting contact angle measurements. The plasma treatments not only create edge surfaces by pit formation, but also completely remove the thin film lamellae by combustion.[6] Several minutes of plasma treating reduces the mechanical properties, because of high oxidation, which creates a high concentration of defects of critical size not seen with scanning electron microscopy (SEM).[4,5,7] The surface is identical in appearance to those of the virgin fibers,[5] but scanning tunneling microscopy (STM) shows the outermost fiber surface seems to have been "burnt."[7] It is also evident that both the high- and low-modulus PAN-based fiber surfaces contain ketones and carboxyl groups after treatment.[8] Air plasma is also capable of introducing a substantial amount of chemical functionality onto pitch-based carbon fibers.[9]

Oxygen Plasma

All carbon fibers become more wettable after oxygen plasma treatment, because the advancing water contact angle is reduced by some tens of degrees.[10] Radio frequency oxygen plasma causes carbon atoms in the hydrocarbon to change into C–OH, C–OOH, C–O–C, and C=O groups; the concentration increases with time but finally reaches a constant level.[2,3,11] Lactone groups also form on the surface but are rare.[11] Oxygen plasma treatment also causes speckling along the longitudinal fiber axis on the surface.[2,3]

The aging behavior of contact angles as a function of aging time depends on the presence of sizing. The wettability of plasma-treated unsized fibers is rather stable in air. Sized fibers lose imparted wettability after aging in air.[10] Sized fibers show the hydrophobic recovery typical of most polymeric materials with a recovery rate that increases as a function of temperature.[12] By contrast, unsized fibers are almost insensitive to aging in air.

Analysis of PAN-based carbon fibers treated by continuous oxygen plasma show increases in carboxylic, alcoholic, and ketone groups on the surface, improved wetting properties, surface striations and surface roughness. The fiber strength decreases very little after the plasma treatments.[13]

Nitrogen and Argon Plasma

XPS analyses show that nitrogen plasma treatment seems to have no significant effect on the topography of carbon fibers. Fibers treated with argon plasma show more significant surface devastation than those undergoing oxygen plasma treatment. Fiber strength decreases, O/C ration increases, and oxygen functionalities such as C–OH and C–O–C are well characterized by XPS on the fiber surface.[3,5] Both argon and nitrogen plasma treatments decrease γ^d_s to make the fiber surface more polar. Again, this correlates to the corresponding increase in the surface oxygen content.

Ammonia Plasma

Ammonia plasma treatment has no evident effect on the fiber surface.[2,5,14] Introduction of nitrogen containing groups (e.g., amine groups) has the potential to facilitate reaction with many polymers, especially epoxy resin.[14] It can successfully introduce C–N functionality without any C–O groups being incorporated; so the gain in surface polarity after ammonia plasma treatment can be correlated to the increase in C–N and C=N moieties.[5] Fiber strength is almost unchanged after a brief period of ammonia plasma treatment.

Unlike air plasma, nitrogen and ammonia plasmas do not have the ability to react with pitch based fibers to any great extent.[9]

Microwave Gas Plasma Treatment

PAN-based carbon fibers treated with a microwave (2.4 GHz) plasma under air and argon give surprising results.[15] More oxygen was introduced with argon plasma treatment of the fibers, presumably by reaction of the active sites created on the surface by the plasma treatment with oxygen, when the fibers, after treatment, are reexposed to air.[16] In comparison, air plasma generates crevices or longitudinal grooves that alter the mechanical properties. Argon plasma offers the advantage of leaving the surface topography unaffected.

Organic Vapor Plasma Treatments

In other experiments, carbon fibers were treated with plasma of acrylonitrile (AN) and styrene (ST) monomers.[17,18] Both monomers produced surfaces that are substantially more polar than the untreated surfaces. XPS and Fourier transformed infrared (FTIR) studies indicate that the plasma polymers contain a high concentration of oxygen in the form of C=O, COOH, C–O–C, and C–OH groups. Also, treated fibers exhibited slightly higher tensile strengths than their untreated counterparts, suggesting that the plasma coatings effectively heal some of the fibers' surface flaws.[17,18]

It has been reported that free radicals remain on the surface after plasma treatment and these radicals possess a long half-life. Reaction of these free radicals with their surrounding can alter the final surface structure. AN plasma-treated fibers are most wettable, followed by those treated in ST plasma, in spite of the higher O/C ratio in ST plasma-treated fibers. Untreated fibers are least wettable with water.

APPLICATIONS

High-performance composites are finding increased applications where high strength, high stiffness, and low weight are important. Plasma treatment provides a relatively simple means of tailoring or designing the fiber surface systematically to accomplish various purposes.

The low adhesion of carbon fibers to a series of thermoplastic polymers is found to be significantly lower when compared with thermosetting epoxy polymers. Amine groups introduced onto surfaces by exposure to ammonia plasma can react with epoxides to form covalent bonds, thereby potentially improving composite performance. The results of PAN-based carbon fibers treated by continuous oxygen plasma show that interlaminar shear strength of carbon-fiber-reinforced epoxy composites increase from 60 MPa to 100 MPa.[13] Fibers that are treated by oxygen and argon plasma embedded in thermoplastic resin poly(phenylene sulfide) (PPS) show more than twice the interfacial shear strength enhancement than do the untreated ones.[3,18]

Plasma polymerization can generate a highly crosslinked, thin layer (400–1000 A) of coatings grafted onto the fiber surface.[17] The effects of polymer deposition by acrylonitrile and styrene plasma that we studied showed that not only was tensile strength preserved, the interfacial shear strength of PAN-based carbon fiber treated with PPS is increased more than 30%. A strong mixed layer resulting from fusion of the plasma-polymerized polymer film and PPS resin resists the interfacial shear stress before failure.

Applying SiC coatings on carbon fibers by rf sputtering process provides optimum conditions for creating high-quality SiC films by varying power input, rf peak voltage, substrate temperature, and gas pressure.[19] SiC, as an adhesion promoter, improves the mechanical property of composites, a result attributed to strong bonding of the whiskers at the graphite substrate and to the increase in bonding between the fiber and resin matrix. However, SiC coatings thicker than 150 nm result in a strong decrease in fiber strength. This fiber damage is initiated from cracks within the SiC layer, which penetrate throughout the fiber and cause it to fail.

The surface treatment of carbon fibers by a low-temperature plasma in an oxygen atmosphere helps the wetting of the carbon fiber surfaces with cement paste and consequently increases the strength of the resultant carbon-fiber-reinforced cements (CFRCs).[20] The treatment of the starting carbon fiber mats by plasma at 50 W increases the three-point bend strength and deflection of the CFRCs prepared—the strength is double that of untreated fibers. Considering the practical applications of CFRCs, the carbon fiber mat has an advantage where high stress is expected.

REFERENCES

1. Veprek, S. *Plasma Chem. Plasma Process.* **1989**, *9(Suppl.)*, 29.
2. Bascom, W. D.; Chen, W. J. *J. Adhesion* **1991**, *34*, 99.
3. Yuan, L. Y. et al. *Compos. Sci. Technol.* **1992**, *45*, 1.
4. Donnet, J. B. et al. *Carbon* **1986**, *24*, 757.
5. Commerson, P.; Wightman, J. P. *J. Adhesion* **1992**, *38*, 55.
6. Bennett, S. C.; Johnson, D. J. *Carbon* **1979**, *17*, 25.
7. Quin, R. Y.; Donnet, J. B. *Carbon* **1994**, *32*, 323.
8. Jones, C.; Sammann, E. *Carbon* **1990**, *28*, 509.
9. Jones, C.; Sammann, E. *Carbon* **1990**, *28*, 515.
10. Morra, M. et al. *Compos. Sci. Technol.* **1991**, *42*, 361.
11. Da, Y.; Wang, D.; Sun, M. *Compos. Sci. Technol.* **1987**, *30*, 119.
12. Morra, M. et al. *J. Colloid. Interface Sci.* **1989**, *132*, 504.
13. Sun, M. et al. *Compos. Sci. Technol.* **1989**, *34*, 353.
14. Loh, I. H.; Cohen, R. E.; Baddour, R. F. *J. Mater. Sci.* **1987**, *22*, 2937.
15. Donnet, J. B. et al. *J. Phys. D.: Appl. Phys.* **1987**, *20*, 269.
16. Donnet, J. B.; Guilpain, G. *Composites* **1991**, *22*, 59.
17. Dagli; Sung, N. H. *Polym. Compos.* **1989**, *10*, 109.
18. Yuan, L. Y.; Lai, J. Y. *Compos. Sci. Technol.* **1992**, *45*, 9.
19. Weisweiler, W. et al. *Thin Solid Films* **1987**, *148*, 93.
20. Inagaki, M. *Carbon* **1991**, *29*, 287.

Carbon Monoxide Copolymers

See: *Carbonylation Polymerization (Palladium-Catalyzed)*
Olefin-Carbon Monoxide Copolymers (Overview)
Olefin-Carbon Monoxide Copolymers, Alternating Palladium-Catalyzed Synthesis (Monomers and Polymers)

CARBON WHISKER (Surface Modification by Grafting)

Norio Tsubokawa, Toshio Yoshihara, and Hiroshi Ueno
Department of Chemistry and Chemical Engineering
Faculty of Engineering
Niigata University

Carbon whiskers (i.e., vapor grown carbon fibers) are produced by depositing a layer of pyrocarbon from vapor phase on a catalytically grown carbon filament.[1–3] These fibers have diameters on the order of submicrons and lengths on the order of microns. Carbon whiskers are heat-resistant, weather proof, chemical proof, and electroconductive and are considered to have a high modulus. Recently, carbon whiskers were shown to be applicable to the formation of high-performance composite materials. Although it is difficult to disperse carbon whiskers into polymer matrices uniformly, the dispersibility of carbon whiskers in solvents and polymer matrices is remarkably improved by the grafting of polymers onto carbon whisker surfaces.

The properties of fiber-reinforced plastics are known to depend not only on the mechanical properties of the fiber and polymer matrices but also on the properties of interfacial regions between the fiber surface and matrix. To transmit stress from the matrix to the fiber, the modification of the fiber surface by grafting polymers was widely investigated.

PREPARATION

Grafting Sites on Carbon Whisker Surface

Carbon whiskers have surface functional groups, such as carboxyl, phenolic hydroxyl, and quinonic oxygen (carbonyl) groups and polycondensed aromatic rings. These functional groups are known to undergo chemical reactions, such as neutralization, esterification, oxidation, and reduction. Therefore, these functional groups and aromatic rings can be used for introduction of initiating groups for graft polymerization.

Grafting Reaction

Radical Grafting From Carbon Whisker Surface

The introduction of azo and peroxyester groups onto the carbon whisker surface was investigated.[4]

The radical polymerizations of vinyl monomers were initiated by surface radicals formed by the thermal decomposition of surface azo groups to give the corresponding polymer-grafted carbon whisker.

The radical graft polymerization of various monomers was also initiated by surface peroxyester groups introduced onto the surface, and the corresponding polymers were effectively grafted onto the surface.

Cationic Grafting From Carbon Whisker Surface

We have reported that the cationic polymerization of *N*-vinylcarbazole and *N*-vinyl-2-pyrrolidone is initiated by surface carboxyl groups on carbon whisker, and a part of the polymer formed is grafted onto the surface.[5]

The polymerization of vinyl monomers such as isobutyl vinyl ether, *N*-vinyl-carbazole, *N*-vinyl-2-pyrrolidone, and styrene was successfully initiated by acylium perchlorate groups on the carbon whisker surface to give the corresponding polymer-grafted carbon whisker.[6]

Acylium perchlorate groups introduced onto the carbon whisker surface were also able to initiate the cationic ring-opening polymerization of cyclic ethers, lactones, and cyclic acetals.[7–9]

Anionic Grafting from Carbon Whisker Surface

We have reported that metallized carbon whisker also initiates the anionic graft polymerization of vinyl monomers to give the corresponding polymer-grafted carbon whisker: the percentage of grafting increased to greater than 100%.[10]

We reported the grafting of polyesters onto the carbon whisker surface by the anionic ring-opening alternating copolymerization of epoxides with cyclic acid anhydrides initiated by potassium carboxylate groups on the surface.[11] The introduction of the initiating groups onto the carbon whisker surface was readily achieved by the neutralization of carboxyl groups with potassium hydroxide.

Reaction of Carbon Whisker with Functional Polymers

Polymers with controlled molecular weights and well-defined structures can be grafted onto the carbon whisker surface by the direct condensation of surface carboxyl groups with functional polymers having terminal hydroxyl or amino groups.[12]

PROPERTIES

Dispersibility

By grafting polymers onto the carbon whisker surface, the dispersibility in solvents and polymer matrices is remarkably improved. Untreated carbon whisker dispersion in THF immediately precipitated, but polymer-grafted carbon whisker gave a stable dispersion in good solvent for grafted chain. Even if carbon whisker was modified by the grafting of polymers, however, carbon whisker precipitated in poor solvents for grafted chain. This indicates that in good solvents for grafted chain the grafted chains on the surface interfere with the aggregation of carbon whisker.[9]

In addition, the wettability of the carbon whisker surface was readily controlled by the grafting of polymers. When polystyrene-grafted and poly(acrylamide)-grafted carbon whisker were dispersed in a two-phase mixture of benzene and water, polystyrene-grafted carbon whisker dispersed only in the benzene phase and no carbon whisker was contained in the water phase. Poly(acrylamide)-grafted carbon whisker, however, dispersed in the water phase and not at all in benzene.

Furthermore, water hardly penetrates a column that is packed with polystyrene-grafted carbon whisker but readily penetrates a column packed with poly(acrylamide)-grafted carbon whisker.

Crosslinking Grafted Polymer Chains on Carbon Whisker

Unsaturated polyester-grafted carbon whisker can be crosslinked with a vinyl monomer by using peroxides as catalysts. For instance, when unsaturated polyester-grafted carbon whisker, which was obtained from the anionic ring-opening copolymerization of styrene oxide with glycidyl methacrylate, was mixed with styrene in the presence of benzoyl peroxide, gel containing carbon whisker was obtained. In such a gel, carbon whiskers were considered to be uniformly incorporated in the polymer matrix with chemical bonds.

REFERENCES

1. Koyama, T. *Carbon* **1977**, *10*, 757.
2. Katsuki, H.; Matsunaga, K.; Rgishira, M.; Kawasumi, S. *Carbon* **1981**, *19*, 148.
3. Tibbetts, G. G.; Devour, M. G.; Rodda, E. J. *Carbon* **1987**, *25*, 367.
4. Ueno, H.; Tsubokawa, N. *Polym. Preprints, Japan* **1993**, *42*, 520.
5. Tsubokawa, N.; Yoshihara, T. *Polym. J.* **1991**, *23*, 177.
6. Tsubokawa, N.; Yoshihara, T. *J. Macromol. Sci., Pure Appl. Chem.* **1993**, *A30*, 517.
7. Tsubokawa, N.; Yoshihara, T. *Colloids and Surfaces A* **1993**, *81*, 195.
8. Tsubokawa, N.; Yoshihara, T. *Polym. Bull.* **1993**, *30*, 421.
9. Tsubokawa, N.; Yoshihara, T. *Composite Interfaces* **1994**, *32*, 71.
10. Tsubokawa, N.; Yoshihara, T.; Sone, Y. *J. Polym. Sci., Part A, Polym. Chem.* **1992**, *30*, 561.
11. Tsubokawa, N.; Yoshihara, T. *J. Polym. Sci.: Part A, Polym. Chem.* **1993**, *31*, 2459.
12. Tsubokawa, N.; Ueno, H.; Yamada, H. *Polym. Preprints, Japan* **1994**.

CARBONYLATION POLYMERIZATIONS (Palladium-Catalyzed)

Robert J. Perry
Imaging Research and Advanced Development
Eastman Kodak Company

In 1974, Heck reported on the formation of amides and esters via palladium-mediated carbonylation and coupling reactions of aromatic halides and amines[1] or alcohols.[2]

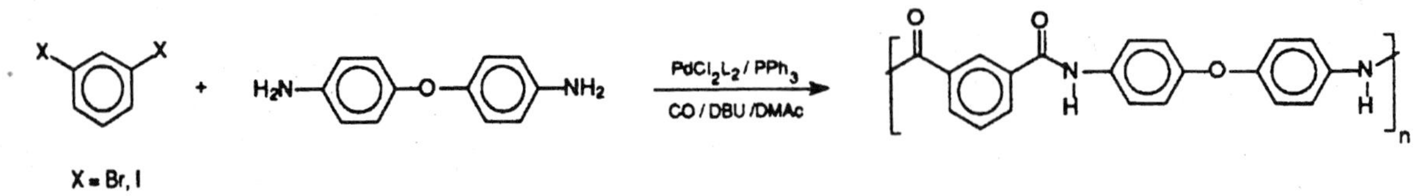

1

Although these were generally clean, high-yielding reactions, it was nearly 15 years before this chemistry was applied to polymer synthesis. In the past several years, a variety of polymeric systems have been prepared using hydrolytically stable aromatic halides under relatively mild conditions. This method complements more conventional approaches that employ acid chlorides or other aromatic acid derivatives.

PREPARATION

Aramids

Aromatic polyamides (aramids) were the first polymers synthesized by palladium-catalyzed carbonylation and condensation reactions. Aromatic diamines and aryldibromides were shown to react under 1 atm carbon monoxide (CO) in the presence of a tertiary amine base, a palladium catalyst, and phosphine ligands in a dipolar aprotic solvent to form aramids of modest molecular weight (**Equation 1**).[3]

Shortly thereafter, it was reported that the use of aromatic diiodides and elevated CO pressures gave aramids of much higher molecular weight in shorter reaction times.[4,5] Increase in the rate-determining carbonylation step of aryldiiodide and a decrease in the rate-determining oxidative addition step of the aryldibromide accounted for this observation. Use of diiodinated aromatics also allowed the formation of a variety of aramids. In both cases, the optimum stoichiometry for achieving maximum molecular weight was to use a slight excess of the diamine. Model studies indicated that this decreased the likelihood of reduction of the aryl–halide bond, which would effectively terminate the polymer chain.[6]

Poly(imide-amide)s (PIAs)

High molecular weight PIAs were also formed by the carbonylation and condensation of aryldiiodides containing one or two preformed imide moieties, diamines, and CO.[7]

Polyimides

Several approaches have been taken to extend carbonylation chemistry to the formation of polyimides. Although variously substituted phthalimides could be easily made,[8] preparation of high molecular weight linear polyimides was more difficult. Reaction of bis(*o*-diiodoaromatics) with diamines and CO gave materials with broad polydispersities, branching, and residual iodide and amide groups.[9]

A different approach to the imide structure incorporated carbonylation chemistry with conventional poly(amide-ester) ring-closing procedures. Bis(*o*-iodoesters) were allowed to react with diamines and CO in the presence of a Pd catalyst. The intermediate poly(amide-ester) then spontaneously cyclized in the presence of base, giving the fully imidized polymer. It was found that *t*-butylesters gave the best results.[10]

Polybenzoxazoles (PBOs)

Another class of fused-ring heterocyclic polymers produced by carbonylation technology were the PBOs. Diiodoaromatics were again used, but in conjunction with bis(*o*-aminophenols).

The phenol groups were sufficiently unreactive under the reaction conditions to allow isolation of the intermediate poly(amide-ol). This PBO prepolymer was soluble in dipolar aprotic solvents and could be precipitated, purified, and stored until needed. Simple thermal or chemical cyclization could then be used to produce the thermally stable PBO.[11]

Polyacylhydrazines

Polymeric precursors to polyoxadiazoles were synthesized from dibromo aromatics and dihydrazines under 1 atm CO. Inherent viscosities from 0.13–0.65 were obtained for a variety of polyacylhydrazides.[12] These prepolymers were not reported to have been cyclized.

Polyesters

Diols also reacted with aromatic dihalides and CO to form polyesters. Both dibromides[13] and diiodides[14] were used to make modest molecular weight materials.

Triflates

Aromatic bis(trifluoromethanesulfonates) (triflates) have also been used instead of aromatic dihalides for polymer syntheses. They are known to undergo oxidative addition and CO insertion reactions and were used in preparing aramids,[15] PIAs,[16] and polyesters.[17] Reactions were performed in a manner similar to the iodo analogues, but only modest molecular weights were achieved.

REFERENCES

1. Schoenberg, A.; Heck, R. F. *J. Org. Chem.* **1974**, *39*, 3327.
2. Schoenberg, A. et al. *J. Org. Chem.* **1974**, *39*, 3318.
3. Yoneyama, M. et al. *Macromolecules* **1988**, *21*, 1908.
4. Perry, R. J. et al. *Macromolecules* **1993**, *26*, 1509.
5. Turner, S. R. et al. *Macromolecules* **1992**, *25*, 4819.
6. Perry, R. J.; Wilson, B. D. . *Macromolecules* **1993**, *26*, 1503.
7. Perry, R. J. et al. . *Macromolecules* **1994**, *27*, 4058.
8. Perry, R. J.; Turner, S. R. *J. Org. Chem.* **1991**, *56*, 6573.
9. Perry, R. J.; Turner, S. R. *Makromol. Chem., Macromol. Symp.* **1992**, *54155*, 159.
10. Perry, R. J. et al. . *Macromolecules* **1995**, *28*, 3509.
11. Perry, R. J.; Wilson, B. D. . *Macromolecules* **1994**, *27*, 40.
12. Yoneyama, M. et al. . *Macromolecules* **1989**, *22*, 4152.
13. Yoneyama, M. et al. . *Macromolecules* **1989**, *22*, 2593.
14. Perry, R. J.; Turner, S. R. U.S. Patent 4 933 419, Eastman Kodak Company, June 12, 1990.

15. Perry, R. J. U.S. Patent 5 210 175, Eastman Kodak Company, May 11, 1993.
16. Perry, R. J. U.S. Patent 5 214 123, Eastman Kodak Company, May 21, 1993.
17. Perry, R. J. U.S. Patent 5 159 057, Eastman Kodak Company, October 27, 1992.

Carboxylated Nitrile Rubber

See: Blends (Interchain Crosslinking)

Carboxymethylcellulose

See: Water-Soluble Polymers (Old Recovery Applications)

CARDO POLYESTERS (Heat Resistant and Photosensitive)

Kazuhiro Watanabe and T. Teramoto
Chemicals Division
Advanced Materials and Technology Research Laboratory
Nippon Steel Corporation

It is well known that polyesters show good heat resistance and mechanical and optical properties. Furthermore, the introduction of bulky structures, one of which is the so-called cardo structure, into the polymer backbone causes excellent heat resistance, good solubility in some organic solvents, and improvement in their moldability.[1–3]

However, these soluble heat-resistant polyesters have disadvantageously poor solvent resistance. The solvent resistance property is an important characteristic for protective coating materials in the micro-electronic field.

We studied the characteristics of unsaturated cardo polyesters whose bis-phenol component is 9,9-bis(4-hydroxyphenyl)fluorene (BPF) with the purpose of introducing thermo- and/or photocurable properties. These unsaturated polyesters were synthesized by an interfacial method and could be made by varying the ratio of saturated/unsaturated acid (terephthalic acid/fumaric acid). Such polyesters had high solubility in common organic solvents and then became insoluble in any solvents due to crosslinking by UV or thermal curing. The cured polymers exhibited a good heat-resistance, solvent resistance, and good adhesion to substrates.

Here, we describe the preparation and the properties of thermal stability of these photosensitive cardo polyesters (**Scheme I**).

RESULTS AND DISCUSSION

Synthesis of Polymers

Interfacial polycondensation was the most convenient method for the preparation of these polyesters, because high molecular weight polymers were obtained.

Characterization of Polymers

These polymers exhibited high heat resistance (T_g>300°C). Before UV irradiation, they were soluble and had an excellent solubility in a variety of solvent (e.g., chloroform, 1,1,2,2-tetrachloroethane, DMF, and NMP). Especially in NMP, the polymer was soluble up to 25 wt%. Even in relatively less polar tetrahydrofuran and dioxane solvents, these polymers were soluble (although to a slight degree) with an increase in the ratio of the fumaric component.

Properties of Films

The films of cardo polyester are easily prepared from a 20% polymer solution in chloroform or tetrachloroethane (TCE) by using the casting method. TCE is an especially good solvent. All the cardo polyester films obtained were transparent.

UV Properties of Photosensitive Polyesters

UV irradiation can cleavage the double bond and give the crosslinked polymer.

The conversion of the double bond in the fumaric unit increased with an increase in irradiation time and amounted to almost 50% within 8 min but no longer increased. This irradiated polyester gradually became insoluble for NMP, effective after 1 min of irradiation. As a result, the photocured coatings made by using a high-pressure Hg-lamp became insoluble in any solvent because of the crosslinking of carbon–carbon double bonds.

CONCLUSION

Heat-resistant, photosensitive cardo polyesters were synthesized from 9,9-bis(4-hydroxyphenyl)fluorene and a mixture of fumaryl dichloride and terephthaloyl dichloride. The polymers indicated good solubility in organic solvents before UV irradiation. Therefore, it was possible to make their films easily, and they showed many excellent properties. The UV-cured films showed good solvent resistance and high transparency and therefore, a wide range of applications is expected.

Photosensitive Cardo polyesters

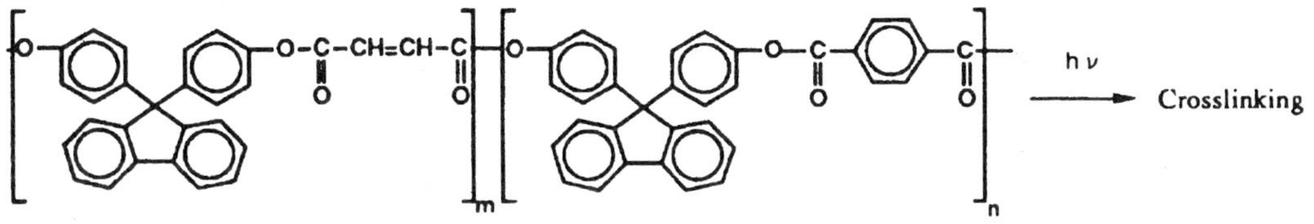

I

REFERENCES

1. Korshak, V. V.; Vinogradovaand, S. V.; Vygodskii, Y. S. *J. Makromol. Sci. Rev. Macromol. Chem.* **1974**, *C11*(1), 45.
2. Usmani, A. M. *J. Makromol. Sci. Chem.* **1982**, *A18*(2), 251.
3. Andoh, H.; Teramoto, T. *Polym. Prep.* **1990**, *31*(2), 677.

Carrageenans

See: Gelling Agents (Agarose and Carrageenan)
Gums (Overview)

CATALYSES (by Polyelectrolytes and Colloidial Particles)

Norio Ise
Central Laboratory
Rengo Company

Hiromi Kitano
Department of Chemical and Biochemical Engineering
Toyama University

The catalytic action of synthetic macromolecules has been studied intensively using various types of chemical reactions, for which numerous reviews were written. To save space, only one review by Ise et al. is referred to here and should be consulted for previous works.[1] The thrust of these investigations has been to determine whether synthetic materials could attain the high efficiency of enzymes. We first confined ourselves to basic polyelectrolyte catalysis of ionic reactions. Later, our interest shifted to ionic polymer latex particles, which serve as crude models of enzymes. Here, important results will be discussed emphasizing recent developments.

POLYELECTROLYTE CATALYSIS

It is easy to observe polyelectrolytes enhancing ionic reactions. For example, the initial rate constant of the $Fe(CN)_6^{4-}$–$S_2O_8^{2-}$ reaction could be increased by a factor of 10^5 in the presence of polybrene, a cationic polymer.[2] Generally, the acceleration factory (k/k^*, where k is the rate constant and * denotes the absence of polyelectrolyte) varies with experimental conditions: it can be quite large for reactions between similarly charged species when oppositely charged macroions are present and the reactants have a high valence and low concentration.

Cationic and anionic polymers decelerate the reactions between oppositely charged species. The simplest case was $NH_4^+ + OCN^- \leftrightarrow (NH_2)_2CO$. As mentioned in an earlier review, this reaction allowed us to quantitatively analyze polyelectrolyte catalysis relative to the Brøstead theory or the primary salt effect.[3]

The most important result from this reaction was discovering mechanistic difference between catalyses by enzymes and polyelectrolytes.[4]

POLYMER RESIN CATALYSIS

Compared with homogeneous polyelectrolyte catalysts, heterogeneous polyelectrolyte catalysts would be useful because products and catalysts can be recovered easily from the reaction solutions. Ion exchange resins were extensively examined as catalysts for various chemical reactions.

Solvolytic Reaction

A solvolytic reaction of α-phenethyl chloride was accelerated by increasing the ionic strength from the stabilized intermediate carbonium ion (ionic strength effect).[5]

Like simple salts and water-soluble macroions, cation exchange resins accelerated the reaction because of the ionic strength effect. In the presence of anion exchange resins, the dissociation of α-phenethyl chloride was hindered by the mass law effect, which decelerated the reaction (common ion salt effect). The deceleration and acceleration effects of ion exchange resins were greater in more polar solvents.

Deamination

The deamination reaction of α-phenethylamine to produce α-phenethyl alcohol was slightly accelerated by cationic resins and decelerated by anionic resins.[5] The reactant was free α-phenethylamine, and the acceleration and deceleration were from destabilizing and stabilizing α-phenethylammonium ion by cationic and anionic polymer resins, respectively. The proton concentration was presumably high in the anionic resin domain, and the reactants concentration was low. The stereochemistry of the product α-phenethyl alcohol in the deamination reaction was affected by the presence of ion exchange resins.

POLYMER LATEX CATALYSIS

Polymer microspheres have been used to carry various catalysts because the catalytic site is more rigidly fixed than in homogeneous polymer systems. To introduce catalytic groups into the microspheres, two methods were used: direct emulsion copolymerization of monomers having a catalytic moiety with other monomers, and modifying reactive groups ($-NH_2$, $-COOH$, and $-OH$, for example) on the polymer microsphere.

Interionic Reactions

Aquation of $Co(NH_3)_5Br^{2+}$ complex induced by Hg^{2+} ion was examined in the presence of anionic latex particles, which were prepared by emulsion copolymerization of styrene and acrylic acid.[6] The reaction was accelerated by a factor of 10^2. However, the reaction between $Co(NH_3)_5Br^{2+}$ and OH^- and that between cationic triphenylmethane dye (crystal violet) and OH^- were decelerated by adding anionically charged latex suspension.[7] Electrostatic interaction dominated the former reaction and the inhibitory effects of the latex particles increased with their surface charge density. These inhibitory effects were attributed to the reactant ions' desolvation near latex particles according to measurements of the activation volume.

Esterolytic Reaction

Latex particles carrying carboxyl groups were prepared by emulsion copolymerization of acrylic acid and styrene. By treating carboxyl groups on the surface with histamine or *L*-histidine in the presence of water-soluble carbodiimide, imidazolyl groups could be introduced easily. Furthermore, *n*-butyl groups were introduced as binding sites for substrates. The latex particles obtained showed a Michaelis-Menten saturation behavior in the hydrolysis of nitrophenyl esters.[8]

Oxidation Reactions

Polymer complexes have been widely used as catalysts in oxidation, reduction, polymerization, and photochemical reactions. After the imidazolyl group was complexed with Cu(II) ion on the latex surface, the microsphere showed catalytic activity for the oxidation of ascorbic acid. The catalytic activity showed a Michaelis-Menten type saturation behavior.[9]

REFERENCES

1. Ise, N.; Okubo, T.; Kunugi, S. *Acc. Chem. Res.* **1982**, *15*, 171.
2. Enokida, A.; Okubo, T.; Ise, N. *Macromolecules* **1980**, *13*, 49.
3. Moore, W. J. *Physical Chemistry*, 3rd ed.; Prentice-Hall: Englewood Cliffs, NJ, 1962.
4. Laidler, K. J.; Bunting, P. S. *The Chemical Kinetics of Enzyme Action*, 2nd ed.; Clarendon: Oxford, England, 1973.
5. Nishiyama, Y.; Moroe, I.; Kitano, H. et al. *J. Polym. Sci., Polym. Chem. Ed.* **1980**, *18*, 3289.
6. Ise, N. *J. Polym. Sci., Polym. Symp.* **1978**, *62*, 205.
7. Ishiwatari, T.; Maruno, T.; Okubo, M. et al. *J. Phys. Chem.* **1981**, *85*, 47.
8. Kitano, H.; Sun, Z.; Ise, N. *Macromolecules* **1983**, *16*, 1306.
9. Sun, Z.; Yan, C.; Kitano, H. *Macromolecules* **1986**, *19*, 984.

CATALYSTS, POLYMERIC (Higher Activity Than Monomeric Analogs)

Shuji Kondo
Department of Materials Science and Engineering
Nagoya Institute of Technology

Polymeric catalysts have been prepared by anchoring monomeric catalysts to soluble or insoluble polymer supports.[1,2] These catalysts have many advantages such as easily separated products and reusable catalysts. However, the catalytic activity is generally lower than monomeric analogs, primarily because the polymer backbone and its steric effects restrict active site mobility.

However, enzymes as natural polymer catalysts display high activity and selectivity mainly because of the polymer's effect. Recently, many polymer catalysts showing higher activity than the monomeric analogs were reported. Here, the preparation and application of these polymer catalysts, except for enzyme-like catalysts, are reviewed briefly.

APPLICATIONS

Although there are many kinds of polymeric catalysts with greater activity than monomeric analogs, here they are classified into three groups: phase transfer catalysts, transition metal complexes, and other catalysts.

Phase Transfer Catalysts

The use of phase transfer catalysts, which promote reactions between organic and aqueous phases, is routine in laboratories where organic synthesis is studied. However, separating products and recovering catalysts such as quaternary onium salts, crown ethers, cryptands, and polyethers are difficult in practice. These problems were overcome by anchoring monomeric catalysts to soluble or insoluble polymers.[4]

Phase transfer reactions catalyzed by polymeric onium salts proceed using the following mechanisms: a counter anion in the catalyst is exchanged with a nucleophile in the aqueous phase, and then the desolvated nucleophile attacks the substrate near active sites in the polymeric catalyst.[5] Therefore, the concentration of onium salt ion sites in a polymer is important.

However, polymer-supported polyethers form complexes with alkali metal cations similar to crown ethers and cryptands. Then, the resulting anion attacks the substrate in organic solvent. Therefore, in these phase transfer catalysts, the functional groups' cooperative effect is the primary factor in catalytic activity.

In fact, many polymeric dipolar aprotic solvents such as *N,N*-dimethylformamide, *N,N*-dimethylacetamide, tetramethylurea, dimethyl sulfoxide, hexamethylphosphoramide, and their analogs displayed catalytic activity, although the monomeric analogs were virtually inactive.[10–12]

Transition Metal Complexes

Functional groups in polymers, especially in insoluble polymers, did not interact effectively because of reduced mobility in the polymer backbone. This feature, known as matrix-isolation of active sites, was used to prepare the polymer-supported metal complex catalysts. For example, polymer-supported titanocene worked as an effective hydrogenation catalyst of olefins, although the activity of the monomeric analog was very low.[14] This acceleration of catalytic activity derived from preventing dimerizaiton of active sites because the dimer had no activity. Similar observations were reported for reducing unsaturated carbonyl compounds with carbon monoxide and water catalyzed by rhodium complexes.[15]

As observed in many oxidation reactions, catalytic activity of transition metal complexes was enhanced by anchoring the ligands to an insoluble polymer.[17]

Other Catalysts

Polymeric acids have been investigated in detail and used for many reactions. Recently, Arai et al.[18] discovered that poly(α-sulfoundecenoic acid) catalyzed the hydrolysis of sucrose and *p*-nitrophenyl acetate more effectively than the monomeric analog. This acceleration resulted from the strong hydrophobic interaction between the substrate and catalyst. The importance of hydrophobicity around the active sites in catalytic activity was also reported for reacting oxiranes with active esters catalyzed by quaternary onium salts.[19]

REFERENCES

1. Sherrington, D. C.; Hodge, P. *Syntheses and Separations Using Functional Polymers*; John Wiley & Sons: Chichester, 1988.
2. Hodge, P. *Rapra Rev. Rep.* **1992**, *5*, 1.
3. Regen, S. L. *Angew. Chem. Int. Ed. Engl.* **1979**, *18*, 421.
4. Ford, W. T.; Tomoi, M. *Adv. Polym. Sci.* **1984**, *55*, 49.
5. Tomoi, M.; Takubo, T.; Ikeda, M. et al. *Chem. Lett.* **1976**, 473.
6. Kondo, S.; Ohta, K.; Inagaki, Y. et al. *Pure Appl. Chem.* **1988**, *60*, 387.
7. Kondo, S.; Furukawa, K.; Tsuda, K. *J. Polym. Sci., Polym. Chem. Ed.* **1992**, *30*, 1503.
8. Bond. W. D.; Brubaker, C. H.; Chandrasekaran, E. S. et al. *J. Am. Chem. Soc.* **1975**, *97*, 2128.

9. Kitamura, T.; Joh, T.; Hagihara, N. *Chem. Lett.* **1975**, 203.
10. Sherrington, D. C. *Pure Appl. Chem.* **1988**, *60*, 401.
11. Arai, K.; Okabe, T. *Polym. J.* **1992**, *24*, 769.
12. Nishikubo, T.; Iizawa, T.; Shimojo, M. et al. *J. Org. Chem.* **1990**, *55*, 2536.

CATALYSTS, POLYMERIC (Pressure Effects)

Yoshihiro Taniguchi and Syoichi Makimoto
Department of Chemistry
Ritsumeikan University

Polymers can have a remarkably large accelerating or decelerating catalytic effect on chemical reactions. The magnitude of these effects is greater than that produced by substances with low molecular weight. One characteristic of a polymeric catalyst is that it incorporates or binds substrate molecules into the polymer with intermolecular interaction. Consequently, polymeric catalysts have three effects: dehydration around the substrate, by which polymeric catalysts incorporate or bind substrate molecules into the polymer; substrates concentrated around the polymeric catalyst; and differences in reaction field atmospheres at the catalytic site of the polymeric catalyst and in bulk water. This last effect can produce a catalytic action that is as efficient as enzymic catalysis according to Michaelis-Menten kinetics. Therefore, to understand how pressure affects a polymer's catalytic action, we must consider two contributions to the polymer catalytic reaction: the substrate binding processes and the catalytic reaction to formed products. Then, the effects of pressure on polymeric catalysis can be used as a model for enzyme catalysis in terms of volume changes for the binding process and the activation volumes of product formation.

PREPARATION AND PROPERTIES

Catalysis and Pressure Theory

The pressure dependence of the polymeric reaction provides information about volume changes occurring in the reaction process, such as intermolecular interaction, hydration at the catalytic site, or the conformational change of the polymeric catalyst upon substrate binding. Therefore, activation or reaction volumes are useful parameters for understanding the polymeric catalysis mechanism.

Pressure Dependence

Polymer

A reaction catalyzed by polyelectrolytes under high pressure was first measured in the spontaneous aquation of $Co(NH_3)_5Br^{2+}$ catalyzed by sodium poly(ethylene sulfonate) (NaPES) shown in **Equation 1**.[1]

$$Co(NH_3)_5Br^{2+} + Pb^{2+} + H_2O \rightarrow Co(NH_3)_5H_2O^{3+} + Pb^{2+} + Br^-$$

Interestingly the spontaneous aquation did not proceed, and the equilibrium lay to the right in the presence of NaPES. Because of pressure dependence from the overall rate constant, the activation volume increased with polymer concentration. The hydrated PB^{2+} was dehydrated by poly(ethylene sulfonate) (PES) anion, which acted as the polymer catalyst so that the PB^{2+} ion easily seized a Br^- ion from $Co(NH_3)_5Br^{2+}$.

The effects of hydrophobic dehydration in polymeric catalysts differed from electrostatic effects. When alkaline faded triphenylmethane dyes catalyzed by polyelectrolytes, $\Delta V^{\neq}$ decreased as the concentration of cationic polyelectrolytes (C_{16}BzPVP: 4-vinyl-*n*-cetylpyridinium bromide or chloride) increased because of the interionic reaction between oppositely charged ionic species.[2] However, despite their opposite charges, $\Delta V^{\neq}$ increased when anionic polyelectrolytes, polymer sulfonic acid (PSS) and sulfonated polystyrene, increased because of hydrophobic dehydration in the dyes during the binding process. The tendency for $\Delta V^{\neq}$ to increase was observed in base hydrolysis of neutral and cationic esters.[3]

Concentration effects, the result of hydrophobic interaction and H-bonding on polymer catalysis, were discussed for the acidic catalyzed hydrolysis of alkylacetate (methyl-, ethyl, *n*-butyl) with PSS and *o*-benzal sulfonated polyvinylalcohol (PVBeS).[4]

Micelles

Micellar catalyzed reactions under high pressure were first measured using acidic hydrolysis of alkyl acetate catalyzed by dodecyl hydrogensulfate micelles.[5] The hydrolysis rate for the esters was markedly accelerated above the critical micelle concentration (CMC) of 1.8 m mol kg^{-1}. The overall rate constant tended to have a minimum at about 100 MPa, which corresponded with the appearance of a maximum in the CMC versus pressure plots at about 100 MPa. Overall activation volume increased with the esters' hydrophobicity because hydrophobic interaction of the ester molecules was incorporated into the micellar phase.

To understand this interesting inversion phenomenon and to analyze the pressure effects quantitatively, the hydrolysis of neutral *p*-nitrophenyl esters and anionic 3-nitro-4-acyloxybenzoic acid, catalyzed by cetyltrimethylammonium bromide (CTAB) micelle, was measured.[6,7]

Cyclodextrins

Cyclodextrins (CDs) are α-1,4-linked cyclic oligomers of D-glucopyranose that include cyclohexaglucopyranose cycloocta-(α–CD), cyclohepta- (β–CD), and cycloocta-(γ–CD). The CDs served as a good model for serine protease as reviewed by Bender et al.[8]

The hydrolysis reactions of substituted phenyl acetates and 2-naphthyl acetate catalyzed by CDs was reported.[9,10]

The tendencies of pressure effects in loose and tight complexes were similar for hydrolysis of *p*-nitro trifluoroacetanilide as an anilide substrate and transacylation of *p*-nitrophenyl (E)-β-ferrocenylacrylate catalyzed by CDs.[11,12]

REFERENCES

1. Ise, N.; Ishikawa, M.; Taniguchi, Y. et al. *J. Polym. Sci., Polym. Lett. Ed.* **1976**, *14*, 667.
2. Ise, N. *Kagaku* **1981**, *36*, 336.
3. Okubo, T.; Ueda, M.; Sugimura, M. et al. *J. Phys. Chem.* **1983**, *87*, 1224.
4. Taniguchi, Y.; Sugiyama, N.; Suzuki, K. *J. Phys. Chem.* **1978**, *82*, 1231.

5. Taniguchi, Y.; Inoue, O.; Suzuki, K. *Bull. Chem. Soc. Jpn.* **1979**, *52*, 1327.
6. Taniguchi, Y.; Makimoto, S.; Suzuki, K. *J. Phys. Chem.* **1981**, *85*, 2218.
7. Makimoto, S.; Suzuki, K.; Taniguchi, Y. *Bull. Chem. Soc. Jpn.* **1983**, *56*, 1341.
8. Bender, M. L.; Komiyama, M. *Cyclodextrin Chemistry*; Springer-Verlag: New York, 1977.
9. Taniguchi, Y.; Makimoto, S.; Suzuki, K. *J. Phys. Chem.* **1981**, *85*, 3469.
10. Makimoto, S.; Suzuki, K.; Taniguchi, Y. *J. Phys. Chem.* **1982**, *86*, 4544.
11. Makimoto, S.; Suzuki, K.; Taniguchi, Y. *Bull. Chem. Soc. Jpn.* **1984**, *57*, 175.
12. le Noble, W. J.; Srivastava, S.; Breslow, R. et al. *J. Am. Chem. Soc.* **1983**, *105*, 2745.

Catalysts, Polymerization

See: *Asymmetric Polymerization (Overview)*
Bisallenes, Polyadditions
Block Copolymers (By Changing Polymerization Mechanism)
Block Copolymers (By Step-Growth Polymerization)
Bulk Polymerization
Butadiene Polymerization (Supported Lanthanide Catalysts)
Butadiene Polymerization, Neodymium-Catalyzed
Carbenoid Polycondensation
Carbonylation Polymerizations (Palladium-Catalyzed)
Condensation Polymers (Using Phosphorus Pentoxide/Methanesulfonic Acid)
Conjugated Polymers (Prepared by Organometallic Processes)
Cyclic Siloxanes (Ring-Opening Polymerization)
Diyne Cycloaddition Copolymerization (Transition Metal Catalyzed)
Electron Transfer Reactions (Novel Polymerization by Divalent Samarium)
Enol Ester Polymers
Functionalized Polyolefins (Using Borane Monomers and Transition Metal Catalysts)
Gas Phase Olefin Polymerization
Glassy Carbon
High Solids Polyurethane Coatings
Ladder Polymers (Cycloaddition Copolymerization of Cyclic Diynes)
Liquid Crystalline Polyacetylene Derivatives
Liquid Crystalline Polymers (Conjugated Chain)
Living Coordination Polymerization
Metallocenel Aluminoxane Catalysts (Orefin Polymerization)
Metallocene Catalysts
Metallocene Catalysts (Cationic Group 4 Metal Alkyl Complexes, Olefin Polymerization)
Metallocene Catalysts (Group 4 Elements, New Polymeric Materials)
Metathesis Degradation, Unsaturated Polymers
Metathesis Polymerization, Cycloolefins
Monomer-lsomerization Polymerization (2-Butene with Ziegler-Natta Catalysts)
Olefin-Carbon Monoxide Copolymers (Overview)
Olefin Polymerization by 2,ω-Linkage (Migratory Nickel-Phosphorane Catalyst)
Olefin Polymerization Catalysts (Experimental and Theoretical Characterization)
Olefin Polymerization Catalysts (Silica-Supported Titanium Tetrachloride)
Olefin Polymerization Catalysts, Soluble (Group 4, Cationic Active Centers)
Oxidative Coupling
Palladium-Catalyzed Synthesis (Monomers and Polymers)
Phenylacetylene (Stereospecific Living Polymerization, Rh Complexes)
Polyacetylene Films
Polyamines (Ru-Catalyzed Polycondensation of Diamines and Diols)
Polybenzimidazoles (Overview)
Poly(ε-caprolactone) (Overview)
Polyelectrolytes, Rigid-Rod [Poly(p-phenylene)-Based]
Polyethylene (Commercial)
Polyethylene (High Density Preparation)
Poly(l,1′-Ferrocene-Arylene)s (Pd -Catalyzed)
Polyimides, Soluble Rigid-Rods (Via Pd-Catalyzed Polycondensation)
3,4-Polyisoprene, Crystalline
Polyolefins (Overview)
Polyphenylacetylene
Polyphenylacetylene-Based Permselective Membranes
Poly(phenylene ethynylene) (Synthesis and Optical Properties)
Poly(phenylene vinylene)s (by the Heck Coupling Reaction)
Polypropylene (Commercial)
Polypropylene (Controlled Degradation)
Polypropylene, Atatic (High Molecular Weight)
Poly(propylene carbonate)
Polypropylene, Hemiisotactic
Poly(propylene oxide)
Polysilylpropynes (Steric Effects on Material Properties)
Polystyrene, Stereoregular
Polystyrene, Syndiotactic
Polyurethane Catalysis
Polyurethane Catalysts
Propylene Polymerization (Stochastic Analysis of the Propagation Mechanism)
Rare Earth-Initiated Polymerization (of Dienes)
Rare Earth Polymerization Initiators
Ring-Opening Coordination Polymerization (by Soluble Multinuclear Alkoxides)
Ring-Opening Metathesis Polymerization (Formation of Cyclic Butadiene Oligomers)
Silacyclobutanes and Related Compounds (Ring-Opening Polymerization)

Catalysts, Reaction

Catenanes

Catheters

Cationic Polymerization

CELL ADHESION-PROMOTING BIOMATERIALS

Yoshihiro Ito* and Yukio Imanishi
Department of Material Chemistry
Faculty of Engineering
Kyoto University

Cell adhesion is a ubiquitous process that influences many aspects of cell behavior. For example, proliferation, migration, secretion, and differentiation are triggered by adhesion to matrix. Extensive investigations have been performed on the effects of surface properties of polymeric materials on cell adhesion. These investigations can be classified into three categories:[1,2] physicochemical (control of surface charge or surface free energy), morphological (control of surface domain or roughness), and biochemical or biological (conjugation with biologically active molecules).

By this last approach it is possible to control complex cellular functions involving adhesion and growth. Many researchers have used proteins such as collagen or fibronectin, which are cell adhesion factors,[3–5] to design and synthesize adhesion-promoting biomaterials.

The discovery of the active peptide sequence Arg-Gly-Asp-Ser (known as RGDS by one-letter abbreviation) in cell adhesive proteins by Pierschbacher and Ruoslahti[6] began a new trend in biomaterials design.[7–12] Various RGDS derivatives, including polyRGDS, cyclic RGDS, and nonpeptidic chemicals have been synthesized.

PROPERTIES

Cell Adhesive Activity

In the adhesion of mouse fibroblast STO cells onto RGDS-immobilized poly(acrylic acid)-grafted polystyrene films, the RGDS-immobilized polystyrene film markedly enhanced cell adhesion.

RGD-OH-immobilized film had no cell adhesion activity. RGD containing an aspartic acid residue with free carboxyl terminal is not active in cell adhesion. The lack of activity of the RGD-OH immobilized film demonstrated that cell adhesion

*Author to whom correspondence should be addressed.

onto RGDS- and fibronectin-immobilized film was based not on physical interactions, but on specific ligand–receptor interactions.

In cell adhesion onto polymerizable RGDS-grafted polystyrene films, the glow discharge treatment increased cell adhesivity. Acrylamide or acrylic acid grafting reduced adhesivity. On the other hand, *N*-(3-trimethylammoniumpropyl)acrylamide chloride-grafted polystyrene film had high adhesivity. These results indicate that hydrophilic polymer grafting reduced cell adhesion, but cationic polymer grafting enhanced it. Film copolymerized with polymerizable RGDS promoted cell adhesion more than did film grafted with cationic polymer, and had cell adhesiveness comparable to that of a fibronectin-coated film.

Stability of Materials

The cell adhesion activity of RGDS- and fibronectin-immobilized polystyrene films decreased with increasing temperature, and the activity of fibronectin-immobilized film decreased more sharply than that of RGDS-immobilized film. The cell adhesion of RGDS- or fibronectin-immobilized film was markedly reduced at high and low pH.

REFERENCES

1. Ito, Y. *Synthesis of Biocomposite Materials*; CRC: Boca Raton, FL, 1992; Chapter 5.
2. Singhvi, R. et al. *Biotech. Bioeng.* **1994**, *43*, 764.
3. Brown, R. A. et al. *Biomaterials* **1994**, *15*, 457.
4. Brunstedt, M. B. et al. *J. Biomed. Mater. Res.* **1993**, *27*, 483.
5. van Wachem, P. B. et al. *J. Biomed. Mater. Res.* **1994**, *28*, 353.
6. Pierschbacher, M. D.; Ruoslahti, E. *Nature* **1984**, *309*, 30.
7. Ito, Y. et al. *Biomed. Mater. Res.* **1991**, *25*, 1325.
8. Ito, Y. et al. *Polymers of Biological and Biomedical Significance*; Shalaby, S. W. et al., Eds. ACS Symposium Series 540, American Chemical Society: Washington, DC, 1994; Chapter 6.
9. Koide, M. et al. *J. Biomed. Mater. Res.* **1993**, *27*, 79.
10. Kasuya, Y. et al. *Biomaterials* **1994**, *15*, 570.
11. Kasuya, Y. et al. *J. Biomed. Mater. Res.* **1994**, *28*, 397.
12. Kugo, K. et al. *J. Biomater. Sci., Polym. Ed.* **1994**, *5*, 325.

Cell-Containing Materials

See: Biodegradable Copolymers [Comprising Poly(L-lactic acid)]
Cell Adhesion-Promoting Biomaterials
Cell Entrapment [Poly(carbamoyl sulfonate) Hydrogels]
Cell Growth-Promoting Biomaterials
Cell Separation Materials
Hybrid Artificial Organs
Immobilized Microbial Cells
Microencapsulation (Artificial Cells)
Radiation Induced Polymerization (Biomedical Applications)
Red Blood Cell Substitutes, Artificial (for Transfusion)
Silk Fibroin (Soft Tissue Compatible Polymer)

CELL ENTRAPMENT [Poly(Carbamoyl Sulfonate) Hydrogels]

K. D. Vorlop, A. Muscat, and A. V. Patel
Institute of Technology
FAL

Immobilized cells have great potential in the laboratory. Their importance is increasing because of their technological advantages over free cells. Among these are continuous process and repeated use, avoiding washout, reaching higher concentration in reactors, protecting sensitive cells, and easy separation. Cells may be immobilized by different methods, such as entrapment, adsorption, and crosslinking. Several authors have reviewed the methods, merits, and use of cell immobilization.[1–3]

Conventional natural materials for cell entrapment (e.g., alginate, carrageenan) possess no toxicity but have a low mechanical stability, are very sensitive to abrasion in stirred reactors, and are biodegradable under nonsterile conditions. Polyurethane (PUR) shows good mechanical and chemical stability,[4] but the raw material (isocyanate prepolymer) is toxic to microorganisms.[5] Furthermore, the short handling time (seconds) during the immobilization process makes it nearly impossible to prepare a large amount of spherical biocatalyst.

Our approach to solve these physiological and engineering problems was based on using chemically blocked isocyanates for the synthesis of polyurethane matrices by making a roundabout in order to lower the toxic effect of the isocyanate groups on the cells.[6]

PREPARATION AND PROPERTIES

Commerical NCO prepolymers react with the blocking agent bisulfite to form poly(carbamoyl sulfonate) (PCS) prepolymer (Figure 1) at room temperature.[6]

This prepolymer forms a PCS hydrosol.

PCS Prepolymer Gelation

When mixed with water, conventional PUR prepolymers crosslink at once (within seconds) to a foam or hydrogel. In contrast to PUR prepolymer, PCS prepolymers show an adjustable "elation time that depends mainly on the pH value of the solution. At a pH of 8.5 an aqueous solution of PCS prepolymers gelled to a PCS-hydrogel in 10 s at room temperature, but at a pH < 5 the solution can be handled up to 10 h.

Molecular Weight Cut Off

We found the molecular weight cut off of the PCS hydrogel membrane fell between that of myoglobin (MW 17,000 g/mol) and albumin (67,000 g/mol) so substances with a MW $>$ 67,000 g/mol cannot diffuse through the PCS hydrogel.

APPLICATIONS

It was now possible to prepare an aqueous PCS solution with a pH of 4–6.5 to support living cells, and to gel this mixture at once by adjusting the pH to 8.5.

Bipolymers like Ca-alginate are normally used to entrap living cells because of their low toxicity. Their disadvantages include poor mechanical stability and biodegradability under nonsterile conditions. Our PCS hydrogel combines the

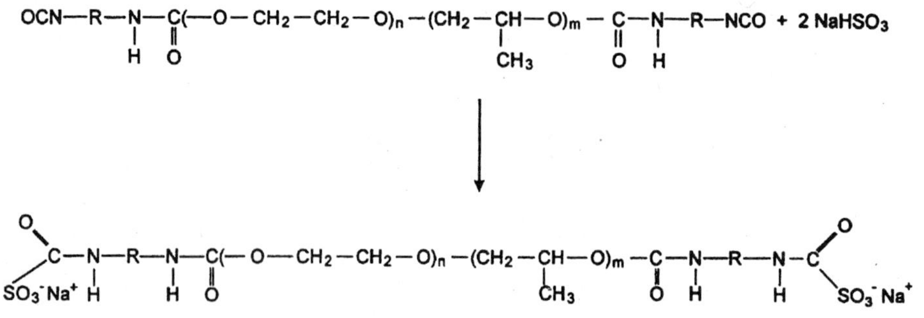

FIGURE 1. Formation of a PCS prepolymer from an isocyanate prepolymer and bisulfite.

advantages of the Ca-alginate (low toxicity) and PUR matrix (high elasticity).

The advantages of PCS hydrogels make biotechnological use in agitated reactors and fluid-bed reactors possible. PCS hydrogel membranes can also be used in biosensors as an immobilization matrix for cell[7] or enzymes.[8]

REFERENCES

1. Klein, J.; Vorlop, K. D. *Comprehensive Biotechnology*; Pergamon: London, 1985; Vol. 2, 203–24.
2. Dervakos, G. A.; Webb, C. *Biotech. Adv* **1991**, *9*, 559–612.
3. Furusaki, S.; Seki, M. *Adv. Biochem. Eng./Biotechnol.* **1992**, *46*, 164–185.
4. Fukui, S. et al. *Methods Enzymol.* **1987**, *135*, 230–52.
5. Klein, J.; Wagner, F. *Appl. Biochem. Bioeng.* **1983**, 4, 11–51.
6. Vorlop, K. D. et al. Ger. Patent 4 217 891, 1992.
7. Muscat, A. et al. *Biosensors Bioelectron.*, submitted 1995.
8. Kotte, H. et al. *Anal. Chem.*, submitted 1995.

CELL GROWTH-PROMOTING BIOMATERIALS

Yoshihiro Ito, Yukio Imanishi, and Ji Zheng
Division of Material Chemistry
Faculty of Engineering
Kyoto University

Jing-Song Li and Takashi Takahashi
Department of Medical Informatics
Faculty of Medicine
Kyoto University

Researchers are interested in the use of biomaterials to control higher level cellular functions, such as growth, differentiation, and movement. Because it is difficult to construct biomaterials using only synthetic substances, biological macromolecules are usually conjugated.[1] Recent research has revealed a number of biosignal molecules regulating cellular functions. Growth factor proteins and cytokines are known to regulate the cellular functions of many processes. These proteins have been noncovalently or covalently incorporated into materials.

Covalent immobilization has several advantages. First, immobilized growth factors are more active than soluble ones. High molecular weight biosignal molecules usually interact with receptors, and the complexes are internalized into the cell to be dissociated and decomposed in lysosome. Immobilization of the biosignal molecule inhibits the down-regulation process initiated by internalization. In addition, immobilization provides a highly localized concentration of biosignal proteins, which should induce multivalent simultaneous stimulation, leading to enhanced complex formation with receptors, crosslinking of the complex, and accelerated signal transduction.

Second, immobilized growth factors can be repeatedly used as a bioreactor support for cell cultures. After the cells are cultured, material immobilized with factor proteins can again be utilized by removing the grown cells from the surface.

PREPARATION

Insulin, epidermal growth factor (EGF), and transferrin, which are usually added in commercial serum-free cell culture media, were immobilized as growth factor proteins on various matrices.[2–11] These included surface-hydrolyzed poly(methyl methacrylate), surface-hydrolyzed poly(ethylene terephthalate), polyurethane-containing amino groups, poly(hydroxyethyl methacrylate)-*co*-poly(ethyl methacrylate) of various compositions, surface-treated polyacrylamide, silane-coupled glass, and collagen beads. Various immobilization reagents, such as water-soluble carbodiimide (WSC), cyanogen bromide, glutaraldehyde, and dimethyl suberimide (DMS) were used.

PROPERTIES

Cell Growth Acceleration

Mouse STO fibroblast cells, mouse fibroic sarcoma cells, mouse hybridomas, bovine endothelial cells, and chinese hamster ovary (CHO) cells were cultured on these biosignal-immobilized materials. The growth of STO cells was accelerated on the insulin-immobilized PMMA membrane. The growth rate on the insulin-immobilized film was higher than that in the presence of free insulin. The immobilized insulin accelerated cell growth 1.8 times as much as nonimmobilized matrix, and 1.2 times as much as free insulin. Such acceleration was observed on other materials, although the degree depended on the immobilization methods, the nature of the matrices, and the nature of cell lines.

Both of the adhesion factors, collagen and PAA, accelerated STO cell growth by themselves. In addition, insulin coimmobilization remarkably enhanced cell growth. The growth rate increased with the amount of immobilized insulin. In both cases, the coimmobilized insulin was more active than free insulin as observed in the insulin-immobilized PMMA membrane. By coimmobilization of insulin with cell adhesion factors, the cell growth rate became comparable to that in serum-containing media.

APPLICATIONS

The biosignal-immobilized materials were applied to *in vivo* and *in vitro* uses. The former was for an artificial organ, the latter for cell culture engineering.

Hybridization of synthetic materials and organs is considered a promising source of high-performance artificial organs. One representative organ is an artificial blood vessel. The vessel is a plastic tube coated with blood endothelial cells. Immobilization of biosignal proteins was used for endothelialization. When we measured the rate of covering a polyurethane tube by endothelial cells, the tube coimmobilized with both insulin and collagen was rapidly covered with endothelial cells, which were stable for more than a year.

Mammalian cell culture is important, not only for fundamental research but also for industrial production of a large quantity of biologically significant materials. Although serum or serum substituents must support cell growth in addition to nutrients, the use of serum is undesirable because of its high cost and the separation complications introduced by the serum proteins. Accordingly, several serum-free media have been developed.

The immobilized biosignal provides another approach for serum-free culture technique. We now have a completely protein-free cell culture medium and recycling culture system.

REFERENCES

1. Ito, Y. *Synthesis of Biocomposite Materials*, CRC: Boca Raton, FL, 1992; Chapter 5.
2. Ito, Y. et al. *Biomaterials* **1991,** *12*, 449.
3. Ito, Y. et al. *Biomaterials* **1992,** *13*, 789.
4. Ito, Y. et al. *Biotechnol. Bioeng.* **1992,** *40*, 1271.
5. Ito, Y. et al. *J. Biomed. Mater. Res.* **1993,** *27*, 901.
6. Ito, Y. et al. *J. Bioact. Comp. Polym.* **1994,** *9*, 170.
7. Liu, S. Q. et al. *Biomaterials* **1992,** *13*, 50.
8. Liu, S. Q. et al. *J. Biophys. Biochem. Method* **1992,** *25*, 139.
9. Liu, S. Q. et al. *Enz. Microb. Tech.* **1993,** *15*, 167.
10. Liu, S. Q. et al. *Int. J. Biol. Macromol.* **1993,** *15*, 221.
11. Zheng, J. et al. *Biomaterials* **1994,** *15*, 963.

CELL SEPARATION MATERIALS

Akihiko Kikuchi
Institute of Biomedical Engineering
Tokyo Women's Medical College

Great progress in immunology and cellular biology has meant significant improvement in cell separation methods. Obtaining sufficient pure and vital cell populations is vital. Novel and effective cell separation techniques are necessary to get cell populations for diagnosis and therapy in clinical medicine, and to produce biotechnological compounds.[1–4]

Blood cells are of special interest because of a great progress in cellular immunology.

Lymphocytes are known to play an important role in immune response, and consist of functionally different subpopulations, B cells and T cells. Separation of B cells and T cells is essential for precise judgment of immunodiseases, cancers, and tissue compatibility by human leukocyte antigen (HLA) typing (donor–recipient matching) in organ transplantation.

Because most cells secrete hormones, growth factors, or enzymes, separation of functional cells and valuable bioactive compounds is necessary in biotechnology.

Cell separation may be based on several parameters: density, volume, charge, hydrophilic/hydrophobic balance (amphiphilic properties of the cell membrane), and biological affinity of membrane components toward specific molecules. Some of these parameters may be combined to increase efficacy. Several methods are based on differential volume and/or density, differential mobility in an electric field, and differential affinity toward solid-phase matrix,[1–3]

Cell separation based on differential affinity toward solid-phase matrices is also used because this method can be operated with relatively simple apparatus and procedures. This method has two categories: the affinity method based on differential biological affinity toward the ligand-immobilized substratum, and adsorption method, based on differences in physicochemical properties of the cell membrane. For both categories, the operating techniques are classified into the panning method and the chromatography method.

I will describe the preparation and application of poly(2-hydroxyethyl methacrylate-*graft*-polymine(HA) copolymers for separation of lymphocyte subpopulations using the adsorption chromatography technique.

PREPARATION

Synthesis of Polyamine Macromonomer

Preparation of polyamine macromonomer was reported by Nitadori and Tsuruta and Nabeshima et al.[12]

Synthesis of HA Copolymer

HA copolymer was synthesized by radical copolymerization of polyamine macromonomer with 2-hydroxyethyl methacrylate (HEMA) in ethanol at 45°C for 1.5 h using 2,2′-azobis(2,4-dimethylpentanonitrile) (V-65) as an initiator.[12,13]

PROPERTIES AND APPLICATIONS

HA Copolymer for Affinity Chromatography Matrix

Although affinity chromatography is recognized as a powerful tool in cell biology research and clinical medicine for cell separation, it occasionally shows a disadvantage of low recovery of attached cells from matrix. Furthermore, ligands are usually expensive and not stable enough for long-term storage.

Kataoka and Tsuruta et al. found that retention behaviors of lymphocytes were dramatically minimized with a certain amount of polyamine in the HA copolymer.[5,16] This phenomenon of minimum cellular retention on a polymer surface with a small amount of cationic amino groups was observed for blood platelets as well as erythrocytes.[17–19] Although cationic groups have strong interaction with negatively charged cellular membranes, the copolymer surface eliminated the nonspecific interaction of these cells. This unique property of HA copolymer could lead to a matrix for affinity separation of lymphocyte subpopulations. That is, a stationary phase with a nonspecific adsorption and simple ligand immobilization through physical adsorption can easily be prepared using HA copolymer.

Kataoka and Tsuruta et al. used HA2, which contains 2 wt % of polyamine in the copolymer, as a matrix because the surface of HA2 showed minimal nonspecific adsorption of lymphocytes.[5,9,10,16,20] A column packed with HA2-coated beads was pretreated with goat and anti-rat IgC in PBS solution to physically adsborb antibody molecules in HA2 surface.

HA Copolymers as Novel Solid-Phase Matrices for Adsorption Chromatography of Lymphocyte Subpopulations

Cell adsorption to matrix surfaces can be described in two distinct but serial processes. The first stage is the physicochemical adsorption process (passive process), in which cells are adsorbed through physicochemical interaction with the substrate surface. The second stage is the active adhesion process, in which cell/material interaction is enhanced through metabolism of the cells accompanied with cell shape changes (active adhesion, contact-induced activation).[14] Functional changes of adhered cells might occur in this process. Although adhesion chromatography, which is based on differential active adhesion of cells, has been widely carried out in the laboratory as well as in the field of clinical medicine, purity and yield of separated cells are not always sufficient and serious problems remain. Among these are damages to cell membrane and function, forced detachment of cells from the adsorbent, restricted separation conditions, and a time-consuming procedure.

By contrast, adsorption chromatography based on differing physicochemical affinity of cells toward adsorbents might overcome the problems of adhesion chromatography because of the minimal contribution of cell metabolisms to separation mechanisms. Kataoka, Maruyama, and Tsuruta et al. first achieved successful resolution of B cells and T cells, based on the difference in physicochemical properties of plasma membrane surface using HA copolymer as adsorbent.[7,14,15]

Separation of B cells from T cells was accomplished at low temperature (i.e., 4°C), indicating that cell retention is based on the physicochemical interaction between cell surface and adsorbent.[7,10]

Medium pH and ionic strength affect the retention behavior of lymphocytes on HA copolymer surfaces,[6,8,10] indicating that electrostatic interaction between HA surfaces and cell membranes of lymphocytes plays an important role. The most influential factor is the protonation of amino groups along polyamine grafts.

The spleen is the most common place to obtain lymphocytes for immunological research. There is a great demand for a method to separate spleen lymphocyte subpopulations. Besides B cells and T cells, spleen lymphocytes contain a considerable amount of another population, called null cells (~ 20% of whole lymphocytes in the spleen). Null cells contain many natural killer cells, which have been recognized as important immune response mediators because of their spontaneous cytotoxicity, and are a good target for a new cell adsorbent.

Separation of T cell subsets, including helper, suppressor, and killer T cells, has become a challenge for clinical usage as well as diagnosis and therapy for immunodiseases. Separation of T cell subsets by immunoaffinity chromatography will help, but some disadvantages remain, such as high cost and unfavorable biological responses of separated cells. Therefore, the development of a synthetic adsorbent with an appropriate affinity toward T cell subsets is a priority.

CONCLUSION

Chromatography for cell separation offers many advantages for separating cell populations by selecting appropriate parameters of interaction between cells and matrices. The major benefits are simplicity of apparatus and short operation time. The development of appropriate matrices that eliminate contact-induced activation of cells and change in cellular functionalities and nonspecific interaction is the most important priority.

REFERENCES

1. Catsimpoolas, N. *Methods of Cell Separation*; Plenum: New York, 1977–80; Vols. 1–3.
2. Pretlow, T. G. II; Pretlow, T. P. *Cell Separation Methods and Selected Applications*; Academic: New York, 1982–87; Vols. 1–5.
3. Sharpe, P. T. *Methods of Cell Separation*; Elsevier: Amsterdam, 1988.
4. Kataoka, K., et al. *J. Biomed. Mater. Res.* **1980**, *14*, 817.
5. Kataoka, K. et al. *Proteins at Interfaces: Physicochemical and Biochemical Studies*; Brash, J. L.; Horbett, T. A., Eds. ACS Symposium Series 343; American Chemical Society: Washington, DC, 1987; Chapter 37.
6. Maruyama, A. et al. *Makromol. Chem., Rapid Commun.* **1987**, *8*, 27.
7. Maruyama, A. et al. *J. Biomed. Mater. Res.* **1988**, *22*, 555.
8. Nabeshima, Y et al. *J. Biomater. Sci. Polym. Ed.* **1989**, *1*, 85.
9. Kikuchi, A. et al. *J. Biomater. Sci. Polym. Ed.* **1992**, *3*, 355.
10. Kikuchi, A. et al. *J. Biomater. Sci. Polym. Ed.* **1994**, *5*, 569.
11. Nitadori, Y.; Tsuruta, *T. Makromol. Chem.* **1979**, *180*, 1877.
12. Nabeshima, Y. et al. *Polym. J.*, **1987**, *19*, 593.
13. Maruyama, A. et al. *Makromol. Chem.* **1986**, *187*, 1895.
14. Kataoka, K. et al. *Makromol. Chem. Supp.* **1985**, *9*, 53.
15. Maruyama, A. et al. *Biomaterials* **1988**, *9*, 471.
16. Kataoka, K. et al. *Biomaterials* **1988**, *9*, 218.
17. Kataoka, K. et al. *Proc. ACS Polym. Mater. Sci. Eng.* **1985**, *53*, 37.
18. Kikuchi. A. et al. *Polym. Adv. Technol.* **1992**, *2*, 245.
19. Kikuchi, A. et al. *Artificial Heart, Heart Replacement*; Akutsu, T.; Koyanagi, H., Eds.; Springer-Verlag: Tokyo, 1993; p 29.
20. Maruyama, A. et al. *Biomaterials* **1989**, *10*, 291.

Cellular Plastics

See: Foamed Plastics
Foaming Agents

Celluloid

See: Cellulosic Materials (Moisture Sorption Properties)

Cellulose

See: Cellulose (Direct Dissolution)
Cellulose (From Fly Cotton Mill Waste)

Cellulose (Overview)
Cellulose (Pyrolysis)
Cellulose (Regioselectively Substituted Esters and Ethers)
Cellulose Acetate Nitrate
Cellulose Derivative-Protein Complex (in Ice Cream)
Cellulose Derivative-Protein Complexes (Whey and Corn Steep Liquor, Enrichment of Macaroni)
Cellulose Derivatives (Reclamation of Proteins from Cheese Whey)
Cellulose Esters, Organic
Cellulose-Filled Composites
Cellulose Grafting (Aspects of the Xanthate Method)
Cellulose Grafting (Ionic Xanthate Method)
Cellulose, Xanthate Grafting
Cellulosic Ethers, Cationic
Cellulosic Liquid Crystals
Cellulosic Materials (Moisture Sorption Properties)
Cotton (Non-formaldehyde Crosslinking Agents)
Cotton Fabric (Water-Repellant Finishes)
High Performance Paper
Hydroxyethyl Cellulose (Infrared Spectra and Dielectric Properties)
Hydroxyethyl Cellulose (Preparation and Use in Protein Precipitation)
Hydroxyethyl Cellulose (Variables Affecting Hydroxyethylation Reaction)
Hydroxypropyl Cellulose (Crosslinked, Cholesteric Liquid Crystal)
Jute
Liquid Crystalline Ethyl-Cyanoethyl Cellulose
Nitrocellulose
Textile Fibers (Structure and Properties)
Vegetal Biomass (1. Monomers and their Polymerization)
Vegetal Biomass (3. Polymers, Derivatives and Composites)
Wood (Chemical Modification)

CELLULOSE (Overview)

Yong Huang and Jianan Chen
Laboratory of Cellulose and Lignocellulocis Chemistry
Guangzhou Institute of Chemistry

Cellulose is the most abundant naturally occurring macromolecular material. Cellulose usually occurs in the cell walls of green terrestrial and marine plants and, up until now, has not been able to be artificially synthesized. Cellulose is never found in a pure form in nature; cotton is probably the purest natural source. Cellulose is generally associated with other substances such as lignin and so-called hemicellulose. Wood contains on a dry basis between 40 and 55% cellulose, 15 and 35% lignin, and 25 and 40% hemicellulose.[1] Generally, cellulose is a linear condensation polymer, consisting of anhydroglucose units joined together by β-1,4-glycosidic bonds. It is thus a 1,4-β–D-glucan.

Cellulose constitutes a ubiquitous and renewable natural resource that has served people's needs for thousands of years. It has been used in a variety of industries such as textiles, paper making, food processing, packaging, building materials, and pharmaceuticals. Sometimes cellulose is used in the form of original fibers, such as cotton fibers and flax and ramie fibers for textile and paper; sometimes it is used in its derivative form such as cellulose xanthogenate for rayon, cellulose nitrate for gunpowder and paint, and carbohydroxyl methyl cellulose (CMC) for a thickener in drill wells and food.

STRUCTURES AND PROPERTIES

Cellulose is a polymer of D-glycopyranose units, and the monosaccharide residues are linked by β-(1 → 4)-bonds and the chains are not branched.[2] The glucose units effectively alternate up and down in the chains because of their β-configuration of the interanhydroglucopyranose. The β-1,4 linkage results in a rigid or semirigid chain polymer that is ideally suited for forming fibrils via inter- and intrachain hydrogen bonding between –OH groups on the glucose residues. Because of very strong inter- and intrachain hydrogen bonding, cellulose is insoluble in water and many organic solvents. However, the polymer can be dissolved in cuprammonium hydroxide solution through complex formation.

Cellulose never occurs in an absolutely pure form in plants, and it is usually embedded in lignocellulose with hemicellulose. The hemicelluloses are polysaccharides, which are mainly heteropolymers and usually branched, of various sugars and some ionic acids that can usually be extracted from the plant substance by an aqueous alkali. Most lignins have to be removed from wood fibers before the cellulose can be used in industry. Delignification is a basic process of chemical pulping for the production of papermaking pulp and cellulose. The mean molecular weight of cellulose is ~ 10^6 if the cellulose samples are carefully extracted and is usually polydisperse.

Cellulose is naturally crystalline, exhibits crystalline polymorphism, and has a highly ordered fibrous morphology.[3] Polymorphology of cellulose crystalline forms and the supermolecular morphology of cellulose fibrous structures are two interesting points in the study of cellulose structures. From the studies of X-ray diffraction and electron diffraction, it is suggested that long-chain cellulose molecules are arranged with their long axes parallel to the length of the elementary fibrils, which are ~ 3.5 nm in width and an indefinite length. The microfibrils associate to form thicker fibrils, which in turn aggregate to cellulose fibers. It is generally believed that the geometry of the arrangement of microfibrils in fiber walls has a marked effect on the physical properties of the fibers and, thus, on the use of these fibers.

Cellulose is a semicrystalline natural polymer. X-ray diffraction and other evidence demonstrate that cellulose is partly crystalline with a regular arrangement of molecules and partly disordered with a nonregular arrangement of molecules, which greatly influences the physical properties of cellulose fibers.

APPLICATIONS

Of the many widely utilized natural substances, cellulose is one of the most important commercial raw materials. It is the parent substance for a large variety of chemical derivatives.

Textile

The most comprehensive use of cellulose is in the textile industry for making clothing, domestic cartons, carpets, and blankets. Cellulose fibers used in textiles are cotton, ramie, and viscous rayon (a man-made fiber).[4] The staple fibers are also widely used for nonwovens, disposables, and sanitary products for general, medical, or surgical applications. The high-tenacity viscose filament yarns or staple fibers are used as tire cord, especially for radial tires, and for belts, tapes, hoses, and coated fabrics.

Paper Making

Paper has been made from wood, straw, and other substances since ancient times. Paper mainly consists of cellulose, which is purified by chemical pulping or biochemical pulping. The higher the content of cellulose in the paper, the higher the paper quality.

Packaging Materials

Paper is one of the most important packaging materials for consumer goods. A film, which is made from cellulose, is also used as packaging material for snacks, cookies, baked goods, and candy and for tobacco and other nonfood uses.

Filter Materials

Cellulose films as filter materials have an important use for artificial kidneys and reverse osmosis. Recently, hollow fibers from many synthetic polymer sources and from cuprammonium have found commercial acceptance. Cellulose acetate is still best overall polymer for use in reverse osmosis.[5]

A large amount of cellulose acetate, which can absorb some harmful substances in smoke, is used for cigarette filters.

Gums and Thickeners

Cellulose ethers are used in laundry detergents, water-based paints, oil wells, cement, plaster, lacquers, foods, textiles, and cosmetics.

Food and Pharmaceuticals

Recently, the development of colloidal microcrystalline cellulose made available much finer particle forms of highly purified crystalline cellulose and more importantly, aqueous suspensions of microcrystalline cellulose particles. These have a smooth texture and unique pseudoplastic properties, including stable viscosity over a wide temperature region. In the food industry, heat stability, the ability to thicken with favorable mouth feel, and flow control are achieved by adding microcrystalline cellulose. Microcrystalline cellulose extends starches, forms sugar gels, stabilizes foam, and controls the formation of ice crystals. A low-calorie dairy food can be made by partially substituting milk (or cream) with colloidal microcrystalline cellulose. Microcrystalline cellulose gel not only has no calories, but also looks like ice cream. In pharmaceuticals microcrystalline cellulose is used mostly to make tablets. It assists in the flow, lubrication, and bonding properties of the tablet ingredients, which improves the stability of the tablet and gives rapid disintegration of the tablet in the stomach.

REFERENCES

1. Nevell, T. P.; Zeronian, S. H. *Cellulose Chemistry and Its Applications*: John Wiley & Sons: New York, 1985; Chapter 1.
2. MacGregor, E. A.; Greenwood, C. T. *Polymers in Nature*; John Wiley & Sons: New York, 1980.
3. Sarko, A. In *Cellulose: Structure, Modification and Hydrolysis*; Young, R. A.; Rowell, R. M., Eds.; Wiley-Interscience: New York, 1986.
4. Mark, H. F.; Atlas, S. M.; Cernia, E. *Man-Made Fibers*; John Wiley & Sons: New York, 1968; Vol. 25.
5. Sourirajan, S. *Reverse Osmosis*; Academic: New York, 1970.

CELLULOSE (Direct Dissolution)

J. A. Cuculo, N. Aminuddin, S. M. Hudson, and A. V. Wilson
Department of Textile Engineering, Chemistry, and Science
North Carolina State University

Cellulose is a natural, high molecular weight polymer. It is the most abundant organic polymer and is renewable and biodegradable. Cellulose, however, has not been exploited, mainly because it neither melts nor dissolves readily in many solvents.

Traditionally, cellulosic fibers have been obtained indirectly by derivatization of the base polymer. This method uses cellulose derivatives, such as cellulose acetate or xanthate. Then, once the final shape has been obtained, the derivative can be converted by various means back to cellulose. These processes are very cumbersome and expensive, and generally have toxic byproducts and polluting effluents. Over the past 20 years, much effort has been devoted to finding direct solvents for cellulose.

We will review direct solvent systems for cellulose. All of these solvents consist of six components: *N*-methylmorpholine-*N*-oxide/water (MMNO/H_2O), lithium chloride/dimethyl acetamide (LiCl/DMAc), trifluoroacetic acid/dichloroethane (TFA/CH_2Cl_2), calcium thiocyanate/water ($Ca(SCN)_2/H_2O$), ammonia/ammonium thiocyanate (NH_3/NH_4SCN), and sodium hydroxide/water (NaOH/H_2O).

Anisotropic solutions have been obtained from semirigid polymers such as cellulose and cellulose derivatives.

The two ways now used to form a high-performance (high modulus and high tenacity) fiber are to eliminate chain folding and to spin fiber from liquid crystal (mesophase) systems.[1] Thus, one of the driving forces behind the study of mesophase solutions of cellulose and its derivatives is the potential of spinning anisotropic cellulosic solutions, or incipient-anisotropic solutions, which will produce high-tenacity, high-modulus cellulosic fibers. Therefore, considerable research is being done in solvent spinning of cellulose solutions and, of course, in searching for new solvents for cellulose. With this in mind, we will discuss the dissolution and spinning of cellulose in various solvents.

N-METHYLMORPHOLINE *N*-OXIDE AND WATER

H. Chanzy et al. reported that cellulose forms a lyotropic mesophase in a mixture of *N*-methylmorpholine *N*-oxide (MMNO) and water.[2] The occurrence of the mesophase depends upon the parameters, such as temperature of the solution, concentration of cellulose, water content of the solvent system, and molecular weight (DP) of cellulose.

In addition to the effects of DP and water content, solution concentration also affects the occurrence of mesophase solution.

Another factor that affects the formation of liquid crystalline solutions is the molar ratio of water in the solvent. The water molar ratio is the number of moles of water per moles of anhydrous MMNO. When the water molar ratio is above 1.0, no mesophase solution forms, regardless of the concentration of the solution.

Chanzy et al. spun the cellulose solutions.[3] The characteristics of the fibers obtained were similar to those of the best viscose rayon, with tenacity of 3.65 g/den (0.5 GPa) and modulus of 153.3 g/den (21 GPa). Previous work with film indicated the cellulose II polymorph.[2] Certain additives in the solution, such as NH_4Cl, could boost the mechanical properties of the fibers.

LITHIUM CHLORIDE AND *N,N*-DIMETHYLACETAMIDE

C.L. McCormick et al. discovered that lithium chloride (LiCl)/*N,N*-dimethylacetamide (DMAc) would dissolve cellulose.[4] He and his co-workers also observed cholesteric lyotropic mesophases of cellulose in this solvent system.[5,6]

Ciferri et al. indicated that the best method of dissolving cellulose in the solvent involves prewetting the cellulose with DMAc.[7]

The viscosity of the cellulose solution exhibits a rapid increase as the concentration of the cellulose increases.

TRIFLUOROACETIC ACID AND CHLORINATED ALKANES

D. L. Patel and R. D. Gilbert demonstrated that mixtures of trifluoroacetic acid (TFA) and chlorinated alkanes, such as 1,2-dichloroethane and methylene chloride, are also good solvents for cellulose and cellulose triacetate.[8,9] In this solvent system, the cellulosic lyotropic mesophases occur at 20% (w/w) concentration and exhibit cholesteric characteristics.

CALCIUM THIOCYANATE AND WATER

J. Dubose found that mixtures of calcium thiocyanate and water would dissolve cellulose.[10] This was the first solvent system for cellulose. There is no evidence, however, of mesophase formation in these solutions, Little work was been done on this solvent system because of its high propensity toward thermal degradation of cellulose. Despite this problem, the solvent demonstrates many properties of other direct solvent systems.

LIQUID AMMONIA/AMMONIUM THIOCYANATE

S. M. Hudson and J. A. Cuculo discovered that ammonia/ammonium thiocyanate (NH_3)/(NH_4SCN) is an excellent solvent for cellulose.[11] They showed that the solvent has several practical advantages, including low cost and readily available components. Another advantage is that a mesophase solution can be obtained at a reasonably low cellulose concentration.

J. J. Cho spun fibers from isotropic and anisotropic solutions.[12] The mechanical properties of the fibers from the isotropic solution were similar to those of normal viscose rayon. The anisotropic solutions produced fibers with slightly better mechanical properties than those from isotropic solutions, as also reported by Bianchi et al.[13] and Glasser et al.[14] Cuculo, Liu, and Smith showed that the modulus of the fiber increased with lower coagulant temperature.[15]

REFERENCES

1. Mark, H. *High-Performance Polymers: Their Origin and Development*; Seymour, R. B.; Kirshenbaum, G. S. Elsevier: New York, NY, 1986.
2. Chanzy, H.; Peguy, A. J. *Polym. Sci., Polym. Phys. Ed.* **1980**, *18*, 1137.
3. Chanzy, H. et al. *Polymer* **1990**, *31*, 400.
4. McCormick, C. L. U.S. Patent 4 278 790, 1981.
5. McCormick, C. L. et al. *Polym. Prepr. (ACS Div. Polym. Chem)* **1983**, *24*(2), 271.
6. McCormick, C. L. et al. *Macromolecules* **1985**, *18*(12), 2394.
7. Terbojevick, M. et al. *Macromolecules* **1985**, *18*, 640.
8. Patel, D. L.; Gilbert, R. D. *J. Polym., Polym. Phys. Ed.* **1981**, *19*, 1231.
9. Patel, D. L.; Gilbert, R. D. *J. Polym., Polym. Phys. Ed.* **1981**, *19*, 1449
10. Dubose, J. *Bull. Rouen* **1905**, *33*, 318.
11. Hudson, S. M.; Cuculo, J. A. U.S. Patent 4 367 191, 1983.
12. Cho, J. J. Ph.D. Thesis, North Carolina State University, 1990.
13. Conio, G. et al. *J. Polym. Sci., Polym. Lett. Ed.* **1984**, *22*, 273.
14. Dave, V.; Glasser, W. G. *J. Appl. Polym. Sci.* **1993**, *48*, 683–699.
15. Liu, C. K. et al. *J. Polym. Sci.: Part B: Polym. Phys.* **1990**, *28*, 449.

CELLULOSE (From Fly Cotton Mill Waste)

Houssni El-Saied* and Mohamed A. El-Taraboulsi
Cellulose and Paper Department
National Research Center and Chemical Engineering Department
Faculty of Engineering
Alexandria University

Cotton mill wastes are a potential source of chemical cotton, that is, cotton linters. There are different grades of cotton mill wastes, collected during opening, picking, carding, roving, and combining operations. The low-grade wastes collected during the first three operations consist mostly of short fibers, a few motes, and some dust. These wastes are of little value except for carding strippings, which are sometimes repressed to make low-grade cotton yarns. Waste varies from 6 to 8% of processed raw cotton, but fly waste represents 1.0 to 1.5% of raw cotton.

We intend to purify Egyptian fly cotton waste and prepare chemical cellulose with low ash and silica. Such cellulose can be useful for the cellulose acetate, cellulose nitrate, cyanoethyl, and carboxymethyl–cellulose industries.

PROPERTIES AND APPLICATIONS

Alkali Digestion and Bleaching

Following preliminary treatment, the fly waste can be easily purified by conventional chemical means, which include alkali digestion (under different conditions) and bleaching.

*Author to whom correspondence should be addressed.

The treated fly waste compares favorably with the purified cotton linters (cellulose 99%, ash 0.1%, ether extractives 0.2%).[1,2]

Cellulose Derivatives from Purified Fly Cotton Waste

Purified and bleached chemical cotton can be considered chemically as suitable for the preparation of cellulose derivatives. Purified cotton has been successfully used as starting material in the preparation of nitrocellulose, primary and secondary cellulose acetate, cyanoethyl cellulose, and derivatives of cellulose acetate.[3,4] Purified chemical cotton is thus a successful raw material for a variety of cellulose industries.

Desalination Membranes from Cellulose Acetate Derivatives

Membranes prepared from Egyptian cellulose acetate are similar to the standard Eastman film, being only 4% lower in flux and salt rejection.

REFERENCES

1. Ott, E., *Cellulose and Cellulose Derivatives*, Interscience: New York, 1943; Vol. V, p 527.
2. Omar, M. et al. *Tappi* **1960**, *43*, 238A.
3. El-Traboulsi, M. A.; El-Saied, H. *Chem. A.R.E.* **1968**, *11*, 373.
4. El-Taraboulsi, M. A., et al. *Carbohyd. Res. J.* **1970**, *13*, 83.

CELLULOSE (Pyrolysis)

Toshimi Hirata
Forestry and Forest Products Research Institute

Cellulose is one of the main chemical components of the cell wall of plants, and in terms of the amount present on the earth it is the most abundant of all organic materials. The mechanical strength, especially the bending and tensile strength, of plants is provided by cellulose. Great quantities of cellulose are used in industries such as publishing, construction, textiles, cosmetic and pharmaceutical products, and food. An enormous number of studies have been carried out on the pyrolysis of cellulose from the standpoint of the protection of cellulose and wood materials from fire and from the standpoint of polymer science.

Cellulose is a linear polymer of repeating glucosan units, all of which are identically linked together through glucosidic bonds. Cellulose molecules form many amorphous and crystalline regions in a material, usually a large part being the latter. Because hydroxyl groups are unstable, the main initial reactions of the pyrolysis in the amorphous region seem to be dehydration and cleavage of glucosidic bonds.[1–4] These reactions appear to be succeeded by the production of glucosans, mostly 1-6 anhydro-β-D glucopyranose(levoglucosan).[5]

On the other hand, as shown in the literature overview by Hirata, it is accepted that the first chemical reaction of pyrolysis in the crystalline region is cleavage of glucosidic bonds followed by the production of glucosan, mostly levoglucosan.[6]

The main route of pyrolysis is assumed to be a course proceeding from the scission of glucosidic bonds to the production of levoglucosan, and dehydration seems to be a side reaction of lesser significance. These reactions might be more or less influenced by the above secondary pyrolysis during the diffusion of active products. Various chemical mechanisms, more than 10, have been proposed as this main route, reflecting the complexity of pyrolysis.[6] Also, several mechanisms for the side reaction and a few mechanisms for the levoglucosan pyrolysis have been proposed.[6] Furthermore, cellulose does not seem to melt with heating. This, in addition to the above multiple and simultaneous reactions makes kinetic analysis difficult. Therefore, it is important to substantiate the appropriate mechanism for the principal route and to build a kinetic model corresponding to it.

Mechanism and Kinetics

To explain both the changes in the mass and dp of cellulose by pyrolysis, a chain reaction mechanism has been presented.[7,8]

The primary mechanism of the pyrolysis of cellulose is described by a chain reaction with the initiation of random scission, the depropagation producing volatile levoglucosan, and the termination of grafting between an activated center and a hydroxyl group. This chain reaction mechanism works as a satisfactory explanation of both the changes in mass and by pyrolysis by means of the kinetic method. The varied orders of reaction reported for mass loss seem to be explainable by relating them to volume changes, although there is no current information pertinent to volume during pyrolysis.

REFERENCES

1. Golova, O. P. *Dokl. Akad. Nauk SSSR* **1957**, *115*, 1122.
2. Golova, O. P.; Kryilova, K. G. *Dokl. Akad. Nauk SSSR* **1957**, *116*, 419.
3. Basch, A.; Lewin, M. *J. Polym. Sci., Polym. Chem. Ed.* **1974**, *12*, 2053.
4. Lewin, M.; Basch, A.; Roderig, C. *Proceedings of the International Symposium on Macromolecules*; Mano, E. B., Ed.; 1974, 226–250.
5. Basch, A.; Lewin, M. *J. Polym. Sci., Polym. Chem. Ed.* **1973**, *11*, 3095.
6. Hirata, T. *Pyrolysis of Celluloses*, An Introduction to the Literature; NBSIR 85-3218; National Institute of Standards and Technology: 1985.
7. Hirata, T. *J. Japan Wood Res. Soc.* **1976**, *22*, 238.
8. Hirata, T. *Bull. Forestry Forest Products Res. Inst.* **1979**, *304*, 77.

CELLULOSE (Regioselectively Substituted Esters and Ethers)

Dieter Klemm, Armin Stein, and Thomas Heinze
Institute of Organic and Macromolecular Chemistry of the Friedrich-Schiller-University of Jena

Burkart Philipp and Wolfgang Wagenknecht
Max-Planck-Institute of Colloid and Interface Research Teltow-Seehof

Cellulose as a polyhydroxylic compound with the chemical structure of a linear 1,4-β-glucan (**Figure 1**) can easily be transformed to esters or ethers in the position at C-2, C-3, and C-6. Examples of these cellulose derivatives produced on a large scale are xanthogenate; cellulose nitrate used as an explosive; and carboxymethyl ether of cellulose, which is widely used as a nontoxic, water-soluble polymer. Product properties of these commercial cellulose derivatives are mostly determined by the degree of polymerization (dp) and the degree of substitution (DS), the latter denoting the average number of OH groups substituted per anhydroglucose unit (AGU) and arising at a value of 3 after complete derivatization of all the OH groups.

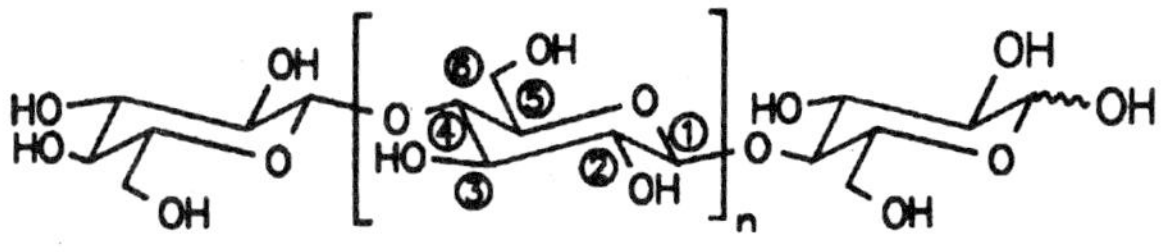

FIGURE 1. Formula of cellulose.

In the case of a partial substitution (DS<3), the uniformity and nonuniformity, respectively, of substituent distribution are within the AGU and along the macromolecule as well as between the polymer chains can exert a strong influence on the product properties.

Little information has been available up to now on the effect of the site or the sites of functionalization within the AGU and in the resulting copolymer patterns on product properties. This is obviously because in a conventional heterogeneous cellulose derivatization process all three types of nonuniformity are superposed and cannot be easily separated with regard to relevance on product properties. In addition, substituent distribution within the AGU was hard to determine by the classical analytical techniques available until the 1960s.

After a brief systematic survey of the reaction routes leading to regioselectively substituted cellulose esters and ethers, we give an overview on synthesis, characterization, and potential application of these "regioselective cellulose derivatives," including for comparison some data and brief comments on substituent distribution within the AGU of commercial cellulose esters and ethers prepared by the conventional heterogeneous course of reaction. Regarding the synthesis of regioselectively substituted products, special emphasis is placed on the previously mentioned route via instable intermediates pursued in a recent research project conducted by the authors.[1]

SYNTHESIS AND STRUCTURE

Regioselectively Substituted Cellulose Esters

Cellulose Esters of Carbonic Acids

A convenient and very versatile route to cellulose esters of aliphatic and aromatic carbonic acids starts from O-trialkylsilyl celluloses by reacting these compounds with an appropriate acyl chloride either in the presence or in the absence of a tertiary amine. In the absence of amine at a temperature of about or above 100°C the silyl ether group is eliminated by reaction to the trimethylsilylchloride, which is distilled off, and the corresponding ester is obtained in a nearly quantitative yield without esterification of free OH groups.[2] This procedure is well suited to prepare long-chain fatty acid esters and numerous aromatic esters of cellulose. But it poses problems with acyl chlorides with low boiling points such as acetyl chloride in which only a partial conversion of the silyl ether group to an acetate group in the 6-position can be achieved.

In the presence of a tertiary amine such as triethyl amine at room temperature, however, the trimethylsilylether group acts as a protecting group. The free OH groups can be esterified by an excess of the appropriate acyl/chloride, resulting in a preferred 2,3-substituted ester of cellulose after desilylation in an acid medium. A more regioselective substituted cellulose acetate or propionate with a DS of 2 in the 2,3-position was synthesized by reacting 6-O-*t*-butyldimethylsilyl or 6-O-thexyldimethylsilyl cellulose.[3]

For a homogeneous acetylation of cellulose with Ac_2O in the DMAc/LiCl system or in *N*-ethylpyridinium chloride, Kamide reported a 3–4 times higher reactivity in the OH groups at C-6 compared with the secondary ones.[4] Another more successful route to a predominately substituted cellulose acetate in the C-6 position consists of a partial deacetylation of a cellulose triacetate at homogeneous conditions in a ternary mixture of DMSO, aliphatic amines such as dimethylamine or hexamethylenediamine, and a defined amount of water.[1]

Miscellaneous Organic Esters of Cellulose

Formation of cellulose xanthogenate Cell-O-C(S)S^-Na^+ relevant as an intermediate in the viscose process, predominantly occurs in the C-2 position up to a DS of ~ 0.5 on reacting Na cellulose with gaseous CS_2, whereas in a "wet xanthation" of slurried Na cellulose with liquid CS_2, a substitution at the C-6 position prevails.[5]

Sulfuric Acid Half-Esters of Cellulose

Regioselectively substituted cellulose sulfate half-esters of the general formula Cell-OSO_3^-(H^+; Na^+) with different patterns of substitution can be prepared by various routes via the choice of the reaction system, the sulfating agent, and the reaction conditions.

In the cellulose trinitrite system prepared by dissolution of the cellulose in a mixture of N_2O_4 and DMF at anhydrous conditions the percentage of C-6 substitution can be regulated within wide limits via the sulfating agent and the temperature of reaction.[6,7]

C-6-substituted or C-6/C-2-substituted Na-cellulose sulfates of high solution viscosity are available from trialkylsilyl celluloses with an appropriate patterns of substitution by sulfation with SO_3 or $ClSO_3H$ in DMSO, DMF, or THF, respectively, using the silylether group as the leaving group and the total DS_S being limited by the DS_{Si} of the starting material.[8]

The free OH groups of commercial cellulose acetates with a DS_{acetyl} between 1.8 and 2.6 can be completely sulfated, too, with an excess of SO_3 or $ClSO_3H$ in DMF, with the acetyl group as a protection group and the substitution pattern of the sulfate being an "inverse image" of that of the acetate.[9]

Miscellaneous Inorganic Esters of Cellulose

Cellulose nitrate formation proceeds quickly to a high DS_N in the conventional ternary system $H_2SO_4/HNO_3/H_2O$, and no route to a regioselectively substituted cellulose nitrate has become known to the authors.

Phosphatation of cellulose with phosphoric acid occurs preferentially in the C-6 position according to Nehls.[10]

Cellulose Ethers

Alkyl Ethers of Cellulose

A practical route to alkyl ethers of cellulose selectively substituted in the C-2/C-3 position is demonstrated by the reaction of 6-O-trialkylsilyl cellulose or 6-O-tritylcellulose with the appropriate alkylhalide in the presence of a strong base and subsequent desilylation or detritylation. Therefore, a tailored optimization of this procedure for each ether in question is required, as demonstrated by the examples of methyl-, benzyl-, and carboxymethyl cellulose.

To introduce the benzyl group into the C-2/C-3 position of the AGU, 6-O-thexyldimethylsilyl celluloses can be recommended as the starting material to be alkylated with benzylchloride in the presence of NaH in THF and subsequently desilylated by means of tetrabutylammoniumfluoride.

Trialkylsilyl Ethers of Cellulose

The OH groups of cellulose can be converted to silylether groups by reacting the polymer with a trialkylchlorsilane in the presence of a weak base such as pyridine or NH_3 in an aprotic solvent such as DMSO, DMF, or *N*-methylpyrrolidone (NMP). The order of reactivity of the OH groups in this reaction is OH-6 > OH-2 > OH-3.

The regioselectivity of a C-6 substitution can be remarkably enhanced by a bulky alkyl residue in the chlorsilane, such as *t*-butyl or a thexyl group, and/or by the reaction conditions.

Ether-Esters of Cellulose

Regioselectively substituted ether-esters were already mentioned in connection with the acylation of OH groups at C-2 and C-3 of 6-O-trialkylsilyl cellulose, with the silyl group acting as a protecting group.

INFLUENCE OF REGIOSELECTIVITY ON PRODUCT PROPERTIES

Experimental data correlating regioselectivity within the AGU and macroscopic product properties are still scarce. Relevant topics are the influence of the position of a functional group within the AGU on its chemical reactivity and on intermolecular interaction in binary or ternary systems, especially in solubility of a cellulose derivative in a low molecular liquid. Furthermore, some sporadic evidence exists on the influence of substitution pattern on solid-state properties of cellulose derivatives.

PROMISING APPLICATIONS

Due to a more expensive process of synthesis, regioselectively substituted esters and ethers of cellulose will not displace the conventionally manufactured derivatization from these established industrial and domestic applications, but they are promising in present and future areas, where the site-specific location of substituents represents a special advantage. Two promising areas are the formation of well-defined, highly ordered supramolecular structures for high-tech applications (e.g., in membranes or sensor devices) and the application in devices and products interacting with living matter and requiring a high level of geometric and energetic specificity.

CONCLUDING REMARKS

The study of regioselective etherification and esterification has broadened the scope of organic chemistry of cellulose and is still opening new routes of research in this area, (e.g., double and triple substitution, reversible substitution with special supramolecular structures obtained after cellulose regeneration, and regioselective introduction of reactive sites into the macromolecule to subsequently build a more complex molecular architecture).

Within polymer material science, site specificity of substitution within the AGU seems to be especially relevant to physical properties, and applications largely determined by geometrically defined intermolecular interactions of the functional groups attached to the polymer, as demonstrated here for solubility, membrane performance, and blood anticlotting activity of some cellulose esters. But more experimental data have to be acquired before the effect of regioselectivity of substitution on material properties of cellulose derivatives can be comprehensively evaluated.

REFERENCES

1. Philipp, B.; Klemm, D.; Wagenknecht, W.; Wagenknecht, M.; Nehls, T.; Stein, A.; Heinze, T.; Heinze, U.; Helbig, K.; Camacho, T. *Das Papier*, **1995**, *49*, 3.
2. Stein, A.; Klemm, D. *Makromol. Chem., Rapid Commun.* **1988**, *9*, 569.
3. Philipp, B.; Klemm, D. *Das Papier*, **1995**, *49*, 58 and 102.
4. Miyamoto, T.; Sato, Y.; Stribata, T.; Inagaki, H. *J. Polym. Sci., Polym. Chem. Ed.* **1984**, *22*, 2363.
5. Kamide, K.; Kowsaka, K.; Okajima, K. *Polym. J.* **1987**, *19*, 231.
6. Wagenknecht, W.; Nehls, I.; Philipp, B. *Carbohydr. Res.* **1992**, *237*, 211.
7. Wagenknecht, W.; Nehls, I.; Philipp, B. *Carbohydr. Res.* **1993**, *240*, 245.
8. Wagenknecht, W.; Nehls, I., Stein, A.; Klemm, D.; Philipp, B. *Acta Polym.* **1992**, *43*, 266.
9. Wagenknecht, W.; Nehls, I.; Kotz, J.; Philipp, B.; Ludwig, J. *Cell. Chem. Technol.* **1991**, *25*, 343.
10. Nehls, I.; Loth, F. *Acta Polym.* **1991**, *42*, 233.

CELLULOSE ACETATE NITRATE

Jun-ichi Kimura
Propellants and Explosives Laboratory
The 1st Research Center
Technical Research and Development Institute
Japan Defense Agency

Cellulose acetate nitrate (CAN) is a cellulose ester of organic and inorganic acid, which is not in great demand. Therefore, the amount of cellulose nitrate produced is small at the present time. However, this unique cellulose derivative is expected to play a prominent part in the fulfillment of special requirements as a specialty polymer.

Desensitization of propellants for military use has been in high demand. Replacement of nitrocellulose is necessary because nitrocellulose is easily ignited and sensitive to shock. Nitrocellulose is only one energetic cellulosic binder having available oxygen in the molecule. CAN is also an energetic binder; the oxygen content of the polymer arbitrarily changed by the selection of the degree of nitration and acetification. From an energetic point of view, CAN is a good candidate for a binder of low vulnerability propellants.

PROPERTIES

CANs are soluble in many organic solvents similar to those for nitrocelluloses. Thus, the conventional wet-process method of compounding can be used in the production of propellants. CANs contain highly polar nitrate groups and the least polar acetate groups in a polymer chain. They have good adhesive power and mechanical properties. Therefore, blending CAN with products such as paint and ink and magnetic tape can improve their quality significantly.[1,2]

The experimental results obtained confirm that CAN is thermally more stable than nitrocellulose and that nitrate groups on the C-6 position are relatively stable compared with those on the C-2 and C-3 positions.

The most striking and unexpected effects of partial acetification of nitrocellulose on the shock sensitivity were obtained; remarkable reduction of shock sensitivity of CAN was observed even in a CAN with a degree of acetification of 0.6.[3]

APPLICATION

Although CANs have been studied since 1905, the polymer has received little attention until now.[4] The growing interest in specialty polymers will lead to an attempt to utilize a more extensive range of CANs to achieve controlled solubility and adhesive power for protective coatings, paint, filters, magnetic tape, and other products.

The apparent lack of interest in CAN in the field of explosives is not difficult to understand. Acetification of even only free hydroxyls causes a decrease in the heat of explosion, which means a loss of performance. Ironically, it is this very property that has, in these several years, led us to apply this polymer to a binder of low-vulnerability (desensitized) gun propellants. CAN can be recognized as desensitized nitrocellulose, improved nitrocellulose, or energetic cellulose acetate. It exhibits less flammability and less shock sensitivity compared with nitrocellulose.

A series of CAN gun propellants has been manufactured at a laboratory scale since 1992 in Japan.[5,6] CAN was also applied to fabricate improved combustible case cartridges. The tensile strength of the case material is 5–10 times higher than that of the conventional material, which is composed of nitrocellulose and craft pulp. High energetic CAN propellants have been successfully developed and a large-scale production plan for CAN is in progress.

REFERENCES

1. Ishino, G.; Taguchi, S. Japanese Patent S56-82849, 1981.
2. Ishino, G.; Taguchi, S. Japanese Patent S63-99276, 1988.
3. Kimura, J.; Arisawa, H.; Shimidzu, T. Technical Report-6532, TRDI; Japan Defense Agency: Tokyo, Japan, 1994.
4. Hausserman, C. *Chem. Ztg.* **1905,** *29*, 667.
5. Nishida, H. *Kunstoffe Vol. 4* **1914**, 141.
6. Kimura, J. Japanese Patent H6-128069, 1992.

CELLULOSE DERIVATIVE-PROTEIN COMPLEXES (in Ice Cream)

Olfat Y. Mansour
Cellulose and Paper Department
National Research Center

Magdy M. El-Sayed
Food Technology and Dairy Science
National Research Center

Soheir H. Ali
Microbiological Chemistry Department
National Research Center

Ice cream mixes used in this study were of the vanilla type prepared from milk-solid powder not fat (MSNF) with cheese whey protein complex (i.e., cheese whey and hydroxyethyl cellulose).

All ice cream mixes were standardized to contain 4% fat, 11.7% MSNF, and 12% sugar. The effect of whey complex solids on some of the properties of ice cream was investigated in a series of five treatments. The first treatment was the control, which contained 11.7% MSNF, without adding whey solid complex. The other four treatments contained 11.7% MSNF, of which 1, 2, 3, and 4% whey solid complex were added, respectively.

Organoleptic Properties of Ice Cream

The effect of whey protein complex content on the flavor of ice cream during storage at –10 and –25°C shows that increased whey protein complex content up to 2$ did not affect the flavor of fresh ice cream kept at –25°C.

Concerning the effect of the whey complex content on the body and texture of ice cream, increasing whey protein complex content up to 2% did not affect the body and texture of fresh ice cream stored at –25 or –10°C. The principal advantages of cellulose–whey protein complex as a replacement for MSNF in ice cream mixes are the low cost compared with other dairy products.

CELLULOSE DERIVATIVE-PROTEIN COMPLEXES (Whey and Corn Steep Liquor, Enrichment of Macaroni)

Olfat Y. Mansour
Cellulose and Paper Department
National Research Center

A. A. E. Yaseen
Food Technology Department
National Research Center

Aziza I. Atwa
Microbial Biotechnology Department
National Research Center

Mathur reported that the current estimated annual world production of fresh liquid whey is ~ 72 billion kg or more.[1] Whey contains ~ 93% water, 5.1% lactose, 0.8–1% protein, 0.5% ash, and 0.3% fat.[2] Protein concentrates of whey show good solubility and functionality and contain from 30 to 60% protein. These products are establishing themselves in standard dairy, bakery, and beverage industries as food supplements.[3–5]

Corn steep liquor contains mineral salts, soluble proteins, carbohydrates, phytates, inositol, and vitamins, particularly from the vitamin B group.[6,7] However, Salem et al. precipitated the protein content of corn steep liquor by pH adjustment in the range of 6–7 by using 6N NaOH and found that the precipitate contained 43% protein.[8]

This study was carried out to investigate the possibility of partial supplementation of semolina flour with precipitated cellulose–protein complex from whey and corn steep liquor in macaroni processing.

SENSORY EVALUATION OF COOKED MACARONI

The addition of precipitated cellulose–protein complex from whey to semolina improved the overall quality at all replacement levels, whereas precipitated cellulose–protein complex from corn steep liquor caused a reduction in overall quality. Flavor and color are major criteria affecting the quality of the cooked macaroni. The addition of 6% CMC–protein from cheese whey to semolina caused ~ 33% and 14% improvement in the color and flavor, respectively. It was found that macaroni prepared from dough mixtures with 6 and 9% of precipitated cellulose–protein complex from corn steep liquor was less acceptable than that prepared from pure semolina.

CHEMICAL COMPOSITIONS

The protein content of the macaroni samples was increased as a result of the addition of precipitated cellulose–protein complex from cheese whey or corn steep liquor.

In conclusion, this study showed that CMC– or HEC–protein complexes from whey and corn steep liquor could be used as a useful, acceptable protein source for macaroni with replacement levels of 9 and 3%, respectively.

ACKNOWLEDGMENT

This article was previously published as "Enrichment of macaroni with cellulose-derivative protein complex from whey and cornsteep liquor" in *Die Nahrung*, **1993**, *37*(6), *544-552*, and appears with permission of the publisher.

REFERENCES

1. Mathur, B. N.; Shabani, K. M. *J. Dairy Sci.* **1979**, *62*, 99.
2. Christensen, V. W. *Diary Industries International* **1976**, *41*, 84.
3. Hidalgo, J.; Camper, E. *J. Dairy Sci.* **1977**, *60*, 1515.
4. Modler, H. W.; Emmons, D. B. *J. Dairy Sci.* **1977**, *60*, *177*.
5. McDonough, F. E.; Hargrove, R. F.; Mattingly, W. A.; Posati, L. R; Alford, J. A. *Dairy Sci.* **1974**, *57*, 1438.
6. Liggett, R. W.; Koffler, H. *Bact. Revs.* **1948**, *12*, 227.
7. Parsons, G. W. U.S. Patent 2 515 157, 1950.
8. Salem, S. A.; Hegazi, S. M.; Foda, M. S.; Badr El-Din, S. M. *Stârke* **1972**, *27*, 290.

CELLULOSE DERIVATIVES (Reclamation of Proteins from Cheese Whey)

Olfat Y. Mansour
Department of Cellulose and Paper
National Research Center

Hoda El-Masry and Aziza I. Atwa
Department of Microbial Biotechnology
National Research Center

Cheese whey is produced as a byproduct from the manufacture of cheese. This liquor is a highly putrescible solution and contains a variety of proteins, lactose, and mineral salts. Great efforts have therefore been made to recover the whey protein either by heating the whey under acidic conditions or alkaline conditions in the range of 50–73%.[1–3] The proteins can also be isolated by using ferric salts, poly(acrylic acid) sodium hexametaphosphate and ultrafiltration, or by adding calcium salts to whey heated to near the boiling point.[4–9]

Carboxymethyl cellulose (CMC), which is anionic, was used as a precipitant for protein from the whey;[1] the product formed is a protein–CMC complex.[10–12] It contained 61.68% protein and more than 95% CMC.[11] The effects of ionic strength, pH, ratio of CMC to protein, and degree of substitution of CMC on the extent of precipitation and complex composition have been determined.[10–12]

The present work aims to precipitate the proteins from cheese whey by using new precipitating agents: water-soluble cellulose ethers such as hydroxyethyl cellulose and hydroxypropyl cellulose.

RESULTS AND DISCUSSION

Effect of the Ethoxyl Content of HEC on the Efficiency of Protein Precipitation

Interaction of HEC with whey proteins led to hydrogels that could not be separated by centrifugation, whereas when using CMC the gel-like mixture produced could be separated by centrifugation and then dried. For trichloroacetic acid, suspended matter formed, was separated by centrifugation, and dried. For HPC, a rubber-like material could be separated.

Effect of Ethoxyl Content of the Prepared Hydroxyethyl Cellulose

HEC with different ethoxyl contents (i.e., 25.95, 28.89, and 30.31 wt.%) were prepared as mentioned elsewhere.[13] These samples were used to precipitate the proteins from fresh cheese whey by dissolving 1% of the HEC in a calculated amount of water and then adding it to the preheated whey. The results obtained showed that a precipitate-like material was formed and could be separated by centrifugation.

REFERENCES

1. Modler, H. W.; Emmons, D. B. *J. Dairy Sci.* **1977**, *60*(2), 177.
2. Cogan, U.; Weiss, J.; Calmanovici, B.; Graff, J.; Gurati, G. *Dairy Cong.* **1978**, *E*, 930.
3. Hul, M. E. *J. Dairy Sci.* **1958**, *41*(2), 330.
4. Watonabe, K.; Matsudo, T.; Nakamurg, R. *Milchwissenschaft* **1985**, *40*(5), 279.
5. Strenberg, M.; Hershberger, D. *Biochim. Biophys. Acta* **1974**, *342*, 195.
6. Mathur, B. N.; Srinivasan, M. R. J. *Food Sci. Technol.* **1979**, *16*(12), 34.
7. Matthews, M. E. J. *Dairy Sci. Technol.* **1978**, *13*, 149.
8. Burton, J.; Skudder, P. UK Pal. Appl. GB2, 188, 526, Appl. 86/8, 604, 074. Feb. 19, 1986.
9. Rozenov, A.; Sokolova, L. *Moloch. Prom.* **1954**, *14*, 37. *J. Dairy Sci., Abs.* **1954**, *16*, 974.
10. Sato, Y.; Hayakawa, S.; Hyakawa, M. *J. Food Technol.* **1981**, *16*(1), 18.
11. Hill, R. D.; Zadow, J. G. *J. Dairy Res.* **1974**, *14*, 373.
12. Hansen, P. M. T.; Hidalyo, J.; Gould, I. A. *J. Dairy Sci.* **1972**, *54*(6), 830.
13. Association of Official Agricultural Chemists *Official Method of Analysis of the Association of Official Agricultural Chemists*; AOAC: Washington, DC, 1965.

Cellulose Esters

See: *Cellulose (Regioselectively Substituted Esters and Ethers)*
Cellulose Acetate Nitrate
Cellulose Esters, Organic
Cellulosic Liquid Crystals
Nitrocellulose
Vegetal Biomass (3. Polymers, Derivatives and Composites)

CELLULOSE ESTERS, ORGANIC

Steven Gedon and Richard Fengl
Eastman Chemical Company

Cellulose is one of nature's most abundant structural materials, providing the primary framework of most plants. Its simplicity lies in the repetitive utilization of the anhydroglucose unit ($C_6H_{10}O_5$) as the building block for chain structure.

Cellulose esters are commonly derived from natural cellulose by reaction with organic acids, anhydrides, or acid chlorides.

Of all commercial cellulose esters, cellulose acetate is by far the most important organic ester because of its broad application in fibers and plastics; as a result, it is currently prepared in multi-ton quantities with degrees of substitution (DS) ranging from that of far-hydrolyzed, water-soluble monoacetates to fully substituted triacetate.[1]

Although cellulose acetate remains the most widely used organic ester of cellulose, its usefulness is restricted by its moisture sensitivity, limited compatibility with other synthetic resins, and relatively high processing temperatures. Cellulose esters using higher aliphatic acids (e.g., C_3 and C_4), however, are able to circumvent these short comings with varying degrees of success, and can be prepared using procedures similar to those used for cellulose acetate. Nevertheless, mixed esters of cellulose containing acetate and either the propionate or butyrate moieties are currently produced commercially only by Eastman Chemical Company in the United States.

Cellulose esters from aromatic or aliphatic mono- and diacids containing more than four carbon atoms are difficult and expensive to prepare because of the poor reactivity of the corresponding anhydrides with cellulose; consequently, little commercial interest has been shown in these esters. Of notable exception, however, is the recent interest in the mixed esters of cellulose succinates and cellulose phthalates.

PROPERTIES

The properties of cellulose esters are determined largely by the number of acyl groups per anhydroglucose unit, acyl chain length, and the dp (molecular weight). By increasing the acyl chain length from C_2–C_6 in the triester, the melting point, tensile strength, mechanical strength, and density generally decrease, whereas the solubility in nonpolar solvents and resistance to moisture increases. On the other hand, fewer acyl groups per anhydroglucose unit (increased hydroxyl content) increase the solubility in polar solvents and decrease moisture resistance. The physical and chemical properties of mixed esters may also vary according to the ratio of the esters used (e.g., acetyl to butyl or acetyl to propionyl). In general, the properties of mixed esters, such as cellulose acetate butyrate, vary as a function of composition in which increasing butyryl (decreasing acetyl) content increases flexibility, moisture resistance, and solubility while decreasing the melting point and density.[2]

HYDROLYSIS

In terms of processability and market value, a true cellulose triacetate (44.8% acetyl) has found few applications. Instead, one finds that a less substituted ester is often preferred due to increases solubility in organic solvents for coatings and fiber applications, and reduced melt temperatures for plastic applications. In order to make this conversion from a cellulose triacetate (44.8% acetyl) to the cellulose diacetate (39.8% acetyl) used in these applications requires some degree of hydrolysis at the end of acetylation. At the end of a typical acetylation reaction, however, the cellulosic dope is anhydrous containing ~ 7% anhydride by weight, which must be quenched through final addition(s) before hydrolysis to a lower acetyl content can begin. But more important than the conversion of acetic anhydride to acetic acid is the conversion of bound cellulose sulfates to free sulfuric acid, which also occurs at this time.

Mixed esters are hydrolyzed by methods similar to those used for hydrolyzing cellulose triacetate. The hydrolysis eliminates small amounts of the combined sulfate ester, which, if not removed, ultimately affects thermal and hydrolytic stability. Several methods of hydrolyzing cellulose esters of higher aliphatic acids are described in References 3 and 4.

STABILIZATION

After the hydrolysis, precipitation, and thorough washing of the cellulose esters to remove residual acids, the esters must be stabilized against thermal degradation and color development, which may occur during processing such as extrusion or injection molding. This thermal instability is caused by the presence of oxidizable substances and small amounts of free and combined sulfuric acid.[5]

ACTIVATION OF CELLULOSE[6]

Native cellulose is a highly crystalline material that typically requires some level of activation before esterification. The amount of activation required, however, depends primarily upon the source of cellulose (cotton linter or wood pulp), purity, and drying history. Normally, water or aqueous acetic acid is the activating agent; however, glacial acetic acid also may be used. Water is more effective because it swells the fibers more than other agents and alters the hydrogen bonding between the polymer chains to provide a greater surface area for reaction. When water or aqueous acids are used, the cellulose must be dehydrated by displacing the water with acetic acid before the start of acetylation.

CATALYSTS FOR ACETYLATION

Sulfuric acid is the preferred catalyst for esterifying cellulose and is the only known catalyst used commercially for this function. The role of sulfuric acid during acetylation is discussed in References 7 and 8. In the presence of acetic anhydride, sulfuric acid rapidly and almost quantitatively forms the cellulose sulfate acid ester.[7] In the absence of anhydride, however, the sulfuric acid may be physically or mechanically retained (sorbed) in the cellulose during activation. The degree of absorption is a measure of the reactivity and accessibility of different celluloses.

More recently, new acetylation technology has emerged which uses a catalytic complex comprised of N,N-dimethyl acetamide (DMAC) and either sulfuric acid or a Lewis acid (i.e., titanium tetraisopropoxide). Due to the toxic nature of these amides, however, this technology has seen little commercial success.[9]

USES

The cellulose esters with the largest commercial value are clearly the acetates, including cellulose triacetate, cellulose diacetates, cellulose acetate butyrate, and cellulose acetate propionate. Cellulose acetate is used in textile fibers, plastics, film, sheeting, and lacquers. The cellulose acetate used for photographic film base is almost exclusively triacetate; some triacetate is also used for textile fibers because of its crystalline and heat-setting characteristics.

Large quantities of secondary cellulose acetate are used worldwide in the manufacture of filter material for cigarettes. Because of its excellent clarity and ease of processing, cellulose acetate film is widely used in display packaging and extruded plastic film for decorative signs. Injection-molded plastics of cellulose acetate are also used in applications such as toothbrush and tool handles.[10]

Low-viscosity cellulose acetate is used extensively in lacquers and protective coatings for paper, metal, glass, and other substrates and as an adhesive for cellulose photographic film because of its quick bonding rate and excellent bond peel strength.[11,12]

Cellulose acetate films, specially cast to have a dense surface and a porous substructure, are used in reverse osmosis to purify brackish water.[13–16] As a filtration device, cellulose acetates are also used as hollow fibers for purification of blood (artificial kidney) and in the fruit drink industry for purifying fruit juices.[17–20]

Eyeglass frames are also made from cellulose acetate plasticized with diglycerol esters, which do not exhibit opaqueness at the frame–lens junction with plastic lenses.[21,22] Biodegradable film, foam molding compositions (e.g., sponges), tobacco substitutes, and microencapsulated drug-delivery systems are potentially new and useful applications for cellulose acetate esters.[23–26]

With the renewed interest in environmentally friendly products, cellulose esters are being re-evaluated as a natural source of biodegradable thermoplastics.

Cellulose acetate propionate and butyrate esters have numerous applications, such as sheeting, molding plastics, film products, lacquer coatings, and melt dip coatings. Cellulose acetate propionates and acetate butyrates are thermoplastic; when properly formulated, these esters are processible by methods such as injection molding and extrusion, or they can be dissolved and cast into forms from a variety of solvents.

Cellulose acetate butyrates with high butyrl content and low viscosity are soluble in inexpensive lacquer resins. They are widely used in lacquers for protective and decorative coatings when applied to automobiles and wood furniture.

Low-viscosity cellulose propionate butyrate esters containing 3–5% butyryl, 40–50% propionyl, and 2–3% hydroxyl groups have excellent compatibility with oil-modified alkyd resins and are used in wood furniture coatings.[27]

In a relatively new decorative-coating technique called wet-on-wet coatings, cellulose acetate butyrate ester as the pigmented basecoat provides good pigment and metal-flake control before applications of a clear topcoat.[28,29] Such coatings provide good appearance and excellent resistance to weathering and are expected to be broadly used in automotive decorative coatings.

Because cellulose acetate propionates have a high melting point, high tolerance for alcohol solvents, low odor, and excellent surface hardness, they make ideal resins for use in printing inks (flexographic and gravure).[30]

FIBERS

Cellulose esters are currently produced in excess of 1 billion pounds annually. Cellulose ester flake production for nonfiber applications represents only a small part of the total U.S. output, with cigarette-filter now and cellulose acetate textile fibers consuming more than 80% of flake production. Production of fiber-grade acetate by both the continuous- or batch-type processes is currently known.

CONCLUSIONS

Through the application of higher order organic acid, mixed acids, and new manufacturing strategies, new cellulose esters will continue to be produced with novel physical properties. From this unique ability to customize products for diverse markets and trends, growth of cellulose ester markets are projected to continue at a modest rate through the 1990s.

REFERENCES

1. Malm, C. J.; Barkey, K. T.; Salo, M.; May, D. C. *Ind. Eng. Chem.* **1957**, *49*, 79.
2. Malm, C. J. *Sven. Kem. Tidskr.* **1961**, *73*, 10.
3. Malm, C. J.; Tanghe, L. J. U.S. Patent 2 801 240, July 30, 1957.
4. Malm, C. J.; Tanghe, L. J.; Herzog, H. M. U.S. Patent 2 816 106, December, 10, 1957.
5. Smirnov, B. P; Tatarnova, A. Z.; Grigoŕeva, T. A. *Sov. Plast.* **1972**, *1*, 71.
6. Malm, C. J.; Barkey, K. T.; May, D. C.; Lefferts, E. A. *Ind. Eng. Chem.* **1952**, *44*, 2904.
7. Malm, C. J.; Tanghe, L. J.; Laird, B. C. *Ind. Eng. Chem.* **1946**, *38*, 77.
8. Rosenthal, A. J. *Pure Appl. Chem.* **1967**, *14*, 535.
9. (a) Bogan, R. T.; Edgar, K. J. U.S. Patent Appl. 367, 025, 1994. (b) Blume, R. C.; U.S. Patent 2 705 710, 1955. (c) Grishen, E. P.; Bonder, V. A.; Mironov, D. P.; Shamolin, A. I. European Patent Appl. 94 904 356.6, Dec. 17, 1993.
10. Patel, D. L.; Gilbert, R. D. *J. Polym. Sci. Polym. Phys. Ed.* **1981**, *19*, 1449.
11. Ueno, W.; Minagaiwa, N. Ger. Offen. 2 104 032, August 5, 1971.
12. Fuji Photo Film, Br. Patent 1 352 605, May 9, 1974.
13. Aptel, P.; Cabasso, I. *Desalination* **1981**, *36*, 25
14. Kesting, R. E. *Proc. Int. Symp. Fresh Water Sea* **1976**, *4*, 73.
15. Ammons, R. D. Gov. Rep. Announce. U.S. **1978**, *79*, 93.
16. Sourirajan, S.; Thayer, W. L.; Kutowy, O. U.S. Patent 4 145 295, March 20, 1979.
17. Daicel Chemical Industries, Japan Kokai Tokkyo Koho JP 82 119 809, July 26, 1982.

18. Kesting, R. E. Ger. Offen. 2 619 250, November 11, 1976.
19. Merson, R. L.; Paredes, G.; Hosaka, D. B. *Polym. Sci. Tech.* **1980**, *13*, 405-13.
20. Baxter, A. G.; Bednas, M. E.; Matsuura, T.; Sounrajan, S. *Chem. Eng. Commun.* **1980**, *4*(4–5), 471–83.
21. Hirotaka, M.; Kenji, W.; Masami, N.; Kimio, I. Japan Kokai Tokkyo Koho JP 53 058 559.
22. Hirotake, M.; Kenji, W.; Masami, N.; Kimio, I. Japan Kokai Tokkyo Koho JP 54 008 654.
23. (a) Penn, B. G.; Stannett, V. T.; Gilbert, R. D. *J. Macromol. Sci. Chem.* **1981**, *A16*(2), 473. (b) Kim, S.; Stannett, V. T.; Gilbert, R. D. *J. Polym. Sci., Polym. Lett.* **1973**, *11*, 731.
24. Fischer, W. Fr. Demande 2 140 454, February 23, 1973.
25. Hiroshi, Y.; Masataka, W.; Kunio, K.; Akio, O. Japan Kokai Tokkyo Koho JP 52 079 096, July 2, 1977.
26. Singh, M.; Bala, K.; Vasudevan, P. *Makromol. Chem.* **1982**, 183, 1897.
27. Brewer, R. J.; Wooten, W. C. U.S. Patent 4 166 809, September 4, 1979.
28. Walker, *K. Farbe + Lack* **1981**, *87*, 198.
29. Walker, K. *Double Liaison-Chim. Print.* **1980**, *27*, 258.
30. Coney, C. H.; Bowen, G. B. *Am. Inkmaker* **1973**, *51*(20), 24.

Cellulose Ethers

See: *Cellulose (Regioselectively Substituted Esters and Ethers)*
Cellulosic Ethers, Cationic
Cellulosic Liquid Crystals
Gums (Overview)
Vegetal Biomass (3. Polymers, Derivatives and Composites)

CELLULOSE-FILLED COMPOSITES

Debesh Maldas*
Centre de Recherche en Pâtes et Papiers
Université du Québec à Trois-Rivières

Cellulosic materials are among the most reliable reinforcing fillers because of the ease with which they can be mixed, the processing advantages they present, their low density, their abundance (~10^{15} lb produced by plant annually), their low cost and use as a renewable resource, and their derivation of carbon from air instead of from petroleum or natural gas.[1] Although high-density conventional fillers (i.e., with a specific gravity of ~2.5) such as calcium carbonate, talc, mica, and glass offer a wide variety of property changes, their use may not result in a cost savings, at least on a volumetric basis. Moreover, cellulosic materials impart added benefits such as weight reduction.

Some of the earlier comprehensive reviews of the commercial uses of cellulosic fillers in polymer industries were by Deanin, Seymour, Galli, and Gillespie.[2–5] Zadorecki and Michell presented a review article on the potential and future prospects for wood cellulose as a reinforcement in polymer composites.[6] Myers and Kolosick compiled a bibliography on composites made from plastics and wood-based fillers.[7] However, very recently, Maldas and Kokta reviewed the current areas where cellulosic materials are used in the polymer industry and detailed the specific end uses of importance.[8] In recent years, short cellulosic fiber-reinforced elastomer composites have gained in practical and economic interest in the rubber industry.[9,10]

CURRENT RESEARCH

The use of cellulosic materials in plastic composites may be classified in a number of ways but, for convenience, here it is divided into three main groups.

Cellulose-Reinforced Thermosets

Cellulosic fibers are used as reinforcement in most of the common thermoset polymers (i.e., amino, phenolic, epoxy, and polyester).[11–13] Decorative laminates and plastic dinnerware based on phenol–formaldehyde, urea–formaldehyde, and melamine–formaldehyde use cellulose fibers. The main benefits of incorporating woodflour in phenolics and paper in amino plastics are decreased shrinkage in the mold and reduced thermal stress during curing of resins. Other benefits are improved impact strength, stiffness, and lower costs to finished products. Both woodflour and paper are suitable for combining with water-soluble resins. The water solution of low molecular weight resin impregnates the cell wall structure of cellulose fibers easily, and during the curing reaction strong bonds between fibers and matrix can be created.

For polyester structural applications, natural cellulosic hard fibers (i.e., jute, flax, ramie, sisal, henequen, and abaca) show similar or even better properties than those of glass fibers but at a fraction of the cost of glass fibers.[14]

In general, polyester composites containing 40% cellulosic pulp show a modulus of about twice that of virgin resin. The strength is equal, but the composites have about one-third the elongation of the virgin polymer.[12] Although cellulosic materials were found to be favorable as a reinforcement for polyester end uses such as pipes, channels, and housing, these composites have poor mechanical properties in wet conditions. As a result, to enhance the interactions of cellulosics with polymers, pretreatment is often required. Cellulosic fibers can be readily grafted to vinyl monomers, as in the case of jute fibers grafted to acrylic acid.

Cellulose fibers treated with different coupling agents (e.g., formaldehyde, dimethylolmelamine, and di- and trichloro-*s*-triazine) have also been evaluated in terms of their reinforcement effect on unsaturated polyesters. The treatment with coupling agents containing double bonds resulted in the formation of covalent bonds between fiber and matrix.

Laminates and Panel Products

There is a wide variety of adhesives, for example, formaldehyde condensation products from urea, melamine, phenol, resorcinol, and a mixture of these resins; poly(vinyl acetate) emulsions or white glues; thermoplastic and thermosetting types; diisocyanates; polyurethanes; epoxides; and hot melts used for bonding cellulosic materials into fabric laminates,

*This article was written and prepared by the author during his vacation and without-pay leave from the University of Québec and before joining the University of Kyoto; therefore, it did not require any permission from any of the institutions.

plywood, particle board, medium-density fiber board, wafer board, and oriented-strand board.[5,6,15–19] However, the great majority of adhesives used in wood-like composite materials are formaldehyde-based resins.

Cellulose-Reinforced Thermoplastics

Plastics and wood often compete sharply in the marketplace, but in the face of a growing shortage of both materials, they are joining forces in a new product: cellulose-reinforced thermoplastic composites. In recent years, there has been growing interest in the field of cellulose-reinforced thermoplastics.[20–26] However, unlike thermoset composites, cellulosic materials are less frequently used in common thermoplastics such as polyethylene, poly(vinyl chloride) and polystyrene because of difficulties associated with surface interactions between hydrophilic cellulosics and hydrophobic thermoplastics Such divergent behavior results in difficulties in compounding these materials and poor mechanical properties. Moreover, commonly used coupling agents in fiber-composite production do not function efficiently in such composite systems and may be too expensive. The most successful, as well as profitable, methods are modification of the polar cellulose surface by grafting with compatable thermoplastic segments or coating with compatibilizing and coupling agents before the compounding step, addition of compatibilizing and coupling agents in the compounding step, and the modification of the matrix polymer with a polar group.

Cellulose-reinforced composite materials have potential applications in the automotive industry, construction materials for low-cost housing, furniture, fencing, road making, packaging materials, grocery bags, garbage cans, flower pots, toys, and many commodity products.

CONCLUSION

The key to the future success of cellulose–plastic composites is the team efforts of chemists and machine tool and die designers. I hope that in the near future more than 50% of cellulosic materials, particularly recycled ones, will be used and more industries will be involved in this area for making various engineering materials and commodity products.

REFERENCES

1. Amin, M. B.; Maasdhah, A. J.; Usmani, A. M. *Polymer Science and Technology;* Carraher, Jr., E.; Sperling, L. H. Eds., Plenum: New York, 1986; Vol. 33, p 29.
2. Deanin, R. D. *Appl. Polym. Symp.* **1975**, *28*, 71.
3. Seymour, R. B. *Popular Plastics* **1978**, *28*, 27.
4. Galli, E. *Plastics Compoundings* **1982**, *5*, 103.
5. Gillespie, R. H. J. *Adhesion* **1982**, *15*, 51.
6. Zadorecki, P.; Michell, R. A. *Polym. Comp.* **1989**, *10*, 69.
7. Myers, G. E.; Kolosick, P C. *Woodfiber Plastic Composite Conf.;* Madison, WI, April, 1991.
8. Maldas, D.; Kokta, B. V. *Trends in Polym. Sci.* **1993**, *1*(6), 174.
9. Flink, P.; Stenberg, B. *Br. Polym. J.* **1990**, *22*, 147.
10. Setua, D. K. *Polymer Science and Technology;* Carraher, Jr., E.; Sperling, L. H., Eds.; Plenum: New York, 1986; Vol. 33, p 275.
11. Hua, L.; Flodin, P.; Rönnhult, T. *Polym. Comp.* **1987**, *8*, 203.
12. Zadorecki, P.; Karnerfors, H.; Lindenfors, S. *Comp. Sci. Technol.* **1986**, 27, 291.
13. Flodin P.; Zadorecki, P. *Composite System from Natural and Synthetic Polymer*; Salmen, L. et al., Eds.; Elsevier Science: Amsterdam, 1986; p 59.
14. Semsarzadeb, M. A. *Plastics Eng.* **1985**, *41*(10), 47.
15. Groah, W. J. *Formaldehyde Release from Wood Products;* ACS Symposium Series 316; American Chemical Society: Washington, DC, 1986; p 17.
16. Zadorecki, P.; Flodin, P. *Polym. Comp.* **1986**, *7*, 170.
17. Irle, M. A.; Bolton, A. J. *Holzforschung* **1988**, *42*, 53.
18. Maldas, D.; Kokta, B. V. *Biores. Technol.* **1991**, *35*, 251.
19. Han, G. S.; Shiraishi, N. *Mokazai Gakkaishi* **1991**, *37*, 39.
20. Myers, G. E.; Clemons, C. M. *Final Report for Solid Waste Reduction and Recycling Demonstration*; U.S. Forest Product Society: Madison, WI, Project No. 91-5; 1993.
21. Felix, J. M.; Gatenholm, P. *Controlled Interphases in Composite Materials*; Ishida, H., Ed.; Elsevier Science: Amsterdam, 1990; p 267.
22. Maldas, D.; Kokta, B. V. *Macromolecules* '92; Kahovec, J., Ed.; VSP International: The Netherlands, 1992; p 349.
23. Maldas, D.; Kokta, B. V. *Wood Fiber/Polymer Composites: Fundamental Concepts, Processes, and Material Options*; Wolcott, M. P., Ed.; U.S. Forest Products Society: Madison, WI, 1993; p 112.
24. Klason, C.; Kubat, J.; Gatenholm, P.; In *Viscoelasticity of Biomaterials*; Glasser, W.; Hatakeyama, H., Eds.; ACS Symposium Series 489; American Chemical Society: Washington, DC, 1992; p 98.
25. Han, G. S.; Shiraishi, N. *Mokuzai Gakkaishi* **1991**, *37*, 241.
26. Harle, M. S. *Globe and Mail*, April 1990.

CELLULOSE GRAFTING (Aspects of the Xanthate Method)

Olfat Y. Mansour, Z. A. Nagieb, and Altaf H. Basta
Cellulose and Paper Department
National Research Center

One of the most effective methods of obtaining hybrid fibers is the simultaneous formation of viscose rayon fibers and their graft copolymerization with synthetic polymers. The redox system proposed for this purpose consists of cellulose xanthate (freshly formed fibers) and hexavalent chromium ions. In this system copolymerization takes place rapidly at relatively low temperatures. In experiments with acrylonitrile a high degree of grafting could be obtained at 20°C within 30–60 s.[1]

A redox system of cellulose xanthogenate–hydrogen peroxide was used by Pavlov et al. to graft polymerize water-insoluble monomers and their mixtures with acrylonitrile to cellulose.[2] The effect of some of the basic parameters was established. The xanthate method could be optimized by properly adjusting the reaction conditions such as the hydrogen peroxide concentration. It is evident that in the studied intervals there exists an increase of grafting efficiency (from 80–90%) and an increase in conversion with H_2O_2 concentration up to 3 g of peroxide (14.28 g/L). Grafting efficiency is not much affected by pH, even though the optimum value seems to be between pH = 4 and 7.[3]

Different parameters were considered in this work on using hydrogen peroxide as a catalyst. Such parameters are hydrogen peroxide concentration, type of monomer, manner of the addition of hydrogen peroxide before or after the addition of monomer, and temperature. The effect of the presence of occluded air in the reaction mixture was also considered.

REFERENCES

1. Morin, B. P.; Starchenko, G. I. Ed., *Allunion Conf. on Chemistry and Physics of Cellulose*, Cromov, V. S., Ed.; Riga: Znatne, **1975**, 88–90.
2. Pavlov, P.; Dimov, K.; Simeonov, N. *J. Appl. Polym. Sci.* **1977**, *21*, 291.
3. Kokta, R. V.; Danneanult, G. *Graft Copolymerization of lignocellulosic Fibers* ACS Symposium Series: American Chemical Society: Washington, DC, 1982, 187, pp 269–283, 291.

CELLULOSE GRAFTING (Ionic Xanthate Method)

Olfat Y. Mansour and Altaf H. Basta
Cellulose and Paper Department
National Research Center

Many unsaturated ketones, esters, and nitriles, undergo addition reactions with active hydrogen-containing compounds. Addition of this general type, known as Michael reactions, requires base catalysts. One of the most powerful catalysts for these reactions is sodium alkoxide. Therefore, sodium cellulosates and alkali cellulose should readily react with certain vinyl compounds. The cyanoethylation of cellulose with acrylonitrile (AN) in the presence of sodium hydroxide is a type of Michael addition reaction. In the case of other vinyl monomers, CH_2–CHX, the tendency to anionic polymerization can be expected to be a function of the electron-withdrawing power of X. Analagous to the anionic polymerization of some vinyl monomers onto alkali cellulose is the use of cellulose xanthate in the presence of sodium hydroxide as an intermediate or substrate for grafting reactions of some vinyl and allyl monomers.

The xanthate method of grafting was invented by Fassinger and Conte.[1] Grafting by this method is based on the introduction of a small number of xanthate groups into the cellulose molecules and then reacting the activated cellulose with polymerizable monomers in the presence of a peroxide catalyst. This type of initiation is based on grafting by the use of the redox system between the cellulose xanthate and the peroxide catalyst.

Previously, radical grafting with the redox system composed of cellulose xanthogenate and hydrogen peroxide was compared with ionic grafting with the xanthate method.[2] In our work different variables of the grafting reaction by use of the ionic xanthate method such as the temperature of the grafting reaction, the degree of xanthation through variation of time of xanthation and the concentration of both alkali and carbon disulfide, sodium hydroxide concentration, liquor ratio, monomer concentrations, and monomer reactivity were considered. Different vinyl and allyl monomers were used.

REFERENCES

1. Fassinger, R. W.; Conte, J. S. Belgian Patent 646 248, October 8, 1964, British Patent 059 641, February 22, 1966, N.S. Patent 3 30 784, September 11, 1967. German Patent 1 468 965, March 6, 1969.
2. Mansour, O. Y.; Nagieb, Z. A.; Basta, A. H. "Aspects of Xanthate Method of Grafting Cellulose" *Acta Polymerica* **1989**, *40*(4), p 251.

Cellulose Xanthate

See: *Cellulose Grafting (Aspects of the Xanthate Method)*
Cellulose Grafting (Ionic Xanthate Method)
Cellulose, Xanthate Grafting
Vegetal Biomass (3. Polymers, Derivatives and Composites)

CELLULOSE, XANTHATE GRAFTING

Altaf H. Basta* and Olfat Y. Mansour
Cellulose and Paper Department
National Research Center

The xanthate method of grafting, known to improve free-radical graft polymerization reactions, also has a positive influence on ionic grafting. Anionic initiation was useful for the graft polymerization of several monomers onto cotton cellulose, including those that cannot be grafted by free-radical mechanisms. The monomers used are acrylonitrile, methyl methacrylate, ethyl acrylate, allyl alcohol, and allyl chloride. The results obtained proved that xanthate groups were essential for the grafting reaction to take place. The grafting yields increased with an increase in the number of xanthate groups (8-number) up to a limit. Maximum graft yields were obtained at a definite temperature (ceiling temperature). Addition reactions due to the presence of a nitrile group or chloride ion in the monomer took place and led to homopolymer formation.

The xanthate method of grafting was invented by Fassinger and Conte.[1] Grafting by this method is based on the introduction of a small number of xanthate groups into cellulose molecules and then reacting the now activated cellulose with polymerizable monomers in the presence of a peroxide catalyst. This type of initiation is based on grafting, by the use of the redox system, between the cellulose xanthate and the peroxide catalyst.

In this work, we considered different variables of the grafting reaction that uses the ionic xanthate method, such as the temperature of the grafting reaction, the degree of xanthation (through variation of time of xanthation and the concentration of both alkali and carbon disulfide), sodium hydroxide concentration, liquor ratio, and monomer concentration. Different vinyl and allyl monomers were used.

ACKNOWLEDGMENT

This article was previously published in *Nordic Pulp & Paper Res. J.*, **1991**, *6*(4), 184 and appears with permission.

REFERENCE

1. Fassinger, R. W.; Conte, J. S. Belgian Patent 646 248, October 8, 1964; British Patent 059 641, February 22, 1966; N.S. Patent 3 330 784, September 11, 1967; German Patent: 1 468 965, March 6, 1969.

*Author to whom correspondence should be addressed.

CELLULOSIC ETHERS, CATIONIC

James V. Gruber*
Amerchol Corporation

R. Lowell Kreeger
Union Carbide Corporation

Cationic cellulosic ethers are a small subset of the larger group of water-soluble cellulosic ethers known today. Cellulose itself is insoluble in water and must be chemically modified to make it water soluble. These chemical modifications result in final polymers that are non-ionic (e.g., hydroxyethyl cellulose), anionic (e.g., carboxymethyl cellulose), or cationic. Water-soluble cellulosic polymers as a whole have a wide range of applications—from coatings, paper making, and water treatment to drilling and personal care. Cationic cellulosics, on the other hand, are unique polymers that are used in the cosmetic and topological drug delivery industries only.

Union Carbide (Danbury, CT) started the commercial development of cationically modified cellulosics with work done by Stone, wherein, the cationization of hydrophilically modified cellulose by various quaternary amine epoxides was carried out.[1] This cellulose chemistry is the basis for Polyquaternium-10.

Union Carbide extended its line of cationic cellulose derivatives with the development of hydrophobically substituted, water-soluble, cationic cellulose.[2] The chemistry for these polymers fell neatly into line with Carbide's existing technology for the Polyquaternium-10. Polyquaternium-24 is a water-soluble, cationic cellulose that has a hydrophobic lauryl group replacing a methyl group in the cationic epoxide used for derivatization of the polymer.

National Starch and Chemical has been very active in the commercialization of water-soluble, cationic celluloses. Iovine patented a process for manufacturing cationic celluloses by graft polymerizing various amine olefins onto water-soluble cellulose.[3] The graft polymerization of diallyldimethyl ammonium chloride onto hydroxyethylated cellulose is the basis for Polyquaternium-4. In addition, National Starch actively manufactures Polyquaternium-10.

Jones has described a group of cationic, hydroxyethylated cellulose derivatives made from quaternary epoxy amines in which one group on the amine is a lipophilic, fatty group.[4] Croda manufactures these water-soluble, cationic celluloses.

PROPERTIES

The functional properties of the cationically modified celluloses can vary considerably depending on the nature of the starting cellulose raw material. Factors such as molecular weight, raw material source, and purity can influence the final product quality and performance. In addition, the level of hydroxalkylation of the cellulose (often called the molar substitution or MS) will strongly affect the water solubility and solution viscosity of the cellulose. The costs and safety issues associated with manufacturing hydroxyethyl cellulose prevent all but the most committed companies from entering this market.

The nature of the cationizing agent and the level of its substitution onto the cellulose will have a profound effect on the behavior of the final polymer. However, keeping in mind that the primary market for the cationically modified celluloses is topological, that is for cosmetics or drug delivery, certain functional parameters must make these polymers attractive to these industries.

In cationic cellulose chemistry, practitioners often speak of the level of cationic substitution (CS) on the cellulose. For hydrophilically modified cationic cellulose, CS can generally be measured by determining the percentage of nitrogen incorporated into the polymer at the completion of the cationization reaction.

For Polyquaternium-10, Stone claimed a CS range of 0.01-1, although in practice the upper limit is probably less than 0.5. For molecular weights, Stone's range probably varied from 50,000 to more than a million. As molecular weight increases, thickening efficiency increases, and as cationic substitution increases, substantivity increases before leveling off. The polymers are essentially Newtonian under low shear conditions but are pseudoplastic at high shear rates.

Solution viscosities for Polyquaternium-4 are generally low. This is in keeping with the intended applications for these polymers as cosmetics and topological delivery agents.

APPLICATIONS

It is well established in the literature that cationic materials can have a strong affinity for anionically charged surfaces.[5–7] Such surfaces might include hair, skin, and mucosal tissue. In the cosmetic industry, this attraction is known as substantivity. Brode recently reviewed the use of cationic polymers, including cationic cellulose, in cosmetics and therapeutics.[8] Goddard has also reviewed Polyquaternium-10 in detail.[9] More recently, a considerable amount of attention has been directed at the interaction of cationically modified polymers and surfactants. This emerging area of interest is a study of the intimate behavior of these cationic polymers in solution. Goddard has exhaustively reviewed the current level of understanding of the interaction of surfactants with, among others, cationically modified polymers, including cationic hydroxethyl cellulose.[10]

Of particular interest in the field of surfactant interaction with cationic cellulose is the extreme viscosity enhancements observed when very low levels of surfactant are added to solutions of cationic cellulose. Leung demonstrated that adding sodium dodecyl sulfate (SDS) to a 1% solution of Polyquaternium-10 could increase the solution viscosity more than 200-fold at the critical micelle concentration (CMC) of the surfactant.[11]

The interest in the interaction of cationic cellulose with surfactants has grown out of the desire by consumers to have shampoo formulations that both clean and condition the hair. The surfactants act to remove oils and dirt from the hair and the cationic cellulose (or other binding polymers) acts to coat the hair, making it soft and easy to comb.

Recently, a new field of application for cationic cellulose has begun to emerge in the delivery of drugs through topical application.

REFERENCES

1. Stone, F. W.; Rutherford, J. M. U.S. Patent 3 472 840, 1969.
2. Brode, G. L.; Kreeger, R. L.; Goddard, E. D.; Merritt, F. M.; Braun, D. B. U.S. Patent 4 663 159, 1987.

*Author to whom correspondence should be sent.

3. Iovine, C. R; Ray-Chandhuri, D. K. U.S. Patent 4 131 576, 1978.
4. Jones, R. T.; Brown, C. A. *Int. J. Cos. Sci.* **1988**, *10*, 219.
5. Reese, G. *Fette Seifen Anstrichmittle* **1966**, *68*, 763.
6. Scott, G. V.; Robbins, C. R.; Barnhurst, J. D. *J. Soc. Cosmet. Chem.* **1969**, *20*, 135.
7. Goddard, E. D.; Hams, W. C. *J. Soc. Cosmet. Chem.* **1987**, *38*, 233.
8. Brode, G. L.; Goddard, E. D.; Hams, W. C.; Salensky, G. A. *Cosmetic and Pharmaceutical Applications of Polymers*; Gebelein, C. G., Ed.; Plenum: New York, 1991; p 117.
9. Goddard, E. D. *Cellulosics: Chemical, Biochemical and Material Aspects*; Kennedy, J. F., Phillips, G. O., Williams, P. A., Eds.; Ellis Horwood: New York, 1993; p 331.
10. Goddard, E. D. *Interaction of Surfactants with Polymers and Proteins;* Goddard, E. D.; Ananthapadmanabhan, K. P., Eds.; CRC: Boca Raton, FL, 1992; p 171.
11. Leung, P. S.; Goddard, E. D. *Coll. and Surf.* **1985**, 13, 47.

Cellulosic Liquid Crystals

See: *Cellulosic Liquid Crystals (Overview)*
Liquid Crystalline Ethyl-Cyanoethyl Cellulose
Lyotropic Polysaccharides
Vegetal Biomass (3. Polymers, Derivatives and Composites)

CELLULOSIC LIQUID CRYSTALS (Overview)

Richard D. Gilbert
Department of Wood and Paper Science
North Carolina State University

The first observation of a liquid crystal solution of a cellulose derivative was by Werbowyj and Gray in 1976.[1]

Since these initial observations the field has expanded rapidly and there are numerous reports of cellulose derivatives that form lyotropic liquid crystals.

Investigators of cellulosic liquid crystals have two prime purposes: to study both lyotropic or thermotropic mesophases from either a scientific or technological viewpoint.[2,3]

The main focus of the technological investigations has been on the potential of preparing high strength/high modulus regenerated cellulose fibers using a less energy-intensive and more environmentally-friendly process than the present day viscose rayon process. Indeed, it is predicted the viscose process will be phased out early in the next century owing to environmental concerns. Another potential use of cellulosic liquid crystal derivatives is in chiroptical filters.[4,5]

LIQUID CRYSTAL SOLUTIONS OF CELLULOSE

Chanzy and Peguy were the first to report that cellulose forms a lyotropic mesophase.[6] They used a mixture of N-methylmorpholine-N-oxide (NMMO) and water as the solvent. Solution birefringence occurred at concentrations greater than 20% (w/w) cellulose. The concentration at which an ordered phase formed increased as the cellulose D.P. decreased. The persistence length of cellulose in NMMO-H_2O is not known, but presumably it has an extended chain configuration in this solvent. Again the question arises as to what is the relevant axial ratio to be used for cellulose.

Simple shearing of an anisotropic solution produced a highly oriented polymer film that, after washing and drying, was shown to have the cellulose II morphology. Long fibers could be pulled from the anistropic solutions. They also had the cellulose II morphology.

Bianchi et al. spun fibers from isotropic and anisotropic solutions of cellulose (D.P. 290) in LiCl (7.8%)–DMAC solutions.[7] The fiber mechanical properties increased through the isotropic-anisotropic transition with elastic moduli as high as 22 GPa (161 g/d) being obtained.

LIQUID CRYSTALLINE CELLULOSE DERIVATIVES

Probably the most widely studied cellulose derivative is hydroxypropyl cellulose (HPC) as the HPC–H_2O system is very tractable. Werbowyj and Gray showed HPC forms ordered solutions in polar organic solvents, such as CH_3OH, C_2H_5OH as well as water.[8] The concentration to form the ordered phase depended on the solvent (from 42–47 wt %) but was relatively insensitive to the molar mass of HPC.

Navard and Haudin[9,10] studied the thermal behavior of HPC mesophases as did Werbowyj and Gray,[8] Seurin et al.,[11] and, as noted above, Conio et al.[9–12] In summary, HPC in H_2O exhibits a unique phase behavior characterized by reversible transitions at constant temperatures above 40°C and at constant compositions when the HPC concentration is above ca. 40%. A definitive paper has been recently published by Fortin and Charlet who studied the phase-separation temperatures for aqueous solutions of HPC using carefully fractionated HPC samples.[13] They showed the polymer-solvent interaction differs in the cholesteric phase (ordered molecular arrangement) from that in the isotropic phase (random molecular arrangement).

Patel and Gilbert showed cellulose triacetate (CTA) forms a mesophase in mixtures of TFA and chlorinated alkanes.[14] Bheda et al. showed that cellulose triacetate forms a mesophase in dichloracetic acid.[15] Zugenmaier and co-workers have published a series of basic studies on the mesophases of various cellulose derivatives in a variety of solvents.[16–22]

THERMOTROPIC CELLULOSE DERIVATIVES

There are now numerous examples of cellulose derivatives that form both lyotropic and thermotropic mesophases. Of course, cellulose itself is unlikely to form a thermotropic liquid crystalline phase because it decomposes prior to melting.

Hydroxypropyl cellulose was shown by Shimamura et al. to form a thermotropic mesophase.[23,24]

Fukuda et al. have recently provided a detailed review of the literature relating to thermotropic cellulose derivatives, including chiral nematic and discotic columnar phases, their dynamic properties, their dielectric relaxation, and the origins of the thermotropicity[3].

REFERENCES

1. Werbowyj, R. S.; Gray, D. G. *Macromolecules* **1976** *34*, 97.
2. Guo, J.-X.; Gray, D. G. In *Cellulosic Polymers Blends and Composites*; Gilbert, R. D., Ed.; Hanser: New York, 1994; Chapter 2.
3. Fukuda, T.; Takada, A.; Miyamoto, T. In *Cellulose Polymers Blends and Composites*; Gilbert, R. D., Ed.; Hanser: New York, 1994; Chapter 3.
4. Marsano, E.; Carpaneto, L.; Ciferri, A. *Mol. Cryst. Liq. Cryst.* **1988**, *158*(B), 267.

5. Charlet, G.; Gray, D. G. *Macromolecules* **1987**, *20*, 33.
6. Chanzy, H.; Peguy, A. *J. Polym. Sci., Polym. Phys. Ed.*; **1980**, *18*, 1137.
7. Bianchi, E.; Ciferri, A.; Conio, G.; Teoldi, A. *J. Appl. Polym. Sci.* **1989**, *27*, 1477.
8. Werbowyj, R. S.; Gray, D. G. *Macromolecules* **1980**, *13*(1), 69.
9. Navard, P.; Haudin, J. M. *Calorim. Anal. Therm.* **1983**, *14*, 207.
10. Navard, P.; Haudin, J. M. In *Polymer Liquid Crystals*; Blumstein, A., Ed.; Plenum: New York, 1983; p 389.
11. Seurin, M. J.; Gilli, J. M.; Fried, F.; Ten, Bosch, A.; Sixou, P. In *Polymer Liquid Crystals*; Blumstein A., Ed.; Plenum: New York, 1983; p 377.
12. Conio, G.; Bianchi, E.; Ciferri, A.; Teoldi, A.; Aden, M. A. *Macromolecules* **1983**, *16*, 1264.
13. Fortin, S.; Charlet, G. *Macromolecules* **1989**, *22*(5), 2286.
14. Patel, D. L.; Gilbert, R. D. *J. Polym. Sci., Polym. Phys. Ed.* **1981**, *19*, 1449.
15. Bheda, J., Fellers, J. E; White, J. L. *Coll. Polym. Sci.* **1980**, *258*, 1335.
16. Zugenmaier, P.; Haurand, P. *Carbohydr. Res.* **1987**, *160*, 369.
17. Vogt, V.; Zugenmaier, P. Ber. *Bunsenges, Phys. Chem.* **1985**, *89*, 1217.
18. Siekmeyer, M.; Zugenmaier, P. *Makromol. Chem., Rapid Commun.* **1987**, *8*, 511.
19. Steinmeier, H.; Zugenmaier, P. *Carbohyd. Res.* **1988**, *173*, 75.
20. Zugenmaier, P.; Haurand, P. *Carbohyd. Res.* **1987**, *160*, 369.
21. Siekmeyer, M.; Zugenmaier, P. *Makromol. Chem.* **1990**, *191*, 1177.
22. Siekmeyer, M.; Steinmeier, H.; Zugenmaier, P. *Makromol. Chem.* **1989**, *190*, 1037.
23. Shimamura, K.; White, J. L.; Feller, J. F. *J. Appl. Polym. Sci.* **1981**, *26*, 2165.
24. Shimamura, K.; White, J. L; Feller, J. F. *J. Polym. Sci., Polym. Lett. Ed.* **1982**, *20*(1), 33.

CELLULOSIC MATERIALS (Moisture Sorption Properties)

Mitsuhiro Fukuda
Textile Materials Science Laboratory
Hyogo University of Teacher Education

Moisture sorbed in polymeric materials shows a dramatic effect on the physical and mechanical properties of the materials. Particularly in the case of the hydrophilic or hygroscopic materials such as cellulose or cellulose derivatives, the degradation of the mechanical properties and dimensional stability by the moisture sorption is critical to practical applications.[1-5] Moisture sorption isotherm is one of the fundamentals to investigate the interaction between water and polymeric materials.

MOISTURE SORPTION PROPERTIES

Moisture Sorption Isotherms of Natural and Regenerated Cellulose

Figure 1 shows the moisture sorption isotherms for natural and regenerated cellulosic fibers at 30°C.[6-10] It is generally accepted that the water molecules interact only with the polar or hydrophilic groups in the noncrystalline region of the polymer.

Cotton and ramie, which have a crystal structure of Cellulose I with higher crystallinity, appears to have a lower moisture uptake than regenerated cellulose fibers. But if we normalize moisture regains in terms of crystallinity, the normalized regains for cotton and ramie are quite the same as those for other regenerated cellulose and amorphous cellulose in the whole range of humidities.

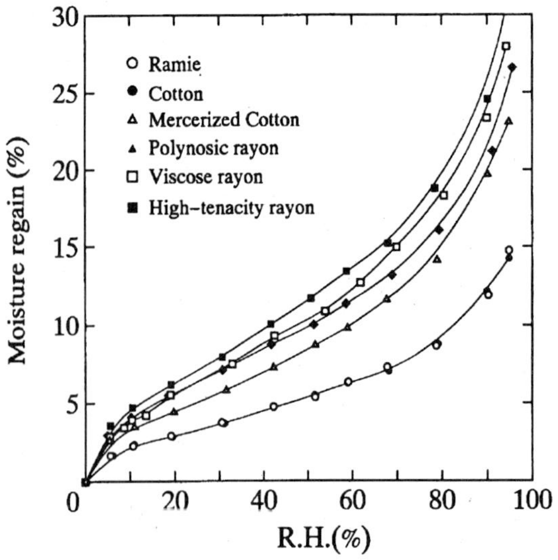

FIGURE 1. Moisture sorption isotherms of natural and regenerated cellulose fibers at 30°C.

Moisture Sorption Isotherms of Other Cellulose Derivatives

HPC is a water-soluble polymer and forms cholesteric liquid crystalline at concentration higher than about 60%(w/w) at room temperature.[11] The 10%(w/w) aqueous solution is optical isotropic and 50%(w/w) solution is optical anisotropic. However, no significant distinction in the moisture regain was observed between the powder and the two cast films from different concentration at less than 80% r.h.

Glass Transition Temperature as a Function of Moisture Regain

Glass transition temperature, T_g, greatly depends on the moisture contents of the polymer.[12] For example, T_g of dried wool of around 180°C is changed to around 0°C at wetness, and that of dried nylon 6 of around 90°C is decreased to below 0°C.[13,14] T_g of native cellulose and regenerated cellulose at dryness is higher than 200°C.[1] T_gs for many cellulose derivatives are ranged from 30°C–180°C.[2] An increase in moisture regain of 1% gives rise to the decrease in glass temperature of about 10°C for many polymers observed.

THE STATE OF WATER IN CELLULOSIC MATERIALS

Heat of Sorption for Cellulosic Materials and Thermodynamical Analysis

Until the 1960s, thermodynamical parameters deduced from the temperature dependence of the sorption isotherms or direct calorimetry of the heat of sorption had been the major approach to interpret the interaction between water and cellulosic materials.[10,15-19] Since then, remarkable development in microcalorimetry such as DSC, in molecular spectroscopies such as

FT–IR and NMR, and in the mechanical relaxation studies enabled us to characterize the nature of the water sorbed in polymers.

Two kinds of heat of sorption, integral heat of sorption and differential heat of sorption, have been employed to interpret the interaction between water and polymer.

Estimation of the State of Sorbed Water from DSC, NMR, and FT-IR Studies

We will summarize the results on the state of the water sorbed in cellulosic materials measured by DSC, NMR, and FT-IR as well as the heat of sorption. Definition of "free water," "bound water," and "nonfreezing water" are reported in DSC study.[20] "Ordered water," "structured water," "strongly interacted water," "weakly interacted water," "tightly bound water," "loosely bound water," "ice-like water," and "immobilized water" have been often used to suggest a kind of bound water, while "noninteracting water" and "mobile water" have been used to suggest the kind of free water in IR and NMR studies. Anyway, they are closely associated the strength of hydrogen bonding and dynamical properties of the water molecules. The state of the water does not change like a stepwise function, but they are continuously changing in the strength of interaction according to the moisture contents. Although a generally acceptable definition of the state of the water seems to be difficult, there are at least two kinds of bound water in dynamics and interactive properties observed in many cellulosic materials. Some authors reported opposite results, namely that the interaction between sorbed water and cellulose derivatives is considerably weaker than that in liquid water, and there is no bound water by DSC or IR studies.[21,22] In this condition, the sorbed water is "dissolved" in Henry's law manner like that in poly(ethylene terephthalate).[23]

REFERENCES

1. Bryant, G. M.; Walter, A. T. *Text. Res. J.* **1959**, *29*, 211.
2. Fujimoto, T.; Inoue, Y.; *Kobunshi Kagaku* (in Japanese) **1960**, *17*, 436.
3. Morton, W. E.; Hearle, J. W. S. *Physical Properties of Textile Fibers*; 2nd ed.; The Textile Institute: London 1975.
4. Nakamura, K.; Hatakeyama, T.; Hatakeyama, H. *Text. Res. J.* **1983**, *53*, 682.
5. Hermans, P. H. *Physics and Chemistry of Cellulose Fibers*; Elsevier: New York, 1949; p 277.
6. Kohata, K.; Miyagawa, M.; Takaoka, A.; Kawau, H. *Sen-i Gakkaishi* **1986**, *42*, T136.
7. Hernádi, A. *Cellulose Chem. Technol.* **1984**, *18*, 115.
8. Pizzi, A. *Cellulosics: Chemical, Biochemical and Material Aspects*; Kennedy, J. F.; Phillips, G. O.; Williams, P. A., Eds.; Ellis Harwood Ltd., 1993; Chapter 30.
9. Zeronian, S. H.; Coole, M. L.; Alger, K. W.; Chander, J. M. *J. Appl. Polym. Sci., Appl. Polym. Symp.* **1983**, *37*, 1053.
10. Fukuda, M.; Ohtani, K.; Iwasaki, M.; Kawai, H. *Sen-i Gakkaishi* **1987**, *43*, 567.
11. Werbowyj, R. S.; Gray, D. G. *Macromolecules* **1980**, *13*, 69.
12. Fuzek, J. F. *Water in Polymers*; Rowland, S. P., Ed.; ACS Symposium Series 127, Washington, 1980; Chapter 31.
13. Wortmann, F. J.; Rigby, B. J.; Phillips, D. G. *Text. Res. J.* **1984**, *54*, 6.
14. Kettle, G. J. *Polymer* **1977**, *18*, 742.
15. Hedge, J. J. *Trans. Farad. Soc.* **1926**, *22*, 178.
16. Guthrie, J. C. *J. Text. Inst.* **1949**, *40*, T489.
17. Rees, W. E. *J. Text. Inst.* **1948**, *39*, T351.
18. Rees, W. E. *Moisture in Textiles*; Hearle, J. W. S.; Peters R. H., Eds.; Butterworths: London, 1960; Chapter 4.
19. Burghoff, H-.G.; Pusch, W. *J. Appl. Polym. Sci.* **1979**, *23*, 473; *Polym. Eng. Sci.* **1980**, *20*, 305.
20. Nakamura, K.; Hatakeyama, T.; Hatakeyama, H. *Text. Res. J.* **1981**, *51*, 608.
21. Toprak, C.: Agar, J. N.; Falk, M. *J. Chem. Soc., Faraday Trans.* **1979**, *73*, 803.
22. Kinard, D. A. *J. Polym. Sci., Polym. Symp.* **1984**, *71*, 183.
23. Fukuda, M.; Kawai, H.; Yagi, N.; Kimura, O.; Ohta, T.; *Polymer* **1990**, *31*, 295.

Cements

See: *Concrete-Polymer Composites*
Dental Polymers (Glass-Ionomer Cements)

Ceramics

See: *Alumina Fibers (from Poly[(acyloxy)aloxane]s)*
Dental Adhesives (for Alloys and Ceramics)
Electrically Conducting Composites
Inorganic Nanostructured Materials
Inorganic/Organic Hybrid Polymers (High Temperature, Oxidatively Stable)
Polyborosiloxanes
Polysilazanes (Through Ring-Opening Polymerization)
Preceramic Polymers
Silacyclobutanes and Related Compounds (Ring-Opening Polymerization)
Silicon Nitride

Charcoal

See: *Carbon Composites*

Charge-Transfer Complexes

See: *Charge-Transfer Polymerization*
Electrically Conducting Composites
Photopolymerization (Initiated by Charge-Transfer Complexes)
Polysulfones
Trisubstituted Ethylene Copolymers

CHARGE TRANSFER POLYMERIZATION

Noriyuki Kuramoto
Department of Materials Science and Engineering
Faculty of Engineering
Yamagata University

Charge-transfer polymerization is a system that involves the charge-transfer complex formed between compounds, such as monomers, solvents, and initiators, present in polymerization systems. Charge-transfer complexes influence the copolymer composition and sequence distribution of the feed composition of polymeric systems. These complexes participate in the initiation, propagation, and termination steps of the polymerization. A typical complex participation system is a polymerization initiated and propagated between electron donor and electron acceptor monomers. The radical copolymerization of a specific electron-accepting monomer with an electron donor monomer can produce an equimolar alternating copolymer. Charge-transfer polymerizations of electron donor monomers with electron acceptor monomers have been extensively investigated.[1]

There have been several studies on the alternating radical copolymerization between a donor monomer and an acceptor monomer. Examples of the acceptor monomer are maleic anhydride (MAH), fumaric esters, vinylidene cyanide, sulfur dioxide, and carbon monoxide. Donor monomers are less specific: styrene (ST), isobutylene, butadiene, vinyl ethers, *p*-dioxene, and vinyl acetate.

Radical alternating copolymerization systems for acrylic monomers such as methyl methacrylate and acrylonitrile were studied in regard to the charge-transfer complex with Lewis acids.[2] The copolymerization can produce a completely alternating copolymer of acrylic monomers and donor monomers in the presence of Lewis acids, for example, zinc chloride and alkylaluminum chloride.

Another charge-transfer complex participation system is a polymerization of electron donor monomers initiated by the charge-transfer complex between the monomer and electron acceptor.[3]

Another complex participation system is a polymerization initiated and propagated through a charge-transfer complex formed between the electron acceptor monomer and electron donor.[4] Polymerizations initiated as a result of a formation of a charge-transfer complex involving an electron acceptor monomer molecule may proceed through the formation of a monomer cation radical by electron transfer from the electron acceptor monomer to the suitable electron donor (i.e., by electron transfer to the monomer from an electron source). Kuramoto et al. reported the copolymerization system of ST with MAH initiated by a charge-transfer complex formed between MAH and the electron donor.[5]

PREPARATION AND PROPERTIES

Measurement of Formation Constants

The charge-transfer complex is composed of a wide variety of intermolecular complexes involving from very weak intermolecular interactions to strong intermolecular interactions, yielding complexes that can be isolated. Ultraviolet spectrophotometry and NMR spectroscopy are most often used, but other techniques such as calorimetry, vapor pressure measurement, and dielectric constant determination have been applied

Nagai et al. studied the copolymerization of ST with chloromaleic anhydride (CMA) initiated by a spontaneous charge-transfer complex and a radical initiator in dioxane.[6,7] CMA copolymerizes spontaneously with ST at a higher rate than MAH does. The copolymerization was strongly accelerated by UV irradiation. An alternating copolymer was obtained by copolymerization over a wide range of feed composition. The initiating radicals for the spontaneous copolymerization are formed through a charge-transfer complex between ST and CMA.

Polymer Composition and Structure

There are several models for explaining polymer composition for charge-transfer polymerization. The first is the Mayo and Lewis model, the so-called terminal model, which allows the copolymer composition and sequence distribution to be related to the monomer feed composition through two reactivity ratios.[8] Bartlett and Nozaki also suggested that there may be an alternative explanation for deviations from the terminal model.[9] They proposed that the charge-transfer complex may participate in the propagation steps of copolymerization, thus accounting for the observed deviations from the terminal model. Some copolymer systems characterized by a high degree of alternation showed no evidence for the formation of donor acceptor complexes between two monomers. Seiner and Litt proposed the extended complex participation model, which allowed for participation of both monomers and the charge-transfer complex in the propagation steps.[10]

Alternating Copolymerization of Acrylic Monomers in the Presence of Lewis Acid

Several mechanisms have been proposed for the alternating copolymerization of acrylic ester and an electron donor monomer in the presence of a Lewis acid, explaining the alternating regulation mechanism in the propagation. It is postulated that a ternary molecular complex forms among the Lewis acids, the acceptor monomer, and the donor monomer. The ternary molecular complex acts as a unit and polymerizes to yield an alternating copolymer.[11,12] The ternary molecular complex mechanism is adequate to explain the specific features of the alternating copolymerization.[13] An acrylic monomer forms a binary complex with a Lewis acid through coordination of the carbonyl or cyano group of the monomer with the central atom of the Lewis acid. The binary complex is fairly stable and can be isolated in the crystalline state.

The alternating copolymerization of acrylic monomers with a donor monomer in the presence of a Lewis acid is possible either by spontaneous initiation, by using a free-radical initiation, by electrochemical initiation, or by photoinitiation. The addition of a radical initiator to the Lewis acid-acceptor, monomer-donor monomer system is effective in all cases for the alternating copolymerization.

Copolymerization of ST with MAH Initiated by Thiol Compounds

The copolymerization system of ST with MAH has been the most thoroughly studied among copolymerization systems for which monomers form charge-transfer complexes to generate copolymerization. Copolymers of MAH with various electron donor monomers exhibit an alternating tendency. Kuramoto et al. reported that thiol compounds were found to be effective initiators for the copolymerization of ST with MAH through the charge-transfer complex between thiol and MAH.[5] Various solvents and thiols such as mercaptoethanol, thioglycolic acid, and *p*-toluenethiol can be used for the copolymerization.

Copolymerization of Furan with MAH Initiated by Thioglycolic Acid

Butler et al. reported the alternating copolymer of furan with MAH was produced regardless of the comonomer composition, if copolymerizations were carried out at 70°C.[14] The presence of a charge-transfer complex between two monomers was observed by using NMR and UV methods. They proposed that the copolymerizations proceeded by homopolymerization of a charge-transfer complex, yielding alternating copolymers. Gaylord and Matyska studied the free-radical copolymerization of MAH with furan.[11]

Kuramoto et al. studied the copolymerization of furan with MAH initiated by thioglycolic acid (TGA).[15]

Copolymerization of Methyl Methacrylate with MAH Initiated by *p*-Toluene Thiol

p-Toluenethiol (PT) initiates the copolymerization of methyl methacrylate with MAH in the presence and in the absence of solvent.[16] Bulk copolymerization initiated by PT occurred at 50°C; the rate of polymerization was high. In the absence of PT or MAH, the yield in the same time period was zero. Homopolymerization of methyl methacrylate does not occur in the presence of PT. The initiation step involves both MAH and PT, which form a charge-transfer complex and initiate the copolymerization. PT was found to be effective as an initiator and as a chain-transfer agent for the copolymerization of methyl methacrylate with MAH and resulted in a marked reduction in the molecular weight of copolymers.

REFERENCES

1. Hill, D. J. T.; O'Donnell, J. J.; O'Sullivan, P. W. *Prog. Polym. Sci.* **1982**, *8*, 215.
2. Hirai, H. *J. Polym. Sci. Macromol. Rev.* **1976**, *11*, 47.
3. Gaylord, N. G. *Macromol. Rev.* **1970**, *4*, 183.
4. Gaylord, N. G.; Dixit, S. S. *J. Polym Sci., Macromol. Rev.* **1974**, *8*, 51.
5. Kuramoto, N.; Iwaki, T.; Satoh, A.; Nagai, K. *J. Polym. Sci., Polym. Chem. Ed.* **1989**, *27*, 367.
6. Nagai, K.; Akiyama, K.; Kuramoto, N. *Makromol. Chem.* **1985**, *186*, 1855.
7. Nagai, K.; Akiyama, K.; Kuramoto, N. *Makromol. Chem.* **1985**, *186*, 1863.
8. Mayo, F. R.; Lewis, F. M. *J. Am. Chem. Soc.* **1944**, *66*, 1594.
9. Bartlett, P. D.; Nozaki, K. *J. Am. Chem. Soc.* **1946**, *68*, 1495.
10. Seiner, J. A.; Litt, M. *Macromolecules* **1971**, *4*, 308.
11. Gaylord, N. G.; Matyska, B. *J. Macromol. Sci., Chem.* **1970**, *A4*, 1507.
12. Gaylord, N. G.; Matyska, B.; Arnold, B. *J. Polym. Sci.* **1970**, *B8*, 235.
13. Gaylord, N. G.; Maiti, S. *J. Macromol. Sci., Chem.* **1972**, *A6*, 1481.
14. Butler, G. B.; Badgett, J. T.; Sharabash, M. *J. Macromol. Sci., Chem.* **1970**, *A4*, 51.
15. Kuramoto, N.; Suda, K.; Nagai, K. *J. Polym. Sci., Polym. Chem. Ed.* **1989**, *27*, 1713.
16. Kuramoto, N.; Iwaki, T.; Shibamura, T.; Nagai, K. *Angew. Makromol. Chem.* **1990**, *175*, 39.

CHARGE TRANSPORT POLYMERS (for Organic Electroluminescent Devices)

Junji Kido
Department of Materials Science and Engineering
Yamagata University

An organic electroluminescent (EL) device is a kind of light-emitting device similar to light-emitting diodes (LEDs) made from semiconductors. Unlike LEDs, organic EL devices have large emitting areas and high brightness levels. This type of device has a luminance of more than 100,000 cd/m^2. It is about 10 times brighter than a common fluorescent lamp.[2] In addition, various colors such as blue, green, red, and white can be obtained.[3–8]

In these devices, organic emitter layers are sandwiched between two electrodes, and electric energy is transformed into light through the excitation of the organic molecules. Excitation mechanisms involve the recombination (reaction) of charge carriers such as electrons (radical anion) and holes (radical cation) that are injected into the organic layers from the electrodes. Hence, it is necessary for the organic component materials to be charge-transporting as well as fluorescent.

Fabrication methods depend on the materials used. For example, in molecular systems, such as fluorescent dyes, organic thin layers can be formed by vacuum deposition. However, the crystallization or aggregation of the vacuum-deposited molecules sometimes causes the destruction of the layered structure and the degradation of the device.[9] In contrast, polymeric materials have stronger mechanical strength and are less crystalline than low molecular weight materials, and thin polymer films can be formed by various costing techniques such as spin coating and dip coating. It is therefore reasonable to use macromolecular systems.

PREPARATION AND PROPERTIES

Poly(*N*-Vinylcarbazole)

Poly(*N*-vinylcarbazole) (PVK) is one of the most well known hole-transporting polymers. In this polymer, hole transport is due to electron hopping between the carbazole side groups. However, the hole drift mobility is low, ~10^{-7} cm^2/Vs at room temperature.[10] The application of a high electric field is, therefore, necessary to transport electrons.

PVK can be synthesized by a radical polymerization of *N*-vinylcarbazole.[11] Cationic polymerization yields polymers with highly isotatic configuration, which have lower drift mobilities than the radically polymerized ones.[11] Therefore, the radical polymerized polymers are more suitable for EL application than are the cationically polymerized polymers.

Polysilane

Another class of well-studied hole-transporting polymers is polysilane. Polysilane is unique in its charge transport properties compared with the saturated carbon polymers. Charge delocalization over the σ Si–Si bond provides low ionization potential and high hole drift mobility, which is ~ 10^{-4}–10^{-5} cm^2/Vs at room temperature.[12,13]

Triphenylamine-Containing Poly(N-Substituted Methacrylamide)

The triphenylamine group has a low oxidation potential, and its derivatives have been used as hole transport materials inorganic EL devices.[1–9] In our laboratory, poly[*N*-*p*-(*N'*,*N'*-diphenylamino)phenyl methacrylamide] (PTPAMA) and poly[*N*-[*p*-*N'*-phenyl-*N'*-(1,1'-biphenyl-4'-[*N'*-phenyl-*N''*-(2-methylphenyl)amino]-4-amino]]phenyl methacrylamide[(PTPDMA) were recently synthesized.[15,16]

Molecularly Doped Polymers

Molecularly doped polymers (MDPs) are binary solid solutions of charge-transporting molecules molecularly dispersed in inert polymeric binders. MDPs have been used as photoconductors in xerography.[10] An advantage of MDPs over the other polymer systems is the flexibility in material design: by selecting a proper dopant molecule and the concentration, it is possible to optimize carrier transport properties.

The MDPs that we have investigated for EL application in our laboratory are *N,N'*-diphenyl-*N,N'*-bis(3-methylphenyl)-1,1'-biphenyl-4,4'-diamine (TPD) in bisphenol-A polycarbonate (PC), poly(methyl methacrylate) (PMMA), and polystyrene (PS). TPD has a high hole drift mobility of 10^{-3} cm^2/Vs, and its hole transport properties in the polymer matrix have been extensively studied.[14] PC, PMMA, and PS are optically and electronically inert and have good film-forming properties with glass transition temperatures of 145, 105, and 90°C, respectively. MDP films can be fabricated by casting the solutions containing polymer and TPD.

Plasma Polymerized Films

Plasma polymerization is an effective method for the preparation of thin, pinhole-free crosslinked films. Common polymerization methods require monomers with polymerizable groups such as a vinyl group. In contrast, in plasma polymerization, it may not be necessary for monomers to meet such requirements because the polymerization proceeds the fragmentation and decomposition of the monomer during the polymerization. The active species, then, react with the another one or the inactive starting monomer, forming highly crosslinked polymers, It is therefore, possible to obtain thin photoconductive pinhole-free films.

APPLICATIONS

Polymer as a Hole Transport Layer

The previously mentioned polymer systems have been used as a hole transport layer in double layer-type organic EL devices that have an electron-transporting emitter layer. Such double-layer-type devices have a layered structure composed of the materials with different carrier transport properties: one transports holes, and the other transports electrons. Therefore, electrons and holes are injected into the organic layers from the electrodes through the corresponding carrier transport layer, and the recombination of the carriers occurs at or near the interface between the two organic layers.

Among the devices, PVK, PTPAMA, PTPDMA, and MDP systems show high luminance.[15–18] For polysilane the device stability is not high, although PMPS possesses a high hole drift mobility.[19]

Polymer as an Emitter Layer

Electroluminescence from PVK

Charge transport polymers can also be used as an emitter layer in organic electroluminescent devices. For example, PVK can be used as an emitter layer in multilayer-type devices in conjunction with appropriate electron transport layers.[20]

Color Tuning in the PVK Device

The EL color of the PVK device can be tuned by doping the PVK layer with fluorescent dyes.[21] The excitation mechanisms of the dopants are not fully understood, but the Förster-type energy transfer from the host to the dopants and/or the direct excitation by the carrier recombination at the dopant sites due to carrier trapping by the dopants may be operative.

Taking advantage of the doping technique, we can use three kinds of dopant at the same time to generate the three primary colors that are combined to yield white color.[22] From the cell with the PVK layer doped with three fluorescent dyes, white emission can be obtained.

Single-Layer-Type Device

Single-layer-type devices that are composed of only one organic layer can be fabricated with polymers. The emitter layer should be designed to transport both electrons and holes so that the recombination can take place in the emitter layer with high efficiencies.

In this device, color tuning can be done easily by adding fluorescent dyes with a different color to the polymer layer. Thus, the fabrication of single-layer-type, polymer-based devices is the simplest among all the device structures.

CONCLUSIONS

Charge transport polymers have been shown to be useful as a component material in organic electroluminescent devices. A hole transport layer and an emitter layer can be made from polymers. These polymer-based devices show different characteristics from the devices based on vacuum-deposited molecular materials. Mechanical strength and processability of the polymer materials are the major advantages over molecular materials.

REFERENCES

1. Tang, C. W.; VanSlyke, S. A. *Appl. Phys. Lett.* **1987**, *51*, 913.
2. Murayama, R.; Kawani, S.; Wakimoto, T.; Sato, H.; Nakada, H.; Namiki, T.; Imai, K.; Nomura, M. *Extended Abstracts (The 54th Autumn Meeting, 1993) The Japan Society of Applied Physics* 1993, p 1127.
3. Adachi, C.; Tsutsui, T.; Saito, S. *Appl. Phys. Lett.* **1989**, *55*, 1489.
4. Hamada, Y.; Adachi, C.; Tsutsui, T.; Saito, S. *Jpn. J. Appl. Phys.* **1992**, *31*, 1812.
5. Kido, J.; Nagai, K.; Ohashi, Y. *Chem. Lett.* **1990**, 657.
6. Kido, J.; Hayase, H.; Hongawa, K.; Nagai, K.; Okuyama, K. *Appl. Phys. Lett.* **1994**, *65*, 2124.
7. Kido, J.; Ohtaki, C.; Hongawa, K.; Okuyama, K.; Nagai, K. *Jpn. J. Appl. Phys.* **1993**, *32*, L917.
8. Kido, J.; Kimura, M.; Nagai, K. *Science*, in press.
9. Adachi, C.; Tsutsui, T.; Saito, S. *Appl. Phys. Lett.* **1990**, *56*, 799.

10. *Electronic Properties of Polymers*; Mort, G.; Pfister, G., Eds.; Wiley-Interscience: New York, 1982; pp 215–265.
11. Itaya, A.; Okamoto, K.; Kusabayashi, S. *Polymer J.* **1985**, *17*, 557.
12. Diaz, A.; Miller, R. D. *J. Electrochem. Soc.* **1985**, *132*, 834.
13. Stolka, M.; Yuh, H. J; McGrane, K.; Pai, D. M. *J. Polym. Sci., Part A, Polym. Sci.* **1987**, *25*, 823.
14. Stolka, M.; Yanus, J. F.; Pai, D. M. *J. Phys. Chem.* **1984**, *88*, 4707.
15. Kido, J.; Harada, G.; Nagai, K. *Kobunshi Ronbunshu* **1995**, *52*, 216.
16. Kido, J.; Komada, M.; Harada, G.; Nagai, K. *Polym. Adv. Technol.* **1995**, *6*, 703.
17. Fujii, T.; Fuijta, M.; Hamada, Y.; Shibata, K.; Tsujino, Y.; Kuroki, K. *J. Photopolym. Sci. Technol.* **1991**, *4*, 135.
18. Kido, J.; Hongawa, K.; Kohda, M.; Nagai, K.; Okuyama, K. *Jpn. J. Appl. Phys.* **1992**, *31*, L960.
19. Kido, J.; Nagai, K.; Okamoto, Y.; Skotheim, T. *Appl. Phys. Lett.* **1991**, *59*, 2760.
20. Kido, J.; Hongawa, K.; Okuyama, K.; Nagai, K. *Appl. Phys. Lett.* **1993**, *63*, 2627.
21. Kido, J.; Hongawa, K.; Nagai, K.; Okuyama, K. *Macromol. Symp.* **1994**, *84*, 81.
22. Kido, J.; Hongawa, K.; Okuyama, K.; Nagai, K. *Appl. Phys. Lett.* **1994**, *64*, 815.

Chelating Polymers

See: *Metal-Chelating Polymers*

Chemical Amplification Resist Materials

See: *Chemical Amplification Resists*
Chemical Amplified Resists (for Deep-UV Lithography)
Chemically Amplified Resists (New Generations)

CHEMICAL AMPLIFICATION RESISTS

Hiroshi Ito
IBM Almaden Research Center

Semiconductor devices are fabricated by a technology known as lithography, in which radiation-sensitive polymeric materials called *resists* are used to produce circuit patterns in substrates such as single crystals of silicon.

For a resist material to be useful in the device fabrication, it must be capable of spin casting from solution into a thin and uniform film that adheres to various substrates such as metals, semiconductors, and insulators. Furthermore, the resist must possess high radiation sensitivity and high resolution capability, and it must withstand extremely harsh environments, for example, high temperature, strong corrosive acids, and plasmas, to be viable for practical use.

CHEMICAL AMPLIFICATION RESISTS

The resist sensitivity plays a key role, especially in the case of high-resolution, short-wavelength lithographic technologies. Quantum yields, expressed as the number of molecules transformed per photon absorbed, characterize the efficiency of photochemical events. The quantum yield of typical diazonaphthoquinones is 0.2–0.3; three to five photons are required to convert a single molecule of the photoactive compound, which places a fundamental limit on the photosensitivity of such systems.

To overcome this intrinsic sensitivity limitation, the concept of "chemical amplification" was proposed in 1982.[1,2] In the chemical amplification scheme, a single photochemical event induces a cascade of subsequent chemical transformations, thus providing a gain mechanism. The chemical amplification concept is based on photochemically induced generation of active species that catalyze numerous subsequent chemical transformations in the resist film. In general, these chemical transformations are accomplished by heating the exposed resist film (postexposure bake or postbake). In principle, the active species could be either ionic or radical. However, the use of photochemical acid generators (PAG) has become a major foundation for the design of an entire family of advanced resist systems.[3–8] Various PAGs for use in chemically amplified lithographic imaging have been reported.[8,9] The choice of PAG depends on a number of factors, including the nature of radiation, solubility, thermal stability, toxicity, and the strength and size of generated acid.

In terms of lithographic processes, chemically amplified resist systems can be divided into positive and negative systems; or UV, electron beam, and X-ray resists; or two- and three-component systems. However, it is more convenient to categorize the chemical amplification resists on the basis of their imaging chemistries. In terms of chemistry, chemically amplified imaging may involve deprotection, depolymerization, rearrangement, polymerization, condensation, and dehydration.

DEPROTECTION

Acid-catalyzed deprotection used to generate a phenol structure was one of the first chemical amplification schemes proposed and has attracted a great deal of attention.[1,2] The greatest research efforts have been placed on this imaging chemistry in recent years to design aqueous base developable positive resist systems for replacement of the diazonaphthoquinone–novolac resist.

A_{AL}-1 Acidolysis of Carbonates and Esters

One of the first chemical amplification resists was based on acid-catalyzed deprotection of poly(4-*t*-butoxycarbonyloxystyrene) (PBOCST).[1,4] As illustrated in **Figure 1**, the phenolic functionality of poly(4-hydroxystyrene) (PHOST) is protected with an acid-labile *t*-butoxycarbonyl (tBOC) group. The lipophilic carbonate polymer is converted upon postexposure bake at ~100°C to PHOST by reaction with photochemically generated acid, releasing carbon dioxide, isobutene, and a proton. The photochemically generated acid is not consumed in one reaction but reproduced to undergo a number of deprotection reactions. The deprotection reaction depicted in Figure 1 results in conversion of a lipophilic polymer to a hydrophilic polymer, providing a large change in the polarity (and hence solubility) of the polymer. This polarity change can be exploited in dual-tone imaging simply by changing the polarity of the developer solvent. The deprotection mechanism has provided a foundation for the design of positive resists that can be developed with aqueous base and of negative resists that are devoid of swelling-induced image distortion.

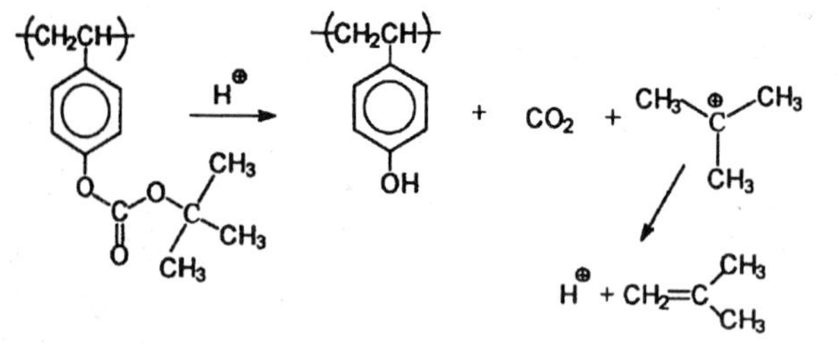

FIGURE 1. tBOC resist imaging mechanism (acid-catalyzed deprotection).

The tBOC resist consisting of PBOCST and 4.75 wt % of triphenylsulfonium hexafluoroantimonate as the acid generator was the first deep-UV resist used in manufacturing.[10] It is also sensitive to electron beams and X-ray radiations.[11,12]

PBOCST can be readily prepared by radical or cationic polymerization of 4-*t*-butoxycarbonyloxystyrene[13] or by treating PHOST with di-*t*-butyl dicarbonate in the presence of a base.[14] PBOCST obtained by radical polymerization has a glass transition temperature (T_g) of ~135°C and undergoes spontaneous thermolysis at ~190°C to produce PHOST, carbon dioxide, and isobutene.[13]

The tBOC group has also been used in the protection of a new class of aqueous-base-soluble polymers, poly[4-(2-hydroxyhexafluoroisopropyl)styrene]. The tBOC-protected polymer can be synthesized by radical polymerization of a protected monomer or by reaction of the fluoro-alcohol polymer with di-*t*-butyl dicarbonate. The base-soluble fluoro-alcohol polymer may be prepared by radical polymerization of the corresponding unprotected monomer or by treating polystyrene with hexafluoroacetone in the presence of a Lewis acid catalyst.[15]

Polymers bearing ester and ether functionalities that undergo A_{AL}-1 acidolysis in the absence of a stoichiometric amount of water are also very useful in dual-tone imaging.

The deprotection mechanism can be extended to all-dry development with oxygen reactive ion etching (RIE).[16,17] The deprotection reaction unmasks the reactive phenolic hydroxyl or carboxylic acid groups, which can then be reacted with a silylating reagent, such as hexamethyldisilazane and dimethylaminotrimethylsilane, while the carbonate or ester groups in the unexposed regions are inert in the silylation reaction. After the gas-phase silylation treatment, the resist film is subjected to oxygen RIE to generate negative-tone images. The surface of the Si-containing exposed area is converted to SiO_x by the oxygen plasma treatment, which prevents further etching, whereas the organic polymer in the unexposed regions is converted to volatile carbon dioxide and water.

Hydrolysis

The deprotection mechanism has been extended to include hydrolysis of trimethylsilyl ether and alcoholysis of tetrahydropyranl ether of phenolic polymers such as PHOST and novolac.[18–21]

Aqueous-Base-Developable Positive Resists

The advanced positive resist systems are almost exclusively based on PHOST partially protected with the above-mentioned acid-labile groups. PHOST is very much transparent at 248 nm in contrast to the novolac resin, and thus highly attractive as a deep-UV matrix resin.[11,22] However, this polymer dissolves too rapidly in aqueous base and it is difficult to control its dissolution rate by adding inhibitors. The partial protection of the OH functionality with lipophilic acid-labile groups results in dramatic reduction of the dissolution rate, whereas the acid-catalyzed deprotection restores the high base solubility, which leads to positive images upon development with an aqueous base. This approach has been extensively studied by many research groups in an attempt to replace the diazonaphthoquinone/novolac resist in the deep-UV application.

Another approach to the design of aqueous-base-developable positive resists is to use a protected dissolution inhibitor, which is converted in a phenolic matrix resin to a base-soluble form by reaction with a photochemically generated acid. Low molecular weight carboxylic acids or phenols are protected in this scheme with acid-labile groups for use as dissolution inhibitors of phenolic resins, especially novolac.[23–25]

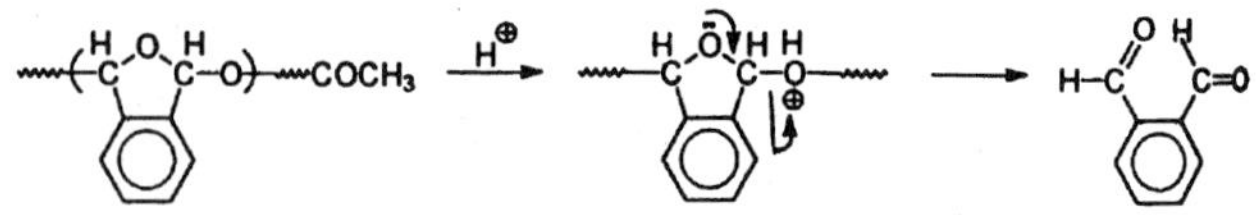

FIGURE 2. Acid-catalyzed depolymerization of polyphthalaldehyde.

DEPOLYMERIZATION

The ultimate form of polymer main-chain degradation is depolymerization. Polymers with low ceiling temperatures (T_c) could undergo depolymerization upon scission of one bond and are potentially useful as sensitive positive resists. Polyaldehydes are good examples that could undergo thermodynamically driven depolymerization after radiation-induced chain scission to volatile monomeric species.[2–4] Another depolymerization mechanism involves repeated catalytic main-chain scission of acid-labile backbone linkages as depicted in the deprotection section.

Thermodynamically Driven Catalyst Depolymerization

Polyphthalaldehyde can be synthesized by anionic or cationic polymerization of phthaladehyde (T_c = ~40°C).[26] Polyaldehydes consist of acetal bonds that are labile toward acids, and the C–O bond of polyphthaldehyde is catalytically cleaved with a photochemically generated acid, which initiates depolymerization of the entire polymer chain and results in spontaneous positive relief image formation by exposure alone (self-development) (**Figure 2**).[2–4]

The polymers that undergo efficient depolymerization can not be expected to withstand dry etching conditions. Two approaches have been taken to overcome the poor dry etch stability of the polyphthalaldehyde resist.[17,27–30] One approach was to use polyphthalaldehyde as a polymeric dissolution inhibitor of a novolac resin, and the other was to incorporate silicon in the polyphthalaldehyde structure for use in oxygen RIE bilayer lithography.

Repeated Main-Chain Acidolysis

Polymers containing tertiary and secondary allylic or secondary benzylic carbonates, esters, or ethers in the backbone form stable carbocations by acid-catalyzed cleavage of the C–O bond when mildly heated at ~100°C. The resulting carbocation undergoes β-proton elimination to form olefins. This reaction is repeated on the polymer chain and results in complete fragmentation to small molecules.[31–33] When the low molecular weight products are volatile, the degradation process can be potentially exploited in thermal development. Polyformals, consisting of secondary allylic and benzylic units, are decomposed upon acidolysis to give volatile products such as an aromatic compound, formaldehyde, and water.[33] Polyethers based on alkoxypyrimidine units undergo an acid-catalyzed tautomeric change to pyrimidone, liberating dienes.[34]

REARRANGEMENT

Acid-catalyzed rearrangement reactions can also be conveniently utilized in the design of chemical amplification resist systems, although such examples are limited.

Pinacol Rearrangement

The pinacol–pinacolone rearrangement involves conversion of *vic*-diols (pinacols) to ketones or aldehydes, with an acid as a catalyst, thus inducing a change of the polarity from a polar to a nonpolar state, whereas acid-catalyzed deprotection results in conversion of a nonpolar polymer to a polar polymer. This polarity change can be exploited in the lithographic imaging.

A polymeric pinacol, poly[3-methyl-2-(4-vinylphenyl)-2,3-butanediol], can be quantitatively converted to a nonconjugated ketone when treated with an acid.[35–37] As a result of the polarity change, the pinacol polymer containing an acid generator functions as a negative resist when developed with alcohol. Alcohol dissolves the polar diol polymer in the unexposed areas but can not dissolve the less polar ketone polymer produced in the exposed regions.

The pinacol rearrangement mechanism can be utilized in the design of aqueous base-developable negative resists. The styrenic pinacol was copolymerized with 4-acetoxystyrene followed by base hydrolysis of the acetoxy group to produce an aqueous base-soluble copolymer.[38]

Claisen Rearrangement

Acidolysis of a cyclohexenyl ether of PHOST results in Claisen rearrangement along with deprotection.[39] Poly(4-phenoxymethylstyrene) is similarly isomerized with acid as a catalyst to produce an *o*-alkylated phenolic structure. The net result of the Claisen rearrangement is a change from a nonpolar to a polar state.

INTRAMOLECULAR DEHYDRATION

Pinacol rearrangement involves acid-catalyzed dehydration to form a carbocation. Tertiary alcohols can undergo intramolecular dehydration by acid catalysis to form olefins, which is another mechanism for changing the polarity from a polar to a nonpolar state.

CONDENSATION

Acid-catalyzed condensation of phenolic resins is the most dominant mechanism used in the design of aqueous-base-developable negative resists and has provided a basis for the first commercialization of chemical amplification resists.[40–49] The condensation resist systems typically use three components: a base-soluble binder resin bearing reactive sites for crosslinking (phenolic resins), an acid generator, and an acid-sensitive latent crosslinking agent. The first condensation resist was based on a novolac–diazonaphthoquinone resist and a *N*-methoxymethylated melamine crosslinker.[40] The indenecarboxylic acid generated by photolysis of the diazoquinone reacts with the melamine to form *N*-carbonium ions, which undergo electrophilic substitution onto the electron-rich benzene ring of the novolac resin, resulting in crosslinking.

The novolac resin has been replaced wit PHOST for deep UV application and the diazonaphthoquinone with chloromethyltriazine as a HCl generator.[43,44,47,48] The rate-determining step in this crosslinking system is the formation of a carbocation from the protonated ether moiety. *O*-Alkylation, as well as the originally proposed *C*-alkylation, of the phenol groups contributes to crosslinking.[48,49]

Self-condensation of silanol compounds in a phenolic matrix resin has been used for aqueous base development of negative images, which is similar to pinacol rearrangement of small *vic*-diol to a dissolution-inhibiting compound and based on a polarity change rather than crosslinking.[50–52] The resist consists of a novolac resin, an onium salt acid generator, and a silanol compound as a dissolution promoter of the phenolic resin in an aqueous base. Base-soluble silanol compounds, such as diphenylsilanediol, undergo acid-catalyzed condensation and are converted to polysiloxanes upon postbake after exposure. Silsesquioxanes have also been used in a similar negative resist system.[53]

CATIONIC POLYMERIZATION

Some of the acid generators (cationic photoinitiators) were originally developed for photochemical curing of epoxy resins. One of the first chemical amplification resists used crosslinking of epoxy resins through cationic ring-opening polymerization of pendant epoxide groups as an imaging mechanism[1–4] Copolymers of styrene with allyl glycidyl ether have been used in deep-UV applications, and polymeric episulfides have been utilized in resist formulation.[54–57]

RECENT PROGRESS

The modern advanced resist systems are exclusively based on the chemical amplification concept; the chemically amplified resists are likely to play a critical role in manufacture of ultra large-scale integrated circuit devices. However, chemical amplification resist systems suffer from a unique problem: the so-called delay effect.[58] Because of the catalytic nature of the imaging mechanisms, a trace amount, on the order of 10 ppb, of airborne basic substances absorbed by the resist film interferes with the desired acid-catalyzed reaction, which results in T-top or skin formation, or line width shift, especially when the coated wafer stands after exposure (before postexposure bake).[58] Several solutions to this contamination problem have been recently proposed:

- purification of enclosing atmosphere by activated carbon filtration,
- application of a protective overcoat to seal off airborne contaminants,
- incorporation of stabilizing additives into resist formulation,
- reduction of activation energy of deprotection for elimination of postexposure bake, and
- reduction of free volume by annealing to minimize contaminant absorption.[58–69]

Significant progress has been made recently in the fundamental aspects of chemical amplification resists such as acid diffusion, kinetics, and modeling.

REFERENCES

1. Ito, H.; Willson, C. G.; Frechet, J. M. J. *Digest of Technical Papers of 1982 Symposium on ULSI Technology*, 1982, 86.
2. Ito, H.; Willson, C. G. *Technical Papers of SPE Regional Technical Conference on Photopolymers*, 1982, 331.
3. Ito, H.; Willson, C. G. *Polym. Eng. Sci.* **1983**, 23, 1012.
4. Ito, H.; Willson, C. G. In *Polymers in Electronics*; Davidson, T. Ed.; ACS Symposium Series 242; American Chemical Society: Washington, DC, 1984; p 11.
5. Iwayanagi, T.; Ueno, T.; Nonogaki, S.; Ito, H.; Willson, C. G. In *Electronic and Photonic Applications of Polymers*; Bowden, M. J.; Turner, S. R., Eds.; Advances in Chemistry 218; American Chemical Society: Washington, DC, 1988; p 107.
6. Reichmanis, E.; Houlihan, F. M.; Nalamasu, O.; Neenan, T. X. *Chem. Mater.* **1991**, 3, 394.
7. Ito, H. In *Radiation Effects on Polymers*; Clough, R. L.; Shalaby, S. W., Eds.; ACS Symposium Series 475; American Chemical Society: Washington, DC, 1991; p 326.
8. Ito, H. In *Radiation Curing in Polymer Science and Technology*; Fouassier, J. P.; Rabek, J. E., Eds.; Elsevier: London, 1993; Vol. 4, Chapter 11.
9. Pawlowski, G.; Dammel, R.; Przybilla, K-J.; Röschert, H.; Spiess, W. *J. Photopolym. Sci. Technol.* **1991**, *4*, 389.
10. Maltabes, J. G.; Holmes, S. J.; Morrow, J.; Barr, R. L.; Hakey, M.; Reynolds, G.; Brunsvold, W. R.; Willson, C. G.; Clecak, N. J.; MacDonald, S. A.; Ito, H. *Proc. SPIE* **1990**, 1262, 2.
11. Ito, H.; Pederson, A.; Chiong, K. N.; Sonchik, S.; Tsai, C. *Proc. SPIE* **1989**, 1086.
12. Seligson, D.; Ito, H.; Willson, C. G. *J. Vac. Sci. Technol.* **1988**, 86(6), 2268.
13. Frechet, J. M. J.; Eichler, E.; Ito, H.; Willson, C. G. *Polymer* **1983**, *24*, 995.
14. Houlihan, F.; Bouchard, F.; Frechet, J. M. J.; Willson, C. G. *Can. J. Chem.* **1985**, 63, 153.
15. Przybilla, K. J.; Roschert, H.; Pawlowski, G. *Proc. SPIE* **1992**, *1672*, 500.
16. MacDonald, S. A.; Schlosser, H.; Ito, H.; Clecak, N. J.; Willson C. G. *Chem. Mater.* **1991**, *3*, 435.
17. Ito, H. *J. Photopolym. Sci. Technol.* **1992**, *5*, 123.
18. Yamaoka, T.; Nishiki, N.; Koseki, K.; Koshiba, M. *Polym. Eng. Sci.* **1989**, *29*, 856.
19. Murata, M.; Takahashi, T.; Koshiba, M.; Kawamura, S.; Yamaoka, T. *Proc. SPIE* **1990**, *1262*, 8.
20. Hesp, S. A. M.; Hayashi, N.; Ueno, T. *J. Appl. Polym. Sci.* **1991**, *42*, 877.
21. Hayashi, N.; Schlegel, L.; Ueno, T.; Shiraishi, H.; Iwayanagi, T. *Proc. SPIE* **1991**, *1466*, 377.
22. Ito, H.; Willson, C. G.; Frechet, J. M. J. *Proc. SPIE* **1987**, *771*, 24.
23. McKean, D. R.; MacDonald, S. A.; Clecak, N. J.; Willson, C. G. *Proc. SPIE* **1988**, *920*, 60.
24. O'Brien, M. J.; Crivello, J. V. *Proc. SPIE* **1988**, *920*, 42.
25. O'Brien, M. J. *Polym. Eng. Sci.* **1989**, *29*, 846.
26. Aso, C.; Tagami, S.; Kunitake, T. *J. Polym. Sci., Part A-1* **1969**, *7*, 497.
27. Ito, H.; Ueda, M.; Schwalm, R. *J. Vac. Sci. Technol.* **1988**, *B6*(6), 2259.
28. Ito, H.; Ueda, M.; Renaldo, A. *J. Electrochem. Soc.* **1989**, *136*, 245.
29. Ito, H. *Proc. SPIE* **1988**, *920*, 33.
30. Ito, H.; Flores, E.; Renaldo, A. *J. Electrochem. Soc.* **1988**, *135*, 2328.
31. Houlihan, F. M.; Bouchard, F.; Fréchet, J. M. J.; Willson, C. G. *Macromolecules* **1986**, *19*, 13.
32. Fréchet, J. M. J.; Eichler, E.; Stanciulescu, M.; Iizawa, T.; Bouchard, F.; Houlihan, F. M.; Willson, C. G. In *Polymers for High Technology*; Bowden, M. J.; Turner, S. R., Eds.; ACS Symposium Series 346; American Chemical Society: Washington, DC, 1987; p 138.
33. Fréchet, J. M. J.; Willson, C. G.; Iizawa, T.; Nishikuko, T.; Igarashi, K.; Fahey, J. In *Polymers in Microlithography*; Reichmanis, E.; MacDonald, S. A.; Iwayanagi, T., Eds.; ACS Symposium Series 412; American Chemical Society: Washington, DC, 1989; p 100.
34. Inaki, Y.; Matsumura, N.; Takemoto, K. In *Polymers for Microelectronics*; Thompson, L. F.; Willson, C. G.; Tagawa, S., Eds.; ACS Symposium Series 537; American Chemical Society: Washington, DC, 1994; p 142.
35. Sooriyakumaran, R.; Ito, H.; Mash, E. A. *Proc. SPIE* **1991**, *1466*, 419.
36. Ito, H.; Sooriyakumaran, R.; Mash, E. A. *J. Photopolym. Sci. Technol.* **1991**, *4*, 319.
37. Ito, H. In *Irradiation of Polymeric Materials*; Reichmanis, E., et al., Eds.; Eds.; ACS Symposium Series 527; American Chemical Society: Washington, DC, 1993; p 197.
38. Ito, H.; Maekawa, Y. In *Polymeric Materials for Microelectronic Applications*; Ito, H., et al., Eds.; Eds.; ACS Symposium Series 579; American Chemical Society: Washington, DC, 1994; p 70.
39. Stöver, H.; Matuszczak, S.; Chin, R.; Shimizu, K.; Willson, C. G.; Fréchet, J. M. J. *Proc. ACS Div. Polym. Mater. Sci. Eng.* **1989**, *61*, 412.
40. Feeley, W. E.; Imhof, J. C.; Stein, C. M.; Fisher, T. A.; Legenza, M. W. *Polym. Eng. Sci.* **1986**, *26*, 1101.
41. Liu, H-Y.; de Grandpre, M. P.; Feeley, W. E. *J. Vac. Sci. Technol.* **1988**, *B6*, 379.
42. Thackeray, J. W.; Orsula, G. W.; Pavelcheck, E. K.; Canistro, D. *Proc. SPIE* **1989**, *1086*, 34.
43. Thackeray, J. W.; Orsula, G. W.; Bohland, J. F.; McCullough, A. W. *J. Vac. Sci. Technol.* **1989**, *B7*(6), 1620.
44. Berry, A. K.; Graziano, K. A.; Bogan, Jr., L. E.; Thackeray, J. W. In *Polymers in Microlithography*; Reichmanis, E., et al., Eds., ACS Symposium Series 412; American Chemical Society: Washington, DC, 1989; p 87.
45. Lingnau, L.; Dammel, R.; Theis, J. *Solid State Technology* **1989**, *32*(9), 105.
46. Lingnau, L.; Dammel, R.; Theis, J. *Solid State Technology* **1989**, *32*(10), 107.
47. Bohland, J. F.; Calabrese, G. S.; Cronin, M. F.; Canistro, D.; Fedynshyn, T. H.; Ferrari, J.; Lamola, A. A.; Orsula, G. W.; Pavelcheck, E.

K.; Sinta, R.; Thackeray, J. W.; Berry, A. K.; Bogan, Jr., L. E.; de Grandpre, M. P.; Feeley, W. E.; Graziano, K. A.; Olsen, R.; Thompson, S.; Winkle, M. R. *J. Photopolym. Sci. Technol.* **1990**, *3*, 355.
48. Thackeray, J. W.; Orsula, G. W.; Rajaratnam, M. M.; Sinta, R.; Herr, D.; Pavelcheck, E. *Proc. SPIE* **1991**, *1466*, 39.
49. Allen, M. T.; Calabrese, G. S.; Lamola, A. A.; Orsula, G. W.; Rajaratnam, M. M.; Sinta, R.; Thackeray, J. W. *J. Photopolym. Sci. Technol.* **1991**, *4*, 379.
50. Ueno, T.; Shiraishi, H.; Hayashi, N.; Tadano, K., Fukuma, E.; Iwayanagi, T. *Proc. SPIE* **1990**, *1262*, 26.
51. Shiraishi, H.; Fukuma, E.; Hayashi, N.; Ueno, T.; Tadano, K.; Iwayanagi, T. *J. Photopolym. Sci. Technol.* **1990**, *3*, 385.
52. Shiraishi, H.; Fukuma, E.; Hayashi, N.; Tadano, K.; Ueno, T. *Chem. Mater.* **1990**, *3*, 621.
53. McKean, D. R.; Clecak, N. J.; Pederson, L. A. *Proc. SPIE* **1990**, *1262*, 110.
54. Stewart, K. J.; Hatzakis, M.; Shaw, J. M. *Polym. Eng. Sci.* **1989**, *29*, 907.
55. Stewart, K. J.; Hatzakis, M.; Shaw, J. M.; Seeger, D. E.; Neumann, E. *J. Vac. Sci. Technol.* **1989**, *B7*, 1734.
56. Dubois, J. C.; Eranian, A.; Datmanti, E. *Proc. Electrochem. Soc.* **1978**, *78–5*, 303.
57. Crivello, J. V. In *Polymers in Electronics*; Davidson, T., Ed.; ACS Symposium Series 242, American Chemical Society: Washington, DC, 1984; p 3.
58. MacDonald, S. A.; Clecak, N. J.; Wendt, H. R.; Willson, C. G.; Snyder, C. D.; Knors, C. J.; Deyoe, N. B.; Maltabes, J.; Morrow, J. R.; McGuire, A. E.; Holmes, S. J. *Proc. SPIE* **1991**, *1466*, 2.
59. Nalamasu, O.; Cheng, M.; Timko, A. G.; Pol, V.; Reichmanis, E.; Thompson, L. F. *J. Photopolym. Sci. Technol.* **1991**, *4*, 299.
60. Kumada, T.; Tanaka, Y.; Ueyama, A.; Kubota, S.; Koezuka, H.; Hanawa, T.; Morimoto, H. *Proc. SPIE* **1993**, *1925*, 31.
61. Oikawa, A.; Santoh, N.; Miyata, S.; Hatakenaka, Y.; Tanaka, H.; Nakagawa, K. *Proc. SPIE* **1993**, *1925*, 92.
62. Röschert, H.; Przybilla, K-J.; Spiess, W.; Wengenroth, H.; Pawlowski, G. *Proc. SPIE* **1992**, *1672*, 33.
63. Funhoff, D. H. J.; Binder, H.; Schwalm, R. *Proc. SPIE* **1992**, *1672*, 46.
64. Przybilla, K. J.; Kinoshita, Y.; Kudo, T.; Masuda, S.; Okazaki H.; Padmanaban, M.; Powlowski, G.; Roeschert, H.; Spiess, W.; Suehiro, N. *Proc. SPIE* **1993**, *1925*, 76.
65. Huang, W.-S.; Kwong, R.; Katnani, A.; Khojasteh, M. *Proc. SPIE* **1994**, *2195*, 37.
66. Hinsberg, W. D.; MacDonald, S. A.; Clecak, N. J.; Snyder, C. D.; Ito, H. *Proc. SPIE* **1993**, *1925*, 43.
67. Ito, H.; England, W. P.; Clecak, N. J.; Breyta, G.; Lee, H.; Yoon, D. Y.; Sooriyakumaran, R.; Hinsberg, W. D. *Proc. SPIE* **1993**, *1925*, 65.
68. Ito, H.; England, W. P.; Sooriyakumaran, R.; Clecak, N. J.; Breyta, G.; Hinsberg, W. D. Lee, H.; Yoon, D. Y., *J. Photopolym. Sci. Technol.* **1993**, *6*, 547.
69. Ito, H.; Breyta, G.; Hofer, D.; Sooriyakumaran, R.; Petrillo, K.; Seeger, D. *J. Photopolym. Sci. Technol.* **1994**, *7*, 433.

CHEMICAL SENSORS

Kurt E. Geckeler and Rongnong Zhou
Institute of Organic Chemistry
University of Tuebingen

Wolfgang Göpel
Institute of Physical and Theoretical Chemistry and Center of Interface Analysis and Sensors

Polymeric materials are gaining increasing interest as sensitive, selective, and stable layers of chemical sensors for the detection of ions, neutral molecules and, in particular, inorganic and organic gases.[1,2] Generally, these materials are based on a variety of functional polymers, preferentially containing a backbone with a high solubilizing power.[3] The main advantage of these materials is the high flexibility for tailoring recognition structures by controlled chemical synthesis and by thin-layer formation.[2–4] In addition, a series of these polymeric materials also has been applied to biosensors.[5]

We summarize recent work on polymers as sensitive layers of chemical sensors for inorganic gases (e.g., CO_2 and NO_2) and organic solvents in air and water. To this end, different transducers based on mass changes, Δm [with quartz microbalance (QMB)], temperature changes, ΔT (with thermopile devices), and capacitance changes, ΔC (with interdigital electrode capacitors) are used.

APPLICATION

QMB Sensors

Detection of CO_2 in Air

Polymers with NH_2 groups can be applied as layers of QMB sensors for the detection of CO_2, according to the reversible acid–base reaction at 70°C.[6]

Detection of NO_2 in Air

The applicability of conducting metal phthalocyanines to monitor NO_2 is well known.[1,7] High sensitivities in the $\mu L\ m^{-3}$ range were reported for conductance sensors with metal phthalocyanines coatings. Polymer-bound metal phthalocyanines containing polystyrene or poly(4-vinylpyridine-*co*-styrene) with covalently bound metal phthalocyaninates of Co(II) and Cu(II) may also be used as chemically sensitive layers of QMB transducers.[7,8] The sensitivities for these QMB sensors are lower than for conductance sensors based on metal phthalocyanines. However, they can be operated at room temperature for the detection of the NO_2 gas in air, whereas the conductance sensors must be operated at high temperatures.

Detection of Organic Solvent Molecules

The bulk dissolution of organic molecules into polymers may be used for their detection in air.[2,6,9,10] Most of the molecule–polymer interactions are completely reversible with response and decay times on the order of minutes at a temperature of 25°C.

Detection of Organic Compounds in Water

Polymer-coated QMB sensors may be used for the detection of chlorinated and aromatic compounds in water.

Capacitive Sensors

Capacitive changes (ΔC) determined with interdigital capacitors contain contributions from changes in the relative dielectric coefficient, ε_3, of the polymer and/or from changes in the effective thickness of the polymer layer, whereas the dielectric coefficients ε_1 and ε_2 of the gas phase and the substrate remain constant.[11] Polar molecules lead to an increased capacitance and nonpolar molecules to a decreased capacitance.

Calorimetric Sensors

Temperature changes during the dissolution of solvent molecules in polymer layers can be determined by using thermopiles as calorimetric transducers.[9,12] Temperature changes are positive during exposure to different concentrations of toluene in air and negative during subsequent exposure to pure air, with negligible analyte concentration. It was found that very polar polymers cannot be used for this type of chemical sensor, because no or only small sensor responses were observed during the exposure of polymer coating to analyte molecules.

REFERENCES

1. *Sensors: A Comprehensive Survey;* Göpel, W., Ed.; VCH: Weinheim, Germany, 1991; Vols. 2 and 3.
2. Göpel, W. *Sensors and Actuators,* B **1995**, *24–25*, 17.
3. Geckeler, K. E.; Pillai, V. N. R.; Mutter, M. *Adv. Polym. Sci.* **1981**, *39*, 65.
4. Schierbaum, K. D.; Göpel, W. *Synth. Met.* **1993**, *61*, 37.
5. Geckeler, K. E.; MŸller, B. *Naturwissenschaften* **1993**, *80*, 18.
6. Zhou, R.; Vaihinger, S.; Geckeler, K. E.; Göpel, W. *Sensors and Actuators B* **1994**, *18–19*, 415.
7. Zhou, R.; Haug, M.; Geckeler, K. D.; Göpel, W. *Sensors and Actuators B* **1993**, *15–16*, 312.
8. Zhou, R.; Tang, L.; Geckeler, K. E.; Göpel, W. *Makromol. Chem. Phys.* **1994**, *195*, 2409.
9. Schierbaum, K. D.; Gerlauch, A.; Haug, M.; Göpel, W. *Sensors and Actuators A* **1992**, *31*, 130.
10. Zhou, R.; Hierlemann, A.; Schierbaum, K. D.; Geckeler, K. E.; Göpel, W. *Sensors and Actuators,* B **1995**, *26–27*, 121.
11. Haug, M.; Schierbaum, K. D.; Enders, H. E.; Drost, S.; Göpel, W. *Sensors and Actuators A* **1992**, *32*, 326.
12. Haug, M.; Schierbaum, K. D.; Gauglitz, G.; Göpel, W. *Sensors and Actuators* B **1993**, *11*, 383.

CHEMICALLY AMPLIFIED RESISTS (for Deep-UV Lithography)

E. Reichmanis and O. Nalamasu
Bell Laboratories
Lucent Technologies

Significant advances are continually being made in microelectronic device fabrication, and especially in lithography, the technique that is used to generate the high-resolution circuit elements characteristic of today's integrated circuits.[1] Today, devices with several million transistor cells are commercially available and are fabricated with minimum features of 0.5 μm or smaller. These accomplishments have been achieved using conventional photolithography (**Figure 1**) (photolithography employing 350–450 nm light) as the technology of choice. Incremental improvements in tool design and performance have allowed the continued use of 350–450 nm light to produce ever smaller features.[2]

The ultimate resolution of a printing technique is governed, at the extreme, by the wavelength of the light (or radiation) used to form the image, with shorter wavelengths yielding higher resolution.[3]

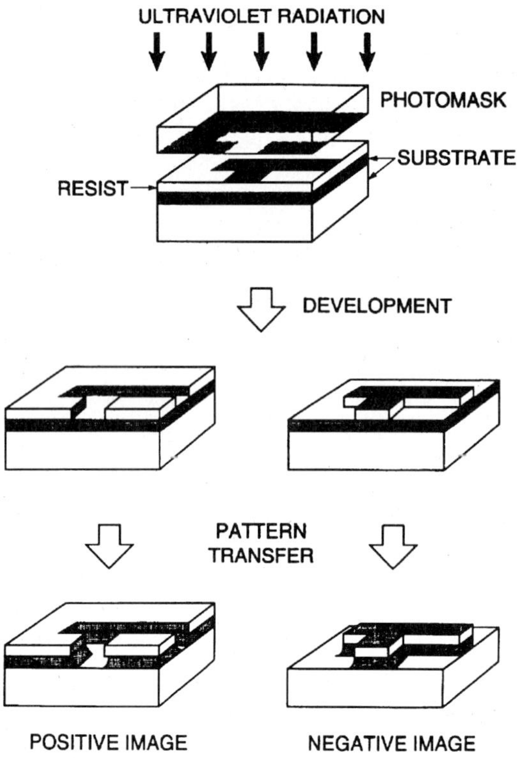

FIGURE 1. The lithographic process.

The introduction of new resist materials and processes will require a considerable lead-time to bring them to the performance level currently realized by conventional positive photoresists.

RESIST DESIGN REQUIREMENTS

These new resists must be carefully designed to meet the specific requirements of each lithographic technology. Although these requirements vary according to the radiation source, device process requirements, and exposure tool design, the following are ubiquitous: sensitivity, contrast, resolution, etching resistance, purity, and manufacturability.[4]

One approach to improving sensitivity involves the concept of chemical amplification,[5,6] which in its most well-known embodiment employs the photogeneration of an acidic species that catalyzes many subsequent chemical events such as deblocking of a protective group or crosslinking of a matrix resin. A chemically amplified resist is generally formulated with three or more elements; a matrix polymer, a photoacid generator, and a moiety capable of effecting differential solubility between the exposed and unexposed regions of the film in a developer, either through a crosslinking reaction or other molecular transformation. These elements may be either discrete molecular entities that are formulated into a multicomponent resist system[5,6] or elements of a single polymer.[7] Additionally, to the extent possible, new resists should be designed around a polymer platform that is extendable to future systems.

CHEMICALLY AMPLIFIED RESIST CHEMISTRY

Acid Generator Chemistry

While some research regarding base catalyzed systems has been reported,[8,9] the predominant chemistry associated with chemically amplified resists involves acidolytic reactions. The dominant ionic photogenerators of acid are a class of materials called onium salts developed by Crivello and co-workers.[5,10–12]

Onium salts are thermally stable (typically >150°C), allow photogeneration of a wide variety of acids (including such strong inorganic acids as hexafluoroarsenic and hexafluoroantimonic acids and the strong organic acid, triflic acid), and may be structurally modified to alter their spectral absorbtion characteristics.

Many systems described in the literature concern the photogeneration of acid from nonionic compounds. Many of these involve the generation of sulfonic acids that are strong organic acids with reasonably low nucleophilicity.

A system that is reported to generate methanesulfonic acid with a high quantum yield when used in conjunction with a novolac resin is 1,3,5-*tris*(methanesulfonyloxy)benzene. UV spectroscopic studies showed that the novolac resin strongly sensitized the substituted benzene towards acid generation, presumably via a charge-transfer intermediate.[13]

Other materials that have been used as photoacid generators in chemically amplified resist compositions include imino sulfonates,[14] 4-nitrobenzenesulfonic acid derivatives,[15] disulfone compounds,[16] sulfonyl substituted diazomethanes,[17] arylmethyl sulfones,[18] and aryl-*bis*-trichloromethyl-*s*-triazines.[19]

Crosslinking Chemistry

Chemical amplification through acid-catalyzed crosslinking for negative working resist applications has been achieved through various mechanisms. These include cationic polymerization, condensation polymerization, electrophilic aromatic substitution, and acid-catalyzed rearrangement. The acid species may be generated from either ionic materials, such as onium salts, or nonionic precursors.

Cationic Polymerization Mechanisms

The first chemically amplified resist systems to be developed were those based on the cationic polymerization of epoxy materials.[5,20] In general, resolution of sub-0.5 μm features in resists based on this mechanism is difficult due to distortion resulting from solvent-induced swelling of the irradiated regions.

Condensation Polymerization Mechanisms

Condensation polymerization mechanisms are probably the most prevalent in the design of chemically amplified negative resists. Such resist systems consist of three essential components: a polymer resin with reactive site(s) (also called a binder) for crosslinking reactions (e.g., a polymer containing a hydroxy functionality); a radiation-sensitive acid generator; and an acid-activated crosslinking agent.[21–25] The photogenerated acid catalyzes the reaction between the resin and crosslinking agent to afford a crosslinked polymer network that is significantly less soluble than the unreacted polymer resin. Additionally, these resists do not show swelling, as the development is done in aqueous-base developers.

Electrophilic Aromatic Substitution Mechanisms

Photo-induced crosslinking can be achieved in styrene polymers that are susceptible to electrophilic aromatic substitution by addition of a latent electrophile, that is, a carbocation precursor and a photoacid generator.[26,27] The photogenerated acid reacts with the latent electrophile during a post-exposure bake step to generate a reactive carbocation that then reacts with an aromatic moiety in the matrix affording a crosslinked network. The latent electrophile may be either an additive or a monomer that is copolymerized into the polymer binder. Examples of latent electrophiles include dibenzylacetate and copolymers of acetyloxymethylstyrene[26,27] and 1,3-dioxlane blocked benzaldehyde.[28]

Deprotection Chemistry

The pioneering work relating to the development of chemically amplified resists based on deprotection mechanisms was carried out by Ito et al.[6] These initial studies dealt with the catalytic deprotection of poly(4-*tert*-butoxycarbonyloxystyrene) (PTBS) in which the thermally stable, acid-labile *tert*-butoxycarbonyl group is used to mask the hydroxyl functionality of polyvinylphenol.[29–31] These resists are sensitive to deep UV and electron-beam irradiation and may be sensitized to longer wavelengths through the addition of appropriate mid- and near-UV dyes.

Alternative resins have been investigated for chemically amplified resist applications. The parent polymer is typically an aqueous-base-soluble high T_g resin. Examples include polyhydroxystyrene,[31] poly(vinyl benzoate),[32] poly(methacrylic acid),[33] *N*-blocked maleimide–styrene resins,[34,35] and poly(hydroxyphenyl methacrylate).[36]

A matrix polymer that combines the advantages of chemically amplified deprotection with conventional chain scission processes is poly(4-t-butoxycarbonyloxystyrene-sulfone) (PTBSS).[37,38] The inclusion of sulfur dioxide into the backbone of the polymer affords a high T_g that gives greater flexibility for processing. When exposed to X-ray irradiation, PTBSS is an effective single-component chemically amplified resist.[7,39] Radiation-induced C–S bond scission leads to generation of either sulfinic or sulfonic acid end groups that subsequently induce the deprotection reaction.

The choice of protective group is equally wide. Clearly this group should be thermally stable and acid labile. Examples of substituents that have been employed include *tert*-butyl,[45] tetrahydropyranyl,[40–45] dihydropyranyl, and α-α-dimethylbenzyl.[33,41,46] Hydrolyzable groups such as trimethyl silyl[47–50] and various acetals and ketals have also been used.[40,41,51,52]

Depolymerization Chemistry

Chemically amplified resists that act through a polymer depolymerization mechanism can be broadly divided into two classes: those that act through a thermodynamically induced depolymerization mechanism and those requiring catalytic cleavage of a polymer backbone. The former process depends upon the use of low ceiling temperature polymers that have been stabilized by suitable end capping. Introduction of a photocleavable moiety either at the end-cap or along the polymer backbone may then allow depolymerization to take place after irradiation and mild heating. A variant of this approach makes use of an end cap, or polymer chain that may be cleaved by photogenerated acid.

The first CA resists that operated by a thermodynamically induced depolymerization were based on polyacetals (or polyaldehydes.[31,32,53] These polymers have very low ceiling temperatures, but can be stabilized by end capping.[54]

The second type of chemically amplified depolymerization resist mechanism depends upon the incorporation of C–O bonds into the polymer backbone, which can be cleaved by either hydrolysis or acidolysis.

CD Sensitivity to Post-Exposure Bake (PEB)

Since the photoacid diffusion distance and the deprotection turnover rate (catalytic chain length) are PEB temperature-dependent, variations in PEB temperature affect the CD (critical dimension) linewidth.

We have previously investigated the PTBSS polymer with nitrobenzyl ester and onium salt PAGs.[55] The sensitivity with a nitrobenzyl ester PAG was a strong function of the PEB temperature. PTBSS with triphenyl sulfonium hexafluorarsenate registered no change in its sensitivity over the 90–130°C PEB range. Resist materials with low or no sensitivity to PEB temperature have been described in literature.[56,57] For example, the ARCH resist material shows no change in linewidth of 0.30 µm features over 12°C PEB range.[58] This performance will become the "norm" as new chemistries are developed for sub-0.25-µm lithographic applications.

SUMMARY

Since the conception of chemical amplification mechanisms for microlithographic applications a decade ago, increasing attention has been given to such processes because they provide advantages in terms of sensitivity and contrast with minimal increase in process complexity. The original work in chemically amplified resists has spawned many research efforts to define chemistries appropriate for matrix materials and photogenerators of catalysts, primarily strong acids. These efforts have led to the development and commercialization of several positive and negative-tone chemically amplified resists. Significant challenges exist today that relate to both fundamental and applied materials chemistry, and to understanding the processes associated with the use of chemically amplified materials.

REFERENCES

1. *Electronic and Photonic Applications of Polymers*; Bowden M. J.; Tumer, S. R., Eds.; Advances in Chemistry Series 218; American Chemical Society: Washington, DC, 1988.
2. McCoy, J. H.; Lee, W.; Varnell, G. L. *Solid State Technol.* **1989,** *32*(3), 87.
3. Thompson, L. F.; Willson, C. G.; Bowden, M. J. In *Introduction to Microlithography*; Symposium Series 219; American Chemical Society Washington, DC, 1983.
4. Thompson, L. F.; Bowden, M. J. In *Introduction to Microlithograph* Symposium Series 219, American Chemical Society: Washington, DC, 1983; pp 162–214.
5. Crivello, J. V. In *Polymers in Electronics*; Symposium Series 242, Davidson, T., Ed.; ACS, Washington, DC, 1984; pp 3–10.
6. Ito, H.; Willson, C. G. In *Polymers in Electronics*; Symposium Series 242; American Chemical Society: Washington, DC. 1984; pp 11–23.
7. Novembre, A. E. *Chem. Mater.* **1992,** *4*, 278.
8. Frechet, J. M. J.; Cameron, J. F. *J. Am. Chem. Soc.* **1991,** *113*, 4252.
9. Winkle, M. R.; Graziano, K. A. J. *Photopolym. Sci. Technol.* **1990,** *3*(3), 419–422.
10. Crivello, J. V.; Lee, J. L.; Conlon, D. A. *Makromol. Chem. Makromol. Symp.* **1988,** *13/14*, 145.
11. Crivello, J. V.; Lam, J. H. W. *Macromolecules* **1977,** *10*, 1307.
12. Crivello, J. V.; Lam, J. H. W. *J. Polym. Sci. Polym. Chem. Ed.* **1979,** *17*, 977.
13. Schlegel, L. et al. *Chem. Mater.* **1990,** 2, 299.
14. Shirai, M. et al. *Makromol. Chem.* **1989,** *190*, 2099.
15. Yamaoka, T. et al. *J. Photopolym. Sci. Technol.* **1990,** *3*, 275.
16. Aoai, T. et al. *Photopolym. Sci. Technol.* **1990,** *3*, 389.
17. Pawlowski, G. *Proc. SPIE* **1990,** *1262*, 16.
18. Novembre, A. E. et al. *Proc. Reg. Tech. Conf. Photopolym.* 1991, October 28–30, 41–50.
19. Buhr, G.; Dammel, R.; Lindley, C. R. *Proc. Polym. Mat. Sci. Eng.* **1989,** *61*, 269.
20. Stewart, K. J. *J. Va. Sci. Technol.* **1989,** *B7*, 1734.
21. Lingnau, J.; Dammel, R.; Theiss, J. *Solid State Technol.* **1989,** *32*(9), 105–112.
22. Lingnau, J.; Dammel, R.; Theiss, J. *Solid State Technol.* **1989,** *32*(9), 107-112.
23. Feely, W. E. Eur. Patent Appl. 232 972, 1980.
24. Buhr, G., U.S. Patent 4 189 323, 1980.
25. Berry, A. K. In *Polymers in Microlithography;* Reichmanis, E.; MacDonald, S. A.; Iwayanagi, T., Eds.; Symposium Series 412; American Chemical Society: Washington, DC 1989; pp 87–99.
26. Reck, B. et al. *Polym. Eng. Sci.* **1989,** *29*, 960–964.
27. Frechet, J. M. J. et al. In *Polymers in Microlithography*; Reichmanis, E.; MacDonald, S. A.; Iwayanagi, T., Eds.; Symposium Series 412; American Chemical Society: Washington, DC 1989; pp 74–85.
28. Schaedeli, U. et al. *Proc. Reg. Tech. Conf. Photopolym.* **1991,** October 28-30, 145 Mid-Hudson Section, SPE.
29. Frechet, J. M. J. et al. *Polym.* **1980,** *24*, 995.
30. Ito, H. et al. *Macromolecules* **1983,** *16*, 1510.
31. Ito, H.; Willson, C. G. *Polym. Eng. Sci.* **1983,** *23*, 1012.
32. Ito, H.; Willson, C. G. *Proc. SPIE* **1987,** *771*, 24.
33. Ito, H.; Ueda, M.; Ebina, M. In *Polymers in Microlithography* Reichmanis, E.; MacDonald, S. A.; Iwayanagi, T., Eds.; Symposium Series 412; American Chemical Society: Washington, DC, 1989; pp 57.
34. Osuch, C. E. et al. *Proc. SPIE* **1986,** *631*, 68.
35. Turner, S. R.; Ahn, K. D.; Willson, C. G. In *Polymers for High Technology*; Bowden, M. J.; Turner, S. R., Eds.; Symposium Series 346; American Chemical Society: Washington, DC, 1987; pp 200–210.
36. Przybilla, K. J. et al. *Proc. Req. Tech. Conf. Photopolym.* **1991,** October 28–30, 131 Mid-Hudson Section SPE.
37. Kanga, R. S. et al. *Chem. Mater.* **1991,** *3*, 66–667.
38. Houlihan, F. M. et al. In *Polymers in Microlithography*; Reichmanis, E.; MacDonald, S. A.; Iwayanagi, T., Eds.; Symposium Series 412; American Chemical Society: Washington, DC, 1989; 39–56.
39. Novembre, A. E. et al. *J. Vac. Sci. Technol.* **1991,** *B9*, 3338.
40. Hayashi, N. et al. *Proc. Polym. Mat. Sci. Eng.* **1989,** *61*, 417.
41. Frechet, J. M. J. et al. *Polym. Bull.* **1988,** *20*, 427.
42. Frechet, J. M. J. et al. In *The Effects of Radiation on High-Technology Polymers*; Reichmanis, E.; O'Donnell, J. H. Eds.; Symposium Series 381; American Chemical Society: Washington, DC, **1989**; pp 155–171.

43. Hesp, S. A. M.; Hayashi, N.; Ueno, T. *J. Appl. Polym. Sci.* **1991**, 42, 877–883.
44. Schlegel, L. et al. *Microelectronic Eng.* **1991**, *14*, 227-236.
45. Taylor, G. N. et al. *J. Vac. Sci. Technol.* **1991**, *B9*, 3348.
46. Ito, H.; Ueda, M. *Macromolecules* **1988**, *21*, 1475.
47. Yamaoka, T. et al. *Proc. Regional Technical Conf. Photopolym.* 1988, October 30–November 2, 27 Mid-Hudson Section SPE.
48. Murata, M. et al. *J. Photopolym. Sci. Technol.* **1992**, *5*(1), 79.
49. Uhrich, K. E. et al. *Chem. Mater.* **1994**, *6*, 287.
50. Schue, E; Giral, L. *Makromol. Chem. Macromol. Symp.* **1989**, *24*, 21.
51. Padmanabhan, M. et al. *Proc. SPIE* **1994**, *2195*, 61.
52. Schwalm, R. et al. *Proc. SPIE* **1994**, *2195*, 2.
53. Ito, H.; Willson, C. G. *Proc. Regional Technical Conf. Photopolym.* **1982**, 331, Mid-Hudson Section, SPE.
54. Vogl, O. *J. Polym. Sci., Polym. Chem. Ed* **1960**, *46*, 261.
55. Nalamasu O. et al. *J. Photopolym. Sci. Technol.* **1991**, *4*, 299.
56. Huang, W-S. et al. *Proc. SPIE* **1994**, *2195*, 37.
57. Houlihan, F. M. et al. *Chem. Mater.* **1991**, *3*, 462.
58. Nalamasu, O. et al. *Proc. Regional Tech. Conf. Photopolym.* 1994, October 30–November 2, Mid-Hudson Section SPE.

CHEMICALLY AMPLIFIED RESISTS (Introduction and Recent Developments)

Graham D. Darling and A. M. Vekselman
Department of Chemistry
McGill University

Advanced microlithographic technologies use *chemically amplified (CA) resists* to produce electronic or photonic circuits and devices, including higher-density components such as I GB DRAMs, 150 MHz microprocessors, or 10 GHz optical couplers.

Resist (photoresist, microresist) has become a general term for a polymer-based material that undergoes a chemical change following brief deep ultraviolet (DUV) or other irradiation. It is typically applied by *spin coating* the solution in *casting solvent* onto a silicon wafer or other substrate, followed by *post-apply baking* (PAB) to produce a thin dry film. This, during *exposure*, captures a briefly-projected radiation image of a circuit pattern (or diffraction grating, etc.) as a *latent image* of altered-unaltered material.

In earlier generations of resists (often based on novolok phenol-formaldehyde polymers), each absorbed quantum of radiation induced on average less than one chemical reaction within the material. In CA resists however, the *primary photochemical event* produces a mobile catalyst that, typically during later *post-exposure baking (PEB)*, goes on to induce a cascade of material-transforming *secondary catalytic events* within a 5–25 nm radius. Such *chemical amplification* thus makes possible an *overall quantum yield*—the number of material reactions divided by number of absorbed photons—of up to several hundred.

Along with their higher *sensitivity* and *contrast* in forming images, CA are also better than earlier resists in being more flexible in design and formulation, versatile in radiation source (electromagnetic or particle beams), and compatible with *dry* (plasma), multilayer, and other advanced pattern-transfer techniques.

The first CA resist for microlithography was reported by the university industry team of Fréchet (University of Ottawa) and Willson and Ito (IBM Inc.),[1] who combined onium salts with the acid-deprotectable poly(4-[*t*-butyloxycarbonyloxy]styrene) (poly-TBOCST), and coined the term *chemical amplification.*

PHOTOCATALYST GENERATORS

The usual photo-supplied catalyst has been strong acid. Triarylsulfonium and diaryliodonium salts have become the standard PAG ingredients in CA resist formulations, because of their generally easy synthesis, thermal stability, high quantum yield for acid (and also radical) generation, and the strength and nonvolatility of the acids they supply.[2] Simple onium salts are directly sensitive to DUV, X-ray and electron radiations, and can be structurally tailored, or mixed with photosensitizers, to also perform well at mid-UV and longer wavelengths. An aromatic polymer matrix alone can also often act as a photosensitizer, presumably through electron transfer within 1.6–2.0 nm.[3] Mechanistic studies have shown onium salt photodecomposition to proceed via both homolytic and heterolytic pathways.[4,5]

Photoradical generators are better known for photopolymerization applications than for microlithographic resists.

CATALYZED REACTIONS

Many reactions can amplify the primary photochemical event of catalyst generation. Besides photopolymerization, whose contrast is often insufficient for microlithography, these include rearrangement, crosslinking, and chain scission/depolymerization.

Rearrangements resulting in solubility changes can further be characterized as nonpolar-to-polar, polar-to-nonpolar, and one of i or ii after the other (*polarity reversal*).[6] Depending on the polarity of a developing liquid, either the exposed or unexposed areas can be finally removed to produce *positive-tone* or *negative-tone* relief images respectively (*dual-tone developability*).

Crosslinkable CA resists have formed negative-tone images mostly through acid-catalyzed electrophilic aromatic substitution reactions. Thus, protonation of small polyfunctional or polymer-bound benzylic ethers, esters, alcohols, or 1,3-dioxolane groups forms carbocations that readily attack polymer phenolic groups.[7–10] Exposed areas thus become heavily crosslinked and insoluble, and sometimes, even unswellable.

Photoacid-catalyzed cleavage of tertiary or allylic ether, ester or carbonate moieties along a polymer backbone leads to chain scission that increases the solubility of exposed areas, for positive-tone images. Photolytic decapping of end-protected polyacetals with low ceiling temperature leads to a depolymerization cascade. Strong acid from irradiated PAG can also lead to simultaneous decapping, mid-chain scission and depolymerization of similar polymers.[11,12]

CONCLUSION AND OBJECTIVES

Whatever new high-resolution radiation sources or techniques will occupy the market in the next decades, CA resists will continue to advance their important idea of *amplifying by catalysis the primary events of irradiation.*[13] Current opportunities for research and development in this area include:

- developing, understanding, and applying new chemistry to CA resists;
- exploring "super amplification" resist designs, where acid-formation and/or resist development can be made auto-accelerating;
- engineering deep- and vacuum-UV, X-ray and E-beam exposure tools, and CA resists that best fit each;
- progressing in dry-developable, multilayer, and top-surface resist techniques;
- developing environment-friendly, water-castable, and water-developable resists;

Along with their mainstream use in microlithography, the relief and functional imaging of CA resists can be useful in any field where submicron features need to be created. Possible novel applications include:

- micro-optical elements for waveguide circuits and other photonics (spatial control of material refractive index, optical rotation, nonlinear optical properties);
- holographic-, photo- and data-recording (photorefractive or photochromic) media, high-resolution color displays (precise positioning of dye spots);
- bioapplications (array biosensors and array biosyntheses);
- templates for space-resolved chemical and physicochemical processes; and
- patterned polymer-inorganic or polymer-organic composite structures.

REFERENCES

1. Frechet, J. M. J.; Eichler, E.; Willson, C. G.; Ito, H. *Polymer* **1983**, *24*, 995.
2. Crivello, J. V. In: *Initiators—Poly-reactions—Optical activity*; Springer-Verlag: Berlin, Advances in polymer science *62*, 1, 1982.
3. Hacker, N. P.; Hofer, D. C.; Welsh, K. M. *J. Photopolym. Sci. Technol.* **1992**, *5*, 35.
4. Dektar, J. L.; Hacker, N. P. *J. Org. Chem.* **1990**, *55*, 639.
5. Dektar, J. L.; Hacker, N. P. *J. Am. Chem. Soc.* **1990,** *112*, 6004.
6. Ito, H. In *Irradiation of Polymeric Materials*; American Chemical Society: Washington, DC, USA, ACS Symp. Ser. 527, 1993; 197.
7. Frechet, J. M. J.; Matuszczak, S.; Lee, S. M.; Fahey, J.; Willson, C. G. *Polym. Eng. Sci.* **1992**, *32*, 1471.
8. Fahey, J. T.; Shimizu, K.; Frechet, J. M. J.; Clecak, N.; Willson, C. G. *J. Polym. Sci.* **1993**, *31*, 1.
9. Kocon, W. W.; Shacham, D. Y.; Frechet, J. M. J.; Fahey, J. *J. Vac. Sci. Technol., B* **1992**, *10*, 2548.
10. Schaedeli, U.; Holzwarth, H.; Muenzel, N.; Schulz, R. *Polym. Eng. Sci.* **1992**, *32*, 1523.
11. Frechet, J. M. J. et al. In *Polymers in Microlithography*; American Chemical Society: Dallas, Texas, USA, ACS Symp. Ser. 412, 1989; 100.
12. Tsuda. M.; Hata, M; Nishida, R.; Oikawa, S. *J. Photopolym. Sci. Technol.* **1993**, *6*, 491.
13. Stix, G. *Scientific American* **1995**, *2722*, 90.

Chemiluminescent Materials

See: Signal Transduction Composites (Biomaterials with Electroactive Polymers)
Stress-Induced Chemiluminescence Imaging

CHEMODEGRADATION (NO_X)

K. L. DeVries and M. C. Perry
University of Utah

The oxidation of polymers has been studied extensively over the years. This review focuses on the study of polymer degradation in NO_x gas. Nitrogen oxides are common by-products associated with fossil fuel combustion and form one of the most significant and costly types of air pollution. Nitric oxide, the primary species formed during combustion, reacts in the atmosphere in the presence of UV radiation to form various oxides of nitrogen, ozone, and acids. These species cause formidable economic and environmental losses that result in a reduced product life for consumers and the increased costs of researching, developing, and producing more resistant materials. With the ever expanding use of polymers in recent years, the costs of degradation are significant.

Nitrogen dioxide, an odd electron molecule, is highly reactive and can couple with itself. At standard temperature and pressure, it is in equilibrium with nitrogen tetroxide, with NO_2 comprising 16 wt%. The decomposition to nitric oxide in diatomic oxygen occurs at higher temperatures, but at standard conditions, NO_2 is more stable. In the atmosphere, the oxides of nitrogen react with other naturally occurring and pollutant gases to play a role in smog, acid rain, and ozone depletion. Ultraviolet radiation decomposes NO_2 to give NO and atomic oxygen. Although the complex of nitrogen oxides are generally associated, the terms NO_x and NO_2 are used interchangeably throughout this review.

Reactions involving water produce nitric and nitrous acids. These reactions are relatively fast with stability favoring acid formation. These reactions may occur in the atmosphere or within the polymer. In either case, the acid products may then be involved further in attacking the polymer, particularly hydrolytically.

Although this treatise is comprehensive, particular attention is devoted to the effects of pollutant gases on polymers, which two independent researchers, their graduate students, and colleagues studied.

Recently, the graduate students of K. L. DeVries studied the effects of pollutant gases, moisture, UV radiation, and stress on various polymers. These researchers focused more on the synergisms between applied mechanical stress and environment, the differences between the degradation of high-strength polymers (aligned molecules) and bulk polymers, and the relationship between molecular occurrences such as chain scission as indicated by electron spin resonance (free radical production) and changes in IR spectrum with macroscopic effects such as strength loss modulus modification and changes in the creep lifetimes.[1–3]

CONCLUSIONS

Two classes of polymers can be distinguished by their reactivity with NO_X gas. Saturated polymers are relatively stable in the presence of NO_X gas, except for those with amide linkages. Cast films and highly drawn nylon materials are especially susceptible to chain scission and a loss of mechanical properties, but the more crystalline Kevlar fiber is affected to a lesser extent. Unsaturated polymers, however, are very susceptible to attack by NO_X gas.

The diffusion of the pollutant gases in the various polymers determines their sensitivity to these gases. Although not considered in detail here, this aspect of the degradation process must be understood to extrapolate the results of studies performed at higher gas concentrations to applications in which the pollutant concentrations are relatively low.[4] Jellinek points out that linear extrapolation from high-concentration test results to low-concentration conditions may be in error because the gas diffusion at lower concentrations may be rate determining.[5] Furthermore, the diffusion behavior will be affected by each polymer's specific crystallinity and morphology.

The combined effects of residual or applied stress and exposure to NO_X gas may be more deleterious to the molecular degradation and mechanical properties than the sum of these effects measured individually. This synergism may be envisioned as a thermally activated bond rupture in which the activation energy required for chain scission is aided by the applied stress and the environmental agent (NO_X gas). Rapid mechano-chemical degradation prevails in high-performance polymers in which high strength and modulus are achieved by aligning covalently bonded chains. The fracture of these materials involves considerable chain scission.[3] Bulk or randomly oriented polymers generally show little synergistic stress-environment effects, because their fracture involves more creep and failure of secondary bonds.

ACKNOWLEDGMENTS

We wish to thank the National Science Foundation (Grant #DMR-9014565 and CM5 9522743) for their support of this research.

REFERENCES

1. Salisbury, D. M. Ph.D. Thesis, University of Utah, 1989.
2. Salisbury, D. M.; DeVries, K. L. Mech. Plast., Plast. Comp. 1989, 191.
3. DeVries, K. L.; Igarashi, M. In *An ESR Investigation of Environmental Effects on Nylon Fibers, In Polymer Characterization by ESR and NMR*; Woodward, A. E.; Bovey, F. A. Eds.; ACS Symposium Series 142; American Chemical Society: Washington, DC, 1980.
4. Jellinek, H. H. G.; Igarashi, S. *J. Phys. Chem.* **1970**, *74*(7), 1409.
5. Jellinek, H. H. G. In *Aspects of Degradation and Stabilization of Polymers*; Jellinek, H. H. G., Ed.; Elsevier Scientific: New York, 1978; pp 431–500.

Chiral Catalysts

See: Asymmetric Polymerization (Overview)
Asymmetric Synthesis (Using Chiral Polymers From (+)-Camphor)
Supported Chiral Catalysts

CHIRAL POLYANILINES

Leon A. P. Kane-Maguire, Mir Reza Majidi, and Gordon G. Wallace
Intelligent Polymer Research Laboratory
Department of Chemistry
University of Wollongong

There has also been considerable interest in the development of chiral-conducting polymers, largely because of their potential as chiral electrodes in electrochemical asymmetric synthesis. Such polymers may also possess specific electromagnetic properties.[1] The first chiral polypyrroles were reported in 1985,[2,3] and subsequent studies have continued to focus on polypyrroles[4–8] and polythiophenes.[4,9–14] These have been generally synthesized by the electropolymerization of monomers in which a chiral substituent, for example an amino acid,[6] has been attached to either the 3-position of the heterocycle ring or to the pyrrole *N*-atom. The high optical activity observed for many of these complexes has been ascribed to the generation of a helical polymer backbone, with a preponderance of one screw sense induced by the optically active side chain.[4,6,8,10] The presence of an optically active side chain does not always lead to main-chain chirality. For example, polypyrroles *N*-substituted with (+)-camphor-sulfonate derivatives only exhibited optical activity due to the chiral substituent.[2]

These routes are generally synthetically tedious and the bulk and electronic properties of the chiral substituents can have a deleterious effect upon the electroactivity, conductivity, and mechanical properties of the polymers.

In 1993 and in subsequent publications we reported the first and remarkably facile synthesis of optically active polyaniline via enantioselective electropolymerization of the achiral monomer, aniline, in the presence of either (+) or (–)-camphorsulfonic acid (HCSA).[15–17] The polyaniline was deposited on ITO-coated glass electrodes as a film of the dark green, conducting polyemeraldine salt **Scheme I**, where A^- is either (+)- or (–)-CSA^-.

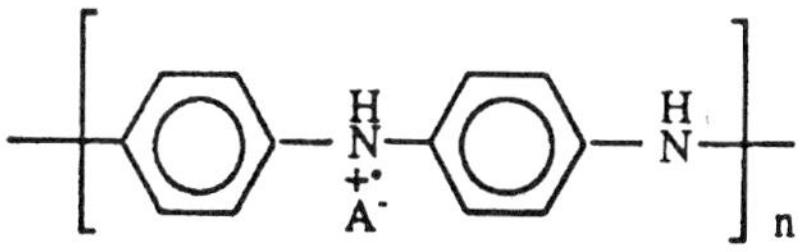

I

We have subsequently found that optically active emeraldine salts of type I can also be readily generated chemically in solution by the enantioselective acid doping of neutral emeraldine base (EB) with either (+)- or (–)-HCSA in various solvents N-methyl-2-pyrrolidinone (NMP), dimethylformamide (DMF), Me_2SO (DMSO), $CHCl_3$.[18,19] We have observed marked changes in the circular dichroism (CD) spectra of such solutions depending on the solvent's nature, which indicate major solvent-mediated conformational changes in the polymer backbone.[19]

We also have CD spectral evidence that other optically active acids of suitable geometry and bifunctionality (such as (+)-tartaric acid) may be used to generate optically active polyaniline salts.[19]

PROPERTIES

Electroactivity

The conductive and electroactive nature of the emeraldine salts (V, A^- = (+)- or (–)- CSA^-) was evident from cyclic voltammetry either during polymer growth or on the deposited polymer films. The well-defined oxidation and reduction processes observed are typical of polyaniline materials.[20] Similar cyclic voltammetric studies of the analogous polytoluidine salts showed that they are considerably less electroactive than the polyanilines.[21]

APPLICATIONS

Our initial search for routes to chiral polyanilines was motivated by our interest in exploring their potential applications as chiral electrodes in asymmetric electrosynthesis and as novel voltage-sensitive stationary phases for the chromatographic separation of racemic chemicals.

We have recently commenced studies of the efficacy of optically active salts (V) as chiral electrodes in the asymmetric electro-oxidation of organo sulfides to the corresponding sulfoxides. We'll continue studies with this and related substrates.

REFERENCES

1. Engheta, N.; Pelet, P. *Opt. Lett.* **1989**, *14*, 593.
2. Salmon, M.; Bidan, G. *J. Electrochem. Soc.* **1985**, *132*, 1897.
3. Elsenbaumer, R. L.; Eckhardt, H. et al. *Mol. Cryst. Liq. Cryst.* **1985**, *118*, 111.
4. Kotkar, D.; Joshi, V.; Ghosh, P. K. *J. Chem. Soc., Chem. Commun.* **1988**, 917.
5. Salmon, M.; Saloma, M. et al. *Electrochim. Acta* **1989**, *34*, 117.
6. Delabouglise, D.; Gamier, F. *Synth. Met.* **1990**, *39*, 117.
7. Kato, T.; Gondaira, M. et al. *Chem. Lett.* **1991**, 713.
8. Moutet, J.-C.; Saint-Aman, E. et al. *Adv. Mater.* **1992**, *4*, 511.
9. Lemaire, M.; Delabouglise, D. et al. In *Recent Advances in Electroorganic Synthesis*; Torii, S., Ed.; Kodansha: Tokyo, Japan, 1987.
10. Lemaine, M.; Delabouglise, D. et al. *J. Chem. Soc., Chem. Commun.* **1988**, 658.
11. Roncali, J.; Garreau, R. et al. *Synth. Met.* **1989**, *28*, C341.
12. Lemaire, M.; Garreau, R. et al. *New. J. Chem.* **1989**, *13*, 863.
13. Andersson, M.; Ekeblad, P. O. et al. *Polym. Commun.* **1991**, *32*, 546.
14. Bouman, M. M.; Meijer, E. W. *Polym. Prepr.* **1994**, *35*, 309.
15. Majidi, M. R.; Kane-Maguire, L. A. P.; Wallace, G. G. *3rd Pacific Polym. Conf.* 1993; pp 835–836.
16. Kane-Maguire, L. A. P.; Majidi, M. R.; Wallace, G. G. Aust. Prov. Patent PM0835, 1993.
17. Majidi, M. R.; Kane-Maguire, L. A. P.; Wallace, G. G. *Polymer* **1994**, *35*, 3113.
18. Majidi, M. R.; Kane-Maguire, L. A. P.; Wallace, G. G. *Polymer* **1996**, *37*, 359.
19. Majidi, M. R.; Kane-Maguire, L. A. P.; Wallace, G. G. *Synth. Met.*, submitted for publication.
20. Genies, E. M.; Tsintavis, C. *J. Electroanal. Chem.* **1986**, *127*, 200.
21. Majidi, M. R.; Kane-Maguire, L. A. P.; Wallace, G. G., unpublished results.

Chiral Polymers

See: Optically Active Polymers

Chitin

See: Chitin (Metal Ion Chelation and Enzyme Immobilization)
Chitin and Chitosan, Graft Copolymers
Chitin and Derivatives
Chitin Chemistry
Chitin Synthesis
Chitinases
Chitosan (Preparation, Structure, and Properties)
Vegetal Biomass (3. Polymers, Derivatives and Composites)
Wound Dressing Materials

CHITIN (Metal Ion Chelation and Enzyme Immobilization)

Riccardo A. A. Muzzarelli
Faculty of Medicine
University of Ancona

The high capacity of chitin and chitosan for transition metal oxyanions (molybdate, tungstate, and vanadate) and the chelating ability for transition metal cations, accompanied by their indifference to alkali and alkali–earth metal ions, Mn(II) and Tl(I), is well documented in terms of collection capacity and distribution coefficients.[1] Chlorocomplexes of the precious metals are collected easily on chitosan and used as silica-supported catalysts for hydrogenation reactions; similarly, photoinduced formation of gold metal films can be obtained.

When using transition metal ions other than oxyanions and chlorocomplexes, chitin and chitosan, as they are produced today, cannot compete with ion-exchange resins, especially in terms of their regeneration, rigidity, solubility, and microbiological spoilage.

Transition metal ion chelates of high crystallinity are obtained with well-crystallized hydrated chitosan polymorph.

The use of chitin or chitosan for metal ion recovery and separation demands functionalization of the polysaccharide to implement the electron dative power of the primary amine and alcohol groups, and optional crosslinking to depress solubility and enhance resistance to bacterial attack. N-Carboxymethyl chitosan, to which a significant freedom of the glycine residues imparts low conformational rigidity, is much more effective than chitosans in binding Cu(II) and Pb(II).

O-CARBOXYMETHYL CHITIN

This is obtained from sodium monochloroacetate under alkaline conditions. Therefore, the groups of interest are primary amine, carboxylate, and alcohol. The carboxyl group can easily approach the C3 hydroxyl group of the neighboring residue and form a site for complexation with a metal ion.

N-CARBOXYMETHYL CHITOSAN

N-Carboxymethylchitosan from crab and shrimp chitosans is obtained in water-soluble form by proper selection of the reactant ratio, i.e., with equimolar quantities of glyoxylic acid and amino groups. The product is in part N-mono-carboxymethylated (0.3), free amine and N-acetylated depending on the starting chitosan.

Temperature and degree of substitution of N-carboxymethylchitosan exert much less effect than does pH on chelation. In all cases the glycine residue is involved in chelation. At least in the case of Cu, both carboxyl and amino groups are involved in the chelate formation. Lowering the pH depresses the binding ability of N-carboxymethylchitosan, especially toward Cu and Pb.

Transition metal ions are chelated by insoluble forms of N-carboxymethylchitosan with formation of colored and generally insoluble products, depending on pH values.

Crosslinking N-carboxymethylchitosan with epichlorohydrin enhances the chelating ability of the insoluble and amorphous product. The chelating ability of the crosslinked N-carboxymethylchitosan can be used to remove transition metal ions from brines, drinking waters, and nuclear plant effluents, some of the most challenging cases in water reclamation. The crosslinked N-carboxymethylchitosan is in fact more effective than ion-exchange resins and activated charcoal where extremely low concentrations of transition metals are involved or predominant concentrations of alkali and alkali-earth metals accompany those traces.

GLUTAMATE GLUCAN

This chitosan, obtained from 2-oxoglutaric acid under reducing conditions, has extended conformationally mobile side chains, regardless of the protonation state of the secondary amine. Instead, the charge state of ionizable groups and chirality of the C atom in the side chain strongly affect the structural features. The chelate initially formed promotes the formation of inorganic aggregates on polymer flakes.

IMMOBILIZED ENZYMES

Chemical linking or adsorption immobilizes enzymes on chitin and chitosan.[2] A number of proteins react with chitosan. The preferred immobilization technique is based on the use of glutaraldehyde. In some cases chitin adsorbs the protein strongly enough without canceling the catalytic effect of the protein; this is the case with α-amylase and glucoamylase.

Magnetic supports are prepared by including magnetite in chitosan with glutaraldehyde. These supports are convenient for the retrieval of the supported enzyme by physical means and its separation from the substrate solution, and for targeting drugs. Crosslinked spherical chitosan beads (Chitopearl®) are also used to immobilize enzymes and recover proteins.

Unspecific Enzymatic Hydrolysis

Many commercial enzyme preparations exert hydrolytic activity on chitosans, water-soluble reacetylated chitin and chitin azure. Several proteases such as pepsin, bromelain, ficin and pancreatin display lytic activities toward chitosans that surpass those of chitinases and lysozyme.

Cellulases, hemicellulases, lipases, and other enzymes are also effective, the presence of a common lytic impurity being ruled out by the high activity and different pH and temperature optima for the various enzyme preparations. Papain and lipase lower the viscosity of chitosan and modified chitosan solutions in a matter of minutes by reducing the size of the longest chitosan chains. In view of the presence of lipases in the digestive fluids, the activity of lipase has important consequences in the application of chitosan in the food area, while for technical purposes lipase, papain, and hemicellulase lend themselves to large scale preparations at pH 3 and 40°C.

REFERENCES

1. Muzzarelli, R. A. A., Ed. *Chitin in Nature and Technology* Plenum: New York, 1986.
2. Muzzarelli, R. A. A., Ed. *Chitin Enzymology*; Alda Tec: Ancona, 1993.

CHITIN AND CHITOSAN, GRAFT COPOLYMERS

Keisuke Kurita
Department of Industrial Chemistry
Faculty of Engineering
Seikei University

Although chitin is structurally similar to cellulose and produced in huge amounts in nature each year, it has remained an almost idle biomass resource owing to its inherent intractable nature. Chitin is an amino polysaccharide having acetamide groups at C-2 and converted into chitosan having free amino groups. It is thus expected to have higher potential than cellulose as a specialty polymeric material in various fields.

With regard to developing advanced functions, much attention had been paid to chemical modifications of chitin. Of the various possible modifications, graft copolymerization is expected to be one of the most promising approaches to a wide variety of molecular designs leading to novel types of tailored hybrid materials, which are composed of natural polysaccharides and synthetic polymers. The properties of the resulting graft copolymers would be widely controlled by the characteristics of the side chains, including the molecular structure, length, and number.

PROPERTIES

Chitin Graft Copolymers

Graft copolymers having poly(sodium acrylate) side chains were soluble in dichloroacetic acid, whereas those having polyacrylamide chains swelled. Both of the copolymers exhibited much enhanced hygroscopicity compared to chitin.[1] Degree of swelling of chitin-*graft*-poly(methylmethacrylate) in *N,N*-dimethylformamide was dependent on the grafting percentage. When the grafting percentage was below 400%, the copolymer showed lower swelling than chitin. Above 400% grafting, however, the swelling became profound, giving transparent gels. Film casting was possible using these highly swollen gels. The glass transition temperature of the copolymer was 130°C as determined by DSC.[2]

Chitin-*graft*-polystyrene prepared from iodo-chitin was almost soluble in aprotic polar solvents, such as dimethyl sulfoxide

(DMSO) and *N,N*-dimethylacetamide containing 5% lithium chloride when the grafting percentage was above 100%.

Graft copolymers of the water-soluble chitin having poly(γ-methyl L-glutamate) chains were soluble in dichloracetic acid and hexafluor-2-propanol. Graft copolymers having poly(sodium L-glutamate) were readily soluble in water, *m*-cresol, and dichloracetic acid.[3]

Chitosan Graft Copolymers

Although chitosan is an effective flocculating agent only in acidic media, the derivatives having poly(acrylic acid) side chains showed high flocculation ability in both acidic and basic regions because of the zwitterionic characteristics.[4] Chitosan-*graft*-polystyrene prepared by the radiation induced method showed higher adsorption of bromine than chitosan. Polystyrene-grafted chitosan films showed less swelling and better elongation in water than chitosan films.[5,6]

APPLICATIONS

Chemistry and biochemistry of chitin and chitosan are advancing quite rapidly as a result of expanding interest in these biopolymers, which have unique characteristics. Applications are being developed in various fields, including water treatment, toiletries, medicine, agriculture, food processing, and separation. The derived graft copolymers are also considered to have high potential in these fields; currently progressing studies on the various modes of graft copolymerization and the relationship between the molecular structure and properties will lead to practical applications in the near future.

REFERENCES

1. Kurita, K.; Kawata, M.; Koyama, Y.; Nishimura, S. *J. Appl. Polym. Sci.* **1991**, *42*, 2885.
2. Ren, L.; Miura, Y.; Nishi, N.; Tokura, S. *Carbohydr. Polym.* **1993**, *21*, 23.
3. Kurita, K.; Yoshida, A.; Koyama, Y. *Macromolecules* **1988**, *21*, 1579.
4. Kim, Y. B.; Jung, B. O.; Kang, Y. S.; Kim, K. S.; Kim, J. I.; Kim, K. H. *Pollino* **1989**, *13*, 126; *Chem. Abstr.* **1989**, *111*, 59884e.
5. Shigeno, Y.; Kondo, K.; Takemoto, K. *J. Macromol. Sci. Chem.* **1982**, *A17*, 571.
6. Shigeno, Y.; Kondo, K.; Takemoto, K. *Makromol. Chem., Rapid Commun.* **1981**, *182*, 709.

CHITIN AND DERIVATIVES

Dang Van Luyen and Dang Mai Huong
Institute of Chemistry
National Center for Natural Science and Technology

Chitin is a widely available biopolymer obtained principally from shrimp and crab shell waste. Chitosan is the main derivative of chitin. Both chitin, chitosan, and their derivatives have many current and potential uses because of the unusual combination of properties they possess, which include toughness, biodegradability, and bioactivity; this combination makes these materials especially attractive.

The chemical structure of chitin and chitosan is very similar to that of cellulose (**Figure 1**), the difference being that instead of the OH-group bonded at C-2 in each D-glucose unit of cellulose, there is an acetylated amino group ($-NHCOCH_3$) in chitin, and an amino group ($-NH_2$) in chitosan.

SYNTHESIS AND CHEMICAL MODIFICATIONS

Sources and Extraction of Chitin

Chitin is produced by removing calcium carbonate and proteins from the shells. In the production of chitin, calcium carbonate is first dissolved by stirring the shells in dilute hydrochloric acid at ambient temperature (demineralization). Proteins are then extracted from the decalcified shells by treating them with dilute aqueous sodium hydroxide (deproteinization); crude chitin is then obtained. Protein, calcium chloride, and carotenoid pigments obtained in the production process as side products may be used further.

Chemical Modifications of Chitin

Chitin is not soluble in most common organic solvents. An introduction of alkyl chains, hydroxyl or carboxyl groups into chitin is likely to improve its solubility, which promotes processing, and thereby widening its practical applications. Chitin can undergo O-substitution reactions to give esters and ethers. Alkali chitin from chitin and concentrated sodium hydroxide is an important intermediate in the synthesis of some chitin derivatives.

Glycol chitin, a partially O-hydroxyethylated chitin, was prepared from alkali chitin with ethylene chlorohydrin $ClCH_2CH_2OH$ and was probably the first chitin derivative to find practical use.[1]

Introducing substituents onto the hydroxyl groups of chitin for obtaining polymers with improved solubility in common organic solvents can be accomplished by reacting chitin with carboxylic anhydrides or carboxylic chlorides, yielding acyl-chitins.

Deacetylation of Chitin to Chitosan

Various techniques of chitin deacetylation have been developed for producing chitosan from chitin. The conventional deacetylation of chitin has been achieved under heterogeneous conditions.

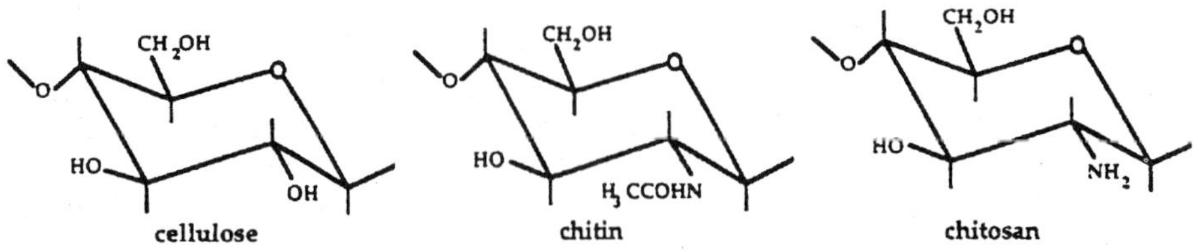

FIGURE 1. Chemical structure of cellulose, chitin, and chitosan.

A special heat treatment has been used to partially destroy the crystalline structure of chitin. After this process, chitin can be more easily converted into chitosan than by following the conventional methods used previously.[2] The deacetylation of chitin at low temperatures can also be achieved by using an alkali impregnation technique and by using a thermo-mechano-chemical technique.[3,4]

Modifications of Chitosan

Derivatives

Chitosan, containing both amine and hydroxyl groups, may be easily modified chemically to produce a wide range of derivatives. Depending on the reaction conditions, O- and N-substitution can take place. Much attention has been paid to the chemical modifications of chitosan to achieve derivatives that are soluble or highly swellable in water or in common organic solvents, for the purpose of creating new materials with high specificity and wide applicability.

N-acylchitosan and *N*-carboxyacylchitosan are examples of two such derivatives. The reaction of chitosan with acid anhydrides or acid chlorides gives *N*-acylchitosan.[5] With dicarboxylic anhydrides the result is *N*-carboxyacylchitosan.[6] The fully acylated chitosan derivatives, especially by acylation with long-chain fatty acyl halides or anhydrides (polyacylchitosan or *N,O*-acylchitosan) are soluble in organic solvents.[7]

N,O-carboxymethylchitosan (NOCC), another chitosan derivative, is formed by carboxymethylation of chitosan and is soluble in water due to the presence of both carboxyl groups and free amino groups.[8]

N-hydroxyalkylchitosan is the reaction product of chitosan with epoxides and is soluble in water.

N-alkylchitosan and *N*-carboxyalkylchitosan are examples of two other chitosan derivatives. The reactions of chitosan with aldehydes and ketones give *N*-alkylidene chitosan, which is converted on hydrogenation (reductive amination) to *N*-alkylchitosan. The reactions of chitosan with aldehydic acids and cetonic acids give *N*-carboxyalkylidene chitosan, which is hydrogenated to *N*-carboxyalkylchitosan.

Quaternized chitosan is obtained by direct alkylation of primary amines in chitosan to their quaternary ammonium salts with organic or inorganic base as catalyst.[9] It is a water-soluble polyelectrolyte with a high charge density that exhibits good properties as a flocculant.

STRUCTURE AND PROPERTIES OF CHITIN, CHITOSAN AND DERIVATIVES

Characterization

Molecular and Crystalline Structure

The solid state, chitin occurs in two principal polymorphic, crystalline forms, namely α- and β-chitin. X-ray diffraction technique has been used intensively to examine the influence of the deacetylation on the crystallinity of chitin derivatives. All the deacetylated products obtained under heterogeneous conditions with a degree of deacetylation up to 71% were crystalline.

The crystal structure of chitosan film has also been studied with X-ray diffraction. The film prepared from chitosan showed low crystallinity. Annealing this film in a solvent, usually in water, and at a high temperature resulted in a higher level of crystallinity. The presence of a high concentration of organic solvent in the chitosan acetate solution altered the crystal structure of the film obtained, rendering this film insoluble in water.[10]

Determination of the Degree of Deacetylation

The degree of deacetylation (DD) can be determined by IR spectroscopy. ^{1}H-NMR spectroscopy, ultraviolet spectrophometry, gel permeation chromatography, colloidal titration, elemental analysis, circular dichroism, and by a picric acid method.[11–15]

Structure Property Relationships

The chain morphology and the molecular structure, for example, mole fraction of *N*-acetylglucosamine or DD, molecular weight, and molecular weight distribution are expected to have a considerable effect on the properties of the polymer.

Chain Morphology and Reactivity Behavior

The two principal polymorphic forms of chitin, α-chitin and β-chitin, differ in the agreement of the molecular chains within the crystal cell. In α-chitin the chains have an antiparallel arrangement, and the reprecipitated α-chitin is more disordered than the original. In β-chitin the chains have a parallel arrangement and reprecipitated β-chitin is more ordered than the original metastable polymorph. In both morphological forms, the chains have interchain hydrogen bonds along the a-axis. In α-chitin the chains also have interchain hydrogen bonds along the c-axis. β-chitin is more reactive than α-chitin (β-chitin is more susceptible to reagents) and swells readily in water to form hydrates, whereas α-chitin does not; β-chitin is soluble in formic acid, α-chitin is not; and β-chitin is soluble in a greater range of formic acid / cosolvent systems than is α-chitin.[16]

Chitosan also occurs in a number of different morphologies. Three forms have been proposed: noncrystalline, hydrated crystalline, and anhydrous crystalline. They have been identified by X-ray crystallography.

The anhydrous crystalline form of chitosan does not form salts with acids, nor complex with transition metal ions, and is not soluble in dilute aqueous acids. The type and extent of the crystalline fraction are important factors in determining the properties of a chitosan sample and both depend on the treatment history of the sample, the degree of deacetylation, and the molecular weight.

Molecular Structure and Properties in Solution

Most samples of chitin and chitosan are copolymers containing two monomer units, *N*-acetyl-D-glucosamine and D-glucosamine. The former unit is predominant in chitin, and the latter is predominant in chitosan. The respective mole fractions of these two monomer units and their distribution pattern along the chain, together with the molecular weight, are major factors in determining the properties of the material.

Chitin is barely soluble, except in some organic solvents containing lithium chloride, such as *N,N*-dimethylacetamide (DMAc)-LiCl, and *N*-methylpyrrolidone-LiCl.

The presence of free amino groups in D-glucosamine units that are capable of being protonated by the acid medium practically enables chitosan to dissolve in dilute acid solutions.

APPLICATIONS OF CHITIN, CHITOSAN, AND DERIVATIVES

The Key Properties Related to Practical Applications

Chitin, chitosan, and their derivatives possess many advantageous properties that make them especially interesting materials. They are nontoxic and can be biodegraded into normal body constituents. Chitosan has both reactive amino groups and hydroxyl groups, which offer many possibilities to chemically tailor its physical and solution properties.

Chitosan is an unbranched cationic polymer with a high charge density in solution and carries a positive (+) charge at pH below 6.5. Many uses of chitosan are based on its "positive charge," which is attracted to negatively (–) charged materials. Most living tissues (skin, bone, hair); polysaccharides (alginat, carboxymethylcellulose, hyaluronate); polyanions; bacteria and fungi; enzymes; and microbial cells are negatively charged. Therefore, chitosan can be used to remove toxic and contaminating bioburden materials, such as proteins and heavy metals for safety reasons. From a biological standpoint, there are several reasons why chitosan is a very attractive material for medicine, pharmacy, and biotechnology. It has been shown that chitosan has many valuable bioactivities. For example, chitosan possesses hemostatic, bacteriostatic, fungistatic, anticancer, and anticholesteremic activity.

Applications of Chitin, Chitosan, and Derivatives

Chitin and its derivatives were originally used as natural flocculants for waste-water treatment. The present trend is toward producing high-value products for medical, pharmaceutical, biotechnological, and cosmetic use. Several high-valued applications have been developed recently.[17]

The use of chitin fabrics as artificial skin cover for burns was approved, and the commercial product is now on sale.

CONCLUDING REMARKS

There are several explanations for the intense worldwide interest in chitin and chitosan research. First, the large quantity of waste by-products originating from seafood processing can be fully utilized and subsequently processed into high-quality materials having interesting properties suitable for biomedical and biotechnological applications. Second, new modification and processing methods can be exploited that result in materials with unique properties, opening up new application possibilities. Finally, the price of chitin, chitosan, and their derivatives will be lowered as demand for the products increases. This contribution is also intended to stimulate further research on chitin and chitosan in order to solve existing problems in creating high-value products, in developing new methods of characterization, and in exploiting novel potential applications of chitin and its derivatives.

ACKNOWLEDGMENT

The authors gratefully acknowledge the assistance of Dr. M. Krenceski in the preparation of this manuscript.

REFERENCES

1. Yamada, H.; Imoto, T. *Carbohydr. Res.* **1981**, *92*, 160.
2. Luyen, D. V., Huong D. M; Mai, P T.; Tuan, L. Q.; Tam, T. D.; Nhung, N. T. Vietnamese Patent 123A 1990.
3. Rao, R. S. V. *Ind. J. Technol.* **1987**, *25*, 194.
4. Pelletier, A.; Lemire, I.; Sygush, J.; Chornet, E.; Overend, R. P. *Biotechnol. Bioeng.* **1990,** *36*, 310.
5. Hirano, S.; Ohe, Y.; Ono, H. *Carbohydr. Res.* **1976**, *47*, 315.
6. Hirano, S.; Moriyasu, T. *Carbohydr. Res.* **1981**, *92*, 323.
7. Fujii, S.; Kumagai, H.; Noda, M. *Carbohydr. Res.* **1980**, *83*, 389.
8. Hayes, E. R. European Patent 265 561, 1988; *Chem. Abstr.* **1988**, *109*, 56927.
9. Domard, A.; Rinado, M.; Terrassin, C. *Int. J. Biol. Macromol.* **1986**, *8*, 105.
10. Grant, S.; Blair, H. S.; McKay, G. *Makromol. Chem.* **1989**, *190*, 2279.
11. Hirai, A.; Odani, H.; Nakajima, A. *Polym. Bull.* **1991**, *26*, 87.
12. Varum, K. M.; Anthonsen, M. W.; Grasdalen, H.; Smidsrod, O. *Carbohydr. Res.* **1991**, *211*, 17.
13. Aba, S.-I. *Int. J. Biol. Macromol.* **1986**, *8*, 173.
14. Domard, A. *Int. J. Biol. Macromol.* **1987**, *9*, 333.
15. Neugebauer, W. A.; Neugebauer, E.; Brzelinski, R. *Carbohydr. Res.* **1989**, *189*, 363.
16. Roberts, G. A. F. *Abstracts of Papers*; Asia-Pacific Chitin and Chitosan Symposium, Bangi, Malaysia, May 1994.
17. Li, Q.; Dunn, E. T.; Grandmaison, E. W.; Goosen, M. F. A. *J. Bioact. Compat. Polym.* **1992**, *7*, 370.

CHITIN CHEMISTRY

Riccardo A. A. Muzzarelli
Faculty of Medicine
University of Ancona

Chitin (CAS Registry No. 1398–61–4) is a linear polymer β(1–4)-linked 2-acetamido-2-deoxy-D-glucopyranose units. Chitin is different from other abundant polysaccharides, because it contains nitrogen in addition to carbon, hydrogen, and oxygen.

The shells of crabs, shrimps, prawns, and lobsters from the peeling machines in a canning factory are suitable for the large-scale preparation of chitin. The isolation includes two steps: demineralization with HCl and deproteination with aqueous NaOH. Lipids and pigments may also be extracted. Fungi, particularly *Absidia coerulea* and *Mucor rouxii*, have also been used for production of chitosan at the laboratory level.

Isolated chitins differ from each other in many ways, including degree of acetylation (i.e., the molar ratio between 2-acetamido-2-deoxy-β-D-glucose units and the totality of sugar units, which is typically close to 0.90); elemental analysis, (the elemental analysis of oligomers is different because there is one water molecule per molecule); and molecular size and polydispersity (the average molecular weight of chitin *in vivo* is on the order of 10^6 D, but chitin isolates have lower values because of partial random depolymerization during isolation and depigmentation steps). The average molecular weights for commercial chitins are 0.5–1.0 MD. Polydispersity may vary depending on treatments and blending of various chitin batches.

Chitin has several properties that define its general behavior and are related to methods of isolation. As a polyacetal it is sensitive to and easily hydrolyzed by acids. It is stable to dilute alkali but is easily oxidized by air in warm concentrated alkali. It is sensitive to heat, even at modest temperatures (50°C), and may denature with prolonged heating. It is a chelating agent for transition-metal ions. Hydrates, alcoholates, and ketonates are formed when chitin is precipitated by appropriate nonsolvents. *N*-Chloro compounds are formed with chlorine bleaches; therefore, peroxide bleaches are preferable. Free amino groups are present in all commonly available chitins. Chitin is a chiral polymer by virtue of both its glucose moieties and its helical structure. In the wet state, chitin is degraded by several microorganisms that produce chitinolytic enzymes or other enzymes with unspecific activity toward chitin. Chitin is promptly degraded by the enzymes present in crustaceans and should be isolated as soon as possible after a catch. It is depolymerized by sonication and microwave treatment. It and many of its derivatives are filmogenic polymers, and it can be blended with other natural and synthetic polymers such as poly(vinyl alcohol).

Chitin is an insoluble biopolymer: *N,N*-Dimethylacetamide (DMA) and 5% LiCl (DMA + LiCl) as well as *N*-methyl-2-pyrrolidione and LiCl are the systems in which it can be dissolved (up to 5%). The viscous chitin solutions in DMA + LiCl, when exposed to moisture, precipitate chitin as a rubber-like solid that swells enormously in water. Films and filaments may be produced from DMA + LiCl solutions, and certain reactions can be carried out in that medium to modify chitin chemically (e.g., chitin is sulfated or reacted with acid chlorides or with isocyanates).

When chitin is partially or totally deacetylated it is usually called chitosan, which indicates a continuum of progressively deacetylated chitins. In general, chitosans have a nitrogen content higher than 7% and a degree of acetylation lower than 0.40. The process by which the acetyl group is removed from 2-acetamido-2-deoxy-glucose units is known as deacetylation and is accomplished by heating the chitin in concentrated NaOH (30–40%) for 2 h or more under nitrogen and in the presence of antioxidants such as sodium borohydride to prevent alkaline depolymerization.

CHITOSANS

The presence of a prevailing number of 2-amino-2-deoxy-glucose units in a chitosan allows the polymer to be brought into solution by salt formation. A chitosan is a primary aliphatic amine that can be protonated by certain selected acids; the p*K* of the chitosan amine is 6.3. The following salts, among others, are water soluble: formate, acetate, lactate, malate, citrate, glyoxylate, pyruvate, glycolate, and ascorbate.

Chitosan behaves as a sphere in aqueous acetic acid solution or as a random coil in urea. The electric charge density on the molecule can be changed by varying pH, ionic strength, concentration, time, and nature of the medium.

Chitosan solutions are difficult to handle because of aging: depolymerization, aggregate formation, microbiological spoilage, and generation of reactive species take place.

REACETYLATED CHITOSANS

Chitosans are reacetylated to chitins with the aid of acetic anhydride in methanol. The degree of acetylation obtained is far from 1.0 and the resulting products are water soluble.

CHITIN SYNTHESIS

Riccardo A. A. Muzzarelli
Faculty of Medicine
University of Ancona

Chitin synthesis is a primitive property of the eukariote cell; the synthetic ability has been lost in some groups like the chordates, emichordates, and echinoderms. Fungi, like animals, have chitin exoskeletons and store glycogen rather than starch.

In the marine environment, chitin is produced by different types of unicellular eukariotes such as diatoms, yeasts, Rhizopoda, Foraminifera, Cnidosporidia, and Ciliata. Molds and fungi are well known chitin producers. In bacteria and blue-green algae, chitin is completely lacking; in the animal kingdom, it is produced by most Coelenterates but not by corals, sea anemones, jelly fishes, or sponges. Chitin is secreted by triblastic invertebrates among which are Bryozoans and Brachiopods. Other chitin-producing invertebrates are mollusks, polychete worms, and arthropods.

The total production of marine chitin has been tentatively calculated, based on data on zooplankton and exuviae production by krill and by large crustacean species, to be at least 2.3×10^9 metric tons per year.

In the terrestrial environment where molds and insects are the main producers, estimates are not quantitative, but the amount is enormous.

Chitin is synthesized at the cytoplasmic membrane and the enzyme, a transmembrane protein, accepts the substrate at the cytoplasmic site while the (1-4)-β-linked N-acetylglucosamine polymer is extruded to the outside. Deposition, crystallization, and fibrillogenesis occur at the outer face of the plasmalemma.

In fungal preparations, chitin synthetase is present as an inactive zymogenic form. Partial proteolysis *in vitro* by trypsin leads to activation of the zymogen, the efficiency of the proteases depending on the fungal species.

Once activated, chitin synthetase becomes rapidly inactivated, and addition of protease inhibitors is ineffective to stop inactivation. Activation and inactivation occur sequentially: the operation of this system *in vivo* requires the constant supply of chitin synthetase to the growing region of the cell for apical growth.

Chitosomes, which convey chitin synthetase to the plasma membrane, have a characteristic lipid composition not necessarily identical to the plasma membrane, and the mere transfer of chitin synthetase from chitosomes to plasma membrane may result in activation of the enzyme. Chitin synthetase can also be activated by the mechanical stretch in the membrane, i.e., the turgor pressure that is the driving force for hyphal extension. The role of metal ions in the activation process is to protect the enzyme from inactivation.

CHITIN MODIFICATIONS IN FUNGI

The product of the reactions described above is a homopolymer of β(1-4)-linked *N*-acetylglucosamine, and very sensitive to the modifying enzymes chitinase and deacetylase.

The modification of newly formed chitin, i.e., deacetylation or linkage to other polysaccharides, interferes with crystallization and hydrogen bonding. In mature hyphal walls seldomly can crystalline chitin be detected.

Crystallization can also be prevented by addition of compounds that adhere to polysaccharides, like the optical brighteners Calcofluor White and Congo Red. These compounds maintain chitin in a reactive form that favors dissolution by chitinase and by acids.

The following modifications may occur after the chitin chains have been extruded into the wall domain: i) linking of the β-glucan to chitin, ii) deacetylation of chitin, iii) hydrolysis of chitin by chitinase.

CHITIN IN INSECT METABOLISM

Arthropods have evolved an orchestrated process of growth based on a series of molts. The molting process is universal among arthropods and like most cells, the epidermal cell forms an envelope around its plasma membrane. Chitin microfibrils make up the bulk of the endocuticular lamellae and the exocuticle is usually sclerotized and melanized.

The synthesis and the deposition of the cuticle take place at the dorsal surface of the plasma membrane of the epidermal cell. The membrane organizes itself in the form of microvilli and secretes fibrous cuticle made of chitin-protein microfibers.

The pesticide diflubenzuron, one of the most important compounds of the class of benzoylphenylureas (trade name Dimilin) acts as an insecticide by inhibiting the deposition of cuticular chitin, thereby disrupting insect growth and development. The inhibitory action is exerted only if the cell membrane remains intact.

CRUSTACEAN CHITIN SYNTHESIS

Subcellular organelles, particularly the Golgi complex, are crucial to the synthesis of chitin in the crustaceans. The crustacean cuticle consists of a protein-polysaccharide complex where chitin is covalently linked to the protein. A primer polypeptide in the rough endoplasmic reticulum is involved in the assembly of a chitin oligosaccharide on a dolichol intermediate. If protein synthesis is blocked by antibiotics, chitin synthesis declines sharply.

The chitoprotein serves as a primer for chitin synthetase, which in the Golgi complex extends the oligosaccharide. Crosslinking of the polypeptides might occur in the Golgi complex as well. The final step is to package the chitin-protein complex into vesicles that move to the apical membrane of the epithelial cells where exocytosis releases the product to the outside. As occurs in bacterial cell wall synthesis, transglycosylases and transpeptidases attach the product to the growing cuticle.

IN VITRO SYNTHESIS

Microfibrils of chitin have been synthesized *in vitro*, taking advantage of both insect and fungal preparations. Early investigations of chitin biosynthesis were performed with crude cell fractions, and the identification of chitin relied on its susceptibility to chitinase, hydrolysis, and alkali insolubility of the product.

Chitin Synthetase

See: *Chitin Synthesis*
Chitinases

CHITINASES

Riccardo A. A. Muzzarelli
Faculty of Medicine
University of Ancona

AUTOLYTIC AND MORPHOGENETIC CHITINOLYSIS

All chitin-containing organisms produce chitinolytic enzymes. Autolytic chitinases also act during sexual reproduction in fungi. Autolytic enzymes accumulate in culture filtrates of senescent fungal cultures.

CHITINOLYSIS IN PATHOGENESIS AND SYMBIOSIS

Pathogens of chitinous organisms characteristically produce chitinases. These assist in penetrating the host and provide nutrients directly in the form of aminosugars and indirectly by exposing other host material to digestion.

Fungal and invertebrate chitin is a considerable part of the diet of many herbivorous and carnivorous animals. There can be three sources of chitinolytic enzymes in the animal digestive system: from the animal itself, from the endogenous gut microflora, and from the food.

CHITINASES

Chitinases (EC 3.2.1.14) are glycosyl hydrolases [poly-1,4-(N-acetyl-β-D-glucosaminide)glycanohydrolases] that catalyze the degradation of chitin.[1–5] The catalytic domains of chitinases can be grouped into two families based on amino-acid sequence similarities. The lack of similarity between two families of proteins suggests that they have different folds.

For insect and crustacean N-acetyl-β-D-glucosaminidases, similarities exist in terms of molecular weight, Michaelis constant (K_m), and number of isoenzymes. These enzymes are part of the defense mechanisms of vertebrates, and elevation of their levels can be obtained by administration of chitosan in a number of ways, including application of chitosan-based wound dressings. Enzyme activities are also elevated in certain acquired or genetic diseases, such as diabetes and leukemia.

The plant chitinases function mainly as a plant defense mechanism, attacking the cell wall of the invading pathogen. In contrast, the bacterial and fungal enzymes are devised to hydrolyze chitin to chitobiose to be used as energy, and a carbon and nitrogen source.

There are analogies between chitinases and cellulases; both act on fibrous, insoluble β-1,4-linked polysaccharides.

OCCURRENCE AND INDUCTION OF PLANT CHITINASES

Chitinases have been reported from over 41 monocotyledonous and dicotyledonous plant species and occur in various tissues including embryos, seeds, leaves and stems, roots, flowers, calli, and protoplasts. Among cultivated crop species. chitinases occur in bean, cabbage, cacao, carrot, celery, corn, cucumber, garlic, melon, oat, potato, pumpkin, rapeseed, rice, soybean, sugar beet, sunflower, tobacco, tomato, wheat, and many others. Noncrop species such as chestnut, petunia, poplar, rubber, spruce, and stinging nettle also contain chitinases.

In fact the plant response to microbial attack involves de novo synthesis of an array of proteins designed to restrict the growth of the pathogen (elicitor). Besides glucanase and chitinase production, defense-related phenomena are induction of enzymes involved in construction or modification of suberin, lignin, wall-bound phenolics, and callose; amylase and protease inhibitors, toxins such as thionins and lectins; and enzymes synthesizing oxidized phenolics, tannins, o-quinones, and phytoalexins. Indeed, enhanced chitinase levels in transgenic plants can reduce the damage caused by pathogens.

CLONING OF CHITINASE GENES

Chitinase genes from bacteria, fungi, and plants have been cloned. The importance of work dealing with this cloning is related to the biocontrol of soil-borne fungal and nematode diseases of crop plants, as chitin is an essential component of fungal walls and nematode egg cases.

Chitinases can be used in a number of ways as biocontrol agents against plant-pathogenic fungi. The free enzyme can be introduced into the irrigation water or incorporated into the seed coating to protect germinating seedlings. Another alternative is to introduce into the soil genetically engineered rhizosphere bacteria or fungi able to express and secrete chitinase.

NUTRITIONAL CHITINOLYSIS

Chitinolytic fungi are present in soils, where they rival the chitinolytic activity of bacteria. Most common are Mucorales, especially *Mortierella* spp., and Deuteromycetes and Ascomycetes, especially the genera *Aspergillus*, *Trichoderma*, *Verticillium*, and *Penicillium*. These fungi have inducible chitinolytic systems.

Chitinolytic bacteria are widespread in all productive habitats. Chitinases are produced by many genera of gram-negative and gram-positive bacteria, but not by Archaeobacteria.

In the sea, vast amounts of chitin are produced, chiefly as carapaces of zooplankton, near the surface. Chitinolytic bacteria are also abundant in fresh water and in soil.

Chitinases are found in many vertebrates' digestive tracts, where they are secreted by the gastric mucosa and, less generally, by the pancreas.

REFERENCES

1. Graham, L. S.; Sticklen, M. B. *Can. J. Bot.* **1994**, *72*, 1057.
2. Henrissat, B.; Bairoch, A. *Biochem. J.* **1993**, *293*, 781.
3. Sahai, A. S.; Manocha, M. S. *FEMS Microbiol. Rev.* **1993**, *11*, 317.
4. Skjak-Braek, G. et al. *Chifin and Chitosan;* Elsevier: Amsterdam, 1989.
5. Spindler, K. D.; Spindler-Barth, M.; Londershausen, M. *Parasitol. Res.* **1990**, *76*, 283.

Chitosan

See: *Chitin (Metal Ion Chelation and Enzyme Immobilization)*
Chitin and Chitosan, Graft Copolymers
Chitin and Derivatives
Chitin Chemistry
Chitosan (Preparation, Structure and Properties)
Dental Polymers (Hydrogels)
Vegetal Biomass (3. Polymers, Derivatives and Composites)
Wound Dressing Materials

CHITOSAN (Preparation, Structure, and Properties)

Attila E. Pavlath, Dominic W. S. Wong, and George H. Robertson
Western Regional Research Center
Agricultural Research Service
U.S. Department of Agriculture

Chitosan is a polysaccharide with a structure similar to cellulose. The pure compound is the polymeric form of 2-amino-2-deoxy-D-glucose with a 1,4-β-glucosidic bonding. Chitin is its N-acetyl derivative. The degree of polymerization can be as high as thousands, with molecular weights in six figures.

Pure chitosan is rarely found in nature, but it is present in *Zygomycetes* cell walls. The most frequent industrial source of chitosan is chitin that is the main component of the exoskeleton of crustaceans, such as crabs, lobsters, shrimp, and crayfish.

The separation of chitin from various crustaceous sources generally follows the same procedures, with only small variations. The starting material is milled first, the proteins are removed by aqueous NaOH, and then the minerals, mostly calcium carbonate are extracted with aqueous HCl treatments. Undesirable color may be removed by solvent extraction or oxidizing agents such as NaOCl and H_2O_2. The degree of acetylation in the final product is generally somewhat less than in the raw material, which may be due to partial removal of the acetyl group during these treatments.

The deacetylation can be carried out in various ways. The most frequently used method is under heterogeneous reaction conditions, using a 50% NaOH solution at 70–100°C for up to 45 hours. Anything at or above 75% deacetylation is considered as suitably pure for a chitosan starting material. A purer chitosan, with only 10% *N*-acetylation, can be obtained if the reaction is carried out at 135°C for 3.5 hours under nitrogen.[1]

Deacetylation is a rapid process at first, but slows and effectively stops before the reaction is complete. The products formed during the high-rate conversion are higher acetylated chitosans. In order to obtain a completely deacetylated product, treatment in a 67% NaOH solution at 110°C for 2 hours under nitrogen atmosphere, with the process repeated twice, is required.[2] Other methods include the use of anhydrous hydrazine–hydrazine sulfate treatments.[3]

A newer method, the so-called thermo-mechano-chemical (TMC) treatment, uses less alkali and a shorter time to convert the chitin to chitosan.

Chitin can also be deacetylated enzymatically, with the advantage of not only providing more random deacetylation, but also minimizing the cleavage of the polymeric chain. However, care has to be used in the selection of the enzyme, since many enzymes will also depolymerize the polymeric chain.[4]

STRUCTURE

Chitosan and cellulose have very similar primary and secondary structures. They can be intimately blended in films prepared from acetic acid and trifluoroacetic acid solvents, as shown by X-ray crystallography and scanning electron micrography (SEM).[5] However, the reactivities of chitosan and cellulose differ. For instance, cellulose was trifluoroacetylated when dissolved in trifluroacetic acid, while chitosan formed only a salt from which chitosan could be easily recovered even after forming a film.[6]

Various structures have been reported for chitosan and its salts. The most frequently occurring polymorph is labeled as "tendon," but other forms such as "form II," "annealed," and "L-2" have been described. The difference between these polymorphs is assumed to be the number of water molecules in the unit cells. The annealed form contains no water.[7]

The structure of chitosan can be sensitive to the degree and agents of crosslinking, interactions with salts and other components, the acids used in neutralization, and the water activity of the matrix. The structural sensitivity may be implied by the results of using non-porous chitosan-based membranes for the pervaporative separation of water from ethyl alcohol-water solutions.

SOLUBILITY

Chitosan can be dissolved in certain inorganic acids, such as HCl and HNO_3, up to 1.1% concentration, but has marginal solubility in sulfuric and phosphoric acids. Pure chitosan is not soluble in water at pH>7, and precipitates from solution if the pH rises above 6. Chitosan is soluble in various organic acids. Aqueous acetic and formic acids are mostly used to make chitosan solutions, which can be formed by stirring from 15 to 30 minutes, depending on the temperature. The partially acetylated chitosan molecules generally are more water soluble. Depending on the degree of acetylation, they may not be precipitated from aqueous solutions by NaOH. Chitosan has a water-binding capacity of 440% (w/w), but the capacities of chitin and microcrystalline chitosan are somewhat less—380% and 340%, respectively.[8]

In order to obtain partially and randomly acetylated chitosan that has good water solubility, chitin is fully deacetylated, then partially acetylated with acetic anhydride in an aqueous acetic acid-methanol–pyridine solution. Best solubility is obtained at about 50% deacetylation.[9]

TOXICITY

Chitosan has no physiological toxicity.

APPLICATIONS

The utilization of chitosan is based on its free amino groups, which can provide valuable polycationic, chelating, film forming, and other properties. Numerous health applications have been reported, including surgical sutures, wound healing, burn treatment, contact lenses, and drugs to control blood cholesterol level and coagulate blood.

Chitosan protects peas from fungal infection. Adding small amounts to the diets of chickens and calves enables these animals to digest lactose. Adding chitosan to paper increases its strength and dyeability. Chitosan can be used for adsorption of nickel, plutonium, and chromate ions.

These are just a few of the many possibilities reported. A thorough discussion of various applications is provided by various review papers.[10–12]

REFERENCES

1. Kurita, K.; Kamiya, M.; Nishimura, S. *Carbohydrate Polymers* **1991**, *16*, 83–92.
2. Mochizuki, A.; Sato, Y.; Ogawara, H.; Yamashita, S. *J. Appl. Polym. Sci.* **1989**, *37*, 3375.
3. Dmitriev, B. A.; Knirel, Y. A.; Kochetkov, N. K. *Carbohydr. Res.* **1975**, *40*, 365.
4. Pantaleone, D.; Yalpani, M.; Scollar, M. *Carbohydr. Res.* **1992**, *237*, 325.
5. Isogai, A.; Atalla, R. H. *Carbohydr. Res.* **1992**, *19*, 25.
6. Hasegawa, M.; Isogai, A.; Onabe, F.; Usuda, M. *J. Appl. Polym. Sci.* **1992**, *45*, 1857.
7. Ogawa, K. *Agric. Biol. Chem.* **1991**, *55*, 2375.
8. Knorr, D. *J. Food Science* **1982**, *47*, 593.
9. Aiba, S. *Int. J. Biol. Macromol.* **1989**, *11*, 249.
10. Muzzarelli, R. A. A. *Carbohydrate Polym.* **1983**, *3*, 53.
11. Knorr, D. *Food Tech.* **1991**, *45*, 114.
12. Knorr, D. *Food Tech.* **1984**, *38*, 85.

CHLORINATED POLYETHYLENE

Gary R. Marchand
The Dow Chemical Company

PREPARATION

Chlorinated polyethylene, or CM as it is designated in ASTM D1418, can be prepared by free-radical substitution the hydrogens of polyethylene [9002-88-4] by chlorine atoms.

Because the reaction is a free-radical substitution, the mechanism is not appreciably affected by the medium and can be conducted in solution or in gas or aqueous slurry suspensions to produce polymers with varying chlorine contents. Molecular chlorine or a chlorine carrier such as $SOCl_2$[1] can be used as the chlorinating agent. The reaction can be initiated thermally or aided by UV light or peroxides.

STRUCTURE AND PROPERTIES

Chlorinated polyethylene structures span a range from a crystalline or rigid thermoplastic to a soft flexible elastomer. The thermoplastic form of chlorinated polyethylene has similar mechanical properties to polyethylene or poly(vinyl-chloride) (PVC) (CAS Registry No. 9002-86-2), depending on the chlorine level, but is better viewed as similar to PVC for processing because the presence of hydrogen and chlorine on adjacent carbons results in the potential for thermally or chemically initiated dehydrochlorination. However, because of their saturated polymer backbones, elastomers made from chlorinated polyethylene are relatively stable to oxidation and UV radiation compared with the more common unsaturated elastomers. Variations in the molecular structure of chlorinated polyethylene that control the resultant properties of the polymer can best be understood in terms of variations in the level and distribution, both inter- and intramolecularly, of chlorine and in the type of polyethylene used.

APPLICATIONS

Applications for chlorinated polyethylene fall into two primary categories: thermoplastic and thermoset elastomer. Chlorinated polyethylene is often the minor component in thermoplastic applications, in which it is used to modify the properties of other polymers such as polyethylene, PVC, and styrenics. In elastomer applications, chlorinated polyethylene is the primary component because of its excellent resistance to oil, chemicals, ozone, and heat. Chlorinated polyethylene is also used in both elastomer and thermoplastic applications in which ignition resistance is required.

Thermoplastics

The single largest application category for chlorinated polyethylene is the modification of PVC, olefinics, and styrenic polymers. Chlorinated polyethylene is used extensively at 2–10 phr to modify rigid PVC for applications such as PVC pipe and vinyl siding and window profiles.[2,3] At higher loadings in PVC blends, compositions from rigid high-impact materials to soft and flexible materials with good mechanical, thermal, and electrical properties can be obtained.[4–6] Chlorinated polyethylene can be added to styrene–acrylonitrile (CAS Registry No. 9003-54-7) and acrylonitrile–butadiene–styrene (CAS Registry No. 9003-56-9) copolymers to improve impact and ignition resistance in the presence of other additives.[7,8]

Elastomers

Chlorinated polyethylene elastomers find use in hydraulic, automotive, and industrial hose applications and in wire and cable jacket materials.[9] Chlorinated polyethylene can be compounded with a variety of reinforcing and nonreinforcing fillers, lubricants, plasticizers, and pigments. It can be cured by organic peroxides to form crosslinked elastomers with properties superior to those of conventional sulfur-vulcanized elastomers.[10]

REFERENCES

1. March, J. *Advanced Organic Chemistry*, 2nd ed.; McGraw-Hill: New York, 1977; pp 631–636.
2. Blanchard, R. R.; Burnell, C. N. *Soc. Plast. Eng. J.* **1968**, *24*(1), 74.
3. Dunn, J. L.; Heffner, M. H. *Soc. Plast. Eng. Tech. Pap.* **1973**, *19*, 483.
4. Pfinster, H. J.; Gruninger, A. E. *Kunstoffe* **1980**, *70*, 556.
5. Frey, H. H. *Kunstoffe* **1959**, *49*(2), 50.
6. Frey, H. H. *Chemiker-Ztg.* **1959**, *83*, 645.
7. Jue, N. O.; Young, W. L. *Soc. Plast. Eng. Tech. Pap.* **1977**, *23*, 19.
8. Grossman, R. F.; Myer, D. J. *Soc. Plast. Eng. Tech. Pap.* **1971**, *17*, 23.
9. *Tyrin® CPE Elastomers*; Dow Chemical Co. Product Literature; Form No. 3006-00094-1190 RJD; Dow Chemical: Midland, MI, 1990.
10. Blanchard, R. R. *J. Elastomers Plast.* **1974**, *6*(1), 3.

Chlorobutyl Rubber

See: *Butyl and Halobutyl Rubbers*
Isobutylene Copolymers (Commercial)

Chloromethylated Polystyrenes

See: *Polystyrene and Derivatives, Photolysis*

Chloroprene Rubber

See: *Chloroprene Rubber (Overview)*
Diene Rubbers, Conventional (Ozone Degradation and Stabilization)
Interpenetrating Polymer Networks, Rubber-Based

CHLOROPRENE RUBBER (Overview)

R. Musch and H. Magg
Bayer AG

Polychloroprene was one of the first synthetic rubbers and has played an important role in the development of the rubber industry as a whole. This fact can be attributed to polychloroprene's broad range of excellent characteristics.

In terms of consumption, polychloroprene has become one of the most important specialty rubbers for nontire applications.

CHLOROPRENE MONOMER PRODUCTION

The modern butadiene process, which is now used by nearly all chloroprene producers, is based on butadiene,[1] which is readily available. Butadiene is converted into monomeric 2-chlorobutadiene-1,3 (chloroprene) via 3,4-dichlorobutene-1, using reactions that are safe and easy to control.

POLYMERIZATION AND COPOLYMERIZATION

The polymerization of chloroprene is exothermic, the enthalpy of polymerization being 65–75 kJ/mol.

In principle it is possible to polymerize chloroprene by anionic, cationic, and Ziegler–Natta catalysis techniques.[2] However, because of the lack of useful properties, production safety, and economic considerations, free-radical emulsion polymerization is used exclusively today. It is carried out on a commercial scale using both batch and continuous processes.

With the aid of radical initiators, chloroprene in the form of an aqueous emulsion is converted into homopolymers or, in the presence of comonomers, into copolymers.[3] The polymerization is stopped at the desired conversion by the addition of a free-radical inhibitor.

Because of the high reactivity of chloroprene, the free-radical initiation and propagation proceed much faster with it than with the other dienes or olefins. Thus chloroprene has a stronger tendency to form homopolymers instead of copolymers.

Comonomers that have been used with success are those with chemical structures similar to that of chloroprene, in particular:

- 2,3-dichlorobutadiene, to reduce the crystallization tendency (the stiffness) of the chain.

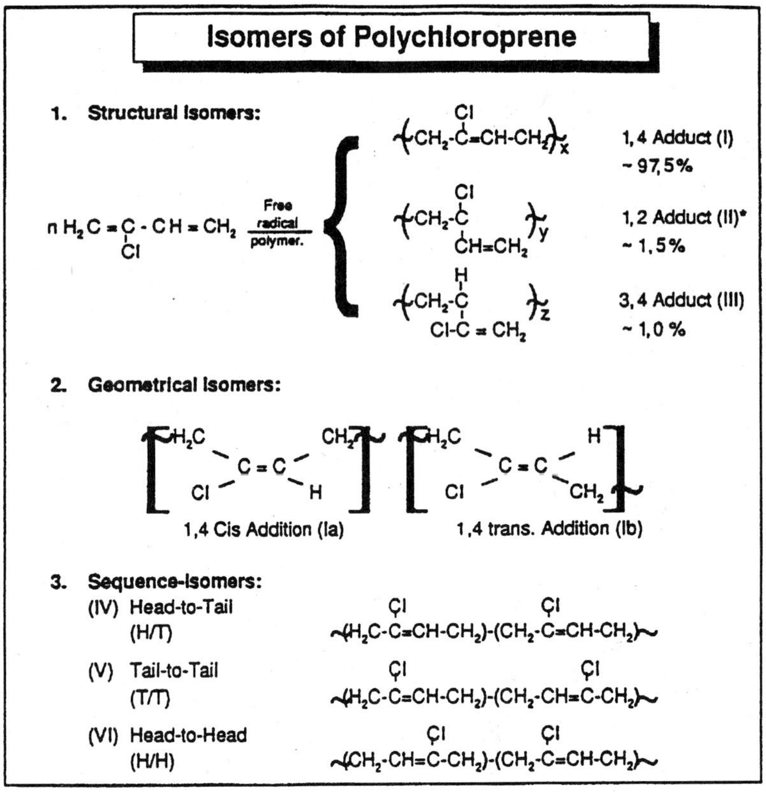

FIGURE 1. Structural units in the polychloroprene chain (typical commercial rubber grade).

- Acrylic or methacrylic acid esters of oligo-functional alcohols, to produce the desired precrosslinked gel polymers.
- Unsaturated acids, for example methacrylic acid, to produce carboxylated polymers.
- Elemental sulfur, to produce polymer chains with sulfur segments in the backbone, facilitating peptization.

Structure and Structural Variables

Polychloroprene is highly regular in structure and consists primarily of *trans*-units; however, there are sufficient *cis*-units to disturb the backbone symmetry and maintain a rubbery state. Therefore, the physical, chemical, and rheological properties of polychloroprene are, to a large extent, dependent on the ability to change the molecular structure, for example the *cis/trans* ratio, long chain branching, and the amount of crosslinking.

Figure 1 compares the structural units of commercially available polychloroprene. In this polymer the 1,4-addition (I), in particular the 1,4-trans-addition (Ib), is dominant. In addition, small proportions of the 1,2-(II) and 3,4-(III) structures are also present. These polymer structures are combined in sequential isomers derived from head-to-tail (IV), head-to-head (VI), and tail-to-tail (V) addition.

The amounts of the different structural units depend on the polymerization temperature.[4]

In addition, the preparation of stereoregular polychloroprene by unusual polymerization conditions has demonstrated that the glass-transition temperature and the melting temperature of the polymer are inverse functions of the polymerization temperature.[5,6]

Using standard polymerization conditions, crystallization is an inherent property of all polychloroprene rubbers.[7]

The crystallinity of polychloroprene makes processing difficult, and the vulcanizate increases in hardness with age. Therefore, polychloroprene polymers are normally produced using high polymerization temperatures (30–60°C) or using additional comonomers that interfere with crystallization. Through such measures, the crystallizing tendency of polychloroprene in both the raw and vulcanized states is reduced.

Variations in microstructure are responsible for significant changes in polymer properties.

COMPOUNDING AND PROCESSING

Chloroprene rubber (CR) vulcanizates can be made using filters, plasticizers, antioxidants, and processing aids commonly used in diene rubber compounding.

Selection of Chloroprene Rubber Grades

To achieve the best compromises in compounds and vulcanizate properties, a proper selection of grades is essential.

Blends with Other Elastomers

Blends of CR and other elastomers are desirable in order to achieve special properties either of CR-based compound or of a compound mainly based on the second component. In many cases general purpose diene rubbers, such as SBR, BR, or NR are also used in order to reduce compound costs.

It is advantageous to select compatible polymers as blending components, in order to form alloys during the mixing process.

With respect to CR this is unlikely in most practical cases. and in addition crosslinking systems are generally of dissimilar reactivity, thus resulting in an inhomogeneous network. Accelerator systems based on thiurams and amines are best for effective co-cure.

Accelerators

CR can be crosslinked by metal oxides alone. Thus, there is a major difference between general-purpose diene rubber and CR. Suitable accelerators help to achieve a sufficient state of crosslinking under the desired conditions.

Zinc oxide (ZnO) and magnesium oxide (MgO) are the most frequently used metal oxides. Lead oxides are used for optimal water/acid/alkaline resistance.

If zinc oxide is used alone, the crosslinking density remains low, although the state of cure is increased. Best results are obtained with a combination of zinc oxide and magnesium oxide. There is a tendency to "marching modulus" characteristics if high dosages of both metal oxides are used.

Various types of lead oxides are used in high dosages, especially if resistance against water, acids, and alkaline solutions is required. Lead oxides make "self-curing" CR compounds possible.

As an organic accelerator, ethylene thiourea (ETU) is widely applied. Different derivatives of thiourea, such as diethyl thiourea (DETU) and diphenyl thiourea (DPTU), are typical ultra-fast accelerators, especially suitable for continuous cure.

Best tear resistance is achieved by a combination of sulfur, thiurams, guanadine-based accelerators, and methyl mercapto benzimidazole (so-called MMBI-system).

A wide variety of other accelerators have been used with CR, but most have not achieved widespread acceptance.

Antioxidants, Antiozonants

Vulcanizates of CR need to be protected by antioxidants against thermal aging and by antiozonants to improve ozone resistance. Some of these ingredients also improve flex-fatigue resistance. Slightly staining antioxidants that are derivatives of diphenylamine, including octylated diphenylamine (ODPA), styrenated diphenylamine (SDPA), and 4,4-bis-(dimethylbenzyl)diphenylamine, are especially effective in CR compounds.

Antioxidant effects are highly dependent on dosage of diphenylamine antioxidants, revealing that a level of 2–4 phr is sufficient for most applications. Similar relationships have been described by Brown and Thompson.[8] Ozone protection is improved if antiozonants are used together with microcrystalline waxes.

Fillers

Generally, CR can be treated as a diene rubber as far as fillers are concerned. Carbon black and fillers based on minerals, either synthetic or natural in origin, can be employed. Active fillers improve physical properties, whereas less active or predominantly inactive fillers are used to reduce compound cost.

Although CR is inherently flame retardant, for certain applications it is necessary to further improve this property. This can be achieved by adding aluminum trihydrate, zinc borate, or antimony trioxide. Silane coupling agents in conjunction with silica fillers are often used to improve this property. Mercaptosilanes or chlorosilanes are preferred.

Plasticizers

Mineral oils, organic plasticizers, and special synthetic plasticizers can be used in typical CR compounds in dosages varying from 5 phr to about 50 phr. These plasticizers can lower glass transition temperature, influence crystallization tendency, and reduce compound cost.

PROPERTIES AND APPLICATIONS

The attraction of CR lies in its combination of technical properties, which is not easily matched by other types of rubber at a comparable price. Properly formulated, CR vulcanizates are capable of yielding a broad range of excellent properties:

- good mechanical properties, independent of the use of reinforcing agents
- good ozone and weather resistance
- good resistance to chemicals
- high dynamic load bearing capacity
- good aging resistance
- favorable flame resistance
- good resistance to fungi and bacteria
- good low temperature resistance
- low gas permeability
- medium oil and fuel resistance
- adequate electrical properties for a number of applications
- vulcanizable over a wide temperature range with different accelerator systems
- good adhesion to reinforcing and rigid substrates, such as textiles and metals

Physical Properties

Polychloroprene vulcanizates possess considerable strength. With optimum formulations, the level is comparable to that of NR, SBR, or NBR. Tear propagation resistance is better than that of SBR. Tear propagation resistance of vulcanizates containing active silica is greater than that of natural rubber. CR vulcanizates show good elasticity, although they do not reach the level of NR.

The abrasion resistance of CR is comparable to that of NBR. The gas permeability is roughly equivalent to NBR of medium acrylonitrile (ACN) content. Thermal conductivity and thermal expansion are comparable to those of other elastomers.

Aging and Heat Resistance

CR vulcanizates, especially those that contain optimized antioxidants and crosslinking systems, display good heat resistance. They neither soften nor harden over a long period of stress, remaining serviceable and elastic. With respect to thermal aging, CR is positioned between NR and CSM in the ASTM D 2000 and SAE J 200 systems.

Low-Temperature Flexibility

Apart from crystallization effects. differential scanning calorimetry reveals that the glass-transition temperature for polychloroprene of around –40°C is practically independent of the type of polymer tested. Compounding ingredients can lower the glass-transition temperature.

The remarkable effect of synthetic low-temperature plasticizers allows CR vulcanizates to exhibit elastic behavior down to temperatures of around –45°C to –50°C, depending on the formulation.

Flame Retardance

CR vulcanizates are inherently flame resistant. The good flame resistance properties of the polymer itself mean that stringent end-user specifications can be fulfilled by appropriate compounding. Limited oxygen index values of 50% can be attained with CR.

However, like all organic substances, CR vulcanizates will decompose at high temperatures like those encountered in open fires. In addition to decomposition products such as carbon dioxide and water, corrosive hydrogen chloride gas is also formed.

Resistance to Fluids

CR possesses medium oil resistance, making the polymer suitable for articles resistant to intermittent oil exposure or exposure to less aggressive oils such as paraffinic/naphthenic oils or the corresponding hydraulic oils. The resistance to regular fuels is limited, and to fuels with high aromatic content it is insufficient. However, properly compounded CR exhibits good resistance to dilute acids and alkaline solutions at moderate temperatures.

Resistance to Fungi and Bacteria

Rubber articles in contact with soil for longer periods of time are liable to attack by soil bacteria and fungi. Rubber-coated underground cables can be destroyed this way. Unlike most other rubber types, CR shows a surprisingly high level of resistance to these microorganisms. This resistance can be further enhanced by means of rubber compounding.

Applications

CR is one of the dominant specialty elastomers and is the basis for a wide variety of technical rubber goods. The estimated worldwide consumption of CR solid rubber, CR adhesive raw materials, and CR latex is about 300,000 tons per year. Approximately two thirds of this consumption is for typical rubber applications.

REFERENCES

1. Bauchwitz, P. S.; Finlay, J. B.; Stewart, C. A. Jr. *Vinyl and Diene Monomers Part II*; Leonhard, E. C., Ed.; John Wiley & Sons: NY, 1971; p 1149.
2. Johnson, P. R. *Rubber Chem. Technol.* **1976**, *49*, 650.
3. Obrecht, W. In *Methoden der organischen Chemie* Bartl, H.; Falke, J., Eds.; Thieme Verlag: Stuttgart, NY, 1987; pp 843–856.
4. Petioud, R.; Tho Pham, Q.; Pol. I. *Sci., Pol. Chem. Ed.* **1985**, *23*, 1333.
5. Garett, R. R.; Hargreaves, C. A.; Robinson, D. N. *J. Macromol. Sci. Chem. A* **1970**, *4.8*, 1679.
6. Aufdermarsh, C. A.; Pariser, R. *J. Pol. Sci.* **1964**, *A2*, 4727.
7. Rohde, E.; Bechen, H.; Mezger, M. *Kautschuk & Gummi. Kunststoffe* **1989**, *42*, 1121. In German.
8. Brown, D. C. H.; Thomson, J. *Rubber World* **1981**, *November, 32.*

CHLOROSULFONATED POLYETHYLENE

Royce E. Ennis
E. I. Du Pont de Nemours and Company, Incorporated
Beaumont Works

Chlorosulfonated olefin polymers, known as chlorosulfonated polyethylene, CSM [68037-39-8], are widely available general-purpose elastomers, coating compositions, and adhesives. CSM originally described chlorosulfonated ethylene homopolymers only, but this family of polymers has been extended to include such base resins as polypropylene and copolymers of ethylene and *alpha*-monoolefins including alkyl esters and carboxylic acids and graft copolymers. CSM elastomers are best known for their excellent combination of heat, oil, and weather resistance, when properly compounded and vulcanized. So they are superior to chloroprene rubber (CR), and inferior to, but more cost effective than silicone elastomers (MQ). They also have a good balance of mechanical properties and resistance to oxidizing chemicals over a broad temperature range. Recent research has concentrated on development of CSM polymers from base resins other than ethylene homopolymers. Thus new grades have been introduced with uniquely temperature-stable vibration dampening properties for use in dynamic applications.[1] Grafted polymers have also been developed for promoting excellent coating and ink adhesion to structural polymers such as polypropylene.[2]

Introduction of linear low density polyethylene resins in the late 1970s and the 1980s paved the way to new CSM polymers with the advantages of both linear and branched polymers, imparting improved low temperature flexibility to the linear based grades.[3] Further extension of chlorosulfonation technology to include other base resins has led to an array of sophisticated polymers ranging in consistency from soft and rubbery to hard and brittle, with a wide variety of specialty applications.

PREPARATION

Chlorosulfonated polyolefins are prepared by interaction of the base resin with chlorine and either sulfur dioxide or sulfuryl chloride in the presence of a radical initiator. Sulfuryl chloride may be used alone as a chlorosulfonating agent, but must be accompanied by a catalytic amount of pyridine or other organic base.

The chlorination and chlorosulfonation reactions are usually carried out simultaneously but may be carried out in stages, that is, the resin may be partially chlorinated with elemental chlorine and then chlorosulfonated to the desired composition with a chlorosulfonating agent. In commercial processes, the reaction is carried out in a homogeneous solution; however, other processes have been disclosed, for example, reaction extrusion,[4] fluidized bed reaction,[5] chlorosulfonation of solvent swollen polymers in aliphatic fluorocarbons,[6] and more.

PROPERTIES

Incorporation of chlorine atoms onto the polyolefin backbone causes sufficient molecular irregularity to break up crystalline chain segments of the base resin. As the chlorine content is increased, the thermoplastic material becomes more amorphous and eventually behaves as an elastomer due to the inherent flexibility of the polyethylene chain in the absence of crystallinity. Pendant chlorine atoms, however, cause interference with chain segment rotation which results in an increase in glass transition temperature with chlorine level. The optimum chlorine content for an amorphous elastomer is the minimum amount required to completely destroy all crystalline segments. Chlorine levels lower than optimum yield products which are partly thermoplastic, whereas higher chlorine levels make them less rubbery and, eventually, brittle at ambient temperatures. Useful products have been produced at chlorine levels ranging from 18% to 65%. A representative polymer at optimum chlorine content for elastomeric uses contains, on average, 18 chlorine atoms and one $-SO_2Cl$ group for each 100 chain carbon atoms.

The chlorine atoms on the polymer backbone not only provide elastomeric properties, but also give useful improvement in oil, chemical, and flame resistance.

Sulfonyl chloride groups on the chain are only there to the extent required to give an adequate number of cure sites for effective vulcanization. The chlorine and sulfonyl chloride groups in CMS undergo predictable reactions from their functions in low molecular weight substances. Thus sulfonyl chloride groups form acids, amides, and esters readily. They also decompose in the presence of heat and organic bases to form polymer free radicals. These reactions, in fact, form the basis for much of the cure chemistry of CSM elastomers.

APPLICATIONS

CSM products are classed as specialty elastomers but enjoy use in a broad spectrum of industrial, automotive, construction, coatings, and miscellaneous applications. In industrial hose CSM is particularly useful because of its combination of heat, oil, and chemical resistance and its outstanding ozone and weather resistance. It is also preferred because of the frequent need for stable bright colors.

In the electrical industry, CSM is preferred as a protective jacketing or sheathing over elastomers with superior insulation resistance but poor weatherability. CSM is also used as an integral insulator for up to 600 volts. This combination of insulation resistance and long-term ozone resistance makes it a clear choice for high quality applications.

Low permeability to moisture and refrigerants makes high chlorine grades ideal for air conditioning hose and tubes.

A variety of miscellaneous end uses take advantage of CSM's oil resistance, colorability, and weather and ozone resistance. Grades based on low density polyethylene are utilized in coating applications such as for awnings, gas hose, escalator handrails, rubber boats, and so on. CSMs are also used for curb pump hose, mining cable jackets, and appliance wire jackets. Grades based on ethylene vinyl acetate polymers are used for traffic and marine paints.

REFERENCES

1. Ennis, R. E. Paper presented at ACS Rubber Div., 1989; Detroit, MI.
2. Brugel, E. G. Hypalon Preliminary Data Sheet, CP 826 and CP 827, 1993; E. I. DuPont de Nemours: Wilmington, DE.
3. Ennis. R. E. Hypalon Bulletin HP 276.12; E.I. DuPont de Nemours: Wilmington, DE.
4. Lorenz, J. C. (DuPont) U.S. Patent 3 347 835, 1967.
5. Riffi, R. M. (Union Carbide) U.S. Patent 4 560 731, 1985.
6. Hoechst, A. G. (Farbewerke) Ger. Patent 1 068 012, 1959.

Chlorosulfonated Polyethylene Rubber

See: *Blends, Interchain Crosslinking*
Chlorosulfonated Polyethylene

Chlorotrifluoroethylene Copolymers

See: *Engineering Thermoplastics (Survey of Industrial Polymers)*
Tetrafluoroethylene Copolymers (Overview)
Vinyl Chloride Copolymers

Chondroitin Sulfate

See: *Chondroitin Sulfate (Overview)*
Glycosaminoglycans (Overview)

CHONDROITIN SULFATE (Overview)

Victor B. Diaz,* Esteban P. Fuentes, Oscar E. Martinez, and Arturo R. Kraglievich
Syntex S. A.

Glycosaminoglycans are polysaccharides which occur widely in the animal kingdom—in vertebrates and invertebrates, terrestrial and marine organisms, and mammals and nonmammals. Five kinds of glycosaminoglycans have been found in the tissues and fluids of vertebrates: chondroitin sulfates, keratan sulfates, dermatan sulfates, hyaluronic acid, heparin, and heparan sulfate.

Chondroitin sulfate is widely found in the connective tissues of animals. It is a linear polymer composed of a repetition of D-glucuronic acid linked alternately through a (1→3) bond to an *N*-acetylgalactosamine molecule. The *N*-acetylgalactosamine holds up an ester sulfate in its C-4 position (originating the chondroitin-4-sulfate, or, chondroitin sulfate A) or in its C-6 position (originating the chondroitin-6-sulfate, or, chondroitin sulfate C). Both sulfation standards exist, keeping between them relatively constant proportions by which they can be defined.[1,2] The acetyl group is always placed as a substituent of the amino sugar nitrogen (**Figure 1**). The disaccharides are in turn linked among themselves following the (1→4) sequence between the *N*-acetylgalactosamine and the uronic acid, which in turn links the following amino sugar through a (1→3) bond.

NATURAL OCCURRENCE AND DISTRIBUTION

Chondroitin-4-sulfate is widely distributed in bovines and porcines. It is found in their nasal and tracheal cartilage, bones, sclera, leucocytes and blood platelets, skin, umbilical cords, and urine. The chondroitin-6-sulfate has been found in skin, fleshy nucleus, umbilical cords, and cardiac valves, showing the same composition but slightly different physical properties from chondroitin-4-sulfate.[3] Both isomers exist in the biological material already mentioned, but the resolution of such a compound is not easy.[3]

*Author to whom correspondence should be addressed.

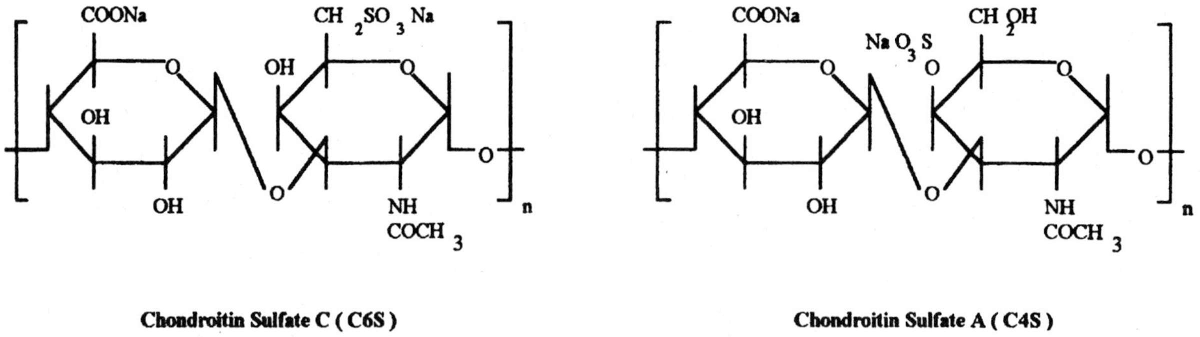

FIGURE 1. Structures of chondroitin-6-sulfate and chondroitin-4-sulfate.

In all cases, chondroitin sulfate may be found as a proteoglycan, but it is never free. As far as we know, hyaluronic acid is the only glycosaminoglycan that is free in fluids and tissues.

ISOLATION AND PURIFICATION

There are two types of methods of isolation and subsequent purification of a glycosaminoglycan. The first primarily isolates polysaccharides from their natural companions—mainly hyaluronic acid and, sometimes, dermatan sulfate. The second, fractionation and purification methods, separate polysaccharides from impurities by using solvent fractionation, ion-exchange chromatography, and combining with quaternary ammonium salts.

IDENTIFICATION AND QUANTIFICATION

Chondroitin sulfate is easily identified by its electrophoretic behavior and by the measure of the specific optical rotation. Electrophoretic methods allow chondroitin sulfate, pure or mixed with other glycosaminoglycans, to be identified. Several methods have been described depending on the support used (i.e., cellulose acetate or agarose gel) or on the developing buffer (i.e., barium acetate, 0.1 M HCl, or pyridine esters.[4–9] The most frequently used method is electrophoresis in barium acetate on cellulose acetate sheets.[10] The quantification is made by using a densitometer or by band elution measuring the absorbance at 605 nm.

Another physical property identifying glycosaminoglycans is the specific optical rotation. The values usually found are heparin (+40/+60), DS(–40/–70), chondroitin-6-sulfate (–10/–15), and chondroitin-4-sulfate)–20/–30), depending on the degree of purification.

The content of sulfur, uronic acids, and hexosamines permits the study of purity and structural integrity. Concerning the latter, the absolute value is as important as the relation between the parameters. The sulfate/carboxylate molar ratio in chondroitin sulfate is ~ 1; in practice, the ratio falls between 0.9 and 1.1, depending on the method of analysis used.

STRUCTURE ANALYSIS

The chondroitin sulfate structure depends essentially on the species and/or tissue of origin and on the treatment used during the manufacturing process. The basic analysis of the structure of glycosaminoglycans, and particularly that of the chondroitin sulfate, includes the study of molecular weights distribution, monosaccharide composition, and polysaccharide chain-linkage region with the protein core in the proteoglycan of origin.

Molecular Weight Determination

The CS molecular weights of chondroitin sulfate depend greatly on the species of origin, on the manufacturing process used, and on the analysis technique.[11] The results found in our laboratory are as follows:

- chondroitin-4-sulfate of bovine trachea has a molecular weight of 20,000 D;
- chondroitin-6-sulfate of shark cartilage has a molecular weight of 45,000–50,000 D; and
- chondroitin-6-sulfate of shark fin has a molecular weight of 36,000 D.

APPLICATIONS

Normal Aspects of Articulations

Articulations are formed by a rounded and a molded bone end. These shapes allow the movement of one bone over the other. The bones are connected by the articular capsule, and this in turn is held and strengthened by ligaments and tendons. The bone ends inside of the articular cavity are covered by articular cartilage. The hyaline cartilage is a specialized connective tissue acting as an elastic damper of the compression forces and as an area of detrition resistance. It is mainly composed of a fibrocartilaginous matrix in which relatively few chondrocites are dispersed. Such extracellular matrix is particularly big and represents more than 90% of the tissue volume. It is composed of a thick net of collagen fibers (type II) soaked in a highly concentrated solution (up to 100 mg/mL) of aggregate proteoglycan molecules.[12]

Proteoglycan Composition

Proteoglycans are a family of compounds with a linear protein core as a basic structure to which is linked, perpendicularly, lateral chains of chondroitin-4-sulfate, chondroitin-6-sulfate, keratan sulfate, and short oligosaccharides. The proteoglycans found in the cartilage have a high molecular weight; Hascall found a range of 0.5×10^6 to 4.0×10^6 daltons.[13] In turn, 20–250 proteoglycan units join a hyaluronic acid chain through their protein core with the help of a link protein to form a proteoglycan aggregate, which has a molecular weight higher than 200 × 10^6 daltons.

Proteoglycans have a rigidly extended spatial structure, because of their ramiform disposition and repulsion among the negative charges of the glycosaminoglycan adjoining the molecules. Their three-dimensional disposition and hydrophilic nature give proteoglycans the following properties: selective permeability, water retention, tissue resilience, and compressive stiffness.

Chondroitin Sulfate Applications

We have outlined the physiological importance of chondroitin sulfate, but a chondroitin sulfate exogenous contribution has also proved to be useful in degenerative articular pathologies.

All over the world, there are different pharmaceutical specialties for human and veterinary use that contain chondroitin sulfate. Such specialties are administered orally or by intramuscular injection. Note that chondroitin sulfate oral bioavailability is discussed in several studies. Palmieri and Conte teach its absorption via oral[14–16] and Baici, covers a lack of absorption.[17] This problem is not easily solved even considering the numerous clinical tests with successful results in exogen chondroitin sulfate therapies.[18–22]

Chondroitin sulfate acts as an antiinflammatory through the inhibition of the complement.[23,24]

The inhibitory effect that chondroitin sulfate exerts on the leukocytic elastase and lysosomal enzymes has been studied *in vitro*.[25–27] These enzymes are responsible for the degradation of the cartilage matrix main components.[28–30] This capacity for chondroitin sulfate to interact, which is related to its molecular weight and its polyanionic characteristic, is important because it allows a regulatory mechanism of the enzymatic activity, thus protecting the articular cartilage by inhibiting its degradation. It has also been suggested that chondroitin-4-sulfate participates in the ossification and calcification processes and that chondroitin-6-sulfate maintains articular integrity.[31,32]

Vacha describes an increase in the cartilage ribonucleic acid metabolism caused by chondroitin sulfate.[33] This work suggests that chondroitin sulfate regulates the formation of a new cartilage matrix by stimulating the chondrocytic metabolism and the synthesis of collagen, intraarticular proteoglycan, and hyaluronic acid.[34,35] Our group, in collaboration with the Facultad de Ciencias Veterinarias de la Universidad del Centro de la Provincia de Buenos Aires, Argentina, performed an activity test on a chondroitin sulfate formulation that was orally administered,[22] based on a model of equine aseptic arthritis caused by intraarticular injection (carpal joint) of 0.5 mL of adjuvant complete Freund (ACF).[36]

The results of the testing showed that chondroitin sulfate had a direct antiinflammatory action on the injury. Differences noted in the response to the strained flexion indicate a positive effect on the arthralgia, thus allowing a fast and better recovery. The inhibiting effect of the enzymatic activity is made evident by the protein content differences found in the synovial fluid. This corroborates the protective effect of a chondroitin sulfate exogen on articular surfaces. The radiographic study of the groups reveals the existence of a subchondral sclerosis and the loss of normal characteristics of the articular capsule, present only in the untreated animals, thus indicating that chondroitin sulfate has an inductive effect on the growth of cartilage injured by ACF.[22]

Chondroitin sulfate has low toxicity. Studies of acute and chronic toxicity via oral intake and intravenous injections were made by administering up to 2 g/kg living weight to mice. No fatal cases or toxic consequences were observed, as it was technically impossible to define an LD50, and no anticoagulant action limiting its use was observed, unlike other glycosaminoglycans and even other modified chondroitin sulfates.[37–39]

REFERENCES

1. Comper, W. D. *Heparin and Related Polysaccharides*; Gordon and Breach Science: New York, 1981; Vol. 7.
2. Chaplin, M. F.; Kennedy, J. F. *Carbohydrate Analysis, A Practical Approach*; IRL: Oxford, England, 1986.
3. Pigman, W.; Horton, D. *The Carbohydrates*; Academic: New York, 1970; pp 597–605.
4. Nanto, V. *Acta Chem. Scand.* **1963**, *17*(3), 857.
5. Jaques, L. B.; Wollin, A. *Anal. Biochem.* **1973**, *52*, 219.
6. Matthews, M. B. *Meth. Carbohydr. Chem.* **1976**, *7*, 116.
7. Wessler, E. *Anal. Biochem.* **1968**, *26*, 439.
8. Wessler, E. *Anal. Biochem.* **1971**, *41*, 67.
9. Gardais, A.; Picard, J.; Tarasse, C. *J. Chromatogr.* **1969**, *42*, 396.
10. Capelleti, R.; Del Roso, M.; Chiarugi, V. P. *Anal. Biochem.* **1979**, *99*, 311.
11. Mathews, M. *Biochem. J.* **1971**, *125*, 37.
12. Hardingham, T. *Biochem. Soc. Trans.* **1981**, *9*, 489.
13. Hascall, V. C.; Sajdera, S. W. *J. Biol. Chem.* **1970**, *245*, 4920.
14. Palmieri, L.; Conte, A.; Giovannini, L.; Lualdi, P.; Ronca, G. *Arzneimittelforschung/Drug Res.* **1990**, *40*, 319.
15. Conte, A.; Palmieri, L.; Segnini, D.; Ronca, G. *Drugs Exptl. Clin. Res.* **1991**, *17*(1), 27.
16. Conte, A.; de Berhardi, M.; Palmieri, L.; Lualdi, P.; Mautone, G.; Ronca, G. *Arzneimittelforschung/Drug Res.* **1991**, *41*, 768.
17. Baici, A.; Hörler, D.; Moser, B.; Hofer, H. O.; Fehr, K.; Wagenhäuser, F. J. *Rheumatol. Int.* **1992**, *12*, 81.
18. Fioravanti, A.; Franci, A.; Anselmi, F.; Fattorini, L.; Marcolongo, R. *Drugs Exptl. Clin. Res.* **1991**, *17*(1), 41.
19. Pagliano, F. *Riv. It. Biol. Med.* **1986**, *6*, 1.
20. Conrozier, T.; Vignon, E. *Litera. Rheum.* **1992**, *14*, 69.
21. L'Hirondel, J. L. *Litera. Rheum.* **1992**, *14*, 77.
22. Videla Dorna, I.; Guerrero, R.; Kraglievich, A. R.; Dománico, R.; Diaz, V. B. presented at the *Jornadas de Argentina, Actualización Técnico Cientificas. Asociación Argentina de Veterinaria Equina, San Isidro*, June 16 and 17, 1994.
23. Raepple, E.; Hill, H. U. *Immunochem.* **1982**, *13*, 251.
24. Biffoni, M.; Paroli, E. *Drugs Exptl. Clin. Res.* **1991**, *17*(1), 35.
25. Avila, J. L.; Convit, J. *Biochem. J.* **1976**, *160*, 129.
26. Marossy, K. *Biochem. Biophys. Acta* **1981**, *659*, 351.
27. Baici, A.; Bradamante, P. *Chem. Biol. Interact.* **1984**, *51*, 1.
28. Baici, A.; Salgam, P.; Cohen, G.; Fehr, K.; Böni, A. *Rheumatol. Int.* **1982**, *2*, 11.
29. Sapolsky, A. I.; Altman, R. D.; Woessner, J. F.; Howell, D. S. *J. Clin. Invest.* **1973**, *52*, 624.
30. Velvart, M.; Fehr, K.; Baici, A.; Sommermeyer, G.; Knöpfel, M.; Cancer, M.; Salgam, P.; Böni, A. *Rheumatol. Int.* **1981**, *1*, 121.
31. Harab, R. C.; Mourão, P. A. S. *Biochem. Biphys. Acta* **1989**, *992*, 237.

32. Mourão, P. A. S. *Arthritis Rheum.* **1988**, *31*(8), 1028.
33. Vácha, J.; Pesáková, V.; Krajicková, J.; Adam, M. *Arzneimittelforschung/Drug Res.* **1984**, *34*(1), 607.
34. Soldani, G.; Romagnoli, J. *Drugs Exptl. Clin. Res.* **1991**, *17*, 81.
35. Adam, M.; Krabcova, M.; Muskova, J.; Pesakova, V.; Brettschneider, I. *Arzneimittelforschung/Drug Res.* **1980**, *30*, 1730.
36. Hamm, D.; Goldman, L.; Jones, E. W. *Vet. Med.* **1984**, *79*, 811.
37. Paroli, E.; Antonelli, I.; Biffoni, M. *Drugs Exptl. Clin. Res.* **1991**, *17*(1), 9.
38. Alvarez, R.; Kraglievich, A. R. *Internal Report*; Syntex S. A.: Buenos Aires, Argentina, 1993.
39. Matera, M.; Scalia, S. *Drugs Exptl. Clin. Res.* **1991**, *17*(1), 65.

Coal

See: Coal (Polymer Structures)
Coal Tar Pitch

COAL (Polymer Structures)

Masashi Iino
Institute for Chemical Reaction Science
Tohoku University

Coal is a complex organic material that consists of aromatic or heteroaromatic rings of various ring sizes with substituents such as hydroxyl, alkyl groups, methylene, ether, and other linkages that connect the rings. Coal generally includes a small percentage of minerals that can be removed by acid washing. When Soxhlet extraction with pyridine, one of the best solvents for coal extraction, is carried out exhaustively on coals of different types, the extraction yields obtained are usually in the range of a few wt % to 40 wt % on the mineral matter free basis.

Recent progress in determining the polymer structure of various forms of coal are given below. Other reviews on polymer structures of coal are also helpful for understanding them.[1,2]

PROPERTIES

Crosslink Structure of Solvent-Insoluble Component

It has been widely accepted that coal consists of covalently bound, crosslinked networks, that are insoluble in any solvents, and solvent-soluble, relatively low molecular weight substances trapped in the networks. This coal structure model, often called the "two-phase model,"[3] does not seem to have definite evidences and comes from the following experimental aspects:

- Coal swells and shows elasticity like rubber when it contacts organic solvents such as pyridine.
- A large portion of coal is always insoluble in any solvent.
- The studies of spin-spin relaxation time (T_2 in ^{1}H-NMR) of solvent-swelled coal show that there are at least two types of protons—immobile protons from the crosslinked network (the immobile phase) and mobile protons from the solvent-soluble substances (the mobile phase).

Brenner[4] carefully made a thin-section sample of coal and measured the stress-strain relationship by the compression mode. For the pyridine-swelled samples, this shows rubber-like elasticity clearly. Recently, viscous compliance observed for the pyridine-swelled thin samples was found to be consistent with the structure of coal being a chain-entangled network or a crosslinked network capable of steady-state bond rupture and re-formation,[5] not a covalently bound crosslinked network as described above.[5]

Carbon disulfide-*N*-methyl-2-pyrrolidinone mixed solvent was found to give very high extraction yields of 50–85 wt % for several bituminous coals, suggesting that covalently crosslinked networks make a small contribution to the structure of coal.[6,7]

A new coal structure model, the "mono-phase model," has been proposed.[8] In this model, coal molecules with a wide molecular weight distribution form a giant aggregate by non-covalent bonds. Further study is needed to clarify whether the crosslinking bond is covalent, non-covalent, or chain-entanglement.

APPLICATIONS

Coal has mainly been used for fuels, cokes, carbon materials, and chemicals. The development of new coal-conversion processes to produce polymer materials with useful functions such as electroconductivity and thermal stability is confidently expected.

REFERENCES

1. Green, T. K.; Kovac, J.; Brenner, D.; Larsen, J. W.; *Coal Structure*; Meyers, R. A., Ed.; Academic: New York, 1982; pp 199.
2. Given, P. H. *Coal Science*, Gorbaty, M. L. et al., Eds.; Academic: New York, 1984; Vol. 3, p 63.
3. Derbyshire, F.; Marzec, A.; Shulten, H-R.; Wilson, M. A.; Davis, A.; Tekeley, P.; Delpuech, J-J.; Jurkiewicz, A.; Bronnimann, C. E.; Wind, R-A.; Maciel, G. E.; Narayan, R.; Bartle, K.; Snape, C. *Fuel* **1989**, *68*, 1091.
4. Brenner, D. *Am. Chem. Soc. Div. Fuel Chem. Prepr.* **1986**, *31*(1), 17.
5. Cody, G. D.; Davis, A.; Hatcher, D. G. *Energy Fuels* **1993**, *7*, 455.
6. Iino, M.; Takanohashi, T.; Ohkawa, T.; Yanagida, T. *Fuel* **1991**, *70*, 1236.
7. Liu, H-T.; Ishizuka, T.; Takanohashi, T.; Iino, M. *Energy Fuels* **1993**, *7*, 1108.
8. Nishioka, M. *Fuel* **1992**, *71*, 941.

COAL TAR PITCH

Akira Tanaka*
Department of Materials Science
The University of Shiga Prefecture

Katsuhisa Tokumitsu and Chiharu Yamaguchi
Osaka Gas Company, Ltd.

Coal tar pitch is a residual product that results from the distillation of coal tar. It is a black, shiny material that is solid and brittle at low temperatures and liquid at high temperatures. Since pitch constitutes over 50% of the crude tar, its use has a major effect on the economics of tar processing.

Coal tar pitch is a mixture of a great number of compounds with different chemical structures and different molecular weights. The number average molecular weight of coal tar pitch

*Author to whom correspondence should be addressed.

ranges from 150 to 4000. Generally, coal tar pitch shows a hydrogen/carbon (H/C) ratio of about 0.5, which compares with a >1.0 ratio for pitch obtained from petroleum, and a 1.0–0.5 ratio for that obtained from organic compounds. The H/C ratio is a measure of the degree of aromatic structure. The lower the ratio, the greater the aromatic structure. The H/C ratio approaches 0.4 when pitch consists mostly of an aromatic structure.

Coal tar pitch is generally categorized according to its softening point (SP); soft (<60°C), intermediate (60–75°C), and hard (>75°C) pitch.

MODIFICATION

Modified pitch may be classified into three types according to structure: isotropic pitch, pseudomesophase pitch, and mesophase (anisotropic) pitch. Isotropic pitch is produced by air blowing and vacuum distillation.[1,2] The pseudomesophase category comprises the premesophase and the dormant mesophase. The premesophase pitch is produced by hydrogen transfer from THQ (1,2,3,4-tetrahydroquinoline) and heat treatment.[3] The dormant mesophase pitch[4] is produced by BenKeser reduction and heat treatment. The premesophase and dormant mesophase pitch have good spinnability and good thermal stability. The as-spun pitch fibers and the infusible fibers made from premesophase and dormant mesophase pitch show isotropic and anisotropic textures, respectively. When spherules meet each other, coalescence occurs to produce larger droplets, leading eventually to bulk mesophase. When the coalesced mesophase is further pyrolyzed, plastic deformation produces two types of fine textures: fibrous and mosaic. This mesophase transformation is essential for the formation of precursors to be carbonized or graphitized, and the fibrous texture of the bulk mesophase is essentially important in connection with graphitizability.[5]

USE OF COAL TAR PITCH

The range of uses for carbon materials has increased measurably. This is due to their many advantageous characteristics, such as chemical resistance, high strength, high elasticity, dimensional stability, heat resistance, sliding performance, resin impregnability, electric conductivity, sound absorbability, and bio-affinity. Besides its application to many kinds of binders, there are many uses of coal tar pitch for high-performance carbon materials, such as needle coke, carbon fiber, absorbent, and high-density isotropic carbon.

Pitch production has tended to drop year by year.[6] But the demand for high performance carbon materials is increasing, as the demand for carbon fiber shows.[7]

REFERENCES

1. Maeda, T. et al. *Carbon* **1993**, *31*, 407.
2. Barr, J. B.; Lewis, I. C. *Carbon* **1978**, *16*, 439.
3. Yamada, Y.; Honda, H. Japanese Patent 58 18421, 1983.
4. Otani, S.; Gomi, S. Japanese Patent 57 100186, 1982.
5. Honda, H. *Carbon* **1988**, *26*, 139.
6. Private communication, Osaka Gas Company Ltd. 1991.
7. Sawanobori, T. "*The 6th Fukugozai Symposium Text*," *Tansosen-i*; Kyokai: 1993, p 13 (in Japanese).

Coatings

See: *Alkyd Emulsions*
Alternating Radical Copolymerization
Antistatic Agents, Polymeric
Electrorheological Materials
Fluoropolymer Coatings (New Developments)
High Solids Coatings (Microspheres from Nonaqueous Polymer Dispersions)
High Solids Coatings (Use of Supercritical Fluids)
High Solids Polyurethane Coatings
Hydrophilic Polymers (for Friction Reduction)
Hyperbranched Aliphatic Polyesters
Hyperbranched Polyesters
Japanese Lacquer: Japan: Urushi (Properties of Urushi Liquid and Urushi Film)
Marine-Antifouling Coating Materials
Methacryloyl Isocyanate and Methacryloylcarbamate
Oriental Lacquer
Paper Coatings
Photoinitiators (for Free Radical Polymerization)
Photoinitiators (for Photocuring)
Poly(ester amide) and Poly(ester imide) Resins (Binders for Air-Drying Protective Coatings)
Poly(imide siloxane)s
Polyimides (Introduction and Overview)
Polypyrrole (Processable Dispersions)
Polysulfides (Prepared from Sulfur Dioxide)
Polyurethanes (Overview)
Polyurethanes, Blocked Copolymer (Reactive Modifiers for Epoxy Resins)
Poly(p-xylylene)s
Poly(p-xylylene)s, Coatings and Films
Powder Coatings (Overview)
Quinone-Diamine Polymers
Resins (Applications in Coatings Industry)
Silicone Release Coatings
Silicone Rubber Latex
Styrene-Maleic Anhydride Copolymer
Surface Modification (Overview)
Ultrathin Films (by Plasma Deposition)
Vinyl Chloride Copolymers
Vinylidene Fluoride-Based Thermoplastics (Applications)
Vinylidene Fluoride-Based Thermoplastics (Blends with Other Polymers)
Water-Borne Coatings (Urethane/Acrylic Hybrid Polymers)

COBALT CARBONYL COMPLEXES (with Main Chain Acetylenic Groups)

William E. Douglas
Unité Mixte CNRS/Rhône-Poulenc/UM II CNRS UMR 44
Université de Montpellier II

It has been known for many years that dicobalt octacarbonyl very readily forms air-stable complexes with acetylenes on simply stirring a solution of the acetylene-containing monomer or polymer with $Co_2(CO)_8$ at room temperature.[1–3]

Dicobalt hexacarbonyl alkyne complexes with organometallic substituents (X,Y = silyl, stannyl, boryl) have been prepared in this way and studied by using multinuclear NMR spectroscopy.[4] The compounds decompose slowly in light and rapidly above 60–70°C. In many cases they can be purified by sublimation or chromatography on alumina. The solutions in chloroform or benzene are stable for several days at RT.

In spite of the wide range of monomeric cobalt carbonyl acetylene complexes that have been prepared and studied, cobalt carbonyl has been complexed to acetylene polymers in only a few cases as discussed below.

PREPARATION AND PROPERTIES

Magnus and Becker synthesized the acetylene cobalt carbonyl polymer **6** starting from the monomeric ethyne cobalt carbonyl complex **2** (**Scheme I**).[5] Complex **2** was treated at –78°C with either $LiNPr_2^i$ or $LiN(SiMe_3)_2$ to give a solution of **3**. The quenching of this solution with water gave rise to a 1:1 mixture of the two dimers **4** and **5**. When **4** was allowed to stand in methanol, the black insoluble polymer **6** was slowly precipitated over a period of 24 h. The microanalytical data indicated that **6** was a polyacetylene of ca. 30 units with each triple bond coordinated to a $Co_2(CO)_6$ group. Preliminary experiments indicated that both **4** and **6** exhibit electrical conductivity.

Soluble ester-containing polydiacetylenes with $Co_2(CO)_6$ units attached to varying proportions of the alkyne bonds have been prepared and the non-linear optical properties investigated.[6]

Matsumoto et al. introduced cobalt carbonyl units into an acetylene polymer containing backbone platinum and silicon groups.[7]

Corriu, Douglas, and Yang incorporated cobalt carbonyl into polycarbosilanes containing main-chain acetylene and aromatic groups.[8]

Polymers containing backbone diacetylene and silicon or germanium groups have been complexed with $Co_2(CO)_6$ units.[9] Thus, polymers **21a–d** were stirred with roughly one equivalent of dicobalt octacarbonyl in THF at RT for 12 h; in the case of **21e** the reaction was performed in toluene at 65°C for 14 h (**Scheme II**). Microanalysis, IR, and multinuclear NMR spectroscopy suggested the overall structures shown in Scheme VII for the resulting polymers **22a–d** with one in two acetylene groups complexed by $Co_2(CO)_6$. In most cases, three ^{29}Si signals were observed and these were assigned to the three silicon environments (α), (β), or (γ) illustrated in Scheme II. It was noted that for **22e**, no cobalt carbonyl units had become attached to the silole rings.

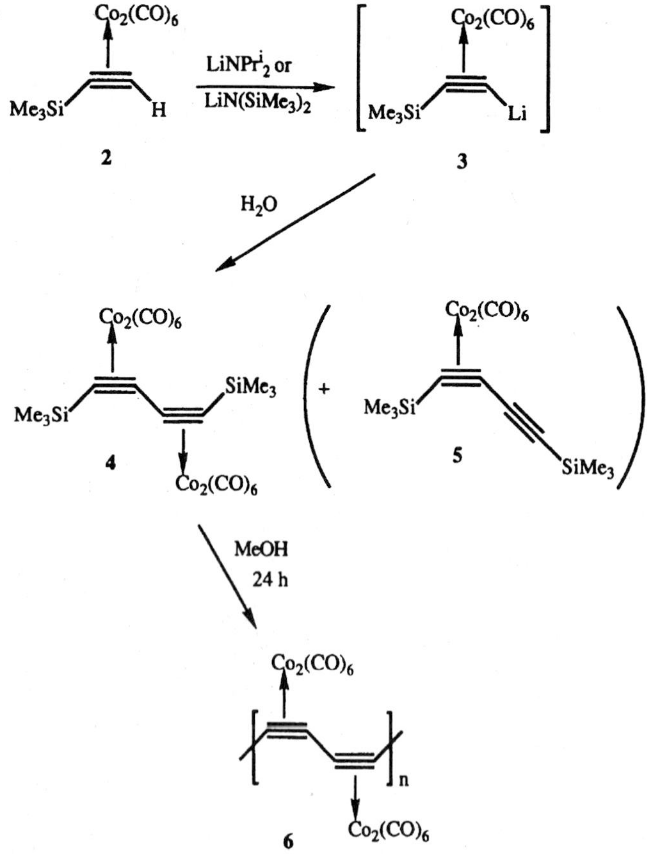

I

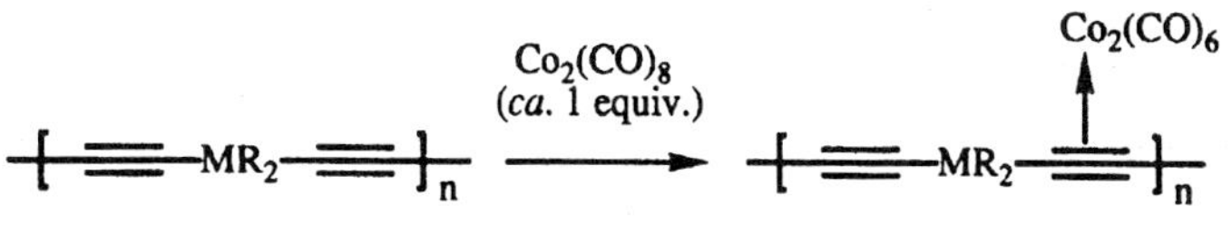

21a-e **22a-e**

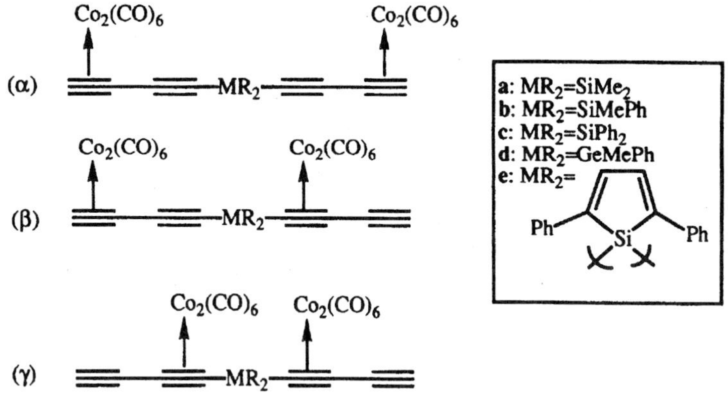

II

APPLICATIONS

Although, as mentioned above, one possible use for acetylene polymers containing cobalt carbonyl units is in electroconductivity,[5,7] it should be noted that in a recent electrochemical study of H C≡C – C_6 H_4 – C≡CH and of its cobalt carbonyl complex $[Co_2(CO)_6]_2$(HC≡C–C_6H_4–C≡CH) no electronic interaction between the two $Co_2(CO)_6$ centers could be detected.[10] It was concluded that π-electron delocalization is hindered by coordination of cobalt carbonyl to the acetylene groups because of the resulting disruption of chain linearity. However, Magnus and Becker pointed out that cobalt carbonyl-containing polymers such as **6** have the potential to act as molecular metals because they are low-dimensional solids containing low-valent metal clusters that can function as electron reservoirs.[5]

As mentioned above, Agh-Atabay et al. have demonstrated the interesting non-linear optical properties of acetylene polymers containing coordinated cobalt carbonyl units.[6] Further developments in this area can be expected.

Finally, polymers such as **22** can be pyrolyzed to give useful mixed ceramics incorporating cobalt phases. The organometallic precursor polymer route to ceramics offers several advantages over classical sintering processes, including initiation of solid-state reactions under milder conditions and also improved control over composition, microstructure, and the final form of the ceramic product.

REFERENCES

1. Greenfield, H.; Sternberg, H. W.; Friedel, R. A.; Wotiz, J. H.; Markby, R.; Wender, I. *J. Am. Chem. Soc.* **1956**, *78*, 120.
2. Kruerke, U.; Hübel, W. *Chem. Ber.* **1961**, *94*, 2829.
3. Dickson, R. S.; Fraser, P. J. *Adv. Organomet. Chem.* **1974**, *12*, 323.
4. Galow, P.; Sebald, A.; Wrackmeyer, B. *J. Organomet. Chem.* **1983**, *259*, 253.
5. Magnus, P.; Becker, D. P. *J. Chem. Soc. Chem. Commun.* **1985**, 640.
6. Agh-Atabay, N. M.; Lindsell, W. E.; Preston, P. N.; Tomb, P. J.; Lloyd, A. D.; Rangel-Rojo, R.; Spruce, G.; Wherrett, B. S. *J. Mater. Chem.* **1992**, *2*, 1241.
7. Matsumoto, T.; Kotani, S.; Shiina, K.; Sonogashira, K. *Appl. Organomet. Chem.* **1993**, *7*, 613.
8. Corriu, R. J. P.; Douglas, W. E.; Yang, Z. *Polymer* **1993**, *34*, 3535.
9. Corriu, R. J. P.; Devylder, N.; Guérin, C.; Henner, B.; Jean, A. *Organometallics* **1994**, *13*, 3194.
10. Osella, D.; Gambino, O.; Nervi, C.; Ravera, M.; Russo, M. V.; Infante, G. *Gazz. Chim. Ital.* **1993**, *123*, 579.

Cobalt-Containing Polymers

See: *Cobalt Carbonyl Complexes (With Main Chain Acetylenic Groups)*
Organometallic Polymers, Cobalt-Containing

COLLAGEN

Hiroshi Itoh and Tereo Miyata
Koken Bioscience Institute

Collagen is the primary connective tissue protein in animals, accounting for about 30% of total body protein in mammals. It serves as an important mechanical component for maintaining body strength and for protecting the body from external stimuli. In addition, collagen also helps determine the shape of the cells and their differentiated function as an extracellular matrix. There are reportedly about 15 different types of collagen. Among them, type I collagen is the most available because of its abundance.

The collagen molecule of type I consists of 3 polypeptide chains, two α 1 chains and one α 2 chain (**Figure 1**). The primary structure of the polypeptide can be repeated by a Gly-X-Y format, where G is glycine, occurring every three residues,

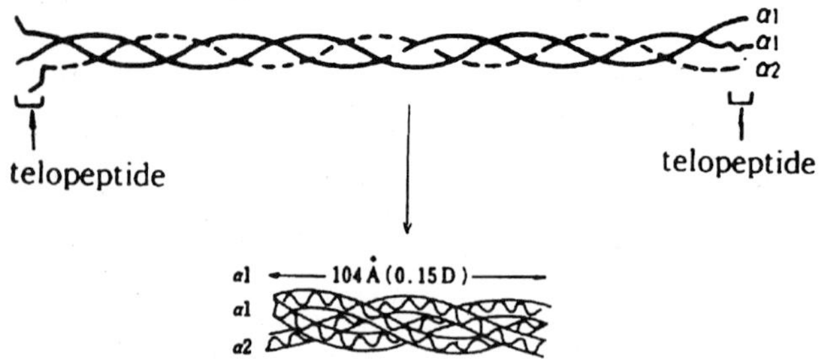

FIGURE 1. A molecule of type I collagen.

and X and Y are other amino acid residues. The polypeptide's secondary structure is a left-handed helix, and in the tertiary structure, the three helices combine to form a smooth, right-handed helix.

The Type I collagen molecule is 300 nm long, 1.5 nm wide, and has a molecular weight of approximately 300,000. In the body, collagen molecules do not exist separately but instead form a Smith's microfibril quaternary structure.

PROPERTIES

The structure of the collagen helix contributes to collagen's levorotation, and the value for the specific optical rotation ($[\alpha]_{-D}$) is usually between –380° and –420°. When an aqueous solution of collagen is heated above 40°C, the helices loosen and form threadlike chains, and the collagen becomes a gelatin. At this point, the levorotation decreases to –130°, and the viscosity also decreases dramatically.

APPLICATIONS

Collagen is currently used widely in industry because of its unique properties. Collagen is marketed worldwide for use in synthetic sausage casings made of bovine hide collagen. Other applications in cosmetics, cell culture matrix, and biomaterials are increasing. Research of collagen as a biomaterial is especially progressing rapidly. Collagen possesses characteristics as a biomaterial distinct from those synthetic polymers, and biopolymers. The most distinct characteristic is collagen's mode of interaction with the body.

In medical products, collagen is used as a hemostatic agent, wound dressing, and for collagen-coated artificial blood vessels. In addition, collagenous native tissue such as blood vessels, pericardiums, or heart valves are used as biomaterials after chemical modification.

Collagen-Based Materials

See: Bioartificial Materials
Collagen
Regeneration Templates, Artificial Skin and Nerves

COLLOIDAL CRYSTALS, POLYMERIC

Tsuneo Okubo
Department of Applied Chemistry
Faculty of Engineering
Gifu University

Colloidal crystals are the suspensions in which colloidal particles distribute regularly as they do in typical crystals like metals and protein crystals.[1–5] Colloidal crystals emit brilliant iridescent light by the Bragg diffraction; they are quite beautiful. When colloidal particles are polymers—for example, polystyrene spheres—their colloidal crystals are called polymer colloidal crystals. The crystal growing process and the morphology of the single crystals of colloidal crystals are quite similar to those of the typical crystals. Furthermore, they undergo phase transition phenomena such as crystallization and melting. Thus, colloidal crystals are ideal systems for model studies of crystals, since we can conveniently analyze with optical techniques and even with the naked eye. Colloidal crystals are true crystals in the suspension (liquid) state, having extremely low rigidity, typically between 10^{-3} to 10^3 Pa (Pascals). Colloidal crystals are easily and sharply distorted by the application of weak external fields such as the gravitational field, shearing forces, an elevated pressure, an electric field, or centrifugal compression.

CRYSTAL STRUCTURE

Crystal structures of monodispersed colloidal spheres in suspensions are generally face-centered cubic (fcc)(or, rarely close-packing) or body-centered cubic (bcc) lattices.[5] From data on the colloidal lattice structures, we may conclude that the fcc lattices are very stable and they transform to the bcc lattice: at low sphere concentrations, in the presence of salts, at elevated suspension temperature, for spheres of high charge densities, and at high pressure. Fcc and bcc structures often coexist. The transformation from the fcc subphases to the bcc lattices is highly related to the increased dead space for particles forming lattice structures; that is, the fcc structure is denser than the bcc form. In other words, crystallike structures are formed most favorably and most stably when the dead space has minimum value and the density of the effective size of spheres including the electrical double layers is largest.

PHASE QUILIBRIA

Phase transition between liquidlike and crystallike structures is discussed by Alder and others with the aid of computer simulation of a finite system of hard spheres.[6,7] They have proposed a realistic *effective hard-sphere model*.[8] In this model, the true diameter of spheres was replaced by an effective diameter of a sphere including thickness of the electrical double layer.[4,5] I have measured phase diagram of the colloidal crystals for the monodispersed polystyrene spheres and the colloidal silica spheres in aqueous suspensions, in salt-containing systems, and in the organic solvents,[9,10] and noted that the phase diagram changes drastically depending on the degree of deionization of the suspension.

VISCOMETRIC, ELASTIC, AND DYNAMIC PROPERTIES

Rheological properties of colloidal suspensions have been studied by many research groups; however, it is very hard to find experimental studies on the deionized and crystallike suspension.[11–15] The reduced viscosities of the deionized suspensions were much larger than those estimated from Einstein's equation for rigid spheres. We ascribe this increase to the important contribution of the electrical double layers. In other words, colloidal spheres move together with the surrounding electrical double layers, and the effective concentration of spheres is much higher than the true concentration of spheres.

EXTERNAL FIELD EFFECT

Elastic moduli of colloidal crystals are very small, in the order of 10^{-2} to 10^3 Pa compared with those of metals, from 10^{10} to 10^{12} Pa. Thus, the colloidal crystal is easily distorted by weak external fields such as gravitational field, shearing forces, elevated pressure, an electric field, centrifugal compression, and so on.[5] The rigidities of the colloidal crystals are conveniently measured using the gravitational compression effect. Colloidal crystals in a rotating disk are also compressed in a centrifugal field.[16] Beautiful color bands appear. The elastic moduli at various sphere concentrations can be obtained from the change in the reflection peak wavelengths in the centrifugal equilibrium.

Electrically induced changes in the colloidal lattices have been studied by several researchers. Here, the interparticle repulsive forces are counter-balanced by the forces from the external electric field. We studied structure of colloidal crystals at high pressures (up to 200 MPa).[17] We observed enhancement of bcc lattice or transformation from fcc to bcc lattice, or both, under elevated pressure. Shearing deformation and melting of colloidal crystals has been studied by Clark and Ackerson,[18–20] Lindsay and Chaikin.[21] Tomita and van de Ven,[22] and Jorand et al.[23] I reported that the fcc and bcc lattices coexisted under flow, although the lattices are deformed and melted at high shear rate.[24] Structural changes under shearing forces have been further studied by several other researchers.[25–29]

APPLICATIONS

Bragg diffraction of colloidal crystals is very sharp, since their *g*-factors are between 0.01 and 0.1, quite similar to those of the typical metals and protein crystals. New nonlinear Bragg diffraction devices have been reported.[30–32] Colors of the colloidal crystals change drastically as the angles of light source and observation change slightly. Then the application in the color display is quite interesting.[33]

Polymer colloidal crystals show promise in the field of laser physics.[34,35] Gels containing the colloidal crystal structure, which will have numerous applications in various areas of technology, have been developed by Asher et al.[36]

REFERENCES

1. Vanderhoff, W.; Van de Hul, H. J.; Tausk, R. J. M. et al. *Clean Surfaces: Their Preparation and Characterization for Interfacial Studies*: Goldfinger, G., Ed.; Marcel Dekker: New York, NY, 1970.
2. Pieranski, P. *Contemp. Phys.* **1983**, *24*, 25.
3. Ottewill, R. H. *Ber. Bunsenges. Phys. Chem.* **1985**, *89*, 517.
4. Okubo, T. *Acc. Chem. Res.* **1988**, *21*, 281.
5. Okubo, T. *Prog. Polym. Sci.* **1993**, *18*, 481.
6. Alder, B. J.; Wainwright, T. E. *J. Chem. Phys.* **1957**, *27*, 1208.
7. Alder, B. J.; Hoover, W. G.; Young, D. A. *J. Chem. Phys.* **1971**, *55*, 1128.
8. Hachisu, S.; Kobayashi, Y.; Kose, A. *J. Colloid Interface Sci.* **1973**, *42*, 342.
9. Okubo, T. *Langmuir* **1994**, *10*, 1695.
10. Okubo, T. *Langmuir* **1994**, *10*, 3529.
11. Okubo, T. *J. Chem. Phys.* **1987**, *87*, 6733.
12. Okubo, T. *Ber. Bunsenges. Phys. Chem.* **1988**, *92*, 504.
13. Okubo, T. *Naturwissenschaften* **1988**, *75*, 91.
14. Okubo, T. *Polym. Bull.* **1988**, *20*, 269.
15. Matsumoto, T.; Okubo, T. *J. Rheol.* **1991**, *35*, 135.
16. Okubo, T. *J. Am. Chem. Soc.* **1987**, *109*, 1913.
17. Okubo, T. *J. Chem. Soc., Faraday Trans.* **1988**, *84*, 1949.
18. Clark, N. A.; Ackerson, B. J. *Phys. Rev. Lett.* **1980**, *44*, 1005.
19. Clark, N. A.; Hurd, A. J.; Ackerson, B. J. *Nature (London)* **1979**, *281*, 57.
20. Ackerson, B. J.; Clark, N. A. *Phys. Rev. Lett.* **1981**, *46*, 123.
21. Lindsay, H. M.; Chaikin, P. M. *J. Chem. Phys.* **1982**, *76*, 3774.
22. Tomita, M.; Van de Ven, T. G. M. *J. Colloid Interface Sci.* **1984**, *99*, 374.
23. Jorand, M.; Koch, A. J.; Rothen, F. *J. Phys. (Paris)* **1986**, *47*, 217.
24. Okubo, T. *J. Chem. Soc., Faraday Trans.* **1988**, *84*, 1171.
25. Hoffman, R. L. *Future Directions in Polymer Colloids*; ElAasser, M. S.; Fitch, R. M., Eds.; Martinus Vijhoff: Dordrecht, The Netherlands, 1987; p 151.
26. Mathis, C.; Bossis, G.; Brady, J. F. *Prog. Colloid Polym. Sci.* **1988**, *76*, 16.
27. Lindner, P.; Markovic, I.; Oberthur, R. C. et al. *Prog. Colloid Polym. Sci.* **1988**, *76*, 47.
28. Bossis, G.; Brady, J. F.; Mathis, C. *J. Colloid Interface Sci.* **1988**, *126*, 1.
29. Mathis, C.; Bossis, G.; Brady, J. F. *J. Colloid Interface Sci.* **1988**, *126*, 16.
30. Asher, S. A. U.S. Patents 4 627 689 and 4 632 517, 1986.
31. Asher, S. A.; Kesavamoorthy, R.; Jagannathan, S.; Rundquist, P. *SPIE* **1992**, *1626*, 238.
32. Spry, R. J. U.S. Patent 4 986, 635, 1991.
33. Fujita, H. *Polymers* **1980**, *29*, 118.
34. Genack, A. Z.; Balachandran, J. M.; Gomes, A. S. L.; Sauvaln, E. *Nature (London)* **1994**, *368*, 400.
35. Lawandy, N. M.; Balachandran, J. M.; Gomes, A. S. L.; Sauvaln, E. *Nature (London)* **1994**, *368*, 436.
36. Asher, S. A.; Holtz, J.; Liu, L.; Wu, Z. *J. Am. Chem. Soc.* **1994**, *116*, 4997.

COLLOIDAL DISPERSIONS, NANOGELS

Derek M. Ölé Kiminta,* G. Braithwaite, and
Paul F. Luckham
Department of Chemical Engineering and Chemical Technology
Imperial College of Science, Technology, and Medicine

Heterogeneous materials, such as cement, paint, ink, milk, blood, and some polymerized synthetic latices belong to a distinct class of materials called colloids. The colloidal state of matter exists according to the shape, size, and form of the individual particles that form the bulk of the material. These materials flow, scatter light, and can also exhibit some solidlike behavior. Colloidal dispersions contain structural entities with at least one linear dimension in the size range of 1 nm to 1 μm. For most monodispense colloids, the particles are normally spherical. Stable aqueous colloidal dispersions of crosslinked poly(*N*-isopropylacrylamide), which is prepared by aqueous dispersion copolymerization of N-isopropylacrylamide (NIPAM) with methylenebisacrylamide (MBAAm), is the main material characterized in this article. Under certain conditions (i.e., below ≈35°C) these particles become swollen by the solvent (water) and each individual particle resembles a gel, therefore at extremely small sizes, they are called "nano or microgels." These microgel dispersions show some of the properties of gels and of colloids. The nature and behavior of colloidal particles which may be in a concentrated dispersion form, a liquid, or in a suspension state, are discussed in this article with emphasis on their inherent physical properties and the immediate influence of the energy contributed from the surfaces.

The most common forms of colloids in our everyday life include ink, milk, paint, fruit juices, and also biocolloids such as blood. The colloidal state is defined as a disperse system in which one dimension is within the range of about 10 nm (10^{-8} m) to about 1000 nm (10^{-6} m), which is the range at which the surface effects predominate. At sizes larger than this, bulk effects (i.e., gravitational forces) are present; at sizes smaller than this, the properties (of the substance) are controlled by molecular effects.[1] There are three phases of matter and eight types of simple colloids. The simplest colloids are those that have only two phases, with the dimensions of the dispersed phase being in the colloidal above-specified range. Naturally occurring colloidal substances include soils (clay or dust), fog, mist, or smoke. There are also many biological structures that are colloidal in nature. These include blood, which is a dispersion of corpuscles in serum, natural fruit juices, the latex obtained from the rubber plant, skeletal bone, which is essentially a dispersion of calcium phosphate embedded in collagen, and many others, which when described in colloidal terms, are appropriately called biocolloids. An important inherent physical property of colloidal dispersions that arises from energy contributions from the surfaces is the tendency of the particles to aggregate. This can be attributed to van der Waals forces of attraction, which exist between neighboring particles.

MICROGELS AS A COLLOIDAL DISPERSION

Perhaps the simplest definition for a microgel can begin by describing the physical structure of a gel. A gel can be defined as a soft, spongy, elastic, deformable solid that is made from a connected assembly of colloidal particles or macromolecules.

When this gel is scaled down to submicron level, and especially if it is formulated through a latex polymerization route, the dispersion formed may be termed a microgel.

Although conventional microgels change only narrowly with the environmental conditions, stimuli-responsive microgels may swell or deswell considerably with relatively small changes of temperature or pH. In this work rheological methods have been used to investigate and characterize these dispersions. Because of their properties and gel-like physical appearance, crosslinked emulsion polymers, such as those based on styrene and divinylbenzene or those based on methyl methacrylate and ethylene diacrylate, display these characteristics. In aqueous solutions, polyNIPAM shows a lower critical solution temperature (LCST), which is in the range of 31–33.5°C.[2,3] The occurrence of this phenomenon is common in aqueous solutions of polar polymers as explained and presented in the works of Malcolm, Dieu, and Meyerhoff.[4–6]

CHEMICAL GELS[7–18]

These are gels with branched networks of covalently crosslinked linearly flexible chains that are surrounded by solvent and are usually formulated from synthetic polymers. The gels that are also in a matrix form are swellable, the extent of the swelling depends on the polymer–solvent interaction.[10,19]

PHYSICAL GELS

Physical gels are formed through physical aggregation and are predominantly disordered, but have regions of local order where physical crosslinking occurs (noncovalent). They can be regarded as reversible because the structures are transient and can reform because they are not permanent. Significant changes occur when the gels are exposed to increasing temperature, pH, and change of solvent qualtiy.[18,19]

AGGREGATED PARTICLE GELS

Gels that are classified as aggregate particles have random (fractals) structures that are dominated by interparticle interactions, which lead to flocculation and phase separation. These gels keep materials in suspension and provide the necessary rheological properties of a formulation, especially in foods such as margarine, yogurt, and other products such as emulsion paints, pharmaceutical and agrochemical dispersions, and so on. In their condition, the random particle gels are not generally thermally reversible.[11,13]

COLLOIDAL GELS

The colloid gels may be considered a special case of flocculation system or aggregated particle gels and are formed when particles flocculate into clusters that interact with each other forming a continuous network of particles before settling occurs. The resulting suspension usually has a very high viscosity and a finite shear modulus.

POLYMER GELS

Polymer gels can also be referred to as colloidal gels, but unlike colloidal gels, described above, which are formed of particle aggregation, polymer gels are more unique and are formed by polymerization and/or crosslinking, both of which are aggregation processes.

SHRINKING AND EXPANDING MECHANISM IN polyNIPAM PARTICLES

Similar to polymer chains, the macromolecular constituent of a gel is not comprised of a single linear strand, but rather a sophisticated topologically complex network. When particles of polyNIPAM microgels are exposed to certain conditions, the response is a change in volume, shape, and overall size of the particle. Ordinarily, the diameter of the particle is used to determine the change in volume of the spherical particles. The mechanism of these changes in expansion and shrinkage of the polymer network involves three forces:

- the rubber elasticity, derived from the elasticity of the individual polymer strands;
- the polymer–polymer affinity; and
- the hydrogen–ion-pressure.

These three forces make up the total pressure acting on the gel. This pressure is known as the osmotic pressure of the gel because it determines whether the gel tends to take up fluid or expel it.

POLYELECTROLYTE GELS

A polyelectrolyte gel is a polymer gel with charged groups present on the polymer gel. The most remarkable characteristic of this type of gels is the swelling and shrinking that can occur with dramatic volume change (about 5-fold).[3] This can be caused by temperature increase, or change of acidity or salinity of the solution with which the gel is in contact.[22]

SWELLING/DESWELLING MECHANISM

The swelling and deswelling of gels is a consequence of the change in the polymer–solvent interaction caused by an increase in temperature.[18–24]

EFFECT OF HEAT ON POLY(*N*-ISOPROPYLACRYLAMIDE)

The thermoresponsiveness characteristic of this polymer is a consequence that results from heating. At room temperature (20–25°C), the microgels maintain their maximum volumetric size, although any time the temperature is increased, the result is deswelling. The swelling characteristics of polymeric materials depend on certain factors directly related to the type and nature of the polymer. Microgel particles are also known to exhibit osmotic characteristics. An example of this is found in poly(*N*-isopropylacrylamide) and styrene-divinylbenzene microgel lattices.[24–26]

Volume change within the microstructure of latex made using poly(*N*-isopropylacrylamide) is caused by various factors of which temperature is the most prominent. The latex bears a hydrophobic characteristic because of the presence of an isopropyl group on and around the polymer chains. Once the polymer is dissolved in water, there is a hydrophobic reaction within the water molecules, causing a cacoonlike layer, which forms around the isopropyl group. As the lower critical solution temperature of 31.5°C is approached, a phase change occurs around this region (between 31.5–33°C). A study done by Fujishige and Ando indicates that the phase change occurring at this particular temperature is independent of polymer molecular weight and concentration.[27] When a particle at room temperature with fully extended macromolecular chains collapses from its original flexible coil, a phase separation occurs. The chain conformation change near this temperature (31.5°C) provides essential information on phase transition.

APPLICATIONS (GELS/MICROGELS)

Gels and microgels of this type have many applications, such as making separation membranes, switching systems, sensors, adsorbents, mechano-transducers, dehydrants, materials for drug delivery systems, and so on. Other uses include the manufacturing polymers, such as rubber gloves, plastics, membranes, and glues. Gels are also used in analytical methods such as electrophoresis and chromatography in which molecules are monitored as they separate according to the speed with which they percolate through the pores of a gel.[13,16,23] Reversible and thermoresponsive microgels are of considerable interest to medical researchers, commercial manufacturers of paints, ink, dyes, and so on. Although most of the uses of microgel technology are only now being fully investigated for industrial applications, some of the more prominent applications that make use of the swelling/deswelling of polyNIPAM occur in the drug industry.

The rapid response of a microgel surface skin temperature changes should enable the application of an "on-off" switch technique for drug permeation and release.[28] One of the more attractive reasons for this is the abrupt volume change in response to temperature.[29] At low temperature a solid skin is formed at the surface of the particle, trapping the drug in the core of the particle, whereas at a higher temperature, the particle structure (skin) collapses allowing the trapped drug to escape.

REFERENCES

1. Tadros, Th. F. *Chem. Ind.* **1985**, April 1, 210.
2. Pelton, R. H.; Chibante, P. *Colloids and Surfaces* **1991**, *58*, 271.
3. Kiminta, D. M. O.; Lenon, S.; Luckham, P. F. *Proceedings of Materials Research Society* Maitland, G., Ed., **1993**, Vol. *289*, p 13.
4. Malcolm, G. N.; Rowlinson, J. S. *Trans. Faraday Soc.* **1957**, *53*, 921.
5. Dieu, H. J. *Polym. Sci.* **1954**, *12*, 417.
6. Uda, K.; Meyerhoff, G. *Makromol. Chem.* **1961**, *47*, 168.
7. Ferry, J. D. *Viscoelastic Properties of Polymers* John Wiley & Sons: New York, 1961, p 391.
8. Ferry, J. D. *Viscoelastic Properties of Polymers* John Wiley & Sons: New York, 1970, p 557.
9. Ferry, J. D. *Viscoelastic Properties of Polymers* John Wiley & Sons: New York, 1980, p 529.
10. Tanaka, T. *Sci. Am.* **1981**, *244*, 110–123.
11. Katayama, S.; Hirokawa, Y.; Tanaka, T. *Macromolecules* **1984**, *17*, 2641.
12. Doi, M.; Edwards, S. F. *The Theory of Polymer Dynamics* Claredon: Oxford, 1986.
13. *Physical Networks: Polymer and Polymer Gels* Burchard, W.; Ross-Murphy, S. R., Eds. Elsevier: New York, 1990.
14. *Synthesis, Characterization and Theory of Polymeric Networks and Gels*; Aharoni, S. M., Ed.; Plenum: New York, 1992; p 423.
15. *Hydrogels for Medical and Related Applications* Andrade, J. D., Ed., American Chemical Society Symposium Series, Washington, DC, 1976; Vol. 31.
16. *Responsive Gels, Volume Transitions I & II*; Dusek, K., Ed.; Advances in Polymer Science, Springer-Verlag: Berlin, 1993; Vol 109/110.
17. Djabourov, M. *Polym. Int.* **1991**, *25*, 135–143.
18. Hoffman, S.; Park, G. J. *Appl. Polym. Sci.* **1994**, *52*, 85–89.
19. Tanaka, T. *Polymer* **1979**, *20*.
20. Rydzewski, R. *Cont. Mech. Thermodyn.* **1990**, *2*.
21. Dickinson, E. *Food Polymers, Gels and Colloids* RSC 1993; *Chem. Industry* **1990**, *Oct. 1*, 595–599.
22. Morro, A.; Müller, I. *Rheol. Acta.* **1988**, *27*.
23. McPhee, W.; Chiu, K.; Pelton, R. J. *Coll. Interface Sci.* **1993**, *156*, 24–30.
24. Okano, T.; Yoshida, R.; Sakurai, Y. *Polymer Gels* DeRossi, D. et al., Eds.; Plenum: New York, 1991.
25. Heskins, M.; Guillet, J. E. *J. Macromol. Sci., Chem.* **1968**, *A2*, 1441.
26. Pelton, R. H.; Chibante, P. *Colloids and Surfaces* **1986**, 20, 247.

27. Fujishige, Ando. *J. Phys. Chem.* **1989**, *93*, 3331.
28. Hoffman, A. S. *Polymer Gels* DeRossi, D. et al., Eds.; Plenum: New York, 1991.
29. Kiminta, D. M. D.; Lenon, S.; Luckham, P. F. Icheme Event: London, 1994; Vol. 2, pp 758–760.

COLLOIDAL METAL

Hidefumi Hirai
Science University of Tokyo

Naoki Toshima
The University of Tokyo

"Colloid" means the dispersed state of the fine particles with the size from 1 nm to 1 μm. Thus, "colloidal metal" is the stable dispersion of small metal particles or metal clusters mainly in liquid.[1] Polymers play an important role in the stabilization of dispersions as well as in the preparation of small particles.

Addition of water-soluble synthetic polymers or surfactants can stabilize the colloidal dispersion of metals. These stabilizers are called "protective colloid." In this case, the protective colloid can adsorb on the surface of colloidal metal particles and the steric repulsion between the protective colloids is considered to stabilize the colloidal dispersions.

The stable colloidal metals protected by polymers are easily prepared by reduction of the corresponding metal ions in the presence of water-soluble polymers.[2] They are applicable to practical catalysts and other purposes, and recognized as "functional metals" as well.

PREPARATION

The preparative methods of colloidal metals can be divided into two categories.[1] One is the preparation of small metal particles by degradation of bulk metals with physical methods such as colloid mill, plasma jet, and electron beam. The other is the preparation of small metal particles by reduction of the corresponding metal ions or metal complexes with chemical methods.

PROPERTIES

Stability and Color

Colloidal dispersions of noble metals usually have transparent dark brown color except gold (red) and silver (yellow), and are stable at room temperature for a long period.[3] Especially PVP-protected colloidal metals are stable over years even under air at room temperature.

Size

Particle size of colloidal metals can be determined by transmission election micrographs (TEM). The size obtained by TEM is that of central metal parts. Average particle size of a central metal part is usually 1 to 8 nm depending on the kind of metal and the preparation conditions. Gold usually has a larger average particle size than the other noble metal particles. TEM photographs can also provide the information of dispersion and coagulation states.

Physical Properties

Colloidal metals bridge between bulk metal and atomic or ionic metal, and have special properties that are quite different from both bulk and atomic metals. These special properties are considered to be due to a quantum size effect.[4]

APPLICATIONS

Colloidal metals are applied to physical, chemical, and biological fields. In the physical field, colloidal metals are the promising advanced materials for application to electronic devices and optical materials. For example, fine gold particles are interesting as the materials for nonlinear optics in third harmonic generation.

In the biological field, colloidal gold has been used for labeling the specific position of cells for the observation by transmission electron micrograph. Colloidal silver is very popular for detection of the specific proteins in the electrophoresis.

The most intensive investigations of applications of colloidal metals are focused on catalyses. Many reports have been published in the application to catalysts.[2,5,6] The catalytic properties of colloidal metals mainly depend on the particle size and the kind of metals as well as the protective colloids.

RECENT TOPICS IN COLLOIDAL METALS

Colloidal Metals in Organic Media

The colloidal metals mentioned above are usually dispersed in water or in an aqueous solution. However, colloidal metal dispersions in organic media are required for the reactions in organic solvents and other purposes.

Recently, methods for transferring colloidal metals from an aqueous medium into an organic medium have been reported. One is the method to use organic ligands like triphenylphosphine and the other is the method to use both surfactants like sodium oleate and inorganic salts like sodium chloride.[7,8]

Another method for the preparation of colloidal metals in organic media is based on the direct syntheses of colloidal metals in organic solvents.[9] For this purpose, metal sources and reductants must be soluble in organic solvents.

Bimetallic Colloids

Bimetallic colloids, where two kinds of metallic elements are contained in one particle, can be prepared by simultaneous reduction of two kinds of metal ions with an alcohol reduction method in the presence of PVP. For example, Pt/Pd, Au/Pd, Au/Pt, and Pt/Rh bimetallic colloids have been prepared by this method.[10]

Colloids of Light Transition Metals

Production of light transition metal colloids requires stronger reducing agents than is the case for noble metal colloids. For example, Cu colloids were prepared by reduction of Cu ions with sodium tetrahydroborate.[11]

Role of Polymers

The colloidal metals usually involve water-soluble polymers such as PVP as a stabilizer. Without the stabilizer, the colloids coagulate to form precipitates. In this case, organic ligands or surfactants are used as the stabilizer. However, polymers are the most useful stabilizer among them.

REFERENCES

1. Everett, D. H. *Basic Principles of Colloid Science*; *Roy. Soc. Chem.*: London, 1988.
2. Hirai, H.; Toshima, N. *Tailored Metal Catalysts*; Iwasawa, Y. Ed.; D. Raidel: Dordredit, 1986; Chapter 2.
3. Henglein, A. *J. Phy. Chem.* **1993**, *97*, 5457.
4. Marzke, R. F. *Catal. Rev. Sci. Eng.* **1979**, *19*, 43.
5. Toshima, N. *Shokubai* **1985**, *27*, 488.
6. Schmid, G. *Chem. Rev.* **1992**, *92*, 1709.
7. Liu, H.-F.; Toshima, N. *J. Chem. Soc., Chem. Commun.* **1992**, 1095.
8. Hirai, H.; Aizawa, H.; Shiozaki, H. *Chem. Lett.* **1992**, 1527.
9. Bönnemann, H.; Brijoux, W.; Brinkmann, R.; Dinjus, E.; Jouben, T.; Korall, B. *Angew. Chem. Int. Ed. Engl.* **1991**, *30*, 1312.
10. Toshima, N. *J. Macromol. Sci.-Chem.* **1990**, *A27*, 1225.
11. Hirai, H.; Wakabayashi, H.; Komiyama, M. *Bull. Chem. Soc. Jpn.* **1986**, *59*, 367.

Colloid Materials

See: *Catalyses (by Polyelectrolytes and Colloidal Particles)*
Colloidal Crystals, Polymeric
Colloidal Dispersions, Nanogels
Colloidal Metal
Colloids, Protective
Conducting Polymer Colloids
Emulsion Polymerization
Natural Rubber Latex (Polymeric Flocculant for Tin Tailings Slurries)
Polyelectrolyte Complexes (Overview)
Surfactants, Polymeric (Colloid Stabilization)
Suspension Polymerization

COLLOIDS, PROTECTIVE

Sadao Hayashi
Faculty of Textile Science and Technology
Shinshu University

We have known for a long time that water soluble proteins protect hydrophobic colloidal dispersion systems. The adsorption of the water soluble polymer to the surface of the hydrophobic colloidal particles is called protective action of a hydrophilic colloid to a hydrophobic colloid. Accordingly, this hydrophilic colloid is called a protective colloid. Although the protective colloids adsorb to the surface of hydrophobic colloidal particles, they are distinguished from emulsifiers because they do not form micelles in a water solution.

Gelatin, until recently, was generally considered one of the best stabilizers for hydrophobic colloids. Currently, many synthetic polymers are used in a similar way, and their protective action equals that of gelatin.[1–4] An interesting problem occurs when the same water soluble polymers act as flocculating agents or as stabilizers, according to their concentration.[5–7] When a small amount of polymer is added, a polymer chain simultaneously adsorbs to two or more colloidal particles, and the polymer bridge craws them together in coagulation. At high polymer concentrations, the colloidal particles become covered with the polymer chains and the resultant coatings prevent the approach of particles.

Synthetic, water soluble polymers are now being employed industrially as protective colloids in emulsion or suspension polymerizations of vinyl monomers, particularly vinyl acetate[8–11] and vinyl chloride.[12,13] The hydrophilic polymers adsorb to the surface of the hydrophobic polymer particles, convert their surfaces from hydrophobic to hydrophilic, and sterically stabilize the polymer particles. This shows that the electric, double–layer effects are not of primary importance when water soluble polymers are present.

To enable the hydrophilic polymers to effectively adsorb to the hydrophobic polymer particles, the stabilizer should be copolymers of hydrophilic and hydrophobic monomers rather than hydrophilic homopolymers. The copolymers produced from the two monomers are divided roughly into three types: random, block, and graft, based on the arrangement and branching of the two monomer units in the polymer chain. With reference to nonionic, surface–active agents, hydrophobic chains in the copolymer will attach to the surface of the hydrophobic colloidal particles and hydrophilic chains in the copolymer will protrude into the water if the copolymer is to impart stability. Judging from the structure, amphipathic block or graft copolymers of the three types can be expected to serve as good stabilizers. Here, the hydrophobic chains in the copolymer attached to the surface of the hydrophobic colloids are known as anchor polymers, and the hydrophilic chains in the copolymer that protruded into water are termed stabilizing moieties.[14]

Grades of partially hydrolyzed poly(vinyl alcohol) with ~10 to 20 percent residual vinyl acetate units are most useful as stabilizers for polymerization. The emulsion polymerization of vinyl acetate that used the partially hydrolyzed poly(vinyl alcohol) as a protective colloid produces a stable poly(vinyl acetate) latex.[8–11] It has been reported that increasing the blockier distribution of the vinyl acetate units in a copolymer produces smaller particles and higher viscosity of the latex, as well as an increase in the content of vinyl acetate units in a copolymer.[15] In this case, if fully hydrolyzed poly(vinyl alcohol) and reacetylated poly(vinyl alcohol) were used as protective colloids, the particle size is larger and the latex viscosity is lower than those of latexes using partially hydrolyzed poly(vinyl alcohol).[16–18] In this manner, it is important to characterize the sequence distribution of the vinyl acetate units in partially hydrolyzed poly(vinyl alcohol) as anchor polymers and stabilizing moieties in controlling adsorption on the surface of the polymer particles.

The graft copolymers with hydrophilic polymers as a backbone are prepared by polymerizing vinyl monomers such as styrene and vinyl acetate in the presence of hydrophilic polymers, such as poly(vinyl alcohol) or poly(ethylene oxide).[19–22] As a result, the graft copolymers were found to be good stabilizers for hydrophobic polymer particles, and were used as a protective colloid for emulsion polymerization.

To produce a more effective stabilizer, new protective colloids that introduce strong chain–transfer activity functional groups into polymer molecules have recently been developed, and were used to prepare graft copolymers of vinyl monomers to the protective colloid as a backbone.

REFERENCES

1. Heller, W.; Pugh, T. L. *J. Chem. Phys.* **1954**, *22*, 1778.
2. Napper, D. H.; Netschey, A. *J. Colloid Interface Sci.* **1971**, *37*, 528.
3. Garvery, M. J.; Tadros, Th. F.; Vincent, B. *J. Colloid Interface Sci.* **1974**, *49*, 57.
4. Hayashi, S.; Nakano, C.; Motoyama, T. *Kobunshi Kagaku* **1995**, *22*, 354.
5. Saunders, F. I.; Sander, J. W. *J. Colloid Sci.* **1956**, *11*, 260.
6. Noda, M. *Nippon Kagaku Kaishi* **1961**, *82*, 1611.
7. Fleer, G. J.; Lyklema, J. *J. Colloid Interface Sci.* **1974**, *46*, 1.
8. Hartley, F. D. *J. Polym. Sci.* **1959**, *34*, 397.
9. Lamont, H. *Adhes. Age* **1973**, *16*, 24.
10. Donescu, D.; Gosa, K.; Ciupitoiu, A. *J. Macromol. Sci.-Chem.* **1985**, *A22*, 931.
11. Abdul–Ghani, R.; Dunn, A. S. *Polym. Com.* **1983**, *24*, 285.
12. Shiraishi, M.; Toyoshima, K. *Brit. Polym. J.* **1973**, *5*, 419.
13. Fabini, M.; Bobula, S.; Rusina, M. *Polym.* **1994**, *35*, 2201.
14. Napper, D. H. *Ind. Eng. Chem. Prod. Res. Develop.* **1970**, *9*, 467.
15. Hayashi, S.; Nakano, C.; Motoyama, T. *Kobunshi Kagaku* **1964**, *22*, 358.
16. Hayashi, S.; Yanagisawa, T.; Hojo, N. *Nippon Kagaku Kaishi* **1973**, 402.
17. Hayashi, S.; Zhang, F. S.; Kurisawa, T.; Hirai, T. *Nippon Setchaku Kyokaishi* **1986**, *22*, 356.
18. Hayashi, S. *Polym.-Plast. Technol. Eng.* **1988**, *27*, 61.
19. Netschey, A.; Napper, D. H.; Alexander, A. E. *J. Polym. Sci., Part B, Polym. Lett.* **1969**, *7*, 829.
20. Chiou, J. J.; Piirma, I. *Polym. Prepr. (Am. Chem. Soc., Div. Polym. Chem.)* **1985**, *26*, 221.
21. Heublein, G.; Hotschansky, P.; Meissner, H. *Acta Polymerica* **1985**, *36*, 246.
22. Hortschansky, P.; Meissner, H.; Heublein, G. *Acta Polymerica* **1985**, *36*, 343.

Colorants

See: Additives (Types and Applications)
Additives (Property and Processing Modifiers)
Colorants (Overview)

COLORANTS (Overview)

Byron G. Hays and Amrit P. Bindra
Engelhard Corporation

The main reason for using a colorant (pigment or dye) in plastic is to increase the visual appeal and market value of the finished part by projecting a desirable color and other optical effects, such as transparency or opacity. This article presents an overview of colorants, their principal requirements and the structures and properties of some of the most important colorants. Articles by Christie and Kaul are excellent current sources for more detailed information on colorants for plastics.[1,2] Webber, like Kaul, lists colorants appropriate for individual plastics, and Ahmed gives details about colorant compounding equipment.[3,4] The *Colour Index* provides much information about all pigments and dyes.[5] Lewis gives additional information about all inorganic and organic pigments.[6] Herbst and Hunger have published an excellent source on all organic pigments.[7]

Pigments differ from dyes in several ways. Although both are finely divided solids, pigments and dyes behave differently when dispersed in molten plastics: pigments are essentially insoluble and remain discrete particles dispersed in the melt, whereas dyes dissolve in the melt, with individual dye molecules associating with the polymer molecules. Pigments can be organic or inorganic, and some are colorless (black, white, metallic, or pearlescent); dyes are organic and colored. Pigments may affect the plastic's physical properties in positive ways such as carbon black's reinforcement of rubber or in negative ways such as the warping of injection-molded polyolefins by copper phthalocyanine blue. Dyes have little, if any, effect on plastics.

REQUIREMENTS

The principal requirement for a pigment or dye is that it imparts the desired optical effects to the finished part. To accomplish this, the colorant must also withstand the processing conditions of plastics and the environment in which the finished part is used and be compatible with the plastic components.

Optical Effects

Materials interact with visible light in two ways: by absorption and by scattering. The absorption of certain wavelengths of visible light allows the remaining wavelengths to reflect as a color. Color can be characterized by hue, chroma, and lightness: hue refers to the shade of color, chroma refers to its tinctorial strength or masstone intensity, and lightness refers to its brightness or cleanliness. Organic pigments and dyes usually have high tinctorial strength (chroma); the more acceptable ones are also clean and bright (high lightness). In contrast, inorganic pigments usually are weak and dull (low chroma and lightness). The degree of scattering of light determines the transparency or opacity. For transparency, the finished part must contain relatively few light scattering points. For opacity, the finished part usually contains an inorganic pigment with a particle size close to or larger than the wavelength of light and with a high refractive index, greatly different from that of the plastic.

Dispersibility

Pigment and dye powders consist of clusters of aggregated particles. Using the aggregates as is would impart relatively little color strength to the finished part, so they must be broken down into smaller aggregates and primary particles. Usually, high shear is required to break down the aggregates; however, pigment manufacturers have made major improvements in the dispersibility (ease of dispersion) of their pigments by using surface treatments that coat the particles with special surface active agents or resins.

Stability

Heat stability or resistance to change in color at high temperatures and long dwell times is an important property during plastics processing. Poor heat stability in the colorant may arise from thermal decomposition of the colorant, a crystal phase change in a pigment or sublimation of a dye from the plastic. In contrast to heat stability, light stability, although unimportant during processing, is very important during use of the finished part. Light stability often depends on the colorant's concentration and the plastic type. Light stability often correlates with weatherability, but sometimes the combination of light and moisture reduces weatherability. Occasionally, stability to alkalies or acids is required (e.g., acid-stable colorants should be used with acid-catalyzed thermoset plastics or with poly(vinyl chloride).

Compatibility

Poor compatibility of the colorant with the plastic and other components in the formulation can manifest itself in several ways. During processing of the plastic, the colorant or other additives may plate out on the metal surfaces of the processing equipment. Some time after the finished part is made, the colorant may diffuse through and bloom on the finished part's surface where the powdery deposit is easily rubbed off. If the finished part has contact with another material, the colorant may mark off on or migrate into and stain the other material. If the finished part is in contact with solvents (e.g., in perfume bottles), the colorant may bleed into the solvent.

Safety

The major safety concerns about colorants pertain to those containing heavy metals such as lead, chromium (hexavalent), and cadmium. The concerns relate mostly to workers' exposure to the heavy-metal dry pigments in the workplace, to possible air pollution while incinerating finished parts and to possible ground water pollution from landfills containing finished parts or incineration ash.

STRUCTURES AND PROPERTIES

Inorganic Pigments

Inorganic pigments are used in plastics for applications that require outstanding resistance to heat, light, weathering, and solvents. Although originally derived from natural sources, most current inorganic pigments (except for some iron oxides) are synthetic.

White

Titanium dioxide (Pigment White)* is the major hiding white pigment. Titanium dioxide imparts high opacity (maximum light scattering) and high whiteness (minimum light absorption).

Colored

The iron oxides are the most important inorganic colored pigments. Red iron oxides (Pigment Red 101 and 102) are primarily anhydrous ferric oxide. Yellow iron oxides (Pigment Yellow 42 and 43) are ferric oxidehydroxide. Brown iron oxides are mixtures of ferrous and ferric oxide. The iron oxides are usually opaque (although some highly transparent products are offered for special applications), low in toxicity, and show excellent weatherability.

Black

Of various black pigments, carbon black (Pigment Black 6 and 7) is definitely the most important and ranks second behind titanium dioxide for volume usage in plastics. Carbon blacks show high tinctorial strength and excellent stability at low cost.

Organic Pigments

Organic pigments are used in plastics for applications where high tinctorial strength, brightness, and usually transparency are needed. Essentially all organic pigments are synthetic.

*Pigment and dye numbers are all *Colour Index* (C.I.) numbers.

Azo

Azo pigments contain one or more azo (-N = N-)** groups and account for most yellow, orange, and red organic pigments.

Heterocyclic

Heterocyclic pigments contain several hydrogen-bonding groups, which impart good properties through strong intermolecular bonding.

Polycyclic

The many aromatic rings in these pigments promote sufficient intermolecular bonding for them to have good properties in plastics; however, many of the polycyclic pigments warp injection-molded plastics.

Dyes

In contrast to pigments, dyes are usually soluble and scatter no light. Thus, they display excellent transparency. Dyes are the only means of coloration available for perfectly transparent products. The advantages of dyes also include high tinctorial strength and brightness, ease of incorporation, and low cost. Dyes generally have poorer lightfastness than pigments, especially when used with opacifiers or in opaque plastics. Modern colorants for plastics include several dyes with significantly improved lightfastness and are suitable for use in more demanding applications. Using ultraviolet absorbers may also significantly enhance the lightfastness of dyes.

Because of their high tinctorial strength, dyes are often combined with pigments to obtain deeper shades, a practice referred to as topping. In a dye pigment combination, the pigment provides the opacity and lightfastness, whereas the dye contributes depth of shade.

Azo

Azo dyes provide a wide range of shades and are useful in coloring polystyrene, phenolics, and rigid PVC.

Anthraquinone

Anthraquinone dyes provide a range of colors covering red, violet, blue, and green shades. Although anthraquinone dyes are more expensive than azo dyes, they offer superior weatherability and heat stability.

Quinoline

Solvent Yellow 33 and Disperse Yellow 54 are the most commonly used quinoline dyes and are characterized by good heat and light stability.

Perinone

Solvent Orange 60 and Solvent Red 135 are the most prominent members of this group. Both dyes have excellent heat stability and lightfastness.

Xanthene

Xanthene dyes display brilliant, fluorescent colors, but have poor heat and light stability.

**Most azo pigments actually exist in the tautomeric hydrazo (-N -N = C) form.

Azine

The high tinctorial strength of nigrosine blacks cannot be matched by using carbon black pigments. Nigrosines are considered complex polymeric dyes containing the phenazine ring system, but their chemical structures are not well defined.

Aminoketone

These dyes are characterized by their brilliant fluorescent colors.

Indigold

The thioindigos function as vat dyes, pigments, and dyes for plastics. They are expensive, but their heat and light stability and clean, bright, slightly fluorescent shade make them valuable colorants.

Phthalocyanine

Several blue dyes are derived from copper phthalocyanine.

Coumarin

Fluorescent Brightener 61 and Fluorescent Brightener 236 are coumarin derivatives used to mask the natural yellowness of plastics such as GPPS.

REFERENCES

1. Christie, R. M. *Polym. Int.* **1994**, *34*, 351.
2. Kaul, B. L. *Rev. Prog. Coloration* **1993**, *23*, 19.
3. Ahmed, M. *Coloring of Plastics*, Van Nostrand Reinhold: New York, 1979.
4. *Colouring of Plastics*, Webber, T. G., Ed.; John Wiley & Sons: New York, 1979.
5. *Colour Index, Pigments and Solvent Dyes*, 3rd ed.; R. L. M. Allen, et al.; H. Charlesworth: Huddersfield, UK, 1982.
6. *Pigment Handbook*, vols. I–III, 2nd ed., Lewis, P. A., Ed.; John Wiley & Sons: New York, 1988.
7. Herbst, W.; Hunger, K. *Industrial Organic Pigments*; VCH: New York, 1993.

COMB-LIKE AMPHIPHILIC POLYMERS (Supramolecular Structure)

Zhishen Mo,* Shanger Wang, Tianru Fang, and Hongfang Zhang
Laboratory of Polymer Physics
Changchun Institute of Applied Chemistry
Chinese Academy of Sciences

A new kind of comb-like amphiphilic polymer, $PAMC_{16}S$, is described. Its supramolecular structure was elucidated by WAXD analysis and its mesophoric character was studied with optical microscopy and differential scanning calorimetry (DSC).

PREPARATION

Monomer

The new amphiphilic monomer, 2-acrylamidohexadecylsufonic acid($AMC_{16}S$), was prepared by reacting acrylonitrile,

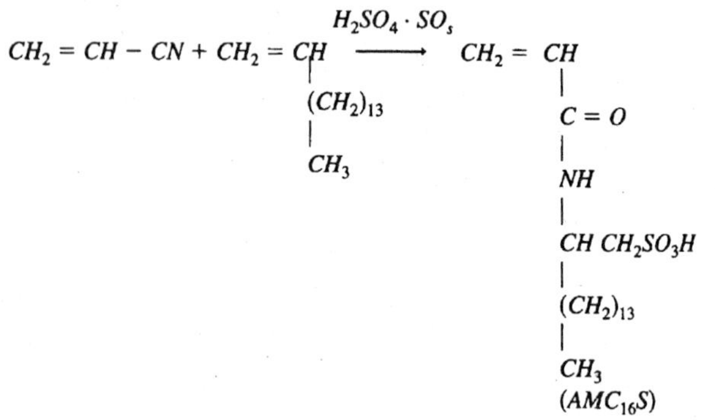

1

1-hexadecylene and oleum (**Equation 1**). $AMC_{16}S$ synthesis and characterization were described in a previous work.[1]

$AMC_{16}S$ not only contained an unsaturated, polymerizable acrylamido group but also had excellent colloidal properties. Its critical micelle concentration (CMC) in water was 2.1×10^{-4} mol/dm^3, and the surface tension of its 1.0 (wt)% aqueous solution was 33.5 mN/m.[1]

Polymerization

$AMC_{16}S$ could be polymerized with a free-radical polymerization mechanism by using AIBN, H_2O_2, or (Ce^{+4} + ROH) redox system as an initiator.

PROPERTIES

General Features

The novel comb-type amphiphilic polymer, $PAMC_{16}S$, was soluble in ethanol, THF, and DMF. Unlike the water-soluble monomer $AMC_{16}S$, however, it was insoluble in water.[2,3]

$PAMC_{16}S$ is an anionic polymer. Viscosity measurements in ethanol revealed its polyion nature. The reduced viscosity, η_{sp}/C, increased markedly with the decrease in concentration. When viscosity was measured in THF solutions, viscosity behavior was normal, and intrinsic viscosities could be obtained according to the Huggins and the Kraemer relationships.[2]

Supramolecular Structure

The evidence from FTIR spectral analysis supported the formation of facing or sideways supramolecular polymeric structures depending on temperature. In our case, the strong hydrophilic groups of sulfonic acid attached to the same side of the polymer backbone with a long lipophilic side chain. These sulfonic groups and lipophilic segments tended to self-assemble forming a layered structure. The sulfonic acid aggregated by hydrogen bonding in a facing way on both sides of the polymer backbone, but the lipophilic moieties segregated from the hydrophilic moieties to construct an aliphatic layer by sandwiching between the polar layers.

Mesomorphic Behavior

Polarized optical microscopy was used to observe the polymer's mesomorphic texture. The polymer melt exhibited high viscosity when the sample, sandwiched between two microscope slides on a hot stage, was sheared. After a pressure was supplied to the sample to make it a uniform thickness, we did not observe distinct birefringence, suggesting that the macroscopically

*Author to whom correspondence should be addressed.

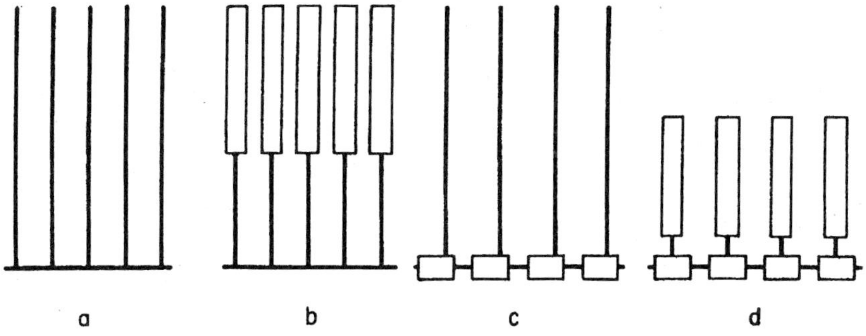

FIGURE 1. Types of comblike polymers: flexible (———) and rigid group (▭).

ordered structure was damaged. The texture then gradually evolved with annealing, and macroscope domains eventually formed. When the shear action was supplied, the birefringent stripes appeared after awhile.

The liquid crystalline phase could be ascribed to a Smectic A mesophase, in which the lamellar is perpendicular to the fiber axis and the short range (4–5Å) order from intermolecular side-chain packing is absent. For the thermotropic mesophase behavior of the comb-like amphiphilic polymer, besides the effects of hydrogen bonding from hydrophilic moieties, the strong lipophilic moieties play a significant role in two ways: by reducing the polymers' melting point so that their liquid-crystalline phases appear before decomposing and isotropizing, and by segregating distinct areas from the polar segments to form a self-aggregated lipophilic layer. Therefore, the length of the lipophilic segments should be intermediate for liquid-crystalline properties.

REFERENCES

1. Fang, T. R.; Shang, Z. P.; Zhu, X. B. et al. *Anal. Chem.* (*Chinese*) **1991**, *19*, 391.
2. Fang, T. R.; Zhu, X. P. *Polym. Bull.* **1991**, *25*, 467.
3. Fang, T. R.; Xu, G. F.; Zhu, X. P. et al. *Thin Solid Films* **1992**, *210/211*, 839.

COMB-LIKE POLYMERS

Yoshio Shibasaki
Saitama University
Department of Chemistry

Comb-like polymers, which have long side chains (see **Figure 1**), are important for liquid-crystalline polymers, polymer surfactants, surface modifiers, and others.

The syntheses, structure, properties, and applications of comb-like polymers have been studied since polymerizing *n*-alkylacrylates.[1] The crystallization of side chains on comb-like polymers is studied in relation to the radiation-induced, solid-state polymerizations for *n*-alkylmethacrylates and *n*-alkylacrylates.[2,3] Many studies of T_g, side-chain crystallization, and mechanical properties were reported for poly(*p*-*n*-alkylstyrene)s, poly(*n*-alkylvinyl ether)s, poly(*N*-*n*-alkylmaleimide)s, poly(*n*-alylkitaconate)s, and poly(*N*-*n*-alkyl-acrylamide)s.[4–8]

Various liquid-crystalline comb-like polymers have been synthesized. The liquid-crystalline properties depend on the structure of mesogenic units and the polymer backbone, and the amount of space between the mesogenic units and the polymer backbone. The research on comb-like polymers has been reviewed extensively.[9–13] Recently, to make thin films with a two-dimensional network structure, Shibasaki et al. studied the photopolymerization in LB films of comb-like polymers containing terminal vinyl groups in the side-chains.[14] The self-organizing behavior of comb-like polymers containing fluorocarbon side chains was also examined.[15]

PROPERTIES

The physical and chemical properties of comb-like polymers depend on the chemical structure of the repeating unit and on a balance of size and polarity for the side chains and backbone. For comb-like polymers composed of flexible side chains and a flexible backbone (Figure 1a) and those composed of rigid side chains and a flexible backbone (Figure 1b), side-chain crystallization was usually observed.[16] Comb-like polymers crystallize in a layered structure, and hence, the side-chain crystallization is influenced by the backbone's stereoregularity.

An isotactic comb-like polymer should not favor side-chain crystallization, although a syndiotactic polymer is preferred for closely packed side chains. This characteristic was verified by comparing the surface pressure areas (π-A) for isotherms of isotactic and atactic poly(octadecyl methacrylate) monolayers spread on the water's surface.[17]

Amphiphilic comb-like polymers with strong polar groups have poor solubility because a good solvent for a hydrophobic side chain is a rather poor solvent for the hydrophilic polymer backbone.

Thermal properties of comb-like polymers were investigated with differential scanning calorimetry (DSC). Comb-like polymers with a flexible backbone and long flexible side chains had higher melting points than those for the corresponding monomer, although the melting points decreased with shorter side chains.[2,3] In the crystals of comb-like polymers with sufficiently long teeth, such as poly(octadecyl methacrylate)s, eight or nine methylene units from the side chains near the main chain were amorphous.[18,19] In addition, for the widely spaced comb-like polymers with teeth (long alkyl groups) appearing on every 4–26 repeating main chain atoms, the extent of crystallization and the crystalline structure were similar to those for conventional comb-like polymers. Flexible main chains can be coiled to allow the side chains to be packed in the crystal lattice.

Comb-like polymers with rigid or less flexible side chains (Figure 1b) crystallized easily in a single- or double-layered structure. A comb-like polymer containing a fluorocarbon chain in the side chains crystallized easily because -$(CF_2)_n$-chains were less flexible and the coagulation force between them, in bulk or in solution, was strong.

Comb-like polymers constructed from teeth having two long hydrocarbon chains were also capable of forming layered structures.[20] In widely spaced comb-like polymers, the long fluorocarbon chains and the double hydrocarbon chains were also easily crystallized as in the conventional comb-like polymers. However, when thin films on solid plates were formed using the Langmuir–Blodgett (LB) method, the monolayer deposited for the widely spaced comb-like polymers was the Z-type (single layer with head-to-tail alignment); for conventional comb-like polymers, it was Y-type (double layer with head-to-head alignment).[21]

Comb-like polymers having rigid groups in side chains ordinarily exhibit mesophases, such as nematic, smectics, and cholesteric.

Comb-like polymers with a rigid backbone and flexible side chains (Figure 1c) exhibited some interesting thermal, mechanical, optical, and electrical properties. For example, *N*-alkylated poly(*p*-phenylene terephthalamide) exhibited good solubility, thermoplastic processability, peculiar mechanical properties, and liquid-crystal formation.[22]

Conjugated polymers have attracted great attention because of their interesting and unusual electrical, electrochemical, and optical properties. Long-chain substituted polythiophenes, polyanilines, polypyrroles, and poly(phenylene vinylene)s were synthesized recently as processable conjugated polymers. Their solubility and stability increased their potential for industrial applications.

Rigid-rod polyaramides, such as Kevlar®, are infusible, only soluble in concentrated sulfuric acid, and incompatible with other polymers.

APPLICATIONS

Polysoaps, for example, polymeric disinfectants, self-assemble at very low concentrations in aqueous solution. Amphiphilic comb-like polymers can be used as dispersing agents for paint and ink. Comb-like polymers having perfluorocarbon chains in their teeth are useful for surface modification of various objects.

Undecyloxazoline homopolymers can be coated on glass slides to form adherent films, 150–200 nm thick, with critical surface energies as low as 21.0 mN m^{-1}.

Poly(2-acrylamidehexadecylsulfonic acid) is a new comb-like polymer, with each repeating unit containing a hydrophilic sulfonic acid group, an amino group, and a long, hydrophobic, aliphatic hydrocarbon chain. This polymer's self-assembled monolayers were formed on gold surfaces and exhibited great adsorption stability during faradaic reactions.[23] The disulfide-bearing, comb-like polymers spontaneously yield two-dimensional, ultrathin polymer films with side-chain-dependent layer thickness of 2.0–4.5 nm by solution adsorption onto freshly deposited gold surfaces.[24] Such ultrathin polymer films should have diverse applications as bound polymeric surface modification reagents.

Immobilizing highly oriented perfluoroalkyl polymers onto porous silica gels can be used for high-performance liquid chromatography.[25]

The polymeric phosphonium salts with different side-chain lengths between the main chain and the active group were prepared, and the effect of the side-chain length on bacterial resistance to *Staphylococcus aureus* and *Escherichia coli* was examined.[26] The antibacterial activity decreased as the side-chain length increased and peaked at an optimum concentration. This activity was closely correlated to the assembly size or the aggregation state of the polycationic biocides near the bacterial cell surfaces.

REFERENCES

1. Rehberg, C. E.; Fisher, C. H. *J. Am. Chem. Soc.* **1944**, *66*, 1203.
2. Shibasaki, Y.; Fükuda, K. *J. Polym. Sci., Polym. Chem. Ed.* **1980**, *18*, 2437.
3. Shibasaki, Y.; Füzuda, K. In *Proceedings of 7th International Conference on Thermal Analysis*; Miller, B., Ed.; John Wiley & Sons: New York; 1982; Vol. II, pp 1517–1523, *Thermal Analysis*.
4. Overberger, C. G.; Frazier, C.; Mandehman, J. et al. *J. Am. Chem. Soc.* **1953**, *75*, 3326.
5. Lal, J.; Trick, G. S. *J. Polym. Sci.* **1964**, *A2*, 4559.
6. Barrales-rieenda, J. M.; Ramos, J. G.; Chaves, M. S. *Br. Polym. J.* **1977**, 6.
7. Cowie, J. M. G.; Henshall, S. A. E.; Macewen, I. J. et al. *Polymer* **1977**, *18*, 612.
8. Jordan, E. F., Jr.; Riser, G. R.; Artymyshyn, B. et al. *J. Appl. Polym. Sci.* **1969**, *13*, 1777.
9. Platé, N. A.; Shibaev, V. P. *J. Polym. Sci., Macromol. Rev.* **1974**, *8*, 117.
10. Platé, N. A.; Shibaev, V. P. *Comb-Shaped Polymers and Liquid Crystals*; Plenum: New York, 1987.
11. *Mesomorphic Order in Polymers and Polymerization in Liquid Crystalline Media*; Blumstein, A., Ed.; ACS Symposium Series 74; American Chemical Society: Washington, DC, 1978.
12. Finkelmann, H. *Polymer Liquid Crystals*; Academic: New York, 1982; Chapter 2.
13. Finkelmann, H. *Liquid Crystallinity in Polymers: Principles and Fundamental Properties*; VCH: New York, 1991; Chapter 8.
14. Shibasaki, Y.; Wen, G.; Nakahara, H. et al. *Thin Solid Films* **1994**, *244*, 732.
15. Shibasaki, Y.; Zhu, Z.-Q.; Nakahara, H. et al. In *proceedings of 46th Symposium on Colloid & Interface Chem.* (*Jpn*); 1993, p 150; J. Nanjin Univ. (Natural Sci: Ed) **1995**, *31*, 335.
16. Wendorff, J. H.; Finkelmann, H.; Ringsdorf, H. *J. Polym. Sci., Polym. Symp.* **1978**, *63*, 245.
17. Duda, G.; Schouten, A. J.; Arndt, T. et al. *Thin Solid Films* **1988**, *159*, 221.
18. Jordan, E. F., Jr.; Feldeisen, D. W.; Wrigley, A. N. *J. Polym. Sci.* **1971**, A1, *9*, 1835.
19. Hsieh, H. W. S.; Post, B.; Morawetz, H. *J. Polym. Sci., Polym. Chem. Ed.* **1976**, *14*, 1241.
20. Schneider, J.; Ringthdorf, H.; Rabolt, J. F. *Macromolecules* **1989**, *22*, 205.
21. Schneider, J.; Erdelen, C.; Ringthdorf, H. et al. *Macromolecules* **1989**, *22*, 3475.
22. Takayanagi, M.; Katayose, T. *J. Polym. Sci., Polym. Chem. Ed.* **1981**, *19*, 1133.
23. Fang, T. R.; Xu, G. F.; Zhu, X. B. et al. *Thin Solid Films* **1992**, *210/211*, 839.

24. Sun, F.; Grainger, D. W. *J. Polym. Sci., Part A: Polym. Chem.* **1993**, *31*, 1729.
25. Hirayama, C.; Ihara, H.; Nagaoka, S. et al. *Polym. J.* **1994**, *26*, 499.
26. Kanazawa, A.; Ikeda, T.; Endo, T. *J. Polym. Sci., Part A: Polym. Chem.* **1994**, *32*, 1997.

COMB-LIKE POLYMERS, WIDELY SPACED

Kenji Yokota
Department of Materials Science and Engineering
Nagoya Institute of Technology

Vinyl polymers with long side chains are called comb-like polymers after their characteristic molecular shape. The long side chains resemble the teeth of a comb, and in most cases, they are *n*-alkyl groups containing 10–20 carbon atoms. The comb-like polymers are crystalline even when the polymer analogues, which have short alkyl side chains, are amorphous. They are crystalline because their long alkyl side chains crystallize like paraffinic hydrocarbons.

The crystalline structure of comb-like polymers has been studied extensively. Excellent reviews of the comb-like polymers were written by Plate and Shibaev.[1,2]

We studied the side-chain crystallization of widely spaced comb-like polymers. We defined widely spaced comb-like polymers as those with long alkyl side chains (teeth) appearing on every three or more main-chain atoms. Conventional comb-like polymers have side chains on every two main-chain carbon atoms. The widely spaced comb-like polymers that we studied were alternating butadiene-alkyl methacrylate copolymers and their hydrogenated products, which had side chains on every six main-chain carbon atoms.[3] Our question was simple: whether the widely spaced side chains could crystallize despite their wide spaces.

PREPARATION

First, we studied alternating butadiene-alkyl methacrylate copolymers and their hydrogenated products.[3] Alkyl substituents were unbranched *n*-alkyl groups containing 12–18 carbon atoms. Alternating copolymerization of butadiene and alkyl methacrylate was carried out in the presence of ethylaluminum sesquichloride and vanadyl chloride using Furukawa's procedure.[4]

The second type of widely spaced comb-like polymers that we studied was a series of α-methyl-substituted and -unsubstituted alternating styrene-alkyl acrylate copolymers.[5] These alternating copolymers were prepared by copolymerizing styrenes and alkyl acrylates in the presence of ethylaluminum sesquichloride.

The third type of widely spaced comb-like polymers was the polyester obtained from 2-alkylpropanediol-1,3 and a diacid chloride in toluene.[6,7]

CONCLUSION

Widely spaced comb-like polymers were defined as those with long alkyl side chains (teeth) appearing on every three or more main-chain atoms. Their alkyl side chains crystallized like those of conventional comb-like polymers despite their wide spaces. The flexible methylene groups of the polymer's main chain allowed the side chains to move closer and crystallize. The rigidness or flexibility of the main chain affected the extent of side-chain crystallization.

REFERENCES

1. Platé, N. A.; Shibaev, V. P. *J. Polym. Sci., Macromol. Rev.* **1974**, *8*, 117.
2. Platé, N. A.; Shibaevk, V. P. *Comb-Shaped Polymers and Liquid Crystals*, Engl. ed.; Plenum: New York, 1987.
3. Yokota, K.; Kougo, T.; Hirabayashi, T. *Polym. J.* **1983**, *15*, 891.
4. Furukawa, J.; Iseda, Y. *J. Polym. Sci.* **1969**, *B5*, 47.
5. Hirabayashi, T.; Kikuta, T.; Kasabou, K. et al. *Polym. J.* **1988**, *20*, 693.
6. Hirabayashi, T.; Kasabou, K.; Yokota, K. *Polym. J.* **1988**, *20*, 911.
7. Yokota, K.; Hirabayashi, T. *Polym. J.* **1986**, *18*, 177.

Comb Polymers

See: *Branched Polymers*
Comb-Like Amphiphilic Polymers (Supramolecular Structure)
Comb-Like Polymers
Comb-Like Polymers, Widely Spaced
Comb Polymers [Poly(ethylene oxide) Side Chains]
Functionalized Polymers (Comb, Rotaxanic, and Dendrimeric Structures)
Hairy Rod-Like Polymers
Poly(ethylene oxide) Macromonomers (Micellar Polymerization)
Stereoregular Polymethacrylate Macromonomer
Supported Lewis Acid Catalysts (on Polypropylene; Recoverable and Reusable)

COMB POLYMERS [POLY(ETHYLENE OXIDE) SIDE CHAINS]

Koichi Ito
Department of Materials Science
Toyohashi University of Technology

Comb polymers are defined here as those with poly(ethylene oxide) (PEO) side chains as their arms or teeth, which are attached regularly and densely to the main chain.

Polymers with varying lengths of the main and side chains should have mixed characteristics because shorter side chains represent a linear polymer and longer side chains indicate a star-like polymer. Long side and main chains indicate a typical comb polymer.

PREPARATION OF MACROMONOMERS

PEO macromonomers, homopolymerized to comb polymers, are listed in **Table 1**.

PREPARATION OF COMB POLYMERS

The PEO macromonomers listed in Table 1 were homopolymerized to produce their corresponding regular comb polymers. Radical initiators such as 2,2′-azobisisobutyronitrile and potassium persulfate were used for (meth)acrylate- or styryl-ended macromonomers in organic and aqueous solution systems, respectively.

TABLE 1. Poly(ethylene oxide) (PEO) Macromonomers Homopolymerized to Comb Polymers

Entry[a]	Structure	References
1	$RO\text{-}[CH_2CH_2O]_n\text{-}C(=O)C(CH_3)=CH_2$; R = CH_3	2,12
	R = tC_4H_9, CH_3	1
	R = H	3
2	$CH_3O\text{-}[CH_2CH_2O]_n\text{-}C(=O)CH=CH_2$	2,11
3	$RO\text{-}[CH_2CH_2O]_n\text{-}CH_2\text{-}C_6H_4\text{-}CH=CH_2$; R = CH_3, C_4H_9, C_8H_{17}, $C_{18}H_{37}$	1,13
	R = H	3
4	$RO\text{-}[CH_2CH_2O]_n\text{-}(CH_2)_m\text{-}C_6H_4\text{-}CH=CH_2$; m = 4, 7	4
5	$RO\text{-}[CH_2CH_2O]_n\text{-}CH=CH_2$; R = CH_3, C_2H_5 ; n = 1-4	5
	R = CH_3 ; n = 7, 12	6
6	$RO\text{-}[CH_2CH_2O]_n\text{-}C(=O)CH=CHC(=O)OR'$; R = CH_3, $Ph_2CHCH_2CH_2$; R' = H, CH_3, PSt	7
7	$CH_3O\text{-}[CH_2CH_2O]_n\text{-}CH_2CH(CH_2SO_3Na)OC(=O)C(CH_3)=CH_2$	8
8	$HOC(=O)CH_2O\text{-}[CH_2CH_2O]_n\text{-}C(=O)C(CH_3)=CH_2$	9
9	$CH_3O\text{-}[CH_2CH_2O]_n\text{-}CH_2C(COOCH_3)=CH_2$	10

[a]Entries are as follows: 1, methacryloyl-ended; 2, acryloyl-ended; 3, *p*-vinylbenzyl-ended; 4, *p*-styrylalkyl-ended; 5, vinyl ether-ended; 6, fumarate-ended; 7, methacryloyl and sulfonate-ended; 8, ω-carboxy PEO methacrylate; and 9, α-(ω-methoxyoligoethylenoxymethyl) acrylate.

Generally, macromonomers polymerized apparently poorly compared with their corresponding low molecular weight monomers because the macromonomers' high molecular weight lowered monomer and initiator concentrations. In fact, at the same molar concentration, the *p*-vinylbenzyl-ended PEO macromonomer polymerized in benzene more rapidly than did lower molecular weight analogs like styrene because of the former's lower rate of diffusion-controlled termination.[1,3]

Interestingly, the polymerization of the *p*-vinylbenzyl-ended macromonomer in water with 4,4′-azobis(4-cyanovaleric acid) was unusually fast and afforded a comb polymer with a very high degree of polymerization.[1,3,13] This result may have occurred because amphiphilic macromonomer molecules formed in the micellar structure.

PROPERTIES AND APPLICATIONS

Solution properties of the comb polymers from *p*-vinylbenzyl-ended monomers were studied using static light scattering, gel permeation chromatography (GPC), and intrinsic viscosity ([η]) measured in tetrahydrofuran.[13] The comb polymers with relatively long PEO arms exhibited compact dimensions compared with those for linear polymers having the same molecular weight. Thus, GPC evaluations calibrated with the linear standard polymers would severely underestimate the comb polymers' molecular weights unless the calibration was performed by measuring the intrinsic viscosity. The plots according to the Mark-Houwink-Sakurada equation ($[\eta] = KM_w^a$) gave very low or nearly zero values for the exponent a and for comb polymers with arm lengths $n = 44$ and 103, suggesting a densely filled, nondraining, rigid, sphere-like conformation. The comb polymers' low dependence on viscosity has potential for applications in coating technology.

Critical micelle concentration of comb polymers from PEO macromonomers was estimated as very low compared with that of the corresponding macromonomers.[11]

The ion-conducting ability of the PEO chains reportedly is promising with comb polymers.[9,14–17]

REFERENCES

1. Ito, K.; Tanaka, K.; Tanaka, H. et al. *Macromolecules* **1991**, *24*, 2348.
2. Gramain, P.; Frère, Y. *Polym. Commun.* **1986**, *27*, 16.
3. Ito, K.; Hashimura, K.; Itsuno, S. et al. *Macromolecules* **1991**, *24*, 3977.
4. Chao, D.; Itsuno, S.; Ito, K. *Polym. J.* **1991**, *23*, 1045.
5. Aoshima, S.; Oda, H.; Kobayashi, E. *J. Polym. Sci., Part A: Polym. Chem.* **1992**, *30*, 2407.
6. Mathias, L. J.; Canterberry, J. B.; South, M. *J. Polym. Sci., Polym. Lett. Ed.* **1982**, *20*, 473.
7. Berlinova, I. V.; Panayatov, I. M. *Makromol. Chem.* **1987**, *188*, 2141; *Makromol. Chem.* **1989**, *190*, 1515.
8. Zhang, S.-S.; Liu, Q.-G.; Yang, L.-L. *J. Polym. Sci., Part A: Polym. Chem.* **1993**, *31*, 2313.
9. Tsuchida, E.; Ohno, H.; Kobayashi, N. et al. *Macromolecules* **1989**, *22*, 1771.
10. Yamada, B.; Kobatake, S.; Aoki, S. *J. Polym. Sci., Part A: Polym. Chem.* **1993**, *31*, 3433.
11. Geetha, B.; Mandal, A. B.; Ramasami, T. *Macromolecules* **1993**, *26*, 4083.
12. Rempp, P.; Lutz, P.; Mason, P. et al. *Makromol. Chem., Suppl.* **1985**, *13*, 47.
13. Ito, K.; Tomi, Y.; Kawaguchi, S. *Macromolecules* **1992**, *25*, 1534.
14. Cowie, J. M. G.; Martin, A. C. S.; Firth, A.-M. *Br. Polym. J.* **1988**, *20*, 247.
15. Khan, I. M.; Fish, D.; Delaviz, Y. et al. *Makromol. Chem.* **1989**, *190*, 1069.
16. Bannister, D. J.; Davis, G. R.; Ward, I. M. et al. *Polymer* **1984**, *25*, 1600.
17. Fish, D.; Xia, D. W.; Smid, J. *Makromol. Chem., Rapid Commun.* **1985**, *6*, 761.

Compatibilizers

See: *Compatibilizers (for Blends)*
Compatibilizers, Polymeric
Compatibilizers, Polymeric (Recycling of Multilayer Structures)
Cyclic Imino Ethers (Ring-Opening Polymerization)
Ethylene-Vinyl Acetate Copolymer (Chemical Modification, Compatibilizer in Blends)
Ionomers (Compatibilization of Blends)
Methacrylic Resins (Improved Heat Resistance)
Organic/Inorganic Polymer Hybrids
Poly(ester imide) Compatibilizer (in Thermotropic Liquid Crystalline Polymer Blends)
Polyolefin Graft Copolymers (Prepared by Borane Approach)
Poly(vinyl chloride), Rigid (Soft Fillers)
Reactive Processing, Thermoplastics
Recycling, Plastics
Surfactants, Polymeric (Compatibilizers for Blends)

COMPATIBILIZERS (for Blends)

David J. Lohse*
Exxon Research and Engineering Company

Sudhin Datta
Exxon Chemical Company

Mixtures of two or more polymers are increasingly common materials despite the general immiscibility of polymer pairs. The compatibility and utility of such blends have been enhanced by interfacially active block or graft copolymers. Here, we outline the principles behind such compatibilization and demonstrate how compatibilizers can improve blend materials.

PRINCIPLES OF COMPATIBILIZATION

The major difficulty in devising a useful polymer blend is the general incompatibility of polymers. There is essentially no entropy of mixing for a blend of high molecular weight polymers. Thus, one major driving force for solubility that is found in mixtures of small molecules is absent in polymer blends. Therefore, one can expect that polymers will be soluble or miscible in one another only in special cases, such as in the presence of specific strong interactions between repeating units. This phenomenon has been outlined in several texts.[1–3]

Consequently, nearly every pair of polymers is immiscible, that is, the mixture will exhibit two phases at essentially all accessible compositions and temperatures. This multiphase morphology is not necessarily a problem when using the blend. Many blends are successful in the marketplace because of this morphology and their corresponding properties. However, the blend's properties will depend on the sizes of the phase domains and the adhesion between them. The purpose of compatibilizers is to control both of these features in polymer blends.

Using block and graft polymers as interfacial agents in polymer blends began almost as soon as these polymers became available.[4,5] The earliest, and still an important, compatibilizer was the use of poly(styrene-*graft*-butadiene) copolymers in a polystyrene–polybutadiene blend to produce high-impact strength polystyrene.[6]

Current understanding of compatibilization can be summarized by a few principles. These rules are not rigorous but have been extracted from the evidence generated by many model and commercial blend systems. The detailed chemistry of how the graft or block compatibilizer is made and how it is introduced into the blend varies from system to system and depends on the nature of the blend components. But the basic physical principles for compatibilization, as they are currently understood, underlie all of these applications, and these rules are summarized below.

LOCALIZATION

The role of the compatibilizer is to control the properties of a multiple phase blend not by converting an immiscible blend into a miscible one but by controlling the size of the phase domains and the adhesion between them. Effective compatibilizers must

*Author to whom correspondence should be addressed.

be located at the interface between the phase domains of a blend.

Compatibilizers have been used to control phase size, the main test for compatibilizing activity.[7] Maximizing the adhesion between the phases is also important, although this has been emphasized less. Recent experimental work has shown the need for the latter. Blends with compatibilizers at the interface show much greater impact strength than uncompatibilized blends made from the same components and having similar particle sizes.[8,9] Moreover, Brown, Kramer, and others have shown the ability of block polymers to improve an interface's mechanical strength.[10–17] Interfacially active compatibilizers provide both particle size control and maximized adhesion between dissimilar phases.

COMPATIBILIZATION: A SPECIFIC EXAMPLE

A commonly used class of polymer blends contains polypropylene (PP) and ethylene-propylene copolymers (EP). These blends are used in a range of composition, from toughened versions of PP containing 5–20% of EP, to thermoplastic elastomers that have a continuous phase of PP and discontinuous EP particles at 60–80%. These blends can be made by mechanical mixing or directly in polymerization reactors. They are versatile materials frequently used in the marketplace.

However, like all immiscible blends, the properties of this blend can be enhanced with compatibilizers, as a series of studies have shown.[18–20] IN these studies, EP-*graft*-PP copolymers were used as compatibilizers for PP–EP blends.

The formation of graft polymers was supported further by their compatibilizing activity in PP–EP blends. The sizes of the disperse EP phases in an 80:20 EP blend were noticeably reduced by adding a small amount of the compatibilizer. Moreover, not only were the sizes smaller in the as-mixed blends, but the phases grew much more slowly in the quiescent melt. By making graft polymers with a small degree of unsaturation, it was also possible to show that the compatibilizers were present at the blend's interfaces. Thus, the control of blend morphology was clearly shown to result from the interfacially active graft polymers.

Adding the graft polymers also improved the toughness of these blends.

From this example and from extensive research completed for various blends, compatibilizers have clearly become an important part of polymer technology. Although only some scientific principles have been firmly established, great progress has been made in understanding compatibilizers so that their use will become even more widespread.

REFERENCES

1. Utracki, L. A. *Polymer Blends and Alloys*; New York, 1990.
2. *Polymer Blends*; Paul, D. R.; Newman, S. Eds., Academic: New York, 1978.
3. *Polymer Blends and Mixtures*; Walsh, D. J.; et al. Eds.; NATO ASI Series E 89; Martinus Nijhoff: Boston, 1985.
4. Baer, M.; Hankey, E. H. U.S. Patent 3 312 756, 1967.
5. Molau, G. E. *J. Polym. Sci.: Part A* **1965**, *3*, 4235.
6. Echte, A. *Rubber-Toughened Plastics*; Riew, C. K., Ed.; Advances in Chemistry Series 222, American Chemical Society: Washington, DC, 1989; Chapter II.
7. Datta, S.; Lohse, D. J. *Polymeric Compatibilizers: Uses and Benefits in Polymer Blends*; Hanser: New York, 1996.
8. Liu, N. C.; Baker, W. E. *Polymer* **1994**, *35*, 988.
9. Lynch, J. C.; Nauman, E. B. *ACS Division Polym. Mater.: Sci. Eng.—Preprints* **1994**, *71*, 609.
10. Brown, H. R.; Deline, V. R.; Green, P. F. *Nature* **1989**, *341*, 221.
11. Brown, H. R.; Char, K.; Deline, V. R. et al. *Macromolecules* **1993**, *26*, 4155.
12. Char, K.; Brown, H. R.; Deline, V. R. *Macromolecules* **1993**, *26*, 4164.
13. Cho, K.; Brown, H. R.; Miller, D. C. *J. Polym. Sci. Phys.* **1990**, *28*, 1699.
14. Creton, C.; Kramer, E. J.; Hadzioannou, G. *Macromolecules* **1991**, *24*, 1846.
15. Creton, C.; Kramer, E. J.; Hui, C.-Y. et al. *Macromolecules* **1992**, *25*, 3075.
16. Washiyama, J.; Creton, C.; Kramer, E. J. *Macromolecules* **1992**, *25*, 4751.
17. Washiyama, J.; Kramer, E. J.; Hui, C.-Y. *Macromolecules* **1993**, *26*, 2928.
18. Datta, S.; Lohse, D. J. U.S. Patent 4, 999, 403, 1991.
19. Datta, S.; Lohse, D. J. *Macromolecules* **1993**, *26*, 2064.
20. Lohse, D. J.; Datta, S.; Kresge, E. N. *Macromolecules* **1991**, *24*, 561.

COMPATIBILIZERS, POLYMERIC

Martha Albores-Velasco, R. I. Gascón, and O. A. Rodríguez
Facultad de Quimica UNAM Circuito Interior Ciudad Universitaria

Polymer blends, which may be defined as "any combination of two or more polymers resulting from a common processing step," have increasing technological importance because they can combine mechanical, thermal, or other useful properties from different polymers to obtain materials that are unavailable from the constituents alone.[1] Blends reduce the time and cost of developing commercial products; therefore, an increased range of commercial blend products is expected in the near future.

Polymeric blends can be homogeneous or heterogeneous, but at a macroscopic level, they should not indicate any phase separation. Compatibility is often used in a thermodynamic sense to be synonymous with miscibility.

A blend's optical transparency depends on the particle size and the refractive indexes of both polymers. Therefore, according to this definition, a lack of transparency does not indicate incompatibility.

The compatibility of polymers varies widely. In fact, most polymer pairs are immiscible or only partially miscible. The thermodynamics of polymer blends plays a major role in the molecular state of dispersion, the morphology of two phase mixtures, or the adhesion between phases. In fact, the enthalpy of mixing should be negative or exothermic for a mixture to be factible. This condition can be achieved if polar groups in the polymer interact in certain ways. These interactions may arise from various mechanisms such as dipole-dipole forces or ionic interactions. Sometimes, two polymers with chemical moieties that have the proper "complementary dissimilarity" can be selected to produce the exothermic heat of mixing. The van der Waal forces between the structure's non-interacting parts, however, help make the reaction endothermic and affect the mixture's properties and applications. The degree of a blend's intermolecular mixing may also be affected as a consequence of mechanical and solvent exposure history.[1]

Few, if any, polymeric blends exist in a thermodynamic equilibrium state. Some thermodynamic compatibility or a physical constraint such as grafting, crosslinking, or an interpenetrating network formation should be present in a stable blend. However, some polymeric species, usually block or graft copolymers, can be used to improve miscibility usually because their segments are identical to those of the polymers being mixed. These species are called compatibilizers.[2] A compatibilizing agent affects the properties of polymer blends and polymer melts and also influences a polymer's dispersion in a nonsolvent.

Compatibilizers can be added to the mixture as a third component or they can be formed *in situ*. *In situ* formation of compatibilizers is achieved by modifying the mixture's components in such a way that one of the polymers reacts with the other.

According to Gaylord, copolymers with a molecular weight greater than 150,000 are poor compatibilizers because the segments have limited access to the polymers.[3]

Solubility is also important for compatibilization. Sometimes, polymers with different structures may be compatible if their solubilities are closely related. For instance, poly(methyl methacrylate), poly(ethyl acrylate), poly(vinyl chloride), and poly(butadiene-co-acrylonitrile) (90/10–60/40) form useful compatible blends because their solubility parameters are closely related.[3]

Recently, we synthesized functionalized polymers with different amounts of polar groups to use as compatibilizers. We introduced the functional groups in commercial polymers by reactions in solution to have a homogeneous distribution of the functional groups instead of graft or block copolymers. According to Paul, functional groups act in the interphase of the polymers being mixed. Block copolymers, however, sometimes have two glass transition (T_g) temperatures. The synthesis and characterization of two types of carboxylated polystyrenes and their use as compatibilizers are reported.[4]

REFERENCES

1. Paul, D. R. *Polymer Blends*; Paul, D. R.; Newman, S., Eds.; Academic: New York, 1978; p. 35.
2. Pouchly, *Polym. Lett.* **1969** *7*, 463.
3. Gaylord, N. G. In *Polymeric Blends*; Paul, D. F.; Newman, S. Eds.; Academic: New York, 1978; Vol. 1, pp. 76–84.
4. Ramírez López, E. G. BsC Thesis. Universidad Nacional Autónoma de México, 1993.

COMPATIBILIZERS, POLYMERIC (Recycling of Multilayer Structures)

Karlheinz Hausmann
DuPont de Nemours International European Technical Center

Plastic waste, especially multilayer film packaging waste, has become an important global issue in recent years. This issue has been triggered by rising concerns over environmental pollution and limited landfill space.[1–3] Plastic multilayer structures are especially targeted because of the perception that they cannot be recycled.

The structure of the packaging contains different layers (structural, barrier, adhesive, and seal), which provide separate functions and are made from different polymers. This structure may present obstacles when the packaging is recycled because melt processing these polymer blends results in poor properties. This problem has led to the perception that multilayer structures are non-recyclable waste.

Polymeric compatibilizers serve—as their name indicates—to make compatible different materials such as those used for plastic multilayer structures in packaging. This problem differs from the recycling of plastic monomaterial packaging, such as poly(ethylene terephthalate) (PET) bottles, which have reasonably clean waste streams to obtain a recyclate with acceptable properties. In this case, compatibilizers are unnecessary. Before discussing the compatibilization of polymer pairs in multilayer film structures, the compatibility of polymer blends and the compatibilizers that are useful for such applications will be described.

PREPARATION AND PROPERTIES

Compatibilizers can be classified as follows: non-reactive and reactive compatibilizers; random, graft, and block copolymers; and the nature of their base polymer. Here, the last category will be focused on, and the first two approaches will be explained only briefly. Compatibilizers based on polyolefins will be considered separately because multilayer structures as they are used in film packaging contain at least one component of a polyolefinic nature, which provides the low sealing temperature and which must to be compatibilized with the remainder of the structure.

Non-Reactive and Reactive

Non-reactive compatibilizers, which compatibilize two polymers, A and B, consist of two parts: the first is soluble in polymer A, and the second is soluble in polymer B. The compatibilizer's effect derives from their solubility. Therefore, the solubility parameters of both parts should be as close as possible to the solubility parameters of the polymer components in the polymer mixture. The solubility parameters are listed by Van Krevelen or have been calculated by Coleman et al.[4,5] An example is a styrene-ethylene butylene-styrene (SEBS) block copolymer compatibilizing polyethylene (PE) and polystyrene (PS).

Block, Graft, and Random Copolymers

A block copolymer contains blocks of the polymer pairs it compatibilizes. These blocks can be reactive or nonreactive polymers and an example is SEBS. Their production process is described in the literature.[6]

In random copolymers, the components, the base polymer B and a comonomer A, are distributed randomly along the polymer chain. Random copolymers only work well as compatibilizers when the comonomer A is reactive. An example is poly(ethylene co-methacrylic acid) (EMMA).

In graft copolymers, either monomers or polymers are grafted onto each other. If only monomers are grafted to the backbone, the monomer should be reactive. An example is PP grafted with maleic anhydride. The exposure of the reactive monomers on the usually non-reactive base polymer backbone makes them more accessible to an attack by other polymers, transforming them into effective compatibilizers. They are produced by three different methods: solution grafting, grafting and reaction of peroxypolymers (macromonomer formation), and reactive extrusion grafting.[7–11]

Nature of the Base Polymer

Modified Polyethylene

Modified polyethylene is the most important category of compatibilizers used in multilayer structures in film packaging because almost every structure contains some kind of polyethylene-based layer that provides flexibility, adhesive strength, or low sealing temperature. There are block copolymers, random copolymers, and graft copolymers. In particular, there are PE-PP block copolymers, and random copolymers based on ethylene. In the last category are also ionomers that are acid copolymers but neutralized with small amounts of metal salts (chosen from zinc, sodium, lithium, and magnesium).

The vast assortment of graft copolymers include a reactive group, mostly maleic anhydride, which is grafted onto the ethylene copolymer backbone. Almost all the non-reactive ethylene copolymers are considered backbone polymers, depending on the type of ethylene copolymer in the base polymer blend.

Modified Polypropylene

Modified polypropylene is the second largest group of compatibilizers, which comprises block or random copolymers of propylene with ethylene such as PP, polypropylene copolymer (COPP), ethylene-propylene rubber (EPR), or EPDM. These copolymers can be used to compatibilize nonpolar components. Grafted polypropylene types are also in this category such as PP grafted with MAH (maleic anhydride) or AA (acrylic acid).

Styrene Based Copolymers

These copolymers are the third largest group of compatibilizers for recycling tasks because PS represents an important part of the packaging waste stream. The most common modified styrene based polymers are SBS, styrene-butylene-styrene terpolymer (SBS), or SEPS block copolymers.

Macromonomers

They are block copolymers that have one block of polymer A grafted onto one block of polymer B. Therefore, they can be used to compatibilize polymers A and B. Examples are polypropylene grafted onto polystyrene (pp-g-ps) and low density polyethylene grafted onto a styrene-acrylonitrile polymer (LLPDE-g-SAN).

APPLICATIONS

Compatibilizing the multilayer structures used in film packaging was considered only recently as governments increasingly imposed quotas to force industries to recycle plastic materials containing virtually incompatible components. The literature began addressing these recycling problems with papers about different multilayers such as PA/PE, PET/PE, and PS/PE.[12–16] This chapter serves to review the most important structures used in film packaging and the most useful compatibilizers of the polymeric components.

Structures

Table 1 lists the most important multilayer structures used in film packaging. This table also gives the functions of these layers and their thickness. Also given are compatibilizers that allow these structures to be recycled and the references that describe these recycling methods.

CONCLUSION

This article reviewed only the most important compatibilizers and how they could compatibilize multilayer structures currently used in film packaging. A standard recipe for each recycling problem is impossible as each case will require a different solution.

The economical viability of recycling has not been discussed in detail because it depends on environmental legislation, which varies with different countries, the availability of inexpensive virgin raw materials, land fill costs, and the individual operational costs of the specific converter. However, commercial, successful applications are known and many others are being developed.[14]

REFERENCES

1. Cheremishoff, N. P. *Poll. Eng.* **1989**, 58.
2. Menges, G.; *Gummi, Kautschuk, Kunststoffe.* **1993**, *46*, 63.
3. Nir, M. M.; Miltz, J.; Ram, A.; *Plastics Eng.* **1993**, *3*, 75.
4. Coleman, M. M.; et al., *Polymer* **1990**, *31*(7), 1187.
5. Van Krevelen, D. W. *Properties of Polymers*; Elsevier: Amsterdam, 1990.
6. Jones, R. U.S. Patent 27145, 1971.
7. Kelusky, E. C. European Patent 370375, 1989.
8. Michel, A.; French Patent 8 122 802, 1981.
9. Minoura, Y.; *J. Appl. Pol. Sci.* **1969**, *13*, 1625–1640.
10. Tatsumi, T.; et al. German Patent 2 023 154, 1970.
11. Yamamoto, T.; *Proc. Compalloy '90 Europe*; Luxemburg, 1990; 55.
12. Grenci, J.; In *Proceedings ANTEC '93*; 1993; 488.
13. Gruetzner, R. E.; *Kunststoffe* **1993**, *83*.
14. Hausmann, K. In *Proceedings 34th International Conference on Advances in Modifiers and Compatibilizers for Polymer Blends and Alloys*; 1994.
15. Jottier, E. *Packaging* **1992**, *6*, 11–15.
16. Nagakawa, M.; Kawachi, H. In *Proceedings SP '91*; Zurich, 1991; 113.
17. Favis, B. D. *Can. J. Chem. Eng.* **1991**, *69*, 619–625.
18. Ducommon, C. B. *Proc. Recycle '91*; Davos, 1991.
19. Akkapeddie, M. K.; et al. *Proc. ACS 1992*; 1992; 317.

Composites

See: *Agro-Fiber/Thermoplastic Blends and Alloys*
Anoxic Polymer Materials
Carbon/Carbon Composites (with Surface Treated Carbon Fibers)
Carbon Composites
Carbon Whisker (Surface Modification by Grafting)
Cellulose-Filled Composites

TABLE 1. Summary of Multilayer Structures and Useful Compatibilizers

Type	Use	Components	Function of components	Composition	Compatibilizer	Reference
PE/PP	Medical	PE-LD[a], PE-LLD	Sealing, flexibility	60	EPR	Dehay, 1993
		PP	Moisture barrier, stiffness	30	SEPS	
		Tie	Adhesion	10		
PA/PE	Pasta, meat	PA6	Oxygen barrier, strength	67	None	14
	Cheese	PE-LD, PE-LLD	Sealing	27	PE-g-MAH	13
	Vegetables, fish	Tie	Adhesion	6	EMAA/Ionomer	15
PA/Ionomer	Pasta, meat	PA6	Oxygen barrier, strength, abrasion res.	67	None	14
	Cheese	Ionomer	Sealing, clarity, abrasion resistance	27	Zn-ionomer	17
		Tie	Adhesion	6	(EMAA)	
PA/EVOH/PE	Sausage casings	PA6	Oxygen barrier, moisture barrier, strength	11	PE-g-MAH	14
	Pate	EVOH	Oxygen barrier	7		
		PE-LD, PE-LLD,	Sealing, flexibility, moisture barrier	78		
		Tie	Adhesion	4		
PP/EVOH/PE	Sausage casings	PP	Moisture barrier	20	PP-g-MAH	14
		EVOH	Oxygen barrier	10	PE-g-MAH	
		PE-LD, PE-LLD	Sealing, flexibility, moisture barrier	60	None	
		Tie	Adhesion	10		
PP/EVOH/PP	Menu portions	PP	Moisture barrier	80	PP-g-MAH	
	Ketchup sauces	EVOH	Oxygen barrier	10		
	Retortable	Tie	Adhesion	10		
PE/EVOH/PE	Milk, juices	PE-LD, PE-LLD	Sealing, flexibility, moisture barrier	77	PE-g-MAH	14
	Purees, sauces	EVOH	Oxygen barrier	10	EPDM-g-MAH	
		Tie	Adhesion	13		
PS/EVOH/PE	Dairy products	PS	Structural layer	75	SEPS-g-MAH	18
	(milk pots)	EVOH	Oxygen barrier	6		Dehay, 1993
		PE-LD, PE-LLD	Sealing, flexibility moisture barrier	12		
		Tie	Adhesion	7		
PET/PE	Liquid detergents	PET	Oxygen barrier	22	E-GMA (Co, Terpol)	19
		PE-LD, PE-LLD	Sealing, flexibility, moisture barrier	77	PE-g-MAH	14
		Tie	Adhesion	6	Ionomer (EMAA)	
PET/PE	Liquid detergents	PET	Oxygen barrier	67	E-GMA (Co, Terpol.)	
		PE-LD, PE-LLD	Sealing, flexibility, moisture barrier	27	PE-g-MAH	
		Tie	Adhesion	6		

[a]Mention of PE-LD also implies the use of EVA with low VA content (0–10%).

Composites (Structure, Properties and Manufacturing)
Composites, Thermosetting Polymers
Concrete-Polymer Composites
Cyclic Imino Ethers (Ring-Opening Polymerization)
Dental Polymers (Overview)
Electrically Conducting Composites
Electroconductive Composites (Emulsion Pathways)
Epoxy Resin-Glass Interphases (Optical Properties)
Epoxy Resins (Overview)
Epoxy Resins (High Performance Composite Applications)
Epoxy Resins (Microwave Processing)
Fibers and Composites (High Modulus and Strength)
Glass-Polymer Melt Blends
Graphite Fiber Composites (Electrochemical Processing)
In Situ Thermoplastic Composites (from Liquid Crystalline Polymers)
Interphases, Ternary Polymer Composites
Liquid Crystalline Ethyl-Cyanoethyl Cellulose
Magnetic Field Processing
Maleic Anhydride Grafted Polypropylene (Solid Phase)
Microwave Absorbing Materials
Molecular Composites (High Modulus Polyimide/Polyimide Films)
Molecular Composites (Liquid Crystalline Polyamide and Amorphous Polyimide)
Molecular Composites (Third Generation Polymers)
Organic/Inorganic Composite Materials
Organic/Inorganic Composites
Organic/Inorganic Polymer Hybrids
Phenolic Composites
Polyamides, Lactam-Based
Polydiacetylenes, Unconventional
Polyimides (Introduction and Overview)
Polyimides (via Interfacial Polyfunctional Condensation)
Polypropylene Blends and Composites
Polysilanes (Bearing Hydroxy Groups and Derivatives)
Preceramic Polymers

Recycling Techniques, Plastic Composites
Solid Polymer Electrolytes
Solid Polymer Electrolytes (Polyether Blends and Composites)
Solid-State Polymerization (Crosslinked Di-p-Ethynyl Substituted Polymers)
Starch-Polymer Composites
Thermoplastic Composites
Unsaturated Polyester Resins (Polyester Composites)
Vegetal Biomass (3. Polymers, Derivatives, and Composites)
Wood (Chemical Modification)
Wood Composites (High Performance)
Wood Graft Copolymers
Wood-Plastic Composites

COMPOSITES (Structure, Properties, and Manufacturing)

J. Karger-Kocsis
Institut für Verbundwerkstoffe GmbH
Universität Kaiserslautern

Composites are defined as materials consisting of two or more distinct phases with a recognizable interface or interphase. This definition is generally restricted in the praxis to materials containing fibrous reinforcements or reinforcements with different length and cross-section dimensions (described by the aspect ratio: e.g., platelet or flake), which are embedded in a continuous matrix material. Incorporating this reinforcement improves mechanical performance. That is the basic difference between reinforced and filled systems in which fillers are used to reduce cost but often degrade instead of improve the mechanical property. The distinction between reinforced and filled systems is sometimes unclear.

Composites can be categorized by their matrix characteristics, including type (metal-, ceramic-, or polymeric-based and inorganic or organic), origin (natural or artificial), and processability (thermoset or thermoplastic). Composites can also be grouped by their fiber properties such as type (glass, carbon, or aramid) and form (discontinuous or continuous) and by their fiber positioning such as alignment (random or undirectional). Still another category is the composite's layout (e.g., laminar with a given stacking sequence of plies). The polymeric composites' build-up of plies, in which the continuous fibers are oriented undirectionally, is often referred to as advanced or high-performance composites.[1–4]

Composites revitalized the research and development of metallic and ceramic materials.

MANUFACTURING

The matrix and reinforcement can be combined in different ways to produce a wide variety of preforms and prefabricates, including granules, pellets, prepregs, and textile fabrics. Their form and appearance determine the processing alternatives for converting them to structural parts.[4,5] **Figure 1** shows the steps for processing thermoplastic-based composites.

The diagram in Figure 1 can easily be modified for thermoset matrix composites. The thermoset matrix exists merely in monomeric or oligomeric form, which wets the reinforcement perfectly because of its low viscosity. This feature can be regarded as the main reason for the further use and growing consumption of thermoset-based composites.

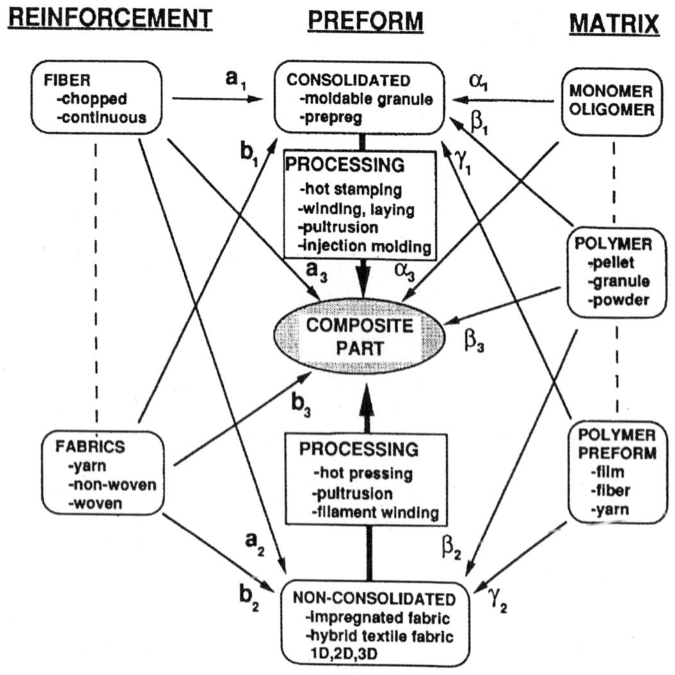

FIGURE 1. An overview of the manufacturing methods for thermoplastic-based polymeric composites.

Figure 1 underlines another difference between composites and other structural materials. The properties of the latter are controlled by the supplier, whereas the composites' properties are defined by the supplier and processor. This difference highlights several important topics related to polymeric composites.

FUTURE TRENDS

The marketing of a third composite, placed between high-volume and high-performing composites, is expected. Its development will borrow techniques used for advanced composites and will affect the thermoset- and thermoplastic-based varieties. Their reinforcements will be based on chopped, long fibers, several millimeters long, or fiber mats. Both will have additional embedding or undirectional fiber or woven fabrics. A slightly higher growth rate can be predicted for thermoplastic-based composites because they are easier to reprocess. If this forecast holds true, manufacturing will focus on developing an efficient sheet-stamping process to replace steel sheets.

REFERENCES

1. *Engineered Materials Handbook*; ASM Int: Metals Park, 1989; Vol. 1, *Composites*.
2. *Concise Encyclopedia of Polymer Science and Engineering*; Kroschwitz, J. I., Ed.; John Wiley & Sons: NY, 1990.
3. *Concise Encyclopedia of Composite Materials*; Kelly, A. et al., Eds.; Pergamon: Oxford, 1989.
4. *International Encyclopedia of Composites*; Lee, S. M., Ed.; VCH: NY, 1991; Vols. 1–6.
5. *Concise Encyclopedia of Polymer Processing and Applications*; Corish, P. I., Ed.; Pergamon: Oxford, 1992.

COMPOSITES, THERMOSETTING POLYMERS

Dorel Feldman
Centre for Building Studies
Concordia University

A composite material may be defined as any substance made by combining two or more materials to produce a multiphase system with different physical properties contained from the constituents. Composite materials may be selected to give unusual combinations of stiffness, strength, weight, high-temperature performance, corrosion resistance, hardness, or conductivity. The elements of a composite are the matrix, the reinforcing, and the component between them, the interphase.

The polymers used in today's advanced composites fall into two groups: thermoplastics and thermosets. Composites can be placed into three categories—particulate, fiber, and laminar—based on the materials' shapes. Concrete, a mixture of cement and gravel, is a particulate composite; fiberglass-reinforced polyester, containing glass fibers embedded in a polymer, is a fiber-reinforced composite; and plywood, having alternating layers of wood veneer, is a laminar composite. The last example is actually a composite of composites because the wood veneer itself is composed of cellulose fibers in a lignin matrix. If the reinforcing particles are uniformly distributed, particulate composites have isotropic properties; fiber composites may be either isotropic or anisotropic; and laminar composites always show anisotropic behavior.

Fiber-reinforced organic polymers are increasingly being considered for structural applications in aerospace, advanced marine systems, construction, electronics, transportation, medicine, and other areas. This emphasis on structural applications is because the composite materials may offer a specific modulus and tensile strength that is high enough for replacing as much metal as possible to reduce weight and save energy. Thermosetting polymers such as phenolformaldehyde (PF), polyesters, epoxies, bis-maleimides, phenolaralkyls, polyimides, poly(styrylpyridine), poly(phenylene quinoxaline), and polybenzimidazoles are widely used as matrices. These polymers offer a combination of high modulus and creep resistance because of their highly crosslinked structures. However, dense crosslinking generally leads to brittle materials that are subject to failure because of internal flaws and progressive crack propagation.

The physical structure of thermosetting polymers is amorphous, and the behavior of most resin matrices is that of rather brittle organic glass.

The matrix serves the following functions: transfers loads in and out of the fibers, prevents continuous crack propagation through the brittle fibers, and protects them from the environment. The matrix properties and the strength of the fiber-matrix bond affect composite characteristics, including temperature, environmental and creep resistance, longitudinal compression strength, transverse compression and tensile strength, interlaminar shear strength, and fatigue strength.[1] Thermosettings come closest to the ideal matrices, at least for applications up to 200°C. **Table 1** shows some properties of a few thermosetting polymers.

Different kinds of fibers, starting with glass and metallic and finishing with synthetic and carbon fibers, can be used in composites with thermosetting matrices. The nature of interfaces in composite materials is very important. The nature of the coupling agents, generally employed to retain a high percentage of bond strength when exposed to environmental factors, continues to receive much attention.[2] The coupling agent-matrix interface is a diffuse boundary where intermixing takes place because the matrix penetrates the chemisorbed coupling agent layers and the physisorbed molecules migrate into the matrix phase. The existence of the matrix interphase is now widely accepted, and its effects on mechanical properties are being studied.[3]

APPLICATIONS

Fiber-reinforced thermosetting polymers are used in many areas such as the aerospace, chemical, and construction industries; the military; sports, and others. Some of their applications include storage bins and floorings of civilian aircraft, doors, fairing, radomes, cockpits, and light load-bearing components in helicopters, small planes, and almost all airplanes. The chemical industry uses these composites for tanks and vessels (pressure and nonpressure), efficient pipelines and fittings, and valves. Because of their corrosion resistance to many fluids and underground water, glass-fiber-reinforced polyester or epoxy tanks are used to store various fluids such as oil and gasoline. These tanks are frequently buried underground where water can damage metal tanks.[6] Composites used in construction are usually classified into two groups: standard products such as flat sheets,

TABLE 1. Properties of the Thermosettings[a]

Properties	PF	MF	EP	BMI	PES
Specific gravity	1.4	1.5	1.11–1.4		1.2
Maximum continuous service temperature °C	110	80	130	400	100–125
Thermal conductivity W/m.k	0.35	0.5	0.88	—	—
Tensile modulus MPa	5,600–12,000	9100	2,150	3700	5,000–11,000
Tensile strength MPa	25	30	28–90	—	50–60
Tensile strength MPa	69–207	241	—	—	70–200
Compressive strength MPa	—	—	—	—	—
Coefficient of thermal expansion °K	0.00003–0.00005	0.00005 —	0.000011–0.000035	—	—
Tg		—	170–206	—	258

[a]PF, phenol-formaldehyde; MF, melamine-formaldehyde; EP, epoxy; BMI, bis-maleimide; and PES, polyester.
Source: References 4 and 5.

corrugated sheets, or sandwich panels; and structures shaped for special applications.

Durability is one of the most important qualities to consider when composites are used externally on buildings because the composites must show good weatherability.[7] Thermosetting composites are suitable for prefabricated units, as cisterns for cold and hot water, for window and door frames, and for concrete frames.[6] In many cold-climate countries, composite deterioration is accelerated by de-icing agents. In most countries, the weight of highway vehicles has steadily increased and now typically exceeds by some forty percent the weight for which many bridges were originally designed. Using polymeric composites to construct new bridges is one avenue that is being explored to reduce the pervasive corrosion problems of steel and reinforced concrete members.

Military applications vary from ordinary helmets to rocket engine cases, and in sports, boron-epoxy composites are used for golf clubs and tennis rackets. Racing yachts and other private boats are examples of applications in which Kevlar fiber is substituted for glass fiber because performance is more important than cost.[8]

REFERENCES

1. Baker, A. A. *Metals Forum* **1983,** *6,* 2, 81.
2. Johanson, O. K.; Stark, F. O.; Vogel, G. E. et al. *Interfaces in Composites* **1969**, ASTM-STP452, p 68 Philadelphia, PA. *452,* 168.
3. Ishida, H. *Polym. Comp.* **1984,** *5,* 2, 101.
4. Daniels, C. A. *Polymers: Structure and Properties*; Technomic: Lancaster, 1989.
5. Feldman, D. *Prog. Polym. Sci.* **1990,** *15,* 603.
6. Feldman, D. *Polymeric Building Materials*; Elsevier Applied Science: London, 1989.
7. Hollaway, L. *Polymer Composites for Civil and Structural Engineering*; Blackie Academic: London, 1993.
8. Chawla, K. K. *Composite Materials*; Springer Verlag: New York, 1987; p 229.

Concrete

See: *Concrete-Polymer Composites*

CONCRETE-POLYMER COMPOSITES

Satish Chandra, H. Justnes, and Y. Ohama
Chalmers University of Technology
SINTEF Nihon University

The use of polymers in concrete is not new; they have been used in their natural form since before Christ. That some of those structures are in good shape today is proof of the polymers durability. The masons of that time knew only that the material worked well, without knowing the reasons or explanations.

Most cement concretes made in the 1930s were based on rubber latexes. Other products based on poly(vinyl acetate) were developed in the 1940s. In the 1960s, latexes based on thermoplastic polymers such as acrylates and poly(vinylidine chloride)s were introduced. Research in this field has continued since then, and new improved latexes are available now for use.

Today, even with advanced technology, the use of polymers is often misunderstood. In some areas, adding polymers is acceptable for significantly improving concrete's durability, a practice that is attracting international attention because of the hostile environmental conditions to which concrete is exposed. Still, polymers are used reluctantly perhaps because of insufficient information about the way polymers are used in concrete.

Using polymers undoubtedly increases the cost of concrete, but considering their linger life span and the cost of repairs, polymers are a lot cheaper in the long run.

CLASSIFICATION AND DEFINITION

Concrete-polymer composites are materials made by replacing part or all of the cement with polymers. The function of a polymer changes with the amount and the technique used. When the polymer is used in small amounts (5%), it modifies the pore structure and does not behave like a binder. When polymer added is more than 5% to the weight of cement, it behaves partially like one. Another factor is whether the polymer is mixed in mortar, concrete, or impregnated. To avoid misconceptions, some classification is needed. Concrete-polymer composites are generally classified into three types based on the amount used and the process technology involved: polymer-modified cement mortar (PCM) and concrete (PCC), polymer mortar (PM) and concrete (PC), and polymer impregnated mortar (PIM) and concrete (PIC).

Polymer-modified cement mortar and concrete are made by partially replacing cement and strengthening its hydrate binders with polymeric modifiers or admixtures such as polymer latexes or dispersions, redispersible polymer powders, water-soluble polymers, liquid resins, and monomers. Polymer-modified mortar and concrete have a monolithic comatrix in which the organic polymer matrix and the cement gel matrix are interwoven.

Polymer mortar and concrete are made by fully replacing the cement with polymeric binders or liquid resins. Thus, unlike conventional mortar and concrete, there is no hydraulic bond; the mortar or concrete is completely chemically bonded and contains no cement hydrate phase.

Polymer-impregnated mortar and concrete are made by impregnating the hardened cement mortar and concrete with monomeric impregnants, which are subsequently polymerized. Vinyl monomers are used as typical monomeric impregnants for the polymer-impregnated mortar and concrete. For PCM, PC, PM, and PC, polymer is mixed in the mortar or concrete, whereas for PIC, concrete is made and then impregnated. This process further strengthens the cement hydrate binders with the resultant polymer. Consequently, the strength of PIC increases.

POLYMER CEMENT MORTAR OR CONCRETE

Adding polymer effects the properties of fresh and hardened polymer-modified mortar and concrete. The factors affecting the properties are polymer type, polymer-cement ratio, air content, and curing conditions. Generally, polymers are easier to work with than ordinary cement mortar and concrete because the ball bearing action of polymer particles, entrained air, and the dispersing effect of the surfactants present in the latexes improve the consistency. The flow of the polymer-modified mortar increases with increasing watercement ratio and polymer-to-cement ratio. The water-cement ratio and polymer-to-cement ratio. The water-cement ratio of the polymer-modified mortar and concrete at a given slump is markedly reduced with the increase in the polymer-cement ratio. This water reduction contributes to the strength development and to drying shrinkage, and durability properties. PCC properties are divided into mechanical and durability properties.

Mechanical Strength

Because cement concrete is an inherently brittle material, the tensile and flexural strengths may be improved by including polymers in the binder. The abrasion process is dominated by tensile forces, and may be strongly influenced by impact as well. Thus, these three mechanical properties are treated separately.

Abrasion Resistance

The abrasion resistance of latex-modified mortar and concrete depends on the type of polymer added, the polymer–cement ratio, and the abrasion or wear conditions. Generally, the abrasion resistance improves considerably with an increase in the polymer–cement ratio.

Impact Resistance

Latex-modified mortar and concrete have an excellent impact resistance compared with conventional mortar and concrete, according to Ohama, because polymers themselves have high impact resistance.[1] The impact resistance generally increases with increasing polymer–cement ratio. However, the data of impact resistance vary markedly between testing methods reported by different workers.

Tensile and Flexural Strengths

Generally, latex-modified mortar and concrete show a significant increase in tensile and flexural strengths but no improvement in compressive strength compared with ordinary concrete or mortar. This difference is measured by the contribution of the high tensile strength of the polymer itself and the overall improvement in the aggregate–binder bond. PCCs' strength properties are influenced by many factors interacting with each other: the nature of the components used (e.g., cement, latex, and aggregate), the control factors for mix proportioning (e.g., polymer–cement, water–cement, and void–binder ratios), curing methods, and test methods.[1]

Durability

Because the polymers interact, the amount of free lime (CH) is significantly reduced. This polymer interaction is described in an earlier paper.[2–4] Moreover, big crystals of preferred orientation do nor form, thereby reducing micro-cracks in concrete and porosity at the past–aggregate interfaces. Subsequently, the penetration of salt solution and gas diffusion are substantially decreased.

Furthermore, polymer addition modifies the pore structure. Polymer addition also reduced the total pore volume. Polymers may not completely prevent the deleterious effects of aggressive agents, but they will certainly delay them.

Freeze-Thaw Resistance

Latex-modified mortar and concrete have improved resistance to freezing and thawing (i.e., frost attack) compared with conventional mortar and concrete. This resistance is partly because of the reduced porosity resulting from a decreased water-to-cement ratio, the filling of pores with polymer, and the entrained air introduced by polymers and surfactants.

Resistance to Chloride Ions and Carbonation

The resistance to chloride ion penetration of polymer-modified mortar and concrete improves with an increase in the polymer–cement ratio. The polymer interaction with cement hydration products, especially calcium hydroxide, significantly reduces its free presence, which subsequently may combine with chlorides and leach out, but because of the filling and sealing of the pores, the permeability and diffusibility are reduced. The pore refinement reduces the transmission of gases and moisture. Because less CO_2 is transmitted, the carbonation resistance of polymer-modified mortar and concrete increases. This increase occurs with an increase in the polymer–cement ratio, which was the case for chlorides, but depends on the polymer type and the CO_2 exposure conditions.

Conclusions

PCC properties depend on the type and the dosage of polymer and the curing conditions. However, polymer additions to mortar or concrete may improve important characteristics such as abrasion resistance, impact resistance, tensile and flexural strength, freeze-thaw resistance, chloride resistance, and reduction in water absorption because of capillary forces. Among the latex polymers, styrene-butadiene rubbers and higher polyacrylic esters generally perform the best. PCCs may be expensive, but they are much more economical in the long run because of their improved longevity.

POLYMER MORTAR AND CONCRETE

Unlike PCC, PC has no cement because the polymer is the sole binder. The polymeric binders used include various thermosetting resins, tar-modified resins, resin-modified asphalts, and vinyl monomers. Polymeric binders themselves cannot set or harden; proper initiators, promoters, or hardeners are selected, and added to the polymeric binders while mixing the polymer mortar and concrete. The working lifetime of polymer mortar and concrete can be controlled by selecting the type and content of the initiators, promoters, or hardeners. PC mixing and batching techniques are the same as for PCC, but the curing methods differ. Dry curing at ambient temperature or heat curing is used. Cast-in-place application systems are usually used for polymer mortar, and the precast application system is used for polymer concrete. The strength, durability, and other properties are governed by the polymeric binder and aggregates. Consequently, it is important to select the most effective mix proportions.

Mechanical Strength and Workability

Generally, PC is more difficult to work with than conventional cement mortar and concrete, with the exception of poly(methyl methacrylate) (PMMA) or acrylic mortars and concrete using a low-viscosity MMA monomer. PC hardens faster compared with CC. PC also strengthens quickly, making the weight reduction of precast products possible by reducing their cross-section. The strength depends primarily on the aggregates' strength and moisture content.

Abrasion Resistance and Other Properties

Most polymer mortar and concrete exhibit good abrasion resistance, impact resistance, and electrical insulating properties compared with conventional cement mortar and concrete. However, these properties depend on the temperature of their use.

Durability

Polymer mortar and concrete are completely monolithic structures made with a chemical binder and filler. They have an almost impermeable microstructure that keeps out water, moisture, or gas. Unlike cement hydration products, they do not contain ingredients that can interact. This feature makes them resistant to many salts (e.g., Cl^-), freezing and thawing, and to gases like SO_2, CO_2, and No_x.

Freeze-Thaw Resistance

Pc freeze-thaw resistance is also superior.[5] The test was performed by rapidly freezing and thawing the samples in water. The changes in weight loss and strength after particular cycles were insignificant. However, polyester mortar and concrete may not be sufficiently water resistant because of their potential degradation from the hydrolysis of the unsaturated polyester resin.

Chemical Resistance

PCs do not contain products (salts) that are susceptible to chemicals like the hydration products of portland cement in conventional concrete. Their microstructure is watertight and impermeable, and they generally have a high chemical resistance. PCs made of epoxy, furan, and acrylics have excellent resistance against acids, alkalies, and salts and good resistance against solvents, oils, and fats, whereas PCs made from polyester are less resistant to alkalies and solvents.

USES AND NEW DEVELOPMENTS

Polymer-modified concrete is widely used for bridge deck overlays and patch repairs in the United States.

In Japan and Germany, polymer-modified pastes or slurries with polymer–cement ratios of 50% or more have been used for waterproofing membrane coatings in apartment houses, silos, and underground tanks.[6–8] The paste consists of Portland cement, high alumina cement, sand, polymer dispersion, other additives, and water, and forms flexible waterproofing membranes with thicknesses of 2–4 mm.

Recently, polymer mortar for underwater construction was developed in Japan.[9] It was produced by using a methyl methacrylate monomer as a binder, and the mortar can be placed and bonded underwater. A recent development in the United States is an electrically conductive polymer concrete overlay used with a cathodic protection system for bridge decks.[10]

At present, polyester concrete is more widely used for various structural and non-structural purposes, and has a few cast-in-place applications. In Europe, precast applications are used in mechanical tool structures, building panels, utility boxes, and underground junction boxes.

POLYMER-IMPREGNATED MORTAR AND CONCRETE

The history of the research and development of polymer-impregnated mortar and concrete is the shortest of the concrete-polymer composites.

Currently, polymer-impregnated mortar and concrete are used infrequently as construction materials and their development is not being pursued actively. The reasons for this are as follows:

- their process technology is complicated, and their process cost is high because considerable thermal energy is consumed in base mortar and concrete drying and in the polymerization processes;
- although they perform similarly to polymer mortar and concrete, the balance between their cost and performance is poor; and
- no accurate, easy methods to measure polymer impregnation depth are known. Their quality control is difficult, and therefore, their reliability is poor in structural applications.

REFERENCES

1. Ohama, Y. *Concrete Admixtures Handbook*; Noyes: Park Ridge, NJ, 1984; Chapter 7.
2. Chandra, S.; Flodin, P.; Berntsson, L. In *Proceedings of the 3rd International Congress on Polymers in Concrete*; College of Engineering, Nihon University: Koriyama, Japan, 1985; p 125.
3. Chandra, S.; Flodin, P. In *Proceedings of the 3rd ACI/CANMET International Conference on Superplasticizers and Other Chemical Admixtures in Concrete*; American Concrete Institute: Detroit, MI, 1989; p 263.
4. Chandra, S.; Ohama, Y. *Polymers in Concrete*; CRC: Boca Raton, FL, 1994; Chapter 7, p 1–47.
5. Ohama, Y. Research Work for the Period, 1st April '69-31st/March '70, Building Research Institute: Tokyo, 1970; 60, p. 304.
6. Ohama, Y. *Bosui J.* **1990**, *21*, 11, 104–106.
7. Volkwein, A.; Petri, R.; Springenschmid, R. *Betonwerk† Fertigteil-Technik* **1988**, *54*, 8, 30.
8. Volkwein, A.; Petri, R.; Springenschmid, R. *Betonwerk† Fertigteil-Technik* **1988**, *54*, 9, 72–78.
9. Ohama, Y.; Demura, K.; Pareek, S. N. et al. Polymer Concrete: American Concrete Institute: Detroit, MI, 1993, pp 93–107.
10. Fontana, J. J.; Reams, W.; Elling, D. *Polymers in Concrete: Advances and Applications*; American Concrete Institute: Detroit, MI, 1989; pp 157–175.

Condensation Polymerization

See: Polycondensation and Step-Growth Polymerization, Addition and Condensation

CONDENSATION POLYMERS (Synthesis from Silylated Diamines)

Yoshiyuki Oishi*
Department of Applied Chemistry and Molecular Science
Iwate University

Yoshio Imai
Department of Organic and Polymeric Materials
Tokyo Institute of Technology

N-Trimethylsilyl-substituted amines are known to be one of the most reactive classes of organosilane compounds. They react with a variety of electrophiles such as carboxylic acid chlorides, acid anhydrides, isocyanates, epoxides, and benzyl chlorides, yielding amine-based compounds such as amides, ureas, and secondary amines.[1] These reactions are applicable for the synthesis of nitrogen-containing condensation polymers.

The *N*-silylated diamine method was successfully applied as a new tool for the synthesis of various types of condensation polymers such as polyamides, polybenzoxazoles, polybenzothiazoles, polyimides, polyureas, polyazomethines, and polyamines.[2]

PREPARATION OF POLYAMIDES

The reaction of *N*-trimethylsilyl-substituted amines with acid chlorides yielded amides with the elimination of trimethylsilyl chloride, and not a combination of *N*-trimethylsilylamides and hydrogen chloride.

The high molecular weight aromatic polyamides (aramids) by the low-temperature solution polycondensation of *N,N'*-bis(trimethylsilyl)-substituted aromatic diamines with aromatic diacid chlorides were successfully prepared (**Structure 1**).

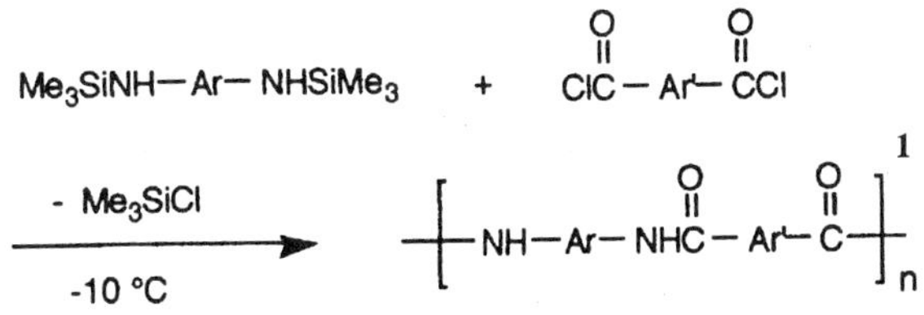

1

We first investigated the progress of the polycondensation through the *N*-silylated diamine method and the conventional diamine route for the synthesis of poly(*p*-phenylene terephthalamide) (DuPont's "Kevlar" molecule). The polycondensation of *N,N'*-bis(trimethylsilyl)-*p*-phenylenediamine with terephthaloyl chloride in a mixture of NMP and hexamethylphosphoramide (HMPA) proceeded much more rapidly and afforded aramid with higher inherent viscosity, compared to the conventional polymerization using parent *p*-phenylenediamine.[3]

Aromatic diamines with low reactivity, which is mainly due to their low basicity, usually produce only low molecular weight polyamides. However, they can be activated markedly by conversion to *N*-silylated diamine derivatives.

Aromatic diamines are usually difficult to purify by crystallization, since they are susceptible to oxidation and give brown-colored compounds. However, the *N*-silylated aromatic diamines were stable to oxidation, keeping them colorless. Thus, the silylation is considered to be an excellent purification method for diamines. *N,N*-Bis(4-aminophenyl)-*p*-perfluoroalkylaniline was one of the promising monomers for the synthesis of fluorine-containing aramids, but this monomer was difficult for purification. The *N*-silylated diamine method revealed that the aramids were easily prepared form the *N*-silylated monomer and aromatic dicarboxylic acid chlorides in NMP.[4]

Aromatic diacid chlorides of considerably lower reactivity than the usual diacid chlorides are also capable of yielding aramids with higher molecular weights by using *N*-silylated diamines.

In addition, the polycondensation was further applicable to that using dicarboxylic acid diphenyl esters which do not have enough high reactivity. The polycondensation of *N*-trimethylsilyl-substituted bis(4-aminophenyl) ether with diphenyl isophthalate proceeded in dimethyl sulfoxide in the presence of cesium fluoride catalyst at 115°C, affording the aramid with moderate inherent viscosity.[5] In the case of the polycondensation of the parent diamine with the diphenyl ester, the aramid having very low viscosity value was obtained only under similar reaction conditions.

The *N*-silylated diamine method has several advantages over the conventional method. First, although aromatic diamines are usually difficult to purify by recrystallization, high-purity *N*-silylated diamines can be obtained simply by distillation. Second, the *N*-silylated diamines are readily soluble in various

*Author to whom correspondence should be addressed.

organic solvents, and hence a variety of solvents can be used as reaction media, ranging from low- to high-boiling and/or nonpolar to polar solvents. Third, the *N*-silylated diamines are more reactive than the parent diamines toward diacid chlorides in polar solvents, and consequently higher molecular weight polyamides can be obtained readily by the polycondensation of these monomer pairs under milder reaction conditions. Fourth, the polycondensation proceeds under neutral reaction conditions with the elimination of volatile trimethylsilyl chloride, and therefore the polymerization solutions, which contain no hydrochlorides, can be used directly for solution casting to prepare pure polymer films. Fifth, the trimethylsilyl chloride recovered from the polymerization system can be recycled as a silylating agent for the diamines.

The *N*-silylated diamine method sounds suitable for the polymerization blending of aramid and other polymers and was extended successfully to the preparation of a compatible blend of the aramid/other polymer system by the polymerization blending.

Aramid synthesis via the silylation method is applicable to a dry process involving vapor deposition polycondensation for the preparation of aramid ultrathin films on appropriate substrates.

PREPARATION OF POLYBENZAZOLES

The reaction of *N,O*-bis(trimethylsilyl)-substituted *o*-aminophenol with benzoyl chloride is also interesting. The acid chloride was found to attack selectively at the *N*-silylated amine site of the aminophenol, giving the corresponding amide product exclusively. This occurred because the amino group of the *o*-aminophenol is activated toward acid chloride attack and the hydroxyl group, in contrast, is deactivated by silylation. This reaction was successfully extended to the polycondensation of the *N,N′,O,O′*-tetrakis(trimethylsilyl)-substituted bis-*o*-aminophenols with aromatic diacid chlorides. The polymerizations were carried out at 5°C in an amide-type solvent such as DMAc, leading to the formation of *O*-silylated poly(*o*-hydroxyamides) of high molecular weights, which were subjected to thermal cyclocondensation affording aromatic polybenzoxazoles. A variety of polybenzoxazoles and their copolymers were prepared by this method.[6,7]

PREPARATION OF POLYIMIDES

Aromatic polyimides are generally prepared starting from aromatic diamines and aromatic tetracarboxylic dianhydrides in two steps: the ring-opening polyaddition of aromatic diamines to aromatic tetracarboxylic dianhydrides yielding soluble poly(amic acid) precursors, and the subsequent thermal cyclodehydration to insoluble polyimides.

N-Trimethylsilyl-substituted amines also react with acid anhydrides in the same fashion. When an *N*-silylated amine is reacted with an anhydride, the intermediate ring-opened adduct, silylated amic acid, can eliminate silanol at high temperature with the formation of an imide compound. The application of this reaction to the synthesis of polyimides was reported by Boldebuck and Klebe in the patent literature.[8] The ring-opening polyaddition of *N*-silylated aromatic diamines to aromatic dianhydrides afforded poly(amic acid trimethylsilyl ester)s, which in turn were converted thermally to polyimides with the elimination of trimethylsilanol.

The *N*-silylated diamine method could be applied successfully to the synthesis of a variety of polyimides, random and block copolyimides, as well as poly(amide imide)s starting from *N*-silylated aromatic diamines and 4-chloroformylphthalic anhydride.

PREPARATION OF POLYUREAS

The addition of *N*-silylated amines to isocyanates gives high yields of *N*-silyl-substituted urea derivatives, which are readily hydrolyzed to the parent ureas. This reaction was utilized for the synthesis of aromatic polyureas by Klebe,[9] who demonstrated the polyaddition of *N*-silylated aromatic diamines to aromatic diisocyanates to yield *N*-silylated polyureas. It is significant that the polyureas derived from the methanolysis of the *N*-silylated polymers have low crystallinity and good solubility in organic solvents, in contrast to the polyureas prepared by the conventional diamine-diisocyanate route.[10]

PREPARATION OF POLYAZOMETHINES

Most aromatic polyazomethines could not be obtained in high molecular weight because of their insoluble and infusible characteristics. Kurosaki and co-workers extensively investigated the preparation of aromatic polyazomethines and found new routes for these polymers by the use of *N*-silylated aromatic diamines.[11,12] One method is the polycondensation of a combination of *N,N′*-bis(trimethylsilyl)-substituted aromatic diamines and aromatic bis(diethyl acetal) compounds. Another method is the polyaddition of *N,N,N′,N′*-tetrakis(trimethylsilyl)-substituted aromatic diamines to aromatic diamines to aromatic dialdehydes in NMP, leading to soluble precursor polymers, followed by thermal conversion to insoluble polyazomethine films.[12]

PREPARATION OF POLYAMINES

The addition of *N*-silylated amines to epoxides gives siloxy-substituted amine compounds. This reaction was utilized for the synthesis of linear polyamines containing siloxy groups by the polyaddition of *N*-silylated amines to bisepoxides.[13]

Therefore, this polyaddition reaction was extended to the curing of epoxy resins using *N*-silylated diamines as curing agents.

Although alkyl halides are less reactive than carboxylic acid chlorides toward aminolysis, the alkyl halides are also capable of attacking *N*-silylated amines. Klebe[14] reported that benzyl chloride and an *N*-silylated amine underwent this substitution reaction and yielded benzylamine and trialkylsilyl chloride quantitatively. The reaction has also been extended to the synthesis of polyamines, which were difficult to obtain in high molecular weight by the conventional dihalide-diamine route because of the occurrence of side reactions.

We found that the aromatic nucleophilic substitution reaction of activated aromatic halides with *N*-silylated aromatic amines leads to aromatic secondary amines. The reaction was effectively catalyzed by fluorides such as cesium and potassium fluorides. This finding has been applied for the first time to the synthesis of aromatic polyamines.

The polycondensation of *N*-silylated aromatic diamine with activated aromatic dihalides in dimethyl sulfoxide proceeded under neutral reaction conditions with the elimination of trimethylsilyl halides, affording high molecular weight polyamine. Removal of the volatile trimethylsilyl halide and the solvent from the reaction system gave a pure polymer containing no byproduct.

REFERENCES

1. Klebe, J. F. *Adv. Org. Chem.* **1972**, *8*, 97.
2. Imai, Y.; Oishi, Y. *Prog. Polym. Sci.* **1989**, *14*, 1723.
3. Oishi, Y.; Kakimoto, M.; Imai, Y. *Macromolecules* **1987**, *20*, 703.
4. Oishi, Y.; Kanazawa, J.; Hirahara, H.; Mori, K. *Polym. Prepr. Jpn.* **1994**, *43*, 3166.
5. Oishi, Y.; Tanaka, M.; Kakimoto, M.; Imai, Y. *Makromol. Chem. Rapid Commun.* **1991**, *12*, 465.
6. Maruyama, Y.; Oishi, Y.; Kakimoto, M.; Imai, Y. *Macromolecules* **1988**, *21*, 2305.
7. Itoya, K.; Kakimoto, M.; Imai, Y. *Polym. Prepr. Jpn.* **1990**, *39*, 787.
8. Boldebuck, E. M.; Klebe, J. F. U.S. Patent 3 303 157, 1967, *Chem. Abstract* **1967**, *66*, 96125f.
9. Klebe, J. F. *J. Polym. Sci. Part B* **1964**, *2*, 1079.
10. Oishi, Y.; Padmanaban, M.; Kakimoto, M.; Imai, Y. *J. Polym. Sci. Part A: Polym. Chem.* **1992**, *30*, 1363.
11. Munetoh, T.; Miyazawa, K.; Matsumoto, T.; Kurosaki, T. *Polym. Prepr. Jpn.* **1987**, *36*, 325.
12. Miyazawa, K.; Munetoh, T.; Matsumoto, T.; Kurosaki, T. *Polym. Prepr. Jpn.* **1987**, *36*, 324.
13. Oishi, Y.; Sekiguchi, H.; Hirahara, H.; Mori, K. *Polym. Prepr. Jpn.* **1993**, *42*, 468.
14. Klebe, J. F. *J. Polym. Sci. Part A* **1964**, *2*, 2673.

CONDENSATION POLYMERS (Using Phosphorus Pentoxide/Methanesulfonic Acid)

Mitsuru Ueda
Department of Materials Science and Engineering
Faculty of Engineering
Yamagata University

The anhydride bond in phosphorus pentoxide (P_4O_{10}) is widely known to be reactive toward nucleophilic attack. However, P_4O_{10} is sparingly soluble in common organic solvents and is difficult to handle because of its high reactivity toward moisture. To alleviate these problems, P_4O_{10} has been chemically modified to products such as poly(phosphoric acid) (PPA),[1] poly(phosphate ester) (PPE),[2] and poly(phosphoric acid trimethylsilyl ester) (PPSE).[3]

In 1973, Eaton et al. recommended phosphorus pentoxide/methanesulfonic acid (PPMA) in a weight ratio of 1:10 as a substitute for PPA, which is difficult to handle. PPMA is a mobile, colorless liquid that can be stirred without difficulty.[4] PPMA is expected to react with a carboxylic acid to give a mixed anhydride of a carboxylic acid and methanesulfonic acid or an acylium ion, which is known to have a high reactivity in nucleophilic reactions. The utility of PPMA as a condensing agent and solvent for the synthesis of various condensation polymers is described below.

SYNTHESIS OF CONDENSATION POLYMERS

Synthesis of Poly(phenylene ether sulfone)

Poly(phenylene ether sulfone)s are a class of materials used as engineering thermoplastics because of their excellent thermal and mechanical properties. The successful synthesis of poly(phenylene ether sulfone) was accomplished by the direct polycondensation of sodium 4-phenoxybenzenesulfonate in PPMA (**Structure 1**).[5]

1

η inh = 0.85 dL/g (DMAc)

SYNTHESIS OF POLY(PHENYLENE ETHER KETONE)

Poly(phenylene ether ketone)s are also typical high-performance engineering plastics because of their unique combination of toughness, stiffness, thermoxidative stability, and retention of physical properties at high temperature. Poly(phenylene ether ketone)s of high molecular weights were prepared readily by the direct polycondensation of aliphatic dicarboxylic acids and aromatic dicarboxylic acids containing phenyl ether structures with diaryl compounds (**Structure 2**).[6,7]

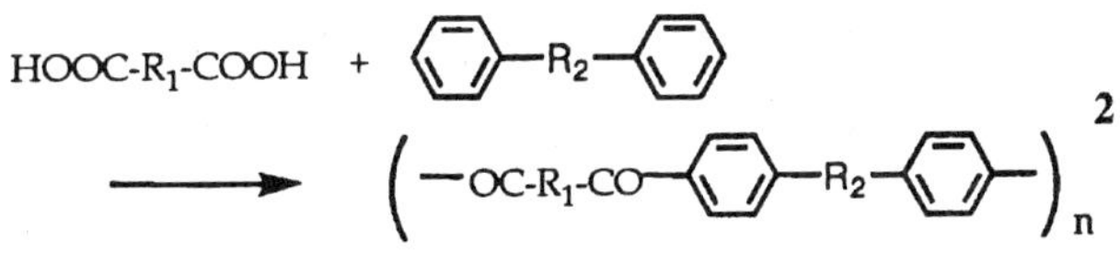

2

η inh = 1.5 dL/g (H_2SO_4)

Taking into consideration the relationship between structure and reactivity, poly(phenylene ether ketone)s were also prepared by the self-polycondensation of 4-(4′-phenoxyphenoxy)benzoic acid and 4-(*o*-methoxyphenoxy)benzoic acid in PPMA.[8]

Many polymers containing various crown ether moieties along the main backbones or side chains have been synthesized and their cation-binding properties have been studied. However, most of the published methods for the incorporation of a crown ether into polymer backbones involve tedious synthetic steps such as the introduction of various functional groups into the crown ether moieties. A simple synthesis of polyketones containing crown ether units by the direct polycondensation of various dicarboxylic acids with dibenzo-18-crown-6 in PPMA was reported.[9]

Synthesis of Polychalcone

Polymers having chalcone structures in their side or main chain are an important class of materials that may be used for photoresists in printing and electronic industries. Polychalcones with high molecular weights were prepared by polycondensation of aromatic dialdehydes with aromatic diacetyl compounds in PPMA.[10]

Synthesis of Polybenzimidazole

The synthesis of 2-substituted benzimidazoles by the reaction of *o*-phenylenediamine with various carboxylic acids was carried out in PPMA at 100°C. The condensations were completed in 30 min and produced the corresponding benzimidazoles in quantitative yields.

The polycondensation of various dicarboxylic acids with 3,3′-diaminobenzidine tetradrochloride monohydrate was carried out in PPMA for several hours at 120–140°C, and produced poly(benzimidazole)s with inherent viscosities up to 5.8 dL/g.[11]

Synthesis of Polybenzothiazole

The direct polycondensation of various dicarboxylic acids with 2,5-diamino-1,4-benzenedithiol dihydrochloride was carried out in PPMA for several hours at 140°C. The polycondensations went on in homogeneous solutions and gave quantitative yields of polybenzothiazoles with high molecular weights.[12]

Synthesis of Polybenzoxazole

Polybenzoxazoles were prepared by the direct polycondensation of aromatic dicarboxylic acids containing a phenyl ether structure with 3,3′-dihydrobenzidine dihydrochloride in PPMA. Polycondensation proceeded smoothly and was completed within 5 h at 140°C and produced polymers with inherent viscosities up to 4.6 dL/g.[13]

Synthesis of Poly(1,3,4-oxadiazole)

The direct polycondensation of hydrazine sulfate with various dicarboxylic acids in PPMA produced poly(1,3,4-oxadiazoles) with inherent viscosities up to 1.4 dL/g. The synthesis of aromatic polybenzazoles from aromatic dicarboxylic acids containing phenyl ether structures usually was carried out at ~140°C.

Synthesis of Polybenzoxazinone

Polybenzoxazinone is also a thermally stable polymer and is prepared by the two-step condensations of diacid chloride with bis(anthranilic acid) in PPA. Polymer was also obtained in PPMA in one step. The direct polycondensation of aromatic acids containing phenyl ether structures with bis(anthranilic acid) in PPMA went on smoothly at 140°C and produced polymers with inherent viscosities up to 2.6 dL/g.[14]

In conclusion, PPMA is an excellent condensing agent and solvent for the synthesis of various kinds of condensation polymers, such as poly(ether sulfone)s, poly(ether ketone)s, and heterocyclic polymers. Furthermore, PPMA is a mobile, colorless liquid that is easy to handle, compared with PPA. It has excellent solvent power for most condensation polymers. Therefore, many new condensation polymers will be developed by using PPMA. The disadvantage of using PPMA is that typical dicarboxylic acids such as iso- and terephthalic acids give low molecular weights of polymers.

REFERENCES

1. Uhlig, F.; Snyder, H. R. *Adv. Org. Chem.* **1960**, *1*, 35.
2. Kanaoka, Y.; Sato, E.; Yonemitsu, O.; Ban, Y. *Tetrahedron Lett.* **1964**, 2419.
3. Imamoto, T. *J. Synth. Org. Chem. Jpn.* **1985**, *43*, 1163.
4. Eaton, P. E.; Carlson, G. R.; Lee, J. T. *J. Org. Chem.* **1973**, *38*, 4071.
5. Ueda, M. *Makromol. Chem. Rapid Commun.* **1985**, *6*, 217.
6. Ueda, M.; Kanno, T. *Makromol. Chem. Rapid Commun.* **1985**, *6*, 833.
7. Ueda, M.; Sato, M. *Macromolecules* **1987**, *20*, 2675.
8. Ueda, M.; Oda, M. *Polym. J.* **1989**, *21*, 673.
9. Ueda, M.; Kanno, T.; Waragai, T.; Sugita, H. *Makromol. Chem. Rapid Commun.* **1985**, *6*, 847.
10. Ueda, M.; Yokoo, T.; Oda, M. *Makromol. Chem., Phys.* **1994**, *195*, 2569.
11. Ueda, M.; Sato, M.; Mochizuki, A. *Macromolecules* **1985**, *18*, 2723.
12. Ueda, M.; Yokote, S.; Sato, M. *Polym. J.* **1986**, *18*, 117.
13. Ueda, M.; Sugita, H.; Sato, M. *J. Polym. Sci. Part-A, Polym. Chem.* **1986**, *24*, 1019.
14. Ueda, M.; Komatsu, S. *J. Polym. Sci. Part-A, Polym. Chem.* **1989**, *27*, 1017.

CONDUCTING POLYMER COLLOIDS

G. M. Spinks,* H. Eisazadeh, and G. G. Wallace
Intelligent Polymer Research Laboratory
University of Wollongong

Conductive polymers offer great promise for the development of a new generation of so-called intelligent products that will utilize the unique dynamic properties of these materials. One of the current challenges in achieving this objective is the development of novel processing routes to produce conductive polymers in the quantities, sizes, and shapes needed for these different applications. The difficulty in meeting this goal arises from the fact that, because they are rather intractable, most of the conductive polymers are difficult to process using conventional polymer processing methods.

One method that has been explored is the preparation of colloidal dispersions of the conductive polymers. These are then "solution processable," although the polymer is not soluble. Additionally, the colloidal particles may be used directly in various applications.

Numerous studies have reported the preparation, characterization, and properties of the colloidal form of various conductive polymers, including polyacetylene,[1,2] polyaniline,[3–6] and polypyrrole.[7–9]

CHEMICAL PRODUCTION OF CONDUCTING POLYMER COLLOIDS

Conducting polymers can be produced by a chemical technique in which the monomer is dissolved in a solvent (commonly water) with an appropriate chemical oxidant.[10–12] Because the polymer is usually insoluble in the solvent, a powdery precipitate forms.[13] The formation of colloidal dispersions is facilitated by the adsorption of stabilizer molecules on the surface of an insoluble phase, in this case the conducting polymer.

Both grafted and nongrafted stabilizers have been used for the formation of conductive polymer colloids.

*Author to whom correspondence should be addressed.

ELECTROCHEMICAL PRODUCTION OF CONDUCTION POLYMER COLLOIDS

Deposition Mechanism During Electrochemical Synthesis

Conventional electrochemical synthesis of conducting polymers, such as polypyrrole and polyaniline, has been shown to involve at least two distinct steps: polymerization, with occurs in solution near the electrode surface, and deposition of the polymer onto the electrode surface.[14] Studies on the mechanism of the deposition step have suggested that it may be possible to prepare conductive polymer colloids via an electrochemical procedure by interrupting this deposition process.[7]

Initial studies examined the effect of steric stabilizers on the electropolymerization of polypyrrole.[14] The time for a fixed charge to be passed ("reaction time") increases with increasing concentration of stabilizer added to the monomer solution. Conductive deposits of polypyrrole (PPy) were still formed during these reactions, although the electroactivity decreased slightly with increasing stabilizer content. Similar behavior was observed during the electropolymerization of polyaniline (PAn) in the presence of steric stabilizers.[15]

A possible explanation for the decrease in reaction is that the polymer is stabilized in the electric double layer by the steric stabilizers, and that this delays the deposition proccss.

PROPERTIES OF CONDUCTIVE POLYMER COLLOIDS

Conductivity

Estimations of the conductivity of colloids have been attempted by drying and compressing the colloid particles into a solid compact. The conductivity is an order of magnitude lower than respective bulk electrochemically prepared polymers. This is attributed to the poor microscopic interparticle contacts in the pellets because of the presence of the insulating adsorbed steric stabilizer.[16]

The conductivity has been found to be sensitive to the reaction conditions. Conductivity decreased with increasing stabilizer concentration[16,17] because of the insulating effect of the adsorbed steric stabilizer. In the case of polypyrrole, increased conductivity has been reported when ferric chloride was used as a chemical oxidant,[18] whereas decreased conductivity was observed when the colloids were produced in nonaqueous media.[19] For polyaniline, conductivity decreases approximately linearly with increasing pH.[20]

Electroactivity

As reported above, conductive polymer colloids of polypyrrole and polyaniline have been found to be electroactive. The type of stabilizer used was found to affect the electroactivity,[8] which is probably due to the different amounts of stabilizer adsorbing on the particle surface in each case.[3] In addition, the electroactivity decreases as the concentration of stabilizer used increases,[15] presumably for the same reason.

Morphology

Particle size (and size distribution) and shape are important for various applications of colloids. As described above, polypyrrole colloids produced electrochemically are spherical with a size range of 10–200 nm. Electrochemically produced polyaniline colloids were either spherical or needle-shaped, depending on the reaction conditions.[21]

Chemically prepared colloids show similar morphologies which can be altered by optimization of the reaction conditions.

APPLICATIONS FOR CONDUCTIVE POLYMER COLLOIDS

A number of applications that directly use conductive polymer colloids have been considered, including coatings,[22–25] novel composites,[21] paints,[26] thermal stabilizers,[27] and others.

Other Applications

The usefulness of the colloidal processing technique has also been demonstrated in a range of other applications, including coating of polymer particles and development of novel composites.

In conclusion, conductive polymer colloids have been shown to be extremely useful in a wide variety of applications. The colloids can be prepared either chemically or electrochemically and retain reasonable conductivities and electroactivities.

Conductive polymer colloids enable the production of unique forms of these materials. They can simply cast to form coatings or membrane films. They can be electrodeposited onto easily corroded substrates, and co-deposited to form unique composite materials. They can be dried and incorporated into other polymer systems such as paints and plastics. These different examples demonstrate the importance of processing technologies to the commercialization of conductive polymer-based products.

REFERENCES

1. Edwards, J.; Fisher, R.; Vincent, B. *Makromol. Chem., Rapid Commun.* **1983**, *4*, 393.
2. Luttinger, L. B. *J. Org. Chem.* **1962**, *27*, 1591.
3. Aldissi, M. *Adv. Mater.* **1993**, *5*, 60.
4. Armes, S. P.; Aldissi, M. *Am. Chem. Soc. Div. Polym. Mater. Sci. Eng. Proc.* **1989**, *60*, 751.
5. Armes, S. P.; Aldissi, M.; Agnew, S.; Gottesfeld, S.; *Langmuir* **1990**, *6*, 1745.
6. Armes, S. P.; Aldissi, M.; Agnew, S.; Gottesfeld, S. *Mol. Cryst. Liq. Cryst.* **1990**, *190*, 63.
7. Eisazadeh, H.; Spinks, G.; Wallace, G. G. *Polymer* **1994**, *35*, 3801.
8. Eisazadeh, H.; Spinks, G.; Wallace, G. G. *Materials Forum* **1992**, *16*, 341.
9. Bjorklund, R. B.; Liedberg, B. *J. Chem. Soc. Chem. Commun.* **1986**, 1293.
10. Machida, S.; Miyata, S.; Techagumpuch, A. *Synthetic Metals* **1989**, *31*, 311.
11. Armes, S. P. *Synthetic Metals* **1987**, *20*, 365.
12. Rapi, S.; Bocchi, V.; Gardini, G. P. *Synthetic Metals* **1988**, *24*, 217.
13. Chao, T. H.; March, J. *Journal of Polymer Science: Pt. A: Polymer Chemistry* **1988**, *26*, 743.
14. Eisazadeh, H.; Spinks, G.; and Wallace, G. G. *Polymer* **1994**, *35*, 1754.
15. Eisazadeh, H. Ph.D. Thesis, University of Wollongong, 1994.
16. Armes, S. P.; Vincent, B. *Synthetic Metals* **1988**, *25*, 171.

17. Armes, S. P.; Miller, J. F.; Vincent, B. *Journal of Colloid and Interface Science* **1987**, *118*, 410.
18. Epron, F.; Henry, F.; Sagnes, O. *Makromol. Chem. Macromol. Symp.* **1990**, *35/36*, 527.
19. Beaman, M.; Armes, S. P. *Colloid and Polymer Science* **1993**, *271*, 70.
20. Vincent, B.; Waterson, J. *J. Chem. Soc. Chem. Commun.* **1990**, 683.
21. Eisazadeh, H.; Spinks, G.; Wallace, G. G. *Electrodeposition of Polyaniline and Polyaniline Composites From Colloidal Dispersions* Polymer International, 1995; Vol. 37, pp 87–91.
22. Cooper, E. C.; Vincent, B.; *J. Phys. D: Appl. Phys.* **1989**, *22*, 1580.
23. Furlong, D. N.; Freeman, P. A.; Lau, A. C. M. *Journal of Colloid and Interface Science* **1981**, *80*, 20.
24. Ge, H.; Wallace, G. G. *Analytical Chemistry* **1989**, *61*, 198.
25. Ge, H.; Wallace, G. G. *Journal of Liquid Chromatography* **1990**, *13*(16), 3245.
26. Eisazadeh, H.; Spinks, G.; Wallace, G. G. *Materials Forum* **1993**, *17*, 241.
27. Eisazadah, H.; Spinks, G.; Wallace, G. G. *Influence of Conductive Polymer Fillers on the Thermal Degradation of Polyethylene*, in preparation.

Conducting Polymers

See: *Batteries (Materials for High Performance)*
Block Copolymers (Anisotropic Conductivity)
Chiral Polyanilines
Conducting Polymer Colloids
Conducting Polymers (for Langmuir-Blodgett Film Fabrication)
Conducting Polymers (Nonconjugated)
Conducting Polymers (Self-Acid-Doped, Conjugated)
Conductive Elastomeric Blends
Conjugated Ladder Polymers
Conjugated Polymers (Insulating and Conducting Forms)
Conjugated Polymers (Prepared by Organometallic Processes)
Coordination Polymers, Dithiooxamides
Electrically Conducting Composites
Electrochemical Polymerization
Electroconductive Composites (Emulsion Pathways)
Electroconductive Paints
Electrorheological Materials
Enzyme-Catalyzed Polymerization
Intelligent Membranes
Ionomers (dc Conduction Properties)
Liquid Crystalline Polyacetylene Derivatives
Metallized Polyimides
Metal-Polymer Complexes
Metallophthalocyanines, Polymeric (Overview)
Oxidative Coupling
Phthalocyanines, Polymeric
Polyacenic Semiconductor Materials (Applications to Rechargeable Batteries)
Polyacetylene Films
Polyaniline (Electrochemical Synthesis)
Polyaniline Network Electrodes (Enhanced Performance of Polymer LEDs)
Polyanilines, Oxidation States
Poly(arylene vinylene)s (Mechanistic Control of a Soluble Precursor Method)
Polybithophenes (Advanced Derivatives)
Poly[(disilanylene) oligophenylene]s
Poly(imide siloxane)s
Polyphenylacetylene
Poly(p-phenylene vinylene)s (Methods of Preparation and Properties)
Polypyrrole (Processable Dispersions)
Polypyrroles (from Isoporous Membranes)
Polysilanes (Overview)
Poly[N,N′-(sulfo-p-phenylene)terephthalamide]
Poly(2,5-thienylene vinylene) and Derivatives
Polythiophenes (Conducting Polymers)
Polythiophenes (Organometallic Synthesis)
Polythiophenes, Substituted (Regioselectivity, Electrical and Optical Properties)
Poly(N-vinylcarbazole)
Rigid-Rod Poly(p-phenylene benzobisthiazole)
Signal Transduction Composites (Biomaterials with Electroactive Polymers)
Small-Bandgap Polymers (Special Class of Organic Conducting Compounds)
Stacked Transition Metal Macrocycles (Semiconductive Properties)
Thiophene Copolymers (Electrical and 3rd-Order NLO Properties)
Topochemical Polymerization (Diacetylenes)

CONDUCTING POLYMERS (For Langmuir-Blodgett Film Fabrication)

L. H. C. Mattoso,* O. N. Oliveira, Jr., and M. Ferreira
Instituto de Física de São Carlos

There has been considerable excitement recently over the use of conducting polymers in a variety of technological applications. Some of these applications have already been commercialized, as in the case of rechargeable batteries, conductive textiles, and electrical displays.[1–6] Sophisticated long-term applications are also envisaged for these materials, including the fabrication of molecular electronic devices,[7] that is, devices in which switching and memory phenomena would be controlled at the molecular level. Two main requirements must be met to achieve such an ambitious goal. First, development of the ability to construct structures whose molecular architecture could be manipulated and appropriately designed. Such structures are most likely to consist of ultra-thin films. Two techniques are available for this purpose, namely, the Langmuir–Blodgett (LB) method,[8,9] which has been investigated extensively for several years, and the more recent Self-Assembly (SA) technique for building multilayers.[10,11]

We are concerned with the first of the above-mentioned aspects, focusing on the LB technique that has been exploited worldwide for fabricating ultra-thin films of a variety of materials, including conducting polymers.[12] Because the formation

*Author to whom correspondence should be addressed.

of LB films requires the spreading of a monolayer onto an aqueous subphase, some prerequisites must be fulfilled by the material of interest. It must be soluble in a reasonably volatile organic solvent and also sufficiently surface-active for spreading to occur. The conducting polymers that were first investigated, namely polyacetylene, polypyrrole, polythiophene, and polyaniline, could not be processed in this manner, and therefore several strategies were developed for overcoming this difficulty.[13]

STRATEGIES FOR BUILDING LB FILMS OF CONDUCTING POLYMERS

Several approaches have been used for processing conducting polymers in the form of LB films. These include mixing the polymer with a surface–active compound, monomer derivatization, polymer synthesis in the Langmuir trough, and polymerization after the LB deposition from a monomer monolayer.

The first approach, mixing, has been widely used for polyalkylthiophenes[14,15] as they are not sufficiently surface-active for spreading on an aqueous surface, even though they are soluble in organic solvents. The amphiphilic nature is conferred to the monolayer by mixing the polythiophene derivative with a surface-active material such as a fatty acid or a plasticizer. The final result may be a film with the polymer molecules embedded into a well-organized matrix of fatty acid molecules.[14]

The second approach, the derivatization of monomers of conducting polymers, was aimed at introducing substituents that make the polymer soluble in an organic solvent. A possibility that leads to good results is the use of preformed derivatized polymers with less bulky substituents. This approach allowed the fabrication of good-quality LB films from poly(*o*-alkoxyaniline)s[16-19] and pure poly(3,4-dibutoxythiophene)[20] with no need for the addition of surface-active materials or processing aids.

The third possibility is the formation of a Langmuir monolayer made from the monomer and the subsequent *in situ* chemical polymerization. In this case, use has been made of amphiphilic derivatized monomers that could readily form a Langmuir monolayer on an aqueous subphase containing a polymerizing agent. Polymerization then occurred by interaction between the agent in the subphase and the film-forming molecules. The polymeric monolayer would be subsequently transferred onto a solid substrate. LB films have been fabricated in this way from polyaniline and polypyrrole derivatives.[21,22]

In the fourth approach, the polymerization may also be carried out after the LB film has been deposited from a monomeric monolayer. The LB film from the polymer precursor is polymerized by thermal treatment, as in the case of poly(*p*-phenylenevinylene)s[23–30] and polyacetylenes,[31] electrochemical oxidation for films of pyrrole derivatives[21] or exposure to UV light.

LB FILMS FROM POLYANILINES

Until very recently, the fabrication of LB films from high molecular weight conducting polymers was considered practically impossible, not only for polyaniline but also for other conducting polymers. The polymer could not be processed, that is, it was not soluble in low-boiling-point solvents as required in the LB technique. Use has therefore been made of polymer derivatization and of mixtures with surfactants to enhance solubility and/or spreadability.[16,32]

Our research group has for some years investigated the fabrication of LB films of poly(*o*-alkoxyaniline)s.[16–18] It was found that more condensed, stable Langmuir monolayers are produced when the polymer is doped in dichloromethane prior to being spread onto an acidic subphase. Doping markedly improves monolayer organization, which also helps in the LB deposition process.

Processing aids have also been employed for fabricating LB films of polyanilines. In some of the most recent work from Rubner's group, polyaniline in its emeraldine-base form (PAni-EB) was mixed with two processing aids for the formation of Langmuir monolayers.[33]

Derivatization as well as mixing with surfactants, however, generally brings disadvantages to the properties of interest in conducting polymers.

The most stable Langmuir films are obtained when the PAni is predoped and the dichloromethane or chloroform solution spread onto an acidic subphase. The monolayers can be transferred onto hydrophilic substrates in the form of Z-type LB films with deposition only on the upstrokes. Deposition as a noncentrosymmetric Z-type film indicates that hydrophilic groups must protrude from the water surface in such a way that the substrate remains hydrophilic after each layer is deposited.

The possibility of depositing good-quality LB films from parent polyaniline may bring a number of welcome features originating from the higher conductivity and more intense electroactivity this polymer possesses in comparison with its derivatives.

PROPERTIES AND POSSIBLE APPLICATIONS OF LB FILMS FROM CONDUCTING POLYMERS

A number of properties of conducting polymers have been exploited such as electrical conductivity, electrochromism, and electroluminescence.[34] The characterization of the LB films is usually made via techniques already employed for other thin films. The optical properties of LB films are usually very similar to those obtained from films produced using different techniques such as spincoating. The electrochromic effects displayed by LB films of polyaniline derivatives are also essentially the same as those of cast films and electrochemically grown films of these materials.[18,19]

Several possible applications have been suggested for LB films from conducting polymers. Electrochromic displays may, in principle, be developed using the electrochromic characteristics of polyaniline LB films.

Gas sensors have long been suggested as one of the possible applications for LB films. This has been suggested for mixed LB films. of poly(3-*n*-alkylthiophenes) ($n = 10$ to 18) and docosanoic acid, whose conductivity increases drastically upon exposure to nitrogen oxides.[35] It is the selectivity in the change in conductivity that makes these LB films potential candidates for gas sensors. Polythiophene derivatives have also been employed in other types of applications. For instance, field-effect transistors have been fabricated using LB films from poly(3-hexylthiophene) and fatty acids.[36]

Various potential applications have been demonstrated for polyimide LB films. To mention just a few, polyimide LB films have been employed as UV resists[37] and have displayed switching and memory phenomena.[38,39]

In conclusion, we have presented several strategies for the fabrication of Langmuir–Blodgett films from conducting polymers, as well as the main features and properties exhibited by such systems. Possible applications for these ultrathin conductive films in several fields, including molecular electronics, have also been described. The main film properties to be exploited relate to the precise control over film thickness and molecular architecture.

ACKNOWLEDGMENTS

The authors acknowledge the financial assistance of FAPESP, CNPq, and Finep (PADCT). They are also grateful to several co-workers, Dr. Roberto M. Faria in particular, who have collaborated in the LB work in São Carlos.

REFERENCES

1. Yang, L. S.; Shan, S. Q.; Liu, Y. D. *Solid State Ionics* **1990**, *40–41*, 967.
2. Gomes, M. A. B.; Gonçalves, D.; Souza, E. C. P.; Valla, B.; Aegerter, M. A.; Bulhões, L. O. S. *Electrochimica Acta* **1992**, *37*, 1653.
3. Genies, E. M.; Lapkowski, M.; Santier, C.; Vieil, E. *Synth. Met.* **1987**, *18*, 631.
4. Chen, S. A.; Fang, Y. *Synth. Met.* **1993**, *60*, 215.
5. Nespurek, S. *Synth. Met.* **1993**, *61*, 55.
6. Roth, S.; Graupner, W. *Synth. Met.* **1993**, *55–57*, 3623.
7. Carter, F. L. *Molecular Electronics Devices II* Marcel Dekker: New York, 1987.
8. Roberts, G. G. *Langmuir–Blodgett Films* Plenum: New York, 1990.
9. Ulman, A. *Introduction to Ultrathin Organic Films from Langmuir–Blodgett to Self-Assembly* Academic: New York, 1991.
10. Decher, G.; Hong, J. D. *Ber. Bunsenges. Phys. Chem.* **1991**, *95*, 1430.
11. Ferreira, M.; Cheung, J. H.; Rubner, M. F. *Thin Solid Films* **1994a**, *244*, 806.
12. Rubner, M. F.; Skotheim, T. A. In *Conjugate Polymers* Brèdas, J. L.; Silbey, R., Eds., Kluwer Academic: Amsterdam, The Netherlands, 1991; p 363.
13. Ferreira, M.; Ph.D. Thesis, Universidade de São Paulo, Brazil, 1994.
14. Watanabe, I.; Hong, K.; Rubner, M. F. *Langmuir* **1990**, *6*, 1164.
15. Pawlicka, A.; Faria, R. M.; Yonashiro, M.; Canevarolo, Jr. S. V.; Oliveira, Jr. O. N. *Thin Solid Films* **1994**, *244*, 723.
16. Gonçalves, D.; Faria, R. M.; Oliveira, Jr. O. N.; Sworakowski, J. *Synth. Met.* **1993**, *57*, 3891.
17. Gonçalves, D.; Bulhões, L. O. S.; Mello, S. V.; Mattoso, L. H. C.; Faria, R. M.; Oliveira, Jr. O. N. *Thin Solid Films* **1994**, *243*, 544.
18. Mattoso, L. H. C.; Mello, S. V.; Riul, Jr. A.; Oliveira, Jr. O. N.; Faria, R. M. *Thin Solid Films* **1994**, *244*, 714.
19. Mello, S. V.; Mattoso, L. H. C.; Faria, R. M.; Oliveira, Jr. O. N. *Synth. Met.* 1995**,** 71, 2039–2040.
20. Callendar, C. L.; Carere, C. A.; Daoust, G.; Leclerc, M. *Thin Solid Films* **1991**, *204*, 451.
21. Duran, R. S.; Zhou, H. C. *Polymer* **1992**, *33*, 4019.
22. Cheung, J. H.; Rosner, R. B.; Watanabe, I.; Rubner, M. F. *Mol. Cryst. Liq. Cryst.* **1990**, *190*, 133.
23. Era, M.; Kamiyama, K.; Yoshiura, K.; Miomi, T.; Murata, H.; Tokito, S.; Tsutsui, T.; Saito, S. *Thin Solid Films* **1989**, *179*, 1.
24. Nishikata, Y.; Kakimoto, M. A.; Imai, Y. *J. Chem. Soc. Chem. Commun.* 15, August 1988, 1040.
25. Nishikata, Y.; Kakimoto, M. A.; Imai, Y. *Thin Solid Films* **1989**, *179*, 191.
26. Nishikata, Y.; Kakimoto, M. A.; Morikawa, A.; Imai, Y. *Thin Solid Films* **1988b**, *160*, 15.
27. Uekita, M.; Awaji, H.; Murata, M. *Thin Solid Films* **1988**, *160*, 21.
28. Lupo, D.; Prass, W.; Scheunemann, U. *Thin Solid Films* **1989**, *178*, 403.
29. Baker, S.; Seiki, A.; Seto, J. *Thin Solid Films* **1989**, *180*, 263.
30. Akatsuka, T.; Tanaka, H.; Toyama, J.; Nakamura, T.; Kawabata, Y. *Synth. Met.* **1991**, *41–43*, 1515.
31. Royappa, A. T.; Saunders, R. S.; Cohen, R. E.; Rubner, M. F. *Mat. Res. Soc. Symp. Proc.* **1991**, *247*, 853.
32. Ando, M.; Watanabe, Y.; Iroda, T.; Honda, K.; Schimidzu, T. *Thin Solid Films* **1989**, *179*, 225.
33. Cheung, J. H.; Rubner, M. F. *Thin Solid Films* **1994**, *244*, 990.
34. *Handbook of Conducting Polymers* Skoteim, T. A., Ed.; Marcel Dekker: New York, 1986; Vols. 1 and 2.
35. Bartkiewicz, S.; Chyla, A.; Sworakowski, J.; Janik, R.; Kucharski, S. *Polish J. Chem.* **1994**, *68*, 1387.
36. Punkka, E.; Rubner, M. F. *Synth. Met.* **1991b**, *41–43*, 1509.
37. Uekita, M.; Awaji, H.; Murata, M.; Mizunuma, S. *Thin Solid Films* **1991**, *205*, 109.
38. Sakai, K.; Matsuda, H.; Kawada, H.; Eguchi, K.; Nakagiri, T. *Appl. Phys. Lett.* **1988**, *53*, 1274.
39. Takimoto, K.; Kawade, H.; Kishi, E.; Yano, K.; Sakai, K.; Hatanaka, K.; Eguchi, K.; Nakagiri, T. *Appl. Phys. Lett.* **1992**, *61*, 3023.

CONDUCTING POLYMERS (Nonconjugated)

Daniel T. Glatzhofer, Shreeram Deshpande,
Gary P. Funkhouser, and Charles J. Neef
Department of Chemistry and Biochemistry and
Laboratory for Electronic Properties of Materials
The University of Oklahoma

The development of materials that combine the desirable properties of polymers with the electronic properties of metals and semiconductors has been an area of considerable growth over the past two decades with several hundred publications now appearing in this area each year. Although low conductivity, poorly defined morphologies, difficult processability, and poor environmental stability have been problems for many conducting polymer systems, several applications of these materials have evolved such as use in low-weight/high-charge-density batteries, polymer-modified electrodes, capacitors, fuses, and solid-state sensors, to name just a few. and considerable progress has been made toward alleviating their drawbacks.

The field of electrically conducting polymers has been largely dominated by the search for higher conductivities, better stabilities, and greater processability. The bast majority of work in the past two decades has focused on linearly conjugated polymers for example, polyacetylene, polypyrrole, polythiophene, poly(phenylene vinylene), and polyaniline, where delocalization or movement of charge carriers along a polymer backbone with contiguous (i.e., conjugated) π-bonds or delocalizable electrons (e.g., lone pairs on heteroatoms) plays a significant role in bulk charge transport.

Less attention has been given to formally nonconjugated polymers, which rely on charge hopping among a high density of charge sites or cofacial overlap of p-orbital systems to establish conductivity on oxidation. In fact, many nonconjugated polymers, when doped, exhibit conductivities in the semiconductor range. Among these are *cis*- and *trans*-polyisoprene and poly(2,3-dimethylbutadiene), which contain isolated double bonds and have conductivities of 10^{-2} to 10^{-1} S/cm when doped with iodine.[1] Another type of polymer in the category of nonconjugated conducting polymers are those that contain [3.2]- or [3.3]paracyclophane units. Certain nonconjugated conducting polymers (e.g., poly(*N*-vinylcarbazole)) have already enjoyed considerable commercial success as photoconducting materials[2,3] for xerography and photoimaging technology. Considerable interest in, and appreciation of, other properties available to this unique class of materials, such as photorefractivity[4] and nonlinear optical effects,[5,6] continues to grow, and new structures are being sought to optimize and modify these useful electronic effects.

CONDUCTING POLYMERS WITH PENDANT AROMATIC GROUPS

Nonconjugated conducting polymers containing electronically active aromatic groups pendant to the polymer backbone are of special interest for a variety of reasons. A large range of possible pendant aromatic groups such as benzene, naphthalene, thiophene, pyrene, and perylene can be used to tune the electronic properties of these types of polymers. They have a wide range of chemical stabilities and mechanical properties (processing) and are used in photoconducting applications.

The classic example of a conducting pendant aromatic group containing polymer is poly(*N*-vinylcarbazole) (PVK), which has enjoyed success as a commercially useful photoconducting polymer in xerography.[7]

Ever since the success of PVK as a commercial organic photoconductor, there have been many reports in which different aromatic amine units have been used pendant to the polymer backbone to obtain electrical and photoconducting polymers.[8,9] For example, poly(4-(*N,N*-diphenylamine)phenylmethyl methacrylate) (PDAPM) was synthesized by free-radical polymerization of the corresponding monomer.[8] On electrochemical doping, this polymer showed good rectification and photovoltaic effects.

Another example of a pendant aromatic group polymer is poly(3-vinylperylene).[10,11] Interest in this polymer is due to the fact that perylene forms highly conducting charge transfer complexes with bromine and iodine.

A novel class of pendant group containing conducting polymers are those with bridged aromatic rings (cyclophanes) pendant to the polymer backbone. Cyclophanes are known to have unusual electronic properties due to transannular interactions between the cofacially held aromatic units in the structure. It was also shown in multilayered cyclophanes that as the number of stacked aromatic rings increases, the absorption spectra of the cyclophanes are red shifted,[12,13] suggesting extended transannular electronic interactions.

POLYORGANOMETALLICS

Research in the area of conductive organometallic polymers has involved the incorporation of numerous metals into polymeric systems that have exhibited a wide range of conductivities.[14] For many of these polymers, it is believed that charge transport is mediated through molecular orbital interactions between the metals and the ligands. Because this interaction can play an important role in the conductivity, it has been observed that doping may adversely affect the conductivity. Another mode of charge mobility that has been observed is a mixed-valent or redox-type between metal centers. In general, these organometallic polymers can be classified into three types: main-chain organometallic polymers, columnar organometallic polymers, and pendant organometallic polymers.

POLYBUTADIENE DERIVATIVES AND MISCELLANEOUS NONCONJUGATED CONDUCTING POLYMERS

In 1988, it was reported that nonconjugated polyisoprene becomes highly conducting (10^{-1} S/cm) on exposure with iodine, and charge transport was proposed to occur by a hopping mechanism.[1] this was quickly confirmed and expanded to include observations of similar behavior by other derivatives of polybutadiene. The main controversy that still exists for these systems concerns the origin of the electrical conductivity and the mechanism of charge transport. There are three main arguments to explain conductivity in these systems.[15] The first is that doping of these polymers with iodine generates a high number of weakly mobile radical cations; the nature of the radical cation is of some debate. The second argument is that the iodine facilitates either oxidation or rearrangement of the poly(butadiene)oid backbone to give domains of highly conducting polyconjugated segments, resulting in a percolation-type of conductivity. The last argument is that conductivity actually occurs majoritively through polyiodides that form in the presence of excess iodine. The most widely accepted view appears to be that iodine forms charge-transfer complexes with polybutadiene repeat units of appropriate redox potential and structural availability. These charge-transfer complexes allow for establishment of an equilibrium concentration of radical cations and polyiodide species. Conductivity then occurs through both species to varying degrees, likely depending on sample preparation and conditions under which measurements are made.

The formation of electronically conducting charge-transfer complexes of polymers with iodine or polyiodides is actually a well-known phenomenon observed in a large number of miscellaneous systems, for example, poly(ethylene oxide) with polyiodide salts[16] and poly(4-vinylpyridine) with iodine.[17]

REFERENCES

1. Thakur, M. *Macromolecules* **1988**, *21*, 661.
2. Mort, J.; Pfister, G. *Electronic Properties of Polymers,* Mort, J.; Pfister, G., Eds.; John Wiley & sons: New York, 1982; p 215.
3. Mylnikov, V. *Polymer Sci.* **1994**, *115*, 1.
4. Moerner, W. E.; *Chem. Rev.* **1994**, *94*, 127.
5. *Nonlinear Optical and Electroactive Polymers*, Prasad, P. N.; Ulrich, D.; Eds.; Plenum: New York, 1988.
6. Burland, D. M.; Miller, R. D.; Walsh, C. A. *Chem. Rev.* **1994**, *94*, 31.
7. Penwell, R. C.; Ganguly, B. N.; Smith, T. W. *J. Poly. Sci. Macromol. Rev.* **1978**, *13*, 63.
8. Shirota, Y.; Noma, N.; Mikawa, H. *Synthetic Metals* **1987**, *18*, 399.

9. Stolka, M.; Pai, D. M.; Renfer, C. S.; Yanus, J. F. *J. Poly. Sci. Polym. Chem. Ed.* **1983**, *21*, 969.
10. Jeon, I.; Noma, N.; Shirota, Y. *Mol. Cryst. Liq. Cryst.* **1990**, *190*, 1.
11. Shirota, Y.; Jeon, I.; Noma, N. *Synthetic Metals* **1993**, *55*, 803.
12. Otsubo, T.; Mizogami, S.; Otsubo, I.; Tozuka, A.; Sakagami, A.; Sakata, Y.; Misumi, S. *Bull. Chem. Soc. Jpn.* **1973**, *46*, 3519.
13. Vögtle, F. *Cyclophane Chemistry*, John Wiley & Sons: New York, 1993; Chapter 6.
14. Biswas, M.; Mukherjee, A. *Adv. Polymer Sci.* **1994**, *115*, 89.
15. Chilkoti, A.; Ratner, B. D. *Chem. Mater.* **1993**, *5*, 786.
16. Hardy, L. C.; Shriver, D. F. *J. Amer. Chem. Soc.* **1986**, *108*, 2887.
17. Lupinski, J. H.; Kopple, K. D.; Hertz, J. J. *J. Polym. Sci., Symp. Ed.* **1967**, *16*, 1561.

CONDUCTING POLYMERS (Self-Acid-Doped, Conjugated)

Show-An Chen,* Mu-Yi Hua, and Gue-Wuu Hwang
Department of Chemical Engineering
National Tsing-Hua University

Conjugated polymers are the polymers alternate in the double and single bonds along the main chain, where the π-electrons can delocalize along a certain range of the chains. They can be doped by oxidation (p-doping) or reduction (n-doping) via chemical or electrochemical methods. The doping, by which some of the π-electrons are removed (p-doping) or external electrons are added (n-doping), is accompanied by an insertion of counterions in the vicinity of charged sites in the polymer chains. These counterions are provided by a reduction or oxidation of the dopants. These counterions can also be bonded to the polymer's main chains; such doping is termed *self-doping* and was first proposed by Heeger and co-workers.[1,2] They prepared the self-dopable polymers, poly[*n*-(3′-thienyl)-alkanesulfonic acids] (P3TPSH) and their sodium salts with alkanes of carbon numbers 2 and 4, in which the dopants are covalently bonded to side chains. Upon electrochemical doping of these polymers in electrolyte solutions, a charge transfer accompanied by a generation of polaron/bipolaron occurs by ejection of protons or metal ions from the bonded dopant and simultaneous ejection of π-electrons from the conjugated main chains, as reflected by an increase of the proton or metal ion concentration in the electrolyte solutions and the presence of free spins during the doping.[1,2] The self-doping concept has been extended to polypyrrole system by Reynolds et al. and Havinga et al.[3–5]

Ikenoue et al. found that poly[3-(3′-thienyl)propanesulfonic acid] (P3TPSH) was actually already self-doped without electrochemical doping, as evidenced by the presence of an additional optical absorption peak at 800 nm.[6,7] Chen and Hua have investigated the structure of this series of polymers and termed this type of doping self-acid-doping, in order to distinguish it from the self-doping previously described.[8]

Another class of conjugated conducting polymers, called polyanilines, can be doped to a conducting state without changing the number of π-electrons through protonation by exposure to an appropriate protonic acid.[9] The protonic acid can also be incorporated to the side chains as prepared by Yue and Epistein for sulfonic acid-ring-substitution and by Chen and Hwang for sulfonic acid-*N*-substitution.[10,11] Other protonic acid substitution and copolymerization of anilines with protonic acid-substituted aniline to yield self-acid-doped polyaniline have also been reported.[12–15]

*Author to whom correspondence should be addressed.

APPLICATIONS OF THE SELF-ACID-DOPED CONDUCTING POLYMERS

Based on the temperature-sensing characteristic, P3TESH film can be used as a temperature indicator for foods and drugs that need to be kept fresh at low temperatures. When food and drugs are exposed for a certain period of time (i.e., about 30 min) to a temperature ~40°C, the P3TESH film would exhibit a permanent color change. P3TESH can also be used as a sensing material for detecting a presence of trace amounts of organic liquid in water, because it will be subject to a color change resulting from solvato-undoping. Because of its moderate conductivity and water solubility, P3TESH can be used as a charge-reduction layer in the electron-beam-lithography process.[16]

The self-acid-doped polyanilines have faster electronic and optical responses to electrochemical potential as compared to its parent polyaniline and thus they are useful for fabrication of electrochemical chromic display.[17] Because of their moderate electrical conductivity, they can be used as a conductor of electricity, for the adsorption of electromagnetic radiation, modulation of electromagnetic beams, the reduction of beam-charging effects in an electron-beam-lithography process, and for a photolithographic technique for patterning sping-coated films.[16–19] The ammonium salts of the self-acid-doped polyanilines can be used as an NH_3 sensor.[20] Furthermore, the self-acid-doped polyanilines can change the pH near the surface of an electrode within 10 ms and they have high charge efficiency compared with other polymer-based pH modulators.[21] It is possible to fabricate an electrically controllable pH modulator as a closed microcell.[22]

ACKNOWLEDGMENT

We thank the National Science Council of ROC for financial aid through project No. NSC 85-2216-E007-047.

REFERENCES

1. Ikenoue, Y.; Votani, N.; Patil, A. O.; Wudl, F.; Heeger, A. J. *Synth. Met.* **1989**, *30*, 305.
2. Patil, A. O.; Ikenoue, Y.; Basescu, N.; Colameri, N.; Chen, F.; Wudl, F.; Heeger, A. J. *Synth. Met.* **1987**, *20*, 151.
3. Reynolds, J. R.; Sundaresan, N. S.; Pomerantz, M.; Basak, S.; Baker, C. K. *J. Electroanal. Chem.* **1988**, *250*, 355.
4. Sundaresan, N. S.; Basak, S.; Pomerantz, M.; Reynolds, J. R. *J. Chem. Soc.* **1987**, 621.
5. Havinga, E. E.; ten Hoeve, W.; Meijer, E. W.; Wynberg, H. *Chem. Mater.* **1989**, *1*, 650.
6. Ikenoue, Y.; Saida, Y.; Kira, M.; Tomozawa, H.; Yashima, H.; Kobayashi, M. *J. Chem. Soc. Chem. Commun.* **1990**, 1694.
7. Ikenoue, Y.; Tomozawa, H.; Saida, Y.; Kira, M.; Yashima, H. *Synth. Met.* **1991**, *40*, 333.
8. Chen, S.-A.; Hua, M.-Y. *Macromolecules* **1993**, *26*, 7108.
9. Chiang, J. C.; MacDiarmid, A. G. *Synth. Met.* **1986**, *13*, 193.
10. Yue, J.; Epstein, A. J. *J. Am. Chem. Soc.* **1990**, *112*, 2800.

11. Chen, S.-A.; Hwang, G.-W. *J. Am. Chem. Soc.* **1994**, *116*, 7939.
12. Chan, H. S. O.; Ho, P. K. H.; Ng, S. C.; Tan, V. T. G.; Tan, K. L. *J. Am. Chem. Soc.* **1995**, *117*, 8517.
13. Nguyen, M. T.; Kasai, P.; Miller, J. L.; Diaz, A. F. *Macromolecules* **1994**, *27*, 3625.
14. Nguyen, M. T.; Diaz, A. F. *Macromolecules* **1994**, *27*, 7003.
15. Nguyen, M. T.; Diaz, A. F. *Macromolecules* **1995**, *28*, 3411.
16. Angelopoulos, M.; Patel, N.; Shaw, J. M.; Labianca, N. C.; Rishton, S. A. *J. Vac. Sci. Technol. B* **1993**, *11*, 2794.
17. Epstein, A. J.; Yue, J. U.S. Patent 5 137 991, 1992.
18. Kondek, C. A.; Poli, L. C. Poli, *SPIE* **1994**, *2194*, 366.
19. Bartolomeo, C. D.; Barker, P.; Petty, M. C.; Adams, P.; Monkman, A. P. *Adv. Mater. Opt. Elect.* **1993**, *2*, 233.
20. Epstein, A. J.; Yue, J. U.S. Patent 5 208 301, 1993.
21. Yue, J.; Epstein, A. J. *J. Chem. Soc., Chem. Commun.* **1992**, 1540.
22. Epstein, A. J.; Yue, J.; Burley, D. R. U.S. Patent 5 250 163, 1993.

CONDUCTIVE ELASTOMERIC BLENDS

Marco-Aurelio De Paoli,* Rita A. Zoppi,
and Maria Isabel Felisberti
Instituto de Química
Universidade Estadual de Campinas

Conductive polymer blends are prepared by combining an insulating polymer or copolymer with an intrinsically conductive polymer. The most commonly used conductive polymers are polypyrrole, polythiophene, and polyaniline — chosen because of their facility of synthesis and environmental stability. Preparation methods include mechanical mixture, casting of a solution containing the components of the blend, or polymerization of one polymer into the other. The latter method can be achieved either chemically or electrochemically, producing blends or interpenetrating networks.

Conductive polymeric blends were first prepared by the electrochemical polymerization of pyrrole into a film of poly(vinyl chloride) coated on a platinum electrode.[1,2] In this method the insulating material is swollen by the electrolyte solution containing the monomer and it polymerizes from the interface of the electrode/polymer to the interface of the polymer/electrolyte. The association of the electronic properties of the conducting phase and the mechanical properties of the insulating phase produce a unique material with conductivities of 10–100 S cm^{-1} and thermoplastic-like mechanical properties.

Chemical polymerization was also used for preparation of polypyrrole/poly(vinyl chloride) blends by the interphase polymerization of pyrrole between aqueous solutions of iron(III) salts and organic solutions of pyrrole separated by poly(vinyl chloride) films.[3] By this method, the structure of the blend could be changed by changing the organic solvent or the oxidant. The reported conductivities of the films were in the range of 0.1 to 10 S cm^{-1}.

BLENDS PREPARATION

Blends Prepared by Electrochemical Synthesis

To obtain blends by the electrochemical method it is necessary to coat the working electrode with a film of the insulating polymer, prior to the anodic deposition of the conducting film. Three conditions must be fulfilled by the insulating polymer film: it must be swollen by the solution containing the monomer and the electrolyte, it must show a certain degree of miscibility with the conducting polymer, and it must be stable in the potential range and in the medium used for the polymerization.

Blends Prepared by Chemical Synthesis

The advantage of the chemical synthesis of conductive polymer blends is that large-scale preparation of conducting materials can be achieved. The synthesis can be carried out in different ways.

APPLICATIONS

Above we describe the variety of ways conductive elastomeric polymer blends are prepared. These have the typical applications envisaged for intrinsically conductive polymers, such as electrochromic displays, electrodes for lightweight batteries, pressure sensors or transducers, and electronic devices. In addition, large-scale preparation methods, such as calendering and producing blends with properties of vulcanized rubber, permit application in other areas, such as anti-static coatings, anticorrosive protections, and electromagnetic shielding.

Blends and semi-IPN of polypyrrole and EPDM rubber absorb 85% of microwave radiation in the 10–13 GHz range.[4] For defense purposes, a larger frequency range should be covered. Surface deposition of polypyrrole on poly(vinyl chloride) produced composites with microwave absorption properties in the 0.1–20 GHz frequency range.[5] These same authors also studied the shielding characteristics of blends of polyaniline and poly(3-octylthiophene) with polystyrene and with the copolymer of ethylene and vinyl acetate.

The large investments in research on new conductive polymer blends reflects the interest and optimism of chemists and technologists to obtain new materials to fill several unexplored application niches.

ACKNOWLEDGMENTS

This work was supported by several grants from the Fundação de Amparo a Pesquisa do Estato São Paulo and FINEP. the authors also acknowledge fellowships from different agencies: MAP, research fellowship, CNPq; RAZ, graduate fellowship, FAPESP.

REFERENCES

1. De Paoli, M.-A. et al. *J. Chem. Soc., Chem. Comun.* **1984**, 1015.
2. De Paoli, M.-A. et al. *J. Polym. Sci., Polym. Chem. Ed.* **1985**, *23*, 1687.
3. Nakata, M.; Kise, H. *Polym. J.* **1993**, *25*, 91.
4. Zoppi, R. A.; De Paoli, M.-A., unpublished material.
5. Olmedo, L.; Hourquebie, P.; Jousse, F. *Adv. Mater.* **1993**, *5*, 373.

Conformational Transitions

See: Hydrophobic Polyelectrolytes (Conformational Transitions)

*Author to whom correspondence should be addressed.

CONJUGATED LADDER POLYMERS

Chyi-Shan Wang and My Dotrong
University of Dayton Research Institute

Ladder polymers originally were prepared to improve the thermal stability of organic polymers for structural applications. However, the various physical and chemical properties of ladder polymers were not extensively characterized until the conjugated benzimidazobenzophenanthroline (BBL)-type ladder polymers were synthesized in late 1960s.[1]

Conjugated polymers, such as polyacetylene with alternating double and single bonds, can lead to extended π-electron delocalization and give rise to interesting electrical, electronic, and optical properties. Conjugated ladder polymers began to be characterized for their solid state properties in the early 1980s.[2–4] It was expected that the planar ladder structure is the preferred structure for π-conjugation. Furthermore, the environmental stability of these polymers is highly desired for device applications. Wang recently reviewed the mechanical, electrical, and nonlinear optical properties of conjugated ladder polymers.[5] A majority of these properties belong to a few processable conjugated ladder polymers. Therefore, it is clear that more synthesis effort is needed to prepare this class of polymers in order to thoroughly characterize their structure and property relationships.

Most ladder polymers are insoluble and infusible because of their rigid molecular structure. Two synthesis approaches have been used to promote the solubility and processability of otherwise insoluble or infusable rigid-chain polymers. The first is to reduce the chain rigidity by introducing flexible bridges (for example, –O–,–CO–, $-CH_2-$ and SO_2–) or asymmetric molecular linkages (in *ortho-* and *meta-* rather than in *para-*position), or by adding long flexible side chains. The second is to prepare a soluble or processable intermediate that, after processing, can be converted to the rigid-chain structure by thermal cyclization, isomerization, or other simple reactions.

SYNTHESIS

Polyquinoxaline (PQL) Ladder Polymers

The formation of quinoxaline by an aromatic *o*-diamine and a 1,2-dicarbonyl is a clean reaction. It leads to an intramolecular cyclization and creates an aromatic nucleus that is *p*-isoelectronic with naphthalene. Thus, the reaction is suitable for synthesizing thermally stable polymers. Stille et al. used 2,5-dihydrozy-*p*-benzoquinone and 1,2,4,5-tetraaminobenzene to synthesize dihydroquinoxaline ladder polymer.[6,7] This polymer does not have a complete aromatic structure because of a low reactivity of hydroxyl groups.

To achieve a fully aromatic polymer backbone, aromatic tetraketones were used to react with tetraaminobenzene where the enolization to the hydroxyketone was restricted.[8–10]

Benzimidazobenzophenanthroline (BBL) Type Ladder Polymers

Poly(7-oxo-7,10H-benz[*de*]imidazo[4′,5′:5,6] benzimidazo[2,1,-*a*]isoquinoline-3,4:10,11-tetrayl)-10-carbonyl] (BBL) is the most characterized conjugated ladder polymer to date because of its processability, heteroaromatic structure, and high molecular weight. It is prepared through the polycondensation in PPA of 1,2,4,5-tetraaminobenzene with 1,4,5,8-naphthalenetetracarboxylic acid or its tetrahydrochloride salt (**Scheme I**).[1] BBL is soluble in strong protonic acids such as methanesulfonic acid (MSA). The BBL molecule illustrated in Scheme I suggests a ribbon-like, quasi-two-dimensional structure. The short-range interaction energies calculated for BBL by a semiempirical method lead to the prediction that the ribbon like BBL possesses an interplanar spacing of 3.3 Å and an alongside spacing of 8.8 Å in the solid state.[11]

Polyquinolino Ladder Polymers

Poly[4-(3-pyridyl),8-methyl,2,3-6,7-quinoline] (PPMQ) is obtained by reacting 2,6-diaminotoluene with 3-pyridine carboxaldehyde in acidic medium.[12] The resulting prepolymer is then condensed in PPA at about 330°C to provide the ladder PPMQ.

Since quinocridone has been shown to be thermally stable and useful in electrophotographic applications, Ruan and Litt synthesized the quinacridone prepolymer by reacting 1,5-diaminonaphthalene with 1,4-cyclohexanedione-2,5-dicarboxylate in *o*-chlorophenol.[13] The prepolymer is then transformed into a ladder structure by condensing in PPA.[13] Poly(1,2-8,9-*trans*-quinocridono) (PQ) is soluble in sulfuric acid.

Phenothiazine and Phenoxozine-Type Polymers

Polyphenothiazine (PTL) and polyphenoxozine (POL) are prepared by polycondensation reaction of 2,5-dichloro-*p*-benzoquinone with 2,5-diamino-1,4-benzene-dithiol or 2,5-diamino-1,4-benzene-diol, respectively.[4] The insoluble black polymers consist of a nearly planar structure of diphenodithiazine units.

Polyphenylene Type Ladder Polymers

Scherf and Muller prepared soluble ladder polymers from 2,5-dihexyl-1,4-phenylenediboronic acid and 2,5-dibromoterephthalic dialdehyde by a two-step process.[14] The Suzuki coupling reaction provides an open-chain prepolymer with a poly(*p*-phenylene) (PPP) backbone. The ring-closure reaction takes place quantitatively within a few seconds with boron trifluoride etherate, leading to a methylene-bridged ladder polymer of polyphenylenes.

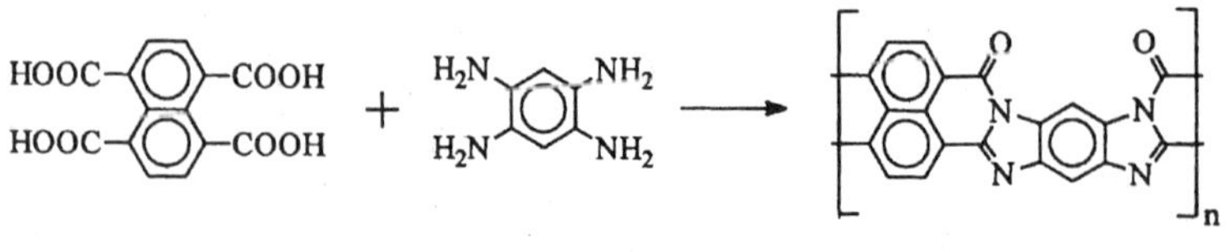

Scheme I Preparation of ladder polymer BBL.

PROPERTIES

Thermal Stability

Conjugated ladder polymers have outstanding thermal and thermooxidative stabilities in comparison with nonladder polymers of a similar structure.

Mechanical Properties

The Young's modulus (E) and the tensile strength (TS) of the cast BBL film compare favorable with those of engineering plastics (E $\cong$ 4.0 GPa and TS $\cong$ 100 MPa). Even the tensile properties of the deposited film are acceptable when one considers its low density.

Chemical and Electrochemical Doping

Conjugated ladder polymers as conducting polymers have advantages over single-stranded conjugated polymers in thermal stability, chemical resistance, mechanical strength, and greater molecular and structural order.

Infrared spectrum of lightly doped BBL suggests that dopants interact with BBL by forming complexes with its imino nitrogens.[2] The n-type doped BBL is very stable in air due to its low redox potentials and its highly fused heteroaromatic rings, which allow extensive delocalization of the polymeric carbanions.[15] The n-type doped BBL film that was electrochemically intercalated with Ag^+ and Ag shows a very high conductivity of 324 S/cm.[15]

Thermo- and Photo-Induced Conductivity

BBL fiber shows a room temperature dc conductivity of 3 $\times 10^{-4}$ S/cm after it is heat-treated at 350°C.[16] This conductivity does not change with the duration of heat treatment and is stable over a period of days, indicating that the thermally induced charge carriers are rapidly excited and extremely long-lived. This thermal excitation of charge carriers in BBL is reversible and the high conductivity can be regained by heat-treating the fiber again. BBL shows a rapid increase in conductivity with temperature to 10^{-2} S/cm at 400°C followed by a sharp decrease due to thermal degradation of the polymer.[17]

Photoconductivity and light-induced electron spin resonance (ESR) have been observed in conjugated ladder polymers such as BBL[18] and the PQL polymers derivatized with long flexible side chains.[19]

Ion Implantation and Pyrolysis

Ion implantation is a process by which a target material is bombarded with a beam of energetic ions that penetrate to a certain depth in the material. It is a well-established technique in the electronic industry for introducing impurities into semiconductors to modify their electronic properties and for fabricating integrated circuits and other microelectronic devices. Ion implantation has been applied to several single-stranded conjugated polymers and has enhanced their electrical conductivity to semiconducting level, 10^{-5} to 1 S/cm. Conjugated ladder polymers are ideal candidates for ion implantation because of their high temperatures stability and fused aromatic of heteroaromatic ring structures, which may withstand the inevitable sample heating during the ion bombardment and minimize the structure damage and bond reforming during the carbonization of the organic matter.

The degree of structural modification and conductivity enhancement resulting from ion implantation depends on ion species, ion energy, current density, and dosage. The conductivity introduced to BBL by low current density ion implantation is due to a synergistic effect of ion doping and structural modification.

Conjugated ladder polymers may be carbonized or graphitized by pyrolysis as well. The ESR susceptibility measured at room temperature for heat-treated BBL increased significantly with heat treatment temperature to about 750°C followed by a sharp decrease until it is no longer measurable.[20] An alternative route to achieve a high conductivity in conjugated ladder polymers is, therefore, through a spatially selective laser annealing. With this method one can form a stable conducting pattern with high intrinsic conductivity of about 30 S/cm.[20] It is suggested by ESR imaging that laser annealing in the ladder polymers primarily produces heat and, therefore, is equivalent to thermal annealing or pyrolysis.[20]

Nonlinear Optical Properties

Technological interest in using organic polymers for nonlinear optical (NLO) applications stems from their large optical nonlinearity and fentosecond response time. In conjugated ladder polymers, the alignment of *p*-orbitals can be further enhanced, leading to a large second hyperpolarizability, γ, of which the third-order nonlinear susceptibility $\chi^{(3)}$ is the macroscopic manifestation.

APPLICATIONS

Conjugated ladder polymers have shown interesting properties in thermal stability, mechanical strength, electrical conductivity, and nonlinear optical susceptibility. They are potential candidates for structure protective coatings because of their environmental stability and chemical resistance. In conjunction with their electrical conductivity, they may be used as electrodes and lightning protection shieldings. Coatings of BBL thin films on platinum electrodes exhibit electrochromism in nonaqueous electrolytes suggesting their potential application in display devices. The spatially selective enhancement of conductivity in ion-implanted or laser-annealed conjugated ladder polymers is useful for fabrication of microelectronic devices and electronic packaging, such as p–n junctions and high-density electronic interconnections. Photoconductivity has been demonstrated in BBL- and PQL-type conjugated ladder polymers. Any material that responds efficiently to the light emitted from the diode lasers used in printers and copiers is of interest for imaging applications. The interesting properties of conjugated ladder polymers will result in their wide application in such long-term generic applications as molecular electronics and optical signal processing and near-term applications as batteries, optical sensors, microelectronic devices and inter-connections, electromagnetic interference shielding, lightning protective structures, solar cells, and displays.

REFERENCES

1. Arnold, F. E.; Van Deusen, R. L. *Macromolecules* **1969**, *2*, 497.
2. Kim, O.-K. *J. Polym. Sci. Polym. Lett. Ed.* **1982**, *20*, 663.
3. Kim, O.-K. *Mol. Cryst. Liq. Cryst.* **1984**, *105*, 161.

4. Kim, O.-K. *J. Polym. Sci. Polym. Lett. Ed.* **1985**, *23*, 137.
5. Wang, C. S. *Trends Polym. Sci.* **1993**, *1*, 199.
6. Stille, J. K.; Mainen, E. L. *J. Polym. Sci.* **1966**, *B*(4), 30.
7. Stille, J. K.; Mainen, E. L. *Macromolecules* **1968**, *1*, 46.
8. Stille, J. K.; Freeburger, M. E. *J. Polym. Sci.* **1966**, *B*(4), 665.
9. Stille, J. K.; Freeburger, M. E. *J. Polym. Sci.* **1967**, *B*(5), 989.
10. Stille, J. K.; Freeburger, M. E. *J. Polym. Sci., Part A-1* **1968**, *6*, 161.
11.. Nayak, K.; Mark, J. E. *Makromol. Chem.* **1986**, *187*, 1547.
12. Ruan, J. Z.; Litt, M. H. *Synth. Met.* **1986**, *15*, 237.
13. Ruan, J. Z.; Litt, M. H. *J. Polym. Sci., Polym. Phys. Ed.* **1988**, *26*, 1483.
14. Scherf, U.; Muller, K. *Macromolecules* **1992**, *25*, 3546.
15. Jenekhe, S. A. *ACS Polym. Mat. Sci. Eng. Proc.* **1989**, *60*, 419.
16. Wang, C. S.; Lee, C. Y-C; Arnold, F. E. *Mat. Res. Soc. Symp. Proc.* **1992**, *247*, 747.
17. Agrawal, A.; Wang, C. S.; Song, H. H. *Mat. Res. Soc. Symp. Proc.* **1994**, *328*, 279.
18. Naryan, K. S. et al. *J. Lumin.* **1989**, *60*, *61*, 482.
19. Belaish, I. et al. *Synth. Met.* **1989**, *33*, 341.
20. Dahm, et al. *Polym. Adv. Tech.* **1990**, *1*, 247.

Conjugated Polymers

See: *Conducting Polymers (Self-Acid-Doped, Conjugated)*
Conjugated Ladder Polymers
Conjugated Polymers (Insulating and Conducting Forms)
Conjugated Polymers (Prepared by Organometallic Processes)
Conjugated Polyradicals
Liquid Crystalline Polymers (Conjugated Chain)
Nonlinear Optical Polymers [Poly(arylene ethynylene) Derivatives]
Organometallic Polymers, Fluoroarylene Bridged (Rigid Rods to Segmented Chains)
Photoluminescent Polymers
Poly(arylene vinylene)s (Mechanistic Control of a Soluble Precursor Method)
Polyazines (Third-Order Nonlinear Optical Properties)
Poly(cyclopentadienylene vinylene)
Poly(phenylene ethynylene) (Synthesis and Optical Properties)
Poly(phenylene vinylene)s (by the Heck Coupling Reaction)
Poly(p-phenylene vinylene)s (Methods of Preparation and Properties)
Poly(2,6-pyridinediylvinylene)
Polyradicals (Synthesis and Magnetic Properties)
Polythiophenes (Conducting Polymers)
Polythiophenes (Organometallic Synthesis)
Polythiophenes, Substituted (Regioselectivity Electrical and Optical Properties)
Small-Bandgap Polymers (Special Class of Organic Conducting Compounds)
Thiophene Copolymers (Electrical and Third-Order NLO Properties)
Topochemical Polymerization (Diacetylenes)

CONJUGATED POLYMERS (Insulating and Conducting Forms)

Daniel J. Sandman
Center for Advanced Materials
Department of Chemistry
University of Massachusetts, Lowell

Considerable attention has been devoted in recent years to polymers that have backbone segments made up of monomeric units in π-conjugation by both polymer scientists and technologists interested in potential applications of these materials based on either or both their electrical and optical properties. This article will, of necessity, be selective in its coverage. First will be a broad discussion of the kinds of synthetic methodologies used to prepare these polymers. Next will be a discussion of the electrically insulating forms of conjugated polymers, materials that are occasionally termed "semiconducting" polymers, with emphasis on electronic and crystal structure as well as linear and nonlinear optical properties and the structure–property relationships associated with them. Finally, the exposure of certain conjugated polymers to oxidizing and reducing agents, a process commonly termed "doping" by analogy to inorganic semiconductors, markedly enhances their dc electrical conductivity. The nature and consequences of this redox process are examined. Potential applications of these materials will be discussed throughout the article, where appropriate.

In contrast to virtually every other field of macromolecular endeavor, the topic of conjugated polymers is blessed with examples of materials that are available as macroscopic polymer single crystals and other highly ordered morphologies. These substances are the polydiacetylenes (PDAs), and their properties have an enhanced degree of definition compared to less-ordered conjugated systems. PDA crystallinity follows from synthesis via a topochemical polymerization of the diacetylene (DA) monomer under thermal, radiative, or mechanical stimuli. The repeat structures of the partially crystalline conjugated polymers and abbreviations that are used herein are given in **Figure 1**.

CRYSTALLOGRAPHY

As noted above, several PDAs are available in the form of macroscopic single crystals, and complete crystal structures are available for these as well as numerous DA monomers.[1] For those systems in which both monomer and polymer structures are available, the crystallographic data allows discussion of the structural changes that accompany the polymerization process. The observed structural data allow assignment of the "en-yne" bond representation for the PDA backbone. Typical values for the double and triple bond lengths are 1.36 and 1.20 Å, respectively. PDA crystal structures typically exhibit as a lattice constant a backbone repeat distance of 4.90 Å.

Since macroscopic single crystals are not available for conjugated polymers other than PDA, the existing crystallographic date are usually obtained from samples that are stretch-aligned films or powders. The polythiophene (PT) and P3AT structures have a trans arrangement of the rings. The conjugated backbones are planar, and there is substantial disorder in the aliphatic group arrangements. For P3AT structures, one dimension

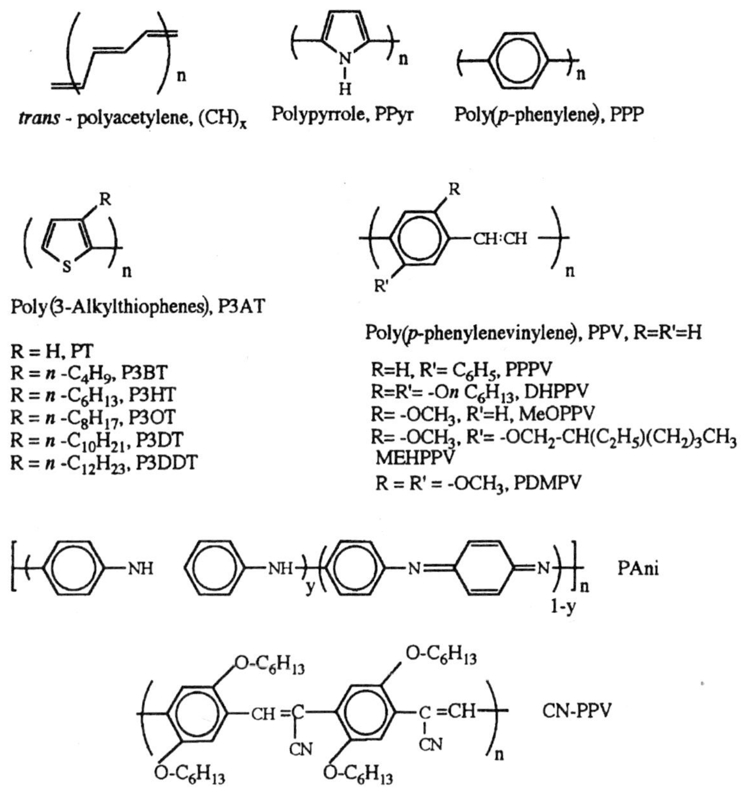

FIGURE 1. Repeat structures and abbreviations for partially crystalline conjugated polymers under discussion.

depends on the length of the alkyl chain, one on the repeat distance along the chain, and one on the van der Waals spacing between chains.

ELECTRONIC STRUCTURE

The energy band structure in conjugated polymers is typically discussed in the tight-binding (Hückel) limit, and that π-electron treatments are useful approximations for the ground state and low-lying excited states. The presence of local levels between the valence and conduction bands is a common feature of energy level diagrams for conjugated polymers.

OPTICAL SPECTROSCOPY

The most detailed optical studies have been carried out on PDA single crystals using polarized specular reflectance. This technique, rather than absorption, is used due to both crystal thickness and intense absorption.[2] A single crystal of PDA from the bis-butoxycarbonylmethylurethane of 5,7-dodecadiyn–1,12-diol (PDA-4BCMU) exhibits the lowest energy optical transition in the direction of the conjugated chain.[3] At the wavelength of maximum absorption, the absorption coefficient exceeds 10^5 cm^{-1}. In contrast, spectra obtained with light polarized perpendicular to the chain are essentially structureless until the lowest energy side-group absorption is reached. The structure in PDA single-crystal spectra is vibrational in origin, and details about the coupling of specific vibrations to the PDA backbone come from excitation profiles in resonance raman spectra.[4] Photoelectronic carriers are typically not produced by irradiation of PDA crystals at their longest wavelengths of absorption. From such observations, there is wide consensus that the lowest energy optical transition is excitonic.[2] PDA crystals typically do not exhibit significant luminescence.

PT, PPV, and their derivatives show easily observed fluorescence that is related to their absorption spectra with relatively small Stokes shifts.[5] The luminescence intensity of P3AT films depends on the size of the hydrocarbon chain: polymers with longer alkyl groups display more intense luminescence.[6]

It is useful to understand that the observed spectrum of a conjugated polymer may be accounted for by a number of repeat units that is significantly less than the number of repeat units found in the polymer chain. The spectra (at liquid helium temperatures) of a growing PDA chain reveal that a chain of about eight repeat units has a spectrum quite similar to that of the completely polymerized material.[7]

SOLID-STATE CHROMISM

Solid-state color changes involving a relatively drastic reversible change in the electronic spectrum have been observed with heat, light, mechanical strains, and chemical effects.[8] Thermochromic materials have applications as color sensors to indicate temperature changes. Thermochromism in PDA was observed in 1976, and has also been reported in P3ATs, polysilanes, arylene vinylenes, and a 1,4-phenylene-2′,5′thiophene copolymer. For all conjugated polymers reported to date, an increase in temperature results in a spectra shift to higher energy. A central issue in these chromoic changes has concerned the structural changes that accompany the spectral shift. Historically, this has been widely interpreted to mean that torsion in the conjugated chain weakened π-electron overlap, resulting in electron localization and a shortened "effective conjugation length."

The best defined and most studied thermochromic polymers are the urethane-substituted PDA from the bis-ethylurethane of 5,7-dodecadiyn-1,12-diol (PDA)-ETCD and PDA from the bis-isopropylurethane of 5,7-dodecadiyn-1,12-diol (-IPUDO) and other closely related substances.[8] Thermochromism in these materials involves an endothermic first-order phase transition in which the crystallographic unit cell increases 3–4% in volume on heating from ambient temperature to above the phase transition. Hydrogen bonding in the urethane group is preserved during the transition, as is the backbone enyne bonding.

THIRD-ORDER NONLINEAR OPTICAL PHENOMENA

In recent years, potential application of ultrafast (subpicosecond) nonlinear optical (NLO) signal processing in computers and communications for functions such as switching, amplifying, and multiplexing has stimulated considerable interest in the measurement of both the magnitude and temporal response of the third-order NLO susceptibility, $\chi^{(3)}$, in conjugated polymers, especially in thin film waveguides. Several schemes for optical limiters to protect sensors, including the human eye, from laser irradiation also use $\chi^{(3)}$, as do several approaches to advanced holographic techniques.

Third-order NLO properties of conjugated polymers have been reviewed, and it is apparent that PDA as a class have values of $\chi^{(3)}$ larger than other classes of conjugated polymers.[9] Quantum chemical approaches to the prediction of third-order properties have been summarized by Dudis et al.[10] Figures of merit for devices based on $\chi^{(3)}$ have been given.[7] In PDA, $\chi^{(3)}$ is highly anisotropic. The value measured in the direction of the conjugated chain is at least a factor of 100 greater than that measured in the perpendicular direction.

ELECTROLUMINESCENCE

Visible light emission under excitation by an electric field is termed electroluminescence (EL), and, at present, its observation in conjugated polymers is an active research field. In EL devices electronic carriers, as electrons and holes, are injected into a layer that emits light. Hence, electrical energy is converted into light, and component materials must be able to transport carriers as well as exhibit fluorescence.

The observation of EL with a quantum efficiency of 0.05% in PPV stimulated current activity in conjugated polymers.[11]

A particularly interesting development is the observation of competition between photoinduced absorption (PA) and stimulated emission (lasing) in PPV.[12] The stimulated emission arises from the same singlet exciton as PL and EL and the PA arises; it has been suggested, from electron-hole pairs bound on adjacent chains. In PPV, it appears that the spectral overlap of PL and PA is an obstacle to developing PPV lasers. However, the challenge to find a material in which lasing dominates remains open.

EL devices based on organic and polymeric materials already have a brightness and efficiency that is attractive for large area displays. Optimization of all device parameters and a better understanding of all device interfaces are necessary if the commercial promise touted for these materials is to come to pass.

CONDUCTING POLYMERS

Exposure of electrically insulating partially crystalline or amorphous conjugated polymers to strong oxidizing and reducing agents (dopants) results in an electron transfer process that creates electronic carriers and significantly enhances the dc conductivity of these materials. Proposed applications for these materials include antistatic elements, batteries, catalysts, controlled release pharmaceutical delivery systems, electrochromic elements, gas separation membranes, electromagnetic shielding, biological and chemical (pH, NH_3, H_2O) sensors, and microelectronic resists, PAni is the most stable and processable of these materials, and it has been introduced commercially. PPyr also is in the marketplace. Typical p-dopants include iodine, $FeCl_3$, and AsF_5. Typical counterions for anodic electropolymerizations include Cl_4^-, PF_6^-, BF_4^-, AsF_6^-, and *p*-toluenesulfonate. While successful n-doping is less common than p-doping, the alkali metals have been used, and quaternary ammonium salts are counterions for cathodic polymerizations.

Since the doping process modifies bond lengths and bond angles, the IR and Raman spectra of doped materials differ from those of the pristine polymers. In addition to the enhancement of the dc conductivity, at least two new electronic optical absorptions that are not present in an undoped polymer appear at energies less than the band gap.

REFERENCES

1. Enkelmann, V. *Adv. Polym. Sci.* **1984**, *63*, 91.
2. Schott, M.; Wegner, G. In *Nonlinear Optical Properties of Organic Molecules and Crystals*; Chemla, D. S., Zyss, J.. Eds.; Academic: London, 1987; Vol. 2, Chapter III-1.
3. Chance, R. R.; Patel, G. N.; Witt, J. D. *J. Chem. Phys.* **1979**, *71*, 206.
4. Batchelder, D. N.; Bloor, D. N. In *Advances in Infrared and Raman Spectroscopy*; Clark, R. J. H.; Hester, R. E., Eds.; Wiley Heyden: Chichester, 1984; Vol. 11, pp 133–209.
5. Bassler, H. In *Optical Techniques to Characterize Polymer Systems*; Bassler, H., Ed.; Elsevier Science: Amsterdam, 1989; pp 181–225.
6. Yoshino, K. et al. *J. Appl. Phys.* **1990**, *68*, 5976.
7. Sixl, H. *Adv. Polym. Sci.* **1984**, *63*, 49.
8. Sandman, D. J. *Trends Polym. Sci.* **1994**, *2*, 44.
9. Nalwa, H. S. *Adv. Mater.* **1993**, *5*, 341.
10. Dudis, D. S.; Yeates, A. T.; Kost, D. *Adv. Mater.* **1994**, *6*, 248.
11. Burroughes, J. H. et al. *Nature* **1990**, *347*, 539.
12. Yan, M. et al. *Phys. Rev. Lett.* **1994**, *72*, 1104.

CONJUGATED POLYMERS (Prepared by Organometallic Processes)

Takakazu Yamamoto
Research Laboratory of Resources Utilization
Tokyo Institute of Technology

Diorganotransition metal complexes such as diorganonickel(II) complexes NiR_2L_n undergo reductive elimination reactions to give R–R, and this coupling reaction on transition metals has been used for C–C coupling reactions.[1]

We have applied the C–C coupling reactions to polycondensation to afford various π-conjugated polymers[2] (**Equations 1**,[3-9] **2**,[10] and **3**.[11-13]

$$n\,X - Ar - X + n\,Metal \xrightarrow{catalyst} -\!(Ar)\!-_n \qquad 1$$

In Equation 1, the metal is Mg, Zn, or Ni(0). The catalyst is ML_n.

$$n\,X - Ar - X + n\,Me_3Sn - Ar' - SnMe_3 \xrightarrow{PdL_m} -\!(Ar - Ar')\!-_n \qquad 2$$

$$n\,X - Ar - X + n\,HC \equiv C - Ar' - O \equiv CH \xrightarrow{PdL_m-NR_3}$$
$$-\!(Ar - C \equiv C - Ar' - C \equiv C)\!-_n \qquad 3$$

PREPARATION

Two examples of the π-conjugated polymers prepared according to Equations 1 and 3 are shown below (**Structures 1 and 2**).

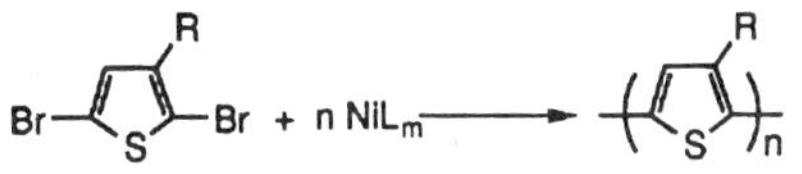

1

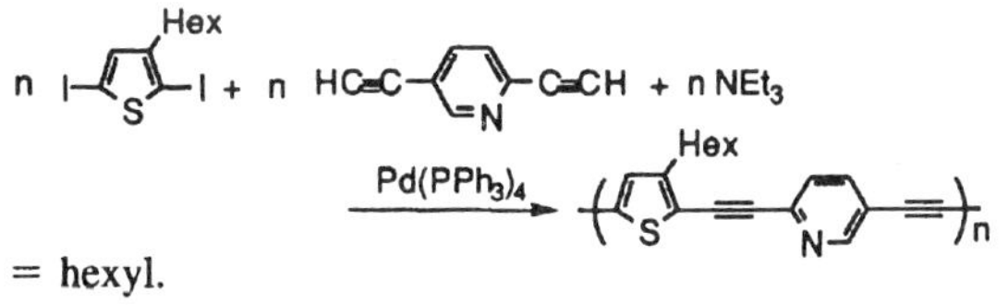

2

Hex = hexyl.

STRUCTURE AND PROPERTIES

Most of the π-conjugated polymers have linear structures.

Doping

Chemical as well as electrochemical oxidation and reduction of the π-conjugated polymers give electrically conductive materials with electrical conductivity (σ) of 10^{-4} – 10^{2} S cm^{-1} as usually measured with compressed powder.

Optical Properties

The π-conjugated polymers give rise to π-π* absorption bands, which are shifted to longer wavelength from those of the recurring aromatic unit; the degree of the shift is in range 3500–13,000 cm^{-1}. They usually exhibit fluorescence with peak positions at onset positions of the π-π* absorption bands. The linear polymers often give excimer emission, presumably due to ease of coagulation of the linear molecules at the excited state. The π-conjugated polymers sometimes show a large optical third-order nonlinear susceptibility $\chi^{(3)}$ of about 5×10^{-11} esu.[13]

Electric and Optical Devices

Since the polymer can be converted into p- and n-conducting properties, they can be used to make electric devices such as electrochromic devices, diodes,[14] electroluminescent devices,[15] and transistors.[16] They are expected to be useful to make a wave guide by applying their large $\chi^{(3)}$ value.

REFERENCES

1. Yamamoto, T.; Wakabayashi, S.; Osakada, K. *J. Organomet. Chem.* **1992**, *428*, 223.
2. Yamamoto, T. *Prog. Polym. Sci.* **1992**, *17*, 1153; *J. Synth. Org. Chem. Jpn.* **1995**, *53*, 999.
3. Yamamoto, T.; Yamamoto, A. *Chem. Lett.* **1977**, 353.
4. Yamamoto, T.; Hayashi, Y.; Yamamoto, A. *Bull. Chem. Soc. Jpn.* **1978**, *51*, 2091.
5. Yamamoto, T. et al. *Makromol. Chem., Rapid Commun.* **1985**, *6*, 671.
6. Yamamoto, T.; Kashiwazaki, A.; Kato, K. *Makromol. Chem.* **1989**, *190*, 1649.
7. Yamamoto, T.; Ito, T.; Kubota, K. *Chem. Lett.* **1988**, 153.
8. Yamamoto, T. et al. *Macromolecules* **1992**, *25*, 1214.
9. Yamamoto, T. et al. *J. Am. Chem. Soc.* **1994**, *116*, 4832.
10. Kanbara, T.; Miyazaki, Y.; Yamamoto, T. *J. Polym. Sci., Polym. Lett. Ed.* **1995**, *33*, 999.
11. Sanechika, K.; Yamamoto, T.; Yamamoto, A. *Bull. Chem. Soc. Jpn.* **1984**, *57*, 752.
12. Yamamoto, T. et al. *J. Chem. Soc., Chem. Commun.* **1993**, 797.
13. Yamamoto, T. et al. *Macromolecules* **1994**, *27*, 6620.
14. Yamamoto, T. et al. *J. Phys. Chem.* **1992**, *96*, 8677.
15. Yamamoto, T.; Inoue, T.; Kanbara, T. *Jpn. J. Appl. Phys.* **1994**, *33*, L250.
16. Yamamoto, T. et al. *Denki Kagaku* **1994**, *62*, 84.

CONJUGATED POLYRADICALS

Paul M. Lahti
Department of Chemistry
University of Massachusetts

The most fundamental notion behind attempts to design new magnetic materials is the exchange interaction between electrons. For example, two independent electrons may be exchange coupled by some mechanism in either a ferromagnetic (FM) manner to give a spin-parallel triplet state, or in an antiferromagnetic (AFM) manner to gave a spin-paired singlet state.

The simple two-electron model may be extended to larger polyradicals and supramolecular assemblages of one or more types of molecules which bear unpaired electron spin. Realistic materials are substantially more complex than the simple two-electron model. One must consider factors such as anisotropy of exchange effects, whereby exchange by one mechanism in a material may be substantially different from exchange by another mechanism.[1]

The summary effect of these mechanisms gives the overall material properties — although one must keep in mind that the bulk ferromagnetism cannot be stronger than the weakest FM exchange coupling mechanism allows. At sufficiently high temperatures, the weakest FM exchange coupling effects are overcome, and isolated spin behavior (paramagnetism, or PM) will result. The temperature at which the transition from FM to PM behavior occurs is T_c, the Curie temperature.

Overall, it is important to recall that magnetism is a bulk property.

Much effort has been put into synthesis and magnetic evaluation of inorganic and organometallic molecular materials.[1–3] This is not surprising, given the outstanding magnetic properties of naturally occurring substances. In terms of magnetic properties, the charge-transfer and other mixed organic/inorganic materials have been some of the most promising to date. A number have been found to exhibit bulk FM coupling, including some with T_c above room temperature.[4] While these systems typically are non-processible, crystalline materials, their behavior shows that the design of desirable bulk magnetic effects from molecular principles can be realized.

Another strategy for designing molecular magnetism has involved identification of pure, stable organic radical and polyradical crystalline materials with FM bulk behavior.[5,6] Many organic stable radicals remain to be studied for potential higher T_c magnetism, as well as mixtures of such species.[7]

The design strategy which we and various others have followed is to make large, conjugated polyradicals that have an appropriate structural connectivity to favor very high spin ground state. A recent review by A. Rajca covers many of the general features of this approach.[7] Basically, conjugated radicals tend to have alternating sites of alpha and beta spin density, which assures the most effective valence bonding arrangement; by convention, an excess of alpha spin exists in a radical. Whenever there is an excess of alpha sites, a high spin state is expected by qualitative spin pairing rules.

It is possible to extrapolate the construction of a π-conjugated spin-bearing system, to give a periodically replicated connectivity pattern where each monomeric unit has at least one excess site of alpha spin density.[8–13]

Rajac's group has done excellent work in the area of making purely organic polyradicals that are based upon the triphenylmethyl radical.[14–19] This group has made the largest organic polyradicals achieved by controlled (controlled structure) design, with up to 31 radical sites. Conformational torsion in these dendritic type systems is considerable, but intramolecular FM coupling between spin sites is maintained in smaller members of the series. In the larger systems of this series, magnetic susceptibility studies have suggested that the full expected number of unpaired spins is not observed, due either to conformational isolation of some spin sites, or to incomplete generation of spin sited by the synthetic methodology.[13] Recently, this group has made efforts to avoid problems of conformation by incorporating their triphenylmethyl spin sites into a cyclic unit, giving a large but conformationally constrained spin domain.[13] Despite these fascinating results, intermolecular coupling in neat solid samples of these materials does not give bulk FM behavior. The spin site stability exhibited by the molecules in this strategy is most impressive, however, giving hope for further development along these lines.

Various groups, including ours, have made polyradicals with pendant stable radical groups attached.[20–35]

Our first work in this area involved the development of semiempirical molecular orbital (MO) methods to describe exchange coupling in model open-shell molecules with two spin sites.[36–39]

As part of our computational studies, we expanded the scope of prediction from two spin site models to larger, conjugated polyradical models.[38,39] We identified a number of structural features as being favoring FM coupling of numerous spin sites in a conjugated chain, as well as some features that we feel should be avoided. For example, the carbonyl group, a synthetically tempting unit of many putative conjugated polymers, was found to be a poor exchange linking unit, while both 1,1- and 1,2-ethenediyl are expected to be effective exchange linkers. Experimental evidence of various types turned out to support this rejection of the carbonyl unit.[40,41] We also noted that conformational mobility in pendant polyradicals could be an impediment to extended FM coupling of a large number of spin sites.[38,39] This latter point, as we shall see, turns out to pose a crucial problem in the design of magnetic materials by the polymeric polyradical method.

Few photochemical or thermal reactions allow the generation of reasonably stable radicals, and fewer still are suited to solid-state work.

Our group was the first to devise unimolecular solid-state reactions appropriate to photochemical generation of phenoxyl and aryloxyl radicals. A similar method was reported subsequently.[42] We aimed at generating aryloxyl radicals, since they have been well studied and are known to exhibit substantial delocalization of spin from the exocyclic oxygen. We have developed oxalate-based methods (AOB, DAO) for making aryloxyl radicals in solution or solid-state,[43–45] or even as radical pairs of interacting radicals.[46] Most recently, we have found that bis(aryloxy)phosphine azides (BAPAs) are substantially more photoefficient solid-state sources of phenoxyl radicals.[47] Work is ongoing to improve further the photoefficiency and generality of these reactions, and to extend their use beyond generation of aryloxyl radicals only.

Another promising area of effort is to make conjugated systems with more inherent conjugation than PPhAs. The poly(1,4-phenylene vinylene) (PPV) system is known to be quite planar in the solid state, although some perturbation occurs with substitution on the rings or the ethylenic positions. Computational modeling studies confirm the potential of PPVs for polyradical design.[39] In addition, a very substantial body of synthetic lore exists for making substituted PPVs, including methods for processing soluble precursors into PPVs.

In addition to the above work, our group is also exploring the possibilities of making doped systems with stable radical cations, making pure intermolecularly FM-coupled radicals, and of using inorganic metal ions to link together stable radical sites into one-dimensional chains and two-dimensional sheets.[48] A variety of synthetic methodologies exists to make polyradicals of desired structure. Still, it has proven difficult to synthesize polyradicals that are sufficiently delocalized to provide effective intramolecular exchange in large systems, while remaining reasonably stable at room temperature. In addition, problems with conformation and morphology remain to be confronted, so that one can design intermolecular exchange effects. Since bulk exchange is no stronger than its weakest exchange interaction, the remaining design problems are considerable. Still, problems confronting the use of conjugated systems as semiconductors, batteries, nonlinear optical devices, and light-emitting diodes are gradually being addressed by improved synthetic and processing strategies. The prognosis for molecular magnetic materials, which are in an earlier stage of development than these other materials, remains promising if processibility or controlled morphology strategies can be elucidated.

APPLICATIONS

It is difficult to improve upon the properties of the iron-based magnetic materials presently used in magnetic recording media. This section limits itself to potential applications of molecular magnetic materials, recalling that these materials are not yet well-defined or obviously achievable.

Magnetic materials constructed mostly of organic subunits have the potential to be processible, insulating, and optically transparent.[3] Such properties are not available in present magnetic materials. Such considerations would have to make up for the inherently low magnetic moment per unit weight of mostly organic materials, which would appear to make the organics uncompetitive with existing materials for existing technologies. Many potential applications could involve combinations of magnetic and optical properties, such as optically switchable magnetic properties and switches. New phenomena may be uncovered as more examples of these materials are synthesized. At present, this field of endeavor is of considerably more academic than industrial interest, which could change drastically if the problems of bulk exchange and processing can be overcome in the near future.

ACKNOWLEDGMENTS

This work could not have been done without the hard work and dedication of the members of my research group, whose work is cited in various places. The field of molecular magnetism is a dynamic and still-growing area, and I have found it to consist of scientists both clever and pleasant with whom to correspond and collaborate. I have had the pleasure of meeting and dealing with many of the people whose work I have cited, and I thank them for sharing their ideas with me. Finally, I acknowledge gratefully financial support at various stages by the Office of Naval Research, the National Science Foundation, and Petroleum Research Fund (American Chemical Society), and the University of Massachusetts Graduate School. Partial industrial support was also provided by the Exxon Education Foundation and by Molecular Simulations, Inc.

REFERENCES

1. Kahn, O. *Molecular Magnetism*; VCH: New York, NY, 1993.
2. Miller, J. S.; Dougherty, D. A., Eds., *Mol. Cryst. Liq. Cryst.* **1989**, *176*, 1.
3. *Magnetic Molecular Materials*; Gatteschi, D. et al., Eds.; Kluwer Academic: Dordrecht, The Netherlands, 1991; Vol. 198E.
4. Manriques, J. M. et al. *Science* **1991**, *252*, 1415.
5. Kinoshita, M. In *Magnetic Molecular Materials*; Gatteschi, D. et al., Eds.; Kluwer Academic: Dordrecht, The Netherlands, 1991; p 87.
6. Iwamura, H.; Miller, J. S.; *Mol. Cryst. Liq. Cryst.* **1993**, *232–233*, 1.
7. Izuoka, A. et al. *J. Am. Chem. Soc.* **1994**, *114*, 2609.
8. Mataga, N. *Theor. Chim. Acta* **1968**, *10*, 273.
9. Ovchinnikov, A. A. *Theor. Chim. Acta* **1978**, *47*, 297.
10. Klein, D. J. et al. *J. Chem. Phys.* **1982**, *77*, 3101.
11. Klein, D. J.; Alexander, S. A. In *Graph Theory and Topology in Chemistry*; King, R. B.; Rouvray, D. J., Eds.; Elsevier: The Netherlands, 1987; Vol. 51, p 404.
12. Klein, D. *J. Pure Appl. Chem.* **1983**, *55*, 299.
13. Misurkin, I. A.; Ovchinnikov, A. A. *Russ. Chem. Res. (Engl.)* **1977**, *46*, 967.
14. Rajca, A. *Chem. Rev.* **1994**, *94*, 871.
15. Rajca, A. J. *Am. Chem. Soc.* **1990**, *112*, 5891.
16. Utamapanya, S.; Rajca, A. *J. Am. Chem. Soc.* **1991**, *113*, 9242.
17. Rajca, A.; Utamapanya, S.; Xu, J. *J. Am. Chem. Soc.* **1991**, *113*, 9235.
18. Rajca, A.; Utamapanya, S. *J. Am. Chem. Soc.* **1993**, *115*, 2396.
19. Rajca, A.; Utamapanya, S.; Smithhisler, D. J. *J. Org. Chem.* **1993**, *58* 5650.
20. Keana, J. F. W. *Chem. Rev.* **1978**, *78*, 37.
21. Braun, D.; Wittig, W. *Makromol. Chem.* **1980**, *181*, 557.
22. Alexander, C.; Feast, W. J. *Polym. Bull.* **1991**, *3*, 598.
23. Abdelkader, M.; Drenth, W.; Meijer, E. W. *Chem. Mater.* **1991**, *3*, 598.
24. Upasani, R. B.; Chiang, L. Y.; Goshorn, D. P. *Mater. Res. Soc. Symp. Proc.* **1990**, *173*, 77.
25. Fuji, A. et al. *Macromolecules* **1991**, *24*, 1077.
26. Inoue, K.; Koga, N.; Iwamura, H. *J. Am. Chem. Soc.* **1991**, *113*, 9803.
27. Iwamura, H. et al. In *Magnetic Molecular Materials*; Gatteschi, D. et al., Eds.; Kluwer Academic: Dordrecht, The Netherlands, 1991; p 53.
28. Miura, Y. et al. *J. Polym. Sci., Polym. Chem.* **1992**, *30*, 959.
29. Nishide, H. *Macromolecules* **1988**, *21*, 3120.
30. Nishide, H. *Macromolecules* **1990**, *23*, 4487.
31. Nishide, H. *Mol. Cryst. Liq. Cryst.* **1993**, *232*, 143.
32. Nishide, H. et al. *Macromolecules* **1994**, *27*, 3082.
33. Rossitto, F. C.; Lahti, P. M. *J. Polym. Sci., Polym. Chem.* **1992**, *30*, 1335–45.
34. Rossitto, F. C.; Lahti, P. M. *Macromolecules* **1993**, *26*, 6308.
35. Rossitto, F. C., Ph.D. Thesis, University of Massachusetts, Amherst, MA, 1994.
36. Lahti, P. M.; Rossi, A. R.; Berson, J. A. *J. Am. Chem. Soc.* **1985**, *107*, 2273.
37. Lahti, P. M.; Ichimura, A. S.; Berson, J. A. *J. Org. Chem.* **1989**, *54*, 958.
38. Lahti, P. M.; Ichimura, A. S. *Mol. Cryst. Liq. Cryst.* **1989**, *176*, 125.
39. Lahti, P. M.; Ichimura, A. S. *J. Org. Chem.* **1991**, *56*, 3030.
40. Hiranol, T. et al. *J. Org. Chem.* **1991**, *56*, 1907.
41. Masters, A. P. et al. *J. Am. Chem. Soc.* **1994**, *116*, 2804.
42. Togo, Y.; Nakamujra, N.; Iwamura, H. *Chem. Lett.* **1991**, *3*, 1201.
43. Modarelli, D. A.; Rossitto, F. C.; Lahti, P. M. *Tetrahedron Lett.* **1989**, *30*, 4473.
44. Modarelli, D. A.; Rossitto, F. C.; Lahti, P. M. *Tetrahedron Lett.* **1989**, *30*, 4477.
45. Modarelli, D. A.; Lahti, P. M. *Chem. Commun.* **1990**, 1167.
46. Modarelli, D. A.; George, C.; Lahti, P. M. *J. Am. Chem. Soc.* **1991**, *113*, 6329.
47. Kalgutkar, R. et al. *Tetrahedron Lett.* **1994**, *35*, 3889.
48. Lung, K. et al. *J. Mater. Chem.* **1994**, *4*, 161.

CONNECTIN (Titin) (Large Filamentous Protein)

Satoru Fujime
Graduate School of Integrated Science
Yokohama City University

Although more than 15 years have passed since its discovery, this long and flexible elastic protein still has two names: "connectin" (connecting filament)[1] and "titin" (titanic molecule).[2] Connectin is a large filamentous protein (single peptide) with a molecular mass of ≈3 MDa, a length of ≈1000 nm and a diameter of ≈3 nm, and resides in striated muscle.[3–6] Recently, a fraction of the molecular structure of connectin was revealed; connectin has repeats of 100 amino acids in the primary structure[7,8] and of 4-nm domains in the tertiary structure.[9] The 4-nm repeats form 43-nm superrepeats consisting of I-I-I-II I-I-II-I-I-II, I and II being type III fibronectin and immunoglobulin C_2 motifs, respectively.[8] The three-dimensional structure of both domains I and II are shown to be a typical β-barrel consisting of seven β-strands.[10,11] A connectin molecule is a chain of these 4-nm domains. Connectin links the thick (myosin) filaments to the Z–line in a sarcomere and produces a passive elastic force when muscle fiber is stretched.[9,12,13] By producing this passive force, connectin would maintain the regular structure of sarcomeres centering the A-band at the center of a sarcomere and would support the tension development.[14–16]

Here our description is limited to an elastic property of isolated connectin studied by dynamic light scattering (DLS).[17,18]

PREPARATION OF β-CONNECTIN

The mother molecule, α-connectin (titin 1) always coexists with its proteolytic fragment, β-connectin (titin 2) in a sodium dodecylsulfate (SDS) extract of rabbit, chicken, and frog skeletal muscle. Electron microscopy with an elegant technique[19] gives the contour length L = 1200 nm for α-connectin and L = 900 nm for β-connectin.[6] Isolation of α-connectin and β-connectin is very difficult at present. β-connectin is soluble in 0.6 *M* KCl and is extractable together with myosin. After the precipitation of myosin by dialysis against 0.1 *M* phosphate buffer at pH 7.0, β-connectin can easily be purified by column chromatography.

PROPERTIES

DLS Results

Qualitative examinations strongly suggest that the connectin filament behaves hydrodynamically as a Gaussian coil. A quantitative analysis follows the method detailed in work by Fujime and Higuchi; g^1 (t)'s were theoretically computed for given parameter values of D = 3.60 and $\langle R^2 \rangle$ from 2.50 to 2.80.[18] The constructed $G^2(mt_m(= 1 + |g^1(mt_m)|^2$ were analyzed by the same program as that used in the experiment. The D_{app} versus K relationships thus simulated are compared with the experimental points. The analysis of the contribution to D_{app} from internal modes of motion suggests again $\langle R^2 \rangle$ = 2.60–2.70. From the observed value of D = 3.60 at 10°C, we have the Stokes radius of $R_m = (k_BT/6\pi\eta D)$ = 44.1 nm. On the other hand, the radius of gyration is estimated to be $R_g = (\langle R^2 \rangle/6)^{1/2}$ = 66.6 nm for $\langle R^2 \rangle$ = 2.66. Then, we have the ratio R_g/R_m = 1.51, which size is characteristic of Gaussian coils, that is, $R_g/R_m = (8/3\pi^{1/2})$ = 1.50 from the Kirkwood formula.

Flexural Rigidity

The flexural rigidity of connectin is an order of magnitude smaller than those of collagen, α-helical polypeptides, and DNA.

Extension of End-to-End Distance Under External Force

A force of a few pico-Newtons stretches a single connectin filament to an almost–straightened state.

Extension of Contour Length Under External Force

When the external force is larger than a few pico-Newtons, an extension of the contour length has to be considered.

The elastic constant k (- k′) = 7 dyn/cm (per L_p = 15 nm) for connectin is an order of magnitude smaller than k′ for collagen and PBLG in DMF.

Higuchi et al. have semiquantitatively discussed a relation between a passive force developed by a stretched muscle fiber and a force $X = k\delta$ by connectin filaments.[17] Folding and unfolding of the β-barrel (4–nm domains) might be responsible for the super-high extensibility of a connectin filament.

REFERENCES

1. Maruyama, K. et al. *J. Biochem. (Tokyo)* **1977**, *82*, 317.
2. Wang, K.; McClure, J.; Tu, A. *Proc. Natl. Acad. Sci. USA* **1979**, *76*, 3698.
3. Wang, K. In *Cell and Muscle Motility*; Plenum: New York, 1985; Vol. 6, pp 315–369.
4. Maruyama, K. *Int. Rev. Cytol.* **1986**, *104*, 81.
5. Trinick, J. *Curr. Opin. Cell Biol.* **1991**, *3*, 112.
6. Maruyama, K. *Biophys. Chem.* **1994**, *50*, 73.
7. Labeit, S. et al. *Nature (London)* **1990**, *345*, 273.
8. Labeit, S. et al. *Eur. Mol. Biol. Organ. J.* **1992**, *11*, 1711.
9. Funatsu, T. et al. *J. Cell Biol.* **1993**, *120*, 711.
10. Leahy, D. H. et al. *Science* **1992**, *258*, 987.
11. Schiffer, M. et al. *Biochemistry* **1973**, *12*, 4620.
12. Fanatsu, T.; Higuchi, H.; Ishiwata, S. *J. Cell Biol.* **1990**, *110*, 53.
13. Wang, K. et al. *Proc. Natl. Acad. Sci. USA* **1991**, *88*, 7101.
14. Higuchi, H.; Umazume, Y. *Biophys. J.* **1985**, *48*, 137.
15. Horowits, R.; Podolsky, R. J. *J. Cell Biol.* **1987**, *105*, 2217.
16. Horowits, R.; Maruyama, K.; Podolsky, R. J. *J. Cell Biol.* **1989**, *109*, 2169.
17. Higuchi, H. et al. *Biophys. J.* **1993**, *65*, 1906.
18. Fujime, S.; Higuchi, H. *Macromolecules* **1993**, *26*, 6261.
19. Nave, R.; Furst, D. O.; Weber, K. *J. Cell Biol.* **1989**, *109*, 2177.

Contact Lens Material

See: Contact Lenses, Gas Permeable
Contact Lenses, Hydrogels

CONTACT LENSES, GAS PERMEABLE

Jay F. Künzler
Department of Chemistry and Polymer Development
Bausch and Lomb Incorporated

There exist three classes of contact lens materials: the hard, high modulus contact lens materials, the soft, low modulus hydrogels, and the soft, low modulus contact lens materials that possess little or no water. The main objective of this research is to design contact lens materials for extended wear applications. This chapter will review the currently available high oxygen permeable contact lens materials and provide a review of the most recent advances in the design of contact lens materials possessing high oxygen permeability.

HARD GAS PERMEABLE CONTACT LENSES

Silicone Acrylates

The class of hard contact lenses consists of transparent, crosslinked polymeric materials that possess a high modulus (> 100,000 g/mm^2) and contain less than 1% water.

Despite its success, the PMMA lens suffered from two major disadvantages: poor initial comfort that required lengthy adaptation wearing times and a low permeability to oxygen. PMMA is essentially a barrier to oxygen with a DK of 0.5 barrers and is not suitable for extended wear. A major advance in the design of hard-gas permeable lenses occurred in 1975 with the introduction of the first siloxane based hard contact lens material developed by Gaylord.[1] Gaylord found that by copolymerizing methyl methacrylate with methacrylate functionalized siloxanes, he could blend the excellent stability and processing characteristics of MMA with the high oxygen permeability characteristics of silicone. Silicone based materials, in fact, possess a permeability to oxygen 1000 times higher than that of PMMA. Specifically, Gaylord discovered that by copolymerizing methyl methacrylate with the silicone acrylate, methacryloylpropyl tris(trimethylsiloxy silane) (TRIS) [17096-07-0], together with methacrylic acid [79-41-4] as a wetting agent and a crosslinker such as ethylene glycol dimethacrylate [97-90-5], a high modulus lens material possessing excellent scratch resistance, wettability, dimensional stability, and oxygen permeability could be obtained.

With the success of the Gaylord material, an extensive amount of research continued on the design of hard contact lenses possessing improved wetting, deposit resistance, and increased levels of oxygen permeability. The majority of this work focused on the synthesis of new TRIS-like silicone acrylate derivatives. Several hard lens materials based on TRIS chemistry are now commercially available. The oxygen permeability values for this class of material are in the 10–30 barrers range.

In addition to the TRIS derivatives, an extensive amount of effort into the design of high DK, rigid materials based on methacrylate end capped polydimethylsiloxane telechelics and methacrylate capped siloxane macromers has been reported with no present commercial application.[2–5]

Fluoroacrylates

The design of contact lenses based on fluoro polymers was first reported by Cleaver, Barkdoll, and Girard at Dupont in the late 1960s.[6,7] The copolymerization of fluoromethacrylates, such as hexafluoroisopropylmethacrylate (HFIM), with TRIS and MMA together with suitable wetting agents and crosslinkers, has resulted in a series of new, dimensionally stable, biocompatible, high DK, hard materials possessing DK values in the 30–160 barrers range.[8–11] Several approaches in the design of fluorine-based hard contact lenses have been pursued including the synthesis of polymers based on fluoro-substituted styrenics, fluorosubstituted itaconates, silicone substituted fumarates with fluoromethacrylates, and silicone-fluorine substituted star polymers prepared by group transfer polymerization.[12–15]

Fluoroethers

A relatively new class of high DK hard materials based on perfluorinated polyethers [25322-68-3] has been developed by Rice and co-workers at 3M.[16] These materials are based on copolymers of methacrylate end-capped, low molecular weight perfluorinated polyethers. The copolymerization of the perfluorinated polyethers with a variety of wetting agents and additional crosslinkers results in contact lenses possessing an excellent resistance to deposition, and oxygen permeability levels exceeding 100 barrers.

Polyacetylenes

The preparation of alkyl substituted acetylenes has been extensively studied by the Higashimura and Masuda research group at Kyota University. They have demonstrated that acetylenes containing bulky substituents such as the monomer 1-trimethylsilyl-1-propyne [6224-91-5] can be effectively polymerized to high molecular weight polymers using the metal halide-based initiators. Solution casting from these polymers has resulted in stable. clear films possessing oxygen permeability levels 10 times that of polydimethylsiloxane.[17,18] It has been shown that the high levels of oxygen permeability decrease with time, presumably owing to polymer rearrangement; however, the permeability levels remain high (200 barrers).

Polyimides

Recent studies have shown that fluorinated polyimides based on hexafluoroisopropylidene diaryldianhydride (6FDA) and isopropylidenedianiline exhibit unexpected high levels of oxygen permeability when compared to other high modulus, transparent polymers. Solution cast films from these polymers show DK levels of 5–16 barrers.[19]

HYDROGELS

High Water Content

Hydrogels are hydrophilic polymers that absorb water to an equilibrium value and are insoluble in water owing to the presence of a three-dimensional network. The hydrophilicity is due to the presence of hydrophilic groups, such as alcohols, carboxylic acids, amides, sulfonic acids, etc. The optimum physical and mechanical properties of a hydrogel for contact lens application include a Youngs modulus between 20 g/mm^2 and 200 g/mm^2, a tear strength greater than 2.0 g/mm, water contents in the 35% to 60% range, and a DK greater than 50 barrers.

In the 1960s Wichterle developed the first hydrogel contact lens based on 2-hydroxyethyl methacrylate (HEMA) [868-77-9] which was commercialized in 1971 by Bausch and Lomb.[20] This 38% water-containing hydrogel lens has been extremely successful. Poly(HEMA) possesses excellent wettability, comfort, and deposit resistance. The only significant limitation of poly(HEMA) is that it has a relatively low permeability to oxygen (10 barrers) and is not suited for long-term, extended-wear application.

There have been several approaches to the design of high water-content hydrogels. The first approach has involved the polymerization of highly hydrophilic, non-ionic monomers such as *N*-vinylpyrrolidinone (NVP) [88-12-0], dimethylacrylamide [2680-03-7], poly(vinyl alcohol) [9002-89-5] and glycerol methacrylate [100-92-5].[20–23] The second method involves the copolymerization of moderately hydrophilic monomers, such as HEMA and hydroxy propyl methacrylate [27813-02-1], with highly hydrophilic, ionic monomers such as methacrylic acid.[24] The ionic functionality in a buffered saline environment dramatically increases the water content of the resultant hydrogel.

There are several basic limitations of high water-content hydrogels. The first is that high water-content materials typically possess poor tear strength. Further, high water-content materials often exhibit a high affinity for protein, particularly for hydrogels possessing an ionic functionality and in dry environments, induce epithelial dehydration, which may cause several adverse physiological responses.[25,26]

Silicone-Based Hydrogels

In an attempt to combine the high oxygen permeability of polydimethylsiloxane materials and the excellent comfort, wetting, and deposit resistance of conventional, non-ionic low water hydrogels, the design of silicone-based hydrogels has been studied for contact lens application. The design of silicone based hydrogels has primarily involved the copolymerization of methacrylate or vinyl functionalized polydimethylsiloxanes with hydrophilic monomers. The biggest limitation in the design of silicone hydrogels is that silicone-based monomers are hydrophobic and insoluble in hydrophilic monomers. There have been a variety of approaches to design transparent silicone hydrogels. One approach has been to synthesize siloxanes containing hydrophilic groups incorporated as polymer end caps, blocks, or side chains.[27–29] In addition, the design of silicone hydrogels based on methacrylate, end-capped polyethylene oxide monomers with methacrylate functionalized silicones has also been pursued.[30,31] Other, more recent, approaches have been the design of hydrogels based on silicone-functionalized urethane prepolymers, methacrylate end-capped macromers, and vinyl functionalized TRIS derivatives, where compatibility in these systems is achieved either by incorporation of a hydrophilic group or by the use of a solubilizing agent.[5,32–37] Finally, the compatibility of hydrophilic monomers with siloxane monomers has been achieved by the trimethylsilyl (TMS) protection of hydrophilic monomers such as HEMA.[39] The TMS protected HEMA is soluble in all proportions with siloxane containing monomers.

Fluorohydrogels

Another approach to prepare oxygen permeable hydrogels has been the design of hydrogels based on fluoropolymers. The preparation of oxygen permeable hydrogels by the copolymerization of fluorinated methacrylates and methacrylate functionalized fluorinated polyethylene oxides with hydrophilic monomers has been reported.[40–47] The oxygen permeability of these hydrogels is in the 30–50 barrers range, significantly less than the silicone based hydrogels; however, it is claimed that the wetting and biocompatibility characteristics of these silicone-free fluorohydrogels are superior to the silicone-based hydrogels.

ELASTOMERS

Low-Water Soft Silicone

The design of low-water (<1%) silicone-based materials has been extensively studied for the past 30 years with limited success. The first series of silicone-based, low-water elastomers was composed of high molecular weight vinyl and hydride functionalized polysiloxanes. These systems were polymerized by a platinum-initiated hydrosilation mechanism, and wettability was achieved by plasma treatment.[48,49] Initial results from the silicone-based elastomers were mixed. Clinical results showed that there was no change in corneal physiology with long-term wear (owing to the extremely high oxygen permeability of these materials (500 barrers); however, the silicone-based elastomer lenses exhibited a high affinity for lipids, poor wetting, and lens adhesion.[50,51]

Recent work has focused on the design of low-water silicone elastomers using methacrylate functionalized siloxanes. Much of the chemistry described for the design of silicone hydrogels has been applied to the design of low water silicone-based lenses.[27–39,52–56]

Poly(ethylene oxide)-Based Elastomers

The design of low-water content, high-oxygen permeable contact lens materials based on poly(ethylene oxide) polymers has been reported by several research groups. These approaches have involved the preparation of methacrylate end-capped polyurethanes containing a poly(ethylene oxide) and propylene oxide soft segments.[57] In addition, the preparation of contact lens materials based on sugars containing methacrylate-functionalized polyethylene oxide grafts has also been reported.[58]

Polyphosphazenes

An extensive amount of research on the design of polyorganophosphazenes membranes for use as gas separation membranes and biomedical polymers has been reported by several research groups.[59,60] The reported oxygen permeability of polytrifluoroethoxyphosphazene [28212-50-2] is 40 barrers and poly(*n*-butyl phosphazene)s is 128 barrers. The major limitation of polyphosphazenes for use as contact lens materials is the difficulty in polymer synthesis and polymer processing. Solution casting is presently the only viable method to obtain shaped bodies.

REFERENCES

1. Gaylord, N. G. U.S. Patent 3 808 178, 1974.
2. Deichert, W. G.; Su, K. C.; Van Buren, M. F. U.S. Patent 4 153 641, 1979.
3. Friends, G. D.; Melpolder, J. B.; Kunzler, J. F.; Park, J. S. U.S. Patent 4 495 361, 1985.

4. Bany, S.; Koshar, R.; Williams, T. U.S. Patent 4 543 398, 1985.
5. Mueller, K. F. U.S. Patent 5 079 319, 1992.
6. Barkdoll, E. U.S. Patent 3 940 207, 1976.
7. Cleaver, C. S. U.S. Patent 3 950 315, 1976.
8. Kawamura, K.; Yamashita, S.; Yokoyama, Y.; Tsuchiya, M. U.S. Patent 4 540 761, 1985.
9. Yamamoto, F.; Suzuki, T.; Ikari, M.; Saito, S.; Ohmori, A.; Yasuhara, T. U.S. Patent 4 684 705, 1987.
10. Stoyan, N. U.S. Patent 4 829 137, 1989.
11. Kossmehl, G.; Fluthwedel, A. *Makromol. Chem.* **1992**, *193*, 157.
12. Falcetta, J. J.; Park, J. S. U.S. Patent 4 690 993, 1987.
13. Ellis, E. J.; Ellis, J. Y. U.S. Patent 4 686 267, 1987.
14. Kawaguchi, T.; Ando, I.; Toyoshima, N.; Yamamoto, Y.; Yoshioka, H.; Yamazaki, T. U.S. Patent 5 250 583, 1993.
15. Seidner, L.; Spinelli, H.; Ali, M.; Weintrab, L. U.S. Patent 5 331 067, 1994.
16. Rice, D.; Ihlenfeld, J. V. U.S. Patent 4 440 918, 1984.
17. Masuda, T.; Isobe, E.; Higashimura, T.; Takada, K. *J. Am. Chem. Soc.* **1983**, *105*, 7473.
18. Masuda, T.; Hamano, T.; Tsuchihara, K.; Higashimura, T. *Macromolecules* **1990**, *23*, 1374.
19. Coleman, M. R.; Koros, W. J. *J. Membrane Sci.* **1990**, *50*, 285.
20. Wichterle, O.; Lim, D. U.S. Patent 3 220 960, 1965.
21. Mancini, W.; Korb, D.; Refojo, M. F. U.S. Patent 3 957 362, 1976.
22. Izumitani, T.; Tarumi, N.; Komiya, S.; Sawamoto, T. U.S. Patent 4 625 009, 1986.
23. Ofstead, R. F. U.S. Patent 4 618 649, 1986.
24. Steckler, R. U.S. Patent 4 036 788, 1977.
25. Minarik, L.; Rapp, J. *Contact Lenses* (*CLOA Journal*) **1989**, *15*, 185.
26. Orsborn, G.; Zantos, S. *Contact Lenses* (*CLOA Journal*) **1988**, *14*, 81.
27. Mueller, K. F.; Kleiner, E. K. U.S. Patent 4 136 250, 1979.
28. Su, K.; Robertson, J. R. U.S. Patent 4 740 533, 1988.
29. Keogh, P. L.; Kunzler, J. F.; Niu, G. C. U.S. Patent 4 260 725, 1981.
30. Lai, Y. C.; Baccei, L. J. *J. Appl. Polym. Sci.* **1991**, *42*, 2039.
31. Chang, S. H.; U.S. Patent 4 182 822, 1980.
32. Harvey, T. B. U.S. Patent 4 711 943, 1987.
33. Tanaka, K.; Takahashi, K.; Kanada, M.; Kanome, S.; Nakajima, T. U.S. Patent 4 139 513, 1979.
34. Ratkowski, D. J.; Lue, P. C. U.S. Patent 4 535 138, 1985.
35. Bambury, R. E.; Seelye, D. E. Eur. Patent 396 364, 1990.
36. Mueller, K. F.; Plankl, W. L. U.S. Patent 5 115 056, 1992.
37. Braatz, J.; Kehr, C. U.S. Patent 4 886 866, 1989.
38. Kunzler, J. F.; Ozark, R. E. U.S. Patent 5 321 108, 1994.
39. Yoshikawa, T.; Shibata, T. U.S. Patent 4 649 184, 1987.
40. Sawamoto, T.; Nomura, M.; Tarumi, N. U.S. Patent 5 008 354, 1991.
41. Mueller K. F. U.S. Patent 4 954 587, 1990.
42. Goldenberg, M. U.S. Patent 4 929 692, 1990.
43. Mueller, K. F. U.S. Patent 5 011 275, 1991.
44. Agou, T.; Sakashita, T.; Shimoda, T.; Sudo, M.; Kuwabara, M.; Tanaka, M. U.S. Patent 5 057 585, 1991.
45. Salamone, J. C. U.S. Patent 4 990 582, 1991.
46. Kossmehl, J.; Volkheimer, J.; Schaefer, H. *Acta Polymer* **1992**, *43*, 335.
47. Futamura, H.; Nomura, M.; Yokoyama, Y. U.S. Patent 5 264 465, 1993.
48. Becker, W. E. U.S. Patent 3 228 741, 1966.
49. Burdick, D. F.; Mishler, J. L.; Polmanteer, K. E. U.S. Patent 3 341 490, 1967.
50. Sweeney, D. F.; Holden, B. A. *Curr. Eye Res.* **1987**, *6*, 139.
51. Josephson, J.; Caffery, B. *Intern. Contact Lens Clinic* **1980**, *7*, 235; Mountford, J. *Aust. J. Optom.* **1978**, *61*, 197.
52. Friends, G. D.; Van Buren, M. F. U.S. Patent 4 254 248, 1981.
53. Chromecek, R. C.; Deichert, W. G.; Falcetta, J. J.; Van Buren, M. F. U.S. Patent 4 276 402, 1981.
54. Keogh, P. L.; Kunzler, J. F.; Niu, G. C. U.S. Patent 4 259 467, 1981.
55. Nakashima, T.; Taniyama, Y.; Sugiyama, A. U.S. Patent 4 814 402, 1989.
56. Schaefer, H.; Kossmehl, G.; Neumann, W. U.S. Patent 4 853 453, 1989.
57. Su, K.; Molock, F. F. U.S. Patent 4 780 488, 1988.
58. Nunez, I. M.; Ford, J. D. U.S. Patent 5 196 458, 1993.
59. Hirose, T.; Mizogucho, K. *J. Appl. Polym. Sci.* **1991**, *43*, 891.
60. Lora, S.; Palma, G.; Bozio, R.; Caliceti, P.; Pezzin, G. *Biomaterials* **1993**, *14*, 430.

CONTACT LENSES, HYDROGELS

Miguel F. Refojo
The Schepens Eye Research Institute
Department of Ophthalmology
Harvard Medical School

Corneal contact lenses are optical devices that are placed on the eye to correct ametropia. the material of construction of a contact lens determines the fit, feel, and physiological interaction of the lens with the eye.[1] Of all the diverse contact lenses on the market, the hydrogel lenses, also called soft and hydrophilic contact lenses, are the most comfortable and most commonly used. The patent literature on contact lenses was reviewed by Refojo, Tighe, and Lai et al.[2–4]

PROPERTIES

Relevant properties required for hydrogel contact lens materials are optimal visible light transmission, sufficient mechanical strength to maintain physical integrity in normal use, oxygen and carbon dioxide permeability (ideally commensurate to the requirement of corneal metabolism), uniform wettability by tears, water retention so that the lens does not dry on the eye, resistance to fouling by deposit formation from the components of tears, as well as resistance to microorganism-binding that may lead to corneal infection. Contact lens transparency and mechanical properties that are adequate for normal use of a lens are obvious and need no further discussion. Improvement in oxygen transmissibility and water retention, as well as decrease fouling and microorganism binding offers the greatest opportunities for the development of newer and better hydrogel contact lenses.

The cornea is an avascular tissue that is rich in cells that need oxygen from the air to maintain their normal metabolism. High oxygen permeability is a property that is desirable for all contact lens materials so that they do not interfere with the normal metabolism of the cornea. The oxidative metabolism of the cornea cells results in production of carbon dioxide that is normally released to the atmosphere and to the anterior chamber of the eye. Accumulation of carbon dioxide on the cornea under a contact lens results in a shift to a non-physiological acidic pH. Therefore, ideally a contact lens must not interfere with the passage of oxygen from the air to the cornea, and with the release of carbon dioxide to the air. The oxygen permeability coefficient of a contact lens material is given by P = D·k, where D is the diffusion coefficient in cm^2/sec, and k is the Henry's Law solubility coefficient in cm^3O_2 (STP)/cm^3·mmHg. The oxygen permeability coefficient of contact lenses is conventionally expressed in the contact lens literature as Dk in barrer units [1 barrer = 1×10^{-11} cm^3O_2 (STP) · cm^2 /cm^3 · sec · mmHg].

The logarithm of the oxygen permeability coefficient of currently available hydrogel contact lens materials with at least 38–40% hydration, is linearly related to the water content of the hydrogel, and extrapolates to the oxygen permeability of water.[5] The oxygenation of the cornea under a given lens is determined by the lens-oxygen transmissibility, Dk/L, where L is the lens thickness in cm. Therefore, to enhance the oxygen transmissibility contact lenses are preferably made as thin as possible. For daily-wear contact lenses that are removed from the eye during sleep, it has been estimated that to avoid corneal hypoxia the oxygen transmissibility of the lenses should be at least 34×10^{-9} cm^3O_2 (STP) · cm/cm^3 · sec·mmHg, and for extended (day and night) wear should be at least 70 to 90 × 10^{-9} cm^3O_2 (STP) · cm/cm^3 · sec · mmHg.[6]

The healthy cornea needs to be continuously covered by the tear film, which consists essentially of an inner mucous layer, a main aqueous layer, and an outer lipid layer, The tear film acts as the lubricant for blinking, as a defensive barrier for microorganisms and other foreign irritants to the eye, and to protect the cornea from drying. Normally, tears from tear glands circulate over the cornea and for the most part drain out of the eye through the tear ducts to the nasal cavity. However, depending on the atmospheric conditions, a substantial portion of the water from the tear film evaporates. Normally, tear evaporation is retarded by the oily layer on the tear film. However, when a contact lens is placed on the eye the normal structure of the tear film is disrupted and water evaporation is enhanced. With a hydrogel contact lens some water evaporates also from the hydrogel itself.

The surface properties of contact lenses are also of the utmost importance because they determine not only comfort, but lens fouling and spoliation. Coatings and deposits on the lenses are formed by the lipids, proteins, and mucus in the tears, and also by external contaminants transferred to the lens from the eye environment or by handling. Contact lens fouling can result in immunological and mechanical irritation of the eye and lids. Contact lenses can also be colonized by microorganisms.

MATERIALS

The contact lens field, previously consisting only of rigid, non-gas permeable lenses made of poly(methyl methacrylate)[PMMA], (CAS-9011-14-7), was revolutionized in the 1960s and early 1970s by the work of the Czech chemists Wichterle and Lim on the hydrogels of the monomethacrylates of ethylene glycol [2-hydroxyethyl methacrylate (HEMA), (CAS-868-77-9)] and its derivatives.[7] Their work resulted in the first commercial hydrogel contact lenses made of poly(2-hydroxyethyl methacrylate) [PHEMA] crosslinked with ethylene glycol dimethacrylate (EGDMA) (CAS-97-90-5).

Optical homogeneous PHEMA hydrogels, owing to the relative low hydrophilicity of PHEMA, have a maximum hydration of about 40% H_2O at equilibrium in water. The early PHEMA lenses were too thick to fulfill the oxygen requirement of the cornea. This was ameliorated by the development of ultrathin lenses, and of newer hydrogels with higher hydration.

It is noteworthy the contribution of Steckler to the use of *N*-vinylpyrrolidone, NVP, (CAS 88-12-0) in the development of the highly hydrated hydrogels contact lenses.[8] To this end, Steckler copolymerized and simultaneously crosslinked a combination of a hydrophilic and a hydrophobic monomer (alkyl acrylate or methacrylate) that resulted in the first high-hydration contact lenses.

Another monomer used to make hydrogel contact lenses is glyceryl methacrylate (GMA, 2,3-dihydroxypropyl methacrylate (CAS-100-92-5)).

Currently the principal hydrophilic monomers used in the manufacture of contact lenses are HEMA, MAA, MMA, NVP, GMA, diacetone acrylamide, and some other monomers used as minor ingredients in the respective formulations. As crosslinking agents, EGDMA is a commonly used one, although several others difunctional or trifunctional methacrylates, allyl methacrylate or divinyl benzene, are used in various formulations.

CLASSIFICATION OF HYDROGEL CONTACT LENSES

Contact lenses are also classified according to the intended use, as daily wear (removed during sleeping time) and extended wear (continuous day and night wear, from 24 hours to one week or more at a time). Disposable lenses can be daily wear, that are discharged after using them during the day for one or more days, with cleaning and sterilization in between uses, and extended wear, which are discharged after about one week of continuous day and night wear. Although highly hydrated materials and very thin lenses are more indicated for continuous lens wear, than the lower hydrated materials or thicker lenses, commercial considerations have resulted in the availability of daily or extended wear lenses of materials from any of the FDA group classifications.

The FDA has classified hydrogel contact lenses into four groups: Group I: nonionic, low-water-content (38–50% water), Group II: nonionic, high-water-content (51–80% water), Group III: ionic, low-water-content (hydration as in Group I), and Group IV: ionic, high-water-content (hydration as in Group II).

RESEARCH MATERIALS FOR HYDROGEL CONTACT LENSES OF ENHANCED OXYGEN PERMEABILITY

As mentioned above, the oxygen permeability coefficient of the standard hydrogel contact lens materials is directly related to their hydration, regardless of their chemical composition. The trend is now toward the development of hydrogels in which the polymer phase will enhance their oxygen permeability compared to standard hydrogels of similar hydration. The challenge to the chemists is to make homogeneous hydrogels by copolymerizing highly hydrophobic monomers with siloxane- and perfluoro-groups that enhance gas permeability, with hydrophilic commoners for water absorption. These hydrogels would have rigid carbon-to-carbon backbones, and the enhanced oxygen permeability would depend, in addition to the water content of the hydrogel, on the bulky siloxane groups, and to a lesser extent on the perfluoro alkyl moieties.

A more likely approach to obtained hydrogels of enhanced oxygen permeability is by the copolymerization of the hydrophilic monomers, with one of several convenient telechelic oxygen-permeable dimethacrylate-end capped polysiloxanes, that is, α,ω-bis(methacryloxypropyl) polydimethylsiloxane.[9]

REFERENCES

1. Ruben, M.; Guillon, M. Eds. *Contact Lens Practice*; Chapman & Hall Medical: London, 1994.
2. Refojo, M. F. *Kirk-Othmer Encyclopedia of Chemical Technology*; 3rd ed.; John Wiley & Sons: New York, NY, 1979; Vol. 6, 720–42. Contact Lenses.
3. Tighe, B. *Optician. Hydrogel Materials: The Patents and the Products* June 2, 17(1989) and July 7, 17(1989).
4. Lai, Y.-C.; Wilson, A. C.; Zantos, S. G. *Kirk-Othmer Encyclopedia of Chemical Technology*, 4th ed.; Contact Lenses. John Wiley & Sons: New York, NY, 1993; 7, 192–218.
5. Refojo, M. F.; Leong, F.-L. *J. Membrane Sci.* **1979**, 4, 415.
6. Holden, B. A. *Optometry and Vis. Sci.* **1989**, 66, 717.
7. Wichterle, O.; Lim, D. *Nature* **1961**, 185, 117.
8. Steckler, R. U.S. Patent 3 532 679, 1970.
9. Künzler, J.; Ozark, R. *Poly. Mat. Sci. and Eng.* **1993**, *69*, 226.

CONTROLLED DRUG DELIVERY SYSTEMS

Joseph Kost
Department of Chemical Engineering
Ben Gurion University

In this chapter the rationale for using polymers in drug delivery systems is discussed and the potential application are presented.

A new development, controlled drug delivery, has evolved from the need for prolonged and better control of drug administration. IN conventional drug delivery modes such as a spray, an injection, or the taking of a pill, each time a person takes medicine, the drug concentration in the blood rises when the drug is taken, then peaks and declines. Because each drug has a plasma level above which it is toxic and below which it is ineffective, the plasma drug concentration in a patient at a particular time depends on compliance with the prescribed routine. The goal of the controlled release devices, which are already available commercially is to maintain the drug in the desired therapeutic range with just a single dose, localize delivery of the drug to a particular body compartment, which lowers the systemic drug level, reduces the need for follow-up care, preserves medications that are rapidly destroyed by the body, and increases patient comfort and/or improves compliance. In general, release rates are determined by the design of the system and are nearly independent of environmental condition.[1]

CLASSIFICATION OF POLYMERIC DRUG DELIVERY SYSTEMS

A convenient classification of controlled-release systems is based on the mechanism that controls the release of the substance in question.

Diffusion-Controlled Systems

The most common mechanism is diffusion. Two types of diffusion-controlled systems have been developed; the first is a reservoir device in which the bioactive agent (drug) forms a core surrounded by an inert diffusion barrier. These systems include membranes, capsules, microcapsules, liposomes, and hollow fibers. The second type is a monolithic device in which the active agent is dispersed or dissolved in an inert polymer. As in reservoir systems, drug diffusion through the polymer matrix is the rate-limiting step, and release rates are determined by the choice of polymer and its consequent effect on the diffusion and partition coefficient of the drug to be released.[1,2]

Chemically Controlled Systems

Chemical control can be achieved using bioerodible or pendant chain systems. The rationale for using bioerodible (or biodegradable) system is that the bioerodible devices are eventually absorbed by the body and thus need not be removed surgically. In a pendant chain system, the drug is covalently bound to the polymer and is released by bond scission owing to water or enzymes.[3,4]

Solvent Activated-Controlled Systems

In these systems the active agent is dissolved or dispersed within a polymeric matrix and is not able to diffuse through that matrix. In one type of solvent-controlled system, as the environmental fluid (e.g., water) penetrates the matrix, the polymer swells and its glass transition temperature is lowered below the environmental (host) temperature.[5] Thus, the swollen polymer is in a rubbery state and it allows the drug contained in it to diffuse through the polymer.

POLYMERS

The polymers used in controlled release systems can be classified to hydrophobic, hydrophilic, biodegradable, and nondegradable. The decision to apply polymer from one of these groups depends upon the route of administration, the type and amount of drug required, and the duration of release required.

The route of administration is a major consideration in the type of polymer selected for the controlled delivery system. For instance, if the drug is unstable in the presence of gastric or intestinal fluids as are many proteins, or if it is quickly metabolized during passage through the liver, then it will have to be delivered parentally or transdermally.

The nature of the drug being delivered also influences the selection of the polymer used. The polymer has to be compatible with the drug and has to be one that does not lead to reaction that will alter either the drug or the polymer. Also, if the drug is unstable under aqueous environment, a more hydrophobic polymer will be preferred.

Nondegradable Hydrophobic Polymers

One of the most common hydrophobic polymers is silicone rubber. Silicones are excellent materials for drug delivery because of their biocompatibility, their ease of fabrication, and high permeability to many drugs. As a result, a number of drug delivery products employing silicones have been commercialized including implants for contraceptive steroids and transdermal patches.

Another widely used nondegradable polymer is ethylenevinyl acetate copolymer. This copolymer also displays excellent biocompatibility, physical stability, biological inertness, and processability. In drug delivery application these copolymers usually contain 30–50 wt% vinyl acetate.

Biodegradable Polymers

Biodegradable polymers in controlled delivery systems have an advantage over other systems in obviating the need to surgically remove the drug-depleted device. In many cases, however, the release is augmented by diffusion through the matrix, rendering the process difficult to control, particularly if the matrix is hydrophilic and thereby absorbs water, promoting degradation in the interior of the matrix. To maximize control over the release process, it is desirable to have a polymeric system that degrades only from the surface and defers the permeation of the drug molecules.

Polyesters

The poly(α-ester)s (ester polymers prepared from α-hydroxy acids), especially poly(lactic acid) (PLA), poly(glycolic acid) (PGA) and their copolymers (PLGA) are the most widely used implantable degradable polymers.

Another polyester, poly(ε-caprolactone), hydrolyzes to 6-hydroxyhexanoic acid.[6] The homopolymer is less biodegradable than PLA and hence is more suitable for long term delivery of drugs. However, its *in vivo* biodegradation can be accelerated by copolymerization with other esters. The polymer is permeable to lipophilic drugs and is therefore useful for the release of steroid drugs. This high permeability of poly(ε-caprolactone) has permitted the development of biodegradable reservoir devices that degrade completely after the drug is depleted.

Poly(ortho esters) were developed to obtain surface-degrading systems. Such systems would result in zero order (constant) drug release rates.

The poly(ortho esters) were evaluated in the developments of bioerodible implants for the release of numerous biological agents such as contraceptives, and cytotoxic and narcotic antagonists.[7]

Polyanhydrides

These polymers are generally synthesized by polycondensation; among the most widely used polyanhydrides are copolymers of sebacic acid (SA) and 1,3-bis(carboxyphenoxy)propane (CPP).

By varying the ratio of the hydrophobic moiety (CPP) and (SA), controlled degradation rates, from days to years, have been achieved for millimeter-thick discs.[8,9]

Polyphosphazenes

Polyphosphazenes consist of a chain of alternating phosphorous and nitrogen atoms, with two side groups attached to each phosphorous.

The physical and chemical properties of these polymers can be controlled by changing the side groups attached. One of the major potentials of this class of polymers is to prepare polymeric drug or polymeric prodrugs by attaching drugs as side chains to the polymer backbone.[10]

Pseudopoly(amino acid)s

The rationale for synthesizing polymers based on amino acids was that these, when degraded, would result in naturally occurring metabolites. L amino acids or dipeptides were polymerized through nonamino (e.g., ester) iminocarbonate bonds, enabling the creation of poly(amino acid)s with tailor-made physical and pharmacological properties.[11]

This approach permits the synthesis of biomaterials that are derived from nontoxic substances, which also have other desirable properties.

HYDROGELS

Hydrogels have special advantages over other synthetic polymers, particularly when the hydrogel and biological surface are in close contact. They resemble in many physical properties biological tissues, having a high water content, being porous and flexible, thus reducing possible injury to the tissue. The nature of the hydrogel-water interface is such that the interfacial tension is low and adsorption onto hydrogel surface is minimized. These properties were the incentive for the frequent use of hydrogels as reservoir and matrix type diffusion-controlled delivery systems.[12–19] The parameters that affect the permeability of drug and hence its release rate are polymer composition, type of releasing agent, water content in hydrogel, and crosslinking agent, water content in hydrogel, and crosslinking density.

Biodegradable Hydrogels Based on Synthetic Polymers

A more desirable hydrogel delivery system that has a high crosslink density so that diffusional release is inhibited and rate of release is mainly controlled by hydrolysis of the hydrogel, was proposed by Heller et al., who synthesized hydrogels by crosslinking a water-soluble polyester from fumaric acid and poly(ethylene glycol) with *N*-vinylpyrrolidone.[20] Chain cleavage takes place by a hydrolysis of ester links with consequent generation of poly(ethylene glycol) and a poly(*N*-vinylpyrrolidone) modified by vicinal carboxylic acid functions.

Biodegradable Hydrogels Based on Natural Polymers

Natural polymers remain attractive primarily because they are natural products of living organisms, readily available, relatively inexpensive, and capable of a multitude of chemical modifications.[21] A majority of investigations of natural polymers as matrices in drug delivery systems have centered on proteins (e.g., collagen, gelatin, and albumin) and polysaccharides (e.g., starch, amylose, dextran, inulin, cellulose, chitin, and hyaluronic acid).

REFERENCES

1. Langer, R. New methods of drug delivery. *Science* **1990**, *249*, 1527–1533.
2. Kost, J.; Langer, R. Controlled release of bioactive agents. *Trends in Biotechnology* **1984**, *2*, 47–51.
3. Heller, J.; Sparer, R. V.; Zenter, G. M. Poly(ortho esters) In *Biodegradable polymers as drug delivery systems*; Chasin, M.; Langer, R., Eds.; Marcel Dekker: New York, 1990.

4. Ron, E.; Langer, R. Erodible systems. 199–224. In *Treatise on controlled drug delivery*; Kydonieus, A., Ed.; Marcel Dekker: New York, 1992.
5. Hopfenberg, H.; Hsu, K. C. Swelling controlled, constant rate delivery systems. *Polymer Eng. Sci.* **1981**, *18*, 18.
6. Pitt, C. G.; Gratzl, M. M.; Kimmel, G. L.; Surles, J.; Schindler, A. Aliphatic Polyesters. II. The degradation of poly(DL-Lactide), poly(ε-caprolactone), and their copolymers in vivo. *Biomaterials* **1981**, *2*, 215.
7. Heller, J.; Maa, Y. F.; Wutrhich, P.; Ng, S. Y.; Duncan, R. Recent developments in the synthesis and utilization of poly(ortho esters). *J. Control. Rel.* **1991**, *16*, 3–14.
8. Leong, K. W.; Brott, B. C.; Langer, R. Bioerodible polyanhydrides as drug-carrier matrices. I. Characterization, degradation and release characteristics. *J. Biomed. Mater. Res.* **1985**, *19*, 941–955.
9. Leong, K. W.; Simonte, V.; Langer, R. Synthesis of polyanhydrides: Melt-polycondensation, dehydrochlorination, and dehydrative coupling, *Macromolecules* **1987**, *20*, 705–712.
10. Allcock, H. R. Polyphosphazenes as new biomedical and bioactive materials. In *Biodegradable Polymers as Drug Delivery Systems*; Chasin, M.; Langer, R., Eds.; Marcel Dekker: New York, NY, 1990; 45, 163–193.
11. Kohn, J.; Langer, R. Polymerization reactions involving the side chains of X-L-Amino acids, *J. Am. Chem. Soc.* **1987**, *109*, 817–820.
12. Peppas, N. A.; Khare, A. R. Preparation, structure and diffusion behavior of hydrogels in controlled release. *Advanced Drug Delivery Rev.* **1993**, *11*, 1–35.
13. Peppas, N. A.; Korsmeyer, R. W. Dynamically swelling hydrogels in controlled release applications. 109–135. In *Hydrogels in Medicine*; Peppas, N., Ed.; CRC: Boca Raton, 1987.
14. Kost, J.; Langer, R. Equilibrium swollen hydrogels in controlled release applications. 95–108. In *Hydrogels in Medicine and Pharmacy*; Peppas, N., Ed.; CRC: Boca Raton, 1987.
15. Gehrke, S. H.; Lee, P. L. Hydrogels for drug delivery systems. 333–392. In *Specialized Drug Delivery Systems*; Tyle, P., Ed.; Marcel Dekker: New York, 1990.
16. Colombo, P. Swelling-controlled release in hydrogel matrices for oral route. *Advanced Drug Delivery Rev.* **1993**, *11*, 37–58.
17. Kamath, K. R.; Park, K. Biodegradable hydrogels in drug delivery. *Advanced Drug Delivery Rev.* **1993**, *11*, 59–84.
18. Yoshida, R.; Sakai, K.; Okano, T.; Sakurai, Y. Pulsatile drug delivery systems using hydrogels. *Advanced Drug Delivery Rev.* **1993**, *11*, 85–108.
19. Base, Y. H.; Kim, S. W. Hydrogel delivery systems based on polymer blends, block co-polymers or interpenetrating networks. *Advanced Drug Delivery Rev.* **1993**, *11*, 109–135.
20. Heller, J.; Helwing, R. F.; Baker, R. W.; Tuttle, M. E. Controlled release of water-soluble macromolecules from bioerodible hydrogels. *Biomaterials* **1983**, *4*, 262–266.
21. Bogdansky, S. Natural polymers as drug delivery systems, In *Biodegradable Polymers as Drug Delivery Systems*; Chasin, M.; Langer, R., Eds.; Marcel Dekker: New York, 1990; 231–259.

Coordination Polymerization

See: Living Coordination Polymerization
Ring-Opening Coordination Polymerization (by Soluble Multinuclear Aldoxides)

COORDINATION POLYMERS, DITHIOOXAMIDES

Anjali Bajpai*
Department of Chemistry
Government Autonomous Science College

Milind Khandwe and U.D.N. Bajpai
Polymer Research Laboratory
Department Studies of Post Graduate and Research in Chemistry
Rani Durgavati University

The metal complexes of dithiooxamide (rubeanic acid) and its derivatives have been known for a long time. Some of their applications include pigments, duplicating inks, vulcanization accelerators, plant growth regulators, rodent and bird repellents, bacterial inhibitors, organic intermediates, and analytical reagents.

In the 1980s the coordination compounds of dithiooxamide were of interest from the point of view of their electronic spectra, magnetic measurement, crystal structure determination, and synthesis of mixed ligand complexes.[1–6] In an acidic medium, dithiooxamide forms low molecular weight mononuclear complexes that are soluble in common solvents, and their full characterization is possible. In a basic medium (e.g., ammoniacal acetone, ethanol, and methanol), dithiooxamide forms polymeric complexes or coordination polymers with 1:1 stoichiometry.[7–9] Some of these coordination polymers are electrically conducting.[10,11] Here, we report on the synthesis, characterization, and possible applications of a new class of coordination polymers of Co(II), Ni(II), Cu(II), Zn(II), Cd(II), and Hg(II) with the following polythiooxamides: poly(ethane thiooxamide) (PETO), poly(butane thiooxamide) (PBTO) (CAS Registry No.: 134847-47-5), and poly(hexane thiooxamide) (PHTO).

INFRARED SPECTRAL STUDIES

The assignments of the IR spectral bands for the ligands and coordination polymers are based on data for dithiooxamide and related compounds and on typical bands generally observed for a *trans*-thioamide group.[7,8,12–17] Coordination polymers were nonelectrolytic in nature and did not respond to the qualitative tests for chloride and acetate ions.

In dithiooxamide there is a great deal of electron delocalization over the thioamide group, and it exists in the tautomeric equilibrium (**Equation 1**). Dithiooxamides are capable of coordination through S and N or S alone. Because the thioamide groups have replaceable hydrogen atoms, metal can form simple salt linkages with nitrogen (through **Structure I**) or sulfur (through **Structure II**).

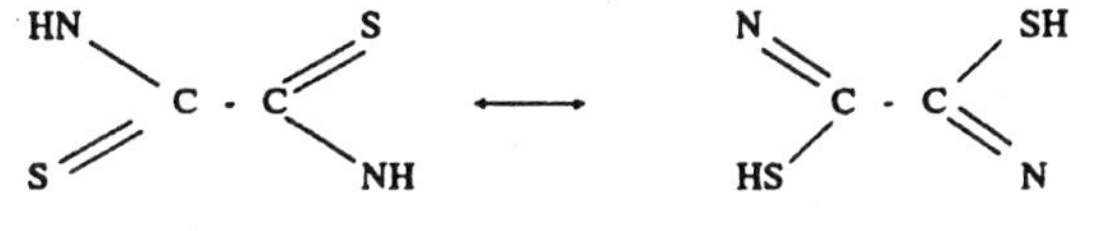

I   II

*Author to whom correspondence should be addressed.

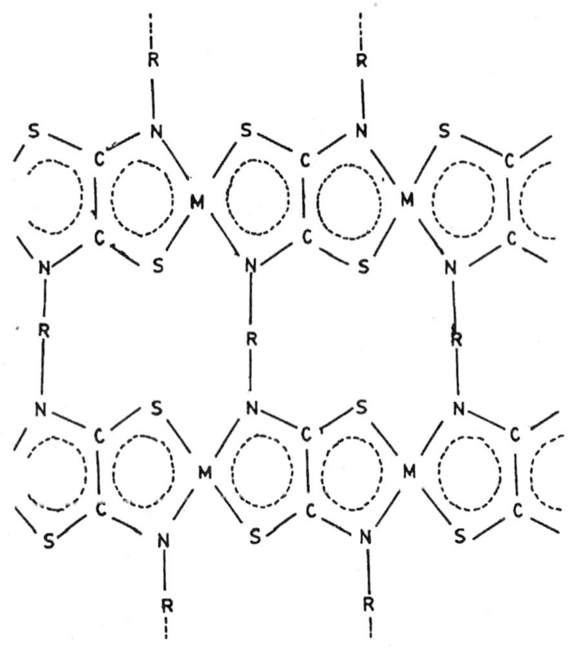

FIGURE 1. Structure representing electron delocalization within chelate: **R** = $(CH_2)_2$, $(CH_2)_4$, $(CH_2)_6$ **M** = Co(II), Ni(II), Cu(II).

ELECTRONIC SPECTRAL STUDIES

Electronic spectra of polymeric and nonpolymeric complexes of dithiooxamides have been extensively studied.[18–20] Because dithiooxamide is used as a reagent for the spectroscopic determination of several metal ions, its electronic transitions are of much interest. In the study of nonpolymeric complexes of dithiooxamide, characteristic assignments were made for the electronic transitions.[2] However, the electronic spectra of polymeric dithiooxamide complexes could not give clear ideas about the geometry of ligand atoms around the metal atom.[18,20] The reasons given for this ambiguous nature were insolubility and structural defects of polymeric complexes. The geometry of polymeric complexes of dithiooxamides can be predicted only on the basis of their colors.

These observations about the geometry of unit structure are very tentative and cannot be explained in the same fashion as for the mononuclear low molecular weight complexes. This suggests that the absorption bands in electronic spectra may be responsible for the transition of electrons within the macromolecular system and not specifically within the electronic levels of metal ions. We have proposed electron delocalization within the five-membered ring, resulting in the formation of an aromatized structure (**Figure 1**).[21]

Electrical Conductivity

Although the conductivity values appear to be very low, they are still significant because the structure of ligands has no specific features, such as conjugated double bonds necessary for conduction. This may be considered to be additional evidence for suggested electron delocalization over macromolecules, resulting in an aromatized structure (Figure 1).

The coordination polymers of the same ligand prepared with different metal ions show different degradation patterns. This suggests that metal ions take an active part in degradation. All the coordination polymers are found to be thermally more stable than the respective ligands.

APPLICATIONS

Coordination polymers are of interest because of the possibilities that they may exhibit thermal stability that is markedly superior to that of available organic polymers while retaining their plasticity. The use of resins to remove, concentrate, and detect metal ions has been practiced for some time. Such reactions are critical analytical, chromatographic polymeric reactions that involve coordination polymers.[22]

Some polymeric complexes of dithiooxamide and its substituted products have been studied for their applications in catalytic processes.[23] The coordination polymers of polythiooxamides exhibit electrical conductivity. We hope that these and similar materials can be used in optoelectronic and molecular electronic devices and can become advanced electronic materials in future.

REFERENCES

1. Hofmans, H.; Desseyn, H. O. *Bull. Soc. Chim. Belg.* **1986**, *95*(2), 83.
2. Green, M. R.; Jurgan, N. B.; Brance, E.; Buch, D. H. *Inorg. Chem.* **1987**, *26*(14), 2326.
3. Antolini, L.; Fabretti, A. C.; Franchimi, G.; Menabue, L.; Dommisse, R.; Hofmans, H. *J. Chem. Soc., Dalton. Trans.* **1987**, *8*, 1921.
4. Geobes, P.; Desseyn, H. O. *Spectrochim. Acta.* **1988**, *44A*(10), 963.
5. Geobes, P.; Slootmakers, B.; Desseyn, H. *Spectroscopy (Ottawa)* **1988**, *6*(5–6), 227.
6. Pilipenko, A. T.; Mellnikova, N. V.; Zubenko, A. I.; Melenevskii, S. V. *Koord. Khim.* **1985**, *11*(2), 244.
7. Voznesenski, S. A.; Pezelski, J.; Tsinn, I. M. *Trans. Inst. Pure Chem. Reagents (USSR)* **1938**, *16*, 98.
8. Odnoralova, U. N.; Kudryavtsev, G. I. *Vysomol. Soedin* **1962**, *4*, 1314.
9. Hofmans, H.; Desseyn, H. O.; Herman, M. A. *Spectrochim. Acta* **1982**, *38A*(11), 1213.
10. Kanda, S. *J. Chem. Soc. Japan* **1962**, *83*, 560.
11. Vozzhennikov, V. M.; Zvonkova, Z. V.; Rukhadze, E. G.; Gulshkova, V. P.; Zhdanov, G. S. *Dokl. Akad, Nauk., SSSR* **1961**, *143*, 1131.
12. Kanda, S.; Suzuki, A.; Ohkawa, K. *Ind. Eng. Chem., Prod. Res. Dev.* **1973**, *12*(9), 88.
13. Hurd, R. N.; De La Mater, G.; McEchney, G. C.; Pieffer, L. V. *J. Am. Chem. Soc.* **1960**, *82*, 4454.
14. Kanekar, C. R.; Casey, A. T. *J. Inorg. Nucl. Chem.* **1969**, *31*, 3105.
15. Pellacani, G. C.; Fabretti, A. C.; Peyronel, G. *Inorg. Nucl. Chem. Lett.* **1973**, *9*, 897.
16. Ray, A.; Sathyanaraya, D. N. *Indian J. Chem.* **1974**, *12*, 1092.
17. Desseyn, H. O.; Aarts, A. J.; Herman, M. A. *Spectrochim. Acta* **1980**, *36A*, 59.
18. Menabue, L.; Pellacani, G. C.; Peyronel, G. *Inorg. Nucl. Chem. Lett.* **1974**, *11*, 176.
19. Peyronel, G.; Pellacani, G. C.; Pignedoli, A. *Inorg. Chim. Acta.* **1971**, *5*(4), 627.
20. Hofmans, H.; Desseyn, H. O.; Donmissee, R.; Herman, M. A. *Bull. Soc. Chim. Belg.* **1982**, *91*, 175.
21. Khandwe, M.; Bajpai, A.; Bajpai, U. D. N. *Macromolecules* **1991**, *24*, 5203.
22. Korkish, J. *Modern Methods for the Separation of Rare Metal Ions*; Pergamon: New York, 1969.
23. Batur, D. G. *Koord. Khim.* **1981**, *7*(1), 49.

Coordination Polymers

See: Metal-Coordinating Polymers

CORNEA, ARTIFICIAL (Hydrophilic Polymeric Sponges)

Traian V. Chirila
Department of Biomaterials and Polymer Research
Lions Eye Institute

The cornea is the first ocular element, located at the interface between the outside world and our eyes. It has a surprisingly complex structure, consisting of five layers (epithelium, Bowman's layer, stroma, Descemet's membrane, and endothelium) or possibly six if the very thin basement membrane between the epithelium and Bowman's layer is considered, all within 600 μm of tissue. This complexity may suggest a multitude of functional demands. Indeed, the cornea probably performs more functions in the eye than any other ocular element. It protects the internal sensory structure of the eye against occasional injury from the outside environment and against microbial invasion. It withstands and contains the intraocular pressure exerted from within the eye. The cornea serves as the most important component of the optical system of the eye by transmitting and refracting the light. About 75% of the dioptic power of the human eye is due to the interface cornea/air. The cornea also absorbs in part UV-B radiation, which is harmful to the retina. The human cornea appears therefore as the ideal evolutionary compromise, able to deal with all functional requirements.[1]

The maintenance of integrity, strength, and transparency of the cornea is essential for normal vision. The cornea maintains its characteristics and good performance through a complex set of processes, and generally it defends itself well most of the time.[2,3] However, owing to injury or pathological conditions, these processes can be seriously disturbed. The resulting degenerated, scarred, or opacified cornea leads to blindness. Drug therapy is less effective in most of the corneal blinding conditions, therefore surgical intervention is generally required. The transplantation of human donor corneal grafts (penetrating keratoplasty) is currently the main surgical rehabilitation procedure and its rate of success is significantly acceptable provided that the patients are not affected by conditions such as chemical burns and certain injuries, ocular pemphigoid, Stevens-Johnson syndrome, trachoma, severe dry-eye syndromes, and recurrent graft rejections. For the latter patients, there is usually no successful outcome to the surgery. Unfortunately, the 200-year-old history of the quest for an artificial cornea (keratoprosthesis) has been plagued by failures caused by long-term complications developed after implantation.[4–9] The most frequent is the spontaneous expulsion ("extrusion") of the implanted prosthetic device, usually preceded by erosive tissue necrosis ("tissue melting") around the prosthetic rim. Other complications include retroprosthetic membrane formation, infection, epithelial downgrowth, glaucoma, retinal detachment, giant papillary conjunctivitis, and poor visual results.

The first polymer to be used as a keratoprosthetic material was poly(methyl methacrylate) [CAS Reg. No. 9011-14-7], PMMA, which is still the preferred material. Remarkably, the idea arose to use a supporting PMMA plate (flange) around the optical cylindrical core.[10] Despite being clinically unsuccessful, this concept led to the so-called "core-and-skirt" keratoprostheses, in which the material in the peripheral zone is usually different from that used in the central transparent zone.

The core-and-skirt keratoprostheses did not reduce significantly the rate of implant extrusion despite a large variety of materials used for skirts, including polymers, ceramics, metals, and preserved autologous tissues. However, there are only certain core-and-skirt keratoprostheses that currently have some success in the long term, such as Cardona "through-and-through" keratoprosthesis, or keratoprostheses in which the PMMA core is surrounded by preserved human tissue (bone, tooth, cartilage, nails).[8,11,12] The attempts to achieve such a tight junction between tissue and implant, in order to avoid the extrusion, led to the use of porous polymers as materials for the skirt.

PHEMA SPONGES: SYNTHESIS AND PROPERTIES

Water is a non-solvent for PHEMA, and it is expected that when used as a diluent in concentrations exceeding a critical value, the effect of the thermodynamically unfavorable interaction between diluent and polymer network will surpass other possible effects. Indeed, when water concentration as a diluent is higher than the equilibrium water uptake of PHEMA, phase separation occurs during the polymerization process. As a result, heterogeneous hydrogels are produced, which are translucent to white materials. The critical diluent concentration for water is probably around 45%, but this value has been variously reported.[13–21] The pore size in PHEMA sponges is much higher than that in macroporous hydrogels. The synthesis and properties of PHEMA sponges have been extensively investigated recently.[22,23]

PHEMA SPONGES: APPLICATION IN ARTIFICIAL CORNEA

Our studies on PHEMA sponges demonstrated first that the pore size, pore morphology, and hydration behavior are readily adjustable by varying the amounts of water, crosslinking agent, or initiator in the initial monomer mixture. However, water concentration is the single most important factor governing the phase separation process, therefore the only effective and also sufficient in designing a material with required characteristics.[22,23]

The manufacture of keratoprostheses based on PHEMA sponges was achieved in a specifically designed mold in which two polymerization processes could be performed successively.[24,25] By synthesizing one of the prosthetic components through polymerization in the presence of the other component, already a fully polymerized material, a permanent joint is created along the boundary between the two components. This polymer combination complies formally with the accepted definition of interpenetrating polymer networks. Using electron microscopy techniques, the existence of an interpenetration at the molecular level between the two polymers was demonstrated.[25] This union is considerably stronger than those previously proposed.

Such a keratoprosthesis, with a core made of a terpolymer HEMA/2-ethoxyethyl methacrylate/acrylamide and having as a skirt a PHEMA sponge produced in 80% water, has been implanted intrastromally into the cornea of rabbits.[26] The joint between core and skirt was always intact. Only minor complications were observed in some cases, such as a nonspecific disruption of the stromal collagen fibrils in front of the prosthesis and transient inflammation. In spite of these promising results, our keratoprosthetic device is still to be improved. To improve the mechanical strength of PHEMA sponges two methods were employed.[27] Copolymerization of HEMA with strengthening comonomers, either hydrophobic or hydrophilic did not have a significant effect. By implanting PHEMA sponges subcutaneously in animals for 6 weeks, the strength increased 3 times, and this method appears valuable for the purpose of improving mechanical characteristics.

ACKNOWLEDGMENTS

The financial support of the National Health and Medical Research Council of Australia through the grant 910167 and of the Australian Foundation for the Prevention of Blindness (Western Australia) Inc. is gratefully acknowledged. The author thanks Yi-Chi Chen, Geoffrey J. Crawford, Ian J. Constable, Sarojini Vijayasekaran, and Paul D. Dalton for their collaboration. The editorial assistance of Ruth Gutteridge is kindly appreciated.

REFERENCES

1. Marshall, J.; Grindle, C. F. J. *Trans. Ophthalmol. Soc. U.K.* **1978**, *98*, 320.
2. Waring, G. O. *Corneal Disorders: Clinical Diagnosis and Management*; W.B. Saunders: Philadelphia, 1984; Chapter 1.
3. Allansmith, M. R. *Trans. Ophthalmol. Soc. U.K.* **1978**, *98*, 361.
4. Tudor Thomas, J. W. *Trans. Ophthalmol. Soc. U.K.* **1955**, *75*, 473.
5. Day, R. *Trans. Am. Ophthalmol. Soc.* **1957**, *55*, 455.
6. Cardona, H. *Am. J. Ophthalmol.* **1962**, *54*, 284.
7. Aquavella, J. V.; Rao, G. N.; Brown, A. C.; Harris J. K. *Ophthalmology* **1982**, *89*, 655.
8. Barron, B. A. *The Cornea*, Churchill Livingstone: New York, 1988; Chapter 30.
9. Barber, J. C. *International Ophthalmol. Clinics* **1988**, *28*, 103.
10. Vanysek, J.; Iserle, J.; Altman, J. *Csl. Oftal.* **1954**, *10*, 108.
11. Donn, A.; Cotliar, A. M. *Oculoplastic, Orbital, and Reconstructive Surgery*; Vol. 2, Williams & Wilkins: Baltimore, MD, 1990, Chapter 128.
12. Casey, T. A. *Proc. Royal Soc. Med.* **1970**, *63*, 313.
13. Rosenberg, M.; Bartl, P.; Lesko, J. *J. Ultrastruct. Res.* **1960**, *4*, 298.
14. Yasuda, H.; Gochin, M.; Stone, W. *J. Polym. Sci. A-1* **1966**, *4*, 2913.
15. Hasa, J.; Janacek, J. *Polym. Sci. C* **1967**, *16*, 317.
16. Wichterle, O.; Chromecek, R.; *J. Polym. Sci C* **1969**, *16*, 4677.
17. Illavsky, M.; Prins, W. *Macromolecules* **1970**, *3*, 425.
18. Kopecek, J.; Lim, D. J. *Polym. Sci. A-1* **1971**, *9*, 147.
19. Wichterle, O. *Encyclopedia of Polymer Science and Technology*; Vol. 15, Wiley-Interscience: New York, 1971; pp 273–291.
20. Warren, T. C.; Prins, W. *Macromolecules* **1972**, *5*, 506.
21. Peppas, N. A.; Moynihan, H. J.; Lucht, L. M. *J. Biomed. Mater. Res.* **1985**, *19*, 397.
22. Chirila, T. V.; Chen, Y. C.; Griffin, B. J.; Constable, I. J. *Polym. International* **1993**, *32*, 221.
23. Chen, Y. C.; Chirila, T. V.; Russo, A. V. *Mater. Forum* **1993**, *17*, 57.
24. Chirila, T. V.; Constable, I. J.; Crawford, G. J.; Russo, A. V. U.S. Patent 5 300 116, 1994a.
25. Chirila, T. V.; Vijayasekaran, S.; Horne, R.; Dalton, P. D.; Constable, I. J.; Crawford, G. J. *J. Biomed. Mater. Res.* **1994**, *28* 745.
26. Chirila, T. V.; Thompson-Wallis, D. E.; Crawford, G. J.; Constable, I. J.; Vijayasekaran, S. *Graefes Arch. Clin. Exp. Ophthalmol.* in press, 1996.
27. Chirila, T. V.; Yu, D. Y.; Chen, Y. C.; Crawford, G. J. *J. Biomed. Mater. Res.* **1995**, *29*, 1029.

Cosmetic Materials

See *Personal Care Polymers*

Cotton

See: *Cellulose (from Fly Cotton Mill Waste)*
Cellulosic Materials (Moisture Sorption Properties)
Cotton (Non-formaldehyde Crosslinking Agents)
Cotton Fabric (Water-Repellant Finishes)
Hydroxyethyl Cellulose (Infrared Spectra and Dielectric Properties)
Hydroxyethyl Cellulose (Variables Affecting Hydroxyethylation Reaction)
Textile Fibers (Structure and Properties)

COTTON (Non-formaldehyde Crosslinking Agents)

B. A. Kottes Andrews
Southern Regional Research Center
Mid South Area Agricultural Research Service
U.S. Department of Agriculture, New Orleans

Cotton and other cellulosic fabrics, unlike their synthetic counterparts, are not inherently resilient and smooth-drying. For this reason they must be chemically treated to impart wrinkle resistance and dimensional stability. The most successful and most economical wrinkle-resistance treatments were based on the cellulose crosslinking capability of the methylolamide class of compounds that have formaldehyde as a starting material.[1]

Recently, the driving force for the development of non-formaldehyde crosslinking agents for cotton and other cellulosics has been a global awareness of our environment and an interest in maintaining its safety. In past years, however, the rationale for elimination of formaldehyde was not based on environmental considerations, but on avoidance of sites for available chlorine from bleach.[2] Free amido nitrogens, from hydrolysis of amidomethyl ether finishes or of unreacted methylolamides, can form chloramides that cause either fabric yellowing of fabric damage when HCl is released on heating.

Limitations to the ready adoption of any new technology by the textile industry are that existing equipment and, preferably, water soluble systems be utilized. A placement process for formaldehyde-based crosslinking systems for wrinkle resistance cotton and cellulosics must be similar to those they replace. The non-formaldehyde cellulose crosslinking agents must be water soluble, suitable for application by mangle padding or foam application and amenable to the high-temperature curing steps during which the crosslinking reaction with cellulose takes place.

Because many of the candidate non-formaldehyde crosslinking agents proved to be more toxic, more expensive, and produced more of a deleterious effect on other functional and aesthetic properties such as strength and fabric color, they have remained laboratory curiosities. This review will present both those non-formaldehyde crosslinking agents that were not adopted by the textile industry as an alternative to formaldehyde based agents for wrinkle resistance and those non-formaldehyde crosslinking agents that have found a place in textile finishing.

VINYL SULFONES AND THEIR DERIVATIVES

One of the first references to cellulose crosslinking by difunctional vinyl compounds described a process for shrink-proofing cotton with divinyl sulfone.[3] Later, crosslinking of cotton with *N,N′*-methylenebisacrylamide by a Michael addition reaction was reported.[4] These agents containing activated vinyl groups are very hazardous and should be handled with caution.

A modification of divinyl sulfone provided an agent without hazard and with no loss of reactivity toward cellulose. Base-catalyzed reaction of bis(2-hydroxyethyl) sulfones with cellulose in the absence of water produced increased resiliency in cotton fabrics.[5,6]

EPOXIDES

Polyepoxides have been claimed to produce crease and shrink resistant fabrics without losses of strength-related properties.[7,8] Acids are the usual catalysts for these reactions with cellulose. The criteria for epoxide reactivity toward cellulose were postulated to be non-volatility, water solubility, and freedom from steric hindrance.[9]

AZIRIDINES

Although aziridine compounds have been proposed for non-formaldehyde durable-press finishing of cotton as well as for flame retardant applications, the aziridine derivatives were found to be toxic.

ISOCYANATES

Both aromatic isocyanates, considered toxic, and the non-toxic cyclohexane diisocyanate have been used successfully to crosslink cotton. The reaction products are carbamic acid esters of cellulose.[10–12] Bisulfic addition products of isocyanates have also been used to crosslink cellulose.[13]

ALKYLATION

Acetals have been studied as crosslinking agents for cellulose. Those most readily available are actually formals, nitrogen free, but not formaldehyde free. Acetals derived from dialdehydes, however, can crosslink cotton to produce improved fabric resiliency.[14,15] Improvements in resiliency, as measured by wrinkle recovery angle, are not as great as that from formaldehyde-containing agents and there are also problems with color and strength loss, as well as cost.[16,17]

ALDEHYDES OTHER THAN FORMALDEHYDE

Dialdehydes can also crosslink cotton. Treatments with reactive aldehydes, however, like formaldehyde itself, usually produce more strength losses in fabric than do those with methylolamides.[18,19] They also have other objectionable properties such as odor, a tendency to discolor fabric, and expense.

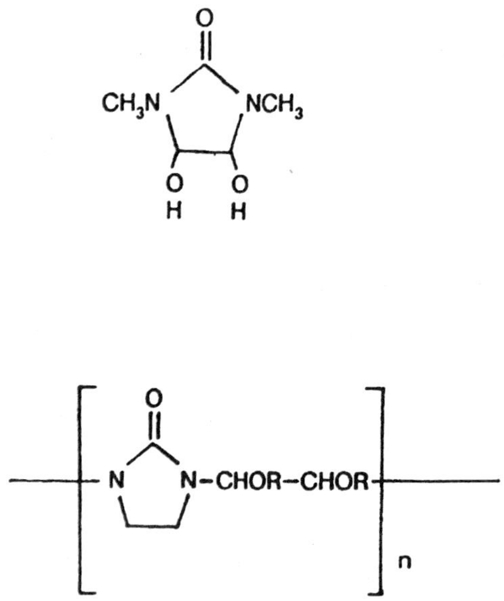

FIGURE 1. Reaction products of glyoxal and amides: top, reaction product of dimethylurea and glyoxal, 1,3-dimethyl 4,5-dihydroxyethyleneurea; bottom, oligomer of reaction product of ethyleneurea and glyoxal.

AMIDE-GLYOXALS

The most commonly used commercial non-formaldehyde crosslinking agents are the amide-glyoxal agents. Two of these are shown in **Figure 1**. Neither the glyoxal-dimethylurea adduct nor the glyoxal-ethyleneurea oligomer produce wrinkle resistance in cotton equivalent to that from dimethyloldihydroxyethyleneurea (DMDHEU).[20–23] In addition, there can be a problem with fabric yellowing if agent preparation is not carefully controlled. Despite the problems, commercial finishing systems have been developed with zinc fluoborate and other highly acid catalysts around these agents. The most widely used non-formaldehyde finish is based on the glyoxal-dimethyl urea adduct, 1,3-dimethyl 4,5-dihydroxyethyleneurea.[24,25] Reaction products of carbamates and glyoxal or gluteraldehyde also serve as crosslinking agents for cotton.[26,27]

POLYCARBOXYLIC ACIDS

Current cellulose esterification crosslinking technology is based on the use of alkali metal phosphites and hypophosphite as catalysts.[28–32] In the presence of these weak base catalysts, the reaction is believed to proceed by an anhydride mechanism. The acids of most interest commercially are cis-1,2,3,4-butanetetracarboxylic acid and citric acid. Refinements of the finishing process produced fabrics with property levels equivalent to those from DMDHEU, the state of the art formaldehyde-based agent.[33–35]

REFERENCES

1. Kottes Andrews, B. A. *Book of Papers, INDA-TEC 93*, **1993**, 113.
2. Reine, A. L.; Reid, J. D.; Reinhardt, R. M. *Am. Dyest. Rep.* **1966**, *55*(19), 91.
3. Schoene, D. L.; Chambers, V. S. U.S. Patent 2 524 399, October 3, 1950.

4. Frick, J. W.; Reeves, W. A.; Guthrie, J. D. *Textile Res. J.* **1957**, *27*, 92.
5. Tesoro, G. C. *Textile Res. J.* **1962**, *32*, 189.
6. Tesoro, G. C.; Pensa, I. *Textile Res. J.* **1964**, *34*, 960.
7. Condo, F. E.; Cerrito, E.; Schroeder, C. W. U.S. Patent 2 752 269, June 26, 1956.
8. Schroeder, C. W.; Condo, F. E. *Textile Res. J.* **1957**, *27*, 135.
9. Steele, R. O. *Textile Res.* J. **1961,** *31*, 257.
10. Ellzey, S. E.; Mack, C. H. *Textile Res. J.* **1962**, *32*, 1023.
11. Ellzey, S. E.; Wade, C. P.; Mack, C. *H. Textile Res. J.* **1962**, *32*, 1029.
12. Verburg, G. B.; Snowden, F. W. *Textile Res. J.* **1967**, *37*, 367.
13. Reich, F. *Textil. Vered.* **1978**, *13*, 454.
14. Frick, Jr., J. G.; Harper, Jr., R. H. *J. Appl. Poly. Sci.* **1984**, *29*, 1433.
15. Frick, Jr., J. G. *J. Appl. Poly. Sci.* **1985**, *30*, 3467.
16. Chance, L. H.; Danna, G. F. U.S. Patent 4 818 243, April 4, 1989.
17. Chance, L. H.; Danna, G. F.; Andrews, B. K. U.S. Patent 4 900 324, Feb. 13, 1990.
18. Frick, Jr., J. G.; Harper, Jr., R. J. *J. Appl. Poly. Sci.* **1982**, *27*, 983.
19. Frick, Jr., J. G.; Harper, Jr., R. J. *J. Appl. Poly. Sci.* **1983**, *28*, 3875.
20. Harper, Jr., R. J.; Frick, Jr., J. G. *Am. Dyest. Rep.* **1981**, *70*(9), 46.
21. Frick, Jr., J. G.; Harper, Jr., R. J. *I&EC Prod. Res. & Devel.* **1982**, *21*(1), 1.
22. Frick, Jr., J. G.; Harper, Jr., R. J. *Textile Res . J.* **1983**, *53*, 660.
23. North, B. F. U.S. Patent 4 285 690, Aug 25, 1981.
24. Vail, S. L.; Murphy, P. J. U.S. Patent 3 112 156, Nov 26, 1963.
25. Petersen, H.; Pai, P. S.; Klippel, F.; Reinert, F. U.S. Patent 4 295 846, Oct. 20, 1981.
26. Frick, Jr. J. G.; Harper, Jr., R. J. *Textile Res. J.* **1981**, *51*, 601.
27. Frick, Jr., J. G.; Harper, Jr., R. J. *Textile Res. J.* **1983**, *53*, 758.
28. Welch, C. M.; Andrews, B. K. U.S. Patent 4 820 307, April 11, 1989.
29. Welch, C. M.; Andrews, B. K. U.S. Patent 4 936 865, June 26, 1990.
30. Welch, C. M.; Andrews, B. K. U.S. Patent 4 975 209, Dec. 4, 1990.
31. Andrews, B. K.; Morris, N. M.; Donaldson, D. J.; Welch. C. M. U.S. Patent 5 221 285, June 22, 1993.
32. Welch, C. M.; Andrews, B. A. Kottes *Textile Chem. Color.* **1989**, *21*, 13.
33. Andrews, B. A. Kottes; Welch, C. M.; Trask-Morrell, B. J. *Am. Dyest. Rep.* **1989**, *78*(6), 15.
34. Andrews, B. A. Kottes *Textile Chem. Color.* **1990**, *22*, 63.
35. Andrews, B. A. Kottes; Collier, B. J. *I&E Chem. Res.* **1992**, *84*, 631.

COTTON FABRIC (Water-Repellent Finishes)

Hiromu Nakamichi
University of Shizuoka

FUNDAMENTAL CONCEPTION

Water repellency is defined by the contact angle for water. In the case of a contact angle larger than 90°, we call it water repellent (not wetting), and in that of less, we call it not water repellent (wetting).

WATER-REPELLENT AGENTS

Water-repellent finishes are practiced by the usage of water-repellent agents for fibers to prevent wetting.

For the water-repellent finishes of fabrics are used wax, aluminum soap, zirconium salt, quaternary ammonium salt, silicone, and fluorine compound.[1]

In water-repellent finishes are often demanded washability and the ability to dryclean. And, in the applications in which durability is required, silicone, quaternary ammonium salts, and fluorine compounds are used.

On the water-repellent finishes that have less information for the consumer, to increase fundamental understanding, Nakamichi tried the following experiments.[2]

The emulsion of fluoroethylene resin, which has been known by the name of Scotch Guard for many years, and the emulsifying agent of which is, chiefly, zirconium salt and the emulsion of reaction type resin degenerated by methylal amide, was used.

REFERENCES

1. Nakao, K. Saishin sen-ikakougizyutu, *The Textile Machinery Society of Japan* 1973, Chapter IV (section 1).
2. Nakamichi, H. *Annual Report of Studies of Shizuoka Women's University* **1986**, *19*, 29.

Counterion Binding

See: *Polyelectrolytes (Association of Hydrophobic Counterions)*
Polyelectrolytes (Chain Models of Polymers)
Polyelectrolytes (Counterion Binding and Hydration)
Polyelectrolytes (Special Properties in Solution)
Polyelectrolytes (Solution Properties)

Coupling Agents

See: *Additives (Property and Processing Modifiers)*
Fillers, Surface Modification
Fluorosilicones (Monomer Synthesis and Applications)
Inorganic Surfaces, Grafting
Silane Coupling Agents (Adhesion Promotors)

Crazing

See: *Deformation and Fracture (Micromechanical Mechanisms)*
Glassy Polymers (Toughening via Dilatational Plasticity)
Glassy Polymers, Crazing
Polystyrene, High Impact
Polystyrene, Rubber Toughened

CROSSLINKED NETWORKS, MACRONETS

A. G. Theodoropoulos* and I. C. Konstantakopoulos
Laboratory of Organic Chemical Technology
Department of Chemical Engineering
National Technical University of Athens

Crosslinked polymers are important in chemical research and practice owing to the wide range of applications. Water treatment, chromatographic techniques, biomedical and biochemical

*Author to whom correspondence should be addressed.

applications, solid-phase organic synthesis, enzyme immobilization, soil conditioning, solvent and ion separations, and various other areas are using crosslinked polymers as the principal substrate.[1–11] Crosslinking is responsible for the three-dimensional network structure that characterizes these materials. Elasticity and swelling properties are attributed to the presence of physical or chemical crosslinks within polymer chains.[12] Thus, it is obvious that the nature of crosslinks would influence strongly the properties of the crosslinked polymer.

In this section we shall focus our interest on the length of crosslinks, or the linear dimension of the crosslinker.

Macronet crosslinked polymers are the networks obtained by copolymerization or post-polymerization that bear bulky molecules as crosslinks. The dimension of divinylbenzene is usually used as a reference in order to distinguish macronet and standard networks.

Some unusual structure characteristics and properties are attributed to macronet networks. High swelling abilities, swelling in poor and non-solvents, permeability of large molecules, and specific network design parameters are referred in this area. During the last decade the recent trends and perspectives of macronets are mostly connected with post-polymerization modifications of linear or grafted polymers.[13]

Another typical advantage of macronets prepared with a post-polymerization technique is the potentiality of applications in polymer recycling.[14] The interest here comes from the fact that already polymerized materials are used as reagents. The important issue and perspective is that by following a post-polymerization route it is possible to control polymer architecture, or even obtain block-order networks, having crosslinks in separate domains.[15] Thus, macronets have been considered as a separate category in polymer synthesis and as a substrate appropriate for various applications in both industrial and laboratory levels.

PREPARATION

Crosslinked macronets are formed from chloromethylated polystyrene or styrene copolymeres because these styrene units are accessible to electrophilic substitution of one of the hydrogen atoms by a chloromethyl group.[16]

Macronets prepared by copolymerization are based on monolefin employment (styrene, acrylic acid, acrylamide, methacrylic acid, etc.) and reactive diolefins as crosslinking agents. Such crosslinking agent application includes bis-acrylamide derivatives (*N,N′*-hexamethylene dimethacrylamides, *N,N′*-decamethylene dimethacrylamides) and copolymers with dimethacrylic esters of 1,12-(*P*-oxyphenyl) dodecane.[17–19]

Macronets prepared by post-polymerization are based on various modifications of linear macromolecules and grafted or block copolymers with multifunctional compounds as crosslinking agents. Some typical crosslinking agents are *p*-dichloromethylbenzene, 4,4′-bis(chloromethyl)biphenyl, adipoyl dichloride, 1,4-diacetoxymethyl-2,5-dimethylbenzene, 1,4 bis(hydroxymethyl)benzene, 1,4-dichloromethyl-2,5-dimethylbenzene and 2,4,6 tris(chloromethyl)mesitylene.[20–25] Uniformity of crosslinking, application on polymer recycling, and maintaining of polymer architecture characterize this category.

Recently macronets have been mostly connected with post-polymerization crosslinking modifications of polymerized materials. A typical solution procedure involves dissolution of the homo- or copolymers in a thermodynamically good solvent and addition of the crosslinker, or a crosslinker solution to the reaction vessel.

MACRONET ION-EXCHANGE RESINS

Introduction of ionic functionality on a macronet substrate is subsequent or simultaneous to crosslinking procedure, in order to obtain macronet ion-exchange resins.

Usually high ion-exchange capacities are observed for macronet ion-exchangers.[26,27] When ionic groups are introduced to crosslinked macronets to a certain degree, usually more than 25% of the available units, the ion-exchange properties are associated with an increase of the hydrophilic character and result in elevated swelling degrees of the polymer in polar media.

PROPERTIES

Macronets may have all the elasticity and swelling properties that derive from their three-dimensional network structure. However, some unusual and more specific properties have been revealed during their investigation. Most of these differences concern swelling characteristics. A macronet polymer is able to swell more significantly compared to a standard network with the same degree of crosslinking.[28] Another interesting fact is that despite the known theories that describe the swelling of crosslinked polymers, an increase in the swelling degree in poor solvents was estimated by increasing the degree of crosslinking.[29–31]

Studies on their thermal properties indicate that polystyrene macronets appeared to elevate glass transition temperatures (T_g) comparatively with their linear analogues. T_g was observed to increase by the extent of crosslinking since segmental motion is restricted by the crosslinks.[32,33]

APPLICATION

The application fields of the macronet polymers do not differ considerably from those of the standard networks. However, the swelling properties that derived from their specific structural characteristics give them a place of prominence in both the laboratory research and the industrial practice when highly and homogeneous, swellable materials are required.

Ionic-functionalized macronets may have all the typical applications of ion-exchange resins. Macronet styrene copolymers derived from chloromethylation were used as substrates for the synthesis of sorbents containing optically active bifunctional and trifunctional α-amino acid groups.[34–37]

Much attention has been directed in recent years to the development of soil conditioning media based on hydrophilic macronets.[38,39] Their action comes primarily from their high swelling ability and secondarily from the availability of ion-exchange sites to various nutrients that are released to plants.

Emphasis should be placed on the fact that post-polymerization procedures for macronet synthesis have been applied by employing various commercial products as raw materials, thus suggesting a novel approach for the recycling of non-biodegradable products.[14]

The high rate of swelling and high swelling ability in water exhibited by some acrylonitrile-containing macronets make them useful as water-retaining media in diapers technology. A transformable soil conditioner is also referred based on a macronet sulfonated styrene-hydrolysed acrylonitrile copolymer that combines properties of water-retaining and soil-aggregating agents.[40]

Regarding the future, macronet polymers of either hydrophobic or hydrophilic type may see increasing use as filters in chromatographic techniques owing to their homogeneous structure. Selective separations of organic solvents can be based on block-order macronets, by altering the proportion of the mid- and end-block phase.[15] Critical phenomena can be applied on macronet hydrogels, suggesting that applications on biomaterials or sensors are possible.[41]

REFERENCES

1. Flemming, H. C. *Water Res.* **1987**, *20*(7), 745.
2. Emerson, D. W.; Shea, D. T.; Sorensen, E. M. *Ind. Eng. Chem. Prod. Res. Dev.* **1978**, *17*(3), 269.
3. Arshady, R. *J. Chromatography* **1991**, *586*, 181.
4. Peppas, N. A.; Mikos, A. G. *Hydrogels in Medicine and Pharmacy,* CRC: 1986; Chapters 1 and 2.
5. Olah, G. A. *Friedel—Crafts and Related Reactions*; 1963; Vol. 1, Chapter 4, 201.
6. Chakrabarti, A.; Sharma, M. M. *React. Polym.* **1993**, *20*, 1.
7. Wiseman, A. *Principles of Enzyme Biotechnology*; John Wiley & Sons: 1986; Chapter 4, 201.
8. Andreopoulos, A. G. *Eur. Polym. J.* **1989**, *25*, 977.
9. Azzam, R. A. I. *Commun. Soil Sci. Plant Anal.* **1980**, *11*, 767.
10. Ladisch, M.; Dyck, K. *Science* **1979**, *205*, 898.
11. Paulson, A. J. *Anal. Chem.* **1986**, *58*, 183.
12. Flory, P. J. *Principles of Polymer Chemistry,* 9th ed.; Cornell Univ.: 1975; Chapter 13, 541.
13. Theodoropoulos, A. G.; Tsakalos, V. T.; Valkanas, G. N. *Polymer* **1993**, *34*, 3905.
14. Theodoropoulos, A. G.; Valkanas, G. N.; Stergiou, D. H.; Vlysidis, A. G. *Macrom. Rep.* **1994**, *A31*, 9.
15. Theodoropoulos, A. G.; Konstantakopoulos, I. C.; Valkanas, G. N. *Eur. Polym. J.* **1994**, *30*, 1375.
16. Belfer, S.; Glozman, R. *J. Appl. Polym. Sci.* **1979**, *24*, 2147.
17. Ergozhin, E. E.; Zhubanov, B. A.; Kushnikov, Y. A.; Prodius, L. N. *Izv. Acad. Nauk. Kaz. Ser. Khim.* **1970**, *20*(3), 44.
18. Musabekov, K. B.; Ergozhin, E. E.; Shapovalova, L. P. *Izv. Acad. Nauk. Kaz. Ser. Khim.* **1972**, *22*(6), 59.
19. Davankov, V. D.; Rogoshin, S. V.; Tsyurupa, M. P. *Ion Exchange and Solvent Extraction*; Marcel Decker: New York, 1977, Vol. 7, Chapters 2 and 29.
20. Grassie, N.; Gilks, J. *J. Polym. Sci. Pol. Chem.* **1973**, *11*, 1531.
21. Wolf, F.; Frederich, K. Ger.(East) Patent 57703, 1967.
22. Bussing, W. R.; Peppas, N. A. *Polymer* **1983**, *24*, 209.
23. Peppas, N. A.; Valkanas, G. N. *Angew. Makromol. Chem.* **1977**, *62*, 163.
24. Barar, D. G.; Staller, K. P.; Peppas, N. A. *J. Polym. Sci. Pol. Chem.* **1983**, *21*, 1013.
25. Iovine, C. P.; Ray-Chaudhuri, D. K. U.S. Patent 4 448 935, 1984.
26. Theodoropoulos, A. G.; Bouranis, D. L.; Valkanas, G. N. *J. Appl. Polym. Sci.* **1992**, *46*, 1461.
27. Tsyurupa, M. P.; Davankov, V. A.; Rogoshin, S. V. *J. Polym. Sci. Symp.* **1974**, *47*, 189.
28. Davankov, V. A.; Rogoshin, S. V.; Tsyurupa, M. P. *Angew. Makromol. Chem.* **1973**, *32*, 145.
29. Davankov, V. A.; Tsyurupa, M. P.; Rogoshin, S. V. *Angew. Makromol. Chem.* **1976**, *53*, 19.
30. Tsyurupa, M. P.; Andreeva, A. I.; Davankov, V. A. *Angew. Makromol. Chem.* **1978**, *70*, 179.
31. Davankov, V. A.; Tsyurupa, M. P. *Angew. Makrom. Chem.* **1980**, *91*, 127.
32. Grassie, N.; Gilks, J. *J. Polym. Sci. Pol. Chem.* **1973**, *11*, 1985.
33. Peppas, N. A.; Bussing, W. R. *Polymer* **1983**, *24*, 898.
34. Davankov, V. A.; Rogoshin, S. V.; Piesliakas, I. I. *Vysokomol. Soedin.* **1972**, *14B*, 276.
35. Davankov, V. A.; Rogoshin, S. V.; Piesliakas, I. I.; Vesa, V. S. *Vysokomol. Soedin.* **1973**, *ISB*, 115.
36. Rogoshin, S. V; Davankov, V. A.; Yamskov, I. A.; Kabanov, V. P. *Zh. Obshch. Khim.* **1972**, *42*, 1614.
37. Rogoshin, S. V; Yamskov, I. A.; Davankov, V. A. *Vysokomol. Soedin.* **1974**, *16B*, 849.
38. Kakoulides, E. R; Papoutsas, I. S.; Bouranis, D. L.; Theodoropoulos, A. G.; Valkanas, G. N. *Commun. Soil Sci. Plant Anal.* **1993**, *24*, 1709.
39. Theodoropoulos, A. G.; Bouranis, D. L.; Valkanas, G. N.; Kakoulides, E. P. *Commun. Soil Sci. Plant Anal.* **1993**, *24*, 1721.
40. Bouranis, D. L.; Theodoropoulos, A. G.; Theodosakis, G. I.; Valkanas, G. N. *Commun. Soil Sci. Plant Anal.* **1994**, *25*, 2273.
41. Tanaka, T. *Polyelectrolyte Gels* Chapter 1, 1 ACS Symposium Series, 1990.

CROSSLINKED POLYMER BRUSHES

Guojun Liu
Department of Chemistry
The University of Calgary

Polymer chains that are crowded at a liquid–sold, liquid–liquid, or liquid–air interface may assume elongated conformations like bristles in a brush. These elongated brush-like polymer structures are referred to as polymer "brushes."[1–4] This paper focuses on polymer brush formation at polymer solution–solid substrate interfaces (**Figure 1**).

Polymer brushes at solution–solid interfaces are currently prepared using either end-functionalized homopolymers or diblock copolymers. If the terminal functional group of a homopolymer binds strongly to a solid substrate, a high surface coverage is possible. Polymer chains in such a crowded layer repel one another and stretch away from the solution–solid interface and this a brush forms. The adsorption of polystyrene with a zwitterion terminal group falls into this category.[5,6]

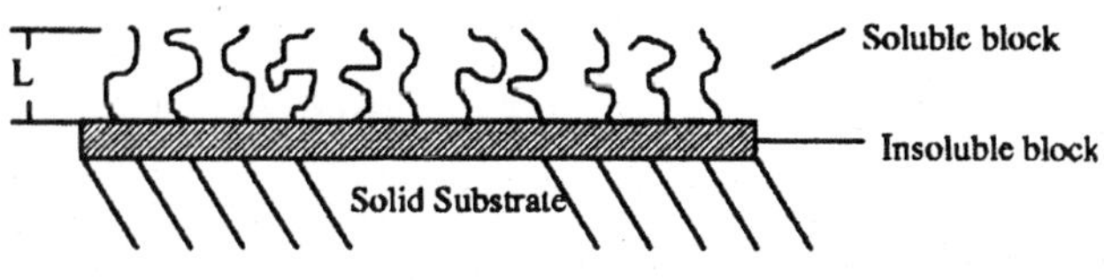

FIGURE 1. A polymer brush.

A diblock copolymer may form a brush on a solid substrate in a block-selective solvent (i.e., one that is poor for one block but good for the other).[1] The driving force for polymer deposition on a solid substrate, in this case, is partially derived from a lowering in enthalpy of the insoluble block when it is removed from the solution phase. If the interaction between the insoluble block and the substrate is favorable, a dense monolayer or a polymer brush forms. In a polymer brush, the insoluble block spreads on the solid substrate like a melt to form the anchoring layer, and the soluble block or so-called top or buoy block/layer stretches into the solution phase roughly perpendicular to the solid and solution interface. The advantage of using a diblock copolymer is that the binding energy between the insoluble block and the solid, which increases with the length of the insoluble block, can be tuned to ensure brush formation.

Polymer brushes have been studied extensively during the past decade. Progress in this field has been recently reviewed.[1–3] This paper focuses on the preparation of crosslinked polymer brushes from diblock copolymers. This can be achieved through the use of copolymers with blocks that are easily crosslinkable.

SYNTHESIS OF DIBLOCK COPOLYMERS

To introduce residual functional groups for crosslinking, we used in one strategy difunctional, especially divinyl, monomers. One functional group of such a monomer would participate in cationic or anionic polymerization, the other survives the polymerization intact. After brush formation, the residual double bond can be crosslinked by another polymerization mechanism such as photoinitiated polymerization. An example in this category is the synthesis of poly(*iso*-butyl vinyl ether)-block-poly[2-(vinyloxy)ethylcinnamate] (PIBVE-*b*-PVEC) (**Structure 1**).[7,8]

The second strategy we took was to prepare a diblock copolymer precursor first.[9,10] Such a precursor can be polystyrene-block-poly(2-hydroxyethyl methacrylate) (PS-*b*-PHEMA) (**Structure 2**)[10]

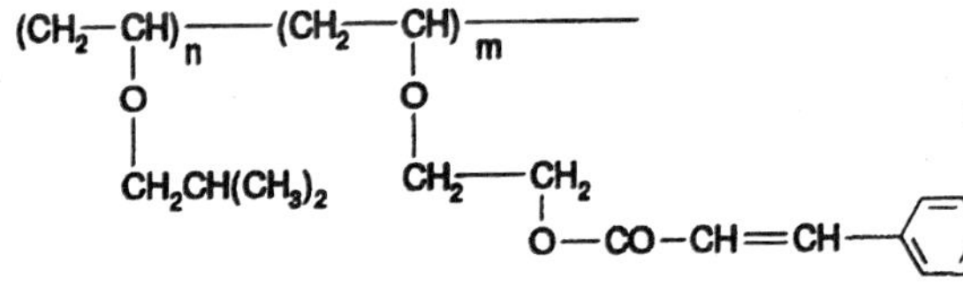

1

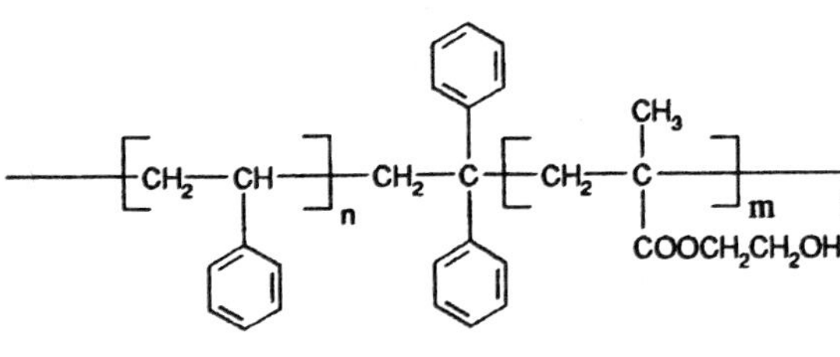

2

PS-*b*-PHEMA

The hydroxyl groups of the PHEMA block were reacted with cinnamic anhydride or cinnamoyl chloride to introduce the photocrosslinkable cinnamate group polystyrene-*b*-poly(cinnamoyethyl methacrylate) (PS-*b*-PCEMA).[10]

PROPERTIES

An uncrosslinked polymer brush is resistant to washing by the solvent from which the brush is built as suggested by Ligoure and Leibler.[11] If rinsed with a solvent that is good for both blocks of a diblock copolymer, a polymer brush may be removed readily. In general, uncrosslinked polymer brushes are not solvent resistant.

Crosslinked polymer brushes, in principle, possess all the properties of uncrosslinked ones and, in addition, are more resistant to solvent attacks.[12]

POTENTIAL APPLICATIONS

Many applications of crosslinked polymer brushes are expected. Among these, a crosslinked brush on the inner wall of a capillary column may function as the stationary phase for gas chromatography. Silica particles covered with crosslinked polymer brushes can be used as HPLC column packing. One can also use a crosslinked brush to immobilize enzymes or to support catalysts. Polymer brush formation can be used for the surface modification of medical devices.

REFERENCES

1. Milner, S. T. *Science* **1991,** *251,* 905.
2. Halperin, A.; Tirrell, M.; Lodge, T. P. *Adv. Polym. Sci.* **1991,** *100,* 31.
3. Gast, A. P. *Scientific Methods for the Study of Polymer Colloids and Their Applications*; Candau, F.; Ottewill, R. H. Eds.; Academic: Amsterdam, 1990.
4. Tuzar, Z.; Kratochvil, P. *Surf Colloid Sci.* **1992,** *15,* 1.
5. Taunton, H. J.; Toprakcioglu, C.; Fetters, L. J.; Klein, J. *Macromolecules* **1990,** *23,* 571.
6. Klein, J.; Kumacheva, E.; Mahalu, D.; Perahai, D.; Fetters, L. J. *Nature* **1994,** *370,* 634.
7. Liu, G.; Hu, N.; Xu, X.; Yao, H. *Macromolecules* **1994,** *27,* 3892.
8. Hu, N.; Liu, G. J. *Macromol. Sci. Pure Appl. Chem.* **1995,** *A32,* 949.
9. Smith, C.; Liu, G.; Hu, N., submitted for publication in *Can. J. Chem.*
10. Liu, G.; Smith, C.; Hu, N.; Tao, J., submitted for publication in *Macromolecules.*
11. Ligoure, C., Leibler, L. *J. Phys. France* **1990,** *51,* 1313.
12. Liu, G.; Xu, X.; Skupinska, K.; Hu, N.; Yao, H. *J. Appl. Polym. Chem.* **1994,** *53,* 1699.

Crosslinking

See: *Benzocyclobutene-Based Monomers (Diels-Alder Ring-Opening Polymerization)*
Benzocyclobutenes (Crosslinking and Related Reactions)
Benzoxazine Monomers and Polymers (New Phenolic Resins by Ring-Opening Polymerization)
Blends (Interchain Crosslinking)
Butyl and Halobutyl Rubbers
Crosslinked Networks, Macronets
Crosslinked Polymer Brushes
Crosslinking (of Poleolefins)
Crosslinking Agents, Phenyltriazenes (for Polyimides and Other Aromatic Polymers)
Cyclomatrix Phosphazene Polymers

CROSSLINKING (of Polyolefins)

Bernt-Åke Sultan
Borealis AB

In their solid state, polyolefins have a semi-crystalline structure with crystalline regions that are mechanically strong and rigid and inaccessible to chemical reagents. The amorphous regions are mechanically weak and fairly open to reagents but give the material its flexibility and plasticity. When crosslinking polyolefins, covalent bonds are formed that link the different regions. The three-dimensional network thus formed dramatically improves important properties such as heat, abrasion, and viscous deformation as well as chemical and stress-cracking resistance. The impact and tensile strength are increased, shrinkage decreased, and the low temperature properties are improved.

Polyolefins are commercially crosslinked by three different methods, peroxide-, irradiation-, and silane-crosslinking. Peroxide crosslinking is today the most widely used crosslinking technique, and peroxide-crosslinked polyethylene has become the dominant insulation material for medium and high voltage power cables. Other important applications are hot-water pipes, shoe soles, and containers for storage of chemically aggressive media.

Silane crosslinking was introduced by the Sioplas and Monosil processes where vinyl silane is grafted onto the polymer during extrusion.[1,2] By addition of a condensation catalyst, crosslinking occurs via formation of a siloxane linkage. Silane crosslinking is of growing importance, especially since the introduction of a copolymerization process.[3]

PEROXIDE CROSSLINKING

Background

In the middle 1950s the most important step was taken in order to be able, in an industrial scale, to produce crosslinkable polyolefins and extrude products thereof. In their patent, Wuestof et al. disclosed the use of tertiary aromatic peroxides like di-α-cumylperoxide (DCP), tert-butyltriphenylmethylperoxide, etc.[4] DCP has a stability that is good enough to safely extrude polyolefins with a melting temperature below 120°C and yet give acceptable curing times. DCP is commonly used for crosslinking Low Density Polyethylene (LDPE), Ethylene-Propylene Rubber (EPR), and Ethylene-Prylene-Diene Rubber (EPDM) as well as polar ethylene copolymers like EVA, Ethylene-Methyl Acrylate (EMA), Ethylene-Ethyl Acrylate (EEA), and Ethylene-Butyl Acrylate (EBA).

Wuesthof's patent meant a breakthrough in the technology of crosslinking of polyolefins, and especially the use of crosslinked LDPE as insulation for medium and high voltage cables. DCP is today by far the most used peroxide for crosslinking of polyolefins.

Crosslinking Chemistry

In presence of a polyolefin the peroxy radicals abstract hydrogens from the hydrocarbon chain. By combination of alkyl radicals, carbon–carbon crosslinks are formed. In order to crosslink LDPE between 1.5–2% DCP is normally needed.

Polyfunctional coagents with a low molecular weight are commonly added to peroxide-crosslinkable compounds, such as triallyl cyanurate.[5]

During the crosslinking reactions these types of substances are grafted to the polymer chain, increasing the number of vinyl groups in the polymer that become the site of similar crosslinking reactions.

Coagents are of large importance for EPR cure systems, where chain scission reactions occur to a greater extent than in PE.[6] Sulfur improves vulcanization by taking part in crosslinking and by inhibiting the chain scission reaction.[7] Polyfunctional coagents such as ethylene dimethacrylate, divinyl benzene, and quinone dioxime, appear to be as good or better than sulfur in preventing chain scission and promoting crosslinking.[7]

More recently, highly crystalline copolymers of propylene, and small amounts of branched diene comonomers, such as 6-methyl-1,6-octadiene and 7-methyl-1.6-octadiene, have been shown to be crosslinkable using peroxides and irradiation.[8]

Applications

The most important application of crosslinked polyolefins is as insulation for Medium (MV) and High (HV) power cables. The good insulating properties of LDPE were utilized in communication cables at an early date.[9]

Modern polyolefin-based MV and HV cable cores, consist of three triple-extruded layers, one inner and one out semiconductive layer, and a thicker insulation layer in the middle. The semiconductive layers are used for homogenizing the electric field and are made semiconductive by addition of carbon black to ethylene copolymers. The insulation layer is most commonly made from LDPE crosslinked using dicumylperoxide. Peroxide-crosslinked EPR is also used as insulation, especially for the production of flexible special cables.

Hot water pipes are the main application for crosslinked polyolefin pipe material. For this application stiffer materials can be used and High Density Polyethylene (HDPE), which compared with LDPE has a higher melting point, and better mechanical properties, is a natural choice. Crosslinking gives improved creep resistance and results in longer-life prices.

Besides how-water pipes, peroxide-crosslinked HDPE is also used for producing containers for storing chemically aggressive media by rotational molding. In injection molding or two-roll mill processes, peroxide-crosslinked and foamed EVA is used for production of shoe soles.

IRRADIATION CROSSLINKING

Background

Charlesby published an extensive study in the field of irradiation crosslinking of LDPE in 1952.[10] He concluded that all forms of high energy radiation would be capable of crosslinking polyethylene. One year later, Lawton and his co-workers showed that LDPE, as predicted by Charlesby, could be crosslinked by high-energy electrons (β-rays) as well.[11] Charlesby further showed that crosslinked LDPE exhibited important improved properties, especially at temperatures over its crystalline melting point, where it was no longer floating but acquired an amorphous structure similar to that of vulcanized rubber at room temperature.[10]

The degree of crosslinking of an irradiated sample is proportional to the absorbed irradiation dose per mass unit of the irradiated object, and measured in terms of rad (radiation absorbed dose, 1Mrad = 10 kJ/kg). Typical doses needed to crosslink polyethylene are in the range of 20–30 Mrad.[12]

Applications

The main advantage of irradiation crosslinking is that high output extrusion can be used because there is no risk of precuring, and it allows a great variety of materials to be crosslinked.

Electron beam irradiation is used for crosslinking of polymeric insulation for wire and cable, for example non-halogen flame retardant polyolefin-insulated cables for low voltage, automotive applications, tubes for hose liner applications, tire components, industrial rubber components, shrinkable sleeves, tubing, and films. Electron beams are also used for sterilizing a wide variety of plastic, medical, disposable items.

SILANE CROSSLINKING

Technologies, Crosslinking Chemistry, and Extrusion Properties

Silane crosslinking is the newest of the important crosslinking techniques. It was introduced in 1968 through the Sioplas process.[1] In this, an unsaturated organosilane (usually vinyltrimethoxysilane), and a small amount of a peroxide, (usually dicumyperoxide) are compounded with polyethylene. Owing to the influence of alkyl radicals formed by the peroxide, the vinylsilane is grafted onto the polymer chain.

During the extrusion step a catalyst masterbatch containing a crosslinking catalyst, normally di-n-butyltin-dilaurate (DBTDL), is added. The presence of the catalyst and water start the crosslinking reaction.

During the 1980s, ethylene-vinyltrimethoxysilane copolymers (EVS) were introduced.[13] These copolymers are produced in the same high pressure reactors as LDPE and other ethylene copolymers such as EVA, EEA, and EBA. Moisture-curable terpolymers consisting of vinylsilane and polar copolymers are also produced by this process.[14]

Compared with the silane-grafted polymer, the copolymer process has many advantages:

- Better storage stability.
- There are no limitations in the choice of antioxidants.
- Production of a polymer with controlled melt viscosity, molecular weight distribution, silane yield, and silane distribution.
- In a copolymerization process, the unreacted silane is recovered by a polymerization unit's gas purification system. In an optimized grafting process, this is equivalent of 20% of the added vinylsilane.

To increase the processing window of moisture-crosslinkable polyethylene to a level more similar to that of thermoplastic compounds, compatible precuring retarders or so called "Scorch Retardant Additives" (SRA) have been developed.[15] This type of additive reacts with water faster than the polymer itself during extrusion, leading to a significant reduced risk for precuring.

Applications

Silane-crosslinked LDPE is mainly used as an insulation for low voltage cables. Other applications are non-halogen, flame-retardant heating and medium-voltage cables.

Silane-grafted HDPE is used for hot water pressure pipes. Moisture-cured LDPE is used for flexible hot water pipes for underground heating, pipes for transport of aggressive media at elevated temperatures, shrink sleeves for cable, and pipe connections.

Silane copolymers are an interesting product for coating applications as they give very high adhesion to different substrates, including aluminum, steel, and glass.[16,17]

REFERENCES

1. U.S. Patent 3 646 155, "Crosslinking of a Polyolefin with a Silane" Scott, H. G. Midlands Silicones, Dow Corning, 1968.
2. U.S. Patent 4 117 195, "Manufacture of Extruded Products" Swarbrick, P.; Green, W. J.; Maillefer, C. BICC Ltd. and Maillefer SA, 1974.
3. U.S. Patent 4 413 066, "Crosslinkable Polyethylene Resin Compositions" Isaka, T.; Ishioka, M.; Shimada, T.; Inoue, T. Mitsubishi Petrochemical Company, 1978.
4. U.S. Patent 2 528 523, "Process for Extruding and Insolubilizing Polymers of Ethylene" Kent, R. E.; Du Pont, 1948.
5. Lazar, M.; Rado, R.; Rychly, J. *Adv. Pol. Sci.* **1990**, *95*, 151.
6. Loan, L. D. *J. Polym. Sci.* **1964**, *A2*, 3053.
7. Fuso, J. V. *Rubber World* **1963**, *5*, 48.
8. Ahlroth, M.; Lappalainen, E.; Malm, B.; Sultan, B-Å; Teimola, R. Internal work, Borealis Polymers Oy.
9. Holmström, A. Doctoral Thesis, "On the Thermal and Thermo-oxidative Degradation of Polyethylene" Chalmers University of Technology, Gothenburg, Sweden, 1973.
10. Charlesby, A. *Proc. Roy. Soc. (London)* **1952**, *A 215*, 187.
11. Lawton, E. J.; Bueche, A. M.; Balwit, J. S. *Nature*, **1953**, *172*, 76.
12. Burns, N. M. IEEE Trans. on Power Apparatus and Systems, Vol. PAS-96, **1977**, *4*, 1196.
13. U.S. Patent 4 413 066; "Crosslinkable Polyethylene Resin Compositions" Isaka, T.; Ishioka, M.; Shimada, T.; Inoue, T. Mitsubishi Petrochemical Co. Ltd., 1979.
14. Sultan, B-Å "Moisture Crosslinkable Polyethylene with Improved Processing and Crosslinking Performance" International Wire and Cable Symposium, Atlanta, 1994.
15. International Patent Application WO 70/07542, "Silane-Crosslinkable Polymer Composition Containing a Silane Compound as a Precuring Retarder" Sultan, B-Å; Ahlstrand, L-E Borealis Holding A/S (Neste Oy), 1988.
16. Ulren, L.; Hjertberg, T. *J. Appl. Pol. Sci.* **1989**, *37*, 1269.
17. Ulren, L.; Hjertberg, T.; Ishid, H. *J. Adhesion* **1990**, *31*, 117.

Crosslinking Agents

See: Additives (Property and Processing Modifiers)
Additives (Types and Applications)
Cotton (Non-Formaldehyde Crosslinking Agents)

CROSSLINKING AGENTS, PHENYLTRIAZINES (for Polyimides and Other Aromatic Polymers)

Aldrich N. K. Lau
Corporate Research and Development
Raychem Corporation

This article serves as a brief review of the synthesis, characterization, and chemistry of phenyltriazene crosslinkers and their applications in crosslinking aromatic polyimides, poly(aryl ether sulftone), poly(arl ether ketone), and fluorinated poly(arlene ether)s.

Aromatic polyimides with high thermal stability find increasing use in electronic applications such as passivation coatings, die-attach adhesives, flexible circuit boards, and interlayer dielectrics.[1–4] However, aromatic polyimides absorb moisture and they are insoluble in common organic solvents.[5] One approach to solve these problems relies on the incorporation of flexible hexafluoroisopropylidene groups, -$(CF_3)_2C$-, into the polymer chain.[6] Aromatic polyimides containing hexafluoroisopropylidene groups exhibit lower dielectric constants that are less sensitive to moisture than the conventional aromatic polyimides.[7] These fluorinated aromatic polyimides, however, are more susceptible to crazing or dissolution upon exposure to organic solvents. Poly(aryl ether sulfone)s, poly(aryl ether ketone)s, and poly(arylene ether)s are also soluble in common organic solvents, preventing them from being used in some electronic applications.

It has been reported that crosslinking improves the resistance to micro-cracking and solvent-induced crazing of fluorinated polyimides.[8,9] Reactive end groups based on polymerizable cycloaliphatics and alkenes also have been used to crosslink polyimides. These groups include norbomene, maleimide, allylphenyl, and benzocyclobutene.[10–13] Unfortunately, the aliphatic residues produced from these crosslinking reactions reduce the thermo-oxidative stabilities of the crosslinked polymers.[14] One approach to improve thermooxidative stability relies on the replacement of these reactive endcaps with [2,2]paracyclophane groups.[14a] Thermal polymerization of the [2,2]paracyclophane endcaps produces a crosslinked network with fewer aliphatic sites. Polyimides can also be crosslinked through the trimerization of cyano pendants.[15] This approach requires a catalyst, a possible source of contamination of electronic devices. Biphenylene units have been incorporated into aromatic polyimides as crosslinking sites.[16]

A desirable crosslinker for thin films of aromatic polymers should be cost effective, processable, and should not impair the stability of the crosslinked polymer.

This article describes the synthesis and characterization of a series of phenyltriazene crosslinkers and their uses in crosslinking aromatic polymers at 200–300°C, through the formation of thermally stable aryl–aryl carbon bonds. Aromatic polymers crosslinked with phenyltriazenes are resistant to organic solvents, without impairing the dielectric constant and thermal stability, therefore qualifying them as potential candidates for many electronic applications, especially the multilayer interconnection in integrated circuits.

HOMOLYTIC AROMATIC SUBSTITUTION WITH PHENYLTRIAZENES

Benzoyl peroxide, *p*-toluenesulfonyl peroxide, benzil, N-phenyl-N′-sulfinylhydrazine, azo compounds, diazonium compounds, N-nitrosoacetanilides, arylcarbonyl-oximes, arylhalides, biphenyls, and triazenes can be used as precursors for phenyl (aryl) radicals.[17–27] The thermolysis of benzoyl peroxide at ~80°C to phenylate benzene and hexafluorobenzene through

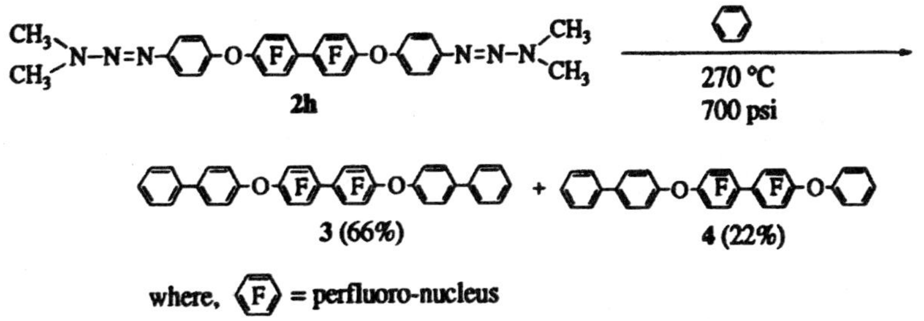

1

homolytic aromatic substitution has been reported.[28,29] Among the above-mentioned precursors for phenyl radicals, phenyltriazenes are preferable because they can be prepared easily with good yields. Phenyltriazenes are also stable at room temperature and in air.

A phenyltriazene compound **2h** reacts with benzene (**Equation 1**) at 270°C and 700 psi, in the absence of a catalyst, to give two major products; **3** produced in 66% yield and **4** in 22% yield.[30]

CROSSLINKING OF AROMATIC POLYIMIDES AND OTHER AROMATIC POLYMERS

A series of aromatic polyimides and other polyaromatics has been crosslinked with various phenyltriazene crosslinkers at 300–400°C under nitrogen. All the cured polymer thin films (10–20 μm thick) are flexible and insoluble in polar organic solvents such as NMP and DMAc. In general, the crosslinked polymers exhibit high gel contents.

The gel content of a crosslinked system is an indication of the crosslinking efficiency of a phenyltriazene crosslinker. For a given aromatic polymer, the efficiency of crosslinking depends on the chemical nature of the phenyltriazene crosslinker.[31]

Hypothetically, a crosslinker with three triazene end-groups should be more efficient than one with only two triazene end-groups.

It is believed that polymer thin films cured at high temperature exhibit residual stress.[32] Solvent swelling releases the residual stress leading to crazing upon drying. By preventing the cured polymer from swelling, chemical crosslinking improves the resistance to solvent-induced stress crazing.

APPLICATION

Phenyltriazene crosslinkers can be used in manufacturing electronic devices. Multilayer interconnection devices comprising two layers of metal conductors and three layers of polymer dielectric on top of a silicon substrate have been fabricated.[33]

REFERENCES

1. Bergen, R. G.; Gregoritsch, A. J. *IEEE Proc. Reliability Physics Symposium,* Las Vegas, Nevada, 1975.
2. Moghadam, F. H. *Solid State Technology* **1984**, *27*, 149.
3. Dupont Technical Information Bulletin, *"Kapton: Summary of Properties"* No. E-50553, 1982.
4. (a) Shah, P.; Laks, D.; Wilson, A.; *IEEE Proc. IEDM,* paper 20.4, pp 465, December, 1979; (b) Wilson, A. M. *Thin Solid Films* **1981**, *83*, 145; (c) Jensen, R. IEEE Trans. Components, Hybrids, and Manufacturing Technology, CHMT-7(4), 1984.
5. (a) Denton, D. D.; Day, D. R.; Priore, D. F.; Senturia, S. D.; Anolick, E. S.; Sceider, D. *J. Electronics Materials* **1985**, *14*, 119; (b) Denton, D. D.; Camou, J. B.; Senturia, S. D. *Proc. Intl. Symp. Moisture and Humidity* p 505, April 15–18, 1985; (c) Beuhler, A. J.; Nowicki, N. R.; Gaudette, J. M. *Polym. Mater. Sci. Eng.* **1988**, *59*, 339.
6. (a) Jones, R. J.; Chang, G. E.; Powell, S. H.; Green, H. E. *NASA Conference Publication,* **1983**, 2385 pp 271; (b) St. Clair, A.; St. Clair, T.; Minfree, W. *Polym. Mater. Sci. Eng.* **1988**, *59*, 28; (c) Goff, D. L.; Yuan, E. L. *Polym. Mater. Sci. Eng.* **1988**, *59*, 186.
7. Polyimides containing hexafluoroisopropylidene groups, for examples, Eymyd-HP™ (Ethyl Corp.) and Sixef-44™ (Hoechst), exhibit lower dielectric constants than the conventional polyimides. However, they are susceptible to solvent-induced stress crazing.
8. Pater, R. H.; Partos, R. D. *Proc. 3rd Intl. Conf. Polyimides,* Ellenville, New York, November, 1988; pp 37.
9. Mercer, F. W. U.S. Patent 4 920 005, 1990.
10. Lubowitz, H. R. *Polym. Prepr.* **1971**, *12*, 329.
11. (a) Stenzenberger, H. D. *Brit. Polym. J.* **1988**, *20*, 383; (b) Kumar, D.; Fohlen, G. M.; Parker, J. A. *J. Polym. Sci., Polym. Chem. Ed.* **1984** *22*, 927.
12. Stenzenberger, H. D.; Konig, P.; Herzog, M.; Romer, W.; Pierce, S.; Canning, M. S. *Proceedings of the 18th International SAMPE Tech. Conference,* 1997, pp 372.
13. (a) Arnold, F. E.; Tan, L. S. *Proc. 31st Intl. SAMPE Symposium* **1986**, *31*, 968; (b) Hahn, S. F.; Towsend, P. H.; Burdeaux, D. C.; Gilpin, J. A. *Polm. Matl.: Sci. Eng.* **1988**, *59*, 190.
14. (a) Baldwin, L. J.; Meador, M. A. B. *Polym. Prepn.* **1988**, *29*, 236; (b) Balde, J. W. *Electronic Materials Handbook*; Vol. 1, Minges, L. Ed. ASM International: 1989, pp 302 and references therein.
15. Hsu, L. C. *ACS Symp. Ser.* **1974**, *4*, 145.
16. (a) Droske, J. P; Gaik, U. M.; Stille, L. K. *Macromolecules* **1984**, *17*, 10; (b) Droske, J. P.; Stille, J. K.; Alston, W. B. *Macromolecules 17*, 14; (c) Takeichi, T.; Stille, J. K. *Macromolecules* **1986**, *19*, 2093.
17. (a) Box, H. C.; Budzinski, E. E.; Freund, H. G. *J. Amer. Chem. Soc.* **1970**, *92*, 5305; (b) Perkins, M. J. *Free Radicals* Vol. II, Kochi, J. K. Ed., John Wiley & Sons: New York, 1973; pp 231.
18. Hisada, R.; Kamigata, N.; Minato, H.; Kobayashi, M. *Bull. Chem. Soc. Jpn.* **1971**, *44*, 3475.
19. Fahr, A.; Stein, S. E. *J. Phys. Chem.* **1988**, *92*, 4951.
20. De Luca, G.; Renzi, G.; Cipollini, R.; Pizzabiocca, A. *J. Chem. Soc. Perkin Trans. I* **1980**, 1901.
21. (a) Koyabashi, M.; Aklyama, E.; Minato, H.; Kito, N. *Bull. Chem. Soc. Jpn.* **1974**, *47*, 1504; (b) Koyabashi, M.; Minato, H.; Watanabe, N.; Kobori, N. *Bull. Chem. Soc. Jpn.* **1970**, *43*, 258; (c) Hey, D. H.; Perkins, M. J.; Williams, G. H. *J. Chem. Soc.* **1965**, 110.
22. Cadogan, J. I. G. *Pure Appl. Chem.* **1967**, *15*, 153.
23. Cadogan, J. I. G. *Advances in Free Radical Chemistry*; Williams, G. H.; Ed. G. H. Heyden: London, 1980 pp 185.
24. Hasebe, M.; Kogawa, K.; Tsuchiya, T. *Tetrahedron Lett.* **1984**, *25*, 3887.

25. (a) Sharma, R. K.; Kharacsh, N. *Angew. Chem. Intl. Ed. Eng.* **1968**, *7*, 36; (b) Grimshaw, I.; De Silva, A. P. *Chem. Soc. Rev.* **1981**, *10*, 181.
26. (a) Recca, A.; Garapon, J.; Stille, J. K. *Macromolecules* **1977**, *11*, 1344; (b) Recca, A.; Stille, J. K. *Macromolecules* **1978**, *11*, 479.
27. Rondestvedt, C. S.; Blanchard, H. S. *J. Amer. Chem. Soc.* **1955**, *77*, 1769.
28. (a) Gill, G. B.; Williams, G. H. *J. Chem. Soc.* **1965**, 995; (b) Bolton, R.; Moss, W. K. A.; Sandall, J. P. B.; Williams, G. H. *J. Fluorine Chem.* **1975**, *5*, 61.
29. (a) Claret, P. A.; Coulson, J.; Williams, G. H. *J. Chem. Soc. (C)* **1968**, 341; (b) Bolton, R.; Sandall, J. P. B. *J. Chem. Soc. Chem. Comm.* **1973**, 286.
30. Lau, A. N. K.; Moore, S. S.; Vo, L. P. *J. Polym. Sci. Poly. Chem.* **1993**, *31*, 1093.
31. Lau, A. N. K.; Vo, L. P. *Macromolecules* **1992**, *25*, 7294.
32. (a)Wachman. E. D.; Frank, C. W. *Polymer* **1988**, *29*, 1191; (b)Kochi, M.; Isoda, S.; Yokota, R.; Mita, I.; Kambe, H. *Polyimides—Synthesis, Characterization, and Applications*; Mittal, K. L. Ed., Plenum: New York, 1984; Vol. 2, pp 671.
33. Mercer, F. W.; Goodman, T. D.; Lau, A. N. K.; Vo, L. P.; Sovish, R. C. U.S. Patent 5 114 780, 1992. Mercer, F. W.; Goodman, T. D.; Lau, A. N. K.; Vo, L. P. U.S. Patent 5 155 175, 1992.

CROWN ETHER ION-EXCHANGE RESINS

Richard A. Bartsch
Department of Chemistry and Biochemistry
Texas Tech University

Due to the strong metal ion-binding properties of crown ethers, considerable attention has been focused upon their incorporation into polymers.[1]

Formaldehyde-type condensation polymers of benzocrown and dibenzocrown ethers were first reported by Blasius and co-workers.[2–6] The resins were usually prepared by condensation of dibenzocrown ethers with formaldehyde in formic acid or of benzocrown ethers with formaldehyde and a crosslinking agent, such as phenol, resorcinol, or xylol, in a mixture of formic acid and sulfuric acid. The formaldehyde-type crown ether polymers are very easy to synthesize and have excellent resistance to heat and to acidic and basic environments.

For the application of such polymers in chromatography, sorption of a metal ion onto the resin from an aqueous solution must be accompanied by concomitant transfer of an anion. Therefore the metal ion sorption is anion dependent.

In 1981, we reported the synthesis of the crown ether carboxylic acid (**Structure 1**) and its application in the solvent extraction of alkali metal cations from aqueous solutions into chloroform.[7] In such proton-ionizable crown ethers, a side arm bearing an acidic group is attached to the crown ether framework.[8] For such proton-ionizable crown ethers, metal ion complexation involves the ion exchange of a metal ion for the proton on the acidic group. Elimination of the need to transfer an aqueous phase anion into the organic medium is of immense importance for potential applications of crown ethers as the next generation of selective metal-ion extractants. For process solvent extraction, the anions normally encountered are chloride, nitrate, and sulfate ions, which are very hydrophilic. The efficiency of metal ion extraction is markedly enhanced for protonionizable crown ethers compared with analogous compounds with non-ionizable side arms.[7,9]

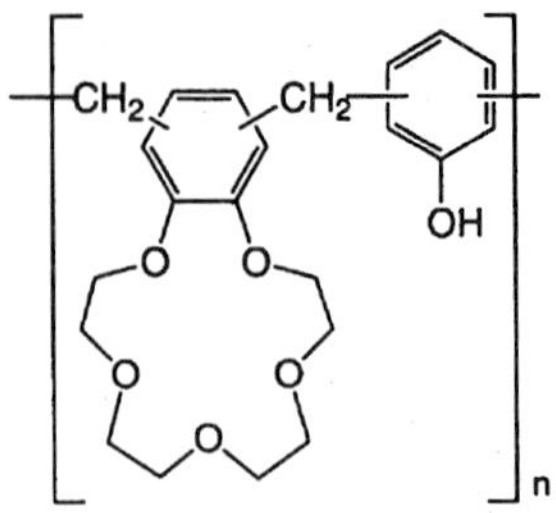

1

It was proposed to prepare formaldehyde-type condensation resins from such proton-ionizable crown ethers. The resultant resins would have both ion-exchange and polyether binding sites for metal ion complexation. It was anticipated that such dual function ion-exchange resins would provide metal ion sorption selectivities which are different from those obtainable with ordinary ion-exchange resins.[10]

SUMMARY

Novel ion-exchange resins have been prepared by condensation polymerization of dibenzocrown ethers having pendent acidic functions with formaldehyde in formic acid. These resins have both ion-exchange and polyether binding sites for metal ion complexation and provide sorption selectivities which cannot be obtained with ordinary ion-exchange resins. New Na^+-selective resins have been formulated from *sym*-(alkyl)-dibenzo-16-crown-5-oxyacetic acids. Conformational positioning of the ion-exchange group in the polyether unit is shown to have an important influence upon the recognition of alkali metal cations. Resins based upon proton-ionizable crown ether resins with phosphonic acid monoethyl ester groups are found to exhibit good sorption selectivity for Pb^{2+} from acidic aqueous solutions over a variety of heavy and transition metal cations, as well as large excesses of alkali metal and alkaline earth cations.

ACKNOWLEDGMENT

This research was supported by the Division of Chemical Sciences of the Office of Basic Energy Sciences of the U.S. Department of Energy (Grant DE-FG03-94ER14416) and the Texas Higher Education Coordinating Board Advanced Research Program.

REFERENCES

1. Smid, J., Sinta, R. *Top. Curr. Chem.* **1984**, *121*, 105.
2. Blasius, E.; Adrian, W.; Janzen, K. P.; Klauthe, G. *J. Chromatogr.* **1974**, *96*, 89.
3. Blasius, E.; Janzen, K. P. *Top. Curr. Chem.* **1981**, *98*, 165.
4. Blasius, E.; Janzen, K. P. *Pure Appl. Chem.* **1982**, *54*, 2115.
5. Blasius, E.; Janzen, K. P.; Keller, M.; Lander, H.; Nguyen-Tien, T.; Scholten, G. *Talanta* **1980**, *27*, 1007.
6. Blasius, E.; Maurer, P. *J. Chromatogr.* **1976**, *125*, 511.
7. Strzelbicki, J.; Bartsch, R. A. *Anal. Chem.* **1981**, *53*, 1894.
8. Bartsch, R. A. *Solvent Extr. Ion Exch.* **1989**, *7*, 829.

9. Bartsch, R. A.; Kim, J. S.; Olsher, U.; Purkiss, D. W.; Ramesh, V.; Dalley, N. K.; Hayashita, T. *Pure Appl. Chem.* **1993**, *65*, 399.
10. Alexandratos, S. D.; Wilson, D. L. *Macromolecules* **1986**, *19*, 280.

Crown Ethers, Polymeric

See: *Crown Ether Ion-Exchange Resins*
Crown Ether, Polymeric (Catalytic Activity)
Ionic Conductivity Switching (Photochromic Crown Compounds)
Molecular Recognition (Macromolecular Ionophore)
Phase Transfer Catalysts, Polymeric (Structure and Activity)
Photodimerizable Poly(crown ether)s
Supported Polythioethers and Polythiacrownethers

CROWN ETHERS, POLYMERIC (Catalytic Activity)

Masao Tomoi
Department of Applied Chemistry
Faculty of Engineering
Yokohama National University

The macrocyclic polyethers such as crown ethers and cryptands, which can complex specifically with alkali or alkaline earth metal ions, have been applied in varied fields of science. (I) solubilization of salts, (ii) phase transfer catalysis, (iii) ion transport (ionophores), (iv) separation of ions (chromatography), (v) sensors (ion-selective electrodes), (vi) enzyme models.[1,2]

The macrocycle units immobilized onto polymer supports may have restricted mobility and be under micro-environments different from those of free macrocycles: these may result in different binding properties of polymer-supported macrocycles or in different reactivity of ion pairs complexed with supported macrocycles.[3]

Soluble polymer-supported macrocycles are applied only as catalysts for limited reactions such as decarboxylation of potassium 6-nitrobenzisoxazole-3-carboxylate.[4] Most macrocyclic polyethers immobilized onto insoluble polymer supports have been extensively used as phase transfer catalysts: the effectiveness of the supported macrocycles was first reported in 1976–77.[5,6]

APPLICATION AS CATALYSTS

Polymer-supported crown ethers and cryptands can complex with alkali or alkaline earth metal salts, and thereby prompt the reactivity of counter anions of the salts.[3] The supported macrocycles, therefore, have been successfully used as phase transfer catalysts.[7–11] The reactions proceed in three phase mixtures consisting of aqueous reagent solution (or solid reagent), organic substrate solution, and swollen polymer catalyst gel and have been called triphase catalysis.[12] The major advantages of the supported macrocycles, compared to soluble macrocycle catalysts, are: (1) Ease of separation from reaction mixtures by simple filtration. (2) Recycling of the catalyst. (3) Exclusion of toxity of macrocyclic compounds. Some macrocyclic compounds have been known to be toxic.[1] (4) Use in flow reactors. The major disadvantages of the supported catalysts are: (1) Higher initial cost, which might be recovered if the catalysts can be used repeatedly. (2) Lower catalytic activity due to diffusional limitations.

The size of macrocycles is an important factor to provide polymer-supported phase transfer catalysts with high activity. It is recognized that cation binding strength for cryptands is considerably larger than that for crown ethers.[1] Crown ethers with 18-membered rings can complex with potassium salts more strongly than do 15-membered crown ethers.[1] Therefore, polymer-supported catalysts with 18-membered crown ethers or cryptands are mainly used for phase transfer reactions.

REFERENCES

1. Gokel, G. *Crown Ethers & Cryptands*; The Royal Society of Chemistry, Cambrige: England, 1991.
2. Starks, C. M.; Liotta, C. *Phase Transfer Catalysis: Principles and Techniques;* Academic: New York, NY, 1978.
3. Smid, J.; Sinta, R. *Top. Curr. Chem.* **1984**, *121*, 105.
4. Smid, J.; Varma, A. V.; Shah, S. C. *J. Am. Chem. Soc.* **1979**, *101*, 5764.
5. Cinquini, M.; Colonna, S.; Molinari, H.; Montanari, F.; Tundo, P. *J. Chem. Soc. Chem. Commun.* **1976**, 394.
6. Molinari, H.; Montanari, F.; Tundo, P. *J. Chem. Soc. Chem. Commun.* **1977**, 639.
7. Tomoi, M.; Ford, W. T. *Syntheses and Separations Using Functional Polymers,* Sherrington, D. C.; Hodge, P. Eds.; John Wiley & Sons: Chichester, England, 1988; Chapter 5.
8. Regen, S. L. *Angew. Chem. Int. Ed. Engl.* **1979**, *18*, 21.
9. Montanari, F.; Landini, D.; Rolla, F. *Top. Curr. Chem.* **1982**, *101*, 147.
10. Ford, W. T.; Tomoi, M. *Adv. Polym. Sci.* **1984**, *55, 49.*
11. Goldberg, Y.; *Phase Transfer Catalysis: Selected Problems and Applications*; Gordon & Breach Sci.: Yverdon, Switzerland, 1992; Chapter 5.
12. Regen, S. L. *J. Am. Chem. Soc.* **1975**, *97*, 5956.

Crytands

See: *Crown Ethers, Polymeric (Catalytic Activity)*

CRYSTALLINE POLYMERS (Structural Aspects)

Kohji Tashiro
Department of Macromolecular Science
Faculty of Science
Osaka University

MORPHOLOGICAL DESCRIPTION

A polymer solid is partially crystalline or noncrystalline.[1] Partially crystalline polymers, called crystalline polymers, are constructed by a complicated aggregation of crystalline and amorphous regions. In the crystalline region the molecular chains are packed parallel together and form a crystallite of finite size, ca. 100–200 Å. A molecular chain is in general far longer than the crystallite size; for example, fully extended polyethylene (PE) chain of molecular weight 50,000 is about

4500Å. Therefore one molecule is considered to pass through many crystalline and amorphous regions successfully.

Crystallites consist of a regular packing of extended chains. these chains are not necessarily fully extended from one end to the other end, but can fold back to the crystallite. In this case, many types of reentry to chains may be possible. Of course such chain folding does not occur for all the polymers but depends on the rigidity of the polymer chain. For example, typical chain folding is observed for flexible PE, while rigid poly(p-phenylene benzobisoxazole) (PBO) cannot fold back easily. PBO is considered to crystallize in a so-called fringed micelle structure in which the polymer chains are arrayed basically parallel and the domains of regular and irregular lateral packings of the chains are repeated along the chain axis.

As stated above, under the usual crystallization condition such as slow cooling from the melt or crystallization from the dilute solution, PE crystallizes in a lamellar form with the chain-folding structure. This type of crystal is called a "folded chain crystal (FCC)." the lamellae consisting of these FCC are aggregated to form a larger scale crystalline system. For example, slow cooling from the melt results in the formation of spherulite, in which lamellar crystals are stacked and grow radially from the center of spherulitic crystal. On the other hand, when the PE sample is crystallized under high pressure at high temperature, it forms an extended chain crystal (ECC) of several μm length, in which the fully extended zigzag chains are gathered to form a giant crystalline block. Ultradawn PE sample prepared by stretching the dried gel by several hundred times the original length is known to form this ECC over a whole fiber. When PE is crystallized in a dilute solution under an agitated stirring, it crystallizes into a form of shish-kebab structure; the shish part consists of ECC and the kebab part of FCC.

The aggregation structure of crystalline and amorphous phases may be changed sensitively and complicatedly depending on the sample preparation condition. By stretching the unoriented sample, the FCC may change into the ECC to form a fringed micelle structure. In the uniaxially drawn sample, the crystallites may distribute with their chain axes oriented along the draw axis but without any preferable orientation of the lateral axes. By rolling this sample, the crystallites may be rearranged with three-dimensional orientation of the crystal axes. This type of sample is called the "double oriented sample."

STRUCTURAL ANALYSIS OF POLYMER CRYSTAL

The crystal structure of polymers is analyzed by the various methods. Among them, the most powerful method may be the X-ray structural analysis. The vibrational spectroscopy, or the infrared and Raman spectroscopy, is also useful for analyzing the molecular and crystal structure of polymers. Solid-state NMR might be another method to analyze the structure as well as the mobility of the chains.

Vibrational Spectroscopy in Structural Investigation

Crystallization-Sensitive Bands

When a polymer sample is annealed at a given temperature for a certain time, several bands in the spectra increase their intensity remarkably. These bands disappear on melting of the crystals. They are called crystallization-sensitive bands, and can be utilized for evaluation of the degree of crystallinity of the sample.

Conformation-Sensitive Bands

Polymers may crystallize into various types of crystal modifications with different molecular conformations. For example, poly(vinylidene fluoride) (PVDF) crystallizes into at least four crystal modifications, depending on the preparation conditions of the sample: form I takes essentially the all-trans zigzag conformation, form II a glide-type $TG T\overline{G}$ conformation, and form III a $TTTGTTT\overline{G}$ conformation.[2] The infrared and Raman spectra of these crystal forms are different from each other and the bands characteristic of *trans* and *gauche* conformers can be detected. By utilizing these conformation-sensitive bands, we may investigate the conformational change occurring in the crystalline region.

Ferroelectric Phase Transition

Vinylidene fluoride-trifluoroethylene (VDF-TrFE) copolymer exhibits a ferroelectric-to-paraelectric phase transition, the first such case for a synthetic polymer.[3] In this transition the electric polymerization and piezoelectric constant of the film disappears above the Curie point T_c. The temperature dependence of the dielectric constant obeys the so-called Curie-Weiss law, $1/\varepsilon \propto T-T_c$. The ferroelectric phase has a polar structure consisting of planar-zigzag chains with the CF_2 dipoles arranged parallel to the b axis. This is essentially the same structure as that of PVDF form I.

Stress-Induced Phase Transition

Natural rubber orients and crystallizes under tension and reverts to its original amorphous state by relaxation.[4] Stress-induced transition is observed in some crystalline polymers, such as poly(tetramethylene terephthalate) (PTMT) and its block copolymers with poly(tetramethylene oxide), PEO, poly-oxacyclobutane, nylon 6, poly(ethylene oxybenzoate) (PEOB), PVDF, polypivalolactone, keratin, and so on.[5–17]

Thermochromic Phase Transition

Electrically conductive poly(3-alkylthiophene) is one typical case; the color changes between red (low temperature) and yellow (high temperature).

Thermochromic transition can be seen also for poly(alkylsilane), [(-SiR_2-), R≡C_6H_{13} etc.] where the ultra-violet absorption spectrum changes its peak position: 370–380 nm at room temperature and 316 nm above 40°C (for R = n-hexyl).[18] The absorption is related to the strong electronic transition of the σ-bonded backbone and thus is sensitive to molecular structural change.

STRUCTURE AND MECHANICAL PROPERTIES OF POLYMER CRYSTALS

Crystallite modulus or the Young's modulus along the chain axis is obtained mainly by measuring the stress-induced shift of the X-ray reflection. It is very sensitive to the molecular conformation.[19] Even if the chemical structure is the same, the modulus is different if the conformation is different.

Such a difference in the Young's modulus among the various chain conformations can be understood by investigating the deformation mechanism or the atomic displacements induced by the external tensile force. The force constants for the bond stretching, bond angle deformation, and internal rotation are lower by one digit to each other, reflecting on the difference in the Young's modulus.

The theoretically derived molecular deformation mechanism can be checked experimentally. Among the various methods, the vibrational spectroscopic method may be the most sensitive to such a slight change in structure and force field caused by a tensile force.

An evaluation of three-dimensional elastic constants is desirable for further understanding of mechanical behaviors of crystalline polymers.

REFERENCES

1. Tashiro, K.; Tadokoro, H. *Encyclo. Polym. Sci. Eng.* **1989**, Suppl., 187
2. Tashiro, K.; Kobayashi, M.; Tadokoro, H. *Macromolecules* **1981**, *14*, 1757.
3. Tashiro, K.; Kobayashi, M. *Phase Transitions* **1989**, *18*, 213.
4. Mandelkern, L. *Crystallization of Polymers*, McGraw-Hill: New York, 1964.
5. Jakeways, R.; Ward, I. M.; Wilding, M. A.; Hall, I. H.; Desborough, I. J.; Pass, M. G. *J. Polym. Sci., Polym. Phys. Ed.* **1975**, *13*, 799.
6. Takahashi, Y.; Osaki, Y.; Tadokoro, H. *J. Polym. Sci., Polym. Phys. Ed.* **1980**, *18*, 1863.
7. Tashiro, K.; Hiramatsu, M.; Ii, T.; Kobayashi, M.; Tadokoro, H. *Sen'i Gakkaishi* **1986a**, *42*, T-597.
8. Tashiro, K.; Hiramatsu, M.; Ii, T.; Kobayashi, M.; Tadokoro, H. *Sen'i Gakkaishi* **1986b**, *42*, T-659.
9. Takahashi, Y.; Sumita, I.; Tadokoro, H. *J. Polym. Sci., Polym. Phys. Ed.* **1973**, *11*, 2113.
10. Tashiro, K.; Tadokoro, H. *Rep. Progr. Polym. Phys. Jpn.* **1978**, *21*, 417.
11. Miyasaka, K.; Ihikawa, K. *J. Polym. Sci., Part A-2* **1968**, *6*, 1317.
12. Takahashi, Y.; Kurumizawa, T.; Kusanagi, H.; Tadokoro, H. *J. Polym. Sci., Polym. Phys. Ed.* **1978**, *16*, 1999.
13. Lando, J. B.; Olf, H. G.; Peterlin, A. *J. Polym. Sci., Part A* **1966**, *4*, 941.
14. Hasegawa, R.; Takahashi, Y.; Chatani, Y.; Tadokoro, H. *Polym. J.* **1972**, *3*, 600.
15. Prud'homme, R. E.; Marchessault, R. H. *Macromolecules* **1974**, *7*, 541.
16. Hearle, J. W. S.; Chapman, B. M.; Senior, G. S. *Appl. Polym. Symp.* **1971**, *18*, 775.
17. Tashiro, K.; Nakai, Y.; Kobayashi, M.; Tadokoro, H. *Macromolecules* **1980**, *13*, 137.
18. West, R. *J. Organomet. Chem.* **1986**, *300*, 327.
19. Tashiro, K. *Progr. Polym. Sci.* **1993**, *18*, 377.

CRYSTALLIZABLE POLYMER BLENDS

Clara Silvestre,* Sossio Cimmino, and Emilia De Pace
Isituto di Ricerca e Tecnologia delle Materie Plastiche
Italian National Council of Research (CNR)

Crystallizable polymer blends represent a commercially important and emerging class of multicomponent polymer systems. It is sufficient, in fact, to consider that roughly one-half to two-thirds of all industrially significant polymers are crystalline or crystallizable and that many of the newly commercialized blends' products involve a crystallizable component.[1,2]

In this paper the up-to-date results on the most important factors that control the miscibility, the crystallization process, the morphology, the phase structure, and the melting behavior of crystallizable polymer blends are reported. The discussion is restricted to binary blends with only one crystallizable component.

CRYSTALLIZATION

As in pure polymers, crystallization in polymer blends can occur in a range of temperatures limited by the glass transition temperature, T_g, and the melting temperature, T_m.[3,4] In the case of miscible blends the range in which the crystallization can occur is altered by the presence of the non-crystallizable component, as T_g and T_m are dependent on composition. Thermodynamic consideration always predicts a depression of the melting temperature as a function of the composition. Therefore, the range of available crystallization temperatures will be expanded or contracted depending on the T_g of the components.

Owing to the presence of the amorphous component, the crystallization process in the blends will be different from that of pure crystalline polymer and will depend greatly on the miscibility between the two components in the melt at T_c.

Several studies are reported in the literature on the crystallization behavior of crystallizable polymer blends. The first studies in the area of crystallization of polymer blends were done by Keith and Padden on atactic polypropylene/isotactic polypropylene and aPS/iPS blends.[5,6]

With few exceptions, the literature reports have indicated a depression of the growth kinetics of the crystallizable component, upon addition of a miscible, non-crystallizable material.[7–19] The entity of the depression is dependent on the nature of the materials used, on the composition, and on a complex balance among the degree of miscibility, the influence of the non-crystallizable material on the nucleation, and on the crystallization regime. Very high depression of the spherulite growth rate owing to the presence of a polymeric diluent is often observed for low values of T_c, whereas when increasing T_c the decrease becomes smaller.

MORPHOLOGY

The phase structure and the morphology of a crystallizable blend depend strictly on the composition, chemical nature of the components, and on the interaction between the components.

The morphologies of the condensed state reported in the literature for several crystallizable polymer blends can be summarized as follows: for blends whose components are miscible in the melt and in the amorphous state no segregated domains of the second component are observed in the melt and after crystallization, indicating that the non-crystallizable material is likely incorporated during crystallization within the interlamellar regions of growing spherulites. The long period increases with the crystallization temperature, and for a given T_c, increases the content of the amorphous component in the blend.

*Author to whom correspondence should be addressed.

The amorphous and the interphase thickness increase with composition. For these blends the structure is constituted of crystalline lamellae separated by amorphous and transition regions containing an homogeneous mixture of the two components.

Many reports can be found in the literature on the morphology of blend systems whose components are immiscible above and below the melting temperature. In the melt and after crystallization the minor component forms its own domains. In the case of blends made by a thermoplastic polymer and an elastomer in the melt, the domains of the dispersed phase are generally spherical. In the condensed state, after crystallization, the material is characterized by the presence of spherulites that include the elastomeric domains in intra- and/or inter-spherulitic regions. The state and mode of dispersion of the rubber particles are strongly dependent on the molecular structure as well as the molecular mass of the rubber component.

The addition of the amorphous component can influence the internal structure of the crystals of semicrystalline polymer.

APPLICATION

The driving force for the development and use of polymer blends relies generally on the cost reduction and synergistic property profiles achievable. In fact, an expensive polymer may be combined with a less costly polymer to provide adequate performance at a significantly reduced price.[20]

Why is it convenient to combine a crystalline polymer and an amorphous polymer? The most used crystalline polymers, such as polyolefins, nylon, and polyacetals, present generally excellent chemical resistance, significant flexural tensile strength, good heat distortion, and mechanical properties, and moreover exhibit low viscosity above their melting point. The most used amorphous polymers in blends, such as ABS, polycarbonate, polysulfone, and polyarylate have generally good dimensional stability, warpage resistance, and impact strength. So when combining a crystalline polymer with an amorphous polymer the resulting blend provides good dimensional stability, ease of processing, chemical resistance, as well as mechanical properties that can be tailored for a specific application.

There is a large body of patent literature and a growing amount of scientific literature on crystallizable polymer blends. Most of the current work in developing engineering blends for automotive industry (car-body panels) relies on the combination of crystalline/amorphous polymers. Blends such as nylon/PPO, poly(butylene terephthalene)/polyarylate and PBT/polycarbonate, provide a combination of warpage resistance for large part molding, chemical resistance to automotive and road chemicals, impact resistance, and the stiffness necessary for application to the panels. This combination of properties is currently unobtainable with a single commercial polymer.

The modification through blending of the poly[D(-)-3-hydroxybutyrate] (PHB), a natural aliphatic polyester produced by bacterial fermentation is interesting. PHB has attracted much interest for medical and agricultural applications because it is biodegradable and highly biocompatible. The use of PHB has some limitations: high cost production, a very narrow processability window, and a relatively low impact resistance. PHB degrades to crotonic acid if it is kept at a temperature of only few degrees above its melting point for a relatively long time; the injection-molded samples have a high crystallinity degree and show brittle behavior, especially at temperatures below the glass transition temperature.

In order to improve the properties of PHB attempts have been made to blend a second polymer component. In literature studies of blends of PHB with PEO, poly(vinyl acetate), polyepichlorohydrin, cellulose esters, and aPMMA have been reported.[18,21–26]

Other groups of crystallizable polymer blends presenting good technological properties include polycarbonate with such crystallizable polyesters as poly(ethylene terepthalate) and poly(butylene terephthalate) and poly(vinylidene fluoride) with a variety of oxygen-containing polymers, typically of the acrylic type and polycaprolactone. The last kind of blends have chemical and solvent resistance that is reduced from that of pure PVF_2, but greatly improved over that of acrylates. The addition of PVF_2 results in materials that are more resistant to environmental stress cracking or crazing. They offer combinations of transparency, toughness, and weatherability and are expected to also find application in the locomotive area as decorative striping and labeling, where clarity, chemical resistance, and adhesion to acrylic paint are required.[27]

The combination of crystallizable polymers with elastomers constitutes another class of crystallizable polymer blends. The crystalline polymers provide strength, stiffness, and thermal resistance, while the elastomers, such as polyurethane, styrene block copolymers, copolyester elastomers, and ethylene-copropylene elastomers provide impact modification. The blending of polypropylene with EPDM for bumper covers is a significant commercial example.[28]

REFERENCES

1. Utracki, L. I. *Polymer Alloys and Blends: Thermodynamics and Rheology*; Hanser: Munich, 1989.
2. Paul, D. R.; Barlow, J. W. *Polymer Sci. Technol.* **1980**, *11*, 239.
3. Mandelkern, L. *Crystallization of Polymers*; McGraw-Hill: NY, 1964.
4. Wunderlich, B. *Macromolecular Physics*; Academic: NY, 1976.
5. Keith, H. D.; Padden, F. J. *J. Appl. Phys.* **1964**, *35*, 1270.
6. Keith, H. D.; Padden, F. J. *J. Appl. Phys.* **1964**, *35*, 1286.
7. Cimmino, S.; Martuscelli, E.; Silvestre, C.; Cecere, A.; Fontelos, M. *Polymer* **1993**, *34*, 1207.
8. Martuscelli, E.; Demma, G. *Polymeric Blends: Processing, Morphology, Properties*; Martuscelli, E.; Palumbo, R.; Kryszeski, M., Eds., Plenum: 1980.
9. Martuscelli, E.; Pracella, M.; Yue, W. P. *Polymer* **1984**, *25*, 1097.
10. Martuscelli, E.; Silvestre, C.; Gismondi, C. *Makromol. Chem.* **1985**, *186*, 2161.
11. Cimmino, S.; Martuscelli, E.; Silvestre, C.; Canetti, M.; De Lalla, C.; Seves, A. *J. Polym. Sci., Polym. Phys. Ed.* **1989**, *27*, 1781.
12. Wang, T. T.; Nishi, T. *Macromolecules* **1977**, *10*, 421.
13. Kwei, T. K.; Patterson, G. D.; Wang, T. T. *Macromolecules* **1976**, *9*, 780.
14. Berghmans, H.; Ovembergh, N. *J. Polym. Sci., Polym. Phys. Ed.* **1977**, *15*, 1757.
15. Robeson, L. M. *J. Appl. Polym. Sci.* **1973**, *17*, 3607.
16. Cimmino, S.; Di Pace, E.; Martuscelli, E.; Silvestre, C. *Polymer* **1993**, *34*, 2799.
17. Yeh, G. S. Y.; Lambert, S. L. *J. Polym. Sci.*, Part A-2, **1972**, *10*, 1183.
18. Greco, P.; Martuscelli, E. *Polymer* **1989**, *30*, 1475.

19. Lotti, N.; Pizzoli, M.; Ceccorulli, G.; Scandola, M. *Polymer* **1993**, *23*, 4935.
20. Kienzle, S. Y. *Advances in Polymer Blends and Alloys Technologies*; Kohudic, M. A., Ed. Technomic: A. G., 1988; Vol. 1.
21. Dubini Paglia, E.; Beltrame, P. L.; Canetti, M.; Seves, A.; Marcandalli, A.; Martuscelli, E. *Polymer* **1993**, *34*, 996.
22. Avella, M.; Martuscelli, E. *Polymer* **1988**, *29*, 1731.
23. Sadocco, P.; Canetti, M.; Seves, A.; Martuscelli, E. *Polymer* **1993**, *34*, 3369.
24. Scandola, M.; Ceccorulli, G.; Pizzoli, M. *Macromolecules* **1992**, *25*, 6441.
25. Canetti, M.; Sadocco, P.; Siciliano, A.; Seves, A. *Polymer* **1994**, *35*, 2884.
26. Siciliano, A.; Seves, A.; De Marco, T.; Cimmino, S.; Martuscelli, E.; Silvestre, C. *Macromolecules* **1995**, *28*, 8065.
27. Murft, S. R.; Barlow, J. W.; Paul, D. R. *Multicomponent Polymer Materials*; Paul, D. R.; Sperling, L. H., Eds.; Adv. Chem. Series, Am. Chem. Soc: Washington, 1986; Chapter XVIII.
28. De, S. K.; Bhowmick, A. K. *Thermoplastic Elastomers from Rubber-Plastic Blends*; Ellis Horwood: 1990.

Crystallization

See: *Crystalline Polymers (Structural Aspects)*
Crystallizable Polymer Blends
Paracrystals
Plasticizers (Polymer–Plasticizer Interactions)
Poly(aryl ether ether ketone) [PEEK] (Bulk Crystallization Kinetics)
Poly(aryl ether ether ketone) [PEEK] (Conformational Order Studied Using FTIR)
Poly(butylene terphthalate) (Annealing Behavior, Thermal Analysis)
Polyethylene (New Aspects of Morphology)
Poly(ethylene terephthalate) (Heat Setting and Thermal Stabilization)
Polyimide, NEW-TPI (Semicrystalline Thermoplastic)
Polymer–Polymer Mixtures (under High Pressure; Physical Properties)
Polypropylene (Crystallization Kinetics)
Polypropylene (Nucleating Agents)
Polypropylene, Isotactic (Polymorphism)
Polypropylene, Isotactic (Supermolecular Structure)
Polystyrene, Stereoregular
Polystyrene, Syndiotactic
Polystyrene, Syndiotactic (Sulfonation and Crystallization)
Poly(vinyl chloride) (Structural Order)
Poly(vinylidene fluoride) (Overview)
Poly(vinylidene fluoride) Phases and Morphology (Effect of Crystallization Conditions)
Poly(p-xylylene)s
Spheres, Poly(butylene terephthalate)

Crystals

See: *Colloidal Crystals, Polymeric*
Crystals, Phase Transitions

CRYSTALS, PHASE TRANSITIONS

Takashi Itoh and Masato Hashimoto
Department of Polymer Science and Engineering
Faculty of Textile Science
Kyoto Institute of Technology

There have been numerous experimental reports for phase transitions in polymer crystals. A polymer generally exhibits several polymorphs according to crystallization condition, mechanical stress, plastic deformation, temperature/pressure change, etc. We introduce several typical and interesting examples of the crystal–crystal or crystal–mesophase phase transitions, classifying the phenomena mainly in terms of the physical parameters (intensive variables) that cause the structural changes in the crystals, as well as the reversibility of the phenomena.

REVERSIBLE PHASE TRANSITIONS WITH TEMPERATURE AT NORMAL PRESSURE

Single Crystal of Nylon 66

Nylon 66 exhibits the characteristic change in the crystal structure with temperature: the 100 reflection and the 010, 110 doublet merge into a single peak and the triclinic unit cell appears to change to a pseudohcxagonal form when the temperature is raised.[1] Similar phenomena have been observed also in nylons 610, 612, 7, etc.[2] New information has been obtained by preparing the single crystals of nylon 66 from the 1,4-butanediol dilute solutions with different concentrations. Electron and X-ray diffraction results have revealed that the first- and second-order phase transitions occur, as well as continuous structural change.[3,4] A theoretical consideration suggests that both of the first- and second-order phase transitions are able to occur in nylon 66 crystal, while they occur with difficulty in nylons 6 and 46, which was also confirmed experimentally.[5–7] The transition mechanism is the order-disorder type, while the phenomenon observed in a bulky sample of nylon 66 is a continuous structural change.[5]

Single Crystal of Poly[2,4-hexadyne-1,6-diol bis(*p*-toulenesulfonate) (PPTS)

A macroscopic single crystal can be formed by solid-state-polymerization of the monomer crystals of diacetylenes with substituents. For example, a macroscopic single crystal of PPTS is obtained from bis(*p*-toluenesulfonate) diacetylene single crystal, which shows the second-order phase transition at 195 K. Temperature dependence of the dielectric-constant tensor was measured using the single crystal of the polymer, and antiferroelectric-type transition mechanism was suggested from the dielectric behavior around the transition point.[8]

Single Crystal of Poly(*p*-xylylene) (PPX)

This polymer crystallizes into the α-form through polymerization, which is irreversibly transformed to hexagonal β-form by heat treatment or drawing.[9,10] The precise feature of phase transition for this polymer became clear with transmission electron microscopy using a single crystal precipitated from 0.5 wt% α-chloronaphthalene solution at 483 K.

Polytetrafluoroethylene (PTFE)

It has been well known that PTFE shows two first-order phase transitions at 292 K and at 303 K under atmospheric pressure.[11] Below 292 K, the chain conformation was considered to be 13/6 helix in phase II crystal but revised to be 2.1598 CF_2 units per turn of the helix on the basis of the electron diffraction result.[11] Above 292 K the polymer is in hexagonal phase IV and above 303 K in hexagonal phase I.

trans-1,4-Polybutadiene (PBD)

PBD crystal changes from form-I crystal (monoclinic, chain conformation is *trans*-ST$\overline{S}$) to form-II crystal (pseudo-hexagonal) at ca. 350 K.[12–15]

Poly(n-alkyl-L-glutamate)s; [Glu(Alkyl)]s (n < 19)

In crystals of [Glu(Alkyl)]s, intramolecular hydrogen bonding gives rigidity to the helical backbone, onto which flexible n-alkyl side groups are appended. These crystals show the transition between 250 and 400 K.[16–18] The transitions seem to be highly affected by the side chains.[16] The transition temperature becomes lower in the case of the polymers with longer side chains owing to the large degree of disorder in the side chain structure.

Ferroelectric Phase Transitions in Poly(vinylidene fluoride) (PVDF) and Vinylidene Fluoride/Trifluoroethylene (VDF/TrFE) Copolymers

There have been numerous reports about crystal structures of PVDF and VDF/TrFE copolymers as well as their ferroelectric properties, and several reviews were already issued.[19–23]

REVERSIBLE PHASE TRANSITIONS UNDER HIGH PRESSURE

PE and PTFE are typical polymers that exhibit new phases and reversible phase transitions at elevated temperatures and pressures. The phase diagrams of both polymer crystals have been reported (PTFE and PE).[24,25] The high-pressure phase (type III) of PTFE is orthorhombic, in which the chain conformation is planar-zigzag.[26] The orthorhombic low-temperature phase of PE transforms to the hexagonal high-temperature phase on raising the temperature under high pressure (triple point: 350 MPa, 488 K).

There are two examples that the order of phase transition changes with increasing pressure. Nylon 66 crystal precipitated from the 0.5 wt% 1,4-butanediol solution at 418 K shows the continuous structural change at atmospheric pressure, while the first-order phase transition was confirmed to occur at pressures higher than 100 MPa. Similar phenomena were also found for VDF/TrFE copolymer with 42% VDF content, that is, the continuous structural change in the crystal at atmospheric pressure changes to the first-order phase transition at pressures higher than 200 MPa.

Unusual pressure-induced phase behavior has been observed in poly(4-methylpentene-1).[27,28] The crystal phase (tetragonal) loses its structural order with increasing pressure isothermally, passing through a continuously varying sequence of mesomorphic phases. This process is reversible and observed in two widely separated temperature regions, which suggests the possibility of reentrant liquid-crystal and amorphous phases for the first time in a single component system.

MESOPHASE

In the high-temperature phases, many polymers exhibit the hexagonal structure as mentioned above except the cases of nylon 66 and PPTS. Such hexagonal phases are considered to be the so-called "mesophase," which includes "CONDIS (conformationally disordered) crystal," "plastic crystal," and "liquid crystal."[29,30]

REVERSIBLE PHASE TRANSITIONS WITH STRESS

α-β Transition of Keratin[31]

The α-helix conformation of keratin, which is included in hair, wool, etc., transforms into β-sheet structure with tensile stress and reverts upon being released from the stress. The transition temperature from α to β decreases with increasing stress and the Clausius-Clapeyron's equation holds true.

Poly(tetramethylene terephthalate) (PTMT) and Poly(pentamethylene terephthalate) (PPMT)[32,33]

Both polymers exhibit the reversible α-β transition induced by drawing and relaxing the sample. The crystal structures of α- and β-forms are triclinic and the chain conformation in β-form is extended; that is, α-β transformation is mainly caused by the expansion and contraction of the methylene group. Under a low stress condition, no intermediate phase appears but α-β forms coexist.

THEORETICAL ANALYSIS

There have not been many theoretical works about the phase transitions in polymer crystals, in contrast to the numerous experimental works; and the theory for phase transitions in polymer crystals has not developed as remarkably in statistical mechanics as has that for polymer solutions, because the molecular chains are strongly restricted in the potential field of polymer crystal and it is extremely difficult to analyze the phase transition mechanism on the basis of the strict estimation for the potential field in the crystal system.

SUMMARY

Several examples of phase transitions in polymer crystals were explained, especially for the phenomena reversible with intensive variables (temperature, pressure, and stress).

No theories have been established to generally explain the molecular mechanism for such phase transitions in polymer crystals as mentioned. Such a situation makes the prediction for the phenomena difficult. Theoretical development is necessary for the molecular design of polymeric material that changes its physical properties drastically through the phase transition. A biological system yields and controls physiological functions utilizing the phase transition phenomena. Applying the properties of each phase of polymer crystals to the field of technology is expected.

REFERENCES

1. Brill, R. *J. Pract. Chem.* **1942**, *161*, 49.
2. Slichter, W. P. *J. Polymer Sci.* **1959**, *35*, 77.

3. Itoh, T.; Konishi, T. *J. Phys. Soc. Jpn.* **1993**, 62, 407.
4. Itoh, T.; Wakaura, M.; Machitani, Y.; Konishi, T. *J. Phys. Soc. Jpn.* **1994**, *63*, 2833.
5. Itoh, T.; Wakaura, M.; Machitani, Y.; Konishi, T. *Polymer Preprnts, Jpn.* **1992**, *41*, 4517.
6. Itoh, T.; Wakaura, M.; Funaki, A.; Konishi, T. *Rept. Progr. Polym. Phys. Jpn.* **1992**, *35*, 217 .
7. Machitani, Y.; Wakaura, M.; Itoh, T.; Konishi, T. *Rept. Progr. Polym. Phys. Jpn.* **1993**, *36*, 171.
8. Orczyk, M. E. *Chem. Phys.* **1990**, *142*, 485.
9. Iwamoto, R.; Wunderlich, B. *J. Polym. Sci. Polym. Phys. Ed.* **1973**, *11*, 2403.
10. Niegisch, W. D. *J. Appl. Phys.* **1966**, *37*, 4041.
11. Weeks, J. J.; Clark, E. S.; Eby, R. K. *Polymer* **1981**, *22*, 1480.
12. Iwayanagi, S.; Sakurai, I.; Sakurai, T.; Seto, T. *J. Macromol. Sci. Phys. Ed.* **1968**, *2*, 163 .
13. Natta, G.; Corradini, P. *Nuovo Cimento, (Suppl. 1)* **1960**, *15*, 9.
14. Tasumi, T.; Fukushima, T.; Imada, K.; Takayanagi, M. *J. Macromol. Sci.* **1967**, *B1*, 459.
15. Finter, J.; Wegner, G. *Makromol. Chem.* **1981**, *182*, 1859.
16. Sasaki, S.; Nakamura, T.; Uematsu, I. *J. Polym. Sci. Polym. Phys. Ed.* **1979**, *17*, 825.
17. Watanabe, J.; Ono, H.; Uematsu, I..; Abe, A. *Macromolecules* **1985**, *18*, 2141.
18. Watanabe, J.; Takashina, Y. *Macromolecules* **1991**, *24*, 3423.
19. Lovinger, A. J. Development in Crystalline Polymers—I; Applied Science: London and New Jersey, 1981; Chapter V.
20. Odajima, A.; Tashiro, K. *J. Cryst. Soc. Jpn.* **1984**, *26*, 103.
21. Furukawa, T. *Phase Transitions* **1989**, *18*, 143.
22. Nalwa, H. S. *Rev. Macromol. Chem. Phys.* **1991**, *C31*, 341.
23. Kepler, R. G.; Anderson, R. A. *Advances in Physics* **1992**, *41*, 1.
24. Matsushige, K.; Enoshita, R.; Ide, T.; Yamauchi, N.; Taki, S.; Takemura, T. *Jpn. J. Appl. Phys.* **1977**, *16*, 681.
25. Hikosaka, M.; Minomura, S.; Seto, T. *Jpn. J. Appl. Phys.* **1980**, *19*, 1763.
26. Nakafuku, C.; Takemura, T. *Jpn. J. Appl. Phys.* **1975**, *14*, 599.
27. Rastogi, S.; Newman, M.; Keller, A. *Nature* **1991**, *353*, 55.
28. Rastogi, S.; Newman, M.; Keller, A. *J. Polym. Sci., Polym. Phys. Ed.* **1993**, *31*, 125.
29. Wunderlich, B. *Adv. Polym. Sci.* **1988**, *87*, 1.
30. Ungar, G. *Polymer* **1993**, *34*, 2050.
31. Chapmann, B. M. *J. Textile Inst.* **1969**, *60*, 181.
32. Yokouchi, M.; Sakakibara, Y.; Chatani, Y.; Tadokoro, H.; Janaka, T.; Yoda, Y. *Macromolecules* **1976**, *9*, 266.
33. Hall, I. H.; Rammo, N. N. *J. Polym. Sci.* **1978**, *16*, 2189.

Curing

See: *Acrylic Elastomer Curing*
Chlorosulfonated Polyethylene
Epoxy Resin-Glass Interphases (Optical Properties)
Epoxy Resins (Curing Reactions)
Epoxy Resins (High Performance Composite Applications)
Epoxy Resins (Microwave Processing)
High Temperature Phthalonitrile Resins
Peroxide Initiators (Overview)
Polysulfides (Use in Sealants)
Polyurethanes, Blocked Copolymer (Reactive Modifiers for Epoxy Resins)
Printing Inks
Quinone-Diamine Polymers
Rubber Additives, Multifunctional
Silicone Release Coatings
Siloxane Elastomers
Unsaturated Polyester Resins (Overview)
Vulcanization (Overview)

CYANATE-BASED RESINS (Polycyanurates)

Monika Bauer and Jörg Bauer
Fraunhofer Institute of Applied Materials Research

Because of its electronic structure, the cyanate group can undergo a multitude of reactions with electrophilic and nucleophilic agents as summarized by Martin.[1]

One of the most remarkable reactions of the cyanate group is its cyclotrimerization into cyanurates. This reaction can be applied successfully to the synthesis of polymers, when using di- or polyfunctional cyanates. Polymers with a high fraction of aromatic structures are obtained by this reaction. Depending on the structure of the monomer, the resulting polycyanurates show a large bandwidth of attractive properties. The main fields of application of polycyanurate thermosets are similar to those of epoxies: binders for laminates for both electronic and structural applications, high-temperature adhesives and casting resins, and binders for miscellaneous substrates.

PREPARATION AND PROPERTIES

Monomers

A universal procedure for the synthesis of cyanates from their corresponding phenols is the cyanogen halide method.[2,3] This method has been widely applied to get mono- and polyfunctional cyanates in good yields and purities on laboratory and production scale. An alternative route for the synthesis of chemically pure cyanates is the thermolysis of thiatriazols.[4,5] However, this method is limited to monofunctional compounds and a laboratory scale.

Polycyclotrimerization

Because of the peculiarity of the cyclotrimerization—three functional groups combine to a junction—cyanate monomers with at least two functional groups are able to form polycyanurates. During this process the cyanate groups are almost completely transformed into trifunctional cyanurate branching points and, therefore, highly crosslinked networks are obtained (**Figure 1**), the branching density of which can cover a wide range through a variation of the monomer structure R and its functionality.

A large variety of substances is able to catalyze the cyclotrimerization, among which metal acetyl acetonates, zinc chloride, and phenols have been of practical importance up to now.[6]

The polycyclotrimerization of di- and polycyanates is a typical step-growth, network-building process: starting from low molecular weight monomers, the system passes a pregel state with a very broad distribution of molar masses, reaches the gel point at a certain value of conversion, and finally arrives at the

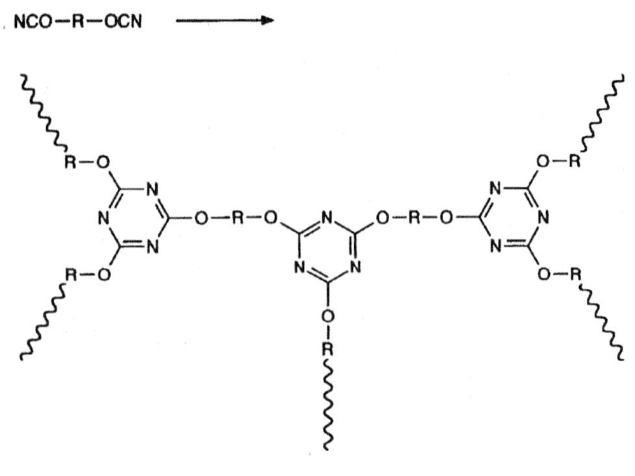

FIGURE 1. Schematic representation of a polycyanurate network synthesized through polycyclotrimerization of dicyanates.

fully cured state, where it consists of one giant macromolecule. For the most widely used monomer, the dicyanate of bisphenol A [DCBA, 2,2-bis(4-cyanatophenyl)propane], it was shown that the network buildup by polycyclotrimerization in bulk is a completely random reaction which can be modeled well by a mean field approach.[7,8]

Copolymers

Copolymerization of polyfunctional cyanates with functionalized comonomers is a valuable tool to design modified polycyanurate networks with special properties.

Besides the use of comonomers with a basic structure similar to that of the cyanate monomer (e.g., bisphenol A, diglycidyl ether of bisphenol A, novolacs, and novolac epoxy resins), functionalized monomers, oligomers, or polymers (e.g., chromophors, rubbers, and thermoplastics) can be chemically bonded to the polycyanurate network. In this way, a wide range of structures and resulting properties can be covered.[9]

Cyanates

The easiest approach to reach a structural modification of the network is the coreaction between poly- and monofunctional cyanates. Because of the possible buildup of two homo- and two mixed-trimer structures, not only branching points from three polyfunctional units but also chain-extending segments, chain ends, and low molecular weight soluble products are formed.

Phenols

According to the mechanism found for the reaction between cyanate and phenolic hydroxyl groups via an iminocarbonic ester intermediate, hydroxyl groups can be exchanged between the two monomers and, therefore, phenolic monomers can be incorporated into the network.[7,8]

In contrast to the monocyanates, the chance to covalently link a phenolic comonomer with the network is strongly dependent on its structure: the higher the pK_a value of the phenol the better it is incorporated into the network because of an improved formation of mixed triazines.[7,10]

Glycidyl Ethers

Copolymers synthesized from polyfunctional cyanates and glycidyl ethers can be regarded as modified polycyanurates and as high-performance epoxies. The main reactions between cyanata and glycidyl ether groups are characterized by consecutive and subsequent steps: cyclotrimerization of cyanates, insertion of oxiranes into the C–O–C bond of the cyanurate, isomerization of trebly alkoxy-substituted triazines, and formation of oxazolidinones.[11–13]

Glycidyl ether groups should be favored to combine the attractive properties of polycyanurates with those of other components such as chromophors, surface-active substances, rubbers, and thermoplastics.

Maleimides

Up to now, there have been conflicting views about whether or not the cyanate group reacts with maleimides under normal polycyclotrimerization conditions. A direct cycloaddition of one maleimide and two cyanate groups has been proposed.[14] However, newer results suggest the formation of interpenetrating polymer networks by separate reactions of both groups. Because the resulting thermosets have attractive properties, work in this field is progressing.

Properties

One of the most important properties of polycyanurate thermosets is their high glass temperature, which lies above that of epoxies derived from monomers with similar structure. Nearly 300°C is obtained using the common DCBA and ~400°C was reached with novolac-polycyanates. Lower values were found for the copolymers with monocyanates, phenols, and glycidyl ethers. Further attractive properties of polycyanurates are their low dielectric constants down to $\varepsilon' \approx 2.6$ at 1 MHz, low moisture absorption in the range of 1%, V-0-flammability of certain mixtures, and many others, which are described elsewhere.[9]

APPLICATIONS

Polycyanurate thermosets are successfully used in applications that take full advantage of their excellent properties. In particular, the high glass temperatures, the low dielectric constants and losses at elevated temperatures, and the high adhesion strength to fibers and metals are utilized for the design of binders for high-performance laminates, encapsulating agents, and high-temperature adhesives for electronics. Further promising applications include structural composites with glass or carbon fibers, casting resins, and binders for miscellaneous organic or inorganic substrates.[9]

REFERENCES

1. Martin, D.; Bacaloglu, R. *Organische Synthesen mit Cyansäureestern*; Akademie-Verlag: Berlin, Germany, 1980.
2. Grigat, E.; Pütter, R. German Patent 1 195 764, 1963.
3. Martin, D.; Bauer, M. *Org. Synth.* **1983**, *61*, 35.
4. Martin, D. *Chem. Ber.* **1964**, *97*, 2689.
5. Jensen, K. A.; Holm, A. *Acta Chem. Scand.* **1964**, *18*, 826.
6. Martin, D.; Bauer, M.; Pankratov, V. A. *Usp. Khim.* **1978**, *47*, 1814.
7. Bauer, M.; Bauer, J. In *The Chemistry and Technology of Cyanate Ester Resins;* Hamerton, I., Ed.; Chapman & Hall: Glasgow, Scotland, 1994.

8. Bauer, J.; Bauer, M. *Acta Polym.* **1987**, *38*, 16.
9. *The Chemistry and Technology of Cyanate Ester Resins*; Hamerton, I., Ed.; Chapman & Hall: Glasgow, Scotland, 1994.
10. Alla, C. Ph.D. Thesis, Université Pierre et Marie Curie Paris, 1992.
11. Bauer, M; Tanzer, W.; Much, H.; Ruhmann, R. *Acta Polym.* **1989**, *40*, 335.
12. Bauer, M.; Bauer, J.; Ruhman, R., Kühn, G. *Acta Polym.* **1989**, *40*, 397
13. Shimp, D. A.; Wentworth, J. E. In *Advanced Materials and Structures from Research to Application*; Brandt, J., et al., Eds.; SAMPE European Chapter: Basel, Switzerland, 1992.
14. Bauer, J.; Bauer, M. *Acta Polym.* **1990**, *41*, 535.

CYANATE ESTER RESINS

C. P. Reghunadhan Nair,* D. Mathew, and K. N. Ninan
Vikram Sarabhai Space Center

A rapidly expanding aerospace industry increasingly demands composites that combine high-strength reinforcements with high-performance and high-temperature-resistant organic matrices.

An ideal system is one that possesses a resin with built-in toughness. Existing thermosetting matrix resins such as phenolics, epoxies, or bismaleimide (BMI) do not possess this quality. Hence, considerable efforts are directed at developing a new matrix system that encompasses all the ideal features, and this goal has led researchers to cyanate ester (CE) systems.

This new generation of thermosetting matrices is polymerized by a simple thermal cyclotrimerization of the cyanato functions to the stable phenolic triazine (PT) network, combining several ideal characteristics of a matrix resin for structural composites. CEs are chosen when thermo-oxidative stability, built-in toughness, and high service temperatures are the criteria for selecting a matrix resin. Here, the chemistry and applications of cyanate ester resins are described.

PREPARATION AND REACTIONS

Cyanates are formed in excellent yields by the reaction of corresponding phenols with cyanogen halides.[1] Most of them are stable.

Reactions with Nucleophiles

Cyanate esters undergo nucleophilic attack. Phenols react with ArOCN to give bisarylimido carbonates in the presence of bases, which generally undergo cleavage with the formation of the more acidic phenol.

Thermal Curing

CEs can undergo thermal or catalytic polycyclo trimerization to give phenolic triazines. The catalysts are usually Lewis acids and transition metal carboxylates.

Reaction with Epoxy Group

A reaction with an epoxy group produces oxazolidinone as the major product.

*Author to whom correspondence should be addressed.

Reaction with Unsaturated Compounds

Unsaturated compounds like maleimide react with cyanate groups. Similar reactions take place with acetylene and benzocyclobutenes.

GELATION AND VITRIFICATION

Because cyanato functional polymers are self-crosslinking, their gelation studies especially interest researchers. The effect of various solvents on the gel formation or 2,2-bis(4-cyanatophenyl)propane was studied, and the conversion was highest when highly polar solvents like nitrobenzene were used.[2] Various catalysts affected the gel conversion to different extents in good solvents. At low monomer concentrations, the conversion at gel point decreased by 15% because the dependence was linear.

COMPARATIVE ADVANTAGES

CEs and PT resins are superior to conventional epoxy, phenolic, and BMI resins in many respects. For example, the inherent brittleness of polyimides makes their processing difficult. The poor hot–wet performance of epoxy resins limits their use as structural components at high temperatures. PT resins are formed by the thermal cyclotrimerization of the cyanato functions. The lack of volatile by-products during cure renders them attractive in void-free composites. Ces combine the processing advantages and the handling convenience of epoxies, the fire resistance of phenolics, and the high-temperature performance of polyimides. All these features clearly indicate that CEs or their blends will replace many existing thermosetting matrices in composites for structural and thermo-structural applications.

APPLICATIONS

Dicyanates of bisphenol derivatives are currently used in composites with established reinforcements such as carbon fiber, glass fiber, silica cloth, and pitch-based graphite fibers. The CE laminates are primarily used in the electronic industry for printed circuit boards. These laminates are characterized by their low dielectric constant, loss factor, and superior copper peel strength. At present, researchers are studying the polymerization, reaction modeling, cure kinetics, and applications of CE especially as matrix resins, either alone or as reactive blends with known systems like epoxies or BMI. Modifications in thermo-mechanical properties are accomplished by various techniques such as blending with epoxies or reactively terminated thermoplastic oligomers and heat treatments. Most of these resins were based on bisphenols as precursors.

Several aryl cyanate esters bearing reactive allyl groups also have been reported. The effect of various additives upon the polymerization behavior of 1-allyl-2 cyanatidobenzene was studied in detail.[3] DSC and NMR analysis indicated a possible co-reaction between the allyl group and reactive additives like maleimide. Since the predominant reaction was cyclotrimerization, no conclusive proof for the formation of bismaleimide–triazine networks was obtained. A recent study showed the possibility for making hard aromatic polycyanurates using high

TABLE 1. Commercially Available Cyanate Esters

Trade name	Producer	Structure		Physical properties
AROCY L-10	Hi-Tek/ Ceiba-Geigy	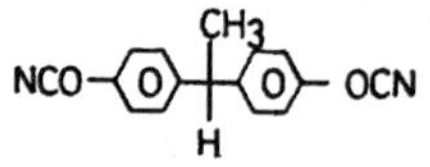	**A**	Liquid
AROCY B-10	"	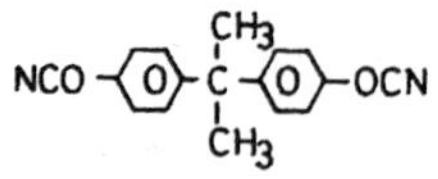	**B**	M.P. 79 °C.
AROCY F 10	"	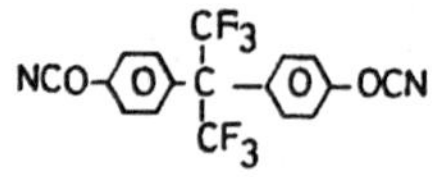	**C**	M.P. 86 °C.
AROCY T 10	"	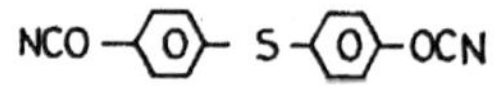	**D**	M.P. 94 °C.
AROCY M 10	"	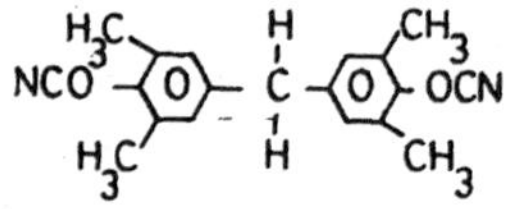	**E**	M.P. 106 °C.
AROCY B-30	"	B-10 oligomer		resinous
AROCY F-40-S	"	AROCY F10 prepolymer		prepolymer solution resin
AROCY M-40S	"	AROCY M 10 prepolymer		prepolymer solution resin
RTx 366	"	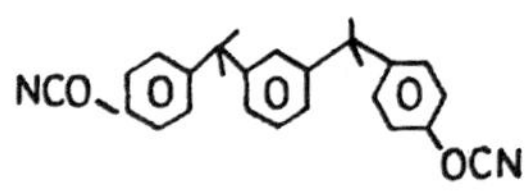	**F**	M.P. 68 °C.
XU.71787.02L	Dow Chemical Co.	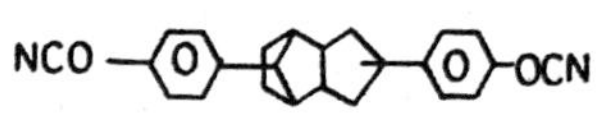	**G**	low melt solid
HX 1553	Hexcel Corp.			
HX 1562		Not known		as prepreg
Primeset PT REX-371	Allied Signal Ceiba–Geigy	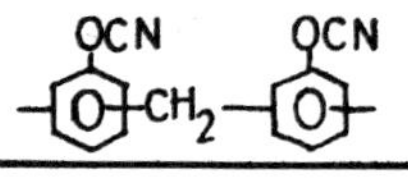	**H**	low melt solid

pressure (200–450 MPa) at 150–280°C while molding dicyanate monomers (**Table 1**).[4]

Cyanate esters generally exhibit better flame retardancy than epoxies. It is possible to do chemical tailoring of these resins to impart the desired degree of flame resistance. A few such resins have been compared with CE-brominated epoxy formulations for their flame retardant characteristics.[5] Such resins for application in printed circuit boards have to pass the stringent UL 94 test specifications. The AroCY cyanate ester family, incorporating elements like fluorine and sulfur possesses inherent flame retardancy (e.g., AroCY-T and AroCY-F laminates). they offer a number of advantages over the heretofore used brominated epoxies. The brominated epoxy-blended AroCY M-405 showed a reduction in onset of thermal decomposition in their E-glass laminates as the concentration of the former increased.

REFERENCES

1. Grigat, E.; Putter, R. *Angew. Chem. Int. Ed.* **1967**, *6*, 206.
2. Korshak, V. V. *J. Polym. Sci. Polym. Chem. Ed.* **1978**, *16*, 1697.
3. Barton, J. M.; Hamerton, I.; Jones, J. R. *Polym. Int.* **1992**, *29*, 145.
4. Itoya, K.; Kakimoto, M.; Imai, Y. *Polymer* **1994**, *35*, 6.
5. Ising, S. J.; Shimp, D. A. Presented at the 34th Intl. SAMPE Symposium, May 1989; paper 1326.

Cyanates

See: Cyanate-Based Resins (Polycyanurates)
Cyanate Ester Resins

Cyanoacrylate Polymers

See: Cyanoacrylates
Poly(2-cyanoacrylates)

CYANOACRYLATES

John Woods
Loctite Corporation

The polymers of alkyl 2-cyanoacrylates are synonymous with "instantly" curable adhesive of "Superglues" which derive their remarkable properties from the unusual chemistry of this polymer group. The commercial products consist mainly of monofunctional monomers, some of whose commonly encountered structures are indicated in **Figure 1**. They are generally low viscosity liquids which exhibit excellent wetting properties. Because of the unique electron withdrawing properties of the adjacent nitrile and carboxylate groups submitted on carbon-2, they undergo rapid anionic polymerization on contact with basic catalysts. The glassy polymers obtained are generally characterized by a high molecular weight, frequently in excess of 10^6. The combination of high molar mass, good wetting properties, and polar effects of the cyano and carboxylate groups provides these materials with their excellent adhesive properties. In addition, they have been found useful as polymeric binding agents in controlled drug delivery systems and as highly sensitive electron beam resists for dry etching processes.

SYNTHESIS

Despite the many synthetic routes to cyanoacrylate which have subsequently been described, the condensation of formaldehyde with cyanoacetate is the most important method for the commercial production of the monomers.

Cyanoacrylates polymerize relatively slowly in the presence of conventional free-radical initiators, but extremely rapidly in the presence of catalytic amounts of anionic and certain covalent bases such as amines and phosphines. Under such conditions they are among the most reactive monomers known. For example, a pure, unstabilized ethyl cyanoacrylate solution in THF (tetrahydrofuran) at 20°C was found to have a half-life of one second following the introduction of 10^{-6} moles/liter Et_3P.[1] Most of the published work on the polymerization of cyanoacrylates relates to anionic polymerization.

Well-defined cyanoacrylate polymers may be prepared in dilute solution of ethereal solvents (THF, dioxane, diethyl ether, or 1,2-dimethoxyethane).[2,3] Preliminary studies suggest that polymerization is more rapid in solvents of high dielectric constant such as nitromethane or acetonitrile,[2] but such solvents are also suspected of acting as polymerization initiators, as has been demonstrated for propylene carbonate.[4]

A wide range of anionic initiators may be employed to affect the polymerization of cyanoacrylates. These include not only conventional initiators such as butyl lithium, but also a variety of organic and inorganic salts including sodium acetate, sodium butyl cyanoacetate, LiBr, KOH, KCN, NaI, and tetrabutyl ammonium iodide.[2] Furthermore, covalent bases such as phosphines, amines, and pyridine are also effective initiators of cyanoacrylate polymerization.[2,5]

PROPERTIES

Kinetics and Mechanism of Cyanoacrylate Polymerization

The anionic solution polymerization kinetics of alkyl 2-cyanoacrylates have been studied in detail by Pepper and co-

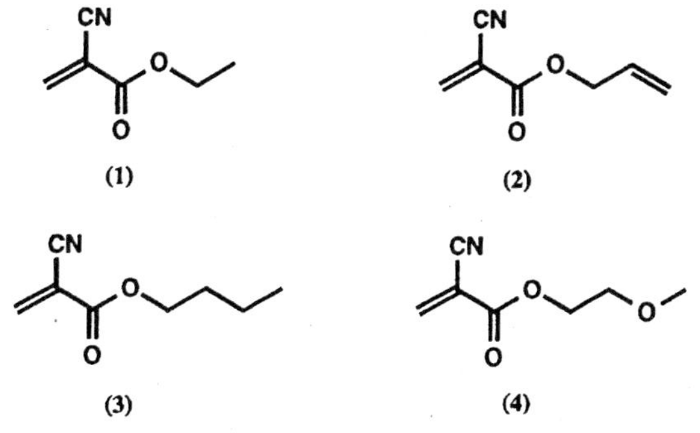

FIGURE 1. The structure of some common cyanoacrylate monomers: (1) ethyl, (2) allyl, (3) n-butyl, and (4) methoxyethyl 2-cyanoacrylate.

workers utilizing adiabatic calorimetry.[1–12] These studies showed that cyanoacrylate polymerization, initiated by base, has no intrinsic termination reaction. The overall kinetics depends on the rate of initiation and may be distinguished between those reactions in which initiators exhibit very rapid rates of initiation (e.g., hydroxyl ions, phosphines) and those where the initiation sequences are relatively slow (e.g., acyclic amines, pyridine). With rapid initiators, nearly ideal living polymerization conditions exist and molecular weights were in close approximation to the theoretical values predicted from the monomer–initiator concentration ratio.[11]

Physical Properties of Poly(alkyl 2-Cyanoacrylates)

Poly(alkyl 2-cyanoacrylates) are hard, colorless, amorphous thermoplastic materials which undergo a clean retropolymerization reaction at temperatures in the region 140–180°C.[13] Several reviews detail the physical properties of cyanoacrylate monomers and polymers.[14–17] Glass transition temperatures (T_g) have been determined by differential scanning calorimetry (DSC), dynamic mechanical analysis (DMA), and dilatometry.[16,18,19]

APPLICATIONS

Adhesives

Adhesives represent the most important application of poly(alkyl 2-cyanoacrylate)s and they are marketed throughout the world for both industrial and consumer use. The typical adhesive product comprises a mixture of cyanoacrylate monomer, stabilizers, and additives, such as polymeric thickeners, rubbers, fillers, and the like, to control the rheology of the uncured adhesive and properties of the cured polymer within the adhesive joint. Comprehensive reviews on cyanoacrylate adhesive technology have been published by Miller and by Lee,[20,21] and more concise accounts by O'Connor and by Coover.[14,15]

The adhesives are typically used in assembly applications where rapid cure and high strength are required, for example, in the assembly of computers, calculators, electric motors, medical equipment, optical components, jewelry, sporting goods, and many others. Cyanoacrylates are also used as surgical adhesives where rapid bonding, a low tendency for abscess or inflammation development, and the elimination of the need for anesthetics make adhesive bonding an attractive alternative to conventional suturing techniques.[22]

The adhesive strengths obtained with cyanoacrylates (CA) match the values for conventional structural adhesives such as epoxies. The great advantage of the CA product is that adhesive strength is developed at room temperature over a period of a few seconds, whereas a comparable single-part epoxy product requires high heat for significant periods (frequently several hours) to cure the adhesive. Cyanoacrylates are therefore particularly suitable for use in high-speed automated assembly operations.

Nanoparticles

Poly(alkyl 2-cyanoacrylate)s are among the most commonly used synthetic polymers for the production of encapsulated colloidal nanoparticles for targeted drug delivery systems.[23–26] Their good encapsulation properties, low toxicity, and satisfactory biocompatibility make them attractive.[23] The preferred polymers for this application are those having four or more carbons in the alkyl side chain, such as *n*-butyl or isohexyl 2-cyanoacrylates.[23]

Photoresists

Poly(alkyl 2-cyanoacrylate)s are useful as photoresists for manufacturing microelectronic devices.[27–31] Cyanoacrylates are particularly suitable for electron beam lithography where they offer better contrast and sensitivity than conventional resists, but they may also be imaged with deep UV light or laser ablation.[25,26] the vapor-deposited material further offers enhanced resistance to plasma etching.[29]

REFERENCES

1. Pepper, D. C.; Ryan, B. *Makromol. Chem.* **1983**, *184*, 383.
2. Donnelly, E. F.; Johnson, D. S. et al. *J. Polym. Sci., Polym. Lett.* **1977**, *15*, 399.
3. Eromosele, I. C.; Pepper, D. C. *Makromol. Chem.* **1989**, *190*, 3085.
4. Eromosele, I. G.; Pepper, D. C. *Makromol. Chem., Rapid Commun.* **1986**, *7*, 531.
5. Pepper, D. C. *Polym. J. (Tokyo)* **1980**, *12*(9), 629.
6. Johnson, D. S.; Pepper, D. C. *Makromol. Chem.* **1981**, *182*, 393.
7. Johnson, D. S.; Pepper, D. C. *Makromol. Chem.* **1981**, *182*, 407.
8. Johnson, D. S.; Pepper, D. C. *Makromol. Chem.* **1981**, *182*, 421.
9. Pepper, D. C.; Ryan, B. *Makromol. Chem.* **1983**, *184*, 395.
10. Cronin, J. P.; Pepper, D. C. *Makromol. Chem.* **1988**, *189*, 85.
11. Pepper, D. C. *J. Polym. Sci., Poly. Symp.* **1978**, *62*, 65.
12. Pepper, D. C. *Makromol. Chem. Macromol. Symp.* **1992**, *60*, 267.
13. Rooney, J. M. *Br. Polym. J.* **1981**, *13*, 160.
14. OÕConnor, J. T. In *Kirk-Othmer Encylopedia of Chemical Technology* 4th ed.; John Wiley & Sons: New York, NY, 1991; Vol. 1, p 344.
15. Coover, H. W.; Dreifus, D. W.; O'Connor, J. T. In *Handbook of Adhesives,* 3rd. ed.; Skeist, I., Ed.; Van Nostrand Reinhold: New York, 1990.
16. Tseng, Y-C.; Hyon, S-H.; Ikada, Y. *Biomaterials* **1990**, *11*, 73.
17. Coover, H. W.; Mclntire, J. M. In *Encyclopedia of Polymer Science* 2nd ed.; John Wiley & Sons: New York, NY, 1985; Vol. 1, 299.
18. Cheung, K. H.; Guthrie, J. et al. *Makromol. Chem.* **1987**, *188*, 3041.
19. Kukarni, R. K.; Porter, H. S.; Leonard, F. *J. Appl. Polym. Sci.* **1973**, *17*, 3509.
20. Millet, G. H. *Structural Adhesives, Chemistry and Technology*; Hartshorn, S. R., Ed.; Plenum: New York, 1986; Chapter 6, 249.
21. *Cyanoacrylate Adhesives-The Instant Adhesives* Lee, H.; Ed.; Pasadena Technology: Pasadena, CA, 1981.
22. Vanholder, R.; Misotten, A. et al. *Biomaterials* **1993**, *14*(10), 737.
23. Lescure, F.; Zimmer, C. et al. *J. Colloid and Interface Sci.* **1992**, *154*(1), 77.
24. Pons, M.; Garcia, M. L.; Valls, O. *Colloid Polym. Sci.* **1991**, 269, 855.
25. Gallardo, M.; Couarraze, G. et al. *Int. J. Pharmaceutics* **1993**, *100*, 55.
26. Chouinard, F.; Kan, F. W. et al. *Int. J. Pharmaceutics* **1991**, *72*, 211.
27. Herlbert, J. N.; Caplan, P. J.; Poindexter, E. H. *J. Appl. Polym. Sci.* **1977**, *21*, 797.
28. Eranian, A.; Datamanti, E. et al. *Br. Polym. J.* **1987**, *19*, 353.
29. Woods, J.; Guthrie, J. et al. *Polymer* **1989**, *30*, 1091.
30. Magan, J. D.; Hogan, M. P. et al. *Chemotronics* **1989**, *4*, 74.
31. Sato, M.; Fujimoto, Y. et al. *Polym. Prepr. Jap.* **1991**, *40*, (1–4), III-7-11, E 209.

CYCLIC BENZO-ALKYNES, ORTHO SUBSTITUTED

Claire A. Tessier, Kyle P. Baldwin, and Wiley J. Youngs
Chemistry Department
University of Akron

Many research groups have been investigating benzo alkynyl molecules in recent years. Several o-cyclynes (**Figure 1,** n=0) have been used as ligands to transition metals.[1] Benzo alkynes have also been used in intramolecular cyclizations, as we and others have shown.[2–5]

Synthesizing carbon tubules and helical π-systems may be possible using o-cyclynes (Figure 1) as the precursor. We then began investigating the synthesis of larger cyclics via stepwise routes with the intent of transforming them into tubules or helical π-systems by using a cyclization reaction. Two types of cyclization reactions may apply to cyclynes. We discovered that reacting TBC (Figure 1, n=0) with lithium in THF yields a product resulting from an intramolecular cyclization.[3] With the discovery that lithium reacts similarly with QBC (Figure 1, n=1), we began synthesizing larger cyclics to make long conjugated π-systems of a helical nature.[4] The second useful cyclization reaction is the Bergman reaction. In the 1970s, Bergman proposed the transformation of *cis*-hexa-3-ene-1,5-diyne into benzene when heated in the presence of an alkane.[6] We have theorized that a tubule may be formed by heating an ortho-substituted cyclyne via a multiple Bergman reaction. One

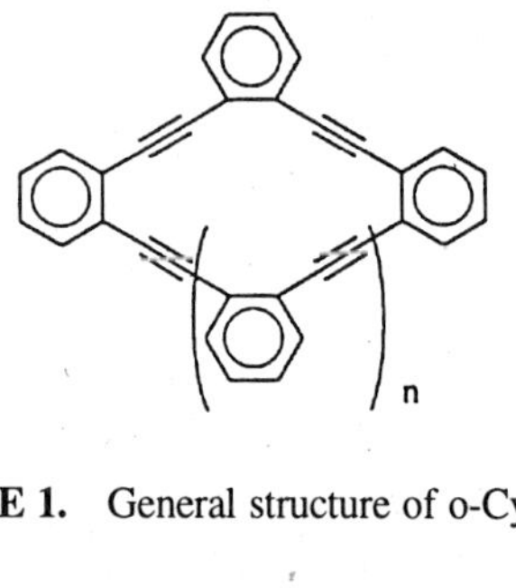

FIGURE 1. General structure of o-Cyclynes.

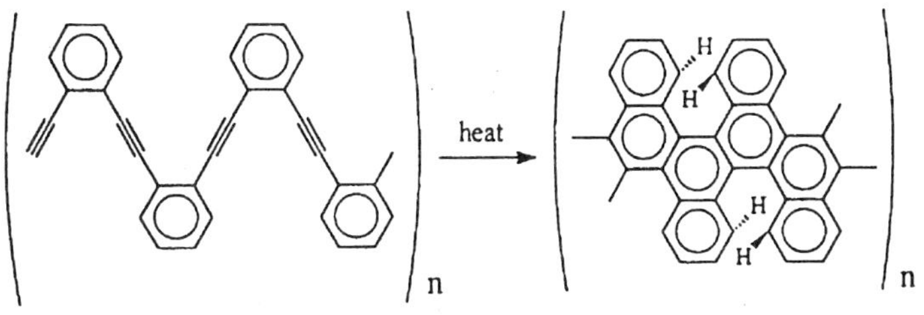

FIGURE 2. Proposed poly-Bergman reaction.

can imagine two radicals propagating around the cyclyne, one in each direction. Then, upon meeting, they would cap each other, resulting in a tubule (**Figure 2**). We wanted to synthesize a cyclyne that would yield a stable tubule. Thus, DBC (Figure 1, n=7) was chosen because cyclized DBC's belt size would have the same diameter as that of C_{60}.

Each of these cyclizations would yield a type of ladder polymer. Ladder polymers, due to an extended π-conjugation through their backbones, make useful conductors and nonlinear optical devices. They also have thermal stability because of multiple backbones.[7] In order for a ladder polymer to degrade, bond scission must occur at multiple sites.[8]

CONCLUSIONS

At present we have shown that our cyclic benzo–alkynes that have been acquired in isolable amounts are large oligomers. At some point, when the properties become constant (i.e., NMR shifts become indifferentiable with the addition of another benzo–alkyne unit) a cyclic polymer will exist. Because the polydispersity index (M_w/M_n) of this type of polymer will equal one, it will be interesting to investigate its mechanical and physical properties.

ACKNOWLEDGMENTS

We would like to thank NASA–Lewis Research Center and NSF for financial support of this research. We would also like to thank Robert Lattimer for running FDMS spectra and Paul Zimmerman and David Hercules for running TOF-SIMS.

REFERENCES

1. Ferrara, J. D.; Tessier-Youngs, C.; Youngs, W. J. *J. Am. Chem. Soc.* **1985**, *107*, 6719. b) Ferrara, J. D.; Tessier-Youngs, C.; Youngs, W. J. *Organometallics* **1987**, *6*, 676. c) Ferrara, J. D.; Tessier-Youngs, C.; Youngs, W. J. *J. Inorg. Chem.* **1988**, *27*, 2201. d) Djebli, A.; Ferrara, J. D.; Tessier-Youngs, C.; Youngs, W. J. *J. Chem. Soc. Chem. Commun.* **1988**, *548*. e) Ferrara, J. D.; Djebli, A.; Tessier-Youngs, C.; Youngs, W. J. *J. Am. Chem. Soc.* **1988**, *110*, 647. f) Ferrara, J. D.; Tanaka, A. A.; Fierro, C.; Tessier-Youngs, C.; Youngs, W. J. *Organometallics* **1989**, *8*, 2089. g) Solooki, D.; Kennedy, V. O.; Tessier, C. A.; Youngs, W. J. *Synlett* **1990**, 427. h) Youngs, W. J.; Kinder, J. D.; Bradshaw, J. D.; Tessier, C. A. *Organometallics* **1993**, *12*, 2406. i) Kinder, J. D.; Tessier, C. A.; Youngs, W. J. *Synlett.* **1993**, 149.
2. a) Youngs, W. J.; Djebli, A.; Tessier-Youngs, C. *Organometallics* **1991**, *10*, 2089. b) Malaba, D.; Djebli, A.; Chen, L.; Zarate, E. A.; Tessier-Youngs, C.; Youngs, W. J. *Organometallics* **1993**, *12*, 1266.
3. Bradshaw, J. D.; Solooki, D.; Tessier, C. A.; Youngs, W. J. *J. Am. Chem. Soc.* **1994**, *116*(8), 3177.
4. Berris, B. C.; Hovakeemian, G. H.; Lai, Y.; Mestdagh, H.; Vollyhardt, K. P. C. *J. Am. Chem. Soc.* **1985**, *107*, 5670.
5. Goldfinger, M. B.; Swager, T. M. *J. Am. Chem. Soc.* **1994**, *116*, 7895.
6. Bergman, R. G. *Acc. Chem. Res.* **1973**, *6*, 25.
7. John, J. A.; Tour, J. M. *J. Am. Chem. Soc.* **1994**, *116*, 5011.
8. a) Bredas, J. L.; Baughman, R. H. *J. Chem. Phys.* **1985**, *83*(3), 1316. b) Yu, L.; Chen, M.; Dalton, L. R. *Chem. Mater.* **1990**, *2*,, 649. c) Wuckel, L.; Lehmann, G.; Mak. Chem., Macromol. Symp. **1990**, *37*, 195. d) Lehmann, G. *Synth. Met.* **1991**, *41–43*, 1615.

Cyclic Carbonate Polymerization

See: *Cyclic Carbonates (Ring-Opening Polymerization)*
Ring-Opening Polymerization, Cationic (with Expansion in Volume)

CYCLIC CARBONATES (Ring-Opening Polymerization)

Hartwig Höcker and Helmut Keul
Lehrstuhl für Textilchemi und Makromolekulare Chemie der Rheinisch-Westfälischen Technischen Hochschule Aachen

Ring-opening polymerization of cyclic carbonates (**Equation 1**) is considered to follow a chain growth mechanism.[1] The active species undergoes an addition or insertion reaction with the monomer. Chain transfer reactions and termination reactions are usually absent; after consumption of the monomer the active site is transferred into an inactive one upon addition of a suitable reagent.

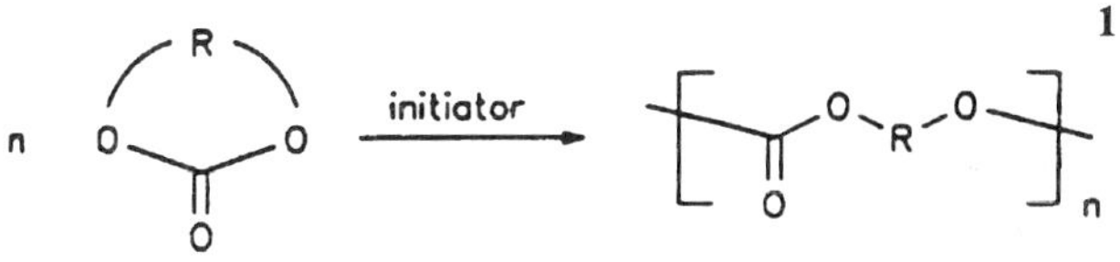

1

The ring-opening polymerization of cyclic carbonates discloses an alternative route to polycarbonates, the typical route to this class of polymers being the polycondensation reaction following a step growth mechanism.[2] The best known technical polycarbonate is the engineering thermoplastic material based on bisphenol-A.

In the following, the results of our laboratory will be presented with respect to homo- and copolymerization of cyclic carbonates initiated with nucleophiles.

CYCLIC CARBONATE MONOMERS

The polymerization of principally two types of cyclic carbonates has been reported in the literature: (i) aliphatic cyclic carbonates and (ii) aromatic cyclic carbonates.

It should be mentioned that five-membered cyclic carbonates, for example ethylene carbonate, upon polymerization with different metal alkoxides result in poly(oxyethylene-stat-ethylene carbonate).[3] Polymers containing low amounts of carbonate groups and polyoxyethylene are formed with anionic (and cationic) initiators. The reason for the inaccessibility of pure poly(ethylene carbonate) is not the low ceiling temperature, but the large positive enthalpy of polymerization.

INITIATION REACTIONS FOR THE NUCLEOPHILIC RING-OPENING POLYMERIZATION OF CYCLIC CARBONATES

Six-membered cyclic carbonates such as 2,2-dimethyl-trimethylene carbonate (DTC) are polymerized anionically, or by insertion in a ring-opening fashion with a variety of initiating systems based on alkali metals (Li, Na, K),[4] earth-alkali metals (Mg),[5] other metals (Al, Zn, Sn),[6,7] and with metal free initiators.[8]

The initiation reaction comprises the nucleophilic attack of the initiator at the carbonyl carbon, followed by an acyl oxygen cleavage and formation of the active species, an alcoholate.[9] The efficiency of the initiation depends on the nucleophilicity of the initiator and on the electron affinity of the monomer; if both are high, initiation is fast and quantitative. Slow initiation implies broadening of the molecular weight distribution and loss of molecular weight control.

CHAIN GROWTH REACTIONS IN THE NUCLEOPHILIC RING-OPENING POLYMERIZATION OF CYCLIC CARBONATES

The polymerization of cyclic carbonates proceeds in two steps. First, in the kinetically controlled regime, the chain propagation reaction is predominant and high molecular weight polymer is formed. Then, in a thermodynamically controlled regime, back-biting reactions occur to form a low molecular weight fraction of cyclic oligomers (M_X).

Alkali Metal Alcoholate and Phenolate Active Site

For alkali metal compounds used as initiators, the active species of polymerization was shown to be an alkali metal alcoholate for the aliphatic cyclic carbonates[9] and an alkali metal phenolate for the aromatic cyclic carbonates.[10] The time necessary to shift from kinetic control to thermodynamic control is around several minutes when the polymerization is performed at 25°C in toluene solution. The larger the alkali-metal counterion the faster is the polymerization and the smaller the selectivity parameter $\beta = k_p/k_b$ (k_p is the rate constant of propagation; k_b, the rate constant of back-biting).

Another point of interest with respect to the active species is the role of the chemical nature of the macroinitiator on the reaction course. Polymerization of DTC with the potassium salt of poly(ethylene oxide) with one terminal hydroxy group (MW 1900) as initiator[11] in toluene at 20°C is a very fast reaction. Already after 30 sec the reaction has passed the regime of kinetic control and is in the regime of thermodynamical control. The enhancement of the rate of propagation and of back-biting has its origin in an interaction between the potassium counterion and the poly(EO) moiety. While the potassium ion is cryptated by the poly(EO) coil the hydroxylate end group of the growing poly(DTC) chain—for incompatibility reasons—is ready to react with the carbonyl group of DTC or poly (DTC) with the result of chain growth, back-biting, or transesterification.

Magnesium, Aluminum, and Zinc Alcoholate Active Sites

With Bu_2Mg as initiator the polymerization of DTC yields a product with a bimodal molecular weight distribution,[5] that is, a high molecular weight polymer and a low molecular fraction consisting of a homologous series of cyclic oligomers. The polymerization of six-membered cyclic carbonates with Al- and Zn-based catalysts shows a high polymer yield at 25°C (95%); even at 80°C in toluene solution the extent of back-biting reactions is not increased.[6] This proves that for Al and Zn in the active species the propagation rate constant is considerably higher than the rate constant for back-biting reactions. Hence, within the scope of a kinetic treatment of the polymerization of DTC with $Al(O\text{-sec-}Bu)_3$ as initiator only the propagation reaction must be considered.[12]

BLOCK COPOLYMERS

Sequential addition of monomers to the initiator is the most obvious procedure for the preparation of block copolymers. Once the first monomer has been polymerized, the resulting living species acts as polymeric initiator for the polymerization of the second monomer. The order in which the monomers are polymerized is determined by the nucleophilicity of the initiator and the electron affinity of the monomer. Some of the examples described in the literature are polystyrene-block-poly(DTC),[13] poly(methyl methacrylate)b-poly(DTC),[14] poly(ε-caprolactone)-b-poly(DTC),[7] polypivalolactone-b-poly(DTC),[15] and poly(DTC)-b-poly(DTC-stat-ECL)-b-poly(DTC).[16] Another strategy for the preparation of block copolymers with B_n–A_m or A_m–B_n–A_m structure is to use terminal functional groups of polymers to prepare the initiator for the ring-opening polymerization, followed by formation of the blocks A_m. Suitable polymers are those with one or two hydroxyl end groups that are transformed into alcoholate groups to initiate the anionic ring-opening polymerization. Some of the examples described in the literature are poly(DTC)-b-poly(ethylene oxide)-b-poly(DTC),[11] poly(DTC)-b-polytetrahydrofuran-b-poly(DTC),[17] and poly(DTC)-b-polydimethylsiloxane-b-poly(DTC).[18]

CONCLUSIONS

Ring-opening polymerization of cyclic carbonaters is a type of polyreaction with a broad scope in the synthesis of polycarbonates and copolymers with carbonate repeating units, especially block copolymers. The fact that aliphatic cyclic carbonates show a volume expansion when polymerized will lead in the future to new and interesting applications.

REFERENCES

1. Höcker, H.; Keul, H. *Adv. Mater.* **1994**, *6*, 21.
2. Freitag, D.; Grigo, U.; Müller, P. R.; Nouvertne, W. In *Encyclopedia of Polymer Science and Engineering,* 2nd ed.; Mark, H. F.; Bikales, N. M.; Overberger, C. G.; Menges, G., Eds.; John Wiley & Sons: 1988; Vol. 2, p 648.
3. Vogdanis, L.; Martens, B.; Uchtmann, H.; Hensel, F.; Heitz, W. *Makromol. Chem.* **1990**, *191*, 465.
4. Keul, H.; Bächer, R.; Höcker, H. *Makromol. Chem.* **1986**, *187*, 2579.
5. Wurm, B.; Keul, H.; Höcker, H.; Sylvester, G.; Leitz, E.; Ott, K.-H. *Makromol. Chem.. Rapid Commun.* **1992**, *13*, 9.
6. Kühling, S.; Keul, H.; Höcker, H. *Makromol. Chem.* **1992**, *193*, 1207.
7. Hovestadt, W.; Keul, H.; Höcker, H. *Polymer* **1992**, *33*, 1941.
8. Hovestadt, W.; Müller, A. J.; Keul, H.; Höcker, H. *Makromol. Chem.. Rapid Commun.* **1990**, *11*, 271.
9. Kühling, S.; Keul, H.; Höcker, H. *Makromol. Chem. Suppl.* **1989**, *15*, 9.
10. Keul, H.; Deisel, F.; Höcker, H.; Leitz, E.; Ott, K.-H.; Buysch, H.-E; Schön, N. *Makromol. Chem., Rapid Commun.* **1991**, *12*, 133.
11. Müller, A. J.; Keul, H.; Höcker, H. *Europ. Polym. J.* **1991**, *27*, 1323.
12. Wurm, B.; Keul, H.; Höcker, H. *Macromol. Chem. Phys.* **1994**, *195*, 34, 85.
13. Keul, H.; Höcker, H. *Makromol. Chem.* **1986**, *187*, 2833.
14. Hovestadt, W.; Keul, H.; Höcker, H. *Makromol. Chem.* **1991**, *192*, 1409.
15. Keul, H.; Höcker, H.; Leitz, E.; Ott, K.-H.; Morbitzer, L. *Makromol. Chem.* **1990**, *191*, 1975.
16. Gerhard-Abozari, E.; Kenl, H.; Höcker, H. *Macromol. Chem. Phys.* **1994**, *195*, 2371.
17. Müller, A. J.; Keul, H.; Höcker, H. *Europ. Polym. J.* **1993**, *29*, 1171.
18. Müller, A. J.; Keul, H.; Höcker, H. *Polym. International* **1994**, *33*, 197.

Cyclic Imino Ethers

See: Oxazoline Polymerization

CYCLIC IMINO ETHERS (Ring-Opening Polymerization)

Keigo Aoi
Department of Applied Biological Science
Faculty of Agricultural Sciences
Nagoya University

Yoshiki Chujo
Division of Polymer Chemistry
Graduate School of Engineering
Kyoto University

Cyclic imino ethers are heterocyclic compounds with an imino ether linkage. Polymerization of cyclic imino ethers generally proceeds via thermodynamically favorable isomerization of the imino ether group to the amide.

Polymerized or copolymerized cyclic imino ethers are classified into two categories, that is, cyclic *endo-* and cyclic *exo-*imino compounds.

Among them, five-membered cyclic *endo*-imino ethers, 2-oxazolines, are important monomers in the field of polymerization chemistry and materials science.[1–7] For more than 20 years, 2-oxazoline polymers have been used. Recent progress in their fundamental chemistry and application are described in this article.

RING-OPENING POLYMERIZATION OF 2-OXAZOLINES

2-Oxazolines are commonly synthesized from the corresponding nitriles or carboxylic acids with modest to good yields. 2-Oxazolines are polymerized with various kinds of cationic initiators, for example, Lewis acids, strong protic acids and their esters, and alkyl halides, to produce the corresponding derivatives of poly(*N*-acylimino)ethylene] (Structure 2) via ring-opening isomerization (**Scheme I**).[2,3,6,8–12] The term poly (*N*-acylethylenimine) is even more common.

1 2 I

The ring-opening polymerization behavior has been widely investigated by using monomers with a variety of 2-substituents, which are, for example, unsubstituted, alkyl, and aryl groups.[2,3,6] The reaction proceeds via two different types of species, that is, ionic and covalent types, depending on the initiator.[2,3,6,7]

Growing species of the polymerization of several 2-oxazolines have been investigated with benzyl chloride, with methyl iodide, with methyl tosylate (MeOTs), and with methyl trifluoromethanesulfonate (MeOTf) as the initiator.[13–18]

The polymerization reaction of 2-oxazolines is not disturbed by chain transfer and termination under the appropriate conditions. The ionic propagating species of a 2-oxazolinium salt is not fragile. Thus, it is conveniently used in the synthesis of end-reactive polymers and block copolymers. The living nature of polymerization of 2-oxazolines is denoted in the previous reviews.[6,7]

POLYMERIZATION OF OTHER CYCLIC IMINO ETHERS

5,6-Dihydro-4*H*-2,3-oxazines (six-membered cyclic imino ethers) are polymerized in a manner similar tot 2-oxazolines. The resulting polymers have a structure of poly[(*N*-acylimino)trimethylene]. Polymerization of several 5,6-dihydro-4*H*-1,3-oxazines has been investigated.[13,19–22]

COPOLYMERIZATION

Copolymerization between 2-*i*-butyl- and 2-(5-hydroxypentyl)-2-oxazoline was examined.[23] Monomer reactivity ratios were determined in the systems of 2-phenyl-2-oxazoline with 2-methyl-, 2-*i*-propyl-, and 2-benzyl-2-oxazolines.[24] Random copolymers of 2-[10-(pentamethyldisiloxanyl)decyl]-2-oxazoline with 2-undesyl- and 2-nonyl-2-oxazolines having a total degree of polymerization of 100 over the whole composition range were made.[25]

AB-type block copolymers are obtained by using living nature of their polymerizations. The "One-Pot Two-Stage Feeding" method, in which two monomers are fed in sequence according to the progress of polymerization, has been used in some of the combinations of 2-oxazolines.[26,27]

Well-defined block copolymers were also obtained from a monomer mixture by using a different reactivity of 2-oxazolines. This methodology, which is used to prepare AB-type block copolymers, is proposed under the name of "One-Shot Block Copolymerization."[28]

Versatile synthetic utility of the living polymerization of cyclic imino ethers has been applied to synthesize various kinds of block copolymers. Other AB-type block copolymers have been prepared with polystyrene, poly(styrene oxide), poly(vinyl ether), and polyoxytetramethylene blocks.[29–33] ABA-type block copolymers consisting of poly(butadiene glycol) and poly(ethylene glycol) as the B blocks were prepared.[34,35] Three-arm star block copolymers were also obtained.[36]

A number of graft copolymers with polyoxazoline side chains have been reported.[37–50] Poly[styrene-*graft*-(*N*-acetylethylenimine)] was prepared by the polymerization of 2-methyl-2-oxazoline with a chloromethylated polystyrene initiator or with an amine-containing polystyrene terminator.[39,40]

Alternating or periodic copolymerization using cyclic imino ethers has been successfully achieved through no-catalyst copolymerization.[51–57]

PROPERTIES AND APPLICATIONS

Polymers of cyclic imino ethers have aroused a great deal of interest in materials science. Poly(*N*-acylethylenimine)s are regarded as a series of polymer homologues of *N*,*N*-dimethyl amides of various carboxylic acids. Thus, the polymer homologues of *N*,*N*-dimethylformamide (DMF) and *N*,*N*-dimethylacetamide (DMAc) are readily synthesized by the polymerization of 2-unsubstituted- and 2-methyl-2-oxazolines, respectively. DMF and DMAc are known as unique solvents because they mix with water freely and dissolve several organic polymers, which are insoluble in common organic solvents. On the basis of the characteristic properties, 2-oxazoline polymers, especially block and graft copolymers, have been studied for their applications.

NONIONIC POLYMER SURFACTANTS

Polymers of 2-methyl- and 2-ethyl-2-oxazoline are hydrophilic and are soluble in water. Hence, they have been used as a hydrophilic segment of nonionic surfactant. On the other hand, the polymers of oxazolines with a longer alkyl substituent are hydrophobic.[26] This different property among the above polymer homologues has successfully been applied to the preparation of nonionic surfactants, which were AB- or ABA-type block copolymers between two different cyclic imino ethers, one giving rise to a hydrophobic block and the other to a hydrophilic block.[26–28,58,59]

NONIONIC HYDROGELS

By utilizing the powerful nonionic hydrophilic property of poly(*N*-acetylethylenimine), hydrogels were prepared by partial hydrolysis of polyoxazoline, followed by the crosslinking with diisocyanate.[60] As an alternative way to prepare oxazoline hydrogels, random copolymerization of 2-methyl-2-oxazoline with a bisoxazoline monomer was carried out.[61,62] The swelling degree depends on the crosslinking density, which is controlled by the feed ratio of the bisoxazoline monomer.

Reversible hydrogels based on poly(*N*-acetylethylenimine) were prepared by means of the following methods: crosslinking by Diels–Alder reaction between diene-modified and dienophile-modified polyoxazolines,[63] crosslinking by the photodimerization of coumarin-modified or anthracene-disulfide-modified polyoxazolines,[64,65] crosslinking by the redox reaction of anthracene-disulfide-modified polyoxazolines,[65,66] crosslinking by the metal coordination of 2,2′-bipyridyl-branched poloxazolines.[67,68]

POLYMER COMPOSITES

With the analogy of DMAc, poly(*N*-acetylethylenimine) has been shown to possess a high miscibility with various commodity polymers.[69–79] It is considered that poly(*N*-acetylethylenimine) is a broadly compatible "polymeric solvent" or compatibilizing agent. Miscibilities of poly(*N*-acylethylenimine)s have been studied with polyamide (Nylon 6), poly(vinyl chloride) (PVC), poly(vinylidene fluoride), polystyrene, poly(acrylonitrile), poly(acrylic acid), and poly(α-hydroxystyrene).[35,69–75]

Poly(*N*-acylethylenimine)s have been applied as an effective compatibilizer ("anchoring") block of surface modifier.[35,77–79]

BIORELATED MATERIALS

A polyoxazoline–enzyme hybrid was synthesized to stabilize an active enzyme function. Carboxylic acid terminated poly(*N*-acetylethylenimine) and poly(*N*-propionylethylenimine) were covalently attached to free amino groups of bovine liver catalase without causing the dissociation of subunits.[80]

Living ends of poly(*N*-acetylethylenimine) and poly(*N*-propionylethylenimine) were terminated by the synthetic sequential peptide recognized by a monoclonal antibody directed against human protein C.[81] The avidity of the antibody for the adducts was characterized with respect to size and hydrophilicity of the oxazoline polymer segments.

Design of sugar-containing polymers by using living polymerization was successfully demonstrated by the ring-opening polymerization of 2-oxazolines with bicyclic *N*-acetyl-D-glucosamine derivatives.[82] Synthetic glycoconjugates are important in the field of biochemical and medical applications, because the carbohydrate moieties possess molecular information, and function as a cell recognition marker.

Graft copolymerization of 2-oxazolines onto a biomass, for example cellulose and its derivatives, has been attained.[48,49]

OTHER APPLICATIONS

Thermotropic liquid crystalline poly(*N*-acylethylenimine) derivatives were synthesized by the ring-opening polymerization of the monomers containing mesogenic groups with flexible spacers.[83]

Langmuir–Blodgett films of poly(*N*-undecanoylethylenimine) were obtained by the vertical deposition method.[84] This polymer forms stable monolayers on the water surface.

Polymers of cyclic imino ethers have been used in application as phase-transfer catalysts, detergents in lubricating oils, oil dispersants, adhesives, fire retardants, and textile colorants.[85–92]

REFERENCES

1. Saegusa, T.; Kobayashi, S. *Macromolecular Science, International Review of Science, Physical Chemistry Series 2*; Butterworth: London, 1975; Vol. 8, Chapter 4.
2. Saegusa, T.; Kobayashi, S. *Encyclopedia of Polymer Science and Technology*; Wiley: New York, 1976; Suppl. Vol. 1, p. 220.
3. Kobayashi, S.; Saegusa, T. *Ring-Opening Polymerization*; Elsevier: Essex, U.K., 1984; Vol. 2, Chapter 11.
4. Kobayashi, S.; Saegusa, T. *Encyclopedia of Polymer Science and Engineering*; Wiley: New York, 1986; Vol. 4, p. 525.
5. Saegusa, T.; Chujo, Y. *Frontiers of Macromolecular Science*; Blackwell Scientific: Oxford, 1989; p. 119.
6. Kobayashi, S. *Prog. Polym. Sci.* **1990**, 15, 751.
7. Chujo, Y.; Saegusa, T. *Ring-Opening Polymerization*; Hanser: Munich, 1993; p. 239.
8. Kobayashi, S.; Uyama, H.; Narita, Y. *Macromolecules* **1990**, *23*, 353.
9. Kobayashi, S.; Uyama, H.; Narita, Y.; Ishiyama, J. *Macromolecules* **1992**, *25*, 3232.
10. Dworak, A.; Schulz. R. C. *Makromol. Chem.* **1991**, *192*, 437.
11. Dworak, A.; Schulz, R. C. *Bull. Soc. Chim. Belge* **1990**, 99, 881.
12. Hrkach, J. S.; Matyjaszewski, K. *Macromolecules* **1992**, *25*, 2070.
13. Saegusa, T.; Ikeda, H.; Fujii, H. *Macromolecules*, **1972**, *5*, 359.
14. Saegusa, T.; Ikeda, H.; Fujii, H. *Polym. J.* **1973**, *4*, 87.
15. Saegusa, T.; Ikeda, H.; Fujii, H. *Macromolecules*, **1973**, *6*, 315.
16. Kobayashi, S.; Tokuzawa, T.; Saegusa, T. *Macromolecules* **1982**, *15*, 707.
17. Miyamoto, M.; Aoi, K.; Saegusa, T. *Macromolecules* **1988**, *21*, 1880.
18. Miyamoto, M.; Aoi, K.; Saegusa, T. *Macromolecules* **1991**, *24*, 11.
19. Miyamoto, M.; Aoi, K.; Morimoto, M.; Chujo, Y.; Saegusa, T. *Macromolecules* **1992**, *25*, 5878.
20. Saegusa, T.; Kobayashi, S.; Nagura, Y. *Macromolecules* **1974**, *7*, 265.
21. Saegusa, T.; Kobayashi, S.; Nagura, Y. *Macromolecules*, **1974**, *7*, 272.
22. Miyamoto, M.; Aoi, K.; Saegusa, T. *Polym. J.* **1992**, *24*, 383.
23. Levy, A.; Litt, M. *J. Polym. Sci., Polym. Chem. Ed.* **1968**, *6*, 1883.
24. Kagiya, T.; Matsuda, T.; Nakato, M.; Hirata, R. *J. Macromol. Sci. Chem.* **1972**, *6*, 1631.
25. Cai, G.; Litt, M. H. *J. Polym. Sci., Polym. Chem. Ed.* **1992**, *30*, 649.
26. Kobayashi, S.; Igarashi, T.; Moriuchi, Y.; Saegusa, T. *Macromolecules* **1986**, *19*, 535.
27. Miyamoto, M.; Aoi, K.; Saegusa, T. *Macromolecules* **1989**, *22*, 3540.
28. Saegusa, T.; Chajo, Y.; Aoi, K.; Miyamoto, M. *Makromol. Chem., Macromol. Symp.* **1990**, *32*, 1.
29. Seung, S. L. N.; Young, R. N. *Polym. Bull.* **1979**, *1*, 481.
30. Seung, S. L. N.; Young, R. N. *J. Polym. Sci. Polym. Lett. Ed.* **1980**, *18*, 89.
31. Seung, S. L. N.; Young, R. N. *J. Polym. Sci., Polym. Lett. Ed.* **1979**, *17*, 233.
32. Liu, Q.; Konas, M.; Davis, R. M.; Riffle, J. S. *J. Polym. Sci., Polym. Chem. Ed.* **1993**, *31*, 1709.
33. Kobayashi, S.; Uyama, H.; Ihara, E.; Saegusa, T. *Macromolecules* **1990**, *23*, 1586.
34. Saegusa, T.; Ikeda, H. *Macromolecules* **1973**, *6*, 805.
35. Miyamoto, M.; Sano, Y.; Saegusa, T.; Kobayashi, S. *Eur. Polym. J.* **1983**, *19*, 955.
36. Percec, V.; Guhanlyogi, S. C.; Kennedy, J. P.; Ivan, B. *Polym. Bull.* **1982**, *8*, 25.
37. Kobayashi, S.; Kaku, M.; Kyogaku, M.; Saegusa, T. *Polym. Prepr. Jpn.* **1984**, *33*, 1315.
38. Kobayashi, S.; Kaku, M.; Mizutani, T.; Saegusa, T. *Polym. Bull.* **1983**, *9*, 169.
39. Saegusa, T.; Kobayashi, S.; Yamada, A. *Macromolecules* **1975**, *8*, 390.
40. Kobayashi, S.; Kaku, M.; Sawada, S.; Saegusa, T. *Polym. Bull.* **1985**, *13*, 447.
41. Kobayashi, S.; Mertesdorf, C.; Tanabe, T.; Saegusa, T. *Polym. Prep. Jpn.* **1986**, *35*, 248.
42. Schulz, R. C.; Schwarzenbach, E. *Makromol. Chem., Macromol. Symp.* **1988**, *13/14*, 495.
43. Saegusa, T.; Yamada, A.; Kobayashi, S. *Polym. J.* **1979**, *11*, 53.
44. Trivedi, P. D.; Schulz, D. N. *Polym. Bull.* **1980**, *3*, 37.
45. Shoda, S.; Masuda, E.; Furakawa, M.; Kobayashi, S. *J. Polym. Sci. Polym. Chem. Ed.* **1992**, *30*, 1489.
46. Sinai-Zingde, G.; Verma, A.; Liu, Q.; Brink, A.; Bronk, J. M.; Marand, H.; McGrath, J. E.; Riffle, J. S. *Makromol. Chem., Macromol. Symp.* **1991**, *42/43*, 329.
47. Chujo, Y.; Ihara, E.; Ihara, H.; Saegusa, T. *Macromolecules* **1989**, *22*, 2040.
48. Ikeda, I.; Kurushima, Y.; Takashima, H.; Suzuki, K. *Polym. J.* **1988**, *20*, 243.
49. Kobayashi, S.; Kaku, M.; Saegusa, T. *Macromolecules* **1988**, *21*, 1921.
50. Aoi, K.; Takasu, A.; Okada, M. *Macromol. Chem. Phys.* **1994**, *195*, 3835.
51. Saegusa, T.; Kobayashi, S.; Kimura, Y.; Ikeda, H. *J. Macromol. Sci., Chem.* **1975**, *9*, 641.
52. Saegusa, T.; Kimura, Y.; Kobayashi, S. *Macromolecules* **1977**, *10*, 236.
53. Odian, G.; Gunatillake, P. A.; Tomalia, D. *Macromolecules* **1985**, *18*, 605.
54. Saegusa, T.; Kobayashi, S.; Kimura, Y. *Macromolecules* **1975**, *8*, 374.
55. Saegusa, T.; Ikeda, H.; Fujii, H. *Macromolecules* **1972**, *5*, 354.
56. Saegusa, T.; Kobayashi, S.; Kimura, Y. *Macromolecules* **1974**, *7*, 139.
57. Rivas, B. L.; del C. Pizarro, G. *Eur. Poly. J.* **1989**, *25*, 231.
58. Litt, M. H.; Chen, T. T.; Hsieh, B. R. *J. Polym. Sci., Polym. Chem.* **1986**, *24*, 3407.
59. Litt, M. H.; Hsieh, B. R.; Krieger, I. M.; Chen, T. T.; Lu, H. L. *J. Colloid Interface Sci.* **1987**, *115*, 312.
60. Chujo, Y.; Yoshifuji, Y.; Sada, K.; Saegusa, T. *Macromolecules*, **1989**, *22*, 1074.
61. Chujo, Y.; Sada, K.; Matsumoto, K.; Saegusa, T. *Polym. Bull.* **1989**, *21*, 353.
62. Chujo, Y.; Sada, K.; Matsumoto, K.; Saegusa, T. *Macromolecules* **1990**, *23*, 1234.
63. Chujo, Y.; Sada, K.; Saegusa, T. *Macromolecules* **1990**, *23*, 2636.
64. Chujo, Y.; Sada, K.; Saegusa, T. *Macromolecules* **1990**, *23*, 2693.
65. Chujo, Y.; Sada, K.; Nomura, R.; Naka, A.; Saegusa, T. *Macromolecules* **1993**, *26*, 5611.
66. Chujo, Y.; Sada, K.; Naka, A.; Nomura, R.; Saegusa, T. *Macromolecules* **1993**, *26*, 883.
67. Chujo, Y.; Sada, K.; Saegusa, T. *Macromolecules* **1993**, *26*, 6315.
68. Chujo, Y.; Sada, K.; Saegusa, T. *Macromolecules* **1993**, *26*, 6320.
69. Sogah, D. Y.; Kaku, M.; Shinohara, K.; Rodriguez-Parada, J. M.; Levy, M. *Makromol. Chem., Macromol. Symp.* **1992**, *64*, 49.
70. Aoi, K.; Takasu, A.; Okada, M. *Macromol. Rapid Commun.* **1995**, *16*, 53.

71. Kobayashi, S.; Kaku, M.; Saegusa, T. *Macromolecules* **1988**, *21*, 334.
72. Galin, M. *Makromol. Chem., Rapid Commun.* **1987**, *8*, 411.
73. Keskkula, H.; Paul, D. R. *J. Appl. Polym. Sci.* **1986**, *31*, 1189.
74. Lin, P.; Clash, C.; Pearce, E. M.; Kwei, T. K. *J. Polym. Sci. Polym. Sci. Polym. Phys. Ed.* **1988**, *26*, 603.
75. Zhu, K. J.; Kwei, T. K.; Pearce, E. M. *J. Appl. Polym. Sci.* **1989**, *37*, 573.
76. Kaku, M.; Hung, M.-H. *Macromolecules* **1993**, *26*, 6135.
77. Yilgor, I.; Ward, R. S.; Riffle, J. S. *Polym. Prepr. (Am. Chem. Soc., Div. Polym. Chem.* **1987**, *28*(2), 369.
78. Yilgor, I.; Steckle, W. P. Jr.; Yilgor, E.; Freelin, R. G.; Riffle, J. S. *J. Polym. Sci., Polym. Chem. Ed.* **1989**, *27*, 3673.
79. Aoi, K.; Miyamoto, M.; Chujo, Y.; Saegusa, T. *Polym. Prepr. Jpn.* 1989, 38, 1596.
80. Miyamoto, M.; Naka, K.; Shiozaki, M.; Chujo, Y.; Saegusa, T. *Macromolecules* **1990**, *23*, 3201.
81. Velander, W. H.; Madurawe, R. D.; Subramanian, A.; Kumar, G.; Sinai-Zingde, G.; Riffle, J. S.; Orthner, C. L. *Biotech. Bioeng.* **1992**, *39*, 1024.
82. Aoi, K.; Suzuki, H.; Okada, M. *Macromolecules* **1992**, *25*, 7073.
83. Rodriguez-Parada, J. M.; Percec, V. *J. Polym. Sci., Polym. Chem. Ed.* **1987**, *25*, 2269.
84. Kaku, M.; Hsiung, H.; Sogah, D. Y.; Levy, M.; Rodriguez-Parada, J. M. *Langmuir* **1992**, *8*, 1239.
85. Kahovec, J.; Jelinkova, M.; Janout, V. *Polym. Bull.* **1986**, *15*, 485.
86. Kelyman, J. S.; Paul, G. A. U.S. Patent 4 152 342, 1979, (to The Dow Chemical Co.).
87. Smith, W. L.; Kelyman, J. S. U.S. Patent 4 120 804, 1978, (to The Dow Chemical Co.).
88. Berg, G.; Bluemel, H.; Seeliger, W. German Patent 1 287 039, 1969, (to Chemische Werke Huls AG).
89. Klemm, E.; Herzok, K.; Hoerhold, H. H. *Acta Polym.* **1980**, *31*, 785.
90. Cai, G.; Litt, M. H. *J. Polym. Sci. Polym. Chem. Ed.* **1992**, *30*, 671.
91. Smith, W. L. U.S. Patent 4 153 466, 1979, (to The Dow Chemical Co.).
92. Evans, G. E. European Patent 30 786, 1981, (to Imperial Chemical Industries).

CYCLIC IMINO ETHERS (Ring-Opening Polymerization of 1,3-Oxazo Monomers)

Michael A. Mitchell and Brian C. Benicewicz
Los Alamos National Laboratory

This contribution attempts to provide an updated review of the synthesis, properties, and emerging applications of polymers derived from 1,3-oxazo monomers, with an emphasis on the commonly studied 2-oxazolines and 2-oxazines.

PREPARATION AND PROPERTIES

Monomer Synthesis

The most commonly studied cyclic imino ethers are the oxazolines, followed by the oxazines. These compounds can be prepared by many methods. The reaction between amino alcohols and nitriles is a very convenient and practical method.[1]

Other preparations of 2-oxazolines and 2-oxazines include the reaction of carboxylic acids with amino alcohols,[2,3] and dehydration of N-(ω-hydroxyalkyl) amides with reagents such as thionyl chloride,[4] sulfuric acid,[5] phosphorotriazolides,[6] and Mitsunobu conditions.[7]

Of particular interest is the synthesis of difunctional monomers, which have an oxazoline moiety.

Polymerization

Both 2-oxazines, and in particular 2-oxazolines, have drawn a great deal of interest for several reasons. First, oxazines and oxazolines are homopolymerized with cationic initiators, which can include alkylating agents. Second, the polymerization is pseudoliving; thus block copolymers and chain-end functionalized macromers can be readily prepared. Third, physical properties can be varied greatly by choice of substitution in the 2-position.

Practical Aspects

Although oxazoline and oxazine polymerizations are more robust than other cationic polymerizations, they are still sensitive to impurities. Oxazolines are usually polymerized between 100°C and 150°C, whereas oxazines often require temperatures of 150°C or higher.[8]

Copolymerization

The attractive features of oxazoline and oxazine polymerizations include the ease with which alternating, block, graft, and random copolymers can be produced.

Alternating/Zwitterionic

Oxazolines and oxazines react with acrylic monomers (acrylic acid,[9–13] α-halo acrylic acids,[14,15] ω-hydroxyalkyl acrylates,[16] and acrylamide[17]) to form alternating copolymers without an initiator. These polymers often have low molecular weights. These polymerizations proceed by formation of a zwitterionic adduct followed by proton transfer.

Copolymerization of 2-oxazoline with β-propiolactone gives 1:1 alternating copolymer via the same intermediate as for the copolymerization of 2-oxazoline and acrylic acid.[18] γ-Propanesultone also gives a 1:1 alternating copolymer.[19] More often, the polymerization of oxazolines with lactones and cyclic anhydrides gives statistical copolymers or polymers rich in the cyclic anhydrides or lactones.[20–29]

Block

Oxazolines have been used to produce both AB and ABA block copolymers, as well as star polymers.[30,31] Oxazoline–oxazoline or oxazoline–oxazine block copolymers are produced by homopolymerization of the first monomer. When the first monomer is consumed, the monomer for the second block is added and the polymerization continues.

Both di- and tri-copolymers have been produced in this manner. Usually the oxazoline copolymers are made with mixtures of hydrophobic and hydrophilic blocks for use as surfactants.[32–36]

Several styrene block copolymers with 2-methyl-2-oxazoline have been reported. For the synthesis of oxazoline-vinyloxazoline triblocks styrene, 2-vinylnaphthalene, and 9-vinylphenanthrene have been used.[37–39]

Triblock copolymers of the structure oxazoline-polyethylene oxide (PEG)-oxazoline have been made by the synthesis of the ditosylate PEG macroinitiator.[40,41] Also, the dichloroformate and the tosyl derivatives of PEG adipate have been used.[42–44]

Block copolymers with siloxanes,[45–47] ε-caprolactone,[48] and styrene oxide[49] have also been synthesized.

Graft

Four methods have been used to synthesize polymers that have oxazoline grafts. One method is the production of macromonomers by the initiation method. In this procedure the polymerization of an oxazoline is initiated by compounds containing a vinyl group. Resulting macromer are then radically polymerized. The second method is macromonomer synthesis by the termination method, which involves terminating an oxazoline polymerization with a compound that contains a polymerizable moiety. The third popular method for making graft copolymers is the polymerization of oxazolines from polymers containing alkylating groups. The fourth method is the grafting of an existing polymer onto another polymer.

Linear Polyamines

The only way to synthesize linear polyamines of the structure $-(CH_2)_n$-NH- is by hydrolysis of poly(2-oxazoline)s[50–52] and poly(2-oxazine)s.[10,53,54] The hydrolysis has been done using either acid or base.

POLYMER APPLICATIONS

Recycling

The degradative thermal and hydrolytic effects of processing recycled poly(ethylene terephthalate) (PET) and polyamides can be compensated by chain extending these polymers with bisoxazolines.[55–57] The addition reaction of bisoxazolines with acid or hydroxyl end groups results in higher molecular weights and lower end-group concentrations. The reaction can be accomplished directly in an extruder and could result in a broader use of recycled PET and nylons. There is extensive worldwide patent literature on using the addition reactions of cyclic imin ethers for thermosetting resins, adhesives, and reaction injection molding (RIM) applications.

Hybrid Organic/Inorganic Composites

Hybrid organic polymer–inorganic composites were made using polyoxazolines.[58,59] These materials were made by blending or reacting polyoxazolines via standard sol–gel techniques to produce novel organic–inorganic composites.

Composite Resins

Kagiya et al. have investigated the addition reactions of bisoxazoline with dicarboxylic acids to give poly(ester amide) resins with high mechanical properties.[60] Culbertson et al. have extended this chemistry to bisoxazoline–phenolic resins and conducted evaluations for aerospace composite applications.[61,62]

Surfactants

Block copolymers have been widely exploited as surfactants in both aqueous and nonaqueous emulsion polymerizations. The living nature of the polymerization of oxazolines and oxazines has been used to synthesize a wide variety of nonionic surfactants with excellent surfactant properties. Copolymers with AB and ABA block copolymer structures containing both hydrophilic and hydrophobic chains show good surfactant properties. The surface-active properties of these polymers have also been explored in applications such as inverse emulsification,[63,64] compatibilization in polymer blends,[65,66] adhesion,[67–70] phase-transfer catalysis,[71] anti-electrostatics,[72] and Langmuir–Blodgett films.[73]

Hydrogels

Hydrogels have found applications in a number of different fields, such as packaging, drying agents, medical devices, and controlled-release agents. In recent years, nonionic hydrogels based on crosslinked polyoxazolines have begun to be explored.[74] Gels produced from lower alkyl-2-oxazolines form stable hydrogels with relatively high swelling properties in both water and 5% aqueous NaCl solutions. Gels from higher alkyl-2-oxazolines form lipogels and those gels made from intermediate alkyl-2-oxazolines form amphigels, which exhibit good swelling characteristics in both water and organic solvents.[75]

REFERENCES

1. Witte, H.; Seeliger, W. *Liebigs Ann. Chem.* **1974**, 996.
2. Vorbruggen, H.; Krolikiewicz, K. *Tet. Lett.* **1981**, *22*(45), 4471.
3. Vorbruggen, H.; Krolikiewicz, K. *Tetrahedron* **1993**, *49*(41), 9353.
4. Hansen, J. F.; Kamata, K.; Meyers, A. I. *J. Heterocyclic Chem.* **1973**, *10*, 711.
5. Levy, A.; Litt, M. *Polymer Lett.* **1967**, *5*, 881.
6. Sund, C.; Ylikoski, J.; Kwiatkowski, M. *Synthesis* **1987**, *9*, 853.
7. Roush, D. M.; Patel, M. M. *Synth. Commun.* **1985**, *15*(8), 675.
8. Mitchell, M. A.; Benicewicz, B. C.; Langlois, D. A.; Thiesen, P. *Polym. Mat. Sci. Eng.* **1993**, *68*(1), 22.
9. Saegusa, T.; Kobayashi, S.; Kimura, Y. *Macromolecules* **1974**, *7*(1), 139.
10. Saegusa, T. *Pure Appl. Chem.* **1974**, *39*(1–2), 81.
11. Saegusa, T.; Kimura, Y.; Kobayashi, S. *Macromolecules* **1977**, *10*(2), 236.
12. Rivas, B.; Canessa, G. S.; Pooley, S. A. *Eur. Polym. J.* **1989**, *25*(3), 225.
13. Odian, G.; Gunatillake, P. A.; *Polym. Prepr.* **1983**, *24*(1), 135.
14. Rivas, B.; Canessa, G. S.; Pooley, S. A. *Eur. Polym. J.* **1992**, *28*(1), 43.
15. Balakrishnan, T.; Periyasamy, M. *Polymer* **1982**, *23*(9), 1372.
16. Saegusa, T.; Kimura, Y.; Kobayashi, S. *Macromolecules* **1977**, *10*(2), 239.
17. Saegusa, T.; Kobayashi, S.; Kimura, Y. *Macromolecules* **1975**, *8*(3), 374.
18. Saegusa, T.; Ikeda, H.; Fujii, H. *Macromolecules* **1972**, *5*(4), 354.
19. Saegusa, T.; Ikeda, H.; Hirayanagi, S.; Kimura, Y.; Kobayashi, S. *Macromolecules* **1975**, *8*(3), 259.
20. Kobayashi, S.; Isobe, M.; Saegusa, T. *Macromolecules* **1982**, *15*(3), 703.
21. Furukawa, J.; Kobayashi, S.; Saegusa, T. *Polym. Bull.* **1989**, *21*(4), 421.
22. Canessa, G. S.; Pooley, S. A.; Parra, M.; Rivas, B. *Polym. Bull.* **1984**, *11*(5), 465.
23. Rivas, B.; Canessa, G. S.; Pooley, S. A. *Polym. Bull.* **1983**, *9*(8–9), 417.
24. Rivas, B.; Canessa, G. S.; Pooley, S. A. *Polym. Bull.* **1985**, *13*(1), 65.
25. Rivas, B.; Canessa, G. S.; Pooley, S. A. *Polym. Bull.* **1985**, *13*(2), 103.
26. Rivas, B.; Canessa, G. S.; Pooley, S. A. *Polym. Bull.* **1985**, *13*(6), 519.
27. Rivas, B.; Canessa, G. S.; Pooley, S. A. *Makromol. Chem.* **1986**, *187*(1), 71.

28. Rivas, B.; Canessa, G. S.; Pooley, S. A. *Makromol. Chem.* **1987**, *188*(1), 149.
29. Rivas, B.; Mena, J.; Pizarro, G. del C.; Tagle, L. H. *Eur. Polym. J.* **1993**, *29*(1), 91.
30. Percec, V; Guhaniyogi, S. C.; Kennedy, J. P.; Ivan, B. *Polym. Bull.* **1982**, *8*(1), 25.
31. Dworak, A.; Schulz, R. C. *Makromol. Chem.* **1991**, *192*(2), 437.
32. Litt, M.; Herz, J. *Polym. Prepr.* **1969**, *10*(2), 905.
33. Litt, M.; Lin, C. H.; Krieger, I. M. *J. Polym. Sci., Part A* **1990**, *28*(10), 2777.
34. Cai, G.; Litt, M.; Krieger, I. M. *Contemp. Top. Polym. Sci.* **1989**, *6*, 139.
35. Hsieh, B. R.; Litt, M. H. *J. Polym. Sci., Part A* **1988**, *26*(9), 2501.
36. Demopolis, T. N.; Cai, G. F.; Irvin, I. M. Litt, M. *Polym. Prepr.* **1988**, *29*(2), 23.
37. Morishima, Y.; Tanaka, T.; Nozakura, S. *Polym. Bull.* **1981**, *5*(1), 19.
38. Ishizu, K.; Fukutomi, T.; Kakurai, T. *J. Polym. Sci., Polym. Lett. Ed.* **1983**, *21*(5), 405.
39. Ishizu, K.; Ishikawa, S.; Fukutomi, T. *J. Polym. Sci., Polym. Chem. Ed.* **1985**, *23*(2), 445.
40. Simionescu, C. I.; Rabia, I. *Polym. Bull.* **1983**, *10*(7–8), 311.
41. Litt, M.; Swamikannu, X. *Polym. Prep.* **1984**, *25*(1), 242.
42. Percec, V. *Polym. Prepr.* **1982**, *21*(1), 301.
43. Simionescu, C. I.; Rabia, I.; Crisan, Z. *Polym. Bull.* **1982**, *7*(4), 217.
44. Percec, V. *Polym. Bull.* **1981**, 5(11–12), 643.
45. Yilgor, I.; Steckle, W. P.; Yilgor, E.; Freelin, R. G.; Riffle, J. S. *J. Polym. Sci., Part A* **1989**, *27*(11), 3673.
46. Liu, Q; Wilson, G. R.; Davis, R. M.; Riffle, J. S. *Polymer* **1992**, *34*(14), 3030.
47. Riffle, S. J.; Sinai-Zingde, G.; Desimone, J. M.; Hellstern, A. M.; Chen, D. H.; Yilgor, I. *Polym. Prepr.* **1988**, *29*(2), 93.
48. Sinai-Zingde, G.; Verma, A.; Liu, Q.; Brink, A.; Bronk, J. M.; Marand, H.; McGrath, J. E.; Riffle, J. S. *Makromol. Chem. Macromol. Symp.* **1991**, *42/43*, 329.
49. Seung, S. N. L.; Young, R. N. *J. Polym. Sci., Polym. Lett. Ed.* **1980**, *18*(2), 89.
50. Saegusa, T.; Ikeda, H.; Fujii, H. *Polymer J.* **1972**, *3*(1), 35.
51. Seeliger, W.; Thier, W.; Kriesten, W. German Patent No. 1720436, July 8, 1971.
52. Tanaka, R.; Ueoka, I.; Takaki, Y.; Kataoka, K.; Saito, S. *Macromolecules* **1983**, *16*(6), 849.
53. Saegusa, T.; Nagura, Y.; Kobayashi, S. *Macromolecules* **1973**, *6*(4), 495.
54. Saegusa, T.; Kobayashi, S.; Ishiguro, M. *Macromolecules* **1974**, *7*(6), 958.
55. Inata, H.; Matusmura, S. *J. Appl. Polym. Sci.* **1986**, *32*(4), 4581.
56. Inata, H.; Matusmura, S. *J. Appl. Polym. Sci.* **1987**, *33*(8), 306.
57. Cardi, N.; Po, R.; Giannotta, G.; Occhiello, E.; Garbassi, F.; Messina, G. *J. Appl. Polym. Sci.* **1993**, *50*(9), 1501.
58. Chujo, Y.; Ihara, E.; Ihara, H.; Saegusa, T. *Macromolecules* **1989**, 22(5), 2040.
59. Chujo, Y.; Ihara, E.; Kure, S.; Suzuki, K.; Saegusa, T. *Macromol. Chem., Macromol. Symp.* **1991**, *42/43*, 303.
60. Kagiya, T.; Narisawa, S.; Maeda, T.; Fukui, K. *J. Polym. Sci., Part B: Polym. Lett.* **1966**, *4*, 257.
61. Culbertson, B. M.; Tiba, O.; Deviney, M. L. *Int. SAMPE Tech. Conf.* **1988**, *20*, 590.
62. Culbertson, B. M.; Tiba, O.; Deviney, M. L.; Tufts, T. A. *Int. SAMPE Symp. Exhib.* **1989**, *34*, 2483.
63. Hsieh, B. R.; Litt, M. H. *Polym. Prepr.* **1986**, 27(2), 122.
64. Litt, M. H.; Hsieh, B. R.; Krieger, I. M.; Chen, T. T.; Lu, H L. *J. Colloid Interface Sci.* **1987**, *115*(2), 312.
65. Sinai-Zingde, G.; Verma, A.; Liu, Q.; Brink, A.; Bronk, J. M.; Allison, D.; Goforth, A.; Patel, N.; Marand, H.; McGrath, J. E.; Riffle, J. S. *Polym. Prepr.* **1990**, *31*(1), 63.
66. Dean, B. D. *J. Appl. Polym. Sci.* **1989**, *37*(6), 1727.
67. Cai, G.; Litt, M. *J. Polym. Sci., Part A: Polym. Chem.* **1992**, *30*(4), 649.
68. Cai, G.; Litt, M. *J. Polym. Sci., Part A: Polym. Chem.* **1992**, *30*(4), 671.
69. Cai, G.; Litt, M. *J. Polym. Sci., Part A* **1989**, *27*(11), 3603.
70. Cai, G.; Litt, M.; Krieger, I. M. *J. Polym. Sci., Part B: Polym. Phys.* **1991**, *29*(7), 773.
71. Kahovec, J.; Jelindova, M.; Janout, V. *Polym. Bull. (Berlin)* **1986**, *15*(6), 485.
72. Miyamoto, M.; Sano, Y. *Eur. Polym. J.* **1983**, *19*(10–11), 955.
73. Kaku, M.; Hsiung, H. Sogah, D. Y.; Levy, M. Rodriguez, Parada, J. M. *Langmuir* **1992**, *8*(5), 1239.
74. Chujo, Y.; Sada, K.; Naka, A.; Nomura, R.; Saegusa, T. *Macromolecules* **1993**, *26*(5), 883.
75. Chujo, Y.; Sada, K.; Matsumoto, K.; Saegusa, T. *Polym. Bull. (Berlin)* **1989**, *21*(4), 353.

Cyclic Ketene Acetal Polymerization

See: Cyclic Ketene Dithioacetals (Polymerization)
Ring-Opening Polymerization, Free Radical

CYCLIC KETENE DITHIOACETALS (Polymerization)

Shiro Kobayashi
Department of Molecular Chemistry and Engineering
Faculty of Engineering
Tohoku University

Jun-ichi Kadokawa
Department of Materials Science and Engineering
Faculty of Engineering
Yamagato University

Cyclic ketene dithioacetals are sulfur analogues of cyclic ketene acetal and have received considerable attention in recent years as important synthetic intermediates.[1] Some cyclic Ketene acetals are known to undergo radical polymerizations.[2–4] The polymerization proceeded involving partial ring-opening of the monomer to give a polymer having a ketene acetal unit and an ester unit.

Radical polymerizations of some cyclic ketene dithioacetals involve two possibilities for the structures of the product polymer, which are a vinylidene unit without ring-opening and a dithioester unit with ring-opening.

RADICAL POLYMERIZATION OF 2-METHYLENE-1,3-DITHIANE (6)[5]

The radical polymerization of 2-methylene-1,3-dithiane (**6** in **Scheme I**), is a six-membered cyclic ketene dithioacetal) is induced by a AIBN initiator to give a polymer.

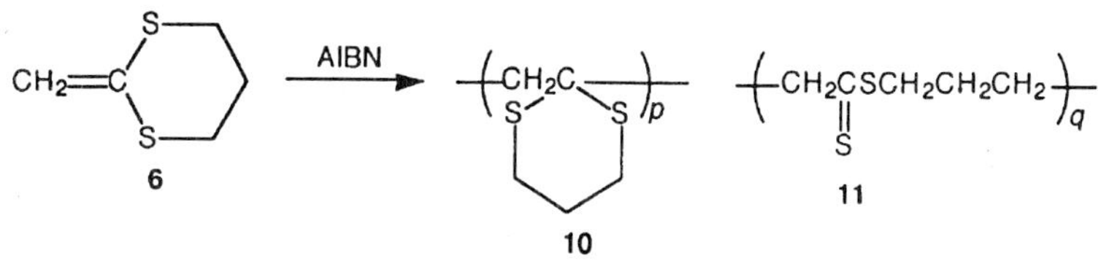

I

The ^{13}C NMR, IR, and UV spectra of the product obtained by solution polymerization supported that the polymer mainly consisted of a dithioacetal unit (**10** in Scheme I) by vinylidene polymerization. The intensities of the absorptions due to C=S group in the spectra indicated that the content of a dithioacetal unit (**11** in Scheme I) in the polymer was low, if present at all.

RADICAL COPOLYMERIZATION OF 2-METHYLENE-1,3-DITHIANE (6) WITH STYRENE AND METHYL METHACRYLATE[6]

Copolymerization of **6** with styrene with AIBN initiator produced a white powdery material. The gel permeation chromatographic (GPC) analysis of the product showed only one peak, indicating that **6** was copolymerized with styrene to form a copolymer.

The ^{1}H NMR, ^{13}C NMR, IR, and UV spectroscopic data were taken to support that the copolymerization proceeded without ring opening of **6**.

REFERENCES

1. Oae, S.; Ohno, A.; Furukawa, N. *Reviews on Heteroatom Chemistry*, MYU K. K., Tokyo, 1988.
2. Bailey, W. J.; Ni, Z.; Wu, S. R. *J. Polym. Sci., Polym. Chem. Ed.* **1982**, *20*, 3021.
3. Endo, T.; Okawara, M.; Bailey, W. J.; Azuma, K.; Nate, K.; Yokono, H. *J. Polym. Sci., Polym. Lett. Ed.* **1983**, *21*, 373.
4. Bailey, W. J. *Polym. J.* **1985**, *17*, 85.
5. Kobayashi, S.; Kadokawa, J.; Shoda, S.; Uyama, H. *Macromol. Reports* **1991**, *A28(Suppl. 1)*, 1.
6. Kobayashi, S.; Kadokawa, J.; Matsumura, Y.; Yen, I. F.; Uyama, H. *Macromol. Reports* **1992**, *A29(Suppl. 3)*, 243.

Cyclic Oligomers

See: *Cyclic Oligomers of Engineering Thermoplastics*
Macrocyclic Aramids
Ring-Opening Metathesis Polymerization (Formation of Cyclic Butadiene Oligomers)

CYCLIC OLIGOMERS OF ENGINEERING THERMOPLASTICS

Daniel J. Brunelle
GE Research and Development Center

The use of ring-opening polymerization for the preparation of engineering polymers has been little explored until recently, due to the difficulty of preparation of the requisite cyclic monomers or oligomers. The use of ROP chemistry for the preparation of engineering thermoplastics is appealing, since low-molecular weight precursors lead to high-molecular weight polymers without formation of by-products. The use of cyclic oligomers would allow processing of the low-viscosity cyclics by many techniques such as pultrusion, resin-transfer molding, melt-filtration, or reaction-injection molding, either concurrent or just prior to polymerization. The macrocyclic polymers discussed in this report have little or no ring strain, and polymerization reactions are driven principally by entropy, leading to completely equilibrated products.

Until recently, most work on cyclic oligomers concentrated on the preparation and characterization of discrete cyclic oligomeric materials, the cyclic carbonate tetramer of bisphenol A (BPA) for example.[1] Over the past ten years, techniques have been devised for the preparation of *mixtures* of cyclic oligomers which avoid tedious separation and purification processes. As these low-molecular weight cyclic polymers became available in multi-gram or even multi-kilogram quantities, more extensive studies of their physical properties and chemistry have become possible. Several techniques have now been established for the preparation of mixtures of cyclic oligomers of polycarbonates, polyarylates, polyesters, poly(ether ketone)s, poly(ether imide)s, and poly(ether sulfone)s, and aramids.

CYCLIC AROMATIC CARBONATES

Brunelle et al. reported that slow addition of a solution of BPA bischloroformate to an efficiently stirred mixture of CH_2Cl_2, triethylamine, and aqueous NaOH would effect a remarkably selective formation of a mixture of cyclic oligomeric aryl carbonates. In high yield (**Equation 1**).[2–4] The product was composed of a mixture of oligomeric cyclics predominantly in the range of dimer to dodecamer, and high-molecular weight polycarbonate, with a ratio of cyclics to polymer of 85/15. The level of linear oligomers present was estimated to be 0.01–0.03%.

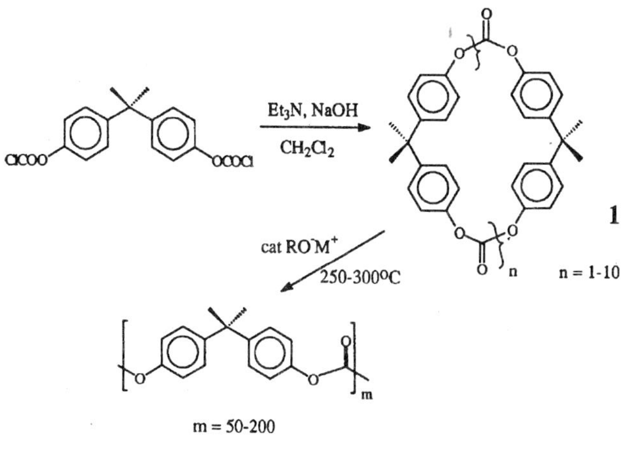

1

Polymerization of BPA Cyclic Carbonate Oligomers[5,6]

Polymerization of BPA oligomeric cyclic carbonates could be achieved in the melt at 200–300°C, or in solution under various conditions, including reaction in DMSO/CH_2Cl_2 at ambient temperature. Polymerization could also be initiated in THF solution with alkyllithiums.[7] A variety of catalysts have been investigated, including Bronsted and Lewis acids and bases, and various metallic compounds. One of the most effective initiators was *n*-Bu_4NBPh_4, an organic-soluble nucleophilic base. Precipitation of the product polycarbonate into acetone and analysis of the soluble portion by HPLC indicated that only about 0.25% cyclics remain after polymerization. The heat of reaction has been measured by differential scanning calorimetry, and has been found to be about −1.2 kJ/mole. This slight exotherm has been correlated to the release of ring strain in opening the cyclic dimer. The exotherm on polymerization was clearly evident in the DSC of pure cyclic dimer.[8]

MACROCYCLIC ARYLATES

A method for the direct formation of cyclic arylate oligomers via phase-transfer catalyzed reaction of iso- and terephthaloyl chlorides with bisphenols has recently been developed.[9–12]

Bisphenol A Poly(Iso/terephthalate)[12]

The mechanism for formation of cyclic polyarylates appears to be quite different from that of aryl polycarbonate formation. Although amine catalysts such as Et_3N could be used to prepare cyclic polyarylates, the yields were low. PTC catalysts such as Adogen (trimethyl-*n*-alkylammonium halides) were much more effective, affording higher yields of cyclics. Since condensation reactions of aroyl halides were somewhat slower than those of bischloroformates, slightly elevated temperatures were preferred. Using these techniques, yields of cyclic polyarylates of up to 85% could be obtained.[13]

The cyclic arylates could be polymerized at elevated temperature (>300°C) in the presence of an anionic initiator.[12] The individual cyclics melted at about 400°C with polymerization occurring, even in the absence of catalyst. Polymerization leads to polyarylates with M_w of about 40–150,000, and the expected glass transition temperatures (bisphenol A isophthalate, T_g = 167°C, spirobiindane isophthalate, T_g = 242°C; mixed BPA iso/terephthalate, T_g = 187°C).

MACROCYCLIC ETHERS AND ETHER SULFONES, ETHER KETONES, AND ETHER IMIDES

A number of aromatic ether and thioether imides, sulfones, and ketones have been formed in cases where a spirobiindane structure has been built into one of the monomers.[14] Using spirobiindane bisphenol (SBI) as a synthetic precursor to dianhydride or diamine has enabled the preparation of a variety of ether polyimide structures via subsequent reaction with various amines or dianhydrides. Cyclic aryl ether ketones have recently been formed with monomers containing the 1,2-dibenzoylbenzene moiety using a high dilution technique.[15]

Preparation of Cyclic Poly(ether sulfone)

Ring-opening polymerization of these cyclic poly(ether imide) structures via a transetherification reaction has been achieved.[16,17] A survey of several potential catalysts has shown that sulfur nucleophiles, such as sodium sulfide or sodium thiophenoxide were effective initiators for the polymerization reaction.

Macrocyclic Aramids

Macrocyclic aramids (aryl aryl amides) have been prepared from SBI-based starting materials such as diamine, via reaction with diacid chlorides in CH_2Cl_2 of THF.[13]

Memeger et al. have recently reported a synthesis of macrocyclic aramids based on terephthaloyl chloride and a substituted *p*-phenylenediamine.[18,19] Reaction of N,N′-diisobutyl-*p*-phenylenediamine with terephthaloyl chloride at elevated temperature in *o*-dichlorobenzene under classical dilution conditions led to good yields (up to 86%) of the macrocyclic aramids.

Ring-opening polymerization of the cyclic aramids was effected in the melt using highly nucleophilic catalysts such as 1,3-dialkylimidazole-2-thiones. Use of an acidic cocatalyst such as phenylphosphinic acid gave complete polymerization, but with comcomitant loss of butyl groups, which gave a desirable increase in T_g from 179 to 218°C. The polymers could be drawn into films. Crystallinity was observed by X-ray in both the cast and drawn films.

REFERENCES

1. Schnell, H.; Bottenbruch, L. *Makromol. Chem.* **1962**, *57*, 1.
2. Brunelle, D. J.; Evans, T. L.; Shannon, T. G.; Boden, E. P. *Polym. Prepr.* **1989**, *30*(2), 569.
3. Brunelle, D. J.; Boden, E. P.; Shannon, T. G. *J. Am. Chem. Soc.* **1990-a**, *112*, 2399.
4. Brunelle, D. J.; Shannon, T. G. *Macromolecules* **1991-a**, *24*, 3035.
5. Evans, T. L.; Carpenter, J. C. *Polym. Prepr.* **1990**, *31*(1), 18.
6. Evans, T. L.; Carpenter, J. C. *Makromol. Chem. Macromol. Symp.* **1991**, *42/43*, 177.
7. Leitz, E.; Bottenbruch, L.; Ott, K. H.; Jung, A.; Grigo, U. Ger. Offen. 3 831 886, 1990.
8. Brunelle, D. J.; Garbauskas, M. F. *Macromolecules* **1993-a**, *26*, 2724.
9. Brunelle, D. J.; Guggenheim, T. L.; Boden, E. P.; Shannon, T. G.; Guiles, J. W. U.S. Patent 4 696 993, 1987.
10. Brunelle, D. J.; Shannon, T. G. U.S. Patent 4 829 144, 1989-b.
11. Guggenheim, T. L.; McCormick, S. J.; Kelly, J. J.; Brunelle, D. J.; Colley, A. M.; Boden, E. P.; Shannon, T. G. *Polym. Prepr.* **1989-a**, *30*(2), 579, 138.
12. Boden, E. P.; Phelps, P. D. U.S. Patent 5 136 018, 1992.
13. Guggenheim, T. L.; McCormick, S. J.; Guiles, J. W.; Colley, A. M. *Polym. Prepr.* **1989-b**, *30*(2), 138.
14. Cella, J. A.; Talley, J. J.; Fukuyama, J. *Polym. Prepr.* **1989-a**, *30*(2), 581.
15. Chan, K. P.; Wang, Yi-feng; Hay, A. S. *Macromolecules* **1995**, *28*, 653.
16. Mullins, M. J.; Galvan, R.; Bishop, M. T.; Woo, E. P.; Gorman, D. B.; Chamberlain, T. A. *Polym. Prepr.* **1992**, *33*(1), 414.
17. Cella, J. A.; Fukuyama, J.; Guggenheim, T. L. *Polym. Prepr.* **1989-b**, *30*(2), 142.
18. Memeger, W.; Lazar, J.; Ovenall, D.; Arduengo, III, A. J.; Leach, R. A. *Polym. Prepr.* **1993-a**, *34*(1), 71.
19. Memeger, W.; Lazar, J.; Ovenall, D.; Leach, R. A. *Macromolecules* **1993-b**, *26*, 3476.

Cyclic Phosphate Polymerization

See: *Poly(alkylene phosphate)s*

Cyclic Phosphite Polymerization

See: *Poly(alkylene phosphate)s*

Cyclic Polymers

See: *Macrocyclic Block Copolymers*
Macrocyclic Polymers (Controlled Dimensions)
Peptide-Based Nanotubes (New Class of Functional Biomaterials)
Tailor-Made Polymers
Uniform Polymers

CYCLIC SILOXANES (Ring-Opening Polymerization)

Julian Choojnowski* and Marek Cypryk
Center of Molecular and Macromolecular Studies of the Polish Academy of Sciences

Polysiloxanes are polymers that have inorganic chain skeletons built of alternatively arranged oxygen and silicon atoms to which organic groups are attached. They are the most important organosilicon polymers, being main components of a variety of commercial products known as "silicones." Methods of synthesis of these polymers are based on two processes: polycondensation of functional silanes, and ring-opening polymerization of cyclic siloxanes. The latter is particularly useful for synthesis of linear polysiloxanes of high molecular weight. Monomers in the polymerization are cyclic oligosiloxanes of structure $(R_1R_2SiO)_n$, $n \geq 3$. (The name "monomer" for oligosiloxanes containing three or more repeating units is not precise as such monomer may contain two or more different siloxane units (see, e.g., **Scheme I**, structure C) and its homopolymerization leads to a copolymer.) Among the numerous polysiloxanes known, polydimethylsiloxane (PDMS) is by far the most important. Thus, rich literature is devoted to the polymerization of $(Me_2SiO)_n$ series, D_n (D denoting a Me_2SiO unit). Among them, D_4 and D_3 (Scheme I, structures A and B, respectively) are the most common.

Apart from conventional siloxane monomers of pure silaacetal structure, such as A, B, and C in Scheme I, there are many other cyclic compounds containing the SiOSi grouping in the ring skeleton. Most of these compounds are able to polymerize, or at least to copolymerize with the cleavage and reformation of the SiOSi bond. Thus, they also may be considered cyclic siloxane monomers. The analogues of cyclic ethers, in which some or all carbon atoms in the skeleton are replaced by silicon, constitute an important group of such monomers. They are exemplified in Scheme I by structures D,[1] E,[2] and F.[3] Siloxazanes having siloxane and silazane grouping (e.g., Structure G)[4] and cycloheterosiloxanes, such as cycloborasiloxane (Structure H),[5] belong to the other class of these monomers. Compounds having two or more siloxane rings may also be regarded as unconventional siloxane monomers. Among them are spirosiloxanes (Structure I), which polymerize to crosslinked polymers,[6] and silsequioxanes (Structure J), which may be polymerized anionically to a ladder polymer.[7]

Cyclic siloxane monomers may be polymerized either on the anionic route when a strong base is used as initiator or by the cationic process initiated with an acid. Both may be performed in bulk, in solution, or in emulsion.[8–10] Polymerization in solid state[11,12] and in heterophase systems[13,14] was explored as well.

The polymer may be obtained with thermodynamic control when the polymerization is carried out to the equilibrium state. This mode of the polymerization is often called equilibration. The alternative way is the polymerization with kinetic control, when the reaction is quenched at a suitable moment to obtain a high polymer yield or to ensure required quality of the product.

THERMODYNAMICS OF THE POLYMERIZATION

Because the polymerization of cyclic siloxanes is often carried out with thermodynamic control, knowledge of the equilibrium state of various siloxane polymerization systems is of particular importance. The equilibrium system is complex because it includes continuous populations of cyclic and linear chain species in equilibrium with each other. Ring-opening equilibria deserve particular attention as they determine the thermodynamic ability of the system to form a polymer.

BASIC PRINCIPLES OF THE THERMODYNAMICALLY CONTROLLED POLYMERIZATION

Results of the thermodynamically controlled polymerization are independent of its route (mechanism, catalysis) and of its initial state. Both anionic and cationic polymerizations are commonly practiced. The initiator should be effective and easily neutralized. "Transient catalysts" are often used. They produce thermolabile propagation centers, such as tetramethylammonium or tetrabutylphosphonium silanolate, which are neutralized by heating.[15]

Triorganosiloxy-ended linear oligosiloxanes are often introduced as the chain stopper which permits control of the molecular weight.

ANIONIC POLYMERIZATION

The anionic ring-opening polymerization of cyclic siloxanes is a convenient method for the equilibration of siloxanes and also provides the unique possibility of precise control of the polymer structure in kinetically controlled synthesis of polysiloxanes. The initiator, which is a strong base such as alkali metal hydroxide, opens the siloxane monomer ring, generating a silanolate group being the active propagating species. If the system is pure and isolated from atmosphere, no termination reaction occurs.

Kinetically Controlled Polymerization and Synthesis of Uniform-Size Linear Telechelic Polysiloxanes

Polymerization with the kinetic control of the product is usually practiced in two cases: when the equilibrium position favors cyclic compounds as it is in the case of poly(γ-trifluoropropylmethylsiloxane) and polydiphenylsiloxane; and when a high degree of control of the polymer structure is required, as

*Author to whom correspondence should be addressed.

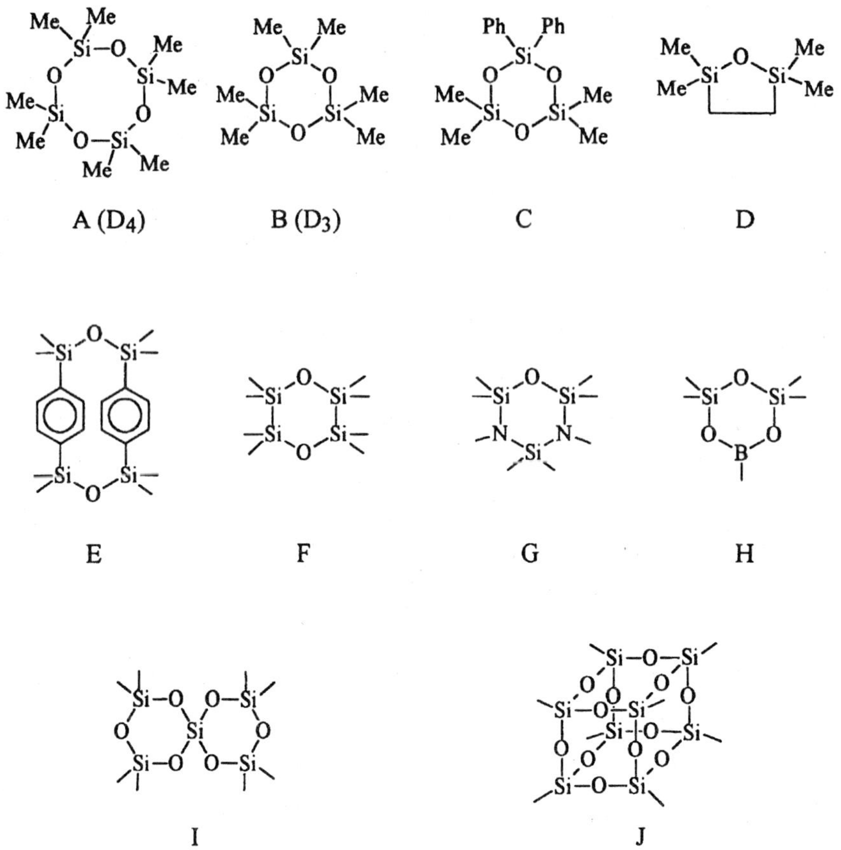

Scheme I Examples of structures of cyclic siloxanes and other monomers capable of ring-opening polymerization via breaking and reforming the SiOSi linkage.

in the synthesis of telechelic polysiloxanes of narrow molecular weight distribution and free of cyclics. Such polymers can be used for the synthesis of block or graft copolymers or for the construction of well-defined polymer networks. Under certain conditions the anionic polymerization of cyclic siloxanes closely approaches living polymerization, giving the possibility of perfect control of molecular weight, heterogeneity, and terminal group structure.[16,17]

there are some rules for performing the kinetically controlled process. First, cyclic monomers of high reactivity, such as cyclotrisiloxanes, must be used. Second, the initiating systems should be selective, that is, they should generate active propagation centers which react much faster with the monomer than with a polymer chain.

Polymer organolithium initiators like polystyryl lithium are often used in the synthesis of organic–siloxane block copolymers.[18,19]

The selective initiator may contain a functional group that is introduced to the end of the polysiloxane chain in the polymerization process. This method of the chain functionalization is called "initiator method."[19–33]

The kinetically controlled polymerization of cyclo siloxanes must be quenched at a proper moment, when the propagation is highly advanced but before back-biting and chain transfer become meaningful. Me_3SiCl is usually used as a terminating agent. The quenching reaction is also utilized for the functionalization of polysiloxane chain ends (the so-called terminator method).

Using a stoichiometric amount of Me_sSiCl_2 to terminate the chain functionalized by the so-called initiator method leads to the chain coupling, giving the linear macromolecule functionalized at both ends.[34] Capping with $MeSiCl_3$ or $SiCl_4$ produces star-branched polymers.[30,33]

The presence of protic contaminants, water in particular, is very detrimental to the control of the polymer structure. Water causes the broadening of molecular weight distribution. This is particularly detrimental in the case of monofunctional initiators.

Effect of the Monomer Structure on the Reactivity

The polymerization rate depends to a considerable extent on the size of the monomer ring. Because the silanolate–siloxane interactions are of a multicenter nature, they promote the polymerization of larger rings. Thus, D_7 polymerizes faster than D_3. this is a purely kinetic effect, and the rates of back-biting are increased to the same extent.[35] Chain-scrambling reactions are also accelerate.[36]

CATIONIC POLYMERIZATION

Cyclic siloxanes can be polymerized using both Brönsted and Lewis acid catalysts.[20,37–40] Cationic equilibration of cyclosiloxanes is a useful method of synthesis of linear polysiloxanes because it may be performed at a suitable rate at ambient temperature and the catalyst may be easily removed from the polymer.[41–43] The cationic route appears to be particularly useful in the synthesis of polysiloxane polymers and copolymers that have substituents that are sensitive to strong bases, such as Si-H.[44–46] The method has recently been applied to the synthesis of organic–siloxane block copolymers[47] and graft copolymers.[44] However, the main disadvantage of the cationic polymerization is the formation of significant amounts of cyclic oligomers since the early state of the reaction.[48–50]

Strained-ring cyclotrisiloxanes are generally much more reactive than larger unstrained cycles. In a series of permethylcyclosiloxanes, the activity of the siloxane bond toward acids decreases in the order $D_3>D_7>D_6>D_5>D_4$.[51] A similar trend was observed in the $(HMeSiO)_n$ series.[45,46]

The Mechanism of Polymerization Initiated by Protic Acids

Protic acids are the most common initiators used in the cationic ring-opening polymerization of siloxanes. Their initiating power increases with the acid strength. Thus, CF_3SO_3H and $HClO_4$ are particularly effective while CF_3CO_2H polymerizes D_3 very slowly. The polymerization is very sensitive to additives.

Mechanism of the Initiation

The initiation of the polymerization involves acidolysis of siloxane bond in the cyclic monomer.

Nonprotic Initiation

Lewis acids are believed to initiate the polymerization in most cases in cooperation with strong protic acids that result from the reaction of those species with traces of water or other protic contaminants present in the system. Indeed, some Lewis acid–protic acid systems are reported to be very effective catalysts.[52] Recent studies, however, performed in high-purity conditions using a proton trap, have shown that some nonprotic systems are themselves able to initiate the polymerization of cyclosiloxanes.[53,54]

Radiation Polymerization

The radiation-induced polymerization of D_3, D_4, and D_5 has been studied in both liquid and solid state.[11,55] The propagation occurs primarily via cationic mechanism. All monomers show very similar reactivities.

REFERENCES

1. Piccoli, W A.; Haberland, G. G.; Merker, R. L. *J. Am. Chem. Soc.* **1960**, *82*, 1883.
2. Dvornic, P. R.; Lenz, R. W. *High Temperature Siloxane Elastomers*; Hüthig and Wepf Verlag: Basel Heidelberg, New York, 1990.
3. Kurjata, J.; Chojnowski, J. *Makromol. Chem.* **1993**, *194*, 3271.
4. Johannson, O. K.; Lee, C. L. *Cyclic Monomers*; Wiley Interscience: New York, 1972; Chapter 6.
5. Liang, M.; Waddling, C.; Honeyman, C.; Foucher, D.; Manners, I. *Phosphorus Sulfur Silicon Rel. Elem.* **1992**, *64*, 113.
6. Andrianov, K. A.; Zachernyuk, A. B. *Pure Appl. Chem.* **1976**, *48*, 251.
7. Brown, J. F.; Vogt, L. H. Jr.; Prescott, P. J. *J. Am. Chem. Soc.* **1964**, *86*, 1120.
8. Graiver, D.; Huebner, D. J.; Saam, J. C. *Rubber Chem. Technol.* **1983**, *56*, 918.
9. Zhang, X.; Yang, Y.; Liu, X. *Goafenzi Tongxun* **1983**, 104; *Chem. Abstract* **1983**, *99*, 140444.
10. Stein, J.; Leonard, T. M.; Smith, J. F. *J. Appl. Polym. Sci.* **1993**, *47*, 667.
11. Chawla, A. S.; St-Pierre, L. E. *J. Polym. Sci.* **1972**, *10*, 2691.
12. Buzin, M. J.; Kvachev, Y. P.; Svistunov, V. S.; Papkov, V. S. *Vysokomol. Soedin. Ser. B* **1992**, *34*, 66.
13. Baglei, N. N.; Bryk, M. T. *Ukr. Khim. J.* **1981**, *47*, 409.
14. Rashkov, I.; Gitsov, I. *J. Polym. Sci. Polym. Chem. Ed.* **1986**, *24*, 155.
15. Gilbert, A. R.; Kantor, S. W. *J. Polym. Sci.* **1959**, *40*, 35.
16. Yilgör, I.; McGrath, J. E. *Adv. Polym. Sci.* **1988**, *86*, 1.
17. Chojnowski, J.; *J. Inorg. Organometal. Polym.* **1991**, *1*, 299.
18. Saam, J. C.; Gordon, D. J.; Lindsay, S. *Macromolecules* **1970**, *3*, 4.
19. Gerharz, B.; Wagner, T.; Ballauf, M.; Fischer, E. W. *Polymer* **1992**, *33*, 3531.
20. Chojnowski, J. *Siloxane Polymers*; Prentice Hall: Englewood Cliffs, NJ, 1993; Chapter 1.
21. Babu, J. R.; Sinai-Zingde, G.; Riffle, J. S. *J. Polym. Sci. Ser. A* **1993**, *31*, 1645.
22. Yu, J. M.; Teyssié, D.; Khalifa, R. B.; Boileau, S. *Polymer Bull.* **1994**, *32*, 35.
23. Gnanou, Y.; Rempp, P. *Makromol. Chem.* **1988**, *189*, 1997
24. Suzuki, T.; Lo, P. Y. *Macromolecules* **1991**, *24*, 460.
25. Suzuki, T.; Yamada, S.; Okawa, T. *Polymer J.* **1993**, *25*, 411..
26. Maschke, U.; Wagner, T. *Makromol. Chem.* **1992**, *193*, 2453.
27. Yin, R.; Hogen-Esch, T. E. *Macromolecules* **1993**, *26*, 6952.
28. Aoyagi, T.; Takamura, Y.; Nakamura, T.; Nagase, Y. *Polymer* **1992**, *33*, 1530.
29. Hunt, M. O.; Belu, A. M.; Linton, R. W.; DeSimone, J. M., *Polym. Prepr.* **1993**, *34*, 445.
30. Dickstein, W. H.; Lillya, C. P. *Macromolecules* **1989**, *22*, 3882.
31. Yoshinaga, K.; Iida, Y. *Chem. Lett.* **1991**, 1057.
32. Kazama, H.; Tezuka, Y.; Imai, K. *Macromolecules* **1991**, *24*, 122.
33. Wilczek, L.; Rubinsztajn, S.; Fortuniak, W.; Chojnowski, J.; Tverdokhlebova, I. I.; Volkova, R. V. *Bull. Acad. Sci. Ser. Sci. Chim.* **1989**, *37*, 91.
34. Kumar, A.; Eichinger, B. E. *Macromolecules* **1990**, *23*, 5358.
35. Laita, Z.; Jelinek, M. *Vysokomol. Soedin.* **1962**, *4*, 1739.
36. Chojnowski, J.; Mazurek, M. *Makromol. Chem.* **1975**, *176*, 2999.
37. Noll, W. *Chemistry and Technology of Silicones*; Academic: New York, 1968.
38. Lichtenwalner, H. K.; Sprung, M. N. *Encyclopedia of Polymer Science and Technology*; Wiley-Interscience: New York, 1970; Vol. 12, p 503.
39. Kendrick, T. C.; Parbhoo, B. M.; White, J. W. *Comprehensive Polymer Science*; Pergamon: Oxford, 1989; Vol. 4, Chapter 25.
40. Kendrick, T. C.; Parbhoo, B. M.; White, J. W. *The Silicon–Heteroatom Bond*; Wiley Interscience: Chichester, 1991; Chapters 3 and 4.
41. Scott, D. W. *J. Am. Chem. Soc.* **1946**, *68*, 2294.
42. Warrick, E. L. U.S. Patent 2 607 792, 1952.
43. Hurd, D. T. *J. Am. Chem. Soc.* **1955**, *77*, 2998.
44. Chujo, Y.; Murai, K.; Yamashita, Y. *Makromol. Chem.* **1985**, *186*, 1203.
45. Graczyk, T.; Lasocki, Z. *Bull. Acad. Pol. Sci. Ser. Sci. Chim.* **1979**, *27*, 181.

46. Gupta, S. P.; Moreau, M.; Masure, M.; Sigwalt, P. *Eur. Polym. J.* **1993**, *29*, 15.
47. Mougin, N.; Rempp, P.; Gnanou, Y. *Makromol. Chem.* **1993**, *194*, 2553.
48. Chojnowski, J.; Ścibiorek, M.; Kowalski, J. *Makromol. Chem.* **1977**, *178*, 1351.
49. Sauvet, G.; Lebrun, J. J.; Sigwalt, P. *Cationic Polymerization and Related Processes*; Academic: London, 1984; p 237.
50. Sigwalt, P.; *Polymer J.* **1987**, *19*, 567.
51. Gobin, C.; Masure, M.; Sauvet, G.; Sigwalt, P. *Makromol. Chem., Macromol. Symp.* **1986**, *6*, 237.
52. Kendrick, T. C.; *J. Chem. Soc.* **1965**, 2027.
53. Jordan, E.; Lestel, L.; Boileau, S.; Cheradame, H.; Gandini, A. *Makromol. Chem.* **1989**, *190*, 267.
54. Sigwalt, P.; Nicol, P.; Masure, M. *Makromol. Chem. Supl.* **1989**, *15*, 15.
55. Naylor, D. M., Stannett, V. T.; Deffieux, A.; Sigwalt, P. *Polymer* **1994**, *35*, 1764.

Cyclic Sulfite Polymerization

See: Ring-Opening Polymerization, Cationic (with Expansion in Volume)

Cycloaddition Polymerization

See: Diyne Cycloaddition Copolymerization (Transition Metal Catalyzed)
Ladder Polymers (Cycloaddition Copolymerization of Cyclic Diynes)
Ladder Polymers (Methods of Preparation)

Cycloaramids

See: Macrocyclic Aramids

CYCLODEXTRIN-POLYMER INCLUSION COMPLEXES

Akira Harada
Department of Macromolecular Science
Faculty of Science
Osaka University

Cyclodextrins (CD) are cyclic molecules consisting of six to eight glucose units linked through α-1-4 linkages. Since the discovery of cyclodextrins, many reports have been published on the formation of inclusion complexes with low molecular weight compounds, ranging from nonpolar hydrocarbons to polar acids and amines.[1] However, there were no reports on the formation of inclusion complexes of CDs with polymers when we started our project in the early 1980s, except for some examples in which a monomer was polymerized *in situ* within CD complexes.

It is of interest to see if CDs form complexes with larger molecules such as oligomers and polymers. Research on the interactions of CDs with various polymers has found that CDs form complexes with some polymers with high selectivity.[2]

PREPARATION

Inclusion Complexes of CDs with Some Water-Soluble Polymers

Aqueous solutions of some nonionic polymers were added to saturated aqueous solutions of CDs to see if insoluble complexes would be formed. Poly(vinyl alcohol) (PVA) and polyacrylamide (PAAm) did not form complexes with any CDs. However, we found that α-CD formed complexes with poly(ethylene glycol) (PEG) to give crystalline compounds in high yields.[3–5] This was the first observation that CD forms complexes with polymers, β-CD did not form complexes with PEG of any molecular weight. However, β-CD formed complexes with poly(propylene glycol) (PPG),[6] which has methyl group on the main chain, while α-CD did not form complexes with PPG of any molecular weight. We found that poly(methyl vinyl ether) (PMeVE), which has the same composition as PPG but has methoxy groups on the main chain, formed complexes with γ-CD,[7] although α- and β-CDs did not form complexes with PMeVE. CDs did not form complexes with polyvinylpyrrolidone. Therefore, there is a clear relationship between the sizes of CDs and the cross-sectional areas of the polymers.

Inclusion Complexes of CDs with Some Water-Insoluble Polymers[8,9]

Recently, we found that CDs form complexes not only with hydrophilic polymers but also with hydrophobic polymers. α-CD formed complexes with oligoethylene, although β- and γ-CD did not form complexes with oligoethylene. β-CD formed complexes with squalan, which has a methyl group on a methylene chain, although α-CD did not form complexes with squalan. γ-CD formed complexes with polyisobutylene,[9] although α-CD formed no complexes with polyisobutylene.

STRUCTURES AND PROPERTIES

Binding Modes

Because the length of two ethylene glycol units is 6.6Å (planer zigzag), which corresponds to the depth of the CD cavity (6.7Å), we proposed that CD includes two ethylene glycol units successively, as shown in **Figure 1**.

In conclusion, CDs were found to form inclusion complexes with various polymers, both hydrophilic and hydrophobic, to give crystalline compounds with high selectivities. These are the first examples of the formation of inclusion complexes between CDs with polymers. There are clear relationships between sizes of CD cavities and the cross-sectional areas of the polymers. CDs also recognize the length of the polymers.

REFERENCES

1. Bender, M. L.; Komiyama, M. *Cyclodextrin Chemistry*; Springer-Verlag: Berlin, 1978.
2. Harada, A.; Li, J.; Kamachi, M. *Proc. Jpn. Acad.* **1993**, *69*, Ser. B, 39.
3. Harada, A.; Kamachi, M. *Macromolecules* **1990**, *23*, 2821.
4. Harada, A.; Li, J.; Kamachi, M. *Macromolecules* **1993**, *26*, 5698.
5. Harada, A.; Li, J.; Kamachi, M. *Macromolecules* **1994**, *27*, 4538.
6. Harada, A.; Kamachi, M. *J. Chem. Soc., Chem. Commun.* **1990**, 1322.
7. Harada, A.; Li, J.; Kamachi, M. *Macromolecules* **1993**, 237.

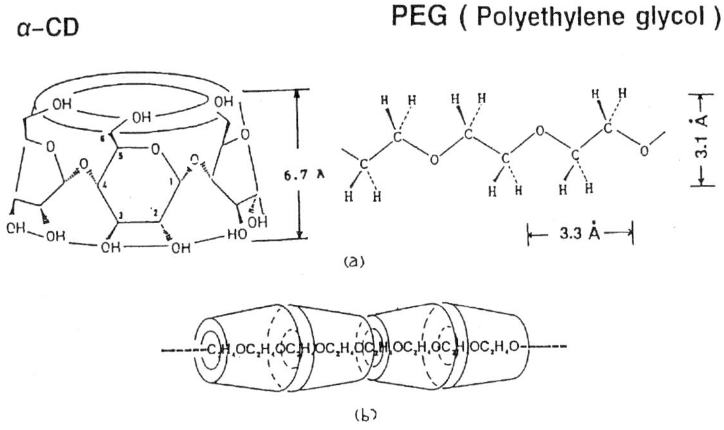

FIGURE 1. Proposed structure of the α-cyclodextrin(α-CD)–poly(ethylene glycol) (PEG) complex.

8. Li, J.; Harada, A.; Kamachi, M. *Bull. Chem. Soc., Jpn.* **1994**, *67*, 2808.
9. Harada, A.; Li, J.; Suzuki, S.; Kamachi, M. *Macromolecules* **1993**, *26*, 5267.

Cyclodextrins

See: *Cyclodextrin-Polymer Inclusion Complexes*
Cyclodextrins (Host-Guest Interactions)
Cyclodextrins (Thermodynamic Functions of Molecular Recognition)
Enzyme Mimics (Cyclodextrins)
Functionalized Polymers (Comb, Rotaxanic, and Dendrimeric Structure)
Host-Guest Chromatography (with Cyclodextrin Derivatives)
Inclusion Complexes (Overview)
Macromolecular Recognition (New Supramolecular Structures)
Metalloenzymes, Artificial (Polyethylenime-Based)
Microcapsules, Molecular Complexes (Dissolution of Drugs by Cyclodextrin)

CYCLODEXTRINS (Host-Guest Interactions)

Tetsuo Osa and Iwao Suzuki
Pharmaceutical Institute
Tohoku University

Cyclodextrins (CDs, or cycloamyloses) are naturally occurring cyclic oligosaccharides composed of D-α-glucopyranose units. At present, hexa-, hepta-, and octaoligomers are commercially available, called α-, β-, and γ-CD, respectively. There have been numerous reports on the complex formation of CDs with guest species, including food components, flavors, drugs, lipids, and engineering materials. Upon complexing, the solubility, absorptivity, and other physicochemical properties of guests are changed. In order to improve the guest binding and other additional properties of CDs, a number of chemically modified CDs have been synthesized and explored on their complexation behavior.[1] Another important aspect of CDs is that they have been investigated as enzyme models.[2,3] Because CDs have the hydrophobic cavity capable of binding a substrate together with the secondary hydroxyl groups activated by intramolecular hydrogen bonds, native CDs are regarded as hydrolytic enzyme models. In this context, a great effort has been made for constructing much better enzyme mimics in which a catalytic moiety was site-selectively introduced. This has corresponded with continual progress in constructing excellent synthetic receptors containing CD structure toward ionic species as well as neutral organic species.[4,5] In addition, a concept of "molecular flask," in which two or more molecules can associate and react with each other, has been proposed for CDs.[6]

PROPERTIES

Detection of Host–Guest Complexes of CDs

The most pronounced property of CDs is the ability to form host–guest complexes with various guest molecules. Usually, the solubility of CD-based host–guest complexes toward water is smaller than that of uncomplexed CDs. Thus, host–guest complexes often precipitate from the saturated CD solution, and the binding strength (association constant, K_a) can be calculated from a phase-solubility diagram. In general, the physiochemical properties of a guest are changed upon complexing with a CD because of the environmental change from polar aqueous phase to a rather apolar CD cavity. Thus, UV–vis and fluorescence spectral variations of a guest can be used to confirm the complexation. Because a CD consists of optically active glucose units by which the inside of the CD cavity possesses a chiral field, circular dichroism is induced when a chromophore guest is inserted into the cavity, even though the guest exhibits no optical activity.

Host–Guest Complexation of Native CDs

Native CDs can accommodate guest species into their hydrophobic cavity in aqueous solutions, and the stability of the resulting host–guest complexes are dependent on the hydrophobicity, size, and shape of guests as well as the cavity size of the

CD. Usually, α- and β-CDs form host–guest complexes with a guest species with 1:1 stoichiometry. However, larger γ-CD can form 1:2 host–guest complexes for aromatic guests. For a guest large in size as compared to the CD cavity, 2:1 or 2:2 host–guest complexation would be formed. If a guest is long enough to capture several CD molecules, resulting host–guest complexes are regarded as polyrotaxanes. For example, α-CD molecules can be threaded with polyethylene glycol.[7]

Host–Guest Complexation by Persubstituted CDs

Relatively simple and readily obtainable per-2,6-O-dimethyl- and per-2,3,6-O-trimethyl-β-CDs are soluble not only in aqueous but also in organic solutions with medium or high polarity, and they can form host–guest complexes in aqueous solutions as performed by a native CD. The methylation is unlikely to change the guest-binding behavior of the CD.

Host–Guest Complexation of CD Duplexes

CD dimers linked with covalent bonds, known as CD duplexes, form extremely stable host–guest complexes for a guest with a compatible shape and size.

APPLICATION

Because native CDs, especially a β-CD, are not so expensive and are essentially nontoxic, they have been used in the food and pharmaceutical industries. Masking odor, making volatile materials nonvolatile, stabilizing unstable materials against moisture or an oxidant, solubilizing undissolved materials to aqueous phase, improving bioavailability, and so forth of a guest species can be achieved by complexing with CD. From the viewpoint of laboratory uses, CD is used for separating enantiomeric molecules in GLC and HPLC.

In conclusion, not only have CDs been attracting great interest both in laboratory uses and as valuable additives in commercial products, CDs and modified CDs both have great potential in selective syntheses, constructing artificial enzymes and molecular indicators, and so forth. Therefore, CDs promise us more sophisticated materials for improving human life. Chemical Abstracts Registry Numbers; α-CD, 10016-20-3; β-CD, 7585-39-9; γ-CD, 17465-86-0.

REFERENCES

1. Croft, A. P.; Barton, R. A. *Tetrahedron* **1983**, *39*, 1417.
2. Tabushi, I. *Acc. Chem. Res.* **1982**, *15*, 66.
3. Breslow, R. *Science* **1982**, *218*, 532.
4. Pregel, M. J.; Jullien, L.; Lehn, J.-M. *Angew. Chem., Int. Ed. Engl.* **1993**, *31*, 1637, for example.
5. Wenz, G. *Angew. Chem., Int. Ed Engl.* **1994**, *33*, 803.
6. Ueno, A.; Osa; T. *Photochemistry in Organized and Constrained Media*; Ramamurty, V., Ed.; VCH: New York, 1991, Chapter 16.
7. Harada, A.; Li, J.; Kamachi, M. *Nature* **1992**, *356*, 325.
8. Tabushi, I.; Kuroda, Y.; Mizutani, T. *J. Am. Chem. Soc.* **1986**, *108*, 4514.
9. Anslyn, E.; Breslow, R. *J. Am Chem Soc.* **1989**, *111*, 5992.

CYCLODEXTRINS (Thermodynamic Functions of Molecular Recognition)

Sadao Takagi,* Maso Fujisawa, and Takayoshi Kimura
Department of Chemistry
Kinki University

Molecular recognition and discrimination may be caused by the interactions among contacting surfaces of colliding molecules in solutions and mixtures. In particular, stereospecific interactions due to neighboring surfaces may play the leading role in, for example, enzyme–substrate reactions, antigen–antibody reactions, some kinds of mechanisms of the senses of smell and taste, and so forth. It is therefore vital to elucidate the role of asymmetric intermolecular interactions due to the stereospecific structure of a molecule in order to understand the mechanisms of reactions in chemistry and biochemistry. To clarify the mechanism of molecular recognition and discrimination in aqueous solutions, Takagi and co-workers have determined systematically thermodynamic functions for the molecular inclusion of monohydric alcohols and some diols into α- and β-cyclodextrin (CD) cavities in dilute aqueous solutions, especially at infinite dilution of the alcohols by the method developed by Takagi and Maeda.[1–8] The importance of the increase in entropy was discovered from the quantitative consideration of these thermodynamic functions.[5]

THERMODYNAMIC PARAMETERS

Discussion

Methanol molecules seem not to distinguish energetically between their circumstances in the cavities of CDs versus bulk water. Energetic stabilization of the molecules due to hydrophobic hydration in some voids formed in clusters of water molecules in bulk water may be weak, because they have only methyl groups as so-called structure-making hydrophobic radicals. In CD cavities, the molecules' energetic stabilization may also be weak, because of their small molecular size.

In aqueous solution, the included methanol molecules might be oscillating and rotating violently with interchanging the positions between the cavities and bulk water. In this case the driving factor for inclusion-complex formation may be the increase in entropy.

The entropy changes on the 1:1 inclusion that have been obtained by Takagi et al. are always positive and large, except hexanol with α-CD. In the cases of α-CD, all enthalpy changes are negative (exothermic), therefore the changes of enthalpy and entropy promote the inclusion, except hexanol. For the cases of ethanol, propanol, butanol, 2-propanol, cyclohexanol, and butanediols with α-CD, and all cases with β-CD, contribution of entropy terms, $-T\Delta_{inc}S$, are predominant. Therefore, in these cases, increase of entropy is an important factor for promoting the inclusion in aqueous solutions, The large increase in entropy might be mainly due to excessive motion of the alcohol molecules in the cavities, especially molecular rotation in the CD cavities, and break of hydrophobic hydration of the alcohol molecules in bulk water.

*Author to whom correspondence should be addressed.

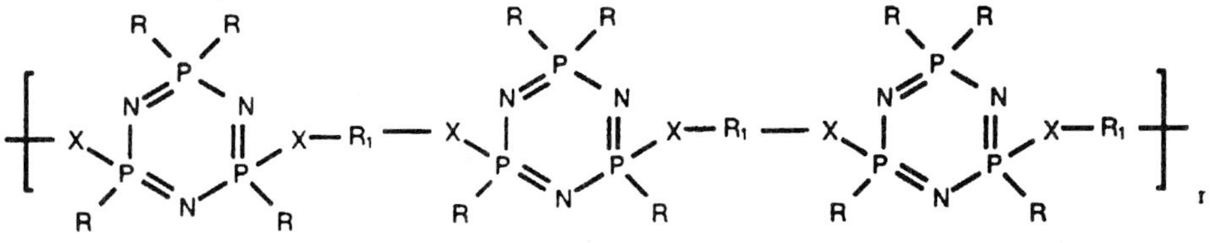

FIGURE 1. Cyclolinear phosphazene polymers. The X and R_1 are organic backbone groups connecting the cyclophosphazene ring containing R as organic substituents.

Finally, molar Gibbs energies of inclusion plotted against N_c tells us as follows: butanol, ethanol, and pentanol can discriminate between α- and β-CDs. However, propanol cannot do so. Cavities of β-CD discriminate among ethanol, butanol, and hexanol. However, ethanol, butanol, and hexanol are only minimally distinguished by α-CD.

REFERENCES

1. Maeda, M.; Takagi, S. *Netsu Sokutei* **1983**, *10*, 43, 103.
2. Takagi, S.; Maeda, M. *J. Inclusion Phenom.* **1984**, *2*, 775.
3. Takagi, S.; Kimura, T.; Maeda, M. *Thermochim. Acta* **1985**, *88*, 247.
4. Fujisawa, M.; Kimura, T.; Takagi, S. *Netsu Sokutei* **1991**, *18*, 71.
5. Takagi, S.; Fujisawa, M.; Kimura, T. *Chem. Express* **1991**, *6*, 93.
6. Fujisawa, M.; Doctoral Dissertation Kinki University, Higashi-Osaka, Japan, 1991.
7. Takagi, S.; Fujisawa, M.; Kimura, T. *Thermochim. Acta* **1991**, *183*, 289.
8. Maeda, M.; Takagi, S. *Nippon Kagaku Kaishi* **1983**, 188 (contains an adequate English summary).

Cyclodiolefin Polymerization

See: Transannular Polymerization

CYCLOLINEAR ORGANOPHOSPHAZENE POLYMERS

Devendra Kumar* and Alka D. Gupta
SETI Institute

Interest is growing in high-performance heat-resistant and fire-resistant polymers to meet the more stringent requirements for applications in microelectronics, membranes, surface coatings, films, and structural materials for space and aerospace usage. Phosphazene polymers possess a number of unusual properties which are expected to dictate the areas of most immediate attention. Cyclolinear phosphazene polymers have an organophosphazene backbone interspaced between inorganic cyclophosphazene rings that are also parts of the polymer backbone (**Figure 1**).

In recent years, our group and others have developed exciting novel methodologies for incorporating cyclotriphosphazenes into the structure of polymers. This approach is based on exploiting the reactivity of organofunctional cyclophosphazenes. The cyclophosphazene ring can easily be substituted by a variety of groups and it is possible to synthesize cyclotriphosphazenes containing suitable functional groups required for addition or condensation type polymerization.

*Author to whom correspondence should be addressed.

CYCLOLINEAR PHOSPHAZENE POLYMERS

Cyclolinear phosphazene polymers are synthesized by linking bifunctional cyclophosphazene rings through bifunctional organic monomers and can have a variety of inorganic-organic polymer backbone structures. The properties of the cyclolinear phosphazene polymers depend upon the nature of the linking group (X), the nature of the substitution (R) on the cyclophosphazene ring, and the polymer backbone structure (R_1) formed by the reaction of the two monomers. The cyclolinear phosphazene polymers generally have thermoplastic characteristics. However, it is also possible to synthesize elastomers by selecting R_1s as mobile and flexible groups. In general, the thermoplastic cyclolinear phosphazene polymers are much more thermally stable than the polyorganophosphazenes and possess a higher glass transition temperature (T_g > 170°C). This can be attributed to the presence of the thermodynamically stable cyclophosphazene ring as an integral part of the polymer chain.

CYCLOLINEAR PHOSPHAZENE POLYIMIDES

A new class of aromatic cyclolinear phosphazene polyimides was recently synthesized from a novel bis(arylenedioxy)spirocyclotriphosphazenediamine.[1] Most of the available commercial linear polyimides degrade almost catastrophically in air above 600°C without leaving any residue. In comparison to other phosphorus-containing linear polyimides,[2,3] the newly developed[5] cyclotriphosphazene-containing polyimide films exhibit higher char yield at 80°C in air. The observation of high char yield may be explained by the higher percentage of phosphorus (in the form of the cyclotriphosphazene ring) per repeat unit of the polyimide chain in comparison to phosphorus-containing polyimides.

CYCLOLINEAR PHOSPHAZENE POLYETHERIMIDES

We have developed novel cyclotriphosphazene-containing poly(ether imide)s by nucleophilic displacement of active nitro groups of tris[4-(3′-nitrophthalimido)phenoxy]tris(phenoxy) cyclotriphosphazene or tris[4-(4′-nitrophthalimido)phenoxy]tris (phenoxy)cyclotriphosphazene and a bisnitrophthalimido derivative by the dianion of bisphenol A.[4]

CYCLOLINEAR PHOSPHAZENE POLYAMIDES

We have synthesized a new class of cyclotriphosphazene-containing polyamides.[5] A key step for the synthesis of these polymers was the molecular design of polymer grade bifunctional monomers that have been synthesized by a simple and direct method in high yields (>80%).

The thermal stability of the synthesized polyamides on dynamic thermogravimetric analysis showed polymer decomposition starting at 380°C.

In 1993, Y. W. Chen-Yang and Y. H. Chaung reported[6] the synthesis of similar cyclotriphosphazene-containing polyamides by polycondensation of a bis(4-aminophenoxy)tetrakis(phenoxy)cyclotriphosphazene (BATPC) with terephthaloyl chloride and sebacoyl chloride.

CYCLOLINEAR PHOSPHAZENE POLYETHERS

In 1974, M. Kajiwara reported on the synthesis of cyclolinear phosphazene polymers by the polycondensation of tetrachlorodiphenylcyclotriphosphazene with aromatic dihydroxy compounds.[7] The cyclolinear phosphazene polyethers derived from tetrachlorodiphenylcyclotriphosphazene and bisphenol-A, or 4,4′-dihydroxydiphenyl sulfone did not soften and exhibited fire resistance.

More than 25 years ago, S. I. Belykh[8] and M. Kajiwara[9] independently reported the synthesis of cyclophosphazene polymers involving transesterification reactions with diphenyldichlorosilane.

Thermal stability data obtained by TGA in air on these systems showed that decomposition of the polymer began below 300°C, with a 10% weight loss occurring at 540°C. Asbestos-containing composite materials made from this type of polymer demonstrated very high thermo-oxidative stability: ~500°C, as determined by isothermal aging.[10]

In 1969, Belykh et al. reported a polymerization reaction of butoxycyclotriphosphazene with diphenyldichlorosilane to yield soluble, glassy polymers.[8] It was presumed that when two butoxy groups are present on each phosphorus, spirocyclic derivatives are formed.

CONCLUSION

The development of cyclolinear phosphazene polymers provides unique polymer backbone structures containing a rigid inorganic cyclotriphosphazene ring synergistically connected with flexible organic groups. This rare combination of rigid inorganic and flexible organic groups provides high thermal and thermo-oxidative stability, high glass transition temperature (usually > 170°C), and thermoplastic processability not available from linear polyorganophosphazenes or any other phosphazene-containing polymers. Cyclolinear phosphazene polymers provide a unique combination of intrinsic properties such as chemical and dimensional stability and resistance to heat and fire, solvents, atomic oxygen, and UV and visible radiation. Cyclolinear phosphazene polymers are particularly suitable for making high performance films and can be processed into composites, laminates, and moldings. Polyimide and poly(ether imide) films have yellow to brown colors but can be tailored to make colorless optically-transparent films. These polyimides and films are potentially useful for applications in membranes for harsh environments, optoelectronics, microelectronics, silicon chip manufacturing, thermal blankets, satellite components, and aircraft interiors and space structures. The cyclolinear polymers are very good for coating applications on surfaces and structures exposed to atomic oxygen and to UV and other radiation.

REFERENCES

1. Kumar, D.; Gupta, A. D. *Macromolecules* **1995**, 6325.
2. Martínez-Núñez, M. F.; Sekhripuram, V. N.; McGrath, J. E. *Polym. Prepr., Am. Chem. Soc., Div. Polym. Chem.* **1994**, *35*(2), 710.
3. Lin, Y. N.; Joardar, S.; McGrath, J. E. *Polym. Prepr., Am. Chem. Soc., Div. Polym. Chem.* **1993**, *34*(1), 515.
4. Kumar, D.; Khullar, M.; Gupta, A. D. *Polymer* **1993**, *34*, 3025–3029.
5. Kumar, D.; Allen, C. W. unpublished work, 1992.
6. Chen-Yang, Y. W.; Chaung, Y. H. *Phosphorus, Sulfur, Silicon, and Rel. Elem.* **1993**, *76*, 261–264.
7. Kajiwara, M. *Angew. Makromol. Chem.* **1974**, *37*, 141–147.
8. Belykh, S. I.; Zhivukhin, S. M.; Kireev, V. V.; Kolesnikov, H. S. *Vysokomol. Soidin., Ser. A* 11, **1969**, *3*, 625.
9. Kajiwara, M.; Sakamoto, A.; Saito, H. *Angew. Makromol. Chem.* **1975**, *46*, 63.
10. Emblem, G. H.; Oxyley, E. C.; Trow, A. S. *Brit. Polym. J.* **1970**, *2*, 83.

CYCLOMATRIX PHOSPHAZENE POLYMERS

Devendra Kumar and Alka D. Gupta
Seti Institute

The field of phosphazene polymers provides a common connecting point for the development of various functional and flame retardant materials. There are several fundamentally different approaches to synthesize phosphazene polymers. **Figure 1** shows the most common approaches used in incorporating the phosphazene unit ($-R_2P{=}N-$) into polymer structures.

CYCLOMATRIX PHOSPHAZENE POLYMERS

Cyclomatrix phosphazene polymers (**Figure 1d**) are synthesized by chain extension or crosslinking of multifunctional cyclophosphazenes and we present them in detail here. These polymers are useful in structural applications in the form of coatings, adhesives, composites, and laminates. Suitably synthesized cyclomatrix phosphazene polymers exhibit useful thermal and chemical properties such as fire-retardancy at higher temperatures, high glass transition temperature, resistance to solvents, atomic oxygen, and oxygen plasma attack. In addition, the synergism of the phosphorus–nitrogen combination in the form of cyclophosphazene offers improved oxidative thermal decomposition and high anaerobic char yields. Also, selection of cyclophosphazene offers flexible synthetic methodologies for the preparation of cyclophosphazenes with various substituents. It is therefore possible to accomplish molecular design and synthesize multifunctional initiators, terminators, polymer precursors, and intermediates.

A key step in the development of cyclomatrix phosphazene polymers is to synthesize a cyclophophazene derivative end-capped with a suitable functional group (usually more than two), polymerizable by thermal, photochemical, or other reactions.

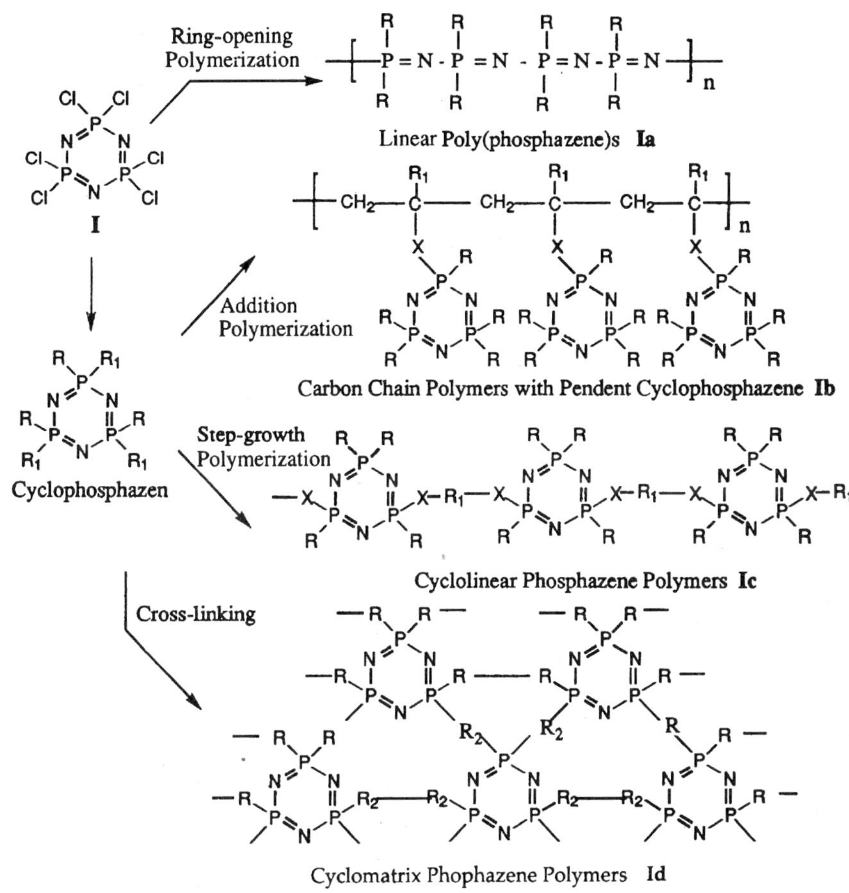

FIGURE 1. Phosphazene-containing polymers derived from the starting material hexachlorocyclotriphosphazene (I). X, A, R, R_1, and R_2 represent an atom or group linking the phosphazene.

The thermal or photochemical addition polymerization of the suitable end-cap is preferred to minimize the formation of volatiles during the thermosetting processing step. The liberated volatiles otherwise may cause voids, resulting in poor thermomechanical properties in the material.

MALEIMIDO-SUBSTITUTED AROMATIC CYCLOPHOSPHAZENES

Several exceptionally improved inorganic–organic cyclomatrix phosphazene polymers of a new class have been developed by thermal polymerization of suitable maleimido end-capped cyclotriphosphazenes.[1–8] The incorporation of cyclotriphosphazene ring into the polymer matrix provided unique properties such as high thermooxidative stability, high glass transition temperature (T_g) (>325°C), high char yields in air (>70% at 800°C), nitrogen (>80% at 800°C) and structural integrity at high temperature. Also, the graphite-fabric composites prepared from these polymers did not burn in pure oxygen (Oxygen Index = 100%) even at 300°C.[1–3] Synthesis of various monomers (namely, cyclotriphosphazene containing hexakisamine, trisamine, bisamine, and their derivatives containing thermally polymerizable end groups) was an important milestone in this research and was notably achieved by the same group.

COMPOSITE FABRICATION

Further improved fire- and heat-resistant cyclomatrix phosphazene polymers have been developed by thermal polymerization of bis-, tris-, tetrakis-(maleimidophenoxy)tris(phenoxy) cyclotriphosphazenes.[1–3] These polymers show char yields of 82–78% at 800°C in nitrogen and 78–71% at 700°C in air. Their synthesis is achieved by reaction of tris(4-aminophenoxy)tris(phenoxy)cyclotrisphosphazene with maleic anhydride alone or in combination with 3,3′,4,4′-benzophenonetetracarboxylic dianhydride (BTDA) and pyromellitic dianhydride (PMDA) followed by curing of the resulting maleimides.

High strength fire- and heat-resistant cyclomatrix phosphazene polymers are obtained by thermally induced melt-polymerization of hexafluoroisopropylidene linked bisimidobis(maleimidophenoxy)-substituted cyclotriphosphazene and bisimidoterakis(maleimidophenoxy)-substituted cyclotriphosphazene. These polymers show good thermal stability and high char yields: 78–80 at 800°C in nitrogen and 60–68 in air at 700°C.

Kumar et al. also prepared a fire- and heat-resistant polymer by thermal polymerization of 2,2-bis(anilino)-4,4,6,6-tetrakis(4-maleimidophenoxy)cyclotriphosphazene.[4,5] The thermal stability of the polymer was evaluated in nitrogen and in air by TGA. The reported polymer is stable up to 345°C and has char yields of 78% at 800°C in nitrogen and 71% at 700°C in air. A heat-resistant polybismaleimide was reported by Kumar utilizing the customary thermal polymerization of bis(maleimidophenoxy)-tetrakis(phenoxy)cyclotriphosphazene.[9] The polybismaleimide is stable up to 340°C and has a char yield of 70% at 800°C in nitrogen and of 60% at 700°C in air.

ALLYL, VINYL, AND NADIMIDO-SUBSTITUTED AROMATIC CYCLOPHOSPHAZENES

At Hercules, heat-resistant cyclophosphazene-containing thermosetting polymers were prepared by Diels–Alder reaction of a bismaleimide-based dienophile and a cyclophosphazene containing two or more aryl or aryloxy groups substituted with allyl or vinyl groups.[10]

METHACRYLATE-SUBSTITUTED CYCLOPHOSPHAZENE

Cyclomatrix cyclophosphazene-based materials have been also developed at Idemitsu Petrochemical Company Ltd. of Japan via the thermal or photochemical curing of hexa(hydroxyethyl methacrylate)cyclotriphosphazene (3-PNC6HEMA).[11] The resulting cured cyclomatrix phosphazene polymer has unique characteristics such as high hardness, optical transparency, abrasion resistance, heat and chemical resistance, and burn-resistance, but actual flame data is not available. The suggested applications are hard coating for plastics and wood.[12]

ETHYNYL-SUBSTITUTED CYCLOPHOSPHAZENES

Improved heat- and fire-resistant thermosetting cyclomatrix phosphazene polymers for composites have been synthesized by the reaction of tris(4-aminophenoxy)tris(phenoxy)cyclotriphosphazene with 4-ethynyl-benzoyl chloride and 4-ethynylphthalic anhydride, respectively.[13] Thermal curing of the synthesized polymer precursors induced crosslinking or chain extension to void-free, heat- and fire-resistant thermosetting cyclomatrix polymers.

EPOXY MATRIX RESINS CONTAINING CYCLOPHOSPHAZENES

Heat- and fire-resistant cyclomatrix phosphazene polymers have been developed for laminates, composites, and adhesives by solvent-less curing of commercially available epoxies with aminophenoxycyclotriphosphazenes.[14,15]

APPLICATIONS

The combination of cyclophosphazene inorganic-ring-systems and polymer chemistry has led to the development of an exciting new class of cyclomatrix phosphazene polymers and materials in the form of composites, adhesives, laminates, protective coatings, matrix resins, semi-permeable membranes, varnishes, and prepregs. The potential applications of the developed cyclomatrix phosphazene polymers and materials lie in such diverse areas as thermooxidative stability, fire retardancy, high-temperature structural integrity, ceramic precursors, separation of effluents, chemical stability, atomic oxygen stability, abrasion resistance, optical transparency, and self-extinguishing burn characteristics. These polymers can be used as such or the cyclophosphazene-based precursors could be co-polymerized with commercial monomers and polymers to improve upon the properties of the latter. Further fine tuning the properties through proper choice of substituents, polymerizable end-capped groups on the phosphorus atoms of the cyclophosphazene ring, and their polymerization can provide new properties and applications of cyclomatrix-phosphazene polymers and materials.

REFERENCES

1. Kumar, D.; Fohlen, G. M.; Parker, J. A. *Macromolecules* **1983**, *16*, 1250.
2. Kumar, D.; Fohlen, G. M.; Parker, J. A. *SAMPE* **1983**, 687.
3. Kumar, D.; Fohlen, G. M.; Parker, J. A. *J. Polym. Sci., Polym. Chem. Ed.* **1983**, *21*, 3155.
4. Kumar, D.; Fohlen, G. M.; Parker, J. A. *J. Polym. Sci., Polym. Chem. Ed.* **1984**, *22*, 927.
5. Kumar, D.; Fohlen, G. M.; Parker, J. A. *J. Polym. Sci., Polym. Chem. Ed.* **1984**, *22*, 1141.
6. Kumar, D. *J. Polym. Sci., Polym. Chem. Ed.* **1985**, *23*, 1661.
7. Kumar, D.; Gupta, A. D.; Khullar, M. *J. Polym. Sci.: Part A: Polym. Chem.* **1993**, *31*, 2739.
8. Kumar, D.; Gupta, A. D.; Khullar, M. J. *Inorg. Organometal. Polym.* **1993**, *3*, 259.
9. Kumar, D.; Fohlen, G. M.; Parker, J. A. U.S. Patent 4 550 177, 1985.
10. Lukacs, A. III. Eur. Pat. Appl. EP 259 803, 1988: U.S. Pat. Appl. 906 000, 1986; *Chem. Abstr.* **1986**, 109(18), 150661z.
11. Yaguchi, A. et al. *Thin Solid Films* **1992**, *216*, 123.
12, Kitayama, M. et al. Presented at the Status and Future of Polyphosphazenes Conference, North Carolina, March 1992.
13. Kumar, D. *J. Polym. Sci.: PartA: Polym. Chem. Ed.* **1993**, *31*, 707.
14. Kumar, D.; Fohlen, G. M.; Parker, J. A. *J. Polym. Sci., Polym. Chem. Ed.* **1986**, *24*, 2415.
15. Kumar, D. *J. Polym. Mater.* **1990**, *7*, 215.

Cycloolefin Polymerization

See: Metathesis Polymerization, Cycloolefins

Cyclophosphazene Inclusion Complexes

See: Inclusion Complexes (Overview)
Inclusion Polymerization

Cyclopolymerization

See: Cyclopolymerization (Overview)
Cyclopolymerization (Cyclization Efficiency and Configuration Control)
Cyclopolymerization (Nonconjugated Dienes)
Cyclopolymerization, Enantioselective Cationic
Group Transfer Cyclopolymerization
Cyclynes
Cyclic Benzo-Alkynes, Ortho Substituted
Molecular Recognition (Macromolecular Ionophore)
Polyimides (for Cyclic Unsaturated Imides)
Ring-Opening-Closing Polymerization

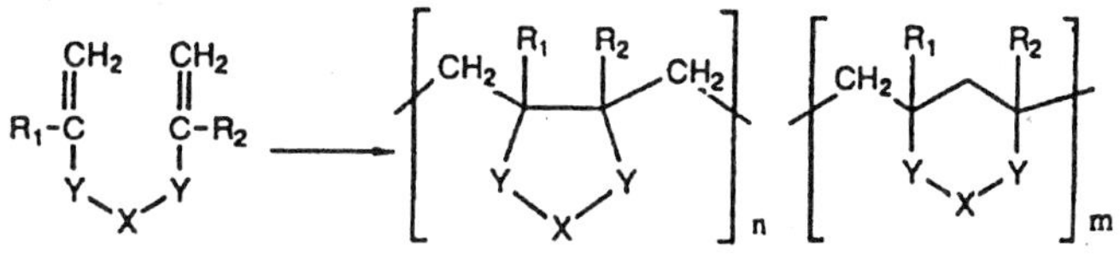

I

CYCLOPOLYMERIZATION (Overview)

George B. Butler
Department of Chemistry
University of Florida

Cyclopolymerization is defined as any type of addition polymerization that leads to the introduction of cyclic structures into the polymer main chain. Cyclopolymerization is a special case of cyclopolymerization in which the developing cyclic structure is derived from two or more separate monomers.

A vast amount of work has now been carried out on cyclopolymerization and cyclocopolymerization.

An important principle established early in the history of modern polymer science is that nonconjugated dienes undergo addition polymerization to form either linear polymers containing unreacted double bonds or, more likely, crosslinked polymers. However, certain exceptions to this principle have been reported in several papers published from 1949 to 1957.[1-7] Polymers produced from diallyl quaternary ammonium salts were soluble in water, not crosslinked, and yet they did not contain residual unsaturation. Both allyl double bonds were involved in the reaction; a mechanism was proposed in which chain growth occurred via alternating intramolecular and intermolecular steps.[8] The presence of cyclic units in these polymers was rigorously established.[9] However, these studies did not establish the size of the ring in the polymers. In this mechanism, cyclization occurs as a characteristic feature of the polymerization process.

More recent investigations have shown that in the case of the quaternary ammonium salts, almost exclusive formation of the kinetically favored five-membered ring occurs instead of the thermodynamically favored six-membered ring (**Scheme I**).[10]

It has been demonstrated that by using appropriate initiating systems, 1,6-dienes can be polymerized to yield soluble, saturated polymers with structures in which rings alternate with methylene groups along the linear chain.[3] Cyclic units in the polymer chain have been obtained by polymerization of monomers containing other unsaturated systems and polymerizable groupings.

RADICAL INITIATED POLYMERIZATION

In addition to the diallylamine salts, a variety of novel monomers have been synthesized. Among these, dimethyl α-α-dimethylenepimelate, a monomer closely related to methyl methacrylate, was polymerized by use of various free-radical initiators.[11] α-α′-Dimethylenepimelonitrile was also cyclopolymerized to produce a soluble polymer that was found to be much more thermally stable than the monoolefinic counterpart, polymethacrylonitrile.

Dimethyl α-α′-dimethylenepimelate copolymerized with acrylonitrile to produce a soluble polymer with no residual unsaturation. These copolymers formed fibers having properties superior to those of polyacrylonitrile.[12]

CATIONIC INITIATED POLYMERIZATION

The number of monomers of the 1,5- or 1,6-heptadiene type that had been subjected to cationic initiation was limited before the synthesis and polymerization of 2,6-diphenyl-1,6-heptadiene had been described.[13] This monomer is unique in that all the known general types of initiation—cationic, conventional anionic, free-radical, metal-coordination, and thermal—led to the same polymer by cyclopolymerization. The cyclopolymer was far superior thermally to its noncyclic analog, poly(α-methylstyrene).

Aliphatic divinyl formals and acetals also have been found to undergo cationic polymerization to yield cyclic polymers containing *m*-dioxane units.[14]

ANIONIC INITIATED POLYMERIZATION

The earliest example of an anionic cyclopolymerization was reported in 1958.[15] A polymer was inadvertently obtained in an effort to synthesize diacrylmethane by a Claisen condensation of methyl vinyl ketone with ethyl acrylate. More recent work has shown the polymer to be a copolymer of diacrylmethane and methyl vinyl ketone.[16] Dimethacrylmethane was also cyclopolymerized anionically.

Polymerization of trimethylenediisocyanate and 1,2-ethylenediisocyanate led to cyclopolymers containing six- and five-membered rings, respectively.[17] Propane-1,2,3-triisocyanate led to a polymer obtained via cumulative ring closures.

COORDINATE POLYMERIZATION

The earliest reported examples of cyclopolymerization by use of metal-coordination catalysts were those of 1,5-hexadiene and 1,6-heptadiene.[18] Both monomers led to soluble, predominately saturated, hydrocarbon polymers containing cyclopentane and cyclohexane rings, respectively. More recent work adds more definitive evidence for the nature of the repeating unit in poly(1,5-hexadiene).[19]

Diallyldimethylsilane and diallyldiphenylsilane were polymerized using a metal-coordination catalyst.[20] Infrared studies are indicative of polymer structures containing 94% of the cyclized monomer units.

COPOLYMERIZATION OF 1,6-DIENES WITH VINYL MONOMERS

Copolymerization of cyclopolymerizable monomers with conventional vinyl monomers has been studied. Both symmetrical and unsymmetrical 1,5- or 1,6-dienes have been investigated.

These encompass formation of copolymers containing carbocyclic rings, nitrogen-containing rings, oxygen-containing rings, and rings containing certain other elements such as sulfur, silicon, and phosphorus. Unsymmetrical monomers lead to copolymers in a similar manner; however, the extent of cyclization, ring size, and other properties may vary. Cyclopolymerizable monomers designed to lead to larger rings have also been studied as comonomers. Examples of such monomers include butanediol dimethacrylate, hydroquinone dimethacrylate, and 4,4′-dihydroxydiphenyl dimethacrylate.[21]

Bis-ethylenically unsaturated amines such as diallylamine, dimethallylamine, and diallylalkylamines have been copolymerized with olefins such as styrene, alkyl acrylates, acrylamide, acrylonitrile, methacrylonitrile, and vinyl acetate.[22] These amines produce copolymers of extremely high molecular weight with acrylamide.

The copolymerization of divinyl acetals with vinyl acetate has been studied.[23]

Crystalline copolymers of ethylene and 1,5-hexadiene over a wide range of compositions have been prepared, using a preformed coordination catalyst.[24]

Copolymers of a wide variety of 1,6-dienes with sulfur dioxide have been prepared.[25] The molar ratio of diene to sulfur dioxide was 1:1, and the copolymers were soluble and essentially saturated.

Copolymerizations of methacrylic anhydride and a variety of common types of vinyl monomers have been studied extensively.[26]

Continued interest in the properties of copolymers of diallyldimethylammonium chloride (DADMAC) and acrylamide has stimulated renewed interest in improving their properties. A novel six-step process for production of high molecular weight copolymers that involves incremental mixing of acrylamide and addition of a chain-transfer agent to the polymerizing system has been described. The resulting copolymers exhibit an intrinsic viscosity of 16 dL/g (25°C, 4% NaCl), cationicity of 1.24 equivalents/g, and 99% conversion of the diallyl monomer.[27]

Radical copolymerization of DADMAC and vinyl acetate has been reinvestigated, and functionalization of the resulting polymers has been studied.

CYCLOPOLYMERIZATION INVOLVING OTHER DOUBLE BONDS

The concept of cyclopolymerization readily extends to symmetrical monomers containing other multiple bonds such as alkynes, aldehydes and ketones, isocyanates, epoxides, and nitriles. Studies have shown that conjugated polymers can be obtained by polymerization of 1,6-dialkynes using appropriate initiators. Dialdehydes have also been converted to soluble, cyclic polymers. A variety of diisocyanates have been studied, confirming the postulate that appropriately substituted compounds would undergo cyclopolymerization. Diepoxides and dinitriles when subjected to polymerization conditions also lead to cyclopolymers.[28]

A variety of monomers having two kinds of polymerizable functional groups appropriately distributed in the molecule have been studied in cyclopolymerization.

Numerous examples of monomers having double bonds between carbon and heteroatoms have been polymerized and their polymers studied. For example, numerous aldehydes have been polymerized through the carbon–oxygen double bond and alkyl isocyanates have been polymerized through the carbon–nitrogen double bond to linear polymers designated as "1 nylons." Difunctional monomers of both the above types of doubly bonded structures have now been prepared and their polymerizations studied.

Polymerization of epoxy compounds has been under study for quite some time. Bifunctional epoxides capable of forming a stable ring system have been cyclopolymerized.

Glycidyl methacrylate and glycidyl acrylate undergo cyclic copolymerization of the methacrylate double bond and the epoxy group by cationic initiation to produce cyclic lactone-ether recurring units in the polymer.[29]

Poly(dihexyldipropargylammonium salt)s were prepared by metathesis polymerization with transition-metal catalysts.[30] Novel side-chain liquid crystalline polymers with a poly(1,6-heptadiene) main chain have been described.[31]

CYCLOPOLYMERIZATION OF UNSYMMETRICAL MONOMERS

In addition to the symmetrical monomers, unsymmetrical diene monomers capable of undergoing cyclopolymerization have been studied. These monomers may be classified according to their hetero atoms as well. These include monomers capable of introducing not only carbon, but nitrogen, oxygen, silicon, sulfur, and phosphorus.

CYCLOPOLYMERIZATION LEADING TO POLYCYCLIC SYSTEMS

Appropriate diene monomers containing preexisting cyclic structures are converted via cyclopolymerization to polymers containing polycyclic systems. Monomers such as *cis*-1,3-divinylcyclopentane and the corresponding cyclohexane compound lead to polymers containing bicyclic systems. Other monomers such as triallylethylammonium bromide and 3-vinyl-1,5-hexadiene, which contain no preformed cyclic unit, undergo a succession of ring closures to yield soluble polymers. The ammonium monomer leads to polymers containing the bicyclo[3.3.1] structure, while the latter monomer leads to a polymer containing 2,6-methylene-linked bicyclo[2.2.1]heptyl rings.[32]

CYCLOPOLYMERIZATION

Previously discussed monomers that undergo cyclopolymerization have involved ring closures in such a manner that all members of the cyclic structure were derived from a single monomer. An example of a bimolecular alternating intramolecular–intermolecular copolymerization has been reported.[33] In 1951, it was shown that divinyl ether (DVE) undergoes copolymerization with maleic anhydride (MA) in a 1:2 molar ratio to yield a soluble, saturated copolymer. It has now been shown that a cyclic structure is formed; the divinyl ether contributes four members and the maleic anhydride contributes the remaining two. The reluctance of either monomer to undergo radical homopolymerization, and the marked tendency of both to undergo copolymerization with each other, support the cyclic structure.

The orderly nature of the copolymerization can be accounted for on the basis of electron donor–electron acceptor interactions, probably between the electron, donor DVE and the electron, acceptor MA. This copolymer (DIVEMA) has been extensively studied for its biological properties; it exhibits a broad spectrum of such activities, including antitumor, antiviral, antibacterial, anticoagulant, and antiarthritic properties, as well as being capable of generating interferon, inhibiting inverse transcriptase, activating macrophages, and eliminating plutonium.[34]

CYCLOPOLYMERIZATION LEADING TO LADDER POLYMERS

Double strand or "ladder" polymers have been sought and their formation has been extensively studied. Cyclopolymerization may be regarded as a halfway step to ladder polymers in that the main chain of a cyclopolymer includes rings. However, a secondary cyclopolymerization reaction on certain preformed polymers can lead to ladder polymers.[35] In fact, the formation of the well-known "Black Orlon" is postulated to occur by such a thermal reaction of polyacrylonitrile.

PRACTICAL SIGNIFICANCE OF CYCLOPOLYMERIZATION

Within 10 years after the discovery and confirmation of cyclopolymerization, poly(DADMAC) was a product of commerce and was being marketed for numerous applications. Also, the copolymer of divinyl ether and maleic anhydride (DIVEMA) was being prepared in semi-commercial quantities for use in further testing of its biological properties. Poly(DADMAC) is useful in water treatment as flocculants and coagulants; as an ionically conducting coating for paper in electrographic reproduction; broadly in the paper and textiles industries; in cosmetics and hair treatment; in biological, medical, and food applications; in the coal, minerals, and glass industries; as membranes; as soil treatment aids; and in a wide and diverse number of miscellaneous applications. Other cyclopolymers or copolymers are useful in water and waste treatment applications; as ion exchangers; as specific catalysts; in resist technology; in the rubber, paper, and textile industries; as thermally stable polymers; as antitumor agents and in other medical applications; as herbicides; in optical applications; in oil well technology; and in printing.[36]

DIVEMA has been investigated broadly as reported in the biological and medical literature for its potential in many applications as an active drug.

The relevance of poly(DADMAC) to flocculation and coagulation has prompted a study of adsorption of the polymer on kaolin.

Microencapsulated *Yarrowia lipolytica* was obtained by immobilization as its polyelectrolyte complex with poly(DADMAC). Other microorganisms and enzymes can be microencapsulated in a similar manner.[37]

An agent found to be effective in increasing petroleum displacement ability contains poly(DADMAC) as an important component.[36]

Water-absorbent materials for use in hygienic products manufacture have been prepared by copolymerizing DADMAC with cross-linking agents.[39]

REFERENCES

1. Butler, G. B. *Cyclopolymerization and Cyclocopolymerization;* Marcel Dekker: New York, NY, 1992; Chapter 1, Reference 1.
2. Butler, G. B. *Cyclopolymerization and Cyclocopolymerization;* Marcel Dekker: New York, NY, 1992; Chapter I, Reference 2.
3. Butler, G. B. *Cyclopolymerization and Cyclocopolymerization;* Marcel Dekker: New York, NY, 1992; Chapter I, Reference 3.
4. Butler, G. B. *Cyclopolymerization and Cyclocopolymerization;* Marcel Dekker: New York, NY, 1992; Chapter 1, Reference 4.
5. Butler, G. B. *Cyclopolymerization and Cyclocopolymerization;* Marcel Dekker: New York, NY, 1992; Chapter 1, Reference 5.
6. Butler, G. B. *Cyclopolymerization and Cyclocopolymerization;* Marcel Dekker: New York, NY, 1992; Chapter 1, Reference 6.
7. Butler, G. B. *Cyclopolymerization and Cyclocopolymerization;* Marcel Dekker: New York, NY, 1992; Chapter 1, Reference 7.
8. Butler, G. B. *Cyclopolymerization and Cyclocopolymerization;* Marcel Dekker: New York, NY, 1992; Chapter 1, Reference 8.
9. Butler, G. B. *Cyclopolymerization and Cyclocopolymerization;* Marcel Dekker: New York, NY, 1992; Chapter 1, Reference 9.
10. Butler, G. B. *Cyclopolymerization and Cyclocopolymerization;* Marcel Dekker: New York, NY, 1992; Chapter 2, Reference 516.
11. Butler, G. B. *Cyclopolymerization and Cyclocopolymerization;* Marcel Dekker: New York, NY, 1992; Chapter 1, Reference 24.
12. Butler, G. B. *Cyclopolymerization and Cyclocopolymerization;* Marcel Dekker: New York, NY, 1992; Chapter 1, Reference 26.
13. Butler, G. B. *Cyclopolymerization and Cyclocopolymerization;* Marcel Dekker: New York, NY, 1992; Chapter 1, Reference 38.
14. Butler, G. B. *Cyclopolymerization and Cyclocopolymerization;* Marcel Dekker: New York, NY, 1992; Chapter 1, Reference 40.
15. Butler, G. B. *Cyclopolymerization and Cyclocopolymerization;* Marcel Dekker: New York, NY, 1992; Chapter 1, Reference 31.
16. Butler, G. B *Cyclopolymerization and Cyclocopolymerization;* Marcel Dekker: New York, NY, 1992; Chapter 1, Reference 41.
17. Butler, G. B. *Cyclopolymerization and Cyclocopolymerization;* Marcel Dekker: New York, NY, 1992; Chapter 1, Reference 44.
18. Butler, G. B. *Cyclopolymerization and Cyclocopolymerization;* Marcel Dekker: New York, NY, 1992; Chapter 1, Reference 52.
19. Butler, G. B. *Cyclopolymerization and Cyclocopolymerization;* Marcel Dekker: New York, NY, 1992; Chapter 1, Reference 53.
20. Butler, G. B. *Cyclopolymerization and Cyclocopolymerization;* Marcel Dekker: New York, NY, 1992; Chapter 1, Reference 54.
21. Butler, G. B. *Cyclopolymerization and Cyclocopolymerization;* Marcel Dekker: New York, NY, 1992; Chapter 7, References 239-271.
22. Butler, G. B. *Cyclopolymerization and Cyclocopolymerization;* Marcel Dekker: New York, NY, 1992; Chapter 1, Reference 57.
23. Butler, G. B. *Cyclopolymerization and Cyclocopolymerization;* Marcel Dekker: New York, NY, 1992; Chapter 1, Reference 58.
24. Butler, G. B. *Cyclopolymerization and Cyclocopolymerization;* Marcel Dekker: New York, NY, 1992; Chapter 1, Reference 60.
25. Butler, G. B. *Cyclopolymerization and Cyclocopolymerization;* Marcel Dekker: New York, NY, 1992; Chapter 1, Reference 61.
26. Butler, G. B. *Cyclopolymerization and Cyclocopolymerization;* Marcel Dekker: New York, NY, 1992; Chapter 1, Reference 66.
27. Gartner, H. A. U.S. Patent 5 110 883, 1992; *Chem. Abstr.* **1992,** *117,* 2747d.
28. Butler, G. B. *Cyclopolymerization and Cyclocopolymerization;* Marcel Dekker: New York, NY, 1992; Chapter 5, References 189–214.
29. Butler, G. B. *Cyclopolymerization and Cyclocopolymerization;* Marcel Dekker: New York, NY, 1992; Chapter 1, Reference 71.

30. Kang, K. L. et al. *Macromolecules* **1993,** *26*(17), 4539; *Chem. Abstr.* **1993,** *119,* 96330g.
31. Jin, A-H. et al. *Macromolecules* **1993,** *26,* 1487.
32. Butler, G. B. *Cyclopolymerization and Cyclocopolymerization;* Marcel Dekker: New York, NY, 1992; Chapter 4, References 171–184.
33. Butler, G. B. *Cyclopolymerization and Cyclocopolymerization;* Marcel Dekker: New York, NY, 1992; Chapter 1, References 16, 87.
34. Butler, G. B. *Cyclopolymerization and Cyclocopolymerization; Marcel* Dekker: New York, NY, 1992; Chapter 8, Reference 2.
35. Butler, G. B. *Cyclopolymerization and Cyclocopolymerization; Marcel* Dekker: New York, NY, 1992; Chapter 9, References 1–187.
36. Butler, G. B. *Cyclopolymerization and Cyclocopolymerization; Marcel* Dekker: New York, NY, 1992; Chapter 12, Reference 476–519.
37. Dautzenberg, H. et al. German Patent DD 285 369, 1990; *Chem. Abstr.* **1991,** *114,* 2433498v.
38. Chalabiev, C. A. et al. U.S.S.R. Patent SU 1 689 596, 1991; *Chem. Abstr.* **1992,** *117,* 194966x.
39. Obayashi, S. et al. U.S. Patent 4 507 438, 1985.

CYCLOPOLYMERIZATION (Cyclization Efficiency and Configuration Control)

Toshiyuki Kodaira
Department of Materials Science and Engineering
Faculty of Engineering
Fukui University

Although the scope of cyclopolymerization is broad, the polymerization of 1,6-dienes is mostly discussed here because they are the monomers that have been studied the most. These investigations revealed that major difficulties in obtaining cyclopolymers arise from imperfect cyclization. The number of reactions changes significantly depending on the monomers and the polymerization conditions, even when favored rings are formed. As a result, controlling the reaction courses, such as the degree of cyclization and the ring size, is a problem characteristic of the difficulties associated with the structural control in cyclopolymerization, together with addition modes (e.g., tacticty of the polymers). These structural controls are important because the structure and properties are closely related. Excellent thermal stability, high glass transition temperatures, and little shrinkage during polymerization compared with conventional linear polymers are advantageous properties of polymers derived from cyclopolymerization. However, the cyclopolymers that have been produced commercially are not used because of these properties. One reason is the difficulty in getting polymers with a well-defined structure. Consequently, here the synthetic approach and recent efforts to control cyclization efficiency and configuration with the polymerization behavior of monomers are discussed.

PREPARATION AND PROPERTIES

Monomers with Non-Polymerizing Functional Groups

Bifunctional monomers with monofunctional counterparts that do not polymerize are likely to result in a highly cyclized polymer, if they can be polymerized at al.[1]

Intramolecular Cyclization Rate

Any interaction between the two functional groups favors rapid cyclization. The formation of a completely cyclized polymer from *N*-allyl-*N*-*t*-butylacrylamide was reported, even though one of its monofunctional counterparts, *N*-propyl-*N*-*t*-butylacrylamide has a high polymerization tendency.[2] The two functional groups are forced into favoring the intramolecular cyclization reaction by the bulky substituent. The effect of these bulky substituents, which enhance cyclization efficiency, has been observed in other monomers, suggesting that, if bulky substituents could be introduced, they would be useful for synthesizing highly cyclized polymers.[3,4]

Ring Structure

There are two possible ring structures formed from the cyclopolymerization of α-ω-dienes. The number of reactions can only be controlled when any changes depend on the reaction conditions. The problem with these reactions concerns the mechanism of a five-membered ring formation from 1,6-denes.

Radical Polymerization

One of the features related to repeating cyclic units derived from the cyclopolymerization of 1,6-dienes is that a five-membered ring is formed in many instances, although a six-membered ring and its radical are considered more stable. For this reason, the mechanism for the five-membered ring formation has been a matter of discussion. The cyclization reaction is characterized by three features.

The first is that the number of five-membered rings increases with an increase in the polymerization temperature. This characteristic is observed in the polymerizations of (meth)acrylic anhydrides.[5]

The second feature is that the number of six-membered rings increases with an increase in the reaction temperature as was observed in the radical cyclization reaction of low molecular weight compounds.[6,7] Kinetically controlled reactions at lower temperatures and thermodynamically controlled reactions at higher temperatures are assumed to explain the results obtained at the cyclization of low molecular weight compounds. This explanation has been applied to cyclopolymerization too.

The third feature is that the ring size formed does not change with the polymerization conditions. Some typical monomers belonging to this group are *N*-substituted dimethacrylamides, *N*-methyldiacrylamide, and *N*-allylmethacrylamide.[8,9] Favorable conformation of monomers for five-membered ring formation is the possible reason assigned to this class of monomers.[9]

Anionic Polymerization

A five-membered ring formation has been reported not only in the radical cyclopolymerization of 1,6-dienes but also in their anionic counterparts. The degree of cyclization of poly(*N*-methyldiacrylamide) and poly(*N*-allyl-*N*-*t*-butylacrylamide) are 100% and 81%, respectively.[10–12] Their repeating cyclic units consisted exclusively of five-membered rings. This observation is extremely odd because five-membered ring formations mean the occurrence of head-to-head and tail-to-tail additions, which have never been observed in the anionic polymerization of vinyl monomers.

Addition Reaction (Asymmetric Induction)

The synthesis of optically active polymers from 1-substituted olefins with structural chirality in the main chain has long been considered improbable. By considering their symmetry, Wulff et al. devised monomer sequences that allowed the formation of optically active polymers with such chirality. Diastereoselective addition was essential at the propagation step to get these sequences. For this purpose, a diene with two monoene units attached stereospecifically to a chiral template molecule, D-mannitol 1,2:3,4:5,6-tris-*O*-[(4-vinylphenyl)boronate, was copolymerized with vinyl monomers such as methyl methacrylate (MMA) or styrene (St.) Poly(St) and poly(St-*co*-MMA) obtained after the template and boronic acid unit split completely showed optical activity.[13,14] The versatility of this method for synthesizing optically active polymers has been proved.[15]

Cyclopolymerization of New Monomers

1,6-Dienes

Since a feasible synthetic route to α-hydroxymethylacrylates was developed, several symmetric and unsymmetric 1,6-dienes were synthesized using them as starting materials, and their cyclopolymerizations were undertaken.[16–20] α-Alkylacrylic esters, the monofunctional counterparts of the former, do not yield high polymers via radical polymerization except for methyl methacrylate, but α-alkyloxymethylacrylates, the monofunctional counterparts of the latter, do yield high polymers.[21,22]

The synthesis and polymerization of several fluorine-containing, unconjugated dienes have been reported. The cyclopolymers obtained are soluble in perfluoro solvents, transparent, and amorphous.[23,24]

α,ω-Dienes

For α,ω-dienes, in which two functional groups are far from each other compared with those of 1,6-dienes, three-dimensional network polymers usually form. However, some α,ω-dienes can yield soluble cyclopolymers.

APPLICATIONS

The cyclopolymers to be manufactured in commercial quantities are polymers derived from diallyl ammonium salt and their copolymers. They have been produced for many diverse commercial applications, including coagulants, electroconductive coatings, antistatic agents, soil conditioners, additives to cosmetics, and anion exchange resins.[25] Cyclopolymers obtained from perfluoro dienes are soluble in perfluoro solvent and transparent amorphous polymers different from conventional fluorine-containing polymers such as polytetrafluorethylene. Although the latter have been used in industry for their unique properties, they are insoluble in any solvent and assume an opaque white color. New perfluoropolymers have excellent properties, and they are used as advanced materials.[23] Polyphthaldehyde has been considered a material for photoresistant, because it can be depolymerized and volatilized by a photochemical reaction.[26] Another interesting cyclopolymer is the copolymer of divinyl ether and maleic anhydride, which has been studied extensively for its antitumor properties.[25]

REFERENCES

1. Kodaira, T.; Aoyama, F. *J. Polym. Sci., Polym. Chem. Ed.* **1974**, *12*, 897.
2. Fukuda, W.; Suzuki, Y.; Kakiuchi, H. *J. Polym. Sci., Part C: Polym. Lett. Ed.* **1988**, *26*, 305.
3. Kodaira, T.; Okumura, M.; Urushisaki, M. et al. *J. Polym. Sci., Part A: Polym. Chem.* **1993**, *31*, 169.
4. Tsuda, T.; Mathias, L. *J. Polym. Prepr.* **1993**, *34*, 339.
5. Matsumoto, A.; Kitamura, T.; Oiwa, M. et al. *J. Polym. Sci., Polym. Chem. Ed.* **1981**, *19*, 2531.
6. Hawthorn, D. G.; Solomon, D. H. *J. Macromol. Sci. Chem.* **1978**, *A10*, 923.
7. Julia, M. *Acc. Chem. Res.* **1971**, *4*, 386.
8. Kodaira, T.; Sakaki, S. *Makromol. Chem.* **1989**, *189*, 1835.
9. Kodaira, T.; Kitagawa, N.; Aoyagi, K. *Koubunshi Ronbunshu* **1989**, *46*, 507; *Chem. Abstr.* **1990**, *112*, 78011e.
10. Kodaira, T.; Tanahashi, H. *Macromolecules* **1989**, *22*, 4643.
11. Kodaira, T.; Tanahashi, H.; Hara, K. *Polym. J.* **1990**, *22*, 649.
12. Kodaira, T.; Urushisaki, M.; Usugaya, M. et al. *Macromolecules* **1994**, *27*, 1320.
13. Wulff, G.; Dhal, P. K. *Angew. Chem., Int. Ed. Engl.* **1989**, *28*, 196.
14. Wulff, G.; Kemmerer, R.; Vogt, B. *J. Am. Chem. Soc.* **1987**, *109*, 7449.
15. Kakuchi, T.; Kawai, H.; Katoh, S. et al. *Macromolecules* **1992**, *25*, 5545.
16. Mathias, L. J.; Kusefoglu, S. H.; Kress, O. A. *Macromolecules* **1987**, *20*, 2039.
17. Mathias, L. J.; Colletti, R. F.; Bielecki, A. *J. Am. Chem. Soc.* **1991**, *113*, 1550.
18. Stansbury, J. W. *Macromolecules* **1991**, *24*, 2099.
19. Mathias, L. J.; Warren, R. M.; Huang, S. *Macromolecules* **1991**, *24*, 2036.
20. Thompson, R. D.; Jarrett, W. J.; Mathias, L. J. *Macromolecules* **1992**, *25*, 6455.
21. Chikanishi, K.; Tsuruta, T. *Makromol. Chem.* **1965**, *81*, 211.
22. Yamada, B.; Satake, M.; Otsu, T. *Makromol. Chem.* **1991**, *192*, 2713.
23. Nakamura, M.; Kawasaki, T.; Unoki, M. et al. *Progress in Pacific Polymer Science*; Springer Verlag: Berlin, 1991, p 369.
24. Uchino, T.; Kojima, G.; Matuso, M. et al. *Polym. Prepr. Jpn.* **1992**, *41*, 99.
25. Butler, G. B. *Cyclopolymerization and Cyclocopolymerization*; Marcel Dekker: New York, 1992.
26. Itoh, H.; Willson, C. G. *Polym. Eng. Sci.* **1983**, *23*, 1012.

CYCLOPOLYMERIZATION (Nonconjugated Dienes)

Akira Matsumoto
Department of Applied Chemistry
Faculty of Engineering
Kansai University

Much work has been carried out on cyclopolymerization and cyclocopolymerization of nonconjugated dienes although the scope of cyclopolymerization is now quite broad and a variety of other multiple-bonded structures, such as alkynes, carbonyls, isocyanurates, epoxides, nitriles, and so forth, also undergo cyclopolymerization.[1–15] We will focus on a mechanistic discussion, including the cyclopolymerizability and ring size of resulting cyclopolymers; this will provide information on the mechanism of the cyclopolymerization of nonconjugated dienes and chain-growth polymerization in general, especially molecular design of polymers in the field of radical polymerization.

RING SIZE

In the radical polymerization of vinyl monomers, head-to-tail (ht) addition for the attack of a growing polymer radical on a monomer is predominant, explained by some qualitative considerations of resonance stability and steric factors leading to the intermediate formation of the more stable free radical and, furthermore, by theoretical treatment based on molecular orbital theory. Most of the early literature on the structure of cyclopolymers obtained in the cyclopolymerization of 1,6-dienes have assumed that a six-membered ring is formed in conformity with expected thermodynamic stability of the secondary or tertiary radical formed via intramolecular ht addition of the uncyclized radical to the internal double bond, Step 2, compared with the required primary radical for head-to-head (hh) addition, Step 3. Thereafter, numerous reports of predominant five-membered ring formation via intramolecular hh addition have been published; in particular, the cyclopolymerization of *N*-substituted dimethacrylamides, diallyl amines and their salts, divinyl acetals, and divinyl phosphonate could be presented as typical examples.

Another approach to the study of the cyclic units formed in these cyclopolymerizations is through the study of radical cyclization reaction of small molecules as selected model compounds. Thus, extensive studies have shown that the five-membered ring is predominant, while new evidence indicates that radical stability exerts a marked influence on the ring size; the correlations between studies on the model compounds and the polymer microstructures are often poor. These conflicting aspects of cyclopolymerization are well-documented, although considerably more investigations will be necessary before definite conclusions can be drawn with respect to the ratio of five- to six-membered rings in the many cyclopolymers already synthesized, and a satisfactory explanation for these extensive variations from one system to another is available.

The factors influencing intramolecular addition modes in the radical cyclopolymerization of some nonconjugated dienes have been investigated in detail and we discuss the selectivity of reaction mode of intramolecular hh or ht addition from the thermodynamical standpoint.[10] For example, in the cyclopolymerization of acrylic anhydride the five-membered ring formation was favored with increased solvent polarity, raised polymerization temperature, and decreased monomer concentration, the ring size of cyclic structure could be controlled freely by choosing appropriate polymerization conditions. We also reduced the rate of polymerization and the molecular weight of the polymer under polymerization conditions where the five-membered ring formation was favored.[16]

APPLICATIONS

Diverse commercial products like cyclopolymers and cyclocopolymers have been developed because they often possess enhanced properties in comparison with their open-chain analogues. Initial hopes of enhanced thermal stability did not lead to any outstanding new materials, but important properties have been discovered, particularly in the quaternary ammonium polymers from diallyl quaternary ammonium salts. Thus, the first cyclopolymer to be manufactured in commercial quantities was poly(diallyldimethylammonium chloride) and its properties are useful in electrographic paper-reproduction processes, where proper functioning of the photoresponsive coating requires rapid dissipation of static electrical charges. Other paper additive uses include antistatic agents, fluorescent whiteners, paperboard reinforcement, and retention aids. In the water-treatment field, quaternary ammonium polymers are used as flocculants and coagulants in potable water, waste water, or sewage sludge treatments. They are applied for the zinc, tin, and lead electroplating industries, as well as in cosmetics, as a biocide in water, as a demulsifier of dispersed oils, and as a detergent additive. Water-soluble cyclocopolymers of dialyl quaternary ammonium salts are also useful; for example, the copolymer of diallyldialkylammonium salts with sulfur dioxide are applied for the coagulation and precipitation of materials suspended in water, and as paper sizes, antistatic agents, surfactants, and thickeners.

The cyclocopolymer of divinyl ether and maleic anhydride, commonly known as Pyran copolymer or DIVEMA, has been extensively investigated as a biologically active product, especially for its antitumor properties; it has been shown to be an interferon inducer, and to possess antiviral, antibacterial, antifungal, anticoagulant, and antiarthritic activity.

Others have found applications as chelators, in chemical association, in composites, as crosslinking agents, in electrode technology, in food application, as herbicides, in oil well technology, in optical applications including synthesis of optically active polymers, and in printing.

The potential for cyclopolymers to compete with their classical counterparts is obvious, although the limiting features generally relate to added monomer cost or other lack of accessibility.

REFERENCES

1. Marvel, C. S. *J. Polym. Sci.* **1960**, *48*, 101.
2. Koton, M. M. *J. Polym. Sci.* **1961**, *52*, 97.
3. Butler, G. B. *Encyclopedia of Polymer Science and Technology*, John Wiley & Sons: New York, NY, 1966; Vol. 4; p 568.
4. Aso, C.; Kunitake, T.; Tagami, S. *Progress in Polymer Science Japan*, Kodansha: Tokyo, Japan, 1971; Vol. 1; p 149.
5. Butler, G. B.; Corfield, G. C.; Aso, C. *Progress in Polymer Science*, Pergamon: London, U.K., 1975; Vol. 4; p 71.
6. Solomon, D. H. *J. Macromol. Sci., Chem.* **1975**, *A9*, 95 and **1976**, *A10*, 855.
7. Butler, G. B. *J. Polym. Sci., Polym. Symp.* **1978**, *64*, 71.
8. Butler, G. B. *Polymeric Amines and Ammonium Salts;* Goethals, E. J., Ed.; Pergamon: New York, NY, 1980; p 125.
9. Butler, G. B. *Anionic Polymeric Drugs;* Doraruma et al., Eds.; John Wiley & Sons: New York, NY, 1980; p 49.
10. Matsumoto, A.; Iwanami, K.; Kitamura, T. et al. *ACS Symp. Ser.* **1982**, *195*, 29.
11. Butler, G. B. *Acc. Chem. Res.* **1982**, *15*, 370.
12. Corfield, G. C.; Butler, G. B. *Development in Polymerization-3*; Haward, R. N., Ed.; Applied Science: Essex, U.K., 1982; p 1.
13. Butler, G. B. *J. Macromol. Sci. Rev., Macromol. Chem. Phys.* **1982**, *22*, 89.
14. Butler, G. B. *Encyclopedia of Polymer Science and Engineering*, John Wiley & Sons: New York, NY, 1985; Vol. 4; p 543.
15. Butler, G. B. *Cyclopolymerization and Cyclocopolymerization*; Marcel Dekker: New York, NY, 1992.
16. Matsumoto, A.; Kitamura, T.; Oiwa, M.; Butler, G. B. *J. Polym. Sci., Polym. Chem. Ed.* **1981**, *19*, 2531.

CYCLOPOLYMERIZATION, ENANTIOSELECTIVE CATIONIC

Osamu Haba* and Kazuaki Yokota
Faculty of Engineering
Hokkaido University

Toyoji Kakuchi
Graduate School of Environmental Science
Hokkaido University

A cyclopolymerization of unconjugate dienes, which produces the polymer consisting of cyclic constitutional units, possesses the capacity to give an optically active polymer as well as the asymmetric polymerization of cyclic olefins. A pioneering work on an enantioselective cyclopolymerization has been demonstrated by Waymouth.[1–3] The optically active polymer was produced by the cyclopolymerization of 1,5-hexadiene using chiral zirconocene catalyst. In this case, the polymer is considered to possess racemo-diisotactic structure, that is to say one of the two *trans* structures formed superiorly as in the case of poly(*N*-phenylmaleimide).

For the polymerization using the metallocene, the initiator participates in the propagation reaction by the coordination to the growing end, and thus the configurational control of the main chain should be expected. Such coordination is found also in cationic living polymerization. This article describes the cyclopolymerization of divinyl ethers using a (+)- or (–)-10-camphorsulfonic acid[(+)- or (–)-**1**] /$ZnCl_2$ initiating system. As a monomer, 1,2-divinyloxybenzene (**2**) and benzaldehyde divinyl acetal (**4**), without asymmetric elements, were used (**Scheme I**).

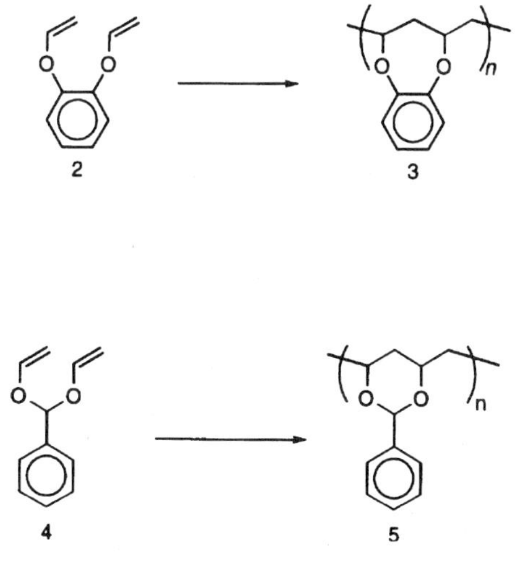

I

*Author to whom all correspondence should be addressed.

PROPERTY

The cyclopolymerization of 1,2-divinyloxybenzene (**2**) has been studied for the relation between the cyclic units in the resulting polymers and the used initiators. Butler reported that **2** polymerized to yield the polymer consisting of the 7-membered ring using cationic initiator.[4] In spite of the heterogeneous polymerization, the resulting polymers were soluble in chloroform and tetrahydrofuran. The polymer obtained with (+)-**1** showed a negative molar optical rotation of –3.4. The observed optical rotation had no influence of **1**, because **1** is soluble in methanol to be removable by reprecipitation and, further, has an opposite sign in optical rotation against the polymer. The polymer obtained with(–)-**1** has a positive molar optical rotation of +5.2, thus being enantiomeric to the polymer with (+)-**1**. Since the polymer has no asymmetric elements in the structure except for the main chain, the enantioselective intramolecular cyclization exactly occurred, resulting in chirality induction in the main chain.

Polymer **5** obtained using $ZnCl_2{\cdot}OEt_2$ was soluble in chloroform and tetrahydrofuran. The polymer structure consisted of only the 6-membered cyclic repeating units.

Although the **1**/$ZnCl_2$ initiating system may be expected to cause living polymerization, lower yield and molecular weight were observed for **5**.

Four possible structures are considered for the cyclic units of **5**, in which only **8** and **9** with a *trans* configuration are chiral. The optical activity of **5** arises from the chiral cyclic units corresponding to **8** or **9**. (4*S*, 6*S*)-(–)- and (4*R*, 6*R*)-(+)-4,6-Dimethyl-2-phenyl-1,3-dioxane[(SS)- and (RR)-**10**, respectively (**Scheme II**), whose configurations are identical to those of **8** and **9**, respectively, were synthesized as model compounds to confirm the absolute configuration of the cyclic units. The absolute configuration of the major, chiral cyclic units in polymer (–)-**5** should correspond to **8**, and that in polymer (+)-**5** to **9**.

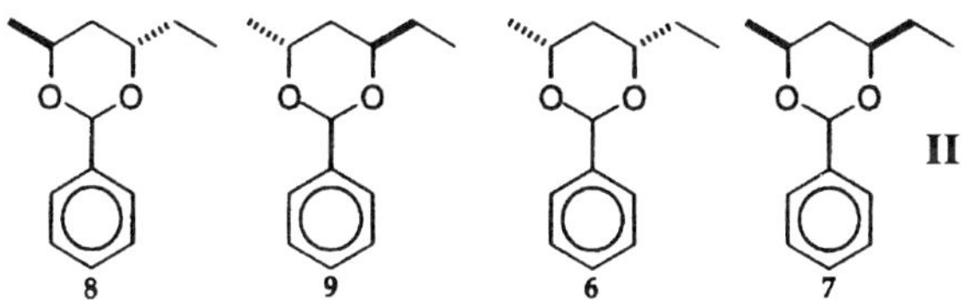

II

REFERENCES

1. Coates, G. W.; Waymouth, R. M. *J. Am. Chem. Soc.* **1991**, *113*, 6270.
2. Cavallo, L.; Guerra, G.; Corradini, P.; Resconi, L.; Waymouth, R. M. *Macromolecules* **1993**, *26*, 260.
3. Coates, G.; Waymouth, R. M. *J. Am. Chem. Soc.* **1993**, *115*, 91.
4. Butler, G. B.; Lien, Q. S. In *Cyclopolymerization and Polymers with Chain-Ring Structures*; Butler, G. B.; Kresta, J. E. Eds.; American Chemical Society: Washington, D.C., 1982; pp 145.

D

Defoaming Agents

See: Additives (Property and Processing Modifiers)

DEFORMATION AND FRACTURE (Micromechanical Mechanisms)

Georg H. Michler
Martin-Luther-Universität Halle-Wittenberg
Institute of Materials Science

Improving the properties of materials is a task of scientific and economic importance. In particular, polymeric materials have a broad inherent capacity for property modification. This is because of their large variety of macromolecular and supramolecular structures (morphology). With a detailed knowledge of structure-property correlations, criteria can be defined for modifying polymer structure and morphology to produce polymers with improved or new properties.

Electron microscopic investigations can provide important information on the role of the different structural details. For polymer studies, several techniques of electron microscopy (EM) can be applied.[1–4]

MICROMECHANICAL PROPERTIES

Rubber-Modified Polymers

Principles of Toughening

By adding relatively small amounts of elastomers to thermoplastics it is possible to increase the fracture toughness up to 1 order of magnitude. This effect was initially utilized in PS and SAN copolymers by blending with butadiene rubber, yielding high-impact polystyrene (HIPS) and acrylonitrile-butadiene-styrene (ABS).

In high-impact polymer systems with disperse structure, the thermoplastic forms the matrix in which the rubber phase is dispersed in particles. In these disperse systems there are two categories of mechanisms: either the energy absorbing step is the preferred formation of crazes at the rubber particles (multiple crazing) as in high-impact PS or numerous ABS grades, or the energy absorption mainly takes place through shear deformation between the modifier particles (multiple shearing) as in impact-modified polyamide (PA) or polypropylene (PP).[2,5–7]

Toughened Polymers with Amorphous Matrix

The fundamental deformation step is the formation of crazes in the stress field around the rubber particles. It can be clearly shown that crazes start in the matrix directly at the interface to the rubber particles. The first step is the stress concentration in the surroundings of the weak rubber particles. In the equatorial zones around the particles (i.e., the zones of highest stress concentration), the amorphous matrix material is transformed by local plastic deformation in the crazes. If the distance between the rubber particles is small enough, crazes are formed between the particles. Crazes starting on particles propagate to neighboring ones, which are located near the propagation direction of the crazes.

HIPS deformed generally by formation of fibrillated crazes, whereas many ABS grades show the formation of fibrillated crazes and homogeneous crazes (homogeneous deformation, shear deformation).

Toughened Polymers with Semicrystalline Matrix

The matrix materials usually applied are PA and PP. In rubber-modified, high-impact PA and PP rubber particles with diameters usually between ~ 100 and 500 nm are homogeneously distributed in the matrix. Results of deformation tests in the high voltage electron microscopy (HVEM) shows rubber particles that are ruptured inside, forming microvoids, which increase the stress concentration between the particles. Therefore, an increased plastic yielding between the particles is initiated. The plastically deformed matrix material is visible between the particles in the form of bright, diffuse zones. This type of deformation differs in principle from the formation of crazes on particles in HIPS and ABS. The plastically deformed matrix material is visible on fracture surfaces in SEM as fibrils.[6]

Results of the investigation of the micromechanical processes by EM are summarized in a three-stage mechanism.[6] Stage 1 is stress concentration. Stage 2 is void formation. Stage 3 is induced shear deformation. The shear deformation process proceeds simultaneously at numerous adjacent matrix bridges and thus takes place in fairly large polymeric volumes.

Particle-Filled Polymers

Particle-filled polymers are often-used examples of combinations of polymers with inorganic components. Besides the spatial distribution of particles, the adhesion or interfacial strength between particles and the polymer matrix is important for the mechanical behavior. Information about the strength at the interface can be gained by microfractographic analysis of fracture surfaces by using scanning electron microscopy (SEM) or by investigating deformed semithin sections in a HVEM.

CONCLUSIONS

Conducting deformation tests of polymers directly inside an electron microscope, including *in situ* tests, is a field in polymer science of increasing importance. By using these possibilities, a better understanding of mechanical properties and a microscopic modelling of the mechanical behavior of polymers can be achieved. Up to now, wider applications of deformation tests in the field of polymers has been hindered by the high sensitivity of polymers to electron irradiation. Here, all existing techniques for protecting the polymeric material from radiation damage should be applied.[4]

Increasing knowledge of the micromechanical processes in dependence on the real existent structure and morphology of a polymeric material makes it possible to define an optimum morphology to obtain a polymer with a desired or new property. In this direction, a new method of polymer manufacturing by "microstructural construction" seems practicable.

ACKNOWLEDGMENTS

The author thanks J. Heydenreich, director of the Max-Planck-Institute of Microstructure-Physics in Halle/S., for providing the opportunity to carry out the deformation tests in the 1000 of kV high-voltage electron microscope and graduate students of M. Enβlen, J. U. Starke, and G. M. Kim for performing *in situ* deformation tests of several polymers.

REFERENCES

1. *Electron Microscopy in Solid State Physics*; Bethge, H.; Heydenreich, J. Eds.; Elsevier Science: Amsterdam, The Netherlands, 1987.
2. Michler, G. H. *Kunststoff-Mikromechanik: Morphologie, Deformations und Bruchmechanismen*; Carl Hanser Verlag: München, Germany, 1992.
3. Michler, G. H. In *Electron Microscopy in Solid State Physics*; Bethge, H.; Heydenreich, J. Eds.; Elsevier Science: Amsterdam, The Netherlands, 1987, Chapter 16, pp 386–407.
4. Michler, G. H. *Appl. Spectros. Rev.* **1993**, *28*(4), 327.
5. Bucknall, C. B. *Toughened Plastics*; Applied Science: London, England, 1977.
6. Michler, G. H. *Acta Polym.* **1993**, *44*, 113.
7. Wu, S. *J. Appl. Polym. Sci.* **1988**, *35*, 549.

Degradation

See: *Anoxic Polymer Material*
Biodegradable Copolymers [Comprising Poly(L-lactic acid)]
Biodegradable Optically Active Polyamides (Stereoregular Polyamides from L-Tartaric Acid)
Biodegradable Polydepsipeptides (Synthesis, Characterization and Degradation)
Biodegradable Polymers (Orthopedic Applications)
Biodegradable Polyesters (Chemically-Induced Surface Degradation)
Biodegradable Polyesters (Theoretical Modeling of Degradation)
Biodegradable Polyesters, Cyclic Ethers
Biomedically Degradable Polymers
Chemodegradation (NOx)
Chitinases
Degradation (Electrical Discharge)
Degradation (Thermal, Polystyrene, and Related Vinyl Polymers)
Degradation (Weatherability)
Diene Rubbers, Conventional (Ozone Degradation and Stabilization)
DNA Degradation (Atomic Target Method using Synchrotron Radiation)
Durability
Environmentally Degradable Polymers
Enzyme-Degradable Hydrogels
Hindered Amine Light Stabilizers, Monomeric
Light Stabilizers (Overview)
Mechano Chemical Polymerization
Mechano Ions (Macro Ionic Products by Mechanical Fracture)
Melamine Resin (Pyrolysis)
Metathesis Degradation, Unsaturated Polymers
Metathesis Polymerization, Cycloolefins
Network Polymers (Chemorheology)
Nitrocellulose
Olefin-Carbon Monoxide Copolymers, Alternating
Photo-oxidation, Polyolefins
Photodegradable Plastics
Polyamide Thermal Stability
Poly(α-amino acids) (Biodegradation, Medical Applications)
Poly(ethylene terephthalate)Degradation (Chemistry and Kinetics)
Polypropylene (Controlled Degradation)
Polysilanes (Overview)
Polystyrene and Derivatives, Photolysis
Polytetrafluoroethylene (Effect of γ-Irradiation)
Poly(vinyl chloride) (Mechanisms of Stabilization)
Reactive Processing, Thermoplastics
Resins and Paints (Analytical Pyrolysis)
Starch-Based Plastics (Measurement of Biodegradability)
Starch Biodegradation (in Starch-Plastic Blends)
Starch-Polymer Composites
Stress-Induced Chemiluminescence Imaging
Thermodegradable Polymers (Azo Groups in Main Chain)
Ultrasonic Degradation (Preparation of Block Copolymers)
Weathering of Polymers (Methodology and Limitations of Accelerated Testing)
Xanthan, Depolymerization
Xanthan Gum (Overview)

DEGRADATION (Electrical Discharge)

Shigeo Kobayashi
Department of Electrical Engineering
Tokyo University of Agriculture and Technology

Sachio Yasufuku
Department of Electrical Engineering
Tokyo Denki University

Synthetic, high polymeric materials play an important role as major electrical insulating materials in high-voltage electrical equipment. This is because they not only have excellent mechanical and electrical properties, but they're also easily available for their applications and versatile for their products. Generally speaking, their properties as insulating materials are apt to deteriorate over time for various reasons. This phenomenon is called electrical insulation degradation. It applies mainly to high-polymer degradation. When degradation falls below the requirements of electric equipment or its parts, the apparatus or parts have reached the end of its useful life. Degradation is classified as physical or chemical deterioration practically, degradation is such that physical and chemical deteriorations take place simultaneously. Factors that can accelerate degradation include thermal, mechanical, electrical, radiative, and miscellaneous. Among these, electrical factors cause degradation when electrical insulation is subjected to voltage application. Electrical factors consist of the following items: (a) electric conduction current, which causes thermal effects as joule heat, as well as an electrochemical effect due to ionic conduction; (b) dielectric loss, which causes a thermal effect in AC voltage application; (c) electromagnetic force/electrostatic force, which is caused by a large, short-circuit current/high voltage and causes mechanical deterioration; and (d) electric discharges, which include partial discharge, treeing, tracking and arcing and causes a thermal action, a chemical action and dielectric breakdown in a solid electrical insulation.

PARTIAL DISCHARGE DETERIORATION

Recent Developments

A basic problem was how partial discharge magnitude can be related to the size, shape and position of a void within a solid dielectric. With the help of Pedersen's theory, the minimum breakdown field for a spherical void was equated with pressure and void radius.

Next, partial discharge deterioration was studied on an epoxy insulation. By use of a CIGRE Method I electrode system, partial discharge was evolved in a rotating, crescent-shaped void of 0.2mm gap that contacted the electrode. At first, a few erosive pits formed due to the void discharge, then a tree evolved at the tip of the largest pit, and finally, total breakdown took place. This is typical of partial discharge in an epoxy resin.[1]

Finally, the partial discharge deterioration of a crosslinked polyethylene was studied and proved that direct discharge synthesis reactions take place in a void through such prime reagents as, CO, CO_2, H_2O and acetophenone to form oxalic acid and other solid, liquid, and gaseous products.[2]

TREEING DETERIORATION

Recent Developments

Until now, numerous contributions for electrical and water-treeing have been presented and many of the findings have been shown, particularly for the mechanisms of evolution and arrival at final electric breakdown. Although it is difficult to establish a perfect demonstration for them, they should be reviewed by summarizing the essentials of the contributions on them.[3]

In electrical treeing in the dielectric of a high voltage crosslinked, polyethylene-insulated power cable at the interface roughness of the semicons or foreign particles included in the polyethylene, very high electrical fields create hot electrons and/or high energetic photons that cleave the molecular bonds and produce radicals. They may lead to double-bond formation and/or crosslinking and to a stepwise degradation. As a result, small voids and tree channels in which gas discharges occur are generated.

Water trees are not formed through a hollow tube, but are bodies of spherical voids that assemble in the direction of the electric fields and partially contain water. In the same dielectric and at the same structure as the above-mentioned electric tree, precipitation of water droplets leads to local mechanical overstressing in the neighboring polyethylene chains and ultimately to their disruption and production of radicals.

To counteract electrical- and water-treeing, keeping the polyethylene clean, any boundaries smooth, and thus, any local field enhancement low, is absolutely important. To counteract water-treeing, it is important to adopt a water-impervious structure above a high-voltage, crosslinked polyethylene power cable, and to prevent the penetration of water and humidity during manufacturing, storage, and construction.[4]

RECENT DEVELOPMENTS

There is another type of tracking phenomena, the division of a conductive path (i.e., the formation of dry belts), and the evolution of gaseous discharges between them to make carbonized products. As in recognizing that when the applied voltage becomes higher, the tracking breakdown becomes hard to evolve in some kinds of electrical insulation, the process of the dry belt formation was investigated. It was proved that in the case of the low applied-voltage, the dry belt is easily formed at the electrode edges, whereas in the case of the high applied voltage it is formed at the middle of the electrode. It was also proved that the voltage dependence of the location at which the dry belt is formed affects the evolved state of gaseous discharge and the tracking breakdown characteristics.[5] Further, in outdoor applications of high voltage apparatus and systems, good maintenance practices were recommended to reduce the possibility of failure from tracking deterioration.[6]

REFERENCES

1. Tanaka, T. *IEEE Trans. Electr. Insul.* **1986**, *EI-12*, 6, 899.
2. Gamez-Garcia, M.; Bartnikas, R.; Wertheimer, M. R. *IEEE Trans. Electr. Insul.* **1987**, *EI-22*, 2, 199.
3. Patsch, R. *IEEE Trans. Electr. Insul.* **1992**, *27*, 3, 532.
4. Fukuda, F. *Trans. IEEJ* **1988**, *108*, 5, 389.
5. Nishida, M.; Yoshimura, N.; Noto, F. *Trans. IEEJ* **1983**, *103*, 11, 593.
6. Kurtz, M. *IEEE Electrical Insulation Magazine* **1987**, *3*, 3, 12.

DEGRADATION (Thermal, Polystyrene, and Related Vinyl Polymers)

Hajime Ohtani and Shin Tsuge
Department of Applied Chemistry
School of Engineering
Nagoya University

Polystyrene (PS) is one of the most widely utilized polymers in industry, because of its easy workability, transparency, and high insulation resistance. It is important to clarify the thermal degradation mechanisms of PS, in view of stability, structural characterization, processing and recycling, and toxicity of decomposed compounds. In this paper, the thermal degradation mechanisms of PS and related vinyl polymers are described on the basis of recent publications.

POLYSTYRENE

It is well known that PS thermally degrades mainly to monomer (ca. 70–80%) with some dimer (2,4-diphenyl-1-butene) and trimer (2,4,6-triphenyl-1-hexene).[3–6,7–14] The formation of higher oligomers such as tetramer and pentamer is only trace, although they give good information about the stereoregularity of PS chains.[15] At higher degradation temperatures the monomer yield increases whereas that of dimer and trimer decreases. The recombination products, 1,2-diphenylethane and 2,4-diphenyl-1,5-hexadiene, also becomes apparent in the dimer region at a higher temperature. However, at much higher temperatures (over ca. 800°C), the yields of non-specific smaller and/or secondary products predominate.[4,7]

The thermal degradation of PS at elevated temperatures is initiated by a random scission of the polymer main chain to give primary and secondary macroradicals.

Both macroradicals chiefly depolymerize to the monomer. The monomer yield generally increases with increase of molecular weight (MW) of PS.[2,7,13] However, difference in the monomer yield becomes negligible in the higher MW region than ~10^4, which is almost equivalent to the zip length of PS.[2a]

Macroradicals can also engage in hydrogen abstraction. However, formation of the previously mentioned dimer and trimer is not reasonably explained through the radical transfer reactions of the primary radical. Instead, the primary macroradical undergoes almost quantitative depolymerization, or converts to the secondary macroradical through elimination of α-methylstyrene. The formation of the dimer and the trimer has been mostly explained through intramolecular 1,3- and 1,5-transfers, respectively, from the terminal groups of the secondary macroradical to the reactive tertiary carbon followed by β-scission.[1,2,4–7,10] Among these, the latter, 1,5-transfer is reasonably accepted as the main path to form the trimer by a back-biting mechanism through a six-membered ring intermediate.

Recently, the thermal degradation of a block copolymer of ordinary and perdeuterated styrene, poly(styrene-b-styrene-d_8) (styrene/styrene-d_8 = ~ 1/1) was investigated by Py-GC/mass spectrometry (MS) and by pyrolysis-field ionization MS (Py-FIMS) to clarify the detailed mechanisms of the dimer formation.[11]

POLYSTYRENE DERIVATIVES

Poly(α-methylstyrene) (P-α-MS) quantitatively depolymerizes into monomers at elevated temperatures, because the tertiary hydrogen atoms in styrene units, which play a very important role for radical transfer reactions to form dimer and trimer, are totally substituted by methyl groups in P-α-MS.[1,16,17]

On the other hand, most of the ring-substituted PS, such as poly(p- and poly(m-chlorostyrene),[18–20] poly(p-methylstyrene),[21–23] poly(p-methoxystyrene), and poly(p-hydroxystyrene),[24] and propionyl derivatives of PS,[25] show basically the same thermal degradation mechanisms as PS. The main pyrolyzates are the corresponding monomers along with the considerable amounts of dimers and trimers.

Polyvinylpyridine (PVP), whose pendant is pyridyl group instead of the phenyl one in PS, is also thermally decomposed in a similar manner to PS.

REFERENCES

1. Madorsky, S. L. *Thermal Degradation of Organic Polymers, Polymer Reviews* Vol. 7, Wiley: New York, 1964.
2. Cameron, G. G.; MacCallum, J. R. *J. Macromol. Sci.- Revs. Macromol. Chem.* **1967**, *C1*, 327.
2a. Tsuge, S.; Okumoto, T.; Takeuchi, T. *J. Chromatogr. Sci.* **1969**, *7*, 250.
3. Sugimura, Y.; Nagaya, T.; Tsuge, *S. Macromolecules* **1981**, *14*, 520.
4. Sousa Pessoa de Amorim, M. T.; Bouster, C.; Vermande, P.; Veron, J. *J. Anal. Appl. Pyrolysis* **1981** *3*, 19.
5. Daoust, D.; Bormann, S.; Legras, R.; Mercier, J. P. *Polym. Eng. Sci.* **1981**, *21*, 721.
6. Ohtani, H.; Tsuge, S.; Matsushita, Y.; Nagasawa, M. *Polym. J.* **1982**, *14*, 495.
6a. Lehrle, R. S.; Peakman, R. E.; Robb, J. C. *Eur. Polym. J.* **1982**, *18*, 517.
6b. Schroeder, U. K. O.; Ebert, K. H.; Hamielec, A. W. *Macromol. Chem.* **1984**, *185*, 991.
6c. Guaita, M. *Brit. Polym. J.* **1986**, *18*, 226.
6d. Costa, L.; Camino, G.; Guyot, A.; Bert, M.; Clouet, G.; Brossas, J. *Polym. Degrad. Stab.* **1986**, *14*, 85.
6e. Mita, I.; Hisano, T.; Horie, K.; Okamoto, A. *Macromolecules* **1988**, *21*, 3003.
6f. Kristina, J.; Moad, G.; Solomon, D. H. *Eur. Polym. J.* **1989**, *25*, 767.
7. Bouster, C.; Vermande, P.; Veron, J. *J. Anal. Appl. Pyrolysis* **1989**, *15*, 249.
8. Dean, L.; Groves, S.; Hancox, R.; Lamb, G.; Lehrle, R. S. *Polym. Degrad. Stab.* **1989**, *25*, 143.
9. Shapi, M. M.; Hesso, A. *J. Anal. Appl. Pyrolysis* **1990**, *18*, 143.
10. McNeill, I. C.; Zulfiqar, M.; Kousar, T. *Polym. Degrad. Stab.* **1990**, *28*, 131.
11. Ohtani, H.; Yuyama, T.; Tsuge, S.; Plage, B.; Schulten, H. R. *Eur. Polym. J.* **1990**, *26*, 893.
12. Atkinson, D. J.; Lehrle, R. S. *J. Anal. Appl. Pyrolysis* **1991**, *19*, 319.
13. Audisio, G.; Bertini, F. *J. Anal. Appl. Pyrolysis* **1992**, *24*, 61.
14. Gardner, P.; Lehrle, R. S. *Eur. Polym. J.* **1993**, *29*, 425.
15. Nonobe, T.; Ohtani, H.; Usami, T.; Mori, T.; Fukumori, H.; Hirata, Y.; Tsuge, S. *J. Anal. Appl. Pyrolysis* **1995**, *33*, 121.
16. Okumoto, T.; Takeuchi, T. *Bull. Chem. Soc. Jpn.* **1973**, *46*, 1717.
17. Guaita, M.; Chiantore, O. *Polym. Degrad. Stab.* **1985**, *11*, 167.
18. Okumoto, T.; Takeuchi, T.; Tsuge, S. *Macromolecules* **1973**, *6*, 922.
19. Okumoto, T.; Tsuge, S.; Yamamoto, Y.; Takeuchi, T.; *Macromolecules* **1974**, *7*, 376.
20. Bertini, F.; Audisio, G.; Kiji, J. *J. Anal. Appl. Pyrolysis* **1994**, *28*, 205.
21. Schroeder, U. K. O.; Ederer, H. J.; Ebert, K. H. *Makromol. Chem.* **1987**, *188*, 561.
22. Nakagawa, H.; Tsuge, S.; Mohanraj, S.; Ford, W. T. *Macromolecules* **1988**, *21*, 930.
23. Luda, M. P.; Guaita, M.; Chiantore, O. *Makromol. Chem., Macromol. Symp.* **1989**, *25*, 101.
24. Still, R. H.; Whitehead, A. *J. Appl. Polym. Sci.* **1977**, *21*, 1199.
25. Venema, A.; Sukkei, J. T. *J. High Res. Chromatogr. Chromatogr. Commun.* **1985**, *8*, 510.

DEGRADATION (Weatherability)

J. R. White
Department of Mechanical Materials and Manufacturing Engineering
University of Newcastle upon Tyne

Polymers often deteriorate rapidly in outdoor applications. The major cause of degradation is solar ultraviolet radiation, and the lifetime of plastics can very considerably according to the climate. Chemical reactions are accelerated by increasing the temperature, plastics are most vulnerable in hot sunny climates. Rainfall can contribute to degradation, especially if the rain reacts with pollutants to form acids. Pollutant gases may cause degradation even in the absence of rain. The chemical reactions that are involved in degradation are complex and different polymers display quite a wide range of weathering sensitivity.

Studies of the chemistry of degradation have led to the development of stabilizing systems that produce spectacular improvements in the service lifetimes of many polymers. Accelerated test procedures are used to assess the weatherability of different polymers and the effectiveness of stabilizing systems, but their reliability is uncertain.

An excellent introduction to the subject is provided in the book by Davis and Sims, which contains a survey of the chemical mechanisms of weathering, relevant climatological information, and discussion of accelerated testing methods.[1] More recent advances have been reviewed by White and Turnbull.[2]

CHEMICAL MECHANISMS OF WEATHER-RELATED DEGRADATION

Polymer degradation promoted by weathering usually involves a complex sequence of chemical reactions. Oxidation is by far the most important type of chemical degradation in polymers. It can occur at elevated temperatures or can be promoted photolytically. Photooxidation is the dominant process in weathering and occupies a central part of this article. Thermooxidation is not of direct concern, but the products of thermooxidation (which may occur during processing, for example) may lead to accelerated photodegradation. For some polymers hydrolysis is a significant hazard.

Photooxidation

In polyolefins and other polymers with hydrocarbon fragments, oxidative degradation proceeds as a radical reaction with branched kinetic chains. Much of the research in this area has been conducted on polypropylene because of its wide application and its sensitivity to weathering. Severini et al. have discussed scission reactions for polypropylene in a recent paper.[3]

Hydrolysis and Other Forms of Attack by Rain

Hydrolysis can cause chain scission and consequent failure in several polymers. Although this is a major hazard with these materials at elevated temperatures, it can proceed slowly at ambient temperatures, with rain as the source of water. Polymers susceptible to this kind of attack include polycarbonate and poly(ethylene terephthalate).

Some polymers absorb water, which leads to other problems. Nylons become plasticized and their Young's modulus can fall by as much as an order of magnitude.[4]

Attack by Pollutants

Degradation of polymers promoted by pollutants has been discussed by Rånby and Rabek.[5] Some pollutants are photolytic, leading to further potential degradants.

STABILIZATION

The inclusion of stabilizers can produce a spectacular improvement in the lifetime of a polymer component in outdoor service.

Photostabilizers

Allen et al. list the methods that offer protection against photooxidation as UV screening, UV absorption, excited state deactivation, free-radical scavenging, and hydroperoxide decomposition.[6]

Thermal Stabilizers

The products of thermal degradation often increase the vulnerability of a polymer to photooxidation and thermal stabilizers are required to limit this type of degradation. Thermal stabilizers do not directly protect the polymer against natural weathering but are of crucial importance in determining weatherability because they limit degradation during processing.

CONCLUSIONS

Unmodified polymers are very vulnerable to degradation when used outdoors, especially in hot, sunny climates. Photooxidation is the main cause of deterioration and the reactions involved in the degradation of polymers when exposed outdoors are fairly well understood, but are quite complex. The reaction rate can be very fast near the surface when it is exposed to high UV intensities and is often limited by lack of oxygen in the interior of thick sections because oxygen is consumed before it can diffuse very far. Extremely effective UV stabilizers have been developed but when they are included the reactions are even more complex. As a result the degradation is very difficult to model. For similar reasons it is difficult to device satisfactory accelerated weathering tests: it is impossible to find a procedure in which acceleration leaves the reaction pathway unchanged. Therefore laboratory results can rarely be extrapolated to provide a reliable guide as to lifetime under service conditions.

REFERENCES

1. Davis, A.; Sims, D. *Weathering of Polymers*; Applied Science: London, 1983.
2. White, J. R.; Turnbull, A. *J. Mater. Sci.* **1994**, *29*, 584.
3. Severini, F.; Gallo, R.; Ipsale, S. *Polym. Degrad. Stab.* **1988**, *22*, 185.
4. Paterson, M. W. A.; White, J. R. *J. Mater. Sci.* **1992**, *27*, 6229.
5. Rånby, B.; Rabek, J. F. In *Effects of Hostile Environments on Coatings and Plastics*; Garner, D. P.; Stahl, G. A., Eds.; ACS Symposium Series 229, American Chemical Society: Washington, DC, 1983; p 291.
6. Allen, N. S.; Chirinos-Padron, A.; Henman, T. J. *Polym. Degrad. Stab.* **1985**, *13*, 31.

Dendrimers

See: Dendrimers, Luminescent and Redox-Active
Dendrimers, Poly(ether ketone)
Dendrimers, Polysiloxanes
Dendritic Polymers, Divergent Synthesis (Starburst Polyamidoamine Dendrimers)
Dendritic Polyradicals
Functionalized Polymers (Comb, Rotaxanic, and Dendrimeric Structures)
Highly Branched Polymers
Hyperbranched Aliphatic Polyesters
Hyperbranched Polyesters
Organometallic Polymers (Transition Metals in Main Chain)
Polyalkylenimines

[1.1.1]Propellanes (New Vistas in their Polymer Chemistry)
Tailor-Made Polymers
Uniform Polymers

DENDRIMERS, LUMINESCENT AND REDOX-ACTIVE

Gianfranco Denti*
Laboratorio di Chimica Inorganica
Istituto di Chimica Agraria
Università di Pisa

Sebastiano Campagna and Scolastica Serroni
Dipartimento di Chimica Inorganica e Struttura Molecolare
Università di Messina

Vincenzo Balzani, Alberto Juris, and Margherita Venturi
Dipartimento di Chimica G. Ciamician
Università di Bologna

In the last 20 years extensive investigations carried out on the photochemical and electrochemical properties of transition metal compounds have shown that Ru(II) and Os(II) complexes of aromatic diimine ligands for example $Ru(bpy)_3^{2+}$ and $Os(bpy)_3^{2+}$ (bpy = 2.2′-bipyridine), exhibit a unique combination of chemical stability, redox properties, excited state reactivity, luminescence, and excited state lifetime.[1–7] Furthermore, all these properties can be tuned within rather broad ranges by changing ligands or ligand substituents. Several hundreds of these complexes have been synthesized and characterized, and some of them have been used as sensitizers for the interconversion of light and chemical energy as well as for other purposes.[1,2]

Essentially, polypyridine complexes of Ru(II) and Os(II) are stable species that feature intense absorption bands in the UV and visible spectral region, a reasonably strong and long-lived luminescence, and an extremely rich redox behavior (reversible one-electron oxidation at the metal center and reversible multielectron reduction at each ligand).[1,2] In these complexes, regardless of the excitation wavelength, the originally populated excited states undergo fast radiationless decay to the lowest triplet ^{3}MLCT, which is luminescent both in rigid matrix at 77 K and in fluid solution at room temperature. This is a very useful property because the occurrence of energy migration can be monitored by luminescence measurements.

Recent research on supramolecular systems has shown that assembly of functionally integrated building blocks into structurally organized arrays is essential to obtain useful functions.[8–13] Modern synthetic chemistry has proven that molecular components indeed can be assembled to give ultralarge supramolecular species of nanometric dimensions. The simplest method for the synthesis of such well-defined, ultralarge supramolecular species is the cascade approach, which leads to tree-like structures (arborols or dendrimers).[14,15]

We have tailored the cascade approach to Ru(II) and Os(II) polypyridine complexes and designed a synthetic strategy to build up luminescent and redox-active dendrimers, where desired metals and ligands can be placed in specific sites of the supramolecular structure. These metal-based dendrimers are of particular interest as they incorporate in their building blocks specific pieces of information, such as electronic excited states and redox levels at accessible energy values.

Thus, it has been possible to obtain species where several important functions are synthetically controlled. In particular, these luminescent and redox active dendrimers can be regarded as efficient artificial antennas for light harvesting, as the direction of electronic energy transfer can be controlled. Moreover, they could play the role of multielectron transfer catalysts since made-to-order control of the number of electrons lost at a certain potential can be obtained.[16]

BUILDING BLOCKS

The reasons for the choice of Ru(II) and Os(II) as metals, and of aromatic diimines as ligands have already been mentioned. To assemble such metal-containing building blocks, we have used the 2,3- and 2,5-bis(2-pyridyl) pyrazine (2,3- and 2,5-dpp) bridging ligands (BL). A particular important point is that it is possible to obtain easily and selectively the methylation of one of the two pyridyl nitrogens of 2,3- and 2,5-dpp, so that one of the two chelating sites can be protected from coordination.[17,18]

GENERAL PROPERTIES

All these dendrimers are soluble in most solvents (e.g., $(CH_3)_2CO$, CH_3CN) and are stable both in the dark and under light excitation. In principle, they can exist as different isomers, depending on the arrangement of the ligands around the metal ions. The polymetallic complexes can also be a mixture of several diastereoisomeric species since each metal center is also a stereogenic center. For these reasons structural investigations on these systems are difficult. However, differences arising from the presence of isomeric species are not expected to be large in the electrochemical and spectroscopic properties.[20]

The species with high nuclearity exhibit a three-dimensional branching structure of the type of those shown by otherwise completely different dendrimers based on organic components.[15,21] Therefore, endo- and exoreceptor properties can be expected, which will be the object of future investigations.[22] We would like to stress that our dendrimers differ from most of those prepared so far for two fundamental reasons: (i) each metal-containing unit exhibits valuable intrinsic properties such as absorption of visible (solar) light, luminescence, and oxidation and reduction levels at accessible potentials; (ii) by a suitable choice of the building blocks, different metals and/or ligands can be placed in specific sites of the supramolecular array, as we have already shown for the tetra-, hexa-, and decanuclear species.[16b,23–25] In other words, our dendrimers are species which can incorporate many *pieces of information* and therefore can be used to perform valuable functions such as light harvesting, directional energy transfer, and exchange of a controlled number of electrons at a certain potential.[16b,19,24b,c,25b,26]

ANTENNA EFFECT

The study of the natural photosynthetic systems has shown that for solar energy conversion purposes one needs supramolecular arrays, which absorbs as much visible light as possible

*Author to whom correspondence should be sent.

and are capable to channel the resulting excitation energy toward a specific site of the array (*antenna devices*).[27]

By locating different building blocks in an appropriate sequence it has been possible to design polynuclear complexes where the electronic energy generated by light absorption is channelled towards a desired direction along the supramolecular structure.[16b,24d,26b]

REFERENCES

1. Juris, A.; Balzani, V.; Barigelletti, F.; Campagna, S.; Belser, P.; von Zelewsky, A. *Coord. Chem. Rev.* **1988**, *84*, 85.
2. Kalyanasundaram, K. *Photochemistry of Polypyridine and Porphyrin Complexes* Academic: London, 1991.
3. Kober, E. M.; Caspar, J. V.; Sullivan, B. P.; Meyer, T. J. *Inorg. Chem.* **1988**, *27*, 4587.
4. Denti, G.; Serroni, S.; Sabatino, L.; Ciano, M.; Ricevuto, V.; Campagna, S. *Gazz. Chim. Ital.* **1991**, *121*, 37.
5. Della Ciana, L.; Dressick, D.; Sandrini, D.; Maestri, M.; Ciano, M. *Inorg. Chem.* **1990**, *29*, 2792.
6. (a) Vlcek, A. A.; *Coord. Chem. Rev.* **1982**, *43*, 39; (b) DeArmond, M. K.; Carlin, C. M. *Coord. Chem. Rev.* **1981**, *36*, 325; (c) Lever, A. B. P. *Inorg. Chem.* **1990**, *29*, 1271.
7. Balzani, V.; Barigelletti, F.; De Cola, L. *Topics Curr. Chem.* **1990**, *158*, 31.
8. Vögtle, F. *Supramolecular Chemistry*; John Wiley & Sons: Chichester, 1991.
9. Schneider, H.-J.; Durr, H. Eds., *Frontiers in Supramolecular Organic Chemistry and Photochemistry*; VCH: Weinheim, 1991.
10. Balzani, V.; Scandola, F. *Supramolecular Photochemistry*; Horwood: Chichester, 1991.
11. Diederich, F. *Cyclophanes*; Royal Society of Chemistry: Cambridge, 1991.
12. Gutsche, C. D. *Calixarenes*, Royal Society of Chemistry: Cambridge, 1991.
13. Balzani, V.; De Cola, L. Eds.; *Supramolecular Chemistry*; Kluwer: Dordrecht, 1992.
14. Buchlein, E.; Wehner, W.; Vögtle, F. *Synthesis* **1978**, *155*.
15. For recent reviews, see: (a) Tomalia, D. A.; Naylor, A. M.; Goddard III, W. A. *Angew. Chem. Int. Ed. Engl.* **1990**, *29*, 138; (b) Newkome, G. R.; Moorefield, C. N.; Baker, G. R. *Aldrichimica Acta* **1992**, *25*, 31; (c) Tomalia, D. A.; Durst, H. D. *Topics Curr. Chem.* **1993**, *165*, 193; (d) Fréchet, J. M. J. *Science* **1994**, *263*, 1710.
16. (a) Serroni, S.; Denti, G.; Campagna, S.; Ciano, M.; Balzani, V. *J. Chem. Soc. Chem. Commun.* **1991**, 944; (b) Denti, G.; Campagna, S.; Serroni, S.; Ciano, M.; Balzani, V. *J. Am. Chem. Soc.* **1992**, *114*, 2944.
17. Serroni, S.; Denti, G. *Inorg. Chem.* **1992**, *31*, 4251, and references therein.
18. Denti, G. unpublished results.
19. Riceuuto, V.; Balzani, V. Submitted for publication.
20. See, e.g., Hage, R.; Dijkhnis, A. H. J., Haasnoot, J. G.; Prins, R.; Reedijk, J.; Buchanan, B. E.; Vos, J. G. *Inorg. Chem.* **1988**, *27*, 2185.
21. For some very recent papers, see: (a) Coffin, M. A.; Bryce, M. R.; Batsanov, A. S.; Howard, J. A. K.; *J. Chem. Soc. Chem. Commun.* **1993**, 552; (b) Van der Made, A. W., Van Leenwen, P. W. N. M.; de Wilde, J.; Brandes, R. A. C. *Adv. Materials* **1993**, *5*, 466; (c) Mülhaupt, R.; Wörner, C. *Angew. Chem. Int. Ed. Engl.* **1993**, *32*, 1306; (d) de Brabander-van den Berg, E. M. M.; Meijer, E. W. *Angew. Chem. Int. Ed. Engl.* **1993**, *32*, 1308; (e) Xu, Z.; Moore, J. S. *Angew. Chem. Int. Ed. Engl.* **1993**, *32*, 1354; (f) Newkome, G. R., Moorefield, C. N.; Keith, J. M.; Baker, G. R.; Escamilla, G. H. *Angew. Chem. Int. Ed. Engl.* **1994**, *33*, 666; (g) Hawker, C. J.; Wooley, K. L.; Fréchet, J. M. J. *J. Chem. Soc. Chem. Commun.* **1994**, 925; (h) Nagasaki, T.; Kimura, O.; Ukon, M.; Arimori, S.; Hamachi, I.; Shinkai, S. *J. Chem. Soc. Perkin Trans.* **1994**, *1*, 75.
22. Tomalia, D. A.; Baker, H.; Dewald, J. R.; Hall, M.; Kallos, G.; Martin, S.; Roeck, J.; Ryder, J.; Smith, P. *Polymer J.* **1985**, *17*, 117.
23. For some relevant, recent papers on the electrochemical and photophysical properties of polynuclear Ru(II) complexes containing the 2,3-dpp bridged ligand, see: (a) Cooper, J. B.; Mac Queen, D. B.; Petersen, J. D.; Wertz, D. W. *Inorg. Chem.* **1990**, *29*, 3701; (b) Kalyanasundaram, K.; Graetzel, M.; Nazeeruddin, Md. K. *J. Phys. Chem.* **1992**, *96*, 5865; (c) Johnson, J. E. B., Ruminski, R. R. *Inorg. Chim. Acta* **1993**, *208*, 231; (d) Richter, M. M.; Brewer, K. J. *Inorg. Chem.* **1993**, *32*, 2827; (e) Roffia, S.; Marcaccio, M.; Paradisi, C.; Paolucci, F.; Balzani, V.; Denti, G.; Serroni, S.; Campagna, S. *Inorg. Chem.* **1993**, *32*, 3003; Richter, M. M.; Jensen, G. E.; Brewer, K. J. *Inorg. Chim. Acta*, in press.
24. (a) Campagna, S.; Denti, G.; Sabatino, L.; Serroni, S.; Ciano, M.; Balzani, V. *J. Chem. Soc. Chem. Commun.* **1989**, 1500; (b) Denti, G.; Campagna, S.; Sabatino, L.; Serroni, S.; Ciano, M.; Balzani, V. *Inorg. Chem.* **1990**, *29*, 4750; (c) Denti, G.; Serroni, S.; Campagna, S.; Ricevuto, V.; Balzani, V. *Inorg. Chim. Acta* **1991**, *182*, 127; (d) Denti, G.; Serroni, S.; Campagna, S.; Ricevuto, V.; Balzani, V. *Coord. Chem. Rev.* **1991**, *111*, 227.
25. (a) Campagna, S.; Denti, G.; Serroni, S.; Ciano, M.; Balzani, V. *Inorg. Chem.* **1991**, *30*, 3728; (b) Denti, G.; Serroni, S.; Campagna, S.; Ricevuto, V.; Juris, A.; Balzani, V. *Inorg. Chim. Acta* **1992**, *198–200*, 507.
26. (a) Denti, G.; Campagna, S.; Sabatino, L.; Serroni, S.; Ciano, M.; Balzani, V. In *Photochemical Conversion and Storage of Solar Energy* Pelizzetti, E.; Schiavello, M.; Eds.; Dordrecht: Kluwer, 1991; p 27; (b) Balzani, V.; Campagna, S.; Denti, G.; Serroni, S. In *Photoprocesses in Transition Metal Complexes, Biosystems, and other Molecules: Experimental and Theory* Kochanski, E.; Ed.; Kluwer: Dordrecht, 1992; p 233.
27. Hader, D. P.; Tevini, M. *General Photobiology*; Pergamon: Oxford, 1987.

DENDRIMERS, POLY(ETHER KETONE)

Atsushi Morikawa
Department of Materials Science
Faculty of Engineering
Ibaraki University

Starburst polymers are highly branched regular molecules that were proposed by Tomalia.[1] They are prepared by stepwise reactions of the building block to the core. The molecules grow concentrically, and their size and shape are completely controlled. Thereby, the starburst polymers are expected to have the different properties compared with corresponding linear polymers. The properties of the molecular such as solubility and affinity with other molecules can be controlled by introduction of functional groups to the periphery. In addition to these possibilities, the convergent synthetic approach, the starburst polymers grow to the inside direction from the peripheral point, and one-step synthetic approach, self-condensation of AB_X type monomers, were proposed so various starburst polymers have been prepared.[2,3] In this article, one-step synthesis and convergent synthesis of starburst poly(ether ketone) dendrimers are described.

PREPARATION AND PROPERTIES

Convergent Synthesis of Starburst Poly(Ether Ketone) Dendrimers

In one-step synthesis, the number of branches, the size, and the shape of the starburst polymers cannot be completely controlled. The starburst poly(ether ketone) dendrimers, possessing the completely regulated structure, are interesting for polymer science. We prepared starburst poly(ether ketone) dendrimers stepwise by the convergent approach.[4] As the building block, 3,5-bis(4-fluorobenzoyl) anisole, which is the protective form of *1*, and as the starting molecule, phenol was selected, respectively. The molecular weight of the starburst poly(ether ketone) dendrimers increased with increasing the number of generations, and the molecular weight distributions were remarkably narrow. These results and NMR spectra indicated that the chemical structure of the starburst poly(ether ketone) dendrimers were obviously elucidated as the expected structure. There are fluorophenyl groups on the periphery in starburst poly(ether ketone) dendrimers prepared by one-step synthesis, but phenyl groups in these starburst poly(ether ketone) dendrimers were prepared by convergent synthesis. Moreover, these starburst poly(ether ketone) dendrimers were soluble in various organic solvents such as amide type solvents, DMSO, THF, dichloromethane, and chloroform. Linear PEEK is known as a high crystalline polymer, and insoluble in common organic solvents, but starburst poly(ether ketone) dendrimers exhibited non-crystalline nature, and were soluble in various organic solvents in spite of possessing similar chemical structure. These phenomena are thought to be caused by their highly branched structure.

REFERENCES

1. Tomalia, D. A.; Nayor, A. M.; Goddard III, W. A. *Angew. Chem. Int. Ed. Engl.* **1990**, *29*, 138.
2. Tomalia, D. A.; Baker, H.; Dewald, J.; Hall, M.; Kallos, G.; Martin, S.; Roeck, J.; Ryder, J.; Smith, P. *Polym. J.* **1985**, *17*, 117.
3. Tomalia, D. A.; Baker, H.; Dewald, J.; Hall, M.; Kallos, G.; Martin, S.; Roeck, J.; Ryder, J.; Smith, P. *Macromolecules* **1986**, *19*, 2466.
4. Tomalia, D. A.; Berry, V.; Hall, M.; Hedstrand, D. M. *Macromolecules* **1987**, *20*, 1167.

DENDRIMERS, POLYSILOXANES

Masa-aki Kakimoto
Department of Organic and Polymeric Materials
Tokyo Institute of Technology

Starburst dendrimer polymers are constructed from various initiator cores, upon which concentric branched layers are built up with geometric processes. They allow precision control of the molecular size as well as disposition of desired functionalities, and are expected to be applied as a microsphere that consists of one molecule. To prepare new highly branched starburst polymers consisting of polysiloxane units, we have developed two synthetic methods: divergent and convergent.

DIVERGENT METHOD

A siloxane bond is generally formed by the reaction of a compound having electrophilic silicon species (Si^+) with a nucleophilic silanol. Since the building block in the present work has both silanol and Si^+ species, one of the functions must be protected to avoid self-condensation. Protection of the Si+ was selected instead of the protection of the silanol in the present work. The reaction starts from the core possessing a synthon of Si^+ such as halosilane, alkoxysilane, and aminosilane. Next, the building block having both Si^+ synthon and silanol is reacted with the core leading to the formation of a siloxane bond. The exterior Si^+ units are again generated after the formation of the siloxane bond, and then react with the silanol moiety of the building block. The polysiloxane starburst polymers are synthesized by repeating these reactions.[1]

Several synthons of Si^+ exists,[2] such as ethoxycarbonylmethylsilanes,[3] allylsilanes,[4,5] and phenylsilanes,[6,7] which are converted to halosilanes by the reaction with hydrogen halides, halogens, or acyl halides.

CONVERGENT METHOD

Recently, Frechet reported a unique synthetic method of dendritic polyethers using a convergent approach in which the molecules were constructed starting at a point that becomes the periphery of the molecules.[8–11] First, the unreactive group of the starting molecule is converted to the reactive group. Next, the building block having two reactive groups and one unreactive group is reacted with it. The reactive groups in the new generation is again generated from the unreactive group, and then reacted with the reactive group of the building block. The starburst dendrimers are convergently synthesized by repeating these steps. Thus, since the starburst dendrimers grow to the inside direction from the peripheral position in the convergent synthesis, the number of the points reacting with the building unit are usually constant, which is only one in most cases.

As mentioned above, the siloxane bond is formed by the reaction of the electrophilic silicon species and silanol. Selection of the synthon of such reactive functions is generally important in the starburst synthesis. Hydrosilylation of chlorosilane to the terminal olefin is used as the key step, where the formation of a new silicon-carbon bond as well as the generation of chlorosilane, the electrophilic silicon species, is achieved at the same time. Next, the siloxane bond formation is carried out by the reaction of aminosilane, which is derived from chlorosilane and the building block which has one terminal olefin and two silanol units.

Siloxane bonds are generally prepared by the reaction between silanol and electrophilic silicon species such as halosilanes or aminosilanes. Terminal olefins were selected as the synthon of electrophilic silicon species in this study.[12,13]

CONCLUSION

The starburst polymers based on the siloxane structure can be successfully prepared by divergent and convergent methods. In the divergent method, the siloxane bond is constructed by the usual siloxane synthetic reaction between an electrophilic silicon species (Si^+) and the nucleophilic silanols, where the combination of phenylsilane and bromine acts as the most suitable synthon of Si^+. On the other hand, Si^+ is generated by the reaction between terminal olefins and chlorosilane in the presence of platinum catalyst.

Because the starburst polymers prepared by the divergent method have phenylsilane units at the exterior position, some functional groups can be easily introduced via the generation of Si^{+} species.[14] The convergent starburst polysiloxanes possess the cyano function in the beginning. One of the potential applications of the starburst polymers could be as drug carriers. Because of the harmless nature of polysiloxanes, the present starburst polysiloxanes could be considered as promising drug microcarriers directly injectable into the body.

REFERENCES

1. Morikawa, A.; Kakimoto, M.; Imai, Y. *Macromolecules* **1991**, *24*, 3469.
2. Birkofer, L.; Stuhl, O. *The Chemistry of Organic Silicon Compounds, Part 1*; Patai, S., Rappoport, Z. Ed.: John Wiley & Sons: New York, 1989, p 724.
3. Gold, J. R.; Sommer, L. H.; Whitmore, F. C. *J. Am. Chem. Soc.* **1948**, *70*, 2874.
4. Sommer, L. H.; Tyler, L. J.; Whitmore, F. C. *J. Am. Chem. Soc.* **1948**, *70*, 2872.
5. Ojima, I.; Kumagai, M.; Miyazawa, Y. *Tetrahedron Lett.* **1977**, 1385.
6. Eaborn, C. *J. Chem. Soc.* **1949**, 2755.
7. McBride, J. J. Jr.; Beachell, H. C. *J. Am. Chem. Soc.* **1952**, *74*, 5247.
8. Hawker, C. J.; Frechet, J. M. J. *J. Chem. Soc., Chem. Commun.* **1990**, 1010.
9. Hawker, C. J.; Frechet, J. M. J. *Macromolecules* **1990**, *23*, 4726.
10. Hawker, C. J.; Frechet, J. M. J. *J. Am. Chem. Soc.* **1990**, *112*, 7638.
11. Wooley, K. L.; Hawker, C. J.; Frechet, J. M. J. *J. Am. Chem. Soc.* **1991**, *113*, 4252.
12. Speier, J. L.; Webster, J. A.; Barnes, G. H. *J. Am. Chem. Soc.* **1957**, *79*, 974.
13. Speier, J. L. *Homogeneous Catalysis of Hydrosilation by Transition Metals*, Advances in Organometallic Chemistry, Academic: New York, 1979; Vol. 17.
14. Morikawa, A.; Kakimoto, M.; Imai, Y. *Polym. J.* **1992**, *24*, 573.

DENDRITIC POLYMERS, DIVERGENT SYNTHESIS (Starburst Polyamidoamine Dendrimers)

Donald A. Tomalia* and Peter R. Dvornic
Michigan Molecular Institute

Dendritic polymers represent the fourth and the most recently discovered class of macromolecular architecture. These polymers are distinguished from all other classical types by containing the ideal one branch juncture per repeating unit. They may be classified into two subgroups: single-trunked, branched arrays called dendrons and multidendron assemblies called dendrimers. Because dendrimers are composed of two or more dendrons, they represent an enhancement in structural complexity within this class of polymers.

The first successful preparation of a well-characterized dendritic polymer was reported in 1984 by one of us.[1–3]

Dendritic macromolecules, including dendrons and dendrimers, may be derived from at least four architectural components. They include initiator cores or focal points; terminal surface groups, which may be chemically reactive or inert; interior branch junctures, possessing various branching characteristics or multiplicities that are equal to or larger than two; and connectors, divalent segments that covalently connect neighboring branch junctures to provide scaffolding upon which terminal surface groups reside. These basic architectural components constitute the branch cells. The resulting dendrimer is defined by a hierarchy of branch cells, including the core branch cell, internal branch cells, and surface branch cells.

These three types of branch cells share one characteristic: they all contain a single-branch juncture point. They may be homogenous or differ in chemical structure or branching functionality (multiplicity). The surface branch cells may contain chemically reactive or passive functional groups. The chemically reactive surface groups may be used for further dendritic growth or for modification of the dendritic surfaces. The chemically passive groups may be used to physically modify the dendritic surfaces (e.g., adjusting hydrophobic–hydrophilic ratios).

PREPARATION

In theory, dendritic polymers may be prepared by two synthetic approaches: convergent and divergent. With the convergent approach, the growth process begins from what will later become the dendron surface and progresses in a radial molecular direction, inward or toward the focal point.[6–10] In contrast, the divergent approach involves dendritic growth from the initiator core or focal point and progresses outward in a radial direction from the core to the surface.[1,2]

The divergent approach can be realized by three different synthetic methods: one-pot synthesis, the protect-deprotect method and the excess monomer method.[1,2,7–29,37]

An important practical advantage of the one-pot synthesis is its relative simplicity to perform. However, this method lacks reaction control, thus leading to more polydispersed products. It is essentially governed by statistical laws, and for that reason, it yields dendritic products with substantial amounts of structural irregularities.

A characteristic feature of the protect—deprotect method is that it provides almost complete control of the dendritic molecular growth process and can be used to produce monodispersed dendritic products. However, lower yields and product loss usually associated with protect-deprotect reaction protocol may severely reduce overall product yields to impractical values after only a few reiterations. In more severe cases, this problem may lead to early extinction of dendritic growth if the reiteration schemes are not nearly quantitative.[30]

The excess monomer method combines the advantages of previous methods. It provides high yields of dendritic products (i.e., nearly quantitative, like the one-pot synthesis), which even after a large number of reiterations, still contain high relative amounts of ideal structures (similar to the protect-deprotect method). For this reason, the excess monomer method currently offers the most promising commercial route to dendrimers with well-defined macromolecular architecture.

*Author to whom correspondence should be addressed.

PREPARATION OF STARBURST

Starburst® polyamidoamine (PAMAM) dendrimers are represented by **Structure 1**:

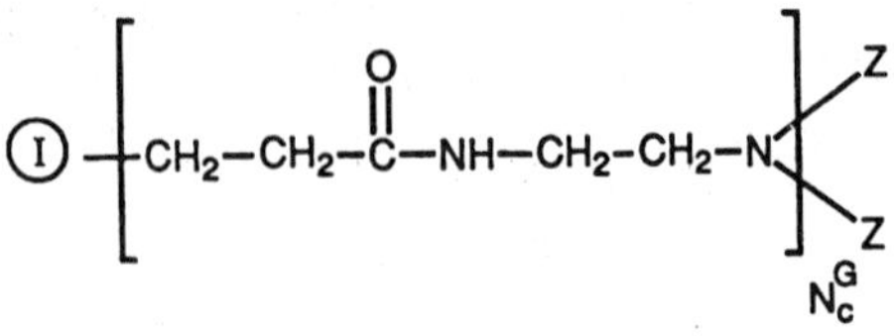

1

where I is the initiator core, which can be $>N-$ (with three bonds), or $>N-(CH_2)_x-N<$ with x = 2,4, ..., or some other suitable amino compound.[1,2] The functionality of I, N_c is 3 if I is $>N-$ (with three bonds) or 4 if I is $>N-(CH_2)_x-N<$, whereas G is the number of generations surrounding the initiator core and containing the repeating units $[CH_2CH_2C(O)N(H)CH_2CH_2N]<$. Z represents the terminal groups, which for PAMAM dendrimers are hydrogen atoms.

The preparation of PAMAM dendrimers involves a typical divergent synthesis.[1,2,20–29] It proceeds via a two-step growth sequence that consists of two reiterating reactions: the Michael addition of amino groups to the double bond of methyl acrylate (MA) followed by the amidation of the resulting terminal carbomethoxy, $C(O)OCH_3$, with ethylene diamine (EDA).

DENDRITIC BRANCHING

The three dendritic architectural components are formed chronologically during the divergent growth process (Figure 1). They develop along the central symmetry axes of each main monodendron. Thus, during a divergent dendritic growth process, each of these axes represents a main reaction coordinate that extends in space (i.e., from the core to the surface) and in time (i.e., from generation to subsequent generation).

When the first layer of branch cells develops around the initiator core, the resulting structures consist of the core cell and the N_c surface cells (generation B of Figure 1), where N_c represents the functionality of the core atom or atomic group. Clearly, this structure does not represent a fully developed dendrimer structure because it is missing one of the three fundamental components required, the interior cells. Therefore, this and other similarly branched structures (generation A of Figure 1) may be considered precursors to dendrimer structures but not dendrimers. The complete development of a dendrimer structure requires at least progression through generation 1.5 (Equation 9 and generation C of Figure 1).

Thus, a fully developed dendrimer structure first appears only after passing through the growth stage above. At that point, the lightly branched structures (generations A and B of Figure 1) transform into fully developed dendrimers that, in the genealogy of the synthesis, contain all three branch cell components (i.e., generation C of Figure 1). At this level of structural complexity, the transition from lightly branched to true dendrimer intermediates takes place. We refer to this transition as the critical degree of branching.[31]

PROPERTIES OF STARBURST

Computer-stimulated modeling related to size and shape have been performed on PAMAM dendrimers. The molecular dynamics calculations based on the force field acting on all atoms in equilibrated structures showed that, with increasing generations, these dendrimers seem to develop by passing through a continuum of molecular shapes ranging from open extended structures to closed globular spheroids. This change in shape apparently occurs because tethered steric constraints are imposed on the developing branches.

Tridendron PAMAM dendrimers appear spherical after about generation 4. Within these spheres, the solvent accessible surface (SAS) seems to increase with generations so that the fraction of the internal molecular surface increases from about 29% of the total SAS for generation 4, to 69% for generation 5, and 124% for generation 6. However, the molecular density (M/V) shows a clear minimum at about generation 4, suggesting that fully developed PAMAM dendrimers have a great deal of accessible internal surface area in a solvent-filled intramolecular free volume, which may consist of internal cavities and channels.[5]

^{13}C NMR measurements of spin-lattice relaxation times of specifically tagged PAMAM dendrimers appear to support this view. They showed considerably reduced mobility in the outer surface groups relative to that in the interior segments.[32–34] This behavior indicates a unimolecular encapsulation type topology in which a relatively soft or spongy interior is enclosed within a considerably harder outer molecular surface or crust.[4,34a]

Our recent rheological and differential scanning calorimetry (DSC) data also appears to support this model.[35] Thus we discovered that PAMAM dendrimer solutions exhibited Newtonian behavior over a wide range of shear rates (from about 150 to about 750–1250 s^{-1}), temperature (from 15 to 40°C), molecular weights (500 to 29,000 or from G = 0 to G = 5), and concentrations (30–75% by weight). In addition, neat dendrimers showed a linear increase in viscosity with molecular weight at temperatures from 90 to 115°C.

Furthermore, DSC measurements on PAMAM dendrimers (from EDA and NH3 cores) showed exponentially increasing glass transition temperatures with their increasing molecular weight reaching an asymptotic value at PAMAM generation 3 or 4.[35]

From these results, PAMAM dendrimers appear to possess outer surfaces that are substantially impenetrable to neighboring dendrimers, while retaining pronounced segmental mobility within the internal volume, possibly around the core and over the first two to three generations.

The molecular dimensions of PAMAM dendrimers have been examined by computer modeling and by size exclusion chromatography.[4,5] In general, excellent agreement between calculated and experimental results was observed. These studies indicate diameters over a range of 1 to 10 nm from generation 0 to generation 8, respectively, with a linear enhancement of 0.7 and 1.2 nm per generation.[4,5] At these nanoscopic dimensions, direct observation of single dendrimer molecules are possible with electron microscopy. These studies showed that dendrimers were observable as highly monodispersed spheroids.[2]

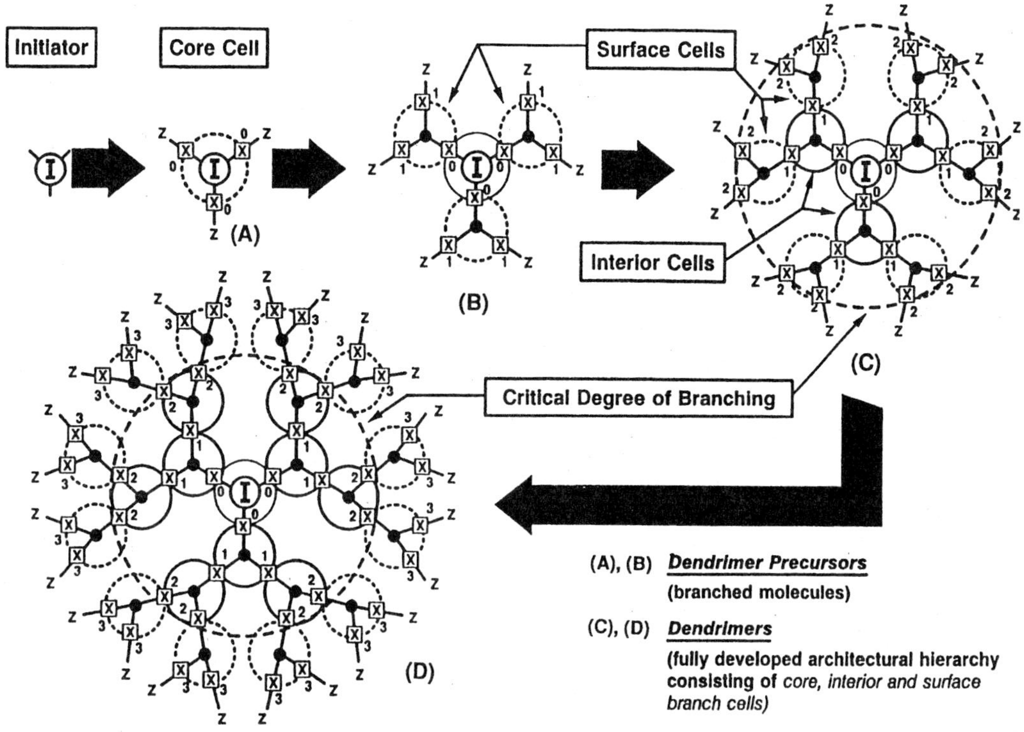

FIGURE 1. Divergent development of dendritic structure. (A and B) dendrimer precursors (i.e., simple branched molecules) (C and D) true dendrimers with fully developed architectural hierarchy consisting of core, interior, and surface branch cells. Note the critical stage of structural development in C.

The dendrimer surface group reactivity changes little until the de Gennes dense-packed stage is approached.

Dendrimer interiors and surfaces can be readily modified to possess chemically reactive or passive functional groups. Dendrimer interiors may contain carbon, nitrogen, oxygen, silicon, phosphorous, metals, or virtually any element found in the periodic table.

Generally, dendrimer surfaces can be modified to contain electrophilic or nucleophilic, hydrophobic or hydrophilic, and cationic or anionic moieties. Thus, dendrimer surface modification allows virtually every known bonding mode to integrate onto the surface of these precisely controlled nanoscopic structures. As such, dendrimers possess highly reactive surfaces that may participate in either classical (sub-nanoscopic chemistry) or novel nanoscopic conversions. In the latter, dendrimers have been validated as reactive nanoscopic building blocks suitable for constructing various megamolecular structures. These building blocks follow basic rules of chemical combination with nano-particles such as proteins, poly(nucleic acids), DNA, RNA, or other dendrimers to produce nanoscopic compounds, clusters, and assemblies.

Many of these new nanostructures exhibit commercial immuno-diagnostic reagents, nano-catalysts, magnetic resonance imaging contrast agents, nano-reactors, and nano-calibrators.[36,38–43] In conclusion, dendrimers should play a significant role in the systematic investigation of nanoscopic chemistry, architecture, and properties in biological and abiotic areas of interest.

REFERENCES

1. Tomalia, D. A.; Dewald, J. R.; Hall, M. J. et al. *Preprints 1st Society Polymer Science Japan International Polymer Conference* Kyoto, Japan, 1984; 65.
2. Tomalia, D. A.; Baker, H.; Dewald, J. R. et al. *Polymer J. (Tokyo)* **1985**, *17*, 117.
3. Tomalia, D. A.; Baker, H.; Dewald, J. R. et al. *Macromolecules*, **1986**, *19*, 2466.
4. Tomalia, D. A.; Naylor, A. M.; Goddard, W. A., III *Angew. Chem. Int. Ed. Engl.* **1990**, *29*, 138.
5. Tomalia, D. A.; Durst, H. D. *Topics in Current Chemistry Vol. 165: Supramolecular Chemistry I—Directed Synthesis and Molecular Recognition*; Weber, E., Ed.; Springer Verlag: Berlin, 1993; p 193–313.
6. Frechet, J. M. J.; Jiang, Y.; Hawker, C. J. et al. In *Proceedings of IUPAC International Symposium on Macromolecules* Seoul, Korea, 1989.
7. Hawker, C. J.; Frechet, J. M. J. *J. Am Chem. Soc.* **1990**, *112*, 7638.
8. Hawker, C. J.; Frechet, J. M. J. *J. Chem. Soc. Chem. Commun.* **1990**, 1010.
9. Hawker, C. J.; Frechet, J. M. J. *Macromolecules* **1990**, *23*, 4726.
10. Miller, T. M.; Neenan, T. X. *Chem. Mater.* **1990**, *2*, 346.
11. Bochkarev, M. N. *Organomet. Chem. USSR* **1988**, *1*, 115.
12. Kim, Y. H.; Webster, O. W. *Polym. Prepr. Am. Chem. Soc.* **1988**, *29*, 310.
13. Kim, Y. H.; Webster, O. H. *J. Am. Chem. Soc.* **1990**, *112*, 4592.
14. Hawker, C. J.; Lee, R.; Fréchet, J. M. J. *J. Am. Chem. Soc.* **1991**, *113*, 4583.
15. Mathias, L. J.; Carothers, T. W. *Polym. Prepr. Am. Chem. Soc.* **1991**, *32*, 633.

16. Mathias, L. J.; Carothers, T. W. *J. Am. Chem. Soc.* **1991**, *113*, 4043.
17. Denkewalter, R. G.; Kole, J. F.; Lukasavage, W. J. U.S. Patent 4 410 688, 1983; *Chem. Abstr. 11*, 103907.
18. Hall, H.; Padias, A.; McConnell, R. et al. *J. Org. Chem.* **1987**, *52*, 5305.
19. Newkome, G. R.; Yao, Z.-Q.; Baker, G. R. et al. *J. Org. Chem.* **1985**, *50*, 2003.
20. Tomalia, D. A.; Dewald, J. R. U. S. Patent 4 507 466, 1985.
21. Tomalia, D. A.; Dewald, J. R. U. S. Patent 4 558 120, 1985.
22. Tomalia, D. A.; Dewald, J. R. U. S. Patent 4 568 737, 1986.
23. Tomalia, D. A.; Dewald, J. R. U. S. Patent 4 587 329, 1986.
24. Tomalia, D. A.; Dewald, J. R. U. S. Patent 4 631 337, 1986.
25. Tomalia, D. A.; Dewald, J. R. U. S. Patent 4 694 064, 1987.
26. Tomalia, D. A.; Dewald, J. R. U. S. Patent 4 713 975, 1987.
27. Tomalia, D. A.; Dewald, J. R. U. S. Patent 4 737 550, 1988.
28. Tomalia, D. A.; Dewald, J. R. U. S. Patent 4 871 779, 1989.
29. Tomalia, D. A.; Dewald, J. R. U. S. Patent 4 857 599, 1989.
30. Buhleier, F.; Wehner, W.; Vogtle, F. *Synthesis* **1978**, 155.
31. Dvornic, P. R.; Tomalia, D. A. *Chem. Br.* **1994**, *30*, 641.
32. Meltzer, A. D.; Tirrell, D. A.; Jones, A. A. et al. *Macromolecules* **1992**, *25*, 4549.
33. Ottaviani, M. F.; Bossmann, S.; Turro, N. J. et al. *J. Am. Chem. Soc.* **1994**, *116*, 661.
34. Gopidas, K. R.; Leheny, A. R.; Caminati, G. et al. *J. Am. Chem. Soc.* **1991**, *113*, 7335.
34a. Tomalia, D. A. *Proceedings of SFC 91, 4e Congres de la Societé Française de Chimie* Strasbourg, France, 1991; p 31.
35. Uppuluri, S.; Dvornic, P. R.; Tomalia, D. A. unpublished results.
36. Haensler, J.; Szoka, F. C., Jr. *Bioconjugate Chem.* **1993**, *4*, 372.
37. Smith, P. B.; Martin, S. J.; Hall, M. J. et al. In *Applied Polymer Analysis and Characterization*; Mitchell, J., Jr., Ed.; Hanser: München/New York, 1987; p 357.
38. Singh, P.; Moll, F., III; Lin, S. H. et al. *Clin. Chem.* **1994**, *40*, 1845.
39. Knapen, J. W. J.; van der Made, A. W.; de Wilde, J. C. et al. *Nature* **1994**, *372*, 659.
40. Tomalia, D. A.; Dvornic, P. R. *Nature* **1994**, *372*, 617.
41. Wiener, E. C.; Brechbiel, M. W.; Brothers, H. et al. *Magnetic Resonance in Medicine* **1994**, *31*, 1.
42. Turro, N. J.; Barton, J. K.; Tomalia, D. A. *Acc. Chem. Res.* **1991**, *24*, 332.
43. Dandliker, P. J.; Diederich, F.; Gross, M. et al. *Angew. Chem. Int. Ed. Engl.* **1994**, *33*, 1739.

DENDRITIC POLYRADICALS

Suchada Rajca and Andrzej Rajca*
Department of Chemistry
University of Nebraska

Properly designed dendritic macromolecules may be viewed as finite fragments of Bethe lattice, which is one of the important models for phenomena related to condensed matter physics and materials science.[2] For example, functionalization with ferromagnetically coupled "unpaired" electrons would allow us to examine some fundamental questions concerning magnetic phenomena in nanometer-size magnetic particles.[3–6] 1,3-Connected triarylmethyl-based polyradicals, which are derived from almost century-old Gomberg triphenylmethyl radical and homologous Schlenk hydrocarbon diradical,[7–9] were selected as the building blocks. Our earlier studies established that an intramolecular ferromagnetic coupling was maintained in a series of polyradicals with up to 10 "unpaired" electrons (e.g., high-spin ground state with spin, $S = 5$ for a decaradical) and polyradicals could be handled in solution at low temperature and inert atmosphere.[10–19]

Here we summarize the synthesis leading to the dendritic polyradical with 31 triarylmethyl sites for "unpaired" electrons.[5]

*Author to whom correspondence should be addressed.

PREPARATION

Synthesis of Polyethers

Preparation of 1,3-connected polyarylmethyl polyethers is based upon a convergent synthetic route, in which branched polymeric arms are built "inward" toward the central core.[1,5,20] We employ three steps: (a) Br/Li exchange, (b) addition of aryllithium to carbonyl compound, and (c) conversion of the alcohol to the corresponding methyl ether. Sequential iterations of steps (a)–(c) yield dendritic polyethers.[5,20]

Generation of Polyradicals

Polyradicals are generated from the corresponding polyethers using the *carbanion method*.[11,16] Unlike the preceding stepwise synthesis of the polyethers, generation of polyradical is carried at all triarylmethyl sites in one pot and without purification.

PROPERTIES

Dendritic polyethers and their synthetic precursors are characterized by Fast Atom Bombardment Mass Spectrometry (FAB MS). Because of the polymeric nature of these dendritic molecules, C–H analyses are similar for all compounds, so they are of little value for structure determination.

Propeller isomerism of polyethers (and their derivatives) adds difficulty to their characterization by NMR spectroscopy.[21] Their NMR spectra are useful only as "fingerprint."[5]

Typically, ESR spectra of sterically congested dendritic polyradicals with 15 and 31 sites are not sufficiently resolved to allow meaningful interpretation.

Magnetization data (SQUID magnetometer) for polyradicals with 15 and 31 sites are interpreted as a mixture of non-interacting spin systems with lower than expected spin values. Such mixtures could form when the spin coupling between a pair of the triarylmethyl sites is interrupted; for example, when at one or more inner sites an "unpaired" electron is not generated. We refer to a site devoid of an "unpaired" electron as a defect.

The key problem with the dendritic connectivity when interaction between many sites has to be maintained is that for any pair of sites there is only one coupling pathway connecting them; therefore, even one defect at an inner site may interrupt the coupling by "dividing" the macromolecule into smaller systems which are uncoupled (or weakly coupled). One of the solutions to this problem are macrocyclic molecules, where two or more pathways are available for coupling.[22]

REFERENCES

1. Rajca, A. *J. Org. Chem.* **1991**, *56*, 2557.
2. Stauffer, D. *Introduction to Percolation Theory* Taylor and Francis: London, U.K., 1985.
3. Rajca, A. *Chem. Rev.* **1994**, *94*, 871.
4. Rajca, A. *Adv. Mater.* **1994**, *6*, 605. (Research News).
5. Rajca, A.; Utamapanya, S. *J. Am. Chem. Soc.* **1993**, *115*, 10688.
6. Rajca, A.; Utamapanya, S. *Liq. Cryst. Mol. Cryst.* **1993**, *232*, 305.
7. Gomberg, M. *J. Am. Chem. Soc.* **1990**, *22*, 757.
8. Gomberg, M. *Ber. Dtsch. Chem. Ges.* **1900**, *33*, 3150.
9. Schlenk, W.; Brauns, M. *Ber. Dtsch. Chem. Ges.* **1915**, *48*, 661.
10. Rajca, A. *J. Am. Chem. Soc.* **1990**, *112*, 5889.
11. Rajca, A. *J. Am. Chem. Soc.* **1990**, *112*, 5890.
12. Rajca, A.; Utamapanya, S.; Xu, J. *J. Am. Chem. Soc.* **1991**, *113*, 9235.
13. Utamapanya, S.; Rajca, A. *J. Am. Chem. Soc.* **1991**, *113*, 9242.
14. Rajca, A.; Utamapanya, S.; Thayumanavan, S. *J. Am. Chem. Soc.* **1992**, *114*, 1884.
15. Rajca, A.; Utamapanya, S. *J. Org. Chem.* **1992**, *57*, 1760.
16. Rajca. A.; Utamapanya, S. *J. Am. Chem. Soc.* **1993**, *115*, 2396.
17. Utamapanya, S.; Kakegawa, H.; Bryant, L.; Rajca, A. *Chem. Mater.* **1993**, *5*, 1053.
18. Rajca, A.; Utamapanya, S.; Smithhisler, D. J. *J. Org. Chem.* **1993**, *58*, 5650.
19. Rajca, A. *Advances in Dendritic Macromolecules*; Newkome, G. R., Ed.; JAI: Greenwich, U.K., 1994; Vol. I. P 133.
20. Rajca, A.; Janicki, S. *J. Org. Chem.* **1994**, *59*, 7099.
21. Mislow, K. *Acc. Chem. Res.* **1976**, *9*, 26.
22. Rajca, A.; Rajca, S.; Padmakumar, R. *Angew. Chem.* **1994**, *33*, 2091.

DENTAL ADHESIVES (for Alloys and Ceramics)

Hideo Matsumura
Department of Fixed Prosthodontics
School of Dentistry
Nagasaki University

The purpose of dental treatment is to recover the anatomic tooth structure and oral function associated with mastication and pronunciation. Once hard tissue is damaged by dental caries, natural healing usually cannot be expected. Various materials, such as polymers, alloys, ceramics, and composites have been used to restore or replace tooth structure. It is vitally important for these materials to be bonded securely to the tooth surface by means of appropriate bonding systems. Lack of bonding between the materials and tooth will induce secondary dental caries or detachment of restorations. Recent dental adhesive techniques are based on chemomechanical bonding between the adhesive and adherend.

Surface conditioning before applying adhesive effectively increases the surface bonding area and enhances the strength of the bonding agent. Tooth enamel is composed of hydroxyapatite ($Ca_{10}(PO_4)_6(OH_2)$). The enamel is etched with an aqueous solution of phosphoric acid.[1] Most current enamel bonding systems employ phosphoric acid in various concentrations and viscosities as an etchant.

Dental casting alloys are categorized into base metals and noble metals. Base metals include cobalt-chromium (Co–Cr), nickel–chromium (Ni–Cr), and titanium (Ti) alloys, while noble metals include gold (Au), silver (Ag) and palladium (Pd) alloys. Although numerous mechanical methods have been proposed, air-abrading with alumina is the most common way to prepare the alloy surface. Electrolytical etching creates micromechanical retention on the alloy surface.[2] Other types of surface modification include electroplating, silica-coating, and ion-coating.[3–5]

Dental ceramics consist mainly of metal oxides of silicon, aluminum, potassium, and sodium. The ceramic surface is roughened extraorally by etching with aqueous solution of hydrofluoric acid.[6]

After surface roughening, the adherend surface is treated with a primer for chemical bonding. Because the composition of dental adhesives is based on methacrylate monomers, the primers for chemical bonding also contain functional monomers. The functional monomer usually consists of a polymerizable methacrylate or vinyl group, hydrophobic intermediate spacer, and a functional group capable of bonding to the adherend surface. Functional groups effective for bonding base metal alloys are the acid anhydride, carboxylic, and phosphoric groups. Thiol derivatives are used to prime noble metal alloys. The dental ceramic surface is primed with silane couplers activated in acetic acid,[7] carboxylic monomers,[8,9] ferric chloride, and/or tri-*n*-butylborane derivative (TBB).[10] Silane couplers have traditionally been used for surface modification of filler particles included in the dental composites.[11]

After chemomechanical surface preparation, the restorative material is bonded to the tooth structure with autopolymerizing adhesive resin. One prerequisite for dental adhesives is an ability to cure in the mouth; this is why methacrylate-based polymer materials are used in dentistry. Initiation systems are benzoyl peroxide (BPO)-tertiary amine or TBB.[12] Base monomers are methyl methacrylate (MMA) or bifunctional methacrylate monomers. Optical opacity, radiopacity, fluoride releasing, etc., are additional requirements for dental adhesives.[13,14] Various inorganic and metal compounds are incorporated to achieve these properties.

APPLICATIONS

In selected cases, restorative materials are bonded directly to the tooth surface with adhesive resins. This minimizes the reduction of healthy tooth substrate in dental treatment. Adhesive resins are also used for bonding orthodontic brackets, amalgam restorative material and many other metal and ceramic materials.[15,16]

REFERENCES

1. Buonocore, M. G. *J. Dent. Res.* **1955**, *34*, 849.
2. Thompson, V. P.; Del Castillo, E.; Livaditis, G. J. *J. Prosthet. Dent.* **1983**, *50*, 771.
3. Yamashita, A.; Kondo, Y.; Fujita, M. *J. Jpn. Prosthodont. Soc.* **1984**, *28*, 1023.
4. Musil, R.; Tiller, H. J. *Dent. Labor* **1984**, *32*, 1155.
5. Tanaka, T.; Hirano, M.; Kawahara, M.; Matsumura, H.; Atsuta, M. *J. Dent. Res.* **1988**, *67*, 1376.
6. Horn, K. R. *Dent. Clin. North Amer.* **1983**, *27*, 671.

7. Paffenbarger, G. C.; Sweeney, W. T.; Bowen, R. L. *J. Am. Dent. Assoc.* **1967**, *74*, 1018.
8. Matsumura, H.; Tanaka, T.; Atsuta, M.; Nakamura, M.; Nakabayashi, N. *J. Jpn. Prosthodont. Soc.* **1987a**, *31*, 1494.
9. Maeda, M.; Mogi, M.; Miura, F.; Nakabayashi, N. *J. Jpn. Orthodont. Soc.* **1987**, *46*, 370.
10. Matsumura, H.; Nakamura, M.; Nakabayashi, N.; Tanaka, T.; Atsuta, M. *Dent. Mater. J.* **1987b**, *6*, 135.
11. Bowen, R. L. *J. Am. Dent. Assoc.* **1963**, *66*, 57.
12. Nakabayashi, N.; Masuhara, E.; Mochida, E.; Ohmori, I. *J. Biomed. Mater. Res.* **1978**, *12*, 149.
13. Yoshida, K.; Taira, Y.; Matsumura, H.; Atsuta, M. *J. Prosthet-Dent.* **1993**, *69*, 357.
14. Matsumura, H.; Sueyoshi, M.; Atsuta, M. *J. Dent. Res.* **1992**, *71*, 2.
15. Miura, F.; Nakagawa, K.; Masuhara, E. *Am. J. Orthodont.* **1971**, *59*, 350.
16. Varga, J.; Matsumura, H.; Masuhara, E. *Dent. Mater. J.* **1986**, *5*, 158.

Dental Materials

See: *Dental Adhesives (for Alloys and Ceramics)*
Dental Polymers (Overview)
Dental Polymers (Glass-Ionomer Cements)
Dental Polymers (Hydrogels)
Dental Sealants
Fluorosilicones (Monomer Synthesis and Applications)
Radiopaque Polymers

DENTAL POLYMERS (Overview)

Lars-Åke Lindén
Polymer Research Group
Department of Dental Biomaterials Science
Karolinska Institute
Royal Institute of Medicine

Polymers in dentistry are used as composite restorative materials, cements and adhesives, cavity liners, and as protective sealants for pits and fissures. Numerous manufacturers and some dentists now advocate the use of composite resins for amalgam replacement. So-called "posterior composites" have gained full acceptance by the American Dental Association.

In addition to the restoration of cavities, polymers are used for the construction of prosthetic devices (such as denture base materials, denture relines, crown and bridge resins, dental impressions and duplicating materials, and plastic teeth) and have also been used to repair anterior teeth, to cement orthodontic brackets, to cover staining, and to repair and combat erosion.

Other devices utilizing plastics include patterns for metal inlays and partial denture-framework castings, contoured impressions trays, orthodontic and periodontic splints, temporary space maintainers and bite plates, obturators for cleft palates and oral implants.

Polymeric materials (plastics) employed in dentistry must meet physical, chemical, biological (enzymatic, bacteriological), and aesthetic requirements (not always fulfilled by currently available products). These requirements include adequate strength; resilience; abrasion resistance; dimensional stability during processing and subsequent use; translucency or transparency to enable the appliance to match the appearance of the mouth tissue that is replaced; good color stability; resistance to mouth (oral) fluids, saliva, and blood, or other substances with which they are in a direct contact; tissue tolerance; low allergenity, toxicity, mutagenicity, cancerogenic responses; and ease of processing into dental appliance. These materials should also be easy to use and inexpensive.

HISTORY

Acrylics processed by compression-molding techniques give dentures which are flexible and moderately strong. Many other polymers such as vinyl-acrylic, polystyrenes, polycarbonates, and polysulfones can be injection-molded to yield dentures with outstanding toughness, high fatigue strength, and low water absorption. Note that epoxy resins are clinically unsatisfactory because of handling problems and dimensional and color instability. Developments in the polymerization (curing) of di- and tri-functional acrylates and methacrylates to highly crosslinked solid (unsoluble) polymers allowed them to be used as restorative materials in early 1950. This "cold curing" polymerization was carried out with redox initiators in an ambient temperature.[1] Significant progress in the application of filling resins was made in 1962 when Bowen introduced a new monomer bis-phenol-*A*-diglycidylether dimethacrylate [2,2′-bis(p-2′-hydroxy-3′-methacryloxypropoxyphenyl)propane] (Bis-GMA) with a ceramic filler added.[2] Application of the silane as a coupling agent between the two components increased the strength characteristics. Most commercially available "composite resins" are based on the Bis-GMA, however many new more sophisticated monomers were lately introduced into these products.

A new class of composite resins based on glass-ionomer cements, invented in 1969 and clinically tested in the early 1970s, are widely used in restorative dentistry.[3,4]

The use of photopolymerization for photocuring of dental resins in the mid-1970s [the first UV-light cured resin was Novalight (L.D. Caulk)] was introduced on the market,[5] and soon after this the ICI Company presented the visible-light-cured polyurethane resin, Fotofil (J&J).

RESTORATIVE MATERIALS[6–9]

Polymeric Composite Filling Materials[6–16]

Composition

The newest polymeric restorative composites are based on di- and tri-functional monomers, mainly modified dimethacrylates. The most common are thermosetting dimethacrylate: 2,2-bis[p-(2′-hydroxy-3′-methacryloxypropoxy)phenyl]propane as (bis-GMA) and triethyleneglycol dimethacrylate (TEGDMA) (**Structures 1 and 2**) used separately or in a mixture.[12,17–21]

The addition of TEGDMA to bis-GMA causes higher contraction at the same conversion, because the concentration of methacrylate groups in the uncured resin is higher.

Free-radical polymerization of resin-based dental materials is initiated chemically by redox systems (cold curing) or photochemically (photocuring, light curing) by visible light (400–500 nm) with photoinitiators. Complete polymerization usually takes place within a few minutes.

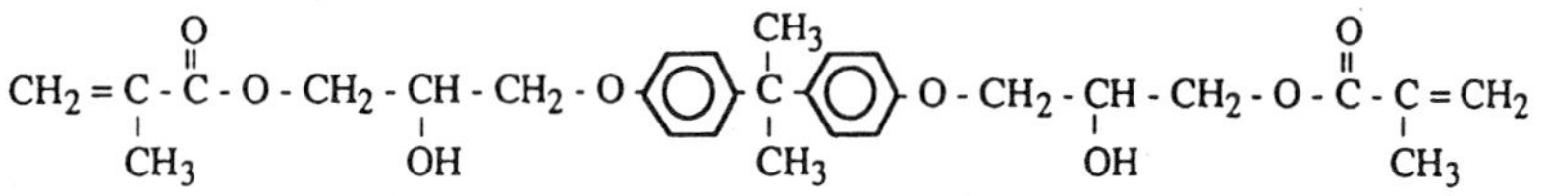

1

(BIS-GMA)

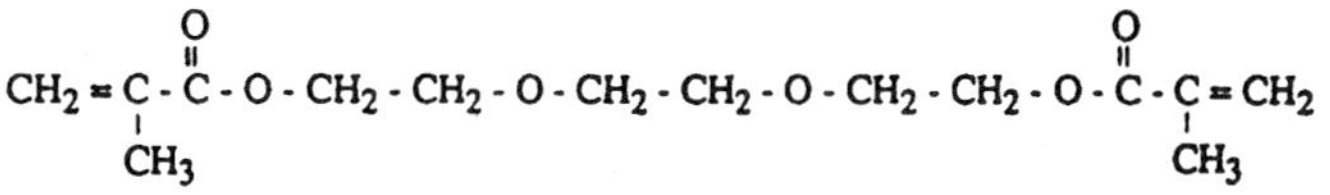

2

(TEGDMA)

Initiator Systems

Redox-initiators systems are based on benzoyl peroxide-tertiary aromatic amines, hydroperoxides-thioureas, peroxides with ascorbic acid or derivatives, or benzoyl peroxide-amine with synergestic activators, such as polythiols.[22–28]

Initiator Systems

Hybrid systems based on camphorquinone (which has a maximum absorption spectrum at 468 nm) are used as photo-initiators and *N,N*-dimethyl-*p*-toluidine or 2-(*N,N*-dimethylamino) ethyl methacrylate are used as coinitiators.[20,29–32]

Toxic Considerations

Unreacted monomer and other low-molecular (nonpolymerized) components gradually diffuse (leach) out of the matrix because they are not compatible with it.[33–36] Some of these substances may be toxic to pulp tissue and surrounding mucous membranes, and they may also be allergens. All methacrylic monomers are toxic (much less than acrylic); amines, with their high toxicity, are also mutagenic and can cause allergic reactions.[37] Polymers, especially crosslinked ones, are nontoxic but their allergenic and mutagenic effects are still not well-known. Many patients and dental personnel show allergic reactions and hypersensitivity in connection with restorative dentistry using polymeric materials.[38–41]

Disadvantages

Polymerization shrinkage (contraction) is a critical limitation for dental composite resins: It results in stresses in restored teeth and also within the resin itself.[42–45] Stresses from shrinkage can cause clinical problems such as postoperative pain, fracture of the tooth, and opening of restoration margins that can result in microleakage and recurrent (secondary) caries.[46–49] The contraction arises from a number of causes: the major factor relates to the fact that the monomers are located at van der Waals' distances from one another, whereas in the corresponding polymer they've moved to within covalent distances. Other factors that influence the amount of shrinkage during polymerization are the change in entropy and the relative free volume in the monomer and polymer.

To obtain low polymerization contraction in conventional dental resin materials, the amount of polymerization methacrylate groups should be limited.[50,51] Another way to minimize shrinkage would be to limit the conversion. However, high conversion is important for good mechanical properties,[52] and for better biocompability.[53]

Resin systems with zero contraction or slight expansion (up to 4%) upon polymerization can be based on expanding monomers.[54]

Addition of ammonia-modified montmorillonite expanding filler particles facilitates constant-volume polymerization.[55]

There are two serious disadvantages of polymeric filling composite materials: the poor resistance of composite resins to mechanical stress (high load, flexural tension, and grinding wear), and limitations due to chemical and biological degradation for their use as posterior restorations. Polymeric materials are also exposed to often rapid thermal changes (from –7 to +70°C). Typical damage from such forces is the formation of microcracks, cavities, and a rough and corrugated surface. This is the result of osmotic pressure built up at the matrix-filler interface due to hydrolytic degradation of the filler.[56] Saliva and fluid foodstuffs may also cause hydrodynamic damage to the surface and so remove material. Such damage surfaces serve as retentive areas for food debris, bacterial plaque, and discolorations.

Cavity Varnishes (Dentin Bonding)[57–68]

The basic purpose of applying varnish (dentin bonding) to the cavity walls is to seal the exposed dentinal tubules and protect the pulp from the irritation of chemicals in the filling materials that may penetrate through the odontoblast processes.[57–60]

Cavity varnishes consist principally of a natural rubber or synthetic polymeric resins dissolved in an organic solvent such as acetone, chloroform, or ether. When the varnish is applied to the cavity preparation, the solvent evaporates and leaves a thin resin film on the surface.

A hydrophilic monomer such as 2-hydroxyethyl methacrylate (HEMA) is a common component of dentin bonding systems. When it is used in conjunction with 4-methacryloxyethyl trimelitate anhydride (4-META)[61–64] or 2-(methacryloxy) ethyl phenyl hydrogen phosphate (phenyl-P),[66,67] or derivatives of phenyl-P,[68] the tensile bond strength increases with time after application. The dominant mechanisms in dentin bonding are diffusion followed by entanglements and micromechanical retention. The components of an efficient bonding system must penetrate the dentin and then polymerize *in situ*.[64,65]

PROSTHETIC MATERIALS[6–8,10,69]

Denture Bases

Nearly all dentures are made of acrylic (methacrylic) resins. Dentures require accurate fit and natural appearance.

A variety of thermoplastic materials such as poly(methylmethacrylate), vinyl–acrylic copolymers, polystyrene, polycarbonates, and polysulfones can be injection-molded. In this processing method, the molded space is filled by injecting the softened resin under pressure. Upon cooling, the thermoplastic resin solidifies in the mold.

A compression-molding process is employed to make the denture base.

Polymerization can also be carried out at room temperature called self-curing, auto-curing, or cold-curing by employing a suitable redox initiator–accelerator system like benzoyl peroxide (2 wt%)–*N*,*N*-dihydroxyethyl-*p*-toluidine or *p*-*N*,*N*-dimethylaminophenethanol (0.4–0.9 wt%).

A serious disadvantage in the processing of denture bases is the polymerization shrinkage of the methacrylate monomer–polymethacrylate dough by 7–10%, whereas shrinkage of pure methyl methacrylate is 21%, and bis-GMA is 5%.[16] The net linear shrinkage of manufactured denture in oral cavity is 0.3–0.4% and this is clinically insignificant, since the tissue on which the denture rests adjusts to such changes.

Denture Reliners

A denture can be readapted to the changing contours of soft tissue by relining it with rigid or resilient materials. Resilient liners reduce the impact of the hard denture bases on soft oral tissues. These materials should provide slow total recovery from deformation, not lose their cushioning effect, possess good wettability, absorb minimum amounts of fluids, not support bacterial or fungal growth, and be easily cleaned.[71–73]

These materials consist of plasticized acrylic (with aromatic plasticizer) or silicones to facilitate usage of the denture.

Crown and Bridge Resins

These materials are used as interim tooth coverage during fabrication of permanent prosthetics. They maintain correct biting relationship, stop drifting of teeth, and protect the prepared tooth against fracture and pulpal irritation. The commercially available materials are based on different alkyl methacrylates.[74]

Plastic Teeth

Plastic acrylic denture teeth are manufactured by injection or transfer molding processes. Acrylic teeth have higher strength than porcelain teeth and break less readily than ceramic teeth. However, they have less resistance to cold flow and have higher water absorption than porcelain teeth. The main disadvantages of acrylic teeth as compared to porcelain teeth are higher wear rate and ultimate loss of occlusional relationship. Plastic teeth exhibit less wear than the supporting structure of the denture prothesis.

Dental Impression and Duplicating Materials[6–8]

Impression materials accurately register to reproduce the dimensions, surface details, and relationships of teeth and soft oral tissues. These materials include rigid gels of reversible hydrogen-bonding type (e.g., agar, which is a biopolymer of mucopolysaccharide (sulfonic acid ester of a galactan complex), irreversible alginate (sodium or potassium salt of anhydro-β-D-mannuronic acid), hydrocolloids, and elastomers such as silicones, polysulfides, or polyethers. Duplicate materials are employed to prepare duplicates of the original cast, as for partial dentures.

Vinyl polysiloxane is considered the most accurate impression material.[75–77]

FUTURE TRENDS

Increasing world-wide demand for dental care will result in a great need for simple and cheap dental applications of polymeric materials. To achieve this, we must stimulate the polymer community to develop joint research programs with dentists; interdisciplinary research will give rapid progress and continued developement to polymeric dental technology that will meet the challenge of the 21st century.

REFERENCES

1. Czapp, A. E.; Schebel, E.; Goelz, A. Ger. Patent 975072, 1941.
2. Bowen, R. L. *J. Am. Dent. Assoc.* **1963**, *66*, 57.
3. Wilson, A. D.; Kent, B. E. *J. Appl. Chem. Biotechnol.* **1971**, *21*, 313.
4. McLean, J. W.; Wilson, A. D. *Aust. Dent. Mater.* **1977**, *22*, 120.
5. Ehrnfors, L. *Swed. Dent. J.* **1983**, *18*, suppl.
6. Brauer, G. M.; Antonucci, J. M. In *Encyclopedia of Polymer Science and Engineering*: Wiley: New York, NY, 1986; Vol. 4, p. 698
7. O'Brien, W. K. *Dental Materials Properties and Selection*; Quintessence: Chicago, IL, 1989. 158, 171.
8. Tesk, J. A.; Antonucci, J. M. et al. In *Kirk-Othmer Encyclopedia of Chemical Technology*; Wiley: New York, NY, 1993; Vol. 7, p 946.
9. Lindén, L.Å. In *Radiation Curing in Polymer Science and Technology*; Fouassier, J. P.; Rabek, J. F., Eds.; Elsevier: London, U.K., 1993; Vol. IV, p 387.
10. Ruyter, I. E. *Acta Odont. Scand.* **1982**, *40*, 359.
11. Lutz, F.; Stecos, J. C. et al. *Dent. Clin. North Am.* **1983**, *27*, 699.
12. McCabe, J. F. *Br. Dent. J.* **1984**, *157*, 440.
13. Ruyter, I. E.; Oysaed, H. *Crit. Rev. Biocompatibil.* **1988**, *4*, 274.
14. Rees, J. S.; Jacobsen, P. H. *Rest. Dent.* **1989**, *5*, 91.
15. Lindén, L.Å. *Proc. Indian Acad. Sci.* **1993**, *105*, 505.
16. Talib, R.; Nihon, J. *Univ. Sch. Dent.* **1993**, *35*, 161.
17. Brauer, G. M.; Dulik, D. M. et al. *J. Dent. Res.* **1981**, *60*, 1966.
18. Ruyter, I. E.; Sjovik, I. J. *Dent. Mater.* **1987**, *3*, 315.
19. Peutzfeld, A.; Asmussen, E. *Acta Odont. Scand.* **1989**, *47*, 229.
20. Yoshida, K.; Greener, E. H. *Dent. Mater.* **1993**, *9*, 246.
21. Venhoven, B. A. M.; De Gee, A. J. et al. *Biomaterials* **1993**, *14*, 871.
22. Asmussen, E. *Acta Odont. Scand.* **1975**, *33*, 129.
23. Antonucci, J. M.; Grams, C. L.; Termini, D. J. *J. Dent. Res.* **1979**, *58*, 1887.
24. Asmussen, E. *Acta Odont. Scand.* **1981**, *39*, 291.
25. Asmussen, E. *Scand. J. Dent. Res.* **1982**, *90*, 490.
26. Brauer, G. M.; Argentar, H. *Am. Chem. Soc. Symp. Ser.* **1983**, *211*, 357.
27. Antonucci, J. M.; Stansbury, J. M.; Dudderar, D. J. *J. Dent. Res.* **1982**, *61*, 270.
28. Ferracane, J. L.; Greener, H. E. *J. Dent. Res.* **1984**, *63*, 1093.
29. Williams, D. F.; Cunningham, J. *Materials in Clinical Dentistry*; Oxford University: Oxford, U.K., 1979; p 175.
30. Brauer, G. M.; Dianne, M. et al. *J. Dent. Res.* **1979**, *34*, 1994.
31. Antonucci, J. M.; Peckoo, R. J. et al. *J. Dent. Res.* **1981**, *60*, 1325.
32. Taira, M.; Urabe, H. et al. *J. Dent. Res.* **1988**, *67*, 24.
33. Ruyter, I. E.; Svenden, S. A. *Acta Odont. Scand.* **1977**, *36*, 75.

34. Thompson, L. R.; Miller, E. G.; Bowles, W. H. *J. Dent. Res.* **1982**, *61*, 989.
35. Tanaka, K.; Tahira, M. et al. *J. Oral Rehabil.* **1991**, *18*, 353.
36. Söhoel, H.; Gjerdet, N. R. et al. *Scand. J. Dent. Res.* **1994**, *102*, 126.
37. Björkner, R.; Niklasson, B.; Persson, K. *Contact Dermatitis* **1984**, *10*, 286.
38. Davidson, W. M.; Sheinis, E. M.; Shepherd, S. R. *Am. J. Orthod.* **1982**, *82*, 502.
39. Hentsen-Pettersen, A.; Örstavik, D.; Wenneberg, A. *Endod. Dent. Traumatol.* **1985**, *1*, 61.
40. Jacobsen, N.; Hentsen-Pettersen, A. *Eur. J. Orthod.* **1989**, *11*, 254.
41. Jacobsen, N.; Aasenden, R.; Hentsen-Pettersen, A. *Comm. Dent., Oral Epidemiol.* **1991**, *19*, 155.
42. Bowen, R. L.; Rapson, J. E.; Dickson, G. *J. Dent. Res.* **1982**, *61*, 654.
43. Davidson, C. L.; De Gee, A. J. *J. Res. Dent.* **1984**, *63*, 146.
44. Fusayama, T. *Quint. Int.* **1993**, *24*, 225.
45. Abdul-Hag Suliman, B. D. S.; Boyer, D. B.; Lakes, R. S. *J. Frosthet. Dent.* **1994**, *71*, 7.
46. Bausch, J. R.; De Lange, K. et al. *J. Prosthet. Dent.* **1982**, *48*, 59.
47. Eick, J. D.; Welch, F. H. *Quint. Int.* **1986**, *17*, 103.
48. Kemp-Scholte, C. M.; Davidson, C. L. *J. Dent. Res.* **1988**, *67*, 841.
49. Tortenson, B.; Oden, A. *Dent. Mater.* **1989**, *5*, 218.
50. Loshaek, S.; Fox, T. G. *J. Am. Chem. Soc.* **1953**, *75*, 3544.
51. Patel, M. P.; Braden, M.; Davy, K. W. M. *Biomaterials* **1987**, *8*, 54.
52. Eliades, G. C.; Vougiouklaiks, G. J.; Caputo, A. A. *Dent. Mater.* **1987**, *3*, 19.
53. Rathbun, M. A.; Craig, R. G. et al. *J. Biomed. Mater. Res.* **1991**, *25*, 443.
54. Sadhir, R. L.; Luck, R. M. *Expanding Monomers*; CRC: Boca Raton, FL, 1992.
55. Liu, C. F.; Collard, S. M.; Armeniades, C. D. *Am. J. Dent.* **1990**, *3*, 44.
56. Söderholm, K. J.; Zigan, M. et al. *J. Dent. Res.* **1984**, *63*, 1248.
57. Causton, B. E. *Br. Dent. J.* **1984**, *156*, 93.
58. Suzuki, T.; Finger, W. *Dent. Mater.* **1988**, *4*, 379.
59. Setcos, J. C. *Am. J. Dent.* **1988**, *1*, 173.
60. Van Meerbeek, B.; Inokoshi, S. et al. *J. Dent. Res.* **1992**, *71*, 1530.
61. Nakabayashi, N.; Takarada, K. *Dent. Mater.* **1992**, *9*, 125.
62. Nakabayashi, N.; Watanabe, A.; Gendusa, N. *J. Dent. Mater.* **1992**, *8*, 259.
63. Takayama, M.; Kashibuchi, N. et al. *J. Jpn. Dent. Appar. Mater.* **1978**, *19*, 179.
64. Nakabayashi, N. *Oper. Dent.* **1992**, *5*, 125.
65. Pashley, D. H.; Horner, J. A.; Brewer, P. D. *Oper. Dent.* **1992**, *5*, 137.
66. Yamaguchi, J. *J. Jpn. Dent. Mater.* **1986**, *5*, 144.
67. Watanabe, I.; Nakabayashi, N. *Dent. Res.* **1993**, *24*, 335.
68. Nakabayashi, N.; Kanda, K. *Kobushi Ronbunshu* **1989**, *45*, 91.
69. *Scientific Aspects of Dental Materials*; Von Frauenhofer, J. A., Ed.; Butterworth: London, U.K., 1975.
70. Smith, L. T.; Powers, J. M. *Am. J. Dent.* **1992**, *5*, 140.
71. Dootz, E. R.; Koran, A.; Craig, R. G. *J. Prosthet. Dent.* **1992**, *67*, 707.
72. Kawano, F.; Dootz, E. R. et al. *J. Prosthet. Dent.* **1992**, *68*, 368.
73. Sinobad, D.; Murphy, W. M. et al. *J. Oral Rehabil.* **1992**, *19*, 151.
74. Ruyter, I. E.; Sjovik, K. *Dent. Mater.* **1987**, *3*, 315.
75. Council on Dental Materials, Instruments and Equipment. *J. Am. Dent. Assoc.* **1990**, *120*, 595.
76. Chee, W. W. L.; Donovan, T. E. *Prosthet. Dent.* **1992**, *68*, 728.
77. De Wald, J. P.; Nakajima, H.; Bell, L. J. *J. Prosthet. Dent.* **1994**, *71*, 394.

DENTAL POLYMERS (Glass-Ionomer Cements)

Lars-Åke Lindén
Polymer Research Group
Department of Dental Biomaterials Science
Karolinska Institute
Royal Institute of Medicine

There are a number of problems in using polymeric composite resins, a major one being its inability to bond to dentine as strongly as to enamel, even with the use of intermediary bonding systems. Attempts to bond resin-based materials to dentine have not been wholly successful, and their adhesive forces do not counteract the inherent polymerization contraction of the composite resin. Consequently, there is no polymeric resin-based material that can provide a perfect seal in a cavity with margins consisting of both enamel and dentin. The new glass-ionomer materials (more recently termed glass-polyalkenoate cements) have been developed to solve problems existing in application of polymeric composite resins. The new materials have found wide application in restorative dentistry.

As for other polymeric materials (plastics) employed in dentistry, glass-ionomer cements must meet strict physical, chemical, biological, and aesthetic requirements. These requirements include adequate strength, resilience, abrasion resistance, dimensional stability during processing and subsequent use. Translucency or transparency is also required to enable the appliance to match the appearance of the mouth tissue that is replaced. Good color stability, resistance to mouth (oral) fluids, saliva and blood flood, or other substances with which they are in direct contact, tissue tolerance, low allergenity, toxicity, mutagenity, carcinogenic responses, and ease of processing into dental appliances are additional requirements. These materials should also be inexpensive and easy to use.

NOMENCLATURE FOR GLASS-IONOMER DENTAL CEMENTS[1]

The term "glass-ionomer cement" is exclusively reserved for a material consisting of an acid-decomposible glass and water-soluble acid, which sets by a neutralization reaction. It covers two sub-groups: the glass-polyalkenoates and the glass-polyphosphonates.

POLYALKENOATE CEMENTS

Glass-ionomer (polyalkenoate) cements are formed by the setting reaction of poly(alkenoic acid)s with the alumino-silicate-fluoride glass.[2–7] The most important of the polyacids used to date have been the poly(acrylic acids)s themselves; copolymers of acrylic and itaconic acids[4] and copolymers[4a] of acrylic and methacrylic acids.[8] There are differences in functionality and acid strength between these polyacids. The polyacid hydrolyzes the glass network, releasing the aluminium and calcium ions and forming a silaceous hydrogel.[9] The poly(alkenoic acid) chains chelate the released cations and become crosslinked, forming a hard ceramic-like cement. A glass-ionomer cement can also be regarded as a glass filled poly(alkenoic acid) that is crosslinked by cations and plasticized by water.

The unique feature of the glasses is their high fluoride content.[10] The fluoride species play an important cariostatic role in dental cements. In addition, the fluoride component facilitates a refractive index match of the glass to the polysalt matrix, enabling translucent cements to be produced. The fluoride ion also causes amorphous phase separation of glass-ionomers to give a very reactive glass droplet phase and less reactive matrix phase.

Glass-ionomer cements have a number of properties which make them ideally suited to application in dentistry as a material for the restoration of anterior (front) teeth,[12] tunnel restoration, a cementing agent for the attachment of stainless steel crowns and bridges, a cavity liner, a base under amalgam and composite restorations, and as a general repair material.[12–22]

RESIN-MODIFIED GLASS-IONOMER CEMENTS

The "resin modified glass-ionomers" (resin modified glass-polyalkenoates) are hybrid cements set partly via an acid-base reaction and partly via a photochemical polymerization (photo-(light-)curing).[25–31]

A simultaneous acid-base reaction and photopolymerization (dual setting system) of resin modified glass-ionomers enhances the physical[30,32] and mechanical (tensile strength)[33,34] properties of the resulting restoration. This technique eliminates the number of steps required in the traditional sandwich restoration procedure, thus reducing technique sensitivity and increasing the efficiency of placement.

SANDWICH RESTORATION

In the sandwich restoration (also known as the glass-ionomer or, polyalkenoate-composite resin laminate technique) dentin is replaced with glass-ionomer cement and enamel with composite resin.[11,24,35–40] The bond formed between the glass-ionomer cement and composite resin can be greater than the strength of the glass polyalkenoate cement itself.[41–48] However, it is controlled by the:[46] tensile bond strength of the polyalkenoate cement; wettability of the resin; the polymerization contraction forces of the composite resin, and the adaptation of the composite resin to the underlying cement.

The bonding of glass-ionomer cement to a polymer composite resin combines the beneficial aspects of both groups of material, particularly the aesthetic properties of composite resin with the dentine and release of fluoride from the glass-ionomer cement.[36] The major disadvantage of sandwich restoration is that this technique is rather time consuming.

THE FUTURE OF GLASS-IONOMER CEMENTS

No other dental filling material has, during the last decades, undergone such a rapid development and gained so much interest in the dental field as glass-ionomer cements. There is every reason to believe that further development of these adhesive cements and clinical improvements in their combination with composite resins under the "sandwich technique," will be of great significance in the 21st century.[49]

REFERENCES

1. McLean, J. W.; Nicholson, J. W.; Wilson, A. D. *Quint. Int.* **1994**, *25*, 587.
2. Crisp, S.; Ferner, A. J.; Lewis, B. G.; Wilson, A. D. *J. Dent.* **1975**, *3*, 125.
3. McLean, J. W.; Wilson, A. D. *Aust. Dent. J.* **1977**, *22*, 31.
4. Kent, B. E.; Lewis, B. G.; Wilson, A. D. *J. Dent. Res.* **1980**, *58*, 1607.

4a. Crisp, S.; Kent, B. E.; Lewis, B. G.; Ferner, A. J.; Wilson, A. D. *J. Dent. Res.* **1980**, *59*, 1055.

5. Wilson, A. D.; Prosser, H. J. *Brit. Dent. J.* **1984**, *157*, 449.
6. Wilson, A. D.; McLean, J. W. *Glass-ionomer cement*; Quint: Chicago, 1988.
7. Wilson, A. D.; Nicholson, J. W. *Acid-Base Cements: Their Biomedical and Industrial Applications*; Cambridge, Univ.: Cambridge, 1993.
8. Schmidt, W.; Purmann, R.; Jochum, P.; Gasser, O. Eur. Patent Appl. 24 056, 1981.
9. Wasson, E. A.; Nicholson, J. W. *Brit. Polym. J.* **1990**, *23*, 179.
10. Wood, D.; Hill, R. *Clin. Mater.* **1991**, *7*, 301.
11. McLean, J. W. *Brit. Dent. J.* **1988**, *164*, 293.
12. Mount, G. J. *An Atlas of Glass-Ionomer Cements: A Clinician's Guide*, 2nd ed.; Martin Dunitz: London, 1994.
13. Garcia-Godoy, F. *Quint. Int.* **1986**, *17*, 617.
14. Garcia-Godoy, F.; Bugg, J. L. *J. Pedodont.* **1987**, *11*, 339.
15. Garcia-Godoy, F.; Marchal, T. M.; Mount, G. J. *Am. J. Dent.* **1988**, *1*, 53.
16. Seeholzer, H. W.; Dasch, W. *J. Clin. Orthod.* **1988**, *22*, 165.
17. Garcia-Godoy, F.; Landry, J. K. *J. Pedodont.* **1989**, *13*, 328.
18. Stratmann, R. G.; Berg, J. H.; Donly, K. J. *Quint. Int.* **1989**, *20*, 43.
19. Forsten, L.; Karjalainen, S. *Scand. J. Dent. Res.* **1990**, *98*, 70.
20. Knight, G. M. *Aust. Dent. J.* **1992**, *37*, 245.
21. Marcushamer, M.; Garcia-Godoy, F.; Chan, D. C. N. *J. Dent. Child.* **1993**, *60*, 300.
22. Wasson, E. A. *Clin. Mater.* **1993**, *12*, 181.
23. Feilzer, A. J.; deGee, A. J.; Davidson, C. L. *J. Prosthet. Dent.* **1988**, *59*, 297.
24. Copenhaver, D. J. *Am. J. Orthod.* **1986**, *89*, 528.
25. Norbo, H. "Glassionomersementer" .3M Dental Nytt.-no. 1, 1990.
26. Croll, T. P.; Killian, C. M. *Quint. Int.* **1991**, *23*, 679.
27. Croll, T. P. *Quint. Int.* **1991**, *22*, 137.
28. Bourke, A. M.; Walls, A. W.; McCabe, J. F. *J. Dent.* **1992**, *20*, 115.
29. Croll, T. P. *Quint. Int.* **1993**, *24*, 109.
30. Knight, G. M. *Quint. Int.* **1994**, *25*, 97.
31. Rasmusson, C. G.; Rasmusson, L. *Tandläkartidn.* **1994**, *86*, 117.
32. Tam, L. E.; McComb, D.; Pulver, F. *Oper. Dent.* **1991**, *16*, 210.
33. Eliades, G.; Palaghias, G. *Dent. Mater.* **1993**, *9*, 198.
34. McCarthy, M. F.; Hondrum, S. O. *Am. J. Orthod. Dentofac. Orthop.* **1994**, *105*, 135.
35. Mitra, S. B. *D. Dent. Res.* **1991**, *70*, 72.
36. McLean, J. W.; Powis, D. R.; Prosser, H. J. *Brit. Dent. J.* **1985**, *158*, 410.
37. McLean, J. W. *Quint. Int.* **1987**, *18*, 517.
38. Mount, G. J. *Aust. Dent. J.* **1989**, *34*, 259.
39. Mount, G. J. *Quint. Int.* **1990**, *21*, 93.
40. Woolford, M. *J. Dent.* **1993**, *21*, 31.
41. Sneed, W. D.; Looper, S. W. *Dent. Mater.* **1985**, *1*, 127.
42. Shimizu, A.; Ui, T.; Kawakami, M. *Dent. Mater.* **1986**, *5*, 225.
43. Hinoura, K.; Moore, B. K.; Phillips, R. W. *J. Am. Dent. Assoc.* **1987**, *114*, 167.
44. Wexler, G.; Beech, D. R. *Aust. Dent. J.* **1988**, *33*, 313.
45. Hinoura, K.; Onosa, H.; Moore, B. K. *Quint. Int.* **1989**, *20*, 31.
46. Fukuda, K.; Katsuyama, S. *J. Dent. Res.* **1989**, *8*, 923.
47. Mount, G. J. *Aust. Dent. J.* **1989**, *34*, 136.
48. Hinoura, K.; Suzuki, H.; Yoshimura, Y. *Dent. Mater.* **1990**, *6*, 94.
49. Lindén, L. Å. *Trends Polym. Sci.* **1994**, *2*, 144.

DENTAL POLYMERS (HYDROGELS)

Lars-Åke Lindén and J. F. Rabek
Polymer Research Group
Department of Dental Biomaterials
Karolinska Institute
Royal Academy of Medicine

Rampant caries and hypersensitive tooth necks have always been a challenge for dentists. To restore, reduce, and prevent these problems is the major task of dentistry worldwide.

The hard tissues of human teeth are penetrated by many microchannels that are remnants from the teeth's embryonic development. The channels communicate and penetrate the tooth's two major hard components, which differ structurally. Enamel is an ectodermal tissue composed of rod-shaped structural units, known as enamel prisms, with a diameter of 4–5 μm, which are mainly built of large hydroxyapatite crystals (calcium phosphate, $Ca_{10}(PO_4)_6$-$(OH)_2$).[2,3] The crystals are embedded in an organic matrix of non-collagenous proteins (0.25–0.45%) and lipids (0.60%). The mesodermal dentin also contains hydroxyapatite (70%). However, the crystals are smaller, and the organic matrix consists of collagen (20%) and water (10%).[4–6] The channels constitute a communication between the interior of the tooth (pulp tissue) and the exterior (saliva) throughout their life. The average diameter of these microchannels is 30– 200Å in enamel and 1–3 μm in dentin. They are filled with a native biohydrogel of a fibrous protein origin.[1]

Biohydrogels are natural polymeric materials that can swell in aqueous biological fluids (e.g., plasma and saliva) and can retain a significant fraction of fluid within the structure. The channels (tubules) play an important role in the physiology and pathology of teeth and serve throughout their life as transport ways for ions and molecules through the dental hard tissues to the pulp, which carries cells and nerve receptors. There is no "biological life" within the channels, but there is a passive ion exchange governed by diffusion and ion concentration equilibria.

To prevent microorganisms from penetrating the channels but still allow the diffusion of ions and molecules, a unique method has been developed in our department for tightening these microchannels with polymeric, ionically crosslinked hydrogels and polymers, which form directly in the channels.

CONCLUSION

The new concept of tightening tooth channels by polymeric, water-penetrable, ionically crosslinked nets, and even by polymers, introduces a new way of clinically treating patients suffering from hypersensitive tooth necks and progressive (rampant) caries. However, more research must be conducted before this method can be used widely in dental practices.

REFERENCES

1. Lindén, L.-Å; Källskog, Ö.; Wolgast, M. *Arch. Oral Biol.* **1995**, *40*, 991.
2. Posner, A. S. In *Adhesive Restorative Dental Materials*; Phillips, R. W.; Ruge, G., Eds.; Spencer, Indianapolis, IN, 1961; p 15.
3. Silverstone, L. M. In *Biocompatibility of Dental Materials*; Smith, D. C.; Williams, D. F., Eds.; CRC: Boca Raton, FL, 1982; Vol. 2, Chapter 2.
4. Ten Cate, A. R.; Torneck, C. In *Biocompatibility of Dental Materials*; Smith, D. C.; Williams, D. F., Eds.; CRC: Boca Raton, FL, 1982; Vol. 1, Chapter 3.
5. Mjor, I. A. In *Dentin and Dentin Reactions in the Oral Cavity*; Thylstrup, A. et al., Eds. Oxford IRL: Oxford, U.K., 1987; p 27.
6. Linde, A. In *Calcification in Biological Systems*; Bonucci, E., Ed.; CRC: Boca Raton, FL, 1987; p 269.

DENTAL SEALANTS

Yasuhiko Tsuchitani
Faculty of Dentistry
Osaka University

Tohru Wada
Kuraray Company Ltd.

Dental sealants are used to seal high caries-susceptible pits and fissures of the deciduous and permanent molars, and also to seal microspaces between the tooth and restorative materials, enabling those materials to adhere firmly both to prepared cavity walls and to other restorative materials. They provide protection from secondary caries and dental pulp involvements.

Most dental sealants are resinous materials, but glass ionomer dental cements have some limited use as sealing materials.

PREVENTIVE DENTAL SEALANTS

Preventive dental sealants used to seal the susceptible areas of teeth are classified into pit and fissure sealants and smooth surface sealants, depending on the part to be sealed. From the viewpoint of material science, pit and fissure sealants can be further classified into resin sealants and glass ionomer cements.

PIT AND FISSURE SEALANTS

Resin Sealants

The chemical nature of resin sealants is a radical-polymerizable monomer mixture that has a viscosity low enough to penetrate into narrow pits and fissures easily and can be cured in them to become a hard and durable sealing material. Bisphenol-A-diglycidyl methacrylate (bis-GMA), which has bisphenol A structure in its molecule, bis-GMA homologues, and urethane dimethacrylate are very popular as the main monomer for resin sealants, and they are usually used together with other monomers that are less viscious in order to increase their penetration ability. Pits and fissures are treated with phosphoric acid, citric acid, or other acidic agents prior to application of sealants. This treatment, often called "acid etching," changes the enamel surface to a micro-rugged structure, which aids penetration of sealants into fissures and results in strong micro-mechanical adhesion between sealants and tooth substances.

Polymerization of sealants is initiated by redox catalyst or light irradiation. In the former case, benzoylperoxide/*t*-amine is most common, and sulfinates also are used when acidic monomers are contained.[1] In the latter case, visible light is irradiated to initiate radical polymerization. Camphorquinone is the most popular photosensitizer and is usually used together with some kinds of amines.[2]

There are some sealants that can release fluoride ion gradually. Fluoride ion converts hydroxyapatite, the major inorganic component of teeth to fluoroapatite, which is more acid-resistant than hydroxyapatite, reducing caries susceptability.

Fluoride release technology, which is increasingly used, requires application of microencapsulated sodium fluoride and application of copolymer of methyl methacrylate and methacrylic fluoride.[3]

Glass Ionomer Cements

Glass ionomer cements basically consist of the powder of fluoro aluminosilicate glass and the aqueous solution of poly(acrylic acid).[4] Typical glass powder contains SiO_2, Al_2O_3, CaO, Na_2O, and CaF_2. Poly(acrylic acid) sometimes contains maleic acid or itaconic acid as a comonomer.[5] Glass ionomer cements are widely used, and their types vary with the dental treatment. Glass ionomer cements for pit and fissure sealing are characterized by low viscosity of the cement paste and fast setting.

SMOOTH SURFACE SEALANTS

Smooth surface sealants are used to seal caries-susceptible tooth surfaces other than pits and fissures in order to prevent them from bacteria, staining, and physical damage. These surfaces are the cervical area, interproximal area, and exposed root surfaces.

Cured sealants are directly exposed to the oral conditions and therefore strong adhesion to enamel and dentin, good physical properties, good wear resistance, chemical stability, biological stability, and thin film formation are required.

RESTORATIVE DENTAL SEALANTS

Amalgam Restoration

Amalgam Sealant

This sealant was developed by Tsuchitani and co-workers.[6] Chemical composition of this material is as follows:

- monomer: diethylene glycol dimethacrylate or *o*-monomethacryloxyethyl phthalate (MEP);
- catalyst: *tert*-butyl hydroperoxide;
- accelerator: *o*-sulfobenzimide; and
- inhibitor: hydroquinone.[6]

A primary characteristic of this sealant is that it is composed of only one liquid, which remains uncured before application, but cures quickly once applied to amalgam restoration.

The second characteristic is that methacryloyloxyethyl acid phthalate (MEP) is combined to enhance penetration of the sealant into microspaces and adhesion both to amalgam and tooth substances.

Amalgam sealant applied along the margin of amalgam restoration penetrates into microspaces by capillary action and cures *in situ* within a few minutes.

Amalgam Bonds

Dental adhesives also are used to seal microspaces in amalgam restoration. These adhesives called amalgam bonds, are of two types: adhesive resin cements, and bonding agents originally developed for composite restorations.

One commonly used adhesive resin cement is Amalgambond (Sun Medical Company, Japan), which consists of poly(methyl methacrylate) (PMMA) powder, methyl methacrylate (MMA) monomer containing 4-methacryloyloxyethyl trimellitic anhydride[9] and tributylborane oxide (TBB-0) as catalyst. TBB-0 is activated by the water on the surface of adherents and promotes graft polymerization of MMA onto collagen in dentin, resulting in strong adhesion.[7]

Another adhesive cement, Panavia 21 (Kuraray Company, Japan) is a sort of composite cement and is characterized by the adhesive component, methacryloyloxydecyl acid phosphate.[8] Panavia 21 exhibits strong adhesion to tooth substances and amalgam, does not deteriorate in wet conditions and provides a reliable seal.

Restorative Dental Sealants in Composite Restoration

Dental composite restoration is a method to repair areas deteriorated by caries with powder-reinforced composite materials that consist essentially of dimethacrylate monomers and silanated inorganic fillers. This dental composite is directly placed into cavities, cures there and serves as a restorative with excellent physical properties. Cured composite is semitranslucent and matches the tooth color well. Wear resistance against tooth brushing and chewing is also very good. Dental composites are the most popular restoratives for anterior and posterior teeth in modern dentistry.

Bis-GMA, Bis-GMA homologues, and urethane dimethacrylate are common as monomers, and quartz, glass, and ceramics are used as fillers.

The sealants used for restorations (called bonding agents) are applied onto the cavity walls prior to the placement of composites and bond the two substances tightly.

REFERENCES

1. Yamauchi, J.; Yamada, K.; Shibatani, K. U.S. Patent 4 182 035, 1980.
2. Dart, E. C.; Nemcek, J. Great Britain Patent 1 408 265, 1975.
3. Kadoma, Y. et al. *Makromol. Chem.* **1981**, *182*, 273–277.
4. Wilson, A. D.; Kent, B. E. *Brit. Dent. J.* **1972**, *132*, 122–135.
5. Crisp, S.; Wilson, A. D. Great Britain Patent 1 484 454, 1977.
6. Fukuda, K. *Jap. J. Cons. Dent.* **1978**, *21*, 595.
7. Nakabayashi, N. et al. *J. Biomed. Mater. Res.* **1978**, *12*, 149–165.
8. Omura, I. et al. U.S. Patent 4 539 382, 1985.
9. Masuhara, E.; Nakabayashi, N.; Takeyama, J. U.S. Patent 4 148 988, 1979.

Deoxycholic Acid Inclusion Complexes

See: Inclusion Complexes (Overview)
Inclusion Polymerization

Deoxyribonucleic Acid

See: DNA

Depolymerization

See: Nylon-6 (Depolymerization)

Dermatan Sulfate

See: Glycosaminoglycans (Overview)

Dewatering Aids

See: Lignin Graft Copolymers

DEXTRAN

Anthony N. de Belder
TdB Consultancy AB

The term "dextran" embraces a family of polysaccharides composed primarily of $1 \rightarrow 6$ linked α-D-glucopyranose units. Many dextrans contain 1- or 2-unit branches attached to O-2, O-3, or O-4 of the backbone chain units.

Dextrans were first recognized through their slime-forming properties in the processing of wines and sugar refining. In the 1940s, however, Ingelman and Grönwall1[1,2] conceived the idea of using partially hydrolyzed dextran fractions as plasma volume expanders. Clinical dextrans are still the plasma volume expander of choice. Throughout the years, an ever-increasing number of applications for dextran fractions in industry and medicine have evolved.

The reviews selected[3–10] represent only a handful of the many review articles that have appeared. They were chosen because they focus on aspects such as biosynthesis, production, structure and properties, etc.

CLINICAL APPLICATIONS

Clinical dextran solutions are still the plasma expander of choice in most countries. These solutions are prepared from clinical dextran fractions of molecular weight 40,000, 60,000, or 70,000 (with the generic names of dextran 40, 60, and 70, respectively). Generally, the higher molecular weight products are recommended for the treatment of shock of impending shock as a result of hemorrhage, burns, or surgery. The lower molecular weight fraction 40,000 was initially developed following the rather surprising observations of its disaggregating and blood flow improvement properties.[11,12]

Dextrans are valuable ingredients for solutions for organ perfusion and preservation, and promising results have been obtained in kidney, liver, lung, and cornea preservation.

OTHER APPLICATIONS

A comprehensive bibliography covering earlier dextran literature and patents is available.[13] Technical dextran fractions are finding growing acceptance in many industrial applications, notably the photographic industry, where the addition of dextran imparts favorable properties to the emulsion.

Dextran fractions have also found wide application in two-phase polymer systems for participating subcellular articles and as stabilizers of sensitive macromolecules (enzymes, proteins) during lyophilization and other drying processes.[14,15]

DEXTRAN DERIVATIVES

The introduction of crosslinked dextran gels (Sephadex®) in 1959 heralded a new era in separation technology. Following the introduction of these neutral gels, a range of substituted gels (ion exchangers) was developed. These gels are now used commercially for downstream purification of proteins and hormones of medicinal value.

Interestingly, a pharmaceutical grade of Sephadex® is available as a wound-cleansing agent (Debrisan®).

Iron-dextran complexes with high relaxativity have been developed, greatly improving the imaging of various tissues (liver, spleen, brain) by magnetic resonance imaging (MRI) techniques.[16,17]

Dextran sulfate, prepared by sulfating dextran with, for example, chlorosulfonic acid in anhydrous media, possesses a wide spectrum of biological properties.

DEAE-dextran is a polycationic derivative of dextran with a molecular weight in excess of 500,000. Perhaps its most useful application is as an agent for enhancing the uptake of nucleic acids and proteins by cells.[18,19] Administered orally, DEAE-dextran decreases plasma cholesterol and low-density lipoprotein cholesterol.[20]

The synthesis and properties of dextran conjugates is providing a fertile area of research.[21,22] Among the benefits to be gained by conjugation of drugs to a biocompatible carrier are increase in solubility of sparingly soluble drugs, increase in plasma half-life, reduced toxicity, and enhanced stability. Fluorescein-labeled dextrans (FITC-dextrans) are used extensively for studying vascular permeability and microcirculation in health and disease.

REFERENCES

1. Grönwall, A.; Ingelman, B. *Nord. Med.* **1944**, *21*, 247.
2. Grönwall, A.; Ingelman, B. *Vox. Sang.* **1984**, *47*, 96.
3. Sidebotham, L. *Adv. Carbohydr. Chem. Biochem.* **1974**, *30*, 435.
4. Jeanes, A. *Encyclopedia of Polymer Science and Technology*; John Wiley & Sons: New York, 1966; Vol. 4, pp 805–824.
5. Jeanes, A. *Am. Chem. Soc. Symp. Ser.* **1977**, *45*, 284.
6. Walker, G. J. *Int. Rev. Biochem.* **1978**, *16*, 75.
7. Alsop, R. M. *Prog. Ind. Microbiol.* **1983**, *18*, 1.
8. Robyt, J. F. *Encyclopedia of Polymer Science and Technology*; John Wiley & Sons: New York, 1985; Vol. 4, pp 752–767.
9. de Belder, A. N. *Dextran*, 2nd ed.; Pharmacia: Uppsala, Sweden, 1990.
10. de Belder, A. N. *Industrial Gums*, 3rd ed.; Academic: New York, 1993; Chapter 14.
11. Thorsén, G.; Hint, H. *Acta Chir. Scand. Suppl.* **1950**, *154*, pp 3–16.
12. Gelin, L. E.; Ingelman, B. *Acta Chir. Scand.* **1961**, *122*, 294.
13. Jeanes, A. *Dextran Bibliography*; Misc. Publication No. 1355, U.S. Dept. of Agriculture, Washington, DC, 1978.
14. Ferris, N. P.; Philpot, R. M.; Oxtoby, J. M.; Armstrong, R. M. *J. Virol. Methods* **1990**, *29*, 43.
15. Schmehl, M. K.; Vazquez, I. A.; Graham, E. F. *Cryobiology* **1986**, *23*, 512.
16. Li, K. C. P.; Quisling, R. G.; Armitage, F. E.; Richardson, D.; Mladinich, C. *Magn. Reson. Imaging* **1992**, *10*, 439.
17. Winter, T. C.; Freeny, P. C.; Nghiem, H. V.; Mack, L. A.; Patten, R. M.; Thomas, C. R.; Elliott, S. *Am. J. Roentgenol.* **1993**, *161*, 1191.
18. Fox, R. M.; Mynderse, J. F.; Goulian, M. *Biochemistry* **1977**, *16*, 4470.
19. Pagano, J. S.; McCutchan, J. H. *Prog. Immunbiol. Standards* **1969**, *3*, 152.
20. Galeone, F.; Giuntoli, F.; Brunelleschi, G.; Saba, P. *Acta Ther.* **1990**, *16*, 313.
21. Larsen, C. *Adv. Drug Delivery Rev.* **1989**, *3*, 103.
22. Virnik, A. D.; Khomyakov, K. P.; Skokova, I. F. *Russ. Chem. Rev.* **1975**, *44*, 1280.

Dextran-Conjugates

See: Anticancer Polymeric Prodrugs, Targetable

Dextran Sulfate

See: Dextran

Diallyl Diglycol Carbonate

See: Allyl Diglycol Carbonate

Diels-Alder Reactions

See: Benzocyclobutene-Based Monomers (Diels-Alder Ring-Opening Polymerization)
Benzocyclobutene Homopolymerization (Chemistry and Applications)
Benzocyclobutenes (Crosslinking and Related Reactions)
Benzocyclobutenes (Diels-Alder Polymerization)
Ladder Polymers (Precursor Route, Fully Unsaturated, All-Carbon)

DIENE RUBBERS, CONVENTIONAL (Ozone Degradation and Stabilization)

Slavtcho Rakovsky
Institute of Catalysis

World industry today produces a tremendous amount of tires and rubber goods. Every year manufacturers of natural and synthetic rubbers increase the total output of these rubbers, which compose the main part of tires and other goods. The technical requirements, storage and exploitation conditions, prolonged use, and environment become more and more complicated. Everywhere in the large cities, particularly at crossings and in traffic jams, a marked growing of ozone concentration is registered. Ozone is a principal enemy of all materials made by diene rubbers because it reacts very rapidly with rubber double bonds. Further, cracks are formed and the materials lose their useful properties. The basic protection against ozone involves application of special additives: antiozonants.

MECHANISM OF THE REACTION OF OZONE WITH DOUBLE BONDS

At present the classic mechanism of the reaction of ozone with isolated double bonds proposed by R. Criegee[4–9,22] is widely accepted.

At the beginning of this century, Harries[1–3] showed that natural rubber reacts rapidly with ozone and that both the molecular weight (MW) and the bromine number decrease. Our investigations were carried out during the last 20 years. We studied the reaction of ozone with rubbers both in solution[23,25–27,29,31,32] and under conditions of the elastic state.[10,12,16,19]

MECHANISM OF THE REACTION OF OZONE WITH RUBBERS IN SOLUTION

We focus on the investigation of conventional rubbers. When very diluted solutions of elastomers are ozonized, the reaction mechanism completely follows Criegee's mechanism.[23,25–27]

MECHANISM OF THE REACTION OF OZONE WITH RUBBERS IN THE ELASTIC STATE

The physicochemical properties of elastomers in the solid state are determined not only by the chemical structure of the molecules, but to a great extent also by the over-molecular structure and defects that occur on the surface.[11,12,16,18–20,30,35,36] The role of the defects is of essential importance because additional energy is concentrated at these spots, the ozone molecule reacting more rapidly with the polymer chain. Thus, one of the most specific phenomena of so-called crack formation and growth takes place.

ANTIOZONANTS

An antiozonant system is very low or high molecular additive to the rubber mixture that increases the rubber's resistance to ozone action. Low molecular antiozonants (AOs)[10,11,28,32–34] are the following classes of organic compounds: *p*-phenylenediamines (*p*-PHDAs), hydroquinolines, *N,N′*-disubstituted hexahydropyrimidines, *N*-substituted-dimethylpyrols, enamines, nitrones, pyrazone derivatives, enolethers, cyclic and acyclic acetals and ketals, other types of heteroatoms containing compounds, dithiocarbamates, and so forth, separately added to the rubber mixtures or in combinations. High molecular antiozonants are polymers or elastomers with attached low molecular antiozonants or saturated chain polymers and elastomers like EPDM,[21,24] fluorine rubbers, etc. To enhance and improve the action of the AO, waxes, paraffins, high molecular lipid acids, metal oxides, and so forth are often added to the rubber mixture.[16,19] The action of antiozonants[10–21,24,35] is complicated, but in general there are two kinds of preservation action: chemical, when an AO reacts with ozone in orders faster than the rubber molecules, and physical, when an AO physically hinders ozone's access to the rubber surface.

There are several general requirements to antiozonant properties:[33] higher reactivity toward ozone than rubber; multifunctional properties; rubber solubility; poison-resistance; water insolubility; lack of color; thermal stability; antiscorching effect during vulcanization. These factors make the selection of compounds very difficult. In addition, no theoretical basis is presently available to satisfy these requirements. In view of this, the search for new compounds with complex properties must continue.

REFERENCES

1. Harries, C. D. *Ber. Dtsch. Chem.* **1903**, *36*, 1933.
2. Harries, C. D. *Untersuchungen Uber das Ozon und Seine Einwirkung auf Organische Verbindungen*; Springer-Verlag: Berlin and New York, 1916.
3. Harries, C. D. *Untersuchungen Uber die Naturlichen und Kunstlichen Kautschukarten*; Springer-Verlag: Berlin und New York, 1919.
4. Criegee, R. *Rec. Chem. Progr.* **1957**, *18*, 111.
5. Criegee, R. *Chem. Ber.* **1975**, *108*, 749.
6. Criegee, R. *Angew. Chemie* **1975**, *87*, 765.
7. Bailey, P. S. *Chem. Rev.* **1957**, *58*, 925.
8. Bailey, P. S. *Ozonation in Organic Chemistry*; Academic: New York, 1978; Vol. 1.
9. Bailey, P. S. *Ozonation in Organic Chemistry*, Academic: New York, 1982; Vol. 2.

10. Parfenov, V. M.; Rakovski, S. K.; Shopov, D. M.; Popov, A. A.; Zaikov, G. E. *Commun. Dep. Chem., Bulg. Acad. Sci.* **1978**, *9*, 180, (in Russian); CA, *90*, 188265a.
11. Parfenov, V. M.; Rakovski, S. K.; Shopov, D. M.; Popov, A. A.; Zaikov, G. E. *Commun. Dep. Chem., Bulg. Acad. Sci.* **1979**, *12*, 51, (in Russian); CA, *92*, 60033x.
12. Popov, A. A.; Parfenov, V. M.; Cherneva, D. R.; Rakovski, S. K.; Shopov, D. M.; Neverov, A. N.; Gurvich, Ya. A.; Zaikov, G. E. *Commun. Dep. Chem., Bulg. Acad. Sci.* **1979**, *12*, 308, (in Russian); CA, *92*, 164941h.
13. Dobreva, R. G.; Rakovski, K. S.; Rakovski, S. K.; Vajarova, A. J. *Coll. Sci. Rep. Chem. Ind., Polymer Ser.*, Central Inst. Chem. Ind. **1979** No. 5, 275, (in Bulgarian).
14. Rakovski, S. K.; Cherneva, D. R.; Vajarova, A. J.; Dobreva, R. G.; Rakovski, K. S. *Coll. Sci. Rep. Chem. Ind., Polymer Ser.*, Central Inst. Chem. Ind. **1979**, No. 5, 283, (in Bulgarian).
15. Nenchev, L. K.; Rakovski, S. K.; Shopov, D. M.; Rakovski, K. S. *Coll. Sci. Rep. Chem. Ind., Polymer Ser.*, Central Inst. Chem. Ind. **1979**, No. 5, 299, (in Bulgarian).
16. Rakovski, S. K.; Shopov, D. M.; Rakovski, K. S.; Dobreva, R. G. *Chem. Ind.* **1980**, *52*, 33, (in Bulgarian); CA *92*, 216514t.
17. Rakovski, S. K.; Vajarova, A. J.; Dobreva, R. G.; Nenchev, L. K.; Rakovski, K. S.; Tzarianska, T. *Coll. Sci. Rep. Chem. Ind., Polymer Ser.*, Central Inst. Chem. Ind., **1982**, No. 15, 9, (in Bulgarian).
18. Rakovski, S. K.; Vajarova, A. J.; Dobreva, R. G.; Cherneva, D. R.; Rakovski, K. S.; Tzarjanska, T. *Coll. Sci. Rep. Chem. Ind., Polymer Ser.*, Central Inst. Chem. Ind. **1982**, No. 15, 17, (in Bulgarian).
19. Rakovski, S. K.; Rakovski, K. S.; Tzarianska, T.; Vajarova, A. J.; Nenchev, L. K. *Coll. Sci. Rep Chem. Ind., Polymer Ser.*, Central Inst. Chem. Ind. **1982**, No. 15, 33, (in Bulgarian).
20. Dobreva, R. G.; Rakovski, K. S.; Tzarjanska, T.; Vajarova, A. J.; Rakovski, S. K. *Coll. Sci. Rep. Chem. Ind., Polymer Ser.*, Central Inst. Chem. Ind. **1982**, No. 15, 83, (in Bulgarian).
21. Dobreva, R. G.; Rakovski, K. S.; Tzarjanska, T.; Vajarova, A. J.; Rakovski, S. K. *Coll. Sci. Rep. Chem. Ind., Polymer Ser.*, Central Inst. Chem. Ind. **1982**, No. 15, 97, (in Bulgarian).
22. Razumovskii, S. D.; Rakovski, S. K.; Shopov, D. M.; Zaikov, G. E. *Ozone Reaction with Organic Compounds*, Monograph, Publ. House, Bulg., Acad. Sci., Sofia, 1983, (in Russian).
23. Anachkov, M. P.; Rakovski, S. K.; Razumovskii, S. D.; Kefely, A. A.; Zaikov, G. E.; Shopov, D. M. *Polym. Deg. Stabil.* **1985**, *10*, 25, CA, *102*, 11487t.
24. Dobreva, R. G.; Rakovski, K. S.; Rakovski, S. K. *Chem. Ind.* (Bulgaria), **1985**, *57*, 448, (in Bulgarian); CA, *104*, 226165k.
25. Anachkov, M. P.; Rakovski, S. K.; Shopov, D. M.; Razumovskii, S. D.; Kefely, A. A.; Zaikov, G. E. *Polym. Deg. Stabil.* **1986**, *14*, 189; CA, *105*, 7714f.
26. Anachkov, M. P.; Rakovski, S. K.; Shopov, D. M. *Commun. Dep. Chem., Bulg. Acad. Sci.* **1985**, *18*, 187; CA, *104*, 35232v.
27. Anachkov, M. P.; Rakovski, S. K.; Razumouskii, S. D.; Zaikov, G. E.; Shopov, D. M. *Commun. Dep. Chem., Bulg. Acad. Sci.* **1985**, *18*, 194; CA, *104*, 35233w.
28. Rakovski, S. K.; Cherneva, D. R.; Rakovski, K. S.; Shopov, D. M. *Commun. Dep. Chem., Bulg. Acad. Sci.* **1985**, *18*, 202, (in Russian); CA, *104*, 20517t.
29. Anachkov, M. P.; Rakovsky, S. K.; Stefanova, R. V.; Shopov, D. M. *Polym. Deg. Stabil.* **1987**, *19*, 293; CA, *108*, 95936v.
30. Parfenov, V. M.; Popov, A. A.; Rakovsky, S. K.; Zaikov, G. E. *VMS-Ser. A*, (Russia), **1989**, *31*, 1250, (in Russian); CA, *111*, 175876c.
31. Anachkov, M. P.; Stefanova, R. V.; Rakovsky, S. K. *Brit. Polym. J.* **1989**, *21*, 429; CA, *111*, 215793g.
32. Rakovski, S. K.; Shopov, D. M.; Zaikov, G. E. *Int. J. Polym. Materials*, **1990**, *14*, 1; CA, *114*, 166058x.
33. Rakovski, S. K.; Cherneva, D. R.; Zaikov, G. E. *Int. J. Polym. Materials*, **1990**, *14*, 21; CA, *114*, 166039s.
34. Rakovski, S. K.; Kasparov, V.; Parfenov, V.; Popov, A. *Int. J. Polym. Materials*, **1990**, *13*, 223; CA, *114*, 123791p.
35. Popov, A. A.; Parfenov, V. M.; Rakovski, S. K.; Shopov, D. M. *Int. J. Polym. Materials*, **1990**, *13*, 123, CA, *115*, 94215s.
36. Rakovski, S. K.; Podmasterev, V.; Razumovskii, S. D.; Zaikov, G. E. *Int. J. Polym. Materials*, **1991**, *15*, 123; CA, *116*, 108050u.

Diethylene Glycol Bis(allyl carbonate)

See: Allyl Diglycol Carbonate

DIISOCYANATES (*In Situ* Generation)

M. Srinivasan
Dr. Bharatram Research and Development Center
SRF Ltd.

Since 1930, diisocyanates, particularly toluene diisocyanate (TDI) and methylene diisocyanate (MDI), have played a significant role in polymer science and technology. Diisocyanates form the basic building units of polyurethanes, poly(urethane urea)s, polyureas, and related polymers, that have properties making them unique for specialty applications such as cryogenic coating. Their versatility in both chemistry and application is attributable at least in part to the diisocyanate.[1]

Diisocyanates as a group are highly toxic, unstable, and corrosive compounds, taking moisture away from whatever they contact. Even under controlled storage, they undergo chain or step-growth polymerization. They can be synthesized by a number of methods.[1] The substituted and flexible aliphatic-rich diisocyanates, however, have not been well documented. But the use of TDI, the most common diisocyanate, leads to polymers whose glass transition temperature, T_g, is rather high, which gives little or no room for property modification, through the rigid [O-C-NH-R-NH-C-O], R being the tolyl unit. In trying to reduce the T_g and hence achieve a useful lower limit for service temperature, several attempts have been made to generate substituted diisocyanates *in situ* to eliminate the handling problems and prepare the polymers by a one-pot reaction.

POLYURETHANES BY *IN SITU* GENERATION OF DIISOCYANATES

Diisocyanates, essential as monomers for production of polyurethanes and polyureas, undergo autopolymerization on standing and decompose as a result of atmospheric moisture. Therefore, attempts[2–11] have been made to generate isocyanates *in situ* during the polymerization process, using the reaction of an alkyl halide with alkali cyanate leading to an alkyl isocyanate.

The reaction of a bis(chloromethyl) compound and potassium cyanate in dimethylformamide (DMF) in the presence of a diol was used to prepare these polyurethanes. In this reaction, the bis(isocyanatomethyl) compound was generated *in situ* and immediately reacted with the diol[12] to yield polyurethane. Some polyurethanes containing acetylenic bonds were also prepared by this method.[13] The bis(chloromethyl) compounds used were 1,4-bis(chloromethyl)-2,5-dimethyl-benzene and 1,5-bichloromethyl-2,4-dimethylbenzene.

Simple glycols, ether glycols, and hydroxy-terminated polyesters have been used as diols to synthesize polyurethanes (PUs), poly(ether urethane)s (POUs), and poly(ester urethane)s (PEUs).

Polyurethanes constitute a very broad spectrum of man-made materials incorporating a variety of monomers.[14–18] Their structural diversity gives them a wide range of properties and thus makes them useful in a variety of applications such as adhesives,[19] coatings,[20] flexible[21] and rigid[22] foams, elastomers,[23] water-proofing agents,[24] thermoset resins,[25] thermoplastic molding compounds,[26] rubber vulcanization,[27] artificial leathers, fibers, and paints.[28,29]

Polyurethanes derived from aromatic diisocyanates generally exhibit considerable crystallinity and are viscoelastic only well above room temperature, whereas introduction of methylene units on the polymer backbone led to a decrease[12] in T_g. Hence, the need to have an approach wherein a flexible isocyanate group is available for reaction, however unstable it may be.

Substituted phenols have been used for the preparation of a diol ester monomer to get lower T_g values. The approach is both simple and convenient. Because the methods for the *in situ* generation of isocyanates are at a moderately elevated temperature, this may make the method an ideal approach for heat-curable polyurethanes, with variable functionality in the polymer backbone.[30]

REFERENCES

1. Allan, G. Ed. *Comprehensive Polymer Science*; Pergamon: New York, 1989; Vol. 5, p 387.
2. Himmel, C. H.; Richards, L. M. U.S. Patent 2 866 801, 1958; *Chem. Abstract* **1958**, *53*, 9145*i*.
3. Graham, B. U.S. Patent 2 866 802, 1958; *Chem. Abstract* **1958**, *53*, 9146*d*.
4. Fukui, K.; Kitano, H. Jpn. Patent 36 4372, 1962; *Chem. Abstract* **1962**, *57*, 12514.
5. Zenner, K. F.; Oertel, G.; Holtschmidt, H. Ger. Patent 1 205 087, 1965; *Chem. Abstract* **1966**, *64*, 19413.
6. Argabright, P. A.; Rider, H. O.; Sieck, R. *J. Org. Chem.* **1965**, *30*, 3317.
7. Ozaki, S. *J. Polym. Sci.* **1967**, *B5*, 1053.
8. Kaiser, D. W. U.S. Patent 2 697 720, 1954.
9. Beachell, H. C.; Son, C. P. N. *J. Appl. Polym. Sci.* **1964**, *8*, 1089.
10. Nagasawa, A.; Kitano, H.; Fukui, K. *Bull. Japan Petrol. Inst.* **1964**, *6*, 72.
11. Miyake, Y.; Ozaki, S.; Hirata, Y. *J. Polym. Sci.* 1969, **A1**, *7*, 899.
12. Ibrahim, A. M. Ph.D. Dissertation, Indian Institute of Technology, Madras, India, 1980.
13. Durairaj, B.; Venkata, K. rao, *Makromol. Chem., Rapid. Commun.* **1980**, *1*, 473.
14. Saunders, J. H.; Frisch, K. C. *Polyurethanes: Chemistry and Technology, Part I*; High Polymer Series; Interscience: New York, 1962, Vol. 16.
15. Saunders, J. H. *Rubber Chem. Technol.* **1960**, *33*, 1259, 1293.
16. Bayer, O. *Angew. Chem.* **1947**, *A59*, 275.
17. Bayer, O.; Rinke, H.; Siefkon, W.; Orthner, L.; Schild, H. Ger. Patent 728 981, 1942.
18. Dombrow, B. A. *Polyurethanes* Reinhold: New York, 1957.
19. Sandler, S. R.; Berg, F. R. *J. Appl. Polym. Sci.* **1965**, *9*, 3909.
20. Wells, E. R.; Hixenbough, J. C. *Amer. Paint. J.* **1962**, *46, No. 47*, 88.
21. Farkas, A.; Mills, G. A.; Erner, W. E.; Maerker, J. B. *Ind. Eng. Chem.* **1959**, *51*, 1299.
22. Darr, W. C.; Gemeinhardt, P. G.; Saunders, J. H. *Ind. Eng. Chem. Prod. Res. Develop.* **1963**, *2*, 194.
23. Anelrood, S. L.; Hamilton, C. W.; Frisch, K. C. *Ind. Eng. Chem.* **1961**, *53*, 889.
24. Hanford, W. E.; Holmes, D. F. Br. Patent 571 975, 1945.
25. Kern, W.; Dall'Asta, G.; Dieck, R.; Kammerer, H. *Makromol. Chem.* **1951**, *6*, 206.
26. Heiss, H. L. *Rubber Age (New York)* **1960**, *88*, 89.
27. Jorczak, J. S.; Fetters, E. M. *Ind. Eng. Chem.* **1951**, *43*, 324.
28. Korshak, V. V.; Strepikheev, Y. A.; Moiseev, A. F. *Sov. Plast.* **1961**, *7*, 12.
29. Rinke, H. *Angew. Chem. Intern. Ed. Engl.* 1962, 1, 419; *Rubber Chem. Technol.* **1963**, *36*, 719.
30. Reddy, T. Ashok; Srinivasan, M. *J. Pol. Sci., Pol. Chem. Ed.* **1989**, *27*, 2805.

DIKETENE POLYMERIZATION

Peter Zarras
Weapons Division
Naval Air Warfare Center

A review on the general subject of ketenes and diketenes has recently appeared in the scientific literature.[1] The purpose of this review is to examine the function of diketenes specifically with respect to their synthesis and polymerization.

REACTIONS OF KETENES

Ketenes can dimerize upon standing at room temperature or below. Wilsmore was one of the first to explore the phenomenon of ketene dimerization.[2] If the ketene $[H_2CCO]$ is left standing, a brown semisolid substance will form. The chief component of this mixture is a colorless liquid. This liquid has a pungent odor which boils at 126–127°C/760 mmHg and has a melting point of –7 to –6°C.

The dimerization of ketenes has been scrutinized to determine its correct structure, while many different structures have been proposed for the dimers.[3,4] The correct structure of the ketene dimer was determined by deuterium-labelling experiments to be the γ-methylene-β-propiolactone (the lactone of 4-hydroxy-3-butenoic acid).

Dimethylketene and other substituted ketene dimers yielded predominantly the cyclobutane-1,3-diones formed by head-to-tail cycloaddition. In addition to the ketene dimerization, diphenylketene also dimerizes to form either a cyclobutane-1,3-dione type dimer or a β-lactone dimer.

Ketenes can be homopolymerized or copolymerized using thermal and/or ionic initiators with varying solvent systems, resulting in regular or irregular structures depending on the solvent and catalyst system employed. The anionic homopolymerization of dimethylketene has been extensively studied by Natta and his group.[5–7] Copolymerization of substituted ketenes with carbonyl compounds such as aldehydes and ketones produced polymers consisting of polyester structures.[8,9]

POLYMERIZATION AND APPLICATIONS

Diketenes can be isolated as stable compounds in good yields and can be purified to polymerization grade. They have been investigated as monomers for condensation polymerization.[1]

Polyamides were obtained from the copolymerization of secondary amines such as piperazine and N,N-dimethyl-1,6-hexamethylenediamine with sebacyl bisketene (which was prepared from sebacyl chloride and triethylamine) at 0, –20, –40, and –60°C in methylene chloride or acetone with corresponding yields greater than 50%.[10] These polyamides were found to be insoluble in the usual polyamide solvents owing to branching and crosslinking.

Polyaddition reactions using diisocyanodiketenes with oligomeric diols and 1,4-butanediol gave linear polyurethanes containing diketene moieties. The diketene moieties could be reacted with ethanol, ammonia, water or acetic acid. The polyurethanes contained β-ketoester, β-ketoamide or ketone groups.[11,12] These materials could potentially be used for the synthesis of polyurethanes containing reactive groups that are susceptible to further reactions and/or the formation of networks from linear poly(diketene-urethane)s.[13]

9,10-Dihydroanthracene-9,10-bisketene can undergo a condensation polymerization with glycols to polyester and with diamines to polyamides.

The isocyanate group is isoelectronic with the ketene group. For decades, polyurethanes, synthesized from diisocyanates with glycols, have been extremely useful polymers. These polymers have one major problem, however. The polyurethanes obtained from aromatic diisocyanates are thermally stable only up to 180°C.[14] In an analogous reaction similar to that of the polyurethane formation; diketenes and appropriate glycols are being studied to produce polyesters with thermal stabilities that might be in excess of 320°C. It has also been proposed[15,16] to use diketenes as chain extenders to increase the molecular weight of hydroxy terminated poly(ethylene terephthalate).

REFERENCES

1. Zarras, P.; Vogl, O. *Prog. Polym. Sci.* **1991**, *16(2/3)*, 173.
2. Wilsmore, N. T. M. *J. Chem. Soc.* 1907, 1938.
3. Johnson, J. R.; Shiner, Jr., V. J. *J. Am. Chem. Soc.* **1953**, *75*, 1350.
4. Huisgen, R.; Otto, P. *J. Am. Chem. Soc.* **1968**, *90*, 5342.
5. Natta, G.; Mazzanti, G.; Pregaglia, G. F.; Binaghi, M. *Makromolek Chem.* **1961**, *44–46*, 537.
6. Natta, G.; Mazzanti, G.; Pregaglia, G. F.; Binaghi, M. *J. Am. Chem. Soc.* **1960**, *82*, 5511.
7. Natta, G.; Mazzanti, G.; Pregaglia, G. F.; Pozzi, G. *J. Polym. Sci.* **1962**, *58*, 1201.
8. Natta, G.; Pregaglia, G. F.; Mazzanti, G.; Zamboni, V.; Binaghi, M. *Eur. Polym. J.* **1965**, *1*, 25.
9. Vogl, O. U.S. Patent 3 775 371, 1973.
10. Garner, D. P. *J. Polym. Sci., Polym. Chem. Ed.* **1982**, *20*, 2979.
11. Garner, D. P.; Fasulo, P. D. *J. Polym. Sci., Polym. Chem. Ed.* **1985**, *23*, 2177.
12. Mormann, W. *Polyurethanes World Congress Proc.* 1981, FSK/SPI 311.
13. Mormann, W.; Hoffmann, S. *Makromol. Chem.* **1988**, *189*, 129.
14. Saunders, K. J. *Organic Polymer Chemistry* 2nd.; Chapman and Hall: New York, 1988, Chapter 16.
15. Vogl, O.; Sustic, A. Makromol. Chem., *Macromol. Symp.* **1987**, *12*, 351.
16. Rosati, L.; Vogl, O. unpublished results.

DIMER ACID-BASED POLYESTERS

U. D. N. Bajpai and Nivedita
Polymer Research Laboratory
Department of Post Graduate Studies and Research in Chemistry
Rani Durgavati University

Dimer acid-based polyesters are mainly the condensation products of di- or polyfunctional hydroxy compounds and dimer acids. These polyesters derive their distinctive physical properties, such as flexibility and resistance to heat, corrosion, chemicals, and so forth, from dimer acids used in their production.

The dimer acids are the polymerized product of C_{18} unsaturated fatty acids or esters such as linoleic acid, ricinoleic acid, and oleic acid derived from vegetable oils for example, dehydrated castor oil, tall oil, and tung oil. Because the major constituent of the product is dibasic fraction having 18 carbon atoms in length per carboxylic group, these are termed dimer acids. In addition to the true dimer fraction, the monomeric fatty acids, the trimeric and the higher polymers are always present in these polymerized "monomers", although in negligible amounts.

Dimer acid (DA), HOOC–D–COOH, is a cyclic dibasic acid where D is a C_{34} divalent hydrocarbon radical with one substituted cyclohexene structure.[1] This structure is formed through Diels-Alder addition reaction between two molecules of dienoic acid. Bicyclic structures have been formed in fatty polymers from trienoic acids and monocyclic predominate in dienoic acids.[2,3] Two of the possible structures for DA can be given as:

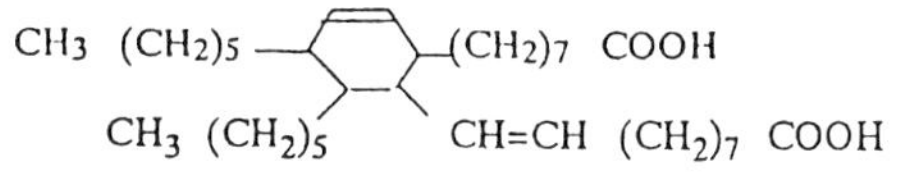

1

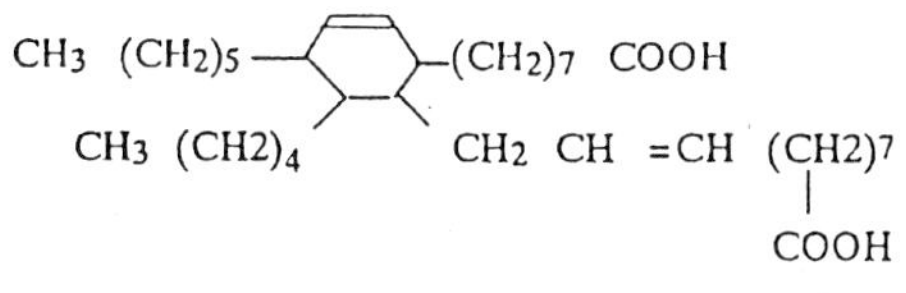

2

Structure 1 is the Δ 9,11-octadecadienoic acid dimer and **Structure 2** is the dimer acid of 9,11-octadecadienoic acid and Δ 9,12-octadecadienoic acid.

The general characteristics, structure, and properties of dimer acids were reviewed earlier.[4,5] These are relatively high molecular weight (~560) viscous liquids at 25°C. The liquidity is due to the presence of many isomers. It does not crystallize and contributes to flexibility in polymers derived from it.

The dimer-acid-based polyesters find a wide variety of applications as hot melt adhesives, films, lubricants, and so forth.[6–8]

The kinetics of the polyesterifications of dimer acid with different polyhydroxy compounds were studied by the present authors. The diols used were ethylene glycol, propanediol, 1,4-butanediol, glycerol, and diethyleneglycol.[12–15]

APPLICATIONS

The dimer-acid-based polyester resin with a broad spectrum of properties can be obtained. Because of good adhesion and flexural properties, dimer-acid-based polyamides are normally used as lubricants and adhesives and form the basis of flexographic inks and heat-seal coatings. They are used as vibration dampers when sandwiched between metal sheets. As a result of their high tensile strength, flexibility, and elongation properties, they are processed by techniques such as injection molding and blow molding.

Various thermoplastic polyester compositions based on dimer acid were prepared.[21–25,32,33] These compositions were heat- and impact-resistant with good toughness. Tomia et al. suggested their usefulness for electric, electronic, automobile, and industrial parts.[16–25]

Curable unsaturated polyester formulations of dimer acid were prepared by using some unsaturated acid anhydride as one of the components and cured either by vinyl toluene or styrene.[10,11] These polyesters find utility as filters for electric devices or as coatings as a result of their good flexibility, adhesion, heat stability, and electric-insulating properties.

A major application for the dimer-acid-based polyester films is in food-packaging materials and shrink labels. These polyester films possess good shrinkage at low temperature and good tensile strength.[9]

Dimer-acid-based hot melt adhesives with good adhesion to metals were used for adhering side seams in tin cans.[6,26–28] Manufacture of lubricating oils by mixing dimer-acid-based polyesters in soybean oil or synthetic oil are other applications of dimer-acid-based polyesters.[8,29,30]

Inking, as well as deinking, agents were prepared by dimer-acid-polyester compositions.

The nontoxic of dimer acid was found useful for preparing personal care products such as conditioners and remoisturizers.[31] These useful polyesters contain residual OH-functional active compounds such as lanolin, cholesterol, vitamins A, D, and E, which are beneficial to skin and hair.

In conclusion, polyesters of dimer acid with desired viscosity, molecular weight, acid, and/or hydroxyl number may be synthesized by condensation with different diols under controlled experimental conditions. These polyesters can find a variety of applications.

REFERENCES

1. Paschke, R. F.; Peterson, L. E.; Wheelar, D. H. *J. Am. Oil Chem. Soc.* **1964**, *41*, 723.
2. Bradley, T. F.; Johnston, W. *Ind. Eng. Chem.* **1940**, *32(6)*, 802.
3. Bradley, T. F.; Johnston, W. *Ind. Eng. Chem.* **1941**, *33*, 86.
4. Leonard, E. C. In *Encyclopedia of Chemical Technology* 3rd ed.; John Wiley & Sons; 1975, Vol. 7.
5. Breurer, T. E. In *Encyclopedia of Chemical Technology* 4th ed.; John Wiley & Sons; 1993, Vol. 8.
6. Jackson, J. W.; Darnell, R. W. U.S. Patent 3 931 073, 1976.
7. Kurome, T.; Tsunashima, K.; Hiraoka, T. Jpn. Kokai Tokkyo Koho, JP 04293985 A2 921019 1992.
8. Sturwold, R. J. Ger. Patent 2 649 684, 1977.
9. Murafuji, Y.; Yamamoto, M.; Makino, T.; Kunimaru, T. Jpn. Kokai Tokkyo Koho Jp 05170944 A2 930709, 1993.
10. Mekjian, A. U.S. 4 535 146 A 850813 1985.
11. Ishikawa, S.; Nagasawa, T. Jpn. Kokai Tokkyo Koho Jp 22797127 A2 901115, 1990.
12. Bajpai, U. D. N.; Hinogorani, S.; Jain, A.; Tiwari, H. P.; Nema, S. K. *Indian J. Chem.* **1988** *27A*(7), 635.
13. Bajpai, U. D. N.; Singh, K.; Nivedita *J. Appl. Polym. Sci.* **1992**, *46*, 1485.
14. Bajpai, U. D. N.; Nivedita, Ms. *J. Appl. Polym. Sci.* **1993**, *50*, 693.
15. Bajpai, U. D. N.; Nivedita, *Polymer Science: Recent Advances* Polymer 94 Conference: Vadodara, India, 1994.
16. Tomita, H.; Pponma, T.; Kishida, Y. Jpn Kokai Tokkyo Koho, JP 05171015 A2 930709, 1993.
17. Tomita, H.; Pponma, T.; Kishida, Y. Jpn Kokai Tokkyo Koho, JP 05117512 A2 930514, 1993.
18. Tomita, H.; Pponma, T.; Kishida, Y. JP 05179128 A2 930720, 1993.
19. Tomita, H.; Pponma, T.; Kishida, Y. JP 05051520 A2 930302, 1993.
20. Tomita, H.; Pponma, T.; Oonishi, T. Jpn Kokai Tokkyo Koho, JP 5214218 A2 930824, 1993.
21. Tomita, H.; Pponma, T.; Oonishi, T. Jpn Kokai Tokkyo Koho, JP 05214219 A2 930824, 1993.
22. Tomita, H.; Pponma, T.; Oonishi, T. Jpn Kokai Tokkyo Koho, JP 05214220 A2 930824, 1993.
23. Tomita, H.; Pponma, T.; Oonishi, T. Jpn Kokai Tokkyo Koho, JP 05214226 A2 930824, 1993.
24. Tomita, H.; Pponma, T.; Oonishi, T. Jpn Kokai Tokkyo Koho, JP 05262965 A2 931012, 1993.
25. Sublett, B. J.; Hilbert, S. D.; U.S. 4 439 598 A 840327, 1984.
26. Eastman Kodak Co., Neth. Appl. NL 7510817 770317, 1977.
27. Imoehl, W.; Drawert, M. Ger. Patent 2 361 486, 1975.
28. Hirakochi, H.; Nakamura, M.; Hachitsuka, T. Jpn. Kokai Tokkyo Koho, JP 04328186 A2 921117, 1982.'
29. Danis, J.; Pare, G. Fr. Patent 2 307 867, 1976.
30. Kenbeck, D.; Poulina, R. R.; Vanderwaal, G. Eur. Pat. Appl. EP 335013 A1 891004, 1989.
31. O'Lenick Jr., A. J. U.S. 5 210 133 A 930511, 1993.
32. Ritter, W.; Bergner, R.; Schaefer, M. Ger. Offen. De 412111 A1 930107, 1993.
33. Manuel, H. J.; Gaymans, R. *J. Polymer*, **1993**, *34*(3), 636.

DIPHENIMIDES (Ring-Opening Polymerization)

Thanun M. Pyriadi
Department of Chemistry
College of Science
University of Baghdad

A commercial polymer made by ring-opening polymerization of seven-membered ring lactam is nylon 6. ϵ-Caprolactam is polymerized by anionic ring opening using a base catalyst.[1,2] Seven-membered cyclic imides have been synthesized and polymerized by Pyriadi and Ahmad.[3] Three of the synthesized *N*-substituted diphenimides were anionically polymerized using *n*-butyl lithium as the initiator,[3] yielding 50% low molecular weight polymer.

PREPARATION AND PROPERTIES

Six monomers of *N*-substituted diphenimides have recently been reported.[3] However, only three have been polymerized anionically by ring-opening polymerization. The monomers are *N*-phenyl-, *N*-(acetoxyphenyl)-, and *N*-(acetoxyethyl)-diphenimides. The seven-membered ring imides were prepared from the corresponding amic acids.[4,5]

Anionic ring-opening polymerization of these substituted diphenimides using *n*-butyl lithium as initiator resulted in the formation of aromatic polyamides possessing high softening point temperatures (300°–350°C).

The importance of polymers obtained from ring opening of *N*-substituted diphenimides is that they are expected to have good thermal and mechanical properties because the produced polymers contain amide linkages with three aromatic rings in the repeating units. However, the molecular weights of the resulting polymers were low. This was attributed to the stability of the monomers.

REFERENCES

1. Frisch, K. C. Ed., *Cyclic Monomers*, Wiley-Interscience: New York, 1972.
2. Sebenda, J. *J. Macromol. Sci. Chem.* **1972**, *A6*, 1145.
3. Pyriadi, T. M.; Ahmad, A. *J. Polym. Sci., Part A, Polym. Chem.* **1993**, *31*, 3199.
4. Cava, M. P. et al. *Org. Syn.* **1961**, *41*, 93.
5. Cotter, R. J. et al. *J. Org. Chem.* **1961**, *26*, 10.

Dispersion Polymerization

See: *Dispersion Polymerization (Overview)*
Macromonomers (Soap Free Emulsion and Dispersion Polymerization)
Monodisperse Polymer Particles (Methods of Preparation)
Precipitation Polymerization (Overview)
Precipitation Polymerization (Process Description and Preparation of Particles)

DISPERSION POLYMERIZATION (Overview)

Harold D. H. Stöver
Department of Chemistry
McMaster University

Kai Li
3M Canada Incorporated

Dispersion polymerization was first developed in the 1950s to meet industrial needs for nonaqueous dispersion coating technologies suitable for automotive paints. The successful development of this technology has since led to many other applications including reprographics, adhesives, encapsulants, colored polymer particles, and coatings. This review concentrates on recent developments of dispersion polymerization in polar organic media, leading to the formation of monodisperse polymer microspheres in the micron size range, thus covering the classical gap in particle size between emulsion and suspension polymerization.[1,2]

Dispersion polymerizations usually start as a homogeneous solution of monomer (i.e., styrene, methyl methacrylate), a radical initiator (azo or peroxide compounds), and a polymeric steric stabilizer in organic solvents, such as hydrocarbons or ethanol. They are thus often called nonaqueous dispersions (NAD) for historical reasons, though water is at times used as cosolvent. The system is designed so that the resulting polymer is insoluble and precipitates out to form a colloidal dispersion, with the polymeric stabilizer added to prevent flocculation. This stabilizer may become adsorbed, or more commonly, grafted on the particle surface. Under favorable conditions, very narrow or monodisperse polymers microspheres can be obtained.

Dispersion polymerization differs from emulsion polymerization largely because an organic polymerization medium is used, a single phase is present at the beginning of the polymerization, and a typically uncharged polymeric stabilizer is used. A central question in dispersion polymerization concerns the role and function of this steric stabilizer, and how its interactions with polymer and solvent determine the final particle size, size distribution, and the polymer molecular weight. Much fundamental work in this area has been done since 1985 at Xerox by Lok, Ober, and Paine, and at Lehigh University by El-Aasser's group.[3–14] Most of the work up to 1990 has been reviewed by Walbridge, Croucher and Winnik, and Ober.[15–17] Here we will present the main aspects of dispersion polymerization, and highlight some recent developments.

PRINCIPLE OF STERIC STABILIZATION

Polymer dispersions prepared by polymerization in absence of stabilizer are unstable and coagulate even as they form (such a process is called precipitation polymerization). Some of the more common stabilizers used in dispersion polymerizations of styrene or methyl methacrylate in polar solvents are hydroxypropylcellulose (HPC), polyvinylpyrrolidone (PVP), and polyacrylic acid (PA). They form small amounts of graft copolymer with the monomer in question, which then serve as the actual steric stabilizer.[4,8,11]

Block copolymers such as poly(styrene-*block*-ethylene-*co*-propylene)[30] and poly(styrene-*block*-ethylene oxide)[31] are being increasingly used as steric stabilizers.[32] In analogy to the graft copolymers described above, block copolymers adsorb selectively with one block onto the growing particle surface, while the other block is solvated by the medium and serves as steric barrier. Due to the absence of a lag period to form the stabilizer *in situ*, block copolymers are usually highly efficient stabilizers.

Another class of steric stabilizers that is receiving increasing attention recently are the macromonomers. These are low to medium molecular weight polymers with high affinity for the polymerization medium, that bear a polymerizable group at one end. These macromonomers function simultaneously as comonomer and as stabilizers and can be used for a wide variety of monomers and reaction media.

TABLE 1. Examples of Dispersion Polymerizations with *In Situ* Graft Stabilization

Monomer	Polymeric Stabilizer	Medium	Reference
Styrene	Poly(acrylic acid-*co*-styrene)	Ethanol	18
Styrene	Poly(acrylic acid)	Ethanol/water	7
Styrene	Poly(acrylic acid)	Iso-propanol/water	19
Styrene	Hydroxypropyl cellulose	Methyl cellosolve/alcohol	4-6,9
Styrene	Poly(*N*-vinylpyrrolidone)	Ethanol	20
Styrene	Poly(vinyl alcohol)	Methanol	21
Styrene	Polyacrylamide	Methyl cellosolve/ethanol	22
Styrene	Poly(2-oxazoline) macromonomer	Ethanol/water	23
Methylmethacrylate	Polydimethylsiloxane	Heptane	24
Methylmethacrylate	Poly(*N*-vinylpyrrolidone)	Methanol	25
Methylmethacrylate	Polyisobutylene	Carbon tetrachloride/ 2,2,4-trimethylpentane	26
Methylmethacrylate	Poly[(propionylimino)ethylene]	Methanol	27
Acrylic monomers	Poly(12-hydroxy stearic acid)	Aliphatic solvents	28,29

POLYMERIZATION AND PARTICLE FORMATION IN DISPERSION POLYMERIZATION

Polymerization

While some dispersion polymerizations have been carried out using anionic and condensation methodologies, most are performed by radical polymerization using common azo or peroxide initiators.[33–38] The initial locus of polymerization is by necessity in the continuous medium, though it may later shift towards the interior of the polymer particles, depending on the solvency of the medium.[5,14] The solvency of the medium plays a crucial role in dispersion polymerization and influences particle size and size distribution, as well as polymer molecular weight.

Provided all other variables are fixed, it is possible to adjust the average particle size by using mixed solvents.

CURRENT DEVELOPMENTS

Much attention is being directed at the formation of dispersion polymer particles containing functional monomer or comonomer for biomedical and medicinal applications.[40] Very promising developments include highly crosslinked acrylic dispersion polymers imprinted with drug molecules, that may point the way to molecule-specific chromatographic resins.[41–43] Recently, chemical modification has been explored as a means to introduce aldehyde and carboxylic acid groups onto the surface of poly(4-methylstyrene) dispersion polymer particles.[44]

Crosslinked polystyrene and poly(methyl methacrylate) microspheres have recently been synthesized by dispersion polymerization. The level of crosslinking agent turned out to have a dramatic effect on both size distribution and morphology of the crosslinked microspheres.

OUTLOOK

Despite the significant advances made in recent years, there continue to be many challenges in both theory and applications of dispersion polymerization. There are still no models that could quantitatively predict particle size, size distribution and molecular weight. In terms of applications, functional dispersion polymer microspheres are already used in biomedicine.[46,47] Other dispersion polymer articles are used to model colloidal interfaces in chemical processes.[29] In timely expectation of more stricter environmental regulations, recycling of solvent and stabilizer in dispersion polymerization has been investigated and dispersion polymerizations in supercritical carbon dioxide are being explored, using fluorinated steric stabilizers.[39,48]

In the future, one may expect more work in the areas of aqueous dispersion polymerizations of polar monomers such as acrolein,[49] preparation of affinity separation resins,[50] magnetic polymer particles,[51] and multiphase dispersion polymer particles prepared using semibatch techniques now common in emulsion polymerization.[22,45]

REFERENCES

1. Corner, T. *Coll. Surf.* **1981**, *3*, 119.
2. Almog, Y.; Reich, S.; Levy, M. *Br. Polym. J.* **1982**, *14*, 131.
3. Lok, K. P.; Ober, C. K. *Can. J. Chem.* **1985**, *63*, 209.
4. Ober, C. K.; Lok, K. P.; Hair, M. L. *Polym. Sci. Polym. Lett. Ed.* **1985**, *23*, 103.
5. Ober, C. K. et al. *Coll. Surf.* **1986**, *21*, 347.
6. Ober, C. K.; Hair, M. L. *Polym. Sci. Part A: Polym. Chem.* **1987**, *25*, 1395.
7. Ober, C. K.; Lok, K. P. *Macromolecules* **1987**, *20*, 268.
8. Paine, A. J. *J. Coll. Interface Sci.* **1990**, *138*, 157.
9. Paine, A. J. *J. Polym. Sci. Polym. Chem.* **1990**, *28*, 2485.
10. Paine, A. J. et al. *J. Coll. Interface Sci.* **1990**, *138*, 170.
11. Paine, A. J.; Luymes, W.; McNulty, J. *Macromolecules* **1990**, *23*, 3104.
12. Paine, A. J. *Macromolecules* **1990**, *23*, 3109.
13. Tseng, C. M. et al. *J. Polym. Sci. Polym. Chem.* **1986**, *24*, 2995.
14. Lu, Y. Y.; El-Aasser, M. S.; Vanderhoff, J. W. *J. Polym. Sci. Polym. Phys.* **1988**, *26*, 1187.
15. Walbridge, D. J. NATO ASI Ser., Ser. E. *Sci. Technol. Polym. Colloids* **1983**, *67*, Vols. 40–50.

16. Croucher, M. D.; Winnik, M. A. In *Scientific Methods for the Study of Polymer Colloids and Their Applications*; Candau, F., Ottewill, R. H., Eds.; Kluwer Academic: Dordrecht, The Netherlands, 1990; pp 35–72.
17. Ober, C. K. *Makromol. Chem. Macromol. Symp.* **1990**, *35/36*, 87.
18. Omni, S. et al. *J. Polym. Sci.: Part A: Polym. Chem.* **1994**, *32*, 571.
19. Tuncel, A.; Kahraman, R.; Piskin, E. *J. Appl. Polym. Sci.* **1994**, *51*, 1485.
20. Paine, A. J.; McNulty, J. *J. Polym. Sci. Polym. Chem. Ed.* **1986**, *24*, 2995.
21. Dawkins, J. V.; Neep, D. J.; Shaw, P. L. *Polymer* **1994**, Vol. 35(24), p 5366.
22. Kobayashi, K.; Senna, M. *J. Appl. Polym. Sci.* **1992**, *46*, 27.
23. Koybayshi, S. et al. *J. Polym. Sci. Polym. Chem.* **1993**, *31*, 3133.
24. Pelton, R. H.; Osterroth, A.; Brook, M. A. *J. Coll. Interface Sci.* **1990**, *137*, 120; **1991**, *147*, 523.
25. Shen, S.; Sudol, E. D.; El-Aasser, M. S. *J. Polym. Sci. Polym. Chem.* **1993**, *31*, 1393.
26. Williamson, B. et al. *J. Coll. Interface Sci.* **1987**, *119*, 559.
27. Kobayashi, S. et al. *Makromol. Chem.* **1992**, *193*, 2355.
28. Kargupta, K.; Rai, P.; Kumar, A. *J. Appl. Polym. Sci.* **1993**, *49*, 1309.
29. Lee, K.-C.; Winnik, M. A.; Jao, T-C. *J. Polym. Sci.: Part A: Polym. Chem.* **1994**, *32*, 2333.
30. Stejskal, J.; Kratochvil, P.; Konak, C. *Polymer* **1991**, *32*(13), 2435.
31. Winzor, C. I. et al. *Eur. Polym. J.* **1994**, *30*(1), 121.
32. Riess, G.; Bahadur, P.; Hurtrez, G. *Encyclopedia of Polymer Science and Engineering*, 2nd ed.; John Wiley & Sons: New York, NY, 1985; Vol. 2, p. 324.
33. Schneider, M.; Mulhaupt, R. *Polym. Bull.* **1994**, *32*, 545.
34. Schwab, F. C.; Murray, J. G. In *Advances in Polymer Synthesis*; Culbertson, B. M., McGrath, J. E., Eds.; Plenum: New York, NY, 1985, pp 381–404.
35. Okay, O.; Funke, W. *Macromolecules* **1990**, *23*, 2620.
36. Jenkins, A. D. et al. *Makromol. Chem., Rapid Commun.* **1992**, *13*, 61.
37. Awan, M. A.; Dimonie, V. L.; El-Aasser, M. S. *Abstracts of Papers*, National Meeting of the American Chemical Society, American Chemical Society: Washington, D.C.; Aug. 21–25, 1994; 208, 403-POLY (PA269).
38. Bunn. A. et al. *Polymer* **1992**, *33*, 3066.
39. Shen, S.; Sudol, E. D.; El-Aasser, M. S. *J. Polym. Sci. Polym. Chem.* **1993**, *31*, 1393.
40. Piskin, E. et al. *J. Biomater. Sci. Polym. Ed.* **1994**, *5*(5), 451.
41. Wulff, G.; Minarik, M. *J. Liq. Chromatogr.* **1990**, *13*, 2987.
42. Li, K. et al. *Macromol. Chem. Phys.* **1994**, *195*, 391.
43. Sellergren, B. *J. Chromatogr.* **1994**, *673*, 133.
44. Li, K. Ph.D. thesis, McMaster University, 1994.
45. Thomson, B.; Rudin, A.; Lajoie, G. *J. Polym. Sci.: Part A: Polym. Chem.* **1995**, *33*, 345.
46. Tuncel, A. et al. *J. Chromatogr.* **1993**, *634*, 161.
47. Ercan, M. T. et al. *J. Microencap.* **1993**, *10*, 67.
48. DeSimone, J. M. et al. *Science* **1994**, *265*, 356.
49. Margel, S.; Wiesel, E. *J. Polym. Sci.: Polym. Chem. Ed.* **1984**, *22*, 145.
50. Sellergren, B. *J. Chromatogr. A.* **1994**, *673*(1), 133.
51. Noguchi, H. et al. *J. Appl. Polym. Sci.* **1993**, *48*, 1539.

Dispersions

See: Polypyrrole (Processsable Dispersions)

1,1-Disubstituted Olefins

See: Captodative Compounds, Polymerization
Captodative Olefins

DIVEMA

See: Anticancer Polymeric Prodrugs, Targetable
Antitumor Active Polymers (DIVEMA-Like)
Cyclopolymerization (Overview)

DIYNE CYCLOADDITION COPOLYMERIZATION (Transition Metal Catalyzed)

Tetsuo Tsuda
Department of Polymer Chemistry
Graduate School of Engineering
Kyoto University

Transition metal-catalyzed 1:1 cycloaddition copolymerization of a diyne, which we recently developed, is a new polymerization reaction.[1] A C≡C bond of the diyne is intermolecularly connected with concomitant formation of a hetero- or carbocycle by a transition metal-catalyzed cycloaddition reaction with a cycloaddition component (Z) to produce a 1:1 copolymer. The transition metal-catalyzed cycloaddition reaction of alkynes[2] has been studied as a tool for organic synthesis. Various compounds may be used as cycloaddition components in the transition metal-catalyzed cycloaddition copolymerization of the diyne: heterocumulenes such as CO_2, an isocyanate, and a carbodiimide; unsaturated compounds such as a nitrile and an olefinic compound; or carbene-type compounds such as carbon monoxide and an isocyanide.[1] Therefore, a variety of new polymers having a repeating unit of a hetero- or carbocycle substituted with a carbon chain or a heteroalkylene chain are expected to be prepared, and transition metal-catalyzed 1:1 cycloaddition copolymerization of the diyne may be a useful synthetic method. The transition metal-catalyzed 1:1 cycloaddition copolymerization of the diyne opens a new aspect of the research field of the transition metal-catalyzed cycloaddition reaction of alkynes.[2]

Here, we describe the nickel(0)-catalyzed diyne/CO_2 copolymerization to a poly(2-pyrone) and the nickel(0)-catalyzed diyne/isocyanate copolymerization to a poly(2-pyridone). Utilization of CO_2 in polymer synthesis by 1:1 CO_2 copolymerization is an attractive approach to efficient CO_2 utilization because 1:1 CO_2 copolymerization is highly selective and the copolymer may be useful as a structural and functional material.[1] Previously, however, the alternating copolymerization of epoxides with CO_2 to polycarbonates[3] is the only example of efficient CO_2 copolymerization. Therefore, the present poly(2-pyrone) synthesis is the first example of the efficient copolymerization of CO_2 with an unsaturated hydrocarbon.[1]

NICKEL(0)-CATALYZED 1:1 CYCLOADDITION COPOLYMERIZATION OF DIYNES WITH CO_2 TO POLY(2-PYRONE)S

3,11-Tetradecadiyne with six methylene groups underwent nickel(0)-catalyzed poly(2-pyrone) formation (**Equation 1**).[5] 3,9-Dodecadiyne with four methylene groups produced bicyclic 2-pyrone as a main product. This finding indicates that diyne/CO_2 copolymerization is controlled by the relative rate of inter- and intramolecular cyclizations of the diyne.

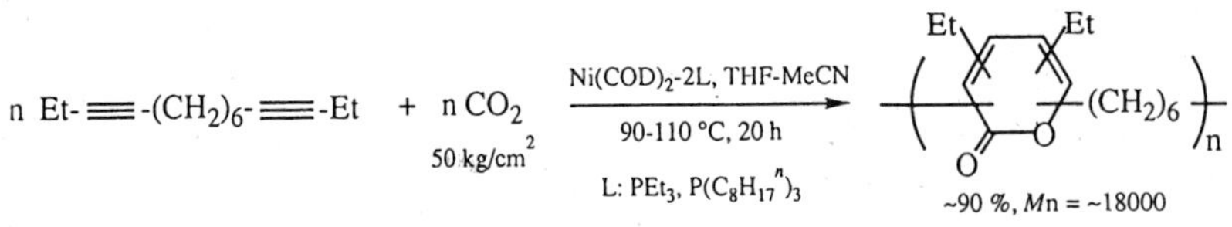

1

A competing side reaction of the poly(2-pyrone) formation is diyne trimerization.[2,4] to produce benzenoid or cyclopentadienoid repeating units in the copolymer.

2,6-Octadiyne with two methylene groups also underwent 1:1 cycloaddition copolymerization with CO_2,[6] but a molecular weight of the poly(2-pyrone) obtained was not high.

If diynes with various structures can be used, diyne/CO_2 copolymerization will have wide application as a new polymerization reaction.

Poly(2-pyrone)s are a new class of polymers. Examination of their physical properties and chemical reactions is interesting. The 2-pyrone ring is known to exhibit a variety of chemical reactivities along with biological activities.[7]

NICKEL(0)-CATALYZED 1:1 CYCLOADDITION COPOLYMERIZATION OF DIYNES WITH ISOCYANATES TO POLY(2-PYRIDONE)

In the nickel(0)-catalyzed cycloaddition copolymerization of diynes, an isocyanate acts as an excellent cycloaddition component to afford a poly(2-pyridone). The isocyanate/diyne cycloaddition copolymerization has features different from those of the CO_2/diyne copolymerization.

3,11-Tetradecadiyne (**1a**), 3,9-dodecadiyne (**1b**), or 2,6-octadiyne (**1c**), which is an aliphatic diyne, copolymerized with phenyl (**2a**), n-octyl (**2e**) or cyclohexyl (**2f**) isocyanate to give the corresponding poly(2-pyridone) **3** (**Equation 2**).[9]

The results thus far obtained on the 1:1 copolymerizability of the diyne with the isocyanate in the nickel(0)-catalyzed diyne/isocyanate copolymerization indicate the following copolymerizability order of the diyne: the aliphatic diyne > 1,4-bis(phenylethynyl)benzene > 1,4-diethynylbenzene. This order may be related to the ease of the formation of metallacycle intermediates generating the 2-pyridone ring from the diyne and the isocyanate components[8] because an order of diyne homopolymerizability was 1,4-bis(phenylethynyl)benzene = 1,4-diethynylbenzene >> the aliphatic diyne. The nickel(0)-catalyzed diyne/isocyanate cycloaddition copolymerization has several significant advantages over the diyne/CO_2 copolymerization: a much higher reactivity of the isocyanate as a cycloaddition component than CO_2, use of various isocyanate cycloaddition components, and control of the yield and the molecular weight of the poly(2-pyridone) by changing the molar ratio of the diyne to the isocyanate. These advantages permit synthesis of a variety of poly(2-pyridone)s. Chemical reactions of poly(2-pyridone)s obtained are a further interesting research subject because various chemical reactions[10] of the 2-pyridone ring such as the Diels-Alder reaction, photochemical reaction, and molecular aggregation are known in addition to its biological activity.[11]

REFERENCES

1. Tsuda, T. *Gazz. Chim. Ital.* **1995**, *125*, 101.
2. Schore, N. E. *Chem. Rev.* **1988**, *88*, 1081.
3. Inoue, S. et al. *J. Polym. Sci., Polym. Lett. Ed.* **1969**, *7*, 287.
4. Inoue, Y. et al. *Bull. Chem. Soc. Jpn.* **1980**, *53*, 3329; Walther, D. et al. *J. Organomet. Chem.* **1987**, *334*, 377.
5. Tsuda, T. et al. *J. Am. Chem. Soc.* **1992**, *114*, 1489.
6. Tsuda, T.; Maruta, K. *Macromolecules* **1992**, *25*, 6102.
7. Staunton, J. *Comprehensive Organic Chemistry*; Pergamon: Oxford, 1979; Vol. 4, p 629; Ellis, G. P. *Comprehensive Heterocyclic Chemistry*; Pergamon: Oxford, 1984; Vol. 3, p 675; Dieter, R. K.; Fishpaugh, J. R. *J. Org. Chem.* 1983, 48, 4439.
8. Hoberg, H.; Oster, B. N. *J. Organomet. Chem.* **1983**, *252*, 359; Hoberg, H.; Oster, B. N. *Synthesis* **1982**, 324.
9. Tsuda, T.; Hokazono, H. *Macromolecules* **1993**, *26*, 1796.
10. Birkofer, L.; Wahle, B., 1983, 116, 3309; Matsushima, R.; Terada, K. *J. Chem. Soc. Perkin Trans. II* **1985**, 1445; Simard, M. et al. *J. Am. Chem. Soc.* **1991**, *113*, 4696.
11. Rigby, J. H.; Balasubramanian, N. *J. Org. Chem.* **1989**, *54*, 224.

DNA

See: *Biopolymers (Overview)*
Biopolymers (Electrochemical Behavior)
DNA (Bending and Twisting Elasticity)
DNA (Intercalation with Drugs)
DNA Degradation (Atomic Target Method using Synchrotron Radiation)
Molecular Shape Recognition
Nucleic Acid Analogs
Nucleic Acid Interactions (with cis-platin)
Nucleohistone Complexes (Protein-Nucleic Acid Interactions)
Polyelectrolyte Complexes (Targeting of Nucleic Acids)

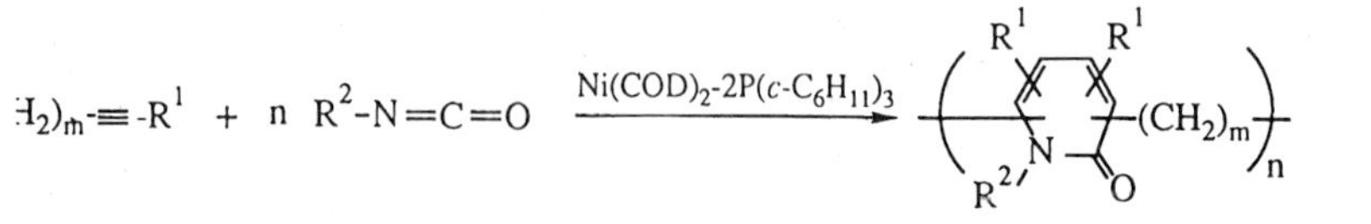

1	2	3
6 (**1a**)	$R^2 = C_6H_5$ (**2a**)	R^1 = Et, m = 6, $R^2 = C_6H_5$ (**3aa**), etc.
4 (**1b**)	$R^2 = C_8H_{17}{}^n$ (**2e**)	
: 2 (**1c**)	$R^2 = c\text{-}C_6H_{11}$ (**2f**)	

2

DNA (Bending and Twisting Elasticity)

John F. Marko
Center for Studies in Physics and Biology
The Rockefeller University

The DNA double helix (B-DNA), in addition to being the genetic information storage element in living things, is a remarkable polymer.

Double-helix DNA is a stiff polymer compared with the typical singly-bonded chains usually studied by polymer scientists. Consequently, its deformations are described by an elastic model quite different from that used to describe singly-bonded main-chain polymers.

DNA BENDING ELASTICITY

Distortion of the B-DNA structure can occur by gentle bends spread over many base pairs. The energy required to make these distortions is described by the classical linear bending elasticity of a thin rod.[1]

A long DNA in room-temperature solution is not straight because thermally excited bends crumple it into a random-coil conformation. Kratky and Porod (KP)[1] provided the first description of this crumpling by using the elastic bending energy E_{bend} as the basis for a statistical–mechanical description of bending fluctuations.

Convincing evidence that the KP model describes room-temperature DNA elasticity was recently obtained from novel micromanipulation experiments that measured the entropic force required to extend a single DNA. The force is entropic because it originates from the disordering tendency of thermal bending fluctuations. Extension of the chain requires it to become more ordered and reduces the entropy of its conformational fluctuations.

DNA TWISTING ELASTICITY

A second type of distortion that can occur is a gentle twist of the double helix, which can be described by the linear twisting elasticity used for rods.[3] This type of polymer elasticity is unusual because commonly studied singly bonded main-chain polymers can be completely rotated around any bond and therefore have no static twisting elasticity. For DNA twist stiffness arises from the ladder-like double-helix structure.[4]

Effects of the twisting elasticity are observed in fluorescence polarization decay experiments (dye molecules can be adsorbed to the base pairs so that their orientation can be sensed using polarized light)[4] and in experiments in which linear DNAs are circularized (chemical reaction of the two ends requires that they meet in a definite orientation requiring some twist).[5]

TWIST-BEND COUPLING

Can B-DNA, which has much less symmetry than a simple rod, really be described using this simple model with decoupled bending and twisting elasticity? Strictly speaking, the answer is no, as it may be shown that deformations of even a highly idealized model of DNA, a straight unstretchable helix with inequivalent major and minor grooves, requires an elastic theory with a coupling between bending and twisting.[6] In that model the twist-bend coupling makes the double helix unwind when it is bent.

Strong bends often occur when proteins bind to DNA *in vivo*. Short regions of B-DNA have been observed to completely untwist when bent sharply (within one helix repeat) during binding to a protein called TBP.[7] TBP plays a role in the startup of transcription, helping the enzymatic machinery that "reads" the DNA to have access to the base pairs.

SUPERCOILING OF DNA

Consider the coiled cord connecting a telephone handset to its base. When the cord is relaxed (the state usually strived for), it hangs in a loose loop from its anchoring points on the handset and on the base. Now, if you take the handset off the base, twist it a few times, and then hang it up, the cord wraps around itself to form an interwound helical supercoil. Circular B-DNA molecules exhibit the same behavior. These structures are referred to as "supercoils" because the superhelix repeat is much larger than the underlying B-DNA double helix repeat of 3.5 nm. Supercoiled DNAs were first observed in 1965,[8] and since then, it has been shown that DNA *in vivo* is often organized into supercoiled loops.[9] DNA supercoiling involves a fascinating interplay of elasticity, topology, and thermal fluctuation.

DNA *IN VIVO*

We have mentioned two recent experiments that studied the entropic elasticity[2] and the supercoiling[10] of DNA molecules. Both these experiments manipulated single DNAs by using proteins that have highly specific interactions with the double helix. Even more dramatic manipulations of DNA *in vivo* involve the action of more highly regulated and specialized protein machines. Consider a single human chromosome that contains one DNA molecule of about 10^8 bp. This ~4 cm-long molecule must be folded up to fit into the ~5-micron diameter nucleus. During DNA replication the two parent strands are completely separated, an operation requiring the removal of 10^7 twists. Such tasks are carried out by the coordinated action of tens of thousands of highly regulated protein machines, and understanding them is the focus of intense research.[11]

CONCLUSION

Understanding DNA's function *in vivo* is the key to understanding much of cell biology. However, DNA should be on the minds of polymer scientists as well. It is a model polymer that can be used to study fundamental questions of polymer statistical mechanics, a simple example being the study of forces needed to stretch single polymers. DNA can play many different roles in experiments thanks to its unique physical properties: short DNAs are rod-like while long ones are random coils; it is a heteropolymer with a tailorable sequence; DNA loops have a rich spectrum of topological states; polyelectrolyte effects can be emphasized or eliminated with solution ionicity. These properties, combined with our capability to enzymatically manipulate DNA nucleotide sequence and topology, point to a future of exciting DNA research at the border of biology and polymer science.

REFERENCES

1. Doi, M.; Edwards, S. F. *The Theory of Polymer Dynamics*; Oxford University: New York, 1984; Sec. 8.8. The KP model is often referred to as the "wormlike" or "persistent" chain.
2. Bustamante, C. et al., *Science* **1994**, *265*, 1599; Fixman, M.; Kovac, J. *J. Chem. Phys.* **1973**, *58*, 1564.
3. Landau, L. D.; Lifshitz, E. M. *Theory of Elasticity*; Pergamon: Oxford, 1986, Chapter II, Sec. 16–20.
4. Barkley, M. D.; Zimm, B. H. *J. Chem. Phys.* **1979**, *70*, 2991; Selvin, P. R. et al. *Science* **1992**, *255*, 82.
5. Taylor, W. H.; Hagerman, P. J. *J. Mol. Biol.* **1990**, *212*, 363; Crothers, D. M. et al. *Meth. Enzymol.* **1992**, *212*, 3.
6. Marko, J. F.; Siggia, E. D. *Macromolecules* **1994**, *27*, 981.
7. Klug, A. *Nature* **1993**, *365*, 486 and references therein.
8. Vinograd, J. et al. *Proc. Natl. Acad. Sci. USA* **1965**, *53*, 1104.
9. Worcel, A.; Burgi, E. *J. Mol. Biol.* **1972**, *71*, 127; Benyajati, C.; Worcel, A. *Cell* **1976**, *9*, 393; Jackson, D. A. et al. *EMBO J.* **1990**, *9*, 567.
10. Boles, T. C. et al. *J. Mol. Biol.* **1990**, *213*, 931.
11. Weintraub, H. *Cold Spring Harbor Symposia on Quantitative Biology*; Cold Spring Harbor Laboratory: Cold Spring Harbor, NY, **1993**, *Vol. LVIII*, p 819.

DNA (Intercalation with Drugs)

Yoshiro Nakata
Department of Biophysics
Faculty of Engineers
Gunma University

Intercalation of planar aromatic molecules between and parallel to adjacent DNA base pairs was first proposed as the result of hydrodynamic and X-ray fiber diffraction studies of DNA in the presence of acridine dyes by Lerman.[1]

Lerman suggested that structural characteristics common to these compounds provide the basis for their mode of binding; they all consist of a system of planar fused (aromatic) rings that have dimensions similar to those of a base pair. In line with observations that these drugs bind with their molecular axes perpendicular to the helix axis, Lerman hypothesized that the planar drug molecules become sandwiched (intercalated) between adjacent base pairs.[2] Model-building studies have shown that intercalation can be accomplished by a stereochemically feasible, localized unwinding of the polynucleotide backbone at the intercalation site, increasing the distance between the base from 3.4 to 6.8 Å. In this situation, rotation around the helical axis is limited, thus intercalating agents not only extend double helices, but extensively unwind them.

The fact that intercalation occurs so readily indicates that it must be energetically favored, with the van der Waals bonds between the inserted molecules and the base pairs being stronger than bonding between stacked base pairs. Intercalation provides additional evidence for the metastability of the double-helical structure. DNA has an ability to temporarily assume a number of inherently unstable configurations that under standard conditions rapidly revert back to the standard B conformation. These metastable variants must include the short stretches of extended chains with gaps separating the adjacent base pairs, into which an intercalating agent may bind.

Owing to DNA's central role in biological replication and protein biosynthesis, modification by such interaction greatly alters cell metabolism, reducing and in some cases terminating cell growth.

The intercalation theory has subsequently contributed significantly to an understanding of both the detailed physico–chemical binding behavior of many nucleic-acid-binding drugs and of their biological properties.

PREPARATION AND PROPERTIES

If an acridine dye is added to a DNA sample and fibers are drawn from X-ray diffraction studies, patterns are obtained in which the reflections resulting from the regular helical structure are blurred or lost, except for equatorial reflections indicative of lateral molecule packing and the strong 3.4-Å meridional reflection cased by base-pair stacking.[1,3,4] In essence, the aromatic 3.4-Å thick drug molecules slide between base pairs without disturbing the overall stacking pattern. However, because base pairs must separate (unstack) vertically to allow for intercalation, the sugar–phosphate backbone is distorted and the regular helical structure is destroyed. According to this model, the DNA will lengthen with the increase in drug concentration. This is indeed observed and can be monitored by measuring the increasing viscosity and reduction in the sedimentation coefficient. These effects also indicate the overall stiffening of the DNA duplex.[1]

The separation of base pairs to make room for the intercalator can be visualized as a combination of pulling along the B-DNA double-helix axis and left-handed unwinding in order not to break the sugar–phosphate backbone.[4]

INTERCALATIVE COMPOUNDS

In general, intercalating compounds have a polarizable, electron-deficient chromophore that is typically composed of three fused six-membered rings. Many compounds are known to act as intercalators, including certain classes of drugs, carcinogens, mutagens, and dyes.

The compounds doxorubicin, mitoxantrone, amsacrine, and actinomycin D are all currently in clinical use. Anthracene-9,10-dione doxorubicin has the widest activity spectrum of any current anticancer drug. A formal positive charge is common in the intercalating drugs, located either on the central chromophore or on attached substituent groups. The presence of a substituent moiety appears to be a necessary precondition for antitumor action, although not for intercalative binding. Thus, unsubstituted acridines are often good intercalators, but have no antitumor activity.

APPLICATIONS

The discovery that several intercalating agents have clinically useful activity has led to a search for superior agents, usually on the presumption that increased DNA affinity is a worthwhile goal. This may be a factor in obtaining improvements in the therapeutic index. Molecular modeling can identify and define key points of molecular interaction and, using a computational approach, can indicate structural modifications that will optimize either general binding affinity or the recognition of a particular nucleotide-binding site.

There is little evidence that intercalating drugs (e.g., adriamycin, actinomycin D, and amsacrine) have a DNA sequence selectivity that extends beyond the 2- or 3-base-pair level. Thus, the sensitivity of some tumor cells to these compounds cannot readily be explained in terms of selective inhibition of tumor-specific DNA sequences (oncogenes). Rather, evidence indicates that the antitumor effects of such agents are related to a drug-induced stabilization of a cleavable DNA complex involving the enzyme DNA topoisomerase II, which results in lethal double-strand breaks.[6] Thus, intercalation is currently considered to be the basis of the anticancer action of these drugs, and is probably the primary recognition event prior to the formation of a ternary drug-DNA-enzyme complex.[7] It has been shown that intercalating agents can be usefully exploited to enhance the value of antisense and antigene oligonucleotides.[8] These exhibit selectivity at the genome level and thus hold considerable promise as totally specific anticancer agents active against defined loci such as oncogene sequences. Rational analysis and application of these findings will be a fruitful area for future molecular modeling studies of drug-DNA intercalation.

REFERENCES

1. Lerman, L. S. *J. Mol. Biol.* **1961**, *3*, 18.
2. Waring, M. J. *Nature* **1968**, *219*, 1321.
3. Neville, D. M. Jr.; Davies, D. R. *J. Mol. Biol.* **1966**, *17*, 57.
4. Fuller, W.; Waring, M. J. *Ber. Bunsenges, Phys. Chem.* **1964**, *68*, 805.
5. Gao, Q. et al. *Proc. Nat. Acad. Sci. USA* **1991**, *88*, 2422.
6. Nelson, E. M.; Tewey, K. M.; Liu, L. F. *Proc. Nat. Acad. Sci. USA* **1984**, *81*, 1361.
7. Denny, W. A. *Anti-Cancer Drug Design* **1989**, *4*, 241.
8. Helene, C.; Toulme, J-J. *Biochem. Biophys. Acta* **1990**, *1049*, 99.

DNA DEGRADATION (Atomic Target Method using Synchrotron Radiation)

Kaoru Takakura
Physics Department
International Christian University

Katsumi Kobayashi
Photon Factory
National Laboratory for High Energy Physics

Although the history of the investigation of DNA degradation induced by radiation is brief, a fair number of reviews on this research exist.[1–7] This is because DNA damage has an important role in the death of cells and carcinogenesis induced by radiation, and the damage significantly effects the heredity of living organisms. These investigations have clarified much concerning the yields of various radioproducts and the mechanisms of the degradation of various types of DNA constituents, *in vivo* and *in vitro*. The radiation sources used for the studies were primarily X-rays from X-ray tubes with metal targets or γ-rays from radioisotopes (such as ^{60}Co or ^{137}Cs). From them it is difficult to isolate monoenergetic X-rays in order to analyze their biological effects, with which we could simulate extreme cases, such as high-LET (linear energy transfer) radiation or low-dose irradiation.

Synchrotron radiation (SR) has some advanced characteristics as a radiation source, and several groups have used it to research DNA degradation.[8–11] It provides an intense photon fluence and has a vast continuous spectrum from the vacuum ultraviolet to the hard X-ray region.

PREPARATION AND PROPERTIES

DNA Degradation in Solid State

Strand breaks in plasmid DNA induced by the K-shell ionization of bromine incorporated in DNA in the form of bromouracil-substituting thymine have been studied by Hieda et al.[11] In this study no X-ray energy-dependent enhancement of DNA damage caused by the K-shell ionization of bromine was observed.

DNA Degradation in Solution

In contrast to the studies concerning DNA degradation in the solid state described above, the effects of K-shell ionization of particular atoms in DNA were clearly observed in solution systems. Takakura has investigated the bromine K-shell edge effect on 5-bromo-2′-deoxyuridine-5′-monophosphate (BrdUMP) in aqueous solution.[12] It was concluded that the selective absorption of X-ray photons by the K-shell in the bromine atom causes an effective bromine-release.

Researchers also have carried out studies concerning the effects of the K-shell photoabsorption of phosphorous on adenosine 5′-triphosphate (ATP) and DNA.[13,14] In this study it was remarkably observed that a slight (0.65%) increase in the photon energy caused a great enhancement of degradation to ATP.

Analyses for single-strand breaks (SSB) and double-strand breaks (DSB) induced in DNA were undertaken using the agarose gel electrophoresis method.[14]

CONCLUSION

Irradiation with monoenergetic X-rays from SR aimed at the atomic target on DNA or DNA constituents can induce both specific degradations and specific radio products, or even an enhancement of the yields of these products. Although the change in the energy of the X-rays used for irradiation was less than 1%, the enhancement in the yields of the radiolytic products were from 13 to 283%. These data demonstrate that this procedure could be used for cancer therapy in the future. Since these remarkable enhancement effects have only been observed in aqueous solution systems, it is supposed that the water molecules surrounding the DNA molecules may play an important role. Further, it is suspected that water molecules may be important not only *in vitro* solution systems, but also *in vivo* biological systems. More research is needed to illuminate the enhancement mechanism, the scavenger effect, and the concentration dependency of the substrate.

ACKNOWLEDGMENTS

We would like to express our gratitude to the staff members of the Photon Factory, the National Laboratory for High Energy Physics (KEK), for their cooperation and technical assistance. We also express our sincere thanks to our colleague, Dr. Ritsuko Watanabe. This work was carried out under the approval of the Photon Factory Advisory Committee (Proposal Nos. 88-153, 90-172 and 92G207), and was partially supported by the Joint Studies Program of the Graduate University for Advanced Studies.

REFERENCES

1. Okada, S. *Radiation Biochemistry*; Academic: New York, NY, 1970; Vol. 1.
2. Blok, J.; Loman, H. *Curr. Top. Radiat. Res. Quart.* **1973**, *9*, 165.
3. Ward, J. F. *Adv. Radiat. Biol.* **1975**, *5*, 181.
4. Bertinchamps, A. J. et al. *Effects of Ionizing Radiation on DNA*; Springer Verlag: Berlin, Germany, 1978.
5. Scholes, G. *Br. J. Radiol.* **1983**, *56*, 221.
6. Hutchinson, F. *Progr. Nucleic Acid Res. Molec. Biol.* **1985**, *32*, 115.
7. Sonntag, C. V. *The Chemical Basis of Radiation Biology*; Taylor & Francis: London, United Kingdom, 1987.
8. Halpern, A. *Uses of Synchrotron Radiation in Biology*; Academic: New York, NY, 1982; p 255.
9. Ito, T.; Saito, M.; Taniguchi, T. *Photochem. Photobiol.* **1987**, *46*, 979.
10. Maezawa, H. et al. *DNA Damage by Auger Emitters*; Taylor & Francis: London, United Kingdom, 1988; 135.
11. Hieda, K.; Ito, T. *Handbook on Synchrotrom Radiation*; Elsevier Science: Amsterdam, 1991; p 431.
12. Takakura, K. *Radiat. Environ. Biophys.* **1989**, *28*, 177.
13. Watanabe, M. et al. *Int. J. Radiat. Biol.* **1992**, *61*, 161.
14. Takakura, K. et al. *Synchrotron Radiation in Biosciences*; Clarendon: Oxford, United Kingdom, 1994; p 756.

DOUBLE ISOMERIZATION POLYMERIZATION

Masatoshi Miyamoto
Department of Polymer Science and Engineering
Kyoto Institute of Technology Matsugasaki

Keigo Aoi
Faculty of Agriculture
Nagoya University Furo-cho

The driving force of the ring-opening polymerization of conventional monomers, such as cyclic ethers, cyclic imines, lactones, and lactams, is the release of ring-strain during the polymerization.[1] So it is well known that cyclic compounds that have no or small ring-strain do not cause ring-opening polymerization. But in some types of cyclic compounds, the driving force for the polymerization is not only the release of ring-strain, but also the energy gain by isomerization.

The cationic polymerization of 2-(1-pyrrolidinyl)-2-oxazoline (**1a**) gives two quite different polymers.[3] One is poly[(*N*-(1-pyrrolidinecarboylimino)ethylene] (**2a**) produced by the usual cationic ring-opening isomerization polymerization of **1a** with an alkyl sulfonate initiator, methyl trifluoromethanesulfonate (triflate) (MeOTf) or methyl *p*-toluenesulfonate (tosylate) (MeOTs), in which the 2-oxazole ring is opened as in the well-known case of cyclic imino ethers.[2] The other is poly[(1,3-diazolidin-2-one-1,3-diyl)tetramethylene] (**3a**), produced by a new mode of cationic ring-opening isomerization polymerization initiated by an alkyl halide, methyl iodide, or benzyl bromide or chloride (**Scheme I**). In the latter polymerization the

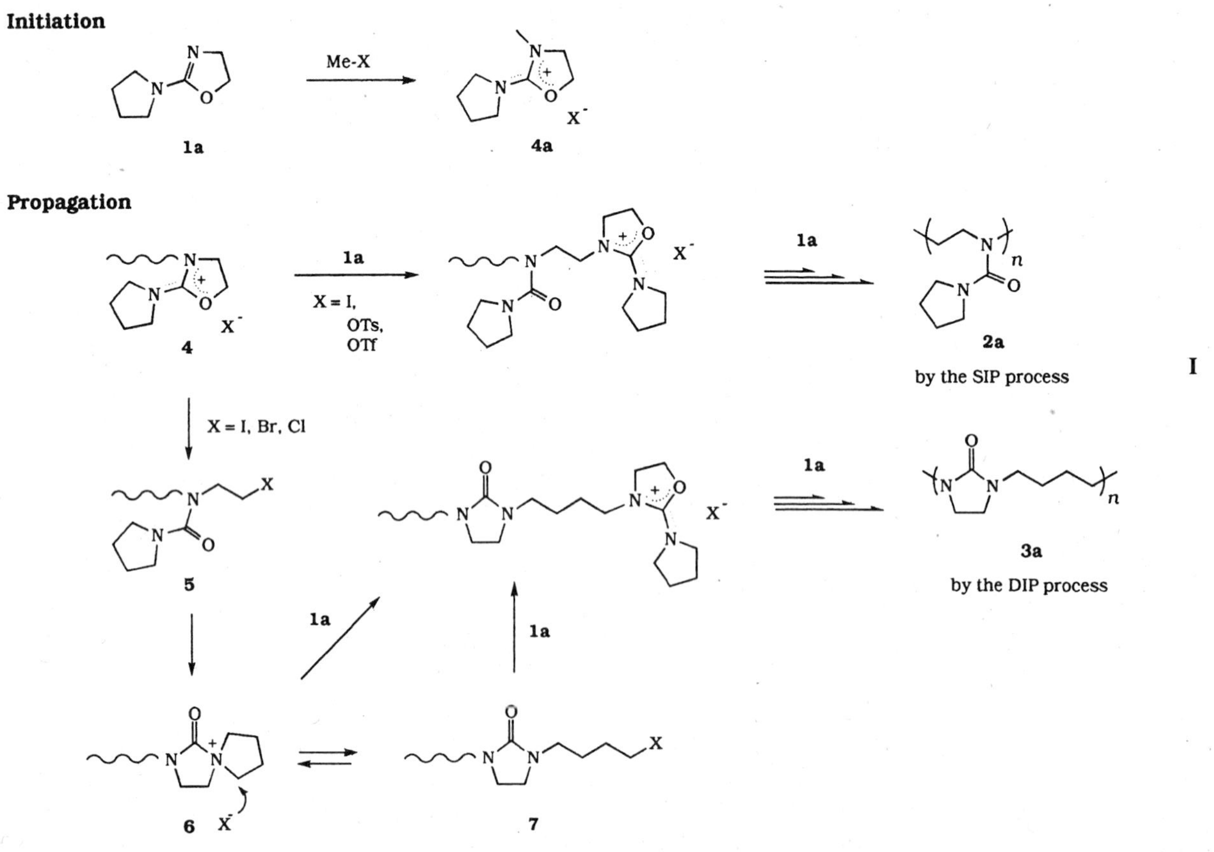

I

2-oxazoline ring in the monomer rearranges to a 5-membered cyclic urea unit and the cyclic imine moiety of the monomer suffers ring-opening since the propagating species isomerizes during the propagation.

Although both modes of polymerization of **1** belong to the same category of isomerization ring-opening polymerization, to distinguish them we will call the first mode to form **2** single isomerization polymerization (SIP) and the second mode, involving the opening of the cyclic imine ring, double isomerization polymerization (DIP), defined as the *isomerization polymerization* accompanying the *isomerization* of propagating species.

PREPARATION AND PROPERTIES

Polymerization Mechanism

To investigate the mechanism of this peculiar polymerization, we prepared 1:1 adducts of monomer **1 a** with MeOTs and with methyl iodide and investigated the isomerization reactions of these species. From these experiments the polymerization mechanisms of the present SIP and DIP processes are explained (**Scheme I**).

DIP of Bis(2-oxazolin-2-yl)piperizine

The present DIP has been successfully applied to a bifunctional monomer, 1,4-bis(2-oxazolin-2-yl)piperazine which yields a linear polyurea, poly[(1,3-diazolidin-2-one-1,3-diyl)ethylene].[4]

Polymerization of Other Pseudoureas

Polymers of 2-alkyl-2-oxazolines can be considered as the models for *N,N*-dimethylformamide or *N,N*-dimethylacetamide. These amide-type polymers have been known to possess high compatibility to commodity polymers and excellent amphiphilic properties, and they have applications as (for example) antielectrostatic agents,[5] hydrogens,[6] and nonionic surfactants.[7,8] Similar developments await the present polyurea, when such studies begin.

REFERENCES

1. Ivin, K. J.; Saegusa, T. *Ring-Opening Polymerization*; Elsevier Applied Science: Barking, Essex, United Kingdom, 1984; Chapter I.
2. Kobayashi, S.; Saegusa, T. *Ring-Opening Polymerization*; Elsevier Applied Science: Barking, Essex, United Kingdom, 1984; Chapter XI.
3. Miyamoto, M. et al. *Macromolecules* **1992**, *25*, 5111.
4. Miyamoto, M. et al. *Macromolecules* **1993**, *26*, 1474.
5. Miyamoto, M. et al. *Macromolecules* **1985**, *18*, 1641.
6. Chujo, Y. et al. *Macromolecules* **1990**, *23*, 1234.
7. Kobayashi, S. et al. *Macromolecules* **1986**, *19*, 535.
8. Litt, M. H. et al. *J. Colloid Interface Sci.* **1987**, *115*, 312.

DRAG-REDUCING POLYMERS

Oh-Kil Kim and Ling-Siu Choi
Chemistry Division
Naval Research Laboratory

Polymer drag reduction (DR) is a reduction of drag in turbulent flow resulting from the addition of a minute amount of a suitable polymer. This allows liquid to flow with less resistance; the ease of pumping a liquid increases, and a significant amount of energy is saved. Alternatively, the flow rate can be raised and a larger pipe can be used with the same pump power.

It has been suggested the DR occurs through the interactions between elastic macromolecules and turbulent flow macrostructures and that in turbulent pipe flow, the region near the wall, comprising the viscous sublayer and buffer layer, plays a major role in drag reduction. Drag reduction is not associated with modification of the surface itself. Several theories based on polymer-flow interactions have been introduced to explain polymer drag reduction. A concept of polymer chain extension by shear flow and a concomitant dissipation (by polymer) of flow energy from turbulent eddies is generally accepted[1–4] and also experimentally verified by flow-visualization photographs.[5] It seems that a drastic increase in the extensional viscosity occurs above a certain elongational strain rate, and this is an important signature of the onset of polymer DR in turbulent flow.[6,7]

Many different types of water-soluble and oil-soluble polymers have been used for DR studies, but most of them are commercial products. As water-soluble polymers, poly(ethylene oxide) (PEO), polyacrylamide (PAM), and poly(acrylic acid) (PAA), and as oil-soluble polymers, poly(methyl methacrylate) (PMMA), polystyrene (PS), and polyisobutylene (PIB) have been most widely studied. Some natural polymers such as guar and xanthan gums are also frequently used. Soluble synthetic polymers and natural polymers are far more effective for drag reduction,[9,10] although some suspended fibers and solid particles are known for DR possessing this characteristic.[11,12] Polymer drag reduction starts to appear generally at a molecular weight of about 100,000, but the polymer structure and solution conditions also contribute to the effect.[13]

Regarding the molecular parameters for DR effectiveness, the linear flexible chain structure and ultra-high-molecular-weight (UHMW) are well recognized to be the most important contributing molecular factors, and solubility is another important parameter that controls the coil dimension.[15–17] Such molecular parameters determine the conformational state of polymers, depending on the external conditions such as polymer concentration and shear rate. The percent drag reduction increases with increasing concentration and shear rate approaching Virk's maximum DR asymptote. This suggests that there exist interactions between polymers and flows. From such a molecular viewpoint, fibers, and other solid particles cannot be as effective as polymers. Colloidal particles of surfactants are also known to have some DR effect.[8,18]

The most serious problem in DR effectiveness of polymers is the chain degradation of polymers by shear strain of turbulent flow. UHMW polymers are more susceptible to shear-induced degradation,[14] and polymers with linear chain-structure are more vulnerable than branched polymers[22,23] and natural gums of semi-rigid structure.[24] The mechanism of shear degradation is assumed to be associated with chain elongation. The chain degradation is often observed when the shear rate is increased to a critical point, after which drag reduction sharply decreases. However, the decrease of drag reduction under high shear is not always the result of chain degradation but of shear-induced interchain association.

PROPERTIES

Molecular Weight Effect

Comparisons of DR effectiveness in terms of molecular weight should be made within the same polymer family and under the same solution conditions, since the hydrodynamic volume of polymers in solution is highly correlated with the onset of drag reduction.[27]

Concentration Dependence

The percent drag reduction increases with increasing polymer concentration, reaching a saturation.[14,29]

The increase of polymer concentration reaches a critical concentration (c*) at which chain overlap occurs, and interchain interactions can be induced. In high shear flows, this can occur at a lesser concentration. Some reports suggest that interchain aggregation facilitates drag reduction of PEO and PAM in an extremely dilute solution.[19,20] In this case, the chain deformation is not a true aggregation but probably is a transient association between elongated chains. A similar observation is also reported with a hydrophobically modified (<1%) PAM.[28] If a true association is responsible for the DR enhancement, the DR should increase even further when the concentration is beyond c^*. Interchain association occurs in a relatively high concentration of PAM and its copolymer solutions at rest.[31,32] Evidence for this is the decrease in viscosity, indicative of interchain H-bonding, which can be prevented by adding a salt or isopropanol. Chang and Meng suggest that there are clear relations between such a viscosity decrease and the decrease in DR effectiveness.[21]

There are interesting observations that the time-dependent conformational transitions of PAA in dilute solution is followed in a high-shear rotating disk flow.[25,26] Since DR measurements in pipe flows take a few seconds, the residence time is too short for macromolecules to form a stable interchain association. Rotating disk flow systems have a merit in that the time-dependent conformational transition can be followed over several minutes under the same flow conditions. In the case of PAA, the initial concentration (before shearing), c_0 (which can be greater than or less than c_0^*), not only determines the onset drag reduction percentage but also determines the process of conformational transitions.[30]

The decreasing drag reduction (as well as viscosity) of PAA (pH 6–8) under the shearing instantaneously and completely recovers, however, by addition of a minute amount of salt (i.e., 0.10 g NaCl per 1 L PAA solution) to the shearing solution.[25] This indicates that the decrease of DR is not due to chain degradation but to the interchain association. The recovery of DR is likely to be associated with a preferential binding of salt cations to polyions, leading to the dissociation of interchain H-bonding.[30] When the added salt exceeds a certain limit, however, PAA molecules are forced to collapse, resulting in a drastic decrease in drag reduction.[33]

APPLICATIONS

Since polymer drag reduction increases the pumpability of fluids, a significant amount of energy can be saved in numerous applications where the transport of fluids is important. Application areas include crude oil and refined petroleum product transport through pipelines, oil-well fracturing operations, firefighting, sewerage and floodwater disposal, and marine and biomedical applications. Among them, the crude oil pipeline application has become increasingly common. Gasoline, diesel fuel, and fuel-oil pipelines are also being studied for DR polymer applications. A similar application is the addition of polymers to oil being pumped from an offshore platform to shore facilities.[34]

One of the first industrial uses of water-soluble polymers was for drastic fracturing of oil wells.[35] In firefighting, application of DR polymers provides many advantages, such as an increase of flow rate, use of smaller hose lines, and increased nozzle pressure, but the practical use of DR polymers for that purpose is still under study.[36] Polymer drag reduction can also provide a cost-effective way to solve flooding problems by increasing the flow capability of storm sewers. Related potential applications are in pipeline transport of coal slurries, large-central heating hot-water installations (having a special benefit of the low heat-transfer coefficient), and irrigation. The efficacy of DR polymers to reduce ship resistance has been reported.[37] The main biomedical application of DR polymers is for improving blood flow to treat circulatory disease. Both *in vivo* and *in vitro* studies have been reported.[38]

ACKNOWLEDGMENTS

The authors wish to express their thanks to Dr. William C. Sanberg of the Naval Research Laboratory for his helpful comments.

REFERENCES

1. Peterlin, A. *Nature* **1970**, *227*, 598.
2. Cottrell, F. R.; Merill, E. M.; Smith, K. A. *Polymer* **1970**, *8*, 287.
3. Keller, A.; Mackley, M. R. *2nd Int. Conf. Drag Red.* 1977, Paper Fl.
4. Berman, N. S. *Ann. Rev. Fluid Mech.* **1978**, *10*, 47.
5. Hoyt, J. W.; Taylor, J. J. *Flow Visualization*; Merzkirch, W., Ed.; Hemisphere: New York, NY 1982; Volume II, p 683.
6. Kulicke, W.-M.; Haas, R. *Ind. Eng. Chem. Fundam.* **1984**, *23*, 308.
7. James, D. F. *The Influence of Polymer Additives on Velocity and Temperature Fluid*; Gampert, B., Ed.; Springer-Verlag: Berlin, Germany, 1985; p 25.
8. Patterson, G. K.; Zakin, J. L.; Rodriguez, J. M. *Ind. Eng. Chem.* **1969**, *61*, 22.
9. Berman, N. S. *Ann. Rev. Fluid Mech.* **1978**, *10*, 47.
10. Sellin, R. H. et al. *J. Hydraulic Res.* **1982**, *20*, 235.
11. Metzner, A. B. *Phys. Fluids* **1977**, *20*, 145.
12. McComb, W. D.; Chen, K. T. J. *Nature* **1981**, *292*, 520.
13. Little, R. C. et al. *Ind. Eng. Chem., Fundam.* **1975**, *14*, 283.
14. Gampert, B.; Wagner, P. *The Influence of Polymer Additives on Velocity and Temperature Fluid*; Gampert, B., Ed.; Springer-Verlag: Berlin, Germany, 1985; p 71.
15. Hoyt, J. W.; Fabula, A. G. *Proc. 5th Symp. Naval Hydrodyn.* **1964**, *112*, 947.
16. Hershey, H.; Zakin, J. L. *Chem. Eng. Sci.* **1967**, *22*, 1847.
17. Peyser, P.; Little, R. C. *J. Appl. Polym. Sci.* **1971**, *15*, 2623.
18. Shenoy, A. V. *Coll. Polym. Sci.* **1984**, *262*, 319.
19. Dunlop, E. H.; Cox, L. R. *Phys. Fluids* **1977**, *20*, 203 (Part II).
20. Kowalik, R. M. et al. *J. Non-Newtonian Fluid Mech.* **1987**, *24*, 1.
21. Chang, H. F. D.; Meng, J. S. *Physicochem. Hydrodyn.* **1987**, *9*, 33.
22. Kim, O-K.; Little, R. C.; Ting, R. Y. *Coll. Interface Sci.* **1974**, *47*, 530.

23. Deshmukh, S. R.; Chaturved, P. N.; Singh, R. P. *J. Appl. Polym. Sci.* **1985**, *30*, 4013.
24. Majumdar, S.; Holey, S. H.; Singh, R. D. *Eur. Polym. J.* **1980**, *16*, 1201.
25. Kim, O-K.; Long, T.; Brown, F. *Polym. Comm.* **1986**, *27*, 71.
26. Kim, O-K. et al. *Polym. Commun.* **1988**, *29*, 168.
27. Virk, P. S. In *Biotechnology of Marine Polysaccharides*; Colwell, R., Parisev, E. R., Sinkey, A. J., Eds.; Hemisphere: Washington, 1985; p 149.
28. Morgan, S. E.; McCormick, C. L. *Eng. Polym. Sci.* **1990**, *15*, 103.
29. Little, R. C. *J. Coll. Interface Sci.* **1971**, *37*, 811.
30. Kim, O-K. et al. *Macromolecules* **1993**, *26*, 379.
31. Kulicke, W-M.; Kniewske, R.; *Makromol. Chem.* **1981**, *182*, 2277.
32. Kulicke, W-M.; Kniewske, R.; Klein, J. *Progr. Polym. Sci.* **1982**, *8*, 373.
33. Kim, O-K.; Choi, L. S. *Makromol. Chem., Makromol. Sym.* **1990**, *39*, 203.
34. Hale, D. *Pipeline Gas J.* **1984**, *211*, 17.
35. Veatch, R. W. Jr. *J. Pet. Technol.* **1983**, *35*, 853.
36. Fabula, A. G. *Trans. ASME* **1971**, *93D*, 453.
37. Canham, H. J. S.; Catchpole, J. P.; Long, R. F. *Naval Architect. J. RINA* **1971**, *2*, 181.
38. Greene, H. L.; Mostardi, R. F.; Nohes, R. F. *Polym. Eng. Sci.* **1980**, *20*, 499.

Drilling Fluids

See: *Ionomers (Overview)*

Drilling Muds

See: *Lignin Graft Copolymers*

Drug Binding

See: *DNA (Intercalation with Drugs)*

Drug Delivery

See: *Anticancer Polymeric Prodrugs, Targetable*
Bioartificial Materials
Biodegradable Copolymers [Comprising Poly(L-lactic acid)]
Bioerodible Polymers A, Controlled Drug Release B. Fracture Fixation
Block Copolymer Micelles (For Drug Targeting)
Block Copolymer Micelles (Microcontainers for Drug Targeting)
Colloidal Dispersions, Nanogels
Controlled Drug Delivery Systems
Cyanoacrylates
Enzyme-Degradation Hydrogels
Glucose Responsive Polymers (Medical Applications)
Hydrogels, Microphase Separated Blends (Biomaterials and Drug Delivery System)
Hydrophobized Polysaccharides (Versatile Functions)
N-Isopropylacrylamide Copolymers (Drug Delivery)
Microcapsules, Molecular Complexes (Dissolution of Drugs by Cyclodextrin)
Microencapsulated Particles
Microencapsulation (Drug Release)
Poly(ethylene oxide) (Applications in Drug Delivery)
Polymeric Drugs
Pseudo-Poly(amino acids)s (Overview)
Radiation Induced Polymerization (Biomedical Applications)
Smart Hydrogels
Stimuli-Responsive Polymers (Thermodynamics in Drug Delivery Technology)
Water-Swellable Polymers (Carbomer Resins)

Drugs, Polymeric

See: *Anticancer Polymeric Prodrugs, Targetable*
Nucleic Acid Analogs
Polymeric Drugs

Drying Oils

See: *Vegetal Biomass (2. Oligomers and their Polymerization)*

DURABILITY

Fujio Ohishi
Faculty of Science
Kanagawa University

POLYMER DEGRADATION

Degradation of polymers takes place under various conditions, just as oxidation of metals does. As such, degradation of polymeric materials is more appropriately expressed by the word "aging." Although polymeric engineering materials are based on particular polymerized molecules, the coexistence of additives such as plasticizers and reinforcers is the rule rather than the exception. Thus, when polymers are exposed not only to additives, but to the service environment (solid, liquid, or gas) and to external energy (light, mechanical force, heat, or electricity),[1] polymer degradation is likely to proceed.

Polymeric materials are classified as viscous elastic bodies. As a result, their performance is greatly affected by temperature and vibration frequency.[2] Defects in polymeric materials vary widely, and therefore, evaluating polymer degradation is not as simple as that for metals. Thus, we proposed fracture analysis procedures for polymer materials.[2]

RELIABILITY AND DURABILITY

Concept

The relationship between durability and reliability for an integrated functional materials system is summarized in Reference 3. Durability contains the following items in general:

- Weatherability: stability against outdoor exposure
- Photostability: stability against sunshine and other light
- Radiation stability: stability against radiation

- Thermal stability
- Electrical stability
- Chemical stability
- Biodeterioration stability
- Dimentional stability
- Creep stability
- Abration stability, wear stability
- Fatigue stability
- Resistance against environment and stress

Typical reaction in degradation of polymers include:

- Main chain rupture: polymers decomposed and depolymerizing with random scission jepar reaction
- Branching
- Oxidation: caused by autoxidation by radical reaction
- Hydrolysis: polymers decomposed by radical chain reaction. In general this reaction is accelerated by acid or alkaline compounds, or high temperature
- Others

Degradative reactions are classified into stepwise reactions such as photodecomposition and chain reactions such as autoxidation, dehydrochloride reaction of PVC, and hydrolysis.

DEVELOPMENT OF DYNAMIC DURABILITY TEST APPARATUS[5]

The durability of material under service conditions, including stress, is defined as dynamic durability. To improve the dynamic durability evaluation of polymeric materials, we first worked to develop fatigue apparatus designed specifically to test plastic materials and then to develop apparatus designed to test elastomers.

Factors Governing Durability of Polymeric Materials

In order to identify the factors that govern durability and the initial properties of plastic, materials research was undertaken making use of the new apparatus devised by us.

Polycarbonate was chosen as the representative engineering plastic, and quantitative analyses were done to determine the influence of the following factors: average molecular weight, unsuitable conditions in the molding process (we cannot observe with the naked eye), preparation methods, and fillers (glass fiber, polyethylene blend).[4] The author summarized the results obtained from this work.[5] From this systematic investigation, the existence of a degree in the intensity of influence of these factors was recognized. It also became apparent that there is an appreciable risk in evaluating dynamic durability on the basis of the static properties of such materials.

REFERENCES

1. Ohishi, F. Durability of Plastics; Jpn Soc. Industrial Survey, 1975.
2. Ohishi, F,; Narusawa, P. *Life of Plastics*, Nikkan Kogyo Shimbun: Tokyo, 1987.
3. Ohishi, F. *Koobunshi (Polymer)* **1986**, *35*, 646.
4. Ohishi, F. et al. *J. Appl. Polym. Sci.* **1976**, *20*, 79.
5. Ohishi, F. *Corrosion Eng.* **1989**, *38*, 655.

Dyes

See: *Colorants (Overview)*
Nonlinear Optical Polymers, Thermoset
Perylene Polymers
Polyimides (From Tricyclic Heteroaromatic Dyes)
Polyimides, Precursors (Dye-Containing Side-Groups)
Silk Fibers (Chemical Modification)
Silk Fibers (Grafting)
Thermosensitive Paper (Thermosensitive Dyes, Thermochromic Compounds)

E

Elastomers

See: *Acrylic Elastomer Curing*
Acrylic Rubber
Acrylonitrile-Acrylic Elastomer-Styrene Terpolymer
Alternating Radical Copolymerization
Chlorosulfonated Polyethylene
Conductive Elastomeric Blends
Contact Lenses, Gas Permeable
Emulsion Polymers (Noncatalytic Diimide Hydrogenation)
Ethylene-Propylene Rubber Modification (Maleic Anhydride and Derivatives)
Ethylene-Propylene Elastomers (Ethylene-Propylene Rubber)
Fluorocarbon Elastomers (Commercial)
Interphases, Ternary Polymer Composites
Ionene Polymers
Ionomers (Overview)
Isobutylene Copolymers (Commercial)
Liquid Crystalline Elastomers
Liquid Crystalline Polymers, Side Chain
Liquid Single Crystal Elastomers
Living Radical Polymerization, Iodine Transfer
Macromonomer Thermoplastic Elastomers
Metallocene Catalysts (Group 4 Elements, New Polymeric Materials)
Peroxide Initiators (Overview)
Photocrosslinking (Overview)
Polypropylene Blends and Composites
Polyurethane Elastomers, Segmented (Non-Hydrogen Bonding Systems)
Polyurethanes (Overview)
Reactive Oligomers (Overview)
Rubber Additives, Multifunctional
Silicone Elastomers
Silicone Rubber Latex
Siloxane Elastomers
Styrene-Butadiene-Styrene Elastomer, Hydrogenated
Styrene-Butadiene-Styrene Triblock Copolymer
Tack, Elastomers
Tetrafluoroethylene Copolymers (Overview)
Thermoplastic Elastomers (Overview)
Unsaturated Polyester Resins (Toughening with Elastomers)
Unsaturated Polyester Resins (Toughened with Liquid Rubber)
Vulcanization (Overview)

Electrical Insulation

See: *Polytetrafluoroethylene*

Electrical Stress

See: *Stress-Induced Chemiluminescence Imaging*

ELECTRICALLY CONDUCTING COMPOSITES

Christoph Kröhnke
Additives Division
CIBA-GEIGY Ltd.

Disadvantages of intrinsically conducting polymers or conducting organic materials in general, such as insufficient mechanical properties and poor environment stability and processability, can be circumvented by preparing composites with stable, nonconducting components, preferentially polymers.[1] Another approach is given by the graft copolymerization.[2] Under appropriate conditions a composite with features of the nonconducting matrix and electrical characteristics of the electroconductor can be achieved.[3,4]

Composites are available by a variety of blending techniques as well as by synthesis of the conducting polymer within a host matrix.[5,6] Reviews describe the carbon black- and metal (Ni, Ag)-filled polymers.[7,8] Synthetic approaches for polymeric intrinsic electroconductors such as polypyrrole, polythiophene, and polyaniline are given by chemical or electrochemical routes.[9–16] The latter has the advantage that the electrical properties of the composite can be changed simply by varying the electrolysis conditions in a controlled way.[17] Moreover, the electrochemical synthesis eliminates the need for strong oxidizing or hazardous agents.

Mechanical properties are usually determined by the matrix of the nonconducting polymer so long as the fraction of the conducting phase is low enough. But the appropriate system consisting of a nonconducting matrix and a conducting phase should be available in such a way that the nonconducting matrix allows the penetration of conducting species or its precursor.

The design of volume conductor composites is limited by the laws of percolation describing the transport of charge carriers between particles.[18,19,67] As part of the rules of percolation, the shape of conducting particles determines the volume fraction to reach the percolation threshold forming a continuous network.[20] According to recent literature, other aspects besides percolation influence the electrical properties of conducting composites, such as physical, chemical, and structural surface properties of individual particles: for example, the inherent electrical properties of matrix polymers and tunneling over small interparticle distances.

Improved structural order, as achieved by stretching, leads to materials with higher conductivity and stability. While intrinsically conducting polymers embedded in a matrix of a nonconducting polymer is a possible method to introduce flexibility and toughness into conducting materials, good mechanical properties can be achieved if the content of the intrinsically conducting polymer is kept as low as possible.

An approach using electrically conductive polymer composites is proposed, taking advantage of the differential etching behavior of the electrically conducting versus the insulating areas under reactive ion etching (RIE) conditions.[21] The electrically conducting areas experience RIE-type etching and therefore etch faster in an oxygen plasma than the insulating sections.

COMPOSITES OF INORGANIC COMPONENTS AND NONCONDUCTING POLYMERS

The addition of inert inorganic substances to polymers and especially to poly(ethylene oxide) based polymers improves their mechanical properties and extends their temperature stability range.

Besides conventional metal-filled polymers, there is a development going on in compounds containing conductive fibers. Metals or metal-surfaced fibers, such as nickel-coated graphite fibers, provide higher levels of conductivity than carbon black, together with useful mechanical properties.[22]

For application as charge-controlling additives for toners, fillers in electrically conductive coatings, conductive synthetic fibers, and copier and telefax drums are claimed mixtures of conducive inorganic particles, such as SnO_2 coated with fluoride ions.[23]

Particle features like shape, size, and arrangement of conducting fillers are the most important parameters to determine the conductivity behavior of a composite together with polymers.[24]

Manufacture of Transparent and Electrically Conductive Plastic Films by Coating Silane Coupler-Coating Acrylate Resins and Metal Oxides

Transparent and electrically conducting polymeric films are manufactured by covering polymers with metal oxides.[26] A two-step procedure results in composites by coating polymeric surfaces (e.g., poly(ether sulfone)s) in the first step with an UV-curable epoxy acrylate containing a silane coupler, curing the system, and coating in a second step with a metal oxide like an ITO membrane.

Thermally conducting composites comprise a polyimide matrix with both aluminium nitride particles and silicon carbide whiskers.

Polyolefin Compositions with Long-Term Antistatic Properties

Molten polyolefins are mixed with antistatic agents and inorganic fillers.[28]

Composites of low density polyethylene and an inorganic powder such as graphite or tin dioxide can be prepared.[29] The morphology and polarity of the particles influence their dispersion and conductivity behavior in low density polyethylene (LDPE).

Propylene can be polymerized on the activated surface of fillers that determine properties like electrical conductivity forming isotactic polypropylene.[30] Compared are the properties as received by this route with properties observed by blending.

Based on polypropylene and polystyrene, conducting composites are obtained with conducting fillers (iron, copper, carbon black) by means of physical blending.[31]

Nanocomposites

Mechanisms and parameters controlling the transport of electrical charges between single conductive particles in organic and polymeric matrices are studied by measuring electrical currents in the picoampère range between different types of electrodes, Au, CuO, Pt, as a function of the relative displacement in the nanometer range.[25] Besides pentadecane and air as separating media thin, *in situ* cured epoxy resin-layers have been used.

Other Conductive Compositions of Polymers with Inorganic Fillers

Claimed are also composites of conductive inorganic fillers dispersed preferentially in amorphous thermoplastic and thermosetting resins.[32]

Ternary conductive composites based on metallized polyester and glass fibers produced by coating nickel alone or double layers of copper and nickel with the aid of electroless plating that are compounded with synthetic resins such as poly(vinyl chloride), acrylic, and polyurethane resins are evaluated for their potential EMI-shielding effectiveness.[33]

Application Approaches: Li-based Ionomers

A complex of poly(ethylene oxide)-grafted poly(methylmethacrylate) with a lithium-salt has been examined as a solid electrolyte of a rechargeable lithium-battery.[27]

COMPOSITES OR INORGANIC COMPONENTS AND CONDUCTING POLYMERS

Various Polypyrrole-Based Composites with Inorganic Components

Charge-discharge capacities of conducting polymer films are limited by the maximum amount of electrolyte anions that are known to be about 0.3 moles and 0.5 moles per mole polypyrrole and polyaniline respectively.[34] These values are too low to apply these electrodes in high energy density batteries. In order to limit the size of the batteries the energy capacity per volume rather than the energy capacity by weight is the important parameter. For a technical improvement, the electrochemical polymerization of pyrrole from a propylene carbonate solution containing additionally β-manganese oxide has been carried out resulting in the incorporation of manganese oxide in the polypyrrole film grown at a glassy carbon anode.

According to their charge-discharge properties, these films can be used as high energy capacity cathodes for rechargeable Li-based batteries with energy capacities almost twice as high as measured for pure polypyrrole.

Pyrrole can be polymerized chemically within montmorrilite clays using Fe(III)- or Cu(II) salts as oxidants.[35] Polypyrrole-clay composites can be also prepared electrochemically. The products show an application potential for both sensor and electrolysis area.

Composites of Nanostructures

Typical technological applications are claimed for nano-dispersed catalyst particles in a matrix of an intrinsically conducting polymer.[37] For example, an electrically conductive polymer film, preferentially polypyrrole, contains colloidal catalyst particles such as platinum with particle sizes < 10nm homogeneously dispersed therein. Advantages of this type of catalysis lies in the ready retrieval of catalytic particles and their efficient utilization as well as in the resistance of the particles to at least high molecular weight poisons, which are hindered to penetrate into the polymer matrix.

Characterization of Polypyrrole-Fiber Composites by TOF Secondary Ion MS and Vibrational Spectroscopy

Of some interest for application reasons are conducting composites consisting of a nonconducting fiber component and an intrinsic electroconductor.[36]

Fiber composites are structured of the underlying nonconducting component covered with an overlayer of the conducting component as confirmed by time-of-flight secondary mass spectroscopy (TOF-SIMS) on polyester-fiber/polypyrrole and quartz-fiber/polypyrrole systems.

COMPOSITES OF CARBON-BASED FILLERS AND POLYMERS

Composites of Carbon-Based Fillers and Nonconducting Polymers

For the past several decades, carbon-based fillers have been applied to modify the properties of polymers. A conventional technical solution from the beginning was to use carbon black that can be produced by the so-called Furnace-process by combustion of oil in air with aid of a carrier gas.[68] Other carbon black types are manufactured by thermal breaking up of acetylene.

Carbon fibers are technologically interesting for special purposes but sensitive against shearing stress. Important is the formation of a polymeric interlayer to improve the interfacial properties of composites.[69–71]

Nonconducting Polymers Modified with Carbon Black

Composites containing a matrix polymer based on a polyester (PE-246) filled with carbon black or with binary fillers containing carbon black together with andesite or graphite have been prepared.[39] The electrical conductivity depends on the filler content and the ratio of components. Increasing temperature and pressure lead to decreasing conductivity owing to thermal expansion of the composite and destruction of the conducting fillers.

Described are applications of heterogeneous compositions of thermoplastic polymers such as poly(ether sulfone), polycarbonate and polyamide-6 with electrically conducting additives such as carbon black powder, carbon black fibers, and steel fibers with low surface resistances used for integrated circuit boards.[40]

Composites with Graphite

Electrically conducting compositions with good resistance to chemicals, water, and heat are claimed mixing polycarbodiimides, silicone modified epoxy-resins, and electrically conducting fillers, e.g., graphite, and heated.[41] The products are described as materials with good peel strength.

Composites with Carbon Fibers

Electrochemical copolymerization can be used to impregnate continuous unidirectional graphite fibers with controlled volume fraction of poly (N-substituted maleimide-costyrene) thermoplastic matrix.[42]

A main purpose for the application of carbon fiber reinforced composites is the combination of high modulus, high strength, and often thermal stability with low density mainly for high-performance equipment.[38] Most of the carbon fibers produced are based on polyacrylonitrile (PAN) fibers; only very limited amounts are based on cellulose fibers (rayon), which are heat-treated to form the final carbon fibers.

Composites use epoxy resins or high temperature thermoplastics as matrix polymer. Those composites are obtainable by successive resin impregnation and carbonization. As for some other conducting composites areas of application are electromagnetic interference (EMI) shielding, antistatic coating, and multiple use as resistance heating elements, for batteries, catalysts, and selective absorbants.

Vapor-growth carbon fibers (VGCF) are also used as fillers to prepare composites with thermoplastic resins, such as ethylene-vinyl acetate copolymer.[43] Compared with conventional carbon filler like carbon black and polyacrylonitrile-based carbon fibers (PAN), VGCF lead to composites with higher conductivity and better stability of the electrical conductivity.

COMPOSITES OF CARBON-BASED FILLERS AND CONDUCTING POLYMERS

Electrochemical Deposition of Polypyrrole on Carbon Fibers for Improved Interfacial Bonding to Epoxy Resins

Improve interfacial bonding is attainable between carbon fibers and epoxy resins after continuous electrochemical deposition of polypyrrole on the carbon fibers, which is characterized by ESCA and scanning electron microscopy.[44]

COMPOSITES OF TWO DIFFERENT ELECTRICALLY CONDUCTING POLYMERS

This chapter deals mainly with the behavior of composites of intrinsically conducting polymers with polyion complexes with different proportions of anionic and cationic groups.

Composites with Polyaniline

Charge balances on composite films composed of polyion complexes as matrix polymers with cationic and anionic functionalities and polyaniline in contact with aprotic media have been studied by FT-IR.[45] So a varying weight-ratio of sodium poly(4-styrenesulfonate) and poly(allylamine hydrochloride) provided large electrochromic changes and improve electrochemical stability of polyaniline in the matrix. These polyaniline composites show a similar electrochemical and electrochromic behavior to homogeneous polyaniline films but improved electrochemical stability due to hybridization with the matrix polymer.

Composites with Polypyrrole

Anionic polyurethane ionomers can be used for a one-step electrosynthesis of conductive composites.[46] A precursor polyurethane prepared from poly(ethylene adipinate), diphenylmethane diisocyanate and butanediol is converted with 1,3-propanesultone/NaH into a water-soluble ionomer with sodium propanesulfonate moieties. This compound acts as polymeric electrolyte and dopant during the electrochemical polymerization of pyrrole forming an electrical conducting polypyrrole/propanesulfonate-composite.

Electrochemical polymerization of pyrrole in the polymer electrolyte poly(ethylene oxide) (PEO)/$LiClO_4$ applying the potential sweep method leads *in situ* to polypyrrole/PEO/$LiClO_4$ junctions.[49]

Application of Polypyrrole/Polymer Anion Composites as Cathodes in Polymer Batteries

Polypyrrole/polymer anion composites are used as the cathode in a conducting rechargeable battery if high energy densities can be reached.[47]

Nafion-Containing Composites of Intrinsic Electroconductors

Nafion composites with polypyrrole can be obtained by polymerizing pyrrole in a diaphragmatic procedure in which a supporting Nafion film was put between the monomer chamber and an oxidant solution chamber to give electrically conductive composite films.[48]

Polyacetylene/Polypyrrole Conducting Composites

Properties characteristic for both of the constituent homopolymers are shown by polyacetylene/polypyrrole conducting composites in which the two components are mixed intimately at molecular level and conduct the current simultaneously.[50]

Composites of Superconducting Ceramics and Nonconducting Polymers

Fe(III)-doped poly(phenylene sulfide) acts as matrix for high-T_c bismuth-containing superconductors to obtain composites with variable properties.[51]

Pores of a structurally weak 3-3 superconducting ceramics $Ba_2YCu_3O_{7-\delta}$ or BaYCO have been impregnated with low viscosity epoxides using simultaneous pressure/push and vacuum/pull techniques.[52]

COMPOSITES OF INSULATING POLYMERS AND CONDUCTING POLYMERS

The physical and electrochemical properties of electrically conducting composites can be adjusted in relatively wide ranges by varying the counterions, introducing substituents at the electrically conducting component, and varying the insulating matrix. The effect of the different adjustments can be used to advantage to custom-tailor the properties of the resulting materials to a specific application.

Several routes are described to prepare those composites in which π-conjugated conductive polymers are combined with nonconducting polymers. Described is an *in situ* procedure for a quasi-composite consisting of polyacetylene and low-density polyethylene (LDPE) by impregnation of the LDPE with Ziegler-Natta catalysts to render the polymerization of acetylene with the LDPE-matrix possible.[53] This approach has been extended to other thermoplastic polymers such as polystyrene, poly(ethylene oxide), poly (methyl methacrylate), and ethylene-vinyl acetate copolymer.[54] Other attempts for a chemical route by sintering poly(p-phenylene) and poly(phenylene sulfide) are described.[55]

From the application point of view, an issue for the preparation of highly conducting polymer blends could be EMI-shielding. Very high levels of shielding performance can be achieved by the use of intrinsically conductive polymers in thermoplastic blends. The most reliable measure of this performance potential is obtained from far-field shielding experiments.

COMPOSITES OF ORGANIC CHARGE-TRANSFER SALTS AND NONCONDUCTING POLYMERS

The thermostable, highly conductive radical cation salt tetraselenotetracene chloride $(TSeT)_2$ Cl was prepared by *in situ* crystallization in a vitrifying polymer matrix e.g., in poly-(bisphenol-A-carbonate), polystyrene, soluble polyimides.[56–58] Highly conductive transparent films (σ = 1–5 S/cm) with networks of $(TSeT)_2$ Cl needle-like crystallites of 1000: 1 aspect ratio (= length/diameter ratio) have been prepared at a very low percolation threshold (< 0.5 wt-%). The conductivity of those films show a metallic behavior in the temperature range from about –200 to > 130°C.

$(TSeT)_2$ halides combine high conductivity at room temperature (σ = 2500 S/cm) with high stability up to over 250°C. The driving force for the formation of a needle network of $(TSeT)_2$ Cl in the polymer matrix is the insolubility of this charge-transfer complex salt. This class of composites was described as the first example of organic composites with real metallic ohmic properties.

Reticulate doping of poly(γ-methyl-L-glutamate) and poly(γ-methyl-DL-glutamate) with 1,3-dimethylimidazolinium-TCNQ complex salt leads to polymeric composites.[59] A high conductivity, even higher than that of pure radical salts measured in pressed pellets, was obtained in poly(γ-methyl-L-glutamate) doped films and was probably due to the α-helical conformation. This high-order confirmation of molecules forces microcrystals of charge transfer salts to grow in long, hairlike shapes distributed regularly in the bulk and in the polymer film.[60]

Reticulate doping of other substrates (polyethylene, polycarbonate) with TCNQ derivatives proves conduction networks that appear to be truly continuous and made of ultrathin needles.[61–66]

REFERENCES

1. Billingham, N. C.; Calvert, P. D. *Adv. Polym. Sci.* Springer-Verlag: Berlin, Heidelberg, **1989**, *90*.
2. Ishizu, K.; Honda, K.; Kanbara, T.; Yamamoto, T. *Polymer* **1994**, *30*(22), 4901.
3. Tassi, E. L.; De Paoli, M. A. *J. Chem. Soc.; Chem. Commun.* **1990**, *155*.
4. Zoppi, R. A.; De Paoli, M. A. *J. Electroanal. Chem.* **1990**, *290*, 275.
5. Hefling, J. R.; Singer, K. D.; Yu, L.; Reynolds, J. R.; Techagumpuch, A. In *"Electroresponsive Molecular and Polymeric Systems"* Skotheim, T. A. Ed.; Marcel Dekker: New York, 1991, pp 1–300.
6. Franck, A.; Biederbick, K. *Plastic Compendium*; 2nd Ed. Vogel; Würzburg, 1988, pp 1–364.
7. Carmona, F. *Ann. Chim. Fr.* **1988**, *13*, 395.
8. Ruschau, R. R.; Yoshikawa, S.; Newnham, R. E. *J. Appl. Phys.* **1992**, *72*(3), 953.
9. Naarmann, H. *Ger. Offen.* DE 3 603 796 (Appl. 1987).

10. Naarmann, H. *Ger. Offen.* DE 3 705 647 (Appl. 1988).
11. Niwa, O.; Kakuchi, M.; Tamamura, T. *Polym. J.* **1987**, *19*, 1293.
12. Tieke, B.; Gabriel, W. *Polymer* **1990**, *31*, 20.
13. Koga, K.; Iino, T.; Ueta, S.; Takayanagi, M. *Polym. J.* **1989**, *21*, 303.
14. Bhat, N. V.; Sunderesan, E. *J. Appl. Polym. Sci.* **1989**, *38*, 1173.
15. Hanack, M.; Roth, S.; Schier, H. *Proceedings of the International Conference on Science and Technology of Synthetic Metals"—ICMS '90; "Synthetic Metals"*; Elsevier Sequoia SA, Subscription Department, Lausanne 1; Switzerland; **1991**, *41*, pp 775–1404.
16. Heeger, A. J.; Smith, P.; Fizazi, A.; Moulton, J.; Pakbaz, K.; Rughooputh, S. *Proceedings of the International Conference on Science and Technology of Synthetic Metals"—ICSM '90 "Synthetic Metals"*; Elsevier Sequoia SA, Subscription Department, Lausanne 1; Switzerland; **1991**, *41*, pp 1027–1032.
17. Diaz, A. F.; Kanazawa, K. K. *"Extended Linear Chain Compounds,"* Miller, J. S. Ed.; Plenum: New York, 1983, Vol. 3, 417.
18. Bhattacharya, S. K.; Chaklander, A. C. D. *Review of Metal-filled Plastics, Part 1, Electrical Conductivity, Polym. Plast. Technol. Eng.* **1992**, *19*(1), 21.
19. Levon, K.; Margolina, A.; Patashinsky, A. Z. *Macromolecules* **1993**, *26*, 4061.
20. Hashin, Z. *J. Appl. Mech.* **1993**, *50*, 481.
21. Bargon, J.; Baumann, R.; Broeker, P. *Proc. Int Soc. Opt. Eng.* 1992, pp 441–447.
22. Maplestone, P. *Modern Plastics Int.* **1989**, *10*, 66.
23. Wedler, M.; Hocken, J.; Rosin, Y.; Hayashi, T. (*Metallgesellschaft AG*), Eur. Pat. Appl. 586 003; September 4, 1992.
24. Yang, L.; Schruben, D. N. *Polym. Eng. Sci.* **1994**, *34*(14), 1109.
25. Zweifel, Y.; Bujard, P.; Hilborn, J. G.; Kausch, H. H.; Plummerand, C. J. G.; Strümpler, R. *Extended Abstract subm. for the PPM Symposium Sion*, September 8–9, 1994.
26. Nakamura, K. (Sumitomo Bakelite Co.), Japan Kokai Tokkyo Koho JP 06 116 425; October 9, 1992.
27. Morita, M.; Fukumara, T.; Motoda, M.; Tsutsumi, H.; Matsuda, Y.; Takahashi, T.; Ashitaka, H. *J. Electrochem. Soc.* **1990**, *137*, 3401.
28. Kubo, K.; Kobayashi, J. (Sumitomo Chem. Co.); Japan Kokai Tokkyo Koho JP 06 128 422; October 16, 1992.
29. Nagata, K.; Kodama, S.; Nigo, H.; Kawasaki, H.; Dek, S.; Mizuhata, M. *Kobunshi Robunshu* **1992**, *49*(8), 677–685.
30. Galashina, N. M. *Vysokomol. Soedin, Ser. A. Ser. B* **1994**, *36*(4), 640.
31. Acosta, J. L.; Rodriguez, M.; Linares, A.; Jurado, J. R. *Polym. Bull.* **1990**, *24*, 87.
32. Baigrie, S.; Chu, E. F.; Park, G. B.; Reddy, V. N.; Rinde, J. A.; Saltman, R. P. (Raychem Corp.); U.S. Patent 5 250 228; November 6, 1991.
33. Shinagawa, S.; Kannbe, T.; Urabe, K.; Suzuki, H. *Proc. of the Asia Pacific Conference on Electromagnetic Compatibility*; Madras, India: SAMEER 1992, pp 136–139.
34. Yoneyama, H.; Kishimoto, A.; Kuwabata, S. *J. Chem. Soc., Chem. Commun.* **1991**, *15*, 986–987.
35. Ma, W.; Lowe, J. A.; Pan, W. P.; Brown, T. *J. Mater. Chem.* **1994**, *4*(5), 771/2.
36. Hearn, M. J.; Fletcher, I. W.; Church, S. P.; Armes, S. P. *Polymer* **1993**, *34*(2), 262.
37. Rajeshwar, K.; Bose, C. S. C. U.S. Patent 5 334 292; August 17, 1992.
38. Ruland, W. *Adv. Mater.* **1994**, *2*, 528.
39. Aneli, D. N.; Gventsadze, D. I.; Shamanauri, L. G. *Plast. Massy (1)* **1993**, 22–25.
40. Quinn, K. R.; McCullough, L. A.; Carreno, C. A. *Annu. Tech. Conf. Soc. Plast. Eng.* **1991**, 557–565.
41. Saito, K.; Okamoto, T.; Osawa, N.; Kitani, M.; Norifune, T.; Kitamura, T. (Isuzu Motors limited), Japan Kokai Tokkyo Koho JP 06 16 907, June 29, 1992.
42. Iroh, J. O.; Bell, J. P.; Scola, D. A.; Wesson, J. P. *Polymer* **1994**, *35*(6), 1306.
43. Katsumata, M.; Yamanashiand, H.; Ushijima, H. *Proc. SPIE-Int. Soc. Opt. Eng.*, 1916 (Smart Materials), **1993**, 140.
44. Chiu, H. T.; Lin, J. S. *J. Mater. Sci.* **1992**, *27*(2), 319.
45. Morita, M. *Makromol. Chem.* **1993**, *194*(5), 1513.
46. Robila, G.; Ivanoiu, M.; Buruiana, T.; Buruiana, E. C. *J. Appl. Poly. Sci.* **1993** *49*(11), 2025.
47. Ohtani, A.; Abe, M.; Higuchi, H.; Shimidzu, T. *J. Chem. Soc., Chem. Commun.* **1988**, 1545.
48. Iyoda, T.; Ohtani, A.; Honda, K.; Shimidzu, T. *Macromolecules* **1990**, *23*(7), 1971.
49. Watanabe, A.; Mori, K.; Takahashi, A.; Nakamura, Y. *Synth. Metals* **1989**, *32*, 201.
50. Mc Donald, R. C.; Pickett, J.; Goebel, F. *IECEC-91: Proc. 26th Intersoc Energy Conversion Eng Conf.* **1991**, pp 74–79.
51. Benlhachemi, A.; Gavarri, J. R.; Musso, J.; Alfred-Duplan, C.; Marfaing, J. *Physica C (Amsterdam)*, **1994**, *230*(3–4), 246.
52. Crosby, B. J.; Unsworth, J.; Du-Moore, J.; Bryant, P. *Adv. Compos. 193, Proc. Int. Conf. Adv. Compos. Mater.*; Chandra, T.; Dhingra, A. K., Eds.; Miner. Met. Mater. Sci., Warrendale, Pa.; **1993**, *1*, 397.
53. Galvin, M. E.; Wnek, G. E. *Polymer* **1982**, *23*, 795.
54. Wnek, G. E. *"Electrically Conductive Polymer Composites, Handbook of Conducting Polymers"* **1986**, pp 205–212.
55. Rueda, D. R.; Cagiao, M. E.; Balta Calleja, F. J.; Pallacios, J. M. *Synth. Met.* **1987**, *22*, 53.
56. Kröhnke, C.; Finter, J.; Mayer, C. W.; Ansermet, J.; Bleier, H.; Hilti, B.; Minder, E.; Neuschäfer, D. *Contemporary Topics in Polymer Science: Vol. 7 (Advances in New Materials)*, Salamone, J. C.; Riffle, J. Eds.; Plenum: New York, 1992, p 191.
57. Bleier, H.; Finter, J.; Hilti, B.; Hofherr, W.; Mayer, C. W.; Minder, E.; Hediger, H.; Ansermet, J. P. *Synth. Met.* **1993**, *57*(1), 3605.
58. Hofherr, W.; Minder, E.; Hilti, B.; Ansermet, J. P. European Patent Appl. EP 521 826; June 25, 1992.
59. Sorm, M.; Dedek, P.; Nespurek, S.; Kotva, R.; Pivkova, H. *Angew. Makromol. Chem.* **1991**, *190*, 201.
60. Sorm, M.; Nespurek, S.; Koropecki, I.; Kubanek, V. *Angew. Makromol. Chem.* **1990**, *174*, 55.
61. Ulanski, J. *Synth. Met.* **1991**, *41*(3), 923.
62. Jeszka, J. K.; Tracz, A.; Kryszewski, M.; Ulanski, J. *Synth. Met.* **1988**, *27*(3–4), B115.
63. Tracz, A.; Shafee, E. E.; Jeszka, J. K.; Kryszewski, M. *Synth. Met.* **1990**, *37*(1–3).
64. Kim, O. K. *Polym. Sci., Polym. Lett. Ed.* **1983**, *2*(17), 575.
65. Imai, A.; Suzuki, T.; Kitano, H. (Matsushita Electric Industrial Comp.), Japan Kokai Tokkyo Koho JP 5 502 681; April 4, 1974.
66. Mizoguchi, K.; Kamiya, T.; Tsuchida, E.; Shinohara, I. *J. Polym. Sci. Polym. Chem. Ed.* **1979**, *17*(3), 649.
67. Weßling, B. *GAK* **1992**, *45*(10), 541.
68. Dixit, S. *GAK* **1984**, *37*, 504.
69. Subramanian, R. V. *Pure Appl. Chem.* **1990**, *52*, 1929.
70. Subramanian, R. V.; Jakubowski, J. J. *J. Polym. Eng. Sci.* **1978**, *18*, 590.
71. Subramanian, R. V.; Jakubowski, J. J. *J. Org. Coat. Plast. Chem.* **1979**, *40*, 688.

Electro-optical Materials

See: *Electroluminescent Polymers (Overview)*
Photopolymer Materials (Development of Holographic Gratings)
Poly(arylene vinylene)s (Mechanistic Control of a Soluble Precursor Method)
Polymer Dispersed Liquid Crystal Display (Driving Circuit for HDTV)

Electroactive Polymers

See: *Batteries (Materials for High Performance)*
Block Copolymers (Anisotropic Conductivity)
Charge Transport Polymers (for Organic Electroluminescent Devices)
Chiral Polyanilines
Conducting Ladder Polymers
Conducting Polymer Colloids
Conducting Polymers (For Langmuir-Blodgett Film Fabrication)
Conducting Polymers (Nonconjugated)
Conducting Polymers (Self-Acid-Doped, Conjugated)
Conductive Elastomeric Blends
Conjugated Ladder Polymers
Conjugated Polymers (Insulating and Conducting Forms)
Conjugated Polymers (Prepared by Organometallic Processes)
Electrically Conducting Composites
Electrochemical Polymerization
Electroconductive Composites (Emulsion Pathways)
Electroconductive Paints
Electroluminescent Polymers (Overview)
Enzyme-Catalyzed Oxidative Polymerization (Aromatic Compounds)
Enzyme-Catalyzed Polymerization
Ferroelectric Polymers (Structural Phase Transitions)
Ferromagnetic Polymers
Glassy Carbon
Intelligent Membranes
Ionomers (dc Conduction Properties)
Langmuir-Blodgett Films (Organic Polymers for Photonics Applications)
Liquid Crystalline Elastomers
Liquid Crystalline Polyacetylene Derivatives
Liquid Crystalline Polymers, Side Chain
Metallized Polyimides
Metal-Polymer Complexes
Metallophthalocyanines, Polymeric (Overview)
Oxidative Coupling
Nonlinear Optical Materials
Nonlinear Optical Polymers [Poly(arylene ethynylene) Derivatives]
Nonlinear Optical Polymers, Thermoset
Photochromic Films, Azo Dyes (Photoinduced Anistropy and Photoassisted Poling)
Phthalocyanines, Polymeric
Polyacenic Semiconductor Materials (Applications to Rechargeable Batteries)
Polyacetylene Films
Polyaniline (Electrochemical Synthesis)
Polyaniline Network Electrodes (Enhanced Performance of Polymer LEDs)
Polyanilines, Oxidation States
Poly(arylene vinylene)s (Mechanistic Control of a Soluble Precursor Method)
Polyazines (Third-Order Nonlinear Optical Properties)
Polybithiophenes (Advanced Derivatives)
Polydiacetylenes, Unconventional
Poly[(disilanylene) oligophenylene]s
Poly(imide siloxane)s
Polyphenylacetylene
Poly(phenylene ethynylene) (Synthesis and Optical Properties)
Poly(p-phenylene vinylene)s (Methods of Preparation and Properties)
Polypyrrole (Processable Dispersions)
Polypyrroles (From Isoporous Membranes)
Polysilanes (Overview)
Polysilanes (Bearing Hydroxy Groups and Derivatives)
Poly[N,N′-(sulfo-p-phenylene)terephthalamide]
Poly(2,5-thienylene vinylene) and Derivatives
Polythiophenes (Conducting Polymers)
Polythiophenes (Organometallic Synthesis)
Polythiophenes, Substituted (Regioselectivity, Electrical and Optical Properties)
Polyureas (Second-Order Nonlinear Optical Properties)
Poly(N-vinylcarbazole)
Poly(vinylidene fluoride) (Overview)
Pyroelectricity
Reversibly Crosslinked Gels
Rigid-Rod Polybenzoxazoles and Polybenzothiazoles (PBZT and PBO)
Rigid-Rod Poly(p-phenylene benzobisthiazole)
Signal Transduction Composites (Biomaterials with Electroactive Polymers)
Small-Bandgap Polymers (Special Class of Organic Conducting Compounds
Solid Polymer Electrolytes
Solid Polymer Electrolytes (Polyether Blends and Composites)
Stacked Transition Metal Macrocycles (Semiconducting Properties)
Supramolecular Self-Assembly (Liquid Crystalline Polymers, Hydrogen Bonding)
Thiophene Copolymers (Electrical and Third-Order NLO Properties)
Topochemical Polymerization (Diacetylenes)
Vinylidene Fluoride-Trifluoroethylene Copolymers (Ferroelectric-to-Paraelectric Phase Transition)

ELECTROCHEMICAL POLYMERIZATION

Ahmet M. Önal
Department of Science Education
Middle East Technical University

It is quite a common analytical technique to add monomers in order to detect the presence of free radicals at an electrode that leads to the polymerization of the monomer. This quite incidental observation brought a new and interesting way of initiating polymerization reactions to a polymer chemist.[1–6] This way of initiating polymerization reaction by species formed in an electrode reaction during electrolysis is known as electrochemical polymerization. In this article, the formation of polymers by the electrolysis of solutions containing monomers will be discussed.

Electrochemical polymerization has not yet led to the synthesis of new polymers that are unattainable by other techniques. However, it is still interesting for a polymer chemist owing to following advantages:

- The amount of initiator can be quickly changed and even eliminated by adjusting the current, which is analogous to photochemical initiation.
- The rate of polymerization may be controlled by controlling the current.
- The molecular weight distribution can be controlled.
- The molecular structure due to adsorption on the electrode surface may be influenced by electrolysis condition.
- New polymeric materials may be obtained through electrochemical coupling reactions.
- Polymer properties may be modified by electrochemically induced crosslinking, grafting, and degradation reactions.
- Copolymers of the desired copolymer composition can be produced by proper selection of the electrode potential.
- A proper design of the electrolysis cell allows one to couple the technique with one of the suitable techniques (i.e., UV-vis, esr, IR) or an electroanalytical technique (i.e., cyclic voltammetry) to elucidate the reaction mechanism and the rate.

APPLICATIONS OF ELECTROCHEMICAL POLYMERIZATION

Electrochemical Polymerization of Vinyl Monomers

The earliest mention of the application of electrochemistry to initiate polymerization is an article published by Szarvasy in 1900.[12] The first review of electrochemical polymerization was given by Fioshin and Tomilov in 1960.[13] Subsequently, Wilson in 1964, Funt and Friedlander in 1966, Yoshizawa in 1969, and Breitenbach in 1972 published their reviews on the subject.[13–18]

In this section, the application of electrochemical polymerization on some common monomers together with recent developments in the field will be given. The interested reader is referred to the excellent review of Funt and Tanner and to most recent reviews of Toppare, and Bezuglyi and Karpinets.[7,19,20]

Methyl Methacrylate (MMA)

In 1962, Funt and Yu published their detailed investigation on the polymerization of MMA.[21] They correlate polymer yield with type of electrolyte, electrolyte concentration, solvent, and electrode material. DMSO and DMF were reported to be the most suitable solvents. Carbon and Pt were found to be the most successful electrode materials. A direct relation between polymer yield and applied current was also noted. In 1963, Tsvetkov and Glotova reported that both the polymer yield and the molecular weight decreased with increasing HCl concentration during the electrochemical polymerization of MMA in aqueous solution.[22] Shelepin, Frumkin, Federova, and Vasina discussed the effect of double layer in the electrochemical polymerization of MMA.[23] Tsvetkov and Koval'chuk initiated the polymerization of MMA by using alternating current.[24] Funt and Bhadani used DMF and quaternary ammonium salts as solvent-electrolyte couple in the electrochemical polymerization of MMA.[25] They found that rate of polymerization was directly proportional to the initial monomer concentration and to the square root of the applied current. An anionic mechanism was suggested by Funt and Bhadani.[25] The kinetics of anodic polymerization of MMA was studied by Tsvetkov and Koval'chuk, and it is noted that polymer yield and molecular weight depend on current density.[26] Breitenbach and his co-workers studied the anodic polymerization of MMA in DMF with alkali metal acetates as supporting electrolyte.[27] The electrochemical polymerization of MMA on carbon fiber electrodes was investigated by Sholnik and Hoecker.[28]

Acrylonitrile (AN)

Electrochemical polymerization of acrylonitrile, AN, has been studied most extensively owing to ease of its polymerization. In 1962, Breitenbach and Srna electrochemically polymerized AN by using tetraalkylammonium salt as the electrolyte and AN itself as the solvent.[9] An anionic mechanism was suggested by Breitenbach and his co-workers during electrochemical copolymerization studies.[29] Funt and Williams postulated that electrochemical polymerization of AN in DMF was initiated via direct electron transfer to the double bond.[30] The constancy of molecular weight at different current densities was explained on the basis of chain transfer by the monomer. The same direct electron transfer mechanism was also suggested by Yamazaki et al.[8] In 1988, Yurttas, Toppare, and Akbulut employed a acetonitrile-tetrabutylammonium tetrafluoroborate solvent-electrolyte couple to polymerize AN electrochemically.[31] It is reported that polymerization proceeds via direct electron transfer to the monomer.

Styrene

Styrene is known to polymerize easily via free radical cationic, or anionic intermediates. Breitenbach and Srna initiated polymerization of styrene by free radicals generated at the anode owing to electrooxidation of the solvent.[9] Funt and co-workers studied the ECP of styrene in a DMF-Bu_4NCl solvent-electrolyte couple.[32,33] High molecular weight polystyrene was obtained during anionic ECP of styrene. It is reported that rate of polymerization is directly proportional to the applied current and monomer concentration. In 1966 Funt, Richardson, and Bhadani studied the ECP of styrene in THF, and they observed living polystyrene anions.[34] Funt and Richardson reported that the concentration of living ends can be changed by varying the current density, the electrolysis time, and the polarity of current, thus allowing molecular weight control.[35]

Electrochemical Polymerization of Allylic Monomers

ECP can also be employed to prepare polymers from allylic monomers, which generally polymerize with difficulty though, and have a wide spectrum of applications.

Electrochemical Ring-Opening Polymerizations

Similar to vinyl polymerizations ring-opening polymerizations involve ionic intermediates; hence polymerization of monomers having ring structures can also be achieved by ECP. The earliest example of such a ring-opening ECP is the polymerization of molten caprolactam.[36] Cationic ECP of some cyclic ethers were reported by Schultz and Strobel.[10] The polymerization of various epoxides was achieved by Onal and his coworkers.[11,37–40]

Conducting Polymers

ECP has recently attracted considerable interest as a synthetic method for conducting polymers. ECP allows easy preparation of conducting films that show reversible redox reactions. One of the most important advantages of utilizing ECP to prepare conducting polymer films is that dopants are incorporated into the polymer film during the polymerization process. Electrochemically prepared conducting polymer films have applications in various fields such as batteries, electronic devices, electrochromism, ion sensors, and modified electrodes.[41–44]

An ideal conducting polymer should possess a high conductivity in its doped state, good reversibility between the two states, good stability in the air, and good mechanical properties.

Most of the work of last decade on ECP is concentrated in the field of preparing conducting polymer films by ECP and improving its properties, namely, conductivity, stability, solubility, and mechanical properties.

Although preparation of polypyrrole, polythiophene, and polyaniline by ECP have attracted the most interest, ECP has also been used to obtain polymer films from furan, carbazole, indole, and other heterocyclic monomers.[45–52] Similar to electrochemical preparation of nonconducting polymers, electrolytic environment (electrodes, solvent, and electrolyte) have great influence on the properties of conducting polymers obtained by ECP.

Polyaniline (PAN)

Among conducting polymers polyaniline, PAN, has been a better studied material because of the ease with which it can be synthesized electrochemically or chemically because of its greater stability. The overall ECP of aniline has been described as a bimolecular reaction involving a radical cation intermediate and a two-electron transfer process for each step of polymerization.

The ECP of aniline to produce PAN was first reported by Adams and his co-workers.[54,55] Hochfeld and his co-workers studied the ECP of aniline and its methyl-substituted derivatives.[56] They noted that the structure of the monomer is an important factor in the determination of kinetics of ECP of aniline. Details of kinetics of ECP of aniline in aqueous H_2SO_4 solution on Pt-electrode is given by Chan and Kim.[57] Osaka et al. used a nonaqueous solvent during the ECP of aniline and investigated the application of PAN in rechargable Li batteries.[58] An increase in both electrochemical activity and stability of electrochemically obtained PAN was observed when electron donating substituents were present on the monomer.[59] Kumar and Trivedi improved electrochemical stability of PAN by electrocopolymerization of aniline with methylaniline.[60] Wei and his co-workers studied ECP of aniline in the presence of small amounts of dimer and observed an increase in the rate of polymerization.[61] Presence of dimer also allows ECP to be conducted at lower potentials, which improves the film quality. Dogan et al. prepared composites of PAN-polycarbonate by conducting ECP of aniline on a polycarbonate-coated electrode to improve its processibility.[62] Preparation of PAN-polystyrene composites with good processibility via ECP was reported by Ruckenstein and Yang.[53]

Polypyrrole (PPy)

Polypyrrole, PPy, is one of the most important conducting polymers because of its high conductivity and environmental stability. PPy can be prepared by oxidative polymerization of pyrrole both chemically and electrochemically. ECP method has the advantage of producing films. Polymerization of pyrolle proceeds via 2-5 coupling as explained by Genies, Bidan, and Diaz.[63]

Mechanisms of ECP of pyrrole and its kinetics was also studied by Lowen and Vandyke.[64] It was found by Imanishi and his co-workers that solvent nucleophilicity has great influence on the quality of electrochemically prepared PPy films.[65] The role of water on the ECP of pyrolle was investigated by Zotti et al.[66] Beck and his co-workers studied the electrodeposition of PPy in aqueous and nonaqueous solutions on different metal electrodes.[67] It is noted that anodic dissolution of the metal competes with the film-forming ECP. Bi and Pei prepared conducting composite films from PPy and polyurethane by ECP of pyrrole in acetonitrile on a Pt electrode precoated with polyurethane.[68] They found that conductivity of composite films range from 1–35 S/cm depending on such conditions as applied potential, reaction time, and type of the electrode.

REFERENCES

1. Wilson, L. C. Rec. Chem. Prog. **1949**, *10*, 25.
2. Dineen, E.; Schwan, T. C.; Wilson, L. C. *Trans. Electrochem. Soc.* **1949**, *96*, 226.
3. Parravano, G. *J. Am. Chem. Soc.* **1951**, *73*, 628.
4. Kolthoff, I. M.; Ferstanding, L. I. *J. Polym. Sci.* **1951**, *6*, 563.
5. Goldschmidt, S.; Stockel, E. *Chem. Ber.* **1952**, *85*, 630.
6. Kern, W.; Quast, H. *Makromol. Chem.* **1953**, *10*, 202.
7. Funt, B. L.; Tanner, J. "Electrochemical Synthesis of Polymers" in Weissberger, A. Ed. Techniques of Organic Chem. Interscience: New York.
8. Yamazaki, N.; Tanaka, I.; Nakahama, S. *J. Makromol. Sci. (Chem.)* **1968**, *6*, 1121.
9. Breitenbach, J. W.; Srna, C. *Pure Appl. Chem.* **1962**, *4*, 245.
10. Schultz, R. C.; Strobel, W. *Monatsch* **1968**, *99*, 1742.
11. Onal, A. M. Ph.D. Thesis, Middle East Technical Univ., 1983.
12. Szarvasy, E. C. *J. Chem. Soc.* **1900**, *77*, 207.
13. Fioshin, M. Y.; Tomilov, A. P. *Plasticheskie Massy* **1960**, *10*, 2.
14. Wilson, L. C. In Encyclopedia of Electrochemistry, Hampel, C. A. Ed., Reinhold: New York, 1964; p 963.

15. Funt, B. L. *Macromol. Rev.* **1966**, *1*, 35.
16. Friedlander, H. *Encyclopedia of Polym. Sci. and Technol.*, 1966; Vol. 5, 629.
17. Yoshizawa, S.; Ogumi, Z.; Fukuhara, K. *Denki Kagaku Oyobi Kogyo Butsuri Kagaku* **1969**, *37*, 740.
18. Breitenbach, J. W. *Advan. Polym. Sci.* **1972**, *9*, 47.
19. Toppare, L. In Cheremisinoff, N. P. ed., Handbook of Polymer Science and Technology, Marcel Dekker: New York, 1989; p 271.
20. Karpinets, A. P.; Bezuglyi, V. D. *Elektrokhimiya* **1992**, *28*(4), 638.
21. Funt, B. L.; Yu, K. C. *J. Polym. Sci.* **1962**, *62*, 359.
22. Tsvetkov, N. S.; Glotova, Z. F. *Vysokomol. Soyed.* **1963**, *5*, 997.
23. Shelepin, I. V.; Frumkin, A. N.; Federova, A. I.; Vasina, S. Y. *Dokl. Akad. Nauk SSSR* **1963**, *154*, 203.
24. Tsvetkov, N. S.; Koval'chuk, E. P. *Visn L'viv Derzh Univ. Ser. Khim.* **1967**, *72*.
25. Funt, B. L.; Bhadani, S. N. *J. Polym. Sci., A* **1965**, *3*, 4191.
26. Tsvetkov, N. S.; Koval'chuk, E. P. *Ser. Khim. Lvov* **1967**, 72.
27. Breitenbach, J. W.; Olaj, O. F.; Sommer, F. *Monatsh. Chem.* **1968**, *99*, 203.
28. Shkolnik, S.; Hoecker, H. *Polymer* **1992**, *33*(8), 1669.
29. Breitenbach, J. W.; Srna, C.; Olaj, O. F. *Makromol Chem.* **1960**, *42*, 171.
30. Funt, B. L.; Williams, F. D. *J. Polym. Sci., A* **1964**, *2*, 865.
31. Yurttas, B.; Toppare, L.; Akbulut, U. *J. Macromol. Sci. (Chem.)* **1988**, *A25*, 219.
32. Funt, B. L.; Laurent, S. W. *Can. J. Chem.* **1964**, *42*, 2728.
33. Funt, B. L.; Bhadani, S. N. *Can. J. Chem.* **1964**, *42*, 2733.
34. Funt, B. L.; Richardson, D.; Bhadani, S. N. *Can. J. Chem.* **1966**, *44*, 711.
35. Funt, B. L.; Richardson, D. *J. Polym. Sci. A1* **1970**, *8*, 1055.
36. Gilch, H.; Michael, D. *Makromol. Chem.* **1966**, *99*, 103.
37. Akbulut, V.; Toppare, L.; Usanmaz, A.; Onal, A. M. *Makromol. Chem. Rapid Commun.* **1983**, *4*, 259.
38. Onal, A. M.; Usanmaz, A.; Akbulut, U.; Toppare, L. *Brit. Polym. J.* **1983**, *15*, 179.
39. Akbulut, U.; Onal, A. M.; Usanmaz, A.; Toppare, L. *Brit. Polym. J.* **1983**, *15*, 187.
40. Onal, A. M.; Usanmaz, A.; Akbulut, U.; Toppare, L. *Brit. Polym. J.* **1984**, *16*, 102.
41. Noufi, R.; Nozik, A. J.; White, J.; Warren, L. C. *J. Electrochem. Soc.* **1982**, *129*, 2261.
42. Inganas, O.; Lundstron, I. *J. Electrochem. Soc.* **1984**, *131*, 1129.
43. Oyama, N.; Ohsaka, T.; Shimuzu, T. *Anal. Chem.* **1985**, *57*, 1526.
44. Kobayashi, T.; Yoneyama, H.; Tamura, N. *J. Electroanal. Chem.* **1984**, *161*, 419.
45. Wernet, W.; Monkenbusch, M.; Wegner, G. *Makromol. Chem. Rapid Commun.* **1984**, *5*, 157.
46. Kazanova, K.; Diaz, A. F.; Gill, W. D.; Giant, P. M.; Street, G. B.; Gardini, G. P.; Kwak, J. F. *Synth. Metals* **1981**, *4*, 119.
47. Towrillan, G.; Garnier, F. *J. Electroanal. Chem. Interfacial Electrochem.* **1984**, *135*, 173.
48. Kaneto, K.; Yashino, K.; Inuishi, Y. *Jpn. J. Appl. Phys. Part 2* **1982**, *21*, L567.
49. Bacon, J.; Adams, R. N. *J. Am. Chem. Soc.* **1968**, *90*, 6596.
50. Garnier, F.; Tourillon, G.; Gazard, M.; DuBois, J. D. *J. Electroanal. Chem. Interfacial Electrochem.* **1983**, *148*, 229.
51. Ambrose, J. F.; Nelson, R. F. *J. Electrochem. Soc.* **1968**, *115*, 1159.
52. Waltman, R. J.; Diaz, A. F.; Bargon, J. *J. Phys. Chem.* **1984**, *88*, 4343.
53. Ruckenstein, Yang, S. *Synth. Metals* **1993**, *53*, 283.
54. Mohilner, D. M.; Adams, R. N.; Argensinger, W. J. *J. Am. Chem. Soc.* **1962**, *84*, 3618.
55. Bacon, J.; Adams, R. N. *J. Am. Chem. Soc.* **1968**, *90*, 6596.
56. Hochfeld, A.; Kessel, R.; Schultze, J. W.; Thyssen, A. *Ber. Bunsenges. Phys. Chem.* **1988**, *92*, 1406.
57. Chan, J. K.; Kim, J. D. *Bull. Korean Chem. Soc.* **1988**, *9*(1), 64.
58. Osaka, T.; Oyama, N.; Ogana, S.; Naoi, K. *J. Electrochem. Soc.* **1989**, *136*, 306.
59. Nonaka, T. *Chem. Express* **1992**, *7*(6), 433.
60. Kumar, D. S.; Trivedi, D. C. *Synth. Metals* **1993**, *60*(1), 63.
61. Wei, Y.; Sun, Y.; Jag, G. W.; Tang, Y. *J. Polym. Sci. Part C, Polym. Lett.* **1993**, *28*, 81.
62. Dogan, S.; Akbulut, U.; Toppare, L. *Synth. Metals* **1992**, *53*, 29.
63. Genies, E. M.; Bidan, G.; Diaz, A. F. *J. Electrochem. Soc.* **1983**, *149*, 101.
64. Lowen, S. W.; Vandyke, T. D. *Abstract of papers on the Am. Chem. Soc.* **1984**, *195*, 194.
65. Imanishi, K.; Yasuda, Y.; Tsushima, R.; Satoh, M.; Aoki, S. *J. Electroanal. Chem. Interfacial Electrochem.* **1988**, *224*, 203.
66. Zotti, G.; Schrovan, G.; Pagani, G.; Berlin, A. *Electrochim. Acta* **1989**, *34*, 881.
67. Beck, F.; Huelser, P.; Michaelis, R. *Bull. Electrochem.* **1992**, *8*(1), 35.
68. Bi, X. J.; Pei, Q. B. *Synth. Metals* **1987**, *22*, 145.

Electrochemical Processing

See: *Graphite Fiber Composites (Electrochemical Processing)*

Electrochemical Sensors

See: *Ion-Selective Electrodes and Biosensors*

Electrochemically-Initiated Polymerization

See: *Antipyryl Acrylamide (Electrochemical Polymerization)*
Electrochemical Polymerization
Polyaniline (Electrochemical Synthesis)
Polyanilines, Oxidation States
Poly(dichlorophenylene oxide)s (by Pyridine/Copper Complex)
Ultrasonically Assisted Polymer Synthesis

Electrochromic Polymers

See: *Enzyme-Catalyzed Oxidative Polymerization (Aromatic Compounds)*

ELECTROCONDUCTIVE COMPOSITES (Emulsion Pathways)

Eli Ruckenstein
Department of Chemical Engineering
State University of New York at Buffalo

The synthesis of polyheterocycles, such as polypyrrole (PPY), polythiophene (PTP) and polyaniline (PANI), has received a great deal of attention in the past decades because of their high electrical conductivity and acceptable environmental stability.[1–3] The conductivity of polyheterocycles has reached values of over 10^2 S/cm without a major decrease over several months of standing in air. However, most conducting polymers have poor processability, being insoluble and infusible, and also poor mechanical properties. For this reason, the conventional methods for polymer processing, such as melt processing and solution-casting, could not be employed for these materials.

A number of methods have been suggested to overcome the processability drawback by combining insulating materials with conducting polyheterocycles. First, such conducting composites have been prepared by incorporating chemically or electrochemically synthesized polyheterocycle molecules in a nonporous polymer matrix.[4–6] Polypyrrole and polythiophene-based conductive composites have been thus prepared. An improvement of the method was achieved by employing a surface active pyrrole in addition to the conventional surfactant in the preparation of the porous medium.[7]

More recently, pathways have been developed in our laboratory in which emulsions have been used to generate the composites. The emphasis of this section is on the latter procedures.

Two emulsion pathways have been developed to produce composites involving polyheterocycles based on pyrrole, aniline, or thiophene and various host polymers. In one of them, an emulsion was generated by dispersing an organic solution containing the host polymer and the conducting monomer in an aqueous solution of a suitable surfactant.[8–11] It is important to emphasize that the phase in which the surfactant is soluble becomes the continuous phase. Subsequently, an aqueous solution of an oxidant and dopant was added dropwise to the emulsion in order to polymerize the conducting monomer and to dope the conducting polymer generated. The composite precipitated by itself or by the addition of methanol. The materials have been processed by cold- or hot-pressing. In the other pathway, an inverted emulsion of water (containing an oxidant and a dopant) in an organic solvent (containing the host polymer, and an oil-soluble surfactant) was first prepared.[12–14] The conducting monomer was dissolved in an organic solvent, which was either the one used as the continuous phase of the emulsion, or another one miscible with it. The latter solution was introduced dropwise in the inverted emulsion. The main advantage of the above two procedures is the good contact between the oxidant solution and the solution containing the conducting monomer and the host polymer. In both cases, the role of the surfactant is to ensure the stability of the emulsion.

O/W EMULSION PATHWAY

This procedure was used to prepare: (a) Polyanilinepolystyrene composites; in this case sodium dodecyl sulfate was used as surfactant, ammonium persulfate as oxidant, HCl as dopant, and benzene as solvent.[8] (b) Polyaniline-poly(alkyl methacrylate) composites; the other ingredients have been as under (a).[9] (c) Polypyrrole-poly(alkyl methacrylate) composites; in this case, sodium dodecylsulfate was used as surfactant, ferric chloride as oxidant and dopant, and chloroform as solvent;[10] (d) Polypyrrole-poly-(ethylene-co-vinyl acetate) or polyethylene composites; in this case sodium dodecylsulfate was the surfactant, ferric chloride the oxidant and dopant, and toluene the solvent.[11]

RESULTS AND DISCUSSION

In summary, polyaniline composites have been prepared by two methods. While in method 1 the polymerization proceeded by introducing an organic solution of aniline into an inverted emulsion containing a host polymer, an oxidant and a dopant, in method 2 it proceeded by adding an aqueous solution of an oxidant and a dopant to an inverted emulsion containing aniline and a host polymer. Method 2 was more successful [the PANI-SBS composites prepared by that method had higher conductivities (in the range of 1.0 to 2.8 S/cm) as well as better mechanical properties (the elongation at break in the range of 95.6% to 23.3%)], probably because the contact between the aqueous solution of oxidant and the monomer is better in procedure 2. This happens because the presence of the aqueous globules of the inverted emulsion allows a more uniform distribution at the microscopic level of the aqueous solution of oxidant and dopant. In addition, the reorganization of the globules because of the additional water introduced, produces mixing at a much smaller scale between the two immiscible phases than any mechanical stirring can generate, which also contributes to the uniform distribution of the additional water. As a result, the contact between the phases is indeed increased. It is important to emphasize that the direct introduction of water (containing the oxidant and dopant) into the organic solvent (containing the surfactant, aniline, and the host polymer) led to electrical conductivities by almost one order of magnitude lower than those provided by method 2.

REFERENCES

1. Skotheim, T. A. Ed., *Handbook of Conducting Polymers* Vol. 1. Marcel Dekker: New York, 1986.
2. Burroughes, J. H.; Bradly, D. D. C.; Brown, A. R.; Marks, R. N.; Mackay, K.; Friend, R. H.; Burnes, P. L.; Holmes, A. B. *Nature* **1990**, *347*, 539.
3. Stevens, T. *Mater. Eng.* **1991**, *108*, 21.
4. Bi, M.; Pei, Q. *Synth. Met.* **1989**, *22*, 145.
5. Yoshikawa, T.; Machida, S.; Ikegami, T.; Techagumpuchm, A.; Miyata, S. *Polym. J.* **1990**, *22*, 1.
6. Li, C.; Song, Z. *Synth. Met.* **1991**, *40*, 23.
7. Ruckenstein, E.; Chen, J. H. *J. Appl. Polym. Sci.* **1991**, *43*, 1209.
8. Ruckenstein, E.; Yang, S. Y. *Synth. Met.* **1993a**, *53*, 283.
9. Yang, S. Y.; Ruckenstein, E. *Synth. Met.* **1993a**, *59*, 1.
10. Ruckenstein, E.; Yang, S. Y. *Polymer* **1993b**, *34*, 4655.
11. Yang, S. Y.; Ruckenstein, E. *Synth. Met.* **1993b**, *60*, 249.
12. Ruckenstein, E.; Hong, L. *Synth. Met.* **1994**, *66*, 249.
13. Sun, Y.; Ruckenstein, E. *Synth. Met.* **1995**, *74*, 261.
14. Ruckenstein, E.; Sun, Y. *Synth. Met.* **1995**, *74*, 107.

ELECTROCONDUCTIVE PAINTS

Kunio Mori
Department of Applied Chemical and Molecular Science
Iwate University

At present, conductive paint types may be listed as dry forming, low-temperature curing, and high-temperature burning types. All of these are mixtures whose three major components are metal powder, binder polymer, and solvent. The concepts and guidelines for proportion have been well established by Mori and co-workers.[1-5]

MECHANISM OF ELECTROCONDUCTION

To date, several models and theories have been proposed regarding conductive mechanisms for paints and plastics.[6-16] These may be summarized as follows: electric charges are transferred when they are assembled or gathered into short chains of particles of conductive materials. It is believed that there is direct electrical contact among the particles. Theories on direct electrical contact specify that particles of conductive material not only contact one another directly but also approach each other at a distance permitted within the scope of tunnel effect experience.

A different point of view maintains that the conductive mechanism is thermal radiation of electronics among the conductive particles. But no matter what theory is adopted, both maintain that the distance between particles is an important factor, that is, electrical contact among particles of conductive materials results in their becoming joined to form short chains. To enable conductive paint film to possess high conductivity, not only is this essential, it is indispensable.

OPTIMAL CONTENT OF CONDUCTIVE MATERIAL

In paints and plastics to which electroconductive materials have been added, conductivity is optional when the conductive agent is added at a certain content.[10,12-16] At this content, there is complete joining of the particles comprising the conductive material and the electrodes are joined, so to speak, through electrical contact.

PACKED STRUCTURE OF PAINT FILM AND CONDUCTIVITY

One may ask what kind of paint should be produced from conductive film to assure high conductivity? Considered on the basis of the conduction mechanism through completion of electrical contact as a result of small-chain formation, the conductivity of the paint film is greater for the nonuniform structure than the uniform structure. Conductive paint is made by mixing three major components—conductive material, binder polymer, and solvent—to prepare a paste that is then applied onto a surface. Following this, the solvent is evaporated off, and the paint film is formed.

To obtain conductive paint film with nonuniform structure and high conductivity, the solvent must be evaporated. But until particles of the conductive material become joined and are sufficiently in contact with each other, the polymer remains dissolved in the solvent and should not adhere to the surface of these particles. It is essential to the formation of a nonuniform structure that the polymer be precipitated as late as possible following establishment of this contact. It thus follows that, for nonuniform structure formation, the solubility of the polymer in the solvent should be adequate.

CONCLUDING REMARKS

Generally, in paint preparations, if dispersibility is satisfactory, few problems are encountered with other features. Even with magnetic paints, if dispersibility is adequate, the magnetic particles are uniformly packed and the magnetic properties are improved. In conductive paints made with alloys, obtaining a high degree of conductivity presents a different situation but is possible when dispersion of the alloy powder does not occur easily. It is in these respects that the behavior and associated phenomena of conductive paints differ from conventional paints. Also, the conductivity of numerous paints remains unclear, and additional research will be required.

REFERENCES

1. Calahorra, A. *J. Coating Technol.* **1988**, *60*, 757.
2. Mori, K. et al. *J. Mater. Sci.* **1989**, *28*, 367.
3. Mori, K. et al. *Kobunshi Ronbunshu* **1992**, *49*, 475.
4. Mori, K. et al. *Kobunshi Ronbunshu* **1992**, *49*, 485.
5. Mori, K. et al. *Kobunshi Ronbunshu* **1992**, *49*, 645.
6. Bulgin, D. *Rubber Chem. Technol.* **1948**, *19*, 669.
7. Polly, M. H.; Boonatra, B. B. *Rubber Chem. Technol.* **1946**, *19*, 669.
8. Grekila, R. B.; Tien, T. Y. *J. Am. Ceram. Soc.* **1965**, *48*, 22.
9. Gurland, R. B. *Trans. Mater. Soc. AIME.* **1966**, *236*, 642.
10. Aheroni, S. M. *J. Appli. Phys.* **1972**, *6*, 2463.
11. Scarisbrick, R. M. *J. Phys.* **1973**, *6*, 2098.
12. Maeda, O. et al. *Kobunshi Ronbunshu* **1975**, *32*, 42.
13. Oono, K. *Nippon Gomu Kyokaishi* **1984**, *57*(3), 178.
14. Asada, Y. *Polymerdaijest* **1984**, *36*, 26.
15. Kotani, T.; Arai, K. *Plastics* **1982**, *33*(2), 51.
16. Miyasaka, K. *Nippon Gomu Kyokaishi* **1985**, *58*(9), 561.

Electroluminescent Materials

See: *Charge Transport Polymers (for Organic Electroluminescent Devices)*
Conjugated Polymers (Insulating and Conducting Forms)
Electroluminescent Polymers (Overview)
Photoluminescent Polymers
Poly(p-phenylene vinylene)s (Method of Preparation and Properties)

ELECTROLUMINESCENT POLYMERS (Overview)

Russell A. Gaudiana and P. Mehta
Polaroid Corporation

Electroluminescence is a phenomenon in which light emission occurs when an electron is injected into a positively charged molecule. In organic materials, the emitting species is the first excited singlet state, that is, a singlet exciton, of a molecule, and hence the emission wavelength and band shape are very similar to those of fluorescence emission from the isolated molecule in solution.

Organic-based light-emitting diodes (LEDs) have received much noteriety and intense interest in academic and industrial research laboratories around the world in recent years. This is due primarily to the belief that these devices represent the potential for large, lightweight, two-dimensional displays that will be used for television and high-density information screens.

Two types of diodes are currently being investigated. The first type[1] is referred to as a "molecular device." The second type, polymeric LEDs, were first realized by the Cambridge University group.[2] The anticipated low cost and general utility of organic-based LEDs originates from the fact that the active components of the cell are extremely thin, are generally easy to synthesize, and most importantly, may be coated on a continuous web.

The electrode serving as the anode must be a high work-function metal or semiconductor; transparent ITO is frequently used. The cathode must have a low work function, and originally Ca, Mg, Mg/Al, and so on were used. More recently, Al has become the metal of choice because of its stability and ease of handling.

Poly(*p*-phenylene vinylene) (PPV) and its homologs were the first polymers to exhibit electroluminescence;[2] the parent polymer is a yellow-green emitter. Soon after these reports, electroluminescence was observed in several other main-chain polymers, including substituted poly(*p*-phenylene)s (PPP),[3] bridged, semiladder poly(*p*-phenylene)s,[4] are blue or blue-green emitters, and more recently ladder poly(*p*-phenylene)s,[5] which peak around 600 nm because of the planar configuration of the phenyl rings.[5,6] Polythiophene exhibits red electroluminescence (630 nm),[7] and a modification of the synthetic procedure generates a polymer containing ≥98% head-to-tail isomer,[8] which sifts the emission wavelength to 650 nm.[9] Hence, emission wavelengths spanning the entire visible spectrum have been observed in main-chain polymers.

Attempts to incorporate predetermined lumophores into main-chain polymers have met with considerable success in several laboratories. Two molecular architectures have been used in this regard. Both require the synthesis of a difunctional lumophore. The functionalities may be diols, diamines, diisocyanates, diacids, etc., which would typically be used in condensation polymerizations. The first type of polymer is classified as a rigid flexible type in that the comonomer contains a flexible, saturated chain that prevents multirepeat unit conjugation. Although the earliest report of electroluminescence in rigid flexible polymers, e.g., poly(ethylene terephthalate),[10] actually predates that of PPV, more recent work on applications of polymeric materials as light-emitting diodes have been performed simultaneously, but independently in three laboratories. Two of these are remarkably similar in their synthetic strategies and resulting structures;[11,12] both are polyesters comprising substituted, short phenylene vinylene lumophores. The emission wavelength (EL) of the second example is 455 nm.

Another blue-emitting, rigid-flexible polymer was obtained via a somewhat more difficult synthetic procedure in which the lumophore was formed during the condensation polymerization. Again in this example the basic structure of the emitter is a phenylene vinylene segment.[13]

Another means of controlling emission wavelength by limiting the extent of conjugation is the incorporation of noncoplanar biphenyls into the polymer backbone of a condensation such as a polyamide or a polyester. The biphenyl rings are forced into a noncoplanar conformation by the placement of substituents on the 2,2′-carbons.[14]

A completely different strategy for controlling the emission wavelength of electroluminescence is the use of polymers containing lumophores attached to a flexible polymer backbone via a flexible spacer. The emitters themselves are well-characterized compounds and many are commercially available. In order to affect attachment to an already-formed polymer or to form a polymerizable monomer, the basic lumophore must typically be functionalized with amine, alcohol, or vinyl groups. Siloxane and methacrylate homopolymers containing a single lumophore and copolymers containing two lumophores in a variety of relative concentrations have been synthesized. Electroluminescent emission wavelength ranging from violet to orange have been demonstrated.[15]

PROPERTIES

Because of the abbreviated nature of this review, it is impossible to discuss in detail the operating characteristics and basic phenomena underlying the mechanisms of organic light-emitting diodes. The reader is referred to several recent references that treat these subjects.[16–18]

APPLICATIONS

Polymeric light-emitting diodes should find utility in a wide range of applications if problems such as efficiency, brightness, power consumption, and operating lifetime can be solved. One of these uses would certainly be flat panel displays; a thin, lightweight, large-area panel has been the dream of television manufacturers for many years. Also, emissive displays will have significant advantages over backlit displays such as LCDs due to the fact that they do not require polarizers, they will have a wide viewing angle, and because they can be coated on a continuous web, manufacturing costs will be lower.

Required brightness levels, and therefore power consumption for a given polymer will depend on the intended use. Currently, polymeric LEDs exhibit brightness levels in the 100–500 cd/m^2 range, which make them suitable for several useful applications. Power consumption for a brightness level of 100 cd/m^2 is about 50 mW/cm^2, which would require 5 W power in a small 10 × 10 cm display. A fairly large display about the size of a laptop computer, 25 × 25 cm would use 30 W. For comparison, laptop LCDs use about 8 W. Hence, in order to make practical flat displays using LEDs, the efficiency and charge transport characteristics must be improved significantly. However, advances are being made very quickly in both materials and devices, and this field holds great promise.

REFERENCES

1. Tang, C. W.; U.S. Patent 4 356 429, 1982; Van Slyke, S. A.; Tang, C. W. U.S. Patent 4 539 507 1985; Tang, C. W.; Chem, C. H. U.S. Patent 4 769 292, 1988.
2. Friend, R. H. et al. U.K. Patent Appl. 8900912 2, PCT Int. Appl. WO 90 13 148, 1989.
3. Grem, G. et al. *Adv. Mater.* **1992**, *4*, 36.
4. Ohmori, Y. et al. *Jpn. J. Appl. Phys.* **1991**, *30*, L1941.
5. Grem, G.; Leising, G. *Synth. Met.* **1993**, *55–57*, 4105.
6. Huber, J. et al. *Acta Polym.* **1994**, *45*, 244.
7. Ohmori, Y. et al. *Jpn. J. Appl. Phys.* **1991**, *30*, L1938.
8. McCullough, R.; Lowe, R. D. *J. Chem. Soc., Chem. Commun.* **1992**, *70*.
9. Chen, F. et al. manuscript in preparation.
10. Kaneto, K. et al. *Jpn. J. Appl. Phys.* **1974**, *13*, 1023.

11. Greiner, A. et al. *Polym. Prepr.* **1993**, *34*, 176.
12. Cumming, W. et al. U.S. Patent 5 376 456 (1995); Cumming, W. et al. *J. Macromol. Sci., Pure Appl. Chem.*, accepted, 1996.
13. Yang, Z. et al. *Macromolecules* **1993**, *26*, 1188.
14. Rogers, H. G. et al. *J. Polym. Sci., Chem.* **1985**, *23*, 2669.
15. Cumming, W. et al. U.S. Patent 5 414 069 (1995); Bisberg, J. et al. *Macromolecules* **1995**, *28*.
16. Baigent, D. R. et al. *Syn. Metals* **1994**, *67*, 3.
17. Heeger, A. J. et al. *Synth. Met.* **1994**, *67*, 23.
18. Schott, M. *Organic Conductors, Fundamentals and Applications*, Farges, J-P. Ed.; Marcel Dekker: New York, 1994; Chapter 12, pp 614–35.

Electron Beam Polymerization

See: Radiation Curing

ELECTRON TRANSFER REACTIONS (Novel Polymerizations by Divalent Samarium)

Takeshi Endo and Ryoji Nomura
Research Laboratory of Resources Utilization
Tokyo Institute of Technology

In recent years, much attention has been paid to lanthanide elements because of their characteristic properties. Divalent and tetravalent lanthanide complexes especially have been established as reducing and oxidizing reagents, respectively, in organic synthesis. Many electron transfer reactions using divalent or tetravalent lanthanide complexes to proceed more efficiency than traditional methods have been reported.[1]

We will outline some new methodologies of polymer synthesis by using electron transfer reaction with samarium (II) iodide (SmI_2). We begin with a report on the novel transformation reaction of the cationic polymerization into an anionic one through electron transfer and then discuss a new method for the polymerization of methacrylates. Finally, we describe novel radical polyaddition of bifunctional carbonyl compounds with bifunctional unsaturated compounds.

RESULTS AND DISCUSSION

Novel Transformation of a Cationic Growing Center into an Anionic One

The direct two-electron reduction of the cationic growing center into an anionic one was achieved, for the first time, using the strong reducing ability of SmI_2.[2] SmI_2 quantitatively reduced the cationic growing center of living poly(tetrahydrofuran) (poly(THF)), (**Scheme I** Structure **1**) into the terminal nucleophile (Structure **2**) in the presence of hexamethylphosphoramide (HMPA) at room temperature (Scheme I). The most probable side reaction during the transformation is the coupling reaction of the prepolymer (Structure 1) with the macroanion (Structure 2). However, the side reaction was found to be negligible by the GPC analysis, in which no significant difference between Structure 1 and Structure 2 was observed.

Because the macroanion (**Scheme II**, Structure **2**) has an organosamarium moiety at the polymer end, it is possible to introduce a variety of electrophiles onto the polymer end by reacting Structure 2 with appropriate electrophiles. These

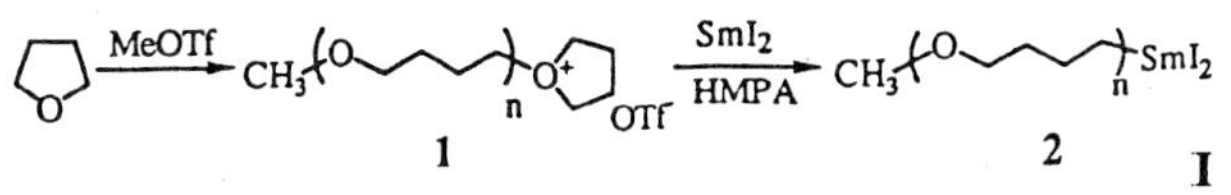

I

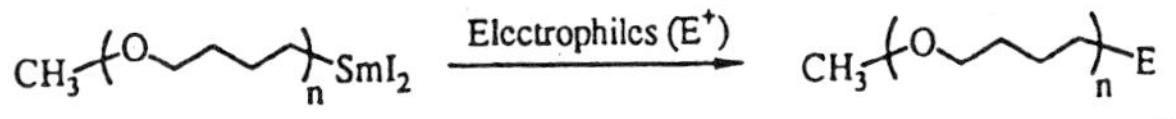

II

results mean that poly(THF), a typical cationic polymer, can be formally endcapped with electrophiles.

As described above, poly(THF)-macroanion (Structure 2) has an organosamarium moiety at the polymer end. Polymerization of anionically polymerizable monomers with Structure 2 is expected to produce block copolymers made of electrophilic and nucleophilic monomers. The scope and limitation of alkylsamarium as an initiator for anionic polymerization was first examined using octylsamarium iodide ($OctSmI_2$).[3]

Polymerization of TBMA that was polymerized by alkylsamarium to afford living poly(TBMA) was attempted by using the macroanion (Structure 2) as an initiator.[4] ^1H NMR analysis supported the formation of the block copolymer of THF with TBMA.

It is well known that trifluoromethanesulfonic anhydride (Tf_2O) polymerizes THF to produce telechelic poly(THF).[5] When Tf_2O is used instead of MeOTf, ABA-type triblock copolymer of TBMA with THF is obtained.[6]

The one-pot ABA-type copolymerization also successively proceeded in diblock copolymerization. The block copolymers were unimodal and exhibited narrow molecular weight distribution. The unit ratio of THF and TBMA was facilely controlled by both the poly(THF)-macroanion and the amount of TBMA, as in the diblock copolymerization.

This method is the first example of one-pot quantitative transformation using electron transfer reaction, and one of the most convenient and simple methods for polarity inversion of growing centers.

New Method for the Polymerization of Methacrylates

Vinyl monomers having electron-withdrawing groups such as methacrylate and methacrylamide are able to undergo the reduction by SmI_2 in the presence of HMPA, leading to homocoupled products.[7] This reaction is initiated by electron transfer from Sm(II) to vinyl compounds to generate anion radicals. The conjugated addition of the anion radical or its protonated form gives the homocoupled product.

The polymerization of methacrylates by the SmI_2/HMPA system proceeded through the dianion, yielding a corresponding polymer having two propagating centers at both polymer ends.[8]

The addition of HMPA was indispensible for the reduction of methacrylates, and the molecular weight distribution of the polymer was affected by the amount of HMPA.

Step Polymerization of Diketones with Bifunctional Unsaturated Compounds Promoted by SmI_2

Ketyl radicals derived from one-electron reduction of carbonyl with SmI_2[9] were found to be available for radical polyaddition to bifunctional olefins or acetylene derivatives.

Polyaddition of diketones with distyryl compounds or bisphenylacetylene derivatives by SmI_2 in the presence of HMPA gave corresponding polyalcohols in good yields.[10]

In summary, we have demonstrated some novel polymerization methods utilizing electron transfer reaction induced by divalent samarium iodide. Electron transfer reaction has a possibility to accomplish unique reactions that cannot be carried out in the traditional way. Appropriate control and application of electron transfer reaction could provide a new route in polymer synthesis.

REFERENCES

1. Molander, G. A. *Chem. Rev.* **1992**, *92*, 29; Inanaga, J. *Yuki Gosei Kagaku Kyokaishi* **1989**, *47*, 200.
2. Nomura, R.; Endo, T. *Macromolecules* **1994**, *27*, 5523.
3. Kikuchi, T. et al. *The 67th Annual Meeting of Chem. Soc. Jpn. Abstr.*, 2M402, 1994.
4. Nomura, R. et al. *Macromolecules* **1994**, *27*, 4853 and 7011.
5. Smith, S.; Hubin, A. J. *J. Macromol. Sci., Chem.* **1973**, *A7*, 1399.
6. Nomura, R. et al. *Macromolecules* **1995**, *28*, 86.
7. Inanaga, J. et al. *Tetrahedron Lett.* **1991**, *45*, 6557.
8. Nomura, R. et al. *Polym. Prep. Jpn.* **1994**, *43*, 158.
9. Otsubo, K. et al. *Tetrahedron Lett.* **1986**, *27*, 5763.
10. Nomura, R.; Endo, T. *Macromolecules* **1994**, *27*, 617 and 1286.

Electrorheological Fluids

See: Electrorheological Fluids (Role of Polymers as the Dispersed Phase)
Electrorheological Materials

ELECTRORHEOLOGICAL FLUIDS (Role of Polymers as the Dispersed Phase)

M. T. Shaw* and R. C. Kanu
Department of Chemical Engineering and Institute of Materials Science
University of Connecticut

For more than a century researchers have studied the effects of electric fields on the deformation and flow properties of liquids and suspensions. Yang and Shing[1] have recently shown that liquid crystals, specifically nematic solutions of poly(*n*-hexyl isocyanate) in *p*-xylene, also show increased viscosity on the application of electric fields. The increases ranged from a factor of 2 to 35. But, for some biphasic materials, viscosity increases by several orders of magnitude when the materials are subjected to external electric fields transverse to their flow direction. These materials are known as electrorheological fluids (ERFs). ERFs were discovered by W. M. Winslow.[2]

ERFs are suspensions of micron-sized polarizable particles dispersed in nonconducting media. ERFs exhibit rapid (on order of milliseconds) liquid-to-solid and solid-to-liquid transitions when an applied electric field is turned on and off, respectively. The transitions of ERFs are accompanied by increases of several orders of magnitude in dynamic modulus (stiffness) and yield stress. This phenomenon is known as electrorheology (ER). The increases in dynamic modulus and yield stress are based on the polarization of the particles, which results in the formation of particle chains between the electrodes.

*Author to whom correspondence should be addressed.

An important aspect of the technological importance of the ER phenomenon is that it permits the direct use of electricity without intermediate transformations. Thus, it has been suggested that ERFs will create a more suitable interface between the electronics and the mechanical parts of devices than is possible by using conventional actuators, such as servo motors and solenoids. ERFs might also reduce or eliminate the number of moving parts, ensuring simplicity of design and manufacture, and equipment efficiency. Other advantages of ERF-based devices are that there is little wearing of surfaces and the components do not require precision machining.

The applications of ERFs are numerous and span a wide range of industries such as pharmaceuticals, power utilities, and automobiles. The automotive devices include clutches,[2–5] shock absorbers,[6–10] engine mounts[11–16] and fuel injectors.[17] Jordan and Shaw[18] suggested that "the greatest promise in realizing the full potential of electrorheology is in the development of new materials."

MATERIALS FOR ERFs

Many materials have been used in formulating ERFs and newer ones are constantly being developed. Block and Kelly[19] have published an extensive list of materials used for ERFs.

On the basis of electrostatic polarization model,[20] the dielectric constant mismatch between the dispersed phase and the continuous phase is responsible for the ER phenomenon. The use of polymers as a dispersed phase has some obvious advantages over materials such as zeolites, semiconducting glasses, dielectric metal oxides, charge transfer salts, and various forms of silica.[21]

POLYMERS FOR ERFs

Polymers tend to be softer; hence they reduce or eliminate the problem of abrasion, which can lead to metallic contamination and electrical failure of the system. Polymers are less dense than minerals, which slows or reduces the rate of sedimentation of ERFs. Perhaps the greatest contribution of polymers to the field is that the chemical and physical properties of a polymer can be changed in order to enhance its dielectric and electrical properties to meet specific requirements.

An illustration of the importance of polymers in ERFs is the work of Kanu and Shaw[22] who studied the role of particle geometry on the response of ERFs using poly(*p*-phenylene-2,6-benzobisthiazole) (PBZT) as the dispersed phase. In their study, they showed that dynamic modulus (stiffness) of the fluid scales is 1.8 times of the geometric aspect ratio (l/d) of the particles.

PBZT is a wholly aromatic, heterocyclic, rigid macromolecule, which frequently forms liquid crystalline as well as crystalline phases.[23] It has favorable dielectric and electrical properties.

As for the future, polymers provide investigators the opportunity to study systematically the variables of the fundamental mechanisms of ER behavior. With better information in hand, it is likely that ERFs with better properties will be developed.

REFERENCES

1. Yang, I.; Shine, D. *J. Rheol.* **1992**, *36*, 1079.
2. Winslow, W. M. U.S. Patent 2 417 850, 1947.
3. Stangroom, J. E. U.S. Patent 4 664 236, 1987.
4. Egyed, M. J. U.S. Patent 5 342 258, 1994.
5. Reuter, D. C. U.S. Patent 5 322 484, 1994.
6. Shtarkman, E. M. U.S. Patent 4 942 947, 1990.
7. Petek, N. K.; Goudie, R. J.; Boyle, F. P. *SAE Technical Paper Series 920275, Int. Congress and Exposition*, Detroit, MI, 1992.
8. Mettner, M. U.S. Patent 5 263 559, 1993.
9. Westkamper, E.; Meschke, J. *Rheology* **1993**, 243.
10. Schuelke, A. U.S. Patent 5 293 968, 1994.
11. Petek, N. K.; Goudie, R. J.; Boyle, F. P. *SAE Technical Paper Series 881785, Passenger Car Meeting and Exposition*, Dearborn, MI, 1988.
12. Doi, K. U.S. Patent 5 145 024, 1992.
13. Morishita, S.; Mitsiu, J. *SAE Technical Paper Series 922290, Int. Fuels and Lubricants, Meeting and Exposition*, San Francisco, CA: 1992.
14. Aoki, H. U.S. Patent 5 236 182, 1993.
15. Shtarkman, E. M. U.S. Patent 5 176 368, 1993.
16. Williams, E. W. *J. Non-Newtonian Fluid Mech.* **1993**, *47*, 221.
17. Hare, N. S. Sr. U.S. Patent 5 019 119, 1991.
18. Jordan, T. C.; Shaw, M. T. *MRS Bull.* **1991**, *XVI*, 38.
19. Block, H.; Kelly, J. P. *J. Phys., D: Appl. Phys.* **1988**, *21*, 1661.
20. Adriani, P. M.; Gast, A. P. *Phys. Fluids* **1988**, *31*, 2757.
21. Zukoski, C. F. *Ann. Rev. Mater. Sci.* **1993**, *23*, 45.
22. Kanu, R. C.; Shaw, M. T. *Am. Chem. Soc. Polym. Prepr.* **1994**, *35*, 337.
23. Bhaumik, D.; Mark, J. E. *J. Polym. Sci.: Polym. Phys. Ed.* **1983**, *21*, 1111.

ELECTRORHEOLOGICAL MATERIALS

Kathleen O. Havelka and Joseph W. Pialet
The Lubrizol Corporation

Electrorheological (ER) fluids, dispersions of polarizable materials in a base fluid, represent a unique class of electroactive materials that exhibit modified flow properties in an AC or DC electric field. In spite of the great potential of ER technology, no large volume commercial ER devices have reached the marketplace. This chapter focuses on ER materials and emphasizes that polymers have emerged as a leading class of materials because they provide flexibility to tailor their physical and chemical properties to maximize the ER effect.

In general, ER fluids consist of a dispersed particulate phase, an insulating base fluid, and often additional components, such as surfactants or dispersants, and in some cases, a polar activator. Although the interactions of all the components are important, the properties of the dispersed particulate phase offer the greatest potential for improving the performance of ER fluids. Based on the emerging role of polymeric ER particulates, the material properties of ER components are the focus of this review.

ELECTRORHEOLOGICAL PHENOMENON

ER fluids are typically composed of electrically polarized particles dispersed in a low dielectric oil. Upon application of an electric field, of typically a few kV/mm, particles become polarized and the local electric field is distorted. The migration of mobile charges to areas with greatest field concentration increases the polarizability of the particle and results in a larger dipole moment. These field induced dipoles attract one another and cause the particles to form chains or fibrillated structures in the direction of the field. These chains are held together by interparticle forces that have sufficient strength to inhibit fluid flow. Subjecting these fibers to a shearing force pulls particles apart, while dipoles on the particles attract replacement particles. An equilibrium is established between the formation and breaking of the chains. The increase in apparent viscosity from particle chain interactions under shear corresponds to the yield stress defined in the Bingham plastic model.[3–5] When the electric field is removed, the particles return to a random distribution and fluid flow resumes.

A widely recognized fact is that the shear stress versus shear rate properties of an ER fluid vary as a function of applied electric field. In the absence of electric field, an ER fluid demonstrates approximately Newtonian behavior.

It is generally assumed that the ER effect is the result of interactions between the viscous forces and the polarization forces. The viscous forces, for a Newtonian fluid, should be a function of the viscosity and the shear rate. The polarization forces should be a function of the field and the permittivity mismatch of the particle and the base fluid.

While many factors influence the ER phenomenon, polarization and structure formation play a leading role in investigating the ER mechanism. The most widely accepted polarization mechanisms involve bulk or interfacial processes.[6–10] Bulk polarization involves creation or alignment of molecular dipoles within the confines of the particulate or continuous phase. Interfacial polarization can involve migration of charge through the particles, along the surface of the particles, or within the double layer region of the dispersion. The above mechanisms may occur at varying magnitudes and rates in ER fluids and will lead to different dynamic particles interactions.

A desired property is for the particles to respond quickly to an applied electric field. In practice, particles do not respond instantaneously to changes in the electric field. The lag time associated with polarization is an important consideration in determining the appropriate frequency and shear rate for testing and device specifications. If an AC field is used, there is a frequency above which the dipole does not have time to reach full strength and the ER stress will decrease. This is typically not a concern in DC fields. However, the shear rate is an issue in both AC and DC fields. Particles can rotate in a shear rate gradient. If the polarization cannot realign fast enough, this rotation will cause misalignment of the dipoles and a decreased ER stress will be observed.[1] In general, particles must have a sufficient rate and magnitude of polarization to be viable ER materials.

MATERIAL PHYSICAL PROPERTIES

The physical properties of the particles impact both dispersion and ER properties. The ability to modify the physical properties of organic materials offer a number of advantages over inorganic materials in ER fluids, including: low density, which decreases settling, and softness, which decreases abrasion and wear. Other important physical properties to consider in formulating, include particle concentration and morphology. Particle-particle separations are a function of particle concentration.

The effects of particle morphology, that is, size and shape, are an area where the synthetic versatility of polymers is being explored. Very small particles limit the strength of the dipole.

The shape of most ER particles in the literature is irregular. The assumption is often made that the particles behave in combination with entrapped solvent as if they were roughly spherical.[1] To examine this, cylindrical particles of poly(*p*-phenylene-2,6-benzobisthiazole) of various aspect ratios (length/diameter) have been synthesized and dispersed in mineral oil.[12] A systematic investigation conducted under low shear rate indicates that there is an increase in the ER response as the aspect ratio of the particle increases. However, since many applications require high shear rates, the study needs to be expanded to higher shear rates to see if the benefit persists. In general, the physical properties of the materials can influence the ER properties; however, the chemical nature of the particles is critically important to the ER effect.

MATERIAL CHEMICAL PROPERTIES

A vast number of materials of very different chemical nature demonstrate ER activity. For simplicity, research in ER particulates can be classified into two general categories: (i) Extrinsically polarizable materials, and (ii) Intrinsically polarizable materials.

Extrinsically Polarizable Materials

Water was originally considered necessary for the ER effect.[13] Extrinsically polarizable ER fluids are composed of hydrophilic particles that require water or some other polar-activator, such as low-molecular weight alcohols or amines, to obtain measurable ER activity. The amount of water required to optimize the ER effect ranges between 1 and 10% and depends on the physical and chemical properties. The ER effect is expected to be predominantly due to interfacial polarization.

Although extrinsically polarizable ER fluids are suitable to demonstrate the ER phenomenon, they have significant limitations at low and high temperatures. An even more serious concern with extrinsically polarizable ER fluids is their very high thermal coefficient of conductance.

Intrinsically Polarizable Materials

The discovery that a polar activator is not necessary to obtain ER activity catalyzed a resurgence of interest in electrorheology.[10,14] There are a number of advantages of intrinsically polarizable materials over extrinsically polarizable materials, including: (i) the system is simpler since a polar activator is not required, and (ii) they have a lower thermal coefficient of conductance, which may facilitate expanding the temperature range of ER activity. A number of intrinsically polarizable materials have been investigated, such as, ferroelectrics, for example $BaTiO_3$ inorganics, for example, anhydrous zeolites, semiconducting polymers, metals, coated conductors, and liquid crystals.

Liquid Crystals

Dispersion and ER properties need to be optimized together. Since micron-size particles tend to settle, and many of the applications may promote settling through vibration or centrifugation of the fluid, dispersion stability is a challenging requirement. A promising candidate for improving both dispersion and ER properties is liquid crystalline polymer solutions.

Liquid crystals have been investigated as both the dispersed and continuous phase. Monomeric liquid crystals have received some attention as potential base fluids with more conventional ER particles as the dispersed phase. However most work has involved liquid crystalline polymer (LCP) solutions, because this one-phase system eliminates particle settling problems inherent to conventional ER fluids.

The need to provide acceptable solubility and freedom for the liquid crystal domains to align has resulted in work to incorporate multiple ligand crystal domains into non-liquid crystal polymers. A current limitation is that liquid crystal domains exist over a limited temperature range. The polarization is probably predominantly bulk polarization. Some encouraging advances have been made in LCP solutions.[15]

BASE FLUIDS

In addition to the particulates, a necessary component in an ER fluid is the base fluid. Information on the effect of the base fluid on ER properties is limited. The electrostatic polarization model predicts that extremely low dielectric constant base fluids should give ideal ER properties.[6]

APPLICATIONS OF ER TECHNOLOGY

Potential applications of ER fluid are numerous and diverse. An applied electric field affects a variety of ER fluid properties, including: changes in the rheological, electrical, optical, thermal, volumetric, and acoustic properties.

Although there have been many industrial applications proposed, most of these involve niche markets. In the past decade, considerable effort has been focused on the development of ER fluids and devices for the more demanding transportation applications, where the market is expected to be substantial. Particular emphasis has been placed on applications that interface with electronic control systems which rapidly sense and respond to environmental inputs in a desired manner. Examples of proposed transportation applications include primary suspension dampers, engine mounts, and auxiliary drive clutches, such as fan clutches, truck cab mounts, torque converters, seat dampers, bushings, hydraulic valves, servoclutches, and brakes.[2,16]

CONCLUSION

This is a pivotal time for ER technology. Considerable progress has been made in ER fluid performance, yet no commercial devices have reached the marketplace. Successful broad commercialization of this promising technology requires improved materials. This is an area where the synthetic versatility of polymers can contribute significantly. Development of novel materials in parallel with innovative device engineering will allow researchers to effectively evaluate and improve ER performance.

REFERENCES

1. Block, H.; Kelly, J. P. *J. Phys. D.* **1988**, *21*, 1661.
2. Coulter, J. P.; Weiss, K. D.; Carlson, J. D. *J. of Intelligent Materials Systems and Structures* **1993**, *4*, 248.
3. Klingenberg, D. J.; Zukoski, C. F. *Langmuir* **1990**, *6*, 15.
4. Deinega, Y. F.; Vinagradov, G. V. *Rheo. Acta* **1984**, *23*, 636.
5. Shulman, Z. P.; Matsepuro, A. D.; Novichenok, L. N.; Demchuk, S. A.; Svirnovskaya, I. L. *J. Eng. Phys.* **1974**, *27*(6), 1569.
6. Gast, A. P.; Zukoski, C. F. *Adv. Coll. Int. Sci.* **1989**, *30*, 153.
7. Klass, D. L.; Martinek, T. W. *J. Appl. Phys.* **1967a**, *38*, 67.
8. Block, H.; Rattray, P. In *Progress in Electrorheology*; Havelka, K. O.; Filisko, F. E., Eds.; Plenum: New York, 1995, p 19.
9. Klass, D. L.; Martinek, T. W. *J. Appl. Phys.* **1976b**, *38*, 75.
10. Block, H.; Kelly, J. P.; Qin, A.; Watson, T. *Langmuir* **1990**, *6*, 6.
11. Havelka, K. O. In *Progress in Electrorheology*; Havelka, K. O.; Filisko, F. E., Eds.; Plenum: New York, 1995; p 43.
12. Kanu, R. C.; Shaw, M. T. In *Proc. XIth Int. Congr. on Rheology*; Moldenaers, P.; Keunings, R. Eds.; Elsevier: New York, 1992; p 766.
13. Stangroom, J. E. *Phys. Technol.* **1983**, *14*, 290.
14. Filisko, F. E.; Radzilowski, L. H. *J. of Rheology* **1990**, *34*(4), 539.
15. Reitz, R. P. W090/00583, 1990.
16. Petek, N. K.; Romstadt, D. J.; Lizell, M. B.; Weyenberg, T. R. SAE Paper 950586 1995.

Emulsifiers

See: Additives (Property and Processing Modifiers)
Emulsion Polymerization

Emulsion Polymerization

See: Emulsion Polymerization (Overview)
Emulsion Polymers (Noncatalytic Diimide Hydrogenation)
Macromonomers (Soap Free Emulsion and Dispersion Polymerization)
Monodisperse Polymer Particles (Methods of Preparation)
Peroxide Initiators (for Water-Based Acrylate Emulsion Polymerizations)
Poly(methyl methacrylate)
Poly(vinylidene fluoride) (Synthesis, Microstructure, and Chain Conformations)
Vinyl Chloride Copolymers

EMULSION POLYMERIZATION (Overview)

Gary W. Poehlein
Georgia Institute of Technology

Conventional emulsion polymerization is a heterogeneous free-radical process that produces polymer particles in the sub-micron size range. This polymer colloid product is commonly called a latex. The products are used for such applications as coatings, adhesives, caulks, and additives for other products. In some cases the polymer is separated from the continuous phase to produce synthetic elastomers and thermoplastics.

Emulsion polymerization is different from suspension polymerization, another important heterogeneous process, in several significant ways. Suspension polymerization proceeds by dispersing the monomer in water with a suspending agent and then converting the dispersed droplets into polymer by reaction with an initiator that is normally soluble in the monomer. The diameters of the final polymer particles can vary from a few microns to several hundred microns.

The monomer droplets are usually more finely dispersed in emulsion polymerization, with diameters ranging from a few microns to > 10 μ. The final polymer particles, however, are commonly < 0.5 μ. Hence, unlike suspension polymerization, the emulsion process does not simply involve polymerization of the monomer droplets. Nucleation and growth of smaller particles is a significant mechanism in emulsion polymerization.

MECHANISMS

A simple recipe for conventional emulsion polymerization is composed of monomer and water in about equal parts. Anionic emulsifier and a water-soluble initiator system are added at about 1 part per 100 parts of monomer. A small amount of buffer is also often used. Recipe variations can be broad, including different types of emulsifiers or stabilizers (polymeric, nonionic, cationic, amphoteric) or mixtures thereof, higher and lower concentrations of all ingredients, seed latex particles, multiple monomers, chain transfer agents, and solvents.

REACTION INGREDIENTS

Initiators

The selection of initiators for free-radical polymerization is determined by the process type and reaction temperature. Initiator activity is often expressed in terms of half-life, which is a function of temperature. In general the initiator should have a half-life of the same magnitude (perhaps smaller) as the reaction time. A proper initiator will be significantly consumed in the reaction but will last long enough to complete the reaction. Persulfate salts (NH_4, Na, and K) are common water-soluble initiators for emulsion polymerization reactions above 50 or 60°C.

Redox initiation systems can be used at temperatures even below 0°C, and they are popular for the production of styrene–butadiene elastomers. The lower temperatures change relative reaction rates and reduce branching and crosslinking, permitting higher conversions before the reaction is stopped. Initiators can also have some influence on product properties by influencing the number of particles formed, the molecular weight and, in emulsion polymerization systems, by the hydrophilic endgroups, which may remain on the surface of the latex particles.

Emulsifiers

Emulsifier selection is probably the most challenging part of emulsion polymer development. The emulsifier needs to permit the formation of a monomer emulsion and to stabilize the polymer particles formed without having an adverse effect on application performance. Anionic oil-in-water emulsifiers are the most widely used in both research and commercial applications, but they can have some disadvantages such as electrolyte sensitivity, lack of freeze-thaw stability, desorption and hence instability in postreaction formulation processes, and detrimental effects on application performance, such as reduced coating adhesion.

Polymeric stabilizers tend to be less sensitive to electrolytes and freeze-thaw situations. They can have other problems, however, such as grafting to the polymer being formed and influencing end-use properties in positive or negative directions. Polymeric stabilizers can also desorb during postreaction formulation processes, and they can influence latex rheology. Polymeric stabilizers, especially when mixed with other emulsifiers, can also promote droplet polymerization or agglomeration of small latex particles.

Cationic emulsifiers are not widely used, but they can have advantages in some applications. Many natural inorganic materials, when immersed in water, have anionic surfaces because the soluble ions are cations. Hence, cationic latexes have a stronger tendency to foul surfaces within reactors, storage vessels, and processing equipment. This can be an advantage, however, if the cationic-anionic interactions can promote the association of different materials; for example coating of an anionic surface by a cationic latex.

Mixed emulsifier systems composed of a classic anionic emulsifier and a highly water-insoluble component can produce very small, often submicron, monomer droplets. The water-insoluble component stabilizes these droplets against diffusion destruction or Oswald ripening. Such processes, often called mini-emulsion polymerization, have been extensively discussed in the literature and will not be considered further here.

Monomers

The key ingredients in any polymerization are the monomers. The basic physical, chemical, and thermal properties of the polymeric materials produced will be largely determined by the monomers used and how they are put together in the macromolecules.

Monomer selection for emulsion polymerization involves several different considerations, that will be outlined briefly here. First, as mentioned earlier, the small particle size and large interfacial area are important characteristics of latexes. Monomers with hydrophilic functional groups such as carboxyl, hydroxyl, sulfate, or sulfonate can influence latex stability, and application performance, and they can be used for post-polymerization reactions. Monomers such as acrylic acid and methyacrylic acid are sometimes used, for example, to enhance adhesion. Such monomers can also be used, via neutralization, to produce gels. Functional monomers are sometimes called polymerizable emulsifiers.

Water solubility is a second important consideration for monomers used in emulsion polymerization because reactions occur in both phases. Monomers that are highly water soluble, such as those mentioned above, can form water-soluble polymer in the aqueous phase. This can be detrimental to the product and a waste of expensive monomer.

Monomers that form polymers of significantly different compositions will generally result in multiphase products. The morphology of such products will depend on thermodynamic driving forces and the mobility of the polymeric molecules under reaction or processing conditions. Particle size manipulation in emulsion polymerization offers the materials scientist a parameter for controlling product morphologies. It should not be surprising, therefore, that recipe selection and reaction engineering are often strongly focused on controlling the morphology of latex particles.

Other Ingredients

Minor accounts of other chemicals are added to emulsion polymerization reactions to achieve specific objectives, or they enter as impurities in commercial-grade reagents. Common additives include buffers, chain transfer agents, and a nitrogen blanket to exclude oxygen. Latex seed particles, produced in a separate reaction, are sometimes used in reaction recipes. Impurities such as inhibitors in the monomers and reaction side-products in the emulsifiers can influence the course of emulsion polymerization and the properties of the latex product.

The least robust or most sensitive part of most emulsion polymerization reactions is particle nucleation. Nucleation rates are generally rapid and can be influenced by many factors, both known and unknown. Variations in properties among the products of different reactions are very likely due to differences in nucleation phenomena. This provides a major motivation for the use of premade seed latex in commercial recipes. One simply accepts variations in nucleation of the seed particles, and then the recipe is adjusted to achieve the desired particle number and size characteristics in the final product. Operational procedures and reactor design offers other methods for manipulating product characteristics.

REACTION ENGINEERING

Reaction engineering covers the entire area of operation and process design, including control strategies. Reaction processes can be classified as batch, semibatch (sometimes called semicontinuous) or continuous.

SUMMARY

Emulsion polymers are manufactured in large quantity and are used in a broad range of applications. The selection of recipe ingredients, reactor system, and process operational procedures permits the control of important product characteristics and thereby the application performance. The fundamental knowledge base in this field is extensive, but significant art remains in the practice of product and process development, scale-up, and commercial production.

GENERAL LITERATURE

1. *Emulsion Polymers and Emulsion Polymerization*; Basset, D. R.; Hamielec, A. E. Eds.; ACS Symposium Series 165: Washington, DC, 1981.
2. *Science and Technology of Polymer Colloids, Vol. I and Vol. II*; Poehlein, G. W. et al. Eds.; NATO ASI Series E: Applied Sciences - No. 67, Matinus Nijhoff The Hague, 1983.
3. *Scientific Methods for the Study of Polymer Colloids and their Applications*; Candua, F.; Ottewill, R. H. Eds.; NATO ASI Series C: Math. And Phys. Sciences - Vol. 303, Kluvier Dordreckt, 1990.
4. *Polymer Colloids*; Buscall, R. et al. Eds.; Elsevier Applied Science: London, 1985.

5. *Emulsion Polymerization and its Applications in Industry*; Elisleva, V. I. et al. Translated from Russian by S. J. Teague; Consultants Bureau: New York, 1981.
6. Blackley, D. C. *Emulsion Polymerization Theory and Practice* Applied Science: Essex, UK, 1975.
7. *Future Directions in Polymer Colloids*; El-Aasser, M. S.; Fitch, R. M. Eds.; NATO ASI Series E: Applied Sciences - No. 138 Martinus Nijhoff: Dordrecht, 1987.
8. *Polymer Latexes, Preparation, Characterization and Applications*; Daniels, E. S. et al. ACS Symposium Series 492; American Chemical Society: Washington, DC, 1991.
9. *Emulsion Polymerization*; Piirma, I.; Gardon, J. L. Eds.; ACS Symposium Series 24; American Chemical Society: Washington, DC, 1976.

EMULSIONS POLYMERS (Noncatalytic Diimide Hydrogenation)

Dane K. Parker
The Goodyear Tire and Rubber Company

CATALYTIC HYDROGENATION OF UNSATURATED POLYMERS

The catalytic hydrogenation of polymers, especially solution-polymerized polydienes, by homogeneous or heterogeneous activation of molecular hydrogen with various transition metals has evolved into a commercially significant process.[1–5] This is particularly true of block copolymers where hydrogenation is a useful technique for broadening their performance range. Hydrogenation not only improves block copolymer thermal and oxidative stability, but may also affect polymer crystallinity and the glass transition temperature.[6–9] For polymers of this type, catalysts prepared by an *in situ* reduction of nickel or cobalt salts with trialkylaluminium or alkyllithium reagents are the more frequently used.[10–13]

Unfortunately however, conventional hydrogenation technology has many potential hazards and costs associated with it. The first of these hazards is hydrogen itself. Hydrogen is a flammable gas and forms explosive mixtures with air or oxygen over a very wide range of concentrations. A second potential hazard is the highly reactive and sometimes pyrophoric nature of active catalyst residues that remain after reduction. Additionally, the solvent(s) used in a catalytic hydrogenation can also pose a serious safety threat. Solvent vapors mixed with air may be inherently explosive. This danger can be made worse by catalysts such as palladium or nickel that can initiate autooxidation of solvent vapors.[14]

Another problem related to the catalyst itself is its efficient removal from the hydrogenated polymer cement. This is not a trivial issue because heavy metals in effluent streams are highly regulated environmental pollutants. Furthermore, transition metal residues, if left in the polymer, can cause undesirable side reactions such as discoloration or oxidation.

Another point of serious concern with catalytic hydrogenation is in the area of selectivity and catalyst poisoning.[15–17] Whereas many hydrogenation catalysts work quite well with hydrocarbon based unsaturated polymers, polymers bearing functional groups such as nitrile, carboxyl, halogen, sulfide, amino groups and the like often tend to retard or even poison catalysts. Additionally, when multiple functional groups are present in a polymer, great care must be exercised in the choice of the appropriate catalyst, solvent, and reaction conditions necessary to achieve the desired hydrogenation selectivity.

Finally, there is the problem of the physical form of the hydrogenated polymers. Currently, hydrogenated polymers are largely available only in a solid form such as bales, pellets, and granules. Unfortunately, many applications that could effectively benefit from the properties of hydrogenated polymers such as adhesives, binders, dips, and coatings require aqueous based emulsion products. Although it is possible to generate a synthetic "inverted latex" from a hydrogenated polymer cement with surfactants, water, and concurrent removal of solvent, this method suffers from high costs and a lack of generality.[18]

Noncatalytic Diimide Hydrogenation of Unsaturated Polymers

Diimide (HN=NH) is a highly reactive, short-lived species known for its ability to hydrogenate multiple bonds. It is also highly selective in its reducing ability. Among olefins, as the degree of alkyl substitution on the double bond increases, diimide's relative reactivity decreases. Additionally, selective olefin reductions with diimide can be readily carried out in the presence of a wide variety of other functional groups—even highly sensitive allylic halides, alcohols, esters, amines, disulfides, and peroxides.[19–26]

Early work on the reduction of polydienes with diimide (generated by the thermolysis of p-toluenesulfonylhydrazide) showed diimide to be an effective alternative to catalytic hydrogenation.

Until about 1984, this was the extent diimide's application in polymer hydrogenation. In that year, however, a new method was first described in a U.S. patent.[27] This new variant of diimide reduction was completely different from all other known diimide-generating systems.

The key discovery in this new method was the fact that diimide reductions could be effectively run in emulsions. Emulsions, by virtue of their large interfacial surface area, enhanced transport, and increased rates of chemical reaction turned out to be the ideal medium for this type of reduction. Each latex particle functions as a microreactor. These emulsion rubber microreactors in turn allow the essential components of hydrazine, catalyst ion (copper), polymer double bonds, carboxylate ion (RCOO-), and oxidant to be efficiently combined in a unique way, resulting in a stable hydrogenated rubber emulsion.[28]

PROPERTIES

To evaluate the effect of hydrogenation on the physical properties of HSBR rubber, SBR latex with 16% bound styrene was reduced by the previously described procedure to yield HSBR latexes with 78, 86, 93, and 98% saturation levels.

The physical properties indicate that HSBR rubber can develop substantial gum-rubber tensile strength upon hydrogenation. The tensile strength increases with increasing degree of hydrogenation. Upon warming these polymers to 100°C however, the tensile strength disappears; only to be regained again upon cooling. This property is the hallmark of a thermoplastic elastomer or TPE.[30–35] The origin of the TPE behavior is derived from the hydrogenation of the longer segments of 1,4-Bd units. These segments develop into a crystalline polyethylene phase while the styrene and hydrogenated 1,2 vinyl Bd units predominantly define the amorphous phase of the polymer.

APPLICATIONS

As a latex, there are numerous applications for high performance elastomers that are resistant to heat, light, ozone, abrasion, and solvents. These applications include water-based latex adhesives, binders for nonwoven fabrics, binders for short fibers (paper and gasketing), foam rubber, latex thread, asphalt additive, latex dipped goods, surface coatings, and paints.[28,29]

For those hydrogenated latexes that can be isolated and processed like conventional rubbers, such as HSBR, additional applications such as hot melt adhesives, blown film, and use as a compatibilizer in polymer blends are also feasible.[29]

REFERENCES

1. Moberly, C. W. *Encyclopedia Polymer Science and Technology—"Hydrogenation"* 1st ed.; pp 557–568.
2. Schulz, D. N. *Encyclopedia Polymer Science and Technology—"Hydrogenation"* 2nd ed.; pp 807–817.
3. Wicklatz, J. *Chemical Reactions of Polymers*; Wiley-Interscience: New York 1964; Chapter 2F.
4. Golub, M. A. *The Chemistry of Alkenes Vol. 2*, Wiley-Interscience: London, 1970; Chapter 9.
5. Schulz, D. N.; Turner, S. R.; Golub, M. A. *Rubber Chem. & Technol.* **1982**, *55*, 809.
6. Halasa, A. F. *Rubber Chem. & Technol.* **1981**, *54*, 627.
7. Noshay, A.; McGrath, J. E. *Block Copolymers—Overview and Critical Survey*; Academic: New York, 1977; pp 232–237.
8. Jones, R. C. (to Shell International Research) U.S. Patent 3 629 371, 1971.
9. Krause, R. L. (to Dow Chemical) U.S. Patent 3 898 206, 1975.
10. Zucas, W. X.; MacKnight, W. J.; Lenz, R. W. *Polymer Bulletin* (Berlin), 2, 2632.
11. Falk, J. C. *Catal. Org. Synth. [Conf.] 5th 1975*; Academic: New York, 1976; pp 305–323.
12. Kang, J. W. (to Firestone Tire and Rubber) U.S. Patent 4 098 991, 1978.
13. Chamberlin, Y.; Pascoult, J. P.; Razzowk, H.; Cheradem, H. *Makromol. Chem. Rapid Commun.* **1981**, *2*, 323.
14. Komarewski, V. I.; Presiz, C. H.; Morritz, F. I. *Technique of Organic Chemistry*, 2nd Ed., Vol. 11, Wiley-Interscience: New York, 1956; Vol. 11, pp 18–93.
15. Chamberlin, Y.; Gole, J.; Pascault, J. P.; Durand, J. P.; Dawans, F. *Makromol. Chem.* **1979**, *180*, 2309.
16. Yokota, K.; Hirabayashi, T. *Poly. J.* **1981**, *13*, 813.
17. Neumann, R.; Sanui, K.; MacKnight, W. J. *Macromolecules* **1975**, *8*, 665.
18. Southwick, J. G.; Raney, K. H.; Borchardt, J. K. (to Shell Oil Co.) U.S. Patent 5 141 986, 1992.
19. Back, R. A. *Reviews of Chemical Internmediates* **1984**, *5*, 293.
20. Miller, C. E. *J. Chem. Ed.* **1965**, *42*, 254.
21. Hunig, S.; Muller, H. R.; Thier, W. *Angew. Chem. Int. Ed. Eng.* **1965**, *4*, 271.
22. Pasto, D. J.; Taylor, R. T. *Organic Reactions* **1991**, *40*, 91.
23. Nelson, D. J.; Henley, R. L.; Yao, Z.; Smith, T. D. *Tet. Lett.* **1993**, *34*, 5835.
24. Nagedrappa, G.; Devaprabhakara, D. *Tet. Lett.* **1970**, 4243.
25. Coxon, J. M.; McDonald, D. O. *Tet. Lett.* **1992**, *33*, 3673.
26. Moriarty, R. M.; Vaid, R. K.; Duncan, M. P. *Syn. Commun.* **1987**, *17*, 703.
27. Wideman, L. G. (to The Goodyear Tire & Rubber) U.S. Patent 4 452 950, 1984.
28. Parker, D. K.; Roberts, R. F.; Schiessl, H. W. *Rubber Chem. & Technol.* **1992**, *65*, 245.
29. Parker, D. K.; Roberts, R. F.; Schiessl, H. W. *Rubber Chem. & Technol.* **1994**, *67*, 288.
30. Holden, G.; Bishop, E. T.; Legge, N. R. *J. Polym. Sci., Part C* **1969**, *26*, 37.
31. Bishop, E. T.; Davison, S. *J. Polym. Sci., Part C* **1969**, *26*, 59.
32. Meier, D. J. *J. Polym. Sci., Part C* **1969**, *26*, 81.
33. Legge, N. R.; Holden, G.; Schroeder, H. E. *Thermoplastic Elastomers: A Comprehensive Review* Macmillan: New York, 1987.
34. Walker, B. M.; Rader, C. P. *Handbook of Thermoplastic Elastomers*, 2nd ed.; Van Nostrand Reinhold: New York, 1988.
35. Harper, C. A.; *Handbook of Plastics, Elastomers and Composites*, 2nd ed.; McGraw, New York, 1992.

Emulsions

See: *Electroconductive Composites (Emulsion Pathways)*
Silicone Polymers, Organo-Modified (Application in Personal Care Products)
Silicone Rubber Latex
Ultrasonically Assisted Polymer Synthesis
Water-Borne Coatings (Urethane/Acrylic Hybrid Polymers)

END-FUNCTIONALIZED OLIGOMERS (by Chain-Transfer Technique)

Elisabetta Ranucci* and Fabio Bignotti
Dipartimento di Chimica e Fisica per i Materiali
Università degli Studi di Brescia

Oligomers are defined by the International Union of Pure and Applied Chemistry as chemical substances composed of molecules containing "a few" of one or more species of atoms, or groups of atoms, repetitively linked to each other and known as constitutional units.[1] The physical properties of oligomers vary with the removal or addition of one or a few constitutional units, making a clear distinction between oligomers and polymers.[2] End-functionalized oligomers have macromolecular chains that end at one (i.e., Category A oligomers) or both sides (i.e., Category B oligomers) with reactive functional groups.

Category B oligomers, with two reactive end groups, are commonly referred to as telechelic polymers.[3] Generally, telechelics can be distinguished as mono-, di-, tri-, and polytelechelics, according to the functionality of the macromolecule. When such groups further participate in polymerization reactions, yielding copolymers or networks, they are called macromolecular monomers or macromonomers.[4] In this work, only Category A and B oligomers (i.e., mono- and ditelechelics) will be reviewed.

*Author to whom correspondence should be addressed.

Rationale for Use of Chain-Transfer Reactions for Preparing End-Functionalized Oligomers by Radical Polymerization

In radical polymerization end-functionalized oligomers can be formed if one of the two reactions leading to the formation of end groups—initiation and chain transfer—introduces reactive functional groups. This first elementary step that permits functionalized residues to be fixed as end groups is initiation, when the initiator's residues, R, contain reactive functions. End-functionalized oligomers can be produced only when large amounts of initiator are used and no side reactions occur with reactive radicals, a requirement that is not easily met in practice. In particular, Category A oligomers are formed when termination proceeds by disproportionation, whereas Category B oligomers are obtained when termination occurs by combination.

Telomerization as a Tool for Obtaining End-Functionalized Oligomers

Strictly speaking, a macromolecule obtained as the result of two chain-transfer reactions with two molecules of the same chain-transfer agent (CTA) is called a telomer. From a practical standpoint, a polymerization process is referred to as telomerization when the large majority of the macromolecules obtained initiate and terminate with a fragment of CTA, the net result being visualized as the insertion of a macromolecular chain in between the CTA molecules, AB, and splitting it into two moieties, A and B, located at the extremities.

SYNTHESIS OF END-FUNCTIONALIZED OLIGOMERS BY USING CHAIN-TRANSFER AGENTS

Synthesis of Category A Oligomers

Substances belonging to several chemical classes have been used for the synthesis of Category A oligomers (monotelechelics) by chain-transfer techniques (**Table 1**). It is apparent, from the data reported in Table 1 that functionalized mercaptans are the CTAs most cited in the literature. This is because the chain-transfer constant of many alkyl mercaptans toward a number of vinyl monomers is not far from one. This makes molecular weight control during the polymerization process particularly easy. However, for monomers with a marked proneness for chain transfer, for which mercaptans are unsuitable as molecular weight regulators because they are too active, compounds belonging to chemical classes (e.g., alcohols and esters) with slower chain-transfer kinetics must be used. Usually these compounds can act as CTAs and as solvents, and are characterized by the presence of hydrogen atoms, which are only moderately mobile toward abstraction by a macroradical (e.g., tertiary hydrogens or hydrogens in the α-position with respect to electron-attracting groups).[5]

Synthesis of Category B Oligomers

Two groups of substances have been extensively used for the synthesis of Category B oligomers (bitelechelics): carbon halogen compounds and disulfides. Aliphatic halogen compounds, especially carbon tetrachloride, have been extensively used as CTAs toward olefins. The abstraction of the halogen atom is followed by the formation of a haloalkyl radical that reinitiates a new chain.

When alkyl and aryl disulfides are used as CTAs, thioether end groups are formed.[21,23] In case of acyl disulfides, the resulting end groups can be cleaved to –SH.

Disulfides have very high transfer constants toward olefins. Alkyl and aryl disulfides as CTAs have reactivities of the same order of magnitude as that of carbon tetrachloride, whereas the reactivities of acyl disulfides is 4 orders of magnitude higher.[24]

END-FUNCTIONALIZED OLIGOMERS BY INIFERTERS

The use of compounds acting both as initiators and CTAs, known as "iniferters" has been reported. Uranek first proposed the use of 3-(3-acetylphenyl diazothio)acetylbenzene as an iniferter.[3]

Category B oligomers (e.g., α,ω-oligostyrene or oligomethylmethacrylate) have been obtained by making use of iniferters such as tetraethyltiuram disulfide or bensoyldisulfide.[22]

APPLICATIONS OF END-FUNCTIONALIZED OLIGOMERS

Category A and B end-functionalized oligomers are, in principle, amenable to a wide variety of interesting applications. Unmodified telomers are often used as additives in various systems to impart special properties, such as high molecular weight plasticizers for improving the processability of plastic materials and the use of coatings of both inorganic and organic materials. Modified telomers, such as Category A oligomers, can be used as precursors for the synthesis of graft copolymers and comblike polymers. Moreover, they can be utilized for the modification of organic and inorganic surfaces.[2,25] In the biomedical field, they can act as drug carriers in controlled drug-delivery systems and have a distinct potential for the modification of therapeutically useful enzymes.[5,13,26–31] Category B oligomers may undergo chain-extension reactions by means of difunctional linking agents and can be used in the synthesis of interpenetrating networks by reacting their reactive groups with multifunctional linking agents. They are among the best starting materials for the synthesis of block copolymers, by reaction of telechelics with different end groups in a polycondensation or polyaddition reaction.[2,25] In particular, condensation block copolymers can be obtained in a stable form only when their constituent blocks are chemically inert toward side reactions such as transesterification and transamidation. These stable blocks commonly comprise carbon chains end capped with one or two reactive groups, and chain-transfer techniques offer unique possibilities for their preparation.

REFERENCES

1. IUPAC Commission on Macromolecular Nomenclature, *Pure Appl. Chem.* **1974**, *40*, 479.
2. Gordon, J.; Loftus, E. *In Encyclopedia of Polymer Science and Technology*; Mark, H. F., et al., Eds.; Interscience: New York, 1992; Vol. 16, pp 533–554.
3. Uranek, C. A.; Hsieh, H. L.; Buck, O. G. *J. Polym. Sci.* **1960**, *46*, 535.
4. Nuyken, O.; Pask, S. In *Encyclopedia of Polymer Science and Technology*; Mark, H. F., et al., Eds.; Interscience: New York, 1992; Vol. 16, pp 495–532.
5. Ranucci, E.; Spagnoli, G.; Bignotti, F.; Sartore, L.; Ferruti, P.; Caliceti, P.; Schiavon, O.; Veronese, F. M. *Macromolecular Chemistry and Physics* **1995**, *196*, 763.

TABLE 1. Examples of Category A Oligomers by Chain Transfer[a]

Chain transfer agent	Functional end group	Initiator	X1	X2	Polymer	References
Diethyl phosphate	$H-[C(X_1)(X_2)-CH_2]_n-P(=O)(OC_2H_5)_2$	$FeCl_3$–benzoin	H	OH	PVA	7 8
2-Iodoacetic acid	$H-[C(X_1)(X_2)-CH_2]_n-CHICOOH$	AIBN	H		PS	9
2-Isopropoxyethanol	$H-[C(X_1)(X_2)-CH_2]_n-C(CH_3)_2-OCH_2CH_2OH$	AIBN	H	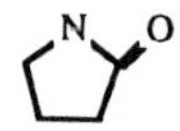	PVP	5
Isopropyl alcohol	$H-[C(X_1)(X_2)-CH_2]_n-C(CH_3)_2-OH$	AIBN			PVP	5
Isopropyl Alcohol/Cyclohexane		γ-Radiation			PS	10
		γ-radiation	CH_3	$COOCH_3$	PMMA	10
		AIBN	H	Cl	PVC	11
		AIBN			PVP	12
2-Mercaptoacetic acid	$H-[C(X_1)(X_2)-CH_2]_n-SCH_2COOH$	AIBN	H	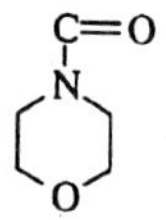	PAcM	13
		AIBN			PMMA	14
		AIBN			PS	15
		AIBN	CH_3	$C(=O)-(CH_2)_2-N(CH_3)_2$	Poly(*N,N*-dimethyl)aminoethyl acrylate	11
2-Mercaptoethanol	$H-[C(X_1)(X_2)-CH_2]_n-SCH_2CH_2OH$	AIBN			PAcM PMMA PS PVP	13 17 15 12
2-Mercaptoethyl amine	$H-[C(X_1)(X_2)-CH_2]_n-SCH_2CH_2NH_2$	AIBN			PAcM	18
3-Mercaptopropionic acid	$H-[C(X_1)(X_2)-CH_2]_n-SCH_2CH_2COOH$	AIBN			PVP Polyacrylamide	19 19
Methylacetate	$H-[C(X_1)(X_2)-CH_2]_n-CH_2COOCH_3$	Peroxide			PE	20

[a]AIBN, azobisisobutylonitrile; PVA: polyvinylalcohol; PS, polystyrene; PVP: poly-N-vinylpyocolidone; PMMA: polymethylmethacrylate; PVC, polyvinylchloride; PAcM: polyactyloylmorpholine; PE, polyethylene.

6. Barson, C. A.; Luxton, A. R.; Robb, J. C. *Trans. Faraday Soc.* **1972**, *68*, 1666.
7. Burczyk, A. F.; O'Driscoll, K. F.; Rempel, G. L. *J. Polym. Sci., Chem. Ed.* **1984**, *22*, 3255.
8. Bauduin, G.; Bondon, D.; Martel, J.; Pietrasanta, Y.; Pucci, B. *Makromol. Chem.* **1981**, *182*, 773.
9. Yamashita, Y.; Ito, K.; Mizuma, H.; Okada, K. *Polym. J.* **1982**, *14*, 255.
10. Altman, O.; Fehrmann, U.; Schnabel, W. *J. Appl. Polym. Sci.* **1979**, *23*, 215.
11. Bockmann, O. C. *J. Polym. Sci.* **1965**, *A3*, 3399.
12. Andreani, F.; Salatelli, E.; Ferruti, P. *J. Bioactive Compat. Polym.* **1986**, *1*, 7.
13. Ranucci, E.; Spagnoli, G.; Sartore, L.; Ferruti, P.; Caliceti, P.; Schiavon, O.; Veronese, F. M. *Macromolecular Chemistry and Physics* **1994**, *195*, 3469.

14. Ito, K.; Usami, N.; Yamashita, Y. *Macromolecules* **1980**, *13*, 26.
15. O'Brien, J. K.; Gornick, F. *J. Am. Chem. Soc.* **1955**, *77*, 475.
16. Chujo, Y.; Shishino, T.; Tsukahara, Y.; Yamashita, Y. *Polym. J.* **1985**, *17*, 133.
17. Ray, K. K.; Pramanik, D.; Palit, S. R. *Makromol. Chem.* **1972**, *153*, 71.
18. Ushakova, V.; Panarin, E.; Ranucci, E.; Bignotti, F.; Ferruti, P. submitted for publication in *Macromol. Chem. Phys.*
19. Akashi, M.; Kirikihara, I.; Miyauchi, N. *Angew. Macromol. Chem.* **1985**, *132*, 81.
20. Fujimoto, T.; Irao, I. *Bull. Soc. Chem. Jpn.* **1974**, *47*, 1930.
21. Pierson, R. M.; Costanza, A. J.; Weinstein, A. H. *J. Polym. Sci.* **1955**, *17*, 221.
22. Otsu, T.; Yoshida, M. *Makromol. Chem. Rapid Commun.* **1982**, *3*, 126.
23. Athey, R. D.; Mosher, W. A.; Weston, N. W. *J. Polym. Sci., Polym. Chem. Ed.* **1977**, *15*, 1423.
24. Young, L. J.; Brandrup, G.; Brandrup, J. In *Polymer Handbook*; Brandrup, J.; Immergut, E. H.; Eds.; Interscience: New York, 1966.
25. Jerome, R.; Henrioulle-Granville, M.; Boutevin, B.; Robin, J. J. *Prog. Polym. Sci.* **1991**, *16*, 837.
26. Ferruti, P.; Ranucci, E. In *High Performance Biomaterials: A Guideline for Biomedical and Pharmaceutical Applications*; Szycher, M.; Ed.; Technomics: Basel, Switzerland, 1991; pp 539–572.
27. Veronese, F. M.; Largajolli, R.; Visco, C.; Ferruti, P.; Miuccio, A. *Appl. Biochem. Biotechnol.* **1985**, *11*, 269.
28. Veronese, F. M.; Visco, C.; Massarotto, L.; Benassi, C. A. *Ann. N. Y. Acad. Sci.* **1987**, *501*, 444.
29. Ferruti, P.; Ranucci, E.; Sartore, L.; Caliceti, P.; Schiavon, O.; Veronese, F. M. Ital. Patent Appl. M192 A0002018, 1992.
30. Ferruti, P.; Ranucci, E.; Sartore, L.; Caliceti, P.; Schiavon, O.; Veronese, F. M. Ital. Patent Appl. M192 A002616, 1992.
31. Sartore, L.; Ranucci, E.; Ferruti, P.; Caliceti, P.; Schiavon, O.; Veronese, F. M. *J. Bioact. Compat. Polym.* **1995**, *10*, 103.

END GROUP ANALYSIS

A. Sezai Saraç and Candan Erbil
Department of Chemistry
Istanbul Technical University

End group analysis for additional polymers with functional end groups prepared by redox systems and for condensation polymers is done with suitable techniques. For linear polymers, determination of end groups gives the number average molecular weight (**Equation 1**):

$$\overline{M}_n = n \times 10^6/m \qquad 1$$

where n is the number of groups that can be determined per macromolecule and m is the end group concentration (in microequivalents per gram).

Titrimetric and instrumental methods are widely used to elucidate polymerization mechanisms and determine the molecular weights of condensation and addition polymers by end group analysis.

In condensation polymers, one or two ends of the polymer have reactive functional groups. A close approximation to the number-average molecular weight, $\overline{\mathbf{M}}_\mathbf{n}$, can be obtained by measuring the total of these groups. The acidic and basic end groups, such as polyesters, polyamides, and polyethers are determined either potentiometrically or conductometrically. Comparison of these methods for determining carboxyl and amine groups shows that the latter is more precise than the former.[1,2] For determining phenolic hydroxyl end groups in polysulfones, potentiometric titration in nonaqueous media has been used.[3]

Spectrophotometric techniques are particularly useful for determining small concentrations of functional groups. Radiochemical methods used to calibrate other analytical methods and to determine molecular weight of stepwise polymers also are used in the study of mechanisms and estimation of polymer structure and end groups, especially for addition polymers. However, small amounts of impurities in the solvent or polymer can cause appreciable errors.

High molecular weight addition polymers have been investigated by end group analysis. Since most commonly used initiators, such as benzoyl peroxide or azobisisobutyronitrile, can not be determined by simple methods, it is necessary to label these catalysts with radioactive groups to obtain good values for $\overline{\mathbf{M}}_\mathbf{n}$ of these polymers if the nature of the termination reaction is known. Isotopic labeling has been applied to poly(methyl methacrylate) and polystyrene.[4–7] This is also a useful technique for insoluble polymers such as tetrafluoroethylene.[8]

For the analysis of the polymer chain end, NMR spectroscopic investigation with isotopic labeling of the end groups has been used widely because of recent advances in NMR technology.[9,10] On the other hand, by using a totally deuterated monomer technique, the fraction of the cyanoisopropyl end group in radical polymerization with azobis(isobutyronitrile) was reported with ^{1}H NMR spectroscopy.[11] Further, high-resolution pyrolysis-gas chromatography and size-exclusion chromatography (SEC) also have been applied to analysis of the polymer end group.[12,13]

Redox systems are used largely as initiators in radical polymerization and more especially in emulsion polymerization.[14] They have the prime advantage of operating at reasonable rates at moderate temperatures, and they are an efficient way to attach a functional group at the end(s) of the polymer chains formed.

Palit and Konar studied general features of the permanganate-oxalic acid initiator system in aqueous polymerization. They wanted to shed light on the nature of the initiating free radical by detecting the end groups of the resulting polymer. Carboxyl and strong acid end groups were detected by dye partition and dye interaction techniques.[15] These tests also are suitable for detecting hydroxyl, amino, quaternary ammonium, and halogen groups. Most of the potentiometric titration techniques permit acid groups of different strengths to be distinguished. Several authors have reported successful measurements by infrared absorption techniques.[16]

Infrared absorption is suitable for quantitative estimation of end groups after being calibrated with polymers of known end group content. Chemical determination according to titration methods would serve as a basis for calibration.

Ceric salts such as the nitrate, perclorate, and sulfate in aqueous acidic solution are used as initiators in vinyl polymerization.[17–19] These ceric salts also form very effective redox systems in the presence of organic reducing agents such as alcohols, aldehydes, carboxylic acids, amino acids, hydroxycarboxylic acids, and polyaminocarboxylic acids.[20–28] In these systems, Ce(IV) ions form an intermediate complex with an organic reducing agent, and this complex acts as a source of free radicals, initiating polymer formation.

When a carboxylic acid reducing agent is used, the same functional group may be present as an end group at the end of the polymer chain. If the molecular weights of these samples are high, accurate determination of functional end groups is difficult because of their very low concentration. Therefore, polymers that have short chain lengths were synthesized. In addition, to obtain vinyl polymers that have different numbers of carboxyl groups at the end of the chains, Ce(IV)-malonic acid (MA), tartaric acid (TA), and citric acid (CA) redox initiated systems were used.[27,30]

Although the potentiometric determination method in nonaqueous media by the glass-calomel electrode system is suitable for acids and bases,[31,32] Pt-glass and glass-calomel electrode systems give similar results in respect to $\overline{\mathbf{M}}_n$ values of low molecular weight polymers.[30] No comparable results were observed with the viscosity-average molecular weights at high molecular weight products. Moreover, good correlation was obtained with $\overline{\mathbf{M}}_n$ and $\overline{\mathbf{M}}_v$ for high molecular weight of PAN polymers by the conductometric method[27] rather than potentiometric results.

The termination mechanism, as well as the polymerization rate and radical production, depend on the choice of experimental parameters. Further, the monomer can be completely soluble (i.e., acrylamide) or not (i.e., acrylonitrile) in the reaction medium.[29,33] Termination mechanisms also are related to the complexity of redox processes involving an organic substrate and an inorganic salt.[24,25] Complexes are formed, but their nature and relative concentrations are affected by changes in experimental conditions. For these reasons, experimental results are usually valid only for the particular conditions defined in that work.

REFERENCES

1. Heidendahl, R. *Dtsch. Textiltech.* **1970**, *20*, 459.
2. Marshall, J. *J. Polym. Sci., Part A-1* **1968**, *6*, 1583.
3. Wnuk, A. J.; Davidson, T. F.; McGrath, J. E. *J. Appl. Polym. Sci. Appl. Polym. Symp.* **1978**, *34*, 89.
4. Arnett, L. M. *J. Am. Chem. Soc.* **1952**, *74*, 2031.
5. Bevington, J. C.; Melville, H. W.; Taylor, R. R. *J. Polym. Sci.* **1954**, *14*, 463.
6. Allen, P. W.; Ayrey, G.; Merrett, F. M.; Moore, C. G. *J. Polym. Sci.* **1956**, *22*, 549.
7. Allen, P. W.; Place, M. A. *J. Polym. Sci.* **1957**, *26*, 386.
8. Berry, K. L. and Peterson, J. H. *J. Am. Chem. Soc.* **1951**, *73*, 5195.
9. Moad, G.; Solomon, D. H.; Johns, S. R.; Willing, R. I. *Macromolecules* **1984**, *17*, 1094.
10. Cywar, D. A.; Tirrell, D. A. *Macromolecules* **1986**, *19*, 2908.
11. Hatada, K.; Kitayama, T.; Masuda, E. *E. Polym. J.* **1986**, *18*, 395.
12. Ohtani, H.; Ishigaro, S.; Tanaka, M.; Tsuge, S. *Polymer J.* **1989**, *21*, 41.
13. Ohtani, H.; Tanaka, M.; Tsuge, S. *J. Anal. Appl. Pyrolysis* **1989**, *15*, 167.
14. Odian, G. *Principles of Polymerization*; McGraw-Hill: New York, 1970; 189.
15. Palit, S. R.; Konar, R. S. *J. Polym. Sci.* **1962**, *57*, 609.
16. Yamadera, R. *J. Polym. Sci.* **1958**, *32*, 323.
17. Saldick, J. *J. Polym. Sci.* **1956**, *19*, 73.
18. Venkatakrishnan, S.; Santappa, M. *Makromol. Chem.* **1958**, *27*, 51.
19. Ananthanaryanan, V.; Santappa, M. *J. Appl. Polym. Sci.* **1965**, *9*, 2437.
20. Lalitha, J.; Santappa, M. *Vijnana Parishad Anusandhan Patrika* **1961**, *4*, 139.
21. Mino, G.; Kaizerman, S.; Rasmussen, E. *J. Polym. Sci.* **1959**, *38*, 393.
22. Ahmed, K. R.; Natarajan, L. V.; Anwaruddin, Q. *Macromol. Chem.* **1978**, *179*, 1193.
23. Subramanian, S. V.; Santappa, M. *J. Polym. Sci. Part A 1* **1968**, *6*, 493.
24. Saraç, A. S.; Basak, H.; Soydan, A. B.; Akar, A. *Angew. Makromol. Chem.* **1992**, *198*, 191.
25. Saraç, A. S.; Erbil, C.; Soydan, A. B. *J. Appl. Polym. Sci.* **1992**, *44*, 871.
26. Erbil, C.; Cin, C.; Soydan, A. B.; Saraç, A. S. *J. Appl. Polym. Sci.* **1993**, *47*, 1643.
27. Erbil, C.; Ustamehmetoğlu, B.; Uzelli, G.; Saraç, A. S. *Eur. Polym. J.* **1994**, *30*(2), 149.
28. Erbil, C.; Soydan, A. B.; Aroğuz, A. Z.; Saraç, A. S. *Die Angew. Makromol. Chem.* **1993**, *213*, 55.
29. Renders, G.; Broze, G.; Jérome, R.; Teysssié, Ph. *J. Macromol. Sci.-Chem.* **1981**, *A16*(8), 1399.
30. Saraç, A. S.; Ustamehmetoğlu, B.; Erbil, C. *Polym. Bull.* **1994**, *32*, 91.
31. Miron, R. R.; Hercules, D. M. *Anal. Chem.* **1961**, *33*, 1771.
32. Onen, A.; Saraç, A. S. *J. Anal. and Appl. Pyrolysis* **1990**, *17*, 227.
33. Samal, R. K.; Nayak, M. C.; Panda, G.; Suryanarayana, G. V.; Das, D. P. *J. Polym. Sci. Polym. Chem.* **1982**, *20*, 53.

Ene Reactions

See: Maleic Anhydride Grafting
Maleic Anhydride-Polybutadiene, Ene Reaction
Natural Rubber (Chemical Modification)

ENGINEERING PLASTICS, MELT SPINNING

Yutaka Ohkoshi
Faculty of Textile Science and Engineering
Shinshu University

The melt processing of so-called engineering plastics, even if it is practicable, has some differences from that of general-purpose polymers. The principal difference is the high processing temperature caused by the high glass transition temperature and/or high melting point. The high processing temperature accelerates the thermal degradation of polymer and decreases the stability of molten polymer. One of the other differences is the viscoelastic behavior of polymer melt. The comparatively long relaxation time due to rigid molecular conformation causes strong elastic behavior, strong nonlinearity of the apparent melt viscosity and a nonlinear stress-optical relation. These differences are shown typically for poly(ether ether ketone) (PEEK), which indicates almost the highest melting temperature of the thermoplastic polymers on the market. Melt spinning[1–5] and drawing,[6,7] of PEEK have been studied by many researchers.

APPLICATIONS

Nylon and poly(ethylene terephthalate) (PET), two engineering plastics, are also typical of synthetic fiber materials used for many purposes, including clothing, medical use, industrial use, etc. Poly(ethylene naphthalene dicarbonate) (PEN) is used in a higher-temperature environment to replace PET, whereas poly(butylene terephthalate) (PBT) is used to replace PET in bathing suits because of its good dyeability and high chlorine resistance.[8] Poly(phenylene sulfide) (PPS) and PEEK are used for the materials of industrial filters and belts, exploiting their high durabilities against high temperature and chemical agents. The fiber/fiber composite method is very flexible for the manufacturing of advanced composite materials (ACM). One can get a suitably shaped ACM by heat compression of cloth that consists of reinforcing fiber and matrix polymer fiber. The matrix fibers are made from high-performance thermoplastic polymers, such as PEEK, PPS, and Poly(ether imide) (PEI). Transparent engineering plastics polycarbonate (PC), Poly(ether sulfone) (PES, etc.) can be the materials of optical fibers for severe environments.

REFERENCES

1. Ohkoshi, Y.; Konda, A.; Kikutani, T.; Shimizu, J. *Sen'i Gakkaishi* **1993**, *49*, 211.
2. Song, S. S.; White, J. L.; Cakmak, M. *Sen'i Gakkaishi* **1989**, *45*, 243.
3. Shimizu, J.; Kikutani, T.; Ohkoshi, Y.; Takaku, A. *Sen'i Gakkaishi* **1987**, *43*, 507.
4. Xu, J. Z.; Kitao, T.; Kimura, Y.; Taniguchi, I. *Sen'i Gakkaishi* **1985**, *41*, T-1.
5. Ohkoshi, Y.; Konda, A.; Ohshima, H.; Toriumi, K.; Shimizu, Y.; Nagura, M. *Sen'i Gakkaishi* **1993**, *49*, 151.
6. Ohkoshi, Y.; Ohshima, H.; Matsushita, T.; Toriumi, K.; Konda, A. *Sen'i Gakkaishi* **1989**, *45*, 509.
7. Ohkoshi, Y.; Ohshima, H.; Matsushita, T.; Miyamoto, N.; Toriumi, K.; Konda, A. *Sen'i Gakkaishi* **1990**, *46*, 87.
8. Yonetani, K. *Engineering Plastics*; Kyouritsu Syuppan.

Engineering Thermoplastics

See: *Acetal Resins*
Acetal Resins (Homopolymers and Copolymers)
Acrylonitrile-Butadiene-Styrene
Bismaleimide Resins (Modification with Engineering Plastics)
Cyclic Oligomers of Engineering Thermoplastics
Engineering Thermoplastics (Survey of Industrial Polymers)
Epoxy Resins (Modification with Engineering Plastics)

ENGINEERING THERMOPLASTICS (Survey of Industrial Polymers)

Rudolph D. Deanin
Plastics Engineering Department
University of Massachusetts

Thermoplastics are discrete, chemically stable, polymer molecules, which can be heated to produce a flowable melt, and the cooled to produce a solid product. This reversible physical equilibrium combines easy processing, recyclability, and favorable economics, so thermoplastics account for 85% of total plastics production. The major thermoplastics, based on ethylene, propylene, styrene, and vinyl chloride, offer the optimum balance of easy processability and moderate properties at low cost for the majority of plastics products. Consequently, they are used in quantities of billions of pounds apiece and are commonly referred to as "commodity thermoplastics."[1]

For more demanding high-performance applications, engineers most often require higher temperature and chemical resistance. In addition, they often have more specialized performance requirements such as creep resistance, adhesion, lubricity, abrasion resistance, dimensional stability, low-temperature toughness, flame retardance, low dielectric constant and loss, and radiation and weather resistance. Because these more sophisticated materials are designed to meet the need of the design engineer, they are commonly referred to as "engineering thermoplastics," typically 3% of total plastics production.

This survey will review the major families of rigid engineering thermoplastics.

FLUOROPOLYMERS[1–6]

Polytetrafluoroethylene (PTFE)

These long rodlike molecules prevent normal melt flow, requiring special processes such as sintering and temporary plasticizers for conversion into finished products. Balanced distribution of the fluorine sheath around the main-chain makes the polymer completely non-polar, giving extreme abhesion/lubricity and electrical insulation (high resistance, low dielectric constant and loss). The non-polarity also offers very little intermolecular attraction, producing high ductility but low strength and serious creep under sustained stress. Finally, the short strong C-F bond gives very high thermal and chemical stability, and the absence of hydrogen gives very high flame-resistance.

These properties lead to high-performance applications in self-lubricating bearings and non-stick cookware, electrical and electronic insulation, and chemical equipment such as gaskets and pumps. The main limitations of PTFE—difficult processing and low mechanical strength—may be overcome by modifying the base polymer, leading to a number of specialty fluoropolymers that have found their niche in special applications.

Poly(vinylidene Fluoride) (PVDF)

PVDF melt processability, modulus, and strength are superior to TFE polymers, but flammability, dielectric constant, and chemical resistance are not as good. It is used mainly in molded products for processing chemicals, paper, nuclear materials, food, and pharmaceuticals, and for handling hot water beyond the limits of the commodity plastics.

Poly(vinyl Fluoride) (PVF)

PVF is much more crystalline than poly(vinyl chloride), and gives films of high strength and excellent weather resistance. These are used for agricultural glazing and as a surface layer for weather-resistant laminates.

Polychlorotrifluoroethylene (PCTFE)

In comparison with PTFE, introduction of a chlorine atom reduces the regularity, lowers the crystallinity, and increases molecular flexibility and intermolecular attraction. This results in easier processability, and higher modulus, strength, and transparency, but lower lubricity, thermal and chemical stability, and higher dielectric constant. Transparent film is used in medical and military packaging, and it also finds use in chemical gasketing and electrical insulation.

POLYOXYMETHYLENE[1,2,7,8]

Acetal (polyacetal) plastics are linear polymers of formaldehyde, either homopolymers or copolymers with ethylene oxide. The copolymers are made by ring-opening copolymerization of trioxane with ethylene oxide or dioxolane, using boron trifluoride etherate in cyclohexane.

Acetals replace metals in gears and bearings, and plumbing fixtures, because of their easier processability, self-lubricating qualities, and corrosion resistance. In competition with nylons, acetals have superior rigidity, creep resistance, fatigue endurance, and water resistance, but are inferior in resistance to impact and abrasion. Modulus is increased still further by addition of glass fibers; lubricity is increased by addition of PTFE powder. Other frequent additives are conductive fillers and UV stabilizers.

POLY(PHENYLENE OXIDE)[1,2,9]

Poly(2,6-dimethyl-*p*-phenylene ether) is produced by oxidative polymerization of 2,6-xylenol. Extended resonance gives a rigid stable molecule with high modulus, strength, creep resistance, dimensional stability, heat deflection temperature, electrical resistance, and water resistance, and low dielectric constant and loss, but very difficult processability. It is commonly polyblended with about equal amounts of polystyrene or high-impact polystyrene, giving easier processing and lower cost, and modulating the other properties so that they are intermediate between the two pure polymers.

AROMATIC ETHER KETONES[1,2,7,10,11]

Aromatic ether ketone polymers are made by condensation polymerization reactions between potassium phenate groups and fluorophenyl groups. Stable functional groups, extended aromatic resonance, and medium crystallinity produce high rigidity, strength, creep resistance, heat deflection temperature, melting point, flame retardance, electrical insulation, radiation resistance, and chemical resistance. Reinforcement by glass and carbon fibers increases modulus, strength, and heat deflection temperature still further. Balance of properties is superior to most other engineering plastics, but the high price limits markets to critical applications, such as aviation and computer insulation, chemical plant equipment, bearings, medical equipment, and small parts in appliances.

POLY(PHENYLENE SULFIDE)[1,2,8,9,12,18]

Condensation polymerization of *p*-dichlorobenzene with sodium sulfide produces linear, stiff, crystalline polymers of high modulus, strength, heat stability, electrical and chemical resistance. They are generally reinforced by glass or carbon fibers, which further increase modulus, strength, and heat deflection temperature; carbon fibers also produce lubricity and semi-conductivity. They find specialty applications where these properties are important in electrical equipment, chemical processing, automotive parts, appliances, and medical equipment.

POLYSULFONES[1,2,7,10]

Aromatic sulfone-aromatic ether polymers are made by polycondensation reaction of sodium phenoxide monomers with chlorophenyl sulfone monomers, producing alternating phenyl-sulfone-phenyl-ether structures. The polymers are amorphous. Aromatic sulfone units give molecular rigidity, and ether links give sufficient molecular flexibility for processing. Aromatic resonance gives high heat deflection temperature, and thermal and chemical stability. Glass fiber reinforcement is frequently added to give high modulus, creep resistance, and dimensional stability. Applications requiring this balance of properties are found in electrical, automotive, and appliance parts, as well as in medical equipment, food processing, chemical processing, and aircraft parts.

POLYCARBONATE[1,2,9,13]

Bisphenol A (BPA) polycarbonate is made by condensation polymerization of BPA with phosgene in the presence of strong base to neutralize the HCl as it is liberated. Benzene rings plus quaternary carbon atoms form bulky stiff molecules, which give rigidity, strength, creep resistance, and high heat deflection temperature. Aromatic resonance, and absence of secondary or tertiary hydrogens, give high heat stability. The bulky chains crystallize with great difficulty, so the polymer is normally amorphous, giving good transparency, but low resistance to solvents and alkali. The bulky amorphous molecules trap considerable free volume, giving high ductility and impact resistance.

Overall balance of properties combines rigidity, high impact strength, high heat deflection temperature, and transparency at a price intermediate between the commodity plastics and the specialized engineering thermoplastics. This makes polycarbonate the second largest volume engineering thermoplastic after polyamides. Major markets, in order of decreasing volume, are unbreakable glazing, auto and aircraft parts, cases and parts for appliances, recreational products, electrical and electronics parts, business equipment, packaging, information storage, and lighting fixtures.

POLYESTERS[1,2,7–9,14,15]

Poly(ethylene terephthalate) is produced from ethylene glycol + terephthalic acid by melt condensation polymerization in vacuum. Flexible ethylene ester groups + rigid p-phenylene groups give bulky chains with good melt processability, but little ability to crystallize directly from the melt. When the amorphous solid is warmed above T_g and stretched, it crystallizes readily up to 40%. This produces high strength, heat and chemical resistance, and impermeability, which have found commodity-scale use, first in fibers, then in film, and finally in blow-molded bottles. There have also been commercial efforts to use it in engineering thermoplastic molding compounds.

The leading polyester engineering thermoplastic molding polymer in commercial practice is based on poly(butylene terephthalate).

A variety of other thermoplastic polyesters are produced by use of cyclohexylene glycol, cyclohexane dimethanol, and bisphenol A, with isophthalic and/or terephthalic acid, to produce crystalline homopolymers or transparent copolymers. These moderately priced engineering thermoplastics find considerable use in automotive parts and electrical/electronic products, with smaller amounts in machinery and appliances.

LIQUID CRYSTAL POLYESTERS (LCP)[1,2,8,10]

Wholly aromatic polyesters, which are also linear and symmetrical, are rod-like molecules that form parallel bundles, and microcrystals, which persist even in the molten state. These give very low melt viscosity and pseudoplastic melt rheology, which result in easy processing and fast molding cycles. In the finished product, these crystalline bundles make the polymers self-reinforcing, thus resulting in high modulus, high strength, dimensional stability, very high heat deflection temperature, chemical resistance, and impermeability. The main-chain aromatic structure also produces good flame-retardance.

Two major problems of these polyesters are high cost of synthesis, and flow orientation in melting processing, which produces very anisotropic structure and properties. These are remedied by addition of short glass fibers, which both lower the cost and help randomize the effects of fiber orientation. The largest use is in microwave cookware, where the high heat deflection temperature and low dissipation factor are extremely important. Other high-performance specialty applications have developed in chemical processing, automotive parts, aerospace, machinery, and electrical and electronic products.

NYLONS[1,2,7,8,14,16]

Aliphatic polyamides which are regular and crystalline were first developed for synthetic fibers, and were so successful that they grew into a multi-billion-pound commodity material. Later they were evaluated as moldable plastics and offered a combination of easy processability, balance of properties, and moderate price. This was superior to the commodity thermoplastics, and gave birth to the concept of engineering thermoplastics. They have a wide and growing variety of applications, and have continued to grow, remaining the leading engineering thermoplastic.

The number of different nylon plastics is greatly magnified by use of additives, particularly reinforcements, fillers, stabilizers, plasticizers, lubricants, flame-retardants, and rubbery impact modifiers. Short glass fibers, in particular, although they make processing more difficult, do provide tremendous increase in modulus, strength, creep resistance, dimensional stability, heat deflection temperature, and moisture resistance.

Major applications, in order of importance, are in auto parts, electrical moldings, film, gears and bearings in machinery, outer sheathing on wire and cable insulation, combs and brushes, and appliances.

POLYPHTHALAMIDES[2,10]

One of the newest engineering thermoplastics is poly(hexamethylene terephthalamide). The flexible hexamethylene segments contribute good melt processability; the rigid resonating terephthalamide units and crystallinity contribute high heat deflection temperature, thermal and chemical stability. Short glass fiber reinforcement enhances these and adds dimensional stability and water resistance. Initial applications are special high-temperature parts in autos, electronics, lighting, and appliances.

POLY(AMIDE IMIDE)[1,2,10]

Condensation of trimellitic anhydride with aromatic diamines (such as p-phenylene diamine, oxydianiline, and/or methylene dianiline) produces highly aromatic polymers with alternating amide and imide links. Resonance, molecular rigidity, and hydrogen-bonding combine to produce high modulus, strength, creep resistance, flame retardance, radiation resistance, and solvent resistance. Fibrous reinforcement increases modulus and dimensional stability still further. Typical applications are in pumps, gears and bearings, electrical, and refrigeration equipment.

POLYIMIDES[1,2,9]

Aromatic polyimides have extreme heterocyclic resonance and molecular rigidity, producing high modulus, creep resistance, dimensional stability, heat deflection temperature, heat resistance, flame retardance, electrical insulation, radiation resistance, and chemical resistance; but their molecular rigidity makes thermoplastic processing almost impossible. Processability may be improved by including some flexible links in the chain. In the best known version, *m*-phenylene-bisphthalimide units are joined by bisphenol A ether units, permitting good thermoplastic processing. Typical applications are circuit breakers, microwave ovens, auto and aerospace parts, heat exchangers, wire and cable insulation and appliances.

REFERENCES

1. Brydson, J. A. *Plastics Materials*; Butterworth Scientific: London, 1982.
2. International Plastics Selector, *Plastic Digest*; D. A. T. A. Business: Englewood, CO, 1995.
3. Ausimont USA, Inc. *Technical Bulletins*, 1994.
4. DuPont, *Technical Bulletins*, 1994.
5. Elf Atochem, *Technical Bulletins*, 1994.
6. Solvay Polymers, Inc. *Technical Bulletins*, 1994.
7. BASF Plastic Materials, *Technical Bulletins*, 1994.
8. Hoechst Celanese, Advanced Materials Group, *Technical Bulletins*, 1994.
9. General Electric Company, Plastics Group, *Technical Bulletins*, 1994.
10. Amoco Performance Products, Inc. *Technical Bulletins*, 1994.
11. Victrex USA, *Technical Bulletins*, 1994.
12. Phillips Petroleum Company, *Technical Bulletins*, 1994.
13. Dow Chemical Company, *Technical Bulletins*, 1994.
14. Allied Signal Inc., Engineering Plastics Division, *Technical Bulletins*, 1994.
15. Eastman Chemical Company, *Technical Bulletins*, 1994.
16. DSM Engineering Plastics, *Technical Bulletins*, 1994.

Enhanced Oil Recovery

See: Water-Soluble Polymers (Oil Recovery Applications)

ENOL ESTER POLYMERS

Kohtaro Osakada, Isao Yamaguchi, and Takakazu Yamamoto
Research Laboratory of Resources Utilization
Tokyo Institute of Technology

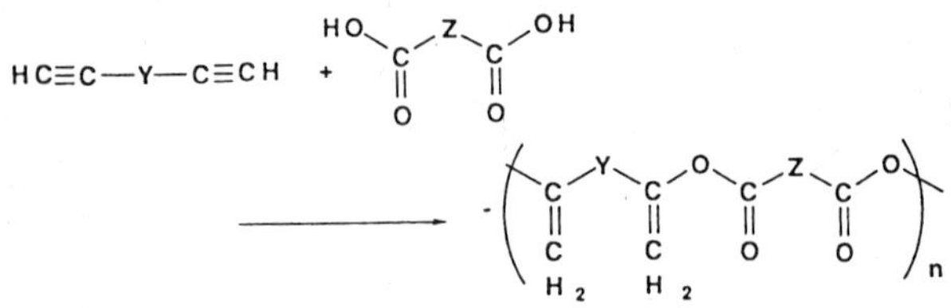

I

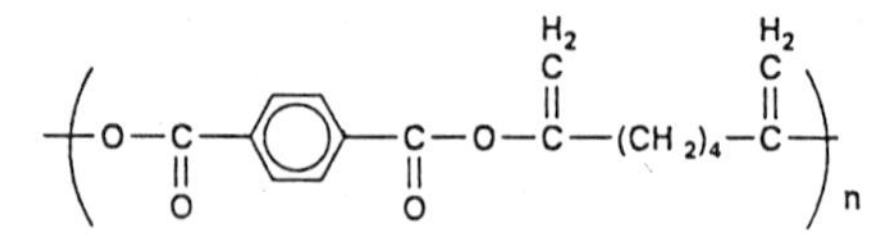

1

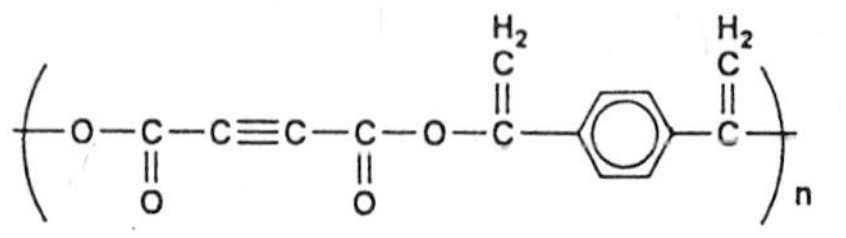

2

II

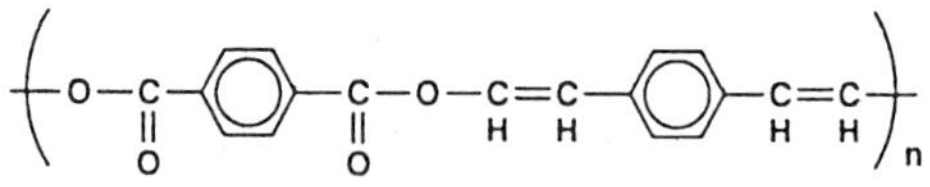

3

Polymers containing the vinylidene group, [-C(=CH_2)-], in the main chain, such as polyallenes[1–4] and polyketene[5–8] have potential utility as precursors of functionalized polymers or crosslinked polymers because they undergo facile H-O, H-S, H-Si bond addition to the C=C double bond or crosslinking caused by addition of radical or cation species.

The vinylidene group bonded to an ester group, -CO-O-C(=CH_2)-, has higher reactivity toward an electrophilic reagent than the vinylidene group in the polyallenes because of the highly polarized C=C double bond. Since an enol ester group with a vinylidene olefin structure is expected to undergo hydrolysis by a protonic acid, introduction of the enol ester group into a polymer chain would provide a new polymer material whose backbone is easily hydrolyzed under acidic conditions. About 5–10 years ago, Mitsudo, Dixneuf, and their respective co-workers[9–14] reported that ruthenium complexes catalyzed the addition of carboxylic acid to 1-alkyene to give enol ester under mild conditions. This reaction in the presence of tributylphosphine (PBu_3) proceeds through Markovnikov addition of the -COOH group to the C≡C triple bond to give the enol ester having the vinylidene (C=CH_2) structure, whereas the reaction in the absence of PBu_3 gives regio-isomer with the vinylene (-CH=CH-) group as the main product.

The Ru$(cod)_2$-PBu_3-catalyzed reaction (where cod is 1,5-η^5-cyclooctadienyl) of dicarboxylic acids with α,ω-diynes would give poly(enol ester)s having the vinylidene group through Markovnikov-type polyaddition, as shown in **Scheme I**.

The regulation of the direction of addition by a catalyst is characteristic of polyaddition reactions catalyzed by transition-metal complexes, including the present reaction of carboxylic acid with diynes catalyzed by the ruthenium complex.

PREPARATION

Three poly(enol ester)s, polymers **1–3** in **Scheme II**, have been prepared by ruthenium-complex-catalyzed reactions of terephthalic acid with 1,7-octadiyne, of acetylene dicarboxylic acid with 1,4-diethynylbenzene, and of terephthalic acid with 1,4-diethynylbenzene, respectively.[5]

Ru$(cod)_2$ is used as the catalyst in the presence of PBu_3 and maleic anhydride according to the procedure reported by Mitsudo's group.[9–11]

PROPERTIES

Molecular Weights and Spectroscopic Results of the Polymers

Polymers **1** and **2** obtained in the above reaction are soluble in polar solvents such as DMF and *N*-methyl-2-pyrrolidone and are not soluble in chloroform, diethyl ether, methanol, and water. The specific viscosity, $[\eta]_{sp}$, of 1 is 0.11 dL/g in DMF solution at 30°C. Gel permeation chromatography (GPC) gives a number-average molecular weight (M_n) of 6.6×10^3 a weight-average molecular weight (M_w) of 1.1×10^4. Polymer **2**, isolated in 55% yield from the reaction mixture by addition of water, shows multimodal GPC curve corresponding to $M_n = 3.3\text{–}7.2 \times 10^3$ and $M_w = 3.3\text{–}9.0 \times 10^3$ (vs. polystyrene). Polymer **3** is sparingly soluble in polar organic solvents such as DMF and dimethyl sulfoxide (DMSO) and practically insoluble in chloroform, diethyl ether, methanol, and water. The DMF-soluble part of **3** (about 10% of the obtained polymer) has $M_n = 1.5 \times 10^3$ and $M_w = 4.7 \times 10^3$ as determined by GPC measurement.

Reaction of terephthalic acid with 1,7-octadiyne catalyzed by Ru$(cod)_2$ in the absence of PBu_3 gives a polymer product that contains the enol ester group with the vinylene (-CH=CH-) structure, whereas the above reaction in the presence of PBu_3 gives exclusively the polymer containing the vinylidene structure.

Thermal Properties of the Polymers

Polymers **1–3** undergo thermal decomposition without melting. Polymer **1** shows a weight loss in the range 270–340°C although **2** and **3** show gradual weight decrease over wider temperature ranges.

REFERENCES

1. Baker, W. P., Jr. *J. Polym. Sci.* **A1963**, *1*, 655.
2. Tadokoro, H.; Takahashi, Y.; Otsuka, S.; Mori, K.; Imaizumi, F. *J. Polym. Sci. B* **1965**, *3*, 697.
3. Otsuka, S.; Nakamura, A. *Polym. Lett.* **1967**, *5*, 973.
4. Mizuya, J.; Yokozawa, T.; Endo, T. *J. Am. Chem. Soc.* **1989**, *111*, 743.
5. Furukawa, G.; Saegusa, T.; Mise, N. *Makromol. Chem.* **1960**, *39*, 243.
6. Hasek, R. H. *J. Org. Chem.* **1962**, *27*, 60.
7. Furukawa, G. *Pure Appl. Chem.* **1962**, *4*, 387.
8. Yamashita, Y.; Nunomoto, S. *Makromol. Chem.* **1962**, *51*, 148.
9. Mitsudo, T.; Hori, Y.; Watanabe, Y. *J. Org. Chem.* **1985**, *50*, 1566.
10. Mitsudo, T.; Hori, Y.; Yamakawa, Y.; Watanabe, Y. *Tetrahedron Lett.* **1986**, *27*, 2125.
11. Mitsudo, T.; Hori, Y.; Watanabe, Y. *J. Org. Chem.* **1987**, *52*, 2230.
12. Devanne, D.; Ruppin, C.; Dixneuf, P. H. *J. Org. Chem.* **1988**, *53*, 925.

13. Philippot, K.; Devanne, D.; Dixneuf, P. H. *J. Chem. Soc., Chem. Commun.* **1990**, 1190.
14. Neveux, M.; Bruneau, C.; Dixneuf, P. H. *J. Chem. Soc., Perkin Trans.* **1991**, *1*, 1197.

ENVIRONMENTALLY DEGRADABLE POLYMERS

Ann-Christine Albertsson and Sigbritt Karlsson
Department of Polymer Technology
The Royal Institute of Technology

Synthetic and natural polymers are susceptible to environmental degradation to a varying extent. Some natural polymers, like proteins and polysaccharides are fairly rapidly degraded by micro- and macroorganisms, whereas a natural polymer like lignin requires many years to be totally degraded. The same is true of synthetic materials; polyanhydrides are sometimes degraded within a couple of hours, but polyethylenes will persist for 50 years or more.

Environmental degradative factors are physical, chemical, mechanical, and biological mechanisms from sun, air, water, pollutions, wind, soil, sand, microorganisms, and macroorganisms.

Attempts to achieve degradable plastics have focused on polymeric blends between degradable components and more stable ones, natural and modified polymers, modifications of synthetic polymers, and synthesis of new degradable polymers. Several factors have to be taken into consideration when designing for degradability: intended mechanisms of degradation, material properties, processability, and cost. With synthesis it is possible to tailor-make materials with specific properties and controlled degradation rates.[1–3] These materials are, however, intended for *in vivo* applications such as implants for which cost may not be the major issue. For materials used in disposable items cost may be important.

One of the simplest ways to render existing polymers more degradable is to accelerate degradation processes already taking place. Based on this, degradable materials can be divided into photodegradable and biodegradable. However, in most cases a combination of mechanisms is responsible for the degradation of these materials; hence they are best referred to as enhanced "environmentally degradable." A material that is designed to have increased sensitivity to environmental degradation may be referred to as "environmentally adaptable." In the following section, we will discuss environmental degradable polymers describing photodegradable, hydrolysable, biodegradable, and starch polymers, which can be classified as environmentally degradable materials.

PHOTODEGRADABLE POLYMERS

For the majority of synthetic polymers, degradation by UV light is the most important mechanism.[4] Natural polymers also degrade under the influence of UV light but are usually more rapidly degraded by microorganisms. For light absorption to take place, the material needs to contain light absorbing groups (chromophores).

Photodegradable plastics are produced by intentionally introducing chromophores into the material, either through copolymerization or by blending them with the polymers. Two major approaches are synthesis of sensitized copolymers and use of photosensitizing additives systems.

Photosensitized Copolymers

Copolymerization of ethylene and carbon monoxide yields UV-sensitive CO groups in the polymer backbone. The result is referred to as E/CO copolymers, first produced on a small scale at Du Pont in 1940.[5] One grade was commercialized in the 1970s and the material is now an established photodegradable polymer available from several manufacturers.[6] The amount of CO does not usually exceed 2% of the material. Since the carbonyl groups are located directly in the polymer backbone, direct chain scission can occur through the Norrish mechanisms. The process is limited to LDPE.

Photodegradable Additive Systems

Additives used to impart enhanced photodegradability in plastics may be photoinitiators or photosensitizers. The former are excited by photoenergy and decompose to give free radicals, whereas the latter absorb energy and transfer it to polymer molecules. Several metal complexes have been found to act as photosensitizers.[7] These are usually divalent transition-metal salts of stearates, dithiocarbamates, or acetoacetonates.[8] A difference between the additive systems and the copolymer systems is that in the former the reactions can continue after a threshold of free radical formation is reached whereas in the latter the reactions stop when the UV source is removed.

HYDROLYZABLE POLYMERS

Hydrolytic degradation takes place when polymers containing hydrolyzable groups, such as polyesters, polyanhydrides, polycarbonates, and polyamides, are exposed to moisture. Hydrolysis proceeds by random hydrolytic chain scission of these linkages. Often the major degradation products are the result of these cleavages, for example hydroxyacids in the case of polyester hydrolysis.[9,10] Hydrolysis also can be achieved enzymatically but is then usually referred to as biodegradation.[11,12]

Microbial polyesters degrade by hydrolysis and the action of microorganisms. Factors such as temperature, pH, composition of copolymers, and processing influence subsequent hydrolysis. Another is release of degradation products. The final degradation product of PHB, D(–)-3-hydroxybutyrate is a physicological compound always present in the human body as a ketone body.

BIODEGRADABLE POLYMERS

Biodegradable polymers have been used to describe materials that are either only part biodegradable or degradable by other mechanisms such as photooxidation. Natural and synthetic polymers are attacked either chemically or mechanically by living organisms. Enzymes are commonly involved in chemical degradation processes. Such reactions can occur along the polymer backbone, resulting in total breakdown, or by attack at the chain ends. A secondary chemical degradation also can result from excretion of substances from microorganisms that might act upon the polymer or alter conditions of the surrounding environment such as pH.[13] Mechanical damage of the polymer may be caused by swelling and bursting due to growing cells of the invading microorganism. It also may be caused by macroorganisms such as insects and larger animals.

Natural polymers such as starch and cellulose are truly biodegradable. A limitation in these as plastics is their high degree of intermolecular interaction resulting in drawbacks in processability and mechanical properties.[14] Of the synthetic polymers, the polyanhydrides contain the most labile bonds.[1] Poly(α-amino acids) and poly(lactic acid) are other examples of completely biodegradable polymers. Polyesters are generally prone to degradation. One example is the polyhydroxyalkanoates, especially homopolymers of hydroxybutyrate (PHB) and its copolymers with hydroxyvalerate (PHB/PHV), which exist as bacterial and synthetic polymers.[15–19]

Biodegradability of most synthetic polymers such as polyethylene, polystyrene, and poly(vinyl chloride) is, however, limited. Biodegradability of synthetic polymers can be enhanced by blending, copolymerizing, or grafting with biodegradable components. Use of starch as the biodegradable component to produce starch-plastic systems are significant examples.

Surface properties have an important influence on degradation in biodegradative mechanism. Biodegradation depends on accessibility of the substrate and, is therefore, interesting to study surface effects such as change from powder to film or addition of a surfactant.[20]

STARCH-PLASTIC SYSTEMS

Starch is one of the least expensive natural polymers, and this, in combination with its biodegradability and ability to withstand processing temperatures, led to exploition of starch as a means of enhancing the biodegradability of otherwise biologically inert plastics.

Use of starch makes it possible to replace part of a synthetic polymer with a renewable resource. Corn starch is commonly used due to its great abundance especially in the USA. Two main techniques can be discerned. One is based on incorporating native granular corn starch into different matrices of synthetic polymers giving rise to what is known as starch-modified plastics. The other technique makes use of gelatinized starch to produce starch-based plastics.

Starch-Modified Plastics

Native granular starch can be incorporated as a filler in polymer matrices. The amount of starch is typically 6–15% (in blown films), and the achieved materials are termed starch-modified plastics. Starch-modified plastics can be processed without modification of the existing processing plant, which is advantageous.

Starch-Based Plastics

Starch-based materials contain more than 40% starch. To incorporate these large amounts while not compromising the mechanical properties, the starch is generally used in its gelatinized or destructurized state; that is, its crystallinity has been destroyed to obtain a thermoplastic melt and thus the polymeric state of starch is used. These two terms refer to basically the same condition of starch and differ merely in the production process. A variety of different compositions has been developed.

REFERENCES

1. Albertsson, A-C.; Lundmark, S. *Brit. Polym. J.* **1990**, *23*, 205.
2. Albertsson, A-C.; Palmgren, R. *J. Makromol. Sci., Pure Appl. Chem.* **1993**, *A30*(12), 919.
3. Mathisen, T.; Albertsson, A-C., *J. Appl. Polym. Sci.* **1990**, *38*, 591.
4. Guillet, J. E. *Pure Appl. Chem.* **1972**, *30*, 135.
5. Statz, R. J.; Dorris, M. C. in *Proceedings of Symposium on Degradable Plastics*; Society of Plastics Industry: Washington, DC, June 10, 1987; p 51.
6. Glover, R. *Internat. Biodeter. Biodegr.* **1993**, *31*, 171.
7. Scott, G. *J. Photochem. Photobiol., A. Chemistry* **1990**, *51*, 73.
8. Gage, P. *Tappi J.* **1990**, *10*, 161.
9. Karlsson, S.; Sares, C.; Renstad, R.; Albertsson, A-C. *J. Chromatogr. A* **1994**, *669*, 97.
10. Löfgren, A.; Albertsson, A.-C. *J. Appl. Polym. Sci.* **1994**, *52*, 1327.
11. Albertsson, A-C.; Ljungquist, O. *J. Macromol. Sci. - Chem.* **1986a**, *A23*, 393.
12. Albertsson, A-C.; Ljungquist, O. *J. Macromol. Sci. Chem.* **1986b**, *A23*, 411.
13. Albertsson, A-C. The Synergism Between Biodegradation of Polyethylenes and Environmental Factors, in *Advances in Stabilization and Degradation of Polymers*; A. Patsis, Ed., Technomic: Lancaster, PA, 1989; Vol. 1.
14. Tao, B. Y. *Overview of Biodegradable Plastic Technology and Research*, International winter meeting of the American Society of Agricultural Engineers, December 1990: Chicago.
15. Bloembergen, S.; Holden, D. A.; Bluhm, T. L.; Hamer, G. K.; Marchessault, R. H. *Macromolecules* **1989**, *22*, 1656.
16. Brandle, H.; Gross, R. A.; Lenz, R. W.; Fuller, R. C. *Appl. Environm., Microbiol.* **1988**, *54*, 1977.
17. Doi, Y.; Kunioka, M.; Nakamura, Y.; Soga, K. *Macromolecules* **1987**, *20*, 2988.
18. Gross, R. A.; DeMello, C.; Lenz, R. W.; Brandle, H.; Fuller, R. C. *Macromolecules* **1989**, *22*, 1106.
19. Kunioka, M.; Kawaguchi, Y.; Doi, Y. *Appl. Microbiol. Biotechnol.* **1989**, *30*, 569.
20. Karlsson, S.; Ljungquist, O.; Albertsson, A-C. *Polym. Degrad. Stab.* **1988**, *21*, 237.

ENVIRONMENTALLY RESPONSIVE GELS

Stevin H. Gehrke and David C. Harsh*
Department of Chemical Engineering
University of Cincinnati

An environmentally responsive gel (or stimulus-sensitive gel) is a gel whose equilibrium solvent sorption is a strong function of its environmental temperature, pH, solvent composition, ionic strength, light, mechanical stress, electric fields, or other specific chemical stimulus. Much of the research on responsive gels has focused on pH- and temperature-sensitive hydrogels ("hydrogels" are gels that absorb water).

The most important property of a gel is the equilibrium degree of swelling, which is the ratio of swollen gel volume to that of the dry polymer. The volume change can be abrupt, displaying a finite volume change over infinitesimal change in stimulus (a phase transition between expanded and contracted gel states), but it also may occur gradually across a broad range of stimulus. Also of interest is the rate of the volume change since many applications require that the volume change occurs in the appropriate time frame. Finally, the permeability of the

*Author to whom correspondence should be addressed.

gel to solutes, both in the swollen and collapsed state, must be controllable for controlled drug delivery devices and chemical separation systems to be viable. Thus wide interest in applications of responsive gels is due in part to the wide range of equilibrium swelling profiles they can display. Numerous applications have been proposed for responsive gels, including artificial muscles, chemical separations, controlled drug delivery, sensors, and actuators. More exhaustive discussions with extensive reference lists of all topics in this review are available in several recently published volumes.[2–4]

SYNTHESIS

In principle, gels can be synthesized by any method used for synthesis of linear polymers. Gels have been synthesized both in solution and in the bulk state; crosslinking can be accomplished during polymerization of monomers, or else linear polymers can be crosslinked by an appropriate multifunctional chemical reagent or by gamma or electron beam irradiation. In some cases, noncovalent crosslinks can be formed via hydrogen bonding, charge complexes, or hydrophobic interactions.

Many environmentally responsive gels are produced by simultaneous copolymerization/crosslinking of the monomer in the presence of a multifunctional crosslinker. Temperature-sensitive gels are frequently synthesized from the N-substituted derivatives of acrylamide, with N-isopropylacrylamide (NIPAAm) the most commonly used monomer.[5–11] Copolymers of acrylamide and acrylic acid or methacrylic acid are widely studied pH-sensitive gels.[11–15]

In addition to chemical crosslinking, linear polymers also can be crosslinked by gamma irradiation, which has been used to produce temperature-responsive gels of poly(vinyl methyl ether)[17] and various polypeptides.[18,19] Linear polymers also can be crosslinked by noncovalent interactions.

EQUILIBRIUM SWELLING

Responsive gels typically show volume changes over a similar range of conditions where corresponding uncrosslinked polymers undergo changes in solvent interaction, whether due to solubility or ionization effects. For example, cellulose ether gels that display temperature-sensitive behaviors in water are synthesized from polymers that exhibit a lower critical solution temperature (LCST). Below the LCST, the polymer is soluble in water due to extensive water-polymer hydrogen bonding, but at temperatures above the LCST, polymer-polymer hydrophobic interactions dominate, polymer-solvent hydrogen bonds are disrupted, and the polymer precipitates from solution.

Studies of polyacrylamide and related gels have directly varied polymer-solvent interaction by studying swelling in various combinations of water and miscible organic solvents like acetone.[1] Some homopolymers, such as poly(N-isopropylacrylamide) (PNIPAAm), display strong temperature dependence of polymer-solvent interaction; thus PNIPAAm gels display temperature dependent swelling behavior. PNIPAAm gels are highly swollen below approximately 32°C, with the degree of swelling depending on the crosslink density. The transition temperature, where the volume transition occurs is relatively unaffected by crosslink density. It is also significant that the volume transition is quite abrupt; the gel changes from highly swollen dilute polymer phase to a shrunken polymer dense phase over a relatively narrow (~5°C) temperature range around the linear polymer LCST. Properties of PNIPAAm gels have been extensively reviewed by Schild.[10]

Gels with ionizable groups may also be sensitive to changes in external pH, the most commonly studied examples being the acrylamide/acrylic acid gels.

Gels that respond to stimuli other than temperature, pH, and solvent concentration have also been described. They include light, mechanical stress, or specific chemical stimuli and generally operate by indirectly altering network ionization or polymer-solvent interaction. For example, by incorporation of chromophores that undergo isomerization upon exposure to ultraviolet light, gels that display volume changes in response to light have been developed.[20] Gels displaying volume changes in response to specific chemical species also have been developed.

Also of interest are reentrant transitions: multiple swollen or collapsed states observed across wide ranges of pH or solvent composition. Gels with both acidic and basic groups may be swollen at extreme pH ranges yet collapsed near neutral conditions in a poor solvent.[11] Conversely, gels exposed to mixtures of good solvents may display reduced swelling relative to that in the pure solvent. This cononsolvency phenomenon has been observed in NIPAAm[11] and cellulose ether[16] gels in aqueous solutions of alcohols, acetone and dimethylsulfoxide.

APPLICATIONS

The volume change of responsive gels has been suggested for use in drug delivery devices, artificial muscles, recyclable absorbents, separations, sensors, and actuators. The way in which these applications use gels can vary greatly. Many designs are based on one of two general schemes: The swelling or shrinking gel is used to drive a mechanism, or the gel is used as a membrane of variable permeability depending on swelling degree.

One of the first proposed applications for environmentally responsive gels was solute concentration by size exclusion.[21] Originally proposed for use with pH-sensitive gels, gel extraction also has been demonstrated with temperature-sensitive gels. This would be useful for solutes such as proteins that could be damaged by exposure to extreme pH.[5]

In addition to isolation of dissolved solutes, dewatering of slurry has been demonstrated for coal slurries, biological sludges, and drilling muds.[6,22] In these cases, the gel is used as a size–selective sponge, removing solvent from the suspended matter.

Perhaps the most intense applications work for environmentally responsive gels has been in drug delivery. Environmentally responsive gels promise ability to create triggered or self-regulating drug delivery systems. The simplest application with an environmentally sensitive gel matrix is to impregnate the gel with the drug to be dispensed. At the appropriate physiological stimulus (e.g., pH), the gel swells allowing entrapped solute to diffuse out of the gel.[7,23] The reverse also has been described: Shrinking the gel would squeeze the drug from an encapsulated reservoir. Another drug delivery design is use of porous membranes coated with the responsive gel.

Artificial muscle applications use volume change to perform useful work. Replacement muscles are required to respond rapidly and have sufficient mechanical strength for the required task. The primary focus of development work on artificial muscles has focused on developing gels that display the desired volume change under appropriate conditions. The mechanical response to changes in environment also can be used as a combined sensor/actuator system.

Many other applications of environmentally responsive gels have been proposed for use in diverse technologies ranging from industrial separations, to medicine and pharmacy, to consumer products, and to agriculture. While responsive gel materials have not yet achieved widespread use, their broad ranges of composition, response, and potential utility implies that commercial use is not far off.

REFERENCES

1. Tanaka, T. *Phys. Rev. Lett.* **1978**, *40*, 820.
2. DeRossi, D.; Kajiwara, K.; Osada, Y.; Yamauchi, A., Eds.; *Polymer Gels: Fundamentals and Biomedical Applications* Plenum: New York, 1991.
3. Dusek, K., Ed., *Advances in Polymer Science, Vol. 109, Responsive Gels: Volume Transitions I*; Springer-Verlag: New York, 1993.
4. Dusek, K., Ed., *Advances in Polymer Science, Vol. 110, Responsive Gels: Volume Transitions II*; Springer-Verlag: New York, 1993.
5. Freitas, R. F. S.; Cussler, E. L. *Chem. Eng. Sci.* **1987**, *42*, 97.
6. Gehrke, S. H. *Advances in Polymer Science, Vol. 110, Responsive Gels: Volume Transitions II*; Springer-Verlag: New York, 1993.
7. Hirotsu, S. *Advances in Polymer Science, Vol. 110, Responsive Gels: Volume Transitions II*; Springer-Verlag: New York, 1993.
8. Hoffman, A. S.; Afrassiabi, A.; Dong, L. C. *J. Cont. Rel.* **1986**, *4*, 213.
9. Saito, S.; Konno, M.; Inomata, H. *Advances in Polymer Science, Vol. 109, Responsive Gels: Volume Transitions I*; Springer-Verlag: New York, 1993.
10. Schild, H. G. *Prog. Polym. Sci.* **1992**, *17*, 163.
11. Shibayama, M.; Tanaka, T. *Advances in Polymer Science, Vol. 109, Responsive Gels: Volume Transitions I*; Springer-Verlag: New York, 1993.
12. Cussler, E. L.; Wang, K. L.; Burban, J. H. *Advances in Polymer Science, Vol. 110, Responsive Gels: Volume Transitions II*; Springer-Verlag: New York, 1993.
13. Gehrke, S. H.; Andrews, G. P.; Cussler, E. L. *Chem. Eng. Sci.* **1986**, *41*(8), 2153.
14. Ilavsky, M. *Polymer* **1981**, *22*, 1687.
15. Park, T. G.; Hoffman, A. S. *J. Appl. Polym. Sci.* **1992**, *46*, 659.
16. Gehrke, S. H.; Cunningham, E.; Fisher, S., Unpublished research, 1992.
17. Huang, X.; Unno, H.; Akehata, T.; Hirasa, O. *J. Chem. Eng. Jap.* **1988**, *21*, 10.
18. Urry, D. W. *J. Protein Chem.* **1988**, *7*, 1.
19. Urry, D. W.; Harris, R. D.; Prasad, K. U. *J. Amer. Chem. Soc.* **1988**, *110*, 3303.
20. Irie, I. M. *Advances in Polymer Science, Vol. 110, Responsive Gels: Volume Transitions II*; Springer-Verlag: New York, 1993.
21. Cussler, E. L.; Stokar, M. R.; Varberg, J. E. *AIChE J.* **1984**, *30*(4), 578.
22. Lyu, L. H. *Dewatering Fine Coal Slurries by Gel Extraction*; M.S. Thesis, University of Cincinnati, 1990.
23. Dong, L. C.; Hoffman, A. S. *J. Cont. Rel.* **1986**, *4*, 223.

ENZYMATIC POLYMERIZATION

Shiro Kobayashi,* Shin-ichiro Shoda, and Hiroshi Uyama
Department of Materials Chemistry
Graduate School of Engineering
Tohoku University

Recently, enzyme-catalyzed polymerization ("enzymatic polymerization") has attracted much attention as a new method of polymer synthesis.[1,2] Specific enzymatic catalysis is expected to synthesize polymers with high selectivity or with novel structures. Here, we define the term "enzymatic polymerization" as a polymerization *in vitro* (in test tubes) via a nonbiosynthetic pathway catalyzed by an isolated enzyme. Until now, there have been papers on the *in vitro* enzymatic syntheses of natural biopolymers as well as non-natural synthetic polymers. This article reviews enzymatic polymerization producing both types of polymers.

POLYSACCHARIDES

The use of an enzyme for glycosylation in polysaccharide synthesis is one of the most promising methods for selective construction of a saccharide unit. The enzymes so far utilized can be classified into glycosidases, glycosyltransferases, and phosphorylases.

Glycosidic bond formation with glycosidases can be achieved by using its reverse reaction of hydrolysis, employing a free sugar as glycosyl donor. The position of the equilibrium can be shifted by increasing substrate concentration, decreasing water, and removing the final product from the reaction system.[3–5]

A glycosyl fluoride also acts effectively as a glycosyl donor.[6] Recently, the first *in vivo* synthesis of cellulose via a nonbiosynthetic pathway has been achieved by an enzymatic polymerization of β-cellobiosyl fluoride as substrate for cellulase.[7]

In addition to the enzymatic polymerizations catalyzed by glycosidases or transferases, a phosphorylase was also found to promote a new enzymatic polymerization. Linear-, star-, and comb-shaped polymers having amylose chains of uniform length were prepared by using D-glucosyl phosphate as substrate for phosphorylase.[8] This method has successfully been applied to the synthesis of poly(dimethylsiloxane-*graft*-(α,1→4)-glucopyranose by using potato phosphorylase, as catalyst. Polysiloxanes with statistically distributed maltoheptaonamide and maltoheptaoside have been employed as polyinitiators.[9]

POLYPEPTIDES

Dipeptidyltransferase (cathespin C) catalyzes the polymerization of dipeptide amides in an aqueous buffer solution.[10] From glycyl-L-phenylalaninamide, the corresponding hexapeptide was mainly formed in about 80% yield.[11] In the case of glycyl-L-tyrosinamide, the hydrolysis of the amide linkage took place during the polymerization, and formation up to an octamer (22% yield) was observed.

Some proteases induced polymerization of esters of α-amino acids to water-insoluble products. Formation of di- or tripeptides

*Author to whom correspondence should be addressed.

was observed in the α-chymotrypsin-catalyzed reaction of isopropyl esters of L-methionine, L-threonine, L-phenylalanine, and L-tyrosine.[12,13] An L-leucine polypeptide, eight to nine units long, was synthesized by the papain-catalyzed polymerization of leucine methyl ester.[14] Papain also catalyzed the polymerization of L-methionine, L-phenylalanine, and L-tyrosine esters to produce water-insoluble oligomers.[15] From L-methionine ethyl ester, an oligopeptide containing mainly an octamer was obtained in yields of up to 85%.[16] The polymerization of diethyl α,γ-L-glutamate catalyzed by papain or α-chymotrypsin was performed.[17] During the reaction, a precipitate composed of five to nine glutamic acid residues was formed in 42% yield.

POLYESTERS, POLYAMIDES, AND POLYCARBONATES

Enzymatic syntheses of polyesters have been attempted by different polymerization methods: polycondensation of oxycarboxylic acids, esters of oxycarboxylic acid, combinations of dicarboxylic acid/glycol and diesters of dicarboxylic acid/glycol, poly(addition-condensation) of cyclic acid anhydride and glycols, and ring-opening polymerization of lactones.

Polymerization of 10 hydroxydecanoic acid in benzene was examined by using poly(ethylene glycol)-modified lipase as a catalyst to give the corresponding oligomer with the degree of polymerization (DP) more than 5 in 48% yield.[18]

Film-forming polyesters with DP up to 100 were obtained by enzymatic polymerization of esters of ω-hydroxycarboxylic acid by using porcine pancreatic lipase (PPL) in hexane at reflux temperature.[19]

A combination of dicarboxylic acid derivatives and glycols undergoes enzymatic polymerizations to polyesters. The polymerization of adipic acid and 1,4-butanediol using lipase catalyst was conducted in a modified reactor in the presence of molecular sieves (4Å).[20]

Enzymatic polymerizations have been extended to a ring-opening polymerization. Lactones of various ring sizes were subjected to lipase-catalyzed polymerization in bulk.[21,22] Medium-size lactones (6- and 7-membered) as well as macrolides (12- and 16-membered) were polymerized to produce polyesters.

Enzymatic synthesis of polyamide and poly(amide ester) was performed by polycondensation of chiral fluorinated compounds having two functional groups (OH or NH_2 and COOH).[23]

An oligocarbonate was enzymatically obtained by the reaction of carbonic acid diphenyl ester with bisphenol-A in the presence of lipase.[24]

POLYAROMATICS

Peroxidases catalyze couplings of a number of phenols and aromatic amines, using hydrogen peroxide as oxidant. Dordick[25] examines this process in a mixture of a water-miscible organic solvent and water to afford a novel class of polyaromatics. The polymerization of *p*-phenylphenol catalyzed by horseradish peroxidase (HRP) in 85% 1,4-dioxane produced a polymer with a molecular weight of 2.6×10^4. From 1-naphthol, insoluble polymeric materials were obtained.

The HRP-catalyzed polymerization of phenol in 1,4-dioxane/phosphate buffer (pH 7.0) (80:20 vol %) produced powdery polymeric materials.[26] The resulting polymer was partly soluble in DMF and dimethyl sulfoxide, and the molecular weight of the soluble part was 3.5×10^4.

Enzymatic synthesis of polyaromatics was conducted in reversed micellar systems.[27]

HRP was also used to catalyze an oxidative polymerization of aniline derivatives. The enzymatic polymerization of *o*-phenylenediamine in an aqueous organic solvent afforded soluble poly(imino-4-aminophenylene) with molecular weight of 2×10^4,[28] which is hard to obtain by a conventional oxidative polymerization.

Lignin is one of the natural polyaromatics and the second most abundant biopolymer after cellulose. Synthetic lignin was claimed to be produced by enzymatic oxidative polymerization of *p*-coumaryl alcohol, coniferyl alcohol, or sinapyl alcohol in an aqueous solution.[29,30]

REFERENCES

1. Ritter, H. *Trends Polym. Sci.* **1993**, *1*, 171.
2. Kobayashi, S.; Shoda, S.; Uyama, H. *Adv. Polym. Sci.* **1995**, *121*, 1.
3. Nilsson, K. *Trends Biotechnol.* **1988**, *6*, 256.
4. Ajisaka, K.; Nishida, H.; Fujimoto, H. *Biotechnol. Lett.* **1987**, *9*, 243.
5. Ajisaka, K.; Nishida, H.; Fujimoto, H. *Biotechnol. Lett.* **1987**, *9*, 387.
6. Barnett, J. E. G.; Jarvis, W. T. S.; Munday, K. A. *Biochem. J.* **1967**, *105*, 669.
7. Kobayashi, S.; Kashiwa, K.; Kawasaki, T.; Shoda, S. *J. Am. Chem. Soc.* **1991**, *113*, 3079.
8. Ziegast, G.; Pfannenmüller, B. *Carbohydr. Res.* **1987**, *160*, 18.
9. Braunmühl, V.; Jones, G.; Stadler, R. *Macromolecules* **1995**, *28*, 17.
10. Fruton, J. S.; Hearn, W. R.; Ingram, V. M.; Wiggans, D. S.; Wintz, M. *J. Biol. Chem.* **1953**, *204*, 891.
11. Heinrich, C. P.; Fruton, J. S. *Biochemistry* **1968**, *7*, 3556.
12. Brenner, M.; Müller, H. R.; Pfister, R. W. *Helv. Chim. Acta* **1950**, *33*, 568.
13. Brenner, M.; Pfister, R. W. *Helv. Chim. Acta* **1951**, *34*, 2085.
14. Sluyterman, L. A. E.; Wijdenes, J. *Biochim. Biophys. Acta* **1972**, *289*, 194.
15. Anderson, G.; Luisi, P. L. *Helv. Chim. Acta* **1979**, *62*, 488.
16. Jost, R.; Brambilla, E.; Monti, J. C.; Luisi, P. L. *Helv. Chim. Acta* **1980**, *63*, 375.
17. Aso, K.; Uemura, T.; Shiokawa, Y. *Agric. Biol. Chem.* **1988**, *52*, 2443.
18. Ajima, A.; Yoshimoto, T.; Takahashi, K.; Tamaura, Y.; Saito, Y.; Inada, Y. *Biotechnol. Lett.* **1985**, *7*, 303.
19. Knani, D.; Gutman, A. L.; Kohn, D. H. *J. Polym. Sci., Polym. Chem. Ed.* **1993**, *31*, 1221.
20. Binns, F.; Roberts, S. M.; Taylor, A.; Williams, C. F. *J. Chem. Soc., Perkin Trans I* **1993**, 899.
21. Uyama, H.; Kobayashi, S. *Chem. Lett.* **1993**, 1149.
22. Uyama, H.; Takeya, K.; Kobayashi, S. *Bull. Chem. Soc. Jpn.* **1995**, *68*, 56.
23. Kitazume, T.; Sato, T.; Kobayashi, T. *Chem. Express* **1988**, *3*, 135.
24. Abramowicz, D. A.; Keese, C. R. *Biotechnol. Bioeng.* **1989**, *33*, 149.
25. Dordick, J. S.; Marletta, M. A.; Klibanov, A. M. *Biotechnol. Bioeng.* **1987**, *30*, 31.
26. Uyama, H.; Kurioka, H.; Kaneko, I.; Kobayashi, S. *Chem. Lett.* **1994**, 423.
27. Rao, A. M.; John, V. T.; Gonzalez, R. D.; Akkara, J. A.; Kaplan, D. L. *Biotechnol. Bioeng.* **1993**, *41*, 531.
28. Kobayashi, S.; Kaneko, Il, Uyama, H. *Chem. Lett.* **1992**, 393.

29. Freudenberg, K. *Science* **1965**, *148*, 595.
30. Tanahashi, M.; Higuchi, T. *Wood. Res.* **1981**, *67*, 29.

ENZYME-CATALYZED OXIDATIVE POLYMERIZATION (Aromatic Compounds)

Masuo Aizawa*
Department of Bioengineering
Faculty of Bioscience and Biotechnology
Tokyo Institute of Technology

Lili Wang
Department of Chemistry
Faculty of Chemistry and Chemical Engineering
Xiamen University

Nonspecificity of some enzymes for substrate is of considerable interest; for example, nonspecific serine exopeptidase Y(CPY) is a potential catalyst for stepwise peptide synthesis.[1–4] In enzymatic polymerization of aromatic and heterocyclic compounds, horseradish peroxidase has been used to catalyze peroxidative couplings of phenol and its derivatives and to produce aromatic polymers.[5–7] High molecular weight poly(*p*-phenylphenol), polybenzidine, and aromatic copolymers have been synthesized by using peroxidase in a hydrogen peroxide-containing mixed solvent of water and dioxane.[5–7]

We have found that bilirubin oxidase (BOX; EC 1.11.1.7) catalyzes oxidative polymerizations of aniline and pyrrole in an aqueous solution.[8,9] This intriguing finding prompted us to take an enzymatic approach toward synthesis of aromatic polymers by exploiting various analogous compounds as substrates of BOX. We also observed enzymatic polymerization of 1,5-dihydroxynaphthalene in a water-organic mixed solvent medium, where it is easily soluble and BOX retains its activity.

BOX, a copper-containing oxidoreductase, catalyzes the oxidation of bilirubin to biliverdin; it also catalyzes, though with less reactivity, the oxidation of *p*-aminophenol, *p*-phenylenediamine, hydroquinone, and some other phenol derivatives.

Polyaniline and polypyrrole have also been synthesized by chemical or electrochemical methods.[10] The unique merit of the enzymatic polymerization of aniline or pyrrole on the aqueous solution/solid matrix interface was that active BOX molecules were self-assembled in polymer thin film-coated solid matrix. Enzymatic polymerization of 1,5-dihydroxynaphthalene produced a new conducting polymer, poly(1,5-dihydroxynaphthalene), that has many of important properties.

PROPERTIES

The properties of enzymatically synthesized polyaniline and polypyrrole are similar to those of chemically or electrochemically synthesized compounds and are briefly discussed.

Solubility

The enzymatically synthesized poly(1,5-dihydroxynaphthalene) is slightly soluble in ethanolamine, pyridine, dimethyl sulfoxide (DMSO), and dimethylformamide (DMF) and insoluble in dioxane, acetone, acetonitrile, formamide, ethanol, benzene, toluene, tetrahydrofuran, chloroform, and tetrachloromethane.

Thermal Property

Thermal stability of the enzymatically synthesized poly(1,5-dihydroxynaphthalene) is much better than that of other enzymatically synthesized phenolic polymer.[6,7]

Electronic Conductivity

The electroconductivity of enzymatically synthesized poly(1,5-dihydroxynaphthalene) depends not only on composition of the reaction medium but also on protonation of the polymer. Electroconductivity of the polymer powder treated with 1 M $HClO_4$ increased one order of magnitude over that of the nontreated one.

Spectroelectrochemistry

The enzymatically prepared poly(1,5-dihydroxynaphthalene) exhibits electrochromic character when the potential of the polymer film-coated ITO/glass electrode is changed from –0.2 V to 0.8 V. More detailed study indicated reversibility and stability of the electrochromism of the poly(1,5-dihydroxynaphthalene) film-coated ITO/glass electrode.[11]

Electroanalysis

The enzymatically synthesized poly(1,5-dihydroxynaphthalene) film-coated electrode has novel catalytic activity for electrochemical oxidation of a number of small organic molecules such as ethanol, methanol, formic acid, and formic aldehyde.[10]

APPLICATIONS

Bilirubin Sensor

The enzymatically synthesized BOX-incorporated polymer film-coated Pt electrode can be directly used to as a bilirubin sensor. When bilirubin was added to a phosphate buffer solution, bilirubin was enzymatically oxidized by BOX entrapped poly(1,5-dihydroxynaphthalene)-coated Pt electrode and produced biliverdin and hydrogen peroxide at the electrode.

The hydrogen peroxide was then electrochemically oxidized at the Pt electrode surface; anodic current rapidly increased in 3 s.

Electrochromic Switching

The electrochromism of enzymatically prepared poly(1,5-dihydroxynaphthalene)-coated ITO/glass electrode can be considered to use as candidate material of electrochromic switching.

Electrolysis of Methanol on Poly(1,5-dihydroxynaphthalene)-Coated Pt Electrode

Electrooxidation of organic fuels on various electrode materials has been a subject of careful investigation in order to develop high-performance organic fuel cells.

The electroanalysis of the enzymatically synthesized poly(1,5-dihydroxynaphthalene) for oxidation of methanol can be considered to be applied to constant potential electrolysis of methanol at 0.65 V vs. Ag/AgCl. In the concentration range from 0 to 1 M, electrolytic current linearly increased with an increase of alcohol concentration.

*Author to whom correspondence should be addressed.

REFERENCES

1. Cramer, S. M.; Horvath, C. *Enzyme Microb. Technol.* **1989**, *11*, 74.
2. Widmer, F.; Johansen, J. T. *Carlsberg Res. Commun.* **1979**, *44*, 37.
3. Widmer, F.; Johansen, J. T. *Carlsberg Res. Commun.* **1980**, *45*, 453.
4. Breddam, K.; Widmer, F.; Johansen, J. T. *Carlsberg Res. Commun.* **1980**, *45*, 237.
5. Dordick, J.; Marletta, M. A.; Kibanov, A. M. *Biotechnol. Bioeng.* **1987**, *15*, 31.
6. Dodick, J. *Enzyme Microb. Technol.* **1989**, *11*, 74.
7. Akkara, J. A.; Sencal, K. J.; Kaplan, D. L. *J. Polym. Sci. Polym. Chem.* **1991**, *29*, 1561.
8. Aizawa, M.; Wang, L.; Shinohara, H.; Ikariyama, Y. *J. Biotech.* **1990**, *14*, 301.
9. Wang, L.; Kobatake, E.; Ikariyama, Y.; Aizawa, M. *Denkikagaku* **1992**, *60*(12), 1050.
10. Skotheim, T. A. *Handbook of Conducting Polymer*; Marcel Dekker: New York, 1986.
11. Wang, L. Ph.D. Thesis, Tokyo Institute of Technology, 1992.

ENZYME-CATALYZED POLYMERIZATION

Joseph A. Akkara and David L. Kaplan
Biotechnology Division
U.S. Army Natick Research Development and Engineering Center

Vijay T. John
Department of Chemical Engineering
Tulane University

Sukant K. Tripathy
Center of Advanced Materials
Department of Chemistry
University for Massachusetts-Lowell

Phenol-formaldehyde polymers, including novolaks and resoles, have a number of applications in coatings, finishes, adhesives, composites laminates, and related areas.[1] These applications require specific polymer functional properties which relate to molecular weight, dispersity, degree of crosslinking, crystallinity, and inter- and intramolecular bonding.[1,2] These properties are difficult to control by oxidative coupling reactions carried out at extremes of temperature, pressure, and pH using inorganic catalysts. Enzyme-catalyzed free radical reactions potentially can overcome some of the deficiencies due their regio- and stereospecificity and the ability to control the environment in which the enzyme functions.

POLYMER SYNTHESIS

Enzyme-catalyzed polymer syntheses can be carried out in monophasic and biphasic solvent systems,[3,4] supercritical fluids,[5] reverse micelles,[6–8] and at the air-water interface in a Langmuir trough.[9,10] Control of the environment in which the enzyme functions, along with control of the organization of the monomers in the reactions, lead to new opportunities in synthesis of novel polymeric products. This chapter will focus on: enzyme-catalyzed synthesis of aromatic polymers (phenols, aniline, and their derivatives), homopolymers and copolymers, on organic solvents with different amounts of water (bulk synthesis), in reverse micelles, and at an air water interface on a Langmuir trough; characterization of polymers synthesized in the three different environments; and applications of polymers synthesized. Many polymers synthesized have useful electronic and nonlinear optical properties due to their aromaticity and conjugated backbone.[8,11] In general, polymerization reactions were carried out with derivatized phenols and aromatic amine compounds as monomers, and the reactions were catalyzed by the enzymes horseradish peroxidase (HRP) or laccase.

Enzymes

Horseradish peroxidase: Peroxidases are widely distributed in nature and they participate in many important physiological functions.[12] HRP (hydrogen-peroxide oxidoreductase; EC 1.11.1.7) is a glycoprotein with a single unbound ferric heme prosthetic group and two calcium ions (Ca^{+2}) bound to one mole of enzyme. The approximate molecular weight of the enzyme is 44,000. The catalytic activity of the enzyme is maintained over a wide range of conditions of pH, ionic strength, and temperature (up to 40°C).[12,13] Simplified reaction steps for the hydrogen peroxide-mediated, free-radical polymerization reaction by HRP are presented in **Figure 1a**.

Laccase (benzenediol:oxygen oxidoreductase; EC 1.10.3.2) is a 'blue oxidase' enzyme and is widely distributed in microorganisms and plants.[14] Laccase is a copper-containing glycoprotein were molecular weight 64,000 to 140,000 Da depending on the source of the enzyme.[15,16] The enzyme in the presence of air catalytically oxidizes phenols and aromatic amines to produce free radicals, which in turn polymerize spontaneously. Laccase requires oxygen as an electron acceptor, but peroxide is not required for catalytic activity: a major difference between laccase and peroxidase. The schematic of the laccase catalyzed reaction is presented in **Figure 1b**.[14–16]

APPLICATIONS

Characterization studies with aromatic polymers prepared from derivatized phenols and anilines by enzyme-catalyzed reactions have shown carbon-carbon aromatic ring to ring linkages with free –OH (with polyphenols) and $-NH_2$ groups (with polyaromatic amines), but without methylene bridges between the aromatic rings. Thus, these polymers have a conjugated backbone, with hydrophilic (–OH and $-NH_2$) and hydrophobic (alkyl and aromatic) side chains. Synthesis of photoresist material from polyphenols prepared by enzyme-catalyzed reactions was reported recently.[17] Synthesis of phenol-formaldehyde-like resins can be achieved with the reactions described here. In addition, electromagnetic applications such as coatings for shielding would also be possible based on the electrical properties of the polymers.

Electrical conductivity and third order nonlinear optical susceptibility studies with fast responses (picoseconds) have indicated potential applications in optical and electrooptic devices such as: optical communication devices to control dispersive properties of optical communication channels waveguides) through nonlinear refractive index and Kerr effect (soliton propagation); image processing through optical correlation and convolution by optically induced hologram formation and erasure; optical computing through ultrafast switching with third order nonlinear optical materials, using electron, hole or exciton (electron-hole) confinement; excitonic and biexcitonic (elec-

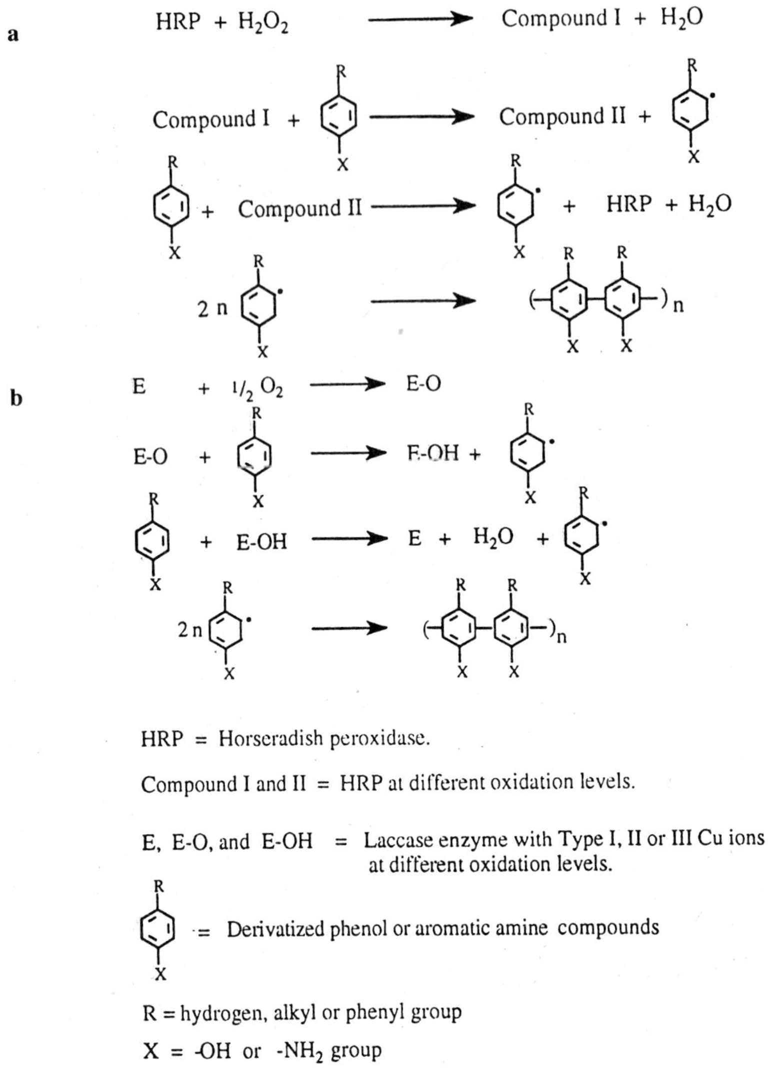

FIGURE 1. Simplified reaction steps for free radical polymerization reaction by enzymes: (a) with horseradish peroxidase in the presence of hydrogen peroxide and (b) with laccase in the presence of oxygen.

tron-hole pair and pair of electron-hole pair) dynamics; large optical nonlinearities in quantum wells, wires, and dots obtained by quantum confinement; optical open air interconnections and spatial light modulation based on hologram formation and modulation using nonlinear optical materials; ultrafast optical multiplexing and serial-to-parallel conversion mediated by third order nonlinear effects using degenerative four-wave mixing geometries; large optical nonlinearities produced by electron delocalization in linear and planar conjugated systems; and parametric processes (sum and difference frequency generation) for wavelength control, oscillation, and amplification.

Polymerization reactions in reverse micelles and in the Langmuir trough combine synthesis and processing into one step within the physical confines of spherical particles and thin films. Polymer spheres prepared in reverse micelles with amphiphilic functional groups and conjugated backbones have many applications, especially for encapsulation of functionalized materials within the spheres. Some examples of these inclusions include ceramic nanoparticles, semiconductor materials, magnetic particles, drugs, nutrients, insecticides, pesticides, affinity column packings, and enzymes. These spherical particles, with inclusions, could function as controlled delivery systems (for drugs, nutrients, insecticides, and pesticides), and as encapsulants for semiconductor nanoparticles stabilized against aggregation and precipitation. Moreover, conductivity and nonlinear optical properties of the polymer could be further enhanced by encapsulation and/or impregnation of electrooptic semiconductor nanoparticles. Polymers have good thermal stability (less than 10% breakdown below 300°C) and have potential application in coatings, finishes, thin films, and fibers with good temperature resistance and performance.

ACKNOWLEDGMENT

The authors thank Dr. D.V.G.L.N. Rao and Mr. Francisco Aranda (Department of Physics, University of Massachusetts, Boston, Boston, MA) for measuring the third order nonlinear optical properties of polymers. V.T.J. thanks the U.S. Army and National Science Foundation for their support.

REFERENCES

1. Kopf, P. W. *Encyclopedia of Polymer Science and Engineering*, 2nd ed.; Mark, H. F.; Bikales, N. M., et al., Eds.; John Wiley & Sons: New York, 1988; Vol. 11.
2. Bogan, L. E. *Macromolecules* **1992**, *25*, 1966.
3. Dordick, J.; Marletta, M. A.; Klibanov, A. M. *Biotech. Bioeng.* **1987**, *30*, 31.
4. Akkara, J. A.; Senecal, K. J.; Kaplan, D. L. *J. Polymer Sci.* **1991**, *29*, 1561.
5. Russell, A. J.; Beckman, E. J.; Chaudhary, A. K. *Chemtech* **1994**, *24*, 33.
6. Rao, A. M.; John, V. T.; Gonzalez, R. D.; Akkara, J. A.; Kaplan, D. L. *Biotechnol. Bioeng.* **1993**, *41*, 531.
7. John, V. T.; Akkara, J. A.; Kaplan, D. L. U.S. Patent 5 324 436, 1994.
8. Akkara, J. A.; Ayyagari, M. et al., *Biomimetic* **1994**, 2, 331.
9. Bruno, F. F.; Akkara, J. A.; et al., *Polymer Preprints* **1991**, *32*, 232.
10. Akkara, J. A.; Kaplan, D. L. et al., U.S. Patent 5 143 828, 1992.
11. Akkara, J. A.; Aranda, F. J. et al., *Frontiers of Polymers and Adv. Materials*; Prasad, P. N., Ed.; Plenum: New York, 1994.
12. Dunford, H. B. *Peroxidases in Chemistry and Biology*; Everse, J.; Everse, K. E.; Grisham, M. B., Eds.; CRC: Boca Raton, FL, 1991; Vol. 2.
13. Poulos, T. L. *Curr. Opin. Biotech.* **1993**, *4*, 484.
14. Malmstrom, B. G.; Andreasson, L.-E.; Reinhammar, B. *The Enzymes*, 3rd ed.; Boyer, P. D., Ed.; Academic: New York, 1975.
15. Fahraeus, G.; Reinhammar, B. *Acta Chem. Scand.* **1967**, *21*, 2367.
16. Reinhammar, B. *Biochim. Biophys. Acta* **1970**, *205*, 35.
17. Liang, R. C.; Pokora, A. R., U.S. Patent 5 212 044, 1993.

ENZYME-DEGRADABLE HYDROGELS

Waleed S. W. Shalaby
College of Medicine
Medical University of South Carolina

Kinam Park
School of Pharmacy
Purdue University

Biodegradable hydrogels are networks that can be solubilized by disrupting interchain or intrachain links. Degradation can be driven by dissociation of electrostatic bonds, chemically induced hydrolysis, or enzyme-catalyzed hydrolysis. Enzyme-degradable hydrogels in particular are well suited as a biomaterial or drug delivery system given their inherent biocompatibility as well as enzyme-controlled biodegradability. A hydrogel can maintain structural integrity to achieve its intended objective and subsequently can be broken down and eliminated by enzymes endogenous to an organ system or cell type. A recent review by Kamath et al.[1] and book by Park, Shalaby, and Park[2] discuss applications for biodegradable hydrogels in drug delivery.

HYDROGELS PREPARED FROM ENZYME-DEGRADABLE POLYMERS

Hydrogels based on polysaccharides or proteins generally require some level of chemical modification that will enable interchain crosslinks to form. Hydrogels can be prepared under aqueous conditions by using activating groups followed by addition of a crosslinking agent or in a single step with the addition or reactive homobifunctional, heterobifunctional, or zero-length crosslinking agents.

Polysaccharide Hydrogels

Polysaccharides can be hetero- or homopolymers consisting of hydroxyl, carboxyl, or amino side chains which can be used for direct crosslinking or to introduce activating groups. The pyranose ring also can be activated through ring-opening oxidation. Much of the work in this area, however, uses a large number of hydroxyl groups to prepare hydrogels.

Polypeptide Hydrogels

The most common reactions used to crosslink polypeptides are alkylation and acylation. Those amino acid residues with amine or carboxyl side chains are primary targets for crosslinking. Reactivity of specific amino acid residues is related to nucleophilicity of the side chain, solvent type, and local steric and electrostatic interactions from the microenvironment.[5]

HYDROGELS PREPARED WITH ENZYME-DEGRADABLE CROSSLINKS

Enzyme-catalyzed hydrolysis at crosslinks is another approach to designing degradable hydrogels. Degradation is characterized by formation and removal of water soluble polymer chains from the network. The crosslink can be tailored to a specific enzyme or designed to degrade over an intended interval of time.

ENZYME-CATALYZED DEGRADATION OF HYDROGELS

Polysaccharide Hydrogels

Dextran

Edman and Sioholom studied modified dextran microspheres as lysosome-directed drug delivery systems.[12–14] The objective was to deliver agents to reticuloendothelial cells with subsequent degradation and drug release by lysosomal enzymes. Microspheres were prepared from glycidyl acrylate-modified dextran. In rats, microspheres were phagocytised predominantly in the liver, spleen, and bone marrow after intravenous injection. The half life of microspheres was directly related to degree of dextran modification. Elimination rates decreased as the degree of dextran modification increased. *In vitro* degradation studies with dextranase provided proof that as the degree of dextran modification increased, the rate of microsphere degradation was slower. These observations illustrate how chemical modification and crosslinking density alter the kinetics of degradation.

Enzyme-degradable hydrogels were prepared from glycidyl acrylate-modified dextran by γ-irradiation.[3] Hydrogel swelling kinetics were used to evaluate dextranase-catalyzed degradation. It was shown that degradation rates decreased as the γ-irradiation dose used for preparation increased.

Starch

Heller et al. developed enzyme-degradable starch hydrogels for "triggered" drug delivery.[4] Soluble starch was functionalized with glycidyl methacrylate under alkaline conditions. All hydrogels tested were susceptible to hydrolysis by α-amylase as measured by gel swelling and dissolution. The rate of hydrogel degradation decreased as the degree of starch modification increased. However, the rate of α-amylase-catalyzed hydrolysis of the uncrosslinked modified starch increased as the degree of modification increased. Thus, although hydrolysis rates increased with increasing chemical modification, corresponding hydrogels were more resistant to degradation. Here the crosslinking density of the network controlled degradation through steric and diffusion constraints on the enzyme even though the modified substrate was more susceptible to hydrolysis.

Artursson et al.[15] and Laakso et al.[16,17] studied degradation properties of modified starch microspheres for lysosomotropic drug delivery.

Chondroitin-6-Sulfate

Modified chondroitin sulfate has been studied as an enzyme-degradable oral drug delivery system since it can be degraded by anaerobic flora of the large intestine.[18,19]

Amylose

Enzyme-catalyzed degradation of epichlorohydrin-crosslinking amylose was studied by Mateescu et al.[20] The rate of enzyme-catalyzed hydrolysis was studied by an iodometric method.

Polypeptide Hydrogels

Albumin

Albumin microspheres have been prepared as injectable delivery systems for a number of drugs.[21] Albumin microspheres can be degraded by collagenase, papain, protease, and trypsin.[21–23] Lee et al. studied the enzyme-degradable properties of progesterone-loaded microspheres.[6]

Willmott et al. studied degradation of adriamycin-loaded glutaraldehyde-crosslinked microspheres.[24] Willmott et al.[22] also studied biodegradibility of ^{125}I-labeled, glutaraldehyde-crosslinked microspheres entrapped in capillary beds of the liver, kidneys, and lungs of rats.

Gelatin

Enzymatic degradation of gelatin hydrogels was studied by Kamath et al.[3] Hydrogels were prepared by exposing solutions of glycidyl acrylate-modified gelatin to γ-irradiation. Once again, increased crosslinking density significantly influenced the rate of enzyme-catalyzed hydrolysis.

Hydrogels Prepared with Degradable Crosslinks

Oligopeptide Crosslinks

The rate of degradation for gels prepared with oligopeptide crosslinks can vary as a function of chain length or subsite-substrate interactions.

Hydrogels Crosslinked by Azo Reagents

The presence of aromatic azo bonds in the crosslinks of hydrogels have been studied for colon-specific drug delivery.[7–9]

Albumin-Crosslinked Hydrogels

Enzyme-degradable hydrogels have been prepared for long-term oral drug delivery by selective retention in the stomach.[10,11,25–30] Hydrogels were susceptible to degradation by pepsin, chymotrypsin, and trypsin. Glycidyl acrylate-modified albumin was used as a polyfunctional crosslinking agent. Albumin-crosslinked hydrogels were prepared by copolymerizing functionalized albumin with acrylic acid, acrylamide, or vinylpyrrolidone.

FUTURE PERSPECTIVES

Interest in enzyme-degradable hydrogels has been significant in the area of pharmaceutics and will likely have unique biomedical and agricultural applications. Although a variety of prototypical hydrogels has been developed for drug delivery, few practical applications have been realized. Currently, colon-specific drug delivery using gels degradable by azoreductase is one of the more promising applications. The advantage is that drug release can be site specific as well as degradation controlled. Drug therapies that may benefit from this approach include chemotherapeutic drugs for the treatment of colon cancer, antiinflammatory agents for inflammatory bowel diseases, or peptide drugs to bypass upper GI degradation. All excellent review by Kopecek et al. examines polymers that may be utilized for colon specific drug delivery.[8] Other applications for biodegradable hydrogels in drug delivery have been recently discussed by Park et al.[2] and Kamath et al.[1]

A recent review by Shalaby discusses use of adjuvants and polymers to overcome physiological and immunological barriers to oral vaccination.[31] There is significant potential for enzyme-degradable hydrogels in the treatment of disease processes. The key to success will be in identifying diseases that cannot adequately be treated by simple medical management.

REFERENCES

1. Kamath, K. R.; Park, K. *Advances Drug Delivery Reviews* **1993**, *11*, 59–84.
2. Park, K.; Shalaby, W. S. W.; Park, H. *Biodegradable hydrogels for drug delivery* Technomic: Lancaster, PA, 1993.
3. Kamath, K. R.; Park, K. *ACS Symposium Series* **1994**, *545*, 55–65.
4. Heller, J.; Pangburn, S.; Roskos, K. *Biomaterials* **1990**, *11*, 345–350.
5. Wong, S. S. *Chemistry of Protein Conjugation and Crosslinking*; CRC: Boca Raton, FL, 1991.
6. Lee, T. K.; Sokoloski, T.; Royer, G. P. *Science* **1981**, *213*, 233–235.
7. Brondsted, H.; Kopecek, J. *Biomaterials* **1991**, *12*, 584–592.
8. Kopecek, J.; Kopecekova, P.; Brondsted, H.; Rathi, R.; Rihova, B.; Yeh, P.; Ihesue, K. *J. Controlled Release* **1992**, *19*, 121–130.
9. Van den Mooter, G.; Samyn, C.; Kinger, R. *Pharm. Res.* **1994**, *11*, 1737–1741.
10. Park, K. *Biomaterials* **1988**, *9*, 435–441.
11. Shalaby, W. S. W.; Park, K. *Pharm. Res.* **1990**, *7*(8), 816–823.
12. Edman, P.; Sjoholm, I. *J. Pharm. Sci.* **1983**, *72*(7), 769–799.
13. Edman, P.; Sjoholm, I.; Brunk, U. *J. Pharm. Sci.* **1983**, *72*(6), 658–665.
14. Edman, P.; Sjoholm, I. *Life Sci.* **1982**, *30*, 327–330.
15. Artursson, P.; Edman, P.; Laakso, T.; Sjoholm, I. *J. Pharm. Sci.* **1984**, *73*(11), 1507–1513.
16. Laakso, T.; Sjoholm, I. *J. Pharm. Sci.* **1987**, *76*(12), 935–939.
17. Laakso, T.; Artursson, P.; Sjoholm, I. *J. Pharm. Sci.* **1986**, *75*(10), 962–967.

18. Rubinstein, A.; Nakar, D.; Sintov, A. *Pharm. Res.* **1992**, *9*(2), 276–278.
19. Sintov, A.; Nakar, D.; Rubinstein, A. *Proceed. Intern. Symp. Control. Rel. Bioact. Mater.* **1991**, *18*, 381–382.
20. Mateescu, M. A.; Schell, H. D. *Carbohydr. Res.* **1983**, *124*, 319–323.
21. Morimoto, Y.; Fujimoto, S. *Crit. Rev. Ther. Drug Carrier Syst.* **1985**, *2*(1), 19–63.
22. Willmott, N.; Chen, Y.; Goldberg, J.; Meardle, C.; Florence, A. T. *J. Pharm. Pharmacol.* **1989**, *41*, 433–438.
23. Mahato, R. I.; Willmott, N.; Vezin, W. R. *Proceed. Intern. Symp. Control Rel. Bioact. Mater.* **1991**, *18*, 375–376.
24. Willmott, N.; Cummings, J.; Stuart, J. F. B.; Florence, A. T. *Biopharm. Drug Dispos.* **1985**, *6*, 91–104.
25. Shalaby, W. S. W.; Peck, G. E.; Park, K. *J. Controlled Release* **1991**, *16*, 355.
26. Shalaby, W. S. W.; Blevins, W. E.; Park, K. *ACS Symposium Series* **1991**, *467*, 484–492.
27. Shalaby, W. S. W.; Blevins, W. E.; Park, K. *ACS Symposium Series* **1991**, *469*, 237–248.
28. Shalaby, W. S. W.; Belvins, W. E.; Park, K. *Biomaterials* **1992**, *13*, 289–296.
29. Shalaby, W. S. W.; Belvins, W. E.; Park, K. *J. Controlled Release* **1992**, *19*, 131–144.
30. Shalaby, W. S. W.; Chen, M.; Park, K. *J. Bioact. Compatible Polym.* **1992**, *7*, 257–274.
31. Shalaby, W. S. W. *Clin. Immunol. Immunopathol.* **1995**, *74*, 127–134.

ENZYME IMMOBILIZATION (Polyion Complexes)

Shinzo Kohjiya*
Institute for Chemical Research
Kyoto University

Hiroko Sato
Department of Polymer Chemistry
Kyoto University

Immobilizing enzymes is useful for biomedical, chemical, analytical, and industrial applications.[1–4] Most enzymes have basic and acidic amino acid residues. Therefore, enzymes are capable of not only immobilizing polyelectrolyte complexes but also mixing with a polyanion or polycation.

The formation of a polyion complex, applied to polyelectrolytes and naturally occurring polymers except enzymes, was reported recently. For example, the complex formation between DNA polyanions and polycations (e.g., copolymer of dimethylaminopropyl acrylamide and acrylamide) affected the transfection of plasmid DNA into the nucleus of eucaryotic cells.[5] Polyion complexes of partially aminated poly(L-glutamic acid) and gum arabic were prepared as antigen-carrier particles with diameters of 2 to 7 µm.[6] These polyion complex particles showed no disturbance of specific reactions, that is, no nonspecific agglutination between antigen and antibody, and were commercialized for diagnoses against hepatitis, human T cell lymphotropic virus, and *Treponema pallidum*.

*Author to whom correspondence should be addressed.

AMIDASE

Margolin et al.[8] studied enzymes immobilized in non-stoichiometric polyion complexes with excess poly(methylacrylic acid) as the polyanion and poly(*N*-ethyl-4-vinyl-pyridinium bromide) as the polycation.[7]

Generally, enzymes allowed to stand in solution have less activity with time or by heating. The relative activity of penicillin amidase, immobilized in non-stoichiometric polyion complexes, increased 7-fold at pH 5.7 and 60°C and 300-fold at pH 3.1 and 25°C, compared with that of native penicillin amidase in solution. The enzyme immobilization of polyion complexes resulted in better thermal stability. Urease, immobilized in a polyion complex as a precipitate or bound with polycations in polyion complexes, was also thermally stable at 70°C.

CATALASE

Catalase is one of many enzymes found in animal or plant cells and microorganisms. Large amounts of this enzyme are contained in hepatic and renal cells where catalase is localized in peroxisomes with other oxidative enzymes.[9] For example, hydrogen peroxide, a substrate of catalase, damages tissues and cells and is produced from the dismutation of superoxide anion radicals, which are also toxic to cells, through the catalysis of superoxide dismutase (SOD).[10] Thus, the catalase immobilization of polyion complexes appears relevant to biological systems. Because the immobilization with polyelectrolytes reduced enzyme activity, we studied the effect of adding inert components to the immobilized catalase.

REFERENCES

1. Silman, I. H.; Katchalski, E. *Adv. Enzymol.* **1971**, *34*, 445.
2. *Immobilized Enzymes*; Chihata, I., Ed.; Kodansha: Tokyo, 1975.
3. Fukui, S. *Seibutsu Yukikagaku*; Kodansha: Tokyo, 1976; Chapter 5.
4. Fukui, S.; Toratani, T. *Seikagaku* **1976**, *48*, 96.
5. Yamaoka, T.; Hamada, N.; Ide, H. et al. *Polym. Prepr. Jpn.* **1994**, *43*, 488.
6. Hirayama, C.; Ihara, H.; Shibata, M. et al. *Polym. J.* **1991**, *23*, 161.
7. Sato, H.; Nakajima, A. *Colloid Polym. Sci.* **1974**, *252*, 294.
8. Margolin, A. L.; Sherstyuk, S. F.; Izumrudov, V. A. et al. *Eur. J. Biochem.* **1985**, *146*, 625.
9. Nicholls, P.; Schonbaum, G. R. *Enzymes* **1963**, *8*, 147.
10. Fridovich, I. *Ann. Rev. Biochem.* **1975**, *44*, 147.

ENZYME IMMOBILIZATION (Porous Polymer Beads)

Toshio Hayashi
Research Institute for Advanced Science and Technology
Osaka Prefecture University

The bioreactor is generally composed of biologically active proteins and their bioinert carriers, and in this system, polymers could become most important in the near future. Many papers have discussed polymeric carriers, especially the immobilization of proteins to the carriers.[1] Because the recovery yield and the reuse of free enzymes as industrial catalysts are limited, enzyme immobilization has greater advantages because it offers the choice of batch or continuous processes for rapidly terminating reactions, controlling production formation, easing enzyme removal from the reaction mixture, and adapting to engineering designs.[2] A concerted or sequential reaction of several enzymes can be obtained by using mixed or stratified beds. Furthermore, interest in immobilized enzymes and their applications in bioprocessing, analytical systems, and enzyme therapy has steadily grown in the past decade.[3–5] Consequently, various methods for preparing water-insoluble enzymes have been explored in recent years to study the enzyme reaction in two-phase systems similar to those existing *in vivo*.[6–8]

The most common method for enzyme immobilization is the covalent attachment to polymer carriers, which yields a strongly bound enzyme. High enzyme loading can be achieved with microspheres or porous bead polymer carriers.

GENERAL FEATURES

Supports

Support materials for immobilizing enzymes usually fall into two categories, inorganic and organic. Organic supports can be in either a particulate or membrane configuration. Inorganic supports are usually in porous form. The advantage of organic supports is the ease with which they form a covalent bond between the enzyme and the support. Because their organic nature allows a selection of many reactive side chains, high enzyme loading is relatively easy to obtain.

Applications

Immobilized enzymes are currently used in industrial, analytical, therapeutic, and for structural applications. Most large-scale industrial applications are in the food industry and in column reactors. Among these applications, the use of immobilized glucose isomerase for producing high fructose corn syrup is the largest.

MATERIALS AND METHODS

Various porous polymer beads were used as carriers. They included polyacrolein microspheres (Matsumoto Yushi Inc.) 0.2 μm in diameter, chitosan porous beads (ChB) (Fuji Spinning Co.) 300 μm in diameter, poly(γ-methyl-L-glutamate) (Ajinomoto Co.) 20 μm in diameter, and poly(vinyl alcohol) (PVA) 120 μm in diameter.[9–17] Each enzyme was covalently immobilized on the porous polymer beads.

Activity

The hydrolytic activity of free and immobilized enzymes was determined using a low molecular weight substrate. The relative activity (RA) of the immobilized enzymes compared to the free enzymes was measured.

CONCLUSION

Enzymes immobilized by covalent bonds on the surface of porous polymer beads with any length of spacer were highly active toward small ester substrates, but far less active toward a high molecular weight substrate. The RA of the immobilized enzyme with spacer gave an almost constant value for the substrate hydrolysis even at low surface concentrations of immobilized enzyme. Compared with free enzymes, immobilized enzymes had higher K_m values but smaller V_m values. The pH, thermal, and storage stability of immobilized enzymes were higher than those for the free enzymes. The initial activity of the immobilized enzyme remained almost unchanged without eliminating or inactivating any enzymes.

REFERENCES

1. Katchalski, E. K. *Enzyme Engineering*; Plenum: New York, 1982; p 12.
2. Axen, R. *Insolubilized Enzymes*; Raven: New York, 1974; p 9.
3. Chibata, I.; Tosa, T.; Sato, T. *Biotechnology, Vol. 7-a: Enzyme Technology*; VCH Verlagsgesellschaft: Weinheim, 1987; p 653.
4. Karube, I. *Biotechnology, Vol. 7-a: Enzyme Technology*; VCH Verlagsgesellschaft: Weinheim, 1987; p 685.
5. Senatore, F.; Bernath, F.; Meisner, K. *J. Biomed. Mater. Res.* **1986**, *20*, 177.
6. Arica, Y.; Hasirci, V. N. *Biomaterials* **1987**, *8*, 489.
7. Sipehia, R.; Chawla, A. S.; Daka, J.; Chang, T. M. S. *J. Biomed. Mater. Res.* **1988**, *22*, 417.
8. Kozhukharova, A.; Kirova, N.; Popova, Y.; Batsalova, K.; Kunchev, K. *Biotech. Bioeng.* **1988**, *32*, 245.
9. Hayashi, T.; Ikada, Y. *Biotech. Bioeng.* **1990**, *35*, 518.
10. Hayashi, T.; Ikada, Y. *Biotech. Bioeng.* **1990**, *36*, 593.
11. Hayashi, T.; Ikada, Y. *J. Appl. Polym. Sci.* **1991**, *42*, 85.
12. Itoyama, K.; Tanibe, H.; Hayashi, T.; Ikada, Y. *Biomaterials* **1994**, *15*, 107.
13. Itoyama, K.; Tokura, S.; Hayashi, T. *Biotech. Progress* **1994**, *10*, 225.
14. Itoyama, K.; Tokura, S.; Seo, H.; Tanibe, H.; Hayashi, T. *Polymer International*, in press.
15. Hayashi, T.; Hirayama, C.; Iwatsuki, M. *J. Appl. Polym. Sci.* **1992**, *44*, 143.
16. Hayashi, T.; Hyon, S. H.; Cha, W. I.; Ikada, Y. *Polymer J.* **1993**, *25*, 489.
17. Hayashi, T.; Hyon, S. H.; Cha, W. I.; Ikada, Y. *J. Appl. Polym. Sci.* **1993**, *49*, 2121.

ENZYME MIMICS (Cyclodextrins)

Kenjiro Hattori
Department of Industrial Chemistry
Tokyo Institute of Polytechnics

An inclusion complex possesses high-stability constants and potentially high stereospecificity because the binding involves multiple interactions. Inclusion complexes resemble enzymatic complexes in this respect. The cyclodextrins (CDs) consist solely of D(+)-glucose units linked through α-1, 4 units in a cyclic array of α-, β-, and γ-cyclodextrins containing six, seven, or eight glucose residues per molecule, respectively.[1–5] They are doughnut-shaped molecules with the glucose units in the C-1 conformation. The primary hydroxyl groups (carbon 6 of the glucose unit) appear on one side of the torus, and the secondary hydroxyls (carbons 2 and 3 of the glucose units) are on the torus' other side. The cavity's interior contains a ring of C-H groups, a ring of glycosidic oxygens, and another ring of C-H groups. Consequently, the interior of the CD torus is relatively apolar compared with water.

The solution complexes are in the main stoichiometric compounds, which contain one molecule of host and one molecule of guest, that is, monomolecular inclusion complexes. CD formation of inclusion complexes in aqueous solution has led to their use as model enzymes.

Various functional groups attached to cyclodextrins can be used to construct enzyme mimics.[6,7] Because the hydrophobic cavity can be used as a site for substrate binding and has the ability to modify the cyclodextrins with catalytic groups, several modified CDs have been studied as enzyme mimics. Many examples of CD-based enzyme models are known and have been reviewed.[1,3,5,8–10]

HYDROLYASE MIMICS

Bender et al.[1,11–13] first reported the hydrolytic catalysis in the presence of a CD. They extensively promoted the chemistry of CDs as enzyme mimics.[11,12] The key that opened the host-guest chemistry was that ester hydrolysis with CDs proceeded in a mechanism like enzyme with inclusion.

CD Catalysis

Imidazole Modified

By introducing an imidazole group into the CD molecule, its rate enhancement may parallel that of the chymotrypsin reaction. The first such attempts at introducing an imidazole group showed only a three-to-four-fold rate enhancement for a combination of CD and imidazole.[14] But a larger effect was found when stereochemistry and binding were considered.[15] γ-CD bearing a histamine was synthesized and used it to treat the cleavage of aryl esters.[19] Bender et al. synthesized a β-CD derivative that had carboxyl, imidazole, and hydroxyl groups.[16]

Non-Inclusion

The hydrolytic mechanism of the nitrophenyl ester has been illustrated as a scheme in which the aryl group is included in the CD cavity at the transition state. Tee et al. showed this representation was not always true.[17–18]

Amide Substrate

There are few reports of amide hydrolysis with CD catalysis. However, new results were recently reported.[2] In a substrate of *p*-nitrotrifluoroacetanilide, there was an acceleration in the presence of α- and β-CDs, and the hydroxypropyl CD.

Asymmetric Selective Mimics

CDs have great potential as tools for asymmetric induction.[21]

CDs, which consists of D-glucose with an α-1,4 linkage, can recognize an enantiomer and can influence the enantioselectivity and reaction rate, especially during the cleavage of carboxylic acid esters.[22] The CD hydroxyl groups and the selection of the appropriate substrate geometry may enhance the rate a million times over.

TRANSAMINASE MIMICS

Breslow et al. described the first example of a coenzyme-attached CD, in which a pyridoxamine residue was covalently linked to a primary carbon of β-CD; the modified CD should catalyze the transamination as an artificial enzyme.[23,24]

Breslow et al. synthesized two transaminase mimics of CD by introducing pyridoxamine on the primary and secondary hydroxyl groups.[25]

Tabushi et al. reported that the artificial vitamin B_6 enzyme, A-(modified B_6)-B-[ω-amino(ethylamino)]-β-CD, accelerated the conversion from keto acids to L-amino acids in a 98/2,98/2,95/5 L/D ratio for phenylglycine, phenylalanine, and tryptophane, respectively.[26]

RIBONUCLEASE MIMICS

Breslow et al. synthesized three isomeric β-CDs by modifying the regiospecificity on two of seven glucoses with imidazole groups.[27]

ASYMMETRIC SELECTIVE CATALYTIC REACTIONS

Sakuraba et al. and Fornasier et al. reported cases of high asymmetric yields for a heterogeneous reaction using the CD inclusion complex in the crystalline state.[20,32,48] In aqueous solution or in an organic solvent, asymmetric reactions in the homogenous phase occur during reactions of addition, oxidation, and reduction.[28–45]

REGIOSELECTIVE CATALYTIC REACTIONS

Because of the CD inclusion, the substituted positioning on the benzene ring was effectively controlled to give ortho-, meta-, and para-oriented products. Breslow et al. first reported the chlorination of benzene.[46,47] In the presence of α-CD, the chlorination of anisole with hydrochlorite gave a para-substituted product with 96% selectivity.

CATALYTIC ANTIBODY MIMICS

A multi-interacting host system prepared by connecting two CD molecules gave higher binding constants and seemed to be a receptor model. In addition, introducing a catalytic site in that system mimicked the behavior of a catalytic antibody. Breslow et al. prepared a CD dimer molecule coordinated with Cu^{2+} on the dipyridyl arm site, which hydrolyzed 220,000 times faster at pH 7.0 and at 37°C than under non-catalytic conditions. Ueno and Toda showed that the CD dimer had an imidazole group hydrolyzed with *p*-nitrophenyl alkanoates.[49] The acceleration was 191 fold with a specific association for substrates.

OTHER ENZYME MIMICS

Models For other enzymes based on CDs were synthesized recently and studied. Nicotinamide adenine dinucleotide (NADH)-dependent enzyme mimics were synthesized by introducing quaternized nicotinamide to the CD molecules. This reduced dihydro derivative was adopted for reduction in aqueous solution.[50–54] Carbonic anhydrase mimics were synthesized. The modified CD enhanced the hydration of CO_2 in the buffer solution containing a zinc ion.[55,56] A glyoxalase mimic was carried out using 2-thioethylamine-CD.[57,58] Flavine attached to the primary or secondary face of the CD was reported.[59] Conjugates of iodosobenzoic acid with β-CD were synthesized, and the turnover catalysis as soman mimics were found.[60]

REFERENCES

1. Bender, M. L.; Komiyama, M. *Cyclodextrin Chemistry*; Springer-Verlag: Berlin, 1979.
2. *Inclusion Compounds*; Atwood, J. L. et al., Eds.; Academic: London, 1984.

3. Szejtli, J. *Cyclodextrin and Their Inclusion Complexes*; Academiai Kiado: Budapest, 1982.
4. *Cyclodextrins and Their Industrial Uses*; Duchene, E.; Ed.; Editions de Santa: Paris, 1987.
5. Szejtli, J. *Cyclodextrin Technology*; Kluwer Academic, Dordrecht, 1988.
6. Croft, A. P.; Bartsh, R. A. *Tetrahedron* **1983**, *39*, 1417.
7. Tabushi, I. *Inclusion Compounds*; Atwood, J. L. et al., Eds.; Academic: London, 1984, Vol. 3, p 446.
8. Breslow, R. In *Inclusion Compounds*; Atwood, J. L. et al., Eds.; Academic: London, 1984; Vol. 3, p 496.
9. Breslow, R. *Acc. Chem. Res.* **1991**, *24*, 159.
10. van Dienst, E. Doctoral Thesis, University Twente, 1994.
11. van Etten, R. L.; Sebastian, J. F.; Clowes, G. A. et al. *J. Am. Chem. Soc.* **1967**, *89*, 3242.
12. van Etten, R. L.; Clowes, G. A.; Sebastian, J. F. et al. *J. Am. Chem. Soc.* **1967**, *89*, 3253.
13. Bender, M. L. *Mechanisms of Homogeneous Catalysis from Protons to Proteins*; Wiley-Interscience: New York 1971; pp 373–382.
14. Cramer, F.; MacKensen, G. *Angew. Chem. Int. Ed. Eng.* **1966**, *5*, 601.
15. Iwakura, Y.; Uno, K.; Toda, F. et al. *J. Am. Chem. Soc.* **1975**, *97*, 4432.
16. D'Souza, V. T.; Bender, M. L. *Acc. Chem. Res.* **1987**, *20*, 146.
17. Tee, O. S.; Hoeven, J. J. *J. Am. Chem. Soc.* **1989**, *111*, 8318.
18. Tee, O. S.; Bozzi, M.; Hoeven, J. J. et al. *J. Am. Chem. Soc.* **1993**, *115*, 8990.
19. Hamasaki, K.; Mochizuki, H.; Ikeda, T. et al. In *Proceedings of the 7th International Cyclodextrins Symposium*; Osan, T., Ed.; Business Center for Academic Societies: Tokyo, 1994; pp 342–345.
20. Granados, A.; de Rossi, R. H. *J. Org. Chem.* **1993**, *58*, 1771.
21. Takahashi, K.; Hattori, K. *J. Inclusion Phenomena Molecular Recognition* **1994**, *17* 1–24.
22. Breslow, R. *Science* **1982**, *218*, 532.
23. Breslow, R.; Overman, L. E. *J. Am. Chem. Soc.* **1970**, *92*, 6869.
24. Breslow R.; Duggan, P. J.; Light, J. P. *J. Am. Chem. Soc.* **1992**, 114, 3982.
25. Breslow, R.; Czarnik, A. W.; Lauer, M. et al. *J. Am. Chem. Soc.* **1986**, *108*, 1969.
26. Tabushi, I. *Pure Appl. Chem.* **1986**, *58*, 1529.
27. Breslow, R. *J. Am. Chem. Soc.* **1978**, *100*, 3227.
28. Schneider, H. J.; Sangwan, N. K. *Angew. Chem. Int. Ed. Eng.* **1987**, *26*, 896.
29. Rao, K. R.; Nageswar, Y. V. D.; Kumar, H. W. S. *Tetrahedron Lett.* **1991**, *32*, 6611.
30. Rao, K. R. *Pure Appl. Chem.* **1992**, *64*, 1141.
31. Czarnik, A. W. *J. Org. Chem.* **1984**, *49*, 924.
32. Fornasier, R.; Scrimin, P.; Tecilla, P. *Phosphorus Sulfur Relat. Elem.* **1988**, *35*, 211.
33. Siegel, B.; Breslow, R. *J. Am. Chem. Soc.* **1975**, *97*, 6869.
34. Drabowicz, J.; Mikolajczyz, M. *Phosphorus Sulfur Relat. Elem.* **1984**, *21*, 245.
35. Sakuraba, H.; Ushiki, S. *Tetrahedron Lett.* **1990**, *31*, 5349.
36. Colonna, S.; Manifredi, A.; Annunziata, R. et al. *J. Org. Chem.* **1990**, *55*, 5862.
37. Weber, L.; Imiolczyk, I.; Haufe, G. et al. *J. Chem. Soc. Chem. Commun.* **1992**, *301*.
38. Baba, N.; Matsumura, Y.; Sugimoto, T. *Tetrahedron Lett.* **1978**, 4281.
39. Hattori, K.; Takahashi, K.; Uematsu, M. et al. *Chem. Lett.* **1990**, 1463.
40. Hattori, K.; Takahashi, K.; Sakai, N. *Bull. Chem. Soc. Jpn.* **1990**, *65*, 2690.
41. Moon, H.; Wadamori, M.; Takahashi, K. et al. *J. Inclusion Phenomena Molecular Recognition* submitting.
42. Takahashi, K.; Wadamori, M.; Hattori, K. *J. Inclusion Phenomena Molecular Recognition* **1994**, *17*, pp 1–24.
43. Saka, W.; Inoue, Y.; Yamamoto, Y. et al. *Bull. Chem. Soc. Jpn.* **1990**, *63*, 3175.
44. Takahashi, K.; Ohtsuka, Y.; Hattori, K. *Chem. Lett.* **1990**, *1990*, 2227.
45. Takahashi, K. *Bull. Chem. Soc. Jpn.* **1990**, *66*, 550.
46. Breslow, R.; Campbell, P. *Bioorg. Chem.* **1970**, *1*, 140.
47. Breslow, R.; Campbell, P. *J. Am. Chem. Soc.* **1969**, *91*, 3085.
48. Komiyama, M.; Sugiura, I.; Hirai, H. *J. Mol. Catal.* **1986**, *36*, 271.
49. Akiike, T.; Nagai, Y.; Yamamoto, Y. et al. *Chem. Lett.* **1994**, 1089.
50. Yoon, C.; Ikeda, H.; Kojin, R. et al. *J. Chem. Soc., Chem. Commun.* **1986**, 1080.
51. Moon, H. T.; Ikeda, T.; Ikeda, H. et al. *Supramol. Chem.* **1993**, *1*, 327.
52. Ikeda, H.; Moon, H. T.; Du, Y. Q. et al. *Supramol. Chem.* **1993**, *1*, 337.
53. Yoon, C. J.; Ikeda, H.; Kojin, R. et al. J. Chem. Soc., *Chem. Commun.* **1986**, *1986*, 1080.
54. Ikeda, H.; Moon, H.-T.; Yoon, C.-J. et al. In *Proceedings of the 7th International Cyclodextrins Symposium*; Osa, T., Ed.; Business Center for Academic Societies: Tokyo, 1994; pp 322–329.
55. Tabushi, I.; Kuroda, Y. *J. Am. Chem. Soc.* **1984**, *106*, 4580.
56. Bonomo, R. P.; Cucinotta, V.; D'Alessandro, F. et al. *Inorg. Chem.* **1991**, *30*, 2708.
57. Tamagaki, S.; Katayama, A.; Maeda, M. et al. *J. Chem. Soc. Perkin Trans.* **1994**, *2*, 507.
58. Tagaki, W.; Ogino, K.; Kimura, Y. et al. In *Proceedings of the 7th International Cyclodextrins Symposium*; Osa, T., Ed.; Business Center for Academic Societies: Tokyo, 1994; pp 334–337.
59. Rong, D.; Ye, H.; Boehlow, T. R. et al. *J. Org. Chem.* **1992**, *57*, 163.
60. Seltzman, H. H.; Lonikar, M. S. In *Proceedings of the 7th International Cyclodextrins Symposium*; Osa, T., Ed.; Business Center for Academic Societies: Tokyo, 1994; pp 285–288.

Enzyme Models

See: *Catalyses (by Polyelectrolytes and Collodial Particles)*
Catalysts, Polymeric (Higher Activity than Monomeric Analogs)
Catalysts, Polymeric (Pressure Effects)
Enzyme Mimics (Cyclodextrins)
Metalloenzymes, Artificial (Polyethylenimine-Based)
Synzymes

ENZYME MODIFICATION (Use in Organic Media)

Yoshihiro Ito* and Yukio Imanishi
Division of Material Chemistry
Kyoto University

Until now, industries have used enzymes in predominantly aqueous environments. However, because many organic compounds have low solubility in water (i.e., fats, oils, fatty acids,

*Author to whom correspondence should be addressed.

alcohols, aromatics, and steroids), undesirable side reactions (hydrolysis, acyl migration, and nucleophilic addition of hydroxide ions) and unfavorable thermodynamic equilibrium, water is a poor reaction medium for most chemical transformations.[1]

To overcome these problems, numerous approaches have been developed. They include using enzymes in aqueous solutions containing a water-miscible organic cosolvent, aqueous-organic mixtures with two phases, reversed micelles, and anhydrous organic solvents. Recently, some groups reported the importance of solvent engineering for controlling enzyme reactions, particularly their regio- and stereoselectivity.[2,3]

Besides these approaches, several researchers have modified the enzyme for use in organic media. The enzyme was made soluble in organic media by a lipid mixture, gene engineering, alkylation, and hybridization with synthetic polymers.[4–9] Hybridization with polymers has the advantage of making the enzyme completely soluble in organic media. Therefore, the reaction rate is usually enhanced 10–100 times, although the final yield in the solvated state is the same as that for suspension. However, the modification polymers have been limited to amphiphilic polymers, such as poly(oxyethylene), carbohydrate-based polymers, and their derivatives. These grafted amphiphilic polymers may provide a water-like microenvironment for the enzymes.

Recently, we devised a new method of synthesizing vinyl polymer and enzyme hybrids, in which various vinyl polymers were selected by using lipoprotein lipase (LPL) or trypsin with an aliphatic azo compound to initiate graft polymerization.[4–9] We then investigated catalysis by placing the hybrid enzyme in organic solvents. This method was applied to various functional polymers, including hydrophobic and hydrophilic polymers.

PREPARATION

We added 5 mg of azo-containing lipase to undiluted styrene (ST), methyl methacrylate (MMA), a toluene solution (60 or 80 vol%) of ST or MMA, or an aqueous solution (5 or 10 vol%) of N-vinylpyrrolidone (VP) or acrylonitrile (AN). After the photoinitiated graft polymerization, the reaction product was recovered and purified.

PROPERTIES

Catalytic Activity

The hybrid LPL's esterification activity in chloroform was measured as follows. We added 5 mg of hybrid LPL to 1 mL of chloroform containing 0.46 M *n*-amyl alcohol and 0.32 M *n*-caprylic acid and the mixture was incubated at 37°C. After various incubation periods, the amount of ester produced in the chloroform solution was determined by high-pressure liquid chromatography.

High catalytic activity in the esterification reaction in chloroform was observed with LPL hybridized with hydrophobic PST or PMMA and amphiphilic PVP. The catalytic activity increased with the content of graft polymers and varied with the kind, number, and length of graft chains for a given amount of graft polymers.

The catalytic activity during the esterification reaction in chloroform of LPL hybridized with various vinyl polymers or with different numbers of PST were plotted against their solubility in chloroform. The catalytic activity increased linearly with the solubility in chloroform. These results indicated that the catalytic activity of LPL enhanced by hybridization with vinyl polymers was not from substrate affinity by the graft chains but from accessibility of the substrate to the enzyme as Zaks and Klibanov noted.[10]

Repeated Use and Thermal Stability

Improved thermal stability by enzymes of modifying them with synthetic polymers was reported.[11] In addition, Zaks and Klibanov reported that the enzyme was more stable in organic media than in water at high temperatures.[12] The method of enzyme modification, which we describe here, effectively improved thermal stability and allowed repeated use. Stabilizing the enzyme by modifying it with synthetic polymers, particularly in organic solvents, may have caused the enzyme's rigid conformation.

REFERENCES

1. Zaks, A. *Biocatalysts for Industry*; Dordick, J. S., Eds.; Plenum: New York, NY, 1991; p 161.
2. Dordick, J. S. *Biotechnol. Prog.* **1992**, *8*, 259.
3. Wescott, C. R.; Klibanov, A. M. *J. Am. Chem. Soc.* **1993**, *115*, 1629.
4. Ito, Y.; Fujii, H.; Imanishi, Y. *Protein Engineering Protein Design in Basic Research, Medicine, and Industry*; Ikehara, M., Ed.; Springer-Verlag: Tokyo, 1990, pp 249.
5. Ito, Y.; Fujii, H.; Imanishi, Y. *Makromol. Chem. Rapid Commun.* **1992**, *13*, 315.
6. Ito, Y.; Fujii, H.; Imanishi, Y. *Biotechnol. Lett.* **1992**, *14*, 1149.
7. Ito, Y.; Fujii, H.; Imanishi, Y. *Biotechnol. Prog.* **1993**, *9*, 128.
8. Ito, Y.; Kotoura, M.; Chung, D. J. et al. *Bioconj. Chem.* **1993**, *4*, 358.
9. Ito, Y.; Fujii, H.; IImanishi, Y. *Biotechnol. Prog.* **1994**, *10*, 398.
10. Zaks, A.; Klibanov, A. M. *Proc. Natl. Acad. Sci. U.S.A.* **1985**, *82*, 3192.
11. Hiroto, M.; Marsushima, A.; Kodera, Y. *Biotechnol. Lett.* **1992**, *14*, 559.
12. Zaks, A.; Klibanov, A. M. *Science* **1984**, *224*, 1249.

ENZYME-POLY(N-ISOPROPYLACRYL-AMIDE) CONJUGATES (Thermoreversibly Soluble/Insoluble)

K. D. Vorlop,* A. V. Patel, and A. Muscat
Institute of Technology
FAL

Enzyme polymer conjugates can be separated from the medium more easily than free enzymes because of the former's high molecular weight. Yet recovering heterogeneous catalysts is easier. Our idea was to combine the advantages of homogeneous and heterogeneous biocatalysts by attaching the enzyme to a water-soluble polymer that could be precipitated and then resolubilized by altering physical parameters such as pH or temperature.

Several enzyme-binding polymers, such as methacrylic acid-methylmethacrylate and alginate, can be precipitated by changing the pH.[1,2] However, the hysteresis between precipitation and

*Author to whom correspondence should be addressed.

resolubilization is quite broad (ΔpH > 1), which can lead to enzyme damage. An alternative to the methods described is using thermoreversible soluble-insoluble polymers. The aims for this method were an adjustable precipitation temperature, narrow hysteresis between solubilization and precipitation, small differences in temperature between the point of first turbidity and complete precipitation, easy attachment of enzymes to the polymer, and facilitated separation of the catalysts by flotation or simple filtration. Such an enzyme polymer conjugate based on poly(*N*-isopropylacrylamide) (PIPAAM) was first designed by Vorlop et al.[3]

PREPARATION AND PROPERTIES

PIPAAM

PIPAAM's most striking characteristic is its thermoreversible precipitation. At temperatures above 33°C, it precipitates.[4,5]

The precipitation temperature can be adjusted by copolymerizing NIPAAM with different hydrophobic or hydrophilic comonomers.

Copolymerizing NIPAAM with HEMA

We wanted to synthesize NIPAAM copolymers, which possess functional groups that allow coupling, thereby immobilizing the enzymes, and simultaneously retain PIPAAM's thermoreversible precipitation properties. Different types of comonomers such as glycidyl methacrylate, methacrylate, 6-heptenate, glycidyl methacrylate and 1,6-diaminohexane, polyethylene glycol monomethacrylate, and hydroxyethylmethacrylate (HEMA) were tested. The best results were obtained with HEMA.[6]

APPLICATIONS

Trypsin Immobilization

A reaction with cyanuric chloride activated HEMA's OH group. We immobilized trypsin type III by dissolving 20 mg of enzyme in 5 mL of phosphate buffer at 4°C. Then, we added 100 mg of activated HEMA–NIPAAM copolymer while stirring.

Conjugate Characterization

The enzyme activity, precipitation characteristics, and long-term stability of the trypsin conjugates were investigated. The conjugate was mostly soluble (94%), and the precipitation point was 34.2°C, 0.6°C higher than pure HEMA–NIPAAM copolymer.

Water-soluble conjugates of enzymes and polymers like PIPAAM, which can be precipitated by a temperature change or by adding small amounts of sodium chloride, offer an alternative to conventional biocatalysts because they allow homogeneous reactions and a simple recovery of the biocatalyst.

REFERENCES

1. Hoshino, K.; Taniguchi, M.; Netsu, Y. et al. *J. Chem. Eng. Jpn.* **1988**, *22*, 54–59.
2. Coughlin, R. W.; Aizawa, M.; Charles, M. *Biotechnol. Bioeng.* **1976**, *18*, 199–208.
3. Vorlop, K. D.; Steinke, K.; Wullbrandt, D. et al. German Patent 3 700 308, 1987.
4. Chiantore, O.; Trossarelli, L.; Guaita, M. *Macromol. Chem.* **1982**, *183*, 2257–2263.
5. Heskins, M.; Guillet, J. E. *J. Macromol. Sci., Chem.* **1968**, *A2*, 1441–1445.
6. Steinke, K.; Vorlop, K. D. In *Proceedings of DECHEMA Biotechnology Conference*; VCH-Verlag: Weinheim, 1990, Vol. 4, pp 889–892.

Enzymes

See: *Biopolymers (Electrochemical Behavior)*
Chitin (Metal Ion Chelation and Enzyme Immobilization)
Chitinases
Enzymatic Polymerization
Enzyme-Catalyzed Oxidative Polymerization (Aromatic Compounds)
Enzyme-Catalyzed Polymerization
Enzyme-Degradable Hydrogels
Enzyme Immobilization (Polyion Complexes)
Enzyme Immobilization (Porous Polymer Beads)
Enzyme Mimics (Cyclodextrins)
Enzyme Modification (Use in Organic Media)
Enzyme-Poly(N-isopropylacrylamide) Conjugates (Thermoreversibly Soluble/Insoluble)
Functional Immobilized Biocatalysts (Stimulus-Sensitive Gels)
Functionalized Polymers (Comb, Rotaxanic, and Dendrimeric Structures)
Globular Proteins (Structures and Functions)
Immobilized Enzymes, Biosensors
Ion-Selective Electrodes and Biosensors
Metalloenzymes, Artificial (Polyethylenimine-Based)
Microencapsulation (Artificial Cells)
Phospholipid-Immobilized Surfaces
Poly(α-amino acids) (Biodegradation, Medical Applications)
Reverse Micelles (Microcontainers for Functional Polymers)
Silk Fibroin (Soft Tissue Compatible Polymer)
Synzymes

EPDM

See: *Ethylene-Propylene Copolymers*

Epichlorohydrin Rubber

See: *Blends, Interchain Crosslinking*

Epipolymerization

See: *Epitaxial Crystallization (of Linear Polymers)*

EPITAXIAL CRYSTALLIZATION (of Linear Polymers)

A. Thierry
Institut Charles Sadron, CNRS

The word epitaxy, introduced by Royer, from the Greek epi, which means "on," and taxos, which means "organized deposition," defines the oriented overgrowth of a crystal upon a foreign crystal that resembles the deposit structurally, for example, similar lattice parameters.[1] Royer set up the lattice matching rules for minerals with a 15% mismatch for the upper limit of corresponding lattice parameters. Pashley and Matthews reviewed the first studies of inorganic and metal epitaxy, and Kern et al. reviewed the first stages of the growth mechanisms.[2–4] More recently, epitaxy has become a standard tool to build inorganic, ordered layers for electronics.

Research in epitaxial crystallization of polymers followed the same steps as that for inorganic materials, namely demonstrating feasibility, analyzing the basic rules, and using epitaxy to tailor macroscopically oriented films with specific properties. Willems et al. and Fischer published the first experiments on polyethylene (PE) epitaxy.[5,6] Polyethylene lamellae can produce oriented overgrowth despite their peculiar chain-folded morphology.[7–9] Many authors have reviewed these early investigations of polymer epitaxy.[10–13] Recently, Wittmann and Lotz reported on polymer epitaxy on organic substrates.[14]

Although the orientation of conventional polymers is usually achieved with standard mechanical deformation, epitaxial interactions are recognized as phenomena that may help control properties via a heterogeneous nucleation by additives or by inducing morphologies that reinforce polymer blends. Moreover, at present, epitaxy is used to design ordered, thin films of specialty polymers which have outstanding non-linear-optical or conducting properties via their highly conjugated chains.[15–17]

FILM PREPARATION

Three procedures are used to deposit the molecules on the substrate: isothermal immersion, recrystallization, and vapor deposition.

SUBSTRATES

The suitability of the substrates depends on three characteristics: crystallographically well-defined faces, a low degree of symmetry of the contact plane, and large sizes.

Single Crystals

Symmetric inorganic crystals (i.e., alkali halides, silicates, mica, silicon wafers, and graphite), which induce multiple orientations, were used first. Mostly alkali halides, were used because of their easy cleavage and facile modulation of lattice dimensions and ionic interactions. Today, graphite basal planes and silicon wafers are preferred for their electronic properties.[19,20]

Because of their chemical similarity to polymers, various low-symmetry organic crystals have been used as templates for polymers.[21] Their cleavage planes are often lined up with rows of phenyl groups producing van der Waals interactions with the polymer.[18] Furthermore, their wettability and melting temperature may be modulated via substitution on the phenyl groups.

Polymers

Bulk crystallization of polymers produces spherulitic morphologies. For well-defined, oriented, crystallographic faces, polymers must be specially organized.

POLYOLEFINS

Polyolefins are the simplest polymers because of their chemical structure, which is neither ionic nor polar. They were the first polymers used in industrial production. Their epitaxial growth was extensively studied from the very beginning.

Inorganic Crystals

Whatever method used—solution growth, melt growth, or vapor deposition (substrate at room temperature)—the stable orthorhombic PE lamellae lie on the (100) plane of alkali halides with their chains parallel to the <110> directions, except for a water-treated substrate.[6,10,23–27]

The crystalline forms of iPP—stable α monoclinic, β hexagonal, and γ orthorhombic—may be ordered on alkali halides.

Organic Crystals

A systematic investigation of PE deposition on low-symmetry organic crystals—aromatic hydrocarbons, benzoic acid, and various organic derivatives—reveals the typical deposition of orthorhombic and monoclinic PE polymorphs with different contact planes.[18,25,28]

The peculiar lamellar branching of the α phase of iPP has been known for some time, but its epitaxial molecular origin was only understood step by step by ascertaining the exact {010} contact planes of the homoepitaxy, the interdigitation of methyl (Me) groups in these planes, and the role of helical handedness.[22,29–31]

MISCELLANEOUS POLYMERS

For apolar inert polymers, the epitaxial relationship relies on geometrical considerations, but for polar chains, there may be more complex criteria. In fact, many cases have been encountered and described, but no general, fully accepted rules have emerged.

APPLICATIONS

The unusual edge-on orientation of epitaxially grown lamellae was used to measure their thickness, determine their structures, and/or supplement X-rays.[32–36] For more practical purposes, polymer epitaxy has found applications in at least three domains: morphology and properties of composites, blends, and laminates; artificial nucleation; and surface decoration.

EPIPOLYMERIZATION

Epipolymerization, short for epitaxial crystallization and polymerization, corresponds to the orientation of monomer deposits on a substrate. Polymerization and crystallization may be simultaneous or not. Epipolymerization is a valuable tool for controlling molecular engineering of polymeric, fold-free, single-crystal-like films with few defects from the topotactic polymerization of an oriented monomer.[37,38] The "epipolymers" are not classified by their growth modes in epipolymerization but by the usual distinction between conventional and speciality polymers.

CONCLUSION

For nonpolar polyolefins and weakly polar polyoxymethylene, the known lattice matching rules applied when the lattice was interpreted broadly as the organization of prominent groups along the chain and/or as interchain distances. However, because of the diverse potential shapes and interactions inherent to organic materials, local molecular interactions may help define and stabilize different forms of order that appear less favorable on purely geometrical grounds.

Although still in its infancy, epitaxial growth of polymers has already demonstrated great potential for various applications, including the control of conventional polymer morphology, and, hence, the enhancement of their mechanical and/or optical properties. Epitaxial growth can also be used to prepare ordered thin films of often untractable conjugated polymer chains. This method should become competitive with common ordering techniques such as ordering in a strong electro-magnetic field, surface alignment of liquid-crystals on rubbed polymers, or ordering by the Langmuir-Blodgett technique, especially as these procedures are limited by their chemical requirements and/or resultant organization.[39–42] None of these restrictions apply to epitaxial growth. Furthermore, the selection of appropriate substrates allows the production of well-defined structures with specific crystallographic orientation(s). These features may place epitaxial crystallization of polymers among the future's more promising techniques. Forthcoming studies should help us understand the initial steps of epitaxial growth and the epitaxial organization of polar or ionic systems with strong interactions.

ACKNOWLEDGMENTS

The author thanks C. Straupé for her help and B. Lotz and J. C. Wittmann for their discussions, which helped shape this paper.

REFERENCES

1. Royer, L. *Bull. Soc. Franç. Miner. Crist.* **1928**, *51*, 7.
2. Pashley, D. W. *Adv. Phys.* **1965**, *14*, 327.
3. Matthews, J. W. *Epitaxial Growth, Part A*; Academic: New York, 1975.
4. Kern, R.; Le Lay, G.; Metois, J. J. *Current Topics in Materials Science*; Kaldis, E., Ed., North-Holland: Amsterdam, NY, 1979; Vol. 3, Chapter 3.
5. Willems, J.; Willems, I. *Experientia* **1957**, *13*, 465.
6. Fischer, E. W. *Kolloid-Z.u.Z. Polym.* **1958**, *159*, 108.
7. Geil, P. H. *Polymer Single Crystals in Polymer Reviews*; John Wiley & Sons: New York, 1963; Vol. 5, Chapters 3 and 4.
8. Till, P. H., Jr. *J. Polym. Sci.* **1957**, *24*, 301.
9. Keller, A. *Phil. Mag. Ser.* **1957**, *8*, 2, 1171.
10. Willems, J. *Experientia* **1967**, *23*, 409.
11. Wunderlich, B. *Macromolecular Physics*, Academic: New York, 1973; Vol. 1, p 266.
12. Mauritz, K. A.; Baer, E.; Hopfinger, A. J. *J. Polym. Sci. Macromol. Rev.* **1978**, *13*, 1.
13. Swei, G. S.; Lando, J. B.; Rickert, S. E. et al. *Encyclopedia of Polymer Science and Engineering*, 2nd ed.; John Wiley & Sons: New York, 1986; Vol. 6, pp 209–224.
14. Wittmann, J. C.; Lotz, B. *Prog. Polym. Sci.* **1990**, *15*, 909.
15. Nalwa, H. S. *Adv. Mater.* **1993**, *5*, 341.
16. Stowell, J. A.; Amass, A. J.; Beevers, M. S. et al. *Polymer* **1989**, *30*, 195.
17. Potember, R. S.; Hoffman, R. C.; Hu, H. S. et al. *Polymer* **1987**, *28*, 574.
18. Wittmann, J. C.; Lotz, B. *J. Polym. Sci., Polym. Phys. Ed.* **1981**, *19*, 1837.
19. Sano, M.; Sasaki, D. Y.; Kunitake, T. *Science* **1992**, *258*, 441.
20. Tanaka, K.; Kubono, A.; Umemoto, S. et al. *Rep. Prog. Polym. Phys. Jpn.* **1987**, *30*, 175.
21. Walton, A. G.; Carr, S. H.; Baer, E. *Polym. Prepr., Am. Chem. Soc. Div. Polym. Chem.* **1968**, *9*(1), 603.
22. Awaya, H. *Nippon Kagaku Zasshi* **1961**, *82*, 1575.
23. Koutsky, J. A.; Walton, A. G.; Baer, E. *J. Polym. Sci. A2* **1966**, *4*, 611.
24. Carr, S. H.; Keller, A.; Baer, E. *J. Polym. Sci., Part A2* **1970**, *8*, 1467.
25. Wittmann, J. C.; Lotz, B. *Polymer* **1989**, *30*, 27.
26. Ghavamikia, H.; Rickert, S. E. *J. Mater. Sci. Lett.* **1983**, *2*, 103.
27. Koutsky, J. A.; Walton, A. G.; Baer, E. *J. Polym. Sci., Part B* **1967**, *5*, 177.
28. Wittmann, J. C.; Hodge, A. M.; Lotz, B. *J. Polym. Sci., Polym. Phys. Ed.* **1983**, *21*, 2495.
29. Padden, F. J. Jr.; Keith, H. D. *J. Appl. Phys.* **1966**, *37*, 4013.
30. Binsbergen, F. L.; De Lange, B. G. M. *Polymer* **1968**, *9*, 23.
31. Lotz, B.; Wittmann, J. C. *J. Polym. Sci., Part B: Polym. Phys.* **1986**, *24*, 1541.
32. Ihn, K. J.; Tsuji, M.; Kawaguchi, A. et al. *Bull. Inst. Chem. Res. Kyoto Univ.* **1990**, *68*, 30.
33. Dorset, D. L. *Macromolecules* **1991**, *24*, 1175.
34. Dorset, D. L.; McCourt, M. P.; Kopp, S. et al. *Acta Cryst.* **1994**, *B50*, 201.
35. Hu, H.; Dorset, D. L. *Acta Cryst.* **1989**, *B45*, 283.
36. Ritcey, A. M.; Brisson, J.; Prud'homme, R. E. *Macromolecules* **1992**, *25*, 2705.
37. Lando, J. B.; Baer, E.; Rickert, S. E. et al. ACS Symposium Series 212; American Chemical Society: Washington, DC, 1983; p 89.
38. Rickert, S. E.; Lando, J. B.; Ching, S. *Mol. Cryst. Liq. Cryst.* **1983**, *93*, 307.
39. Findlay, R. B. *Mol. Cryst. Liq. Cryst.* **1993**, *231*, 137.
40. Takeno, A.; Okui, N.; Kitoh, T. et al. *Thin Solid Films* **1991**, *202*, 213.
41. Yokoyama, H. *Mol. Cryst. Liq. Cryst.* **1988**, *165*, 265.
42. Miyashita, T. *Prog. Polym. Sci.* **1993**, *18*, 263.

Epitaxial Films

See: *Epitaxial Crystallization (of Linear Polymers)*
Epitaxial Films (Epitaxy During Polymerization)

EPITAXIAL FILMS (Epitaxy During Polymerization)

Masahito Sano
π-Electron Materials Project - JRDC

Polymerization-induced epitaxy (PIE) is the epitaxial growth of polymers on a solid surface during polymerization in solution. During a conventional polymerization, when a crystalline solid is present in the reaction mixture, polymers crystallize on the solid surface as ultra-thin crystalline.[1–5] As an epitaxial film, polymer chains should be oriented on the surface with a definite crystallographic relationship to the substrate lattice. Even when such a relationship cannot be verified, an oriented growth of crystalline polymers on a substrate surface is often referred to as an epitaxial growth in polymer science.[6] Therefore, PIE represents a surface analog of bulk crystallization of polymers during polymerization, although the mechanisms of surface and bulk crystallization may differ.[7]

Other common techniques of polymer epitaxy include epitaxial polymerization and epitaxial crystallization. Submicron film thicknesses have been attained by these methods. Epitaxial crystallization is the ordinary recrystallization of polymers dissolved in solution on a freshly prepared solid surface. Epitaxial crystallization has been reviewed in detail.[8] Epitaxial polymerization differs from PIE in that it involves solid-state topochemical polymerization of epitaxially grown monomer crystals, whereas PIE uses solution polymerization of dissolved monomers. A general review of epitaxial polymerization has been published.[9]

Various polymers can form ultra-thin crystalline films by PIE. Current studies suggest that PIE does not involve any surface-catalyzed reactions, but the crystalline films that are formed consist of polymers that are chemically identical to those produced in solution. These polymers are physiosorbed on the surface. Furthermore, the reaction conditions indicate that it is neither an ordinary crystallization process nor a spontaneous adsorption of polymers. Although the PIE mechanism is not well understood, various polymers with different molecular structures have been grown epitaxially using this technique. Epitaxial films form independently of the polymerization type and mechanism, implying that the vast knowledge of solution polymerization can be applied to epitaxial growth. Therefore, the diverse structures and chemical reactions make PIE a versatile technique for polymer epitaxy.

REFERENCES

1. Sano, M.; Sasaki, D. Y.; Kunitake, T. *Macromolecules* **1992**, *25*, 6961.
2. Sano, M.; Sasaki, D. Y.; Kunitake, T. *Science* **1992**, *258*, 441.
3. Sano, M.; Sasaki, D. Y.; Yoshimura, S. *Faraday Discuss.* **1994**, *98*, 307.
4. Sano, M.; Sasaki, D. Y.; Kunitake, T. *Langmuir* **1993**, *9*, 629.
5. Sano, M.; Sandberg, M. O.; Yamada, N.; Yoshimura, S. *Macromolecules* **1995**, *28*, 1925.
6. Wittmann, J. C.; Lotz, B. *Prog. Polym. Sci.* **1990**, *15*, 909.
7. Wunderlich, B. *Fortschr. Hochpolym.-Forsch.* **1968**, *5*, 568.
8. Mauritz, K. A.; Baer, E.; Hopfinger, A. J. *J. Polym. Sci. Macromol. Rev.* **1978**, *13*, 1.
9. Swei, G. S.; Lando, J. B.; Rickert, S. E. et al. *Encyclopedia of Polymer Science and Technology*; Wiley-Interscience: New York, 1986, Vol. 6.

EPM

See: Ethylene-Propylene Copolymers

EPOXIDE–AMINE ADDITION POLYMERS, LINEAR

Joachim E. Klee
De Trey Dentsply

Hans-Heinrich Hörhold
Institut für Organische Chemie und Makromolekulare Chemie
Universität Jena

In 1934, Paul Schlack developed the conception of linear epoxide–amine addition polymerization as the reaction between bisepoxides and amines having two active hydrogen atoms.[1]

Since 1934, numerous experiments have been performed, obtaining high molecular weight, linear addition polymers. Low molecules weight polymers, 2000–6000 g/mol, were obtained in solution.[4–9] The bulk polymerization of epoxides and amines showed a contradictory picture. With special polymerization conditions and amines such as aniline, amino ethanol, *N,N'*-dimethylethylenediamine, diazacrownether, or special aliphatic and aromatic liquid-crystalline amines, thermoplastic linear addition polymers were obtained.[3,7,8,10–15] However, a series of aliphatic amines (piperazine, butyl amine, hexylamine, decylamine, and ethanol amine) probably forms linear addition polymers that react further to insoluble and nonmelting network polymers.[6,16–18] These contradictory results were clarified by an investigation of conditions, raising the possibility that linear epoxide–amine addition polymerization exists.[19]

ADDITION POLYMERIZATION

Primary Monoamines

Addition polymerization of stoichiometric amounts of primary monoamines and DGEBA led to epoxide–amine addition polymers having molecular weights of 10,000–20,000 g/mol at bulk polymerization during 10–70 h at 20–130°C (**Scheme I**).[19–23]

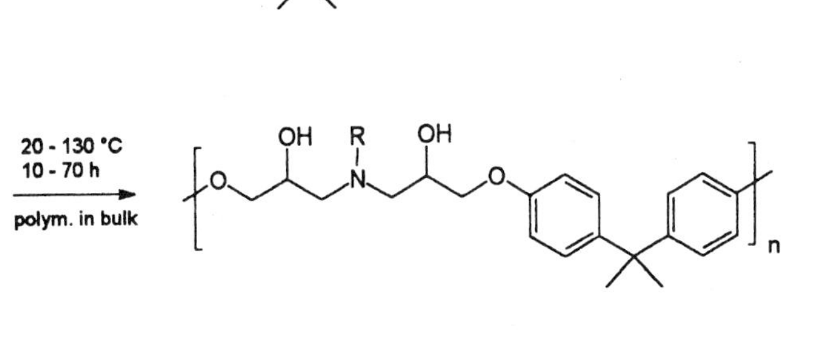

I

1

The high molecular weight epoxide–amine addition polymers were soluble in organic solvents such as pyridine, tetrahydrofuran (THF), 2-methoxy ethanol, and chloroform-methanol mixtures. In pyridine solution, the addition polymers reacted with acidic anhydride to form O-diacetyl-derivatives.

With phenylisocyanate, urethane-derivatives were obtained. The reactions of some polymers with alkylating agents such as benzylchloride led to N-quaternized polymers. These polyelectrolytes were soluble in polar organic solvents and formed transparent films that adhered well to glass and metal surfaces.

Carboxylic Acid Hydrazides

Because the NH_2 group is not incorporated in the mesomery of the amide group, carboxylic acid hydrazides have a higher nucleophily and react as primary monoamines with DGEBA to high molecular weight addition polymers.[25]

α-Amino Carboxylic Esters

Stoichiometric amounts of α-amino carboxylic esters and DGEBA polymerize in bulk at 60–100°C during 75 h to form high molecular weight, soluble addition polymers.

Disecondary Diamines

Similar to primary monoamines, disecondary diamines polymerize with DGEBA to form high molecular weight addition polymers while being heated at 20–120°C for 10–100 h.[19–21,24,26–30,33,34]

Ditertiary Diamines

Recently, linear addition polymers were obtained by the addition polymerization of ditertiary diamines, DGEBA, and SO_2. For example, 1,4-diazabicyclo[2,2,2]octane, DGEBA, and SO_2 were polymerized in solvents such as 2-methoxy ethanol or DMF to form addition polymers with M_n 20,000 g/mol.[32] These ionic polymers were soluble in methanol.

PROPERTIES AND APPLICATIONS

Linear epoxide–amine addition polymers are increasingly more interesting because they combine advantageous mechanical, optical, and thermal properties of commonly known crosslinked epoxide–amine networks with thermoplasticity and solubility. Currently, linear epoxide–amine addition polymers are used in the optical industry and in dentistry.[28–31,33,34] Further investigations concern their use in electrophotographic recording systems, membranes, materials for second-order nonlinear optical processes and liquid crystalline polymers.[2,12–16,35] To obtain liquid-crystalline side-chain polymers, mesogenic amines were used.[12–15] The liquid-crystalline phase of the addition polymer was tentatively identified as nematic from optical microscopy.

SUMMARY

The principle of synthesizing linear epoxide–amine addition polymers is realized, provided that the monomers are used in stoichiometric amounts, and selected aromatic, araliphatic, cycloaliphatic, or substituted aliphatic primary monoamines, disecondary diamines either *N-N′*-dibenzyl-substituted or aromatic, hydrazides, or α-amino carboxylic esters are applied. Furthermore, the reaction temperature must be in the range of the glass transition temperature of the corresponding addition step (20–130°C).

New applications of linear epoxide–amine addition polymers are possible because they combine good mechanical, electrical, and optical properties with solubility and thermoplasticity. Interesting applications for epoxide–amine addition polymers are in the development of new photoconductors, optical adhesives, dental filling materials, and in liquid-crystalline polymers and polymers having NLO properties.

REFERENCES

1. Schlack, P. DR Patent 676117; *Chem. Abstr.* **1939**, *33*, 6619.
2. Renner, A.; Michaelis, P. *J. Polym. Sci., Polym. Chem. Ed.* **1984**, *22*, 249.
3. Vinnik, M.-I.; Vinnik, R. M.; Grinev, V. E. et al. *Russ. Chem. Bull.* **1993**, *42*, 843.
4. Iwakura, Y.; Matsuzaki, K. *Kobunshi Kagagu* **1960**, *17*, 703; *Chem. Abstr.* **1961**, *55*, 24080f; *Chem. Zbl.* **1964**, *135*, 19-0942.
5. Canonica, L.; Cappuccio, V.; Caldo, C. et al. *Makromol. Chem.* **1968**, *116*, 158.
6. Jones, F. B. U.S. Patent 3 554 956; *Chem. Abstract* **1971**, *74*, 77005r.
7. Bormann, S.; Brossas, J.; Franta, E. et al. *Tetrahedron (London)* **1975**, *31*, 2791.
8. Gramain, P.; Frére, Y. *Macromolecules* **1979**, *12*, 1038.
9. Gramain, P.; Frére, Y. *Polymer* **1980**, *21*, 921.
10. Johnson, W. E.; Henson, W. A. U.S. Patent 3 317 471; *Chem. Abstract* **1967**, *67*, 3351j.
11. Griffith, J.R. U.S. Patent 3 592 946; *Chem. Abstract* **1971**, *75*, 130456n.
12. West, J. L. U.S. Patent 5 093 471, 1989.
13. Zentel, R. *Polymer* **1992**, *33*, 4040.
14. Zentel, R.; Jungbauer, D.; Twieg, R. J. et al. *Makromol. Chem.* **1993**, *194*, 859.
15. Chien, L.-C.; Lin, C.; Fredley, D. S. et al. *Macromolecules* **1992**, *25*, 133.
16. Silvis, H.-C.; White, J. E.; Crain, S. P. *J. Appl. Polym. Sci.* **1992**, *44*, 1751.
17. Hiltpold, R.; Fisch, W. German Patent 1 137 863; *Chem. Abstract* **1963**, *58*, 7000e.
18. Huang, J. C.; Gu, F. *Adv. Polym. Technol.* **1992**, *11*, 213.
19. Hörhold, H.-H.; Klee, J.; Bellstedt, K. *Z. Chem.* **1982**, *22*, 166.
20. Hörhold, H.-H.; Klee, J.; Flammersheim, H.-J. *Makromol. Chem. Rapid Commun.* **1981**, *2*, 113.
21. Hörhold, H.-H.; Opfermann, J.; Raabe, D. et al. *Isvest. Akad. Nauk Kazachchskoj SSR, Ser. Chim.* **1981**, *(6)*, 23.
22. Klee, J.; Hörhold, H.-H.; Schütz, H. *Acta Polym.* **1987**, *38*, 293.
23. Klee, J.; Hörhold, H.-H.; Schütz, H. et al. *Angew. Makromol. Chem.* **1989**, *170*, 145.
24. Klee, J. Universität Jena, 1990.
25. Hörhold, H.-H.; Bellstedt, K. DD Patent 226451, 1985.
26. Klee, J.; Hörhold, H.-H.; Flammersheim, H.-J. *Angew. Makromol. Chem.* **1990**, *178*, 63.
27. Hörhold, H.-H.; Klemm, D.; Klee, J. *J. Prakt. Chem.* **1980**, *322*, 445.
28. Hörhold, H.-H.; Klemm, D.; Bellstedt, K. et al. DE-OS Patent 3 010 247, 1980.
29. Hörhold, H.-H.; Klemm, D.; Bellstedt, K. et al. DD Patent 141677, 1980.

30. Hörhold, H.-H.; Klemm, D.; Bellstedt, K. et al. British Patent 2 045 269, 1980.
31. Klee, J. E. EP 0673 673, 1996.
32. Nippon Point Co. Japanese Patent 80135138; *Chem. Abstract* **1981**, *94*, 104104u.
33. Hörhold, H.-H.; Klemm, D.; Bellstedt, K. et al. U.S. Patent 4308085, 1981.
34. Hörhold, H.-H.; Klemm, D.; Bellstedt, K. et al. CS Patent 227363, 1984; *Chem. Abstract* **1981**, *94*, 157803e.
35. Mandal, B. K.; Jeng, R. J.; Kumar, J. et al. *Makromol. Chem., Rapid Commun.* **1991**, *12*, 607.

EPOXIDIZED NATURAL RUBBER

Ian R. Gelling
Malaysian Rubber Producers' Research Association
Tun Abdul Razak Laboratory

Epoxidized natural rubber (ENR) is a derivative of natural rubber produced by chemical modification. It was not until the mid 1980s that pure samples of ENRs were prepared and their properties fully recorded.[1–3]

PREPARATION

Natural rubber (NR)[CAS.9006.04.06] is a 1,4-polyisoprene with the molecules having an all *cis* configuration except for a sequence of three chain-end *trans* units that are believed to be the remnants of starter molecules involved in the rubber's biosynthesis.[4] The chemistry of epoxidizing unsaturated compounds employing peroxyacids is well-documented in the literature and has been applied to NR to produce a range of ENRs [CAS.138009.59.3].[5,6] Currently, 50 mol % epoxidized natural rubber (ENR-50) is commercially available, and other epoxidation levels can be manufactured on request.[7] The process employs an *in situ* technique based on hydrogen peroxide and formic acid to epoxidize NR latex.[8]

Epoxidation reactions are stereospecific, and thus, ENR is a *cis*-1,4-polyisoprene (**Figure 1**) with epoxide groups randomly situated along the polymer backbone.[9]

APPLICATIONS

ENRs have a unique set of properties because of their ability to strain crystallize and because of the increase in the glass transition temperature and solubility parameter with epoxidation. These properties have been used for various rubber goods, such as high-damping engineering mounts, acoustic devices, and adhesives.

The greatest potential for ENRs is in tires, although this use has not yet progressed for commercialization. The profile of tire treads based on 25 mol % epoxidized natural rubber show high wet traction and low rolling resistance.[1,11] The partial replacement of carbon black by silica in these tread compounds further reduces rolling resistance and hence, improves fuel economy. ENRs have also been investigated as replacements for butyls in tire inner tubes and liners with excellent results.[12] This work was performed in Malaysia, but a stiffening of ENR–50-based inner liners has been observed in regions with low ambient temperatures because of ENR–50's relatively high glass transition temperature.

The epoxide groups in ENRs have also been investigated as routes to new crosslinking systems and rubber-bound antidegradants and as intermediates for further chemical modification.[10–12]

REFERENCES

1. Baker, C. S. L.; Gelling, I. R.; Newell, R. *Rubber Chem. Technol.* **1985**, *58*, 67.
2. Gelling, I. R. *Rubber Chem. Technol.* **1985**, *58*, 86.
3. Gelling, I. R.; Morrison, N. J. *Rubber Chem. Technol.* **1985**, *58*, 243.
4. Tanaka, Y. *Int. Rubber Conf., Harrogate, UK.* 4A/1, 1987.
5. Swern, D. *Organic Peroxides*; Wiley Interscience: New York, 1971; Vol. 2, Chapter V.
6. Trahanovsky, N. S. *Oxidation in Organic Chemistry*; Academic : New York, 1978, Vol. 5, Part C, Chapter III.
7. Kumpulan Guthrie Berhad, Jalan Sungai Ujong, 70200 Seremban, Malaysia.
8. British Patent 2 113 692, 1982.
9. Witnauer, L. P.; Swern, D. *J. Amer. Chem. Soc.* **1950**, *12*, 3364.
10. Teik, L. C. *Int. Rubber Conf., Kuala Lumpur* 368, 1985.
11. Baker, C. S. L.; Gelling, I. R.; Palmer, J. *Int. Rubber Conf., Kuala Lumpur* 336, 1985.
12. Loh, P. C.; See Toh, M. S. *Int. Rubber Conf., Kuala Lumpur* 312, 1985.

EPOXIES, RUBBER-MODIFIED

Roberta J. J. Williams
Institute of Materials Science and Technology
University of Mar del Plata and National Research Council

Epoxies constitute a most important class of thermosetting polymers used in adhesives, matrices for fiber-reinforced composites, and coatings. Epoxies have excellent mechanical and thermal properties such as high strength, elastic modulus, and glass transition temperature (T_g). These properties result from the chemical nature of the starting monomers and the high crosslink density of the final materials. However, this structure also leads to low resistance to crack initiation and propagation so that methods to increase toughness are needed. When rubbery domains in the micrometer range are randomly dispersed in the epoxy matrix, the fracture energy can be greatly increased. Common toughening mechanisms are matrix shear-yielding, particle cavitation, and rubber bridging.[1] There are

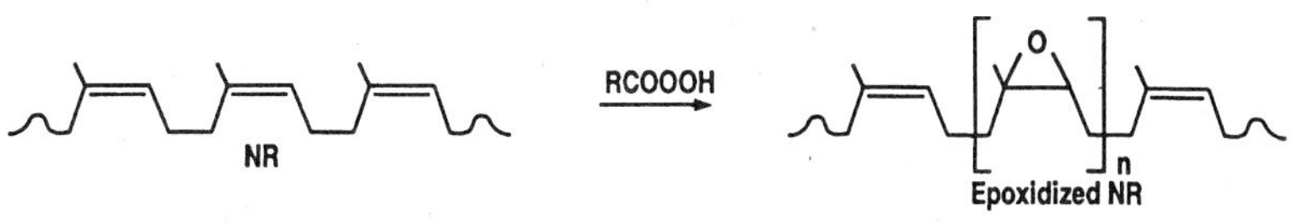

FIGURE 1. Epoxidation of natural rubber with peroxycarboxylic acid.

two main procedures for dispersing rubbery particles: phase separation during polymerization of an initial homogeneous solution (reaction-induced phase separation), and a two-phase initial formulation by either dispersing elastomeric particles (usually the core shell type) in the mixture of monomers or by carrying out an emulsion polymerization in the epoxy monomer.

The phase separation method, which is older, started with the work of McGarry and his colleagues.[2–4] Epoxies based on diglycidylether of bisphenol A (DGEBA) were toughened by adding low molecular weight, liquid, carboxyl-terminated poly(butadiene-acrylonitrile) copolymers (CTBN). Rubbery domains precipitated *in situ* during cure yielding toughened epoxy materials. Numerous investigations have been devoted to this area in the last two decades, and several books give a global vision of the field.[5–7] Although poly(butadiene-acrylonitrile) copolymers with different terminal groups (i.e., epoxy, carboxyl, amine, and non-functional) still constitute the most important group of modifiers used in propietary rubber-modified epoxies, several other modifiers also have been proposed: polysiloxanes, fluoroelastomers, carboxyl-terminated acrylate oligomers, polyepichlorohydrin rubbers, and vegetal oils.[8–15] In the 1980s, a novel approach to toughen epoxies, consisting of replacing the rubber with an engineering thermoplastic, was developed.[16]

Preparing rubber-modified epoxies from reaction-induced phase separation has the advantage of processing initial homogeneous solutions with relatively low viscosity. The main disadvantage is that the morphologies generated (i.e., volume fraction of dispersed phase, distribution of particle sizes, concentration, and composition of particles, etc.) and the resulting properties depend on the initial amount of modifier, cure cycle, and the presence of other additives in the formulation.[17–21] However, this disadvantage can be an advantage if the phase separation process is effectively controlled. For example, bimodal (eventually multimodal) particle size distributions or "sandwich" structures may be generated using an appropriate cure cycle with different temperature vs. conversion trajectories.[14,22,23]

The second technique for preparing rubber-modified epoxies consists of introducing or generating performed particles in the initial epoxy formulation. Systems based on dispersed acrylic elastomers and particles consisting of a rubbery core embedded in a compatibilizing thermoplastic shell have been reported.[24–31] The dispersion is stabilized by adjusting the shell composition or by copolymerizing the acrylic monomers with an adequately functionalized epoxy (i.e., a vinyl ester obtained by reacting the epoxy monomer with methacrylic acid).[25,26,28] This procedure's main advantage is the initial control of the dispersed phase's volume fraction and composition. Also important is the potentially accurate studies of the interfacial adhesion's influence on resulting properties.[28] Disadvantages are the relatively high initial viscosity of the dispersions and the potential for particle agglomeration or macroscopic phase separation during storage or processing. Some of these systems reportedly perform poorly in the presence of high inorganic filler contents.[32]

APPLICATIONS

Rubber-modified epoxies are used in formulations for adhesives, fiber-reinforced composites, and coatings among others.

For adhesives formulated with toughened epoxies, a strong dependence of adhesive fracture behavior on bond thickness was observed.[33,34] Optimum bond thickness depends on the temperature and loading rate. For composites, the interlaminar fracture energy increases with the resin fracture energy, the relative effect being more important for low values of toughness.[1,35] Elastomer-modified epoxies are not applied in high-performance structural composites, which require high values of T_g and modulus. In this case, modification with engineering thermoplastics constitutes a better approach. Hybrid-particulate composites based on an epoxy-matrix, a rubber modifier, glass beads, or other fillers offer an interesting alternative to obtain convenient values for elastic modulus and yield strength, combined with a significant increment of fracture toughness. Adhesives based on a diglycidylether had significantly improved mechanical properties when formulated as hybrid materials.[36]

REFERENCES

1. Huang, Y.; Hunston, D. L.; Kinloch, A. J. et al. *Toughened Plastics I: Science and Engineering*; Advances in Chemistry 233; American Chemical Society: Washington, DC, 1993; Chapter 1.
2. McGarry, F. J.; Willner, A. M. *Org. Coat. Plast. Chem.* **1968**, *28*, 512.
3. Sultan, J. N.; Laible, R. C.; McGarry, F. J. *J. Appl. Polym. Sci.* **1971**, *6*, 127.
4. Sultan, J. N.; McGarry, F. J. *Polym. Eng. Sci.* **1973**, *13*, 29.
5. *Rubber-Modified Thermoset Resins*; Riew, C. K.; Gillham, J. K., Eds.; Advances in Chemistry 208; American Chemical Society: Washington, DC, 1984.
6. *Rubber-Toughened Plastics*; Riew, C. K., Ed.; Advances in Chemistry 222; American Chemical Society: Washington, DC, 1989.
7. *Toughened Plastics I: Science and Engineering*; Riew, C. K.; Kinloch, A. I., Eds.; Advances in Chemistry 233; American Chemical Society: Washington, DC, 1993.
8. Yorkgitis, E. M.; Tran, C.; Eiss, Jr., N. S. et al. *Rubber-Modified Thermoset Resins*; Advances in Chemistry 208; American Chemical Society: Washington, DC, 1984; Chapter 10.
9. Mijovic, J.; Pearce, E. M.; Foun, C. C. *Rubber-Modified Thermoset Resins*; Advances in Chemistry 208; American Chemical Society: Washington, DC, 1984; Chapter 18.
10. Banthia, A. K.; Chaturvedi, P. N.; Jha, V. et al. *Rubber-Toughened Plastics*; Advances in Chemistry 222; American Chemical Society: Washington, DC, 1989; Chapter 16.
11. Jackson, M. B.; Edmond, L. N.; Varley, R. J. et al. *J. Appl. Polym. Sci.* **1993**, *48*, 1259.
12. Ruseckaite, R. A.; Williams, R. J. J. *Polym. Int.* **1993**, *30*, 11.
13. Ruseckaite, R. A.; Hu, L.; Riccardi, C. C. et al. *Polym. Int.* **1993**, *30*, 287.
14. Ruseckaite, R. A.; Fasce, D. P.; Williams, R. J. J. *Polym. Int.* **1993**, *30*, 297.
15. Frischinger, I.; Dirlikov, S. *Toughened Plastic I: Science and Engineering*; Advances in Chemistry 233; American Chemical Society: Washington, DC, 1993; Chapter 19.
16. Pearson, R. A. *Toughened Plastics I: Science and Engineering*; Advances in Chemistry 233; American Chemical Society: Washington, DC, 1993; Chapter 17.
17. Manzione, L. T.; Gillham, J. K.; McPherson, C. A. *J. Appl. Polym. Sci.* **1981**, *26*, 889.
18. Manzione, L. T.; Gillham, J. K.; McPherson, C. A. *J. Appl. Polym. Sci.* **1981**, *26*, 907.
19. Williams, R. J. J.; Borrajo, J.; Adabbo, H. E. et al. *Rubber-Modified Thermoset Resins*; Advances in Chemistry 208; American Chemical Society: Washington, DC, 1984; Chapter 13.
20. Montarnal, S.; Pascault, J. P.; Sautereau, H. *Rubber-Toughened Plastics*; Advances in Chemistry 222; American Chemical Society: Washington, DC, 1989; Chapter 8.

21. Verchere, D.; Sautereau, H.; Pascault, J. P. et al. *Toughened Plastics I: Science and Engineering*; Advances in Chemistry 233; American Chemical Society: Washington, DC, 1993; Chapter 14.
22. Chen, T. K.; Jan, Y. H. *J. Mater. Sci.* **1992**, *27*, 111.
23. Fang, D. P.; Frontini, P. M.; Riccardi, C. C. et al. *Polym. Eng. Sci.* **1995**, *35*, 1359.
24. Hoffman, D. K.; Arends, C. B. U.S. Patent 4 708 996, 1987.
25. Ho, T. H.; Wang, C. S. *J. Appl. Polym Sci.* **1993**, *50*, 477.
26. Hoffman, D. K.; Ortiz, C.; Hunston, D. L. et al. *Polym. Mater. Sci. Eng.* **1994**, *70*, 7.
27. Ortiz, C.; McDonough, W.; Hunston, D. L. et al. *Polym. Mater. Sci. Eng.* **1994**, *70*, 9.
28. Sue, H. J.; Garcia-Meitin, E. I.; Pickelman, D. M. et al. *Toughened Plastics I: Science and Engineering*; Advances in Chemistry 233; American Chemical Society: Washington, DC, 1993; Chapter 10.
29. Riew, C. K.; Siebert, A. R.; Smith, R. W. et al. *Polym. Mater. Sci. Eng.* **1994**, *70*, 5.
30. Qian, J. Y.; Pearson, R. A.; Dimonie, V. L. et al. *Polym. Mater. Sci. Eng.* **1994**, *70*, 17.
31. Maazouz, A.; Sautereau, H.; Gerard, J. F. *Polym. Bull.* **1994**, *33*, 67.
32. Mülhaupt, R.; Buchholz, U. *Polym. Mater. Sci. Eng.* **1994**, *70*, 4.
33. Bascom, W. D.; Cottington, R. L.; Jones, R. L. et al. *J. Appl. Polym. Sci.* **1975**, *19*, 2545.
34. Bascom, W. D.; Cottington, R. L. *J. Adhes.* **1976**, *7*, 333.
35. Hunston, D. L. *Compos. Technol. Rev.* **1984**, *6*, 176.
36. Maazouz, A.; Sautereau, H.; Gerard, J. F. *J. Appl. Polym. Sci.* **1993**, *50*, 615.

EPOXY ACRYLATE-BASED RESINS*

Peter J. Burchill and P. J. Pearce
Aeronautical and Maritime Research Laboratory

Epoxy acrylate-based resins or vinyl esters, as they are more commonly known, have various industrial applications from primary structures and containers to adhesives and dentistry. These resins were introduced commercially in 1965, and since then, their use has been firmly established.[1] Because epoxy acrylates have better chemical resistance than the cheaper polyester resins, especially toward water and most aqueous solutions, the use of polyester resins is being challenged. Similarly, the greater control over cure rate and conditions has enabled epoxy resin markets to be penetrated. Developments in the chemistry and applications of epoxy acrylate-based resins have included attempts to improve their toughness.

SYNTHESIS

Epoxy acrylate resins are produced by a reaction between epoxy compounds and acrylic or methacrylic acid. The most important commercial member of these chemicals (**Figure 1**) is made from the diglycidylether of bisphenol-A (DGEBA) and methacrylic acid.[2]

The simple synthesis (Figure 1) enables considerable structural variation of the final product and allows much latitude for adjusting various conflicting properties, from the liquid resin's viscosity to the cured resin's toughness and heat distortion temperature.[3–7] For example, a reactive liquid polymer (RLP) may replace some of the methacrylic acid.

*The Commonwealth of Australia holds patent rights.

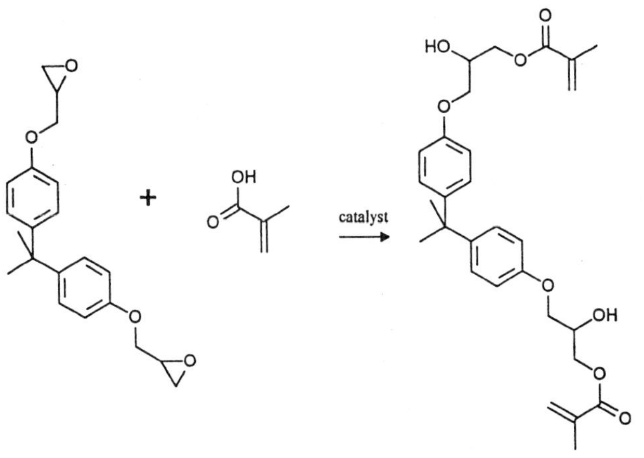

FIGURE 1. Reaction of a bisphenol-A glycidylether with methacrylic acid. Other reactants that may be present include bisphenol-A, maleic anhydride, and carboxy-terminated butadiene acrylonitrile copolymer.

CURE AND PROPERTIES

Epoxy acrylates are polymerized by a free-radical reaction, which a thermal or catalytic decomposition of peroxides or a decomposition of a photo initiator usually initiates. The most commonly used epoxy acrylate, monomer 1 in Figure 1, gives a brittle, highly crosslinked network with a glass transition temperature of about 220°C and is generally used diluted with other vinyl monomers such as styrene or other acrylates or methacrylates.[8]

Toughening

Not as much attention has been paid to improving the toughness of epoxy acrylate-based resins compared with epoxy resins.[9–12] Nevertheless, toughness has improved for DGEBA-based resins by incorporating a carboxy terminated RLP in the synthesis.

Adhesive properties of modified epoxy acrylate-styrene copolymers also improve when the resins are blended with RLPs.[13]

APPLICATIONS

DGEBA methacrylate-based resins have excellent chemical resistance compared with polyester resins and low water absorption. These properties and the flexibility allowed in the resin's cure underlie many of its applications, which range from electro-refining tanks to swimming pools. Novolac epoxy-based resins with their higher glass transition temperature are used in chemical storage vessels and chemical reactors. Modified DGEBA resins are used wherever improved adhesion, wear resistance, and fracture toughness are required.

An epoxy acrylate-based carbon fiber composite has been used to reinforce-prone superstructures in ships.

REFERENCES

1. May, C. A.; Burge, R. E.; Christie, S. H. *Soc. Plastics Engrs. J.* **1965**, *21*(9), 1106.
2. Doyle, T. E.; Fekete, F.; Keenan, P. J. et al. U.S. Patent 3 317 465, 1967.

3. White, W. D.; Puckett, P. M.; Blankenship, L. T. et al. European Patent 315 101, 1989.
4. Najvar, D. J. U.S. Patent 3 892 819, 1975.
5. Waters, W. D. U.S. Patent 3 928 491, 1975.
6. Takano, H.; Kunitomi, T.; Okada, S. et al. U.K. Patent 2 158 830, 1985.
7. White, M. N.; Puckett, P. M.; Blankenship, L. T. European Patent 315 086, 1989.
8. Yilgor, I.; Yilgor, E.; Banthia, A. K. et al. *Polym. Comp.* **1983**, *4*, 120.
9. Pham, S.; Burchill, P. J. *Polymer* **1995**, *36*, 3279.
10. Grabovac, I.; Pearce, P. J.; Camilleri, A. W. *IUPAC International Symposium Polymer* **1991**, *Vol. 91*, p 202.
11. Baker, A. W.; Martin, P. H. U.S. Patent 4 824 919, 1989.
12. Siebert, A. R.; Guiley, C. D.; Egan, D. R. SPI 47th Annual Conference Composites Institute, 1992, 17-C, 1.
13. Peare, P. J.; Siebert, A. R.; Egan, D. R. et al. *J. Adhesion* **1995**, *49*, 245.

EPOXY-IMIDE AND PHOSPHORYLATED EPOXY-IMIDE POLYMERS

Wei-Kuo Chin*
Department of Chemical Engineering
National Tsing Hua University

Min-Da Shau
Department of Applied Chemistry
Chia Nan Junior College of Pharmacy

Research on the structural modifications of epoxy to enhance its thermal and oxidative stability has recently received great attention. Polymers such as polyimides, which contain aromatic or heterocyclic groups, show good thermal stability, and their thermal decomposition temperatures are ~ 400–500°C. Therefore, using imide groups to modify the structure of epoxy has become the focus of research.[1–20] In these studies, imide-modified epoxy polymers can be found as either thermoplastic epoxy–imide polymers[1–7] or thermosetting epoxy–imide polymers.[8–20]

THERMOPLASTIC EPOXY–IMIDE POLYMERS

In synthesized thermoplastic epoxy–imide polymers, the imide groups are introduced into the structure in either the backbone or the side chain (3, 4). Generally, epoxy–imide polymers that contain aromatic and/or heterocyclic rings have good thermal properties. The maximum decomposition temperatures of these polymers in air are ~ 400°C.

THERMOSETTING EPOXY–IMIDE POLYMERS

The imide-modified thermosetting epoxy systems can be grouped into three types: those created by modifying the backbone structure of an epoxy with an imide group; by using a curing agent that would contain imide groups; or by a combination of the first two.

The thermal properties of polymers are improved through the introduction of imide groups into the structure of epoxy resins and curing agents. The thermal resistance and oxidative stability of the crosslinked epoxy–imide polymers depend on several factors; that is, type of epoxy resin or epoxy–imide resin used, type of curing agent or imide-modified curing agent used, molar ratio of resin and curing agent, curing conditions, and resultant crosslinking density.

*Author to whom correspondence should be addressed.

PHOSPHORYLATED EPOXY–IMIDE POLYMERS

Epoxy has been found to have extensive industrial applications. However, the common epoxy system cannot satisfy the flame-resistance requirements for these field applications. The imide group has been proved able to significantly enhance the thermal resistance of epoxy, whereas its improvement in flame resistance is relatively low. In order to improve the flame resistance of an epoxy–imide polymer, we use phosphorus groups to modify the structure of the diamine curing agent and the epoxy–imide resin.[21]

APPLICATIONS

In field applications such as aerospace and microelectronics that require the material to sustain both a high service temperature and thermal oxidative stability, the imide-modified epoxy resins and curing agents have great potential. Epoxy–imide resins are used in composites in structural parts; they are also used as organic insulators, as padding and encapsulating compounds, and as adhesives.

REFERENCES

1. Martínez, P. A.; Cádiz, V.; Mantecón, A.; Serra, A. *Angew. Makromol. Chem.* **1985**, *133*, 97.
2. Shau, M-D.; Chin, W-K. *J. Polym. Sci., Polym. Chem. Ed.* **1993**, *31*, 1653.
3. Serra, A.; Cádiz, V.; Martínez, P. A.; Mantecón, A. *Angew. Makromol. Chem.* **1986**, *138*, 185.
4. Serra, A.; Cádiz, V.; Martínez, P. A.; Mantecón, A. *Angew. Makromol. Chem.* **1986**, *140*, 113.
5. Mantecón, A.; Cádiz, V.; Serra, A.; Martínez, P. A. *Angew. Makromol. Chem.* **1987**, *148*, 149.
6. Serra, A.; Cádiz, V.; Mantecón, A. *Angew. Makromol. Chem.* **1987**, *155*, 93.
7. Sasaki, S.; Hasuda, Y. *J. Polym. Sci., Polym. Letters Ed.* **1987**, *25*, 377.
8. Ichino, T.; Hasuda, Y. *J. Appl. Polym. Sci.* **1987**, *34*, 1667.
9. Martínez, P. A.; Cádiz, V.; Serra, A.; Mantecón, A. *Angew. Makromol. Chem.* **1985**, *136*, 159.
10. Mantecón, A.; Cádiz, V.; Serra, A.; Martínez, P. A. *Eur. Polym. J.* **1987**, *23*, 481.
11. Martínez, P. A.; Cádiz, V.; Mantecón, A.; Serra, A. *Eur. Polym. J.* **1987**, *23*, 961.
12. Soler, H.; Cádiz, V.; Serra, A. *Angew. Makromol. Chem.* **1987**, *152*, 55.
13. Mantecón, A.; Cádiz, V.; Serra, A.; Martínez, P. A. *Angew. Makromol. Chem.* **1988**, *156*, 37.
14. Takeda, S.; Hakiuchi, H. *J. Appl. Polym. Sci.* **1988**, *35*, 1351.
15. Rao, B. S. *J. Polym. Sci., Polym. Letters Ed.* **1988**, *26*, 3.
16. Rao, B. S. *J. Polym. Sci., Polym. Letters Ed.* **1989**, *27*, 133.
17. Bucknall, C. B.; Gibert, A. H. *Polymer* **1989**, *30*, 213.
18. Ochi, M.; Shiba, J.; Takeuchi, H.; Yoshizumi, M.; Shimbo, M. *Polymer* **1989**, *30*, 1079.

19. Misaki, T.; Hirohata, T. *J. Appl. Polym. Sci.* **1989**, *37*, 2617.
20. Chin, W-K.; Hu, J. J.; Shau, M. D., 4E-07, Nov., International Union of Pure and Applied Chemistry (IUPAC) Polymer Symposium Preprints, Taipei, 1994, Taipei, 1994.
21. Shau, M-D.; Ph.D. Thesis, National Tsing Hua University, Hsinchu, Taiwan, Republic of China, 1992.

EPOXY RESIN-GLASS INTERPHASES (Optical Properties)

Michael Hess*
Department of Physical Chemistry
Gerhard-Mercator University
and
Department of Material Science
University of North Texas

Robert Kosfeld
Department of Physical Chemistry
Gerhard-Mercator University

The behavior of polymer composites strongly depends on the properties of the interphase between the polymer and the reinforcing material. Provided there is a perfect adhesion between, for example, glass fibers and an epoxy resin, optimum composite solidity can be achieved. Applied forces can be completely exchanged between the resin matrix and the reinforcing phase, using the whole stability range of both the fiber and the polymer. However, certain time-dependent factors such as aging, relaxation processes, mechanical stress, temperature effects, and chemical reactions generally affect the properties of the interphase. Interphase conditions are governed by complex interactions of adhesion, adsorption, wetting, viscosity, friction, and other physical effects, as well as by chemical interaction.

Surfaces and interphases may be characterized by analysis of contact angles[1,2] or by measurements of electrical conductivity along the fibers.[3] These investigations, however, are time consuming and not always adequately sensitive.[2] Monitoring the optical transmission (transparency) during the curing process, or even thereafter, may serve as a fast and sensitive alternative, but it is restricted to samples with a reasonable optical transparency. The principle introduced by Ehrenstein[4] and by Krolikowski and co-workers[2,5] and applied to the analysis of glass-fiber-filled polyesters has been developed and modified.[6]

MATERIALS AND METHODS

The glass used was an alumino–boro–silicate weaving provided by Fa. Interglas, Ulm, Germany. The surfaces of the fibers were finished with a Cr(II)-modified methacryl complex, covered with textile size, to obtain a surface with minor quality, or the surface was the free glass, obtained after thermally removing the textile size. The textile size was a hardly defined mixture of starches, oils, and so forth, which is necessary to process the fibers.

*Author to whom correspondence should be addressed.

RESULTS AND DISCUSSION

Transmission During Impregnation and Wetting

Three main factors control the transparency of a composite having zero or negligible absorbance at the observation wavelength:

- the difference in refractive index of polymer and fiber;
- inhomogeneities in the polymer matrix; and
- inhomogeneities in the interphase polymer/glass.

When a glass fabric is dipped into a liquid polymer, impregnation and wetting[7,8] take place. In impregnation, the liquid is forced to penetrate the weaving because of its weight, the use of rollers, or other external forces. In wetting, the glass/air interphase is replaced by the liquid resin, a process that is governed by the surface tension of the components. A good wetting is an important precondition to achieving good resin-glass adhesion and henceforth a high-performance composite.

Because there are large differences in the refractive index of air ($n_D^{20} = 1.003$) and the commonly used glass ($n_{D,glass}^{20} \approx 1.552-1.557$), the system shows a quite low transparency. This is also true for air and resin ($n_{D,resin}^{20} \approx 1.537-1.549$). After a good wetting of the glass by the liquid resin, the difference between the refractive indices of these two phases is small; therefore, during wetting an increase of transparency is expected.

Transmission During the Crosslinking Process

In contrast to neat resins, the behavior of a resin that contains a curing agent and in which the crosslinking reaction starts simultaneously with mixing the components or at least when the mixing is poured into the cuvette, is completely different. In this case, impregnation and wetting are governed by the influence of the progress of the chemical reaction.

In curing systems, the changing refractive index during the crosslinking reaction plays an important role. The measurements could not be followed beyond the gel-point. During the reaction, the refractive index of the curing resin exceeds the corresponding value of the glass. As long as the refractivity of the resin is lower than that of the glass, transmission of the composite is low but increasing. Maximal transparency is reached when the refractive indices are equal. Further growth of refractive index of the resin again results in a decreasing transparency of the composite.

Effects of Surface Treatment

The best transparency—the highest value of transmission—was observed in composites with finished glass surfaces, supporting the interpretation that a finished surface shows a strong resin/glass interaction giving rise to only a small number of defects that act as scattering centers. The lowest transparency is observed if there is textile size on the surface, whereas a clean glass surface from which any surface modifier has been removed thermally obtains an intermediate position. A finished surface shows the best results.

Nonstoichiometric Systems

The course of the transmission depends strongly on the stoichiometry of the reaction. It is thus possible to shift transparency to higher final values by increasing the amount of hardener within a certain range.

Increasing the amount of curing agent above the stoichiometric ratio results in a highly branched matrix with lower crosslink density and shorter polymer segments; the matrix becomes softer, and smaller segments have a better adhesion to glass surfaces because of the reduced stress in these interphases.[1] The result is a more homogeneous interphase with a smaller amount of scattering centers. This is demonstrated by a higher transparency of the composite.

Influence of Curing Temperature

Monitoring the optical transmission of a curing epoxy-glass composite is clearly also suitable to determine the end of the curing process. Temperature variation strongly influences the time course of transparency; the initial slope and the time of the maximum in transparency are especially affected. The adhesion process occurs more rapidly at higher temperatures. After cooling down to room temperature, a comparison of samples cured at different temperature shows that the final transparency decreases with increasing curing temperature (or rate).

ACKNOWLEDGMENTS

The investigations have been supported by a grant from the Arbeitsgemeinschaft Industrieller Forschungsförderer (AIF), Cologne, Germany, grant 7020. The authors thank Th. Marzi and U. Schroeder for carrying out the experiments; The Bakelite, A.G., Duisburg, Germany, for providing epoxy systems and hardeners; and Fa. Interglas, Ulm, Germany, for providing glass weavings.

REFERENCES

1. Selden, P. H. *Glasfaserverstärkte Kunststoffe*; Springer: Berlin, 1967.
2. Krolikowski, W.; Izbicka, J.; Miklaszewics, J. *Conference on Reinforced Plastics*; Karlovy Vary: 1972; Conference Proceedings, p 24.
3. Anderson, A. C.; Healey, J. H. *13th ATC-SPJ, Sect. 3-B* **1958**.
4. Ehrenstein, G. H., Ph.D. Thesis, Hannover, 1967.
5. Willax, H. O.; Ehrenstein, G. W. *Kunststoffe-German Plastics* **1981**, *62*, 887.
6. Krolikowski, W.; Czech, Z. *Kunststoffe-German Plastics* **1981**, *71*, 442.
7. Hansen, D., Ph.D. Thesis, Duisburg, 1985.
8. de Gennes, P. G. *Adv. in Colloid and Interface Sci.* **1989**, *27*, 189.

Epoxy Resins

See: *Epoxide-Amine Addition Polymers, Linear*
Epoxidized Natural Rubber
Epoxies Rubber-Modified
Epoxy Acrylate-Based Resins
Epoxy-Imide and Phosphorylated Epoxy-Imide Polymers
Epoxy Resin-Glass Interphases (Optical Properties)
Epoxy Resins (Overview)
Epoxy Resins (Curing Reactions)
Epoxy Resins (High Performance Composite Applications)
Epoxy Resins (Microwave Processing)
Epoxy Resins (Modification with Engineering Plastics)
Ethylenimine-Modified Polymers
Fluoropolymers, Acrylates, and Epoxies (With Low Dielectric Constants)
Magnetic Field Processing
Novolak-type Epoxy Resins
Ordered Epoxy Networks
Polysulfides (Use as Modifiers in Epoxy Systems)
Polyurethanes, Blocked Copolymer (Reactive Modifiers for Epoxy Resins)
Powder Coatings (Overview)
Reaction Injection Molding (RIM) Materials
Reactive Liquid Rubbers (Epoxy Toughening Agents)
Resins (Applications in Coatings Industry)
Resins and Paints (Analytical Pyrolysis)
Thermosets (Main Chain Liquid Crystalline Polymers)
Unsaturated Polyester Resins (Toughened with Liquid Rubbers)

EPOXY RESINS (Overview)

M. S. Bhatnagar
Retired

Epoxy resins are polyether resins containing more than one epoxy group capable of being converted into the thermoset form. These resins, on curing, do not create volatile products in spite of the presence of a volatile solvent.

The simplest epoxy resin is prepared by the reaction of bisphenol A (BPA) (80-05-7) with epichlorohydrin (ECH) (106-89-8) (**Scheme I**). The value of *n* varies from 0 to 25. This determines the end-use applications of the resin.

Apart from innumerable patents, several books,[1–4] review articles,[5–18] and technical bulletins have been published on various aspects of epoxy resins.

Applications for epoxy resins are extensive: adhesives, bonding, construction materials (flooring, paving, and aggregates), composites, laminates, coatings, molding, and textile finishing. They have recently found uses in the air- and spacecraft industries.

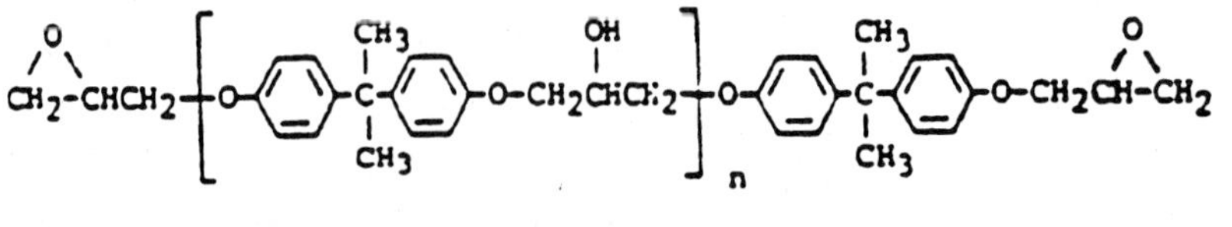

I

APPLICATIONS

Foam

Epoxy resins are used to form rigid, lightweight, foamy structures with good insulation properties.

Adhesives

The versatile properties of epoxy resins make them valuable as adhesives in civilian and military applications.[19–21]

Construction

Epoxy resins are now used as binders in materials for construction.[22]

MOLDING

Molding can be broadly classified as simple molding and embedding processes.[23,24]

Epoxy Powders

The most commonly used epoxy powders are based on B-stage resins with aromatic amines or methane diamine. Dicyanodiamide or guanidine are also used, and inert fillers include silica, clay, and calcium carbonate. These resins are also used in foundry molding.

Composites and Laminates

Composites and laminates are materials produced by reinforcing the polymers with continuous fibers. These can be fabricated into structural materials. This industry consumes 27.6% of the epoxy resins produced.[25,26]

Laminates

Laminated epoxy insulations are largely used as sheets, rods, and tubes. Laminated epoxy structures are used in building construction for concrete molds, honeycomb cores, facing for foams, wood, metal assemblies, and reinforced pipes.

Composites

The fibers that dominate the field of advanced composites are carbon, graphite, glass, aluminum, boron, and aromatic polyamides.[25,26] These fibers possess the desirable properties of low density [1.44–2 4.27 g/cm^2] and extremely high strength (80–550 GPa). When combined with the resin binder and laminated to support applied load, they provide mechanical properties.

Pipes

Epoxy resins are used in filament-bound, glass-reinforced pipes and tubes.

COATINGS

The coatings industry is the biggest consumer of epoxy resins. These resins are used mostly as chemical and special purpose coatings. Epoxy resins provide thin-layer durable coatings having mechanical strength and good adhesion to a variety of substrates. They are resistant to chemicals, corrosion, and solutions. They find applications in washing machines and appliances, ships and bridges, pipelines and chemical plants, automobiles, farm implements, containers, and floor coatings. Epoxy coating formulations are available as liquid resins, solid resins, high molecular weight thermoplastic resins, multifunctional resins, radiation curable resins, and special purpose resins.[27–33]

High-solid coating solution formulations attain maximum film properties (adhesion, appearance, and freedom from defects). These are based on liquid epoxy resin acrylic adducts with epoxy resins. These adducts have proved useful in automotive primers.

Waterborne coatings are made by dispersing or emulsifying the resins with surfactants. High-solid coatings have an additional advantage, as they are useful on steel, brass, metal furniture, buildings, and miscellaneous products. Application of powders is accomplished by electrostatic spray fluidized-bed coating and electrostatic fluidized-bed coating. For marine use, epoxy resins that cure under water or are resistant to seawater have been developed.

REFERENCES

1. Lee, H.; Neville, K. *Handbook of Epoxy Resins*, McGraw Hill: New York, NY, 1982.
2. May, C. A.; Tanaka, Y. *Epoxy Resins—Chemistry and Technology*; Marcel Dekker: New York, NY, 1987.
3. *Adhesion Polymers*, Japan Welding Society: Tokyo, 1993.
4. Bhatnagar, M. S. *Epoxy Resins*; Universal Book: Bombay, India, 1996.
5. *Encyclopedia of Polymer Science and Technology*, John Wiley & Sons: New York, NY, 1986, Vol. 6; pp 208–271; 322–382.
6. *Encyclopedia of Polymer Science and Engineering*, John Wiley & Sons: New York, NY, 1986, Volume 6; pp 208–271; 322–382.
7. *Ullmann's Encyclopedia of Industrial Chemistry*, VCH: London, United Kingdom, 1987, Volume A9.
8. *Encyclopedia of Chemical Process and Design*, Marcel Dekker: New York, NY, 1983, Volume 19.
9. Bhatnagar, M. S. *Polym.-Plastic Techol. Eng.* **1993**, *32*, 53.
10. Bauer, R. S. *Applied Polymer Science*; American Chemical Society: Washington, DC, 1985.
11. Petrosyan, V. A. et al. *Arin. Khim. Zh.* **1989**, *42*, 54.
12. Polyakov, V. A. et al. *Mekh. Kompoz. Mater.* **1989**, *2*, 218.
13. Shode, L. G. et al. *Lakrokras. Mater. Ikh. Primen.* **1989**, *1*, 60; *3*, 31; *2*, 48.
14. Kharakha, Zho G. et al. *Polimery* **1989**, *34*, 204.
15. Czvikovszky, T. *Kem. Kozl.* **1989**, *6*, 4.
16. Singh, J. *Paintindia* **1989**, *4*, 1.
17. Bhatnagar, M. S. *Pop. Plast.* **1987**, *32*, 20, 29.
18. Shibalovich, V. et al. *Khim. Prom. St. Rubezhok* 1988; Volume 3, p. 28.
19. Shiruishi, N. *ACS Symp. Ser.* **1989**, *397*, 488.
20. Manfred, H. *Adhesion* **1989**, *33*, 21.
21. Matsuura, M. *Nippon Setchaku Gokkaishi* **1994**, *30*, 24.
22. Hinderwaldner, R. *Adhesion* **1989**, *33*, 16.
23. Kharakhus, O. G. et al. *Polimery* **1989**, *34*, 204.
24. Kinjo, N. et al. *Adv. Polym. Sci.* **1989**, *88*, 1.
25. Czvikovozky, T. *Kem. Kozl.* **1989**, *69*, 45.
26. Wyatt, B. S. *Australian Corrosion Association*; Wavehampton, Australia, 1993.
27. Puranik, V. V. *Paintindia* **1989**, *39*, 21.
28. Kochenova, Z. A. et al. *Lakrokras. Mater. Ikh. Primen* **1989**, *2*, 82.
29. Pesat, V. *Kowze. Ochz. Mater.* **1989**, *33*, 1.
30. Klien, D. et al. *Verfkromik* **1989**, *62*, 140.
31. Thankachan, C. J. *Coat. Technol.* **1989**, *61*, 39.

32. Hashim, A. S. *Kompoz. Polim. Mater.* **1991**, *48*, 16.
33. Hudrlik, F. P.; Hudrlik, M. A. *Adv. Silicon Chem.* **1993**, *2*, 1.

EPOXY RESINS (Curing Reactions)

Shigeo Nakamura* and Masao Arima
Department of Applied Chemistry
Faculty of Engineering
Kanagawa University

Epoxy resins contain two or more epoxy groups in a molecule. Their backbone may be aliphatic, cycloaliphatic, or aromatic. The epoxy groups react with curing agents to yield insoluble and infusible 3-D networks. Curing agents have two or more reactive groups in a molecule, which can react with epoxy groups. The properties of cured epoxy resins are determined by the epoxy prepolymer and the curing agent. Because of the versatility of their properties, epoxy resins are used in a variety of applications such as coatings, laminates, composites, bonding, and adhesives.[1]

Because it has a strained three-membered ring structure, an epoxy group can react with many nucleophilic and electrophilic reagents. However, for curing of epoxy resins, compounds with active hydrogen atoms, for example, amines, phenols, alcohols, thiols and carboxylic acids, and acid anhydrides are used for their workability and availability. Except for acid anhydrides, the conventional curing agents for epoxy resins leave pendant hydroxyl groups in the cured resins.

To fulfill a demand for high-performance epoxy resins, novel curing agents are still being sought.

PROPERTIES

Various properties of epoxy resins are affected by the epoxy prepolymer, curing agent, and additives. Above all, the optimum curing temperature and the thermal stability of epoxy resins depend on the type of curing agent. The curing temperature and thermal stability are the most important factors in determining the application of epoxy resins. The curing temperature and thermal stability of epoxy resins decrease as follows, depending on the curing agent: acid anhydride > phenol = aromatic amine > cycloaliphatic amine > aliphatic amine.

Amine-cured epoxy resin finds a wide variety of applications such as casting, laminates, adhesive, and bonding because its properties are remarkably affected by the backbone structure of multifunctional amines. Aliphatic primary amines are so reactive that curing reaction occurs at room temperature. The cured resins are stiff and resistant to inorganic acids; they are used for adhesives and laminates. Although aromatic primary amines are less reactive than aliphatic ones, the cured resins are more thermally stable and chemically resistant and are used for substrates of carbon fiber-reinforced plastics. Hindered amines have recently been used for preparation of prepreg of print board.

The curing conditions for anhydrides are more severe than those for amines. However, pot life is longer, the exothermic heat of curing is lower, and shrinkage is smaller. The cured resins have excellent electric, chemical, and mechanical properties. Therefore, anhydrides are used for electric and electronic applications.

Because the curing rate of thiols is high at lower temperatures, they are used as low-temperature bonding and coating for construction, even at 0 to –20°C by adding tertiary amines as accelerators.

Phenol-formaldehyde resin (novolac) is used for semiconductor packaging materials because of the high thermal stability and low water uptake that results from its structure. Carboxylic acids are used only for coating.

The properties of epoxy resins cured with active esters were examined in detail.[2–4]

Active esters are expected for application that requires low dielectric constant and low water uptake such as electric and electronic use. In these fields, anhydrides are now used. However, they liberate free acids during storage, which may impair the mechanical properties of the cured resins, whereas active esters are very stable during storage and have very long pot life.

*Author to whom correspondence should be addressed.

REFERENCES

1. McAdams, L.; Gannon, J. A. *Encyclopedia of Polymer Science and Engineering* 2nd ed.; Mark, H. F.; Bikales, N. M.; Overberger, C. G.; and Menges, G., Eds. John Wiley & Sons: New York, 1986; Vol. 6, p 322.
2. Arima, M.; Ibe, H.; Nakamura, S. *J. Appl. Polym. Sci.*, in press.
3. Nakamura, S.; Arima, M. *Int. J. Polym. Anal. Characterization* **1995**, *1*, 75.
4. Arima, M.; Nakamura, S. *J. Appl. Polym. Sci.*, in press.

EPOXY RESINS (High Performance Composite Applications)

Anthony Bosch
Fiberite, Incorporated

High-strength, lightweight, composite materials have helped transform many creative ideas into technical realities. The application of these materials has affected a wide range of industries including transportation, communications, construction, and leisure goods. In the aerospace sector alone the impact of these materials can be seen in the space program, rocket motors, stealth military aircraft, commercial transports, and so forth. Although most of these benefits are derived from the excellent properties of high-strength reinforcing fibers, the resin matrices play an equally important role in defining these advantages and determining the proper applications. More than simply binding the fibers together, the polymer matrix reins must help transfer loads between fibers and offer good processability, thermal resistance, environmental protection, toughness, and fatigue resistance. Of the various polymer matrices in use, the epoxy resins have occupied a dominant position in the manufacture of high-performance composites, primarily because of their overall performance balance relative to processing, properties, and cost.

PREPARATION AND AVAILABILITY OF EPOXY RESINS

Because the chemistry of epoxy resins is well documented in various literature sources[1–7] the following will serve only as a brief review.

Although the diglycidylether of bisphenol A (DGEBPA) its brominated derivatives, and other difunctional resins are widely used in the composites industry, their use in aerospace applications is limited by the maximum attainable glass transition temperature (T_g) of approximately 175°C (347°F). For many high-performance applications it is therefore necessary to use epoxy resins with higher functionalities that yield more highly crosslinked, 3-D polymer networks with increased T_g values. Tri- and tetrafunctional epoxy resins derived from p-aminophenol [5026-74-4], tris (*p*-hydroxyphenyl) methane [66072-38-6], and 4,4′-methylenedianiline [28768-32-3] are therefore used extensively, even though at considerably higher cost. Similar to the DGEBPA resins, the tetraglycidyl derivatives of 4,4′-methylenedianiline (TGMDA) are available in several viscosity ranges, from a low of 3–6 Pascal second (3000–6000 cps) to 17–19 Pa.s (17,000–19,000 cps) at 50°C.

CURING AGENTS

Although a great variety of curing agents (hardeners) based on amines, amides, phenols, thiols, carboxylic acids, and acid anhydrides exist, the number of hardeners available for high-performance applications and prepreg manufacture in particular is limited. The curing agent must have latent reactivity in order for the resulting resin and prepreg to possess acceptable outtime (good tack and drape, generally for a minimum of 10–14 days at ambient temperature) and for the resulting cured system to have both a high T_g and a maximized resin modulus. The most latent hardener, dicyandiamide [461-58-5], in conjunction with accelerators such as substituted ureas, is still widely used for 121–177°C (250–350°F) curing systems. However, for optimal properties the application of aromatic diamines is preferred. Although *m*- and *p*-phenylenediamine [108-45-2, 106-50-3] and 4,4′-methylenedianiline [101-77-9] have been used extensively, these products are losing favor because of toxicity concerns. The reagent of choice is 4,4′-diaminodiphenyl sulfone (4,4′-DDS, [80-08-0]), which provides the best balance of desired properties and processability although requiring a two-hour cure at 177°C (350°F) for complete reaction.

FORMULATED EPOXY RESINS/PREPREGS: AVAILABILITY AND APPLICATION

High-performance or advanced composites are manufactured using a variety of matrix materials including organic polymers, ceramics, carbon, and metals. Of these matrices, the organic polymers have been the most widely used, whereas within this group epoxy resins have been the more dominant category, easily outpacing other resin systems such as bismaleimides, cyanate esters, polyimides, and thermoplastics. Although advanced composites are finding increased application in areas as varied as sporting goods, biomedical devices, electronics, transportation, and civil engineering structures, the major usage has been in the aerospace industry and will likely remain there in the near future despite the significant downturn in the defense business. Main driving forces behind the use of composites have been the weight reduction resulting from the high strength/weight and high modulus/weight ratios, reduction in number of parts, chemical and corrosion resistance, and excellent fatigue properties.

RESIN TRANSFER MOLDING (RTM)

RTM is a process wherein a dry fiber preform is placed in a two-part, matched metal mold with a controlled gap in between. After the mold is closed, resin is injected under pressure at as low a temperature as possible, thus permeating the fiber reinforcement.[8–10] Curing can be accomplished in the mold or separately once sufficient integrity and strength has been obtained to allow demolding of the part.

Resin Film Infusion (RFI) is a closely related but alternate method to RTM. Instead of being injected, the resin is placed directly in the mold in film or slab form and permeates the preform under vacuum or pressure as the temperature is increased. Because resin infusion is limited to the thickness of the part, this technology appears to be advantageous in the manufacture of large parts.

PREPREGS, FIBERS, AND FABRICATION

Although some applications exist in processes such as wet filament winding, RTM/RFI, reaction injection molding (RIM), and pultrusion, most high-performance epoxy resins are not marketed as neat resins and are available only in prepreg form. Preimpregnated fibers (prepregs) are manufactured by either a solution coating or a hot-melt process[11,12] in three forms: roving (single-fiber tow), unidirectional tape, and woven fabrics (broadgoods).

Most high-performance epoxy resins can be cured in two hours at temperatures ranging from 121–177°C (250–350°F) and at pressures of 550–690 kPa (80–100 psi), although parts manufactured using honeycomb are limited to 310 kPa (45 psi) pressure. High-strength, low-weight composites can be made in this fashion having typical specific gravities of 1.6 with carbon fiber, 1.9 with glass fiber, and 1.4 with aramid fiber (cf. aluminum 2.7, steel 7.8).

SAFE HANDLING

In view of the reactive oxirane and amino groups present in high-performance epoxy matrices, due care must be taken in the handling of these resins and their resulting prepregs.[11]

REFERENCES

1. Lee, H.; Neville, K. *Handbook of Epoxy Resins*; McGraw-Hill: New York, 1967.
2. McAdams, L. V.; Gannon, J. A. *Encyclopedia of Polymer Science and Engineering*; John Wiley & Sons: New York, 1986; Vol. 6, p 322.
3. Ilschner, B.; Lees, J. K.; Dhingra, A. K.; McCullough, R. L. *Ullmann's Encyclopedia of Industrial Chemistry*, 5th ed.; VCH Verlagsgesellschaft mbH: Weinheim, Germany, 1986; Vol. A7, p 369.
4. May, C. A. *Engineered Materials Handbook (Composites)*; ASM International: Metals Park, Ohio, 1987; Vol. 1, p 66.
5. Muskopf, J. W.; McCollister, S. B. *Ullmann's Encyclopedia of Industrial Chemistry*, 5th ed.; VCH Verlagsgesellschaft mbH: Weinheim, Germany, 1987; Vol. A9, p 547.
6. Bauer, R. S.; Stewart, S. L.; Stenzenberger, H. D. *Kirk-Othmer Encyclopedia of Chemical Technology*, 4th ed.; John Wiley & Sons: New York, 1993; Vol. 7, p 33.
7. Gannon, J. *Kirk-Othmer Encyclopedia of Chemical Technology*, 4th ed.; John Wiley & Sons: New York, 1994; Vol. 9, p 730.
8. Becker, W. *Aerospace Composites and Materials* **1989**, *1*, 12.

9. Stover, D. *Advanced Composites* **1990**, *5*, 60.
10. Stover, D. *Advanced Composites* **1992**, *7*, 39.
11. Kantz, M. R. *Applied Industrial Hygiene, Special Issue* December 1989, p 1.
12. McConnell, V. P. *Advanced Composites* **1991**, *6*, 53.

EPOXY RESINS (Microwave Processing)

Henri Jullien* and Michel Delmotte
Laboratoire de Microstructures et Mécanique des Matériaux
Center National de la Recherche Scientifique, Research Unit
Ecole Nationale Superiour d'Arts et Méticrs

STATE OF THE ART

Microwave energy can be used as a heat source for dielectric materials containing polar groups, the most efficient of them being the water molecule; many processes for drying miscellaneous substances have been described. But polar dielectrics that do not contain water, like macromolecular compounds such as poly(vinyl chloride), polyurethanes, polyesters, and polyimides, can also be heated by microwaves. Since 1970 many patents have described microwave processes and devices for curing polymeric materials such as synthetic rubbers, melamine and wood laminates, and so forth. In the field of epoxy resins, after the early experiment of Williams,[1] who described qualitatively the microwave cure of glass-reinforced epoxy pipes, a scientific approach started with the work of Wilson and Salerno,[2] who developed a mathematical model for the microwave heating of epoxy samples as functions of the dielectric constant, the electric field, and material parameters, in multimode microwave applicators such as 2.45 GHz kitchen ovens. A few years later, Strand[3] studied temperature-time profiles of neat or reinforced epoxies, finding that the microwave cure time was considerably reduced in comparison to conventional cure.

Loos and Springer[4] and Lee and Springer[5,6] proposed a model for the microwave curing of either glass-fiber or graphite-fiber-reinforced epoxy matrix composites, yielding the resin flow-out of the composites,[4] the thermochemical aspects,[5] and the electromagnetic aspects[6] of the process. The authors presented experimental results that seemed to be in good agreement with the model.

In the same time a number of papers started to be published, dealing with temperature-time profiles in microwave experiments. Working in a 2.45 GHz monomode waveguide system, Gourdenne et al.[7] showed a correlation between the temperature and the absorbed power in samples made of an epoxy resin (diglycidylether of bisphenol-A, DGEBA, cured with diaminodiphenylmethane), neat,[7,8] or reinforced with glass fibers,[9] or aluminum powder,[10] or carbon black powder.[11] Gourdenne et al. also used pulsed microwaves (as previously experimented by Jullien and Valot[12] for polyurethane networks, and by Thuillier et al.[13] for epoxy networks), and found correlation between cure time, temperature, network structure, and absorbed power in samples.[14,15]

*Author to whom correspondence should be addressed.

To obtain a pertinent comparison between conventional and microwave mode of heating, Jow et al.[16] used an on-off control to monitor the temperature-time profile and to maintain a constant temperature inside the sample cured in the same single-mode cylindrical resonator as Singer, also providing dielectric data; they concluded that microwave heating provides a better control of the temperature profile in the sample.

The kinetic effect of a microwave electric field on the crosslinking reaction kinetics is still being debated. As did Strand,[3] Lewis et al.[17] found for a DGEBA-DDS system a significant increase of reaction rate; Mijovic and Wijaya[18] used an on-off control of the temperature-time profile similar to Jow's system,[16] for the cure of small DGEBA-DDS samples in a single-mode rectangular resonator. The cured products were characterized by glass-transition temperature and degree of cure determined by Differential Scanning Calorimetry (DSC), showing slight differences between conventionally cured and microwave-cured samples. A few years later, for a polysulfone-modified epoxy network, Liptak et al.[19] also found significant differences in morphology between microwave-cured and conventionally cured samples; at the same time, the microwave crosslinking of DGEBA by metaphenylene-diamine was found to be faster than the conventional curing by Wei et al.[20] But Urro et al.[21] studied the benzyldimethylamine-catalyzed crosslinking reaction of DGEBA with dicyandiamide (DDA) in a rectangular resonator, with either conventional or continuous microwave heating, providing also dielectric constant-temperature profiles. The chemical structure of products was analyzed by Fourier Transform Infrared Spectroscopy FT-IR, and no difference was found in samples which had followed the same temperature-time profiles: and for the cure of a phenylglycidylether-aniline model system, analyzed by FT-IR spectroscopy and High Performance Liquid Chromatography (HPLC), Mijovic et al.[22] found no difference between the two modes of cure, so that the kinetics of reaction and consequently the properties of cured networks would not differ.

However, Jullien and Petit[23] studied the same model system, the reagents being fixed on inorganic supports, heated either in a multimode oven or in a conventional oil bath. They observed kinetic effects essentially due to supports: nevertheless, differences appeared in the chemical structures of products, studied by proton and ^{13}C NMR spectroscopy, which they attributed to microwave-induced changes in the relative kinetics of side reactions, epoxy-amine, and etherification reactions, respectively,[24] as previously observed by Thuillier et al.[13] in a DGEBA-DDS system under pulsed microwaves.

Lastly Outifa et al.[25–28] have proposed a new approach for the microwave processing of large epoxy composite samples, which includes chemical, wave propagation, dielectric, thermal, and heat science considerations.

REFERENCES

1. Williams, N. H. *J. Microwave Power* **1967**, *2*, 123.
2. Wilson, L. K.; Salerno, J. P. *AVRADCOM Report no 78-46*, 1978.
3. Strand, N. S. *Prepr. 35th Conf. Of Reinforced Plastic/Composites Inst., The Society of Plastics Industry Inc.* **1980**, *Sec. 24C*, 1.
4. Loos, A. C.; Springer, G. S. *J. Compos. Mater.* **1983**, *17*, 135.
5. Lee, W. I.; Springer, G. S. *J. Compos. Mater.* **1984**, *18*, 357.
6. Lee, W. I.; Springer, G. S. *J. Compos. Mater.* **1984**, *18*, 387.

7. Gourdenne, A.; Maassarani, A. H.; Monchaux, P.; Aussudre, S.; Thourel, P. *Polym. Prepr.* **1979**, *20*(2), 471.
8. Van, Quang Le; Gourdenne, A. *Eur. Polym. J.* **1987**, *23*, 777.
9. Gourdenne, A.; Van, Quang Le *Polym. Prepr.* **1981**, *22*(2), 125.
10. Baziard, Y.; Gourdenne, A. *Eur. Polym. J.* **1988**, *24*, 873, 881.
11. Bouazizi, A.; Gourdenne, A. *Eur. Polym. J.* **1988**, *24*, 889.
12. Jullien, H.; Valot, H. *Polymer* **1985**, *26*, 506.
13. Thuillier, F. M.; Jullien, H.; Grenier-Loustalot, M. F. *Polym. Communic.* **1986**, *27*, 206.
14. Beldjoudi, N.; Bouazizi, A.; Douibi, D.; Gourdenne, A. *Eur. Polym. J.* **1988**, *24*, 49.
15. Beldjoudi, N.; Gourdenne, A. *Eur. Polym. J.* **1988**, *24*, 53, 265.
16. Jow, J.; DeLong, J. D.; Hawley, M. C. *SAMPE Quart.* **1989**, *20*(2), 46.
17. Lewis, D. A.; Hedrick, J. C.; Ward, T. C.; McGrath, J. E. *Polymer Prepr.* **1987**, *28*(2), 330.
18. Mijovic, J.; Wijaya, J. *Macromolecules* **1990**, *23*, 3671.
19. Liptak, S. C.; Gungor, A.; Ward, T. C.; McGrath, J. E. *Polym. Mater. Sci. Eng.* **1992**, *66*, 387.
20. Wei, J.; Hawley, M. C.; DeMeuse, M. T. *Polym. Mater. Sci. Eng.* **1992**, *66*, 478.
21. Urro, P.; Jullien, H.; Duhot, V. *Ann Composites* **1990**, *merged issues of January and February*, 137.
22. Mijovic, J.; Fishbain, A.; Wijaya, J. *Macromolecules* **1992**, *25*, 979, 986.
23. Jullien, H.; Petit, J. *Polym. Mater. Sci. Eng.* **1992**, *66*, 378.
24. Jullien, H.; Petit, J.; Mérienne, C. *Microwave and High Freq. 1993 Intern. Congress*; SIK: Göteborg, Sweden, Vol. 1 (Proceedings), D2, September 28–30, 1993.
25. Outifa, L.; Jullien, H.; Delmotte, M. *Polym. Mater. Sci. Eng.* **1992**, *66*, 424.
26. Outifa, L.; Ph.D. Dissertation, University Pierre et Marie Curie, Paris, France, 1992.
27. Outifa, L.; Delmotte, M.; Jullien, H. *Microwave and High Freq. 1993 Intern. Congress*; SIK: Göteborg, Sweden, Vol. 2 (Posters), 1:6, September 28–30, 1993.
28. Outifa, L.; Guyonvarch, G.; Delaunay, D.; Moré, C. *Microwave and High Freq. 1993 Intern. Congress*; SIK: Göteberg, Sweden, Vol. 1 (Proceedings), E4, September 28–30, 1993.

EPOXY RESINS (Modification with Engineering Plastics

Takao Iijima
Department of Applied Chemistry
Faculty of Engineering
Yokohama National University

Epoxy resins are composed of epoxy oligomers and hardeners. A major drawback of these resins is that they are brittle and have poor resistance to crack propagation.

Toughness may be defined as resistance to crack propagation. A fracture toughness (or stress intensity) factor (K_{IC}) or a fracture energy (or a strain energy release) rate (G_{IC}) is used as a measure of toughness of materials.[1]

The toughness of cured epoxy resins has been increased with loss of their mechanical and thermal properties by blending with reactive liquid rubbers such as carboxyl-terminated butadiene acrylonitrile rubbers (CTBN).[2] Epoxide-containing acrylic elastomers are also effective modifiers for epoxies.[3] These elastomers, however, are not always effective modifiers for improving toughness of highly crosslinked epoxy matrices.[4,5] Engineering thermoplastics are interesting materials as modifiers for epoxy resins from the viewpoint of maintaining mechanical and thermal properties of the matrix resins. Recently various types of ductile thermoplastics have been used as alternatives to reactive rubbers for improving toughness of epoxy resins.[6–8]

PREPARATION AND PROPERTIES

At first commercial poly(ether sulfone) (PES) (Victrex 100P, ICI) was used as a modifier in the modification of polyfunctional epoxies such as *N,N′*-tetraglycidyl diaminodiphenylmethane (TGDDM) (MY 720, Ciba Geigy) using 4,4′-diaminodiphenyl sulphone (4,4′-DDS) or dicyandiamide, but less effective owing both to their highly crosslinking structures and to poor interfacial adhesion between two uncompatibilized phases in the cured resin.[9] Researchers also used PESs (Victrex 100P and 300P, ICI) as modifiers for the TGDDM (MY 720)/aromatic acid anhydride system.[10,11] Terminally functionalized polysulfones (PSF) are more effective modifiers than commercial grades of PES.[12,13]

The reaction-induced phase separation in the DGEBA/4,4′-DDS/PES modification system has been investigated, and the relation between the curing conditions and the microstructure of the cured resins has been examined by Inoue et al.[14]

Poly(aryl ether ketone)s (PEEK) have poor compatibility with epoxies, so it is difficult to prepare the modified resin without solvents such as dichloromethane. Recently, both bisphenol A-type and *t*-butylhydroquinone-type PEEKs with terminal amine groups were used as modifiers without solvents.[15]

Poly(etherimide)s (PEI) are interesting and effective modifiers. Toughening can be achieved because of ductile drawing and tearing of the PEI-rich phases and crack-bridging of the PEI phases.

PEI (Ultem 1000) is also effective in the modification of the DGEBA and/or triglycidyl *p*-aminophenol (TGAC)/dicyandiamide system.[16]

Poly(butylene terephthalate) (PBT) is used as a crystalline polymer and effective in the modification of Epon 828/aromatic diamine (mainly containing 4,4′-diaminodiphenyl methane) system.[17]

Poly(phenylene oxide) (PPO, mol wt 34,000) has been reported as a modifier for the DGEBA/piperidine systems.[18] The incorporation of PPO into the networks is stabilized by inclusion of a styrene-maleic anhydride copolymer and the modified resin has a particulate morphology. The hybrid modifier composed of PPO and styrene–butadiene–styrene triblock copolymer also has been examined.

N-Phenylmaleimide-styrene copolymers (PMS) were effective modifiers for the Epikote 828/4,4′-DDS system.[19]

Polycarbonate (PC) (M_n 12,000, Rubber and Plastics Research Association) is used in the modification of the phenol novolak epoxy resin (DEN 438, Dow Chemical Company)/nadic methyl anhydride system.[20,21]

Researchers used five kinds of pre-formed Nylon 12 particles (5, 10, 20, 40, and 50 (μm)) as modifiers to toughen a crosslinkable epoxy thermoplastic resin developed by Dow.[22]

Toughening mechanisms in the modification by engineering thermoplastics are different from those in the rubber modification system. In the latter systems toughening can be attained based on the particulate structure, but cocontinuous or inverted-phase structures are desirable in many cases of the former systems, where the ductile drawing and tearing of the thermoplastic-rich phases absorb the fracture energy in the course of the crack propagation, and toughening can be achieved.

REFERENCES

1. Williams, J. G. *Adv. Polym. Sci.* **1978**, *27*, 69.
2. Yee, A. F.; Pearson, R. A. *J. Mater. Sci.* **1986**, *21*, 2462, 2476 and references cited therein.
3. Iijima, T.; Horiba, T.; Tomoi, M. *Eur. Polym. J.* **1991**, *27*, 1231 and references cited therein.
4. Pearson, R. A.; Yee, A. F. *J. Mater. Sci.* **1989**, *24*, 2571.
5. Iijima, T.; Yoshioka, N.; Tomoi, M. *Eur. Polym. J.* **1992**, *28*, 573.
6. Hedrick, J. C.; Patel, N. M.; McGrath, J. E. *Toughened Plastics*; American Chemical Society: Washington, DC, 1993; Vol. I, Chapter 11.
7. Pearson, R. A. *Toughened Plastics*; American Chemical Society: Washington, DC, 1993; Volume I, Chapter 17.
8. Iijima, T.; Tomoi, M. *Kobunshi Kako* **1994**, *43*, 21.
9. Bucknall, C. B.; Partridge, I. K. *Polymer* **1983**, *24*, 639.
10. Raghara, R. S. *J. Polym. Sci., Part B, Polym. Phys.* **1987**, *25*, 1017.
11. Raghara, R. S. *J. Polym. Sci., Part B, Polym. Phys.* **1988**, *26*, 65.
12. Hedrick, J. H. et al. *Polym. Bull.* **1985**, *13*, 201.
13. Hedrick, J. H. et al. *Polymer* **1991**, *32*, 2020.
14. Yamanaka, K.; Inoue, T. *Polymer* **1989**, *30*, 662.
15. Bennett, G. S.; Farris, R. J.; Thompson, S. A. *Polymer* **1991**, *32*, 1663.
16. Murakami, A. et al. *Nippon Setchaku Gakkaishi* **1991**, *27*, 364.
17. Kim, J.; Robertson, R. E. *J. Mater. Sci.* **1992**, *27*, 161, 3000.
18. Pearson, R. A.; Yee, A. F. *Polymer* **1993**, *34*, 3658.
19. Iijima, T. et al. *Eur. Polym. J.* **1992**, *28*, 1539.
20. Martuscelli, E. et al. *Angew. Makromol. Chem.* **1993**, *204*, 153.
21. Martuscelli, E. et al. *Angew. Makromol. Chem.* **1994**, *217*, 159.
22. Shi, Y-B.; Groleau, M. R.; Yee, A. F. *Polym. Mater. Sci. Eng.* **1994**, *70*, 110.

Ethyl-Cyanoethyl Cellulose

See: Liquid Crystalline Ethyl-Cyanoethyl Cellulose

Ethylene-Acrylic Acid Copolymer

See: Powder Coatings (Overview)

Ethylene-Acrylic Rubber

See: Blends, Interchain Crosslinking

Ethylene-Carbon Monoxide Copolymers

See: Olefin-Carbon Monoxide Copolymers (Overview)
Olefin-Carbon Monoxide Copolymers, Alternating

Ethylene-Chlorotrifluoroethylene Copolymers

See: Engineering Thermoplastics (Survey of Industrial Polymers)

Ethylene-Propylene Copolymers

See: Ethylene Propylene Rubber Modification (Maleic Anhydride and Derivatives)
Ethylene-Propylene Elastomers
Ionomers (Overview)
Metallocene Catalysts
Viscosity-Index Improvers

ETHYLENE-PROPYLENE ELASTOMERS

Stephen C. Davis, Walter Von Hellens, and Hayder A. Zahalka
Bayer Rubber Incorporated

Klaus-Peter Richter
Bayer AG

Ethylene-propylene elastomers refer to a family of synthetic rubbers that are prepared by polymerization of ethylene, propylene, and optionally, a nonconjugated diene. Unlike the majority of synthetic elastomers, these polymers contain a saturated backbone. There are two classes: ethylene-propylene copolymers (EPM) and ethylene-propylene-diene terpolymers (EPDM). Typical chemical structures are shown in **Figure 1a.** Here the term EP(D)M will be used in examples that apply to both EPM and EPDM. Extensive reviews of EP(D)M technology have been published.[1–3]

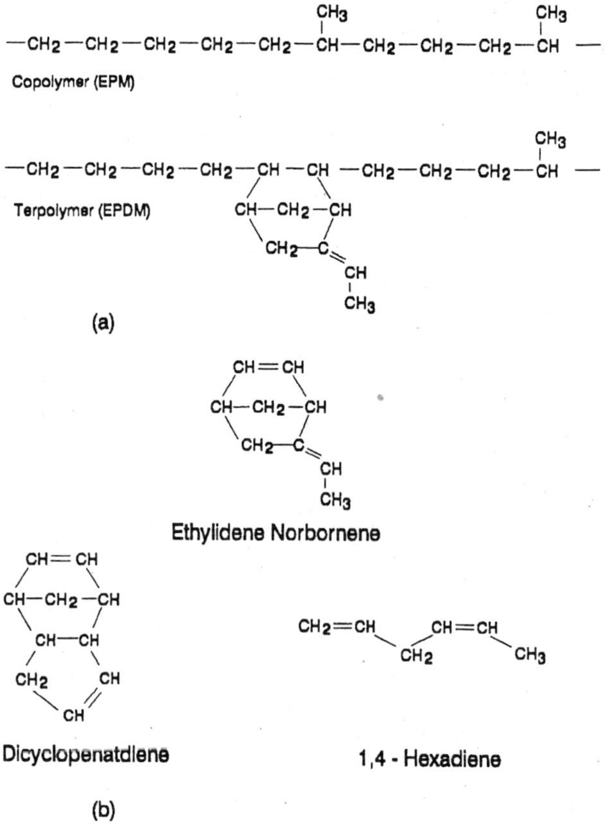

FIGURE 1. The structure of: (a) EPM and EPDM, and (b) dienes employed.

Degradation of synthetic elastomers often occurs by reactions involving backbone double bonds (attack by ozone, oxygen, ultraviolet, etc.), which result in bond cleavage and loss of physical properties. The saturated backbone found in EPM and EPDM is fairly resistant to chemical attack. This fact gives these elastomers excellent weatherability and heat resistance. In use, EPM and EPDM will not develop cracks due to reactions with ozone, oxygen, and light as readily as other elastomers. Due to the nonpolar hydrocarbon nature of the polymers, EPM and EPDM have excellent water resistance but poor oil resistance.

The elastomeric nature and chemical resistance of EPM was recognized in the 1950s.[4–5] Copolymers were synthesized initially. However, in order to obtain adequate elastomeric properties, EPM required crosslinking using peroxides. Peroxide crosslinking was not widely accepted in the rubber industry at that time. To enable the use of sulfur vulcanization, researchers used terpolymerization with dienes. This resulted in EPDM containing a saturated backbone and pendant double bonds for crosslinking. Screening of third monomers led to three types of commercial use. The three types of ethylidene norbornene (ENB), dicyclopentadiene (DCPD), and 1,4-hexadiene (HD) (**Figure 1b**).

Recent introduction of elastomers produced by copolymerization of ethylene with longer chain olefins (butene, hexene, octene) are expanding the family of ethylene-olefin elastomers beyond ethylene-propylene.[6]

APPLICATIONS

Traditional Rubber Applications

The inherent properties of EPM and EPDM that result in the widespread application of these polymers are: good ozone, oxygen, and heat resistance; low swell in water; low electrical conductivity; and the ability to accept high loadings of fillers and oils. Typical applications where these material characteristics are used are: automotive window, door and trunk seals; roofing membranes; gaskets and seals; hose and tubing; wire and cable insulations; tire sidewalls; and footwear.[3]

EPDMs also find some utility in blends with highly unsaturated polymers.[7] In the blend situation the EPDM polymer is frequently used to enhance the ozone and weathering resistance of the blend.

Plastics Modification Applications

In recent years an increasingly important applications of EPDM has been in blending with plastics to obtain enhanced material properties. Because of their compatability, EPMs, and EPDMs are particularly suited for blending with polyproplene and polyethylene. There are two classes of plastic/rubber blends. TPOs (thermoplastic olefins) and TPEs (thermoplastic elastomers) as defined by Coran.[8] TPOs are essentially rigid at the service temperature but the brittleness has been reduced by the addition of EPDM to produce a soft plastic, thus increasing their utility. TPEs exhibit rubbery properties at the service temperature but will process as thermoplastics. Both classes of polymer have found their major application to date in automotive applications such as bumper covers, decorative facia, and air ducting; however, wire and cable and hose applications are showing substantial rates of growth.

REFERENCES

1. Baldwin, F. P.; VerStrate, P. *Rubber Chem. Technol.* **1972**, *45*, 709.
2. Cesca, S. *Macromol. Rev.* **1975**, *10*, 1.
3. Easterbrook, E. K.; Allen, R. D. *Rubber Technology*, 3rd ed.; Morton, M., Ed.; 1987, Chapter 9.
4. Natta, G. *Rubber Plast. Age* **1957**, *38*, 495.
5. Natta, G. *Rubber Plast. Age* **1957**, *40*, 437.
6. *Metcon '94 Proceedings*; Catalyst Consultants: Springhouse, PA, 1994.
7. McDonel, E. T.; Baranwal, D. C.; Andries, J. C. *Polymer Blends*; Academic: London, United Kingdom, 1978; Vol. 2, Chapter 19.
8. Coran, A. Y.; Patel, R. P. Presented at the American Chemical Society Rubber Division Meeting, Nashville, TN November 1992, Paper 41.

ETHYLENE-PROPYLENE RUBBER MODIFICATION (Maleic Anhydride and Derivatives)

Fernanda M. B. Coutinho* and Maria Inês P. Ferreira
Instituto de Macromoléculas Professora Eloisa Mano
Universidade Federal do Rio de Janeiro

SYNTHETIC ROUTES

Introducing succinic groups (SAH) in ethylene-propylene (EPM) rubbers is a useful method for improving their adhesion characteristics and for obtaining new materials designed for applications in which unmodified hydrocarbon rubber performs poorly. Depending on the elastomer's end use, two main synthetic routes have been used to produce (ethylene-propylene-dieme rubbers) (EPDM)-g-SAH copolymers: solution and bulk techniques.

APPLICATIONS

The development of EPDM-g-SAH has received considerable attention from researchers because of the succinic groups' high reactivity in successive reactions.[4] One important application for maleated ethylene-propylene rubbers is in the area of multipurpose lubricating oil additives.[6–9] Whereas elastomer's hydrocarbon molecular improves the oil viscosity index, the succinic groups can post-react with amines and alcohols to provide dispersant and detergent characteristics to the additive.[6–9] Researchers found that the use of EPDM-g-SAH containing a high-propylene-grade rubber also provides better shear stability to the additive compared with the unmodified rubber.[5]

Another EPDM-g-SAH that involves the post-reaction of succinic groups is the production of thermoplastic elastomers.[3,10,11] In blends of EPDM and polar plastics, the usual adhesion promoter is EPDM-g-SAH.[12–19] Total or partial substitution of EPDM by the maleated rubber improves the rubber's dispersion in the plastic matrix because smaller particle sizes are obtained. Consequently, the blend's brittle-tough transition shifts to lower temperatures.[15]

Thus, succinic modified EPDM and the *in situ* formation of the compatibilizing agent can increase the impact resistance of different thermoplastics such as poly(butylene terephthalate) (PBT), poly(ethylene terephthalate) (PET), poly(phenylene oxide) (PPO), poly(vinyl chloride) (PVC), and polyamides.[1,2,4,5a,12–21] In the last system, succinic units of the modified rubber react with

*Author to whom correspondence should be addressed.

the polyamide's -NH_2 end groups making it possible to produce the supertough nylons.[22,23]

REFERENCES

1. Lielo, V. D.; Malinconico, M.; Martuscelli, E. et al. *Die Ang. Makromol. Chem.* **1990**, *141*, 141.
2. Greco, R.; Malinconico, M.; Martuscelli, E. et al. *Polymer* **1987**, *28*, 1185.
3. Caywood, S. W. U.S. Patent 4 010 223, 1973.
4. Greco, R.; Maglio, G.; Musto, P. V. *J. Appl. Polym. Sci.* **1987**, *33*, 2513.
5. Ferreira, M. I. P. D. Sc. Thesis, Instituto de Macromoléculas, 1994.

5a. Okamoto, A.; Okita, T.; Okino, E. Japanese Patent 87 053 304, 1987.

6. Kiovsky, T. E. U.S. Patent 4 235 731, 1976.
7. Hayashi, K. U.S. Patent 4 357 250, 1978.
8. Papay, A. G.; O'Brien, J. P. European Patent 050 994, 1980.
9. Papay, A. G.; O'Brien, J. P. European Patent 061 889, 1982.
10. Rees, R. W.; Heinhardt, H. G. U.S. Patent 3 997 487, 1976.
11. Scott, C.; Ishida, H.; Maurer, F. H. J. *Rheol. Acta.* **1988**, *27*, 272.
12. Xanthos, M. *Polym. Eng. Sci.* **1988**, *28*, 1392.
13. Cimmino, S.; Greco, R.; Maglio, G. et al. *Polym. Eng. Sci.* **1984**, *24*, 48.
14. Borggreve, R. J. M.; Gaymans, R. J.; Ingen Housz, J. F. *Polymer* **1987**, *28*, 1489.
15. Borggreve, R. J. M.; Gaymans, R. J.; Luttmer, A. R. *Makromol. Chem. Makromol. Symp.* **1988**, *16*, 195.
16. Borggreve, R. J. M.; Gaymans, R. J. *Polymer* **1989**, *30*, 63.
17. Borggreve, R. J. M.; Gaymans, R. J.; Schuijer, J. 1989, 30, 71.
18. Borggreve, R. J. M.; Gaymans, R. J.; Eichenwald, H. M. *Polymer* **1989**, *30*, 78.
19. Martuscelli, E.; Riva, F.; Sellitti, C. et al. *Polymer* **1985**, *26*, 270.
20. Boon, W. H.; Henderson, J. N. European Patent 119 150, 1983.
21. Yates, J. B.; Ullman, T. J. U.S. Patent 4 745 157, 1986.
22. Gaylord, N. G. *Chemtech* **1988**, 435.
23. Chen, D.; Kennedy, J. P. *Polym. Bull.* **1987**, *17*, 71.

Ethylene-Tetrafluoroethylene Copolymers

See: Engineering Thermoplastics (Survey of Industrial Polymers)

Ethylene-Vinyl Acetate Copolymer

See: Alternating Ethylene-Vinyl Alcohol Copolymer
Controlled Drug Delivery Systems
Ethylene-Vinyl Acetate Copolymer (Chemical Modification, Compatibilizer in Blends)
Ethylene-Vinyl Alcohol Copolymer (Overview)

ETHYLENE-VINYL ACETATE COPOLYMER (Chemical Modification, Compatibilizer in Blends)

Ailton de Souza, Gomes, Bluma G. Soares, and Ronilson V. Barbosa
Instituto de Macromoléculas
Universidade Federal do Rio de Janeiro

Several strategies for blend compatibilization exist, including the introduction of nonreactive graft or block copolymers, reactive polymers, and low molecular weight coupling agents. The nature of the components in the polymer blend dictate which method is best.

When the polymer components of a blend contain reactive groups, the compatibilization may be easily performed during the melt processing.[1,2] Such reactive groups can be incorporated through a variety of chemical reactions. A typical example of this approach is the well-known free-radical-catalyzed grafting of maleic anhydride onto polyolefins to be used in nylon-based polymer blends.[3]

Chemical modification of preformed polymers may be very profitable when functional groups are already present along the backbone. That is the case of poly(ethylene-covinyl acetate) (EVA). The incorporation of vinyl acetate in the ethylene chain imparts flexibility, toughness, and clarity as compared to low-density polyethylene. Other important features include toughness at low temperature and intrinsically good processability. Because of these important properties, EVA copolymers have found several applications in sheeting, wire and cable coatings, flexible tubing, shoe soles, and food packaging.[4] They also represent a good partner for polymer blends as an impact modifier.[5]

Besides their good properties, EVA copolymers can be easily converted into compatibilizing agents for EVA-based polymer blends. Using simple organic reactions, one can transform the acetate groups along the backbone into active centers for a series of grafting reactions and a series of other organic reactions. In addition, the presence of acetate groups facilitates the generation of free radicals along the backbone, which enables promotion of grafting reactions in the presence of several monomers (such as vinyl chloride and styrene).[6,7] Such graft copolymers can be prepared *in situ* during the blend preparation.

Chemical modification of EVA by introducing a functional group that can act as a chain transfer agent is another elegant pathway for the synthesis of EVA-based graft copolymers.

Considering the versatility of EVA conversion into several other reactive polymers and the behavior of mercaptan groups as chain transfer agent, a new functional polymer derived from EVA was developed with a simple esterification technique between hydrolyzed EVA and mercaptoacetic acid.[8]

APPLICATIONS IN POLYMER BLENDS

One important application of SH-modified EVA consists of its use as reactive compatibilizing agent in polymer blends. SH groups along the EVA backbone can join several other functional groups through specific interactions, hydrogen bonding or even covalent bonding, and promote graft copolymerization *in situ* during the polymer processing.

CONCLUSIONS

The use of mercaptan-modified EVA alone has demonstrated to be a versatile method for blend compatibilization involving EVA or LDPE with other polymers. In addition, it can be used as a precursor for graft copolymerization to be employed as a nonreactive compatibilizing agent for polymer blends. Both EVALSH preparation and graft copolymerization involve simple synthetic routes without the use of special drying techniques as anionic polymerization requires. The large number of monomers able to polymerize using the free-radical technique increases the possibility of synthesis of new graft copolymers that can be used as compatibilizing agents for several other polymer blends.

ACKNOWLEDGMENTS

The authors would like to thank the following coworkers: E. F. Silva, P. J. Oliveira, M. A. R. Moraes, and M. Pinto. This work has been sponsored by CNPq, CAPES, CEPG-UFRJ.

REFERENCES

1. Xanthos, M.; Dagli, S. S. *Polym. Eng. Sci.* **1991**, *31*, 929.
2. Gaylord, N. G. J. *Macromol. Sci., Macromol. Chem.* **1989**, *A26*, 1211.
3. Liu, N. C.; Baker, W. E. *Adv. Polym. Technol.* **1992**, *11*, 249.
4. Andrews, G. D.; Dawson, R. L. In *Encyclopedia of Polymer Science and Engineering*; John Wiley & Sons: New York, NY, 1986; Vol. 6, p 383.
5. Gupta, A. K.; Ratnam, B. K.; Srinivasan, K. R. *J. Appl. Polym. Sci.* **1992**, *45*, 1303.
6. Itazawa, H.; Isuo, B. *Bull. Japan Soc. Sci. Fish.* **1979**, *45*, 323.
7. Pilz, D. *Plaste Kautschuk* **1979**, *21*, 647.
8. Barbosa, R. V.; Soares, B. G.; Gomes, A. S. *J. Appl. Polym. Sci.* **1993**, *47*, 1411.

Ethylene-Vinyl Alcohol Copolymer

See: *Alternating Ethylene-Vinyl Alcohol Copolymer*
Ethylene-Vinyl Acetate Copolymer (Chemical Modification, Compatibilizer in Blends)
Ethylene-Vinyl Alcohol Copolymer (Overview)

ETHYLENE-VINYL ALCOHOL COPOLYMER (Overview)

Yukio Ozeki and Jason Kim
EVAL Company of America

Ethylene-vinyl alcohol copolymer (EVOH) is a thermoplastic material obtained by hydrolyzing the copolymerization product of ethylene and vinyl acetate. This polymer possesses outstanding gas barrier properties and excellent resistance to chemicals such as solvents and hydrocarbons. It also offers an excellent barrier to odor and flavor permeance. EVOH resins are generally more thermally stable than other commercial barrier resins; consequently, they are more easily processable.

Since the commercial introduction of EVOH resin in 1972, these materials have been and continue to be used primarily to food package because they provide an excellent oxygen barrier. Food packaging containing EVOH has replaced many glass and metal containers. The unique properties of EVOH to preserve fragrance and resist chemicals and solvents account for its growing use in other commercial applications.

The current worldwide producers of EVOH and their registered trade names are EVAL Company of America, USA EVAL®); Kuraray Company, Ltd., Japan (EVAL®); and Nippon Gohsei, Japan (Soarnol®).

APPLICATIONS

Existing Applications

EVOH is used in flexible and rigid packaging. Barrier structures containing EVOH resins or films are used in most types of food processing, including aseptic, hot-fill, and retort processes. Products packaged in structures containing EVOH include sauces, juices, meat and cheese products, and processed fruit. Nonfood applications include packaging of solvents, chemicals, and medical products.

New Applications

New EVOH grades have improved thermal stability, oxygen barrier, and thermoformability properties. New applications include automotive fuel systems, protective clothing, and advanced insulation systems. EVOH-based fuel tanks meet the high standards required by more stringent emission control regulations. Its use in protective clothing continues to grow. The use of EVOH for vacuum-panel insulation to replace traditional foam and fiberglass insulation is under development. Other potential applications are PVDC replacement in flexible film, retortable flexible pouches, medical applications requiring barrier films, refrigerator liners, and wallpaper and textile applications.

Ethylenimine-Containing Polymers

See: *Aziridine-Based Polymers*

ETHYLENIMINE-MODIFIED POLYMERS

Ichimoto Akasaki
Nippon Shokubai Company, Ltd.

Ethyleneimine (EI) is so reactive that it can add to a carboxylic group with an opening-chain reaction at low temperature. This reaction is called aminoethylation and allows anionic polymers to be easily converted to cationic polymers (**Structure 1**).

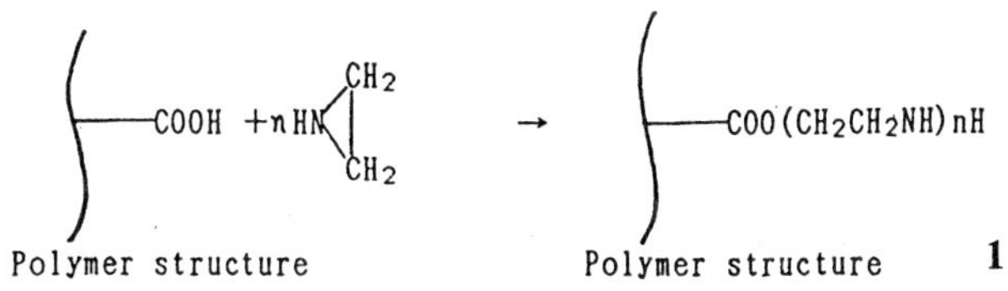

1

Polymers containing carboxylic groups can be modified with EI and gain amine functionality.

This technology produces mostly primary amine group, but secondary and tertiary amine groups are also possible. EI modified polymer has high reactivity to many functional groups, such as epoxy and acryroyl, and good adhesive properties to many substances, because the primary amine group has active hydrogen and a loan pair of electronics on its nitrogen atom.

EI MODIFIED POLYMER

There are three kinds of polymers: solvent-type EI modified polymer, waterborne EI modified polymer, and emulsion-type EI modified polymer. Waterborne EI modified polymer can be produced by neutralizing solvent-type EI modified polymer with acids. Emulsion-type EI-modified polymer can be manufactured by modifying emulsion with EI.

REFERENCES

1. Usala, H. U.S. Patent 3 705 076.
2. Akasaki, I., JP Koukai Tokkyo Kouhou S63-056549.
3. Masuda, M., JP Koukai Tokkyo Kouhou H01-121338.
4. McFadden, R. T. U.S. Patent 3 634 372.
5. Martin, P. H. U.S. Patent 3 719 629.
6. McFadden, R. T. U.S. Patent 4 272 621.
7. Young, R. G. *J. Coat. Technol.* **1977**, *49*, 623.
8. McDowell, M. J. U.S. Patent 3 261 797.
9. Werner, E. R. Jr. U.S. Patent 3 282 879.
10. McDowell, M. J. U.S. Patent 3 309 331.
11. Mesec, K. J. U.S. Patent 3 386 939.
12. Dye, J. I. U.S. Patent 3 455 727.
13. Pinkney, P. S. U.S. Patent 3 547 845.
14. Saeki, K. U.S. Patent 5 082 881.

EUCOMMIA ULMOIDE GUM

Rui-Fang Yan
Institute of Chemistry
Academia Sinica

Eucommia ulmoide gum (EU gum) is a natural polymer whose chemical name is *trans*-polyisoprene. It is known as gutta-perch in Southeast Asia and balata in South America.

EU gum is an isomer of NR, the high elastomer that plays an important role in the rubber industry. But EU gum has no elasticity, and is only durable in plastics form. Many scientists have tried to change EU gum into the high elastomer, but without success.

In 1984, I obtained the German patent on preparing high-elastic *trans*-polyisoprene.[1] I then made high-elastic EU gum and tires of blend EU gum/PB rubber.[2,3] And moreover, I have advanced a series of new theories, for example, the three stages of the vulcanizing process of EU gum and its critical conversion;[2,4] the three kinds of materials corresponding to the three stages; the competitive principle between the opposite factors controlling the critical conversion;[5] the duality of plasticity and elasticity of EU gum and its transition characteristic;[6,7] the stage character of the interrelated materials on the basis of the transition principle, (for example, EU gum, 3 stages; NR, 2 stages; PE, 2 stages); and the systematic material engineering of EU gum.

The new materials derived from these theories and developed from EU gum cover a wide range of materials, from improved functional plastics, thermoplastic elastomers, and thermoelastomers to the high elastomers. This means that research and development on EU gum has entered an entirely new stage.

APPLICATIONS

General Uses

The traditional uses of EU gum are submarine cables, golf balls, HF vessels, dental filler, chewing gum filler, molder, and adhesive. These are all plastic products. Since the advent of cheap synthetic plastics, many uses of EU gum have been replaced, so research and development of EU gum has almost stopped for a long time.

New Application Development Under the Guide of the Three Stages Principle

My group has made the first batch of tires from blend EU gum/PB rubber (1:1). They have been used on motorcycles for two years, and their contour and cross-section are undamaged.

Shape Memory Function of B-Stage EU Gum and Its Applications

B-Stage EU gum is a partial crystalline network polymer. It has different mechanical behaviors below and above melting point. Below melting point, it is a plastic, but above melting point, it becomes a thermoelastomer. Its elastic deformation can be frozen by cooling it to crystallize in the presence of an external force. The deformation can be kept after the elimination of the force. This means that the sample has obtained shape memory; it can return to its original shape again after being heated.

Many new materials have been made using this principle, for example, temperature-controlling switches, multiple functional connectors, new sealing material for stopping leaks, and so on.[8–11]

Material for transmitting radar waves,[12] artificial limb sockets, and protective vests for sportsmen[9] have also been made from EU gum.

REFERENCES

1. Yan, R. F. Ger. Patent DE 3227757.
2. Yan, R. F. *Polymer. Bull.* (Chinese) **1989**, *2*, 39.
3. Yan, R. F. *Preprints, China-Japan Bilateral Symposium on Polymer and Materials*, Guang Zhou, China, 1990, p E15.
4. Yan, R. F. et al. Presented at *Annual Meeting, PPS* (Japan), 1989, p 259.
5. Yan, R. F. et al. Presented at *International Rubber Conference* Orlando, FL, 1993, Session A 60.
6. Yan, R. F. Presented at *International Rubber Conference* Beijing, 1992, p 304.
7. Yan, R. F. *Rubber Ind.* (Chinese) **1992**, *10*, 620.
8. Yan, R. F. et al. *China Patent* CN 88207365.6.
9. Yan, R. F. et al. *China Patent* CN 88103742.7.
10. Yan, R. F. et al. *China Patent* CN 92114761.9.
11. Yan, R. F. et al. *China Patent* CN 93118760.5.
12. Yan, R. F.; Lin, C. L. *China Patent* CN 90101267X.

Expansion in Volume

See: *Bicyclic and Spirocyclic Monomers, Oxygen-Containing (Ring-Opening Polymerization)*
Dental Polymers (Overview)
Ring-Opening Polymerization, Anionic (with Expansion in Volume)
Ring-Opening Polymerization, Cationic (with Expansion in Volume)
Ring-Opening Polymerization, Radical (with no Shrinkage in Volume)

Extrusion Coating Polymers

See: *Polyethylene (Commercial)*

F

Fabrics

See: Cotton Fabric (Water-Repellent Finishes)
Polyester Fabrics (Surface Modification by Low Temperature Plasma)
Polytetrafluoroethylene

FERROCENE-BACKBONE POLYMERS

Ian Manners,* Daniel A. Foucher, and
John K. Pudelski
Department of Chemistry
University of Toronto

A resurgent interest in polymers with skeletal transition metal atoms is driven by their potentially novel electrical, electrochemical, optical, and magnetic properties.[1-7] In addition, polymers with transition elements in the backbone may serve as precursors to metal-containing ceramic materials.[7,8] The most extensive research in transition metal-containing polymers has involved the synthesis of ferrocene-based polymers whose low cost, radiation resistance, and high thermal stability provide many potential uses as advanced materials.[1,2,5] Our focus in this article is on the synthesis, characterization, and properties of representative well-defined polymers containing skeletal ferrocene units rather than attempting to review the vast number of poorly characterized materials.

Most ferrocene-backbone polymers have been prepared via condensation routes. These syntheses include simple salt elimination/coupling reactions or Lewis acid-catalyzed condensation reactions. Because of the difficulty in achieving the stringent reaction conditions required for the formation of high molecular weight polymer, most ferrocene-containing polymers produced via condensation routes have low molecular weights.[5,9] Free-radical initiated reactions, more commonly associated with the preparation of high molecular weight polyvinylferrocene[10] containing pendant ferrocene groups, have also been used to prepare relatively low molecular weight polyferrocenylene.[11,12] In 1992, Rauchfuss et al.[13] described the elegant atom-abstraction induced polymerization to form high molecular weight polyferrocenylpersulfides consisting of an alternating backbone of ferrocenyl moieties and disulfide bridges. A few months later n the same year, a thermal ring-opening polymerization (TROP) route to high molecular weight polyferrocenylsilanes was also reported (Scheme I).[14] This discovery has led to the synthesis of analogous ferrocene-backbone polymers with germanium,[15] phosphorus,[16] and hydrocarbon spacers.[17]

*Author to whom correspondence should be addressed.

SYNTHESIS OF OLIGOMERS AND POLYMERS WITH FERROCENE GROUPS IN THE MAIN CHAIN

Ring-Opening Routes

Atom-Abstraction Induced Ring-Operating Polymerization

Poly(ferrocenylene persulfides), $[Fe(\eta\text{-}C_5H_3R)_2S_2)]_n$ (**16**) (where R = H, *n*-Bu, *t*-Bu), were recently prepared by Rauchfuss et al.[13,28] These well-characterized polymers were prepared from the novel sulfur abstraction reaction with one equivalent of $n\text{-}Bu_3P$. The desulfurization of the doubly strapped [3,3] ferrocenophane $Fe(\eta\text{-}C_5H_3)_2(S_3)_2$ with $n\text{-}Bu_3P$ results in the formation of an insoluble three-dimensional polymeric network, which was characterized by (CP/MAS) ^{13}C NMR spectroscopy. Soluble, exceptionally high molecular weight ($M_w = 3.0 \times 10^6$, $M_n = 8.5 \times 10^5$) networks have been prepared by the selective alkylation of the monomer.

TROP OF [1]Ferrocenophanes with Si, Ge, or P Atoms in the Bridge

Ring-opening polymerization routes to high molecular weight, soluble polyferrocenylsilanes were first reported by Manners et al. in 1992.[14] These polymers, $[Fe(\eta\text{-}C_5H_4)_2SiRR']_n$ consisting of an alternating backbone of ferrocene and alkyl- or arylsilane groups are readily prepared from the correspondingly strained ferrocenophanes, $[Fe(\eta\text{-}C_5H_4)_2SiRR']$.[14,23,29-33] The ring strain, as is evident from the high degree of ring-tilt found in the structures of silicon bridged [1]ferrocenophanes, is believed to be the driving force in these polymerizations.[32] Spontaneous, exothermic, and quantitative thermally induced ROP reactions are achieved by heating the silicon-bridged [1]ferrocenophanes in the melt at elevated temperatures in evacuated, sealed Pyrex tubes (Scheme I).

Several polyferrocenylsilanes with symmetrical[14,29] (R = R′:) and unsymmetrical[33] (R ≠ R′) side groups have been prepared. Even silicon-bridged [1]ferrocenophanes with large or bulky substituents such as *n*-octadecyl, 5-norbornyl, or ferrocenyl, readily undergo TROP.

The simplicity and ease of preparing high molecular weight polyferrocenylsilanes from the thermal ring-opening polymerization of their respective monomers has led to parallel investigations of analgous polymers with bridging atoms other than Si, such as the polyferrocenylgermanes.[15,34,35]

The synthesis of high molecular weight polyferrocenylphosphines via ring-opening polymerization has also been reported by Manners et al.,[16,35] and other similar polymers were first prepared by Pittman[26] and Seyferth[27] by condensation routes.

TROP of Hydrocarbon-Bridged [2]Metallocenophanes

The extension of the TROP strategy to [2]metallocenophanes (**19**) has also proven successful.[17] The hydrocarbon-bridged [2]ferrocenophanes, which possess ring tilts similar to those observed for silicon-bridged ferrocenophanes, undergo TROP at 300°C. Similar polymerization behavior has been observed for hydrocarbon-bridged [2]ruthenocenophanes.[36]

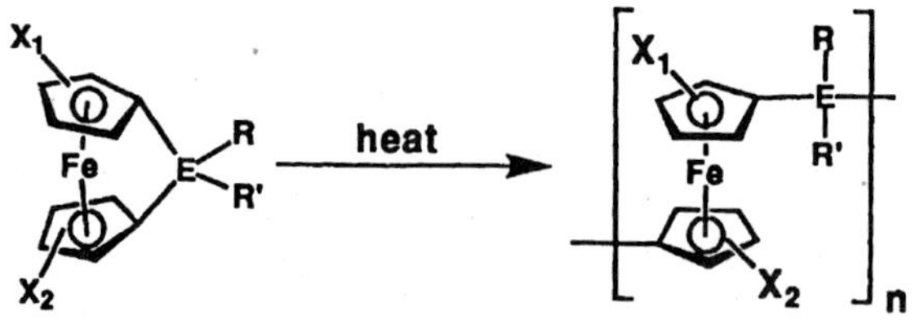

17

I

PROPERTIES OF FERROCENE-BACKBONE OLIGOMERS AND POLYMERS

Electronic Properties and Polyferrocenes

UV-Vis and Near-IR Spectroscopy

Polyferrocenylsilanes possess an interesting main chain that comprises potentially conjugated σ, pπ, and dπ units. Solution UV-vis spectra of polyferrocenylsilanes in the 200-900 nm range were obtained for solutions of the polymers in THF.[23,29,30,32,33] The resulting spectra were similar to those of ferrocene and, in particular, to monomeric ferrocenes with organosilicon groups attached to each cyclopentadienyl ring.

Partially doped bi,[37,40-42,45] tri-, and tetraferrocenes,[43] aside from displaying the expected visible ligand to metal charge transfer (LMCT) band ($^2E_{2g} \rightarrow {}^2E_{1u}$) between 540 and 800 nm, have also been shown to exhibit intervalance electronic transitions (IVET) in the near-IR. Very recent work by Yamamoto et al. on a band description of the electronic states in polyferrocenylene indicates that electronic delocalization in the polymer is intermediate between a highly delocalized π-conjugated polymer and an insulating system.[46]

Electrochemical Studies

Electrochemical studies carried out on ferrocene-containing polymers such as polyvinylferrocene[38,47] and polyferrocenylphosphazenes[39] indicate that pendant ferrocenyl groups do not interact extensively, if at all, with each other, whereas two or more waves are found for polymers in which the ferrocene is part of the polymer backbone, such as polyferrocenylene-persulfides[13] or polyferrocenylsilanes.[14,33,48]

Conductivity Studies

The conductivities for several partially doped ferrocene-containing polymers have been reported.[9,10,18,20,44,49] The virgin ferrocene polymers are insulating, with conductivities of 10^{-14} Scm^{-1}, whereas conductivities for mixed-valence polymers containing pendant ferrocene groups are as high as 6.0×10^{-3} Scm^{-1} for a polyvinylbiferrocene derivative when tetracyanoquinodimethane (TCNQ) is used as an oxidant.[49]

Thermal Transition Behavior

Glass and melt transition data for ferrocene-backbone polymers have, until very recently, remained noticeably rare in the literature. Sharp melt transitions between 55 and 80°C had been reported for a series of polyferrocenylsiloxanes prepared by Pittman and co-workers.[24] No glass transitions were observed for these materials. More recently, the thermal transition behavior of several polyferrocenylsilanes[29,30,33,50] and polyferrocenylgermanes[34] has been investigated by manners et al.

The T_g values for the symmetrical polyferrocenylsilanes generally decreased as the length of the organic side group increased.[23,50] This is a well-established trend for many polymer systems and can be attributed to the generation of additional free volume by pushing the polymer main chains further apart from one another.[94]

The thermal behavior of the polyferrocenylgermanes was investigated by DSC, which showed weak yet identifiable T_gs in each case.[34]

Thermal Stability

The thermal stabilities of several ferrocene-containing polymers have been reported.[8,13,19,21,22,26]

REFERENCES

1. *Advances in Organometallic and Inorganic Polymer Science*; Carraher, C. E. et al., Eds.; Marcel Dekker: New York, **1982**.
2. *Metal-Containing Polymer Systems*; Carraher, C. E. et al., Eds.; Plenum: New York, **1985**.
3. Davies, S. J. et al. *J. Chem. Soc., Chem. Commun.* **1991**, 187.
4. Fyfe, H. B. et al. *J. Chem. Soc., Chem. Commun.* **1991**, 188.
5. Manners, I. *Adv. Organomet. Chem.* **1995**, 37, 131.
6. Zeldin, M. et al. *Inorganic and Organometallic Polymers;* ACS Symposium Series 360; American Chemical Society: Washington, DC, 1988.
7. Peuckert, M. et al. *Adv. Mater.* **1990**, *2*, 398.
8. Tang, B. Z. et al. *J. Chem. Soc., Chem. Commun.* **1993**, 523.
9. Tanaka, M.; Hayashi,T. *Bull Chem. Soc. Jpn.* **1993**, *66*, 334.
10. Pittman, C. U. et al. *Macromolecules* **1970**, *6*, 746.
11. Korshak, V. V. et al. *Dokl. Akad. Nauk. SSSR* **1960**, *132*, 360.
12. Nesmeyanov, A. N. *Dokl. Akad. Nauk. SSSR* **1961**, *137*, 1370.
13. Brandt, P. F.; Rauchfuss, T. B. *J. Am. Chem. Soc.* **1992**, *114*, 1926.
14. Foucher, D. A. et al. *J. Am. Chem. Soc.* **1992**, *114*, 6246.
15. Foucher, D. A.; Manners, I. *Makromol. Chem., Rapid Commun.* **1993**, *14*, 63.
16. Honeyman, C. H. et al. *Poly. Prepr., Am. Chem. Soc. Div. Polym. Chem.* **1993**, *34*(1), 330.
17. Nelson, J. M. et al. *J. Am. Chem. Soc.* **1993**, *115*, 7035.
18. Yamamoto, T. et al. *Inorg. Chim. Acta* **1983**, *75*, 75.
19. Neuse, E. W. *J. Organomet. Chem.* **1966**, *6*, 92.
20. Cowan, D. O. et al. *J. Am. Chem. Soc.* **1972**, *94*, 5110.
21. Abd-Alla, M. M. et al. *J. Appl. Polym. Sci.* **1993**, *47*, 323.
22. Seyferth, D.; Withers, H. P. *J. Organometallics* **1982**, *1*, 1275.
23. Foucher, D.; Ph. D. Thesis, University of Toronto 1993.

24. Patterson, W. J. et al. *J. Polym. Sci.; Polym. Chem. Ed.* **1974**, *12*, 837.
25. Neuse, E. W.; Chris, G. J. *Macromol. Sci. Chem.* **1967**, *3*, 371.
26. Pittman, C. U. *J. Polym. Sci., Polym. Chem. Ed.* **1967**, *5*, 2927.
27. Withers, H. P. et al. *Organometallics* **1982**, *1*, 1283.
28. Compton, D. L.; Rauchfuss, T. B. *Am. Chem. Soc. Polym. Prepr.* **1993**, *34*, 351.
29. Foucher, D. A. et al. *Macromolecules* **1993**, *26*, 2878.
30. Manners, I. *J. Inorg. Organomet. Polym.* **1993**, *3*, 185.
31. Nguyen, M. T. et al. *Chem. Mater.* **1993**, *5*, 1389.
32. Finckh, W. et al. *Organometallics* **1993**, *12*, 823.
33. Foucher, D. A. et al. *Macromolecules* **1994**, *27*, 3992.
34. Foucher, D. A. et al. *Organometallics* **1994**, *13*, 4959.
35. Honeyman, C. H.; Manners, I. *Organometallics* **1995**, *14*, 5503.
36. Nelson, J. M. et al. *Angew. Chem. Int. Ed. Engl.* **1994**, *33*, 989.
37. LeVanda, C. et al. *J. Am. Chem. Soc.* **1997**, *99*, 2964.
38. Flanagan, J. B. et al. *J. Am. Chem. Soc.* **1978**, *100*, 4248.
39. Saraceno, R. A. et al. *J. Am. Chem. Soc.* **1988**, *100*, 7254.
40. LeVanda, C. et al. J. *Am. Chem. Soc.* **1974**, *96*, 6788.
41. LeVanda, C. et al. J. *Org. Chem.* **1976**, *41*, 2700.
42. Dong. T. Y. et al. *J. Organomet. Chem.* **1990**, *391*, 377.
43. Delgado-Pena, F. et al. *J. Organomet. Chem.* **1983**, *253*, C43.
44. Pittman, C. U.; Surynarayanan, B. J. *Am. Chem. Soc.* **1974**, *96*, 7916.
45. Kramer, J. A.; Hendrickson, D. N. *Inorg. Chem.* **1980**, *19*, 3330.
46. Seki, K. et al. *Chem. Phys. Lett.* **1991**, *178*, 311.
47. Smith, T. W. et al. *J. Polym. Sci.* **1976**, *14*, 2433.
48. Foucher, D. A. et al. *Angew. Chem. Int. Ed. Engl.* **1993**, *32*, 1709.
49. Pittman, C. U. et al. *Inorganic Compounds with Unusual Properties;* American Chemical Society: Washington, DC **1976**, *150, 46.*
50. Seker, F. et al. unpublished results, 1994.
51. Sperling, L. H. *Physical Polymer Science*; Wiley-Interscience: Toronto, 1986.

Ferrocene-Containing Polymers

See: Ferrocene-Backbone Polymers
Germanium-Containing Polymers (Overview)
Metal-Containing Silicon-Based Polymers
Organometallic Polymers (Transition Metals in Main Chain)
Poly(1,1′-Ferrocene-Arylene)s (Pd-Catalyzed)
Polysilanes (Ring-Opening Polymerization of Cyclotetrasilanes)

Ferroelectric Polymers

See: Ferroelectric Polymers (Structural Phase Transitions)
Liquid Crystalline Elastomers
Liquid Crystalline Polymers, Side Chain
Polyamides, Alicyclic and Aromatic (Ferroelectric Hystersis and Pyroelectricity)
Poly(vinylidene fluoride) (Overview)
Supramolecular Self-Assembly (Liquid Crystalline Polymers, Hydrogen Bonding)
Vinylidene Fluoride-Trifluorethylene Copolymers (Ferroelectric-to-Paraelectric Phase Transition)

FERROELECTRIC POLYMERS (Structural Phase Transitions)

Kohji Tashiro
Department of Macromolecular Science
Faculty of Science
Osaka University

FERROELECTRICITY IN POLYMERS

Polymer materials exhibiting high piezo- and pyroelectricity include poly(vinylidene fluoride) (PVDF), vinylidene fluoride-trifluoroethylene (VDF-TrFE) copolymers, vinylidene fluoride-tetrafluoroethylene (VDF-TFE) copolymers, copolymers of vinylidene cyanide with *n*-fatty acid vinyl ester, odd nylons, polyureas, poly(L-lactic acid), and so on. In particular, PVDF and its copolymers containing the crystal form I show the largest piezo- and pyroelectric effects among the synthetic polymers.[1-5] In the form I crystal, planar-zigzag chains are packed together with their CF_2 dipoles parallel along the b axis, giving a large polarity to the whole crystal.[6-8]

From the early stage of X-ray structural analysis, PVDF form I has been speculated to be a ferroelectric material. If a material is ferroelectric, the dipoles of the crystal should inverse reversibly without any change in the resulting structure when the sign of the external electric field is inverted.[9] The measurements of *D-E* hysteresis loop (*D*: electric displacement and *E*: electric strength) and inversion current, the direct measurements of rotation of chain dipoles by Ir/Raman spectroscopy, NMR, X-ray diffraction, and so on have confirmed that PVDF form I is a ferroelectric polymer material. One of its significant properties is the phase transition between ferroelectric and paraelectric phases through the Curie transition temperature (T_c).

Another remarkable ferroelectric transition is that of ferroelectric liquid crystalline polymers (FLCP). These so-called ferroelectric liquid crystals (FLC) have been applied widely in industry. FLCP is the polymer containing these chiral mesogenic groups as side chains or in the main chain.

FERROELECTRIC PHASE TRANSITION IN FLUORINE POLYMERS

VDF Content Dependence of Transitional Behavior

Transitional behavior is affected largely by the VDF content.[11-13] Tashiro et al. measured the temperature dependencies of IR absorbance and X-ray diffraction intensity during heating of PVDF form I and found trans-to-gauche conformational exchange immediately below the melting point (~ 172°C).[14,15]

As pointed out in several papers, the data presented by Tashiro et al. might be interpreted in another way: that the "center" of T_c locates above the melting temperature and only the beginning of the transition could be detected because a broad transition might overlap even partly with a broad melting,[16] or that the melt of form I crystal and recrystallization to form III crystal might occur in the temperature region close to the melting point.[17] But at any rate, a possibility of ferroelectric phase transition of PVDF form I was presented, and the T_c was proposed to be in the same temperature region as the melting phenomenon.

Ferroelectric Phase Transitions in VDF-TFE Copolymers

Vinylidene fluoride-tetrafluoroethylene (VDF-TFE) copolymers exhibit ferroelectric phase transitions similar to those of VDF-TrFE copolymers, but their behavior differs depending on the VDF content.[16,18-22]

OTHER TYPES OF FERROELECTRIC POLYMERS

Vinylidene Cyanide Copolymers

The cyanide group is highly polar. Poly(vinylidene cyanide) (PVDCN), which has a chemical structure of $-[-CH_2C(CN)_2-]-$, is another candidate to demonstrate ferroelectric properties. This homopolymer is not chemically stable, so the copolymers between VDCN and *n*-alkylfatty acid ester have been synthesized.[23,24] These copolymers with different alkyl chain length *n* were reported to show characteristic electric behavior.[25-27] In particular, an alternating copolymer of $n = 2$ or VDCN-vinyl acetate (VDCN-VAc) copolymer[28-30] attracts much attention because of its high piezoelectricity, comparable to that of PVDF form I.

Odd Nylons

Since the crystal structure of odd nylon (e.g., nylon 11) was proposed by X-ray diffraction analysis, many investigators have speculated on the ferroelectricity of this crystal.[31] The unit cell structure is polar as a whole; essentially all-trans zigzag chains are packed with the amide C=O groups parallel to each other. Piezoelectric constant, dielectric constant, and so on have been measured for a series of odd nylons.[32-36] Until recently, however, no clear proof of ferroelectric behavior of odd nylons had been presented. Scheinbeim et al. prepared the oriented films by melt quenching and drawing at room temperature. They showed a clear *D-E* hysteresis loop.[35-37] The *D-E* hysteresis loop indicates that odd nylons may be ferroelectric.

Ferroelectric Liquid Crystalline Polymers

Liquid crystalline molecules having chiral carbon atoms as the linkage between mesogenic-group and end-alkyl chains may exhibit ferroelectric behavior in the smectic C phase (Sm C*).[10] In this phase the molecular axis tilts from the normal to the layer plane and the electric dipoles of C=O orient into a common direction to each other, that is, the layers are polar in the lateral direction. On the other hand, the smectic A phase (Sm A) is paraelectric because the molecules are packed in the layer with their axes set vertically and they rotate in a liquidlike fashion around the molecular axis. The ferroelectric-to-paraelectric phase transition is observed between Sm C* and Sm A phases. If such liquid crystalline molecules are introduced into a polymer system as a component of the skeletal chain or as a side group sticking out of the main chain, a ferroelectric liquid crystalline polymer (FLCP) is obtained.

REFERENCES

1. Hayakawa, R.; Wada, Y. *Progr. Polym.* Sci. **1973**, *11*, 1.
2. Wada, Y.; Hayakawa, R. *Jpn. J. Appl. Phys.* **1976**, *15*, 2041.
3. Tashiro, K. et al. *Macromolecules* **1980**, *13*, 691.
4. Tashiro, K. et al. *Ferroelectrics* **1981**, *32*, 167.
5. Furukawa, T. *Phase Transitions* **1989**, *18*, 143.
6. Lando, J. B. et al. J. *Polym. Sci., Part A-1* **1966**, *4*, 941.
7. Hasegawa, R. et al. *Polym. J.* **1972**, *3*, 600.
8. Tashiro, K.; Tadokoro, H. *Macromolecules* **1983**, *16*, 961.
9. Lines, M. E.; Glass, A. M. *Principles and Applications of Ferroelectrics and Related Materials*; Oxford University: London, 1977.
10. Meyer, R. B. et al. *J. Phys. (Paris) Lett.* **1976**, *36*, L-69.
11. Tashiro, K. et al. *Ferroelectrics* **1984**, *57*, 297.
12. Lovinger, A. J. et al. *Polymer* **1983**, *24*, 1225.
13. Lovinger, A. J. et al. *Polymer* **1983**, *24*, 1233.
14. Tashiro, K. et al. *Polymer* **1983**, *24*, 199.
15. Tashiro, K. et al. *Polym. Bull.* **1983**, *10*, 464.
16. Lovinger, A. J. *Macromolecules* **1982**, *16*, 1529.
17. Takahashi, Y.; Miyamoto, N. *J. Polym. Sci., Polym. Phys. Ed.* **1985**, *23*, 2505.
18. Lovinger, A. J. et al. *J. Appl. Phys.* **1984**, *56*, 2412.
19. Murata, Y.; Koizumi, N. *Polym. J.* **1985**, *17*, 1071.
20. Lovinger, A. J. et al. *Macromolecules* **1988**, *21*, 78.
21. Tashiro, K. et al. *Polymer* **1992**, *33*, 2915.
22. Tashiro, K. et al. *Polymer* **1992**, *33*, 2929.
23. Kishimoto, M. et al. *Polym. Prepr. Jpn.* **1988**, *37*, 2249.
24. Tasaka, S. et al. *Polymer* **1989**, *30*, 1639.
25. Zou, D. et al. *Polymer* **1990**, *31*, 1888.
26. Wang, T. T.; Takase, Y. *J. Appl. Phys.* **1987**, *62*, 3466.
27. Seo, I. et al. *Nonlinear Optics of Organics and Semiconductors* Kobayashi, T. Ed.; Springer-Verlag: Berlin, 1989; p. 196.
28. Inoue, Y. et al. *Polymer* **1989**, *30*, 698.
29. Mirau, P. A.; Heffner, S. A. *Polymer* **1992**, *33*, 1156.
30. Tasaka, S. et al. *Ferroelectrics* **1984**, *57*, 267.
31. Hasegawa, R. K. et al. *Polym. Prepr. Jpn.* **1974**, 713.
32. Wu, G. et al. *Polym. J.* **1986**, *18*, 51.
33. Newman, B. A. et al. *J. Appl. Phys.* **1980**, *51*, 5161.
34. Scheinbeim, J. I. *J. Appl. Phys.* **1981**, *52*, 5939.
35. Lee, J. W. et al. *J. Polym. Sci.: Part B: Polym. Phys.* **1991**, *29*, 273.
36. Lee, J. W. et al. *J. Polym. Sci.: Part B: Polym. Phys.* **1991**, *29*, 279.
37. Takase, Y. *Macromolecules* **1991**, *24*, 6644.

FERROMAGNETIC POLYMERS

Alexander A. Ovchinnikov
Russian Academy of Sciences

Val. N. Spector
Department of Electronics of Organic Materials
Russian Academy of Sciences

Magnetism has always been a subject of fascination and is still not completely understood even in such traditional substances as iron. Nevertheless, extended *d*- and *f*-electronic orbitals play a decisive role in ferromagnetic ordering of nonpaired electrons under strict requirements to crystalline structure.

In the mid 1980s the theory of organic ferromagnetic substances[1] found its realization in synthesis of ferromagnetic polymers.[3-6] Despite many controversies[2,7,8] the existence of experimentally reproducible ferromagnetism in organic materials became an indisputable fact.

PREPARATION OF FERROMAGNETIC POLYMERS

High-Multiplicity Polyradicals

The first most evident approach to increasing the length of polymer chains with structurally ordered stable free radicals was used by Japanese chemists, who finally prepared polymeric molecules with spins as high as $^{11}/_2$.[9,10]

The major breakthrough in synthetic preparation of ferromagnetic polymers took place when the first quasi-one-dimensional magnetic polymer was obtained[4] and later reproduced in a number of independent laboratories.[4-6] It was prepared by an explosive polymerization of a monomer of the following structure (Structure 1).[11]

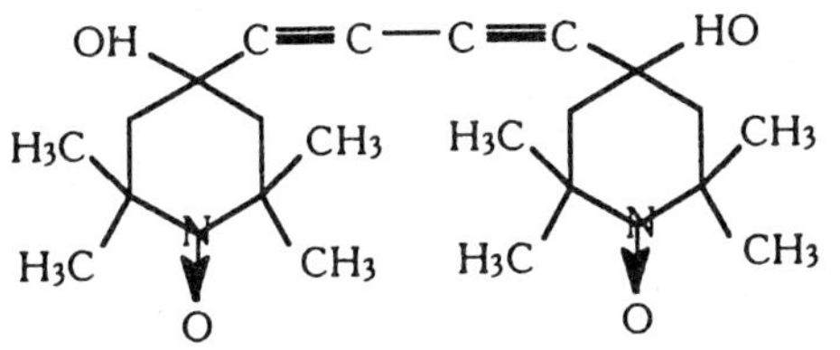

1

Quasi-Two-Dimensional and Bulk Ferromagnetic Polymers

In a few months after publication of our paper[3] a new two-dimensional ferromagnetic polymer was announced.[13] It was obtained by an oxidative condensation of sym-tri-aminobenzene, and the best run was characterized by the following magnetic properties: $Ms<1.6$ cm^{3}G g^{-1} and H_c <100 Öe. The inescapable logic of this paper, that irreversibility of magnetism on a decomposition of the polymer is a clear indication of its organic origin, is applicable to ferromagnetic polydiacetylenes.

Higher demensionally of ferromagnetic polymers is desirable, as it diminishes the role of random intermolecular interaction. In pursuit of this goal, researchers have undertaken pyrolysis of several polymers with structurally ordered sequences of pendant groups, such as polyacrylonitrile (PAN),[14] polypropylene (PP),[15] and polyaniline (PA).[16] One very interesting variation was worked out that led to milder condensation conditions (and better structural control) and a new class of ferromagnetic thermostable and thermosetting COPNA resins with triarylmethane structure.[17] The variation originated from polycyclic aromatic compounds by heating them with aromatic aldehydes under acidic catalyst.

All new ideas for synthesis of ferromagnetic polymers are, in one way or another, attempts to get away from topochemical polymerization, which implies too-narrow gates for entry into the ferromagnetic polymer realm. All avenues, whether more stable radicals for further polymerization,[12,18] or the generation of ordered paramagnetic sequences in diamagnetic prepregs by either transformation into radical state at reactive sites[16,17,19-21] or decomposition of unstable molecular moieties with formation of radical states,[10a,22] seem equally promising.

CONCLUSION

Polymer (organic) ferromagnetic materials are now well established. Independent groups have prepared more than 50 samples of magnetic polymers and products of polymeranalogous transformations with intrinsic ferromagnetism. All these substances are characterized by more or less the same magnetic and physicochemical properties. Quasi-one-dimensional ferromagnetic polymers have low but stable (around 1 cm^3 G g^{-1}) saturation magnetization, coercive force between 100 and 800 Öe, and an expressed tendency to form spin glass. Quasi-two-dimensional ferromagnetic products of oxidative condensation of polymers with an ordered sequence of side-chain substituents usually have higher values of saturation magnetization, around 5 cm^3 G g^{-1}, lower coercive force (as a rule < 100 Öe), and tendency to temporal and environmental degradation. Substances from both these groups usually do not possess a Curie point, as this temperature is normally above the limit of their thermal decomposition or degradation.

Second, there is an established, however unpleasant, fundamental difficulty connected with contemporary synthetic approaches to preparation of ferromagnetic polymers and magnetic products of their polymer-analogous transformations. As a rule, yield of a magnetic fraction in any existing techniques is dismally low (<5% mass). The same may be said about reproducibility of the results in many such techniques, which is still rather poor.

The third conclusion concerns knowledge of chemical structures of ferromagnetic polymers and magnetic products of polymer transformations. Two ferromagnetic organic compounds with an established structure are known[18c,d,23] and they may be models for other polymers.[24] The other approach, which has not been instantly successful, is connected with topochemical polymerization. The most promising way to get better insight into ferromagnetic polymer chemical structure is instrumental studies of local moiety structures.

The difficulties on the way to ferromagnetic polymers are many, but these polymers promise new magnetic materials, new properties, and new applications. Some of these applications are based on such anticipated properties as low density, flexibility, and processibility at low temperature.[25] Expected applications include bulk permanent magnets, magnetic shielding of low-frequency magnetic fields, novel memory media, optical disks, magnetic dielectric materials, gas sensors, and photosensitive materials. Efforts and expenses in this field are certain not to be wasted and may bring a manifold return in the near future.

ACKNOWLEDGMENTS

We are grateful for helpful discussion to R. Baughman, K. Bozhenko, J. Cao, L. Dulog, A. Epstein, R. Friend, F. Garnier, H-H. Horhold, K. Itoh, H. Iwamura, M. Kinoshita, D. Klein, K. Kimura, V. Krivnov, O. Lazareva, J. Lobanovsky, B. Lubentzov, J. Miller, K. Müllen, Y. W. Park, J. Ribo, I. Shamovsky, A. Shchegolikhin, T. Sagawara, N. Tyutyulkov, P. Wang, G. Wegner, F. Wudl, K. Yamaguchi, and many other colleagues and collaborators.

REFERENCES

1. (a)Ovchinnikov, A. A. *Rep. Ac. Sci. USSR* **1977**, *236*, 928; (b)Ovchinnikov, A. A. *Theoret Chim. Acta* **1978**, *47*, 297.
2. Miller, J. S. et al. *J. Am. Chem. Soc.* **1987**, *109*, 769.

3. (a)Korshak, Y. V. et al. *JETP Lett.* (Russ.) **1986**, *43*, 399; (b)Korshak, Y. V. et al. *Nature* **1987**, *326*, 370; (c)Ovchinnikov, A. A.; Spector, V. N. *Synth. Met.* **1988**, *27*, B615; (d)Ovchinnikov, A. A. et al. *Rep. Ac. Sci. USSR* **1988**, *302*, 634; (e)Shchegolikhin, A. N. et al. *Synth. Met.* **1995**, *71*, 1825; (f)Lee, W. P. et al. *J. Kor. Phys. Soc.* **1992**, *25*, 432.
4. (a)Sugano, T. et al. *Abstr. Symp. Mol. Struct.* **1987**, 504-505; (b) Sugano, T. et al. Preprint of ISSP "Magnetic Properties of Polydiacetylene Polymers" Kanazawa (Japan), 1987.
5. (a)Cao, J. et al. *Solid State Commun.* **1988**, *68*, 817; (b)Cao, J. et al. *Synth. Met.* **1988**, *27*, B625; (c)Cao, J. et al. *Academia Sinica Preprint* "Chemical and Magnetic Characterization of Organic Ferromagnet–PolyBIPO" Beijing: 1988.
6. (a)Sugawara, T. et al. *J. Am. Chem. Soc.* **1988**, *108*, 368; (b)Kimura, K. private communication (measurement data for a sample presented by A. Ovchinnikov, ref. 4.)
7. (a)Dulog, L. *Nachr. Chem. Tech. Lab.* **1990**, *38*(4), 445; (b)Dulog, L; Lute, S. *Liebigs Ann. Chem.* 1991, p. 971; (c)Dulog, L.; Lute, S. *Macromol. Chem. Rapid Commun.* **1993**, *14*, 147.
8. Ed.: Organizing Committee. *Abstracts of International Conference on Science and Technology of Synthetic Metals* Santa Fe, NM: 1988.
9. (a)Itoh, K. *Pure Appl. Chem.* **1978**, *50*, 1251,; (b)Itoh, K. *Chem. Phys. Lett.* **1967**, *1*, 235; (c)Teki, Y. et al. *J. Am. Chem. Soc.* **1986**, *108*, 2147.
10. (a)Iwamura, H.; Izuoka, A. *Nippon Kagaki Kaisi* **1987**, *#4*, p. 595; (b)Sugawara, T. et al. *J. Am. Chem. Soc.* **1984**, *106*, 6449; (c) Sugawara, T. et al. *J. Am. Chem. Soc.* **1986**, *108*, 368; (d)Teki, Y. et al. *J. Am. Chem. Soc.* **1983**, *105*, 3722; (e)Teki, Y. et al. *J. Am. Chem. Soc.* **1986**, *108*, 2117; (f)Iwamura, H. *Pure Appl. Chem.* **1986**, *58*, 187; (g)Murata, S.; Iwamura, H. *J. Am. Chem. Soc.* **1991**, *113*, 5547.
11. (a)Rozantzev, E. G. *Theor. I Exper. Khimia (Russ)* **1966**, *2*, 415; (b)Rozantzeu, E. G. Free Nitroxyl Radicals; Plenum Press, New York/London, 1970.
12. (a)Hosokoshi, Y. et al. *Proc. TODAI Symp. 4-th ISSP Int. Symp.* "Frontiers in High Magnetic Fields" ISSP, Tokyo, Japan 1993; (b) Nakazawa, Y. et al. *Phys. Rev. B* **1992**, *46*, 8906; (c)Shiomi, D. et al. *J. Phys. Soc. (Jpn)* **1993**, *62*, 289; (d)Hosokoshi, Y. et al. *ISSP Technical Rep. Series A* # 2791; (e)Turek, P. et al. *Chem. Phys. Lett.* **1991**, *180*, 327; (f)Tamura, M. et al. *Mol. Cryst. Liq. Cryst.* **1993**, *232*, 45; (g)Le, L. P. et al. *Chem. Phys. Lett.* **1993**, *206*, 405; (h)Kinoshita, M. *Mol. Cryst. Liq. Cryst.* **1993**, *232*, 1; (I)Kinoshita, M. *Synth. Met.* **1993**, *55-57*, 3285; (j)Kinoshita, M. et al. *Chem. Lett. (Jpn)* **1991**, 1225; (k)Shiomi, D. et al. *Mol. Cryst. Liq. Cryst.* **1993**, *232*, 109; (l)Kanno, F. et al. J. *Phys. Chem.* **1993**, *19*, 13267.
13. Torrance, J. B. et al. *Synth. Met.* **1987**, *19*, 709.
14. Ovchinnikov, A. A. et al. *Rep. Ac. Sci. USSR* **1988**, *302*, 885.
15. (a)Smirnova, S. G. et al. *JETP Lett.* **1988**, *48*, 212; (b)Smirnova, S. G. et al. *Vysokomol. Soed. (Russ)* **1989**, *318*, 323; (c)Grigorov, L. N. et al. *Macromol. Chem. Macromol. Symp.* **1990**, *37*, 177.
16. Madjarova, G. et al. *Macromol. Theory Simul.* **1994**, *3*, 803.
17. (a)Ota, M.; Otani, S. *Chem. Lett. (Jpn)* **1989**, p. 1179; (b)Tanaka, H. et al. *Chem. Lett. (Jpn)* **1990**, 1813; (c)Ota, M. et al. *Abstracts of Papers, Annual Meeting of the American Chemical Society* Dallas, TX: 1989, p. 33.
18. (a)Dulog, L.; Wang, W. *Adv. Mater.* **1992**, *4*, 349; (b)Dulog, L.; Wang, W. *Leibigs Ann. Chem.* **1992**, 301; (c)Dulog, L.; Kim, J. S. *Korea Polym. J.* **1993**, *1*, 96; (d)Dulog, L.; Kim, J. S. *New Scie.* **1990**, *1727*, 29; (e)Batz, C. et al. *Liebigs Ann. Chem.* **1994**, 739; (f)Zierer, D.; Dulog, L. *Liebigs Ann. Chem.* **1993**, 691; (g) Dulog, L.; Gittinger, A. *Mol. Cryst. Liq. Cryst.* **1992**, *213*, 31; (h)Dulog, L.; Gittinger, A. *Mol. Cryst. Liq. Cryst.* **1992**, *237*, 235.
19. (a)Baumgarten, M. et al. *Proc. ICSM'94* Seoul, Korea, 1994, *Synth. Metals* **1995**, *70*, 1389; (b)Baumgarten, M.; Mullen, K. *Topics Curr. Chem.* **1994**, *169*, 1; (c)Karabunarliev, S. et al. *Chem. Phys.* **1994**, *179*, 421; (d)Baumgarten, M. et al. *Chem. Phys. Lett.* **1994**, *221*, 71.
20. (a)Horhold, H.-H. et al. *Proc. ICSM'94* Seoul, Korea: 1994 *Synth. Metals* **1995**, *69*, 525; (b)Helbig, M.; Horhold, H.-H. *Electronic Properties of Polymers*; Springer Series in Solid State Sci. Kuzmany, H. et al. Ed. Springer-Verlag: Berlin-Heidelberg, 1992, p. 321; (c) Horhold, H.-H.; Helbig, M. *Macromol. Chem. Macromol. Symp.* **1987**, *12*, 229; (d) Raabe, D.; Horhold, H.-H.; *Acta Polym.* **1992**, *43*, 275.
21. Lubentzov, B. V. et al. *In: Abstr. Int. Conf. Sc. Techu. Synth. Met.* 1994, Seoul, Korea, ICSM'94, Ed., Organizing Committee, p. 488.
22. (a)Iwamura, H. et al. In *Abstract of Papers, IUPAC CHEMRAWN VI* Tokyo, Japan, 1987, IC05; (b)Fujii, A. et al. *Macromolecules* **1991**, *24*, 1077.
23. Wudl, F. *Proc. Noble Symp.* Lulleo, Sweden, 1992, NS-81.
24. (a)Friend, R. H. *Nature* **1987**, *326*, 368; (b)Alexander, C. et al. *J. Mater. Chem.* **1992**, *2*, 459.
25. Miller, J. S. *Adv. Mater.* **1994**, *6*, 322.

Fertilizers

See: Ionomers (Overview)

Fiber Reinforcement

See: Composites (Structure, Properties, and Manufacturing)
Composites, Thermosetting Polymers
Thermoplastic Composites

Fiberboard

See: Wood Composites (High Performance)

Fibers

See: Alumina Fibers (from Poly[(acyloxy)aloxanes]s)
Aramid Fibers and Nonwovens
Aramid Fibers, High Strength
Aramids (Processable Precursors)
Biospinning (Silk Fiber Formation, Multiple Spinning Mechanisms)
Carbon/Carbon Composites (with Surface Treated Carbon Fibers)
Carbon Fibers (Overview)
Carbon Fibers (Plasma Surface Treatments)
Carbon Whisker (Surface Modification by Grafting)
Cellulose (Overview)
Cellulose (Direct Dissolution)
Cellulosic Materials (Moisture Sorption Properties)
Composites (Structure, Properties, and Manufacturing)
Composites, Thermosetting Polymers
Fibers (Moisture Sorption Properties)
Fibers (Stainproofing)
Fibers and Composites (High Modulus and Strength)
Fibers, Synthetic
Fibers, Synthetic (Unspun Natural Pulp-Like)
Flame Retardant Finishing
Functional Fibers
High Performance Paper

FIBERS (Moisture Sorption Properties)

Takashi Ohtsubo
Matsuyama Shinonome Junior College

With a few exceptions, such as polyethylene or polypropylene fibers characterized by nonpolar chemical structure, most textile fibers can absorb moisture from the atmosphere. Fibers with a high ratio of length to thickness and small section absorb moisture faster, and moisture sorption of a fiber attains equilibrium more rapidly than general plastic moldings.

TABLE 1. Polymer, Chemical Groups, and Moisture Regain of Major Fibers

Fiber	Polymer	Chemical groups	Moisture regain (70 °F, 65% R.H.)
Cotton	Cellulose	-OH, $-CH_2OH$	8.5
Wool	Keratin	-NHCO-, $-NH_2$ etc.	15.0
Silk	Fibroin	-NHCO-, $-NH_2$ etc.	11.0
Viscose rayon	Cellulose	-OH, $-CH_2OH$	12.0
Acetate	Secondary cellulose acetate	-OH, $-COOCH_3$, $-CH_2COOCH_3$	6.5
Triacetate	Cellulose triacetate	$-COOCCH_3$, $-CH_2COOCH_3$	3.5
Nylon	Poly(hexamethylene adipamide)	-NHCO-, $-NH_2$, -COOH	4.5
Polyester	Poly(ethylene terephthalate)	-COO-, -OH, -COOH	0.4
Acryl	Polyacrylonitrile	$-CH_2(CN)-$	2.0
Polypropylene	Polypropylene	$-CH_2-CH(CH_3)-$	0

The amount of water within a fiber exerts a profound influence on its rheological and other properties. Moreover, moisture-absorbing properties will significantly affect the usefulness of fibers for textures. For example, a fiber that permits some moisture absorption is comfortable to wear, especially in hot weather.

PROPERTIES

A fiber's ability to absorb moisture is a direct reflection of its chemical structure. Chemical groups in the polymer molecule of a fiber are divided into hydrophilic groups and hydrophobic groups. The major hydrophilic groups are hydroxyl (–OH), nitryl (–CN), amino ($-NH_2$), carboxyl (–COOH), carbonyl (>CO), ester (–COO–) and amide (–CONH–). On the other hand, methyl ($-CH_3$) and methylene ($-CH_2-$) are hydrophobic groups.

All natural vegetable and animal fibers have chemical groups in their molecules that attract moisture. One the other hand, most synthetic fibers contain few if any moisture-attractive groups, which accounts for their low moisture absorption.

The cellulose polymer found in cotton and viscose rayon is highly reactive because of the three free hydroxyl groups ($-OH-CH_2OH$) on each anhydroglucose unit, the basic monomeric repeating unit of cellulose. The three hydroxyl groups in cellulose serve as principal adsorption sites for water molecules by hydrogen bonding.

Moisture regain is affected not only by fiber's chemical groups but also by other factors such as crystallinity. For example, the moisture regain of cellulosic fibers is linearly related to the amorphous fraction of a fiber,[1] and the higher moisture regain of viscose rayon fiber compared with that of cotton is due to rayon's lower crystallinity.

Wool is a hygroscopic fiber. At standard humidity and temperature, it has a regain of approximately 15%. Wool belongs to a class of proteins known as keratin, and the polypeptide chains of keratin contain at least 19 amino acids. Once basic disadvantage of conventional polyester (PET) fibers is their low absorbency, reflecting their lack of sorption sites and their rigid backbone polymer, which restricts chain mobility and accessibility. The low moisture absorbency of PET fibers make them uncomfortable to wear. Therefore, a variety of techniques have been developed to improve absorbency. One method that has received considerable attention is reacting PET fibers with base. The most hydrophilic groups in PET are carboxyl (–COOH) and hydroxyl (–OH). For example, the caustic-soda treatment for PET fibers increased the carboxyl end-groups and consequently increases moisture regain.[2]

The most important polymeric linkage for nylons is amide groups, which also serve as the site for moisture absorption in nylon fibers; this makes them reasonably hydrophilic.

REFERENCES

1. Jeffries, R. *J. Appl. Polym. Sci.* **1964**, 3, 1213.
2. Hendrix, H. *Zeitschr. Gesamte Text.* **1961**, 63, 962.

FIBERS (Stainproofing)

Yashavanth K. Kamath,* X. X. Huang, S. B. Ruetsch, and H.-D. Weigmann
TRI/Princeton

Fabric staining is an important consumer concern, one that has led to the development of several stain-release products.

Most staining substances are acidic food colors. These are relatively small molecules capable of diffusing into the constituent fibers of the fabric and binding with the functional groups of the polymer. Poly(ethylene terephthalate)–(PET), polypropylene–(PP), and polyacrylonitrile–(PAN)-based fibers are more resistant to diffusion because of lower swelling. However, natural polyamides such as silk and wool, synthetic polyamides like nylon, and cellulosic fibers such as cotton are more susceptible to adsorption and diffusion of stains because of the presence of strong polar groups and their high degree of swelling in aqueous environments. The ionic nature of the stains plays a significant role in the staining of polyamides because of their ability to bind through salt linkages.

Without diffusion, staining involves the adsorption of stain molecules on the fiber surface. Therefore, surface energy of the fiber, which is responsible for adsorption, plays an important role in staining. The surface energy of the fiber, which is a function of the chemical composition of the surface, can be divided into dispersive and acid-base components.[1] Fibers based on hydrocarbon polymers have low surface energies because they possess only the dispersive type of surface energy.[2] Aqueous stains with surface tensions higher than the critical surface tension of the fiber surface do not spread and penetrate (wick) fabrics made from these fibers.[3] They adsorb on the surface of these fibers by van der Waal's forces, which are weak, and therefore can easily be removed. Fibers based on polar polymers, on the other hand, possess both dispersive and acid-base components of surface energy.[4] Because of their higher surface energies, aqueous stains penetrate and spread easily through fabrics of these fibers. Also, the staining molecules chemisorb on these surfaces, rendering them more difficult to dislodge.

*Author to whom correspondence should be addressed.

Staining of synthetic polyamides nylon 6 and 66 is especially important because of their extensive use in the manufacture of carpets. Because carpets are expensive and cannot be replaced frequently, stainproofing technology has been selectively aimed at these products. Attempts are being made to extend this technology to other products, such as wool carpets, home furnishings, and even apparel.

PREPARATION OF SYNTAN STAIN BLOCKERS

Syntans are generally prepared by condensing formaldehyde with sulfonic acid-substituted aromatic phenols.[5,6] These can be phenolsulfonic acids, naphtholsulfonic acids, or a sufonated mixture of phenol, naphthol, cresol, and bisphenol. Sometimes thiophenolics, such as monothiobisphenols and dihydroxydiphenylsulfones, are also used. The cost of raw materials is the guiding factor in selecting a particular phenolic component. The preferred condensation polymer is one having a M_n of 1000 or less.

Stain-blocking polymers are generally bound to nylon through electrovalent bonds between the sulfonate anion and the protonated amine end groups. Formation of covalent linkages through methylol groups cannot be completely ruled out. Instability of the salt linkage in aqueous environments, often encountered in the shampooing of carpets, leads to loss of the stain blocker. This problem has been alleviated by recent work reported in the patent literature involving coupling agents to bind the stain blocker covalently to specific functionalities in the substrate. The patent[8] claims significant improvements in the washfastness of the stain blocker.

Stain-Blocking Mechanisms

The double layer repulsion at the fiber surface is the most commonly invoked mechanism for the effectiveness of the stain blocker.[7] According to this mechanism, the negatively charged syntan adsorbed on the fiber surface establishes an electrical double layer with a negative zeta potential. This prevents the adsorption of dye anions on the fiber surface, a prerequisite for diffusion into the fiber.

EFFECT OF TEMPERATURE ON STAIN RESISTANCE

Although stain blockers are effective at room temperature, they lose their stain resistance at higher temperatures. The effect is reversible unless the fiber is subjected to temperatures high enough to disrupt the crosslinked barrier membrane. Carpet manufacturers take advantage of this fact when dyeing carpets treated with stain blocker.

Stain Repellancy

One finish extensively used in the protection of carpets against soiling is based on the polymer developed by Sherman.[9] It is a block copolymer of hydrophobic fluorocarbon polymer blocks and the hydrophilic polyethylene glycol blocks. While dry, hydrophobicity dominates and therefore the surface is soil repellant. In washing or steam cleaning, hydrophilic polyethylene glycol blocks dominate and prevent soil redeposition by osmotic repulsion.[10] If the hydrophilic blocks are charged polymers, then the soil redeposition is prevented by an electrostatic repulsion mechanism.[11]

REFERENCES

1. Fowkes, F. M. *J. Phys. Chem.* **1962**, *66*, 382.
2. Good, R. J.; Choidhury, M. K. *Fundamentals of Adhesion*; Plenum: New York, 1991; Chapter 3.
3. Shafrin, E. G.; Zisman, W. A. *J. Phys. Chem.* **1960**, *64*, 519.
4. Good, R. J.; Choudhury, M. K.; van Oss, C. J. *Fundamentals of Adhesion*; Plenum: New York, 1994; Chapter 4.
5. Ucci, P. A.; Blyth, R. C. U.S. Patent 4 501 591, February 26, 1985.
6. Blyth, R. C.; Ucci, P.A. U.S. Patent 4 680 212, July 13, 1987.
7. Harris, P. W.; Hangey, D. A. *Textile Chem. Color.* **1989**, *21*, 25.
8. Sargent, R. R.; Williams, M. S. U.S. Patent 5 316 850, May 31, 1994.
9. Sherman, P. O.; Smith, S.; Johannesen, B. *Textile Res. J.* **1969**, *39*, 449.
10. Fong, W.; Lundgren, H. P. *Textile Res. J.* **1953**, *23*, 769.
11. Fong, W.; Ward, W. H. *Textile Res. J.* **1954**, *24*, 881.

FIBERS AND COMPOSITES (High Modulus and Strength)

Roger S. Porter
Polymer Science and Engineering Department
University of Massachusetts, Amherst

Already, commercial polymers, their fibers and composites excel in such properties as toughness and impact strength. Advanced polymers now overlap the high impact properties of metals and even go beyond, for energy absorption, even providing entirely reversible deformations.[1] The highest impact listed is for the DuPont Lycra-type segmented urethane fiber compositions which, of course, are also the basis of sleek, athletic body suits. In addition, the tremendous strain-energy capacity of such fibers, suitably assembled, can in the future provide a low-cost, lightweight, totally resilient bumper component for automobiles.

FIBER CONCEPTS

Polymers comprise the full range of stiffness. Extremely soft polymers lattices have been synthesized in orbit to reduce their distortion by gravity. At the high modulus end, the polymer called diamond is the current zenith.

FIBERS IN TENSION

Currently the strongest fibers available are made by the thermal cyclization of polyacrylonitrile, or PAN, as by Nippon Steel Company. The second envelope of carbon fiber properties comes from liquid crystal technology, starting with selected pitch bottom stock from petroleum.

The molecular weight, morphology, and draw conditions for polyethylene have been established to achieve high uniaxial draw and high modulus. This has been accomplished directly from selected reactor powders by compaction followed by calendering, drawing, and splitting. This process is being developed by the Nippon Oil Company, and is based on solid-state-forming processes originated at the University of Massachusetts.

FIBERS IN COMPRESSION

For use in composites, the goal is to load reinforcing fibers only in tension. This is because organic fibers buckle under compression. An understanding of this vulnerability is emerging.

COMPOSITE CONCEPT

The development of effective macro composites is intriguing. The utilization of such continuous fiber reinforcement provides new opportunities. One is to co-weave a thermoplastic (matrix) fiber with reinforcing fibers before heat treatment. Another is the art of injecting a thermosetting matrix over a reinforcing preform, as in resin transfer molding.

Beyond the simplest method of filament winding, high levels of structural integrity can be achieved by weaving and braiding technologies. These methods can provide high concentrations and controlled orientations of reinforcing fibers. Although space-filling, three-dimensional weaving processes have been developed. However, such construction becomes impossibly complex. Jim Sidles at BF Goodrich has suggested a simple, versatile, and alternative three-dimensionally reinforced structure for large composites, reminiscent of bonded Velcro. Application of a thermoset to a shaped Velcro can thus produce a 3-D reinforced composite.

The concept of a molecular Velcro has also been demonstrated by Professor Anna Balazs, of Pittsburgh University, with Dennis Pfeiffer and co-workers. This clever concept provides a bond at a composite interface which, at a molecular level, can be called compatibilization. The idea is to use graft copolymers, with the grafts acting as tines for Velcro interpenetration.[2]

SMART FIBERS AND COMPOSITES

Continuous fiber reinforcement in composites offers an opportunity for on-line, real-time evaluation of both stress exposure and composite life expectancy. Moreover, responses detected through the composite fibers may be utilized to adjust properties to the environment, that is, to enable the material to perform as a smart composite.[3]

CONCLUSION

With major success to date, there assuredly are many bright prospects for both the volume and diversity of new systems of polymer fibers and composites.

Not to be forgotten is our recognized collective responsibility to measure and manage the recycling and lifetime of organic materials in society. The safety net, with few exceptions, is the retained combustion energy of organic materials including polymers, just as in the parent petroleums, from which polymers generally have been derived.

REFERENCES

1. Farris, R. J.; Chien, J. C. W.; MacKnight, W. J. U.S. Patent 4 357 041, 1982.
2. Gersappe, D.; Irvine, D.; Balazs, A. C.; Liu, Y.; Sokolov, J.; Rafailovich, M.; Schwarz, S.; Peiffer, D. G. *Science*, **1994**, *265*, p 1072.
3. Porter, R. S. *Plastic Engineering*, May 1994, p 67.

FIBERS, SYNTHETIC

J. George Tomka
Department of Textile Industries
University of Leeds

Fibers produced from synthetic polymers belong to a broader category of man-made (manufactured) fibers. The first man-made fibers were based on a natural polymer, cellulose.

In commerce, synthetic fibers are classified by generic names, which use chemical names in a more restricted way that is common in polymer science.

Originally, the synthetic fibers were viewed as cheaper substitutes for natural fibers, particularly silk. However, it quickly became apparent that synthetic fibers can in many aspects surpass the natural fibers. Thus, synthetic fibers not only replaced successfully the natural fibers in the traditional end-uses, such as clothing and furnishing, but also enhanced the range of fiber applications in the technical sector.[1]

FIBER PROPERTIES

Each specific end-use requires a different balance of properties. It is always important to take into consideration the in-use conditions such as the composition of the environment and the temperature. Furthermore, due attention must be paid to fiber durability related to its thermal/thermo-oxidative stability, hydrolytic stability, solvent resistance, and biostability.

For many noncritical end-uses, fibers are selected from the available commodity products on the basis of minimum cost for acceptable performance. However, high-value traditional applications stimulated innovations resulting in the upgrading of commodity fibers. Furthermore, a range of specialty fibers have been developed for critical technical applications and for special effects.[2]

FACTORS DETERMINING FIBER PROPERTIES

The key issue in both innovation and development of new fibers is the understanding of factors determining their properties. In common with other shaped articles, the properties of fibers are determined by their geometry (i.e., shape and dimensions) and by the properties of the constituent material or materials; these are in turn determined by the composition (*which* molecules are used), and by the physical structure (*how* these molecules are arranged with respect to each other and to the fiber axis).[3]

Whilst the fiber composition is essentially determined by the choice of the starting material, the fiber geometry and physical structure are controlled by the fiber formation processes. Generally, production of synthetic fibers involves several stages: spinning (also called "extrusion") and drawing, optionally followed by texturing, heat setting, and cutting.[4,5]

FIBER COMPOSITION

The commodity fibers are suitable for the majority of end-uses. However, there is a demand for specialty fibers belonging to the following main categories: fibers with enhanced chemical stability and combustion resistance; high-strength, high-modulus fibers; and fibers for special effects.

Obviously, the synthetic fibers of the first two categories must be competitive with the established inorganic fibers (glass, carbon, and ceramic). Many polymers, typically those containing aromatic rings and/or heterocycles joined by stable linkages, have been considered for production of fibers belonging to the first category.[6] Commercially, the most important flame-resistant fibers are those produced from poly(*m*-phenylene isophthalamide), but aromatic polyamide-imide and copolyimide fibers are also commercially available.[7]

We now turn to the second category of specialty fibers, namely high-strength, high-modulus (HSHM) fibers.[8,9] It has been established that one of the prerequisites for obtaining the required fiber strength and stiffness is the elimination of chain folding. To achieve this, two types of approach, classified respectively as chemical and physical, are used. The chemical approach utilizes linear rigid-chain polymers.[10]

The physical approaches are directed to realizing fully the potential of the linear flexible-chain polymer. The most successful physical approach is based on gel spinning of high molecular weight ($\overline{M}_w > 10^6$) polyethylene.[11]

A number of fibers for special effects have been developed.[2] Particularly important are fibers of high elasticity, especially those based on segmented polyurethanes (generic name: spandex (U.S.) or elastane).[12,13] Although these fibers were commercialized in the 1950s (Lycra®), their performance has been much improved and their range of applications is growing rapidly, particularly in leisurewear and sportswear.[14]

CONCLUSIONS

The success of commodity fiber produced in large quantities from a few mass-produced polymers is due to their cost-effectiveness combined with versatility, which is based on our capability to manipulate fiber geometry and physical structure. Minor modifications of composition (e.g., skillful utilization of additives) are also used for this purpose. While product and process innovations will continue, it is highly unlikely that a completely new commodity fiber will appear in the foreseeable future. In contrast, there is still scope for the development of new specialty fibers for critical end-uses.

REFERENCES

1. Brody, H. Ed., *Synthetic Fibre Materials*; Longman: Harlow, 1994.
2. Hongu, T.; Phillips, G. O. *New Fibres*; Ellis Horwood: Chichester, 1990.
3. Tomka, J. G.; Johnson, D. J.; Karacan, I. *Advances in Fibre Science*; Mukhopadhyay, S. K., Ed.; The Textile Institute: Manchester, 1992; Chapter 8.
4. Ziabicki, A. *Fundamentals of Fibre Formation*; John Wiley & Sons: London, 1976.
5. McIntyre, J. E.; Denton, M. J. *Encyclopedia of Polymer Science and Engineering, Vol. 8*, Mark, H. F. et al., Eds.; John Wiley & Sons: New York, 1987; 802.
6. Mohajer, A. A.; Ferguson, W. *J. Textile Progress* **1976**, *8* (No. 1), 97.
7. Desitter, M. G.; Cassat, R. *Synthetic Fibre Materials*; Brody, H., Ed.; Longman: Harlow, 1994; Chapter 11.
8. Jiang, H.; Adams, W. W.; Eby, R. K. *Materials Science and Technology, Vol 12–Structure and Properties of Polymers*; Thomas E. L., Ed.; VCH: Weinheim, 1993; Chapter 13.

9. Ward, I. M.; Cansfield, D. L. M. *Advances in Fibre Science*; Mikhopadhyay, S. K., Ed.; The Textile Institute: Manchester, 1992; Chapter 1.
10. Yang, H. H. *Aromatic High-Strength Fibers*; John Wiley & Sons: New York, 1989.
11. Lemstra, P. O.; Kirschbaum, R.; Ohta, T.; Yasuda, H. *Developments in Oriented Polymers, Vol. 2*; Ward, I. M., Ed.; Elsevier: London, 1987; Chapter 2.
12. Couper, M. *Handbook of Fiber Science and Technology, Vol. III, Part A*; Lewin, M.; Preston, J. Eds.; Marcel Dekker: New York, 1985; Chapter 2.
13. Ultee, A. J. *Encyclopedia of Polymer Science and Engineering, Vol. 6*; Mark, H. F. et al., Eds.; John Wiley & Sons: New York, 1987; 733.
14. Holme, I. *Synthetic Fibre Materials*, Brody, H. Ed. Longman: Harlow, 1994; Chapter 5.

FIBERS, SYNTHETIC (Unspun Natural Pulp-Like)

Byoung Chul Kim
Division of Polymer Researches
Korea Institute of Science and Technology

Fibers are slender, threadlike materials characterized by a high ratio of length to diameter. Fibers typically range from 5 to 25 μm in diameter. Up to now, almost all synthetic fibers have been manufactured by spinning and subsequent longitudinal drawing, just like silkworms spin cocoons.[1] On the other hand, most naturally occurring cellulose fibers, such as cotton and flax, are believed to be formed by lateral aggregation of cellulose microfibrils, although controversy as to the mechanisms still exists.[2,3] Consequently, natural cellulose fibers have much greater surface area than synthetic fibers because of their more finely divided fibrillar structures.[1]

The history of unspun synthetic fibers dates back to the early 1960s, when a method of fibrillation without spinning was reported.[4,5] It was based on the concept that an extrusion of incompatible polymer blends leads to elongational deformation of the dispersed phase if the processing conditions are properly chosen.[6,7] This principle has enabled one to obtain very thin submicron synthetic fibers, as reviewed by Robeson et al.[8] Recently, thermotropic liquid crystalline polymer (TLCP) proved appropriate for producing unspun fibers because of the anisotropy and rigidity of the polymer chains.[9] An extrusion of incompatibile TLCP polymer blends also yields very thin TLCP fibers under specific processing conditions.[10-12]

In the early 1980s and 1990s, novel processes of preparing unspun synthetic fibers were reported for *p*-aramid[13-16] and acrylic polymers.[17-20] The processes are based on *in situ* pulping technology, an orientation and fibrillation during polymerization and extrusion, respectively. The cardinal point of *in situ* pulped unspun fibers lies in their structural similarity to naturally occurring cellulose fibers.

PREPARATION OF NATURAL PULP-LIKE SYNTHETIC FIBERS

p-Aramid Fiber by Growth Packing Synthesis

Natural pulp-like *p*-aramid fiber is prepared by *in situ* pulping during polymerization via growth packing synthesis; that is, orientation and fibrillation of growing polymer chains during the polycondensation of terephthaloyl chloride (TPC) and *p*-phenylene diamine (PDA) in a tertiary amide solvent containing pyridine and a metal halide salt.[13]

Two important prerequisites for this *in situ* pulping process are[14] (1) the rate of polymerization should be high enough to obtain high molecular weight gel within a few seconds, and (2) the viscosity of the resultant gel should be high enough to allow the molecular orientation in the gel to proceed to form microfibrils.

Acrylic Fiber by *In Situ* Pulping Extrusion

An incorporation of water makes acrylic polymers with high acrylonitrile contents melt via hydration.[22] This fact that water acts as a plasticizer for acrylic polymers has been repeatedly used to develop a melt spinning process of acrylic polymers, but it has not been successful commercially.[23,24] However, this principle was adapted to the *in situ* pulping extrusion process of acrylic polymers in the early 1990s by Yoon et al.[17,18] They found that an extrusion of the hydrated acrylic melt at the supercooled state produces a microfibrillated extrudate. Opening the extrudate yields the natural pulp-like acrylic fiber.

In the *in situ* pulping extrusion process of the supercooled hydrated acrylic melt, the molecular orientation of the mesophase is very sensitive to processing temperature and shear.[17] If the extrusion temperature exceeds the upper limit of supercooling, the extrudate develops little molecular orientation, but a cellular foamed extrudate results.

PROPERTIES OF NATURAL PULP-LIKE SYNTHETIC FIBERS

p-Aramid Fiber

The structure of the *p*-aramid fiber prepared by the growth-packing synthesis is distinctively different from conventional spun fibers.[15] There is a close similarity between natural asbestos and the pulped *p*-aramid fiber. That is, natural asbestos and aramid fiber exhibit uniform and finely fibrillated structure. In the commercial *p*-aramid pulp, however, an unground rod-like filament structure is observed.

A closer examination of the *in situ* pulped *p*-aramid fiber reveals that it has a physical form that resembles naturally occurring wood pulp fibers. That is, the aramid fiber is composed of a plurality of individual unit fibers that have the shape of an elongated ribbon with irregular cross-sections and the same needle point-like ends.[14] This structure has been observed in naturally ocurring wood pulp fibers.

It is reported that the *in situ* pulped *p*-aramid fiber does not exhibit the Maltese cross that predominently appears in conventional spun fibers when the cross-section of the fiber is observed by a polarizing microscope.[14] When a given point of the cross-section of the fiber specimen is rotated through 360° around the rotating axis of the specimen, bright and dark positions are alternately detected at every rotation angle of 45°. This indicates that the aramid fiber is in the form of a parallelogram network in the direction perpendicular to the axes of fibers.

Acrylic Fiber

The microfibrillated extrudate obtained by extruding the hydrated acrylic melt at the supercooled phase can be easily divided in a longitudinal direction by mechanical beating or opening. A mechanical beating of the chopped extrudate yields short fibers,[18] whereas mechanical opening of the continuous extrudate yields an endless white acrylic fiber.[19-20] Both of the *in situ* pulped acrylic fibers have a high degree of orientation and a unique fiber structure similar to naturally occurring cellulose fibers. Unlike the conventional spun synthetic fibers, the acrylic fiber exhibits a microfibrillated structure, which means that the fiber is composed of parallel fibrils. This is because of the different mechanisms of fiber formation. In other words, most orientation in the fiber spinning process is developed by a longitudinal drawing of the amorphous as-spun fiber, whereas fiber orientation is obtained by predetermined mesophase structure in the *in situ* pulping extrusion process. As a consequence, the *in situ* pulped acrylic fiber exhibits a laminated fibrillar structure rather than a continuous smooth-filament structure.[17]

As can be imagined, the physical properties of the acrylic fiber vary with the chemical composition; increasing the acrylonitrile content increases the tensile modulus, tensile strength, and resistance to chemicals. It is another merit of this extrusion process that the chemical composition can be freely changed because the polymer need not be dissolved in organic solvents.[17] Even the homopolymer of acrylonitrile without any soft comonomer can be readily fabricated to fibers. In consequence, the *in situ* pulped acrylic fiber has a wide range of physical properties depending on the chemical composition of the acrylic polymer: a tensile strength of 10 to 70 kg/mm^2, an elastic modulus of 300 to 1500 kg/mm^2, an elongation of 5 to 20%.[18]

APPLICATIONS OF NATURAL PULP-LIKE SYNTHETIC FIBERS

The *in situ* pulping process has several advantages over conventional manufacturing processes. First of all, a considerable cost reduction is feasible as a result of much simplified pulping process.[14,18,21] Particularly with the acrylic *in situ* pulping process, replacement of toxic polar solvents such as dimethylformamide and aqueous sodium thiocyanate solution with water may be helpful to solve the environmental problems of the acrylic fiber industries.[18] In addition to the benefits in cost and ecology, the morphological advantage of *in situ* pulped synthetic fibers, fineness and high ratio of surface to weight can lead to a wide range of applications, particularly in the separation and insulation fields.

Owing to their excellent physicochemical and morphological properties, the *in situ* pulped *p*-aramid fiber can substitute for asbestos, which has turned out to be a carcinogenic.[16] The aramid fiber fabricated as sheet form is suitable for electric and sonic insulators, and for structural materials such as honeycombs and helmets. On carbonization, the fiber gives a high carbon yield (> 50%) and large surface area, which are desirable in active carbon materials. The carbonized materials can be used as filter media for separation purposes.

In situ pulped acrylic fiber can be used in construction materials as reinforcing fillers because of its excellent alkaline resistance and desirable morphology.[17] The acrylic fiber is suitable for apparel purposes as well. The opened endless acrylic fiber may be cut to desired staple lengths for blending with other natural cellulose fibers. The acrylic fiber is easily fabricated to a synthetic paper by a conventional wet spraying method. Its good miscibility with timber pulps may lead to applications in the papermaking industry as an antishrinkage agent. If the fiber is heat treated, resistance to heat and chemicals is greatly improved.[25] In addition, the acrylic fiber can be used in manufacturing an active carbon paper.

REFERENCES

1. Moncrieff, R. W., *Man-Made Fibers, 6th ed.*, Newnes-Butterworth, London, **1975**.
2. Muller, S. C.; Brown, R. M., *J. Cell Biol.*, **1980**, *84*, 315.
3. Grout, B. W. W., *Planta Berlin*, **1975**, *123*, 275.
4. U.S. Patent 3 097 991 (to Union Carbide), 1963.
5. U.S. Patent 3 099 067 (to Union Carbide), 1963.
6. Chin, H. B.; Han, C. D., *J. Rheol.*, **1979**, *23*, 557.
7. Han, C. D.; Kim, Y. W., *J. Appl. Polym. Sci.*, **1974**, *18*, 2589.
8. Robeson, L. M. et al. *Polym. News*, **1994**, *19*, 167.
9. U.S. Patent Appl. No. 08/303,875 (to KIST), 1994.
10. Carfagna, C. et al., *J. Appl. Polym. Sci.*, **1991**, *43*, 839.
11. Mantia, F. P. L. et al., *Rheol. Acta*, **1989**, *28*, 417.
12. Hong, S. M. et al. *Polym. Eng. Sci.*, **1993**, *33*, 630.
13. Yoon, H. S., *A Fundamental Study on the Manufacturing of Aromatic Polyamide Pulp*, KIST Annual Report, April 1983.
14. U.S. Patent 4 511 623 (to KIST), 1985.
15. Yoon, H. S., *Nature*, **1987**, *326*, 580.
16. Yoon, H. S., *Mat. Res. Soc. Symp. Proc.*, **1990**, *174*, 187.
17. Yooh, H. S.; Kim, B. C., *A Study on the Manufacturing of Acrylic Pulp*, KIST Annual Report, May 1992.
18. U.S. Patent 5 219 501 (to KIST), 1993.
19. U.S. Patent Appl. No. 08/446,287 (to KIST), 1996.
20. U.S. Patent Appl. No. 08/128,657 (to KIST), 1993.
21. U.S. Patent 5 028 372 (to Du Pont), 1991.
22. Throne, D. J. et al. *Fiber Sci. Technol.*, **1970**, *3*, 119.
23. U.S. Patent 4 219 523 (to American Cynamid Co.), 1980.
24. U.S. Patent 4 271 056 (to American Cynamid Co.), 1981.
25. U.S. Patent Appl. No. 08/064,345 (to KIST), 1993.

Fibrin

See: Bioartificial Materials

Fibroin

See: Silk (Its Formation, Structure, Character, and Utilization)
Silk (Physico-Chemical Properties)
Silk Fibroin (Soft Tissue Compatible Polymer)

Field Responsive Polymers

See: Electrorheological Fluids (Role of Polymers as the Dispersed Phase)
Electrorheological Materials

Filaments

See: Superconducting Filaments, High Tc (Suspension and Solution Spinning from PVA Solution)

Fillers

See: *Additives (Property and Processing Modifiers)*
Additives (Types and Applications)
Carbon Black
Carbon Black (Graft Copolymers)
Cellulose-Filled Composites
Dental Polymers (Overview)
Electrically Conducting Composites
Fillers (Glass Bead Reinforcement)
Fillers and Reinforcing Agents
Fillers, Propropylene
Fillers, Surface Modification
Interphases, Ternary Polymer Composites
Polypropylene (Commercial)
Polypropylene (Crystallization Kinetics)
Poly(vinyl chloride), Rigid (Soft Fillers)
Poly(vinyl chloride), Soft (Calcium Carbonate Filled)
Powder Coatings (Overview)
Reaction Injection Molding (RIM) Materials
Silicone Rubber Latex
Silicone Sealants
Siloxane Elastomers
Thermoplastic Composites
Unsaturated Polyester Resins (Overview)
Unsaturated Polyester Resins (Polyester Composites)
Unsaturated Polyester Resins (Toughening with Elastomers)
Vulcanization (Overview)

FILLERS (Glass Bead Reinforcement)

Charles L. Beatty and Mostafa A. Elrahman
Department of Materials Science and Engineering
University of Florida

Reinforced polymers constitute an excellent mechanical and tribological system owing to their specific strength, stiffness, and their properties that are available through controlled combination of reinforced materials and matrices. Hollow and solid glass microspheres have become increasingly important for filler applications. Spheres have the lowest surface area to volume ratio of any particle shape, and therefore have low resin requirements to wet the filler surface. Also, spherical glass beads are particularly amenable to analysis of a model system of the particulate fillers. In addition, the glass beads, owing to their geometry, do not increase the viscosity of the melt as much as do glass fibers or mineral particles. Furthermore, glass beads reinforcement reduces shrinkage, offers improved compressive strength, hardness, creep of base polymer, and costs less than glass fibers.[1] In the field of the tribomaterials (self-lubricated materials), the glass beads combined with thermoplastics, poses an excellent tribological behavior owing to their abrasive action and ability to modify the counterface surface topography.[2]

MECHANICAL BEHAVIOR

Early work was done by Lewis et al. that measured the dynamic mechanical properties, shear modulus, damping, and glass transition temperature of glass bead/epoxy composites.[3] They concluded that in general when a filled system is compared with unfilled material, shear modulus increases and damping decreases. Also, a small increase in the glass transition temperature was observed for both the glass bead volume fraction and glass bead size. Recently, Vratsanos and Farris have proposed an alternative approach to predict the mechanical behavior of particulate composites with good success.[4,5] The effect of glass bead size and volume fraction upon the fracture behavior of epoxy resins containing spherical glass beads has also been considered.[6]

The mechanical properties of polystyrene composites containing various percentages of glass microspheres over a wide range of temperature have been studied.[7,8] This study concluded that the presence of glass microspheres filler increases the elastic modulus and ultimate elongation while decreasing the strength of polystyrene composites. Also, the overall area under the stress/strain curve, failure energy, is enhanced by addition of glass microspheres. For polypropylene composites, an investigation according to LEFM (linear elastic fracture mechanics) principles has been carried out to determine the impact resistance of polypropylene containing calcium carbonate in comparison with glass spheres and talc fillers.[9] Both homopolymer polypropylene and that containing talc showed low elongation at break. Specimen fracture occurs a short time after the maximum load, while with products containing calcium carbonate and glass spheres see an increase in ductility. In this case the addition of filler causes the maximum load to decrease.

Recently, a series of polystyrene/glass beads composites were studied by using dynamic mechanical spectrometry.[10] This study pointed out that the magnitude of the experimental main relaxation of the composite is significantly reduced with increasing the amount of glass beads. In glass bead filled ionic polyurethane, the effect of glass bead content (7.4, 16.4, 35.4, and 53.8%) and glass bead size (10, 38, and 150 μm) on the mechanical properties of polyurethane was investigated. This investigation showed that there is an overall increase in failure stress for all microspheres studied with increasing concentration, followed by a decrease in the failure stress. The data suggest that depending upon the size of the filler, there is an optimum concentration below which composite reinforcement occurs and above which the filler behaves in a non-reinforcing fashion.

In foamed material, it is known that low density foams deform primarily by bending of cell edges. Therefore their mechanical properties could be improved by increasing the flexural rigidity of the wall.[11] In this field the elastic modulus of low density foams filled by hollow glass sphere was investigated.[12]

GLASS BEAD/MATRIX ADHESION

In polymers containing rigid particles it is not always necessary to have good particles/matrix adhesion for the material to possess the optimum mechanical properties. Some mechanical properties are improved while others are made worse as the level of adhesion increased.[6] The degree of adhesion in rigid-particulate composite is generally controlled by employing an appropriate surface pretreatment for the particles, prior to their addition to the matrix. For example, there are many silane-based coupling agents now available for either primary (chemical) bonding or weaker, secondary (e.g., van der Waals) bonding across the particle/matrix adhesion interface. In the case of glass beads/matrix, adhesion is found to have a significant effect upon the fracture strengths of both epoxy and polyester resins reinforced with glass beads.[13]

TRIBOLOGICAL BEHAVIOR

It is known that the wear behavior of polymer composites depend on the ability of filler material to modfy the initial roughness of the counterface surface.[14] From this point of view, the tribological behavior of the recycled glass bead filled HDPE was studied.[2] The effect of glass bead size and various concentrations of the glass beads were investigated. According to this investigation, the specific wear rate decreased as a result of increasing the glass bead concentration by up to 10% by weight. However, the coefficient of friction between the sliding surfaces decreased continuously with the glass bead concentration. Also, this study showed that the smallest bead size tested (75 μm average diameter) has better enhancement of the wear rate and coefficient of friction of the composite.

REFERENCES

1. Seymour, R. *Additives for Plastics*; Academic: New York, 1978.
2. Elraham Mostafa, A. *Wear and Sliding Properties of Polymer Composites*; Ph.D. Thesis, Minia University, Egypt, granted in June 1995.
3. Lewis, T.; Nielsen, L. *Dynamic Mechanical Properties of Particulate-Filled Composites; Journal of Applied Polymers Science* **1970**, *14*, 1449-1471.
4. Vratsanos, L.; Farris, R. *A Predictive Model for the Mechanical Behavior of Particulate Composites. Part 1: Model Derivation*; Polymer Engineering Science **1993**, *33*, 1458-1465.
5. Vratsanos, L.; Farris, R. *A Predictive Model for the Mechanical Behavior of Particulate Composites. Part II: Comparison of Model Predictions to Literature Data; Polymer Engineering Science*, November **1993**, *33*, 1466-1474.
6. Kinloch, A.; Young, R. *Fracture Behavior of Polymers*; Elseveir Applied Science; 1983.
7. Lavengood, R.; Nicolais, L.; Narkis, M. *A Deformational Mechanism in Particulate-Filled Glassy Polymers; Journal of Applied Polymer Science* **1973**, *17*, 1173-1185.
8. Nicolais, L.; Guerra, G.; Migliaresi, C.; Nicodemo, L.; Bendetto, A. *Mechanical Properties of Glass-Bead Filled Polystyrene Composites*; Composites, January 1981, 33-37.
9. Bramuzzo, M.; Savadori, A.; Bacci, D. *Polypropylene Composite: Fracture Mechanics Analysis of Impact Strength; Polymer Composites*, January **1985**, *6*, 1-8.
10. Agbossou, A.; Bergeret, A.; Benzarti, K.; Alberola, N. *Modelling of the Viscoelastic Behavior of Amorphous Thermoplastic/Glass Beads Composites Based on the Evaluation of the Complex Poisson's Ratio of the Polymer Matrix; Journal of Materials Science* **1993**, *28*, 1963-1972.
11. Gibson, L.; Ashby, M. *Cellular Solids: Structure and Properties*; Pergaman: Oxford, 1988.
12. Huang, J.; Gibson, L. *Elastic Moduli of Composite of Hollow Spheres in a Matrix; Journal of Mechanics Physics Solids* **1993**, *41*, 55-75.
13. Sahu, S.; Broutman, L. *Mechanical Properties of Particulate Composites; Polymer Engineering and Science* **1972**, *12*, 91-100.
14. Anderson, J. *The Wear and Friction of Commercial Polymers and Composite Materials*; Friedrich, Ed.; New York, 1986.

FILLERS AND REINFORCING AGENTS

Dennis G. Sekutowski
Engelhard Corporation

Cost reduction has historically been the primary purpose for a filler in a resin composite. The filler is less costly than resin and essentially acts as a diluent, for example, wood flour in phenolic molding compound. However, nearly all fillers do more than just take up space and extend a polymer resin. Fillers will always modify the flow characteristics and mechanical properties of a composite, often improving such properties as stiffness and temperature resistance. The fillers that benefit composite properties are called reinforcing agents or reinforcing fillers, for example, fiber glass in nylon. In coating applications, the word *extender* rather than filler is often used. Fillers, extenders, and reinforcing agents are finely divided solid particles which are added to polymer systems to enhance properties or reduce costs. More detailed information can be found in many of the excellent reviews, standard references and handbooks on minerals, fillers, and pigments.[1-7]

COMPOSITE CONCEPTS

All three phases–the filler, the resin, and the interphase between the two–have an influence upon final properties. The interphase is the least understood of the three, making composite formulation still very much an art.

SHAPE AND SIZE

Particles come in a variety of shapes and sizes. Most filler applications require particles smaller than 40 microns (μm). Fine particles, less than 5 μm, are needed when mechanical properties are important. In general, the finer the particle the better the mechanical properties, but the poorer the viscosity and the more difficult a composite is to process.

The shape of a particle has a great influence on a composite or coating's mechanical and surface properties (gloss and sheen) and permeability. In general, a high aspect ratio material in a composite is more difficult to process and yields higher anisotropy, but better mechanical properties, than a low aspect ratio particle.

CALCIUM CARBONATE

Calcium carbonate is the most abundant white mineral in the earth's crust; consequently, it is widely exploited around the world. Calcium carbonate is the largest-volume mineral filler in the plastic, sealant, rubber, paper, coating, adhesive, textile, and carpet backing industries. Carbonate's wide utility is due to its whiteness, low abrasion, ready availability, wide particle-size range, and low cost.

SILICA

Four types of commercial silica gel are available: ground, precipitated, fumed, and gel. All are chemically silicon dioxide. The major commercial markets for silica are abrasives, ceramics, glass, chemicals, foundry uses, paint, paper, ink, rubber, plastic, pharmaceutical, cosmetic, greases, and catalyst supports. Many large volume uses, however, use grades that are much too coarse for filling applications.

Silica is available with a variety of surface treatments, for example, organic silicones, silanes, and waxes. High treatment levels are needed for the synthetic grades due to their inherent high surface areas.

KAOLIN

Kaolin is a white crystalline aluminum silicate mineral. It is often referred to as clay even though its physical and chemical properties greatly differ from other commercial clay silicates such as bentonites or fire clays. Kaolin is one of the most finely divided and highly refined naturally occurring minerals. The primary market for kaolin is the paper industry, where it is used for filling and coating applications because of its optical properties and platy character. Kaolin also has important industrial niches in the plastics, rubber, paint, fertilizer, ceramic, ink, adhesive, textile, cement, and pharmaceutical industries.

MICA

Mica is a generic name given to a family of hydrous potassium aluminum silicates having similar properties. Muscovite and phlogopite are the two most common commercial filler varieties. The flake morphology of mica is its most important attribute, yielding high reinforcing properties even at low loading levels. The most important markets for mica are joint cements, roofing, well drilling muds, paints, plastics and rubber.

TALC

Talc is a white to pale green platy hydrated magnesium silicate. Talc is the softest filler and has a slippery texture. The major markets for talc are ceramics, paint, paper, plastics, cosmetics, and roofing materials.

WOLLASTONITE

Wollastonite is a white to brownish acicular mineral consisting of calcium metasilicate. Low aspect ratio wollastonite (3-5) is used in ceramics and metallurgical fluxes, whereas high aspect ratio (3-20) wollastonite is used in reinforcing applications.

Wollastonite is largely used in engineering thermoplastics, phenolics, and polyester SMC because of its acicular nature and low oil absorption. Like fiber glass it increases fexural modulus and tensile strength. Since asbestos is being phased out because of health concerns, wollastonite is being used as a replacement, in phenolic brake linings and floor tile backing, for example. Surface-treated grades are used in molding nylon products.

FELDSPAR AND NEPHELINE

Feldspar and Nepheline are similar aluminum silicate minerals. Their physical properties are very similar and thus they are used for similar purposes. As fillers they are unique in that their refractive index of 1.53 is similar to that of most polymers. Hence these minerals appear translucent in resin films and coatings. Both are abrasive and are coarse (8-12 μm). In addition to their optical properties they impart good weatherability and abrasion resistance to coatings.

BEADS

Both hollow and solid glass microspheres are available as fillers. With an aspect ratio of 1, spheres do not contribute to the mechanical strength of a composite. However, they will have little effect on melt viscosity and can improve flow properties, composite uniformity, and surface effects. Glass microspheres are used in many specialty thermoplastic and thermoset applications.

ORGANIC FILLERS

Organic fillers are produced from process residues that would otherwise be discarded. These fillers are used primarily to reduce cost, but they are nonabrasive and can contribute improved processibility, stiffness, warpage, and dimensional stability. Temperature is a limitation; 380°F is an upper limit. Organic fillers are hygroscopic and may contain up to 10% moisture. In thermosets this is not a concern, but in other applications a redrying becomes necessary.

GLASS FIBER

Glass fiber is the most commonly used reinforcing filler. Nearly all glass fiber for reinforcement is produced from E-(electrical) glass. Chopped strand is the largest volume form of fiber glass used in plastic composites.

Glass fiber is used in nearly all types of plastics under all types of processing conditions. The highest volume for chopped strand is in thermoplastic, SMC, and BMC application.

Because of the high aspect ratio, glass fiber imparts great mechanical strength to a polymer composite. Its drawbacks are poor flow characteristics and high anisotropy of mechanical properties in end-product performance.

CARBON BLACK

Carbon black is not usually considered a filler even though it does provide reinforcing properties, particularly in rubber compounds. Carbon black is used as an ultraviolet (UV) stabilizer thermal antioxidant, antistat, polymerization modifier (in unsaturated polyesters), conductive filler, and colorant.

ALUMINA TRIHYDRATE

Alumina trihydrate (ATH) is produced as an intermediate in the manufacture of aluminum metal. ATH is primarily used as a flame retardant or smoke suppressor, especially in PVC applications.

BARIUM SULFATE

Barium sulfate fillers, the heaviest nonmetallic fillers, are white and very inert.

SUMMARY

New sources and new applications of fillers are continually being found. A current example is plastic-lumber which contains recycled plastic and glass scrap and is being promoted for deck construction applications. Few new polymer or filler materials are being invented, but many new composites are being made by modifying processing methods and reformulating or combining different additives and fillers to improve composite properties for new applications.

REFERENCES

1. *Pigment Handbook* Patton, T.C. Ed.; Wiley: New York, 1973; Volumes I, II, III.
2. *Handbook of Fillers for Plastics* Milewski, J. V.; Katz, H. S., Eds.; Van Nostrand Reinhold: New York, NY, 1987.

3. *Handbook of Reinforcements for Plastics* Milewski, J. V. and Katz, H. S., Eds.; Van Nostrand Reinhold: New York, NY, 1987.
4. *Plastics Compounding Redbook,* Advanstart Communications Inc., 1994-95.
5. *Industrial Minerals and Rocks* Lefond, S. J. Ed.; American Institute of Mining, Metallurgical and Petroleum Engineers; Port City: Baltimore, MD, 1975.
6. Stoy, W. S.; Washabaugh, F. J. *Encyclopedia of Polymer Science and Engineering,* Wiley: New York, NY, 1986; Vol. 6.
7. Sekutowski, D. G. *Plastics Additives and Modifiers Handbook.* Van Nostrand Reinhold: New York, NY, 1992; Chapters 35-42.

FILLERS, POLYPROPYLENE

Robert A. Shanks and Long Yu
Applied Chemistry and Cooperative Research Centre for Polymer Blends
Royal Melbourne Institute of Technology

Fillers dispersed in polymers give higher melt viscosity, stiffening, higher tensile strength, low heat distortion, and greater density. Most fillers are inorganic materials, which have ionic or polar surfaces; therefore they need modification to enhance interaction with a polymer. Modification is specific to a particular filler but usually involves coating the surface with a nonpolar additive. Polypropylene can be filled with almost all fillers.

Polypropylene (PP) is a semi-crystalline polymer with moderate melting temperature (165°C) and low glass transition temperature (T_g) (–25°C). PP has low intermolecular forces, so its strength depends on regularity. The isotactic index of a PP often determines the characteristics of the material. An increase in isotactic index of a PP will sharply increase the yield strength of the material. Hardness, stiffness, and tensile strength also increase. However, elongation and impact strength will decrease as isotactic index increases. The supramolecular structure of PP is spherulitic with a number of crystal modifications.[1]

In world production of filled thermoplastics, filled PP is the new leader. Each year, its production grows by 15 to 20 percent.[2] There are several reasons for this: the low density of the matrix polymer, its advantage in heat stability to deformation at the time of filling, low cost, high chemical resistance, and ease of processing. Although it is necessary to consider the interaction between the PP and fillers, (for example, their interface), it is the properties of the filler and not filled PP which form the major content of this section.

CHARACTERIZATION OF VARIOUS FILLERS

Talc

Talc is a very soft mineral with a greasy feel. Due to its plate-like form, or high aspect ratio, talc is considered as a reinforcing filler in many plastic applications. Composites filled with plate-like talc always exhibit a higher stiffness and creep resistance.

Talc is also utilized as medium filler of average whiteness in plastics where improvement in electrical insulation, heat and moisture resistance, chemical inertness, and good machinability of molding is required.

Talc is among the reinforcements and fillers for PP with the highest growth rates. It is the classic filler that imparts higher stiffness, an increase in the heat deflection temperature, and dimension stability. One of the major disadvantages of talc-filling PP is reduced impact strength at low temperature.

Calcium Carbonate

As a filler, $CaCO_3$ has unique advantages as follows: it is nontoxic, odorless, white, soft, it allows for particle size and size distributions, it is easily coated in dry form for modification, easily mixed into formulations, etc. However, as a polar, reactive substance, $CaCO_3$ has several possible disadvantages: chemical reaction with acid, little reinforcing action, and moisture sensitivity. To improve impact strength of filled PP, $CaCO_3$ has been used with talc as a mixed filler.

Calcium carbonate, like talc, is one of the most popular fillers used in PP. It has distinguishing advantages: white color, easier dispersion, better surface of finished parts, suitability for contact with food, and so forth.

Mica

Mica has high strength and stiffness, good chemical resistance, low solubility in water, low coefficient of friction, and low hardness.

The main traditional use for mica as a filler in the plastic industry is in electrical applications, where its excellent electric strength, low power factor, and high dielectric constant make it eminently suitable. Mica is the most effective mineral filler for reducing warpage and increasing stiffness and heat deflection temperature. However, mica is difficult to wet with olefins, which may produce a product with poor mechanical properties. When the mica is used in nonpolar polymers, the surface of the mica must be treated.

In the automotive industry, mica is also used as an inexpensive substitute for glass fibers in PP parts.

Kaolin

Kaolins generally possess a high degree of whiteness, are nonconductors of electricity, are highly resistant to chemicals, and have good resistance toward acids, even strong ones. Kaolins have relatively high surface area, which promotes high melt viscosities.

As fillers, in general, kaolins offer improved chemical resistance and electrical properties, and reduced water absorption. The finer grades of kaolin are particularly effective for increasing tensile strength and modulus flow of low T_g thermoplastics, especially surface-treated types, without attendant severe loss of elongation and impact strength. It also lowers the crack formation tendency in the finished article and improves shock resistance and surface quality.

Kaolin is mainly used as a filler for insulating materials since it improves the electrical properties.

Glass Fiber

Glass fibers have a high tensile strength and are different from natural fibers in that they have no inner cellular structure and therefore do not absorb moisture internally. They do absorb moisture on the surface, however, and can be wetted with organic liquids.

Property improvements through fiber glass reinforcements are increases in tensile, compressive, and flexural strengths, impact strength, stiffness and flexural modulus, and dimensional stability. Disadvantages of fiber glass include a somewhat rougher surface of product, painting difficulty, and high abrasion from processing machines.

Glass fiber-reinforced PP is one of the most important reinforcements for thermoplastics. The main advantages lie in the high tensile strength over a wide temperature range, the high stiffness, improvement in creep behavior, and more.

Glass Microspheres

Glass microspheres are classified into solid and hollow spheres. Spheres clearly have no aspect ratio to provide reinforcement properties, but their geometry affords important advantages in processing. They have the smallest surface area of any shapes and therefore they have the smallest influence on viscosity and melt flow. The unique shape of glass spheres also decreases high stress concentrations associated with regularly shaped and sharp-edged particulate fillers that result in severely reduced mechanical properties.

In PP, glass spheres are used less frequently, but they result in higher stiffness at elevated temperature, better compressive strength, and dimension stability.

Carbon Fiber

The use of carbon fibers permits the manufacture of composites which are extremely lightweight but possess high strength and stiffness.

Although most of the early industrial activity on carbon fiber composites involved thermoset resins, we anticipate the predominance of thermoplastics composites.

Asbestos

Asbestos is a generic name for a number of fibrous, hydrated Mg and Na silicate materials. The advantages of asbestos in thermoplastics include higher modulus of elasticity, higher tensile strength, and higher dimensional stability under heat. But, asbestos does impart higher melt viscosity, lower impact strength, and it has higher stabilizer requirements.

PP filled by asbestos offers excellent impact strength and stiffness at temperatures up to 120°C. Other favorite properties include low coefficient of friction, low heat absorption, excellent chemical resistance, and good dielectric properties.

Asbestos is a hazardous material, known to cause various types of cancer and asbestosis under certain conditions. The hazards associated with exposure to asbestos dust led to the development of nondusting grades that represent a much lower health risk.

Organic Fillers

Organic fillers generally can be classified into natural and synthetic materials. Natural materials include various wood flours, shell fiber, cotton fiber, vegetable fiber, and miscellaneous materials. Synthetic materials mainly include various fibers, beads, regenerated cellulose, rubber dust, and so forth. The properties of these fillers differ significantly.

Organic materials have been useful as additives for plastics throughout the history of the industry. PP filled with wood flour exhibits good acoustic properties, but has low elongation and impact strength. Synthetic fibers, such as polyamide or polyeser fibers increase the tensile stress at yield and the impact strength of PP.

Metal and Magnetic Fillers

Metallic powders consisting of aluminum, bronze, zinc, copper, nickel, and some metallic oxides are used in thermoplastics if products with very high thermal or electrical conductivity are required. Heavy metal powders additionally increase resistance to neutron and gamma radiation. The principal advantages in the use of powdered metals for fabricating parts lie in low scrap losses, low labor input, high volume production, and improved strength factors by alloying during fabrication.

The magnetic phenomena found in many metallic and ceramic materials can be utilized to produce useful plastic magnets when fine magnetic particles are dispersed in polymer materials. Plastic magnets have an advantage over metal and ceramic magnets in their shape versatility in conjunction with dimensional accuracy, and their low cost.

Fire Retardant Fillers

The use of flammable materials such as plastics is restricted by a great and growing number of codes, regulations, and standards. Some special fillers, like antimony oxide and alumina trihydrate, and others, are used as flame-retardant fillers.

Synergism between antimony oxide and halogens accounts for the widespread use of antimony oxide in a variety of plastics. Antimony oxide (5-10 wt%) and high melting temperature halogen are used in PP to reduce flammability.

Mixed Fillers

In practice many fillers are used as mixtures; e.g., glass fiber has been used in talc-PP composites to improve their mechanical strength and alumina hydrate has been used to reduce the flammability of $CaCO_3$-PP composites.

Other Fillers

Theoretically, any substance which can disperse in a polymer matrix can be used as a filler. Two other groups of filler for polymers are wastes and gas-fillers. PP can be filled by almost all of the fillers. Some small quality fillers, such as pigments and stabilizers, can also have an effect on various properties of filled polymers.

SURFACE TREAMENT OF FILLERS

Coupling Agents

Coupling agents have been widely used to treat the surface of fillers. They enhance adhesion between the filler and the polymer and they create a strong bond between the two at their interface. Most common coupling agents are silanes and organotitanates.

Surface-Active Agents

Surface-active agents are used to favor interaction between a hydrophilic filler and a hydrophobic polymer. Their effect may be simply described as an increase in the wetting of filler surface by polymer molecules and a resultant improvement in the mechanical properties.

Crosslinking Agents

Peroxides are one of the typical crosslinking agents used in the polymer industry. With certain peroxides, chemical bonds can be formed between a polymer and filler. Peroxides can also be considered coupling agents.

Polymerization on Fillers

To improve their compatibility with a polymer matrix, filler particles can be modified by the grafting of organic polymer chains to their surfaces. Polymerization filling considerably increases the area of contact between the matrix and the filler in comparison with conventional mixing techniques, because the monomer can penetrate the tiniest cracks and pores in the filler, therefore the bonds between the components are strengthened.

The polymerization of PP on a porous filler was accompanied by formation of microfibrils with macromolecules orientation normally in relation to the filler surface.[3] PP synthesized on different fillers showed a strength about equal to that of compounded samples, but the impact strength was exceptionally higher than that of normal filled PP, and had a superior frost resistance.[4,5]

EFFECT OF FILLERS ON POLYPROPYLENE

The effect of filler on PP is determined by the characteristics of the filler, such as shape of the particles, particle size and size distribution, surface area, particle packing, chemical composition, and so forth.

Morphologies

Morphologies of filler-PP composites depend not only on the characteristics of the filler but also on the processing.

Thermal Properties

Because PP has very low T_g (–25°C), its melting temperature, T_m, relative to crystallization temperature, T_c, allows crystallization to take place rapidly after processing. The solidification rate of the crystalline PP is increased by fillers which increase the number of nuclei.

MECHANICAL PROPERTIES OF FILLED POLYPROPYLENE

Generally, fillers in PP produce the following beneficial increases in mechanical properties: stiffness, modulus of elasticity, flexural modulus, bend-creep modulus, hardness, dimension stability, tensile stress at break and tensile strength, and compressive strength. However, filler may reduce elongation, impact strength, and creep tendency.

REFERENCES

1. Varga, J. *J. Mat. Sc.* **1992**, *27*, 2557.
2. Enikolopian, N. S.; Volfson, S. A. *Handbook of Polyolefins* Vasile, C.; Seymour, R. B., Eds.; Marcel Dekker: New York, NY, 1993; pp 772.
3. Maksimova, T. V. et al. *Plast. Massy* **1986**, *6*, 50.
4. Wypyeh, G. Fillers, Chemtec: Toronto, 1993, p 109.
5. Khunova, V.; Zamorsky, Z. *Polym-Plast. Technil. Eng.* **1993**, *32*(4), 289.

FILLERS, SURFACE MODIFICATION

Kazuta Mitsuishi and Hitoshi Kawasaki
Industrial Technology Center of Okayama Prefecture

Modification of polymers and fillers is performed for various industrial uses. With a view to giving multifunction and high performance to polymer composites. So, control of polymer-filler interface is becoming a significant task for composite material design. We show here the characteristics of modifiers and the physical or chemical properties of modified particle-filled polymers. The modification agents for various fillers can be classified as silane, titanate, aluminate, fatty acids, fats and oils, olefin wax, and surface active agent.

SILANE COUPLING AGENT

Silane coupling agent is applied for various industrial and household articles such as plastics, elastomers, paints, and adhesive agents.[1]

J. G. Marden described the interaction between polymer matrix and silane coupling agent: covalent bond, formation of interphase communicating load, affinity of polymer-filler interface, prevention of water from polymer surface, and compatibility.[2,3] The action of silane coupling agent on inorganic filler, however, is suspected to be covalent bond, hydrogen bond, physical adhesion, bridge formation, and prevention of water from filler surface.

These results show promise for the effect of silane coupling agent. We expect it to improve mechanical and electric properties, filler dispersion, thermal stability, formation, waterproof stability, high loading of filler, uniformity of material properties, and reduction of materials costs.

TITANATE COUPLING AGENT

Silane coupling agent is used for glass, silica, alumina, talc, and mica to reinforce the mechanical properties of inorganic particle-filled polymer, whereas titanate coupling agent is utilized in electronics, automobiles, construction, paints, plastics and so on for improving filler dispersion in process formation, high loading, reflux rate, elongation at break, impact strength and flexibility of composites, adhesion to metal and viscosity reduction.

FATTY ACIDS

Since precipitated calcium carbonate ($CaCO_3$) was developed in 1927, various types of modified $CaCO_3$ have been used for industrial products and household articles.

POLY(ETHYLENE GLYCOL) DERIVATIVES

Poly(ethylene glycol) derivatives and poly(ethylene glycol) are modifiers for $CaCO_3$.[4,5]

PHOSPHATE COUPLING AGENT

Phosphate coupling agent has been developed as a modifier for $CaCO_3$.[6] In this report we conclude that the existence of a hydrophobic group of alkyl dihydrogen phosphate on the $CaCO_3$ surface promotes flexibility of PP composites. The reason for the improvement of adhesion at the PP-$CaCO_3$ interface and of mechanical properties is mainly attributed to the following factors: restriction of aggregation of filler, relaxation of stress concentration surrounding filler, and increase of wettability at the polymer-filler interface.[7]

REFERENCES

1. Yoshioka, H.; Ikeno, M. *Surface* **1983**, *21*, 157.
2. Marden, J. G. *Appl. Polym. Symp.* 1970, p 14.
3. Erickson, P. W. *Composite Materials*, Academic: New York, NY, 1974; Vol. 6.
4. Mitsuishi, K.; Kodama, S.; Kawasaki, H. *Kobunshi Ronbunshu* **1985**, *42*, 129.
5. Kalinski, R.; Galeski, A.; Kryszewski, M. *J. Appl. Polym. Sci.* **1981**, *26*, 4047.
6. Nakatsuka, T. et al. *J. Appl. Polym. Sci.* **1982**, *27*, 259.
7. Mitsuishi, K. et al. *Die. Angew. Makromol. Chemia* **1991**, *189*, 13.

Films

See: Aramid Film
Aramids (Processable Precursors)
Compatibilizers, Polymeric (Recycling of Multilayer Structures)
Electrically Conducting Composites
Ethylene-Vinyl Alcohol Copolymer (Overview)
Hydroxypropyl Cellulose (Crosslinked, Cholesteric Liquid Crystal)
Liquid Crystalline Polymers (Heat-Strengthened, Thermotropic)
Liquid Crystalline Polymers (Polyesters)
Molecular Composites (High Modulus Polyimide/Polyimide Films)
Molecular Composites (Polyimide/Polyimide Laminate Processability)
Multilayer Films (Polyelectrolytes)
Polyacetylene Films
Polyethylene (Commercial)
Polyethylene Film Gels (Photografting of Hydrophilic Monomers)
Polyethylene, Multicomponent
Polyimides, High Performance Films
Polystyrene, Syndiotactic
Poly(vinyl alcohol) (High Strength and High Modulus)

Filters

See: Cellulose (Overview)

Finishes

See: Flame Retardant Finishing

FIRE RETARDANTS, ANTIMONY-CONTAINING

N. Khalturinskij and A. Antonov
Institute of Synthetic Polymeric Materials

Antimony trioxide in most cases is used as a fire retardant. This compound is a so-called synergistic inhibitor; that is, it is used with other substances to lower flammability. As a rule, it is used in combination with halogen-containing compounds. Most reports concerned with the influence of antimony trioxide on polymer flammability point out its extremely low effectiveness when used without halogens. The formation of intermediate compounds that inhibit burning in the gaseous phase, as halogens do, may be responsible for a possible synergetic mechanism between antimony and halogens.

In 1947, Little was the first to propose the retardant capabilities of antimony trioxide and halogen containing compounds.[1] Little concluded that the formation of the antimony oxyhalide that inhibits burning in the gaseous phase causes the synergism. Other authors, such as Lions and Hastie, confirmed this proposition.[2,3]

Bendow and Cullis considered that the appearance of an antimony halide in the burning zone causes the formation of aerosol Sb_2O_3 and atomic halogen.[4] Antimony trioxide acts as a catalyst of free-radical recombination and atomic halogen causes the formation of a hydrogen halide known as a burning inhibitor. However, Martin and Price theorized that halogens were only transport mechanisms for the antimony transference from condensed to gaseous phases.[5]

Thus, experimental data currently available are discrepant and do not allow us to make a final conclusion concerning the inhibition properties of antimony-containing fire retardants. Our aim here is to investigate some characteristic properties of antimony compounds and refine the combustion inhibitory mechanism of fire retardants.

RESULTS AND DISCUSSION

Based on the obtained data we can concern ourselves with the extremal relationship between the oxygen index of composition and antimony trioxide percentage in it to two acting factors: the introduction of antimony trioxide increased the heat loss from the "cap" surface, but, the time needed for the beginning of pyrolsis decreased at the expense of the elevation of composition optical density.

Thus, we studied in detail the influence of antimony compounds on physical processes occurring in polymer pyrolysis and burning and showed the following: Gas-phase inhibition by antimony compounds is ineffective and does not substantially reduce epoxide composition flammability. Antimony trioxide alone can be an effective burning inhibitor. The inhibitory mechanism of polymer composition flammability was defined which consists of changing the optical characteristics of polymer compositions and coke and gives rise to the elevation of heat loss from the burning surface. To put this mechanism in action in the case of soot flames we must seek the production of coke that has a high reflection coefficient, especially in the spectrum range where the radiation maximum of flame occurs. Since full reflection is virtually unachievable, the coke must be opaque to radiant heat in addition to having high reflection properties.

REFERENCES

1. Little, R. W. *Flameproofing Textile Fabrics*; American Chemical Society Monograph No. 104; New York, 1947.
2. Lions, J. W. *The Chemistry and Uses of Fire Retardants*; Wiley-Interscience: New York, 1970.
3. Hastie, J. W. *J. Res. Nat. Bur. Stand. Sect. A* **1973**, *77*(6), 733.

4. Bendow, A. W.; Cullis, C. F. *Combustion Institute European Symposium*; Academic: New York, 1973, p 183.
5. Martin, F. J.; Price, H. R. *J. Appl. Polym. Sci.* **1968**, *12*, 143.

FLAME-RETARDANT FINISHING

Shizuo Kubota
Industrial Technology Center of Wakayama Prefecture

Dense urban populations, and building regulations and codes, are contributing factors to the increased market demand for flame-retardant finished textile products. These products are the most effective fire-fighting countermeasures because a fire can be extinguished in its initial stage, and prevented from spreading. Many countries are regulating textile products like sofas and bedding, curtains, carpets, decorative wallpaper, tents, sheets for working, etc., as well as thc interior and fabrics used in airplanes and automobiles for fire-safety reasons.

Fibers such as cotton, polyester, etc., are termed flame retardant according to their after-treatment with phosphorus halogen-containing flame retardants. This paper mainly reports on the flame-retardant finishing of cotton, polyester, and wool fibers.

FLAME-RETARDING MECHANISM

Flame-retardant finishing utilizes a chemical mechanism to obstruct the heat decomposition of fibers by radical scavenging, or endothermic reaction, or to change the heat decomposition reaction in its mechanism to prevent the generation of flammable gases, and it uses a physical mechanism by which heat and air are shielded, flammable gases are diluted, etc.

Empirically, it is known that elements having a flame-retarding effect are magnesium (Mg), boron (B), aluminum (Al), nitrogen (N), phosphorus (P), antimony (Sb), sulfur (S), and halogens.

Phosphorus acts as a catalyst for dehydration of cellulose, and lowers the decomposition temperature and simultaneously promotes the synthesis of polyene carbide residues, caused by β-elimination, so that the cellulose is converted into a flame-retardant substance. This action largely depends on the chemical structure of the phosphorus compounds.

Among the halogens, bromine is more flame retardant than chlorine. Halogens interfere with the chain decomposition reaction caused by free radicals in burning gas with flame-retarding effects. Moreover, halogen-containing compounds, when mixed with emitted gases, are burn-inhibiting and are effective for cellulose in their gaseous or solid states, promoting the production of carbide residues.

The combinations of phosphorus with nitrogen, antimony with halogen, and phosphorus with halogen produce synergistic effects for flame retardation. Antimony halides achieve the carbide formation reaction in the solid state and react to scavenge free radicals in the gaseous state, making them flame retardant.

FLAME-RETARDANT FINISHING OF COTTON FIBERS

Transitory Flame-Retardant Finishing

Products made from fibers, such as curtains that are not required to be washable, are finished with orthophosphoric acid, ammonium phosphate, ammonium sulfate, ammonium sulfamate, ammonium bromide, borax-boric acid, thiourea phosphate, phosphoric acid carbamate, sulfuric acid carbamate, guanylurea methylolated phosphate, etc.

Ammonium phosphate is inexpensive, and the finishing can be easily carried out. However, it is not water-resistant and cannot be washed. Fabric materials finished with ammonium phosphate tend to be brittle with white powder showing on their surface. Fabric materials having moisture-adsorptive properties are sticky and are not suitable, because they may cause corrosion of metal fasteners and hooks.

Durable Flame-Retardant Finishing

This finishing is used for work materials, bed sheets, and similar fabrics. Several methods follow.

Tetrakis(hydroxymethyl)phosphonium Salt (Proban-Finishing Method)

This method, using THPC [tetrakis(hydroxymethyl)-phosphonium chloride $(HOCH_2)_4\ P^+Cl^-$] together with urea and trimethylolmelamine, can be applied practically.[1] However, fabrics made using this finish have a relatively stiff hand and have a considerably low tearing strength.

N-hydroxymethyl-3-dimethlphosphonopropionamide $(CH_3\ O)P(=O)CH_2CH_2C(=O)NHCH_2OH$, Pyrovatex CP (Ciba Geigy Co.)

The *N*-hydroxymethyl group is capable of reacting with cellulose. In addition, trimethylolmelamine is used to enhance its washability.

Vinylphosphonate Oligomer

This oligomer, called Fyrol 76, was developed by Stauffer Chemical Co. The oligomer is synthesized by the condensation copolymerization of bis(2-chloroethyl)vinylphosphonate and methylphosphonate.[2]

The finishing agent is applied along with *N*-methylolacrylamide and a potassium persulfate catalyst. Examination of finishing quality is made using electron beams.[3]

Aminophosphazene

We finished cotton fibers with purified aminophosphazene, using an acid catalyst. Flame-retardant finished pieces of cloth were obtained having both good washability and durability and eliminating time-dependent changes. Use of the acid catalyst in an insufficient or excessive amount caused deterioration of the endurance. When finished with the proper amount of the acid catalyst, we found satisfactory flame-retardant properties, after placing them in 60°C temperature and 95% humidity for one week, then washing them. The finished fabrics were laundered after storage for two years and had flame-retardant properties.[5,6]

Finishing with a purified aminophosphazene has the following characteristics: formaldehyde is not released, the finished pieces of cloth have a soft hand, the tear strength retention ratio is good, that is, 90% in filling. We obtained durability and flame-retarding properties. Moreover, no toxic halogen gases and no hydrogen halide gases are emitted during the time of burning, since no halogen is present.

FLAME-RETARDANT FINISHING OF POLYESTER FIBERS

The amounts of elements required for flame-retardant finishing of polyester fibers are listed below.[7]

- ~5% P(organophosphorus compounds)
- 1% P and 15-20% Cl
- 2% P and 6% Br
- 25% Cl
- 12-15% Br
- 2% Sb_4O_6 and 16-18% Cl
- 2% Sb_4O_6 and 8-9% Br

The following compounds are available: Tricresyl-phosphate $O{=}P(OC_6H_4CH_3)_3$, tris(tribromonopentyl)phosphate $O{=}P(OCH_2\text{-}C(CH_2Br)_3)_3$, a cyclic methylphosphonate ester,[8] and tetrabromobisphenol A ethylene oxide adduct, tetrabromocyclooctane, hexabromocyclododecane.

With these compounds, polyester fibers can be finished by the pad-dry-cure method or by the absorption method from a dye liquid (hexabromocyclododecane).

FLAME-RETARDANT FINISHING OF WOOL

Wool fibers, having an ignition temperature of 570-600°C, a combustion heat of 4.9 Kcal/g, and LOI of 25.2 are resistant to burning, but are not acceptable as "flame-retardant fibers."

Friedman and Tillin studied the flame-retardant finishing of wool by the graft-copolymerization of bis(2-chloroethyl) vinylphosphonate or by the Michel addition method.[9]

REFERENCES

1. Reeves, W. A.; Guthrie, J. D. *Text. Res. J.* **1953**, *23*, 527.
2. Weil, E. D. *JFF/Fire Retardant Chemistry* **1974**, *1*, 125.
3. Weider, A. et al. *Text. Chem. Color* **1977**, *9*, 109.
4. Kubota, S.; Taniguchi, H.; Maeda, R. *Sen'i Gakaishi* **1987**, *43*, 263.
5. Kubota, S.; Tamura. S. *Pacific Polymer Preprints* **1989**, *1*, 145, Hawaii.
6. Kubota, S.; Horiuchi, Y. *Proc. Int. Symp. Dyeing & Finishing Text.* Fukui: Japan, 1994; p 168.
7. Lyons, J. W. *The Chemistry and Uses of Fire Retardants* Robert E. Krieger: Florida, 1987, p 398.
8. Johnston, B. E.; Jurewicz, A. T.; Ellison, T. *Proc. Symp. Tex. Flammability* **1977**, *5th*, 209.
9. Friedman, M.; Tillin, S. *Text. Res. J.* **1970**, *40*, 1045.

Flame Retardant Materials

See: *Additives (Property and Processing Modifiers)*
Additives (Types and Applications)
Epoxy-Imide, Phosphorylated Epoxy-Imide Polymers
Fire Retardants, Antimony-Containing
Flame-Retardant Finishing
Flame Retardants (Overview)
Flame Retarding Polymers
Flammability (Char Formation)
Phenolic Composites
Poly(ethylene terephthalate) Copolymers
Unsaturated Polyester Resins (Overview)
Unsaturated Polyester Resins (Polyester Composites)

FLAME RETARDANTS (Overview)

J. R. Ebdon and M. S. Jones
The Polymer Centre
School of Physics and Chemistry
Lancaster University

Although synthetic polymers have become widely accepted in almost every application imaginable, many are extremely flammable and, given sufficient heat and oxygen, will burn readily and rapidly. The flammability and the resulting destruction of property are not the only problem. Fire fatalities are essentially due to the evolved smoke and toxic gases, exacerbated in some cases by the poisonous fumes emitted from some synthetic organic polymers. This has led to the introduction of stricter legislation and safety standards concerning flammability, and extensive research into the area of flame retardants for polymers has been the result.

THERMAL DECOMPOSITION AND FLAME RETARDANTS

The understanding of flammability and flame retardancy requires an understanding of the thermal decomposition processes of polymers.[1] The most important step in the burning of a polymer is the fuel production stage, in which an external heat source causes an increase in temperature, resulting in the dissociation of chemical bonds and the evolution of volatile fragments. These fragments diffuse into the surrounding air to create a flammable mixture and combustion is initiated when this mixture reaches the ignition temperature.

Flame retardants are classified into two main categories: *additives*, which are mechanically blended with the polymeric substrate, and *reactives*, which are chemically bound to the polymer either by simple copolymerization or by modification of the parent polymer. Both additives and reactives can interrupt the burning cycle of a polymer in several ways: by altering the thermal decomposition mechanism of the polymer, by quenching the flame, or by reducing the heat transferred from the flame back to the decomposing polymer.[4] These methods can involve, among others, the use of endothermically decomposing heat sinks, the generation of free radical inhibitors, and the formation of nonflammable protective coatings.

TESTING PROCEDURES

The most widely used laboratory test is the limiting oxygen index (LOI) ASTM-D-2683 technique. The LOI value is the minimum amount of oxygen in a mixture of oxygen and nitrogen, expressed as a volume percentage, that will just support flaming combustion. This value therefore enables the combustibility of a polymer to be expressed and compared with that of other materials. Other attractive aspects of the LOI value are that it correlates well with other flammability parameters such as flame spread, heat of combustion, char yield, and flame-retardant concentration.[7-11]

HALOGEN FLAME RETARDANTS

Although in the last few years there has been an increase in the number of heteroelements used in flame retardants, the commercial market is still dominated by compounds containing halogens, notably bromine and chlorine. They have exceptional efficiency and may be incorporated as either additives, such as ethylene bis(tetrabromophthalimide), decabromobiphenyl, hexabromocyclododecane, and Dechlorane-plus, or as reactives, for example, tetrabromobisphenol-A, vinylidene chloride, chlorostyrene and pentabromobenzyl acrylate.[15-21]

Bromine compounds are generally more effective than their chlorinated analogues, but are more expensive and less thermally and photochemically stable as a result of the weak carbon-bromine bond.[22] A halogenated flame retardant may act in the condensed phase but detailed research indicates that they are primarily vapor-phase retardants, interfacing with the free radical reactions involved in flame propagation.[2,4,6,12,22]

Halogen compounds are frequently used with antimony trioxide, Sb_2O_3, or antimony pentoxide, Sb_2O_5, synergists, creating a flame-retardant effect greater than the sum of the individual effects.[6] Antimony trioxide is an inert filler with no flame-retardant effect of its own, but forms antimony halides and oxyhalides on heating with organohalogens.[4] These halides primarily act in the gas phase, removing $HO^\bullet$ and $H^\bullet$ radicals and subsequently poisoning the flame.

High loadings of halogenated flame-retardants and antimony oxide fillers often improve the physical properties and long term aging effects of polymers, but at the same time have deleterious effects on the tensile and impact strengths. The best method to reduce the flammability of advanced composites, with minimal effect on the mechanical properties is currently to use a brominated reactive such as 2,2′,6,6′-tetrabromobisphenol-A with antimony oxide fillers.[19,24]

The use of halogenated compounds and antimony oxides has one major drawback: they increase the amounts of smoke and toxic decomposition products evolved during polymer combustion. Consequently, they have been the subject of environmental concern for some time.[15,25,26] The use of halogenated materials also gives rise to the additional hazard of strongly acidic gases, for example, HCI and HBr, being liberated on heating. These gases can not only cause lung damage but can also corrode electrical equipment rendering them inoperative. Therefore there has been increasing research into innovative, environmentally friendly, halogen-free flame retardants.

PHOSPHORUS-BASED FLAME RETARDANTS

Phosphorus-based flame retardants can be inorganic, organic, or elemental (red phosphorus), can be active in the vapor phase or in the condensed phase, and sometimes may operate simultaneously in both phases.[27]

The condensed phase mechanism arises as a consequence of thermal generation of phosphorus acids from the flame retardant, for example, phosphoric acid or poly(phosphoric acid). These acids act as dehydrating agents altering the thermal degradation of the polymer, promoting char formation at the expense of flammable volatiles. The water vapor liberated through the dehydration acts as a diluent of the flammable volatiles and cools the flame. The polymers concerned are usually oxygen containing species, often polyols or cellulosics.[12,23]

Char formation may also occur via acid catalyzed elimination of water from alcohol. These mechanisms provide the basis for intumescence, the formation of a foamed char protective coating.

Ammonium polyphosphate (APP), and compounds based on APP, are usually used as the inorganic acid source, forming the very strong phosphorylating agent polyphosphoric acid (PPA) on decomposition; but almost any phosphorus containing species is capable of decomposing to form the acid catalyst.[1,28] Occasionally other acids, such as sulphuric and boric acids, are used.[27a,29]

The flame retardance of polymers is at present dominated by the use of additives, but increasing research activity is now being directed at the chemical modification of polymers to impart flame retardance. It is believed that low levels of modification may have comparable effects to those achieved with relatively high loadings of flame-retardant additives. The modification of a polymer can be carried out effectively in two ways: by altering the structure of the parent polymer with the introduction of a potential flame-retardant species by modification of a functional group or by a simple addition reaction to an unsaturated polymer, and by relatively straightforward copolymerization techniques.

There are many other interesting methods used to modify a variety of polymers to reduce flammability. Four recent examples are: the preparation of phosphonated polystyrenes by modification of either the polystyrene homopolymer or poly(4-chloromethylstyrene-*co*-styrene) copolymers, the phosphorylation of polyethlyene with PCl_3/O_2 and water, the preparation of highly flame-retardant polyethylene by electron beam radiation grafting of vinyl phosphonate oligomers, and the epoxidation and phosphorylation of natural rubber with performic acid and dibutylphosphate.[9,30-33]

The more conventional approach to the incorporation of phosphorus into a polymer to impart flame retardance is by copolymerization.

PHOSPHORUS-NITROGEN SYSTEMS

Although some naturally occurring nitrogen-containing organic polymers have low flammabilities, for example, silk and wool, most nitrogen-containing synthetic polymers, such as polyurethanes, are extremely flammable.[2] Occasionally, nitrogen-containing organic compounds, such as melamine salts, are used as additives, for example, in polyamides to increase dripping and heat removal during decomposition.[34,35] Nitrogen-containing organic compounds, for example urea and isocyanates, can also be used as reactives within certain polymers.[36]

The main application of nitrogen compounds in flame retardance is in conjuction with phosphorus to give P-N synergistic effects. It has been known for some time that mixtures of nitrogen bases and phosphoric acids provide greater flame retardance with cellulosics than would be expected on the basis of the combined individual effects.[2] The nitrogen is believed to facilitate the phosphorylation of cellulose by phosphoric acid and therefore to assist the production of intumescence.[37]

The most interesting area of P-N research is in the development of phosphazenes as flame retardant materials. Monomeric phosphazenes have been used as flame retardant additives in a number of applications: polystyrene blends, PET, polyurethanes, and cyanoacrylate adhesives.[39-42] The flame-retardant effect in each case is associated with vapor phase activity.[38]

The use of polymeric phosphazenes as flame retardants is a much more exciting application. The phosphazene polymers may be classified into three categories: linear polyphosphazenes with an alternating P-N backbone, systems based on carbon chain polymers with cyclophosphazene substituents as pendant groups, and the cyclolinear/cyclomatrix polymeric phosphazenes, linked via exocyclic groups to give linear polymer chains or crosslinked polymer networks.[38]

PHOSPHORUS-HALOGEN SYNERGISM

The possibility of phosphorus-halogen synergism has been studied for a number of years.[2,13,23,43,44] The mechanism is believed to involve the formation of phosphorus halides and oxyhalides acting as free radical interceptors, terminating the radical chain reactions involved in flame propagation. The phosphorus-halogen system not only provides improved flame retardancy, hence reducing the amount of flame retardant required, but also removes the need to use antimony oxides, as often P-Sb systems exhibit antagonistic effects.[45]

Although phosphorus-halogen synergism undoubtedly provides excellent flame retardants, there is again the problem of toxicity of bromine compounds and brominated decomposition products.

METAL-BASED FLAME RETARDANTS

At present the most widely used acid-free and halogen-free flame retardant/smoke suppressant is aluminium trihydroxide (ATH). ATH operates in several ways as a flame retardant:[46]

1. The endothermic decomposition absorbs thermal and radiative energy from the flame. The endothermic cooling is believed to favor dehydrogenation reactions and protective char formation.[47]
2. The evolution of water not only has a cooling effect but also blankets the flame and acts as a diluent.
3. The aluminium oxide layer, formed by dehydration, has an exceptionally high surface area. This high surface area alumina adsorbs smoke and flammable molecules, causing a reduction in the amount of carbon dioxide evolved.
4. It has been suggested that ATH acts as an electron donor to terminate radical reactions, by giving rise to a relatively unreactive inorganic free radical, which is incapable of initiating further radical reactions.[48]

Some current applications of ATH are as a flame retardant filler/additives for EVA, LDPE, and LLDPE cable insulation and jackets, for polypropylene, and for a variety of thermosets.[49-51]

Magnesium hydroxide and a hydromagnesite filler have recently attracted growing interest as potential flame retardants/smoke suppressants.[52] Magnesium hydroxide endothermically decomposes at 340°C, higher than ATH, and can be used with the polymers requiring high temperature processing. The higher cost of the magnesium hydroxide in comparison to that of ATH, and the requirement of higher loadings, are drawbacks to its widespread acceptance.

Molybdenum, iron, tin, and zinc compounds have all shown potential as flame retardants/smoke suppressants for organic polymers. Molybdenum compounds are well-known flame retardants for cellulosics and over the last few years have shown promise as flame retardants for a variety of other polymers.[53,54] Molybdenum trioxide and zinc and ammonium molybdates all act as replacements and partial replacements for antimony trioxide with halogenated polymers such as PVC.[53] Unfortunately, molybdenum compounds have limited use because they are expensive and colored, but used as partial replacements for antimony oxides they are excellent flame retardants and smoke suppressants.[55]

Organometallic and inorganic iron compounds, for example, ferrous oxides, ferric oxides, and ferrocene, have been widely used as flame retardants and smoke suppressants, dramatically increasing the resistance of various polymers to ignition and burning.[56,57]

The powerful Lewis acid, $FeCl_3$, is thought to be the key intermediate for smoke suppression and char formation with PVC containing systems. Many other metal-based flame retardants also have shown potential as flame retardants and smoke suppressants, including chromium, zinc, and tin compounds.

BORON-CONTAINING FLAME RETARDANTS

Borax ($Na_2B_4O_7 \cdot 10H_2O$) and boric acid have been well-established flame retardants for cellulosics for some considerable time.[58] Boric acid is a good afterglow suppresser while borax is a good flame retardant. Therefore, mixtures of the two are frequently used, giving a combined vapor and condensed phase active flame retardant.

Boric acid and the alkali metal borates are water soluble and cannot be used for synthetic polymers, so alternatives have to be found. The zinc borates have emerged as partial, and cheaper, replacements for antimony oxides as synergists in halogenated polymers and halogen additive/polymer systems.[22] Another major attraction of the zinc borates is that they have no adverse effects on the strength, elongation, and thermal aging of many polymers. The zinc also catalyzes HCl loss from PVC and subsequent crosslinking reactions, leading to char. The borate also forms a viscous, vitreous layer on the polymer surface, which smothers the flame and prevents oxidation of the underlying carbonaceous char. As well as having been used for PVC, zinc borate is an effective synergist/flame retardant with other halogen-containing systems, e.g., brominated polyesters, nylons, and epoxy resins.[58-61] The zinc borates have also been employed as flame retardants/smoke suppressants on their own in silicone-based polymers used in electrical cables.[62]

CONCLUSIONS FOR THE FUTURE

As more legislative actions are taken in an attempt to minimize the environmental impact of flame retardants, the balance between fire safety issues and environmental issues becomes more of a problem. Adequate flame-retardants are mandatory requirements, for example in applications of polymers in the mining, building, and transport industries, but there is a need also to minimize pollution and toxicity risks. Thus the use of the polybrominated diphenyl ethers, which are excellent flame retardants but which termally degrade to give highly toxic products, becomes more problematic in these industries. This type of problem may lead to dramatic changes in the design of flame retardants and/or result in the wider introduction of flame-retardant free, thermally stable polymers.

REFERENCES

1. McNeil, I. C. *Makromol. Chem., Macromol. Symp.* **1993**, *74*, 11.
2. Cullis, C. F.; Hirschler, M. M. *The Combustion of Organic Polymers*; Oxford, 1981.
3. Camino, G.; Costa, L.; Luda di Cortemiglia, M. P. *Polymer Degradation and Stability* **1991**, *33*, 131.
4. Grassie, N.; Scott, G. *Polymer Degradation and Stabilisation* Cambridge University: 1985.
5. Aseeva, R. M.; Laikov, G. E. *Advances in Polymer Science* **1985**, *70*, 171.
6. Gann, R. G.; Dipert R. A.; Drews, M. J. *Encyclopaedia of Polymer Science and Engineering*, 2nd ed.; Wiley-Interscience: New York, 1987; Vol. 7, p 154.
7. Komamiya, K. J. *Fire and Flammability* **1973**, *4*, 82.
8. Johnson, P. R. *J. Appl. Polym. Sci.* **1974**, *18*, 491.
9. Ebdon, J. R.; Banks, M.; Johnson, M. *Proceedings from the Flame Retardants '94 Conference* The British Plastics Federation, Interscience Communications Limited, London, 1994; p 183.
10. Van Krevelen, D. W. *Polymer* **1975**, *6*, 39.
11. Larson, E. R. *J. Fire and Flammability/Fire Retardant Chem.* **1975**, *2*, 209.
12. Camino, G.; Costa, L. *Polymer Degradation and Stability* **1988**, *20*, 271.
13. Hindersinn, R. R.; Witschard, G. *Flame Retardancy of Polymeric Materials*; Kuryla, W. C.; Papa, A. J. Eds.; Marcel Dekker: New York, 1978; Vol. 4.
14. Pagliari, A.; Cicchetti, O.; Bevilacqua, A.; Van Hees, P. *Proceedings from the Flame Retardants '92 Conference* The Plastics and Rubber Institute, Elsevier Applied Science, London, 1992; p 41.
15. Troitzsch, J. *Makromol. Chem., Macromol. Symp.* 1993, 74, 125.
16. Vandenberk, B. *Proceedings from the Flame Retardants '94 Conference* The British Plastics Federation, Interscience Communications Limited, London, 1994; p 163.
17. Valange, B. M.; Calewarts, S. E.; Bonner, G. A.; Pettigrew, F. A.; Schleifstein, R. A. *Proceedings from the Flame Retardants '90 Conference* The British Plastics Federation, Elsevier Applied Science: London, 1990; p 67.
18. Markezich, R. L. *Proceedings from the Flame Retardants '92 Conference* The Plastics and Rubber Institute, Elsevier Applied Science, London, 1992; p 187.
19. Kidder, R. C. *Proceedings from the Flame Retardants '92 Conference* The Plastics and Rubber Institute, Elsevier Applied Science: London, 1992; p 21.
20. Tesoro, G. C. *J. Polym. Sci.* **1978**, *13*, 283.
21. Georlette, P.; Gramse, Y.; Peled, M.; Simons, J.; Utevskii, L. *Fire Retard Technol. Mark., Int. Conf. Fire Saf.* Fire Retard. Chem. Assoc.: Lancaster, PA, 1990; p 43.
22. Landrock, A. *Handbook of Plastic Flammability and Toxicology* Park Ridge Noyes, 1983.
23. Papa, A. J. *Flame Retardancy of Polymeric Materials* Vol. 3, Kuryla, W. C.; Papa A. J. Eds.; Marcel Dekker: New York 1975.
24. Kunz, D. H. E. *Makromol. Chem., Macromol. Symp.* 1993, 74, 155.
25. Rappe, C. *Proceedings from the Flame Retardants '92 Conference* The Plastics and Rubber Institute, Elsevier Applied Science: London, 1992; p 133.
26. McAllister, D. L. *Proceedings from the Flame Retardants '92 Conference* The Plastics and Rubber Institute, Elsevier Applied Science: London, 1992; p 149.
27. Green, J. *J. Fire Sciences* **1992**, *10*, 470; (a)Cicchetti, O.; Bertelli, G.; Bevilacqua, A.; Cipolli, R.; Pagliari, A. *Proceedings from the Flame Retardants '94 Conference* The British Plastics Federation, Interscience Communications Limited: London, 1994; p 129.
28. Becker, W. *Makromol. Chem., Macromol. Symp.* 1993, 74, 225.
29. Lyons, J. W. *The Chemistry and Uses of Fire Retardants* Wiley-Interscience: New York, 1970.
30. Banks, M.; Ebdon, J. R.; Johnson, M. *Polymer* **1993**, *34*, 4547.
31. Catala, J.-M.; Brossas, J. *Progress in Organic Coatings* **1993**, *22*, 69.
32. Kaji, K.; Yoshizawa, I.; Kohara, C.; Komai, K.; Hatadu, M. *J. Appl. Polym. Sci.* **1994**, *51*, 841.
33. Derguet, D.; Radhakrishnan, N.; Brosse, J.-C.; Boccaccio, G. *J. Appl. Polym. Sci.* **1994**, *52*, 1309.
34. Gilleo, K. B. *Advances in Fire Retardant Textiles (Prog. Fire Retardancy Ser.)*; Techonomic: Westport, 1975; Vol. 5, p 165.
35. Spepniczka, H. E. *Advances in Fire Retardant Textiles (Prog. Fire Retardancy Ser.)*; Techonomic: Westport, 1975; Vol. 5, p 409.
36. Kuryla, W. C. *Flame Retardancy of Polymeric Materials*; Kuryla, W. C.; Papa, A. J. Marcel Dekker: New York, 1973, Vol. 1.
37. Troitzsch, J. H. *Progress in Organic Coatings* **1983**, *11*, 41.
38. Allen, C. W. *J. Fire Sciences* **1993**, *11*(4), 320.
39. Benzoari, M. D. *US486047* 1990. {Chem. Abstract, *113*: 24242h}.
40. Laszkiewicz, B.; Kotek, R. *Hung. J. Ind. Chem.* **1989**, *17*, 221. {Chem. Abstract, *112*: 8647t}.
41. Kurach, Y.; Okuyama, T.; Oohasi, T. *J. Mater. Sci.* **1989**, *24*, 2761.
42. Nagasawa, K.; Nakada, C. Jpn. Kokai Tokyo Koho, *61/40386* A2, 1986. {Chem. Abstract, *105*: 98812a}.
43. Lyons, J. W. *J. Fire and Flammability*, **1970**, *1*, 302.
44. Hilado, C. J. *Flammability Handbook for Plastics* Technomic, Stamford, CT, 1969.
45. Green, J. *ACS Symp. Series* **1989**, *425*, 259.
46. Mead, N. G.; Bown, S. C. *Preprints from the Inorganic Fire Retardants–All Change? Conference* Royal Society of Chemistry–Industrial Division: London, January 13, 1993.
47. Vandersall, H. L. *J. Fire and Flammability* **1971**, *2*, 97.
48. Hosaka, H.; Megurio, K. *Prog. Org. Coat.* **1973**, *2*, 315.
49. Claus, Jr., W. D. *Proceedings from the Flame Retardants '92 Conference* The Plastics and Rubber Institute, Elsevier Applied Science: London, 1992, p 66.
50. Kirschbaum, G. *Proceedings from the Flame Retardants '94 Conference* The British Plastics Federation, Interscience Communications Limited: London, 1994; p 169.
51. Brown, S. C.; Herbert, M. J. *Proceedings from the Flame Retardants '92 Conference* The Plastics and Rubber Institute, Elsevier Applied Science: London, 1992; p 100.
52. Doyle, M.; Clemens, M.; Lees, G.; Briggs, C.; Day, R. *Proceedings from the Flame Retardants '94 Conference* The British Plastics Federation, Interscience Communications Limited: London, 1994; p 193.
53. Skinner, G. A.; Haines, P. J. *Fire and Materials* **1986**, *10*, 63.
54. Church, D. A.; Moore, F. W. *Plast. Eng.* **1975**, *31*, 36.
55. Murfitt, P. S.; Moore, F. W.; Musselman, L. L. *Proceedings from the Flame Retardants '92 Conference* The Plastics and Rubber Institute, Elsevier Applied Science: London, 1992; p 176.
56. Carty, P.; White, S. *Preprints from the Inorganic Fire Retardants–All Change? Conference* Royal Society of Chemistry–Industrial Division: London, January 13, 1993.
57. Carty, P.; White, S. *Appl. Organomet. Chem.* **1991**, *5*, 51.

58. Arthur, L. T.; Quill, K. *Proceedings from the Flame Retardants '92 Conference* The Plastics and Rubber Institute, Elsevier Applied Science: London, 1992.

59. Quill, K. *Preprints from the Inorganic Fire Retardants–All Change? Conference* Royal Society of Chemistry–Industrial Division, London, January 13, 1993.

60. Woods, W. G.; Bower, J. G. *Mod. Plast.* 140, June 1970.

61. Shen, K. K. *Plastics Compounding* **1985**, *8*(5), 66.

62. Cella, J. A.; O'Neil, E. A.; Williams, D. A. GB 2,193,216A.

FLAME-RETARDING POLYMERS

Joseph Green
FMC Corporation

The objectives in flame-retarding polymers are to increase ignition resistance and reduce the rate of flame spread (slow burning). The product does not become noncombustible, but a flame retardant may prevent a small fire from becoming a major catastrophe. Organic flame retardants, either additive or reactive, include halogen and phosphorus compounds. Chlorine and bomine are the only halogen compounds used commercially, the carbon-fluorine bond being too stable and the carbon-iodine bond being too unstable for processing. Aliphatic halogen is more effective than aromatic halogen; alicyclics are in between. Generally, halogen needs antimony as a flame-retardant synergist, as in polyolefins and styrenic resins. Phosphorus and halogen combinations are perhaps more effective[1] and can be synergistic, as in polyurethanes and engineering polymers.[2] Inorganic compounds containing a high concentration of water of hydration, such as alumina trihydrate, magnesium hydroxide, and magnesium carbonate, are effective in specific polymer systems when used at concentrations of 60-65%. Additives are used in thermoplastics and thermosets; reactives are used mainly in thermosets. Zinc borate and molybdenum compounds function as smoke suppressants in polyvinyl chloride and perhaps in unsaturated polyesters.

Halogen is believed to function in the gas phase by a free-radical trap mechanism, and phosphorus can function either in the gas or condensed phase,[3,4] depending on the thermal stability and volatility of the phosphorus compound. Imtumescent agents form voluminous quantities of porous char that protects the virgin polymer.

The addition of flame retardants may severely degrade the polymer and may present processing problems. Flame retardants that plasticize reduce thermal properties such as heat distortion under load (HDUL), whereas nonmelting solid additives may severely degrade impact properties. Furthermore, the flame retardant may have a limited thermal stability and restrict the processing temperature, or discolor because of high shear. Halogenated additives can have poor UV stability and restrict the use of the final product. The use of flame retardants differs from other additives that may improve the properties of the polymer, such as antioxidants or UV stabilizers; flame retardants degrade properties and are a compromise.

POLYSTYRENE

General-purpose polystyrene is conventionally flame-retarded with alicyclic bromine compounds such as hexabromocyclododecane (HBCD). These compounds have limited thermal stability (to ~200°C), but this is adequate for polystyrene applications.

The major applications for flame-retarded general purpose polystyrene are extruded foamed insulation board, expandable polystyrene bead board insulation, and extruded construction profiles.

HIGH-IMPACT POLYSTYRENE (HIPS)

High-impact polystyrenes differ from general-purpose polystyrene in that rubber is added as an impact modifier. This makes the polymer more difficult to flame-retard. Applications that require flame retardancy include television receiver cabinets, business machine housings, and structural foam.

The flame-retardant used almost exclusively is decabromodiphenyl oxide (decabrom) plus antimony oxide. Decabrom contains 83% bromine, the highest level of any commercially available flame retardant. The key properties for this product are Izod impact resistance and heat distortion under load. Decabrom absorbs in the sunlight region and therefore has very poor UV stability.

ACRYLONITRILE-BUTADIENE-STYRENE (ABS)

Acrylonitrile-butadiene-styrene (ABS) is an impact resin with higher Izod impact and higher heat distortion temperature than HIPS. The impact resistance of ABS degrades rapidly with the addition of filler and, as a result, the flame retardants of choice are melt-blendable materials, which presumably plasticize the resin. Three aromatic bromine-containing additives are used commercially, namely, bis-(tribromophenoxy) ethane (FF-680/Great Lakes Chemical), octabromodiphenyl oxide, and tetrabromobisphenol A.

MODIFIED POLY(PHENYLENE OXIDE) (PPO)

The flame retardants in HIPS and ABS are bromine additives with antimony oxide as synergist. Modified PPO uses phosphate esters as flame retardants; antimony oxide is not needed. The additives used in HIPS and ABS are all solids, but the phosphate esters are liquid.

Alkylated phosphate esters are thermally stable at processing temperatures, but are volatile and can "juice" during injection molding. If the molded part is wetted in high stress areas the part will stress-crack. These flame retardants are therefore being replaced with resorcinol diphosphate (RDP/Akzo, FMC), which is higher in molecular weight and less volatile.

Modified PPO can be considered the third in the series of impact resins. It has an Izod impact about twice that of ABS and a higher heat distortion temperature under load. The HDUL of the modified PPO resins depends upon the PPO content; the higher the PPO the higher the HDUL.

POLYCARBONATE

Polycarbonates have Izod impacts of ~14. They are inherently flame resistant, having an oxygen index of ~25 and a UL-94 self-extinguishing rating of V-2. To get a V-0 rating, a low concentration of bromine or bromine-phosphate compounds is added.

The commercial flame retardants include a polycarbonate oligomer of tetrabromobisphenol, chain capped with phenol for a bromine content of 51.3% (BC-52/Great Lakes Chemical), a similar oligomer chain capped with tribromophenol for a bromine content of 58% (BC-58/Great Lakes Chemical), and a brominated phosphate containing 60% bromine and 4% phosphorus (Reoflam PB-460/FMC Corporation).

Another method of flame-retarding polycarbonate resin is to add an alkali metal salt of organosulfonate. These work apparently by lowering the decomposition/charring temperature.

POLY(BUTYLENE TEREPHTHALATE) (PBT)

Poly(butylene terephthalate) has poor impact resistance and a low heat distortion temperature. It is generally used with reinforcing fillers such as glass or minerals. Flame-retardant addivites include many aromatic bromine compounds such as decabromodiphenyl oxide, brominated polystyrene (Pyro-Chek 68PB/Ferro), brominated polycarbonate oligomers BC-52 and BC-58 (Great Lakes Chemical), brominated acrylic (Ameribrom/Dead Sea Bromine), and brominated epoxy (Ameribrom/Dead Sea Bromine). Antimony oxide is required. A brominated phosphate Reoflam PB-460 (FMC Corporation) is very effective because of the combination of bromine and phosphorus. With mineral-filled PBT, no antimony oxide is needed with brominated phosphate for a V-0 rated product.[5] In glass-filled PBT, the brominated phosphate improves processibility and impact resistance, distinct advantages over the brominated flame retardants. Approximately 10% flame retardant plus 3% antimony oxide is used.[6]

POLY(ETHYLENE TEREPHTHALATE) (PET)

Glass-reinforced poly(ethylene terephthalate) is flame retarded with brominated polystyrene and sodium antimonate. Antimony oxide cannot be used because it degrades the polymer. The brominated phosphate does not require an antimony synergist, and the sodium antimonate can be replaced with additional brominated phosphate.[5]

POLYCARBONATE/ABS BLEND (PC/ABS)

Polycarbonate resin can be difficult to process, but polycarbonate/ABS blends are used to mold computer housings. Resins with high polycarbonate content are used because they are easier to flame-retard. Bromine, phosphorus, and brominated phosphates can be used without the need for antimony synergist.

POLYCARBONATE/POLY(ETHYLENE TEREPHTHALATE) BLEND (PC/PET)

Polycarbonate/poly(ethylene terephthalate) blends can be flame retarded with bromine compound solutions, all-phosphorus compound, a mechanical blend of the two, or with a brominated phosphate in which both the bromine and the phosphorus are in the same molecule.

POLYPROPYLENE

Nondripping flame-retarded V-0 polypropylene requires about 40% loading of additives, making it expensive and very high in specific gravity because halogen compounds and antimony oxide have high specific gravity. Polypropylene can be made V-2 with only 5-6% of total flame retardant, maintaining resiliency, gloss, and hinge properties.

Flame retardants for V-2 products include bromine compounds such as Saytex BN-451 (Albemarle), PE-68 (Great Lakes Chemical), and Reoflam PB-370 (FMC).

Antimony oxide is required to flame-retard polypropylene at about a 2:1 weight ratio of flame retardant/antimony oxide. Polypropylene can also be flame-retarded by using intumescent additives.

POLYETHYLENE

Brominated compounds such as decabromodiphenyl oxide and Saytex BT-93 (Albemarle) with antimony oxide are used to flame-retard polyethylene. The concentration required is highly dependent on the melt index of the resin. The higher the melt index of the resin, the less flame retardant is required to obtain a V-2 rating. Crosslinked polyethylene can also be flame retarded, but it takes significantly more retardant.[7]

POLYAMIDES

Chlorinated compounds such as Dechloane Plus (Occidental) and brominated compounds such as BT-93 (Albemarle) are used with antimony oxide. The use of various phosphorus compounds is also reported.

POLY(VINYL CHLORIDE) (PVC)

PVC contains 57% chlorine and is inherently flame retardant. Rigid PVC requires stabilizers and processing aids, but additional flame retardant is not required. Flexible PVC contains high concentrations of organic plasticizers (40-90 parts), and hence is flammable.

Many additives are used to flame-retard plasticized PVC, including antimony oxide, phosphate ester plasticizers, chlorinated paraffin plasticizers, alumina trihydrate, and zinc and barium borates. As measured by oxygen index, the most effective are antimony oxide, which functions as a synergist for the chlorine in the polymer, and phosphate esters, which replace in whole or in part the flammable organic plasticizer.[8]

Flame-retarded transparent film and sheet is made by replacing the organic plasticizer in whole or in part, depending on the flame-retardant requirement, with phosphate ester plasticizers. These are classified into three types: triaryl, alkyldiaryl, and trialkyl. The triaryl phosphate is the most effective and trialkyl the least. The alkyldiaryl phosphate ester, while not the most effective in flame retardancy, is a more efficient plasticizer than the triaryl phosphates, and gives products with better low-temperature flexibility and lower smoke.

UNSATURATED THERMOSET POLYESTERS

Unsaturated polyesters are condensation polymers, and flame retardancy can be obtained by replacing the monomers with halogenated diols or dicarboxylic acids. Thermoset polymers also have the ability to accept high concentrations of fillers, and both inert and active fillers are used for flame retardancy. Halogenated monomers used include tetrabromophthalic anhydride (to replace in part, or in whole, the phthalic anhydride), tetrachlorophthalic anhydride, chlorendic anhydride, dibromoneopentyl glycol (to replace the ethylene or propylene glycol), and post-bromination of resin-containing tetrahydrophthalic anhydride. Alicyclic bromine is the most effective of these and aromatic chlorine the least effective.

EPOXY RESIN

Epoxy resins are condensation polymers that use 54 parts of tetrabromobisphenol A as a monomer to attain flame retardancy. This is the largest volume organic flame retardant in the industry. The major application is epoxy glass laminate circuit boards. Phosphine oxides are highly efficient in this resin, although it is not used commercially.[9]

POLYURETHANE FOAM

Polyurethanes fall into three types: rigid foam, flexible foam, and thermoplastic elastomer. The flame-retardant requirements for rigid foam are now met with polyisocyanurates, which are inherently flame-resistant.

Flexible polyurethane foam uses chlorinated phosphates at a concentration of about 12 parts per hundred of polyol.

MARKETS

The markets for flame retardants by polymer application are shown below. The large market shown for rubber is due to the large volume of alumina trihydrate used in carpet backing.

Product	% of total market
Plastics	66
Textiles	5
Coatings/adhesives	3
Rubber	24
Wood/paper	2

REFERENCES

1. Lyons, J. W. *The Chemistry and Uses of Fire Retardants*, Wiley-Interscience: New York, 1970; pp 20-24.
2. Green, J. *J. Fire Sci.*, **1994**, *12*, 257-267.
3. Hastie, J. W.; McBee, C. L. *National Bureau of Standards/IR* August 1975, p 75-741.
4. Carnahan, J., et al., Fourth Int'l. Conf. of Flammability and Safety, San Francisco, January 1979.
5. Green, J. *J. Fire Sci.*, **1994**, *12*, 388-408.
6. Green, J. *J. Fire Sci.*, **1990**, *8*, 254-265.
7. Green, J. *Flame Retardant Polymeric Materials*, Vol. 3, Lewin, M., et al., Plenum: New York, 1982; pp 1-37.
8. Green, J. *Plast. Compound*, November/December, 1984.
9. Fretz, E. R.; Green, J. *Printed Circuit Fabrication*, **1983**, *6*, 55-63.

Flammability

See: Ignition Temperatures

FLAMMABILITY (Char Formation)

Peter Carty
University of Northumbria at Newcastle

The combustion of polymers is a very complex, rapidly changing chemical system which is not yet fully understood. Organic polymers undergo degradation processes, depending on their structures, and form volatile organic compounds when they are heated above certain critical temperatures.[1] If the gaseous mixture of the volatile organics with air is heated to a temperature greater than its ignition temperature, combustion begins.

The burning characteristics of polymers cannot be thoroughly evaluated by determining a few simple fire parameters in the laboratory. Fire and combustion are very complex and at least four major features interact during the overall burning process: ignition, rate of spread of flame, rate of heat development and release to the surroundings, and formation of smoke and gases. These processes all contribute to combustion and the hazards associated with burning polymers.

FLAME-RETARDANTS

The purpose of adding flame retardants to polymers is to slow down or stop the burning process. All flame retardants function during the early stages of fire by interrupting the self-sustaining combustion cycle. It must be recognized that no organic polymer can be made fireproof, even a polymer such as polytetrafluoroethylene (PTFE) which has a limiting oxygen index (LOI) value of 95 will burn in a real fire.

CHAR FORMATION AND FLAMMABILITY

In the mid-1970s van Krevelen clearly established that carbonaceous char formation during polymer combustion is an important aspect of flame retardancy.[2] He showed that there was a clear correlation between LOI values and the amount of char a polymer forms when it burns. Polymers with high LOI values tend to be more resistant to ignition and burning than those with lower LOI values, and the higher the LOI value, the less flammable the material.

Char formation is important to flame retardancy, but little progress has been made since these early studies to improve the flame resistance of polymers without significantly modifying their fundamental structure. Intumescent systems are gaining interest.[3] When heated, these systems are cellular foamed char on the surface, which is thermally stable and protects the underlying polymer from the action of the flames. The major disadvantages of most intumescent systems is that they require high loading (20-30%) to be effective. This char-forming mixture of compounds does not become an integral part of the polymer matrix, so physical and mechanical properties of the polymer can be adversely affected. Kroenke has found that incorporating low-melting sulfate glasses into rigid poly(vinyl chloride) induces char formation and crosslinking processes, and reduces smoke production.[4]

CHAR FORMATION IN POLYMER BLENDS

Blending (or alloying) polymers has been long recognized as an important route to novel and commercially successful polymer materials.

Unplasticized PVC (UPVC) is suitable for blending with other polymers and is ideal in flame-retardant applications because it is inherently nonflammable.[5] Its high LOI value (49.8) is an indication of its inability to burn freely in the air.

UPVC can be successfully blended with other polymers such as ABS and polypropylene (PP) without serious deterioration in physical properties.

ABS is very different from UPVC: it is a very flammable material, its low LOI value (18.3) ensures that it readily burns in air. Unlike UPVC it is easily ignited, burns with a hot flame, and produces large quantities of thick black smoke.

Polypropylene, like ABS, is easily ignited (LOI = 17.6); an aliphatic polymer, it does not produce a great deal of smoke when it burns in air.

Carbonaceous char, which is formed during the burning of polymers, depends basically on the structure of the polymer itself. Char-forming, condensed-phase reactions are generally restricted to polymers which contain aromatic groupings in the structural unit. Typically, polystyrene, polyester, phenolic resins, poly(phenylene oxide), polycarbonates, polysulfones, etc., yield quite significant amounts of char when heated because they tend to condense to aromatic chars. Aliphatic polymers such as polyethylene and polypropylene form little or no char as they are totally decomposed to monomers when heated. Although loosely classed as an aliphatic polymer, UPVC is a noticeable exception and forms quite significant amounts of char across a range of temperatures. High char-forming polymers tend to have high LOI values because they produce fewer gaseous flammable breakdown products when decomposed.

The formation of char during the thermal decomposition of polymers is a result of crosslinking reactions taking place in the solid phase. One of the beneficial effects of char formation is that polymer carbon is retained in the solid phase and as a consequence is not available for the formation of oxides of carbon, volatile organic compounds, and of course smoke. Hence, substances which react with the polymer to increase char formation during heating or burning are in fact smoke suppressants and only a few elements in the periodic table are known to react in this way. Zinc, molybdenum, iron, and possibly tin compounds are known smoke suppressants and have been available commercially for use in PVC formulations. However, each of these has problems associated with its use in plastic materials. Zinc oxide has very effective char-forming and smoke-suppressing activity but tends to decompose PVC at elevated temperatures. Molybdenum oxides are good smoke suppressants in PVC, but the high cost of molybdenum has prevented them from being used commercially in substantial quantity. Many compounds of iron are excellent char formers in a wide range of PVC formulations and it is only the color of these substances that has limited their use. Adding iron to the blended formulations has very significant effects on char formation even at 800°C.

The char-forming ability of iron compounds appears to depend on their ability to form active chemical species which promote crosslinking reactions and smoke-suppressing activity by a series of Lewis acid catalyzed reactions. The overall effect is that polymer carbon is converted into a stable char. The smoke-suppressing activity of iron compounds is thus a chemical effect which has great potential for use in a wide range of polymer systems.

REFERENCES

1. Cullis, C. F.; Hirschler, M. M. *The Combustion of Organic Polymers*; Clarendon: Oxford, 1981.
2. Van Krevelen, D. W. *Polymer* **1975**, *16*, 615.
3. Camino, G. In *Atmospheric Oxidation and Antioxidants*; Scott, G., Ed.; Elsevier: The Netherlands, 1993; Chapter 10.
4. Kroenke, W. J. *Journal of Materials Science* **1986**, *21*, 1123.
5. Elliott, A. R.; Hien, M. D.; Warry, D. L. *Journal of Vinyl Technology* **1993**, *15* (2), 76.

Flocculants

See: Flocculants (Organic, Overview)
Flocculants, Cationic
Lignin Graft Copolymers
Natural Rubber Latex (Polymeric Flocculant for Tin Tailings Slurries)

FLOCCULANTS (Organic, Overview)

Sun-Yi Huang and David W. Lipp
Cytec Industries

Synthetic organic flocculants are water-soluble polymers which are prepared by vinyl addition, condensation, ring-opening, and cyclopolymerization. In this review article, we emphasize cationic and anionic polymers. In addition, we mention poly(alkylene oxide)s, polyvinylpyrrolidinones, poly(vinyl alcohol)s, and polyampholytes.

Synthetic organic flocculants are a family of polymeric materials which have been developed commercially and have been extensively studied scientifically. There are two primary reasons for this growth. First, there has been an increasing need for more efficient synthetic organic flocculants in major areas such as water treatment, mineral processing, paper manufacturing, and food processing. Second, there have been increasingly tightened government regulations because of environmental concerns. These materials have become some of the most important products in our modern society.

Because of the very large number of applications for water soluble polymers and the extensive literature,[1-7] this article will be a broad overview of the state of the art. To provide a perspective on the types of synthetic organic flocculants, we organize them according to chemical functionality: cationic, anionic, nonionic, and ampholytic compositions. The most versatile and useful types of flocculants are cationic, followed by anionic, and then by nonionic. Structurally similar polymers are grouped into broad classes. We will discuss synthetic methods and commercial processes.

Low to medium molecular weight polymers are prepared by polycondensation or addition polymerization, whereas high molecular weight polymers are prepared by vinyl addition polymerization. These flocculants are supplied to the user in a variety of forms: dry powders, water-in-oil emulsions, beds, aqueous solutions, gels, or dispersions. Each of these forms has combinations of economic and technical advantages and disadvantages related to ease of handling, storage stability, and shipping.

SYNTHETIC ORGANIC FLOCCULANTS: SYNTHESIS AND CHEMISTRY

Cationics

Polymers and Copolymers of Quaternized Tertiaryaminoalkyl Esters or Amides with Acrylamide (AMD)

Homopolymers and copolymers of aminoalkylacrylates, aminoalkylmethacrylates, aminoalkylacrylamides, and aminoalkylmethacrylamides represent a very important class of organic flocculants used for waste water treatment. Although there are many cationic monomers, only a very few are commercially important.[8]

The cationic esters are generally synthesized by transesterification of methyl esters using *N,N*-dimethylaminoethanol. Traditional catalysts for these reactions are metal alkoxides; magnesium diisopropoxide, aluminum triisopropoxide or titanium (IV) tetraisopropoxides.[9-11] Methyl chloride, dimethylsulfate and benzyl chloride are the common quaternization agents. The most widely used cationic esters are acryloyloxyethyltrimethylammonium chloride (AETAC) [44992-01-0] and methacryloyloxyethyltrimethylammonium chloride (MAETAC) [5039-78-1]. The quaternary amide monomers widely used in synthetic organic flocculants are synthesized by Michael addition of *N,N*-dimethyl-1,3-propanediamine to methylacrylic or acrylic esters, and transesterification to the same esters, followed by cracking of the resultant aminopropionamide.[12,13] The aminoamide monomers are then quaternized, using methyl chloride, to methacrylamidopropyltrimethylammonium chloride (MAPTAC) [51410-72-1] and acrylamidopropyltrimethylammonium chloride (APTAC) [45021-77-0].

The hydrolytic stabilities of cationic polymers have been determined. High molecular weight cationic copolymers are readily produced by copolymerizing AMD with various proportions of cationic monomers. For applications, the best ratio of AMD and quaternary monomer depends on the cost and performance relationship. For most of the applications in waste water treatment, the cationic charge is in the range of 10-60 mol%. Cationic ester monomers are less expensive than cationic amide monomers. However, these esters in cationic polyacrylamides are very susceptible to base-catalyzed hydrolysis under mild polymerization conditions.[14-16]

Cationic Carbamoyl Polymers

In aminomethylation of polyacrylamide, formaldehyde and a secondary amine react via the Mannich reaction with the amide group to form a polytertiary amine.[17-20] Because of the simplicity of the process, low capital investment in manufacturing equipment, and low raw materials cost, these cationic polymers represent a commercially important group of synthetic organic flocculants.

Cationic Ring-Opening and Stepwise Condensation Polymers: Polyalkyleneimine and Polyhydroxylalkyleneamine

Ring-opening (ring-to-chain) polymerizations are intermediate in character between condensation and addition polymerizations. Polymer growth occurs exclusively by reaction of the cyclic reactant with the intermediate-stage polymer chain. Polyethlenimine (PEI) [26913-06-4] and poly(2-hydroxypropyl-1,1-*N*-dimethylammonium chloride) [39660-17-8] are commercially important flocculants which are formed by ring-opening polymerizations. Etyleneimine (aziridine) [151-56-4] polymerizes very exothermically at elevated temperatures in the presence of catalytic amounts of acids.[21-23] The polymer is highly branched with an average of one tertiary amine group per three chain nitrogen atoms.[24] The branched PEI is amorphous. Solution properties of branched PEI indicate compact and brush-like structures which tend to aggregate due to hydrogen-bonding, whereas linear PEI is a flexible semipermeable coil and more prone to aggregation.[25] PEI and modified PEI[26,27] are very effective flocculants for industrial wastewater,[28] municipal wastewater,[29] and in the petroleum industries,[30] food industries,[31] and coal concentration.[32]

Methylamine reacts with epichlorohydrin to form poly(2-hydroxypropyl-1-*N*-methylammonium chloride) [31568-35-1]. This polymer is a commercial flocculant. However, the cationic charge of this product varies with pH. To overcome the pH sensitivity, industry uses quaternary polyamines. These polymers are obtained by either alkylation (using alkyl halides or alkyl sulfates) of polyalkyleneamine or polyhydroxyalkyleneamine,[33,34] or by direct synthesis from dialkylamines and difunctional epoxy compounds under alkaline conditions.[35,36] Quaternary polyamines are excellent as flocculating agents for raw water clarification,[36] filter aids in treating coal washings,[37] treatment of sludge,[38] and color removal.[39,40]

Cycloaddition Polymers: Poly(diallyldimethylammonium chloride) (DADMAC) and Related Polymers

Nonconjugated diene monomers would normally be expected to form crosslinked networks. Diallyldialkylammonium salts, however, polymerize under radical initiation to form water soluble linear highly cationic polymers.[41] This polymerization is known as a cyclopolymerization and has been studied extensively since the first published report.[42] Poly(DADMAC) [26062-79-3] consists of chains containing five membered pyrrolidine rings with a cis to trans ratio of 6 to 1.[43]

There are many copolymers of cationic quaternary monomers with AMD. One of the most important is the copolymer of DADMAC and AMD[26590-05-6], poly(DADMAC-*co-AMD)*.

Cyclic Amindine Polymers

High molecular weight polymers possessing the cyclic amindine moiety have a unique structure and a very high cationicity. These flocculants have a high dewatering efficiency for water clarification.[45-54]

Poly(1-aminoethylene) and Related Polymers

The simplest polybasic structure, next to polyethyleneimine, is polyvinylamine (PVAM) [26336-38-9]. Since PVAM cannot be prepared directly from vinylamine monomer, the synthesis of this polymer with varying degrees of amino composition must be indirect.

Polyvinylamine hydrochloride[29792-49-2] is manufactured commercially using the acid hydrolysis of polyvinylacetamide.[55]

Quaternized polyvinylamines have been used for various applications as de-inking loop clarification aids, retention aids, clarification agents, and water treatment agents.[59-63]

Cationic Dicyandiamide Polymers

Many patents describe dye-fixation agents based on the condensation products of dicyandiamide[461-587-5] and formaldehyde,[56-58] sometimes modified with polyfunctional amines or ammonium salts.[64-71] Other polymeric condensation products are prepared from amino building blocks including urea[57-13-6],[44] melamine,[72,73] guanidine,[74] cyanamide,[75] and related nitrogen-containing compounds.[76]

Cationic dicyandiamide polymers are very effective flocculants when used alone, with inorganic flocculants, or with high molecular weight anionic flocculants for removal of organic dyes from waste water.[75-77] Paper de-inking process waters are clarified with cationic polymeric flocculants, followed by flotation.[78,79] These flocculants include epichlorohydrin polyamines [106-89-8D], dicyandiamide polymers[461-58-5D], polyethylenimine[9002-98-6D], poly(diallyldimethylammonium chloride) [26062-79-3] and poly(acrylamide-*co*-diallydimethylammonium chloride) [26590-05-6].

Melamine-Formaldehyde Resins Acid Colloids

Melamine-formaldehyde amine resins are prepared at a specific mole ratio of melamine[108-78-1] to formaldehyde[30525-89-4] in a mildly alkaline medium.[80-83] The acid colloid formed is very effective as a better additive for paper manufacturing, and as a flocculant for potable water. Because of the low cost of acid colloids, paper-making industries and waste water management use them extensively. However, the free formaldehyde content is an environmental issue.

Anionics

Acrylic Acid Copolymers and Salts

Poly(acrylamide-*co*-sodium acrylate) [25085-02-3] and poly(acrylamide-*co*-ammonium acrylate) [26100-47-0] are flocculants of considerable practical importance (AMD/NaAA and AMD/NH_4AA). They can be prepared by hydrolysis of polyacrylamide or by copolymerization of comonomers. For most industrial applications a polymer with 30 mol% hydrolysis is sufficient.

Acrylamide can be copolymerized with acrylic acid or acrylic salts by free-radical initiation in solution, inverse emulsion,[84] or in inverse microemulsion.[85] Reactivity ratios vary with pH.[84] At low pH the reactivity ratio for the acrylic acid is higher, but at high pH the reactivity ratio for acrylamide is higher.

Hydroxamated Polyacrylamide

Polyacrylamide reacts with hydroxylamine to form poly(hydroxamic acid). Stable inverse emulsions of hydroxamated polymers can be prepared by adding an inverse emulsion of a hydroxylamine salt, sodium thiosulfate, and excess NaOH to a PAM precursor inverse emulsion.[86-88]

Sulfonate Polymers, Copolymers, and Salts

Since 1981, 2-acrylamido-2-methylpropanesulfonic acid (AIBS) and the monosodium salt (NaAIBS) have been available in commercial quantities. AIBS can be prepared from acrylonitrile, isobutene, and sulfuric acid.[89] AIBS can also be prepared by reacting *tert*-butyl alcohol with a sulfonating agent under anhydrous conditions, and then with acrylonitrile and water.[90] Pure AIBS can be obtained by recrystallization in dry methanol.[91] The monomer may also be purified as a salt by crystallization in water in the presence of Group IA metals (Na preferred), or Group IIA metals (Mg preferred), or amines.[92] AIBS and NaAIBS can be polymerized with most other water-soluble vinyl monomers using azo compounds, organic peroxides, inorganic peroxides, redox initiators, and photoinitiators.

Nonionics

Acrylamide Polymers

Polyacrylamide-based polymers play a very dominant role in the synthetic organic flocculants market. A few comments on their unusual properties are pertinent.

Kulicke et al. have summarized a wide variety of polymerization methods such as azo-compound initiated polymerization, redox initiated polymerization, photo-polymerization, radiation-induced polymerization, electroinitiated polymerization, ultrasonic polymerization, and others.[85]

The polymerization of acrylamide in inverse microemulsions is a significant advance in the state-of-the-art.[93-98]

Poly(ethylene oxide)

Nonionic ethylene oxide polymers[25322-68-3] with molecular weights greater than 3×10^6 can be prepared by coordinate ring-opening polymerization using special catalysts. These polymers are used in the mineral processing industry, particularly for coal slurry dewatering, because they form flocs of unique character.

Poly(N-vinylpyrrolidione)

Poly(*N*-vinylpyrrolidione) [9003-39-8] is prepared by free-radical addition polymerization. Flocculant usage is specialized. Toxicity is very low. Examples include clarification of fruit juices, beer, and wine, and other food processes.[99] The polymer can be blended with other flocculants such as polyethylenimine to improve settling times of dispersed inorganic solids.[100,101]

Poly(vinyl alcohol)

Poly(vinyl alcohol) [9002-89-5] is produced on an industrial scale by alkaline alcoholysis of poly(vinyl acetate) [9003-20-7].

Polyampholytics

Polyampholytes

Flocculants containing cationic and anionic groups are fundamentally interesting because of the ability of the cationic group to flocculate particles by a charge-patch mechanism and the ability of the anionic group to interact with multivalent cations. Mannich reaction products of polyacrylamide with low levels of carboxyl groups are claimed to be highly efficient for dewatering sewage sludges.

REFERENCES

1. Hamza, H. A.; Picard, J. L. *Index of Commercial Flocculants* Energy Research Program, Energy Research Lab, Canmet Report 77-78, 1975.
2. Alverson, F.; Panzer, H. P. *Kirk-Othmer: Encyclopedia of Chem. Tech.*, 3rd ed.; Wiley-Interscience: New York, NY 1980; Vol. 10, p 489.
3. Heitner, H. I. *Kirk-Othmer: Encyc. of Chem. Techn.*; 4th Ed.; Wiley-Interscience: New York, NY 1994; Vol. 11, p 61.
4. Heitner, H. I.; Foster, T.; Panzer, H. P. *Encyc. of Polym. Sci. and Eng.*; Wiley-Interscience: New York, NY, 1987; Vol. 9, p 824.
5. Thomas, W. M.; Wang, D. W. *Encyc. of Polym. Sci. and Eng.*; 2nd ed.; Wiley-Interscience: New York, NY, 1985; Vol. 1, p 182.
6. Mortimer, D. A. *Polym. Int.* **1991,** *25,* 29.
7. Lipp, D. W.; Kozakiwicz, J. J. *Kirk-Othmer: Encyc. of Chem. Tech.*; 4th ed.; Wiley-Interscience: New York, NY, 1991; Vol. 1, p 266.

8. Luskin, L. S. In *Functional Monomers* Yocum, R. H.; Nyquist, E. B. Eds.; Marcel Dekker: New York, NY, 1974; Vol. 2, p 555.
9. Farrar, D. (Allied Colloids, Ltd., U.K.) U.S. Patent 4 609 755, 1986.
10. Rehberg, C. E.; Fisher, C. H. *J. Org. Chem.* **1947,** *12,* 226.
11. Rohm and Haas Co. Br. Patent 820 560, 1959.
12. Moss, P. H.; Gipson, R. M. (Jefferson Chem. Co.) U.S. Patent 3 878 247, 1975.
13. Barron, B. G. (Dow Chem. Co.) U.S. Patent 3 652 671, 1972.
14. Tanaka, H. *J. Polym. Sci., Polym. Chem. Ed.* **1986,** *24,* 29.
15. Aksberg, R.; Wagberg, L. *J. Appl. Polym. Sci.* **1989,** *38,* 297.
16. Iafuma, F.; Durand, G. *Polym. Bull.* **1989,** *21,* 315.
17. Suen, T. J.; Schiller, A. M. (Amer. Cyanamid Co.) U.S. Patent 3 171 805, 1965.
18. Wisner, R. L. (Dow Chem. Co.) U.S. Patent 3 539 535, 1970.
19. McDonald, C. J.; Beaver, R. H. *Macromolecules* **1979,** *12,* 203.
20. Agababyan, A. G.; Gevorgyan, G. A.; Mndzhoyan, O. L. *Russian Chem. Rev.* **1982,** *51(4).*
21. Dermer, O. C.; Ham, G. E. *Ethylene Imine and Other Aziridines*; Academic: New York, NY, 1969.
22. Ethyleneimine Dow: Midland, MI, 1965.
23. Ham, G. E. *Kirk-Othmer: Encycl. of Chem. Tech.*; John Wiley & Sons: New York, NY, 1981; vol. 13, p 142.
24. Dick, C. R.; Ham, G. E. *J. Macromol. Sci.* **1970,** *A4,* 1301.
25. Bekturov, E. A.; Bakauova, Z. K. *Synthetic Water-Solubule Polymers in Solution* Huthig & Wepf; New York, NY, 1986.
26. Ise, N.; Okubo, T. *Polymer* **1972,** *13,* 552.
27. Akagae, K.; Allen, G. G. Skikizai Kyokaish **1973,** *46,* 239.
28. Naletskaya, G. N. et al. *Chem. Abstract* **1978,** *89,* 117201.
29. Butseva, L. N. et al. *Chem. Abstract* **1978,** *89,* 117200u.
30. Veitser, Y. I. et al. *Chem. Abstract* **1978,** *89,* 17199A.
31. Aleksandrova, L. D. et al. *Chem. Abstr.* **1978,** *89,* 200108R.
32. Aleksandrova, L. D.; Borts, M. A.; Stepanora, D. I. *Chem. Abstr.* **1978,** *89,* 200108.
33. Coscia, A. T., *Kirk-Othmer: Encyc. of Polym. Sci. and Chem. Tech.*; John Wiley & Sons: New York, NY, 1969; Vol. 10, p 616.
34. Garms, D. C. (Dow Chem. Co.) U.S. Patent 3 275 588, 1966.
35. Panzer, H. P.; Rabinowitz, R. (American Cyanamid Co.) U.S. Patent 3 725 312, 1973.
36. Panzer, H. P.; Dixon, K. W. (American Cyanamid Co.) U.S. Patent 3 894 944, 1975.
37. Panzer, H. P.; Dixon, K. W. (American Cyanamid Co.) U.S. Patent 3 894 946, 1975.
38. Panzer, H. P.; Dixon, K. W. (American Cyanamid Co.) U.S. Patent 3 880 753, 1975.
39. Shkut, M. V. et al. USSR SU 1772159, 1992; *Izobreteniya* **1992,** *85,* (40).
40. Finck, M. R.; Reed, P. E.; Shetty, C. S. (Nalco Chem. Co.) U.S. Patent 5 209 854, 1993.
41. Butler, G. B.; Ingley, F. L. *J. Am. Chem. Soc.* **1951,** *71,* 3120.
42. Butler, G. B. *Acc. Chem. Res.* **1982,** *15,* 370.
43. Lancaster, E.; Baccei, L.; Panzer, H. P. *J. Polym. Sci., Polym. Lett.* **1976,** *14,* 549.
44. Cramm, J. R.; Kravitz, F. K. (Nalco Chem. Co.) U.S. Patent 5 248 744, 1993.
45. Annand, R. R.; Redmore, D.; Rushton, B. M. (Petrolite) U.S. Patent 3 509 046, 1970.
46. Redmore, D. (Petrolite) U.S. Patent 3 576 741, 1971.
47. Hurwitz, M. J.; Park, E.; Aschkenasy, H. (Rohm and Haas Co.) U.S. Patent 3 406 139, 1968.
48. Panzer, H. P.; O'Conner, M. N.; Baccei, L. J. (American Cyanamid Co.) U.S. Patent 4 006 247, 1977.
49. Panzer, H. P.; O'Conner, M. N.; Baccei, L. J. (American Cyanamid Co.) U.S. Patent 4 007 200, 1977.
50. Panzer, H. P.; Acholonu, K. U. (American Cyanamid Co.) U.S. Patent 4 137 416, 1979.
51. Panzer, H. P.; Acholonu, K. U. (American Cyanamid Co.) U.S. Patent 4 137 415, 1979.
52. Panzer, H. P.; Acholonu, K. U. (American Cyanamid Co.) U.S. Patent 4 157 442, 1979.
53. Panzer, H. P.; O'Conner, M. N.; Rothenberg, A. S. *Polym. Mater. Sci. Eng.* **1987,** *57,* 130.
54. Sawa, N. *Nippon Kagaku Zasshi* **1968,** *89,* 780.
55. Gless, R. D.; Dawson, D. J.; Wingard, E. (Dynapol Co.) U.S. Patent 4 018 826, 1977.
56. Technical Bulletin, Air Products and Chemicals, Inc.
57. Chang, Y.; McCormick, C. L. *Macromolecules* **1993,** *26,* 4814.
58. Kathmann, E. E. L.; McCormick, C. L. *Macromolecules* **1993,** *26,* 5249.
59. Harrington, J. C.; Chen, J. C.; Chen, F. (Betz Lab) U.S. Patent 5 269 942, 1993.
60. Harrington, J. C.; Chen, J. C.; Chen, F. (Betz Lab) CA 2 110 365-AA, 1994.
61. Harrington, J. C.; Chen, J. C.; Chen, F. (Betz Lab) CA 2 110 366-AA, 1994.
62. Chen, J. C. et al. (Betz Lab) CA 2 110 455-AA, 1994.
63. Chen, F.; Bair, K. A.; Devore, D. CA 2 110 458-AA, 1994.
64. Stettler, H.; Hofer, K. (Sandoz Ltd.) Ger. Offen. 1 917 050, 1969.
65. Hasek, J. et al. *Litsty Cukrov.* **1977,** *93(5),* 104.
66. Hahn, E. et al. (BASF A. -G., Germany) EP 309 908 A2, 1989.
67. Seeholzer, J.; Von Seyerl (SKW Trostberg A. -G. Germany) EP 320 986 A2, 1989.
68. D'Elia, M.; Romano, A. Braz. Pedido PI BR 8 900 194 A, 1989.
69. Schwarz, R.; Hennig, T. Ger. Offen. DE 1 243 646, 1967.
70. Odum, J. J.; Shumaker, T. P.; Bloomquist, P. R. (Reichold Chemicals, Inc.) U.S. Patent 3 484 837, 1969.
71. Odum, J. J.; Shumaker, T. P.; Bloomquist, P. R. (Kuxita Ind., Co., Ltd.) Jpn Kokai Tokkyo Koho JP 71/10 408, 1971.
72. Renner, A. (Ciba-Geigy, Switzerland) U.S. Patent 3 716 483, 1973.
73. Renner, A. (Nippon Carbide Kogy KK) Jpn. Patent 50 105 575, 1978.
74. Breuer, W.; Riese-Meyer, L. (Henkel KGaA, Germany) Ger. Offen. DE 4 236 530 A1, 1994.
75. Kzozlov, V. V. et al. USSR SU 1782938 A1, 1992; *Izobretenya* **1992,** *87,* (47).
76. Charannes, J. P. (Santoz-GmbH, Germany) Ger. Offen. DE 4 301 364 A1, 1993.
77. Avideera, E. I. et al. USSR SU 1733395, 1992; *Izobreteniya* **1992,** *86,* (18).
78. Langley, J. G. et al. PCT Int. Appl. WO 93/02 967 A1, 1993.
79. Langley, J. G.; Peter, L.; Broughton, R. PCT Int. Appl. WO 93/02 966 A1, 1993.
80. Dickson, C. *Paper Trade J.* **1948,** *November 11.*
81. Yates, R. W. (Brit. Ind. Plastics, Ltd., England) U.S. Patent 3 721 651, 1973.
82. Yates, R. W. (Borden Inc.) JP 47/001 890 A, 1972.
83. Yates, R. W. (Borden Inc.) CA 884 931A, 1971.

84. Frank, S.; Coscia, A. T.; Frisque, A. J. (American Cyanamid Co.) U.S. Patent 4 439 332, 1984.
85. Kulicke, W. M.; Kniewske, R.; Klein, *J. Prog. Polym. Sci.* **1982,** *8,* 373.
86. Heitner, H. I.; Ryles, R. G. EPA 514 648, 1992.
87. Heitner, H. I. EPA 514 647, 1992.
88. Heitner, H. I. (American Cyanamid Co.) U.S. Patent 5 256 331, 1993.
89. Hoke, D. (Lubrizol Co.) Brit. Patent GB 1 341 104, and Ger. Patent DE 2 105 030, 1978.
90. Benn, G.; Farrar, D.; Flesher, P. (Allied Colloids Ltd.) U.S. Patent 4 876 047, 1989.
91. Cahalan, P. T.; Coury, A. J.; Jeyne, A. H. U.S. Patent 4 650 614, 1987.
92. Burk, W. M. et al. (Lubrizol Co.) World Patent Appl. WO 92/21652, 1992.
93. Speiser, P.; Birrenbach, G. (Forsch, Switzerland) U.S. Patent 4 021 364, 1977.
94. Candau, F. In *Polymer Association Structures: Microemulsions and Liquid Crystals* El-Nokaly, M. A. Ed.; ACS Symp. Ser.: **1984,** 384.
95. Candau, F.; Buchert, P. (Norsolor) World Patent WO 88/10274, 1988.
96. Candau, F.; Leong, Y. S.; Fitch, R. M. *J. Polym. Sci., Polym. Chem. Ed.* **1985,** *23,* 193.
97. Carver, M. T. et al. *J. Polym. Sci., Polym. Chem. Ed.* **1989,** *27,* 2161.
98. Candau, F. In *Polymerization in Microemulsion, Polymerization in Organized Media*; Gordon and Breach: Philadelphia, PA, 1992; Chapter 4.
99. Ryznar, J. W. (Nalco Chem. Co.) U.S. Patent 3 492 224, 1970.
100. Azorlosa, J. L.; Williams, E. P. (GAF Co.) U.S. Patent 3 835 084, 1974.
101. Azorlosa, J. L.; Williams, E. P. (GAF Co.) U.S. Patent 3 951 792, 1976.

FLOCCULANTS, CATIONIC

Truis Smith-Palmer
Chemistry Department
St. Francis Xavier University

Cationic polyelectrolytes are of great importance in many industrial applications. They are used as flocculants in the clarification of drinking water and in the clean up of industrial wastes, sewage, and sludges, and as retention aids in the paper-making industry.

Much of the material required to be flocculated (e.g., coalfines, clays, many minerals, and cellulose-derived substances) has a negative surface charge and is electrostatically attracted to positively charged polyelectrolytes. In many situations, such as the clarification of water, it has been established empirically that the most important feature of the flocculant is its cationic charge density, whereas for the treatment of sewage sludges both a high molecular weight and cationicity are required.[2] Some surfaces will adsorb only cationic polyelectrolytes and show no affinity for polyacrylamide.[3-5]

One commonly used cationic flocculant is the copolymer of acrylamide and 2-trimethylammonioethyl acrylate chloride (coTMAEA) (CAS Registry No. 69418-26-4).

These polymers have been synthesized by several methods and are available commercially under a variety of trade names. Other similar polymers have also been used, including the copolymer of acrylamide and 2-trimethylammoniopropyl acrylate chloride (coTMAPA) (CAS Registry No. 7150-29-7).

When the cationic moieties are joined to the backbone by ester linkages, which are easily hydrolyzed, aqueous solutions of the polymers should be used immediately unless they are prepared in a pH 4 buffer, in which they are indefinitely stable.

The first step in the interaction of a polymer with a particle is adsorption. Adsorption isotherms are generally measured by adding a known amount of polymer to a suspension, allowing flocculation to finish, and then determining the residual polymer concentration in the supernatant. Several methods have been used to determine cationic polymers, including charge-density titrations, refractometry, suppression of a polarographic maximum, organic carbon analysis, fluorescent labeling, and tensammetry.[1,3,7-11]

Adsorption isotherms for cationic polyelectrolytes onto suspended particles generally start with a region where the uptake of polymer is almost complete and then round off to a plateau. The rounding off of the isotherm is attributed to the polydispersity of the molar masses of the polymers involved.[12]

In many flocculations, the subsidence rates are inversely proportional to the quantities of polymer adsorbed. For example, where we obtained very high settling rates when kaolin was flocculated with coTMAEA, we found a correspondingly very small degree of polymer adsorption.[6] When lower settling rates were obtained by using partially hydrolyzed polymer, the adsorption increased. This inverse relationship can be explained by the fact that when flocculation occurs, surfaces rapidly become sterically unavailable as particles are incorporated into flocs.

A portion or portions of a polymer may interact with a particle surface, forming various loops, tails, and trains (i.e., portions lying flat on the particle surface) or adsorbing almost completely flat onto the particle surface. Because of the attraction of opposite charges, researchers have speculated that large portions of cationic polymers are adsorbed onto the surface of particles, "neutralizing" the charges on the particles and leading to flocculation, hence the term "charge neutralization" to describe this mechanism. A modification of this description involves adsorbed cationic polymers interacting with negatively charged regions on other particles, giving rise to the term "charge-patch" mechanism.[13,14] The mechanism by which cationic polyelectrolytes produce flocculation thus has an extra dimension when compared with the mechanism of flocculation by anionic and nonionic polymers. These latter polymers bind to surfaces mostly by hydrogen bonding, and if the polymers are of high enough molecular weight, parts of them will adsorb onto different molecules and "bridges" will be formed, leading to flocculation. This bridging mechanism is expected to yield larger flocs and faster settling rates than the charge-patch or charge-neutralization mechanisms and occurs when cationic flocculants of high molar mass are used.[15,16]

In conclusion, for large cationic polyelectrolytes, both the charge density and the molecular weight are important in the flocculation process. Electrostatic attractions will predominate with polymers of low molar mass, and bridging will be at least partially responsible for flocculation with polymers of high molar mass. Other important factors include the nature of the suspended particles, the pH, and the ionic strength. The surface conformation of polymers during the adsorption and flocculation process is still not well defined.

REFERENCES

1. Mabire, F.; Audebert, R.; Quivoron, C. *J. Coll. Interface Sci.* **1984,** *1,* 120.
2. Lockyear, C. F.; Jackson, P. J.; Warden, J. H. *Polyelectrolyte Users Manual, Technical Report TR 184*; Water Resource Centre: England, Marlow/Bucks, U.K. SL7 2HD, 1983.
3. Wang, T. K.; Audebert, R. *J. Coll. Interface Sci.* **1988,** *121,* 32.
4. Eriksson, L.; Alm, B.; Stenius, P. *Colloids Surfaces A: Physicochem. Eng. Aspects* **1993,** *70,* 47.
5. Ishimaru, Y.; Lindström, T. *J. Appl. Poly. Sci.,* **1984,** *29,* 1675.
6. Smith-Palmer, T.; Campbell, N.; Bowman, J.; Dewar, P. *J. Appl. Polym. Sci.* **1994,** *52,* 1317.
7. Wang, L. K.; Wang, M. H.; Kao, J-F. *Water, Air, Soil Pollution* **1978,** *9,* 337.
8. Gill, R. I. S.; Herrington, T. M. *Colloids Surfaces* **1986,** *22,* 51.
9. Wentzell, B. R.; Smith-Palmer, T.; Donini, J. C. *Can. J. Chem.* **1987,** *65,* 557.
10. Tanaka, H.; Ödberg, L.; Wågberg, L.; Lindstrom, T. *J. Coll. Interface Sci.* **1990,** *134,* 229.
11. Smith-Palmer, T.; Roberts, C. *Can. J. Chem.* **1991,** *69,* 1516.
12. Vander Linden, C.; Van Leemput, R. *J. Coll. Interface Sci.* **1978,** *67,* 63.
13. Gregory, J. *J. Coll. Interface Sci.* **1973,** *42,* 448.
14. Kasper, D. R. Ph.D. Thesis, California Institute of Technology, Pasadena, 1971.
15. Hogg, R. In *Proc. Int. Symp.*; Somasundarun, P., Ed.; Fine Particle Processing, Proceedings International Symposium; AIME: New York, NY, 1980; Vol. 2, p 990.
16. Gill, R. I. S.; Herrington, T. M. *Colloids Surfaces* **1987,** *28,* 41.

FLOOR FINISHES

Daniel J. Gross* and James F. Hermann
S. C. Johnson Wax

Floor finishes have evolved from natural wax-based, solvent dispersion waxes of the past to modern acrylic polymer based floor finishes. The terms floor wax, floor polish, or floor finish are all used interchangeably to describe products that provide a thin, temporary protective coating to a floor substrate. Floor finishes provide a sacrificial wear layer that prevents traffic damage to the flooring substrate, seals the floor and makes it easier to clean, and gives a glossy appearance. Floor finishes are formulated to provide a nonhazardous, predictable walkway. Surface damage to the floor finish caused by foot traffic can be renewed easily through cleaning, spray buffing, burnishing, or recoating. When traffic wear is severe enough so that the gloss can not be renewed, then the old worn floor finish is chemically stripped. A new coating of floor finish is applied to the stripped floor and a new floor maintenance cycle is begun. Floor finishes are relatively easy to remove chemically in contrast to wood floor coatings and concrete floor coatings that are usually chemically crosslinked and therefore must be removed by sanding or other mechanical means.

*Author to whom correspondence should be addressed.

HISTORY

The first floor polishes were solvent dispersions of natural waxes (carnauba, beeswax, and paraffin). They needed to be buffed after application to produce a shine. In the 1960s, detergent resistant floor finishes were introduced. These polishes were based on the zinc crosslinking technology patented in 1967.[1] Self-stripping floor polishes were developed and marketed in the 1970s. They served the dual role of cleaning and polishing in one step. These polishes are based on alkali-soluble acrylic polymers[2] and pressurized wax emulsions.

In the industrial and institutional marketplace, the 1980s saw development of the ultrahigh speed (UHS) burnishing method of floor maintenance, which uses a UHS buffing machine that spins a floor finish buffing pad at 1500-2500 rpms. In the 1990s, regulations have driven the floor finish developments. California has passed regulations limiting the VOC content in floor finishes to 7 wt.% for resilient floor finishes and 10 wt.% for nonresilient finishes.[3] New polymer technology[4] has been developed to formulate zinc-free floor finishes. Most major finish suppliers now offer zinc-free floor finishes to cover these special cases.

CURRENT MARKET

High-performance floor finishes used in the industrial and institutional marketplace as well as self-stripping consumer polishes are based primarily on acrylic and styrene-acrylic emulsion polymers.

INGREDIENTS

Floor finishes are formulated by combining several polymeric materials. The major component is an acrylic or styrene-acrylic emulsion polymer, which accounts for 50-90% of the solids of most floor finishes. A low molecular weight polyethylene or polypropylene wax that has been functionalized to make it emulsifiable comprises 5-20% of the solids. An alkali-soluble, low molecular weight polymer usually makes up from 3-15% of the floor finish film. In the floor finish industry this material is referred to as leveling resin. The other major component is the plasticizer. Plasticizer can make up from 0-25% of the floor finish solids. A minor component is a divalent metal ion, usually zinc. The zinc ion forms an ionic crosslink with the carboxyl functionality of the emulsion polymer and the leveling resin. This crosslink provides detergent resistance, improves recoat properties, and gives improved traffic durability. Coalescing solvent and wetting/leveling agents also are used to get good application properties and provide tough, durable finishes.

REFERENCES

1. Rogers, J. R.; Sesso, L. M. (to S. C. Johnson & Son) U.S. Patent 3 308 078, 1967.
2. Dwyer, S. G.; Hackbarth, D. J. (to S. C. Johnson & Son) U.S. Patent 4 013 607, 1977.
3. California Clean Air Act of 1988.
4. Gray, R. T.; Owens, J. M.; Killam, H. S. (to Rohm and Haas Co.) European Patent Appl. EP 438 216, 1991.

Floor Waxes

See: Ionomers (Overview)

Flow Modifiers

See: Polytetrafluorethylene

FLUORESCENT POLYMERS (2-Vinylquinoline Hydrogels and Soluble Polymers)

Inés F. Piérola,* Elena Morales, and
M. Rosa Gómez-Antón
Universidad Nacional de Educación a Distancia

In recent years, fluorescent polymers have found extensive applications.[1] One of them is in luminescent sensors both for analytes (pH, ions, ...) and for biological membranes. In the investigation of biological systems the polymeric structure has the advantage of increasing the sensor selectivity, relative to low molecular weight compounds. In the determination of analytes, the role of the polymer has usually been to support the luminescent sensor, although a very interesting system has been proposed based on fluorescence quenching by the analyte of the optical signal generated in a fluorescent polymeric optical fiber.[1]

Sensors are usually placed at one end of an optical fiber, and their luminescence is carried to a detector at the opposite end. Fiber optic sensors show important advantages compared with electrical devices.[2] They can be used in monitoring many different optical properties of the sensor such as changes in absorbance, reflectance, energy transfer, fluorescence intensity, or fluorescence lifetime.

Several sources for fiber optic devices, based on the fluorescence characteristics of an immobilized chromophore, have been developed in the last decade.[2] The aim of this work is to develop self-supported polymeric pH sensors in which polymer is both the matrix and the emitting chromophore.

Several difficulties preventing practical applications of fluorescent sensors[3] have already been overcome.[4]

Polymers bearing heterocycles show no excimer emission, as is usual for homocyclic polymers.[5] They have a fluorescent spectrum formed by two bands whose relative intensity is pH dependent as described by Piérola, et al.,[6] Gómez-Antón, et al.,[7] and Morales et al.[8] The fluorescent polymers here studied contain quinoline groups covalently joined to the polymer skeleton. We will try to study their fluorescent properties and their availability as self-supported fluorescent sensors.

PROPERTIES

Fluorescence spectra of 2-methylquinoline (2MQ) and the atactic or isotactic homopolymers of 2-vinylquinoline (2VQ) show two bands whose relative intensity changes with pH. They have been ascribed to the emissions of the neutral and protonated forms of the heterocycle. The spectral positions of the two bands depend on the system, and polymer effects have also been observed in the basicity of the quinoline in both the ground (pK_a) and the excited (pK_a) states. It is well known that the pK_a of a heterocycle joined to a polymer chain depends on its degree of ionization (a),[6,9] whereas it is constant for a low molecular weight compound like 2MQ.

The most important conclusion from these results is that the dissociation constant of the chromophore, pK_a, which is of prime importance in determining the pH range applicability of the sensor,[2] depends very much on the matrix or support.

Since pH is measured in aqueous solution and poly(2-vinylquinoline) (P2VQ) is water insoluble and unswellable, it is necessary to develop hydrophilic systems containing the quinoline group. Thus, we have prepared water-soluble copolymers of 2VQ and four nonfluorescent comonomers and we have studied the dependence of their fluorescent characteristics on pH, ionic strength, temperature, and quinoline molar fraction.

CONCLUSIONS

The consequence of the results is that one can design a self-supported fluorescent sensor based in quinoline, valid to measure pH with the desired precision and in the desired pH range, by changing the comonomer and the composition of the hydrogel.

In any of the systems here studied the emission of quinoline is in the visible region and requires as excitation wavelength which is also in the range of the emission of simple tungsten lamps. These are clear advantages for the use of these systems as pH fiber optic sensor devices.

Two more advantages should be remarked upon: the dual emission allows use of FRs, which are better than fluorescence intensities, and the versatility of the system, which can be "tailor made" for any particular experimental situation.

ACKNOWLEDGMENTS

We express our thanks to the Comisión Interministerial de Ciéncia y Tecnologia for financial support under Grant MAT93-1067. Also we thank Dr. A. P. Dorado for technical assistance with computers.

REFERENCES

1. Barashkov, N. N.; Gunder, O. A. *Fluorescent Polymers*; Ellis Horwood: New York, **1994**.
2. Cámara, C. M.; Moreno, C.; Orellana, G. In *Biosensors with Fiber Optics*; Wingard, L. G., Jr.; Wise, D. L., Eds.; Humana: Clifton, NJ, 1990.
3. Seitz, W. R. *Anal. Chem.* **1984,** *56,* 16A.
4. Jordan, D. M.; Walt, D. R.; Milanovich, F. P. *Anal. Chem.* **1987,** *59,* 437.
5. Cáceres, P.; Piérola, I. F.; Horta, A. *Makromol. Chem., Rapid Commun.* **1987,** *8,* 573.
6. Piérola, I. F.; Turro, N. J.; Kuo, P. L. *Macromolecules* **1985,** *18,* 508.
7. Gómez-Antón, M. R.; Rodriguez, J. G.; Piérola, I. F. *Macromolecules* **1986,** *19,* 2932.
8. Morales, E.; Gómez-Antón, M. R.; Piérola, I. F. *Makromol. Chem., Macromol. Symp.* **1994,** *84,* 227.
9. Kirsh, Y. E.; Komarova, O. P.; Lukoukin, G. M. *Eur. Polym. J.* **1973,** *9,* 1405.

*Author to whom correspondence should be addressed.

FLUORINATED HIGH PERFORMANCE POLYMERS (Imide and Hexafluoroisopropylidene Groups)

Maria Bruma
Institute of Macromolecular Chemistry

John W. Finch and Patrick E. Cassidy
Chemistry Department
Southwest Texas State University

High-performance polymers maintain their structural integrity and physical and mechanical properties in application demanding service at enhanced temperature. The last two decades have witnessed the spectacular growth of research in high-performance polymers containing hexafluoroisopropylidene (6F) groups. The incorporation of 6F units into the macromolecular chain increases the solubility, thermal stability, flame and oxidation resistance, glass transition temperature (T_g), adhesion, optical transparency, and environmental stability, while decreasing the crystallinity, dielectric constant, water adsorption, and color.[1] This combination of excellent properties is a result of the disruption of the crystallinity and conjugation provided by the bulky but highly thermally stable 6F group.

In 1989, the literature on polymers derived from hexafluoroacetone was thoroughly reviewed.[2]

POLYIMIDES

6F-containing polyimides are synthesized via a typical AA-BB-type polycondensation, with component AA consisting of a dianhydride and component BB consisting of a diamine, one or both of them containing 6F groups.

Much work has been reported on polyimides based on 6F diphthalic anhydride (6FDA). Polycondensation of 6FDA with a series of aromatic diamines[3] yielded high-performance thermoplastic polyimides, **Structure 1**, that had T_g values of about 350°C and that formed tough, transparent flexible films when cast from *N*-methylpyrrolidinone solutions or when compression-molded above T_g. These polyimides exhibited higher hydrolytic stability than commercially available polyimide films, which do not contain the 6F group.[4]

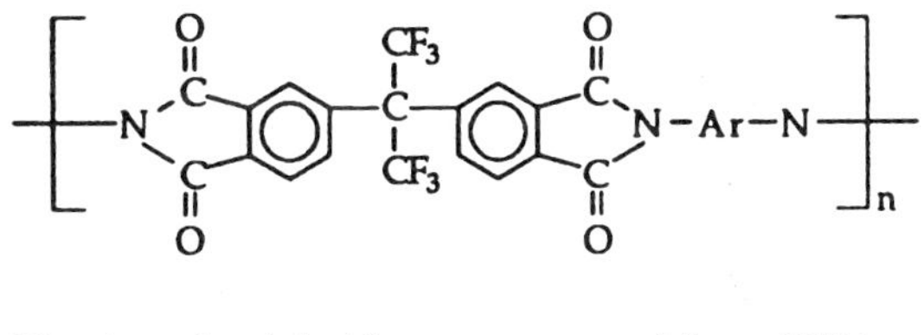

1

Fluorinated polyimides were prepared from 6FDA and 2,2-bis(trifluoromethyl)-4,4′-diaminobiphenyl. They exhibited low dielectric constant, low refractive index, and high optical transparency, whereas a polyimide prepared from pyromellitic dianhydride and the same diamine showed a low thermal expansion coefficient.[5]

Polyimides obtained from 6FDA and 4,4′-diaminodiphenylmethane or 4,4′-diaminodiphenyl ether have been used for gas separation membranes.[6] Polymers based on 6FDA and 4,4′-diaminodiphenyl ether had the best gas permeability and gas permeation selectivity for CO_2/CH_4 separation.

Mixtures of 6FDA and ordinary, nonfluorinated aromatic dianhydrides reacted with aromatic diamines to give polymers that were soluble in toluene, *N*-methylpyrrolidinone, *N,N*-dimethylacetamide, and *m*-cresol and that had high thermal stability, with decomposition temperatures above 440°C.[7]

Various polyimides were synthesized by reacting fluorinated diamines with ordinary aromatic dianhydrides. Polymers based on 2,2-bis[4(4-aminophenoxy)-phenyl]hexafluoropropane and pyromellitic dianhydride could be processed by compression molding, showing a high T_g (after post-cure), 449°C, and high thermal stability, 500°C.[8] Heat-resistant sheets manufactured by extrusion of such polyimides showed good mechanical properties and low water adsorption. Other polyimides, prepared from cyclobutantetracarboxylic dianhydride and 2,2-bis([4-(4-aminophenoxy)phenyl]hexafluoropropane or its mixtures with other aromatic diamines, have been found useful for electric and electronic devices, especially for semiconductor devices and liquid crystal displays.[9] Photoreactive polyimides based on benzophenonetetracarboxylic dianhydride and 2,2-bis(2-amino-3-methylphenyl)hexafluoropropane showed high photosensitivity and excellent thermal stability and electrical properties.[10]

Polyimide-propylene oxide block copolymers have been prepared by copolymerization of pyromellitic diester diacid chloride with 2,2-bis[(4-aminophenoxy)phenyl]hexafluoropropane and amine-endcapped polyoxypropylene oligomers.[11]

Highly fluorinated polyimides have been synthesized by polycondensation of 6FDA with various fluorinated diamines.

Ultrathin films have been deposited on silver or on highly oriented graphite by the Langmuir-Blodgett technique using a polyamic acid prepared from a mixture of 6FDA with pyromellitic dianhydride and 2,2-bis[4(4-aminophenoxy)phenyl]hexafluoropropane.[12]

Polyimides prepared from 6FDA and fluorinated diamines have been processed into sheets and used in laminates and multilayer circuit boards.

Fluorinated polyimides having low energy loss in the near-IR region can be used for optical waveguides in optical communication components. They are introduced as more advanced materials.[13] Thus, an optical waveguide made from a polyimide based on 2,2′-bis(trifluoromethyl)benzidine and 6FDA or mixtures of 6FDA with pyromellitic dianhydride showed high heat resistance and low optical loss, the light transmittance loss (1.3 μm) being 0.3 dB/cm. Optical waveguides made from polyimides based on 6F dianiline showed high quality due to molecular ordering up to the thermal decomposition temperature of the material.

Highly fluorinated polyimides have been synthesized by polycondensation of 6FDA with various aromatic diamines containing fluorine substituents on benzene rings or other fluorinated diamines.[14] Systematic studies showed that fluorinated polyimides offered up to 20% reduction in dielectric constant in integrated circuit packaging over standard polyimides, which translated to faster signal speed throughout the package. They also showed less than one-third the moisture absorption of standard polyimides. The differences in water absorption were related to ease of formation of polar resonance structures.

POLYIMIDES CONTAINING AMIDE, SULFONE, ETHER, OR OTHER FLEXIBLE GROUPS

Aromatic poly(imide amide)s have attracted considerable attention because they bring together the superior mechanical properties associated with amide groups and the high thermal stability associated with the imide ring. The all-around properties of poly(imide amide)s represent a very desirable combination even though they do not exhibit the same level of thermal stability or performance at elevated temperatures as polyimides.

Fluorinated poly(imide amide)s have been prepared by polycondensation of aromatic diamines with aromatic dianhydrides and aromatic tricarboxylic acid monoanhydrides, one or two of them containing 6F.[15] These materials possessed high T_g, useful mechanical properties, and outstanding thermoplastic flow behavior that rendered them readily melt processable into fibers, films, sheets, and other molded articles. They were soluble in many organic solvents and were thus amenable to solution casting techniques. They also showed excellent resistance to thermooxidative degradation and had low moisture uptake. Poly(imide amide)s prepared by reaction of aromatic diamines with 2,2-bis(4-chlorocarbonylphenyl)hexafluoropropane and 6FDA showed a T_g of 300°C and a decomposition temperature of 500°C, and were easily processable as films, moldings, and fibers.[16] Other poly(imide amide)s, made by reacting diaminobenzanilide with 6FDA or mixture of 6FDA, oxydiphthalic anhydride, benzophenontetracarboxylic dianhydride, and isophthaloyldiphthalic anhydride, showed high T_g, high thermal oxidative stability, and good solvent resistance.

Fluorinated poly(imide amide)s containing pendant cyano groups were prepared by reacting bis(4-aminophenoxy)benzonitriles with fluorinated diacid chlorides containing imide rings.[17,18] Cyano groups were incorporated in these polymers as latent crosslinking sites.

Fluorinated poly(imide sulfone)s were prepared by reacting 6FDA with diamino diphenyl sulfones and were processed into transparent films having a whiteness index of 91, light transmission (500 nm) of 88%, and moisture absorption of 0.7% compared with 57, 72%, and 1.9%, respectively, for a film of a related nonfluorinated polyimide based on pyromellitic dianhydride.[19] Fluorinated poly(imide sulfone amide)s were prepared by reacting diaminodiphenyl sulfones with diacid chlorides containing 6F and imide groups.[20] Thin, flexible, colorless films showed T_g values of 279-359°C, decomposition temperatures of 464-479°C, and dielectric constants of 3.49-3.68. The expected effect of 6F groups on the decrease of dielectric constant seemed to be offset by the presence of polar sulfonyl and amide groups.

Fluorinated poly(imide ether)s have been prepared by reacting 6FDA with aromatic diamines containing ether groups and various content of fluorine.[21] Systematic studies on the relationship between fluorine content in these polymers and their dielectric constant and moisture absorptions showed that the dielectric constant and moisture absorption could be reduced to about 2.7 and 0.79%, respectively, by incorporating both a fluorinated diamine and a fluorinated dianhydride.

Fluorinated poly(imide ester)s were obtained by solution polycondensation in a high-boiling-point solvent of diphenols containing 6F and preformed imide groups with diacid chlorides containing ester bridges.[22] Colorless, flexible films showed T_g values in the range 170-270°C, decomposition temperatures above 450°C, and dielectric constants of 2.8-3.21.

Fluorinated poly(imide siloxane)s were prepared from aromatic dianhydrides and aromatic diamines, either or both containing 6F groups, and/or diaminosiloxanes.[23] Application and thermal treatment of the polyamic solutions gave interlayer insulators for multilayered wirings, with low dielectric constant (2.7 at 15 GHz), low water absorption, and good chemical resistance.

Fluorinated polyimide containing phosphorus in the main chain was prepared by reacting 6FDA with bis(3-aminophenyl)phenylphosphine, showing a T_g of 282°C and a decomposition temperature of 503°C[23a] Such a polymer is of interest because of its good fire resistance which is usually characteristic to phosphorus-containing polymers, and its ability to complex with metals.

Fluorinated poly(imidine imide)s were synthesized by reaction of 3,3′-dibenzilidenepyromellitimide with xylylenediamine or diaminodiphenyl ether, followed by reaction with 6FDA and cyclodehydration of the intermediate polyamic acid; the resulting polymers showed high thermal stability and good solubility and processability into tough, flexible films.[24,25]

ADDITION-TYPE POLYIMIDES AND POLYIMIDE BLENDS

Addition-type polyimides prepared from bis(maleimide)s and aromatic diamines, one or both of them containing 6F bridges,[26,27] showed high T_g, 200-290°C; low moisture absorption, 0.14-0.4%; and low dielectric constant, 2.6-3.15. Thermosetting resins have been obtained through the synthesis of amine-terminated polymides derived from 6FDA and fluorinated diamines that were then reacted with maleic anhydride and then with other aromatic diamines with or without 6F groups.[28] These polymers had good processability and were used in the preparation of laminated circuit boards having low dielectric constant, 3.35; good heat resistance with a decomposition temperature of 495°C and a T_g of 325°C; good dimensional stability; good flexibility; and through-hole bonding reliability.

POLYIMIDES CONTAINING VARIOUS HETEROCYCLES

Copolymers containing imide and other heterocyclic units—such as oxadiazole, benzoxazole, benzimidazole, quinoxaline, and benzoxazinone—have been synthesized and studied in order to attain an even better balance of physical, chemical, and mechanical properties, based on the particular contribution of each of these heterocycles.

Fluorinated poly(1,3,4-oxadiazole imide)s were prepared by reaction of dihydrazide of hexafluoroisopropylidene bis(benzoic acid) with diacid chlorides containing preformed imide rings, followed by cyclodehydration in a film state to give the oxadiazole structure.[29] They had a decomposition temperature above 400°C, exhibited no T_g, and were unchanged after isothermal aging at 300°C in air.

Fluorinated poly(benzoxazole imide)s prepared by condensation of aromatic dianhydrides with diaminobenzoxazoles containing 6F groups or by reaction of 6FDA with fluorinated bis(aminophenol)s containing amide groups[30] were used to make composites with glass fibers showing weight loss 8.6% after 100 h at 650°C.

Poly(benzimidazole imide)s were prepared by reaction of diaminobenzimidazoles with diacid chlorides containing 6F and preformed imide moieties. The resulting polymers were soluble in polar aprotic solvents and gave flexible films with decomposition temperatures above 450°C and low dielectric constants (3.1-3.4).[31]

CONCLUSIONS

Research on polymers containing 6F groups has expanded in the past 5 years, particularly on heterocyclic polymers, and among them polyimides were the most commonly studied. The benefits realized–mostly in processability and electrical, optical, and surface properties–make these polymers attractive for various high-performance applications. Increased numbers of publications and patents appearing from one year to another certifies that these polymers are continuing to be of interest to both synthetic chemists and application engineers.

ACKNOWLEDGMENTS

It is a great privilege to acknowledge the financial support provided to Maria Bruma by the Robert A. Welch Foundation, Houston, Texas (Grant AI-0524).

REFERENCES

1. Cassidy, P. E. *J. Macromol. Sci., Rev. Macromol. Chem. Phys.* **1994,** *C34,* 1.
2. Cassidy, P. E.; Aminabhavi, T. M.; Farley, J. M. *J. Macromol. Sci., Rev. Macromol. Chem. Phys.* **1989,** *C29,* 365.
3. McGrath, J. E.; Rogers, M. E.; Arnold, C. A.; Kim, Y. J.; Hedrich, J. C. *Macromol. Chem., Macromol. Symp.* **1991,** *51,* 103.
4. Laius, L.; Zhukova, T.; Kuznetsov, N.; Kudryavtsev, V.; Svetlichnyi, V.; Simakov, B. V.; Rastorgueva, N.; Nikiforova, G. *Vysokomol. Soedin. Ser. B.* **1991,** *33,* 851.
5. Sasaki, S.; Matsuura, T.; Ando, S.; Nishi, S. *N.T.T.R. & D.* **1991,** *40,* 967, *Chem. Abstract* **1992,** *116,* 130626h.
6. Matsumoto, K.; Xu, P. *Maku* **1992,** *17,* 395; *Chem. Abstract* **1993,** *118,* 16896n.
7. Park, J.; Lee, M.; Liu, J.; Chang, J.; Rhee, S. *Macromolecules* **1994,** *27,* 3459.
8. Jones, R. J.; Silverman, E. M. *Int. SAMPE Tech. Conf.* **1988,** *20,* 542; *Chem. Abstract* **1989,** *111,* 11641n.
9. Komasa, N.; Kobayashi, T. Jpn. Kokai Tokkyo Koho, Jpn. Patent 02 219 827, 1990; *Chem. Abstract* **1991,** *114,* 103078h.
10. Omote, T.; Koseki, K.; Yamaoka, T. *Polym. Eng. Sci.* **1989,** *29,* 945.
11. Lakshmanan, P.; McGrath, J. E. *Polym. Prepr., Am. Chem. Soc. Div. Polym. Chem.* **1994,** *35,* 713.
12. Tsai, W. H.; Cave, N. G.; Bocrio, F. J. Langmuir **1992,** *8,* 927.
13. Reuter, R.; Wagner, D.; Franke, H. *Proc. SPIE, Int. Soc. Opt. Eng.* **1990,** *1147,* 271; *Chem. Abstract* **1991,** *114,* 124031c.
14. Hougham, G.; Tesoro, G.; Shaw, J. *Macromolecules* **1994,** *27,* 3642.
15. Chen, P.; Glick, M.; Cooper, W.; Jaffe, M. *High Perf. Polym.* **1990,** *2,* 39.
16. Chen, P.; Vora, R. H. U.S. Patent 4 962 183, 1990; *Chem. Abstract* **1991,** *114,* 144140r.
17. Mercer, F.; McKenzie, M.; Bruma, M.; Schulz, B. *Polymer International* **1994,** *33,* 399.
18. Bruma, M.; Mercer, F.; Schulz, B.; Dietel, R.; Fitch, J.; Cassidy, P. *High Perf. Polym.* **1994,** *6,* 183.
19. Morita, N.; Nakayama, T. Jpn. Kokai Tokkyo Koho, Jpn. Patent 02 185 582, 1990; *Chem. Abstract* **1991,** *114,* 25529q.
20. Bruma, M.; Mercer, F.; Fitch, J.; Cassidy, P. *J. Appl. Polym. Sci.* **1995,** *56,* 527.
21. Mercer, F. W.; Goodman, T. D. *High Perf. Polym.* **1991,** *3,* 297.
22. Bruma, M.; Sava, I.; Negulescu, I.; Mercer, F.; Daly, W.; Fitch, J.; Cassidy, P. *High Perf. Polym.* **1995,** *7,* 411.
23. Arnold, C. A.; Chen, Y. P.; Rogers, M. E.; Graybeal, J. D.; McGrath, J. E. *Int. SAMPE Electron. Conf.* **1989,** *3,* 198; *Chem. Abstract* **1991,** *114,* 229512p. (a)Martinez-Nunez, M. F.; Sekharipuram, V. N.; McGrath, J. E. *Polym. Prepr., Am. Chem. Soc. Div. Polym. Chem.* **1994,** *35,* 709.
24. Farley, J. M.; Cassidy, P. E. *Macromolecules* **1988,** *21,* 3372.
25. Arai, M.; Cassidy, P. E.; Farley, J. M. *Macromolecules* **1989,** *22,* 989.
26. Misra, A. C.; Tesoro, G. *Polymer* **1992,** *33,* 1083.
27. Nagai, A.; Takahashi, A.; Mukoh, A. *J. Appl. Polym. Sci.* **1992,** *44,* 159.
28. Shirai, M.; Hayashi, S. Jpn. Kokai Tokkyo Koho, Jpn. Patent 04 114 035, 1992; *Chem. Abstract* **1993,** *118,* 60710w.
29. Thaemlitz, C. J.; Weikel, W. J.; Cassidy, P. E. *Polymer* **1992,** *33,* 3278.
30. Khanna, D. N.; Lee, W. R. Eur. Patent Appl. 387 060, 1990; *Chem. Abstract* **1991,** *114,* 82719.
31. Hamcuic, E.; Bruma, M.; Mercer, F. *Angew. Makromol. Chem.* **1993,** *210,* 143.

FLUORINATED PLASTICS, AMORPHOUS

Ming-H. Hung, Paul R. Resnick, Bruce E. Smart, and Warren H. Buck
DuPont Fluoroproducts
DuPont Central Research and Development
Experimental Station

The discovery of polytetrafluoroethylene (PTFE) by R. Plunkett of DuPont in 1937 ushered in the era of commercial fluoroplastics, and PTFE has remained the largest volume fluoropolymer. The processability of PTFE, however, is limited by its high melt viscosity (high crystallinity), and melt-processable copolymers such as Teflon® FEP (tetrafluoroethylene-hexafluoropropylene copolymer) and Teflon® PFA [tetrafluoroethylene-perfluoro (propyl vinyl ether) copolymer] that contain tetrafluoroethylene and modifying fluoromonomers were subsequently developed by DuPont. All of these fluoroplastics are crystalline and certain of their properties, such as optical transparency or solubility, therefore are generally poor. These limitations prompted industry's search for "amorphous" fluoroplastics, which culminated in the first commercial offerings by DuPont and Asahi Glass in 1989.

The new classes of amorphous perfluoropolymers with extraordinary properties were developed and commercialized by DuPont as Teflon® AF and by Asahi Glass as Cytop®. These polymers combine the characteristic high thermal stability, excellent chemical resistance, and low surface energy of the traditional crystalline polymers with superior physical and optical properties. A unique property among perfluoropolymers is their solubility in selected fluorinated solvents, which allows

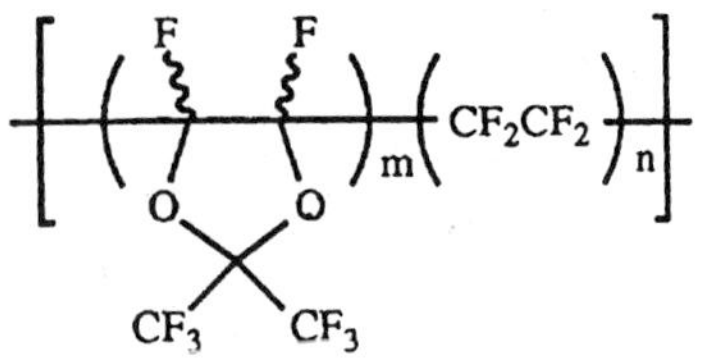

Teflon AF

1

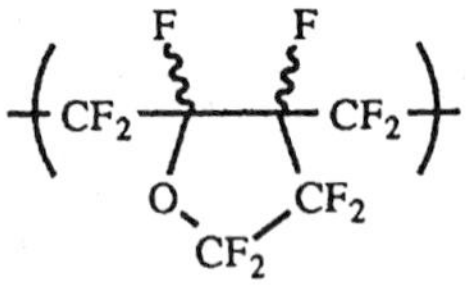

Cytop

2

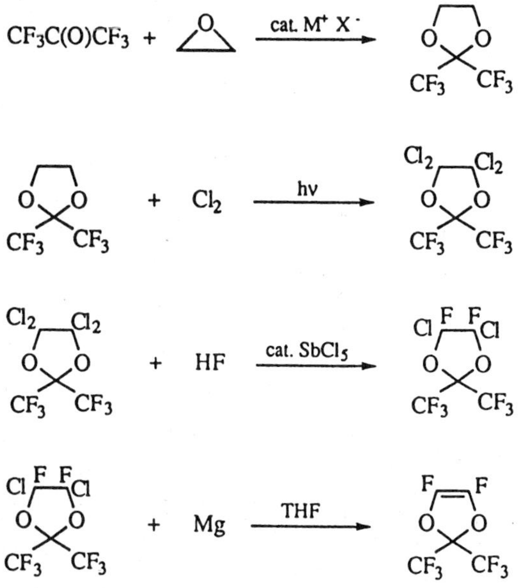

I

1a

their solution processing and the preparation of thin cast films. These enhancements provide the opportunity to use fluoropolymers in many new applications (Structures 1 and 2).

PREPARATION

Teflon® AF is a family of copolymers of 2,2-bis(trifluoromethyl)-4,5-difluoro-1,3-dioxole (**1a**) and tetrafluoroethylene (**Scheme I**) where **1a** is the principal monomer.[1-5]

PROPERTIES

The homopolymer of **1a** exhibits a T_g at 330°C, which is one of the highest among amorphous plastics.[6] Cytop polymer also is totally amorphous but has a substantially lower T_g of 108°C.[7] One rationale for the high glass transition temperature of Teflon AF is that steric interactions involving the two trifluoromethyl groups in the dioxole homopolymer lead to a highly congested chain structure with very limited mobility. This hypothesis is supported by the following observations.

For the copolymers of **1a** with tetrafluoroethylene, the T_gs drop rapidly as the percentage of tetrafluoroethylene (TFE) increases. Moreover, the T_g of the dioxole polymers is highly sensitive to the structure of the dioxole monomer. Indeed, replacing the trifluoromethyl group at the 2-position of dioxole ring with either fluorine or larger fluoroalkyl groups dramatically affects polymer T_g and the polymerization reactivity of the monomer.[8]

2,2-Bis(trifluoromethyl)-4,5-difluoro-1,3-dioxole (**1a**) copolymerizes readily with other fluorinated olefins besides TFE. Typical examples include chlorotrifluoroethylene, hexafluoropropylene, perfluorovinyl ethers (R_f-O-CF=CF_2), vinyl fluorides, vinylidene fluorides, trifluoroethylene, perfluoroalkylethylene (R_f-CH=CH_2), and functional fluoromonomers.

Teflon AF polymers are totally amorphous and no polymer melting point (T_m) is observed over a wide temperature range. The glass transition temperature of Teflon AF depends on its composition: T_g decreases with increasing content of TFE. For most applications, the T_g range is 80-250°C, which corresponds to ~ 20-90 mol% of **1a** in the polymer chain. The homopolymer of **1a** is difficult to process since it decomposes at temperatures required for melt flow. Its solubility is also very limited.

Teflon AF and Cytop polymers combine enhanced optical, electrical, and mechanical properties with the traditional properties of fluoropolymer resins (**Table 1**). The thermal stabilities of Teflon AF and Cytop are comparable to those of other perfluoroplastics. Like other perfluoropolymers, Teflon AF and Cytop are highly chemical resistant.

The dielectric constants of Teflon AF polymers are the lowest of any organic polymer and they are almost unaffected by humidity.

Traditional fluoroplastics like PTFE have unexceptional optical transmission properties due to the crystallinity interference. Teflon AF and Cytop polymers, on the contrary, have excellent optical clarity as a result of their totally amorphous character.

The densities of Teflon AF and Cytop polymers are lower than other common fluoroplastics, presumably due to either their noncrystalline structures or the poor polymer chain-packing resulting from their bulky ring molecular frameworks.

TABLE 1. Comparison of Telfon AF, Cytop, and Teflon PFA Properties

Property	Teflon AF 1600 (2400)	Cytop	Teflon PFA
Morphology	Amorphous	Amorphous	Semi-crystalline
Density (g/cm^3)	1.78 (1.67)	1.84	2.12–2.17
Optical clarity	>95%	95%	Translucent to opaque
Abbe no.	92 (113)	90	—
Thermal stability (°C)	360	400	380
Linear thermal expansion (ppm/°C)	74 (81)	—	150
Water absorption (%)	<0.01	<0.01	<0.01
Flame resistance (LOI)	95%	95%	95%
Tensile strength (MPa)	27.0 (24.6)	32	28–31
Tensile modulus (MPa)	1550 (1540)	1170	271–338
Creep resistance	High	High	Low
Water contact angle (°)	105 (104)	110	115
Refractive index	1.31 (1.29)	1.34	1.34–1.35
Dielectric constant	1.93 (1.89)	2.1–2.2	2.1

Perfluorodioxole polymer thin films are unusually permeable to several gases and their high permeability is matched only by some polysiloxanes. The gas permeability increases with the dioxole monomer content, and Teflon AF homopolymer exhibits an extraordinary oxygen permeability of 154,000 cB and a nitrogen permeability of 81,000 cB.[9-11] By contrast, Cytop and closely related structures are unexceptional with permeabilities only slightly greater than those for PTFE films.[12]

Properties such as high gas permeability, low refractive index, low dielectric constant, and low density suggest the presence of "microvoids" or considerable free volume in the Teflon AF polymer structure. This has been substantiated by positron annihilation lifetime experiments.[13]

The amorphous perfluoropolymers have improved mechanical properties compared to other fluoropolymer resins. They are stiffer, tougher materials as reflected by their considerably higher tensile modulus (**Table 1**).

A property of Teflon AF and Cytop that is unique to perfluoroplastics is their solubility in perfluorinated solvents at ambient temperature. Many perfluorinated solvents can dissolve Teflon AF and Cytop polymers. Examples are hexafluorobenzene, perfluorohexane, perfluoro(methylcyclohexane), perfluorodecalins, perfluoroalkyldecalin, perfluorooctane, perfluoro(*n*-alkyltetrahydrofuran) (Fluorinert® FC-75; 3M Company), and perfluoro(alkyl amine) (Fluorinert FC-40; 3M Company). Crystalline fluoropolymer resins, by contrast, are insoluble at room temperature and dissolve only slightly in octafluoronaphthalene at temperatures close to the polymer melting points. In general, the solubility of Teflon AF decreases with increasing perfluorodioxole monomer content.

PROCESSING AND APPLICATIONS

Amorphous perfluoropolymers can be processed by several methods. They are melt-processable just like Teflon PFA and Teflon FEP, so conventional methods such as extrusion or compression molding and injection molding can be used to fabricate polymers into shaped articles. Since the polymers can be dissolved in fluorinated solvents at ambient temperature, a wide variety of solution processing methods is also available. Other fabrication methods for Teflon AF include laser ablation and vacuum pyrolysis, which provide extremely thin films without using solvents.[14-16]

EXPERIMENTAL POLYMERS

Since the commercialization of Teflon AF and Cytop resins, industrial laboratories worldwide have increased their efforts to develop new, more cost-effective amorphous fluoroplastics. Much of this research seeks to develop structure-T_g correlations aimed at designing high T_g polymers that retain the desirable properties of Teflon AF. Because fluorinated dienes offer the greatest opportunity for structural variation in the cyclopolymer, in both ring size and substituents, research on these systems has been particularly active.

REFERENCES

1. Resnick, P. R. U.S. Patent 3 865 845, 1975.
2. Resnick, P. R. U.S. Patent 3 978 030, 1976.
3. Hung, M.-H.; Resnick, P. R.; Squire, E. N. *Proceedings of the First Pacific Polymer Conference*; Maui, Hawaii, Dec. 1989; p 331.
4. Resnick, P. R. *Polym. Prepr., Am. Chem. Soc. Div. Polym. Chem.* **1990,** *31,* 312.
5. Hung, M.-H. U.S. Patent 4 908 461, 1990.
6. Buck, W. H.; Resnick, P. R. *"Teflon® AF Technical Information Bulletin"*; DuPont: Wilmington, DE, 1993.
7. *"Cytop® Technical Bulletin"*; Asahi Glass: Yokohama, Japan, 1988.
8. Hung, M.-H. *Macromolecules* **1993,** *26,* 5829.
9. Nemser, S. M. U.S. Patent 5 051 113, 1991.
10. Nemser, S. M. U.S. Patent 5 053 059, 1991.
11. Nemser, S. M.; Roman, I. C. U.S. Patent 5 051 114, 1991.
12. Nakamura, H. et al. JP Tokukai 63-23811, 1988.
13. Gregory, R. B. *J. Appl. Phys.* **1991,** *70,* 4665.
14. Blanchet, G. B. *Appl. Phys. Lett.* **1993,** *62,* 479.
15. Nason, T. C.; Moore, J. A.; Lu, T. M. *Appl. Phys. Lett.* **1992,** *60,* 1866.
16. Grieser, J.; Swisher, R.; Phipps, J.; Pelleymounter, D.; Hildreth, E. *Optical Surfaces Resistant to Severe Environments; Proceedings of the SPIE,* **1990,** *1330,* 111.

Fluorination

See: Fluropolymers (Preparation by Fluorination of Polymers)

FLUORINE-CONTAINING POLY(AMINO ACID)S (Conformation and Gas Permeability)

Yoshiyuki Kondo* and Yoshiro Ogoma
Department of Functional Polymer Science
Faculty of Textile Science and Technology
Shinshu University

The atomic size of fluorine is nearly equal to that of hydrogen, but introduction of fluorine to a molecule results in endowment of electronic characteristics because of greater electronegativity. Fluorinated compounds thus have been applied to medicine and engineering, such as artificial blood[1] and a functional membrane with higher O_2 permselectivity.[2-4] Poly(amino acid)s have good biocompatibility as an analog of protein, and application in biomaterial, such as artificial skin, has been reported.[5] Therefore, fluorine-containing poly(amino acid) is expected to have some of the characteristics of both poly(amino acid) and fluorine.

We synthesized the new fluorine-containing poly(amino acid)s, poly(γ-4-fluorobenzyl-L-glutamate) (F-PBLG) and poly(γ-4-trifluoromethylbenzyl-L-glutamate) (CF_3-PBLG), and examined the effect of fluorine on the secondary structure of polymers in solid state and in solution. We also examined properties such as contact angle and O_2 as well as N_2 permeabilities of their films.

PROPERTIES

Conformation of Fluorine-Containing Polyamino Acids in the Solid State

X-ray diffraction photographs were taken with Rigakudenki Geiger flex (CuKα). Strong spacings at 13.1-13.4 and around 5.2 Å in all polymers indicated that the polymers had α-helical structure in solid state and that fluorine had little influence on the backbone structure of the polymers.

*Author to whom correspondence should be addresed.

Conformation of Fluorine-Containing Polyamino Acids in Solution

CD spectra were obtained with a JASCO J-40A automatic recorder. The result showed that a large negative CD band was observed at 222 nm in all samples, indicating that the polymers had a right-handed α-helical structure in chloroform.

Contact Angle and Gas Permeability

The contact angles of PBLG, F-PBLG, and CF_3-PBLG were 74°, 70°, and 74°, respectively, indicating that hydrophobicities of the polymers are almost the same. Nevertheless, introduction of fluorine to a molecule generally increases its hydrophobicity.

O_2 and N_2 permeabilities were measured by JASCO Gasperm-100 using cast films from chloroform. However, only CF_3-PBLG cast film did not succeed because of lower solubility to chloroform. O_2 and N_2 permeabilities of F-PBLG were 85.9 and 90.4 cm^3STP·cm/cm^2·sec·cmHg, respectively, and these values were much higher than those of PBLG. These results suggest that F-PBLG and CF_3-PBLG are useful for optical materials for ophthalmology such as contact lenses and artificial corneas.

REFERENCES

1. Wesseler, E. P.; Iltis, R.; Clark, L. C. Jr. *J. Fluorine Chem.* **1977,** *9,* 137.
2. Kajiyama, T.; Washizu, S.; Ohmori, Y. *J. Membrane Sci.* **1985,** *24,* 73.
3. Terada, I.; Haraguchi, T.; Kajiyama, T. *Polym. J.* **1986,** *18,* 529.
4. Higashi, N.; Kunitake, T.; Kajiyama, S. *kobunshi ronbunshu,* in Japanese **1986,** *43,* 761.
5. Lin, S. D. et al. *Trans. Am. Soc. Artifi. Intern. Organs,* **1981,** *27,* 522.

Fluorine-Containing Polymers

See: *Comb-Like Polymers*
Contact Lenses, Gas Permeable
Crosslinking Agents, Phenyltriazenes (for Polyimides and Other Aromatic Polymers)
Cyclolinear Organophosphazene Polymers
Engineering Thermoplastics (Survey of Industrial Polymers)
Ferroelectric Polymers (Structural Phase Transitions)
Fluorinated High Performance Polymers (Imide and Hexafluoroisopropylidene Groups)
Fluorinated Plastics, Amorphous
Fluorine-Containing Poly(amino acid)s (Conformation and Gas Permeability)
Fluorine-Containing Polymers (Polycarbonates and Polyformals)
Fluoro-Acrylate Adhesive Blends (Surface Segregation Behavior)
Fluoroacrylic Acid Oligomers
Fluorocarbon Elastomers (Commercial)
Fluoropolymer Coatings (New Developments)
Fluoropolymers (Preparation by Aromatic Fluoroalkylation)
Fluoropolymers (Preparation by Fluorination of Polymers)
Fluoropolymers (Surface Modification by Eximer-Laser Irradiation)
Fluoropolymers (Trifluoromethylated PES and PEEK)
Fluoropolymers, Acrylates and Epoxies (with Low Dielectric Constants)
Fluorosilicon Oligomers
Fluorosilicones (Monomer Synthesis and Applications)
Ionomeric Membranes, Perfluorinated
Living Radical Polymerization, Iodine Transfer
Maleimides, Fluorine-Containing
Perfluorinated Ionomers (Overview)
Perfluorocyclobutane Aromatic Ether Polymers
Polyfluoroacrylates (Effects of Side-Chain H/F Variations)
Poly(fluoroalkyl α-fluoroacrylate)s
Poly(hexafluoro-1,3-butadiene)
Polyphosphonates and Polyphosphates (Low Temperature Solution Polycondensation)
Polytetrafluoroethylene
Poly(vinylidene fluoride) (Overview)
Poly(vinylidene fluoride) (Synthesis, Microstructure, and Chain Conformations)
Poly(vinylidene fluoride) Phases and Morphology (Effect of Crystallization Conditions)
Tetrafluoroethylene Copolymers (Overview)
Vinylidene Fluoride-Based Thermoplastics (Overview and Commercial Aspects)
Vinylidene Fluoride-Based Thermoplastics (Applications)
Vinylidene Fluoride-Based Thermoplastics (Blends with Other Polymers)
Vinylidene Fluoride-Based Thermoplastics (Homopolymerization and Copolymerization)
Vinylidene Fluoride-Trifluoroethylene Copolymers (Ferroelectric-to-Paraelectric Phase Transition)

FLUORINE-CONTAINING POLYMERS (Polycarbonates and Polyformals)

Yasuo Saegusa,* Minoru Kuriki, Akihiro Kawai, and Shigeo Nakamura
Department of Applied Chemistry
Faculty of Engineering
Kanagawa University

Due to the outstanding properties of thermal stability, transparency, impact resistance, and self-extinguishment, aromatic polycarbonates, such as that derived from 2,2-bis(4-hydroxyphenyl)propane (Bisphenol A), are ranked as high-performance engineering thermoplastics.[1] Fluorine-containing polycarbonate also has been synthesized from 2,2-bis(4-hydroxyphenyl)-1,1,1,3,3,3-hexafluoropropane (Bisphenol AF) and phosgene.[2] However, no detailed properties of this polymer, except thermal ones, were disclosed. Structurally related aromatic polyformal has been obtained from Bisphenol A and dichloromethane (DCM).[3-5] This polymer also has a number of prominent physical properties comparable to those of Bisphenol A-based polycarbonate.[4,5]

*Address to whom correspondence should be addressed.

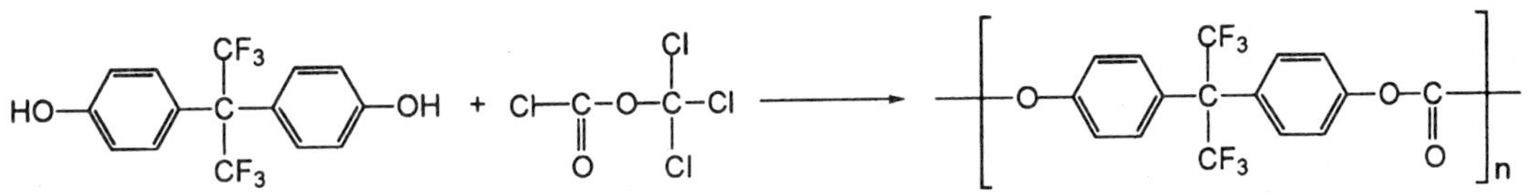

1

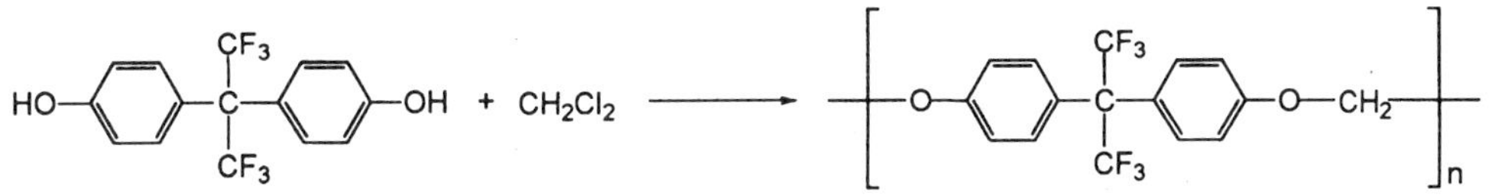

2

PREPARATION

Polycarbonates

Commercial polycarbonates have been produced mainly through a low-temperature solution polycondensation technique using phosgene as a carbonylation agent.[6] In view of phosgene vapor's extreme toxicity, this preparative method is not favorable in a laboratory scale of polycarbonate syntheses. Formation of high molecular weight polycarbonates also has been achieved by using *N,N'*-carbonyldiimidazole as a condensing agent under mild conditions.[7,8] Bisphenol AF-based homopolycarbonate was synthesized by the two-phase phase-transfer-catalyzed polycondensation of Bisphenol AF with TCF in an organic solvent-aqueous alkaline solution system with a variety of quaternary ammonium salts, such as tetra-*n*-butylammonium chloride (TBAC), tetra-*n*-butylammonium bromide (TBAB), benzyltriethylammonium chloride (BTEAC), and benzyltriethylammonium bromide (BTEAB) at room temperature. The carbonylation agent TCF is a liquid and can be handled more easily than phosgene for polycarbonate preparations. Chlorinated hydrocarbons, such as DCM, chloroform (CF), tetrachloromethane (TCM) and 1,2-dichloroethane (DCE), and nitrobenzene (NB) were used as organic media (**Equation 1**).

Judging both from the yield and reduced viscosity of the polymers obtained, use of TBAB as a phase-transfer catalyst, sodium hydroxide as a base, and DCE as an organic medium were found to be suitable for preparing a high molecular weight polycarbonate in high yield.

A series of fluorine-containing random copolycarbonates and Bisphenol A-based homopolycarbonate was also synthesized by reacting the respective mixtures of Bisphenol AF/Bisphenol A feed ratios of 80/20, 60/40, 50/50, 40/60, and 20/80, and 0/100 with TCF under the most suitable conditions: in the DCE-aqueous sodium hydroxide two-phase system with TBAB as a catalyst.

Viscosity values of the copolycarbonates decreased markedly with increasing feed ratio of Bisphenol AF to Bisphenol A, although the yields are still high. This is probably due to the lower nucleophilicity of Bisphenol AF induced by strongly electron-withdrawing trifluoromethyl groups at the para position than that of Bisphenol A.

Polyformals

Bisphenol AF-based homopolyformal was synthesized by solution polycondensation of Bisphenol AF with DCM in highly polar cosolvents in the presence of potassium hydroxide at 75°C for 4 h, according to conditions used by Hay et al.[3-5] for the preparation of Bisphenol A-based homopolyformal. Polar aprotic solvents such as *N,N*-dimethylacetamide (DMAc), *N,N*-dimethylformamide (DMF), dimethyl sulfoxide (DMSO), hexamethylphosphoramide (HMPA), and *N*-methyl-2-pyrrolidone (NMP) were used as comedia (**Equation 2**).

Use of a large excess of DCM to Bisphenol A is required for production of polyformal with high molecular weight.[4,5]

A series of fluorine-containing copolyformals and Bisphenol A-based homopolyformal also was synthesized by reacting the respective mixtures of Bisphenol AF/Bisphenol A feed ratios of 75/25, 50/50, 25/75, and 0/100 with DCM under the most preferable conditions.

PROPERTIES

Polycarbonates

Introduction of a fluorine atom into Bisphenol A-type of polycarbonate successfully improved the physical properties of the resulting polycarbonates such as solubility, mechanical, surface and optical properties, and thermal behavior.

Polyformals

Introduction of a fluorine atom into Bisphenol A-type of polyformal made them highly soluble, less wettable, and more thermally stable without a decrease in mechanical properties.

REFERENCES

1. Freitag, D.; Grigo, U.; Muller, P. R.; Nouvertne, W. *Encyclopedia of Polymer Science and Engineering*; 2nd ed.; Wiley-Interscience: New York, 1995.
2. Knunyants, I. L.; Chen, T.-S. et al. *Zh. Vses. Khim. O-va.* **1960,** *5,* 114.
3. Hay, A. S.; Williams, F. J. et al. *Polym. Prepr. Div. Polym. Chem. Am. Chem. Soc.* **1982,** *23,* 117.
4. Hay, A. S.; Williams, F. J. et al. *J. Polym. Sci. Polym. Lett. Ed.* **1983,** *21,* 449.
5. Hay, A. S.; Williams, F. J. et al. *J. Macromol. Sci. Chem.* **1984,** *A21,* 1065.
6. Fox, D. W. *Macromolecular Syntheses*, Coll. Vol. 1; Wiley-Interscience: New York, 1977.
7. Wada, Y.; Ito, T.; Suzuki, S. *Jpn. Kokai Tokkyo Koho* 55, 526 1982a.
8. Wada, Y.; Wada, M. et al. *Jpn. Kokai Tokkyo Koho* 57, *121,* 031, 1982b.

FLUORO-ACRYLATE ADHESIVE BLENDS (Surface Segregation Behavior)

Saburo Akiyama* and Yoshihisa Kano
Faculty of Technology
Tokyo University of Agriculture and Technology

In recent decades, aided by the development of various surface analyses, such as X-ray photoelectron spectroscopy (XPS),[1] secondary ion mass spectrometry (SIMS),[2] attenuated total-reflection Fourier-transform infrared spectroscopy (ATR-FTIR),[3] and photoacoustic step-scan FTIR (PAS-FTIR),[4] the surfaces and interfaces of polymer alloys have been examined. In particular, surface segregation behavior and functionally gradient material are found in polymer blends, block copolymers, and graft copolymers by the above-mentioned surface techniques. In polymer blends, surface segregation occurs as a low surface tension component, is preferentially enriched on the surface of samples according to differences between surface tensions of components. On the other hand, in functionally gradient material, the concentrations of a mixture gradually change from surface to bottom side (surface in contact with substrate) of film samples. Recently, knowledge of surface segregation phenomena has been applied to surface modification,[5-7] medical material,[8] and the hot melt adhesives.[9] It is thought that the appearance of surface segregation (functionally gradient material) is affected by miscibility, surface tension, density, rate of solvent casting, and convection in solution if the blend film was prepared by the solvent casting method. Surface segregation phenomena would be very interesting to the pressure-sensitive adhesive (PSA) industry if tackiness of adhesives could be controlled easily.

We and our colleagues[10-12] have found surface segregation behavior (functionally gradient material) for blends of acrylate PSA copolymer with fluoro-copolymer by using XPS, ATR-FTIR, and PAS-FTIR analyses. The thermodynamics of surface segregation behavior were investigated by miscibility, surface tension, and density. Finally, the tack properties to apply in PSAs were evaluated in terms of dynamic mechanical properties.

PREPARATION

Poly(2-ethylhexyl acrylate-*co*-acrylic acid-*co*-vinyl acetate) [P(2EHA-AA-VAc)] was selected as an acrylate PSA copolymer that possesses excellent tackiness and is used in commercial PSA tape. Its tackiness is confirmed by evaluation of various properties, such as fluorescence,[13,14] wettability,[15] dynamic mechanical properties, and peeling morphology.[16] On the other hand, poly(vinylidene fluoride-*co*-hexafluoroacetone) [P(VDF-HFA)] was noted as a low surface tension component, since the surface tension of P(VDF-HFA) is lower than that of poly(vinylidene fluoride) (PVDF).

APPLICATIONS

In general, tackiness is dependent on dynamic mechanical properties, surface tension, and surface morphology. As P(2EHA-AA-VAc)/P(VDF-HFA) blends exhibited surface segregation (gradient structure), their tackiness might be different from that of commercial PSA tape. In these blends, the tackiness of the surface in contact with the substrate (the bottom side) was evaluated by the J. Dow ball tack method.

We judge the difference of J. Dow ball tack between surface side and bottom side is influenced by the difference of the dynamic mechanical property (surface component) based on surface segregation (gradient structure). We expect P(2EHA-AA-VAc)/P(VDF-HFA) blends can be utilized as a new type of PSA tape, using no backing.

*Author to whom correspondence should be addressed.

REFERENCES

1. Thomas, H. R.; O'Malley, J. J. *Macromolecules* **1981,** *14,* 1316.
2. Chujo, R.; Nishi, T.; Sumi, Y.; Adachi, T.; Naito, H.; Frentzel, H. *J. Polym. Sci., Polym. Lett. Ed.* **1983,** *21,* 487.
3. Cowie, J. M. G.; Devlin, B. G.; McEwen, I. J. *Polymer* **1993,** *34,* 501.
4. Dittmar, R. M.; Chao, J. L.; Palmer, R. A. *Appl. Spectrosc.* **1991,** *45,* 1104.
5. Green, P. F.; Russell, T. P. *Macromolecules* **1991,** *24,* 2931.
6. Pertsin, A. J.; Gorelova, M. M.; Levin, V. Y.; Makarova, L. I. *J. Appl. Polym. Sci* **1992,** *45,* 1195.
7. Chen, X.; Gardella, Jr., J. A. *Macromolecules* **1994,** *27,* 3363.
8. Takahara, A.; Tashita, J.; Kajiyama, T.; Takayanagi, M.; MacKnight, W. J. *Polymer* **1985,** *26,* 978.
9. Tachikake, M.; Nakamura, M.; Sakurai, Y.; Mitoh, M. *Nippon Setchaku Gakkaishi* **1994,** *30,* 200.
10. Kano, Y.; Ishikura, K.; Kawahara, S.; Akiyama, S. *Polym. J.* **1992,** *24,* 135.
11. Kano, Y.; Akiyama, S.; Kasemura, T. *J. Appl. Polym. Sci.* **1993,** *50,* 1619.
12. Kano, Y.; Akiyama, S.; Kasemura, T. *Polym. J.* **1995,** *27,* 339.
13. Ushiki, H.; Kano, Y.; Akiyama, S.; Kitazaki, Y. *Eur. Polym. J.* **1986,** *22,* 381.
14. Akiyama, S.; Ushiki, H.; Kano, Y.; Kitazaki, Y. *Eur. Polym. J.* **1987,** *23,* 327.
15. Kano, Y.; Akiyama, S.; *Eur. Polym. J.* **1993,** *29,* 1099.
16. Kano, Y.; Akiyama, S.; Ushiki, H. *J. Adhesion* **1993,** *43,* 223.

FLUOROACRYLIC ACID OLIGOMERS

Hideo Sawanda
Department of Chemistry
Nara National College of Technology

Acrylated or methacrylated polymers containing perfluoroalkyl groups are known to exhibit excellent physical, thermal and surface properties.[1] Especially, polymers possessing long-chain (C>5) perfluoroalkyl groups have been widely used in various fields such as polymeric fluorinated surfactants. Efforts have focused on synthesis and surfactant properties of long-chain fluorinated amphiphiles such as perfluoroalkyl-polyoxyethylenes in which the hydrophobic tail is a chemically relatively nonreactive perfluoroalkyl chain, and the hydrophilic head is a polyoxyethylene unit.[2] Usually perfluoroalkyl groups are introduced into these polymers through the ester bond, and these materials are generally unstable under acid or alkaline conditions owing to the ester moieties. Hence, development of more efficient synthetic methodology for direct introduction of

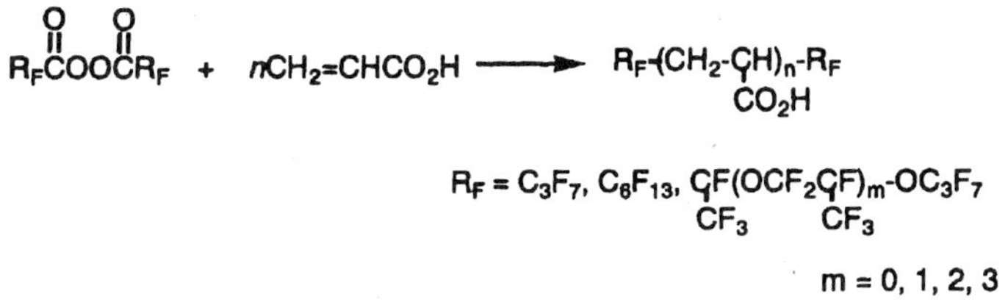

$$\underset{R_F}{\overset{O}{\parallel}}COO\underset{}{\overset{O}{\parallel}}CR_F + nCH_2=CHCO_2H \longrightarrow R_F\text{-}(CH_2\text{-}\underset{CO_2H}{CH})_n\text{-}R_F$$

$$R_F = C_3F_7,\ C_6F_{13},\ \underset{CF_3}{CF}(OCF_2\underset{CF_3}{CF})_m\text{-}OC_3F_7$$

$$m = 0, 1, 2, 3$$

I

$$\text{-}[\underset{}{\overset{O}{\parallel}}CR_F\overset{O}{\parallel}COO]_p\text{-} + q\,pCH_2=CHCO_2H \longrightarrow \text{-}\{R_F\text{-}[CH_2\text{-}CH(CO_2H)]_q\}_p\text{-}$$

$$\text{-}R_F\text{-} = \text{-}(CF_3)CF[OCF_2(CF_3)CF]_n\text{-}O(CF_2)_5O\text{-}[CF(CF_3)CF_2O]_mCF(CF_3)\text{-}$$

$$(n + m = 3)$$

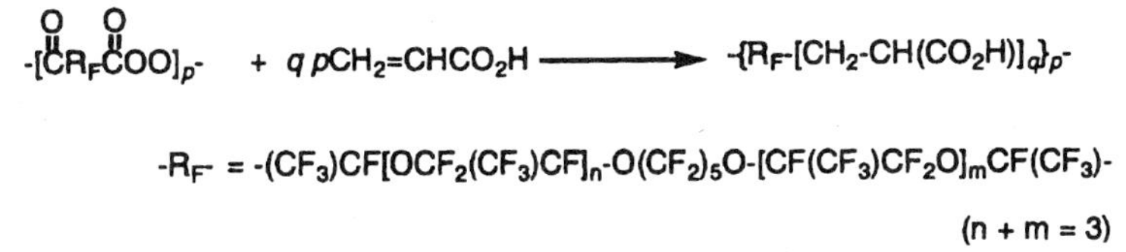

II

perfluoroalkyl groups with carbon-carbon bond formation into these organic molecules is desirable. In general, perfluoroalkyl iodides are convenient tools for introduction of perfluoroalkyl groups with a radical process[3] or copper-induced Ullmann-type reactions.[4,5] However, there have been few reports of preparations of fluoroalkylated polymers (or oligomers) with carbon-carbon bond formation except for those of perfluoroalkylated polystyrenes[7,8] or perfluoroalkylated polyimides.[9-11] We have succeeded in preparing acrylic acid oligomers containing two fluoroalkylated end groups by using fluoroalkanoyl peroxides via a radical process (**Scheme I**).[12,13]

These fluoroalkylated acrylic acid oligomers and fluoroalkylated methacrylate or acrylate oligomers indicated surface active properties typical of fluorinated amphiphiles, although they are oligomeric (high molecular mass) materials.[7,9,11,12,14-16] In this account, we show synthesis and properties of fluoroalkylated acrylic acid, methacrylate, and acrylated oligomers with carbon-carbon bond formation by using fluoroalkanoyl peroxides. We also demonstrate the direct introduction of a perfluoro-oxaalkylene unit into acrylic acid oligomers by using novel polymeric perfluoro-oxaalkane diacyl peroxide and the properties of these new acrylic acid oligomers (**Scheme II**).[17]

PROPERTIES

We found that acrylic acid oligomers containing two fluoroalkyl end-groups [R_F-$(CH_2CHCO_2H)_n$-R_F] obtained by the reactions with fluoroalkanoyl peroxides are easily soluble in water and water-soluble polar solvents such as methanol, ethanol, and tetrahydrofuran. Thus, these fluoroalkylated acrylic acid oligomers are applicable as new fluorinated surfactants.

Perfluoroalkylated and perfluoro-oxaalkylated acrylic acid oligomers were effective in reducing the surface tension of water. Acrylic acid oligomers possessing perfluoro-oxaalkyl groups were more effective in reducing the surface tension of water to 10 dyn/cm levels. This property should depend on the flexibility of perfluoro-oxaalkyl groups as a result of the ether linkage and a particular number of trifluoromethyl groups in a perfluoro-oxaalkyl chain. Thus, our fluoroalkylated, especially perfluoro-oxaalkylated acrylic acid oligomers, are applicable for new fluorinated surfactants though these oligomers possess only two fluoroalkyl end groups in one oligomeric molecules. From this point of view, perfluoro-oxaalkylated compounds are

expected to have various unique properties such as extremely low surface tensions, which cannot be achieved by the corresponding perfluoroalkylated ones.

Many new fluorinated amphiphiles can be prepared readily by use of fluoroalkanoyl peroxides. We found that perfluoro-oxaalkylated oligomers containing hydroxylated ethylene oxide units {R_F-$[CH_2CRCO_2(CH_2CH_2O)_qH]_x$-$R_F$} exhibit surface properties typical of amphiphiles from the contact angle measurements of water and dodecane.

The surface behavior of perfluoro-oxaalkylated oligomers containing hydroxylated ethylene oxide units is of great interest because the contact angles of these oligomers, especially those of oligomers bearing long ethylene oxide chains, exhibit strong hydrophilic properties despite the fact that these oligomers possess perfluoro-oxaalkyl groups.

REFERENCES

1. Fielding, H. C. *Organofluorine Chemicals and Their Industrial Applications*; Banks, R. E. Ed.; Ellis Horwood Ltd.; London, 1979.
2. Selve, C.; Schiiefu, S. *J. Chem. Soc., Chem. Commun.* **1990,** 911 and references cited therein.
3. Kunitake, T. *Angew. Chem. Int. Ed. Engl.* **1992,** *31,* 709.
4. Brace, N. O. *J. Org. Chem.* **1962,** *27,* 3023, 3027, 4491.
5. McLoughlim, V. C. R.; Thrower, J. *Tetrahedron* **1969,** *25,* 5921.
6. Kobayashi, Y.; Kumadaki, I. *Tetrahedon Lett.* **1969,** 4095.
7. Sawada, H.; Mitani, M. et al. *Polym. Commun.* **1990,** *31,* 63.
8. Thrower, J.; Hewins, M. A. H. *Royal Aircaaft Establishment Technical Report 70056,* April 1970.
9. Matsuura, T.; Ishizawa, M. et al. *Polymer Preprints, Japan* **1988,** *38,* 434.
10. Labadie, J. W.; Hedrick, J. L. *Macromolecules* **1990,** *23,* 5371.
11. Sawada, H.; Mitani, M. et al. *Kobunshi Ronbunshu* **1992,** *49,* 141.
12. Sawada, H.; Gong, Y.-F. et al. *J. Chem. Soc., Chem. Commun.* **1992,** 537.
13. Sawada, H.; Minoshima, Y.; Nakajima, H. *J. Fluorine Chem.* **1993,** *65,* 169.
14. Sawada, H.; Minoshima, Y. et al. *J. Jpn. Oil. Chem. Soc.* **1992,** *41,* 649.
15. Sawada, H.; Gong, Y.-F. et al. *Chem. Lett.* **1992,** 531.
16. Sawada, H.; Komoto, K. et al. *Kobunshi Ronbunshu* **1993,** *50,* 983.
17. Sawada, H.; Sumino, E. et al. *J. Chem. Soc., Chem. Commun.* **1994,** 143.

FLUOROCARBON ELASTOMERS (Commercial)

Werner M. Grootaert
Specialty Fluoropolymers Department
3M Company

Fluorocarbon elastomers are fluorine containing, crosslinked amorphous polymers with a carbon-carbon backbone. Their presence in the marketplace today stems from a military interest in the early 1950s for elastomers with excellent thermal and chemical resistance.

Today, a variety of fluorocarbon elastomers are available from several suppliers. Major manufacturers of these materials are 3M (**Fluorel**™), DuPont (**Viton**™), Ausimont (**Tecnoflon**™), Daikin (**Dai-el**™), and Asahi Glass (**Aflas**™).

Fluorocarbon elastomers are available with or without incorporated cure systems. They are used in a wide variety of applications that demand both chemical and high temperature resistance, and are used, for example, in the aerospace, military, automotive, and chemical processing industries.

This article deals with the manufacture, cure chemistry, and physical properties of various classes of fluorocarbon elastomers.

MANUFACTURE OF FLUOROCARBON ELASTOMERS GUMS

The vast majority of fluoroelastomer gums are made by free-radical, aqueous-emulsion polymerization.[1-5] Because most of the monomers used are gaseous under ambient conditions, the polymerization is conducted under pressure. The two most common manufacturing processes are semi-batch and continuous emulsion polymerization.

DIFFERENT TYPES OF FLUOROCARBON ELASTOMERS

By far the most abundant fluorocarbon elastomer in the industry today is a copolymer of about 78 mole% vinylidene fluoride (VDF) and 22 mole% hexafluoropropylene (HFP) (CAS 9011-17-0). Its glass transition temperature, measured by DSC, is around −18°C.

Other widely used fluoroelastomers are copolymers of vinylidene fluoride, hexafluoropropylene, and tetrafluoroethylene (TFE) (CAS 25190-89-1).[6] Their composition can vary over a broad range, but the HFP content remains usually around 15-20 mole% in order to assure their amorphous nature.

These types of fluoroelastomers are commonly cured by a nucleophilic cure system. In some cases small amounts of cure site monomers are incorporated in the polymer backbone in order to allow a peroxide cure. They are typically iodine- or bromine-containing fluoro olefins, such as 1-bromo-2,2-difluoroethylene or 4-bromo-3,3,4,4,-tetrafluorobutene-1.[7-14]

Specialty fluoroelastomers include the copolymers of vinylidene fluoride and perfluoroalkyl vinyl ethers (e.g., perfluoromethyl vinyl ether) with or without tetrafluoroethylene.[15-22] The replacement of hexafluoropropylene with perfluoroalkyl vinyl ethers results in polymers with improved low-temperature performance.

Another important fluoroelastomer is the copolymer of tetrafluoroethylene and propylene (P) (CAS 27029-05-06).[23-26]

Perfluoroelastomers are another important class of specialty fluorocarbon elastomers. These polymers are copolymers of tetrafluoroethylene, perfluoroalkyl vinyl ethers, and a small amount of a cure site monomer.

CURE CHEMISTRY OF FLUOROELASTOMERS

The cure chemistry of the amorphous copolymers of vinylidene fluoride and hexafluoropropylene has been the subject of several studies.[27-34]

PROPERTIES OF FLUOROCARBON ELASTOMERS

When fully cured and postcured, fluorocarbon elastomers typically exhibit tensile strengths of 10-20 Mpa, with an elongation at break of 100-500%. The hardness ranges between 50-95 Shore A depending on the type and amount of filler used. Fluorocarbon elastomers exhibit excellent thermal stability.

REFERENCES

1. Rexford, D. R. U.S. Patent 3 051 677, 1962.
2. Honn, F. H.; Hoyt, S. M. U.S. Patent 3 845 024, 1962.
3. Weaver, S. D. U.S. Patent 3 845 024, 1974.
4. Logothetis, A. L. *Prog. Polym. Sci.* **1989,** *14,* 251-296.
5. Grootaert, W. M.; Millet, G. H; Worm, A. T. *Kirk-Othmer Encyclopedia of Chemical Technology Fourth Ed.*; John Wiley & Sons, Inc. **1993,** *8,* pp 990-1005.
6. Pailthrop, J. P.; Schroeder, H. E. U.S. Patent 2 968 649, 1961.
7. Apotheker, D.; Krusic, P. J. U.S. Patent 4 035 565, 1977.
8. Yamabe, M. et al. U.S. Patent 4 418 186, 1983.
9. Barney, A. L.; Keller, W. J.; Van Gulick, N. M. *J. Polym. Sci. A-1* **1970,** *8,* 1091.
10. Barney, A. L.; Kalb, J. H.; Kahn, A. A. *Rubber Chem. Technol.* **1971,** *44,* 660.
11. Kalb, J. H.; Barney, A. L.; Kahn, A. A. *Am. Chem. Soc. Div. Polym. Chem.* **1972,** *13,* 490.
12. Ogintz, S. M. *Lubric. Eng.* **1978,** *34,* 327.
13. Tomoda, M.; Ueta, Y. U.S. Patent 4 251 399, 1981.
14. West, A. C. U.S. Patent 4 263 414, 1981.
15. Apotheker, D.; Finlay, J. B.; Krusic, P. J.; Logothetis, A. L. *Rubber Chem. Technol.* **1982,** *55,* 1004.
16. Apotheker, D.; Krusic, P. J. U.S. Patent 4 035 565, 1977.
17. Apotheker, D.; Krusic, P. J. U.S. Patent 4 214 060, 1980.
18. Finlay, J. B.; Hallenbeck, A.; MacLachlan, J. D. *J. Elastomers Plast.* **1978,** *10,* 3.
19. Yamabe, M.; Kojima, G.; Wachi, H.; Kodama, S. U.S. Patent 4 418 186, 1983.
20. Yamabe, M.; Kodama, S.; Kojima, G. U.S. Patent 4 619 983, 1986.
21. Yamabe, M.; Kodama, S.; Kojima, G. U.S. Patent 4 368 308, 1983.
22. Yamabe, M.; Kodama, S.; Kojima, G. U.S. Patent 4 654 394, 1987.
23. Yabata, Y.; Ishiguri, K.; Sobue, H. *J. Polym. Sci. A-2* **1964,** 2235.
24. Kojima, G.; Tabata, Y. *J. Macromol. Sci. Chem.* **1971,** *A5(6),* 1087.
25. Ishigure, K.; Tabata, Y.; Oshima, K. *Macromolecules* **1973,** *6,* 584.
26. Kojima, G.; Tabata, Y. *J. Macromol. Sci. Chem.* **1972,** *A6(3),* 417.
27. Schmiegel, W. W. *Kautsch. Gummikunst.* **1971,** *31,* 137.
28. Schmiegel, W. W. *Angew. Makromolek. Chem.* **1979,** *76/77,* 39.

29. Venkateswarlu, P. et al. Presented at the 136th ACS Rubber Division Meeting, Detroit, MI, October 1989, p 123.
30. Ardella, V. et al. Presented at the 140th ACS Rubber Division Meeting, Detroit, MI, October 1991, p 57.
31. Smith, J. F. *Rubber World* **1960,** *142,* 103.
32. Smith, T. L. *J. Polym. Sci. A-2* **1972,** *10,* 133.
33. Apotheker, D.; Finlay, J. B.; Krusic, P. J.; Logothetis, A. *Rubber Chem. Technol.* **1982,** *55,* 1004.
34. Schmiegel, W. W.; Logothetis, A. *Polymers for Fibers and Elastomers*; ACS Symposium Series 260; American Chemical Society: Washington, DC, **1984,** Vol. 10, p 159.

Fluoroelastomers

See: Fluorocarbon Elastomers (Commercial)
Living Radical Polymerization, Iodine Transfer
Tetrafluoroethylene Copolymers (Overview)

FLUOROPOLYMER COATINGS (New Developments)

John Scheirs
ExcelPlas Australia

The ongoing demand for more durable surface coatings has encouraged the continued development of fluoropolymer coatings. Fluoropolymers have exceptional properties, such as outstanding outdoor durability (up to 30 years), excellent chemical resistance, and low surface energy.[1] Significant quantities of fluoropolymers are presently used for the coating of facades of large buildings and structures as well as for the coil coating of sheet metal.[1] New markets are growing as new uses for fluoropolymers are discovered: automotive coatings, antigraffiti paints, anticorrosion coatings, specialty coatings to upgrade the service life of thermoplastics, and the preservation of historic facades.[2-10] In civil engineering and in architecture the use of durable, maintenance-free coatings based on fluoropolymers are becoming increasingly important because they retain their aesthetic and protective qualities for decades. As top coats, fluoropolymers are finding an increasing application as weather-resistance coatings for other materials whereby a UV absorber in the fluoropolymer screens out damaging UV rays and thus protects the more sensitive substrate.

Improvements in polymer synthesis and molecular design have enabled the development of coatings which overcome the solvency and high-temperature curing, problems that exist with conventional fluoropolymers. The chain extension of a fluorinated backbone makes it possible to tailor the structure of the coating for a specific application and to achieve desired end-use properties.[1]

VDF-BASED COATINGS

Fluoropolymer coatings based on vinylidene fluoride (VDF) are extensively used as surface coatings on architectural aluminium cladding as well as on postformed steel sheets (also known as metal coil). They have gained widespread acceptance in the building industry because of their low maintenance properties, such as outstanding retention of color and gloss and their self-cleaning characteristics (such as their ability to repel dust).[1]

Over the past decades the main fluoropolymer that has been successfully exploited in the coatings field is poly(vinylidene fluoride) (PVDF). The outstanding durability of architectural PVDF coatings allows designers to add on to a building, up to 20 years later, without objectionable color variance in the facade. PVDF coatings resistance to weathering as well as atmospheric pollutants and acid rain has led to their endorsement by the American Aluminium Manufacturers Association (AAMA).[12] The major advantages of PVDF coatings are their outstanding gloss retention and their ability to resist chalking with outdoor exposure.[13] Furthermore, PVDF coatings have excellent tensile strength and elongation, which allows them to be bent and roll-formed without damage to the coatings or loss of protection of the metallic substrate.[14] This property is particularly important in the metal coil-coating industry.

The major shortcoming of PVDF-based coatings is that they are low in gloss.[15] A further disadvantage is that conventional PVDF coatings cannot be cured at ambient temperatures, thus the object being coated is restricted by the size of the curing oven available. Other limitations of PVDF include poor solubility in most solvents, poor pigment wettability, propensity for pinhole formation, inferior marring resistance compared with polyester coatings, and a high content (about 60 wt%) of organic solvents (usually a mixture of isophorone and dimethyl phthalate.[11,16]

A number of coating systems based on copolymers of PVDF have been reported. By copolymerizing VDF with fluorinated comonomers it is possible to reduce the crystallinity of the polymer and thereby increase its solubility in solvents. The most common comonomers polymerizable with VDF are hexafluoropropylene (HFP), tetrafluoroethylene (TFE), and chlorotrifluoroethylene (CTFE).[18,19] In addition to copolymers, terpolymers of PVDF have also been developed by Atochem (Kynar™ ADS), Dainippon Ink and Chemical Co., and Ausimont (Fluorobase™ T).[17,20,21]

The newly developed Fluorobase T system is quite unique in that it is a waterborne fluoro-terpolymer comprised of VDF, HFP, and TFE.[21] The polymer is cured using amine-based curing agents to produce a coating with a rubber-like surface and possess exceptional chemical resistance. The high fluorine content of Fluorobase T (68 wt%) imparts high chemical resistance and thermooxidative stability.[21]

VDF terpolymers also have up to double the gloss of PVDF-based systems. This is especially advantageous in the powder-coating field where the coating is applied as a powder and sintered by heat.

TFE-BASED COATINGS

The most well-known coating based on tetrafluoraethylene (TFE) is PTFE or Teflon™. Its use in the coatings field has, however, generally been restricted to nonstick saucepans and the waterproofing of textiles.[22] This is due primarily to the high rigidity of the main chain in PTFE, which limits processability, solubility, and miscibility. In order to preserve its excellent chemical and environmental resistance while inducing backbone flexibility, propylene segments can be inserted into the molecular chain of PTFE to give poly(tetrafluorethylene propylene) (PTFEP). This TFE-propylene alternating copolymer contains a low concentration of glycidyl vinyl ether. It has excellent chemical resistance and can withstand continuous service at 200°C.[23]

VINYL ETHER AND VINYL ESTER-BASED COATINGS

The first such fluoropolymer to be commercialized and the most widely recognized is Lumiflon™, which is developed by Asahi Glass Co. Lumiflon is a soluble, room-temperature curing fluoropolymer resin with properties that are ideal for high-performance coatings,[15,24] known chemically as poly(fluoroethylene-vinyl ether) (PFEVE). It consists of fluoroolefin units, [(namely chlorotrifluoroethylene (CTFE) or tetrafluoroethylene)], various vinyl ether units, and pendant polar groups (e.g., hydroxyl groups). The polymer chain of PFEVE-type polymers is comprised of alternating vinyl-based units and fluoroolefin groups.[15,24]

PFEVE is an amorphous copolymer that is readily soluble in aromatic solvents, such as xylene, or esters, such as butyl acetate, and curable at ambient temperature with isocyanate-based hardeners. Its unique properties include high gloss, excellent chemical resistance, excellent outdoor weathering performance, and a good adhesion to a variety of substrates. PFEVE has been patented extensively by Asahi Glass.[25-27] Modified PFEVE (known as Cefral) with vinyl ester in place of vinyl ether has been patented by Central Glass Co.[28] PFEVE can be used as a weather-resistant top coat for automobiles, as well as for coating hardwood, masonry, engineering polymers, and fiber glass plastic sheets.[27-31]

A distinct advantage of PFEVE resins is that they can be cured at room temperature, which makes them useful for restoration of existing coatings on large buildings as applying a heat source to the coating *in situ* in order to effect cure would pose obvious problems.[32] Furthermore, PFEVE coatings can be used to coat heat-sensitive materials such as plastics, and in doing so can upgrade their durability.[15] PFEVE coatings also exhibit high gloss (double that of PVDF coatings), are very compatible with the majority of commercial hardeners, and provide good interfacial adhesion to already cured PFEVE finishes, enabling them to be used for recoating and overcoating.[31,32]

PFPE-BASED COATINGS

Perfluorinated polyether (PFPE) (Fluorobase Z™ 1000) polymers are thermosetting resins that can be crosslinked at room temperature with isocyanates. Fluorobase Z 1000 is obtained by chain extension of a copolymer consisting of a random distribution of perfluorooxyethylene and perfluoroxymethylene.[2,33] Backbone flexibility is imparted by the ether linkages, which possess very high chemical stability compared to ether bonds in hydrogenated polymers because each ether unit is flanked by two-CF_2-groups.[2] Advantages of these new resins include high gloss (double that of PVDF), retention of gloss and color on aging even under severe environmental conditions, good adhesion, and good hardness. Other unique properties of this polymer are transparency to UV light, a low refractive index, and very low surface energy (18 dyne/cm).[2] The high concentration of perfluoropolyether blocks in the coating improves properties like cleanability and surface lubrication effects.[2]

KEY PROPERTIES OF FLUOROPOLYMER COATINGS

Gloss

The gloss value of a fluorinated coating is dependent on the mode of film formation, the type of crosslinker used, the intrinsic chemistry of the polymer, and the surface topography of the coating. Gloss is the main criteria by which the weathering of polymer coatings is assessed.

Fluorine Content

The key performance properties of fluoropolymer coatings are their resistance to weathering and chemical attack. The outdoor durability and chemical resistance of a fluorinated coating is directly related to its fluorine content.[2]

It is not only the concentration of fluorine but also the position at which it is incorporated in the polymer chain that is important when considering the environmental stability of polymer coatings. The fluorine must be contained in the polymer main chain in order to improve durability.

Tensile Properties

Tensile strength and elongation are important properties of a fluoropolymer coating because they allow coated metal panels to be bent and post formed without the coating cracking or loss of protection of the metallic substrate.[14]

Surface Energy

Nonstick fluorinated coatings have found broad usage in applications such as stain-resistant fabrics, nonfouling release surfaces for ship hills, and peelable substrates for adhesive labels.

It was recently discovered that a low-energy surface is best achieved by closely packing and orienting perfluoroalkyl groups.[34] A number of companies have been focusing their research efforts in this direction by producing coatings based on perfluorinated prepolymers.

REFERENCES

1. Scheirs, J.; Burks, S.; Locaspi, A. *Trends Polym. Sci.* **1995,** *3,* 74.
2. Lenti, D.; Locaspi, A.; Tonelli, C. *Abstracts of Papers, Proceedings of 25° ISATA Jubilee Conference*; Florence, paper #920664, 1992.
3. Schmidt, D. L.; Doburn, C. E.; McCrakin, P. J. U.S. Patent 5 006 624, 1991.
4. Schmidt, D. L.; Coburn, C. E.; DeKoven, B. M.; Potter, G. E.; Meyers, G. F.; Fischer, D. A. *Nature* **1994** 368, 39.
5. Daikin Co. Japanese Patent 02 298 544, 1991.
6. Anon, *Polym. Paint Col. J.* **1992,** 182, 541.
7. Strassel, A. *Kunststoffe* **1988,** *78,* 9.
8. Ellison, T. M.; Shimanski, M. A. *ANTEC '90* **1990,** 788.
9. Moggi, G. *Proceedings of XVth International Conference in Org. Coat. Sci. and Tech.*, Athens **1989,** p 283.
10. Moggi, G. *Abstracts of Papers, Proceedings of XVIth International Conference in Org. Coat. Sci. and Tech.*, Athens **1990,** p 251.
11. Norwood, D. S. D. *Pigment Resin Tech.* March, 1987, p 4.
12. Chrisman, N. *Abstracts of Papers, Proceedings of FINISHING '85 Conference*, Detroit, 1985, paper #8.

13. Kondo, T.; Fujishiro, H.; Kokubo, M. In *Durability of Building Materials and Components*; Chapman and Hall: New York, 1993; Vol. 6.
14. McCann, J. *Surface Coatings Aust.* **1990,** *27,* 8.
15. Symes, W. R.; Conti-Ramsden, J. H. *Paint Resin* October 20, 1986.
16. Munekata, S. *Prog. Org. Coat.* **1988,** *16,* 113.
17. Tournut, C.; Kappler, P.; Perillon, J. L. *Abstracts of Papers, Proceedings of the 2nd International Conference Fluorine in Coatings*; Salford, paper #11 1994.
18. Daikin Co. European Patent 435 330, 1993.
19. Nippon Oils and Fats Japanese Patent 61 238 863, 1987.
20. Mitani, T.; Hihata, I. U.S. Patent 4 640 966 (to Dainippin Ink and Chemicals) 1987.
21. Poggio, T.; Lenti, D.; Masini, L. *Abstracts of Papers, Proceedings of the 2nd International Conference Fluorine in Coatings*; Salford, paper #24 1994.
22. Miller, W. A. *Chem. Eng.*; New York, April 1993, 163.
23. Kojima, G.; Yamabe, M.; Munekata, S.; Abe, T. *Abstracts of Papers, Proceedings of XIth Int. Conference in Org. Coat. Sci. and Tech.*; Athens, **1985,** p 113.
24. Khan, A. K.; Saxena, M. S.; Chandra, S. *Pigment and Resin Tech.* **1992,** 4.
25. Asahi Glass Co. European Patent 312 834, 1989.
26. Asahi Glass Co. Japanese Patent 02 294 309, 1991.
27. Asahi Glass Co. Japanese Patent 04 018 473, 1992.
28. Central Glass Co. Japanese Patent 04 039 311, 1992.
29. Motohashi, K.; Deno, T. In *Durability of Building Materials and Components*; Nagataki, T. et al. Eds.; New York, 1987; Vol. 4, 121.
30. Motohashi, K.; Miyazaki, N. In *Durability of Building Materials and Components*; Nagataki, T. et al. Eds.; New York, 1987; Vol. 4, 317.
31. Yamabe, M.; Higaki, H.; Kojima, G. In *Organic Coatings-Science and Technology*; Parfitt, G. D.; Patsis, A. V. Eds.; Marcel Dekker: New York, 1984, p 25.
32. Fluorobase™ C Technical Brochure, Ausimont USA, 1992.
33. Sianesi, D.; Pasetti, A.; Fontanelli, R.; Bernardi, G. C.; Caporiccio, G. *Chimica E L'Industri* **1973,** *55,* 208.
34. Brady, R. F. *Nature* **1994,** 368, 16.

FLUOROPOLYMERS (Preparation by Aromatic Fluoroalkylation)

Hideo Sawada
Department of Chemistry
Nara National College of Technology

We have been studying the reaction behaviors of perfluoroalkanoyl and perfluoro-oxaalkanoyl peroxides [$R_FC(=0)OOC(=0)R_F$- = $F(CF_2)_n$-, $C_3F_7O[CF(CF_3)CF_2O]_mCF(CF_3)$-, n = 1, 3, 6, 7; m = 0, 1, 2, 3],[1] and found that these peroxides are useful reagents for direct introduction of fluoroalkyl groups into aromatic and heteroaromatic compounds such as benzene, toluene, thiophene, furan, and pyrrole via a single electron transfer reaction from the HOMO of substrate to the LUMO of O-O bond of the peroxide as shown in **Scheme I.**[2]

PROPERTIES

We measured glass transition temperatures: T_g (by DSC) and the refractive indices (n_D) of a series of perfluoroalkylated polystyrenes (PSt-R_F) obtained by using perfluoroalkanoyl peroxides. T_g of the PSt-R_F was raised from 89°C to about 106°C by perfluoroalkylation. Moreover, n_D of the PSt-R_F was clarified to depress drastically compared with that of the parent polystyrenc. For example, the n_D of the PSt-C_3F_7 (R_F ratio = 69%) is close to that of polytetrafluoroethylene (1.35).[3]

In general, perfluoroalkanoic and perfluoroalkanesulfonic acids, and perfluoroalkyl acrylate polymers possessing long-chain alkyl groups [$F(CF_2)n$-; n>5], have been widely used as fluorinated surfactants. However, these straight perfluoroalkyl chains and their derivatives are known to be hard and inflexible and to give low solubilities in various solvents. Therefore, it is interesting to develop novel surfactants bearing other fluoroalkyl groups such as perfluoro-oxaalkyl. We measured the critical surface tensions of perfluoro-oxaalkylated polystyrenes in addition to those of perfluoroalkylated ones for comparison.[4-6]

Critical surface tensions (γ_c) of the perfluoro-oxaalkylated polystyrene films are diminished dramatically in comparison with those of perfluoroalkylated polystyrenes. The γ_c of both perfluoro-oxaalkylated and perfluoroalkylated polystyrene films decreased with an increase in the R_F ratio, and remained constant above a specific R_F ratio value.

REFERENCES

1. a) Sawada, H. *J. Fluorine Chem.* **1993,** *61,* 253; Sawada, H. *Reviews on Heteroatom Chemistry*; Oae, S., Ed.; Myu: Tokyo 8, 205, 1993; Sawada, H.; Matsumoto, T.; Nakayama, M. *Yuki Gosei Kagaku Kyokaishi* **1992,** *50,* 592.
2. a) Yoshida, M.; Amemiya, H. et al. *J. Chem. Soc., Chem. Commun.* **1985,** 234; Sawada, H.; Yoshida, M. et al. *Bull. Chem. Soc. Jpn.* **1986,** *59,* 215.
3. Sawada, H.; Mitani, M. et al. *Polymer Commun.* **1990,** *31,* 63.
4. Ogino, K.; Abe, M. et al. *J. Jpn. Oil Chem. Soc.* **1991,** *40,* 1115.
5. Abe, M.; Morikawa, K. et al. *J. Jpn. Soc. Colour Mater.* **1992,** *65,* 475.
6. Morikawa, K.; Abe, M. et al. *J. Jpn. Soc. Colour Mater.* **1992,** *65,* 612.

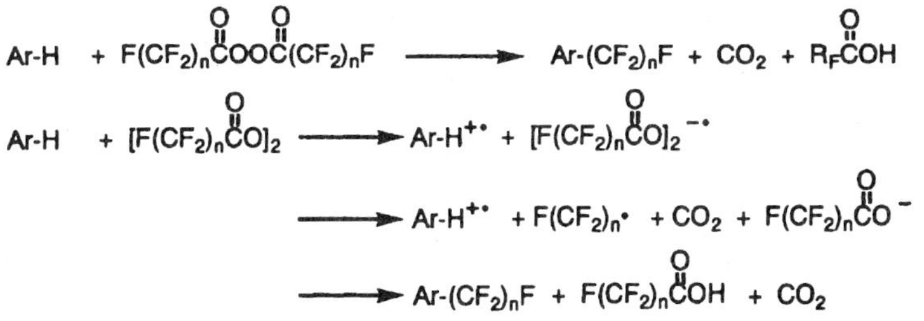

I

FLUOROPOLYMERS (Preparation by Fluorination of Polymers)

Sukumar Maiti, Partha S. Das, and Basudam Adhikari
Materials Science Center
Indian Institute of Technology Kharagpur

Fluoropolymers find numerous applications due to their outstanding chemical and solvent resistance, excellent electrical insulation, weather resistance, and very desirable thermal, mechanical, and non-stick properties.[1] In general, fluoropolymers are made by conventional polymerization of fluoromonomers, such as tetrafluoroethylene, vinyl fluoride, and so on.[1] However, the manufacture of fluoromonomers and their subsequent polymerizations are hazardous and difficult. The fabrication of fluoropolymers is also difficult and expensive. Attempts have, therefore, been made for alternative routes to obtain fluoropolymers.

METHODS OF FLUORINATION

Fluorination of polymers has been shown to be a successful new route to fluoropolymers. Polymers are fluorinated either by direct or indirect way. In direct fluorination, highly active fluorinating agents such as fluorine, hydrogen fluoride, or sulfur tetrafluoride convert the polymeric material completely to a fluorocarbon polymer.

Direct Fluorination

By Fluorine

Fluorine is a highly active fluorinating agent due to its low dissociation energy. It forms extremely stable bonds with carbon.[2]

Fluorination of polymers by fluorine may be divided into types: bulk fluorination and surface fluorination.

Bulk fluorination. *Lagow and Margrave*[3-5] reported the 'LaMar' process for direct fluorination of various hydrocarbon polymers. Polyethylene (PE), polypropylene (PP), polystyrene (PS), polyacrylonitrile (PAN), polyacrylamide (PAA), PF-resin, polyisobutylene, poly(*p*-xylylene) and ethylenepropylene copolymers have been successfully fluorinated to perfluoropolymers.

Different grades of fluorinated polymers may be obtained from a single raw material by controlling the extent of fluorination, or by regulating the density and structure of the starting hydrocarbon polymer. Another physical advantage in the LaMar process is the surface fluorination of solid objects to a depth of at least 0.2 mm.

Surface fluorination. The propagation of fluorination into bulk of PE and other hydrocarbon polymers is diffusion controlled and requires greater pressure of F_2 as the depth of fluorination increases.[6-8] High concentration of F_2 promotes crosslinking and degradation during fluorination. Hence, the surface fluorination of polymers in the form of fabricated articles is convenient and attractive.[9-12]

The commercial fluorination process of polymers by F_2 is limited mainly due to the following drawbacks.

i) Fluorination process involves highly reactive, toxic, and hazardous F_2 gas, which requires special and usually costly reaction vessels.
ii) Difficulty in manufacturing, handling, and storage fluorine gas.
iii) Fluorine reacts with hydrocarbons vigorously with possibility of extensive degradation and fragmentation of starting materials.
iv) To control the fluorination process, fluorine gas is often diluted with a large quantity of costly inert gases (argon, helium, or nitrogen).
v) Fluorination of polymer is diffusion controlled.
vi) Simultaneous crosslinking, oxidation, hydration, and hydrocarbon contamination take place during fluorination.

Fluorination by Hydrogen Fluoride

Although HF is extensively used as an industrial fluorinating agent, it is unsuitable for fluorination of organic polymers.[2]

Fluorination by Sulfur Tetrafluoride

SF_4 is a selective agent for introducing fluorine to organic compounds having hydroxyl, carbonyl, and carboxyl groups. This idea was extended to polymers for making fluoropolymers. SF_4 replaces the hydroxyl groups of polyvinyl alcohol at low temperature. The extent of fluorination depends on the concentration of SF_4.[14,15]

Various carbonyl-containing polymers like poly(methyl vinyl ketone)s, unsaturated polyacrolein, and polymethacrolein were fluorinated by SF_4.[16-18]

Polymers containing carboxyl and anhydride groups also undergo fluorination with SF_4 to give fluoropolymers having an acid fluoride group.[19,20]

Indirect Fluorination of Polymers

In an effort to overcome the disadvantages of conventional fluorinating agents such as F_2, HF, or SF_4, nontoxic fluorocarbons, chlorofluorocarbons, and sulfur hexafluoride are used. These gases cannot be used directly as fluorinating agents. However, when exposed to high-energy environments such as plasma, glow discharge, or γ-radiation, they generate active fluorinating agents.[21]

PROPERTIES OF FLUORINATED POLYMERS

Completely fluorinated polymers obtained by fluorination with F_2 show similar physical properties to known structurally related fluorocarbon polymers obtained by polymerization of fluorocarbon monomers.[3]

A dense product is usually obtained from the LaMar process than from normal polymerization of fluoromonomer possibly due to extensive crosslinking. The latter is a loosely packed and flaky polymer.[3] Thermal stability of polymers was significantly improved due to fluorination. The fluorinated polymers show decreased solubility or complete insolubility in the solvents in which nonfluorinated polymers were soluble. These are due possibly to formation of fluorinated polymer, as well as crosslinking of the polymer during fluorination.

In most of the cases the bulk properties of the surface fluorinated polymers are unaltered and only the surface properties get affected.

The surface energy of the plastics is improved due to fluorination.[22] This results in improved polymer surface for good storage stability. Crosslinking of polymer chains during fluorination improves mechanical strength and thermal stability of the polymer.

APPLICATION OF FLUORINATED POLYMERS

Fluorinated plastic surface is impervious to most solvents and has good chemical, solvent and hydrological resistances.[4,9-12,22] Various plastic bottles, containers, and tanks are fluorinated to handle safely chemicals and solvents. Hence, the fluorinated plastic containers are found in use as containers for gasoline, paint, turpentine, motor oil, and varnish.[23,24]

Since fluorination of plastics reduces permeation as much as 98%, fluorinated plastics are successful used in the manufacture of plastic fuel tanks for the automobile industry.[25] It removes all the restrictions on fuel tank shape and at the same time achieves substantial weight savings.

PE coated razor blades are fluorinated to have a thin fluorinated layer which protects the metal surface.[26] Antistatic packing materials are made by partial fluorination of PE, polystyrene and polyurethane.[27] The fluorination of poly(methyl methacrylate) and polycarbonate provides an anti-reflecting layer on surface for plastic lens, optical fiber, and sheet applications.[6]

Polyester fabrics are fluorinated for improving dyeability and light fastness.[28] Low-temperature plasma fluorinated woven fabric of polyethylene terephthalate is used as a soiling resistant fabric.[29] These fabrics have good water and oil resistant properties.

Fluorinated polyacetylenes are useful as electrodes for a battery. These are also used in conductors and in solar cells.[13] To modify the surface for electrical wire insulation paper, textiles, and polyethylene are plasma treated with CF_4 under electrodeless discharge.[30,31]

REFERENCES

1. Grayson, M.; Ed. *Kirk-Othmer Encyclopedia of Chemical Technology Vol. 11, Ed. 3*; Wiley-Interscience: New York, 1980; pp 1-74.
2. Sheppard, W. A.; Sharts, C. M. *Organic Fluorine Chemistry* New York: Benzamin, 1969; pp 10-14, 53-66.
3. Lagow, R. J.; Margrave, J. L. *J. Polym. Sci. Polym. Lett. Edn.* **1974,** *12,* 177.
4. Lagow, R. J.; Margrave, J. L. *Chem. Eng. News* **1970,** *48,* 40.
5. Otsuka, A. J.; Lagow, R. J. *J. Fluorine Chem.* **1974,** *4,* 371.
6. Shimade, J.; Hoshino, M. *J. Appl. Polym. Sci., Polym. Chem. Edn.* **1979,** *18,* 157.
7. Clark, D. T.; Feast, W. J.; Musgrave, W. K. R.; Ritchie, I. *J. Polym. Sci., Polym. Chem. Edn.* **1975,** *13,* 857.
8. Anand, M.; Cohen, R. E.; Baddour, R. F. *Polymer* **1981,** *22,* 361.
9. Schonhorn, H.; Hansen, R. H. *J. Appl. Polym. Sci.* **1968,** *12,* 1231.
10. DAMW Associates, Fr. Patent, 2 243 207, 1975.
11. Margrave, J. L.; Lagow, R. J. U.S. Patent 3 758 450, 1973; *Chem. Abstract* **1974,** *80,* 60546.
12. Woytekand, A. J.; Gentilcore, J. F. A New Blow Molding Process to Reduce Solvent Permeation of Polyolefin Containers, Paper No. 13. Presented at *Advances in Blow Molding Conference* Rubber and Plastics Institute, London, December 6, 1977.
13. Asahi Chemical Ind. Co. Ltd., Jpn Kokai Tokkyo Koho, Jap. Patent, 58 219 202, 1983.
14. Bezsolitsen, V. P.; Gorbunov, B. N.; Nazarov, A. A.; Khardin, A. P.*Vysokomol. Soedin. Ser A* **1972,** *14,* 950; *Chem. Abstract* **1972,** *77,* 75710.
15. Bezsolitsen, V. P.; Gorbunov, B. N.; Khardin, A. P. *Khim Khim Technol.* **1988,** 3.
16. Protopopov, P. A.; Kuznetsov, A. A. *Vsb Funkts Org. Soedin Polym* **1975,** 153; *Chem. Abstract* **1976,** *85,* 124588.
17. Khardin, A. P.; Protopopov, P. A.; Kuznetsov, A. A. *Vysokomol Soedin, Ser B* **1976,** *18(7),* 502; *Chem. Abstract* **1976,** *85,* 109251.
18. Gorbunov, B. N.; Protopopov, P. A.; Khardin, A. P. *USSR* **1974,** 443, 879; *Chem. Abstract* **1975,** *82,* 98890.
19. Solomina, T. I. *Funkts Org. Soedin. Polym.* **1972,** 66; *Chem. Abstract* **1974,** *80,* 146598.
20. Berkova, G. A.; Gorbunov, B. N.; Nazarov, A. A. *Funkts Org. Soedin. Polym.* **1972,** 84; *Chem. Abstract* **1974,** *81,* 13950.
21. Strobel, M.; Corn, S.; Lyons, C. S.; Corba, G. A. *J. Polym. Sci., Part A Polym. Chem.* **1987,** *25,* 1295.
22. Oberhonnefield, R. M.; Lauterbach, B. M. *Kunststoffe* **1992,** 82(10), 978.
23. Gentilcore, J. F.; Trialo, M. A.; Waytek, A. *J. Plast. Eng.* **1978,** *34(9),* 40.
24. Frankfurt, A. G. *Gas Aktuell* **1986,** *32,* 17.
25. Kalichemie, A. G.; Hans Bockler-Alee 20, Hannover, Germany *Modern Plastics International* September 1989; p 19.
26. Dodd, C. G. U.S. Patent 4 330 576, 1982.
27. Milligam, D. J.; Roland, P. U.S. Patent 4 595 706, 1986.
28. Dixon, D. D. Ger. Patent, 2 247 615, 1973.
29. Akagi, T.; Sakamoto, I.; Yamaguchi, Y. Jpn Kokai Tokkyo, Jap. Patent, 62 104 965, 1987.
30. Manion, J. P.; Davies, D. J. U.S. Patent 3 740 256, 1973; *Chem. Abstract* **1973,** *79,* 55190.
31. Manion, J. P.; Davies, D. J. U.S. Patent 3 740 325, 1973; *Chem. Abstract* **1973,** *79,* 54321.

FLUOROPOLYMERS (Surface Modification by Eximer-Laser Irradiation)

Shunichi Kawanishi
Osaka Laboratory for Radiation Chemistry
Japan Atomic Energy Research Institute

Fluoropolymer materials have many excellent properties such as thermal stability and solvent resistance with poor adhesiveness and wettability because of the chemical inactiveness. Therefore, the improvement of the surface of the fluoropolymers while keeping the excellent properties is very important from a viewpoint of the enlargement of applications. At present, surface modifications of fluoropolymers (e.g., improvement of the adhesion and endowment with wettability) has been carried out by an alkaline metallic solution[1] and a plasma discharge,[2] but the properties endowed by these treatments could not be satisfied completely with these effects. Therefore, we have attempted to proceed efficiently the surface modification of typical fluoropolymers using an excimer laser cooperated with Kurabo Ltd.[3-5] and Gunze Ltd.[6]

We chose polytetrafluoroethylene (PTFE) and tetrafluoroethylene-perfluoroalkyl vinyl ether copolymer (PFA) as fluoropolymers. PTFE is a typical fluorine resin that has excellent heat resistance, chemical resistance, and electrical properties. However, because of the chemical inactiveness, PTFE has poor adhesive properties. Therefore, the improvement of the surface properties of PTFE is very important from a viewpoint of the enlargement of application, for example, in which PTFE is laminated with various materials. On the other hand, PFA has advantages of the molten processing with excellent properties the same as PTFE. Development of an effective method for the endowment with wettability has become a very important subject for the enlargement of applications such as medical-use.

In this study, we report the chemical reaction induced on the fluoropolymer, such as ethylene-tetrafluoroethylene copolymer (ETFE), and also report the effective improvement of adhesiveness and wettability for the surface of PTFE and PFA, respectively, using an intense UV-radiation from excimer lasers. These treatments are very convenient and the effects are better than ordinary engineering treatment, such as chemical etching.

CONCLUSION

The adhesive strength of PTFE film was enhanced remarkably by KrF-laser irradiation in air when a small amount of aromatic polymer was blended with PTFE. The enhancement of adhesive properties are attributable both to the chemical effect owing to the formation of the polar groups and the physical effect owing to the formation of the fine unevenness by the intense laser irradiation.

Tetrafluoroethylene-perfluoroalkyl vinyl ether copolymer was endowed with high wettability by ArF-laser irradiation in water-dissolved carbon monoxide. The enhancement of wettability was the formation of the polar groups with the release of fluorine atoms.

These results show that the surface of fluoropolymers can be conveniently modified by excimer laser, and the modified properties are more excellent than ordinary engineering treatments such as chemical etching.

REFERENCES

1. Nelson, E. R.; Kilduff, T. J.; Benderly, A. A. *Ind. Eng. Chem.* **1958**, *50*, 329.
2. Schonhorn, H.; Hansen, R. H. *J. Appl. Polym. Sci.* **1967**, *11*, 1161.
3. Nishii, M.; Sugimoto, S.; Shimizu, Y.; Suzuki, N.; Nagase, T.; Endo, M.; Eguchi, Y. *Chem. Lett.* **1992**, *1992*, 2089.
4. Nishii, M.; Sugimoto, S.; Shimizu, Y.; Suzuki, N.; Nagase, T.; Endo, M.; Eguchi, Y. *Chem. Lett.* **1993**, *1993*, 1063.
5. Nishii, M.; Sugimoto, S.; Shimizu, Y.; Suzuki, N.; Kawanishi, S.; Nagase, T.; Endo, M.; Eguchi, Y. *Polym. Mater. Sci. Eng.* **1993**, *68*, 165.
6. Okada, A.; Negishi, Y.; Shimuzu, Y.; Sugimoto, S.; Nishii, M.; Kawanishi, S. *Chem. Lett.* **1993**, *1993*, 1637.

FLUOROPOLYMERS (Trifluoromethylated PES and PEEK)

James H. Clark,* James Denness, and David Wails
Department of Chemistry
University of York

Since the commercialization of polytetrafluoroethylene (PTFE) more than 40 years ago, there has been considerable interest in the use of fluorine to modify the properties of polymers, especially in terms of improved chemical stability, low dielectric constant, high hydrophobicity and low surface energy. Difficulties in processing PTFE and related materials have been a major obstacle to their more widespread application. It would be desirable to combine the useful physical and chemical properties of fluoropolymers with the useful mechanical properties of polyaromatic thermoplastics, to wit the significant recent interest in fluorine-containing polyaromatics, although much of the work to date has been limited to the easily synthesized $C(CF_3)_2$-bridged polymers.[1-3] Other materials in this category include $(CF_2)_6$-bridged poly(aryl ether)s[4,5] and fluoroalkoxy ring-substituted polyimides.[6] These materials show many useful properties, including a low dielectric constant, good solubility and reduced water uptake and, in some cases, good thermal stability.

*Author to whom correspondence should be addressed.

PREPARATION

Several methods for synthesis of the target monomers were attempted (see **Structures 1-3**). The most efficient route was the reaction of $SOCl_2$ with trifluoromethyl aryl Grignard reagents followed by oxidation with CrO_3.[7]

$$2ArMgBr + SOCl_2 \longrightarrow ArSOAr \longrightarrow ArSO_2Ar$$

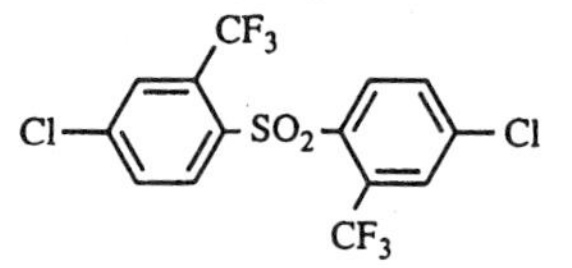

1

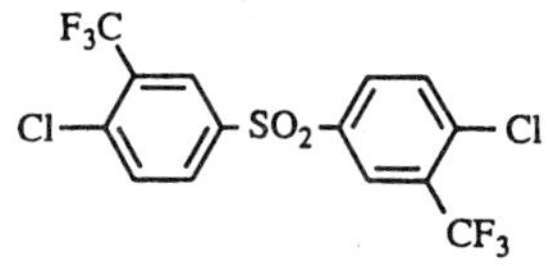

2

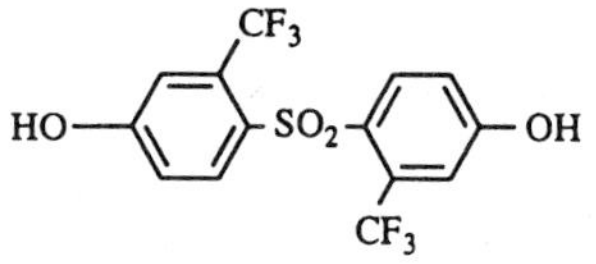

3

A range of trifluoromethylated copolymers was prepared by replacing some of the 4,4′-difluorodiphenylsulfone used in the standard PES preparation with either 1 or 2. A polymer containing CF_3 on every ring was made by reaction of Structure 1 with Structure 3.

CONCLUSIONS

The effects of trifluoromethylation of the properties of PES and PEEK polymers are marked and unexpected. Trifluoromethyl groups ortho to the ether bond in PES polymers are subject to a decomposition process during polymerization, leading to a reduction in molecular weight and brittle polymer films. Remarkably, trifluoromethyl groups ortho to the ether bond in PEEK polymers are stable to polymerization. High molecular weight, tough PES polymers possessing trifluoromethyl groups ortho to the sulfone can be made.

There is another striking contrast between the effects of trifluoromethylation on PES and PEEK polymers in the crystallinity of the materials, which is increased in one case and reduced in the other. Significantly, trifluoromethylation consistently reduces water update, suggesting that even with selective fluorination, material hydrophobicity can be improved.

REFERENCES

1. Cassidy, P. E. *Am. Chem. Soc. Div. Polym. Chem.* Polym. Prep. **1990,** *31,* 338.
2. Cassidy, P. E.; Aminabhavi, T. M.; Farley, J. M. *J. Macromd. Sci.-Rev. Macromol. Chem. Phys.* **1989,** C29, 338.
3. Farnham, A. G.; Johnson, R. N. U.S. Patent 3 332 909, 1967.
4. Labadie, J. W.; Hedrick, J. L. *Macromolecules* **1990,** *23,* 5371.
5. Labadie, J. W.; Hedrick, J. L. *Am. Chem. Soc. Div. Polym. Chem.* Polym. Prep. **1990,** *31,* 334.
6. Ichino, T.; Sasaki, S.; Matsura, T.; Nishi, S. *J. Polym. Sci.; Poly. Chem. Ed.* **1990,** *28,* 232.
7. Clark, J. H.; Denness, J. E.; Wynd, A. J. *J. Fluorine Chem.* in press.

FLUOROPOLYMERS, ACRYLATES AND EPOXIES (With Low Dielectric Constants)

Henry S.-W. Hu
Geo-Centers, Incorporated

James R. Griffith
Naval Research Laboratory

Materials exhibiting dielectric constants (DEs) below 3 are in increasing demand in the aerospace and electronic circuit industries.[1] DE is defined as a measure of the ability of a dielectric to store an electric charge. A dielectric is a nonconducting substance or an insulator. DE is directly proportional to the capacitance of a material, which means that the capacitance is reduced when the dielectric constant of a material is reduced. The DE of insulators may vary with changes in frequency for some commonly used polymers. For high-frequency, high-speed digital circuits, the capacitances of substrates and coatings are critical to the reliable functioning of the circuits.

With recent trends toward microminiaturization and utilization of very thin conductor lines, close spacings, and very thin insulation, greater demands are being placed on the insulating layer. Reductions in such parasitic capacitance can be achieved in a number of ways through the proper selection of materials and the design of circuit geometry.[4]

Polytetrafluoroethylene (PTFE), which is also known by DuPont's tradename "Teflon", is a solid at room temperature, and has a DE in the range of 2.00-2.08, while its monomer, tetrafluoroethylene, is a gas at room temperature.[2] PTFE is exceptionally chemically inert, has excellent electrical properties, has outstanding stability, and retains mechanical properties at high temperatures. The problem with PTFE is that even at high temperature it is soluble only in perfluorinated cycloalkanes such as $C_{14}F_{24}$ and slightly soluble in perfluorinated kerosene, and therefore has limited processability.[5] A family of commercial polymeric materials known as "Teflon AF" is believed to be a terpolymer of tetrafluoroethylene, hexafluoropropylene, and 2,2-bis(trifluoromethyl)-4,5-difluoro-1,3-dioxole (a derivative of hexafluoroacetone). It is reported to have a DE in the range of 1.89-1.93 and to be more processable than PTFE.[6] Teflon FEP has a dielectric constant of 2.1 and is a copolymer of tetrafluoroethylene and hexafluoropropylene,[7] while Teflon PFA has a DE of 2.06 and is a copolymer of tetrafluoroethylene and perfluoroalkyl vinyl ether.[8]

The highly polar nature of the C-F bond has been used to provide some high-performance characteristics in comparison with their hydrogen-containing or other halogen-containing analogues. Fluorine-containing epoxies or acrylics generally exhibit resistance to water penetration, chemical reaction, and environmental degradation; they also show unusual properties of surface tension, friction coefficient, optical clarity, refractive index, vapor transmission rate, and electromagnetic radiation resistance.

Recently we[9-12] reported the preparation of a series of processable heavily fluorinated acrylic and methacrylic homo- and copolymers which exhibit DEs as low as 2.06, very close to the minimum known values of 2.0-2.08 for Teflon and 1.89-1.93 for Teflon AF.[6]

This task represents a continuation of efforts to maximize the hydrophobicity of acrylic, epoxy, and other polymeric systems for resistance to water penetration and environmental degradation, and to minimize the DE and improve the processability for adhesives and coatings.

PREPARATION

Fluorinated Alcohols

One of the major problems in research of this type is that of preparing a series of hexafluoroacetone-based polyfunctional alcohols. This was carried out in multistep routes according to the reported procedures.[13-17]

Acrylic Resins

Monomers

The synthesis of acrylic and methacrylic esters (**Scheme I**) was carried out in a fluorocarbon solvent such as Freon® 113 by the reaction of the respective fluorinated alcohol with acryloyl chloride or methacryloyl chloride and an amine acid acceptor such as triethylamine.

Epoxy

Monomers

In **Scheme II**, it is shown that heavily fluorinated aromatic di- and triglycidyl ethers can be obtained in high yields by the standard procedures.

LOWER DEs

DEs of these materials can be further lowered by known means such as by incorporating air bubbles in the materials or by inhibiting crystallization. A difference of a couple of hundredths in the DE value may be important when one is at the low extremes thereof.

The amorphous Resnick Teflon AF terpolymer is reported with the lowest refractive index in the range of 1.29-1.31 and also with the lowest DE in the range of 1.89-1.93. The optical refractive index n and the DE are theoretically related by the Maxwell relation: DE = n^2.[18] In view of the good agreement between the refractive index and the DE in amorphous organic polymers, the opportunity to obtain DEs in the range of 1.89-2.06 by modifying our synthesis into higher fluorine content and ether linkage is high.

In summary, the factors which affect the DE and thermal stability are the fluorine content, the polymer type, and the molecular architecture. Lower DEs are obtained as fluorine contents on

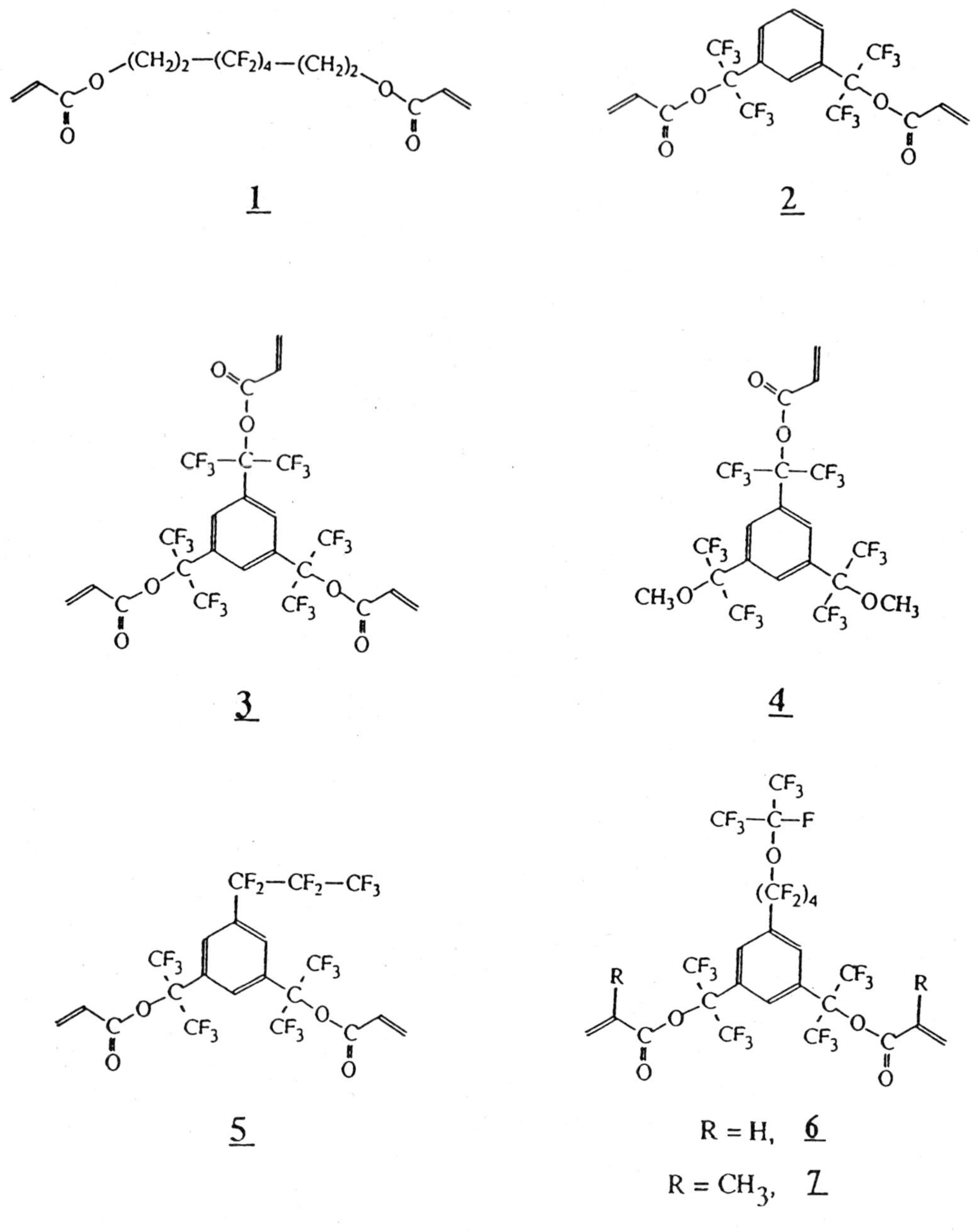

SCHEME I. Acrylic and methacrylic monomers.

the polymer backbone or side chain increase, when acrylate is replaced by methacrylate, when ether linkages are present, and when the aromatic structure is symmetrically *meta*-substituted.[19]

REFERENCES

1. Dorogy, W. E. Jr.; St. Clair, A. K. *Polym. Mater. Sci. Eng.* **1991,** *64,* 379.
2. Licari, J. J.; Hughes, L. A. *Handbook of Polymer Coating for Electronics, 2nd ed.* Noyes: Park Ridge, NJ, 1990; p 114.
3. Grove, A. S. *Physics and Technology of Semiconductor Devices*; John Wiley & Sons: New York, 1967; pp 254-255.
4. St. Clair, A. K.; St. Clair, T. L.; Winfree, W. P. *Polym. Mater. Sci. Eng.* **1988,** *59,* 28.
5. Tuminello, W. H.; Cavanaugh, R. J. *PCT Int. Appl. WO* 93 05,100; *Chem. Abstract* **1993,** 119, 161635b.
6. Resnick, P. R. *Polym. Prepr.* **1990,** *31(1),* 312.
7. Gangal, S. V. In *Kirk-Othmer Encyclopedia of Chemical Technology*; Wiley: New York, 1980; see p 26 for FEP.
8. Johnson, R. L. In *Kirk-Othmer Encyclopedia of Chemical Technology*; Wiley: New York, 1980; see p 45 for PFA.
9. Hu, H. S.-W.; Griffith, J. R. *Polym. Mater. Sci. Eng.* **1992,** *66,* 261.
10. Hu, H. S.-W.; Griffith, J. R. *Polym. Prepr.* **1993,** *34(1),* 401.
11. Hu, H. S.-W.; Griffith, J. R. In *Fluoropolymers: Syntheses and Properties*; Davidson, T., Ed.; Plenum: New York, in press.
12. Griffith, J. R.; Hu, H. S.-W. U.S. Patent 5 292 927, 1994.
13. Griffith, J. R.; O'Rear, J. G. In *Biomedical and Dental Applications of Polymers*; Gebelein, C. G.; Koblitz, F. F., Eds.; Plenum: New York, 1981; pp 373-377.
14. Griffith, J. R.; O'Rear, J. G. *Polym. Mater. Sci. Eng.* **1985,** *53,* 766.

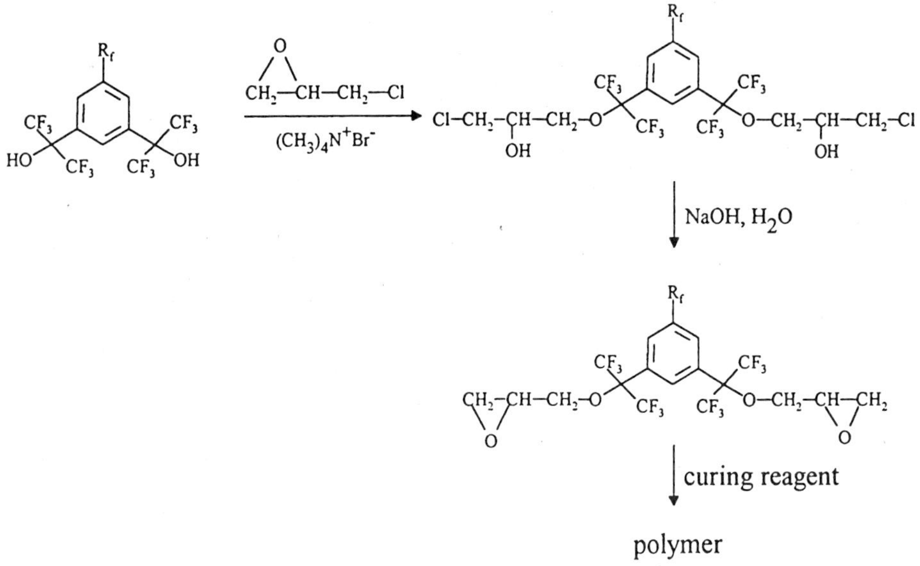

SCHEME II. Typical preparation of epoxy monomers and polymers.

15. O'Rear, J. G.; Griffith, J. R. *Org. Coatings Plastics Chem.* **1973**, *33*, 657.
16. Soulen, R. L.; Griffith, J. R. *J. Fluorine Chem.* **1989**, *44*, 195.
17. Soulen, R. L.; Griffith, J. R. *J. Fluorine Chem.* **1989**, *44*, 203.
18. Lin, L.; Bidstrup, S. A. *J. Appl. Polym. Sci.* **1994**, *54*, 553.
19. Hu, H. S.-W.; Griffith, J. R. In *Polymers for Microelectronics* ACS Symposium Series; Wilson, G. et al., Eds.; American Chemical Society: Washington, DC; 1994; Vol. 537, p 507.

FLUOROSILICON OLIGOMERS

Hideo Sawada
Department of Chemistry
Nara National College of Technology

Very recently we have found that a series of fluorine-containing organosilicon compounds are easily obtained under mild conditions by using fluoroalkanoyl peroxides. In this account, I describe the synthesis and properties of a series of organosilicon compounds containing perfluoroalkyl or perfluoro-oxaalkyl groups.

PREPARATION

A typical procedure for the preparation of fluorine-containing organosilicon compounds (fluorine-containing silicon oligomers obtained by the reaction of perfluorobutyryl peroxide and trimethoxyvinylsilane) is as follows.[1,2]

A solution containing perfluorobutyryl peroxide (117 mmol) and trimethoxyvinylsilane (234 mmol) in Freon® 113 ($CF_2ClCFCl_2$; 500 ml) was stirred at 30°C for 10 h under nitrogen. The mixture was then distilled under reduced pressure to give a colorless oligomer.

Both perfluoropropylated and perfluoro-oxaalkylated organosilicon oligomers were obtained in excellent to moderate yields under very mild conditions.

Furthermore, 3-methacryloxypropyltrimethoxysilane (MMA-Si) an 3-acryloxypropyltrimethoxysilane (AC-Si) were shown to react smoothly with the peroxides under mild conditions to give fluorine-containing silicon oligomers in good yields via a radical process.[3]

In contrast, allylsilanes such as trimethoxyallylsilane and trimethylallylsilane were found to react with fluoroalkanoyl peroxides to afford 1:1 adducts $[R_F\text{-}CH_2CH(OCOR_F)CH_2Si(R)_3]$ in good yields without oligomer formation.[2]

PROPERTIES

Organosilicon compounds containing perfluoroalkyl and perfluoro-oxaalkyl groups obtained by the reactions of VM-Si, MMA-Si, and AC-Si with fluoroalkanoyl peroxides are useful as surface-active materials.

Fluorosilicon co-oligomers, $R_F\text{-}[CH_2CHSi(OMe)_3]_x\text{-}[CH_2\text{-}CMeCO_2(CH_2CH_2O)_pH]_y\text{-}R_F$, are suggested to be useful as surface-active substances, since these co-oligomers possess a reactive group such as a methoxy group on a silicon atom. In fact, these co-oligomers were shown to have good surface activity and to be a new type of amphiphilic fluorosilane coupling agent.[4]

The surface properties of these fluoroalkylated silicon co-oligomers were evaluated by surface tension measurement of their aqueous and *m*-xylene solutions. These amphiphilic oligomers, $R_F\text{-}(CH_2CHCO_2H)_x\text{-}(CH_2CHSiMe_3)_y\text{-}R_F$, were found to decrease the surface tension of both water and *m*-xylene from 72.3 and 28.8 dyn/cm, respectively.

In the case of aqueous solutions of fluorinated silicon co-oligomers, the co-oligomers possessing longer perfluoro-oxaalkyl chains were not so effective in reducing the surface tension of water. However, shorter perfluoro-oxaalkyl chains were effective in reducing the surface tension of water to 20.6 dyn/cm. In contrast, the degree of reduction in surface tension of *m*-xylene depends on the length of the fluoroalkyl group in co-oligomers as well as the usual fluorinated surfactants. Thus, longer fluoroalkyl chains, especially, longer perfluoro-oxaalkyl chains, were found to reduce the surface tension of *m*-xylene more effectively. Further studies of these unique surface behavior's of fluorosilicon co-oligomers are in progress.

REFERENCES

1. Sawada, H.; Nakayama, M. *J. Chem. Soc., Chem. Commun.* **1991,** 677.
2. Sawada, H.; Gong, Y.-F.; Matsumoto, T.; Nakayama, M.; Kosugi, M.; Migita, T. *J. Jpn. Oil Chem. Soc.* **1991,** *40,* 730.
3. Sawada, H.; Minoshima, Y.; Matsumoto, T.; Nakayama, M.; Gong, Y.-F.; Kosugi, M.; Migita, T. *J. Fluorine Chem.* **1992,** *59,* 275.
4. Sawada, H.; Mitani, M.; Nishida, M.; Gong, Y.-F.; Kosugi, M.; Migita, T.; Kawase, T. *J. Jpn. Oil Chem. Soc.* **1994,** *43,* 65.

FLUOROSILICONES (Monomer Synthesis and Applications)

Norio Yoshino
Science University of Tokyo

Recently, dramatic developments have occurred in the utilization of organofluoro compounds as new high-performance materials. The syntheses of fluoroalkyl and fluoroaryl titanates[1] as well as fluoroalkyl silicates[2] have been reported. Although these alkoxides were expected to be stable against moisture from the general water repellency of the fluorinated carbon structure, all the products were very sensitive to water and readily decomposed. The syntheses and characterization of titanium butoxide tris(polyfluoroalkanoate)s[3] and dimethoxysilanediyl bis(polyfluoroalkanoate)s[4] have also been reported. The TiOR bond was hydrolyzable in the titanium compounds,[3] but the TiOCORf (Rf = polyfluoroalkyl group) bond was relatively stable in water. However, in the silicon compounds[4] the moieties of the SiOCORf and the SiOR were hydrolyzed with exactly the same ease, and these compounds were decomposed easily by atmospheric moisture. Thus, these silicon compounds could not be applied to surface modification.

In general, silane coupling agents to promote surface modification and/or adhesion are the most familiar example of the R_nSiX_{4-n} (n=1, 2, and 3) class of organosilane materials having two types of substituents, where R is a nonhydrolyzable organic group that may be relatively inert in one case, such as a hydrocarbon radical, or may be reactive to particular organic systems in another case. The X functionality is a hydrolyzable group; often an alkoxyl group.[5]

PROPERTIES

Surface Modification of Glass[6-9]

A few plates of glass were allowed to react with the fluorosilicones in 100 cm^3 1,1,2-trichloro-1,2,2-trifluoroethane (F-113) at various concentrations. The solvent, F-113, is a good solvent against these fluorosilicones. After the glass was treated thermally in an oven at 150°C for 2 h, the contact angles of water and oleic acid were measured against these glass samples. The contact angles on the modified glass surface depend on the fluorocarbon chain length. From the modification of the glass surface using fluorosilicones, it was found that the hydrophilic glass surface could be covered effectively by the longer fluorocarbon chains.

In conclusion, the density of fluorocarbon, which displayed water and oil repellency, was increased on the top surface of the modified glass with the longer fluorocarbon chain. The largest contact angles for water on the modified glass surface were 118°. This contact angle was larger than that of polytetrafluoroethylene (110°).[10] In regard to water and oil repellency on the modified glass surface, $CF_3(CF_2)_9CH_2CH_2Si(OCH_3)_3$ displayed the highest repellency.[6] The contact angles for water on the glass surface modified with $CF_3(CF_2)_9CH_2CH_2SiCH_3(OCH_3)_2$, which contains the same fluorocarbon chain length, were 110°.[7] The difference in contact angles may be caused by the structural difference of the silane coupling agents. The glass surface modified with $CF_3(CF_2)_9CH_2CH_2Si(OCH_3)_3$ has no Si-CH_3, which cannot increase the contact angles in comparison with the fluorocarbon chain, and the fluorocarbon chains tend to pack together to construct the siloxane networks. In other words, the top-surface modified with $CF_3(CF_2)_9CH_2CH_2SiCH_3(OCH_3)_2$ was more enriched with fluorocarbon chain than that modified with $CF_3(CF_2)_9CH_2CH_2SiCH_3(OCH_3)_2$.

Measurement of Oxidation Resistance against Modified Glass Surface[6-9]

From the ESCA measurements of the glass surface modified with $CF_3(CF_2)_9CH_2CH_2SiCH_3(OCH_3)_2$ and its oxidized surface, the following facts were observed.[7] The atomic ratios of O/Si: C/Si: F/Si: F/C = 1.97: 1.25: 0.87: 0.70 were obtained from the modified surface (non-oxidized) and the values of O/Si: C/Si: F/Si: F/C = 1.99: 0.75: 0.56: 0.74 were observed from the oxidized surface. This data showed that the atomic content of fluorine and carbon on the glass surface was reduced by the oxidation. From these results, two possibilities of the mechanism of the contact angle reduction can be discussed. One possibility is that the oxidation occurs in the neighborhood of hydrocarbon moiety, which exists in the inside-surface of the modified glass and is easily oxidized in comparison with the fluorocarbon chain. This is because, the C-C chain of the fluorocarbon is shielded by compactly arranged fluorine atoms, and additionally, C-F bond energy (484 kJ mol^{-1}) is larger than that of C-H (416 kJ mol^{-1}). The second possibility is that some of the Si-O-Si bondings formed by condensation reaction of silane coupling agents with the glass surface are hydrolyzed by the reaction with water in hot concentrated nitric acid, such as -Si-O-Si-Glass → -Si-OH + HO-Si-Glass.

After treatment of the modified glass in concentrated hydrochloric acid, which has no oxidation ability, in the same way as the treatment in nitric acid, the contact angles for water were not reduced at all. This result supports the idea that the oxidation of the hydrocarbon moiety in these fluorosilicones led to reduction of the contact angle. The high oxidation resistance of the modified surface treated with fluorosilicones having longer fluorocarbon chains must result from the oxidation-protection ability against the inside-surface by fluorocarbon chains which possess high C-F bond energy.

APPLICATIONS

Application for Dentistry[11]

Some dental materials were modified using $CF_3(CF_2)_9$ — $CH_2CH_2Si(OCH_3)_3$, and the contact angles of the modified surfaces were measured with water and oleic acid. On the basis of these results, the fluorosilicone was applied to the surface modification of a denture to provide stain-protecting ability.

It was found that a fluorosilicone, $CF_3(CF_2)_9CH_2CH_2Si(OCH_3)_3$, had a high degree of surface modification ability for these dental materials. The modified denture teeth displayed a heightened stain-protecting ability. The modified composite materials for dentistry also displayed a reduction of coloration and an increment of detachment ability of experimental plaque. It is expected that these results will be useful for the health of our oral cavity.

The metal surfaces modified with $CF_3(CF_2)_9CH_2CH_2Si(OCH_3)_3$, such as metals for general use (Fe, Al, Ni, Cu, SUS-316 and SUS-304) and the metals for dentistry (Au, Ti, Co-Cr alloy, and Au-Ag-Pd alloy) also displayed high water and oil repellence. These surfaces have high abrasion resistance and their thermal stabilities are > 250°C.

REFERENCES

1. Yoshino, N.; Tomita, J.; Hirai, H. *Bull. Chem. Soc. Jpn.* **1989,** *62,* 2208.
2. Yoshino, N.; Terajima, H.; Hirai, H. *Synth. React. Inorg. Met.-Org. Chem.* **1990,** *20(6),* 729.
3. Yoshino, N.; Shiraishi, Y.; Hirai, H. *Bull. Chem. Soc. Jpn.* **1991,** *64,* 1648.
4. Yoshino, N.; Tominaga, S.; Hirai, H. *Bull. Chem. Soc. Jpn.* **1991,** *64,* 2735.
5. Owen, M. J.; Williams, D. E. *J. Adhesion Sci. Technol.* **1991,** *5,* 307.
6. Yoshino, N.; Yamamoto, Y.; Hamano, K.; Kawase, T. *Bull. Chem. Soc. Jpn.* **1993,** *66,* 1754.
7. Yoshino, N.; Yamamoto, Y.; Seto, T.; Tominaga, S.; Kawase, T. *Bull. Chem. Soc. Jpn.* **1993,** *66,* 472.
8. Yoshino, N.; Nakaseko, H.; Yamamoto, Y. *Reactive Polymer* **1994,** *23,* 157.
9. Yoshino, N.; Sasaki, A.; Seto, T. *J. Fluorine Chem.* **1995,** *71,* 21.
10. Ishikawa, N. *Fussokagobutu no Gosei to Kino,* 1st ed., CMC Co. Ltd.: Tokyo, Japan, 1987.
11. Yoshino, N.; Yamamoto, Y.; Teranaka, T. *Chem. Lett.* **1993,** 821.

Foam Control Agents

See: Antifoaming Agents

FOAMED PLASTICS

Rudolph D. Deanin
Plastics Engineering Department
University of Massachusetts, Lowell

When tiny gas bubbles are dispersed in a solid polymer, the product is referred to as a foamed plastic, or a cellular plastic or plastic foam. U.S. production may be conservatively estimated at 4.7 billion pounds/year, or 6% of total plastics production, for the two leading materials, polyurethane and polystyrene; but most of the other commercial polymers are also foamed in some smaller, unreported amounts, so the total must be considerably higher.[1]

FOAM PROCESSES IN GENERAL

Foamed plastics may be produced by a great variety of processes, but most of them involve three basic steps: the liquid system, fine gas bubbles, and solidification. The liquid system may be produced by using reactive oligomers (for example, polyol + polyisocyanate), by melting thermoplastics (for example, polystyrene), by dispersing polymer in plasticizer [poly(vinyl chloride) (PVC) plastisol], or by using aqueous (rubber) latex. The gas bubbles may be produced by using permanent gases (air or nitrogen), by boiling volatile liquids (pentane), or by chemical reaction (isocyanate + water, or thermal decomposition of azo compounds). Solidification is accomplished by polymerization and crosslinking of thermosetting oligomers, by cooling of thermoplastic melts, by fusion and gelation of plastisols, or by coagulation of latexes.

GENERAL TYPES OF FOAMS

Most foams are blown to low densities, typically 0.01–0.2 g/cc. The gas bubbles may be open (interconnecting) cells, producing a continuous gas phase; or they may be closed (unicellular) cells that isolate the gas within each cell. Polystyrene and rigid polyurethane are typically unicellular, flexible polyurethane is open cell, and PVC plastisol is generally a mixture of open and closed cells. Density and open/closed cell structure are the major factors controlling foam properties.

UNIQUE PROPERTIES

Foamed plastics have many unique properties that distinguish them from solid polymers and are particularly useful in practical applications.[3] Perhaps most obvious is light weight, which produces economy in general and buoyancy in closed-cell foams. Mechanical properties are characterized by rigidity in closed-cell foams and sandwich structures, softness and flexibility in open-cell foams, and impact cushioning in both types. Open-cell foams provide acoustic absorption. Even though polymers themselves are thermal and electrical insulators, the gas phase in foams provides much more insulation than the polymer itself. And open-cell foams provide high permeability and sponge absorption/release behavior.

STRUCTURE AND PROPERTIES

The critical structural features of foamed plastics may be itemized as follows: polymer, gas, density, open/closed-cell ratio, cell size, anisotropy, and structural foam/sandwich structure.[3]

POLYURETHANES

Polyol + polyisocyanate is a stepwise addition polymerization reaction, with viscosity increasing rapidly as it proceeds from liquid to solid. If the blowing agent releases gas at the proper liquid viscosity, this produces polyurethane foams.[2,4-7] When the polyol is a high-molecular-weight triol (1000-2000), and the gas ruptures the walls of the bubbles before they solidify, flexible foams are produced. When the polyol is a low-molecular-weight high-functionality polyol (100-200), and the cell walls solidify before the gas can rupture them, rigid foams are produced. These are both low-density products. When polyol + polyisocyanate are mixed and injected into a closed mold with a smaller amount of blowing agent, the process is called reaction injection molding, and the resulting products are medium-to-high density moldings.

General Foam Ingredients

Polyols are either polyethers, which give the best flexibility, hydrolysis-resistance, and economics; or polyesters, which give the best strength. Polyisocyanates are generally toluene diisocyanate (TDI) for flexible foams, methylene diphenylisocyanate (MDI) for higher strength, or polymeric MDI (PMDI) for rigid foams. Polymerization catalysts are generally organotins and/or tertiary amines.

Flexible Foams

When high-molecular-weight polyoxypropylene triols are mixed with toluene diisocyanate, in the presence of catalyst, surfactant, and water, flexible polyurethane foams are produced, generally 1-6 pounds/cubic foot (PCF) density.

Rigid Foams

Rigid polyurethane foams are generally made from polymeric methylene diphenylisocyanate plus low-molecular-weight high-functionality polyols, typically tetrahydroxy polyethers with a molecular weight of ~500.

Reaction Injection Molding (RIM)

RIM requires little heat or pressure, and produces large strain-free parts rapidly. Adding moderate amounts of blowing agents produces structural foams with solid skins and foamed cores, thus combining the rigidity, strength, and surface finish of the skins with the low density and additional rigidity of the cores. This has been a major force in the growth of plastics in the auto industry.

Polyurethanes have proven to be the ideal materials for RIM. In addition to meeting the above requirements, they also provide all the high performance of polyurethanes in general, and also permit any desired degree of foaming. There are two types of polyurethane RIM: elastomer and structural foam.

Polyurethane RIM elastomers are particularly useful in auto bumpers, which must withstand an impact of 6 mph, and also in total auto facias.

POLYSTYRENE

Processes

Most polystyrene foam is produced by extrusion, melting the polymer, injecting pentane blowing agent, and extruding in the form of billets up to a 12" × 24" cross section, boards up to a 4" × 48" cross section, or hollow tubing, which is then slit to produce film and sheet.[2] A somewhat smaller amount is produced by swelling 5-8% pentane into 1-mm beads, heating them with steam to expand them to the desired final density ("prepuff"), and then molding the foamed beads at low pressure to bond them into the desired shape.

Properties

Polystyrene is generally a closed-cell foam that is rigid, water resistant, and excellent electrical insulation.

Applications

The major uses of polystyrene foam are in packaging, for thermal insulation, and for the protection of delicate products; it is also used in building, for thermal insulation.

Extruded foam is used mainly in residential construction (roofs, walls, foundations) and commercial buildings (refrigeration, freezers, livestock). Extruded sheet is used for food trays. Molded bead is used primarily for hot and cold drinking cups. Molded board is used for commercial insulation, flotation, toys, and novelties.

Structural Foam

Wood-like products (structural foam) are made by injection molding high-impact polystyrene containing sodium bicarbonate or an azodicarbonamide blowing agent.[8,9] Densities are typically 0.6-0.7, but can be as low as 0.3. The major applications of structural foam are in furniture, housewares, and toys.

POLYOLEFIN FOAMS

Processes

Low-density polyethylene is extruded into closed-cell foams by several processes.[2,10]

Properties

Foams range from soft and flexible at low density to semirigid at high density. Electrical insulation and water resistance are excellent.

Applications

The major uses of polyethylene foam are in the packaging of delicate products; construction (sealants, roof insulation, noise reduction); sports equipment; flotation devices; wire and cable insulation; pipe insulation; and automotive panels, padding, and insulation.

POLY(VINYL CHLORIDE)

Flexible (Plasticized) PVC

Emulsion-polymerized spray-dried PVC is stirred into liquid plasticizer to form a viscous liquid plasticol.[2,11-13] This can then be foamed, most commonly by thermal decomposition of azodicarbonamide, less often by whipping in air to form a froth. Fusion and cooling then produce flexible vinyl foams of medium to low density, containing a mixture of open and closed cells. Because air is less viscous than liquid plasticizer, it is actually an extremely efficient softener.

Whereas solid plasticized PVC is firm and fairly stiff, foaming gives it a soft, luxuriant feel like high-quality leather, which is widely used in upholstery, winter clothing, pocketbooks, and other accessories. Other major applications include cushioning under vinyl flooring and carpet backing, weather stripping, and gasketing in bottle caps.

Rigid PVC

Extrusion of molten PVC, containing azodicarbonamide (AZDN) or sodium bicarbonate blowing agent, produces medium-to-high density structural foam. Typical uses of rigid PVC foam are in pipe, tubing, and conduit, and in wood-like applications such as trim and molding.

OTHER POLYMERS

Foaming of most commercial polymers has been mentioned in the literature.[2,14-16]

REFERENCES

1. *Mod Plastics* **1995,** *72,* 63-67.
2. Berins, M. L. *SPI Plastics Engineering Handbook*; Van Nostrand Reinhold: New York, 1991; Chapter 19.
3. Deanin, R. D. In *Applied Polymer Science*; Tess, R. W.; Poehlein, G. W., Eds.; American Chemical Society: Washington, DC, 1985; Chapter 20.
4. Bailey, F. E., Jr. In *Handbook of Polymeric Foams and Foam Technology*; Klempner, D.; Frisch, K. C., Eds.; Hanser: Munich, 1991; Chapter 4.
5. Backus, J. K. In *Handbook of Polymeric Foams and Foam Technology*; Klempner, D.; Frisch, K. C., Eds.; Hanser: Munich, 1991; Chapter 5.
6. Ashida, K. In *Handbook of Polymeric Foams and Foam Technology*; Klempner, D.; Frisch, K. C., Eds.; Hanser: Munich, 1991; Chapter 6.
7. Tabor, R.; Turner, R. B. In *Handbook of Polymeric Foams and Foam Technology*; Klempner, D.; Frisch, K. C., Eds.; Hanser: Munich, 1991; Chapter 7.
8. Suh, K. W. In *Handbook of Polymeric Foams and Foam Technology*; Klempner, D.; Frisch, K. C., Eds.; Hanser: Munich, 1991; Chapter 8.
9. Howard, M. J. *Foams*; Cordura; **1980,** p 409.
10. Park, C. P. In *Handbook of Polymeric Foams and Foam Technology*; Klempner, D.; Frisch, K. C., Eds.; Hanser: Munich, 1991; Chapter 9.
11. Brathun, R.; Zingsheim, P. In *Handbook of Polymeric Foams and Foam Technology*; Klempner, D.; Frisch, K. C., Eds.; Hanser: Munich, 1991; Chapter 10.
12. Deanin, R. D.; Kapasi, V. C.; Georgacopoulos, C. N.; Picard, R. *J. Polym. Eng. Sci.* **1974,** *14(3),* 193.
13. Deanin, R. D.; Agarwal, P. *Am. Chem. Soc. Polym. Mater. Sci. Eng.* **1985,** *53,* 818.
14. Benning, C. J. *Plastic Foams*; John Wiley & Sons: New York, 1969.
15. Frisch, K. C.; Saunders, J. H. *Plastic Foams*; Marcel Dekker: New York, 1972.
16. Klampner, D.; Frisch, K. C. In *Handbook of Polymeric Foams and Foam Technology*; Klempner, D.; Frisch, K. C., Eds.; Hanser: Munich, 1991.

FOAMING AGENTS

Brendan J. Geelan
Uniroyal Chemical Company, Incorporated

Foaming agents are materials which are used to create a foamed or cellular structure within a given polymer system. Generally speaking, they do this by causing a gas to be generated at the appropriate time during polymer processing. This gas is then captured by the polymer matrix in the form of a separate phase, ideally consisting of a large number of small voids. These may be in the form of individual bubbles (closed cell) or an interconnected series of voids (open cell). The physical properties of a foamed polymer may be extremely different from those of the unfoamed material and will be a function of the amount of gas captured, and the morphology of the cellular structure.

There are a wide variety of applications wherein the properties of a foamed polymer are desirable. Some of these include the following: flotation devices take advantage of the reduction in bulk density; very light weight foams have excellent thermal insulation properties; wire and cable electrical insulation is improved with a foam layer between the wire and jacketing; plastic pipe can be made lighter in weight and use less material per foot by having a foamed core; and the embossed pattern in many vinyl floors is achieved via the use of foaming agents.

Structural foam-molded parts have improved rigidity at equal weight as compared to solid parts. Soft but resilient padding, packaging, and weather proofing materials utilize foaming technology. Extruded plastic profiles for wood replacement are foamed to improve their "nailability" and wood-like feel. These are only some of the more common examples.

FOAMING THEORY

In order to have a better appreciation of the different types of foaming agents, a brief review of the foaming process itself is helpful. The generic description of a polymeric foaming process consists of the following steps: (1) mixing of the polymer and foaming agent, (2) bringing the polymer phase to a fluid state, (3) cell nucleation (generating a separate gas phase), (4) cell growth, and finally (5) stabilization of the cellular structure (solidification of the polymer phase). The order of steps 1 and 2 may be reversed, depending upon the foaming agent and process type.

FOAMING AGENT TYPE OVERVIEW

Foaming agents may be divided into two principal categories according to their mode of gas generation: Physical foaming agents (PFAs) generate gas via a physical change of state–they change from a liquid to a gas without changing their chemical identity during the foaming process. On the other hand, chemical foaming agents (CFAs) generate gas via a chemical reaction. Most often, this chemical reaction is a thermally induced decomposition of the chemical foaming agent wherein one or more of the major products of decomposition is a non-condensible gas.

Within each of these major categories, further sub-groupings are possible. Physical foaming agents may be compressed or liquefied simple gases such as nitrogen and carbon dioxide (the only commercially noteworthy examples of this sub-class), or low boiling liquids.

Within the chemical foaming agent category, there are two main sub-types: organic and inorganic. The organic types in general are based on hydrazine chemistry and decompose exothermically to produce nitrogen gas. In general, the inorganic foaming agents are based on carbonate and bicarbonate chemistry and decompose endothermically to produce primarily carbon dioxide gas. In all cases for chemical foaming agents, there will be some solid residue in addition to the gas produced (which of course is not the case with physical foaming agents).

FOAMING AGENT APPLICATIONS

Physical Foaming Agents

Compressed Gases

Historically, the use of compressed gas as a foaming agent has had a very limited and specialized area of applicability. Their use requires that they be metered into the polymer melt under high pressure, and that the pressure on the melt itself be maintained high enough that the gas will remain dissolved in the polymer until foaming is to occur. This will happen when the pressure is released. Processes have been developed to apply this principle to injection molding and extrusion applications.

Volatile Liquids

The use of this type of physical foaming agent has been extremely effective in the attainment of very low density (less than five pounds per cubic foot) foam polystyrene and polyolefin sheeting materials. These foams have a wide applicability in the food packaging and construction industries and a result of their thermal insulation characteristics. The unique foaming activity of this type of PFA is particularly well suited to this process and product.

The production of very low density, non-crosslinked foam sheet is usually a two-extruder process. The first extruder is used to flux the resin and to obtain a thorough mix of resin, foaming agent, and nucleant. In the second extruder, the mixture is cooled to the point where the viscosity is suitable for foam development.[1]

Another major application area for PFAs has been with rigid (thermoset) polyurethane foams. In this case, it is the heat of reaction which volatilizes the foaming agent rather than any reduction in pressure.

Chemical Foaming Agents

At the risk of oversimplification, the preceding discussion indicates that PFAs tend to be used in highly specialized processes where the finished polymer product bears little resemblance to its solid counterpart. On the other hand, CFAs are more apt to be used in a fashion analgous to plastic and rubber additives. That is to say, they are used as an ingredient in a more or less standard polymer processing technique, and the finished product often bears a fairly strong resemblance to the unfoamed polymer.

The variety of different chemicals which have been used to foam polymers over the past century is staggering. Based on the collective experiences of CFA users over that time, however, the list today has dwindled to just a few which have commercial significance. These are the ones which have excelled in the basic criteria for CFA suitability, including the following: (1) generation of significant volumes of non-condensable gas within a well-defined temperature range; (2) storage stability; (3) lack of toxicity, odor, color, environmental, or related concerns; (4) absence of adverse interactions with the polymer, and (5) good balance of cost and performance.[2]

The chemicals and chemical types, which have survived the test of time and account for the vast majority of the worldwide CFA usage today, are as follows: azodicarbonamide (either pure or blends), sulfonyl hydrazides, *n*-nitroso compounds (especially dinitroso pentamethylene tetramine), semicarbazides, tetrazoles, carbonates, bicarbonates, and blends of carbonates and bicarbonates with organic acids.

Azodicarbonamide

Referred to as "AZ" and "ADCA," this is the most commonly used CFA in the world today. It measures up very well against the previously listed criteria for CFA suitability. Additionally, its normal operating or decomposition temperature range encompasses the processing temperature range of most high-volume thermoplastics, such as PE, PP, ABS, and HIPS. Furthermore, the decomposition temperature of AZ can be easily modified downward via the use of "activators." In this way it can be made suitable for lower temperature processing materials such as flexible vinyl and thermoplastic and thermoset rubber formulations. The ability to customize the decomposition temperature of AZ to suit a wide variety of polymers and processes is perhaps its most unique and valuable asset.

It is apparent from the preceding why AZ is such a popular CFA: it yields a large volume of gas per unit weight; the solid residue left behind is non-colored and is relatively non-toxic; it is produced via simple chemistry from common, inexpensive raw materials, so that its cost-performance is excellent; the shelf life is practically unlimited under normal conditions; and the rate of decomposition can be custom tailored to many applications.

While zinc oxide is perhaps the most commonly used activator for AZ decomposition, a wide variety of materials may be used. As a class, metal salts in general are effective. The metal ions that have shown the most activity are lead, zinc, barium, cadmium, and tin. Other materials which are often used to activate AZ are urea, certain organic acids, and triethanolamine.

Other (Non-AZ) CFAs

In spite of the broad range of applicability of AZ, there are instances where other CFAs are preferred. These generally involve foaming where the process temperature is either below 150°C (300°F) or above 220°C (425°F), or where the alternative CFA has some unique feature not available with AZ. The following are some of the more common alternative CFAs with typical applications.

Sulfonyl Hydrazides. These CFAs find use in very low-temperature plastisols, such as those used for "puff" inks. They are also commonly used in some expandable sealants and glues where low processing temperatures are necessary. Room temperature foaming is possible with peroxide-cured polyester systems in conjunction with metallic activation. The sulfonyl hydrazides can be activated like AZ, but are sensitive to far fewer materials, and generally to a lesser extent. As a class, amines are useful–most notably triethanolamine.

***p*-Toluene Sulfonyl Semicarbazide.** The foaming of some high-viscosity styrenics and polyolefins, such as high heat ABS and low-melt index PP, sometimes requires processing temperatures which are too high for AZ (>225°C). For these applications, TSSC has proven to be successful.

Phenyl Tetrazole. The foaming of some glass-filled nylons and polycarbonate requires still higher processing temperatures (240-250°C). In addition to this, these polymers may be attacked by the minor amounts of ammonia released along with the nitrogen from both AZ and TSSC. For both these reasons, the tetrazoles have found applicability with these polymers. Most notably, 5-Phenyl tetrazole has a gas yield, working temperature range, and gas composition (100% N_2), which has made it a favorite with PC processors.

Dinitroso Pentamethylene Tetramine. This CFA has a very long history of usage, particularly in low-temperature applications such as silicones. While it is not very low in decomposition temperature itself, it is usually used in blends with activators such as urea. In this form it gives large volumes of gas rapidly at low temperature.

Inorganic CFAs. Historically, the simple carbonate and bicarbonate salts are perhaps the longest lived of the CFAs. The very first polymer foams were produced using sodium bicarbonate. While they have been replaced largely by the more efficient and adaptable CFAs, the inorganics continue to maintain an application niche. Their basic chemistry has not changed over the years, but means have been found to lessen their tendency to absorb water on storage, and also to make their decomposition residues less harmful to processing equipment. The fact that these materials produce carbon dioxide as the foaming gas, and that they are endothermic (absorb heat) during decomposition sets them apart from most organic CFAs. Since CO_2 migrates through most polymer matrices more rapidly than N_2, parts foamed with inorganic CFAs need less "degassing" time prior to painting. Their endothermic decomposition also tends to produce parts with smoother surfaces. In many injection molding applications where these factors are important, inorganic CFAs dominate.

REFERENCES

1. Berins, M. L. *Plastics Engineering Handbook ed. 5*; Van Nostrand Reinhold: New York, 1991; Chapter 19.
2. Morgan, P. *Plastics Progress*; Philosophical Library: New York, **1955,** p 55.

Foams

See: Foamed Plastics
Foaming Agents
Polyethylene Foams
Polyimides (Introduction and Overview)
Polyurethanes (Overview)
Rigid Polyurethane Foams

FOOD POLYMERS

H. N. Cheng
Research Center
Hercules Incorporated

A large number of polymers are allowed for direct food use; these are mostly of natural origin, sometimes modified, and sometimes made through microbial means.[1-8] Most of them are polysaccharides or their derivatives, to which the term "gums" and "hydrocolloids" are often applied.

Food polymers are usually grouped by their origin. The following classification includes most of the common polymers (**Table 1**).

PROPERTIES AND USES

In general, food polymers are tasteless, odorless, colorless, and non-toxic materials. Many of them (e.g., polysaccharides with β-glycosidic linkages) are non-caloric and can be used in reduced-calorie food formulations and as dietary fibers. Two key properties of the food gums render them attractive in food preparation: their thickening and their gel-forming tendency. Almost all food gums are water-soluble or water-dispersible.[9]

Examples of gel-forming gums are agar, alginate, carrageenan, furcellaran, gelatin, pectin, and starch. The rheology and gelling properties of the solutions or dispersion depend on the polymers involved, pH, concentration, temperature, salt and other solutes present, and other factors.

Food polymers serve many uses in food and food formulations. The choice of a food polymer for a given application depends on a large number of factors: price, use level, physical properties (solubility, viscosity, gelation, crystallization, etc.), chemical reactivity (sensitivity to heat, moisture, pH, browning, stability, etc.), special functional properties, appearance (color, cloudiness, body, etc.), texture and mouthfeel, flavor/odor, compatibility and synergism with other food components, and nutritive content (or lack thereof). In general, if a thickener is needed, then a polymer that produces high viscosity at low concentration is preferred. If a binder or protective colloid is desired, then a polymer that gives low viscosity at high solids may be chosen.[1,2]

STRUCTURE AND FUNCTION

The properties of the gums and their functional uses are related to their chemical structures and physical characteristics. The key structural parameters are molecular weight, branching, homopolymer or copolymer, random or block microstructure (for copolymer), and electric charge.

In general, the higher the molecular weight, the higher the viscosity of the polymer solution.

Linear uncharged polysaccharides, particularly homoglycans (e.g., cellulose and amylose) tend to associate and to crystallize. As a result, they dissolve poorly or not at all. In contrast, branched polymers tend to give more stable solutions

TABLE 1. Common Food Polymers

Seed gums	Tuber and root gums	Plant exudates	Plant extracts
Guar	Konjac mannan	Ghatti	Agar
Locust bean gum	Potato starch	Gum arabic	Alginate
Corn starch	Tapioca starch	Karaya	Carrageenan
Wheat starch		Tragacanth	Furcellaran
			Pectin
Animal source	**Microbial**	**Modified gums**	
Casein	Dextran	Carboxymethylcellulose (CMC)	
Gelatin	Gellan	Methylcellulose (MC)	
	Pullulan	Hydroxypropylcellulose (HPC)	
	Xanthan	Hydroxypropylmethylcellulose (HPMC)	
		Modified starches	

which will not retrograde or precipitate. Examples are amylopectin, gum arabic, guar, and locust bean gum.

Many polysaccharides contain ionic functional groups and these modify the properties of the polymer. For example, pectin and alginic acid contain carboxylic acids, and carrageenan contains sulfonic acid groups which can form gels with suitable polyvalent cations. Other examples are modified cellulosics and starches, xanthan, gellan, and furcellaran. These are noted for their ready solubility in water and their affinity for cationic salts.

Alginic acid is an example of a polymer with random and block comonomer structures. It is composed of mannuronic acid (M) and guluronic acid (G) where the M and the G exist as all M or all G segments as well as M/G alternating segments. Each of these three segments has a different geometry and a different tendency towards gel formation (using calcium ions).

REFERENCES

1. Glicksman, M. *Food Hydrocolloids*; CRC: Boca Raton, FL, Vol. 1, 1982; Vol. 2, 1983; Vol. 3, 1986.
2. Whistler, R. L.; BeMiller, J. N., Eds.; *Industrial Gums*, 3rd ed.; Academic: New York, 1991.
3. Phillips, G. O.; Wedlock, D. J.; Williams, P. A., Eds.; *Gums and Stabilizers for the Food Industry*; Pergamon: Oxford, Vol. 1, 1982; Vol. 2, 1984; Elsevier: London, Vol. 3, 1986; IRL: Oxford, Vol. 4, 1988; Vol. 5, 1990; Vol. 6, 1992.
4. Alexander, R. J.; Zobel, H. F., Eds.; *Developments in Carbohydrate Chemistry*; Amer. Assoc. Cereal Chem.: St. Paul, MN, 1992.
5. *Plant Polymeric Carbohydrates*; Spec. Pub. Roy. Soc. Chem.; Vol. 134, 1993.
6. Imeson, A., Ed.; *Thickening and Gelling Agents for Food*; Blackie Academic: London, 1992.
7. Dziezak, J. D. *Food Technol.* March **1991,** 116.
8. Davidson, R. L., Ed.; *Handbook of Water Soluble Gums and Resins*; McGraw-Hill: New York, 1990.
9. For example, Levine, H.; Slade, L., Eds.; *Water Relationships in Foods*; Plenum: New York and London, 1991.

FRACTURE IMMOBILIZATION POLYMERS

Robert E. Richard
Johnson and Johnson Professional Incorporated

The technology of plaster-based fracture management products is relatively simple. The chemical process involves the conversion of gypsum ($CaSO_4 2H_2O$), a hard, rock-like substance, to plaster of Paris (POP) ($CaSO_4 0.5H_2O$), a soft and powdery material. The conversion takes place by heating the former to approximately 128°C. In application of a cast, the POP roll is dipped into water to begin the rehydration of the softer POP to the hard gypsum.

The advantages associated with POP-based products are their low cost, excellent moldability, low allergic response, and good shelf life. The disadvantages cited for POP are its lack of strength and durability, poor water resistance, poor radiolucency (transmission of X-rays), along with its heavy weight. It is these disadvantages that have prompted effort to develop fracture immobilization products based on polymeric materials.

APPLICATIONS

POP/Polymer Composites

One approach designed to address the above-stated shortcomings of POP was a POP/polymer composite that utilized 5-30% of a water-dispersible melamine formaldehyde resin.[1,2] In this system a polymerization catalyst was incorporated into the bandages and was activated to induce polymerization of the melamine resin once the bandage was wetted for application. This resulted in an interpenetrating polymer network within the POP matrix. This development provided the first truly waterproof cast, and at the same time significantly improved the cast strength as well.

Polymer-Based Fracture Management Products

The driving forces behind the development of polymer-based fracture products stems from the desire to improve the comfort of patients while portions of their skeletal system are under repair. By designing polymer-based systems, a lighter weight and more water-resistant product is achieved. In addition, the ability to X-ray the fracture site without removing the cast is also a major advantage. Other benefits associated with polymer-based systems are improved air and moisture permeability, which lowers the potential of skin maceration. The approaches taken to design such products can be broadly classified as follows:

- Plasticized polymers
- Heat-softened polymers
- Radiation-activated resins
- Chemically activated resins

THE FUTURE OF FRACTURE IMMOBILIZATION

Efforts will continue to develop improved products for fracture management, especially due to the active lifestyle of our society. Areas of focus are patient comfort and speed of recovery. Recently a new technology for bone repair has been receiving significant attention in the medical industry. This technology, developed by researchers at the Norian Corporation, involves implantation of a bone-like substance at the fracture site.[3] The material consists of a carbonated form of apatite (inorganic calcium and phosphate salts) known as dahllite. It is applied to the fracture site by direct injection as a surgical paste and hardens in minutes. It achieves significant compressive strength after only 10 min and after 12 hrs the strength is greater than cancellous bone. It reportedly serves as an internal fixation while the healing process occurs, and is continually remodeled with living bone over time. The day may come when a bone fracture can be healed in only a fraction of the time it once took.

REFERENCES

1. Foglia, A. J.; Smith, D. F.; Detwiler, E. B. U.S. Patent 2 842 120, Johnson and Johnson: 1958.
2. Billings, O. B.; Brickman, L.; Eberl, J. J. U.S. Patent 2 842 138, Johnson and Johnson: 1958.
3. Constantz, B. R. et al. *Science* **1995,** *267,* 1796.

FREE RADICAL INIFERTERS

C. P. Reghunadhan Nair
Polymers and Special Chemicals Division
Vikram Sarabhai Space Center

Although free-radical method addresses to many of the technological challenges posed by other techniques, it is still not the one of choice when it concerns synthesis of polymers with narrow dispersity at high conversion and synthesis of well-defined block copolymers. The free-radical method has now gained importance for graft copolymers thanks to the advancement in macromonomer concept, and the same is true for telechelics, owing to extensive research in this domain.

The current innovative research in free-radical domain is directed toward addressing some of these problems. Access to molecular architecture by way of controlling the chain initiation, transfer, and termination is the motive for the current interest in this domain, which is centered about "living" radical polymerization, radical addition-fragmentation, and radical iniferters.

WHAT ARE INIFERTERS?

Following the lines of Kennedy's[1] "inifers" for the controlled synthesis of telechelic and block copolymers by carbocationic polymerization, the terminology "iniferter" was given by Otsu[1a] to a class of compounds acting as radical initiators, chain-transfer agents, and chain terminators. It was recognized that tetraalkyl thiuram disulfides act in these three ways during polymerization of vinyl monomers in their presence. The result is that all chain ends are invariably end-capped with the dithiocarbamate moieties.

There are only limited organic compounds that exhibit simultaneously these three properties. Apart from thiuram disulfides and its analogues,[2,3] the tritylazobenzene class of compounds have been shown to manifest the "initer" property.[4] No transfer is encountered in this case.

Because the trityl radical is too stable to initiate polymerization on its own, it participates in the radical termination process. The trityl-terminated polymer species is capable of undergoing reversible thermal dissociation, opening a way for living radical polymerization facilitating synthesis of block copolymers as claimed by Otsu.[4]

TETRAPHENYL ETHANE DERIVATIVES

Tetraphenyl ethane derivatives also fall into this category. Thermal, vinyl polymerization in their presence proceeds in two phases.

THIURAM-DISULFIDE-BASED INIFERTERS

Thiuram disulfides are known thermal and photochemical initiators for polymerization for electron-rich vinyl monomers.

In radical polymerization induced by thiuram disulfides, the three principal mechanisms operative are given in **Scheme I**. Except for the termination process, all sequences of the polymerization process are the same as in a conventional one, and the chain length and functionality are decided by the three phenomena.

APPLICATION OF THERMAL INIFERTERS IN POLYMER SYNTHESIS

Telechelics

Thermal iniferters can be successfully used for synthesis of telechelic and block copolymers; synthesis of telechelic polymers warrants design of a functional iniferter. Thus thiuram disulfides, bearing groups like hydroxy, carboxy, phosphate, etc., have been used to derive perfect telechelics with respect to that function.[9,10]

TriBlock Copolymers

If instead of simple functions, polymer segments are located on the alkyl group of the thiuram disulfide, the resulting polymer would have a triblock structure.

This technique has been adopted to realize triblock copolymers of various types[11,12] such as poly(ethyl acrylate)-PMMA and poly(ethyl acrylate)-polystyrene. Amphiphilic polymers with central water-soluble polymethacrylic acid and terminal styrene blocks have been made by polymerizing methacrylic acid in the presence of polystyrene (PS)-based macroiniferters.[5]

Multiblock Copolymers

The free-radical route is now increasingly sought to realize this type of block copolymer, as it permits combination of a large variety of blocks. Iniferter technique can be used successfully if one of the segment-forming polymers constitutes the polymeric iniferter. This way, multiblock copolymers such as

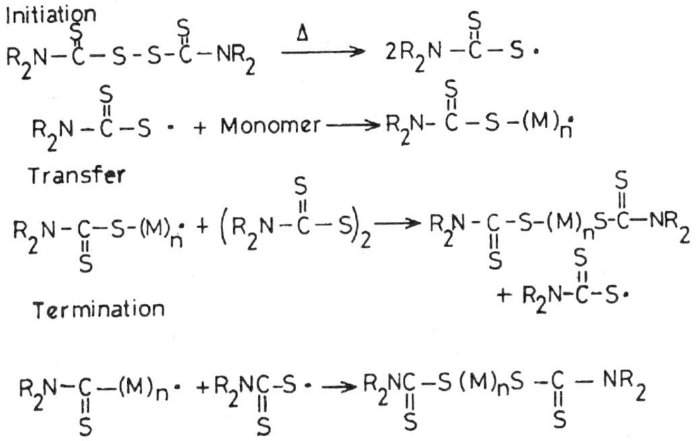

I

phenyl phosphonamide-*b*-PMMA,[10] (PDMS-*b*-PS), polydimethyl siloxane-*b*-polystyrene[11] and polyoxyethylene-*b*-PS polystyrene[8] having different characteristics have been prepared.

Although it is possible to control the length of the vinyl sequence inserted across the disulfide linkage, the absolute number of A-B blocks per chain is beyond precise control. It normally increases with increase in monomer conversion.

PHOTO INIFERTERS AND THEIR APPLICATIONS

The above-described systems constitute thermal iniferters in the sense that polymerization is initiated by their thermally induced decomposition. Dialkyl dithiocarbamates, however, can induce radical polymerization in the presence of UV light. Dithiocarbamate-terminated polymers are easily formed by vinyl polymerization in the presence of thiuramdisulfides.

The method can be extrapolated to the design of a variety of block copolymers, the blocks of which are constituted by either homo- or random copolymers in fairly high yield.

GRAFT COPOLYMERS VIA PHOTOINIFERTERS

Polymers bearing pendant dithiocarbamate groups can be made via polymerization of vinyl monomers bearing them. Such polymers used as polymeric photoiniferters give rise to graft copolymers.[12]

PHOTOINIFERTERS FOR STAR POLYMER SYNTHESIS

Star polymers and star block copolymers are made using multifunctional photoiniferters, for example, 1,2,4,5-tetrakis(N,N diethyl dithiocarbamoyl methyl) benzene was used as a tetrafunctional iniferter. The polymer grows from each arm.[13]

LIMITATION OF THERMAL AND PHOTOINIFERTERS

Thermal iniferters have only a relatively mild initiating ability, and the efficiency decreased for polymeric iniferters. For photoiniferters it is claimed that the counter-dithiocarbamate radical and its photo-degraded products can induce homopolymerization of the monomer. These are some of the problems that need immediate attention in this domain.

REFERENCES

1. Kennedy, J. P.; Marechal, E. "Carbocationic Polymerization" Wiley-Interscience: New York, 1982.
1a. Otsu, T.; Yoshida, L. M. *Makromol. Chem. Rapid. Commun.* **1982,** *3,* 127.
2. Staudner, E.; Beniska, J. *Eur Polym. J. Suppl.* **1969**, 537.
3. Niwa, M.; Matsumoto, T.; Izumi, H. *J. Macromol. Sci. Chem. Ed.* **1987**, *24,* 567.
4. Otsu, T.; Tazaki, T. *Polym. Bull.* **1986**, *16*, 277.
5. Nair, C. P. R.; Chaumont, P.; Clouet, G. In *Macromolecular Design, Concept and Practice;* Mishra, M. K., Ed.; Polymer Frontier International: New York, 1994, Chapter 11.
6. Nair, C. P. R.; Clouet, G.; Chaumont, P. *J. Poly. Sci. Polym. Chem.* **1989**, *27*, 1795.
7. Nair, C. P. R.; Clouet, G.; *Makromol. Chem.* **1989**, *190*, 1243.
8. Nair, C. P. R.; Chaumont, P.; Clouet, G. *J. Macromol. Sci. Chem.* **1990**, *A27*, 791.
9. Nair, C. P. R.; Richou, M. C.; Clouet, G. *Makromol. Chem.* **1991**, *192*, 579.
10. Nair, C. P. R.; Clouet, G. *Polymer* **1988**, *29*, 1909.
11. Nair, C. P. R.; Clouet, G. *Macromolecules* **1990**, *23*, 1361.
12. Otsu, T.; Yamashita, K.; Tsuda, K. *Macromolecules* **1986**, *19*, 287.
13. Kuriyama, A.; Otsu, T. *Polym. J.* **1984**, *16*, 511.

FREE RADICAL INITIATORS, BIFUNCTIONAL

Irja Piirma
The University of Akron

Bifunctional and multifunctional free radical initiators have been synthesized and studied in many applications. Many are commercially available (Lucidol-Penwalt, Akzo Chemie and Atochem, for example). Their one important and successful use has been in the synthesis of block copolymers. In bifunctional diperoxides, the two peroxide groups can have different thermal stabilities and thus be activated to add various monomers sequentially. Hence, these diperoxides can be used to synthesize diblock copolymers, while some azodiperesters would provide the means for a triblock synthesis.

Bifunctional free radical initiators also have interesting applications in free radical-initiated bulk and solution polymerizations. With customary monofunctional initiators, initiation and termination rates are interdependent and therefore restrict the possibility of increasing the polymerization rate without lowering the molecular weight of the polymer. Use of diperoxides with sequential decomposition kinetics allows faster rates without lowering the molecular weight.

PREPARATION

Since commercial availability of the multifunctional free radical initiators is good, a considerable number of investigators have carried out block copolymer synthesis and polymerization kinetic studies with these materials. However, some interesting initiators have been synthesized by researchers such as Simionescu et al.,[1,2] Önen and Yagči,[3] and others.

APPLICATIONS

Preparation of Block Copolymers Using Bifunctional Initiators

Di- and trifunctional initiators have been used to create di- and triblock copolymers, depending on the primary mode of termination of the polymerizing chain. Use of these multifunctional initiators for formation of block copolymers is facilitated by the possibility of sequential activation of the initiating radicals. This allows formation of a polymeric chain which would retain the other azo- or peroxy group (polymeric initiator), ready to be activated in the presence of another monomer for formation of the block copolymer.

Multifunctional free radical initiators have been used with increasing success in block polymerization.

A sequential free radical initiator, di-*t*-butyl-4,4-azobisiso(4-cyanoperoxy valerate) has thermal stabilities of the azo group and the perester groups different enough so the block copolymer synthesized can be carried out in two separate stages of polymerization. Depending on the types of monomers used and their predominant modes of termination, it is either the azo group or the perester groups that can be activated first, followed by a second-stage polymerization with the other monomer. Polymerizations with styrene and methyl methacrylate as monomers using the above reaction scheme have resulted in formation of a mixture of di- and triblock copolymers with some homopolymer formation.[4]

In an alternate approach with the azo-diperester initiator, the first monomer could be initiated by activating the perester groups, thus creating a polymeric initiator with an azo group in the polymer backbone. This azo group could then be activated thermally to polymerize the second monomer.[5] This approach has an added advantage that very little homopolymer is formed during the block copolymerization stage. Triblock yields as high as 74 wt% have been obtained in the preparation of poly(methyl methacrylate)- block-polystyrene-block-poly(methyl methacrylate) triblock.[6]

Diperoxides as Homopolymerization Initiators

The authors developed a model for the kinetics of styrene bulk polymerization using three different commercially available bifunctional initiators. The experiments for different polymerization conditions gave reaction rate, molecular weight and molecular weight distribution data that agreed with the model predictions. A comparison of the results of bifunctionally initiated systems with the mono-functional initiated (benzoyl peroxide) ones showed that reduction of polymerization time of up to 75% could be achieved with bifunctional initiators for a wider range of polymerization conditions without significantly affecting the molecular weight and molecular weight distribution of the final polymer.

REFERENCES

1. Simionescu, C.; Sik, K. G.; Comanita, E.; Dimitriu, S. *Eur. Polym. J.* **1984**, *20*, 467.
2. Simionescu, C.; Comanita, E.; Dumitriu, A.; Petrovici, A.; Shaikh, A. S. *Rev. Chim.* **1982**, *33*, 423.
3. Onen, A.; Yagci, Y. *J. Macromol. Sci.* **1990**, *A27*, 743.
4. Piirma, I.; Chou, L. H. *J. Appl. Polym. Sci.* **1979**, *24*, 2051.
5. Gunesin, B. Z.; Piirma, I. *J. Appl. Polym. Sci.* **1981**, *26*, 3103.
6. Su, J. S. N.; Piirma, I. *J. Appl. Polym. Sci.* **1987**, *33*, 727.

Free Radical Polymerization

See: *Alternating Radical Copolymerization*
Azo Initiators (Function and Application)
Block Copolymers (By Changing Polymerization Mechanism)
Bulk Polymerization
Captodative Compounds, Polymerization
Captodative Olefins
Cyclic Ketene Dithioacetals (Polymerization)
Cyclopolymerization (Overview)
Emulsion Polymerization
End Functionalized Oligomers (by Chain-Transfer Technique)
Free Radical Iniferters
Free Radical Initiators, Bifunctional
Free Radical Polymerization (Recent Developments)
High Pressure Radical Polymerization
Inhibition Kinetics, Radical Polymerization
Inhibition, Radical Polymerization
Living Radical Polymerization
Living Radical Polymerization, Iodine Transfer
Magnetic Field Polymerization
Mechano Chemical Polymerization
Network Structure Formation (Radical Polymerization)
Poly(methyl methacrylate)
Polystyrene Manufacture (Using Bulk Free Radical Polymerization)
Redox Polymerization
Ring-Opening Polymerization, Free Radical
Ring-Opening Polymerization, Radical (With No Shrinkage in Volume)
Styrenesulfonic Acid and its Salts (Polymerization and Applications)
Suspension Polymerization
Trisubstituted Ethylene Copolymers
Ultrasonically Assisted Polymer Synthesis
Vesicles, Polymerization in
Vinyl Polyperoxides (Oxygen Copolymerization)

FREE RADICAL POLYMERIZATION (Recent Developments)

David H. Solomon and Grace Y. N. Chan
Department of Chemical Engineering
The University of Melbourne

Even with the idealized mechanism of free radical polymerization, the resultant polymers inherently have structural groups that vary in stability. While these structural anomolies represent only a small percentage of the total polymer, their presence becomes significant when we relate these structures to mechanisms of polymer degradation.

In recent years, much of the development in free radical polymerization has been concerned with the importance of these defect groups within a polymer, and with correlating the type of initiator, and the conditions used in the synthesis, with polymer structure and properties.[1,2]

Furthermore, it has been suggested that the propagation steps (and hence polymer formation) may be controlled by the use of additives, chain transfer agents, and iniferters to produce highly regular functionalized polymers and oligomers. Functionalized chain transfer agents in particular are quite versatile, and may also participate in controlling polymer functionality arising from termination reactions. The termination process in itself may also be studied in closer detail in order to further our understanding of the complex mechanisms involved in free radical polymerization.

INITIATION

Extensive studies in the last decade have shown that the mechanism of initiation is far from simple, and that for a given polymerization, initiation depends markedly on the particular radical species, monomer, and total polymerization conditions.[1,2]

Tail addition of the initiating radical to the monomer is rarely the only reaction pathway, as head addition and abstraction often occur at competitive rates, and different radical species possess differing specificities for a given monomer. This may be illustrated by considering two common oxygen-centered radicals, the *tert*-butoxy (*t*-BuO) and benzoyloxy ($PhCO_2$) radicals, and reaction with each of two common monomers, styrene and methyl methacrylate.

Pathways of initiation can be deduced by using the technique of radical trapping. In the case of the initiators, *t*-BuO and $PhCO_2$, the use of the nitroxide radical trap 1,1,3,3,-tetramethylisoindolinyl-2-oxyl (O-T) enabled the trapping of the alkyl radicals produced from the initiation reaction, as their corresponding alkoxyamines.[3-5] From these types of experiments, one can determine the relative proportions of alkyl radicals produced, and this enables some insight into how steric and polar effects may influence the mode of initiation.

PROPAGATION

The challenge of controlling the growth step in free radical reactions has been of interest to researchers for many years. This is desirable in order to minimize the formation of structural defects such as head-to-head and tail-to-tail linkages within a polymer, as well as to control the nature of the end groups, molecular weight of a polymer, monomer incorporation, and tacticity.

One manner in which researchers are attempting to control polymer formation is through the use of chain transfer agents. The most commonly employed chain transfer agents are derivatives of thiols, such as thioglycolic acid ($RSH = HSCH_2COOH$).[9] These compounds may participate in hydrogen atom transfer with a propagating radical to give a polymer chain with the desirable fully saturated chain end. The newly generated thiol radical is then capable of initiating another chain, to produce a functionalized polymer.

Functionalized chain transfer agents such as allylic sulfides also enable the formation of low molecular weight oligomers bearing specific functional groups.[10]

In the choice of suitable chain transfer agents, it is of interest to consider the chain transfer constant (C_T). C_T is defined as the ratio of the rate constant of propagation (k_p) to the rate constant of the reaction of propagating radical with the transfer agent (k_{tr}), that is, $C_T = k_p / k_{tr}$. Ideally, we would like C_T to have a value of unity, such that the rates of propagation and chain transfer are equal, and a narrow polymer molecular weight distribution is obtained.

Another method by which researchers are attempting to control polymer formation is through the use of complexing agents such as square planar cobaloximes and cobalt porphyrins which also act as chain transfer agents.[11-13] Unlike the thiols, these molecules are not consumed during their reaction with radicals, and because of this regenerative ability, have been termed catalytic chain transfer agents.

The concept of living free radical polymerization through the use of "iniferters" (*ini*tiation-trans*fer* agent-*ter*mination) was first proposed by Otsu in 1982.[14] Polymerization performed in the presence of iniferters resulted in a polymer with two iniferter fragments, one at each chain end. Under appropriate conditions, these fragments were once again able to dissociate to two radical and initiate further polymerization. In this way, living radical polymerization may be visualized as an insertion of monomers between the iniferter end group and the backbone polymer.[15,16]

Two types of iniferters have been described; photoiniferters and thermal iniferters. An example of a photoiniferter is *p*-xylylene *N,N*-diethyldithiocarbamate (XDC). XDC is known as a difunctional iniferter. In other words, it possesses two dithiocarbamate end groups. XDC has previously been employed by Otsu in the synthesis of ABA block copolymers of styrene and methyl methacrylate.[17]

Nair and Clouet have recently developed thermal iniferters with some remarkable properties. Their studies with thiuram disulphides has made possible the formation of unusual polymer-terminated macroiniferters used for the synthesis of ABA block copolymers.[18] The method developed by Nair and Clouet has the advantage that amphiphilic block copolymers composed of usually incompatible segments such as poly(ethylene oxide) (POE) and poly(styrene) may be formed by an addition-condensation mechanism.

While much research has been directed toward understanding solvent effects in free-radical polymerization, controversy still exists concerning their modes of action, and many aspects of solvent influence in radical systems remain unclear. A further challenge for chemists striving for propagation control lies in producing highly stereoregular polymers through the control of tacticity. A readily accessible industrial process for the synthesis of polymers with controlled tacticity by free radical processes would be highly desirable.

TERMINATION

In polymer synthesis, the termination mechanism is usually represented as a radical-radical reaction that leads to either combination or disproportionation.

Ideally, if termination occurs by combination, two initiator fragments may be detected. If disproportionation is the preferred termination pathway, only one initiator fragment may be seen. However, it should be noted that the free radical systems are not always ideal, and where initiation occurs by abstraction of a hydrogen from the monomer to give an initiating radical, a polymer terminal initiator fragment will not be present.

The advent of NMR has provided a valuable tool for studying mechanisms of polymerization. NMR may be used to distinguish between a number of different initiator residues, such as those incorporated into the polymer chain, the unchanged initiator fragment, and end groups formed by transfer to the initiator, or by copolymerization of initiator by-products.[6,8]

As mentioned, combination and disproportionation are the most common pathways of termination for a free radical reaction. However, a number of competing termination pathways are also known to exist. These include chain transfer, primary radical termination, and transfer to initiator.

Chain transfer reactions through the use of thiol compounds have already been discussed previously as a means of controlling polymer terminal functionality. However, one further advantage of these chain transfer agents is that termination occurs through hydrogen transfer, to form a saturated chain end, and it is therefore possible to avoid the undesirable head-to-head and unsaturated chain ends that contribute to polymer instability.

Primary radical termination involves the reaction of a propagating radical with an initiator-derived radical. This method of termination gives products that may originate from either of three reaction pathways: combination of the two radical species, hydrogen abstraction from the propagating radical by the initiator-derived radical to give a polymer with an unsaturated chain end, or hydrogen abstraction from the initiator-derived radical by the propagating radical to a polymer with a saturated chain end.

Transfer to initiator involves the termination of a propagating radical by reaction with a molecule of initiator. The resulting polymer possesses two initiator-derived end groups, and in the process, a new primary radical is formed, which may initiate a new chain.

CONCLUSION

The last decade has seen great advances in our understanding of the factors that influence free radical polymerization. Whereas radical reactions were once considered nonselective and governed by thermodynamic considerations, modern approaches have enabled great control and selectivity in radical processes to be achieved. The developments in our understanding of the initiation and termination processes have resulted in more scientific selection of reaction conditions for polymerization, and significant advances in the functionalization and molecular weight control of polymers. Further challenges in determining methods for controlling the propagation process to suit particular applications remain.

REFERENCES

1. Moad, G.; Solomon, D. H. *Aust. J. Chem.* **1990,** *43,* 215.
2. Chung, R. P–T.; Solomon, D. H. *Prog. Org. Coat.* **1992,** *21,* 227.
3. Griffiths, P. G.; Rizzardo, E.; Solomon, D. H. *Macromolecules* **1982,** *15,* 909.
4. Griffiths, P.G.; Rizzardo, E.; Solomon, D. H. *J. Macromol. Sci. Chem.* **1982,** *A17,* 45.
5. Moad, G.; Rizzardo, E.; Solomon, D. H. *Aust. J. Chem.* **1983,** *36,* 1573.
6. Moad, G. et al. *Macromolecules* **1984,** *17,* 1094.
7. Krstina, J.; Moad, G.; Solomon, D. H. *Polym. Bull.* **1992,** *27,* 425.
8. Moad, G. et al. *Macromolecules* **1982,** *15,* 1188.
9. Farina, M. *Makromol. Chem. Makromol. Symp.* **1987,** *10/11,* 3122.
10. Meijs, G. F.; Rizzardo, E.; Thang, S. H. *Macromolecules* **1988,** *21,* 3122.
11. Burczyk, A. F.; O'Driscoll, K. F.; Rempel, G. L. *J. Polym. Sci. Polym. Chem. Ed.* **1984,** *22,* 3255.
12. Enikolopyan, N. S. et al. *J. Polym. Sci. Polym. Chem. Ed.* **1981,** *19,* 879.
13. Gridnev, A. A. et al. *Organometallics* **1993,** *12,* 4871.
14. Otsu, T.; Yoshida, M. *Makromol. Chem. Rapid. Commun.* **1982,** *3,* 127.
15. Otsu, T. et al. *Polym. J.* **1989,** *25,* 643.
16. Otsu, T.; Yoshida, M.; Kuriyama, A. *Polym. Bull.* **1982,** *7,* 45.
17. Otsu, T.; Kuriyama, A. *Polym. Bull.* **1984,** *11,* 135.
18. Nair, C. P. R.; Clouet, G. *J. Macromol. Sci. Chem. Phys.* **1991,** *C31,* 341.

Fullerene-Containing Polymers

See: Fullerene-Functionalized Polymers
Supramolecular Mesomorphic Polymers (Assembling in Ultrathin Films)

FULLERENE-FUNCTIONALIZED POLYMERS

Abhimanyu O. Patil
Corporate Research Laboratory
Exxon Research and Engineering Company

One of the most rapidly growing areas of science is the study of fullerenes (often referred to as buckyballs in the literature). These large molecules, made entirely of carbon, have structures resembling geodesic domes. The smallest member of the fullerene family, C_{60} or Buckminsterfullerene [99685-96-8], resembles the pattern traced out by the seams on a soccer ball.

Although C_{60} was observed earlier[1] the fullerenes remained laboratory curiosities until 1990, when Wolfgang Kratschmer and Donald Huffman discovered that the fullerenes could be made in tangible quantities by consumption of graphite in an electric arc.[2] They found that when an electric arc is struck between carbon rods, a soot is formed from which the fullerenes can be extracted by organic solvents. With this development, scientists around the world were able to begin the study of the chemical and physical properties of fullerenes. Here we focus on research on fullerene functionalized polymers.

FULLERENES AS A NEW CARBON FORM

Fullerenes are an interesting class of molecules that have an even number of carbon atoms arranged over the surface of a closed hollow cage. Each atom is linked to three nearest neighbors by bonds that define a polyhedral network. This network consists of 12 pentagons and n hexagons, where n can be any number except one.

Fullerenes are produced by high-temperature reactions of elemental carbon or of carbon-containing molecules. They are also reported to be produced from coal.[4] It has also been shown that partial combustion of benzene in an argon–oxygen mixture produces fullerenes.[5] All fullerene-producing processes typically lead to mixtures of various all-carbon molecules. The portion of the product that is soluble in organic solvents such as toluene is commonly referred to as fullerenes (fullerite) or crude fullerenes; this product typically contains mostly C_{60}, lesser amounts of C_{70} [115383-22-7], and much smaller amounts of many different fullerenes of higher carbon number.

FULLERENE CHEMISTRY

Fullerenes are composed of sp^2-hybridized carbon atoms, like those in alkenes and aromatics. Initially envisaged as rather unreactive, aromaticlike molecules, the fullerenes undergo a great number of reactions associated with poorly conjugated and electron-deficient alkenes. Some recent reviews summarize the current state of knowledge about the chemical reactions of fullerenes.[3,6-8]

FULLERENE FUNCTIONALIZED POLYMERS

Because fullerenes are interesting materials with unique structures, their attachment to polymers would provide functional polymers with a variety of technologically important applications such as emulsifiers, compatibilizers, adhesion promoters, adhesives, additives, membranes, and coatings. Moreover, the incorporation of reactive fullerene functionality would provide sites for further functionalization, including functionalization by methods that would fail for the original polymer.

The facile reactivity of fullerenes with organic reagents suggests that fullerenes could be incorporated into polymers to create unique structures that bring different attributes to the polymers.

One of the first reports on fullerene-containing polymers involved the synthesis of a copolymer of C_{60} and *p*-xylylene [140853-30-1].[10] The copolymer was prepared by a free radical polymerization leading to a brown insoluble crosslinked adduct with a xylylene:C_{60} ratio of approximately 3.4:1.

The reaction of fullerene-C_{60} with polymers having amino groups in their side chain is an elegant method for the synthesis of polymer-bound C_{60}.[11,12] Fullerenes attached to the amino group of polymers could be prepared.

The acid-catalyzed alkylation reaction, or fullerenation, of aromatics has been reported by Olah et al.[13]

The reaction of living polystyrene with C_{60} has also been reported, in which a wide distribution of reaction products is obtained, with the number of attached polymer chains ranging from 1 to 10.[14] They designate these polymers as "flagellenes" because they comprise several flexible polymer chains covalently attached to a fullerene core. The products are fusible and soluble in organic solvents, and can be spin-coated, solvent-cast, or melt-extruded to give films and fibers having a high concentration of the fullerenes covalently bound to the polymer matrix. Transmission electron microscopy (TEM) of the products showed nanophase-separated morphologies with fullerene-rich domains anologous to those observed to occur spontaneously in diblock copolymers prepared from chemically distinct monomers. Researchers also reported fullerene dications as initiators of polymerization of 1,3-butadiene in the gas phase.[15]

The Santa Barbara group reported the novel approach of making the methanofullerenes, fulleroids C_{61} (inflated fullerenes), which can be used to make pendant chain polymers.[9,16,17]

Polymers containing fulleroid units are prepared by polycondensation of a dihydroxyphenyl fulleroid with sebacoyl chloride or hexamethylene diisocyanate [144767-55-5].

Fullerenes undergo Diels–Alder cycloaddition with a variety of reactive dienes.[17] This concept was extended to make a cyclopentadiene-functionalized polymer that reacts with C_{60} at room temperature to give fullerene-containing polymers. This addition was found to be reversible, allowing recovery of C_{60} upon heating the polymer adduct.[18]

C_{60}: A NOVEL BRIDGING GROUP

Fullerene-grafted olefin polymers were reacted with amines such as primary-tertiary diamines to give novel multiarm amine-functionalized polymers.

In contrast to reaction with a hydrocarbon polymer with an average of one amine per chain, when C_{60} was reacted with poly(ethylenimine) [9002-98-6] (50% aq. solution, average 50,000-60,000 mol wt), under similar reaction conditions, an insoluble crosslinked polymer was obtained.

Reaction of C_{60} with amine-containing polymers has been extended to dendrimers with amine functionality.[19]

APPLICATIONS

Possible applications of fullerenes range from semiconductors to medical applications to new materials. Some of the proposed applications include a new class of lubricants,[20,21] rechargeable batteries,[22] semiconductors,[23] superconductors,[24] chemical feedstocks,[21] and catalysts.[25] One barrier to application is cost. Fullerenes are currently produced only in small quantities for research purposes. Because of this, they are expensive.

The unique structure of fullerenes, their solubility, and facile reactivity has raised expectations that these novel materials will find commercial applications. In consequence, many patents concerning preparation and potential uses of fullerenes and fullerene derivatives have been filled worldwide. Although the true potential of fullerenes remains unknown, fullerene-based materials will undoubtedly play an important role in many applications.

ACKNOWLEDGMENT

I would like to thank my co-workers, Drs. G. W. Schriver, R. D. Lundberg, and S. J. Brois for useful suggestions, and S. M. Kaback for literature search help.

REFERENCES

1. Kroto, H. W. et al. *Nature* **1985**, *318*, 162.
2. Kratschmer, W. et al. *Nature* **1990**, *347*, 354.
3. Taylor, R.; Walton, D. R. *Nature* **1993**, *363*, 685.
4. Pang, L. S. K.; Vassalo, A. M.; Wilson, M. A. *Energy & Fuels* **1992**, *6*, 176.
5. Howard, J. B. et al. *Nature* **1991**, *352*, 139.
6. Kroto, H. W.; Allaf, A. W.; Balm, S. P. *Chem. Rev.* **1991**, *91*, 1213.
7. Hirsch, A. *Angew. Chem. Int. Ed. Engl.* **1993**, *32*, 1138.
8. *Fullerenes: Synthesis, Properties and Chemistry of Large Carbon Clusters*; Hammond, G. S.; Kuck, V. J., Eds.; American Chemical Society: Washington, DC, 1992.
9. Wudl, F. *Acc. Chem. Res.* **1992**, *25*, 157.
10. Loy, D. A.; Assink, R. A. *J. Am. Chem. Soc.* **1992**, *114*, 3977.
11. Patil, A. O. et al. *Polym. Bull.* **1993**, *30*, 187.
12. Geckeler, K. E.; Hirsch, A. *J. Am. Chem. Soc.* **1993**, *115*, 3850.
13. Olah, G. A. et al. *J. Am. Chem. Soc.* **1991**, *113*, 9387.
14. Samulski, E. T. et al. *Chem. Mater.* **1992**, *4*, 1153.
15. Wang, J. et al. *J. Am. Chem. Soc.* **1992**, *114*, 9665.
16. Suzuki, T. et al. *J. Am. Chem. Soc.* **1992**, *114*, 7301.
17. Shi, S. et al. *J. Am. Chem. Soc.* **1992**, *114*, 10656.
18. Guhr, K. I.; Greaves, M. D.; Rotello, V. M. *J. Am. Chem. Soc.* **1994**, *116*, 5997.
19. Hawker, C. J.; Wooley, K. L.; Frechet, J. M. J. *J. Chem. Soc. Chem. Commun.* **1994**, 925.
20. Stoddart, J. F. *Angew. Chem. Int. Ed. Engl.* **1991**, *30*, 70.
21. Selig, H. et al. *J. Am. Chem. Soc.* **1991**, *113*, 5475.
22. Cox, D. M. et al. *J. Am. Chem. Soc.* **1991**, *113*, 2940.
23. Haddon, R. C. et al. *Nature* **1991**, *350*, 320.
24. Hebard, A. F. et al. *Nature* **1991**, *351*, 600.
25. Chai, Y. et al. *J. Phys. Chem.* **1991**, *95*, 7564.

FUNCTIONAL FIBERS

Hiromu Takeda* and Sei-ichi Manabe
Faculty of Human Environmental Science
Fukuoka Women's University

Advances in fiber science and technology have made possible the production of functional fibers with special characteristics such as temperature resistance, ion exchangeability, electroconductivity and antielectrostatic properties, optical transparency, magnetic susceptibility, water absorption and reversible gel properties, immobilization of enzymes and microorganism cells, biodegradability, self-microfibrillation, separation activity, ultra-fineness of fibers, and anti-bacterial properties. Here we describe some functional fibers.

PREPARATION, PROPERTIES, AND APPLICATIONS

Hollow-fiber Membrane for Particle Separation

Hollow fibers have found wide application in purification and the concentration processes. Their uses can be divided roughly into five areas, based on mean pore size: reverse osmosis membranes, dialysis membranes, ultrafiltration membranes, nanofiltration membranes, and microfiltration membranes. All types are commercially available. Studies using field emission scanning electron microscopy (FE-SEM), wherein the specimen is prepared using a novel method,[1-3] teach us that these membranes fall into two categories: those that undergo phase separation and those that do not, judging from the membrane preparation method. The supermolecular structure for the hollow-fiber membrane obtained through the first method is explained by the aggregation model composed of the fine particles generated through the microphase separation.[4-6] Here the particle to be separated includes a molecule itself in addition to small particles such as colloids.

Preparation

The Microphase Separation Method

Kamide and Manabe discuss the microphase separation method for preparing porous polymeric membranes.[4] In this method, the pores in a finished membrane are generated from the polymer lean phase or from the space between neighboring fine particles after the phase separation.

By contrast, in droplet phase separation a pore generates from inside the fine particle (i.e., the droplet), where the nonsolvents and swelling agents are localized and the interfacial boundary of the particle is composed of polymer chain.[7] After the particles collide with each other, they form into hexagonal packing, and the continuous evaporation of solvents or extraction from inside the particles causes them to break into tetrahedral-type networks, resulting in straight-through pores.

Detailed EM investigations of this mechanism of solidification to a membrane supports the explanation of microphase separation mechanism, since most of the samples collected in the intermediate stage of the process preparing the membrane show the fine particles of the polymer-rich phase.[4,5]

At the initial stage of the phase separation, it is not yet clear which process is dominant: nucleation of the nuclei or spinodal decomposition.[8] In the case of wet spinning, although nucleation followed by growth is confirmed, spinodal decomposition may proceed when the phase separation occurs with decreasing temperature (which also occurs in the melt-spinning process).

This microphase separation method is also used in the case of cuprammonium-regenerated cellulose hollow-fiber such as Bemberg® microporous membrane (BMM) for virus removal.[1-3] The tension created when the hollow fiber is taken up gives rise to breaks in the continuous body of secondary particles, and the void-type pores tend to link. This void linkage should be minimized in order to get high removability of viruses, by lowering the stress on the hollow fiber after coagulation.

Researchers have observed aggregation of fine particles in the case of hollow fibers used for dialysis and of those prepared by using dry and wet spinning. Transport properties are controlled by the total amount of particles, particle size, and the degree of the amalgamation of particles.[9]

Phase separation usually occurs at the heterogeneous area in the composition, such as the interfacial plane between the feed solution and the coagulant solution in the case of wet spinning. Along with diffusional transport of the solvent molecules, phase separation proceeds from the surface to the inside of the feed solution. Progress of the phase separation is marked by two characteristics: the interfacial plane shows more porosity than the inside of the membrane, and the membrane shows the opposite gradient in porosity. This phenomenon is explained by the change in the polymer concentration of the solution in the interfacial boundary.[5]

The Drawing and Annealing Method[10]

Some kinds of melt-spun polymers can be cold drawn, resulting in the generation of fine cracks that appear as slits. The cracks are stabilized as pores by annealing after drawing. This method is applied to crystalline polymers such as polyethylene, isotactic polypropylene, Nylon 6, polyethylene terephthalate, polytetrafluoroethylene, and polyvinylidene chloride.

Properties

Characteristics of the supermolecular structure fall into two categories, observed in both plane and hollow fibers and influenced by the individual preparation. First considered are pore characteristics (such as pore size and porosity), and second, the orientation of a molecular chain and a crystal plane.

Applications

Particle Separation Using a Hollow Fiber

It has been more than 20 years since reverse osmosis with a hollow fiber membrane was used for water desalinization. These membranes now find applications in many fields, among the organic solvent separation. When the mean pore size of the membrane decreases <50 nm, the mechanism of particle separation is influenced by the intermolecular interaction between the raw materials of the membrane and the particles undergoing separation.

Commercially available BMMs with mean pore sizes ranging from 10 to 75 nm can effectively remove viruses while recovering proteins at a high rate.

*Author to whom correspondence should be addressed.

Concentration of Proteins and Viruses

Regenerated cellulose hollow fiber for an artificial kidney has been used to concentrate proteins produced through a biotechnological process. The virus removal filter BMM can concentrate retroviruses without disturbing their activity.[11] Use of these fibers for concentration is likely to increase with future developments in this field.

ION EXCHANGE FIBERS (CONJUGATED FIBERS)

It is well known that ion-exchange fibers are made from modified acrylic copolymers and their derivatives, poly(vinyl alcohol) derivatives, and polystyrene-based derivatives. But these fibers are less in demand, because their ion-exchange capacity and chemical stability rapidly decline with repeated operations, and further, because their mechanical properties are insufficient for industrial materials.

A recent significant development is the production of an ion-exchange fiber made from conjugated fiber, formed by the morphological structure of a "multi-islands" ingredient dispersed in a "sea" ingredient.[12] The fiber's mechanical properties are derived from the multi-islands ingredient (polypropylene); its ion-exchange properties come from the sea ingredient–polystyrene-based derivatives–which are sulfonated and aminated after crosslinking. This ion-exchange fiber is characterized by a high surface-to-weight ratio and by a reduced pressure drop of liquid passing in column.[12,13] The ion exchange rate of this fiber is from 10 to 100 times that of particle-type ordinal ion-exchange resins.[12] This fiber is used for hydrolysis catalyst,[14] immobilization of enzymes,[15] separation of bacterias and microorganism cells,[16] absorption and separation of heavy metal ions and dyes,[13] preparation of ion-free water for electronics use, and of virus- and pyrogen-free water for medical use.

SELF-MICROFIBRILLATION OF FIBERS (POLYMER ALLOY FIBERS)

Polymer alloys and multicomponent polymer systems possess properties and functions that monocomponent polymer materials cannot. Most polymer alloys consist of polymer blends combined with graft polymers or block polymers. The science of those polymer systems–that is, statistic thermodynamics, compatibility, phase separations and morphological structures, polymer–polymer interfaces, and reological properties of polymers–must continue to be expanded to yield new useful polymer alloys. These self-microfibrilable fibers, which find uses as artificial leather and ultra-fine fibers, are produced by combining polymer alloys with advanced spinning technology and fiber science.[17]

HIGH-PERFORMANCE FIBERS

The demand for high-performance fibers that have high strength, high modulus, high temperature resistance, fire resistance, irradiation resistance, low meltability, and high dimension stability has intensified with increasing public awareness of safety standards and sophisticated technology. Since the 1950s, considerable attention has been paid to fibers made from aromatic polyamides,[18,19] polybenzimidazoles,[20,21] polyimides,[22-24] polybenzothiazoles,[25-28] and polybenzoxazoles,[28-32] and many fibers have been investigated. These polymers, which have strong intermolecular forces, are not melted but are dissolved in specific solvents such as sulfuric acid, methane sulfonic acid, DMAc, DMF, and DMSO. Concentrated polymer solutions prepared with those solvents have appropriate viscosity and stability, and are therefore suitable for spinning dope.

These fibers are found in textile fabrics, knitted fabrics, nonwoven materials, paper materials, and reinforcement materials used in industrial, aerospace, environmental, household, sports, and military applications.

OTHER FUNCTIONAL FIBERS

Fibers for absorbing and regenerating solar energy have been significantly improved by sheath-core conjugated fiber, which is made from polyester (sheath) and zirconium carbide-polyester (core). This fiber is produced using high-speed conjugated spinning. To obtain good quality fiber, submicron particle sizes of zirconium carbide (ZrC) in the core are necessary.[33,34] ZrC efficiently absorbs light energy and is transformed into heat energy; the heat energy is then accumulated in ZrC. Textiles and fabrics made from this fiber thus create warmth from sunlight and are ideal for winter sportswear.

Another fiber rapidly absorbs water 500-2,000 times its own weight. This fiber is made from polymers such as slightly crosslinked poly(acrylic acid) (PA) and its salts, PVA-PA polymer alloy, modified PVA, and starch derivatives (SD) graft polymerized with vinyl monomers.

REFERENCES

1. Turumi, T. et al. *Polym. J.* **1990**, *22*, 304.
2. Turumi, T. et al. *Polym. J.* **1990**, *22*, 751.
3. Turumi, T. et al. *Polym. J.* **1990**, *22*, 1085.
4. Kamide, K.; Manabe, S. *Materials Science of Synthetic Membranes*; Lloyd, D. R., Ed.; American Chemical Society: Washington DC, 1985; 197.
5. Manabe, S. et al. *Poly. J.* **1987**, *19*, 391.
6. Fujioka, R.; Manabe, S. *Bull. Home Life Sci.* **1995**, *26*, 13 (Fukuoka Women's University).
7. For example, Kesting, R.; Manafee A. *Kolloid Z.* **1969**, *230*, 341.
8. Baumgartner, A.; Heerman, D. W. *Polymer* **1986**, *27*, 1776.
9. Fujii, Y. et al. *Polym. Prepr. Jpn.* **1987**, *36*(7), 1917.
10. For example, Druns, M. L. U.S. Patent 3 801 404, 1974.
11. Makino, M. et al. *Arch. Virol.* **1994**, *139*, 87.
12. Yoshioka, T.; Shimamura, M. *Bull. Chem. Soc. Jpn.* **1983**, *56*, 3726; Japan Patent 56 8046, 1981, and U.S. Patent 4 693 828, 1987.
13. Yoshioka, T.; Shimamura, M. *Bull. Chem. Soc. Jpn.* **1985**, *58*, 2618; Japan Patent 59 23852, 1984.
14. Yoshioka, T.; Shimamura, M. *Bull. Chem. Soc. Jpn.* **1984**, *57*, 334; Japan Patent 58 9699, 1983.
15. Yoshioka, T.; Shimamura, M. *Bull. Chem. Soc. Jpn.* **1986**, *59*, 399; Japan Patent 57 20408, 1982.
16. Yoshioka, T.; Shimamura, M. *Bull. Chem. Soc. Jpn.* **1986**, *59*, 77; Japan Patent 59 50313, 1984.
17. Nakagawa, K.; Itho, M. *Fiber Sci.* **1986**, *28*, 27; Japan Patent 59 59920, 1984, and U.S. Patent 4 514, 459, 1985.
18. U.S. Patents 3 671 542, 1972, 3 767 756, 1973, 3 817 941, 1974, 4 075 269, 1978.
19. Hancock, T. A.; Spruell, J. E.; White, J. L. *J. Appl. Polym. Sci.* **1977**, *21*, 1227.

20. Wang, W. F. et al. *J. Macromol. Sci. Phys.* **1977**, *B22*, 231.
21. Moelter, G. M.; Tetreault, R. F.; Hefferon, M. *J. Polym. News* **1983**, *9*, 138.
22. Raush, K. W., Jr.; Farrissey, W. J.; Sayigh, A. A. R. preprints, at the 28th Annual Technical Conference. Reinforced Plastics/Composites Institute, the Society of the Plastic Industry: North Haven, Conn. **1973** Section 11-E, p 1.
23. Technical Report, 1984; Chemifaser Lenzing A. G.: Oesterreich.
24. Wada, A. *Sen'i Gakkaishi* **1994**, *50*, 119.
25. Allen, S. R. et al. *Macromolecules* **1981**, *14*, 1138.
26. Chenevey, E. C.; Timmons, W. J. *Preprints,* Fall Meeting of the Materials Research Society, Materials Research Society: Boston, 1988, p 300.
27. Cohen, Y.; Thomas, E. L. *Polym. Sci. Eng.* **1985**, *25*, 1093.
28. Fratini, A. V.; Resh, T. *Preprints,* Fall Meeting of the Materials Research Society, Materials Research Society: Boston, 1988, p 309.
29. Choe, E. W.; Kim, S. N. *Macromolecules,* **1981**, *14*, 920.
30. Takeda, H. *Proc. Int. Symp. Fiber Sci. Tech.,* **1994**, 305.
31. Im, J. H. *Proc. Inter. Symp. Fiber Sci. Tech.* **1994**, 33.
32. Im, J. H. *Preprints,* Fall Meeting of the Materials Research Society, Materials Research Society: Boston, 1988, p 308.
33. Furuta, T. *Sen'i Gakkaishi* **1989**, *45*, 38.
34. Japan Patents H1 132816, 1989, H1 219048, 1989, H1 220798, 1989, H1 256759, 1989, H1 221468, 1989.

FUNCTIONAL IMMOBILIZED BIOCATALYSTS (Stimulus-Sensitive Gels)

Etsuo Kokufuta
Institute of Applied Biochemistry
University of Tsukuba

The immobilizaton of enzymes or microbial cells generally provides the following advantages: continuous operation becomes practical; biocatalysts can be recovered and reused after reactions; biocatalysts can be formed into shapes, such as membranes or beads, required for fitting to specific reaction processes; and in some cases, biocatalysts become stable with regard to changes in temperature, pH, and inhibitor concentration.

A renewed interest in this field may lead to the construction of "functional" immobilized biocatalysts that surpass the conventional definition, or usually credited advantages, of immobilized biocatalysts–for example, immobilized enzyme systems in which an enzymatic process can be controlled by externally applied stimuli such as light, electric fields, pH, temperature, or mechanical force.[1] In such cases, what is crucial in system construction is not to rely on a possible alteration in the property of the biocatalyst (e.g., an enhanced thermal stability) that has frequently been expected as a result of immobilization, but to impose a new capability on a biocatalyst system as the result of a process of rational design. As the principal means of reaching this goal, polymer chemists and scientists have made important contributions to developing supporting matrices using several functional polymers such as pH-sensitive microcapsules,[2-4] thermosensitive polymer gels,[5] and reversibly soluble polyelectrolyte complexes.[6,7]

In some cases the concept of biomimetic engineering-stimulating natural biofunctions in the design and construction of functional immobilized biocatalysts–has also been taken into consideration. On the basis of this idea, a model immobilized enzyme system capable of promoting the pore diffusion not only of a low molecular-weight substrate but also of a polymeric substrate was designed and prepared.[8,9] Polymer gels were used as the support into which enzymes were co-entrapped with carriers having a binding affinity towards the substrates. However, gels have for the most part played the role only of the supports "maintaining" or "holding" the enzymes and carriers (sugar-binding proteins in this case), little attention having been paid to the application of the functional capabilities of gels in the preparation of immobilized biocatalysts.

Recent years have seen dramatic developments in the field of polymer gels, the majority of which have involved the successful synthesis of stimulus-sensitive polymer gels that undergo reversible, discontinuous, or continuous volume changes–often as large as several hundred times–in response to small variations in the conditions surrounding the gel.[5] This phenomena is called the "volume phase transition" of polymer gels. Variables that trigger volume phase transitions include temperature, solvent composition, pH, ion concentration, small electric fields, and light. This property of polymer gels may be one of the most interesting and important subjects in the development of functional immobilized biocatalysts.[10] Here we describe functional immobilized biocatalysts which can be designed and constructed using stimulus-sensitive polymer gels.

Three types of functional immobilized biocatalysts have been prepared using stimulus-sensitive polymer gels:

- the biocatalysts capable of absorbing substrate solutions through thermally controlled volume changes in gel supports;
- the biocatalysts whose catalytic activity can be terminated reversibly through thermally controlled collapse of gel supports; and
- the biocatalysts capable of converting biochemical energy into mechanical work through enzymatically induced volume changes in the gel supports (i.e., biochemomechanical systems).

PROPERTIES

Enhancement of Substrate Diffusion

When immobilizing biocatalysts within polymer gels using physical entrapment methods, we may take advantage of the great resistance to the diffusion of macromolecular substances due to the gel porosity. However, this limited diffusion within the gel phase also causes a reduced mass transfer rate for low molecular-weight compounds, which diminishes the chances for collision and reaction between the molecules lying within the gel and coming from the external bulk phase. One way of applying stimulus-sensitive gels in the area of functional immobilized biocatalysts involves the utilization of their capability for enhancing the diffusion of a substrate from the exterior to the interior of a gel, based on the possible absorption of the surrounding medium by the gel during a thermally controlled swelling process. Contrarily, the shrinking process squeezes from inside the gel the absorbed medium which contains products resulting from the immobilized enzyme reaction along with any remaining substrate. Therefore, repetition of these swelling and shrinking processes enhances mass transfer rates within the gel and elevates immobilized enzyme activity.

Initiation-Termination Control of Enzyme Reaction

A completely collapsed gel acts as a permeability barrier, causing a reduced mass transfer rate for molecules lying within the gel phase and for those passing from the external bulk phase. This is the case both when the gel collapse occurs either at the surface or throughout the whole body of the gel. Therefore, the initiation and termination of enzyme reactions can be regulated through the collapsed and swollen states of stimulus-sensitive gels with immobilized enzymes.

Biochemo-Mechanical Functions

Almost all living systems convert biochemical energy into mechanical work; a good example of this is muscle contraction. It would be interesting to develop gels containing immobilized enzymes that undergo changes in shape due to contractile or expansive forces created as a result of enzymatic reactions. An immobilized enzyme of this sort would be regarded as a "biochemo-mechanical system" that converts biochemical energy into mechanical work through swelling and shrinking. With respect to the purpose of research, therefore, such immobilized biocatalysts are distinct from more usual immobilized systems that have been studied to utilize their functional abilities in chemical conversion.

One approach for constructing biochemo-mechanical systems may be the enzymatic control of the phase transition of gels. Among the many variables that trigger phase transitions, solvent composition and solution pH seem to be controlled enzymatically. It also seems that the phase transition occurring as the result of a change in the balance between the ionic (repulsive) and hydrophobic (attractive) forces can be regulated through an enzyme-induced change in pH or solvent composition.

APPLICATIONS

The immobilized biocatalysts described above, which are capable of enhancing substrate diffusion and also of regulating biochemical reations, could have potential applicability in the design of biochemical conversion processes in which it would be desirable to control enzymatic activities using a small temperature change. In contrast, the enzymatically driven gel (biochemo-mechanical system) provides technological applications outside the field of chemical conversion processes. The gels undergoing reversible swelling and shrinking changes in response to biochemical stimuli in the form of enzyme reactions may serve as drug delivery devices. In fact, the release of a protein such as insulin from the gel system in question occurred when the substrate was added to an aqueous bulk solution to initiate the enzyme reaction within the gel.[11]

In addition, the functional capability of an enzymatically driven gel has been demonstrated to serve as a biochemo-mechanical valve, when a sintered glass filter was prepared with pores that were filled with *N*-isopropylacrylamide/acrylic acid gel containing immobilized glucose oxidase.[12]

REFERENCES

1. Kokufuta, E. *Prog. Polym. Sci.* **1992**, *17*, 647.
2. Kokufuta, E.; Sodeyama, T.; Katano, K. *J. Chem. Soc., Chem. Commun.* **1986**, 641.
3. Kokufuta, E.; Shimizu, N.; Nakamura, I. *Biotechnol. Bioeng.* **1988**, *32*, 289.
4. Kokufuta, E. *ACS Symp. Ser.* **1993**, *532*, 85.
5. Tanaka, T. et al. *Mechanics of Swelling*; NATO ASI Series; Springer-Verlag: Berlin, Germany, Vol. H 64, 1992; pp 683-703.
6. Zezin, A. B.; Izumrudov, V. A.; Kabanov, V. A. *Makromol. Chem., Macromol Symp.* **1989**, *26*, 249.
7. Margolin, A. L. et al. *Biotechnol. Bioeng.* **1982**, *24*, 237.
8. Kokufuta, E.; Shimohashi, M.; Nakamura, I. *Biotechnol. Bioeng.* **1988**, *31*, 382.
9. Kokufuta, E.; Jinbo, E. *Macromolecules* **1992**, *25*, 3549.
10. Kokufuta, E. *Adv. Polym. Sci.* **1993**, *110*, 159.
11. Kokufuta, E. et al. *ACS Symp. Ser.* **1994**, *548*, 507.
12. Kokufuta, E. In *Industrial Biotechnological Polymers*; Gebelein, C. G.; Carraher, C. E., Eds.; Technomic: Lancaster, PA, 1995; Chapter 18, pp 283-295.

Functional Monomers

See: Dental Adhesives (for Alloys and Ceramics)
Functional Polymaleimides, N-Protected
Living Anionic Polymerization (Protected Functional Monomers)

FUNCTIONAL POLYMALEIMIDES N-PROTECTED

Kwang-Duk Ahn
Functional Polymer Laboratory
Korea Institute of Science and Technology

Maleimides are known to be radically polymerizable ethylenic monomers providing good thermal properties. Polymers of maleimides can be classified into polyimides, important high-performance engineering plastics, because of their five-membered succinimide ring structure.[1] The monomers of maleimide (MI) **1** with a substituent at the imino nitrogen, however, exhibit much different polymerization behaviors from the polyimide condensates. The *N*-substituted maleimides (RMI) **2** belong to 1,2-disubstituted ethylenic monomers having a five-membered ring structure (**Structure 1**). The polymerization behaviors of variously substituted RMIs and thermal properties of the polymers have been extensively studied, primarily by Matsumoto and Otsu.[2] They found some RMIs to be quite reactive in radical polymerizations, giving high molecular weights,[3,4] even though 1,2-disubstituted ethylenes are known to react reluctantly in radical homopolymerization.[5]

Despite sluggish homopolymerization, RMIs copolymerize with other vinyl monomers quite smoothly in many cases.[6-10] RMIs are known to produce alternating copolymers when copolymerized with electron-rich monomers such as styrene and vinyl ether derivatives.[11,12] The tendency of alternating copolymerization of electron-deficient RMIs with electron-rich monomers is explained by charge-transfer complexation of the system.[8,13]

RMI polymers are used primarily as high-temperature structural materials, for example as thermosetting resin compositions based on aryl bismaleimides,[1,14] in reactive extrusion of styrene-maleic anhydride copolymers,[15] and as linear polymers of maleimides having benzocyclobutene moieties.[16] Such materials are ideally suited to high-temperature applications because of their excellent thermal stability and high glass transition temperatures (T_g).

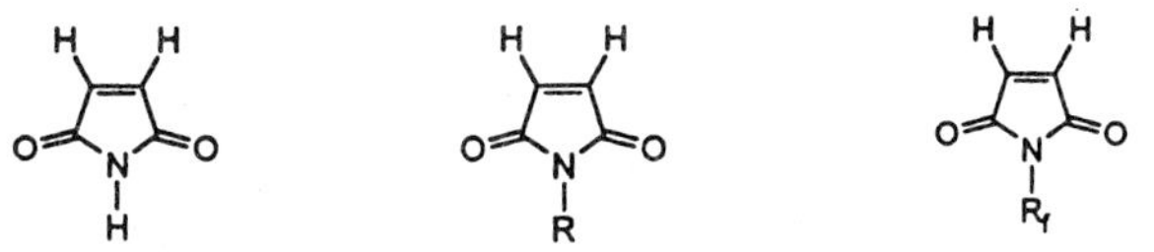

1 (MI) 2 (RMI) 3 (R_fMI) **1**

R = alkyl, aryl
R = desired functionality

FUNCTIONAL MALEIMIDE MONOMERS

Great strides in research have been made on *N*-substituted functional polymaleimides. Advances in research on the functional polymers of RMIs are continuously increasing; three factors account for this progress: First, functional maleimide monomers (R_fMI) **3** are easily prepared by facile and safe introduction of desired functional groups into the imino nitrogen (**Structure 1**). Second, copolymerizations of monomers **3** with electron-rich monomers readily produce high molecular weight alternating copolymers with high conversions within short periods of time. Homopolymerizations or copolymerizations of RMIs with common vinyl monomers are also possible. Third, polymaleimides are thermally stable with high T_gs over 200°C, which are not commonly obtained from conventional vinyl polymers.

Various functional groups are advantageously introduced into the imino nitrogen of **1** (MI) as a substituent, since the ethylenic double bond of **2** (RMI) is less reactive than that of common vinyl monomers toward some useful chemical reactions.

Here we describe a new type of tailor-made functional polymaleimides having readily removable protecting groups. These polymers can bring facile structural modification so that one can attain desired specific changes in polymer properties. Several *N*-protected functional maleimides (R_fMI) that have appropriate protecting groups as *N*-substituents were prepared and radically copolymerized with various styrene monomers. Various aspects of deprotection behaviors and characteristic changes in properties and applicability in photoimaging were investigated for the *N*-protected functional polymaleimides.

PREPARATION OF *N*-PROTECTED FUNCTIONAL MALEIMIDES

Table 1 is a list of (*N*-substituted) functional maleimides **3** (R_fMI) used as a monomer for investigating the characteristic properties of the resulting functional polymaleimides. Several R_fMIs in Table 1 (such as *t*-BOCMI **15**, *t*-BuOMI **16**, iPOCMI **17**, TsOMI **18**, *t*-BOCOMI **19**, and *t*-BOCOPMI **22**) were prepared to obtain a new type of functional polymaleimide that can be modified to attain desired properties through facile deprotection of the polymer side-chain protecting groups. These maleimide monomers were chosen to synthesize *N*-protected functional polymaleimides because they possess radical polymerizability with styrene monomers in an alternating structure as well as facile deprotection capabilities; one can obtain the desired properties in addition to high thermal stability. Most of the *N*-protected R_fMIs discussed here were prepared by retro-Diels–Alder reaction pathways–known to be a useful method for the synthesis of reactive maleimide compounds.[17,23]

N-PROTECTED FUNCTIONAL POLYMALEIMIDES

Polymerization

Unblocked maleimides such as MI **1**, HOMI **4** and HOPMI **21** are known to be reluctant in radical homopolymerizations. The *N*-protected maleimides (R_fMIs) are still reluctant to radical homopolymerizations, producing only low molecular weight polymers in low conversions. But the radical copolymerizations of R_fMIs with styrene monomers yield high molecular weight alternating copolymers in very high conversions over 90% within a short period of reaction time (based on a 1:1 molar feed ratio). The electron-rich styrene monomers (X-St) used are *p*-methylstyrene (MeSt), *p*-chlorostyrene (ClSt) *p*-acetoxystyrene (AcOSt), and *p*-trimethylsilylstyrene (SiSt).

TABLE 1. *N*-Protected Functional Maleimides (R_fMI) **3** and Physical Properties

Compound No.	R_f	R_fMI	mp (°C)	T_{dp}[a] (°C)	Reference
1	H	maleimide (MI)	92	—	
4	OH	*N*-hydroxymaleimide (HOMI)	125	—	
15	CO_2-Bu-*t*	*N*-(*t*-butyloxycarbonyl)maleimide (*t*-BOCMI)	62–63	163	18
16	O-Bu-*t*	*N*-(*t*-butoxy)maleimide (*t*-BuOMI)	92	207	19
17	OCO_2-$CH(CH_3)_2$	*N*-(isopropyloxycarbonyloxy)maleimide (iPOCMI)	70–72		20
18	OTs	*N*-(*p*-toluenesulfonyloxy)maleimide (TsOMI)	148	—	21
19	OCO_2-Bu-*t*	*N*-(*t*-butyloxycarbonyloxy)maleimide (*t*-BOCOMI)	84	102	22
20	OCO_2-Ph	*N*-(phenyloxycarbonyloxy)maleimide (PhOCMI)	98	—	23
21	C_6H_4-OH	*N*-(*p*-hydroxyphenyl)maleimide (HOPMI)	184	—	24
22	$C_6H_4OCO_2$-Bu-*t*	*N*-[*p*-(*t*-butyloxycarbonyloxy)phenyl]maleimide (*t*-BOCOPMI)	142	168	22
23	$Si(CH_3)_3$	*N*-(trimethylsilyl)maleimide (TMSMI)	—[b]		25
24	*t*-Bu	*N*-(*t*-butyl)maleimide (*t*-BuMI)	—[c]		36

[a] T_{dp} is a deprotection temperature of protecting groups.
[b] bp 75 °C (5 mmHg).
[c] bp 54 °C (3.5 mmHg).

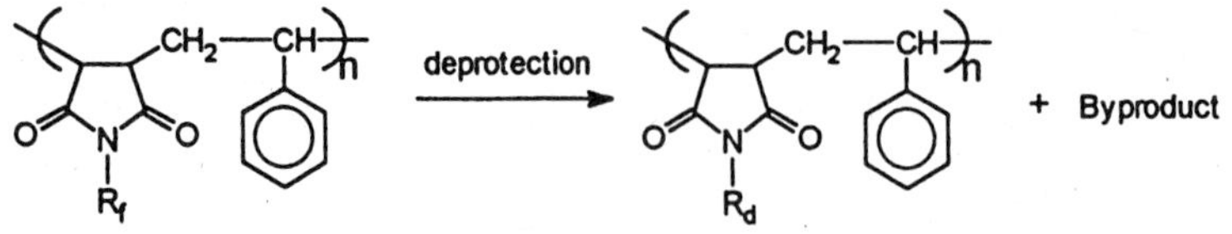

P(R_fMI/St) 32-36 P(R_dMI/St) 42-44

R_f = t-BOC, P(t-BOCMI/St) 32 → R_d = H, P(MI/St) 42 + isobutene + CO_2

R_f = t-BuO, P(t-BuOMI/St) 33 → R_d = OH, P(HOMI/St) 43 + isobutene

R_f = O-CO_2-Pr-i, P(iPOCMI/St) 34 → R_d = OH, P(HOMI/St) 43 + propene + CO_2

R_f = O-t-BOC, P(t-BOCOMI/St) 35 → R_d = OH, P(HOMI/St) 43 + isobutene + CO_2

R_f = C_6H_4-O-t-BOC, P(t-BOCOPMI/St) 36 →

R_d = C_6H_4-OH, P(HOPMI/St) 44 + isobutene + CO_2

I

Thermal Deprotection Behaviors

Scheme I describes thermal deprotection of the side-chain protecting groups of *N*-substituted functional polymaleimides P(R_fMI/St) to P(R_dMI/St). The thermal deprotection behaviors are well understood from thermal analysis techniques with differential scanning calorimetry (DSC) and thermogravimetric analysis (TGA). In a TGA thermogram of P(*t*-BOCMI/St) **32**, it is stable up to 130°C, but above 150°C it undergoes a precipitous mass loss by the complete deprotection of *t*-BOC groups, and then a plateau persists until the main-chain decomposition temperature at about 370°C.[18] P(*t*-BOCMI/St) **32** is converted to P(MI/St) **42** by deprotection, releasing isobutene and carbon dioxide. The DSC thermograms of P(*t*-BOCMI/St) reveal an endothermic event during the deprotection of *t*-BOC groups at 152°C, defined as a thermal deprotection temperature (T_{dp}) in the first run, and show the glass transition temperature (T_g) at 245°C in the second run.[18] The glass transition temperature observed in the second run corresponds to that of the deprotected polymer P(MI/St) **42** and the temperature is confirmed to be the same as the glass transition temperature of P(MI/St), obtained by the direct copolymerization of MI and St monomers without deprotection.

Upon deprotection, the *N*-protected maleimide units are converted to the deprotected forms as follows: *t*-BOCMI to MI, *t*-BuOMI to HOMI, iPOCMI to HOMI, *t*-BOCOMI to HOMI, *t*-BOCOPMI to HOPMI, and *t*-BuMI to MI. The *t*-BOC protecting groups release isobutene and carbon dioxide while the iPOC protecting group releases propene and carbon dioxide, and the *t*-BuO group only releases isobutene.

Properties and Application

The alternating copolymers P(R_fMI/X-St) are white powders having a good film-forming property. They are very soluble in common organic solvents such as acetone, chloroform, toluene, anisole, dioxane, and DMF, but insoluble in methanol, hexane, and aqueous base solutions. On the contrary, the deprotected polymers P(HOMI/St), P(MI/St), P(HOPMI/St) and P(R_dMI/HOSt) are soluble in dioxane, dimethylformamide, and aqueous base solutions, but insoluble in common organic solvents due to the large polarity changes in the side-chains. The significant change in solubility by deprotection is a useful property for the application of the polymers, as resist materials in lithography or image-making processes.

A novel technology based on the chemical amplification concept was established to afford a drastic modification of polymer properties through deprotection of side-chains by thermal and photochemical acidolytic processes. During the past decade, the chemically amplified resists have attracted a great deal of attention because of their high sensitivity and versatility in lithographic applications.[27-29] Enormous efforts have been made to develop novel polymeric photoresist materials that can provide high-resolution images that possess high thermal stability with high glass transition temperatures and aqueous base developing capabilities. The *N*-protected functional polymaleimides P(R_fMI/X-St) are potential candidates by virtue of their unique thermal stability and polar functionality.

Improvement of resist properties in applications of *N*-protected functional polymaleimides is ongoing; the polymers possess versatile usefulness for modification of desired properties based on their unique protection-deprotection behaviors. In applications as resist materials utilizing other *N*-protected maleimide polymers, researchers report generation of high-resolution image patterns.[30-32]

ACKNOWLEDGMENTS

The authors are deeply grateful to MOST and KIST for the financial support on the MI-polymer research, to C. G. Willson for the initial contribution, and to the coauthors Chan-Moon Chung, Deok-Il (Jae-Sun) Koo, Jong-Hee Kang, and colleagues in the Functional Polymer Laboratory.

REFERENCES

1. Takekoshi, T. *Advances in Polymer Science* Springer-Verlag: Heidelberg, Germany, **1990**, Vol. 94, p 1.
2. a) Matsumoto, A.; Oki, Y.; Otsu, T. *Eur. Polym. J.* **1993**, *29*, 1225; b) From Ref. 61 as 2.b.
3. Matsumoto, A. et al. *Chem. Lett.* **1991,** 1141.
4. Otsu, T.; Matsumoto, A.; Tatsumi, A. *Polym. Bull.* **1990,** *24*, 467.
5. Minoura, Y. In *Structure and Mechanism in Polymerization* Tsuruta, T., O'Driscoll, K. F., Eds., Marcel Dekker: New York, NY, **1969,** p 190.
6. Tawney, P. O. et al. *J. Org. Chem.* **1961,** *26*, 15.
7. Van Paesschen, G.; Timmerman, D. *Makromol. Chem.* **1964,** *78*, 112.
8. Olson, K. G.; Butler, G. B. *Macromolecules* **1984,** *17*, 2480.
9. Sandreczki, T. G.; Brown, I. M. *Macromolecules* **1990,** *23*, 1979.
10. Iwatsuki, S. et al. *Macromolecules* **1991,** *24*, 5009.

11. Barrales, J. M.; DeLa Campa, J. I. G.; Ramos, J. G. *J. Macromol. Sci.-Chem.* **1977,** A11, 267.
12. Turner, S. R.; Anderson, C.; Kolterman, K. M. *J. Polym. Sci., Polym. Lett.* **1989,** *27,* 253.
13. Mohamed, A. A.; Jebrael, F. H.; Elsabee, M. Z. *Macromolecules* **1986,** *19,* 32.
14. Chisholm, M. S. et al. *Polymer* **1992,** *33,* 838, 842, 847.
15. Vermeesch, I. M.; Groeninckx, G.; Coleman, M. M. *Macromolecules* **1993,** *26,* 6643.
16. Kirchhoff, R. A.; Bruza, K. J. *Chemtech* **1993,** *23,* 22.
17. Narita, M.; Teramoto, T.; Okawara, M. *Bull. Chem. Soc. Jpn.* **1971,** *44,* 1084.
18. Ahn, K-D.; Lee; Y-H Koo, D-I.; *Polymer* **1992,** *33,* 4851.
19. Ahn, K-D; Koo, D-I. In *Polymers for Microelectronics* Thompson, L. F.; Willson, C. G., Tagawa, S., Eds.; ACS Symposium Series 537; American Chemical Society: Washington, DC, **1994,** Chapter 9, p 124.
20. Ahn, K-D.; Willson, C. G. *Bull. Korean Chem. Soc.* **1995,** *16,* 443.
21. Ahn, K-D.; Chung, C-M.; Koo, D-I. *Chem. Mater.* **1994,** *6,* 1452.
22. Ahn, K-D.; Koo, D-I.; Willson, C. G. (Ref. 44) *Polymer* **1997,** *36,* 2621
23. Akiyama, M. et al. *J. Chem. Soc., Perkin I* **1980,** 2122.
24. Park, J. O.; Jang, S. H. *J. Polym. Sci., Polym. Chem.* **1992,** *30,* 723.
25. Matsumoto, A.; Oki, Y.; Otsu, T. *Polym. J.* **1991,** *23,* 201.
26. Matsumoto, A.; Kubota, T.; Otsu, T. *Polym. Bull.* **1990,** *24,* 459.
27. MacDonald, S. A.; Willson, C. G.; Fréchet, J. M. *J. Acc. Chem. Res.* **1994,** *27,* 151.
28. Reichmanis, E. et al. *Chem. Mater.* **1991,** *3,* 394.
29. *Polymers for Microelectronics* Thompson, L. F.; Willson, C. G.; Tagawa, S., Eds.; ACS Symposium Series 537; American Chemical Society: Washington, DC, 1994.
30. Osuch, C. E. et al. *Proc. SPIE* **1986,** *631,* 68.
31. Leuschner, R. et al. *Polym. Eng. Sci.* **1992,** *32,* 1558.
32. Schaedeli, U. et al. *Proc. SPIE* **1994,** 2195, 98.

Functional Polymers

See: *Aminosiloxanes*
Benzocyclobutene-Based Monomers (Diels-Alder Ring-Operating Polymerization)
Benzocyclobutene Homopolymerization (Chemistry and Applications)
Benzocyclobutenes (Crosslinking and Related Reactions)
End Functionalized Oligomers (by Chain-Transfer Technique)
Fullerene-Functionalized Polymers
Functional Fibers
Functional Polymaleimides, N-Protected
Functional Polymers (Advanced Functions by Chemical and/or Physical Modification)
Functional Polymers (By Carbocationic Polymerization)
Functional Polymers (Telechelics and Macromonomers)
Functional Polysiloxanes
Functionalized Polymers (Comb, Rotaxanic and Dendrimeric Structures)
Functionalized Polymers (Telechelics and Macromonomers, via Cationic Polymerization)
Functionalized Polyolefins (Using Borane Monomers and Transition Metal Catalysts)
Living Carbocationic Polymerization, Olefins
Living Coordination Polymerization
Microemulsion Polymerization (Functionalization of Microlatices)
Photoreactive Langmuir-Blodgett-Kuhn Assemblies (Functionalized Liquid Crystalline Side Chain Polymers)
Poly(aryl ether ketone) (Functionalization)
Poly(aryl ether ketone)s, Amine-Terminated
Polyesters [Bis(Cyclic Ether) and Diacyl Chloride Polyaddition]
Poly(terephthalic acid anhydride) (as a Latent Monomer)
Polyvinylamine (Overview)
Polyvinylamine (Preparation, Properties and Applications)
Propellanes (New Vistas in their Polymer Chemistry)
Reactive Oligomers (Overview)
Reverse Micelles (Microcontainers for Functional Polymers)
Styrene-Maleic Anhydride Copolymer
Supported Lewis Acid Catalysts (on Polypropylene; Recoverable and Reusable)
Tailor-Made Polymers
Uniform Polymers

FUNCTIONAL POLYMERS (Advanced Functions by Chemical and/or Physical Modification)

Yasuhiko Kurusu
Department of Chemistry
Faculty of Science and Technology
Sophia University

Natural high polymers such as jute yarn, silk, and cotton show "function," that is, they keep the human body at the proper temperature and protect our skin from the environment. Synthetic high polymers are now more common than natural polymers, because of the stable chemical, mechanical, electrical, and thermal properties. Synthetic polymers may be considered functional polymers in a broad sense. We will discuss common polymers modified by chemical reactions and synthesized polymers designed to display advanced function.[1] Functional polymers are classified by chemical reactivity, catalysts, energy transfer, transportation, and transmission of information. These kinds of polymers are also known as electron-transfer polymers.[2]

COMMON POLYMERS

Functional Polymer Indicating Reactivities

Polyethylene (PE), polypropylene (PP) and polystyrene (PS) are chemically stable. In order to improve this character, functional groups are connected to these polymers by chemical reactions. Polymers obtained by this method were used for polymer-type chemical reagents.

Other methods to obtain functional polymers were then used. Functional polymers containing a halogen group at the terminal position in the alkyl chain were reported.[3]

Natural Polymer

By the formation of ester or ether linkage from the hydroxy group of cellulose, the solubility of the cellulose increased. This modification greatly improves the character of the organic solvents. The value of the cellulose increases because it is "reactive." Polysaccharide derivatives in nature (for example, dextrin, chitin, and so on) are also functionalized by this method.

Inorganic Polymer

Silica is hydrophilic and stable in organic reaction conditions. The introduction of organic functional groups improves the reactivity and affinity of silica to organic solvents and leads to a new application. In order to introduce a functional group, silane coupling reagents ($(RO)_3SiR'X$; X is a reactive group such as—NH_2 and halogen) were used. This modified compound also supports organic reactions. The modified silica has been widely used for packing in chromatographic separations.

FUNCTIONAL POLYMER WITH CATALYTIC ACTIVITY

Macromolecular Ion

The solvation of polyethylenimine hydrochloric acid, polystyrenesulfonic acid sodium salt, and poly(vinylpyridinium salt) accelerates ionic reactions. In this reaction, the larger the molecular weight, the faster the reaction. These polymers could be used for ester hydrolysis, esterification, dehydradion, Knoevenagel reaction, etc. Mathur has summarized these reactions with catalyst polymers.[1b]

Polymer Containing a Nucleic Acid Base

When a monomer containing adenine(A) was polymerized by a radical initiation in the presence of a polymer obtained from a monomer containing chimine(T), the rate of polymerization increased with concentration of the polymer. This is because the hydrogen bond between A and T accelerates the polymerization.

Polymer with Metal Complex

Metal complex polymer are synthesized by metals reacting with polymer functional groups that act as ligands. For example, mixed metal complex polymer of Fe(III) and Cu(II) was synthesized by reacting poly(vinyl alcohol) with Cu(II) and Fe(II). This complex polymer was effective for decomposing of hydrogen peroxide and acceleration was reported.

Poly(dimethyl-*co*-isocyanopropyl-*p*-methylsiloxane) was synthesized and metal vapors were reacted with the polymer. Different oxidation states for the metals were observed, and novel redox-active polymers may be possible.[4]

Immobilization at linear polymer enhances the solubility of the chelate in organic solvents and water. Phthalocyanines (Pc) exhibit excellent photocatalytic properties. Polymer binding of such sensitizers can enhance the photocatalytic properties because of hindered intermolecular interaction of Pc molecules and the stabilizing effect of the surrounding polymer.[5]

Many oxidation catalysts were also reported. The activity of the catalyst was estimated by singlet oxygen acceptor. Rose bengal was binded to polystyrene, which was applied to photooxidation.[6] Cobalt(II) phthalocyanine tetrasodium sulfonate was attached to polyvinylamine. In the oxidation of thiols, this functional polymer showed 50-fold enhancement compared to polymer-free cobalt(II) phthalocyanine.[7]

Polystyrene-bound phenylseleninic acid was used for oxidation of olefins, ketones, and aromatic compounds.[8]

Heterogeneous and Homogeneous Polymer in Organic Reactions

Many studies have been made of ion-exchange resin followed by redox polymer, and many reactive polymers have appeared. We will now summarize ongoing work in this field.

Oxidation

The introduction of thioanisol to insoluble polymer was useful for Moffatt-type oxidation.[9] The oxidation of OH group was carried out by poly(vinylpyridine dichromate).[10]

Reduction

Poly(4-vinylpyridine borane) was reported as a polymeric reducing agent.[11,12] This reagent was effective for the selective reduction of alkynes to *z*-alkenes.[13]

Halogenation

Ammonium perbromate was introduced to the polymer-containing ammonium group and was effective for the α-bromination of ketones. This reaction occurs mainly from the less hindered side of the molecule.[14]

Acylation

Polystyrene-bound-4-hydroxy-3-nitrobenzoyl group has been reported as a polymer-type acylating reagent. This polymer was acylated by acid chloride acid anhydride, or carboxylic acid in the presence of DCC.

Polymer Containing Enzymelike Function

One approach is the attachment of a unique performance group. For example, cyclam strongly accepts Cu^{2+} ion, and diethylenetriamine cyclodextrine forms selectively additive products with benzene and naphthalene. Introducing functional substances makes the functional polymer show higher performance.

ENERGY TRANSFER POLYMER

Polymer that Transfers from Chemical Energy to Mechanical Energy

Polymer electrolytes vary solubility and conformation according to pH, concentration of salt, and dielectric constant of the solvent. When the polymer is solid state, the size alters.

Polymer Exchangeable from Thermal Energy to Mechanical Energy

The complex formation of poly(ethylene glycol) and poly(methacrylic acid) is dependent on temperature. The volume change of the polymer coming from the complex formation and dissociation applies to energy transfer.

Polymer Exchangeable from Photo Energy to Mechanical Energy

The reversible transformation polymer material by photoreaction makes possible the conversion of photo energy to mechanical energy. The method involves changing molecular volume by photoisomerization of the spirobenzopyrane group.

Polymer Exchangeable from Photo Energy to Electro Energy

The p-n contact is the connection of polyacetylene sensitized by an acceptor and polyacetylene sensitized by a donor. Polymer composed by this method can be used for a photo battery.

LIQUID TRANSPORTATION THROUGH POLYMERIC FILM

Acetylcellulose film can separate water and ions such as Na^+ and Cl^-. Perfluorosulfonic acid film is an example of such a film. The separation is made possible by film with different charge ion.

REFERENCES

1. (a) *Iwanami Gendai Kagaku* vol. 20; Functional Polymers, Ise, N.; Tabushi, I. Ed. Iwanami Schoten: Tokyo 1980. (b) Mathur, N. K. et al. *Polymers as Aids in Organic Chemistry,* Academic 1980. (c) Hodge, P. *Polymer-Supports Reactions in Organic Synthesis* Sherrington, D. C. Ed.; John Wiley & Sons: New York, 1980. (d) *Polymer Reagents and Catalysts* Ford, W. T. Ed. ACS Symposium Series 308; American Chemical Society: Washington, D. C., 1986. (e) Hodge, P. *Syntheses and Separation Using Functional Polymers* Sherrington, D. C. Ed.; John Wiley & Sons: New York, 1988.
2. Manecke, G.; Storck, W. *Encyclopedia of Polymer Science and Engineering* 2nd ed.; John Wiley & Sons: New York, **1986,** vol. 5.
3. McManus, S. P.; Olinger, R. D. *J. Org. Chem.* **1980,** *45,* 2717.
4. Albers, M. O.; Conille, N. J. *J. Chem. Soc., Chem. Commun.* **1980,** 1039.
5. Wohrle, D.; Krawczyk, G. *Polym. Bull.* **1986,** *15,* 193.
6. Blossey, E. C. et al. *J. Am. Chem. Soc.* **1987,** *95,* 5820.
7. Brouwer, W. M. et al. *J. Mol. Catal.* **1985,** *29,* 335.
8. Taylor, R. T. *J. Org. Chem.* **1983,** *48,* 5160.
9. Crosby, G. A. et al. *J. Am. Chem. Soc.* **1975,** *97,* 2232.
10. Frechet, J. M. J. et al. *J. Org. Chem.* **1981,** *46,* 1728.
11. Domb, A.; Avny, Y. *J. Macromol. Sci.-Chem.* **1985,** A22, 167.
12. Hallensleben, M. L. *J. Polym. Sci., Symposium No. 47,* **1974,** 1.
13. Kini, A. D. et al. *Tetrahedron Lett.* **1994,** *35,* 1507.
14. Cacchi, S. et al. *Synthesis* **1979,** 64.

FUNCTIONAL POLYMERS (By Carbocationic Polymerization)

Alain Deffieux
Laboratoire de Chimie des Polymères Organiques
ENSCPB-CNRS
Université Bordeaux-l

The recent discovery that alkenyl monomers could polymerize by cationic processes that are close to living polymerizations has drastically increased interest in this technique for the preparation of well-controlled polymer structures. Among these, end-functional and side chain-functional polymers prepared by cationic polymerization are an important class of materials.

END-FUNCTIONALIZED POLYMERS: MONOTELECHELICS

Synthesis by Functional Initiation

Depending on the monomer and polymerization conditions, various approaches have been developed for the introduction of functional groups at the chain head (noted Y at the α-end). They involve chain precursors bearing functional groups that serve, after activation, as polymerization initiators. The prerequisite is that the Y function remains inert during the polymerization; otherwise it should be introduced in a protected form.

α-Telechelic Poly(vinyl ether)s

A typical method consists of a vinyl ether bearing the Y function as a chain precursor.[4] Most functional vinyl ethers have been prepared from chloroethyl vinyl ether by chloride substitution.[5-8]

Another approach is the formation of an α-halogeno ether derivative by reaction of a carbonyl or an acetal compound with trimethylsilyl iodide (TMSI).[10-15] In a similar way, hydroxyl headgroups have been introduced into poly(vinyl ether)s by starting at the polymerization from cyclic acetal derivatives.[12]

α-Telechelic Polystyrenes

Head-functional polystyrenes with a halide atom at the α-end were prepared, more than 20 years ago, by Kennedy using $Cl_2/AlMe_3$[16] and $Br\text{-}C(CH_3)\ (CH_2)_3\text{-}C(CH_3)_2\text{-}Cl/AlEt_3$[17] as initiator.

Recently Higashimura and co-workers have shown that α-halogeno ether derivatives are excellent initiators of styrene derivatives polymerization. A series of α-functional poly(*p*-methoxy- and *p*-*tert*-butoxystyrene)s were obtained by initiation of the polymerization with a functional vinyl ether-hydrogen iodide adduct ($Y\text{-}CH_2\text{-}CH_2\text{-}O\text{-}CHI\text{-}CH_3$) in conjunction with ZnI_2.[18,19]

In the same way, α-functional polystyrenes and poly(α-methyl styrene)s were obtained by the association of a functional vinyl ether-hydrogen iodide adduct with $SnCl_4$ and an ammonium salt.[20]

α-Telechelic Polyisobutylene

Functional initiation has also been used to introduce headgroups in polyisobutylene and relevant alkenyl monomers.[2,3]

Chloride headgroups, stable under isobutylene polymerization conditions, have been introduced using $Cl_2/AlMe_3$.[16] In this case, however, chain termination may occur by chlorination. In a recent study, the functionalization of polyisobutylene chains by haloboration initiation has been reported.[21,22]

Preparations of polyisobutylene macromonomers with dicyclopentadienyl or styrenyl endgroups, starting respectively from dicyclopentadienyl chloride/$AlEtCl_2$,[23] *para*-chloromethylstyrene/BCl_3,[24] and *p*-2-bromoethylstyrene/$AlMe_3$[25] followed by dehydrobromination, have also been described.

As illustrated above, the synthesis of head-functionalized polymers by controlled initiation of living carbocationic polymerization is extremely versatile, making available a broad range of new α-functionalized polymers.

Synthesis by Functional Termination

Endcapping of living polymer chains or inifer systems by functional terminating agents may be considered as both an alternative route to monotelechelics and as a complementary method for the synthesis of di- and multifunctional polymers. This functionalization procedure is of course much more dependent on the livingness of the polymerization (absence of transfer and true termination) and, as we will see, on the reactivity of polymer ends that exist both in active and dormant forms.

ω-Telechelic Poly(vinyl ether)s

In vinyl ether polymerization, the living chains can be directly quenched by a nucleophile (Nu–Z) bearing a functional group Z or its protected form.[2] In most cases the nucleophile directly adds to the living end, leading to ω-end-functionalized chains.

Alcohols readily react with α-iodo ether polymer ends to form acetal terminals.[26] The latter can be used to restart the polymerization of another vinyl ether[13] or styrene,[15] thus leading to block copolymers. Hydroxyl- and carboxyl-terminated poly(alkyl vinyl ether)s have also been prepared by quenching polymer chains with substituted anilines (H_2N-C_6H_4-Z, Z= $CO_2C_2H_5$, CO_2H).[29]

Poly(isobutyl vinyl ether)s with stable carboxylic acid endgroups have been prepared by endcapping the living chains with sodiomalonic ester ($NaCH(CO_2Et)_2$) and subsequent hydrolysis of the malonate by alkali and thermal decarboxylation.[8]

Vinyl ether macromonomers of isobutyl and 2-benzoyloxyethyl vinyl ethers have been synthesized using the same technique. The chain endcapping agent was $NaC(CO_2Et)_2CH_2CH_2OCH{=}CH_2)$.[30]

ω-Telechelic Polystyrenes

Functional termination of living polystyrenes by nucleophiles is more difficult because of the high nucleophilic character of the halide terminal. Most of the time it remains attached to the chain end, leading to *sec*-benzylic halide (often chloride) termini. An intermediate situation is observed with *p*-alkoxystyrenes. They can be polymerized with HI/ZnI_2 initiating systems[18,31] the iodide is not tightly attached to the terminal benzylic carbon. Indeed, poly(*p*-alkoxystyrene)s can be endcapped with functional primary alcohols ($HOCH_2CH_2$-Z with Z = acetate, acrylate, and methacrylate) yielding ω-end functional polymers, but tertiary alcohols and sodiumalonate are ineffective as endcapping agents.

ω-Telechelic Polyisobutenes

Important efforts have been made to develop simple ω-functionalization techniques for polyisobutylene chains prepared by controlled polymerization. However, as mentioned for some polystyrene derivatives, as soon as a chlorine is present either in the initiator or in the catalyse (BCl_3, $TiCl_4$), most attempts to introduce another nucleophile at the chain end have failed[33] and yielded polyisobutylenes with *tert*-butyl chloride termini.

Direct end-quenching of polyisobutylene living ends was achieved only with allyltrimethylsilane and allyltrimethylstannate, leading to allyl-capped polymers.[34,35]

Though direct ω-functionalization of polyisobutylenes by endcapping is limited, a large series of telechelics have been prepared by post-polymerization reactions either from tertiary chlorine or from unsaturated terminated oligomers. The procedures have been described in detail by Kennedy and Ivan.[3]

DI- AND PLURITELECHELICS

Linear difunctional, branched, and star plurifunctional macromolecules bearing functional endgroups are important mainly as precursors of multiblock structures and networks. Endfunctional multiarmed polymers are also interesting for their spatial shape, which can afford a particular distribution of functional termini at the polymer/surrounding interface.

Synthesis by Multifunctional Initiation

Owing to the fast exchanges that take place between active and dormant species in the modern cationic polymerization of alkenyl monomers, the synthesis of di- and multiendfunctional polymers from di- or plurifunctional initiators appears as an appropriate procedure.

ω-Di- and ω-Pluritelechelic Poly(vinyl ether)s

The preparation of ω-difunctional poly(vinyl ether)s by bifunctional initiation has been achieved using divinyl ether/HX[28] or diacetal compound/TMSI systems[27] as chain precursors. A broad series of divinyl ether molecules, most of them prepared by substitution reactions of 2-chloroethyl vinyl ether with diols, are available because divinylethers are used in industry as reagent in cationic photopolymerization.[36-40]

Poly(vinyl ether)s end-quenching by functional molecules lead to the corresponding ditelechelic poly(vinyl ether)s.

An alternative route to ditelechelic poly(vinyl ether)s, based both on monofunctional initiation and functional termination, has also been used. Though there is, in principle, no particular advantage to the latter approach for homofunctional ditelechelics, a relatively broad series of homo ditelechelic poly(vinyl ether)s have been prepared by this route.[7-9,41]

ω-Di- or ω-Pluritelechelics Polystyrenes

Only a few examples of di- or multitelechelic polystyrenes have been prepared so far using multifunctional initiators. The preparation of dichloro-telechelic poly(*p*-chlorostyrene) using dicumyl chloride/BCl_3 has been described by Zsuka.[32] A hexa-armed polystyrene with chloride termini was recently obtained from a chain precursor bearing six 1-phenylethyl chloride groups and with $SnCl_4/nBu_4NCl$ as catalytic system.[42]

ω-Di- and ω-Pluritelechelic Polyisobutenes

The use of di- and multifunctional initiators for the synthesis of multitelechelic polyisobutylenes has been extensively studied.[3] Alkyl, alkenyl, and aromatic derivatives[43-53] bearing *tert*-butyl chlorine, *tert*-butyl alcohol, *tert*-butyl ester, or *tert*-butyl ether groups as chain initiators with BCl_3, $TiCl_4$, or an aluminum alkyl catalyst have usually been used. Di-, tri- and tetratelechelic polyisobutylenes have been obtained by the (so-called) inifer technique or by living polymerization.

Synthesis by Chain End-Coupling

The synthesis of di- and multiend-functional polymers can be achieved by a combination of monofunctional initiation and coupling of the growing polymer ends with appropriate multifunctional monomers or molecules. At the end of the polymerization, addition of a second monomer with a functionality higher than 2, divinyl benzene for example, yields to the formation of spherical polymers in which a number of arm chains are attached to a micro gel core of the second multifunctional monomer. In the second approach the number of arm chains is determined by the functionality of the coupling agent.

SIDE-CHAIN FUNCTIONALIZED POLYMERS

Side-Chain Functionalized Poly(vinyl ether)s

A great variety of vinyl ether derivatives bearing halide, ester, ether, and other polar substituents, which were found to act as chain transfer or terminating agents in conventional cationic polymerizations, have been polymerized by a living-type mechanism using HX-vinyl ether adducts as chain precursors and I_2 or a weak Lewis acid (ZnX_2,...) as catalysts.[1,2] Under such mild conditions, poly(functional vinyl ether)s with controlled molecular weights and narrow molecular weight distributions were synthesized.

An important series of random and block copolymers made up of several functional and nonfunctional vinyl ethers has been synthesized. Various block copolymers containing one or several functional blocks were synthesized by the sequential controlled polymerization of alkyl vinyl ethers and functional vinyl ethers[55-58] or aromatic monomers.[59,60]

Amphiphilic star-shaped poly(vinyl ether)s with separate sets of hydrophilic and hydrophobic arms, or with arms constituted of hydrophilic and hydrophobic segments placed either in inner or outer positions, have been synthesized by the micro gel core technique. Also based on a controlled cationic polymerization, the synthesis of poly(chloroethyl vinyl ether)s with a macrocyclic[9a,64] or plurimacrocyclic chain architecture[65] has been achieved. Some of these star[61-63] and cyclic[54,66] polymers exhibit interesting complexing properties toward organic substrates.

Side-Chain Functionalized Polystyrenes

As indicated in previous sections, the controlled carbocationic polymerization of styrene and of its *p*-methyl, *p*-methoxy, *p*-chloro, and *p*-fluoro derivatives has been accomplished under different conditions. The presence of a methoxy or halide group on the phenyl ring induces significant variation in the glass transition temperature and presents a particular interest for flame retardancy.

REFERENCES

1. *Comprehensive Polymer Science*; Eastmond, G. C. et al., Eds.; Pergamon: Oxford, 1989; Vol. 3, Chapter 39-44.
2. Sawamoto, M. *Prog. Polym. Sci.* **1991,** *16,* 111.
3. Kennedy, J. P.; Ivan, B. *Comprehensive Polymer Science*; Hanser: Munich, 1991.
4. Aoshima, S. et al. *Macromolecules* **1985,** *18,* 2097.
5. Aoshima, S. et al. *Polym. Bull.* **1985,** *14,* 425.
6. Hashimoto, T. et al. *J. Polym. Sci., Polym. Chem.* Ed. **1987,** *25,* 1073.
7. Sawamoto, M. et al. *Makromol. Chem., Macromol. Symp.* **1988,** *13/14,* 513.
8. Sawamoto, M. et al. *Macromolecules* **1987,** *20,* 1.
9. Shohi, H. et al. *Polym. Bull.* **1989,** *21,* 357. (a) Schappacher, M.; Deffieux, A. *Makromol. Chem., Rapid Commun. Symp.* **1991,** *12,* 447.
10. Kamigaito, M. et al. *J. Polym. Sci., Polym. Chem. Ed.* **1991,** *29,* 1909.
11. Kamigaito, M. et al. *Makromol. Chem.* **1993,** 194, 727.
12. Meirvenne, D. V. *Polym. Bull.* **1990,** *23,* 185.
13. Haucourt, N. et al. *Makromol. Chem., Rapid Commun.* **1992,** *13,* 329.
14. Goethals, E. J. et al. *Macromol. Symp.* **1994,** *85,* 97.
15. Cramail, H. et al. *Polym. Adv. Technol.* **1994,** *5,* 568.
16. Kennedy, J. P.; Sivaram, S. *J. Macromol. Sci. Chem.* **1973,** A-7, 969.
17. Kennedy, J. P.; Melby, E. *J. Polym. Sci., Polym. Chem.* Ed. **1975,** *13,* 29.
18. Shohi, H. et al. *Macromolecules* **1992,** *25,* 53.
19. Shohi, H. et al. *Macromol. Chem.* **1992,** 193, 1783.
20. Miyashita, K. et al. *Macromolecules* **1994,** *27,* 1093.
21. Kennedy, J. P.; Carter, J. D. *Polym. Prepr., Am. Chem. Soc. Div. Polym. Chem.* **1986,** *27(2),* 29.
22. Balogh, L. et al. *Prepr., 35th IUPAC on Macromolecules*, Akron, OH, 1994.
23. Kennedy, J. P. et al. *Polym. Bull.* **1983,** *9,* 268.
24. Kennedy, J. P.; Frisch, K. C. U.S. Patent 4 327 201, 1982; *Prepr., Macro Florence* **1980,** *2,* 162.
25. Kennedy, J. P. *Polym. Prepr., Am. Chem. Soc. Div. Polym. Chem.* **1982,** *23(2),* 99.
26. Miyamoto, M. et al. *Macromolecules* **1984,** *17,* 2228.
27. Bennevault, V. et al. *Macromol. Chem. Phys.* **1995,** 196, 3075.
28. Miyamoto, M. et al. *Macromolecules,* **1985,** *18,* 123.
29. Sawamoto, M. et al. *Polym. Bull.* **1987,** *18,* 117.
30. Sawamoto, M. et al. *Polym. Bull.* **1986,** *16,* 117.
31. Higashimura, T. et al. *Makromol. Chem.* **1989,** *15,* 127.
32. Zuska, M.; Kennedy, J. P. *J. Polym. Sci., Polym. Chem.* Ed. **1991,** *29,* 875.
33. Faust, R.; Kennedy, J. P. *J. Polym. Sci., Polym. Chem.* Ed. **1987,** *25,* 1847.
34. Ivan, B.; Kennedy, J. P. *Polym. Mater. Sci. Eng.* **1988,** *58,* 866.
35. Ivan, B.; Kennedy, J. P. *J. Polym. Sci., Polym. Chem.* Ed. **1990,** *28,* 89.
36. Mathias, L. J. et al. *J. Polym. Sci., Polym. Lett.* Chem. Ed. **1982,** *20,* 473.
37. Crivello, J. V. et al. *J. Radiat. Curing* **1983,** *10,* 6.
38. Butler, G. B.; Lien, Q. S. *Polym. Prepr., Am. Chem. Soc. Div. Polym. Chem.* **1981,** *22(1),* 54.
39. Crivello, J. V.; Conlon, D. A. *J. Polym. Sci., Polym. Chem.* Ed. **1983,** *21,* 1785.
40. Gallucci, R. R.; Going, R. C. *J. Org. Chem.* **1983,** *48,* 342.
41. Shohi, H. et al. *Polym. Prepr.,* Jap. Engl. Ed. **1990,** *31(1),* 590.
42. Cloutet, E. et al. *J. Chem. Soc., Chem. Commun.* **1994,** 2433.
43. Kennedy, J. P.; Chung, D. Y. L. *J. Polym. Sci., Polym. Chem.* **1981,** *19,* 2729.
44. Mah, S. et al. *Polym. Bull.* **1987,** *18,* 433.
45. Kennedy, J. P.; Smith, R. A. *Polym. Prepr., Am. Chem. Soc. Div. Polym. Chem.* **1979,** *20(2),* 316.
46. Kennedy, J. P.; Smith, R. A. *J. Polym. Sci., Polym. Chem.* Ed. **1980,** *18,* 1523.
47. Santos, R. et al. *J. Polym. Sci., Polym. Chem.* Ed. **1984,** *22,* 2685.
48. Santos, R. et al. *Polym. Bull.* **1984,** *11,* 261.
49. Santos, R. et al. *Polym. Bull.* **1984,** *11,* 341.
50. Kennedy, J. P. et al. *Polym. Bull.* **1981,** *4,* 67.
51. Kennedy, J. P. et al. *Polym. Bull.* **1981,** *5,* 5.
52. Pask, S. D.; Nuyken, O. *Polym. Bull.* **1982,** *8,* 457.
53. Huang, K. J. et al. *Polym. Bull.* **1988,** *19,* 43.
54. Deffieux, A.; Schappacher, M. *Macromol. Rep.* **1994,** *A31(6-7),* 699.
55. Higashimura, T. et al. *J. Polym. Sci., Polym. Chem.* Ed. **1989,** *27,* 2937.
56. Minoda, M. et al. *Macromolecules* **1987,** *20,* 2045.
57. Minoda, M. et al. *Macromolecules* **1990,** *23,* 1897.

58. Minoda, M. et al. *J. Polym. Sci., Polym. Chem.* Ed. **1990,** *28,* 1127.
59. Kojima, K. et al. *Polym. Bull.* **1990,** *23,* 149.
60. Kojima, K. et al. *Macromolecules* **1991,** *24,* 2658.
61. Kanaoka, S. et al. *Macromolecules* **1993,** *26,* 254.
62. Kanaoka, S. et al. *Makromol. Chem.* **1993,** 194, 2035.
63. Kanaoka, S. et al. *J. Polym. Sci., Polym. Chem.* Ed. **1993,** *31,* 2513.
64. Deffieux, A. et al. *Polymer* **1994,** *35,* 4562.
65. Schappacher, M.; Deffieux, A. *Macromolecules* **1992,** *25,* 6744.
66. Deffieux, A.; Schappacher, M. *Pol. Adv. Tech.*, in press.

FUNCTIONAL POLYMERS (Telechelics and Macromonomers)

Oskar Nuyken
Lehrstuhl für Makromolekulare Stoffe
Technische Universität München

Functional polymers are reactive prepolymers that allow further reactions. More specifically, they are capped by functional terminal groups. In modern terminology they are called telechelics (oligomers with one, two, or even more functional end groups such as –OH, –COOH, –NCO, –SH, $-NH_2$, etc.) or macromonomers (oligomers with at least one end group that allows homopolymerization, for example $-CH_2{=}CH_2$ or 2-oxazoline).

Both telechelics and macromonomers are of great interest, owing mainly to their ease of processing. Because of their low molecular weights they have low melt viscosities and can be processed without solvent. They are also less toxic and have lower vapor pressure than their corresponding monomers.

These functional polymers can be converted into high polymers, including networks, block, and graft copolymers. These extension reactions and network formations proceed with little shrinkage.

Although the terms "telechelics" and "macromonomers" have been known for years,[1-3] it was not until 1989 that one could search for these keywords in *Chemical Abstracts*. A measure of their new-found importance is the great number of references in *Chemical Abstracts* in recent years (1500 references in 1990-1994) as more and more reviews on this topic have become available.[3-8]

TELECHELICS

Telechelics via Polycondensation

Polycondensation (step growth reaction) is probably the oldest and easiest way to telechelics. This concept has been applied successfully in the addition of thiols to olefins.[9,10]

Excess of norbonadiene yielded olefinic end groups and excess of 1,3-dimercapto benzene resulted in telechelics with SH-functions.

Examples of telechelics by step growth process on the basis of epoxide/amine and epoxide/dithiols have recently been reviewed.[11]

Telechelics by Radical Polymerization

Two pathways have been used for the synthesis of telechelics by a radical mechanism:[3,12] large amount of initiator (dead end)[13] and transfer. Functionalized initiators, either azo compounds[14] or alkanes with labile C–C bonds[15] open a suitable pathway to telechelics.

A more universal pathway is polymerization in the presence of effective functionalized transfer agents. Suitable transfer agents are either solvents or mercapto compounds and disulfides.[15,16]

Telechelics by Anionic Polymerization

Living anionic polymerization is the method of choice for the synthesis of telechelics. Termination with CO_2 or oxirane yields polymers with –COOH or CH_2OH end groups.

Initiation with bifunctional anionic initiators yields telechelics.[17]

A new telechelic poly(methyl methacrylate) having OH and COOH in the chain was prepared by anionic initiation with a functional initiator made by the reaction between 4-[2-*tert*-butyldimethylsiloxyethyl]isopropenylbenzene and sodium naphthalene, followed by termination with carbon dioxide and hydrolysis of the $OSi(CH_3)_2C(CH_3)_3$ group.[18]

Telechelic poly(methyl methacrylate) (PMMA) with two different terminal groups (OH and COOH) was prepared by anionic polymerization using the same system. However, the bifunctional initiator is replaced by LiBu.[19]

Living anionic polymerization of polar monomers became possible by protection of the polar groups.[20,21]

Another important discovery was group transfer polymerization.[22,23]

Telechelics by Cationic Polymerization

This topic has been reviewed recently.[24,25] One of the milestones in the field is the development of the INIFER technique for the polymerization of isobutylene.[26-29] Bi- and trifunctional initiators yield telechelics with two or three functional end groups.

Another highlight in living cationic polymerization is the development of HI/I_2 initiated polymerization of vinylethers.[30-33] An alternative pathway was developed in our laboratories.[34-36] Nucleophilic substitution of the iodine functionalized chain ends opens a variety of possible end groups. Bifunctional initiators yield telechelics with functional groups at each end. Sequential addition of monomers or conventional linkage of preformed telechelics opens the door to block- and star-shaped polymers.[33] This route also allows the synthesis of poly(isobutyl vinyl ether-*b*-vinylcarbazole).[37]

Telechelics by Chain Scission

Chain scission reactions are normally uncontrolled and therefore not suitable for the synthesis of telechelics. However, reductive chain scission of polysulfides has become a very important technical process (liquid Thiokol®).[38]

Telechelics by Criss-Cross Cycloaddition

Although many structures are available via criss-cross reaction, this reaction type has not been used to synthesize telechelics until recently.[39]

MACROMONOMERS

Definition

Macromonomers are oligomers with at least one polymerizable group at the chain end. Polymeric monomers are important for preparing tailored graft copolymers with well-defined structure and composition. For this application it is essential to have synthetic routes to allow the control of molar masses and functionalities.[8]

Macromonomer Syntheses

Two strategies are described in the literature for introducing unsaturated head or tail groups into polymers: via initiators[8,40,41] and via termination agents (end capping).[42,43]

CONCLUSION

Telechelics and macromonomers rank among the most important polymer chemistry developments of the past 10 to 15 years. Some of these polymers are commercialized, and others will have a great future. However, there are still some problems to be solved. These include

- purity,
- reaction time,
- coloration,
- restrictions in the monomers and solvents, and
- expense of the procedure.

REFERENCES

1. Uranek, C. A. et al. *J. Polym. Sci.* **1960,** *46,* 535.
2. Milkovich, R. *Am. Chem. Soc. Symp. Ser.* **1981,** 166, 41.
3. *Telechelic Polymers: Synthesis and Application*; Goethals, E. J., Ed.; CRC: Boca Raton, FL, 1989.
4. Nuyken, O.; Pask, S. D. *Encyclopedia of Polymer Science and Engineering* 2nd Ed.; John Wiley & Sons: New York, 1989; Vol. 16, p 494.
5. Percec, V. et al. *Comprehensive Polymer Science*; Pergamon: Oxford, 1989; Vol. 6, p 281.
6. Tezudky, Y. *Progr. Polym. Sci.* **1992,** *17,* 471.
7. *Macromolecular Design: Concept and Practice*; Mishra, M. K., Ed.; Polymer Frontiers International: Hopewell Jct., NY; 1994.
8. *Chemistry and Industry of Macromonomers*, Yamashita, Y. Ed., Hüthig & Wepf: Basel, 1993.
9. Nuyken, O. et al. *Makromol. Chem.* **1991,** 192, 1959.
10. Nuyken, O.; Völkel, T. *Makromol. Chem.* **1990,** 191, 2465.
11. Klee, J. E. *Acta Polym.* **1994,** *45,* 73.
12. Boutevin, B. *Adv. Polym. Sci.* **1990,** *94,* 70.
13. Tobolsky, V. V. *J. Am. Chem. Soc.* **1958,** *80,* 5927.
14. Bledzki, A. et al. *Makromol. Chem.* **1983,** 184, 745.
15. Heitz, W. In *Telechelic Polymers: Synthesis and Application*; Goethals, E. J., Ed.; CRC: Boca Raton, FL, 1989; p 77.
16. Freidlina, R. K.; Chukovskaya, E. C. *Synthesis* **1974,** 477.
17. Richards, D. H. et al. *Telechelic Polymers: Synthesis and Application*; CRC: Boca Raton, FL, 1989; p 33.
18. Ohata, M. et al. *Macromolecules* **1993,** *26,* 5339.
19. Ohata, M. et al. *Macromolecules* **1992,** *25,* 5131.
20. Hirao, A.; Nakahama, S. *Makromol. Chem.* **1985,** 186, 1157.
21. Hirao, A.; Nakahama, S. *Macromolecules* **1988,** *21,* 561.
22. Hertler, W. R. et al. *Macromolecules* **1984,** *17,* 1415.
23. Sogah, D. Y.; Webster, O. W. *J. Polym. Sci., Polym. Lett.* Ed. **1983,** *21,* 927.
24. Kennedy, J. P.; Ivan, B. *Designed Polymers by Carbocationic Macromolecular Engineering*; Hanser Verlag: München, 1992.
25. Nuyken, O. *Polymeric Materials Encyclopedia*; CRC: Boca Raton, FL, 1996.
26. Kennedy, J. P.; Smith, R. A. *J. Polym. Sci., Polym. Chem. Ed.* **1980,** *18,* 1523.
27. Kennedy, J. P. et al. *Polym. Bull.* **1981,** *4,* 67.
28. Kennedy, J. P. et al. *Polym. Bull.* **1981,** *5,* 5.
29. Kennedy, J. P.; Hiza, M. *J. Polym. Sci., Polym. Chem.* Ed. **1983,** *21,* 1033.
30. Miyamoto, M. et al. *Macromolecules* **1984,** *17,* 265 and 2228.
31. Higashimura, T.; Sawamoto, M. *Makromol. Chem. Suppl.* **1985,** *12,* 153.
32. Higashimura, T. et al. *Makromol. Chem., Macromol. Symp.* **1988,** *13/14,* 457.
33. Sawamoto, M. Presented at the *35th IUPAC Symposium on Macromolecules*, Akron, OH, 1994; *Abstract Book* p 51.
34. Nuyken, O.; Kröner, H. *Makromol. Chem.* **1990,** 191, 1.
35. Nuyken, O. et al. *Makromol. Chem., Macromol. Symp.* **1990,** *32,* 181.
36. Nuyken, O. et al. *Polym. Bull.* **1990,** *24,* 513.
37. Nuyken, O. et al. *Macromol. Reports*, submitted 1994.
38. Cheradame, H. *Telechelic Polymers: Synthesis and Application*; CRC: Boca Raton, FL, **1989,** p 14 and p 154.
39. Maier, G. et al. *Makromol. Chem.* **1993,** 194, 2413.
40. Rempp, P.; Franta, E. *Adv. Polym. Sci.* **1984,** *58,* 1.
41. Kobayashi, S. et al. *Polym. Bull.* **1983,** *9,* 169.
42. Milkovich, R.; Chiang, M. T. Invs. Ger. Offen. 2.500.494, (1975), *Chem. Abstract* **1975,** *83,* 165083f(1975).
43. Rempp, P. et al. *Makromol. Chem. Suppl.* **1984,** *8,* 3.

FUNCTIONAL POLYSILOXANES

Valeria Harabagiu, Mariana Pinteală, Cornelia Cotzur, and Bogdan C. Simionescu*

"P. Poni" Institute of Macromolecular Chemistry

Department of Macromolecules

"Gh. Asachi" Technical University

Polysiloxanes, especially polydimethylsiloxane (PDMS), possess extremely low glass transition temperature (T_g = −123°C), low surface tension and surface energy, and low solubility parameters and dielectric constants. In addition, PDMS is transparent to UV light, very resistant to ozone and corona discharge, and stable against atomic oxygen. These properties show only a small variation over a wide range of temperature. Polysiloxanes (PSs) also have film-forming ability, high gas permeability, release action, chemical and physiological inertness, and high hydrophobicity.

The introduction of organic functionalities into siloxane backbones determines the conservation or the diminishing of these characteristics, depending on the nature, number, or position of the functionality.

SILICOFUNCTIONAL POLYSILOXANES (SPSS) ~O–SI–X

SPSs are polysiloxanes whose functional group is directly linked to the silicon atom. The main route to SPSs is the controlled (co)hydrolysis of organohalogen- or organoalkoxysilanes.

*Author to whom correspondence should be addressed.

The process is quite complex, involving the hydrolysis of Si–X linkage and condensation reactions and gives rise to a mixture of linear and cyclic polysiloxanes. The ratio between the two types of species depends on the reaction condition, nature and concentration of the reagents, and solvent (miscible, partially miscible, or unmiscible with water).

Other functionalities directly bonded to the silicon atom are usually obtained by nucleophilic substitution in the Si–Cl unit[1,2] or by electrophilic displacement of hydrogen in the relatively polar Si–H bond.[3]

Detailed studies on the preparation and chemical characteristics of Si–X functionalized siloxanes area available.[1,4,5]

ORGANOFUNCTIONAL POLYSILOXANES (OPSS) ~O–SI–R–X

OPSs are starting materials for the synthesis of block and graft copolymers. They have important advantages over their silicofunctional homologues. They are more stable and easier to store and manipulate, they provide unreactive Si–C linkages in derived copolymers, and their miscibility with organic media can be improved by changing the nature of (R) radicals. There are several reviews of the synthesis, properties, and applications of OPSs.[6-8]

SYNTHESIS OF ORGANOFUNCTIONAL POLYSILOXANES

Equilibration of Cyclic Siloxanes

The equilibration or redistribution of cyclic siloxanes (especially three- or four-membered species) are anionic or cationic processes involving continuous breaking and reforming reactions of the siloxane bonds until the system reaches it thermodynamic equilibrium (linear polymer $\Leftrightarrow$ cyclosiloxanes).[9] Starting from cyclic siloxanes and organofunctional disiloxanes (end blockers), α,ω-organofunctionally terminated polysiloxanes are obtained. Polysiloxanes with organofunctional groups in lateral position can be prepared by copolymerization of functional and unfunctional cyclosiloxanes.

Living Anionic Polymerization

Organofunctional well-defined polysiloxanes were obtained by anionic polymerization of cyclosiloxanes, especially hexamethylcyclotrisiloxane (D_3). This is a very powerful method for the preparation of macromonomers[10] or of OPSs containing other terminal functionalities (such as γ-trifluoropropyl[11] or arylamine).[12] The reactive groups can be introduced by blocking the macromolecular anion with an organofunctional halogenosilane,[13,14] by the initiator,[15] or can be recovered in both ways.[16]

Hydrosilation

Hydrosilation is the term used for the addition of Si–H functional group to unsaturated organic reagents in the presence of transition metal compounds (H_2PtCl_6 or Pt, Pd, Rh complexes), peroxides, or bases. Hydrosilation gives rise to a mixture of α- and β- isomers or 1,4-adducts in unsaturated esters. The nature and the ratios of the addition products, as well as the extent of side reactions, depend on the type of catalyst; the structures of olefinic compound and hydrosilane, respectively; temperature; solvent nature; and the presence of water or other active hydrogen-containing compounds.[17-19]

Chemical Transformation

Organic modification of functional polysiloxanes broadens the range of OPSs. The main restriction of this method is the sensitivity of the siloxane backbone to acidic or basic reaction condtions.

PROPERTIES OF ORGANOFUNCTIONAL POLYSILOXANES

Physical Properties

The typical physical properties of PDMSs are more or less modified by the displacement of methyl groups with different functional units. Two types of polymers can be distinguished:[7] slightly modified polymers (containing 5 mol% functional groups) and highly modified polymers (with >5 mol% functional groups). PDMSs possess a very low solubility parameter ($\delta = 7.3$–7.5 $(cal/cm^3)^{0.5}$); therefore they are soluble in nonpolar solvents like petroleum ether, hexane, and so on. The introduction of polar groups in the siloxane chain makes it soluble in highly polar solvents (DMF, DMSO, alcohols)[6,24] or even in water.[26] Because of their high flexibility, polysiloxanes have very low bulk and solution viscosities. Depending on the number and the nature of the Si–R–X groups, this behavior can be dramatically changed.[27,28]

The very low surface tension of PDMSs is highly modified by the presence of polar ionic or nonionic groups.[26] Polysiloxanes of this type are miscible with both organic and aqueous phases and are used as emulsifiers. They present high gliding ability and antistatic properties.

The very low glass transition temperature of PDMS is slightly changed by the presence of final organic functionalities[6] whereas the substitution in a greater extent of the methyl groups with organic functionalities such as aminopyridyl[25] or carbohydrates[21] increased the T_g to positive values. The thermal behavior of siloxanes is certainly influenced by the introduction of mesogen units, which provide liquid crystalline polymers. Also, the presence of photoactive units induces specific properties.[22,23]

Chemical Properties

The reactivity of a Si–R–X group depends on the nature of (R) radical and on the type and position of (X) functionality. Changing the nature of the (R) radical from aliphatic to aromatic also has a major influence on the chemical reactivity of the (X) group. The functional units attached to the siloxane chain display reactivities typical of the organic groups.

APPLICATIONS OF ORGANOFUNCTIONAL POLYSILOXANES

Because of their chemical reactivity, OPSs are starting materials in the synthesis of block and graft copolymers or of different types of siloxane[7] or siloxane-containing networks.[6] Also, OPSs have found applications in textile treatments,[7,29] in photolithography,[30,31] as cosmetic[32] and biomedical[20,33] products, emulsifiers,[26] and polymer electrolytes.[34]

REFERENCES

1. Noll, W. *Chemistry and Technology of Silicones*; Academic: New York, 1968.
2. Kossmehl, G. et al. *Makromol. Chem.* **1992,** *193,* 157.
3. Barton, T. J.; Boudjouk, P. *Adv. Chem. Ser.* **1990,** *224,* 3.
4. Eaborn, C. *Organosilicon Compounds*; Butterworth: London, 1960.
5. Andrianov, K. A. *Heteroorganic Chemistry Methods. Silicon*; Nauka: Moscow, 1968.
6. Yilgor, I.; McGrath, J. E. *Adv. Polym. Sci.* **1988,** *86,* 1.
7. Kendrick, T. C. et al. *The Chemistry of Organic Silicon Compounds*; John Wiley & Sons: New York 1989; Chapter 21.
8. Pinteala, M. et al. *Roum. Chem. Quart. Rev.*, in press.
9. Voronkov, M. G. et al. *The Siloxane Bond*; Consultants Bureau: New York, London, 1978.
10. Kawakami, Y. et al. *Polym. J.* **1985,** *17,* 1159.
11. Kawakami, Y. et al. *Polym. Commun.* **1985,** *26,* 133..
12. Sysel, P. et al. *Polym. Prepr.* **1992,** *33(2),* 218.
13. DeSimone, J. M. et al. *Macromolecules* **1991,** *24,* 5330.
14. Villar, M. A. et al. *J. Macromol. Sci., Pure Appl. Chem.* **1992,** A29, 391.
15. Li W.; Huang, B. *Makromol. Chem.* **1989,** 190, 2373.
16. Suzuki, T.; Lo, P. Y. *Macromolecules* **1991,** *24,* 460.
17. Pomerantseva, M. G.; Voronkov, M. G. *J. Organomet. Chem.* **1977,** *5,* 1.
18. Yu, J. M. et al. *Polym. Bull.* **1992,** *28,* 435.
19. Harabagiu, V. et al. *Synth. Polym. J.*, **1994,** *1,* 259.
20. Aoyagi, T. et al. *Makromol. Chem.* **1992,** 193, 2821.
21. Jonas, G.; Stadler, R. *Makromol. Chem., Rapid Commun.* **1991,** *12,* 625.
22. Riedel, J. H.; Hocker, H. *J. Appl. Polym. Sci.* **1994,** *51,* 573.
23. Yagci, Y. et al. *Turkish J. Chem.* **1994,** *18,* 101.
24. Pinteala, M. et al. *Polym. Bull.* **1994,** *32,* 173.
25. Nemoto, N. et al. *Polym. Commun.* **1990,** *31,* 65.
26. Goldschmidt Catalogue: *Oligomers as Synthetic Components and Active Agents*, 1991.
27. Rogler, W. et al. *Polym. Prepr.* **1988,** *29(1),* 528.
28. Harabagiu, V. Ph.D. Thesis, "Gh. Asachi" Technical University, Jassy, Romania, 1992.
29. White, J. W. et al. U.S. 4 599 438, 1986.
30. Reichmanis, E. et al. *Adv. Chem. Ser.* **1990,** 224, 265.
31. Puyenbroek, R. et al. *Polym. Prepr.* **1993,** *34(1),* 238.
32. Roidl, J. *Parfuem. Kosmet.* **1986,** *67,* 232.
33. Minoura, N. *Biomedical Applications of Polymeric Materials*; CRC: Boca Raton, Ann Arbor, London, Tokyo, 1993; Chapter 3.2.
34. Smid, J. et al. *Adv. Chem. Ser.* **1990,** 224, 113.

FUNCTIONALIZED POLYMERS (Comb, Rotaxanic, and Dendrimeric Structures)

Helmut Ritter
Bergische Universität-GH Wuppertal
FB 9
Organische Chemie und Makromolekulare Chemie

Functionalized and reactive polymers have been of great scientific and commercial interest since the beginning of macromolecular chemistry.

Functional groups cannot only be attached within a linear polymer chain but can also be combined with a branched or comb-like polymer (**Scheme I; A-D**). Typically, comb-like polymers contain a linear main chain of different chemical types and at least four C-atoms in the side chains. Although comb polymers were originally investigated because of their thermal and solution properties, up to now only a moderate knowledge exists about functionalized combs.[1] We recently have synthesized comb polymers containing noncovalently anchored cyclodextrins around linear and branched side chains (Scheme I; **E** and **F**). This new class of polyrotaxanes is characterized by a supramolecular[2] architecture. Finally, polymerizable chiral dendrimers containing functional groups have been prepared recently (Scheme I; **G**).

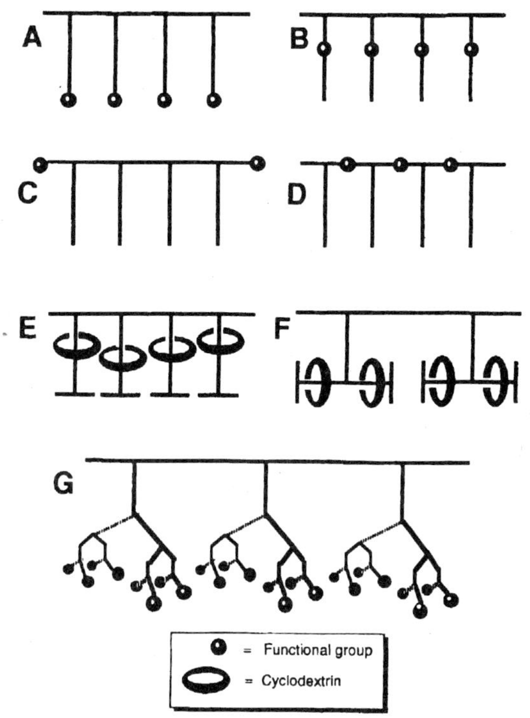

SCHEME I. Structures of functionalized comb-polymers.

COMB-LIKE POLYMERS WITH PHOTOREACTIVE GROUPS

Comb-like polymers containing azobenzene and cinnamic acid moieties in the side chains have recently been synthesized.[3] The E/Z-isomerization of the azogroup of the homopolymer is much faster in solution than in the condensed phase. Therefore, an amorphous copolymer with methyl methacrylate (MMA) was prepared and a significant increase in mobility of the azobenzene function in the solid phase was observed.[2]

We have also synthesized aromatic polyamides containing cinnamic acids in the side chains to produce asymmetric photocrosslinkable membranes. We found the mechanism of the photoreaction is controlled by the position of the cinnamic double bonds.[4]

A new class of photosensitive polymers containing mesoionic groups in the side chains was prepared by radical polymerization of a corresponding styrene derivative. Additionally, synthesis of photosensitive polymers containing mesoionic groups in the main chain was successfully achieved.[5]

DIELS–ALDER AND ENE-REACTIONS

Modification of 2,2′-azoisobutyronitrile (AIBN) with furan derivatives was performed to produce new radical initiators that were used for the synthesis of telechelic comb polymers containing furan endgroups (Scheme I; **C**). In a similar way, C,C-splitting initiators were also used to prepare oligomeric systems with furan endgroups. As an example, stearyl acrylate was polymerized with this modified C,C-labile initiator, yielding bifunctional oligomers with comb structure. These telechelic systems were used for crosslinking of unsaturated polyesters prepared from maleic anhydride and diols via Diels–Alder additions.[6] In this connection, comb polymers were prepared with furan groups at the end of the side chains (Scheme I; **A**).

Unsaturated comb polymers (Scheme I; **D**) were obtained via ene-reaction of stearylacrylate with polybutadiene. The oxidative crosslinking reaction was found to increase significantly above the melting point of the side chains.[7]

ENZYMES

Many enzymes have been extensively applied as selective catalysts in low molecular weight organic chemistry. Also, the degradation of suitable polymers, such as peptides, polyesters, or cellulose with enzymes is well known. In contrast, only a few papers have appeared describing the anchoring of low molecular weight compounds at synthetic polymers.[8]

Recently, the chemoenzymatic epoxidation of methacryl monomers containing chalcone components was realized in the presence of glucose, glucose oxidase, and oxygen in a preparative scale. The similar system was used to degrade water soluble polymers.[9]

ROTAXANES AND DENDRIMERS

Recently, comb-like polymers containing rotaxanes in the side chains have been produced (Scheme I; **E**). A barrier group N-(11-aminobutanoyl)-4-triphenyl methylaniline was shown to give a stable complex with 2,6-dimethyl-β-cyclodextrin. The aminogroup in the semirotaxane is still nucleophilic and can react with an activated comb polymer. The structure of the resulting polyrotaxane was proved by a characteristic broadening and shift of the NMR signals of the 2,6-aniline protons.[10]

Recently, the synthesis of a "tandem polyrotaxane" according to Scheme I; **F** was performed via similar polymer analgous reaction. Enzymatically catalyzed degradation of the cyclodextrine rings from the polyrotaxane was demonstrated, yielding the pure guest polymer.[11]

The synthesis of chiral (meth)acrylic monomers with a dendrimeric structure bearing four or even eight ester groups produces polymers with highly branched side chains. It was demonstrated that these types of monomers can be polymerized by radical mechanism.

REFERENCES

1. (a)Kapellen, K. K.; Stadler, R. *Polym. Bull.* **1994,** *32,* 3-10; (b) Volkov, V. V. et al. *Polym. Bull.* **1994,** *32,* 193-200; (c) Daly, W. H. *Progr. Polym. Sci.* **1994,** *19,* 79-135; (d) Grutke, S. et al. *Macromol. Chem. Phys.* **1994,** 195, 2875-2885; (e) Ritter, H.; Stock, A. *Macromol. Rapid Commun.* **1994,** *15,* 271-277.
2. Vögtle, F. *Supramolekulare Chemie* 2d. ed.; B. G. Teubner: Stuttgart, 1992.
3. (a) Ciardelli, F. et al. *B. Polym.* **1989,** *21,* 97-106; (b) Niemann, M.; Ritter, H. *Makromol. Chem.* **1993,** 194, 1169; (c) Stumpe, J. et al. *Makromol. Chem.* **1992,** 193, 1567.
4. (a) Ommer, H.; Ritter, H. *Makromol. Chem.* **1993,** 194, 767-776; (b) Ommer, H.; Ritter, H. *Macromol. Chem. Phys.* **1996,** 197, in press.
5. (a) Ritter, H. et al. *Macromol. Chem. Phys.* **1994,** 195, 3823-3834; (b) Deutschmann, T.; Ritter, H., unpublished results.
6. (a) Edelmann, D.; Ritter, H. *Makromol. Chem.* **1993,** 194, 2375-2384; (b) Ritter, H. et al. *Makromol. Chem.* **1993,** 194, 1721-1931.
7. Luchtenberg, J.; Ritter, H. *Macromol. Chem. Phys.* **1994,** 195, 1623-1632.
8. Ritter, H. *Trends Polym. Sci.* **1993,** *6,* 171-173.
9. (a) Goretzki, C.; Ritter, H. *Macromolecular Reports, A 32 (Suppls. 1&2)* **1995,** *237-245*; (b) Cho, M. D.; Okamoto, Y. *Macromol. Rapid Commun.* **1994,** *15,* 629-631.
10. Born, M. et al. *Polym. Acta.* **1994,** *45,* 68-72.
11. Born, M.; Ritter, H. *Angew. Chem. Jut. Ed. Engl.* **1995,** *34,* 309-311.

FUNCTIONALIZED POLYMERS (Telechelics and Macromoners via Cationic Polymerization)

Oskar Nuyken
Lehrstuhl für Makromolekulare Stoffe Technische Universität München

Although macromonomers and telechelics are oligomers, all these terms can be distinguished from one another by the functionality of their chain ends and by the products resulting from the reactions of their chain ends. Telechelics are oligomers containing at least one reactive endgroup. Macromonomers are oligomers that contain at least one homopolymerizable endgroup. In contrast to telechelics and macromonomers, oligomers do not necessarily contain any reactive endgroups.[1]

This topic has been (partly) reviewed several times.[1-6] We will examine the synthesis and applications of functionalized oligomers via cationic polymerization.

TELECHELICS BY CATIONIC TECHNIQUES

Poly(vinyl ether)s

The initiation of vinyl ether polymerization with HI/I_2 yields polymers with extremely narrow molecular weight distribution.[7-9]

The reaction of bifunctional monomers with HI/I_2 yields α,ω-iodine telechelics or, after precipitation in methanol, acetal-terminated oligomers.

Living cationic polymerization can be obtained not only by activation with I_2 but also by $ZnCl_2$, $ZnBr_2$, ZnI_2,[10,11] $SnCl_2$, and SnI_2.[11] Living polymerization of vinyl ethers has also been observed using HI / R_4NClO_4 as initiator.[12-14]

In the presence of inert nucleophiles such as esters and ethers (ethyl acetate; tetrahydrofuran; diethyl ether),[15,16] $EtAlCl_2$ can also yield narrow distributed poly(vinyl ether). In the absence of nucleophiles these systems revert to rapid conventional chain-transfer dominated polymerizations.[15]

End-functional polymers can be prepared by terminating a living polymerization with suitable functionalized agents.

Aliphatic amine endgroups are available via terminating agents containing an imide function after their deprotection with base.[17] Telechelic PVE can be prepared by endcapping of living polymerization induced by bifunctional initiators such as 1,4-bis(1-iodoethoxy)butane/I_2 or 1,4-bis(2-vinyl-oxyethoxy) benzene-HI adduct/I_2.[18]

Several new functionalized polymers have been synthesized via living cationic polymerization including different types of block copolymers, end-functionalized polymers, sequence-regulated oligomers and amphiphilic polymers. It is now possible to control not only the molar masses but also specific functionalizing and three-dimensional shapes of the polymers.[19-23]

Polydivinylbenzene

Linear unsaturated polymers have been obtained from a reaction between 1,4-divinylbenzene (DVB) and cationic initiators such as CF_3SO_3H and $AcClO_4$.[24-27] A sequence of protonation, propagations, and transfer steps yields telechelics with α,ω-divinyl endgroups.

Both the terminal and the internal double bonds of the resulting polymers can be selectively reacted to provide polymers containing a variety of functional groups such as OH or oxirane functions.

Polyindane

1,4-Diisopropenylbenzene can be converted into polyindane under certain conditions.[28-33]

Telechelics are available by copolymerization of 1,4-diisopropenylbenzene with functionalized isopropenylbenzene.

Polyisobutylene

Hydrosilylation of olefinic telechelics provides a pathway for telechelics with Si–Cl endgroups.[35] Amino-group-terminated telechelics are available via either Gabriel synthesis[36] or nitration of aromatic endgroups followed by a reduction step.[37]

The next generation of initiators contained tertiary ester and ether functions.[38-40] Living polymerization was also observed for isobutylene with *tert* alcohol/BCl_3.[41] The modern view on living cationic polymerization of isobutylene is supported by fundamental studies,[42-45] spectroscopy intermediates and conductivity measurements,[46] and kinetic simulations on the basis of experimental results.[47-49]

Telechelics by Ring-Opening Polymerization

The synthesis of telechelics by ring opening has attracted great interest because of the commercial potential of the resulting products, such as polyethers, polyols, etc. In many cases, the endgroup can be introduced by initiation, endcapping, or transfer. Preparation of telechelics of THF, oxiranes, cyclic acetals, amines and siloxanes has been extensively reviewed.[50]

Tetrahydrofuran (THF)

Under proper conditions, THF polymerization is living in the sense that the concentration of oxonium ions remains constant. The active ends are converted into functional groups by endcapping.

Cyclic Acetals

Several cyclic acetals with varying ring size have been polymerized using cationic initiators. 1,3-Dioxolane (DXL) and 1,3-dioxepane (DXP) were observed living when triflic acid or triflic anhydride were applied as initiators.[51,52]

Cyclic Amines

Aziridines and azetidines can be polymerized only by a cationic mechanism.[53] Monofunctional initiators such as methyltriflate[54] produce monofunctional telechelics from *t*-butylaziridine. Bifunctional living PTHF can initiate *t*-butylaziridine, resulting in bifunctional telechelics, which can also be considered as ABA-block copolymer.[55]

Cyclic Siloxanes

Cationic polymerization is initiated by strong protonic acids. The polymerization is characterized by backbiting during polymerization, creating an equilibrium between open chains and rings.[56,57] Bifunctional polysiloxanes with a variety of endgroups are obtained by the application of end blockers with different substituents R.[58,59]

2-Oxazolines

The cationic ring-opening polymerization of 2-oxazolines is well established and has been applied several times for the synthesis of telechelics.[60-62]

MACROMONOMERS

Macromonomers are often an ideal starting material for well-defined graft monomers.[1-6,34,63]

Unsaturated functional groups are introduced either via initiator or by endcapping.

REFERENCES

1. Percec, V. et al. Allen, G.; Bevington, J. C., Eds.; In *Comprehensive Polymer Science*, Pergamon: Oxford, 1989; Vol. 6, p 281.
2. Rempp, P.; Franta, E. *Adv. Polym. Sci.* **1984**, *58*, 1.
3. *Telechelic Polymers: Synthesis and Application*; Goethals, E. J., Ed. CRC: Boca Raton, FL, 1989.
4. Nuyken, O.; Pask, S. D. In *Encyclopedia of Polymer Science and Engineering*, 2nd ed.; John Wiley & Sons: New York, **1989**, Vol. 16, p 494.
5. *Chemistry and Industry of Macromonomers*; Yamashita, Y., Ed. Hüthig & Wepf: Basel, 1993.
6. *Macromolecular Design: Concept and Practice*; Mishra, M. K. Ed. Polymer Frontiers, International: Hopewell Jct., NY, 1994.
7. Miyamoto, M. et al. *Macromolecules* **1984**, *17*, 265.
8. Miyamoto, M. et al. *Macromolecules* **1984**, *17*, 2228.
9. Higashimura, T.; Sawamoto, M. *Makromol. Chem. Suppl.* **1985**, *12*, 153.
10. Kojima, K. et al. *Macromolecules* **1989**, *22*, 1522.
11. Sawamoto, M. et al. *Macromolecules* **1987**, *20*, 2693.
12. Nuyken, O.; Kröner, H. *Polym. Prepr. Am. Chem. Soc. Div. Polym. Chem.* **1988**, *29(2)*, 87.
13. Nuyken, O. et al. *Makromol. Chem., Rapid. Comm.* **1988**, *9*, 671.
14. Nuyken, O.; Kröner, H. *Makromol. Chem.* **1990**, *191*, 1.
15. Aoshima, S. et al. *Polym. Bull.* **1985**, *14*, 425.
16. Aoshima, S. et al. *Polym. Prepr. Jpn.* **1985**, *34*, 170.
17. Higashimura, T. et al. *Makromol. Chem., Macromol. Symp.* **1988**, *13/14*, 457.

18. Heroguez, V. et al. *Makromol. Chem., Macromol. Symp.* **1990,** *32,* 199.
19. Sawamoto, M. Presented at the *35th IUPAC Symposium on Macromolecules* Akron, OH, 1994*: Abstract Book,* p 51.
20. Kanaoka, S. et al. *J. Polym. Sci., Polym. Chem. Ed.* **1993,** *31,* 2513.
21. Minoda, M. et al. *J. Polym. Sci., Polym. Chem. Ed.* **1993,** *31,* 2789.
22. Hashimoto, T. et al. *Makromol. Chem.* **1993,** 194, 2323.
23. Cramail, H.; Deffieux, A. *Makromol. Chem.* **1992,** 193, 2793.
24. Hasegawa, H.; Higashimura, T. *Macromolecules* **1980,** *13,* 1350.
25. Aoshima, S.; Higashimura, T. *J. Polym. Sci., Polym. Chem. Ed.* **1984,** *22,* 2443.
26. Higashimura, T. et al. *Macromolecules* **1982,** *15,* 1221.
27. Higashimura, T.; Sawamoto, M. *Adv. Polym. Sci.* **1984,** *62,* 49.
28. Dittmer, et al. *Makromol. Chem.* **1989,** 190, 1755 and 1771.
29. Nuyken, O. et al. *Makromol. Chem.* **1991,** 192, 1969.
30. Nuyken, O. et al. *Makromol. Chem.* **1991,** 192, 3071.
31. Nuyken, O. et al. *Makromol. Chem.* **1992,** 193, 487.
32. Nuyken, O. et al. *Makromol. Chem., Macromol. Symp.* **1992,** *60,* 57.
33. Mauer, G. et al. *Polym. Bull.* **1992,** *28,* 135.
34. Kennedy, J. P.; Ivan, B. *Designed Polymers by Carbocationic Macromolecular Engineering*; Hanser: Munich, 1991.
35. Fang, T. R.; Kennedy, J. P. *Polym. Bull.* **1983,** *10,* 82.
36. Percec, V. et al. *Polym. Bull.* **1983,** *9,* 27.
37. Kennedy, J. P.; Hiza, H. *J. Polym. Sci., Polym. Chem. Ed.* **1983,** *21,* 3573.
38. Faust, R.; Kennedy, J. P. *Polym. Bull.* **1986,** *15,* 317.
39. Nagy, A. et al. *Polym. Bull.* **1985,** *13,* 97.
40. Mishra, M. K.; Kennedy, J. P. *J. Macromol. Sci. Chem.* **1987,** A24, 933.
41. Mishra, M. K. et al. *Polym. Bull.* **1989,** *22,* 455.
42. Hadjikyriacou, S.; Faust, R. Presented at the 35th IUPAC Symposium on Macromolecules, Akron, OH 1994; *Abstract Book,* p 96 (98), p 98.
43. Freyer, C. V. et al. *Angew. Macromol. Chem.* **1986,** 145/146, 69.
44. Nuyken, O. et al. *Makromol. Chem., Macromol. Symp.* **1986,** *3,* 129.
45. Nuyken, O. et al. *Makromol. Chem.* **1983,** 184, 553.
46. Dittmer, T. et al. *Makromol. Chem.* **1992,** 193, 2547, 2555.
47. Puskas, J. E. et al. *Macromolecules* **1991,** *24,* 5278.
48. Freyer, C. V.; Nuyken, O. *Makromol. Chem., Macromol. Symp.* **1988,** *13/14,* 319.
49. Freyer, C. V. et al. *Macromol. Theory Simul.* **1994,** *3,* 845.
50. Goethals, E. J. *Makromol. Chem., Macromol. Symp.* **1986,** *6,* 53.
51. Chwialkowska, W. et al. *Makromol. Chem.* **1982,** 183, 753.
52. Kubisa, P. et al. In *Cationic Polymerization and Related Processes*; E. J. Goethals, Ed.; Academic: London, 1984; p 395.
53. Goethals, E. J. In *Ring-Opening Polymerization*; Irvin, H. J.; Saegusa, T., Eds.; Elsevier: Barking, UK, **1984,** p 715.
54. Munir, A.; Goethals, E. J. *J. Polym. Sci., Polym. Chem. Ed.* **1981,** *19,* 1985.
55. Van de Velde, M.; Goethals, E. J. *Makromol. Chem., Macromol. Symp.* **1986,** *6,* 271.
56. Kantor, S. W. et al. *J. Am. Chem. Soc.* **1954,** *76,* 5190.
57. Beevers, M. S.; Semlyen, J. A. *Polymer* **1971,** *12,* 373.
58. Sermani, P. M. et al. In J. E. McGrath (Ed.), *Ring-Opening Polymerization*; McGrath, J. E., Ed.; ACS Symposium Series 286, American Chemical Society: Washington, DC, 1985; p 147.
59. Yilgor, I. et al. In *Reactive Oligomers*, Harris, F. A.; Spinelli, H. J. Eds.; ACS Symposium Series 282, American Chemical Society: Washington, DC, 1985; p 161.
60. Kobayashi, S. et al. *Macromolecules* **1989,** *22,* 2878.
61. Miyamoto, M. K. et al. *Macromolecules* **1989,** *22,* 1604.
62. Chujo, Y. et al. *Polym. Bull.* **1993,** *31,* 311.
63. Schwarzenbach, E. Ph.D. Thesis, Mainz, 1987.

FUNCTIONALIZED POLYOLEFINS (Using Borane Monomers and Transition Metal Catalysts)

T. C. Chung
Department of Materials Science and Engineering
The Pennsylvania State University

Polyolefins, especially polyethylene, polypropylene, poly(l-butene) and their copolymers, are used in a wide range of applications since they incorporate an excellent combination of mechanical, chemical, and electronic properties and processability.[1] Nevertheless, deficiencies, such as the lack of reactive groups in the polymer structure, have limited some of their end uses, particularly those in which adhesion, dyeability, paintability, printability, or compatibility with other functional polymers is paramount. Accordingly, the chemical modification of polyolefins has been an area of increasing interest as a route to higher value products, and various methods of functionalization[2-5] have been employed to alter their chemical and physical properties.

It is well known that the Ziegler-Natta process is the most important method for preparing polyolefins,[6] but the direct polymerization of functional monomers by this method is normally very difficult because of catalyst poisoning and other reactions.[7] Some attempts at using post-polymerization processes, modifying the preformed polyolefins, suffer from other problems such as the degradation[8] of polymer backbone. It is clear that there is a fundamental need to develop new chemistry that can address the challenge of preparing functionalized polyolefins with controllable molecular weight and functional group concentration.

In the past few years, we have been investigating a new approach by using borane-containing monomers in Ziegler-Natta polymerizations. The initial idea of using borane monomers, ω-alkenylboranes,[9-14] was based on three considerations. (a) The stability of borane to transition metal catalyst. (b) The solubility of borane monomers and polymers in the hydrocarbon solvents (hexane and toluene) used in Ziegler-Natta polymerization. (c) The versatility of borane groups, which can be transformed to a remarkably fruitful variety of functionalities, as shown by Brown.[15] Many new functionalized polyolefins with various molecular architectures have been obtained based on this chemistry. Most of them would be very difficult to obtain by other existing methods.

PREPARATION

Equation I illustrates the general scheme of copolymerization reactions between borane monomers and α-olefins using Ziegler-Natta catalysts.

The copolymerization between borane monomer (5-hexenyl-9-BBN) and various α-olefins, such as ethylene, propylene, 1-butene, and 1-octene was carried out in an inert gas atmosphere

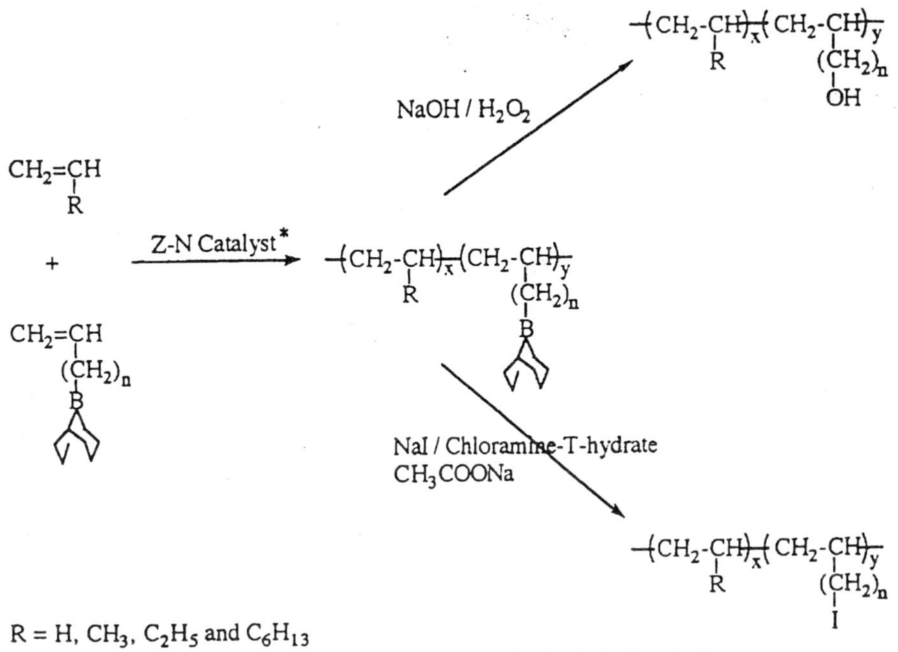

I

R = H, CH_3, C_2H_5 and C_6H_{13}

* Z-N Cat. = $TiCl_3/EtAlCl_2$, Cp_2ZrCl_2/MAO and $C_2H_4[Ind]_2ZrCl_2$/MAO

using both heterogeneous catalysts, such as $TiCl_3 \cdot AA/Et_2AlCl$ and homogeneous metallocene catalysts such as Cp_2ZrCl_2 and $Et(Ind)_2ZrCl_2$ with MAO.

The borane groups in polymers were converted to the corresponding hydroxy groups by reacting with $NaOH/H_2O_2$ reagents at 40°C for 3 hours. It is very interesting to note that the resulting functionalized polyethylene (LLDPE-OH) is structurally similar to that of linear low-density polyethylene (LLDPE).

PROPERTIES

Morphology

DSC (Differential Scanning Calorimetry) and Polarized Optical Microscopy were used to study the crystalline structure of copolymers. As expected, both the melting point (T_m) and crystallinity (χc) of copolymers decrease with the side chains. The higher the density of the side chain,[5] the lower the T_m and χc.

The comparison of DSC curves of polypropylene homopolymer and hydroxylated polypropylene copolymers, obtained from a continuous monomer feed and hetergeneous $TiCl_3AA$ and Et_2AlCl catalyst system, showed all samples have similar melting points (~160°C) despite the difference in concentration of hydroxy groups, indicating the long sequences of propylene units in the polymer backbone that form crystalline phases similar to those in pure isotactic polypropylene. While the polypropylene segments crystallize, the functional groups located at the end of side chains are expelled from the crystalline phases into the amorphous regions.

Thermal Stability

Thermal stability of copolymers was examined by TGA technique. The thermal stability of PP-OH is higher than that of pure isotactic polypropylene. The decomposition temperature of hydroxy polypropylene is above 280°C in argon and about 205°C in air. A slightly better resistance in decomposition may be contributed from the relatively high thermal stability of polyhexenol. This thermal stability stems from the excellent stability of the primary alcohol and the fact that the functional group is pendant on a side chain.

Co-crystallization

The co-crystallization of PP and PP-OH was revealed by DSC analysis.

APPLICATIONS

Adhesion Studies of PP/Al and PP/Glass Laminates

The bondings between PP to glass or aluminum are of both scientific and commercial interest. The hydroxylated polypropylene was used as surface modifier to improve PP adhesion. The flexibility of hydroxy groups located at the ends of side chains in PP-OH may enhance the interaction of PP-OH to substrates. On the other hand, high crystallinity of PP-OH (similar T_m as PP) may be co-crystallized with PP, which provides strong adhesion. Both drawn and undrawn PP films (commercial products) were laminated with PP-OH treated substrates. PP/Al laminates bonded by PP-OH were found to exhibit an extraordinary 7–10 fold increase in peel strength over acid-etched samples. It is clear that cohesive failure occurs and gives rise to the high peel strengths observed for these PP/Al laminates. When peeling Al from well bonded PP, the failure path appeared to propagate within the PP layers. It also helps to explain why the peel strength of drawn PP/Al laminates are greater than those of undrawn PP sample.

The same results were observed in the PP/Glass laminates.[16] Most of the speciments have strong peel strength. Especially, the drawn PP/acid-etched E-glass laminates, with PP-OH interfacial agent and hot-press process, have considerably high peel strength (about 1200 to 1500 N/M).

It is believed that the strong adhesion between PP-OH and glass surface is primarily due to the chemical bonding. The reflection IR studies provide the experimental evidence of the chemical reaction between free Si-OH groups on glass surface and hydroxy groups in PP-OH.

ACKNOWLEDGMENT

The authors would like to thank the financial support from the Polymer Program of the National Science Foundation.

REFERENCES

1. Baijal, M. D. *Plastics Polymer Science and Technology*; John Wiley & Sons: New York, 1982.
2. Carraher, E. C. Jr.; Moore, J. A. *Modification of Polymers*; Plenum: Oxford, 1982, pp 33-52.
3. Pinazzi, C.; Guillaume, P.; Reyx, D. *J. Eur. Poly.* **1977**, *13*, 711.
4. Chung, T. C.; Raate, M.; Berluche, E. et al. *Macromolecules* **1988**, *21*, 1903.
5. Chung, T. C. *J. Polym. Sci., Polym. Chem. Ed.* **1989**, *27*, 3251.
6. Boor, J. Jr. *Ziegler-Natta Catalysts and Polymerizations*; Academic: New York, 1979.
7. Purgett, M. D. Polymer and Copolymers from ω-Functionally-Substituted α-Olefins, Ph.D. Thesis, University of Massachusetts, 1984.
8. Ruggeri, G.; Aglietto, M.; Petragnani, A. et al. *Eur. Polymer J.* **1983**, *19*, 863.
9. Chung, T. C. U.S. Patent 4 734 472 and 4 751 276, 1988.
10. Chung, T. C. *Macromolecules* **1988**, *21*, 865.
11. Chung, T. C.; Ramakrishnan, S.; Kim, M. W. *Macromolecules* **1991**, *24*, 2675.
12. Chung, T. C.; Chasmawala, M. *Macromolecules* **1991**, *24*, 3718.
13. Ramakrishnan, S.; Berluche, E.; Chung, T. C. *Macromolecules* **1990**, *23*, 378.
14. Ramakrishnan, S.; Chung, T. C. *Macromolecules* **1990**, *23*, 4519.
15. Brown, H. C. *Organic Synthesis via Boranes*; Wiley-Interscience: New York, **1975**, pp 77-121.
16. Lee, S. H.; Li, C. L.; Chung, T. C. *Polymer* **1994**, *35*, 2980.

Fungicidal Polymers

See: Antibacterial Resins

Furan Polymers

See: Furan Resins (2-Furfuryl Alcohol Based)
Vegetal Biomass (1. Monomers and their Polymerization)

FURAN RESINS (2-Furfuryl Alcohol Based)

Zsuzsa László-Hedvig, and M. Szesztay
Central Research Institute for Chemistry
Hungarian Academy of Sciences

Furan resins are the linear condensation products of furfuryl alcohol (FA) or other self- or co-condensates containing furan rings, catalyzed by acids and promoted by heat. They belong to the group of renewable polymeric adhesives derived from agricultural waste products.[1] The linear, soluble resins are crosslinked at ambient or elevated temperatures in an acid- or oxygen-catalyzed process. The most important industrial furan resins are based on 2-furfuryl alcohol. The acid-catalyzed self-condensation reaction leads to formation of linear oligomers, containing methylene and methylene-ether linked furan-ring sequences.[2,3]

Along with self-condensates, the reaction products of FA and formaldehyde (F) co-condensation are widely used in industrial practice (FA/F resins). They are often modified in the course of preparation by urea or phenol (primarily for lowering free F-content of the resin) and are crosslinked similarly to FA resins.

Formaldehyde elimination from methylol groups or from methylene-ether linkages connecting two furan rings is negligible during acid-catalyzed resinification in the range of 60–100°C. Both FA self-condensation and FA/F co-condensation methylene linkages dominate over methylene-ether linkages. This statement is valid for both linear polycondensates and cured (crosslinked) resins.[4-7,9]

The reason for the darkening observed during condensation (and curing) of furan resins was not found until recently, when some data have been published[8] that suggests that increasing methylene-linked furan ring concentration in both self- and co-condensation systems is responsible for this phenomenon. This structure, referred to as a "homoconjugated"[10] system, broadens the range of UV absorption and shifts its maxima toward lower frequency as compared with similar structures of simple aromatic rings. The main route of polycondensation and chain growth is the formation of 5-furfuryl-furfuryl-alcohol type oligomers according to **Equations 1** and **2**:

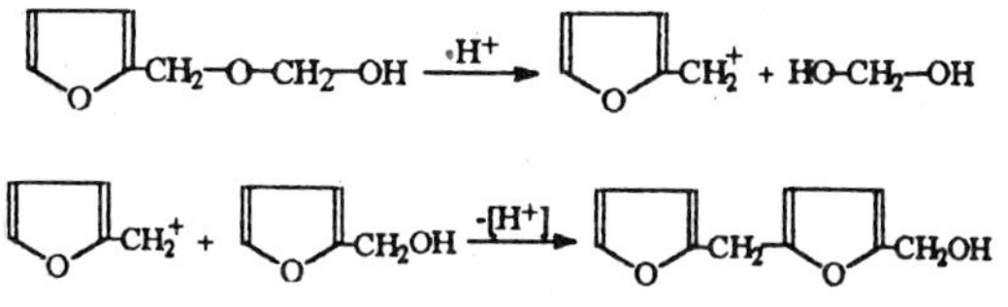

1

2

In FA self-condensation HF formation is almost negligible because F elimination during the early stage of polycondensation is negligible, too.

It was established that in FA/F acid-catalyzed polycondensation, the overall rates of FA and HF consumption are the same at a given ratio of substrates. The activation energies for FA and HF consumption are the same, too (~93 kJ/mol).

Besides equilibrium and consecutive reactions, parallel reactions take place in FA/F co-condensation, namely, self-condensation of FA. A uniform reaction mechanism seems to be valid for both FA self-condensation and FA/F co-condensation.[11,12]

The crosslinking (curing) of FA-based resins is catalyzed by acids and proceeds at ambient temperatures. It is accelerated by heat, oxygen, and addition of such compounds as phenol and urea. The crosslinking mechanism also depends on the atmosphere during the reaction.[13] The extent of crosslinking was followed by IR spectroscopy,[13] differential scanning calorimetry (DSC),[14] and solid-phase ^{13}C-MAS-NMR.[6]

FURAN RESINS: PROPERTIES AND APPLICATION

Furan resins are used mainly in the foundry industry, as sand binders for casting molds and cores.

Molds and cores made of furan resin-covered sands have unusually good thermal stability and chemical resistance. Interest has also been awakened because of their rapid curing.[7] Their invention also significantly decreased foundry casting-cycles and increased production accuracy.[18] Furans are relatively more expensive than other binders, but the possibility of sand reclamation[16] in the casting cycle decreases costs. Furan resins are often used in combination with other resins; for example, they may be applied as binders for foundry sands along with phenolic acid and urea resins.

Modification of furan-based resins with additions such as urea/formaldehyde or phenol/formaldehyde (novolac) also has economic reasons. The amount of these additions depends on the applied technology.

FA-based resins have outstanding resistance against bases and strong acids (hydrochloric, phosphoric, sulfuric, acetic and its chloroderivatives), and they are not attacked even by boiling organic solvents (chloroform, toluene, chlorobenzene, acetone, ethers, or esters). The resins are attacked only by strong oxidizing agents.

Furan resins are applied wherever corrosion resistance is needed: in mortars and cements, in acid-brick flooring, in waste and sewer disposal systems, in tank-lining topcoats, and in impervious graphite production. When the resins are reinforced with glass fibers, corrosion resistance is supplemented with improved mechanical properties; FA-resin-glass fiber systems are used in piping.

Because of their high strength, even when incompletely cured, FA-based polymers are also used as rapid-repair materials. With their chemical resistance and thermal stability, their application is mentioned for wells-plugging in oilfield operation.[19]

Because of their controllable curing process, FA resins are widely used as adhesives and binders, especially in wood technology, to make particle boards and other water- and chemical-resistant wood composites.[15,20]

REFERENCES

1. McKillip, W. J. ACS Symposium Series, *Adhesives from Renewable Resources*, Chapter 29, Chemistry of Furan Polymers *385*, American Chemical Society: Washington, DC, 1989; p 408.
2. Dunlop, A. P.; Peters, F. N. *The Furans*; Reinhold: New York, 1953.
3. Gandini, A. *Adv. Polym. Sci.*, Springer **1977,** *25,* 47.
4. Fawcett, A. H.; Dadamba, W. *Macromol. Chem.* **1982,** *183*, 2799.
5. Lászlo-Hedvig, Zs. et al. *Angew. Makromol. Chem.* **1982,** *107*, 61.
6. Chuang, I. S. et al. *Macromolecules* **1984**, *17,* 1087.
7. Gonzales, R. et al. *Macromol. Chem.* **1992,** *193*, 1.
8. Szesztay, M. et al. *Angew. Macromol. Chem.* **1989,** *170*, 173.
9. Barr, J. B.; Wallon, S. B. *J. Appl. Polym. Sci.* **1971,** *15,* 1079.
10. Ferguson, L. N.; Nandi, J. C. *J. Chem. Education* **1965,** *42,* 529.
11. Szesztay, M. Ph.D. *A furfuril-al-kohol-formaldehid polikonden zá ciós reakció kez deti szakaszának KinetiKai vizsgálata GPC modszerrel,* dissertation presented at the Hungarian Academy of Sciences, Budapest, February 20, 1991.
12. László-Hedvig, Zs. Doct. Sci. Thesis, Hungarian Academy of Sciences (HAS), 1994.
13. Conley, R. T.; Metil, I. *J. Appl. Polym. Sci.* **1963,** *7,* 37; 7, 1083.
14. Chanda, M.; Dinesh, S. R. *Angew. Macromol. Chem.* **1978,** *69,* 85.
15. Philippou, J. L.; Zavarin, E. *Holtzforschung* **1984,** *38,* 119.
16. *Foundry Management and Technology, Sand/Binders/Sand Preparation/Core Making,* Section D-3, December 1991, January 1993, January 1994.
17. Buell, W. H. *Foundry* **1961,** *89(2),* 64.
18. Dorfmueller, A. *Foundry* **1962,** *90(4),* 54.
19. Hess, P. H. *J. Petrol. Technol.* **1980,** *10,* 1834.
20. Johns, W. E. et al. *Holzforschung* **1978,** *32,* 163.

Fuzzy Rods

See: Hairy Rod-Like Polymers

G

GAS BARRIER POLYMERS

William J. Koros* and Maryam Moaddeb
Department of Chemical Engineering
The University of Texas at Austin

Polymeric materials whose molecular structure and segmental packing hinder permeation of gases and vapors such as oxygen, carbon dioxide, and water are classified as barrier polymers. Light weight, transparency, processability, shatter resistance, and low cost make polymers very attractive in the packaging industry and strongly competitive with more traditional materials such as glass, metal, and paper. Food packaging is one of the largest markets for polymeric materials.[1]

Water vapor permeation into or out of a package can damage the integrity of a product or damage the product indirectly by increasing the permeation of an undesirable gas-like oxygen in hydrophilic polymers such as ethylene-vinyl alcohol copolymers.

Because the tolerable oxygen gain for many products is typically small (as low as 1 to 5 ppm for some food products), the strategy for product preservation has been to eliminate any type of undesirable permeation, especially oxygen permeation, by engineering polymers that are excellent barriers.[2,3] Clearly, engineering a single polymer with the right balance of physical, mechanical, and barrier properties that also meets government requirements is extremely challenging. Therefore, much research has been directed toward improving existing polymers by blending alloying, laminating, and by coextrusion to achieve the desired properties.[4]

Advances in technology, resulting from years of research, now make it possible to tackle increasingly challenging problems. Although for most applications the permeable nature of polymers is a drawback, so-called smart structures that control the atmosphere by selectively admitting gases and vapors into and out of a package represent a goal for some specialty applications. In food packaging, controlled atmosphere packages (CAPs) keep the product naturally fresh from supplier to consumer by establishing optimum levels of oxygen and carbon dioxide.

Concerns about health care costs and epidemics such as AIDS have urged manufacture of safe disposable packages (for handling blood) and biomaterials that promote reduction of health care cost by shortening hospital stays. Exxaire™ is a "smart" porous polyolefin film for wounds introduced by Exxon Chemical. This material promotes healing by allowing air and water vapor through while keeping microorganisms and liquids such as water, alcohol, and blood out. Polyurethane wound dressings are another product replacing sutures or staples for closing surgical incisions. These can be peeled off by patients at home four days after the surgery, thereby reducing health care costs by eliminating follow-up visits to the hospitals.[5]

The microelectronics industry also depends heavily on the advancement of plastics technology for encapsulation of electronic integrated circuit (IC) devices. Moisture barrier properties are essential to prevent electrooxidation.[6]

Environmental concerns are at the forefront of the packaging industry's problems. These days, whether a package is recyclable is becoming as important for consumers as convenience, visual appeal, and other design criteria.[7] A surge of research during the past decade has improved recycling possibilities. In the past, the technology did not exist to break multi-layer packages into their individual recyclable resins.[8] Today, compatibilizers can allow blending of otherwise unstable components. Reactive units, for instance, can bond to units such as the hydoxyl groups in ethylene-vinyl alcohol copolymers; other segments in the compatibilizer may be miscible with purely hydrocarbon components such as polypropylene or polyethylene. The main drawback of such an approach is the cost of the compatibilizer. More recently, the possibility of blow molding laminates from a blend of ethylene-vinyl alcohol and 100% recycled multilayer bottles[9] and formation of ethylene-vinyl alcohol copolymer laminates with reground waste-containing layers with good gas barrier and heat resistance properties has been reported.[10] Searches for creative solutions to the recycling problem are expected to receive considerable effort over the next decade or so.

FACTORS INFLUENCING TRANSPORT PROPERTIES

Chemical Nature of Polymer and Penetrant

The permeability of a gas molecule in a polymer is a direct function of its diffusivity and solubility. Both constituents are influenced by the chemical nature of the polymer and penetrant. Interchain attraction, mobility of the chains, and penetrant size affect diffusivity. The solubility coefficient is primarily dependent on the condensability of the gas as determined by its critical temperature, and on polymer-penetrant interactions. Previous work has demonstrated that for most glassy polymers changes in diffusivity can result in order of magnitude changes in permeability; manipulation of the solubility coefficient does not typically affect permeability significantly.[12] Therefore, most efforts to engineer new polymeric structures focus on manipulation of factors affecting diffusivity.

High barrier polymers typically contain hydroxyl, cyano, halogen, ester, or amide groups, which create strong chain-to-chain forces that restrict movement of the chains and hinder diffusion.[1] Having all these structures in the repeat unit of a polymer results in low oxygen permeability, but only groups such as –Cl or –F also yield low water permeability. Water is readily attracted to structures with which hydrogen bonding can occur. The high solubility coefficient of water in such structures results in plasticization-induced increases in the diffusion coefficient.

One of the lowest water permeabilities among polymers is reported for polychlorotrifluoroethylene (PCTFE), which has a reasonably good oxygen permeability (in the range of nylon) as well. PCTFE is primarily used in blister (strip) packaging of pharmaceutical tablets, where a high degree of protection from water vapor is required.[12a]

*Author to whom correspondence should be addressed.

Most high barrier polymers are relatively poor water barriers. Copolymerization and blending often result in polymers with improved properties. For example, ethylene-vinyl alcohol is a copolymer of ethylene and vinyl alcohol. Poly(vinyl alcohol) is an exceptional gas barrier for O_2, N_2, and CO_2 in the dry state, but is swellable at high relative humidities and difficult to process because of its instabilities. Copolymerization with ethylene yields polymers with high gas and improved moisture barrier properties that are typically referred to as ethylene–vinyl alcohol copolymers $[(-(CH_2CH_2)_m-(CH_2\underset{\underset{OH}{|}}{C}H)_n-)]$.

Copolymers of carbon monoxide and olefins, known as polyketones, are formed by polymerization of the two structures in the presence of a metal catalyst. Polyketones are attractive because of $[-(CH_2-CH_2-CO-]_n-[-(P)-CO-]_m$ (P is the propylene unit) their processability and competitive gas barrier properties. In some cases excellent water barrier properties are also exhibited (Carilon™ Polyketone).

Blending polymers or additives with polymers may result in enhanced or reduced barrier properties depending on the chemical nature of the polymers and additives used.

Additives and modifiers are often blended with polymers to enhance processability. Barrier properties may be lost in such cases.

Transport of penetrants in polymers is influenced by polymer-penetrant interactions (which affect the solubility coefficient) as well as penetrant shape and size (which affect the diffusion coefficient). At higher sorption levels, these two effects cease to be neatly separable; polymer-penetrant interactions that promote high solubilities can also promote chain segment motion and facilitate penetrant diffusion. This phenomenon is usually referred to as plasticization.[13]

Polymer Morphology

Regular arrangement of polymer chains achieved through crystallization, orientation, or a combination of the two almost invariably improves barrier properties. The morphology of polymers typically consists of crystalline domains dispersed among amorphous regions, because most polymers never completely crystallize. The crystallinity of a polymer depends on interchain forces, chain stiffness, and the thermal history of the polymer.[10] Orientation of the semicrystalline polymers organizes both crystalline and amorphous backbone regions, and often yields higher degrees of crystallinity because of stress-induced crystallization. Higher degrees of crystallinity and orientation increase tortuosity and reduce diffusivity of polymer, significantly improving the barrier properties of the oriented material.[11] Many commercial processing techniques impose biaxial orientation on a polymer to achieve higher degrees of orientation.

Amorphous polymers are optically clear and are often utilized in applications requiring this quality. In such cases, orientation is the only possible method of improving barrier properties while maintaining optical clarity.[10]

With noted exceptions involving poly(4-methyl-1-pentene)[14] crystallinity decreases permeation by reducing both solubility and diffusivity of molecules. Gas molecules are essentially insoluble in the crystalline regions, and their solubility in a semicrystalline polymer is a linear function of the amorphous fraction.[10] Crystallinity affects diffusivity by providing impermeable blocking domains and by inhibiting the mobility of amorphous chains restrained between the crystalline regions.

Since the 1970s a class of polymers known as liquid crystalline polymers (LCPs) has attracted the attention of researchers, primarily for their excellent mechanical properties. However, recent characterization of their gas transport properties points to attractive gas barrier properties as well.[10]

Liquid crystalline polymers have highly ordered, uniaxially oriented rodlike chains, or thin platelets with biaxial order that allow very close packing of the chains.

An oxygen permeability of 47×10^{-5} Barrers at 35°C is reported for Vectra™, a commercial LCP.[10] This value is lower than oxygen permeability of polyacrylonitrile (PAN) at the same temperature ($P_{oxygen} = 54 \times 10^{-5}$ Barrers).

Surface Treatments

Chemical modification of surface of polymers is an attractive method of improving barrier characteristics of polymers, which are otherwise considered ideal materials for packaging. With the exception of low gas barrier properties, polyolefins are extremely attractive because of their low cost, toughness, processability, and excellent water barrier properties. Surface treatment is ideal for such polymers, because they are easily processed and made into better barriers by surface modification either during processing or afterward.[1,15] Modifications using chlorine, fluorine, and sulfur have been examined. Fluorine attaches to the polymer near the surface and, because of its bulkiness and polar nature, improves gas and nonpolar liquid barrier properties.[16] However, fluorinated materials may not maintain their barrier properties after repeated flexing. The chlorination reaction is too slow and not practical, but it results in good barrier properties with more resistance to flexing. Surface sulfonation yields excellent gas barrier properties under dry conditions, is relatively simple, and does not affect the mechanical stability of the polymer.[15]

Surface fluorination is used to improve the barrier properties of the inner surface of polyethylene during the blow molding process for formation of bottles. Polyethylene is nonpolar and, therefore, a poor barrier to nonpolar hydrocarbons. Such treatments with highly polar fluorine or sulfur significantly improve its barrier properties. Surface-fluorinated containers are commonly used for gasoline, herbicides, pesticides, and other products that normally penetrate polyethylene.[3]

PROJECTIONS

The importance of environmental concerns in determining future developments in barrier polymers will continue to increase. Polymers often are not considered "environmentally friendly" materials because they are nonbiodegradable. Nevertheless, the food packaging, medical, microelectronics, and other industries have become extremely dependent on polymeric materials because of their excellent properties. The chief issue is the development of recycling or disposal techniques for barrier polymers and structures without loss of their primary barrier and processability properties. Research in this area has significantly improved the situation over the past decade, and researchers have even been able to affect the rate of biodegradation of some polymers. Water soluble poly(vinyl alcohol) (PVA) is recognized as a biodegradable material and the water solubility of PVA, in turn, is a function of crystallinity and its degree of hydrolysis. Recent research has shown that addition of polyhydroxy compounds significantly affects the rate of biodegradation of water-insoluble PVA.[17]

Applications of barrier polymers will further increase as new developments in recycling possibilities change consumers' perception of these materials as "environmentally unfriendly" to "environmentally friendly." Controlled atmosphere packages allow reaching markets that are geographically far from the point of packaging for sensitive materials; however, the actual size of the market for such materials is not well defined. Indeed, the sophistication of such packages will increase costs and the trade-off with alternative technologies such as refrigeration will likely be determined by balancing convenience and economics. For standard, large-volume barrier materials and structures, the balance between demands for environmentally friendly structures, low cost, and good processability is likely to be evolving. Consumer perceptions are likely to be as significant as barrier performance in determining new approaches to high-performance packaging. There are many opportunities and pitfalls, and care is advised at each step of product development.

REFERENCES

1. Salame, M. *The Wiley Encyclopedia of Packaging Technology*; Bakker, M. Ed.; John Wiley & Sons: New York, 1986.
2. Koros, W. J. *Barrier Polymers and Structures*; ACS Symposium Series 423; American Chemical Society: Washington, DC, 1990; Chapter 1.
3. Leaversuch, R. D. et al. *Mod. Plast.* **1985,** *62(8),* 41.
4. Combellick, W. A. *Encyclopedia of Polymer Science and Engineering*; Kroschwitz, J. I., Ed.; John Wiley & Sons: New York, 1985.
5. Leaversuch, R. D. et al. *Mod. Plast.* **1993,** *70(2),* 40.
6. Myers, J.; Moore, S. *Mod. Plast.* **1994,** *71(3),* 60.
7. Leaversuch, R. D. *Mod. Plast.* **1994,** *71(4),* 79.
8. Holusha, J. *Austin American-Statesman* December 21, 1989, p-E4.
9. Umeyama, H.; Ono, K. *Jpn. Kokai Tokkyo Koho JP* 04,363,228 **1992.**
10. Imaizumi, M.; Kuriyama, M. *Jpn. Kokai Tokkyo Koho JP* 03,262,642 **1991.**
11. Graham, T. *Philos. Mag.* **1866,** *32,* 401.
12. Pixton, M. R.; Paul, D. R. *Polymeric Gas Separation Membranes*; CRC: Boca Raton, FL, 1994; Chapter III. (a) Briston, J. H. *The Wiley Encyclopedia of Packaging Technology*; Bakker, M. Ed.; John Wiley & Sons: New York, 1986.
13. Koros, W. J.; Hellums, M. W. *Fluid Phase Equilibria* **1989,** *53,* 339.
14. Puleo, A. C. et al. *Polymer* **1989,** *30,* 1357.
15. Walles, W. E. *Barrier Polymers and Structures*; ACS Symposium Series 423; American Chemical Society: Washington, DC, 1990; Chapter XIV.
16. Hobbs, J. P. et al. *Barrier Polymers and Structures*; ACS Symposium Series 423; American Chemical Society: Washington, DC, 1990; Chapter XV.
17. Loomis, G. L.; Flammino, A. *Polym. Mater. Sci. Eng.* **1992,** *67,* 292.

Gas Permeable Polymers

See: *Contact Lenses, Gas Permeable*
Contact Lenses, Hydrogels
Fluorine-Containing Poly(amino acids) (Conformation and Gas Permeability)
Gas Barrier Polymers
Gas Separation Membranes
Polyimides (via Interfacial Polyfunctional Condensation)
Polyphenylacetylene-Based Permselective Membranes
Polysilylpropynes (Steric Effects on Material Properties)

GAS PHASE OLEFIN POLYMERIZATION

Kyu Yong Choi
Department of Chemical Engineering
University of Maryland

Gas phase polymerization processes are widely used in the polymer industry to manufacture polyolefins such as polypropylene and ethylene homo- and co-polymers. In a gas phase polymerization process, gaseous monomers are polymerized over high-activity solid-transition metal catalysts known as Ziegler-Natta type catalysts. In industrial gas phase olefin polymerization processes, continuous flow reactors such as fluidized beds and mechanically stirred bed reactors are used. Since no liquid phase is present in the reactor, the gas phase polymerization process is conceptually simpler than liquid slurry or solution processes where liquid diluent or solvent must be separated, purified, and recycled.

Although the idea of polymerizing olefins in gas phase was introduced in late 1950s, it was by the Union Carbide Corporation in late 1960s that the first commercial gas-phase ethylene polymerization process was developed. A key to the successful development of commercial gas-phase olefin polymerization process was the high-activity organo-chromium catalyst. Currently, there are several companies that have developed their own gas-phase polymerization technology using different types of proprietary catalysts and reactor systems. The polyolefins manufactured by gas phase technology include polypropylene, high-density polyethylene (HDPE, 0.931–0.970 g/cm^3); low density polyethylene (LDPE, 0.910–0.930 g/cm^3), linear low-density polyethylene (LLDPE), and recently developed special grade polyethylenes.

A large number of highly active and stereospecific catalysts have been developed recently and used successfully for commercial production of olefin polymers. Catalysts that are inexpensive and that have improved controllability of polymer properties are persistently pursued as witnessed in the hundreds of patents granted in past decades. For example, the metallocene catalyst that offers single-site activity is emerging as a next-generation catalyst. It has been claimed that metallocene catalyst can provide a narrow molecular weight distribution and more precise control of polymer properties than other olefin polymerization catalysts. Reviews on the recent progress in transition metal catalyzed olefin polymerization processes and catalysts can be found in the literature.[1-12]

One of the requirements for the catalyst in gas phase polymerization is high activity. The gas phase olefin polymerization catalyst must have activity such that no catalyst residue removal or atactic polymer extraction step (in polypropylene process) is required. It is well known that solid catalyst particles disintegrate into smaller sized particles shortly after polymerization is started. New active catalytic sites buried in the transition metal crystals are exposed by disintegration, resulting in the enhancement of catalytic activity. Regardless of the type of catalyst being used, the shape of the polymer particle produced is a close replica of the original catalyst particle. This implies that polymer particle morphology can be controlled to some extent by the catalyst particle morphology.

PRODUCTION OF OLEFIN POLYMERS

α-Olefin polymers are manufactured in commercial gas phase polymerization processes by using three different types of reactors: fluidized beds, vertical stirred beds, and horizontal stirred multicompartmental reactors.

REFERENCES

1. Pino, P.; Rotzinger, B. *Makromol. Chem. Suppl.* **1984,** *7,* 41.
2. Choi, K. Y.; Ray, W. H. *J. Macromol. Sci–Rev. Macromol. Chem. Phys.* **1985,** C25(1), 1.
3. Choi, K. Y.; Ray, W. H. *J. Macromol. Sci–Rev. Macromol. Chem. Phys.* **1985,** C25(1), 57.
4. Karol, F. J. *Catal. Rev. Sci. Eng.* **1985,** *26(3-4),* 557.
5. Hsieh, H. L. *Catal. Rev. Sci. Eng.* **1985,** *26(3-4),* 631.
6. McDaniel, M. P. *Adv. Catal.* **1985,** *33,* 47.
7. Barbe, P. C.; Cecchin, G.; Moristi, L. *Adv. Polym. Sci.* **1987,** *81,* 1.
8. Seppala, J. V.; Auer, M. *Progr. Polym. Sci.* **1990,** *15,* 147.
9. Nowlin, T. E. *Progr. Polym. Sci.* **1989,** *11,* 29.
10. James, D. E. *Encyclopedia of Polym. Science and Engineering*; John Wiley & Sons: New York, NY, 1986; Vol. 6, 429.
11. Beach, D. L.; Kissin, Y. V. *Encyclopedia of Polym. Science and Engineering*; John Wiley & Sons: New York, NY, 1986; Vol. 6, 454.
12. Xie, T.; McAuley, K. B.; Hsu, J. C. C.; Bacon, D.W. *Ind. & Eng. Chem. Res.* **1994,** *33,* 449.

Gas Phase Polymerization

See: Bulk Polymerization
Gas Phase Olefin Polymerization
Polyethylene (High Density Preparation)

Gas Sensors

See: Amphiphilic Gas Sensors (Based on Polymer-Coated Piezoelectric Crystals)

GAS SEPARATION MEMBRANES

Lloyd M. Robeson
Air Products and Chemicals Incorporated

Separation of gas mixtures is a significant field in commercial technology with myriad applications that clearly impact the worldwide economy as well as our standard of living. Cryogenic distillation, adsorption, and chemical separation are well-established technologies in this field. More recently, membrane separation of gases has joined this list of available technologies. Membrane separation offers the specific advantages of low energy use, simplicity, ease of operation, limited space requirements, as well as economic benefits. Although membrane separation is not expected to completely replace the established technologies, it offers significant benefits and will be a competitive system in specific existing applications as well as in new and unique applications that cannot be performed using established technologies. Membrane separation is used for N_2 generation to blanket the storage of flammable materials as well as controlling the atmosphere to extend the life of perishable products. The initial use of membrane separation involved H_2/N_2 separation in ammonia production. Additional applications to be detailed later include CO_2/CH_4 separation for natural gas and biogas separation, water dehydration, He recovery, H_2/hydrocarbons (petrochemical processes), and production of medical oxygen. These are a few of the present and future applications that demonstrate the applicability of this rapidly expanding technology.

POLYMERS USED/PROPOSED FOR GAS MEMBRANE SEPARATION

There has been a wide variety of polymers used or proposed for gas separation. Except for modest utility of silicone rubber (for O_2 enrichment), the polymers used are glassy polymers.

Polymers that have either been used or proposed for gas separation membranes are listed in **Table 1** along with their permselective properties.

Specific polymer classes that are of interest for gas separation membranes include polycarbonates, polysulfones, polyarylates, and polyimides.

Polycarbonates, of which Bisphenol A polycarbonate has achieved significant recognition as an engineering polymer, offer many variations based on the choice of the Bisphenol.

Aromatic polyesters (also termed polyarylates) are based on the condensation of aromatic diacids and bisphenols. As such, the property balance is similar to that of aromatic polycarbonate including permselectivity.

Polysulfone (based on the condensation polymer of Bisphenol A and 4,4′ dichlorodiphenylsulfone) has been widely used in membrane systems (e.g., ultrafiltration) and was the first polymer utilized in large, commercial-scale gas separation for H_2/N_2 (ammonia purge gas). Studies reported in the literature include methyl substituted Bisphenols, biphenol-based polysulfones, effect of structural symmetry and polyethersulfone.[3-7]

More recently, polyimides have received primary attention in both the patent and open literature. Polyimides have been shown to exhibit permselectivity characteristics for most of the gas separations of interest (O_2/N_2, CO_2/CH_4, H_2/N_2, H_2/CH_4). A series of publications from Yamaguchi University (Japan) has reported on the permselectivity of many polyimide variants.[8,9] Stern and co-workers (of Syracuse University) have also reported on many additional polyimide variants.[10,11]

Although the classes of polymers noted above have been the primary focus of researchers and comprise many of the present and emerging commercial membranes, other polymers have generated both academic and commercial interest. Poly(phenylene oxide) (PPO) has been of interest because of its high permeability and good permselectivity (relative to other commercially available polymers). Permeability data on PPO has been reported by Paul and co-workers.[2,12] Poly(4-methyl-1-pentene)

TABLE 1. Polymers Proposed/Utilized for Gas Separation

	$P(O_2)$	$P(N_2)$	P(He)	$P(CO_2)$	$P(CH_4)$	T(°C)	References
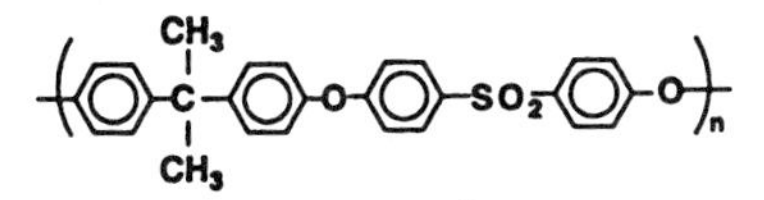	1.2	0.2	11.0	4.9	0.21	35	23
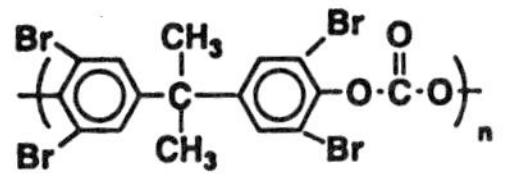	1.36	0.182	17.6	4.23	0.126	35	16
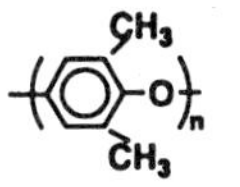	14.6	3.5	82.3	65.5	4.1	35	12
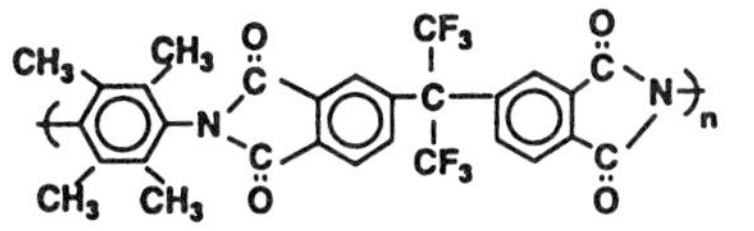	122	35.6	—	440	28.2	35	24
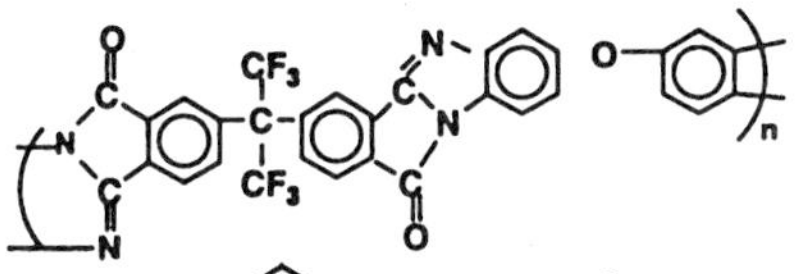	7.9	1.22	89.0	27.6	0.54	35	18
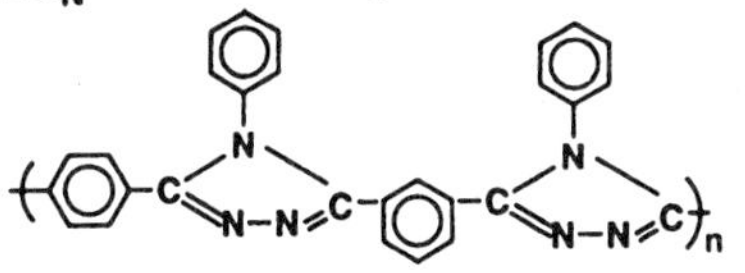	1.2	0.133	14.0	5.5	0.10	41	17
Cellulose acetate 2.45 DS	0.82	0.15	16.0	4.75	0.15	35	19
Ethyl cellulose 47.9% ethoxy content	19.4	5.8	66	116	12.4	35	20
Poly(4-methyl-1-pentene)	27.0	6.7	95.4	84.6	14.9	35	25
Silicone rubber	960	460	580	4500	1300	35	21
Polytrimethylsilylpropyne	9710	6890	6750	37,000	18,400	25	13
Syndiotactic PMMA	0.105	0.00957	0.013	0.415	0.0064	35	22

has had commercial activity as a melt-spun fiber for O_2/N_2 separation.

Cellulose acetate and ethyl cellulose have been used for CO_2/CH_4 and O_2/N_2 separations. Cellulose acetate has been widely employed for reverse-osmosis applications, facilitating the translation of fabrication technology to gas separation. Silicone rubber has had only limited commercial use for gas separation, it has been used primarily for medical oxygen.

Recently, one of the more widely studied polymers has been polytrimethylsilylpropyne (PTMSP) and its variants. PTMSP exhibits the highest permeability of known polymers, however, it has only limited separation capabilities. It was noted in many publications that PTMSP shows a major flux decline with time. This was initially attributed to densification (collapse of free volume). However, a recent summary of publications indicated that the observed flux decline can be attributed to other factors, including contamination (compressor oil, vacuum pump oil, permeability equipment, gaskets, etc.) and the immersion of samples in alcohol prior to testing.[13] There does appear to be a consistent flux decline dependent on the original casting solvent, however. Recently Chen et al. observed only limited flux decline on PTMSP membranes stored in air for 6 months.[14] In addition to the references cited, Stern has compiled a comprehensive review of polymers that are proposed or already used for gas separation.[15]

MEMBRANE FABRICATION

The key discovery in membrane fabrication allowing for the production of thin, dense separating layers was made by Loeb and Sourirayan.[1] The controlled coagulation process yielded asymmetric membranes comprised of a very thin dense skin supported by a porous substrate. This procedure and variations thereof are employed in most of the membrane systems presently in use.

The two most common membrane-separation systems are spiral-wound flat sheet and hollow-fiber assemblies.

Although asymmetric film and fibers comprise many of the important commercial membrane systems, other constructions have utility. Thin-film composites involve the coating of a thin dense film onto a porous substrate. A variation of this technique involves coating a porous substrate with a very permeable polymer, followed by coating with a thin film of a separating polymer to allow for pinhole free constructions. Another variant of membrane fabrication involves surface modification to improve selectivity.

APPLICATIONS

During the 1980s, commercialization of membranes for nitrogen generation resulted in several significant applications. These included nitrogen blanketing for inerting flammable ship-bound cargoes (LNG/oil) and offshore platform inerting, and atmosphere control for storage of perishable vegetables, flowers, and fruits. Other nitrogen-generating applications involve blanketing for process industries (metallurgical heat treatment, glass manufacture, chemical production) and silo use to prevent dust explosion. Presently, nitrogen generation is the largest use of polymeric membranes for gas separation.

CO_2/CH_4 separation applications are quite feasible with membrane technology because of the high separation factors achieved (>20) along with generally high CO_2 flux. Natural gas enrichment to remove CO_2, H_2S, and H_2O is an application used presently that holds significant promise for the future as supplies become more limited and lower-grade sources are more commonly tapped for production. Additional uses include recovery of CH_4 from landfills or agricultural wastes. Enhanced oil recovery employing CO_2 flooding yields CO_2 contamination of the natural gas coproduced in this process. Enrichment of the natural gas allows for a salable fuel and recovery of CO_2 allows for reinjection.

H_2/CH_4 (and higher hydrocarbons) separation in refining operations is an important application involving polymeric membrane systems. Specific applications include hydrocracker purge, reformer off-gas streams, catalytic cracker purge recovery, and hydrotreater purge H_2 recovery. The high H_2/CH_4 separation factors for relevant polymeric membranes (>50) allow for efficient hydrogen recovery in these systems.

The water-permeation rate through polymers is generally much higher than that of air or CH_4 (500 to 1000 times), thus membrane separation can be efficiently employed for dehydration of air or natural gas streams. Applications include dry compressed air production and natural gas dehydration.

Other applications include helium recovery, medical oxygen, gasoline vapor recovery, and various pervaporation separations.

The primary advantages for membranes vs. other methods of gas separation include ease of operation (no moving parts), unit size, low capital investment, low energy requirements, and the overall economies of low-scale operation. Limitations could include high temperature use, inability to handle specific multicomponent feed streams (solvent-induced membrane failure, contamination sensitivity), and economics vs. established techniques for large-scale operations.

REFERENCES

1. Loeb, S.; Sourirayan, S. *Adv. Chem. Ser.* **1963**, *38*, 117.
2. Toi, K.; Morel, G.; Paul, D. R. *J. Appl. Polym. Sci.* **1982**, *27*, 2997.
3. McHattie, J. S.; Koros, W. J.; Paul, D. R. *Polym.* **1991**, *32*, 840.
4. McHattie, J. S.; Koros, W. J.; Paul, D. R. *Polym.* **1992**, *33*, 1701.
5. Aitken, C. L.; Koros, W. J.; Paul, D. R. *Macromolecules* **1992**, *25*, 3424.
6. Aitken, C. L.; Koros, W. J.; Paul, D. R. *Macromolecules* **1992**, *25*, 3651.
7. Kumazawa, H.; Wang J.-S.; Sada, E. *J. Polym. Sci. Part B: Polym. Phys.* **1993**, *31*, 881.
8. Tanaka, K.; Kita, H.; Okamoto, K.; Nakamura, A.; Kusuki, Y. *J. Membr. Sci.* **1989**, *47*, 203.
9. Okamoto, K.; Tanaka, K.; Kita, H.; Ishida, M.; Kakimoto, M.; Imai, Y. *Polymer J.* **1992**, *24*, 451.
10. Mi, Y.; Stern, S. A.; Trohalaki, S. *J. Membr. Sci.* **1993**, *77*, 41.
11. Stern, S. A.; Liu, Y.; Feld, W. A. *J. Polym. Sci. Part B: Polym. Phys.* **1993**, *31*, 939.
12. Aguilar-Vega, M.; Paul, D. R. *J. Polym. Sci. Part B: Polym. Phys.* **1993**, *31*, 1577.
13. Robeson, L. M.; Burgoyne, W. F.; Langsam, M.; Savoca, A. C.; Tien, C. F. *Polymer* **1994**, *35*, 4970.
14. Chen, G.; Griesser, H. J.; Mau, A. W. H. *J. Membr. Sci.* **1993**, *82*, 99.
15. Stern, S. A. *J. Membr. Sci.* **1994**, *94*, 1.
16. Muruganandam, N.; Koros, W. J.; Paul, D. R. *J. Polym. Sci. Part B: Polym. Phys.* **1987**, *25*, 1999.
17. Gebben, B.; Mulder, M. H. V.; Smolders, C. A. *J. Membr. Sci.* **1989**, *46*, 29.
18. Walker, D. R. B.; Koros, W. J. *J. Membr. Sci.* **1991**, *55*, 99.
19. Puleo, A. C.; Paul, D. R.; Kelley, S. S. *J. Membr. Sci.* **1989**, *47*, 301.
20. Houde, A. Y.; Stern, S. A. *J. Membr. Sci.* **1994**, *92*, 95.
21. Stern, S. A.; Shah, V. M.; Hardy, B. J. *J. Polym. Sci. Part B: Polym. Phys.* **1987**, *25*, 1263.
22. Min, K. E.; Paul, D. R. *J. Polym. Sci. Part B: Polym. Phys.* **1988**, *26*, 1021.
23. Barbari, T. A.; Koros, W. J.; Paul, D. R. *J. Membr. Sci.* **1989**, *42*, 69.
24. Tanaka, K.; Okano, M.; Toshino, H.; Kita, H.; Okamoto, K.-I. *J. Polym. Sci. Part B: Polym. Phys.* **1992**, *30*, 907.
25. Mohr, J. M.; Paul, D. R. *Polym.* **1991**, *32*, 1236.

GASEOUS INITIATOR POLYMERIZATION (Nitrogen Dioxide and Sulfur Dioxide)

Suraj N. Bhadani* and Sumanta K. Sen Gupta
Department of Chemistry
Ranchi University

NITROGEN DIOXIDE AS GASEOUS INITIATOR

NO_2 is a red-brown gas with a pungent smell. It is an odd-electron molecule (seventeen valence electrons) and is, therefore, paramagnetic.[1,2] It has an angular structure with O–N–O bond angle 132° and bond distance N–O is 1.19 A°. Thus, the N-atom is sp^2-hybridized and σ-bonds are formed by the electrons of molecular orbitals localized between nitrogen and oxygen atoms. Besides ten electron pairs, there are two *p*-electrons, which form a three electron bond joining all three atoms. The following resonating are assigned to NO_2 molecule (**Structure 1**):[1,3-9]

1

*Author to whom correspondence should be addressed.

Being an odd electron molecule, NO_2 causes a number of free-radical reactions. It has a propensity to dimerize and undergo association reactions with other free radicals. It adds to double bonds in unsaturated molecules, and thus it causes the polymerization of vinyl monomers.[1]

SULFUR DIOXIDE AS GASEOUS INITIATOR

SO_2 is a colorless gas with a suffocating smell, m.p. 75.5°C, and b.p. 10.02°C. The magnitude of the bond angle O–S–O 199.5° indicates that the sulfur atom is in the trigonal valence state. The reaction of SO_2 with ethyl magnesium bromide in ethereal solution to give ethyl sulfinic acid indicates the presence of double bonds in SO_2. The S–O double bonds originate from pπ–dπ bonding due to the lateral overlap of *p*-orbitals of oxygen with d-orbitals of sulfur. The bond angle 119.5° clearly agrees with the structure predicted from sp^2-hybridization. Two resonating forms are shown in **Structure 2.**[3,8,23]

2

Therefore, liquid SO_2 itself copolymerizes with a number of monomers, such as alkenes and dienes, to form polysulfones in presence of free radical.[24] It further functions as a diluent in the polymerization. However, SO_2 gas in low concentration also acts as initiator for the polymerization of some vinyl monomers.[25,26] Aqueous SO_2 or its alkali metals salt also initiates aqueous vinyl polymerization.

DISCUSSION

NO_2 Initiated Polymerization of Acrylamide in Nonaqueous Solvents

When NO_2 is passed into dimethylformamide (DMF), quite a number of species seem to be formed. Some of them may have short lives. Thus, the fresh solution shows a spectrum of multiple absorption bands, among them a broad one with maximum 590 nm. When acrylamide is added to NO_2 solution in DMF, the intensity of the absorption peaks decrease with time and finally all peaks vanish, indicating an interaction between NO_2 and the monomer. Both the blue and air exposed solutions of NO_2 initiate polymerization.

NO_2 Initiated Polymerization of Acrylamide in Water

NO_2 also initiates the polymerization of acrylamide in water.[10] Unlike its polymerization in organic solvents, polymerization of acrylamide with NO_2 in water yields a viscous homogenous reaction mixture. During the course of polymerization, the reaction mixture becomes one solid glassy gel.

NO_2 Initiated Copolymerization

Acrylamide copolymerizes with methyl methacrylate in the solution of NO_2 in DMF and forms high molecular weights of copolymers.[11]

NO_2 Initiated Polymerization of *N*-Vinylcarbazole

N-vinylcarbazole (N-VCZ) is easily polymerized by several methods such as free radical,[12] cationic,[13] Zeigler type,[14] charge transfer for solution or melt,[15] and high energy radiation in solid state.[12-18] It is also initiated by gaseous initiators such as NO_2, SO_2, and Cl_2 in solution and solid state.[17,18] Bhandani et al. investigated the polymerization of N-VCZ in the solution of NO_2 in 1,2-dichloroethane.[19] The rate of polymer formation increases rapidly with increasing NO_2 concentration, giving a maximum 75% monomer conversion to polymer within a few minutes of polymerization. The free-radical inhibitor, 2,2-diphenyl-1-picryl-hydrazyl does not inhibit the polymerization of N-VCZ with NO_2. The polymer formation does not occur in basic solvents, such as dimethylformamide. A characteristic feature of the N-VCZ polymerization is the initiation by electron accepting compounds. The electron releasing effect of nitrogen in the carbozole ring may induce high nucleophilicity of the vinyl group and cationic propagation species may be stabilized by conjugation with carbazole group.

NO_2 Initiated Polymerization of Some Other Vinyl Monomers

NO_2 initiates also successfully the polymerization of methyl methacrylate, acrylic acid, and acrylonitle in organic solvents via free-radical mechanism.[20-22] In all cases, polymer yield increases with increasing the concentrations of monomer and NO_2. The polymerization rate follows the first order dependence on monomer concentration and square root of NO_2 concentration.

SO_2 Initiated Polymerization of Some Vinyl Monomers

Driscoll et al. reports that SO_2 initiated polymerization of methyl methacrylate, styrene, methyl acrylate, acrylic acid, acrylonitrile, and acrylamide in non-aqueous solvents.[26]

SO_2 Initiated Copolymerization

SO_2 initiated copolymerization of methyl methacrylate and styrene at various feed ratios was carried out and the compositions of the resultant copolymers were determined spectrophotometrically.[27]

Polymerization of N-Vinylcarbazole by SO_2

Similar to NO_2, SO_2 also causes the successful polymerization of N-VCZ in 1,2-dichloroethane but with a relatively slow rate.[28] The rate increases with increasing SO_2 concentration. SO_2 is also a charge transfer electron acceptor like NO_2, and initiates the polymerization through a charge transfer mechanism.

REFERENCES

1. Gray, P.; Yoffe, A. D. *Chem. Rev.* **1955,** *55,* 1069.
2. Cotton, F. A.; Wilkinson, G. *Advanced Inorganic Chemistry*; Wiley Eastern Limited: New Delhi, 1967; Chapter XXI.
3. Dutta, R. L. *Inorganic Chemistry Chemistry*; The New Book Stall: Calcutta, 1981; Chapter XIX.
4. Orville-Thomas, W. *Chem. Rev.* **1957,** *57,* 1179.
5. Tanaka, J. *Nippon Kagaku Zasshi* **1957,** *78,* 1647.
6. Sutherland, G.; Penney, W. *Nature* **1935,** 136, 146.

7. Sutherland, G.; Penney, W. *Proc. Roy. Soc.* **1936,** A156, 678.
8. Pauling, L. *The Nature of the Chemical Bond*; Cornell University: New York, 1967; Chapter X.
9. McEven, K. *J. Chem. Phys.* **1961,** *32,* 1801.
10. Bhadani, S. N.; Prasad, Y. K. *J. Polym. Sci.* **1978,** B16, 639.
11. Fineman, F.; Ross, S. D. *J. Polym. Sci.* **1956,** *15,* 259.
12. Davidge, H., *J. Appl. Chem.* **1959,** *9,* 553.
13. Reppe, W.; Keyssner, E.; Dorrer, E. U.S. Patent 2 072 465.
14. Soloman, O. F. *J. Polym. Sci.* **1961,** *52,* 205.
15. Scott, H.; Miller, G. A.; Labes, M. M. *Tetrahedron Letters* **1963,** 1073.
16. Cornish, E. H. *Plastics* **1962,** *27,* 132.
17. Haldon, R. A.; Hay, J. N. *J. Polym. Sci.* **1968,** A-1, *6,* 951.
18. Meyers, R. A.; Christman, E. M. *J. Polym. Sci.* **1968,** A-1, *6,* 945.
19. Bhadani, S. N.; Sen Gupta, S. K.; Sinha, A. K. *J. Appl. Polym. Sci.* **1992,** *44,* 173.
20. Bhadani, S. N.; Mishra, M. K. *J. Polym. Sci.* **1983,** A-1, *21,* 3601.
21. Mishra, M. K.; Bhadani, S. N. *Makromol. Chem.* **1983,** 184, 955.
22. Mishra, M. K.; Bhadani, S. N. *J. Appl. Polym. Sci.* **1983,** *28,* 2005.
23. Durrant, P. J.; Durrant, B. *Introduction to Advanced Inorganic Chemistry*; Norfolk: ELBS and Longman Group Limited, 1975; Chapter XXII.
24. Florjanczyk, Z. *Inst. Polym. Sci.* **1991,** *16,* 509.
25. O'Driscoll, K. F. *Structure and Mechanism in Vinyl Polymerization*; Tsurta, T.; O'Driscoll, K. F. Eds.; Marcel Dekker: New York, 1969; Chapter III.
26. Ghosh, P.; O'Driscoll, K. F. *J. Polym. Sci.* **1966,** *B4,* 519.
27. Eisenberg, A.; O'Driscoll, K. F.; Tobolsky, A. V. *Anal. Chem.* **1959,** *31,* 203.
28. Bhadani, S. N.; Kundu, S. *Makromol. Chem., Rapid Commun.* **1981,** *2,* 139.

Gel Fibers

See: Polyacrylonitrile Gel Fibers

Gel-like Spherulite Press Method

See: Polyethylene, High Performance (Gel-Like Spherulite Press Method and Ultradrawing)

Gel Spinning

See: Poly(vinyl alcohol) (High Strength and High Modulus)
Poly(vinyl alcohol) Fiber (High Strength and High Modulus)
Ultra-Strong Polyethylene Fibers

Gelan Gum

See: Welan Microbial Polysaccharide

Gelatin

See: Gelatin (Molecular Weight Distributions)

GELATIN (Molecular Weight Distributions)

Chi Wu
Department of Chemistry
The Chinese University of Hong Kong

Gelatin constitutes a class of proteinaceous substances derived from a naturally occurring protein–collagen–through certain processes mainly involving destruction of collagen's secondary structure. Gelatin is well known for its property of forming elastic gels at room temperature at relatively low concentrations: a few percent of gelatin in water. Gelatin is widely and extensively used in the food, photographic, and pharmaceutical industries as an important stabilizer ingredient (e.g., for pulverulent formulations of carotenoids and vitamin A).[1,2]

Much experimental research has been done both on the practical uses of gelatin and on fundamental aspects of the gelation process.[3-5] Much less attention, on the other hand, has been given to gelatin's molecular characterization. This is due mainly to its polyelectrolyte and polydisperse nature.

At room temperature, there are two kinds of intermolecular actions in aqueous gelation solution. One such interaction is the electrostatic activity in the presence of electric charges on the polypeptide chain. Another is the formation of hydrogen bond between the amino acid units. These interactions have to be eliminated in order to identify the true molecular characteristics of gelatin. Adding salts (as in solutions of synthetic polyelectrolytes) is one way to screen the electrostatic interaction and suppress the polyelectrolyte effect. However, most types of salts added in aqueous gelatin solution at room temperature will not prevent the hydrogen bonding, which is reflected in either gelation or aggregation, depending on the gelatin concentration. Only certain salts, such as KSCN and LiBr, are capable of screening out electrostatic interactions while preventing the formation of the hydrogen bond.

When salts are present in aqueous gelatin solution, it is known that there is a problem of preferential sorption of the salt in the domain of gelatin molecules, just as synthetic polyelectrolytes in aqueous salt solutions. Thus, in order to determine the true molecular parameters, the apparent molecular parameters are extrapolated to infinitely dilute salt concentration. However, this does not give the true molecular parameters, as the behavior of gelatin in aqueous solution with and without salts could be completely different.

Considering this preferential sorption, some researchers have used non-aqueous solvents, such as formamide, glycerol, and trifluoroethanol, to dissolve gelatin[6] with an assumption that those non-aqueous solvents can suppress the above two types of intermolecular interactions.

As an absolute method, laser light scattering (LLS) has been used extensively to characterize polymers. Current-model LLS spectrometers can perform both static and dynamic measurements. In static LLS, the angular and concentration dependence of absolute scattered intensity are measured; the weight-average molecular weight (M_w), the z-average radius of gyration ($<R_g^2>_z^{1/2}$), and the second virial coefficent (A_2) can be determined from the measured absolute scattered intensities. In dynamic LLS, the intensity-intensity time correlation function is measured. By making a Laplace inversion, we can convert the measured correlation function into a characteristic line-width distribution (G(Γ)) which could be further reduced to a translational diffusion coefficient distribution (G(D)) or even to a molecular weight distribution (MWD) if we can establish a calibration between D and M.

PREPARATION

When gelatin is extracted from collagen, the entire structure of the collagen fiber is lost. This extraction process requires both chemical and thermal treatments, such as in acidic or basic environment and heating. In order to establish a general methodology to characterize the molecular weight distribution of gelatin, two special, laboratory-prepared gelatins were used. One is an A-type, the other a B-type–termed hereinafter gelatin-A and gelatin-B, respectively. Analytical grade formaldehyde was used as solvent. Both gelatins and formamide were used without further purifications.

APPLICATIONS

Since we have confirmed that formamide is a good solvent for dissolving gelatin at room temperature, our experimental results suggest that the gelatin chain in formamide at room temperature is flexible and has a random-coil conformation. Both the effects of polyelectrolytes and hydrogen bonding in gelatin molecules are eliminated in formamide so that it can be used as a solvent in other solution methods to characterize gelatin.

It should be noted that the above-listed calibration between D and M, i.e., K_D and α_D are independent on a particular laser light scattering instrument as long as formamide as solvent and room temperature (25°C) are used, which means that with this calibration, dynamic light scattering becomes a good, absolute, and fast experimental method for characterizing the molecular weight distribution of gelatin even if it does not have the same resolution as SEC.

In addition, in the future, with the known values of k_d and f, we should be able to determine the molecular weight distribution of a given gelatin sample by measuring *only* one dilute gelatin solution at *only* one scattering angle.

REFERENCES

1. Ward, A. G.; Courts, A. *The Science and Technology of Gelatin* Academic: London, 1977.
2. Veis, A. *The Macromolecular Chemistry of Gelatin*; Academic: London, 1964.
3. Djabourov, M. *Contemp. Phys.* **1988**, *29(3)*, 273.
4. Newcomer; Jones, T. A.; Aqvist, J.; Sundelin, J.; Eriksson, U.; Rask, L.; Peterson, P. A. *EMBO J.* **1984**, *3(7)*, 1451.
5. Quellet, C.; Eicke, H. F.; Gehrke, R.; Sager, W. *Europhys. Lett.* **1989**, *9(3)*, 293.
6. Umberger, J. Q. *Photographic Sci. & Engineering* **1967**, *11(6)*, 385.

GELATION, PHYSICAL (Atactic Polystyrene)

Jeanne François and Dominique Sarazin
Institut Charles Sadron/ULP

For a long time, physical gelation in polymer solutions was believed to be limited to crystallizable polymers. Discovery by Tan et al.[1] of the ability of atactic polystyrene (aPS) to exhibit this phenomenon in several solvents was first considered astounding and suspicious but was confirmed. Among the various solvents used, carbon disulfide (CS_2) has attracted attention because it exhibits the highest temperature gelation.

PHASE DIAGRAMS: SOL-GEL TRANSITION

Tan et al.[1] showed that atactic polystyrene with molecular weights ranging from $4*10^3$ to $2*10^6$ are capable of gelation in many solvents. For a sample of a given molecular weight, the sol-gel transition temperature (T_{gel}) varies with the polymer concentration (c_p). Below T_{gel}, which is an increasing function of c_p, the polymer forms a clear gel without syneresis.

Effect of Branching

The phase diagrams for model starlike aPS in CS_2 has been described by Leshaini et al.[3] When compared to their linear homologs, a significant increase of $T_{m,gel}$ with the number of branches is observed. Since small changes in the tacticity play an important role in the gelling properties of aPS, a ^{13}C RMN study was performed which confirms that the aPS stars exhibit the same tacticity as the linear aPS. Thus, no ambiguity exists on the origin of the higher stability of the star gels.

DIFFERENTIAL SCANNING CALORIMETRY DSC

The first finding of a gelation exotherm in the aPS-CS_2 system was published by Tan et al.[2] Later, a more systematic DSC study brought in clear evidence of a gelation exotherm and a melting endotherm. Important conclusions have been drawn:

- The occurrence of exotherms and endotherms suggests a first-order transition.[5]
- The gelation phenomenon is completely different from the glass transition.
- The exotherms and endotherms are very broad for all samples and lie in the same temperature range independent of the molecular weight.
- Enthalpies of formation and melting plotted as a function of polymer concentration do not differ significantly and pass through a maximum for cp = 500 g/L. Such a behavior has been interpreted in terms of formation of a type of solvent-polymer complex. The solvent is assumed to participate in the crosslink formation by intercalation between polymer chains. This may explain why gelation is a solvent-dependent phenomenon. This is also consistent with the concentration dependence of melting and crystallization enthalpies of various solvents, ΔH_{cs} and ΔH_{ms}, in aPS solutions.[6]

It has further been proposed from DSC experiments performed on polystyrene of variable tacticity that the syndiotactic sequences are mainly involved in structures forming the crosslinks.[7]

LIGHT SCATTERING

Isotropic Enhanced Low Angle Scattering (ELAS)

ELAS seems to depend on the method of preparation of the solutions. The general question for semidilute polymer solutions in water or organic solvents is relative to the amount of polymer lost in centrifugation and filtration. Gan et al.[8] showed that if a solution that exhibits ELAS is centrifuged, the effect disappears but the concentration becomes significantly lower. For solutions where ELAS is absent, centrifugation does not provoke change of concentration. ELAS reveals without ambiguity the presence of heterogeneities in the polymer solutions, but it is never clear if they really reflect particular polymer-polymer interactions and have a physical significance or if they result from artifacts related to a bad preparation of the solutions or samples.

Anisotropic Light Scattering

A recent anisotropic light-scattering study showed progress in the characterization of the structure of crosslinked domains. For interacting molecules, the excess isotropic scattering is affected by the correlations in position, while the anisotropic scattering is affected by the orientational correlations.

NEUTRON SCATTERING

Izumi et al.[4] performed neutron scattering experiments of aPS-CS_2 over a temperature range of –45 to 27 °C. The authors have concluded that the aPS at the vicinity of the gel point behaves like a branched polymer with an increasing crosslink functionality as the temperature decreases. For the lower temperature, the number of branches in a model of star has been found to reach 109 for monodisperse stars or 310 for polydisperse stars. Gathering of such a high number of chains in a crosslinked domain should result in a syneresis, but this effect was never encountered with this system.

CONCLUSION

aPS solution in CS_2 is able to form physical gels. The thermodynamic gelation properties can be interpreted in terms of the Ferry-Eldridge approach. Crosslinks seem to be related to formation of a type of polymer-solvent complex where syndiotactic sequences are preferentially involved and solvent molecules are intercalated in between polymer chains. Nevertheless, there are now two descriptions of the structure of these crosslinks. Lehsaini et al.[9] proposed a model of association of limited number of polymer chains stiffened along a number of segments, which decreases with increasing concentration. Izumi et al.[4] discovered formation of stars of up to 100 branches. The first model seems to be in better agreement with the formation of optically clear gels without syneresis.

REFERENCES

1. Tan, H. M.; Moet, A.; Hiltner, A.; Baer, E. *Macromolecules* **1983,** *16,* 28.
2. Tan, H. M.; Chang, B. H.; Baer, E.; Hiltner, A. *Eur. Polym. J.* **1983,** *19,* 1021.
3. Lehsaini, N.; Boudenne, N.; Zilliox, J. G.; Sarazin, D. *Macromolecules* **1994,** *27,* 4353.
4. Izumi, Y.; Katano, S.; Funahashi, S.; Furusaka, M.; Harai, M. Private Communication, 1991.
5. Xie, X. M.; Tanioka, A.; Miyasaka, K. *Polymer* **1991,** 32, 479.
6. Gan, Y. S.; François, J.; Guenet, J. M. *Macromolecules* **1986,** *19,* 173.
7. François, J.; Gan, Y. S.; Sarazin, D.; Guenet, J. M. *Polymer* **1988,** *29,* 893.
8. Gan, Y. S.; François, J.; Guenet, J. M. *Makromol. Chem.* **1985,** *6,* 225.
9. Lehsaini, N.; Muller, R.; Weill, G.; François, J. *Polymer* **1994,** *35,* 2180.

GELLING AGENTS (Agarose and Carrageenan)

Mineo Watase
Applied Biological Chemistry
Faculty of Agriculture
Shizoha University

AGAROSE

Agarose has been studied as a model gelling substance in biology and in the food industry. Many investigations have been carried out to clarify the gelation mechanism and gel properties of agarose.[1-5] Since the gelation and gels of agarose are influenced strongly by the cosolute, the effects of cosolutes such as salt,[6,7] polyols,[8] or sugars,[9] have also been studied.

Agarose is extracted from a red seaweed. It consists of D-galactose and 3,6-anhydro-L-galactose, and it does not contain sulfate groups. It is a main component of agar-agar and governs the mechanical properties of agar-agar gels.

The gelling ability of agarose is not influenced by alkali metal ions or alkali earth ions, but it is decreased by guanidine hydrochloride. The gelling mechanism of these polysaccharides is considered to be as follows: (i) single or double helices are formed from random coils when the temperature of the solution is lowered; (ii) these helices aggregate to form an ordered structure which is called the junction zone or the crystalline region; and (iii) as a result, the three-dimensional network is formed.

The Effects of Molecular Weight [10]

The dynamic Young's modulus E' of 3% agarose gels with molecular weight higher than 1.81×10^5 slightly increased with increasing temperature up to a certain temperature, T_0, and then decreased. T_0 shifted to higher temperature with increasing molecular weight. Even at temperatures lower than the gel-to-sol transition temperature, some polymer segments are released from junction zones into flexible chains, as was described previously by a reel-chain model.[11]

The structure of the agarose gel network is believed to consist of ordered junction zones and flexible chains connecting these junction zones.[2,12]

Junction zones consist of aggregated double helices. An increase in molecular weight of agarose may increase the contour length of the flexible chains and/or the number of junction zones and/or the size of junction zones. Since E' increased with increasing molecular weight, the number of elastically active chains should increase, hence the number of junction zones should also increase. It has been suggested that contour length of flexible chains increases because both breaking stress and strain increase with increasing molecular weight of agarose.[13]

Effects of Sugars and Polyols on the Gel–Sol Transition of Agarose

Effects of various sugars and polyols such as ethylene glycol, glycerol, ribose, glucose, fructose, mannose, galactose, sucrose, maltose, and raffinose on the gel–sol transition of agarose gels are studied by differential scanning calorimetry (DSC). Melting temperature T_m and setting temperature T_s shift to lower temperatures upon the addition of polyols and ribose. All the other sugars examined in this study shift T_m and T_s to higher temperatures. The shift of T_m and T_s to higher temperature is explained by the increase of the number of zippers created by hydrogen bonds between hydroxyl groups in sugars and agarose. The shift of T_m and T_s is linearly related to the dynamic hydration number or the number of equatorially attached hydroxyl groups in sugars or polyols.[14]

The elastic modulus of agarose gels was increased by the addition of ethylene glycol and glycerin. The elastic modulus of thermoreversible gels was proposed to be a function of the number of junction zones, the bonding energy, the number of segments liberated from the junction zone, and the ceiling number, which is the upper limit number of segments which can be liberated from the junction zone before the gel-to-sol transition occurs. The addition of polyols seems to increase the number of junction zones, which is equivalent to the number of zippers in a zipper model approach.[15] The shift of T_m or T_s to lower temperature, however, suggests the structure of agarose gels became less thermally stable. This may be mainly due to the increase of the rotational freedom of parallel links consitituting a zipper upon the addition of polyols.

The elastic modulus of agarose gels increased and T_m and T_s shifted to higher temperatures with increasing concentrations of added sugars.

The result that the number of zippers increases with increasing sucrose concentration is consistent with the rheological result that the molecular weight of the junction zone decreased and the rigidity increased upon the addition of sucrose.[5,12]

CARRAGEENAN

Since κ-carrageenan and agarose have been used as texture modifiers and gelling agents in the food industry, there have been many investigations on the mechanism of gelation and rheological and thermal properties of these materials. κ-carrageenan consists of D-galactose 4-sulfate and 3,6-anhydro-D-galactose. Rees, Morris, and their co-workers[16,17] proposed that the gelation of κ-carrageenan occurs in two steps. (i) When aqueous κ-carrageenan solutions are cooled, some random-coil molecules of κ-carrageenan transform into double-helix molecules. (ii) Further cooling or introduction of metal ions promotes the aggregation of these double-helix molecules, and the aggregates form crosslinking junctions. Recently, these workers have emphasized the point that the junction in κ-carrageenan gels consists of cation-mediated aggregates of double helices.

On the other hand, Smidsrod, Grasdalen, and their coworkers[18,19] did the osmometry, light scattering, and viscometry to examine the gelation mechanism of κ-carrageenan. Since the intrinsic viscosity increased while the molecular weight was not increased in the formation of ordered conformations, a stiffening of chains was suggested. These results, together with the correlation between ion binding and gel strength measurements, led these authors to the "cation-specific, nested, single-helix model" for the gelation of κ-carrageenan.

Although the details of the gelation mechanisms proposed by these two research groups are different, the essential point is that the κ-carrageenan gels consist of junction zones connected by some kind of long flexible chains.

Alkali metal ions or alkaline-earth metal ions promote the gelation of κ-carrageenan,[16-22] although excessive addition of these cations can inhibit gelation.[23] However, addition of excess guanidine hydrochloride has also been shown to inihibit gelation.[24] Ethylene glycol or glycerin also promotes the gelation; however, once again, addition of an excess of these poly-hydric alcohols has been shown to inhibit gelation.[25]

Effect of Alkali Metal Ions

The effect of the addition of the monovalent cations Li^+, Na^+, K^+, and Cs^+ on the gelation of κ-carrageenan and agarose aqueous gels has been studied by the measurement of longitudinal vibration. The dynamic Young's modulus E' of κ-carrageenan and agarose gels containing the alkali metal salts LiCl, NaCl, KCl, or CsCl at various concentrations has been measured at various temperatures. When the alkali metal salt is added, the value of E' for agarose gels is influenced only slightly, while for κ-carrageenan E' is increased substantially. κ-carrageenan has many sulfate groups. The addition of the alkali metal ions screens the electrostatic repulsions between these groups. As a result, the helical structure of κ-carrageenan is stabilized and the helices may form densely packed aggregates, so increasing E'. In contrast, agarose has a naturally stable molecular structure and, therefore, the structure and hence E' is not sensitive to added ions. The K^+ and Cs^+ ions increase E' more than Li^+ and Na^+ for κ-carrageenan gels.[24]

REFERENCES

1. Rees, D. A. *Pure Appl. Chem.* **1981,** *53,* 1.
2. Clark, A. H.; Ross-Murphy, S. B. *Adv. Polym. Sci.* **1987,** *83,* 57.
3. Tokita, M.; Hikichi, K. *Phys. Rev. A* **1987,** *35,* 4329.
4. Hayashi, A.; Kanzaki, T. *Food Hydrocolloids* **1987,** *1,* 317.
5. Watase, M.; Nishinari, K.; Clark, A. H.; Ross-Murphy, S. B. *Macromolecules* **1989,** *22,* 1196.
6. Watase, M.; Nishinari, K. *Rheol. Acta* **1982,** *21,* 318.
7. Picullel, L.; Nillsson, S. *J. Phys. Chem.* **1989,** *93,* 5596.
8. Nishinari, K.; Watase, M. *Agric. Biol. Chem.* **1987,** *51,* 3231.
9. Watase, M.; Nishinari, K.; Williams, P. A.; Phillips, G. O. *J. Agric. Food Chem.* **1990,** *38,* 1181.
10. Watase, M.; Nishinari, K. In *Symposium on Polymer Gels and Networks*, Tsukuba, Japan **1993,** p 10.
11. Nishinari, K.; Koide, S.; Ogino, K. *J. Phys. (France)* **1985,** *46,* 793.
12. Nishinari, K.; Watase, M.; Kohyana, K.; Nishinari, N.; Oakenfull, D.; Koide,S.; Ogino, K.; Williams, P. A.; Phillips, G. O. *Polym. J.* **1992,** *24,* 871.
13. Watase, M.; Nishinari, K. *Rheol. Acta* **1983,** *22,* 580.
14. Watase, M.; Kohyama, K.; Nishinari, K. *Thermochim. Acta* **1992,** 206, 163.
15. Nishinari, K.; Koide, S.; Williams, P. A.; Phillips, G. O. *J. Phys. (France)* **1990,** *51,* 1719.

16. Morris, E. R.; Rees, D. A.; Robinson, G. O. *J. Mol. Biol.* **1980,** 138, 349.
17. Rees, D. A.; Morris, E. R.; Thom. D.; Madden, J. K. In *The Polysaccharides*; Aspinall, G. O., Ed.; Academic: New York, 1982; Vol. 1.
18. Smidsrod, O.; Grasdalen, H. *Hydrobiologia* **1984,** 116/117, 19.
19. Paoletti, S.; Smidsrod, O.; Grasdalen, H. *Biopolymers* **1984,** *23,* 1771.
20. Rochas, C.; Rinaudo, M. *Biopolymers* **1984,** *23,* 735.
21. Piculell, L.; Hakansson, C.; Nilsson, S. *Int. J. Biol. Macromol.* **1987,** *9,* 297.
22. Norton, I. T.; Goodall, D. M.; Morris, E. R.; Rees, D. A. *J. Chem. Soc. Faraday Trans. 1* **1983,** *79,* 2475.
23. Watase, M.; Nishinari, K. *Rheol. Acta* **1982,** *21,* 318.
24. Watase, M.; Nishinari, K. *Food Hydrocolloids* **1988,** *2,* 25.
25. Nishinari, K.; Watase, M. *J. Agric. Food Chem.* **1990,** *38,* 1188.

Gels

See: *Acryloyl-L-Proline Alkyl Ester Hydrogels*
Agarose (Overview)
Bridged Polysilsesquioxanes (Highly Porous Hybrid Organic-Inorganic Materials)
Colloidal Dispersions, Nanogels
Dextran
Environmentally Responsive Gels
Functional Immobilized Biocatalysts (Stimulus-Sensitive Gels)
Gelatin (Molecular Weight Distributions)
Gelation, Physical (Atactic Polystyrene)
Gelling Agents (Agarose and Carrageenan)
Gels (Overview)
Gels, HPLC Packing Materials (For Structural Characterization)
Hydroxypropyl Cellulose (Crosslinked, Cholesteric Liquid Crystal)
N-Isopropylacrylamide Copolymers (Drug Delivery)
Lyotropic Polysaccharides
Microgels, Macrogels (by Photocrosslinking)
Network Polymers (Vinyl-Type)
Polyethylene Film Gels (Photografting of Hydrophilic Monomers)
Poly(N-isopropylacrylamide) Hydrogels
Polysaccharide-Polysaccharide Interactions
Poly(vinyl alcohol)-Ion Complex Gels
Reversibly Crosslinked Gels
Solid Polymer Electrolytes
Starch Graft Copolymers
Starch Polymers, Natural and Synthetic
Supramolecular Formation and Structure (PEO-PPO-PEO Block Copolymers)
Thermoreversible Gelation (Polyacrylonitrile Solutions)
Thermoreversible Gels (Isolactic, Syndiotactic, and Atactic Polystyrene)
Water-Soluble Polymers (Oil Recovery Applications)
Water-Swellable Polymers (Carbomer Resins)
Welan Microbial Polysaccharide
Wood, Chemically Modified (Thermoplasticization and Solubilization)

Gels (Overview)

Yoshihito Osada* and J. P. Gong
Division of Biological Sciences
Graduate School of Science
Hokkaido University

A polymer gel consists of an elastic crosslinked network and a fluid filling the interstitial space of the network. The network of long polymer molecules holds the liquid in place and so gives the gel what solidity it has. These gels are wet and soft and look like a solid material but are capable of undergoing large deformation. Living organisms are largely made of gels. Except for bones, teeth, nails, and the outer layers of skin, mammalian tissues are highly aqueous gel materials largely composed of protein and polysaccharide networks, in which water contents range up to 90% (blood plasma). This enables the organism to transport ions and molecules more easily and effectively while keeping its solidity.

There are a variety of ways to classify the gel: natural gel or synthetic gel, hydrogel or organogel according to the liquid medium the polymer network is swollen in, chemical gel or physical gel according to their crosslinkage, etc.

PREPARATION

A polymer gel is a network of flexible chains. Structures of this type can be obtained by chemical or physical processes. Some gels are crosslinked chemically by covalent bonds (chemical gel), whereas other gels are crosslinked physically by weak forces such as hydrogen bonds, van der Waals forces, or hydrophobic and ionic interactions (physical gel). Physical gelation processes are usually reversible and called sol-gel transitions. The final gel structures and properties are sensitive to the preparation methods.

PROPERTIES

Solvent Property and Phase Transition

Gels can be assumed to be constructed of a macromolecular network consisting of a number of small cavities. Water molecules, especially for those in hydrophilic and polyelectrolyte gels, are restricted in their motion compared with common water molecules. The restriction of the motion of water arises from the extensive association of water molecules with the network. The existence of an ordered structure at the water/macromolecule interface has been generally accepted. These water molecules possess certain preferred orientations and cannot move independent of their neighbor molecules. In this sense, water molecules near macromolecules are structured (bound water), those sufficiently far from the macromolecule have bulk water structure (bulk water), and those in between have decreasing order as a function of distance from the macromolecule.[1] The bound water in the gel is not frozen even when the temperature of the gel sample gets below the freezing point of the bulk water.[2] The presence of bound solvent was also observed in organic gels.[3]

Drastic changes in the state of the gel can be brought about by small changes in the external conditions. Under some

*Author to whom correspondence should be addressed.

conditions, swelling or shrinking is discontinuous and, therefore, a minute change in temperature causes a large change in volume. Phase transition in polymer gels was theoretically predicted on the base of the coil-globule transition observed in solution by Dusek and Patterson[4] and later experimentally confirmed by Tanaka.[5]

ELECTRICAL PROPERTIES

When water-swollen polyelectrolyte gel is interposed between a pair of electrodes in the air and direct current is applied, the gel undergoes electrically induced contraction at one electrode and concomitant water exudation at another electrode.[6] Electrically induced contraction is attributed to electrokinetic mechanisms.[7] The applied electric field induces the migration of hydrated counterions toward the oppositely signed electrode (electrophoresis) together with the water, thereby transporting water to the electrode.

APPLICATIONS

Gels have applications in numerous fields, including the food industry, medicine, biotechnology, chemical processing, agriculture, civil engineering, and electronics.

A polymer gel can absorb solvent up to several thousand times its original weight, depending on the chemical structure of the gel. Disposable diapers, feminine napkins, and perfumes used in everyday life are typical examples of the applications of the highly water-absorbing property of hydrogels.

There is considerable interest and activity in the application of synthetic and biological polymer gels in medicine. Interest has focused on the utilization of the bulk or the surface properties of hydrogels for biomedical applications. The bulk property of swelling is of particular interest for "swelling implants," which can be implanted in a small dehydrated state via a small incision and which then swell to fill a body cavity and/or to exert a controlled pressure. The swelling of synthetic and natural gels may also help elucidate swelling and osmotic mechanisms in biological tissues.

A number of biomedical applications for hydrogels have been mentioned in the literature.[8] The wide range of biomedical applications for hydrogels can be attributed both to their satisfactory performance upon *in vivo* implantation in either blood-contacting or tissue-contacting situations and to their ability to be fabricated into a wide range of morphologies.

Modulation of swelling forces in gels by chemical or physical stimuli enables dynamic control of the gel hydration and, thereby, the effective diffusibility and permeability of solutes can be obtained. Drug delivery systems, permselective membranes for selective extraction, and chemical valves are examples of applications of stimuli-responsive polymer gels.

A polymer gel with electrically driven motility first realized the artificial chemomechanical system.[9-11] The concept of a chemomechanical system was first proposed by Kuhn[12] and Katchalsky.[13] It refers to thermodynamic systems capable of transforming chemical energy directly into mechanical work or, conversely, of transforming mechanical work into chemical potential energy. Simple though it is, this system exhibits the essential characteristics that set "soft" chemomechanical systems apart from mechanical devices made of more rigid materials. In contrast to conventional motors and pumps, gels are gentle and flexible, and their movement is more reminiscent of muscle than that of metallic machines. Gels have potential uses as actuators and power sources under water, in space, and sometimes even in the human body, where any power supply is limited or difficult to chain.

REFERENCES

1. Drost-Hansen, W. *Indust. Eng. Chem.* **1969,** *61,* 10.
2. Quinn, F. X.; Kampff, E.; Smyth, G.; McBrierty, J. *Macromolecules* **1988,** *21,* 3191.
3. Nishide, T.; Gong, J. P.; Yasunaga, H.; Nishi, N.; Osada, Y. *Macromolecules* **1994,** *27,* 7877.
4. Dusek, K.; Patterson, D. *J. Polym. Sci.*, Part 2 **1968,** *6,* 1209.
5. Tanaka, T. *Phys. Rev. Lett.* **1978,** *40,* 820.
6. Osada, Y.; Hasebe, M. *Chem. Lett.* **1985,** 1285.
7. Gong, J. P.; Nitta, T.; Osada, Y. *J. Phys. Chem.* **1994,** *98,* 9583.
8. Ratner, B. D.; Hoffman, A. S. In *Hydrogels for Medical and Related Applications*; Andrade, J. D., Ed.; ACS Symposium Series 31; American Chemical Society: Washington, DC, 1976.
9. Osada, Y.; Okuzaki, H.; Hori, H. *Nature* **1992,** 355, 242.
10. Osada, Y.; Ross-Murphy, S. *B. Sci. Amer.* **1993,** 268(5), 82.
11. Okuzaki, H.; Osada, Y. *Macromolecules* **1994,** *27,* 502.
12. Kuhn, W. *Experientia* **1949,** *5,* 318.
13. Katchalsky, J. *J. Polym. Sci.* **1952,** *1,* 393.

GELS, HPLC PACKING MATERIALS (for Structural Characterization)

Yasuyuki Tanaka and Hasaya Sato
Faculty of Technology
Tokyo University of Agriculture and Technology

High-performance liquid chromatography (HPLC)–including size exclusion chromatography (SEC) or gel permeation chromatography (GPC)–has been widely applied to the separation, purification, and characterization of compounds having low molecular weight as well as high molecular weight polymers. The separation is in principle based on the mutual interaction between the stationary phase, mobile phase, and solute. Polymer packing materials have been used as an important stationary phase in HPLC. One of the first polymer packing materials as diethylaminoethyl derivatives of polysaccharide, used for the adsorption and desorption of proteins by changing the ionic strength of the eluent.[1] Crosslinked polydextran gels have been applied to the separation of water-soluble compounds by molecular weight.[2] These types of polymer packings are lightly crosslinked polymer gels. Moore synthesized macroporous beads by suspension copolymerization of styrene and divinylbenzene in the presence of a porogen or diluent, a compound in which monomers are soluble but the copolymer is not.[3] One characteristic of these polymer gels is the ability to control pore size almost independently from the crosslinking density, which has enabled preparation of porous beads with a highly crosslinked tough structure. Various types of polymer gels have been similarly synthesized as HPLC packings by considering bead properties such as polarity, solubility parameters, and rigidity.

PREPARATION OF POLYMER GELS

Polymer packing materials for HPLC take the form of 5–10-μm-diameter beads with a narrow size distribution and with controlled pore size. Preparation of polymer beads can be performed by suspension polymerization and so-called seed polymerization of vinyl monomers, or microspherization techniques.

Suspension polymerization is used for many lipophilic vinyl monomers and creates porous polymer beads by copolymerization of divinyl or trivinyl monomers in the presence of porogen. The particle size can be controlled by regulating the size of monomer droplets, usually by selecting suitable stirring conditions for the suspension prior to the polymerization. One can control pore size mainly by the type and amount of porogen. Typical packing materials prepared by suspension polymerization are styrene-divinylbenzene gels,[3,4-6] alkyl methacrylate gels,[7] hydroxyethyl methacrylate gels,[8-10] glycidyl methacrylate gels,[11,12] vinyl pyridine gels,[13] and fluorine containing gels.[14,15] Vinyl alcohol gels can be prepared by hydrolysis of vinyl acetate-triallylisocyanate gels prepared by suspension polymerization.[16]

Conventional suspension polymerization cannot be applied to highly hydrophilic vinyl monomers that are insoluble in organic solvents, such as acrylamide and sodium acrylate. In this instance, inverse suspension polymerization is used (i.e., a suspension of monomers and water in organic solvent).[18]

There exists a distribution of particle size of polymer beads prepared with conventional suspension or inverse suspension polymerization, and so, fractionation of beads by sieves, wind classifiers, or decantation is necessary to obtain polymer beads having a narrow particle size distribution. Very narrow particle size distribution can be obtained by the seed polymerization technique–a kind of suspension polymerization in monodispersed droplets synthesized by swelling of seeds with monomer, divinyl monomer, porogen, and a radical initiator. Monodispersed seeds of about 1 μm in diameter are prepared by emulsifier-free emulsion polymerization.[19] or dispersion polymerization.[20,21] Practically monodispersed styrene gels,[22-26] glycidyl methacrylate gels,[27] and oligoethylene dimethacrylate gels[28] have been prepared by the seed polymerization technique. An emulsification technique using uniform microporous glass membrane shirasu porous glass (SPG) also provides polymer beads with extremely narrow particle size distribution.[29]

APPLICATION TO POLYMER CHARACTERIZATION

HPLC has become an indispensable tool to characterize polymers using size exclusion, fractional dissolution, and adsorption mechanisms. If interaction between the sample and the stationary phase is negligible, the size exclusion mechanism governs the separation. In HPLC of polymers, SEC is most frequently used to determine the molecular weight distribution. If additional information is obtained about the effluent, some structural changes depending on the molecular weight can be detected.

It is expected that polymers can be separated by properties other than molecular volume, if they are separated by adsorption or phase separation. However, adsorption of a polymer to the stationary phase or occurrence of phase separation (i.e., precipitation) makes isocratic elution impossible. With a gradient elution increasing the composition of a good solvent for the sample, copolymers are expected to be separated reflecting their chemical composition.[17,30-35]

Adsorption HPLC has been applied to separate homopolymers by tacticity.

REFERENCES

1. Sober, H. A.; Peterson, E. A. *J. Am. Chem. Soc.* **1954,** *76,* 1171.
2. Porath, J.; Flodin, P. *Nature* (London) **1959,** 183, 1657.
3. Moore, J. C. *J. Polym. Sci., Part A* **1964,** *2,* 835.
4. Tanaka, Y.; Takeda, J.; Noguchi, K. U.S. Patent 4 338 404, 1982.
5. Sederel, W. L.; DeJong, G. J. *J. Appl. Polym. Sci.*1973, *17,* 2835.
6. Lloyd, L. L. *J. Chromatogr.* **1991,** 544, 201.
7. Tanaka, Y. et al. *J. Chromatogr.* **1987,** 407, 197.
8. Coupek, J.; Krivakova, M.; Pokorny, S. *J. Polym. Sci., Part C* **1973,** *42,* 185.
9. Mikes, O.; Strop, P.; Coupek, J. *J. Chromatogr.* **1978,** 153, 23.
10. Coupek, J. et al. Czech. Patent Application 6207 85, 1985.
11. Tennikova, T. B. et al. *J. Chromatogr.* **1988,** 435, 357.
12. Tennikova, T. B. *J. Chromatogr.* **1989,** 475, 187.
13. Sugii, A.; Harada, K. *J. Chromatogr.* **1991,** 544, 219.
14. Hirayama, C. et al. *J. Chromatogr.* **1989,** 465, 241.
15. Jurgen, H.; Heitz, J. W. *Makromol. Chem.* **1993,** 194, 963.
16. Noguchi, K.; Kasai, M. European Patent 0058381, 1982.
17. Tanaka, Y. *Handbook of Polymer Science and Technology*; Cheremisinoff, N. P., Ed.; Marcel Dekker: New York, 1989; Vol. 1, p 677.
18. Ogino, K.; Sato, H. *Kobunshi Ronbunshu* **1989,** *46,* 667.
19. Smigol, V. et al. *Angew. Macromol. Chem.* **1992,** 195, 151.
20. Paine, A. J.; Luymes, W.; McNulty, J. *Macromolecules* **1990,** *23,* 3104.
21. Tseng, C. M. et al. *J. Polym. Sci., Polym. Chem. Ed.* **1986,** *24,* 2995.
22. Kulin, L. I. et al. *J. Chromatogr.* **1990,** 514, 1.
23. Ellingsen, T. et al. *J. Chromatogr.* **1990,** 535, 147.
24. Wang, Q. C. et al. *Anal. Chem.* **1992,** *64,* 1232.
25. Hosoya, K.; Frechet, J. M. J. *J. Polym. Sci., Part A* **1993,** *31,* 2129.
26. Ogino, K. et al. *J. Chromatogr.* **1995,** *699,* 59.
27. Smigol, V.; Svec, F. *J. Appl. Polym. Sci.* **1992,** *46,* 1439.
28. Ogino, K. et al. *J. Chromatogr.* **1995,** *669,* 67.
29. Omi, S. et al. *J. Appl. Polym. Sci.* **1994,** *51,* 1.
30. Glockner, G. *Gradient HPLC of Copolymers and Chromatographic Cross-Fractionation*; Springer-Verlag: New York, NY, 1991.
31. Sato, H.; Takeuchi, H.; Tanaka, Y. *Macromolecules* **1986,** *19,* 2613.
32. Sato, H. et al. *J. Polym. Sci., Polym. Chem. Ed.* **1991,** *29,* 1073.
33. Sato, H. et al. *Makromol. Chem. Rapid Commun.* **1984,** *5,* 719.
34. Sato, H.; Takeuchi, H.; Tanaka, Y. *Proc. Int. Rubb. Conf.* **1985,** 596.
35. Asada, N. et al. *Rubber Chem. Technol.* **1990,** *63,* 181.

Gene Targeting

See: Polyelectrolyte Complexes (Targeting of Nucleic Acids)

Genetic Engineering

See: Spider Silk (Production of Polypeptide Polymers)

Germanium-Containing Polymers

See: Alternating Radical Copolymerization
Ferrocene-Backbone Polymers
Germanium-Containing Polymers (Overview)

GERMANIUM-CONTAINING POLYMERS (Overview)

Shiro Kobayashi* and Shin-ichiro Shoda
Department of Materials Chemistry
Graduate School of Engineering
Tohoku University

Organometallic polymers also have become interesting substances from the viewpoint of materials science.[1] Based on progress in silicon chemistry, attention has been paid mostly to syntheses and applications for silicon-containing polymers,[2,3] and few reviews on germanium-containing polymers have appeared so far. Here we review the syntheses and properties of germanium-containing polymers where a germanium atom is incorporated into the polymer main chain.

PREPARATION

From Tetravalent Germanium Compounds

The most popular method for preparation of polymers having a germanium atom in the main chain is based on the concept of polycondensation, starting from a tetravalent germanium species. These chemical processes normally require elimination of a small molecule as a result of an acid reaction or a dehydration reaction. For example, in the case of polycondensation between a germanium dichloride and a diol, two hydrogen chloride molecules are generated in the course of the polycondensation.

A linear germanium-containing polymer of μ-oxotype has been prepared. Researchers have achieved synthesis of germanium-containing polysters by the interfacial technique.[4-8] In general, organometallic dihalides in organic solvents are added to stirred solutions of diacid salts in water. The interfacial condensation technique allows synthesis of oligomeric germanium-containing ferrocene polyesters (**Scheme I**).[6]

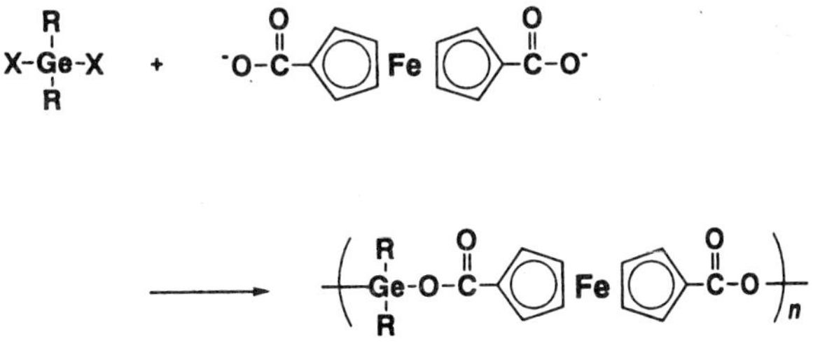

I

From Divalent Germanium Compounds

Among the divalent chemical species of group 14 elements, carbenes and silylenes are normally quite reactive and very unstable. Germylenes are more stable and can be isolated by distillation, provided that an appropriate ligand is attached to the germanium atom.[9-12] These chemical species which possess a strong reducing ability are potentially useful for construction of various polymers because they can form two new chemical bonds via α,α-addition on the germanium atom by changing its valency state from two to four.

Germylenes add to alkenes, alkynes, or conjugated dienes.[9-12] Germylenes react with ethylene and various substituted ethylenic compounds with formation of organogermanium polymers.[13,14] These reactions may be rationalized in terms of initial formation of an unstable, three-membered heterocycle.

The addition of germylenes to acetylenic derivatives usually leads to either dimers or polymers, via unstable germacyclopropenes.[15-17]

Recently, a series of novel copolymerization reactions using these stable germylenes have been developed. Germylenes were found to copolymerize with *p*-benzoquinone derivatives to give copolymers having an alternating germanium unit and a *p*-hydroquinone unit in the main chain (**Scheme II**).[18] The resulting copolymers are of relatively high molecular weight and soluble in organic solvents.

Five- and six-membered cyclic germylenes react with a *p*-benzoquinone derivative under similar conditions, giving rise to the copolymer having a 2:1 periodic structure with a molecular weight higher than 10^5.[19]

Germylene and cyclic α,β-unsaturated ketones having *s*-trans structure give alternating copolymers having a germanium enolate structure in the main chain.[20]

Germylene was copolymerized with acetylene monomers in the presence of a rhodium catalyst, giving rise to copolymers with a structure having acetylene unit in excess.[21] Among catalysts used, a rhodium norbornadiene complex was found to be most effective for the copolymerization.

High molecular weight polygermanes have been prepared by Wurz coupling reactions of dichlorogermanes with sodium metal or by a method using diiodogermylene and Grignard reagents or organolithiums.[22,23]

A polymerization reaction for the synthesis of polygermane via a ligand substitution polymerization of germanium dichloride with organolithium compounds has been reported (**Scheme III**).[24]

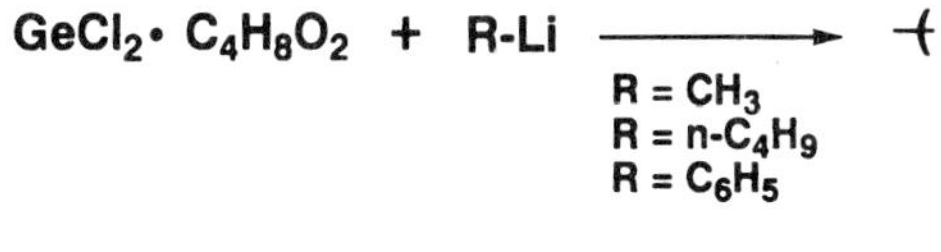

III

PROPERTIES

Polygermane shows very strong UV absorptions owing to the Ge–Ge σ-bond conjugation. For instance, the polygermane prepared by the ligand substitution polymerization is photosensitive, especially in the solution form. When the polymer solution was exposed to daylight, the molecular weight decreased, and all of the higher molecular weight portion disappeared after

*Author to whom correspondence should be addressed.

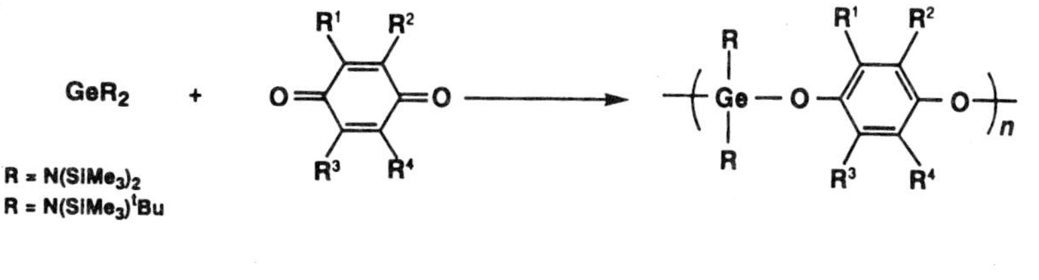

II

a 2 h exposure. The polymer chain scission behavior was also observed in the solid form.

REFERENCES

1. Zeldin, M.; Wynne, K. J.; Allcock, H. R. Eds.; *Inorganic and Organometallic Polymers*; American Chemical Society: Washington, DC, 1988.
2. Stark, F. O.; Falender, J. R.; Wright, A. P. In *Comprehensive Organometallic Chemistry*; Pergamon: Oxford, United Kingdom, 1982; Chapter 9.3.
3. West, R. In *Comprehensive Organo-metallic Chemistry*; Pergamon: Oxford, United Kingdom, 1982; Chapter 9.4.
4. Carraher, C. E. Jr.; Dammeier, R. L. *Makromol. Chem.* **1971,** 141, 245.
5. Carraher, C. E. Jr.; Dammeier, R. L. *J. Polym. Sci.*, Part A-1 **1972,** *10,* 413.
6. Carraher, C. E. Jr.; Jorgensen, S.; Lessek, P. J. *J. Appl. Polym. Sci.* **1976,** *20,* 2255.
7. Carraher, C. E. Jr. et al. In *Biological Activities of Polymers*; American Chemical Society: Washington, DC, 1982; pp 13-25.
8. Andersen, D. M. et al. In *Biological Activities of Polymers*; American Chemical Society: Washington, DC, 1982; pp 27-33.
9. Petz, W. *Chem. Rev.* **1986,** *86,* 1019.
10. Lappert, M. F.; Rowe, R. S. *Cood. Chem. Rev.* **1990,** 100, 267.
11. Barrau, J.; Escudié, J.; Satgé, *J. Chem. Rev.* **1990,** *90,* 283.
12. Neumann, W. P. *Chem. Rev.* **1991,** *91,* 311.
13. Satgé, J.; Massol, M.; Rivière, P. *J. Organomet. Chem.* **1973,** *56,* 1.
14. Rivière, P.; Satgé, J.; Castel, A. *C. R. Hebd. Seances Acad. Sci., Ser C* **1976,** 282, 971.
15. Nefedov, O. M.; Manakov, M. N. *Angew. Chem.* **1966,** *78,* 1039.
16. Harris, D. H. et al. *J. Chem. Soc., Dalton Trans.* **1976,** 945.
17. Scibelli, J. V.; Curtis, M. D. *J. Organomet. Chem.* **1972,** *40,* 317.
18. Kobayashi, S. et al. *J. Am. Chem. Soc.* **1990,** 112, 1625.
19. Kobayashi, S.; Iwata, S.; Hiraishi, M. *J. Am. Chem. Soc.* **1994,** *116,* 6047.
20. Kobayashi, S. et al. *J. Am. Chem. Soc.* **1992,** 114, 4929.
21. Kobayashi, S.; Cao, S. *Chem. Lett.* **1993,** 25.
22. Mochida, K.; Masuda, S.; Hayada, Y. *Chem. Lett.* **1992,** 2281.
23. Mochida, K. et al. *Organometallics* **1993,** *12,* 586.
24. Kobayashi, S.; Cao, S. *Chem. Lett.* **1993,** 1385.

Glass-Containing Materials

See: Epoxy Resin-Glass Interphases (Optical Properties)
Fillers (Glass Bead Reinforcement)
Glass-Polymer Melt Blends
Inorganic Fibers (Surface Modification by Grafting)

Glass-Ionomer Cements

See: Dental Polymers (Glass-Ionomer Cements)
Dental Sealants

GLASS MAT REINFORCED THERMOPLASTICS

Jozsef Karger-Kocsis
Institut für Verbundwerkstoffe GmbH
Universität Kaiserslautern

Glass mat reinforced thermoplastic (GMT) sheets were first introduced in the early 1970s to produce economically large parts by hot stamping and pressing techniques.[1] Their mechanical performance and cost has enabled GMTs to fill the gap between the injection-moldable chopped glass fiber (GF), reinforced thermoplastics, and advanced thermoplastic composite laminates. Since large parts of high surface quality are also produced from thermoset sheet-molding compounds (SMC), the number of competing materials including processing technologies is further increased.[2] Though GMTs are available with different matrices grouped into high-volume (e.g., polypropylene [PP]) and engineering thermoplastics, such as poly(ethylene terephthalate) (PET), poly(butylene terephthalate) (PBT), and various polyamides (PA), the overwhelming majority of the market is controlled by GMT-PP at present (90-95%).

Since low-cost production cannot be achieved with advanced thermoplastic composites and SMC hardly meets the requirements of economic recycling (although "particle recycling" is practiced at present), increased competition is likely between the chopped long GF-reinforced injection moldable compounds and GMTs in the future.[3]

USE AND MARKET TRENDS

GMTs are increasingly replacing steel and SMC in the automotive industry. They are widely used for the production of noise shields, driver and passenger seat shells, bumpers and construction front ends. Further applications that may suit GMTs include inner door panels and spare wheel covers. Another use for GMTs may be in materials handling and transportation (e.g., replacing wood pallets). Other applications will emerge in sports equipment (helmets, snow boards).

OUTLOOK

Further development of GMTs or GMT-like materials depends on several factors. Property improvements such as scratch-resistant class A surface products may trigger further (automotive) applications. High innovation potential is due to structural reaction injection molding (S-RIM) processes. In this case a low-viscosity pumpable prepolymer mixture is transferred into the mold containing the previously positioned reinforcement. After wetting-out the reinforcement, the prepolymer polymerizes *in situ* in the mold. This process works with the Nyrim® system (nylon block copolymer) of DSM.[4,5]

A recently announced new on-line process that integrates the manufacturing and processing (forming) steps of long (≈12 mm) GF-reinforced thermoplastics may become a severe competitor for GMT processors. Economics can be expected to regulate the future market for these thermoplastics.

REFERENCES

1. Bigg, D. M. *Polypropylene: Structure, Blends and Composites*; Karger-Kocsis, J., Ed.; Chapman and Hall: London, United Kingdom, 1995; Vol. 3, Chapter 7.
2. Kia, H. G. *SMC, Sheet Molding Compounds*; Hanser: Munich, Germany, 1993.
3. Karger-Kocsis, J.; Harmia, T.; Czigány, T. *Compos. Sci. Technol.* **1995,** *54,* 287.
4. Macosko, C. W. *RIM, Fundamentals of Reaction-Injection Molding*; Hanser: Munich, Germany, 1989.
5. Karger-Kocsis, J. *J. Appl. Polym. Sci.* **1992,** *45,* 1595.

GLASS-POLYMER MELT BLENDS

Candace J. Quinn, Paul Frayer, and George Beall
Science and Technology Group
Corning Incorporated

In recent years, researchers at Corning have searched for durable low-temperature glasses for compounding with high-temperature polymers to produce dimensionally stable, stiff injection-moldable materials for high-temperature applications. By far the best glasses discovered to date, in terms of glass stability, chemical durability, and low working temperature, are mixed-alkali zinc-pyrophosphate compositions.

THE ROLES OF GLASS COMPONENTS

It is useful to comment on the role of specific chemical species and their effects on the properties of alkali zinc pyrophosphate glasses.

Phosphate

P_2O_5 is the primary glass former in this composition region. Its tendency to form linear molecular structures, in contrast to the 3-D networks typical in silicate glasses, offers an explanation for the low melting and glass transition temperatures of these glasses. In silicate glasses sufficient thermal energy to break chemical bonds must be supplied to dissociate the network for viscous flow to occur. It seems that linear phosphate molecules, include $(P_2O_7)^{-4}$ dimers, can move past one another, allowing viscous flow at temperatures well below 400°C, and at sufficiently low glass viscosities to co-form with polymers.

Alkali

Alkali oxides are modifiers in phosphate glasses, as they are in other glass systems. Alkali phosphates form stable glasses of moderate to poor chemical durability. The importance of mixed alkali effects on glass stability is pronounced. Sodium and mixed alkali zinc pyrophosphate glasses are quite stable. The best glasses for durability over wide pH ranges are obtained from relatively high lithia levels (~10 mol%). However, devitrification during compounding and molding can result. Lower lithia levels (~6 mol%) produced stable glasses with some compromise in durability.

Divalent Cations

The combinations of ZnO and alkali in pyrophosphate compositions has been the key to developing useful glasses with working temperatures as low as 350°C, good glass stability for compounding and molding processes, and water durability generally equivalent to (or in some cases better than) soda lime glass. The best chemical durability seems to occur with zinc levels above 40 mol%.

PbO and SnO in partial replacement of zinc are effective for increasing refractive index, slightly improving durability, and lowering T_g in zinc pyrophosphate glasses.

Alkaline earth oxides also can be substituted for ZnO with increases in durability and T_g. Some beneficial mixed alkaline earth effects on glass stability have been noted.

Multivalent Cations

Alumina is a key minor glass former that improves durability, particularly in resistance to boiling water, steam, and acid. About 1-2 mol% appears optimum. Higher alumina increases the glass transition temperature and promotes crystallization; lower levels introduce long-term hydration effects.

Special additions to zinc pyrophosphate glasses include rare-earth oxides, titania, zirconia, and molybdenum oxide. The rare-earth oxides have a strong effect in improving alkali resistance. Cerium and praseodymium oxides appear to have the most beneficial effect, probably because of their tendency to be stable in both trivalent and the preferred quadravalent condition.

Titania helps both alkali and acid durability, but is highly insoluble in pyrophosphate glasses. Moreover, it has a very strong tendency to increase T_g, and is therefore of limited use. Zirconia is helpful in creating dense opals in the glass. Zirconia, like titania, is very insoluble but (unlike titania) forms an emulsion or finely dispersed particulate suspension of zirconium phosphate in the melt, thus creating opacity in the glass.

Silica, although normally considered a glass former, may well be a modifier in pyrophosphate glasses. Levels up to 3 mol% have proven useful in reducing devitrification and preventing coalescence of glass droplets in polymer blends. These effects may be due to surface tension influences.

Optimal Glass Compositions

The preferred glass composition area for glass-polymer blends (as outlined in two U.S. patents)[2,3] is: P_2O_5, 30–36 mol%; ZnO, 30–49 mol%; SnO, 0–10 mol%; alkali as M_2O, 12–25 mol% (where M_2O consists of at least two alkali metal oxides of the indicated proportions 3–12% Li_2O, 4–13% Na_2O and 0–12% K_2O); Al_2O_3, 1–3.5 mol%; SiO_2, 0–2%; Re_2O_3, .5 to 3%; and 0–5 mol% of the combination CaO + MgO + F. The preferred range of sulfophosphate glasses described in another patent[4] is 21–33% P_2O_5, 9–17% SO_3, 35–51% ZnO, and 10–20% M_2O.

GLASS STRUCTURE

The structure of phosphate glasses is clearly important in determining their chemical resistance to various aqueous solutions as well as their bonding or reactive properties with high-temperature polymers. The results of Raman spectra studies show a consistent pattern of cation-dependent depolymerization of P_2O_5.[1] Binary zinc phosphate glasses show a broader distribution of structural species in comparison to sodium phosphate glasses of equimolar proportions.

This broad speciation of phosphate units in binary zinc glasses probably accounts for the unique glass-forming behavior of up to 70 mol% ZnO in this particular binary phosphate system. The observation that substitution of alkalis for zinc increases water durability while lowering T_g can also be explained from the Raman spectral data.

GLASS-POLYMER MELT BLENDS

Although some results are given for other polymers, here we discuss primarily the preparation, properties, and applications for composites made with a liquid crystalline polymer (LCP) and a low T_g phosphate glass wherein the glass is in the dispersed phase.

COMPOUNDING

The glass-polymer melt blends are made by conventional twin screw extrusion compounding and pelletizing.

COMPOSITE RHEOLOGY

Polymer composites with hard inorganic fillers such as E-glass at 50–60 vol% (70–80 wt%) often cannot be injection-molded because of their high viscosity. Noteworthy exceptions are highly filled thermosets and lower molecular weight, low-viscosity thermoplastic resins.

Creating the dispersed particles *in situ* in the melt phase where the glass is soft and deformable during the compounding step circumvents the well-known difficulties in viscous polymer melts of adequately wetting out and dispersing inorganic fillers. The composite viscosity is dramatically different between hard and soft dispersed particles.

LCP resins are known to be very shear thinning. The low viscosity of an LCP allows one to make composites with E-glass beads up to about 70 wt% (56 vol%), but the LCP molecular weight is reduced, the mechanical properties are poor, and viscosity comparisons are less meaningful.

MICROSTRUCTURE DEVELOPMENT

A composite in which the glass is well melted and dispersed consists of very small glass beads where most particles are about 5 μm in diameter or less. These are labeled as "F" or "fine" composites.

Despite the smaller than conventional size filler particles, the phosphate glass beads do not readily agglomerate or coalesce. Another significant difference is that the glass particles are soft and nonabrasive towards the melt processing equipment including the mold.

The relatively high glass viscosity allows one to make reinforcing microstructures so that stronger composites can be made despite the high glass loading. Thus, one can produce resin matrix composites with mixtures of small beads, fine diameter fibers, or ribbons and plates. Also these fibers can have relatively high aspect ratios as they meander throughout the composite resin matrix phase. Materials that contain mainly fibers and beads are conveniently called "S" for structured composites; those that also contain ribbons are designated "M" for macro-composites.

PROPERTIES

Normally, polymer composites containing such high levels of an inorganic phase give materials that are not only very difficult to process, but also have poor mechanical properties. The subject phosphate glass composites are exceptions, because the glass is soft during melt processing so that injection molding is possible. Also, the glass is well dispersed and wetted easily by the polymer melt because the structures are formed *in situ.*

These composites are also unusual because the glass composition is adjusted to give good interfacial bonding without the use of conventional coupling agents. The interfacial bond strength along with the glass particle size and shape distributions determine the mechanical properties.

An LCP composite made with 56 vol% (70 wt%) E-glass beads was found to have a tensile strength of only 46 MPa. These new "Fine" composites are about 30% stronger. LCP composite made with 40% E-glass beads and 30% E-glass fibers had a flexural strength of only 75 MPa while the "structured" phosphate glass composites at similar volume fraction filler exhibit a flexural strength about 60% higher!

APPLICATIONS

The materials described in this article are not yet commercially available, but they have been studied by a number of companies around the world. Three general areas of interest have been identified: high dimensional stability, high precision tolerances, and high heat resistance. Materials exist that combine all three features.

High levels of glass improve a number of properties some of which are discussed below. Dimensional stability, molding to high precision tolerances, and high heat resistance are all maximized by maximizing the amount of inorganic phase. But, of course, there are trade-offs with other properties such as strength. Therefore, the optimal range of glass was selected to be 50–60 vol%.

The high inorganic content increases resistance to creep and stress relaxation and decreases the volumetric expansion coefficient to about 80 ppm. Therefore, the dimensional stability is excellent. Also, these materials do not burn even in 100% oxygen. Other possible applications requiring heat resistance and high dimensional stability are integrated circuit burn-in sockets and carrier trays.

Another area of interest has been high-precision molded parts, which are generally small, but volumes can be quite high such as in automotive or consumer electronics. The goal with glass-polymer blends is to directly mold parts without the need for machining. One example is fuel injection pump impellers; another is optical fiber connectors. Other potential applications are possible and the search for these continues.

ACKNOWLEDGMENTS

The authors wish to thank Experimental Melting, PRC Operations, and Analytical Services Departments for their cooperation and support. Particular appreciation goes to the Cortem technical team for their enthusiasm, determination, and hard work in developing this new technology.

REFERENCES

1. Quinn, C. J.; Dickinson, J. E.; Beall, G. H. *Alkli Phosphate Glasses for Polymer Blends*; Spanish Society of Ceramics and Glass: Madrid, 1992; Vol. 4 (31-C), pp 79-*84 (Bull. SSCG, Int. Cong. Glass).*
2. Beall, et al. U.S. Patent 4 940 677, 1990.
3. Beall, et al. U.S. Patent 4 996 172, 1991.
4. Beall, et al. U.S. Patent 5 328 874, 1994.

GLASS TRANSITION (Theoretical Aspects)

Hans Adam Schneider
Institut für Makromolekulare Chemie
"Hermann-Staudinger-Haus" and
Freiburgrer Materialforschungszentrum FMF
der Universität

The glass transition is considered the most important characteristic in any decision to use amorphous polymers. Although the glass transition is generally a kinetically determined "freeze-in" process, there are also serious hints of a thermodynamic background. The kinetic nature of the glass transition is supported by the observation that any liquid can be transformed in a glass if its cooling rate overcomes the crystallization rate. In polymers, however, an additional structure-dependent ordering possibility has to be considered, which is connected with the tacticity of the macromolecule. The thermodynamic background of the glass transition is supported by the "Kauzmann paradox." Kauzmann[1] has shown that, if any thermodynamic property of a material observed in the liquid state is extrapolated through and below the glass temperature, absurd results, such as smaller specific volume of the glass compared with that of the crystal, or even negative entropies, are obtained. The glass transition is also characterized by changes specific to a thermodynamic second-order transition, such as shoot-ups in expansion coefficient, compressibility, and heat capacity.

Taking into account this dual aspect of the glass transition, two main models have been proposed for the theoretical interpretation of the glass transition phenomenon: "free volume" and thermodynamic models (**Table 1**).

According to the first theory[2] the molecular mobility is controlled by "free volume," or that part of the "excess" volume which can be randomly redistributed with no energy requirement. The glass transition is the result of the temperature-dependent decrease of this free volume of the amorphous phase below some small characteristic value, which prevents any further molecular rearrangement. Accordingly, below the glass temperature, T_g, further molecular rearrangements of the polymer are forbidden, so any mobility of the polymer is "frozen in." In the free volume theory the system is described by a "P–V–T" equation of state and the glass is considered a "frozen-in" metastable state of the matter. Because of this kinetic aspect of the glass transition, polymer glasses are characterized by "relaxation" and "aging" phenomena.[3]

Unfortunately, there is no way to verify experimentally the validity of these two theories. On the one hand, because of differing definitions, the free volume responsible for the glass transition cannot be quantified. The estimated values of the critical free volume differ by several orders of magnitude.[4] On the other hand, the zero conformational entropy can be calculated, but the transition temperature, T_2, is experimentally inaccessible because of the kinetic freeze-in phenomenon. Both theories share this dilemma. Further details concerning glass formation, glassy behavior and glass temperatures are given elsewhere.[5]

TABLE 1. Theories of the Glass Transition

"Free Volume" Model	"Conformational Entropy" Model
"Free volume" controls mobility	Conformational changes are controlled by "flexible" bonds
"P-V-T"-equation of state	*"S-V-T"- equation of state*
Glass metastable "frozen in" state	*Glass thermodynamic stable fourth state of matter*
Kinetic controll of glass transition, T_g (relaxation and aging)	Second-order "phase transition," $T_2 \sim T_g$-50 K for "zero"-conformational entropy

ATTEMPTS TO PREDICT THE GLASS TEMPERATURE OF POLYMERS

Taking into account the theoretical and technological importance of knowing the glass transition, different methods have been presented for predicting the glass temperature of polymers, either by knowing the chemical structure of the repeating unit or relating the glass temperature to specific properties of the polymer, for instance, density[6] or specific volume.[7]

One of these semiempirical methods scales the volume contributions to T_g of the different moieties in the chemical structure of the repeating unit of polymers.[8]

Hopfinger et al.[9] have elaborated a more sophisticated model for estimating glass temperatures of linear polymers by conformational flexibility and mass moments of the polymer. These characteristics were evaluated in terms of torsion angle units composing the polymer. The model has been extended by Koehler and Hopfinger to account for intermolecular energetic contributions.[10]

GLASS TEMPERATURES OF RANDOM COPOLYMERS

Random copolymers as well as compatible polymer blends are characterized by a single, composition-dependent T_g. Both the free volume theory and the thermodynamic model were consequently extended to explain the composition dependence of the glass temperature of random copolymers and compatible polymer blends.

To explain the composition dependence of the glass temperature of random copolymers, the free volume models start with the Fox equation.[11] As Schneider et al.[12] have shown, the Fox equation results in a first approximation from the Gordon-Taylor expression, which was deduced in the supposition of additivity of the specific volumes.[13] As Kovacs[14] has explicitly demonstrated, the supposition of specific volume additivity requires that the respective free volumes be additive.

DiMarzio and Gibbs have extended the thermodynamic model of the flexible bond contribution to T_g for copolymers[14] in the supposition of additivity of the flexible bond contributions of the monomeric units to the T_g of the copolymer.

REFERENCES

1. Kauzmann, W. *Chem. Revs.* **1948,** *43,* 219.
2. Doolitle, A. K. *J. Appl. Phys.* **1951,** *22,* 1471; Hirai, N.; Eyring, H. *J. Polym. Sci.* **1959,** *37,* 51; Turnbull, D.; Cohen, M. H. *J. Chem. Phys.* **1961,** *34,* 120; Simha, R.; Boyer, R. F. *J. Chem. Phys.* **1962,** *37,* 1003; Simha, R.; Smocynscky, T. *Macromolecules* **1969,** *2,* 342.
3. Robertson, R. F. et al. *Macromolecules* **1984,** *17,* 911.
4. Haward, R. N. In *The Physics of Glassy Polymers*; Haward, R. N. Ed.; Wiley: New York, 1973; pp 25-41.
5. Kovacs, A. J. *Fortschr.Hochpolym.-Forsch.* **1963,** *3,* 394; Rehage, G.; Borchard, W. *In The Physics of Glassy Polymers*; Haward, R. N. Ed.; John Wiley & Sons: New York, 1973; pp 55-102; Eisenberg, A. In *Physical Properties of Polymers*; American Chemical Society: Wash-

ington, DC, 1984; Chapter 2, pp 57-95; McKenna, G. B. In *Comprehensive Polymer Science*; Booth, C.; Price, C., Eds.; *Polymer Physics*; Pergamon: Oxford, 1989; Vol. 2, Chapter 10, pp 311-362.
6. Askadskii, A. A. *Polym. Sci. USSR* **1967,** *9,* 417.
7. Rogers, S. S.; Mandelkern, L. *J. Phys. Chem.* **1957,** *61,* 985.
8. Wiff, D. R. et al. *J. Polym. Sci., Polym. Phys. Ed.* **1985,** *23,* 1165.
9. Hopfinger, A. J. et al. *J. Polym. Sci. Polym. Phys. Ed.* **1988,** *26,* 2007.
10. Koehler, M. G.; Hopfinger, A. J. *Polymer* **1989,** *30,* 116.
11. Fox, T. G. *Am. Phys. Soc.* **1965,** *1,* 123.
12. Schneider, H. A.; Leikauf, B. *Thermochim. Acta* **1987,** 114, 165; Brekner, M-J. et al. *Polymer* **1988,** *29,* 78.
13. Gordon, M.; Taylor, J. S. *J. Appl. Chem.* USSR **1952,** *2,* 493.
14. DiMarzio, E. A.; Gibbs, J. H. *J. Polym. Sci.* **1959,** *40,* 121.

GLASSY CARBON

Thomas X. Neenan*
AT&T Bell Laboratories

Olivier J. A. Schueller, Wenhua H. Huang,
Nicolas L. Pocard, Richard L. McCreery, and
Matthew R. Callstrom*
Department of Chemistry
The Ohio State University

The widespread use of graphitic carbons in electrochemical systems is due to their high electrical conductivity, good corrosion resistance, and reasonable mechanical and dimensional stability. Specific applications of graphitic carbons include electrodes, electrocatalyst supports, and bipolar electrode separators. Because of the high ash content of natural graphites, most graphitic carbons used in electrochemical applications are synthesized from hydrocarbons. A unique feature of synthetic graphites is the wide variety of microstructures accessible, depending upon the source of the carbon precursors and the method of preparation.

Of these various forms of carbon, glassy or vitreous carbon (GC) has been widely used as an electrode. Glassy carbon derives its name from exhibiting fracture behavior similar to glass, from having a disordered structure over large dimensions (although it contains a graphitic microcrystalline structure), and because it is a hard shiny material capable of high polish. Glassy carbon is particularly useful in electrochemical applications because of its low electrical resistivity, impermeability to gases, and high chemical resistance, and because it has the widest potential range observed for carbon electrodes.[2] A model for the structure of glassy carbon has been proposed by Jenkins and Kawamura.[1] They concluded that glassy carbon (**1**) consists of long microfibrils that twist, bend, and interlock to form interfibrillar bonds, and that these microfibrils are randomly oriented.

Although certain ambiguities exist about the precise molecular structure of glassy carbon, this model does take into account the relatively low density of glassy carbon, ~1.5 g/cm^3 vs. 2.25 g/cm^3 for graphite (which suggests the existence of voids), the impermeability of glassy carbon to gases (which suggests that these voids are not connected), and the isotropic conductivity observed for glassy carbon.

The current approach to the conventional preparation of glassy carbon solids involves the careful pyrolysis of one of a variety of polymeric materials, including poly(vinyl chloride) (**2**), poly(vinylidene chloride) (**3**), cellulose (**4**), phenolic resin (**5**), poly(furfuryl alcohol) (**6**), and polyacrylonitrile (**7**).[1,3]

The processing requirements for the preparation of glassy carbon are quite exacting. It is important to note that all conventional approaches to synthetic graphites, including glassy carbon, require extensive heat treatment (1000–3000°C).

We have pursued an alternative approach to the preparation of modified glassy carbon materials by the synthesis of new carbon solids with elements other than carbon.[4-10] These dopants are incorporated into the carbon solid and significantly modify the structure and surface energy of the electrode surface, which ultimately controls the interaction of small molecules with the surface. These new doped glassy carbon materials exhibit excellent behavior for electrocatalysis.[5,6] We have found that the incorporation of dopants, including halogens and metals, on the molecular level in a carbon precursor, a poly(arylene diacetylene), (**8**), followed by thermolysis at relatively low temperatures (<600°C), results in a conductive, dimensionally stable carbon matrix containing the dopant (**9**) (**Scheme I**).[5]

We have also found that thermolysis of carbon precursor systems doped with metals results in the formation of metal particles of controlled composition, size, and catalytic activity.[6-10]

PREPARATION AND PROPERTIES

As outlined above, we have recently discovered a novel synthesis of glassy carbon that requires relatively low (<600°C) thermal treatment for its preparation. This low-temperature synthesis allows, for the first time, the preparation of heteroatom-doped glassy carbon (HGC) materials in which the dopant is homogenously dispersed within the carbon matrix. The synthesis of HGC materials involves the preparation of diacetylenic oligomer or polymer precursors, molding of the powder to form a disk, or alternatively, cast as a film, and thermally treated to form glassy carbon.[4,5,7]

APPLICATIONS

Our method of preparation allows for the preparation of novel electrodes in which precious metal particles are dispersed within the carbon matrices. The application of these materials to the preparation of practical fuel cell electrodes is described in this section.

Microcrystalline Platinum Doped Glassy Carbon

Our initial experiments designed to incorporate platinum in glassy carbon involved the dispersion of commercially available platinum(IV) oxide in the parent carrier polymer. X-ray photoelectron spectroscopy analyses of these pellets confirmed the incorporation of platinum as a dopant in the glassy carbon solid (PtO_2-GC). A study of the binding energy of the Pt($4f_{7/2}$) and Pt($4f_{5/2}$) electrons found that there was a mixture of different oxidation states of platinum [Pt(IV), Pt(II), Pt(0)]. The partial reduction of the platinum(IV) oxide to lower oxidation states is not surprising given the fact that a relatively labile oxide is

*Author to whom correspondence should be addressed.

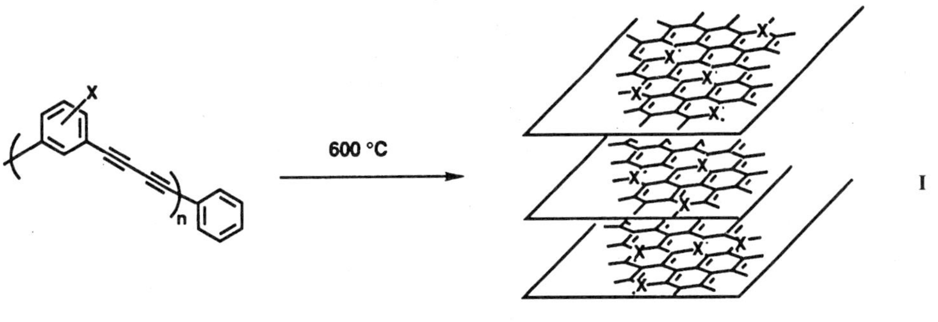

being heated in a strongly reducing atmosphere during the formation of the glassy carbon matrix. In contrast with nondoped glassy carbon electrodes, where no response was recorded upon the application of a reducing potential, the immobilization of 0.25 and 1 atom% of platinum in glassy carbon led to efficient hydrogen evolution. These platinum HGC electrodes exhibited excellent stability and no evidence of degradation was found after several thousand reduction cycles. However, an overpotential of 50–100 mV, in comparison with a bulk platinum electrode, was found for the reduction.

Synthesis of Nanoscale Clusters of Platinum in Glassy Carbon

In an effort to prepare glassy carbons with a more intimate platinum-carbon relationship, we sought chemistries that would allow the dispersion of platinum species at the atomic level in glassy carbon precursors. Coordination of platinum to the carbon-carbon triple bond of poly(phenylene diacetylene) offered a convenient method of dispersing platinum at an atomic level along the glassy carbon precursor, and, under the most optimistic of scenarios, offered a method of dispersing platinum at the atomic level in the glassy carbon solid itself.

Conversion of the platinum containing oligomers to platinum-doped glassy carbon at 600°C was accomplished as either a film on conventional glassy carbon or in the form of a disk. X-ray photoelectron spectroscopic and microprobe analyses of the materials confirmed the incorporation of approximately 0.5–1 atom% of platinum(0) in glassy carbon, with no evidence for the presence of other oxidation states of platinum.[11]

Following a similar approach (synthesis of an organometallic polymer precursor followed by controlled thermolysis), we were able to incorporate other metallic dopants such as cobalt,[12] iron,[13] nickel,[14] palladium,[15] and ruthenium[16] into a glassy carbon matrix. Bimetallic systems were also studied.[16]

REFERENCES

1. Jenkins, G. M.; Kawamura, K. *Polymeric Carbons-Carbon Fibre, Glass and Char*; Cambridge University: Cambridge, 1976.
2. For a review of the use of glassy carbon in electrochemical applications, see: van der Linden, W. E.; Dieker, J. W. *Anal. Chim. Acta* **1980,** 119, 1.
3. Fitzer, K. et al. In *Chemistry and Physics of Carbon*; Walker, P. L. Jr., Ed.; Marcel Dekker: New York, 1971; Vol. 7, p 237.
4. Neenan, T. X. et al. Br. *Polym. J.* **1990,** *23,* 171.
5. Callstrom, M. R. et al. *J. Am. Chem. Soc.* **1990,** *112,* 4954.
6. Pocard, N. L. et al. *J. Am. Chem. Soc.* **1992,** *114,* 769.
7. Pocard, N. L. *J. Mater. Chem.* **1992,** *2,* 771.
8. Schueller, O. J. A. et al. *Chem. Mater.* **1993,** *5,* 11.
9. Hutton, H. D. et al. *Chem. Mater.* **1993,** *5,* 1110.
10. Hutton, H. D. et al. *Chem. Mater.* **1993,** *5,* 1727.
11. Hartley, F. R. *Chemistry of Platinum and Palladium*; Applied Science: London, **1973,** p 398.
12. Neenan, T. X. et al. *Polym. Prepr. Am. Chem. Soc. Div. Polym. Chem.* **1993,** *34,* 356.
13. Pudelski, J. K. Ph.D. Thesis, The Ohio State University, 1993.
14. Baumgard, K. D. M.S. Thesis, The Ohio State University, 1993.
15. Diener, M. R. M.S. Thesis, The Ohio State University, 1993.
16. Schueller, O. J. A. Ph.D. Thesis, The Ohio State University, 1995.

Glassy Polymers

See: *Deformation and Fracture (Micromechanical Mechanisms)*
Glassy Polymers (Toughening via Dilatational Plasticity)
Glassy Polymers, Crazing
Photochromic Reactions, Polarization-Induced (in Polymer Solids)

GLASSY POLYMERS (Toughening via Dilatational Plasticity)

Stephen H. Spiegelberg
Division of Applied Sciences
Harvard University

Robert E. Cohen
Department of Chemical Engineering
Massachusetts Institute of Technology

Glassy polymers comprise a sizable group of materials used in the plastics industry. These materials are found in many structural applications because of their high tensile strength, ease of processing, low cost, and optical transparency. An unfortunate quality of glassy polymers, however, is their brittle behavior under tensile loads. Polymers such as polystyrene and poly(methyl methacrylate) typically fracture after being stretched to a few percent strain; such behavior obviously limits their use. This observation is the motivating factor for the extensive work that has been devoted to the toughening of glassy polymers.

Considerable effort has been devoted to the enhancement of a phenomenon termed "crazing" observed during tensile tests. Crazing is a localized cavitational mode of deformation whereby appreciable levels of inelastic strain may be realized without a change in cross-sectional area of the test section; thus crazing is a lenticular dilatational transformation. As the craze grows, polymer chains are drawn out of the matrix to form the craze tufts, which then lengthen until the craze reaches a final thickness between 1 and 10 μm, depending on the nature of the polymer material.[2]

It is the long-chain nature of polymers that allows them to exhibit crazing behavior. Experimental evidence suggests that chain entanglement density plays a strong role in the crazing behavior of a polymer. As the chain entanglement density (or, similarly, chemical crosslink density) increases, it becomes more difficult to draw polymer chains out into craze tufts. Thus, high molecular weight glassy polymers and thermosetting polymers will not craze. At the other end of the spectrum, polymers with chain entanglements below a critical limit exhibit little craze tuft stability because of chain pull-out and thus fracture readily.[3-5]

PRINCIPLES OF TOUGHENING

Having recognized the dilatational property of crazing, it became apparent to researchers that crazing might be a valuable tool for toughening glassy polymers. To prevent fracture, a sample must be able to absorb the amount of energy placed into it by an external excitation. This energy absorption usually manifests itself as a plastic deformation. After the initial elastic deformation, crazes begin growing in bulk when the material reaches its craze flow stress σ_c, and continue to grow until the sample reaches its fracture strain.

In the following sections, we describe three techniques that have been used to exploit the dependence of tensile toughness on craze front density and craze velocity. Each technique uses a different mechanism of toughening and can be described quantitatively from a theoretical viewpoint based on the sample morphology. It will be shown that the stress response of the material and the final craze structure is dependent on the toughening technique utilized.

BLOCK COPOLYMERS

The extensive attention devoted to block copolymer morphology has made these materials a valuable tool in studying toughness in glassy polymers. The large degrees of freedom associated with the manufacture of these materials, such as choice of copolymer, homopolymer molecular weight, and block molecular weight, give the researcher a vast array of morphological possibilities. The dependence of the equilibrium morphology on the relative volume fractions in a heterogeneous copolymer system is well known.[6]

The role of block copolymer morphology in dilatational toughening was not readily understood initially. The characteristic length scale of block copolymers is on the order of 100 Å, which is to be compared with a craze thickness of 0.1–0.5 μm for polystyrene. Hence, the stress-induced displacement misfit for these materials would be too low to achieve craze initiation. Mechanical studies of these materials, however, showed measurable increases in toughness from their glassy homopolymer counterparts. Clearly some toughening mechanism was occurring.

TOUGHNESS OF GLASSY POLYMERS CONTAINING COMPOSITE PARTICLES

Toughness in glassy polymers is ultimately limited by fracture arising from the breakdown of craze matter under an applied stress. Typically fracture results from micron-sized flaws, such as inorganic heterogeneities, entrapped in the bulk of the sample. Although these particles can act as potent craze initiators, their lack of cohesion to the matrix material quickly transforms them into superficial flaws, which results in premature fracture.

One successful approach to this goal is the distribution of large, compliant heterogeneities in a glassy matrix polymer by either synthesis routes or blending schemes. The two most common materials using this approach are high impact polystyrene (HIPS) and acrylonitrile-butadiene-styrene (ABS).[8-13]

These materials typically contain 1–10% particles with sizes in the range of 1–10 μm. The particles themselves are composites, often containing 80 vol% of glassy occlusions in a continuous rubber phase.[6] The role of the particles is to initiate crazes at relatively low stress levels throughout the sample volume and thereby increase the craze front density.

The craze-initiating potency of the compliant composite particles will lower the stress at which crazing occurs. Thus stress concentrations at less compliant inorganic inclusions will never reach the levels necessary to initiate crazes, which results in a reduced chance of premature craze rupture and subsequent macroscopic failure. Additionally, the high craze front density leads to substantial dilatational strain.

The efficacy of craze initiation from composite particles depends not only on their composition but also their morphology. An analysis of four model particles was discussed by Argon and Cohen.[6]

The mechanical responses of the blends containing these three particles suggest that HIPS and CSS particles are effective in initiating crazes while the K-resin particles were not; TEM analysis confirmed this hypothesis. The high Young's modulus of the K-resin particle (2.4 GPa) versus that of PS (3.25 GPa) results in a rather small modulus misfit.[7] The CSS and HIPS particles have lower moduli on the order of 0.1–0.3 GPa, which produces a modulus misfit substantial enough to initiate crazes.

TOUGHENING BY DILUENT-INDUCED LOCALIZED PLASTICIZATION

Although the result was completely unexpected in view of the previously described mechanisms of toughening, Gebizlioglu et al.[14] found that the addition of >0.4 wt% of low molecular weight polybutadiene (PB) (2,760 g/mol) resulted in a gradual rise in tensile toughness up to a maximum PB level of 1.1 wt%, after which the toughness dropped abruptly. The low rate stress-strain plots showed the characteristic decrease in craze flow stress accompanied by an increase in strain-to-fracture; the maximum toughness at 1.1% PB was roughly four times that of pure polystyrene, an enhancement roughly equivalent to that introduced by some HIPS particles at 10% rubber content.

Argon et al.[6] developed a mechanistic model describing the dependence of the craze growth rate on PB content. This model is based on the interface convolution mechanism developed for craze growth in glassy polymers.[1]

This model was compared with results from isolated craze growth experiments both at room temperature[16] and at reduced temperature.[15]

The craze velocity for all blends increases with applied stress; a higher stress corresponds to an increased rate of deformation. It is more relevant to state that for a higher imposed strain rate, a greater stress must be applied in order to achieve a higher craze growth rate.

The addition of a few weight percent of PB results in an elevation of the observed craze velocity by several orders of magnitude from the velocities recorded for pure PS at the same applied stress. The craze velocities for these materials were on the same order of magnitude as those found for the diblock copolymer blends exhibiting cavitation. Note that the opacity of the HIPS and ABS materials precludes craze velocity measurements.

CONCLUSIONS

Three mechanisms for toughening glassy polymers have been discussed. All these mechanisms rely on dilatational plasticity, or crazing, to achieve their high levels of toughness. Blends containing block copolymers typically craze by cavitation of the phase-separated morphology and thus achieve high craze densities. Blends containing particles, such as HIPS, also achieve a high number of crazes, but accomplish this from the craze-initiating potency of the added particles. Finally, glassy polymers containing a few weight percent of a low molecular weight PB achieve high toughness through higher-than-expected craze velocities, a result of the plasticizing action of the PB. The latter mechanism relies on surface flaw-initiated crazes, and is therefore effective only in thin samples.

REFERENCES

1. Argon, A. S.; Salama, M. M. *Phil. Mag.* **1977,** *36,* 1217.
2. Michler, G. H. *J. Mater. Sci.* **1990,** *25,* 2321.
3. Gent, A. N.; Thomas, A. G. *J. Polym. Sci. Part A-2* **1972,** *10,* 571.
4. Spiegelberg, S. H. et al. *J. Appl. Polym. Sci.* **1995,** *58,* 85.
5. Yang, A. C. M. et al. *Macromolecules* **1986,** *19,* 2010.
6. Argon, A. S.; Cohen, R. E. In *Advances in Polymer Science*; Springer-Verlag: Berlin, Heidelberg, 1990; Vol. 91/92; p 302.
7. Argon, A. S. et al. In *Mechanical Behaviour of Materials-V*; Pergamon: Beijing, 1987; p 3.
8. Bucknall, C. B.; Clayton, D. *J. Mater. Sci.* **1972,** *7,* 202.
9. Bucknall, C. B.; Stevens, W. W. *J. Mater. Sci.* **1980,** *15,* 2950.
10. Chen, C. C.; Sauer, J. A. *J. Appl. Polym. Sci.* **1990,** *40,* 503.
11. Cigna, G. et al. *J. Appl. Polym. Sci.* **1992,** *44,* 505.
12. Hall, R. A. *J. Mater. Sci.* **1990,** *25,* 183.
13. Okamoto,Y. et al. *Macromolecules* **1991,** *24,* 5639.
14. Gebizlioglu, O. S. et al. *Macromolecules* **1990,** *23,* 3968.
15. Spiegelberg, S. H. et al. *J. Appl. Polym. Sci.* **1994,** *53,* 1251.
16. Spiegelberg, S. H. et al. *J. Appl. Polym. Sci.* **1993,** *48,* 85.

GLASSY POLYMERS, CRAZING

Makoto Kawagoe
Department of Mechanical Systems Engineering
Faculty of Engineering
Toyama Prefectural University

Polymeric materials are used in many fields. To use them in structural components, it is necessary to recognize their mechanical strength. Two types of plastic deformation of amorphous polymers have been observed. One is shear flow followed by the formation of a localized deformed region, called shear band. The other is yielding caused by a number of fine fissures, called craze. This type of deformation is significant in media such as organic liquids. Although to the naked eye a craze is similar to a crack, it is distinguished by its microscopic structure, which consists of many elongated fibrils and voids. Scission of the fibrils, however, turns the craze into a crack and leads the polymer to brittle fracture. Because crazes lead to both plastic deformation and polymer fracture, they exert a great influence on polymer strength. The phenomenon of crazing, therefore, has been extensively studied.[1-5]

MICROSTRUCTURE AND MICROMECHANICS

The microstructure and micromechanics of isolated crazes have drawn the attention of many investigators,[6-12] probably because of the similarity in shape between crazes and cracks. The researchers revealed that there are two different mechanisms of craze thickening: the surface-drawing mechanism and the fibril creep mechanism.

In surface drawing, which is general for crazes grown in air, new fibrillar matter is drawn in from the surrounding unoriented polymer through a thin strain-softened layer called an active zone. In fibril creep, generally found in environmental crazes, the fibrils in a craze are further elongated. Particularly in air crazes, the fibrils are formed by disentanglement or scission of molecular chains in the active zone.

CRITERIA FOR CRAZING

Crazing mechanisms and criteria have long been studied, but are still not established, probably because of limited experimental results under multiaxial stress and considerable differences in crazing behavior under different conditions. Theories of craze nucleation mechanisms and criteria may be divided into two categories. One theory asserts that crazing is a cavitational plastic deformation at a localized region and thus the dilatational component of applied stress is essential to crazing.

On the other hand, the second theory asserts quite the opposite. Under conditions such as high pressure, where the specimens are protected from contact with the pressure medium,[13] and in liquid environments,[14] crazing can occur even in a wide stress field where the hydrostatic component is compressive. Kawagoe and Kitagawa[15] reexamined data on biaxial crazing stress in cylindrical specimens of glassy polymers and revealed that no crazing under pure shear is due not to vanishing of the dilatational stress component, but to development of shear flow.

CRAZING AT A CRACK TIP

Analyzing the stress and displacement of a craze formed at a crack tip is important for understanding brittle fracture. Recently, optical interference measurements have progressed in response to this problem. Döll, and Döll and Könczöl[16,17] have measured craze opening displacement by this technique and obtained the craze contour stress based on the Dugdale model of fracture mechanics. They showed the stress-strain relation in the craze zone at the crack tip and also demonstrated jumps in craze displacement under cyclic loading. They have also explained mechanisms of retarded crack growth in fatigue processes. Direct measurements of craze contour displacement by scanning electron microscopy (SEM) have also been developed recently.[18]

CRAZE-INDUCED DEFORMATION

As mentioned earlier, crazes act as both crack precursors and sources of plastic strain. In practical applications,[19] crazes increase the toughness of polymer alloys such as high-impact polystyrene (HIPS) by forming crazes at the interfaces between polymer components and absorbing applied energy under tensile loading.

Several researchers have studied the effect of crazing on fatigue failure. Sauer and co-workers[20,21] have shown that craze formation enhances the energy loss in the hysteresis loop of stress and displacement under cyclic loading. They also noted that the fatigue lifetime of HIPS is less than that of homo-PS. This means that dispersed rubber particles in HIPS produce craze formation much earlier during the fatigue process, and therefore crazes are not always beneficial.

EFFECTS OF ENVIRONMENTS

Crazing is facilitated under the action of some organic media. Most studies have examined the effects of liquid environment on crazing, with particular regard to the solubility of the liquid in the polymer. Kawagoe and co-workers,[22,23] however, have demonstrated that for PMMA exposed to *n*-alkanes and alcohols, the critical crazing stress increases with molar volume of the reagents and is not correlated with solubility parameter, equilibrium solubility, opposing the above results of the crazing strain.

CONCLUDING REMARKS

Crazing is localized plastic deformation and causes both plastic strain, contributing to toughness enhancement, and crack formation, leading to brittle fracture. Although the phenomenological features of crazing have been extensively studied, the interpretations of their results are not always established. For instance, is the crazing process at the nucleation stage cavitational deformation dependent on dilatational stress? In addition, why are different criteria applied to air crazing and environmental crazing? These questions must be clearly answered in order to establish criteria for crazing. Correlation between crazing and shear flow should be more precisely investigated, with special reference to the effects of environmental reagents.

One reason these problems are difficult is that crazing falls between two extremes: the molecular level at nanometers or less and the macroscopic level allowing continuum mechanics such as fracture mechanics. The progress of knowledge and experimental techniques at the molecular level will enable these problems to be solved in the future.

ACKNOWLEDGMENTS

The author thanks Dr. J. Qiu, T. P. U., for valuable discussions on the content of this paper. K. Nakamura and Y. Doi are acknowledged for their help in preparing the manuscript.

REFERENCES

1. Kambour, R. P. *J. Polym. Sci., Macromol. Rev.* **1973,** *7,* 1.
2. Kramer, E. J. In *Developments in Polymer Fracture*; Andrews, E. H., Ed.; Applied Science: London, 1979; Chapter 3.
3. Kinloch, A. J.; Young, R. F. *Fracture Behaviour of Polymers*; Applied Science: London, 1983; Chapter 5.
4. Kausch, H. H., Ed.; *Adv. Polym. Sci.* **1983,** *52/53,* 1.
5. Kausch, H. H., Ed.; *Adv. Polym. Sci.* **1990,** *91/92,* 1.
6. Beahan, P. et al. *Proc. R. Soc.* **1975,** A343, 525.
7. Kambour, R. P.; Holik, A. S. *J. Polym. Sci.* **1969,** A2(7), 1393.
8. Kambour, R. P.; Kopp, R. W. *J. Polym. Sci.* **1969,** A2(7), 183.
9. Lauterwasser, B. D.; Kramer, E. J. *Phil. Mag.* **1979,** A39, 369.
10. Kramer, E. J.; Berger, L. L. *Adv. Polym. Sci.* **1990,** *91/92,* 1.
11. Plummer, C. J. G.; Kausch, H. H. *Polymer* **1993,** *34,* 305.
12. Plummer, C. J. G.; Donald, A. M. *J. Polym. Sci., Polym. Phys. Ed.* **1989,** *27,* 327.
13. Matsushige, K. et al. *J. Mater. Sci.* **1975,** *10,* 833.
14. Kawagoe, M.; Kitagawa, M. *J. Polym. Sci., Polym. Phys. Ed.* **1981,** *19,* 1423.
15. Kawagoe, M.; Kitagawa, M. *J. Mater. Sci.* **1988,** *23,* 3927.
16. Döll, W. *Adv. Polym. Sci.* **1983,** *52/53,* 105.
17. Döll, W.; Könczöl, L. *Adv. Polym. Sci.* **1990,** *91/92,* 136.
18. Brown, N.; Wang, X. *Polymer* **1988,** *29,* 463.
19. Bucknall, C. B. *Toughened Plastics*; Applied Science: London, 1977.
20. Sauer, J. A.; Chen, C. C. *Adv. Polym. Sci.* **1983,** *51/52,* 167.
21. Sauer, J. A.; Hara, M. *Adv. Polym. Sci.* **1990,** *91/92,* 69.
22. Kawagoe, M.; Kitagawa, M. *J. Mater. Sci.* **1987,** *22,* 3000.
23. Kawagoe, M.; Morita, M. *J. Mater. Sci.* **1994,** *29,* 6041.

GLOBULAR POLYBASES, MICROEMULSION-LIKE

Michel Vert
Faculty of Pharmacy
University of Montpellier I

For synthetic and natural polyelectrolytes, the presence of hydrophobic residues close to the hydrophilic ionized ones is a critical factor. It can cause collapse from extended-coil to compact-coil conformation with either the formation of polysoap monomolecular globules, dispersed aggregates, or macroscopic precipitation, depending on the relative amounts of hydrophobic and hydrophilic moieties within the same molecules.[1]

During the past decade, it was reported that bifunctional basic polyelectrolytes of the partially quaternized poly[thio-l-(*N,N*-diakylaminomethyl)ethylene][Q-P(T,N-R_1-N-R_2,AE)-X] type, where X is the percentage of quaternary ammonium groups (X = 100 m/m + p) where m and p are the mole fractures in quaternary ammonium and tertiary amine repeating units, respectively (**Structure 1**), exhibit unusual polyelectrolytical characteristics in aqueous media. In particular, some members of the family can take on a pH-dependent globular conformation at neutral pH and can undergo a globule-to-coil cooperative transition in a narrow range of pH values.[2] Such macromolecules are actually copolymers. These copolymers were obtained by ring-opening polymerization of *N,N*-dialkyl-*N*-(thiirane-2-ylmethyl) amine followed by partial *N*-methylation of some of the tertiary amine side groups present in poly[thio-1-(N-R_1-N-R_2-aminomethyl) ethylene] molecules.

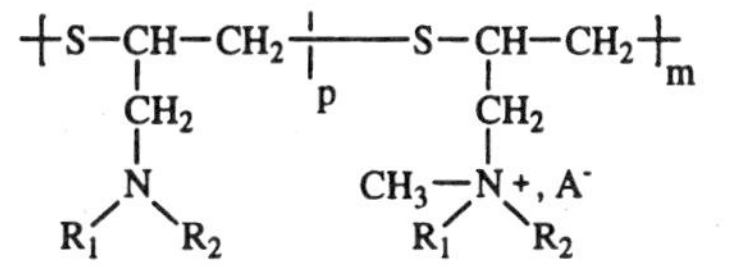

1

Q – P (T, N–R1–N–R2, AE) X

After identification and study of the composition-dependent physicochemical behaviors of these copolymer systems in aqueous media, the effects of factors such as ionic strength, presence of chiral centers, nature of the main chain, and nature of alkyl substituents were investigated.[2-5] The data showed the formation of protonation-dependent polysoap-like globular structure that appeared later on to be able to solubilize and transport water-insoluble compounds temporarily entrapped in the core of the globules. This is similar to what usually happens in the core of a micelle of amphiphiles, with the difference that the core of the macromolecular globule did not undergo any exchange of equilibrium with isolated amphiphilic small molecules. The microemulsion-like globule-based microphase was first considered to be suitable for drug transport and sustained-release drug delivery. The system worked correctly from the physicochemical viewpoint and *in vivo* after intramuscular injection.[6] However, blood toxicity precluded any effective therapeutic applications.[7] More recently, the potential of the globules to act as macromolecular microreactors was investigated. An enzyme-like activity of globular bifunctional polybases was demonstrated.[8] The catalytic potential of the globule-forming bifunctional polybases appeared versatile and seems to be important with respect to organic chemistry in water.

PROPERTIES

The originality of partially quaternized tertiary polyamine bifunctional polyelectrolytes comes from the presence of both alkylated quaternary ammonium groups and weakly basic tertiary amine groups on the same backbone. These pendant groups are more or less randomly distributed along the polymer chain. The former are always ionized regardless of the counterion, which can be either OH^- to give a strong base form or an A^- anion to give a neutral salt form. The latter can be neutral (deprotonated) or charged (protonated) depending on the solvent medium. When X>33, Q-P(T,N-R_1-N-R_2,AE)X macromolecules behave as well-solvated hydrophilic polyelectrolytes.[2] Because of the difference in basicity between strong base quaternary ammonium groups and weak base tertiary amine groups, the addition of an acid to a solution of dibasic Q-P(T,N-R_1-N-R_2,AE)X first causes an ion exchange at quaternary ammonium sites; next, protonation of tertiary amine residues occurs.

In the globular state, the core of each globule forms a hydrophobic microdomain that is stabilized by the permanent quaternary ammonium charges grouped at the surface, as for globular proteins. When macromolecules are highly charged because of the protonation of tertiary amine groups, they are in a well-solvated open-coil state. The addition of small amounts of acids to a solution of globular Q-P(T,N-R_1-N-R_2,AE)X molecules causes protonation of tertiary amine residues and triggers an all-or-none globule-to-coil cooperative conformational transition, which is reversible upon addition of a base and occurs almost instantaneously over a narrow range of pH values.

Another amazing property of globular Q-P(T,N-R_1-N-R_2,AE)X molecules is that, because of the very small size of the globules (~ 80Å diameter), the viscosity of aqueous solutions is very low, even for high molecular weights and high polymer concentration. In contrast, the viscosity of the same solution becomes high if the globules are turned to the solvated extended-coil conformation.

The microemulsion-like property of globular Q-P(T,N-R_1-N-R_2,AE)X is characterized by the ability of the core of the globules to accommodate water-insoluble lipophilic compounds and thus to temporarily entrap them–a phenomenon called "molecular encapsulation."

The core of the globules can also be used to temporarily accommodate water-insoluble reactive chemicals and to catalyze their chemical modification in an unusual surrounding medium. The catalytic activity is based on the dissolution of a lipophilic water-insoluble organic reagent in the hydrophobic core of the globular macromolecules and its transformation to a product, as already observed is microemulsion and microemulsion-like polymeric systems.[9]

APPLICATIONS

The ability of globular Q-P(T,N-R_1-N-R_2,AE)X to entrap lipophilic water-insoluble compounds and transport them in aqueous media was first used for sustained-release drug delivery. It has been shown that a steroid, such as radioactive ^{75}Se-cholestenol can be entrapped in globular Q-P(T,N-R_1-N-R_2,AE)X to be injected as an aqueous solution *in vivo* via the various parenteral routes, that is, subcutaneously, intramuscularly, intraperitoneally, and intravenously. The carrier system worked satisfactorily when given intramuscularly with sustained release for more than 35 days.[6] In contrast, a dramatic lethal effect was observed after intravenous injection, probably because of strong electrostatic interactions with anionic-type proteins and anionic membranes of blood platelets and cells.[7,10] This lethal effect can be minimized by using infusion at low rates and concentrations. So far, effective therapeutic uses in humans have appeared too risky, even through intramuscular injections.

The catalytic properties of globular Q-P(T,N-R_1-N-R_2,AE)X are currently being investigated. The observed rate increases should be extendible to several other chemical reactions with suitable characteristics. The catalytic activity of the globular macromolecules might open the door to organic chemistry in water because the very small size of the catalytic device allows easy and fast diffusion of the substrate and the product. An experimental apparatus for online chemical modifications in water was tested and has remained functional for several hours without decrease of catalytic activity.[11] Moreover, the characteristics of the globular macromolecules and of the globule-to-coil transition might also be of interest in the case of oil recovery. The oil-trapping microphase would have a very low viscosity, and the recovery could be achieved almost instantaneously by acidification before returning the polymer to its globular form. Whether such a system would be workable and cost-effective is still unknown. The economy of using globular Q-P(T,N-R_1-N-R_2,AE)X as an oil solubilizer and temporary carrier has not been evaluated.

ACKNOWLEDGMENTS

I am indebted to my former co-workers, J. Huguet and J. P. Couvercelle for collaboration on the primary work.

REFERENCES

1. *Microdomains in Polymer Solutions, Polymer Science and Technology*; Dubin, P., Ed.; Plenum: New York, 1985; Vol. 30.
2. Vallin, D.; Huguet, J.; Vert, M. *Polym. J.* **1980,** *12,* 113.
3. Vallin, D.; Huguet, J.; Vert, M. *Polymeric Amines and Ammonium Salts*; Goethals, E. J., Ed.; Pergamon: Oxford, England, 1980; p 219.
4. Huguet, J.; Vallin, D.; Vert. M. *Polym. J.* **1982,** *14,* 335.
5. Huguet, J.; Vert. M. In *Microdomains in Polymer Solutions*; Dubin, P. Ed.; *Polymer Science and Technology*; Plenum: New York, 1985; Vol. 30, Part 1, Chapter 3.
6. Illum, L.; Huguet, J.; Vert, M.; Davis, S. S. *J. Contr. Rel.* **1986,** *3,* 77.
7. Illum, L.; Huguet, J.; Vert, M.; Davis, S. S. *Int. J. Pharm.* **1985,** *26,* 113.
8. Couvercelle, J. P.; Huguet, J.; Vert, M. *Macromolecules* **1991,** *24,* 6452.
9. Couvercelle, J. P.; Huguet, J.; Vert, M. *Macromolecules* **1993,** *26,* 5015.
10. Vert, M.; Huguet, J. *J. Control. Rel.* **1987,** *6,* 159.
11. Couvercelle, J. P. Ph.D. Thesis, University of Rouen, France, 1990.

GLOBULAR PROTEINS (Structures and Functions)

Tamaichi Ashida* and Atsuo Suzuki
Department of Biotechnology
School of Engineering
Nagoya University

Proteins are classified by their molecular shapes into two large groups: globular proteins and fibrous proteins. The molecules of fibrous proteins, including collagen, keratin, and fibroin, are long and slender, and their biological roles are mainly physical (i.e., structural, protective, connective, and motive). They have simple repetitive amino acid sequences and are mostly biochemically inactive. They are hardly soluble to solvents, and they are not crystallizable. The globular proteins are biochemically active and control almost all the biochemical reactions in living organisms. They work as enzymes, carriers, antibodies, hormones, and toxins. Although their molecular shapes are not necessarily globular, some are far from being globular and are very irregular. They have unique three-dimensional molecular structures and will crystallize under appropriate conditions. The X-ray crystal structure analysis method displayed the molecular structures of many globular proteins, and the investigation of the structure–function relationship of globular proteins is one of the most important fields for biomolecular science and its application in biological technology.

*Author to whom correspondence should be addressed.

CHEMICAL STRUCTURE

Major components of proteins are unbranched polypeptide chains consisting of L-α-amino acids linked by the peptide bonds between the α-carboxyl group of one residue and the α-amino group of the next residue. Only 20 kinds of amino acids are involved in naturally occurring proteins. Some proteins have metal ions or small organic molecules as cofactors, and these are essential for the activity of the proteins. An enzyme lacking the essential cofactor is called an apoenzyme, and an intact enzyme with the bound cofactor is called a holoenzyme. Some kinds of proteins play specific roles in the form of oligomer consisting of chemically identical or different subunits not covalently linked together.

Peptide Bond

The CO–NH bond (C represents the C atom of the carbonyl group) is the peptide bond. The C′–N bond has a partial double-bond character of about 40%,[2] hence the six atoms of $C^{\alpha}CONHC^{\alpha}$ are coplanar, and the $C^{\alpha}C'NC^{\alpha}$ chain is usually restricted to be trans ($\omega \approx 180°$). Generally only prolyl residue can make a cis peptide bond ($\omega \approx 0°$) with its N atom with a probability of about 5%.

In addition to the peptide bonds formed between adjacent amino acid residues, many extracellular proteins have covalent bridging disulfide bonds made within the same polypeptide chain or between different polypeptide chains. It often is possible to reduce and break these disulfide linkages in a single polypeptide chain and then reversibly oxidize to reform the linkages. Such linkages play an important role of stabilizing three-dimensional structures of many proteins. Protein engineering with a site-directed mutagenesis has been used to make proteins more stable by introducing stereochemically well-designed unstrained disulfide bridges in protein molecules.

HIERARCHY OF PEPTIDE CHAIN STRUCTURE

The four levels of protein structures, primary, secondary, tertiary, and quaternary structures, build a hierarchy of the peptide chain structures.

The primary structure of a protein molecule, coded by the nucleotide sequence of its gene, determines its three-dimensional molecular structure and biological function.

The secondary structure is the local regular structures of the main chain without regard to the conformation of the side chain. They are α helices, β sheets, and β turns.

Tertiary structure is the three-dimensional arrangement of all atoms in a molecule, a subunit, or a domain. Many proteins consist of two or more domains even if the proteins are made of single polypeptide chains. The tertiary structure consists of secondary structures linked with loop regions (random coil) of various lengths and shapes, including β turns.

Quaternary structure is the assembled structure of molecules, subunits, and/or domains into an integrated biologically active unit. In many enzymes the active site locates in the crevice made between two domains.

BUILDING BLOCKS

The most important building blocks in protein structures are the α helix and the β sheet, which were proposed by Linus Pauling in 1951.[1] They have been found in fibrous proteins and in globular proteins.

The α helices observed in proteins are right handed. The left-handed α helix of L-amino acids is significantly less favorable than the right-handed helix due to the steric repulsions among nonbonded atoms. The helix has 3.6 residues per turn with a pitch of 5.4 Å, a rise along the helix axis of 1.5 Å and a rotation of 100° around the helical axis per residue. In globular proteins the α helices can vary considerably in length ranging from four residues to more than 40 residues, but most of them are rather short with the average measuring approximately 10 residues in three turns of length of about 15 Å.[2]

The β sheet, another major building block in globular proteins, is made of several parallel or antiparallel β strands of different regions of polypeptide chains. Mixed β sheets consisting of a mixture of parallel and antiparallel β strands are also found in globular proteins. The β strands are usually five to ten residues long with an average of six residues, and most β sheets contain less than six strands.[2] Almost all β sheets found in proteins are twisted in a right-handed sense when viewed down the peptide chain axis. Conformational energy calculations have shown that the extended peptide chains prefer to be slightly twisted in a right-handed sense. This twisting behavior of the β sheets is important, because twisted β sheets frequently constitute the backbones of protein structures such as the saddle shaped surfaces in flavodoxin, carboxypeptidase, and β barrels in triosephosphate isomerase and taka-amylase.

The β turn (or β bend)[3] is a frequently found small building block made of four consecutive residues. It changes inversely the direction of the peptide chain to fold back a polypeptide chain.

ASSEMBLY OF BUILDING BLOCKS

Most protein structures consist of combinations of a helices and/or β strands forming core regions. These blocks are connected by loop regions of various lengths and shapes and often by β turns. These connecting units are mostly at the surface of a molecule and are rich in polar and charged hydrophilic residues. The loop regions frequently participate in forming the active sites of enzymes.

STRUCTURE AND FUNCTIONS OF ENZYMES

More than 1500 enzymes are known, but a few simplifying features help classify structures and functions of enzymes and indicate molecular evolution. An overview of the structures, functions, and evolutions of a family of the serine proteases will serve as an example.[4]

Serine proteases constitute an enzyme family with extensively studied structures and functions. They have a unique reactive serine residue in the catalytic triad, Asp–His–Ser, which makes an acyl-enzyme as an intermediate of a hydrolysis of the substrate. The mammalian pancreatic serine proteases, trypsin, chymotrypsin, and elastase, are homologous with each other. They essentially have the same peptide backbone structure and the same disposition of the catalytic triad, with their sequence homologies of amino acids being about 50%. They differ in the specificity for substrates: Trypsin hydrolyzes the peptide bonds following the long basic side chains, chymotrypsin hydrolyzes the peptide bonds following aromatic side chains, and elastase hydrolyzes those following small side chains. The differences are due to the sizes and natures of the substrate recognition pockets. In contrast, thrombin, one of the plasma serine proteases, has the same backbone structure and reactive site structure, but its sequence homology between bovine trypsin is slightly less than 40%. These four mammalian serine proteases appear to represent a typical divergent evolution of protein molecules from a common ancestral enzyme to the proteins having different specificities.

ROLE OF WATER MOLECULES FOR THE STRUCTURE AND FUNCTION OF PROTEINS

Water molecules play an important role for the structures and function of globular proteins. They fill crevices of molecular surfaces and open inner spaces of protein molecules, making hydrogen bond networks with polar groups of proteins. Furthermore, one or two layers of water molecules are bound by hydrogen bonds at the protein surface. Many of these molecules behave as parts of protein molecules and take part in maintaining tertiary structures. If these bound water molecules are removed from proteins, proteins are denatured and lose their unique three-dimensional structure and biological functions. For several homologous proteins, water molecules occupy nearly the same sites, and these are important for the structure and function of the proteins. Further, such waters are conserved.

REFERENCES

1. Pauling, L.; Corey, R. B.; Branson, H. R. *Proc. Natl. Acad. Sci. U.S.A.* **1951,** *37,* 205-211.
2. Schultz, G. E.; Schirmer, R. H. *Principles of Protein Structure*; Springer-Verlag: New York, 1979; p 66-78.
3. Venkatachalam, C. M. *Biopolymers* **1968,** *6,* 1425-1436.
4. Fersht, A. *Enzyme Structure and Mechanism*; W. H. Freeman: New York, 1985; 2nd edition, p 17-21.

GLUCOSE RESPONSIVE POLYMERS (Medical Applications)

Joseph Kost
Department of Chemical Engineering
Ben-Gurion University of the Negev

Diabetes mellitus is a chronic disease with major vascular and degenerative complications. It is treatable inasmuch as the acute metabolic disorder can be kept under control. There is, however, no cure for diabetes; nor are we capable of fully preventing its long-term complications, which are believed to result from chronically elevated blood glucose levels. Thus, more than 90% of all diabetics eventually develop eye diseases;

in 30-40%, this advances to proliferative retinopathy, which is among the most common causes of blindness. In about 40% of patients end-stage renal failure develops, requiring dialysis and finally kidney transplantation. The leading cause of nontraumatic amputation in the west is diabetes, and the risk of cardiovascular accidents is 5-10 times higher in diabetics.[1]

None of the present modes of treatment including insulin pumps fully mimics the physiology of insulin secretion. Therefore even the most compliant patient who checks his blood glucose 6-7 times daily fails to achieve a completely normal glucose metabolism. The realization that blood sugar levels must be kept in normal limits if complications are to be avoided has led to self-regulated insulin delivery systems[2-4] capable of mimicking the natural pattern, in which insulin release is in response to glucose levels in the blood. Several approaches have been devised: immobilized glucose oxidase in pH-sensitive polymers, competitive binding, and polymer complex systems.

IMMOBILIZED GLUCOSE OXIDASE IN pH SENSITIVE POLYMERS

Heller et al.[5] were the first to use immobilized enzymes to alter local pH and thus cause changes in polymer erosion rates. Responsive systems based on pH-sensitive polymers have been performed along three different approaches: pH-dependent swelling, degradation, and solubility.

pH-Dependent Swelling

The systems consist of immobilized glucose oxidase in a pH-responsive polymeric hydrogel enclosing a saturated insulin solution (reservoir configuration). As glucose diffuses into the hydrogel, glucose oxidase catalyzes its conversion to gluconic acid, thereby lowering the pH in the microenvironment of the membrane and causing swelling (the hydrogel swells in acidic pH).

Horbett and co-workers[6-13] immobilized glucose oxidase in a crosslinked hydrogel made from *N,N*-dimethylaminoethyl methacrylate (DMA), hydroxyethyl methacrylate (HEMA) and tetraethylene glycol dimethacrylate (TEGDMA).

The great advantage of reservoir systems is the ease with which they can be designed to produce constant-release-rate kinetics. Their main advantage is leaks, which are dangerous as the incorporated drug could be rapidly released. To overcome this limitation Goldraich and Kost[14] evaluated a matrix system in which the drug and enzyme are uniformly distributed throughout a solid polymer. The matrices displayed faster and higher swelling and release rates at lower pH or higher glucose concentrations. Swelling and release kinetics were also responsive to step changes in glucose concentration in the physiological range.

pH-Dependent Degradation

Heller et al.[15-17] suggested a system in which insulin is immobilized in a pH-sensitive bioerodible polymer prepared from 3,9-bis(ethylidene 2,4,8,10-tetraoxaspiro) (5,5) undecane and *N*-methyldiethanolamine, which is surrounded by a hydrogel containing immobilized glucose oxidase. When glucose diffuses into the hydrogel and is oxidized to gluconic acid, the resultant lowered pH triggers enhanced polymer degradation and release of insulin from the polymer in proportion to the concentration of glucose.

pH-Dependent Solubility

Glucose-dependent insulin release was proposed by Langer and co-workers,[18,19] based on the fact that insulin solubility is pH dependent.

COMPETITIVE BINDING

The basic principle of competitive binding, first presented by Brownlee and Cerami[20,21] suggests the preparation of glycosylated insulins that are complementary to the major combining site of carbohydrate binding proteins such as Concavalin A (Con A). Con A is immobilized on sepharose beads. The glycosylated insulin, which is biologically active, is displaced from the Con A by glucose in response to, and proportional to, the amount of glucose that competes for the same binding sites. Kim et al.[22-31] found that the release rate of insulin also depends on the binding affinity of an insulin derivative to the Con A and can be influenced by the choice of saccharide group in glycosylated insulin. By encapsulating the glycosylated insulin-bound Con A with a suitable polymer that is permeable to both glucose and insulin, the glucose influx and insulin efflux would be controlled by the encapsulation membrane.

REFERENCES

1. Dubernanrd, J. M.; Sutherland, D. E. *International Handbook of Pancreas Transplantation*; Kluwer Academic: Dordrecht, The Netherlands, 1989.
2. Hrushesky, W. J. M. et al. *Ann. N.Y. Acad. Sci.* **1991,** Vol. 618.
3. Kost, J. *Pulsed and Self-Regulated Drug Delivery*; CRC Press: Boca Raton, FL, 1990.
4. Kost, J.; Langer, R. *Adv. Drug. Delivery Res.* **1990,** *6,* 19-50.
5. Heller, J.; Trescony, P. V. *J. Pharm. Sci.* **1979,** *68,* 919-21.
6. Horbett, T. A. et al. *Am. Chem. Soc. Div. Polym. Chem.* **1983,** *24,* 34-35.
7. Horbett, T. A. et al. In *Polymers as Biomaterials*; Shalaby, S. et al. Eds.; Plenum: New York, 1984; p 193-207.
8. Kost, J. et al. *J. Biomed. Mater. Res.* **1985,** *19,* 1117-33.
9. Albin, G. et al. *J. Controll. Rel.* **1985,** *2,* 153-64.
10. Albin, G. et al. *J. Controll. Rel.* **1987,** *6,* 267-91.
11. Albin, G. In *Pulsed and Self-Regulated Drug Delivery*; Kost, J., Ed.; CRC Press: Boca Raton, FL, 159-85.
12. Klumb, L. A.; Horbett, T. A. *J. Controll. Rel.* **1992,** *18,* 59-80.
13. Klumb, L. A.; Horbett, T. A. *J. Controll. Rel.* **1993,** *27,* 95-114.
14. Goldraich, M.; Kost, *J. Clin. Mater.* **1993,** *13,* 135-42.
15. Heller, J. et al. *Proceed. Intern. Symp. Control. Rel. Bioact. Mater.* **1989,** *16,* 155.
16. Heller, J. In *Pulsed and Self-Regulated Drug Delivery*; Kost, J., Ed.; CRC Press: Boca Raton, FL, 1990.
17. Heller, J. *Adv. Drug Delivery Rev.* **1993,** *10,* 163-204.
18. Fischel-Ghodsian, F. *Proc. Natl. Acad. Sci. USA* **1988,** *85,* 2403.
19. Brown, L. *Proceed. Intern. Symp. Control. Rel. Bioact. Mater.* **1988,** *15,* 166.
20. Brownlee, M.; Cerami, A. *Science* **1979,** 206, 1190.
21. Brownlee, M.; Cerami, A. *Diabetes* **1983,** *32,* 499.
22. Kim, S. W. et al. In *Recent Advances in Drug Delivery Systems*; Anderson J. W.; Kim, S. W., Eds.; Plenum: New York, 123.
23. Jeong, S. Y. et al. *J. Controll. Rel.* **1984,** *1,* 57.
24. Sato, S. *J. Controll. Rel.* **1984,** *1,* 67.
25. Jeong, S. Y. et al. *J. Controll. Rel.* **1985,** *2,* 143.

26. Sato, S. et al. *Pure. Appl. Chem.* **1984,** *56,* 1323.
27. Seminoff, L. *Proceed. Intern. Symp. Control. Rel. Bioact. Mater.* **1988,** *15,* 160.
28. Seminoff, L.; Kim, S. W. In *Pulsed and Self-Regulated Drug Delivery*; Kost, J., Ed.; CRC Press: Boca Raton, FL, 1990.
29. Kim, S. W. et al. *J. Controll. Rel.* **1990,** *11,* 193-201.
30. Makino, K. et al. *J. Controll. Rel.* **1990,** *12,* 235-39.
31. Pai, C. M. et al. *J. Pharm. Sci.* **1992,** *81,* 532.

Glue Resins

See: Urea-Formaldehyde Glue Resins

Glued-Laminated Timber

See: Wood Composites (High Performance)

GLUTEN PLASTIC, BIODEGRADABLE

Mitsuo Yasui, Takashi Domae, and Iwao Yamashita
Hyogo Prefectural Institute of Industrial Research

Plastic materials have caused serious environmental problems, because ecological systems alone cannot dispose of their wastes. Therefore, development of biodegradable plastics became a very important goal for the good of the global environment.

Several approaches exist to develop biodegradable plastics from such materials as starch, chitosan, and others. On the other hand, developing plastic from a protein resource has not been studied, because the heating procedure during processing is liable to cause a structural change (gelation). This would make protein difficult to plasticize.

We tried to utilize "gluten," a globular protein made from wheat, to develop biodegradable plastics. Gluten is a general name of protein contained in wheat, and it consists of gliadin, glutenin, and the other polypeptides that form a macromolecule linked by S–S bonds, hydrogen bonds, and other bonds. Ordinarily, gluten contains 10 to 12% carbohydrates and 8 to 12% lipids. Gluten is rigid and brittle when it contains little water, but is soft and rubber-like with higher water content. Therefore, we investigated the dynamic mechanical properties of gluten containing a polyol instead of water and found we could plasticize gluten with a simple procedure. Also the obtained plastic is moldable and biodegradable.

PLASTICIZATION AND MOLDING OF GLUTEN

Gluten powder was preplasticized by adding a little water and then blending the mixture with a plasticizer such as glycerin, ethylene glycol, or diethylene glycol by using a roll-mixing procedure. The added water was removed from the compound system by a drier set to 40°C.

BIODEGRADABILITY OF GLUTEN PLASTICS

A burial test of a sheet sample was performed in soil to evaluate the biodegradability of gluten plastic. The sheet sample's thickness was about 300 mμ, and samples with sulfur and without sulfur were used. Several pieces of both samples were buried individually about 10 cm under the surface of inclined ground. The samples were extracted each week for observation. The two kinds of samples degraded equally under this environmental condition.

APPLICATION OF GLUTEN PLASTIC

Gluten plastic might not cover the same field of use as conventional plastics. Gluten plastic's mechanical properties are weakened because it contains much water. However, the change of properties of plastic with moisture in the air was not observed when the sample sat in a room for four years. Moreover, compression and injection molding were the most suitable for the molding of gluten plastic.

Gluten plastic is low in cost compared with current biodegradable plastics. Therefore, this plastic would be expected to applicate to golf tees (injection molding), trays and tablewares, cushion material for packing (forming product), and forming products and eatable films.

REFERENCES

1. Yasui, M.; Domae, T.; Yamashita, I. *Polym. Prepr. Japan* **1990,** *39(7),* 227.
2. Yasui, M.; Domae, T.; Yamashita, I. *Polym. Prepr. Japan* **1991,** *40(3),* 906.
3. Yasui, M.; Domae, T.; Yamashita, I. *Polym. Prepr. Japan* **1991,** *40(8),* 2665.
4. Yamashita, I.; Yasui, M.; Domae, T. *Proco. 5th Elastomer Symp.* **1991,** p 2.

GLYCOPROTEINS (Synthesis by NCA Method)

Masahiko Okada
Department of Biological Sciences
School of Agricultural Sciences
Nagoya University

A class of polymeric substances called glycoproteins are widely distributed in nature. Found in higher animals, plants and microorganisms, glycoproteins contain a protein chain of any of the 20 naturally occurring L-α-amino acid units. Covalently attached to this protein backbone and pendant to it are glycochains that, in most cases, are branched hetero-oligosaccharides consisting of less than 10 sugar residues. The molecular weights of glycoproteins range from 15,000 up to millions. Some glycoproteins have only one glycochain in a molecule, whereas other glycoproteins possess many glycochains in a molecule.[1-4]

It would be extremely difficult to artificially construct glycoproteins that are the same as natural glycoproteins through chemical synthesis alone. Rather, it seems unnecessary to do so. The principal purposes of synthesizing glycoprotein conjugates are to synthesize adequate models for understanding sophisticated functions displayed by naturally occurring glycoproteins on a molecular level and to create artificial polymeric materials possessing characteristic properties of both glycochains and proteins.

Three methods exist for synthesizing simple glycopeptides that mimic naturally occurring glycoproteins (**Scheme I**). The first is a synthesis of α-amino acid derivatives containing a

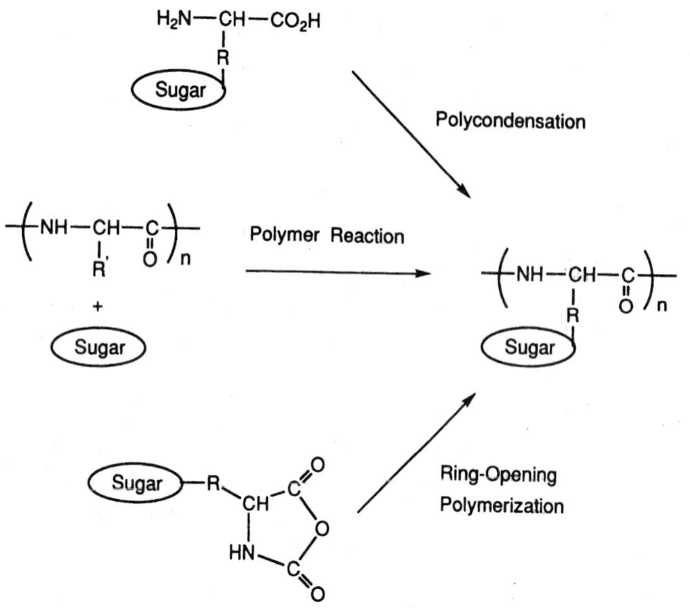

I

sugar moiety followed by their successive condensation.[4-9] The second is a synthesis of polypeptides and their chemical modification by sugar-containing reagents, and the third method is a synthesis of sugar-containing *N*-carboxy α-amino acid anhydrides (NCAs) and their ring-opening polymerization.[10-14]

To synthesize complex glycopeptides that have glycochains with well-defined structures and controlled sequences of α-amino acid residues, the stepwise condensation method commonly must be used for peptide synthesis. However, it is extremely tedious and difficult to synthesize glycopeptides of high molecular weights by this method. The chemical modification method of synthetic polypeptides creates high molecular weight polypeptides relatively easily, but it has difficulty in precisely controlling the introduction of pendant glycochains. The NCA method has the disadvantage of complex monomer syntheses that generally require a sequence of several reaction steps. However, under properly selected reaction conditions, anionic ring-opening polymerization of NCAs proceeds without transfer and termination and, therefore, glycopeptides of well-defined structures and of controlled chain lengths can be produced. However, this method cannot synthesize glycopeptides with controlled sequences of different α-amino acid residues in a peptide chain.

GLYCOPEPTIDE SYNTHESIS BY RING-OPENING POLYMERIZATION OF NCAS

The NCA method is particularly effective for synthesizing glycopeptides of simple structures. It generally applies to a variety of sugar-containing NCAs, including *O*-acetylated *O*-(β-glycopyranosyl)-L-serine derivatives of D-galactose, lactose, cellobiose, and *N*-acetyl-D-glucosamine. Although the biological functions of the synthetic glycopeptides are subjects of future investigation, "cluster effects" originating from the presence of glycochains in high density should be expected. Because of specific functions such as molecular recognition and cell recognition that are displayed by sugar moieties, it is expected that the synthetic glycopeptides could be a new class of biofunctional polymeric materials, particularly applicable to biomedical purposes.

CHEMICAL SYNTHESIS OF RELATED GLYCOCONJUGATES BY THE NCA METHOD

This section looks at the synthesis of some related glycoconjugates by ring-opening polymerization of NCAs.

Cationic ring-opening polymerization of 2-oxazolines proceeds with isomerization to yield poly[*N*-acylimino)ethylene]s.[15,16] Poly [*N*-acylimino)ethylene]s can be regarded as pseudopolyamides. Those bearing pendant acetyl or propionyl groups are hydrophilic, wherease those having longer acyl groups or benzoyl groups are hydrophobic.

Poly(2-methyl-2-oxazoline) having a terminal amino group was prepared by cationic ring-opening polymerization of 2-methyl-2-oxazoline followed by termination with ammonia. By using this as a macroinitiator, serine NCA with a β-D-glucose residue was polymerized in dichloromethane at 27°C for three days to yield a diblock copolymer in quantitative yield.[17]

Graft copolymers of cellulose and chitin with polypeptides can be included in the category of glycopeptide conjugates in a broader sense. Cellulose-*graft*-poly(γ-benzyl L-glutamate) has been prepared by polymerizing γ-benzyl L-glutamate NCA in dimethyl sulfoxide with a cellulose derivative containing a small amount of aminoethyl groups (degree of substitution, 0.05) as a macroinitiator.[18] The solubility tests of the reaction products showed that all aminoethyl cellulose molecules were grafted with polypeptide chains.

Next to cellulose, chitin is the second largest biomass produced in the world. Chitin consists of *N*-acetyl-D-glucosamine repeating units. It is widely distributed in the shells of Crustacea (e.g., crabs and shrimps) and the cuticles of insects and also in the cell walls of some fungi and microorganisms. However, its utilization has been limited because of difficulties in isolation, purification, and solubilization. Since a simple procedure to dissolve it in water was developed, its chemical modification has been undertaken in various ways.[19] Water-soluble chitin is a partially (about 50%) and randomly deacetylated chitin prepared by homogeneous hydrolysis. Because it has free amino groups, it is capable of initiating the ring-opening polymerization of NCAs by the nucleophilic addition mechanism.

REFERENCES

1. Sharon, N. *Complex Carbohydrates: Their Chemistry, Biosynthesis and Function*; Addison-Wesley: Reading, MA, 1975.
2. Kennedy, J. F.; White, C. A. *Bioactive Carbohydrates: In Chemistry, Biochemistry and Biology*; Ellishorwood: Chichester, England, 1983; Chapter 9.
3. Bahl, O. P. In *Glycoconjugates*; Allen, H. J.; Kisailus, E. C., Eds.; Marcel Dekker: New York, **1992,** Chapter 1.
4. Lee, Y. C.; Lee, R. T. In *Glycoconjugates*; Allen, H. J.; Kisailus, E. C., Eds.; Marcel Dekker: New York, **1992,** Chapter 6.
5. Garg, H. G.; Jeanioz, R. W. *Adv. Carbohydr. Chem. Biochem.* **1985,** *43,* 135.
6. Kunz, H.; Birnbach, S. *Angew. Chem., Int. Ed. Eng.* **1986,** *25,* 360.
7. Kunz, H. *Angew. Chem., Int. Ed. Eng.* **1987,** *26,* 294.
8. Cohen-Ansfield, S.; Lansbury, P. T. *J. Am. Chem. Soc.* **1993,** 115, 10531.
9. Wong, C. H.; Schuster, M.; Wang, P.; Sears, P. *J. Am. Chem. Soc.* **1993,** 115, 5893.
10. Stowell, C. P.; Lee, Y. C.; *Adv. Carbohydr. Chem. Biochem.* **1980,** *37,* 225.
11. Kuroyanagi, Y.; Kobayashi, H.; Seno, M.; Ishida, M.; Tominaga, N.; Akaike, T.; Sakamoto, M.; Ebert, G. *Int. J. Biol. Macromol.* **1984,** *6,* 266.
12. Kuroyanagi, Y.; Kubota, T.; Miyata, T.; Seno, M. *Int. J. Biol. Macromol.* **1986,** *8,* 52.
13. Kobayashi, K.; Zhou, S. H.; Sumitomo, H.; Okada, M.; Akaike, T. *Kobunshi Ronbunshu* **1991,** *48,* 253.
14. Aoi, K.; Tsutsumiuchi, K.; Okada, M. *Macromolecules* **1994,** *27,* 875.
15. Kobayashi, S.; Saegusa, T. R. In *Ring-Opening Polymerization*, Ivin, K. J.; Saegusa, T., Eds.; Elsevier Applied Science: London, 1984; Vol. 2, Chapter 11.
16. Kobayashi, S.; Saegusa, T. In *Encyclopedia of Polymer Science and Engineering, 2nd ed.*; Mark. H. F. et al., Eds.; John Wiley & Sons: New York, 1986; Vol. 4, p 525.
17. Tsutsumiuchi, K.; Aoi, K.; Okada, M. *Polym. Reprint. Japan*, **1993,** *42,* 3442.
18. Miyamoto, T.; Takahashi, S.; Tsuji, S.; Ito, H.; Inagaki, H. *J. Appl. Polym. Sci.* **1986,** *31,* 2303.
19. Sannan, T.; Kurita, K.; Iwakura, Y. *Makromol. Chem.* **1976,** 177, 3589.

Glycosaminoglycans

See: Chondroitin Sulfate (Overview)
Glycosaminoglycans (Overview)

GLYCOSAMINOGLYCANS (Overview)

Nicola Volpi
Department of Biologia Animale (Animal Biology)
University of Modena

Glycosaminoglycans (i.e., hyaluronic acid, keratan sulfate, chondroitin sulfates, heparan sulfate, and heparin) are linear, complex, polydisperse polysaccharides.[1-7] With the exception of keratan sulfate, they consist of alternating copolymers of uronic acids and amino sugars; their structures are commonly represented by characteristic disaccharide sequences.[1,3,4,6] With the exception of hyaluronic acid, glycosaminoglycans are sulfated heteropolysaccharides with different degrees of charge density due to sulfate groups in varying amounts and linked in different positions.[6,8] They are very heterogeneous polysaccharides in terms of relative molecular mass, charge density, physicochemical properties, and biological and pharmacological activities. These heteropolyacids are a class of macromolecules of great importance in the fields of biochemistry, pathology, and pharmacology.

With the exception of hyaluronic acid, glycosaminoglycan chains are covalently attached at their reducing end through an *O*-glycosidic linkage to a serine residue or through an N linked to asparagine (for keratan sulfate) in a core protein; the resulting macromolecules are called proteoglycans.[6,8-11] They are localized at cellular (intracellular granula and membrane-associated proteoglycans) and extracellular levels where they play structural and regulating roles because of their interaction with several proteins.[9-15]

Most glycosaminoglycans are derived from animal origins by extraction and purification processes.[16-23] Recently, these natural substances have been chemically modified, and synthetic analogues have been developed.[2] Currently, glycosaminoglycan-based drugs represent natural or depolymerized hyaluronic acid, heparin, dermatan sulfate, and chondroitin sulfates and mixtures of these polysaccharides.[2,3,24-28] Derivatives of heparin devoid of anticoagulant action (with less *N*-sulfate and larger amounts of *N*-acetyl groups more closely resembling the structure of heparan sulfate) are also utilized for pharmaceuticals.[29] Source material, manufacturing processes, procedures of derivatization, presence of contaminants, and many other factors contribute to the overall biologic and pharmacologic actions of these agents.

Hyaluronic acid is a linear polysaccharide composed of alternating residues of the monosaccharides D-glucuronic acid and *N*-acetyl-D-glucosamine linked by β(1->3) bonds in repeating units.[6] Disaccharides are linked to each other by β(1->4) bonds.

Keratan sulfate chains are based on a repeating *N*-acetyllactosamine sequence of -β(1–3)-[D-galactose-β(1->4)-*N*-acetyl- D-glucosamine]-β(1–3)-, which is usually sulfated at the C-6 position of acetylglucosamine, and further sulfate groups may be present at the C-6 position of galactose.[30]

Chondroitin sulfates are heteropolysaccharides composed of alternate sequences of D-glucuronic acid and differently sulfated residues of *N*-acetyl-D-galactosamine linked by β(1->3) bonds.[1,2,6]

Polysaccharide chains of dermatan sulfate (chondroitin sulfate B) consist of a prevailing disaccharide unit, [(1->4)-*O*-(idopyranosyluronic acid)-(1->3)-*O*-(2-acetamido-2-deoxy-D-galaytopyranosyl 4-sulfate)].[2,5]

Heparan sulfate and heparin have a heterogeneous structure due to the presence of variously sulfated regions distributed along the chains.[31-33] They are polysaccharides composed of alternate sequences of differently sulfated residues of uronic acid (D-glucuronic acid L-lduronic acid) and *N*-acetyl-D-glucosamine or *N*-sulpho-D-glucosamine linked by α(1->4)bonds.[1-4] Sulfate groups can be O-linked in the 2 position of uronic acids, in the 6 position of *N*-acetyl-D-glucosamine, and in the 3 and 6 positions of *N*-sulpho-D-glucosamine. Heparan sulfate chains consist of a high percentage of glucuronic acid, *N*-acetyl-D-glucosamine, and low-charged disaccharides, whereas heparin chains are formed by a high percentage of iduronic acid, *N*-sulfate-D-glucosamine, and high-charged disaccharides.[32,33] However, specific kinds of haparan sulfate molecules are constituted by sequences typical of heparin that impart heparin-like properties, such as anticoagulant capacity, to these polysaccharides.[34-36]

PROPERTIES

Relative Molecular Mass

The molecular mass of glycosaminoglycans affects their biological and pharmacological properties. They are polydisperse depending on the species and source of preparation. Several physical methods are currently used to estimate their molecular mass. The values of number-average molecular mass (M_n), weight-average molecular mass (M_w), and polydispersity of heteropolysaccharides are generally obtained by high-performance size-exclusion chromatography (HPSEC).[37-40] Molecular mass has also been estimated by using viscosimetry, ultracentrifugation, and light-scattering methods.

Degree of Epimerization

One important feature making the structure of glycosaminoglycans complex is the presence of alternating disaccharide and oligosaccharide sequences containing either glucuronic or iduronic acid; the glucuronic/iduronic ratio of a sulfated polysaccharide depends on the source. This is a key factor because several biological and pharmacological activities of glycosaminoglycans have been related to the conformational properties of iduronic acid.[41]

Specific Oligosaccharide Sequences Inside the Polysaccharide Chains

Several biological and pharmacological activities of glycosaminoglycans depend on the specific interactions between oligosaccharide sequences localized inside the polysaccharide chains and proteins, such as the interaction between heparin and antithrombin III, heparin/heparan sulfate and basic fibroblast growth factor, and dermatan sulfate and heparin cofactor II.[12,42,43] The characterization of these peculiar oligosaccharide sequences requires isolation, after chemical or enzymatical polysaccharide degradation, by gel permeation and anion-exchange chromatography, and analysis by NMR, fast atom bombardment mass spectrometry (FABMS), and disaccharide evaluation after bacterial lyases treatment.

APPLICATIONS

Hyaluronic Acid

The rheological properties used as lubricating and shock absorbing agents of high molecular mass hyaluronic acid provide for its use as an adjunct in ophthalmic surgery and in facilitating manipulation of ocular tissues.[24] In animal models, hyaluronic acid protected corneal endothelium from damage caused by synthetic lenses.

The rationale for the use of hyaluronic acid in the treatment of patients with rheumatoid or osteoarthritis is based on the increase of lubricating properties of synovial fluid after exogenous supplementation. Of great interest is the finding that high molecular mass hyaluronic acid has some ability to repair or protect cartilage fqrm the pathophysiological processes of arthritis in animal models and in small numbers of patients with osteoarthritis.[6]

Heparin

Anticoagulant, Antithrombotic, and Thrombolytic Properties

Heparin has long been known to arrest blood clotting. The mechanism of this anticoagulant action requires a cofactor in plasma, the proteinase inhibitor antithrombin III. Although the anticoagulant and antithrombotic actions of heparin have been mainly attributed to the inhibition of thrombin generation and potentiation of antithrombin III activity, it has also been suggested that this compound has a profibrinolytic effect inhibiting the growth of a preexisting thrombus. This most probably is caused by the release of tissue plasminogen activator or to the inhibition of fibrin(ogen) accumulation on thrombi. Heparin is a valuable drug for the treatment of deep venous thrombosis and as a prophylactic agent to prevent and limit venous and arterial thromboembolism.

Effects on Lipolysis

The appearance of lipolytic activity in plasma, following heparin injection, is well known. Lipases are normally located on the heparin sulfate proteoglycans of the cell membrane of endothelial cells. Heparin competes for the heparan sulfate binding site of the lipases, releasing into the circulation the lipolytic enzymes triglyceride lipase and lipoprotein lipase. This leads to a rapid increase of plasma-free fatty acid levels.[1-4]

Regulation of Growth Factor Activities

The affinity of polypeptides for immobilized heparin has led to the purification, amino acid sequencing, and molecular cloning of several growth factors, collectively known as heparin-binding growth factors (e.g., basic and acidic fibroblast growth factor).[12,15] The specific interaction of heparin (and derivatives) with growth factors leads to the capacity of this glycosaminoglycan to regulate the fundamental cellular functions, such as cellular proliferation and differentiation, nerve cell development and regeneration, cellular adhesion and aggregation, induction of chemotaxis *in vitro*, induction of DNA synthesis and protein expression, regulation of angiogenesis and possible influence on tumor growth, and metastasis development.

Heparin-Coated Surfaces

Heparin has been immobilized on several supports to obtain blood-compatible materials on which favorable and specific events such as preferential albumin adsorption, antithrombogenicity, platelet adhesion, and cell attachment were promoted or, in general, to produce biocompatible materials.[44-50]

Chondroitin Sulfates

Mixtures of chondroitin sulfates, containing mostly chondroitin sulfate A and C, are currently used as chondroprotective drugs in the treatment of osteoarthritis.[51-54] Moreover, these polysaccharides are generally able to inhibit certain enzymes present in the synovial fluid (e.g., elastase and hyaluronidase), which may damage joint cartilage, and they have also been shown to act as anti-inflammatory drugs. In addition, chondroitin sulfates can be used in osteoarthritis therapy because this drug is absorbed by the body and concentrated in the cartilage and produces no toxic or teratogenic effects. In clinical studies performed to date, chondroitin sulfates have always yielded clinical improvement of painful symptoms due to their proven anti-inflammatory activity.

Dermatan Sulfate

Dermatan sulfate shows anticoagulant capacity by inhibiting thrombin activity via activation of a protease inhibitor, heparin cofactor II. Several studies indicate that dermatan sulfate prevents venous thrombosis and inhibits the growth of a preexisting thrombus.

REFERENCES

1. Ofosu, F. A.; Danishefsky, I.; Hirsh, J. *Heparin and Related Polysaccharides, Structure and Activities*; New York Academy of Science: New York, 1989; Vol. 556.
2. Mammen, E. F. In *Seminars in Thrombosis and Hemostasis*; Theme Medical: New York, 1991; Vol. 17, Suppl. 1-2.
3. Lane, D. A.; Lindahl, U. Heparin, *Chemical and Biological Properties. Clinical Applications*; Edward Arnold: London, England, 1989.
4. Lane, D. A.; Björk, I.; Lindahl, U. *Heparin and Related Polysaccharides, Advances in Experimental Medicine and Biology*; Plenum: New York, 1992; Vol. 313.
5. Scott, J. E. *Dermatan Sulfate Proteoglycans*; Portland: London, England, 1993.
6. Kuettner, K. E.; Schleyerback, R.; Peyron, J. G.; Hascall, V. C. *Articular Cartilage and Osteoarthritis*; Raven: New York, 1992.
7. Crescenzi, V.; Dea, I. C. M.; Paoletti, S.; Stivala, S. S.; Sutherland, I. W. *Biomedical and Biotechnological Advances in Industrial Polysaccharides*: Gordon and Breach Science: New York, 1989.
8. Wight, T. N.; Mecham, R. P. *Biology of Proteoglycans*; Academic: New York, 1987.
9. Kjellen, L.; Lindahl, U. *Ann. Rev. Biochem.* **1991,** *60,* 443.
10. Ruoslahti, E. *Ann. Rev. Cell. Biol.* **1988,** *4,* 229.
11. Heinegard, D.; Oldberg, A. *FASEB J.* **1989,** *3,* 2042.
12. Jackson, R. L.; Busch, S. J.; Cardin, A. L. *Physiol. Rev.* **1991,** *71,* 481.
13. Cardin, A. D.; Weintraub, H. J. R. *Arteriosclerosis* **1989,** *9,* 21.
14. Bernfield, M.; Kokenyesi, R.; Kato, M.; Hinkes, M. T.; Spring, J.; Gallo, R. L.; Lose, E. J. *Ann. Rev. Cell Biol* **1992,** *8,* 365.
15. Burgess, W. H.; Maciag, T. *Ann. Rev. Biochem.* **1989,** *58,* 575.
16. Volpi, N. *Carbohydr. Res.* **1993,** *247,* 263.
17. Volpi, N. *Carbohydr. Res.* **1994,** *255,* 133.
18. Volpi, N. *Carbohydr. Res.* **1994,** *260,* 159.
19. Volpi, N.; Bolognani, L.; Conte, A.; Petrini, M. *Leukemia Res.* **1993,** *17,* 789.
20. Poblacion, C. A.; Michelacci, Y. M. *Carbohydr. Res.* **1986,** *147,* 87.
21. Linhardt, R. J.; Al-Hakim, A.; Liu, S. Y.; Kim, Y. S.; Fareed, J. In *Seminars in Thrombosis and Hemostasis*; Mammen, E. F., Ed.; Theme Medical: New York, 1991; Vol. 17, p 15-22.
22. Heinegard, D.; Sommarin, Y. *Methods in Enzymology* **1987,** *144,* 319.
23. Roden, L.; Baker, J. R.; Cifonelli, J. A.; Mathews, M. B. *Methods in Enzymology* **1972,** *28,* 73.
24. Goa, K. L.; Benfield, P. *Drugs* **1994,** *47,* 536.
25. Linhardt, R. J.; Wang, H-M.; Ampofo, S. A. *Advances in Experimental Medicine and Biology*; Plenum: New York, 1992; Vol. 313, 37-47.
26. Linhardt, R. J.; Loganathan, D. *Biomimetic Polymers*; Plenum: New York, 1990; 135-173.
27. Linhardt, R. J.; Desai, U. R.; Liu, J.; Pervin, A.; Hoppensteadt, D.; Fareed, *J. Biochem. Pharm.* **1994,** *47,* 1441.
28. Cristofori, M.; Mastacchi, R.; Barbanti, M; Sarret, M. *Arzneim. Forsch.* **1985,** *35,* 1513.
29. Gervasi, G. B.; Bartoli, C.; Farina, C.; Catalini, R.; Carpita, G.; Pistelli, N. *Arzneim. Forsch.* **1991,** *41,* 410.
30. Lauder, R. M.; Huckerby, T. N.; Nieduszynski, I. A. *Biochem. J.* **1994,** 302, 417.
31. Gallagher, J. T.; Lyon, M.; Steward, P. *Biochem. J.* **1986,** 236, 313.
32. Gallagher, J. T.; Turnbull, J. E.; Lyon, M. *Int. J. Biochem.* **1992,** *24,* 553.
33. Gallagher, J. T.; Turnbull, J. E. *Glycobiology* **1992,** *2,* 523.
34. Nader, H. B.; Dietrich, C. P.; Buonassisi, V.; Colburn, P. *Proc. Natl. Acad. Sci. USA* **1987,** *84,* 3565.
35. Lindblom, A.; Bengtsson-Olivecrona, G.; Fransson, L-A. *Biochem. J.* **1991,** *279,* 821.
36. Horner, A. A. *Biochem. J.* **1991,** *280,* 393.
37. Nielsen, J. I. *Thromb. Haem.* **1992,** *68,* 478.
38. Ahsan, A.; Jeske, W.; Mardiguian, J.; Fareed, J. *J. Pharm. Sci.* **1994,** *83,* 197.
39. Maderich, A. B.; Sugita, E. T. *J. Chromatogr.* **1993,** *622,* 278.
40. Volpi, N.; Bolognani, L. *J. Chromatogr.* **1993,** *630,* 390.
41. Ferro, P. R.; Provasoli, A.; Ragazzi, M.; Casu, B.; Torri, G; Bossennec, V.; Perly, B.; Sinay, P.; Petitou, M.; Choay, J. *Carbohydr. Res.* **1990,** *195,* 157.
42. Maccarana, M.; Casu, B.; Lindahl, U. *J. Biol. Chem.* **1993,** *268,* 23898.
43. Tollefsen, D. M.; Peacock, M. E.; Monafo, W. J. *J. Biol. Chem.* **1986,** *261,* 8854.
44. Hlady, V.; Andrade, J. D.; Ho, C-H.; Feng, L.; Tingey, K. *Clin. Mater.* **1993,** *13,* 85.
45. Llanos, G. R.; Sefton, M. V. *Biomaterials* **1992,** *13,* 421.
46. Collis, J. J.; Embery, G. *Biomaterials* **1992,** *13,* 548.
47. Ito, Y.; Iguchi, Y.; Imanishi, Y. *Biomaterials* **1992,** *13,* 131.
48. Ma, X.; Mohammed, S. F.; Kim, S. W. *J. Biomed. Mater. Res.* **1993,** *27,* 357.
49. Ho, C-H.; Hlady, V.; Nyquist, G.; Andrade, J. D.; Caldwell, K. D. *J. Biomed. Mater. Res.* **1991,** *25,* 423.
50. Nadkarni, V. D.; Pervin, A.; Linhardt, R. *J. Anal. Biochem.* **1994,** *222,* 59.
51. Paroli, E.; Antonilli, L.; Biffoni, M. *Drugs Exptl. Clin. Res.* **1991,** *17,* 9.
52. Paroli, Ed. *Int. J. Clin. Res. Pharm.* **1993,** *13,* 1.
53. Setnikar, I. *Int. J. Tiss. Reac.* **1992,** *14,* 253.
54. Pipitone, V. R. *Drugs. Exptl. Clin. Res.* **1991,** *265,* 18263.

GRADIENT POLYMERS

Carlos F. Jasso-Gastinel
Department of Chemical Engineering
University of Guadalajara

Because of the unusual polymer–polymer incompatibility, polymer blending has become an art based on obtaining stable materials with tailor-made properties. Two phase materials can have superior properties compared with homogeneous materials of equivalent chemical composition, if phase separation is maintained at the microscopic level. That synergistic effect can be enhanced by preparing physical polymer blends or interpenetrating polymer networks (IPNs) that have a gradient in composition. Shen and Bever defined the gradient polymers and "systems in which structure and properties vary continuously in space."[2] Conceptually they discussed what a gradient polymer structure could offer for heterogeneous materials and even for single phase materials in which a crosslinking degree gradient can modify solubility properties, for example.

The gradient idea can be applied to any constituent, state, or phase morphology of a particular system or polymeric product.

PREPARATION

Gradient profiles of different geometries can be prepared by physical or chemical methods. Xie et al. used the former method after blend processing.[5] Nevertheless, until now most researchers have used sequential polymerizations to obtain gradient polymers.

Following the technique used by Akovali et al., Jasso et al. prepared sheets of slightly crosslinked poly(methyl methacrylate) (PMMA) by photopolymerization; then chloroethyl acrylate monomer (with a small amount of crosslinker) was allowed to diffuse into the host PMMA matrix at 60°C for various periods of time.[3,4] When a specific desired amount of monomer was absorbed, a second photopolymerization was immediately carried out.

To prepare traditional IPNs, thermodynamic equilibrium swelling has to be reached after monomer sorption. The composition profiles of the gradient polymer samples were obtained by chlorine combustion analysis of layers obtained by mechanical sectioning. The average content for the second polymer prepared in this way was determined by gravimetry. This characteristic is important for the final properties of the material. Depending on the chemical and morphological structure of the systems used and on the conditions for diffusion and polymerization, the desired gradient may or may not be reached. Dror et al. prepared a gradient of polyacrylamide by using poly(ether urethane) (PEU) as a matrix.[6] They used solvent-monomer solutions for diffusion to swell the "soft segments" (or "soft" matrix) of the PEU to interlock the polyacrylamide to be formed. They also included the crosslinker as a variable in the respective diffusion solution, taking swelling up to equilibrium.

They determined that even though the crosslinker concentration nonmonotonically influenced the amount of polyacrylamide in the matrix, the presence of solvent limited the gradient to some extent. This was because solvent penetration kinetics was much faster than that of the monomer, limiting monomer diffusion and affecting polymerization rates.

If a semicrystalline polymer is used as a host matrix, along with diffusion time, temperature, and polymerization initiation method, the amount, size, and possibly geometry of the polymer crystals will influence monomer diffusion, which in turn will affect gradient profile.

One slight variation for addition sequential polymerizations is to diffuse the second component while the first polymerization is still running. By doing this, there is some monomer interdiffusion, which at the end will produce some copolymer regions. This method is used for the preparation of optical gradient elements (gradans). Usually, cylindrical rods with radial composition variation are prepared by using this three-stage method.[8]

In general, for all the gradient polymer systems mentioned, diffusion control is critical to achieve desired profiles.

DIFFUSION PHENOMENA

Based on the thermodynamics of irreversible processes, DeGroot indicated that the flow rate of a substance in a multicomponent system depends on the gradients of chemical potential of each component.[9] Frisch found that systems below a "glass-like" transition temperature will possess history-dependent diffusion coefficients and be subject to time-dependent surface concentrations.[10] Rogers noted that the permeability of a polyethylene-graded graft copolymer film was highly dependent on the sample preparation method.[11]

STRUCTURE MORPHOLOGY AND TRANSITIONS

In general, thermodynamic polymer–polymer incompatibility leads to phase separation in polymer blends. When doing sequential polymerization by diffusing the second monomer (or monomers) into a thermoplastic or crosslinked polymer matrix, followed by the second polymer synthesis, separation is expected at the microphase level. It is this microphase separation that gives polymer blends their technological importance. Domain size-distribution, and degree of separation depend specifically on blend preparation.[12] By using transmission electron microscopy (TEM), Elsabee et al. found an increase in domain size when the overall polyacrylamide concentration increased in the gradient IPN.[7]

Jasso et al. have observed by using scanning electron microscopy (SEM) the difference in domain size and dispersion uniformity over the surfaces of polystyrene poly(butyl acrylate) slabs coated with gold for traditional and gradient IPNs.[13] Local free volume differences in the amorphous polystyrene matrix conduct to nonuniform monomer diffusion. Because of rapid *in situ* polymerization to form the gradient composition in the bulk, such differences arise in the product.

PROPERTIES AND APPLICATIONS

A suitable continuous gradient in a composition causing a continuous variation in properties can be of great value in many applications in various fields. Because most of the patents on polymer blends are focused on mechanical properties, studies on those properties are of interest for gradient polymers.[1]

After Akovali et al. presented yielding behavior in samples (expected to have a gradient in composition) with a low content of the rubbery component while retaining a high Young's modulus, Jasso et al. clearly demonstrated the tensile performance superiority of the gradient polymers compared with traditional IPNs of similar average composition.[3,4,14]

Interpretation of the remarkable mechanical performance of gradient polymers was presented by following two different hypothetical mechanisms.[14] Because the gradient polymers can be taken as IPNs with an infinite number of layers of varying composition and because uniform tensile deformation was observed throughout the samples, one can expect that the harder layers must bear greater stresses than the layers richer in elastomer. Such a stress-biased mechanism requires the redistribution of the load among supporting elements, which in turn leads to an increase in yield. An alternative mechanism for high tenacity may be a reduction in surface-layer imperfections as a result of the presence of some regions rich in elastomer, which can support high deformation before cracking. Continuing the work in this chemical system, Martin et al. concluded by microscopic observation of impact fractured samples that the presence of rubbery layers on the surface of the gradients may inhibit crack growth and increase the stress for craze intiation.[15] These results support the second mechanism.

In addition to superior mechanical properties, Dror et al. obtained water absorption by using a hydrogel in their gradient absorption gradient absorption capability.[6] This is important for biomedical applications.

Elaboration of optical gradient elements, stands for application on optical rays spreading along curved trajectories when refractive index is changing in the medium.

Gradient compositions in polymer blends, whether crosslinked or not, may be useful in many fields. However, understanding the behavior remains to be fully clarified. Composition gradient and layer microphase separation level greatly depend on the compatibility of components, which affects properties.

REFERENCES

1. Utracki, L. A. *Polymer Alloys and Blends*; Hanser-Gardner: Munich, 1989.
2. Shen, M.; Bever, M. B. *J. Mater. Sci.* **1972**, *7*, 741.
3. Akovali, G.; Biliyar, K.; Shen, M. *J. Appl. Polym. Sci.* **1976**, *20*, 2419.
4. Jasso, C. F.; Hong, S. D.; Shen, M. *ACS Polym. Prepr.* **1978**, *19(1)*, 63.
5. Xie, X. M.; Matsuoka, M.; Takemura, K. *Polymer* **1992**, *33(9)*, 1996.
6. Dror, M.; Elsabee, M. Z.; Berry, G. C. *J. Appl. Polym. Sci.* **1981**, *26*, 1741.
7. Elsabee, M. Z.; Dror, M.; Berry, G. C. *J. Appl. Polym. Sci.* **1983**, *28*, 2151.
8. Kosjakov, V. I.; Ginzburg, L. I. *Diffusion Phenomena in Polymers*; Chernogolovka: 1985; p 86.
9. DeGroot, S. R. *Thermodynamics of Irreversible Processes*; North-Holland: Amsterdam, 1952.
10. Frisch, H. L. *J. Chem. Phys.* **1964**, *41*(12), 3679.
11. Rogers, C. E. *Polym. Sci., Part C* **1965**, *10*, 93.
12. Jordhamo, G. M.; Manson, J. A.; Sperling, L. H. *Polym. Eng. Sci.* **1986**, *26(8)*, 517.
13. Jasso, C. F.; Castellanos, E.; Sánchez, N.; Laguna, O. Unpublished results.
14. Jasso, C. F.; Hong, S. D.; Shen, M. *Multiphase Polymers; Advances in Chemistry*; American Chemical Society: Washington, DC, 1979; Chapter 23, p 443.
15. Martin, G. C.; Ensanni, E.; Shen, M. *J. Appl. Polym. Sci.* **1981**, *26*, 1465.

GRAFT AND NETWORK COPOLYMERS (by Macromolecular Ion Coupling Reactions)

Yasuyuki Tezuka
Department of Organic and Polymeric Materials
Tokyo Institute of Technology

Tomoo Shiomi
Department of Material Science and Technology
Nagaoka University of Technology

A graft copolymer is a branched macromolecule in which different polymer segments are covalently linked to each other, and relates to a block copolymer, which maintains a linear structure. The term "graft copolymer" refers to a specific polymer molecule whereas "graft (or grafted) polymer" refers to specific polymer material commonly prepared through chemical or physical treatments on the surface of bulk polymer material. Hence, care should be taken to avoid confusing these two terms.

A multicomponent polymer network (MPN) is a macromolecular network comprising different polymer segments, covalently linked. Thus, it is an extension of a graft copolymer into a three-dimensional structure. So the MPN relates to the interpenetrating polymer network (IPN), which is the entanglement of different polymer networks–an extension of a polymer blend to a three-dimensional structure.

PREPARATION OF GRAFT AND NETWORK COPOLYMERS

Graft Copolymers

Graft copolymers are synthesized by three principal techniques: a "grafting from" method, a "grafting onto" method, and a "macromonomer" method.

In the "grafting from" method, reactive species are generated either chemically or physically on the prepolymer main chain, and preferably at specific positions. The reactive groups introduced along the main chain can initiate the polymerization reaction of a specific monomer to produce a graft segment covalently linked to the main chain segment.

In the "grafting onto" method, the reaction between the different prepolymers is involved, namely, one containing reactive groups along the main chain and another containing a single reactive group at the chain end.

The "macromonomer" method is a technique in which a prepolymer having a reactive (polymerizable) end group, that is, a macromonomer, is subjected to the copolymerization reaction with a low molecular weight monomer.[1] This process can provide a graft copolymer with well-defined graft segment components with respect to the chain length and the distribution density along the main chain.

Network Copolymers

Network copolymers are synthesized by two principal techniques; first, a crosslinking reaction through the polymerization of a low molecular weight monomer in the presence of a prepolymer containing multiple polymerizable groups along the main chain or at both chain ends (bifunctional macromonomer). The second is a crosslinking reaction between two different prepolymers, one containing reactive groups along the main chain and another having reactive groups at both chain ends (telechelic polymer).

The former method was traditionally employed in the production of commercial unsaturated polyester resins, in which the polymerization of styrene in the presence of poly or oligoesters containing unsaturated groups in the main chain produced the crosslinked material.[2]

Although the bifunctional macromonomer method is versatile and convenient, the crosslinking reaction proceeds in a stepwise fashion and the reaction will be retarded after the gel point of the system. The loss of the fluidity of the reaction system will significantly suppress the interdiffusion of polymer molecules and may lead to such structural defects as free branch segments in the final network product.

The crosslinking reaction between prepolymers, however, is generally less efficient in producing an MPN because the repulsive interaction between different polymer segments prevents the efficient coupling between the two reactive groups located on different polymer molecules. Therefore, this system is rarely applied for the synthesis of network copolymers, although it can, in principle, provide a well-defined MPN from precharacterized segment components.

GRAFT AND NETWORK COPOLYMERS BY TELECHELICS HAVING CYCLIC ONIUM SALT GROUPS

Telechelic polymers having moderately strained cyclic onium salt groups are efficient and useful prepolymers in various polymer reactions, including the synthesis of graft and network copolymers.[3] We prepared a series of monofunctional telechelic poly(THF)s having cyclic onium salt groups and subjected them to the ion-coupling reaction with poly(styrene-*co*-sodium acrylate) containing defined carboxylate contents up to ca. 5 mol%. During this procedure, the ion-exchange reaction between the two ionic groups located on the separate prepolymers took place by expelling sodium triflate into the surrounding aqueous medium to combine the prepolymers by an ionic interaction. The subsequent ring-opening reaction either at ambient condition or at elevated temperatures, depending on the nature of the cyclic onium salt groups, transformed the bond between the prepolymers from ionic to covalent by the ring-opening reaction through the nucleophilic attack of the carboxylate counter anion on the cyclic onium salt end groups.

A gel permeation chromatographic (GPC) study on the products demonstrated the formation of the ion coupling product, that is, polystyrene-*graft*-poly(THF), by the evident increase in the molecular weights of the coupling product compared to those of the starting prepolymers. In addition, the higher the carboxylate content in the prepolymer (styrene–sodium acrylate copolymer), the higher the molecular weight of the product, indicating the increase in the number of graft segments in the product.

SWELLING PROPERTY OF NETWORK COPOLYMERS

A two-component copolymer network (MPN) is a single network consisting of the two polymer segments and is distinct from the interpenetrating polymer network (IPN), which is formed from the entanglement of two kinds of polymer networks. Hence, the two-component copolymer network may exhibit features characteristic of both block or graft copolymers and of polymer networks, to provide a unique opportunity to realize unconventional gel properties for a wide area of applications.

Swelling behavior is an important fundamental property of polymer networks. In the two-component copolymer network in the swollen state, multiple combinations of different interactions between the different polymer segments and solvent components can operate on each other. Consequently, the two-component MPN is expected to show a unique swelling behavior, particularly in mixed solvents.

REFERENCES

1. Tsukahara, Y. *Macromolecular Design: Concept and Practice*; Polymer Frontiers International: New York, NY, 1994, Chapter 5.
2. Tezuka, Y.; Hoshino, C. *J. Polym. Sci., Part A: Polym. Chem.* **1994,** *32,* 897.
3. Tezuka, Y. *Progr. Polym. Sci.* **1992,** *17,* 471.

Graft Copolymers

See: *Acrylonitrile-Butadiene-Styrene*
Antistatic Agents, Polymeric
Benzocyclobutenes (Crosslinking and Related Reactions)
Block Copolymers (From Macroinitiators)
Carbon Black (Graft Copolymers)
Carbon Whisker (Surface Modification by Grafting)
Cellulose Grafting (Aspects of the Xanthate Method)
Cellulose Grafting (Ionic Xanthate Method)
Cellulose, Xanthate Grafting
Chitin and Chitosan, Graft Copolymers
Compatibilizers (for Blends)
Compatibilizers, Polymeric (Recycling of Multilayer Structures)
Cyclic Imino Ethers (Ring-Opening Polymerization)
Cyclic Imino Ethers (Ring-Opening Polymerization of 1,3-Oxazo Monomers)
Ethylene-Propylene Rubber Modification (Maleic Anhydride and Derivatives)
Ethylene-Vinyl Acetate Copolymer (Chemical Modification, Compatibilizer in Blends)
Graft and Network Copolymers (by Macromolecular Ion Coupling Reactions)
Guar (Graft Copolymerization)
Hydrophilic Polymers (for Friction Reduction)
Hydrophilic Surfaces (Plasma Treatment and Plasma Polymerization)
Inorganic Fibers (Surface Modification by Grafting)
Inorganic Surfaces, Grafting
Ion-Chelating Polymers (Medical Applications)
Ion-Exchange Membranes
Lignin Graft Copolymers
Living Anionic Polymerization
Living Radical Polymerization
Macromonomers (Preparation, Polymerizability, and Applications)
Macromonomers (Synthesis, Polymerization, and Utilization)
Maleic Anhydride Grafted Polypropylene (Solid Phase)
Maleic Anhydride, Grafting
2-Oxazolines, Polymerization
Pervaporation Membranes (for Separating Organic Solvents)
Pervaporation Membranes (Materials)
Poly(N-acylethylenimine)s
Polyester Fabrics (Surface Modification by Low Temperature Plasma)
Polyethylene (Surface Functionalization)
Polyethylene Film Gels (Photografting of Hydrophilic Monomers)

Polyolefin Graft Copolymers (Prepared by Borane Approach)
Polyphosphazene Block Copolymers
Reactive Oligomers (Overview)
Reactive Processing, Thermoplastics
Silk Fibers (Grafting)
Siloxane-Containing Polymers
Starch-Based Plastics (Measurement of Biodegradability)
Starch Graft Copolymers
Stereoregular Polymethacrylate Macromonomer
Stimuli-Responsive Polymers (Thermodynamics in Drug Delivery Technology)
Surface Modification (by Low Energy Electron Beam)
Surfactants, Polymeric (Overview)
Surfactants, Polymeric (Colloid Stabilization)
Tailor-Made Polymers
Vegetal Biomass (3. Polymers, Derivatives, and Composites)
Vinyl Chloride Copolymers
Vinylidene Fluoride-Based Thermoplastics (Homopolymerization and Copolymerization)
Wood Graft Copolymers
Wood Plastic Composites
Wool Grafting

Graphite-Containing Materials

See: *Glassy Carbon*
Graphite Fiber Composites (Electrochemical Processing)
Magnetic Field Processing

GRAPHITE FIBER COMPOSITES (Electrochemical Processing)

Jude O. Iroh
Department of Materials Science and Engineering
University of Cincinnati

Advanced composites consist of a polymeric matrix and fibrous reinforcement combined in such a way that the dispersion of the latter in the matrix is controlled in order to optimize their properties. The properties of advanced composites are a result of the combined properties of the fiber, the matrix, and the interface between the fiber and the matrix. The rigidity and strength of polymers are enhanced by the incorporation of high-strength, high-rigidity, and low-density reinforcing fibers in accordance with the rule of mixture.[1-3]

Advanced composites have excellent specific properties; they are, therefore, the preferred choice in applications in which weight savings are highly desired.

The matrix used for advanced composites is viscoelastic and continuous. It supports the rigid and elastic fibers that carry much of the applied load. The fibers are separated from the matrix by the interface. A wide range of polymer matrices ranging from thermoplastics, thermosets, and semiinterpenetrating resins have been used in composite technology.

The transmission of the applied load through the matrix to the fibers depends on the state of the bonding between the matrix and the fibers. For short-fiber composites, the load applied to the matrix is transferred to the fibers by shear at the interface.[4-7] The stronger the interfacial bonding, the more efficient the transfer of the applied load through the interface to the fibers. Fiber lengths must be greater than the critical length of the fiber in order to create an interfacial area large enough to permit the fiber to carry much of the applied load.

ELECTROCHEMICAL PROCESSING OF COMPOSITES: HISTORICAL BACKGROUND

A variety of techniques are used to infiltrate graphite fiber preforms with polymeric resins. They include solvent infiltration, solution coating, hot-melt infiltration, powder coating, and comingling of the reinforcing fibers with the polymeric matrix extruded as fibers. Some of these techniques are well suited for processing high-melt viscosity thermoplastic matrix composites, whereas the others are used to process reactive thermosetting resins. Although these techniques have been used for composite processing for about two decades, no significant innovation has been made. Some of the problems associated with these traditional resin infiltration techniques include the inability to adequately wet individual graphite fiber filaments contained in a graphite fiber bundle of about 3–12*K*. Traditional resin infiltration techniques are slow, cumbersome, lack proper control and cannot be effectively automated. The practice whereby the volume fractions of the resin and the fibers are determined after the consolidation of the prepreg, and in some cases after the mechanical evaluation of the composite laminate, is inefficient and outdated.

The difficulty in adequately controlling the coating process and coating uniformity add weight to the argument for innovative approaches to resin infiltration. New and novel techniques of prepregging are also needed in order to keep pace with the advancement in the other areas of composite technology. The drawbacks of the traditional resin infiltration techniques can be overcome by applying the matrix directly onto graphite fibers, using either an electropolymerization or electrodeposition technique. The electrochemical technique allows for easy control of resin infiltration via the control and manipulation of processing parameters, such as current density, monomer concentration, pH of the solution, electrolyte concentration, and reaction time.

ELECTRODEPOSITION

In electrodeposition, the resin is applied directly onto the graphite fiber preform from an emulsion of the polymer salt. Electrodeposition involves the migration of charged particles to oppositely charged electrodes under the influence of an applied electric field. It consists of three stages: the transport of the charged particles to the electrode (electrosphores), electrochemical redox reaction at the electrode surface (electrolysis), and the squeezing out of the solvent and precipitant from the film (electroendosmosis).[8] Both cathodic (positively charged polymer ions depositing on the cathode) and anodic (negatively charged polymer ions depositing at the anode) deposition can be performed from aqueous and nonaqueous media.[9] The formation of *in situ* polyimide-graphite fiber composites has been demonstrated in our laboratory.[10] The rate of impregnation of the graphite fiber preforms and the volume fraction of the polyimide resin increases with increasing applied current, carboxylic acid/TEA (triethylamine) ratio, and precipitant/solvent (P/S) ratio.[10,11] Further investigation is actively being pursued.

PROCESSING OF ELECTROPOLYMERIZED COMPOSITES

Electropolymerization has been performed in a **3**-compartment electrochemical cell.[12,13] A preparation of ~0.2 M solution of the comonomers (styrene and maleimide) is prepared in a 1:1 DMAc:dilute sulfuric acid ~0.02 M solution and placed in the middle chamber of the cell. Electropolymerization is started when the cell's on/off button is switched on. The weight gain of the fibers, represented by the difference between the weight of the high-temperature (250°C) dried coated fibers and the weight of the uncoated fibers, is determined as a function of the electropolymerization time, other reaction parameters remaining constant.

CONCLUSIONS

The electrochemical processing of *in situ* polymer matrix–graphite fiber composites is highly attractive because of the capability of fine control, automation, and reduced cost. Graphite fiber reinforcements are well suited for this process because of their excellent electrical and thermal conductivity. Continuous glass-fiber preforms coated with conducting carbon powder can also be used. Formation of *in situ* polyimide matrix-graphite fiber prepregs with a resin content of ~10 $^w/o$ to 90 $^w/o$ has been accomplished by electrodeposition. Graphite fibers preimpregnated with ~35 $^w/o$ to 40 $^w/o$ thermoplastic resins, by electropolymerization, were used in composite fabrication. The electropolymerized matrix–graphite fiber composites have excellent flexural strength ~1800 MPa and impact strength ~230 KJ/m^2 at 40 volume fraction of the matrix. The mechanical performance of the electropolymerized composites compare favorably with the state-of-the-art graphite fiber–epoxy composites. The interlaminar shear strength of ~60 MPa suggests the possibility of further improvement in the shear strength. It is expected that prior surface treatment and functionalization of the graphite fiber surfaces should further enhance the adhesion between the reinforcements and the matrix. Reduction of the numbers of graphite fiber bundle layers in the preform from five to one results in composites with improved mechanical properties.

REFERENCES

1. Agarwal, B. D.; Broutman, L. J. *Analysis and Performance of Fiber Composites*; Wiley-Interscience: New York, 1980; p 211.
2. Hull, D. *An Introduction of Composite Materials* 3rd ed.; Cambridge University: Cambridge, 1978; p 200.
3. Kelly, A.; Tyson, W. R. *High Strength Materials*; Zackay, V. F., Ed.; John Wiley & Sons: New York, 1965; p 578.
4. Rosen, R. W. *Strength of Unixial Fibrous Materials*; Pergamon: New York, 1970.
5. Pitt, W. G.; Lakenan, J. E.; Fogg, D. M.; Strong, A. B. *SAMPE Q* **1991,** *23,* 39.
6. Tyson, W. R.; Davis, G. J. *Brt. J. Appl. Phys.* **1965,** 16.
7. Broutman, L. J.; Agarwal, B. D. *Polym. Eng. Sci.* **1974,** *14,* 581.
8. *Encyclopedia of Polymer Science Engineering,* 2nd ed.; Mark, H. F.; Bikales, N. M.; Overberger, C. G.; Merger, G; Kroschwitz, Eds.; John Wiley & Sons: New York, 1985-1989; Vol. 3, p 645.
9. Scala, L. C.; Alvino, W. M.; Fuller, T. J. In *Polyimide: Synthesis, Characterization and Applications*; Plenum: New York, 1984; pp 1081.
10. Yuan, W.; Iroh, J. O. Electrodeposition of Polyimides onto Graphite Fibers, *J. Appl. Polym. Sci.* **1996,** *59,* 737.
11. Yuan, W. Master of Science Thesis, University of Cincinnati, 1994.
12. Iroh, J. O.; Bell, J. P.; Scola, D. A. *J. Appl. Polym. Sci.* **1990,** *41,* 735.
13. Iroh, J. O. Ph.D. Thesis, University of Connecticut, 1990.

GROUP TRANSFER ALTERNATING COPOLYMERIZATION

Shiro Kobayashi*
Department of Molecular Chemistry and Engineering
Faculty of Engineering
Tohoku University

Jun-ichi Kadokawa
Department of Materials Science and Engineering
Faculty of Engineering
Yamagata University

In the series of the studies on alternating copolymerizations via zwitterion intermediates several monomers having an acidic hydrogen such as acrylic acid, acrylamide, α-keto acids, ethylene sulfonamide, and vinylphosphonic acid monoethyl ester, have been employed in conjunction with cyclic phosphorus (III) compounds.[1-12] The copolymerization using these monomers proceeded to produce 1:1 alternating copolymers by a hydrogen-transfer process.

In addition polymerization termed "group-transfer polymerization (GTP)," a trimethyl silyl group is transferred from an initiator to the incoming monomer or from a monomer to an initiator.[13-26] Here, a novel alternating copolymerization is described, which involves a trimethylsilyl group transfer step, that is, "group-transfer alternating copolymerization (GTAC)." Such a new copolymerization (GTAC) has been reported in the monomer combination of 2-phenyl-1,3,2-dioxaphosphorinane (**1**) and acrylic ester having trimethylsilyl carboxylate or trimethylsilyl sulfonate and proceeded via the zwitterion intermediate.[27,28] Therefore, GTAC belongs to the series of the zwitterionic-type alternating copolymerization. In GTAC, the hydrogen-transfer copolymerization has been extended to the trimethylsilyl group-transfer copolymerization. The reaction of GTAC is based on the nature of the trimethylsilyl group so called "feeble proton" or "super proton."[29]

GTAC USING TRIMETHYLSILYL 3-(ACRYLOYLOXY)PROPIONATE (2) AS COMONOMER[27]

When an equimolar mixture of six-membered cyclic phosphonite **1** and trimethylsilyl 3-(acryloyloxy)propionate (**2**) in a solvent was heated without catalyst, the copolymerization

*Author to whom correspondence should be addressed.

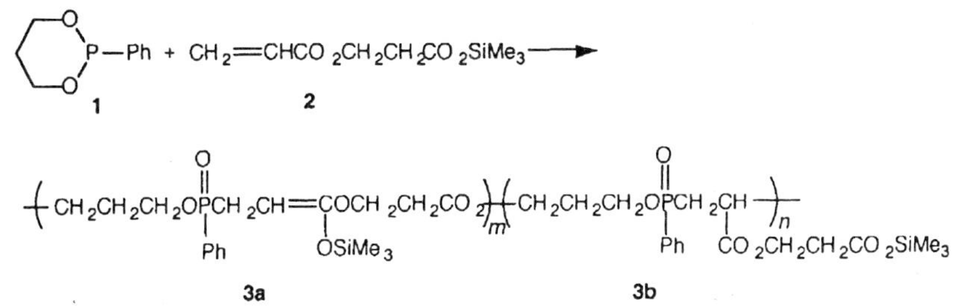

1

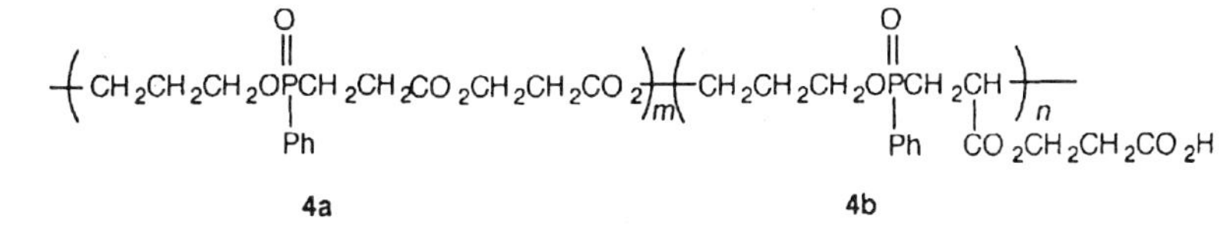

2

occurred to afford alternating copolymer **3**, as shown in **Structure 1**.

The ^{1}H and ^{13}C NMR spectra of the copolymerization mixture before work-up showed that the copolymer consists of two units, **3a** and **3b**, where the former was formed by the trimethylsilyl group-transfer process and the latter was afforded without the group-transfer process.

During the isolation procedures, hydrolysis of copolymer **3** took place; labile trimethylsilyl groups were cleaved off, and units **3a** and **3b** were converted into units **4a** and **4b**, respectively, by hydrolysis with moisture (see **Structure 2**).

REFERENCES

1. Kobayashi, S.; Saegusa, T. *Alternating Copolymers*; Plenum: New York/London, 1985; Chapter V.
2. Saegusa, T.; Kimura, Y.; Ishikawa, N.; Kobayashi, S. *Macromolecules* **1976**, *9*, 724.
3. Saegusa, T.; Yokoyama, T.; Kobayashi, S. *Polym. Bull.* **1978**, *1*, 55.
4. Saegusa, T.; Kobayashi, S.; Furukawa, J. *Macromolecules* **1978**, *11*, 1027.
5. Kobayashi, S.; Huang, M. Y.; Saegusa, T. *Polym. Bull.* **1982**, *6*, 389.
6. Suzuki, M.; Mizuno, K.; Sakaya, T.; Otani, K.; Kobayashi, S.; Saegusa, T. *Polym. Bull.* **1989**, *22*, 435.
7. Kobayashi, S.; Okawa, M.; Saegusa, T. *Macromolecules* **1990**, *23*, 1873.
8. Saegusa, T.; Yokoyama, T.; Kimura, Y.; Kobayashi, S. *Macromolecules* **1977**, *10*, 791.
9. Kobayashi, S.; Yokoyama, T.; Saegusa, T. *Polym. Bull.* **1980**, *3*, 505.
10. Kobayashi, S.; Yokoyama, T.; Kawabe, K.; Saegusa, T. *Polym. Bull.* **1980**, *3*, 585.
11. Kobayashi, S.; Kadokawa, J.; Yen, I. F.; Shoda, S. *Macromolecules* **1989**, *22*, 4390.
12. Kadokawa, J.; Yen, I. F.; Shoda, S.; Uyama, H.; Kobayashi, S. *Polym. J.* **1992**, *24*, 1205.
13. Webster, O. W.; Hertler, W. R.; Sogah, D. Y.; Farnham, W. B.; RajanBabu, T. V. *J. Am. Chem. Soc.* **1983**, 105, 5706.
14. Sogah, D. Y.; Webster, O. W. *J. Polym. Sci., Polym. Lett. Ed.* **1983**, *21*, 927.
15. Hertler, W. R.; Sogah, D. Y.; Webster, O. W.; Trost, B. M. *Macromolecules* **1984**, *17*, 1415.
16. Webster, O. W.; Hertler, W. R.; Sogah, D. Y.; Farnham, W. B.; RajanBabu, T. V. *J. Macromol. Sci-Chem.* **1984**, A21, 943.
17. Sogah, D. Y.; Webster, O. W. *Macromolecules* **1986**, *19*, 1775.
18. Sogah, D. Y.; Hertler, W. R.; Webster, O. W.; Cohen, G. M. *Macromolecules* **1987**, *20*, 1473.
19. Hertler, W. R.; RajanBabu, T. V.; Ovenall, D. W.; Reddy, G. S.; Sogah, D. Y. *J. Am. Chem. Soc.* **1988**, 110, 5841.
20. Hertler, W. R.; Reddy, G. S.; Sogah, D. Y. *J. Org. Chem.* **1988**, *53*, 3532.
21. Hirabayashi, T.; Ito, T.; Yokota, K. *Polym. J.* **1988**, *20*, 1041.
22. Hirabayashi, T.; Kawasaki, T.; Yokota, K. *Polym. J.* **1990**, *22*, 287.
23. Quirk, R. P.; Ren, J. *Macromolecules* **1992**, *25*, 6612.
24. Sumi, H.; Hirabayashi, T.; Inai, Y.; Yokota, K. *Polymer* J. **1992**, *24*, 669.
25. Charleux, B.; Pichot, C. *Polymer* **1993**, *34*, 195.
26. Mori, Y.; Sumi, H.; Hirabayashi, T.; Inai, Y.; Yokota, K. *Macromolecules* **1994**, *27*, 1051.
27. Kobayashi, S.; Kadokawa, J.; Uyama, H. *Macromolecules* **1991**, *24*, 4475.
28. Kobayashi, S.; Kadokawa, J. *Macromol. Chem. Phys.* **1995**, 196, 2113.
29. Fleming, I. *Chem. Soc. Rev.* **1981**, *10*, 83.

GROUP TRANSFER CYCLOPOLYMERIZATION

Sam-Kwon Choi
Department of Chemistry
Korea Advanced Institute of Science and Technology

Yang-Bae Kim
Korea Chemical Company

Weon-Jung Choi
Polymer Laboratory
Inchon Research Center
Yokong Ltd.

Kwon-Taek Lim
Pusan National University of Technology

Cyclopolymerization, established by Butler, is a field that can give ring-fused structure in polymer main chain.[1-5] The discovery of cyclopolymerization helped in broadening of the scope of monomer-to-polymer systems.

However, DuPont announced a new method of a living polymerization called "Group Transfer Polymerization (GTP)" in 1983.[6] (Meth)acrylic monomers can be polymerized by a trimethyl silyl ketene acetal initiator in the presence of nucleophilic or electrophilic catalyst to give the living polymers with controlled molecular weight and molecular weight distribution.[7,8]

Because of the living characteristics of GTP, it is important to extend the monomers capable of GTP in preparing new low polydisperse polymers and/or (meth)acrylate based block copolymers. In this review, the results of group transfer cyclopolymerization are discussed.

PREPARATION AND PROPERTIES

Characteristics of Group Transfer Polymerization

Although group transfer cyclopolymerization was reported only a few years ago, the applications of group transfer polymerization to the cyclopolymer have some advantages compared with other cyclopolymerization techniques such as free radical, cationic, and anionic methods. As far as our knowledge goes, group transfer cyclopolymerization of dimethacrylic derivatives has these advantages: control of ring size, enhanced reactivity to the cyclopolymer at higher monomer concentration, fairly good molecular weight control, and block copolymerization with other methacryl monomers.

Cyclopolymerization is proceeded through an alternating "intermolecular–intramolecular" reaction. The same reaction mechanism can be applied to the group transfer cyclopolymerization. A silyl ketene acetal initiator reacts with a methacryl group of dimethacryl monomer by a Michael reaction. During the reaction, the silyl group transfers to the methacryl group, generating a new silyl ketene acetal. The new silyl ketene acetal reacts with another methacryl group of the same monomer to give a living cyclic silyl ketene acetal. A proposed mechanism of the group transfer cyclopolymerization is represented in **Scheme I.**

It has been known that in numerous cases of free radical or anionic cyclopolymerization of 1,6-dienes, the ring size of the cyclized polymer depends on the orientation of the second vinyl group and stabilization of reaction intermediate at the intramolecular propagation step, head-to-head or head-to-tail.

However, group transfer cyclopolymerization allows only a head-to-tail mechanism although a controversy exists whether cyclization results directly from a hypervalent intermediate to the second methacryl group or an enolate.[11-13] Therefore, group transfer cyclopolymerization can offer a novel method for the facile formation of a cyclic polymer with the control ring size compared with other techniques.

Group Transfer Cyclopolymerization of Dimethacrylic Monomer

The first example of group transfer cyclopolymerization of dimethacryl derivatives with livingness was reported by Choi et al.[9] Bis(2-carbomethoxy allyl)methyl amine (DCMA) can be cyclopolymerized to give cyclicpolymer piperidine moieties.

2,6-Dicarbomethoxy-1,6-heptadiene (DCHD) could also be cyclopolymerized by GTP conditions to give 6-membered cyclopolymerization.[10] This reaction is cyclopolymerization through an alternating "intermolecular–intramolecular" reaction, which is expected to produce recurring cyclohexane units by the fact that GTP allows only a head-to-tail mechanism.

Even at 35 wt% of monomer concentration, cyclopolymerization proceeded without insoluble product obtained by the crosslinking. The polymer yield was quantitative and the polydispersity (D = $\overline{M}w/\overline{M}n$) was estimated to be 1.23–1.28.

In order to investigate the life of this cyclopolymerization, an ABA triblock copolymer was successfully prepared by adding methyl methacrylate as the second monomer to a completely polymerized DCHD solution. This block copolymerization resulted in quantitative yield and polymers with fairly low dispersity ($\overline{M}w/\overline{M}n$ = 1.33–1.37).

APPLICATIONS

Cyclopolymers are known to possess many advantageous properties such as high glass-transition temperature, excellent thermal stability, and light weight in comparison with noncyclic linear polymers. But the use of cyclopolymers is limited to only a few cases because of economic aspects.[14,15] Therefore cyclopolymers will be used in specialty applications where polymers with promising properties are used. Several examples shown in this review suggest that group transfer cyclopolymerization offers a powerful tool for the cyclopolymers with well-defined polymer structure and molecular weight. Because group

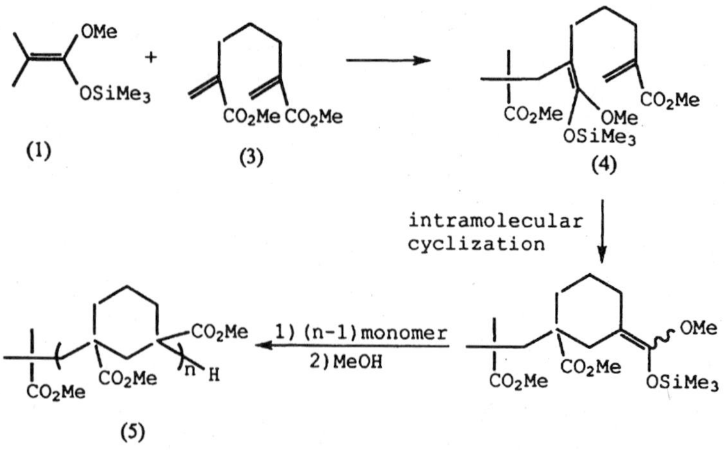

I

transfer cyclopolymerizations proceed with life, there are considerable merits in the glass-transition temperature of the target system by choosing the type and amount of particular monomers. For many industrial applications, continuous studies of group transfer polymerization are still required.

REFERENCES

1. Butler, G. B.; Bunch, R. L. *J. Am. Chem. Soc.* **1949,** *71,* 3210.
2. Butler, G. B.; Ingley, F. L. *J. Am. Chem. Soc.* **1991,** *73,* 895.
3. Butler, G. B.; Goette, R. L. *J. Am. Chem. Soc.* **1952,** *74,* 1339.
4. Butler, G. B. *Encyclopedia of Polymer Science and Engineering,* 2nd ed.; John Wiley & Sons: New York, 1986; Vol. 4.
5. Butler, G. B. *Comprehensive Polymer Science,* 1st ed.; Pergamon: London, UK, 1989; Vol. 4.
6. Webster, O. W.; Hertler, W. R.; Sogah, D. Y.; Farnham, W. B.; Rajanbabu, T. V. *J. Am. Chem. Soc.* **1983,** 105, 5706.
7. Sogah, D. Y.; Hertler, W. R.; Webster, O. W.; Cohen, G. M. *Macromolecules* **1987,** *20,* 1473.
8. Webster, O. W.; Sogah, D. Y. *Comprehensive Polymer Science,* 1st ed.; Pergamon: London, UK, 1989; Vol. 4.
9. Choi, W. J.; Kim, Y. B.; Lim, K. T.; Choi, S. K. *Macromolecules* **1990,** *23,* 5365.
10. Kim, Y. B.; Choi, W. J.; Choi, B. S.; Choi, S. K. *Macromolecules* **1991,** *24,* 5006.
11. Sogah, D. Y.; Farnham, W. B. *Organosilicon and Bioorganosilicon Chemistry: Structure, Bonding, Reactivity, and Synthetic Applications*; Sakurai, H. Ed.; John Wiley & Sons: New York, 1985; Chapter 20.
12. Quirk, R. P.; Bidinger, G. P. *Polym. Bull.* **1989,** *22,* 63.
13. Quirk, R. P.; Ren, J. *Macromolecules* **1992,** *25,* 6612.
14. Butler, G. B. In Goethals, E. J. Ed., *Polymeric Amines and Ammonium Salts*; Pergamon: Elmsford, New York, 1980; pp 125-142.
15. Butler, G. B. In Donaruma, L. G.; Ottenbrite, R. M.; Vogl, O., Eds.; *Ammonium Polymeric Drugs*; John Wiley & Sons: New York, 1980; pp 49-141.

Group Transfer Polymerization

See: *Block Copolymers (By Changing Polymerization Mechanism)*
Group Transfer Alternating Copolymerization
Group Transfer Cyclopolymerization
Group Transfer Polymerization (Overview)

GROUP TRANSFER POLYMERIZATION (Overview)

Walter R. Hertler, retired
Experimental Station
E. I. DuPont De Nemours and Company

In 1983 Webster[1,2] reported the successful application of the silicon-mediated Michael addition reaction of Mukiyama,[3] to polymer synthesis providing, for the first time, a practical process to take advantage of the benefits of living anionic polymerization of acrylics–the ability to control polymer architecture, number average molecular weight ($\overline{M}_n$), and molecular weight distribution (MWD)–but at the reflux temperatures of polymerization solvents, rather than at the low temperatures usually required by anionic polymerization processes (**Equation 1**). This silicon-mediated polymerization of acrylic monomers is termed group transfer polymerization (GPT). The ability to prepare polymers by GTP at elevated temperatures opened up the possibility for practical commercial-scale synthesis of acrylic polymers with controlled architecture. GTP is referred to as a "living polymerization" because, through manipulation of reaction conditions, rates of transfer and termination can be made diminishingly small relative to the rate of propagation.[4,5] In GTP a silyl ketene acetal (**1**) undergoes a nucleophilic anion-catalyzed Michael-type addition to an α,β-unsaturated ester, such as methyl methacrylate (MMA) to give an adduct (**2**) which is also a silyl ketene acetal. In the presence of additional MMA, (**2**) undergoes repeated additions of monomer until the MMA is depleted, and the living polymer (**3**) results. At this point, addition of a different methacrylate monomer would lead to formation of a block copolymer. To deactivate the living polymer, a proton source, such as water or alcohol R 'OH, is added to give the polymethacrylate (**5**) and a trimethylsilyl ether (**4**). Since the silyl ether (**4**) is usually volatile, the polymer can readily be obtained free of impurities such as the alkali metal salts which accompany the polymers prepared by classical anionic polymerization. The only contaminant in a polymer prepared by anion-catalyzed GTP is the catalyst, which generally is present at a level of only about 1 mol% of the silyl ketene acetal initiator.

GTP is common with other living polymerization processes, is capable of preparing polymethacrylates with very narrow MWD and with DP (degree of polymerization) equal to the

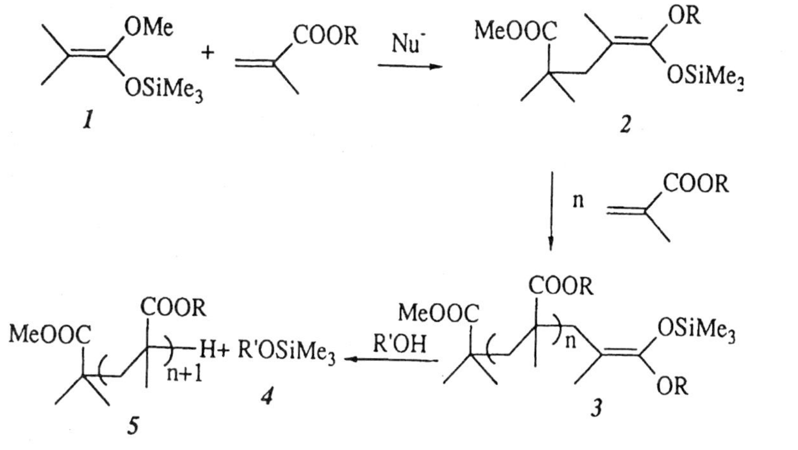

1

molar ratio of monomer to initiator (**1**). A principal use of GTP is the preparation of block copolymers, for which livingness is critical. Müller et al. provided an analysis of the limitations of GTP in synthesis of block copolymers.[6] Although GTP is not limited to methacrylates, this class of monomer gives the best control of $\overline{M}_n$ and narrowest MWD.

Onium salts of nucleophilic anions, such as sulfonium bifluorides, quaternary ammonium carboxylates,[7-9] and quaternary ammonium cyanide are the catalysts most commonly used for GTP, but Lewis acids,[10] Lewis bases,[11,12] Zeolites,[13] mercuric iodide,[14-17] and high pressure (without catalyst)[18,19] have also been used. It is important to note that silyl ketene acetals, including the living polymer (**3**), are stable over long periods of time in the absence of catalyst. But in the presence of catalyst they are highly reactive, and in the absence of monomer undergo a variety of decomposition reactions.[20-22] Replacement of the silicon of silyl ketene acetals with germanium, tin,[23] titanium,[24,25] hafnium,[24] zirconium,[24,26] and lanthanides, notably samarium,[27-29] has been shown to support GTP-like processes for polymerization of acrylic monomers.

There are several general reviews of GTP,[12,30-34] reviews with a particular emphasis on the mechanism of GTP,[35,36] and a review of polymer architecture with emphasis on block polymers.[37]

SCOPE

Monomers

The class of monomers most successfully polymerized by anion-catalyzed GTP is methacrylates. Acrylates are best polymerized by mercuric iodide-catalyzed GTP. A variety of other kinds of monomers have been polymerized by GTP, for example, *N,N*-dialkylacrylamides,[23,38] acrylonitrile,[14,23,39-41] methacrylonitrile,[42,43] α-methylenebutyrolacetones,[23,44,45] *N*-substituted maleimides,[46] and alkyl pentadienoates.[44,46]

Monomers containing functional groups are of particular importance in synthesis of functional polymers. Some functional groups cannot survive GTP without causing termination, while others are compatible with GTP. Functions with active hydrogen atoms, such as carboxylic acids, alcohols, and phenols must first be protected before they can be used in GTP. Aldehydes, ketones, nitro compounds, primary and secondary amides are not compatible with GTP.

APPLICATIONS

The practicality of GTP for the large-scale synthesis of acrylic polymers with controlled structure is attested by the successful scale up of the process by the DuPont Company to commercial reactors capable of producing several million pounds of block polymers per year. The accessibility of polymers prepared by GTP has led to their evaluation in numerous applications. Well-controlled block copolymers have been prepared for use as pigment dispersants for finishes,[48-51] and inks.[52-56] Block polymers prepared by GTP have also been used to stabilize dispersions of poly-(pyrrole).[57]

In the area of coatings, polymers with controlled structures prepared by GTP have found various applications. These polymers offer advantages over conventional resins in the application and appearance of coatings and in final film properties. Some of these advantages are high solids, improvements in the hardness–flexibility balance, control of the rheology and application of coatings, and better pigment dispersions.[58]

The use of GTP for the synthesis of acid-labile acrylic polymers with narrow MWD has been applied in photoresists[59,60] and in color proofing materials.[61-65]

REFERENCES

1. Webster, O. W. (DuPont) U.S. Patent 4 417 034, 1983; *Chem. Abstract* **1984,** *100,* 86327e.
2. Webster, O. W.; Hertler, W. R.; Sogah, D. Y. et al. *J. Am. Chem. Soc.* **1983,** *105,* 5706.
3. Narasaka, K.; Saigo, K.; Aikawa, Y. et al. *Bull. Chem. Soc. Jpn.* **1976,** *49,* 779.
4. Matyjaszewski, K. *Proc. Am. Chem. Soc. Div. Polym. Mater., Sci. Eng.* **1993,** *68,* 58.
5. Matyjaszewski, K. *Macromolecules* **1993,** *26,* 1787.
6. Muller, M. A.; Augenstein, M.; Dumont, E. et al. *New Polym. Mater.* **1991,** *2,* 315.
7. Dicker, I. B.; Cohen, G. M.; Farnham, W. B. et al. *Polym. Prepr., Am. Chem. Soc. Div. Polym. Chem.* **1987,** *28(1),* 106.
8. Dicker, I. B.; Cohen, G. M.; Farnham, W. B. et al. *Macromolecules* **1990,** *23,* 4034.
9. Dicker, I. B. et al. (DuPont) U.S. Patent 4 588 795, 1986; *Chem. Abstract* **1986,** *105,* 98135.
10. Hertler, W. R.; Sogah, D. Y.; Webster, O. W. et al. *Macromolecules* **1984,** *17,* 1415.
11. Hertler, W. R. (DuPont) U.S. Patent 4 605 716, 1986; *Chem. Abstract* **1986,** *105,* 227555v.
12. Hertler, W. R. In *Silicon in Polymer Synthesis* Kricheldorf, Ed.; Springer-Verlag: Heidelberg, Germany, in press.
13. Corbin, D. R.; Sormani, P. M. E. (DuPont) U.S. Patent 5 162 467, 1992; *Chem. Abstract* **1993,** *118,* 39608.
14. Dicker, I. B. (DuPont) U.S. Patent 4 732 955, 1988; *Chem. Abstract* **1988,** *109,* 38426a.
15. Dicker, I. B. *Polym. Prepr. Am. Chem. Soc. Div. Polym. Chem.* **1988,** *29(2),* 114.
16. Dicker, I. B. (DuPont) U.S. Patent 4 866 145, 1989.
17. Muller, A. H. E. *Macromolecules* **1994,** *27,* 1685.
18. Sogah, D. Y. U.S. Patent 4 957 973, 1990; *Chem. Abstract* **1991,** 114, 43757S.
19. Sogah, D. Y.; Hertler, W. R.; Dicker, I. B. et al. *Makromol. Chem., Macromol. Symp.* **1990,** *32,* 75.
20. Martin, D. T.; Bywater, S. *Makromol. Chem.* **1992,** 193, 1011.
21. Schmalbrock, U.; Bandermann, F. *Makromol. Chem.* **1993,** *194,* 2543.
22. Steinbrecht, K.; Bandermann, F. *Makromol. Chem.* **1989,** *190,* 2183.
23. Sogah, D. Y.; Hertler, W. R.; Webster, O. W. et al. *Macromolecules* **1987,** *20,* 1473.
24. Farnham, W. B.; Hertler, W. R. (DuPont) U.S. Patent 4 728 706, 1988; *Chem. Abstract* **1988,** *109,* 55438y.
25. Jenkins, A. D. *Polym. Bull.* **1988,** *20,* 243.
26. Collins, S.; Ward, D. G. *J. Am. Chem. Soc.* **1992,** *114,* 5460.
27. Yasuda, H.; Ihara, E.; Morimoto, M. et al. *Polym. Prepr., Am. Chem. Soc. Div. Polym. Chem.* **1994,** *35(2),* 532.
28. Yasuda, H.; Yamamoto, H. et al. *Macromolecules* **1993,** *26,* 7134.
29. Yasuda, H.; Yamamoto, H.; Yokota, K. et al. *J. Am. Chem. Soc.* **1992,** *114,* 4908.
30. Eastmond, G. C.; Webster, O. W. In *New Methods of Polymer Synthesis*; Ebdon, J. R., Ed.; Blackie and Son: Glasgow, Scotland, 1991; p 22.

31. Webster, O. W. In *Encyclopedia of Polymer Science and Engineering*; Kroschwitz, J. I., Ed.; Wiley-Interscience: New York, NY, 1987; p 580.
32. Webster, O. W. *Makromol. Chem., Macromol. Symp.* **1990,** *33,* 133.
33. Webster, O. W.; Anderson, B. C. In *New Methods for Polymer Synthesis*; Mijs,W. J., Ed.; Plenum: New York, NY, 1992; p 1.
34. Webster, O. W.; Sogah, D. Y. In *Recent Advances in Mechanistic and Synthetic Aspects of Polymerization*; Fontanille, M.; Guyot, A., Eds.; Reidel: Dordrecht, The Netherlands, 1987; p 3.
35. Brittain, W. *J. Rubber Chem. Technol.* **1992,** *65,* 580.
36. Bywater, S. *Makromol. Chem., Macromol. Symp.* **1993,** *67,* 339.
37. Hertler, W. R. In *Macromolecular Design of Polymeric Materials*; Hatada, K. et al., Eds.; Marcel Dekker: New York, NY, in press.
38. Eggert, M.; Freitag, R. *J. Polym. Sci., Part A., Polym. Chem.* **1990,** *32,* 803.
39. Hertler, W. R.; Sogah, D. Y.; Beottcher, F. P. *Macromolecules* **1990,** *23,* 1264.
40. Pan, R.; Xia, H. *Xiamen Daxue Xuebao, Ziran Kexueban* **1987,** *26(4),* 474.
41. Xia, H.; Zou, Y.; Pan, R. *Gaofenzi Xuebao* **1991,** *2,* 225.
42. Bandermann, F.; Speikamp, H. D. *Makromol. Chem., Rapid. Commun.* **1985,** *6,* 335.
43. Catagil, H.; Jenkins, A. D. *Eur. Polym. J.* **1991,** *27,* 651.
44. Hertler, W. R.; RajanBabu, T. V. et al. *J. Am. Chem. Soc.* **1988,** 110, 5841.
45. Suenaga, J.; Sutherlin, D. M.; Stille, J. K. *Macromolecules* **1984,** *17,* 2913.
46. Saito, A.; Tirrell, D. A. *Polym. J.* **1994,** *26,* 169.
47. Laschewsky, A.; Ward, M. D. *Polymer* **1991,** *32,* 146.
48. Harper, L. R.; McDonnell, G. S. (DuPont) U.S. Patent 5 206 295, 1993; *Chem. Abstract* **1993,** *119,* 119612.
49. Hutchins, C. S.; Shor, A. C. (DuPont) U.S. Patent 4 656 226, 1987; *Chem. Abstract* **1987,** *107* 23910p.
50. Webster, O. W. *Makromol. Chem., Macromol. Symp.* **1992,** *53,* 307.
51. West, M. J. W. (DuPont) U.S. Patent 4 755 563, 1988; *Chem. Abstract* **1989,** 110, 39568f.
52. Dicker, I. B. et al. (DuPont) U.S. Patent 5 219 945, 1993; *Chem. Abstract* **1993,** 119, 250807b.
53. Ma, S. H. et al. (DuPont) Eur. Patent Appl. 556649, 1993; *Chem. Abstract* **1994,** 120, 79539k.
54. Ma, S. H. et al. (DuPont) Eur. Patent Appl. 556650, 1993; *Chem. Abstract* **1994,** 120, 33116s.
55. Ma, S. H. et al. (DuPont) U.S. Patent 5 272 201, 1993.
56. Ma, S. H. et al. (DuPont) U.S. Patent 5 085 698, 1992; *Chem. Abstract* **1992,** 116, 237559r.
57. Beadle, P. M.; Rowan, L. et al. *Polymer* **1993,** *34,* 1561.
58. Simms, J. A.; Spinelli, H. J. *J. Coatings Technol.* **1987,** 589, 125.
59. Hertler, W. R.; Sogah, D. Y.; Taylor, G. N. (DuPont and AT&T) U.S. Patent 5 206 317, 1993.
60. Taylor, G. N.; Stillwagon, L. E. et al. *Chem. Mater.* **1991,** *3,* 1031.
61. Chen, G. Y. Y. T. et al. (DuPont) U.S. Patent 5 071 731, 1991.
62. Chen, G. Y. Y. T. et al. (DuPont) U.S. Patent 5 093 221, 1992.
63. Hertler, W. R. et al. (DuPont) U.S. Patent 5 229 244, 1993; *Chem. Abstract* **1994,** 120, 257549y.
64. Raymond, F. A.; Hertler, W. R. *J. Imaging Sci. Technol.* **1992,** *36,* 243.
65. Simmons, H. E., III.; Hertler, W. R.; Sauer, B. B. *J. Appl. Polym. Sci.* **1994,** *52,* 727.

Growth Factors

See: Cell Growth-Promoting Biomaterials

GUAR (Graft Copolymerization)

U. D. N. Bajpai
Polymer Research Laboratory
Department of Post Graduate Studies and Research in Chemistry
Rani Durgavati University

Guar is an edible carbohydrate polymer.[1] It is found in the seeds of leguminous plants like *Cyanmposis tetragonolobous* and *psoraloids.* Guar is a galactomannan that consists of a straight chain of mannose units joined by β-D (1–4) linkages having α-D galactopyranose units attached to this linear chain by (1–6) linkages.[2-4] In pure polysaccharide guar, the ratio of D-galactose to D-mannose units is 1:2. The molecular weight of guar is approximately 2.2×10^5.[5,6] The structure of guar is shown in **Figure 1**. Guar gum is produced from guar seeds.

Guar gum is a cold water swelling polymer. In its powdered commercial form, the rate of thickening and the final viscosity reflect the process history of the product. Guar gum is one of the most highly efficient water thickening agents. Very low concentration of guar gum when dissolved in water forms highly viscous solutions. The viscosity of the guar gum solution can be measured with rotational shear-type viscometers. The viscosity of fully hydrated 1% guar gum solution varies almost directly with the temperature (20-80°C). Guar gum is stable over a wide pH range because of its non-ionic nature. Hydrated guar gum converts into gel by crosslinking.[7] Guar gum is used as a viscosity builder and water binder in many industries: explosives,[8-13,58,59] foods,[14,15] cosmetics and pharmaceuticals,[16,17] mining,[18,19] pulp and paper,[20] tobacco,[21] and textiles.[22]

Although the viscosity of aqueous solution of guar gum is advantageous, it is difficult to control because of its quick biodegradation. Polysaccharides like guar gum are very susceptible to biodegradation.[23-25] Huang has observed that within a few hours of solution preparation, biodegradation starts by the action of micro-organisms or enzymes.[26] The hydrolyzable linkages along the main chain in guar gum are hydrolyzed by an attack of enzymes, followed by oxidation. This limits its applications. This drawback can be avoided either by adding suitable preservatives and biocides or by taking measures which check or inhibit microbiol growth. Guar gum can be made less susceptible to biodegradation by derivatization of its hydroxyl

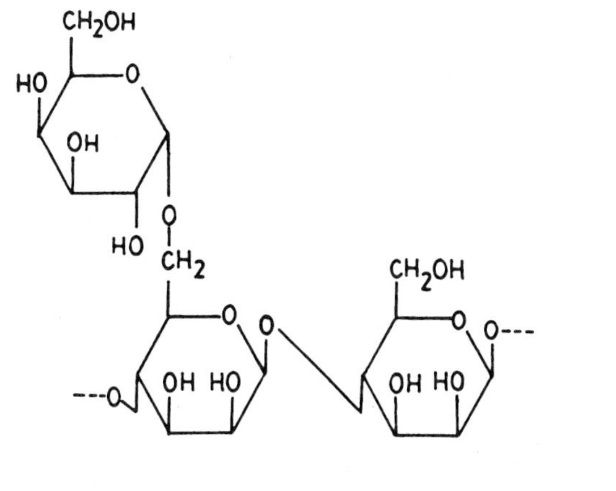

FIGURE 1. Structure of guar gum.

groups and graft copolymerization with vinyl polymers. Vinyl polymers are resistant to biodegradation since they do not contain hydrolyzable linkages along the chain.

Modifications of guar gum by derivatization include carboxymethylation,[27,28] ethylation,[29] oxidation,[30] phosphation[31] and sulfation.[32] These modified guar polymers have better properties and applicability than unmodified or natural guar gum. Sufficient literature is available on the modification of guar gum through derivitization but available information about the modification of guar gum by graft copolymerization is rare.[33-36]

Chemical modification of guar gum through graft copolymerization of vinyl monomers onto it is useful for altering chemical and physical properties to give this natural polymer new applications. Vinyl graft copolymerization through chemical initiation is an easier process than thermal, photochemical, and other methods of initiation because activation energy for the redox initiation is comparatively low.[37-41]

PREPARATION

Guar-grafted polymers have been prepared mostly by free-radical-initiated graft copolymerization.[37,42] A free radical is generated onto guar gum which reacts with the monomer to form a graft copolymer.

Mino and Kaizerman suggested the generation of free radical onto polysaccharide backbone by its interaction with Ce (IV) salts.[43] Deshmukh and Singh prepared guar-grafted polyacrylamide using ceric ammonium nitrate in nitric acid solution.[25] Deshmukh and Singh did not discuss the mechanism of graft copolymerization,[25] but analgous to the interaction of Ce (IV) with starch, I suggest that guar gum first forms a complex with Ce (IV), which breaks into guar gum macroradicals. These macroradicals then initiate polymerization of acrylamide yielding guar-grafted polyacrylamide.

Chain transfer reactions have also been employed to generate free radicals onto guar gum. A variety of oxidation–reduction pairs which generate transient free radicals by single electron transfer reaction have been used frequently. These free radicals then abstract hydrogen atoms from guar gum. Free radicals generated by redox reaction instead of by hydrogen abstraction reaction may also initiate the homopolymerization of vinyl monomer to be grafted. Formation of a significant amount of homopolymer along with graft copolymer suggests that H-abstraction reaction and vinyl polymerization occur simultaneously. Graft copolymerization of guar gum with a hydrogen peroxide–oxygen system have also been studied.

Raval et al. obtained guar-grafted with poly(methyl methacrylate) (PMMA) by the graft copolymerization of methyl methacrylate (MMA) onto guar gum in an aqueous slurry with H_2O_2 as initiator.[44] The same group of authors synthesized a series of guar-grafted polymers by the graft copolymerization of various vinyl monomers such as acrylonitrile, MMA, acrylamide, styrene, methacrylic acid, and acrylic acid onto guar gum in aqueous slurry using H_2O_2 as initiator.[45] The reaction products were characterized by grafting parameters such as percent grafting, grafting frequency, and grafting efficiency.

My co-workers and I modified guar gum by the graft copolymerization of acrylamide onto guar gum using a potassium permanganate–oxalic acid redox initiating system in nitrogen atmosphere.[46]

Modification of guar gum by grafting onto it water-soluble polymers using low cost initiating systems has always been of prime importance. Guar gum grafted with polyacrylamide has better properties like lower susceptibility to biodegradation. Usually atmospheric oxygen plays an inhibiting role in radical polymerization but in a few systems it plays a cocatalytic role. Graft polymerization of acrylamide onto guar gum has been carried out with a potassium bromate–thiomalic acid redox system which works well in the presence of oxygen. This avoids the use of pure inert gases to remove dissolved oxygen. This system was very effective for the homopolymerization of acrylamide[40] and methyl methacrylate.[47]

Guar grafted with polyacrylonitrile by the graft copolymerization of acrylonitrile guar gum using persulfate and ascorbic acid initiating system has also been synthesized in my laboratory.[48] We carried out this preparation in the presence of oxygen. Saline and water retention capacities of these guar-modified polymers reduced with increase in percent add-on.

APPLICATIONS

Singh et al. have studied turbulent drag reduction by guar-grafted copolymers in detail.[24] Graft copolymers are fairly shear-stable in drag reduction.[49,50] Because it is shear-stable, guar gum alone can be an effective drag reducer but its fast biodegradation restricts its application. Guar grafted with polyacrylamide is a good shear-stable drag reducing agent.

Guar-grafted with polyacrylamide having long grafted chains was found to be more shear-stable than the small-branch graft copolymers. However, pure guar gum was more shear-stable than the grafted copolymers because of the incorporation of polyacrylamide. No graft copolymer solution showed any loss of viscosity up to ten days.

Guar-grafted polyacrylamides also act as very effective flocculating agents for industrial effluents containing metallic and nonmetallic contaminants.[51] Guar gum and its modified polymers have also been used for the removal of toxic metals from water, process water, and waste water in the form of biomass or chemical extractants immobilized in porous insoluble polysulfone beads. The metals are removed from the beads using dilute mineral acids.[52]

Guar gum and its modified polymers, along with inorganic or organic powdery fertilizer with a soluble polymer, gelation agent, and inorganic additives (like acidic clay) have been used in the manufacture of powdery fertilizer that produces a stable slurry or paste with water.[53]

Water-soluble guar-modified polymers with water absorbing resin having uronic acid or its salt have been used by the shield process in the preparation of drilling-mud composition for use in construction work. This composition is a good water thickener and has high resistance to salt and alkali.[54]

Lokhande et al. synthesized a new guar gum-based superabsorbant using gamma radiation as a soil additive. The evaluation of this new polymer as an agent to enhance the water-holding capacity of soil is promising. The new product also improves drainage. The material has shown a promising application: to conserve the available water in the soil and then release it slowly to the surrounding soil.

Hydrophilic–hydrophobic polygalactomannan-bearing amino groups have been used as thickeners for aqueous compositions.

Mixtures of modified guar gum and acrylic polymers can be used in the printing of polyester and cotton–polyester fabrics with disperse dyes. Modified guar gum alone gives good results.[55]

A physical mixture of crosslinked, water-like sodium salt of acrylic acid-methylene bisacrylamide and their modified polymers with high absorption capacities for water and body fluids are being used as swelling agents and absorbants in hygienic goods and soil amendments.

The membranes of guar gum and its modified polymers along with other ingredients have also been used as synthetic skin substitutes.[56]

Guar-grafted polymers have been used in improving the tensile and internal strength of paper while retaining reasonable tear strength.[57]

REFERENCES

1. Whistler, R. L. *The Encyclopedia of Polymer Science and Technology*; John Wiley & Sons: New York, NY, 1969; Vol. 11, p 416.
2. Hirst, E. L.; Jones, J. K. N. *J. Chem. Soc.* **1948,** 1278.
3. Whistler, R. L.; Durso, D. F. *J. Am. Chem. Soc.* **1951,** *73,* 4189.
4. Whistler, R. L.; Durso, D. F. *J. Am. Chem. Soc.* **1952,** *74,* 5140.
5. Boggs, A. D.; M.S. Thesis, Perdue University, Lafayette, Indiana, 1949.
6. Hoyt, J. W. *J. Polym. Sci., Part B* **1966,** *4,* 713.
7. Chrisp, J. D. U.S. Patent 3 301 723, 1967; CA 67, 92485w, 1967.
8. Taylor, J. W. U.S. Patent 2 654 666, 1953.
9. McIrvin, J. U.S. Patent 3 108 917, 1963.
10. Clarlevado, H. U.S. Patent 3 129 126, 1964; 3 214 307, 1965.
11. Logan, J. S.; Zalowsky, J. A. U.S. Patent 3 235 423, 1966.
12. Ferguson, J. D. U.S. Patent 3 235 423, 1966.
13. Lyerly, W. U.S. Patent 3 355 336, 1967.
14. Goldstein, A. M.; Alter, E. N. *Industrial Gums*; Academic: New York, NY, 1959; p 321.
15. Werbin, S. J. U.S. Patent 2 502 397, 1950.
16. Hutchins, H. H.; Singiser, R. E. *J. Am Pharm. Asst. Pract. Pharm. Ed.* **1955,** *16,* 226.
17. Eartherton, L. E.; Platz, P. E.; Cosgrove, F. P. *Drug Stand,* **1955,** *23,* 42.
18. Atwood, G. E.; Bourne, D. J. *Mining Eng.* **1953,** *5,* 1099.
19. Atwood, G. E.; Bourne, D. J. U.S. Patent 2 696 912, 1954.
20. Swanson, J. J. *Tappi* **1956,** *39,* 257.
21. Samfield, M. M.; Block, B.A.; Locklair, E. E. U.S. Patent 2 708 175, 1955.
22. Prakash, A.; Kapoor, R. C. *Cell. Chem. Tech.* **1984,** *18,* 207.
23. Kenis, P. R.; Hoyt, J. W. *Friction Reduction by Algal and Bacterial Polymers* Naval Undersea Research and Development Centre, San Diego, USA Report No. NUC TP **1971,** 204.
24. Deshmukh, S. R.; Chaturvedi, P. N.; Singh, R. P. *J. Appl. Polym. Sci.* **1985,** *30,* 4013.
25. Deshmukh, S. R.; Singh, R. P. *J. Appl. Polym. Sci.* **1987,** *33,* 1963.
26. Huang, S. J. In *Encyclopedia of Polymer Science Eng.,* 2nd ed.; Mark, H. F. et al., Eds.; Wiley-Interscience: New York, NY, 1985; pp 220-243.
27. Gulrajani, M.; Choudhry, L.; Roy, A. K. *Colourage* **1981,** 28(10), 3.
28. Boonstra, J. D.; Bekker, A. Ger. Patent 244 012, 1975.
29. Nordgren, R. U.S. Patent 3 303 184, 1967.
30. Keen, J. L.; Ward, W. J.; Swanson, R. R.; Dunning, H. N. U.S. Patent 3 239 500, 1966.
31. Tiefenthaler, H. O.; Wyass, U. Eur. Patent 30 4443, 1981.
32. Kenneth, G. B. *Am. Chem. Soc. Symp. Ser.* **1978,** *77,* 148.
33. Cottrell, I. W.; Alan, E. R.; Racciato, J. S. Ger. Patent 2 756 870, 1978.
34. Nordgren, R. U.S. Patent 3 22 028, 1966.
35. Wheeler, D. H.; Keen, J. L. U.S. Patent 2 321 559, 1966.
36. Japan Patent 80 165 909, 1981.
37. Misra, G. S.; Bajpai, U. D. N. *Prog. Polym. Sci.* **1982,** *8,* 61.
38. Bajpai, U. D. N.; Bajpai, A. K. *J. Macromol. Sci. Chem.* **1983,** *A19,* 487.
39. Bajpai, U. D. N.; Bajpai, A. K. *J. Polym. Sci., Polym. Chem. Ed.* **1984,** *22,* 1803.
40. Bajpai, U. D. N.; Ahi, A. *Rev. Roumaine. Chim.* **1984,** *29,* 605.
41. Bajpai, U. D. N.; Bajpai, A. K. *Macromolecules* **1985,** *18,* 2113.
42. Bhattacharya, S. N.; Maldas, D. *Progr. Polym. Sci.* **1984,** *10,* 171.
43. Mino, G.; Kaizerman, S. *J. Polym. Sci.* **1958,** *31,* 242.
44. Raval, D. K.; Patel, R. G.; Patel, V. S. *Starch* **1988,** *40,* 66.
45. Raval, D. K.; Patel, R. G.; Patel, V. S. *J. Appl. Polym. Sci.* **1991,** *35(8),* 2201.
46. Bajpai, U. D. N.; Rai, S. *J. Appl. Polym. Sci.* **1988,** *35(5),* 1169.
47. Shukla, J. S.; Shukla, S. K.; Tiwari, R. K.; Sharma, G. K. *J. Macromol. Sci. Chem.* **1983,** A20, 13.
48. Parashar, P.; Datt, S. C.; Jain, A.; Bajpai, U. D. N. *Polym. Test.* **1992,** *11(3),* 185.
49. Little, R. C.; Hauseh, R. J.; Hunston, D. L. et al. *Ind. Eng. Chem. Fund.* **1975,** *14,* 283.
50. Singh, R. P.; Majumdar, S.; Reddy, G. V. *Preprints IUPAC-MACRO '82* University of Massachusetts: Amherst, 1982.
51. Singh, R. P.; Jam, S. K.; Lan, N. *Polymer Science: Contemporary Themes*; Sivaram, S., Ed.; Tata-McGraw-Hill: New Delhi, India, 1991; p 716.
52. Seidel, D.; Jeffers, T. U.S. Patent 429 326, 1990; CA 113(8), 64740a, 1994.
53. Aibe, T.; Onodera, Y. Japan Patent 05105570 A2, 1993.
54. Aoyama, K.; Adachi, T.; Mita, K. et al. Japan Patent 05098249 A2 930420 Heisei, 1993.
55. Teli, M. D.; Ramani, V. Y. *Am. Dyest. Rep.* **1992,** *81(5),* 55.
56. Nagia, A.; Hung, C. T. Eur. Patent 489206, 1992.
57. Seiko Chemical Industry Co. Ltd., Jpn. Patent 02026994, 1990; CA 112(24): 219089a, 1994.
58. Cook, M. A.; Farnam, H. U.S. Patent 2 930 685, 1960.
59. Barnhart, R.; Sawyer, F. U.S. Patent 3 072 509, 1963.

Guar Gum

See: Guar (Graft Copolymerization)
Gums (Overview)

Gum Arabic

See: Gums (Overview)

Gums

See: Food Polymers
Guar (Graft Copolymerization)
Gums (Overview)
Polysaccharide-Polysaccharide Interactions
Welan, Microbial Polysaccharide

Xanthan, Depolymerization
Xanthan Gum (Overview)
Xanthan Gum (Conformation and Solution Properties)

GUMS (Overview)

James N. BeMiller
Purdue University

Gums are polymers that produce viscous solutions or gels in a solvent or swelling agent. As such, the term gum can be applied to a variety of polymers that produce viscous, gummy, sticky, or slimy molecular dispersions or gels. Most are specific chemical substances. Some, especially the ancient plant gums like balsam and gum rosin, contain several components.

Because the term gum can be applied to a variety of substances with gummy, sticky, viscous, or mucilaginous characteristics, it cannot be defined precisely. As employed in industry, however, the term most often refers to hydrophilic polymers that cause water to thicken or gel at low concentrations. More specifically, the term gum most often refers to water-soluble polysaccharides and derivatized polysaccharides. When used in foods, they are sometimes referred to as hydrocolloids. (Gelatins are also classified as hydrocolloids.)

Since polysaccharides are abundant, come from renewable sources, are relatively inexpensive, are safe (nontoxic), and are amenable to both chemical and biochemical modification, it is not surprising that they find widespread and extensive use. Industrial gums, that is the water-soluble or dispersible polysaccharides of commerce, are most often used because of their ability to thicken or gel aqueous systems at low concentration, but these general behaviors are manifested in a variety of ways because polysaccharides have a variety of chemical and three-dimensional structures.

CLASSIFICATIONS

Polysaccharide gums can be classified by origin, solubility, the number of different glycosyl units in their structures, shape, charge, and source.

GENERAL PROPERTIES

Solutions of polysaccharide gums exhibit the characteristics of solutions of hydrated polymer molecules or, because intermolecular interactions are common, associations of polymer molecules. The usefulness of industrial gums stems from a variety of functional properties, but the common characteristic is the ability to modify flow characteristics or to bind and hold water. The rheology of gum solutions is non-Newtonian.

Polysaccharides in nature are heterogeneous with respect to molecular size, that is, any preparation from any source contains molecules in a range of molecular weights. Polysaccharides can be rather easily depolymerized by acid- and enzyme-catalyzed hydrolysis of the glycosidic (acetal) linkages joining the monomeric (saccharide) units, by treatment with an oxidant followed by a base, and sometimes by physical means such as the application of shear forces. Gum preparations of reduced average molecular weights have reduced thickening capacity. Industrial gums are usually available in a range of viscosity types for different applications.

The physical properties of gums, which determine their functional properties, are determined by the chemical nature of the gum molecules; the nature and amount of solvent, almost always an aqueous solution, including its pH and the presence of dissolved salts, sequestrants, and other solutes; and their tendency to interact with other molecules of the same gum or with other molecules in the system, especially other polymer molecules. Other controlling variables that effect the properties gums impart to aqueous systems include concentration, shear rate during dispersion, shear rate during use, and temperature.

Because all gums are thickeners and modifiers of the rheology of aqueous systems, the choice of a gum for a particular application often depends on its other characteristics: those that allow gums to provide body and bulk, to form gels, to absorb and bind water, and to be used in coatings and as adhesives, binders chelaters, crystallization inhibitors, clarifying agents, cloud agents, emollients, emulsifying agents, emulsion stabilizers, encapsulating agents, film formers, flocculating agents, foam stabilizers, humectants, mold-release agents, protective colloids, suspending agents, suspension stabilizers, swelling agents, syneresis inhibitors, texturing agents, and whipping agents.

Gums are tasteless, odorless, and colorless. Except for the starches and starch derivatives, they are, for practical purposes, noncaloric. All gums are subject to microbiological attack.

SOURCE, PREPARATION, PROPERTIES, AND APPLICATIONS OF SPECIFIC GUMS

The most common non-starch-based water-soluble gums are described briefly, in alphabetical order.

Algins

Algins are salts (generally sodium [9005-38-3], ammonium [9005-34-9], or potassium [9005-36-1]) or esters (propylene glycol [9005-37-2]) of alginic acid [9005-32-7], a generic term for copolymers of D-mannuronic acid and L-guluronic acid units. Algins are extracted from brown algae.

Alginate molecules in aqueous solutions are highly hydrated linear polyelectrolytes in extended conformation. An important and useful property of alginates is their ability to form gels by reaction with calcium ions. Sodium alginate is used extensively in textile printing pastes made with reactive dyes and in gelled food products.

Carrageenans, Agars, and Furcellarans

Carrageenan is a generic term applied to polysaccharides extracted from a number of closely related species of red algae. Agar [9002-18-0] and furcellaran [9000-21-9] are also red seaweed extracts and are members of the same larger family. All polysaccharides in this family are derivatives of linear galactans. Commercial carrageenans are composed primarily of three types of polymers, κ-, ι-, and λ-carrageenan.

A useful property of the red seaweed extracts is their ability to form water and milk gels. Carrageenans are blended and standardized to provide products that will form a wide variety of gels: clear and turbid gels, rigid and elastic gels, tough and tender gels, heat stable and thermally reversible gels, gels that undergo syneresis and those that do not.

Agars are the least soluble of this class of polysaccharides. By far the greatest use of agar in the United States is in the preparation of microbiological culture media. Agar is also used in bakery icings. Agarose, the linear component of agar, is used in making gels for electrophoresis, in size-exclusion chromatography, and in several biotechnological applications.

Guar and Locust Bean Gums

Guaran, the purified polysaccharide from guar gum [9000-30-0], is a polymer of D-galactose and D-mannose. Locust bean (carob) gum [9000-40-2] is, like guaran, a galactomannan [11078-30-1], and has a similar structure.

Commercial guar gum is the ground endosperm of guar seeds. Guar gum forms very high viscosity, pseudoplastic solutions at low concentrations. Guar gum is used in food products, in textile printing pastes, and to thicken and gel blasting agents and explosive slurries, but the largest uses are in water and water-methanol-based fracturing fluids for oil and gas wells and as processing aids in the separation of certain minerals from their ores.

Like guar gum, commercial locust bean gum (LBG) is the ground endosperm of the seeds of the locust bean (carob) tree. Locust bean gum has low cold-water solubility and is generally used when delayed viscosity development is desired. LBG is used primarily in food products, often making use of its ability to form gels with κ-carrageenan or xanthan.

Gum Arabic

Of the gums of ancient commerce, which were dried, gummy exudations collected by hand from various trees and shrubs, only gum arabic [9000-01-5], also called gum acacia and acacia gum, is still in significant use. Gum arabic is unique among gums because of its high solubility and the low viscosity and the Newtonian flow of its solutions. Its main uses are in the preparation of flavor oil emulsions and dry powders made by spray-drying such emulsions, and in coating certain confections.

Pectins

Pectins [9000-69-5, 16048-08-1, 58128-44-2] are mixtures of polysaccharides that originate from plants, and contain poly(α-D-galactopyranosyluronic acid)s [9046-38-2, 84149-03-1, 25249-06-3] in a partial methyl ester form and various degrees of neutralization as the major components. The commercial importance of pectin is predominantly the result of its unique ability to form spreadable gels (jams, jellies, preserves, etc.) in the presence of a solute that competes for water of hydration (almost always sugar) at pH ca. 3, or in the presence of calcium ions.

Xanthan Gum

Xanthan (known commercially as xanthan gum) [11138-66-2] has a main chain identical to the sructure of cellulose molecules. However, in xanthan every other β-D-glucopyranosyl unit in the main chain is substituted with a trisaccharide unit. The molecular weight is probably in the order of 2×10^6, although much higher figures have been reported.

Xanthan solutions are extremely pseudoplastic and have high yield values, making them almost ideal for the stabilization of aqueous dispersions, suspensions, and emulsions. Whereas other polysaccharide solutions decrease in viscosity when they are heated, xanthan solutions containing a small amount of salt (0.1%) change little in viscosity over the temperature range 0–95°C. Although xanthan is anionic, pH has almost no effect on the viscosity of its solutions over the range pH 1–12.

Xanthan is used in various aspects of petroleum production, including oil well drilling, hydraulic fracturing, workover, completion, pipeline cleaning, and enhanced oil-recovery fluids. It also finds application in a variety of food, consumer, and agricultural chemical products.

Cellulose Derivatives

Cellulose [9004-34-6] can be derivatized to make both water-soluble gums and hydrophobic polymers. Carboxymethylcellulose (CMC) [9004-42-6] is the carboxymethyl ether of cellulose.

Viscosity building is the single most important property of CMC. Solutions of CMC are either pseudoplastic or thixotropic, depending on the type.

CMC is useful as a warp size, in a broad spectrum of food products, in cosmetic and personal care products, in oil and gas well-drilling fluids, and in a variety of other applications.

Hydroxyalkylcelluloses are cellulose ethers prepared by reacting alkali cellulose with ethylene oxide (to prepare hydroxyethylcellulose [HPC, 9004-64-2]).

Both HEC and HPC form clear, smooth, uniform solutions. Both gums, like other cellulose derivatives, are produced in a wide range of viscosity grades, including high-viscosity types. Also, like other cellulose ethers, their solutions are Newtonian at low shear rates and become pseudoplastic at higher shear rates.

Clear, water-soluble, oil- and grease-resistant films of moderate strength can be cast from HEC solutions. Flexible, nontacky, heat-sealable packaging films and sheets can be produced from HPC by conventional extrusion techniques. Both gums can be used in the formulation of coatings, and to form edible films and coatings. The primary use of HEC is for thickening latex paint. HPC is used as a secondary stabilizer in suspension polymerization of vinyl chloride and as a granulating agent for pharmaceutical tablet and capsule mixes.

Methylcellulose [9004-67-5] contains methoxyl groups in place of some of the hydroxyl groups along the cellulose molecule. Hydroxyalkylmethylcelluloses contain, in addition to methoxyl groups, hydroxyalkoxyl groups in place of some of the hydroxyl groups. The properties of methyl- and hydroxyalkylmethylcelluloses are also a function of the type(s) of derivatization, the amount of each type of substituent group, the molecular weight distribution, and to some extent, the physical nature of the product (e.g., fibrous vs. powdered, mesh size, surface treatment). Because these variables can be controlled to some degree, the members of this methylcellulose family are tailor-made products, as are other starch and cellulose derivatives.

The most interesting property of these nonionic products is thermal gelation. Solutions of members of this family of gums that are soluble in cold water decrease in viscosity when heated, like solutions of other polysaccharides. However, unlike other gums, when a certain temperature is reached, the solution viscosity increases rapidly and the solution gels.

The main use of methylcellulose is in the preparation of tape joint compounds for gypsum board paneling, in gypsum spray plaster, and in ceramic tile adhesive, grout, and mortar formulations. It is also used in ceramics, food products, wallpaper adhesives, shampoos, and a variety of other applications.

GENERAL REFERENCE

Industrial Gums, 3rd ed.; Whistler, R. L.; BeMiller, J. N., Eds.; Academic: San Diego, CA, 1993.

Gutta-Percha

See: Eucommia Ulmoide Gum

H

HAIR AND SKIN CARE BIOMATERIALS (Cosmetology and Dermatology)

Bernard R. Gallot
Laboratoire des Materiaux Organiques Proprietes Specifiques
Centre National de la Recherche Scientifique

Cosmetology and dermatology are interrelated and complementary in many ways although they have different purposes. Cosmetology is concerned with hygiene, beauty enhancement, and the care of hair and skin, whereas dermatology is concerned with preventing and curing diseases of the skin (including the scalp) and its appendages.[1] In this paper we will focus on polymeric biomaterials used in hair and skin applications. The main areas of application of polymeric biomaterials are hair and scalp hygiene (shampoos), hair treatment and care, hair-setting, hair-coloring, and skin-care, including makeup.

COSMETIC PRODUCTS INCORPORATING POLYMERS

Shampoos

A shampoo is a product available as a clear, opaque, or opalescent liquid, gel, cream, foam, aerosol, or powder that dissolves in use. A shampoo is formulated using surface-active agents (surfactants and polymers), and exhibits detergent, emulsifying, and foaming properties to clean hair and to make hair supple and glossy, and easy to style and disentangle.[1] Therefore, shampoos contain many ingredients, such as detergents, foam stabilizers and softeners, thickeners, opacifiers and agents providing pearlescence, special care products, sequestering agents, preservatives, perfumes, and coloring agents.

Hair-Care Products

Hair-care products restore the natural beauty of hair to give it lightness, volume, spring control, suppleness, softness, and sheen.

Basic Compounds

Basic compounds of hair-care products are organic acids, fatty compounds and their derivatives, vitamins, protein derivatives, cation-active surfactants, cationic polymers, and hair strengtheners.

Formulation

Different formulations are used to fulfill the various requirements of hair-care products: conditioners and deep conditioners, lotions or fluid gels, aerosol foams, hair oils, conditioning shampoos, and hair tonics.

Hair-Setting Products

Products for Maintaining Hair Styles

These are applied to wet hair, after shampooing but before combing. They form a flexible film on each hair, provide an overall stiffening of the hair, and retard moisture uptake.

The essential components of products for maintaining hairstyle are film-forming polymers; plasticizers for the polymers; disentangling, softening and glossening agents; solvents; perfumes; and colorants.[1]

Hair Sprays

Hair sprays are applied after drying the styled hair. An aerosol formula contains the following components: polymers, plasticizers, softening and glossening agents (often lanolin derivatives), perfume solvents (ethyl alcohol is the best), and propellants (the nature of which has evolved with environmental legislation and greatly influenced the nature of the polymers used).

Skin-Care Products

Skin-care products protect, lubricate, and clean the skin. They incorporate a variety of polymers, including polyethylene, poly(ethylene glycol)s, ethylene-propylene oxide copolymers, vinyl and acrylic polymer resins, hydrophilic polymers, and silicone. Polymers are used as thickeners, film-formers, binders, lubricants, water-resistant agents,[2] and in specialized delivery systems such as liposomes, "microsponge", hydrogels.[3]

Eye Makeup Products

The main makeup products are mascaras, shadows, liners, brow makeup, creams, false eyelash adhesives, and makeup removers.[4] In these products polymers are used as binders, thickeners or film-formers and may provide water resistance, waterproofing, and gloss.[5] The main families of polymers used here are cellulose and derivatives, polyethylenes, ethyleneoxide homo and copolymers, vinyl polymers, and acrylic polymers such as carbomers[6] and acrylic/acrylates copolymers.

POLYMERS USED IN COSMETOLOGY

The number of characteristics of polymers used in cosmetology and dermatology are rapidly increasing. To put some order in this crowded family of polymers, we have divided them into six classes: natural polymers and derivatives, vinyl and acrylate polymers, cationic polymers, ozone debate polymer bond, water compatible polymers, and siloxanes.

Natural Polymers and Derivatives

Five main types of natural polymers have been used in cosmetics: shellac, cellulose derivatives, xanthan gum, hyaluronic acid, and protein derivatives.

Shellac

Shellac consists primarily of three complex hydroxycarbonic acids: aleuric, shellolic, and jalaric acid.[7] Bleached and dewaxed shellac has good holding ability and excellent weather resistance, but poor solubility in propellants and solvents. It is also very difficult to brush and shampoo away and its films powder easily.

Cellulose Derivatives

Cellulose derivatives sodium carboxymethyl, methyl, hydroxypropylmethyl, hydroxyethyl and hydroxypropyl ether are used in concentrations between 0.1% and 5% in cosmetology, mainly as polymeric thickeners.

Xanthan Gum

Xanthan gum is a polysaccharide used as an emulsion stabilizer and suspending agent in moisturizing lotions and creams.[8]

Hyaluronic Acid

Hyaluronic acid is a mucopolysaccharide that can be used as a transdermal delivery system as it forms a matrix on the skin allowing for increased skin penetration due to skin hydration.[9]

Protein Derivatives

Protein molecules are too large to penetrate hair and fix onto hair keratin, so they are used in the form of partial hydrolysates with molecular weights higher than 1000. Many benefits have been claimed for inclusion of proteins in hair-care formulations.

VINYLIC AND ACRYLIC POLYMERS

Polyvinylpyrrolidone (PVP)

Among vinylic polymers the best known and most widely used is polyvinylpyrrolidone (PVP). It has been developed for hair sprays because it exhibits conditioning properties if applied via shampoo. It forms transparent films on hair, providing smoothness and luster. PVP exhibits other useful properties in shampoos: it improves foam stabilization, increases viscosity, reduces eye irritation and is easily removed. Apart from its ability to act as a thickener, dispersing agent, lubricant, or binder, it has a strong affinity for dyes that makes it suitable for use in coloring shampoos and in eye-makeup products. PVP applications in hair and skin formulations result from the following properties: complete solubility in water; high solubility in a wide variety of organic solvents, especially in ethanol propellant mixtures; and absence of odor, taste, and toxicity.

Unfortunately, PVP also has drawbacks, such as hygroscopycity (its films tend to become sticky, dull, and tacky as humidity is absorbed). To overcome such problems formulators introduced small amounts of shellac or silicon oil in their formulations. A more efficient way to solve the problem was the copolymerization of the hydrophilic vinylpyrrolidone with a hydrophobic monomer.

Copolymers

The first copolymers synthesized were statistic copolymers of vinyl acetate and vinylpyrrolidone, but they were soon followed on the market by many other copolymers.

CATIONIC POLYMERS

Cationic polymers were developed to overcome the disadvantages of cationic surfactants (absence of setting properties, incompatibility with most anionic surfactants used in formulating shampoos, eye irritation, etc.) They are revolutionary in the way that they are substantive for hair and able to cover the hair surface with a continuous film and thereby protect hair against external attack.[1]

Following the CTFA classification cationic polymers are called "Polyquateriums," as they are polymers that contain a quaternary ammonium radical. Cationic polymers can be divided into four groups: polysaccharide derivatives, acrylamide derivatives, vinylpyrrolidone derivatives, and ionenes.

WATER COMPATIBLE POLYMERS

An elegant way to fulfill the aerosol regulations consists of replacing aerosol hair sprays by aqueous gels. Aqueous gels are obtained by adding polymers exhibiting one or more of the following properties: thickening, gelling, suspending, emulsifying, and thixotropic.[10]

Classical Polymers

Three types of classical thickening agents are particularly important. They are cellulose derivatives, such as hydroxypropylmethyl cellulose; carbomers obtained by precipitation polymerization of acrylic acid and used in hand and body lotions and creams and hair styling gels;[6,11] and PVM/MA decane crosspolymers used in styling gels, hand and body lotions, and antidandruff shampoos.[12]

Lipopeptide-Based Polymers

A different type of polymer, lipopeptide-based block and comb copolymers, can be used to obtain aqueous gels and stable emulsions.

Block Copolymers

Block copolymers are amphiphilic lipopeptides Cn(AA)p, formed by a hydrophobic lipidic chain, Cn, containing 12 to 22 carbon atoms linked through an amide bond to a hydrophilic peptidic chain, (AA)p, with a number average degree of polymerization, p, between 1 and 70, where the hydrophilic amino acids are: sacrcosine, serine, glutamic acid, lysine, N–hydroxyethylglutamine, or N-hydroxypropylglutamine.[13–15] Lipopeptides with $p = 1$ are prepared by coupling the amino acid by its α-carboxylic function to the end of a fatty amine. Lipopeptides with p higher than 1 are prepared by polymerization of the N–carboxyanhydride (NCA) of the amino acid, or of the side-chain-protected amino acid at the end of a fatty amine.[13–15]

Lipopeptides exhibit liquid crystalline gels in water solution, and the structure of the gels goes from lamellar to cylindrical hexagonal and to cubic when the degree of polymerization of the amino acid increases.[13,14] Lipopeptides with Cn equal to or greater than C16, and low degrees of polymerization, p, exhibit very interesting emulsifying properties. They give very stable oil in water (O/W) emulsions for very small amounts of lipopeptide (1%–2%) with a wide variety of the oils used in the cosmetic industry such as isopropylmyristate, butylstearate, dodecane, migliol, cosbiol, wheat germ oil, vaseline oil, ricin oil, and silicone oils, as well as stable mini-emulsions of the same oils if 1% to 3% of cetyl alcohol is added.[16–18] Lipopeptides in C12 with degrees of polymerization between 8 and 20 exhibit good foaming and disentangling properties. Moreover, lipopeptides are highly biocompatible.[18]

Comb-Like Polymers

Comb-like polymers consist of a polyvinyl main chain of polyacrylamide, polymethacrylamide or polystyrylamide, and lipopeptide side chains.

As a result of their physical and chemical properties, polysiloxanes are non-toxic, non-irritating, and classified as environmentally friendly.[18a]

Polydimethylsiloxane

Polydimethylsiloxane (CTFA name: Dimethicone) are used both for skin- and hair-care. Skin-care products containing Dimethicone are used for protective hand and body lotions and creams, cleansing creams, suntan lotions, sun creams, and aerosol shave lathers. Dimethicones are incorporated in hair-care formulations to improve luster and sheen.[2,19]

Organopolysiloxanes

Organopolysiloxanes have been developed to overcome the drawbacks of Dimethicone, such as its lack of solubility in water, alcohol, and common solvents. They are prepared by chemically modifying Dimethicone.

The main organopolysiloxanes are: alkyl, alkoxy, ether, amino, betaine, amidoquaternary, and phosphate esters polysiloxanes.

Alkylpolysiloxane Copolymers

Their CTFA name is Alkyldimethicone. When the alkyl chains contain more than 10 carbon atoms they are called silicon waxes and are mostly used in makeup products such as lipsticks and mascaras where they improve the spreadability as well as luster and distribution of pigments.[19] When they contain two types of side chains randomly distributed they enhance the smoothness and softness of the skin and are used in water-in-oil skin creams, water-in-oil protection creams and oil-in-water moisturizing lotions for sun protection.[21]

Polyetherpolysiloxanes Copolymers

Their CFTA name is Dimethiconecopolyol. They are soluble in water when they contain mostly ethylene oxide and soluble in oils when they contain mainly propylene oxide. In these products the polydimethylsiloxane skeleton brings gloss and silky feel, while the ethylene oxide chains increase adherence on polar surfaces. They are used as surface-tension depressants, wetting agents, emulsifiers, or foam builders in shampoos, liquid soaps, foaming bath, shower gels, skin care creams, shaving lather, deodorants, and antiperspirants sticks.[19,20,22]

Polyalkylpolyetherpolysiloxane Copolymers

Very efficient water-in-oil emulsifiers are obtained because polyether chains are dissolved in water and alkyl chains are dissolved in the oil phase while the polysiloxane skeleton fixes the total molecule at the interface.

They also give water in silicon emulsions that are more substantive than the corresponding water-in-oil emulsions.[19,23]

Polyaminopolysiloxane Copolymers

They are divided in two families in the CFTA classification: Trimethylsylilamodimethicone (TSA) and Amodimethicone.

Silicones with amino functions are strongly substantive to hair, and are ideal conditioners conferring luster and enhancing wet and dry combing.[20,22] Amodimethicones exhibit higher substantivity than TSA, because their α,ω–OH groups allow chain extension and crosslinking.[23]

Polyorganobetainepolysiloxane Copolymers

They are amphoteric surfactants that can exist with a positive or a negative charge or both in aqueous solution as a function of the pH of the solution. They are of two types: carboxy or phosphobetaines.

Aminoquaternarycompoundspolysiloxane Copolymers

Aminoquaternary compounds can be of the comb or terminal types. Their CFTA name is Silicon Quaternary.

Silicone Quaterniums exhibit antistatic properties, improved wet combing and gloss, skin conditioning, and can be formulated into gels, mousses, and virtually any desired product form. They are used at low concentrations (1 to 2%) in conditioning shampoos, hair conditioners, and bubble baths, where they provide silky feel, conditioning properties, and foam stabilization.[24]

Polyphosphatesterspolysiloxane Copolymers

Polyphosphate esters are anionic surfactants of the terminal or comb type.

Silicone-based phosphate esters are emulsifiers producing oil-in-water emulsions. They are used in emollient lotions, light body lotions, sunscreen lotions, and creams at concentrations of about 3%. Incorporated in hair conditioners, they improve curl retention and combability.[25]

REFERENCES

1. Zviak, C. *The Science of Hair Care*; Marcel Dekker: New York and Basel, 1986.
2. Idson, B. *Cosmet. Toilet., Polymers in Skin Cosmetics* **1988**, *103*, 63.
3. Gans, E. H. *Cosmet. Toilet., Polymer Developments of Cosmetic Interest* **1988**, *103*, 94.
4. Wetterhahn, J. *Eye Makeup in Cosmetics Science and Technology*; Bolsam, M. S.; Sagarin, E. Eds.; Wiley-Interscience: 1972; Vol. 1.
5. Kapadia, Y. M. *Cosmet. Toilet., Use of Polymers in Eye Makeup* **1984**, *99*, 53.
6. Amjad, Z.; Hemker, W. J.; Malden, C. A.; Rouse, W. H.; Sauer, C. E. *Cosmet. Toilet., Carbomer Resins: Past, Present and Future* **1992**, *107*, 81.
7. Tannert, U. *S.O.F.W. Journal, Shellac: A Natural Polymer for Hair Care Products* **1992**, *118*, 1079.
8. Rieger, M. *Cosmet. Toil., The Role of Water in Performance of Hydrophilic Gums* **1987**, *102*, 101.
9. Balazs, E. A.; Bond, P. *Cosmet. Toilet. Hyaluronic Acid: Its Structure and Use* **1984**, *99*, 65.
10. Lochhead, R. Y. *Cosmet. Toilet., The History of Polymers in Hair Care* **1988**, *103*, 23.
11. Barzaghi, M.; Tadini, G. *Cosmet. Toilet. Synthalin: The New Carbomers of Highest Purity* **1991**, *106*, 100.
12. Kopalov, 1993.
13. Douy, A.; Gallot, B. *Makromol. Chem. New amphipatic lipopeptides: 1) Synthesis and mesmorphic structures of lipopeptides with polysarcosine peptidic chains* **1986**, *187*, 465.
14. Gallot, B.; Douy, A.; Haj Hassan, H. Mol. Cryst, Liq. Cryst. *Synthesis and structural study by X-ray diffraction of lyotropic lipopeptidic block copolymers with polylysine and poly(glutamic acid) peptidic chains* **1987**, *153*, 347.

15. Gallot, B.; Haj Hassen, H. *Mol. Cryst. Liq. Cryst., Lyotropic lipo-amino-acids: synthesis and structural study* **1989**, *174*, 89.
16. Gallot, B.; Haj Hassan, H. ACS Series, *Liquid crystalline phases and emulsifying properties of block copolymers with hydrophobic aliphatic chains and hydrophilic peptidic chains* **1989**, *384*, 116.
17. Gallot, B. ACS Series, *Liposarcosine based polymerizable and polymeric surfactants* **1991**, *448*, 103.
18. Gallot, B. *Polymers in medicine: biomedical and pharmaceutical applications*; Ottenbrite, A. M.; Chiellini, E., Eds.; Thecnomic: Lancaster, Basel, 1992; Chapter XVI.
18a. Chandra, G.; Disapio, A.; Frye, C. L.; Zellner, D. *Cosmet. Toilet. Silicones for Cosmetics and Toiletries: An Environmental Update* **1994**, *109*, 63.
19. Schaefer, D. *Tenside Surf. Det., Silicone Surfactants: Organo-Modified Polydimethyl Siloxanes as Surface Active Ingredients in Cosmetic Formulations* **1990**, *27*, 154.
20. Roidl, J. *Parfumerie und Kosmetik, Anwendung Der Silicon-polymere und Silicone mit Funktionellen Gruppen in der Kosmetik* **1986**, *67*, 232.
21. Powell, V.; Thimneur, R. *Cosmet. Toilet., New Developments in Alkyl Silicones* **1993**, *108*, 87.
22. Roidl, J. *Seifen-Ole-Fette-Wachse, Bedeutung der Silicone in der Modernen Kosmetik* **1988**, *114*, 51.
23. Disapio, A.; Fridd, P. *Internat. J. Cosmet. Sci. Silicones: Use of Substantive Properties On Skin and Hair* **1988**, *10*, 75.
24. O'Lenick, A. J.; Parkinson, J. K. *Cosmet. Toilet., Silicone Quaternary Compounds* **1994**, *109*, 85.
25. O'Lenick, A. J.; Parkinson, J. K. Cosmet. Toilet., *Silicone quaternary compounds* **1994**, *109*, 85.

HAIR AND SKIN COSMETICS

Sanjeev Midha* and Raymond E. Bolich, Jr.
The Procter and Gamble Company

Natural and synthetic polymers represent a significant share of raw materials used in the cosmetic industry. Hair and skin care (including nail care) chemicals dominate total U.S. cosmetic chemicals sales. In 1993 sales of polymers and resins for cosmetics and toiletry products were estimated at $191 million.[1]

Polymers and resins used in personal care products can be natural, synthetic, or semi-synthetic. Natural polymers have unique properties, but they are sometimes used simply because of the growing consumer demand for natural products. They do, however, vary in color, odor, and purity, and are relatively expensive compared to common synthetic polymers used in the industry. Synthetic and semi-synthetic polymers (chemically modified natural polymers) have been developed to match, improve, and offer unique formulatibility advantages over natural polymers. This article attempts to show the benefits of polymers in hair and skin care cosmetics.

GENERAL FUNCTIONS

Conditioning/Lubrication

Particularly in shampoos and conditioners, polymers are used to provide softness, suppleness, and volume to hair. Either cationic water soluble or silicone polymers are used to condition hair. Silicone gums/fluids are dispersed in shampoo/conditioning products and may deposit onto hair by interfiber entrapment of silicone droplets.[2] These droplets spread easily on the hair fiber surface and provide a non-oily, smooth feel.

Foam Boosting

The high surfactant content of shampoos can irritate skin and eyes. Using mild surfactants or less anionic surfactant significantly decreases foaming. Cellulosic polymeric surfactants are used to enhance lathering.[3]

Rheology Modification (Thickening)

Thickening and gelling the product provides mechanical stability and coherence to the product matrix.[4] Rheology modification is also very important to stabilize emulsions and suspending active ingredients, such as antidandruff compounds in shampoos. Polymers modifying rheology through their high molecular weights, chain entanglements, and polymer-solvent interactions. A few other mechanisms have also been utilized to achieve gelation and thickening of products. Anionic polymers, such as crosslinked acrylic acid of various molecular weights, have been used extensively as gelling and suspending agents. Hydrophobically modified, water soluble polymers (for example, hydrophobically modified (hydroxyethyl)cellulose) enhance solution viscosity through intermolecular association of hydrophobic groups.[7]

Welding and Gluing

In hairstyling products, polymers are used to weld and glue hair fibers. Hair spray polymers maintain hairstyle through bonding and friction. Styling products work primarily by bonding hair, as the hair is not distributed after product application. Water/alcohol-soluble, semi-synthetic polymers, and synthetic polymers are used to provide welding and gluing.

Emulsification

Amphiphatic non-ionic graft and block copolymeric surfactants are used as surface-active polymers to emulsify cosmetic ingredients. Polymeric surfactants are primarily used to enhance spreading properties and to emulsify oils and perfumes. Block copolymers of propylene and ethylene oxides, depending on chain length and ratio, exhibit good emulsification and gel-forming properties.

Opacification

Opacification of products to provide milkiness is done to give an impression of softness and gentleness. Dispersions of styrene/vinylpyrrolidone copolymer in water are used to opacify shampoos and conditioners.

Substantivity Aids

Polymers can enhance the deposition of active ingredients. These polymers trap the active ingredient on the hair surface. Ethoxylated polyethyleneimine and cationic polymers can be used to improve the deposition of the antidandruff compound, zinc pyrithione, on the scalp through a shampoo.[8,9]

Film Formation

Film-forming, water/alcohol-soluble polymers have been used to enhance the gloss and shine of hair. Film-forming polymers are used in styling products to provide hold and as binders in hair coloring products to resist ruboff or disintegration of coating.[10]

*Author to whom correspondence should be addressed.

MISCELLANEOUS

Hair Restrengthening

Low molecular weight proteins, such as hydrolyzed animal collagen, penetrate heavily damaged hair and occupy voids in the hair fiber.[15] Hair strengthening probably results from the hydrogen bonding of the hydrolyzed collagen with adjacent polypeptides within the hair.

Anti-Resoiling Agents

Sebaceous glands in the scalp secrete sebum, a complex mixture of lipids. The sebum climbs the hair shaft due to capillary forces and coats the hair fiber. Low T_g-fluorinated polymers can be deposited onto hair from shampoo/rinse formulations to prevent sebum from moving up the hair shaft.[12]

Shampoo/conditioners and styling products are the two major product categories in which polymers are used.

APPLICATIONS OF POLYMERS

Shampoos and Conditioners

A variety of polymers are used in shampoos and conditioners. Their primary functions are product thickening and opacification. Benefits include neutralization of hair static charge, hair body/fullness, gloss, and ease of combing. Cationic polymers are most commonly used for conditioning. Cationic polymers have been used in shampoos to provide body and fullness to hair through polymer–surfactant complex (coacervate) deposition on hair during shampooing and conditioning.[6] However, excessive deposition of coacervate during continuous product use leads to buildup, resoiling, and limpness. Synthetic polymers of lower cationic charge density can be used to decrease excessive polymer buildup on the hair. Despite these drawbacks, use of cationic conditioning polymers is increasing.

Silicone polymers (for example, polydimethylsiloxanes) condition hair without the significant tradeoffs associated with the cationic polymers. They are excellent lubricants and at low levels provide hair with non-oil feel.[5] They lower the surface energy of hair and thereby lower the wettability of hair due to increased surface hydrophobicity. Dewetting the hair surface results in reduced hair tangling and contributes to easier wet combing.[6]

Natural polymers, such as hydrolyzed collagen, and various synthetic polymers in latex emulsions, such as styrene/acrylate copolymer, vinylpyrrolidone/vinyl acetate, have also been used in shampoos to provide conditioning and hair-holding properties.[13]

Hair Styling Products

A broad range of polymers are used in styling products, mainly to weld and glue hair fibers. These polymers provide hold and protect hair from mechanical deformation, wind, and humidity. Usually, aqueous or aqueous-alcoholic solutions of polymers are applied. The polymer properties that determine performance are solubility, strength, adhesion to hair surface, glass transition temperature, moisture resistance, and shampoo removeability. Polar, high T_g high molecular weight, film-forming polymers are mainly used in styling products. The following are the types of polymers used in hair styling products.

Anionic Polymers

Anionic polymers are most commonly used in hair sprays. Anionic polymers are also used as thickeners; for example, crosslinked, neutralized acrylic acid is used primarily as a thickener in water-based hairstyling gels.

Cationic Polymers

Most cationic polymers are water soluble and are used most commonly in "mousses" and "gels." Cationic polymers have high water-sensitivity, and the amount of hold depends on humidity. Most of the commercially available cationic polymers have quaternary amine-containing units in the backbone. The level of quaternization determines the film tackiness, product viscosity, and shampoo removeability of the polymer.

Nonionic Polymers

Nonionic polymers are compatible with both anionic and cationic polymers. These polymers are used both in aqueous and aqueous-alcoholic formulations. The most common nonionic polymer is polyvinylpyrrolidone (PVP).

POLYMERS IN SKIN CARE PRODUCTS

Polymers perform multiple functions in cosmetics. They are used as pigment dispersants, skin conditioners and lubricants, water-proofing agents in sunscreens, rheology modifiers in gels and lotions, emulsifiers, and controlled-release agents.[20,21] Some of the functional properties of polymers are due to large size, the ability to adsorb at interfaces, the tendency to form coascervates with charged surfactants, and the ability to build three-dimensional networks in aqueous media.

GENERAL FUNCTIONS

Film Formation/Substantivity

Many film-forming polymers are used in skin-care creams and lotions to protect the skin. These polymers form a film that prevents or retards transdermal water loss. Many oil- and water-soluble polymers are used. Water-soluble polymers, such as sodium hyaluronate and polyvinylpyrrolidone, are believed to help moisturize skin because of their water holding and absorbing properties. Oil-soluble polymers act primarily by forming an occlusive film on the skin. In addition, these film-forming polymers are used to enhance the rub-off resistance of cosmetic products, enhance water repellence, and increase fragrance longevity.[16] Examples of commercially available polymers are polyethylenes, oxidized polyethylenes and functional groups containing copolymers of polyethylenes, acrylates/tert-octylpropenamide copolymers polyvinylpyrrolidone, and polydimethylsiloxanes.

Conditioning

Polymers can act as emollients or skin conditioning agents. Emollients are substances that modify skin texture. Examples of liquid polymers used in skin-care formulations are hydrogenated polyisobutenes, polydecenes, and polydimethylsiloxanes. In addition, water soluble polymers, such as poly(vinylpyrrolidone), carboxylic acid salts of chitosan, and poly(ethylene glycol)s (MW 3000–6000), have been used to impart a smooth-skin feel.[17]

Cationic polymers can also be used to make skin-washing products mild.

Emulsification and Stabilization

Crosslinked polymers of acrylic or methacrylic acid and long chain alkyl acrylates are used as primary emulsifiers for skin lotions and creams based on oil/water systems. These hydrophobically modified polyelectrolytes act by adsorbing on the oil/water interface.

Controlled Release

Polymeric entrapment has been explored to enhance the sustained release formulation actives. By sustaining the release, polymeric entrapment agents help deliver low levels of potential skin irritants on a continuous basis.[18]

Thickening and Gelation

Many neutral, semi-synthetic and synthetic polymers are used to thicken skin-care products. These polymers also help provide a moisturized, silky, after-feel to the products and sometimes help suspend active ingredients in formulations. Examples of polymers used for thickening and gelling are crosslinked acrylic acid copolymers, crosslinked acrylamide, xanthan gum, and various cellulose gums.

FUTURE OF POLYMERS IN COSMETIC PRODUCTS

Although polymers have been used in cosmetics for many years, their full potential is only now being realized. What were once thought of mainly as thickening agents in cosmetics or as gluing materials are now key components providing cosmetics with powerful and varied new consumer benefits. Additionally, polymers, by virtue of their molecular weight and size, have low toxicity; many functionalized polymers are safer than their nonpolymeric counterparts.

New synthetic methods have opened up routes to the synthesis of block and graft copolymers that could not be synthesized previously. Graft and block copolymers are being designed with multiple functional attributes.[19] New environmental regulations on volatile organic solvents and biodegradability issues are keeping the cosmetic industry in constant flux. These requirements will have a major impact on all facets of the cosmetic industry but will be an enormous challenge for the polymer industry.

REFERENCES

1. Ainsworth, S. J.; *C&E News* **1993**, *April 26*, 36.
2. Bolich, R. E.; Williams, T. B. U.S. Patent 4 788 006, Nov. 29, 1988.
3. Conklin, J. R. *Cosmetic & Pharmaceutical Applications of Polymers*; Gebekein, C. G.; Cheng, T. C.; Yang, V. C., Eds.; Plenum: New York, 1991; p 15.
4. Silberberg, A. *Polymers in Aqueous Media*; Glass, J. E., Ed.; American Chemical Society: Washington, DC, 1989; p 3.
5. Hunting, A. L. L. *Cosmetic & Toiletries* **1982**, *97*(3), 53.
6. Lochhead, R. Y.; Davidson, J. A.; Thomas, G. M. *Polymers in Aqueous Media*; Glass, J. E., Ed.; American Chemical Society: Washington, DC, 1989; p 113.
7. Sau, A. C.; Landoll, L. M. *Polymers in Aqueous Media*; Glass, J. E., Ed.; American Chemical Society: Washington, DC, 1989; p 344.
8. Parran, J. J. U.S. Patent 3 761 471, September 25, 1973.
9. Glover, D. A. U.S. Patent 4 557 928, December 10, 1985.
10. Quisling, S. U.S. Patent 2 154 822, February 8, 1935.
11. Jones, R. T. *Croda Inc. Technical Report* August 1983.
12. Parks, L. R.; Builder, S. E. U.S. Patent 3 959 462, May 25, 1976.
13. Tsushima, R.; Kondo, A. *Cosmetic & Pharmaceutical Applications of Polymers*; Gebekein, C. G.; Cheng, T. C.; Yang, V. C. Eds.; Plenum: New York, 1991; p 33.
14. Lochhead, R. Y. *Cosmetic & Toiletries* **1992**, *107*(9), 131.
15. Siciliano, A. A.; Szycher, M.; Brooks, G. *Cosmetic & Toiletries* **1992**, *107*(9), 123.
16. Guth, J.; Martino, G.; Pasapane, J.; Patel, D. *HAPPI* **1991**, *28*(5), 80.
17. Idson, B. *Cosmetic & Toiletries* **1988**, *103*(12), 63.
18. Pugliese, P.; Hines, G.; Wielinga, W. *Cosmetic & Toiletries* **1990**, *105*(5), 105.
19. Garbe, J. E.; Kantner, S. S.; Kumar, K.; Mitra, S. U.S. Patent 5 209 924, May 11, 1993.

HAIRY ROD-LIKE POLYMERS

Henning Menzel
Institut für Makromolekulare Chemie
Universität Hannover

"Hairy rod" polymers are comb-shaped polymers with a "rigid rod-like" backbone and more or less flexible side chains. The combination of main chain rigidity and side chain flexibility results in a new class of materials with interesting properties. Hairy rod polymers are believed to be useful for membranes and sensors and in the growing field of optoelectronics.

HAIRY RODS WITH A HIGH DENSITY OF FLEXIBLE SIDE CHAINS

The properties of rigid, rod-like polymers are influenced by the length of the alkyl side chains if their density is high enough for interaction. Several investigations of this influence have been performed with substituted polyglutamates. Due to their α-helical structure, polyglutamates have a rod-like backbone. The synthesis and properties of these polymers have been reviewed recently by Daly et al.[1] Robinson revealed that solutions of polypeptides form liquid crystalline (cholesteric) phases above a certain concentration.[4] The thermotropic LC phase of hairy rod polymers can be explained as a solution of rigid polypeptides in a liquid matrix of the side chains ("attached solvent"). Besides the expected cholesteric phases, a columnar hexagonal liquid crystalline phase has been found for an octadecyl side chain.[2,3,5]

The strong influence of the nature and length of the alkyl chain on the structure of the thermotropic LC phase also has been demonstrated with a series of aromatic polyesters and polyamids equipped with alkyl side chains. In short side chains, "normal" nematic melts are formed, but if the side chains are longer than 8 carbon atoms, a new layered structure is found.[6,7]

Liquid crystalline phases also form if oligo ethylene oxide chains are attached to the polyglutamate backbone. As in the case of the alkyl chains the structure of the LC phase depends on the length of the side chain. The side chains are very mobile above their melting temperature ($<0°C$) and can form complexes with $LiClO_4$. The polymers, therefore, show a high ionic conductivity.[8]

FUZZY RODS

Hairy rod polymers sparked particular interest after Wegner et al. found that phthalocyaninato-polysiloxanes unsymmetrically substituted with short and long alkyl chains from monomolecular films at the air/water interface, which can be deposited on solid substrates using the Langmuir-Blodgett (LB) technique.[9–12] The LB films of hairy rod polymers are superior to that of classical amphiphiles in terms of thermal and mechanical stability and homogeneity.[13,14] The stability is due to the polymeric nature of the molecules and the rigid polymer backbone prevents a diffusion of the molecules within the LB film. The great homogeneity of the LB films from hairy rod polymers is due mainly to the two-dimensional, nematic structure of the monolayer at the air/water interface.[15]

Polysilanes

A further example of hairy rod polymers are polysilanes. These materials are particularly interesting because of the delocalization of the σ-orbitals of the silicon-atoms which causes interesting spectral and non-linear optical properties, as well as radiation sensitivity. Embs et al. synthesized and investigated a series of polysilanes with various side groups, including linear and branched alkyl chains, alkylaryl moieties and alkoxyaryl groups.[19] Within this series only polysilanes with butoxyphenyl-butylphenyl or bis(butoxyphenyl) substituents form stable monolayers. It was concluded that two substituent phenyl rings are essential, because they cause the conformational restrictions necessary for chain stiffness.

Cellulose and Amylose Derivatives

Cellulose has a relative stiff backbone and some hydrophilic sites. Therefore, we can expect cellulose derivatives also to form monomolecular films. There are several starting materials for synthesizing hairy rod polymers with a cellulose backbone. Esters and ethers can be synthesized from cellulose itself, the soluble hydroxypropyl- and hydroxyethylcellulose can also be esterified, and cellulose acetate has been successfully converted into cellulose ethers.[21–24]

Poly(1,4-phenylene)s and Related Compounds

Polymers with a conjugated π-system are also very stiff, making them potential candidates for backbones in hairy rod polymers. Several reports on polyanilines, polythiophenes and polypyrroles with alkyl chains have been reviewed recently.[25,26]

BRISTLY RODS

As shown by Wegner and co-workers, a fluid matrix around the polymer backbone is essential for obtaining high-quality LB films.[13,17] For this reason the degree of substitution should not be too high and/or a branched alkyl chain should be used. In light of these results it is noteworthy that high-quality LB films can be obtained from 100%-substituted polyglutamates with stiff azobenzene moieties in the side chain.[27–29] This behavior is attributable to the occurrence of a special liquid crystalline phase in which the main chains are embedded in a nematic matrix of the side chains.

APPLICATIONS

LB films have been proposed for applications in advanced optical, electronic, and bio-related systems.[30] For most technical applications the requirements of mechanical, thermal, and chemical stability are decisive. By using polymers, the stability of the LB films can be dramatically enhanced.[31,32] Furthermore, LB films of hairy rod polymers have been proven to be very homogeneous and to have excellent optical properties.[33] Hairy rod polymers are therefore promising materials for such LB film applications.

Data Storage

Due to the high homogeneity of LB films from polyglutamates, small differences in the optical properties, which can be induced by photoisomerization of a photochromic dye admixed to the monolayer, can be read out using surface plasmon microscopy (SPM). This process can be used for persistent optical storage.[34] Problems can arise from segregation phenomena due to the mobility of the dye, which can lead to inhomogeneities in the LB film and distort the correct read-out. This can be avoided by tethering the chromophores covalently to the polymer backbone as in copolyglutamates.[18]

Waveguides

As mentioned previously, the LB films from polyglutamates are of high optical quality and therefore can be used as waveguides. The waveguides show low losses between 2.6 and 5.5 dB/cm, which probably can be further reduced by optimizing the materials and process parameters.[35,36] These values are significantly lower than those of classical amphiphiles. This is attributable to the special structure of the hairy rod polymers, which behave like a molecular-reinforced liquid and therefore exhibit less grain boundaries in the LB films than the polycrystalline systems of classical amphiphils.

Orientation Layers

Conventional isotropic organic surface coatings in the micrometer range play a very important economic role. Particularly in the field of orientation layers for liquid crystal displays, LB films may have some advantages over the classical, spin-coated layers. The hairy rod polymers are of particular interest because, they show molecular axis orientation in dipping direction.[16] It has been shown that nematic liquid crystals are oriented along this molecular axis in the case of polysilanes, polyglutamates, and phthalocyaninato-polysiloxanes.[20,37] The orienting power of the LB film was stronger than that of an underlying, rubbed polymer surface or fine stripe pattern.[37]

Sensor Applications

LB films promise to be useful in the design of sensors and biosensors. This is because they can mimic biological membranes. Biological membranes not only separate different compartments but also control transport and recognition processes in a very specific way. However, where biosystems are thermodynamically open and show self-healing phenomena, artificial systems are thermodynamically closed and thus have limited stability. In this respect LB films of hairy rod polymers are favorable because of their excellent stability compared to classical amphiphils. Besides their function as biomimetic membranes, LB films of hairy rod polymers can have special optical or electrical properties.

Photoresists (Crosslinking)

For many applications of LB films stabilizing the structure by crosslinking is useful. Beside this stabilization, the possibility of pattern formation in LB films by crosslinking a part of the film and removing the uncrosslinked portion is very important. To obtain crosslinkable LB films of hairy rod polymers, reactive groups have to be incorporated into the side chains.

CONCLUSIONS

Combining a rigid polymer backbone with flexible side chains has opened a new class of materials. In particular, if these hairy rod polymers are subjected to the Langmuir-Blodgett technique, several applications are possible. The outstanding properties of the LB films of these polymers are caused by the two-dimensional liquid crystalline phase at the air/water interface and the fact that the films can be seen as a molecular-reinforced liquid. It has been shown that the original concept of rigid rods embedded in a fluid matrix can be extended to systems with stiffer side chains that form a liquid crystalline phase in the side chain region, without losing the outstanding properties of the LB films. This extension opens new possibilities to design materials because several interesting groups can be incorporated in the hairy rod polymer.

REFERENCES

1. Daly, W. H.; Poche, D.; Negulescu, I. I. *Prog. Polym. Sci.* **1994**, *19*, 79.
2. Watanabe, J.; Ono, H.; Uematsu, I.; Abe, A. *Macromolecules* **1985**, *18*, 2141.
3. Watanabe, J.; Gato, M.; Nagase, T. *Macromolecules* **1987**, *20*, 298.
4. Robinson, C. *Mol. Cryst.* **1966**, *1*, 467.
5. Watanabe, J.; Takashima, Y. *Macromolecules* **1991**, *24*, 3423.
6. Ballauf, M.; Schmidt, G. F. *Makromol. Chem., Rapid Commun.* **1989**, *8*, 93.
7. Watanabe, J.; Harkness, B. R.; Sone, M.; Ichimura, H. *Macromolecules* **1994**, *27*, 507.
8. Watanabe, M.; Aoki, S.; Samui, K.; Ogata, N. *Polym. Adv. Technol.* **1993**, *4*, 179.
9. Blodgett, K. B. *J. Am. Chem. Soc.* **1935**, *57*, 1007.
10. Gaines, G. L. *Insoluble Monolayers at Liquid-Gas Interfaces*; Wiley-Interscience: New York, 1966.
11. Ullman, A. *An Introduction to Ultrathin Organic Films–From Langmuir-Blodgett to Self-Assembly*; Academic: San Diego, 1991.
12. Orthmann, E.; Wegner, G. *Angew. Chem. Int. Ed. Engl.* **1986**, *25*, 1105.
13. Wegner, G. *Thin Solid Films* **1992**, *216*, 105.
14. Tsukruk, V. V.; Foster, M. D.; Reneker, D. H.; Schmidt, A.; Knoll, W. *Langmuir* **1993**, *9*, 3538.
15. Tredgold, R. H.; Jones, R. *Langmuir*, **1989**, *6*, 531.
16. Schwiegk, S.; Vahlenkamp, T.; Xu, Y.; Wegner, G. *Macromolecules* **1992**, *25*, 2513.
17. Duda, G. Ph.-D. Thesis, Johannes-Gutenberg-Universität Mainz, Germany.
18. Menzel, H.; Hallensleben, M. L. *Polym. Bull.* **1991**, *27*, 89.
19. Embs, F.; Wegner, G.; Neher, D.; Albouy, P.; Miller, R. D.; Wilson, C. G.; Schrepp, W. *Macromolecules* **1991**, *24*, 5068.
20. Kani, R.; Yoshida, H.; Nakano, Y.; Murai, S.; Mori, Y.; Kawata, Y.; Hayase, S.; *Langmuir* **1993**, *9*, 3045.
21. Itoh, T.; Suzuki, H.; Matsumoto, M.; Miyamoto, T.; in *Cellulose, Structure and Functional Aspects*; Kennedy, J. F.; Phillips, G. O.; Williams, P. A., Eds.; Ellis Horwood: London, 1989; Chapter 54.
22. Minx, C.; Menzel, H. unpublished results.
23. Mao, L.; Ritcey, A. M. *Thin Solid Films* **1994**, *242*, 263.
24. Kano, T.; Gray, D. G. *J. Appl. Polym. Sci.* **1992**, *45*, 417.
25. Rubner, M. F.; Skotheim, T. A. in Bredas, J. L.; Shilbey, R. Eds., *Conjugated Polymers*; Kluwer: Dordrecht, 1991; 363.
26. Ando, M.; Watanabe, Y.; Iyoda, T.; Honda, K.; Shimidzu, T. *Thin Solid Films* **1989**, *179*, 225.
27. Menzel, H.; Weichart, B.; Hallensleben, M. L. *Polym. Bull.* **1992**, *27*, 637.
28. Menzel, H.; Weichart, B.; Hallensleben, M. L. *Thin Solid Films* **1993**, *223*, 181.
29. Menzel, H.; Weichart, B. *Thin Solid Films* **1994**, *242*, 56.
30. Fuchs, H.; Ohst, H.; Prass, W. *Adv. Mater.* **1991**, *3*, 10.
31. Erdelen, C.; Laschewesky, A.; Ringsdorf, H.; Schneider, J.; Schuster, A. *Thin Solid Films* **1989**, *180*, 153.
32. Tippmann-Krayer, P.; Riegler, H.; Paudler, M.; Möhwald, H.; Siegmund, H-U.; Eickmanns, J.; Scheunemann, U.; Licht, U.; Schrepp, W. *Adv. Mater.* **1991**, *3*, 46.
33. Wegner, G.; Mathauer, K. *Mat. Res. Symp. Proc.* **1992**, *247*, 767.
34. Hickel, W.; Duda, G.; Wegner, G.; Knoll, W. *Makromol. Chem., Rapid Commun.* **1989**, *10*, 353.
35. Hickel, W.; Duda, G.; Jurick, M.; Kröhl, T.; Rocheford, K.; Stegemann, G. I.; Swalen, J. D.; Wegner, G.; Knoll, W. *Langmuir* **1990**, *6*, 1403.
36. Mathy, A.; Mathauer, K.; Wegner, G.; Bubeck, C. *Thin Solid Films* **1992**, 98.
37. Suzuki, M.; Ferencz, A.; Iida, S.; Enkelmann, V.; Wegner, G. *Adv. Mater.* **1993**, *5*, 359.

Halobutyl Rubber

See: Butyl and Halobutyl Rubbers
Isobutylene Copolymers (Commercial)

Hardeners

See: Epoxy Resins (Modification with Engineering Plastics)

HEAD-TO-HEAD POLYMERS

Stephen F. Hahn
Central Research and Development
The Dow Chemical Company

The polymerization of monosubstituted olefinic monomers usually give the head-to-tail configuration.[1–7] Ionic forms of polymerization, in which stabilization of a particular propagating ionic species is critical to chain growth, are not known to give rise to head-to-head sequences in several ways. Inverse monomer addition, although uncommon, can give rise to head-to-head sequences in some polymers. Side reactions, such as disproportionation and recombination of growing polymer radicals, can give head-to-head sequences in small amounts.

Synthesis of Polymers with Head-to-Head Configuration

Interest in head-to-head polymeric structures has led to a variety of synthetic efforts to prepare these polymers.[8–10] The primary interest in these structures has been to understand how the head-to-head configuration influences polymer properties, especially thermal degradation. Two distinct approaches have been used: polymerization of specially designed monomers that already have the required head-to-head substituents appropriately placed, and the chemical modification of existing, high molecular weight polymers.

Synthesis of Head-to-Head Polymers by Polymerization of Designed Monomers

The preparation of head-to-head polymers has often been approached by synthesizing 2,3-disubstituted butanes or butadienes and polymerizing those monomers through the 1 and 4 carbons. Synthesis of head-to-head polymers by the polymerization of 2,3-disubstituted butadienes has been studied in some detail due to the ease of synthesis of the precursor monomers and the variety of polymerization techniques available that give high polymer from these monomers.

Head-to-Head Polystyrene

Head-to-head polystyrene[11–19] has been prepared by the radical polymerization (bulk and emulsion) of 2,3-diphenyl-1,3-butadiene followed by hydrogenation of the residual backbone unsaturation. This polymer has also been hydrogenated to give the head-to-head equivalent of poly(vinyl cyclohexane).[20]

Head-to-Head Polypropylene

Head-to-head polypropylene has been prepared by the alternating copolymerization of ethylene and *cis*-2-butene using a VCl_3/alkyl aluminum catalyst system.[21,22] This polymer has also been prepared by anionic[23,24] emulsion,[25,26] and $TiCl_4$/aluminum alkyl catalyzed polymerization of 2,3-dimethyl-1,4-butadiene followed by hydrogenation of the backbone olefin.[27,28]

Head-to-Head Polyisobutylene

Polyisobutylene has been prepared by the cationic polymerization of 2,5-dimethyl-2,4-hexadiene followed by hydrogenation of the residual olefin, although complete hydrogenation was not achieved.[29]

Head-to-Head Acrylates

The head-to-head equivalent polymers of acrylate monomers have been prepared by the alternating copolymerization of maleic anhydride with ethylene or 1,2-disubstituted olefins followed by esterification of the anhydride moieties.

Head-to-Head Polymers by Chemical Reaction of High Polymer

A second common synthetic approach to the synthesis of head-to-head polymers involves the chemical modification of high molecular weight polymers. Bromine and chlorine add readily to the backbone unsaturation in polybutadiene to provide vicinal dihalides; this experiment is usually performed by simply adding the halogen to a methylene chloride or chloroform solution of the polymer.[30–40]

The chlorination of polybutadiene in the presence of nucleophiles has been used to prepare head-to-head structures with a variety of additional functional groups. Early chlorination work suggested that when methanol was present some of it was incorporated as a methoxide substituent on the polymer chain. Further work has shown that alcohols,[34,38] cyclic ethers,[41] carboxylic acids,[42] phenols, nitriles, and ketones[43] can participate in this reaction. These structures are equivalent to head-to-head copolymers of vinyl chloride and vinyl ethers (for alcohol participation), vinyl acetates (carboxylic acid participation), etc.

Physical Properties

Much of the interest in head-to-head polymers has been to determine the effect this orientation has on the physical properties of the polymer. Although the quantity of head-to-head structure is typically small, some physical properties that are sensitive to these structures can be influenced. Crystallization in polymeric systems, which is influenced by regularity of polymer structure, is one such property. When present even at low levels in largely head-to-tail polymer structures (less than 5%), head-to-head units can act as flaws in the chain structure, affecting crystallization.

The positioning of substituents on adjacent carbons can have a profound effect on the spectral and chemical behavior of the polymer. Attempts to elucidate the relationship between the quantity of head-to-head structures and thermal stability has been responsible for much of the research performed in this area. It has been thought that the presence of substituents on adjacent carbons could, in some instances, lead to stabilization of intermediates created by thermal degradation reactions, which would then initiate thermal degradation of the bulk polymer.

Glass transition temperatures have been determined for a number of head-to-head polymers. While the formation of polymers with substituents on adjacent carbons might be expected to lead to differences in T_g due to an increase in the energy required for chain rotation with respect to the head-to-tail versions, the relative flexibility of the ethane tail-to-tail coupling minimizes the measured differences for most nonpolar polymers. In fact, for hydrocarbon head-to-head polymers, T_g tends to decrease. A notable exception is polyisobutylene, in which the glass transition temperature of the head-to-head form is 160°C higher than that of the head-to-tail form; in this case, the adjacent dimethyl substituted carbons provide a significant impediment to chain rotation. For polymers with polar substituents, glass transition temperatures of the head-to-head structures tend to be higher than those of the head-to-tail versions.

Perhaps the most thoroughly studied of the purely head-to-head polymers is polystyrene. The two different forms of this polymer have similar glass transition temperatures, but a comparison of dilute solutions properties shows meaningful differences. The theta temperature of head-to-head polystyrene in cyclohexane is significantly lower than that of the head-to-tail polymer, and the size of the unperturbed polymer coil is much larger for the head-to-head polymer (similar to that of isotactic polystyrene).

Rheological study of head-to-head and head-to-tail polystyrenes of similar molecular weight show virtually identical melt viscosities, although the viscosity of the head-to-head polymer deviates from Newtonian behavior at lower frequencies than head-to-tail polystyrene.[44]

REFERENCES

1. Flory, P. J. *Principles of Polymer Chemistry*; Cornell University: Ithaca, New York, 1953; Chapter 6.
2. Mark, H. U.S. Patent 3 081 282.
3. Lenz, R. W. *Organic Chemistry of Synthetic High Polymers*; John Wiley & Sons: New York, 1967; pp 343–344.
4. Staudinger, H.; Steinhofer, A. *Ann.* **1935**, *517*, 35.
5. Marvel, C. S.; Sample, J. H.; Roy, M. F. *J. Am. Chem. Soc.* **1939**, *61*, 3241.
6. Marvel, C. S.; Levesque, C. L. *J. Am. Chem. Soc.* **1938**, *60*, 280.
7. Marvel, C. S.; Denson, C. E. *J. Am. Chem. Soc.* **1938**, *60*, 1045.
8. Vogl, O.; Malanga, M.; Berger, W. *Contemporary Topics in Polymer Science*, ACS Symposium Series, American Chemical Society: Washington, DC, 1983; Vol. 4, p 35.
9. Vogl, O.; Bassett, W. Jr.; Grossman, S.; Kawaguchi, H.; Kitayama, M.; Kondo, T.; Malanga, M.; Xi, F. *J. Macromol. Sci. Chem.* **1984**, *A21*, 1725.
10. Vogl, O.; Jaycox, G. D.; Hatada, K. *J. Macromol. Sci.-Chem.* **1990**, *A27*(13+14), 1781.
11. Inoue, H.; Helbig, M.; Vogl, O. *Macromolecules* **1977**, *10*, 1331.
12. Corley, L. S.; Inoue, H.; Helbig, M.; Vogl, O.; Seki, K.; Tirrell, D. A. *Macromol. Synth.* **1985**, *9*, 31.
13. Starzielle, C.; Benoit, H.; Vogl, O. *Eur. Polym. J.* **1978**, *14*, 331.
14. Marchal, E.; Benoit, H.; Vogl, O. *J. Polym. Sci., Polym. Phys. Ed.* **1978**, *16*, 949.
15. Laupretre, F.; Monnerie, L.; Vogl, O. *Eur. Polym. J.* **1978**, *14*, 981.
16. Weill, G.; Vogl, O. *Polym. Bull.* **1978**, *1*, 181.
17. Lindsell, W. E.; Robertson, F. C.; Soutar, I. *Eur. Polym. J.* **1981**, *17*, 203.
18. Torkelson, J. M.; Lipsky, S.; Tirrell, M. *Macromolecules* **1981**, *14*, 1603.
19. Foldes, E.; Deak, G.; Tudos, J.; Vogl, O. *Eur. Polym. J.* **1993**, *29*, 321.
20. Helbig, M.; Inoue, H.; Vogl, O. *J. Polym. Sci.: Polym. Symp.* **1978**, *63*, 329.
21. Natta, G.; Dall'Asta, G.; Mazzanti, G.; Pasquon, I.; Valvassori, A.; Zambelli, A. *J. Am. Chem. Soc.* **1961**, *83*, 3343.
22. Natta, G.; Allegra, G.; Bassi, I. W.; Corradini, P.; Ganes, P. *Makromol. Chem.* **1962**, *58*, 242.
23. Prud'homme, J.; Roovers, J. E. L.; Bywater, S. *Eur. Polym. J.* **1972**, *8*, 901.
24. Khlok, D.; Deslandes, Y.; Prud'homme, J. *Macromolecules* **1976**, *9*, 809.
25. Arichi, S.; Pedram, M. Y.; Cowie, J. M. G. *Eur. Polym. J.* **1979**, *15*, 107.
26. Arichi, S.; Pedram, M. Y.; Cowie, J. M. G. *Eur. Polym. J.* **1979**, *15*, 113.
27. Grossman, S.; Yamada, A.; Vogl, O. *J. Macromol. Sci.* **1981**, *A16*(4), 897.
28. Grossman, S.; Stolarczyk, A.; Vogl, O. *Monatsh, Chem.* **1981**, *112*, 1279.
29. Malanga, M.; Vogl, O. *J. Macromol. Sci.* **1985**, *A22*(12), 1623.
30. Cantarino, P. J. *Ind. Eng. Chem.* **1957**, *49*, 712.
31. Bailey, F. E. Jr.; Henry, J. P.; Lundberg, R. D.; Whelan, J. M. *J. Polym. Sci., Part B* **1964**, *2*, 447.
32. Pinazzi, C. P.; Gueniffey, H. *Makromol Chem.* **1966**, *93*, 109.
33. Drehfahl, G.; Horhold, H. H.; Hess, E. *J. Polym. Sci.: Part C* **1967**, *16*, 965.
34. Murayama, N.; Amagi, Y. *J. Polym. Sci., Part B* **1966**, *4*, 119.
35. Kawaguchi, H.; Loeffler, P.; Vogl, O. *Polymer* **1985**, *26*, 1257.
36. Kawaguchi, H.; Sumida, Y.; Muggee, J.; Vogl, O. *Polymer* **1983**, *23*, 1805.
37. Takeda, M.; Endo, R.; Matsuura, Y. *J. Polym. Sci., Part C* **1968**, *23*, 487.
38. Bevington, J. C.; Ratti, L. *Polymer* **1975**, *16*, 66.
39. Mitani, K.; Ogata, T.; Awaya, H.; Tomari, Y. *J. Polym. Sci., Polym. Chem. Ed.* **1975**, *13*, 2813.
40. Crawley, S.; McNeill, I. C. *J. Polym. Sci., Polym. Chem. Ed.* **1978**, *16*, 2593.
41. Dreyfuss, M. P.; Nevius, M. R.; Manninen, P. R. *J. Polym. Sci.: Part C: Polym. Letters* **1987**, *25*, 99.
42. Tarshiani, Y.; Dreyfuss, M. P. *J. Polym. Sci.: Part A: Polym. Chem.* **1990**, *28*, 205.
43. Hahn, S. F.; Dreyfuss, M. P. *J. Polym. Sci.: Part A: Polym. Chem.* **1993**, *31*, 3039.
44. Jacovic, M.; Vogl, O. *Polym. Eng. Sci.* **1978**, *18*, 875.

Heart Valves

See: Blood Compatible Polymers

Heat Stabilizers

See: Additives (Property and Processing Modifiers)
Organo-Tin Polymers

Helical Polymers

See: Asymmetric Polymerization (Overview)
Nylons 3, Helical [poly(α-alkyl-β-L-aspartates)
Polypeptide Restricted Backbone Conformation (Cα,α-Symmetrically Disubstituted Glycines)

Helicates

See: Metallosupramolecular Helicates

Helicenes

See: Heterohelicenes

Hemicellulose

See: Hemicelluloses
Jute

HEMICELLULOSES

Ronald D. Hatfield
U.S. Dairy Forage Research Center
USDA Agricultural Research Service

James L. Minor (retired)
Forest Products Laboratory
USDA Forest Service

Hemicelluloses [9034-32-6] are a diverse class of plant polysaccharides found in close physical association with cellulose, but chemically unrelated to cellulose. The name was originally proposed for polysaccharides that could be extracted from plants with aqueous alkali, but it now includes all non-cellulosic, cell-wall structural polysaccharides of land plants.

Wood is a convenient source of hemicelluloses. They may be used in the intact wood or large quantities may be obtained as byproducts of pulp for papermaking or for the production of cellulose derivatives. Hemicelluloses from wood vary with species and location within the tree, but the types of predominant hemicelluloses are characteristic of gymnosperms (softwoods) or angiosperms (hardwoods).

PREPARATION

Biosynthesis of Structural Polysaccharides

Although oligosaccharides have been chemically synthesized, making it theoretically possible to chemically synthesize polymers, isolating hemicellulose polysaccharides from plant cell walls is far more efficient. Studying the biosynthetic pathways of polysaccharides is important for future control and modification of polysaccharide syntheses both *in vivo* and *in vitro*.

PROPERTIES AND USES

Hemicelluloses play an important, but poorly understood biological role that is the subject of current research. They are known to form an amorphous gel embedding cellulose microfibrils. It is also believed that they may regulate cellulose biosynthesis.[1] The complexity of structural features, including polydispersity (molecular mass), makes it difficult to determine the absolute chemical structure of a given hemicellulosic polysaccharide. However, these features control the physicochemical properties of the molecule and dictate the type, strength, and abundance of interactions with other wall components. Such interactions control the structural and functional roles of a given wall and influence degradation of the matrix, utilization by ruminants, and properties as papermaking fibers.

Hemicelluloses are largely responsible for hydration and development of bonding during beating of chemical pulps. Pulps containing high percentages of hemicellulose typically have weak tearing strength but develop high bonding strengths. It is important to learn how to modify these properties if chemical pulp yields are to be maximized. Hemicelluloses have been proposed as additives in papermaking.[2]

Hemicelluloses are used as a feedstock for the production of sugars that can then be converted to other products, such as xylitol, furfural from xylose, or fuel alcohol from fementable sugars. Xylitol is used as a non-caloric sweetener. Considerable research has been expended to improve the fermentation of the pentan hemicelluloses.[3]

Because hemicelluloses are found in grains and other food plants for animals and humans, they have various uses in the food industry.

Hemicelluloses are important constituents of bowel-movement-improving products incorporating water-soluble and slightly water-soluble types of dietary fibers.

Because hemicelluloses are susceptible to enzymatic action and are non-toxic, they can be used in making sustained-release composition. Also delayed-release pharmaceutical tablets may be formed by incorporating hemicellulose and a matrix-forming adjuvant that does not disintegrate.[4]

Oxidation products of hemicelluloses may be added to surface-active detergents or cleaning agents.[5] Strong, water-soluble, non-hygroscopic adhesives may be made from hemicelluloses.

The above list of uses is not conclusive, but it does indicate how hemicelluloses are put to practical use.

REFERENCES

1. Atalla, R. H.; Hackney, J. M.; Uhlin, I.; Thompson, N. S. *Int. J. Biol. Macromol.* **1993**, *15*, 109.
2. El-Ashmawy, A. E.; Mobarak, F.; Fahmy, Y. *Cellulose Chem. Technol.* **1973**, *7*, 315.
3. Jeffries, T. W.; Kurtzman, C. P. *Enzyme Microb. Technol.* **1994**, *16*, 992.
4. Wegleitner, K.; Korsatko, W.; Korsatko-Wabnegg, B. U.S. Patent 5 151 273, 1992.
5. Cornelissens, E. G.; Zielke, R.; Diamantoglou, M.; Magerlein, H. U.S. Patent 4 056 400, 1977.
6. Reintjes, M.; Starr, L. D. U.S. Patent 3 832 313, 1974.

Hemiisotactic Polypropylene

See: *Metallocene Catalysts*
Polypropylene, Hemiisotactic

Hemoglobin

See: *Blood Substitutes*
Hemoproteins
Red Blood Cell Substitutes, Artificial (for Transfusion)

HEMOPROTEINS

Yasuhiko Yamamoto
Department of Chemistry
University of Tsukuba

Hemoproteins are perhaps the most extensively studied group of proteins. This is due not only to their wide distribution and abundance, but also to their many unique spectral properties. The unusual spectral properties of hemoproteins stem from the presence of the heme group, usually iron protoporphyrin IX. This molecule is found as the prosthetic group of a number of proteins possessing quite different functions: electron transport proteins (cytochromes), oxidase enzymes (peroxidase catalase), and oxygen transport/storage proteins (hemoglobin (Hb)/myoglobin (Mb)). Since the proteins all contain the same prosthetic group, their functional differences are attributed to differences in the way the protein interacts with the heme and with potential substrates. The heme in the electron transport hemoproteins is exceptional in that it has two' strong axial ligands and generally does not bind exogenous ligands. The heme in all the other hemoproteins has one accessible coordination site at the heme, which allow the ligand to bind or react with peroxidase and other substrates.

Cytochromes are found in all organisms except a few types of obligate anaerobes.[1] Each group of cytochromes contains a different heme derivative. The b-type cytochromes contain protoheme IX. The heme group of c-type cytochromes differs from protoheme IX in that its vinyl groups are reacted with cysteine sulfhydryls to form thioether linkage to the protein. Heme *a* contains a long hydrophobic tail of isoprene units attached to the porphyrin, as well as a formyl group in place of a methyl substituent. The axial ligands to the heme iron also vary with the cytochrome type. In cytochromes a and b, both ligands are histidine (His) residues, whereas in cytochrome c one is His and the other is methionine. Among the different types of cytochromes occurring in the cell, the soluble c-type cytochromes are most abundant and have been used extensively to examine relations of evolutionary variations in the protein structure to the phylogeny of the species.[3]

Peroxidases and catalases contain ferric protoheme IX.[4,5] Peroxidases primarily oxidize molecules using hydrogen peroxide, and catalase mainly catalyzes the disproportionation reaction of hydrogen peroxide to water and molecular oxygen and decomposition of organic peroxides. These proteins, including a variety of other heme enzymes, also occur in most living systems.

Hb and Mb are possibly the most-studied of all hemoproteins.[6–9] Their reactions with various ligands have been studied in detail.[6] Furthermore, sperm whale Mb was the first protein crystal structure to be determined.[10]

PROPERTIES

Structure

Mb is a monomeric hemoprotein consisting of a single polypeptide chain (globin) and a prosthetic group (heme). The protein moiety in sperm while Mb is composed of 153 amino acid residues (its molecular weight is about 17,000) and is largely helical (eight helices in sperm whale Mb are labeled A to H from the amino to the carboxyl terminus). Despite large differences in the sequence, the three-dimensional structures of globins are remarkably similar.

Human Hb (called Hb A) is a globular protein consisting of two pairs of polypeptide chains, usually called subunits, two α subunits and two β subunits, and the tertiary structure of the subunits resembles that of Mb. The four Mb-like subunits are assembled in a tetrahedral array to form a roughly spherical molecule. The arrangement of the subunits is called quaternary structure. The quaternary structure is stabilized by non-covalent bonds, such as van der Waals contacts, hydrogen bonds, and salt bridges between unlike subunits. There are a few salt bridges between the α subunits, whereas no direct interaction exists between the β subunits.

Heme iron in Hb and Mb generally is in either the ferrous (+2) or ferric (+3) state. The number of electrons in 3d orbitals of ferrous and ferric iron are six and five, respectively. The spin quantum number S is integer and half-integer in ferrous and ferric hemes, respectively. The spin state of hemoprotein depends on the chemical nature (or field strength) of the ligand.

Function

Functional properties of Hb and Mb have been thoroughly investigated.[6] The function of Mb is to carry oxygen from Hb to the electron transfer system in mitochondria of muscle cells. The principal functions of Hb are to transport oxygen from the lungs to the tissues and to facilitate the transport of carbon dioxide from the tissues to the lungs. Since heme iron in each subunit binds oxygen, an Hb molecule carries four oxygen molecules. The binding site of carbon dioxide is not the heme, but the amino acid terminals of individual polypeptide chains of Hb.

Heme-Modified Protein

Reconstitution reaction of hemoprotein allows chemical modification of the heme.[9–15] Chemical modification of the heme peripheral side chains and substitution of the central metal ion with unusual metals can be easily carried out to prepare reconstituted heme-modified protein. Comparison of functional properties between native and heme-modified proteins provides the relative importance of heme electronic structure and heme-protein contacts in determining the functional properties of the proteins. Heme-modified proteins often exhibit quite striking differences from native protein with respect to such properties as redox potential, oxygen affinity, and ligand binding kinetics.

Control of Ligand Binding in Mb

Molecular mechanisms operating in ligand binding in Mb have been exhaustively investigated using both wild-type proteins, such as mollusk *Aplysia limacina* Mb, Asian elephant Mb and mollusk *Dolabella auricularia* Mb, and genetic mutants.[16–45]

REFERENCES

1. Dickerson, R. E. *Sci. Am.* **1972**, *226*(4), 58.
2. Saunders, B. C.; Holmes-Siedle, A. G.; Stark, B. P. *Peroxidases*; Butterworths: London, 1964.
3. Schonbaum, G. R.; Chance, B.; *The Enzyme*; Boyer, P. D., Ed.; Academic: New York, 1976; p 363.
4. Antonini, E.; Brunori, M. *Hemoglobin and Myoglobin in Their Reactions with Ligands*; North-Holland: Amsterdam, 1971.
5. *Methods in Enzymology* Vol. 76, *Hemoglobin*; Antonini, E.; Rossi-Bernardi; L.; Chiancone, E. Eds.; Academic: New York, 1981.
6. Imai, K. *Allosteric Effects in Haemoglobin*; Cambridge University: Cambridge, 1982.
7. Dickerson, R. E.; Geis, I. *Hemoglobin: Structure, Function, Evolution and Pathology*; Benjamin/Cummings: Menlo Park, CA, 1983.
8. Kendrew, J. C.; Dickerson, R. E.; Strandberg, B. E.; Hart, R. G.; Davies, D. R.; Phillips, D. C.; Shore, V. C. *Nature (London)* **1960**, *185*, 422.
9. Yonetani, T.; Asakura, T. *J. Biol. Chem.* **1968**, *243*, 4715.
10. Makino, R.; Yamazaki, I. *Arch. Biochem. Biophys.* **1973**, *157*, 356.
11. Makino, R.; Yamazaki, I. *Arch. Biochem. Biophys.* **1974**, *165*, 485.
12. Yamada, H.; Makino, R.; Yamazaki, I. *Arch. Biochem. Biophys.* **1975**, *169*, 344.
13. Sono, M., Asakura, T. *J. Biol. Chem.* **1976**, *250*, 5227.
14. Inubushi, T.; Yonetani, T. *Methods Enzymol.* **1981**, *76*, 88.
15. Makino, R.; Iizuka, T.; Sakaguchi, K.; Ishimura, Y. *Oxygenases and Oxygen Metabolism*; Nozaki, M.; Yamamoto, S.; Ishimura, Y.; Coon, M. J.; Ernster, L.; Estabrook, W. R., Eds.; Academic: New York, 1982; p 467.
16. Giacometti, G. M.; Da Ros, A.; Antonini, E.; Brunori, M. *Biochemistry* **1975**, *14*, 1584.

17. Giacometti, G. M.; Ascenzi, P.; Brunori, M.; Rigatti, G.; Giacometti, G.; Bolognesi, M. *J. Mol. Biol.* **1981**, *151*, 315.
18. Giacometti, G. M.; Ascenzi, P.; Bolognesi, M.; Brunori, M. *J. Mol. Biol.* **1981**, *146*, 363.
19. Rousseau, D. L.; Ching, Y.-C.; Brunori, M.; Giacometti, G. M. *J. Biol. Chem.* **1989**, *264*, 7878.
20. Bolgnesi, M.; Cannillo, E.; Ascenzi, P.; Giacometti, G. M.; Merli, A.; Brunori, M. *J. Mol. Biol.* **1982**, *158*, 305.
21. Bolgnesi, M.; Coda, A.; Gatti, G.; Ascenzi, P.; Brunori, M. *J. Mol. Biol.* **1985**, *183*, 113.
22. Bolognesi, M.; Onesti, S.; Gatti, G.; Coda, A.; Ascenzi, P., Brunori, M. *J. Mol. Biol.* **1989**, *205*, 529.
23. Bolognesi, M.; Coda, A.; Frigerio, F.; Gatti, G.; Ascenzi, P.; Brunori, M. *J. Mol. Biol.* **1990**, *213*, 621.
24. Mattevi, A.; Gatti, G.; Coda, A.; Rizzi, M.; Ascenzi, P.; Brunori, M.; Bolognesi, M. *J. Mol. Recogn.* **1991**, *4*, 1.
25. Conti, E.; Moser, C.; Rizzi, M.; Mattevi, A.; Lionetti, C.; Coda, A.; Ascenzi, P.; Brunori, M.; Bolognesi, M. *J. Mol. Biol.* **1993**, *233*, 498.
26. Pande, U.; La Mar, G. N.; Lecomte, J. T. J.; Ascoli, F.; Brunori, M.; Smith, K. M.; Pandey, R. K.; Parish, D. W.; Thanabal, V. *Biochemistry* **1986**, *25*, 5638.
27. Pyton, D. H.; La Mar, G. N.; Pande, U.; Ascoli, F.; Smith, K. M.; Pandey, R. K.; Parish, D. W.; Bolognesi, M.; Brunori, M. *Biochemistry* **1989**, *28*, 4880.
28. Qin, J.; La Mar, G. N.; Ascoli, F.; Bolognesi, M.; Brunori, M. *J. Mol. Biol.* **1992**, *224*, 891.
29. Bartnicki, D. E.; Mizukami, H.; Romero-Herrera, A. E. *J. Biol. Chem.* **1983**, *258*, 1599.
30. Kerr, E. A.; Yu, N.-T.; Bartnicki, D. E.; Mizukami, H. *J. Biol. Chem.* **1985**, *260*, 8360.
31. Krishnamoorthi, R.; La Mar, G. N.; Mizukami, H.; Romero, A. *J. Biol. Chem.* **1984**, *259*, 265.
32. Krishnamoorthi, R.; La Mar, G. N.; Mizukami, H.; Romero, A. *J. Biol. Chem.* **1984**, *259*, 8826.
33. Yamamoto, Y.; Osawa, A.; Inoue, Y.; Chûjô, R.; Suzuki, T. *Eur. J. Biochem.* **1990**, *192*, 225.
34. Yamamoto, Y.; Chûjô, R.; Suzuki, T. *Eur. J. Biochem.* **1991**, *198*, 285.
35. Yamamoto, Y.; Iwafune, K.; Chûjô, R.; Inoue, Y.; Imai, K.; Suzuki, T. *J. Mol Biol.* **1992**, *228*, 343.
36. Yamamoto, Y.; Iwafune, K.; Nanai, N.; Chûjô, R.; Inoue, Y.; Suzuki, T. *Biochem. Biophys. Acta* **1992**, *1120*, 173.
37. Yamamoto, Y.; Iwafune, K.; Chûjô, R.; Inoue, Y.; Imai, K.; Suzuki, T. *J. Biochem.* **1992**, *112*, 414.
38. Yamamoto, Y.; Chûjô, R.; Inoue, Y.; Suzuki, T. *FEBS Lett.* **1992**, *310*, 71.
39. Yamamoto, Y.; Suzuki, T.; Hori, H. *Biochim. Biophys. Acta* **1993**, *1203*, 267.
40. Yamamoto, Y.; Suzuki, T. *Biochim. Biophys. Acta* **1993**, *1163*, 287.
41. Rohlfs, R. J.; Mathews, A. J.; Carver, T. E.; Olson, J. S.; Spriner, B. A.; Egeberg, K. D.; Sligar, S. G. *J. Biol. Chem.* **1990**, *265*, 3168.
42. Cutruzzolá, F.; Allocatelli, C. T.; Ascenzi, P.; Bolognesi, M.; Sligar, S. G.; Brunori, M. *FEBS Lett.* **1991**, *282*, 281.
43. Rajarathnam, K.; La Mar, G. N.; Chu, M. L.; Sligar, S. C.; Singh, J. P.; Smith, K. M. *J. Am. Chem. Soc.* **1991**, *113*, 7886.
44. Rizzi, M.; Bolognesi, M.; Coda, A.; Cutrozzolá, F.; Allocatelli, C. T.; Brancaccio, A.; Brunori, M. *FEBS Lett.* **1993**, *320*, 13.
45. Allocatelli, C. T.; Cutruzzolá, F.; Brancaccio, A.; Brunori, M.; Qin, J.; La Mar, G. N. *Biochemistry* **1993**, *32*, 6041.

Heparin

See: *Glycosaminoglycans (Overview)*
Heparin (Overview)
Heparinoids, Artificial (Biological Activities)
Ion-Chelating Polymers (Medical Applications)
Lung, Artificial

HEPARIN (Overview)

K. B. Johansen and Anni Larnkjaer
Heparin Research Laboratory
Leo Pharmaceutical Products

Heparin belongs to the group of polysaccharides known as glycosaminoglycans (GAGs), and is composed of alternating, 1–4–linked hexuronic acid and D–glucosamine (GlcN).[1] Both the hexuronic acid and the glucosamine residues are sulfated in a complex pattern. L–Iduronic acid (IdoA) is the predominant uronic acid and D–glucuronic acid (GlcA) is present in minor amounts. The D–glucosamine units are either *N*-sulfated or *N*-acetylated. The anomeric configurations of the glucosamine and the iduronic acid are α, whereas the glucuronic acid is in the β form. As a consequence of the variation in the building blocks, the heparin chains display extensive structural variability. However, all GAGs consist of repeating disaccharide units, which account for the largest fraction of these polysaccharides. For heparin the predominant repeating unit is the trisulfated disaccharide (1–4)–α–L–iduronic acid 2–sulfate–(1–4)–α–D–glucosamine *N*,6–disulfate, and it represents up to 75% and 85% of the structure in pig mucosal and beef lung heparin, respectively.[2,3]

Essentially as a result of its high negative-charge density, heparin is able to interact with clusters of basic amino acids on numerous proteins and cell membranes, such as coagulation proteinases, serine protease inhibitors,[4,5] growth factors,[6–8] lipoprotein and hepatic lipase,[9,10] apolipoproteins B and E,[11,12] adhesive matrix proteins,[13] platelets,[14,15] and endothelial cells.[16–18] Although many of these interactions are not of biological significance, the effect of heparin on blood coagulation is exploited clinically as an anticoagulant and antithrombotic drug.

Besides conventional heparin, also called standard heparin, a second generation of heparin drugs, the low molecules weight (LMW) heparins, has been developed. The LMW heparins differ in molecular weight and in coagulation profiles (anti-XA and anti-IIa activities) from standard heparin. In addition the LMW heparins show improved pharmacokinetic properties, resulting in clinical advantages.[19–22]

SYNTHESIS

The Manufacture of Heparin

Previously, bovine lung and bovine liver were the most common sources for preparation of commercial unfractionated (standard) heparin. However, due to the relatively low yield of heparin from lungs and the alternative use of lungs in the pet food industry, procine intestinal mucosa is the raw material of choice. The mucosa is a side product of manufacturing sausage casings and most heparin used today is of porcine origin.[23]

Heparin production takes place in two major steps; the first involves extraction and isolation of crude heparin, and the subsequent step consists of purifying the crude heparin.[24]

Low molecular weight (LMW) heparins are derived from conventional heparin by different means of partial depolymerization combined with size fractionation. Some of the principal depolymerization methods comprise deaminative cleavage by nitrous acid; enzymatic cleavage with heparinase; alkaline hydrolysis of heparin esters; and peroxidative depolymerization.[25] The non-reducing and reducing terminal residues will differ depending on the method of depolymerization used.

PROPERTIES

Physico–Chemical Properties

Appearance and solubility–Purified heparin appears as an odorless white to creamy-white hygroscopic powder. Heparin's high content of sulfate groups makes it the most acidic compound in nature. It dissolves readily in water and is practically insoluble in organic solvents.[24]

Molecular weight–Heparin preparations are mixtures of polyanions in a relatively wide range of molecular weights. The polydispersity is in the range of 1.5. The mean molecular weight generally ranges between 12,000 and 15,000 and 4000 to 6000 Da for standard heparin and LMW heparin, respectively.[21]

Conformation–Whereas the conformations of D–glucuronic acid and D–glucosamine are well-established as $^4C_1$ chair conformation the conformation of L–iduronic acid is complicated by well-documented conformational mobility of this residue.[26] When L–iduronic acid is present in saccharide sequences two conformations, the $^1C_4$ chair and $^2S_0$ skew boat, are accessible. This equilibrium is displaced towards the $^1C_4$ form when it is a non-reducing residue whereas the $^4C_1$ form also contributes to the equilibrium when the iduronic acid is at the reducing end. This flexibility might be exploited in heparin's interaction with other molecules and contribute to the binding properties and biological versatility of heparin.[27]

Biological Properties

Heparin mediates a variety of biological processes by interacting with proteins. These interactions may affect biochemical reaction in different ways. It may lead to changes of the conformation of a protein with a subsequent increase or reduction of its activity; it can act as a template to bring proteins together and accelerate their interaction; it may displace proteins from their ligands at cell surfaces; and it may bind to cell receptors and, hence, potentially prevent proteins from binding to their receptors.[28] Traditionally two kinds of heparin/protein interactions have been recognized–positive cooperative binding and specific binding.[29]

APPLICATIONS

Heparin is used as an anticoagulant in circumstances where the normal propensity for clotting must be overcome, such as in hemodialysis and hemofiltration. It is also used to prevent and treat deep venous thromboembolism. LMW heparins have a number of advantages over standard heparin. The plasma-life of LMW heparins is two to four times longer and less dependent on dosage than the half-life of unfractionated heparin. Additionally, the anticoagulant response to weight-adjusted doses of LMW heparins is less variable compared to unfractionated and their bioavailability is improved. These properties allow LMW heparin to be administered once daily and without laboratory monitoring, which is a critical advantage in the clinical setting.[30–32] These dissimilarities in pharmacokinetic properties occur because of different binding characteristics to plasma proteins and endothelial cells.

FUTURE

The clinical use of heparin as anticoagulant and antithrombotic agents is now well-established in practice and future directions might be non-anticoagulant uses of heparin. Heparin interacts with many biological systems and a possible role in the treatment of atherosclerosis and restenosis,[33–36] angiogenesis and tumor metastasis,[35,37] human retroviral infections,[38,39] and exercise-induced asthma[40] has been speculated. However, much work and research remain before a non-anticoagulant derivative of heparin reaches the market.

REFERENCES

1. Jackson, R. L.; Busch, S. J.; Cardin, A. D. *Physiological Reviews* **1991**, *71*, 481.
2. Casu, B. In *Heparin, Chemical and Biological Properties, Clinical Applications*; Lane, D. A.; Lindahl, U., Eds.; Arnold: London, 1989; Chapter II.
3. Casu, B. *Ann. NY. Acad. Sci.***1989**, *556*, 1.
4. Bourin, M.-C., Lindahl, U. *Biochem. J.* **1993**, *289*, 313.
5. Pratt, C. W.; Church, F. C. *Blood Coa. Fibrinol.* **1993**, *4*, 479.
6. D'Amore, P. A. *Haemostasis* **1990**, *20*, 159.
7. Klagsbrun, M. *Curr. Opin. Cell. Biol.* **1990**, *2*, 857.
8. Burgesand, W. H.; Maciag, T. *Ann. Rev. Biochem.* **1989**, *58*, 575.
9. Hata, A.; Ridinger, D. N.; Sutherland, S.; Emi, M.; Shuhua, Z.; Myers, R. L.; Ren, K.; Cheng, T.; Inoue, I.; Wilson, D. E.; Iverius, P. H.; Lalouel, J-M. *J. Biol. Chem.* **1993**, *268*, 8447.
10. Olivecrona, T.; Bengtsson-Olivecrona, G.; Østergaard, P. B.; Liu, G.; Chrevreuil, O.; Hultin, M. *Haemostasis* **1993**, *23*, 150.
11. Cardin, A. O.; Weintraub, H. J. R. *Arteriosclerose* **1989**, *9*, 21.
12. Mahley, R. W.; Hussain, M. M. *Curr. Opin. Lipidol.* **1991**, *2*, 170.
13. Biswas, C. *Trend. Glycosci. Glycotech.* **1992**, *4*, 53.
14. Lane, D. A.; Pejler, G.; Flynn, A. M.; Thompson, E. A.; Lindahl, U. *Biochem. J.* **1986**, *261*, 3980.
15. Messmore, H. L.; Griffin, B.; Fareed, J.; Coyne, E.; Seghatchian, J. *Ann. NY. Acad. Sci.* **1989**, *556*, 217.
16. Barzu, T.; Molho, P.; Tobelem, G.; Petitou, M.; Caen, J. *Biochem. Biophys. Acta* **1985**, *196*, 845.
17. Vannucchi, S.; Pasquali, F.; Porciatti, F.; Chiarugi, V.; Magnelli, L.; Bianchini, P. *Thromb. Res.* **1988**, *49*, 373.
18. Larnkjaer, A.; Østergaard, P. B.; Flodgaard, H. *J. Thromb. Res.* **1994**, *75*, 185.
19. Fareed, J.; Walenga, J. M.; Hoppensteadt, D.; Huan, X.; Racanelli, A. *Haemostasis* **1988**, *183*.
20. Vinazzer, H. *Semin. Thromb. Hemost.* **1991**, *17*, 385.
21. Hirsh, J.; Levine, M. M. *Blood* **1992**, *79*, 1.
22. Plessas, C. T.; Plessas, S. T. *Epitheor. Klin. Farmakolo. Farmakokin. Int.* **1992**, *6*, 5.
23. Coyne, E. *Chemistry and Biology of Heparin*; Lundblad, R. L.; Brown, W. V.; Mann, K. G.; Roberts, H. R., Eds; 1981; p 9.

24. Nachtmann, F.; Atzl, G.; Roth, W. D. *Analytical Profiles of Drug Substances* **1983**, *12*, 215.
25. Barrowcliffe, T. W. *Adv. Exp. Med. Biol.* **1992**, *313*, 205.
26. Gatti, G.; Casu, B.; Hamer, G. K.; Perlin, A. S. *Macromolecules* **1979**, *12*, 1001.
27. Ferro, D. R.; Provasoli, A.; Ragazzi, M. *Carbohydr. Res.* **1990**, *195*, 157.
28. Lane, D. A.; Adams, L. *N. Engl. J. Med.* **1993**, *329*, 129.
29. Zhou, F.; Höök, T.; Thompson, J. A.; Höök, M. *Adv. Exp. Med. Biol.* **1992**, *313*, 141.
30. Hirsh, J. *N. Z. J. Med.* **1992**, *22*, 487.
31. Young, E.; Wells, P.; Holloway, S.; Weitz, J.; Hirsh, J. *Thrombos. Haemost.* **1994**, *71*, 300.
32. Friedel, H. A.; Balfour, J. A. *Drugs* **1994**, *48*, 638.
33. Berk, B. C.; Gordon, J. B.; Alexander, R. W. *J. Am. Coll. Cardiol.* **1991**, *17*, 111B.
34. Currier, J. W.; Pow, T. K.; Haudenschild, C. C.; Minihan, A. C.; Faxon, D. F. *J. Am. Coll. Cardiol.* **1991**, *17*, 118B.
35. Hyslop, S.; de Nucci, G. *Semin. Thromb. Hemost.* **1993**, *19*, 89.
36. Lövqvist, A.; Emanuelsson, H.; Nilson, J.; Lundqvist, H.; Carlsson, J. *J. Int. Med.* **1993**, *233*, 215.
37. Thorpe, P. E.; Derbyshire, E. J.; Andrade, S. P.; Press, N.; Knowles, P. P.; King, S.; Watson, G. J.; Yang, Y-C.; Rao-Betté, M. *Cancer Res.* **1993**, *53*.
38. Holondiny, M.; Kim, S.; Katzenstein, D.; Konrad, M.; Groves, E. T.; Merigan, T. C. *J. Clin. Microbiol.* **1991**, *29*, 676.
39. Spear, P. G.; Shieh, M-T.; Herold, B. C.; WuDunn, D.; Koshy, T. I. *Adv. Exp. Med. Biol.* **1992**, *313*, 341.
40. Ahmed, T.; Girrigo, J.; Danta, I. *N. Engl. J. Med.* **1993**, *329*, 90.

HEPARINOIDS, ARTIFICIAL (Biological Activities)

Kenichi Hatanaka
Department of Biomolecular Engineering
Faculty of Bioscience and Biotechnology
Tokyo Institute of Technology

Sulfated polysaccharides are polyanions having polysaccharide backbone with N–sulfate, O–sulfate, and/or carboxyl groups, and they exhibit many kinds of biological activity. Natural sulfated polysaccharides (glycosaminoglycans) are classified as heparin, heparan sulfate, chondroitin sulfate, dermatan sulfate, keratan sulfate, and so on, according to the type of glycosidic linkage, the composition of amino sugar and hexuronic acid, and degree of sulfation. The anticoagulant activity of heparin is a well-known biological activity and other glycosaminoglycans also show important biological activities. Natural and artificial sulfated polysaccharides, except heparin, are called heparinoids.

There are two kinds of interactions between heparin or heparinoids and other biomolecules, specific and nonspecific. Two examples are strong recognition of a specific chemical structure of heparin and heparinoids, and simple electric interaction with negative charges of polyanions.

This section describes preparation and biological functions of artificial heparinoids. Some artificial heparinoids have higher biological activity than natural heparin or heparinoids. For example, lipemia clearing activity, which depends heavily on the degree of sulfation, of highly sulfated dextran sulfate is much stronger than that of heparin. Low molecular weight dextran sulfate, which has low toxicity, is therefore used therapeutically.

PROPERTIES

Anticoagulant Activity

It is well known that sulfated polysaccharides, such as heparin, bind to antithrombin III (heparin cofactor I) and heparin cofactor II to inhibit blood coagulation. The anticoagulant activity of artificial heparinoids depends not only on the kinds of saccharide and linkage, but also on the molecular weight and the degree of sulfation. Generally, the anticoagulant activities of furanan heparinoids are higher than those of the corresponding sulfated pyranan polysaccharides.[1] Moreover, the anticoagulant activity of each artificial heparinoid increases molecular weight and degree of sulfation.

Heparin and natural heparinoids have linear polysaccharide backbone structure, whereas sulfated branched polysaccharides can be synthesized artificially. The sulfated branched polysaccharide, such as sulfated (1→6)–α–D–glucopyranan, having glucopyranosyl branching units shows higher anticoagulant activity than the sulfated linear polysaccharide.[5]

Anti-HIV Activity

It is well known that AIDS (acquired immune deficiency syndrome) is a contagious disease that shows high mortality and is caused by HIV (human immunodeficiency virus).

Recently, it has been reported that heparin, dextran sulfate, and other sulfated polysaccharides inhibited HIV-induced cytopathic effects *in vitro*.[6–8] Lentinan sulfate,[9,10] curdlan sulfate,[3,4] and cellulose sulfate[11] prepared by sulfating natural polysaccharides strongly inhibit HIV-induced cytopathic effects *in vitro*. Furthermore, sulfated synthetic pentofuranans such as ribofuranan sulfate[2,9] and xylofuranan sulfate[9] also prevented HIV-induced cytopathic effects at low concentration.

Sulfated polysaccharides with low sulfations and polysaccharides having carboxyl groups[12] show no activity on HIV-induced cytopathic effects. Therefore, strength and density of the anion on the polysaccharide influence the anti-HIV activity. There may be specific and nonspecific interactions; i.e., strong recognition of a specific carbohydrate chain structure of sulfated polysaccharide, and simple electric interaction with negative charges of sulfated polysaccharide.

Interaction with Growth Factors

Acidic fibroblast growth factor (aFGF) and basic fibroblast growth factor (bFGF) are heparin-binding growth-factors, whose activities are changed by binding to heparin and heparan sulfate. Many researchers have studied the effective heparin fragment or heparan sulfate fragment that can be prepared by digestion with heparinase, heparanase, heparitinase, or other enzymes. Recently, growth factor binding fragment of heparin or hepan sulfate is gradually coming out, and there must be a specific interaction. However, the minimum structure necessary to show the non-specific interaction is also investigated.

APPLICATIONS

Although artificial heparinoids also exhibit as many kinds of biological activities as heparin, heparin and natural heparinoids are preferred for medical treatment because of their safety. However, because the density of sulfate groups influences lipemia clearing activity, dextran sulfate sodium salt with high sulfation is already produced as an oral medicine for hypercholesterolemia and arteriosclerosis. Recently, a column that contains porous cellulose gel coupling with dextran sulfate and adsorbs low density lipoprotein in the blood was developed by Kaneka Corporation, Japan, and is used to treat hyperlipemia.

REFERENCES

1. Hatanaka, K.; Yoshida, T.; Miyahara, S.; Sato, T.; Ono, F.; Uryu, T.; Kuzuhara, H. *J. Med. Chem.* **1987**, *30*, 810.
2. Hatanaka, K.; Nakajima, I.; Yoshida, T.; Uryu, T.; Yoshida, O.; Yamamoto, N.; Mimura, T.; Kaneko, Y. *J. Carbohydr. Chem.* **1991a**, *10*, 681.
3. Yamamoto, I.; Takayama, K.; Gonda, T.; Matsuzaki, K.; Hatanaka, K.; Yoshida, T.; Uryu, T.; Yoshida, O.; Nakashima, H.; Yamamoto, N.; Kaneko, Y.; Mimura, T. *British Polym. J.* **1990**, *23*, 245.
4. Yoshida, T.; Hatanaka, K.; Uryu, T.; Kaneko, Y.; Suzuki, E.; Miyano, H.; Mimura, T.; Yoshida, O.; Yamamoto, N. *Macromolecules* **1990**, *23*, 3717.
5. Yoshida, T.; Arai, T.; Mukai, Y.; Uryu, T. *Carbohydr. Res.* **1988**, *177*, 69.
6. Ito, M.; Baba, M.; Sato, A.; Pauwels, R.; Clercq, E.; Shigeta, S. *Antiviral Res.* **1987**, *7*, 361.
7. Nakashima, H.; Yoshida, O.; Tochikura, T. S.; Yoshida, T.; Mimura, T.; Kido, Y.; Motoki, Y.; Kaneko, Y.; Uryu, T.; Yamamoto, N. *Jpn. J. Cancer Res.* **1987a**, *78*, 1164.
8. Nakashima, H.; Kido, Y.; Kobayashi, N.; Motoki, Y.; Neushul, M.; Yamamoto, N. *Antimicrob. Agents Chemother* **1987b**, *31*, 1524.
9. Yoshida, O.; Nakashima, H.; Yoshida, T.; Kaneko, Y.; Yamamoto, I.; Uryu, T.; Matsuzaki, K.; Yamamoto, N. *Biochem. Pharmacol.* **1988**, *37*, 2887.
10. Hatanaka, K.; Yoshida, T.; Uryu, T.; Yoshida, O.; Yamamoto, N.; Mimura, T.; Kaneko, Y. *Jpn. J. Cancer Res.* **1989**, *80*, 95.
11. Yamamoto, I.; Takayama, K.; Honma, K.; Gonda, T.; Matsuzaki, K.; Hatanaka, K.; Uryu, T.; Yoshida, O.; Nakashima, H.; Yamamoto, N.; Kaneko, Y.; Mimura, T. *Carbohydr. Polym.* **1991**, *14*, 53.
12. Mizumoto, K.; Sugawara, I.; Ito, W.; Kodama, T.; Hayami, M.; Mori, S. *Jpn. J. Exp. Med.* **1988**, *58*, 145.

HETEROAROMATIC POLYMERS

Gerhard Maier
Technische Universität München

Originally, reasons to incorporate heteroaromatic rings into polymers were (1) the exceptionally high thermal stability of the resulting polymers; (2) the solubility and, hence, processability (films, fibers) of these polymers in contrast to similar, purely phenyl-based systems; and (3) their good mechanical properties, which were retained at temperatures up to more than 300°C. Polybenzimidazole, Polybenzobisoxazole and Polybenzobisthiazole are examples of high-temperature-resistant polymers that can be spun into fibers or cast into films directly from their polymerization solution. These materials are soluble only in strongly acidic systems, such as sulfuric acid, methane sulfonic acid, polyphosphoric acid, or Lewis-acid containing systems such as $GaCl_3$/nitromethane.[1,4–6] The 1,3–azole rings present in these polymers are electron deficit compared to phenyl rings, and are therefore less prone to oxidation. Another reason for the thermooxidative stability of such systems may be the reduced number of hydrogen atoms, when CH–groups in aromatic rings are replaced by heteroatoms because hydroxyl- and hydroperoxide-radicals produced by hydrogen abstraction are the chain carriers in combustion reactions.[3,7,8] The presence of nitrogen atoms allows for protonation or complexation and solubilization of the resulting polyions or complexes. In the case of polybenzimidazole (PBI), the mobility of the backbone, which is introduced by the bisbenzimidazolyl- and the 1,3-phenylgroups, resulting in enough chain flexibility to allow solubility in DMAc.[1,2]

In the recent literature, several trends can be observed in the field of polymers with five-membered heteroaromatic systems:

- Incorporation of heteroaromatic units into known polymers, such as polyamides, polyimides, polyesters, and polyaryl ethers, is used to affect temperature resistance, solubility, and processability.
- Attempts to improve the mechanical properties of fibers and films for high temperature applications.
- Improvement of solubility, increase of the range of applicable solvents.
- Improvement of synthesis processes.
- Rigid rod polyelectrolytes.
- Exploitation of the possible conjugation of the backbones of heteroaromatic groups in polymers leads to studies of electrical properties such as conductivity, electroluminescence, and electrochromic effects as well as second- or third-order non-linear optical properties.

INCORPORATION OF FIVE-MEMBERED HETEROAROMATIC RINGS INTO POLYMERS OF VARIOUS STRUCTURES

Poly(heteroarylene ether)s

Heteroaromatic rings can be incorporated into the backbone of poly(arylene ether)s in four different ways: (1) The formation of the heteroaromatic unit can be used as the polymer-forming reaction; the ether group is present in the monomers; (2) the heteroaromatic system can have two hydroxy or hydroxyphenyl groups, which can react with activated aromatic dihalides such as 4,4–difluorobenzophenone or 4,4′-difluorodiphenyl sulfone; (3) the heteroaromatic ring can be used as the activating group for halogen displacement. In this case the heteroaromatic system contains two halogen or halophenyl substituents in appropriate positions, which can react with bisphenols; and (4), Friedel–Crafts acylation of diphenyl ether with diacid chlorides.

Thiophene,[9–16] oxazole,[17–22] isoxaxole,[23] thiazole,[18] pyrazole,[24–26] imidazole,[27–30] 1,3,4-oxadiazole,[31–37] 1,3,4-thiadiazole,[38] 1,2,4-triazole,[35,39,40] benzoxazole,[35,41–44] benzothiazole,[35,45] and benzimidazole[42,46–53] containing poly(aryl ether)s have been prepared using nucleophilic halogen displacement.

POLYESTERS

A poly(ester benzoxazole) was prepared by Sirigu et al.[54] from terephthalic acid and the diacetate of 2–(4′-hydroxyphenyl)–6–hydroxybenzoxazole. This material exhibits a nematic phase between 408 and 518°C; however, the onset of weight loss coincides with the second transition temperature. The same authors also studied liquid crystalline poly(ester benzimidazole)s derived from the above mentioned benzoxazole monomer and aliphatic dicarboxylic acids.[55] Kricheldorf et al.[56–58] studied polyesters with similar structures, in which the orientation of the ester bonds was varied to determine the influence of this isomerism on polymer properties such as structure of the mesophases.

A polyester with 2,2′-biimidazole units in the main chain was prepared by Lister and Collier who studied the copper binding properties of the biimidazole group.[59]

POLYAMIDES

Polyamides based on 1,4–phenylene rings can exhibit lyotropic liquid crystalline phases, and can therefore be spun into high modulus fibers such as Kevlar or Twaron.

POLYIMIDES

The thermally stable heteroaromatic rings are ideal candidates for modifying the properties of polyimides. Without extensive loss in thermal stability, solubility of the fully imidized polymers can be achieved in some cases. In addition, the glass transition temperatures can be adjusted to the desired range.

IMPROVEMENT OF PROPERTIES OF POLYBENZOBISAZOLES

Fiber Properties

Polybenzobisoxazoles (*cis*-PBO) and polybenzobisthiazoles (*trans*-PBZT) are rigid rod polymers with superior thermooxidative stability. They can be prepared in high molecular weights by condensation of 1,3-diamino-4,6-dihydroxybenzene or 1,4-diamino-2,5-dimercaptobenzene and terephthalic acid in polyphosphoric acid. These materials can be processed into very strong fibers from lyotropic liquid crystalline solutions. It is generally assumed that PBO and PBZT behave like stiff, rod-like molecules in solution.

While the tensile properties of these fibers are comparable to carbon fibers, PBO and PBZT fibers lack sufficient compressive strength for application in truly high performance composites. This triggered considerable efforts to increase the mechanical properties along the axes perpendicular to the orientation axis. It is generally felt that the compressive strength of such fibers is compromised by the fact that the individual polymers can slide along each other, as the interactions between them are not sufficient to prevent sliding. The attempts to increase chain interaction and prevent sliding include increase of lateral order, crosslinking of the fibers in an additional step after spinning, and further stiffening of the chains by introducing a ladder-like structure.

Molecular Composites

Materials in which rigid-rodlike polymer molecules are dispersed on a molecular basis in a flexible polymer, which serves as a matrix, are referred to as molecular composites. In these materials, the reinforcement is thought to be maximized, if the rod-like polymer molecules can indeed be dispersed on a molecular level. The main problem in this respect is that rod-like molecules tend to aggregate due to their shape anisotropy.

Optical Clarity

High temperature, transparent structural materials are required by the U.S. Air Force, and the approach which was used to progress in this direction was the synthesis of semi-aliphatic PBZT's with rigid-rodlike aliphatic moieties.

IMPROVEMENT OF THE PROCESSABILITY OF POLYBENZAZOLES

Aside from the incorporation of benzazole units into other high temperature polymers, many attempts have been made recently to improve the processability of benzazole-containing polymers. The aim was especially to achieve thermoplastic behavior of such materials, and to some extent also to obtain solubility in non-acidic polar solvents such as NMP or DMAc. In general, the formation of the heterocycle is used as the polymer-forming reaction. The difference is mainly that less rigid structure elements are incorporated to leave the rigid-rodlike geometry.

The most popular flexibilizing unit is the hexafluoroisopropylidene group, (or 6F–group). Instead of the hexafluorisopropylidene group, the isopropylidene group[61] or the trifluorophenylethylidene group[62] (3F–group) have also been used for limited flexibilization.

Another interesting feature of the hexafluoroisopropylidene unit is that fluorinated groups in polymers tend to decrease the dielectric constant of these materials. This is important for the production of microchips, as a lower dielectric constant means better electrical insulation and therefore higher integration density on the chip.

Other approaches to enhance processability involve the functionalization of polybenzimidazoles at their NH groups with various side groups[63–65] and the incorporation of aromatic units that are not based on phenyl rings into polybenzobisoxazoles and polybenzobisthiazoles, such as thiophene,[60,66] bithiophene or terthiophene,[67] thianthrene,[68] and phenothiazine.[69] It is interesting to note that again the thiophene unit does not prevent the formation of the lyotropic phase, despite its angular geometry.[60,66] The phenothiazine and oligothiophene units were incorporated into benzobisthiazole-containing polymers in order to achieve high third-order susceptibilities for NLO applications.

IMPROVEMENT OF SYNTHETIC PROCESSES FOR POLYBENZAZOLES

The most common way to prepare polybenzazoles involves the use of polyphosphoric acid as solvent, to which P_2O_5 is added after dissolution of the monomers to aid solubilization of the polymer and to ensure complete cyclodehydration. This method was developed by Wolfe et al.[1,70] However, it cannot be used for all combinations of monomers.

For the synthesis of the more flexible polybenzazoles, a two-step process in an amidic solvent can be employed. McGrath for example used the formation of a poly(hydroxyamide) from diaminodiphenols with diacid chlorides in NMP/pyridine at low temperatures, followed by ring closure on heating the solution formed in the first step to achieve removal of reaction water as an azeotrope with *o*-dichlorobenzene.[61] However, when enough flexibilizing links are not present, there remains the problem of the insolubility of the precursor polyhydroxy–, polymercapto–, or polyaminoamides. Hattori et al.[71–73] addressed this problem by preparing rigid-rod precursor polymers having functionalized alkyl substituents at their mercapto groups. Heating to 360–400°C removes the alkyl groups and effects cyclization.

RIGID ROD POLYELECTROLYTES

Aside from studying the solution properties of polyelectrolytes in general, the interest in rigid-rod polyelectrolytes also comes partly from the possible application in the preparation of molecular composites. It is assumed that rigid-rod polyelectrolytes would not be able to form aggregates within the polymer matrix because there would be considerable Coulombic repulsion between charged chains.[74–77]

Truly rigid-rod polyelectrolytes were also prepared, based on sulfonated polybenzobisthiazole.[78,79]

CONCLUSION

Despite the many variations of polyheteroaromatics that have been prepared so far, the number of representatives of this class of polymers that is commercially available is very small. Many possible applications that have been proposed, especially for structural materials, such as fibers, films, composites (including molecular composites), high temperature thermoplastics, etc. However, between the purely phenyl-based polyimides, poly(ether ketone)s, poly(ether sulfone)s, and poly(ether imide)s, there is little room left for the heteroaromatic polymers in the field of high temperature applications. In most cases, the heteroaromatic polymers are more expensive than phenyl-based polymers, usually because the monomer synthesis is more expensive. Thus, polyheteroaromatics can find applications as structural materials only if their manufacturing costs can be reduced; e.g., by developing new polymerization processes that allow the use of cheaper monomers. The only other case that allows a market for these polymers is when their properties considerably exceed those of their competitors. This is why polybenzimidazole fiber has found its application in fire protective garments. PBI fiber is not only inflammable, but compared to conventional polyaramids, it shows better dimensional stability when exposed to flames, to form a flexible chair, and its subjective wear comfort is similar or superior to cotton.[80]

Nevertheless, the variety of different heteroaromatic ring systems allows a very subtle fine tuning of the geometry of the polymer backbone, as well as its polarity. The tailoring of free volume, chain stiffness, and chain interactions certainly opens up interesting perspectives for applications as functional materials, such as separation membranes for applications in hostile environments. Conceivable examples are treatment of waste gas at high temperatures, or water contaminated with aggressive pollutants such as acids, bases, or oxidizing agents.

Some of the high T_g poly(heteroaromatics)s are promising candidates for backbones of polymers with second order NLO properties. In such materials, the NLO chromophores (usually π-conjugated systems with an electron acceptor on one end and an electron donor on the other end) are aligned parallel in an electric field.

Despite the amount of work that has been done already, the field of polyheteroaromatics will continue to grow as it finds more and more possible applications in the future.

REFERENCES

1. Wolfe, J. E. *Polybenzothiazoles and Polybenzoxazoles*, In *Encyclopedia of Polymer Science and Engineering*; 2nd ed., Mark, H. F.; Bikales, N. M.; Overberger, C. G.; Menges, G., Eds.; John Wiley & Sons: New York, 1988; Vol. 11.
2. Rossbach, V.; Oberlein, G. *Thermostable Polyheterocyclics*, In *Handbook of Polymer Synthesis*; Kricheldorf, H. R., Ed.; Marcel Dekker: New York, 1992; Chapter XIX.
3. Arnold, C. *J. Polym. Sci., Macromol. Rev.* **1979**, *14*, 265.
4. Jenekhe, S. A.; Johnson, P. O.; Agrawal, A. K. *Macromolecules* **1989**, *22*, 3216.
5. Jenekhe, S. A.; Johnson, P. O.; Agrawal, A. K. *Macromolecules* **1990**, *23*, 4419.
6. Roberts, M. F.; Jenekhe, S. A. *Polymer Commun.* **1990**, *31*, 215.
7. Kim, Y.-K.; Thurber, E. L.; Rasmussen, P. G. *Polym. Prepr.* **1991**, *32*(3), 590.
8. Allan, D. S.; Rasmussen, P. G. *J. Polym. Sci. Polym. Chem. Ed.* **1992**, *30*, 1413.
9. DeSimone, J. M.; Stompel, S.; Samulski, E. T. *Polym. Prepr.* **1991**, *32*(2), 172.
10. Brennan, A. B.; Wang, Y. Q.; DeSimone, J. M.; Stompel, S.; Samulski, E. T. *Polym. Prepr.* **1992**, *33*(1), 485.
11. DeSimone, J. M.; Sheares, V. V.; Samulski, E. T. *Polym. Prepr.* **1992**, *33*(1), 418.
12. DeSimone, J. M.; Stompel, S.; Samulski, E. T.; Wang, Y. Q.; Brennan, A. B. *Macromolecules* **1992**, *25*, 2546.
13. Sheares, V. V.; DeSimone, J. M. *Macromolecules* **1992**, *25*, 4235.
14. Sheares, V. V.; Berndt, S. C.; DeSimone, J. M. *Polym. Prepr.* **1992**, *33*(2), 170.
15. Promislov, J. H.; Preston, J.; Samulski, E. T. *Macromolecules* **1993**, *26*, 1793.
16. Yi, M. H.; Choi, K.-Y.; Jung, J. C. *J. Polym. Sci., Part A: Polym. Chem.* **1989**, *27*, 2417.
17. Maier, G.; Hecht, R.; Nuyken, O.; Helmreich, B.; Burger, K. *Macromolecules* **1993**, *26*, 2583.
18. Maier, G. et al. *Polym. Prepr.* **1993**, *34*(1), 429.
19. Maier, G. et al. *Makromol. Chem. Macromol. Symp.* **1993**, *75*, 199.
20. Maier, G.; Schneider, J. M.; Nuyken, O. *Macromol. Reports* **1994**, *A31* (Suppls. 1&2), 179.
21. Maier, G. et al. *Makromol. Chem. Macromol. Symp.* **1994**, *82*, 143.
22. Carter, K. R.; Hedrick, J. L. et al. *Polym. Prepr.* **1994**, *35*(1), 529.
23. Bass, R. G. et al. *Polym. Prepr.* **1994**, *35*(1), 383.
24. Bass, R. G., Srinivasan, K. R. *Polym. Prepr.* **1991**, *32*(1), 619.
25. Bass, R. G.; Srinivasan, K. R.; Smith, Jr., J. G. *Polym. Prepr.* **1991**, *32*(2), 160.
26. Bass, R. G. et al. *Polym. Prepr.* **1993**, *34*(1), 441.
27. Connell, J. W.; Hergenrother, P. M. *J. Polym. Sci., Part A: Polym. Chem.* **1990**, *29*, 1667.
28. Connell, J. W.; Hergenrother, P. M. *High Perf. Polym.* **1990**, *2*, 211.

29. Connell, J. W.; Croall, C. I. *Polym. Prepr.* **1991**, *32*(2), 162.
30. Connell, J. W. et al. *Polym. Prepr.* **1994**, *35*(1), 543.
31. Connell, J. W.; Hergenrother, P. M.; Wolf, P. *Polym. Mat. Sci. Eng. Proc.* **1990**, *63*, 366.
32. Connell, J. W.; Hergenrother, P. M.; Wolf, P. *Polymer* **1992**, *33*, 3507.
33. Connell, J. W.; Hergenrother, P. M. et al. *Polym. Prepr.* **1993**, *34*(1), 525.
34. Hedrick, J. L. *Polym. Bull.* **1991**, *25*, 543.
35. Hedrick, J. L.; Carter, K. R. et al. *Polym. Prepr.* **1992**, *33*(1), 388.
36. Hedrick, J. L. *Polym. Prepr.* **1992**, *33*(1), 1016.
37. Hedrick, J. L.; Twieg, R. *Macromolecules* **1992**, *25*, 2021.
38. Şaegusa, Y., Iwasaki, T.; Nakamura, S. *J. Polym. Sci., Part A: Polym. Chem.* **1994**, *32*, 249.
39. Hedrick, J. L.; Carter, K. R. et al. *Polym. Prepr.* **1993**, *34*(1), 481.
40. Carter, K. R.; Miller, R. D.; Hedrick, J. L. *Macromolecules* **1993**, *26*, 2209.
41. Hedrick, J. L.; Hilborn, J.; Labadie, J. W.; Volksen, W. *Poly. Prepr.* **1990**, *31*(1), 446.
42. Hilborn, J. G.; Labadie, J. W.; Hedrick, J. L. *Macromolecules* **1990**, *23*, 2854.
43. Hergenrother, P. M. et al. *Polym. Prepr.* **1991**, *32*(1), 647.
44. Smith, J. G.; Connell, J. W.; Hergenrother, P. M. *Polymer* **1992**, *33*, 1742.
45. Hedrick, J. L. *Macromolecules* **1991**, *24*, 6361.
46. Smith, Jr., J. G.; Connell, J. W.; Hergenrother, P. M. *Polym. Prepr.* **1991**, *32*(3), 193.
47. Connell, J. W.; Hergenrother, P. M.; Smith, Jr., J. G. *Polym. Prepr.* **1992**, *33*(1), 411.
48. Hergenrother, P. M. et al. *Polym. Prepr.* **1992**, *33*(1), 1098.
49. Smith, Jr., J. G.; Connell, J. W.; Hergenrother, P. M. *Polymer* **1993**, *34*, 856.
50. Smith, J. G.; Connell, J. W.; Hergenrother, P. M. *J. Polym. Sci., Part A: Polym. Chem.* **1993**, *31*, 3099.
51. Kane, J. J.; Gao, F. *Polym. Prepr.* **1994**, *35*(2), 631.
52. Hedrick, J. L. et al. *Macromolecules* **1988**, *21*, 1883.
53. Takekoshi, T. et al. *J. Polym. Sci., Polym. Chem. Ed.* **1981**, *19*, 1635.
54. Caruso, U.; Centore, R.; Roviello, A.; Sirigu, A. *Macromolecules* **1992**, *25*, 2290.
55. Centore, R.; Fresa, R.; Roviello, A.; Sirigu, A. *Polymer* **1993**, *34*, 4536.
56. Kricheldorf, H. R.; Thomsen, S. A. *J. Polym. Sci., Polym. Chem. Ed* **1991**, *29*, 1752.
57. Kricheldorf, H. R.; Thomsen, S. A. *Makromol. Chem.* **1993**, *194*, 2063.
58. Kricheldorf, H. R.; Domschke, A. *Polymer* **1994**, *35*, 198.
59. Lister, R. L.; Collier, H. L. *Polym. Prepr.* **1993**, *34*(1), 360.
60. Promislow, J. H.; Samulski, E. T.; Preston, J. *Polym. Prepr.* **1991**, *32*(2), 211.
61. Joseph, W. D.; Mercier, R.; Prasad, A.; Marand, H.; McGrath, J. E. *Polymer* **1993**, *34*, 866.
62. Joseph, W. D.; Abed, J. C.; Mercier, R.; McGrath, J. E. *Polym. Prepr.* **1993**, *34*(1), 397.
63. Gan, L.; Gan, Y.; Weber, W. P.; Amis, E. J. *Polym. Prepr.* **1993**, *34*(2), 386.
64. Reynolds, J. R. et al. *Polym. Prepr.* **1989**, *30*(1), 151.
65. Arnold, F. E. *Mat. Res. Soc. Symp. Ser.* **1990**, *134*, 117.
66. Promislow, J. H.; Preston, J.; Samulski, E. T. *Macromolecules* **1993**, *26*, 1793.
67. Dotrong, M.; Tomlinson, R. C.; Sinsky, M.; Evers, R. C. *Polym. Prepr.* **1991**, *32*(1), 85.
68. Johnson, R. A.; Mathias, L. J. *Polym. Prepr.* **1994**, *35*(2), 653.
69. Scherf, U.; Müllen, K. *Polym. Prepr.* **1994**, *35*(1), 200.
70. Wolfe, J. F.; Sybert, P. D. U.S. Patent 4 533 693, 1985.
71. Hattori, T.; Akita, H.; Kakimoto, M.; Imai, Y. *Macromolecules* **1992**, *25*, 3351.
72. Hattori, T.; Akita, H.; Kakimoto, M.; Imai, Y. *J. Polym. Sci., Polym. Chem. Ed.* **1992**, *30*, 197.
73. Hattori, T.; Kagawa, K.; Kakimoto, M.; Imai, Y. *Macromolecules* **1993**, *26*, 4089.
74. Gieselmann, M. B.; Reynolds, J. R. *Macromolecules* **1990**, *23*, 3188.
75. Gieselmann, M. B.; Reynolds, J. R. *Polym. Prepr.* **1992**, *33*(1), 931.
76. Gieselmann, M. B.; Reynolds, J. R. *Polym. Prepr.* **1992**, *33*(1), 1056.
77. Gieselmann, M. B.; Reynolds, J. R. *Macromolecules* **1992**, *25*, 4832.
78. Dang, T. D.; Arnold, F. E. *Polym. Prepr.* **1992**, *33*(1), 912.
79. Reynolds, J. R.; Lee, Y.; Kim, S.; Bartling, R. L.; Gieselman, M. B.; Savage, C. S. *Polym. Prepr.* **1993**, *34*(1), 1065.
80. Buckley, A.; Stuetz, D. E.; Serad, G. A. *Polybenzimidazoles*, In *Encyclopedia of Polymer Science and Engineering*; 2nd ed.; Mark, H. F.; Bikales, N. M.; Overberger, C. G.; Menges, G., Eds.; John Wiley & Sons: New York, 1988; Vol. 11.

HETEROHELICENES

Hiroshi Kawazura,* Hiroko Nakagawa, and
Koh-ichi Yamada
Faculty of Pharmaceutical Sciences
Josai University

Heterohelicenes refer to orthocondensed aromatic compounds in which benzene and heteroaromatic rings are angularly annulated to give helically shaped molecules. Thus far, heterohelicenes of fifteen rings having three decks have been reported.[1] These molecules are asymmetric chromophores themselves and manifest interesting structural, spectral, and optical behaviors. This paper mainly describes thiaheterohelicenes (helicenes consisting of benzene and thiophene rings) because of the detailed and systematic reports available. Abbreviations according to the literature are occasionally used for the notation of heterohelicenes: the number of constituting aromatic rings, between brackets, followed by the symbols B = benzene, S = thiophene, NH = pyrrole, N = pyridine, and O = furan.[2]

PREPARATION

Most thiaheterohelicenes have been prepared by photocyclization of the corresponding precursor ethenes in a dilute solution in the presence of an oxidizing agent, such as oxygen or iodine, as shown in **Scheme I.**

PROPERTIES

Chiroptical Properties

Optical Resolution of Enantiomers

The first optical resolution of heterohelicene was achieved by crystallizing racemic [7]SBSBSBB from the chiral solvent, (–)-α-pinene, which gave the crystals of (+)-antipode excess.[3]

*Author to whom correspondence should be addressed.

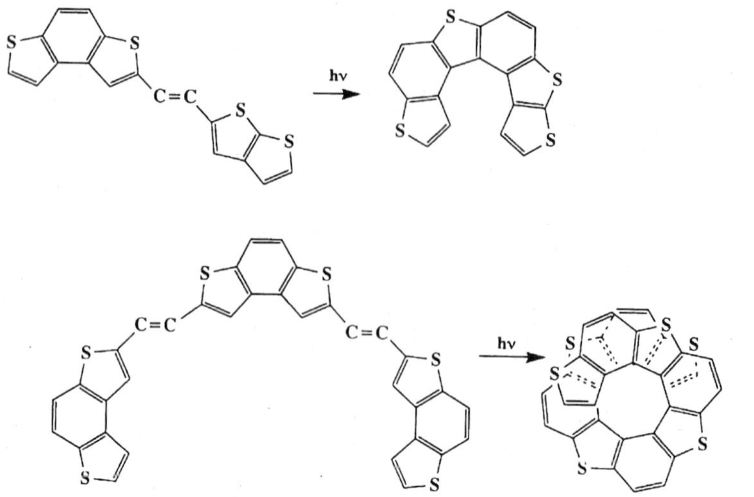

I

A more effective method to resolve enantiomers of heterohelicenes seems to be high performance liquid chromatography (HPLC) equipped with columns coated with or bonded to chiral charge-transfer (CT) complexing agents. 2–(2,4,5,7–tetranitro–9–fluorenylidene aminooxy)propionic acid (TAPA) and its congeners (TABA, TAIVA, and TAHA) have been examined as a chiral resolving agent for helicenes.[4–6] However, TAPA shows a better resolving capacity than any other congeners possessing larger alkyl groups at the chiral center, the bulkiness of the substituent determining the ease of optical resolution of helicenes. It turns out that the right handed *P*-enatiomer of thiaheterohelicene interacts strongly with *S*-TAPA in the case of HPLC.

Some examples have been reported on the use of aluminum oxide coated with TAPA to resolve heterohelicenes such as [7]BOBBBOB.[7] Various chiral selectors have been studied for their abilities to separate enantiomers of heterohelicene. Binaphthyl-2,2′-diyl-hydrogenphosphate shows a fairly good separation of enantiomers of thiaheterohelicene provided with a greater electro-donating capacity by the substitution of bromine.[5,6] The n-dodecyl ester of N-(2,4-dinitrophenyl)-L–alanine is able to discriminate between the enantiomers of l-aza-[6]-heterohelicene, when used as a chiral dopant on a reversed phase column.[8] Several naturally occurring selectors have been successfully reported for the resolution of helicenes such as riboflavin,[9] adenosine, 3′-AMP, and guanosine.[10]

Optical Rotation

Absolute configurations of heterohelicenes have been confirmed by calculations and X-ray analyses,[11] which demonstrated that the dextrorotatory enantiomer (+) of thiaheterohelicene possesses a right-handed (*P*) helicity. The specifications of heterohelicenes increase on the whole with increasing the number of aromatic rings. However, we also observed that the magnitude of the rotation depends upon the sort and the aligning order of the constituent rings in a molecule.

THERMAL RACEMIZATION

Reports on racemization of carbohelicenes were surprisingly in light of the flexibility of angularly fused benzene rings.[12] Thermal racemization of heterohelicenes has also been studied from the perspective of flexibility of heteroaromatic rings compared with a benzene ring. The racemization rate depends on the winding degree of the helix, which can be estimated with the internal angle (IA) value of a heterohelicene molecule.

APPLICATION

Some applications of heterohelicenes have been attempted on the basis of the following properties that make heterohelicene molecules conspicuous: a large π-electron density in the highly delocalized system, molecular chirality arising from the helical structure, and the characteristics of the heteroatoms involved.

Fukumi determined second molecular hyperpolarizabilities γ of a series of thiaheterohelicenes and their methyl derivatives composed of alternating thiophene and benzene rings in benzene solution by a degenerate four-wave mixing method.[13] This γ value is promising to apply thiaheterohelicenes as non-linear optical material because the γ is deemed to come mainly from electronic origin.

As asymmetric transformation (AT) in which racemic [5]SBSBS is converted into a single enantiomer was found on the diastereomeric CT complex formation with chiral TAPA by taking advantage of a rapid racemization of [5]SBSBS at room temperature in solution.[14] It was suggested that [5]SBSBS may become a sensitive probe for measuring the extent of chiral recognition ability of various albumins, other material-transport proteins, and enzymes by using the intense absorptions in CD spectrum of the heterohelicene.

Bell synthesized expanded azaheterohelicenes with partially saturated structures show a molecular coil and rapidly interconvert between enantiomers in solution.[15] These compounds considered as a class of preorganized, monohelical polypyridine ligand, can wrap around guests such as sodium ion. The complexation with metal ions increases the barrier to helix inversion. The authors described that these expanded helicenes are a new type of ligands and may have a potentially useful application by utilizing the optical response to the conformational change.

REFERENCES

1. Yamada, K.; Ogashiwa, S.; Tanaka, H.; Nakagawa, H.; Kawazura, H. *Chem. Lett.* **1981**, 343.
2. Laarhoven, W. H.; Prinsen, W. J. C. *Top. Curr. Chem.* **1984**, *125*, 63.
3. Wynberg, H.; Groen, M. B. *J. Am. Chem. Soc.* **1970**, *92*, 6664.
4. Mikes, F.; Boshart, G.; Gil-Av, E. *J. Chromatgr.* **1976**, *122*, 205.
5. Mikes, F.; Boshart, G. *J. Chromatgr.* **1978**, *149*, 455.
6. Mikes, F.; Boshart G. *J. Chem. Soc. Chem. Comm.* **1978**, 174.
7. Numan, H.; Helder, R.; Wynberg, H. *Rec. Trav. Chim. Pays-Bas* **1976**, *95*, 211.
8. Lochmüller, C. H.; Jensen, E. C. *J. Chromatgr.* **1981**, *216*, 333.
9. Kim, Y. H.; Tishbee, A.; Gil-Av, E. *J. Am. Chem. Soc.* **1980**, *102*, 5915.
10. Kim, Y. H.; Tishbee, A.; Gil-Av, E. *Science* **1981**, *213*, 1379.
11. Stulen, G.; Visser, G. J. *J. Chem. Soc. Chem. Comm.* **1969**, 965.
12. Martin, R. H.; Marchant, M. J. *Tetrahedron* **1974**, *30*, 347.
13. Fukumi, T.; Sakaguchi, T.; Miya, M.; Nakagawa, H.; Yamada, K.; Kawazura, H. *Rev. Laser Engineering* **1994**, *22*, 409.
14. Nakagawa, H.; Yamada, K.; Kawazura, H. *J. Chem. Soc. Chem. Comm.* **1989**, 1378.
15. Bell, T. W.; Jousselin, H. *J. Am. Chem. Soc.* **1991**, *113*, 6283.

Heterophane Polymerization

See: *Vapor Deposition Polymerization ([2.2]Paracyclophane and Analogues)*

Hevea brasiliensis

See: *Natural Rubber (Hevea brasiliensis)*
Natural Rubber Latex (Polymeric Flocculant for Tin Tailings Slurries)

High Density Polyethylene

See: *Polyethylene (Commercial)*
Polyethylene (High Density Preparation)
Polyethylene, Recycled
Polyolefins (Overview)

High Impact Polystyrene

See: *Bulk Polymerization*
Polystyrene, High Impact
Polystyrene, Rubber Toughened
Styrenic Resins

High Modulus Polymers

See: *Carbon/Carbon Composites (with Surface Treated Carbon Fibers)*
Carbon Fibers (Overview)
Fibers and Composites (High Modulus and Strength)
Liquid Crystalline Polymers (Polyesters)
Molecular Composites (High Modulus Polyimide/Polyimide Films)
Molecular Composites (Polyimide/Polyimide, Laminate Processability)
Poly(vinyl alcohol) (High Strength and High Modulus)
Poly(vinyl alcohol) Fiber (High Strength and High Modulus)
Poly(vinyl alcohol) Fibers (High Modulus; by Vibrating Zone-Drawing)
Textile Fibers (Structure and Properties)

High Molecular Weight Polymers

See: *Ultrahigh Molecular Weight Polymers*

HIGH PERFORMANCE PAPER

Yoshinari Kobayashi
Shikoku National Industrial Research Institute

Just as human bodies are composed of muscles and bones, which form figures, and brains, nerves and sense, which allow human activity, all materials can be roughly classified into two groups: "structural materials", corresponding to skeletons, and "functional or high performance materials", which are equivalent to nerves and sense. Because paper is a material, it can also be categorized into "commodity paper", equal to "structural materials in paper" and "high performance or functional paper (H.P.P.)". "Commodity paper" can be called a mass-produced cellulosic paper that is widely used but has lower added-value.

High-performance paper (H.P.P.) is described as "a group of new functional and/or high-performance papers made up not only of non-cellulosic, short-cut fibers, such as synthetics and ceramics, but also of cellulosic fibers".[1]

H.P.P. also includes fine Japanese paper that consists of long bast fibers and is commonly used for arts and crafts. According to this definition, some of the high performance papers composed of cellulosic fibers overlap with so-called "specialty papers". In other words, high performance papers cover a wider range of higher added-value sheets than simply specialty papers. These papers are mainly applied in the fields of advanced technology or high-technology, e.g., information, electronics, medical, and environmental conservation. This term can also be interchanged with the expressions "engineered papers" and "advanced paper", by the analogy of "engineered fabrics", the alias for nonwovens and "advanced materials", respectively.[2]

Paper has generally been considered as the two-dimensional product composed of plant fibers, i.e., natural cellulosic fibers or cellulosic pulps. This is so-called "paper" in common usage and is the "first group" of paper. In this group, plant cellulose comprises the short-cut fibers. The "second group" of paper is termed "chemical fiber paper", "man-made fiber paper", or "artificial fiber paper", in which chemical (or man-made or artificial) fiber represents a wider range that includes regenerated fibers.

Therefore, high-performance papers comprise a part of the first group of paper which has especially high performance and/or functionality and also the whole of the second group.

TECHNOLOGICAL TRENDS IN CLASSIFIED HIGH PERFORMANCE PAPERS

The classifications of functionality and/or high performance for papers are grouped into 10 characteristics as follows: (1) mechanical, (2) thermal, (3) electric, electronic, and magnetic,

(4) optical, (5) acoustic, (6) chemical, (7) biological and biochemical, (8) human feelings, (9) hybridized or composite functions, and (10) paper-analogous products.[3] The importance of functionality and/or high performance for paper was ranked according to the frequency of responses for each group. It was found that in commercialized high performance papers, heat-resistant paper was ranked first, electrical insulating paper was ranked second, and chemically processed papers having adhesive, tacking, bonding, and releasing characteristics were ranked third.

Heat-resistant papers were grouped into inorganic paper and organic papers. Inorganic papers include glass-fiber paper, alumina-fiber paper, ceramic-fiber paper, asbestos paper, carbon-fiber paper, and inorganic-pigment-filled paper. Their applications vary widely and include heat-insulating sheets, nonflammable construction sheets, heat-resistant industrial materials, gaskets for combustion instrumentation, heat-protecting clothes, packaging for sterilization, exothermic sheet composites-backing sheet materials for lining high-temperature furnaces; pipe-thermal insulation paper, welding spark shielding mat, and heat-resistant sheet carriers for catalysts. Examples of commercialized organic heat-resistant paper are: fire-proofing vinylon-fiber paper, aramid-fiber paper, fluorocarbon-fiber paper, heat-retardant polyester fiber paper, sheet consisting of synthetic pulp (fibrid) with heat-resistance and fire-protection, and phenolic resin-filled paper. Their usages include flame-retardant insulation for articles of daily living; backing material for wallpaper, interior architectural material; packing sheets, packaging sheets, substrate sheets for printed circuit boards; gasket sheets; and fire-protective cable wraps. As these examples show heat-resistant papers are widely applied in the fields of construction, industry, daily living, and medical usage.

Electrical insulation paper can be represented by coil-insulating paper, cable-insulating paper, battery-separator sheets, condenser paper for dielectric, interlayer paper for transformer coils, and paper used for printed circuits. The fibers of electrical-insulation papers are classified as inorganic fiber, cellulosic pulp, polyolefin fiber, polyester fiber, aramid fiber, etc.

In contrast with the above high-performance papers, low-tack-pressure-sensitive paper is completely different in production process because higher value is added in the secondary paper-converting process instead of in papermaking. This is a high-performance paper which is born by combining two different technologies, papermaking and post-chemical application, that can safely be called "high-processed paper". Low tack paper is used widely in temporary labels, stickers, postcard form envelopes provided with low tack seal, transparent paper, tape with easy release characteristics, separate paper, and masking paper consisting of two-faced tack tape.

METHODS FOR PROVIDING FUNCTIONALITY AND/OR HIGH PERFORMANCE TO PAPER

As illustrated in alginate fiber paper,[4] there are three stages in which paper is provided with functionality and/or high performance: (1) in the selecting fibers as the main constituent of paper, (2) in actual paper-making, and (3) in paper-converting or paper-processing.

FRONTIER FIELD OF PAPER INDUSTRY

High performance papers can be said to be directed toward pioneering a new boundary between the fields of nonwoven, film, membrane, synthetic leather, and its related fields by applying high technology materials and methods to the basic papermaking process.

ACKNOWLEDGMENT

The authors wish to express the deepest appreciation to Mrs. Patricia and Dr. Karl V. Kraske for their kindnesses in critical reading and elaborating the manuscript.

REFERENCES

1. Kobayashi, Y.; Fujiwara, K.; Inagaki, H. *Kinohshi Sohran '93* ("Overview of High Performance Paper, Japan '93"), Kakohgijutsu Kenkyu Kai, Tokyo, 1993, in Japanese.
2. Kobayashi, Y. *Kami Parupu Gijutsu Times* (Paper & Pulp Times: Japanese Journal of Paper Technology), **1994**, *37*(4), 9, in Japanese.
3. Kobayashi, Y.; Fujiwara, K. *Kami Pa Gikyo Shi* (Japan Tappi), **1992**, *46*(10), 1215, in Japanese.
4. Kobayashi, Y. "*Introduction to Applied Phycology*"; Akatsuka, I., Ed.; SPB Academic: The Hague, The Netherlands, 1990; pp 407–427, in English.

High Performance Polymers

See: *Acetylene-Terminated Monomers (Catalyzed Cure)*
Anoxic Polymer Material
Aramid Fibers and Nonwovens
Aramid Fibers, High Strength
Aramid Film
Aramids (Processable Precursors)
Aramids and Aromatic Polyimides, Soluble (Bulky Substituents)
Carbon Whiskers (Surface Modification by Grafting)
Cellulose (Direct Dissolution)
Cyclolinear Organophosphazene Polymers
Epoxy Resins (High Performance Composite Applications
Fibers, Synthetic
Fluorinated High Performance Polymers (Imide and Hexafluorisopropylidene Groups)
Fluorinated Plastics, Amorphous
Fluorine-Containing Polymers (Polycarbonates and Polyformals)
Fluoropolymer Coatings (New Developments)
Functional Fibers
Heteroaromatic Polymers (Five-Membered Rings in Main Chain)
Ladder Polymers (Methods of Preparation)
Liquid Crystalline Polymers (Heat-Strengthened Thermotropic)
Liquid Crystalline Polymers (Polyesters of Bibenzoic Acid)
Liquid Crystalline Polymers (Thermotropic Polyesters)
Perfluorocyclobutane Aromatic Ether Polymers
Poly(alkylene naphthalate)
Poly(amide imide)s (with Improved Melt Processability)

Poly(aryl ether ketone ketone)s and Copolymers (Synthesis, Structure, Morphology, and Properties)
Poly(aryl ether ketone)s (Overview)
Poly(aryl ether ketone)s (Including Thioether Ketones)
Poly(aryl ether ketone)s Amine-Terminated
Poly(aryl ether sulfone)s
Poly(arylene-1,3,4-oxadiazole)s
Polybenzimidazoles (Overview)
Poly(benzobisthiazole-imide)s
Poly(ether sulfone)s (Properties, Applications, and Trends)
Poly(ether sulfone)s and Poly(ether ketone)s (Methyl-Substituted Aromatic)
Polyethers (1,3,4-Thiadiazole-Containing)
Polyethylene High Performance (Gel-Like Spherulite Press Method and Ultradrawing)
Polyimide Blends
Polyimides (Arylene Ether Diamines having Cyclic Side Groups)
Polyimides (From Tricyclic Heteroaromatic Dyes)
Polyimides, High Performance Films
Polyisoimides
Polyisophthalamides (Water-Uptake Enhancement)
Polyoxyarylenesiloxanes
Polyoxymethylene Fiber, Superdrawn
Poly(p-phenylene sulfone)
Poly(vinylidene fluoride) (Overview)
Rigid-Rod Polybenzoxazoles and Polybenzothiazoles (PBZT and PBO)
Rigid-Rod Poly(p-phenylene benzobisthiazole)
Siloxane Elastomers
Spider Silk (Production of Polypeptide Polymers)
Thermosets (Main Chain Liquid Crystalline Polymers)
Ultra-Strong Polyethylene Fibers
Whiskers of Rigid-Rod Polymers

HIGH PRESSURE POLYCONDENSATION (for High Temperature Polymers)

Yoshio Imai
Department of Organic and Polymeric Materials
Tokyo Institute of Technology

A variety of methods are known for the synthesis of condensation polymers, however, the application of high pressure has seldom appeared in the literature to date. Quite recently, we investigated systematically the high-pressure polycondensation leading to the formation of polyimides,[1–6] and polybenzoxazoles.[4,7,8]

AROMATIC POLYIMIDES

The final goal of the high-pressure synthesis of polyimides is to develop a new method of polymerization and simultaneous hot pressing for intractable wholly aromatic polyimides directly from the salt monomers.[2] It has been shown that aromatic diamines also formed salt monomers with pyromellitic acid diester, which led to the formation of wholly aromatic polyimides through thermal polycondensation.[9] We also prepared the salt monomer, 44PME, with the reaction of pyromellitic acid diethyl ester with 4,4′-oxydianiline.

The polycondensation proceeded under high-pressure conditions, however, the inherent viscosity of the resultant polymer was 0.19 dL/g in concentrated sulfuric acid, indicative of low molecular weight. The polyimide oligomer thus formed had high crystallinity.

When this oligomer was subjected to postpolymerization further polycondensation readily proceeded, resulting in the polyimide P-44PM which has a high inherent viscosity of 1.11 dL/g. These results clearly indicated that to obtain polyimide 44PM, with a high-glass transition temperature of 420°C, a high temperature of 400°C was essential to enhance the mobility of the polymer chain ends of the growing polyimide molecule, which is necessary for the progression of polycondensation in the solid state. The high-molecular-weight polyimide thus formed was also crystalline.

AROMATIC POLYBENZOXAZOLES

The high-pressure polycondensation through the *o*-aminophenol–aliphatic nitrile reaction was extended to the aromatic nitrile counterpart for the synthesis of aromatic polybenzoxazoles.[7] When 3-amino-4-hydroxybenzonitrile hydrochloride was subjected to high-pressure polycondensation under 200 MPa at 400°C for 40 h, we obtained the aromatic polybenzoxazole, which had an inherent viscosity of 0.40 dL/g in concentrated sulfuric acid, along with the liberation of ammonium chloride. The postpolycondensation of this polymer at 400°C for 1 h under a thin stream of nitrogen afforded the polybenzoxazole with a higher viscosity value of 0.75 dL/g, which is indicative of high molecular weight.

The other aromatic polybenzoxazoles with inherent viscosities ~0.4 dL/g were also prepared by the high-pressure polycondensation from the combinations of AHB or AHB–HCl and aromatic dinitriles under similar reaction conditions.

REFERENCES

1. Itoya, K.; Kumagai, Y.; Kakimoto, M.; Imai, Y. *Polym. Prepr.* **1992**, *41*, 2131.
2. Kumagai, Y.; Itoya, K.; Kakimoto, M.; Imai, Y. *Polymer* **1995**, *36*, 2827.
3. Itoya, K.; Kumagai, Y.; Kakimoto, M.; Imai, Y. *Macromolecules* **1994**, *27*, 4101.
4. Imai, Y. *A. C. S. Polym. Prepr.* **1994**, *35*, 399.
5. Imai, Y.; Inoue, T.; Watanabe, S.; Kakimoto, M. *Polym. Prepr. Jpn.* **1994**, *43*, 371.
6. Imai, Y.; Inoue, T.; Kakimoto, M. *Polym. Prepr. Jpn.* **1994**, *43*, 3178.
7. Imai, Y.; Sawada, H.; Kakimoto, M. *Polym. Prepr. Jpn.* **1994**, *43*, 372.
8. Itoya, K.; Sawada, H.; Kakimoto, M.; Imai, Y. *Macromolecules* **1995**, *28*, 2611.
9. Bell, V. L. *J. Polym. Sci., Part B, Polym. Lett.* **1967**, *5*, 941.

High Pressure Polymerization

See: *High Pressure Polycondensation (for High Temperature Polymers)*
High Pressure Polymerization Processing (High Temperature Thermosetting Polymers)
High Pressure Radical Polymerization

Nylons (By High Pressure Solid-State Polycondensation)
Peroxide Initiators (Overview)

HIGH PRESSURE POLYMERIZATION PROCESSING (High Temperature Thermosetting Polymers)

Yoshio Imai
Department of Organic and Polymeric Materials
Tokyo Institute of Technology

Thermosetting polymers that are resistant to high temperatures such as aromatic polycyanurates have high glass-transition temperatures, which makes melt processing of these aromatic polymers very difficult. A possible alternative approach for the polymerization and simultaneous processing is the application of high pressure in order to get high-temperature thermosetting moldings directly starting from the monomeric reactants.

We have investigated the high pressure polymerization processing through polyaddition reactions from the monomeric reactants. The reactions are subdivided into two categories. The first is the Michael-type polyaddition of an aromatic diamine to a bismaleimide, which produces polyaminoimide.[1] The second is the cycloaddition polymerizations of the monomers, such as aromatic dicyanantes, diisocyanates, and a nitrile *N*-oxide, leading to the formation of polycyanurates, polydiazetidinediones, polyisocyanurates, and poly(1,2,4-oxadiazole), respectively.[2–5] These three types of monomers are isomeric with regard to each other.

HIGH-PRESSURE MICHAEL-TYPE POLYADDITION

Polyaminoimide resins obtained by the Michael-type polyaddition of aromatic diamines to aromatic bismaleimides are known as a class of high-temperature thermosetting addition-type polyimides, and are used as matrices for composites in electronics and aerospace industries.[6–8] We carried out the high pressure polymerization processing for the polyaminoimide derived from 4,4′-bismaleimidodiphenylmethane (BMI) and 4,4′-methylenedianiline (MDA) without a solvent and a catalyst by using a piston-cylinder type hot-pressing apparatus in a Teflon® capsule.[1]

The crosslinked polyaminoimide obtained under high pressure was transparent, amber-colored, and a very hard resinous mass with Vickers hardness of 330–360 MPa, it also had high modulus (>1.5 GPa (giga-Pascal)), compared with the linear polymer prepared under atmospheric pressure. The hardness value was nearly comparable to that of the thermosetting phenolic resin. The crosslinked polymer had a glass-transition temperature (T_g) and decomposition temperature (in nitrogen) ~220°C and 280°C, respectively; these temperatures were almost the same as those of the ordinary-pressure polymerized polymer.

HIGH PRESSURE CYCLOADDITION POLYMERIZATION

We have studied the high pressure cycloaddition polymerization of three isomeric functional monomers composed of CNO elements, such as cyanate (-O-C≡N), isocyanate (N=C=O), and nitrile N-oxide (-C≡N-O).

REFERENCES

1. Itoya, K.; Kumagai, Y.; Kanamaru, M.; Sawada, H.; Kakimoto, M.; Imai, Y.; Fukunaga, O. *Polym. J.* **1993**, *25*, 883.
2. Itoya, K.; Kakimoto, M.; Imai, Y. *Polymer* **1994**, *35*, 1203.
3. Itoya, K.; Kakimoto, M.; Imai, Y. *Polym. Prepr. Jpn.* **1991**, *40*, 1821.
4. Itoya, K.; Kakimoto, M.; Imai, Y. *Macromolecules* **1994**, *27*, 7231.
5. Itoya, K.; Kakimoto, M.; Imai, Y.; Fukunaga, O. *Polym. J.* **1992**, *24*, 979.
6. Crivello, J. V. *J. Polym. Sci., Polym. Chem. Ed.* **1973**, *11*, 1185.
7. Cassidy, P. E. *Thermally Stable Polymers*; Marcel Dekker: New York, 1980.
8. Wilson, D.; Stenzenberger, H. D.; Hergenrother, P. M. *Polyimides*; Blackie: New York, 1990.

HIGH PRESSURE RADICAL POLYMERIZATION

A. L. Kovarskii and Yu M. Sivergin
Semenov Institute of Chemical Physics
Russian Academy of Sciences

Over the past two to three decades the application of high pressure (HP) for polymer synthesis grew considerably. Along with hydrostatic pressure, the interest in high pressure arose with respect to the investigation of polymer synthesis resulting from shock waves and the combined action of HP and shear deformation.[1,2]

The term "high pressure" is conventional to a great extent. The range of pressure may be divided into at least two regions: pressure that causes the reduction of intermolecular distance only (less than 1 GPa) and the higher pressures in which the changes of energetic characteristics of particles may occur. The latter range is apparently typical for plastic deformation by way of shock waves and shear deformations.

The available data are concerned mainly with the radical polymerization as reflected in the title of this article. However, the authors decided to include some results on other types of polymer-formation processes that may be of interest to the reader.

POLYMERIZATION UNDER VOLUME COMPRESSION

An intensive study of monomer polymerization under pressure began in the late 1920s to 1930s. A number of researchers made great contributions to the development of this subject.[3–12]

The application of pressure makes it possible to polymerize monomers that do not allow polymerization under atmospheric pressure, with the pressure changing the structural and spatial direction of the reaction; to increase the polymerization rate and the degree of conversion; to produce polymers with a larger molecular mass; to obtain information about the structure of the activated complex, its polarity, and the degree of solvation, and of the mechanism polymerization. Whether or not polymerization proceeds to resulting products is influenced by material transport, topochemistry steric hinderance, and other physicochemical factors.[8]

Behavior of Monomers Under Pressure

When pressure is applied to a monomer or a polymerizable oligomer, the following effects take place:[1,4–12]

- The increase of the concentration of reacting molecules is caused by the decrease of its volume.
- The diffusion rate of molecules changes because the reaction system becomes denser and its viscosity increases.
- As the pressure increases, the relaxation-time spectrum of kinetic units shifts in the direction of a larger relaxation time, the selection of conformers that provide denser packing of molecules occurs, the angle of internal rotation decreases, and the solubility of components and phase states change.
- At high pressure, the deformation of molecules occurs changing the shape of electron clouds, the effectivity and frequency of the collision of reacting particles, the occurrence of orienting effects, and the ability to overcome sterical obstacles become possible, and so on.[8]

Under hydrostatic pressure, monomers and polymerizable oligomers behave like crystallizable or glass-like substances. For example, methyl methacrylate (MMA), *n*-butyl methacrylate, *n*-hexyl methacrylate, styrene, glycol dimethacrylates, and so on have compression isotherms specific for crystallized monomers.[8,13]

The viscosity of the polymerization medium is known to substantially influence the velocity of polymerization and, first, the chain-termination reaction. The higher the pressure, the lower the degree of conversion at which the gel effect may be observed.[8,12]

In certain reaction systems, chemical equilibrium at atmospheric pressure in the monomer to polymer system takes place at low temperatures and thus accomplishes polymerization.

Finally, a very important factor is the variation of the solubility of forming macromolecules in the monomer when the pressure is enhanced, which can be used for fractionating, but may cause unwanted fractionating products.

COPOLYERMIZATION

The monomer's behavior at copolymerization is caused by steric and polaric effects, resonance stability of monomers and their active particles, and so on. Pressure exercises influence on these factors, ultimately detecting the composition and the structure of copolymers. The constants of copolymerization r_1 and r_2 and the polarity parameter *e* depends on the pressure. As *P* increases, the product r_1r_2 and the value *e* may enhance, decrease, or remain independent from *P* according to the chemical nature of monomers M_1 and M_2.

POLYMERIZATION UNDER COMBINED ACTION OF HIGH PRESSURE AND SHEAR DEFORMATION

As shown in the previous section, high pressure can substantially affect the polymerization rate as well as polymer structure and molecular mass. However, the reaction mechanism remains unchangeable, that is, initiators and catalysts are to be used and temperature is to be increased in some cases, this involves the same elementary stages of polymerization. The pattern strongly changes when shear deformation is applied to a solid monomeric substance in a compressed state. In this case polymerization occurs at extremely high rates with low temperatures in the absence of initiators; the dependence of polymerization kinetics on energetic characteristics of deformation (compressive load and deformation angle) being its peculiar feature. Thus, the chemical reaction of polymer formation from a monomer is replaced by the mechanochemical process. Among the most important features of this process is the possibility of transforming polymers from monomers that do not polymerize or that polymerize with great difficulty under normal conditions. This feature widens the scope of synthesis in polymer chemistry.

The characteristic features of polymerization under HP + SD are as follows:

- High reactivity, high rates of polymerization, and large polymer yield are observed.
- Polymerization occurs in the absence of solvents, initiators, and catalysts.
- Polymerization of monomers that do not polymerize or polymerize with difficulty under normal conditions and polymerization of insoluble and unmelting monomers is realized.
- Homogeneous component mixtures are formed during copolymerization.
- It is possible to regulate the selectivity of a chemical process owing to the fact that the rate of parallel reactions is sometimes comparable to that of the main reaction and can even exceed it.
- The reaction rate can be varied using matrices differing in shear stress.
- Polymerization can take place at low temperatures (e.g., the temperature of liquid nitrogen).
- The fundamental kinetic features are dependence of the polymerization kinetic parameters on the value of deformation and shear stress; low values of activation parameters; and applicability of formal kinetics methods and techniques.

POLYMERIZATION UNDER SHOCK WAVES

Among the various actions of high pressure on substances is the shockwave (SW) method. The detection of macromolecules in organic compounds exposed to SW made this method attractive to the specialists in polymer and biopolymer science.[14–25] Their interest has been primarily related to the fact that such experiments model the substance transformations in extreme conditions (e.g., explosions, tectonic processes, falling meteorites).

In order to produce SWs in various materials, detonation of condensed-phase explosives is usually applied. This generates jumps in pressure, temperature, and density propagating at supersonic speed. The material to be exposed to SW is placed in special capsules of different designs (so-called recovery capsules).[25,26] The detonation velocity, *D*, and particle, velocity, *U*, are the main parameters to be measured by special sensors and an oscillograph.

SW is characterized by pressure values much higher than the maxima achievable under static conditions, high-rate plastic deformations (10^{-10}–10^{-6} s), great shear stresses, and pressure gradients exceeding a GPa in the 100s.[25] These conditions determine the extremely high rates of active center generation. The analysis of shock adiabats indicates that chemical transformations occur 10^7–10^{14} times quicker than in static pressure conditions. This is explained by the fact that chemical reactions are primarily influenced by plastic microdeformation processes observed under SW. It is also known that the shock-compressed specimen is then exposed to relief waves, which also provoke chemical and physical transformations.

Contrary to the HP + SD conditions, the SW action on specimen is much shorter in duration but causes a sharper increase in temperature. Therefore, the significant thermal and posteffects are expected (and in most cases observed). Polymerization under SW is shown to proceed in three ways.[25]

- Polymerization at the compression stage under SW can proceed within a few microseconds in the mode of self-accelerated adiabatic reaction.
- Postpolymerization occurs at small compression amplitudes and is most noticeable when the experimental temperature is 77 K and the obtained specimen is then stored at room temperature.
- Postpolymerization in the "thermal explosion" mode is observed at high compression amplitudes and T_{res}.

These three modes can be realized in a single experimental explosion, however, the share of polymer forming via this or that mechanism is condition-dependent.

REFERENCES

1. Jenner, J. In *High Pressure Chemistry* Kelm, H., Ed.; Reidel: Dordrecht, 1977; p 218.
2. *High Pressure Chemistry and Physics of Polymers*: Kovarskii, A. L., Ed.; Boca Raton, FL, 1994; 430.
3. Weale, K. E. *Q. Rev.* **1962**, *16*, 267.
4. Weale, K. E. *Chemical Reactions at High Pressures*; Spon: London, 1967; p 349.
5. Imoto, T. *Kagaku Kogaku* **1968**, *32*, 224.
6. Gonikberg, M. G. *Chemical Equilibria and Reactions Rates at High Pressures*; Khimiya: Moscow, 1969; p 427.
7. Ehrlich, P.; Mortimer, G. A. *Adv. Polym. Sci.* **1970**, *7*, 386.
8. Sivergin, Yu M. *High Pressure Chemistry and Physics of Polymers*; Kovarskii, A. L., Ed.; Boca Raton, FL, 1994; p 195.
9. Zhulin, V. M. *Rev. Phys. Chem. Jap.* **1980**, *50*, 217.
10. Ogo, Y. *J. Macromol. Sci., Macromol. Chem. Phys.* **1984**, *C24*, 1.
11. Zabisky, R. C. M.; Chan, W. M.; Gloor, P. E.; Hamielec, A. E. *Polymer* **1992**, *33*, 2243.
12. Sivergin, Yu M. *Polymerization under Pressure* (Part 2), Deposit VINITI: Moscow, N1549-B93 Dep., 1993; p 101.
13. Sasuga, T.; Takehisa, M. *J. Macromol. Sci., Chem.* **1978**, *A12*, 1307.
14. Duvall, G. E.; Fowles, G. R. In *High Pressure Physics and Chemistry*; Bradley, R. S., Ed.; Academic: New York, 1963; Vol. 2, p 209.
15. Adadurov, G. A.; Barkalov, I. M.; Goldanskii, V. I., et al. *Polymerization in a shock wave, Polymer Sci. USSR* **1965**, *7*, 196.
16. Adadurov, G. A.; Barkalov, I. M.; Goldanskii, V. I., et al. *Acad. Sci. USSR, Proc. Phys. Chem. Sect.* **1965**, *165*, 835.
17. Barkalov, I. M.; Adadurov, G. A.; Dremin, A. M., et al. *J. Polym. Sci.* **1967**, *C16*, 2567.
18. Adadurov, G. A.; Goldanskii, V. I. *Russian Chem. Rev.* **1981**, *50*, 948.
19. Dodson, B. W.; Graham, R. A. In *Shock Waves in Condensed Matter–1981*; Nellis, W. J. et al., Eds.; American Institute of Physics: New York, 1982; p 42.
20. Graham, R. A.; Morosin, B.; Dodson, B. W. The Chemistry of Shock Compression: A Bibliography Sandia National Labs. Albuquerque, New Mexico, 1983: Rep. SAND 83-1887.
21. Anonymous *Report of the Committee on Shock Compression Chemistry in Materials, Synthesis and Processing*; National Materials Advisory Board, National Research Council. NMAB-414, National Academy: Washington, DC, 1984; Publication NMAB-414.
22. Graham, R. A.; Morosin, B.; Venturini, E. L.; Carr, M. J. *Annu. Rev. Mater.* **1986**, *16*, 315.
23. Adadurov, G. A. *Uspekhi Khimii* **1986**, *55*, 555.
24. Graham, R. A. In *Shock Waves in Condensed Matter–1982*; Schmidt, S. C.; Holmes, N. C., Eds.; Elsevier: Amsterdam, 1988; p 11.
25. Gustov, V. W. In *High Pressure Chemistry and Physics of Polymers*; Kovarskii, A. L., Ed.; Boca Raton, FL, 1994; p 430.
26. Graham, R. A. In *High Pressure Explosive Processing of Ceramics*; Graham, R. A.; Savaoka, A. B., Eds.; Transaction Technical Publication, 1987; p 31.

High Solids Coatings

See: High Solids Coatings (Microspheres from Nonaqueous Polymer Dispersions)
High Solids Coatings (Use of Supercritical Fluids)
High Solids Polyurethane Coatings
Hyperbranched Polyesters
Peroxide Initiators (for High Solids Polyacrylate Resins)

HIGH SOLIDS COATINGS (Microspheres from Nonaqueous Polymer Dispersions)

Motoshi Yabuta, Yasushi Nakao, and Akira Tominaga
Technical Research Laboratory
Kansai Paint Company, Ltd.

The investigation and development of nonaqueous polymer dispersion (NADs) has a history extending over almost three decades. Twenty years have passed since the publication of the comprehensive review of NADs by Barrett in 1975, and a variety of NADs and their applications have been proposed during these years.[1] NADs have been extensively used, particularly in the field of high solids metallic finishes for automobiles. In the case of high solids coatings, which are mainly composed of low T_g, low molecular weight polymers and are thus not free from several inherent defects, NADs are in indispensable component for improving their rheology, mechanical properties, durability, etc.[2-4]

The NAD coatings composed of a mixture of NAD/solution acrylic polyol/melamine formaldehyde resin (MFR) give a typical two-phase structured system not only in liquid enamel, but also in crosslinked films with discrete polymer particles, which brings out specific features in the coating performance.

In this chapter, the functions of polymeric microspheres from NAD in high solids coatings will be reviewed including rheological and optical properties of the coatings.

PREPARATION OF NAD AND CLEAR ENAMEL FORMULATION

Up to now, various kinds of dispersants and NADs stabilized by them have been proposed and patented. They include comb-like dispersants from the graft-copolymerization of poly(12 hydroxystearic acid) macromonomer and MMA, alkyletherized MFRs, alkyl methacrylate copolymers, oil-modified alkyd resins, alkoxysilane-containing resins, and fluoropolymers.[5–10]

The type of polymeric dispersant of NAD plays a decisive role in the performances of the coatings both in liquid paint and in dried films. In the development of NAD coatings for automotive metallic finishes, NADs, which are stabilized with acrylic dispersants, are the most useful because the acrylics have a large degree of freedom in modulating the molecular weight, T_g, solubility parameter (SP) and the functional groups of the dispersant polymer. The obtained NADs are easily combined with various kinds of solution polymers and crosslinkers, which will form the continuous phase of the coatings.

The miscibility between the dispersant and the continuous phase polymers is an important factor and should first be considered when selecting NADs to formulate them in the coatings. Also, acrylic dispersants are easy to use for modulating the miscibility by changing the monomer composition.

APPLICATION TO METALLIC FINISHES FOR AUTOMOBILES

Two-phase Morphology of the Crosslinked Films

The TEM image shows the typical two-phase morphology of the crosslinked film; evidently each particle of NAD remains without coalescing with each other and is dispersed almost randomly in the crosslinked matrix.

The concept of two-phase morphology is accepted as the general way for designing high solids automotive metallic finishes both in clearcoat and basecoat enamel, and hence the two-phase structure exhibits many advantages in the coating performances, mechanically, optically, etc.

At around room temperature, the magnitude of the internal stress of the two-phased films are almost one-half that of the solution type film.

From the results, it is considered that in the case of the two-phase films, the stress relaxation is more rapid following the decrease in temperature compared to the homogeneous structured film. It should be concluded that NAD particles have a "Particle Reinforcement Effect" similar to the rigid fillers together with a "Stress Relaxation Effect" like rubber particles exhibit in ABS and HIPS resins.

In general, it is said that cracks and adhesion failure of the coatings during outdoor exposure or from the accelerated weathering test would be caused by internal stress, which was repeatedly generated due to cycling of dry-wet and heat-cool along with chemical decomposition by UV, oxygen and water.

Optical Properties of the Films Containing NAD Particles

To produce a top clearcoat containing NAD particles for the metallic system, the refractive indices of the particle and continuous phase polymer, along with the particle size and its distribution, must be modulated to attain completely transparent film.[11]

In order to analyze the optical properties of the composite, the ratio of acrylic/MFR along with styrene content, which were thought to domonate the refractive index of the composite, were examined. The refractive indices linearly increase with an increase in MFR ratio and the effect of styrene contents of the acrylic copolymer are also remarkable.

Due to high optical density of triazine rings in MFR as well as benzene rings in acrylics, the contents of these moieties in the composites dominate the refractive index, and hence determine the optical properties, especially transparency of the two-phase coating films.

SUMMARY

The functions of NAD in high solids coatings are demonstrated in this chapter. Recently in the field of automotive finishes, new performances such as acid-etch resistance and mar resistance, are demanded in the clearcoat, and further decreasing the VOC using HAPS-free solvents without sacrifice of any performance of the coatings.

Beside the traditional combination of hydroxy/MFR, other crosslink chemistries such as acid/epoxy, hydroxy/blocked isocyanate and alkoxysilanes are actively introduced in order to answer these demands.[13–15] Also, various kinds of polymers such as fluoropolymers, polydimethylsiloxane and polyurethanes, are combined with acrylics as film forming binders. NADs have to be designed to meet these new chemistries and polymers, which include miscibility, functionalities, and refractive index.

In developing these NADs, acrylic dispersants still have the largest degree of freedom in modulating their composition as is described, but great effort should be made to extend the dispersant menu and the particle formation methods.[9,10] Through these efforts, NADs will be advanced and can open new windows to coatings world.

REFERENCES

1. Barett, K. E. J. *Dispersion Polymerization in Organic Media*, John Wiley & Sons: London, 1975.
2. Yabuta, M.; Sasaki, Y. *Organic Coatings, Science and Technology*, Marcel Dekker: New York, 1986; Vol. 8, p 329.
3. Yabuta, M.; Sasaki, Y. *13th International Conference in Organic Coatings Science and Technology*, Athens, 1987; p 443.
4. Yabuta, M.; Sasaki, Y. *Polymeric Materials Science and Engineering*, Washington, DC, Fall meeting, 1990; Vol. 63, p 767. Nakao, Y.; Yabuta, M.; Tominaga, A. *Progress in Organic Coatings* **1992**, *20*, p 369.
5. British Patent 1 122 397 1964; Imperial Chemical Industries.
6. U.S. Patent 3 736 279, 1964; Ford Motor Company.
7. U.S. Patent 3 232 903, 1961; Rohm & Haas.
8. Japanese Patent 73-60180, 1973; Kansai Paint.
9. Japanese Patent 89-95116, 1989; Kansai Paint.
10. Japanese Patent 87-25105, 1987; Kansai Paint.
11. Yabuta, M.; Sasaki, Y. *Polymeric Material Science and Engineering*, Washington, DC: Fall meeting, 1990; Vol. 63; p 772.
12. Backhouse, A. J. *J. Coat. Technol.* **1982**, *54* [698], p 83; U.S. Patent 4 180 489, 1979, Imperial Chemical Industries.
13. U.S. Patent 4 650 718, 1987; PPG Industries; U.S. Patent 96 282, 1980; Du Pont de Nemours.
14. U.S. Patent 5 025 060, 1991; Kansai Paint.
15. U.S. Patent 5 063 114, 1991; Kanegafuchi Kagaku Kogyo.

HIGH SOLIDS COATINGS (Use of Supercritical Fluids)

Kenneth A. Nielsen
Union Carbide

Use of environmentally benign supercritical fluids for the spray application of high solids coatings is a new pollution prevention technology first commercialized in 1991. The supercritical fluid is used to thin the coating formulation to a sprayable viscosity instead of using volatile organic solvents that cause air pollution.[1] However, unlike liquid solvents, the supercritical fluid provides energy for spray formation by decompression and expansion. This produces a fundamentally new type of atomization called a decompressive spray.[2] It has favorable characteristics for the application of coatings when compared to conventional spray methods.[3] This is an important factor in reducing application cost and improving coating appearance and performance.

Carbon dioxide is the preferred supercritical fluid because it is environmentally compatible, nonflammable, inexpensive, readily available and generally has high solubility. The process actually reduces carbon dioxide emissions because natural oxidation of organic solvents generates several times more carbon dioxide. Ethane also can be used because it does not generate ozone in the atmosphere. However, ethane is flammable, costs more and generally has lower solubility. Nitrous oxide has properties similar to carbon dioxide and decomposes to nitrogen and oxygen, but it is generally avoided at supercritical conditions because it can become unstable.

Conventional and high solids coating formulations often can be reformulated for spray application with supercritical carbon dioxide by adjusting the solvent level and blend, with little or no change made to the polymer system and pigments.[4] Typically, 60% to 80% of the organic solvent is removed. The reformulated coating is called a coating concentrate because it has much higher viscosity and higher solids level than normal high solids coatings.

The process is broadly applicable to most coating polymers, including acrylics, alkyds, polyesters, melamines, formaldehydes, cellulosics, silicones, vinyls, epoxies, phenolics, ureas, urethanes and waxes. It has been used with air-dry, thermoplastic, thermosetting, catalyzed, ultraviolet light cured, and two-component reactive systems to apply clear, pigmented, and metallic coatings.

SOLUBILITY AND PHASE CHEMISTRY

Supercritical carbon dioxide is dissolved into the coating concentrate under pressure and elevated temperature to form a spray solution that has low viscosity and is capable of forming a decompressive spray. The solubility, and therefore the amount of carbon dioxide that can be used for spraying, depends on pressure, temperature, polymer type and solids level and the solvent blend.[5]

Supercritical carbon dioxide brings a new dimension to the chemistry of polymer-solvent interactions because its density and therefore solubility parameter can vary markedly with pressure and temperature.[6] This produces new degrees of freedom because solvation effects can be modified without changing solvent composition.

The important property is not how well carbon dioxide dissolves polymers, but how well polymers dissolve carbon dioxide, that is, the solubility of carbon dioxide in polymers and polymer solutions, which is much higher. In thermodynamic terms, carbon dioxide is a poor solvent but a good solute. Carbon dioxide solubility depends on its compatibility with coating polymers, in addition to pressure and temperature.

A useful parameter in correlating the solubility behavior of carbon dioxide (or ethane) in coating formulations is called the critical bubble pressure.[7] It is defined as the critical pressure for a mixture of carbon dioxide and the solvent blend of a coating formulation (excluding the polymer), at a given temperature above the critical temperature for pure carbon dioxide. At temperatures below the critical temperature, the critical bubble pressure becomes the vapor pressure of carbon dioxide. The critical bubble pressure curve is useful because supercritical carbon dioxide is liquid-like above it and gas-like below it. Further, it is a reasonably good indicator of the minimum pressure required to form the liquid–liquid region so it helps predict the pressure at which the cloud point curve intersects the bubble point curve. Further, the critical bubble–pressure curve is insensitive to solvent composition; it pivots very little about the critical point for pure carbon dioxide as the solvent composition changes.

Solubility studies[6] demonstrate that polymers can be designed to have relatively high carbon dioxide solubility, which allows coating concentrates to be sprayed at very low solvent levels. Important factors are polymer molecular weight, polydispersity, solubility parameter, functionality, hydrogen bonding, and structure. Significant increases in solubility have been obtained by including fluorine, silicon, and bulky substituent groups in the polymer structure. Further, the data suggest that a variety of liquid polymers can be developed to give solvent-free coating formulations with sufficiently high carbon dioxide solubility for spraying. Three types of solvent-free coatings already have been sprayed. Therefore, the supercritical fluid spray process has potential for becoming a liquid analog of powder coating.

For many polymers, solubility also depends on interactions between polar groups and on hydrogen bonding. As hydrogen bonding between polymer chains increases, carbon dioxide solubility decreases. Hence polymers with carboxylic groups (strongest hydrogen bonding) have the lowest solubility. Hydroxyl groups also lower the carbon dioxide compatibility of a copolymer, while epoxy groups (no hydrogen bonding) have no effect on solubility relative to a copolymer with no functional groups.

APPLICATION

Commercial application studies[4] generated the unforseen result that the supercritical fluid spray process, due to its atomization characteristics, can improve coating appearance and performance, despite removing most of the solvent from the coating formulation to prevent pollution. Material and operating costs also have been reduced by eliminating application steps and reducing the amount of coating solids sprayed per part. Dry film thicknesses applied in one coat have ranged from very thin (5 microns) to very thick (> 125 microns). Thinner films have been applied with equally good appearance and thicker films have been applied with no sag when compared to coatings applied by conventional spray methods. Solvent pop tolerance has also been increased. Corrosion resistance was greatly increased in several coating systems for reasons that are not yet understood.

The supercritical fluid spray process can be used in nearly all industrial applications where coatings are spray applied. Applications include plastics, automotive topcoats and components, adhesion promoters, general industrial, metal finishing, wood furniture, adhesives, release surfaces, aircraft, marine and appliances.

REFERENCES

1. Nielsen, K. A.; Busby, D. C.; Glancy, C. W.; Hoy, K. L.; Kuo, A. C.; Lee, C. *J. Oil & Color Chemists Assoc.* **1991**, *74*(10), 362.
2. Semerjian, H. G. *Proceedings of the 5th International Conference on Liquid Atomization and Spray Systems* NIST Special Publication 813, Gaithersburg, MD, 1991; paper 37.
3. Senser, D. W. *Proceedings of the 7th Annual Conference on Liquid Atomization and Spray Systems*, ILASS, Bellevue, WA, 1994.
4. Storey, R. F.; Thames, S. F. *Proceedings of the 20th Water-Borne, Higher-Solids, and Powder Coatings Symposium*, University of Southern Mississippi, Hattiesburg, MS, 1993.
5. Kiamos, A. A. *High-Pressure Phase-Equilibrium Studies of Polymer-Solvent-Supercritical Fluid Mixtures*, M. S. Thesis, The Johns Hopkins University, Baltimore, MD, 1992.
6. Storey, R. F.; Thames, S. F. *Proceedings of the 21st Water-Borne, Higher-Solids, and Powder Coatings Symposium*, University of Southern Mississippi, Hattiesburg, MS, 1994.
7. Nielsen, K. A.; Lear, J. J.; Argyropoulos, J. N. U.S. Patent 5 312 862, 1994.

HIGH SOLIDS POLYURETHANE COATINGS

Edward P. Squiller,* Douglas A. Wicks, and Philip E. Yeske
Bayer Corporation

The reduction of volatile organic compound (VOC) emissions associated with coatings is the major driving force in the paint industry today. Increasingly, stringent environmental and safety regulations coupled with consumer demands for ecologically friendly materials are eliminating many traditional coatings systems and forcing radical changes in the surviving systems. These changes are necessary to overcome the dependence of traditional coatings on high solvent content, which is the primary source of VOCs. The most graphic and often cited example of such change is in the consumer house paint market where, in the last 30 years, low VOC latex paints have virtually eliminated solvent-borne, drying oil alkyd systems.

Equally dramatic technology changes are now taking place in non-consumer and industrial markets such as automotive refinish, maintenance, architectural, and Original Equipment Manufacturer (OEM) applications. Many of these industries demand high performance in durability, solvent-chemical resistance, or weatherability, and high solids polyurethanes are often chosen for these reasons over traditional lacquers and low solids alkyds. High solids two-component polyurethanes represent a robust, viable route for significantly reducing emissions while still yielding high-performance coatings. With these products, a reactive crosslinking coating system is formed by adding a polyisocyanate to a co-reactant just before application to the substrate.

*Author to whom correspondence should be addressed.

Here, we will focus on the chemistries used to achieve high solids polyurethane coatings along with some limitations of existing technology. In addition, we will discuss some emerging products for the future.

Raw Materials

Many isocyanates are commercially available to the urethane industry, including aliphatic or aromatic di- and polyisocyanates. In general, aromatic isocyanates cost less and are faster than aliphatic isocyanates. The choice of isocyanate can affect key properties in a polyurethane coating. For example, aliphatic isocyanates provide better hydrolytic stability and weatherability than aromatic isocyanates in a polyurethane coating. Additionally, coatings based on aliphatic isocyanates are more color stable in outdoor coating applications.

Except for bis(4-4′-isocyanatophenyl)methane (MDI), isocyanates are not used directly in polyurethane coatings applications. Instead, various chemical modifications are made to reduce vapor pressure and to increase the functionality.

Common co-reactants containing hydroxyl groups and used in the coatings industry include the following:

- linear and branched poly(ether glycols) such as poly(propylene glycol)s (PPG), poly(ethylene glycol)s (PEG), and poly(tetramethylene ether glycol)s (PT-MEG, PTHF) with molecular weights ranging from 1000 to 4000;
- polyester diols and triols such as poly(alkylene) glycol adipates with molecular weights ranging from 500 to 5000;
- poly(carbonate)diols; and
- hydroxy functional polyacrylates.

When combined with polyisocyanates, these polyols (or co-reactants) yield high-performance coatings, exhibiting a wide range of protective properties. In addition, high solids coatings can be formulated from these materials. Additional information about co-reactant raw materials for polyurethanes can be found in the *Polyurethane Handbook* and in *Polyurethanes.*[1,2]

Types of Coatings

Polyurethanes are often referred to as reactive coatings because they are functionalized, low molecular weight resins until they are cured on the substrate. This aspect of polyurethane coatings is especially conducive to designing and formulating high solids coatings. Polyurethane coatings are used in various ways, as one- or two-component coating systems, as ambient temperature or heat-cured systems, and as reactive coatings or completely reacted resins.

One-component polyurethane coatings are characterized by their ready use directly out of the container.

Two-component polyurethane coatings differ from one-component in that the two major portions of the coating binder system are mixed immediately before use. One portion contains the polyol and additives, and the second contains the polyisocyanate. Thus, two-component systems have a limited pot life.

LIMITING FACTORS OF FORMULATIONS

The viscosity of the reactive resins and the formulator's ability to control the rate of the crosslinking reactions limit one from achieving very low VOC content in reactive coating systems. What follows is a description of how these factors limit the formulation of low VOC coatings and how formulators are pushing the limits of this technology with catalysts and solvents.

Polyisocyanates

An analysis of the components' solvent requirements in a two-component urethane coating makes it apparent that the polyisocyanate is already the high solids component of the coatings. Commercially available isocyanurate and biurets of hexamethylene diisocyanate (HDI) are being pushed to lower viscosities, with the practical limitations of these products being 1000 and 2000 mPa, respectively. At these viscosities their solvent demand is only 10–20% by weight on resin to achieve reasonable application viscosities.

New types of polyisocyanate-containing resins are pushing the viscosity envelope below 500 mPa.

Polyols

For most applications acrylic polyols have been favored because of their superior weathering, low cost, and low isocyanate demand. In the past, these products have been of relatively high molecular weight (10,000–20,000 g/mol) with hydroxyl equivalent weights of 1000 g/equivalent or more. The biggest drawback of high molecular weight acrylics is their high solvent demand, which is 60–70% by weight solvent and corresponds to only 30–40% solids at application viscosities.

The first step toward higher solids polyurethane coatings is to reduce the acrylic's molecular weight and functionality. Reducing the equivalent weight by including more hydroxy functional monomers reduces the number of nonfunctional chains but can lead to brittleness and substantially increases the resin cost. Products of this type can be produced so that the solvent demand is reduced to 35–45% by weight, thus yielding resins that are 55–65% solid at application viscosity.

Accelerated Reactions from Increasing Solids

The effect of concentration can be seen easily in a model system with the same polyols and polyisocyanates at 50% and 70% solids. Assuming no change in density, the change from 50% to 70% solids increases the reactant's relative concentration by a factor of 1.4, thus doubling the reaction rate. This increased reaction rate results in a faster molecular weight build-up with the accompanying viscosity increases. One way to reduce the reaction rate in the solution is to lower the catalyst level.

The change in resin solids is not the only factor increasing the reaction rate at higher solids. In fact, this change is compounded by the movement toward lower equivalent weight co-reactants. Again, assuming no changes in solution density, the change from 1000 to 500 equivalent weight at 70% solids almost doubles the reactant concentrations and triples the reaction rate. Combining the effects, the overall change from 50% to 70% solids while reducing equivalent weight from 1000 to 500 g/equivalent more than doubles the reactive group concentration so that the reaction rate *increases more than five times.*

Modification of Catalysis

Changing the relative rate of reactions before and after spraying can be accomplished in several ways. First, a blocked or hindered catalyst system that has a low catalytic value in solution but can be activated after application can be used. A second method is through the choice of cosolvents and other volatile additives that are included in the formulation and volatilize after application.

NEW RAW MATERIALS

To meet the demanding requirements of high solids polyurethane coatings, many new types of raw materials have been developed. They include polyisocyanates and co-reactants that help reduce the solvent demand, the requirement that defines high solids coatings.

CONCLUSIONS

Over the years, most solvent reduction in coatings was achieved by using thermoset coatings, which are inherently higher in solids than the thermoplastic solution resins (e.g., polyurethane lacquers). State-of-the-art thermoset polyurethane coatings use relatively low molecular weight resins that undergo additional crosslinking reactions on the substrate to give high molecular weight polymers with high-performance properties. Continual increases in the solids content of polyurethane coatings are now achieved by continual decreases in the resins' viscosity, which for the most part are obtained by lowering the molecular weight. There are limitations, however, to how low the molecular weight and viscosity can go and still maintain the properties needed for good coatings. Researchers are working to circumvent these limitations by developing new resins, reactive diluents, and resins synthesized by different methods that allow for lower molecular weight distributions. The commercially successful developments will be cost effective and will maintain high performance levels.

REFERENCES

1. *Polyurethane Handbook: Chemistry-Raw Materials-Processing-Application-Properties*; Oertel, G., Ed.; Hanser: New York, 1985; pp 510–546.
2. Encyclopedia of Polymer Science and Engineering 2nd ed.; Backus, J. K., Ed.; John Wiley & Sons: New York, 1988; Vol. 13, pp 288–294.

High Strength Polymers

See: Aramid Fibers, High Strength
Carbon Fibers (Overview)
Carbon Fibers (Plasma Surface Treatments)
Fibers and Composites (High Modulus and Strength)
Polyacrylonitrile Gel Fibers
Poly(vinyl alcohol) (High Strength and High Modulus)
Poly(vinyl alcohol) Fiber (High Strength and High Modulus)
Poly(vinyl alcohol) Fibers (High Modulus; by Vibrating Zone-Drawing)

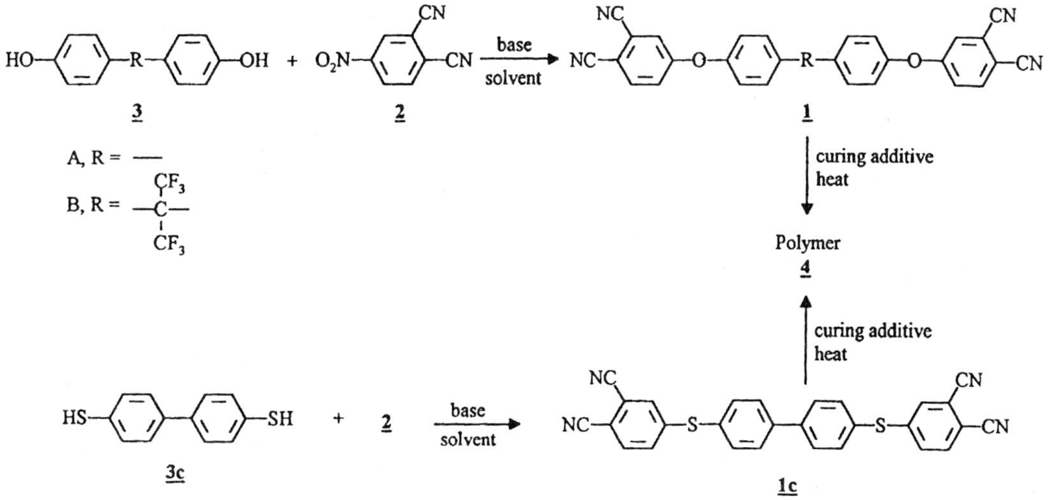

I

HIGH TEMPERATURE PHTHALONITRILE RESINS

Teddy M. Keller
Materials Chemistry Branch
Naval Research Laboratory

Phthalonitrile-based polymers are a class of high-temperature polymers having a wide range of potential uses such as composite matrices, adhesives, and electrical conductors.[1–6] Aromatic ether, thioether, sulfone, and imide linkages have been incorporated successfully between terminal phthalonitrile units.[7–16] Polymerization occurs through the cyano groups to afford heterocyclic crosslinked products.[17,18] Curing agents that have been used to cure phthalonitrile resins include organic amines, strong organic acids, strong organic acid–amine salts, metallic salts, and metals.[15–19] Only a minute quantity of curing additive is needed to initiate the polymerization reaction. The polymerization reaction occurs through the cyano groups of the phthalonitrile units by an addition mechanism and appears to propagate through multiple reaction pathways involving polytriazine, polyimine, and polyphthalocyanine formation. These heterocyclic crosslinked products exhibit good thermal and oxidative stability. The preferred reaction pathway probably depends on the curing agent used. The polymerization rate is easily controlled as a function of the concentration of curing additive and curing temperature, which enhances the importance of the phthalonitrile-based resins as a matrix material for resin transfer molding (RTM). The following discussion focuses on the development of aromatic ether- and aromatic thioether-linked phthalonitrile resin systems.

SYNTHESIS

Bisphenol-Based Phthalonitriles

The phthalonitrile monomers (**1A**, **1B**, and **1C**) are synthesized in high yield by a simple nucleophilic displacement reaction (see **Scheme I**).[10,19]

Oligomeric Phthalonitriles

In a series of low molecular weight oligomeric phthalonitriles, the average molecular weights were varied by reacting different molar ratios from an excess of 4,4′-dihydroxybiphenyl (3a) or 4,4′-(hexafluoroisopropylidene)-bisphenol (3b) with 4,4-difluorobenzophenone (6), yielding 7 and end-capping with 2. This series (5) was prepared in a two-step one-pot reaction (**Scheme II**). This synthetic method provides the ability to obtain resins with a crosslink density that can be readily controlled. By varying the distance between the polymerization centers, the effect of changing the characteristics in the interconnecting linkages on the corresponding polymer's physical and mechanical properties can be ascertained.

POLYMERIZATION

The phthalonitrile monomers are readily converted to highly crosslinked thermosetting polymers (4) in the presence of thermally stable organic amines, phenols, strong organic acids, mineral acids, metals, or metallic salts. Most of our polymerization studies involved using an aromatic amine such as 3,3′-bis(3-aminophenoxy)benzene (APB) and, to a lesser extent, a strong organic acid such as *p*-toluenesulfonic and as a curing additive. Polymerization occurs through the terminal phthalonitrile moieties to afford void-free polymeric materials. The polymerization of phthalonitrile monomers appears to occur by competing reactions involving polyimine, phthalocyanine and triazine polymerization products.[6–8]

After adding the curing additive to the melt of these monomers, the reaction mixture rapidly converts from a crystalline to an amorphous phase (prepolymer). Moreover, the polymerization reaction can be performed in one step to gelation or can be advanced to any viscosity desired (B-staged prepolymer), quenched, and stored at room temperature until ready to convert into a plastic, thermosetting material.

THERMAL STABILITY

The thermal and oxidative stabilities of the phthalonitrile resins have been assessed under isothermal and dynamic conditions. The

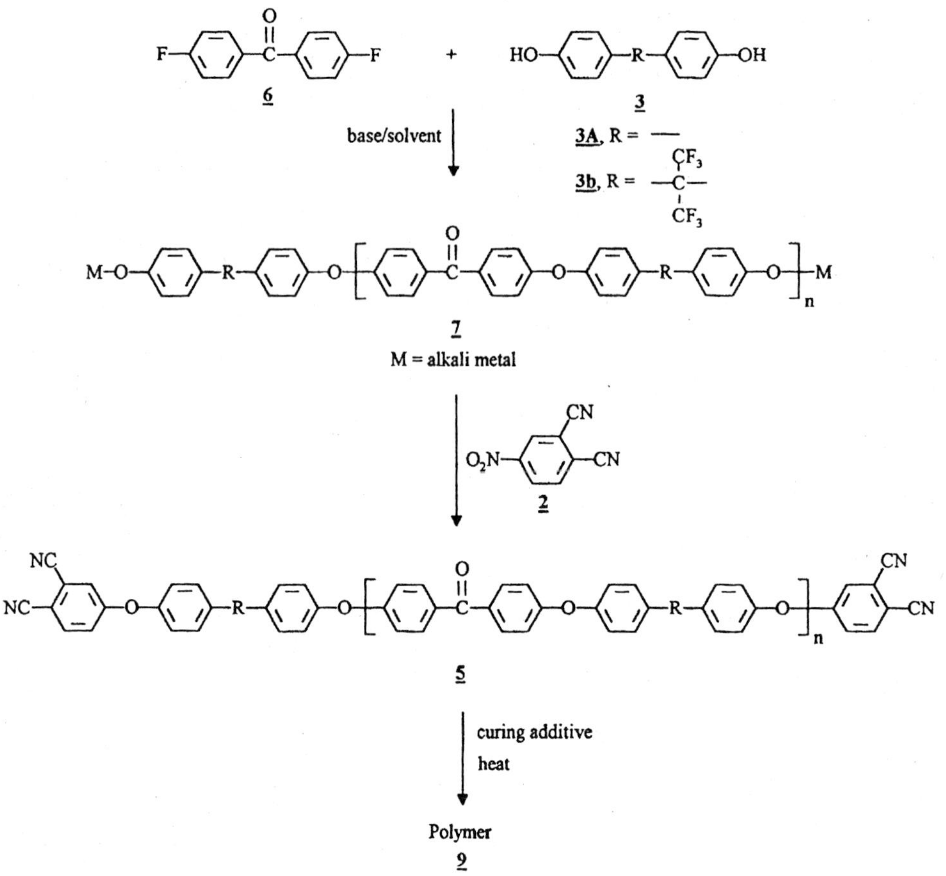

II

thermal stability was a function of the heat treatment. Upon curing and postcuring to 375°C under inert conditions, the polymers show excellent thermal stability and a char yield exceeding 70%. When the polymers were pyrolyzed to higher temperatures, improvements in the char yield and oxidative stability were observed. The thermal and oxidative properties were found to be enhanced as the polymers were postcured at elevated temperatures.

MECHANICAL PROPERTIES

The phthalonitrile-based polymers exhibit excellent mechanical properties.[20] Compared with other state-of-the-art thermosetting polymers, the phthalonitrile polymers display superior tensile strength values and retain these values when aged at 315°C for extended periods in an oxidizing environment.

The phthalonitrile polymers also exhibit a fracture toughness similar to the values of unmodified epoxy resins.

ELECTRICAL CONDUCTIVITY

Several polymers (e.g., polyacrylonitrile, dianil phthalonitrile resins, and poly[*N,N′*-(*p,p′*-oxydiphenylene)pyromellitimide] have exhibited high conductivity upon pyrolysis.[14–19] Moreover, the conductivity can be changed in a controlled manner by a precise pyrolytic procedure without adding external chemical dopants. Interest in pyropolymers for practical application does appear to be growing because of their excellent conductive stability in air and water.[21] The electrical properties of pyrolyzed polymers can probably be attributed to the buildup and ordering of polycondensed rings in the developing conductive matrix, which allows current to be transported by charge hopping or tunneling between carbonized islands.[14]

The excellent thermal stability displayed by the phthalonitrile polymers contributed to further pyrolytic studies. The electrical behavior can be systematically changed from an insulator to a semiconductor and made to approach metallic regions by controlling the thermal processing temperature.

CONCLUSION

Because of their unique properties, monomers 1 and 5 are potential candidates as matrices for numerous composite applications. The monomers are easily made into shaped, void-free components, possess high thermal and oxidative stability, and have high char yields. When postcured at high temperatures, the polymers do not exhibit a T_g. Upon pyrolysis to 900°C, pyrolysates from 4A and 4C exhibited electrical conductivity ranging from an insulator to a semiconductor and approaching metallic behavior. Potential applications for the polymers are as a matrix material for advanced composites and as a molding material to make electronic devices because the polymers are thermally stable at the soldering temperatures required for mounting components on a circuit board surface.[22]

ACKNOWLEDGMENT

The author wishes to thank the Office of Naval Research for their financial support of this work.

REFERENCES

1. Delvigs, P.; Hsu, L. C.; Serafini, T. T. *J. Polym. Sci. B* **1970**, *8*, 29.
2. Bilow, N.; Landis, A. L.; Aponyl, T. J. *SAMPE Symp.* **1975**, *20*, 618.
3. Wilson, D. *Br. Polym. J.* **1988**, *20*, 405.
4. Keller, T. M. *Polym. Commun.* **1991**, *32*, 2.
5. Keller, T. M. *Polymer* **1993**, *34*, 952.
6. Burchill, P. J. *IPAC Int. Symp. Polym. Mater. Prepr.* **1991**, 186.
7. Burchill, P. J. *J. Polym. Sci., Part A: Polym. Chem.* **1994**, *34*, 1.
8. Snow, A. W.; Griffith, J. R. *Macromolecules* **1984**, *17*, 1614.
9. Keller, T. M. *J. Polym. Sci., Part C: Polym. Lett.* **1986**, *24*, 211.
10. Critchley, J. P.; Wright, W. W. *Rev. High Temp. Mater.* **1979**, *4*, 107.
11. Truong, V. T. *Polymer* **1990**, *31*, 1669.
12. Kinloch, A. J.; Shaw, S. J.; Tod, D. A. et al. *Polymer* **1983**, *24*, 1341.
13. Baughman, R. H.; Ivory, D. M.; Miller, G. G. et al. *Org. Coat. Plast. Prepr.* **1979**, *41*, 139.
14. Bruck, S. D. *Polymer* **1965**, *6*, 319.
15. Grassie, N.; McNeill, J. C. *J. Polym. Sci.* **1958**, *27*, 207.
16. van Beek, L. K. H. *J. Appl. Polym. Sci.* **1965**, *9*, 553.
17. Walton, T. R.; Griffith, J. R.; Reardon, J. P. *J. Appl. Polym. Sci.* **1985**, *30*, 2921.
18. Brom, H. B.; Tomkiewicz, Y.; Aviram, A. et al. *Solid State Comm.* **1980**, *35*, 135.
19. Lin, J. W.-P.; Epstein, A. J.; Dudek, L. P. et al. *ACS Org. Coat. Plast. Prepr.* **1980**, *43*, 482.
20. Warzel, M. L.; Keller, T. M. *Polymer* **1993**, *34*, 663.
21. Keller, T. M. *J. Polym. Sci., Part A: Polym. Chem.* **1987**, *25*, 2569.
22. Frisch, D.; Ciccarone, R. *Circuits Manufacturing*; Benwill: 1977.

High Temperature Polymers

See: *Aramids (Processable Precursors)*
Aromatic Hydrocarbon-Based Polymers
Cardo Polyesters (Heat Resistance and Photosensitive)
High Temperature Phthalonitrile Resins
Inorganic/Organic Hybrid Polymers (High Temperature, Oxidatively Stable)
Polybenzimidazoles (Overview)
Polyimides, Processable (Modified with other Rigid Groups)
Polyisophthalamides (Thermal Stability Enhancement by Pendent Groups)
Polyoxadiazoles (Overview)
Poly(p-phenylene sulfide) (Synthesis by p-Dichlorobenzene and Sodium Sulfide)
Poly(thioether ketone and sulfone)s
Poly(1,2,4-triazole)s
Quinone-Diamine Polymers
Solid-State Polymerization (Crosslinked Di-p-Ethynyl Substituted Polymers)

High Temperature Superconductors

See: *Superconducting Filaments, High Tc (Suspension and Solution Spinning from PVA Solution)*

HIGHLY BRANCHED POLYMERS

Young H. Kim
Du Pont Central Research and Development Experimental Station

Highly branched dendritic polymer is one of the new architectural polymers which delineate from conventional linear polymer architecture. These polymers have an intrinsically globular shape, and show promise as additives or crosslinking agents, taking advantage of their unique structure. Another characteristic of these polymers is that, in an ideal stepwise addition of an AB_2 type monomer to a highly branched polymer, each new tier would almost double the molecular weight of the polymer.

HIGHLY BRANCHED POLYMERS

Polyamide

Rao synthesized the highest molecular man-made protein having a dendritic structure, utilizing the fast molecular weight growth of the divergent approach.[2]

Tomalia prepared dendritic polyamidoamines by alternating Michael reactions and amidation on NH_3 using methyl acrylate and ethylenediamine, respectively.[3]

Bayliff prepared monodispersed dendritic aromatic polyamides using convergent growth.[4]

Polyesters

Polyester is one of the most-exploited polymers with highly branched structure. Arborol is a series of cascade oligomeric polyesters which have a three-dimensional structure with the outer surface covered with polar functional groups. There are various types of core material and branching repeat units.

Dumbbell-type 2-directional arborol, for example, [9]-n-[9]arborols,$[(HOCH_2)_3CNHCO]_3C(CH_2)_nC[CONHC(CH_2OH)_3]_3$(n = 3-10), prepared by ethoxycarbonylation of the appropriate $Br(CH_2)_nBr$ with $NaC(CO_2Et)_3$ followed by amidation with $H_2NC(CH_2OH)_3$ showed a self aggregation to form a gel structure.[5,6] More recently, Newkome reported structurally more complex four-dimensional arborols using adamantane derivatives,[7,8] or $[Br(CH_2)_3]_4C$ with 4-amino-4-(3-acetoxypropyl)-1,7-diacetoxyheptane and di*tert*-Bu 4-amino-4-[2-(*tert*-butoxycarbonyl)ethyl]heptanedioate as the building blocks.[9] New structural arborols containing metal chelating ligands,[10] or azacrown ether[11] were also exploited.

There are aliphatic and aromatic polyesters having some AB_x monomers, such as $MeO_2C(CH_2)_2N(-(CH_2)_2OH)_2$ and 3,5-dihydoxy benzoic acid diacetate. Glass transitions (T_gs) of the polymers depend on the polymer backbone and terminal group.[12]

Polyamine

Palladium catalysed ring-opening polymerization of 5,5-dimethyl-6-ethenylperhydro-1,3-oxazin-2-one with primary or secondary amine initiator gave a hyperbranched polyamine.[13]

Polyether

Padias synthesized an aliphatic dendritic polyether with a pentaerythritol core or a tris(hydroxymethyl)ethane core and a hydroxymethyl bicyclic orthoformate as the building block.[14]

Polyphenyl ether from 1,4,6-tribromophenol is one of the earliest-known hyperbranched polymers.[15,16] Kim recently reported higher molecular weight of the same polymer, and homopolymer of 1,4-dibromophenol, which accompanied C-C coupling in addition to C-O coupling.[17]

Polyphenylene

Dendritic polyphenylenes were convergently prepared by a coupling of arylboronic acids to 3,5-dibromo-1-trimethylsilylbenzene using Pd(0) catalyst.

Similarly, Kim synthesized hyperbranched polyphenylenes from AB_2 type monomers, for example (3,5-dibromophenyl)boronic acid and 3,5-dihalophenyl Grignard reagents, by using Pd(0) and Ni(II)-catalyzed aryl–aryl coupling reactions.[18] He obtained polymers with M_n 5000–35,000 and polydispersity less than 1.5. ^{13}C NMR indicated about 70% branching efficiency.[19]

Silanes and Siloxanes

Lower solution viscosity of branched polysilanes could be beneficial for conventional silicone applications. Because the silicone is used as the branching agent, the silane monomers are typically AB_3 type monomers, thus the surface congestion can occur at low generation.

Hyperbranched polysiloxanes were prepared by coupling of hydroxyl containing silicones with chlorosilane,[22] hydrosilyation of allyl silane,[23] or vinyl tris(dimethylsiloxy) silane and tris(dimethylvinylsiloxy)silane.[24]

PROPERTIES AND APPLICATIONS OF HIGHLY BRANCHED POLYMERS

The physical properties and utility of these polymers are just being exploited. The most intriguing properties are unusual viscosity and glass transition.

Viscosity

One of the most peculiar properties of the dendritic polymer is their solution or α viscosity. Some polymers are known to exhibit a very low constant,[1,20,21,25] a characteristic of nondraining spheres, but others exhibit a maximum in the Mark–Houwink plot.[26,27]

Glass Transition

Even though complete understanding of the glass transition phenomena of dendritic polymers remains elusive, Wooley successfully used a modified chain-end free volume theory to explain the changes of T_g with molecular weight and number of chain end.[28] In general, the T_g increased with end-group polarity.[29] More careful dielectric relaxation measurement study maybe able to further elucidate the relaxation phenomena of these polymers.[30]

REFERENCES

1. Aharoni, S. M.; Crosby, C. R.; Walsh, E. K. *Macromolecules* **1982**, *15*(4), 1093.
2. Rao, C.; Tam, J. P. *J. Am. Chem. Soc.* **1994**, *116*(15), 6975.
3. Tomalia, D. A. et al. *Polym. J. (Tokyo)* **1985**, *17*(1), 117.
4. Bayliff, P. M.; Feast, W. J.; Parker, D. *Polym. Bull.* Berlin **1992**, *29*(3-4), 265.
5. Newkome, G. R. et al. *J. Chem. Soc., Chem. Commun.* **1986** (10), 752.
6. Newkome, G. R. et al. *J. Am. Chem. Soc.* **1990**, *112*(23), 8458.
7. Newkome, G. R. et al. *J. Org. Chem.* **1991**, *56*(25), 7162.
8. Newkome, G. R. et al. *J. Org. Chem.* **1992**, *57*(1), 358.
9. Newkome, G. R. et al. *J. Org. Chem.* **1993**, *58*(4), 898.
10. Newkome, G. R. et al. *J. Chem. Soc., Chem. Commun.* **1993**, (11), 925.
11. Nagasaki, T. et al. *J. Chem. Soc., Chem. Commun.* **1992**, (8), 608.
12. Figuly, G. D. U.S. Patent 5136014, 1992.
13. Suzuki, A.; Ii, A.; Saegusa, T. *Macromolecules* **1992**, *25*(25), 7071.
14. Padias, A. B. et al. *J. Org. Chem.* **1987**, *52*(24), 5305.
15. Hunter, W. H.; Woollett, G. H. *J. Am. Chem. Soc.* **1921**, *43*, 135.
16. Staffine, G. D.; Price, C. C. *J. Am. Chem. Soc.* **1960**, *82*, 3632.
17. Kim, Y. H. *Polymer Preprint* **1993**, *34*(1), 56.
18. Kim, Y. H.; Webster, O. W. *J. Am. Chem. Soc.* **1990**, *112*(11), 4592.
19. Kim, Y. H.; Webster, O. W. *Macromolecules* **1992**, *25*(21), 5561.
20. Morikawa, A.; Kakimoto, M.; Imai, Y. *Macromolecules* **1991**, *24*(12), 3469.
21. Morikawa, A.; Kakimoto, M.; Imai, Y. *Macromolecules* **1992**, *25*(12), 3247.
22. Uchida, H. et al. *J. Am. Chem. Soc.* **1990**, *112*(19), 7077.
23. Mathias, L. J.; Carothers, T. W. *J. Am. Chem. Soc.* **1991**, *113*(10), 3994.
24. Rubinsztajn, S. J. *Inorg. Organomet. Polym.* **1994**, *4*(1), 61.
25. Aharoni, S. M.; Murthy, N. S. *Polym. Commun.* **1983**, *24*(5), 132.
26. Mourey, T. H. *Macromolecules* **1992**, *25*(9), 2401.
27. Tomalia, D. A.; Naylor, A. M.; Goddard, W. A. *Angew. Chem. Int. Ed. Engl.* **1990**, *29*, 138.
28. Wooley, K. L. et al. *Macromolecules* **1993**, *26*(7), 1514.
29. Kim, Y. H.; Beckerbauer, R. *Macromolecules* **1994**, *27*(7), 1968.
30. Malmstroem, E. et al. *Polym. Bull.* **1994**, *32*(5-6), 679.

Hindered Amine Light Stabilizers

See: Additives (Agents for Sustaining Properties)
Antioxidants
Hindered Amine Light Stabilizers, Monomeric
Light Stabilizers (Overview)

HINDERED AMINE LIGHT STABILIZERS, MONOMERIC

Wayne W. Y. Lau
Department of Chemical Engineering
National University of Singapore

Pan Jiang Qing
Institute of Chemistry
Academia Sinica

Hindered amine light stabilizers (HALS) usually refer to derivatives made from 2,2,6,6-tetramethylpiperidine TMP (**Structure 1**). These hindered amine derivatives have proven to be the most effective photostabilizers; their effectiveness is two to six times higher than that of nickel chelete photostabilizers.[1,2] With HALS, photostability of polymeric materials, especially polypropylene, has increased.

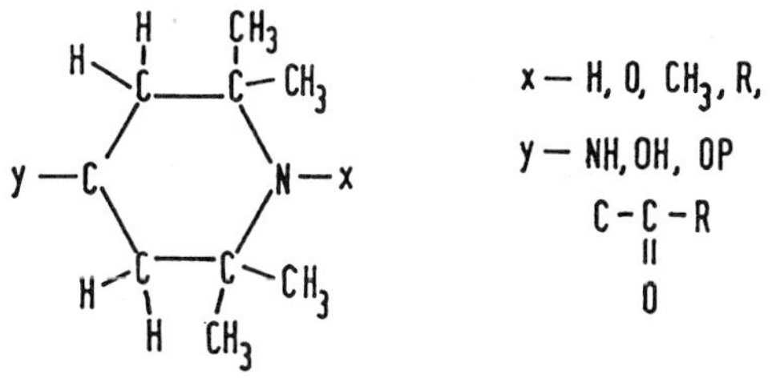

1

MECHANISM OF PHOTOSTABILIZATION

Aging of polymeric materials exposed to sunlight is caused by photooxidation through weathering. The free radicals formed in this process trigger chain reactions which lead to polymer chain scissions and degradation. To increase the weatherability of polymeric materials, various types of photostabilizers have been developed.

Screening

Screeners act by shielding the substrate from light. They belong to first generation photostabilizers, of which carbon black is an effective UV screener.

Absorption of Harmful UV Light

Among popular UV absorbers are *o*-hydroxybenzophenone (UV-531) and *o*-benzotriazole (UV 327).

On absorption of UV light, these molecules are raised to an excited state and they can dissipate added energy harmlessly by a mechanism of intramolecular proton transfer between the keto and the enol tautomers.

Quenching Excited State

When a polymer at ground state absorbs light energy, it may be raised to an excited state. A quencher is able to absorb the excited–state energy from the excited macromolecule to let it return unharmed to its ground state. Nickel cheletes are effective quenchers.

Scavenging Free Radicals

Free radical scavengers act by scavenging free radicals formed by auto-oxidation in a polymer matrix.

Low molecular weight (MW) HALS have higher volatility and extractability by liquids. Therefore, low MW HALS are generally not durable.[3,4] Durability in a substrate can be increased by an increase in MW. To increase a HALS MW to an optimal level is a major objective in HALS research and development. Homopolymerization and copolymerization of monomeric HALS are effective routes to increase the MW of HALS. Synthesis of monomeric HALS, therefore, is an important first step toward making HALS of higher MW.

SYNTHESIS OF MONOMERIC HALS

The key functional group in a HALS is TMP. Accordingly, synthesis of piperidines is a crucial first step, which is followed by conversion to monomers.

PROPERTIES OF MONOMERIC HALS

Physical Properties

In general, monomeric HALS are colorless solids, but those made from epoxides are yellowish oil-like liquids. UV-absorption spectra of monomeric HALS indicate that these HALS do not absorb UV light of wavelength longer than 300 nm, probably due to the fact that these HALS are aliphatic, not aromatic amines. Research using ESR[4,6,10] showed that in the oxidation and photostabilizing actions of these HALS against polymer degradation, these are strong signals attributable to the presence of stablc nitroxy-free radicals. It is believed, therefore, that the photostabilization mechanism of monomeric HALS, like other HALS, is via scavenging alkyl-free radicals by HALS to form stable nitroxy radicals.

Chemical Properties

In an additional reaction through a vinyl group, polymeric HALS can be made through free-radical homopolymerization, copolymerization and grafting of HALS carrying vinyl functionality.[4–9,11–13]

APPLICATIONS OF MONOMERIC HALS

As Effective Photostabilizers

Low MW HALS usually have low thermal stability and high volatility. Many low MW piperidinyl HALS, therefore, cannot be used as effective photostabilizers. A recent study by Lau et al.[14] indicated that 1,2,2,6,6-pentamethyl-4-piperidinyl-m-isopropenyl-α,α-dimethyl benzyl carbamate and its 2,2,6,6-tetramethyl counterpart are effective HALS comparable to Tinuvin-770 due to their moderately high MW.

These two monomeric HALS exhibit lower thermal stability and higher extractability than their polymeric counterparts, but their high photostabilizing capability can land them in applications in which high-temperature processing is not involved, such as surface coating.

Surface Modification

It has been reported[13] that resistance to photodegradation of polypropylene (PP) film was markedly increased after a monomeric HALS was grafted onto the film, which effectively distributed HALS over the surface of PP film where photooxidation is most active.

Polymer-Bound HALS

Macroradicals exist in the latter part of a free-radical polymerization and through chain scissions in high-temperature polymer processing such as melt extrusion. Monomeric HALS with vinyl functionality can be added to these macroradicals such that the HALS become bound to the polymer.

Preparation of Star Oligomer HALS

Monomeric HALS with a vinyl group can be added to silicon via hydrosilylation[15] to make star oligomer HALS of optimal MW and for better MW control.

Preparation of Polymeric HALS

Monomeric HALS are building blocks of polymeric HALS. The higher MW of polymeric HALS has given them higher thermal stability and resistance to extraction.

Higher thermal stability and resistance to extraction of polymeric HALS make them applicable for high-temperature processing and long exposure to weathering, such as in polymer melt extrusion and outdoor usage of synthetic fibers, films, and pipes.

The higher MW of polymeric HALS reduces their mobility. Thus there arises a problem of an optimal MW for a HALS.[4,10–12]

FUTURE DEVELOPMENT

As a result of their lower mobility due to higher MW, their photostabilization effectiveness is lower than to monomeric HALS.[3,4,13,14,16] Hence the question of an optimal MW for a HALS arises. It is generally believed that a HALS should have MW between 1000 and 2000, and many commercial photostabilizers are oligomers of MW less than 3000.

Surface modification by grafting HALS onto polymers is an attractive proposition, which still must be viewed along with other problems.

It can be expected that copolymerization of a monomer carrying hindered phenol with a monomeric HALS would yield stabilizers with dual capabilities: there of an antioxidant and a photostabilizer.

REFERENCES

1. Klemchuk, P. P. *Polymer Stabilization and Degradation* ACS Symposium Series **280**, Washington, DC, 1985.
2. Pan, J. Q. *Polymer Bulletin* **1992**, *3*, 138.
3. Chmela, S.; Hrdlovic, P. *Polym. Degrad. Stab.* **1985**, *11*(4), 339.
4. Pan, J. Q.; Lau, W. W. *Polym. Degrad. Stab.* **1994**, 44, 85.
5. Rabek, J. F. *Photostabilization of Polymers*; Elsevier Science: New York, 1990.
6. Chmela, S.; Hrdlovic, P. *Chem. Zvesti.* **1984**, *38*, 199.
7. Fen, H. B. *Gaofenzi Tongxin* **1981**, (2), 38.
8. Lee, C. S.; Lau, W. W. Y.; Goh, S. H. *J. Polym. Sci. Chem.* **1992**, *30*(6), 983.
9. Kurosaki, T.; Lee, K. W.; Okawara, M. *J. Polym. Sci. Chem.* **1972**, *10*, 3295; **1974**, *12*(7), 1407.
10. Pan, J. Q.; Lau, W. W. Y. *Polym. Degrad. Stab.* **1994**, *46*(1), 45.
11. Chmela, S.; Hrdlovic, P. *Polym. Degrad. Stab.* **1985**, *11*(3), 233.
12. Gugumus, F. *Res. Discl.* **1981**, *209*, 357.
13. He, M.; Hu, X. *Polym. Degrad. Stab.* **1987**, *18*(4), 321.
14. Lau, W. W. Y.; Pan, J. Q. *J. Appl. Polym. Sci.* **1993**, *50*(3), 403.
15. Pan, J. Q.; Lau, W. W. Y. *J. Polym. Sci. Chem.* **1994**, *32*(5), 997.
16. Pan, J. Q.; Lau, W. W. Y. *Polym. Degrad. Stab.* **1993**, *41*(3), 275.

Histones

See: Nucleohistone Complexes (Protein-Nucleic Acid Interactions)

Hollow Fiber Membranes

See: Functional Fibers

Hollow Particles

See: Multihollow Particles
Vesicles, Polymerization in

Holographic Gratings

See: Photopolymer Materials (Development of Holographic Gratings)

HOST-GUEST CHEMISTRY [Using α-Helical Poly(L-Lysine)]

Hirotaka Ihara, Atsushi Matsumoto, Masaaki Shibata, and Chuichi Hirayama
Department of Applied Chemistry and Biochemistry
Faculty of Engineering
Kumamoto University

Many researchers have great interest in "host-guest" chemistry using synthetic compounds because of its important role in understanding biofunctions at the molecular level. These studies have also led to various developments in biomimetic applications such as artificial receptors for sensors, organic media for separation, and transducers for chemical signals. Although a great number of host compounds have been discovered during the past half-century, almost all of these compounds are restricted to low-molecular cyclic compounds such as cyclodextrins, calixarenes, crown ethers, cryptands, cyclic polyamines, cyclophanes and cyclic dipeptides, or their polymer-supported materials. In these cases, the polymers do not play a main role. However, we know that specificities of biofunctions in enzymes and DNAs are derived from their three-dimensional configurations constructed by secondary structural polypeptides and polysaccharides, respectively.

Therefore, in this study, we aim to prove that α-helical synthetic polymers would be useful as host materials. As poly(L-lysine) is the simplest class of synthetic polymers that can produce chiral secondary structures spontaneously, we selected it as a host polymer and investigated its enantioselectivity in host-guest interaction. We focused especially on its α-helical conformation, in which the molecules are rather rigid and the residual amino groups assume identical position. However, random-coiled molecules are too heterogeneous to use for a molecular recognition. Unfortunately, we encountered two serious problems in this investigation: first, ionic property of the residual ammonium groups in useful as a driving force for selective binding, but charged poly(L-lysine) usually forms random coils in water. It is necessary to find a special condition in which charged poly(L-lysine) forms an α-helical conformation. Second, it is very difficult to detect enantioselective binding with chiral substances, because the interaction is not usually accompanied by a spectrophotometrically detectable response.

In this study, we avoided these problems by selection of methanol as a solvent, and by establishment of a new evaluating method for detecting the interaction, respectively. The latter technique is based on the fact that an achiral cyanine dye NK-2012 bound to polycations shows remarkable changes in the visible and circular dichloism (CD) spectra due to dissociation of the polycation-dye complexes induced by the interaction between polycations and anionic guest molecules.[1–4]

Here we describe how α-helical poly(L-lysine) acts as an enantioselective host molecules for *N*-benzyloxycarbonyl α-amino acids as guest molecules and also on the molecular recognition mechanism.

RESULTS

Formation of pLL-Dye Complexes

Poly(L-lysine hydrobromide) (pLL·HBr) was dissolved in water (pH 7) to show a CD spectrum with 3000 deg cm^2 $dmol^{-1}$ at 222 nm. This spectrum agrees with that in random coiled pLL.

We confirmed that 10 m*M* of pLL·HBr could be readily dissolved in methanol. This methanol solution provided a typical CD pattern (-34×10^3 deg cm^2 $dmol^{-1}$ at 222 nm) belonging to right-banded α-helical conformation. This CD strength indicates that the content of α-helix is almost 100%. The random coil-to-α-helix transition of the pLL main chain is due to lowering of electrostatic repulsion among the residual ammonium groups of pLL·HBr caused by using methanol as a solvent. This finding is very useful for investigation on the function of α-helical polymer because no additive is used.

Enantioselective Dissociation of the Complexes by Amino Acid Derivatives

The pLL-dye complexes in a methanol solution provide extremely strong exciton couplings at the absorption band of dimeric dyes. We confirmed that the CD strengths were very sensitive to additional ions.[1,3,4] For example, the values decreased remarkably with addition of *N*-benzyloxycarbonyl L-phenylalanine triethylamine salt (Cbz-L-Phe). These results indicate that the pLL-dye complexes are dissociated by Cbz-L-Phe to produce monomeric dyes.

The enantioselectivity decreased remarkably in the following case: when *N*-tert-butyloxycarbonyl derivatives of phenylalanine (Boc-L-Phe and Boc-D-Phe) were examined instead of the Cbz derivatives, we observed no enantioselectivity. This result indicates that the Cbz group contributes to the enantioselective interaction.

CONCLUSIONS

In this study we have proved that α-helical poly(L-lysine) is useful as a host polymer for molecular recognition and that the secondary structure plays in an important role. We have also clarified that poly(L-lysine) shows geometrical selectivity for dicarboxylic acids as reported elsewhere.[3,4] These successful findings have been supported by the establishment of a new method for detecting selective binding behavior between poly(L-lysine) and the guest molecule, using induced chirality due to bound cyanine dyes. This technique could have an extended range of applications in various highly ordered systems, for example, lipid bilayer membrane,[5] and ionic polysaccharides.[6]

REFERENCES

1. Ihara, H.; M. Shibata Hirayama. C. *Chem. Lett.* **1992**, 1731.
2. Shibata, M.; Ihara, H.; Hirayama, C. *Polymer* **1993**, *34*, 1106.
3. Ihara, H. et al. *Mater. Sci. Eng. C.* **1994**, C2, L1.
4. Ihara, H. et al. *Proc. 3rd Pac. Polym. Conf.* **1993**, 557.
5. Arimura, T.; Ihara, H.; Hirayama, C. *Anal. Sci.* **1993**, *9*, 401.
6. Shiratsuchi, J. et al. *Polym. Preeprints Jpn.* **1995**, *44*, 689.

HOST-GUEST CHROMATOGRAPHY (with Cyclodextrin Derivatives)

Keiko Takahashi
Department of Industrial Chemistry
Faculty of Engineering
Tokyo Institute of Polytechnics

Host-guest chromatography is the selective separation of substances by the formation of the host-guest complexes (inclusion compounds).[1] A primary condition for chromatographic separation is the extent of interaction between the substance forming the stationary phase and the eluent.

Cyclodextrins (CDs), which are doughnut-shaped oligosaccharides composed of six or more (α)-D-gluocpyranose units linked by α-1,4-glycoside bonds, are among the most extensively investigated biomimetic enzyme models and are employed in many different technological applications.[2,3] A fascinating property of CDs is their ability to selectively incorporate other organic compounds into their cavity, both in the solid state and in solution (especially in water). Such host-guest complexes must be separated from the clathrates. CDs are also powerful resolving agents for racemic mixtures,[4] their recognition ability has been recently examined using circular dichroism spectra, ^1H, ^{13}C or ^{19}F NMR spectroscopy and molecular dynamics simulation.[5–11] Moreover, the properties of CDs, such as solubility, polarity, the size or shape of the hydrophobic cavity, etc., can easily change by chemical modification of the many hydroxyl groups (three groups per one glucose unit) on both rims of the hydrophobic cavity. All of their properties suggest that CDs are attractive host compounds and have been used in multi-stage chromatographic processes. CDs and CD derivatives can be classified into two large groups: insoluble CD and soluble CD. The former group, which contains the CD polymer and "immobilized" CDs, can be used as the chiral stationary phase (CSP). The latter group, the native CD (α, β, γ-CD, and branched CD) and modified CD, can be used as additives to the mobile phase (as the chiral selector) or as the separation mediator in HPLC, TLC, and capillary electrophoresis (CE), and as CSP in GC.

APPLICATIONS

Direct Separation with Host-Guest Complex Formation (Separation Using CD as Stationary Phase)

High-Performance Liquid Chromatography

The CD polymer has been used as the stationary phase in LC, where nucleic acids and aromatic amino acids separated. The CD polymer can separate enantiomers of vinca alkaloids and methyl mandelate. The behavior of the CD polymer indicates the possibility of a CD as a chiral stationary phase (CSP). However, the CD polymer is not suitable for HPLC use because it lacks physical and chemical stability and CD cavity density. The first CD-CSP for HPLC use was reported as immobilized CD on silica gel.[10] Presently, three CD-CSPs and nine modified CD-CSPs are commercially available. Different types of CD-CSPs have achieved the separation of hundreds of enantiomers.[11–13] There have been very few reports for inclusion with molecular recognition in an organic solvent by CD. Recently, Takahashi and Hatori reported the asymmetric reduction and regioselective Wittig reaction in organic solvents with CDs.[14] The achievement of separation under normal-phase conditions increased the possibility of CD chromatography and a new type of CD complex formation in organic solvents.

Gas Chromatography[15]

Native CD and modified CD without immobilizing the stationary phase can also be used as a stationary phase in GC. As of now, various chemically modified CDs have been reported, and some of the chemically modified CD capillary columns are commercially available in many forms from a variety of manufacturers.

Some immobilized type CD-CSPs on silica gel have also achieved enantiomeric separation in GC and supercritical fluid chromatography (SFC).[16–18]

Cyclodextrin Mediated Separation (Separation Using CD as Mobile Phase Additive)

High-Performance Liquid Chromatography

The number of separations using some types of CDs as the mobile-phase additive in TLC, HPLC, and CE has dramatically increased. For reasons not fully appreciated, the efficiency of such separations significantly exceed those employing CDs as the stationary-phase component. The utility of CDs as mobile-phase additives is sometimes limited by their solubility. This problem can sometimes be alleviated using an eluent system or additives.

Capillary Electrophoresis (CE)

Most of the interest in the resolution of enantiomers by CE has focused on the use of CD buffer additives.[19–21] This separation depended on the type and the concentration of CD, pH, temperature, and the concentration of organic solvent. Now, more than 50 enantiomers have been successfully separated.

REFERENCES

1. Hattori, K.; Takahashi, K. *Kobunshi* **1987**, *36*, 84.
2. Bender, M. L.; Komiyama, M. *Cyclodextrin Chemistry*; Springer Verlag: Berlin, 1979.
3. Szejtli, J. *Cyclodextrin and Their Inclusion Complexes*; Academiai Kaido: Budapest, 1982.
4. Armstrong, D. W.; Ward, T. J.; Armstrong, R. D. *Science* **1986**, *232*, 1132.
5. Li, S.; Purdy, W. C. *Anal. Chem.* **1992**, *64*, 1405.
6. Qi, Z. H. et al. *J. Org. Chem.* **1991**, *56*, 1537.
7. Hanazawa, M. et al. *J. Chem. Soc., Chem. Commun.* **1992**, 206.
8. Brown, S. E. et al. *J. Chem. Soc., Faraday Trans* **1991**, *87*, 2669.
9. Köhler, J. E. H. et al. *Angew. Chem. Int. Ed. Engl.* **1992**, *31*, 319.
10. Armstrong, D. W.; Demond, D. *J. Chromatgr.* **1984**, *22*, 411.
11. Armstrong, D. W.; Hans, S. M. *CRC Crit. Rev. Anal. Chem.* **1988**, *19*, 175.
12. Watt, A. P.; Rathbone, D. *J. Liq. Chromatgr.* **1993**, *16*, 3423.
13. Zukowski, J.; Pawlowska, M. *J. High Resolut. Chromatgr.* **1993**, *16*, 505.
14. Takahashi, K.; Hattori, K. *J. Incl. Phenom.* **1994**, *17*, 1.
15. Shurig, V.; Nowotny, H. P. *Angew Chem. Int. Ed. Engl.* **1990**, *29*, 939.
16. Yi, G. et al. *J. Org. Chem.* **1993**, *58*, 4844.
17. Yi, G. et al. *J. Org. Chem.* **1993**, *58*, 2561.
18. Jung, M.; Schurig, V. *J. High Resolut. Chromatgr.* **1993**, *16*, 289.
19. Ward, T. J. *Anal. Chem.* **1994**, *66*, 633A.
20. Novotny, M. et al. *Anal. Chem.* **1994**, *66*, 646A.
21. Riley, C. M.; Lough, W. J.; Wainer, I. W. Ed. *Pharmaceutical and Biomedical Applications of Liquid Chromatography* Pergamon: Oxford, England, 1994.

Host-Guest Complexes

See: Cyclodextrins (Host-Guest Interactions)
Host-Guest Chemistry [using ∝ Helical Poly(L-Lysine)]
Host-Guest Chromatography (with Cyclodextrin Derivatives)
Inclusion Complexes (Overview)
Inclusion Polymerization
Molecular Shape Recognition
Phenolics (Linear and Cyclic Oligomers)

Hot Melt Adhesives

See: Polyterpene Resins

HPLC Materials

See: Gels, HPLC Packing Materials (For Structural Characterization)
Host-Guest Chromatography (with Cyclodextrin Derivatives)
Nucleic Acid Analogs
Optically Active Polyamides, Coumarin Dimer (Chiral Stationary Phases for HPLC)

Hyaluronic Acid

See: Bioartificial Materials
Glycosaminoglycans (Overview)
Hyaluronic Acid (Macroionic, Rheological, and Lubricity Properties)

HYALURONIC ACID (Macroionic, Rheological, and Lubricity Properties)

Akira Takahashi, Mibuko Kaburaki, and Yukihiro Oka
Department of Chemistry for Materials
Mie University

Hyaluronic acid, also called hyaluronan or sodium hyaluronate (abbreviated as HA, CAS Registry Numbers 9067-32-7), is a glycosaminoglycan and is composed β(1→4)-linked disaccharide units that consist of D-glucuronic acid and *N*-acetyl-D-glucosamine linked by a β(1→3) bond (**Figure 1**). It is an

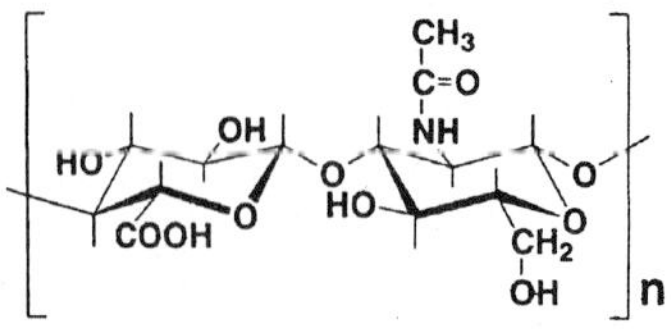

FIGURE 1. Disaccharide unit of hyaluronic acid.

$(A\text{-}B)_n$ type alternative copolymer and also an anionic polyelectrolyte. HA is composed of 150 to 25,000 disaccharide units ($M = 5 \times 10^4$ to 8×10^6). It is found in vitreous body, various connective tissues, skin, cartilage and synobial fluid. The nomenclature "hyaluronic acid" is derived from "hyalos" in Greek, which means glass-like or transparent, and uronic acid. Its physiological function in tissues is the attraction of a large amount of water and the resultant swelling of the tissue matrix and thereby facilitates cell migration during morphogenesis and repair.[1] HA solution exhibits viscoelastic properties, especially shear thinning. This behavior makes the HA excellent shock absorbers and lubricants.

PREPARATION

HA is extracted for commercial purposes from rooster comb or synthesized by microbial fermentation using streptococcus.

PROPERTIES

Solution Properties

Hyaluronic acid is a weak and linear polyelectrolyte with carboxyl and acetoamide functional groups. Under physiological conditions, it exists as sodium hyaluronate and is soluble in water. HA binds Ca^{2+} tightly. X-ray fiber analysis of the Ca^{2+} hyaluronate indicates that HA anion forms an extended left-handed single strand helix with three disaccharide units per turn, which is stabilized by intramolecular hydrogen bonds.[2]

HA is a semirigid worm-like polyelectrolyte; nevertheless, it adopts highly extended random coil conformation, thereby occupying a large volume relative to the dry state. This is the basis of the moisturization and water retention capabilities of HA.

Rheological Properties

In increasing polymer concentration, one can define the dilute, semidilute, and concentrated regimes. In the semidilute regime, although it is very limited, the rheological behavior is governed by the degree of overlapping of polymer chains. Above the critical overlap concentration, chains interpenetrate and form a highly entangled network. The viscosity of HA solutions is shear dependent, and above a certain critical rate of shear sec^{-1} (γ_c) shear thinning occurs.[3] The onset of shear thinning is found to be correlated with the inverse of the longest relaxation time of HA in solution.[4] The semirigid HA molecules probably tend to line up under the flow.

LUBRICITY

It is now recognized that cartilage consists of collagen fibrils and proteoglycans. The proteoglycan–hyaluronic acid complex is a highly organized supramacromolecular assembly resembling bottle brushes and consists of hyaluronic acid, link protein, core protein, chondroitin sulfate, and keratan sulfate.[5] It is also known that synovial fluid contains HA and proteins. Nevertheless, mechanisms of the friction and lubrication at joints have not yet been explained from the molecular point of view.

Takahashi and Kozaki (and Klein et al.) have focused attention on the studies of lubrication between surfaces bearing adsorbed polymers or polymer brushes, or both, namely, grafted or tethered chains.[6,7] Kato, Kozaki, and Takahashi measured the static friction coefficient μ_s of thin liquid films of polymer solutions between glass surfaces.[8]

APPLICATION

Hyaluronic acid is biocompatible and biodegradable. Highly purified HA is nonpyrogenic, nontoxic and nonimmunogenic.[9] Its unique lubricant property is used in the treatment of inflammatory and degenerative joint diseases and in viscosurgery.[10,11] HA is used in a ophthalmic surgery.[12] The moisturization and water retention capabilities have application in cosmetics.[13] Shah and Barnett showed that crosslinked HA gels reversibly swell about 15 times in water at a crosslinking ratio of 1.27.[14]

REFERENCES

1. Laurent, T. C.; Fraser, J. R. E., *In Functions of the Proteoglycans, Ciba Foundation Symposium 124* Evered, D.; Whelan, J. W., Eds.; John Wiley & Sons: New York, 1986; pp 9–29.
2. Winter, W. T.; Arnott, S. *J. Mol. Biol.* **1977**, *117*, 777.
3. Takigawa, T.; Ishida, M.; Masuda, T. *Polymer Preprints Japan* **1989**, *38*, Number 10, 3599.
4. Fouissac, E.; Milas, M.; Rinaudo, M. *Macromolecules* **1993**, *26*, 6945.
5. Caplan, A. I. *Sci. Am.* **1984**, *251*(4), 87.
6. Takahashi, A.; Kozaki, N. *Polymer J.* **1987**, *19*, 945.
7. Klein, J. et al. *Macromolecules* **1993**, *26*, 5552.
8. Kato, T.; Kozaki, N.; Takahashi, A. *Polymer J.* **1986**, *18*, 111 and 189.
9. Larson, N. et al. *ACS Polymeric Materials Science and Engineering Fall Meeting Proceedings* 1990, Vol. 63, p 34.
10. Asheim, A.; Lindbald, G. *Acta Vet Scand.* **1979**, *17*, 379.
11. Namiki, O.; Toyoshima, M.; Morisaki, N. *Clin. Orthop* **1982**, *146*, 260.
12. Miller, D.; Stegman, R., Eds.; *Healon, A Guide to its use in ophthalmic surgery*; John Wiley & Sons: New York, 1983.
13. Balazs, E. A.; Band, P. *Cosmet. Toiletries* **1984**, *99*, 65.
14. Shah, C. B., Barnett, S. M. In *Polyelectrolyte Gels*; Harland, R. S.; Prud'homme, Eds.; ACS Symposium Series 480, ACS: Washington, DC, 1992; Chapter 7.

HYBRID ARTIFICIAL ORGANS

Hiroo Iwata and Yoshito Ikada
Research Center for Biomedical Engineering
Kyoto University

Most natural organs metabolize various substances, synthesize biologically important molecules, and secrete them in response to biological signals, but artificial organs made from nonviable materials cannot replace these synthetic and metabolic functions. Hybrid artificial organs are defined as medical devices made by combining viable cells and nonviable materials.[1] The viable cells take the synthetic and metabolic functions, while the nonviable materials afford a suitable environment for the cells to perform their functions. The nonviable materials used for the hybrid artificial organs form four categories, according to their functions: resorbable templates for tissue regeneration, nondegradable scaffolds for tissue regeneration, immuno-isolation membranes, and bioreactors for temporary use (**Table 1**).

TABLE 1. Examples of Hybrid Artificial Organs

Organs	Tissues or cells	Devices	Materials	References
		Template for tissue regeneration		
Skin	Fibroblast-Keratinocyto	Gel	Collagen	3
	Keratinocyte	Sponge	Collagen and čondroitin-6-sulfate	4
	Fibroblast	Nonwoven mesh	Poly(glycolic-*co*-lactic acid) (Vicryl™)	2
Cartilage	Chondrocytes	Nonwoven mesh	Poly-(L-lactic acid) or Poly(glycolic acid) and their copolymers	5
	Embryonic chondrocytes	Gel	Extracellular matrix	6
	chondrocytes	Sponge	Collagen with a fibrobrast growth factor	7
Esophagus	Mucosal cells	Sponge	Collagen	8
		Scaffold for tissue regeneration		
Blood vessel	Endothelial cells	Knitted	Poly(ethylene terephthalate)	9
	Endothelial cells	Expanded	Polytetrafluoroethylene	10
	Endothelial cells and smooth muscle cells	Knitted	Poly(ethylene terephthalate)	11
		Immuno-isolation		
Pancreas	Allogeneic islets	Diffusion Chamber	Microfilter made of cellulose acetate nitrate (Millipore®)	12
	Xenogeneic islets	Hollow fiber	Ultrafilter made of random copolymer of vinyl chloride and acrylonitrile (Amicon® XM-50)	13
	Xenogeneic islets	Microcapsule	Polyion complex membrane made of Alginate-poly-(L-lysine)	14
	Xenogeneic islets	Microcapsule	Hydrogel made of Alginate crosslinked with barium ion	15
	Xenogeneic islets	Microcapsule	Hydrogel made of Agarose	16
	Xenogeneic islets	Microcapsule	Hydrogel made of Agarose/Poly(styrene-sulfonic acid)	17
	Allo- or xenogeneic islets	Intravascular capillary unit	Ultrafilter made of random copolymer of vinyl chloride and acrylonitrile (Amicon® XM-50)	18
		Bioreactor for temporary use		
Liver	Xenogeneic hepatocytes	Extracorporeal bioreactor	Microfilter fibers and hepatocytes on collagen-coated dextran microcarriers	19
	Allogeneic hepatocytes	Extracorporeal bioreactor	Collagen-coated borosilicate glass plates	20
	Hepatocytes	Tissue flask	Poly-(*N*-*p*-vinylbenzyl-D-lactonamide)	21
	Human hepatoblastoma cell line.	Extracorporeal bioreactor	Hemodialysis fibers made of cellurose acetate	22

SUMMARY

The advent of mammalian cell culture technology and the development of various biomaterials opened the door for hybrid artificial organs. Some of them have been successfully applied to clinical treatments, but the polymers used in preparation of the hybrid artificial organs, such as polyesters, cellulose derivatives, and collagen, are quite limited. They were initially developed by industries for nonmedical purposes or for medical purposes that did not involve hybrid artificial organs. Thus one can say that the hybrid artificial organ is still in its infancy. Polymers and polymer-based composites that can be used for high-performance hybrid artificial organs will be molecularly designed and developed after accumulation of experimental and clinical experiences.

REFERENCES

1. Williams, D. E., Ed., *Definitions in Biomaterials*; Elsevier Science: Amsterdam, The Netherlands, 1987.
2. Halberstadt, C. et al. *Mat. Res. Soc. Symp. Proc.* **1992**, *252*, *323*.
3. Bell, E. et al. *Science* **1981**, *211*, 1052.
4. Boyce, S. T.; Hansbrough, J. F. *Surgery* **1988**, *103*, 421.
5. Puelacher, W. C. et al. *Biomaterials* **1994**, *15*, 774.
6. Itay, S.; Abramovici, A.; Nevo, Z. *Clin. Orthop.* **1987**, *220*, 284.
7. Ikada, Y.; Fujisato, K.; Sajiki, T. *Biomaterials* **1995**, in press.
8. Natsume, T. et al. *ASAIO Trans.* **1990**, *36*, M435.
9. Herring, M. G. et al. *Ann. Surg.* **1979**, *190*, 84.
10. Graham, L. M. et al. *Surgery* **1982**, *91*, 550.
11. Weinberg, C. B.; Bell, E. *Science* **1986**, *231*, 397.

12. Strautz, R. L. *Diabetologia* **1970**, *6*, 306.
13. Archer, J.; Kaye, R.; Mutter, G. J. *J. Surg. Res.* **1980**, *28*, 77.
14. Lim, F.; Sun, A. M. *Science* **1980**, *210*, 908.
15. Zekorn, T. et al. *Acta Diabetol.* **1992**, *29*, 99.
16. Iwata, H. et al. *J. Biomed. Mater. Res.* **1994**, *28*, 1003.
17. Iwata, H. et al. *J. Biomed. Mater. Res.* **1994**, *28*, 1201.
18. Sullivan, S. J. et al. *Science* **1991**, *252*, 718.
19. Rozga, J. et al. *Biotechnol. Bioeng.* **1994**, *43*, 645.
20. Uchino, J. et al. *ASAIO Trans.* **1989**, *34*, 972.
21. Kobayashi, A. et al. *Makrom. Chem. Rapid Commun.* **1986**, *7*, 645.
22. Sussman, N. L. et al. *Artif. Organs* **1994**, *18*, 390.

Hydroboration Polymerization

See: Organoboron Main Chain Polymers

Hydrocolloids

See: Acetan
Food Polymers
Gelatin (Molecular Weight Distributions)
Gums (Overview)

HYDROGEL BIOMATERIALS (HEMA-Based)

Kuo-Huang Hsieh
Department of Chemical Engineering
National Taiwan University

Tai-Horng Young
Center for Biomedical Engineering
College of Medicine
National Taiwan University

Hydrophilic gels are a very important class of polymeric materials with extensive applications as biomedical products. The first prepared and described polymer hydrogels for biomedical application were proposed by Wichterle and Lim.[1,2] They used 2-hydroxyethyl methacrylate (HEMA) as the monomer in their development. The most widely used hydrogel in hydroxyalkyal methacrylates or acrylates is poly(2-hydroxyethyl methacrylates) (PHEMA).

SYNTHESIS OF PHEMA

The structure of the monomer HEMA is (**Structure 1**): To form a gel that is hydrophilic but insoluble in water, the HEMA must be copolymerized with a crosslinking agent in aqueous solution. Refojo and Yasuda, Ranter and Hoffman, and Gregonis et al. described the preparation and properties of PHEMA.[3–5] PHEMA is usually prepared by free radical solution polymerization. Simultaneous copolymerization and crosslinking reaction in solution is the preferred method to prepare PHEMA since the polymerization rate is fast and the shape of gels can be controlled.[3] The free radicals for the polymerization of HEMA are generated by chemical initiators, ionizing radiation, or photochemical initiations. Volume contraction occurs when the initial soft matrix is formed. As the network forms completely, it becomes more rigid with no more contraction.[6,7]

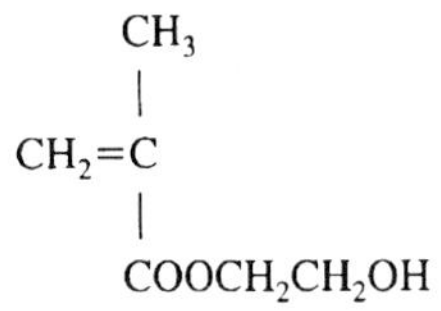

1

Ethylene glycol dimethacrylate (EGDMA) is the most often used crosslinking agent. Other crosslinking agents are 3-oxapentamethylene dimethacrylate, 2,3-dihydroxytetramethylene dimethacrylate, and trimethylolpropane trimethacrylate. The critical properties of hydrogels, such as sorption and desorption, mechanical behavior, swelling properties, etc., can be controlled by network characteristics, that is, the degree of crosslinking and the density, distribution, and length of crosslinks.

Many solvents have been employed in the preparation of PHEMA and their concentrations determine the homogeneous or heterogeneous structure of the resultant gel.[3,4,6–8] Impurities found in commercial PHEMA include EGDMA, diethylene glycol dimethacrylate (DEGMA), methacrylic acid (MAA) and acrylic acid (AA).

PHEMA with ordered structure is polymerized through highly specific reaction and exhibits good biocompatibility.[9] Ordered PHEMA is obtained in the syndiotactic or isotactic form.[10]

PROPERTIES OF PHEMA

PHEMA exhibits good chemical stability and resistance to acid hydrolysis and reaction with amines.[11] Acrylates are less often used alone due to their brittle property, but PHEMA gives the flexible hydrogels.[12] The swelling property of PHEMA was evaluated.[6,11] The maximum water content of PHEMA can be regarded as the equilibrium swelling.

Any factors changing the water content will cause a change in dimension of the hydrogels. PHEMA is a very stable hydrogel and its water content is not easily influenced by the change of pH value, temperature, and tonicity.

A lot of comonomers can be copolymerized with HEMA to change its resultant properties. To increase the tensile strength of PHEMA, we use copolymerization of HEMA with various hydrophobic monomers. Hydrophobic comonomers are used to improve the tensile strength of PHEMA for use in contact lenses and other implants.[13–15]

The crosslinking agents also plays an important role in controlling hydrophilicity and mechanical properties. Ethylene glycol dimethacrylate (EGDMA) is hydrophobic to control mechanical property. 3-Oxapentamethylene dimethacrylate with hydrophilic and hydrophobic properties through its ethoxy groups is suitable for contact lenses.[16]

PHEMA and its related copolymers have been developed as materials for contact lenses. For the biomedical applications, the diffusion of gases through the polymers, especially oxygen through contact lenses, is important to diminish hypoxia or edema problems. In general, the factors affecting the degree of swelling also affect the permeation property of the hydrogel.

The disadvantage of PHEMA used in implants is calcification. Comonomers with anionic fixed charges introduced into the PHEMA can minimize the calcification.[17,18]

APPLICATIONS

Polymeric materials based on HEMA monomer have been used extensively, especially in biomedical applications. PHEMA is interesting as a biomaterial because it will maintain the precise geometric configuration and sufficient mechanical strength. It has appropriate compliance in the application to retain its integrity and to minimize mechanical and frictional irritation to the surrounding tissue.[19,20] Furthermore, the hydrogel network can be easily obtained[21] and it has low interfacial tension to minimize the driving force for protein adsorption or cell adhesion, which is an important property for a biocompatible material.[22–25]

Contact Lenses

Contact lenses are the successful biomedical applications of the PHEMA, while other applications are employed less. This may be because the problems of biocompatibility in the eyes are slightly low compared to those in contacting with blood.

Blood Compatible Materials

PHEMA and its copolymer have both hydrophilic and hydrophobic properties in each molecule leading to their antithrombogenic properties, so they are developed as materials for blood contact such as hemodialysis membranes,[26–28] catheters,[29,30] vascular grafts,[31] by regulating the relative hydrophilic–hydrophobic balance of the polymer surface containing microdomain structure.

Drug Delivery Systems

Equilibrium swelling measurements are very important in using hydrogels as drug delivery systems because there is a possible correlation between the rate of drug release and the water-swelling property of hydrogels. The resultant diffusional properties are the important determinants for the use of PHEMA copolymers as drug delivery systems.

Combined with Collagen

PHEMA represents the well-known artificial biomaterials which are recognized as inert toward host organisms. However, normal animal or human cells are not able to adhere to their surfaces. Stol found that this characteristic of PHEMA can be totally changed by the simple addition of collagen.[32] The collagen-PHEMA hydrogels were found to be well-tolerated, nontoxic and highly biocompatible materials.[33]

REFERENCES

1. Wichterle, O.; Lim, D. *Nature* **1960**, *185*, 117.
2. Wichterle, O.; Lim, D. Patent, 2 976 576, 1961.
3. Refojo, M. F.; Yasuda, H. *J. Appl. Polym. Sci.* **1965**, *9*, 2425.
4. Ranter, B. D.; Hoffman, A. S. In *Hydrogels for Medical and Related Applications*, Andrade, J. D., Ed.; ACS Symp. Ser. 31, American Chemical Society: Washington, DC, 1976.
5. Gregonis, D. E.; Chen, C. M.; Andrade, J. D. In *Hydrogels for Medical and Related Applications*, Andrade, J. D., Ed.; ACS Symp. Ser. 31, American Chemical Society: Washington, DC, 1976.
6. Wichterle, O.; Chromecek, R. *J. Polym. Sci.* **1969**, *C16*, 1677.
7. Kopecek, J.; Lim, D. *J. Polym. Sci. A1* **1971**, *9*, 147.
8. Hasa, J.; Janacek, J. *J. Polym. Sci.* **1967**, *C16*, 317.
9. Ranter, B. D. *Biocompat. Clin. Implant Mater.* **1981**, *2*, 145.
10. Gregonis, D. E. et al. *Polymer* **1978**, *19*, 1279.
11. Sevick, S. et al. *J. Polym. Sci.* **1967**, *C16*, 821.
12. Refojo, M. F.; Liu, H. S. *Ophthalmic Surg.* **1978**, *9*, 43.
13. Sedlacek, B.; Overberg, C. J.; Mark, H. F., Eds., In *J. Polym. Sci. Polym. Symp. Number 66*; John Wiley & Sons: New York, NY, 1979.
14. Ranter, B. D. In *Biocompatibility of Clinical Implant Materials*, Williams, D. F., Ed.; CRC: Boca Raton, FL, 1981; Vol. II.
15. Refojo, M. F. *Surv. Ophthalmology* **1982**, *26*, 257.
16. Gustafson, R. U.S. Patent 3 728 315, 1973.
17. Klomp, G. F. et al. *J. Biomed. Mater. Res.* **1983**, *17*, 865.
18. Pinchuk, L. et al. *Appl. Polym. Sci.* **1984**, *29*, 1749.
19. Kardos, J. L. et al. *Biomater. Med. Devices, Artif. Organs* **1974**, *2*, 387.
20. Lyman, D. J. et al. *Trans. Am. Soc. Artif. Organs* **1977**, *23*, 253.
21. Homsy, C. A. *J. Biomed. Mater. Res.* **1970**, *4*, 341.
22. Andrade, J. D. *Med. Instrum.* **1973**, *7*, 110.
23. Andrade, J. D. et al. *J. Polym. Sci. Polym. Symp.* **1979**, *66*, 313.
24. Zdrahala, R. J. et al. *J. Appl. Polym. Sci.* **1979**, *24*, 2041.
25. Sun, Y. et al. *Polym. Prepr. Am. Chem. Soc., Div. Polym. Chem.* **1987**, *28*, 292.
26. Luttinger, M.; Cooper, C. W. *J. Biomed. Mater. Res.* **1967**, *1*, 67.
27. Ranter, B. D.; Miller, I. F. *J. Biomed. Mater. Res.* **1973**, *7*, 353.
28. Muzykewicz, K. J. et al. *J. Biomed. Mater. Res.* **1975**, *9*, 487.
29. Levowitz, B. S. et al. *Trans. Am. Soc. Artif. Organs* **1968**, *14*, 82.
30. Kaganov, A. L.; Stamberg, J.; Synek, P. *J. Biomed. Mater. Res.* **1976**, *10*, 1.
31. Ranter, B. D.; Hoffman, A. S. *Synth. Biomed. Polym.: Concepts Appl.* **1980**, *1*, 133.
32. Stol, M. J. *J. Bioact. Compat. Polymers* **1991**, *6*, 308.
33. Jeyanthi, R.; Rao, K. P. *Biomaterials* **1990**, *11*, 238.

Hydrogels

See: Acryloyl-L-Proline Alkyl Ester Hydrogels
Agarose (Overview)
Amphoteric Latex
Bioartificial Materials
Cell Entrapment [Poly(carbamoyl sulfonate) Hydrogels]
Contact Lenses, Gas Permeable
Contact Lenses, Hydrogels
Controlled Drug Delivery Systems
Cyclic Imino Ethers (Ring-Opening Polymerization)
Cyclic Imino Ethers (Ring-Opening Polymerization of 1,3-Oxazo Monomers)
Dental Polymers (Hydrogels)
Environmentally Responsive Gels
Enzyme-Degradable Hydrogels
Fluorescent Polymers (2-Vinylquinoline Hydrogels and Soluble Polymers)
Gels (Overview)
Glucose Responsive Polymers (Medical Applications)
Hydrogel Biomaterials (HEMA-Based)
Hydrogels, Microphase Separated Blends (Biomaterials and Drug Delivery System)
Intelligent Polymers (in Medicine and Biotechnology)
N-Isopropylacrylamide Copolymers (Drug Delivery)

Nonthrombogenic Materials
Poly(α-amino acids) (Biodegradation, Medical Applications)
Poly(ethylene oxide) (Applications in Drug Delivery)
Poly(N-isopropylacrylamide) Hydrogels
Poly(vinyl alcohol) Fiber (High Strength and High Modulus)
Smart Hydrogels
Superabsorptive Polymers (from Natural Polysaccharides and Polypeptides)
Vitreous Body, Artificial

HYDROGELS, MICROPHASE SEPARATED BLENDS (Biomaterials and Drug Delivery System)

Kishore R. Shah
ConvaTec

Covalent and ionically crosslinked networks of hydrophilic polymers, commonly referred to as hydrogels in their hydrated state, have been known for a long time and have found a number of biomedical applications such as soft contact lenses, wound management, and controlled drug delivery. The methods of preparation of these hydrogels are subject to the constraints imposed by their thermosetting nature, and consequently they do not enjoy the benefits of thermoplastic processing. In addition, such hydrogels are mechanically weak, which further limits their usefulness. However, the hydrophobic/hydrophilic domain polymer systems that hydrate to form hydrogels are of special interest on account of their unique morphological features that greatly influence their properties. Such polymer systems include block copolymers, graft copolymers, and polymer blends that exhibit hydrophilic/hydrophobic microphase domain structures. The hydrophobic domains in such systems behave as thermally labile pseudocrosslinks. An approach to the preparation of blends, which are compatible on an optical scale and exhibit microphase separation by virtue of a controlled extent of interpolymer hydrogen bonding interactions, has been reported earlier.[1] Such blends are characterized by their unique physical properties arising from their two-phase morphology. In particular, blends of poly(*N*-vinyl-2-pyrrolidone) (PVP) (and those of water-soluble *N*-vinyllactam copolymers) with certain acidic–group-containing water-insoluble copolymers form stable, optically transparent hydrogels upon equilibration in water.[2]

INTERPOLYMER INTERACTIONS AND MICROPHASE SEPARATION

Infrared spectroscopic studies have shown that optical transparency and apparent compatibility of *N*-vinyl-2-pyrrolidone (VP) polymers and copolymers with acidic–group-containing copolymers result from interpolymer hydrogen bonding interactions between the pyrrolidone carbonyl groups and the carboxyl groups.

POLYMER BLEND COMPOSITIONS

Compatibility, as monitored by optical transparency and mechanical integrity, of the polymer blend both before and after hydration, is necessary for the formulation of stable hydrogels. Although blends containing major fraction of VP homopolymer or a water-soluble VP copolymer with an acidic–group-containing water-insoluble copolymer are transparent and coherent in the dry state, most of them become opaque and incoherent when placed in water. Some of them may even disintegrate into crumbs by the effect of hydration, which seems to dramatically decrease the interfacial adhesion. In order to retain the interfacial adhesion even after hydration, the water-insoluble copolymer of the blend must have a certain hydrophobic–hydrophilic balance. Thus, for example, blends of PVP with different water-insoluble copolymers containing a minor proportion of a hydrophilic comonomer retain coherency and optical transparency upon equilibration in water to form strong and stable hydrogels.

In the case of the polymer blend hydrogels, the domains of the water-insoluble copolymer constituting the dispersed phase seem to act as multiple crosslinks and prevent dissolution of the water-soluble polymeric continuous phase in water. The polymeric constituents of the blend being uncrosslinked, the blend is soluble in suitable organic solvents and possesses the ability to be repeatedly shaped or formed under the influence of heat and pressure. Thus, in the dry state, the polymeric blend exhibits thermoplasticity, but it forms a swollen gel when equilibrated in water. The thermoplasticity of these hydrogel-forming blends confers on them a special processing advantage over covalently crosslinked synthetic hydrogels. The polymer blend hydrogels differ in an important way from the block and graft copolymers, in that the hydrophobic and hydrophilic phases in the former are not connected to one another by covalent bonds.

The equilibrium water content of the hydrogels depends upon the proportion of the water-insoluble copolymer in the blend.

Optimum interphase compatibility and mechanical properties of the polymer blend hydrogels are controlled by comonomeric composition of the polymeric components. The water-soluble polymeric component must contain 25 to 100 mole% of VP units in the chain for interpolymer hydrogen bonding, in the absence of which coherency upon hydration is lost. Likewise, composition of the water-insoluble copolymer is also critical for optimizing properties of the hydrogels.

BIOMEDICAL APPLICATIONS

High degree of tissue and blood compatibility of the polymer blend hydrogels makes them potentially useful in diverse biomedical applications, such as soft contact lenses, burn and wound dressings, catheters, and controlled drug delivery.

The hydrophobic/hydrophilic two-phase nature of the polymer blend hydrogels and their excellent biocompatibility make them particularly suitable for controlled release of bioactive agents, such as drugs. The principle of this delivery system is to utilize the microphase domains of the dispersed polymeric phase as "depots" or microreservoirs for the drug to be released.[3] A drug, when incorporated into the polymer blend matrix, would be distributed between the dispersed and the continuous phases depending upon its relative solubilities (partition coefficient) in the hydrophobic and the hydrophilic phases. A relatively hydrophilic drug would be mainly concentrated in the hydrophilic phase. In an aqueous environment, the hydrated continuous phase can serve as a rate controlling membrane for release of the drug into the surrounding medium.

REFERENCES

1. Shah, K. R. *Polymer* **1987**, *28*, 1212.
2. Shah, K. R. U.S. Patent 4 300 820 November 17, 1981.
3. Shah, K. R. U.S. Patent 4 693 887, September 15, 1987.

HYDROGEN-BONDED BLENDS

Issa A. Katime and Carlos C. Iturbe
Grupo Nuevos Materiales
Departamento de Química Física
Universidad del País
Vasco Lejona España

Polymer blends are an important class of materials that may exhibit a variety of mechanical, thermal, optical, or electrical properties depending on their blend composition and the characteristics of the homopolymers. Although chemical synthesis is the more common method for developing new polymeric materials, the physical mixture of two or more polymers is acquiring a great relevance in this field. The mixtures of polymers, also known as polyblends or polymer alloys, suggest a potential similar to that of alloys in the field of metallurgy. Although single-phase materials (miscible mixtures) present the most possibilities, components constituting two or more phases (compatible mixtures) offer interesting practical advantages.

It is well known that polymer blends are generally not miscible. This is due to the very low entropy of mixing of long polymer chains and the unfavorable enthalpy of mixing of most polymer pairs due to van der Waals interactions.

Miscible polymer blends generally require some type of specific interaction between the two components to provide a negative enthalpy of mixing. This is called complementary dissimilarity.[1]

The concept of specific interactions is wide but in general includes intermediate forces between the dispersive or van der Waals forces and the chemical bonds. We can emphasize among them hydrogen bonding, charge transfer, and ion–ion or ion–dipole interactions. This article refers exclusively to hydrogen bonding, which provides negative enthalpies of mixing.

Hydrogen bonding has no universally accepted definition, but it arises through interactions between electron-deficient hydrogen atoms and atoms of high-electron density. The hydrogen bonds have an accused directional character and their bond energies are relatively low, ≈ 13–25 kJ/mol. However, the hydrogen bond should be considered a dynamic equilibrium in which bonds are constantly forming and breaking. The temperature of the system plays an essential role in this equilibrium. Hydrogen bonding progressively decreases when the temperature increases, the equilibrium frequently is recovered when the temperature of the system decreases again.

Although it is possible to find a wide variety of situations that lead to the formation of hydrogen bonds, this interaction usually involves at least two functional groups in the same or in different molecules (intra- or intermolecular hydrogen bonding). One of these groups behaves like a proton donor (acid group) and the other as an electron acceptor (basic group). In most of the cases the proton is donated by a hydroxyl, carboxyl, amine, or amide group. Protons close to other electronegative atoms like halogens and sulfur (including carbon) can also produce hydrogen bonding, but it is much weaker. Normally, electrons donors are oxygen atoms located on carbonyl, ether, or hydroxyl groups; nitrogen atoms on amines and heterocyclic compounds; and in some cases halogens in molecules with a very particular environment.

Given this information, one can predict that the pairs of polymers that contain complementary donor and acceptor groups in their chains will establish intermolecular interactions by hydrogen bonding of miscible mixtures. At present, a wide variety of miscible blends with these characteristics are known. In **Figure 1** we have several miscible hydrogen-bonded polymer blends. One method to increase polymer–polymer miscibility consists in introducing a small proportion of complementary chemical groups in the components, by means of copolymerization or modification.[2–6] This strategy is efficient and effective.

Likewise, the presence in a polymer of remaining donor or acceptor groups derived from conditions in which their synthesis has been carried out could increase their capacity of forming miscible mixtures.

When the interactions between donors and acceptors are particularly intense and acquire a cooperative character they could form polymer-polymer complexes also called interpolymers.[7,8] Polymer-polymer complexes have compositions that involve a simple stoichiometries (1:1 or 1:2) of repetitive units of the employed homopolymers. The complex depends on factors such as temperature, concentration, or solvent. Most of the polymer–polymer complexes involve carboxylic polyacids [or other polymers with strong acid character like poly(4-vinylphenol)] and polymers with a strong basic character. Examples are the polymer–polymer complexes obtained between the poly(carboxylic acid)s and polyvinylpyrrolidone,[9–12] poly(ethylene oxide),[13,14] poly(vinyl alcohol),[15,16] polyvinylpyridines,[17] polyacrylamide,[18] poly(ethyl oxazoline).[19]

DETECTION AND CHARACTERIZATION OF HYDROGEN BONDING IN BLENDS

Techniques based on spectroscopic methods are most often used in the characterization and study of hydrogen bonding in polymer blends. Although fluorescence or nuclear magnetic resonance studies have been carried out, infrared spectroscopy is the spectral technique most extensively used in the study of hydrogen bonding in polymer blends. Several spectral modes corresponding to typical donor or acceptor groups are notably affected by hydrogen bonding, in their intensity and positions. Frequently the functional groups involved in hydrogen bonding in each polymer exhibit spectral contributions that do not overlap, which simplifies the spectral analysis.

HOW THEY AFFECT THE PROPERTIES OF THE MIXTURES

The experimental criterion of miscibility is the presence of a sole T_g in the mixture, that usually depends on the composition and intermediates temperatures between pure components. Obviously, this criterion is also valid in mixtures with hydrogen bonding. However, here the behavior of the T_g with respect to the composition of the mixture is particular. Usually, in the case of mixtures with strong specific interactions, such as hydrogen bonding, we observe clear deviations of the evolution of the T_g

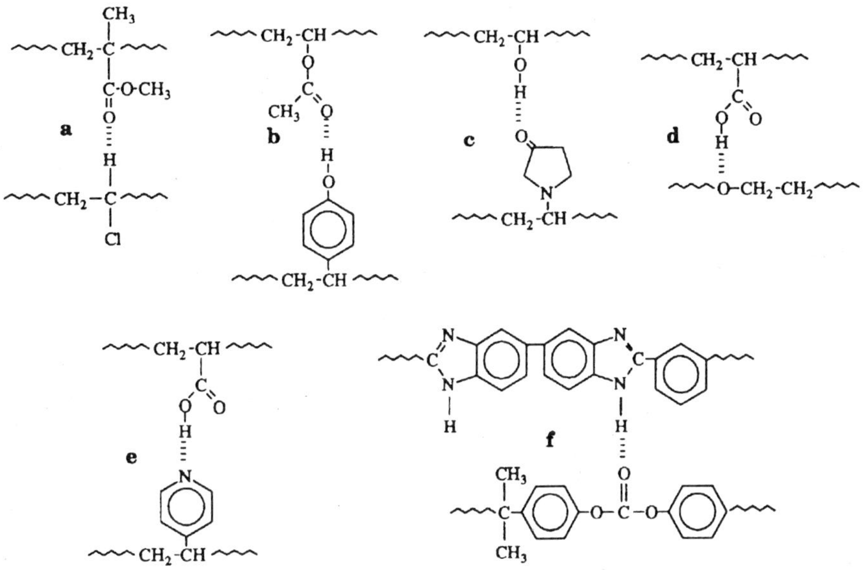

FIGURE 1. Several hydrogen bonded miscible polymer blends: (a) poly(methyl methacrylate)/poly(vinyl chloride);[29] (b) poly(vinyl acetate)/poly(4-vinylphenol); (c) poly(vinyl alcohol)/poly(*N*-vinylpyrrolidone);[21] (d) poly(acrylic acid)/poly(ethylene oxide);[30] (e) poly(methacrylic acid)/poly(2-vinylpyridine);[17] (f) polybenzimidazole/poly(bisphenol-A carbonate).[20]

with the composition compared to the weight average value of each composition's pure components.

PREDICTION OF MISCIBILITY IN MIXTURES WITH HYDROGEN BONDING

Two different approaches have been developed to consider the thermodynamic of mixtures with strong interactions. Sanchez and Balasz have incorporated the directional specific interactions to a compressible lattice model, without taking into account the possibility of interactions in the pure components.[22] Coleman, Graf, and Painter have developed an association model for polymer blends in which hydrogen bonding is present.[23] This model is notably successful in the prediction of miscibility in this type of systems.

The association model is also useful for the study of mixtures involving copolymers. This is particularly easy if one of the comonomers behaves like an "inert" diluent for the processes of association via hydrogen bonding. However, the model is also applied to homopolymer/copolymer blends in which the three repetitive units are susceptible of forming hydrogen bonds.[24] It has also been applied to the study of mixtures constituted by three homopolymers.[25,26]

Determining the enthalpy of mixing by means of the association model allows us to incorporate the effect of hydrogen bonding on the dependence of the T_g with the composition of the mixture.[27] The model has been also applied to the analysis of the melting point depression in hydrogen-bonded blends in which one of their components is semicrystalline.[28]

ACKNOWLEDGMENTS

The authors acknowledge the CICYT (Project MAT 464/92-C02), CYTED and Vicerrectorado de Investigación de la Universidad Del País Vasco for its financial support.

REFERENCES

1. Olabisi, O. *Macromolecules* **1975**, *40*, 324.
2. Moskala, E. J. et al. *Macromolecules* **1984**, *17*, 1671.
3. Dkjadoun, S.; Goldenberg, R. N.; Morawetz, H. *Macromolecules* **1977**, *10*, 1015.
4. Pearce, E. M.; Kwei, T. K.; Min, B. Y. *J. Macromol. Sci., Chem.* **1984**, *A21*, 1181.
5. Shah, K. R. *Polymer* **1987**, *28*, 1212.
6. Chu, E. Y. et al. *Makromol. Chem. Rapid Commun.* **1991**, *12*, 1.
7. Bekturov, E. A.; Bimendina, L. A. *Adv. Polym. Sci.* **1981**, *41*, 99.
8. Tsuchida, E.; Abe, K. *Adv. Polym. Sci.* **1982**, *45*, 1.
9. Bimendina, L. A. et al. *J. Polym. Sci., Polym. Symp.* **1979**, *66*, 9.
10. Ohno, H.; Abe, K.; Tsuchida, E. *Makromol. Chem.* **1978**, *179*, 755.
11. Cesteros, L. C. et al. *Polym. Commun.* **1990**, *31*, 152.
12. Cesteros, L. C.; Isasi, J. R.; Katime, I. *J. Polym. Sci.: Part B; Polym. Phys.* **1994**, *32*, 223.
13. Smith, K. L.; Winslow, A. E.; Petersen, D. E. *Ind. Eng. Chem.* **1959**, *51*, 1361.
14. Bailey, F. E.; Lundberg, R. D.; Callard, R. W. *J. Polym. Sci. Part A* **1964**, *2*, 845.
15. Ohkrimenko, I. S.; D'yakonova, E. B. *Vysokomol. Soyed. Ser. A* **1964**, *6*, 1891.
16. Hughes, L. J.; Britt, G. E. *J. Applied Polym. Sci.* **1961**, *5*, 337.
17. Piérola, I. F. et al. *Eur. Polym. J.* **1988**, *24*, 895.
18. Eustace, D. J.; Siano, D. B.; Drake, E. N. *J. Appl. Polym. Sci.* **1988**, *35*, 707.
19. Lin, P. Y. et al. *J. Polym. Phys. Ed.* **1988**, *26*, 603.
20. Musto, P. et al. *J. Polym.* **1991**, *32*, 3.
21. Ping, Z.; Nguyen, Q. T.; Neel, J. *Makromol. Chem.* **1988**, *189*, 437.
22. Sanchez, I. C.; Balazs, A. C. *Macromolecules* **1989**, *22*, 2325.
23. Coleman, M. M.; Graf, J. F.; Painter, P. C. *Specific Interactions and the Miscibility of Polymer Blends*; Technomic: Lancaster, PA, 1991.
24. Coleman, M. M. et al. *Macromolecules* **1992**, *25*, 4414.

25. Le Menestrel, C. et al. *Macromolecules* **1992**, *25*, 7101.
26. Pomposo, J. A.; Cortázar, M.; Calahorra, E. *Macromolecules* **1994**, *27*, 252.
27. Painter, P. C.; Graf, J. F.; Coleman, M. M. *Macromolecules* **1991**, *24*, 5630.
28. Painter, P. C. et al. *Macromolecules* **1991**, *24*, 5623.
29. Jager, H.; Borenkamp, E. J.; Challa, G. *Polym. Commun.* **1983**, *24*, 290.
30. Aubin, M.; Woyer, R.; Prud'homme, R. E. *Makromol. Chem. Rapid Commun.* **1984**, *5*, 419.

Hydrogenated Polymers

See: *Emulsion Polymers (Noncatalytic Diimide Hydrogenation)*
Styrene-Butadiene-Styrene Elastomer, Hydrogenated
Viscosity-Index Improvers

Hydrolytic Degradation

See: *Degradation (Weatherability)*
Environmentally Degradable Polymers

HYDROPHILIC POLYMERS (For Friction Reduction)

Y. L. Fan
Union Carbide Corporation

Nearly all water-soluble polymers with high molecular weight exhibit some degree of lubricity on hydration. This friction-reduction phenomenon is believed to be associated with a large amount of water held by the hydrophilic polymeric molecules and the resulting high degree of freedom of the polymeric chains in the hydrated state. This property has been used extensively for friction reduction on the surfaces of medical devices.

Hydrophilic surface lubricity is a common phenomenon in marine life and in the living tissues of humans and animals. This surface lubricity serves to reduce friction in motion, and to protect tissue damage, or both. It is therefore not surprising that such a hydrophilic lubricious surface has become highly desirable for medical devices intended for either transient or permanent implant in the human body.[1]

Synthetic and natural-occurring polymeric materials, such as polyurethane, acrylic resin, vinyl resin, polyolefin, nylon, and rubber are widely used for making catheters, contact lenses, implant devices, heart valves, intrauterine devices, peristaltic pump chambers, endotracheal tubes, gastroenteric feed tubes, and arteriovenous shunts. It is highly desirable that these materials can be fabricated to provide surfaces that are not only hydrophilic, but that also have low coefficients of friction when in contact with an aqueous fluid, such as a body fluid. For ease of handling, it is even more desirable if the surface exhibits a normal plastic feel when dry but becomes slippery only on exposure to the aqueous fluid.

A variety of approaches have been undertaken to develop hydrophilic lubricious surfaces for medical devices that would provide good handling characteristics when dry, but become instantaneously slippery on exposure to an aqueous body fluid.[2–9] This field was reviewed recently by this author.[10,11] A more general discussion on this subject was provided in a recently published monograph by Technomic Publishing Company.[12]

HYDROPHILIC POLYMERS FOR FRICTION REDUCTION OF SURFACES

Both physical and chemical methods have been developed recently to immobilize a variety of hydrophilic polymers to the surfaces of either polymeric or metallic devices for friction reduction in the hydrated state. For most applications, surfaces that offer a fast hydration rate coupled with a very low coefficient of friction in water and a durable wet coating are desired. The performance characteristics of coatings produced by these methods may vary greatly depending on the nature of the hydrophilic polymer used, as well as the process by which the polymer is deposited on the device. The current methods may be categorized into five groups and are discussed individually below.

Physical Attachment of Hydrophilic Polymers

A technique that is often used to impart a hydrophilically slippery surface to a medical device, such as surgical gloves, catheters, and sutures, is the application of an exterior coating containing a suitable hydrophilic polymer. The latter may be either water soluble or water swellable. The coating may be applied using any suitable coating process, but a dip-coating process is probably used most often.[13]

Poly(*N*-vinylpyrrolidone) or poly(ethylene oxide) is an effective hydrophilic polymer for providing a coated surface with a low coefficient of friction when wet, which retains a normal plastic feel when dry.[13–19] Coatings made from these polymers have been accepted by the medical communities because of their proven processability, biocompatibility, some antithrombogenicity, protein repellent properties, and commercial availability.[20,21]

Other synthetic polymers useful for providing a friction-reduction coating include poly(alkylene oxalate), poly(vinyl alcohol), ionene (ionic amine) polymers, and caprolactone copolymers.[23–26] Caprolactone copolymers constitute a class of biodegradable, lubricating polymers, however, the homopolymer is water insoluble.[27]

Chitin [poly(N-acetyl-D-glucosamine)] and its derivatives have been used to impart lubricity on items necessary for surgery, for example, tubing, catheters, drains, and gloves.[28]

Certain hydrophilic lubricious coatings are formulated intentionally to be leachable or absorbable, so that the hydrophilic polymer is removed from the device in a controlled fashion.[22,27,29] This type of coating is often used on surgical sutures and shaving devices.

Hydrophilic coatings with somewhat improved durability, adhesion, or both have been obtained by forming either a polymeric blend or complex with the substrate material or an additive.

A lubricious coating consisting of a noncrosslinked, crystallized polymer or polymer blend has been used to modify the surfaces of diagnostic probes, catheters, and guide wires.[30] A fatty acid–polymer mixture has been used to develop wet lubricity in nasogastric intubation devices on exposure to stomach fluids. Lubricity is developed when the fatty acid is converted to its salt.[31]

The hydrophilic lubricity of many of the coating compositions described in this group, although good initially, may exhibit questionable durability. The hydrophilic polymers are often removed from the substrate surfaces through leaching or mechanical dislocation because of a lack of adhesion. Leaching is improved by crosslinking the hydrophilic polymer in the finished coating; mechanical dislocation is often improved through either complexation with another polymer or surfactant, or blending the hydrophilic polymer with a structure polymer of the substrate, such as polyurethane.

Immobilization of Hydrophilic Polymers Through Formation of Interpenetrating Polymeric Networks

The need for greater adhesion between the lubricious coating and the substrate has led to the development of techniques that form a surface layer of an interpenetrating polymeric network (IPN) in which the hydrophilic polymer is anchored permanently with the substrate molecules. In some instances, however, the hydrophilic polymer present in the coating composition is in reality a thermoplastic rather than a thermosetting polymer, therefore the resultant compositions may not be true IPNs, but are mixtures in which the long-chain hydrophilic polymers intertwine physically within the network of the polyurea, polyurea–polyurethane, or polyurea–polyacrylate depending on the nature of the coating composition and substrate used in the device. Recently, this approach has received a great deal of attention and a variety of IPN-type coatings have been developed using either chemical-, photo-, or radiation-induced polymerization techniques.

The hydrophilic polymer often cited for this application is poly(*N*-vinylpyrrolidone). It is usually used in conjunction with a multifunctional isocyanate with or without the presence of a polyurethane.[32–35] Poly(ethylene oxide) has been applied in a very similar fashion with a polyisocyanate to produce a network structure through a mostly physical entanglement.[35a]

Hyaluronic acid and its salts, chondritic sulfate and agarose, are another class of biocopolymers used for producing a hydrophilic lubricious coating on medical devices. The dimensional stability of such a coating was significantly improved by the incorporation of albumin.[36]

Chemical Attachment of Hydrophilic Polymers Using a Reactive Primer

Although good adhesion does not necessarily involve chemical bonding, the presence of chemical bonding in an adhesive joint usually ensures a more durable bond. Following this thinking, a number of systems have been devised in which the hydrophilic polymers are capable of chemically bonding to the substrate, usually through a reactive primer system.

Water-soluble polymers, such as cellulosic derivatives, poly(maleic anhydride), polyacrylamide, or water-soluble nylon have been claimed to chemically bond to a polymeric substrate through a reactive primer, such as an aldehyde, epoxy, or isocyanate.[37] Depending on the chemical nature of the substrate material, however, the reactive primer may or may not bond chemically to the substrate. Polysaccharides, such as hyaluronic acid, have been bonded to acrylic polymers using a polyisocyanate to afford improved hydrophilicity and lubricity.[38,39]

The hydrophilic lubricious coatings have been claimed to be durable and not leachable. A common feature claimed by the patentees of this group was that the hydrated hydrophilic coating was abrasion resistant and would survive either repeated abrasions during a surgical procedure, a long period of implanting in the human body, or both. The success of chemical bonding is affected not only by the reactivity of coating chemicals used and the chemical nature of the substrate to be coated, but also the coating conditions employed. Unless all these required elements are met, this approach may not necessarily lead to a high-quality lubricious coating. Because of the chemical reactivity of the hydrophilic polymers and primers used in these reactive-coating systems, a two-step coating process is usually employed. Good process control is essential for a successful application of the chemically reactive coatings.

In Situ Polymerization or Grafting

Instead of solution-coating hydrophilic polymers, they can be polymerized *in situ* from the corresponding monomer(s) or prepolymer(s) directly onto the surfaces of a medical device. A free-radical polymerization process is usually employed.

A radiation-grafting process of hydrophilic monomers onto organic polymeric substrates in the presence of cupric or ferric ions was described by Ratner and Hoffman.[40]

Plastic contact lenses have been treated with electric-glow discharge to render the surface hydrophilic without altering the optical characteristics or the physical dimensions of the lenses.[41] A similar glow-discharge polymerization technique was employed to impart a hydrophilic, optically clear, impermeable barrier to soft corneal contact lenses.[42] Wettability of polymeric materials were improved by plasma treatment in the presence of allene.[43] A sulfonation process was reported recently for increasing the wettability and thrombo-resistance of a wide range of plastic devices.[44]

An advantage of the free-radical-initiated grafting process is that a thin and conforming coating can be readily obtained, regardless of the contour and shape of the medical device to be coated. However, this process may require a higher capital investment on equipment and a more elaborate control process than those of a conventional solution-coating operation. Furthermore, one needs to be concerned with residual monomers in the coated article, such as acrylamide, which may be highly toxic. When plasma polymerization is employed, the degree of lubricity of the modified surfaces may be limited because of an excessive crosslinking of the resultant hydrophilic polymer. This technique holds promise, however, for modifying surfaces of intricate medical devices that may not lend themselves readily to conventional solution-coating processes.

Hydrophilic Bulk Polymers

Besides surface modification, lubricious medical devices may be fabricated directly from a hydrophilic polymer. The latter is produced by either copolymerization or chemical modification of a base resin. (Ethylene–vinyl acetate) copolymers have been radiation-graft copolymerized with acrylic acid to impart a greater hydrophilicity. The neutralized resin was suitable for making surgical devices with high water-swelling characteristics and slipperiness.[45,46]

SUMMARY

A variety of physical and chemical modification methods are currently available for friction reduction of polymeric or metallic surfaces in an aqueous environment. This property has found a broad application for imparting hydrophilic lubricity to medical devices. Among these methods, the formation of lubricious surfaces using either a chemically reactive hydrophilic polymeric coating, a chemical grafting of hydrophilic monomer(s) or prepolymer, or the formation of an IPN composition appear to offer the most satisfactory performance characteristics. The coated medical device exhibits a normal plastic feel when dry, but becomes instantly lubricious on exposure to bodily fluids. The performance of a satisfactory hydrophilic lubricious coating has been interpreted using a double-layer model in which the outer layer consists of fast-hydrating and mobile polymeric chains, whereas the inner layer is a matrix of hydrophilic polymer molecules firmly anchored to the substrate through either the coupling action of a reactive primer or direct chemical grafting or entanglement with the substrate matrix molecules. Hydrophilic lubricious coatings meeting these criteria have been shown to offer good handling characteristics, quick lubricity in bodily fluids, and improved coating durability in the hydrated state.

REFERENCES

1. Mardis, H. K.; Kroeger, R. M. *Urol. Clin. North Am.* **1988**, *15*, 3.
2. Shook, D. R. et al. *J. Biomech. Eng.* **1986**, *108*(2), 168–74.
3. Triolo, P. M.; Andrade, J. D. *J. Biomed. Mater. Res.* **1983**, *17*, 149–65.
4. Cohen, A. *J. Hosp. Infect.* **1985**, *6 (Suppl. A)*, 155–61.
5. Pearce, R. S. et al. *Am. J. Surg.* **1984**, *148*, 687–91.
6. Harrison, L. H. *J. Urol.* **1980**, *124*, 347–9.
7. Kikuchi, Y. et al. *Cardiovasc. Intervent. Radiol. (U.S.)* **1989**, *12*(2), 107–9.
8. Rodeheaver, G. T. et al. *Surg. Gynecol. Obstet. (U.S.)* **1987**, *164*(1), 17–21.
9. Buter, H. K.; Kunin, C. M. *J. Urol. (U.S.)* **1965**, *100*(4), 560–6.
10. Fan, Y. L. *Amer. Chem. Soc., Polym. Mater. Sci. Eng.* **1990**, *63*, 709–716.
11. Fan, Y. L. *Polymernews* **1992**, *Vol. 17*, No. 3.
12. Ikada, Y.; Uyama, Y. *Lubricating Polymer Surfaces*; Technomic: Lancaster, PA 1993.
13. Schwartz, A.; Graper, J.; Williams, J. U.S. Patent 4 589 873, 1986.
14. Poly(vinyl pyrrolidone) USP grades are available from GAF and BASF, and poly(ethylene oxide) NF grades are available from Union Carbide.
15. Althans, W.; Throne, J. European Patent 321 679, 1989.
16. Becker, L. F. et al. U.S. Patent 4 835 003, 1989.
17. Podell, D. L. U.S. Patent 3 813 695, 1974.
18. American Cyanamid Co., European Patent Application 558965-A2, 1992.
19. Boston Scientific Corp., World Organization Patent Application 9107-200A, 1991.
20. Goldstein, A. U.S. Patent 4 482 577, 1984.
21. Kim, S. W.; Feijen, J. *Critical Reviews in Biocompatibility*; CRC: Boca Raton, **1985**, *Vol. 1*, 229–260.
22. Casey, D. J. et al. U.S. Patent 4 716 203, 1987.
23. Shaluby, S. W.; Jamiolkows, D. U.S. Patent 4 105 034, 1978.
24. Miller, R. A. British Patent 2 179 258, 1987.
25. Schaper, R. J. U.S. Patent 4 166 894, 1979.
26. Schaper, R. J. U.S. Patent 4 075 136, 1978.
27. Messier, K. A.; Rhum, J. D. U.S. Patent 4 624 256, 1986.
28. Casey, D. J. U.S. Patent 4 068 757, 1978.
29. Gillette Co., The Netherland Patent 7 904 061, 1979.
30. Ofstead, R. F. U.S. Patent 4 977 901, 1990.
31. Etheredge, R. W.; Charkoudian, J. C. U.S. Patent 4 983 170, 1991.
32. Lambert, H. R. U.S. Patent 4 666 437, 1987.
33. Micklus, M. J.; On-Yang, D. T. U.S. Patent 4 100 309, 1978.
34. ibid U.S. Patent 4 119 094, 1978.
35. PCT Patent Application WO-9 002 579-A, 1990.
35a. Lambert, H. R. U.S. Patent 4 487 808, 1984.
36. Markle, R. A.; Brusky, P. L.; Baker, J. H. U.S. Patent 4 943 460, 1990.
37. Takemura, N.; Tanabe, S. European Patent 166 998, 1986.
38. Beavers, E. M. U.S. Patent 4 663 233, 1987.
39. Halpern, G. et al. U.S. Patent 4 801 475, 1989.
40. Ratner, R. D.; Hoffman, A. S. U.S. Patent 3 909 049, 1976.
41. Gesser, H. D.; Warriner, R. E. U.S. Patent 3 925 178, 1975.
42. Peyuman, G. A.; Koziol, J. E.; Yasuda, H. U.S. Patent 4 312 575, 1982.
43. Thurm, S.; Noreiks, U.; Schwabe, P. German Patent Application 4235300-A1, 1994.
44. Medical Product Manufacturing News, Vol. 6 No. 5, September/October 1990, p 6.
45. Fydelor, P. et al. U.S. Patent 4 785 059, 1988.
46. Fydelor, P. et al. European Patent 179 839, 1986.

Hydrophilic Sponges

See: Cornea, Artificial (Hydrophilic Polymeric Sponges)

HYDROPHILIC SURFACES (Plasma Treatment and Plasma Polymerization)

Norihiro Inagaki
Faculty of Engineering
Shizuoka University

When polymeric materials are exposed to plasma, the plasma can interact with the surface of the materials causing chemical and physical changes at the polymer's surface. The changes include etching, implantation of atoms, and radical formation. This process is called plasma treatment. When organic molecules rather than polymeric materials interact with the plasma, polymers deposit directly onto the surface of the materials. This reaction is called plasma polymerization. Plasma treatment and plasma polymerization are effective ways of modifying the surface of polymeric materials so they become hydrophilic. Each of these processes is briefly described.

IMPLANTATION

The implantation process that occurs in plasma treatment is one of the most effective methods of surface modification of polymeric materials. Gas molecules, such as oxygen and nitrogen, are activated by the plasma. The activated species interact with the polymer's surfaces, and then special functions, such as hydroxyl, carbonyl, carboxyl, amino, and amido groups are formed at the surface of the polymers. As a result, the implantation reactions lead to large changes in the surface properties of the polymer, for example, the polymers change from hydrophobic to hydrophilic. "Plasma treatment" is frequently used for the improvement of adhesion and wettability of polymeric materials. Carbon oxide, carbon dioxide, nitrogen oxide, nitrogen dioxide, and ammonia are used as plasma gases for hydrophilic surface modification. Polypropylene, polyester, polystyrene, rubber, polytetrafluoroethylene and so on, rather than polyethylene are successfully modified by the plasma treatment. The details of the implantation process are reviewed in the literature.[1]

One needs to consider what species in the plasma initiates the implantation. Electrons, ions, and radicals are active species within the plasma. The radicals are neutral in electrical properties, and the electrons and ions are charged negatively or positively. Therefore, the radicals are easily isolated from the electrons and ions in the plasma by the electric potential.[2]

We should recognize that the plasma treatment is a result of interactions between the plasma and the polymeric material. As a result, the modification depends not only on the plasma operating conditions, such as the kind of plasma gas, rf power, treating time, and pressure; but it also depends on the nature of polymer materials.

PLASMA-GRAFT COPOLYMERIZATION

When polymeric materials are exposed to plasma, the plasma creates radicals in the polymer chains. When polymer materials make contact with monomers in a liquid or gas phase, polymerization reactions of the monomers are initiated from the radical sites. Graft polymers are then formed on the surface of the polymeric materials. In a chemical sense, this polymerization process is the same as the process of graft copolymerization induced by radiation such as X-rays, gamma rays, and accelerated electrons.

The surface modification by plasma-graft copolymerization is applied in many polymer systems to improve properties such as adhesion, printability, dyeability, gloss, tribology, antifogging, and biocompatibility. Acrylamide, acrylic acid, methyl methacrylic acid, styrene, 4-vinylpyridine, maleimide, and glycidyl acrylate are monomers that are used for grafting.[3–12] Typical polymer systems involved in this process are poly(ethylene terephtalate) (PET)/acrylic acid (AA), PET/acrylamide, polyethylene (PE)/acrylamide, PE/AA, PE/methacrylic acid, PE/2-hydroxyethyl methacrylate (HEMA), polypropylene/glycidyl methacrylate, poly(vinyl chloride)/AA, cotton/HEMA, silk/HEMA, poly[1-(trimethylsilyl)-1-propylene]/AA, and so on.[13–23]

PLASMA POLYMERIZATION

Plasma polymerization is a thin-film-forming process. In this process, the growth of low-molecular-weight molecules (monomers) into high-molecular-weight molecules (polymers) occurs with the assistance of the plasma's energy. Polymers are deposited directly onto the surface of substrates without any fabrication to create a new surface with a coating of the deposited plasma polymers. In this sense, the plasma polymerization is a complicated process involving both polymerization and coating process. The reaction mechanism of the polymer-forming process in plasma polymerization is very different from that of conventional polymerizations. The basic concept of plasma polymerization was proposed by Yasuda.[23] Monomers in plasma gain high energy from the plasma and are fragmented into activated small fragments, in extreme cases they are fragmented into atoms. These activated fragments are recombined, and sometimes rearranged, and the molecules become molecules with high molecular weight. The repetition of activation, fragmentation, and recombination leads to the formation of polymers. Yasuda's concept means that plasma polymerization is not a chain reaction but rather a stepwise reaction, and the chemical composition of the deposited plasma polymers is determined by how the used monomer molecules are fragmented in the plasma. Yasuda emphasizes that a key factor in controlling the chemical composition of the deposited plasma polymers is to input energy through electric power to maintain a glow discharge. This indicates that plasma polymerization is an operational condition-dependent process. The operational conditions, especially the level of the electric input power, determine the chemical composition of the deposited plasma polymers. Readers should recognize that the plasma polymerization at two different levels of electric input power will yield two plasma polymers with completely different chemical compositions, even if the same monomer is used for the plasma polymerization. Characterization of the plasma polymers requires the description of the operational conditions used for the plasma polymerization.

Hydrophilic plasma polymers can be easily formed through plasma polymerization. Organic oxygen- or nitrogen-containing compounds are used as monomers for the plasma polymerization. Alcohols, carboxylic acids, esters, ethers, and so on are typical monomers of the oxygen-containing compounds; amines, amides, nitriles, nitros, etc. are specific monomers of the nitrogen-containing compounds. A functional group of the monomer used for the plasma polymerization, for example, a hydroxyl group, is not always incorporated into the deposited plasma polymer, because the monomer in plasma is necessarily fragmented. The optimum operational conditions to minimize the degradation of the functional group in plasma are chosen for the preparation of hydrophilic plasma polymers.

Ethylene oxide (EO) and nitro compounds are good monomers to use for the deposition of hydrophilic plasma polymers.[24] The plasma polymerization of EO deposited colorless and hydrophilic films. The surface energy of the deposited films showed a strong dependence on the reaction conditions operating the plasma polymerization.

Nitro compounds involving nitroethane ($C_2H_5NO_2$), 1-nitropropane ($C_3H_7NO_2$), and 2-nitropropane ($(CH_3)_2\,CHNO_2$) produce high hydrophilic plasma polymer films.[24] These compounds were plasma polymerized to deposit light-yellow transparent films that possessed high surface energy.

REFERENCES

1. Inagaki, N. *Plasma Surface Modification and Plasma Polymerization*; Technomic: Lancaster, PA 1996.
2. Inagaki, N.; Yamamoto, K. *Nippon Kagaku Kaishi* **1987**, *302*, 1987.

3. Allmér, K.; Hult, A.; Rånby, B. *J. Polym. Sci., Polym. Chem. Ed.* **1988**, *26*, 2099.
4. Allmér, K.; Hult, A.; Rånby, B. *J. Polym. Sci., Polym. Chem. Ed.* **1989**, *27*, 1641.
5. Allmér, K.; Hult, A.; Rånby, B. *J. Polym. Sci., Polym. Chem. Ed.* **1989**, *27*, 3405.
6. Allmér, K.; Hult, A.; Rånby, B. *J. Polym. Sci., Polym. Chem. Ed.* **1989**, *27*, 3419.
7. Allmér, K.; Hilborn, J.; Larsson, P. H.; Hult, A.; Rånby, B. *J. Polym. Sci., Polym. Chem. Ed.* **1990**, *28*, 173.
8. Yao, Z. P.; Rånby, B. *J. Appl. Polym. Sci.* **1990**, *40*, 1647.
9. Yao, Z. P.; Rånby, B. *J. Appl. Polym. Sci.* **1991**, *43*, 621.
10. Yao, Z. P.; Rånby, B. *J. Appl. Polym. Sci.* **1990**, *41*, 1459.
11. Yao, Z. P.; Rånby, B. *J. Appl. Polym. Sci.* **1990**, *41*, 1469.
12. Edge, S.; Feast, W. J.; Pacynko, W. F.; Preston, L.; Walker, *Polym. Bull.* **1992**, *27*, 441.
13. Suzuki, M.; Kishida, A.; Iwata, H.; Ikada, Y. *Macromolecu* **1986**, *19*, 1804.
14. Hirotsu, T. *J. Appl. Polym. Sci.* **1987**, *34*, 1159.
15. Hirotsu, T. *Ind. Eng. Chem. Res.* **1987**, *26*, 1287.
16. Hirotsu, T.; Nakajima, S. *J. Appl. Polym. Sci.* **1988**, *36*, 1.
17. Hirotsu, T.; Isaya, M. *J. Membr. Sci.* **1989**, *45*, 137.
18. Uchida, E.; Uyama, Y.; Iwata, H.; Ikada, Y. *J. Polym. Sci., Polym. Chem. Ed.* **1990**, *28*, 2837.
19. Masuda, T.; Kotoura, M.; Tsuchihara, K.; Higashimura, T. J. *Appl. Polym. Sci.* **1991**, *43*, 423.
20. Vigo, F.; Uliana, C.; Traverso, M. *Eur. Polym. Sci.* **1991**, *27*, 779.
21. Hirotsu, T.; Asai, N. *J. Macromol. Sci., Chem.* **1991**, *A28*, 461.
22. Hsieh, Y.-L.; Wu, M. *J. Appl. Polym. Sci.* **1991**, *43*, 2067.
23. Inagaki, N.; Tasaka, S.; Horikawa, Y. *Polym. Bull.* **1991**, *26*, 283.
24. Inagaki, N.; Suzuki, K. *J. Polym. Sci., Polym. Chem. Ed.* **1987**, *25*, 1633.

HYDROPHOBIC POLYELECTROLYTES (Conformational Transitions)

Shintaro Sugai
Department of Bioengineering
Faculty of Engineering
Soka University

The hydrophobic polyelectrolytes (HP), which have ionizable and hydrophobic groups, have been known to form local micelles in aqueous solutions. Research indicates that poly(acrylic acid) (PAA) and poly(methacrylic acid) (PMA) are typical polyelectrolytes; however, some conformational behaviors of PMA differ from those of PAA.[3–5] PMA assumes compact-globule form at low degrees of ionization of the carboxyl groups and the pH-induced reversible globule-to-coil transition; such a globule form has been considered to be a result of the (hydrophobic) interaction between the methyl groups. This has made various derivatives of PMA and copolymers of methacrylic acid (MA) worthy of notice.[4–9] Polysoaps have been recognized as functional polymers in the industrial and medical fields.[13,14] As a result of these fields of interest, new functional hydrophobic polyelectrolytes have been designed, and the new polymers attempt to extend their application with special attention placed on photochemistry.[15–20]

Some physicochemical research groups of HP have pointed out the similarity of their globule-to-coil transition to the denaturation process of globular protein.[21,22] Recently, the globular protein has been shown to assume the equilibrium (or the kinetic) intermediate(s) on the pathway of denaturation, which is in the partially denatured and fluctuated, but compact (molten globule), form.[23–25] Globule HP is similar to the molten globule protein. Therefore, the conformational transition of HP can be viewed as a denaturation model of globular protein.

HYDROPHOBIC POLYELECTROLYTES

PMA and its Derivatives

The polysoap characteristics of poly(weak acid)s seem to depend not only on the hydrophobicity of the repeating unit, but also on various additional factors. Poly(ethacrylic acid), (PEA) is the HP, and its globule form is more stable than that of PMA.[26,27] Comparison of the stability of the globule form between PMA and PEA is consistent with the fact that the main interaction stabilizing the globule form in poly(α-alkyl acrylic acid) is hydrophobic.[27]

Hydrophobic Polyelectrolytes with Sulfonic Acid

Various copolymers with sulfonic acid have recently been used as functional HP; these include copolymers of 2-acrylamide-2-methylpropanesulfonic acid (AMPS) with cholestanyl methacrylate, lauryl methacrylate, 3-acrylamideo-3-methylbutanoic acid, 2-acrylamido-2-methylpropanedi (or tri)-methylammonium hydrochloride, or styrene.[28–30] Some of these can assume the hydrophobic globule form.

Copolymers of Maleic Acid (MA)

Various alternating copolymers of MA with alkyl vinyl ethers $(MA\text{-}AVE)_n$, assume the globule form at acid pH, and the pH-induced conformational transitions have been observed for the copolymers with longer side chains than butyl.[10–12,32–35] The globule form of the copolymers of very long alkyl vinyl ethers are very stable, and even in the fully ionized state they remain in the globule form.

An alternating copolymer of MA with styrene $(MA\text{-}St)_n$ assumes the acid globule form, and the pH-induced conformational transition has been studied by various titrations.[36–42] The kinetics of the conformational transition of $(MA\text{-}St)_n$ have been analyzed with its optical properties.[41] We prepared three derivatives of $(MA\text{-}St)_n$-poly(MA-*alt*-α-methyl, -*m,p*-methyl-, and perdeutrostyrene), $(MA\text{-}\alpha MSt)_n$, $(MA\text{-}m,pMSt)_N$ and $(Ma\text{-}DSt)_n$, respectively, and compared the pH-induced transition profile among them.[43–45]

Hydrophobic Polybases

The conformational properties of quaternized derivatives of poly(4-vinyl- or 2-vinylpyridine) and of various partially N-alkylated derivatives of poly(tertiary amines), such as poly(thio-1-(*N,N*-diethyl) aminoethylene), which has a polythioether backbone and a tertiary amine pendant group, and partially quaternized derivatives of poly(*N*-sec-butyl *N*-methyl aminoethyl ethylene) with a polyolefin type backbone, were studied.[46,47] Recently, new hydrophobic polybases with quaternary ammonium and methylene groups in backbone structures have been worthy of notice because of their conformational analysis and technical applications.[48] The charge density and hydrophobicity on a repeating unit can easily be adjusted in this type of ionene.

Block-Type HP

The block HP, such as polystyrene-b-PMA and poly(methylmethacrylate)-b-poly(1-methyl-4-vinylpyridinium) assumes local micelles in aqueous media.[49,50] Some of them form the reverse micelles in organic solvents including a small amount of water, and such reverse micelles are appropriate as a model for the enzyme reaction of water-insoluble substrates.

CONFORMATIONAL TRANSITION OF HYDROPHOBIC POLYELECTROLYTES

Poly(α-alkyl acrylic acid)s and Their Derivatives

Some data on PAA hints at the existence of a hydrophobic domain even in acidic PAA.[51] The transitional behavior of aqueous hydrophobic poly(α-alkyl acrylic acid)s have clearly been shown in terms of intrinsic viscosity. During ionization of carboxyl groups, the molecules transform from a very compact to an extended-coil form. That organic solvents expand the globule PMA and eliminate the hydrophobic domain is consistent with its hydrophobic stabilization.[52,53] Also, solubilities of water-insoluble substances in the polyacids show the existence of the hydrophobic globule form.

Alternating Copolymers of Maleic Acid (MA)

During ionization of carboxyl groups in MA, the abnormal pH-titration curve of poly(MA-*alt*-butyl, -pentyl, -hexyl, or -octyl vinyl ether) $(MA\text{-}BVE)_n$, $(MA\text{-}PVE)_n$, $(MA\text{-}HEV)_n$, and $(MA\text{-}OVE)_n$, respectively has been observed. The results are very similar to those obtained for PMA or PEA.[1–3,32–35]

Other Hydrophobic Polyelectrolytes

There are few reports of studies on the conformation of polybases. Rather, some interesting reports of the conformation of an aqueous strong polyacid, polystyrene-sulfonate, which is considered to be a normal polyelectrolyte, are found. The hydrophobic interaction between the sulfonated phenyl groups forms the hydrophobic microdomain.[54] Such studies are valuable because many HPs of AMPS have recently been used as functional polymers.[26–31]

REFERENCES

1. Leyte, J. C.; Mandel, M. *J. Polym. Sci.* **1964**, *A2*, 1879.
2. Mandel, M.; Leyte, J. C.; Stadhouder, M. G. *J. Phys. Chem.* **1967**, *71*, 603.
3. Michaeli, I. *J. Polym. Sci.* **1968**, *C16*, 4169.
4. Jager, J.; Engerts, J. B. F. N. *Eur. Polym. J.* **1987**, *23*, 295.
5. Bottiglione, V.; Morcellet, M.; Loucheux, C. *Makromol. Chem.* **1980**, *181*, 469.
6. Bottiglione, V.; Morcellet, M.; Loucheux, C. *Makromol. Chem.* **1980**, *181*, 485.
7. Morcellet-Sauvage, J.; Morcellet, M.; Loucheux, C. *Makromol. Chem.* **1981**, *182*, 949.
8. Vorreux, G.; Morcellet, M.; Loucheux, C. *Makromol. Chem.* **1982**, *183*, 711.
9. Morcellet-Sauvage, J.; Morcellet, M.; Loucheux, C. *Makromol. Chem.* **1982**, *183*, 821, 831, 839.
10. Dubin, P.; Strauss, U. P. *J. Phys. Chem.* **1967**, *71*, 2757.
11. Dubin, P.; Strauss, U. P. *J. Phys. Chem.* **1970**, *74*, 2842; **1973**, *77*, 1427.
12. Straus, U. P.; Schlesinger, M. S. *J. Phys. Chem.* **1978**, *82*, 571.
13. Ottenbreite, R. *Biological Activities of Polymers*; Canaher, C. E.; Gebelein, C. G., Eds.; American Chemical Society: Washington, DC, 1982.
14. Goldberg, E. P.; Nakajima, A. *Biomedical Polymers–Polymeric Material and Pharmaceuticals for Biomedical Use*; Academic: New York, 1980.
15. Morishima, Y.; Kobayashi, T.; Furui, T.; Nozakura, S. *Macromolecules* **1987**, *20*, 1707.
16. Slama-Schwok, A.; Rabani, J. *Macromolecules* **1988**, *21*, 764.
17. Delaire, J. A.; Barrie, M. S.; Webber, S. E. *J. Phys. Chem.* **1988**, *92*, 1252.
18. Morishima, Y.; Furui, T.; Nozakura, S.; Okada, T.; Mataga, N. *J. Phys. Chem.* **1989**, *93*, 1643.
19. McCormick, C. L.; Hoyle, C. E.; Clark, M. D. *Macromolecule* **1991**, *24*, 2397.
20. Mumick, P. S.; Welch, P. M.; Salazar, L.; McCormick, C. L. *Macromolecules* **1994**, *27*, 323.
21. Anufreiva, E. V.; Birstein, T. M.; Nekrasova, T. N.; Ptitsyn, O. B.; Sheveleva, T. V. *J. Polym. Sci.* **1968**, *C16*, 3519.
22. Sugai, S. *Adv. Colloid Interface Sci.* **1986**, *24*, 247.
23. Kuwajima, K. *Proteins Struct. Funct. Genet.* **1989**, *6*, 87.
24. Ptitsyn, O. B. In *Protein Folding*; Creighton, T. E., Ed.; W. H. Freeman: New York 1992; p 243.
25. Sugai, S.; Ikeguchi, M. *Adv. Biophys.* **1994**, *30*, 37.
26. Joyce, D. E.; Kurucsev, T. *Polymer* **1981**, *22*, 415.
27. Sugai, S.; Nitta, K.; Ohno, N.; Nakano, H. *Colloid & Polymer Sci.* **1983**, *261*, 159.
28. McCormick, C. L.; Middleton, J. C.; Cummins, D. F. *Macromolecules* **1992**, *25*, 1991.
29. McCormick, C. L.; Salazar, L. C. *Polymer* **1992**, *33*, 4384.
30. Morishima, Y.; Tsuji, M.; Seki, M.; Kamachi, M. *Macromolecules* **1993**, *26*, 3299.
31. McCormick, C. L.; Middleton, J. C.; Grady, C. E. *Polymer* **1992**, *33*, 4184.
32. Strauss, U. P.; Vesnaver, G. *J. Phys. Chem.* **1975**, *79*, 1558.
33. Strauss, U. P.; Schlesinger, M. *J. Phys. Chem.* **1978**, *82*, 1627.
34. Martin, P. J.; Morss, L. R.; Strauss, U. P. *J. Phys. Chem.* **1980**, *84*, 577.
35. Strauss, U. P. In *Microdomains in Polymer Solutions*; Dubin, P., Ed.; Plenum: New York, 1985; p 1.
36. Ohno, N.; Nitta, K.; Makino, S.; Sugai, S. *J. Polym. Sci. Phys. Ed.* **1973**, *11*, 413.
37. Sugai, S.; Ohno, N.; Nitta, K. *Macromolecules* **1974**, *7*, 961.
38. Okuda, T.; Ohno, N.; Nitta, K.; Dugai, S. *J. Polym. Sci. Phys. Ed.* **1977**, *15*, 749.
39. Ohno, N.; Okuda, T.; Nitta, K.; Sugai, S. *J. Polym. Sci. Phys. Ed.* **1978**, *16*, 513.
40. Sugai, S.; Ohno, N. *Biophys. Chem.* **1980**, *11*, 387.
41. Sugai, S.; Nitta, K.; Ohno, N. *Polymer* **1982**, *23*, 238.
42. Sugai, S.; Nitta, K.; Ohno, N. In *Microdomains in Polymer Solutions*; Dubin, P., Ed.; Plenum: New York, 1985; p 13.
43. Ohno, N. *Polymer J* **1981**, *13*, 719.
44. Ohno, N.; Sugai, S. *Macromolecules* **1985**, *18*, 1287.
45. Ohno, N. *JOSMER Japan* **1990**, *3*, 49.
46. Joyce, D. E.; Kurucsev, T. *Polymer* **1980**, *21*, 1457.

47. Huguet, J.; Vert, M. In *Microdomains in Polymer Solutions*; Dubin, P., Ed.; Plenum: New York, 1985; p 51.
48. Zhoumei, L.; Xuexin, Z.; Yuanpei, C.; Yuanzhen, Z. *Macromolecules* **1992**, *25*, 450.
49. Prochazka, K.; Kiserow, D.; Ramireddy, C.; Tuzar, Z.; Munk, P.; Webber, S. E. *Macromolecules* **1992**, *25*, 454.
50. Bekturov, E. A.; Kudaibergenov, S. E.; Frolova, V. A.; Khamzamulina, R. E. *Makromol. Chem. Rapid Commun.* **1991**, *12*, 37.
51. Bednar, B.; Trenena, J.; Svoboda, P.; Vajda, S.; Fidler, V.; Prochazka, K. *Macromolecules* **1991**, *24*, 2054.
52. Priel, Z.; Silberberg, A. *J. Polym. Sci.* **1970**, *A-2*, 689, 705, 713.
53. Sedlak, M.; Konak, C.; Labsky, J. *Polymer* **1991**, *32*, 1688.
54. Yamagishi, A. In *Microdomains in Polymer Solutions*; Dubin, P., Ed.; Plenum: New York, 1985; p 67.

Hydrophobic Polymers

See: *Amphiphilic Polymers (Binding Properties for Small Molecules)*
Amphiphilic Polymers (Fluorescence)
Hydrophobic Polyelectrolytes (Conformational Transitions)
Hydrophobized Polysaccharides (Versatile Functions)

HYDROPHOBIZED POLYSACCHARIDES (Versatile Functions)

Junzo Sunamoto,* Kazunari Akiyoshi, Shigeru Deguchi, and Takehiro Nishikawa
Department of Synthetic Chemistry and Biological Chemistry
Graduate School of Engineering
Kyoto University

Naturally occurring polysaccharides are one type of the abundant and diverse families of biopolymers. The polysaccharides, which are biodegradable polyhydroxyl compounds, have unique properties such as formation of hydrogel or liquid crystal.[1] Therefore, they have been widely utilized in pharmaceutical applications.[1–3] Especially, the polysaccharide derivatives with long alkyl chains have been prepared for a variety of applications such as hydrophobic gel chromatography[4] enzyme immobilization,[5,6] polymer drugs,[7] and coating of cytoplasma membrane surface.[8–10] Solution viscosity drastically increases in the presence of a hydrophobized polysaccharide such as hydroxypropyl cellulose derivatives that are partly substituted by long hydrocarbon chain.[11] This is due to the intramolecular and/or intermolecular association of hydrophobic groups. The property as rheological modifier is applied to many commercial materials such as paints, inks, cosmetics, and pharmaceuticals.

Since 1982 we have been developing several hydrophobized polysaccharides and have utilized them in biotechnology and medicine. This article gives an overview of the physicochemical characterization and the function of several naturally occurring polysaccharides partly hydrophobized, especially cholesterol-bearing pullulan (CHP).

*Author to whom correspondence should be addressed.

PREPARATION

Synthesis of Hydrophobized Polysaccharides

Cholesterol-bearing polysaccharides were first synthesized *via* aminoethylcarboxymethyl derivative of pullulan followed by condensation with cholesterylchlorofomate.[12] In this procedure, however, a trace amount of carboxylic acid and aminoethylcarboxymethyl groups remained as unreacted in the final product. In order to synthesize a nonionic hydrophobized polysaccharide, the reaction between polysaccharide and cholesteryl *N*-(6-isocyanatehexyl)carbamate (1) was proposed.[13] Various hydrophobized polysaccharides such as pullulan, dextran, mannan, and amylose were synthesized by the same procedure.

PROPERTY

Solution Property of Hydrophobized Polysaccharide in Water

Self-assembly of polymer amphiphile in water is of growing interest with respect to biological importance and medicinal or biotechnological application, and especially the solution property of block copolymer micelles or self-aggregates of hydrophobized polymers in water has been extensively studied.[14–17] Polysaccharides have been used as the hydrophilic backbone of the polymer amphiphile. The addition of surfactants dramatically affects the rheology of the aqueous hydrophobized polysaccharide solution.[18–20] We also found that the cholesterol-bearing polysaccharides such as pullulan, dextran, mannan, and amylose form an intermolecular self-aggregate in water upon hydrophobic association of cholesterol groups and/or polysaccharide skeletons. Especially, cholesterol-bearing pullulan (CHP) formed stable hydrogel nanoparticles upon self-aggregation.

Complexation of CHP Self-Aggregate with Hydrophobic Substances

The CHP self-aggregate complexes with various hydrophobic substances.[13] The complexation would be primarily driven by hydrophobic association between the two because the binding constant closely related to the hydrophobicity of the probe employed. We can estimate the micropolarity around these probes in the self-aggregate from their emission maxima.

The CHP self-aggregate is expected to provide a chiral binding site because it has many chiral centers within the particle. Bilirubin (BR) is useful to probe the existence of the chiral environment. In fact, induced circular dichroism (ICD) of BR upon enantioselective complexation has been observed in various systems such as serum albumin,[24] acyclic oligosaccharide,[25] nucleoside,[26] or cyclodextrins,[27,28] and an aqueous deoxycholate micelle.[29] BR strongly complexed with the CHP self-aggregate and exhibited a bisignate ICD Cotton effect from negative to positive.[21–23] Not only CHP but also CHD (cholesterol-bearing dextran) and CHM (cholesterol-bearing mannan) showed ICD in the BR binding.

Complexation with Soluble Proteins

The CHP self-aggregates strongly interact with various soluble proteins and enzymes.[30–32] The maximum amount of the protein bound to the CHP nanoparticle depended on the molecular weight (or the size) of protein. The CHP nanoparticle seems to have a limited capacity for a suitable size of protein. The CHP self-aggregate has a hydrophilic domain of polysaccharide skeleton, and also a hydrophobic cholesterol domain. Such an amphiphilic property of the nanosize hydrogel plays an important role for complexation with soluble proteins that have both hydrophobic and hydrophilic patches on their surface.

Complexation with Other Molecular Assemblies

The cholesterol-bearing polysaccharide derivatives strongly interact with other molecular assemblies such as liposome, lipid monolayer, and black lipid membrane. Surface pressure and surface potential isotherms of eggphosphatidylcholine monolayers were greatly modified by addition of cholesterol-bearing polysaccharides to the aqueous subphase. The results showed that the cholesterol moieties of polysaccharide derivatives penetrate the lipid monolayer.[33,34] The stability of black lipid membranes was significantly improved by coating them with the polysaccharide derivatives.[35]

The cholesterol-bearing polysaccharides effectively coat the liposomal surface.[36–40] The polysaccharide-coated liposomes are physicochemically stable against the external stimuli such as pH, ionic strength, and *in vivo* biodegradation by lipases, lipooxidases, and serum proteins compared with the conventional liposome.[39]

APPLICATION

Effective delivery of drugs to a target cell or tissue largely diminishes the toxic side effect and increases the pharmacological activity. Targeting of drugs to a specific cell or tissue is the most important problem for developing useful drug delivery system (DDS). The colloidal particles have been widely accepted to be a possible drug carrier.[42] Various kinds of device such as saccharides, glycoproteins, glycolipids, lectins, peptide hormones, antigens, and antibodies have been proposed to provide a cell specificity of the drug carrier.[43,44] The hydrophobized polysaccharide nanoparticle itself, the polysaccharide-coated liposome, and o/w-emulsion are expected to be an excellent drug carrier. Since 1982, we have been extensively studying the polysaccharide-coated liposome as a cell-specific drug carrier.[12,45,46] Recently, we have shown that o/w-emulsion stabilized by CHP is promising to a systemic administration of a lipophilic and antitumor drug such as α-linolenic acid,[41] hydrogel nanoparticle complexed and stabilized antitumor agents such as adriamycin, or bioactive proteins such as insulin.[21–23]

REFERENCES

1. Yalpani, M. *Polysaccharides Studies in Organic Chemistry*; Elsevier Science: New York, 1988.
2. Schacht, E.; Vercauteren, R.; Vansteekiste, S. *J. Bioact. Compat. Polym.* **1988**, *3*, 72.
3. Sezaki, H.; Takakura, Y.; Hashida, M. *J. Bioact. Compat. Polym.* **1988**, *3*, 2796.
4. Butler, L. G. *Arch. Biochem. Biophys.* **1975**, *171*, 645.
5. Caldwell, K. D.; Axen, R.; Porath, J. *J. Biotechnol. Bioeng.* **1976**, *18*, 433.
6. Sandberg, M.; Lundahl, P.; Greijer, E.; Belew, M. *Biochim. Biophys. Acta* **1987**, *924*, 185.
7. Suzuki, M.; Mikami, T.; Matsumoto, T.; Suzuki, S. *Carbohydr. Res.* **1977**, *53*, 223.
8. Tsumita, T.; Ohashi, M. *J. Exp. Med.* **1964**, *119*, 1017.
9. Hämmerling, U.; Westphal, O. *Eur. J. Biochem.* **1967**, *1*, 46.
10. Wolf, Đ. E.; Henkart, P.; Webb, W. W. *Biochemistry* **1980**, *19*, 3893.
11. Landoll, L. M. *J. Polym. Sci., Part A: Polym. Chem.* **1982**, *20*, 443.
12. Sato, T.; Sunamoto, J. *Prog. Lipid Res.* **1992**, *31*, 345.
13. Akiyoshi, K.; Deguchi, S.; Moriguchi, N.; Yamaguchi, S.; Sunamoto, J. *Macromolecules* **1993**, *26*, 3062.
14. Ringsdorf, H.; Schlarb, B.; Venzmer, J. *Angew. Chem. Int. Ed. Engl.* **1988**, *27*, 113.
15. Shalaby, S. W.; McCormick, C. L.; Butler, G. B., Eds.; *Water-Soluble Polymers*; American Chemical Society: Washington, DC, 1991.
16. *Macromolecular Complexes in Chemistry and Biology*; Dubin, P.; Bock, J.; Davies, R. M.; Schultz, D. N.; Thies, C., Eds.; Springer-Verlag: Berlin, 1994.
17. Morishima, Y. *Trends Polym. Sci. (TRIP)* **1994**, *2*, 31.
18. Dualeh, A. J.; Steiner, C. A. *Macromolecules* **1991**, *24*, 112.
19. Lindman, B.; Thalberg, K. *Polymer-Surfactant Interaction*; Goddard, D. E.; Anathapadmanabhan, K. P., Eds.; CRC: Boca Raton, 1991; 203.
20. Kästner, U.; Hoffmann, H.; Donges, R.; Ehrler, R. *Colloids and Surfaces* **1994**, *82*, 279.
21. Akiyoshi, K.; Deguchi, S.; Tajima, H.; Nishikawa, T.; Sunamoto, J. *Proc. Japan Acad.* **1995**, *71, Ser. B*, 15.
22. Akiyoshi, K.; Nishikawa, T.; Shichibe, S.; Mitsui, Y.; Miyata, T.; Kodama, M.; Sunamoto, J. submitted to *Macromolecules*.
23. Akiyoshi, K.; Nishikawa, T.; Shichibe, S.; Sunamoto, J. *Chem. Lett.* **1995**, 707.
24. Lightner, D. A.; Reisinger, M.; Landen, G. L. *J. Biol. Chem.* **1986**, *261*, 6034.
25. Kano, K.; Yoshiyasu, K.; Hashimoto, S. *J. Chem. Soc., Chem. Commun.* **1988**, 801.
26. Kano, K.; Yoshiyasu, K.; Hashimoto, S. *Chem. Lett.* **1990**, 21.
27. Kano, K.; Yoshiyasu, K.; Yasuoka, H.; Hata, S.; Hashimoto, S. *J. Chem. Soc. Perkin Trans.* **1992**, *2*, 1265.
28. Lightner, D. A.; Gawronski, J. K.; Gawronska, K. *J. Am. Chem. Soc.* **1985**, *107*, 2456.
29. Perrin, J. H.; Wilsey, M. *J. Chem. Soc., Chem. Commun.* **1971**, 769.
30. Akiyoshi, K.; Sunamoto, J. *Organized Solutions*; Marcel Dekker: New York, 1991, 289.
31. Akiyoshi, K.; Nagai, K.; Nishikawa, T.; Sunamoto, J. *Chem. Lett.* **1991**, 1727.
32. Nishikawa, T.; Akiyoshi, K.; Sunamoto, J. *Macromolecules* **1994**, *27*, 7654.
33. Baszkin, A.; Rosilio, V.; Albrecht, G.; Sunamoto, J. *Chem. Lett.* **1990**, 299.
34. Baszkin, A.; Rosilio, V.; Albrecht, G.; Sunamoto, J. *J. Colloid Interface Sci.* **1991**, *145*, 502.
35. Moellerfeld, J.; Prass, W.; Ringsdorf, H.; Hamazaki, H.; Sunamoto, J. *Biochim. Biophys. Acta* **1986**, *857*, 265.
36. Takada, M.; Yuzuriha, T.; Katayama, K.; Iwamoto, K.; Sunamoto, J. *Biochem. Biophys. Acta* **1984**, *802*, 237.
37. Sunamoto, J.; Iwamoto, K. *CRC Crit. Rev. Therapeutic Drug Carrier Systems* **1986**, *2*, 117.
38. Sunamoto, J.; Sato, T. In *Multiphase Biomedical Materials*; Tsuruta, T.; Nakajima, A., Eds.; VSP: The Netherlands, 1989; 167.
39. Sunamoto, J.; Sato, T.; Taguchi, T.; Hamazaki, H. *Macromolecules* **1992**, *25*, 5665.
40. Sato, T.; Sunamoto, J. *Liposome Technology 2nd Edition Vol. II*; Gregoriadis, G., Ed.; CRC: Boca Raton, 1993; 179.

41. Fukui, H.; Akiyoshi, K.; Sato, T.; Sunamoto, J.; Yamaguchi, S.; Numata, M. *J. Bioactive and Compatible Polym.* **1993**, *8*, 305.
42. *Microcapsules and Nanoparticles in Medicine and Pharmacy*; Donbrow, M., Ed.; CRC: Boca Raton, 1992.
43. Poznansky, M. J.; Juliano, R. L. *Pharmacological Review* **1984**, *36*, 277.
44. Machy, P.; Leserman, L. In *Liposomes in Cell Biology and Pharmacology*: John Libbey Eurotext: London, Paris, 1987.
45. Sunamoto, J.; Iwamoto, K.; Takada, M.; Yuzuriha, T.; Katayama, K. *Polymers in Medicine*; Plenum: New York, 1983; 157.
46. Sunamoto, J.; Goto, M.; Iida, T.; Hara, K.; Saito, A.; Tomonaga, A. In *Receptor-Mediated Targeting of Drugs*; Gregoriadis, G.; Poste, G.; Senior, J.; Trouet, A., Eds.; Plenum: New York, 1984; 359.

HYDROXYAPATITE-BIOPOLYMER INTERACTIONS

Saburo Shimabayashi* and Tadayuki Uno
Faculty of Pharmaceutical Sciences
The University of Tokushima

Hydroxyapatite, $Ca_{10}(PO_4)_6(OH)_2$ (HAP), is a basic calcium phosphate that is formed through a transformation from amorphous calcium phosphate in an aqueous phase at pHs higher than ~5. Biological HAP is a major inorganic constituent of the hard tissues of animals, that is, teeth, bones, and pathological calculi. Synthesized HAP is being developed as a biomedical material for artificial bones and teeth. It is also used in HAP chromatography for separation and purification of biopolymers from their mixture or raw extract by elution with a phosphate buffer in a laboratory because the HAP prepared for this purpose adsorbs various biopolymers with various affinities.

The interaction between HAP and biopolymers must be thoroughly known in order to understand the mechanisms of formation of biological hard tissues and pathological calculi, the mechanisms of implantation with high affinity for biological body, and the mechanisms of separation and purification of biopolymers. The interaction between HAP and biopolymers and compounds closely related to biopolymers will be briefly reviewed.

MECHANISMS FOR INTERACTION AND ADSORPTION

Electrostatic Attraction

The adsorption amount of a nonionic polymer is usually very low. This suggests that van der Waals interaction is weak as a driving force for the adsorption. However, various kinds of biopolymers (i.e., polyelectrolytes or polyampholytes) are strongly, and often irreversibly, adsorbed by HAP, where the adsorption amount is affected with a concentration of an indifferent electrolyte. This means that the dominant driving force for the adsorption is caused by electrostatic attraction between the charge of the adsorbate polymer and that of opposite sign on the surface of the HAP. The latter is donated by the surface ions and sites defective in lattice ions on the surface.

*Author to whom correspondence should be addressed.

Formation of the Intermolecular Complex on the Surface of HAP

The apparent amount of adsorption of sodium chondroitin sulfate increases with a concentration of added $CaCl_2$ for two reasons: first, Ca^{2+} adsorbed on HAP makes the adsorption sites increase in number. Second, Ca^{2+}, which is bound to chondroitin sulfate forming a complex of Ca^{2+}-chondroitin sulfate, causes shrinkage of the polymer coil because of the depression of intersegment electric repulsion. The decrease in an occupied area per molecule is the increase in the adsorption amount.[1]

Albumin at pHs lower than its isoelectric point forms a complex with chondroitin sulfate on the surface of HAP. Therefore, the adsorption amount of one polymer increases in the presence of the other.[2]

Although nonionic polyvinylpyrrolidone is not adsorbed directly onto the surface of HAP, it is adsorbed when HAP is pretreated with sodium dodecyl sulfate. This is explained in terms of a surface complex formed with sodium dodecyl sulfate on HAP. Dodecyl sulfate ion is adsorbed via its sulfate group, whereas its hydrophobic carbon chain protrudes into an aqueous phase. The hydrophobic chains capture the polymer segments in an aqueous phase through hydrophobic interaction, resulting in the formation of a surface complex.[3]

Thus, modification of the surface by an ionic amphiphile can enhance the amount of adsorption of nonionic and hydrophobic polymers. This suggests that, in biological systems, nonionic and hydrophobic compounds/groups, which rarely interact directly with HAP, still contribute to the formation of biological hard tissues through hydrophobic interaction with surface amphiphilic compounds adsorbed on HAP. As a matter of fact, hard tissues are a composite material of organic and inorganic compounds.

CONCOMITANT EFFECTS OF THE ADSORPTION

The concentration of phosphate ion released from HAP increases with the amount of adsorption of dodecyl sulfate or chondroitin sulfate. This is because of isomorphous substitution between phosphate ion on HAP and the ester sulfate of the adsorbates.

The dissolution rate of HAP decreases with adsorption. This is explained in terms of the diffusion barrier on the surface by the adsorbed layer against acids toward the surface and against dissolved ions toward the bulk aqueous phase.

Crystal growth and precipitate formation of HAP from a supersaturated solution are inhibited by various kinds of biopolymers such as saliva and enamel proteins. Condensed phosphates also inhibit them.[4] These inhibiting effects are significant in preventing pathological calculi and in regulating the formation of hard tissues in animal bodies. The effect of these inhibitors is explained in terms of adsorption of inhibitors on the crystal nuclei and/or crystal surface either generally or at the active growth sites.

The mean size of secondary particles of ripened HAP becomes small also in the presence of chondroitin sulfate or condensed phosphates. The small amount of phosphate ion coexisting together with these polymers amplifies the degree of dispersing effect.[5] These facts are explained in terms of electrostatic and steric repulsion between polymers adsorbed on the surface of HAP.

Dispersion/aggregation depends on molecular weight of the polymer even though the polymer is the same species. The other factors affecting the degree of dispersion/aggregation are polymer species, the amount of adsorption, the size of the polymer loop protruding into an aqueous phase, pH, temperature, and the surface potential and charge of the particles covered with the adsorbed layer.

REFERENCES

1. Shimabayashi, S.; Itoi, K. *Chem. Pharm. Bull.* **1989**, *37*, 1437.
2. Shimabayashi, S.; Hirao, K.; Bando, J.; Shinohara, C. *J. Macromol. Chem.* **1994**, *A-31*, 65.
3. Shimabayashi, S.; Yoshida, Y.; Arima, K.; Uno, T. *Phosphorus Research Bulletin* **1994**, *4*, 89.
4. Shimabayashi, S.; Nakagaki, M. *Chem. Pharm. Bull.* **1985**, *33*, 3589.
5. Shimabayashi, S.; Sumiya, S.; Aoyama, T.; Nakagaki, M. *Chem. Pharm. Bull.* **1984**, *32*, 1279.

Hydroxyethyl Cellulose

See: Cellulose Derivative-Protein Complex (in Ice Cream)
Cellulosic Materials (Moisture Sorption Properties)
Hydroxyethyl Cellulose (Infrared Spectra and Dielectric Properties)
Hydroxyethyl Cellulose (Preparation and Use in Protein Precipitation)
Hydroxyethyl Cellulose (Variables Affecting Hydroxyethylation Reaction)
Water-Soluble Polymers (Oil Recovery Applications)

HYDROXYETHYL CELLULOSE (Infrared Spectra and Dielectric Properties)

I. Z. Selim
Department of Physical Chemistry
National Research Center

A. H. Basta and O. Y. Mansour
Department of Cellulose and Paper
National Research Center

A. I. Atwa
Department of Microbial Biotechnology
National Research Center

Hydroxyethyl celluloses with relatively high ethoxyl contents and low activation energies were prepared.[1] The IR spectra, the crystallinity, and the asymmetry index of these compounds have been discussed and related to these properties of native cotton linters, along with the dielectric properties at different temperatures and frequencies. The dielectric behavior of cellulose and cellulose derivatives greatly depends on the nature of the side group, the degree of hydrogen bonding between the different chains, and the micropores of the fiber.[2,3]

The dielectric relaxation of hydroxyalkyl cellulose below 180°C in films of different ethoxyl contents has been studied.[4] It was found that even large variations in the content of the hydroxyalkyl groups have little effect on the dielectric constant at room temperature.

RESULTS AND DISCUSSION

Hydroxyethyl celluloses with different ethoxyl contents (wt %) were prepared from cotton linter. There was no relation between the intrinsic viscosities and the ethoxyl contents and consequently the molar substitution.

Dielectric constant values, ε', of native cellulose and hydroxyethylated celluloses of different ethoxyl content have been measured at various frequencies and different temperatures; ε' increased with increasing temperature at all frequencies (0.2–10 MHz) due to decrease of the dipole–dipole interactions. Such a decrease enhances the ease of rotation and polarizability of the side groups and the other flexible portions of the cellulose chains.[5] The increase in ε' with temperature may be due also to the excitation of charge carriers.[6] Thus, the expansion of the polymeric network contributed to the addition of many hydroxyethyl groups to the cellulose backbone and consequently to interchain spaces which show no hindrance effect. This means that most of the hydrogen bonds of the hydroxyethylated cellulose are weak or that more amorphous regions are formed on hydroxyethylation.

The crystallinity index of samples containing 25.95% ethoxyl is lower than that of samples containing 28.95% and approaches that of native cotton cellulose.

REFERENCES

1. Mansour, O. Y; Basta, A. H.; Atwa, A. I. Presented at the Cellulose 91 Conference, New Orleans, LA, December 1991; poster xx.
2. Hanna, A. A. *Acta Polymerica* **1984**, *35*, 325.
3. Shinouda, H. G.; Hanna, A. A. *J. Appl. Polym. Sci.* **1977**, *21*, 1479.
4. Domkin, V. S.; Shishenkova, T. E.; Katalevskaya, I. V. *Vysomol. Soed.* **1973**, *15A*, 1625.
5. Boutros, S.; Hanna, A. A. *J. Polym. Sci., Chem. Ed.* **1978**, *16*, 1443.
6. Beam, W. R. *Electronics of Solids*; McGraw Hill: New York, 1965; p 385.

HYDROXYETHYL CELLULOSE (Preparation and Use in Protein Precipitation)

Altaf H. Basta* and Olfat Y. Mansour
Department of Cellulose and Paper
National Research Center

Ibrahim Z. Selim
Department of Physical Chemistry
National Research Center

Aziza I. Atwa
Department of Microbial Biotechnology
National Research Center

Much work has been carried out and many patents have been obtained on the preparation of hydroxyethyl cellulose. But little work has been done on the mechanism of this reaction, and more or less contradictory results have been obtained.[1–5] In this work, different variables are considered. Dielectric properties and IR-spectra of hydroxyethyl cellulose with different ethoxyl contents were studied. The possibility of using hydroxyethyl cellulose as a new precipitating agent of proteins from cheese

*Author to whom correspondence should be addressed.

whey compared with conventional precipitants also was evaluated.

PREPARATION AND MEASUREMENT OF HYDROXYETHYL CELLULOSE (HEC) SAMPLES

Formation of hydroxyethyl cellulose from cotton linters was carried out in a two-stage process. First was formation of alkali cellulose (mercerization stage) and second was etherification.[6]

Precipitation of proteins from cheese whey using HEC with different ethoxyl contents and other precipitating agents was carried out by the method described by Mansour et al.[7]

PHYSICAL PROPERTIES OF PREPARED HEC[7]

Measurements of IR-spectra of native cotton linters and hydroxyethyl cellulose samples of different ethoxyl contents 23.34–30.31% indicate that weaker bonded OH groups are present in hydroxyethylated cellulose of ethoxyl contents of 23.34–28.89%. However, at ethoxyl content 30.31%, the crystallinity index increased; that is, a higher state of order is formed.

The dielectric constant, ε', of the aforementioned samples was measured at different temperatures and frequencies. The dielectric constant increased with increasing temperature at all frequencies, 0.2–10 MHz, for native and hydroxyethylated samples. This can be attributed to decrease of dipole-dipole infractions. Such decrease enhances ease of rotation and polarizability of the side groups and other flexible portions of the cellulose chains.[9] It is also clear that decrease in the crystallinity index and degree of hydrogen bond strength, on hydroxyethylation at EtO,% 23.34–25.95%, resulted in a greater chance for the disordered chain to achieve random disorder; therefore, a higher dielectric constant was achieved compared to native cellulose. However, blocking the free OH groups on the cellulose molecule by a substituted group of a relatively less polar character (O $-CH_2-CH_2OH$), especially in the case of 28.89–30.31% ethoxyl content, would hinder absorption, and therefore a decrease in the ε' occurs compared with HEC of ethoxy content 25.59 wt%. In other words, measurements of IR-spectra and the dielectric constant are correlated.

HYDROXYETHYL CELLULOSE FOR RECLAMATION OF PROTEINS FROM CHEESE WHEY[8]

Hydroxyethyl cellulose of different ethoxy contents wt.%, 25.59, 28.89, and 30.31, were prepared. These samples were used to precipitate proteins from fresh cheese whey. The results obtained showed that the efficiency of protein recovery is related to the percent of ethoxyl content of prepared HEC. The highest value is obtained at 28.89% ethoxyl content.

REFERENCES

1. Broderik, A. E. (to Carbide and Carbon Chemicals Corp.), U.S. Patents 2 173 470 and 2 173 47 1939; CA, 34, 621, 1941.
2. Klug, E. D.; Tennent, H. G. (to Hercules Powder Company), U.S. Patent 2 572 039 1951, CA, 46, 1256, 1954.
3. Battista, O. A.; Tasker, C. W.; Cornwell, R. T. K. *Svensk Papperstidn* **1958**, *61*, 272.
4. Claus, B.; Hartwig, R.; Kords, V.; Redener, C.; Detmor (Bayer, A-G, Wolff Walsrode A-G), *Ger. Offen.* DE3, 044, 696 ((c) CO811/20) 24 June 1982; Appl. 27, Nov. 1980, pp 13.
5. Reuben, J.; Casti, T. E. *Carbohyd. Res.* **1987**, *163*, 91.
6. Mansour, O. Y.; Basta, A. H.; Atwa, A. *Polym. Plast. Technol. Eng.* **1993**, *32*(5), 415.
7. Selim, I. Z.; Basta, A. H.; Mansour, O. Y.; Atwa, A. *Polym. Plast. Technol. and Eng.* **1994**, *33*(2), 173.
8. Mansour, O. Y.; Basta, A. H.; Atwa, A. *Cellulose e Carta* **1993**, *4*, 37.
9. Boutros, S.; Hanna, A. A. *J. Polym. Sci. Polym. Chem. Ed.* **1978**, *16*, 1443.

HYDROXYETHYL CELLULOSE (Variables Affecting Hydroxyethylation Reaction)

Olfat Y. Mansour and Altaf H. Basta
Department of Cellulose and Paper
National Research Center

Aziza I. Atwa
Department of Microbial Biotechnology
National Research Center

A tremendous amount of work has been carried out and many patents have been obtained on the preparation of hydroxyethyl cellulose. On the other hand, little work has been done on the mechanism of this reaction, and more or less contradictory results have been obtained.[1–14]

RESULTS AND DISCUSSION

Effect of Concentration

The ethylene oxide builds up a polymer chain on the hydroxyl group; i.e., a graft polymerization reaction may take place when ethylene oxide contacts alkali cellulose.

Effect of Time, Temperature, and Activation Energy

Three reaction temperatures were chosen–30, 50, and 75°C–to reveal the effect of these variables on the course of the hydroxyethylation reaction.

The rate of hydroxyethylation reaction as well as the rate of polymerization increased when the reaction temperature was increased from 30°C to 50°C; this can be attributed to an increase in the kinetic energies of the molecules and, hence in the rate of formation of ethoxyl groups on the cellulose backbone. However, due to the exothermic nature of the hydroxyethylation reaction, increasing the reaction time and temperature led to progressive degradation of the formed ethoxyl groups,[14] and hence the polymer load decreased or leveled off. At 75°C, the rates of hydroxyethylation and polymerization decreased compared with those of 50°C. However, when 14.5% sodium hydroxide was used, the rate of hydroxyethylation was higher at 75°C than at 50°C, which may be attributed to a decrease in the degradation effects. Hence the rate of hydroxyethylation overcame the rate of progressive degradation.

REFERENCES

1. Broderik, A. E. U.S. Patent 2 173 470 and 2 173 471 to Carbide and Carbon Chemicals Corp. 1939; *Chem. Abstract* **1940**, *34*, 621.
2. Klug, E. D.; Tennent, H. G. U.S. Patent 2 572 039 to Hercules Powder Company 1951; *Chem Abstract* **1954**, *46*, 1256.
3. Yokota, H. *Cell Chem. Technol.* **1986**, *20*, 487.

4. Khin, N. N.; Prokofeva, M. V.; Shairkov, Y. V.; Bogaslosku, V. E.; Golubev, V. M.; Andreev, Y. R. *Khim. Technol. Priozvood. Tsellyol.*; U.S.S.R., 1971; from *Zh. Khim. Abstract* **1972**, *No. 9*, 460.
5. Stratta, J. J. *Tappi* **1963**, *46*, 717.
6. Seneker, S. D. Ph.D. Thesis, North Dakota State University, 1986; p 190.
7. Glass, J. E.; Lowther, R. G. (Union Carbide Corp.) Ger. Offen. 2 751 411 (Cl. CO8B11/08), May 24, 1978, US Appl. 742, 885, November 18, 1976; p 45.
8. Glass, J. E.; Buthner, A. M.; Lowther, R. G.; Young, C. S.; Cosby, L. A. *Carbohyd. Res.* **1980**, *84*(2), 245.
9. DeMeber, J. R.; Taylor, L. D.; Trummer, S.; Rubin, L. E.; Chiklis, C. K. *J. Appl. Polym. Sci.* **1977**, *21*, 621.
10. Reuben, J.; Casti, T. E. *Carbohyd. Res.* **1987**, *163*, 91.
11. Claus, B.; Hartwig, R.; Kords, V.; Redener, C.; Detmor (Bayer, A-G, Wolff Walsrode A-G), Ger. Offen. DE3 044 696 (Cl. C0811/20), June 24, 1982; Appl. November 27, 1980; p 13.
12. Khin, N. N.; Egorova, E. V.; Rotenberg, I. M.; Petropavlvskii, G. A.; Larina, E. I.; Prokofeva, M. V. *Khim. Drev.* **1980**, *5*, 25.
13. *Encyclopedia of Polymer Science and Technology*; John Wiley & Sons: New York, 1965; Vol. 3, 469.
14. Battista, O. A.; Tasker, C. W.; Cornwell, R. T. K. *Svensk Papperstidn* **1958**, *61*, 272.

Hydroxypropyl Cellulose

See: Hydroxypropyl Cellulose (Crosslinked, Cholesteric Liquid Crystal)
Magnetic Field Processing

HYDROXYPROPYL CELLULOSE (Crosslinked, Cholesteric Liquid Crystal)

Shinichi Suto
Department of Materials Science and Engineering
Yamagata University

Hydroxypropyl cellulose (HPC) is a water-soluble, crystalline cellulose derivative. HPC was the first cellulose derivative discovered that formed a lyotropic cholesteric liquid crystal.[1] This discovery was the driving force behind the increasing interest in the chemistry, physics, and engineering of cellulosic liquid crystals.[2–5]

The cholesteric liquid-crystalline order (CLCO) of HPC cannot be used effectively as fiber but can be used in film applications.[6] Some researchers have tried to retain the cellulosic CLCO in the solid films.[6–9] HPC is a good model for preparing such films because it is soluble in many solvents, the texture is easily controlled, and it forms thermotropic liquid crystals. From an industrial viewpoint, however, the application of HPC films is less likely because of its tremendous solubility in many solvents, including water.

One way to overcome this drawback is to crosslink HPC films. Crosslinking widens the fields for HPC applications. Of great interest is retaining CLCO in the solid films through crosslinking. This film serves as a good model for studying the mechanical and swelling behavior of isotropic and anisotropic crosslinked polymeric materials. There is also the potential for gas or liquid separation from the film, drug delivery, and optical filters.[6,10,11]

PREPARATION

There are reportedly two kinds of crosslinking processes: gamma or UV irradiation[12–15] and chemical crosslinking.[10,16–27] Despite their advantages, gamma and UV irradiation need a chemical crosslinker and catalyzer. Therefore, the chemical crosslinking process offers more promising prospects for preparing crosslinked films or gels.

After the crosslinking conditions are determined, the type of solvent and the polymer concentration must be decided. The solvent type affects the efficiency of the crosslinking reaction.

The polymer concentration is the chief determining factor for preparing crosslinked films or gels retaining CLCO.

CONTROLLING PHASE

As noted earlier, HPC is a crystalline polymer. To distinguish the liquid-crystalline characteristics of crosslinked films, the film retaining no texture acts as a good basis for comparison. The liquid-crystalline gels with low crosslinking density can transform from an anisotropic to an isotropic phase as the temperature increases.[25] The phase transformation occurs reversely.

We suggest another method for preparing films with no texture: controlling the concentrations of polymer and crosslinker.[28]

CHARACTERISTICS

Swelling in Solvent

The most remarkable behavior of crosslinked films retaining CLCO was their preferential swelling in water or alcohols. The films swelled in the thickness direction, which was greater than width- or lengthwise swelling by a factor of 3, depending on the crosslinker type and concentration.[29] In films retaining CLCO, the molecules for each layer aligned parallel to the films surface, and therefore, the preferential swelling direction was perpendicular to the layers.

Another feature of the film's swelling behavior was its great dependence on time. This dependence fell into three categories: rapid swelling, gradual swelling, and equilibrium.[29]

Tensile Properties

There have been conflicting reports of the tensile properties of crosslinked films with CLCO, probably because the films were prepared under different conditions.[21,22,24,27]

Dynamic Mechanical Properties

The temperature dependence of strage modulus for crosslinked films cast from different solvents or crosslinkers systems was reported, and no correlation between the properties and cast conditions was observed at given temperatures.[16,19,21,22,24]

The glass transition temperature (T_g) was the most useful information obtained from studying the temperature dependence of dynamic mechanical properties. Rials et al. described the T_g as shifting to a higher temperature with greater crosslinker concentrations.[16]

Tensile Creep Behavior

The creep behavior provides not only information about the film's ability to be drawn, but also provides the Eyring-activated volume, which reflects the liquid-crystalline domain behavior during deformation.[21,30] The Eyring-activated volume decreased with increasing crosslinker concentration, and the crosslinking reduced the creep rate under constant loads.[30]

PROBLEMS TO SOLVE IN FUTURE

The molecules falling into order during the liquid-crystalline phase are somewhere between the amorphous and crystalline phases. Usually, crosslinking occurs in the amorphous and not in the crystalline phase. However, it is unclear whether the crosslinking occurs in the liquid-crystalline phase because nearly all cellulose derivatives are considered semirigid. Furthermore, it is unclear whether any interdomain crosslinking occurs because HPC films consist of polydomains.[31]

Thermotropic Liquid Crystals

HPC films form lyotropic and thermotropic crystals. At temperatures near the melting point, HPC exhibits CLCO. When an appropriate crosslinkers is added or gamma and UV rays are irradiated, thermotropic CLCO will be crosslinked. It will be interesting to compare the characteristics of crosslinked films retaining the lyotropic and thermotropic CLCO.

De-Crosslinking

HPC is eatable and nontoxic to humans. Thus, crosslinked films retaining CLCO could be used as biomembranes in human patients. Controlling de-crosslinking will need to be investigated further except for main-chain scissions and biodegradations by specific enzymes.

Gas Permselectivity

The permselectivity of oxygen and nitrogen gases for ethyl cellulose films improves when CLCO is retained in the films.[32] The same improvement was observed for HPC films by comparing their permeability data with amorphous films.[33] Crosslinking HPC films that retain CLCO may induce permselectivity because there are fewer disordered regions in the films. Accordingly, further improvement in selecting gases or liquids is effected by using crosslinked films retaining CLCO.

Self-Coloring Films

HPC liquid-crystalline solutions are self-coloring because of CLCO. This self-coloring is related to the selective reflection of circular polarized light in the visible portion of the spectrum. If the color in the solution can be retained in solid films, colored films will be obtained. Furthermore, if the cholesteric pitch can be controlled homogeneously, the hue, value, and chroma of the films can be change freely. Crosslinking may even allow the films to retain their color and prevent them from fading. Such an investigation is already in progress.[28]

CONCLUSIONS

We summarized the preparation, properties, and applications of crosslinked HPC solid films retaining CLCO. HPC must be investigated satisfactorily for biomaterial applications, and crosslinking blends with HPC may provide other interesting applications.

REFERENCES

1. Werbowyj, R. S.; Gray, D. G. *Mol. Cryst. Liq. Cryst. (Lett.)* **1976**, *34*, 97.
2. Gray, D. G. *J. Appl. Polym. Sci., Appl. Polym. Symp.* **1983**, *37*, 179.
3. Gilbert, R. D.; Patton, P. A. *Prog. Polym. Sci.* **1983**, *9*, 115.
4. Gray, D. G. *Faraday Discus. Chem. Soc.* **1985**, *79*, 257.
5. Gilbert, R. D. In Agricultural & Synthetic Polymers: Biodegraded Utilization; ACS Symposium *American Chemical Society*; Series 433; Washington, DC, 1990; p 259.
6. Suto, S.; Kudo, M.; Karasawa, M. *J. Appl. Polym. Sci.* **1986**, *31*, 1327.
7. Nishio, Y.; Yamane, T.; Takahashi, T. *J. Polym. Sci., Polym. Phys. Ed.* **1985**, *23*, 1053.
8. Onogi, Y. In *Cellulose: Structural and Functional Aspects*; Kennedy, J. F. et al., Ed.; Ellis Horwood: Chichester, 1989; Chapter 49.
9. Giasson, J.; Revol. J.-F.; Ritcey, A. M. et al. *Biopolymer* **1988**, *27*, 1999.
10. Harsh, D. C.; Gehrke, S. H. *J. Controlled Release* **1991**, *17*, 175.
11. Charlet, G.; Gray, D. G. *Macromolecules* **1987**, *20*, 33.
12. Bhadani, S. N.; Gray, D. G. *Mol. Cryst. Liq. Cryst. (Lett.)* **1984**, *102*, 255.
13. Giasson, J.; Revol. J.-F.; Gray, D. G. et al. *Macromolecules* **1991**, *24*, 1694.
14. Song, C. Q.; Litt, M. H.; Zloczower, I. M. *J. Appl. Polym. Sci.* **1991**, *42*, 2517.
15. Song, C. Q.; Litt, M. H.; Zloczower, I. M. *Macromolecules* **1992**, *25*, 2166.
16. Rials, T. G.; Glasser, W. G. *J. Appl. Polym. Sci.* **1988**, *36*, 749.
17. Suto, S. *J. Appl. Polym. Sci.* **1989**, *37*, 2781.
18. Suto, S.; Tashiro, H. *Polymer* **1989**, *30*, 2063.
19. Suto, S.; Tashiro, H.; Karasawa, M. *Sen-i Gakkaishi* **1990**, *46*, 56.
20. Suto, S.; Tashiro, H.; Karasawa, M. *J. Mater. Sci. Lett.* **1990**, *9*, 768.
21. Suto, S.; Tashiro, H.; Karasawa, M. *J. Appl. Polym. Sci.* **1992**, *45*, 1569.
22. Suto, S.; Yoshinaka, M. *J. Mater. Sci.* **1993**, *28*, 4644.
23. Suto, S.; Suzuki, K. *J. Appl. Polym. Sci.* **1995**, *55*, 139.
24. Yanagida, N.; Matsuo, M. *Polymer* **1992**, *33*, 996.
25. Mitchell, G. R.; Guo, W.; Davis, F. J. *Polymer* **1992**, *33*, 68.
26. Yang, Y.; Kloczkowski, A.; Mark, J. E. et al. *Colloid Polym. Sci.* **1994**, *272*, 284.
27. Mark, J. E.; Yang, Y.; Kloczkowski, A. et al. *Colloid Polym. Sci.* **1994**, *272*, 393.
28. Suto, S.; Takata, H. *Rept. Prog. Polym. Phys. Jpn.* **1994**, *37*, 243; Sen-i Gakkaishi **1995**, 51, 368.
29. Suto, S.; Suzuki, K. *Rept. Prog. Polym. Phys. Jpn.* **1992**, *35*, 283.
30. Suto, S.; Suzuki, K. *Rept. Prog. Polym. Phys. Jpn.* **1993**, *36*, 281.
31. Asada, T. *J. Soc. Rheol. Jpn.* **1982**, *10*, 51.
32. Suto, S.; Sugiura, T. *Rept. Prog. Polym. Phys. Jpn.* **1991**, *34*, 139.
33. Suto, S.; Kobayashi, T. *Rept. Prog. Polym. Phys. Jpn.* **1995**, *38*, 187.

HYPERBRANCHED ALIPHATIC POLYESTERS

A. Hult,* E. Malmström, and M. Johansson
Department of Polymer Technology
Royal Institute of Technology

A hyperbranched polymer resembles a dendritic, but linear defects, such as not fully branched repeating units, are allowed in the structure. Hyperbranched materials are believed to exhibit

*Author to whom correspondence should be addressed.

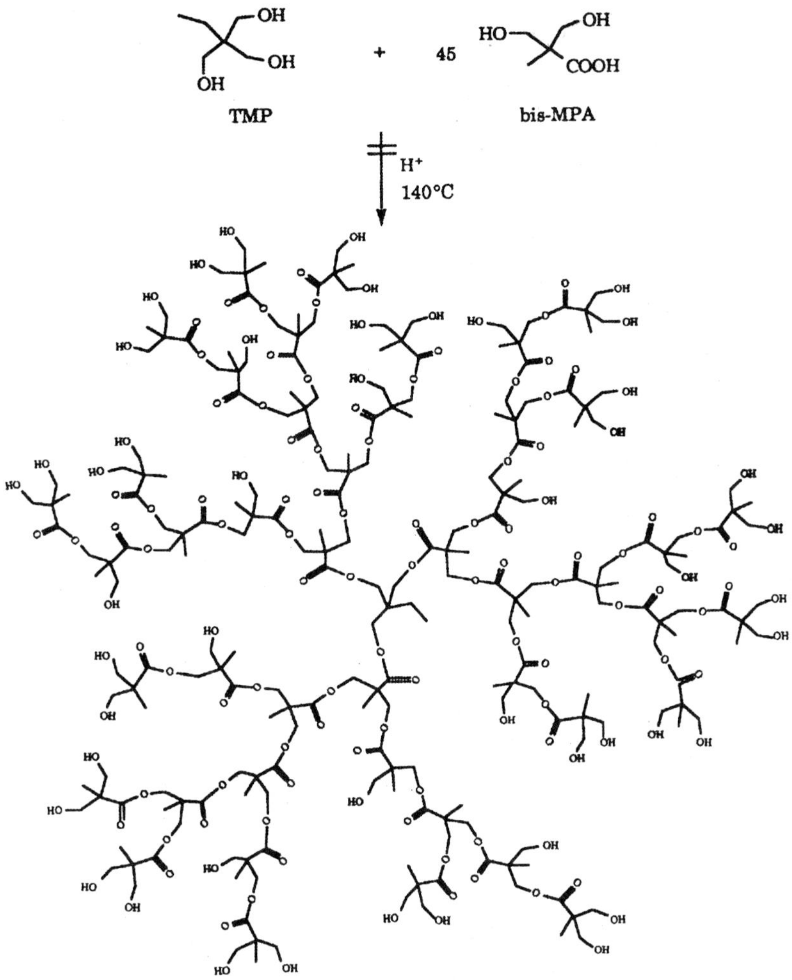

FIGURE 1. Idealized reaction scheme illustrating the synthesis of hyperbranched polyester no. 3 with a degree of branching around 80%.

properties resembling those of dendritic polymers[1] but can be produced more easily on a large scale at reasonable cost.

The architecture of these polymers is believed to reduce viscosity because they have few or no entanglements. Consequently, one possible application for hyperbranched polymers is as resins in coatings and lacquers, especially when VOCs must be reduced. Polyesters are often used in coatings, and therefore, hyperbranched polyesters are of interest.

A few papers concerning dendritic and hyperbranched polyesters have been published. Kwock et al.,[2] Miller et al.,[3] and Fréchet et al.[4] synthesized monodisperse aryl ester dendrimers, the latter group synthesizing hyperbranched aromatic polyesters via a one-step procedure.[5,6] A patent by Figuly describes the preparation of hyperbranched, functional polyesters by self-condensation of AB_x-type monomers, and in another patent, Baker et al. discuss the polycondensation of polyhydroxy monocarboxylic acids.[7,8] The hyperbranched polyesters whose pseudo one-step synthesis is described here have been evaluated for use as coatings when functionalized with allyl ether-groups.[9,10] These polyesters have also been studied for their relaxation processes with dielectric spectroscopy and thermal analysis.[11]

PREPARATION

Synthesis

Hyperbranched aliphatic polyesters of theoretically calculated molar masses 1200–44300 have been synthesized. The polyesters are based on 2,2-bis(hydroxymethyl)propionic acid (bis-MPA) for the AB_2-monomer and 2-ethyl-2-(hydroxymethyl)-1,3-propanediol (TMP) for the core moiety. When the reaction was performed without the core moiety, the resulting polymer was less soluble indicating the importance of the central core.

The pseudo one-step procedure involved adding the monomer in several portions, each one corresponding to the stoichiometric amount of the next theoretical generation. The purpose of this procedure was to increase the probability of an unreacted bis-MPA monomer reacting with a hydroxyl group on the hyperbranched skeleton instead of another free bis-MPA unit. This favored the growth of the hyperbranched skeleton instead of just growing dendrons. An idealized reaction scheme appears in **Figure 1**.

Some hyperbranched polyesters were prepared using 1,1,1-tris(4-hydroxyphenyl)ethane as the core moiety instead of TMP so that the core's effect on the polymer's properties could be studied. The first step in synthesizing these polyesters was a low-temperature esterification in which bis-MPA was attached to the core. Further polyester growth resulted from the pseudo one-step synthesis.

The synthetic procedure made it possible to synthesize hyperbranched polyesters on a large scale and from a commercial point of view, produced more interesting materials.[12]

End-Capping

The large number of hydroxyl groups on the terminal units provides numerous possibilities for tailoring the polymer to a certain application. The glass transition temperature for hyperbranched materials was shown to depend primarily on the polarity of the end groups;[13] the more polar the groups were the higher the glass transition was. The hyperbranched material can be made miscible or nonmiscible with another polymer simply by changing the polarity of the end groups. If non-reactive groups are attached to the surface, the polymers become thermoplastics. If the hydroxyl groups are end-capped with reactive groups, the resulting polymer could be crosslinked and used as a thermosetting resin.

PROPERTIES

The glass transition temperature, as determined by differential scanning calorimetry (DSC) for the hydroxy functional polyesters, showed only a slight increase with molar mass whereas the polarity of the terminal groups had greater effect. The T_g decreased from 35°C for terminal hydroxyl groups to 15°C for benzoates and –20°C for propionates. The thermal stability was examined with thermo gravimetric analysis (TGA) and was good for these polyesters. They lost about 3.5 wt% up to 340°C, the point at which thermal degradation starts.

Rheology

Rheological studies of hyperbranched polyesters with different end groups or with a different core molecule show that the most important structural part determining the properties is the polarity of the end groups.

APPLICATIONS

The hyperbranched polyesters described here have been used mostly for resin applications in coatings. Alkyd resins based on aliphatic hyperbranched polyesters exhibit a lower resin viscosity and simultaneously a reduced drying time[14] compared with conventional high-solid alkyds.[12] The polyesters have also been used for other types of resins such as unsaturated polyesters[10] and acrylates.[12]

REFERENCES

1. Chu, F.; Hawker, C. J. *Polym. Bull.* **1993**, *30*, 265.
2. Kwock, E. W.; Neenan, T. X.; Miller, T. M. *Chem. Mater.* **1991**, *3*, 775.
3. Miller, T. M.; Kwock, E. W.; Neenan, T. X. *Macromolecules* **1992**, *25*, 3143.
4. Hawker, C. J.; Fréchet, J. M. J. *J. Chem. Soc., Perkin Trans.* **1992**, *1*, 2459.
5. Hawker, C. J.; Fréchet, J. M. J. *J. Am. Chem. Soc.* **1991**, *113*, 4583.
6. Wooley, K. L.; Hawker, C. J.; Lee, R. et al. *J. Polymers* **1994**, *26*, 187.
7. Figuly, G. D. U.S. Patent 5 136 014, 1992.
8. Baker, A. S. U.S. Patent 3 669 939, 1972.
9. Malmström, E.; Johansson, M.; Hult, A. *Macromolecules* submitted for publication.
10. Johansson, M.; Malmström, E.; Hult, A. *J. Polym. Sci., Polym. Chem. Ed.* **1993**, *31*, 619.
11. Malmström, E.; Liu, F.; Boyd, R. H. et al. *Polym. Bull.* **1994**, *32*, 679.
12. Sörensen, K.; Johansson, M.; Malmström, E. et al. Swedish Patent 9200564–4, 1993.
13. Kim, Y. H.; Beckerbauer, R. *Macromolecules* **1994**, *27*, 1968.
14. Pettersson, B.; Sörensen, K. Proceedings of the *Waterborne, Higher-Solids, and Powder Coatings Symp.* 1994, 753–764.

HYPERBRANCHED POLYESTERS

Brigitte I. Voit* and S. Richard Turner
Technische Universität München
Lehrstuhl für Makromolekulare Stoffe

Hyperbranched polymers are macromolecules obtained in one-step polycondensation reactions of multifunctional A_xB monomers. Based on the high degree of branching, which results in a highly functional, globular-shaped macromolecules, hyperbranched polymers can be classified as dendritic macromolecules. However, in contrast to the well-defined, monomolecular, symmetrical perfect dendrimers, these hyperbranched structures are irregular and exhibit broad molecular weight distributions. The main reason for these differences is the method of synthesis.

The field of hyperbranched polymers is rapidly growing, paralleling progress in perfect dendritic polymers. Hyperbranched structures cover not only polyesters but a broad variety of condensation polymers.[1]

Here, we will discuss the synthesis, properties, and potential applications of hyperbranched polyesters, an intensively studied class of hyperbranched polymers.

PREPARATION

Hyperbranched polymers are prepared in a one-pot polycondensation reaction of A_xB monomers. The first examples of synthetic hyperbranched polymers were aromatic polyethers and aliphatic polyesters synthesized by a nucleophilic displacement reaction and by common melt polycondensation, respectively.[2,7] Since then, many hyperbranched structures have been reported in the literature: polyethers, polyphenylenes, poly(ether ketone)s, poly(ether sulfone)s, poly(siloxy silane)s, polyureas, LC polymers, metal-containing structures, and several polyesters.[3,4,6,8–19] Usually, A_2B monomers were used, and only in a few cases, A_3B monomers.[2,3] The polycondensation can be carried out in solution or in melt using the same catalysts and promotes that are used for linear polycondensates.

PROPERTIES

Like linear polymers, the properties of hyperbranched polyesters depend on the molecular weight and monomer structure. Rigid monomers produce polymers with higher T_g than flexible

*Author to whom correspondence should be addressed.

monomers and the solubility is reduced somewhat for high molecular weight samples. However, two other characteristics, the degree of branching and the nature of the terminal groups, are significantly responsible for the unique properties of hyperbranched polyesters and hyperbranched polymers in general.

The degree of branching can vary widely. Values from 15% to 90% have been reported for different types of hyperbranched polymers.[4,15,22,24] For most hyperbranched polyesters, the degree of branching ranges from 50% to 60%[14–22] but after the procedure described by Hult et al., hyperbranched polyesters could be synthesized so that a degree of branching as high as 80 or 90%[23–25] is possible.

The special branched structure of hyperbranched polymers affects their properties. First, hyperbranched polymers are highly soluble compared with their linear analogs. Second, the high solubility is combined with low solution viscosities and low Mark–Houwink constants, usually below 0.4.[15] This low value results from the dense, globular structure of the macromolecules. However, the molecular weight–viscosity dependence is linear and never reaches a maximum as observed for perfect dendrimers.[1,28,29]

The terminal groups and the globular shape influence not only the solubility but also the thermal behavior of hyperbranched polymers. Hyperbranched polymers show no crystallinity, even for structures with highly crystalline linear analogs. However, it has been possible to prepare hyperbranched polyethers and polyamides with liquid-crystalline behavior.[5,12] The thermal stability and especially glass transition temperatures depend heavily on the nature of the terminal groups.

The globular structure of hyperbranched polymers could be demonstrated by molecular modeling[27] and their solution viscosity behavior. Especially of interest is how the branching influences the internal mobility of the branched polymer backbone and the intermolecular interactions of the polymer molecules. The strong influence of the end groups on the glass transition temperature indicates that mostly end group mobility and end group interaction are observed at that point.

The results from dynamic mechanical measurements of hyperbranched polyesters with aliphatic and aromatic structures indicate no or few entanglements: no rubber plateau is observed.[19,23,26] This result agrees with the observation that all hyperbranched polymers are brittle compared with their linear analogs.

Hyperbranched polymers show lower solution and melt viscosities and enhanced solubility compared with linear structures. The polymers share several characteristics with perfect dendrimers: high solubility, low solution viscosities, large numbers of terminal functional groups, and a strong dependence of T_g on those groups. However, in contrast to the monodisperse dendrimers, hyperbranched polymers exhibit broad molecular weight distributions and imperfect branching, resulting from the nonselective but easy one-step synthetic process.

APPLICATIONS

As interest in perfect dendrimers has increased, the research of hyperbranched polymers in academic and industrial laboratories has also increased. Although dendrimers can be produced commercially,[28] the applications for perfect dendrimers will likely be limited to high value products such as pharmaceuticals because the step-wise synthetic process costs more. However, for hyperbranched polymers, the relative ease of synthesis and lower costs of the final polymers provides more opportunities to exploit their characteristics when perfect architecture and monodisperse molecular weights are not needed. For example, the high solubility, low solution viscosity, and high functionality suggest potential applications as functional oligomers in coatings or as carrier molecules for controlled release.

Acrylated hyperbranched polyesters were used in radiation-curable systems, and resins with high hardness and high glass transition temperatures were obtained.[30]

Another possible application for hyperbranched polymers may be as polymer modifiers to reduce melt viscosity or to improve the mechanical properties of linear polymers. Hyperbranched polyesters with functional polar groups were miscible with various linear polyesters and polyamides, and even in an immiscible system, such as aromatic hyperbranched polyester with linear polycarbonate, an increase of the tensile and compression modulus was observed.[31] However, these dense and stiff polyesters acted more or less as hard spheres, similar to fillers, and therefore, the samples became brittle at a high content of hyperbranched polymer. The brittleness of all hyperbranched polymers from the absence of entanglements will be a problem for some applications.

Hyperbranched polymers are a relatively new class of macromolecules, and the research in this area is accelerating. Clearly, considerable work is needed to characterize and evaluate the properties of these unique materials. Because of the ease of synthesis and the unusual properties, it is likely that hyperbranched polymers, especially hyperbranched polyesters, will find future commercial applications.

REFERENCES

1. Tomalia, D. A.; Durst, H. D. *Top. Curr. Chem.* **1993**, *165*, 194.
2. Hunter, W. H.; Woollett, G. H. *J. Am. Chem. Soc.* **1921**, *43*, 135.
3. Mathias, L. J.; Carothers, T. W. *J. Am. Chem. Soc.* **1991**, *113*, 4043.
4. Kim, Y. H.; Webster, O. W. *Macromolecules* **1992**, *25*, 5561.
5. Kim, Y. H. *J. Am. Chem. Soc.* **1992**, *114*, 4947.
6. Uhrich, K.; Hawker, C. J.; Fréchet, J. M. J.; Turner, S. R. *Macromolecules* **1992**, *25*, 4583.
7. Baker, A. S.; Walbridge, D. J. U.S. Patent 3 669 939; German Patent 2 129 994, 1972; *Chem. Abstract 76*, 128968.
8. Chu, F.; Hawker, C. J. *Polym. Bull.* **1993**, *30*, 265.
9. Miller, T. M.; Neenan, T. X.; Kwock, E. W.; Stein, S. M. *Macromol. Symp.* **1994**, *77*, 35.
10. Spindler, R.; Fréchet, J. M. J. *Macromolecules* **1993**, *26*, 1453.
11. Kumar, A.; Ramakrishnan, S. *J. Chem. Soc., Chem. Commun.* **1993**, 1453.
12. Percec, V.; Kawasumi, M. *Macromolecules* **1992**, *25*, 3843.
13. Bochkarev, M. N. *Organomet. Chem. (USSR)* **1988**, *1*, 115.
14. Hawker, C. J.; Lee, R.; Fréchet, J. M. J. *J. Am. Chem. Soc.* **1991**, *113*, 4583.
15. Turner, S. R.; Voit, B. I.; Mourey, T. H. *Macromolecules* **1993**, *26*, 4617.
16. Turner, S. R.; Walter, F.; Voit, B. I.; Mourey, T. H. *Macromolecules* **1994**, *27*, 1611.
17. Kricheldorf, H. R.; Stöber, O. *Macromol. Rapid Commun.* **1994**, *15*, 87.

18. Johansson, M.; Malmström, E.; Hult, A. *J. Polym. Sci. Part A: Polym. Chem.* **1993**, *31*, 619.
19. Connolly, M.; Wedler, E.; Ma, B. et al. Presented at the 35th IUPAC International Symposium on Macromolecules, Akron, OH, July 1994, Conference p 328.
20. Wooley, K. L.; Hawker, C. J.; Lee, R. et al. *Polym. J.* **1994**, *26*, 187.
21. Turner, S. R.; Voit, B. I. Eastman Kodak: Rochester, unpublished results.
22. Hawker, C. J.; Kambouris, P.; Chu, F. Presented at the 35th IUPAC International Symposium on Macromolecules, Akron, OH, July 1994, Conference p 324.
23. Johansson, M.; Hult, A.; Manson, J.-A. E. Presented at the 35th IUPAC International Symposium on Macromolecules, Akron, OH, July 1994, Conference p 333.
24. Malmström, E.; Johansson, M.; Hult, A. *Macromolecules* **1995**, *28*, 1698.
25. Malmström, E.; Liu, F.; Boyd, R. H. et al. *Polym. Bull.* **1994**, *32*, 679.
26. Voit, B. I.; Turner, S. R.; Colby, R. Eastman Kodak: Rochester, unpublished results.
27. Voit, B. I.; Shnidman, Y. Eastman Kodak: Rochester, unpublished results.
28. de Brabander-van den Berg, E. M. M.; Meijer, E. W. *Angew. Chem.* **1993**, *105*, 1370; *Int. Ed. (Engl.)* **1993**, *32*, 1308.
29. Mourey, T. H.; Turner, S. R.; Rubinstein, M. et al. *Macromolecules* **1992**, *25*, 2401.
30. Shi, W.; Rånby, B. J. *J. Appl. Polym. Sci.* **1996**, *59*, 1951.
31. Massa, D. J.; Shriner, K. A.; Turner, S. R.; Voit, B. I. *Macromolecules* **1995**, *28*, 3214.

Hyperbranched Polymers

See: *Highly Branched Polymers*
Hyperbranched Aliphatic Polyesters
Hyperbranched Polyesters

I

IGNITION TEMPERATURES

Masahide Wakakura
Kanagawa Industrial Research Institute

Kougaku Komamiya
Laboratory of Urban Safety Planning

The high energy of combustion and generation of toxic gases of plastics have caused serious damage in case of fire accident. In the investigation of fire and explosion, the importance of the ignition temperature is ascertainable. Ignition characteristics of several materials are measured by a variety of tests based on different properties.

High pressure thermal analysis (using DSC, DTA, DTA-TG) is the conventional technique used to measure the practical ignition temperature of organic compounds. The ignition temperature of coals, activated carbons, and aircraft engine oils has been measured using these apparatus.

We also measured the ignition temperature of polymers using a high pressure DTA-TG (HDTA-TG).

RESULT

In the case of plastics, thermal analysis under high-pressure air conditions provided not only the ignition temperature, but the sequence of combustion. Alkyl-based plastics (PP, PE, etc.) burned rapidly, but the burning rate of aromatic based plastics (PC, PET, etc.) was low or did not ignite. Halogenous plastics ignited after elimination of halogens or smokeless burning.

REFERENCES

1. Zabetakis, M. G.; Scott, G. S. Bureau of Mines, U.S. Dept of the Interior, RI6112 1965.
2. Ocrates, G. S. *Const. & Build. Mater.* **1988**, *2*, 131.
3. Wirtz, H. W.; Alfred, Z. *Thermchim. Acta* **1988**, *112*, 297.
4. Wakakura, M. J. *Japan Petrol. Inst.* **1981**, *24*, 385.

Immobilized Cells

See: *Cell Adhesion-Promoting Biomaterials*
Cell Entrapment [Poly(Carbamoyl sulfonate) Hydrogels]
Immobilized Microbial Cells
Microencapsulation (Artificial Cells)

Immobilized Enzymes

See: *Chitin (Metal Ion Chelation and Enzyme Immobilization)*
Enzyme Immobilization (Polyion Complexes)
Enzyme Immobilization (Porous Polymer Beads)
Functional Immobilized Biocatalysts (Stimulus-Sensitive Gels)
Glucose Responsive Polymers (Medical Applications)
Immobilized Enzymes, Biosensors

IMMOBILIZED ENZYMES, BIOSENSORS

M. Helena Gil* and A. P. Piedade
Department of Chemical Engineering
University of Coimbra

J. T. Guthrie
Department of Colour Chemistry
University of Leeds

In order to have a biosensor that has a high degree of specificity, good stability, good retention of the biological activity, good reproducibility, and good resistance to the contamination present in the sample, one has to extensively analyze the immobilization reaction of the biocatalyst and the support characteristics. In the present article, we will report some of the work that we have been doing in this area.

The major portion of the research and development work carried out to date on biosensor technology has been concerned with enzymes. Enzymes are biological catalysts that convert a specific substrate into a product without being consumed as a consequence of the reaction. Therefore, they are suitable for use in biosensors that function as continuous monitors of analyte concentration.

In the literature, potentiometric sensors are reported as being suitable for the preparation of biosensors. They have been used for the determination of urea, penicillin, creatinine, and aspartame among other substrates.[4–11] In addition to electrochemical transduction systems, there is now significant and increasing interest in optical biosensors based on immobilized enzymes. Fiber optical technology, when linked to a biological compound, has great potential value in sensor applications and is the subject of many studies.[12,13]

The rapid growth in the development of new enzyme electrode applications is related to advances in immobilization enzyme technology and the availability of a variety of transducers. The different methods of immobilizing enzymes, cells, and other biological compounds have been extensively reviewed in the literature.[14–16] The selection of the support depends on the potential use of the system and on the enzyme being immobilized.

We have been interested in the use of graft copolymers as supports for the binding of enzymes through covalent links. These supports offer several advantages, because they can be prepared in an enormous variety of forms, allowing the choice of the microenvironment for the enzyme on the support and the control of the hydrophobicity and mechanical properties of the copolymer. In addition it is possible to use a great variety of reactive groups through which it is possible to bind the biocatalysts.

In our studies of enzyme immobilization on graft copolymers, we have used nylon, cellulose, polyethylene, pectin, polytetrafluoroethylene, as substrates for grafting, and polyacrylic acid, poly(hydroxyethyl methacrylate), and poly(maleic anhydride/styrene) as grafted branches.[17–22] We have attempted to

*Author to whom correspondence should be addressed.

immobilize onto these supports the following enzymes: acid phosphatase, glucose oxidase, trypsin, α-chymotrypsin, lipases, and cholesterol oxidase.[1,3,20,23–26]

Based on these procedures, poly(ethylene-*g*-*co*-acrylic acid) and poly(tetrafluoroethylene-*g*-*co*-acrylic acid) membranes were prepared and used for the immobilization of catalase in order to make a single membrane hydrogen peroxide sensor.[2] This sensor was designed to function on the oxygen electrode principle with significantly improved characteristics with respect to the double membrane version in terms of sensitivity and response time.[27]

We also developed an electrochemical bienzyme sensor for free cholesterol.[3] Catalase and cholesterol oxidase were covalently immobilized onto poly(ethylene-*g*-*co*-acrylic acid). The single membrane containing the two enzymes was characterized and the activity and stability of the enzyme studied. This amperometric biosensor for the determination of free cholesterol showed good reproducibility.

More recently, we have been interested in the use of cellulose derivatives as the membranes for biosensors. We reported the use of these polymers in potentiometric sensors for urea determination and amperometric sensors for ascorbic acid analysis.[5,28]

CHARACTERIZATION OF POLYMERIC MEMBRANES

Poly(tetrafluoroethylene-g-co-acrylic acid)

Poly(tetrafluoroethylene) (Teflon™) was grafted with acrylic acid using a ^{60}Co source and the amount of grafted branches was quantified.[7] Catalase was covalently linked to the graft copolymer by using 1-cyclohexyl-3-(2-morpholinoethyl)carbodiimide metho-p-tolueno-sulfonate (CMC) as indicated earlier.[7]

The results indicated that the introduction of -COOH groups could provide a good environment for enzyme immobilization; the enzymatic activity can be affected by too low hydrophilicity; and the grafting process increases the wettability of the support surface, providing a better reaction to the support with the activating agent and the enzyme.

Cellulose Derivatives

The three cellulose derivatives in the study were cellulose acetate (CA), cellulose acetate butyrate (CAB), and ethyl cellulose (EC).

The activation procedure involved three steps: the production of aldehyde groups on the cellulose chain via periodate reaction, the linkage of an extensor arm of hexamethylene diamine to the produced aldehyde groups, and finally the activation of the amino groups with a bifunctional agent–glutaraldehyde. The urease immobilization occurred between the aldehyde groups of glutaraldehyde and the free amino groups of the enzyme.[28]

Hydrophilic/Hydrophobic Characteristics

We can assess the variations induced in the hydrophobic/hydrophilic characteristics of polymeric systems by the activation and immobilization procedures. In all cases the water wets the polymers with difficulty owing to their low value of critical wetting tension (γ_c) when compared with the surface tension of water (7.3N m-1). For all the supports the activation procedure provides a better surface for wetting and thereby a better chance to react with the enzyme aqueous solution.

These results are reinforced by the values of the sorption capacity of the polymeric membranes that show that the activation and immobilization procedures also increase the sorption capacity of the cellulose derivatives.

Zeta-Potential ($\Psi\xi$)

The value of this parameter can provide information about the surface charge of a polymeric membrane when in contact with a given aqueous solution.

CAB has a different behavior from the other two cellulose derivatives. This can be due to a low density of the surface charge in this polymer, which means that for low values of applied pressure the charge will be removed and the membranes will collapse between them during the determination.

Diffusion of NH_4^+ Ions Through the Polymeric Membranes

As already stated, this study of characterization of cellulose derivatives was carried out to establish the best polymer to be used in a urea biosensor. As the NH_4^+ is the cation that results from the degradation of urea by urease, and it is crucial for the biosensor function and performance that the cation will diffuse through the polymeric membrane and reach the inner part of the transducer, the study of this parameter is of great importance.

From all the results reported we can suggest that the behavior of the polymeric supports used in the enzyme immobilization for the applications in biosensors are greatly changed with the activation and immobilization procedures; the behavior of the immobilized biocatalyst is dependent on the physical and chemical characteristics of the microenvironment provided by the polymeric support; and the choice of a given polymeric support must be made after assessing all the possible characteristics of the support before and after the activation procedure, if it is the case.

APPLICATIONS

As already mentioned, the polymers characterized in the previous section were used, with very promising results, in the development of amperometric and potentiometric biosensors.

ACKNOWLEDGMENTS

The authors would like to thank Mr. A. Kazlauciunas for all the SEM analysis, and Mr. Ricardo Lagoa for his contribution in the literature research. We also thank JNICT-Portugal which partially supported this work.

REFERENCES

1. Alves da Silva, M.; Gil, M. H.; Redinha, J. S.; Brett, A. M.; Pereira, J. L. C. *J. Polym. Sci. Part A* **1991**, *29*, 275.
2. Alves de Silva, M.; Gil, M. H.; Piedade, A. P.; Redinha, J. S.; Brett, A. M.; Costa, J. M. *J. Polym. Sci. Part A* **1991**, *29*, 269.
3. Brett, A. M.; Gil, M H.; Piedade, A. P. *Bioelectrochemistry and Bioenergetics* **1992**, *28*, 105.
4. Hulanicki, A.; Glab, S.; Ingman, F. *Pure and Appl. Chem.* **1991**, *63*, 1247.
5. Gil, M. H.; Alegret, S.; Alves da Silva, M.; Alegria, A. C.; Piedade, A. P. *Cellulosics: Materials for Selective Separations and Other Technologies*; Polymer Science and Technology, 1993; 163.

6. Bergveld, P.; *IEEE Trans. Biomed. Eng.* **1970**, *17*, 70.
7. Guilbault, G. G.; Montalvo, J. G. *Nature* **1969**, *91*, 2164.
8. Nilsson, H.; Akerlund, A.; Mosbach, K. *Biochim. Biophys. Acta* **1973**, *320*, 529.
9. Tor, R.; Freeman, A. *Anal. Chem.* **1986**, *58*, 1042.
10. Guilbault, G. G.; Chen, S. P.; Kuan, S. S. *Anal. Lett.* **1980**, *13*, 1607.
11. Guilbault, G. G.; Lubano, G. J.; Kauffmann, J. M.; Patriache, G. J. *Anal. Chim. Acta* **1988**, *206*, 369.
12. Arnold, M. A. *Anal. Chem.* **1992**, *21*, 1015.
13. Wollfbeis, O. S.; Trettnak, W. *Biosensors–Applications in Medicine, Environmental Protection and Process Control*; Schmid, R. D.; Scheller, F., Eds., VCH: New York, 1989; 213.
14. Zaborsky, O. R. *Immobilized Enzymes* CRC: Cleveland, Ohio, US, 1973.
15. Chibata, I. *Immobilized Enzymes* Kodausha, Tokyo, John Wiley & Sons: London, New York, 1979.
16. Chaplin, M. F.; Bucke, C. *Enzyme Technology*; Cambridge University: 1990.
17. Abedel, Hay, F. I.; Beddows, C. G.; Gil, M. H.; Guthrie, J. T. *J. Polym. Sci., Polym. Chem. Ed.* **1983**, *21*, 2463.
18. Beddows, C. G.; Gil, M. H.; Guthrie, J. T. *Polym. Photochem.* **1986**, *7*, 213.
19. Beddows, C. G.; Gil, M. H.; Guthrie, J. T. *Appl. Polym. Sci.* **1988**, *35*, 135.
20. Beddows, C. G.; Gil, M. H.; Guthrie, J. T. *Polym. Bull.* **1984**, *11*, 1.
21. Beddows, C. G.; Gil, M. H.; Guthrie, J. T. *Biotechnol. Bioeng.* **1982**, *24*, 1371.
22. Beddows, C. G.; Gil, M. H.; Guthrie, J. T. *Biotechnol. Bioeng.* **1985**, *27*, 199.
23. Alves da Silva, M.; Gil, M. H.; Guiomar, J.; Lapa, E.; Machado, E.; Moreira, M.; Guthrie, J. T.; Kotov, S. *Radiat. Phys. Chem.* **1990**, *36*, 589.
24. Alves da Silva, M.; Gil, M. H.; Guthrie, J. T.; Guiomar, J. *J. Appl. Polym. Sci.* **1990**, *41*, 1629.
25. Ramos, M. C.; Gil, M. H.; Garcia, F. A.; Cabral, J. M.; Guthrie, J. T. *Biocatalyst* **1992**, *6*, 223.
26. Rocha, J. M.; Gil, M. H.; Garcia, F. A. P. *Biocatalyst* **1994**, *9*, 157.
27. Aizawa, M.; Karube, I.; Suzuki, S. *Anal. Chim. Acta* **1974**, *69*, 431.
28. Gil, M. H.; Piedade, A. P.; Alegret, S.; Alonson, J.; Martinez-Fabregas, E.; Orellana, A. *Biosensors and Bioelectronics* **1992**, *7*, 645.

Immobilized Growth Factors

See: Cell Growth-Promoting Biomaterials

IMMOBILIZED MICROBIAL CELLS

Nariyoshi Kawabata
Department of Chemistry and Materials Technology
Faculty of Engineering and Design
Kyoto Institute of Technology

Crosslinked poly(*N*-benzyl-4-vinylpyridinium halide) prepared in a pulverized form was found to show strong affinity for cells of microorganisms and capture them alive on the surface during contact with them.[1,2] Crosslinked copolymer of *N*-benzyl-4-vinylpyridinium halide with styrene and other vinyl monomers prepared in a bead form is an advanced material.[3] Non-woven cloth coated with a small amount of linear (not crosslinked) copolymer of *N*-benzyl-4-vinylpyridinium halide with styrene and other hydrophobic vinyl monomers is a further improved material.[4,5] These water-insoluble pyridinium-type polymers are used for immobilization of microbial cells on the surface, and for wide applications in the fields of biotechnology and water purification.[6] For example, the immobilized microbial cells are used for bioreactors and biosensors.[3,4,7,8] The procedure of immobilization is used for effective removal of microorganisms and viruses from water without addition of toxic chemicals to the treated water as well as the remainder of toxic chemicals in the treated water.[1,5,9,10] In addition, incorporation of the pyridinium group into a synthetic hydrophobic polymer makes the polymer biodegradable, probably due to the immobilization of microbial cells on the polymer surface exerted by the presence of the pyridinium group in the polymer chain, followed by acclimatization of the immobilized cells for the degradation of the synthetic hydrophobic polymer.[11]

On the other hand, in the case of linear (not crosslinked) and water-soluble poly(*N*-benzyl-4-vinylpyridium halide), the strong affinity for microbial cells appears as a strong antibacterial activity, and coagulation and sedimentation of microbial cells in aqueous suspension.[12,13]

PROPERTIES

The above described polymers containing *N*-benzyl-4-vinylpyridinium halide exhibit a strong affinity for microbial cells and viruses. Water-insoluble polymers capture microbial cells alive on the surface.[1,2] The capture is irreversible and equally effective both for viable cells and also for the cells killed by autoclaving.

Affinity of the polymer for microbial cells increases with content of the pyridinium group in the polymer chain.[2] Among the pyridinium groups examined, *N*-benzyl-4-vinylpyridinium chloride and bromide were most effective for the capture. Pyridinium groups containing a long alkyl group and other hydrophobic functional groups are much less effective for the capture. In addition, introduction of styrene and other hydrophobic monomer units into the polymer chain exerts a marked reduction of the affinity for microbial cells. On the other hand, enhancement of the character of swelling in water by decrease of the degree of crosslinking through reduction of the content of divinylbenzene in the polymer chain remarkably increases the affinity of the polymer for microbial cells. Thus, water appears to play an important role in the capture.

Affinity of the pyridinium-type polymer for microbial cells depends on the species of microorganisms, although the affinity does not appear to depend on the gram-positive or gram-negative character. The electrostatic interaction between positive charge of the polymer surface and negative charge of the cell surface plays an important role in the capture, but the interaction does not exhibit the definitive contribution.[14] The importance of other factors such as hydrophobic interaction between the polymer surface and the cell surface, as well as solvent-mediated and hydrodynamic forces, was suggested for the capture.

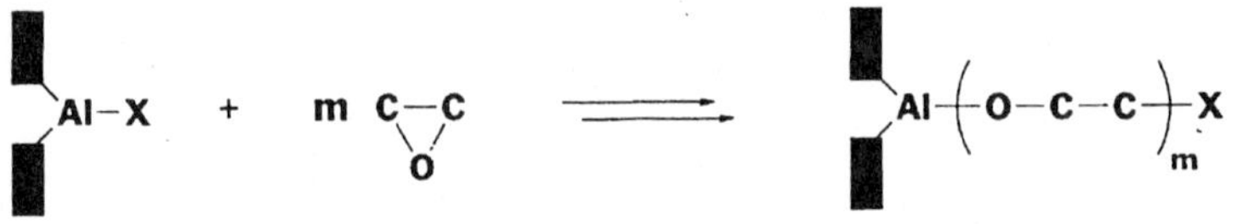

1 6 4_{alive} 1

Linear (not crosslinked) and water-soluble polymers exhibit strong bacteriocidal activity, and coagulation and sedimentation of microbial cells suspended in water.[12,13]

Application to Make Synthetic Polymers Biodegradable

Synthetic polymers, especially hydrophobic polymers with exclusively carbon–carbon bonds in the main chain, are highly resistant to microbial degradation or deterioration, although oligomers of these polymers exhibit biodegradability to some extent. However, incorporation of a small amount of *N*-benzyl-4-vinylpyridinium chloride into the polymer chain makes poly(methyl methacrylate) degradable when treated with activated sludge from the aeration tank of a domestic sewage works.[11] Biodegradation of the synthetic hydrophobic polymer appears to be facilitated by enhancement of the affinity for microbial cells. Immobilization of microbial cells on the polymer surface exerted by the pyridinium group would facilitate acclimatization of the immobilized cells for the degradation of the polymer.

REFERENCES

1. Kawabata, N.; Hayashi, T.; Matsumoto, T. *Appl. Environ. Microbiol.* **1983**, *46*, 203.
2. Kawabata, N.; Hayashi, T.; Nishikawa, M. *Bull. Chem. Soc. Jpn.* **1986**, *59*, 2861.
3. Kawabata, N.; Nishimura, S.; Yoshimura, T. *Biotechnol. Bioeng.* **1990a**, *35*, 1000.
4. Kawabata, N.; Nakagawa, K. *J. Ferment. Bioeng.* **1991**, *71*, 19.
5. Kawabata, N.; Inoue, T.; Tomita, H. *Epidemiol. Infect.* **1992**, *108*, 123.
6. Kawabata, N.; *Prog. Polym. Sci.* **1992**, *17*, 1.
7. Kawabata, N.; Nakamura, N.; Sato, M. *Sen-i Gakkai Symp. Biotechnol. Prepr.* **1986**, 21.
8. Kawabata, N.; Teramoto, K.; Ueda, T. *J. Microbiol. Methods* **1992**, *15*, 101.
9. Kawabata, N.; Hashizume, T.; Matsumoto, T. *Agric. Biol. Chem.* **1986**, *50*, 1551.
10. Kawabata, N.; Yamazaki, K.; Otake, T.; Oishi, I.; Minekawa, Y. *Epidemiol. Infect.* **1990**, *105*, 633.
11. Kawabata, N.; Uchihori, D.; Fukuda, S.; Funahashi, H. *J. Appl. Polym. Sci.* **1994**, *51*, 33.
12. Kawabata, N.; Nishiguchi, M. *Appl. Environ. Microbiol.* **1988**, *54*, 2532.
13. Kawabata, N.; Takagishi, K.; Nishiguchi, M. *Reactive Polym.* **1989**, *10*, 269.
14. Kawabata, N.; Ueno, Y.; Torii, K.; Matsumoto, T. *Agric. Biol. Chem.* **1987**, *51*, 1085.

IMMORTAL POLYMERIZATION

Shohei Inoue
Department of Industrial Chemistry
Faculty of Engineering
Science University of Tokyo

Takuzo Aida
Department of Chemistry and Biotechnology
Faculty of Engineering
The University of Tokyo

PRINCIPLE OF IMMORTAL POLYMERIZATION

"Immortal" polymerization is the catalytic version of "living" polymerization, where a narrow molecular weight distribution (MWD) polymer is formed when the number of the polymer molecules exceeds that of the initiator molecules.[1] The principle of immortal polymerization has been discovered in the course of our studies on polymerization of epoxides (**6**) initiated with metalloporphyrins.[2] Aluminum porphyrins (**1**) initiate the living polymerization of epoxides via a (porphinato)aluminum alcoholate ($\mathbf{4}_{alive}$) as the growing species (**Equation 1**).[3] Polymerization of epoxides (**6**) initiated with aluminum porphyrins (**1**) is not terminated by protic compounds, and polymerization with this character is known as immortal. Its immortal character is due to the unusual reactivity of the aluminum atom–axial ligand bond in **1**.

Polymerization involving chain transfer reactions usually gives a broad MWD polymer. On the other hand, immortal polymerization enables formation of a narrow MWD polymer, owing to the much more rapid exchange reaction than the chain growth. Thus, immortal polymerization can be regarded as the catalytic version of living polymerization in the sense that a narrow MWD polymer is formed with the number of polymer molecules exceeding that of initiator molecules.

MONOMERS, INITIATORS, AND CHAIN TRANSFER AGENTS FOR IMMORTAL POLYMERIZATION

The monomers confirmed to undergo living polymerization by metalloporphyrin initiators are listed in **Table 1**.[1a]

Selection of chain transfer agents for immortal polymerization is important. For immortal polymerization of **6** with **1**, a wide variety of protic compounds, such as alcohol, water, phenol, carboxylic acid, and hydrogen chloride, are usable as chain transfer agents[4] since the aluminum porphyrins (**1**) formed by the reaction of the alcoholate growing species (**4**) with these protic compounds all possess enough nucleophilic reactivity to reinitiate polymerization. On the contrary, in polymerization of lactones (**10–12**), the situation is different depending on the size of the lactone ring.

TABLE 1. Monomers for Living Polymerization with Metalloporphyrins

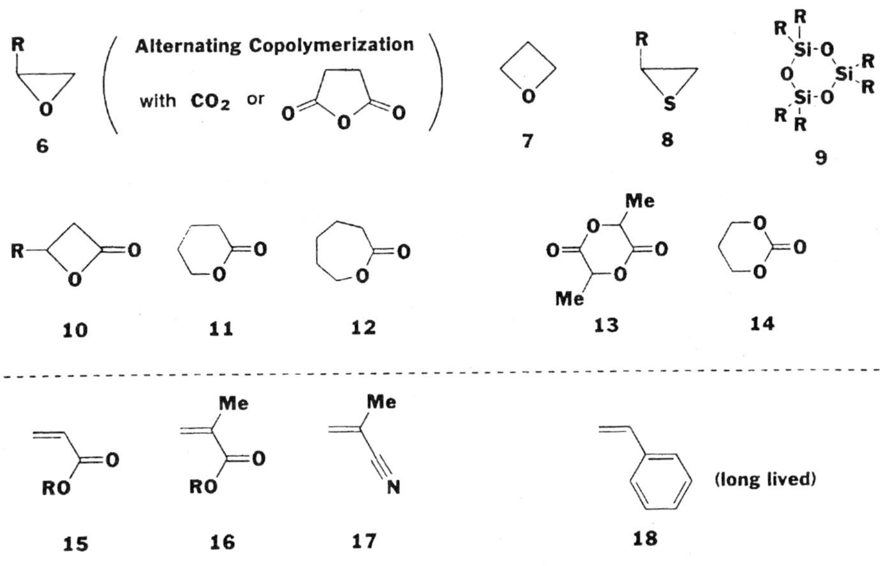

APPLICATION OF IMMORTAL POLYMERIZATION

In immortal polymerization, a group originating from the protic chain transfer agent is quantitatively introduced into the polymer terminal. By taking advantage of this, a variety of telechelic polymers and oligomers with narrow MWD can be synthesized. Use of unsaturated protic chain transfer agents such as acrylic acid, methacrylic acid, and 2-hydroxyethyl methacrylate for immortal polymerization leads to narrow MWD polymers carrying (meth)acrylate and styryl end groups, which are usable as macromonomers.[6] Use of benzyl alcohol as a chain transfer agent for immortal polymerization of 1,2-epoxypropane (**6**, R = Me) results in formation of a polyether having a $PhCH_2$-O moiety. Multifunctional protic chain transfer agents such as diols and triols enable synthesis of telechelic and star-shaped polymers.

Immortal polymerization also can be applied to the tailored synthesis of block copolymers.[5]

In summary, immortal polymerization is an attractive method for controlled polymer synthesis from fundamental and practical viewpoints. In this regard, an aluminum porphyrin bound to crosslinked polystyrene has been developed as an initiator for immortal polymerization of epoxides, where the initiator is easily separable from the polymerization mixture and can be used repeatedly without loss of activity.[7]

REFERENCES

1. (a) Aida, T. *Prog. Polym. Sci.* **1994**, *19*, 469. (b) Inoue, S.; Aida, T. Controlled Polymer Synthesis with Metalloporphyrins In: *Molecular Design of Polymeric Materials*; Vogl, O.; Hatada, O and K. Eds., Dekker: New York, in press.
2. Asano, S.; Aida, T.; Inoue, S. *J. Chem. Soc., Chem. Commun.* **1985**, 1148.
3. Aida, T.; Inoue, S. *Macromolecules* **1981**, *14*, 1166.
4. Aida, T.; Maekawa, Y.; Asano, S.; Inoue, S. *Macromolecules* **1988**, *21*, 1195.
5. Endo, M.; Aida, T.; Inoue, S. *Macromolecules* **1987**, *20*, 2982.
6. Inoue, S.; Aida, T. *Makromol. Chem., Macromol. Symp.* **1986**, *6*, 217.
7. Uno, H.; Tanaka, K.; Mizutani, Y. *Reactive Polymers* **1991**, *15*, 121.

Immunologically Active Polymers

See: Polyelectrolyte Complexes (in Immunology)

Impact Modifiers

See: Additives (Property and Processing Modifiers)
Polypropylene Blends and Composites

In Situ Composites

See: In Situ Thermoplastic Composites (from Liquid Crystalline Polymers)
Poly(ester imide) Compatibilizer (in Thermotropic Liquid Crystalline Polymer Blends)

IN SITU THERMOPLASTIC COMPOSITES (from Liquid Crystalline Polymers)

Donald G. Baird
Department of Chemical Engineers
Virginia Polytechnic Institute and State University

The general method by which composites based on thermotropic liquid crystalline polymers (TLCPs) and thermoplastics are generated is to subject the blend to an extensional flow during processing. The TLCP drops are deformed into fibrils, which are then oriented and frozen to lock in the morphology and orientation. Because the fibril reinforcement is generated during processing, they are referred to as "*in situ* composites."

The most frequently used processes for generating *in situ* composites are injection molding, fiber spinning, strand extrusion and drawing and film drawing.

In situ composites have been produced by using a wide variety of materials and several different processing methods. One of the first studies of *in situ* composites was published by Kiss.[1] In this study a thermotropic copolyester and a copoly(ester amide) were blended with various thermoplastics, including polycarbonate (PC), poly(butylene terephthalate) (PBT), polyarylate (PAR), poly(ether imide) (PEI), and poly(ether ether ketone) (PEEK). Since this work, subsequent studies have been performed with blends of polypropylene (PP),[2,3] poly(ethylene terephthalate) (PET),[4,5] polystyrene (PS),[6,7] PC,[1,8] PEI,[8,9] and various TLCPs. The addition of a TLCP to a thermoplastic generally results in an increase in tensile properties. Additionally, the highest properties appear to be generated in fiber spinning and strand extrusion.

The properties that can be achieved in an *in situ* composite are dependent on the processing technique. For example, drawn strands and fibers are seen to show a greater improvement in tensile properties than injection-molded parts of the same TLCP composition. To understand why this occurs, one must look to the way that the morphology is developed in each of these processes.

To begin to understand how morphology is developed in a two-phase system, it is necessary to look to studies performed specifically on the behavior of a dispersed phase in a liquid medium. Many studies of this type have been performed on both Newtonian and non-Newtonian droplet/medium systems.[12,13] These studies have shown that the droplets deform over a greater range of viscosity ratios in elongational flow fields than in shear flow fields. This is an important result in terms of *in situ* composites because it provides an idea of what is necessary to create the desired fibrillar reinforcement in TLCP/thermoplastic blends.

In a study by Blizard and Baird,[14] the development of morphology in capillary dies and in shear fields was studied for nylon 6/6 and PC blends with 60% HBA/PET.

Blizard et al.[8] attempted to provide a means of predicting the relationship between component rheology and processing conditions on subsequent blend morphology.

Another means of producing *in situ* composite strands is the dual extruder mixing method. This method may be unique in that it is the only method that does not rely on droplet deformation and breakup to form TLCP fibrils.[15] Due to the nature of the static mixer used to combine the matrix and TLCP phase, the fibrils are formed in the matrix by repeated splitting of the component streams. As a result of this method of fibril formation, no skin-core morphology is found in the composite strands. An unoriented core can prevent *in situ* composites from achieving their maximum properties. Additionally, it is also the only method by which continuous TLCP fibrils can be created in the composite at any composition of TLCP.

PROPERTIES

The major issue of concern at this point is the potential ability of TLCPs to reinforce thermoplastics. In the case of *in situ* composites, the reinforcing fiber can be one of many TLCPs available commercially. Additionally, the modulus of the neat TLCPs is dependent on the processing technique used to produce the sample.

To obtain the maximum modulus of *in situ* composites, it is necessary to achieve the maximum modulus in the TLCP fibrils.

As the properties of *in situ* composites are highly dependent on the properties of the TLCP fibrils, the upper limit of the *in situ* composite properties should be reflected in the as-spun TLCP fiber properties.

It was determined that the TLCP containing the highest level of HBA, and, therefore, the highest molecular stiffness, had the highest orientation and modulus.

Several authors[10,11,17,18] have studied the compression molding of drawn *in situ* composite strands, fibers, and films. It is believed that by using materials containing high-modulus TLCP fibrillar reinforcement in compression molding the fiber properties can be maintained, and the resulting compression-molded part will be less anisotropic because of lay-up options during the consolidation step. Bassett and Yee[11] attempted to use *in situ* composite fibers to produce compression composite strands, fibers, and films, It is believed that by using materials containing high-modulus TLCP fibrillar reinforcement in compression molding the fiber properties can be maintained, and the resulting compression-molded part will be less anisotropic because of lay-up options during the consolidation step.

Handlos and Baird[17] performed a similar study using the dual-extruder mixing method to produce *in situ* composite strands for compression molding experiments. The advantage of using this method over the method used by Bassett and Yee[10] is that strands made with the dual-extruder mixing system have no skin-core morphology and contain TLCP fibrils of infinite aspect ratio which can be generated at any TLCP level, and the improved control over the heat transfer in the system can be used to eliminate freezing of the TLCP during drawing. Furthermore, there is no limit in the amount of TLCP that can be used.

Pre-extruded thermoplastic/TLCP sheets, strands, and fibers can be used to produce composite materials with biaxial properties. Clearly, the properties of the compression-molded composites are dependent on the properties of the sheets, strands, and fibers used to make the composite. If the properties of the pregenerated materials can be maximized, the properties of the compression-molded materials can also be maximized. It is apparent that one of the best ways to obtain the optimum reinforcing effect from TLCPs is to start with highly drawn composite fibers or strands.

REFERENCES

1. Kiss, G. *Polym. Eng. Sci.* **1987**, *29*, 410.
2. Datta, A.; Chen, H. H.; Baird, D. G. *Polymer* **1993**, *34*, 759.
3. Chapleau, N.; Carrecu, P. J.; Peleteiro, C.; Lavoie, P. A.; Malik, T. M. *Polym. Eng. Sci.* **1992**, *32*, 1876.
4. Seppala, J.; Heino, M.; Kapanen, C. *J. Appl. Polym. Sci.* **1992**, *44*, 1051.
5. Joseph, E. G.; Wilkes, G. L.; Baird, D. G. In *Polymer Liquid Crystals* Blumstein, A., Ed.; Plenum: New York, 1985.
6. Zhuang, P.; Kyu, T.; White, J. L. *Polym. Eng. Sci.* **1988**, *28*, 1095.
7. Crevecoeur, G.; Groeninckx, G. *Polym. Eng. Sci.* **1990**, *30*, 532.
8. Blizard, K. G.; Federick, C.; Federico, O.; Chapoy, L. L. *Polym. Eng. Sci.* **1990**, *30*, 1442.

9. Carfagna, C.; Amendoc, E.; Nicolais, L.; Acieno, D.; Rancescangeli, O.; Yang, B.; Rustichelli, P. *J. Appl. Polym. Sci.* **1991**, *43*, 839.
10. Bassett, B. R.; Yee, A. F. *Polym. Composites* **1990**, *11*, 10.
11. Crevecoeur, G. Ph.D. Thesis, Katholieke Universiteit Leuven, 1991.
12. Bentley, B. J.; Leal, L. G. *J. Fluid Mech.* **1986**, *167*, 241.
13. Utracki, L. A.; Shi, Z. H. *Polym. Eng. Sci.* **1992**, *32*, 1824.
14. Blizard, K. G.; Baird, D. G. *Polym. Eng. Sci.* **1987**, *27*, 653.
15. Baird, D. G.; Bafna, S. S.; de Souza, J. P.; Sun, T. *Polym. Comp.* **1993**, *14*, 214.
16. Bafna, S. S.; de Souza, J. P.; Sun, T.; Baird, D. G. *Polym. Eng. Sci.* **1993**, *33*, 808.
17. Handlos, A. A.; Baird, D. G. *SPE Tech. Pap. (ANTEC 93)* **1993**, 1170.
18. Handlos, A. A. Ph.D. Thesis, Virginia Polytechnic Institute, 1994.

Inclusion Complexes

See: *Cyclodextrin-Polymer Inclusion Complexes*
Cyclodextrins (Host-Guest Interactions)
Enzyme Mimics (Cyclodextrins)
Host-Guest Chromatography (with Cyclodextrin Derivatives)
Inclusion Complexes (Overview)
Inclusion Polymerization
Macromolecular Recognition (New Supramolecular Structures)
Molecular Complexes [Poly(ethylene oxide) and Urea]
Poly(vinyl alcohol)-Iodine Interactions

INCLUSION COMPLEXES (Overview)

Kazuaki Suehiro
Department of Applied Chemistry
Faculty of Science and Engineering
Saga University

Inclusion complexes are molecular complexes in which atoms or molecules (the "guests") are confined in cavities constructed by other chemical species (the "hosts") in a definite proportion. Polymer inclusion complexes can be classified into three categories: the guest is a polymer, the host is a polymer, and both are polymers. However, little is known about the last case. There are three types of host compounds: macrocyclic monomolecular hosts having holes in the center that can include guest compounds (e.g., cyclodextrins), polymolecular hosts that make a cavity due to the assembling of a number of host molecules (e.g., urea), and macromolecular hosts.

The shape of the cavity built up by host compounds is usually cage- or channel-like in the crystalline state. However, a polymer guest cannot be accommodated within a cage-like cavity in terms of size, and so all inclusion complexes with a polymer guest have only a channel structure.

INCLUSION COMPLEXES OF UREA AND THIOUREA

Polyesters

Urea is well known to form inclusion complexes, so-called adducts, with various low molecular weight linear compounds such as *n*-alkanes, and is perhaps the host compound that was first found to form inclusion complexes with polymers. The adducts have the same crystal structures. The unit cell is hexagonal and contains six urea molecules.[3] The hydrogen-bonded host urea molecules form honeycomb-like channels with cavities of about 5 Å in diameter. The guest molecules are located within the hexagonal channels with their long axis parallel to, but statistically around the *c* axis.

Polyethylene

An inclusion complex of urea with polyethylene, which can be regarded as a high molecular weight homologue of *n*-alkanes, was prepared for the first time by substitution of polyethylene for *n*-alkane by heating a urea–*n*-alkane complex in a xylene solution of polyethylene.[1] This complex also has hexagonal unit cell dimensions similar to those of urea–*n*-alkane complexes.

Polyethers

Prior to these investigations, compounds containing a surface-active agent of the polyoxyethylene (POE) type which are fluid were converted to a solid by adding urea and holding the mixture at room temperature.[4] The products were excellent detergents. Later, Parrod and Kohler[5] and Bailey and France[6] revealed that POE forms a crystalline complex with urea or thiourea. The X-ray diffraction patterns of the urea-POE complex are different from those of the urea–*n*-alkane, and a trigonal unit cell was proposed.[7]

Polyethers $(CH_2)_m$–O–$]_n$ ($m \geq 3$), also form inclusion complexes with urea.[8] The number of urea molecules complexing with a monomeric unit of polyethers changes linearly with the number of CH_2 groups in the monomeric unit. These complexes have a hexagonal unit cell very similar to the urea–*n*-alkane complexes. Therefore, it could be assumed that the polyether chains take on a planar zigzag conformation.

Other Polymers

Inclusion complexes of some diene polymers have been prepared through inclusion polymerization–that is, solid-state polymerization of inclusion complexes of corresponding monomers by irradiation with γ-rays.

Complexes of poly(2,3-dichlorobutadiene) and poly(2,3-dimethylbutadiene) with thiourea have also been prepared by radiation polymerization of their monomer complexes.

INCLUSION COMPLEXES OF DEOXYCHOLIC ACID

Deoxycholic acid (3α, 12α-dihydroxy-5β-cholan-24-oic acid, DCA) is a natural product present in the bile of vertebrates in the form of conjugated bile acid. DCA has the ability to form inclusion complexes with a wide variety of small organic compounds.[9] DCA was found to form complexes with many unsaturated monomers.[29]

Later, inclusion polymerization was also attained in the channels of apocholic acid (ACA), which is closely related to DCA.[11]

Direct preparation of polymer inclusion complexes of DCA was examined with various polyethers by slow evaporation from ethanol solution.[12,13] Inclusion complexes were obtained with polyethers $[-(CH_2)_{m-1}CHR–O–]_n$ (R = H, $m \geq 2$; and $R = CH_3$, $m = 2$), poly(1,3-dioxolane) and several POE derivatives. All crystals were orthorhombic, similar to those most commonly observed for the DCA complexes with small molecules.

INCLUSION COMPLEXES OF PERHYDROTRIPHENYLENE

Polyethylene, *cis*- and *trans*-1,4-polybutadiene, and POE were found to give rise to very stable inclusion complexes with the *trans*, anti-*trans*, anti-*trans* isomer of perhydrotriphenylene (PHTP), $C_{18}H_{30}$, either by crystallization from a solution or by melt-cooling of a mixture of the two components.[14] In several cases, inclusion complexes of polymers were obtained through canal polymerization of the monomer inclusion complexes.[5] A high degree of stereoregularity was observed for some of the resultant polymers.

INCLUSION COMPLEXES WITH SPIROCYCLOTRIPHOSPHAZENES

Tris(*o*-phenylenedioxy)cyclotriphosphazene forms inclusion complexes with a variety of molecules. Butadiene, vinyl chloride, isobutylene, divinylbenzene, 4-bromostyrene, and acrylic monomers (acrylic acid, acrylic anhydride, acrylonitrile, methyl methacrylate, methyl vinyl ketone) were polymerized in the channel of the complex by γ-irradiation.[41–43] Tris(2,3-naphthalenedioxy)cyclotriphosphazene was also used as a host compound for styrene and 4-bromostyrene. However, little attention has been devoted to the resultant polymer inclusion complexes with these host compounds.

INCLUSION COMPLEXES OF CYCLODEXTRINS

CDs can form monomolecular inclusion complexes with a great number of low molecular weight compounds of appropriate sizes in water as well as in the solid state.

Inclusion polyamines prepared by treating a diamine with β-CD have been reacted with acid chlorides to give the inclusion polyamides.[19] Polymer inclusion complexes of CDs were also prepared by inclusion polymerization of vinylidene chloride within β-CD by radiation or a radical initiator.[20,21] Inclusion copolymerization of a disubstituted butadiene system was also studied in the channel of β-CD.[22] Threading CD rings on polymer chains results in polyrotaxane compounds.

α-CD was found to form crystalline complexes with poly(ethylene glycol) (PEG)–i.e., POE.[23,24]

INCLUSION COMPLEXES OF AMYLOSE

Amylose, a linear polysaccharide, is the principal constituent of starch together with branched amylopectin. The iodine coloration reaction, which has long been used to detect starch, is based on the inclusion of iodine molecules within the helical amylose chain. The iodine–starch reaction is also observed in solution. It is considered that the amylose chain retains its helical structure even in solution and that a monomolecular inclusion complex is formed with iodine molecules. Amylose can form crystalline inclusion complexes with a variety of small organic molecules such as alcohols, fatty acids, ketones, aldehydes, phenols, nitro compounds, hydrocarbons, and lysolecithin, in addition to iodine.

Structures of several amylose complexes were reviewed by Sarko and Zugenmauer.[25]

REFERENCES

1. Monobe, K.; Yokoyama, F. *J. Macromol. Sci., Phys.* **1973**, *B8*, 277.
2. Farina, M.; Natta, G.; Allegra, G.; Loffelholz, M. *J. Polym. Sci., Part C* **1967**, *16*, 2517.
3. Smith, A. E. *Acta Crystallogr.* **1952**, *5*, 224.
4. Barker, G. E.; Ranauto, H. J. *J. Am. Oil Chem. Soc.* **1955**, *32*, 249.
5. Parrod, J.; Kohler, A. *C. R. Hebd. Seances Acad. Sci.* **1958**, *246*, 1046.
6. Bailey, F. E., Jr.; France, H. G. *J. Polym. Sci.* **1961**, *49*, 397.
7. Tadokoro, H.; Yoshihara, T.; Chatani, Y.; Murahashi, S. *J. Polym. Sci., Part B* **1964**, *2*, 363.
8. Suehiro, K.; Urabe, A.; Yoshitake, Y.; Nagano, Y. *Makromol. Chem.* **1984**, *185*, 2467.
9. Herndon, W. C. *J. Chem. Educ.* **1967**, *44*, 724.
10. Miyata, M.; Takemoto, K. *J. Polym. Sci., Polym. Lett. Ed.* **1975**, *13*, 221.
11. Miyata, M.; Kitahara, Y.; Takemoto, K. *Polym. Bull.* **1980**, *2*, 671.
12. Suehiro, K. *J. Incl. Phenom.* **1988**, *6*, 9.
13. Suehiro, K.; Kuaramori, M. *J. Macromol. Sci., Phys.* **1994**, *B33*, 1.
14. Farina, M.; Allegra, G.; Natta, G. *J. Am. Chem. Soc.* **1964**, *86*, 516.
15. Finter, J.; Wegner, G. *Makromol. Chem.* **1979**, *180*, 1093.
16. Allcock, H. R.; Ferrar, W. T.; Levin, M. L. *Macromolecules* **1982**, *15*, 697.
17. Allcock, H. R.; Levin, M. L. *Macromolecules* **1982**, *18*, 1324.
18. Saenger, W. *Angew. Chem. Int. Ed. Engl.* **1980**, *19*, 344.
19. Ogata, N.; Sanui, K.; Wada, J. *J. Polym. Sci., Polym. Lett. Ed.* **1976**, *14*, 459.
20. Maciejewski, M. *J. Macromol. Sci., Chem.* **1979**, *A13*, 77.
21. Maciejewski, M.; Gwizdowski, A.; Peczak, P.; Pietrzak, A. *J. Macromol. Sci., Chem.* **1979**, *A13*, 87.
22. Schneider, C.; Greve, H. H.; Bohlman, H. P.; Rehbold, B. *Angew. Makromol. Chem.* **1986**, *145/146*, 19.
23. Suehiro, K. Presented at the Joint Symposium of The Chugoku-Shikoku and Kyush Branches of the Chemical Society of Japan, Tokushima, November 1986, paper 138.
24. Harada, A.; Kamachi, M. *Macromolecules* **1990**, *23*, 2821.
25. Sarko, A.; Zugenmaier, P. In *Fiber Diffraction Methods*; American Chemical Society: Washington, DC, 1980; Chapter XXVIII.

INCLUSION POLYMERIZATION

Mikiji Miyata
Department of Material and Life Science
Faculty of Engineering
Osaka University

Inclusion polymerization proceeds in molecular-level spaces called channels and interlayers, which inclusion compounds provide for monomers. These spaces play a decisive role in determining reaction behaviors of the monomers. Thus, the polymerization serves as a space-dependent polymerization on a molecular-level, enabling us to observe a real feature of low-dimensional polymerizations. This is because the monomers form any low-dimensional molecular assemblies in the low-dimensional spaces. Accordingly, the polymerization presents a strong method for preparing low-dimensional materials.

Historically speaking, studies on inclusion polymerization started in an application to stereoregular polymerizations. Since steric conditions are advantageous for polyconjugated monomers but not for vinyl monomers, the polymerization produced stereoregular diene polymers that could not be prepared by other methods. For example, the polymers were completely 1,4-*trans* tactic and in some cases optically active, leading to an application to asymmetric polymerization.[1–7]

Development of boundary researches, particularly host-quest chemistry, called attention to a basic aspect of the polymerization. Introduction of a general concept, molecular-level spaces, forced us to change the conventional view that inclusion polymerization is a special way to prepare stereoregular polymers. The proposed replacement for that view is that features of inclusion polymerization are intermediate between those of bulk or solution polymerization and solid-state polymerization. The key is that space effects are negligible in conventional polymerizations but not in inclusion polymerization.[8,9]

POLYMERIZATION PROCESSES

Polymerization Procedure

Inclusion polymerization consists of three steps: (1) formation of the inclusion compounds with the monomers; (2) polymerization; and (3) separation of the resulting polymers from the hosts, if necessary.

The inclusion compounds are formed by two methods, intercalation and recrystallization. The first method is widely applicable for preparing the inclusion compounds of inorganic hosts, because some inorganic polymers spontaneously absorb the guest components to yield the compounds. Similarly, some organic, layered assemblies can absorb the monomer molecules spontaneously to yield the compounds. Other organic assemblies are formed by the second method, recrystallization.

The polymerization consists of the following elementary reactions: an initiation, a propagation, a termination, and a chain-transfer reaction, as in conventional polymerizations. There are several methods for initiating the polymerization. Among them, γ-ray irradiation from a ^{60}C source is preferable, since γ-rays can transmit into the solids homogeneously.

Hosts and Monomers

Channels

Strict complementarity between the spaces and the resulting polymers is required for the one-dimensional inclusion polymerization. In general, conjugated dienes and trienes rather than vinyl monomers are preferable, since size ratios of pendent groups to polymeric chains are important.

A pair of hosts, urea and thiourea, was first used for inclusion polymerization. However, since the hosts are less versatile for the monomers, the polymerization was successfully applied only to limited monomers. A flexible host, perhydrotriphenylene, extended the scope of polymerizable monomers to many methyl-substituted butadiene derivatives. These monomers gave polymers with highly 1,4-*trans* tactic structures. Optical resolution of the host brought about the first asymmetric inclusion polymerization of 1,3-pentadiene. The great difficulty of the resolution directed us to a natural compound, deoxycholic acid, as well as its derivative, apocholic acid, as chiral hosts. These steroidal acids are well known to form channel-type inclusion compounds with a variety of organic substances.

Interlayers

Montmorillonite and graphite-potassium layered compound were used, respectively, for the polymerization of vinyl monomers such as methyl methacrylate and styrene.

POLYMERIZATION BEHAVIORS

Space Size Effects and a Suitable Sets of Hosts

We expect the same monomers to display different reaction behaviors in different sizes of channels, although they have the same electronic structures. This may be called a space-size effect. This effect is common in reactions of small molecules in the case of macrocyclic compounds, such as a set of α-, β-, γ-cyclodextrins. For inclusion polymerization, it was difficult to find a suitable pair of hosts. For example, urea and thiourea are not suitable for observing the effect, because the same monomers cannot be included in both hosts. However, deoxycholic acid and apocholic acid serve as a suitable pair of hosts. The former forms a slightly smaller channel than the latter. In addition, their related compounds, such as cholic acid and its derivatives, were found to form the inclusion compounds involving the channels with similar environments. Their extensive inclusion abilities led to the first inclusion of the same monomers in the different channels.

Propagating Radicals

After γ-irradiation, any radicals are induced through unknown processes in both the host components and the monomers. The resulting radicals are readily observed by ESR spectroscopy.

Polymerizabilities and Kinetics

The motions of molecules are smaller in channels than in solution at identical temperature, resulting in a depression of the polymerization rates in the channels. In fact, the polymerization often proceeds very slowly in spite of radical species.

Structural Analysis

X-ray crystallography is a powerful tool for carrying out structural studies of crystalline inclusion compounds and provides direct evidence that the monomers polymerize in the channels.

APPLICATIONS

Stereoregular Polymers

Regiospecific Addition

Since the monomer molecules lie along the channels, the propagating reactions usually take place between a propagating-chain end and the other end of the neighboring monomer to give α,ω-addition. Therefore, even conjugated trienes can polymerize only by 1,6-addition. However, the selectivity may depend on the relative sizes of the monomers and the channels.

Similarly, *trans*-addition rather than *cis*-addition mostly takes place in the one-dimensional polymerization. The contents also depend on the relative sizes of monomers and channels.

Tacticities

Regarding vinyl monomers, the inclusion compounds of urea gave more syndiotactic poly(vinyl chloride) and more isotactic polyacrylonitrile than usual. For diene monomers, the poly(1,3-pentadiene) was highly isotactic in the channels of perhydropriphenylene, while the poly(1,3-pentadiene) obtained was prevailingly isotactic (meso:racemic = 2:1) in the channels of deoxycholic acid.

2,4-Hexadiene polymerized in channels of perhydrotriphenylene, deoxycholic acid, and apocholic acid to yield polymers with *erythro-* rather than *threo*-rich structure. The polymerization preferentially proceeded in the channels of deoxycholic acid through *trans* opening to give *erythro* diisotactic structure.

Asymmetric Induction

Inclusion polymerization enabled us to obtain optically active polymers from many prochiral diene monomers with a substituent at 1-position. Such a polymer was prepared from *trans*-1,3-pentadiene by using optically active prehydrotriphenylene. Optically, active polymer was obtained from *cis*-1,3-pentadiene by using a natural chiral host, decoycholic acid. Later, asymmetric inclusion polymerization of many prochiral monomers was performed by using both deoxycholic acid and apocholic acid. The predominant absolute configurations were constantly (*R*), and the optical yield amounted to a maximum of 36%.

Molecular Composites

One-Dimensional Composites

Inclusion polymerization might lead to a novel method of obtaining a potential molecular wire. The wire is attractive, since nanometer composite materials will be indispensable in future technology.

We expect that ionic channels of inorganic hosts will differ from organic hosts. Inclusion polymerization of pyrrole and aniline is based on a redox reaction between the host and guest components. For example, the composite of polypyrrole was synthesized in the channels of zeolite Y, mordenite, and $[(Me_3Sn)_3Fe(CN)_6]_n$.

Two-Dimensional Composites

The spontaneous redox reaction was utilized to prepare the composite materials of conductive polymers with layered inorganic hosts, such as the composite of FeOCl layered compounds with pyrrole, layered $V_2O_5nH_2O$ xerogels or MoS_2 with aniline. The latter composite material displays enhanced electrical properties with a metal-to-insulator transition at 9 K. The inclusion polymerization of acrylonitrile between the interlayers gave an intercalation compound of montmorillonite with polyacrylonitrile, which works as a precursor of graphite. Ring-opening polymerization of ε-caprolactam intercalated into montmorillonite produced a hybrid of nylon 6 with clay, which has excellent physical properties compared to nylon 6 itself.

Polymer Aggregates

In general, we can isolate polymer aggregates from the composites after the polymerization. The resulting aggregates consists of elongated polymer chains without folding.

REFERENCES

1. Chatani, Y. *Prog. Polym. Sci. Japan* **1974**, *7*, 149.
2. Farina, M. *Makromol. Chem. Suppl.* **1981**, *4*, 21.
3. Farina, M.; Di Silvestro, G.; Sozzani, P. *Mol. Cryst. Liq. Cryst.* **1983**, *93*, 169.
4. Farina, M. In *Inclusion Compounds*; Atwood, J. L.; Davies, J. E. D.; MacNicol, D. D., Eds.; Academic: London, 1984; Vol. 3, pp 297–329.
5. Farina, M.; Di Silvestro, G. *Encyclopedia of Polymer Science and Engineering*; 2nd ed.; Mark, H. F.; Bikales, N. M.; Overberger, C. G.; Menges, G. Eds.; John Wiley & Sons: New York, 1988; Vol. 12, pp 486–504.
6. Takemoto, K.; Miyata, M. *J. Macromol. Sci.-Rev. Macromol. Chem.* **1980**, *C18*, 83.
7. Takemoto, K.; Miyata, M. *Advances in Supramolecular Chemistry* Gokel, G. W., Ed.; JAI: Greenwich, 1993; Vol. 3, pp 37–63.
8. Miyata, M. In *Polymerization in Organized Media* Paleos, C. M., Ed.; Gordon and Breach: New York, 1992, pp 327–367.
9. Miyata, M. *Comprehensive Supramolecular Chemistry*; Lehn, J.-M., Ed.; Pergamon: Oxford, 1996; Vol. 10; Chapter 19.

Indene Polymerization

See: Living Carbocationic Polymerization, Olefins

INHIBITION KINETICS, RADICAL POLYMERIZATION

F. Tüdõs
Central Research Institute for Chemistry
Hungarian Academy of Sciences and
Department of Chemistry Technology
Eötvös Loránd University

KINETICS OF POLYMERIZATION IN THE PRESENCE OF ADDITIVES

Under the usual conditions of radical polymerization, three elementary reactions should be considered: initiation, chain propagation, and bimolecular termination of the polymer radicals.

If an additive capable of participating in the radical reaction is added to the system, then, depending on its reactivity, the additive reacts with the propagating radical. This reaction proceeds either by addition or by transfer mechanism.

The additive changes the macrokinetics of the system, depending on the rate of the reaction between the intermediate radical and the monomer molecule.

This reaction produces radicals in which the odd electron is localized on the part of the molecule originating from the monomer unit. The reactivity of the molecule is practically equal to that of primary propagating radicals (so-called chain regeneration).

Intermediate radicals can, irrespective of the actual mechanism, participate only in the termination reaction, the additive finally decreases the concentration of chain-carrier radicals. This effect leads to a decrease in the rate of polymerization, that is, an inhibition effect can be observed macroscopically.

KINETICS OF INHIBITED POLYMERIZATION

During the inhibition period the polymerization proceeds at a much lower rate. As the inhibitor is being consumed, the rate of the process gradually increases and, after the consumption of the whole inhibitor, it reaches the rate of the noninhibited process. The inhibition period increases with the inhibitor concentration.

The length of the inhibition period is proportional to the initial inhibitor concentration and the stoichiometric coefficient and inversely proportional to the rate of initiation. In other words, the inhibitor is consumed in a zero order reaction; the rate of its consumption is independent of its concentration.

ACKNOWLEDGMENT

We express our thanks for copyright permission to Elsevier Science Ltd. in using the review paper published in *Progress in Polymer Science*, **1989**, *14*, 717.

INHIBITION, RADICAL POLYMERIZATION

F. Tüdõs
Central Research Institute for Chemistry
Hungarian Academy of Sciences and
Department of Chemical Technology
Eötvös Loránd University

The inhibition of radical polymerization is important from both the theoretical and practical point of view. Inhibitors are used for decreasing the rate of polymerization or for stopping the process. Polymerization is generally, with the exception of a well-defined polymerization procedure carried out to produce a desired polymer, an undesirable process. Under uncontrolled conditions, polymerization may cause loss of monomer, or produce polymer of the wrong molecular mass and mass distribution, spoiling the quality of the product. During storage or, in some technological steps where heat transfer conditions are inadequate, polymerization may accelerate autocatalytically and, in extreme cases, can lead to thermal explosion.

In order to avoid monomer loss and technological disturbance, different inhibitors are used. According to their functions, these can be classified as follows:

- storage inhibitors:
- inhibitors that can be used during technological operations (e.g., rectification);
- inhibitors for quick stopping of the polymerization (e.g., in the case of any disturbance–the so-called short stop inhibitors);
- inhibitors to regulate the polymerization (e.g., partial or complete elimination of gel effect, regulation of the molecular mass, etc.);
- inhibitors used for stopping the polymerization and increasing stability of the polymer (e.g., antioxidants).

The first three types stop the polymerization almost completely; therefore inhibitors of high reactivity should be applied in relatively small concentrations. The last two types involve the application of moderately or slightly reactive inhibitors (retarders) in relatively high concentration. Secondary aspects should also be considered when choosing inhibitors (for example, the stability and possible side reactions of the inhibitor itself, the temperature regime, the color and reaction products of the inhibitor, its price, toxicity, possible deposition/elimination of its side products, etc.).

Inhibited radical polymerization is an excellent model for study of kinetics and mechanism of radical reactions and for the investigation of the relationship between structure and radical reactivity of the reacting molecules.

DETERMINATION OF THE RATE OF INITIATION

One of the central problems of inhibition kinetics is determining the rate of initiation. For this purpose we can use only those inhibitors that participate with one electron in the reaction with the polymer radical, assuming that the inhibition reaction consists only of one elementary step (e.g., recombination). Such inhibitors include the transient metals[1] (e.g., Fe^{3+}) and the stable-free radicals.

For determining the rate of initiation, the Banfield radical[2] is usually used. The picryl group of DPPH often causes strong secondary retardation,[3,4] therefore the 1,1-diphenyl-2-(2,6-dinitrophenyl)hydrazyl (DPDH) having the same stability is more suitable for inhibition kinetic investigations.[4]

INVESTIGATION OF MOLECULAR INHIBITORS

If the electron structure of the inhibitor consists of an even number of electrons, that is, if it is a diamagnetic molecule, then it can furnish diamagnetic molecules unable to undergo further radical reactions only if it reacts with two free radicals containing unpaired electrons.

The mechanism of inhibition can sometimes be complex, however.

VINYL MONOMERS AS INHIBITORS

The polymerization of a mixture of two monomers can generally be considered copolymerization. If, however, one of the radicals formed in the reaction is much less reactive (e.g., owing to delocalization) than the propagating radical, the process becomes inhibition from a kinetic point of view. For example, styrene inhibits the polymerization of vinyl acetate in a very low concentration.

Because of the release of considerable delocalization energy of the resonance-stabilized radical formed, the elementary inhibition reaction is very strongly exothermal.[5]

AROMATIC HYDROCARBONS

The condensed aromatic ring systems are favored models for the investigation of radical reactivity, because these compounds can easily be studied by quantum chemical methods.

Only a few of these compounds exert measurable retarder effect on styrene polymerization; simultaneously, more or less chain regeneration can be observed. Inhibition is caused only by tetracene.

QUINONES

Quinones are the most extensively studied inhibitors of radical polymerization. Szwarc,[6] investigating methyl affinities of numerous benzo- and naphtoquinones, has concluded that methyl radicals attack the C=C double bond of quinones. Other very reactive radicals (e.g., phenyl radicals) are likely to react by a similar mechanism.

The actual mechanism of the reaction is more complicated. Nevertheless, in the elementary inhibition reaction the quinoidal π-electron structure will be transformed to a benzoidal one, that is, to an aromatic structure.

The very extensive studies of Tüdõs and co-workers suggested that, considering their reactivity in styrene polymerization, the quinones can be divided in three groups:

- Benzoquinone and its substituted derivatives,[7–9]
- Halogen-substituted derivatives of benzoquinone,[10] and
- Quinones with condensed ring systems.[11]

AROMATIC NITRO COMPOUNDS

The first inhibition kinetic investigations with nitro compounds were carried out as early as the 1940s. More thorough kinetic study, however, was done by Bartlett and co-workers.[12–14] They found that in the polymerization of vinyl acetate and styrene, the propagating radicals attach at the oxygen atom of the nitro group. Further studies[15,16] revealed that electron-accepting substituents increase the reactivity of the nitro group.

The polymerization of vinyl acetate proved especially suitable for the investigation of the substituent effect in radical reactions, because no side reaction takes place.

NITROSO COMPOUNDS AND NITRONES

Owing to their high reactivity, aromatic nitroso compounds are suitable *par excellence* for practical inhibition.

INHIBITION BY TRANSFER MECHANISM

Effect of Phenols

The inhibition period brought about by oxygen dissolved in styrene is lengthened substantially by the presence of phenols,[18] whereas in oxygen-free systems the rate of polymerization is unaffected by hydroquinone for example. For a long time, phenols were believed to react only with peroxy radicals[19] and not with carbon radicals. This phenomenon was explained by the transformation of phenols into quinones in the presence of oxygen, followed by inhibition of radical polymerization.[20] Most such investigations were unfortunately performed with styrene. The polystyrene radical is relatively unreactive, and phenols of polar character are only slightly soluble in styrene. These two factors together restricted the effects of simple phenols in oxygen-free systems to the limit of detection.

Effect of Aromatic Amines

The polymerization of vinyl acetate was inhibited or retarded by substituted aniline derivatives. The reactivities of the 23 compounds investigated[21] vary within a range of 2 orders of magnitude.

The *N*-heterocyclic compounds containing N–H bonds also belong to this group.

ACKNOWLEDGMENT

We express our thanks for copyright permission to Elsevier Science Ltd. in using the review paper published in *Progress in Polymer Science* **1989**, *14*, 717.

REFERENCES

1. Bamford, C. H. et al. *Proc. R. Soc.* **1957**, *A239*, 214; Bamford, C. H. et al. *The Kinetics of Vinyl Polymerization by Radical Mechanism*; Butterworths: London, 1958.
2. Banfield, F. H.; Kenyon, J. *J. Chem. Soc.* **1926**, 1612.
3. Tüdõs, F.; Smirnov, N. I. *Acta Chem. Hung.* **1958**, *15*, 389.
4. Tüdõs, F. et al. *Acta Chim. Hung.* **1960**, *24*, 91.
5. Tüdõs, F. D.Sc. Thesis, LTI im. Lensovieta, Leningrád, 1964.
6. Rembaum, A.; Szwarc, M. *J. Am. Chem. Soc.* **1955**, 4468; Buckley, R. P. et al. *J. Chem. Soc.* **1958**, 344.
7. Tüdõs, F.; Simándi, T. L. *Vysokomol. Soedin.* **1962**, *4*, 1271.
8. Tüdõs, F.; Simándi, T. L. *Vysokomol. Soedin.* **1962**, *4*, 1425.
9. Simándi, T. L.; Tüdõs, F. *Eur. Polym. J.* **1985**, *21*, 865; Simándi, T. L. et al. *J. Polym. Sci.* **1969**, *C16*, 4607.
10. Tüdõs, F. et al. *Vysokomol. Soedin.* **1962**, *4*, 1431.
11. Tüdõs, F. et al. *Eur. Polym. J.* **1970**, *6*, 1321.
12. Bartlett, P. D.; Kwart, H. *J. Am. Chem. Soc.* **1950**, *72*, 1051.
13. Bartlett, P. D.; Kwart, H. *J. Am. Chem. Soc.* **1952**, *74*, 3969.
14. Hammond, G. S.; Bartlett, P. D. *J. Polym. Sci.* **1950**, *5*, 617.
15. Sinitsina, Z. A.; Badasaryan, H. S. *Zh. Fiz. Khim.* **1958**, *32*, 2614; *32*, 2663.
16. Tüdõs, F. et al. *J. Polym. Sci.* **1958**, *A1*, 1353; *1369*, 1963.
17. Foord, S. G. et al. *J. Chem. Soc.* **1940**, *48*.
18. Breitenbach, J. *Ber.* **1938**, *71*, 1438.
19. Wallig, C.; Briggs, E. R. *J. Am. Chem. Soc.* **1946**, *68*, 1141.
20. Dolgoplosk, B. A.; Korotkina, D. Sh. *Zh. Obshch. Khim* **1957**, *27*, 2226; Dolgoplosk, B. A.; Parfenova, G. A. *Zh. Obshch. Khim* **1957**, *27*, 3083.
21. Simonyi, M.; Tüdõs, F. In *Kinetics and Mechanism of Polyreactions*; Hungarian Academy of Sciences: Budapest, 1969; Vol. 3, p 119.

Inhibitors

See: *Inhibition Kinetics, Radical Polymerization*
Inhibition, Radical Polymerization

Iniferters

See: *Captodative Compounds, Polymerization*
Free Radical Iniferters
Living Radical Polymerization

Initiators

See: *Anionic Initiators (Lithium Aggregate Effects, Anionic Polymerization)*
Asymmetric Polymerization (Overview)
Azo Initiators (Function and Application)
Block Copolymers (By Changing Polymerization Mechanism)
Block Copolymers (From Macroinitiators)
Cyclic Carbonates (Ring-Opening Polymerization)
Cyclopolymerization, Enantioselective Cationic

Initiators, Polymeric

INJECTION MOLDINGS (Residual Stresses)

J. R. White
Department of Mechanical
Materials and Manufacturing Engineering
University of Newcastle upon Tyne

Residual stresses, sometimes called molding stresses or internal stresses, are almost always present in injection-molded articles. They can have a significant influence over the properties of the molding, and it is important to know how to control them. The stresses are usually fairly weak and tensile in the interior of the molding and compressive and sometimes quite strong near the surface. If the stresses are not symmetrically distributed across a molded section, warping occurs. If an environmental stress-cracking agent comes into contact with the molding it may penetrate into the interior, where the tensile stresses may be sufficient to promote crazing or cracking even without the application of an external force. The stresses will usually cause the material to be birefringent; this may be detrimental to the optical performance of the article in some applications. On the other hand, strong compressive stresses near the surface will inhibit the formation and propagation of cracks. This is beneficial because fracture from a surface flaw is a common mode of failure. This kind of stress distribution is deliberately introduced into inorganic glasses to toughen them. Another benefit is that residual compressive stresses may retard photodegradation near the surface.

We will examine the formation and control of residual stresses, methods of measurement, and the changes in residual stresses that occur after molding as a consequence of aging or conditions that promote physical or chemical changes that alter the stress distribution.

FORMATION AND CONTROL OF RESIDUAL STRESSES

The injection-molding process consists essentially of injecting a polymer melt into a cold mold and allowing the material to set by cooling under pressure. The material enters the mold through a narrow constriction ("gate") or a series of gates. The injection system maintains pressure after the mold is full. More melt enters the mold cavity ("topping up") to compensate for thermal shrinkage during the early part of the cooling phase, but this can no longer happen once the gate freezes. The material that is in contact with the mold cavity wall sets rapidly; then the solidification front moves progressively inward as the interior cools. Because polymers are poor thermal conductors there is a large temperature gradient present when the gate freezes. The material in the interior continues to cool and shrink, but the material that has already solidified opposes the shrinkage. This produces hydrostatic tensile stresses in the interior and may cause voiding or sinking. The tensile stresses must be balanced by compressive stresses near the surface.

In addition to these "thermoelastic" stresses are "flow stresses," or shear and extensional stresses that are generated in the melt during mold filling and become partly frozen-in if the cooling rate is too high to permit their complete relaxation.

Theoretical methods to predict residual stress distribution have been reviewed by Haworth et al.,[1] Isayev and Crouthamel,[2] White,[3] and Baaijens and Douven.[4]

Crystallizing polymers present yet another problem. A large volumetric change occurs on crystallization with most polymers, but there are no satisfactory theories to predict the depth variation in crystallinity in injection moldings with sufficient accuracy to allow for its effect. Furthermore, the elastic properties of injection-molded material vary through the depth of the molding, and this is not taken into account, partly because this would magnify still more the complexity of the analysis and partly because there is a shortage of data on this property upon which to base calculations.

The simplest theories predict that the thermoelastic stress distribution should be parabolic, with the tensile maximum at the center of the section. The magnitude of the compressive maximum, located at the surface, is double that of the tensile maximum. Stress distributions predicted by the more complex theories show the same basic features but depart from the exact parabolic form. It is generally agreed that the flow-induced stresses are smaller than the thermoelastic stresses; their magnitudes are different parallel and transverse to flow.

The theories predict that the stress magnitudes should increase as the temperature of the mold decreases, and this has been confirmed by experimental investigations.

Injection moldings possess large variations in morphology through depth. A recent introduction to the literature of this field has been presented by Chen and White.[5] These characteristics have a strong influence on the properties of the moldings and are sensitive to the molding conditions. Therefore, if a change is made in the molding conditions in order to produce a change in the residual stress distribution, there will generally be changes in other characteristics of the molding, and it is not easy to determine the source of the consequent property changes.

Residual stress is called "internal stress" by some authors, and this term is also used for molecular orientation. Although it is expected to be present whenever flow stresses occur, molecular orientation alone does not imply an active stress is present at temperatures below the glass transition temperature. Molecular orientation can exist even in the absence of residual stress.

CHANGES IN RESIDUAL STRESSES AFTER MOLDING

Residual stresses may change after molding for a variety of reasons. Stress relaxation or secondary crystallization may occur as slow room-temperature aging phenomena. These processes may be accelerated by applying an elevated temperature; they may produce unbalanced stresses and warping if promoted by a temperature gradient. Absorption of a penetrant modifies the residual stress distribution if the distribution of penetrant is nonuniform; an overall reduction in the stress magnitudes occurs if the penetrant causes a reduction in the stiffness of the polymer. These effects are often much larger than those that can be made by altering the molding conditions.

EFFECT OF RESIDUAL STRESS ON PROPERTIES

Many attempts have been made to determine the influence of residual stresses on properties of polymer moldings. Determining this is not as simple as it appears because any action taken to change the stresses, whether by changing the molding conditions or by applying a post-molding conditioning treatment, will cause other important changes within the molding in addition to changes in residual stresses. Changes in molding conditions produce changes in molecular orientation, crystallinity, and density, all of which can be expected to influence the properties. Aging and annealing cause "physical aging" which manifests itself in a variety of ways, including densification, and leads to changes in property.[6] Secondary crystallization causes property changes additional to those that may follow from the consequent changes in residual stress. Absorption of water or other penetrants may lead to crazing or the formation of other flaws that may then nucleate cracks. Weathering causes serious chemical degradation, most commonly chain scission by photooxidative reactions.

CONCLUSIONS

Residual stresses are always present in injection moldings and are often large enough to have a significant influence on mechanical and optical properties and on dimensional accuracy. They are sometimes detrimental and sometimes beneficial. There is some scope for controlling them through adjustment of the molding conditions, but this causes changes in other characteristics, such as molecular orientation and crystallinity, that may be equally important. Residual stresses may change significantly during service, and this may alter the properties.

REFERENCES

1. Haworth, B.; Hindle, C. S.; Sandilands, G. J.; White, J. R. *Plast. Rubb. Proc. Applics.* **1982**, *2*, 59.
2. Isayev, A. I.; Crouthamel, D. L. *Polym. Plast. Technol. Eng.* **1984**, *22*, 177.
3. White, J. R. *Polym. Testing* **1984**, *4*, 165.
4. Baaijens, F. P. T.; Douven, L. F. A. *Appl. Sci. Res.* **1991**, *48*, 141.
5. Chen, Z.; White, J. R. *Plast. Rubb. Compos. Proc. Applics.* **1992**, *18*, 289.
6. Struik, L. C. E. *Physical Aging in Amorphous Polymers and Other Materials*; Elsevier: Amsterdam, 1978.

Inks

See: Printing Inks

Inorganic-Based Polymers

See: Antitumoral and Antiviral Polyoxometalates (Inorganic Discrete Polymers of Metal Oxide)
Charge Transport Polymers (for Organic Electroluminescent Devices)
Contact Lenses, Gas Permeable
Controlled Drug Delivery Systems
Cyclolinear Organophosphazene Polymers
Cyclomatrix Phosphazene Polymers
Magnetic Field Processing
Metal-Containing Silicon-Based Polymers

Organo-Tin Polymers
Phosphorus-Containing Polymers (Overview)
Polyphosphazene Block Copolymers
Polyphosphazenes (Solution Properties)
Polysilanes (Overview)
Polysilanes (Bearing Hydroxy Groups and Derivatives)
Polysilanes (Ring-Opening Polymerization of Cyclotetrasilanes)
Polysilazanes (Through Ring-Opening Polymerization)
Preceramic Polymers
Silicon Nitride
Ultrasonically Assisted Polymer Synthesis

Inorganic Fibers

See: *Alumina Fibers (from Poly[(acyloxy)aloxane]s)*
Inorganic Fibers (Surface Modification by Grafting)
Preceramic Polymers

INORGANIC FIBERS (Surface Modification by Grafting)

Norio Tsubokawa
Department of Chemistry and Chemical Engineering
Faculty of Engineering
Niigata University

Kazuhiro Fujiki
Division of Life and Health Sciences
Joetsu University of Education

Inorganic fibers possess the desirable properties of high temperature resistance, high modulus, extremely high strength, low density, and chemical resistance. Starting with glass fiber, a large family of inorganic fibers has been developed. Over the past 20 years, carbon fibers have emerged as the main reinforcement for high-performance composite materials.

Surface grafting of polymers onto fibers induced by chemical (radical and ionic)[1–3] and physical (plasma and radiation)[4–6] processes has recently been used to enhance adhesiveness between the components. By using these polymer-grafted fibers, it is possible to improve the mechanical properties of composite materials, especially their interlaminar shear strength.

PREPARATION

Grafting of Polymers onto Carbon Fiber Surface

The cationic polymerization of *N*-vinylcarbazole, *N*-vinyl-2-pyrrolidone, and isobutyl vinyl ether can be initiated by carbon fiber, and part of the resultant polymers graft onto the surface.[7] The initiating sites of the polymerizations are considered to be carboxyl groups of the carbon fiber because the blocking carboxyl groups, initiating ability disappears.

Tsubokawa et al. reported that the anionic ring-opening copolymerization of epoxides with cyclic acid anhydrides was initiated by potassium carboxylate (COOK) groups introduced onto the carbon fiber surface by the reaction of carboxyl groups with aqueous solution of potassium hydroxide.[8]

Anionic graft polymerization of vinyl monomers onto the carbon fiber surface is also initiated by metallized carbon fiber, which is prepared by the reaction of polycondensed aromatic rings of the carbon fiber surface with *n*-butyllithium (BuLi) in *N,N,N′,N′*-tetramethylethylenediamine (TMEDA).

It is reported that radical grafting of vinyl polymers onto the carbon fiber was achieved by the chain transfer reaction of growing polymer radicals to mercapto or unsaturated groups introduced onto the carbon fiber surface. The mercapto groups were attached to the surface by the reaction of ethylene sulfide with surface hydroxyl groups of the carbon fiber, which were oxidized with 20% nitric acid solution.[9] Introduction of unsaturated groups onto the surface was carried out by the reaction of surface hydroxyl or carboxyl groups of the oxidized carbon fiber with epoxy groups of glycidyl methacrylate using boron trifluoride etherate as a catalyst.[10]

Grafting of Polymers onto Glass Fiber Surface

All glass fibers used in the manufacture of glass fiber-reinforced plastics are treated by coupling agents to promote fiber-matrix adhesion. The most common coupling agents for glass fibers are organosilanes, which have the form R_3–Si–X. The R is an alkoxy group that can react with silanol groups on the glass fiber surface to form siloxane linkages. The X is an alkyl group that is varied to optimize reactivity with the polymer matrix. When X has a terminal functional group such as amino, epoxy, or unsaturated, these coupling agents are useful for introducing surface functional groups to act as anchors for grafting.

Silanol groups on the glass fiber surface can also be used to introduce lithium silanolate (Si–OLi) groups, which are capable of initiating anionic polymerization, by treating the glass fiber with BuLi.[11]

Grafting of Polymers onto Alumina and Silicon Carbide Fiber Surface

Surface hydroxyl groups of alumina and silicon carbide fibers as well as glass fiber can be transformed into alkoxy lithium (OLi) groups by treatment with BuLi, and these OLi groups can initiate anionic polymerization.

PROPERTIES

Properties of Polymer-Grafted Carbon Fiber

Tensile strength and breaking elongation tend to increase with an increase in the percentage of grafting, whereas Young's modulus tends to decrease with increasing percentage of grafting.

Mechanical Properties of Composite

Carbon fibers are principally used as reinforcement in various matrices, including synthetic thermosetting resins such as epoxides, bismaleimides, polyimides, vinyl esters, and a number of thermoplastic resins. Glass fibers are also extensively used as a plastic reinforcement because they are inexpensive to produce and possess high strength, low density, good chemical resistance, and good insulating characteristics.

As mentioned before, the mechanical properties of composite are markedly influenced by the surface interaction of fiber with matrix polymer.

REFERENCES

1. Tanaka, A. et al. *J. Polym. Sci., Polym. Chem. Ed.* **1980**, *18*, 2267.
2. Hashimoto, K. et al. *J. Appl. Polym. Sci.* **1982**, *27*, 4529.
3. Yosomiya, R.; Fujisawa, T. *Polym. Bull.* **1985**, *13*, 7.
4. Garnet, J. L.; Phuoc, D. H. *J. Macromol. Sci. Chem.* **1976**, *A10*, 709.
5. El-Assy, N. B. *J. Appl. Polym. Sci.* **1991**, *42*, 885.
6. Mori, M. et al. *J. Polym. Sci., Part A, Polym. Chem.* **1994**, *32*, 1683.
7. Tsubokawa, N. et al. *J. Macromol. Sci. Chem.* **1988**, *A25*, 171.
8. Tsubokawa, N. et al. *Polym. Plast. Technol. Eng.* **1989**, *28*, 201.
9. Fujisawa, T. et al. *Makromol. Chem.* **1982**, *183*, 2923.
10. Yosomiya, R. et al. *Angew. Makromol Chem.* **1983**, *118*, 133.
11. Fujiki, K. et al. *Kobunshi Kako* **1994**, *43*, 534.

INORGANIC NANOSTRUCTURED MATERIALS

Kenneth E. Gonsalves and Xiaohe Chen
Polymer Science Program
Institute of Materials Science and Department of Chemistry
University of Connecticut

Nanostructured materials (also referred to as nanocrystalline, nanophase, nanosized, or nanoscale materials) are defined as materials having an average phase or grain size or < 100 nm.[1] In a broader sense, any material that contains grains or clusters below 100 nm, or layers or filaments of that thickness, can be considered nanostructured.[2] Because of the small size of the structural unit (particle, grain, or phase) and the high surface-to-volume ratio, nanostructured materials exhibit unique behavior compared to conventional materials with micron-scale structures.[3] Extensive fundamental and applied investigations have been conducted on nanoscale materials in the past decade.[4–9]

Nanostructured materials have properties different from those of bulk systems. Structural factors, including dimensionality, composition, and interface, influence their performance. The first of these, dimensionality, includes the size and size distribution, as well as the shape and morphology of the structural domains. The chemical composition of the constituent phases in nanostructured materials is directly related to the performance of end-use materials, which usually have been consolidated or processed into specific form. The third important structural factor consists of the interfaces (more specifically the grain boundaries, the heterophase interfaces, or the free surface) created between constituent phases and the nature of the interaction across the interfaces, because the number of interfaces in nanostructured materials is large.

The preparation of nanostructured materials requires consideration of the above factors. Studies on inorganic nanostructured materials generally cover the range from metals and ceramics (oxide, nitrides, silicides, carbides) to biominerals. Many synthesis and processing methods have been used to fabricate inorganic nanostructured materials.[10,11,14–18,20] These methods can be classified into physical and chemical techniques[3,8,9] although many preparation methods are based on more than one principle and the boundary between these methods might be diffuse.

Physical methods, such as gas-phase condensation,[10] sputtering deposition,[11] laser ablation,[12] and plasma synthesis,[13] are widely used. Chemical synthetic methods offer several attractive features: tailored synthesis via assembly of atomic or molecular precursors, controlled stoichiometry, mixing of constituent phases at the molecular level, and cost-effective production of bulk quantities of materials. The most frequently used chemical methods include sol–gel conversion,[14] chemical vapor deposition,[15] prepolymer pyrolysis,[16] thermolysis of organometallics,[17] laser pyrolysis,[18] and borohydride reduction.[19] Newer preparation methods for nanostructured materials, such as sonochemical synthesis[20] and mechanosynthesis,[21] have recently been reported.

TABLE 1. Properties and Applications for Inorganic Nanostructured Materials

Property	Application	Examples of materials
Surface area	Catalysts IR sensors	Ni, Co, Fe, MoS_3[26] Au[7]
Superior hardness and wear resistance	Materials with enhanced mechanical properties	WC, WC-Co[29]
Single magnetic domain	Magnetic recording superparamagnetism Magnetic fluids	Fe/Co/Ni[7] Fe, Co, Ni, Fe_2O_3, Fe_3N[24] Fe_2O_3, Fe_3N[22]
Quantum confinement	Nonlinear optical material	Au colloid[25]
Nanosized filler	Nanocomposite with ultrahigh refractive index	PbS/polyoxyethylene[28]
Nanosized pores	Inorganic membranes	TiO_2, ZrO_2, SiO_2[23]

The properties exhibited by nanostructured materials synthesized so far show a marked improvement over those of the coarser-grained counterparts of the same chemical composition.[2–9] Industrial applications of nanostructured materials appear to be increasingly promising.[7,22,23,24–26,29,51] **Table 1** delineates a number of unique properties as well as existing or potential applications.

The search for advanced materials with novel architecture at the nanometer scale covers almost all areas in material science. The major goal is to explore newly designed nanostructured architecture with tailored properties to fit specific applications.

PROPERTIES AND APPLICATIONS

Inorganic nanostructured materials possess many unique features. It is these advanced features that have caused a flourish of activity toward their fabrication and application. As these properties originate from the nature of the materials, the practical performance of the final products are strongly dependent on their physical and chemical characteristics.

The list of applications for nanostructured materials is enormous and keeps growing. The use of their large surface area for catalysts[26] is an obvious example. Another early application is the fabrication of arrays of particles for devices based on quantum confinement.[27] The use of nanoscale ceramic powders provides many advantages for processing, such as greatly reduced sintering temperature.[28] Potential applications for consolidated materials include very hard, wear-resistant coatings:[29] superparamagnetic materials for magnetic refrigeration;[24] net shape forming via superplastic deformation;[30] and transparent ceramic windows.[31]

Many inorganic nanostructured materials show promise for practical applications. The future of nanostructured materials will depend on their ability to improve materials by controlling their structures and producing them in commercially viable quantities.

REFERENCES

1. Gleiter, H. *Adv. Mater.* **1992**, *4*, 474; Gleiter, H. *Nanostruct. Mater.* **1992**, *1*, 1; Birringer, R.; Gleiter, H. *Encyclopedia of Materials Science & Engineering, Suppl. Vol. 1*; Cahn, R. W. Ed.; Pergamon: Oxford, UK, 1988; p 339.
2. Siegel, R. A. *Nanostruct. Mater.* **1993**, *3*, 1; Dagani, R. *Chem. Eng. News* **1992**, *72*(47), 18.
3. Andres, R. P. et al. *J. Mater. Res.* **1989**, *4*(3), 704; Gleiter, H. *Prog. Mater. Sci.* **1989**, *33*, 223; Birringer, R. et al. *Suppl. Trans. Jpn. Inst. Met.* **1986**, *27*, 43; Feynman, F. *Miniaturization*; Gilbert, H. D. Ed.; Reinhold: New York, 1961.
4. Andrievski, R. A. *J. Mater. Sci.* **1994**, *29*, 614; Siegel, R. W. *Mater. Sci. and Eng.* **1993**, *B19*, 37–43; Shull, R. D.; Bennett, L. E. *Nanostruct. Mater.* **1992**, *2*(1), 53; Suryanaraya, C.; Froes, F. H. *Metall. Trans.* **1992**, *23A*, 1071; Whitesides, G. M. et al. *Science* **1991**, *254*, 1312.
5. Gonsalves, K. E. et al. In MRS Symposium Proceedings, Vol. 351, San Francisco, April, 1994.
6. Yacaman, M. J. et al. *Nanostruct. Mater.* **1993**, *3*(1–6), Special Proceedings Vol.
7. Ichinose, N. et al. *Superfine Particle Technology*, (Translated from Japanese); Springer-Verlag: London, 1992.
8. Gonsalves, K. E.; Chow, G. M. In *Nanostructured Materials: Synthesis, Properties; and Uses*; IOP Publishing Ltd. of Techno House: Bristol, UK, 1995.
9. Chakravorty, D.; Giri, A. K. In *Chemistry for the 21st Century: Chemistry of Advanced Materials*; Rao, C. N. R. Blackwell Scientific: London, 1993; p 217.
10. Siegel, R. W. In *Materials Science and Technology*; Cahn, R. W., Ed.; VCH: Weinheim, 1991; Vol. 15, p 583; Uyeda, R. *Progr. Mater. Sci.* **1991**, *35*, 1; Hahn, H.; Averback, R. S. *J. Appl. Phys.* **1990**, *67*, 1113.
11. Aita, C. R. *Nanostruct. Mater.* **1994**, *4*(3), 257.
12. Balkus, K. J. Jr., et al. In *Molecularly Designed Ultrafine/Nanostructured Materials*; Gonsalves, K. E. et al. Eds.; MRS Symposium Proceedings, 1994, Vol. 351, p 437; Liu, Y. et al. *J. Chem. Phys.* **1986**, *85*, 7434.
13. Kameyama, T. et al. *J. Mater. Sci.* **1990**, *25*, 1058; Ishizaki, K. et al. *J. Mater. Sci.* **1989**, *24*, 3553.
14. Phalippou, J. In *Chemical Processing of Ceramics*; Lee, B. I.; Pope, E. J. A.; Marcel Dekker: New York, 1994; p 265; Brinker, C. J.; Schener, G. *Sol-Gel-Science, the Physics and Chemistry of Sol-Gel Processing*; Academic: Boston, 1990.
15. Chang, W. et al. *Nanostruct. Mater.* **1994**, *4*(3), 345; Hahn, H.; Averback, R. S. *J. Appl. Phys.* **1990**, *67*, 1113.
16. Gonsalves, K. E.; Xiao, T. D. In *Chemical Processing of Ceramics*; Lee, B. I.; Pope, E. J. A. Eds.; Marcel Dekker: New York, 1994; p 359; Gonsalves, K. E. et al. In *Inorganic and Organometallic Polymers II: Advanced Materials and Intermediates*; ACS Symp. Ser., 572, Wisian-Neilson, D. P. et al. Eds.; American Chemical Society: Washington, DC, 1994; p 195.
17. Gonsalves, K. E. et al. *Nanostruct. Mater.* **1994**, *4*(2), 139; Gonsalves, K. E. U.S. Patent 4 842 641, 1989.
18. Gonsalves, K. E. et al. *J. Mater. Sci.* **1992**, *27*(12), 3231; Xiao, T. D. et al. *J. Mater. Sci.* **1993**, *28*, 1334.
19. Rivas, J. et al. *J. Magnet. Magnet. Mater.* **1993**, *122*(1-5), 1; Glavee, G. N. et al. *Nanostruct. Mater.* **1993**, *3*(1-6), 391.
20. Suslick, K. S. et al. In *Molecularly Designed Ultrafine/Nanostructured Materials*; Gonsalves, K. E. et al., Eds.; MRS Symposium Proceedings, 1994, Vol. 351, p 201 and p 443.
21. Koch, C. C. *Nanostruct. Mater.* **1993**, *2*, 109; Matterzzi, P. et al. *Nanostruct. Mater.* **1993**, *2*, 217; Dogan, C. P. et al. *Nanostruct. Mater.* **1994**, *4*(6), 631.
22. Bashtovoy, V. G. et al. *Introduction to Thermomechanics of Magnetic Fluids*; Berkovsky, B. M.; Rosenweig, R. E., Eds. Springer-Verlag: Berlin, 1988; Bibette, J. *J. Magnet. Magnet. Mater.* **1993**, *122*, 37.
23. Xu, Q.; Anderson, M. A. *J. Am. Ceram. Soc.* **1994**, *77*(7), 1939; Hyun, S. H.; Kang, B. S. *J. Am. Ceram. Soc.* **1994**, *77*(12), 3093.
24. Gonsalves, K. E. et al. *Adv. Mater.* **1994**, *6*(4), 291; Shull, R. D. et al. *Nanostruct. Mater.* **1993**, *2*, 205; Ziolo, R. F. et al. *Science* **1992**, *257*, 219.
25. Bloemer, M. J. et al. *J. Opt. Soc. Am.* **1990**, *7B*, 5, 790; Olsen, A. W.; Kafafi, Z. H. *J. Am. Chem. Soc.* **1991**, *113*(20), 7758.
26. Boakye, E. et al. *J. Colloid. Inter. Sci.* **1994**, *163*, 120.
27. Arakawa, Y. *Solid State Electronics* **1994**, *37*(4–6), 523.
28. Chen, D. J.; Mayo, M. J. *Nanostruct. Mater.* **1993**, *2*, 469.
29. Kear, B. H.; McCandlish, L. F. *J. Adv. Mater.* **1993**, *10*, 11.
30. Higashi, K. *Mater. Sci. Eng.* **1993**, *A166*, 109.
31. Gallas, M. R. et al. *J. Am. Ceram. Soc.* **1994**, *75*(8), 2109.

INORGANIC/ORGANIC HYBRID POLYMERS (High Temperature, Oxidatively Stable)

Teddy M. Keller* and David Y. Son
Materials Chemistry Branch
Naval Research Laboratory

Inorganic-organic hybrid polymers are an emerging technology that holds great promise for extending the temperature stability of polymeric materials. Such materials provide novel important engineering tools for technological advances.

The search for polymeric materials with new and improved high-temperature properties led us to investigate inorganic-organic hybrid polymers containing carborane-siloxane, siloxane, and acetylene segments in the backbone of linear polymers. These polymers behave as precursors to thermosets and ceramics, which have exceptional thermal and oxidative stabilities. Our approach combines the unique properties of inorganic (high temperature and oxidative stability) and organic (processability) units into the same polymeric systems. Poly(carborane-siloxane) and polysiloxane elastomers lack functional groups for crosslinking purposes. The siloxane units provide thermal and chain flexibility to polymeric materials. Poly(carborane-siloxane) elastomers exhibit superior thermal and oxidative properties and retarded depolymerization[1,2] at elevated temperatures relative to polysiloxanes. Upon exposure to air, carborane units are oxidized to boron oxide. It has been recognized that attachment of a carborane unit to a polymeric backbone does not enhance thermal properties. However, considerable improvements in thermal properties were achieved when the carborane unit was situated in the polymeric backbone.[3] In our studies, the chemistry involved in synthesizing polysiloxanes and poly(carborane-siloxane)s has been modified to accommodate inclusion of an acetylenic segment in the backbone. The presence of the acetylenic linkage within the backbone provides an opportunity to convert the initially formed liquid or low melting

*Author to whom correspondence should be addressed.

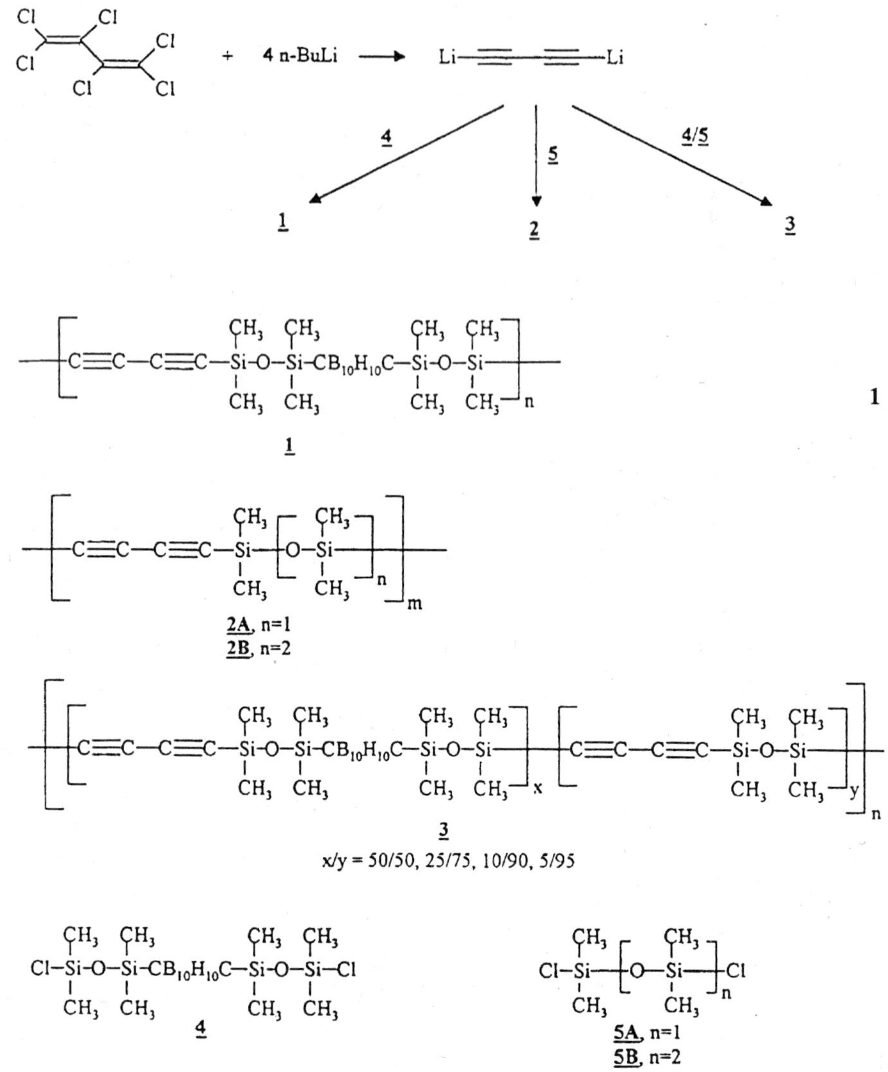

1

linear polymers into thermosets. The acetylenic functionality provides many attractive advantages relative to other crosslinking centers. The acetylene group remains inactive during processing at lower temperatures and reacts either thermally or photochemically to form conjugated polymeric crosslinks without evolution of volatiles.

RESULTS AND DISCUSSION

Synthesis

Synthesis of **1**, **2**, and **3** is a one-pot, two stage reaction (**Scheme 1**). Dilithiobutadiyne was not isolated and was reacted with an equimolar amount of the 1,7-bis(chlorotetramethyldisiloxyl)-*m*-carborane **4**, disiloxyl dichlorides **5**, or combinations of **4** and **5** to afford dark brown viscous polymers **1**, **2**, or **3** respectively in high yield (90–97%).

Polymerization

Crosslinking of **1**, **2**, and **3** occurs by thermal or photochemical means through acetylenic units to afford thermosetting polymers **6**.[4–6]

Poly(carborane-siloxane-acetylene)s

The poly(carborane-siloxane-acetylene)s **1** and **3** are readily converted in sequence into high-temperature and oxidatively stable thermosets and ceramic-based materials.[7]

The importance of the carborane unit in stabilizing the thermal properties is apparent. The char yield upon heat treatment of 3 to 1000°C decreases as the percentage of carborane is diminished in the copolymer. Decreasing the amount of carborane also lowers the exothermic maximum temperature. Apparently, a minimum carborane composition of approximately 10% will provide thermo-oxidative stability of 1000°C.

Increasing the amount of carborane in the copolymer decreases the thermoset weight loss during heating to 1000°C.

Poly(siloxane-acetylene)s[8]

Most conventional organic polymers lack the ability to withstand temperatures in excess of 200–250°C for extended periods in an oxidizing environment. In addition, they tend to have poor hot-wet mechanical properties. The thermo-oxidative degradation of a polysiloxane is a free radical chain reaction involving attack at the methyl group.[9,10] While most polymers containing carbon-carbon (C–C) single bond main chain units began to decompose above 250°C, pure polysiloxanes are stable under high vacuum or in an inert atmosphere to at least 350–400°C. This difference is attributed to the considerably greater strength of the Si-O bond relative to the C–C bond. At elevated temperatures in an inert atmosphere, linear polysiloxanes degrade by a process that yields a mixture of volatile, low-molecular-weight products.[10]

A polymer combining siloxyl and organic units within the backbone should exhibit enhanced thermal and oxidative stability relative to an organic polymeric material. Siloxyl groups are a logical choice for inclusion in these polymers as they possess good thermal and oxidative stability and high hydrophobicity.[9] Further, their flexibility should contribute favorably to the processability of the resulting polymers.

The thermal and oxidative properties of the linear poly(siloxane-acetylene)s were determined to 1000°C. As observed previously for linear polysiloxanes,[10] the thermal stability is diminished as the size of the siloxyl segment increases.

CONCLUSION

Poly(carborane-siloxane-acetylene)s **1** and **3** and polysiloxaneacetylenes **2** are precursors for high-temperature thermosetting polymers and ceramic-based materials that exhibit outstanding thermal and oxidative properties. Linear polymers are either liquids or low melting solids at room temperature, which enhances their importance for structural and electronic applications. The desirable features of inorganics (high thermal and oxidative stability) and organics (processability) are incorporated into these novel polymeric materials. Our studies have shown that relatively modest amounts of carborane in diacetylene-siloxane polymers can greatly enhance the thermal and thermo-oxidative stabilities of the resulting thermosets and ceramics formed during heat treatments at elevated temperatures. Further studies are underway to evaluate and exploit polymers **1**, **2**, and **3** as matrix materials for high temperature composites and carbon/ceramic composites.

ACKNOWLEDGMENT

The authors thank the Office of Naval Research for financial support of this work. The authors are grateful to Dr. Pehr Pehrsson of the Surface Branch of the Chemistry Division for the XPS studies, Dr. Tai Ho of George Mason University for his assistance with the GPC measurements, and Dr. Leslie J. Henderson for the original synthesis of poly(butadiyne-1,7-bis(tetramethyl- disiloxyl)-*m*-carborane.

REFERENCES

1. Papetti, S.; Schaeffer, B. B. et al. *J. Polym. Sci.* **1966**, *4*(A-1), 1623.
2. Critchley, J. P.; Knight, G. J.; Wright, W. W. *Heat-Resistant Polymers*, Plenum: New York, 1983 ed.
3. Williams, R. E. *J. Pure Appl. Chem.* **1972**, *29*, 569.
4. Rutherford, D. R.; Stille, J. K. *Macromolecules* **1988**, *21*, 3530.
5. Neenana, T. X.; Callstrom, M. R. et al. *Macromolecules* **1988**, *21*, 3525.
6. Callstrom, M. R.; Neenan, T. X.; Whitesides, G. M. *Macromolecules* **1988**, *21*, 3528.
7. Henderson, L. J.; Keller, T. M. *Macromolecules* **1994**, *27*, 1660.
8. Son, D. Y.; Keller, T. M. *Polym. Mat. Sci. & Eng.* **1994**, *71*, 305.
9. Goldovskii, E. A.; Kuzminskii, A. S. *Polym. Sci. Tech.* **1979**, *6*(4), 75.
10. Dvornic, P. R.; Lenz, R. W. *High Temperature Siloxane Elastomers* Hüthig & Wepf: Heidelberg, 1990.

Inorganic/Organic Polymer Hybrids

See: Organic/Inorganic Hybrid Polymers

INORGANIC SURFACES, GRAFTING

Walter Caseri
Eidgenössische Materialprüfungs-und-Forschungsanstalt EMPA (Abt. 136)

Ulrich W. Suter
Eidgenössische Technische Hochschule ETH
Institut für Polymere

Modifying the surfaces of inorganic substrates plays an essential role in the technology of composites and adhesives. Among the methods employed to this end, the grafting of polymers to inorganic surfaces is enjoying increasing attention.[1–5] The inorganic surfaces are usually modified with organic substances before grafting is attempted.

The most intense interest is in grafting sites, the point at which the polymeric material is attached to the substrate. On the surface, there is a certain density of potential sites available to the grafting reaction, with a certain spatial distribution. The availability might change during the grafting process,[6] and one should expect it to change because the reaction zone is gradually being covered by the reaction product. For instance, a monolayer of grafted coils might make further grafting difficult and lower the amount of bound polymer from that anticipated by analysis of the early grafting reaction.[7] The attachment proceeds most frequently through a sequence of covalent bonds, although no conclusive spectroscopic proof has been put forth for such a situation. When a polymer cannot be removed from the inorganic substrate by a typical solvent, it is frequently claimed as evidence for grafting, but one must consider that polymers readily adsorb on surfaces and that polymers often cannot be removed from the adsorbed state, even by prolonged extraction with a solvent.

Macromolecules with many functional groups capable of significant interaction with the substrate are most susceptible to this artifact. This is unfortunate, because the most widely used grafting method is copolymerization of a vinyl monomer to a surface modified previously with C=C double bonds. Such systems are not well defined.[7] Better defined systems can often be obtained by grafting via endgroups, and we limit this discussion to terminal grafting.

We will distinguish different grafting situations on the basis of the process used. The basic distinction in this context, used to structure this review, is that between grafting of a polymer onto a solid substrate, and a grafting polymerization starting from a solid surface.[6,7]

ORGANIC SURFACE MODIFIERS (COUPLING AGENTS)

In most cases, surfaces are modified with silanes as coupling agents before grafting. The most common agents are compounds of the formula $(RO)_3SiX$,[8] such as 3-aminopropyltriethoxysilane

(APS),[3,9,10] aminophenyltrimethoxysilane (APTS),[2] *p*-(chlormethyl)phenyltriethoxysilane[8] or vinyl triethoxysilane.[8,11] p-Vinyl-benzyl trichlorosilane can also be used to introduce reactive groups on silica.[7]

The morphology and chemical composition of the inorganic material's surface may depend on the preparation method, that is, the tendency for formation of bonds of a specific type between silane coupling agents and inorganic surfaces might vary for a material with the same bulk composition.

GRAFTING INITIATED BY SURFACE-BOUND RADICAL INITIATORS

Grafting to carbon black was reported starting from a surface-bound peroxide group.[6] It appears that when the amount of bound initiator is increased, the mass coverage of polymer on the surface and the molecular weight reach a maximum. Monomer concentration and reaction temperature also influence the polymer coverage on silica.[2]

A phenyldiazo-(2-naphthyl) thioether (PANT) was attached to silica via a reaction involving chlorination of the silica surface followed by phenylation and further reactions, supposedly via a covalent bond.[14] The grafted mass increases with increasing reaction temperature, monomer concentration (in toluene), and number of initiator molecules in the system.[14]

Typically about 10% of the bound initiator molecules initiate a polymerization.[14] Because only one of two radicals formed after the initiator decay is attached to the surface, "free" (nongrafted) polymer is also formed in the solution.[14] M_η of the free polymer is ~ 20–100% higher than that of the grafted polymer and, accordingly, the mass of the nongrafted is larger than that of the grafted polymer.[14]

GRAFTING INDUCED BY SURFACE-BOUND IONIC INITIATORS

When silica modified to bear styryl groups is treated with butyl lithium anionic polymerizations can be initiated from the surface, yielding PS that is attached to the surface.[7] The amount of grafted PS increases linearly with increasing monomer concentration, as expected.[7] Anionic polymerization initiated from the surface offers the possibility of block copolymer grafting: with this technique, poly(styrene-*block*-isoprene-*block*-styrene), poly(styrene-*block*-methyl methacrylate), and poly(styrene-*block*-2-vinylpyridine) were prepared.

GRAFTING OF POLYMERS ONTO MODIFIED SURFACES

Grafting of carbon black was reported as a result of the attack of polymers with anionic endgroups to surface-bound ester groups.[6] Similarly, when poly(ethylene oxide) (PEO) with two amino endgroups and a molecular weight of 1000 and 5000 was reacted with a silica surface previously modified with aldehyde groups, grafted chains were observed; it seems that most polymer chains react only with one endgroup.[8]

GRAFTING TO SURFACE-BOUND CHAIN TRANSFER REAGENTS

Silica surfaces can be modified with mercapto trialkoxysilanes.[8,15] When such surfaces are treated with vinyl benzyl chloride or styrene in the presence of azobisisobutyronitrile (AIBN) or dibenzoyl peroxide (BPO), grafting occurs via chain transfer, while polymer is produced simultaneously in the solution.[8,15]

GRAFTING TO MODIFIED MICA SURFACES VIA AN UNEXPECTED MECHANISM

The radical initator azobis(isobutyramidine hydrochloride) (AIBA) strongly adsorbs on mica containing lithium surface ions via ion exchange;[12] two lithium ions are released per AIBA molecule adsorbed.

After treatment of AIBA-modified mica with styrene at 60°C, PS is bound to the surface.[13,16,17] If nonmodified mica is kept at 60°C in the presence of styrene or if AIBA-modified mica is suspended in a 5% w/w PS solution in toluene, no surface-bound polymer is created.[16] These experiments suggest that the polymer is bound to the surface either by one or by two ionic endgroups, depending on the type of chain termination reaction.[16]

The grafted polymer is probably not formed by polymerization initiated from the surface.[16] The polymer attached to mica is formed by reaction of thermally initiated polystyryl radicals with decomposition products of the bound AIBA. Indeed, an increase of polystyryl radicals in the solution by addition of AIBN strongly accelerates the rate of bound polymer.

NANOPHASES CONSISTING OF INDIVIDUAL GRAFTED POLYMER MOLECULES

Although the average thickness of grafted polymers can be calculated from the grafted mass per surface area, it is usually not known if the molecules from uniform layers, "islands," or other structures.

GRAFTING ONTO NONMODIFIED SURFACES

Relatively few studies deal with grafting to nonmodified surfaces. For example, PS (M_n = 9200) with a dimethyl-methoxysilyl endgroup can be bound to silica,[18] and poly(ethylene glycol) with a trimethyl ammonium endgroup can be grafted to mica by ion exchange.[19]

CONCLUSIONS

Although the grafting of polymer chains to the surfaces of inorganic substrates is of great importance in the technology of composites and adhesives, the detailed structure of the systems thereby obtained is still vague. Some progress has been made in the analysis of the simplest case, that of end-grafted chains. However, the type of interaction that attaches the functional group of the polymer to the grafting site at the surface is often unknown, the most notorious case being that of polymers on silica surfaces. Nevertheless, work on carefully chosen, well-defined systems has allowed significant progress and promises deeper insight into more complex technological situations.

REFERENCES

1. Heublein, G. et al. *Makromol. Chem.* **1989**, *190*, 9.
2. Boven, G. et al. *Polymer* **1990**, *31*, 2377.
3. Yu, X, D. et al. *J. Mater. Sci.* **1990**, *25*, 3255.
4. Widmann, B. et al. *Kunststoffe* **1992**, *82*, 1185.
5. Tishin, S. A. et al. *J. Mater. Sci.* **1993**, *28*, 325.

6. Donnet, J.-P.; Papirer, E. *J. Polym. Mater.* **1982**, *9*, 73.
7. Tsubokawa, W.; Kogure, A. *J. Polym. Sci. A, Poly. Chem.* **1991**, *29*, 697.
8. Guiot, A. et al. Makromol. Chem., Markomol-Symp. **1993**, 70/71, 265.
9. Kopeckovà et al. *New Polym. Mater.* **1990**, *4*, 289.
10. Jiang, M. et al. *Mater. Sci. Lett.* **1990**, *9*, 1239.
11. Chaimberg, M. et al. *J. Appl. Polym. Sci.* **1989**, *37*, 2921.
12. Shelden, R. A. et al. *J. Colloid Interface Sci.* **1993**, *157*, 318.
13. Shelden, R. A. et al. *Polymer* **1994**, *35*, 1571.
14. Laible, R.; Hamann, K. *Makromol. Chem.* **1975**, *48*, 97.
15. Tsubokawa, N. et al. *Colloid Polym. Sci.* **1993**, *271*, 940.
16. Meier, L. P. et al. *Macromolecules* **1994**, *27*, 1637.
17. Shelden, R. A. et al. *Acta Polym.* **1993**, *44*, 206.
18. Milling, A. et al. *Colloids Surf.* **1991**, *57*, 185.
19. Shelden, R. A. et al. unpublished results, 1994.

Intelligent Materials

See: *Fibers and Composites (High Modulus and Strength)*
Intelligent Membranes
Intelligent Polymers (in Medicine and Biotechnology)
Poly(N-isopropylacrylamide) Hydrogels
Smart Hydrogels
Stimuli-Responsive Polymers (Thermodynamics in Drug Delivery Technology)

INTELLIGENT MEMBRANES

W. E. Price, A. Mirmohseni, C. O. Too, G. G. Wallace, and H. Zhao
Intelligent Polymer Research Laboratory
Department of Chemistry
University of Wollongong

In the past five years a new generation of membranes has been developed based on conducting electroactive polymers. In these the dynamic nature of a conducting polymer gives a system that can adapt and respond to its environment, with properties that are tunable *in situ* during operation. These systems are capable of sophisticated separations and have been dubbed "intelligent membrane systems." The idea behind an intelligent membrane is that characteristics can be modulated by electrical signals in response to changes in the environment.

Conducting polymers are organic polymeric materials that are able to conduct electrons by dint of extended conjugated systems. We will briefly review conducting polymers in the context of intelligent systems and membranes, using polypyrrole (PPy) as an illustration.

The dynamic nature of conducting electroactive polymers makes them well suited to membrane separation and provides the possibility of producing membrane systems with tunable, controllable separation behavior.

PREPARATION

Conducting polymers may be produced either chemically or electrochemically.[1] Electrochemical synthesis has a number of advantages, including more precise control of polymer growth and therefore film quality and thickness. The electrochemical procedures for making PPy membranes are very similar to those for making thin PPy films for applications such as sensors.[3] A suitable material (e.g., platinum) is used as the working electrode and the polymer is grown on the surface under either controlled potential or controlled current conditions.

However, in order to make a free-standing film the polymer has to be removed from the substrate electrode, and therefore the choice of substrate in terms of film adhesion is important. In the Intelligent Polymer Research Laboratory (IPRL), we have extensively investigated the problem of substrate-polymer adhesion. It was found that highly polished stainless steel plates of an appropriate size gave optimal results. We have succeeded in producing a wide range of high-quality freestanding PPy membranes with large areas and uniform thickness. These include membranes with aromatic sulfonate counterions,[5] surfactants,[6] and other polymers such as PVS and Nafion in composite structures.

Polyaniline films are best prepared by chemical means because electrochemically produced films have much lower conductivity and poor mechanical properties. Extensive investigations on optimizing the polymerization of polyaniline have been carried out, notably by MacDiamid's group.[7,8]

PROPERTIES

The properties of conducting polymers made on electrodes or cast as films have been investigated for many years. We will consider only the chemical and physical properties of the conducting polymer membranes. In particular, those parameters most relevant to use as intelligent membrane materials are highlighted. These are conductivity, mechanical strength, electroactivity, and the environmental stability of the material.

The conductivity of the material is intrinsic to its use as a membrane. Composite PPy membranes made with either poly(vinyl sulfonate) or Nafion were found to have low conductivities (10–15 S cm^{-1}),[11] but are sufficient for their use in applications. Even more subtle changes in the counterion can have some effect on the inherent conductivity of the membranes.

For a film to be used in a membrane system it must be mechanically robust. Free-standing conducting polymer membranes have adequate mechanical properties. PPy membranes have been prepared routinely with tensile strengths of 80–120 MPa using sulfonated aromatic counterions such as pTS.[9,10] Previous work with coated polypyrrole films showed that the counterion plays an important role in the mechanical properties.[12] Polyaniline films have also been shown to have adequate mechanical properties for use as membranes.[10]

If the membranes are to be used in an intelligent separation system, they need to retain their electroactivity after repeated cycling of the applied potential. Much work has been done on the reversible electroactivity of coated films of conducting polymers. Membranes made from these materials exhibit similar excellent characteristics.[9–11] The electroactivity is the basis of the mechanism for intelligent separations.

A final property of these materials that is important to their use as membranes is morphology and porosity. Porosity measurements indicate a subnanometer pore size.[2] Interestingly, there is a marked difference between the two surfaces for an electrochemically synthesized film. The side in contact with the solution shows greater contrast in topology than the face in contact with the electrode. In addition, the versatility of conducting polymer membranes is highlighted by the fact that different (surface) chemistries may be written into a membrane by changing the polymer.

APPLICATIONS

Initial Developments of Intelligent Polymer Membranes

Most work to date using conducting polymers as intelligent membranes has used PPy materials. Burgmayer and Murray in the early 1980s were first to realize the potential of PPy as a membrane material.[4]

It was, however, the IPRL group that took the unique properties of conducting electroactive polymers and incorporated them into a dynamic adaptive membrane system. The crucial development was to use a membrane as a working electrode in a three-electrode electrochemical cell. This meant that the polymer could be oxidized and reduced. This would induce ion movement in and out of the polymer and initiate net transport across the membrane.

CONCLUSION

The work described here is the basis of a new generation of membrane materials based on conducting polymers capable of performing sophisticated separations. The scope to manipulate the molecular architecture of these systems, through varying the polymer structure or using copolymers and changing the counterions present, is vast. The next few years should see continued growth in this area, in applications in areas such as metal refining, water purification, and environmental/effluent control.

REFERENCES

1. Skomheim, T. A. *Handbook of Conducting Polymers*; Marcel Dekker: New York, 1986.
2. Mulder, M. *Basic Principles of Membrane Technology*, Kluwer: Dordrecht, The Netherlands, 1991.
3. Wang, E.; Anhua, L. *Anal. Chim. Acta* **1991**, *252*, 53.
4. Burgmayer, P.; Murray, R. W. *J. Phys. Chem.* **1984**, *88*, 2515.
5. Huijun, Z. et al. *J. Membrane Sci.* **1994**, *87*, 47.
6. Zhao, H. et al. *Reactive Polymer* **1994**, *23*, 213.
7. McDiarmid, A. G. et al. *Mol. Cryst. Liq. Cryst.* **1985**, *121*, 173.
8. Angelopoulos, M. et al. *Synth. Met.* **1987**, *21*, 21.
9. Huijun, Z. et al. *J. Electroanal. Chem.* **1992**, *334*, 111.
10. Huijun, Z. et al. *J. Intelligent Material Sys. Struc.* **1994**, *5*, 605.
11. Mirmohseni, A. et al. *J. Intell. Material Sys. Struc.* **1993**, *4*, 43.
12. Buckley, L. J. et al. *J. Polym. Sci. Part B* **1987**, *251*, 2179.
13. Ansari, R. et al. *Polymer*, in press (1995).
14. Ansari, R., Wallace, G. G. et al. *Polymer* **1994**, *35*, 2372.

INTELLIGENT POLYMERS (in Medicine and Biotechnology)

Allan S. Hoffman
Center for Bioengineering
University of Washington

One can define "intelligent" polymers as those polymers that respond with large property changes to small physical or chemical stimuli. These polymers may be in various forms, such as in solution, on surfaces, or as solids. One may also combine "intelligent" aqueous polymer systems with biomolecules, to yield a large family of polymers that respond "intelligently" to physical, chemical, or biological stimuli. This article overviews such interesting and versatile polymer systems.

The term "intelligent polymers" refers to soluble, surface-coated, or crosslinked polymer systems that exhibit relatively large and sharp physical or chemical changes in response to small physical or chemical stimuli. Although the well-known glass and melting transitions of solid polymers can fit within this definition, most of the recent interest in "intelligent" polymer systems focuses on aqueous polymer solution, interfaces, and hydrogels. "Intelligent" polymers are also sometimes called "smart", "stimuli-responsive", or "environmentally sensitive" polymers.

Many different properties of the polymer system may change when sharp responses to stimuli occur. For example, when a soluble polymer is stimulated to precipitate, it will be selectively removed from solution, which will become cloudy. When such polymers are grafted or coated onto a solid support, then one may reversibly change the water absorption into the coated polymer, thus changing the wettability of the surface. When a hydrogel is stimulated to collapse, it will squeeze out its pore water, turn opaque, become stiffer, and shrink in size. One may take advantage of one or more of these "signals" or phase changes for different end-uses. A significant amount of research on these interesting systems has been carried out over the past 10 to 15 years, producing many publications and patents.[1–15]

There are a number of possible molecular mechanisms that can cause sharp, sometimes discontinuous transitions in polymer systems. Water is involved in most of these mechanisms. There are numerous publications describing and discussing such mechanisms in both natural and synthetic polymers.[16–21]

TEMPERATURE-SENSITIVE POLYMERS AND COPOLYMERS

Temperature-sensitive, "smart" polymers have been extensively studied over the past 5 to 10 years.[1–3,6–15,22] There are many polymers which exhibit a cloud point (CP) or lower critical solution temperature (LCST) in aqueous solutions. One property which is common to these water soluble-insoluble polymers is that they each have a balance of hydrophilic and hydrophobic groups. The main mechanism of a thermally induced phase separation is the release of hydrophobically bound water. This is the mechanism of precipitation as well as of physical adsorption of a soluble LCST polymer onto a solid polymer substrate.[23–25] If one increases or decreases the relative hydrophilic content of the temperature-sensitive polymer, this will usually cause an increase or decrease, respectively, in the LCST, and it will have a similar effect on its tendency to physically adsorb onto a particular solid polymer substrate.[25–27]

We have recently been studying intelligent copolymer systems with more than one stimulus-response component. These copolymers can exhibit very interesting properties, with many new and novel applications. If one combines temperature-sensitivity with pH-sensitivity in the same "smart" polymer, then the LCST of the copolymer may be especially sensitive to the pH, due to the strong hydrophilic character of the ionized state of the pH-sensitive component.

In order to retain both sensitivities, we have synthesized novel "hybrid intelligent" graft and block copolymer structures where the backbone polymer and grafted or block polymer chains each independently exhibit and retain a different stimulus-response sensitivity, in contrast to the random copolymers.[29]

COMBINING BIOMOLECULES AND INTELLIGENT POLYMERS

A large number of biologically active molecules ("biomolecules") may be combined with intelligent polymer systems. The biomolecules may be conjugated to pendant groups along a polymer backbone, or to one or both terminal ends of the polymer. In either case, the smart polymer may be (a) a soluble polymer, (b) a polymer grafted to a solid support, (c) a physically adsorbed polymer on a solid support, or (d) a polymer chain segment within a hydrogel.

One of the important aspects of the conjugation of a biomolecule to a polymer molecule is the possibility of conjugating many biomolecules to the same polymer molecule thereby providing the opportunity for significant amplification of the biological activity. The biomolecule may also be physically entrapped within a hydrogel, either permanently, as in the case of a large protein, or temporarily, as in the case of a small drug molecule. There are many diverse biomedical and biotechnological applications of environmentally sensitive "smart" polymeric biomaterials, whether or not they contain immobilized biomolecules.

STIMULI-RESPONSIVE POLYMERS IN AQUEOUS SOLUTIONS

Soluble, environmentally sensitive polymers in aqueous solutions can be precipitated at specific environmental conditions. Such systems can be useful as temperature or pH indicators, or as "on-off" light transmission switches.

When soluble smart polymers are mixed with liposome or cell suspensions, they may be phase-separated by a stimulus, and may interact with liposomal or cell membranes by hydrophobic interactions. This can cause dramatic changes, such as lysis.[31] We and others have conjugated cell or liposomal membrane components (e.g., phospholipids) and cell surface receptor ligands (e.g., the RGD peptide) to a temperature-responsive polymer.[32–36] After the conjugate interacts with the membranes and is caused to phase-separate, a gel may be formed, and cells may be reversibly cultured on surfaces.[35,36]

Another kind of "intelligent" polymer molecule in solution is based on the random incorporation of the key chemical groups of a known recognition sequence from a natural biomolecule along a water-soluble polymer backbone, and in the same ratio as in the natural recognition molecule.[37] If the polymer is a stimuli-responsive polymer, one might then be able to stimulate this polymer to change its conformation, enhancing the possibility of recognition.

STIMULI-RESPONSIVE POLYMERS ON SURFACES

Stimuli-responsive polymers can also be chemically grafted or physically adsorbed onto solid polymer supports, and then one can rapidly change surface film thickness, wettability, or surface charge in response to small changes in stimuli such as solution temperature, pH, or specific ionic concentrations.[25,38] These responses can be much faster than for solids as hydrogels since the surface coatings can be very thin. Permeation "switches" can be prepared by depositing "intelligent" polymers onto the surfaces of pores in a porous membrane, and stimulating their swelling (to block the pore flow) or collapse (to open the pore to flow).[39–42]

If proteins or cells are exposed to "intelligent" polymer surfaces that are in the swollen or collapsed state, they usually will preferentially adsorb on the more hydrophobic surface compositions.[43,44] One potential use of such a system is to cycle the temperature and thereby reduce fouling by adsorbed proteins or cells on surfaces. Another interesting application is to reversibly culture cells on a chemically grafted LCST polymer surface.

STIMULI-RESPONSIVE HYDROGELS

The most extensive work on "intelligent" polymers has been carried out on stimuli-responsive hydrogels, particularly those based on pH-sensitive monomers and thermally sensitive monomers.[1–3,6–14,22,28,45–50] Stimuli derived from changes in solvent mixtures, specific ions or solutes, pH, temperature, electric fields, or electromagnetic radiation have all been used to effect collapse or swelling of stimuli-responsive hydrogels. This action has led to application such as desalting and/or dewatering of protein solutions, microrobotics and artificial muscles, and "on-off" immobilized enzyme reactors.[2,3,13,51,52]

Delivery of drugs has been one of the most extensively studied application areas for stimuli-responsive hydrogels.[4–6,11–14,28,30,35,48,50] Application of cyclic temperature or electric field stimuli have resulted in cyclic delivery of physically incorporated drug molecules.[11,14,53]

KINETIC CONSIDERATIONS

The kinetics of the various "intelligent" responses described in this article will be very sensitive to the speed of the stimulus and the dimensions of the system being stimulated.

CONCLUSIONS

There are many diverse possibilities for the design of compositions, molecular structures and physical properties of "intelligent" polymers and hydrogels. Such diversity provides great opportunities for many diverse and novel applications in medicine and biotechnology.

ACKNOWLEDGMENTS

I would like to acknowledge the "stimulation" I received from the many "intelligent" and creative contributions of my students and colleagues over the past 10 to 15 years while working in this exciting field. I would also like to note that the reference list is relatively brief and therefore is only illustrative and not comprehensive, and I apologize for all omissions. This article was previously published in *Macromolecules*, Vol. 98, pp 645–664, 1995 and appears with permission.

REFERENCES

1. Tanaka, T. *Sci. Amer.* **1981**, *244*, 124.
2. Cussler, E. L. et al. *AIChE J.* **1984**, *30*, 578.

3. Cussler, E. L. U.S. Patent 4 555 344, 1985.
4. Ishihara, K. et al. *Polymer J.* **1984**, *16*, 625.
5. Heller, J. *Med. Dev. Diag. Ind.* **1985**, *7*, 34.
6. Hoffman, A. S. *J. Contr. Rel.* **1987**, *6*, 297.
7. Peppas, N. A.; Korsmeyer, R. W., Ed. "*Hydrogels in Medicine and Pharmacology*" CRC: Boca Raton, FL, 1987.
8. Bae, Y. H. et al. *Makromol. Chem.-Rapid Comm.* **1987**, *8*, 481.
9. Hoffman, A. S.; Monji, N. U.S. Patent 4 912 032, 1990.
10. Kost, J., ed. "*Pulsed and Self-Regulated Drug Delivery*" CRC: Boca Raton, FL, 1990.
11. Okano, T. et al. *J. Contr. Rel.* **1990**, *11*, 255.
12. Hoffman, A. S. *Mater. Res. Soc. Bull.* **1991**, *XVI*(9), 42.
13. DeRossi, D. et al., Eds. "*Polymer Gels*," Plenum: New York, 1991.
14. Kwon, I. C. et al. *Nature* **1991**, *354*, 291.
15. Osada, Y. et al. *Nature* **1992**, *355*, 242.
16. Dusek, K.; Patterson, D. *J. Polym. Sci., Part A-2* **1968**, *6*, 1209.
17. Verdugo, P. *Biophys. J.* **1986**, *49*, 231.
18. Tanaka, T. *Nature* **1987**, *325*, 796.
19. Tanaka, T. In: Nicolin, C. Ed. "*Structure and Dynamics*" Amsterdam: Nijhoff, 1987; p 237.
20. de Gennes, P. G. *Phys. Lett A38* **1972**, *339*.
21. Urry, D. et al. *JACS* **1988**, *110*, 3303.
22. Schild, H. G. *Progr. Pol. Sci.* **1992**, *17*, 163.
23. Heskins, H.; Guillet, J. E. *J. Makromol. Sci.-Chem, A2* **1968**, *1441*.
24. Monji, N. et al. *Biochem. Biophys. Res. Comm.* **1990**, *172*, 652.
25. Miura, M. et al. *J. Biomat. Sci.-Polym. Ed.*, **1994**, *5*, 555.
26. Taylor, L. D.; Cerankowski, L. D. *J. Polym. Sci., Polym. Chem.* **1975**, *13*, 2551.
27. Priest, J. H. et al. In: Russo, P., Ed. "*Reversible Polymeric Gels and Related Systems*" *ACS Sympos. Series* 350; ACS: Washington, DC, 1987, p 255.
28. Dong, L. C.; Hoffman, A. S. *J. Contr. Rel.* **1991**, *15*, 141.
29. Chen, G. H.; Hoffman, A.S. *Nature* **1995**, *373*, 49–52.
30. Chen, G. H.; Hoffman, A. S. *Macromol. Rapid Comm.* **1995**, *16*, 175.
31. Park, T. G.; Hoffman, A. S. Unpublished results; Seattle, WA: University of Washington, 1990.
32. Nightingale, J. A. S. et al. *Proc. Soc. for Biomat.* **1987**, *56*.
33. Ringsdorf, H. et al. *Macromol.* **1991**, *24*, 1678.
34. Schild, H. G.; Tirrell, D. A. *Langmuir* **1991**, *7*, 1319.
35. Wu, X. S. et al. *J. Intell. Mtls. Syst. and Struct.* **1993**, *4*, 202.
36. Miura, M.; Hoffman, A. S. Unpublished results; Seattle, WA: University of Washington, 1992.
37. Tardieu, M. et al. *J. Cell Physiol.* **1992**, *150*, 194.
38. Uenoyama, S.; Hoffman, A. H. *Radiat. Phys. Chem.* **1988**, *32*, 665.
39. Tirrell, D. A. *J. Contr. Rel.* **1987**, *6*, 15.
40. Iwata, H. et al. *J. Membr. Sci.* **1991**, *55*, 119.
41. Osada, Y. O. et al. *J. Membr. Sci.* **1986**, *27*, 327.
42. Yoshida, M. et al. *Rad. Eff. and Defects in Solids* **1993**, *126*, 409.
43. Kawaguchi, H. et al. *Colloid Polym. Sci.* **1992**, *270*, 53.
44. Okano, T. et al. *J. Biomed. Mater. Res.* **1993**, *27*, 1243.
45. Kuhn, W. et al. *Nature* **1950**, *165*, 514.
46. Ilavsky, M. *Macromol.* **1982**, *15*, 782.
47. Yu, H.; Grainger, D., Ph.D. Thesis, Yu, H. Beaverton, Oregon: Oregon Graduate Institute, 1994.
48. Siegel, R. A. et al. *J. Contr. Rel.* **1988**, *8*, 179.
49. Nakamae, K. et al. *Makromol. Chem.* **1992**, *193*, 983.
50. Dong, L. C. et al. *J. Contr. Rel.* **1992**, *19*, 171.
51. Park, T. G.; Hoffman, A. S. *Appl. Biochem. and Biotech.* **1988**, *19*, 1.
52. Park, T. G.; Hoffman, A. S. *Biotech. Bioeng.* **1990**, *35*, 152.
53. Bae, Y. H. et al. *ACS Sympos. Series* **1994**, *545*, 98.

Intercalated Complexes

See: DNA (Intercalation with Drugs)

Interchain Copolymerization

See: Reactive Processing, Thermoplastics

Interfacial Polycondensation

See: Cardo Polyesters (Heat Resistant and Photosensitive)
Organo-Tin Polymers
Polyimides (via Interfacial Polyfunctional Condensation)

Interpenetrating Polymer Networks

See: Blends (Silanol Functional Groups)
Hydrophilic Polymers (for Friction Reduction)
Interpenetrating Polymer Networks (Overview)
Interpenetrating Polymer Networks, Rubber-Based
Reaction Injection Molding (RIM) Materials
Silicone Elastomers
Unsaturated Polyester Resins (Overview)
Unsaturated Polyester Resins (Polyester Composites)

INTERPENETRATING POLYMER NETWORKS (Overview)

L. H. Sperling
Materials Research Center
Lehigh University

Vinay Mishra
Department of Chemical Engineering
Center for Polymer Science Materials Research Center
Lehigh University

Interpenetrating polymer networks (IPNs) form a special class of polymer blends in which both polymers generally are in network form. The two networks share the same region of space, that is, the macroscopic volume of the sample. This blend of two network polymers is "physical" in nature; ideally, there are no chemical grafts between them. Yet IPNs must be made "chemically," since physically blending two networks is not possible without breaking chemical bonds. So at least one polymer must be polymerized and/or crosslinked in the immediate juxtaposition of the other.[1]

Most IPNs phase separate just as their corresponding (linear) blends do because of their low entropy of mixing and often positive heats of mixing. However, multiple catenane type structures restrict the domain size significantly in IPNs. Thus, where two linear immiscible polymers may not form a useful blend owing to coarse phase separation, their corresponding IPN may have a morphology controlled by the crosslink densities of the two networks. This allows for better combinations of the useful properties of two immiscible polymers, resulting in materials with improved tensile and impact strength, flexibility, chemical and solvent resistance, weatherability, flammability resistance, reduced creep and novel biomedical applications. Another result of incomplete phase separation and fine morphologies in IPNs is that they can have a single broad glass transition, ranging from the glass transition of one homopolymer to another. Such IPNs form excellent sound and vibration damping materials over a wide temperature range. Very often, IPNs are clear or translucent against light despite phase separation because the domain sizes are smaller: of the order of the visible light wavelength.

The single most common system found in the scientific and engineering literature involves *cross*-polyurethane-*intercross*-poly(methyl methacrylate) (PU/PMMA IPNs). This system constitutes an excellent model for understanding the various aspects of IPN technology and is therefore frequently chosen throughout this article as an example.

IPN CLASSIFICATION BY SYNTHESIS

There are two general methods of IPN synthesis: sequential (resulting in the so-called sequential IPNs, or IPNs) and simultaneous (resulting in simultaneous interpenetrating networks, or SINs).[1–7]

MORPHOLOGY AND GLASS TRANSITION BEHAVIOR

Frisch and co-workers[10] studied IPNs of the thermodynamically miscible polymer combination of poly(2,6-dimethyl phenylene ether) and polystyrene (PPO-PS). These IPNs were single phased in all compositions and exhibited single glass transitions, varying with IPN compositions. Such systems are truly interpenetrating on the molecular and segmental scales but also are very rare. Most IPNs phase separate, although the degree of dispersion is much higher compared with corresponding linear blends. In sequential IPNs, network I tends to be continuous, with network II forming the dispersed phase whose dimensions are of the order of the distance between adjacent crosslinks of network I. In SINs, the network gelling first tends to form the more continuous phase.

SOUND AND VIBRATION DAMPING WITH IPNS

By selecting the proper sequence of the three important events (gelation of polymers I and II and phase separation of I from II), many IPNs can be prepared with high degrees of phase dispersion and accompanying broad glass transitions. In the special case of a microheterogeneous morphology, when domains are of the order of 10–20 nm, the whole system is virtually all interphase material. Subsequently, a single glass transition results, which is broadened over the temperature range between the glass transitions of the two pure polymers. Such IPNs form excellent sound and vibration damping materials.[11] One such known material is based on vinyl-phenolic IPNs.[11a] Yao[7b] recently reviewed the IPN technology devoted to achieving high damping behavior over a wide temperature. Klempner and co-workers extensively studied damping properties of IPNs of polyurethanes and a variety of other polymers.[1g,5a] Other contributors in this area are Greenhill and Hourston,[12] Chen et al.,[13] and Chang et al.[14]

ENGINEERING AND PHYSICAL BEHAVIOR

Because of their high and controllable degrees of phase dispersions, several IPN systems have high tensile and impact strengths and other advantages as engineering materials. For example, epoxy-poly(*n*-butyl acrylate) SINs were tougher than the pure epoxy with some sacrifice in the initial modulus.[9] In PU-PMMA SIN system, tensile strength and ultimate elongation were increased considerably over corresponding linear blends as well as semi- and pseudo-IPNs. In general, by selecting polymers of widely different glass transitions, useful materials ranging from reinforced elastomers to toughened plastics can be prepared, depending on which phase is continuous.

Control in IPN morphology also provides subsequent control in the permeability coefficients of several gases and vapors through IPN membranes. Some interesting applications resulting from this are in pervaporation and medical wound dressings. Pervaporation is a membrane separation process that can be used to separate two liquids from their mixture.

PHASE SEPARATION BEHAVIOR

The morphology of an IPN depends on the interplay among several factors: reaction kinetics, miscibility and dynamics of phase separation.[1i] Lipatov and co-workers[1j,16] have pointed out the nonequilibrium conditions under which phase separation occurs and morphology develops in IPNs. Specifically, phase separation may proceed simultaneously with polymerization due to increased thermodynamic immiscibility.

APPLICATIONS

Many commercial applications of IPN-based materials exist. Some prominent examples are ion-exchange resins, damping materials and automotive parts, the latter including bumpers, tires, and hoses. Other include artificial teeth, medical devices and sheet molding compounds. IPN applications in membrane separation of gases and liquids is being investigated in Korea.[8,15] Pervaporation is potentially useful where conventional distillation techniques are difficult to use, such as fractionation of azeotropic or isomeric mixtures.

Ion-exchange resins are another interesting application of IPNs.

Several other potential applications of IPN materials exist. IPNs prepared with a renewable resource material, such as castor oil or vernonia oil, are economically attractive, especially in countries rich in these resources, for example China, India and Colombia. In particular, numerous investigations on IPNs from a castor oil-based polyurethane and a rigid polymer like polystyrene or PMMA have been carried out over the last decade.[17,18] These materials have excellent engineering properties and make tough abrasion-resistant materials. The outstanding abrasion resistance (to flying sand) of IPNs based on castor oil-polyurethanes and styrenic-acrylic copolymers has led to coatings for the Ge Zhou Ba Hydroelectric Power Station in the People's Republic of China.[11]

ACKNOWLEDGMENT

The authors thank the 3M company for their financial support in the form of a doctoral fellowship for V. M.

REFERENCES

1. Klempner, D.; Sperling, L. H.; Utracki, L. A. Eds. *Interpenetrating Polymer Networks* ACS Books, Advances in Chemistry 239, American Chemical Society: Washington, DC, 1994. *Chapters cited a.* Eschbach, F. O.; Huang, S. J. Chapter 9 *b.* Bauer and Briber, Chapter 7 *c.* Ma, S.; Tang, X. Chapter 20 *d.* Hsieh, K. H.; Lee, S. T.; Liao, D. C.; Wu, D. W.; Ma, C. C. M. Chapter 21 *e.* Holdsworth, Hourston Chapter 22 *f.* Liucheng, Z.; Xiucuo, L.; Tianchang, L. Chapter 18 *g.* Sophiea, D.; Klempner, D.; Sendijarevic, V.; Suthar, B.; Frisch, K. C. Chapter 2 *h.* Dillon, M. Chapter 19 *i.* Utracki, L. A. Chapter 3 *j.* Lipatov, Y. S. Chapter 4 *k.* Sperling, L. H. Chapter 1 *l.* Tan, P. Chapter 29.
2. Sperling, L. H. *Interpenetrating Polymer Networks and Related Materials* Plenum: New York, 1981.
3. Sperling, L. H. in Paul, D. R.; Sperling, L. H. Eds. *Multicomponent Polymer Materials* ACS Books, Advances in Chemistry 211, American Chemical Society: Washington, DC, 1986.
4. Klempner, D.; Frisch, K. C. Eds. *Advances in Interpenetrating Polymer Networks*; Plenum: Lancaster, PA, 1989; Vol. I. *a.* Silverstein, M. S.; Narkis, M. p. 117.
5. Klempner, D.; Frisch, K. C. Eds. *Advances in Interpenetrating Polymer Networks* Plenum: Lancaster, PA, 1990; Vol. II. *a.* Klempner, D.; Muni, B.; Okoroafor, M.; Frisch, K. C. p. 1. *b.* Suthar, B. p 281.
6. Klempner, D.; Frisch, K. C. Eds. *Advances in Interpenetrating Polymer Networks* Plenum: Lancaster, PA, 1992; Vol. III.
7. Klempner, D.; Frisch, K. C. Eds. *Advances in Interpenetrating Polymer Networks* Plenum: Lancaster, PA, 1994; Vol. IV.
8. Lee, D. S.; Kang, W. K.; An, J. H.; Kim, S. C. *J. Membr. Sci.* **1992**, *75*, 15.
9. Touhsaent, R. E.; Thomas, D. A.; Sperling, L. H. in *Toughness & Brittleness of Plastics*; Deanin, R. D.; Crugnola, A. M. Eds.; Adv. Chem. Ser. No. 154, ACS: Washington, DC, 1976.
10. Frisch, H. L.; Klempner, D. et al. *Macromolecules* **1980**, *13*, 1016.
11. Corsaro, R. D.; Sperling, L. H. Eds. *Sound and Vibration Damping with Polymers*; ACS Symposium Series 424, American Chemical Society: Washington, DC, 1990. *a.* Yamamoto, K.; Takahashi, A. (Chapter authors).
12. Greenhill, D. A.; Hourston, D. J. *Polym. Mater. Sci. Eng.* **1989**, *60*, 644.
13. Chen, Q.; Ge, H. et al. *J. Appl. Polym. Sci.* **1994**, *54*, 1191.
14. Chang, M. C. O.; Thomas, D. A.; Sperling, L. H. *J. Polym. Sci.: Polym. Phys.* **1988**, *26*, 1627.
15. Kim, S. C. *Makromol. Chem. Macromol. Symp.* **1991**, *51*, 79.
16. Lipatov, Y. S.; Alekseeva, T. T. et al. *Polymer* **1992**, *33*(3), 610.
17. Song, M.; Donghua, Z. *Plastics Industry* **1987**, *2*, 42.
18. Patel, P.; Shah, T.; Suthar, B. *J. Appl. Polym. Sci.* **1990**, *40*, 1037.

INTERPENETRATING POLYMER NETWORKS, RUBBER-BASED

Harry L. Frisch and Yongpeng Xue
Department of Chemistry
State University of New York at Albany

Interpenetrating polymer networks (IPNs) represent a mode of blending two or more crosslinked polymer networks to produce a polymer alloy[1,2] in which phase separation is limited due to permanent entanglements occurring between the topologically linked networks with no covalent bonds.[3,4] The compatibility of IPNs can be enhanced because the polymers are interlocked in a three-dimensional network structure during polymerization (crosslinking) before phase separation occurs. There are several ways to prepare IPNs to achieve a unique combination of properties from the component polymers.[5] IPNs generally can be made by sequential or simultaneous crosslinking of the component networks; in the later case the resulting IPNs are designated as SINs. A pseudo-IPN (or semi-IPN) is a polymer alloy in which only one of the polymers is crosslinked. For the weakly crosslinked interpenetrating networks, the statistical theory of rubber networks suggests that sequentially formed networks are comparatively less stable than simultaneously crosslinked ones;[6] it also states that under certain conditions single phase IPNs can be formed from incompatible linear polymers.

One of the attractions of SINs is the possibility of making systems that are potentially easy to process compared with most types of IPN.[5,7] SINs allow for some degree of control of mutual miscibility of the constituent polymer chains, and they can produce miscible IPNs at all compositions even from incompatible chains, such as poly(2,6-dimethyl-1,4-phenylene oxide) (PPO) with polybutadiene (PB) (cis and trans),[8] high molecular weight natural rubber (NR) with poly(carbonate-urethane) (PCU) at large average molecular weight between crosslinks (M_c) of crosslinked NR,[9] etc. The thermodynamic forces leading to phase separation are so powerful that these IPNs show microphase separation at all compositions.[10] In recent years a number of miscible or partially miscible simultaneous rubber-based IPNs were prepared from NR and some commercially available synthetic rubbers in our laboratory.

NR-BASED IPNS

A series of simultaneous IPNs (including full IPNs and pseudo-IPNs) containing high molecular weight NR from Brazilian *Manihot glaziovii* and another plastic or elastomeric component were prepared and characterized.[9,11–14] The high molecular weight NR was always crosslinked by a free radical mechanism by some vinyl crosslinkers when it was in either a pure network form or IPNs. Two papers on SINs or NR and PPO[11,12] were devoted to the phase morphology investigations for these materials. PPO is a well-known commercial plastic resin. In the SINs the PPO network is crosslinked by the condensation reaction of suitable diamines with methyl brominated PPO.[15] A single phase morphology is found in full IPNs with either high or low weight percent NR. The intermediate composition full IPNs, pseudo-IPNs, and linear blends all exhibited microphase separation.

Frisch and DeBarros[13] and Frisch and Chen[14] have made and studied a number of single-phased pseudo-IPNs from linear high molecular weight NR and crosslinked PCU network. One of the interesting properties of these single-phased pseudo-IPNs is their electrical conductivity upon doping with iodine. The authors believe that in these pseudo-IPNS the linear NR chains are fully entangled within the meshes of the crosslinked PCU network. Upon iodine doping, the clear pseudo-IPN materials darken to a violet black and develop a metal-like cluster. The electrical dc conductivity, σ, of the pseudo-IPN rises on I_2

doping by eight orders of magnitude to ~10^{-4}S·cm^{-1}. I_2 doping of the pure NR causes a rise in σ also by eight orders of magnitude to 10^{-4} S·cm^{-1}. The iodine-doped pure crosslinked PCU has a conductivity an order of magnitude smaller than this. A simple theory based on the assumption that conduction occurs essentially along the linear NR chains, composing a percolation cluster,[14,16,17] in the iodine-doped pseudo-IPNs of NR and PCU accounts for the observed electrical conductivity dependence on temperature, iodine molarity, and weight fraction of NR at or near room temperature.

POLYBUTADIENE RUBBER (PBR)-BASED IPNS

Sequential IPNs based on PBR with another glassy polymer such as PS or a copolymer have been well documented in the literature.[18–24] All of the above IPNs prepared by a sequential mode show microphase separation even though the domain sizes and domain distribution may vary from specimen to specimen. Ghosh and Ray[24] studied the sequential IPN's of PBR with PS and found that the full IPNs exhibited higher tensile strength and modulus, while the pseudo-IPNs exhibited greater toughness. They suggest that the higher mechanical properties of the full IPNs as compared with those of the corresponding pseudo-IPNs are apparently linked to the presence of finer PBR domains in the former.

Frisch and Hua[8] synthesized and investigated the physical properties and morphology of SINs of PB (cis and trans) and PPO. We have also prepared and studied a simultaneous IPN based on hydroxyl-terminated liquid PBR and PPO in which both polymer networks are crosslinked by condensation reactions.[25] The phase morphology of the PPO/PB (diol) SINs is highly dependent on the mass ratio of the two polymer components. A single-phase morphology of the SINs is achieved when the PB (diol) content is <40 wt % in the initial composition.

POLYCHLOROPRENE RUBBER (CR) BASED-IPNS

CR has the distinction of being the earliest commercially successful synthetic rubber. The vulcanizates of CR possess excellent oil resistance, chemical resistance, weather resistance, and nonflammability.

We have synthesized and characterized a novel simultaneous IPN based on high molecular weight CR (M_w = 2.9 × 10^5, Aldrich) and PCU.[26] Here CR was crosslinked by a free radical reaction under mild temperature conditions. All CR/PCU SIN materials are tough and flexible. A single-phase morphology of the IPNs was achieved when the content of the CR component was below 50 wt %. The microphase separation of the component networks in the IPNs occurred in samples whose weight percentage of the CR component was 50% and higher. We have also reported on an interesting pseudo-SIN based on high molecular weight CR and cis-PBR.[27] We expect that novel combination properties from high molecular weight CR with other high molecular weight diene polymers can be produced by using the pseudo-IPN approach.

SUMMARY

Completely miscible, single-phase rubber-based SINs can be made from components whose linear polymer chains are wholly immiscible [e.g., PB (cis and trans)–PPO, NR–PCU (with high M_c of NR), PB–PCU]. In some cases, single-phase SINs that are miscible over only a portion of the composition range [e.g., NR–PPO, CR–PCU, PB (diol)–PPO], whereas other IPNs are phase separated over the whole composition range [e.g., PB(cis)–PPO]. Generally, the investigated rubber-based pseudo-SINs are phase separated. Only in the example of NR/PCU pseudo-IPNs in which linear NR possess an exceptionally high molecular weight are they single phased.

ACKNOWLEDGMENT

This work was supported by the National Science Foundation, Grant DMR 9023541.

REFERENCES

1. *Polymer Alloys* Klempner, D.; Frisch, K. C., Eds.; Plenum: New York, 1977.
2. *Polymer Alloys II* Klempner, D.; Frisch, K. C., Eds.; Plenum: New York, 1980.
3. Frisch, H. L.; Klempner, D.; Frisch, K. C. *Polym. Lett.* **1969**, *7*, 775.
4. Klempner, D.; Frisch, H. L.; Frisch, K. C. *J. Polym. Sci. A-2* **1970**, *8*, 921.
5. Sperling, L. H. *Interpenetrating Polymer Networks and Related Materials* Plenum: New York, 1981.
6. Binder, K.; Frisch, H. L. *J. Chem. Phys.* **1984**, *81*, 2126.
7. Touhsaent, R. E.; Thomas, D. A.; Sperling, L. H. *J. Polym. Sci., Part C* **1974**, *46*, 175.
8. Frisch, H. L.; Hua, Y. *Macromolecules* **1989**, *22*, 91.
9. DeBarros, G. G.; Frisch, H. L.; Travis, J. L. In *Elastomeric Polymer Networks, A Memorial to Eugene Guth* Mark, J. E.; Erman, B., Eds.; Prentice Hall: Englewood Cliffs, NJ, 1992; pp 313–326.
10. Frisch, H. L.; Xue, Y. unpublished results.
11. DeBarros, G. G.; Huang, M. W.; Frisch, H. L. *J. Appl. Polym. Sci.* **1992**, *44*, 255.
12. Xue, Y.; Chen, Z.; Frisch, H. L. *J. Appl. Polym. Sci.* **1993**, *51*, 1835.
13. DeBarros, G. G.; Frisch, H. L. *J. Polym. Sci., Part A* **1992**, *30*, 937.
14. Frisch, H. L.; Chen, Z. *J. Polym. Sci., Part A* **1994**, *32*, 1317.
15. Fyvie, T. J.; Frisch, H. L.; Semlyen, J. A.; Mark, J. E. *J. Polym. Sci., Part A* **1987**, *25*, 2503.
16. Frisch, H. L.; Chen, X. *J. Polym. Sci., Part A* **1993**, *31*, 3307.
17. Forsyth, M.; Shriver, D. F.; Ratner, M. A.; DeGroot, D. C.; Kannewurf, C. R. *Chem. Mat.* **1993**, *5*, 1073.
18. Curtius, A. J.; Covitch, M. J.; Thomas, D. A.; Sperling, L. H. *Polym. Eng. Sci.* **1972**, *12*, 101.
19. Fernandez, A. M.; Wignall, G. D.; Sperling, L. H. *Polym. Mater. Sci. Eng.* **1984**, *51*, 478.
20. Kim, J. W.; Cho, Y. G.; Huh, Y.; Kim, S. K.; Cho, C. S. *Pollino (Korea)* **1985**, *9*, 368.
21. An, J. H.; Fernandez, A. M.; Sperling, L. H. *Polym. Mater. Sci. Eng.* **1986**, *55*, 169.
22. An, J. H.; Fernandez, A. M.; Sperling, L. H. *Macromolecules* **1987**, *20*, 191.
23. Burford, R.; Chaplin, R.; Mai, Y. M. *Contemp. Top. Polym. Sci.* **1989**, *6*, 699.
24. Ghosh, P.; Ray, P. *J. Mater. Sci.* **1991**, *26*, 6104.
25. Xue, Y.; Frisch, H. L. *J. Polym. Sci., Part A* **1994**, *32*, 257.
26. Xue, Y.; Frisch, H. L. *J. Polym. Sci., Part A* **1993**, *31*, 2165.
27. Xue, Y.; Chen, Z.; Frisch, H. L. *J. Appl. Polym. Sci.* **1994**, *52*, 1833.

INTERPHASES, TERNARY POLYMER COMPOSITES

Michael Hess
Center for Material Science
University of North Texas

Robert Kosfeld
Department of Physical Chemistry
Gerhard-Mercator University

Thermoplasts are usually modified with fillers to improve mechanical stability and modulus. Elastomers are added to increase impact resistance, especially at low temperatures. Unfortunately, the effects of these modifiers counteract one another, so that a mineral filler generally decreases impact resistance and an elastomer phase decreases the modulus. Consequently, the question arises whether it is possible to optimize these properties by proper choice of the filler type and amount and of the amount of elastomer.

Polypropylene (PP) is–after low density polyethylene (LDPE) and poly(vinyl chloride) (PVC)–one of the most widely used thermoplasts, mainly in the automotive industry and as insulating materials in electrical engineering. Hence, binary systems of the type described above and with PP as the main component have been investigated intensively. The number of investigations of ternary systems, however, is significantly lower,[4–25] and only a few authors have focused on the correlation between mechanical properties and morphological structure and the existence of an interphase. In binary systems, this interphase consists of a polymer layer around the filler particles.[26] Its properties differ in general from those of the bulk polymer.[1–3,23–25] In the following article, the properties and morphological features of interphases and their effect on macroscopic thermal and mechanical behavior are discussed with respect to optimization of impact resistance and modulus.

RESULTS AND DISCUSSION

Analysis of the Interphase

According to Lipatov,[26] and as verified by Maurer et al.,[1–3] a filled polymer may be modeled as a three-phase system consisting of the polymer matrix, an interlayer of a certain thickness, which is an interphase rather than an interface, and the filler particle itself. At the surface of the filler, interactions with the polymer lead to a more or less tightly bound layer of polymer, which can only partially be removed by a solvent. The formation of these interphases has been discussed by de Gennes.[27–29] The residual polymer layer on the filler is now defined as the interphase.

Mechanical Properties of the Interphase

Several existing models describe the torsional modulus $G'(T)$ of a composite. Linear and inverse mixing rules give only upper and lower limits, respectively, between which the actual behavior of the composites is found. The suspension model[30,31] takes into account the volume fractions but not the values of the modulus of the components. The Takanayagi model[32] works quite well below the glass transition temperature T_g but deviates at higher temperatures. A satisfactory description of $G'(T)$ of the composites–not only at temperatures below T_g–is given by a modified[1] van der Poel model.[33–37] This turned out to be applicable for ternary systems as well.[25]

Mechanical Properties of Ternary Composites: Influence of Morphology

Compared with the PP–filler system, significant differences are observed when EPDM is also present. The reason for this behavior is that the interphases in ternary composites differ markedly from those of the binary PP–filler composites. EPDM–filler interactions depend on their relative strength. The question of whether the presence of the filler results in a partial miscibility of PP and EPDM is still open.

According to Pukansky et al.,[21] the surface interaction with EPDM is good when the surface energy of the filler is high. When it is low, no EPDM interphase is built, and the elastomer is dispersed in the matrix separated from the filler. When the dispersion is more homogeneous, much better mechanical values are to be expected. Calculations of the torsional modulus of the interphase show significant differences between binary PP–filler systems and ternary PP-filler–EPDM systems.

Calorimetric experiments confirm that the observed material properties are not due to effects in crystallinity: the specific enthalpy of fusion of PP is found to be nearly independent of the composition. In contrast, the crystallinity of EPDM increases with the filler content and does not depend on the filler type.

ACKNOWLEDGMENTS

The authors wish to thank Dr. Theodor Uhlenbroich, Sachtleben Chemie, Duisburg, Germany, for helpful discussions and for providing fillers and compounding; the Hüls A. G., Marl, Germany, for providing the polymers; and Dr. K. Schaefer for carrying out the measurements. Dr. Burchard and Dr. Klar, Institut für Rasterelektronenmikroskopie der RWTH Aachen, Germany, and Prof. Dr. Wenig, Physics Department, University of Duisburg, Germany, provided us with SEM investigations and measurements of the X-ray crystallinity, respectively. Finally, the Arbeitsgemeinschaft Industrieller Forschungsförderer e.V. (AIF), Cologne, Germany, is gratefully acknowledged for financial support.

REFERENCES

1. Maurer, F. H. J. Ph.D. Thesis, University of Duisburg, 1983.
2. Maurer, F. H. J.; Kosfeld, R.; Uhlenbroich, T.; Bosveliev, L. G. Proceedings of the 27th Annual Symposium on Macromolecules, Strasbourg, 1981; Vol. 2, p 1251.
3. Maurer, F. H. J. *Progr. Sci. Eng. Comp. ICCM-IV* Hayashi, T, Ed.; North Holland: Tokyo, 1984.
4. Serafimov, B. *Plaste u. Kautschuk* **1986**, *33*, 331.
5. Serafimov, B. *Plaste u. Kautschuk* **1985**, *32*, 299.
6. Serafimov, B. *Plaste u. Kautschuk* **1982**, *29*, 598.
7. Serafimov, B. *Plaste u. Kautschuk* **1979**, *26*, 39.
8. Chu, K. C.; Wright, A. N.; Woodhams, R. T. Presented at the Annual Conference on Reinforced Plastics/Composites Institute; The Society of Plastics Industries, 1986, 1/Sess. 21-C.
9. Scott, C.; Ishida, H.; Maurer, F. H. J. *J. Mater. Sci.* **1987**, *22*, 3963.
10. Scott, C.; Ishida, H.; Maurer, F. H. J. In *Composite Interfaces* Ishida, H.; Koenig, J. Eds.; North Holland: London, 1986, p 58.

11. Scott, C.; Ishida, H.; Maurer, F. H. J. Presented at the 42nd Annual Conference on Reinforced Plastics/Composites Institute; The Society of Plastics Industries, 1987; 1/Sess. 19-E.
12. Scott, C.; Ishida, H.; Maurer, F. H. J. *Rheol. Acta* **1988**, *27*, 273.
13. Stamhuis, J. E. *Polym. Comp.* **1984**, *5*, 292.
14. Stamhuis, J. E. *Polym. Comp.* **1988**, *9*, 72.
15. Stamhuis, J. E. *Polym. Comp.* **1988**, *9*, 280.
16. Faulkner, D. L. *J. Appl. Polym. Sci.* **1988**, *36*, 467.
17. Fernando, P. L. *Polym. Eng. Sci.* **1988**, *28*, 806.
18. Pukansky, B.; Tüdos, F.; Kelen, T. *Polym. Comp.* **1986**, *7*, 106.
19. Zerjal, B.; Musil, V.; Pegrad, B.; Malavasic, T. *Thermochim. Acta* **1988**, *134*, 139.
20. Kolarik, J.; Pukansky, B.; Lednicky, F. In *Interfaces in Polymer, Ceramic and Metal Matrix Composites*; Ishida, H. Ed.; Elsevier: New York, 1988; p 453.
21. Pukansky, B.; Tüdos, F.; Kolarik, J.; Lednicky, F. *Polym. Comp.* **1990**, *11*, 104.
22. Kolarik, J.; Lednicky, F.; Jancar, J.; Pukansky, B. *Polym. Comm.* **1990**, *31*, 201.
23. Theisen, A.; Hess, M.; Kosfeld, R. In *Integration in Fundamental Polymer Science and Technology Vol. V*; Lemstra, P.; Kleintjens, L. Eds.; Elsevier Science: New York, 1991; p. 72.
24. Schaefer, K.; Hess, M.; Kosfeld, R.; Uhlenbroich, T. *Kunststoffe-German Plastics* **1990**, *80*, 1363.
25. Schaefer, K.; Theisen, A.; Hess, M.; Kosfeld, R. *Polym. Eng. Sci.* **1993**, *33*, 1009.
26. Lipatov, Y. S. *Adv. Polym. Sci.* **1977**, *22*, 1.
27. de Gennes, P. G. *Rev. Mod. Phys.* **1985**, *57*, 827.
28. de Gennes, P. G. *Adv. Polym. & Interface Sci.* **1987**, *27*, 189.
29. de Gennes, P. G. *Macromolecules* **1980**, *13*, 1069.
30. Nielsen, L. E. *Mechanical Properties of Polymers and Composites Volume II*; Marcel Dekker: New York, 1974.
31. Einstein, A. *Ann Physik.* **1906**, *19*, 289.
32. Takanayagi, M.; Nemura, S.; Minami, S. *J. Polym. Sci.* **1964**, *C5*, 113.
33. van der Poel, C. *Rheol. Acta* **1958**, *1*, 198.
34. Kerner, E. H. *Proc. Phys. Soc.* **1956**, *B69*, 802.
35. Lewis, T. W.; Nielsen, L. E. *J. Appl. Polym. Sci.* **1970**, *14*, 1449.
36. Dickie, P. *J. Appl. Polym. Sci.* **1973**, *17*, 45.
37. Halpin, J. C. *J. Comp. Mater.* **1969**, *3*, 732.

Interpolymer Interactions

See: Molecular Recognition (Interpolymer Interactions)

INTERPOLYMERS

I. I. P. E. Vointseva and A. N. Nesmeyanov
Institute of Organo-Element Compounds
Russian Academy of Sciences, Moscow

High-molecular polymers containing reactive functional groups as their low-molecular analogs can enter various chemical reactions. Topological structure and the properties of the resulting polymers will essentially depend on the amount and the arrangement of functional groups in the macromolecule.

When the functional groups are arranged at the ends of the reacting macromolecules, the chemical interaction of polymers leads to formation of block copolymers with a linear arrangement of blocks. When one of the two reacting polymers contains functional groups in the repeating units of chain and the other ones at the chain ends, the reaction results in graft copolymers.

But if the functional groups are arranged in the repeating units of each of the reacting polymers, the chemical reaction between them leads to the formation of interpolymers (IPs).

(IPs) are two-strand branched polymacromolecular compounds in which the macromolecules of different types are connected along the chains with covalent bonds.

The number of resulting interchain covalent bonds characterizes the conversion degree of IP reaction (θ) and may differ, depending on the amount of the reacted functional groups of each polymer.

The IP polymacromolecule can contain different numbers of chemically combined initial macromolecules of different types.

The chemical reaction in solution between two different polymers containing functional groups in each repeating unit (with the exception of abnormal units) may occur as follows: as a result of initial collision between different macromolecular coils, one or a small number of chemical bonds develop between different macromolecules, and a hybrid polymer of branched structure is formed.

In each repeating unit of the different macromolecules that form the IP, there are reactive functional groups. This offers a principal possibility of subsequent plural chemical interactions of already bonded macromolecules. In this case, if the reacting polymers are incompatible, their chains are repelled, hindering the following intramolecular interaction of functional groups of already bonded macromolecular coils.

But if there is affinity between the functional groups of polymers (for instance, donor–acceptor interaction), the following interaction of functional groups will be facilitated and the IP reaction can proceed cooperatively, as in the case of IP complex formation.

After formation of the first chemical bond, the process can develop intra- or intermolecularly.

The ability of the polymers to enter into IP reaction depends initially on the reactivity of their functional groups; this can be evaluated by the ability of each of the polymers to enter the corresponding polymer analogous reactions.

It is also necessary to take into consideration the presence of a common solvent for reacting polymers and their compatibility in solution.

Only in the solution, can such conditions be selected that cause predominantly an intramolecular process development and lead to the formation of soluble IPs with a maximum number of chemical links between macromolecules ("a regime of intramolecule interactions").

Polymer compatibility in solution affects the ability of reacting macromolecular coils to mutual penetration–that is, the accessibility of functional groups for the reaction.

POLYMERS THAT HAVE A COMMON SOLVENT, BUT ARE INCOMPATIBLE IN SOLUTION

The interaction of polystyrene (PS) with poly(1,1,2-trichlorobutadiene) (PTCB) and poly(vinyl chloride) (PVC) in the presence of $AlCl_3$ (the Friedel–Crafts IP reaction) were studied as a model reaction of this type.[1–3]

The conversion of functional groups in the IP reactions between PTCB or PVC with PS usually varies from 1–10%. In the corresponding polymer analogous reactions between PTCB or PVC and benzene, this value reaches 85% and 50%, respectively. Apparently, in polymer analogous reactions, the molecules of low molecular reagent (serving simultaneously as solvent) easily penetrate deeply into the swollen macromolecular coil. In the IP reaction with participation of incompatible polymers, the mutual penetration of reacting macromolecular coils is restricted, and most of their functional groups (90–99%) remain inaccessible for interaction; the reaction occurs only in the limited zone of mutual penetration.

The rate ratios of each stage of the IP reactions depend on the conditions of its conduction.

Under mild conditions of the "regimes of intramolecular interactions," the rate of intramolecular interactions of functional groups of the bonded coils is higher than the intermolecular interaction of unbounded coils.

In contrast, under severe conditions, called as the "regime of crosslinking", the intermolecular interaction on nonbonded macromolecular coils predominates.

The influence of the molecular mass of initial polymers on the structure and properties of the interpolymers formed is reported.

If the degrees of polymerization ($\overline{P}_w$) of the reacting polymers are close, $\overline{M}_{IP}$ is lower or equal to the sum of M_w of the reacting polymers. Most of the particles of such IP are formed in the interaction of two or three initial macromolecules.

If $\overline{P}_w$ of the reacting polymers differs by several factors, then $\overline{M}_{IP}$ is essentially higher than the sum of $\overline{M}_w$ of the initial polymers–that is, the particles of the soluble IP are being formed from a large number of initial macromolecules.

For each pair of polymers with the chosen values of $\overline{M}_w$ conditions corresponding to the regime of "intramolecular interactions" can be easily selected by means of parameter $[\eta]_m C_0$, where $[\eta]_m$ is the intrinsic viscosity of the initial polymer mixture in the same solvent in which the reaction is conducted. This parameter takes into account the combined effect of the molecular weights and the ratio of the units of reacting polymers, polymer concentration in solution, and the thermodynamic characteristic of the solvent on the process course.

If the concentration of the polymer mixture in solution is lower than crossover ($[\eta]_m C_0 < 1$), fully soluble IPs are formed.

If the concentration of the polymer mixture in solution is higher than crossover ($[\eta]_m C_0 > 1$), at some time a gel is formed in the system.

The optimum concentration of the initial polymer mixture in a "poor" solvent (nitrobenzene, dichloroethane) corresponds to crossover ($[\eta]_m C_0 = 1$).

Thus, by changing the molecular weights of initial polymers and the conditions of the IP reaction, the topological structure and the properties of both soluble and network IPs can be varied.

The IPs maintain specific properties characteristic of each of the initial homopolymers.

POLYMERS THAT ARE COMPATIBLE AT THE COST OF SPECIFIC INTERACTION OF FUNCTIONAL GROUPS BUT HAVE NO COMMON SOLVENT

This reaction is exemplified by the interaction of PTCB with polyhexamethyleneguanidine (PHMG).[4]

POLYMERS THAT ARE INCOMPATIBLE AND HAVE NO COMMON SOLVENT

The polymers such as chlorinated and chlorosulfonated polyethylene (CSPE) contain the mobile chlorine atoms but have no double bonds in the macromolecule.

These polymers enter into the IP reaction with polyethylenimine (PEI) and PHMG, but there are no features of a specific interaction of functional groups of polymers, since no polyconjugation effects arise in the reaction course.

The main field of application of IPs based on PHMG and chlorine-containing polymers is associated with the biocide properties of PHMG. This polymer has a broad region of bactericidal activity: antimicrobe, antivirus, antifungus activity; it is a substance of low toxicity without color and odor, but its water solubility hinders it prolonged fixation on the surface.

The IP reactions allow fixation of PHMG on the surface as water-insoluble protecting coatings.

As a result, on the basis of IPs, the coatings are capable of preserving, for long periods of time, the reservoirs with water (vessels, swimming baths) from bioencrustation and make water antiseptic in a vessel, even on a rather severe microbe attack.

CONCLUSION

Performance of irreversible IP reactions in solution offers wide possibilities of preparation of modified polymeric materials with specific properties (conductivity, bactericide, ion-exchange). These materials combine the properties of polymers taken into reaction and reveal enhanced strength characteristics.

Formation of chemical bonds between different macromolecules allows for overcoming many drawbacks of the polymer mixture caused by incompatibility of most of the polymer pairs.

The synthesis of IPs can be especially useful in the case when hybrid polymers cannot be obtained from monomers by copolymerization or copolycondensation. New advances in preparing modified polymers can be expected when secondary reactions of paired polymers leading to the formation of triple IPs are studied.

REFERENCES

1. Korshak, V. V.; Suprun, A. P.; Slonimskii, G. L.; Askadskii, A. A.; Birshtein, T. M.; Vointseva, I. I.; Mustafaeva, B. B.; Nikolskii, O. G. *Makromol. Chem.* **1986**, *187*, 2153.
2. Askadskii, A. A.; Vointseva, I. I. *Visokomolekularnye soedinenia* **1987**, *29A*, 2654.
3. Vointseva, I. I.; Askadskii, A. A. *Sov. Sci. Rev. B. Chem.* **1991**, *16*, 2.
4. Vointseva, I. I.; Rhranina, T. I.; Gembitskii, P. A. *Lakokrasochnye materialy* **1994**, *1*, 22.

Inverse-Emulsion Polymerization

See: Acrylamide, Inverse Emulsion Polymerization (in Supercritical Carbon Dioxide)
Flocculants (Organic, Overview)
Inverse-Emulsion/Suspension Polymerization

INVERSE-EMULSION/SUSPENSION POLYMERIZATION

David J. Hunkeler and José Hernández-Barajas
Department of Chemical Engineering
Vanderbilt University

The first water-in-oil polymerization was described in 1962 by Vanderhoff and was termed an "inverse-emulsion," analogous to the direct oil-in-water process.[1] In Vanderhoff's paper, inverse-micelles were postulated, although not yet detected. Similar processes where inverse-micelles were absent and nucleation proceeded in the monomer droplets have been termed inverse-suspensions or inverse-microsuspensions, with the prefix "micro" added to denote the nominal particle size.[2] Most real water-in-oil systems share elements of emulsion and suspension behavior, and an international group of researchers in the field have categorized both of these polymerizations as "inverse-macroemulsions."[3]

The principal advantage of inverse-macroemulsion polymerization over other processes is the simultaneous attainment of high molecular weights and high reaction rates. Two disadvantages of inverse-macroemulsion polymerization are: (i) Water-in-oil polymeric macroemulsions require conversion, either into an aqueous solution or a dry powder, prior to use. (ii) Large amounts of coagulum are formed in the reactor during the synthesis of homo- or copolymers.

CATEGORIZATION AND COMPARISON OF WATER-IN-OIL POLYMERIZATIONS

Water-in-oil polymerizations can be categorized as heterophase polymerizations due to the differences in the solubility of the phases involved at the outset of the reaction. Four mutually exclusive domains have been distinguished in heterophase polymerization: (I) macroemulsion; (II) microemulsion; (III) inverse macroemulsion; and (IV) inverse microemulsion. Two primary criteria exist for categorizing these domains: a surface-tension driving force and a stability threshold.

Subdomains I and II include systems produced at surfactant levels below a critical concentration at which micelles exist (CMC). The term "suspension polymerization" is used to describe the systems with [E] < CMC. The distinction between emulsions and suspensions is not ambiguous. It is generally agreed that for a polymerization to be considered an emulsion/inverse emulsion, two criteria must be satisfied: (i) The kinetics, as defined by the average number of macroradicals per particle ($\bar{n}$), must not be significantly larger than one. (ii) In an emulsion or inverse-emulsion polymerization, the mechanism of polymer particle nucleation resides outside the monomer droplets. These mechanisms require the initiator to be insoluble in the monomer phase, such as a water-soluble initiator and an organically soluble monomer, or vice versa. In contrast, if the initiator is soluble in the monomer phase, all the components of the reaction are contained in the dispersed phase, and the continuous phase serves only to decrease the viscosity and dissipate heat. These polymerizations are considered inverse-suspensions.[5]

MECHANISM OF INVERSE-MACROEMULSION POLYMERIZATION

Oil-Soluble Initiators

The first general mechanism for any type of inverse-macroemulsion polymerization was proposed in 1987 by Hunkeler et al. for the "inverse-microsuspension" polymerization of acrylamide.[4] It was subsequently expanded and developed into a kinetic model in 1989.[2] The mechanism consisted of a series of reactions in the continuous organic phase, including initiator decomposition, and the reaction of primary radicals with emulsifier, hydrocarbon, and monomer molecules. Primary- and oligo-radical transfer from the organic phase to the aqueous phase was also included. In the aqueous dispersed phase, a series of reactions that are common to all free-radical polymerizations (propagation, transfer to monomer, and termination) was also considered for two populations of macroradicals. This mechanism included the following unusual steps: (i) a reaction between a macroradical with the hydrophilic moiety of an interfacial emulsifier molecule, (ii) a long-chain branching reaction with radically active functional groups on the emulsifier, and (iii) a chain-length dependent mass transfer of primary radicals and oligoradicals from the continuous to the dispersed phase. In particular, this mechanism was explicitly derived for fatty acid esters of sorbitan and aliphatic continuous phases in order to explain the experimental findings such as the unusual dependence of the polymerization rate with emulsifier concentration ($R_p = f([E]^{-0.2})$) and the dependence of the polymerization rate on the initiator concentration ($R_p = f([I]^{1.0})$).

It is obvious that the particle size and distribution will have a significant influence on the interfacial reactions. Therefore, the emulsifier has both a physical and chemical role. In addition, factors that influence the particle diameter, such as the rate of agitation and the impeller size and type, will also have a minor influence on the rate of the reaction. Since the temperature also affects the interfacial tension, it too will have a physical role, although it is much smaller than the role of temperature on the initiator decomposition.

The mechanism has been subsequently applied to additional, more commercially relevant, emulsifier systems involving block copolymeric surfactants.[5]

The optimization of the linear molecular weight of polymers produced by inverse macroemulsion requires an analysis of the radically active functional groups on the stabilizers. The ideal emulsifier, in addition to providing a protective sheath, should either be without radically functional groups or multifunctional.

The inverse-microsuspension polymerization mechanism derived for homopolymers has been extended for copolymerizations of acrylamide with quaternary ammonium cationic monomers by considering the unique reactivities of the two monomers and four types of macroradicals.[6]

Water-Soluble Initiators

The mechanism of inverse-macroemulsion polymerization is complicated by the chemical nature of the emulsifier and the heterophase mass transfer of small molecules. These are particularly exacerbated when an oil-soluble initiator is utilized, because the chain-initiating species must traverse a condensed interfacial boundary. One can alleviate these mass-transfer and degradative chain-transfer limitations by employing a water-soluble initiator.

However, these are not usually utilized commercially since oil-soluble initiators provide a more efficient product in applications such as flocculation (higher molecular weight), as well as a safer reactor operating strategy.

A "hybrid complex-cage" mechanism has been proposed for the persulfate-initiated polymerizations of acrylic water-soluble monomers.[7] This mechanism involves bonding, either hydrogen or electrostatic, between the monomer and initiator that ultimately leads to association. This monomer-initiator associate (or monomer swollen cage) decomposes via a donor-acceptor intermediate formed between the amide, acid, or quaternary ammonium group and the persulfate. This has been developed into a kinetic model that accounts for the high rate order with respect to monomer concentration observed for the polymerization of acrylamide, acrylic acid, and quaternary ammonium cationic monomers.[8–10]

This mechanism is essentially equivalent to the elementary reaction scheme observed for aqueous solution polymerization, with the continuous organic phase having no chemical influence and serving primarily to dissipate the heat and reduce the viscosity.

KINETIC MODELING

Since the pioneering work of Vanderhoff over three decades ago there have been several kinetic investigations on inverse-suspension and inverse-emulsion polymerizations.[1] The kinetics differ for each particular investigation due to the unique chemistries employed, in particular the broad range in organic phases, initiators, and emulsifiers utilized. For example, micellization is influenced by both the interfacial composition and nature of the dispersion media.[11–13] These water-in-oil polymerizations are also characterized by transitions from inverse-suspension to inverse-emulsions, concomitant with a change in the particle nucleation mechanism, for minor increases in emulsifier concentration. The solubility of the initiator in the organic phase also strongly influences the polymerization mechanism.

With respect to the solubility, one must define the phase of primary solubility because neither the initiator nor the monomer are absolutely insoluble in either aqueous or organic media, and both partition to some extent.

The dependence of the physical properties of these inverse-macroemulsions on the chemical nature of the system renders the generalization of water-in-oil polymerization processes more difficult than for direct oil-in-water polymerizations.

COMMERCIAL PRODUCTION

A commercial inverse suspension/emulsion recipe for the production of water-soluble homo- and copolymers basically consists of a continuous aliphatic organic phase, a mixture of emulsifiers to achieve an overall HLB between 4–6, monomer(s), deionized water, chemical initiator(s), and additives. The organic phase is a mineral oil, kerosene, or a specific solvent with a high boiling point (>200°C) and flash point. Monomers include acrylamide, anionic species such as acrylic and methacrylic acids, and quaternary ammonium cations. The aqueous-to-organic phase ratio varies between 2–4:1. The emulsifiers used are too extensive to list but include sorbitan esters of fatty acids of low HLB and block copolymers.

REFERENCES

1. Vanderhoff, J. W.; Bradford, E. B.; Tarkowski, H. L.; Shaffer, J. B.; Wiley, R. M. *Adv. Chem. Ser.* No. 34, 1962; American Chemical Society: Washington, DC; 32.
2. Hunkeler, D. J.; Hamilec, A. E.; Baade, W. *Polymer* **1989**, *30*, 127.
3. Hunkeler, D. J.; Candau, F.; Pichot, C.; Hamielec, A. E.; Xie, T. Y.; Barton, J.; Vaskova, V.; Guillot, J.; Dimonie, M. V.; Reichert, K. H. *Adv. Polym. Sci.* **1994**, *112*, 115.
4. Hunkeler, D. J.; Baade, W.; Hamielec, A. E. *Polym. Mat. Sci. Eng.* **1987**, *57*, 854.
5. Hernandez-Barajas, J.; Hunkeler, D. J. *Proceedings of the Third International Symposium on Copolymers in Dispersed Media* Lyon, France, April 1994, 175.
6. Hunkeler, D. J.; Hamielec, A. E. *Polymer* **1991**, *32*, 2626.
7. Hunkeler, D. J. *Macromolecules* **1991**, *24*, 9, 2160.
8. Riggs, J. P.; Rodriguez, F. *J. Polym. Sci., Polym. Chem. Ed.* **1967**, *5*, 3151.
9. Manickman, S. P.; Venkatatao, K.; Subbaratnam, N. R. *Eur. Polym. J.* **1979**, *15*, 483.
10. Jaeger, W.; Hahn, M.; Seehaus, M.; Reinisch, G. *J. Macromol. Sci. Chem.* **1984**, *A21*, 593.
11. Bhattacharyya, D. N.; Kelkar, R. Y. *J. Colloid Interface Sci.* **1983**, *92*, 260.
12. Baade, W.; Reichert, K. H. *Makromol. Chem. Rapid Commun.* **1986**, *92*, 260.
13. Bartelt, G.; Reichert, K. H. In *Polymer Reaction Engineering* Reichert, K. H.; Geiseler, W., Eds., VCH: Berlin, 1989.

Inverse-Suspension Polymerization

See: Inverse-Emulsion/Suspension Polymerization

Iodine Transfer Polymerization

See: Living Radical Polymerization, Iodine Transfer

ION-CHELATING POLYMERS (Medical Applications)

Paolo Ferruti
Dipartimento di Chimica Organica e Industriale
Università di Milano

Ion-chelating polymers are usually defined as polymers capable of acting as chelating (i.e., multidentate) ligands.[1] This definition includes either polymers containing multidentate ligand groups attached to discrete points by covalent linkages, or polymers that although endowed with "simple" rather than multidentate groups, are nevertheless capable for sterical reasons of forming multidentate complexes through the cooperation of two or more groups in different positions. Most exciting ion-chelating polymers, including commercial resins, fall into the broad category of polyelectrolytes, and therefore ion chelation can often be regarded as a particular aspect of polyelectrolyte chemistry.

When dealing with the potential of ion-chelating polymers for medical applications, it is probably better to accept the broadest possible definition, considering as ion chelation the formation of polymer–polymer complexes via multiple ionic

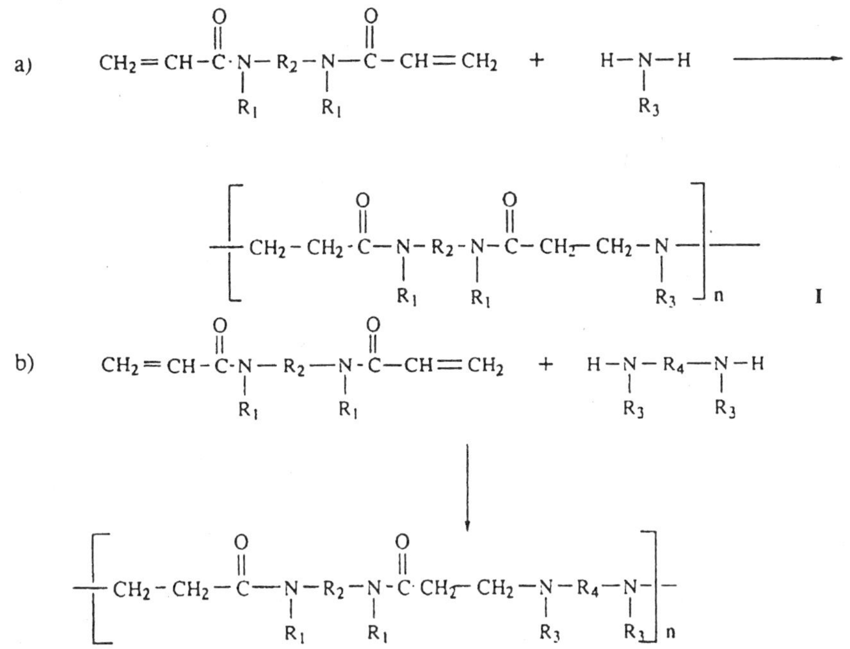

a)
$$CH_2=CH-\overset{\overset{O}{\|}}{C}\cdot\underset{\underset{R_1}{|}}{N}-R_2-\underset{\underset{R_1}{|}}{N}-\overset{\overset{O}{\|}}{C}-CH=CH_2 \quad + \quad H-\underset{\underset{R_3}{|}}{N}-H \longrightarrow$$

$$\left[-CH_2-CH_2\cdot\overset{\overset{O}{\|}}{C}-\underset{\underset{R_1}{|}}{N}-R_2-\underset{\underset{R_1}{|}}{N}-\overset{\overset{O}{\|}}{C}-CH_2-CH_2-\underset{\underset{R_3}{|}}{N}-\right]_n \qquad \mathbf{I}$$

b)
$$CH_2=CH-\overset{\overset{O}{\|}}{C}\cdot\underset{\underset{R_1}{|}}{N}-R_2-\underset{\underset{R_1}{|}}{N}-\overset{\overset{O}{\|}}{C}-CH=CH_2 \quad + \quad H-\underset{\underset{R_3}{|}}{N}-R_4-\underset{\underset{R_3}{|}}{N}-H$$

$$\downarrow$$

$$\left[-CH_2-CH_2-\overset{\overset{O}{\|}}{C}-\underset{\underset{R_1}{|}}{N}-R_2-\underset{\underset{R_1}{|}}{N}-\overset{\overset{O}{\|}}{C}\cdot CH_2-CH_2-\underset{\underset{R_3}{|}}{N}-R_4-\underset{\underset{R_3}{|}}{N}-\right]_n$$

interactions. Accordingly, we will consider as medical applications of ion-chelating polymers all applications of polymers in the medical field that involve interactions with ionic species, including high molecular weight ones.

We will review synthesis, properties, and medical applications of ion-chelating polymers obtained by stepwise polyaddition of compounds bearing mobile hydrogens to compounds bearing carbon–carbon double bonds activated by electron-attracting groups in α position. Most of the discussion will be devoted to polyamidoamines (PAAs). Other polymers structurally related to PAAs will also be considered, and reference will be made, when opportune, to structurally unrelated polymers with similar medical potential.

POLYAMIDOAMINES

Structure and Synthesis

PAAs are synthetic polymers characterized by the presence of amido and tertiary amino groups regularly arranged along the macromolecular chain.

Linear PAAs are obtained by polyaddition of primary monoamines of *bis*-secondary amines to *bis*-acrylamides (**Scheme I**).

PAAs carrying additional functions as side substituents can be easily obtained, starting from the appropriate monomers. In fact, hydroxy groups, carboxy groups, tertiary amino groups, allyl groups, and so on, if present in monomers, do not interfere with the polymerization process. In the case of amino acids the polyaddition can be performed in the presence of a base, for instance triethylamine. In the absence of added base no reaction occurs in practice, under normal conditions, between neutral amino acids and *bis*-acrylamides in water. As a rule, natural α-amino acids other than glycine give low molecular weight products even in the presence of bases, but the polyaddition proceeds reasonably well with all amino acids bearing no substituents in

α position, as well as with peptides, provided the first amino acid residue of the latter fulfills the same condition.

Oligomeric compounds carrying either two secondary or one primary amino group, such as α,ω-*sec*-amino-polyoxyethylenes or α-methyl-ω-aminooxyethylenes, can be used as monomers in the same way as simple amines, giving block[7] or graft copolymers,[8] respectively.

From a practical standpoint, the overall monomer concentration in the reacting mixture should be kept as high as possible, if high molecular weight polymers are desired.

Under nonselective conditions *bis*-primary amines usually give crosslinked products on reaction with *bis*-acrylamides, because they react tetrafunctionally. However, it has recently been reported that under special conditions, including low reactant concentrations, low initial temperatures, and an excess of *bis*-amines, on a molar basis, soluble PAAs carrying secondary amino groups in their main chain can be obtained. These PAAs have been found to provide suitable soluble carriers for drug attachment.[9]

Polyamidoamines Related Polymers

Polymer structures related to PAAs will be obtained by substituting *bis*-acrylic esters or divinylsulfone[2] for *bis*-acrylamides or hydrazine,[10] or phosphines[11] for amines. Other polymers, of poly(amino-ketone) structure, can be obtained by polycondensation of ketonic *bis*-Mannich bases with *bis*-amines.[12-14] These polymers can be reduced to poly-γ-aminoalcohols.[15] Similarly, the polycondensation of *bis*-secondary amines with quaternarized phenolic *bis*-Mannich bases gave polyaminophenols,[16] and the polycondensations of *bis*-secondary amines with *bis*-Mannich bases of *bis*-carboxyamides gave poly(N-aminomethyleneamides).[17]

The ring-opening addition reaction of *bis*-secondary amines, or primary monoamines, to ethylene sulfide gave a new family of monomers, 2,2'-alkylidenediiminodiethanethiols.

These monomers could be used in polyaddition reactions with *bis*-acrylamides, *bis*-acrylic esters, or divinylsulfone, in a way formally similar to *bis*-secondary amines.[18–20] They are, however, more reactive, and relatively high molecular weight polymers could also be obtained in aprotic solvents.

Moreover, the polyaddition also proceeded with *bis*-methacrylic esters[21] and *bis*-methacrylamides,[22] although methacrylic derivatives, as a rule, are inactive to polyaddition with *bis*-amines under the usual conditions of PAA synthesis. The same *bis*-mercapto amines gave high molecular weight *ter*-amino polymers by polyoxidative coupling.[23]

Properties

General Properties

Most PAAs synthesized so far are soluble in, or at least swollen by, water. Water-insoluble PAAs are soluble in aqueous acids. PAAs are also soluble in chloroform, lower alcohols, and many other organic solvents. PAAs bearing carboxy groups as side substituents, however, tend to dissolve only in water.

The intrinsic viscosities of PAAs in organic solvents or aqueous media range from about 0.15–1 dL/g. The number-average molecular weights of some PAAs with intrinsic viscosities of 0.1–0.4 dL/g were determined by vapor-pressure osmometry,[24,25] and found to be in the range $2 \times 10^3 - 1 \times 10^4$. The number-average molecular weights of polymers having intrinsic viscosities higher than 0.4 dL/g were not determined.

Molecular weight distributions of water-soluble PAAs can be conveniently studied by size-exclusion chromatography (SEC) in aqueous buffers at slightly alkaline pHs.

Owing to their regular structure, many PAAs are partially crystalline in the solid state.[3] The thermal stability of PAAs is not very high. In most instances decomposition begins at about 140°C in air and 170–180°C under inert atmosphere.[3]

PAAs as Polyelectrolytes

Protonation studies All PAAs contain ionizable aminic nitrogens in their main chains, and therefore can be classified as polyelectrolytes. However, in all the PAAs studied, except those deriving from amino acids, potentiometric titrations clearly show that the basicity of the aminic nitrogens $\rangle$N–R of each repeating unit does not depend on the degree of protonation of the whole macromolecule.[26–38] Consequently, "real" basicity constants can be determined. The number of basicity constants is always equal to the number of the aminic nitrogens present in the repeating unit, and these are similar to those of the corresponding non-macromolecular models.

PAAs deriving from amino acids, and therefore carrying carboxylate groups as side substituents, behave in a different way from all other PAAs.[39] In fact, they exhibit a typical polyelectrolyte behavior.

Heavy metal ions complexing behavior. Many PAAs are able to give coordination complexes with heavy metal ions, such as Cu^{2+}, Ni^{2+}, Co^{2+}. "Real" stability constants could also be determined for PAAs in complex formation. To give an example, the coordinating ability of some PAAs, and their models, with respect to copper(II) ion in aqueous solution were studied.[26,27,32,33,35,38,41] Leaving apart the case of PAAs deriving from amino acids, the complex formation ability pertains to PAAs containing at least two aminic nitrogens per repeating unit that do not belong to the same cyclic structure and are not separated by more than three methylene groups. For most other PAAs, insoluble hydroxides precipitate prior to the formation of a significant amount of complex. These results show that increasing the chelating ring size from five to six leads to a distinct reduction in the (CuL) (L = ligand) complex stability or even to the inability to chelate copper(II) to give stable complexes.

As regards the non-macromolecular models, the (CuL) stability constants are always slightly higher than those of the corresponding polymers. This is presumably related to different entropy effects.

The electronic and electron paramagnetic resonance (EPR) spectra of both polymeric and nonpolymeric complexes are similar, and consistent with an octahedral tetragonally distorted structure. The fact that the EPR and electronic data of both families of complexes are quite similar is in agreement with the substantial independence of the repeating units of the polymer in the complex formation process.

Degradation Behavior of Polyamidoamines in Aqueous Media

All PAAs, having hydrolyzable amidic bonds in their main chain, are in principle degradable in the presence of water. This property in biomedical applications in a sense is double edged, being advantageous when PAAs are considered as soluble carriers for preparing polymeric prodrugs, but disadvantageous for all applications requiring the absence of leachable products, such as de-heparinizing resins or heparinizable biomaterials.

The degradation of several PAAs was studied.[42,43] We found that the structure of both the aminic and amidic moieties has an influence on the degradation rate.

As a final observation, we have demonstrated that, other conditions being the same, the degradation rate is strongly influenced by pH but does not seem to be affected by tritosomal enzymes at pH 5.5.[44]

POLYAMIDOAMINES MACROMONOMERS AND MEDICAL APPLICATIONS OF POLYAMIDOAMINES AND POLYAMIDOAMINE-BASED MATERIALS

Until recently, PAAs were considered for medical applications only in relation to their ability to form complexes with heparin. However, it is being recognized that PAAs have a definite potential as drug carriers. Biomedical and pharmacological applications of PAAs will be treated separately.

Polyamidoamines in the Biomedical Field as Heparin-Complexing Agents

Structure and Properties of Heparin

Heparin is a well-known anticoagulant agent, widely used in clinical practice. It is a mucopolysaccharide containing carboxyl- and sulfonic groups. In aqueous solution, it behaves as a polyanion with a high density of negative charges.

Soluble Polyamidoamines as Heparin-Neutralizing Agents

In many cases, the anticoagulant activity of heparin must be inhibited when no longer needed. This is usually done by administering protamine sulfate. Protamine is a highly basic natural polymer of polypeptide structure. It is itself a powerful anticoagulant when administered alone, or in excess over heparin. Moreover, it has several untoward side effects, and the neutralization of heparin by protamine may be followed by the so-called heparin rebound.[45]

In a previous study, Marchisio found that several PAAs are able, in a linear form, to neutralize the anticoagulant activity of heparin, much as protamine sulfate does.[4–6]

There is little doubt that the mechanism of action was related to polyelectrolyte complex formation with heparin.

Polyamidoamines as Macromonomers

By considering the synthetic process leading to PAAs, we see that in principle the endgroups of the products are either sec-, amino-, or acrylamido groups. By performing the reaction with an excess of one of the two monomers. PAAs prevailingly or totally terminated with either of the two groups can be obtained.

These can be employed for preparing block and graft copolymers, as well as crosslinked resins, and therefore can be regarded as macromonomers.

Polyamidoamine-Based Crosslinked Resins

The simplest way for obtaining crosslinked PAAs is to partly substitute α,ω-*bis*(aminoalkanes) for difunctional aminic monomers in the polymerization process. According to the general scheme of PAA synthesis, *bis*-primary amines, having four mobile hydrogens, act as tetrafunctional monomers.[3,46]

Alternative methods can be used. For instance, acrylamide-terminated PAAs can be post-polymerized with radical initiators, either alone or in the presence of a vinyl comonomer, for instance *N*-vinylpyrrolidinone.

Crosslinked PAAs are clearly insoluble in water. In aqueous media, however, they swell to a large extent. Those obtained by the standard method give brittle gels when swollen. The other methods, as a rule, lead to gels with better mechanical properties. Moreover, their chemical resistance, for instance to hydrolytical degradation, is apparently much higher.

Ion-Chelating Properties and Medical Interest of Polyamidoamines-Based Crosslinked Resins

Crosslinked polyamidoamines as heavy metal ions absorbers. The ion-complexing behavior of linear PAAs in aqueous media is peculiar in the polyelectrolyte domain, because their complexes have well-defined stoichiometries and sharp stability constants can be determined in aqueous solution. Crosslinked PAAs still have ion-complexing ability.

Heparin absorbing resins. Crosslinked PAAs in aqueous media have protonation and heavy metal ion complexation similar to their linear counterparts. Correspondingly, crosslinked resins based on PAAs capable of interacting with heparin in their linear form exhibit remarkable heparin-absorbing capacity, even from very dilute solutions.

The best use of a heparin-absorbing resin with negligible side effects is as the bioactive component of a deheparinizing filter for achieving regional heparinization in extracorporeal circuits, especially hemodialysis.

Patients needing chronic hemodialytic treatment for irreversible renal failure must in many cases undergo three 4-h treatments per week for life. In order to avoid blood coagulation on contact with foreign surfaces in the dialysis machine, heparin is commonly administered. In the normal practice, the whole patient's blood is heparinized, which can result in morbidity because of a direct anticoagulant effect, delayed thrombogenic effects, and interference with lipid metabolism. Moreover, about 5% of the patients can be considered "risk patients," owing to the presence of latent diseases such as ulcers, adversely affected by temporary disorders of the coagulative system. The insertion of a heparin-absorbing device before blood reenters the patient's body, coupled with continuous administration of heparin to the patient's blood before entering the extracorporeal circuit, would constitute the best solution of the problem. To achieve this goal, the bioactive component of a heparin-absorbing device should have high capacity; high selectivity; no adverse effects on normal blood components, including blood cells; and last but not least, reasonable cost. In all these respects, PAAs seem to offer at present the best combination of properties.

Polyamidoamine-Modified Heparinizable Materials

Most non-physiological materials induce thrombus formation when placed in contact with blood. An interesting approach to the development of non-thrombogenic surfaces is to coat synthetic polymers with heparin.

The first heparinized material was a graphite-benzalkonium-heparin coating.[48] Heparin was electrostatically bonded through its anionic groups to the detergent benzalkonium chloride, which had been previously adsorbed on a graphite-impregnated base.

Novel poly(aminoether urethane urea)s containing tertiary amino groups either in side chains or in the main chain were synthesized, quaternarized, and heparinized.[49,50]

When the material is not swollen by water to a large extent, the ability to absorb heparin and to become nonthrombogenic is essentially a surface property. It follows that grafting heparin-complexing PAAs on the surface of a given material, not necessarily an organic polymer, would eventually lead to the same result, with the additional advantage of avoiding all risks connected with the potentially toxic quaternary ammonium groups.

Surface grafting of polyamidoamines onto various materials. Several methods have been devised for surface-grafting PAAs on plasticized poly(vinyl chloride),[51] poly(ethylene terephthalate),[52] polyurethanes,[40] and glass.[47] The PAA-grafted materials were able to absorb and tenaciously retain heparin.

Polyamidoamines-based block and graft copolymers. PAA macromonomers provide an excellent starting point for the preparation of block and graft copolymers capable, in principle, of combining the mechanical properties of conventional polymers with the ability to absorb large amounts of heparin. The first approach[53–56] involved the use of a vinyl-terminated PAA (V-PAA). Styrene was radically polymerized in DMF solution in the presence of this macromonomer.

The addition of a chain-transfer agent and the use of a fairly high molecular weight sample of V-PAA prevented extensive crosslinking. A structural and mechanical characterization of the resulting starlike styrene-PAA block copolymers was performed. The product was hard and somewhat brittle. Most subsequent work on PAA block and graft copolymers made use of *sec*-amino-terminated macromonomers (A-PAA).

In principle, three different synthetic pathways involving A-PAAs can be followed. The first one, leading to graft copolymers, involves the functionalization of a known, often commercial polymer that subsequently enters into a coupling reaction with the selected A-PAA. The second pathway is, in a sense, the reverse of the first one. The amino endgroups of A-PAA are transformed into a new reacting function capable of giving a coupling reaction with chemical groups already present in a conventional polymer. The third pathway involves the reaction of A-PAA with a prepolymer of conventional type carrying reactive functions, that is, a complementary macromonomer.

Polyamidoamines as drug carriers. The potential of PAAs as soluble drug carriers has been only recently recognized.[9,57,58] Polycations, particularly poly(L-lysine), are being explored as drug carriers for cancer chemotherapy,[59] and more recently as vehicles for gene and oligonucleotide delivery, for example in liver-directed gene therapy[60] and antisense nucleotide delivery.[61] However, polycations are in general very toxic. For instance, the IC_{50} of poly-L-lysine *in vitro* was found to be in the range 1–60 g/mL, depending on the cell type and incubation conditions.[62] Many of them have proved to be much less toxic *in vitro* than polylysine, their ID_{50} being higher by 2 or 3 orders of magnitude or even more in some cases, and therefore they appear to have definite potential as drug carriers.[57] Several PAAs have been found to be active in reducing number and average weight of metastases in mice infected with cancer cells.[63]

A poly(amidoamine) derivative capable of pH-dependant cell lysis. PAAs bearing two aminic nitrogen in their repeating unit show a marked conformational change during movement from neutral to acidic pH as a result of the modification of their average charge, this effect being more pronounced when the aminic nitrogens were separated by only two methylene groups.[36,38]

This property provides, in principle, the possibility of designing polymer–drug conjugates that are, following intravenous administration, relatively compacted and thus protect a drug payload in the circulation, where the pH is 7.4. Following pinocytic internalization into an acidic (pH ≈ 5.5) intracellular compartments unfold, permitting pH-triggered intracellular drug delivery.

REFERENCES

1. Hodgkin, J. H. *Encyclopedia of Polymer Science and Technology*; Mark, H. F.; Gaylord, N. G.; Bikales, N. M. Eds. Interscience: New York, 1992; Vol. 3, pp 363–381.
2. Danusso, F. et al. *Chimica e Industria* **1967**, *49*, 826.
3. Danusso, F.; Ferruti, P. *Polymer* **1970**, *11*, 88.
4. Marchisio, M. A. et al. *Eur. Surg. Res.* **1971**, *3*, 240.
5. Marchisio, M. A. et al. *Eur. Surg. Res.* **1972**, *4*, 312.
6. Marchisio, M. A. et al. *Experientia* **1973**, *29*, 93.
7. Ranucci, E.; Ferruti, P. *Macromolecules* **1991**, *24*, 3747.
8. Vansteenkiste, S. et al. *Makromol. Chem.* **1992**, *193*, 937.
9. Caldwell, G. et al. *J. Appl. Polym. Sci.* **1993**, *50*, 393.
10. Ferruti, P.; Brzozowski, Z. *Chimica e Industria* **1968**, *50*, 441.
11. Ferruti, P.; Alimardanov, R. *Chimica e Industria* **1967**, *49*, 831.
12. Andreani, F. et al. *J. Polym. Sci. Polym. Lett.* **1981**, *19*, 443.
13. Angeloni, A. S. et al. *Polym. Commun.* **1983**, *24*, 87.
14. Galli, G. et al. *Makromol. Chem. Rapid Commun.* **1983**, *4*, 681.
15. Angeloni, A. S. et al. *Polymer* **1982**, *23*, 1693.
16. Ghedini, N. et al. *Makromol. Chem. Rapid Commun.* **1984**, *5*, 181.
17. Angeloni, A. S. et al. *Polym. Commun.* **1984**, *25*, 119.
18. Ferruti, P. et al. *Makromol. Chem., Rapid Commun.* **1988**, *9*, 807.
19. Ferruti, P. et al. *Polym. Commun.* **1989**, *30*, 157.
20. Ferruti, P.; Ranucçi, E. *Polym. J.* **1991**, *23*, 541.
21. Ranucci, E.; Ferruti, P. *Polymer* **1991**, *32*, 2876.
22. Ranucci, E. et al. *Polym. J.* **1993**, *25*, 625.
23. Ranucci, E.; Ferruti, P. *Chimica e Industria* **1990**, *72*, 49.
24. Tanzi, M. C. et al. *Biomaterials* **1984**, *5*, 357.
25. Tanzi, M. C. et al. *Biomaterials* **1992**, *13*, 425.
26. Ferruti, P. In *Polymeric Amines and Ammonium Salts*; Goethals, E. J. Ed.; Pergamon: Oxford and New York, 1980; p 305.
27. Ferruti, P. et al. *J. Chem. Soc. Dalton Trans.* **1981**, *539*.
28. Ferruti, P.; Barbucci, R. *Advanc. Polym. Sci.* **1984**, *58*, 57.
29. Barbucci, R. et al. *Polymer* **1978**, *19*, 1329.
30. Barbucci, R. et al. *Atti Acc. Naz. Lincei VIII* **1978**, *64*, 481.
31. Barbucci, R.; Ferruti, P. *Polymer* **1979**, *20*, 1061.
32. Barbucci, R. et al. *Polymer* **1979**, *20*, 1298.
33. Barbucci, R. et al. In *Polymeric Amines and Ammonium Salts* Goethals, E. J. Ed.; Pergamon: Oxford and New York, 1980; p 263.
34. Barbucci, R. et al. *Polymer* **1980**, *21*, 81.
35. Barbucci, R. et al. *J. Chem. Soc. Dalton Trans.* **1980**, *253*.
36. Barbucci, R. et al. *Macromolecules* **1981**, *14*, 1203.
37. Barbucci, R. et al. *J. Polym. Sci., Polym. Symp.* **1981**, *69*, 49.
38. Barbucci, R. et al. *Gazz. Chim. It.* **1982**, *112*, 105.
39. Barbucci, R. et al. *Macromolecules* **1986**, *19*, 37.
40. Barbucci, R. et al. **1985**, *26*, 1353.
41. Barbucci, R. et al. *Polymer* **1982**, *23*, 148.
42. Ferruti, P. et al. *Biomaterials* **1994**, *15*, 1235.
43. Ferruti, P. et al. *Biomater. Sci. Polym. Ed.* **1994**, *6*, 833.
44. Ranucci, E. et al. *Polymer* **1991**, *32*, 2876.
45. Frick, P. G. *Surgery* **1996**, *59*, 721.
46. Ferruti, P. et al. *Polymer* **1985**, *26*, 1336.
47. Ferruti, P. et al. *Biomaterials* **1983**, *4*, 217.
48. Gott, V. L. et al. *Science* **1963**, *142*, 1297.
49. Yoshiro, I. et al. *J. Biomed. Mater. Res.* **1986**, *20*, 1017.
50. Shibuta, R. et al. *J. Biomed. Mater. Res.* **1986**, *20*, 971.
51. Ferruti, P. et al. *Biomaterials* **1982**, *3*, 33.
52. Barbucci, R. et al. *Biomaterials* **1985**, *6*, 107.
53. Ferruti, P.; Provenzale, L. *Transplant. Proc.* **1976**, *VIII*(1), 103.
54. Ferruti, P. et al. *Chimica e Industria* **1976**, *58*, 539.
55. Ferruti, P. et al. *J. Polym. Sci.* **1977**, *15*, 2151.
56. Martuscelli, E. et al. *Chimica e Industria* **1976**, *58*, 542.
57. Ranucci, E. et al. *J. Biomater. Sci. Polym. Ed.* **1991**, *2*, 303.
58. Duncan, R. et al. *J. Drug Targeting* **1994**, *2*, 341.
59. Chu, B. C. F.; Howell, S. B. *J. Pharmacol. Exp. Ther.* **1981**, *219*, 389.
60. Wu, G. Y.; Wu, C. H. *Adv. Drug. Rel. Rev.* **1993**, *12*, 159.
61. Citro, G. et al. *Br. J. Cancer* **1994**, *69*, 463.
62. Sgouras, D. et al. *J. Mater. Sci.: Mater. Med.* **1990**, *1*, 61.
63. Ferruti, P. et al. *J. Med. Chem.* **1973**, *16*, 497.

Ion-Conducting Polymers

See: *Ion-Exchange Membranes*
Ionomers (dc Conduction Properties)
Perfluorinated Ionomers (Anisotropic Ionic Conductivity)
Poly[N,N′-(sulfo-p-phenylene) terephthalamide]
Solid Polymer Electrolytes
Solid Polymer Electrolytes (Polyether Blends and Composites)

Zwitterionic Polymers (Ionic Conductivity)

Ion-Exchange Materials

See: *Azidation Polymer*
Crown Ether Ion-Exchange Resins
Functional Fibers
Ion-Exchange Membranes
Ion-Selective Reagents (Dual Mechanism Bifunctional Polymers)
Metal Ion-Binding Resins (Synthesis Analytical Properties)
Polyimides (from Cyclic Unsaturated Imides)

ION-EXCHANGE MEMBRANES

Yres Hervaud and B. Boutevin
Laboratoire de Chimie Appliquée
Ecole Nationale Supérieure de Chimie de Montpellier

C. Gavach
Laboratoire des Matériaux et Procédés Membranaires
Centre National de la Recherche Scientifique

In recent years, ion-exchange membranes (IEM) have seen great development, first to solve industrial, medical, or environmental requirements; second because we have a better knowledge of their physicochemical characteristics. IEMs have functional groups that allow ionic exchanges. They seem similar in certain ways to the ion-exchange resins (IER), but their nature and fabrication are different because they are generally used as film shapes.

From the fabrication point of view, we must distinguish two kinds of IEMs: heterogeneous membranes and homogeneous membranes. The first are IER grains supported by a polymer, the second are polymer films containing the ion-exchange groups (IEG). Heteropolar membranes are a combination of the two types and, depending on the IEG positions, may be amphoter, bipolar, or mosaic. IEMs may also be classified by the nature of the exchanged ions into anion exchange membranes (AEM), cation exchange membranes (CEM), and amphoter membranes.

IEM SYNTHESIS

Considering their structure–property relationships, IEMs are found as functionalized polymer shapes and can be represented in **Structure 1**.

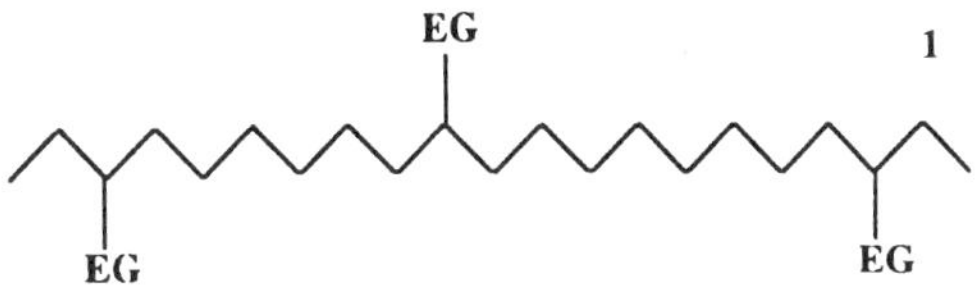

1

EG-ionic exchange group (anionic or cationic)

Polymer Synthesis Generalities

The main factors that directly affect polymer or copolymer formation are the monomer's nature, the initiation mode, the presence or absence of solvent, the concentrations, the temperature, the pressure, the reaction time, and the later treatments. The IEMs are obtained from polymers or copolymers on which different chemical or radiochemical modifications are made before or after their processing.

Synthesis of IEM from Grafted Polymers

The polymers used most often are polyolefins (PE and PP) and fluoropolymers (PTFE, for example). The grafts are often acrylic (acid or ester) or pyridinic (4-vinyl pyridine).[1–16]

The EG groups coming from the grafts are more often cation exchangers. When the graft comes from another monomer we obtain grafted copolymers. We have two very important kinds of grafting: radiation and ozonization.

Different methods lead to the creation of active radical sites on a polymer chain; the radiochemical ones[17] seem to be the most advantageous.

The ionizing radiations mostly used are γ-rays and electrons. Nevertheless, several examples of grafting by a chemical method using ozone have been described.[1,2,5,18]

Perfluorinated IEM Synthesis

The general characteristics of fluoropolymers include thermal resistance, weather resistance, good electrical properties, good surface properties (oil and water repellency, low surface energy, low stickiness) and low refractive index.[19]

Some Examples of Particular IEM

IEMs can appear in different forms, so we find such IEMs as porous composite amphoteric, mosaic, or bipolar membranes.

Porous IEMs are generally obtained by addition of inorganic fillers (e.g., MgO, Z_nO, Ca CO_3) to polymers or copolymers.[20,21] Some IEMs are used to support catalyst particles either inside the material or on its surface. They are called composite IEMs and are obtained by a radiolytic process.[22]

When looking at bifunctional membranes, we note three classes that differ only in the positions of the different exchanger groups EG.[23] The amphoteric membranes contain the two EG on the same support with equal quantities of opposite-sign electrical charges, and they are without apparent order arrangement.

The mosaic membranes process both cation exchange regions and anion exchange regions that are adjacent. Electroneutrality is maintained by the simultaneous passage of cations and anions through the membrane.

Bipolar membranes (BPM) are formed by an anion exchange layer coated with a cation exchange layer.

IEM APPLICATIONS AND UTILIZATIONS

IEMs have several large-scale uses. In the H_2–O_2 fuel cell, the CEM is placed between the porous electrodes. The transfers of protons and water, in opposite directions through the membrane, link the two electrode reactions.

IEMs are used as separators in electrolysis cells, where they prevent the mixing of the gas produced by the electrolysis reactions. In water electrolysis, the CEM is a proton conductor. For the chloralkali electrolysis, the CEM is highly permeable to sodium and not to hydroxyl ions. Only perfluorinated membranes have a chemical resistance to the drastic conditions of chloralkali electrolysis (oxidative contact on the anodic side, concentrated NaOH on the cathodic, high temperature). Conventional electrodialysis involves a couple of CEMs and AEMs that are alternately placed and form a stack with the diluate and the concentrate circuits. Because of the combined effects of the permselectivity of the membrane and the migration of ion under an applied electrical field, the electrolyte is removed from the diluate and reconcentrated in the concentrate circuit. Therefore this electromembrane separation technique is used for the elimination of electrolyte (desalting of brackish waters, demineralization of whey) and also for the reconcentration of salts (production of brine from seawater). These three applications are the most important for electron desposition (ED).

Conventional ED is used in three large-scale applications but the electromembrane process is also used, at a lower existent, in other fields: metallurgy, surface treatment, chemical industries, biotechnology, pharmacy, food industries, and water treatment.

NEW PROSPECTS

A new orientation of IEM development is surface modification, which improves specific permeability. Commercial CEMs are already produced to make brine by ED. A positively charged polyelectrolyte is fixed at the surface of the membrane. Electrostatic repulsions hinder the transmembrane transport of bivalent cations such as Ca^{2+} and Mg^{2+} that are present in seawater.

AEMs are permeable to protons because their polymeric ion exchanging material is swollen and water is a proton conductor. New AEMs have been developed for reducing the proton leakage. With this kind of membrane, ED can be applied to acid solutions and especially to the treatment of used acids (acid solutions containing metal salts). Combining an AEM that exhibits a reduced proton leakage and a surface-modified CEM with a very low permeability to bivalent cations, ED is used to treat used acid. The recycled acid is purified and reconcentrated.

This new application of IEMs is symbolic of the penetration of IEM technology into the environment field. Many kinds of effluent-containing electrolytes can be treated by IEM technology, not only for the elimination or the reconcentration of pollutant ions but also for recycling.

REFERENCES

1. Elmidaoui, A.; Sarraf, T.; Gavach, C.; Boutevin, B. *J. Appl. Polym. Sci.* **1991**, *42*(9), 2551.
2. Elmidaoui, A. et al. *J. Polym. Sci., Part B, Polym. Phys.* **1991**, *29*(6), 257.
3. Hsiue, G. H.; Huang, W. K. *J. Chin. Inst. Chem. Eng.* **1985**, *16*(3), 257.
4. Ishigaki, I. et al. *Radiat. Phys. Chem.* **1981**, *18*(5–6), 899.
5. Gineste, J. L. Thesis, University 2 of Montpellier, France, 1992.
6. Chakravorty, B. et al. *Int. J. Environ. Stud.* **1986**, *27*(3–4), 173.
7. Chakravorty, B. et al. *J. Membr. Sci.* **1989**, *41*, 155.
8. Chakravorty, B. et al. *Desalination* **1983**, *46*, 353.
9. Omichi, H.; Okamoto, J. *J. Polym. Sci., Polym. Chem. Ed.* **1982**, *20*(6), 1559.
10. Omichi, H.; Okamoto, J. *J. Appl. Polym. Sci.* **1985**, *30*(3), 1277.
11. Ishigaki, I. et al. *Polym. J.* **1978**, *10*(5), 513.
12. Kostov, G. K.; Atanassov, A. N. *J. Appl. Polym. Sci.* **1993**, *47*, 1269.
13. Omichi, H.; Okamoto, J. *J. Polym. Sci., Polym. Chem. Ed.* **1984**, *22*(7), 1775.
14. Omichi, H.; Okamoto, J. *J. Polym. Sci., Polym. Chem. Ed.* **1982**, *20*(2), 521.
15. Denamganai, J. Ph.D. Thesis, University 2 of Montpellier, France, 1988.
16. Munari, S. et al. *Ric. Sci.* **1967**, *37*(7–8), 641.
17. Maorgan, P. W.; Corelli, J. D. *J. Appl. Polym. Sci.* **1983**, *28*, 1879.
18. Elmidaoui, A., Ph.D. Thesis, University 2 of Montpellier, France, 1988.
19. Yamabe, M. *Macromol. Chem., Macromol. Symp.* **1992**, *64*, 11.
20. Bryjak, M. et al. *Angew. Makromol. Chem.* **1992**, *200*, 93.
21. Bryjak, M. et al. *Angew. Makromol. Chem.* **1993**, *205*, 131.
22. Platzer, O. et al. *J. Phys. Chem.* **1992**, *96*(5), 2334.
23. Mulder, M. *Basic Principles of Membrane Technology*; Kluwer: The Netherlands, 1991.

Ion-Selective Electrodes

See: *Ion-Selective Electrodes and Biosensors*
Ion-Selective Microchemical Sensors (Cardiology)

ION-SELECTIVE ELECTRODES AND BIOSENSORS

M. Koudelke-Hep, P. van der Wal, D. J. Strike, and N. F. de Rooij
University of Neuchâtel
Institute of Microtechnology

In recent years electrochemical sensors and biosensors have been the subject of growing interest arising from their selectivity and simplicity, together with the possibility of real-time analysis. We will define biosensor as a device incorporating a sensing element of biological origin;[1] all other devices fall in the chemical sensor category. Whereas most ion-selective electrodes (ISEs) cannot be classified as biosensors, polymer membrane-based ISEs incorporating antibiotic ionophores from living organisms might be included in the biosensor family.

Basically, most electrochemical biosensors of either potentiometric or amperometric types rely on a combination of a solid-state transducer and one or more organic or inorganic membranes. The most common examples are pH and other ISEs for potentiometric devices and the Clark-type oxygen sensor for amperometric devices. The polymer membranes that complete the transducer fulfill widely different purposes such as simple protection of the device, immobilization matrices for sensing elements, permselective membranes based on size or charge exclusion, and so on. Accordingly, the polymer materials used are numerous, ranging from the classical PVC through various hydrogels to the more recent conducting polymers.

For different analytical applications, miniaturization of sensors can be either mandatory (e.g., measurements *in vivo*) or, at least beneficial (e.g., use of smaller sample volumes).

ISEs based on polymer membranes have, since their introduction in 1970[2], been commercialized rapidly and are now among the most widely used sensors. ISEs are typically based on plasticized PVC, with an ionophore to assure selectivity, and can measure the activity of many different ions in various, often complex, solutions. Since the introduction of the ion-sensitive field-effect transistor (ISFET),[3] which may be considered as an extension of ISEs, there has been intensive research on the modification of ISFETs with these ion-selective polymer membrane materials. Besides the miniaturization and mass production, the possibility of making a multi-ion sensor is of particular interest.

For mass production of these potentially low-cost miniaturized sensors, the development of an IC-compatible membrane deposition technique is necessary. Therefore, photopolymerizable materials are of particular interest because they allow on-wafer deposition and patterning with photolithography. This approach also enables the realization of multiple sensing devices.

Several studies reported[4,5] on the use of alternative membrane materials to achieve better adhesion of thin solvent cast membranes to ISFET surfaces, such as modifications of PVC, polyurethanes, and polysiloxanes.

In our work we have doped well-defined polysiloxane structures using the plasticized PVC membranes normally used to make ion-selective electrodes. Because adhesion between the PVC and the polysiloxane is poor, the PVC can be totally removed to leave a clean polysiloxane surface. The use of viscous THF-based solutions of PVC, plasticizer, and ionophore present little risk of spreading to neighboring polysiloxane membranes. This technique was used in our laboratory to make four-function multi-ion sensors for K^+ (with valinomycin as the ionophore), Ca^{2+} (with ETH 5234), NO_3^- (with a tetradodecyl ammonium ion exchanger) and H^+ (uncovered ISFET).

The performance characteristics of the sensors were comparable to devices fabricated individually,[6] having nearly Nernstian slopes and good selectivities.

Among different amperometric biosensors, those that measure glucose have received the greatest attention. This is due of course to the importance of glucose measurements in human health care and bioprocess control. In most cases, a combination of the enzyme glucose oxidase (GOx) with an amperometric detector measuring either oxygen consumption or hydrogen peroxide formation is used. The polymers used in amperometric enzyme electrodes based on hydrogen peroxide detection can be divided into three categories by function: outer diffusion limiting membranes, enzyme immobilization matrices, and inner permselective membrane.

Polyurethane and Nafion® are probably the most commonly used materials for the outer diffusion-limiting membrane.[7–10] It is important to control the outer membrane diffusional properties to obtain a favorable ratio of oxygen over glucose diffusion. The reproducibility of the whole sensor is frequently determined by the outer membrane.

The main requirements of enzyme immobilization are retention of enzyme activity upon immobilization, good stability or low immobilized enzyme leakage and, for enzyme immobilization on miniaturized transducers, good spatial control. Immobilization procedures range from a simple adsorption through entrapment in a gel or conducting polymer matrix to chemical crosslinking using bifunctional reagents.

The inner permselective membrane prevents interference by other electrochemically active species that are oxidized at the same potential as hydrogen peroxide. The interference problem is especially acute in physiological samples (blood, serum, and urine), in which the presence of several compounds (e.g., uric acid, paracetamol, and ascorbate) compromises the selectivity of the enzymatic system. Thus, a permselective membrane should permit only hydrogen peroxide to reach the electrode surface while screening out all other interfering species. Use of cellulose acetate,[8] polyethersulfone,[13] 1,2-diaminobenzene[14] and polypyrrole[11,12] has been proposed. The latter two are examples of an electrochemical approach for making spatially well-controlled permselective membranes that offer a convenient and versatile way to vary membrane exclusion properties.

Another interesting application of polymers is their use as enzyme stabilizers because, except for glucose oxidase, most enzymes are not stable enough to allow the production of biosensors useful over long periods. Using a combination of a positively charged polymer, diethylaminoethyl (DEAE)-dextran, and a sugar alcohol, several enzymes can be stabilized at a considerable increase in lifetime.[15]

CONCLUSION

Many of the techniques and materials we have described are also used in classical macro devices. Although these are well established, they cannot be directly applied to silicon-based devices. Miniaturized invasive sensors, bedside discrete multiparameter analyzers, and on-line monitoring with real-time information in clinical chemistry and bioprocess control are all exciting possibilities. But development of polymeric membrane deposition technology and new materials that present good adhesion to silicon substrates must continue before silicon-based sensors can fulfill the growing expectations of the past few years.

ACKNOWLEDGMENT

Funding by the Swiss National Science Foundation and Committee for the Promotion of Applied Scientific Research is gratefully acknowledged.

REFERENCES

1. Turner, A. P. F. et al. *Biosensors – Fundamentals and Applications*; Oxford Science, 1987.
2. Moody, G. J. et al. *Analyst* **1970**, *95*, 910.
3. Bergveld, P. *IEEE Trans. Biomed. Eng.* **1970**, *BME-17*, 70.
4. van der Wal, P. D. et al. *Anal. Chim. Acta* **1990**, *231*, 41.
5. Cha, G. S. et al. *Anal. Chem.* **1991**, *63*, 1666.
6. van der Wal, P. D. et al. *Sens. Actuators* **1994**, *B(18-19)*, 200.
7. Shichiri, M. et al. *Diabetologia* **1993**, *24*, 179.
8. Bindra, D. S. et al. *Anal. Chem.* **1991**, *63*, 1692.
9. Turner, R. F. B. et al. *Sens. Actuators* **1990**, *B1*, 561.
10. Urban, G. et al. *Biosens. Bioelectron.* **1991**, *6*, 555.
11. Centoze, P. et al. *Fresenius J. Anal. Chem.* **1992**, *342*, 729.
12. Hämmerle, M. et al. *Sens. Actuators* **1992**, *B6*, 106.
13. Mutlu, S. et al. In ACS Symposium Series 556, American Chemical Society: Washington, DC, 1994; 71.
14. Sasso, S. V. et al. *Anal. Chem.* **1990**, *62*, 1111.
15. Gibson, T. D. et al. *Biosens. Bioelectron.* **1992**, *7*, 701.

ION-SELECTIVE MICROCHEMICAL SENSORS (Cardiology)

Richard P. Buck, Miklós Erdösy, Timothy A. Johnson, and Vasile V. Cosofret
Departments of Chemistry and Medicine
University of North Carolina

Ernö Lindner
Institute for General and Analytical Chemistry
Technical University of Budapest

Ion-selective potentiometric sensors, based on neutral carriers, have become routine tools in clinical chemistry and have gained acceptance in clinical analyzers and instrumentation since 1970.[1–3] They have been used to measure the extracorporeal bloodstream during open-heart surgery,[4] continuous monitoring in intensive care units,[5] and during hemodialysis.[6] An important aspect of cardiology research is the measurement of biopotentials in and on the myocardium and epicardium. Knowing the spatial and temporal changes in these potentials is essential to understanding the mechanism of cardiac dysrhythmias and developing models of cardiac function. Cardiovascular physiologists would like to use flexible multielectrode probes with well-defined geometries in order to describe the potential and the ionic distribution in the heart as a function of time. Ischemic events have ionic markers, usually hydrogen and potassium ions, that change magnitude over time following brief periods of ischemia. Consequently, a number of investigators have used chemically sensitive sensors to measure changes in myocardial extracellular K^+ and H^+ in the intact heart.[7–9]

Acute myocardial ischemia (deprivation of oxygen to the heart) often leads to impulse conduction slowing, cardiac arrhythmias and ventricular fibrillation and, eventually, to sudden cardiac death. Chemical consequences of fibrillation and defibrillation also require *in vivo* monitoring of ionic species, especially H^+, K^+, Na^+, and Ca^{2+}. Our main efforts have been directed to design fabrication of planar, solvent–polymeric ion-selective microsensors for cardiovascular applications using the advantages of existing semiconductor and microelectronics technologies. Physiological considerations require that the polymer-based sensor be robust, small, flexible, and made of biocompatible materials. We recently produced planar microelectrochemical sensors, based on flexible polyimide (Kapton®, DuPont [25036-53-7]) substrate, that meet these requirements.[10,11]

In addition, we reduced the size of the sensors by changing the classic membrane electrode construction (i.e., internal reference electrode-internal buffer solution–polymeric selective membrane) to a planar sensor arrangement without a large volume of liquid internal contact. Several problems were brought to light concerning the stability, reproducibility, response time, and lifetime of these sensors, especially considering the stringent requirements in this field. Improved adhesion, biocompatibility, and low resistance were found to be decisive in polymeric membrane optimization.[12,13]

RESULTS AND DISCUSSION

We intend to show that pH, pK, pNa, and pCa membranes can be optimized in order to fabricate planar microchemical sensors for these ionic species with applications in cardiology.

Poly(vinyl chloride) was introduced as a polymeric matrix for chemical sensors in 1967.[14–17] The polymer traps the organic solvent (plasticizer) used in liquid membranes, and improves the flexibility of the membrane in the physical sense as well as in the ease of its application.[1] The solvent (plasticizer) is a swelling agent for the polymer that is held in place by solvation forces. The polarity of the solvent and its concentration in the membrane significantly influence the ion partition coefficients and mobilities, and therefore the analytical properties, of the sensor. Generally, in polymeric ion-selective membrane electrodes, one part polymer and two parts plasticizer (1:2) is the most commonly used composition.

Membrane Materials Selection

The optimal membrane composition was systematically studied using macroelectrodes with massive internal solution contact (typical membrane sizes of several millimeters). In the optimization procedure, besides the general analytical performances of the electrodes, some crucial membrane characteristics (i.e., selectivities, response time resistance) were followed over time.[11,18] Our goal was to adapt existing semiconductor and microelectronics technologies to produce closely spaced sensor arrays on a flexible substrate (e.g., Kapton) that moves with the muscle fibers within the wall of the beating heart. The adhesion of the polymeric membrane to this substrate becomes crucial.

Adhesion is considerably better for the modified PVC samples (PVC-NH_2 ~ PVC-COOH > PVC-OH >> HMW-PVC). Unfortunately, aminated PVC (PVC-NH_2) cannot always be used as a membrane matrix because of its pH sensitivity.[13,18–20] The properties of ionophore-free H^+ selective membranes (PVC-NH_2) are dependent on the chemical structure of the amine as well as on several parameters related to the synthesis process.[18,19]

Carboxylated PVC (PVC-COOH) is known to be a good membrane substrate for ion-selective electrodes.[18,21–24] Plasticized PVC-COOH is surprisingly responsive to H^+, Na^+, K^+, and Ca^{2+} in much the same way that glass electrodes containing specific compositions show enhanced cation response.

For our membrane casting, we used Tecoflex polyurethane as a membrane matrix because of its excellent biocompatibility,[25,26] and on the basis of the encouraging preliminary results of the Cha et al.[27] and D'Orazio et al.[28] This material shows good adhesive properties to polyimide-coated Kapton substrate and low membrane resistance.[12] Unfortunately, the electroanalytical properties of H^+, K^+, and Na^+ sensors obtained with this matrix were inferior to sensors made with PVC-COOH.

Analytical Performances of Planar Microchemical Sensors

For measuring ions of cardiovascular interest the relevant properties of the membrane sensors are selectivity, stability, and response time.

All efforts to design new ionophores for membrane fabrication take into consideration the required selectivities for blood, plasma, and serum applications.[1] Because the ionophore does not solely determine the selectivity of a membrane, a well-selected plasticizer and lipophilic salt additive may enable the selectivity to reach the required value.[1]

The new miniature pH, K^+, Na^+, and Ca^{2+} sensors developed by our group have overcome one of the principal limitations of small electrochemical sensors for biomedical applications, stability.

CONCLUSION

Microelectronically fabricated pH, K^+, Na^+, and Ca^{2+} planar sensors based on polyimide substrate (Kapton) and using various polymeric membrane compositions demonstrate good analytical performances and extended lifetime (up to at least three months of continuous soaking and use). Single pH and K^+ sensors were used to monitor the fall in pH and K^+ activity, during acute ischemia at several sites in a beating porcine heart. We hope to use these data to develop practical devices with good analytical and polymeric membrane properties.

ACKNOWLEDGMENTS

This work was supported in part by the NSF Engineering Research Center (grant number CDR-8622201) and NIH grants R01 HL49818 and P01 H127430.

REFERENCES

1. Oesch, Y. et al. *Clin. Chem.* **1986**, *32*, 1448.
2. Czaban, J. D. *Anal. Chem.* **1985**, *57*, 345A.
3. Meier, P. C. et al. *Medical and Biological Applications of Electrochemical Devices*; John Wiley & Sons: Chichester, 1980; Chapter 2.
4. Osswald, H. F. et al. *Clin. Chem.* **1979**, *25*, 39.
5. Treasure, T. *Intensive Care Med.* **1978**, *4*, 83.
6. Haase, E. A. et al. *GIT Labor-Medizin* **1992**, *15*, 84.
7. Johnson, T. A. et al. *Am. J. Physiol.* **1990**, *258*, H1224.
8. Cosofret, V. V. et al. *Am. Chem. Soc.* **1993**, *70*, 137.
9. Cosofret, V. V. et al. *Talanta* **1994**, *41*, 931.
10. Lindner, E. et al. *J. Chem. Soc. Faraday Trans.* **1993**, *89*, 361.
11. Lindner, E. et al. *Fresenius J. Anal. Chem.* **1993**, *346*, 584.
12. Lindner, E. et al. *J. Biomed. Mater. Res.* **1994**, *28*, 591.
13. Cosofret, V. V. et al. *J. Electroanal. Chem.* **1993**, *345*, 169.
14. Moody, G. J. et al. *Analyst* **1970**, *95*, 910.
15. Craggs, A. et al. *J. Chem. Educ.* **1974**, *51*, 541.
16. Bloch, R. et al. *Biophys. J.* **1967**, *7*, 865.
17. Moody, G. J.; Thomas, J. D. R. *Ion-Selective Electrode Rev.* **1979**, *1*, 3.
18. Lindner, E. et al. *Talanta* **1993**, *40*, 957.
19. Cosofret, V. V. et al. *Analyst* **1994**, *119*, 2283.
20. Buck, R. P. et al. *Anal. Chim. Acta* **1993**, *282*, 273.
21. Anzai, J. et al. *Bull. Chem. Soc. Jpn.* **1987**, *60*, 4133.
22. Satchwill, T.; Harrison, D. J. *J. Electroanal. Chem.* **1988**, *202*, 75.
23. Ma, S. C. et al. *Anal. Chem.* **1988**, *60*, 2293.
24. Lindner, E. et al. *Anal. Chem.* **1988**, *60*, 295.
25. Coury, A. J. et al. *Biomater. Appl.* **1988**, *3*, 130.
26. Lelah, M. D.; Cooper, S. L. *Polyurethane in Medicine*; CRC: Boca Raton, FL, 1986.
27. Cha, S. C. et al. *Anal. Chem.* **1991**, *63*, 1666.
28. D'Orazio, P. et al. *Presented at Electrolytes, Blood Gases, and Other Critical Analytes: The Patient, The Measurement, and The Government*; International Symp.; Chatham, MA, 1992; 21-34.

ION-SELECTIVE REAGENTS (Dual Mechanism Bifunctional Polymers)

Spiro D. Alexandratos
Department of Chemistry
University of Tennessee

The synthesis and characterization of polymer-supported reagents with high binding affinities for selected metal ions is important for many applications, including catalysis, chromatography, and environmental separations.[1] Numerous ligands have been immobilized onto crosslinked polymer supports (sulfonates, carboxylates, thiols, polyamines, glycols, oximes, imidazoles, etc.) and their metal ion affinities have been determined.[2] Much of our understanding of the affinities is embodied within hard–soft acid–base theory.[3]

Dual-mechanism bifunctional polymers (DMBPs) are designed as substrate-selective reagents with optimal selectivity and kinetics. DMBPs consist of a crosslinked network on which are immobilized two different types of ligands, one of which operates through an access mechanism and one through a recognition mechanism. The access mechanism allows for a rapid entry of substrates such as metal ions into the network, enabling them to come into close proximity to the site responsible for the recognition mechanism and, in turn, for the observed selectivity. Most studies have centered on polystyrene crosslinked with divinylbenzene as the inert support and ion exchange as the access mechanism because of the rapid kinetics associated with that process.

The recognition mechanism defines the three classes of DMBPs: Class I polymers are the ion exchange–redox resins in which reduction of the M^{n+} substrate to the free metal is the recognition mechanism;[4] Class II polymers are the ion exchange–coordination resins in which different coordinative ligands are responsible for the observed selectivity;[5] and Class III polymers are the ion exchange/precipitation resins in which precipitation of the metal ion as the metal salt is the operative recognition mechanism.[6] This phosphinic acid polymer is an example of Class I DMBPs (1); phosphonate monoester/diester polymers are an example of Class II DMBPs (2); and phosphonic acid–quaternary amine polymers are an example of Class III DMBPs (3).

Under highly acidic conditions, the phosphonic acid ligand is unable to ion exchange but is able to coordinate with substrates through the phosphoryl oxygen. A bifunctional sulfonic acid/phosphonic acid resin (**Structure 1A**) was prepared and characterized as another example of a Class II DMBP.[7] A sulfonic acid–diphosphonic acid resin was also prepared and, by displaying extraordinarily high affinities at short contact times, is now available commercially as Diphonix™ resin (Structure **1B**) through Eichrom Industries.[8]

1

(A) (B)

The concept of dual mechanism (access/recognition) bifunctional polymers reaches an important point with the advent of Diphonix. It was originally developed for the recovery of actinide ions. In a representative example, the distribution coefficient for Am(III) from a 1 M HNO_3/4 M $NaNO_3$ solution is 49 for a sulfonic acid ion exchange resin and 5600 for Diphonix.™ Chelation through the geminal diphosphonate ligands is probably critical to its high distribution coefficients, because a comparable monophosphonic acid resin has a distribution coefficient for Am(III) of 24. It is currently being evaluated for the recovery of transition metal ions.

ACKNOWLEDGMENTS

We are grateful to the Department of Energy, Office of Basic Energy Sciences, for continuing support of this research (DE-FG05-86ER13591). We have greatly benefited from our interaction with Dr. E. Philip Horwitz of the Argonne National Laboratory. This research is also dependent on the dedication of a very able group of postdoctoral associates and graduate students, currently consisting of Robert Beauvais, Latiff Hussain, Cheryl Laughter, Dorothy Miller, Vijay Patel, Kelly Ripperger, Chris Shelley, and Dr. Andrzej Trochimczuk.

REFERENCES

1. D. C. Sherrington and P. Hodge Eds., *Syntheses and Separations Using Functional Polymers*, John Wiley & Sons: New York, 1988.
2. Warshawsky, A. *Ion Exchange and Sorption Processes in Hydrometalurgy*, M. Streat and D. Naden, Eds., Critical Reports in Analytical Chemistry, Vol. 19, John Wiley for Soc. Chem. Ind., 1987.
3. R. G. Pearson, Ed., *Hard and Soft Acids and Bases*, Dowden, Hutchinson and Ross, Inc., PA, 1973.
4. Alexandratos, S. D.; Wilson, D. L. *Macromolecules* **1986**, *19*, 280.
5. Alexandratos, S. D.; Quillen, D. R.; Bates, M. E. *Macromolecules* **1987**, *20*, 1191.
6. Alexandratos, S. D.; Bates, M. E. *Macromolecules* **1988**, *21*, 2905.
7. Trochimczuk, A.; Alexandratos, S. D. *J. Appl. Poly. Sci.* **1994**, *52*, 1273.
8. Horwitz, E. P.; Chiarizia, R.; Diamond, H.; Gatrone, R. C.; Alexandratos, S. D.; Trochimczuk, A.; Crick, D. W. *Solv. Extr. Ion. Exch.* **1993**, *11*, 943.

IONENE OLIGOMERS

Hiromichi Noguchi
Department of Chemistry and Biotechnology
Graduate School of Engineering
The University of Tokyo

Ionene oligomers are defined as low–molecular weight compounds that are produced directly or as byproducts in reactions that, at least formally, afford ionene polymers, which are polymers having positive hetero atoms in their main chains.

Two types of ionene oligomers are known: cyclic ionene oligomers and linear ionene oligomers. Although both have been known for a long time, no particular attention had been paid to them until Brown et al.[1] published an epoch-making paper on the efficient template-directed synthesis of a macrocyclic ionene oligomer, cyclobis(parquat-*p*-phenylene).

Their fascinating work since then[2–7] as well as the work of several other research groups[8–11] has resulted in remarkable development of the synthetic chemistry of pseudorotaxanes, rotaxanes, and catenanes based on ionene oligomers and polymers. This, as well as concurrent progress in constructing other types of polyrotaxanes and polycatenanes, is no doubt paving an avenue to a new field of polymer science.[12–16]

Many reactions can produce ionene polymers. However, only the Menschutkin reactions yield ionene oligomers directly. In this synthesis, the reaction path and product(s) depend on several factors: structure of reactants, molar ratio of reactants, reactant concentration, solvent, and temperature. In the Stoddart method for the synthesis of macrocyclic ionene-oligomers, a template molecule also influences the reaction. In both cases, the syntheses have seldom been investigated under catalyzed conditions.

SYNTHESIS

Conventional Methods

Linear Ionene Oligomers

Oligomers of the structure of **3** [**Scheme I**: Z = Br; (x,y) = (4,2), (6,1) and (6,2)], presumably the first linear ionene oligomers with well-defined structures, have been prepared by 1:2 addition in the reactions of N,N,N′,N′-tetramethyl-α-ω-diaminoalkanes (**1**) with α,ω-dihaloalkanes (**2**), called x, y, Z reactions for short, which were carried out at a molar ratio of 1:1.[17] Linear ionene oligomers **4**, could be prepared by 2:1 addition only when the molar ratio of **1** to **2** was 2:1[18] or when a large excess of **1** was used.[19]

Thus, reactions of **3** and **4** with a large excess of **1** and **2**, respectively, under homogeneous conditions are expected to produce linear oligomers with four positive nitrogens. Repeating these procedures alternately should in principle give higher oligomers of well-defined structures. In fact, Gulyayeva et al. reported that they prepared linear ionene oligomers with up to eight N^+s from **1** (x = 2) and **2** (y = 5, Z = Br).[20] Monodisperse linear ionene oligomers with more than eight N^+s have not been reported.

Cyclic Ionene Oligomers

Cyclic ionene oligomers have been known much longer than linear ionene oligomers. The x-y, Br reactions with (x, y) values of (2,1), (2,2), (2,3), and (3,2) afford five- to seven-membered cyclic ionene oligomers (**5**) by 1:1 addition.[17,28]

Recently, I have shown that even the 2-3 reactions (Z = Cl, Br) when carried out in acetonitrile at a molar ratio of 1:1 also give linear diammonium salts **4** by 2:1 addition together with the seven-membered cyclic diammonium salt by 1:1 addition.[18]

In contrast to the 2-4,Br reaction at a molar ratio of 1:1, which gives an ionene polymer, 2,4-ionene bromide,[17] the reaction of **1** (x = 2) with *o*-xylylene dibromide affords an eight-membered cyclic diammonium salt by 1:1 addition,[23] indicating that the product type can be altered by the rigidity of reactants.

Macrocyclic ionene oligomers having four or six N^+s have been prepared by Tabushi et al,[24] Murakami et al.,[25] Kawakami et al.,[26] Winkler et al.,[27] and Schneider et al.[28,29] by stepwise

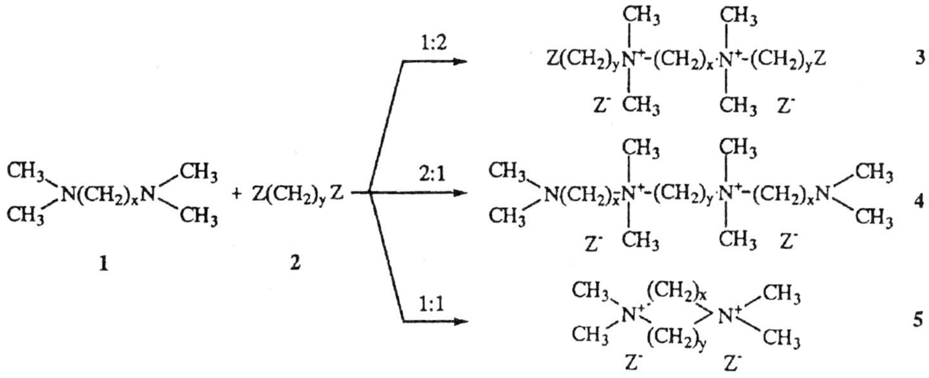

SCHEME I. Synthesis of linear and cyclic ionene oligomers from **1** and **2**.

methods, in all cases construction of the basic macrocyclic skeletons being independent of the Menschutkin reaction.

PROPERTIES AND APPLICATIONS

Reactions of ionene oligomers have been much less investigated than those of ionene polymers. The main reactions, excluding skeletal degradation, are (i) counterion exchange reactions, (ii) reactions of terminal groups, and (iii) dequaternization reactions by removal of side groups. The second reaction is, of course, limited to linear ionene oligomers.

The counterion exchange reactions of ionene oligomers has been investigated little. One application of this reaction is to control the solubility of oligomers, enabling a wider variety of investigations.

Few papers have been published on the application of ionene oligomers. Rembaum has reported that, although the toxicity of some ionene oligomers, **3** and **4**, was higher than that of most x,y-ionene bromides, the former exhibit much lower antimicrobial and antifungal activities than the latter.[32] Considering the well-known pharmacological activity of N^+-containing compounds such as tubocrarine, decamethonium iodide, and 6,3-ionene bromide (Polybrene®), activity of photocleavage of DNA of a linear tetracation having 2,7-dipyrenium cations at both terminals,[33] the biological activity of a naturally occurring ionene polymer[21] and naturally occurring macrocyclic ionene oligomers,[34] isolated from different marine sponges, or the ubiquitous presence of polyamines in nature, much potential is expected for the biological, pharmacological, and medical applications of ionene oligomers. It should also be noted that a macrocyclic per-N-alkylated tetraammonium salt has been found to interact with the DNA double helix at its major groove much more strongly than do linear tetraammonium salts[34] and that ionene polymers form complexes with DNA as well.[32]

Macrocyclic ionene oligomers having four N^+s and S^+s, have been known to incorporate hydrophobic compounds in their cavities. This inclusion phenomenon has been applied to promote various reactions such as ester hydrolysis,[24] nucleophilic substitution,[28] transamination,[27] radical cholorination,[36] and Diels–Alder reaction.[37]

Ionene polymers with high charge density, in particular 2,4-ionene, have been found to be effective promoters for the autoxidation of the thiols catalyzed by Co(II)phthalocyanine tetrasodiumsulfonate, $CoPc(NaSO_3)_4$, namely the so-called Merox oil sweetening process.[38]

No practical applications of pseudorotaxanes and catenanes have so far been reported. Stoddart thinks that these supramolecular assemblies could lead to molecular computers.[2,3]

CONCLUDING REMARKS

Supramolecular chemistry using ionene and other types of oligomers and polymers is being rapidly developed, particularly in the past few years. A variety of pseudopolyrotaxanes and polyrotaxanes have already been prepared.[16,30] In contrast, synthesis of linear polycatenanes remains at an oligomeric stage, and appears to require a dramatic breakthrough in the synthetic method.[31] The physical properties and applications of these novel types of polymers will be of greater importance in the future.[14–16] Applications of ionene oligomers in the fields of biology, pharmacology, medicine, and catalysis will also be promising.

ACKNOWLEDGMENT

The author wishes to thank Ms. Satoko Nishiyama for her help in preparing the schemes and tables.

REFERENCES

1. Brown, C. L.; Philip, D.; Stoddart, J. F. *Synlett.* **1991**, 462.
2. Stoddart, J. F. *Host-Guest Molecular Interactions: From Chemistry to Biology*; Ciba Foundation Symposium 158; John Wiley & Sons: New York; 1991; pp 5–22.
3. Stoddart, J. F. *Chem. Br.* **1991**, *27*, 714.
4. Philp, D.; Slawin, A. M. Z.; Spencer, N.; Stoddart, J. F.; Williams, D. J. *J. Chem. Soc., Chem. Commun.* **1991**, 1584.
5. Philp, D.; Stoddart, J. F. *Synlett.* **1991**, 445.
6. Anelli, P. L.; Ashton, P. R.; Ballardini, R.; Balzani, V.; Delgado, M.; Gandolfi, M. jT.; Goodnow, T. T.; Kaifer, A. E.; Philp, D.; Pietraszkiewicz, M.; Prodi, L.; Reddington, M. V.; Slawin, A. M. Z.; Spencer, N.; Stoddart, J. F.; Vincent, C.; Williams, D. J. *J. Am. Chem. Soc.* **1992**, *114*, 193.
7. Amabilino, D. B.; Stoddart, J. F. *Chem. Mater.* **1994**, *6*, 1159.
8. Vögtle, F.; Müller, W. M.; Müller, U.; Bazuer, M.; Rissanen, K. *Angew. Chem. Int. Ed. Engl.* **1993**, *32*, 1295.
9. Benniston, A. C.; Harriman, A. *Angew. Chem. Int. Ed. Engl.* **1993**, *32*, 1459.
10. Gunter, M. J.; Hockless, D. C. R.; Johnston, M. R.; Skelton, B. W.; White, A. H. *J. Am. Chem. Soc.* **1994**, *116*, 4810.

11. Wylie, R. S.; Macartney, D. H. *Supramol. Chem.* **1993**, *3*, 29.
12. Harada, A.; Li, J.; Kamachi, M. *J. Am. Chem. Soc.* **1994**, *116*, 3192.
13. Harada, A.; Li, J.; Kamachi, M. *Macromolecules* **1994**, *27*, 4538.
14. Gibson, H. W.; Marand, H. *Adv. Mater.* **1993**, *5*, 11.
15. Gibson, H. W.; Bheda, M. C.; Engen, P. T. *Prog. Polym. Sci.* **1994**, *19*, 843.
16. Wenz, G. *Angew. Chem. Int. Ed. Engl.* **1994**, *33*, 803.
17. Noguchi, H.; Rembaum, A. *J. Polym. Sci., Part B, Polym. Lett.* **1969**, *7*, 383.
18. Noguchi, H. Presented at Second Pacific Polymer Conference, Otsu, Japan, Preprints, November 26–29, 1991, p 72.
19. Schipper, E. T. W. M.; Roelofs, A. H. C.; Piet, P.; German, A. L. *Polym. Int.* **1994**, *33*, 79.
20. Gulyayeva, Zh. G.; Zansokhova, M. F.; Razvodovskii, F.; Yefimov, V. S.; Zezin, A. B.; Kabanov, V. A. *Polym. Sci. U.S.S.R.* **1983**, *25*, 1436.
21. Coleman-Davis, M. T.; Faulkner, D. J.; Dubowchik, G. M.; Roth, G. P.; Polson, C.; Fairchild, C. *J. Org. Chem.* **1993**, *58*, 5925.
22. Rembaum, A.; Noguchi, H. *Macromolecules* **1994**, *94*, 1183.
23. Noguchi, H.; Uchida, Y. *J. Polym. Sci., Part B, Polym. Lett.* **1975**, *13*, 773.
24. Tabushi, I.; Kimura, Y.; Yamamura, K. *J. Am. Chem. Soc.* **1978**, *100*, 1304.
25. Murakami, Y.; Nakano, A.; Miyata, R.; Matsuda, Y. *J. Chem. Soc., Perkin 1* **1979**, 1669.
26. Kawakami, H.; Yoshino, O.; Odashima, K.; Koga, K. *Chem. Pharm. Bull.* **1985**, *33*, 5610.
27. Winkler, J.; Coutouli-Argyropoulou, E.; Leppkes, R.; Breslow, T. *J. Am. Chem. Soc.* **1983**, *105*, 7198.
28. Schneider, H.-J.; Busch, R. *Angew. Chem. Int. Ed. Engl.* **1984**, *23*, 911.
29. Schneider, H.-J.; Philippi, K.; Pohlmann, J. *Angew. Chem. Int. Ed. Engl.* **1984**, *23*, 908.
30. Ashton, P. R.; Philp, D.; Spencer, N.; Stoddart, J. F. *J. Chem. Soc., Chem. Commun.* **1991**, 1677.
31. Amabilino, D. A.; Ashton, P. R.; Reder, A. S.; Spencer, N.; Stoddart, J. F. *Angew. Chem. Int. Ed. Engl.* **1994**, *33*, 1286. See also: *Chem. Eng. News.* August 29, 1994, p 28.
32. Rembaum, A. *Appl. Polym. Symp.* **1973**, *22*, 299.
33. Blacker, A. J.; Jazwinski, J.; Lehn, J.-M.; Wilhelm, F. X. *J. Chem. Soc., Chem. Commun.* **1986**, 1035.
34. Fusetani, N.; Asai, N.; Matsunaga, S.; Honda, K.; Yasumuro, K. *Tetrahedron Lett.* **1994**, *35*, 3967.
35. Schneider, H.-J.; Blatter, T. *Angew. Chem. Int. Ed. Engl.* **1992**, *31*, 1207.
36. Schneider, H.-J.; Philippi, K. *Chem. Ber.* **1984**, *117*, 3056.
37. Schneider, H.-J.; Sangwan, N. K. *J. Chem. Soc., Chem. Commun.* **1986**, 1787.
38. Brouwer, W. M.; Piet, P.; German, A. L. *J. Mol. Cat.* **1965**, *31*, 169.

IONENE POLYMERS

Hiromichi Noguchi
Department Chemistry and Biotechnology
Graduate School of Engineering
The University of Tokyo

Polymers having positive nitrogen atoms in their main chains were first prepared by Gibbs et al. (**Equation 1**),[1] followed by

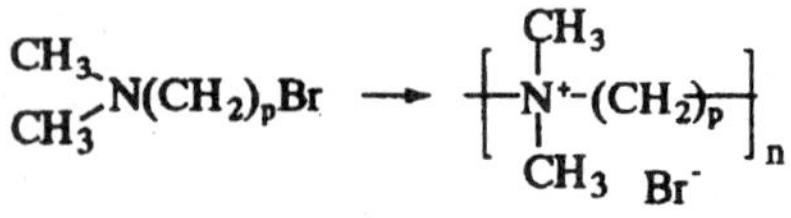

1

1 (p = 3, p > 6) **2**

Kern et al.,[2] based on the Menschutkin reactions. The investigations of this type of polymers had remained sporadic until Rembaum et al.[3] published the epoch-making paper in which they coined the name *ionenes*, for such polymers, based on the fact that they can be synthesized through the ionization of amines. A series of papers by the Rembaum group since then obviously aroused the interest of many scientists in various fields, recently resulting in about seventy papers a year.[4–13] During the developing period, naturally occurring ionene oligomers and polymers have also been isolated from some marine sponges.

Recent developments on this type of polymers have revealed that synthesis and properties of polymers having other positive hetero atoms such as P^+s and S^+s in the main chains are analogous to those of conventional ionenes. Therefore, apart from the original definition, the author proposes to extend formally the original definition of ionene polymers to polymers having in their main chains positive *hetero* atoms other than positive nitrogens.

PROPERTIES

Reactivity

Reactions of ionene polymers can be classified into several categories: counterion exchange, reactions of the terminal groups, dequaternization by removal of the side groups, dequaternization and degradation of the backbone chain, oxidation of the positive hetero atoms, and unusual reactions of phosphonium ionenes.

The most well-known reaction of ionene polymers is counterion exchange, leading to various ionene polymers with different properties. Both organic and inorganic atoms can be introduced by this method. It is well known that replacement of halide ions of ionene polymers with TCNQ radical anions, i.e., $TCNQ^-$, particularly in the presence of neutral TCNQ, affords conductive organic materials.[12] Similarly, 6,10-ionene glasses having organic dye counterions that exhibit linear and non-linear optical properties[20,21] and photochromic 6,10-ionenes with spiropyran sulfonate as counterions[22] have been prepared. Further, Simmrock et al.[23] have prepared various ionene polymers having mono- or divalent counter anions such as BPh_4^-, $ZnBr_4^{2-}$, $CuBr_4^{2-}$, HgI_4^{2-} and so forth by ion exchange. Some of these were of high refractive indices with low optical dispersion. Polymeric anions can also be introduced by this method as counterions of ionene polymers.

The terminal NMe_2 groups of ionene polymers **2** have been utilized to prepare A-B-A triblock polymers [A: 3-ionene, B: polyoxyethylene or PTMO].[24,25] However, preparation of block polymers consisting of ionene segments and totally hydrophobic segments is in general quite difficult because of extremely different solubilities of the two components.

Dequaternization of ionene polymers by removal of side groups is useful to prepare poly(*tert*-amine)s with controlled distribution of tertiary nitrogens in the polymer backbone and, subsequently, different ionene polymers by requaternization.

Formation of Polyelectrolyte Complexes

For the quaternized poly(4-vinylpyridine) [QPVP]/poly (sodium styrene sulfonate) [PNaSS] system, the yield of PEC reaches a maximum (ca. 100%) at a molar ratio of 1:1, regardless of the mixing order, indicating that a water-soluble equimolar complex is always obtained.[7,26] In marked contrast, the mixing order exerts a vital influence on the PEC yield in the reactions of ionenes with polyanions of the same type.

Solution Viscosity and Molecular Weight

It is well known that an ionene polymer with high charge density in pure water behaves as a typical polyelectrolyte, and its solution viscosity fits the Fuoss equation.[27] On addition of a simple electrolyte its intrinsic viscosity ([η]) can be estimated in the usual manner.

The MWs have been determined in most cases by the light scattering method and, by membrane osmometry. The MWs of various ionene halides have also been determined by the analytical ultracentrifugation method.[18,19,28,29] If low (MW < ca. 15000), the MWs of aliphatic ionene polymers can be estimated by end group determination after pretreatment(s).

Surface Tension and Formation of Micelle-Like Structure

While an ionene polymer with short spacers such as 3,4-ionene bromide forms a multilayer structure only when the counterion is a long one like dodecyl sulfate,[31] ionene polymers with long organic spacers between positive hetero atoms exhibit micelle-like behaviors in an aqueous solution even when the counterions are small, short ones.

Catalytic Activity

Alkaline hydrolysis of *p*-nitrophenyl acetate is accelerated in the presence of 3,y-ionene bromides as in the presence of cetyltrimethylammonium bromide (CTAB).[30]

Biological Activity

Early works on the biological activity of ionene polymers and oligomers have been reviewed by Rembaum.[8] It has been found that polymeric nature is an indispensable requisite for a high antimicrobial or antifungal activity, because the ionene oligomers and model components are inactive. Further, the structure of the ionene polymers plays an important role in the antimicrobial activity. Additionally, ionene polymers have been found to be more efficient ganglionic blockers with longer duration effects than their oligomeric model compounds.

All these results concerning the antibacterial activity of ionene polymers suggest that the mechanism of their action at least involves their adsorption to, and subsequent disruption of, the bacterial cytoplasmic membranes.

Hygroscopicity, Solubility, and Crystallinity

The hygroscopicity and solubility of ionene polymers widely vary depending on their chemical structure, counterions, and charge density. Those that have a low volume fraction of organic components and hence a high charge density exhibit typical polyelectrolyte behaviors. They are relatively hygroscopic, particularly when the counterions are Br^- and Cl^- and the primary structure is irregular. In general, ionene polymers with higher hygroscopicity have higher solubility in water and those with lower hygroscopicity exhibit higher tendency to dissolve in aprotic organic solvents. It should be noted, however, that the crystallinity of ionene polymers also influences their hygroscopicity and solubility.

The crystallinity of ionene polymers also depends on the chemical structure, counterions, and charge density. It also depends more or less on the history and water content of the sample as well as on the preparation conditions.[33,34] For instance, some amorphous ionenes turn into crystalline polymers upon absorption of water.[32] Ionene polymers that have a high charge density are prone to readily crystallize to highly or perfectly crystalline materials, especially when the primary structure is highly regular. In contrast, ionene polymers with a high volume fraction of organic components or with a low charge density are hydrophobic, and partially crystalline or even amorphous. In particular, when the polymer backbone chain is irregular, crystallization of the polymer is hindered, giving rise to *ionene glasses* of high quality. When the charge density is extremely low, they behave like ionomers,[35] and hence belong to *cationomers*.

Thermal Properties

The thermal properties of ionene polymers such as glass transition temperature (T_g), melting point (T_m), and stability of the melt markedly depend on the chemical nature and length of the organic spacers between the ionic sites, the size of side groups, and the type and size of counterions.

Molecular Motions (NMR Spectroscopy)

The mechanical and transport properties of solid polymers largely depend on molecular motions. Therefore, characterization of the motion at different parts of a polymer is fundamentally important. Solid-state ^2H and ^{13}C NMR spectroscopy have been shown to be excellent means for this purpose.

Elastomeric Ionenes

Ionene polymers having rubbery soft segments, exhibit large elongation and high tensile strength[35,37–41] and hence are called *elastomeric ionenes.* Crosslinking is not necessarily indispensable, because, if properly constructed, ionene polymers exhibit ionomeric properties owing to ion-clustering of the hard segments, namely N^+s. Thus, the ion clustering works as physical crosslinking points. Therefore, such ionene polymers can also be regarded as a sort of ionomers, namely *ionene cationomers.*

The excellent tensile properties of elastomeric ionenes have been considered to result from a high degree of microphase separation or ionic domain formation.[39,41,42]

Liquid Crystalline Ionenes

Ionene polymers having mesogenic groups in either side or main chain have been prepared that exhibit thermotropic and/or lyotropic LC phases.

Interestingly, some of the aliphatic ionene polymers having a long alkyl side chain, N^+-x,y-Me,C_pH_{2p+1}-Z (x, y = 3–10, p = 14–20; Z=Br, BF_4), which are called *polysoaps*[15] or *comb-like ionenes*,[16] also exhibit smectic phases,[43] even though they do not have mesogenic groups. The mesomorphism is attributed to the packing of the long aliphatic side chains and the absence of crystallinity of the main chains.

Electrical Conductivity

Since the early work of Rembaum et al.[3,44] the electrical conductivity of $TCNQ^-$/TCNQ complexes of ionene polymers has been most extensively studied[12,45] and is beyond the scope of this article.

Construction of Pseudopolyrotaxanes

Pseudopolyrotaxanes and polyrotaxanes, which are molecular composites comprising macrocycles threaded by linear polymers not having and having, respectively, large stopper groups at terminals, have recently been attracting much interest [see also *Ionene Oligomers*]. Recently, Gibson et al.[46,53,54] prepared a series of segmented elastomeric poly(urethane rotaxane)s.

Similarly, Marsella et al.[47] prepared pseudopolyrotaxanes from a poly(pyridinium vinylene) and crown ethers.

APPLICATIONS

Polyelectrolytes including ionene polymers have been used in various industrial fields.[48] Ionene polymers, which can be produced from cheap starting materials, have been used in relatively large quantities for the purpose of flocculation, waste water treatment, paper making, and so forth. Many applications of ionene polymers are based on the direct interaction of their positive charges with the negative charges of the materials to be treated.

Medical Use

Rather surprisingly, medical application of 6,3-ionene bromide, also called *hexadimethrine bromide* or Polybrene® (Abbott Laboratories, U.S.A.), had been investigated before the term ionene appeared.[49] It neutralizes the effects of heparin, an anionic polymer, and hence prevents its anticoagulant action. The antagonistic capacity would be due to the PEC formation between the ionene and heparin. Polybrene can also be used as an antispasmolytic drug.

Microbiocides

Some ionene polymers have been found to be highly effective microbiocides.[50]

Flocculation

Ionene polymers are effective for the flocculation of a dilute clay suspension.[51] Various types of ionene polymers are commercially available for water treatment.[17]

Miscellaneous Applications

Other potential or suggested applications of ionene polymers include: diagnostic and chemotherapeutic applications of ionene polymers,[14] applications in integrated optics,[36] use of elastomeric ionene for electrochromic display,[11] and for high speed and high density optical recording,[52] control of the mechanical properties of elastomeric ionene by light,[38] and medical use of elastomeric ionene as an adhesive and artificial skin that exhibits high adhesive property and high permeability of water vapor, oxygen, and nitrogen.[11,37]

CONCLUDING REMARKS

Most of the earlier and some of the recent developments not described here can be consulted with the aid of reviews[4–13] as well as of the recent papers cited here.

Summarizing the above-mentioned, such interest in ionene polymers would be due to the following facts: Syntheses of ionene polymers are in general straightforward and can be carried out under mild conditions, even in the absence of catalyst. Various positive hetero atoms can be incorporated into the polymer backbones. The distribution of positive hetero atoms in the polymer chain as well as the charge density can be controlled without difficulty. The groups between positive hetero atoms can be widely varied and hence it is relatively easy to prepare ionene polymers with enhanced functionality. The side groups pendant from positive hetero atoms can also be varied relatively easily either by using appropriate monomers or, if applicable, by removing the side groups from ionene polymers and then requaternizing. The counterions of ionene polymers can be readily replaced with other ions including polyanions. These features enable one to systematically prepare a huge variety of ionene polymers and hence to readily clarify the relationship between the polymer structures and their properties. The functionalities of ionene polymers are matchlessly diversified compared with any other types of polymers. And finally, the properties of special kinds of ionene polymers can be modified even by the formation of pseudopolyrotaxanes using them as thread molecules.

In view of the recent developments as well as above-mentioned diversified features, the author expects that still more interesting properties and applications of ionene polymers as well as of ionene oligomers will be found and devised. To further specify this conjecture, the following would be worth commenting: Ubiquitous presence of polyamines in plants and animals, the known biological activity of (poly)ammonium salts, and recent isolation of naturally occurring ionene oligomers and polymers with biological activities suggest that biological and medical applications of ionene polymers as well as of ionene oligomers should further be examined. The investigations on ionene polymers having positive hetero atoms other than positive nitrogens are still at the embryonic stage, and hence unique properties and applications would be expected for them. Also, it should be noted that a new field of polymer science concerning synthesis and properties of pseudopolyrotaxanes, which one might call supramolecular polymer science, is being rapidly developed. As exemplified here by pseudopolyrotaxanes having ionene polymers as thread molecules, such supramolecularly assembled polymers appear to be promising and important in the future from various viewpoints including modifying biological, chemical, and physical properties of the polymers.[53,54]

ACKNOWLEDGMENT

The author wishes to thank Ms. Satoko Nishiyama for her help in the preparation of the equations, tables, and figures.

REFERENCES

1. Gibbs, C. F.; Littmann, E. R.; Marvel, C. S. *J. Am. Chem. Soc.* **1993**, *55*, 753.
2. Kern, W.; Brenneisen, E. *J. Prakt. Chem.* **1941**, *159*, 194 and 219.
3. Rembaum, A.; Baumgartner, W.; Eisenberg, A. *J. Polym. Sci., Part B, Polym. Lett.* **1968**, *6*, 159.
4. Hoover, M. F. *J. Macromol. Sci.-Chem.* **1970**, *A4*, 1327.
5. Hoover, M. F.; Butler, G. B. *J. Polym. Sci., Polym. Symp.* **1974**, *45*, 1.
6. Tsuchida, E.; Sanada, K. *Kagaku no Ryoiki (J. Jpn. Chem.)* **1971**, *25*, 732 and 864.
7. Tsuchida, E. *J. Macromol. Sci., Pure Appl. Chem.* **1994**, *A31*, 1.
8. Rembaum, A. *J. Appl. Polym. Sci., Appl. Polym. Symp.* **1973**, *22*, 299.
9. Tsutsui, T. *Developments in Ionic Polymers–2*; Wilson, A. D.; Prosser, H. J., Ed.; Elsevier Appl. Sci.: London, 1986; Chapter 4, p. 163.
10. Meyer, W. H.; Dominguez, L. *Polymer Electrolyte Reviews–2*; MacCallum, J. R.; Vincent, C. A., Ed.; Elsevier Sci.: London, 1989; Chapter 6, p 191.
11. Kohjiya, S.; Yamashita, S. *Kautsch. Gummi Kunst.* **1991**, *44*, 1128.
12. Kryszewski, M.; Pecherz, J. *Polym. Adv. Technol.* **1994**, *5*, 146.
13. Han, H.; Bhowmik, P. K. *Trends Polym Sci.* **1995**, *3*, 199.
14. Rembaum, A.; Senyei, A. E.; Rajaraman, R. *J. Biomed. Mater. Res. Symp.* **1977**, *8*, 101.
15. Sonnessa, A. J.; Cullen, W.; Ander, P. *Macromolecules* **1980**, *13*, 195.
16. Wang, J.; Meyer, W. H.; Wegner, G. *Acta Polym.* **1995**, *46*, 233.
17. Tsuda, M.; Yoshida, A.; Yuki, T. *J. Synth. Org. Chem. Jpn.* **1973**, *31*, 703.
18. Bortel, E.; Kochanowski, A.; Gozdecki, W.; Kozlowska, H. *Makromol. Chem.* **1981**, *182*, 3099.
19. Bortel, E.; Kochanowski, A. *Makromol. Chem.* **1987**, *188*, 2019.
20. Meyer, W. H.; Pecherz, J.; Mathy, A.; Wegner, G. *Adv. Mater.* **1991**, *3*, 153.
21. Pecherz, J.; Kryszewski, M.; Wegner, G.; Meyer, W. H. *Mol. Cryst. Liq. Cryst.* **1992**, *216*, 79.
22. Uznanski, P.; Pecherz, J.; Kryszewski, M. *Mol. Cryst. Liq. Cryst.* **1994**, *246*, 351.
23. Simmrock, H.-U.; Mathy, A.; Dominguez, L.; Meyer, W. H.; Wegner, G. *Angew. Chem. Int. Ed. Engl. Adv. Mater.* **1989**, *28*, 1122.
24. Takahashi, A.; Kawaguchi, M.; Kato, T.; Kuno, M.; Matsumoto, S. *J. Macromol. Sci.-Phys.* **1980**, *B17*, 747.
25. Kawaguchi, M.; Oohira, M.; Tajima, M.; Takahashi, A. *Polym. J.* **1980**, *12*, 849.
26. Tsuchida, E.; Abe, K. *Adv. Polym. Sci.* **1982**, *45*, 1.
27. Casson, D.; Rembaum, A. *Macromolecules* **1972**, *5*, 75.
28. Schmir, M.; Rembaum, A. *Polym. Sci. Technol., Vol. 2: Water-Soluble Polymers*; Bikales, N. M., Ed.; Plenum: New York, 1973; p 327.
29. Timofejeva, G. J.; Pavlova, S. A.; Wandrey, C.; Jaeger, W.; Hahn, M.; Linow, K.-J.; Görnitz, E. *Acta Polym.* **1990**, *41*, 479.
30. Soldi, V.; Erismann, N. Magalhães, Quina, F. H. *J. Am. Chem. Soc.* **1988**, *110*, 5137.
31. Harada, A.; Nozakura, S. *Polym. Bull.* **1984**, *11*, 175.
32. Tsutsui, T.; Tanaka, R.; Tanaka, T. *J. Polym. Sci., Polym. Phys. Ed.* **1976**, *14*, 2273.
33. Dominguez, L.; Meyer, W. H. *Solid State Ionics* **1988**, *28–30*, 941.
34. Dominguez, L.; Enkelmann, V.; Meyer, W. H.; Wegner, G. *Polymer* **1989**, *30*, 2030.
35. Feng, D.; Venkateshwaran, L. N.; Wilkes, G. L.; Leir, C. M.; Stark, J. E. *J. Appl. Polym. Sci.* **1989**, *38*, 1549.
36. Mathy, A.; Simmrock, H.-U.; Bubeck, C. *J. Phys. D: Appl. Phys.* **1991**, *24*, 1003.
37. Kohjiya, S.; Ohtsuki, T.; Yamashita, S.; Taniguchi, M.; Hashimoto, T. *Bull. Chem. Soc. Jpn.* **1990**, *63*, 2089.
38. Hashimoto, T.; Kohjiya, S.; Yamashita, S.; Irie, M. *J. Polym. Sci., Part A, Polym. Chem.* **1991**, *29*, 651.
39. Hashimoto, T.; Sakurai, S.; Morimoto, M.; Nomura, S.; Kohjiya, S.; Kodaira, T. *Polymer* **1994**, *35*, 2672.
40. Feng, D.; Wilkes, G. L.; Leir, C. M.; Stark, J. E. *J. Macromol. Sci.-Chem.* **1989**, *A26*, 1151.
41. Feng, D.; Wilkes, G. L.; Lee, B.; McGrath, J. E. *Polymer* **1992**, *33*, 526.
42. Venkateshwaran, L. N.; Leir, C. E.; Wilkes, G. L. *J. Appl. Polym. Sci.* **1991**, *43*, 951.
43. Meyer, W. H.; Wang, J.; Wegner, G. *Polym. Prepr.* **1995**, *36*(1), 558.
44. Rembaum, A.; Hermann, A. M.; Stewart, F. E.; Gutmann, F. *J. Phys. Chem.* **1969**, *73*, 513.
45. Mostovoy, R. M.; Sukharev, V. Ya; Berendyayev, V. I.; Kotov, B. V.; Myasnikov, I. A. *J. Phys. Chem. Solids* **1989**, *50*, 541.
46. Loveday, D.; Wilkes, G. L.; Bheda, M. C.; Shen, Y. X.; Gibson, H. W. *J. Macromol. Sci., Pure Appl. Chem.* **1995**, *A32*, 1.
47. Marsella, M. J.; Swager, T. M. *Polym. Prepr.* **1994**, *35*(1), 275.
48. Mortimer, D. A. *Polym. Int.* **1991**, *25*, 29.
49. Preston, F. W.; Hohf, R.; Trippel, O. *Quart. Bull. Northwestern Univ. Med. School* **1956**, *30*, 138.
50. May, O. W.; Morgan, D. E. *Appita* **1979**, *32*, 466.
51. Casson, D.; Rembaum, A. *J. Polym. Sci., Part B, Polym. Lett* **1970**, *8*, 773.
52. Nagamura, T.; Isoda, Y.; Sakai, K. *Polym. Int.* **1992**, *27*, 125.
53. Gibson, H. W.; Marand, H. *Adv. Mater.* **1993**, *5*, 11.
54. Gibson, H. W.; Bheda, M. C.; Engen, P. T. *Prog. Polym. Sci.* **1994**, *19*, 843.

Ionenes

See: Ionene Oligomers
Ionene Polymers
Liquid Crystalline Polymers (Viologen: Thermotropic and Lyotropic)
Matrix Polymerization
Polyelectrolytes, Cationic
Viologen Polymers (Electron-Transfer Catalysts)

IONIC CONDUCTIVITY SWITCHING (Photochromic Crown Compounds)

Keiichi Kimura
Chemical Process Engineering
Faculty of Engineering
Osaka University

Polymer electrolytes have recently received considerable attention.[1–3] However, few organic materials are able to undergo ionic-conductivity switching by external stimulation, although photoresponsive ion-conducting organic materials are especially useful for designing memory, display, and printing devices.

Macrocyclic polyethers, which are called crown compounds, can contribute to ion migration in solids or quasisolids as well

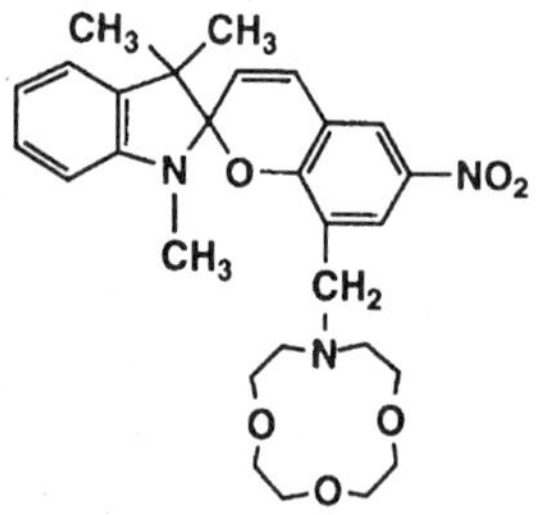

1

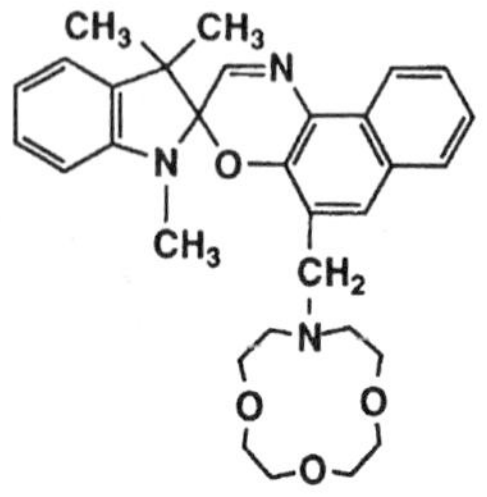

2

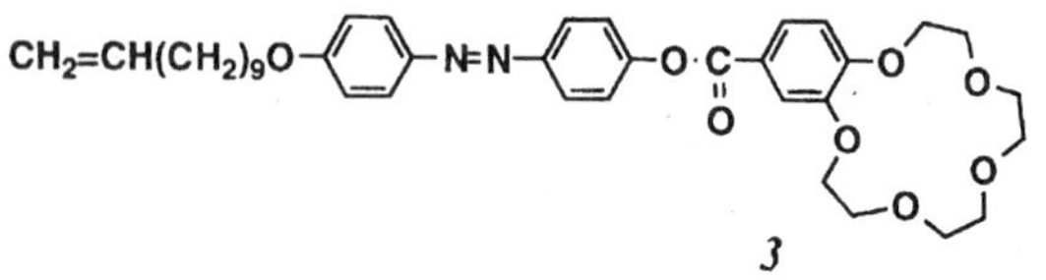

3

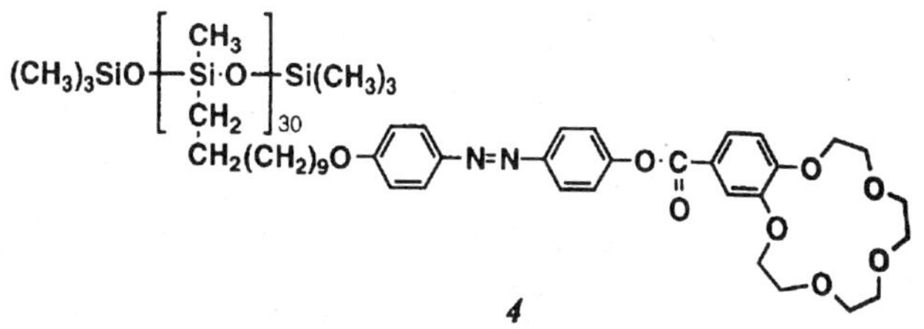

4

as in solution[4] and are promising materials for ionic conduction. One possibility for photoresponsive ion-conducting organic materials is crown compounds that can isomerize by photoirradiation, that is, photochromic crown compounds. We have designed several such compounds and have studied their behavior in photoinduced ionic conductivity switching in polymer films.

Photochromic crown compounds include spirobenzopyran and spironaphthoxazine derivatives incorporating a crown ether moiety (crowned spirobenzopyran and spironaphthoxazine), **1** and **2**, which undergo photochemical ionic conductivity switching by molecular control of cation binding.

Another type of compound includes crowned azobenzene derivative **3** and its polymer **4**, which realize marked phase transition of the compounds by photoisomerization, which in turn causes ionic conductivity switching (**Structures 1–4**).

PHOTOCHEMICAL SWITCHING BEHAVIOR OF IONIC CONDUCTIVITY

Ionic Conductivity Switching Based on Molecular Control of Cation Binding

Crowned spirobenzopyran **1** isomerizes to its corresponding zwitterionic merocyanine form by UV irradiation even in the presence of a lithium salt. In the merocyanine – Li^+ complex, the metal ion complexed by its crown ring can be trapped more strongly than in the spiropyran complex by intramolecular interaction between the phenolate anion and metal ion.[5] The ensuing visible-light irradiation produces isomerization back to the corresponding spiropyran form, thus resulting in some Li^+ release. This molecular cation binding control affects mobility and concentration of the metal ion significantly.

The crowned spirobenzopyran was first applied to a bi-ionic conducting system, in which both the cation and anion are able to participate in ionic conduction.[6] UV light accelerates isomerization to the merocyanine form. This indicates that the photoisomerization of crowned spirobenzopyran is quite reversible even in the composite film. Similar ionic conductivity changes in the composite film can be attained with alternate turning on and off of visible light, which increases and decreases the ionic conductivity, respectively. This photoirradiation cycle has an advantage over alternate irradiation of UV and visible lights for practical application of photochemically switchable ion-conducting systems, because UV irradiation may cause deterioration of the film components, especially the spirobenzopyran derivative.

Because both Li^+ and its counter anion take part in ion conduction in bi-ionic conducting composite films, the ionic conductivity decrease (or increase) through photoinduced cation binding enhancement (or alleviation) may be canceled by the resulting anion contribution. Crowned spirobenzopyran was therefore applied to a single-ionic conducting system, in which only a cation participates in ion conduction. Efficient photoinduced switching in ionic conductivity can be expected in a single-ionic conducting system containing a polyanion as the Li^+ counter anion, because a cation transport number of nearly 1 is attainable through the extremely low mobility of the polyanion.

Crowned spironaphthoxazine **2**, which undergoes photoisomerization similar to that of the spirobenzopyran derivative, generally shows excellent light-fatigue resistance and can therefore by expected to be a more reliable and durable photochromic control system than the corresponding spirobenzopyran derivative.

Ionic Conductivity Switching Based on Phase Transition

Another type of photochromic crown compounds, crowned azobenzenes **3** and **4**, undergo photochemical ionic conductivity switching based on phase transition induced by photoisomerization.[7,8] Photoisomerization of the crowned azobenzenes from *trans* to *cis* forms brings about significant changes in their phase transition behavior, which in turn affects their ionic conducting behavior in composite films.

Crowned azobenzene polymer **4** is able to form a highly oriented mesophase, smectic liquid crystal, whereas its corresponding monomer **3** forms only a nematic liquid crystal. UV-induced isomerization of the azobenzene moiety of **4** from *trans* to *cis* forms disturbs its smectic state. In the smectic state of crowned azobenzene polysiloxane, its adjacent side chains, consisting of lipophilic crowned azobenzene moiety, are closely oriented to one another.

POSSIBLE APPLICATIONS

Ion-conducting composite films containing photochromic crown compounds have many potential applications. One of them is electrostatic imaging.[9] By irradiation of an appropriate light on a photoresponsive ion-conducting composite film through an exposure image, the photoirradiated region of the film becomes relatively ion-conducting while the unirradiated region remains poorly ion-conducting. The ionic conductivity difference between the photoirradiated and unirradiated regions forms a latent image, which can be easily read by corona charging on its surface; the surface charges in the photoirradiated region are immediately compensated because of the region's high ionic conductivity. The surface of the unirradiated region, however, is charged stably by corona charging. The subsequent toner development enables duplication of the image. Other applications of photoresponsive ion-conducting materials might be in photo-writable electrochromic displays and photorefractive imaging.

REFERENCES

1. Fenton, D. E. et al. *Polymer* **1973**, *14*, 589.
2. Armand, M. *Solid State Ionics* **1983**, *9/10*, 745–54.
3. Ratner, R. A.; Shriver, D. F. *Chem. Revs.* **1988**, *88*, 109–124.
4. Inoue, Y.; Gokel, G. W. *Cation Binding by Macrocycles*; Marcel Dekker: New York, 1990.
5. Kimura, K. et al. *J. Phys. Chem.* **1992**, *96*, 5614–17.
6. Kimura, K. et al. *J. Chem. Soc., Perkin Trans. 2* **1992**, 613–19.
7. Tokuhisa, H. et al. *Chem. Mater.* **1993**, *5*, 989–93.
8. Tokuhisa, H. et al. *Macromolecules* **1994**, *27*, 1842–46.
9. Kimura, K. et al. *J. Chem. Soc., Chem. Commun.* **1989**, 1570–71.

IONOMERIC MEMBRANES, PERFLUORINATED

Charles W. Martin,* Pramod J. Nandapurkar, and Sanjeev S. Katti
The Dow Chemical Company

The perfluorinated ionomer membranes as a class exhibit a unique combination of properties, including excellent chemical and thermal stability, high ionic conductivity with permselectivity, and super acid strength.[1–3] The acidity of traditional functional groups such as sulfonic, carboxylic, and phosphonic acids is significantly increased by the fluorocarbon stabilization of anionic charge. Other nontraditional acid functional groups such as bis-sulfonimides behave as super acids whereas sulfonamides and tertiary alcohols have acidities similar to acetic acid.[4–6] Despite the high cost of these materials, their unique properties have led to their commercialization as permselective membrane separators in mature electrochemical processes such as chlor-alkali, and they are the membranes of choice in the rapidly developing proton-exchange membranes (PEM) fuel cell industry.[7,8] These complex materials have been the subject of numerous reviews and monographs.[9–11]

*Author to whom correspondence should be addressed.

POLYMERIZATION

Fluoropolymers, with functional groups containing exchangeable ions or a precursor, can be prepared via solution, bulk, or emulsion polymerization. An excellent review discussing polymeric fluorocarbon acid preparation appears in Ukihashi et al.[12]

STRUCTURE–PROPERTY RELATIONSHIPS

The structure–property relationships for the perfluorinated sulfonic acid (PFSA) polymers have been well documented in the literature.

In commercial and scientific literature, the properties of fluorocarbon ionomers are typically presented as a function of equivalent weight (EW) or its inverse, ion-exchange capacity (IEC). However, properties such as crystallinity are more appropriately correlated with mole percent functional monomer. The relationship between ionic EW and mole percent comonomer is nonlinear.

The ion-containing perfluoropolymers exhibit a complex, three-phase morphology consisting of crystalline and amorphous fluoropolymer phases and a separate clustered ionic phase. Furthermore, thermal analysis via differential scanning calorimetry (DSC) shows two widely separated crystalline melt endotherms. It has been proposed that two distinct crystalline forms coexist in these polymers.[14] The properties of a given polymer are influenced by the interplay of ionic species (functional group and salt form) and ionic content (EW) with the ratio of crystalline to amorphous phase. The crystalline/amorphous ratio in turn is influenced by the side-chain length and the mole percent of the functional monomer.

The first detailed model of the morphology of the perfluorinated ionomers was proposed by Hsu and Gierke.[15] This "cluster-network" model envisioned roughly spherical, inverted micellular clusters connected by narrow channels. The model was consistent with data from small-angle X-ray scattering (SAXS) studies and with transmission electron microscopy (TEM) studies. It provided the first explanation of the unusual permselectivity of these materials. The fixed anionic charges on the surface of the narrow channels allow cations to pass easily but block the transport of mobile anions such as chloride or hydroxide.

APPLICATIONS

Perfluorinated ionomers were first developed as membrane separators for electrochemical applications. Although this general use remains the most important, a wide variety of membrane and several nonmembrane uses have been developed.[16] Amongst the major applications are brine and water electrolysis, PEM fuel cells, superacid catalysis, gas separation, electrodialytic applications, and electroorganic syntheses. These polymers have excellent thermal and chemical stability and are inert in several environments. However they are expensive as well and the application spectrum is narrowed as a result.

Brine Electrolysis

This is the most important and widely practiced commercial application of the perfluorinated ionomeric membranes. In this process, caustic soda and chlorine are produced by electrolysis of brine. It is by far the largest industrial electrolytic process in the world.[13]

PEM Fuel Cell

The perfluorinated sulfonic ionomer membranes were used as the electrolyte in PEM fuel cells long before the commercial development of the membrane chlor-alkali process.

Water Electrolysis

Functionally, a PEM electrolyzer is the reverse operation of a fuel cell (i.e., DC power is used to split water to produce hydrogen and oxygen). It is typically used for on-site production of hydrogen for small-scale applications.[17]

Heterogeneous Acid Catalysis

PFSA polymers exhibit catalytic activity in a wide variety of chemical reactions such as alkylation, olefin hydration, alcohol dehydration, isomerization, nitration, and condensation. In general, any reaction system that is catalyzed by a protic acid may benefit from the use of the PFSA catalyst. Extensive studies have been summarized in several reviews.[3,18,19]

Membrane Separation Processes

The sulfonic ionomer membranes have been used for separating water vapor from a variety of gases and for concentrating inorganic acids.[20,21]

Electrodialysis

Because perfluorosulfonate ionomers are available as both anion and cation selective membranes, they find use in applications involving colloids, such as dialysis and electrodialysis.[22,23] Some of the applications are desalination of brackish water, waste water treatment by electrodialysis, and diffusion dialysis.

Hydrogen Peroxide Synthesis

Reduction of humidified oxygen to hydrogen peroxide can be carried out in a SPE electrochemical reactor.[24]

REFERENCES

1. Martin, C. R.; Feldheim, D. L.; Lawson, D. R. *NTIS Order Number: AD-A258 542/0/XAD.*
2. Yeager, H. L.; In *Perfluorinated Ionomer Membranes* Eisenberg, A.; Yeager, H. L., Eds. ACS Symposium Series 180 American Chemical Society; Washington, DC, 1982; Chapters 3 and 4.
3. Olah, G. A.; Iyer, P. S.; Prakash, G. K. S. *Synthesis* **1986**, *7*, 513.
4. Koppel, I. A.; Taft, R. W.; Anvia, F.; Zhu, S-Z.; Hu, L-Q.; Sung, K-S.; DesMarteau, D. D.; Yagupolskii, L. M.; Yagupolskii, Y. L. *J. Am. Chem. Soc.* **1994**, *116*(7), 3047.
5. Trepka, R. D.; Harrington, J. K.; Belisle, J. W. *J. Org. Chem.* **1974**, *39*(8), 1094.
6. Kuopio, R.; Kivinen, A.; Murto, J.; Juhani *Acta Chem. Scand., Ser. A* **1976**, *A30*(1), 1.
7. Burney, H. S.; Talbot, J. B. *Report of the Electrochemical Industry* IEEE Division of The Electrochemical Society; Electrochemical Society: 1990; pp 3–11.
8. Burney, H. S.; Talbot, J. B. *Report of the Electrochemical Industry* IEEE Division of The Electrochemical Society; Electrochemical Society: 1990; p 83.
9. Pourcelly, G.; Gavach, C. *Chem. Solid State Mater.* **1992**, *2*, 294.
10. Eisenberg, A.; Yeager, H. L., Eds. *Perfluorinated Ionomer Membranes* ACS Symposium Series 180; American Chemical Society: Washington, DC, 1982.
11. Pineri, M.; Eisenberg, A. *Structure and Properties of Ionomers* NATO ASI Series; Reidel: Dordrecht, The Netherlands, 1987.
12. Ukihashi, H.; Yamabe, M.; Miyake, H. *Prog. Polym. Sci.* **1986**, *12*, 229.
13. Smith, P. J. In *Electrochemical Science and Technology of Polymers* Lindford, R. G., Ed.; Elsevier Applied Science: 1987, p 293.
14. Tant, M. R.; Lee, K. D.; Darst, K. P.; Martin, C. W. In *Multiphase Polymers: Blends and Ionomers* Utracki, L. A.; Weiss, R. A., Eds.; ACS Symposium Series, 395 American Chemical Society: Washington, DC, Chapter 15.
15. Hsu, W. Y.; Gierke, T. D. *Proc. Electrochem. Soc.* **1982**, *82*(10), pp 158–177.
16. Yeager, H. L. *Polym. Prepr.* **1988**, *29*(2), 440.
17. Coker, T. G.; LaConti, A. B.; Nutall, L. J. In *Proceeding of the Symposium on Membranes and Ionic and Electronic Conducting Polymers*; Yeager, E. B., et al., Eds.; The Electrochemical Society: 1983; Vol. 83(3), p 191.
18. Waller, F. J. In *Polymer Reagents and Catalysts*; Ford, W. T., Ed. ACS Symposium Series 308 American Chemical Society: Washington, DC, 1986; p 42.
19. Sondheimer, S. J.; Bunce, N. J.; Fyfe, C. A. *J. Macromol. Sci., Rev. Macromol. Chem. Phys.* **1986**, *C26*(3), 353.
20. Kipling, B. In *Perfluorinated Ionomer Membranes*; Eisenberg, A.; Yeager, H. L., Eds.; ACS Symposium Series 180 American Chemical Society: Washington, DC, 1982; p 475.
21. Ping, Z. H.; Ding, Y. D.; Chang, Y. X.; Cheng, W. H.; Nguyen, Q. T.; Neel, J. *Recents Prog. Genie Procedes* **1992**, *6*(21), 325.
22. Bahador, S. K. *Makromol. Chem., Macromol. Symp.* **1990**, *37*, 129.
23. Strathmann, H. *Makromol. Chem., Macromol. Symp.* **1993**, *70/71*, 363.
24. Tatapudi, P.; Fenton, J. M. *J. Electrochem. Soc.* **1993**, *140*(4), 155.

Ionomers

See: *Dental Polymers (Glass-Ionomer Cements)*
Electrically Conducting Composites
Ionomeric Membranes, Perfluorinated
Ionomers (Overview)
Ionomers (Compatibilization of Blends)
Ionomers (dc Conduction Properties)
Ionomers (Mechanical Properties)
Ionomers (Solution Behavior)
Perfluorinated Ionomers (Overview)
Perfluorinated Ionomers (Anisotropic Ionic Conductivity)
Poly(ethylene terephthalate) (Modified with Ionic Groups)
Polystyrene, Syndiotactic (Sulfonation and Crystallization)
Polyurethane Ionomers
Polyzwitterions (Overview)
Water-Borne Coatings (Urethane/Acrylic Hybrid Polymers)

IONOMERS (Overview)

Adi Eisenberg* and Joon-Seop Kim
Department of Chemistry
McGill University

Although the word "ionomer" is only 30 years old, it was recognized much earlier that ionic forces can exert major effects on polymer properties.[1] The work before 1957 is summarized in a review by Brown, which centers on elastomeric materials.[2] Most of the recent work started in the 1960s with the styrene ionomers, and later came the advent of commercial ethylene ionomers.[1,3–7] Since then, the interest in ionomers has grown steadily, mainly because of their effects on polymer properties.

There is no unanimous agreement on the definition of the word "ionomer." Historically, it referred to thermoplastic materials with low dielectric constants containing a relatively small percentage of ionic groups.[1] This is a useful definition; however, it gives rise to some ambiguities, for example, ionomers defined this way can possess polyelectrolyte-like properties. Therefore, another definition, based on the behavior of materials and the reasons for that behavior, was recently proposed.[8] Thus, ionomers are defined as polymers in which the bulk properties are governed by ionic interactions in discrete regions of the material (i.e., ionic aggregates), whereas polyelectrolytes are defined as polymers in which solution properties in solvents of high dielectric constant are governed by electrostatic interactions over distances larger than typical molecular dimensions.

To illustrate the magnitude of the changes that can be induced by the presence of ionic groups, it is useful to look at three examples of the effect of ionic interactions on properties. The first example comes from the area of glass transitions and involves an inorganic polymer in which no decomposition takes place in the region of interest.[9] The glass transition temperature (T_g) of polymetaphosphate in the acid form, $(HPO_3)_x$ is –10°C. Neutralization with Na^+ raises the glass transition to 280°C, whereas neutralization with Ca^{2+} raises it further to 520°C. This is strictly an effect of ionic forces because, in all other respects, the three polymers are identical. It has been shown that the glass transition temperature can be related to the strength of the ionic interaction through a simple q/a effect, where q is the cation charge and a is the distance between centers of charge of anion and cation, which suggests that the glass transition temperature is a function of the work needed to separate the anion from the cation.[10] The magnitude of effect (from –10 to 520°C) is larger than the range of glass transition temperatures of most of the common organic polymers and illustrates the importance and extent of ionic interactions on the glass transition.

Another useful example is the behavior of the modulus as a function of temperature for the poly(styrene-*co*-sodium methacrylate) [P(S-co-MANa)] copolymers of different ion contents. Pure polystyrene behaves like a normal thermoplastic. The addition of 4.5 mol % of the ionic comonomer changes the behavior of this material to one resembling a phase-separated blend or block copolymer system in which the volume fraction of the high T_g phase is on the order of 50%, the T_g of that phase being ~200°C.

*Author to whom correspondence should be addressed.

For the third example, it is useful to look at the viscosity of ionomers as a function of ion content.[12] Unmodified polystyrene, the melt viscosity of which is 4×10^3 poise (P) at 220°C, was sulfonated and carboxylated to various extents. At 2% carboxylation, the viscosity of the carboxylated sample of identical molecular weight is 4×10^5 P, whereas that of the sulfonated sample is 3×10^8 P (i.e., an increase of ~ 5 orders of magnitude). Naturally, the viscosity also increases with increasing ion content.

As a result of these drastic changes accompanying the incorporation of ions into polymers, the study of ionomers has been of great industrial and academic interest. In this article, we offer a brief overview of ionomer properties.

MORPHOLOGY

Ionic aggregation is one of the major features of the study of ionomers and will form the basis of the discussion of morphology.

The aggregation of ionic groups into entities termed "multiplets" frequently gives rise to a pronounced peak in the small angle region of the X-ray scattering profile.[13] The existence of this peak in almost all ionomers, at least under some conditions, provided that the ionic species has a high enough atomic number, which gives the scattering centers a sufficient electron density. The interpretation of this ionomer peak, as can be imagined, it not straightforward, because a single feature of this type can be ascribed either to a distance between scattering centers or to intraparticle features.

The Eisenberg-Hird-Moore (EHM) model, which incorporates both morphological and rheological features appeared in 1990.[18] Multiplets (i.e., primary aggregates of ionic groups) form the starting point for the description.[13] It is assumed in the model that the small-angle X-ray peak is a result of interparticle scattering, along the lines suggested by Cooper and coworkers.[15,16] The size of the multiplet is limited by a number of factors; the most important are steric effects, which lead to an upper limit on the size. These steric effects arise because each ion pair is attached through primary bonds to a polymer chain or to a spacer and because all the ionic groups must be inside the ionic multiplet, while the organic segments must be outside the multiplet.[13] Naturally, a number of other factors also influence the formulation and size of multiplets, such as the size of the ion pairs and the distance from the ionic group to the backbone.[19,20]

As the concentration of ionic groups in the polymer increases, the concentration of multiplets also increases, and more and more regions of restricted mobility start overlapping. Eventually, as the ion concentration increases still further, the overlapping regions of reduced mobility reach dimensions of 50 to 100 Å, above which independent phase behavior becomes possible. These large regions of reduced mobility are called clusters.[16] This overlapping of reduced mobility regions results in the appearance of a second glass transition temperature at a temperature higher than that of the normal T_g, despite the very low volume fraction of ionic groups in the region of restricted mobility.

EFFECT OF IONS ON THE GLASS TRANSITION TEMPERATURE

As was pointed out previously, two glass transitions are observed in ionomers especially when studied by dynamic mechanical techniques. These techniques, which allow for the measurement of the loss tangent, are particularly suited for the determination of these glass transitions in ionomers. The two T_g values will be referred to as the matrix glass transition and the cluster glass transition in order of increasing temperature.

Matrix Glass Transition Temperature

Two cases can be distinguished in connection with the glass transition temperature of the matrix phase (unclustered area). In one case, the glass transition temperature can be strictly due to a crosslinking effect in the absence of any ion hopping. This would occur under circumstances in which the ionic multiplet is extremely stable (i.e., the temperatures at which the polymer chains undergo their glass transition temperature are too low for any ion hopping to occur). In this case, no *q/a* effect would be expected.[20,21] However, one can envisage a situation in which the crosslinking effect still persists, but the glass transition is now due to a combination of chain mobility and ion hopping resulting from that mobility. In this case, a *q/a* effect would be observed.[22,23]

The glass transition temperature does not always increase with increasing ion content. In the studies of the glass transition temperature of ionomers, one has to keep in mind that the clustered or unclustered regions can, under some circumstances, be very small.[11]

Cluster Glass Transition Temperature

Although cluster T_g values have been observed in many ionomer systems, the number of detailed investigations of the positions of the cluster T_g as a function of ion content is relatively small.

Although the mechanism of the matrix glass transition is clear and involves the same mechanism as that in non-ionic materials, the mechanism of the cluster glass transition is not nearly as clear. Several studies have suggested that in the region of the glass transition of the clusters the multiplets are subject to ion hopping and that this process is intimately associated with the glass transition.[21] Activation energies for ion hopping have also been determined. The activation energies for ion hopping in the methacrylates generally tend to be higher than in the sulfonates in spite of the opposite trend in the interaction energies. This suggests that the strength of the ionic interaction influences the position of the second glass transition temperature but is not related to the activation energies for the ion hopping process.

MECHANICAL PROPERTIES OF STYRENE IONOMERS

The styrene ionomers appear to be the best understood ionomer family from the point of view of mechanical properties and their relation to various molecular parameters, both because they are noncrystalline and because the parent polymer has been investigated extensively.

Stress Relaxation

Stress relaxation in P(S-*co*-MANa) ionomers was investigated by Eisenberg and Navratil.[25] It was found that below ~ 6 mol % of ions, time–temperature superposition was followed, but that above that ion concentration, time–temperature superposition failed.

Dynamic Mechanical Properties

For the ionomers of intermediate ion content, two inflection points are seen in the plot of modulus versus temperature. The lower inflection point is dependent of the ion content, whereas the positions of the higher modulus are a function of the ion content. The latter modulus is associated with the crosslinking or filler effect of the ionic groups and has been called the "ionic" modulus.

Filler Effect

Because the clustered regions obviously have a much higher glass transition temperature than the unclustered surroundings, the value of the ionic modulus might be interpretable in terms of a filler effect. With the volume fraction of clustered material obtained from the areas under the loss tangent curves, one can calculate the value of ionic modulus, assuming that the filler effect is in operation. Such a study has, indeed, been performed.[11] This suggests that clusters act as a filler in the temperature range between the two T_g values at concentrations below the percolation threshold.[11]

Percolation Behavior

Given the morphological picture previously presented, it seems reasonable to suggest that the percolation concept should be useful in these materials, and an attempt was made to fit the ionic modulus of the P(S-*co*-MANa) system by percolation theory.[11,36–38]

Type of Pendant Ions

In the previous discussion, a brief comparison was already made between the mechanical properties of the styrene ionomers based on methacrylate and those based on sulfonate pendant ions. Another group that has been investigated in this content is the vinylpyridinium ions.[29] In contrast to the sulfonate or methacrylates, the vinylpyridinium ionomer does not show an ionic inflection point. However, the matrix glass transition temperature increases with ion content at about the same rate as it does in the sulfonates and methacrylates. This suggests that although multiplets do form and act to increase the glass transition temperature, they do not survive above the matrix glass transition temperature. Therefore, above the glass transition temperature, only the entanglement modulus governs the properties.

OTHER IONOMERS

Ethylene Ionomers

The ethylene ionomers have been studied very extensively since their appearance in 1965 under the trade name Surlyn®; these materials are flexible, transparent, and tough and can be melt processed. Some early investigators were Rees and Vaughan, MacKnight and co-workers, Ward and Tobolsky,

Longworth and Vaughan, Otocka and Kwei, and Bonotto and Bonner, among others.[1,6,7,19,24,30–32] Extensive recent studies come from the group of Yano.[33–35]

Perfluorinated Ionomers

Systematic studies of the mechanical properties of perfluorosulfonate ionomers were performed by Eisenberg and coworkers.[36–40] A comparison was also made between the nonionic precursor, (i.e., the material that contains $-SO_2F$ terminal groups of the side chain rather than SO_3^- groups) and the ionomer of an equivalent weight of 1155.[37] Perhaps because of the small amount of crystallinity that is present, time–temperature superposition still appears to be applicable for both the precursor and the acid, although the original stress relaxation curves were measured over only 4 orders of magnitude of time.[37]

Ethylene–Propylene–Diene Terpolymers

A completely different example in terms of mechanical behavior is that of the EPDM-based rubbers.[41] Ethylene and propylene form the base elastomer, and the exo-double bond of 5-ethylidene-2-norbornene, which had been copolymerized with the ethylene–propylene rubber, forms the site that can be sulfonated. Only ~ 0.6 mol % of Zn-sulfonate groups enable the polymer to remain mechanically intact up to ~ 150°C. As was found in the other ionomers, the height of the rubbery plateau increases with increasing ion content. Recall that in this particular material, the ion content is very low, so no extensive clustering behavior is to be expected, merely ion association leading to crosslinking via simple multiplet formation.

Block Copolymers

Although ionic block copolymers in micelle form are under extensive study at this time in both aqueous and nonaqueous solutions, the viscoelastic properties of ionic block copolymers, in which the ionic block represents the minority component, have not received extensive attention.[42–46] One example of a brief dynamic mechanical study of this system has been performed by Desjardins and Eisenberg.[46] It was shown first that the viscoelastic properties of diblocks were not of particular interest. The short ionic diblocks associate to form micelle-like aggregates, and the viscoelastic properties are very similar to those of stars or polymers of low molecular weight in general, because a rapid drop in the modulus is seen in the glass transition temperature region with only a hint of a rubbery plateau.

Triblocks, by contrast, are much more interesting because an extended rubbery plateau is seen. The height of the rubbery plateau increases with increasing length of the ionic block.

Telechelics and Stars

An extensive study of the telechelic materials (i.e., those in which ions are placed at the chain ends), has been performed by Teyssié, Jérôme, and co-workers.[47–52] Most of the studies were devoted to solution properties, although some attention was also paid to bulk properties.[47,51,52] As might be expected, the viscosities of the bulk telechelics are higher than those of the parent polymers because of the presence of end-group association. Differences between cations are appreciable, because they control the degree of association of end groups and thus influence viscoelastic properties.[51]

PLASTICIZATION

It was recognized very early that it should be possible to plasticize independently the ionic regions and the regions of low polarity in ionomers. Lundberg et al. explored this dual plasticization possibility in 1980.[53] In that study, sulfonated polystyrene containing 1–5 mol % of ions was plasticized in the ionic regions by using a highly polar plasticizer such as glycerol, which weakened the ionic interactions and thus reduced the viscosity of the material. A material such as dioctyphthalate, which has a relatively low polarity in contrast to glycerol, is known to interact with the nonpolar regions of ionomers and thus lowers the glass transition temperature of both the cluster and matrix regions. This effect is due to the relatively even distribution of plasticizers of low polarity in both the cluster and matrix regions. The proper use of these two types of plasticizers allows the production of materials, starting with sulfonated polystyrene, which have many of the properties of plasticized poly(vinyl chloride).

TYPICAL APPLICATIONS OF IONOMERS

The range of applications of ionomers is very broad; however, within the framework of the present treatment, only a few selected applications can be described.

Membranes

One of the most interesting applications of ionomers is their use as membrane materials, of which the most important are the perfluorinated ionomers, which are used as superpermselective membranes.[54]

Floor Waxes

Another application of ionomers is in the area of floor waxes.[57]

Plastics

To the general public, perhaps the best known single application of ionomers is golf balls.[58] In addition to golf balls, however, a very wide range of other solid plastic items has been made out of these ionomers. These include bowling pin covers, body side moldings and bumper guards of cars, shoe parts (e.g., toe puffs, heel counters, and top lifts), bottle stoppers, and ski boot shells. Still other applications of the ethylene ionomers include use as packaging materials, impact modifiers, rheology modifiers, modifiers for glass-reinforced thermoplastics, ionomer forms, pipe coatings, and coatings on fluorescent light tubes.

Elastomers

Self-vulcanizing elastomers are based on ionomers.[56]

Magnetic Recording Aids

Water-dispersible polyurethane ionomers are used as dispersing agents and binders for magnetic particles in magnetic recording media, and several patents have been issued to BASF on this technology.[59,60]

Catalysis of Catalytic Supports

Ionomers have proven useful in the field of catalysis. The earliest work is perhaps that of Olah on superacid catalysis in the perfluorosulfonates (PFSA).[61–63] However, various ions in the PFSA ionomers can also be used catalytically. Use of the ionic aggregates as microreactors suggests that a wide range of possibilities exists in this area, especially involving catalytically active ions or metals that can be used as catalysts and are produced in the phase-separated regions of the ionomers.[64]

Drilling Fluids

An unusual application of ionomers is as a component of drilling fluid.[55,65] Sulfonated polystyrene and sulfo-EPDM have served in that capacity.

Fertilizers

The use of slow-release coating technology in agricultural fertilizers results in a reduction of the total cost and an improvement in performance. Very recently, a new controlled-release coating technique for fertilizers has been developed using Zn sulfo-EPDM, which takes advantage of the unique bulk and solution properties of ionomers.[66]

OMITTED TOPICS

This article has been devoted to an overview of ionomers, especially a brief discussion of the effect of ionic aggregation on the morphology and properties of ionomers and of some of the typical applications of these materials. Because of space limits, a wide range of topics had to be omitted. Some of these topics were excluded because they are mentioned in other articles; others were skipped because they do not fit into the present article, even though they are important. For information on these topics, including solution properties,[67–74] engineering properties,[75–76] coil dimensions,[78–83] orientation effects,[84–86] and synthesis,[87,88] see the original literature.

REFERENCES

1. Rees, R. W.; Vaughan, D. J. *Polym. Prepr., Am. Chem. Soc. Div. Polym. Chem.* **1965**, *6*, 287.
2. Broen, H. P. *Rubber Chem. Technol.* **1957**, *30*, 1347.
3. Edri, N. Z.; Morawetz, H. *J. Colloid Sci.* **1964**, *19*, 708.
4. Fitzgerald, W. E.; Nielsen, L. E. *Proc. R. Soc. Ser. A.* **1964**, *282*, 137.
5. Rees, R. W. U.S. Patent 3 264 272, 1966.
6. MacKnight, W. J.; McKenna, L. W.; Read, B. E. *J. Appl. Phys.* **1967**, *38*, 4208.
7. Ward, T. C.; Tobolsky, A. V. *J. Appl. Polym. Sci.* **1967**, *11*, 2403.
8. Eisenberg, A.; Rinaudo, M. *Polym. Bull.* **1990**, *24*, 671.
9. Eisenberg, A.; Farb, H.; Cool, L. G. *J. Polym. Sci. Part A-2* **1966**, *4*, 855.
10. Eisenberg, A. In *Physical Properties of Polymers*; Mark, J. E. et al. Eds.; ACS Professional Reference Book; American Chemical Society: Washington, DC, 1993; Chapter 2.
11. Kim, J-S.; Jackman, R. J.; Eisenberg, A. *Macromolecules* **1994**, *27*, 2789.
12. Lundberg, R. D.; Markowski, H. S. *Ions in Polymers*; Eisenberg, A. Ed.; Advances in Chemistry Series 187; American Chemical Society: Washington, DC, 1980; Chapter 2.
13. Eisenberg, A. *Macromolecules* **1970**, *3*, 147.
14. Marx, C. L.; Caulfield, D. F.; Cooper, S. L. *Macromolecules* **1973**, *6*, 344.
15. Yarusso, D. J.; Cooper, S. L. *Macromolecules* **1983**, *16*, 1871.
16. Eisenberg, A.; Hird, B.; Moore, R. B.; *Macromolecules* **1990**, *23*, 4098.
17. Wollmann, D.; Williams, C. E.; Eisenberg, A. *Macromolecules* **1992**, *25*, 6775.
18. Moore, R. B.; Bittencourt, D.; Gauthier, M.; Williams, C. E.; Eisenberg, A. *Macromolecules* **1991**, *24*, 1376.
19. Bonotto, S.; Bonner, E. F. *Macromolecules* **1968**, *1*, 510.
20. Yang, S.; Sun, K.; Risen, Jr, W. M. *J. Polym. Sci., Part B, Polym. Phys.* **1990**, *28*, 1685.
21. Hird, B.; Eisenberg, A. *Macromolecules* **1992**, *25*, 6466.
22. Eisenberg, A. *Adv. Polym. Sci.* **1967**, *5*, 59.
23. Matssura, H.; Eisenberg, A. *J. Polym. Phys. Ed.* **1976**, *14*, 1201.
24. Otocka, E. P.; Kwei, T. K. *Macromolecules* **1968**, *1*, 401.
25. Eisenberg, A.; Navratil, M. *Macromolecules* **1973**, *6*, 604.
26. Zallen, R. *The Physics of Amorphous Solids*; John Wiley & Sons: New York, 1983.
27. Deutscher, G. et al. Eds.; *Percolation Structures and Processes*; Annals of the Israel Physical Society; Israel Physical Society: Jerusalem, Israel, 1983; Vol. 5.
28. Stauffer, D. *Introduction to Percolation Theory*; Taylor and Francis: London, 1985.
29. Gauthier, S.; Duchesne, D.; Eisenberg, A. *Macromolecules* **1987**, *20*, 753.
30. MacKnight, W. J.; McKenna, L. W.; Read, B. E.; Stein, R. S. *J. Phys. Chem.* **1968**, *72*, 1122.
31. McKenna, L. W.; Kajiyama, T.; MacKnight, W. J. *Macromolecules* **1969**, *2*, 58.
32. Longworth, R.; Vaughan, D. J. *Polym. Prepr., Am. Chem. Soc. Div. Polym. Chem.* **1968**, *9*, 525.
33. Tadano, K.; Hirasawa, E.; Yamamoto, H.; Yano, S. *Macromolecules* **1989**, *22*, 226.
34. Tachino, H.; Hara, H.; Hirasawa, E.; Kutsumizu, S.; Tadano, K.; Yano, S. *Macromolecules* **1993**, *26*, 752.
35. Tachino, H.; Hara, H.; Hirasawa, E.; Kutsumizu, S.; Yano, S. *Macromolecules* **1994**, *27*, 372.
36. Yeo, S. C.; Eisenberg, A. *J. Appl. Polym. Sci.* **1977**, *21*, 875.
37. Hodge, I. M.; Eisenberg, A. *Macromolecules* **1978**, *11*, 289.
38. Kyu, T.; Eisenberg, A. In *Perfluorinated Ionomer Membranes*; Eisenberg, A. and Yeager, H. L., Eds.; ACS Symposium Series 180, American Chemical Society: Washington, DC, 1982; Chapter 6.
39. Kyu, T.; Hashiyama, M.; Eisenberg, A. *Can. J. Chem.* **1983**, *61*, 680.
40. Kyu, T.; Eisenberg, A. *J. Polym. Sci., Polym.* **1984**, *71*, 203.
41. Agarwal, P. K.; Makowski, H. S.; Lundberg, R. D. *Macromolecules* **1980**, *13*, 1679.
42. Moffitt, M.; Khougaz, K.; Eisenberg, A. submitted for publication in *Acc. Chem. Res.*
43. Gauthier, S.; Eisenberg, A. *Macromolecules* **1987**, *20*, 760.
44. Long, T. E.; Allen, R. D.; McCrath, J. E. In *Chemical Reaction on Polymers*; Benham, J. L. and Kinstel, J. F., Eds.; ACS Symposium Series 364, American Chemical Society: Washington, DC, 1988, Chapter 19.
45. Weiss, R. A.; Sen, A.; Pottick, L. A.; Willis, C. L. *Polymer* **1991**, *32*, 2785.
46. Desjardins, A.; Eisenberg, A. *Plast. Rubber. Compos. Process. Appl.* **1992**, *18*, 161.
47. Jérôme, R.; Broze, G. *Rubber Chem. Technol.* **1984**, *58*, 223.

48. Broze, G.; Jérôme, R.; Teyssié, P. In *Polymer Science and Technology*; Dubin, P., Ed.; Plenum: New York, 1985; Vol. 30.
49. Jérôme, R. In *Telechelic Polymers: Synthesis and Applications*; Goethals, E., Ed.; CRC: Boca Raton, Fl, 1989; Chapter 11.
50. Vanhoorne, P.; Jérôme, R. In *Physical Chemistry of Ionomers–A Monograph*; Schlick, S., Ed.; CRC: Boca Raton, Fl., in press.
51. Boze, G.; Jérôme, R.; Teyssié, P.; Marco, C. *J. Polym. Sci., Polym. Phys. Ed.* **1983**, *21*, 2205.
52. Charlier, P.; Jérôme, R.; Teyssié, P.; Utracki, L. A. *Macromolecules* **1992**, *25*, 617.
53. Lundberg, R. D.; Makowski, H. S.; Westerman, L. In *Ions in Polymers*; Eisenberg, A., Ed.; ACS Symposium Series 187, American Chemical Society: Washington, DC, 1980; Chapter 3.
54. *Perfluorinated Ionomer Membranes*; Eisenberg, A.; Yeager, H. L., Eds.; ACS Symposium Series 180; American Chemical Society: Washington, DC, 1982.
55. Lundberg, R. D. In *Structure and Properties of Ionomers*; Pineri, M. and Eisenberg, A., Eds.; NATO ASI Series, Series C: Mathematical and Physical Sciences, D. Reidel: Dordrecht, 1987; Vol. 198, p 279.
56. Lundberg, R. D. In *Structure and Properties of Ionomers*; Pineri, M.; Eisenberg, A., Eds.; NATO ASI Series, Series C: Mathematical and Physical Sciences, D. Reidel: Dordrecht, 1987; Vol. 198, p 429.
57. Rogers, J. R.; Randall, F. J. *Polym. Prepr., Am. Chem. Soc. Div. Polym. Chem.* **1988**, *29*(2), 432.
58. Statz, R. J. *Polym. Prepr., Am. Chem. Soc. Div. Polym. Chem.* **1988**, *29*(2), 435.
59. Lehner, A.; Hartmann, H.; Bachmann, R.; Kohl, A.; Spoor, H.; Mahler, K.; Balz, W. U.S. Patent 4 310 565, 1982.
60. Kohl, A.; Balz, W.; Melzer, M.; Schneider, N.; Koester, E.; Lehner, A. Ger. Offen. DE 3 248 327, 1984.
61. Olah, G. A.; Kaspi, J.; Bukala, J. *J. Org. Chem.* **1977**, *42*, 4187.
62. Olah, G. A.; Meider, D. *Synthesis* **1978**, 358.
63. Olah, G. A.; Laali, K.; Mehrotra, A. K. *J. Org. Chem.* **1983**, *48*, 3360.
64. Risen Jr, W. M.; Drake, E. N. *Physical Chemistry of Ionomers - A Monograph*; Schlick, S., Ed.; CRC: Boca Raton, FL, to be published.
65. Portnoy, R. C.; Lundberg, R. D.; Welein, E. R. Presented at the IADC/SPE Drilling Conference, February 10–12, 1986; paper 14795.
66. Drake, E. N. *Polym. Prepr. Am. Chem. Soc., Polym. Chem.* **1994**, *35*(2), 14.
67. Lundberg, R. D. In *Structure and Properties of Ionomers*; Pineri, M.; Eisenberg, A., Eds.; NATO ASI Series, Series C: Mathematical and Physical Sciences, D. Reidel: Dordrecht, 1987; Vol. 198, p 387.
68. Lantman, C. W.; MacKnight, W. J.; Higgins, J. S.; Peiffer, D. G.; Sinha, S. K.; Lundberg, R. D. *Macromolecules* **1988**, *21*, 1339.
69. Gabrys, B.; Higgins, J. S.; Lantman, C. W.; MacKnight, W. J.; Pedley, A. M.; Peiffer, D. G.; Rennie, A. R. *Macromolecules* **1989**, *22*, 3746.
70. Pedley, A. M.; Higgins, J. S.; Peiffer, D. G.; Burchard, W. *Macromolecules* **1990**, *23*, 1434.
71. Wang, J.; Wang, Z.; Peiffer, D. G.; Shuely, W. J.; Chu, B. *Macromolecules* **1991**, *24*, 790.
72. Hara, M.; Lee, A. H.; Wu, J. *J. Polym. Sci., Part B, Polym. Phys.* **1987**, *25*, 1407.
73. Hara, M.; Wu, J-L.; Lee, A. H. *Macromolecules* **1988**, *21*, 2214.
74. Chu, B. *Langmuir* **1995**, *11*, 414.
75. Hara, M.; Jar, P-Y. *Macromolecules* **1988**, *21*, 3183.
76. Hara, M.; Sauer, J. A. *Macromolecules* **1990**, *23*, 4465.
77. Hara, M.; Jar, P.; Sauer, J. A. *Macromolecules* **1990**, *23*, 4964.
78. Pineri, M.; Duplessix, R.; Gauthier, S.; Eisenberg, A. In *Ions in Polymer*; Eisenberg, A., Ed.; Advances in Chemistry Series 187: American Chemical Society: Washington, DC, 1980; Chapter 18.
79. Earnest, T. R. Jr.; Higgins, J. S.; Handlin, D. L.; MacKnight, W. J. *Macromolecules* **1981**, *14*, 192.
80. Forsman, W. C.; MacKnight, W. J.; Higgins, J. S. *Macromolecules* **1984**, *17*, 490.
81. Squires, E.; Painter, P.; Howe *Macromolecules* **1987**, *20*, 1740.
82. Register, R. A.; Cooper, S. L.; Thiyagarajan, P.; Chakapani, S.; Jérôme, R. *Macromolecules* **1990**, *23*, 2978.
83. Register, R. A.; Pruckmayr, G.; Cooper, S. L. *Macromolecules* **1990**, *23*, 3023.
84. Zhao, Y.; Bazuin, C. G.; Prud'homme, R. E. *Macromolecules* **1989**, *22*, 3788.
85. Fan, X-D.; Bazuin, C. G. *Macromolecules* **1993**, *26*, 2508.
86. Bazuin, C. G.; Fan, X-D.; Lepilleur, C.; Prud'homme, R. E. *Macromolecules* **1995**, *28*, 897.
87. Eisenberg, A.; Hoover, M. F. In *Ion-Containing Polymers*; ACS Short Courses; American Chemical Society: Washington, DC, 1972.
88. Storey, R. F. Presented at the 20th Annual Water-Borne, Higher-Solids and Powder Coatings Symposium, New Orleans, LA, Feb. 22–23, 1993; Ion-containing Polymer Short Course.

IONOMERS (Compatibilization of Blends)

C. Geraldine Bazuin
Université Laval

Ionomers, by definition, contain a small number of ionic groups which are chemically incorporated into a (normally) hydrocarbon matrix. It is these ionic groups that are exploited in the use of ionomers to compatibilize polymer blends. This can be done in several ways. One way is to blend two ionomers possessing complementary ionic groups that interact strongly with the ionic group of one component acting as the counterion for the other. Another is to blend ionomers with nonionic polymers (generally polar) so that there are electrostatic attractive forces between the ionic groups of the ionomer and the polar groups of the nonionic polymer. Since ionomers are composed principally of nonionic material, a certain level of compatibilization can also take place between this material and a homopolymer with which it is normally miscible; in this case, the ionic groups act to limit the compatibilization. Finally, ionomers can be used as ternary components, or as emulsifiers, to compatibilize two otherwise immiscible polymers.

In all cases, we must keep in mind the particular morphology of ionomers. Ionomers are frequently biphasic. One phase (often called the matrix or soft phase) is composed mainly of the nonionic matrix material, and is more or less depleted of ionic groups. The other phase (often called the cluster or hard phase) is the ion-rich phase; it is composed of ionic aggregates, or multiplets, surrounded by nonionic material whose mobility is reduced as a consequence of the rigidity of the multiplets. The cluster phase thus has a higher glass transition temperature (T_g) than the matrix phase. Their relative volumes in a particular ionomer depend on various factors, one of the more important being ion content. In some cases, the cluster phase becomes the continuous phase at rather low ion contents. When ionomers are blended with other polymers, two extreme situations can occur. In the first, the interactions between the blend components can strongly modify the biphasic structure of the ionomer(s) involved. In the other, the biphasic morphology can strongly reduce the compatibility between the blend components.

BLENDS OF IONOMERS WITH IONOMERS

In these blends, both components are either ionomers to begin with or ionomer precursors that become ionomers during the blend process. The ionic groups in the two components can, in principle, either be of the same type (leading to ion pair–ion pair interactions) or be complementary (leading to ion–ion interactions).

It has been more usual to prepare ionomer–ionomer blends from components with mutually neutralizing ionic groups, leading to ion–ion interactions or ionic crosslinks. Frequently, this has involved blending an acid-containing ionomer precursor [for example, poly(styrene-*co*-styrene sulfonic acid) (PS-SSA)] with a base-containing ionomer precursor [for example, poly(ethyl acrylate-*co*-4-vinylpyridine) (PEA-VP)], so that proton transfer from the acid to the base occurs; this leads to an ion pair (in this case, also hydrogen-bonded) in which the ion of one component is the counterion for the oppositely charged ion of the other component, or polymeric ion pairs.[1] The concept has been applied to a variety of blends, including PS–isoprene,[3] PEA–Nafion® (a polytetrafluoroethylene ionomer),[4] and PS–polydiacetylene,[5] and is often based on SSA/VP interactions.

Zhang and Eisenberg have also prepared ion–ion blends from pairs of already-neutralized ionomers.[6]

Because the ionic crosslinks in these blends are necessarily interchain interactions, and thus force dissimilar chain segments within each other's vicinity, they are very effective in enhancing the miscibility of the blends. The extent of exhanced miscibility will depend strongly on the ion content: the higher the ion content, the greater the number of contacts between the dissimilar chains, and therefore, the smaller the phase sizes.[7]

The most widely investigated ionically crosslinked blend is the PS-SSA–PEA-VP system mentioned above.

The type or strength of mutual interaction also plays a role in the extent of miscibility achieved.

The strong ion–ion interactions have also been used to prepare blends of end-functionalized polymer chains with complementary end-functionalized chains (blends of telechelic ionomers) or with appropriate statistically functionalized polymers (blends of a telechelic ionomer with a random ionomer or ionomer precursor). In the first case, noncovalent block copolymer-like materials may be formed,[8–12] and, in the second, noncovalent graft copolymer-like materials.[13,14]

In general, when the chains are sufficiently long (and well-defined), we obtain microphase-separated structures. Russell et al. and Iwasaki et al. have shown that blends of complementary bifunctional telechelic ionomers give rise to lamellar structures resembling multiblock copolymers.[9,11]

BLENDS OF IONOMERS WITH POLAR POLYMERS

In these blends, only one of the components is an ionomer. The other is generally a polymer (often a homopolymer, but it can also be a copolymer) possessing a polar repeat unit; this polymer is usually immiscible with the ionomer matrix. Numerous examples of such blends have been reported in the literature. Smith et al. and Natansohn et al. refer to many.[2,15]

In general, the interactions enhancing miscibility in these systems are between the ion pairs of the ionomer and the polar units in the polymer.[2,16] These are dipolar in nature, and thus significantly weaker than the interactions in the mutually interacting (ion–ion) ionomer blends.

Partial miscibility is often observed. In favorable cases, we can obtain complete miscibility over the entire composition range, according to the criteria of DSC and DMA. This is generally only for ionomers with higher ion contents and with small counterions. The miscibility is also temperature-dependent, and we may observe lower critical solution temperature (LCST) behavior.

BLENDS OF IONOMERS WITH OTHER POLYMERS

A few investigations have been reported of ionomer–homopolymer blends where the intercomponent interactions occur between the homopolymer and the matrix segments of the ionomer. A typical example is the polystyrene ionomer–poly(2,6-dimethyl-1,4-phenylene oxide) homopolymer (PPhO).[17–20]

TERNARY BLENDS WITH IONOMERS

Because strong interactions can take place between ionomers and other polymers, ionomers may also be useful as (ternary component) compatibilizers, or emulsifiers, in immiscible or incompatible blends. Several such blends have been reported in the literature, although to date none have yet been studied in depth. They usually involve ionomers which have weak (matrix) interactions with one component and strong (electrostatic or chemical) interactions with the other.

In one series of studies, Surlyns® in both their Na and Zn forms are used to compatibilize polyolefin and polyamide blends.[21,22] Generally, the ionomer decreases the size of the dispersed phase, until there is presumably interfacial saturation.[23]

REFERENCES

1. Smith, P.; Eisenberg, A. *J. Polym. Sci.: Polym. Lett. Ed.* **1983**, *21*, 223.
2. Smith, P.; Hara, M.; Eisenberg, A. In *Current Topics in Polymer Science Volume II* Ottenbrite, R. M.; Utracki, L. A.; Inoue, S. Eds.; Hanser: New York, 1987; Chapter 6.3.
3. Zhou, Z.-L.; Eisenberg, A. *J. Polym. Sci.: Polym. Phys. Ed.* **1983**, *21*, 595.
4. Murali, R.; Eisenberg, A. *J. Polym. Sci.: Part B: Polym. Phys.* **1988**, *26*, 1385.
5. Eisenbach, C. D.; Hofman, J.; MacKnight, W. J. *Macromolecules* **1994**, *27*, 3162.
6. Zhang, X.; Eisenberg, A. *J. Polym. Sci.: Part B: Polym. Phys.* **1990**, *28*, 1841.
7. Brereton, M. G.; Vilgis, T. A. *Macromolecules* **1990**, *23*, 2044.
8. Horrion, J.; Jérôme, R.; Teyssié P. *J. Polym. Sci.: Part C: Polym. Lett.* **1986**, *24*, 69.
9. Russell, T. P. et al. *Macromolecules* **1988**, *21*, 1709.
10. Charlier, P.; Jérôme, R.; Teyssié, P. *Macromolecules* **1992**, *25*, 2651.
11. Iwasaki, K.; Hirao, A.; Nakahama, S. *Macromolecules* **1993**, *26*, 2126.
12. Fleischer, C. A.; Morales, A. R.; Koberstein, J. T. *Macromolecules* **1994**, *27*, 379.
13. Weiss, R. A.; Sasongko, S.; Jérôme, R. *Macromolecules* **1991**, *24*, 2271.
14. Plante, M. Ph.D. Thesis Laval University, Quebec, 1993.
15. Natansohn, A.; Murali, R.; Eisenberg, A. *Makromol. Chem., Macromol. Symp.* **1988**, *16*, 175.

16. Lim, J-C.; Park, J-K.; Song, H-Y. *J. Polym. Sci.: Part B: Polym. Phys.* **1994**, *32*, 29.
17. Bazuin, C. G. et al. *J. Polym. Sci.: Part B: Polym. Phys.* **1993**, *31*, 1431.
18. Register, R. A.; Bell, T. R. *J. Polym. Sci.: Part B: Polym. Phys.* **1992**, *30*, 569.
19. Hseih, D-T.; Peiffer, D. G. *Polymer* **1992**, *33*, 1210.
20. Tomita, H.; Register, R. A. *Macromolecules* **1993**, *26*, 2796.
21. Fairley, G.; Prud'homme, R. E. *Polym. Eng. Sci.* **1987**, *27*, 1495.
22. Willis, J. M.; Favis, B. D. *Polym. Eng. Sci.* **1988**, *28*, 1416.
23. Favis, B. D. *Polymer* **1994**, *35*, 1552.

IONOMERS (dc Conduction Properties)

Shoichi Kutsumizu* and S. Yano
Faculty of Engineering
Gifu University

Ionomers are ion-containing polymers, but they are electrical insulators in the dry state because the ions in ionomers are tightly bound to the side groups. The direct current (dc) conductivity of ionomers is, in general, as low as ~10^{-18} S cm^{-1} at room temperature.[1] To date only a few studies have been done on the dc conduction of ethylene and styrene ionomers. It is generally recognized that the dc conduction of ionomers above the glass transition temperature (T_g) mainly comes from ionic conduction; under a dc electric field, ionic carriers migrate in the amorphous regions in ionomers, which are found in most insulating polymers. Therefore, the dc conduction of ionomers is expected to depend on the state of conduction paths, which is closely connected with the states of amorphous regions and ionic aggregates in ionomers.

dc CONDUCTION PROPERTIES OF ETHYLENE IONOMERS

Temperature dependence of the dc conductivities for ethylene ionomers was first investigated by Hirota.[2] The ionomers used were Na and Zn(II) salts of poly(ethylene-*co*-methacrylic acid) (EMMA) containing 5 or 7 mol % methacrylic acid (MAA). He showed that the conduction behavior of the Na salts shows a critical temperature (T_c) at 322–303 K, and above and below T_c, each temperature dependence of conductivity (σ) was of the Arrhenius type. The T_c corresponded to a glass–rubber transition temperature of ionic aggregates, which was determined by dielectric and mechanical measurements.[1] He observed that the conductivity values at $T_c + 20$ decrease with an increase in the degree of neutralization and concluded that the most plausible carriers for these ionomers are protons and not Na ions (i.e., the neutralizing cation). He also investigated the effect of heat treatment on conduction behavior. He pointed out that the growth of ionic aggregates by heat treatment enhances the conductivity.

Recently, Kutsumizu et al. have extensively studied EMAA ionomers neutralized with various metal cations [i.e., Na, K, Mg, Ca, Co(II), Cu(II), Mn(II), and Zn(II)] and transition-metal(II) complex cations with 1,3-bis(aminomethyl)cyclohexane (BAC).[3–5] The results indicated that the conduction behavior is sensitive to the state of microstructure of these ionomers, for example, the state of the polyethylene crystalline region.

The temperature dependence of the conductivity was also investigated for the alkali–metal and alkaline–earth metal salts of EMAA. All the samples exhibited a σ peak at 319–327 K in the σ-$1/T$ curve. A good relationship between the appearance of the σ peak and microphase structure was also demonstrated.

The dc conduction in EMAA ionomers above T_i was mainly attributed to ionic conduction, but it is difficult to determine the ionic carriers. The most probable carriers are dissociated protons and/or an unavoidable amount of ionic impurities.

To date, two studies have surveyed the potential applications of the K salts of EMAA to ion-conductive or antielectrostatic ionomers because the reduction of electrostatic accumulation is important when ionomers are used for packaging of electronic parts.[6,7] Takeoka et al. recently reported that the ionic conductivity of the K salts increases with an increase in the content of polyoxyethylene introduced.[6] Tachino et al. showed that the method of mixing two K salts of EMAA containing high and low MAA contents is effective in reducing the magnitude of moisture absorption to the extent of practically acceptable levels while maintaining the anti-electrostaticity, where antielectrostaticity requires the magnitude of surface resistivity less than 10^{12} Ω under an atmosphere of 60% relative humidity.[7]

dc CONDUCTION PROPERTIES OF STYRENE IONOMERS

Arai and Eisenberg studied the dc conductivities for poly(styrene-*co*-methacrylic acid)s (SMAA) and the Na salts.[8]

REFERENCES

1. *Ionic Polymers*; Holliday, L., Ed.; Applied Sciences: London, 1975.
2. Hirota, S. *Rep. Prog. Polym. Phys. Jpn.* **1973**, *13*, 437.
3. Kutsumizu, S.; Hashimoto, Y.; Yano, S.; Hirasawa, E. *Macromolecules* **1991**, *24*, 2629.
4. Kutsumizu, S.; Hashimoto, Y.; Hara, H.; Tachino, H.; Hirasawa, E.; Yano, S. *Macromolecules* **1994**, *27*, 1781.
5. Kutsumizu, S.; Hashimoto, Y.; Sakaida, Y.; Hara, H.; Tachino, H.; Hirasawa, E.; Yano, S. *Macromolecules* **1994**, *27*, 5457.
6. Takeoka, S.; Sakai, H.; Shin, H.; Ohta, T.; Tsuchida, E., submitted for publication in *Polym. Adv. Tech.*
7. Tachino, H.; Hara, H.; Hirasawa, E.; Kutsumizu, S.; Yano, S. *Polym. J.* **1994**, *26*, 1170.
8. Arai, K.; Eisenberg, A. *J. Macromol. Sci., Phys.* **1980**, *B17*, 803.

IONOMERS (Mechanical Properties)

Masanori Hara and J. A. Sauer
Department of Mechanics and Materials Science
Rutgers University

Ionomers are long-chain polymers that contain ionic groups, usually occurring as side-chain substituents attached to some of the monomeric units of the non-ionic backbone chains. They can be prepared from a wide variety of polymers and copolymers, including rubbery ones such as polybutadiene, glassy ones such as polystyrene (PS) and poly(methyl methacrylate) (PMMA), and partially crystalline ones such as polyethylene

*Author to whom correspondence should be addressed.

(PE) and polypropylene (PP). In many of the ionomers, the ionic groups are distributed randomly along the hydrocarbon chain, but in some, such as the telechelic ionomers, the ionic groups are situated only at the chain ends. The concentration of ionic groups generally falls within the range of 1–15 mol %.

The specific nature of the ionic groups may vary, but frequently they consist of negative sulfonate or carboxylate ions fully, or partially, neutralized with positive metallic ions to form sulfonate or carboxylate salts. Interaction between the ionic groups modifies and strengthens the intermolecular forces and leads to a physical type of crosslinking. The resulting network structure causes the mechanical properties of the ionomer to differ from the properties of the original non-ionic polymer, and a broad range of properties can be realized by controlling the ion content and other variables.[1] For example, modulus, yield strength, (tensile strength), and clarity can be enhanced in partially neutralized ionomers based on poly(ethylene-*co*-methacrylic acid); i.e., Surlyn® resins successfully introduced into the commercial market in the mid-1960s.

INFLUENCE OF ION CONTENT

Ion content is one of the principal variables that affects the deformation modes and the mechanical properties of ionomers. Deformation modes can be assessed by transmission electron microscopy (TEM) examination of strained thin films. There are as yet few such studies that have been carried out on thin films of ionomers.

The influence of ion content on the tensile strength of bulk samples of Na-SPS ionomers has been studied.[3] The energy to fracture, or toughness, was also measured, and it followed a similar trend to the tensile strength. The introduction of ionic groups has led to appreciable increases in the average tensile strength (and toughness), particularly for ion contents in the range of 6–8 mol %. These beneficial changes in mechanical properties with increasing ion content are believed to arise from changes in the microstructure. At low ion contents, there is a small increase in strength due to an increase in the entanglement strand density. Then as the ion content approaches the critical value, at which the cluster phase begins to dominate, there is a more significant rise in tensile strength (and toughness). Evidently, the ionic clusters provide a more effective type of crosslinking and, because they constitute a second phase, they may also act to some extent as a reinforcing filler.

In elastomeric ionomers, where the T_g is below room temperature, it may be anticipated that increases in tensile strength with ion content would be appreciably greater than for glassy ionomers where the T_g is well above room temperature. Here, too, the enhancement of mechanical properties is attributed to the development of an ionic crosslinking network and to the possible effects of the ionic cluster phase acting as a reinforcing filler.

In all ionomers, one may expect an increase in modulus and stiffness with increasing ion content as a direct result of the ionic interactions between the macromolecular chains. The PE-based ionomers develop transparency and toughness, as well as enhanced strength and stiffness, and these characteristics make them attractive candidates for many industrial applications.

INFLUENCE OF DEGREE OF NEUTRALIZATION

In many ionomers all of the acid groups are converted to salt groups to obtain optimum performance. However, optimum properties, based on strength, stiffness, toughness, and fatigue durability, may be realized in partially neutralized ionomers or in ionomers containing an excess of neutralizing agent. Surlyn-type ionomers, based on poly(ethylene-*co*-methacrylic acid), are good examples of ionomers in which only partial neutralization is desired. In these PE-based ionomers, good mechanical properties, as well as transparency, can be reached at relatively low degrees of neutralization.

In some other ionomers, as in the Zn salts of poly(butyl acrylate-*co*-acrylic acid), as the percent conversion increases, a linear rise in tensile strength occurs until the material is fully neutralized, but a maximum strength is reached only when ~50–70% excess neutralizing agent is added.

INFLUENCE OF TYPE OF COUNTERION

The nature, valence, and size of the counterions can have some influence on mechanical properties, depending on the type of ionomer. In general, ionomers based on poly(ethylene-*co*-methacrylic acid), in which the percent conversion is low, show relatively little influence of the counterion on properties such as strength and stiffness. However, in other types of ionomers, such as SPS ionomers and telechelic polyisoprene ionomers, the nature of the counterion can have an appreciable effect on properties and deformation modes.

Tests made on telechelic, polyisoprene ionomers show that the stress–strain characteristics of these ionomers are significantly affected by the nature of the counterions present. In particular, the size of the counterion was found to be very important.

INFLUENCE OF BLEND COMPOSITION

By blending two different polymers, properties not held by either component can be obtained. When the component polymers are compatible, the properties of their blends are intermediate between those of each component. When the component blends are incompatible, as are most blends of ionomers with unmodified polymers, microphase separation occurs and synergistic effects may be attained in which the property of certain blends is well above values given by the rule of mixtures and may exceed values of that particular property of either components. Studies of the mechanical properties of ionomer blends are relatively new, but the field is a promising one and several examples illustrate the enhancement in properties that can be attained by blending.

INFLUENCE OF OTHER VARIABLES

Although the principal variables affecting the mechanical properties of ionomers appear to be ion content, extent of conversion, and nature of the counterion, other factors such as processing, thermal treatment, and the presence of other ingredients such as plasticizers or additives can play a part. Aging can also have a significant effect on deformation modes induced in ionomers by straining.

Polar plasticizers, such as DMF and glycerol, essentially penetrate and destroy the ionic clusters. They also lower, to some extent, the temperature of the principal loss peak, or modulus drop, that is associated with the multiplet-containing matrix. Nonpolar plasticizers, such as dioctyl phthalate (DOP) and dibutyl (DBP), have a much more significant effect on the lower temperature loss peak but have relatively little effect on the higher temperature transition associated with the ion-rich cluster phase. The presence of plasticizer in an ionomer can also have a significant effect on mechanical properties measured at ambient temperature.

The tensile strength and stiffness of ionomers can be reduced by the presence of plasticizers, but they can be increased in some ionomers by the presence of certain additives. For example, the tensile strength and toughness of Zn salts of a sulfonated butyl rubber ionomer can be appreciably increased by the addition of zinc stearate.

APPLICATIONS

From the many experimental studies on ionomers of widely different types, it is evident that appreciable enhancement in mechanical properties can be achieved by the introduction of ionic groups into non-ionic polymers and copolymers. Also, by control of the many variables that are involved, such as ion content, degree of neutralization (conversion), nature of the counterion, thermal treatment, extent of blending, and presence of plasticizing agents or other types of additives, it is possible to prepare ionomers that have a very wide range of mechanical properties. As a result, ionomers are used in a wide variety of industrial applications, such as molded thermoplastic products, thermoplastic elastomers, films, compatibilizing agents, and membranes for various industrial uses, and it is anticipated that many new industrial applications will be forthcoming as the unique properties of these materials become more widely known.

REFERENCES

1. Hara, M.; Sauer, J. A. *Rev. Macromol. Chem. Phys.* **1994**, *C34*, 325.
2. Bellinger, M.; Sauer, J. A.; Hara, M. *Macromolecules* **1994**, *27*, 6147.

IONOMERS (Solution Behavior)

Masanori Hara
Department of Mechanics and Materials Science
Rutgers University

Ion-containing polymers with a relatively small number of ionic groups (up to 10–15 mol %) in nonionic backbone chains are referred to as *ionomers*. Because of significant changes in physical properties caused by incorporation of ionic groups into polymer chains, much work has been devoted to elucidating the structure–property relationship of ionomers in the solid state.[1-3] Compared with the work conducted on solid state behavior, relatively little has been done to study the structure–property relationship of ionomer solutions. This is in marked contrast to the situation of another class of ion-containing polymers, polyelectrolytes, where major interest has been concentrated on (aqueous) solution properties.[4,5]

It is now well established that ionomers show two types of behavior depending primarily on the polarity of the solvent;[10] one is *polyelectrolyte behavior* arising from electrostatic interactions among ions (fixed ions and counterions) in a polar solvent, such as dimethylformamide (DMF) (ε=37); another is *aggregation behavior* arising from dipolar attractions of ion pairs in a nonpolar or low-polarity solvent, such as toluene (ε = 2.4) or tetrahydrofuran (THF) (ε = 7.6); here, ε is the dielectric constant of the solvent.

So far, two types of ionomers in terms of the molecular architecture have been widely used for studying ionomer solutions: *random ionomers* and *telechelic ionomers.* In random ionomers, such as partially sulfonated polystyrene (SPS) (sodium salt), ionic groups are randomly distributed along backbone chains. In contrast, in telechelic ionomers, ionic groups are located only at the chain ends: monotelechelic (also called monochelic or semitelechelic) ionomers have only one ionic group at the chain end of a linear chain, ditelechelic ionomers have two ionic groups at the chain ends of a linear chain, and tritelechelic ionomers have ionic groups at the chain ends of three-arm star polymers, etc.[9] Therefore, the distance between ionic groups, as well as the number and the location of ionic groups, is well defined for telechelic ionomers, and thus these ionomers are considered simple model systems for more complex random ionomers.

We describe results on both polyelectrolyte behavior and aggregation behavior of random ionomers separately, followed by the description of both polyelectrolyte and aggregation behavior of telechelic ionomers; we focus more on the polyelectrolyte behavior because of our research interests.

PROPERTIES

Polyelectrolyte Behavior

Viscosity measurements are probably most widely used to study the characteristic polyelectrolyte behavior of ionomer solutions. The random ionomers used were sodium salts of SPS having various ion contents, and the solvent used was a polar solvent DMF. The following characteristics are observed: the reduced viscosity, η_{sp}/c, increases markedly with decreasing polymer (ionomer) concentration; the higher the ion content, the greater the upturn of the viscosity curve at low polymer concentration; and even for an ionomer sample with very low ion content (0.94 mol %), there is an upturn is reduced viscosity at low polymer concentration. The first two points are well-known for polyelectrolyte solutions.[4,5] Other aspects of viscosity behavior for ionomer solutions are commonly observed for polyelectrolyte solutions: first, the viscosity data from ionomer solutions follow the Fouss equation, that is, the reciprocal reduced viscosity, c/η_{sp}, increases linearly with $\sqrt{c}$. Second, the method of so-called isoionic dilution can be used to obtain the linear plot of η_{sp}/c versus c by adjusting the total ionic concentration of the solution. Third, the addition of simple salts can suppress the upturn of the reduced viscosity: as the concentration of added salts increases, the reduced viscosity decreases significantly, then a maximum appears in the η_{sp}/c versus c curve and finally, straight lines that are reminiscent of neutral polymer solutions are obtained.

Small-angle neutron scattering (SANS) and small-angle X-ray scattering (SAXS) experiments were conducted on ionomers in a polar solvent, showing typical scattering behavior observed for polyelectrolytes in aqueous solution: a broad single peak in scattered intensity versus scattering vector.[11,12]

Osmotic pressure measurements for ionomer solutions generally show behavior similar to polyelectrolyte solutions.[13]

In summary, all of the observations for ionomer (nonaqueous) solutions parallel those for polyelectrolyte (aqueous) solutions: a detailed review of the behavior of polyelectrolytes and ionomers in nonaqueous solution is available.[7] It should be stressed that ionomer solutions can be used as a good model system for studying complex salt-free polyelectrolyte behavior because of various advantages that they possess, as already described.

Aggregation Behavior

Next, we focus on the aggregation behavior of random ionomers in a low-polarity or nonpolar solvent. In contrast to ionomers dissolved in a polar solvent, where most of the counterions are dissociated, in ionomer solutions of a nonpolar solvent most of the counterions are associated to form ion pairs (ionic dipoles) that further interact with each other to form molecular aggregates in solution. Most initial studies of aggregation behavior have been conducted with low-polarity solvents, such as THF ($\varepsilon = 7.6$).[8,10,14] Recently, there have been additional reports with nonpolar solvents, such as toluene and xylene ($\varepsilon = 2.4$).[15–17] While these solvents require ionomer samples having lower ion contents to form a homogeneous solution, the ionomer electrical environment is closer to that found in the solid state (for example, $\varepsilon = 2.5$ for polyethylene and polystyrene); therefore, information obtained in these nonpolar solvents may be used to understand the nature of aggregates in solid ionomers, which are still not well understood.

Initial experimental studies on the aggregation behavior were conducted mainly by viscosity measurements.[8] At low concentrations, the reduced viscosity of ionomer solutions is smaller than that of the starting polymer, suggesting that the average size of the coil is smaller than that of the starting polymer; this was interpreted as arising from dominant intramolecular interactions that cause chain contraction. As concentration is increased, the reduced viscosity increases markedly, surpassing that of the starting polymer, and eventually forming gels at higher concentrations. This was interpreted as arising from dominant intermolecular interactions that lead to formation of large aggregates.

Low-angle light scattering results obtained at very low concentration, where the reduced viscosity of ionomer solutions is smaller than that of the parent polymer, show higher molecular weights in ionomer solutions, suggesting that intermolecular aggregation persists down to very low concentrations.[10,18]

In addition to light scattering, SANS has been used to study ionomer solutions.[14,15] Studies on SPS ionomers in deuterated THF show that both M_w and R_g values of the ionomer aggregates increase with increasing ionomer concentration.

Dynamic light scattering has also been used to study ionomer aggregates in solution. Initial studies by the method of cumulants have shown that the hydrodynamic radius value increases with ionomer concentration up to a certain point, above which the value becomes almost constant.[18] It has also been found that the quality factor, a measure of polydispersity of molecular aggregates, increases with increasing ionomer concentration.[16] In addition, the dynamic scattering data suggest the existence of single collapsed coils, but mostly at concentrations below those where initial viscosity measurements suggested the existence of collapsed coils due to intramolecular interactions.

In summary, aggregation due to ion-pair (dipolar) attractions causes characteristic solution behavior, which is of interest to various areas, including theoretical consideration of polymer chains with stickers[19] and gelation due to rather strong physical crosslinks.

Telechelic Ionomer Solutions

The basic behavior of telechelic ionomer solutions is similar to that of random ionomer solutions, which are described above; however, we will discuss them separately because of their unique advantageous features. In the early 1980s, ditelechelic ionomers with very sharp molecular distribution were synthesized by anionic polymerization and used for studying aggregation behavior.[20] Also, in the late 1980s, polyelectrolyte behavior of telechelic ionomers in a polar solvent was reported.[21] Although extensive work has been conducted on aggregation behavior of telechelic ionomers, here we focus on the polyelectrolyte behavior of telechelic ionomers in a polar solvent; aggregation behavior is mentioned only when it is related to the discussion of polyelectrolyte behavior.

Telechelic ionomers do indeed develop characteristic polyelectrolyte behavior. We have already described similar behavior for random ionomers with low-charge density in nonaqueous solution and for polyelectrolytes with high-charge density in aqueous solution. Since no intramolecular electrostatic interactions are available for this ionomer, the *intermolecular electrostatic interactions* should be responsible for the behavior. The essential factor causing this behavior is intermolecular electrostatic interactions. More recently, another characteristic "polyelectrolyte" behavior is noted:[22] i.e., negative angular dependence of the reciprocal scattering in static light scattering and an appearance of the fast mode in dynamic light scattering. These results indicate that *only single charge per chain* is sufficient to produce *intermolecular electrostatic interactions*, which is responsible for such characteristic behavior.

Ditelechelic ionomers in toluene behave more or less like random ionomers, because both intra- and intermolecular associations between ion pairs may still be available. In contrast, the monotelechelic ionomer in toluene shows that the reduced viscosity values are always higher than those of the ionomer precursor at all concentrations. This is due to the fact that intramolecular interactions are not available for this type of ionomer. Therefore, the lower viscosity of ionomers compared with that of the parent polymer is closely related to intramolecular associations, as we pointed out for random ionomer solutions. At higher concentrations, the reduced viscosity increases slightly and the curve is concave-down, suggesting the much smaller degree of aggregation compared with ditelechelic ionomers, which show even gel formation. We stress that the experiments conducted to distinguish the role of intra- and intermolecular interactions in aggregation behavior of ionomers are completely parallel to those in polyelectrolyte behavior of ionomers.

APPLICATIONS

Ionomers dissolved in a polar solvent will be a good model system for studying more complex salt-free polyelectrolyte solutions whose structures are still controversial. Ionomers dissolved in a nonpolar solvent will offer a good model system for studying aggregation (solution) behavior of polymers with stickers, which have attracted much attention, both for theoretical considerations and for studying solid-state structures of ionomers, which are still not well understood. Since aging effects make solid ionomers difficult to handle, whereas concentrated ionomer solutions may reach an equilibrium easily, the study of concentrated ionomer solutions should produce useful information.

Although the description given above is focused on basic studies of ionomer solutions, various applications of ionomer solution can be found. For example, ionomers that can be soluble in oils are used as viscosity modifiers.[6] SPS ionomers are used as a viscosifier for oil-based drilling mud, in which the SPS ionomer, having outstanding high-temperature viscosification characteristics, keeps the components in the oil dispersed by distributing itself at the oil-water interface.

REFERENCES

1. Holliday, L., Ed. *Ionic Polymers*; Applied Science: London, 1975.
2. Eisenberg, A.; King, M. *Ion Containing Polymers*; Academic: New York, 1977.
3. MacKnight, W. J.; Earnest, T. R. *J. Polym. Sci., Macromol. Rev.* **1981**, *16*, 41.
4. Rice, S. A.; Nagasawa, M. *Polyelectrolyte Solutions* Academic: New York, 1961.
5. Oosawa, F. *Polyelectrolytes*; Marcel Dekker: New York, 1971.
6. Lundberg, R. D. Review: Ionomer Solution Behavior, In *Structure and Properties of Ionomer*; Pineri, M.; Eisenberg, A.; Eds., NATO ASI Ser. **1987**, Volume 198, p 387.
7. Hara, M. In *Polyelectrolytes: Science and Technology*; Hara, M. Ed.; Marcel Dekker: New York, 1993; Chapter 4.
8. Lundberg, R. D.; Phillips, R. R. *J. Polym. Sci., Polym. Phys. Ed.* **1982**, *20*, 1143.
9. Goethals, E. J. *Telechelic Polymers: Synthesis and Applications*; Goethals, E. J., Ed., CRC: Boca Raton, FL, 1989; Chapter 1.
10. Hara, M.; Wu, J. *Macromolecules* **1988**, *21*, 402; **1986**, *19*, 2887.
11. Wang, J.; Wang, Z.; Peiffer, D. G.; Shuely, W. J.; Chu, B. *Macromolecules* **1991**, *24*, 790.
12. Lantman, C. W.; MacKnight, W. J.; Sinha, S. K.; Peiffer, D. G.; Lundberg, R. D.; Wignall, G. D. *Macromolecules* **1988**, *19*, 1344.
13. Rochas, C.; Domard, A.; Rinaudo, M. *Polymer* **1979**, *20*, 76.
14. Lantman, C. W.; MacKnight, W. J.; Higgins, J. S.; Peiffer, D. G.; Sinha, S. K.; Lundberg, R. D. *Macromolecules* **1988**, *21*, 1339.
15. Pedley, A. M.; Higgins, J. S.; Peiffer, D. G.; Rennie, A. R. *Macromolecules* **1990**, *23*, 2494.
16. Pedley, A. M.; Higgins, J. S.; Peiffer, D. G.; Burchard, W. *Macromolecules* **1990**, *23*, 1434.
17. Hara, M.; Wu, J.; Lee, A. *Macromolecules* **1988**, *21*, 2214.
18. Lantman, C. W.; MacKnight, W. J.; Peiffer, D. G.; Sinha, S. K.; Lundberg, R. D. *Macromolecules* **1987**, *20*, 1096.
19. Cate, M. E.; Witten, T. A. *Macromolecules* **1986**, *19*, 732.
20. Broze, G.; Jérôme, R. J.; Teyssié, Ph. *Macromolecules* **1981**, *14*, 224: **1982**, *15*, 920: **1982**, *15*, 1300.
21. Wu, J.; Hara, M. *Macromolecules* **1994**, *27*, 923.
22. Kupperblatt, G.; Hara, M.; Vanhoorne, P.; Jérôme, R. J. *Polymer* in press.

Ionophores

See: Molecular Recognition (Macromolecular Ionophore)
Phenolics (Linear and Cyclic Oligomers)

ISO 9000 (An Industrial Management Tool)

Charles E. Carraher, Jr.
Department of Chemistry
Florida Atlantic University

Shawn M. Carraher
Department of Management
Indiana State University

While ISO 9000 is largely considered a management tool, it has affected all levels of industrial polymer science, including synthesis, manufacturing, fabrication, and sales. Its influence is increasing.

ISO

The International Organization for Standardization (ISO) is an organization with members from about 100 countries. It often works closely with the International Electrotechnical Commission (IEC) as a single unit to develop common global standards.

The ISO 9000 series is a living process in which every five years, the "standards" are revised.

Most major and many secondary international companies must have ISO certification if they want to do business in Europe. Many other companies and governmental entities are also favoring, and some also requiring, ISO 9000 certification before products and components are accepted. Thus, it is critical for polymer-based companies to be aware of, and for many to receive, ISO 9000 certification.

ISO 9000 is a series of five quality standards, two of which focus on guidance and three on contract standards.

The ISO 9000 series encompass the entire product development sequence from strategic planning to customer service. ISO 9000 is entitled "Quality Management Standards: Guidelines for Selection and Use." It defines terms and sets general principles. It introduces the other four major standards.

ISO 9001 is the most comprehensive of the standards. It describes the design, production, installation and servicing of products, and development with respect to quality assurance necessary to achieve registration under 9001. ISO 9002 is similar to 9001 but it does not include design elements. ISO 9002 does provide a model for quality assurance in installation as well as production and servicing. ISO 9003 is the least as well as comprehensive, dealing only with inspection and testing. ISO 9004 provides guidelines for developing and applying internal Total Quality Management (TQM) aspects. It also examines in more detail the elements covered in ISO 9001–9003.

ISO 9000 certification is generally obtained to promote a company's quality, to provide supplier control, and to promote management practices (typically TQM). It acts as a worldwide standardizing "tool" with respect to business and industry in its broadest sense, including food stores, plastic parts, manufacturing units, governmental service agencies, and banks.

Third Party Certification

ISO 9000 calls for third party assessment. "Third party" refers to an outside reviewer that certifies that the "first party" has satisfied ISO 9000 procedures. "First party" refers to the supplier company that requests ISO 9000 certification. "Second party" refers to the customer whose needs have been met by the "first party" through the use of quality management procedures achieved through ISO 9000 compliance.

ISO 9001–9003 are written from a "second party" standpoint, while ISO 9004 is written from a "first party" perspective. Thus, ISO 9001–9003 are not quality management standards, but quality assurance standards.

Because individual industries, such as the chemical industry, have peculiar concerns, specialized companies have emerged to do the required "registration audits". Specialized consulting groups have also been formed to assist units in their registration process.

Registration

Registration takes time, creativity, and money. Initial registration is estimated to cost an average unit between $250,000 and $500,000 and requires six months to several years to achieve, depending on type of unit, unit size and complexity, prior experience with TQM, etc.

Total Quality Management

ISO 9000 is often referred to as Total Quality Management applied to industry. In truth, many of the concepts included ISO 9000 are TQM concepts. Two major themes of both TQM and ISO 9000 are the importance of human resource management and customer satisfaction.

European Community

One of the major driving forces of ISO 9000 was the desire of the European Community, EC, to create a certified quality system as a method of assuring that quality goods freely move within the EC. Thus, it has acted and been used as a unifying factor with respect to the EC.

The EC divides products into two categories–regulated and nonregulated. Regulated products are those that have an effect on health, safety, or the environment.

For units desiring to do work in the EC it has become almost mandatory to have ISO certification. Some USA governmental agencies are also requiring ISO 9000 (or equivalent) registration.

While the EC and most of the world certify both products and quality systems, the USA certifies products and register quality systems.

ISO 14000

ISO 14000 is an international series of standards intended to assist a unit in managing the impact of its products, operations, and services on the environment. It addresses the need to have one internationally accepted environmental management system. While it has been devised to be a voluntary set of standards, as in the case of ISO 9000, it will be market-driven and may become a requirement for doing business in the international marketplace.

ISOBUTYLENE COPOLYMERS (Commercial)

Anthony Jay Dias
Butyl Product Development
Exxon Chemical Company

Commercially available isobutylene copolymers consist of four families of products: butyl rubber, halobutyl rubber, branched butyl rubber, and poly(isobutylene-*co*-4-bromomethylstyrene). These polyisobutylene-based elastomers offer a unique combination of attributes including superior air retention, good barrier properties, unique damping properties, good aging characteristics, and ozone and flex resistance. This subject has recently been reviewed with an emphasis on chemistry[1] and properties.[2]

POLY(ISOBUTYLENE-*co*-ISOPRENE): BUTYL RUBBER

Butyl rubber resulted from early synthetic rubber research, which had been conducted in Germany, England, Russia, and the United States. Though isobutylene was initially oligomerized in the 1870s, it was not until the 1930s that a method of polymerization was developed which produced high molecular weight polyisobutylene.[4] These polymers exhibited rubber-like properties by virtue of their high molecular weight; however, they lacked suitable functionality for vulcanization. These homopolymers are commercially available today under the trade names of VISTANEX® or OPPANOL® and their chief applications include adhesives, caulks and sealants, and chewing gum.

Continued research in the 1930s led to the discovery of a vulcanizable isobutylene rubber by copolymerizing a small amount of a diene.[3] This copolymerization first introduced the concept of limited unsaturation for vulcanization in an otherwise saturated polymer. Vulcanizates of the new copolymer, butyl rubber, were found to have unique and desirable properties such as low gas permeability, high hysteresis, and good resistance to heat, ozone, and chemical attack.

HALOGENATED BUTYL RUBBER: CHLOROBUTYL AND BROMOBUTYL

Butyl rubber, with its limited unsaturation, represents a compromise between reactivity and stability. The low levels of unsaturation used in commercial butyl rubber grades gives butyl a stable backbone when compared to highly unsaturated general-purpose rubbers like polybutadiene or natural rubber. The introduction of a halogen to the butyl molecule, almost stoichiometrically with the amount of unsaturation present, yields a very reactive allylic halogen site for vulcanization. This increased reactivity resulted in a dramatic broadening of the crosslinking window and vulcanization rates. A direct result of this enhanced reactivity is the ability to produce halobutyl compounds which cure as fast as the higher unsaturation general-purpose elastomers. The better cure properties and greater opportunity to covulcanize with other elastomers enhanced tire manufacturers' ability to build tires using halobutyl innerliners.

BRANCHED POLY(ISOBUTYLENE-*co*-ISOPRENE)

Butyl rubbers have unique processing behavior which results largely from their molecular structure. The high molecular weight between entanglements and lack of crystallization on extension combine to give low green strength and very little creep resistance. Maintaining the typical physical properties of butyl while enhancing its processability is one of the main objectives for producing branched butyl polymers. Two methods of branching butyl rubber are commercially practiced: one involves the use of a difunctional monomer, divinylbenzene. The other introduces a multifunctional polymeric branching agent into the reactor during the polymerization, yielding a star-like branched fraction of butyl rubber.

The very high molecular weight "star-branched" fraction contributes to increased green strength. The improved green strength-stress relaxation balance facilitates functions like tire building where maintaining shape is important. The physical properties after vulcanization are equivalent to those of conventional butyl and halobutyl polymers.

POLY(ISOBUTYLENE-*co*-4-BROMOMETHYLSTYRENE)

Poly(isobutylene-*co*-4-bromomethylstyrene) was developed by Exxon Chemical (commercialized under the name Exxpro™) to further improve the aging, ozone, and flex resistance of isobutylene-based rubber compounds. This family of isobutylene copolymers no longer has backbone unsaturation: the saturated backbone is inert to degradation by ozone. Instead, crosslinking takes place through the highly reactive benzylic bromide sites.

The poly(isobutylene-*co*-4-methylstyrene) copolymer [134737-24-9] has a completely saturated backbone with pendant 4-methyl substituted rings. High molecular-weight linear copolymers can be prepared over a broad composition range of isobutylene and 4-methylstyrene. These copolymers cannot easily be directly vulcanized. Instead, the pendant 4-methyl substituted rings are used as sites for bromination. The pendant 4-methyl sites provide the ideal substrate for selective free-radical bromination.

Chemical Properties

Polyisobutylene has the chemical properties of a saturated hydrocarbon. Poly(isobutylene-*co*-methylstyrene) and poly(isobutylene-*co*-4-bromomethylstyrene), Exxpro™, exhibit in saturated hydrocarbon properties, for example, resistance to ozone and chemical attack and a completely saturated backbone. These copolymers are reactive at the pendant benzylic hydrogens and bromine. The benzylic bromine is especially versatile and can easily be converted to a number of derivatives and graft copolymers.

Physical Properties

Polyisobutylene copolymers have a number of physical properties that make them unique: low gas permeability, low permeability to water and oxygenated solvents, broad vibration damping response, and good mechanical properties. The most important of these physical properties are those exhibited in the vulcanized compounds. Polyisobutylene copolymers have a glass transition (T_g) of –70°C. In the uncrosslinked state, polyisobutylene copolymers are tacky and self-adhere quite well. They pack well and have densities of 0.92 g/cm^3. This dense packing further increases an already large backbone-bond rotational barrier. This combination of good packing and sluggish mobility combine to give poly(isobutylene) copolymers their unique permeability properties.[5]

Compounding

As with most thermoset elastomers, the ultimate physical properties are determined by compounding and vulcanization. Compounding is specifically done for each application area and involves a vast array of fillers, processing aids, plasticizers, tackifiers, stabilizers, and curatives.

Applications

Polyisobutylene copolymers have a unique combination of chemical and physical characteristics which make them useful for a wide variety of applications. Tire inner liners and tubes are two of the prime applications which take advantage of the low permeability to air. Tire-curing bladders require the long-term heat stability of poly(isobutylene) copolymer vulcanizates. The unique damping characteristics of isobutylene copolymers are used widely in vibration isolation applications like automotive bumpers, exhaust hangers, and body mounts. The unique viscoelastic response of isobutylene copolymers can also be utilized to improve the property balance for tire tread compounds. In addition, the chemical resistance and ozone stability of isobutylene copolymers are advantages in tire sidewall blends.

Tire Treads

The unique viscoelastic response of polyisobutylene copolymers imparts high damping properties as well as good compliance. These copolymers have a high potential for improving the coefficient of friction of tire tread compounds with a variety of surfaces and road conditions. The use of halobutyl as a component in tread general-purpose rubber-blend compounds dramatically improves wet skid resistance as well as traction. This improved performance often comes with a sacrifice in tread life and rolling resistance. The blend system is currently under development using Exxpro™ in general-purpose rubber blends.

REFERENCES

1. Kresge, E.; Wang, H. C. *Kirk-Othmer Encyclopedia of Chemical Technology, Fourth Edition*; John Wiley & Sons: New York, NY 1993; Vol. 8.
2. Fusco, J. V.; Hous, P. *Rubber Technology, Third Edition*; Van Nostrand Reinhold: New York, NY, 1987; Chapter 10.
3. Thomas R. M.; Sparks, W. J. U.S. Patent 2 356 127, 1937.
4. Plesch, P. H. *Cationic Polymerization and Related Complexes* Academic: New York, NY, 1953.
5. Krishna Pant, P. V.; Boyd, R. H. *Macromolecules* **1993**, *26*, 679.

Isoelectric Points

See: Amphoteric Latex

Isomer Separation

See: Polymerization Separation (Ring-Opening Metathesis Polymerization)

Isomerization Polymerization

See: *Cyclic Imino Ethers (Ring-Opening Polymerization)*
Cyclic Ketene Dithioacetals (Polymerization)
Double Isomerization Polymerization
Macromonomers (Through Isomerization Polymerization)
Monomer-Isomerization Polymerization (2-Butene with Ziegler-Natta Catalysts)
Nylon 3
Olefin Polymerization by 2, ω-Linkage (Migratory Nickel-Phosphorane Catalyst)
Poly(hexafluoro-1,3-butadiene)
Tautomer Polymerization
Vinylsilanes (Isomerization Polymerization)

Isoporous Membranes

See: *Polypyrroles (From Isoporous Membranes)*

N-ISOPROPYLACRYLAMIDE COPOLYMERS (Drug Delivery)

You Han Bae and Sung Wan Kim
Center for Controlled Chemical Delivery
Department of Pharmaceutics and Pharmaceutical Chemistry
University of Utah

N-isopropylacrylamide (NiPAAm) based thermosensitive polymeric systems, which are currently being used for controlled drug delivery applications, include random copolymers with hydrophilic or hydrophobic comonomers, or both, full or semi-interpenetrating polymer networks (IPN), graft copolymers, and conjugates of polyNiPAAm with other molecules.

The unique thermosensitive properties of polyNiPAAm are a result of its changing molecular shape in an aqueous environment. As the temperature is raised above a critical temperature–lower critical solution temperature (LCST), the polymer chains transform from a solvated expanded coil to a collapsed globule. This transition is attributed to temperature-dependent hydrogen bonding and hydrophobic interactions among polymer segments and water molecules. This transition is known as an endothermic process associated with the reduction of ice-like water structure surrounding hydrophobic moieties along polymer chains. Breaking hydrogen bonds among water molecules by heat absorption at the transition temperature is accompanied by an increase in entropy.

This coil–globule transition of polyNiPAAm was applied for thermoregulation of solute permeability by grafting polyNiPAAm onto a porous nylon 2,12 capsule membrane.[1]

Wu et al. synthesized an erodible, gel-like, polyNiPAAm-lipid composite and studied it for insulin loading and release.[2]

The major mechanism for water soluble solute diffusion in a swollen gel phase is based on the "free volume theory" and thus, the solute diffusion in the hydrogels can be controlled by adjusting water content. Swelling levels can be manipulated, for conventional hydrogels, by changing the structure of the repeat unit, random copolymerization, and crosslink density. For crosslinked polyNiPAAm, the degree of water swelling decreases abruptly, even discontinuously in some cases, around the LCST with increasing temperature. The end point for the gel deswelling process is related to the LCST of the linear polymers and will be referred to as the gel collapse temperature (GCT) hereafter. The temperature dependency of polyNiPAAm gel can be optimized for efficient drug loading and release.

Another advantage of NiPAAm-based hydrogels is their unique swelling properties to control drug loading. Aqueous drug loading into a hydrogel at low temperatures may be advantageous over conventional hydrogel systems, where organic solvents or an organic solvent/water mixture have commonly been used for the loading process. The heparin loading content in NiPAAm copolymer gels was linearly proportional to the degree of swelling in the aqueous drug solutions and thus the loading content was easily manipulated by temperature.[3,4]

Several concurrent mechanisms may be responsible for the unique drug release properties of NiPAAm-based polymer gels.

Historically, research on drug delivery has focused on achieving zero-order (time-independent) release kinetics to minimize the peak-trough profile of drug plasma concentration.

For some drugs, the concept of constant release rate may not be useful: rather, a pulsatile or timed release will be beneficial to minimize drug tolerance (receptor down-regulation) or to supply drugs when required, such as insulin for Type I diabetic patients. As one approach, a drug could be released from a device in response to internal or external signals, such as physical or chemical stimuli. When internal stimuli are used for a feed-back mechanism, it is called closed-loop (or self-regulating) drug delivery. For an externally modulated system, temperature can be a candidate signal for easy control. Thus, NiPAAm-based copolymers have been utilized for modulated release by temperature fluctuation.

We (and others) fabricated drug-loaded polymer gels composed of NiPAAm and hydrophobic comonomer.[5–8] When the temperature was raised above the GCT during release, an outer surface dense layer was formed before bulk shrinkage, as previously described, accompanied by a small degree of drug squeezing. This layer prevented solute from further permeation. When the temperature was lowered below the GCT, the dense outer layer started to reswell, allowing solute release. This resulted in an "on–off" release of a drug. The on–off release was repeated with several cycles by fluctuating the temperatures around GCT.

We obtained the opposite on–off release behavior (drug release above GCT) by placing drug-loaded gel particles, composed of NiPAAm and a hydrophilic comonomer, in a container having a release orifice.[9]

Because macromolecular drugs, such as polypeptides and proteins, are available by virtue of advanced genetic engineering and recombinant technology, the methodology for macromolecular drug delivery is important. Hydrogels and biodegradable polymers have been considered as macromolecular drug carriers. However, many problems are associated with the drug loading procedure due to the labile properties of macromolecular drugs. For instance, when the protein drug is exposed to organic solvents or unfavorable conditions, the protein may become denatured, losing its biological activities. Thus, mild conditions are required for formulation or fabrication of polymeric drug dosage forms. The swelling difference of crosslinked polyNiPAAm gel at low and high temperatures has been used for high molecular weight drug loading and release.

For the purpose of easy, safe, and uniform loading and controlled release of polypeptide drugs, Kim et al. have synthesized soluble terpolymers of NiPAAm, BMA, and AA.[10] When an aqueous solution of NiPAAm copolymer and insulin kept at a low pH (<pH4) and below LCST (4°C) was dropped into a warm oil phase, the solution droplet surface was solidified (skin formation), forming a capsule. These capsules were then isolated from the oil phase by filtration at room temperature, washed, and carefully dried to produce solid beads. The loaded polypeptide (insulin) was not released at gastric pH and protected from gastric enzymatic attack *in vivo* and released in a controlled fashion at high pH. The released insulin preserved its conformation and bioactivity confirmed by a circular dichroism spectrum and blood glucose depression after intravenous injection to normal rats. These facts indicate that such polymeric carriers could be used for oral delivery of polypeptide drugs if appropriate formulations are developed.

In summary, NiPAAm copolymers in various forms could be applied to molecular separation based on molecular size, zero-order release kinetics, on–off modulation of release, and enteric coating materials or enteric carriers for labile drugs, such as proteins.

REFERENCES

1. Okahata, Y.; Noguchi, H.; Seki, T. *Macromolecules* **1986**, *19*, 493.
2. Wu, X. S.; Hoffman, A. S.; Yager, P. *J. Intel. Mat., Syst. & Structures* **1993**, *4*, 202.
3. Gutowska, A. et al. *J. Control. Rel.* **1992**, *22*, 95.
4. Gutowska, A. et al. *Macromolecules* **1994**, *27*, 4167.
5. Bae, Y. H.; Okano, T.; Kim, S. W. *Pharm. Res.* **1991**, *8*, 531.
6. Yoshida, R. et al. *J. Biomater. Sci. Polym. Edn.* **1991**, *3*, 155.
7. Bae, Y. H.; Okano, T.; Kim, S. W. *Makromol. Chem., Rapid Commun.* **1987**, *8*, 481.
8. Mukae, K. et al. *Polym. J.* **1991**, *23*, 1179.
9. Bae, Y. H.; Kim, S. W.; Valuev, L. I. U.S. Patent 5 226 902, 1993.
10. Kim, Y.-H.; Bae, Y. H.; Kim, S. W. *J. Control. Rel.* **1994**, *28*, 143.

Isotactic Polypropylene

See: Polypropylene (Commercial)
Polypropylene, Isotactic (Polymorphism)
Polypropylene, Isotactic (Supermolecular Structure)

Isotactic Polystyrene

See: Polystyrene, Stereoregular
Styrene, Stereospecific Polymerization
Thermoreversible Gels (Isotactic, Syndiotactic, and Atactic Polystyrene)

J

JAPANESE LACQUER: JAPAN: URUSHI (Properties of Urushi Liquid and Urushi Film)

Masato Kasamori, Makoto Sakamoto, Kaoru Awazu, Tachio Ichikawa, and Toshiro Egashira
Industrial Research Institute of Ishikawa Prefecture

Lacquer ware is a varnish that was developed in the Orient, particularly Japan. Lacquer ware is representative of traditional artistic handicrafts of Japan.[1] Here, we use the Japanese word for lacquer ware, urushi. A major component of urushi is urushiol.

Urushi ware, famous for its beauty and luster, is not only art; it is a daily necessity. Urushi is made by various processes from the sap that exudes from under the bark of the lacquer tree (*Rhus vernicifera*). A layer or urushi is coated on the surface of an object and is allowed to dry in moist air under the action of the laccase enzyme to form a film with a network structure.[2] The urushi film obtained is highly adhesive and resistant to corrosion.

Studies have been conducted on improving the refining process, creating colorful urushi, and improving measurement techniques.[5–26]

PREPARATION AND PROPERTIES

Properties of Urushi Liquid

The urushi consists of urushiol, gummy substances containing an enzyme (laccase), polysaccharides, nitrogenous substances, and water. Urushiol is a major component of urushi that has catechol derivatives with long hydrocarbon chains.[3,4,21]

The viscosity and weight-average molecular weight of urushiol extracted from refine urushi increase as refining time increases.

The drying time of urushi changed with the refining time and temperature. The water content, viscosity, and drying time are closely related to the emulsion state of urushi. The emulsion state or urushi can be divided into three types.[13]

Properties of Urushi Film

Changes of gloss in urushi films are caused by changes in the refining process. The longer these processes are, the glossier the urushi film becomes. Regarding the diffuse transmittance by the measurement, of haze, the longer the refining time (i.e., the longer the stirring time), the lower the diffuse transmittance and the lower the haze.[19,20] The gloss and transmittance of urushi films relate to the stirring time only, not to the heating and stirring process, heating time, or temperature. Consequently raw urushi becomes more homogeneous during the stirring process because gummy substances, nitrogenous substances, and the spherical size of water become finer.

Changes in the dynamic mechanical properties are a result of changes in the stirring time. At high temperatures, storage modulus, E', becomes lower as the stirring process proceeds. Also, the magnitude of the loss tangent, δ, maxima becomes larger. Furthermore, the temperatures of the maxima shift somewhat lower. As a result, more stirring time results in less crosslinking in urushi film. The shorter the heating and stirring time or the higher the maxima temperature of the process results in a higher storage modulus, E', and a lower magnitude of the loss tangent, δ. These results suggest that the longer the stirring times or the higher the temperature of the refining process becomes, the lower the activity of laccase seems to be.

Quality Improvement of Urushi

The properties and quality of urushi films are strongly influenced by the refining process. To improve the quality of urushi, a homogenizer and an evaporator were utilized for the stirring process and the heating and stirring process, respectively.[15] These methods reduced the refining time to one-fifth that of present methods and prevented the rise of urushi viscosity, added gloss without the use of lustering agents, increased transparency, and improved light resistance (residual gloss).

Studies have been done on improving light resistance of urushi film through ion implantation.[22,24] Ion implantation induces the surface carbonization of urushi films to inhibit the change in gloss and haze by photoirradiation.

Because black urushi products, especially bowls, are frequently exposed to severe conditions (e.g., hot water and washing), discoloration of the surface occurs with use. The effects of coloring agents on the physical properties of urushi and urushi film and the prevention of the discoloration of black urushi film were investigated.[16] Discoloration can be prevented by decreasing the content of gummy substances in urushi film by adding urushiol, making gummy substances insoluble by baking, or by adding carbon-black. The addition of urushiol also improves the quality of urushi (e.g., it increases transparency and improves light resistance).[25]

We have worked on improving urushi ware and on developing colorful urushi.[5–9,11,17] We hope to proceed with further research on urushi.

REFERENCES

1. JGC; *The Wajimanuri, A Series of Stories on Japan's Traditional Handicrafts, No. 5*; JGC, 26, Tokyo, August, 1980.
2. Kumanotani, J.; Achiwa, M.; Oshima, R.; Adachi, K. In *Proc. 2nd IS-CRCP, Cultural Property and Analytical Chemistry* **1979**, 51. Attempted to understand Japanese lacquer as a super durable material.
3. Kumanotani, J. In *7th Inter. Conf. On Org. Coat. Sci. Tech.* 1981, 265.
4. Kumanotani, J.; Tanaka, S.; Matsui, T. In *Proc. 12th Inter. Conf. In Org. Coat. Sci. Tech.*, 1986; p 195.
5. Nishimura, K.; Ichikawa, T. *Report Ind. Research Inst. Ishikawa* **1980**, *29*, 113.
6. Nakagawa, M. *Report Ind. Research Inst. Ishikawa* **1980**, *30*, 109.
7. Nishimura, K.; Takano, S.; Kuwata, K.; Ichikawa, T. *Report Ind. Research Inst. Ishikawa* **1981**, *30*, 127.
8. Nishimura, K.; Takano, S.; Kuwata, K.; Ichikawa, T. *Report Ind. Research Inst. Ishikawa* **1981**, *30*, 131.
9. Nishimura, K.; Takano, S.; Kuwata, K.; Ichikawa, T. *Report Ind. Research Inst. Ishikawa* **1982**, *31*, 155.
10. Ichikawa, T. *J. Jpn. Soc. Colour Material* **1984**, *32*, 25.
11. Sakamoto, M.; Ichikawa, T.; Takano, S.; Nishimura, K. *Report Ind. Research Inst. Ishikawa* **1985**, *33*, 51.

12. Ichikawa, T.; Sakamoto, M.; Ishibashi, Y. *Report Ind. Research Inst. Ishikawa* **1985**, *33*, 59.
13. Ishibashi, Y.; Kuwamura, T.; Kasamori, M.; Takano, S. *Report Ind. Research Inst. Ishikawa* **1989**, *37*, 37.
14. Egashira, T.; Sakamoto, M.; Ichikawa, T. *Report Ind. Research Inst. Ishikawa* **1990**, *38*, 29.
15. Kasamori, M.; Egashira, T.; Kuwata, K. *Report Ind. Research Inst. Ishikawa* **1990**, *38*, 35.
16. Sakamoto, M.; Egashira, T.; Ichikawa, T. *Report Ind. Research Inst. Ishikawa* **1991**, *39*, 9.
17. Sakamoto, M.; Ichikawa, T.; Egashira, T. *Report Ind. Research Inst. Ishikawa* **1993**, *41*, 23.
18. Mizuno, H.; Takahashi, T.; Shiho, M.; Sakamoto, M.; Ichikawa, T.; Sakikawa, K. *Report Ind. Research Inst. Ishikawa* **1993**, *41*, 39.
19. Sakamoto, M. *Report Ind. Research Inst. Ishikawa* **1994**, *42*, 27.
20. Kasamori, M.; Egashira, T. *Polym. Prepr., Jpn.* **1989**, *38*, E1128.
21. Kasamori, M.; Egashira, T.; Kuwata, K. In *First Pacific Poly. Conf. Preprints*: Anderson, B. C., Ed.; 1989; Vol. 1, p 417.
22. Kasamori, M.; Egashira, T.; Sakamoto, M. In *The 1989 Inter. Chem. Conf. Pacific Basin Societies Abstracts, Part I*; 1989.
23. Awazu, K.; Nishimura, Y.; Ichikawa, T.; Sakamoto, M.; Watanabe, H.; Iwaki, M. *Nucl. Instr. Meth.* **1993**, *B80/81*, 1332.
24. Egashira, T.; Ichikawa, T.; Sakamoto, M.; Ogawa, T.; Kumanotani, J. *Inter. Symp. on Oriental Lacquers* **1993**, 9.
25. Awazu, K. *IONICS* **1984**, *20*(3), 27.
26. Egashira, T.; Ogawa, T. *Polym. Prepr., Jpn.* **1994**, *43*(1), E731.
27. *The Merck Index*, 10th ed.; Merck & Co.; Rahway, NJ, 1983; p 1414.

JUTE

Premamoy Ghosh
Department of Polymer Science and Technology
University of Calcutta

P. K. Ganguly
Jute Technological Research Laboratory
Indian Council of Agricultural Research

Jute is a ligno-cellulosic bast fiber obtained from the bark of the two cultivated species of the genus *Corchorus (C), C. capsularis* (white jute) and *C. olitorius* (tossa jute) of the family Tiliaceae.[1]

Jute textiles are mainly used as packaging materials because of low cost, high strength, and stiffness. Backings of tufted carpets and cheap floor coverings and mattings are other important traditional uses of jute. Because of stiff competition from synthetic substitutes based on polyolefins during the past decades, we have research and development efforts to improve the quality of jute as a textile material and to find nontraditional applications for the fiber. However, the continuous escalation in the prices of petroleum products and petrochemicals, the energy intensive nature of their production, and their resistance to biodegradability have led to renewed interest in jute. Jute has the advantages of being both renewable by agro-efforts and environmentally friendly due to biodegradability. Nontraditional applications envisaged for jute include decorative and furnishing fabrics, floor coverings, woven and nonwoven blankets, primary nonwoven air-filtration media, woven and nonwoven geotextiles, nonwoven thermal and sound insulation media, and jute-reinforced plastics and composites.

PHYSICAL STRUCTURE

Jute, unlike cotton, is a multicellular fiber. A single fiber of jute is a bundle of overlapping cells forming one of the links of the tubular mesh (often termed a reed) extracted from one plant.

CHEMICAL COMPOSITION

Jute has three principal chemical constituents, namely α-cellulose, hemicellulose, and lignin. In addition, it contains minor constituents such as fats and waxes, inorganic (mineral) matters, nitrogenous matter and traces of pigments like β-carotene and xanthophyll.

Interconstituent Linkages in Jute

Jute fiber may be considered a composite with the anisotropic cellulose microfibrils acting as the load-bearing entity in an isotropic lignin matrix, and with hemicellulose acting as the coupling agent.[2]

Acid Nature of Jute

Jute is mildly acidic and defatted jute can possess an acid value of 3.6 milli-equivalents per 100 g which increases to about milli-equivalents per 100 g on rendering jute mineral free using 0.1 M HCl.[5]

PROPERTIES OF JUTE

Jute is a stiff fiber with very low extensibility. Its color ranges from pale cream to golden brown to dirty grey depending upon the variety and quality of the fiber. It is lustrous in appearance and generally has a rough feel.

Jute fibers (or filaments) contain a variable number of cells along their length. Hence, the value of filament strength within a sample varies widely. Appreciably higher moisture regain value of jute compared to that of cotton is attributed to the presence of hemicellulose in jute. Transverse swelling of jute increases appreciably on removal of either lignin or hemicellulose. Moisture absorption reportedly brings about a reduction in the degree of crystallinity value of jute.[6]

In addition to being mildly acidic, jute is chemically reducing in nature and copper number of jute varies roughly between 1.5 and 4.[3]

The presence of lignin in jute makes it more resistant to the action of acid than cotton or ramie.

Jute is more susceptible to alkali, particularly NaOH, than cotton or ramie because of its hemicellulosic constituents. Action of aqueous alkali solutions on jute is generally associated with loss in weight due mainly to some loss of hemicellulosic component, and it depends on temperature, the concentration of the alkali, and the duration of treatment.

On continuous exposure to strong sunlight, jute darkens in color and loses strength due to photodegradation of its lignin constituent.

Jute can be dyed with all the dyes used for cotton and wool. However, dyed jute, particularly pale shades, lacks photostability; discoloration occurs on prolonged exposure to light. Lignin and hemicellulose constituents have significant effects on jute's ability to accept dye.[8]

CHEMICAL MODIFICATION OF JUTE

Chemical modification of jute is mainly based on reactions involving the reactive hydroxyl (OH) groups present in all three major chemical constituents of jute: cellulose, hemicellulose, and lignin. Cellulose contains both primary and secondary hydroxyl groups; hemicellulose contains secondary hydroxyl, aldehydic, and some free carboxyl groups; and lignin bears both alcoholic and phenolic OH groups.

Bleaching

Jute can be bleached to various degrees of whiteness ranging from pale cream to milk white by treating with alkaline solutions of Ca and Na hypochlorites, $KMnO_4$, H_2O_2, and acidic $NaClO_2$ solution in aqueous medium.

Poor photostability of bleached jute (yellowing of the fabric on exposure of light) has been attributed to the carbonyl and cinnamaldehyde types of linkages present in the lignin moieties.[4]

Esterification and Etherification

Jute has been methylated, acetylated, cyanoethylated, and maleated. It is commonly crosslinked using methylolated resins such as urea-formaldehyde (U/F) resins or dimethylol dihydroxy ethylene urea (DMDHEU) resin.[9–12] All these modifications are associated with various degrees of strength loss.

Rot-Resistance and Fire-Retardance

Treatment of jute with copper naphthenate and basic copper chromate improves its resistance to weathering and rotting and the treatments are fairly fast to leaching.[13]

Treatment with diammonium phosphate (DAP), urea–DAP combination, DAP–ammonium sulfate/sulfamate, and Rochelle salt render jute fire-retardant.[14–16]

Graft Copolymerization

Many studies present graft copolymerization of vinyl monomer such as acrylonitrile, methyl methacrylate, acrylamide, styrene, maleic anhydride, and methacrylic acid to jute by γ-irradiation,[17,18] photoactivation (UV rays),[19,20] and chemical activation.[7,21–25] Removal of lignin and treatment with aqueous NaOH significantly enhance the graft copolymerization reactions.[7,24] Improvements to such properties as moisture-resistance, rot-resistance, thermal stability, light-fastness, ability to accept dye, and tensile strength and modulus values of jute and modified jute have been reported on vinyl grafting.[7,23]

JUTE FIBER REINFORCED PLASTIC (JRP) COMPOSITES

Lower cost, higher specific modulus, lower density, lower energy input requirement of jute production compared to glass,[27,29] and its renewability make jute suitable and attractive for use as a reinforcing agent in JRP composites using different matrix resins such as the phenolics, unsaturated polyester (USP), and epoxy resins.

Various investigators have highlighted the potential of use of jute as a reinforcing agent, either alone or in combination with glass fiber.[26–31] They suggest that significant improvements in moisture-resistance of jute and in its compatibility with hydrophobic synthetic resins (such as USP) are important preconditions for improving this application. Attempts at improving properties or performance of JRP composites by pretreating or modifying jute with polyesteramide polyols,[30] titanate coupling agent,[31] silane coupling agents,[31,32] diluted USP resin solution,[31] toluene di-isocyanate,[31] phenol-formaldehyde (PF) and resorcinol–formaldehyde (RF) resins,[26] treatment with NaOH, or on vinyl grafting[26,33] resulted in various degrees of improvements in strength, modulus, and water-resistance of the JRP composite.

REFERENCES

1. Kundu, B. C.; Basak, K. C.; Sarkar, P. B. In *Jute in India*, Indian Central Jute Committee: Calcutta, India, 1959; pp 1, 129, 337–360.
2. Navell, T. P.; Zeronian, S. H., Eds.; *Cellulose Chemistry and Its Applications* Ellis Harwood: Chichester, England, 1985.
3. Sarkar, P. B.; Majumder, A. K.; Pal, P. B. *J. Textile Inst.* **1948**, *39*, T44.
4. Bag, S. C. et al. *Cellulose Chem. Technol.* **1984**, *18*, 149.
5. Sarkar, P. B. and Majumdar, A. K. *Text. Res. J.* **1955**, *25*, 1016.
6. Ray, P. K. *J. Appl. Polym. Sci.* **1976**, *20*, 1765.
7. Ghosh, P.; Ganguly, P. K. *Polymer* **1994**, *35*, 383.
8. Ganguly, P. K.; Chanda, S. *Indian J. Fiber and Textile Res.* **1994**, *19*, 38.
9. Callow, H. J. *J. Textile Inst.* **1952**, *43*, T423.
10. Zahn, H.; Das, P. C. *Gesamite Textile Ind.* **1965**, *67*, 353, 550.
11. Som, N. C.; Bagchi, A.; Mukherjee, A. K. *Indian J. Textile Res.* **1987**, *12*, 78, 126.
12. Reddy, S. S.; Bhaduri, S. K.; Pandey, S. N. *J. Appl. Polym. Sci.* **1993**, *47*, 73.
13. Pal, P. N.; Ghosh, B. L.; Macmillan, W. G. *Indian J. Technol.* **1964**, *2*, 311.
14. Mondal, S. P.; Roy, A. *Bombay Textile Res. Assocn. (India) Silver Jubilee Monograph*, No. 12; 1981.
15. Samajpati, S.; Ganguly, P. K.; Bag, S. C. *J. Appl. Polym. Sci.* **1993**, *47*, 747.
16. Sharma, U. *Colourage* **1986**, *33*, 19.
17. Majumdar, S. K.; Rapson, W. H. *Textile Res. J.* **1964**, *34*, 1007, 1015.
18. Ghosh, P.; Bandyopadhyay, A. R.; Das, S. *J. Macromol. Sci. Chem.* **1983**, *A19*, 1165.
19. Ghosh, P.; Paul, S. K. *J. Macromol. Sci. Chem.* **1983**, *A20*, 169.
20. Ghosh, P.; Biswas, S.; Datta, C. *J. Material Sci.* **1989**, *24*, 205.
21. Trivedi, I. M.; Mehta, P. C. *Cellulose Chem. Technol.* **1973**, *7*, 401.
22. Abou-Zeid, N. Y.; Higazy, A.; Hebeish, A. *Angew. Makromol. Chem.* **1984**, *121*, 69.
23. Ghosh, P.; Ganguly, P. K. *J. Appl. Polym. Sci.* **1994**, *52*, 77.
24. Ghosh, P.; Ganguly, P. K.; Bhaduri, S. K. *Eur. Polym. J.* **1994**, *30*, 749.
25. Mohanty, A. K. *JMS-Rev. Macromol. Chem. Phys.* **1987–88**, *C27*, 593.
26. Ghosh, P.; Ganguly, P. K. *Plastics Rubber and Composites Processing and Applications* **1993**, *20*, 171.
27. Bowen, D. H. "The Use of Vegetable Fibers for Plastic Reinforcement", *Proc. of Int. Symp. of Energy and Composite Matter*, Venice, Italy, 1981.
28. Chawla, K. K.; Bastos, A. C. *Proceedings, 3rd Intl. Conf. on Mech. Behavior of Materials* Toronto, Canada, 1979.
29. Roe, P. J.; Ansell, M. P. *J. Mater. Sci.* **1985**, *20*, 4015.
30. Mukherjee, R. N.; Pal, S. K.; Sanyal, S. K. *J. Appl. Polym. Sci.* **1983**, *28*, 3029.
31. Verma, I. K.; Ananthakrishnan, S. R.; Krishnamoorthy, S. *Composites* **1989**, *20*, 383.
32. Phani, K. K.; Bose, N. R. *J. Mater. Sci.* **1987**, *22*, 1929.
33. Varma, D. S.; Murali, S. *Indian J. Textile Res.* **1989**, *14*, 9.

K

Keratin

See: *Keratins*
Wool Keratin

KERATINS

David A. D. Parry
Department of Physics
Massey University

The mammalian epidermis and its wonderfully diverse derivatives provide an animals first line of defense against its external environment and potential predators. As such, the mechanical properties of the keratins play a large part in determining whether or not an animal will survive and flourish in its natural habitat.[1] The keratins are commonly divided into three groups, largely on the basis of the structures adopted by their various protein constituents rather than on their physical attributes. These are (1) hard α-keratins that include hair, nails, claws, beaks, quills, hooves, baleen, and horns, and the (soft) epidermal keratins that form the *stratum corneum*, corns, and callouses; (2) β-keratins, which are not a naturally occurring form but which are produced by the action of pressure and temperature on native α-keratins; (3) feather keratins, which include feathers, scales, and also parts of beaks and claws. Greatest attention has been focused on the first group, primarily because of the commercial importance of wool and the medical significance of hair and skin.

In order to become familiar with the keratinous family of proteins, it is important first of all to gain an appreciation of their molecular and filamentous structures. Only then is it possible to gain some understanding of their *raison d'être.*

HARD AND EPIDERMAL α-KERATIN

X-ray diffraction patterns of hard and epidermal α-keratin have been interpreted in terms of filamentous assemblies of highly oriented molecules with conformations based on the right-handed α-helix. The structure of the individual intermediate filament (IF) molecule, the assembly of these molecules into oligomers, and the incorporation of oligomers into the intact IF have now been characterized in large part.[2]

Hard α-keratin IF are embedded in matrix proteins. The matrix consists of three families of proteins: those rich in cysteine residues (the high-sulphur proteins), those extremely rich in cysteine residues (the ultra high-sulphur proteins) and those rich in glycine and tyrosine residues (the high-tyrosine proteins). The content and composition of the matrix proteins vary with source, age, and nutrition.[1,3] Although the matrix proteins as a group play an important role mechanically, especially in compression, it is the number and disposition of the covalent disulphide bonds formed within and between the matrix proteins and the terminal domains of the IF molecules that dominate the physical attributes of hard α-keratin. There is no matrix as such in the epidermal keratins, but IF-associated proteins (IFAP) such as filaggrin are responsible for aggregating IF into larger assemblies that act as the functional unit *in vivo.*[2]

β-KERATIN

β-Keratin does not occur naturally but can readily be produced when mammalian hard α-keratin is stretched by about 100% in steam over the course of five to six hours.

Over the years the structure of β-keratin proved to be controversial, even though it was commonly agreed that the basic protein conformation was that of hydrogen-bonded sheets of β-strands packed together to form a β-crystallite. Some evidence, however, was taken to indicate that the chains were parallel, whereas other data showed that they were antiparallel. The problem was solved by Fraser et al.[4] The analysis showed that the β-strands were arranged in a regular antiparallel β-pleated sheet, but that the stacking of the sheets displayed disorder. In fact, the β-crystallites contained only two to three sheets, and in each of these there were only ten chains.

β-Keratin provides a neat structural link between the α-keratins and the feather structure. It has also provided insights into the relationship between the chain and molecular structure of epidermal keratin and its filamentous aggregate (IF).

FEATHER AND SCALE KERATIN

The hard keratin in avian epidermal appendages such as feathers, beaks, and claws can largely be accounted for by a single protein species with a molecular weight of about 10.4 kDa (Dalton). While this feature is markedly dissimilar to that seen in the α-keratins, where distinct families of filament-forming and matrix proteins were identified, electron microscopy has shown that feather keratin also consists of filaments embedded in a "matrix".[5] The feather keratin protein is unique, however, in that it forms both the filament and the "matrix". The filament consists of an antiparallel pair of intertwined helices.

Interestingly, the amino acid sequences of feather and scale keratin display a high degree of homology. Whereas scale keratin molecules have a much higher molecular weight than those from feather keratin, this difference lies almost *in toto* with the addition of a four-fold repeat of a 13 residue motif in scale keratin.

PHYSICAL PROPERTIES

Each of the keratinous tissues has unique features that enable it to function optimally. For example, the thermal insulation attributes of hair necessitate that it must be flexible and moderately extensible. In addition, water sorption properties are important. In contrast, it is easily seen that the hard α-keratin of rhinoceros horn (which is actually hair) and the epidermal keratin that forms the surface layer of skin must have quite different physical attributes. In contrast yet again, feather keratin must be light and inextensible without being brittle.[1]

Of course, the physical properties of keratins do not depend solely on the composite nature of the material. Other aspects that are important include the water content, the lipid content (especially in the epidermal keratins) and the composition and content of the proteins that constitute either the matrix or the proteins associated with the IF (the IFAP). The orientation of the IF in the aggregates that form is also crucial. The leading role played by IF orientation in specifying mechanical properties is often not fully appreciated.

Bendit and Feughelman have collected a great quantity of physical data for the α-keratins.[6] However, we now understand one of the key physical properties of native hard α-keratins relating specifically to the disposition of water within the IF-matrix composite. In turn, this would lead to a decrease in the swelling of the fiber in water and would cause the wet matrix to become stiffer.

Fraser et al. have provided data on the site of water deposition within α-keratin.[1] In particular, they showed that when hair swells in water there is only about 1% change in length but a 16% change in diameter. Furthermore, in quill the IF were shown to be relatively impermeable to water in comparison to the matrix, which increased in volume by about 53%.[7] Naturally, these figures change from one type of α-keratin to another, but the overall trend is maintained absolutely.

In contrast to the highly disulfide-bonded structure of the hard α-keratins the epidermal keratins present a very flexible barrier to their environment.

There has been little work done on the physical properties of feather keratin. Nonetheless, we know that the water sorption properties of feather keratin and hard α-keratin are remarkably similar in spite of the large differences in the conformations of the constituent molecules. This implies that molecular structure *per se* is not a critical feature in water sorption though it is clearly important in determining other physical attributes, such as tensile properties. Feather keratin can be elongated by only about 6% before rupturing, whereas α-keratin can be extended by more than 30%. This feature, at the least, is very clearly related to the nearly fully extended molecular conformation in feather keratin and the much more highly folded structure in hard α-keratin.

REFERENCES

1. Fraser, R. D. B.; MacRae, T. P.; Rogers, G. E. *Keratins: Their Composition, Structure and Biosynthesis*: Charles C Thomas, Springfield, IL, 1972.
2. Parry, D. A. D.; Steinert, P. M. *Intermediate Filament Structure*; R. G. Landes: Austin, TX 1995.
3. Gillespie, J. M. *Cellular and Molecular Biology of Intermediate Filaments*: Goldman, R. D.; Steinert, P. M., Eds., Plenum: New York, NY, 1990; p 95.
4. Fraser, R. D. B. et al. *Polymer* **1969**, *10*, 810.
5. Rogers, G. E.; Filshie, B. K. *Ultrastructure of Protein Fibers*; Borasky, R., Ed., Academic: New York, NY, 1963; p 123.
6. Bendit, E. G.; Feughelman, M. *Encyclopedia of Polymer Science and Technology*; John Wiley & Sons: New York, NY, 1968; Vol. 8, p 1.
7. Fraser, R. D. B. et al. *Appl. Polym. Symp. No. 18* **1971**, 65.

Ketene Polymerization

See: Diketene Polymerization

L

Lacquer

See: *Japanese Lacquer: Japan: Urushi (Properties of Urushi Liquid and Urushi Film)*
Oriental Lacquer

Lactam Polymerization

See: *Bicyclic Lactams, Ring-Opening Polymerization*
Nylon 6 (Depolymerization)
Nylon 3, Helical [Poly(a-Alkyl-b-L-Aspartates)]
Polyamide Fibers, Aliphatic (Nylon 66 and Nylon 6)
Polyamides, Lactam-Based
Polyisopeptides

Lactone Polymerization

See: *Bicyclic Lactones, Ring-Opening Polymerization*
Polyesters (Synthesis from Lactones)
Ring-Opening Coordination Polymerization (by Soluble Multinuclear Aldoxides)
Ring-Opening Polymerization, Anionic (with Expansion in Volume)

Ladder Polymers

See: *Conjugated Ladder Polymers*
Cyclolinear Organophosphazene Polymers
Cyclopolymerization (Overview)
Ladder Polymers (Cycloaddition Copolymerization of Cyclic Diynes)
Ladder Polymers (Methods of Preparation)
Ladder Polymers (Precursor Route, Fully Unsaturated, All-Carbon)
Ladder Polyphenylsilsesquioxane
Silsesquioxane-Based Polymers
Small-Bandgap Polymers (Special Class of Organic Conducting Compounds)
Template Polymerization (Overview)
Topochemical Polymerization (Diacetylenes)

LADDER POLYMERS (Cycloaddition Copolymerization of Cyclic Diynes)

Tetsuo Tsuda
Department of Polymer Chemistry
Graduate School of Engineering
Kyoto University

A new generation of the ladder polymer has been brought forth by the synthesis of soluble ladder polymers using two improved approaches: a repetitive Diels–Alder reaction.[1] and a polymer-analogous cyclization of linear prepolymers.[2,3] This successful synthesis of soluble ladder polymers including π-conjugated ladder polymers is stimulating further interest in exploitation of new synthesis of soluble ladder polymers and their application to electronic and optical materials.

We describe here facile and efficient synthesis of soluble ladder poly(2-pyridone)s and poly(2-pyrone)s by the nickel(0)-catalyzed 1:1 cycloaddition copolymerization of cyclic diynes with isocyanates and CO_2, respectively. The present synthetic method of the soluble ladder polymer is new and is characterized by transition metal-catalyzed one-step cycloaddition of the acetylene with the heterocumulene as an elementary reaction.

Equation 1 shows the copolymerization[8] of three cyclic diynes: 1,7-cyclotridecadiyne (1a), 1,7-cyclotetradecadiyne (1b), and 1,8-cyclopentadecadiyne (1c), with three isocyanates: phenyl (2a), cyclohexyl (2b), and *n*-octyl (2c) isocyanates, carried out in tetrahydrofuran (THF) at 60°C in the presence of a nickel(0) catalyst generated from bis(1,5-cyclooctadiene)nickel ($Ni(COD)_2$) (10 mol %) and 2 equivalents of a tricyclohexylphosphine ligand in a manner similar to the acyclic diyne-isocyanate copolymerization.[5–6] Nine ladder poly(2-pyridone)s (Equation 4:3) with molecular weights of ca. 15000–65000 were obtained in high yield by purification with CH_2Cl_2/Et_2O or CH_2Cl_2/MeOH. They were soluble in CH_2Cl_2, $CHCl_3$, and acetic acid.

The present nickel(0)-catalyzed cycloaddition copolymerization of the cyclic diyne with the isocyanate provides a new synthetic method of the soluble ladder polymer and has two significant advantages: synthesis of a variety of ladder poly(2-pyridone)s by changing a structure of the ioscyanate cycloaddition component along with that of the cyclic diyne; and control of the 1:1 copolymerizability of the copolymerization reaction, which is related to the solubility of the poly(2-pyridone), by changing the feed ratio of the isocyanate to the cyclic diyne.

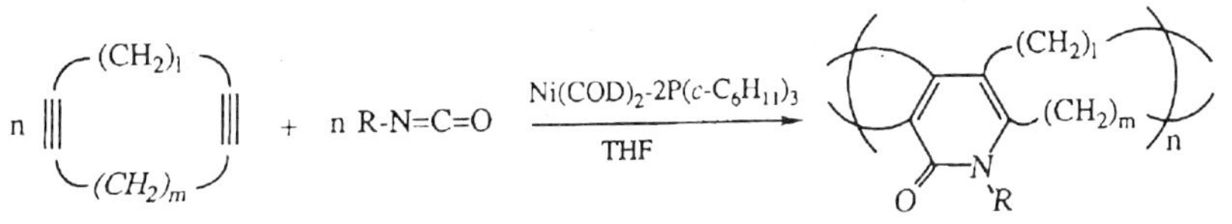

(1)

1	2	3
l,m = 4,5 (**1a**)	R = C_6H_5 (**2a**)	l,m = 4,5; R = C_6H_5 (**3aa**)
l,m = 4,6 (**1b**)	R = *c*-C_6H_{11} (**2b**)	l,m = 4,6; R = *c*-C_6H_{11} (**3bb**)
l,m = 5,6 (**1c**)	R = $C_8H_{17}{}^n$ (**2c**)	l,m = 5,6; R = $C_8H_{17}{}^n$ (**3cc**) etc.

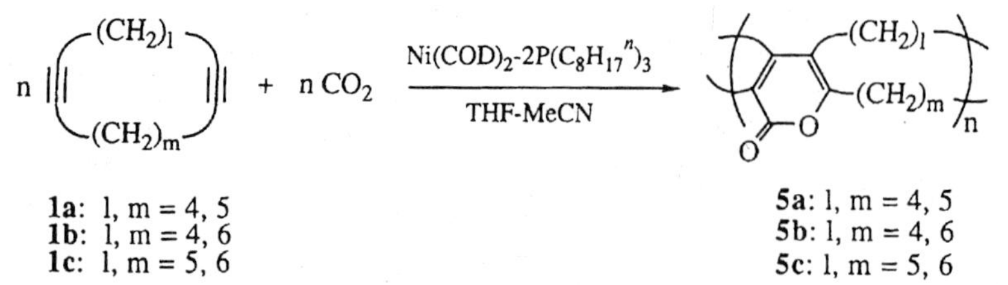

2

LADDER POLY(2-PYRONE) SYNTHESIS

A nickel(0) catalyst generated from $Ni(COD)_2$ and two equivalents of tri-*n*-octylphosphine effected the 1:1 cycloaddition copolymerization of cyclic diynes (1) with CO_2 to ladder poly(2-pyrone)s (5) in a mixed solvent of THF/MeCN (1/1) (**Equation 2**).[4,9] The (1a)–CO_2 copolymerization quantitatively afforded ladder poly(2-pyrone) (5a) with a molecular weight of ca. 7000 by purification with CH_2Cl_2/Et_2O. The copolymer was white powder and soluble in CH_2Cl_2 and $CHCl_3$. Cyclic diynes (1b) and (1c) also produced ladder poly(2-pyrone)s (5b) and (5c) in high yield.

REFERENCES

1. Schlüter, A-D. *Adv. Mater.* **1991**, *3*, 282.
2. Scherf, U.; Müllen, K. *Polymer* **1992**, *33*, 2443.
3. Tour, J. M.; Lamba, J. J. S. *J. Am. Chem. Soc.* **1993**, *115*, 4935.
4. Tsuda, T.; Maruta, K.; Kitaike, Y. *J. Am. Chem. Soc.* **1992**, *114*, 1498.
5. Tsuda, T.; Hokazono, H. *Macromolecules* **1993**, *26*, 1796.
6. Tsuda, T.; Tobisawa, A. *Macromolecules* **1994**, *27*, 5943.
7. Tsuda, T.; Tobisawa, A. *Macromolecules* **1995**, *28*, 1360.
8. Tsuda, T.; Hokazono, H. *Macromolecules* **1993**, *26*, 5528.
9. Tsuda, T. et al. *Macromolecules* **1995**, *28*, 1312.

LADDER POLYMERS (Methods of Preparation)

Niyazi Biçak and A. Sezai Saraç
Department of Chemistry
Instanbul Technical University

Polymers with repeating units that link to each other at two connecting points are called "ladder" or "double strand" polymers. Ideal ladder polymers consist of fully uninterrupted series of rings. Free rotation cannot occur around the linkages between rings unless the bond breaks. Ladder polymers only degrade if at least two bonds on the same ring are broken. The probability of breaking two bonds on the same ring is far less than a single bond rupture, and ladder polymers should exhibit superior thermal, mechanical, and chemical stability when compared to single chain polymers. However, purity and maintaining a perfect ladder form are mainly responsible for the thermal stabilities.

METHODS OF PREPARATION

Methods of preparation of ladder polymers are classified into 5 different categories.

1. Ring closure on linear polymers
2. Multifunctional polycondensation
3. Cycloaddition
4. Ring forming-Electrooxidative coupling
5. Ladder polymers from polyelectrolytes

RING CLOSURE ON LINEAR CHAIN POLYMERS "ZIP UP" PROCEDURE

This method makes a linear polymer containing reactive pendant groups along the chain and links these groups by adding or eliminating reactions to form a ladder structure. Cyclization of poly(methyl vinyl ketone) is the first example in this area.[1]

Upon heating to 300°C, water is given off, leaving a glossy red solid ladder polymer containing fused cyclohexene rings.

The second and most studied example is pyrolyzed polyacrylonitrile (PAN), the so called "Black Orlon." It is obtained by simultaneous cyclization and oxidation at 160°–300°C (**Structure 1**).[2,3]

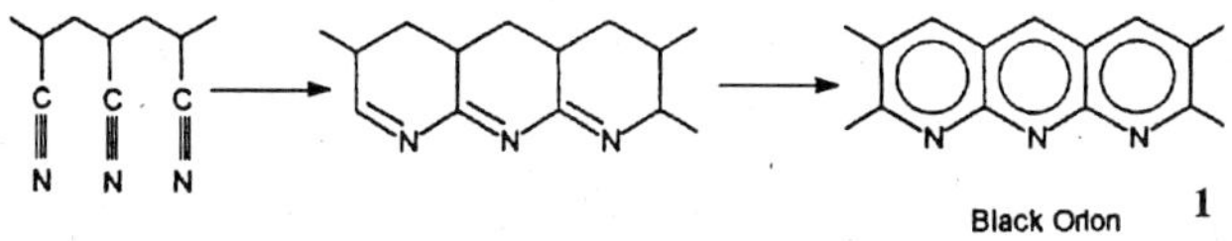

1

The "zip-up" procedure has been applied to poly(allyl methacrylete)[4] and poly(vinyl alcohol) that is esterified by methacrylic acid[5] to give partial (80%) ladder structures.

MULTIFUNCTIONAL POLYCONDENSATION

Polypyrones derived from aromatic tetraamines with tetracarboxylic acids or its anhydrides are the most studied class of heteroarylene polymers obtained by polycondensation. When pyromellitic dianhydride is heated in poly(phosphoric acid) (PPA) with 1,2,4,5-tetraminobenzene, the polypyrone formed is soluble in sulfuric acid. No weight loss is observed on heating up to 600°C.[6] However, some rings in the structure remain open.[7]

POLYQUINOXALINES

o–Phenylene diamine, reacts with 1,2-diones to give quinoxalines. When this procedure is extended to tetramines and tetrones, ladder polyquinoxalines are obtained.[8–11]

POLYPHENOXAZINES AND POLYPHENYLTHIAZINES

Two step condensation of 2,5-dihydroxybenzoquinone and its substituted derivatives with 1,3-diamino-4,6-dihydroxybenzene also creates polyphenoxazines in ladder form.[8,12] However, thermally cyclized materials show poor thermal stability due to incomplete ladder formations.

LADDER POLYMERS BY CYCLOADDITION

The Diels–Alder reaction is often used for simple cycloaddition reactions that, in turn, lead to ladder polymers.

The Diels–Alder reaction involving 1,4-addition of a carbon–carbon double bond to a 1,3-diene synthesized a ladder polymer with six membered rings.

RING-FORMING ELECTROOXIDATIVE COUPLING

Only two examples of ring-forming electrooxidative coupling leading to ladder polymers exist in the recent literature. Electropolymerization of 2-aminophenol gives semiconducting thin films of polyphenoxazine on an electrode surface.[13]

Also, electrolysis of 1,2-diaminobenzene in aqueous acid solutions results in a ladder polymer with phenazine rings.[14]

LADDER POLYMERS FROM POLYELECTROLYTES

It is presumed that two oppositely charged polymers will interact to form ladder polymers with ionic interlinkages.[15]

An IR spectrum of the polymeric salt obtained by matrix polymerization of 4-vinylpyridine in the presence of polymethacrylic acid is almost identical to that of the mixture of poly(4-vinylpyridine) and poly(methacrylic acid). A ladder form has been proposed for their structures. A similar report is based on the study of solid films obtained by poly(acrylic acid) and linear polyethyleneimine.[16]

However, no evidence has been presented to assign a ladder form for their structures. Even though two polymer chains are partially paired to form ionic ladders, the possibility of perfect ladder form has not been accepted by many authors.

LADDER POLYMERS IN INTERAMIDATION

The directional regularity in poly(acrylic acid) (PAA)-polyvinylamine (PVA) complex salt has been investigated.[17] For this purpose PAA-PVA complex salt was prepared by template polymerization of acrylic acid monomer onto PVA matrix.

If all amino and carboxyl groups in the polysalt are directed toward each other, its thermal dehydration must result in formation of a ladder structure that is covalently interlinked.

The resulting interamidation product is soluble in 2-methoxyethanol-water (1:1). This means that the interamidation product must be in a double-stranded ladder form.

INTERAMIDATION OF POLY(ACRYLIC ACID) WITH PIPERAZINE

When an aqueous solution of PAA (MW 220.000) interacts with equimolar amounts of piperazine, a white precipitate forms at the beginning. While stirring continuously, the precipitate disappears and a clear solution is obtained.

A quantitative amidation can be ascribed to the completely oriented polysalt structure. Quantitative amidation and solubility of the resulting interamidation product in ethylene glycol-water (1:1) indicate that the amidation product should be in ladder form.

REFERENCES

1. Marvel, C. S.; Levesque, C. L. *J. Am. Chem. Soc.* **1938**, *87*, 2671.
2. Houtz, R. *Textile Res. J.* **1950**, *20*, 786.
3. Grassie, N.; McNeill, I. C. *J. Chem. Soc.* **1956**, 3929.
4. Sawamura, S.; Sato, H.; Tanaka, Y. *Kobunshi Ronbunsku* **1978**, *35* (2), 95.
5. Jantas, R.; Polowinski, S. *J. Polym. Sci. Pol. Chem. Ed.* **1986**, *24*, 1819.
6. Dawans, F.; Marvel, C. S. *J. Polym. Sci. Part A* **1965**, *3*, 3549.
7. Arnold, F. E.; Van Deusen, R. L. *Macromolecules* **1969**, *3*, 3549.
8. Stille, J. K.; Mainen, E. L. *J. Polym. Sci. Part B* **1966**, *4*, 39.
9. Stille, J. K.; Freeburger, M. E. *J. Polym. Sci. Part B* **1967**, *5*, 989.
10. Stille, J. K.; Mainen, E. L. *Macromolecules* **1968**, *1*, 36.
11. Stille, J. K. *J. Macromol. Sci. Chem.* **1969**, *3* (6), 1043.
12. Stille, J. K.; Mainen, E. L. *J. Polym. Sci. Part B* **1966**, *5*, 665.
13. Kunimura, S.; Ohsaka, T.; Oyama, M. *Macromolecules* **1988**, *21*, 894.
14. Chiba, K.; Ohsaka, T.; Ohnuki, Y.; Oyama, N. *J. Electroanal. Chem. And Interfacial Electrochem* **1987**, *219*, 117.
15. Watson, J. D.; Crick, F. H. C. *Natural* London, **1953**, *171*, 737, 964.
16. Komarov, V. S.; Bogacheva, V. B.; Bezzubov, A. A.; Zezin, A. B. *Vysokomol. Soedin. Ser. B* **1976**, *18*(10), 784.
17. Biçak, N.; Koza, G.; Sarac, A. S.; Afay, T.; Senkal, F. *J. Polym. Sci. Chem. Ed.*, submitted in 1994.

LADDER POLYMERS (Precursor Route, Fully Unsaturated, All-Carbon)

Arnulf-Dieter Schlüter,* M. Löffler, B. Schlicke, and H. Schirmer

Freie Universität Berlin

Institut für Organische Chemie

We have tried to synthesize structurally perfect, fully unsaturated double-stranded (ladder) polymers for some time.[1] Such polymers should have a planar, board-like structure. We applied the precursor concept that has proven very successful in the synthesis of insoluble and yet well-defined polymers.[2–4] We used Diels–Alder (DA) polymers, which are soluble and therefore characterizable, as precursors. The reactions during the polymer analogous transformation of the precursors into the targets were first examined with the help of model reactions. The knowledge obtained from these model studies helped generate the title polymers. The following gives a brief and up-to-date description of these developments focusing on the synthesis of appropriate precursor polymers, the model reactions performed, and the chemically induced conversion of a number of precursors into the target polymers.

During the past six years a variety of DA ladder polymers have been prepared and studied in great detail. Investigations into constitutional, configurational, and conformational aspects, including thorough molecular weight determination, showed these polymers had perfect double-stranded structures. Their backbones consist of linear or angular sequences of six-membered rings.[1,5] Recent studies also paid attention to those ladder polymers with backbones consisting of linear sequences of six-

*Author to whom correspondence should be addressed.

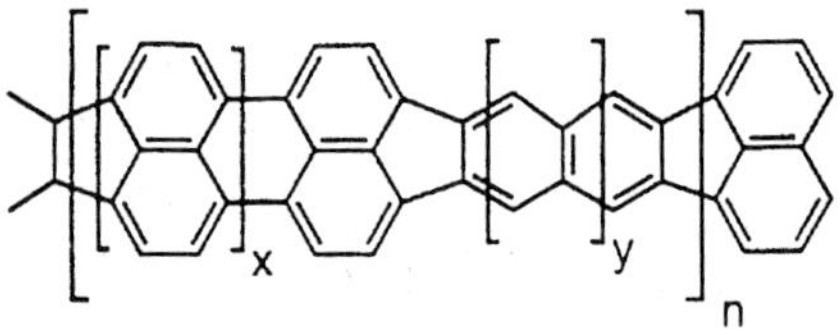

1

a X = 0 Y = 0

b X = 0 Y = 1

c X = 0 Y = 2

d X = 1 Y = 1

and five-membered rings. Because these polymers can transform into the unsaturated targets, they are of major interest. **Structure 1** is a comprehensive representation of the target polymers' structures. They closely resemble the hypothetical open-chain, polymeric analogue of the belt-region of the icosahedric C_{60}. Besides the synthetic challenge, the interest in such polymers originates in their potential usefulness as materials for electroluminescence and photovoltaics applications.

OUTLOOK

Three main goals are easily identified. First, the unsaturated ladder polymers have to be prepared in a form to allow physicists to study their properties. Second, the aromatic boards have to be investigated for their properties. Despite their extended π-conjugation, the polymers are stable toward oxygen and may therefore be of interest for electroluminescence and photovoltaics applications.[6] Because of the relatively mild reaction conditions required for their generation and their double-stranded structure, our polymers are expected to have fewer interruptions of the π-conjugation than the single-stranded poly(phenylene vinylene), presently the material of prime interest for the active elements in the light-emitting diodes.[7,8] Last, there remains the challenge to make unsaturated polymers from all six-membered rings (graphite ribbons in contrast to peeled buckyballs), the goal we originally set out to reach. The success story described in this paper nourishes hopes that such distant goals will be within reach some day.

ACKNOWLEDGMENTS

We are thankful to Prof. H.-H. Limbach, Berlin, and his coworkers for recording several CPMAS ^{13}C NMR spectra for us. Financial support by the Deutsche Forschungsgemeinschaft and the Fonds der Chemischen Industrie is gratefully acknowledged.

REFERENCES

1. Schlüter, A.-D. *Adv. Mater.* **1991**, *3*, 282.
2. Feast, W. J.; Edwards, J. H. *Polymer* **1980**, *21*, 595.
3. Wessling, R. A. *J. Polym. Sci., Polym. Symp. Vol.* **1985**, *72*, 55.
4. Lenz, R. W.; Han, C.-C.; Stenger-Smith, J.; Karasz, F. E. *J. Polym. Sci., Polym. Chem.* **1988**, *26*, 3241.
5. Scherf, Y.; Müllen, K. *Synthesis* **1992**, *23*.
6. Schwoerer, M. *Phys. Bl.* **1994**, *50*, 52.
7. Burroughes, J. H. et al. *Nature* **1990**, *347*, 539.
8. Braun, D.; Heeger, A. *J. Appl. Phys. Lett.* **1991**, *58*, 1982.

LADDER POLYPHENYLSILSESQUIOXANE

Akemi Ueyama* and I. Karino
Materials and Electronic Devices Laboratory
Mitsubishi Electric Corporation

S. Yamamoto and H. Adachi
Central Research Laboratory
Mitsubishi Electric Corporation

Classically, polymers were divided into two groups: linear polymers and randomly crosslinked network polymers. However, Brown reviewed the general characteristics of synthetic procedures that give ordered rather than random structures.[1] He discussed the characteristics of double-chain polymer systems, including the polyphenylsilsesquioxane (PPSQ). PPSQ, a trifunctional silicone polymer, is considered to have a "ladder"-like linear network (**Structure 1**) and it is called "ladder polymer". Andrianov et al., Zhang and Shi also studied synthesis and characterization of PPSQ.[2,3]

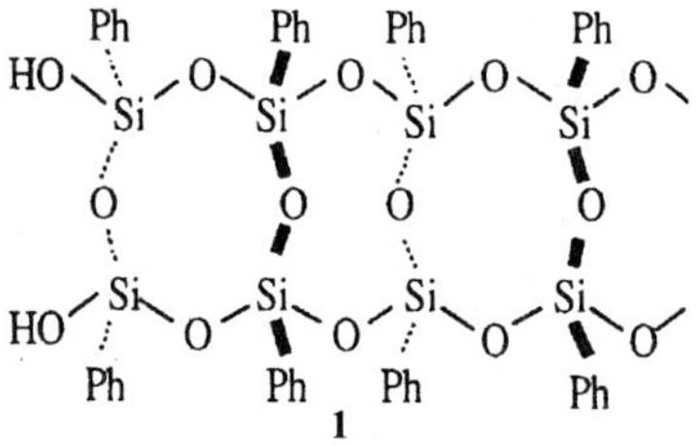

1

Polyorganosilsesquioxanes receive attention because they, especially PPSQ, are highly heat-resistant, a feature of ladder polymers. PPSQ is a high-molecular-weight soluble thermoplastic resin and displays a strong affinity for inorganic films due to the Si-O bonds in its main chain. Therefore, PPSQ is one of the most promising heat resistant material for semiconductor devices. Our group first characterized the thermal and mechanical properties of PPSQ.[4] Then we used PPSQ as a stress buffering film for a mold resin sealed semiconductor device, and as surface protective films for color filters of light-receiving elements and display element.[5,6]

PREPARATION

PPSQ are synthesized from phenyltrichlorosilane, phenyltrialkoxysilane, or *cis*-(1,3,5,7-tetrahydroxy)-tetraphenyl cyclotetrasiloxane.[1–4,7,8,11,12]

PROPERTIES

High molecular weight PPSQ has excellent properties. It can be dissolved in solvents such as benzene, toluene, methoxybenzene, and chloroform. The casted PPSQ film is clear and colorless. It has high heat resistance, low shrinkage during cure, low stress and high adhesiveness.

Thermogravimetric analysis shows that the thermal decomposition temperature and the 5% weight loss temperature of the cured PPSQ films are 500°C and 550°C, respectively.

*Author to whom correspondence should be addressed.

APPLICATIONS

Because PPSQ has excellent properties, it can be used as a heat resistant silicone polymer for coating materials, buffer coating and interlayer insulation for electronic devices.

We used it for a surface protective film for a color filter of light-receiving elements and display elements.[5] A silicone ladder resin, which can be cured at low temperature about 200°C, is obtained by substituting a small amount of phenyl group in PPSQ to the alkenyl group such as a vinyl and an allyl group and adding a trace amount of aromatic diazido compound as a catalyst.

We also investigated a mold resin sealed semiconductor device with excellent moisture resistance and high reliability.[6] The stress buffering film in the device is formed of organosilicone ladder polymer having a hydroxyl group at its end. A chemical reaction (dehydration) occurs between the hydroxyl group on the substrate surface and the hydroxyl group of PPSQ. Accordingly, this enhances the adhesive property of a PPSQ film to the underlying substrate.

Nonogaki's group developed a positive photoresist with O_2-RIE resistance, which is a mixture of a commercially available positive photoresist, a sensitizer and PPSQ oligomer, as a top-layer for two-layer resist systems.[9,10]

Furthermore, Hurwitz et al. described that pyrolyzed polysilsesquioxane offers promise for a variety of applications, including matrices, infiltrants and coating in ceramic matrix composites.[11]

REFERENCES

1. Brown, Jr., J. F. *J. Polymer Sci. C* **1963**, *1*, 83.
2. Andrianov, K. A.; Bushin, S. V.; Vitovskaya, M. G.; Yemel'yanov, V. N.; Lavrenko, P. N.; Makarova, N. N.; Muzafarov, A. M.; Nikolayev, V. Y.; Kolbina, G. F.; Shtennikova, I. N.; Tsvetkov, V. N. *Polymer Sci. USSR* **1977**, *19*, 536.
3. Zhang, X.; Shi, L. *Chinese J. Polym. Sci.* **1987**, *5*, 197.
4. Adachi, H.; Adachi, E.; Yamamoto, S.; Kanegae, H. *Mat. Res. Soc. Symp. Proc.* **1991**, *227*, 95.
5. Adachi, H.; Adachi, E.; Yamamoto, S. "*Silsesquioxane films for protection of color filters*", German Patent 4138180 A1, 1992.
6. Adachi, E.; Adachi, H.; Mochizuki, H. "*Semiconductor device sealed with mold resin*", U.S. Patent 5 278 451, 1994.
7. Ueyama, A.; Yamamoto, S.; Adachi, H.; Karino, I. *Proceedings ACS; Polym. Mat. Sci. Engn.* **1992**, *67*, 246.
8. Yamamoto, S.; Ueyama, A.; Karino, I.; Adachi, H. *Proceedings ACS; Polym. Mat. Sci. Engn.* **1994**, *71*.
9. Toriumi, M.; Shiraishi, H.; Ueno, T.; Hayashi, N.; Nonogaki, S.; Sato, F.; Kadota, K. *J. Electrochem. Soc.* **1987**, *134*, 936.
10. Hayashi, N.; Ueno, T.; Shiraishi, H.; Nishida, T.; Toriumi, M.; Nonogaki, S. *ACS Symp. Ser.* **1987**, *346*, 211.
11. Hurwitz, F. I.; Heimann, P.; Farmer, S. C.; Hembree Jr., D. M. *J. Mat. Sci.* **1993**, *28*, 6622.
12. Sprung, M. M.; Guenther, F. O. *J. Polymer Sci.* **1958**, *28*, 17.

Laminated Veneer Lumber

See: Wood Composites (High Performance)

Laminates

See: Cellulose-Filled Composites
Epoxy Resins (Overview)
Microwave Absorbing Materials
Molecular Composites (Polyimide/Polyimide, Laminate Processability)

LANGMUIR-BLODGETT FILM POLYMERIZATION

Tokuji Miyashita
Institute for Chemical Reaction Science
Tohoku University

Langmuir–Blodgett (LB) films have received much attention recently. These functional ultrathin films have been extensively investigated. Because LB films are built up from condensed monolayers on a water–air interface by multiple depositions onto a solid support, the thickness of the LB film is controlled by the size of the monolayer compound used and the number of layers deposited. Moreover, the LB technique can provide organized molecular assemblies with well-defined molecular orientation and an ordered layer structure that mimics biological membranes. The functionality and property of the organized multilayer assemblies can be designed from a molecular level by selecting the corresponding monolayer and building up different monolayers. Therefore, LB technique is expected to be a potential candidate for molecular-architecture technology related to the concept of "Molecular Devices."

Most conventional LB multilayers are prepared from compounds with low molecular weight, such as long alkyl chain fatty acids. It has been established that these LB films have poor thermal and mechanical stabilities or poor resistance to dissolution by organic solvents, which are major obstacles to practical applications. The preparation of polymer LB films has been employed to improve this poor stability. The polymerization in Langmuir–Blodgett Films is expected to produce a new class of functional polymer films. This article describes some examples of Langmuir–Blodgett film polymerization.

SCHEME FOR PREPARATION OF POLYMER LB FILMS

As shown in **Scheme I** there are three methods in the preparation of polymer LB films. Langmuir–Blodgett film polymerization is represented by the process shown in Scheme I(A).

POLYMERIZATION OF LANGMUIR–BLODGETT FILM

The polymerization of LB multilayers has been studied mainly with amphiphiles containing double or triple bonds (unsaturated monomers), although a few polycondensations in LB multilayers or monolayers were carried out using long-chain esters of amino acids and octadecyl urea.[1–5]

Vinyl Amphiphiles

The polymerization of vinyl stearate, octadecyl acrylate, octadecyl methacrylate, α-octadecyl acrylamide, and octadecyl fumaric and maleic esters in the LB multilayers were studied.[6–15] These earlier works focused on confirming the polymerization and characterization of the polymers by spectroscopic methods.

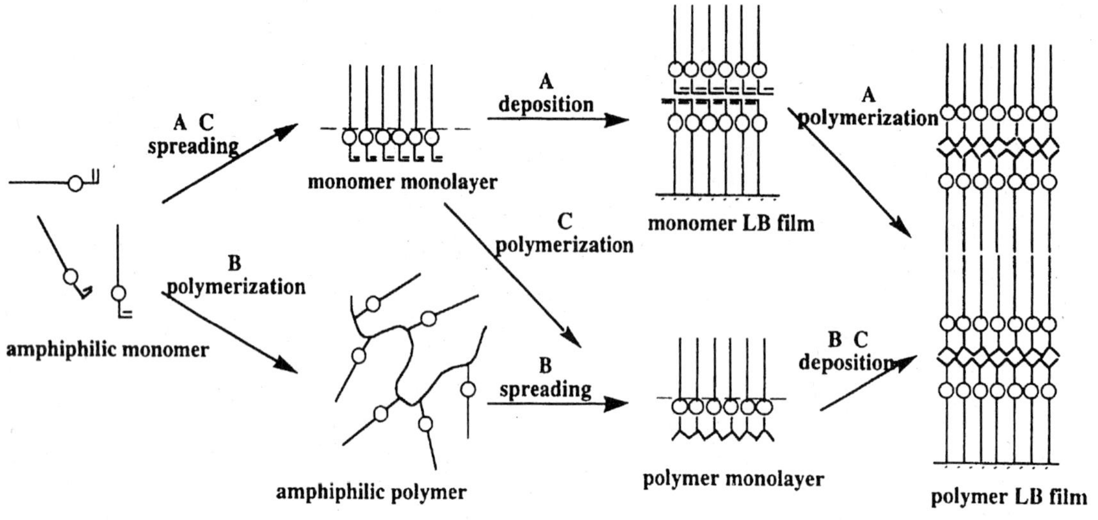

SCHEME I. Preparation of polymer Langmuir–Blodgett films.

Functionality was not considered. We have investigated the polymerization of *N*-alkyl-acrylamide LB multilayers systematically.[16–23]

Dienic Amphiphiles

There is some research available on the polymerization of LB films with diene derivatives.[24–27] The LB films of diene derivatives yield disorder structures and defects when polymerized. No research is available regarding their functionality.

Diacetylenic Amphiphiles

A variety of diacetylenic amphiphiles have been systematically synthesized and studied. The polymerization of diacetylenes were first studied in the solid state by Wagner.[28,29]

The diacetylene structure, which yields close packing and a high rate of polymerization, has been incorporated not only into LB-forming compounds, but also into biomembrane mimetic amphiphiles forming bilayers or liposomes.[31,32] Ringsdorf et al. have prepared various polymerizable lipids containing a diacetylene unit and have investigated the spreading behavior and polymerization on water.[30,31]

The diacetylene polymer LB films were used as insulator thin film for electronics devices such as metal/insulator/metal (MIM) or metal/insulator/semiconductor (MIS) structures.[33] Larkins, Jr. et al. reported that the metal-LB film-semiconductor field-effect transistor (MLSFET) devices would be a useful replacement for a similar Si-SiO_2 metal-oxide-semiconductor field-effect-transistor (MOSFET).[34] Because the diacetylene polymers have a one-dimensional conjugated π-electron system, their use in nonlinear optics has been proposed, especially for third-harmonic generation (THG).[35–37]

Diacetylene polymer LB films have two colored phases (a red and a blue phase) and are applicable to photomemory and optical devices.[38,39]

Other Research

Electrochemical polymerization of LB films of pyrrole and aniline derivatives, which are expected to have high conductivity, have been carried out. The resulting polymer LB films exhibited anisotropic conductivity because of the alternating layered structure of conducting polymer layers and insulating alkyl chain layers.[40]

REFERENCES

1. Fukuda, K.; Shibasaki, Y.; Nakahara, H. *J. Macromol. Sci. Chem.* **1981,** *A15,* 999.
2. Fukuda, K.; Shibasaki, Y.; Nakahara, H. *Thin Solid Films* **1989,** *179,* 103.
3. Folda, T.; Gros, L.; Ringsdorf, H. *Makromol. Chem. Rapid Commun.* **1982,** *3,* 167.
4. Miyasaka, T.; Nishikawa, N.; Orikasa, A.; Ono, M. *Chem. Lett.* **1991,** 969.
5. Rossilio, C.; Ruaudel-Teixier, A. *J. Polym. Sci., Polym. Chem. Ed.* **1975,** *13,* 2459.
6. Cemel, A.; Fort, T.; Lando, J. B. *J. Polym. Sci. Polym. Chem. Ed.* **1972,** *10,* 2061.
7. Puterman, M.; Fort, T.; Lando, J. B. *J. Colloid. Interface Sci.* **1974,** *47,* 705.
8. Fukuda, K.; Shibasaki, Y. *Thin Solid Films* **1980,** *68,* 55.
9. Fukuda, K.; Shibasaki, Y.; Nakahara, H. *Thin Solid Films* **1983,** *99,* 87.
10. Hatada, M.; Nishi, M. *J. Polym. Sci. Polym. Chem. Ed.* **1977,** *15,* 927.
11. Banerjie, A.; Lando, J. B. *Thin Solid Films* **1980,** *68,* 67.
12. Banerjie, A.; Lando, J. B. *Org. Coat. Plast. Chem.* **1978,** *38,* 634.
13. Naegele, D.; Lando, J. B.; Ringsdorf, H. *Macromolecules* **1977,** *10,* 1339.
14. Rabe, J. P.; Rabolt, J. F.; Swalen, J. D. *J. Chem. Phys.* **1986,** *84,* 4096.
15. Laschewsky, A.; Ringsdorf, H.; Schmidt, G. *Polymer* **1988,** *29,* 448.
16. Miyashita, T.; Yoshida, H.; Murakata, T.; Matsuda, M. *Polymer* **1987,** *28,* 311.
17. Miyashita, T.; Yoshida, H.; Matsuda, M. *Thin Solid Films* **1987,** *155,* L11.
18. Miyashita, T.; Matsuda, M. *Thin Solid Films* **1989,** *168,* L47.
19. Miyashita, T.; Sakaguchi, K.; Matsuda, M. *Polymer* **1990,** *31,* 461.

20. Miyashita, T.; Yoshida, H.; Itoh, H.; Matsuda, M. *Nippon Kagaku Kaishi* **1987**, 2169.
21. Miyashita, T.; Sakaguchi, K.; Matsuda, M. *Langmuir* **1992**, *8*, 336.
22. Miyashita, T.; Suwa, T. *Langmuir* **1994**, *10*, 3387.
23. Miyashita, T.; Itoh, Y. *Thin Solid Films* **1995**, *260*, 217.
24. Fukuda, K.; Shibasaki, Y.; Nakahara, H. *Thin Solid Films* **1985**, *133*, 39.
25. Schupp, H.; Hupfer, B.; Van Wagenen, R. A.; Andrade, J. D.; Ringsdorf, H. *J. Colloid. Polym. Sci.* **1982**, *260*, 262.
26. Laschewsky, A.; Ringsdorf, H.; Schmidt, G. *Thin Solid Films* **1985**, *134*, 153.
27. Laschewsky, A.; Ringsdorf, H. *Macromolecules* **1988**, *21*, 1936.
28. Wegner, G. *Z. Naturforsch.* Teil B **1969**, *24*, 824.
29. Wegner, G. *Pure Appl. Chem.* **1977**, *49*, 443.
30. Ringsdorf, H.; Schlarb, B.; Venzmer, J. *Angew. Chem. Int. Ed. Engl.* **1988**, *27*, 113.
31. Bader, H.; Dorn, K.; Hupfer, B.; Ringsdorf, H. *Adv. Polym. Sci.* **1985**, *64*, 1.
32. Lando, J. B.; Hansen, J. E. *Thin Solid Films* **1989**, *180*, 141.
33. Kaneko, F.; Shibata, M.; Inaba, Y.; Kobayashi, S. *Thin Solid Films* **1989**, *179*, 121.
34. Larkins, G. L.; Fung, C. D.; Rickert, S. E. *Thin Solid Films* **1989**, *180*, 217.
35. Kajzar, F.; Messier, J.; Zyss, J.; Ledoux, I. *Opt. Commun.* **1983**, *45*, 13.
36. Kajzar, F.; Messier, J. *Thin Solid Films* **1985**, *132*, 11.
37. Okada, S.; Matsuda, H.; Nakanishi, H.; Kato, M. *Thin Solid Films* **1989**, *179*, 423.
38. Grunfeld, F.; Pitt, C. W. *Thin Solid Films* **1983**, *99*, 249.
39. Kanetake, T.; Tokura, Y.; Koda, T. *Solid State Commun.* **1985**, *56*, 803.
40. Ando, M.; Watanabe, Y.; Iyoda, T.; Honda, K.; Shimidzu, T. *Thin Solid Films* **1989**, *179*, 225.

Langmuir-Blodgett Films

See: *Comb-Like Polymers*
Conducting Polymers (for Langmuir-Blodgett Film Fabrication)
Cyclic Imino Ethers (Ring-Opening Polymerization)
Hairy Rod-Like Polymers
Langmuir-Blodgett Film Polymerization
Langmuir-Blodgett Films (Organic Polymers for Photonics Applications)
Monolayer and Langmuir-Blodgett Films (of Protein)
Nonlinear Optical Materials
Photochromic Poly(a-aminoacid)s
Photoredox Langmuir-Blodgett Films (for Light Energy Conversion)
Polysilanes (Bearing Hydroxy Groups and Derivatives)
Pyroelectricity
Signal Transduction Composites (Biomaterials with Electroactive Polymers)
Supramolecular Mesomorphic Polymers (Assembling in Ultrathin Films)
Ultrathin Films (Self-Assembled)

LANGMUIR-BLODGETT FILMS (Organic Polymers for Photonics Applications)

Paras N. Prasad and W. M. K. P. Wijekoon
Photonics Research Laboratory
Department of Chemistry
The State University of New York at Buffalo

The method of transferring molecular monolayers from the air–water interface onto a substrate bears the names of its inventors and is known as the Langmuir–Blodgett (L–B) technique.[1–3]

L–B films are popular in nonlinear optics (NLO) investigations for two major reasons: First, they provide a means to build well-oriented multilayer films without introducing an inversion center. The absence of an inversion center is a prerequisite for observation of second-order NLO effects. Second, L–B films allow precise control film thickness at the molecular level. In recent years, the L–B films of a variety of amphiphiles, such as organic polymers, organometallics and organic dyes, have enjoyed a resurgence in popularity in NLO research.[2,3] This paper describes the progress of NLO research of polymeric L–B film assemblies and covers only the L–B films of polymeric materials for second- and third-order NLO applications.

NLO PROPERTIES OF L–B FILMS

Although many monomeric amphiphiles have been tested for their NLO activity in the form of L–B films, from the viewpoint of device application a polymeric structure is highly preferable. To meet the stringent thermal, environmental, and chemical stabilities necessary for device uses, many attempts have been made to fabricate L–B films of novel polymeric materials. In this respect, preformed polymers containing various structural features have been investigated by several research groups. Hall et al. synthesized a polymer with a hydrophilic polyether backbone containing 50% of the hemicyanine chromophore as the side groups.[4] This particular polymer orients on the water surface in a manner in which its backbone is on the water surface and the NLO chromophores extend away from the water surface. The L–B layers prepared in the Y-mode exhibited an SHG response quadratically dependent on the film thickness.

It has been found that the $\chi^{(2)}(-2\omega;\omega,\omega)$ of some polymeric L–B films can be permanently modified photochemically irradiating the film with UV radiation.[5] This type of spatially modified NLO response may be useful in the application of guided waves because such a grating along the guiding direction can act as an NLO source for phase-matched SHG.

Optical waveguides capable of guiding blue light with low loss (2–6 dB cm^{-1}) over several centimeters have also been fabricated from fluorocarbon NLO polymers by using the L–B technique.[6]

In these L–B films it is necessary to alternate the NLO polymer layer with an inert polymer layer to achieve the noncentrosymmetric arrangement of the NLO chromophores. Although this procedure is effective, the number density of the NLO-active molecules in such a film is reduced because of the presence of inert layers. Therefore, approaches have been made to replace the passive polymeric layer with another NLO-active polymer layer that has NLO chromophores oriented opposite the other NLO-active polymer. Such polymers allow fabrication of L–B multilayers with the use of the Y-type deposition, without introducing an inversion center in the film.[7] An L–B multilayer deposited from such polymers shows quadratic enhancement of the SHG intensity as a function of the number of bilayers.

Some authors have investigated the L–B films of new azobenzene functionalized dimethylamino (DMA) copolymers for their SHG efficiency activity.[8] Monolayer films of these materials can be deposited onto hydrophilic glass substrates via Z-type deposition. However, multilayer films peel off from the substrate; therefore, these authors deposited only a bilayer film.

Generally, in amorphous polymers a high degree of long-range molecular order is not entropically favorable. Some attempts have been made to reduce chromophore mobility through chemical bonding while retaining the chemical processability. To achieve this goal, some authors have incorporated NLO chromophores into polymer backbones in head-to-head and tail-to-tail (i.e., syndioregic) manners while introducing spacer groups to achieve the necessary amphiphilic properties.[9] These syndioregic polymers form stable monolayers on the water surface that can be transferred onto a substrate in either the Y-mode or the Z-mode. Both the Y-type and the Z-type multilayer L–B films result in noncentrosymmetric structures, which are SHG active. The noncentrosymmetry in Y-type layers is attained possibly through regular chain folding. However, the SHG activity in these films decays over a period of days. The SHG response does not follow the expected quadratic behavior as the film thickness increases. Also, the Y-type films seem to be more noncentrosymmetric than the Z-type films.

Some authors have investigated the orientations of pendent NLO group-attached polyamic acid L–B films by optical SHG.[10] These dye monolayers show SHG activity on the water surface. However, after being transferred onto a substrate, the SHG activity disappears because of reorientation of the molecules on the substrate. Also, the L–B film formation has been reported for polymers in which NLO chromophores are attached to the side chains of polymethacrylate and poly(vinyl alcohol) backbones.[11]

The Langmuir–Schaefer (LS) technique has also been used to prepare multilayers of both NLO and mesogenic moieties containing polysiloxane copolymer.[12] Optical and X-ray data of these LS films clearly show that individual layers are organized in a head-to-tail fashion is characteristic of the X-ray deposition. The NLO groups in an LS film are oriented in a direction almost perpendicular to the film plane, exhibiting a highly polar structure.

The polarization behavior of the SHG signal of L–B films suggest that one end of the molecule extends into the air and the other end is attached to the surface of the substrate.

The fundamental structural requirements necessary for the observation of a third-order NLO process are different from that of a second-order process. The π-electron conjugated polymers are favorable structures for third-order NLO applications.[3] The first third-order NLO investigation in an organic polymeric material was demonstrated in 1976 by Stuteret and co-workers in polydiacetylene solutions.[13] The measured $\chi^{(3)}$ value of this polymer was 8.5×10^{-10} esu. The $\chi^{(3)}$ value in bulk single crystals of this polymer (in the direction parallel to the polymer chains) is approximately 13 times higher than that of GaAs. This exciting result created an intense interest in using conjugated organic polymers for the third-order NLO investigations.

A group of soluble polydiacetylene (poly-*n*-dibutyl 4,19-dioxo-5,18-dioxa-3,20-diaza-10,12-docosadiynedioate [BC-MU]) has also been manipulated in the monolayer film to control the polymer conformation by Biegajski et al.[14,15] These authors demonstrated that a monolayer of poly-3-BCMU (yellow) transforms to a bilayer (blue) and a monolayer of poly-4-BCMU (yellow) becomes a bilayer (red) as the surface pressure of the monolayer is varied. This indicates the change in the conjugation length in these monolayer films.

Optical THG has also been applied to investigate the third-order NLO behavior of centrosymmetric L–B films that were fabricated with polar dye molecules such as stilbazolium and merocyanine.[16] By analyzing the THG Maker fringes, a $\chi^{(3)}(-3\omega,\omega,\omega,\omega)$ value of 1.4×10^{-12} esu at 1.064 μm and a value of 0.9×10^{-12} esu at 1.907 μm has been obtained for these centrosymmetric multilayer L–B films. Because it is possible to control film thickness very accurately with the L–B technique, the authors were able to determine the phase of $\chi^{(3)}(-3\omega,\omega,\omega,\omega)$ in these L–B layers. In other studies, Casstevens et al. and Prasad used multilayer LS films of an axially substituted siliconphthalocyanine for DFWM studies.[17,18] These LS films generated a DFWM signal proportional to the square of the number of layers (or the square of the film thickness), which should be expected for a phase-matched NLO process with no depletion of the interacting beams. It was evident that the nonlinearity of this particular phthalocyanine compound was large enough to obtain a DEFW signal even in a monolayer film. These authors simultaneously monitored the first- and second-order (i.e., $\chi^{(3)}$ and $\chi^{(5)}$) diffractions from the transient gratings and concluded that the dynamics of the excitons determining the nonlinearity was governed mainly by the bimolecular exciton–exciton interaction. The DFWM measurement of the LS layers has resulted in a $\chi^{(3)}(-\omega;\omega,\omega,-\omega)$ value of 2×10^{-9} esu at 602 nm.

REFERENCES

1. Gaines, G. L., Jr. *Insoluble Monolayers at Liquid-Gas Interfaces*; John Wiley & Sons: New York, 1966.
2. *Langmuir-Blodgett Films*; Roberts, G., Ed.; Plenum: New York, 1990.
3. Prasad, P. N.; Williams, D. J. *Introduction to Nonlinear Optical Effects in Molecules and Polymers*; John Wiley & Sons: New York, 1992.
4. Hall, R. C.; Lindsay, G. C.; Anderson, B.; Kowel, S. T.; Higgins, B. G.; Stroeve, P. *Mater. Res. Soc. Proc.* **1988**, *109*, 351.
5. Hsiung, G. *Appl. Phys. Lett.* **1991**, *59*, 2495.
6. Clays, K.; Armstrong, N. J.; Ezenyilimba, M. C.; Penner, T. L. *Chem. Mater.* **1993**, *5*, 1032.
7. Motschmann, H. R.; Penner, T. L.; Armstrong, A. J.; Ezenyilimba, M. C. *J. Phys. Chem.* **1993**, *97*, 3933.
8. Tamada, T.; Yokohoma, S.; Kajikawa, K.; Ishikawa, K.; Takezoe, H.; Fukuda, A.; Kakimoto, M.; Imai, Y. *Thin Solid Films* **1994**, *244*, 754.
9. Hoover, J. M.; Henry, R. A.; Lindsay, G. A.; Nadler, M. P.; Nee, S. F.; Seltizer, M. D.; Stenger-Smith, J. D. *ACS Symposium Series* **1992**, *493*, 94.
10. Verbiest, T.; Persoon, A.; Samyn, C. *Proc. SPIE* **1991**, *1560*, 353.
11. Takahashi, T.; Chen, Y. M.; Rahaman, A. K.; Kumar, J.; Triphaty, S. K. *Thin Solid Films* **1992**, *210/211*, 202.

12. Ou, S. H.; Mann, J. A.; Lando, J. B.; Zhou, L.; Singer, K. D. *Appl. Phys. Lett.* **1992**, *61*, 2284.
13. Satuteret, C.; Herman, J. P.; Frey, R.; Fradere, F.; Ducuing, J.; Daughman, R. H.; Chance, R. R. *Phys. Rev. Lett.* **1976**, *36*, 956.
14. Biegajski, J.; Burzynski, R.; Cadenhead, D. A.; Prasad, P. N. *Macromolecules* **1986**, *19*, 2457.
15. Biegajski, J.; Cadenhead, D. A.; Prasad, P. N. *Langmuir* **1988**, *4*, 689.
16. Kajzar, F.; Girling, I. R.; Peterson, I. R. *Thin Solid Films* **1988**, *160*, 209.
17. Casstevens, M. K.; Samoc, M.; Pfleger, J.; Prasad, P. N. *J. Chem. Phys.* **1990**, *92*, 2019.
18. Prasad, P. N. *Mat. Res. Symp. Proc.* **1992**, *255*, 247.

Langmuir-Blodgett-Kuhn Assemblies

See: *Photoreactive Langmuir-Blodgett-Kuhn Assemblies (Functionalized Liquid Crystalline Side Chain Polymers)*

Laser-Induced Polymerization

See: *Photoinduced Polymerization*

Latent Monomer

See: *Poly(terephthalic acid anhydride) (as a Latent Monomer)*

Latex

See: *Amphoteric Latex*
Emulsion Polymerization
Emulsion Polymers (Noncatalytic Diimide Hydrogenation)
Microemulsion Polymerization (Overview)
Microemulsion Polymerization (Functionalization of Microlatices)
Microemulsion Polymerization (Oil-in-Water)
Natural Rubber (Chemical Modification)
Natural Rubber (Hevea Brasiliensis)
Natural Rubber Latex (Polymeric Flocculant for Tin Tailings Slurries)
Peroxide Initiators (for Water-Based Acrylate Emulsion Polymerizations)
Vinyl Chloride (Minisuspension Polymerization)

Lewis Acid Catalysts

See: *Supported Catalysts (Lewis Acid and Ziegler-Natta)*
Supported Lewis Acid Catalysts (on Polypropylene; Recoverable and Reusable)

Ligament Replacement

See: *Ligament Replacement, Artificial*
Ligament Replacement Polymers (Commercial Products)
Ligament Replacement Polymers (Technological and Design Aspects)

LIGAMENT REPLACEMENT, ARTIFICIAL

J. V. Wening,* A. Katzer, and K. H. Jungbluth
Department of Trauma and Reconstructive Surgery
University Hospital Hamburg-Eppendorf (UKE)

S. Schulz
Institute of Organic Chemistry
University of Hamburg

M. Dauner and H. Planck
Institut für Textil und Verfahrenstechnik
Denkendorf Forschungsbereich Blomedizintechnik

First repairs of primary cruciate ligaments were described in 1895 by Mayo Robson and 1898 by Battle who tried to suture the injured structures using silk or catgut for refixation. The long term clinical results were surprisingly promising and other surgeons were encouraged to follow this example.[1,2]

Progress in anorganic chemistry, surgical knowledge about foreign body reactions in man under physiological conditions, asepsis, sterilization, and immunology led to the first alloplastic prosthesis for anterior cruciate reconstruction after World War II in 1949; Supramid (Rüther) was a thermoplastic polyamide synthesized from hexamethylenediamine and aditine acid.[3]

Bucheron and Pessereau reported in 1956 about their experiences with Nylon ligamentous implants. Contact of synovial fluid led to degeneration of these polyamides, and infections with fistula and rejection of the implant made explanation inevitable. Furthermore, publications by Oppenheimer and Nothdurft in 1958 about carcinogenic side effects and foreign body sarcomas after implantation of polymers induced reluctant acception of these new materials.[4–6]

Nevertheless, further investigation as well as chemical modification and invention of other polymers introduced more and more new products into the medical market.

SYNTHESIS AND PROPERTIES

Polymers, still available for cruciate ligament replacement or ligament augmentation, are mainly polyesters, such as poly(ethylene terephthalate) (PETP), polyester composites, polypropylene, polyglactine 910, and polyparadioxanones, and for experimental items poly(phenylene terephthalamide) (PPPT) as well as poly(ether ether ketone) (PEEK) and poly(L-lactide) (PLLA).

Poly(ethylene terephthalate) (PETP) (CAS 25038-59-9)

Despite its good resistance against many chemicals, PETP is somewhat degraded when used as an implant. For the production of strands, fibers are formed from the melt oriented along their axis by a stretching process, which yields a highly firm strand.[9,11]

Carbon Fiber (CAS 7440-44-0) Composites

A polyester carbon fiber composite is intended to combine complementary material qualities. Adversely to obviously delete effects after intra-articular implantation, biocompatibility

*Author to whom correspondence should be addressed.

of carbon fibers is claimed to be good and they are used to reinforce other polymers.[8,11,12]

Polypropylene (CAS 9003-07-0)

Only the isotactic form is suitable for fiber production. After long-term implantation into the human body the elasticity modulus of polypropylene has shown to be altered.[14]

Polyglactine 910 (CAS 26780-50-7)

Polyglactine 910 (VICRYL), a biodegradable polyester, is a copolymer, consisting of 90% glycolic acid and 10% lactic acid. The material undergoes slow degradation in organisms, by hydrolysis rather than by enzymatic attack.

Poly(L-lactide)

This biodegradable polyester is similar to polyglactine 910 and consists solely of L-lactide-acid monomers. It can be synthesized according to the procedures described for polyglactine 910 and has similar material properties.

Poly(*p*-dioxanone) (PDS) (CAS 31852-84-3)

Poly(*p*-dioxanone)s are resorbable polyesters synthesized by catalytic polycondensation of 1,4-dioxanone.

In water as well as in the human body degradation of poly(*p*-dioxanone) takes place by hydrolysis of the ester bond.[15,16]

Poly(*p*-phenylene terephthalamide) (PPPT) (CAS 24038-64-5)

Chemically, poly(*p*-phenylene terephthalamide) is extremely inert and resists oxidation. The material is not expected to lose its properties after implantation into the human body despite existence of two theoretical degradation pathways for aromatic polyamides *in vivo*, either by hydrolysis or, in minor quantity, by monooxygenase and, in presence of microorganisms, by dioxygenase. Subsequently degradation by monooxygenase cytochromoxidase P450 may theoretically influence formation of carcinogenic benzoloxyde. PPPT macromolecules are oriented along the fiber axis and linked by hydrogen bonds resulting in extremely high strength and flexibility. The material is available with different elasticity modulus (KEVLAR-29, KEVLAR-49; duPont de Nemours Int. S.A., Geneve).[17–24]

Poly(ether ether ketone) (PEEK) (CAS 57947-42-9)

The material is a semi-crystalline thermoplast with a density of 1.30 g/cm^3, which shows a good combination of strength, stiffness, toughness, and environmental resistance. Biocompatibility has to be proved in further *in vitro* and long-term *in vivo* investigations.[25–27]

APPLICATIONS

Ligament prostheses can be characterized by their application as a total replacement of a certain anatomical structure (e.g., anterior or posterior cruciate ligament), by the proposed implantation technique and by the intended, unlimited time of function. Therefore, biocompatibility as well as permanent and lasting mechanical resistance against stress, strain, creep, fatigue, and lack of degeneration after contact with synovial fluid make basic claims on polymer implants.

While in the 1970s and early 80s many prostheses made from polymers were implanted for total ligament replacement (TLR), the poor overall performance of most implants led to the use of autologeous tendons, which were for certain applications intended to be supported by a ligament augmentation device (LAD).

From the surgical point of view, augmentation devices are intended for ligamentous reconstruction using transosseous sutures and autologeous ligament support or replacement, respectively, to allow early continuous passive or even active motion including weight bearing after operation. The polymeric implants are designed as long-term support for natural ligaments and are, guided parallel to–in case of ligament reconstruction–or through the center of the transplant–in case of autologeous transplantation.

In contrast to ligament prostheses and augmentation devices support materials (SM) in ligament surgery are resorbable and degradable without residue in a more or less well defined period. At present, only polydioxanone (PDS) cords and bands meet this demand.

The classification criteria mentioned above overlap sometimes, because some implants are used for total ligament replacement as well as for ligament augmentation, others are implanted isometrically and over-the-top, respectively, and even long-term biostable materials for total ligament replacement are degraded by organisms.

FUTURE ASPECTS FOR POLYMERS IN LIGAMENT SURGERY

Major problems in artificial ligament replacement consisting of non-resorbable polymers are related with mechanical properties (stiffness, elasticity, creep, fatigue, strength), biostability (degradation behavior in organism), tissue reaction, as well as intraosseous fixation.

Almost all prostheses made from polymers are designed as a linear one-bundle cord of different cross-sections consisting of various amounts of single fibers woven, knitted, or manufactured in a way that allows soft tissue ingrowth and surface coverage by soft tissue. In contrast to nature, where the large number of fiber bundles of the anterior cruciate ligament are arranged in three parts (anteromedical, posterolateral, and intermedial bundle), which guarantees low friction and low tension during all ranges of motion, industrial products generally imitate only the macroscopic aspect of the anterior cruciate ligament structure, forming a simple cord inserted through drill holes punctiform.[28,30]

By and large all materials used as implants are called biomaterials despite that they do not possess any regeneration capacity. Therefore, they are subject to changes in their mechanical behavior as a result of continuous or changing long-term stress after implantation. In the knee joint, load of the cruciate ligaments predominately takes place in redirection areas, e.g., the edges of drill channels in the bone, anchorage points, and during rapid and uncontrolled movements and extreme flexion and extension. Even physiological guidance of artificial ligaments in open knee surgery or controlled by arthroscopy with notch-plasty and so-called biological fixation devices (bone plugs) could not overcome these problems. In contrast with biomechanical data obtained by *in vitro* testing, long term performance of all polymers approved for medical application leads to elongation and creep with reappearance of instability combined with clinical signs of inflammatory effusion and painful symptomatic giving-way of the knee joints after anterior cruciate replacement using polymers.[7,21,22,28,29] Cellular and humeral induced enzymatic and hydrolytic reactions may result in lymphytic accumulation of debris as one of the first steps to mechanical failure and prosthesis demarcation.

From experimental results regarding poly(p-phenylene terephthalamide) (PPPT) as an example we were able to work out general basic biochemical demands for biomaterials to be used in ligament surgery.

Collagen-linked surface reaction allowing soft tissue ingrowth without interface development could improve biocompatibility of polymers. Thus, we would consider enzyme-linked modification by either introduction of a carboxy group or a glucose spacer, thus allowing covalent linkage to human collagen, a decisive progress in development of an "ideal" ligament polymer.[22,24,29]

Furthermore, direct radiologic imaging of polymers currently used in ligament surgery is neither possible by conventional X-ray examinations nor by computer tomography (CT), and magnetic resonance imaging (MRI), respectively. Contrast opacification with surrounding tissue would be essential to prove the prosthesis' integrity and continuity in clinical follow-ups or generally in case of knee joint instability complaints after alloplastic replacement of the anterior cruciate ligament.[22,31,32]

REFERENCES

1. Battle, H. W. A case of open section of the knee joint for irreducible traumatic dislocation. *Clin. Soc. London Trans.* **1990**, *33*, 232–233.
2. Robson, A. W. Ruptured crucial ligaments and their repair by operation. *Ann. Surg.* **1902**, *37*, 716–718.
3. Rüther, H. Die Verwendung von Supramid in der Orthopädie. *Z. Orthop.* **1949**, *78*, 151–163.
4. Boucheron, A.; Pessereau, R. Sur le traitment des laxites du genou. *Rev. Orthop.* **1956**, *42*, 564–566.
5. Nothdurft, H. Tumorerzeugung durch Fremdkörperimplantation. *Abh. Dtsch. Acad. Wiss.* **1960**, *3*, 204–213.
6. Oppenheimer, B. S.; Oppenheimer, E. T.; Stout, A. P. The latent period in carcinogenesis by plastics in rats and its relation to the presarcomatous stage. *Cancer II* **1958**, 204–314.
7. Munzinger, U. Heutiger Stand des künstlichen Ersatzes. In: Jakob, R. B.; Stäaube, H. E., Eds.; Knlegelenk und Kreuzbänder. Springer-Verlag: Germany, 1990.
8. Rushton, N.; Dandy, D. J.; Naylor, C. P. E. The clinical, arthroscopic and histological findings after replacement of the anterior cruciate ligament with carbon fibre. *J. Bone Jt. Surg.* **1983**, *65-B*, 308–309.
9. Frazier, C. H. Tendon repairs with Dacron vascular graft suture: a follow up report. *Orthopaedics* **1981**, *4*, 539–540.
10. Park, J. P.; Grana, W. A.; Clitwood, J. S. A high-strength Dacron augmentation for cruciate ligament reconstruction: a two-year canine study. *Clin. Orthop.* **1985**, *196*, 175–185.
11. Turner, R. J.; Hoffmann, H. L.; Weinberg, S. L. Knitted Dacron double velour grafts. In: Stanley, J. C., Ed.; Biologic and synthetic vascular protheses: Grune & Stratton: New York, 509–522.
12. Helbing, G.; Wolter, D.; Neugebauer, R.; Odewey, J. The reaction of tissue to friction particles of LT I carbon and carbon fibre reinforced prothesis in the knee joint of the rat. Trans. 3rd Ann. Meeting Soc. Biomat., New Orleans, 1:26, 1977.
13. Wolter, D. Biocompatibility of carbon fibre and carbon microparticles. In: Burri/Claes, Ed.; Akt. Probl. Chir. Orthop. 26: Alloplastic knee ligament replacement: 28–36.
14. Schwertassek, A.; Dvorak, J. *Faserforschung und Textiltechnik* **1972**, *23*, 66.
15. Dociou, N.; Hein, P. PDS ein neues chirurgisches Nahtmaterial. In: Ethicon OP Forum 4: 108 1981.
16. Ray, J. A.; Doddi, N.; Regula, D.; Williams, J. A.; Melveger, A. Polydioxanone (PDS), a novel monofilament synthetic absorbable suture. *Surg. Gynaec. Obstet.* **1981**, *153*, 497–507.
17. Allen, S. R. Deformation behaviour of Kevlar aramid fibres. *Polymer* **1989**, *30*, 996–1003.
18. Brinkmann, O. A.; Müller, K. M. What's new in introperitoneal test of Kevlar (Asbestos substitute) *Path. Res. Pract.* **1989**, *185*, 412–417.
19. Dunnigan, J.; Nadeau, D.; Paradis, D. Cytotoxic effects of aramid fibres on rat pulmonary macrophages: Comparison with chrysotile Asbestos. *Toxicol. Lett.* **1984**, *20*, 277–282.
20. Katzer, A. Vorderer Kreuzbandersatz durch eine Kevlar-Kunststoffprothese. Eine tierexperimentelle Untersuchung am Schaf. Inaug-Dissertation, Universität Hamburg, Germany, 1992.
21. Wening, J. V.; Hahn, M.; Vogel, M.; Katzer, A.; Jungbluth, K. H. Ossäre Reaktion im Bohrkanal belasteter Bandprothesen am Beispiel Aramid. *Hefte zu der Unfallchirurg* **1993**, *230*, 1226–1230.
22. Wening, J. V.; Katzer, A.; Lorke, D. E.; Jungbluth, K. H. Kevlarfaser–eine Alternative fü den alloplastischen, vorderen Kreuzbandersatz? In: Rahmanzadeh, R.; Scheller, E. E., Eds.: Alloplastische Verfahren und mikrochirurgische Massnahmen: 520–524, 1994; Einhorn-Presse-Verlag: Reinbek, Germany.
23. Wening, J. V.; Lorke, D. E. A scanning electron microscopic (SEM) investigation of aramid (Kevlar) fibers after incubation in plasma. *Clinical Materials* **1992**, *9*, 1–5.
24. Wening, J. V.; Marquardt, H.; Katzer, A.; Jungbluth, K. H.; Marquardt, H. Cytotoxicity and mutagenicity of Kevlar: an in vitro evaluation. *Biomaterials* **1995**, *16*(4), 337–340.
25. Heidel, P. Fasem aus Polyetherketon. Chemiefasem/Textilindustrie 38, 90 Jahrgang, 1988.
26. Jones, D. P.; Leach, D. C.; Moore, D. R. Mechanical properties of poly(ether-ether-ketone) for engineering applications. *Polymer* **1985**, *26*, 1385–1393.
27. Williams, D. F.; McNamara, A.; Turner, R. M. Potential of polyetheretherketone (PEEK) and carbon-fibre-reinforced PEEK in medical applications. *J. Materials Sci. Letters* **1987**, *6*, 188–190.
28. Wening, J. V.; Katzer, A.; Tesch, C.; Jungbluth, K. H. Kunstbandfixation im Knochen durch biomorphologisches Prothesendesign–eine experimentelle Untersuchung, Poster: Chirurgen-Kongress München, April 1993, Germany.
29. Wening, J. V.; Tesch, C.; Katzer, A.; Francke, W.; Jungbluth, K. H. Polymers in artificial ligament replacement–How does the ideal material look? Poster presentation: The Polymer Conference, University of Cambridge, U.K., July 1993.
30. Grood, E. S.; Noyes, F. R. Cruciate ligament prothesis: Strength, creep and fatigue properties. *J. Bone Jt. Surg.* **1976**, *58-A*(No. 8), 1083–1088.
31. Katzer, A.; Wening, J. V., unpublished data, 1995.
32. Wening, J. V.; Katzer, A.; Nicolas, V.; Hahn, M.; Jungbluth, K. H. Zur bildgebenden Darstellung alloplastischen Bandmaterials. Eine in-vivo und in-vitro Untersuchung am Beispiel Kevlar. *Unfallchirurgie* **1994**, *20*(2), 61–65.

LIGAMENT REPLACEMENT POLYMERS (Biocompatability, Technology, and Design)

Martin Dauner* and Heinrich Planck
Institut für Textil und Verfahrenstechnik
Denkendorf Forschungsbereich Blomedizintechnik

Today all ligament prostheses are produced from fibers, that combine the advantage of low diameter, meaning low resulting bending stresses, with the high strength of fibrous materials and uncontested flexibility.

With that view, the prostheses are designed according to their nature. Based on the observations of Prockop, Kastelic, and others, who have shown that down to ranges of 10 to 100 nm the microstructure of the ligaments is formed by fibrils and finally by microfibrils of peptide helices, one can compare that structure with twisted and bundled staple fibers of polymeric origin.[1–3]

CLASSIFICATION

Ligament prostheses can be characterized by their application as a replacement or as a supporting structure, by their proposed way of implantation, and by their intended time of function.

In the following discussion the ligament prostheses will be described by "ligament replacement," and the supporting structures are named "augmentation devices."

Depending on the application, and also on the cross-section of the implant and the philosophy of the inventors, the devices will be implanted "quasi-isometrical" or "over-the-top."

While a total replacement usually needs a nondegradable, long-term biostable biomaterial, for augmentation a degradable and preferable resorbable material can be used. Some authors consider the self-restoring potency of the organism and propose a long-term, degradable material for "guided tissue regeneration."[4,5]

All these possible classifications cannot be applied strictly, because some implants are used for replacement as well as for augmentation, some are implanted quasi-isometrically and over-the-top, and the biostability of some materials is in question.

REQUIREMENTS ON PROSTHESES AND AUGMENTATION DEVICES

Different requirements must be considered for ligament replacements and for augmentation devices with respect to their application, for long-term prostheses, and for degradable augmentations.

Biocompatibility

The general requirements for the biocompatibility of any implanted material must be extended for ligaments to their wear particles, which often are produced at the intraarticular entrance of the bore channels.

Biocompatibility, mild early inflammatory response and no signs of cancerogenicity, and mutagenicity are essentially not only attributed to the implant material and its wear products, but also to the possible degradation products. For totally resorbable materials the observation period can be terminated by the time of total elimination of the material and its degradation products from the body. For other materials the chemical long-term stability must be secured. The aggressive and oxidative potency of the intra-articular synovial fluid must be regarded for evaluation of the biostability; subcutaneous implantation tests don't give conclusive evidence.

Fiber materials are generally finished by a preparation to enhance the manufacturing conditions, to reduce electrostatic effects, and to adhere the single fibers together. The biocompatibility of the respective preparation has to be established as well, or the preparation has to be extracted below the measurable quantity.

Mechanical Properties

It follows that a high water resistance of the implant material is advantageous. With the use of high-strength materials, the amount of implanted allogenic material and consequently the size of the prosthesis can be kept small. Unfortunately high modulus materials mostly have a low transversal strength and, therefore, the theoretical longitudinal strength can be used for the prosthesis design in parts only. High creep resistance and low visco-elastic effects are provisions for replacements. For augmentation devices a well determined creep can be part of the functional design to share the load increasingly with the transplant.

For the prosthesis design, the following two principles may be considered: the implant must stabilize the respective joint enabling the physiological range of motion over the intended time of use, and in any potential case of dysfunction it should not harm the body tissue.

Following these principles and looking at the performance of the natural ligaments in more concrete terms, the mechanical requirements of ligament prostheses are:

- Strength of ligament replacements about 1500 N
- Initial strength of augmentation devices of 500 to 1000 N
- Strength of resorbable augmentations beyond 500 N over the first two months
- Progressive stress–strain curve[6]
- Elongation according to the implantation and the implant size
- Stress relaxation asymptotically to 20–25% over 3 hours according to natural ligaments[7]
- Augmentation devices can relax to a high degree, but very slowly (in the range of months) in order to share the load increasingly with the restored or replaced ligament

For prostheses often an "isoelastic behavior" is claimed. This behavior is coupled strongly to the "isometric" implantation site, to its position over the implantation period, and to the integrity of the prosthesis.[3]

Design Aspects

From a commercial view the number of sizes and shapes for different application sites of the ligament prostheses should be kept low. The most frequent application of a replacement as well as of an augmentation device is the restorage of the anterior

*Author to whom correspondence should be addressed.

cruciate ligament. But often a combined ACL/MCL-rupture or other combined rupture requires versatility of the implant.

With a universally applicable ligament the different sizes can more easily be handled, yet a more sophisticated design regarding fixation and mechanical behavior cannot be considered. A small cross-section of the prosthesis generally facilitates the implantation, enables in isometric position, and enlarges versatility.

MANUFACTURING TECHNOLOGIES

The mechanical requirements of the ligament prostheses and the example of the natural ligaments suggest textile processing for the implants. Accordingly, only few devices were produced from bulk material. The above mentioned POLYFLEX-prosthesis [U.S. Patent 503990-12331] is the only one that was considered for clinical application. Experimental designs with a core of PTFE, silicone, and polyurethane, reinforced with fiber materials such as aramide, carbon, and polyester were tested in animal or in preliminary clinical studies.[8–11] Creep of PTFE and abrasion between fibers and the elastomeric materials disqualified the prototypes for further use.

Similar problems must be expected when covering the fibers with elastomeric materials. Wear particles generally provoke synovitis or other adverse tissue reactions. Therefore, further discussion will concentrate on textile-processing methods.

Of five principal textile processing methods, the latter three are used for ligament prostheses: nonwovens, weft knits, warp knits, woven fabric, and braids. The interfibrillar strength of a nonwoven is very poor, so that it could be used only as a covering structure. The elastic recovery and the form stability of weft knits are poor. More details for design and production of textile, structures are reported by the authors elsewhere.[3,7,12–14]

REFERENCES

1. Prockop, D. J.; Guzman, N. A. *Hosp Pract* **1977**, *12*, 61.
2. Kastelic, J.; Galeski, A.; Baer, E. *Connective Tissue Res* **1978**, *6*, 11.
3. Dauner, M.; Planck, H.; Syre, I.; Dittel, K. K. In: Planck, H.; Dauner, M.; Renardy, M., Eds.; *Medical Textiles for Implantation*; Springer: Berlin Heidelberg, Germany, 1990; Chapter II, 8.
4. Alexander, H.; Weiss, A.; Parsons, J. R. *Bulletin of the Hospital for Joint Diseases Orthopaedic Institute* **1986**, *46*(2), 155.
5. Leandri, J.; Dahhan, P.; Tarragano, O.; Cerol, M.; Rey, P.; Geiger, D. In: de Putter; de Lange, G. L.; deGroot, K.; Lee, A. J. C., Eds.; *Implant Materials in Biofunction*; Elsevier: Amsterdam, The Netherlands, 1988; p 113.
6. Butler, D. L.; Grood, E. S.; Noyes, F. R. *Exercise and Sports Science Review 6*, 1978, p 125 ff.
7. Dauner, M.; Planck, H.; Brüning, H. J. In: Claes, L., Ed. *Die wissenschaftlichen Grundlagen des Bandersatzes*; Springer: Berlin Heidelberg, Germany, 1994; p 25 ff.
8. James, S. L.; Woods, G. W.; Homsy, C. A.; Prewitt, J. M.; Slocum, D. B. *Clinical Orthop. and Related Research* **1979**, *143*, 90.
9. Woods, G. W.; Homsy, C. A.; Prewitt, J. M.; Tullos, H. S. *Am. J. Sports Med* **1979**, *7*(6), 314.
10. Trembley, G. R.; Laurin, C. A.; Drovin, G. *Clinical Orthop. and Related Research* **1980**, *147*, 88.
11. Peterson, C. J.; Donachy, J. H.; Kalenak, A. *JBMRBG* **1985**, *19*, 589.
12. Planck, H. In: Planck, H.; Dauner, M.; Renardy, M., Eds. *Medical Textiles for Implantation*; Springer: Berlin Heidelberg, Germany, 1990; Chapter I.1.
13. Planck, H. In: Planck, H., Ed. *Kunststoffe und Elastomere in der Medizin*; Kohlhammer: Stuttgart, 1993; Chapter 6.1.
14. Dauner, M.; Planck, H. In: Planck, H., Ed. *Kunststoffe und Elastomere in der Medizin*; Kohlhammer: Stuttgart, 1993; Chapter 6.3.1.

LIGAMENT REPLACEMENT POLYMERS (Commercial Products)

Martin Dauner and Heinrich Planck
Institut für Textil und Verfahrenstechnik
Denkendorf Forshungsbereich Biomedizintechnik

Of the many prostheses reported on in the research literature, only a few have been brought to clinical use. Of those, most replacements have been retracted from the market in recent years due to what manufacturers cite as decreasing turnover.

In this chapter, only the prosthesis still available is discussed. An overview on ligament prostheses is given in **Table 1**.

FUTURE ASPECTS IN PROSTHESES DEVELOPMENT

The ligaments and especially the often damaged anterior cruciate ligaments are high strength materials with high stretchability and perfect elasticity. The natural tissue has the outstanding feature of self-maintenance healing capacity. Any synthetic or transplant replacement can only be an imperfect solution. In particular, the twisted two or three bundle ACL cannot be appropriately designed as a prosthesis that is universally implantable. The requirements for a perfect insertion of that prosthesis also seems unreachable.

Improper materials and improper design are the technical reasons for the failure of so many prostheses. During the operation, the deviation from the isometric insertion or the improper insertion of the prosthesis and an unsufficiant notch plasty also contribute to failure.

Total ligament replacements are of limited efficiancy unless these methods are brought successfully to the operation. Prostheses mostly fail due to abrasion or loosening, which is related to some degree to the insertion.

Autologeous transplants have the ability to adapt to the actual situation to some degree. In the first months they need a substantial support which can be provided by augmentation devices.

A totally resorbable augmentation device may protect from late foreign tissue reaction when it is totally resorbed. In the stage of resorption, the large amount of (lactic) acid set free intra-articularly and intra-osseously may provoke strong adversive reactions. A step-by-step degradation using different fiber materials with different degradation and resorption kinetics may bring a more retarded release of acids. At the same time, the load sharing to the transplant will be more steady. Today much research is focused in that direction.

TABLE 1. Overview on Ligament Prostheses

Product	Manufacturer/ official supplier	Appli- cation	Material	Structure	Rupture force (N)	Elonga- tion (%)
ABC	Surgicraft UK	REPL	Polyester and Carbon fibers	Axial fibers and braid	3.300	n.i.
LEEDS KEIO	Neoligaments Ltd. UK	REPL	Poly(ethylene-terephthalate)	Woven net	840 2.000	n.i.
PHP 00 to PHP 74	Cendis Médical France	REPL	High Modulus Polyethylene	Different structures	up to 4.900	diverse
STRYKER	Meadow Medical USA	REPL	Poly(ethylene-terephthalate)	Woven and velour	3.000	17
Trevira-Band	Hoechst/Telos Germany	REPL /LA	Poly(ethylene-terephthalate)	Woven	3.600	17
ABC	Surgicraft UK	LA	Polyester	Woven	1.400	n.i.
Kennedy LAD	3M Company USA	LA	Polypropylene	Flat braid	1.730 (8mm) 1.500 (6mm)	n.i.
PROLAD	Sulzer/Protek Switzerland	LA	Poly(ethylene-terephthalate)	Braid	1.700	n.i.
Tetra-L3	Hoechst/Telos Germany	LA	Poly(ethylene-terephthalate)	Flat braid	1.050	17
VICRYL	Ethicon USA	RES	Polyglactin 910	Woven or braid	700 (woven of 10 mm)	n.i.
PDS	Ethicon USA	RES	Polydioxanon	Woven or braid	500 (braid of diam. 2mm)	n.i.

(REPL = replacement; LA = ligament augmentation; RES = resorbable:) (n.i.: no information given)

Light Emitting Diodes

See: *Electroluminescent Polymers (Overview)*
Polyaniline Network Electrodes (Enhanced Performance of Polymer LEDs)
Poly(arylenevinylene)s (Mechanistic Control of a Soluble Precursor Method)
Poly(p-phenylenevinylene)s (Method of Preparation and Properties)

Light Energy Conversion

See: *Light Harvesting Polymers*
Photoredox Langmuir-Blodgett Films (for Light Energy Conversion)

LIGHT HARVESTING POLYMERS

Marye Anne Fox* and Diana M. Watkins
Department of Chemistry
University of Texas at Austin

Light harvesting polymers (LHPs) may be defined as strategically designed macromolecular assemblies bearing an array of light absorbing moieties and one or more electron transfer relays and traps. The function of such polymers is to facilitate the transfer of light energy across macroscopically observable distances and to facilitate electron transfer between pendant groups. The end result of such long distance energy and electron transfers is formation of long-lived, charge-separated species capable of redox chemistry, optical switching, photoconductivity, or solar energy conversion. The potential uses of these systems in microelectronics and photolithography, for example, make them particularly interesting.

The need to understand the basic processes that constitute natural photosynthesis has stimulated researchers to construct synthetic models that combine the essential components into an organized array along a polymer backbone.[1] Polymers are ideal as a "molecular template" upon which several components with individual specific functions can be assembled in an organized manner to work synergistically toward a single desired function.

Designs for functionalized photoactive polymers generally follow a common strategy. Each assembly contains a single unit, or a series of photoactive functional groups, to absorb light; a series of intervening units to act as an electron transfer relay; and ultimately an electron acceptor to trap the electron.

To effectively absorb and convert light, an LHP must have rigid backbone to which pendant functional groups are attached and on which the relative orientations of pendant donors and acceptors is constrained to an appropriate position to facilitate electron and/or energy transfer.[1] The flexible nature of most organic polymers, however, has made the preparation of such rigid systems a synthetic challenge. Furthermore, much work must still be done to resolve the conflicting tendencies toward rigidity and solubility induced by the introduction of highly functionalized groups. This article describes some recent efforts toward the design, preparation, and characterization of physical

*Author to whom correspondence should be addressed.

properties, and toward defining the general utility of light harvesting polymers.

POLYMERIZATION METHODS

The optimal method for preparing a specific LHP depends on the design of the individual array. Hence, a variety of different techniques can be utilized.[2,3] The high loading of substituents in most LHPs also requires that the method of polymerization be chemically controllable, producing well defined polymers in which individual components can be placed at desired positions along the polymer chain. These methods include living anionic/cationic polymerization, metal-catalyzed polymerization, group transfer polymerization, dendrimer polymerization, biopolymerization, and what we have called a "covalent assembly of polymeric arrays".[1–7] Clearly, this is not an exhaustive list, but it represents the many methods that have been used to prepare LHPs.

Vinyl Polymerizations with Anionic, Cationic, or Radical Initiators

Both anionic and cationic polymerizations have been used to prepare LHPs, as both methods form stereoregular polymers in which the monomer possesses an asymmetric center.[8–15]

Free radical polymerization has been used in many of the early preparations of LHPs.[16–21] LHPs prepared via free radical intermediates exhibit complicated photophysical properties because of these large variations in chain lengths and chromophore/quencher compositions. In light of these complications, living anionic and cationic methods of polymerization have largely replaced radical polymerizations, because they allow for much better control over molecular weight distributions and chain transfer reactions of the polymers.

Metal-Catalyzed Polymerization

Much better properties are attained in LHPs prepared using transition metal-catalyzed polymerizations. *Ring opening metathesis polymerization* (ROMP) and *isocyanide polymerization*, in particular, produce well-defined polymers with unique stereoregular structures and interesting photophysical properties.[22,23]

The highly regular helical structure of functionalized polyisocyanides makes them particularly interesting as LHPs.[1]

Group Transfer Polymerization

Recently group transfer polymerization (GTP) has been applied to the preparation of LHPs, since the step-wise progress of this technique provides narrow molecular weight distributions and several derivatizable centers for the attachment of redox or photoactive groups.[24]

Biopolymers

Biologically derived polymers, e.g., polypeptides[25] or polymerizable lipids,[26] are also useful backbones for LPHs.[25,26] Polypeptides in particular can exist as well-defined helices that provide extremely important assistance in establishing distance and orientation between pendant donors and acceptor groups.

The use of polymerizable lipids in the preparation of LHPs is a unique biological approach.[26] O'Brien and co-workers have used lipid bilayers to pre-organize a membrane-bound acceptor (cyanine dye) and donor (porphyrin) by electrostatic association with charged head-groups providing supramolecular arrays produced by self-assembly on which efficient energy transfer can occur.[26]

Dendrimers

The preparation of LHPs based on dendrimer assemblies is currently being investigated.[27]

Summary

While the array of methods seems quite expensive, a pressing need still exists for even more strictly controlled methods than those listed, so that polymers possessing greater functionality and rigidity can be produced routinely.

FUNCTION AND PROPERTIES OF LHPS

Function

The desired function of an LHP is the "capture of a photon by one chromophore on a polymer backbone, followed by transfer of this energy to an intrapolymer energy trap."[28] Following this event, the energy transferred to the trap can be utilized to sensitize other photochemical reactions, to activate optical switches, etc. These functions ensue from the series of photophysical events that occur between adjacent chromophores pendant to, or contained within, the polymer chain.[29]

Properties

Controlling photophysical properties by adjusting polymer conformational rigidity has clearly been the key to building successful light harvesting polymers. The photophysical properties of LHPs derive largely from the attached chromophores and the interactions among these groups following the absorption of a photon. A variety of photophysical properties ensue from these interactions, the two most important being energy transfer (exchange of excitation energy between one chromophore and another) and electron transfer (passage of an electron from a donor to an acceptor).[29]

INTRAMOLECULAR ENERGY TRANSFER AND MIGRATION IN POLYMERS

An ultimately goal of controlled energy transfer in polymer systems involves the transport of excitation energy through a series of chromophores to a center where redox chemistry can take place. This process occurs through a sequence of isoenergetic excitation transfers between chromophores either pendant to, or contained within, the polymer backbone. A variety of possible mechanisms for energy transfer have been discussed.[29]

INTRAMOLECULAR ELECTRON TRANSFER THROUGH A CONJUGATED OR NONCONJUGATED POLYMER BACKBONE

Electron transfer between a donor and acceptor appended to a rigid polymer and can occur via two pathways: through solvent or through the π- or σ-network that separate them.[30] With conjugated polymer backbones, electronic coupling is likely to take place through the π-bond system. Conversely, when the backbone is saturated, electron coupling can take place through the σ-network or through the solvent separating the donor and acceptor. As with energy transfer, vectorial electron transfer requires backbone rigidity to minimize back electron transfer.

Photoconductivity studies by Meier and Albrecht reveal that electron migration is also possible in bridged polymeric phthalocyanato metal complexes $[PcML]_n$.[31]

ENERGY STORAGE IN POLYMERS WITH PHOTORESPONSIVE GROUPS

Another desirable characteristic of LHPs is the potential for storage of light energy. Toward this objective, polymer-bound photoactive groups can store absorbed light as chemical energy by forming new high energy bond or by valence isomerization to a more highly strained isomer.[32] Upon reversion to the original state (breaking the newly formed bonds), perhaps in contact with a catalyst, thermal energy is released.[33–35] The utility of these polymers derives from their applicability to photoswitching and optical data storage. Energy storage has been achieved by Nishikubo and co-workers in a polymer incorporation photoisomerizable norbornene-quadricyclane moieties.[34,35] Irradiation at 311 nm (into the norbornadiene (NBD) groups) appended to poly[2–[[(3–phenyl–2,5–norbornadienyl)–2–carbonyl]oxy]ethyl vinyl ether] results in the formation of the higher energy (but thermally stable) isomer, quadricyclane (QC). Photochemical reversion of the QC isomer to its original NBD state is achieved either thermally, catalytically, or by irradiation at 248 nm, producing the release of ~92 kJ/mol of usable energy.

Photoresponsive polymers bearing photoisomerizable 2,4,6–triisopropylbenzophenone groups have been investigated by Frechét and co-workers as potential vehicles for optical energy storage.[36] Upon irradiation at 313 nm, Norrish Type II photocyclization yields a benzocyclobuentol-functionalized polymer, which stores the incident light in a form from which energy is released upon heating. These light-induced isomerizations induce changes in the optical density and/or the refractive index of the polymer, permitting image formation.

APPLICATIONS

The design and synthesis of multicomponent polymers able to separate charges over defined distances is a field with growing technological importance. Although most initial work in this field was done with the aim of investigating polymer stereochemistry and macromolecular conformation,[36] the unique properties of LHPs have made them prime candidates for many optical applications. In fact, their use in redox chemistry,[37] in optoelectronic devices, and in information storage[32] is bringing molecular science to the rational design of usable materials. Although many scientific questions about their structures remain unanswered, significant progress has been made toward correlating microscopic structure and macroscopic properties in these materials. Accordingly, the ability to effect synthetic changes at the molecular level promises access to new families of interesting polymer materials.

Direct applications of these polymers materials generally fall into three categories: 1) solar energy conversion; 2) photoconduction; and 3) information storage. We have discussed a number of polymers capable of capturing and transferring light energy. These properties are relevant to solar energy conversion in which absorbed light is directed (via energy migration) toward a site where it is converted into useful chemical potential. This potential can be transformed, for example, into redox chemistry[37,38] as is encountered in photosynthesis, where charge separated radical ions are used to reduce CO_2 and oxidize H_2O.[37,38]

In addition, LHPs are useful in photoconductive devices as components that provide extended π- or σ-networks through which electrons can flow (as electrical current) following the absorption of light and the application of a potential.[55] Photoconductive polymeric materials are useful as luminescence detectors in spectrophotometric devices, optical computing applications, or in xerographic films. Photoconduction in polymeric materials is also responsible for converting light into electrical signals, as is required for optical switches and optical waveguides.[40]

Another highly useful application of LHPs is in the field of nonlinear optics (NLO).[41] NLO specifically requires the alignment of photoactive species which induce a net-positive dipole that interact with linearly polarized light. The result is a nonlinear output of light that is directly applicable in new lasers.[42]

LHPs can also be used for energy storage.[43] Storage of large amounts of information can be most readily accomplished using LHPs because multiple light collection units (photoisomerizable groups) can be appended to a single polymer strand. The polymers bearing photoisomerizable groups can store information (or energy) as chemical bonds. That information can easily be released by a variety of different mechanisms (i.e., light, heat, or catalysis). The liberation of this stored energy can readily be applied as a means of optical recording. Optical information storage, effected by laser irradiation of an optically active material, can be used to induce changes in the macroscopic refractive index of a polymeric material.[44] This process finds important application in optical recording.

In short, LHPs are useful in a variety of known applications. Furthermore, advances in polymer research to permit easy synthesis of highly controlled and organized polymers are required if greater flexibility in the fabrication of molecular devices and molecular technology is to be attained. The preparation and study of these highly functionalized photoactive polymers will also lead to a greater understanding of the key photophysical events that occur in macromolecular systems, and ultimately how we can control them.

REFERENCES

1. Fox, M. A.; Jones, W. E.; Watkins, D. M. *Chem. & Engr. News* **1992,** *70*, 38–48.
2. Allcock, H. R. *Contemporary Polymer Chemistry*; Prentice Hall: New Jersey, 1990.
3. Mijs, W. J.; Addink, R. *Recl. Trav. Chim. Pas-Bas* **1991**, *110*, 526–542.
4. Fréchet, J. M. J. *Science* **1994**, *263*, 1710–1720.
5. Caminati, G.; Turro, N. J.; Tomalia, D. A. *J. Am. Chem. Soc.* **1990**, *112*, 8515–8522.
6. Hawker, C. J.; Fréchet, J. M. J. *J. Am. Chem. Soc.* **1992**, *114*, 8405–8413.
7. Arshady, R. *Functional Polymers for Engineering Technologies*, "Light Harvesting Polymers", in press.
8. Lin, J.; Fox, M. A. *Macromol.* **1994**, *27*, 902–907.
9. Sowash, G. G.; Webber, S. E. *Macromol.* **1988**, *21*, 1608–1611.

10. Cao, T.; Munk, P.; Ramireddy, C.; Tuzar, Z.; Webber, S. E. *Macromol.* **1991**, *24*, 6300–6305.
11. Hargreaves, J. S.; Webber, S. E. *Macromol.* **1984**, *17*, 235–240.
12. Webber, S. E. *Chem. Rev.* **1990**, *90*, 1469–1483.
13. Bai, F.; Chang, C.-H.; Webber, S. E. *Macromol.* **1986**, *16*, 2484–2494.
14. Cameron, J. F.; Fréchet, J. M. J. *Macromol.* **1991**, *24*, 1088–1095.
15. Nishikubo, T.; Kameyama, A.; Kishi, K.; Kawashima, T.; Fujiwara, T.; Hijikata, C. *Macromol.* **1992**, *25*, 4469–4475.
16. Baxter, S. M.; Jones, W. E.; Danielson, R.; Worl, L.; Strouse, G. Younathan, J.; Meyer, T. J. *Coord. Chem. Rev.* **1991**, *111*, 47–71.
17. Meyer, T. J. *Acc. Chem. Res.* **1989**, *22*, 164–170.
18. Anufrieva, E. V.; Tcherkasskaya, O. V.; Krakovyak, M. G.; Ananieva, T. D.; Lushchik, V. B.; Nekrassova, T. N. *Macromol.* **1994**, *27*, 2623–2627.
19. Shand, M. A.; Rodgers, M. A. J.; Webber, S. E. *Chem. Phys. Lett.* **1991**, *177*, 11–16.
20. Chatterjee, P. K.; Kamioka, K.; Batteas, J. D.; Webber, S. E. *J. Phys. Chem.* **1991**, *95*, 960–965.
21. Jones, G., II; Rahman, M. *Chem. Phys. Lett.* **1992**, *200*, 241–250.
22. Schrock, R. R. *Accts. Chem. Res.* **1990**, *23*, 158–165.
23. Deming, T. J.; Novak, B. M. *J. Am. Chem. Soc.* **1993**, *115*, 9101–9111.
24. Fox, M. A.; Fox, H. H. *J. Org. Chem.* **1994**, submitted.
25. Meier, M. S.; Fox, M. A.; Miller, J. R. *J. Org. Chem.* **1991**, *56*, 5380–5383.
26. Armitage, B.; Klekotka, P. A.; Oblinger, E.; O'Brien, D. F. *J. Am. Chem. Soc.* **1993**, *115*, 7920–7921.
27. Fox, M. A.; Stewart, G.; Galoppini, E. *J. Am. Chem. Soc.* **1996**, *118*, in press.
28. Webber, S. E. *Chem. Rev.* **1990**, 1469–1482.
29. Guillet, J. *Polymer Photophysics and Photochemistry* Cambridge Univ: New York, 1985.
30. Balzani, V. *Tetrahedron* **1992**, *48*, 10443–10514.
31. Meier, H.; Albrecht, W. *Mol. Cryst. Liq. Cryst.* **1991**, *194*, 75–83.889.
32. Mittal, K. *Polymers in Information Storage Technology*; Plenum: New York, 1988.
33. Kamagawa, H.; Yamada, M. *Macromol.* **1988**, *21*, 918–922.
34. Nishikubo, T.; Kameyama, A.; Kishi, K.; Kawashime, T. *Macromol.* **1992**, *25*, 4469–4475.
35. Nishikubo, T.; Kawashima, T.; Inomata, K.; Kameyama, A. *Macromol.* **1992**, *25*, 2312–2318.
36. Cameron, J. F.; Fréchet, J. M. J. *Macromol.* **1991**, *24*, 1088–1095.
37. Wrighton, M. *Comm. Inorg. Chem.* **1985**, *4*, 269–294.
38. Fox, M. A. *Accts. Chem. Res.* **1992**, *25*, 569–574.
39. Lessard, R. A. *Proc. Int. Soc. Opt. Eng.* **1991**, *1599*, 1–499.
40. Schildkraut, J. S. *Appl. Phys. Lett.* **1991**, *58*, 340–342.
41. Chemala, D. S.; Zyss, J. Eds.; *Nonlinear Optical Properties of Organic Molecules and Crystals*; Academic: Boston, MA, 1987; Vol. 1.
42. Shen, Y. R., Ed.; *The Principles of Nonlinear Optics*; John Wiley & Sons: New York, NY, 1984.
43. Mittal, K. *Polymers in Information Storage Technology*; Plenum: New York, 1988.
44. Kumar, G. S. *Azo Functional Polymers*; Technomic: Hyderabad, India, 1992.

Light Stabilizers

See: *Additives (Agents for Sustaining Properties)*
Antioxidants (Overview)
Hindered Amine Light Stabilizers, Monomeric
Light Stabilizers (Overview)

LIGHT STABILIZERS (Overview)

James P. Galbo
Ciba Additives

Photooxidation causes changes in the physical and mechanical properties of plastic materials. Examples are loss of gloss and cracking of automotive coatings, yellowing of poly(vinyl chloride) and polyurethane, and embrittlement of polypropylene and polyethylene during exposure to the sun or artificial light sources. Light stabilizers are additives that are incorporated into polymers to inhibit the detrimental effects caused by ultraviolet light. The development of light stabilizer technology has been a critical factor in the growth of the plastics industry.

Polymers degrade when sufficient energy is added to cause bonds to rupture. The source of this energy may be heat, mechanical stress, or light. Antioxidants such as hindered phenols and phosphites are used to inhibit degradation that results when polymers are exposed to oxygen at elevated temperatures during manufacture, processing into finished articles, or end-use.[2–4] Light stabilizers must be able to protect polymers during long periods of exposure to sunlight.

An effective light stabilizer additive must be physically and chemically compatible with the polymer it protects, and must not be readily transformed into any species that will promote degradation or alter the physical and mechanical properties or the appearance of the polymer. From an economic standpoint, synthetic feasibility and production cost are important considerations in determining what structures are suitable stabilizers for specific applications.

PHOTODEGRADATION[1,5–6]

The absorption of ultraviolet light by a polymer that contains a chromophore may lead to bond cleavage from an excited state. For example, photolysis of an aromatic polyurethane, polyamide, polycarbonate, or polyester can produce a radical pair that recombines to form an aromatic amine or phenol. This reaction is known as the photo-Fries rearrangement.[1,5,6a,7] The amine or phenol can undergo further chemistry.

LIGHT STABILIZERS

Light stabilizers are conveniently separated into different classes based on the principal mechanism by which they operate.[1,5,8,9] However, a stabilizer molecule may protect a polymer by more than one mechanism.

Ultraviolet Absorbers (UVAs)

UVAs are compounds that have very high extinction coefficients in the ultraviolet portion of the solar spectrum. They function by transforming harmful short-wavelength-high-energy sunlight into less harmful heat energy. In the process, the excited state UVA returns to its ground state chemically unchanged and can subsequently absorb another photon. UVAs are typically used at concentrations of 0.1 to 3 percent. Because total light absorbance is a function of path length. UVAs are less effective in protecting thin films and polymer surfaces. The most widely used UVAs are 2-hydroxybenzophenones and 2-(2′-hydroxyphenyl)benzotriazoles.

Different substituents can be used to change the absorption spectrum and molecular weight of the UVA. For example, 5-sulfur substituted 2-(2′-hydroxyphenyl)benzotriazoles have enhanced absorption at wavelengths over 350 nm.[10,11] Hydroxyphenyltriazines are a recent development in this class of light stabilizers.[12] Less common UVAs are oxanilides and cyanoacrylates.

Energy Quenchers

Energy quenchers accept energy from an excited polymer molecule before any harmful chemical transformations can occur. The energy transfer can occur by a dipole-dipole interaction (Förster mechanism) or by a contact mechanism. The energy quencher must have a more excited state of lower energy than the excited polymer molecule. Once energy is transferred, the quencher returns to the ground state by dissipating the excess energy by fluorescence, phosphorescence, or conversion to heat.

The difference between UV absorbers and energy quenchers is that UVAs protect the polymer by preferentially absorbing incident light, while energy quenchers deactivate excited polymer molecules. In reality, UVAs may also function as energy quenchers, and energy quenchers may also act by absorbing light.[1]

Hydroperoxide Decomposers

Hydroperoxides are one of the most harmful species to polymers. They are formed by oxidation during processing and during later exposure of plastics to heat and light in the presence of air. Because of the weak O–O bond, hydroperoxides are very susceptible to homolysis. The presence of transition metal ions accelerates the decomposition of hydroperoxides. Photolysis of hydroperoxides in polyolefins can lead to the formation of carbonyl compounds and/or alkoxy radicals. Efficient hydroperoxide decomposers include nickel sulfur-containing complexes.[1] These thiosynergists can also function as UV absorbers and energy quenchers.[13] Phosphites, such as Irgafos® 168, react with hydroperoxides and hindered amines react with peroxy acids.[9]

Radical Scavengers

Traditional antioxidants, such as hindered phenols, usually do not inhibit photooxidation effectively enough during long term polymer exposure. In polyolefins, some benzophenone UV absorbers are able to trap radicals, but the UVA is consumed in the process.[9] Hindered amine light stabilizers (HALS or HAS) are very effective polymer stabilizers. They are typically incorporated at levels of 0.1 to 2 percent. The mechanism of polymer stabilization by hindered amines is complex.[14–17]

STABILIZER PERMANENCE

One of the important factors that determine the effectiveness of a light stabilizer is its ability to remain in the polymer substrate. The additive should have sufficient solubility in the polymer to minimize extraction or migration to the surface where it can be easily lost or affect the appearance or properties of the plastic material. One way to increase stabilizer permanence is to use higher molecular weight or polymeric additives. These molecules are less likely to volatilize at elevated temperatures or be extracted. Studies have shown that there is an optimum molecular weight range for stabilizers.[18,19]

Another way to increase stabilizer permanence is to anchor the additive in the polymer by a chemical reaction. Recent examples include vinyl substituted HALS, HALS azo derivatives, HALS substituted by hydroxyl and amino groups and reactable UVA stabilizers.[20–24]

STABILIZER PACKAGES AND SYNERGISM

Different classes of polymer stabilizers are often used in combination to maximize protection of the polymer. For example, the use of UVAs with HALS often provides better photostability than either class used alone.[25,26] Light stabilizers are also used in combination with antioxidants such as hindered phenols and processing stabilizers such as phosphites.[1,27] Different stabilizers can be added separately to a polymer formulation or chemically combined to form a new stabilizer molecule.

ANTAGONISM

Hindered amines may be less effective in specific applications because of chemical reactions with other stabilizers or the substrate. The use of hindered amines with thiosynergists can result in the formation of HALS-sulfonamides which interfere with the free radical scavenging of the HALS and the hydroperoxide decomposing ability of the sulfide.[28]

There are often ways to avoid or minimize these antagonistic interactions. Placing bulky substituents on the *ortho* and *para* positions of hindered phenolic antioxidants reduces the antagonism with HALS. Synergistic effects can be observed.[28] Hydroxybenzyl sulfides interact less strongly with HALS than do hydroxyphenylene sulfides.[28]

CONCLUSIONS

Light stabilizers are used to protect a wide variety of polymers from photodegradation caused by exposure to sunlight. The light stabilizer industry will have to continue to keep pace with the development of new polymers, changes in manufacturing processes, the demand for longer lifetimes for many plastic materials, and stricter environmental regulations in the chemical industry. Better understanding of polymer degradation chemistry and photostabilizing mechanisms, combined with the refinement of accelerated test methods will allow the development of more effective light stabilizer packages.

REFERENCES

1. Gugumus, F. *In Plastics Additives Handbook*, 3rd ed.; Gächter, R.; Müller, H., Eds.; Hanser: Munich, 1990; Chapter 3.
2. Gugumus, F. In *Plastics Additives Handbook*, 3rd ed., Gächter, R.; Müller, H., Eds.; Hanser: Munich, 1990; Chapter 1
3. Pospíšil, J.; Klemchuk, P. O. *Oxidation Inhibition in Organic Materials*; CRC: Boca Raton, FL, 1990; Vols. 1 and 2.
4. Dexter, M. *In Kirk-Othmer Encyclopedia of Chemical Technology* 4th ed.; John Wiley & Sons: New York, 1992; Vol. 3, p 424.
5. Rabek, J. F. *Mechanisms of Photophysical Processes and Photochemical Reactions in Polymers*; John Wiley & Sons: Chichester, 1987.

5a. Kelen, T. *Polymer Degradation*: Van Nostrand Reinhold: New York, 1983.
6. Schnabel, W. *Polymer Degradation*; Macmillan: New York, 1981.
6a. Decker, C.; Moussa, K.; Bendaikha, T. *J. Polym. Sci.: Part A: Polym. Chem.* **1991**, *29*, 739.
7. Hoyle, C. E.; Shah, H.; Nelson, G. L. *J. Polym. Sci.: Part A: Polym. Chem.* **1992**, *30*, 1525.
8. Shlyapintokh, V. Y. *Photochemical Conversion and Stabilization of Polymers*; Hanser: Munich, 1984.
9. Gugumus, F. *Polym. Degrad. Stab.* **1993**, *39*, 117.
10. Winter, R. A. E.; von Ahn, V. H.; Stevenson, T. A.; Holt, M. S.; Ravichandran, R. U.S. Patent 5 278 314, 1994.
11. Winter, R. A. E.; von Ahn, V. H.; Stevenson, T. A.; Holt, M. S.; Ravichandran, R. U.S. Patent 5 280 124, 1994.
12. Stevenson, T. A.; Holt, M. S.; Ravichandran, R. U.S. Patent 5 354 794, 1994.
13. Carlsson, D. J.; Wiles, D. M. *J. Polym. Sci., Polym. Chem. Ed.* **1974**, *12*, 2217.
14. Gugumus, F. *Polym. Degrad. Stab.* **1993**, *40*, 167.
15. Klemchuk, P. P.; Gande, M. E. *Polym. Degrad. Stab.* **1988**, *22*, 241.
16. Gijsman, P.; Hennekens, J.; Tummers, D. *Polym. Degrad. Stab.* **1993**, *39*, 225.
17. Shlyapintokh, V. Y.; Ivanov, V. B. In *Developments in Polymer Stabilization-5*; Scott, G., Ed.; Applied Science: London, 1992; Chapter 3.
18. Friedrich, H.; Jansen, I.; Rühlmann, K. *Polym. Degrad. Stab.* **1993**, *42*, 127.
19. Malík, J. Hrivík, A.; Alexyová, D. *Polym. Degrad. Stab.* **1992**, *35*, 125.
20. Ravichandran, R.; Schirmann, P. J.; Mar, A. U.S. Patent 4 983 737, 1991.
21. Ravichandran, R.; Patel, A. R. U.S. Patent 5 380 828, 1995.
22. Cortolano, F. P.; Seltzer, R.; Patel, A. R. U.S. Patent 5 004 770, 1991.
23. Galbo, J. P.; Ravichandran, R.; Schirmann, P. J. Mar, A. U.S. Patent 5 145 893, 1992.
24. DeBergalis, M.; O'Fee, R. P. U.S. Patent 4 495 325, 1985.
25. Gugumus, F. *Polym. Degrad. Stabil.* **1989**, *24*, 289.
26. Schirmann, P. J.; Dexter, M. In *Handbook of Coatings Additives* Calbo, L. J., Ed.; Marcel Dekker, Inc.: New York, 1987.
27. Berner, G. U.S. Patent 4 426 471, 1984.
28. Lucki, J.; Jian, S. Z.; Rabek, J. F.; Rånby, B. *Polym. Photochem.* **1986**, *7*, 27.

Lignin

See: *Jute*
Lignin-Based Polymers
Lignin Graft Copolymers
Vegetal Biomass (2. Oligomers and their Polymerization)

LIGNIN-BASED POLYMERS

Wolfgang G. Glasser
Department of Wood Science and Forest Products
Virginia Polytechnic Institute

LIGNIN DEFINITION AND STRUCTURE

Lignin is a naturally occurring aromatic polymer that comprises between 20 and 35% of all perennial land-growing plants.[1,2] Lignin serves the tree as sealant, as fiber-bonding matrix and adhesive, and an antioxidant. Its repeat units are variations of 1-(4-hydroxyphenyl)propene-ol-3, which vary in the extent of methoxylation in positions three or five of the aromatic ring, ortho to the phenolic OH group. Lignin's non-uniform and irregular structure is a result of the existence of the phenoxide radical in several isomeric (mesomeric) forms that result from a delocalization of the free electrons.

In addition to differences between lignins in native plant tissue, lignins also vary in relation to their method of isolation. Whereas lignins isolated by means of aqueous alkali typically are high in phenolic OH content and low in ether bonds, lignins isolated under acidic conditions, in the presence of sulfite ions, typically are lower in phenolic OH content and higher in alkyl-aryl ether bonds.[3,4] In addition, isolated lignins often have functional groups and substituents that were introduced during the process of isolation, and that are absent in native lignin. Lignins are commercially isolated from industrial pulp and paper processes as "kraft" lignin (with SH groups) or as lignin sulfonates (with $SO_3^{\ominus}$-groups).

The physical structure of lignin in wood reveals phase separation from carbohydrates at the nano-level.[5,6] Lignin in wood appears to undergo a glass-to-rubber transition between 50 and 100°C and this varies in relation to the moisture content.[7] Isolated lignins have glass-transition temperatures between just under 100 and 200°C, and these transitions are often unpronounced because of high poly-dispersity and secondary bonding.[2]

Molecular weights of isolated lignins typically range from the tetrameric structure to the hundreds of thousands, with weight average molecular weights varying in relation to method of isolation between approximately 3,000 and 20,000.[8]

Isolated lignins are dark brown-colored powders that often resist thermal softening, as well as dissolution in organic solvents. Chemical modification, the preparation of lignin derivatives, has been demonstrated to be a useful technique for improving lignin's handling characteristics.[9] Lignin esters and ethers have been prepared in the laboratory, and they have been shown to possess significantly improve thermal, solubility, and molecular weight characteristics.[10,11] Lignin esters and hydroxy alkyl ethers have been the subject of extensive investigations in both thermoplastic and thermosetting polymer systems.

LIGNIN IN THERMOSETS

Many thermosetting polymer systems have benefitted from the incorporation of particulate lignin in the form of fillers. Lignin-filled thermosets, especially phenolic resins and reinforced rubber formulations, benefit from lignin's glassy (i.e., high modulus) nature, and from its antioxidant properties.[9] The inclusion of lignin in network structures by covalent bonding has been achieved by means of chemical modification. Covalent linking of lignin to phenolic resins has been mediated by phenolation; polyurethanes and polyamines resulted from the crosslinking of hydroxylalkyl lignin derivatives with diisocyanates and melamine, respectively; polyacrylates on lignin basis were formed from acrylated lignin derivatives; and cured epoxy systems resulted from both carboxylated lignins crosslinked with multifunctional oxiranes, and from glycidyl ether-modified lignins crosslinked with diamines or anhydrides of dicarboxylic acids.[13–23]

LIGNIN IN THERMOPLASTICS

Multiphase material systems in which lignin assumes the role of the discontinuous phase have been studied extensively in relation to processibility and phase morphology.[9]

The addition of lignin and thermoplastic lignin derivatives to polyethylene and ethylene-vinyl acetate copolymers has produced evidence for the capacity of lignin to form polyblends in which the phase morphology is dictated by secondary (hydrogen) bonds.[24] Factors qualified to enhance phase compatibility were found to include reduced molecular weight and chemical modification, especially modification of phenolic hydroxy groups.[12,25] Blends of lignin and lignin derivatives were reported with poly(vinyl alcohol), poly(vinyl acetate), polyethylene, poly(ethylene-co-vinyl acetate), poly(methyl methacrylate), poly(caprolactone), polystyrene, poly(vinylchloride), and cellulose ethers and esters.[24-34] Star-shaped copolymers of lignin with caprolactone had remarkable phase compatibility with poly(vinyl chloride).[31] The interaction of lignin and lignin derivatives with cellulose and cellulose derivatives demonstrated partial miscibility and the formation of a well-mixed amorphous phase.[25,32,33]

Transformation of lignin by chemical modification into star-like copolymers with aliphatic ethers, aliphatic esters (caprolactone), or cellulose ester blocks produced low molecular weight, thermoplastic substances on lignin basis with excellent processing characteristics, and with phase behavior ranging from the highly compatible to the highly incompatible.[35]

CONCLUDING REMARKS

Lignin-based polymer systems follow the normal rules governing resin cure behavior (i.e., thermosets) and phase separation and morphology (i.e., thermoplastics). Cure behavior and phase separation characteristics are dictated by parameters related to chemical structure and molecular fractionation.

Although lignin derivatives can be prepared with greatly reduced light absorption characteristics, all lignin containing polymer and materials systems have characteristically brown pigmentation.

REFERENCES

1. Sarkanen, K. V.; Ludwig, C. H., Eds. *Lignins-Occurrence, Formation, Structure and Reactions*; Wiley-Interscience: New York, 1971.
2. Glasser, W. G.; Kelly, S. S. *Lignin, Section in Encyclopedia of Polymer Sci. Eng.*; John Wiley & Sons: New York, 1987; Vol. 8, pp 795–852.
3. Glasser, W. G.; Barnett, C. A.; Sano, Y. *Appl. Polym. Symp.* **1983**, *37*, 441–460.
4. Glasser, W. G.; Barnett, C. A.; Muller, P. C.; Sarkanen, K. V. *J. Agric. Food Chem.* **1983**, *31*(5), 921–930.
5. Atalla, R. H.; Agarwal, U. P. *Science* **1985**, *227*, 636–638.
6. Agarwal, U. P.; Atalla, R. H. *Planta* **1986**, *169*, 325–332.
7. Kelley, S. S.; Rials, T. G.; Glasser, W. G. *J. Materials Sci.* **1987**, *22*, 617–624.
8. Glasser, W. G.; Davé, V.; Frazier, C. E. *J. Wood Chem. Technol.* **1993**, *13*(4), 545–559.
9. Glasser, W. G.; Sarkanen, S., Eds., *Lignin: Properties and Materials*; ACS Symp. Ser. No. 397 American Chemical Society: Washington, DC, 1989.
10. Glasser, W. G.; Jain, R. K. *Holzforschung* **1993**, *47*(3), 225–233.
11. Jain, R. K.; Glasser, W. G. *Holzforschung* **1993**, *47*(4), 325–332.
12. Kelley, S. S.; Ward, T. C.; Rials, T. G.; Glasser, W. G. *J. Appl. Polym. Sci.* **1989**, *37*, 2961–2971.
13. Muller, P. C.; Glasser, W. G. *J. Adhesion* **1984**, *17*(2), 157–173.
14. Muller, P. C.; Kelley, S. S.; Glasser, W. G. *J. Adhesion* **1984**, *17*(3), 185–206.
15. Saraf, V. P.; Glasser, W. G. *J. Appl. Polym. Sci.* **1984**, *29*(5), 1831–1841.
16. Rials, T. G.; Glasser, W. G. *Holzforschung* **1984**, *38*(4), 191–199.
17. Newman, W. H.; Glasser, W. G. *Holzforschung* **1985**, *39*(6), 345–353.
18. Yoshida, H.; Mörck, R.; Kringstad, K. P.; Hatakayama, H. *J. Appl. Polym. Sci.* **1987**, *34*, 1187–1198.
19. Naveau, H. P. *Cell. Chem. Technol.* **1975**, *9*, 71.
20. Glasser, W. G.; Wang, H.-X. *Derivatives of Lignin and Ligninlike Models with Acrylate Functionality* In *Lignin: Properties and Materials*; Glasser, W. G.; Sarkanen, S., Eds.; ACS Symp. Ser. No. 397, Washington, DC, 1989; Chapter 41, pp 515–522.
21. Tomita, B.; Kurozumi, K.; Takemura, A.; Hosoya, S. *Ozonized Lignin-Epoxy Resins: Synthesis and Use*; In *Lignin: Properties and Materials*; Glasser, W. G.; Sarkanen, S., Eds.; ACS Symp. Ser. No. 397, Washington, DC, 1989; Chapter 38, pp 488–495.
22. Hofmann, K.; Glasser, W. G. *J. Wood Chem. Technol.* **1993**, *13*(1):73–95.
23. Hofmann, K.; Glasser, W. G. *Macromol. Chem. Phys.* **1994**, *195*, 65–80.
24. Glasser, W. G.; Knudsen, J. S.; Chang, C.-S. *J. Wood Chem. Technol.* **1988**, *8*(2), 221–234.
25. Rials, T. G.; Glasser, W. G. *Polymer* **1990**, *31*, 1333–1338.
26. Ciemniecki, S. L.; Glasser, W. G. *Polymer* **1988**, *29*, 1030–1036.
27. Ciemniecki, S. L.; Glasser, W. G. *Polymer* **1988**, *29*, 1021–1029.
28. Kelley, S. S.; Ward, T. C.; Glasser, W. G. *J. Appl. Polym. Sci.* **1990**, *41*, 2813–2828.
29. de Oliveira, W.; Glasser, W. G. *Macromolecules* **1994**, *27*, 5–11.
30. de Oliveira, W.; Glasser, W. G. *J. Wood Chem. Technol.* **1994**, *14*(1), 119–126.
31. de Oliveira, W.; Glasser, W. G. *J. Appl. Polym. Sci.* **1994**, *51*, 563–571.
32. Rials, T. G.; Glasser, W. G. *J. Appl. Polym. Sci.* **1989**, *37*, 2399–2415.
33. Rials, T. G.; Glasser, W. G. *Wood and Fiber Science* **1989**, *21*(1), 80–90.
34. Davé, V.; Glasser, W. G.; Wilkes, G. L. *Polymer Bulletin* **1992**, *29*, 565–570.
35. de Oliveira, W.; Glasser, W. G. *Comparison of Some Molecular Characteristics of Star-block Copolymers with Lignin;* In *Cellulosics: Chemical, Biochemical and Materials Aspects*; Kennedy, J. F.; Phillips, G. O.; Williams, P. A., Eds.; Ellis Horwood Ltd.: 1993; Chapter 37, pp 263–272.

LIGNIN GRAFT COPOLYMERS

John J. Meister
Department of Chemistry
New Mexico Institute of Mining and Technology

Lignin [8068-00-6] is a natural product produced by all woody plants. As a commodity forest product, lignin has a long history as a waste product for which functional uses are sought. The most common use of lignin is as a fuel for the bioprocessing operations that produce it. However, as oil and gas are depleted, the grafting of lignin will become an important industrial process for producing the chemicals and materials that society needs. Woody plants synthesize lignin from trans-4-coumaryl alcohol (grasses), trans-coniferyl alcohol (pines), and trans-sinapyl alcohol (deciduous) by free radical crosslinking initiated by enzymatic dehydrogenation.[1]

Lignin has been grafted with ethenylbenzene (styrene),[7,8] 4-methyl-2-oxy-3-oxopent-4-ene (methyl methacrylate),[9,10] 2-propenamide (acrylamide),[11] 2-propene nitrile (acrylonitrile),[12] cationic monomers,[2,3] anionic monomers,[4] and propenoic acid ethoxylates.[5,6]

Lignin is grafted by using an initiation system that preferentially attacks repeat units in lignin to create a site for polymer chain growth. The reaction appears general and works on almost all ethene monomers. By this reaction, lignin is converted into process polymers for industrial use or thermoplastics for use in consumer items.

PROPERTIES

The properties of the grafted products are a combination of the properties of lignin and the properties of the grafted side chain. Lignin is a deep brown, fluffy powder which can be thermoformed into hard, brittle solids when heated above its glass temperature. Further, the deep brown color is a product of free radicals in the lignin which, if bleached away, will slowly reform and react with atmospheric oxygen.[14] This behavior can be a major drawback for applications of lignin to consumer products. Added to these difficulties are the variations in lignin caused by the crosslinking reaction that creates it, the different alcohols used by the vegetation that synthesizes it, and the chemical alteration produced in the extraction processes that recover it.

These properties of lignin appear in the graft copolymer weighted by mole fraction of lignin and side chain repeat units in the product. Thus, grafting a side chain of poly(1-amidoethylene), a water-soluble polymer, onto lignin will make the copolymer progressively more water soluble as the mole fraction of 1-amidoethylene repeat units increases.

Poly[lignin-*g*-(1-amidoethylene)

Molecules of poly[lignin-*g*-(1-amidoethylene)] are nonionic, small in size, readily adsorbed on silica surfaces, and prone to complex di- and trivalent metal ions from aqueous solution.[15,16]

Anionic and Cationic Copolymers

Poly[lignin-*g*-((1-amidoethylene)-*co*-(sodium 1-2(methylprop-2N-yl-1-sulfonate)amidoethylene))] and poly[lignin-*g*-((1-amidoethylene)-r-(1-methyl-1-(1-oxo-2-oxybutylenetrimethylammonium)ethylene methylsulfate))] are anionic, or cationic graft copolymers, respectively.[3,4] The synthesis, characterization, and testing of these products shows them to be water soluble polymers that are effective dispersing, flocculating, and surface active agents.

Poly[lignin-*g*-(1-phenylethylene)]

When grafted with ethenylbenzene, 2-propene nitrile, 1,3-butadiene, or 2-methyl-2-oxy-3-oxopent-4-ene, lignin forms a thermoplastic.

The thermoplasticity of the poly[lignin-*g*-(1-phenylethylene)] graft copolymers can be verified by measurements of glass transition temperature. The T_g of lignin is 115 to 194°C while that of pure poly(1-phenylethyene) is 90 to 109°C.[17] The copolymers have T_gs between 94 and 102°C. These thermoplastics also occupy surfaces on wood and act to alter the wetting properties of the plant matrials.[18] Copolymers give smooth, adherent surface coatings on the wood with contact angles against water of 90 to 110°, a 50° contact angle increase from that of the original wood. These data show that copolymers of lignin are surface active, preferentially orienting the lignin portion of the product towards wood while the plastic sidechain is oriented outward to create a new surface with different wetting properties.[19] Thus, these copolymers are surface-active, coupling agents which can bind wood to hydrophobic phases such as plastic.

The coupling of wood to plastic by a graft copolymer increases the binding strength of the plastic to the wood.

APPLICATIONS

The nonionic polymers and their hydrolysis products are effective thinners and suspending agents for drilling mud formulations.[21]

Replacing portions of the 2-propenamide in the copolymerization reactions with 2,2-dimethyl-3-imino-4-oxohex-5-ene-1-sulfonic acid produces anionic polymers. These strongly anionic polymers are thickening agents for fluid flow control, particularly in strong brines.[4] These properties are useful in oil recovery processes such as polymer flooding.[22,23]

Cationic graft copolymers of lignin have been found to be effective dewatering aids and flocculating agents for sewage.[2]

Lignins grafted with 1-phenylethylene, 2-propene nitrile, or 2-oxy-3-oxo-4-methylpent-4-ene have been shown to be thermoplastics, coupling agents for wood and plastic, and biodegradable plastics.[13,18–20] By grafting thermoplastic chains onto lignin, a surface active agent is created that will wet the surface of wood and convert it into a hydrophobic material. The lignin/plastic copolymer acts as a coupling agent to connect the wood and plastic phases together, increasing tensile strength and slightly decreasing modulus when added to a wood/plastic composite.

Polymers have been reinforced with many fibers, including glass, asbestos, and Kevlar.[24] Wood, however, offers many advantages as a reinforcing fiber.

Wood/plastic composites coupled by copolymers are being developed as materials for consumer products.

Bioconversion and degradation of lignin-(1-phenylethylene) graft copolymer allows these copolymers to be used in disposable consumer items that can be composted by white rot fungi rather than being landfilled or becoming litter. Copolymerization of synthetic side chains onto naturally occurring backbones is being developed as a way of producing compounds that are more easily degraded in the environment. In particular, grafting of lignin with synthetic side chains such as poly(1-phenylethylene) will form a much more biodegradable material than synthesis of a polymer from pure, petroleum-derived hydrocarbons.

REFERENCES

1. Kirk, T. K.; Higuchi, T.; Chang, H. *Lignin Biodegradation: Microbiology, Chemistry, and Potential Applications*, CRC: ISBM 0-8493-5459-5 **1980**, *1*, 5.
2. Meister, J. J.; Li, C. T.; Tewari, K. K.; Simoliunas, S. *Am. Chem. Soc. Meet. Abstracts*, **April, 1990**, *199*, 418.
3. Meister, J. J.; Li, C. T. *Macromolecules* **1992**, *25*(1), 611–616.
4. Meister, J. J.; Patil, D. R.; Augustina, C.; Lai, J. Z. *Lignin, Properties and Uses*. S. Sarkanen and W. Glasser, Editors, American Chemical Society Symposium Series, American Chemical Society: Washington, D.C., **1989**, *397*, 294–310.
5. Meister, J. J. Soluble or Crosslinked Graft Copolymers of Lignin Based on Hydroxyethylmethacrylate and Acrylamide, U.S. Patent 5 121 801, June 16, 1992.
6. Meister, J. J.; Patil, D. R.; *J. Timber Assoc. India* **1989**, *35, #3*, 9–20.
7. Koshijima, T.; Muraki, E. *Zairy O.* **1967**, *16, #169*, 834–838.
8. Phillips, R. B.; Brown, W.; Stannett, V. T. *J. Appl. Poly. Sci.* **1971**, *15*, 2929–2940.
9. Koshjima, T.; Muraki, E. *Nihon Mokuzai Gakkaishi* **1964**, *10, 44, #3*, 110–115.
10. Koshjima, T.; Muraki, E. *Nihon Mokuzai Gakkaishi* **1964**, *10, #3*, 116–119.
11. Meister, J. J.; Patil, D. R.; Channell, H. *Ind. Eng. Chem. Prod. Res. Dev.* **1985**, *24, #2*, 306–313.
12. Simionescu, Cr.; Ceratescu-Asandei, A. *Stoleru, Cellulose, Chem. Tech.* **1975**, *9, #4*, 363–380.
13. Meister, J. J.; Chen, M. J. *Macromolecules* **1991**, *24, #26*, 6843–6848.
14. Sarkanen, K. V.; Ludwig, C. H. *Lignins; Occurrence, Formation, Structure, and Reactions*, ISBN 0-471-75422-6, John Wiley & Sons: New York, **1971**, 1.
15. Meister, J. J.; Patil, D. R.; Field, L. R.; Nicholson, J. C. *J. Polym. Sci., Polym. Chem. Ed.* **1984**, *22*, 1963–1980.
16. Meister, J. J.; Patil, D. R.; Channell, H. *J. Appl. Polym. Sci.* **1984**, *29*, 3457–3477.
17. Immergut, E. H.; Brandrup, J. *Polymer Handbook*, 3rd ed., Wiley-Interscience, New York, **1989**, *Chapter VI*, p 227.
18. Gunnells, D. W.; Gardner, D. J.; Chen, M. J.; Meister, J. J. *Am. Chem. Soc. Meet. Abstracts: 204*, American Chemical Society: Washington, D. C., **August, 1992**, 125.
19. Gunnells, D. W.; Gardner, D. J.; Chen, M. J.; Meister, J. J. *Procd. Am. Chem. Soc., Div. Poly. Mater.: Sci. Eng.* **1992**, *67*, 227.
20. Meister, J. J.; Chen, M. J.; Wang, A.; Aranha, A.; Proceed. Symp. Biodeg. Polym. Packaging, June 16–18, 1992, Natick, Massachusetts, Published as *Biodegradable Polymers and Packaging*, C. Ching, D. L. Kaplan, E. L. Thomas, Eds., Technomic: Lancaster, PA, **1993**, 75–96.
21. Meister, J. J.; Patil, D. R. *Macromolecules* **1985**, *18*, 1559–1564.
22. Martin, F. D.; "Laboratory Investigations of the Use of Polymers in Low Permeability Formations", 49th Annual Fall Meeting, SPE-AIME, SPE-5100, Houston, **1974**.
23. Mungan, N. Programmed Mobility Control in Polymer Flooding, *Can. J. Chem. Eng.* **1971**, *V. 49*, 32–37.
24. Maldas, D.; Kokta, B. V. "Effect of Fiber Treatment on the Mechanical Properties of Hybrid Fiber Reinforced Polystyrene Composites: III. Use of Mica and Sawdust as Hybrid Fiber", *Journal of Reinforced Plastics and Composites*, January, 1991, *Vol. 10*.

Linear Low Density Polyethylene

See: Metallocene Catalysts
Polyethylene (Commercial)
Polyolefins (Overview)

LIPID MEMBRANES (High Stability, from Archaebacterial Extremophiles)

Kiyoshi Yamauchi
Department of Bioapplied Chemistry
Faculty of Engineering
Osaka City University

Archaebacteria proliferate in extreme environments such as hot springs and salt lakes. It is believed that these microorganisms appeared in the primeval biosphere, maintaining unique lines to the present age in the spots which are reminiscent of the early earth.[1,2]

The lipid structure of archaebacteria differs from ordinary organisms. Whereas eubacteria and eukaryotes have glycerolipids, which consist chiefly of fatty acids and glycerol with an ester bonding between them, archaebacterial lipids are composed of saturated isoprenoid (isopranyl) groups of 20, 25, or 40 carbon atoms and polyol (glycerol or its derivative) joined by an ether bonding. Recently it was reported that the major ether-type lipid structure of *S. acidocaldarius* had a polyol backbone of 2-hydroxymethyl-1-(2,3-dihydroxypropoxy)2,3,4,5-cyclopentane tetraol.[11] Furthermore, the stereochemistry at the sn-2 position of the glycerol is opposite to that of the conventional glycerolipids (**Figure 1**). Conceivably, the bacteria needed these unique lipids to survive the extreme habitats. One may speculate that their membranes might function as a stable barrier at high temperatures and in highly salty and acidic environments. Isolation, biosynthesis and spectroscopic constants of the lipids have been reviewed.[2–7]

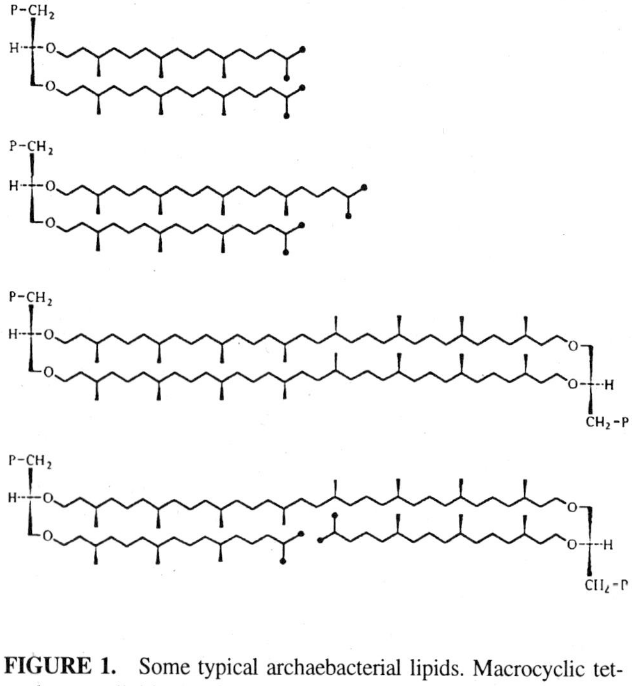

FIGURE 1. Some typical archaebacterial lipids. Macrocyclic tetraether-type lipids are often isolated from thermophiles. The P groups include H, OPO_3H_2, sugars phosphosugars, etc.

PREPARATION

Lipids

The extraction of archaebacterial cells provides a wide variety of lipids. The amphiphiles are often hydrolyzed into a simple mixture of the universal hydrophobic cores. One may design seminatural lipids by chemically binding a polar head to the core.[8]

Membranes

The morphology of lipid molecular assemblies is determined primarily by lipid structure. It seems that lipids having large polar groups readily afford the curved assemblies, such as liposomes, while lipids bearing the polar group similar in a cross-sectional area to the hydrocarbon chains are good for preparing planar assemblies like Langmuir membranes and black membranes.

THERMAL PHASE TRANSITION

Molecular assemblies of lipids generally undergo a gel-to-lipid crystalline phase transition to absorb the transition energy. A sonicated dispersion of bipolar lipids from *Thermoplasma acidophiles* was found to have a transition at –20°C.[9] The very low transition temperature or the lack of transition to the gel phase, could be explained by the steric hindrance of the methyl groups to ordered packing of the hydrocarbon residues, or by a small energy difference between the trans and gauche rotamers. This suggests that archaebacterial lipid-assemblies are in a liquid crystalline-like phase, and do not take a gel phase at an ordinary temperate condition (–10 ~ 50°C).

MECHANISM OF STABILITY

Aqueous suspensions of liposomes made of archaebacterial lipids are stable for more than a year at ambient temperatures. This stability may be ascribed to the hindered fusion liposomal membranes due to impermeability of the membrane against the bulky isopranyl chains of a counter membrane, and the inability of polar head groups to cross the thick hydrophobic membranes.

The archaebacterial lipids are also stable chemically and biologically. The isopranyl chains in particular are more resistant to oxidation than the unsaturated fatty acid chains of conventional lipids. The ether linkages of archaebacterial lipids are much more resistant to hydrolysis by acid and alkali than the ester linkages of conventional lipids.

Low permeability of various substances over a wide temperature range and high salt tolerance could be ascribed to the greater intramembrane steric resistance of the branched isopranyl chains to diffusing materials (permeants or the lipid molecules of a counter membrane). It has also been reported that the branched molecules diffuse via a lipid pathway more slowly than do small and simple compounds.[10]

REFERENCES

1. Woese, C. R. *Microbiol. Rev.* **1987,** *51*, 221.
2. Langworthy, T. A. *The Bacteria*; Academic: Orlando, FL, Vol. 8, p 459–497.
3. Koga, K.; Ishihara, M.; Morii, H.; Akagawa-Matsushita, M. *Microbiol. Rev.* **1993**, *57*, 164.
4. Balch, W. E.; Fox, G. E.; Magrum, L. J.; Woese, C. R.; Wolfe, R. S. *Microbiol. Rev.* **1979**, *43*, 260.
5. Langworthy, T. A.; Holzer, G.; Zeikus, J. G.; Tornabene, T. G. *System. Appl. Microbiol.* **1983**, *4*, 1.
6. Kates, M. *Prog. Chem. Fats Other Lipids* **1978**, *15*, 301–342.
7. De Rosa, M.; Gambacorta, A.; Gliozzi, A. *Microbiol. Rev.* **1986**, *50*, 70.
8. Lazrak, T.; Milon, A.; Wolff, G.; Albrecht, A. M.; Miehe, M.; Ourisson, G.; Nakatani, Y. *Biochim. Biophys. Acta* **1987**, *903*, 132.
9. Blöcher, D.; Gutermann, R.; Henkel, B.; Ring, K. *Biochem. Biophys. Acta* **1984**, *778*, 74.
10. Wright, E. M.; Diamond, J. M. *Proc. R. Soc.* London, 1969, B 172, 203, 227, and 273.
11. Sugai, A.; Sakuma, R.; Fukada, I.; Kurokawa, N.; Itoh, H. Y.; Kon, K.; Ando, S.; Itoh, T. *Lipid* **1995**, *30*, 339.

Lipids

See: Lipid Membranes (High Stability from Archaebacterial Extermophiles)
Phospholipid Bilayers
Phospholipid-Immobilized Surfaces

Liquid Crystal Display

See: Polymer Dispersed Liquid Crystal Display (Driving Circuit for HDTV)

LIQUID CRYSTALLINE AROMATIC POLYESTERS

Jung-Il Jin*
Department of Chemistry and Advanced Materials
Chemistry Research Center
Korea University

Jin-Hae Chang
Department of Polymer Science and Engineering
Kum-Oh University of Technology

Formation of liquid crystalline (LC) states by low molar mass compounds has been known for more than 100 years, since Reinitzer's discovery in 1888.[1] However, the LC states of polymers were found only 45 years ago by Elliott and Ambrose for poly(γ-benzyl L-glutamate) dissolved in chloroform.[2] Nevertheless, there is no doubt that the discovery of the aramids such as poly(*p*-phenylene terephthalamide) (Kevlar) (1) and poly(*p*-benzamide) (2) by DuPont de Nemours Company in the 1970s has spawned unparalleled scientific and industrial interest for the past 25 years. These aromatic polyamides form LC states when dissolved in solvents (lyotropic) such as sulfuric acid. In addition, more recent commercialization of aromatic copolyesters [e.g., Xydar (3) and Vectra (4)] that form LC states in melt (thermotropic) sparked unabated growth in the field of LC polymers (LCPs).

There are two types of thermotropic polymers: main-chain and side-chain liquid crystalline polymers.[3] A spacer between

*Author to whom correspondence should be addressed.

mesogenic groups in main-chain LCPs, or between the backbone and mesogenic units in side-chain LCPs, is sometimes necessary to form the LC phase. Among these varieties, we will treat here only the main-chain thermotropic, aromatic polyesters. The references provide a more thorough review on thermotropic LCPs.[4–7]

The thermotropic aromatic polyesters have either long stretches of linear rigid units or an alternating sequence of rigid units and flexible or rigid spacers along the main-chain.[5,8] The mesogenic units generally contain either two or three aromatic (or cycloaliphatic) rings connected in para positions by ester links which maintain the linear alignment of the aromatic rings. These mesogenic structures have been synthesized using mainly terephthalic acid, 2,6-naphthalenedicarboxylic acid, *p*-hydroxybenzoic acid, hydroquinone,and *p,p'*-biphenol in either their unsubstituted or substituted forms. We can include low levels of bent aromatic units such as isophthalic acid and resorcinol moieties as long as their presence does not destroy liquid crystallinity.[9]

Wholly aromatic LCPs, especially when they are homopolymers, are highly crystalline, very high melting, and often interactable materials. Two representative examples are poly(*p*-hydroxybenzoate) (5) and poly(*p*-phenylene terephthalate) (6). The other class of wholly LCPs contains bent structural moieties such as resorcinol and bisphenol-A in addition to para-linked aromatic structures. Inclusion of asymmetrically substituted units such as chloro- or methylphenylene units can further depress melting temperature. The LCPs belonging to aliphatic-aromatic LCPs consist of mesogenic groups sequenced in a regular fashion by flexible aliphatic segments of different lengths. The spacer in the repeating units plays a fundamental role in decoupling the intramolecular interactions among the mesogens, thus allowing for substantial decreases in melting and clearing temperatures, accompanied by appreciable solubility in conventional organic solvents.

High molecular weight is essential to achieve good mechanical properties.

PROPERTIES

Thermal Properties

Two representative *para*-linked aromatic polyesters, poly(*p*-phenylene terephthalate) and poly(*p*-oxybenzoate), show melting point temperature (T_ms) of about 600°C or above, competitive with their degradation temperature. Highly rigid and rod-like LCPs are difficult to handle because of their melting points. In order to reduce the high transition temperature, several structural modifications have been adopted: introduction of flexible spacers to separate rod-like mesogenic groups in the main chain from each other, unsymmetrical substitution of the aromatic rings, addition of an element of dissymetry to the main chain by copolymerizing mesogenic units of different structures, and introduction of rigid kinks into the straight polymer chains.

There are many studies on the nature of the liquid crystalline phase of polymers with flexible methylene spacers.[11–14] The type of flexible spacer has been restricted in most cases to polymethylene chains from 2 to 14 carbon atoms in length. Poly(ethylene oxide) and polysiloxane segments are also commonly investigated as spacers.[15–18] In most cases, polymers containing an even number of methylene units have higher transition temperatures (melting, T_m; and isotropization, T_i) and entropy changes than those with odd units.

Unsymmetrical attachment of a ring substituent depresses the melting and clearing temperatures of LC aromatic polyesters. Symmetric disubstitution, however, can elevate T_m, as compared to the monosubstituted polymer. Generally speaking, the melting temperature as well as clearing (or isotropization) point decreases with the size of the substituent.

We emphasize that the thermal transition values are dependent on the molecular weight and thermal history of the samples. Regioregularity of substituent position also greatly influences transition temperatures, particularly, the melting temperatures. In addition, thermal stability of the polymers is highly influenced by the presence of substituents.

Among the different ways to reduce the T_m of aromatic LCPs, the development of random copolymers is the most important.[10a,19]

Several researchers have reported investigations on the effects of inclusion of meta isomers, that is, rigid kinks.[5,9,10,20] Replacement of *p*-hydroxybenzoic acid (PHB), hydroquinone (HQ), and terephthalic acid (TPA) units in LC polyesters by *m*-hydroxybenzoic acid (MHB), resorcinol (RE), and isophthalic acid (IPA) units, respectively, results in low melting point polymers. An excessive level of meta isomers can destroy the LC phase.[19]

Rheological Properties

LCPs are easily oriented by external force fields such as electromagnetic fields or mechanical stresses. The degree of orientation then affects the melt viscosity with rheological properties depending on the texture of the mesophase. The rheological behavior of LCPs in the isotropic phase is similar to that of conventional polyesters. In the LC phase, however, a drastically lower viscosity is observed even though the temperature of LC phases is lower than the temperature of the isotropic melt.

The behavior of LCPs differs from that of ordinary polymers in at least one important aspect: a very strong dependence of melt viscosity on shear rate. Data reported by Denn and coworkers show a shear thinning behavior over as much as eight decades of shear rate.[19] Under normal processing conditions, in general, the viscosity values of thermotropic melt of wholly aromatic LC polyesters can be two or more orders of magnitude lower than those of an isotropic melt such as PET.[21]

LCP melts are very efficiently oriented in extensional flow and, as a result, extensional stresses at the entrance to a capillary influences the shear flow in the capillary to a much greater extent than is usually found with non-LC polyesters. One of the attractive properties of LCPs is the intrinsic anisotropy of the mesophase to create molecular orientations during the processing. One obvious application of this concept is in fiber spinning where orientation in the fiber direction is created by the elongational flow prevailing in the spinning line. High modulus and strength of fibers are achieved largely relying on the degree of molecular alignment.

Mechanical and Other Properties

LC aromatic polyesters produce fibers with tenacity and modulus values comparable to those of aramid fibers.

APPLICATIONS

LCPs can be processed into fibers, films, and sheets and development of new applications is ongoing. LCPs based on aromatic polyesters are being used mainly in plastic applications.

Polymer blends containing LCPs and thermoplastics and thermoset matrix polymers have received considerable attention in recent years.[22,23] Since LCPs form fibrous structures and have low melt viscosity, we expect them to act as reinforcing agents and processing aids when added to matrix polymers, and these blends should be easier to process than the pure matrix polymers.

Since aromatic LCPs exhibit good solvent- and chemical-resistance, they can be molded into various shapes of packing materials for distillation and absorption towers. The excellent vibration damping properties of aromatic LCPs are already advantageously employed in the manufacture of speaker cones, and are expected to make LCPs a major competitor in tennis racket and ski-goods production. Vibration damping characteristics together with excellent thermal stability, oil-resistance, and dimensional stability will lead aromatic LCPs to new uses in under-the-hood automobile parts such as fuel-pump impellers, cylinder-head covers, locker covers, and sensors. The same is true for various aerospace vehicle parts. Coating of optical fibers, printed circuit boards, encapsulating IC and condensers, and extrusion of thin films are now under development.

Fibers based on Vectra, Ekonol, and Xydar show excellent strength and modulus and very low water absorption. They also exhibit remarkable abrasion and chemical resistance, as well as antithermal aging and self-extinguishing properties. Combination of these properties will make the fibers competitive against aramid fibers in many applications such as belts, ropes, cords, rubber reinforcing, asbestos substituents, and reinforcers for plastic and cement.

Aromatic LCPs are often called third generation engineering plastics and their consumption is increasing rapidly because of their excellent processing characteristics, physical and chemical properties, and mechanical strength.

REFERENCES

1. Reinitizer, F. *Monatsheft Fur Chemie* **1888**, *9*, 421.
2. Elliott, A.; Ambrose, E. J. *Disc. Farad. Soc.* **1950**, *9*, 246.
3. Helfrich, W.; Heppke, G. *Liquid Crystals of One- and Two-Dimensional Order*; Springer-Verlag: Berlin, Germany 1984; Chapter IX.
4. Chapoy, L. L. *Recent Advances in Liquid Crystalline Polymers*; Elsevier: New York, NY, 1985.
5. Ober, C. K.; Jin, J.-I.; Lenz, R. W. *Adv. Polym. Sci.* **1984**, *59*, 103.
6. La Mantia, F. P. *Thermotropic Liquid Crystal Polymer Blends*; Technomic: Lancaster, Pennsylvania, 1993.
7. Blumstein, R. B.; Thomas, O.; Gauthier, M. M. et al. In *Polymer Liquid Crystals*; Blumstein, A., Ed.; Plenum: NY, 1985.
8. Shibaev, V. P.; Plate, N. A. *Adv. Polym. Sci.* **1984**, *60/61*, 173.
9. Lenz, R. W.; Jin, J.-I. *Macromolecules* **1981**, *14*, 1405.
10. Lenz, R. W. *Organic Chemistry of Synthetic High Polymers*; Interscience: New York, NY 1967.

10a. Blaschke, F.; Ludwig, W. U.S. Patent 3 395 119, 1968.

11. Griffin, A. C.; Havens, S. J. *J. Polym. Sci., Polym. Lett. Ed.* **1980**, *18*, 259.
12. Griffin, A. C.; Havens, S. J. *J. Mol. Cryst. Liq. Cryst. Liq. Cryst.* **1979**, *49*, 239.
13. Roviello, A.; Sirigu, A. *Eur. Polym. J.* **1979**, *15*, 61.
14. Krigbaum, W. R.; Asrar, J.; Tariumi, H. et al. *J. Polym. Sci., Polym. Lett. Ed.* **1982**, *20*, 109.
15. Meurisse, P.; Noel, C.; Monnerie, L.; Fayolle, B. *Brit. Polym. J.* **1981**, *13*, 55.
16. Fayolle, B.; Noel, C.; Billard, J. *J. Phys.* **1979**, *40*, C3-485.
17. Jo, B.-W.; Jin, J.-I.; Lenz, R. W. *Eur. Polym. J.* **1982**, *18*, 233.
18. Ringsdorf, H.; Schneller, A. *Brit. Polym. J.* **1981**, *13*, 43.
19. Jackson, Jr., W. J. *Brit. Polym. J.* **1980**, *12*, 154.
20. Chung, T. S. *Polym. Eng. Sci.* **1986**, *26*, 901.
21. Jerman, R. E.; Baird, D. G. *J. Rheol.* **1981**, *25*, 275.
22. Dutta, D.; Fruitwala, H.; Kohli, A.; Weiss, R. A. *Polym. Eng. Sci.* **1990**, *30*, 17, 1005.
23. Roetting, O.; Hinrichsen, G. *Adv. Polym. Tech.* **1994**, *13*, 57.

LIQUID CRYSTALLINE ELASTOMERS

Rudolf Zentel and Martin Brehmer
Institute of Organic Chemistry
Universität Mainz

Liquid crystalline (LC) polymers generally combine the properties of polymers with that of liquid crystalline phases.[1–5] From this combination, new property combinations result.[6]

LC-elastomers combine the properties of LC phases with a very typical polymer property: rubber elasticity.[7,8] They are prepared by chemically linking mesogenic groups into slightly crosslinked polymer networks that add rubber elasticity. Alternatively, materials with similar properties can be prepared from hybrid materials between densely crosslinked LC-resins and not chemically linked low molar mass liquid crystals.[9–12] These hybrid materials are two phasic, but they can display properties of a homogenous material on a macroscopic level. For reviews on LC-elastomers including hybrid materials, see the references.[7,8,13–18]

SYNTHETIC ROUTES TO LC-ELASTOMERS

The preparation of LC-elastomers can be done in three possible ways (see **Figure 1**). The synthesis of the LC-polymer and the crosslinking can be done in one step (see Figure 1a), or a preformed LC-polymer can be crosslinked in a second step (see Figure 1b). Alternatively the crosslinking of the LC-polymer can be done by a second polymerization process leading to two types of intertwined polymer chains (see Figure 1c).

PROPERTIES OF LC-ELASTOMERS

General Properties

The LC phase is retained for the crosslinked elastomers. X-ray measurements performed with the uncrosslinked and the slightly crosslinked polymers show the same LC phases.[14,19,21–24] Thus the crosslinking reaction, which transforms a soluble polymer into a soft solid, does not influence the LC order as long as the crosslinking density is low. In this limit of low crosslinking density also the phase transition temperatures are not shifted significantly.[7,8,19,20,23,24]

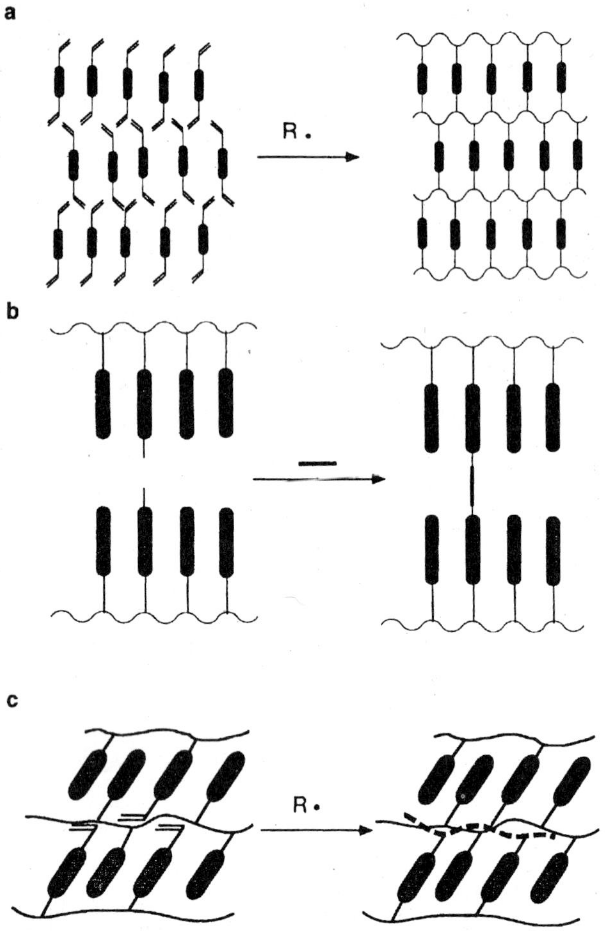

FIGURE 1 Different ways to LC-elastomers: (a) from bifunctional monomers, (b) reaction of a polymer with a bifunctional crosslinker, (c) crosslinking by a second polymerization reaction.

The most prominent property of all LC elastomers is their rubber-like elasticity and their mechanical orientability.[7,8,16,23–27] The mechanical orientability of the LC elastomers is due to the anisotropic conformation, which the polymer chains adopts in the LC phase.[21,28–30]

Properties of Chiral LC-Elastomers

The combination of elastomer and LC-properties becomes especially interesting when liquid crystalline phases with ferroelectric properties (chiral smectic C* phase = S_C*) are encountered. These are used in applications in ferroelectric liquid crystal displays.[31]

Ferroelectric LC-Elastomers

LC-elastomers prepared from ferroelectric LC-polymers are interesting materials for two reasons.[8,32] From a materials point of view, researchers are interested in their ferroelectric, piezoelectric, and pyroelectric properties. From a scientific point of view they are fascinating because they allow a study of the interplay of electric and mechanic forces in a rubber material. This happens because the reorientation of the mesogenic groups in the electric field creates stress in the network of the polymer chains.

Composite Materials of LC-Thermosets and Monomers

An alternative approach to LC-networks with ferroelectric and piezoelectric properties uses composite materials of LC-thermosets filled with LC-monomers. While the pure LC-thermosets are interesting optical materials, with nearly temperature independent properties (up to the decomposition temperature), the composite materials allow director reorientations.[35] They are thus interesting for display application and piezoelements.[17,32]

Ferroelectric LC-Elastomers for Nonlinear Optics

Since chiral smectic C* phases possess a polar structure, they have the required geometry for second order nonlinear optical effects (frequency doubling or the pockets-effect) in the helix unwound state. The potential of ferroelectric LC-polymers for these types of applications may be proven soon.[36,37] The use of LC-elastomers is interesting to stabilize the polar and helix unwound state mechanically. Two possible ways to perform this goal are apparent. Researchers will either attempt stabilization of the polar state by photo-crosslinking,[33] or try a mechanical untwisting of the helical superstructure. This last possibility, using only mechanical forces to prepare a polar sample, was first demonstrated in 1989.[8,38] Very recently the nonlinear optical effects associated with this effect were determined.[39] These systems are interesting because no relaxation of the polar order is possible as long as the sample is held mechanically. In addition phasematching is possible in chiral smectic C* phases.[34]

REFERENCES

1. Cantow, H. J.; Platé, N. A. "Polymer Liquid Crystals II/III", *Adv. Polymer. Sci. 60/61*, Aufl., Springer-Verlag: Berlin, 1984.
2. Ciferri, A. *Liquid Crystallinity In Polymers* VCH: New York, 1991.
3. Kowlek, S. L.; Morgan, P. W.; Schaefgen, J. R. *Encyclopedia of Polymer Science and Engineering*; John Wiley & Sons: New York, 1987; Vol. 9.
4. McArdle, C. B. *Side Chain Liquid Crystal Polymers*, Blackie and Son: Glasgow, 1989.
5. Weiss, R. A.; Ober, C. K. "Liquid-Crystalline Polymers", American Chemical Society, *ACS Symposium Series*, Washington, DC, 1990; Vol. 435.
6. Kapitza, H.; Poths, H.; Zentel, R. "Functionalized LC-polymers towards NLO-applications and piezosensors", *Macromol. Chem., Macromol. Symp.* **1991**, *44*, 117.
7. Gleim, W.; Finkelmann, H. "Side Chain Liquid Crystalline Elastomers" in 4, Chapter 10.
8. Zentel, R. "Liquid Crystalline Elastomers", Angew. Chem. Adv. Mat. **1989**, *101*, 1437; *Angew. Chem. Int. Ed. Engl. Adv. Mat.* **1989**, *28*, 1407.
9. Bhadani, S. N.; Gray, D. G. "Crosslinked cholesteric network from the acrylic acid of (hydroxyproply)cellulose", *Mol. Cryst. Liq. Cryst.* **1984**, *120*, 255.
10. Broer, D. J.; Gossink, R. J.; Hikmet, R. A. M. "Oriented polymer networks obtained by photopolymerization of liquid-crystalline monomers", *Angew. Makromol. Chem.* **1990**, *183*, 45.
11. Hikmet, R. A. M. "From Liquid Crystalline Molecules to Anisotropic Gels" *Mol. Cryst. Liq. Cryst.* **1991**, *198*, 357.

12. Strzelecki, J.; Liebert, L. "Synthesis and polymerization of a new mesomorphic monomer", *Bull. Soc. Chim. Fr.* **1973**, 597.
13. Brehmer, M.; Zentel, R. "Liquid-Crystalline Elastomers; characterization as networks", *Mol. Cryst. Liq. Cryst.* **1994**, *243*, 353.
14. David, F. J. "Liquid-crystalline elastomers", *J. Mater. Chem.* **1993**, *3*, 551.
15. Finkelmann, H. in: *Liquid Crystallinity in Polymers*, A. Ciferri (Hrsg.), VCH: Weinheim, 1991.
16. Meier, W.; Finkelmann, H. "Liquid Crystalline Elastomers", *Condens. Mat. News* **1992**, *1*, 15.
17. Broer, D. J. "Creation of super-molecular, thin film architectures with liquid crystalline networks" at International Liquid Crystall Conference, Budapest, 1994; abstract, Vol. 1, p 498.
18. Hikmet, R. A. M. "Anisotropic Gels in Liquid Crystal Devices", *Adv. Mat.* **1992**, *4*, 679.
19. Zentel, R.; Schmidt, G. F.; Meyer, J.; Benalia, M. "X-ray Investigations of linear and crosslinked LC-main chain and combined polymers", *Liq. Cryst.* **1987**, *2*, 651.
20. Zentel, R.; Reckert, G. "Liquid crystalline elastomers based on liquid crystalline side group main chain and combined polymers", *Makromol. Chem.* **1986**, *187*, 1915.
21. Davidson, P.; Noirez, L.; Cotton, J. P.; Keller, P. "Neutron scattering study and discussion of the backbone conformation in the nematic phase of a side chain polymer. *Liq. Cryst.* **1991**, *10*, 111.
22. Degert, C.; Richard, H.; Mauzac, M. "Polymorphism and thermoelastic behavior of some crosslinked mesogenic polysiloxanes", *Mol. Cryst. Liq. Cryst.* **1991**, *214*, 179.
23. Mitchell, G. R.; Davis, F. J.; Ashman, A. "Structural studies of side-chain liquid-crystal polymers and elastomers", *Polymer* **1987**, *28*, 639.
24. Zentel, R.; Benalia, M. "Stress induced orientation in lightly crosslinked liquid-crystalline side-group polymers", *Makromol. Chem.* **1987**, *188*, 665.
25. Bräuchler, M.; Boeffel, C.; Spiess, H. W. "Dynamic Fourier transform infrared spectroscopy of LC-elastomers", *Makromol. Chem.* **1991**, *192*, 1153.
26. Mitchel, G. R.; Davis, F. J.; Guo, W.; Cywinski, R. "Coupling between mesogenic units and polymer backbone in side-chain liquid crystal polymers and oligomers", *Polymer* **1991**, *32*, 1153.
27. Schätzle, J.; Kaufhold, W.; Finkelmann, H. "Nematic elastomers: The influence of external mechanical stress on the liquid-crystalline phase behavior", *Makromol. Chem.* **1989**, *190*, 3269.
28. Davidson, P.; Levelut, A. M. "X-ray diffraction by mesomorphic comb-like polymers", *Liq. Cryst.* **1992**, *11*, 469.
29. Kriste, R. G.; Ohm, H. G. "The conformation of liquid crystalline polymers as revealed by neutron scattering," *Makromol. Chem. Rapid Commun.* **1985**, *6*, 179.
30. Moussa, F.; Cotton, J. P.; Hardouin, F.; Keller, P.; Lambert, M.; Pepy, G.; Mauzac, M.; Richard, H. "Conformational anisotropy of a liquid-crystalline side chain polymer: a neutron scattering study", *J. Phys.* (Les Ulis, Fr.) **1987**, *48*, 1079.
31. Escher, C.; Wingen, R. "Ferroelectric liquid crystalls in high information containing displays", *Adv. Mater.* **1992**, *4*, 189.
32. Hikmet, R. A. M. "Piezoelectric networks obtained by photopolymerization of liquid crystal molecules", *Macromolecules* **1992**, *25*, 5759.
33. Brehmer, M.; Wiesemann, A.; Zentel, R.; Siemensmeyer, K.; Wagenblast, G. "LC-elastomers by chemical and physical crosslinking", *Polymer Preprints* **1993**, *34*(2), 708.
34. Kocot, A.; Wrzalik, R.; Vij, J. K.; Zentel, R. "Pyroelectric and electro-optical effects in the Sm C* phase", *J. Appl. Phys.* **1994**, *75*, 728.
35. Broer, D. J.; Heynderickx, I. "Three dimensionally ordered polymer networks with a helicoidal structure", *Macromolecules* **1990**, *23*, 1021.
36. Ozaki, M.; Utsumi, M.; Yoshino, K.; Skarp, K. "Second harmonic generation in a ferroelectric liquid crystalline polymer", *Jpn. J. Appl. Phys.* **1993**, *32*, 852.
37. Wischerhoff, E.; Zentel, R.; Redmond, M.; Mondain-Monval, O.; Coles, H. "Ferroelectric liquid crystalline polysiloxanes designed for high second-order nonlinear susceptibilities," *Macromol. Chem. Phys.* **1994**, *195*, 1593.
38. Zentel, R.; Reckert, G.; Bualek, S.; Kapitaz, H. "LC-elastomers with cholesteric and chiral smectic C* phases", *Makromol. Chem.* **1989**, *190*, 2869.
39. Benné, I.; Semmler, K.; Finkelmann, H. Second harmonic generation on mechanically oriented sC*-elastomers at 23. Freiburger Arbeitstagung Flüssigkristalle, Freiburg, 1994; abstract, No. 6.

LIQUID CRYSTALLINE ETHYL-CYANOETHYL CELLULOSE

Yong Huang
Laboratory of Cellulose and Lignocellulosics Chemistry
Guangzhou Institute of Chemistry

Cellulose and its derivatives are semirigid chain polymers and can be dissolved in many organic and inorganic solvent systems. Cellulose and its derivatives can form lyotropic liquid crystals in the appropriate solvent systems and some of cellulose derivatives can also form thermotropic liquid crystals. Following the first observation of a liquid crystalline solution of hydroxypropyl cellulose in water, many cellulose derivatives have been reported to be able to form liquid crystals either in solution or in bulk.[1–3] The liquid crystals of cellulose and its derivatives are generally cholesteric. The cellulose backbone is chiral and the chiroptical properties are exhibited in specifically substituted cellulose derivatives, from which the relationship between the molecular structure of cellulose derivatives and their liquid crystalline behavior has been investigated.

The chain stiffness depends on the nature of substituents on cellulose derivative backbone, the degree of substitution, and solvents. Consequently, the liquid crystalline behavior of cellulose derivatives is greatly influenced by their molecular structures such as the chemical characteristics of substituents, the degree of substitution and its distribution, the molecular weight and its distribution, and the solvent systems.

Ethyl-cyanoethyl cellulose [(E-CE)C] is a cellulose derivative with two kinds of substituents on its main chain. One substituent is ethyl and the other is cyanoethyl. It can be dissolved in many organic solvents and form cholesteric liquid crystalline phase.

PROPERTIES

Like many cellulose derivatives, (E-CE)C exhibits lyotropic liquid crystalline behavior in many organic solvent systems, such as dichloroacetic acid (DCA), acetic acid (AA), trifluoroacetic acid (TFA), benzene, ethyl cyanide, and when the concentration of (E-CE)C is above the critical value. In general, (E-CE)C lyotropic liquid crystals are cholesteric. When the concentration is under the first critical value, C_1, the (E-CE)C solution is isotropic and when it is above the second critical value, C_2, the (E-CE)C solution is completely anisotropic. In the concentration region between C_1 and C_2, the isotropic phase and the anisotropic co-exist in the solution, which means that the solution is biphasic in this concentration region.

The molecular structures of (E-CE)C, such as the molecular weight, the degree of substitution, and the distribution of the substituent, markedly influence its lyotropic behavior.[4]

The lyotropic behavior of (E-CE)C also depends on the temperature of the system. The solution can transform from liquid crystalline state to the isotropic state when the solution is heated to above the critical temperature $T_{C\text{-}I}$, and when the solution is cooled to below the $T_{I\text{-}C}$, the liquid crystalline phase can appear in the solution again.

In the (E-CE)C lyotropic liquid crystalline solutions, the mesophase generally exhibits multi-texture behavior, which means that the mesophase texture can vary with the concentration and the temperature of the solutions.[5]

In a certain concentration region, the (E-CE)C lyotropic cholesteric liquid crystalline phase can show the planar texture with vivid colors. The vivid colors of liquid crystalline solutions can be easily observed because the mesophase selectivity reflects visible light in a certain wavelength.

The cholesteric order of (E-CE)C lyotropic liquid crystalline solutions can be frozen by rapid polymerization of the solvent in the solutions.[6]

REFERENCES

1. Werbowyji, R. S.; Gray, D. G. *Mol. Cryst. Liq. Cryst. (Lett)* **1976**, *34*, 97.
2. Gray, D. G. *Appl. Polym. Symp.* **1983**, *37*, 179.
3. Gilbert, R. D. In *Agricultural and Synthetic Polymer, Biodegradability and Utilization*; ACS Symp. Ser. 433; Glass, J. E.; Swift, G., Eds.; American Chemical Society: Washington, DC, 1990; Chapter 22.
4. Jiang, S. H.; Huang, Y. *Polym. Bull.* **1995**, *34*, 203.
5. Huang, Y. *J. Macromol. Sci.-Phys.* **1989**, *B28*, 13.
6. Jiang, S. H.; Huang, Y. *J. Appl. Polym. Sci.* **1993**, *49*, 125.

LIQUID CRYSTALLINE POLYACETYLENE DERIVATIVES

Kazuo Akagi* and Hideki Shirakawa
Institute of Materials Science
University of Tsukuba

Polyacetylene is the most common conducting polymer. Owing to its insolubility and infusibility, however, polyacetylene is hard to process in solution, as in casting, and also to evaluate explicitly its molecular weight and conjugation length. Introduction of alkyl or aromatic substituents into a polyacetylene chain makes the polymer soluble in organic solvent when the alkyl chain length is long enough; i.e., hexyl, pentyl, and octyl groups. Nevertheless, the substituted polyacetylenes have lower electrical conductivities, compared with those of nonsubstituted polyacetylene.

If the substituent is a liquid crystalline group, the polymer would not only be soluble in organic solvents, but could also be aligned by spontaneous orientation of the liquid crystalline group. Besides it can be macroscopically oriented by an external perturbation, such as shear stress, electric field or magnetic force field. As a result, the main chain is aligned following the orientation of the liquid crystalline side chain. This means that a mono domain structure of the liquid crystalline phase is constructed at the macroscopic level. Under such a circumstance, the polymer is expected to show a higher electrical conductivity compared with that of random orientation. At the same time, one can control the macroscopic orientation and, hence, the electrical conductivity of the polymers by the external force.

*Author to whom correspondence should be addressed.

Liquid crystalline polyacetylene derivatives recently developed will be reviewed[1–9] here as a new conducting polymer with liquid crystalline function in the side chain. We synthesized acetylene derivatives substituted with liquid crystalline groups and then polymerized them with Ziegler–Natta and metathesis catalysts. The liquid crystalline substituent of the monomer is composed of a mesogen core of phenylcyclohexy (PCH) moiety, methylene chain linked with an ether oxygen atom, $-(CH_2)_3O$, as a spacer, and an alkyl chain, $(-C_nH_{2n+1}$, n = 2, 3, 5 ~ 8), as a terminal group. This monomer is abbreviated as PCHn03A, where A represents the acetylene segment. Another kind of substituent consists of a biphenyl (BP) mesogen, a spacer of a methylene chain linked with oxygen, and *n*-pentyl group as a terminal moiety. This type of monomer is abbreviated as BP503A. The monomers were polymerized by using Ziegler–Natta and metathesis as PPCHn03A and PBPn03A as shown below (**Scheme I**).

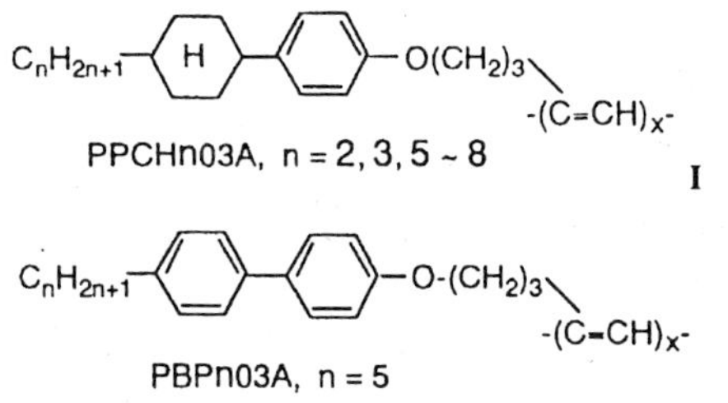

Liquid Crystalline Conducting Polymers

PROPERTIES

Liquid Crystalline Phases of Monomers and Polymers

All the PCHn03A monomers except PCH203A and PCH303A showed Schlieren textures in the polarizing optical microscope, characteristic of the nematic phase. However, all PCH and BP polymers, regardless of the catalyst used in the polymerizations, exhibited fan-shaped textures characteristic of smectic liquid crystalline phases. Meanwhile all polymers showed smectic phases, which have a higher order than the nematic phase. This can be rationalized with a polymerization effect; i.e., a higher order in molecular arrangement and/or packing was generated by the polymerization of liquid crystalline molecules.

MOLECULAR ORIENTATION AND ELECTRICAL CONDUCTIVITY

Orientation by Shear Stress

It is evident that an optical bâtonet texture, characteristic of a smectic liquid crystal, is aligned along the direction of the shear stress. From these results, we can confirm that the side-chain, liquid–crystalline–conjugated polymers are macroscopically aligned by the shear stress as the external force.

CONCLUSION

We synthesized two liquid–crystalline, polyacetylene derivatives using both a Fe-based, Ziegler–Natta catalyst and Mo-based, metathesis catalyst. The polymers showed not only an enantiotropic smectic A phase, by virtue of spontaneous orientation of liquid crystalline side chain composed of phenylcyclohexyl or biphenyl mesogenic moiety, but also an electrical conductivity upon iodine doping.

Electrical conductivity of the polymers was 10^{-8}–10^{-7} S/cm upon iodine doping to the cast films. The alignment of the polymer chain, accompanied with the side-chain orientation using an external magnetic force of 0.7 ~ 1.0 Tesla, enhanced the electrical conductivity up to 10^{-6} S/cm and gave rise to a notable electrical anisotropy.

REFERENCES

1. Oh, S.-Y.; Akagi, K.; Shirakawa, H.; Araya, K. *Macromolecules* **1993**, *26*, 620.
2. Akagi, K.; Oh, S.-Y.; Shirakawa, H.; Nishizawa, T. *Polym. Prep. Jpn.* **1993**, *42*, 698.
3. Akagi, K.; Oh, S.-Y.; Goto, H.; Kadokura, Y.; Shirakawa, H. *Trans. Mat. Res. Soc. Jpn.* **1994**, *15A*, 259.
4. Shirakawa, H.; Kadokura, Y.; Goto, H.; Oh, S.-Y.; Akagi, K.; Araya, K. *Mol. Cryst. Liq. Cryst.* **1994**, *255*, 213.
5. Akagi, K.; Goto, H.; Kadokura, Y.; Shirakawa, H.; Oh, S.-Y.; Araya, K. *Synth. Met.* **1995**, *69*, 13.
6. Akagi, K.; Goto, H.; Shirakawa, H.; Nishizawa, T.; Masuda, K. *Synth. Met.* **1995**, *69*, 33.
7. Goto, H.; Akagi, K.; Shirakawa, H.; Oh, S.-Y.; Araya, K. *Synth. Met.* **1995**, *71*, 1899.
8. Akagi, K.; Goto, H.; Iino, K.; Shirakawa, H.; Isoya, J. *Mol. Cryst. Liq. Cryst.* **1995**, *267*, 277.
9. Yoshino, K.; Kobayashi, K.; Kawai, T.; Ozaki, M.; Akagi, K.; Shirakawa, H. *Synth. Met* **1995**, *69*, 49.

LIQUID CRYSTALLINE POLYMER-BASED BLENDS (Thermotropic)

F. P. La Mantia
Dipartimento di Ingegneria Chimica dei Processi e dei Materiali
Università di Palermo, viale delle Scienze

P. L. Magagnini
Dipartimento di Ingegneria Chimica
Chimica Industriale e Scienza dei Materiali
Università di Pisa

A well-established cost-effective route to differentiating and tailoring polymeric materials, for a wider variety of uses, is to combine two or more commercially available polymers through alloying or blending.[1] In particular, the addition of relatively small amounts of thermotropic liquid crystalline polymers (LCPs) into commercial thermoplastics has been shown to lead to advantages, including a strong reduction of the melt viscosity[2,4] and an enhancement of the mechanical properties of the finished articles.[3–5] The easier processing and the improved engineered performances of appropriate designed LCP/polymer blends may outbalance the economic outlay of using relatively expensive LCPs. Therefore, it is not surprising that in recent years considerable attention has been paid to the study of the structure-property-application relationships in these LCP/polymer blends.[3,4]

PHASE BEHAVIOR

Since the two components of the LCP/polymer blends generally show no or incomplete miscibility, most such blends consist of biphasic systems. The rheological and mechanical properties of a biphasic polyblend depend not only on the properties of the individual components but also to a considerable extent on the phase interactions.

DSC is particularly suited to provide information on the miscibility of LCPs with flexible polymers. The observation of one composition dependent glass transition temperature (T_g) is generally taken as a demonstration that the two components of a polyblend form a single, homogeneous amorphous phase.[1] On the contrary, two composition independent T_gs indicate that the two components are immiscible. Partial miscibility is generally invoked to explain the presence of two composition dependent T_gs. This criterion has been used to characterize several LCP/polymer blends and, in particular, those based on the two most widely investigated LCPs Vectra-A and (PET/H). Vectra-A (VA), by Hoechst-Celanese, is a wholly aromatic copolyester composed of 4-hydroxybenzoic (H) and 2-hydroxy-6-naphthoic (N) units. (PET/H), first synthesized by Eastman Kodak, consists of poly(ethylene terephthalate) (PET) copolymerized with different proportions of H.

VA was shown to be immiscible with most thermoplastic polymers, including PET,[6] and poly(butylene terephthalate) (PBT),[7] poly(ether ether ketone) (PEEK),[8] and high density polyethylene (HDPE).[7] The behavior of the VA blends with bisphenol-A polycarbonate (PC) is less clear. For the LCP/polymer blends, miscibility should involve the amorphous phase of the matrix polymer and the organized, nematic phase of the LCP and this is probably unfavorable on a thermodynamic ground.

The behavior of the blends containing PET/H is much more intricate because this LCP is considered to be biphasic itself. Thus, PBT,[9] PET,[10] and PC[10–12] were found to be miscible with the PET-rich phase of the LCP and immiscible with the H-rich phase, whereas polystyrene (PS)[10] and other polymers were shown to be completely immiscible with this LCP.

MORPHOLOGY

As already noted, most LCP/polymer blends are biphasic in nature. SEM examination of fracture surfaces generally reveals that the LCP dispersed phase, in the form of spherical or elongated particles, is partially pulled out of the matrix; these are signs of poor interfacial adhesion.

For LCP/polymer blends characterized by poor interfacial adhesion (which is the most common event) reasonable toughness and strength may still be obtained if the LCP particles have sufficiently high aspect ratio, i.e., if they possess a fibrillar geometry. The higher the aspect ratio, the higher the gripping force which opposes fiber pulling-out and the lower the need for adhesion.

The morphology of LCP/polymer blends is determined by the characteristics of the individual components (for example the relative viscosities), by the volume fraction of the LCP, and especially by the conditions adopted for the mixing and processing stages.

Under uniform shear field conditions, the LCP droplets appear generally spherical, with no preferential alignment in the flow direction.[13,14]

RHEOLOGY AND PROCESSING

Shear Flow

The more important effect of blending small amounts of TLCP with flexible polymers is the drastic reduction of the viscosity.[3,4] The viscosity of the flexible matrix is remarkably reduced at low concentration of TLCP.

The lower viscosity improves the processing of the polymers. Moreover it allows materials very difficult to process to be processed or makes it possible to reduce the processing temperature and avoid dangerous degradation phenomena. Two systems show two different behaviors. In the first case, the viscosity of the blends is intermediate between those of the polymer parents. In the second case, the flow curves of the blends are below those of the two components. The viscosity is intermediate where the viscosity of the flexible matrix is higher than that of the rigid dispersed phase.[16,17]

Elongational Flow

The reduction of the shear viscosity also causes a drastic decrease of the elongational viscosity of the blends. These materials could not then be used in processing operations where the elongational flow is involved. The few data in non-isothermal elongational flow reported in the literature[4,15] confirm this only in part.

MECHANICAL PROPERTIES

The reinforcing effect of a second phase in a blend depends on the mechanical properties, the size, and the shape of the particles of the dispersed phase. In particular, high aspect ratio values give rise to remarkable enhancement of many mechanical properties. For this reason, fiber-filled polymers exhibit mechanical properties better than those of powered-filled polymers.

The effects of the TLCPs on the properties of blends with flexible polymers can be summarized as follows:

- Improvement of the modulus and, in some cases, of the tensile strength
- Reduction of elongation at break and impact strength

In order to obtain the best mechanical properties, it is necessary to stipulate:

- Viscosity of the LCP phase similar or slightly lower than that of the matrix
- Appropriate cooling behavior
- Elongation or convergent flow

REFERENCES

1. Paul, D. R.; Barlow, J. W. *J. Macromol. Sci., Rev. Macromol. Chem.* **1980**, *C18*, 109.
2. Cogswell, F. N.; Griffin, B. P.; Rose, J. B. U.S. Patents 4 386 174, 1983; 4 433 083, 1984; 4 438 236, 1984.
3. Dutta, D.; Fruitwala, H.; Kohli, A.; Weiss, R. A. *Polym. Eng. Sci.* **1990**, *30*, 1005.
4. La Mantia, F. P., Ed., *Thermotropic Liquid Crystal Polymer Blends* Technomic: Lancaster, 1993.
5. Kiss, G. *Polym. Eng. Sci.* **1987**, *27*, 410.
6. Silverstein, M. S.; Hiltner, A.; Baer, E. *J. Appl. Polym. Sci.* **1991**, *43*, 157.
7. Harada, T.; Tomari, K.; Hamamoto, A.; Tonogai, S.; Sakaura, K.; Nagai, S.; Yamaoka, K. *SPE Antec* **1992**, 376.
8. Mehta, A.; Isayev, A. I. *Polym. Prepr.* **1989**, *30/2*, 548.
9. Kimura, M.; Porter, R. S. *J. Polym. Sci., Polym. Phys. Ed.* **1984**, *22*, 1697.
10. Zhuang, P.; Kyu, T.; White, J. L. *SPE Antec* **1988**, 1237.
11. Huang, S.; Griffin, A. C.; Porter, R. S. *Polym. Eng. Sci.* **1989**, *29*, 55.
12. Nobile, M. R.; Amendola, E.; Nicolais, L.; Acierno, D.; Carfagna, C. *Polym. Eng. Sci.* **1989**, *29*, 244.
13. Blizard, K. G.; Baird, D. G. *Polym. Eng. Sci.* **1987**, *27*, 653.
14. La Mantia, F. P.; Valenza, A.; Paci, M.; Magagnini, P. L. *Polym. Eng. Sci.* **1990**, *30*, 7.
15. La Mantia, F. P.; Valenza, A.; Scargiali, F. *Polym. Eng. Sci.* **1994**, *34*, 799.
16. La Mantia, F. P.; Valenza, A. *Makromol. Chem. Macromol. Symp.* **1992**, *56*(151).
17. La Mantia, F. P.; Valenza, A. *Polym. Networks Blends* **1993**, *3*, 125.

Liquid Crystalline Polymerization

See: *Solid-State Polymerization (Crosslinked Di-p-Ethynyl Substituted Polymers)*

Liquid Crystalline Polymers

See: *Block Copolymers (By Changing Polymerization Mechanism)*
Cellulosic Liquid Crystals
Comb-Like Amphiphilic Polymers (Supramolecular Structure)
Comb-Like Polymers
Cyclic Imino Ethers (Ring-Opening Polymerization)
Electrorheological Materials
Engineering Thermoplastics (Survey of Industrial Polymers)
Epoxide-Amine Addition Polymers, Linear
Ferroelectric Polymers (Structural Phase Transitions)
Glass-Polymer Melt Blends
Hairy Rod-Like Polymers
Hydroxypropyl Cellulose (Crosslinked, Cholesteric Liquid Crystal)
In Situ Thermoplastic Composites (from Liquid Crystalline Polymers)
Ionene Polymers
Liquid Crystalline Aromatic Polyesters
Liquid Crystalline Elastomers
Liquid Crystalline Ethyl-Cyanoethyl Cellulose
Liquid Crystalline Polyacetylene Derivatives
Liquid Crystalline Polymer-Based Blends (Thermotropic)
Liquid Crystalline Polymers (Conjugated Chain)
Liquid Crystalline Polymers (Heat-Strengthened Thermotropic)

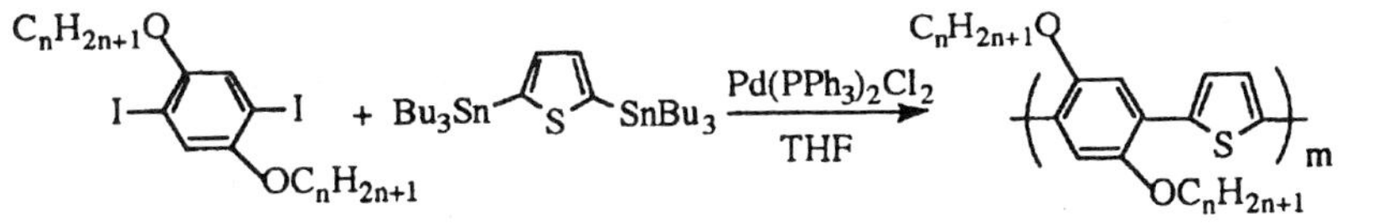

Polymer **1** n = 4, 5, 6, 7, 8, 9, 12, 16

SCHEME I. Synthesis of poly(*p*-phenylene thiophene) from Stille reaction.

LIQUID CRYSTALLINE POLYMERS (Conjugated Chain)

Luping Yu and Zhenan Bao
Department of Chemistry
The University of Chicago

The conjugated polymer chain is obviously a rigid rod which could act as a mesogenic unit to manifest the LC properties if the conjugated polymers are fusible.[1,2] Compared to the normal main-chain LC polymers, such as polyesters and polyamides, many special physical properties associated with the delocalized π-electron system in a conjugated LC polymer add new dimensions to the studies of liquid crystalline materials. These physical properties can be enhanced due to the orientation effect induced by the LC ordering. For example, the conductivity of the electroactive polymers and the third order nonlinear optical (NLO) susceptibility strongly depend upon the orientation of the conjugated polymer chains.[3–7] After the polymer is aligned by mechanical stretching, the conductivity parallel to the stretching direction is much larger than in the perpendicular direction. However, the conductivities measured in stretched materials are still far below the expected intrinsic conductivity of conjugated polymers. Among the reasons is the inefficient alignment of the polymer chain. If a conjugated polymer possesses liquid crystallinity, the alignment of the conjugated LC polymer chain can be more effectively achieved through the field effect of the liquid crystalline materials. Consequently, the anisotropy of physical properties can be enhanced and new physical properties may occur.

To make this rationale work, the key step is the preparation of conjugated polymers soluble and fusible under an appropriate temperature range.

POLY(*p*-PHENYLENE-1,4-THIOPHENE)S (PPT)[8–10]

Synthesis

The Stille coupling reaction is utilized to synthesize polymer **1**.[8,10] The polymerization was carried out easily in THF in the presence of a catalytic amount of $Pd(PPh_3)_2Cl_2$ (5% mol eq.) under a nitrogen atmosphere as shown in **Scheme I**. All of these polymers were soluble in common organic solvents and fusible under certain temperature range. Their basic physical properties are reported in Reference 9.

Liquid Crystallinity Studies

For polymer **1**, polarized microscopy studies revealed a birefringent fluid melt above their melting temperatures.[9] A characteristic shear opalescence for nematic phase was observed in every polymer sample.

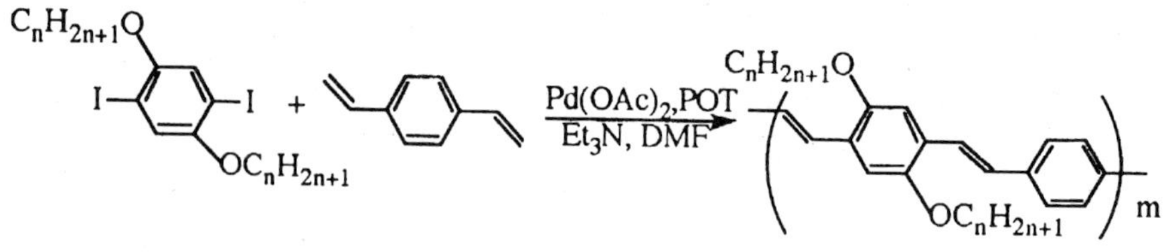

Polymer **2** n = 4, 7, 9, 12, 16

SCHEME II. Synthesis of poly(*p*-phenylene vinylene) from Heck reaction.

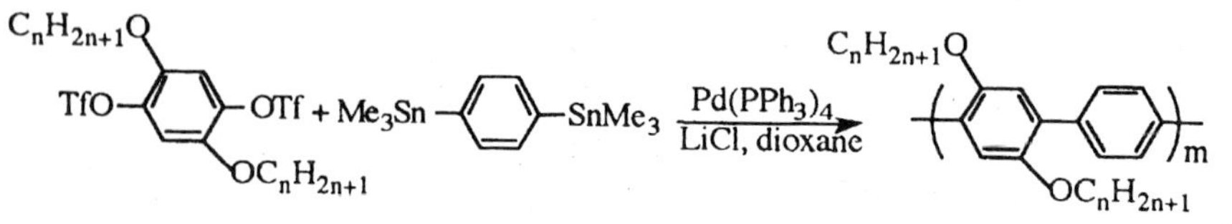

Polymer **10** n = 6, 8, 11, 16

SCHEME III. Synthesis of poly(*p*-phenylene)s from Stille reaction.

POLY(*p*-PHENYLENE VINYLENE)S (PPV)[11]

Synthesis

Poly(phenylene vinylene) can be synthesized from a number of methods, such as the McMurry reaction and the Wittig reaction, providing the elimination of bis-sulfonium salt from a precursor polymer.[12–17] The Heck reaction was developed recently for synthesizing soluble and fusible polymers with well-defined structures.[11,18,19]

Poly(2,5-dialkoxyphenylene vinylene)s were synthesized according to **Scheme II**.

Liquid Crystallinity Studies

The *trans*-poly(phenylene vinylene) backbone is a rigid rod which can act as the mesogenic unit. Poly(phenylene vinylene) with side chain substitutes has a similar mesogenic structure to rigid rod polyester and polyamide with side chains. Therefore, it should be able to manifest liquid crystalline properties.

POLY(*p*-PHENYLENE)S (PPP)[20]

Synthesis

A number of synthetic methods have been developed to synthesize PPPs.[21–23] Recently, the Pd(O) catalyzed coupling reaction of 2,5-dialkyl-*p*-dibromobenzene with benzenediboronic acids has been successfully applied to synthesize high molecular weight PPPs.[24] PPPs were also synthesized from the pyrolysis of the precursor polymer from a transition metal-catalyzed polymerization of heteroatom-functionalized cyclohexadienes and from nickel-catalyzed coupling of bistriflates of substituted hydroquinones.[25,26]

Polymers **10** were synthesized following **Scheme III** where the bistriflates of *p*-dialkoxy-substituted hydroquinone was synthesized from 2,5-dihydroxy-*p*-benzoquinone via three steps.[27–29]

Recently, Wegner et al. reported the observation of liquid crystallinity in 2,5-disubstituted poly(*p*-phenylene)s.[30]

The liquid crystalline properties in methyl-substituted oligo(*p*-phenylene)s were reported by Heitz et al.[31,32]

PHYSICAL PROPERTIES

It is widely known that conjugated polymers exhibit unusual electronic and optical properties. In polymer systems **1**, **2**, and **10** discussed above, many physical properties have been investigated. These polymer systems all exhibit strong fluorescence when they are excited by a proper light source, clearly demonstrating the potential for the electroluminescent (EL) applications.[8,10,11,20] It is known that many conjugated polymers exhibit EL activity, but the low quantum yield is the major problem.[33] If those materials also possess liquid crystallinity, the polymer chain can be aligned effectively and the EL quantum yield can be enhanced due to a closer chain packing.

Both polymer system **1** and polymer system **2** exhibit reasonably large third-order optical nonlinearity. After the polymer chain is aligned by the magnetic field, we can expect that some of the third-order nonlinear optical susceptibility tensor components can be enhanced. Furthermore, these polymers are found to be photoconducting; the photoconductivity can also be enhanced by the chain alignment in LC state. These are several examples of physical properties which can gain advantages from the LC properties of the materials.

REFERENCES

1. Skotheim, T. A. *Handbook of Conducting Polymers*; Marcel Dekker: New York, 1986, Vols. 1 and 2.
2. Skotheim, T. A. *Electroresponsive Molecular and Polymeric Systems*; Marcel Dekker: New York, Vol. 2, 1991.
3. Bradley, D. D. C.; Friend, R. H.; Lindenberger, H.; Roth, S. *Polymer* **1986**, *27*, 1709.
4. Gagnon, D. R.; Capistran, J. D.; Karasz, F. E.; Lenz, R. W. *Polym. Bull.* **1984**, *12*, 293.
5. Murase, I.; Ohnishi, T.; Noguchi, T.; Hirooka, M. *Polymer. Comm.* **1984**, *25*, 327.
6. Murase, I.; Ohnishi, T.; Noguchi, T.; Hirooka, M. *Synth. Met.* **1987**, *7*, 639.
7. Machado, J. M.; Denton, R. R.; Schlenoff, J. B.; Karasz, F. E.; Lahti, P. M. *J. Polym. Sci. Polym. Phys.* **1989**, *27*, 199.

8. Bao, Z. N.; Chan, W. K.; Yu, L. P. *Chem. Mater.* **1993**, *5*, 2.
9. Yu, L. P.; Bao, Z. N.; Cai, R. B. *Angew. Chem.* **1993**, *105*, 1392.
10. Bao, Z. N.; Cai, R. B.; Yu, L. P. *Polym. Prepr.* **1993**, *34*(2), 749.
11. Bao, Z. N.; Chen, Y. M.; Cai, R. B.; Yu, L. P. *Macromol.* **1993**, *26*, 5281.
12. Rajaraman, L.; Balasubramanian, M.; Nanjan, M. J. *Curr. Sci.* **1980**, *49*, 101.
13. Feast, W. J.; Millichamp, I. S. *Polym. Commun.* **1983**, *24*, 102.
14. McDonald, R. N.; Campbell, T. N. *J. Am. Chem. Soc.* **1960**, *82*, 4669.
15. Gooding, R.; Lillya, C. P.; Chien, J. C. W. *J. Chem. Soc. Chem. Commun.* **1983**, *151*.
16. Kossmehl, G.; Wallis, J. *Makromol. Chem.* **1982**, *183*, 331.
17. Gagnon, D. R.; Capistran, J. D.; Karasz, F. E.; Lenz, R. W. *Polym. Bull.* **1984**, *12*, 293.
18. Greiner, A.; Heitz, W. *Makromol. Chem. Rapid Commun.* **1988**, *9*, 581.
19. Martelock, H.; Greiner, A.; Heitz, W. *Makromol. Chem.* **1991**, *192*, 967.
20. Yu, L. P.; Bao, Z. N. *Adv. Mater.* **1994**, *6*, 156.
21. Feast, W. J. *Handbook of Conducting Polymers*; Skotheim, T. A., Ed.; Dekker: New York, 1986.
22. Reynolds, J. R.; Pomerantz, M. *Electroresponsive Molecular and Polymeric Systems*, Vol. II; Skotheim, T. A., Ed.; Dekker: New York, 1988.
23. Tour, J. M.; Stephens, E. B. *J. Am. Chem. Soc.* **1991**, *113*, 2309 and references therein.
24. Martina, S.; Schluter, A. D. *Macromolecules* **1992**, *25*, 3607 and references therein.
25. Gin, D. L.; Conticello, V. P.; Grubbs, R. H. *J. Am. Chem. Soc.* **1992**, *114*, 3167.
26. Percec, V.; Okita, S.; Weiss, R. *Macromolecules* **1992**, *25*, 1816.
27. Colletti, R. F.; Stewart, M. J.; Taylor, A. E.; MacNeill, N. J.; Mathias, L. J. *J. Polym. Sci.: Part A: Polym. Chem.* **1991**, *29*, 1633.
28. Berg, S.; Krone, V.; Ringsdorf, H. *Makromol. Chem., Rapid Commun.* **1986**, *7*, 381.
29. Echavarren, A. M.; Stille, J. K. *J. Am. Chem. Soc.* **1987**, *109*, 5478.
30. Witteler, H.; Lieser, G.; Wegner, G.; Schulze, M. *Makromol. Chem. Rapid Commun.* **1993**, *14*, 471.
31. Heitz, W. *Makromol. Chem. Macromol. Symp.* **1989**, *26*, 1.
32. Heitz, W. *Makromol. Chem. Macromol. Symp.* **1991**, *47*, 111.
33. Burroughes, J. H.; Bradley, D. D.; Brown, A. R.; Marks, R. N.; Mackay, K.; Friends, R. H.; Burns, P. L.; Holmes, A. B. *Nature* **1990**, *347*, 539.

LIQUID CRYSTALLINE POLYMERS (Heat-Strengthened, Thermotropic)

Robert R. Luise
Hytem Consultants, Incorporated

Main-chain thermotropic liquid crystalline polymers (TLCPs) represent a class of high performance condensation polymers composed mainly of *para*-extended aromatic polyesters.[1]

Often-quoted advantages of TLCPs include high thermal, electrical, and solvent resistance combined with excellent (unparalleled) orienting flow characteristics and dimensional stability–ideal for injection-molded applications such as thin-walled electrical insulators with close tolerances. In film form, excellent barrier properties are afforded by the sense uniaxial LC morphology and it is anticipated that medical and food packaging opportunities in addition to electronics will emerge rapidly coincident with lower resin costs and expanded TLCP production.[2]

Perhaps the outstanding TLCP characteristics, certainly the one which attracted major industrial attention to these materials in the 1970s, is its ability in oriented fiber form to dramatically strengthen by postheat treatment due to solid-phase polymerization.[3] The latter afforded unprecedented tensile properties for melt-spun polyesters seen only before in "Kevlar" solution-processed polyamide fibers.

While main-chain TLCPs have been covered extensively in a number of reviews, the historical perspective on these materials remains somewhat clouded, largely because essentially all of it lies in the often-overlooked patent literature.[4–7] Moreover, much of the work related to post-heat strengthening of these materials, particularly in nonfiber form, remains embedded in patent literature. The present paper seeks to further clarify the historical perspective on these materials based on the published literature and is also intended as a review of the unique post-strengthening attributes of these materials observed not only in fibers, but also in other forms such as films, sheets, and pultruded and molded shapes.

HEAT STRENGTHENING

As shown in early DuPont disclosures heat strengthening consists of solid-phase polymerization principally of axially-oriented as-spun fibers in a continuously purged inert atmosphere.[8–12] Purging is required to remove post-polymerization by-products and drive the equilibrium polycondensation reaction forward.[3] The atmosphere should be inert or *in vacuo* to minimize undesirable side reactions, which may cause branching (e.g., Fries rearrangement in polyesters) and retard chain length growth. Preferred temperatures are near, but below, the melting and heating point stages since the melting point may increase as much as 30°C in the process. It is important not to exceed the melting point in heating to prevent deorientation.

Orientation is critical for strength enhancement; retention of orientation despite heating in the unconstrained state is a novel feature of the rigid extended-chain uniaxial morphology which imparts exceptional dimensional stability (little or no change in fiber length observed).[3]

While heat strengthening has been applied mainly to fibers, often overlooked are other TLCP materials discussed below that also respond to post-treatment such as films, uniaxial sheets, molded or extruded shapes, pultruded shapes, foams, pulps, and pressed fiber webs.

Fibers

Fiber heat strengthening has been covered most extensively in a number of review and research investigations.[4–7] Early DuPont disclosures describe the process as outlined briefly above, including the use of inert coatings (e.g., talc or graphite) to reduce interfilament sticking and packaging options for yarns such as skeins, soft-covered bobbins, and piling in perforated baskets.[8–12] Other reported investigations include studies of the effects of spinning variables and structure on heat-treatment response by Krigbaum and co-workers and a tenacity model proposed by Yoon based on uniaxial chain length growth.[14,15,17] Krigbaum et al. examined the effect of heat-treatment response of "Vectra"-type TLCPs, documenting the importance of spin-stretch and as-spun orientation for tenacity improvement. They also observed increases in molecular weights, crystallinity, and crystalline melting point on heat treatment, consistent with other investigators.[16] Elongation, increases, consistent with molecular weight growth, were also reported.[14]

Tenacity Model

Yoon's uniaxial tenacity model corroborated experimental observations that tenacity increase on heat treatment is primarily due to an increase in chain length.[3,5,8,17]

"Vectran" Fiber

Presently, the only commercial TLCP fiber is Hoechst-Celanese' "Vectran", introduced in 1988. The fiber is based on the "Vectra" HBA/HNA 73/27 copolyester, and is produced in both as-spun and heat-strengthened forms, the latter by Kuraray in Japan.

Films

Heat-strengthened TLCP film was first disclosed in DuPont's 1975 patent application.[8] A U.S. patent to Celanese in 1982 claims improved flow orientation and mechanical properties for "Vectra"-type extruded film using a grid of cone-shaped holes in the die.[18] Improved orientation with the grid resulted in a larger heat-strengthening response with MD tensile strengths to .83 GPa. This compares with a much greater tenacity response observed for more highly oriented fibers.[12,17]

Biaxial Films

Since high MD film orientation leads to poor transverse (TD) and tear mechanical properties, blown film techniques have been utilized to improve the property balance.

Barrier Properties

Excellent permeability resistance to gases (e.g., oxygen, water vapor) have been reported for TLCP films.[19]

Uniaxial Sheets

TLCP fibers can be used as building blocks for larger uniaxial shapes because of their ease of self-adhesion and retention of orientation when heated in aligned bundles under pressure near (but below) the melting point.[20–23]

Sheets can be heat-strengthened similar to that of a fiber to yield a novel 100% fiber resinless advanced composite, with tensile strengths to 1.9 GPa and moduli in the 50–60 GPa range.[3] The transparency is retained in the heat-strengthening step, another indication that the uniaxial morphology (i.e., interchain distance) is preserved on heat treatment, consistent with solid-phase polymerization. Alternatively, high strength sheets can be laminated into multi-ply structures with conventional epoxy of TLCP film adhesives, or as previously mentioned, be converted to biaxial film by transverse stretching on heating due to good lateral ductility.[3,20]

Pultruded Shapes

In a manner similar to uniaxial sheets, a U.S. patent to DuPont shows that rovings of TLCP as-spun yarn can be pultruded by conventional methods into larger uniaxial shapes (e.g., rods, tapes) without a matrix binder.[23] Significant improvements in tensile and shear mechanical properties are observed after heat strengthening by methods similar to fibers.[3] Retention of fiber orientational properties is good.

Molded Articles

Injection-molded and extruded articles can also be heat strengthened as discussed in another U.S. patent to DuPont.[24,25] Improvements in tensile and flexural strengths (20–90%) are noted for standard plaques and flex bars up to 3.2 mm thickness, and also for 1.2 mm extruded rods of copolyesters and polyazomethines. Toughness (or energy-to-break) and impact strength are most improved (100–200%) due to increased elongation-to-break, indicating the effectiveness of solid-phase polymerization even in these larger shapes. Strength improvements are generally lower than those of other more highly oriented uniaxial materials discussed above.

Mechanical Anisotropy

A number of injection-molding studies of TLCPs (mainly copolyesters) have noted the anisotropy of properties dependent on extensional flow-induced orientation.[13,25–30]

Other Materials

A European patent application to Celanese discloses rigid TLCP foams with improved strength and thermal, flame and solvent stability by heat strengthening.[31]

SUMMARY

Although heat strengthening has been applied mainly to fibers and appears to be the basis for the "Vectran" high-tenacity TLCP fiber manufactured by Kuraray and Hoechst-Celanese, other TLCP product forms exist where heat strengthening has been effectively utilized. The latter includes film, uniaxial transparent sheets, and pultruded shapes prepared from fibers, injection-molded and extruded articles, and foams. While strength response is generally orientation-dependent (fibers > films > moldings), other benefits such as enhanced thermal and solvent resistance due to increased melting point and molecular weight should have value in many non-fiber applications. Molded parts, in particular, should be ideal for large-scale heat strengthening. In addition, larger uniaxial shapes such as pressed sheets and pultruded materials utilizing fibers as the basic building blocks should have important future reinforcement applications (e.g., infrastructures), particularly as TLCP costs decline and lateral bonding techniques are utilized to improve sheer properties. Transparent sheets may also have optical display applications.

In terms of fundamentals, while a number of heat-strengthening studies discussed in this work tend to support a basic strengthening mechanism of post-polymerization (e.g., Yoon's tenacity model; unaffected lateral spacing indicated by film permeability and uniaxial sheet data) there is still a need for basic kinetic data for the solid-phase polymerization process, including transesterification rates for copolyesters and the effects of thickness and orientation that are often difficult to separate on by-product diffusion and conversion rate. It is hoped that these will appear as TLCP commercial acceptance grows.

REFERENCES

1. Jackson, Jr. W. J. *The British Polymer Journal* **1980**, December, 154.
2. Thomas, L.; Roth, D. D. "Films from Liquid Crystals," *Chemtech* **1990**, September, 546.

3. Luise, R. R. "Liquid Crystalline Condensation Polymers," In *Integration of Fundamental Polymer Science and Technology - 5*, Lemstra, P. J.; Kleintjens, L., Eds.; Elsevier: 1990; p 207.
4. Kwolek, S. L.; Morgan, P. W.; Schaefgen, J. R. "Liquid Crystal Polymers," In *Encyclopedia of Polymer Science and Engineering*, 2nd ed.; John Wiley & Sons: 1987; Vol. 9, p 1.
5. Calundann, G. W.; Jaffe, M. Anisotropic Polymers: Their Synthesis and Properties," Proceedings of the Robert A. Welch Conference on Chemical Research, 1982, Synthetic Polymers, Vol. XXVI.
6. Economy, J.; Volkson, W. "Structural Properties of Aromatic Polyesters," In *The Strength and Stiffness of Polymers*, Zachariades, A. E.; Porter, R. S., Eds.; M. Dekker: 1983, p 293.
7. Chung, T-S. *Polymer Engineering and Sciences*, **1986**, *Vol. 36*, 13, 901.
8. Kleinschuster, J. J.; Pletcher, T. C.; Schaefgen, J. R.; Luise, R. R. (DuPont), German Offen. 2 520 820, November 27, 1975; Belgian Patents 828 935 and 828 936, November 12, 1975.
9. Pletcher, T. C. (DuPont), U.S. Patent 3 991 013, November 9, 1976.
10. Kleinschuster, J. J. (DuPont), U.S. Patent 3 991 014, November 9, 1976.
11. Schaefgen, J. R. (DuPont), U.S. Patent 4 118 372, October 3, 1978.
12. Luise, R. R. (DuPont), U.S. Patent 4 183 895, January 15, 1980.
13. Jackson, W. J. Jr.; Kuhfuss, H. F.; Gray, T. F. Jr. *30th Anniversary Technical Conference, Reinforced Plastics/Composites Institute*, The Society of the Plastics Industry, Inc., 1975, Section 17-D, pp 1–4.
14. Miramatsu, H.; Krigbaum, W. R. *Macromolecules* **1986**, *19*, 2850.
15. Yang, D. K.; Krigbaum, W. R. *J. Polym. Sci. B. Polym. Physics* **1989**, *27*, 1837.
16. Sarlin, J.; Tormala, P. *J. Polymer Sci. B. Polym. Physics* **1991**, *29*, 395.
17. Yoon, H-N. *Colloid Polym. Sci.* **1990**, *268*, 230.
18. Ide, Y. (Celanese) U.S. Patent 4 332 759, June 1, 1982.
19. Thomas, L. D.; Roth, D. D. *Chemtech* **1990**, *September*, 546.
20. Luise, R. R. (DuPont) U.S. Patent 4 786 348, November 22, 1988.
21. Luise, R. R. (DuPont) U.S. Patent 4 939 026, July 3, 1990.
22. Luise, R. R. (DuPont) EP 0354285, February 14, 1990.
23. Luise, R. R. (DuPont) U.S. Patent 4 832 894, May 23, 1989.
24. Luise, R. R. (DuPont) U.S. Patent 4 247 514, January 27, 1981.
25. Huynh-Ba, G.; Cluff, E. F. "Structural Properties of Rigid and Semirigid Liquid Crystalline Polyesters," In *Polymeric Liquid Crystals*; Blumstein, A., Ed., 1985, p 217.
26. Snokoc, E.; Sarlin, J.; Tormala, P. *Mol. Cryst. Liq. Cryst.* **1987**, *153*, 515.
27. Sweeney, J.; Brew, B.; Duchett, R. A.; Ward, I. M. *Polymer* **1992**, *33*, 4901.
28. Boldizar, A. *Plastics and Rubber Processing and Applications* **1988**, *10*, 73.
29. Bangert, H. *Kunstoffe German Plastics* **1989**, *79*, 1327.
30. Meges, G.; Schacht, T.; Becker, H.; Ott, S. *Int. Polymer Processing* **1987**, *2*, 281.
31. Ide, Y. (Celanese) EP 70658, October 1, 1985.

LIQUID CRYSTALLINE POLYMERS (Main-Chain, Thermotropic)

W. A. MacDonald
ICI Films

Low molecular weight liquid crystalline compounds have been known for about 100 years but main-chain thermotropic liquid crystal polymers have attained prominence only in the last 15 years. The first well-characterized description of a polymer exhibiting thermotropic behavior appeared in the mid 1970s when Jackson described a series of copolymers prepared by the acidolysis of poly(ethylene terephthalate) with *p*-acetoxybenzoic acid, which exhibited the phenomenon of opaque melts, low-melt viscosities and anisotropic properties.[1] There has since been considerable research in both industry and academia and a recent survey of the liquid crystal polymer (LCP) field estimated that 40 companies are active in LCP research internationally with further work being carried out in smaller specialized companies.[2]

DESIGN OF THERMOTROPIC LIQUID CRYSTAL POLYMERS

Completely rigid rod-like molecules such as poly(4-oxybenzoyl) or poly(*p*-phenylene terephthalate) tend to be highly crystalline and intractable with melting points above the decomposition temperature of the polymers (>450°C). The problem of thermotropic LCP design is to disrupt the regularity of the intractable *para*-linked aromatic polymers to the point at which mesomorphic behavior is manifest below the decomposition temperature and the materials can be processed in fluid, yet ordered, states. However, the disruption must not be taken to the stage, where conventional isotropic fluid behavior is preferred. These requirements that the polymer retain some rod-like nature, but at the same time be melt-processable below 400–450°C have limited thermotropic LCPs mainly to polymers based on the linear ester or ester/amide bonds.

PROPERTIES OF THERMOTROPIC LIQUID CRYSTAL POLYMERS

Rheology

In general, the shear viscosity of LCPs is much lower than that of conventional polymers at a comparable molecular weight and the transition from the isotropic state to the liquid crystalline state is usually accompanied by a significant decrease in melt viscosity. At the onset of nematic behavior, the melt viscosity of the LCPs is three decades' less order of magnitude than that of a similar but nonmesogenic polymer.[3]

Molecular orientation occurs readily during melt flow and it has been demonstrated that elongational/extensional flow is primarily responsible for orientation of LCPs during melt processing.[5,6]

From a commercial point of view, the low melt viscosity of LCPs at high shear rates relative to conventional polymers is one of the key features of LCPs and enables their use.[7]

Mechanical

Mechanical properties, particularly tensile strengths and stiffness, depend upon the degree of orientation. This is limited to some extent by the fabrication method and type of article produced.

As the level of elongational flow increases, the mechanical properties increase to the very high mechanical moduli and tensile strengths demonstrated by LCP fibers.[4,8]

LCPs exhibit low elongation to break, typically around 5–10%, due to the ordered structure of the solid which resembles that of fiber reinforced conventional plastics. Adding glass fiber to LCPs further reduces the elongation to break.

LCPs are tough and a benign failure is experienced on impact, similar to that exhibited by long fiber-reinforced polymers or natural wood, i.e., the failure is neither ductile nor brittle and the moldings generally do not shatter.

LCPs exhibit a significant drop in modulus with temperature, and although this can be influenced by control of the component parts of the LCP, this is obviously undesirable in an engineering resin.[9,10] However, this is partly offset by the exceptionally high stiffness and strength of LCPs at room temperature and the presence of useful mechanical properties at high temperatures (>200°C).

Fabricated LCP articles are anisotropic and the anisotropy ratio, i.e., the difference in properties along and across the flow direction, increases with orientation and so is highest in fibers.

LCPs absorb low levels of moisture (typically less than 0.2% on immersion in water) and therefore the change in dimension of moldings of the LCPs due to moisture absorption is very low. The coefficient of linear thermal expansion of LCPs is much lower than that for conventional polymers (even when glass fiber reinforced) and is comparable to those for metals.

LCPs also exhibit very low mold shrinkage and minimal sinkage and warpage compared to conventional isotropic polymers, permitting precision molding of components.

This is largely related to the unusual molecular morphology of LCPs where there is very little change in the general configuration of the molecules before and after melting.[9–11]

Miscellaneous Properties

LCPs have excellent resistance to a wide range of organic solvents and exhibit very good hydrolysis resistance. The retention of properties in both acidic and basic environments is also very good.[4] As in gas barrier properties the liquid crystalline order is likely responsible for this excellent chemical resistance.

APPLICATIONS

The main application for LCPs will be in areas that exploit combinations of the key properties such as strength, easy flow, excellent dimensional stability, the ability to incorporate high levels of fillers and excellent chemical resistance. Examples of these can be seen in applications in the electronic industry, such as surface mount units and printed wiring boards, where the similarity in thermal expansion for metal and LCPs is expected to result in good component integrity and minimal strain when components containing metals (e.g., solder) and LCPs contact and are subject to thermal cycling or shock. In addition, the easy flow and low warpage will allow precision molding of complicated components and the ability to withstand strong solvents, increasingly used to clean between ever smaller places, will be increasingly important.

Applications outside the electronic industry include LCP tower packing saddles for formic acid plants that replace ceramics. Here, the better chemical resistance and breaking strengths are the critical properties. Interestingly the main application areas identified so far exploit the thermal and chemical properties rather than mechanical strength, the original property that LCPs were pushed for.

LCPs offer a unique combination of properties and the market for them will grow as new applications exploiting this range of properties are developed.

REFERENCES

1. Jackson, W. J. and Kuhfuss, H. F. *J. Polym. Sci. Chem. Ed.*, **1976**, *14*, 2043.
2. Technology Catalyst Inc, Liquid Crystal Polymer Technology Technical Review, 1988.
3. Griffin, B. P.; Cox, M. K. *Br. Polym. J.* **1980**, *12*, 147.
4. Calundann, G. W.; Jaffe, M. 'Robert A Welch Conferences in Chemical Research Proc. Synth. Polymers', 1982, 247.
5. Ide, Y.; Ophir, Z. *Polym. Engng. Sci.* **1983**, *23*, 261.
6. Viola, G. G.; Baird, D. G.; Wilkes, G. L. *Polymer Engng. Sci.* **1985**, *25*, 888.
7. Cox, M. K. *Mol. Cryst. Liq. Cryst.* **1987**, *153*, 415.
8. Prevorsek, D. C. In *Polymer Liquid Crystals*; Ciferri, A.; Krigbaum, W. R.; Meyer, R. B., Eds.; Academic: 1982.
9. MacDonald, W. A. In *Liquid Crystal Polymers: From Structures to Applications*; Collyer, A. A., Eds.; Elsevier Applied Science: 1992.
10. MacDonald, W. A. In *High Value Polymers* (Special Publication No. 87); Fawcett, A. H., Ed.; The Royal Society of Chemistry: 1990.
11. Blundell, D. J. *Polymer* **1982**, *23*, 359.

LIQUID CRYSTALLINE POLYMERS (Polyesters)

Toshihide Inoue* and Shigeru Okita
Plastics Research Laboratories
Toray Industries, Incorporated

In Japan, the Research Association for Basic Polymer Technology, as a part of R & D of Basic Technology for Future Industries sponsored by MITI, started a research program with the aim of developing high-modulus polymeric materials that can replace metals. Toray participated in this program and found various types of high-modulus LCPs.[1] We also started a project to develop a new LCP independently and finally succeeded to commercialize our original LCP under the trademark Siveras® in 1994.

In this paper, recent advances of LCPs, including the results of our research and development performed at Toray, will be described.

PREPARATION

Polyesters based on HBA or hydroquinone (HQ)/TPA do not exhibit a liquid-crystalline state due to their high melting temperature. To lower the melting temperature of the polymers based on HBA, three approaches had been used.[2]

- incorporation of unsubstituted, rigid rod-like segments such as 4,4′-biphenylene and naphthalene ring into the main chain.
- incorporation of flexible aliphatic units such as 2,2′-dioxyethylene into the main chain, and
- incorporation of rigid kinks such as 1,3-phenylene ring into the main chain.

*Author to whom correspondence should be addressed.

PROPERTIES

General Properties

Many fibrils exist in the cross-section of tensile-fractured as-spun fiber of LCP. These fibrils act as reinforcements and, therefore, LCP is often referred to as a self-reinforced polymer.[4] However, this anisotropy can become a problem when LCPs are used as injection-molded parts, because LCPs produce oriented moldings with anisotropic mechanical and thermal properties. Properties along the flow direction tend to be superior to those across the flow direction. We found this disadvantage was overcome by modifying the monomer composition. LCPs from substituted HQ and 4,4′-diphenyl-dicarboxylic acid (BB) exhibited reduced anisotropy while their moduli were high.[1] Fillers are often added to LCPs to reduce the anisotropy because the addition of any filler disrupts the alignment of the LCP molecules.[5]

This is the reverse of what happens with conventional thermoplastics. Filled LCPs are, therefore, recommended for injection molding. Filled LCPs have excellent processability that leads to short cycle times because of the low-heat fusion and the high flowability in thin sections due to low melt viscosity under high shear. Filled LCPs exhibit little or no flash due to the low injection pressure and the shear sensitivity of viscosity. Molded parts exhibit very little warpage and shrinkage and provide high dimensional stability.[6]

Mechanical Properties

Filled LCPs possess exceptionally high strength and modulus compared with conventional thermoplastics. However, the weldline strength of LCP is very low compared with that of other conventional thermoplastics. Therefore, new technologies for optimizing molding conditions and mold/shear design such as the gate relocating method, the overflow cavity method and the cavity vibration-pressure method have been developed to improve the weldline strength.[3,7]

Thermal Properties

Most LCPs exhibit crystallinity even if they are copolymers. Some of the LCPs, such as methylhydroquinone(Me-HQ)/BB/1,2-bis(2-chlorophenoxyethane)-4,4′-dicarboxylic acid (Cl-PEC), show crystallinity in all compositions.[8]

DuPont discovered amorphous LCPs (amorphous means that crystallinity is not high enough to be detected by differential scanning calorimetry).[9] We found that a bulk substituent such as phenyl hydroquinone (Ph-HQ) synthesized with amorphous LCPs prevented the polymer chains from packing into crystal lattice.[10]

Filled LCPs exhibit a high deflection temperature under load (DTUL) and superior solder resistance.

APPLICATIONS

Injection Molding

The primary target to LCPs is electric, and electric and electronic parts account for 70 to 80% of the total demand. Recently, demand for LCPs for superminiaturizing design and surface mount technology (SMT) is increasing. In addition to excellent heat resistance, outstanding flowability also has been demanded for polymeric materials. LCP is the only material exhibiting both characteristics.[11]

The second largest market for LCPs is office equipment and precise machines based on the LCPs' reduced mold shrinkage and excellent dimensional stability due to their low thermal expansion and high modulus. Commercial applications include floppy disk drive (FDD), hard disk drive (HDD), and printer and camera parts. Recently, LCPs have been used in audio and video parts because of their excellent damping property, which is attributed to their ability to absorb the vibrational energy at the interface of the clear skin/core sandwich layer.

Molded interconnector devices provide chemically plated circuits on molded electronic parts that are expected to replace metal lead frames. Metallizing grades of LCPs are being used for this purpose due to their good soldering resistance, low thermal expansion, and superior flowability. Other commercial applications include light-emitting devices, photointerruptors, solar sensors, isolators, and double balanced mixers.

Fibers

Kuraray succeeded in developing a LCP fiber that was trademarked Vectran® in 1989. Because Vectran provides excellent wear resistance, dimensional stability, electric insulation, strength, and vibrational damping, it is applied to fishing mesh for globe fish and Spanish mackerel, industrial rope, cord, tension members of heaters, optical-fiber tension members, protection clothes, and sports goods such as golf clubs and tennis rackets.

Others

LCPs also are expected to be used as base sheet of printed circuits and Integrated circuit (IC) packaging materials.

POLYMER ALLOYS

Extensive studies for polymer alloys continue with industry, government, and academia working in close cooperation to improve moldability, mechanical properties, heat resistance, dimensional stability, and gas barrier properties. Alloys of LCP with various types of polymers are under development.

REFERENCES

1. Inoue, T. *Proceedings of the second Pacific Polymer Conference and Progress in Pacific Polymer Science 2*, Springer-Verlag; 1992, p. 261.
2. Jackson, Jr. W. J. *Br. Polym. J.* **1980**, *12*, 154.
3. Iijima, M. M. *Proc. Third Japan International SAMPE Symposium* **1993**, *1*, 754.
4. Inoue, T.; Okamoto, M.; Hirai, T. *Kobunshi Ronbunshu* **1986**, *43*, 253.
5. Duska, J. J. *Plastics Engineering* **1986**, *42*(12), 39.
6. Kaslusky, A. *Advanced Materials & Processes* **1993**, *144*(6), 38.
7. Miyazaki, *Plastics Age* **1993**, *39*(3), 175.
8. Inoue, Y.; Yamanaka, T. *Kobunshi Ronbunshu* **1988**, *45*, 783.
9. Inoue, T. *Japan Plastics Industry Annual* **1992**, *35*, 138.
10. Inoue, T.; Yamanakà, T. *Kobunshi Ronbunshu* **1989**, *46*, 75.
11. Edwards, P. *Modern Plastics International* **1989**, *19*(3), 32.

LIQUID CRYSTALLINE POLYMERS (Polyesters of Bibenzoic Acid)

Ernesto Pérez
Instituto de Ciencia y Tecnología de Polímeros
Consejo Superior de Investigaciones Científicas

Since the discovery of the liquid crystalline, or LC, state in 1888 by Reinetzer,[1] there has been enormous development in the field of low molar mass liquid crystals,[2] encouraged by the special applications of these substances in optoelectronic and other devices.

Research on liquid crystalline polymers is more recent, but they have also received considerable attention during the past decades,[3] because they combine the spontaneous anisotropic behavior of liquid crystals with the excellent properties of macromolecules.

Focusing attention on main-chain thermotropic systems, all-aromatic rod-like polymers offer the attraction of very good chemical and thermal stability, but transforming them is difficult, due to their high melting points and limited solubility. Different methods have been used to lower the transition temperatures.[4,5] One is to utilize flexible spacers so that a typical thermotropic system includes a mesogenic group connected to a flexible spacer through a functional group. Many of these structures have been extensively reviewed.[4,6] The simplest of these mesogenic groups is a biphenyl, provided by polycondensation of either 4,4′-dihydroxybiphenyl with a dicarboxylic acid or 4,4′-biphenyldicarboxylic acid (*p,p*'-bibenzoic acid) with a diol.

This article reviews that the mesophase behavior and general properties of thermotropic polyesters derived from bibenzoic acid. The structural formula of these polybibenzoates, or PBs, is the following (**Structure 1**): where A is the spacer. Depending on the spacer, some of these polyesters present easily accessible transition temperatures, good solubility in some common solvents and mesophases stable in a wide range of temperatures, while other polybibenzoates display properties that make them suitable for high-performance polymers.

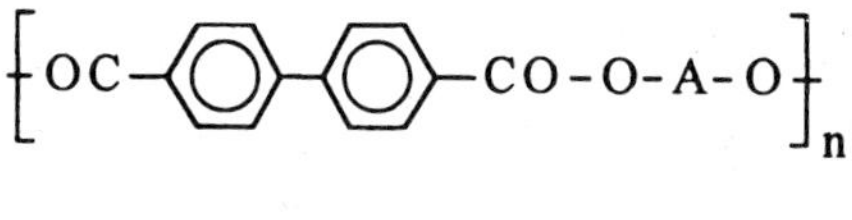

1

Several works have been carried out[8–28] on the thermotropic behavior and properties of PBs, and they show the ability of the biphenyl group to produce mesophase structures. Similar to other main-chain thermotropic polymers, the transition temperatures and the nature and stability of these mesophases are significantly affected by the structure of the spacer. Thus, the role of the spacer is more complex than simply serving to decouple the mesogenic groups.[4]

The spacers included in the reported PBs can be divided into three main categories:

- linear all-methylene spacers;
- linear oxyalkylene spacers, and
- branched spacers

The properties derived from these three groups differ significantly.

In all cases, the polybibenzoates have been synthesized by melt transesterification of the dimethyl or diethyl ester of *p,p*'-bibenzoic acid and the corresponding diol, using a titanium compound as catalyst.

Polybibenzoates with Linear All-Methylene Spacers

Several PBs with linear all-methylene spacers, have been reported,[8–13,15,18] with a number, n, of methylene groups ranging from 2 to 10. In the following, these polymers will be named as PBn.

The transition temperatures from the various authors do vary. This can be explained by the different molecular weights. Thus, the inherent viscosities of these polyesters cover a wide range: from about 0.2 to 1.3 dl g^{-1}. From this article, a molecular weight of 60,000 corresponds approximately with an inherent viscosity of unity.

Up to four thermal transitions have been reported in the melting of PBs with low viscosities.[9,10] On the contrary, the DSC melting curves for the case of higher viscosities seem to be less complicated,[13] as only two endotherms are reported (in the second and subsequent heating runs) for the lower members of the series and only one when n is seven or more (and also for PB3). However, two exotherms are usually obtained in all the cases when cooling from the melt.

The transition temperatures in PBs (as in thermotropic polyesters) depend on the number of methylene units in the spacer and on the odd or even character of that number. Real melting temperatures (transition crystal-isotropic melt) are obtained only for $n \leq 7$ and $n = 3$ (monotropic behavior), while for the other members T_m actually represents the transition crystal-mesophase. In any event, the effect of the number of methylene units in the spacer is important, and a decrease of about 200°C is obtained on passing from PB2 to PB10.

Polybibenzoates with Linear Oxyalkylene Spacers

Oxyalkylene spacers have been widely used in thermotropic polymers.[4] The most common are ethylene oxide oligomers and several works report on PBs with these oxyethylene spacers. Moreover, the dimer of trimethylene oxide has been also used.

A general characteristic of these PBs with ether groups in the spacer is the considerable decrease in the rate of mesophase-crystal transformation. The advantage is that the mesophase of these polymers can usually be investigated at room temperature.

Spacers Derived from Trimethylene Oxide

Several articles[14,17,20,23,25,26] deal with the phase behavior and properties of polyoxybis(trimethylene) *p,p*'-bibenzoate, PDTMB, where the spacer is the dimer of trimethylene oxide, i.e., the structure of PDTMB is the same as that of PB7, but with the central methylene replaced by an oxygen atom.[14] The mesophase of PDTMB was found to be stable at any temperature (below its isotropization point) for considerable time. However, at very long annealing times, this mesophase transforms to the crystal,[23] and two endotherms are obtained in the DSC melting curve of an annealed sample of PDTMB; i.e., this polymer exhibits enantiotropic behavior, contrary to the case of PB7, the all-methylene analog.

Spacers Derived from Ethylene Oxide

Ethylene oxide oligomers have also been used as spacers in polybibenzoates.[16,18,19,24,28] These spacers consist of the dimer, trimer and tetramer of ethylene oxide, and the corresponding PBs will be named as PDEB, PTEB and PTTEB, respectively (the polymer derived from the monomer is PB2, which can be included both in this group and in the all-methylene spacers class).

It has been reported for other families of thermotropic polymers that the phase transitions of the polymers with oxyethylene spacers do not differ very much from those of the all-methylene counterparts.[4] For instance, this is the case of polyesters with a terphenyl group as mesogen.[8]

However, the behavior of PBs differs dramatically in several aspects. First, and similar to PDTMB, the rate of the mesophase-crystal transformation is rather slow, and only an exothermic peak is obtained when cooling PDEB and PTEB from the melt.[16,19,24] Even more, PTTEB appears to be an amorphous polymer when cooled from the melt,[16] and only the glass transition was obtained in the melting of this polymer. Second, the isotropization temperatures of PDEB and PTEB (192° and 114°C, respectively) are significantly lower than those for the corresponding all-methylene homologues. Finally, the type of mesophase is also different.

The behavior of PBs with oxyethylene spacers is, however, similar to that of polyesters derived from *trans*-4,4′-stilbenedicarboxylic acid.[13,30] Only the dimer and trimer have been used as spacers with the stilbene group as mesogen. The transition temperatures for the stilbene series are always higher than those for PBs; i.e., the stilbene group displays a better efficiency as mesogen.[8]

The mesophase of PTEB experiences a very slow transformation into the crystal.

The phase behavior of PDEB is similar to that of PTEB, but with much higher transition temperatures. The glass transition of PDEB occurs at 52°C,[24] and therefore the mesophase of this polymer is stable at room temperature.[19]

The liquid crystalline character is lost for the polybibenzoate with only four ethylene oxide units in the spacer (PTTEB), while in the case of a terphenyl as mesogen, even the polymer with a spacer containing ten such units has been found to be liquid crystalline.[8] This gives an idea of the low effectiveness of the biphenyl group for promoting mesophase properties.

Polybibenzoates with Branched Spacers

The properties of several PBs with branched spacers have been also reported.[20,22,27] These spacers can be divided into two series: I) derived from trimethylene oxide, and II) derived from 1,4-butanediol. None of the polybibenzoates of series I has been found to give liquid-crystalline structures.[20,27] In the PBs studied of series II, all but one of them were found to be liquid crystalline.[22]

SOLUBILITY OF POLYBIBENZOATES

One problem with rod-like polyesters is their very low solubility, which is partly responsible for the fact that very few lyotropic systems have been described.[6]

In general, all PBs are soluble in mixtures of phenols and tetrachloroethane and many of the viscosity determinations have been carried out in such mixtures.[8,11,13,22]

We have also reported a more complete analysis of the effect of ether groups and branches on the solubility behavior of several PBs, by determining the solubility spectra and solubility parameters.[16,20] It was found that the range of solubility is significantly widened by an increase in the number of oxyethylene units in the spacer and by the presence of branches.

In conclusion, most of the polybibenzoates display easily accessible transition temperatures and good solubility in several common solvents.

MECHANICAL PROPERTIES AND VISCOELASTIC RELAXATIONS

The study of macroscopic properties in thermotropic polymers is of great importance, considering the potential applications of these materials. Concerning PBs, several works have been published about different mechanical properties.[13,19,31] as well as about the viscoelastic relaxations of these polymers.[17,24,28,32]

APPLICATIONS

Although we are not aware of any industrial production of polybibenzoates, several patents showing some potential applications have been issued on these polyesters. First, some PBs display high tensile strengths and improved toughness, and are suitable for injection molding.[33,34]

A second potential application is found in optically active PBs. Moreover, such polybibenzoates can be also employed in layered color compensation plates for liquid crystal displays.[35]

Another potential application of PBs is to thermoset polyesters suitable for powder coating. Furthermore, some PBs (and polyesters derived from 3,4′-biphenyldicarboxylic acid) were suggested for gas barriers.[36] Another patent has been issued on compositions of PBs (or PET) with lactone derivatives, moldable at low temperature and displaying low molten viscosities.[37] Finally, blends of smectic liquid crystalline polymers, including PBs, are found to improve the mechanical properties of thermoplastics like polyoxymethylene.[38]

In conclusion, liquid crystalline polybibenzoates present potential applications for high-performance and high-technology polymers.

REFERENCES

1. Reinitzer, F. *Monatsh. Chem.* **1888**, *9*, 421.
2. Demus, D. *Mol. Cryst. Liq. Cryst.* **1988**, *165*, 45.
3. *ACS Symp. Ser.* **1990**, 435.
4. Ober, C. K.; Jin, J.; Fhou, Q.; Lenz, R. W. *Adv. Polym. Sci.* **1984**, *59*, 103.
5. Nöel, C. *Makromol. Chem., Macromol. Symp.* **1988**, *22*, 95.
6. Kwolek, S. L.; Morgan, P. W.; Schaefgen, R. J. *Encyclopedia of Polymer Science and Engineering*, Wiley: New York, 1987; Vol. 9, p 1.
7. Ober, C. K.; Weiss, R. A. *ACS Symp. Ser.* **1990**, *435*, 1.
8. Meurisse, P.; Nöel, C.; Monnerie, L.; Fayolle, B. *Br. Polym. J.* **1981**, *13*, 55.
9. Krigbaum, W. R.; Asrar, J.; Toriumi, H.; Ciferri, A.; Preston, J. *J. Polym. Sci. Poly. Lett. Ed.* **1982**, *20*, 109.
10. Krigbaum, W. R.; Watanabe, J. *Polymer* **1983**, *24*, 1299.
11. Watanabe, J.; Hayashi, M. *Macromolecules* **1988**, *21*, 278.
12. Watanabe, J.; Hayashi, M. *Macromolecules* **1989**, *22*, 4083.
13. Jackson, W. J.; Morris, J. C. *ACS Symp. Ser.* **1990**, *435*, 16.

14. Bello, A.; Pérez, E.; Marugán, M. M.; Pereña, J. M. *Macromolecules* **1990**, *23*, 905.
15. Pérez, E.; Bello, A.; Marugán, M. M.; Pereña, J. M. *Polymer Commun.* **1990**, *31*, 386.
16. Pérez, E.; Benavente, R.; Marugán, M. M.; Bello, A.; Pereña, J. M. *Polymer Bull.* **1991**, *25*, 413.
17. Pereña, J. M.; Marugán, M. M.; Bello, A.; Pérez, E. *J. Non-cryst. Solids* **1991**, *131–133*, 891.
18. Pérez, E.; Riande, E.; Bello, A.; Benavente, R.; Pereña, J. M. *Macromolecules* **1992**, *25*, 605.
19. Pérez, E.; Pereña, J. M. Benavente, R.; Bello, A.; Lorenzo, V. *Polymer Bull.* **1992**, *29*, 233.
20. Marugán, M. M.; Pérez, E.; Benavente, R.; Bello, A.; Pereña, J. M. *Eur. Polym. J.* **1992**, *28*, 1159.
21. Watanabe, J.; Kinoshita, S. *J. Phys. II* **1992**, *2*, 1237.
22. Watanabe, J.; Hayashi, M.; Kinoshita, S.; Niori, T. *Polymer J.* **1992**, *24*, 597.
23. Bello, A.; Riande, E.; Pérez, E.; Marugán, M. M.; Pereña, J. M. *Macromolecules* **1993**, *26*, 1072.
24. Benavente, R.; Pereña, J. M.; Pérez, E.; Bello, A. *Polymer* **1993**, *34*, 2344.
25. Pérez, E.; Marugán, M. M.; VanderHart, D. L. *Macromolecules* **1993**, *26*, 5852.
26. Pérez, E.; Marugán, M. M.; Bello, A.; Pereña, J. M. *Polymer Bull.* **1994**, *32*, 319.
27. Bello, A.; Pereña, J. M.; Pérez, E.; Benavente, R. *Macromol. Symp.* **1994**, *84*, 297.
28. Benavente, R.; Pereña, J. M.; Pérez, E.; Bello, A.; Lorenzo, V. *Polymer* **1994**, *35*, 3686.
29. Watanabe, J.; Komura, H.; Niiori, T. *Liq. Cryst.* **1993**, *13*, 455.
30. Jackson, W. J.; Morris, J. C. *J. Appl. Polym. Sci., Appl. Polym. Symp.* **1985**, *41*, 307.
31. Benavente, R.; Pereña, J. M.; Pérez, E.; Bello, A. *Proceedings of the 9th International Conference on Deformation, Yield and Fracture of Polymers* Cambridge, U.K., 1994.
32. Pérez, E.; Zhen, Z.; Bello, A.; Benavente, R.; Pereña, J. M. *Polymer* **1994**, *35*.
33. Morris, J. C.; Jackson, W. J. (to Eastman Kodak Co.) Eur. Patent Appl. 378 031, July 18, 1990.
34. Morris, J. C.; Jackson, W. J. (to Eastman Kodak Co.) U.S. Patent 4 904 747, February 27, 1990.
35. Iida, S.; Toyooka, T.; Takiguchi, Y.; Enomoto, T. (to Nippon Oil Co., Ltd.; Ricoh Co., Ltd.) Jpn. Kokai Tokkyo Koho JP 91 28 822, February 7, 1991.
36. Mang, M. N.; Brewbaker, J. L. (to Dow Chemical Co.) U.S. Patent 5 138 022, August 1, 1992.
37. Saiki, N. (to Teijin Ltd.) Jpn. Kokai Tokkyo Koho JP 91 200 859, September 2, 1991.
38. Watanabe, J.; Naka, M.; Hijikata, K. (to Polyplastics Co., Ltd.) Eur. Patent Appl. 323 127, July 5, 1989.

LIQUID CRYSTALLINE POLYMERS (Polypeptides)

Satoshi Itou
Aichi College of Technology

A number of polymeric substances can form a liquid crystal, which has practical applications and offers a model of the molecular aggregation state. Technological advances in the fabrication of temperature-resistant and ultrahigh-strength fiber materials depend upon a liquid crystal organization of macromolecules, e.g., poly(*p*-phenylene terephalamide) (Kevlar). Measurements of the basic liquid-crystal properties of polymeric liquid crystals is useful both as a basis for understanding their characteristics and as a guide to the development of theory relating their molecular and macroscopic properties. Understanding their anisotropic properties also is important, including the chain length-to-diameter ratio, L/d, the degree of chain flexibility, and the volume fraction, *v*, of polymer in solution.

Poly(γ-benzyl L-glutamate) (PBLG) solution is particularly well suited for polymeric liquid-crystal studies for two reasons. First, the intrinsic molecular parameters of PBLG are well known. Secondly, the preparation method of PBLG is established and well-controlled PBLG on molecular weight is easily synthesized.

PROPERTIES

Liquid Crystal Formation

Robinson[1] has reported that concentrated solutions of PBLG, a rod-like molecule, form liquid crystals.

Structures of PBLG Liquid Crystals

Robinson[1] found that concentrated solutions of PBLG in appropriate solvent behave in some aspects as cholesteric liquid crystals. When the solutions are viewed in the polarization microscope, the cholesteric phase is characterized by regions of regular striations arranged in a fingerprint pattern.

The cholesteric pitch varies with the polymer concentration. The cholesteric twist sense of poly(γ-benzyl D-glutamate) (PBDG) solution is opposite that of the PBLG solution.[2] PBLG retains the right-handed α-helical conformation, and PBDL retains the left-handed one. In this case, the molecular structure, L and D, directly influences the molecular assemble structure, that is, the cholesteric twist sense.

The structure of PBLG liquid crystal is strongly influenced by solvents.[1–3] The cholesteric pitch drastically varies with solvents. Moreover, the cholesteric twist sense is transformed by the change of solvent from dioxane to dichloromethane.[4] The compensated state (nematic) has been observed in the mixtures of these two solvents.

In many cases, the inverse cholesteric pitch varies linearly with the inverse temperature.

Mechanical Properties of PBLG Liquid Crystals

Mechanical properties of polypeptide liquid crystals are characterized by the Frank elastic constants and the Leslie viscosity coefficients in the same method as low molecular ones.[5] In the elastic constants, three types of deformation, splay, twist, and bend, occur in nematics. These constants have been determined by many methods. The most important ones are light scattering and Freedericksz transition, methods that are applicable to the lyotropic polymer liquid crystals.

Light Scattering Method

The turbidity of liquid crystalline samples is one of their most characteristic features. An oriented liquid crystal has strong depolarized scattering of light (several orders of magnitude stronger than in the isotropic phase), which is caused by orientational fluctuations of the director around its equilibrium value.

Freedericksz Transition

PBLG liquid crystal has the anisotropy of the diamagnetic and dielectric susceptibilities. Therefore, the free energy of an ensemble of liquid-crystalline molecules in an external magnetic or electric field has a minimum for a well-defined orientation of the molecular axes relative to the field. The director tends to align itself along the field. If the directions of the field and the director of the liquid crystal do not correspond to the condition of minimum free energy, a reorientation of the director will occur with a strong field, which overcomes the elastic forces of the liquid crystal. This effect is called the Freedericksz transition.

Experimentally, the Freedericksz transition is applicable to the lyotropic polymer systems. However, the ramp rate in the polymer systems is slower than that in the small molecular liquid crystals. In the polymer systems, we need to devise the cell to avoid solvent evaporation.

APPLICATIONS

Liquid crystalline polymers are widely applied for their temperature-resistant and extremely high strength properties on display, memory, sensor, and optical devices. For PBLG liquid crystals, an attempt has been made to use temporary records of an electron microscope instead of photograph.[6]

REFERENCES

1. Robinson, C. *Trans. Faraday Soc.* **1956**, *52*, 571.
2. Robinson, C.; Ward, J. C.; Beevers, R. B. *Disc. Faraday Soc.* **1958**, *25*, 29.
3. Patel, D. L.; DuPre, D. B. *J. Chem. Phys.* **1980**, *72*, 2515.
4. Robinson, C. *Tetrahedron* **1961**, *13*, 219.
5. de Gennes, P. G. In *The Physics of Liquid Crystals*; Oxford University: Oxford, 1974.
6. Matsuda, J.; Kiyohara, T.; Sawada, H.; Akahane, T.; Moriito, N. *Oyo Buturi* **1985**, *54*, 492.

LIQUID CRYSTALLINE POLYMERS (Rheology)

Francesco Paolo La Mantia
Dipartimento di Ingegneria Chimica dei Processi e dei Materiali
Università di Palermo

The peculiar rheological behavior of liquid crystalline polymers (LCP) is strictly connected with their molecular conformation.

The conformation of the macromolecules of flexible polymers is a random coil with many entanglements. At rest, the molecular segments are randomly oriented (isotropic). During flow, however, the macromolecules become oriented (anisotropy). This anisotropy is lost during relaxation after the flow stops. The LCP molecules, both rigid and semirigid, are anisotropic at equilibrium (nematic phase) and are parallel one to another. The anisotropy is not due to the flow, but it is an intrinsic characteristic of the macromolecules that is determined by their geometry.

In spite of thermal motions, the rigid molecules in nematic phase maintain a common orientation, indicated by the vector "director". The average orientation determined by a director is called "domain." In a very few cases the director is uniform in the space and the material becomes a "monodomain." The presence of defects caused, for example, by non uniform boundary conditions or by external forces, implies that all the LCPs are "polydomain."

The imposition of a flow can alter this texture giving rise to more oriented or to monodomain structures. When the flow stops, the equilibrium structure is restored. The time required to restore the equilibrium structure is usually very long, extending beyond the relaxation times of the molecules of the flexible polymers.

The unusual rheological behavior of liquid crystalline polymers is a consequence of this anisotropic structure.

The flow behavior of liquid crystalline polymers has become a research field, and many points, experimental and theoretical, are far from being well assessed. Interesting reviews already exist on this topic, and the rheological modelling of rigid polymers, although very young, is able to predict many unusual shear flow properties of these materials.[1–13]

MAIN RHEOLOGICAL CHARACTERISTICS OF LCPS

Change of Viscosity in the Nematic Transition Region

By decreasing the temperature, the viscosity of flexible polymers (and of all the ordinary liquids) is steadily increased. As far as the polymer solutions is concerned, the viscosity increases by increasing the polymer concentration. In the system of PBA in DMAA + 3% LiCl, at low polymer concentrations the system exists in isotropic phase and the viscosity rises continuously up to the transition. In the nematic phase, the viscosity drops dramatically. The anisotropic phase shows a very low viscosity with respect to the isotropic one. The isotropic-nematic transition is a function of temperature, molecular weight, and shear rate.

Shape of the Flow Curve

The typical flow curve of LCPs is schematically depicted in **Figure 1**. This three-region curve, proposed by Asada et al. has been frequently reported for solutions of LLCPs and, in some cases, for TLCPs.[15]

The most important difference with the flow curve of flexible polymers is the lack of a plateau (Newtonian viscosity) at very low shear rates. In this zone, rather, rapid upturn accompanies decreasing shear rate followed by a region of almost constant viscosity and, at high shear rates, a drastic shear thinning region.

Flexible polymers do not show the first region of the above-reported flow curve (Figure 1), which is, instead, typical of filled or crosslinked polymers.

Normal Stresses

Normal stresses in shear flow arise from the viscoelastic nature of the polymers. For flexible polymers the first normal stress difference, N_1, is invariably a positive quantity whereas the second normal stress difference, N_2, is negative. Both are a steadily increasing function of the shear rate.

Some relationships between viscosity and normal stresses have been observed. Roughly, the maximum of N_2 occurs at about the same shear rate of the minimum of N_1, and the negative values of N_1, and approximately the positive values of

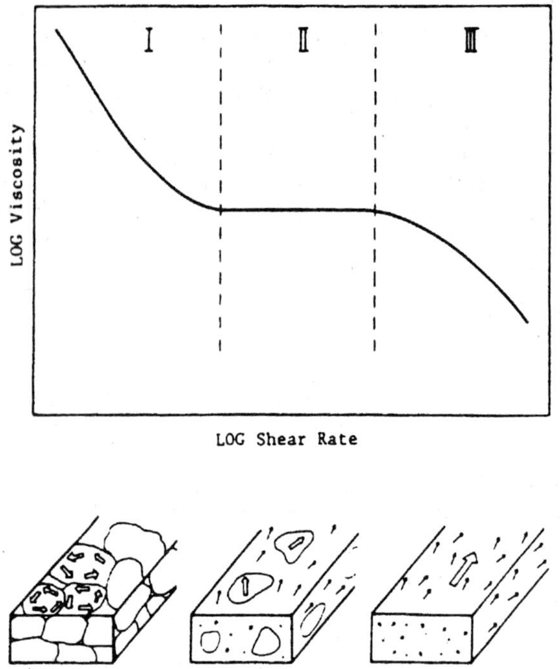

FIGURE 1. Typical flow curve of LCPs.

N_2, occurs in the shear rates range when the region III of the viscosity begins.

This behavior has been often revealed for lyotropic systems,[18] while both negative and positive N_1 values have been found for thermotropics.[19–21]

Transient Stress Behavior

The transient stress growth curves of LCPs, both shear and normal stresses, exhibit interesting distinctions from the analogous curves of the flexible polymers. The transient curves of these latter ones show two distinct behaviors: monotonic growth up to the steady state value at low shear rates and the presence of overshoots before reaching steady value at high shear rates.

The transient curves of LCPs present two or more maxima and minima (or better several oscillations) before it reaches the steady state.[16,17,21,22]

Die-Swell

At the exit of a capillary the polymeric extrudate undergoes a significant increase of the diameter with respect to that of the capillary. The phenomenon is due to the elasticity of the material: the melt tends to recover its dimensions before flowing through the capillary. The amount of the recovery depends on the relaxation times of the polymer, on the shear stress and on the length of the capillary. For all the flexible polymers, the die-swell, B, defined as the ratio between the diameter of the extrudate and that of the capillary, is larger than unity. That does not hold true for LCPs. Die-swell values of a rigid TLCP sample are reported as a function of the apparent shear rate.[14]

FLOW-INDUCED ORIENTATION

As previously stated, the shear flow cannot orient the rigid rod molecules along the flow direction. The tumbling nature of the LCPs is, on the contrary, irrelevant in elongational flow, which easily orients the molecules in the flow direction. In this flow no polydomain structure is observed because the flow induces uniform orientation of the directors. The high level of orientation gives rise to very high values of mechanical properties.

REFERENCES

1. Baird, D. G. *Liquid Crystalline Order in Polymers* Blumstein, A., Ed.; Academic: New York, 1978.
2. Wissbrun, K. F. *J. Rheol.* **1981**, *25*, 619.
3. Asada, T.; Onogi, S. *Polym. Eng. Rev.* **1983**, *3*, 323.
4. Marrucci, G. *Liquid Crystallinity in Polymers*, Ciferri, A., Ed.; VCH: New York, 1991.
5. Marrucci, G.; Greco, F. *Advances in Chemical Physics*, Prigogine, I.; Rice, S., Eds.; John Wiley & Sons: 1993; Vol. 86.
6. Marrucci, G. *Thermotropic Liquid Crystal Polymer Blends*, La Mantia, F. P., Ed.; Technomic: Lancaster, 1993.
7. Kulichikhin, V. G.; Volkov, V. S.; Platé, N. A. *Comprehensive Polymer Science First Supplement*, Aggarwal, S. L.; Russo, S., Eds.; Pergamon: Oxford, 1992.
8. Doi, M.; Edwards, S. F. *J. Chem. Soc. Faraday Trans.* **1978**, *74*, 560.
9. Doi, M. *J. Polym. Sci. Polym. Phys.* **1981**, *19*, 229.
10. Marrucci, G.; Maffettone, P. L. *Macromolecules* **1989**, *22*, 4076.
11. Marrucci, G.; Maffettone, P. L. *J. Rheol.* **1990**, *34*, 1217, 1231.
12. Marrucci, G.; Maffetone, P. L. *J. Rheol.* **1990**, *34*, 1231.
13. Larson, R. G. *Macromolecules* **1990**, *23*, 3983.
14. La Mantia, F. P.; Valenza, A. *Polym. Eng. Sci.* **1989**, *29*, 625.
15. Onogi, S.; Asada, T. *Rheology*, Astarita, G.; Marrucci, G.; Nicolais, L., Eds.; Plenum: New York, 1980.
16. Grizzuti, N.; Cavella, S.; Cicarelli, P. *J. Rheol.* **1990**, *34*, 1293.
17. Cocchini, F.; Nobile, M. R.; Acierno, D. *J. Rheol.* **1991**, *35*, 1171.
18. Kiss, G.; Porter, R. S. *Mol. Cryst. Liq. Cryst.* **1980**, *60*, 267.
19. Prasadarao, M.; Pearce, E. M.; Han, C. D. *J. Appl. Polym. Sci.* **1982**, *27*, 1343.
20. Gotsis, A. D.; Baird, D. G. *J. Rheol.* **1985**, *29*, 539.
21. Guskey, S. M.; Winter, H. H. *J. Rheol.* **1991**, *35*, 1191.
22. Baird, D. G.; Gotsis, A.; Viola, G. *Liquid Crystalline Order in Polymers*, Blumstein, A., Ed., Academic: New York, 1978.

LIQUID CRYSTALLINE POLYMERS (Thermotropic Polyesters)

Garret D. Figuly
DuPont Central Research and Development

Four primary classes of liquid crystalline polymers may be identified as a) rigid main chain, b) flexible spacer main chain, c) side chain, and d) discotic. The features of these primary structural polymers can be mixed and matched to produce at least 20 different structural classes.[3] As with small molecules, these polymers may exhibit liquid crystalline behavior (mesomorphism) in solution (lyotropic) or upon melting (thermotropic). Those which produce a liquid crystalline phase directly upon heating from a solid phase are classified as enantiotropic, while those which only exhibit a liquid crystalline phase upon supercooling from an isotropic liquid phase below the melt temperature are monotropic.[2] Thermotropic polyesters of pri-

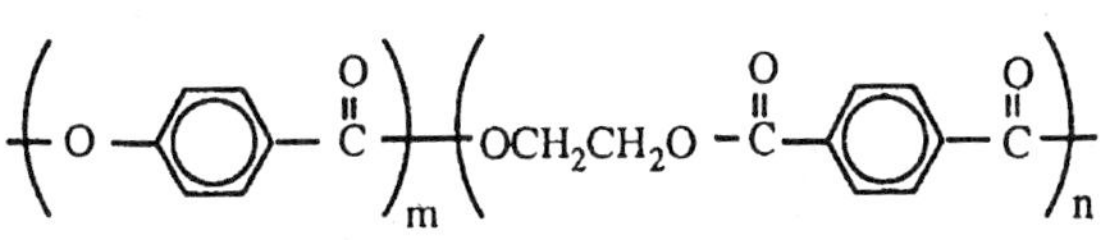

1

mary importance are rigid main chain liquid crystalline polymers (LCPs).

PREPARATION

Initial explorations eventually leading to melt processible LCPs focused on identification of melt processible polymers with exceptionally high thermal stability (> 300°C). It was generally recognized that aromatic structures could provide the required thermal properties and initial work was carried out on the readily available *p*-hydroxybenzoic acid.

Eastman's work toward a polyester with high thermal stability led to the first aromatic-aliphatic polyester. The resulting random copolymer was the first thermotropic polyester to be injection molded, melt spun into fiber, and characterized as being liquid crystalline.[4,5] The aromatic-aliphatic copolymer is formed via a melt phase acidolysis of pre-formed poly(ethylene terephthalate) (PET) with *p*-acetoxybenzoic acid (**Equation 1**).

PET copolymers containing greater than 40% *p*-hydroxybenzoic acid exhibit distinct liquid crystalline character, including melt birefringence, anisotropic mechanical, and optical properties, and significant lowering of melt viscosity at a number of different shear rates.[6–8] Whereas, the aromatic components are directly responsible for liquid crystalline character, the flexible ethylenedioxy components produce a dramatic decrease in melt viscosity resulting from lowering the polymer melt temperature. The lower melt temperature allows accessibility to the nematic phase at process temperatures.

Much of the recent synthetic work has focused on identification of readily processible LCPs to provide high performance characteristics (e.g., high melting points, high strength and modulus, processibility, etc.). In addition to melting point lowering by copolymerization, powerful alternative techniques have been developed through the use of non-linear, "crankshaft", or asymmetrically substituted ingredients.

Crankshaft units cause irregularities within the polymer chains that disrupt crystal packing forces and allow lower melt temperatures. Typically, a melting point depression of 200–300°C is observed for compositions containing significant amounts of naphthalene-based crankshafts.[1] The melting point can be adjusted by altering the ratio of crankshaft ingredient. The addition of other comonomers can also reduce melting points. Much of DuPont's work has centered on the production of melt processible A-A/B-B systems.[9–14] With the use of melt acidolysis techniques, the introduction of comonomers, asymmetrically substituted aromatic rings, and swivel point (nonlinear) monomers such as diphenyl ethers, sulfides, ketones, or meta-substituted aromatics has been accomplished. All of these ingredients disrupt crystal packing and lead to a lowering of crystalline melting points. The use of highly unsymmetrical phenylhydroquinone (PhHQ) leads to a lowered melting point of at least 250°C. A combination of unsymmetrical chlorohydroquinone (ClHQ) with a nonlinear benzophenone derivative (HBBP) drops the melting point 300°C. In extreme cases, no crystalline structure will develop at all and glassy, amorphous LCPs are produced.[15,16]

Virtually all preparative methods for thermotropic liquid crystalline polyesters, other than melt acidolysis, require solvents and may produce undesirable by-products which are difficult or impossible to remove to recover. Therefore, the preferred method of preparation on a commercial scale usually makes use of melt acidolysis done under an initial nitrogen purge.

Synthetic methodology has evolved toward the production of melt processible polymers with high performance properties through techniques that lower the crystalline melting point and allow nematic phases to be reached. However, the ability to produce thermotropic liquid crystalline polyesters with well defined polymer architecture remains elusive. The synthetic methodology generally used for these polymers only affords production of completely random copolymers. Thus, synthetic control to influence thermotropic liquid crystalline polyester architecture beyond random configurations remains a primary challenge.[17]

PROPERTIES

In general, thermotropic liquid crystalline polyesters (LCPs) offer a number of high performance properties that have a high degree of orientational order and high temperature stability. Unlike all other engineering resins (isotropic melts), which are difficult to process due to high melt viscosities at desirable molecular weights, LCPs flow more readily to fill very intricate molds because their ordered nematic structures are subject to less resistance to flow (especially in the direction of flow).[18–23]

Bulky chain substituents, "crankshafts", "swivels", flexible units, and comonomers used to lower in LCP's melting point are also used to modify the properties of an LCP. Typically, as more adjustments are made to deviate from complete polymer linearity, a decrease in the melt temperature, increasing T_g, decreasing tensile properties, decreased hardness, better toughness, a quicker onset of an isotropic melt phase (a lower nematic to isotropic transition temperature), and somewhat reduced anisotropic character (more balanced physical properties parallel and perpendicular to melt flow) can be observed.

The relationships between structure and properties of main-chain polyesters has been studied extensively.[24–30] In general, LCP properties result from the synthetic chemist's ability to alter both the aspect ratio (L/W ratio) of a constituent mesogen and the persistence length of the polymeric chain.

Mechanical properties of thermotropic liquid crystalline polyesters derive from aspect ratio and persistence length. Since rigidity is constant for aromatic ester links, overall molecular stiffness, and mechanical properties, are controlled by the frequency and size of paraaromatic units in the backbone, molecular weight and its distribution, and comonomer ratio. The frequency of paraaromatic units in an LCP structure may provide a measure of polymeric rigidity.[4]

Polymeric molecular weight can have a very large effect on mechanical properties. This is especially true for LCP tensile properties. Very broad molecular weight distributions tend to be detrimental to mechanical properties. This is due to the fact that the solid state properties of oriented LCPs are dependent on interchain interactions, and large disparities in chain lengths act as defects (stress concentrators) which diminish strength and related properties. Strength improvements in LCPs may be achieved by appropriate heat-treatment.[30a] Heat-treatment of LCP fibers is also known to produce higher melting polymer (sometimes melt processibility is lost), improved chemical stability, increased tensile property levels at elevated temperatures, and may also provide annealing to heal defects in fiber morphology. Heat-treated LCP fibers provide very high levels of tensile properties and are comparable with other high performance fibers.

The primary use of LCPs is molding resins. Using the compositional variables previously outlined, researchers have developed a very wide range of LCPs that exhibit a broad spectrum of resin properties.

APPLICATIONS

Thermotropic liquid crystalline polyesters offer many high performance properties including low melt viscosity, high orientation along the machine direction (in shaped objects this leads to high strength and modulus), high thermal stability (low coefficient of thermal expansion and high heat distortion temperature), dimensional stability (low mold shrinkage), solvent resistance, low moisture regain, low oxygen permeation, and flame resistance. LCP resins may be processed containing talc, particulates, minerals, graphite flake, glass fibers to enhance desired properties and reduce costs. The ease of melt processing LCPs give ready access to injection molded parts, fibers, films, coatings, and melt blends with other polymers or materials. To date, primary applications for thermotropic LCPs have derived from injection molding parts, often incorporating fillers.

Examples of applications include: cookware (dual ovenable), electronic connectors [including those for surface mount technology (SMT)], molded interconnection devices, coil bobbins (all of which take advantage of the LCP's high flow, good toughness, low warpage, and high strength in thin sections), packing elements for distillation columns, chemical pumps, automotive parts (including sensors, bearings, power train elements, and ignition systems), aircraft parts (fuselage rivets, aerodynamic fins), audio-visual parts (speaker covers, headphone cases, CD pick-up bodies), fiber optic couplers, office equipment parts (floppy disc carriages, copy machine parts), medical and dental instrumentation and minimally invasive surgical devices.[5,31–33]

Films and coatings derived from thermotropic liquid crystalline polyesters could have an impact on such substrates for recording media, flexible circuit board substrates, barrier layers in packaging construction, encapsulation of components and adhesives.[34,35] LCPs are currently under investigation as coating media for encapsulation of electronic components and fiber optic wires. A very exciting potential application is that of a barrier resin. As a film, LCPs can provide a barrier to both oxygen and moisture up to 10 times more effectively than conventional barrier resins (Saran) in layers as thin as 0.1 mil (0.0001 inch).[36]

Thermotropic liquid crystalline polyesters are a rapidly growing family of high performance melt processible polymers. Their melt flow capabilities combined with their dimensional stability, high strength and stiffness, and excellent electrical, heat, and flame resistance promise to provide high performance polymeric resins, fibers, and films well into the next century. Some difficulties remain around processing (anisotropic properties), cost effective heat treatment processes and ingredient costs. Despite these difficulties, conservative estimates show by the year 2000, approximately 25 million pounds of LCP polymer will be sold into high performance resin, fiber, and film applications.[37]

REFERENCES

1. Priestley, E. B.; Wojtowicz, P. J.; Sheng, P., Eds.; *Introduction to Liquid Crystals* Plenum: New York and London, 1975.
2. Blumstead, A., Ed.; *Polymeric Liquid Crystals* Plenum: New York and London, 1985.
3. Brostow, W. *Polymer* **1990**, *31*, 979.
4. Calundann, G. W.; Jaffe, M. *The Robert A. Welch Foundation Conference on Chemical Research, XXVI. Synthetic Polymers* Nov. 15–17; 1982, 247.
5. Jackson, Jr., W. J. *Mol. Cryst. Liq. Cryst.* **1989**, *169*, 23.
6. Jackson, Jr., W. J.; Kuhfuss, H. F. *J. Polym. Sci., Polym. Chem.* **1976**, *14*, 2043.
7. Kuchfuss, H. F.; Jackson, Jr., W. F. **Eastman Kodak Co.** U.S. Patent 3 778 410, 1973.
8. Kuchfuss, H. F.; Jackson, Jr., W. F. **Eastman Kodak Co.** U.S. Patent 3 804 805, 1974.
9. Schaefgen, J. R. **E. I. du Pont** U.S. Patent 4 118 372, 1978.

9a. Payet, C. R. **E. I. du Pont** U.S. Patent 4 159 365, 1979.

10. Irwin, R. S. **E. I. du Pont** U.S. Patent 4 269 965, 1981.
11. Irwin, R. S. **E. I. du Pont** U.S. Patent 4 381 389, 1983.
12. Siemionko, R. K. **E. I. du Pont** U.S. Patent 4 412 058, 1983.
13. Harris, Jr., J. F. **E. I. du Pont** U.S. Patent 4 294 955, 1981.
14. Irwin, R. S. **E. I. du Pont** U.S. Patent 4 335 232, 1982.
15. Connolly, M. S. **E. I. du Pont** U.S. Patent 4 664 972, 1987.
16. Choe, E. W. **Hoechst Celanese Corp.** U.S. Patent 5 298 591, 1994.
17. McCullagh, C. M.; Blackwell, J.; Jamieson, A. M. *Macromolecules* **1994**, *27*, 2996.
18. Wissbrun, K. E. *Journal of Rheology* **1981**, *25*(6), 619.
19. Wissbrun, K. F. *Faraday Discuss. Chem. Soc.* **1985**, *79*, 161.
20. White, J. L. *J. Appl. Polym. Sci.* **1985**, *41*, 241.
21. Baird, D. G. *Poly. Sci. Technol.* **1985**, *28*, 119.
22. Muir, M. C.; Porter, R. S. *Mol. Cryst. Liq. Cryst.* **1989**, *169*, 83.
23. Kim, S. S.; Han, C. D. *Macromolecules* **1993**, *26*, 6633.
24. Lenz, R. W. *J. Polym. Sci.: Poly. Symp.* **1985**, *72*, 1.
25. Lenz, R. W.; Furukawa, A.; Wu, C. N. Relationships Between Polymer Structure and Liquid Crystalline Phase Behavior for Thermotropic Polyesters *IUPAC - Polymers for Advanced Technologies* 1987, 491.
26. Blackwell, J.; Biswas, A. *Dev. Orient. Polym.* **1987**, *2*, 153.
27. Calundann, G. W. *ACS Proceedings, New York* April 15–18; pp 235–249, Seymour, R. B.; Kirshenbaum, G. S.; Elsevier: New York, 1986.
28. Huynh-ba, G.; Cluff, E. F. *Polymeric Liq. Crystals - Polym. Sci. Technol.*; Blumstein, A. Plenum: **1985**, *28*, pp 217–238.
29. Jackson, Jr., W. J.; Kuhfuss, H. F. *J. Polym. Sci. Polym. Chem.* **1976**, *14*, 2043.
30. Jackson, Jr., W. J. *J. Appl. Polym. Sci.* **1985**, *41*, 25.

30a. Warner, S. B.; Lee, J. *J. Pol. Sci: Part B, Poly. Phys.* **1994**, *32*(10), 1759 and references therein.

31. See: *Polymer News* **1994**, *19*(5), p 146.

32. Fujiwara, K. "LCP," *Plastics*, **1990**, *41*(1), 145.
33. Okada, T. "Trend of Development of LCP," *5th JRP-C Seminar* **1991**, pp 18–21.
34. Susko, J. R.; Volksen, W.; Wheater, R. A. Economy, J. **IBM Corp.** EP 139924 1988.
35. Cobbs, Jr., W. H. **Nordson Corporation** U.S. Patent 4 560 594, 1982.
36. *Modern Plastics*, 1988, May, 52.
37. McChesney, C. E.; Dole, J. R. "Higher Performance in Liquid Crystalline Polymers," *Modern Plastics*, 1988, January, 112.

LIQUID CRYSTALLINE POLYMERS (Viologen: Thermotropic and Lyotropic)

Pradip K. Bhowmik and Haesook Han
Department of Chemistry
University of Detroit Mercy

Rafil A. Basheer
Polymer Department
GM Research and Development Center

Liquid crystalline polymers (LCPs) are an interesting class of polymers, because they offer a unique combination of properties as high performance materials for versatile applications in modern technology. They can be classified into three principal groups: one exhibits liquid crystallinity in a heat-induced melt and is known as thermotropic; the second exhibits in solution and is termed as lyotropic; and the third exhibits liquid crystalline both in the melt and in the solution and is termed amphotropic. Among the thermotropic polymers, one can generally find both wholly aromatic polyesters and semiflexible polyesters.[1] Wholly aromatic polyamides and poly(γ-benzyl-L-glutamate), PBLG, are notable examples of lyotropic polymers.[1] An amphotropic class of polymers are less well-known. However, the family of polyisocyanates having aliphatic or certain aralkyl side chains of appropriate lengths, some derivatives of PBLG, some derivatives of cellulose, and a few aromatic copolyesters exhibit amphotropicity.[1–4]

The 1,1′-dialkyl-4,4′-bipyridinium salts are commonly called viologens. Viologen polymers are an important class of materials, which exhibit a wide range of properties because of the presence of viologen moiety. They exhibit electrical conductivity either doping with an electron acceptor such as 7,7,8,8-tetracyanoquinodimethane (TCNQ) or without a dopant, photochromism–color change with light, electrochromism–color change electrochemically with an applied reduction potential, thermochromism–color change with temperature and photomechanical behavior–change of mechanical properties with light.[5–9] Furthermore, they have other interesting properties for applications to photochemical conversion, storage of solar energy and electron transport across a polymeric membrane.[10,11] They are also known as ionene polymers because of the presence of 4,4′-bipyridinium ions in the polymer backbone.

In this chapter, we describe the preparation, both thermotropic and lyotropic LC properties, and potential applications of a series of viologen polymers. The structures and designations of the viologen polymers are shown in **Scheme I**.

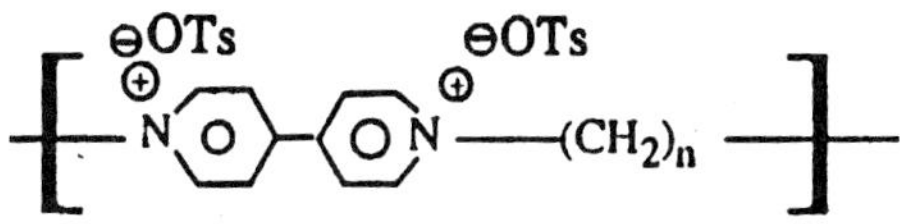

I

n = 10 (I-1), 11 (I-2) and 12 (I-3).

PROPERTIES

Dilute Solution Properties and Molecular Weights

In contrast to an analogous series of viologen polymers containing bromide as a counterion, all of these polymers were sparingly soluble in water, but highly soluble in methanol.[12,13] Their low solubility in water may presumably arise because of the large hydrophobic effect of tosylate when compared to bromide ion. As expected, each of them showed a polyelectrolyte behavior in methanol, that is, its inherent viscosity, IV, increased markedly with decreasing concentration.

Thermal Properties

Polymer I-1 showed two endotherms in the first heating cycle of the DSC thermogram. In the subsequent cooling cycle, there was only an exotherm. Correspondingly, in the second heating cycle and in the second cooling cycle there were an endotherm and an exotherm, respectively. The first endotherm corresponded to the crystal-to-LC phase transition, T_m, at 86°C and the second endotherm to the LC-to-isotropic transition, T_i, at 204°C as confirmed with the PLM studies. The endotherm in the second heating cycle was related to T_i, and the exotherm in each of the cooling cycles was related to the isotropic-to-LC phase transition. Polymer I-2 showed a T_g at 99°C and three endotherms in the first heating cycle of the DSC thermogram. In the subsequent cooling and heating cycles there was neither an exotherm nor an endotherm, respectively. The intermediate endotherm was presumably related to LC-to-LC transition. Polymer I-3 showed several endotherms in the first heating cycle and one endotherm in the second heating cycle, respectively. The intermediate endotherms were possibly related to LC-to-LC transitions, but the nature of these transitions could not be determined by PLM studies. In each of the cooling cycles there was only an exotherm, which corresponded to the isotropic-to-LC phase transition.

Optical Textures

Each of these ionene polymers formed a turbid viscous melt above its T_m that exhibited strong stir opalescence, which was taken as the first indication of its liquid crystalline behavior.

Morphology

All of the "as-made" viologen polymers were examined by WAXD method at room temperature. Their WAXD patterns contained several sharp symmetric circular rings, which were indicative of semicrystalline polymers.

Thermal Stability

The thermal stability of each of these polymers was determined in nitrogen by thermogravimetric analysis, TGA. They had thermal stabilities with a narrow range of 330–334°C, which were well above the T_m, LC-to-LC transition, and T_i values of these polymers.

Lyotropic Properties

The formation of lyotropic phase of polymers is quite complex and determined by several key requirements.[14,15] The polymers, generally, should have a rod-like structure with an extended chain character to facilitate the alignment of the polymer chain along a particular direction. The solubility needs to be sufficiently high to exceed the critical concentration at which the formation of a biphasic solution occurs, that is, where a LC phase coexists with an isotropic phase. The degree and nature of interaction between the polymer chains themselves and that between the polymer chains and solvent molecules can also provide a major influence on the phase behavior of a polymer. The solubility and chain stiffness of a polymer are themselves affected by the polymer microstructure, molecular weight, polymer–polymer and polymer–solvent interactions and temperature. Therefore, it appears that the ionic groups present in these viologen polymers might increase polymer–solvent and also polymer–polymer interaction, hence the solubility and the prospect of formation of a lyotropic phase in common polar organic solvents.[16–18] Polymer I-1 formed an isotropic solution at a low concentration. At an intermediate concentration, a biphasic solution occurred where a LC phase coexisted with an isotropic phase; at a high concentration a lyotropic phase occurred. Similarly, each of the polymers I-2 and I-3 formed an isotropic solution at a relatively higher concentration when compared to polymer I-1, but there was no occurrence of a biphasic solution for these two polymers, which was in contrast to polymer I-1.

APPLICATIONS

As viologen polymers exhibit a number of interesting properties, they can be used for versatile applications in modern technology. Their major applications include: (a) photochromic, thermochromic, electrochromic, and photomechanic materials; (b) in the preparation of modified electrodes and electron-transport membranes; (c) electron-transfer catalysts in organic synthesis; and (d) in the photocatalyzed water cleavage for solar energy harvesting. Furthermore, an additional interesting property, the thermotropic and lyotropic LC behavior in common organic solvents of this class of polymers, may provide a new avenue for applications in the fields of films, coatings, barrier membranes, polymer blends and composites. Particularly, from the viewpoint of thermodynamics of polymer blends, it is very unlikely that a blend of a LCP and a random-coil polymer exhibits miscibility because of the unfavorable low entropy of mixing of the two macromolecules, even for an athermal mixture. Additionally, the phase separation between the components is enhanced by the preference of a LCP component for an anisotropic phase. This general problem of limited compatibility of two macromolecules can be overcome by increasing the affinity between the two blend components or through specific interactions to provide a negative enthalpy of mixing. Therefore, viologen polymers containing ionic groups in conjunction with their liquid crystallinity may offer an interesting class of materials, which would enhance the miscibility with other appropriate macromolecules through the ionic interactions.

ACKNOWLEDGMENT

P. K. B. wishes to acknowledge the Donors of The Petroleum Research Fund, administered by the American Chemical Society, for partial support of this research.

REFERENCES

1. Donald, A. M.; Windle, A. H. *Liquid Crystalline Polymers*; Cambridge University: Cambridge, 1992.
2. Iizuka, E.; Abe, K.; Hanabusa, K.; Shirai, H. *Current Topics in Polymer Science*; Ottenbrite, R. M.; Utracki, L. A.; Inoue, S., Eds.; Hanser: New York, 1987; 1, 235.
3. Ogata, N.; Sanui, K.; Zhao, A.-C.; Watanabe, M.; Hanaoka, T. *Polymer J.* **1988**, *20*, 529.
4. Heitz, W. *Makromol. Chem., Macromol. Symp.* **1991**, *47*, 111.
5. Merz, A.; Reitmeier, S. *Angew. Chem. Int. Ed. Engl.* **1989**, *28*, 807.
6. Simon, M. S.; Moore, P. T. *J. Polym. Sci., Polym. Chem. Ed.* **1975**, *13*, 1.
7. Hashimoto, T.; Kohjaya, S.; Yamashita, S.; Irie, M. *J. Polym. Sci., Polym. Chem. Ed.* **1991**, *29*, 651.
8. Sato, H.; Tamamura, T. *J. Appl. Polym. Sci.* **1979**, *24*, 2075.
9. Moore, J. S.; Stupp, S. I. *Macromolecules* **1986**, *19*, 1815.
10. Nambu, Y.; Yamamoto, K.; Endo, T. *Macromolecules* **1989**, *22*, 3530.
11. Willner, I.; Riklin, A.; Lapidot, N. *J. Am. Chem. Soc.* **1990**, *112*, 6438.
12. Yu, L.-P.; Samulski, E. T. In *Oriented Fluids and Liquid Crystals*; Griffin, A. C.; Johnson, J. F., Eds.; Plenum: New York, 1984; Vol. 4, 697.
13. Bhowmik, P.; K.; Han, H. *Polym. Prepr. (ACS, Div. Polym. Chem.)* **1994**, *35*(2), 617.
14. Kwolek, S. L.; Morgan, P. W.; Schaefgen, J. R. In *Encyclopedia of Polymer Science and Engineering*, 2nd ed.; Mark, H. F.; Kroschwitz, J. I., Eds.; Wiley-Interscience: New York, 1988; Vol. 9, 1.
15. Preston, J. *Angew. Makromol. Chem.* **1982**, *109/110*, 1.
16. Bhowmik, P. K.; Han, H. *J. Polym. Sci., Polym. Chem. Ed.* **1995**, *33*, 1745.
17. Han, H.; Bhowmik, P. K. *Polym. Prepr. (ACS, Div. Polym. Chem.)* **1994**, *35*(2), 856.
18. Han, H.; Bhowmik, P. K. *Polym. Prepr. (ACS, Div. Polym. Chem.)* **1994**, *35*(2), 619.

LIQUID CRYSTALLINE POLYMERS (Wholly Aromatic Polyesters)

Pradip K. Bhowmik* and Haesook Han
Department of Chemistry
University of Detroit Mercy

Wholly aromatic, thermotropic polyesters are an interestingly class of liquid crystalline polymers (LCPs) because of their ease of processing in the nematic LC phase to obtain either high strength fibers or engineering thermoplastics. In general, however, they have high crystal-to-nematic transitions, T_m, because of their high enthalpy change and low entropy change at these transitions. Moreover, they demonstrate very low solubility in all but aggressive organic solvents.

Several structural modifications can decrease the T_m values of this class of polymers to a convenient level to prevent thermal

*Author to whom correspondence should be addressed.

degradation during nematic melt processing.[1–5] Several structural modifications also exist to increase their solubility in common organic solvents.[6–9] Their solubility is highly desirable for the ease of characterization, for the ease of preparation of films by solution casting, and for the study of polymer blends.

A large number of studies have been devoted to wholly aromatic, thermotropic polyesters based on substituted hydroquinones (HQs), substituted terephthalic acids (TAs), and substituted 4,4′-biphenols (BPs), but a few studies have been performed to prepare polymers based on 3,4′-benzophenone dicarboxylic acid (3,4′-BDA).[10–17] Skovby et al. reported the results of a series of thermotropic copolyesters based on phenylhydroquinone (PhHQ) with mixtures of TA and 3,4′-BDA.[18] All except the copolymer containing 90 mol% of 3,4′-BDA exhibited nematic melts in the temperature range of 185–372°C. This suggests that the 3,4′-benzophenone dicarboxylate unit was capable of decreasing the T_m values considerably with the increased amount of substitution for TA. Recently, we reported the results of a series of thermotropic homopolyesters of 3,4′-BDA with various aromatic diols.[19] These diols were HQ, 2,6-, 1,4-, 1,5-, 2,3-, and 2,7-naphthalenediol isomers. All the homopolymers of 3,4′-BDA, except 2,7-naphthalenediol, formed nematic phases in the temperature range of 305 to 360°C.

In this chapter, we describe the preparation, thermotropic LC properties and applications of a new series of homopolyesters of 3,4′-BDA with BP, 3,3′,5,5′-tetramethyl-4,4′-biphenol (TMBP), 3-phenyl-4,4′-biphenol (MPBP), 3,3′-bis(phenyl)-4,4′-biphenol (DPBP), and 1,1′-binaphthyl-4,4′-diol. Two copolyesters of 3,4′-BDA with either BP or BND and 30 mol% of 6-hydroxy-2-naphthoic acid (HNA) also were included.

PREPARATION

Wholly aromatic, thermotropic polyesters usually are prepared by a step-growth or condensation polymerization reaction with or without a catalyst. The most common polycondensation reaction is carried out in the melt at a relatively high temperature from the diacetate derivatives of aromatic diols and aromatic dicarboxylic acids or acetate derivatives of hydroxy-aromatic acids.

Solubility

All the polyesters prepared by melt polycondensation reactions, with the exception of polymers I-1, I-6, and I-7, were soluble in *p*-chlorophenol. Their inherent viscosity (IV) values ranged from 0.58–1.88 dL/g in this solvent at 50°C despite the inclusion of a flexible 3,4′-benzophenone dicarboxylate moiety as a nonlinear component. These values suggest the polymers have sufficiently high molecular weights, and their thermal properties, optical textures, and other properties can be compared, neglecting the effect of molecular weight on these properties.

Optical Textures

All polyesters of 3,4′-BDA formed turbid melts that exhibited stir opalescence. This property can be taken as the first indication of their liquid crystalline behavior. For further characterization of their melt morphology, they were evaluated by visual observations with polarizing light microscope. All polyesters above their T_m/T_f values exhibited a typical nematic appearance with the so-called polished marble texture or a threaded texture depending on the thickness or temperature of the sample.

Thermal Stability

The thermal stability of each thermotropic polyester was determined by TGA either in air or in nitrogen. Generally, the thermal stability of a polyester in air was lower than that in nitrogen. The temperature at which 5% weight loss occurred varied from 465 to 511°C in nitrogen, indicating that the polymers had good thermal stabilities for nematic melt processing at elevated temperatures.

APPLICATIONS

Wholly aromatic, thermotropic polyesters generally offer a unique combination of properties as high performance materials for versatile applications in modern technology. These properties include low nematic melt viscosity, fast cycle times, very low mold shrinkage, low thermal expansion coefficient, excellent mechanical properties, good chemical resistance, excellent solder resistance, low flammability and low water absorption. Their low-melt viscosity enables one to create injection molding components with long or complex flow profiles and thin sections and to use these as a processing aid. The major applications include high performance composites, ballistic protection devices, electronic and telecommunication devices, hostile environments use, and other specialty devices where cost is not critical. Finally, the combination of excellent barrier properties, tolerance for high temperature and all practical means of sterilization may find applications in medical and food packaging industries.

ACKNOWLEDGMENT

P. K. B. wishes to acknowledge the Donors of the Petroleum Research Fund, administered by the American Chemical Society, for their support of this research.

REFERENCES

1. Griffin, B. P.; Cox, M. K. *Br. Polym. J.* **1980**, *12*, 147.
2. Ober, C. K.; Jin, J.-I.; Zhou, A.-F.; Lenz, R. W. *Adv. Polym. Sci.* **1984**, *59*, 103.
3. Erdemir, A. B.; Johnson, D. J.; Tomka, J. G. *Polymer* **1986**, *27*, 441.
4. Yang, H. H. *Aromatic High Strength Fiber* John Wiley & Sons: New York, 1989.
5. Cai, R.; Preston, J.; Samulski, E. T. *Macromolecules* **1992**, *25*, 563.
6. Sinta, R.; Gaudiana, R. A.; Minns, R. A.; Rogers, H. G. *Macromolecules* **1987**, *20*, 2374.
7. Heitz, W.; Schmidt, H.-W. *Makromol. Chem. Macromol. Symp.* **1990**, *38*, 149.
8. Navarro, F.; Serrano, J. L. *J. Polym. Sci., Polym. Chem. Ed.* **1992**, *30*, 1789.
9. Bhowmik, P. K.; Han, H. *Macromolecules* **1993**, *26*, 5287.
10. Jackson, Jr., W. J. *Contemporary Topics in Polym. Sci.* **1984**, *5*, 177.
11. Krigbaum, W.; Hakemi, H.; Kotek, R. *Macromolecules* **1985**, *18*, 965.
12. Jin, J.-I.; Choi, E.-J.; Jo, B.-W. *Macromolecules* **1987**, *20*, 934.
13. Ballauff, M.; Schmidt, G. F. *Mol. Cryst. Liq. Cryst.* **1987**, *147*, 163.

14. Bhowmik, P. K.; Atkins, E. D. T.; Lenz, R. W. *Macromolecules* **1993**, *26*, 440.
15. Bhowmik, P. K.; Atkins, E. D. T.; Lenz, R. W. *Macromolecules* **1993**, *26*, 447.
16. Bhowmik, P. K.; Lenz, R. W. *J. Polym. Sci., Polym. Chem. Ed.* **1993**, *31*, 2115.
17. Irwin, R. S.; Logullo, Sr., M. U.S. Patent 4 500 699, 1985.
18. Skovby, M. H. B.; Heilmann, C. A.; Kops, J. in *Liquid Crystalline Polymers*, Weiss, R. A.; Ober, C. K., Eds.; ACS Symposium Series 435, American Chemical Society: Washington, DC, 1990, p 46.
19. Bhowmik, P. K.; Han, H.; Garay, R. O. *J. Polym. Sci., Polym. Chem. Ed.* **1994**, *32*, 333.

LIQUID CRYSTALLINE POLYMERS, SIDE CHAIN

J. M. G. Cowie
Department of Chemistry
Heriot-Watt University

The liquid crystalline (LC) phase is a genuine, thermodynamically stable state of matter, exhibiting the properties of both a crystalline solid and a liquid.

Only certain molecules can self assemble and form a liquid crystal phase. These tend to be either long lathe-like molecules with high aspect ratios of disc-like units. While the earlier work was carried out on small molecule mesogens it was found that these could be incorporated into polymeric structures with retention of the liquid crystal phase. Such materials not only display liquid crystal phases, but also possess the unique properties associated with macromolecules, which is a desirable combination for many applications.

SIDE-CHAIN LIQUID CRYSTAL POLYMERS (SCLCPS)

The formation of lyotropic liquid crystal phases in polymer systems was first predicted theoretically by Flory in 1956, and confirmed experimentally by Kwolek in 1965 for solutions of polyaromatic amides (aramids).[1,2] The discovery of thermotropic main-chain liquid crystal polymers followed in 1976, and in 1978 Finkelmann et al. described the synthesis of the first SCLCP.[3,4] In this paper, a fundamentally important concept is introduced, which opens up this field of investigation, viz., that the mesogenic unit should be attached to the main polymer chain by a flexible spacer unit. This serves to decouple the random motion of the flexible polymer backbone chain and allows the mesogens to enter into long-range ordering and the formation of a mesophase. If the mesogens are linked directly to the backbone chain, the liquid crystal phases are less likely to form because the self-assembly process requires substantial distortion of the backbone from its normal random coil conformation to occur.

While there are now several examples of SCLCPs exhibiting mesophages in which the mesogen is directly attached to the backbone chain,[5] this guiding principle has helped to establish the field and a wide range of side-chain (comb-branched) structures have been synthesized. This growing interest in SCLCPs has generated extensive literature on the subject which has now been comprehensively reviewed. The major work in this area, edited by McArdle, covers work up to 1989.[6] Subsequent publications contain chapters covering the advances made since then, and a review devoted to chiral SCLCPs has recently appeared.[6–11]

Effect of the Backbone Polymer Chain

The nature of the backbone chain does not normally influence the type of mesophase formed, but it can affect the stability range. Percec has concluded that the more flexible the backbone chain, the broader the temperature range of mesophase stability and that the nature of the least-ordered phase is governed mainly by the length of the spacer unit.[12]

Effect of Spacer Length

The majority of SCLCPs (but not all) are prepared by attaching the mesogenic unit to the backbone via a flexible spacer of varying length, comprising typically $-(CH_2)-_m$, $-(CH_2CH_2O)-_m$, $-(Si\text{-}O)-_m$ sequences, with m varying from 2-15. The length of the spacer affects the nature of the mesophases formed quite profoundly. For amorphous polymers, the increase in the spacer length results in a decrease in T_g, the classical internal plasticizing effect seen in most comb-branched polymers, and this is accompanied by a corresponding lowering of the temperature range for the smectic mesophase formed.

The ordering in the mesophase is also governed by the spacer length. There is a general tendency for a nematic phase to form when the spacer is short, and, as this increases, smectic phase formation is induced.

Mesogen Placement

The majority of SCLCPs have the mesogen attached longitudinally to the chain with the spacer unit attached to one end. An alternative approach is to attach the spacer to the center of the mesogen making it oriented laterally to the main chain. This reduces the tendency for formation of smectic phases, and nematic phases are normally obtained irrespective of the spacer length.

Polymers with disc-like mesogens attached pendant to the main chain have also been prepared both as homo- and copolymers.[13]

CHIRAL SIDE-CHAIN LIQUID CRYSTALLINE POLYMERS

The introduction of chirality into SCLCPs has led to an important class of materials that has tremendous potential for use in emerging optical technologies. The unique properties of these polymers are usually derived from either of two mesophase types; the chiral nematic (cholesteric) phase N*, or the chiral smectic C phase (S_C*).

The chiral centers can be located either in the tail unit of the mesogen, or in the spacer unit, or both.

Other properties include high sensitivity of light reflection to temperature changes (thermochromic materials) and an extremely high optical activity.

Of special significance is the ability to "freeze" the liquid crystalline structure into the glassy state of a polymer, thereby retaining the optical characteristic, this makes N* SCLCPs ideal candidates for optical information storage materials.

COPOLYMERIZATION

Statistical copolymerization of mesogenic monomers with a comonomer, which may or may not be mesogenic itself, results in dilution of the primary mesogen but rarely causes a change in the nature of the mesophase.

Copolymerization of a mesogen with a nonmesogenic monomer will of course result in dilution and eventual loss of the mesophase as the content of nonmesogen increases.

PROPERTIES AND APPLICATIONS

Side-chain liquid crystal polymers have considerable potential, much of which is still to be realized, in areas such as optical information storage, nonlinear optics, and chromatography, where the properties of facile alignment and the ability to scatter light from the fine textures formed are particularly useful.

Liquid Crystalline Elastomers

Elastomeric materials can be prepared by crosslinking SCLCPs with sufficiently low values of T_g, to ensure the elastomeric phase is retained at ambient temperatures. If the crosslink density is low, the liquid crystalline phase of the uncrosslinked material is retained.

If the elastomers are chiral nematic or chiral smectic systems then the possibilities arise of observing piezoelectric and ferroelectric behaviour. The latter arises because the helical structure of the mesophase does not possess a mirror plane and has been observed in siloxane elastomers and in combined main chain and side chain LC structures that have the S_C* phase.[14,15]

Optical Information Storage

The properties described in the previous section can be used to produce a laser addressed information storage system with write and erase capacity.

If a chiral SCLCP is used, that displays selective reflection of visible light, this can act as a base on which to impress data using a thermal head, rather than a laser, to disrupt the regions of interest. This approach can produce selective filters and reflectors or to act as a holographic storage unit.

Ferroelectric SCLCPs

One disadvantage of polymeric systems is that their response times are too slow for use as devices that depend on field induced molecular orientation for their operation. This has led to increased interest in polymers with S_C* phases with a high spontaneous polarization (ferroelectricity).

Non-Linear Optics (NLO)

NLO active comb-branch polymers can be used as frequency modulators, as a material capable of second-order harmonic generation, or for frequency doubling applications while those with a large $\chi^{(3)}$ term exhibit frequency tripling capabilities. Molecules that have high NLO activity should be noncentro symmetric, have a large fixed dipole, conjugation along the length of the molecules for easy movement of delocalized electrons, and should also be long and straight. If the material is an SCLCP, it should have a nematic phase to assist poling which is an essential requirement for good alignment. Good results are often obtained by mixing into a SCLCP matrix an efficient NLO chromophore, such as 4,4′-N,N′-dimethylaminonitrostilbene (DANS), which has an electron withdrawing group at one end and an electron donating group at the other to give the necessary "push-pull" effect on the electrons.

Alternatively, the functionalized NLO chromophore can be incorporated as a monomer by copolymerization with a suitable chiral monomer.

Second harmonic generation, without resorting to external poling, has been observed in liquid crystalline polysiloxane elastomers, exhibiting the S_C* phase. As LC elastomers are very sensitive to mechanical orientation, there is no need to apply an external electric field, as the effect on orientation is much weaker than that produced by the mechanical field.[16]

OUTLOOK

The great potential of SCLCPs has not yet been fully realized. Their anisotropic nature and ease of orientation in shear, magnetic, or electric fields, combined with the ability to capture this in the glassy state of the polymer, make them attractive in information technology and electronic device manufacture. However, more quality control is now necessary and well-designed, narrow molar-mass distribution samples are needed. More effort is required to effect control of these parameters in order to ensure that further development is not impeded, and must not be neglected in the drive to discover new materials and new applications.

REFERENCES

1. Flory, P. J. *Proc. Roy. Soc. London, Series A* **1956**, *234*, 66.
2. Kwolek, S. L. DuPont U.S. Patent 3 600 350, 1971.
3. Jackson, W. J.; Kuhfuss, H. F. *J. Polym. Sci. Chem. Ed.* **1976**, *14*, 2043.
4. Finkelmann, H.; Ringsdorf, H.; Wendorff, J. H. *Makromol. Chem.* **1978**, *179*, 273.
5. Pugh, C.; Percec, V. *Chemical Reactions on Polymers*; Benham, J. L.; Kinstle, J. F., Eds.; Amer. Chem. Soc. Symp. Ser. 364, Washington, DC, 1988; Chapters 8, 9.
6. McArdle, C. B., Ed., *Side Chain Liquid Crystal Polymers*; Blackie: Glasgow, 1989.
7. Collyer, A. A., Ed., *Liquid Crystal Polymers* Elsevier Applied Science: London, 1992.
8. Donald, A. M.; Windle, A. H. *Liquid Crystalline Polymers*; Cambridge Univ.: Cambridge, 1992.
9. Cifferi, A., Ed., *Liquid Crystallinity in Polymer: Principles and Fundamental Properties*: VCH: New York, 1991.
10. Platé, N. A. *Liquid Crystal Polymers*; Plenum: New York, 1992.
11. Cowie, J. M. G.; Hinchcliffe, T. *Chiral Side Chain Liquid Crystal Polymers* in *Functional Polymers for Emerging Technologies*; Arshady, R., Ed.; Amer. Chem. Soc., in press.
12. Percec, V.; Tomazos, D. *Contemporary Topics in Polymer Science* Vol. 7; Salamone, J. C.; Riffle, J., Eds.; Plenum: New York, 1992.
13. Kreuder, W.; Ringsdorf, H. *Makromol. Chem. Rapid Commun.* **1983**, *4*, 807.
14. Meier, W.; Finkelmann, H. *Makromol. Chem. Rapid Commun.* **1990**, *11*, 599.
15. Vallerieu, S. U.; Kremer, F.; Fischer, H. W.; Kapitza, H.; Zentel, R.; Poth, R. *Makromol. Chem. Rapid Commun.* **1990**, *11*, 593.
16. Benné, I.; Semmler, K.; Finkelmann, H. *Makromol. Chem. Rapid Commun.* **1994**, *15*, 295.

LIQUID CRYSTALLINE POLYMERS, SIDE CHAIN (Structure–Property Relationships)

Corrie T. Imrie
Department of Chemistry
University of Aberdeen

LIQUID CRYSTAL POLYMERS

The mesogenic groups responsible for the observation of liquid crystallinity in low molar mass compounds can be incorporated into a polymeric architecture in one of two ways: as part of the polymer backbone or as a pendant attached onto the polymer.[3] The former arrangement, in which the mesogenic units are inserted directly into the polymer main chain, gives rise to main-chain liquid crystal polymers, and these can be subdivided into rigid and semi-flexible classes.[4]

The second major class of liquid crystal polymers are side-chain liquid crystal polymers in which the mesogenic units are linked to the polymer backbone via a flexible spacer. This article focuses upon structure–property relationships in thermotropic side-chain liquid crystal polymers containing calamitic mesogenic groups.

SIDE-CHAIN LIQUID CRYSTAL POLYMERS

Side-chain liquid crystal polymers have attracted considerable research interest since their discovery in the late seventies, because of their considerable application potential in a range of advanced electro-optic technologies, including optical information storage[6] and nonlinear optics,[7] and they challenge our understanding of the molecular factors that promote self-organization in polymeric systems.[8] Side-chain liquid crystal polymers comprise three essential structural components: a polymer backbone, a mesogenic unit, and a flexible spacer. The flexible spacer plays a critical role because it decouples, to some extent, the ordering tendencies of the mesogenic units from those of the backbones to adopt random coil conformations. Its presence endows upon the polymer a unique duality of properties. Thus, side-chain liquid crystal polymers exhibit macromolecular characteristics, such as ease of processability and mechanical integrity, coupled with the electro-optic properties of low molar mass mesogens, albeit on a much slower time-scale. It is this unique combination of properties that forms the basis of the proposed applications.[5]

Design strategies explored by polymer chemists have included laterally attached mesogens,[9] discotic mesogens,[10] copolymers containing differing mesogenic groups[11] or mesogenic and nonmesogenic units,[12] block copolymers,[13] graft copolymers,[14] and self-assembling supramolecular systems.[15] This flexibility, however, comes at a price, and the molecular engineer must first understand the structure–property relationships in homopolymers if this richness of structure is to be exploited in a rational manner.

STRUCTURE-PROPERTY RELATIONSHIPS

Molecular Weight and Polydispersity

In general, the transition temperatures of side-chain liquid crystal polymers are considered to be independent of molecular weight when DP $\geq$ ($\approx$20–30). The enthalpies of transition exhibit only a moderate or essentially no molecular weight dependence.[17,23–25,27–30]

Komiya et al. have shown that polydispersity does not effect either the transition temperatures or the width of the biphasic region provided that the component polymers are in the molecular weight independent regime.[23]

Mesogenic Core Structure

The mesogenic core in a calamitic system is normally comprised of phenyl rings separated via short unsaturated linkages. Attached to this are two terminal substituents. In the case of side-chain liquid crystal polymers, one of these is the flexible spacer, while the other is chosen to control, for example, the phase behavior, dielectric properties, or transition temperatures. In general, the rules developed relating properties to structure in low molar mass systems can be applied to the mesogenic unit in side-chain polymers.[1,2,16,31] There are, however, exceptions. For example, in low molar mass systems the cyano group normally is more effective in enhancing clearing temperatures than a methoxy group. When attached to a polymethacrylate backbone, this order is actually reversed, and methoxybiphenyl-based polymers exhibit considerably higher clearing temperatures than the analogous cyanobiphenyl-based materials.[32,33] This result may be rationalized in terms of the difference in the packing of the side-chains in the smectic phase.

Length and Parity of the Flexible Spacer

In order to establish the effects of varying the length and parity of the flexible spacer on the thermal behavior of side-chain polymers, we must consider the properties of homologous series containing ten or more members in which the only variable is the number of methylene units in the spacer. Such complete studies have been performed only for a limited number of systems: poly(vinyl ether)s,[18,19,21,22,34-38] polymethacrylates,[32,33] polynorbornenes,[23–25] and polystyrenes.[29,39–43]

Backbone Flexibility

We now explore the effect of backbone flexibility on the thermal properties of side-chain liquid crystal polymers. To make such a comparison we require materials in which only the nature of the backbone is varied. The largest set of such data exists for polymers containing 4-cyanobiphenyl as the mesogenic unit. This has been attached to a wide range of polymer backbones: poly(vinyl ether),[18,19,21,22,34–38,44] polystyrene,[29,42] polynorbornene,[25] polyacrylate,[26,45–55] polysiloxane,[20,56–62] and polymethacrylate.[26,33,50,54,55,63]

There is also the dependence of the clearing temperatures on the length of the flexible spacer for each series. For short spacer lengths there is a complex dependence of the clearing temperature on backbone flexibility but, as the spacer length is increased, a pattern emerges in which the more flexible the backbone is, then the higher the clearing temperature.

SUMMARY

This article provides an overview of our current understanding of structure–property relationships in side-chain liquid crystal polymers. It also highlights several areas in which our knowledge is either inadequate to establish trends or dependent on a rather limited number of systems. To verify the generality of the relationships discussed, many more complete homologous series of polymers must be synthesized and extensively characterized. The relationships outlined, however, will contribute to the rational design of new polymers having targeted properties.

REFERENCES

1. Gray, G. W. *The Molecular Physics of Liquid Crystals*, Luckhurst, G. R., Gray, G. W., Eds.; Academic: London, 1979; Chapter 1.
2. Gray, G. W. *The Molecular Physics of Liquid Crystals*, Luckhurst, G. R., Gray, G. W., Eds.; Academic: London, 1979; Chapter 12.
3. Finkelmann, H. *Thermotropic Liquid Crystals*, Critical Reports on Applied Chemistry, Gray, G. W., Ed.; John Wiley & Sons: 1987; Vol. 22, Chapter 6.
4. Ober, C. K.; Jim, J.-I.; Lenz, R. W. *Adv. Polym. Sci.* **1984**, *59*, 103.
5. Attard, G. S. *Trends Polym. Sci.* **1993**, *1*, 79.
6. Bowry, C.; Bonnett, P. *Opt. Comput. Proc.* **1991**, *1*, 13.
7. Möhlmann, G. R.; van der Vorst, C. P. J. M. *Side Chain Liquid Crystal Polymers*, McArdle, C. B., Ed.; Blackie and Sons: Glasgow, 1989; Chapter 12.
8. *Side Chain Liquid Crystal Polymers*, McArdle, C. B., Ed., Blackie and Sons: Glasgow, 1989.
9. Hessel, F.; Finkelmann, H. *Polym. Bull.* **1985**, *14*, 375.
10. Kreuder, W.; Ringsdorf, H. *Makromol. Chem., Rapid Commun.* **1983**, *4*, 807.
11. Schleeh, T.; Imrie, C. T.; Rice, D. M.; Karasz, F. E.; Attard, G. S. *J. Polym. Sci. Polym. Chem. Ed.* **1993**, *31*, 1859.
12. Percec, V.; Tomazos, D. *Adv. Mater.* **1992**, *4*, 548.
13. Bohnert, R.; Finkelmann, H. *Macromol. Chem. Phys.* **1994**, *195*, 689.
14. Hefft, M.; Springer, J. *Makromol. Chem., Rapid Commun.* **1990**, *11*, 397.
15. Imrie, C. T. *Trends Polym. Sci.* **1995**, *3*, 22.
16. Percec, F.; Pugh, C. *Side Chain Liquid Crystal Polymers*, McArdle, C. B., Ed.; Blackie and Sons: Glasgow, 1989; Chapter 3.
17. Percec, V.; Tomazos, D.; Pugh, C. *Macromolecules* **1989**, *22*, 3259.
18. Percec, V.; Lee, M. *Macromolecules* **1991**, *24*, 2780.
19. Percec, V.; Lee, M. *Macromolecules* **1991**, *24*, 1017.
20. Percec, V.; Lee, M. *J. Macromol. Sci.-Chem.* **1991**, *A28*, 651.
21. Percec, V.; Lee, M.; Ackerman, C. *Polymer* **1992**, *33*, 703.
22. Percec, V.; Lee, M.; Jonsson, H. *J. Polym. Sci., Part A: Chem. Ed.* **1991**, *29*, 327.
23. Komiya, Z.; Pugh, C.; Schrock, R. R. *Macromolecules* **1992**, *25*, 3609.
24. Komiya, Z.; Pugh, C.; Schrock, R. R. *Macromolecules* **1992**, *25*, 6586.
25. Komiya, Z.; Schrock, R. R. *Macromolecules* **1993**, *26*, 1393.
26. Kostromin, S. G.; Talroze, R. V.; Shibaev, V. P.; Platé, N. A. *Makromol. Chem., Rapid Commun.* **1982**, *3*, 803.
27. Hahn, B.; Percec, V. *Macromolecules* **1987**, *20*, 2961.
28. Hsu, C. S.; Percec, V. *Makromol. Chem., Rapid Commun.* **1987**, *8*, 331.
29. Imrie, C. T.; Karasz, F. E.; Attard, G. S. *J. Macromol. Sci.-Pure Appl. Chem.* **1994**, *A31*, 1221.
30. Percec, V.; Hahn, B. *Macromolecules* **1989**, *22*, 1588.
31. Imrie, C. T.; Schleeh, T.; Karasz, F. E.; Attard, G. S. *Macromolecules* **1993**, *26*, 539.
32. Craig, A. A.; Imrie, C. T. *J. Mater. Chem.* **1994**, *4*, 1705.
33. Craig, A. A.; Imrie, C. T. *Macromolecules* **1995**, *28*, 3617.
34. Heroguez, V.; Schappacher, M.; Papon, E.; Deffieux, A. *Polym. Bull.* **1991**, *25*, 307.
35. Johsson, H.; Sundell, P.-E.; Percec, V.; Gedde, U. W.; Hult, A. *Polym. Bull.* **1991**, *25*, 649.
36. Percec, V.; Lee, M. *Polym.* **1991**, *32*, 2862.
37. Percec, V.; Lee, M. *Polym. Bull.* **1991**, *25*, 131.
38. Percec, V.; Lee, M. *Macromolecules* **1991**, *24*, 4963.
39. Attard, G. S.; Dave, J. S.; Wallington, A.; Imrie, C. T.; Karasz, F. E. *Makromol. Chem.* **1991**, *192*, 1495.
40. Imrie, C. T.; Karasz, F. E.; Attard, G. S. *Macromolecules* **1992**, *25*, 1278.
41. Imrie, C. T.; Karasz, F. E.; Attard, G. S. *Macromolecules* **1993**, *26*, 545.
42. Imrie, C. T.; Karasz, F. E.; Attard, G. S. *Macromolecules* **1993**, *26*, 3803.
43. Imrie, C. T.; Karasz, F. E.; Attard, G. S. *Macromolecules* **1994**, *27*, 1578.
44. Kostromin, S. G.; Cuong, N. D.; Garina, E. S.; Shibaev, V. P. *Mol. Cryst. Liq. Cryst.* **1990**, *193*, 177.
45. Bormuth, F. J.; Biradar, A. M.; Quotschalla, U.; Haase, W. *Liq. Cryst.* **1989**, *5*, 1549.
46. Dubois, J.-C.; Decobert, G.; LeBarny, P.; Esselin, S.; Friedrich, C.; Noël, C. *Mol. Cryst. Liq. Cryst.* **1986**, *137*, 349.
47. Gubina, T. I.; Kise, S.; Kostromin, S. G.; Talroze, R. V.; Shibaev, V. P.; Plate, N. A. *Liq. Cryst.* **1989**, *4*, 197.
48. Ikeda, T.; Kurihara, S.; Karanjit, D. B.; Tazuke, S. *Macromolecules* **1990**, *23*, 3938.
49. Kostromin, S. G.; Shibaev, V. P.; Diele, S. *Makromol. Chem.* **1990**, *191*, 2521.
50. Kostromin, S. G.; Sinitsyn, V. V.; Tal'roze, R. V.; Shibayev, V. P. *Poly. Sci. U.S.S.R.* **1984**, *26*, 370.
51. Kurihara, S.; Ikeda, T.; Tazuke, S. *Macromolecules* **1991**, *24*, 627.
52. Kurihara, S.; Ikeda, T.; Tazuke, S. *Macromolecules* **1993**, *26*, 1590.
53. Le Barny, P.; Dubois, J.-C.; Friedrich, C.; Noël, C. *Polym. Bull.* **1986**, *15*, 341.
54. Piskunov, M. V.; Kostromin, S. G.; Stroganov, L. B.; Shibaev, V. P.; Platé, N. A. *Makromol. Chem., Rapid Commun.* **1982**, *3*, 443.
55. Shibaev, V. P.; Kostromin, S. G.; Plate, N. A. *Eur. Polym. J.* **1982**, *18*, 651.
56. Gemmell, P. A.; Gray, G. W.; Lacey, D. *Mol. Cryst. Liq. Cryst.* **1985**, *122*, 205.
57. Hsu, C. S.; Percec, V. *Polymer Bull.* **1987**, *18*, 91.
58. Hsu, C. S.; Rodriguez-Parada, J. M.; Percec, V. *J. Polym. Sci., Polym. Chem. Ed.* **1987**, *25*, 2425.
59. Kalus, J.; Kostromin, S. G.; Shibaev, V. P.; Kunchenko, A. B.; Ostanevich, Y. M.; Svetogorsky, D. A. *Mol. Cryst. Liq. Cryst.* **1988**, *155*, 347.
60. Krücke, B.; Scholssarek, M.; Zaschke, H. *Acta Polym.* **1988**, *39*, 607.
61. Nestor, G. N.; White, M. S.; Gray, G. W.; Lacey, D.; Toyne, K. J. *Makromol. Chem.* **1987**, *188*, 2759.
62. Ringsdorf, H.; Schneller, A. *Makromol. Chem., Rapid Commun.* **1982**, *3*, 557.
63. Shibaev, V. P.; Platé, N. A. *Pure and Appl. Chem.* **1985**, *57*, 1589.

LIQUID CRYSTALLINE POLYURETHANES

Piotr Penczek and Barbara Szczepaniak
Industrial Chemistry Research Institute

Liquid crystalline polyurethanes have been synthesized by incorporating the rigid aromatic ring moieties into the main chain or into the side chains. High concentration of the built-in mesogenic moieties results in a high melting point, usually exceeding the decomposition temperature of the polyurethane. In addition, such polyurethanes exhibit a limited solubility. This is an effect of high molecular cohesion and low mixing enthropy of liquid crystalline polymers.

To decrease the phase transition temperatures, to broaden the temperature range of the mesophase, and to improve the solubility, flexible chain spacers are inserted between rigid mesogenic segments in the main chain, or between the pendant mesogenic groups and the main chain of the polyurethane.

Other commonly used methods consist of incorporation of a pendant flexible chain or a voluminous substituent, as well as in the introduction of unsymmetric comonomers, which disorder the linearity of the molecule.

Liquid crystalline polyurethanes were described first by Iimura et al. using the mesogenic component, 3,3′-dimethyl-4,4′-biphenyldiyl diisocyanate.[1] Oligoethylene glycols (di-, tri-, and tetraethylene glycol) were applied as the diols, also playing the role of flexible spacers. In addition, aliphatic α,ω-alkanediols (2–12 C atoms) were also used. Thermal properties only of the polyurethanes were reported. In a similar series of experiments, 4,4′-azobenzene diisocyanate was used.[2] Liquid crystalline properties were found only if using triethylene glycol.

Besides that of the diisocyanate component, mesogenic diols also affect transition temperatures. The influence of 2-hydroxyethoxy and 6-hydroxyhexoxy derivatives of biphenyl, azobenzene, and benzophenone was compared.[5] Segmented polyurethanes with liquid crystalline polyester hard segments were investigated.[7] Liquid crystalline urethane ionomers were synthesized using ionic chain extenders.[8] The polymers are water emulsifiable.

Amphiphilic polyurethanes, containing hydrophobic long alkyl chains in the side chains, and amino or quaternary ammonium groups in the main chain exhibit smectic structure.[4] Cholesteryl groups, which were built into a side chain of an amphiphilic polyurethane, imparted liquid crystalline properties.[10]

PREPARATION

Liquid crystalline polyurethanes are usually synthesized by the polyaddition of a diisocyanate with a mesogenic diol in solution. The diol can be partially replaced by a polyether diol (flexible spacer).

PROPERTIES

Investigations of properties of liquid crystalline polyurethanes involve spectrometric methods as well as the molecular weight distribution measurements, which do not differ from the chemical composition studies being performed using common polyurethanes.

In addition, there are methods of investigation which are specific for liquid crystalline polyurethanes. The most important one is the determination of transition temperatures (melting and isotropic transition).

In addition to the thermal transitions, morphology of liquid crystalline polyurethanes is investigated. In order to investigate the sorption behavior of liquid crystalline polyurethanes with hydrocarbons, sorption constants, diffusion coefficients, and selectivity factors are determined using polyurethane films.

Structure–properties relationships for a series of liquid crystalline polyurethanes were studied.[11,12,15–17] The polyurethanes were synthesized from 4,4′-bis(2-hydroxyethoxy)biphenyl BHBP (the mesogenic component), poly(tetramethylene ether)diol PTMO (flexible spacer) and 2,4-tolylene diisocyanate (2,4-TDI).

The effect of the method of synthesis (one-step or two-step) and the content of mesogenic agent on the properties of liquid crystalline polyurethanes was investigated.[11]

Polyurethanes prepared by the one-step method usually exhibit higher T_m and T_i values than the polymers obtained by a two-step polyaddition. Moreover, the higher the PTMO content, the lower the phase transition temperatures for the obtained liquid crystalline polyurethanes, independent of the polyaddition method applied.

The conclusion can be drawn that the crystalline properties of polyurethanes are connected with the occurrence of the BHBP–2,4-TDI sequence. The crystallinity was more pronounced by increasing the BHBP–2,4-TDI sequences, as well as for polyurethanes obtained by the one-step method. The one-step method favors the formation of longer BHBP–2,4-TDI sequences, where the two-step method produces molecules with a more regular arrangement of the BHBP–2,4-TDI and PTMO–2,4-TDI sequences.

The effect of the flexible spacer length on some properties of liquid crystalline polyurethanes was investigated for a series of polyurethanes, which differ in the flexible spacer length and BHBP content.[15] Poly(tetramethylene ether) diols of different molecular weights (PTMO, M_n = 250, 650, 1000, 2000) were used as flexible spacers. The isotropization temperatures for all polyurethanes containing 75 mol% BHBP are similar, independent from the flexible spacer length. Decrease in the amount of the mesogenic agent results in a reduction of the melting temperature, as well as the isotropization temperature. Increase in the molecular weight of the flexible spacer at a constant content of the mesogenic agent exerts the same effect. Both melting and isotropization temperatures are less sensible to the changes of PTMO spacer length at higher BHBP content.

APPLICATIONS

Expected applications of liquid crystalline polyurethanes were mentioned in some articles dealing with synthesis, morphology, and thermotropic properties of the polyurethanes. As an example, elastomers and fibers, in general, and melt processed fibers and films are given.[6,14] Also, fibers spun from solutions were mentioned.[3]

Liquid crystalline urethane ionomers which were crosslinked with a melamine resin could find applications as films and coatings.[8]

In the field of urethane elastomers, blends of liquid crystalline polyurethanes with common segmented polyurethanes should become an object of future applications.[13]

In any proposed application, increased mechanical strength and modulus of elasticity is taken into account.

Membranes cast from solutions of liquid crystalline polyurethanes could find application for organic liquids separations. It was expected that incorporation of mesogenic groups into a polyurethane chain should enhance the durability of the membrane and influence its transport properties.[17] Attention should also be paid to the potential application of amphiphilic liquid crystal polyurethanes.[9]

REFERENCES

1. Iimura, K.; Koide, N.; Tanabe, H.; Takeda, M. *Makromol. Chem.* **1981**, *182*, 2569.
2. Kossmehl, G.; Hirsch, B. *Acta Polym.* **1990**, *41*, 597.
3. Tanaka, M.; Nakaya, T. *Makromol. Chem.* **1986**, *187*, 2345.
4. Tanaka, M.; Nakaya, T. *Macromol. Chem.* **1988**, *189*, 771.
5. Onouchi, Y.; Inagaki, Sh.; Okamoto, H.; Furukawa, J. *Kenkyu Hokoku–Asahi Garasu Zaidan* **1990**, *57*, 241; Ch. Abstract **1992**, *116*, 21571.
6. Gerbi, D. J.; Macosko, Ch. W.; Mormann, W.; Benadda, S. *Polym. Prepr. (Am. Chem. Soc., Div. Polym. Chem.)* **1992**, *33*, 1109.
7. Mormann, W.; Benadda, S. *Polym. Prepr. (Am. Chem. Soc., Div. Polym. Chem.)* **1993**, *34*, 739.
8. Lorenz, R.; Els, M.; Haulena, F.; Schmitz, A.; Lorenz, O. *Angew. Makromol. Chem.* **1990**, *180*, 51.
9. Tanaka, M.; Nakaya, T. *Makromol. Chem.* **1989**, *190*, 3067.
10. Tanaka, M.; Nakaya, T. *Adv. Urethane Sci. Technol.* **1993**, *12*, 1.
11. Penczek, P.; Frisch, K. C.; Szczepaniak, B.; Rudnik, E. *J. Polym. Sci., Polym. Chem. Ed.* **1993**, *31*, 1211.
12. Szczepaniak, B.; Frisch, K. C.; Penczek, P.; Rudnik, E. *J. Polym. Sci., Polym. Chem. Ed.* **1994**, *32*, 2559.
13. Gähde, J.; Pohl, M.-M.; Mix, R.; Dany, R.; Hinrichsen, G. *Polymer*, in press.
14. Pollack, S. K.; Shen, D. Y.; Hsu, S. L.; Wang, Q.; Stidham, H. D. *Macromolecules* **1989**, *22*, 551.
15. Szczepaniak, B.; Frisch, K. C.; Penczek, P.; Mejsner, J.; Leszczyńska, I.; Rudnik, E. *J. Polym. Sci., Polym. Chem. Ed.* **1993**, *31*, 3223.
16. Szczepaniak, B.; Frisch, K. C.; Penczek, P.; Rudnik, E.; Cholińska, M. *J. Polym. Sci., Polym. Chem. Ed.* **1993**, *31*, 3231.
17. Wolińska-Grabczyk, A.; Cabasso, A. I.; Szczepaniak, B.; Penczek, P. *J. Membr. Sci.*, to be submitted.

LIQUID CRYSTALLINE STATE POLYMERIZATION

John Tsibouklis
Advanced Polymers and Composites Research Group
University of Portsmouth

Low molecular mass liquid crystals containing polymerizable moieties react in the liquid-crystalline state to give polymers that may exhibit liquid crystallinity. The first example of such a reaction was reported by Herz et al.[1] Four years later, Amerik and Krentsel published a comparative study of the radiation-induced polymerization of vinyl oleate in the solid, mesomorphic, and isotropic phases.[2] Since then, liquid-crystalline state polymerization have been reported for a number of reactive mesogens, including acrylates and methacrylates, vinyl ethers and other vinylic compounds, diacetylenes, and epoxides.

Two systems have attracted particular attention, namely:

- the thermal polymerization of liquid-crystalline diacetylenes; and,
- the photopolymerization of liquid-crystalline compounds containing reactive double bonds.

PREPARATION AND PROPERTIES

Thermal Polymerization of Liquid-Crystalline Diacetylenes

The observation that mesogenic diacetylene monomers can form thermotropic liquid-crystal phases that can undergo polymerization on thermal annealing, dates back to the middle of the last decade, when three independent groups demonstrated the feasibility of the reaction. Garito et al. demonstrated the liquid-crystalline state polymerization of substituted divinyldiacetylenes, but the structure of the resulting polymers was believed to be different from that obtained from topotactic solid-state reactions.[6] This work was extended by Schen and Tsibouklis et al. who showed that polydiacetylenes, with structures analogous to those obtained by solid-state reactions, could be prepared by the liquid-crystalline state polymerization of both symmetrically and unsymmetrically disubstituted diacetylene monomers.[4,5] Since then, the reaction has been shown to occur with a large number of diacetylenic compounds.[4,5,10–12] The liquid-crystalline state polymerization process can be summarized as shown in **Scheme I**.

monomer crystal ⇌ (heat / cool) liquid crystal monomer → (heat or irradiation) polymer **I**

The polymers formed do not normally exhibit liquid-crystalline properties, but a number of polymeric materials, in which the liquid-crystalline order can be preserved, have been prepared. The preservation of liquid crystallinity postpolymerization is normally achieved if the monomer-to-polymer conversion is low and/or suitably designed liquid-crystalline mixtures are employed where the monomer acts as a low-concentration solute in a liquid-crystalline solvent.[4,7] As a rule, however, the liquid-crystalline order is preserved in systems that conform to the structural features required for the formation of liquid-crystalline polymers in general.[8,9]

The Polymerization of Liquid Crystalline Compounds Containing Reactive Double Bonds

The liquid crystalline state polymerization reaction of monomers containing reactive double bonds dates back to the 1960s. Amerik and Krentsel, during their pioneering work on the photopolymerization of vinyl oleate, found that the polymer morphology was strongly dependent on the phase in which the reaction had occurred.[12] Further work involving a comparative investigation of the polymerization of p-methacrylobenzoic acid in p-cetyloxybenzoic acid (a liquid crystalline solvent) and in dimethyl formamide established that a higher molecular weight polymer was obtained in the former medium.[13] Following these first experiments, a number of monomers containing reactive double bonds were prepared.[3,16,20–22]

The criteria for the development of liquid crystallinity in polymers were established by two independent groups that showed that polymers containing mesogenic moieties will exhibit liquid crystallinity if the motion of the mesogen is decoupled from the main chain by the use of appropriate spacer groups.[8,9,14] Thus, polymers derived from monomers that adopt on polymerization the now well-established main-chain or side-chain structures will normally exhibit liquid crystalline properties.[15] The phenomenon of liquid crystallinity in polymers is therefore a consequence of the propensity of polymers to self-organize rather than a function of the mode of polymerization adopted for their preparation.

More recently, a significant research effort has been directed towards the study of the liquid-crystalline state polymerization of mono- and di-acrylates. Thus, Broer et al. have reported the photoinitiated polymerization of 4-biphenylyl, 4-(6-acryloyloxyhexyloxy)benzoate.[3] The monomer, which exhibited a series of smectic and nematic mesophases was oriented in polyimide-coated glass cells and characterized by birefringence measurements. It was found that both the birefringence and the liquid-crystalline transition temperatures increased during polymerization, suggesting that the orientation imposed by the monomer structure could be retained in the polymer.

The same group also investigated a series of 1,4-phenylene bis4-(ω-acryloyloxy)alkyloxybenzoates and found that for these materials an increase in spacer length results in a decrease of the observed isotropic transition temperatures and a stabilization of the smectic phases.[17,18] The effect of variable spacer lengths on the order parameters was found to be minimal.

APPLICATIONS

The preparation of polymeric materials with controlled orientation is one of the major challenges facing modern polymer science. There are, of course, many techniques available to construct such structures, but the liquid-crystalline state polymerization of mesogenic monomers represents one of the most easily accessible methods. It is best for engineering materials or fabricating novel device structures. The advantages of this approach stem from the basic principle that the oriented state of a liquid crystalline monomer can be frozen-in by *in-situ* polymerization.

The wide range of possible applications of polymerization reactions in the liquid crystalline state becomes apparent when the above features and the ease with which these reactions can be carried out are considered. The highly anisotropic mechanical behavior in the direction of the molecular orientation may be of interest to workers in the areas of polymeric fibers and filaments. The anisotropic thermal behavior (with a low thermal expansion coefficient in the direction of the molecular orientation) would be expected to find uses in structures for which thermal stresses must be avoided, such as in the fabrication of polymeric coatings or the encapsulation of inorganic substances in polymer networks. Finally, the large difference in refractive index and optical absorption, between the principal directions of polymeric arrays constructed by the method, is already being actively investigated by a number of industrial researchers as a possible means for the fabrication of active elements for optical and optoelectronic devices, confirming that the technique is ripe for commercial exploitation.[19]

REFERENCES

1. Herz, J.; Reiss-Husson, F.; Rempp, P.; Luzzatti, V. *J. Polym. Sci., Part C* **1963**, *4*, 1275.
2. Amerik, Y. B.; Krentsel, B. A. *J. Polym. Sci., Part C* **1967**, *16*, 1383.
3. Broer, D. J.; Finkelmann, H.; Kondo, K. *Makromol. Chem.* **1988**, *189*, 185.
4. Tsibouklis, J.; Shand, A. J.; Werninck, A. R.; Milburn, G. H. W. *Liquid Crystals* **1988**, *3*(10), 1393.
5. Schen, M. A. *SPIE* **1988**, *971*, 178.
6. Garito, A. F.; Teng, C. C.; Wong, K. Y.; Zammani'Khamiri, O. *Mol. Cryst. Liq. Cryst.* **1984**, *106*, 219.
7. Tsibouklis, J.; Petty, M.; Pearson, C.; Petty, M. C.; Feast, W. J. *Materials for Non-Linear and Electro-Optics, IoP Conf. Ser. No. 103*; IoP: Bristol, 1989, pp 187–192.
8. Wendorff, J. H.; Finkelmann, H.; Ringsdorf, H. *J. Polym. Sci., Polym. Symp.* **1978**, *63*, 73.
9. Wendorff, J. H.; Finkelmann, H.; Ringsdorf, H. *J. Polym. Sci., Polym. Symp.* **1978**, *63*, 245.
10. Milburn, G. H. W.; Campbell, C.; Shand, A. J.; Werninck, A. R. *Liquid Crystals* **1990**, *8*(5), 623.
11. Izuoka, A.; Okuno, T.; Ito, T.; Sugawara, T. *Mol. Cryst. Liq. Cryst.* **1993**, *226*, 201.
12. Attard, G. S.; West, Y. D. *Liquid Crystals* **1990**, *7*(4), 487.
13. Amerik, Y. B.; Konstantinov, I. I.; Krentsel, B. A. *J. Polym. Sci., Part C* **1968**, *23*, 231.
14. Shibaev, V. P.; Plate, N. A.; Freidzon, Y. A. *J. Polym. Sci., Polym. Chem. Ed.* **1979**, *17*, 1655.
15. Dobb, R. G.; McIntyre, J. E. *Advances in Polymer Science*; Springer-Verlag: New York, 1984; Vol. 60/61.
16. Broer, D. J.; Boven, J.; Mol, G. N.; Challa, G. *Makromol. Chem.* **1989**, *190*, 2255.
17. Broer, D. J.; Hikmet, R. A. M.; Challa, G. *Makromol. Chem.* **1989**, *190*, 3201.
18. Broer, D. J.; Mol, G. N.; Challa, G. *Makromol. Chem.* **1991**, *192*(1), 59.
19. Heynderickx, I.; Broer, D. J. *Mol. Cryst. Liq. Cryst.* **1991**, *202*, 113.
20. Perplies, V. E.; Ringsdorf, H.; Wendorff, J. H. *Ber. Bunsenges. Phys. Chem.* **1974**, *78*, 921.
21. Magagnini, P. L. *Makromol. Chem. Suppl.* **1991**, *4*, 223.
22. Finkelmann, H.; Ringsdorf, H.; Wendorff, J. H. *Makromol. Chem.* **1978**, *179*, 273.

Liquid Rubbers

See: Reactive Liquid Rubbers (Epoxy Toughening Agents)

LIQUID SINGLE CRYSTAL ELASTOMERS

Heino Finkelmann, Stefan Disch, and Claudia Schmidt
Institut für Makromolekulare Chemie der Universität Freiburg

A liquid single crystal elastomer (LSCE) is a stable monodomain of a weakly crosslinked liquid-crystalline polymer–a so-called liquid-crystalline elastomer (LCE).[1] Like any LCE, it combines the properties of a liquid crystal (LC) and an elastomer. It shows high molecular mobility combined with anisotropic order, which are the characteristics of the LC phase, and it exhibits the rubber elasticity of a weakly crosslinked polymer network.[2,3] If the concentration of the crosslinks is increased, the elastic properties are lost and duromers are obtained.[4–7] The third important aspect of an LSCE that distinguishes it from a simple LCE is that the LSCE is a LC monodomain–the LC analogue of a single crystal. Therefore, the LSCE shows macroscopically uniform order throughout the sample. This feature, and the fact that LSCE samples of considerable size can be prepared, makes LSCEs interesting objects for both fundamental investigations and technological applications. Here we consider only side-chain LSCEs, for which the mesogens are attached as side groups to the polymer backbone, in contrast to main-chain systems, where the mesogenic units are incorporated in a linear fashion into the polymer backbone.

We will give an overview on nematic LSCEs, on which most studies have been performed. The nematic phase lacks any long-range positional order and shows only a one-dimensional orientational order of the moieties that cause the formation of the LC phase, the mesogens. The average mesogen direction is called the director. LSCEs with LC phases of higher order will be treated in the third section.

NEMATIC LIQUID SINGLE CRYSTAL ELASTOMERS

Synthetic Concepts

These steps are required to synthesize stable monodomain LCEs: preparation of an adequate precursor compound; alignment by a suitable external field to form the monodomain; and chemical fixation of the monodomain state. Two concepts for the synthesis of LSCEs have been developed based on these steps. In the first case, alignment is induced by a magnetic field, which acts on the mesogens. In the second approach, a mechanical field, which acts on the polymer network, is employed. The first concept was realized by Legge, Davis, and Mitchell.[8] They use a linear nematic side-chain prepolymer as a starting point. The linear LCP is macroscopically aligned by an external magnetic field.

The other concept was developed by Küpfer et al.[1] This approach is known as the two-step crosslinking concept because two crosslinking reactions with different reaction rates are used.

The LSCEs obtained by these methods are optically transparent and show a macroscopically uniform director orientation. Similar to single crystals in solid-state physics, which allow the study of the physical properties of crystals, LSCEs allow study of the nematic phase on a macroscopic scale. The anisotropic macroscopic properties, like anisotropic elastic modulus, anisotropy of swelling and, in particular, birefringence, make LSCEs interesting materials for technical applications.

Thermal Expansion

One example of the unusual properties of LSCEs, which is observed on the macroscopic scale, is the dependence of the dimensions of the elastomer on temperature. At the nematic-isotropic phase transformation, the dimensions of the LSCE change drastically: the length of the sample in the direction of the internal mechanical field, that is, parallel to the director, decreases by more than 20 percent when the LSCE is brought from the nematic into the isotropic state.[9]

Phase Behavior

For low molar mass and polymeric liquid crystals, the nematic–isotropic phase transformation and its dependence on external fields can be described by the phenomenological Landau–de Gennes theory.[10–12] The theory predicts that at a critical field strength the transformation becomes second order with an order parameter decreasing continuously with increasing temperature. Beyond the critical point, there is no phase transformation. However, for electrical or magnetic fields, the critical field strength is larger than experimentally accessible, so that the critical point of the nematic–isotropic transformation has not been detected for low molar mass or polymeric LCs.

The application of mechanical fields to nematic elastomers, however, has shown an increase of the order parameter in the LC phase and a shift of the nematic–isotropic transformation temperature of up to 10 K, in agreement with the predictions of the Landau–de Gennes theory.[13,14]

State of Order and Mechanical Behavior

Two order parameters are used to analyze the state of order of LSCEs. The nematic order parameter S_N describes the degree of orientation of the mesogens with respect to the director. The director order parameter S_D takes into account the degree of orientation of the local directors. When an external mechanical field is applied to a nematic elastomer, the polydomain structure with isotropic director distribution ($S_D = 0$) is converted into the monodomain state where S_D equals one for a sample with ideal uniform director orientation.[13] In LSCEs, an internal mechanical field is permanently present. This internal mechanical field is introduced by the mechanical deformation of the elastomer before the second crosslinking step takes place.

Mechanical Reorientation

As an example of reorientation effects of nematic systems in external fields, a director reorientation is observed when an external mechanical field is applied perpendicularly to the director of an LSCE. LSCEs offer the unique possibility to follow the director reorientation process experimentally because a stress applied perpendicularly to the original director induces a certain degree of reorientation that can be analyzed experimentally.

Dynamic Behavior

Although various aspects of nematic LSCEs are well understood, investigations on the dynamic behavior are small in number and all fairly recent. Dynamic stress-optical measurements of LCEs were carried out in the isotropic phase.[15] The relaxation frequencies and the relaxation strength show critical behavior in qualitative agreement with the Landau–de Gennes theory. The strong crosslinking density dependence of the relaxation strength shows that the crosslinking points locally disturb the LC order.

LSCES OF HIGHER LIQUID CRYSTALLINE ORDER

The columnar phases of discotic LCs show long-range positional order of the columns formed by the stacking of the disk-shaped mesogens. Macroscopically oriented discotic phases are interesting as systems with rapid directed charge migration.[16]

The two-step crosslinking concept was successfully applied to discotic side-chain elastomers.[17] By using a discotic crosslinker, a discotic mesophase that is stable up to high temperature could be obtained.[18] The quantitative investigation of the network anisotropy of a macroscopically ordered discotic elastomer by thermal expansion measurements revealed an anisotropy of the radius of gyration of 1.4.[18]

The concept of LSCEs is not only applicable to thermotropic liquid crystals where the formation of LC phases depends on temperature, but also to lyotropic liquid crystals that are formed in the presence of a solvent. It has been shown that macroscopically aligned lyotropic polymer networks can be obtained by using mechanical fields. An example of an aligned hexagonal phase was given by Löffler, who obtained the orientation of the cylindrical aggregates by anisotropic swelling of the crosslinked sample in a cylindrical tube.[19] An aligned lamellar phase was obtained by compressing the sample uniaxially.[20]

PERSPECTIVE

The development of LSCEs provides materials with novel, interesting physical properties and a large variety of potential applications. Due to their optical properties, nematic LSCEs turned out to be suitable materials for bifocal contact lenses. LSCEs can act as membranes for separation processes, as they show both high selectivity and high permeability.[21] S^*_c LSCEs are new materials for nonlinear optics and could be applied as piezoelectric sensors because their elastic moduli are much smaller than those of the common piezoelectric elements.

REFERENCES

1. Küpfer, J.; Finkelmann, H. *Makromol. Chem. Rapid Commun.* **1991**, *12*, 717.
2. Finkelmann, H.; Rehage, G. *Adv. Polym. Sci.* **1984**, *60/61*, Springer-Verlag, Berlin.
3. Davis, F. J. *J. Mater. Chem.* **1993**, *3*, 551.
4. Broer, D. J.; Boven, J.; Mol, G. N.; Challa, G. *Makromol. Chem.* **1989**, *190*, 2255.
5. Broer, D. J.; Hikmet, R. A. M.; Challa, G. *Makromol. Chem.* **1989**, *190*, 3202.
6. Broer, D. J.; Mol, G. N.; Challa, G. *Makromol. Chem.* **1991**, *192*, 59.
7. Küpfer, J.; Nishikawa, E.; Finkelmann, H. *Polym. Adv. Technol.* **1994**, *5*, 110.
8. Legge, C. H.; Davis, F. J.; Mitchell, G. R. *J. Phys. II France* **1991**, *1*, 1253.
9. Küpfer, J.; Finkelmann, H. *Macromol. Chem. Phys.* **1994**, *195*, 1353.
10. de Gennes, P. G. *Mol. Cryst. Liq. Cryst.* **1971**, *12*, 193.
11. de Gennes, P. G. *C. R. Seances Acad. Sci. Ser. B.* **1975**, *28*, 101.
12. de Gennes, P. G. *Polymer Liquid Crystals*; Ciferri, A.; Krigbaum, W. R.; Myer, R. B., Eds.; Academic: New York, 1982.
13. Schätzle, J.; Kaufhold, W.; Finkelmann, H. *Makromol. Chem.* **1989**, *190*, 3269.
14. Kaufhold, W.; Finkelmann, H.; Brand, H. R. *Makromol. Chem.* **1991**, *192*, 2555.
15. Sigel, R.; Stille, W.; Strobl, G.; Lehnert, R. *Macromolecules* **1993**, *26*, 4226.
16. Adam, D.; Closs, F.; Funhoff, D.; Haarer, D.; Ringsdorf, H.; Schuhmacher, P.; Siemensmeyer, K. *Phys. Rev. Lett.* **1993**, *70*, 457.
17. Bengs, H.; Finkelmann, H.; Küpfer, J.; Ringsdorf, H.; Schuhmacher, P. *Makromol. Chem. Rapid Commun.* **1993**, *14*, 445.
18. Disch, S.; Finkelmann, H.; Ringsdorf, H.; Schuhmacher, P. *Macromolecules* **1995**, *28*, 2424.
19. Löffler, R.; Finkelmann, H. *Makromol. Chem. Rapid Commun.* **1990**, *11*, 321.
20. Fischer, P.; Schmidt, C.; Finkelmann, H. *Macromol. Rapid Commun.* **1995**, *16*, 435.
21. Reinecke, H.; Finkelmann, H. *Makromol. Chem.* **1992**, *193*, 2945.

Lithographic Materials

See: *Chemical Amplification Resists*
Chemically Amplified Resists (for Deep-UV Lithography)
Chemically Amplified Resins (New Generation)
Novolak Resins (for Microlithography)

LIVING ANIONIC POLYMERIZATION

Michel Fontanille
ENSCPB
Université Bordeaux-l

The importance of living systems is crucial in macromolecular engineering; it turns out that anionic polymerization has been by far the most often applied mechanism for making polymers of well-defined structures.

To achieve living conditions, sites–other than the polymerizable group–that are sensitive to nucleophilic attack, should not be present. Only a few monomers and solvents compel such a requirement. Reactivity and basicity of propagating species being closely related, the propensity of a growing anion to react with an electrophilic site can be estimated from the pK_a of its conjugated acid.

Moreover, provided the chosen anionic initiator reacts with a total efficiency, the molar mass of the resulting polymers is closely related to the initiator concentration. To obtain high molar masses, low concentrations of initiator are required that necessitate extremely pure reactants.

VINYL AND RELATED MONOMERS

Monomers which are prone to polymerize via an anionic mechanism necessarily contain a polymerizable group that is sensitive to nucleophilic addition.[1] This requires that substituents with an electron-withdrawing character be present to activate the double bond towards a nucleophilic attack by carbanions.

HETEROCYCLES

A wide variety of heterocyclics can be opened and polymerized by nucleophilic substitution.

The above requirements regarding the presence of an electrophilic site also apply to heavy types; this considerably limits the number of solvents that can be used in anionic polymerization; two types are generally distinguished:

- hydrocarbons (either aliphatic or aromatic) which act as diluents,
- ethers (cyclic or not) which act not only as diluents but also as solvating agents and sometimes as dissociating media. In the latter case, they play an important role in the reactivity of the species.

STRUCTURE OF ACTIVE SPECIES

Carbanionic species arising from vinyl and related monomers were found to exist under various structures; similar behaviors have been observed for species derived from heterocyclic monomers.

Indeed, carbanionic active centers exist under a wide variety of structures whose proportion in the reaction medium are a function of the monomer considered, the counterion, the solvent, the temperature, and the possible presence of additives.

When various organo-lithiums are simultaneously present in the reaction medium, they lead to mixed aggregates whose disaggregation equilibrium constant is generally higher than that of the homo-aggregates.[2] An increase of the overall reactivity can thus be obtained. The addition of ethers (tetrahydrofurane, dimethoxyethane, dioxane) in amounts comparable to that of organo-lithium species provokes an acid–base interaction with Li^+ cation. Depending on the structure of the additive, this coordination leads to solvated ion-pairs whose structure can take different forms. Such species are non-aggregated. For example, N,N,N′,N′-tetramethylethylenediamine forms externally solvated ion-pairs.[3]

The intrinsic reactivity is, in this case, lower than that of non-aggregated, non-solvated species.[4] Crown-ethers and crown-amines are known to bring about loose ion-pairs.

An increased interionic distance often results that leads to a strong increase in the reactivity.[5]

VARIOUS TYPES OF INITIATORS

There are two ways to initiate the anionic polymerization depending on the nature of the solvent used:

- initiation by electron transfer,
- initiation by nucleophilic attack.

For both industrial and (often) laboratory purposes, initiation of anionic polymerization is performed via nucleophilic addition. In the case of vinyl and related monomers, fast and complete nucleophilic addition requires initiators with high nucleophilicity.

In hydrocarbon solvents, lithium alkyls are soluble due to their partial covalency and they are often used as initiators. Butyllithium (n-, sec-, tert-) is the most common anionic initiator.

The relative reactivity of various isomers is determined by both the value of the apparent disaggregation constant and the electronic effect of the substituents. The reactivity decreases in the following sequence with isoprene in hexane.[6] ter-BuLi > sec-BuLi > n-BuLi but a different order of reactivity may be observed with other solvents and monomers. Other lithium derivatives (even if less reactive) are used to initiate strongly electrophilic monomers; they are obtained either by nucleophilic addition of butyllithium on nonpolymerizable monomers.[7]

REACTIONS OF LIVING ENDS

The high reactivity of propagating species is responsible for several side reactions that are either spontaneous or result from the presence of reactive functional groups in the reaction medium. The living character is obtained only if such reactions may be neglected during the propagation step.

Because of possible side reactions and low concentration in active sizes, the perfect purification of the reaction medium has to be performed with great care, particularly if high molar masses have to be reached.

When performed after the quantitative monomer conversion, the reaction of the end-standing active sites with various electrophilic compounds is used to end-functionalize the polymeric chains.

If the polymerization is initiated by a bifunctional initiator (transfer from radical-ion or dilithium derivative), end-functionalization leads to telechelics, which can be used either for chain extension or as precursors of networks.

ANIONIC POLYMERIZATION AND MACROMOLECULAR SYNTHESIS

Even though limited to a few monomers because of the above requirements, living anionic polymerization is a particularly useful tool of macromolecular engineering. Due to both the persistence of reactive chain-ends and the low dispersity in chain length, anionic polymerization can be used to synthesize complex and well-defined architectures.

The most important application for which anionic polymerization cannot be matched is the preparation of block copolymers. Addition of a monomer B to living poly A after completion of A polymerization, leads to diblock copolymers.[8]

Macromonomers which are interesting intermediates in polymer synthesis are commonly prepared by anionic processes.[9]

Graft copolymers with relatively well-defined structure can be prepared by anionic copolymerization of macromonomers.

Two other methods are used to synthesize graft copolymers with nearly monodisperse grafts. The first, "grafting from," consists of the generation of anionic active sites on the poly-A backbone, to anionically initiate the polymerization of monomer B.

The second method of grafting is called "grafting onto," and consists of the reaction of living chains of polymer B onto a backbone bearing reactive groups.

Anionic polymerization is also used to prepare star polymers. The two previous principles may again be used.

Other types of macromolecular architectures are accessible via anionic polymerization: macrocycles, plurimacrocycles, dendrimers, hetero-arm star polymers, and others whose synthesis is described in this Encyclopedia.

REFERENCES

1. Worsfold, D. J.; Bywater, S. *Can. J. Chem.* **1960**, *38*, 1891.
2. Guyot, A.; Vialle, J. *J. Macrom. Sci.* **1970**, *4*, 79.
3. Hélary, G.; Fontanille, M. *Eur. Pol. J.* **1978**, *14*, 345.
4. Fontanille, M.; Hélary, G.; Szwarc, M. *Macromolecules* **1988**, *21*, 1532.
5. Hélary, G.; Fontanille, M. *Pol. Bull.* **1980**, *3*, 159 (a) Mélary, G.; Lefèvre-Jenot, L.; Fontanille, M.; Smid, J. *J. Organomet. Chem.* **1981**, *205*, 139.
6. Roovers, J. E. L.; Bywater, S. *Macromolecules* **1975**, *8*, 251.
7. Fowells, W.; Schuerch, C.; Bovey, F. A.; Steiner, E. C. *J. Amer. Chem. Soc.* **1967**, *89*, 1396.

8. Vanzo, E. *J. Polym. Sci. A-1* **1966**, *4*, 1727.
9. Rempp, P.; Franta, E. *Adv. Polym. Sci.* **1984**, *58*, 1.

LIVING ANIONIC POLYMERIZATION

Seiichi Nakahama and Akira Hirao
Department of Polymer Chemistry
Tokyo Institute of Technology

The living polymerization system is currently the only method that leads to polymers with well-regulated chain lengths and a variety of molecular architectures. In the last decade, a variety of living polymers have been developed based on anionic, cationic, coordination, and even radical mechanisms for achieving well-regulated and defined polymer synthesis. The most established method among these living polymerization systems is currently the anionic living polymerization of styrene and 1,3-dienic monomers.[1] Unfortunately, only a limited range of vinyl monomers are amenable to this anionic living polymerization system. In addition to styrene and 1,3-dienes, similar living polymers are obtained from a few styrene derivatives substituted with alkyl, aryl, and alkoxy functions. Under certain conditions, 2- and 4-vinylpyridines, alkyl methacrylate, *tert*-butyl acrylate, and some heterocyclic monomers have been found to undergo anionic living polymerization.[2] The main drawback of this method is that very useful functional groups, such as hydroxy, amino, mercapto, and carbonyl, are generally not compatible with anionic species at the chain ends of living polymers derived from the above-mentioned monomers. This means that anionic living polymerization of the monomers with these functional groups is difficult. To overcome this obstacle, Nakahama and Hirao have developed a new strategy in which the concept of protective group is introduced into anionic living polymerization.[3] The strategy involves protecting the functional group during living polymerization, followed by deprotection to regenerate the original functional group from the resulting polymer. Styrene is selected as a basic monomer skeleton, since an excellent class of living polymer is obtained from styrene itself. The key necessity is to find a suitable protecting group for a functional group which is completely stable during anionic living polymerization and is quantitatively removable without decomposing and crosslinking of resulting polymer chain. If the route is realized, the resulting polymer should retain the desirable characteristics of living polymers; i.e., a well-controlled molecular weight and narrow molecular weight distribution, and a functional group in all monomer units.

PREPARATION AND PROPERTIES

tert-Butyldimethylsilyl-protected 4-vinylphenol was the first successful example of anionic living polymerization of a protected functional monomer.[4,5] By protecting the phenolic hydroxy group of 4-vinylphenol with *tert*-butyldimethylsilyl group, the anionic living polymerization of the silyl protected 4-vinylphenol can be achieved satisfactorily under typical conditions. Accordingly, the poly(4-vinylphenol)s thus obtained should be linear and well-regulated structures with known molecular weights and narrow molecular weight distributions. By employing a similar method, a variety of novel anionic living polymers have been produced from styrene derivatives by protecting functional groups during polymerization. Functional polystyrenes with regulated chain lengths and narrow molecular weight distributions were obtained after deprotection. At present, the functional groups cover most useful functionalities that involve hydroxy (phenolic and alcoholic), amino, mercapto, formyl, acetyl, carboxy, ethnyl, and silanol.[5–20]

Anionic living polymers are also produced from 1,3-diene monomers and alkyl methacrylates. Therefore, the same ideas used for the anionic living polymerization of the functionalized styrene derivatives may be extended to the dienes and methacrylates with functional groups.[21–26]

APPLICATION

The protection and anionic living polymerization of functional monomers can be applied to the syntheses of block copolymers containing functional groups and of the telechelic polymers with regulated molecular weights and very narrow molecular weight distributions. A number of functional block copolymers are produced from various combinations of the usual monomers and the protected functional monomers. Here are three typical cases.

Poly(2-hydroxyethyl methacrylate) (PHEMA)-*block*-polystyrene-*block*-PHEMA and polystyrene-*block*-PHEMA are generated by the sequential block copolymerization of styrene and (2-trimethylsilyloxy) ethyl methacrylate followed by deprotection of the protecting group from the resulting copolymers.[21,22] The block copolymers of styrene and HEMA show excellent antithrombogenic activity and are being applied to various medical devices.[27–29]

The block copolymer of poly(1,3-butadiene)-*block*-poly-[2-(trialkoxysilyl)-1,3-butadiene] prepared by anionic living polymerization was hydrogenated to afford polyethylene-*block*-poly[2-(trialkoxysilyl)-1,3-butadiene] with precisely regulated block lengths. The hydrogenated block copolymer plays an important role in adhering polyolefins and inorganic materials like silica.[30]

A number of telechelic polymers with very narrow molecular weight distributions are produced by the reactions of the anionic living polymer. These include living polystyrene, with α-haloalkanes ω-substituted with protected functions. Thus, telechelic polystyrenes having amino, carboxy, and aminoxy groups at the chain terminals are included in this group.[31–33]

The nearly monodisperse polystyrenes and polymethacrylates having alkoxysilyl groups produced by anionic living polymerization show great resistance against plasma etching as a photoresist with very high resolution in the lithography process.[34]

REFERENCES

1. Morton, M. *Anionic Polymerization: Principles and Practice*, Academic, London, 1983.
2. Rempp, A.; Franta, E.; Herz, J-E. *Adv. Polym. Sci., Macromolecular Engineering by Anionic Methods*, **1988**, *86*, 145.
3. Nakahama, S.; Hirao, A. *Prog. Polym. Sci. Protection and Polymerization of Functional Monomers; Anionic Living Polymerization of Protected Monomers*, **1990**, *15*, 299.
4. Hirao, A.; Yamaguchi, K.; Takenaka, K.; Suzuki, K.; Nakahama, S.; Yamazaki, N. *Makromol. Chem., Rapid. Commun., Synthesis of Poly(4-vinylphenol) by Means of Anionic Living Polymerization*, **1982**, *3*, 941.

5. Hirao, A.; Takenaka, T.; Packirisamy, S.; Yamaguchi, K.; Nakahama, S. *Makromol. Chem., Studies on Anionic Living Polymerization of 4-(tert-Butyldimethylsilyloxy)styrene*, **1985**, *186*, 1803.
6. Hirao, A.; Yamamoto, A.; Takenaka, K.; Yamaguchi, K.; Nakahama, S. *Polymer, Anionic Living Polymerization of 4-[2-(trialkyl)silyloxyethyl]styrene as Protected 4-(2-hydroxyethyl)styrene*, **1987**, *28*, 303.
7. Yamaguchi, K.; Hirao, A.; Suzuki, K.; Takenaka, K.; Nakahama, S.; Yamazaki, N. *J. Polym. Sci., Polym. Lett. Ed., Anionic Living Polymerization of p-N,N-Bis(trimethylsilyl) aminostyrene*, **1983**, *21*, 395.
8. Suzuki, K.; Yamaguchi, K.; Hirao, A.; Nakahama, S. *Macromolecules, Synthesis of Well-defined Poly(4-aminostyrene) by Means of Anionic Living Polymerization of 4-(N,N-Bis(trimethylsilyl)amino)styrene*, **1989**, *22*, 2607.
9. Suzuki, K.; Hirao, A.; Nakahama, S. *Makromol. Chem., Anionic Living Polymerization of 4-[N,N,-Bis(trimethylsilyl) aminomethyl]styrene and 4-[N,N-Bis(trimethylsilyl)aminoethyl]styrene*, **1989**, *190*, 2893.
10. Hirao, A.; Shione, H.; Wakabayashi, S.; Nakahama, S.; Yamaguchi, K. *Macromolecules, Anionic Polymerization of 4-Vinylphenyl t-Butyldimethylsilyl Sulfide and 2-(4-Vinylphenyl)ethyl t-Butyldimethylsilyl Sulfide*, **1994**, *27*, 1835.
11. Hirao, A.; Ishino, Y.; Nakahama, S. *Makromol. Chem. Synthesis of Linear Polymerization of 1,3-Dimethyl-2-(4-vinylphenyl)imidazolidine and Subsequent Hydrolysis*, **1989**, *187*, 141.
12. Hirao, A.; Nakahama, S. *Macromolecules, Synthesis of a Well-defined Poly(4-vinylbenzaldehyde) by the Anionic Living Polymerization of N-[(4-Ethenylphenyl)methylene]cyclohexamine*, **1987**, *20*, 2968.
13. Ishizone, T.; Kato, R.; Ishino, Y.; Hirao, A.; Nakahama, S. *Macromolecules, Anionic Living Polymerization of 2-(3-Vinylphenyl)-1,3-dioxolane and the Related Monomers*, **1991**, *24*, 1449.
14. Hirao, A.; Kato, K.; Nakahama, S. *Macromolecules, Synthesis of Well-defined Poly(vinylacetophenone)s by Means of Anionic Living Polymerization of tert-Butyldimethylsilyl Enol Ethers of Vinylacetophenone*, **1992**, *25*, 535.
15. Ishino, Y.; Hirao, A.; Nakahama, S. *Macromolecules, Anionic Living Polymerization of 2-(4-Vinylphenyl)-4,4-dimethyl-2-oxazoline*, **1986**, *19*, 2307.
16. Hirao, A.; Ishino, Y.; Nakahama, S. *Macromolecules, Synthesis of Well-defined Poly(4-vinylbenzoic acid) by Means of Anionic Living Polymerization of 2-(4-Vinylphenyl)-4,4-dimethyl-2-oxazoline*, **1988**, *21*, 561.
17. Ishizone, T.; Hirao, A.; Nakahama, S. *Macromolecules, Anionic Living Polymerization of tert-Butyl 4-Vinylbenzoate*, **1989**, *22*, 2602.
18. Ishizone, T.; Hirao, A.; Nakahama, S.; Kakuchi, T.; Yokota, K.; Tsuda, K. *Macromolecules, Anionic Living Polymerization of 4-(Trimethylsilyl)ethynylstyrene*, **1991**, *24*, 5230.
19. Hirao, A.; Nagawa, T.; Hatayama, T.; Yamaguchi, K.; Nakahama, S. *Macromolecules, Anionic Living Polymerization of (4-Vinylphenyl)dimethyl-2-propoxysilane*, **1985**, *18*, 2101.
20. Hirao, A.; Nagawa, T.; Hatayama, T.; Yamaguchi, M.; Yamaguchi, K.; Nakahama, S. *Macromolecules, Anionic Living Polymerization of 4-(Alkoxysilyl)styrene*, **1987**, *20*, 242.
21. Hirao, A.; Kato, H.; Yamaguchi, K.; Nakahama, S. *Macromolecules, Synthesis of Poly(2-hydroxyethyl methacrylate) with a Narrow Molecular Weight Distribution by Means of Anionic Living Polymerization*, **1986**, *19*, 1294.
22. Mori, H.; Wakisaka, O.; Hirao, A.; Nakahama, S. *Makromol. Chem. Phys., Synthesis of Well-defined Poly(2-hydroxyethyl methacrylate) by Means of Anionic Living Polymerization of Protected Monomers*, **1994**, *195*, 3213.
23. Mori, H.; Hirao, A.; Nakahama, S. *Macromolecules, Anionic Living Polymerization of (2,2-Dimethyl-1,3-dioxolan-4-yl) methyl Methacrylate*, **1994**, *27*, 35.
24. Ozaki, H.; Hirao, A.; Nakahama, S. *Macromolecules, Anionic Living Polymerization of 3-(Tri-2-propoxysilyl)propyl Methacrylate*, **1992**, *25*, 1391.
25. Takenaka, K.; Hattori, T.; Hirao, A.; Nakahama, S. *Macromolecules, Anionic Polymerization of 2-(Trialkoxysilyl)-1,3-butadiene*, **1989**, *22*, 1563.
26. Takenaka, K.; Hattori, T.; Hirao, A.; Nakahama, S. *Macromolecules, Anionic Polymerization of 2-Silyl-substituted 1,3-Butadienes with Mixed Substituents*, **1992**, *25*, 96.
27. Nojiri, C.; Okano, T.; Grainger, D.; Park, K. D.; Nakahama, S.; Suzuki, K.; Kim, S. W. *Trans. ASAIO, Evaluation of Nonthrombogenic Polymers in a New Rabbit A-A Shunt Model*, **1987**, *10*, 596.
28. Nojiri, C.; Okano, T.; Koyanagi, H.; Nakahama, S.; Park, K. D.; Kim, S. W., *J. Biomater. Sci., Polymer Edn, In vivo protein adsorption on polymers: visualization of adsorbed proteins on vascular implants in dog*, **1992**, *4*, 75.
29. Nojiri, C.; Nakahama, S.; Senshu, K.; Okano, T.; Kawagoshi, N.; Kido, T.; Sakai, K.; Koyanagi, H.; Akutsu, T. *ASAIO J, A New Amphiphilic Block Co-polymer with Improved Elastomeric Properties for Application in Various Medical Devices*, **1993**, *39*, 322.
30. Takenaka, K.; Kato, K.; Hirao, A.; Nakahama, S. *Macromolecules, Catalytic Hydrogenation of Poly(2-silyl-substituted-1,3-butadiene)s*, **1990**, *23*, 3619.
31. Ueda, K.; Hirao, A.; Nakahama, S. *Macromolecules, Reaction of Anionic Living Polymers with a-Halo-w-aminoalkanes with a Protected Amino Functionality*, **1990**, *23*, 939.
32. Hirao, A.; Nagahama, H.; Ishizone, T.; Nakahama, S. *Macromolecules, Synthesis os Polymers with Carboxy End Groups by Reaction of Polystyryl Anions with Trimethyl 4-Bromoorthobutyrate*, **1993**, *26*, 2145.
33. Yoshida, E.; Ishizone, T.; Hirao, A.; Nakahama, S.; Takata, T.; Endo, T. *Macromolecules, Synthesis of Polystyrene having an Aminoxy Terminal by the Reactions of Living Polystyrene with an Oxoaminium Salt and with the Corresponding Nitroxy Radical*, **1994**, *27*, 3119.
34. Yamaguchi, K.; Ozaki, H.; Hirao, A.; Hirose, N.; Harada, Y.; Uzawa, Y.; Sekine, M.; Yoshimori, S.; Kawamura, M. *J. Electrochem. Soc., New Positive EB Resist with Strong Resistance to Plasma Damage*, **1992**, *139*, L33.

LIVING CARBOCATIONIC POLYMERIZATION, OLEFINS

Rudolf Faust
Chemistry Department
University of Massachusetts, Lowell

Living polymerization is a chain polymerization process that proceeds in the absence of chain transfer to monomer and irreversible termination. The irreversibility of termination is emphasized because, in contrast to conventional living anionic polymerization, in living cationic polymerizations the concentration of active species–i.e., the cations–is often very small. Most of the chain ends are in a dormant form arising by reversible termination. When initiation is relatively fast compared to propagation, polymers with molecular weights controlled by the [monomer]/[initiator]) ratio and narrow molecular weight distribution (MWD) are obtained ($M_w/M_n \sim 1.0$). However, it is important to note that narrow MWD is not a criterion of living polymerization. It may indicate living polymerization, but broad MWD does not necessarily imply nonliving behavior.

INITIATION OF LIVING CARBOCATIONIC POLYMERIZATIONS

The central question in cationic polymerization is the nature, activities, and concentration of the active species. They are determined in most systems by the mechanisms of initiation, which in turn will determine that head- and end-groups. They control the molecular weights by determining chain transfer and termination reactions. Cationic polymerization may be induced by a variety of chemical and physical methods; however, the most important initiating system from a scientific and practical perspective is the cation donor (initiator)/Friedel–Crafts acid (coinitiator) system.

The first example of living carbocationic polymerization, of isobutyl vinyl ether with a mixture of hydrogen iodide and iodine (HI/I_2), was discovered in 1984.[1] The scope has since been rapidly expanded to different vinyl ethers, propenyl ethers and other cationically highly reactive monomers, such as *N*-vinylcarbazole and *p*-methoxystyrene, and to other initiating systems based on weak Lewis acids.[2]

While weak Lewis acids such as Zn halides may be necessary to effect living polymerization of the more reactive vinyl ethers, they are ineffective to induce polymerization of the less reactive monomers, such as isobutylene (IB) and styrene (St). In the first reported living cationic homo- and copolymerization of IB, St, and St derivatives, the much stronger BCl_3 and $TiCl_4$ were used in conjunction with organic esters or ethers as initiators.[3]

When using a cation source in conjunction with a Friedel–Crafts acid, the concentration of growing centers is most often difficult to measure and remains unknown. By using stable carbocation salts, the uncertainty of the concentration of initiating cations may be eliminated. However, their use is limited to cationically fairly reactive monomers (e.g., *N*-vinylcarbazole, *p*-methoxystyrene, alkyl vinyl ethers, α-methylstyrene, etc.) as they are too stable and, therefore, ineffective initiators of less reactive monomers, such as IB. Selection of an initiator providing fast initiation compared to propagation is important in living cationic polymerization that usually needs to be determined for each monomer. In general, because initiation involves two subsequent events, ion generation and cationation, species that form ionic species slowly and/or in extremely low concentration (primary or secondary alkyl halides), or species that would form ions in high concentration, but are too stable to cationate the monomer (triphenyl methyl halides), are less active, maybe completely inactive.

BCl_3 alone can initiate the polymerization of IB, St, and St derivatives using polar solvents in the presence of a proton trap to prevent initiation by protic impurities.[5,6] Although initiation by metal halides has been postulated before, direct initiation has been proved for the first time. Kinetic and mechanistic studies supported the proposed new initiation mechanism via haloboration and explained the apparent livingness of the polymerization.[7]

MONOMERS

Isobutylene

Isobutylene is the most studied monomer that can polymerize only by a cationic mechanism. Living polymerizations to date are based on BCl_3 and $TiCl_4$ coinitiators. The activity of the BCl_3-based system is greatly solvent dependent; i.e., sufficient activity occurs only in polar solvent.

A wide variety of initiators, organic esters, halides, ethers, and alcohols have been used to initiate living polymerization of IB at temperatures up to –10°C.[6] The true initiating entity with ethers and alcohols is the chloro derivative arising by fast chlorination. The polymerization involving the BCl_4^- counter anion is very slow, measured in hours, compared to the fast polymerization by protic impurities, and, in the absence of proton scavengers, the monomer is consumed mainly by this process. In the presence of a proton trap or EDs, similar rates, controlled molecular weights, and narrow MWDs (M_w/M_n ~ 1.2) are obtained.[4]

Styrene

The living carbocationic polymerization of St was first achieved under well-defined conditions by the 1-*p*—methylphenyl)ethyl acetate/BCl_3 system in CH_3Cl at –30°C.[8] It was determined that side reactions such as intra- and intermolecular alkylation were absent even in the relatively polar CH_3Cl and at moderately high temperature. Polymers with theoretical molecular weights but with broad MWDs were obtained, due most likely to the presence of protic impurities and/or slow exchange between dormant and active species.

Living polymerization and PSt with narrow MWD were obtained using $SnCl_4$, a milder Lewis acid as co-initiator in conjunction with 1-phenylethyl halides as initiators in a nonpolar solvent ($CHCl_3$) and solvent mixtures or in a polar CH_2Cl_2 in the presence of nBu_4NCl.[9,10]

Substituted Styrenes

The living polymerization of *p*-methylstyrene (pMeSt) and that of 2,4,6-trimethylstyrene were reported by the organic ester/BCl_3 initiating systems in CH_3Cl or CH_2Cl_2 at up to –30°C.[11,12] While polymers with M_w/M_n ~ 1.1–1.2 were obtained with 2,4,6-trimethylstyrene, PpMeSt exhibited broad MWD (M_w/M_n ~ 2.5).

Indene

Polyindene (PInd) of theoretical molecular weight up to at least 13,000 and narrow MWD (M_w/M_n ~ 1.2) can be reportedly obtained by the cumyl chloride/BCl_3 initiating system in CH_3Cl at –80°C.[14] Living polymerization is also claimed with the TMPCl/$TiCl_4$ initiating system using CH_3Cl:Hex 40:60 v:v[14] and CH_3Cl:MeChx 40:60 v:v[15] mixed solvents at –80°C.

α-Methylstyrene

The living polymerization of α-methylstyrene (αMeSt) was first achieved with the vinyl ether–HCl adduct/$SnBr_4$ initiating system in CH_2Cl_2 at –78°C.[16] The obtained M_ns were consistent with the calculated ones, assuming that one polymer chain is formed per initiator. Polymers with M_ns up to 110,000 were obtained with M_w/M_n ~ 1.1.

BLOCK COPOLYMERS BY SEQUENTIAL MONOMER ADDITION

The emergence of living polymerization provided the simplest and most convenient method for the preparation of block copolymers by sequential monomer addition. PIB-based block copolymer thermoplastic elastomers (TPEs), predicted to have superior properties, have been the focus of interest.

The synthesis of PSt-*b*-PIB-*b*-PSt TPE has been accomplished by many research groups.[17–20] The synthesis invariably involved sequential monomer addition using di- or trifunctional initiator in conjunction with $TiCl_4$ in moderately polar solvent mixture at low (–70° to –90°C) temperatures.

A novel scheme was recently developed to synthesize block copolymers by living carbocationic sequential block copolymerization when the second monomer is more reactive than the first.[13,21] It involves capping with a highly reactive but nonpolymerizable monomer, such as DPE, forming stable cations. This is followed by tailoring the Lewis acidity to the reactivity of the second monomer by the addition of titanium(IV) alkoxide $\{Ti(OR)_4\}$, replacing the Lewis acid with a weaker one or by the use of a common ion salt. This approach is potentially applicable to all living cationic polymers.

FUNCTIONAL POLYMERS BY *IN SITU* FUNCTIONALIZATION OF THE LIVING ENDS

Success remains limited in the synthesis of functional polymers by *in situ* functionalization of the living carbocationic chain ends. The lack of success is due to the nature of living cationic polymerization. In contrast to living anionic polymerization, in living cationic polymerizations the concentration of active species, i.e., the cations, is often very small since most of the chain ends are in a dormant form.

In situ functionalization of the living ends by a variety of nucleophiles was only recently realized using DPE capping followed by end-quenching.[22]

REFERENCES

1. Miyamoto, M.; Sawamoto, M.; Higashimura, T. *Macromolecules* **1984**, *17*, 265.
2. For a recent review, see Sawamoto, M. *Trends Polym. Sci.* **1993**, *1*, 111.
3. Faust, R.; Kennedy, J. P. *Polymer Bulletin* **1986**, *15*, 317, and publications therein. For a review see Kennedy, J. P. and Ivan, B., Designed Polymers by Carbocationic Macromolecular Engineering, Hanser: 1991.
4. Balogh, L.; Faust, R. *Polymer Bulletin* **1992**, *28*, 367.
5. Balogh, L.; Faust, R.; Wang, L. *Macromolecules* **1994**, *27*, 3453.
6. Wang, L.; Svirkin, J.; Faust, R. *Polymeric Materials Science and Engineering, Preprint* 1995.
7. Balogh, L.; Fodor, Zs.; Kelen, T.; Faust, R. *Macromolecules* **1994**, *27*, 4648.
8. Faust, R.; Kennedy, J. P. *Polymer Bulletin* **1988**, *19*, 21.
9. Kwon, O.-S.; Kim, Y.-B.; Kwon, S.-K.; Choi, B.-S.; Choi, S.-K. *Makromol. Chem.* **1993**, *194*, 251.
10. Higashimura, T.; Ishihama, Y.; Sawamoto, M. *Macromolecules* **1993**, *26*, 744.
11. Faust, R.; Kennedy, J. P. *Polymer Bulletin* **1988**, *19*, 29.
12. Faust, R.; Kennedy, J. P. *Polymer Bulletin* **1988**, *19*, 35.
13. Fodor, Zs.; Faust, R. *J. Macromol. Sci.* **1994**, *A31*(12), 1983.
14. Tsunogae, Y.; Majoros, I.; Kennedy, J. P. *J. Macromol. Sci.* **1993**, *A30*(4), 253.
15. Kennedy, J. P.; Midha, S.; Keszler, B. *Macromolecules* **1993**, *26*, 424.
16. Higashimura, T.; Kamigaito, M.; Kato, M.; Hasebe, T.; Sawamoto, M. *Macromolecules* **1993**, *26*, 2670.
17. Kaszas, G.; Puskas, J.; Kennedy, J. P.; Hager, W. G. *J. Polymer Sci.* **1991**, *A29*(1), 427.
18. Everland, H.; Kops, J.; Nielsen, A.; Ivan, B. *Polymer Bulletin* **1993**, *31*, 159.
19. Storey, R. F.; Chisholm, B. J.; Choate, K. R. *J. Macromol. Sci.* **1994**, *A31*(8), 969.
20. Gyor, M.; Fodor, Zs.; Wang, H.-C.; Faust, R. *Polymer Preprints* **1993**, *34*(2), 562 *J. Macromol. Sci.* **1994**, *A31*(12), 2053.
21. Fodor, Zs.; Hadjikyriacou, S.; Li, D.; Faust, R. *Polymer Preprint* **1994**, *35*(2), 492, *Makromol. Chem. Macromol. Symp.*, in press.
22. Hadjikyriacou, S.; Fodor, Zs.; Faust, R. *J. Macromol. Sci.* **1995**, *A32*(6).

LIVING COORDINATION POLYMERIZATION

Kazuo Soga
Japan Advanced Institute of Science and Technology

Satoshi Ueki and Masahide Murata
Tonen Corporate Research and Development Laboratory
Tonen Chemical Corporation

Recently, several examples of living coordination polymerizations by means of transition metal catalysts have been reported. This section reviews progress in the living polymerizations of olefins and the syntheses of tailor-made polymers utilizing living coordination polymerization.

PREPARATION AND PROPERTIES

Syntheses of Living Polymers

Living Polymerization of Ethylene

The $[Cp*_2LuH]_2$ ($Cp*: C_5(CH_3)_5$) catalyst is reported to produce polyethylene with a relatively narrow molecular mass distribution (MMD) ($M_w/M_n = 1.4$).[1] Such diene complexes of tantalum as Cp*(butadiene)$TaMe_2$ and Cp(butadiene)$TaCl_2$ ($Cp: C_5H_5$) combined with methylalumoxane (MAO) also can catalyze the living polymerization of ethylene below 20°C, and the molecular weight of polyethylene produced with these tantalum catalyst systems increases directly in proportion to the polymerization time and its distribution is very narrow ($M_w/M_n < 1.3$).[2]

Similar diene complexes of niobium have given almost monodispersed polyethylene ($M_w/M_n = 1.05$) with a much higher activity.[3]

Living Polymerization of Propylene

In 1967, a homogeneous catalyst system composed of VCl_4 and AlR_2Cl promoted the syndiospecific living polymerization of propylene.[5] Both the yield and molecular weight of polypropylene increased proportionally to the polymerization time at a temperature as low as –78°C. This might be the first discovery of the living olefin polymerization with the use of Ziegler-Natta catalysts. Since then, the vanadium-based catalyst system has been markedly extended.

The structure of living polypropylene chain ends has been discussed. The vanadium–carbon bond of living polypropylene reacts instantaneously with I_2 to give an iodine-bonded polypropylene in approximately 100% yield.

The heterogeneous catalyst system composed of $TiCl_3$ and $Cp^*{}_2Ti(CH_3)_2$ also catalyzes the living polymerization of propylene.[7] The aluminium free $TiCl_3$ is more preferable. In contrast to the vanadium systems, the catalyst yields highly isotactic living polypropylene.

Living Polymerization of Higher α-Olefins

A yttrium catalyst is reported to produce the living polymers of higher α-olefins like 1–pentene and 1–hexene.[8]

Living Polymerization of Substituted Acetylenes

The living polymerization of substituted acetylenes takes place with some molybdenum-based catalysts essentially composed of $MoOCl_4$ or $(MoCl_5)$, n–Bu_4Sn and ethanol.[9] Living polymers of 1–chloro–1–octyne, *o*–Me_3Si–, *o*–CF_3– and *p*–n–Bu–*o,o,m,m*–F_4–phenylacethylenes have been obtained. The polydispersities (Mw/Mn) of these polymers range from 1.05 to 1.3.

Terminally Functionalized Polypropylenes

Various functional groups can be incorporated in the living polypropylene chain end.[10,11] The addition of I_2 to the vanadium-catalyzed propylene polymerization system yields the terminally-iodized polypropylene. Polypropylene in the terminal amino group is easily synthesized from the reaction between the iodine-bonded polypropylene and ethylenediamine in aqueous alkaline medium.[12] The reaction of living polypropylene with carbon monoxide generates polypropylene having the terminal aldehyde group.[13] Introductions of vinyl, phenyl, and hydroxyl groups to the living polypropylene chain end result from using butadiene, styrene, and propylene oxide as reactants.[14] Further treatment of living polypropylene having the terminal hydroxyl group with trimethylsilylchloride yields trimethylsilyl functionalized polypropylene. Introduction of ester group also is possible with the addition of methyl methacrylate to the vanadium-catalyzed propylene polymerization system.[15] Other terminally-functionalized polypropylenes are synthesized with the use of methacrylate derivatives.[16,17]

Block Copolymers Containing Polypropylene Unit

The addition of a small amount of ethylene during the living polymerization of propylene prepares triblock copolymer of PP–EP–PP. After the consumption of ethylene monomer, the living polymerization of propylene continues, resulting in the formation of a syn·PP–EP–syn·PP triblock copolymer.[6] Some examples of the syntheses of diblock copolymers containing polyethylene (PE) unit have also been reported. For example, diblock copolymers such as PE–PMMA and PE–PEA(EA:ethyl acrylate) are obtained by successive polymerization of corresponding monomers using the $Cp^*{}_2SmCH_3$ complex.[4]

REFERENCES

1. Jeske, G.; Lauke, H.; Mauermann, H.; Sweptson, P. N.; Schumann, H.; Marks, T. J. *J. Am. Chem. Soc.* **1985**, *107*, 8091.
2. Mashima, K.; Fujikawa, S.; Nakamura, A. *J. Am. Chem. Soc.* **1993**, *115*, 10990.
3. Mashima, K.; Fujikawa, S.; Urata, H.; Tanaka, E.; Nakamura, A. *J. Chem. Soc. Chem. Commun.* **1994**, 1623.
4. Yasuda, H.; Ihara, E.; Yoshioka, S.; Nodono, M.; Morimoto, M.; Yamashita, M. *Catalyst Design For Tailor-Made Polyolefins*; Sofa, K., Ed.: Kodansha, Japan, 1992, p 237.
5. Zambelli, A.; Natta, G.; Pasquon, I.; Signorini, R. *J. Polym. Sci. Pt-C* **1967**, *16*, 2455.
6. Doi, Y.; Keii, T. *Adv. Polym. Sci.* **1986**, *73/74*, 201.
7. U.S. 4408019 (1983), Hercules Inc., invs.: H. W. Blunt; *Chem. Abstr.* 100, 7417 (1984); Japan (open), 57-111307 (1982), Hercules Inc., inv.: G. A. Lock; *Chem. Abstr.* 101, 132232r (1984).
8. Yasuda, G.; Ihara, E. *Koubunshi* **1994**, *43*, 534.
9. Masuda, T. *Yuki Gousei Kagaku* **1991**, *49*, 138.
10. Doi, Y.; Ueki, S.; Keii, T. *Coordination Polymerization*: Price, C. C.; Vandenberg, E. J., Eds., Plenum: New York-London, 1983; p 249.
11. Doi, Y.; Watanabe, Y.; Ueki, S.; Soga, K. *Macromol. Chem., Rapid Commun.* **1983**, *4*, 533.
12. Doi, Y.; Nozawa, F.; Soga, K. *Makromol. Chem.* **1985**, *186*, 2529.
13. Doi, Y.; Murata, M.; Soga, K. *Makromol. Chem., Rapid Commun.* **1984**, *5*, 811.
14. Doi, Y.; Hizal, G.; Soga, K. *Makromol. Chem.* **1987**, *188*, 1273.
15. Doi, Y.; Koyama, T.; Soga, K. *Makromol. Chem.* **1985**, *186*, 11.
16. Furuhashi, H.; Murakami, N.; Ueki, S.; Doi, Y. *Polymer Preprint Jpn.* **1991**, *40*, 1775.
17. Furuhashi, H.; Murakami, N.; Ueki, S.; Doi, Y. *Preprint of 4th SPSJ International Polymer Conference* **1992**, *329*.

Living Polymerization

See: *Anionic Initiators (Lithium Aggregate Effects, Anionic Polymerization)*
Block Copolymers (By Changing Polymerization Mechanism)
Cyclic Carbonates (Ring-Opening Polymerization)
Cyclic Imino Ethers (Ring-Opening Polymerization)
Free Radical Iniferters
Functional Polymers (Telechilics and Macromonomers)
Functional Polymers (By Carbocationic Polymerization)
Immortal Polymerization
Living Anionic Polymerization
Living Anionic Polymerization (Protected Functional Monomers)
Living Carbocationic Polymerization, Olefins
Living Coordination Polymerization
Living Polymerization, Fast (Vinyl Ethers)
Living Radical Polymerization
Living Radical Polymerization, Iodine Transfer
Macrocyclic Block Copolymers
Metathesis Polymerization, Cycloolefins
Miktoarm Star Polymers
Molecular Weight Distribution (Ionic Polymerization)
Monodisperse Polymers
Phenylacetylene (Stereospecific Living Polymerization, Rh Complexes)
Polyalkylenimines
Polyisocyanides
Rare Earth Polymerization Initiators

LIVING POLYMERIZATION, FAST (Vinyl Ethers)

Eiichi Kobayashi
Department of Industrial Chemistry
Science University of Tokyo

Conventional cationic polymerization of 2-methylpropene was industrialized by the effort of Thomas and Sparks. However, the variety of initiation, propagation, transfer, and termination reactions limited subsequent development for carbocationic polymerization industries. Milestones in the history of carbocationic polymerizations and a modern controlled synthesis of living carbocationic polymerization of vinyl ethers[1] and olefins[2] are extensively studied.[3]

Living cationic polymerization of vinyl monomers is achieved by a nucleophilic stabilization of the growing carbocations. This is accomplished with suitable nucleophilic counteranions or with added proper Lewis bases. For example, HI adds quantitatively to vinyl ethers and yields a 1 : 1 adduct. Living polymerization started when I_2 is added to the system containing monomer.[4]

FAST LIVING POLYMERIZATION

Research progressed from the seedtime to extensive objectives with such new useful material preparations as high molecular weight polymers and the controlled synthesis of multiblock copolymers. The first aim was a systematic study of added bases. The author expects that basicity and sterical structure of added bases are important controlling factors. The apparent polymerization rate constant k of isobutyl vinyl ether (IBVE) increases with increase of the alkyl substituent R. $\overline{M}_w/\overline{M}_n$ was used as a measure of polydispersity of molecular weights.[5] To determine a critical condition for fast living polymerization, IBVE was polymerized in the presence of methyl acetate and related ester compounds. Methyl chloroacetate turned out to be effective for the fast living polymerization. The apparent polymerization rate constant k was larger by a factor of 400~500 than that in the presence of methyl acetate.[6]

CYCLIC FORMAL ADDITIVE

Basicity and sterical structure controlling factors also were recognized in the living polymerization of IBVE by 1-(isobutoxy)ethyl acetate/$EtAlCl_2$ in the presence of cyclic ethers, cyclic formals, and acyclic ethers with oxyethylene units.[7] The apparent polymerization rate constant k depended on the number of oxygen's and ring size, concerning the basicity and sterical structure. That is, polymerization in the presence of cyclic formals was much faster than that of cyclic ethers.

PROPERTIES

Stereoregularity

The stereostructure of polymers controls the physical and chemical properties. The stereostructure may be concerned with an ether group of vinyl ether monomers, the structure of added bases, counteranions, and the polymerization temperature. Counteranion and added base in the living cationic polymerization of IBVE did not affect the stereostructure of polymers.

Thermally Induced Phase Separating Polymers

Thermally induced phase separating polymers have attracted much attention by reason of practical and theoretical interests.[8] However, the systematic investigations are limited, since the syntheses of polymers with controlled molecular weight and sequence distribution of monomers are generally difficult. For such well-controlled polymer synthesis, the living polymerization of vinyl ethers is a versatile method. The phase separation temperature (T_{ps}) was measured by visual observation or by transmittance at 500nm light beam. The phase separation turned out to be quite sensitive (ΔT_{ps} = 0.3 ~ 0.5°C) and reversible toward heating and cooling cycles. The T_{ps} of random copolymers depended mainly on the monomer composition. However, the phase separation of block copolymers showed multi-stage behavior. For example, triblock copoly(EOVE/MOVE/EOVE) exhibits a phase separation starting at about 20°C corresponding to the phase separation temperature of poly(EOVE) and a complete phase separation at about 50°C.[9] At moment, steep increase in viscosity of 5 wt % aqueous solution of triblock copoly($EOVE_{50}/MOVE_{400-800}/EOVE_{50}$) has been observed at about 40 ~ 50°C. The value was more than 1200 cp. Here, suffix number means number of monomeric units.[10]

REFERENCES

1. Higashimura, T.; Sawamoto, M. *Adv. Polymer Sci.* **1984**, *62*, 49.
2. Faust, R.; Kennedy, J. P. *Polymer Bull.* **1986**, *15*, 317.
3. Kennedy, J. P.; Iván, B. *Designed Polymers by Carbocationic Macromolecular Engineer: Theory and Practice*, Carl Hanser: Verlag, D-8000, München, 1992.
4. Higashimura, T.; Miyamoto, M.; Sawamoto, M. *Macromolecules* **1985**, *18*, 611.
5. Aoshima, S.; Shachi, K.; Kobayashi, E.; Higashimura, T. *Makromol. Chem.* **1991**, *192*, 1749.
6. Aoshima, S.; Shachi, K.; Kobayashi, E. *Makromol. Chem.* **1991**, *192*, 1759.
7. Aoshima, S.; Fujisawa, T.; Kobayashi, E. *J. Polymer Sci. Part A. Polymer Chem.* **1994**, *32*, 1719.
8. Kobayashi, E.; Aoshima, S. *Chemistry and Chemical Industry* **1993**, *46*, 1426.
9. Aoshima, S.; Oda, H.; Kobayashi, E. *Kobunshi Ronbunshu* **1992**, *49*, 937.
10. Ueda, K.; Kobayashi, E.; Aoshima, S. *Polymer Prep. Japan* **1995**, *44*, 1G13.

LIVING RADICAL POLYMERIZATION

Ezio Rizzardo and Graeme Moad
CSIRO Division of Chemicals and Polymers

Living polymerization mechanisms offer polymers of controlled architecture and molecular weight distribution. They provide a route to narrow polydispersity homopolymers, high purity block copolymers, and end-functional polymers. Traditional methods of living polymerization, based on ionic, coordination, or group transfer mechanisms, are limited for economic reasons (due to stringent requirements on reaction conditions and monomer purity) and by the relatively small number of monomers that are amenable to these methods.[1]

The use of living radical polymerization (also known as quasi- or pseudo-living radical polymerization) offers the potential to overcome many of these problems. Living radical polymerization has been the subject of several previous reviews.[2–6]

There are three types of radical polymerization called "living", but this chapter considers systems with reversible termination where the bond formed by primary radical termination or chain transfer is kinetically unstable under the reaction conditions. According to Quirk's criteria,[7] these polymerizations should be considered as true living polymerizations. It is possible to make block copolymers and narrow polydispersity resins, since all chains grow, though not simultaneously. The success of the mechanism, illustrated in **Scheme I**, depends on the choice of a suitable initiator (R–T).

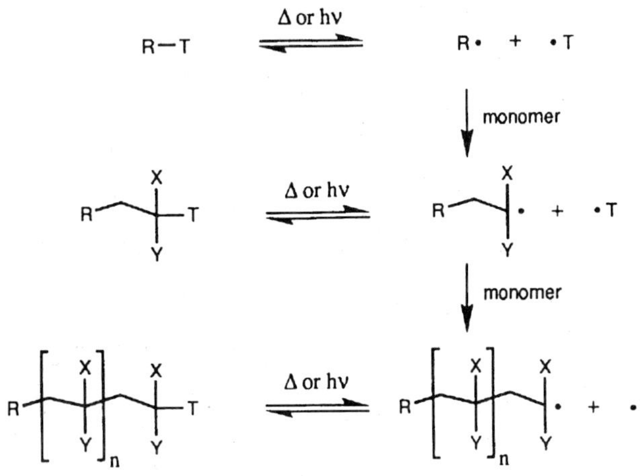

I

Four main types of initiator have been used for this form of living radical polymerization: organosulfides, tri- and diarylmethyl derivatives, alkoxyamines, and organocobalt derivatives.

ORGANOSULFIDES

The potential of organosulfides as initiators of living radical polymerization was first recognized by Otsu and Yoshida.[8] The C–S bonds of certain organosulfides are photolabile (e.g., (**1**), **Scheme II**). Living polymerization initiated by these species then involves monomer insertion into the C–S bond.

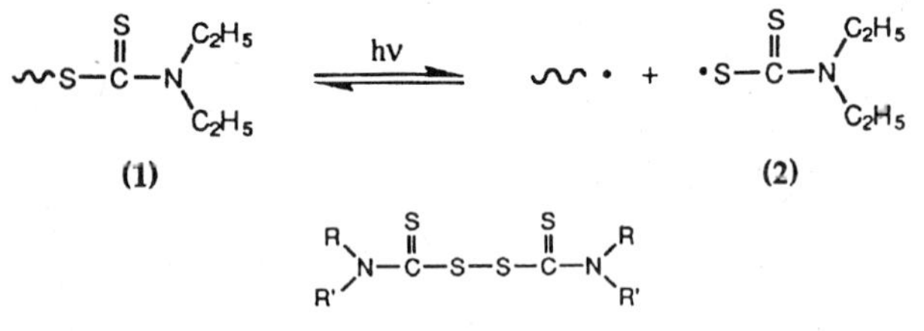

(1) (2) (3)

The formation of block (or graft) copolymers by this chemistry generally requires a two-step process. First a macroinitiator (the A-block) is prepared either by radical polymerization with a disulfide initiator or by other chemistry. This is then dissociated photolytically in the presence of a second monomer to give an A–B block copolymer. Only the second step should be regarded as a true living polymerization.

Disulfide initiators used in this context include diaryl disulfides,[9,10] dibenzoyl disulfide,[8] dithiuram disulfides,[8,11–15] and bis(isopropylxanthogen) disulfide.[16] The most effective of these appears to be the dithiuram disulfides (Structure 3).

DI- AND TRIARYLMETHYL DERIVATIVES

Di- and triarylmethyl radicals are also effective in reversibly coupling with certain propagating radicals (**Scheme III**).

III

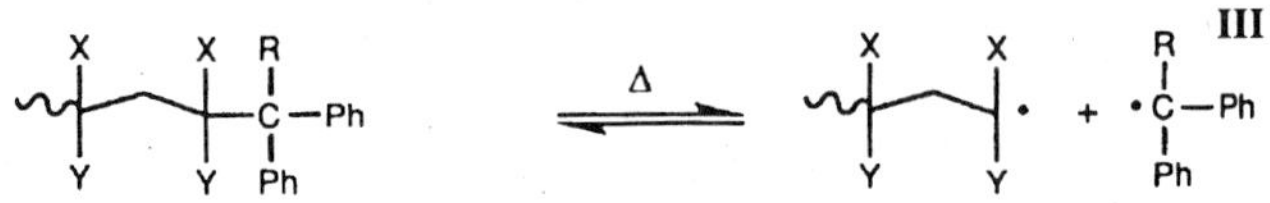

Otsu and Tezaki[17] reported on the use of triphenylmethylazobenzene as an initiator of living polymerization. The phenyl radical initiates polymerization and the triphenylmethyl radical reversibly terminates chains. Acar and Yagci[18] also used this initiator to prepare methacrylate block copolymers.

ALKOXYAMINES

Rizzardo et al.[19,20] pioneered the use of alkoxyamines as initiators of living radical polymerization. The C–O bond of certain alkoxyamines is relatively weak and undergoes reversible homolysis on heating to yield a reactive carbon-centered radical and a stable nitroxide (**Scheme IV**). The carbon-centered radical initiates polymerization. Nitroxides are inert towards monomer under normal reaction conditions, but they react with the propagating radicals by primary radical termination to form a new oligo- or polymeric alkoxyamine initiator. The result is monomer insertion into the C–O bond. The method is applied to make block and graft copolymers[19,21–23] and functional and narrow polydispersity homopolymers.[24–26] The method is most suited for the synthesis of polymers based on acrylate ester and styrenic monomers.

IV

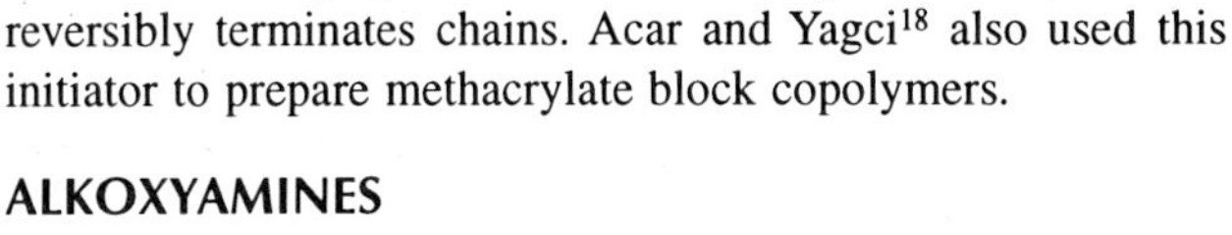

ORGANOCOBALT COMPOUNDS

Certain square planar cobalt porphyrin complexes have been reported to reversibly inhibit butyl acrylate polymerization.[27–30] Derivatives of these compounds may function as initiators of quasi-living polymerization.[31,32]

REFERENCES

1. Webster, O. *Science* **1991**, *251*, 887.
2. Greszta, D.; Mardare, D.; Matyjaszewski, K. *Macromolecules* **1994**, *27*, 638.
3. Moad, G.; Rizzardo, E.; Solomon, D. H. *Comprehensive Polymer Science*, Eastmond, G. C.; Ledwith, A.; Russo, S.; Sigwalt, P., Eds.; Pergamon: London, 1989; Vol. 3, p 141; Moad, G.; Solomon, D. H. *The Chemistry of Free Radical Polymerization*, Pergamon: London, 1995; p 335–346.
4. Kuchanov, S. I. *Comprehensive Polymer Science*; Agarwal, S. L.; Russo, S., Eds.; Pergamon: London, 1992; Vol. Suppl. 1, p 23.
5. Nair, C. P. R.; Clouet, G. *J. Macromol. Sci., Rev. Macromol. Chem. Phys.* **1991**, *C31*, 311.
6. Georges, M. K.; Veregin, R. P. N.; Kazmaier, P. M.; Hamer, G. K. *Trends Polym. Sci.* **1993**, *2*, 66.
7. Quirk, R. P.; Lee, B. *Polym. Int.* **1992**, *27*, 359.
8. Otsu, T.; Yoshida, M. *Makromol. Chem., Rapid Commun.* **1982**, *3*, 127.
9. Otsu, T.; Kuriyama, A. *J. Macromol. Sci., Chem.* **1984**, *A21*, 961.
10. Shefer, A.; Grodzinsky, A. J.; Prime, K. L.; Busnel, J. P. *Macromolecules* **1993**, *26*, 2240.
11. Otsu, T.; Yoshida, M. *Polym. Bull. (Berlin)* **1982**, *7*, 197.
12. Turner, S. R.; Blevina, R. W. *Macromolecules* **1990**, *23*, 1856.
13. Haque, S. A. *J. Macromol. Sci., Chem.* **1994**, *A31*, 827.
14. Liu, F.; Cao, S.; Yu, X. *J. Appl. Polym. Sci.* **1993**, *48*, 425.
15. Nair, C. P. R.; Clouet, G.; Chaumont, P. *J. Polym. Sci., Part A: Polym. Chem.* **1989**, *27*, 1795.
16. Niwa, M.; Matsumoto, T.; Izumi, H. *J. Macromol. Sci. Chem.* **1987**, *A24*, 567.
17. Otsu, I.; Tazaki, T. *Polym. Bull. (Berlin)* **1986**, *16*, 277.
18. Acar, M. H.; Yagci, Y. *Macromol. Rep. (Suppl. 2)* **1991**, *A28*, 177.
19. Solomon, D. H.; Rizzardo, E.; Cacioli, P. Eur. Pat. Appl. EP135280 *(Chem. Abstr.)*. **1985**, *102:221335q*.
20. Rizzardo, E. *Chem. Aust.* **1987**, *54*, 32.
21. Rizzardo, E.; Chong, Y. K. *2nd Pacific Polymer Conference, Preprints*; Pacific Polymer Federation: Tokyo, 1991; p 26.
22. Yoshida, E.; Ishizone, T.; Hirao, A.; Nakahama, S.; Takata, T.; Endo, T. *Macromolecules* **1994**, *27*, 3119.
23. Yoshida, E.; Nakamura, K.; Endo, T. *J. Polym. Sci., Part A: Polym. Chem.* **1993**, *31*, 1505.
24. Johnson, C. H. J.; Moad, G.; Solomon, D. H.; Spurling, T. H.; Vearing, D. J. *Aust. J. Chem.* **1990**, *43*, 1215.
25. Georges, M. K.; Veregin, R. P. N.; Kazmaier, P. M.; Hamer, G. K. *Macromolecules* **1993**, *26*, 2987.
26. Veregin, R. P. N.; Georges, M. K.; Kazmaier, P. M.; Hamer, G. K. *Macromolecules* **1993**, *26*, 5316.
27. Oganova, A. G.; Smirnov, B. R.; Ioffe, N. T.; Enikolopyan, N. S. *Bull. Acad. Sci. USSR* **1983**, *1837*.
28. Oganova, A. G.; Smirnov, B. R.; Ioffe, N. T.; Kim, I. P. *Bull. Acad. Sci. USSR* **1984**, 1154.
29. Oganova, A. G.; Smirnov, B. R.; Ioffe, N. T.; Enikopyan, N. S. *Doklady Akad. Nauk SSR (Engl. Transl.)* **1983**, *268*, 66.
30. Morozova, I. S.; Oganova, A. G.; Nosova, V. S.; Novikov, D. D.; Smirnov, B. R. *Bull. Acad. Sci. USSR* **1987**, 2628.
31. Wayland, B. B.; Poszmik, G.; Mukerjee, S. L.; Fryd, M. *J. Am. Chem. Soc.* **1994**, *116*, 7943.
32. Smirnov, V. R. *Polym. Sci. USSR (Engl. Transl.)* **1990**, *32*, 583.

LIVING RADICAL POLYMERIZATION

Krzysztof Matyjaszewski and Daniela Mardare
Department of Chemistry
Carnegie-Mellon University

The latest developments in living polymerizations of vinyl and cyclic monomers are based on the reversible protection of the growing chains, in the form of inactive "dormant" species, that exchange reversibly with the active growing chains (**Scheme I**).

$$P_n\text{-}X \underset{K}{\overset{\text{(cat.)}}{\rightleftharpoons}} P_n^{\bullet} + X \xrightarrow{M} P_n\text{-}M^{\bullet} + X \underset{K}{\overset{\text{(cat.)}}{\rightleftharpoons}} P_n\text{-}M\text{-}X$$

"dormant" active

cat. = Nu^-, Lewis acid (GTP), hν, ΔT
X = protective group
P_n = polymer chain
M = monomer
K = equilibrum constant

SCHEME I. Living polymerizations involving active and "dormant" species.

In a living process the number of growing chains usually is the same as the number of initiating molecules, and it does not change during propagation until the total monomer consumption. If the exchange reactions between dormant and active species occur fast enough, polymers with low polydispersities can be formed. However, some polymerization systems provide well-defined polymers in which chain breaking reactions cannot be completely avoided. We call such systems controlled or "living" since they do not fulfill criteria of living polymerization because of the presence of either transfer of termination. Radical polymerization is one of them.

The synthesis of well-defined polymers by controlled polymerization should occur in systems with a low momentary (stationary) concentration of growing radicals that should be reversibly deactivated to provide a relatively large number of macromolecules.

Controlled Radical Polymerization–Types of Initiators and Radical Scavengers

All systems developed so far are based on a low stationary concentration of growing radicals and suppression of the termination process. This can include either physical methods (precipitation, emulsions, inclusion complexes, template polymerization, stiff chains, and viscous media) or chemical ones such as reversible termination with scavengers.

Some systems are far from being ideal: initiation is slow, scavengers react with monomers and reversibility is not observed except under photochemical conditions, and some side reactions lead to decomposition products. Nevertheless, in some systems block copolymers have been prepared but, as will be discussed later, probably not via chain extension reactions.

Systems Based on Reversible Recombination of Growing Radicals with Scavenging Radicals–Iniferters, Tetra(hexa)substituted Ethanes, Alkoxyamines

In 1982, Otsu reported a living radical polymerization of alkenes initiated by tetraalkythiuram disulfide and described its action as inifer or iniferter, which means that it acted as initiator, transfer agent, and terminator.[1–6]

Another class of thermal initiators with growing radicals that are reversibly scavenged by stabilized radicals is based on tetraarylethanes[7,8] and phenylazotriphenylmethane.[9]

Another class of initiators, generated from hyponitrite and either arenediazoate or cyanate anions reacting with such electron-acceptors as arenediazonium ions or activated alkyl halides, provide long-lived scavenging oxygen-centered radicals.[10] These polymerizations partially demonstrated their "living" nature by an increase in DP_n with conversion and by synthesis of block-copolymers poly(methyl methacrylate)–poly(butyl acrylate). However, broad polydispersities (2.0–3.0) and low conversions were obtained.

Some of the best controlled polymers obtained by radical polymerization are prepared with iniferters based on alkoxyamines, which can be prepared by decomposing an azo- or peroxy-initiator in the presence of a nitroxyl radicals.[11]

Radical Polymerizations in the Presence of Organometallic Compounds

Reversible reactions of carbon-based radicals with organometallic compounds play an important role in biological processes. For example, cobaloximes react reversibly with methyl radicals. Coderivatives have served as very efficient chain transfer reagents to regulate molecular weights of various polymers synthesized by radical processes. It was reported that alkylcobaloximes with generation of radicals used for both carbon–carbon and carbon–heteroatom bond-forming reactions.[12–15]

Enikolopian[16–18] and O'Driscoll[19] reported use of various cobaloximes and coporphyrins for catalyzed chain-transfer in the free radical polymerization of methyl methacrylate.

Recently, Harwood showed that free radical polymerization of acrylates, initiated by alkyl cobaloximes as photoinitiators, proceeded almost quantitatively, and molecular weight increased linearly with conversion.[20] The mechanism of propagation is believed to involve the cleavage of the cobalt–carbon bond at the chain-end, the monomer addition, and a rapid recombination. Also, organometallic derivatives of cobalt tetramesitylporphyrin initiate and control the polymerization of acrylates to form homopolymers and block-copolymers.[21] The linear increase in average molecular weight with monomer conversion and relatively low polydispersities (1.1–1.3) are indicative of an effective living radical process.[21]

Systems Based on Degenerative Transfer

The concept of degenerative transfer is based on the reversible reaction of a transfer agent, R' – X, which reacts with a propagating radical, P, to form a dormant polymer chain, P–X. Since the new radical R' can then reinitiate polymerization, the number of polymer chains is equal to the concentration of the transfer agent. Subsequently, the newly formed polymer chain, P', can react with the dormant polymer chain, P–X, in order to form P'–X and P. Optimally, the exchange reaction between the dormant and active radical species should be a thermodynamically neutral reaction. Therefore, the transfer agent should resemble the polymer chain end.

The basic requirement for the successful degenerative transfer is the sufficient thermal stability of P–R and fast exchange with P in comparison with propagation. Therefore, in order to prepare well-defined polymers, especially when $[M]_o >> [P\text{–}R]_o$ the optimal values for transfer coefficients should be larger than 1. Additionally, the reactivity of the P_1–R species in the initially added transfer agent should be similar to or higher than that in the macromolecular species P_n–R. Thus, P_1 should have a structure similar to the growing radical P_n.

The radical polymerization of styrene and butyl acrylate in the presence of degenerative transfer agents followed such typical behavior of controlled polymerization as low polydispersities and a linear evolution of molecular weight with conversion, even though it was initiated by classic radical initiators. Moreover, addition of a new portion of monomer or addition of another monomer extends chain growth leading to block copolymers as in the case of polystyrene and poly(butyl acrylate).

REFERENCES

1. Otsu, T.; Yoshida, M.; Kuriyama, A. *Polym. Bull. (Berlin)* **1982**, *7*, 45.
2. Otsu, T.; Yoshida, M.; Tazaki, T. *Makromol. Chem. Rapid Commun.* **1982**, *3*, 133.
3. Otsu, T.; Kuriyama, A. *J. Macromol. Sci., Chem.* **1984**, *A21*.
4. Otsu, T.; Matsunaga, T.; Kuriyama, A.; Yoshioka, M. *Eur. Polym. J.* **1989**, *25*, 643.
5. Otsu, T.; Yoshida, M. *Makromol. Chem. Rapid Commun.* **1982**, *3*, 127.
6. Otsu, T.; Matsumoto, A.; Yoshioka, M. *Kikan Kagaku Sosetsu* **1993**, 244.
7. Bledzki, A.; Braun, D. *Makromol. Chem.* **1981**, *182*, 1047.
8. Bledzki, A. *Polimery (Warsaw)* **1990**, *35*, 349.
9. Otsu, T.; Tazaki, T. *Polym. Bull. (Berlin)* **1986**, *16*, 277.
10. Druliner, J. D. *Macromolecules* **1991**, *24*, 6079.
11. Solomon, D. H.; Waverly, G.; Rizzardo, E.; Hill, W.; Cacioli, P. U.S. Patent 4 581 429, 1986.
12. Thomas, G.; Giese, B. *Helv. Chim. Acta* **1992**, *75*, 1123.
13. Giese, B.; Hartung, J.; Jianing, H.; Hutter, O.; Koch, A. *Angew. Chem. Int. Ed. Engl.* **1989**, *28*, 3.
14. Branchaud, B. P.; Yu, G. X. *Organometallics* **1993**, *12*, 4262.
15. Branchaud, B. P.; Meier, M. S.; Malekzadeh, M. N. *Org. Chem.* **1987**, *52*, 212.
16. Enikolopian, N. S.; Smirnov, B. R.; Ponomarev, G. V.; Belgovskii, I. M. *J. Polym. Sci., Polym. Chem. Ed.* **1981**, *19*, 879.
17. Smirnov, B. R.; Marchenko, A. P.; Plotnikov, V. O.; Kuzayev, A. I.; Enikolopian, N. S. *Polymer Sci. USSR* **1981**, *23*(5), 1169.
18. Smirnov, B. R.; Marchenko, A. P.; Korolev, G. V.; Bel'govskii, I. M.; Enikolopian, N. S. *Polymer Sci. USSR* **1981**, *23*(5), 1158.
19. Burczyk, A. F.; O'Driscoll, K. F.; Rempel, G. L. *J. Polym. Sci., Polym. Chem. Ed.* **1984**, *22*, 3255.
20. Arvanitopoulos, L. D.; Greuel, M. P.; Harwood, H. J. *35th IUPAC International Symposium on Macromolecules*; Akron, July 11-15, 1994; p 8.
21. Wayland, B. B.; Poszmik, G.; Mukerjee, S. L.; Fryd, M. *J. Am. Chem. Soc.* **1994**, *116*, 7943.

Initiation:	radical source(eg. peroxide) $\rightarrow$ $R'\cdot$ (radical)		(1)
	$R'\cdot + RI \rightarrow R\cdot + R'I$		(1')
Propagation:	$R\cdot + \mathbf{M} \rightarrow R\mathbf{M}\cdot$		(2)
	$(R'\cdot + \mathbf{M} \rightarrow R'\mathbf{M}\cdot)$	minor	
	$R\mathbf{M}\cdot + n\mathbf{M} \rightarrow R\mathbf{M}_{n+1}\cdot$		(2')
	$(R'\mathbf{M}\cdot + n\mathbf{M} \rightarrow R'\mathbf{M}_{n+1}\cdot)$	minor	
Transfer:	$R\mathbf{M}_n\cdot + RI \rightarrow R\mathbf{M}_nI + R\cdot$		(3)
	$(R\mathbf{M}_n\cdot + R'I \rightarrow R\mathbf{M}_nI + R'\cdot)$	minor	
	$R\mathbf{M}_m\cdot + R\mathbf{M}_nI \rightarrow R\mathbf{M}_mI + R\mathbf{M}_n\cdot$	$(m>n)$	(3')
Termination:	$2R\cdot \rightarrow R\text{-}R$		(4)
	$R\mathbf{M}_n\cdot + R\mathbf{M}_m\cdot \rightarrow R\mathbf{M}_{n+m}R$, etc.		(4')

[RI : alkyl iodide, **M** :monomer]

SCHEME I. Telomerization steps.

LIVING RADICAL POLYMERIZATION, IODINE TRANSFER

Masayoshi Tatemoto
Chemical Division
Daikin Industries Ltd.

CONCEPTUAL ARCHITECTURE

Iodine Transfer Polymerization (ITP) is one of the radical living devices developed since the 1970s.[1] It originates from the radical telomerizations developed by M. S. Kharasch[2] and R. N. Haszeldine.[3]

In the case of ITP of olefinic monomer, the terminal active bond is always C–I originating from the initial iodine-containing chain transfer agent (telogen) and monomer.

Scheme I indicates typical radical telomerization reaction steps to obtain low molecular weight telomers. High boiling residues in this reaction were useless by-products. But if the standard telomerization steps, especially step (3) are considered, then living radical polymerization becomes possible. Such successive chain transfers through the reaction (3′) are often observed in the case of iodine-containing telogens, and are especially common in fluorocarbon chemistry because of the lowest level of C–I bond dissociation energies in fluorocarbon compounds and telomeric species.[4] Therefore, such chain transfer reactions extending to a higher molecular weight range are the key controlling factors in obtaining favorable results in ITP. The name "Iodine Transfer Polymerization" comes from such considerations. The situation could be improved by reducing the amount of initiator in step (1), and would contribute to delay possibilities of steps (4) and (4′). Another improvement is also possible by using di-iodide, poly-iodide, and in some cases unsaturated iodide as a chain transfer agent (RI). These possibilities will be discussed later.

POLYMERIZATION PRACTICES IN FLUOROELASATOMERS

ITP was first applied to the copolymers of vinylidene fluoride (VDF) with hexafluoropropylene (HFP), and occasionally with tetrafluoroethylene (TFE), the most widely used fluoroelastomers ever developed. The polymerization was carried out by persulfate-initiated emulsion polymerization using perfluorinated surfactants as emulsifiers, and perfluoroalkyl and perfluoroalkylene iodides as chain transfer agents. These fluoroelastomers produced by ITP will be abbreviated as I-FKM below.

CROSSLINKING BY IODINE TRANSFER POLYMERIZATION

These fluoroelastomers obtained by ITP are easily cured by the peroxide curing system using multi-functional, unsaturated compounds, such as triallyl isocyanurate as crosslinking coagents. Crosslinking only occurs at the terminal iodine. This is an interesting and industrially useful application of the polymers.[5] Other practical effects of the molecular purity concerning terminal iodines were observed in this peroxide-curing process.

The peroxide-curable fluoroelastomers produced by ITP are now available as Daiel®[6] and are being developed for a wide variety of uses in automobiles, chemical plants, semiconductors, and others.

The peroxide curing system involves some care in the curing process. Because of the radical nature of the process, any radical scavengers, including air, must be strictly excluded during the process except after treatment to remove volatiles. Cured elastomers are stable up to 200°C, but unstable at temperatures higher than 200°C.

IODINE TRANSFER POLYMERIZATION IN FLUOROPLASTICS

ITP can be applied not only to elastomers but also to thermoplastic polymers and their crosslinking.

ITP was also applied to poly(vinylidene fluoride) (PVDF). The number average molecular weight of 1300 and M_w/M_n ratio of 1.02 were obtained at the front edge fraction of chain growth.

THERMOPLASTIC ELASTOMERS

The most remarkable use of living polymerization is to prepare block polymers. Among the most useful products in block polymers are thermoplastic elastomers, the most famous of which is polystyrene-*b*-polybutadiene-*b*-polystyrene. Such approaches are also possible for ITP as illustrated in **Scheme II**.

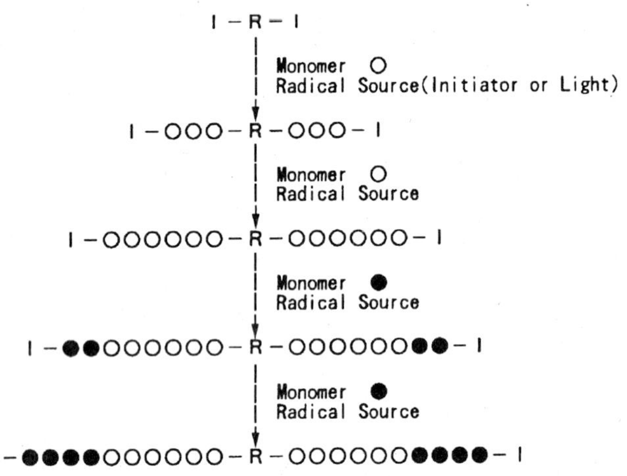

SCHEME II. Block polymer manufacture by ITP.

TELECHELICALLY REACTIVE LIQUID POLYMERS

Crosslinking using the reactivity of terminal iodines is characterized as "telechelic" in nature and should be extended to various reactive liquid polymers. Although various low molecular liquid polymers originating from ITP are crosslinked easily by the peroxide curing process, terminal iodines are also convertible to other functional groups such as -OH, -COOH, $-CH_2-\underset{\backslash / \atop 0}{CH-CH_2}$,$-CH_2CH_2SI(OR)_3$, etc.

The telechelic nature and the narrower molecular weight distribution up to higher molecular weight range in ITP make possible obtaining favorable prepolymers and thus excellent telechelic liquid polymer products without additional purification. Many future uses are anticipated especially for fluorinated products, in view of their heat stability, chemical and weather resistance, anti-humidity, low refractive index, lower surface energy, and other qualities. So-called "hybridized silicones" composed of fluoropolymers and silicones are expected to be most promising.[7]

OTHER BLOCK POLYMERS BY ITP

Modification of Fluorinated Oil with PTFE

Some perfluoropolyethers have been developed as heat and chemically resistant lubricating oils.[8] Greases are prepared by combining microfine polytetrafluoroethylene wax as the thickening agent with these oils. The block polymer of perfluoropolyether with PTFE can be used as a more effective thickening agent in those greases. Smaller amounts of PTFE, accounting for the weight fraction of PTFE segment of the block polymer, than PTFE wax itself, are sufficient to prevent oil from separating from the grease at temperatures of 200°C or so. The effect comes from the micro-phase separating properties of the block polymer itself or in oil. The typical properties of these block-polymer greases resemble that of classical metal soap greases with respect to such micelle formation.[9]

Mosaic Structures in Water-Soluble Polymers

Perfluorovinylacetic acid (FVA) easily copolymerizes with vinylidene fluoride (VDF) to produce water-soluble copolymers. This copolymer was prepared by ITP and then post-polymerized with vinylidene fluoride to yield a block polymer illustrated as PVDF-b-(FVA-co-VDF)-b-PVDF. The block polymer specimen casted from solution showed linear reversible size transitions against electrolyte concentrations in water.

OTHER PROBLEMS TO BE SOLVED IN ITP

ITP offers a general process for radical living polymerization of fluorocarbon and hydrocarbon monomers. The characteristic features of ideal ITP are summarized below:

1. Numbers of polymer molecules are essentially equal to the molar amount of iodide;
2. Average degree of polymerization is determined by the molar ratio of consumed monomers to iodide;
3. Molecular weight distributions are sharply controlled;
4. Stepwise chain growth, including block-polymer synthesis, is possible;
5. Useful in producing telechelically reactive polymers, including telechelic crosslinking;
6. Orientational selectivity of addition of unsymmetrical olefins seldom occurs at the front edge in stepwise additions of the monomers as a result of the difference in chain transfer ability of the terminal iodines (Iodine Transfer Polymerization in Fluoroplastics); and
7. Initiation by photons upon iodide is preferable.

All items except No. 7 are much the same as other living polymerization systems. But the most remarkable problem is that the iodide itself cannot act as initiator differing from others, and the present ITP must be initiated by another radical initiator that is the source of impure polymer species containing no iodine(s). Item 7 will become an important target in the future.

REFERENCES

1. Tatemoto, M. *Kobunshi Ronbunshu* **1992**, *49*(10), 765–783.
2. Kharasch, M. S.; Jensen, E. V.; Urry, W. H. *Science* **1945**, *102*, 128; Hanford, W. E. and Joyce, Jr., R. M. U.S. Patent 2 440 800, 1948.
3. Haszeldine, R. N. *J. Chem. Soc.* **1949**, *1949*, 2856–61.
4. Starks, C. M. *Free Radical Telomerization*, 1st ed., Academic: New York, 1974.
5. Oka, M.; Tatemoto, M. *Contemporary Topics in Polymer Science* Plenum: 1984, *4*, p 763.

6. Daikin Ind., Ltd., Fluroelastomer Catalogue.
7. Tatemoto, M. U.S. Patent 5 081 192, 1992, European Patent 0343526, 1993.
8. Kasai, P. H. *Macromolecules* **1992**, *25*(25), 6791–9.
9. Kirk-Othmer, *Encycl. Chem. Tech.*, 2nd ed.; 7, p 283.

Locust Bean Gum

See: *Gums (Overview)*

Low Density Polyethylene

See: *Polyethylene (Commercial)*
Polyethylene (Low Density Preparation)
Polyolefins (Overview)

Low Dielectric Polymers

See: *Fluoropolymers, Acrylates and Epoxies (with Low Dielectric Constants)*
Perfluorocyclobutane Aromatic Ether Polymers

Low Profile Additives

See: *Additives (Property and Processing Modifiers)*

Lube Oil Additives

See: *Viscosity-Index Improvers*

Lubricants

See: *Additives (Types and Applications)*
Hydrophilic Polymers (for Friction Reduction)
Polytetrafluoroethylene

Luminescent Polymers

See: *Dendrimers, Luminescent and Redox-Active*
Electroluminescent Polymers (Overview)
Photoluminescence Polymers

LUNG, ARTIFICIAL

Tatsuo Nakamura* and Yasuhiko Shimizu
Department of Artificial Organs
Research Center for Biomedical Engineering
Kyoto University

An artificial lung is an apparatus that substitutes for the functions of the lung. In general, "artificial lung" is synonymous with "oxygenator." The term "oxygenator" is used strictly for the gas-exchanger part of an artificial heart-lung machine (pump-oxygenator), which is used for cardiopulmonary bypass (CPB) during open heart surgery. In a broader sense "extracorporeal pulmonary support devices" and "intravascular blood gas exchange machines," which are used in patients with respiratory failure, are included in the artificial lung category.

Among the many physiological functions of the natural lung, such as blood–gas exchange, maintaining acid-base balance, controlling circulatory blood volume, filtering blood, defending against infection, and assisting in the immune response, contemporary artificial lungs carry out only blood–gas exchange, namely oxygen delivery to, and CO_2 removal from the patients' blood. The most important technical issue in artificial lungs is creating a large contact area between the blood and the gas. With regard to blood–gas contact, the methods employed in artificial lungs can be broadly classified into two types. In one, the gas makes contact directly with the blood, while in the other, the gas and blood are divided into separate compartments and gas exchange is performed through a septum with either a limited or an absent gas–blood interface. The former includes the vertical screen type, disc type, and bubble type oxygenator, and the latter, the membrane oxygenator.

TYPES OF MEMBRANES

Oxygenator membranes are classified according to their microstructures as dense membranes, porous membranes, and combined membranes. A typical dense membrane is silicone, which offers excellent blood and gas separation phases, thus promoting gas exchange by diffusion through a semipermeable barrier. As a result, blood damage is minimal. The disadvantage of silicone membrane is its low tear strength.

Polypropylene with a thickness of 22 to 50 μm is now widely used as a porous membrane. In porous membranes, blood and gas contact at the pores. Due to the hydrophobicity of the membrane and surface tension of the blood, the blood does not leak. Porous membranes provide a higher gas-exchange efficiency than do dense membranes. Unfortunately, the gas-exchange efficiency of porous membranes decreases with long-term use.

To overcome this problem, new membranes have been designed to incorporate the advantages of dense and porous membranes. These are called combined membranes. For example, microporous polypropylene membrane is covered with thin silicone and micro-porous polypropylene is sandwiched with porous polyethylene. These methods, however, are not yet in wide use.

TYPES OF MEMBRANE OXYGENATORS

Membrane oxygenators can be classified by their structures into parallel plate-type; the coil-type; and the hollow-fiber (capillary)-type.

Heparin-Coated Membrane Oxygenator

A current trend is to coat the membrane and circuit with heparin. Since the contact between blood and nonendothelialized "foreign" surfaces is inevitable in extracorporeal circulation, it is necessary to add the anticoagulant heparin to the patients' circulating blood. The heparin coating significantly reduces intrinsic clotting activity and minimizes the need to add heparin to circulating blood. In addition, it provides high biocompatibility and reduces the aggregation of granulocytes, complements, and platelets.

*Author to whom correspondence should be addressed.

ARTIFICIAL LUNGS IN THE BROAD SENSE: APPLICATION OF MEMBRANE OXYGENATORS FOR EXTRACORPOREAL LIFE SUPPORT

Recently developed membrane oxygenators produce minimal blood damage, and continuous use over several days is now possible. Hence, they have been used not only in cardiac operations but also in patients with pulmonary insufficiency in hospital wards. Typical systems are extracorporeal membrane oxygenation (ECMO).[1]

For the patient with chronic respiratory failure, CO_2 retention and increase in blood is a major problem. For such patients, removal of CO_2 has been tried using an extracorporeal circulation via veno–arterial (V–A) or veno–venous (V–V) bypass. Since the concentration of bicarbonate ion (HCO_3) in blood is normally 20 times higher than that of CO_2, the driving potential for bicarbonate ion transfer thus is much larger than that for CO_2. This extracorporeal CO_2 removal is called $ECCO_2R$. This system employs a hemodialyzer, namely a liquid–liquid blood–gas exchanger.[2] Liquid–liquid type artificial lung systems are also designed with a hemodialyzer, in which an artificial blood (parfluorotri-n-butylamine emulsion) is mixed with the perfusate to enhance oxygen transfer.[3]

Currently, for emergency cardiac patients, hollow-fiber membrane oxygenators are used for percutaneous cardiopulmonary support (PCPS). The femoral approach with this method is highly practical for rapid cannulation of emergency patients and is providing promising results.[4]

NEW IMPLANT-TYPE ARTIFICIAL LUNG

Beside the above devices, the intravascular blood–gas exchange system has been developed as an implantable artificial lung.

OUTLOOK

The development of cardiopulmonary bypass has undoubtedly made open-heart surgery possible. The history of the oxygenators involved represents the development of increasingly physiological devices, as early disc and bubble oxygenators gave way to the membrane oxygenators. Moreover, advances in membrane oxygenators have paved the way for treatment of respiratory failure, although they are now applied mainly as temporary support measures. In the future, artificial lungs will be employed as an alternative breathing system, for chronic respiratory failure.

REFERENCES

1. Gattinoni, L.; Pesenti, A.; Mascheroni, D. *JAMA* **1986**, *256*, 881.
2. Matsunobe, S.; Isobe, J.; Mizuno, H.; Shimizu, Y. *American Society of Artificial Internal Organs* **1987**, *10*, 441.
3. Mizuno, H.; Isobe, J.; Matsunobe, S.; Nakamura, T.; Shimizu, Y.; Hitomi, S. *Int. J. Artif. Organs* **1994**, *17*, 609.
4. Hill, J. G.; Bruhn, P. S.; Cohen, S. E.; Gallager, M. W.; Manart, F.; Moore, C. A.; Seifert, P. E.; Askari, P.; Banchieri, C. *Ann. Thorac. Surg.* **1992**, *54*, 699–704.

Lyotropic Liquid Crystalline Polymers

See: *Cellulosic Liquid Crystals*
Liquid Crystalline Ethyl-Cyanoethyl Cellulose
Liquid Crystalline Polymers (Rheology)
Liquid Crystalline Polymers (Viologen: Thermotropic and Lyotropic)
Lyotropic Polysaccharides

LYOTROPIC POLYSACCHARIDES

Peter Zugenmaier
Institute of Physical Chemistry
Technical University Clausthal

The term lyotropic derives from the Greek words lyo, for dissolved, and tropic, meaning directional behavior. It expresses an alignment of particles or molecules in a solvent. Although the term is used mostly to describe liquid crystals, recent studies in the polysaccharide field suggest the molecules may align and form clusters and similar other structures already in dilute and semi-dilute solutions. Polysaccharide interaction with surfactants and inorganic ions may also lead to certain structural features in solution. Many polysaccharides and their derivatives adapt solvent in the solid state, resulting in crystalline complexes with solvent located on lattice points or attached to the polysaccharide chain.

LYOTROPIC LIQUID CRYSTALS

Most research on lyotropic liquid crystals (lc) of polysaccharides; that is, polysaccharides in highly concentrated solutions, has been carried out on cellulosics, although the lc behavior of xanthan and scleroglucan and some other species have been studied.

A structural model for the cholesteric lc phase, the one observed in almost all cases, is depicted in **Figure 1**. The basic units of the supermolecular helicoidal structure consist of the clusters introduced in the previous section, which are twisted from one sheet to the next. A pitch p or helicoidal period can be defined, if a 2π twist is reached.

The pitch p depends on the substituents of the chain backbone, the degree of substitution, the solvent temperature, concentration, and also on the molar mass of the samples. Other structural quantities, such as the domain sizes of the cholesteric phases, have yet to be determined.

Pitch lowering occurs as molecular mass increases in aqueous hydroxypropyl cellulose (HPC) liquid crystals, whereas mesophases of cellulose triphenylcarbamate (CTC) in diethylene glycol monoethyl ether (DEME) provide increasing pitch values as the molecular mass rises and $M_w > 150{,}000$ g/mol levels off asymptotically.

The solubility of cellulosics plays an important role in the formation and existence of highly concentrated lc solution. In this discussion of the molecular shape in semi-dilute solution, it is clear that the lc state will not occur if single molecules exist at high concentration, nor will it occur if the highly concentrated system cannot be realized because of gelation. The lc state represents a delicate balance of polysaccharide molecule-solvent interaction.

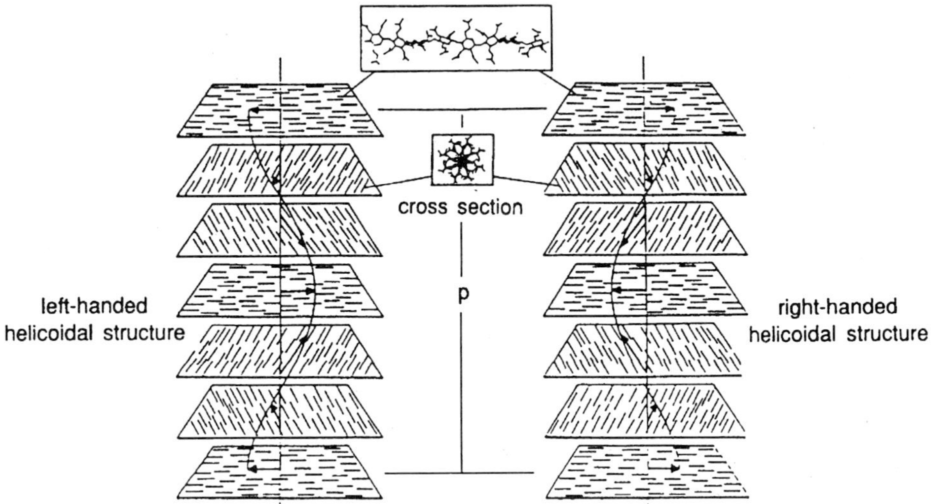

FIGURE 1. Schematic representation of a cholesteric liquid crystal of a cellulose derivative with pitch p.

PHASE TRANSITION

It has been shown that the basic behavior and properties of polymeric systems can be modeled with the assumption of a few features of the macromolecules, such as chain stiffness, chain length, and their space fillings. Various theories have been developed to match the observed anisotropic-isotropic transition which adapt the basic experimental parameters of the polymeric chains. Although these theories have been applied successfully to a number of materials, it is difficult to describe lyotropic trisubstituted cellulose derivatives.

CONCLUDING REMARKS

Lyotropic polysaccharides represent a broad field of research where experimental tools can now provide new insights. Only some basic ideas are addressed here which might be covered and overlapped by special structural features of the chain molecules, such as the existence of double and triple helices, helix coil transitions by temperature, or ionic strength variation. In studying polysaccharides, it should be remembered that these macromolecules belong to a unique class of polymeric materials which are created for particular functions and organized in various structural appearances, from insoluble solids to molecular dispersed chains in solution.

GENERAL LITERATURE

1. *The Polysaccharides*, Aspinall, G. O., Eds.; Academic: London, 1985; Vol. 1–3.
2. Rees, D. A. *Polysaccharide Shapes*; Chapman and Hall: London, 1977.
3. *Polysaccharide*, Buchard, W., Ed.; Springer Verlag: Berlin, 1985.
4. Gray, D. G.; Harkness, B. R. *Liquid Crystalline and Mesomorphic Polymers*; Shibaev, V. P.; Lam, L., Eds.; Springer Verlag: New York, 1994; p 298.
5. *Cellulosic Polymers*, Gilbert, R. D., Ed.; Carl Hanser Verlag: München, 1994.
6. Gilbert, R. D. *Agricultural and Synthetic Polymers* ACS Symp. Ser. 433; Glass, J. E.; Swift, G., Eds.; American Chemical Society: Washington, DC, 1990; p 259.

M

Macrocycles

See: *Bicyclic Lactones, Ring-Opening Polymerization*
Calixarenes (as Supramolecular Cyclic Oligomers)
Crown Ethers, Polymeric (Catalytic Activity)
Cyclic Benzo-Alkynes, Ortho Substituted
Cyclic Oligomers of Engineering Thermoplastics
Ionene Oligomers
Macrocyclic Aramids
Macrocyclic Block Copolymers
Macrocyclic Polymers (Controlled Dimensions)
Ring-Opening Metathesis Polymerization (Formation of Cyclic Butadiene Oligomers)
Tailor-Made Polymers
Uniform Polymers

MACROCYCLIC ARAMIDS

Wesley Memeger, Jr.
Du Pont Central Research and Development
Experimental Station
Du Pont Company

Macrocyclic aliphatic amides have been known since the early work of Carothers on aliphatic polyamides.[1]

Macrocyclic aromatic amides (aramids) are also known, but these do not include the simplest aromatic amide paracyclophane analogs (2*n*-aza[2_n]paracyclophane-2*n*-ones) (**Structure 1**) corresponding to cyclic oligomers of the rigid linear aramid poly(*p*-phenylene terepthalamide) (Kevlar® aramid substrate). That the cycloaramids in Structure 2 are not apparently formed during the polymerization of *p*-phenylenediamine and terephthaloyl chloride is believed to be due to the *trans*-conformation of the amide bond in the growing polymer chain.[2a] However, as flexible and/or kinked linkages are introduced into the intermediates, the propensity toward macrocyclization increases, especially under dilute reaction conditions. It is worth noting at this point that there are virtually no reports, with the exception of the work by Elhadi et al. discussed below, on designed macrocyclizations leading to formation of aromatic amide macrocycles.

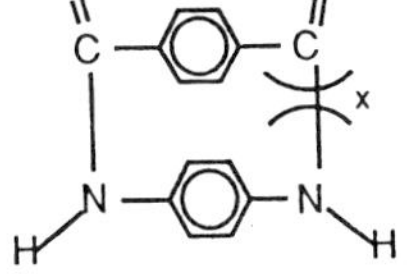

1

EFFECT OF FLEXIBLE LINKAGES AND KINKS ON MACROCYCLIZATION

This section illustrates some potential difficulties in synthesizing cycloaramids. **Table 1** shows a number of macrocyclic amide structures synthesized under dilute reaction conditions with the Schotten–Baumann reaction employing amines and acid chlorides in solvents such as benzene, chloroform, and tetrahydrofuran. One exception to the Schotten–Baumann reaction to prepare the macrocycle was in case entry 3, where the Schmidt reaction was employed.

CYCLOARAMIDS AND EFFECTS INFLUENCING THEIR FORMATION

Entries 9–13 represent true cycloaramids where both moieties adjacent to the amide group are aromatic. Here, except for entry 13, the overall yields are much lower than for the macrocycles containing alkylene linkages. No entry is shown in Table 1, wherein only *p*-phenylene or other all-*p*-oriented all-aromatic moieties are linked only by amide groups. The reason is the rectilinear nature of the growing chain imposed by the *transoid*-amide bonds inhibit cyclizatiion.

In contrast to the results of attempting to synthesize all *para*-oriented cycloaramids, *N*-alkyl substitution of the amide bond encourages the *cisoid*-conformation. This, in turn, facilitates macrocyclization. The presence of *cis*-amide bonds in a growing chain seems to predispose the system toward formation of macrocycles, as the chain would tend to fold back upon itself. The proposed steric guiding by *cis*-amide bonds towards formation of macrocycles compares to the *gem*-dimethyl effect exerted in certain intramolecular Diels–Alder reactions where large rate enhancements have been observed.[11]

The *cisoid*-amide effect emerged in Memeger's preparation of aromatic polyamide (aramid) foams[12] via the pyrolysis of *n*-alkyl-substituted aromatic polyamides. In addition to the linear poly(*N,N'*-di-sec-butyl-*p*-phenylene terephthalamide), a surprisingly high level (10%) of a low molecular weight fraction consisting of the *N,N'*-di-s-butyl macrocyclic amide homologues with up to 13 repeating units was formed during the high temperature solution polymerization of *N'N'*-di-s-butyl-*p*-phenylenediamine and terephthaloyl chloride in *o*-dichlorobenzene at relatively high concentration (1.1 molar).

In some recent, unpublished work, Memeger expanded work on aramid-like structures to include new melt-processible poly(ether ketone amide)s which are analogs of an all-*para* poly(ether ketone ketone) PEKK. These ordered polymers were prepared by a condensation polymerization, but should also be available from a ring-opening polymerization, which could be advantageous vs. PEKK in the fabrication of fiber-reinforced composites.

In another recent study, Lorenzi, Tomasic and Suter reported on *N*-methylated oligo(*p*-phenylene terephthalamide)s with a high propensity for intramolecular cyclization.[13]

RING-OPENING POLYMERIZATIONS, POLYMER PROPERTIES, AND APPLICATIONS

Since the macrocyclic aramid class is relatively new, work on their ring-opening polymerization is modest. In entry 13, no attempt was reported on the homopolymerization of the mixture of the high melting macrocycles using a procedure analogous to that employed in the ROP of a mixed aliphatic/aromatic macrocycle, entry 6 (x = 12, y = 1), which involved heating without a catalyst in the melt phase.[7] Homopolymerization under anionic catalytic conditions employed successfully with ε-caprolactam[14]

TABLE 1. Macrocyclic Amides via Dilute Reaction of Diacid Halides with Diamines

		x	y	m.p., °C	yield, %	Reference
1	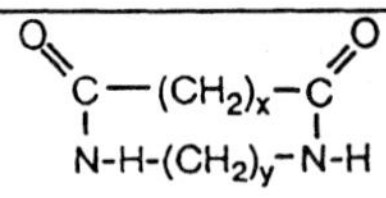	4	6	245	76	**3**
2	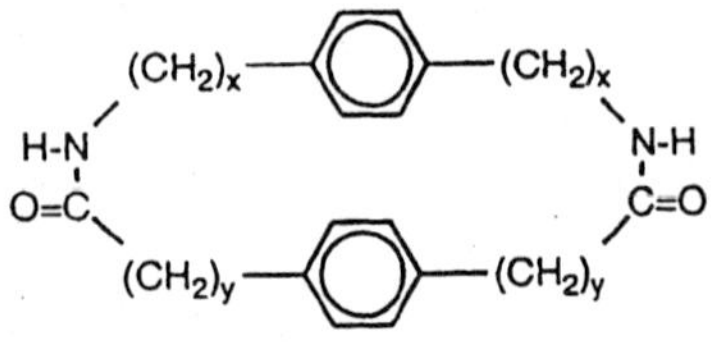	1	1	308–10	30	**4**
		1	2	208	54	**4**
3	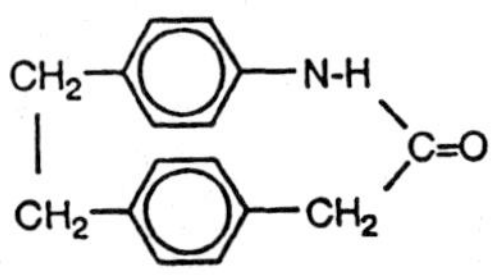	—	—	> 217 °C (dec.)	—	**5**
4	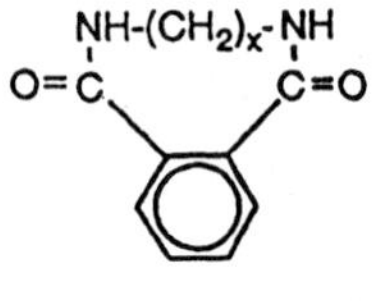	2	—	178	32	**4**
		4	—	171	52	**4**
		6	—	224	65	**4**
5	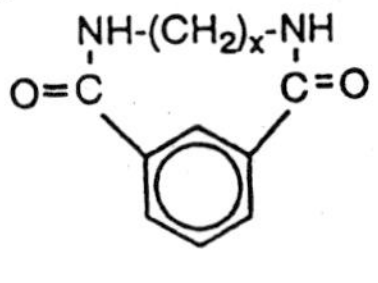	4	—	365–370	7.4	**4**
		6	—	325	19	**4**
6	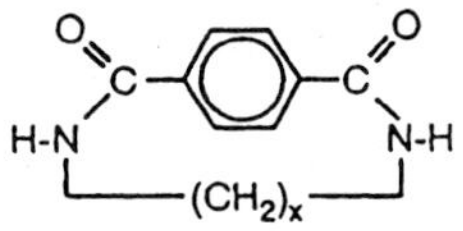	5		407	20	**6**
		6		296 (dec.)	6	**4**
		8		290 (dec.)	30	**4**
		10		275 (dec.)	57	**4**
		12		273	50	**7**
7	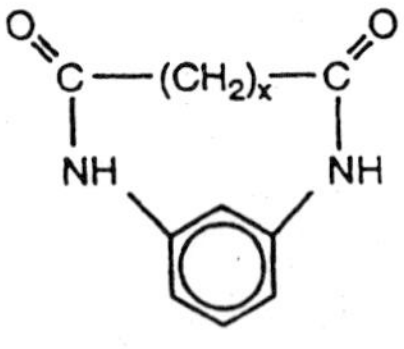	4		215 (dec.)	17	**8**
		6		264 (dec.)	77	**8**
		8		245	88	**8**
8	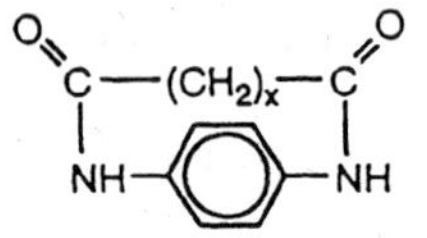	6		320 (dec.)	13	**8**
		8		310 (dec.)	35	**8**
		10		208	66	**8**

met with only limited success.[10] For example, heating the macrocycles from 275–375°C under nitrogen, neat or in solution, in the presence of catalytic amounts of sodium hydride, gave a polymer with modest MW (M_n = 10,000) but very dark color, indicating decomposition. However, when the macrocycles were dissolved in molten ε-caprolactam at a level 50 wt.% (15 mole%) in the presence of a catalytic amount of sodium hydride and held for 12 min. at 265°C, a copolymer was produced with an M_n = 10,000 and M_w of 47,000. The latter has potential utility as a molding resin with better properties than those of nylon 6.

TABLE 1. Macrocyclic Amides via Dilute Reaction of Diacid Halides with Diamines (continued)

		x	y	m.p., °C	yield, %	Reference
9	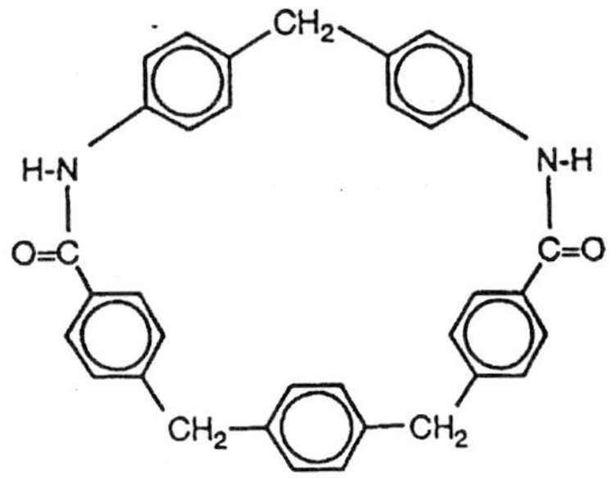	—	—	> 400	5	**3**
10	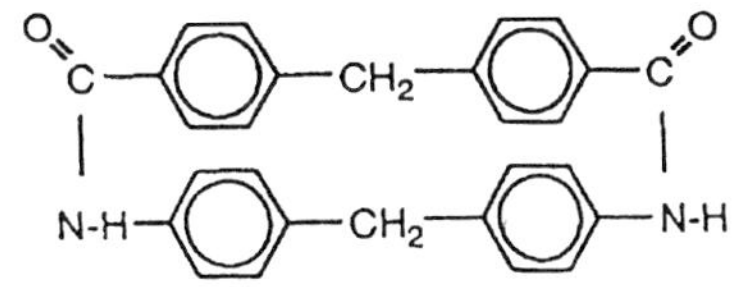	—	—	> 420	6	**3**
11	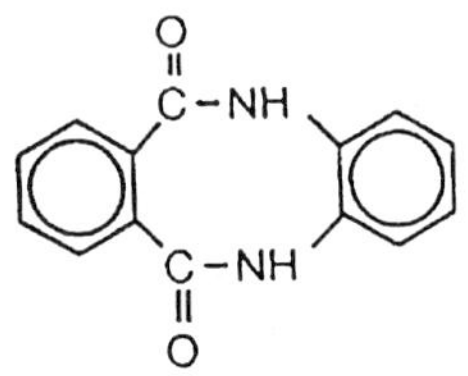	—	—	300	15	2
12	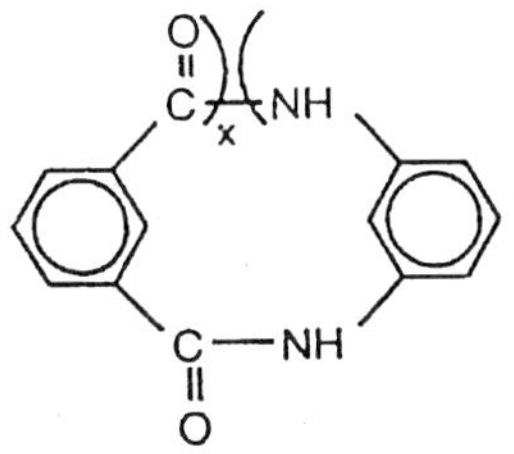	—	—	—	v. low	**9**
13	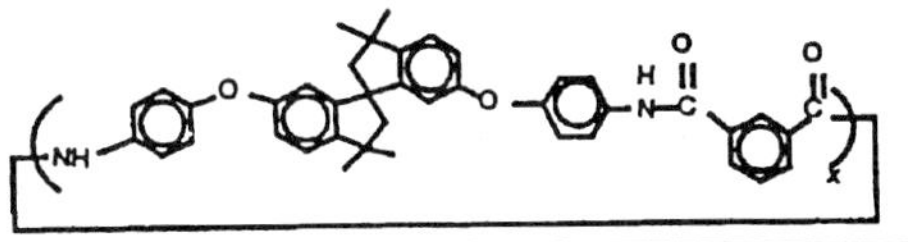	1–15[a]	—	245–285	80	**10**

[a]Ratio of major HPLC fractions: 1:2:3:4:5:6 = 78:28:8:4:2:1.

Unfortunately, attempts were not made to ring-open polymerize the *N*-methylated macrocyclic aramids prepared by Lorenzi and co-workers for, unlike the *N-s*-butyl substituted macrocycles, they would be thermally more stable and would resist deakylation during the ring-opening polymerization, making them potentially more suitable for high temperature resin applications.

REFERENCES

1. Carothers, W. H. U.S. Patent 2 071 253, 1937; Carothers, W. H., U.S. Patent 2 130 948.
2. Mori, S.; Fururawa, M.; Tacheuchi, T. *Anal. Chem.* **1970**, *42*, 661.
2a. Itai, A.; Toriumi, Y.; Tomioka, N.; Kagechika, H.; Azumaya, I.; Shudo, K. *Tet. Lett.* **1989**, *30*, 6177.
3. Stetter, H.; Marx, J. *Liebigs Ann.* **1957**, *607*, 59.
4. Stetter, H.; Marx-Moll, L.; Rutzen, H. *Chem. Ber.* **1958**, *91*, 1775.
5. Hertler, W. R.; Sharkey, W. H.; Andersen, B. C. *Macromolecules* **1976**, *9*, 523.
6. Livingston, H. K.; Gregory, R. L. *Polymer* **1972**, *13*, 297.
7. Glons, J. H.; Akkapeddi, M. K. *Macromolecules* **1992**, *25*, 5526.
8. Stetter, H.; Marx-Moll, L.; Rutzen, H. *Macromolecules* **1958**, *91*, 677.
9. Guggenheim, T. L.; McCormick, S. J.; Guiles, J. W.; Colley, A. M. *Polymer Preprints* **1989**, *30*(2), 138.
10. Guggenheim, T. L.; McCormick; Guiles, J. W.; Colley, A. M. U.S. Patent 4 868 279, 1989.

11. Jung, M. E.; Gervay, J. *Tet. Lett.* 2988, **1988,** *29,* 2429.
12. (a) Memeger, W. U.S. Patent 4 178 419, 1979; (b) Memeger, W., U.S. Patent 4 226 949, 1980.
13. Lorenzi, G. P.; Tomasic, L.; Suter, U. W. *Macromolecules* **1993**, *26*, 1183.
14. Sorenson, W. R.; Campbell, T. W. *Preparative Methods of Polymer Chemistry*, 2nd ed., Wiley-Interscience; 1968; Chapter V, p 238.

MACROCYCLIC BLOCK COPOLYMERS

Jingjing Ma
3M Adhesive Technology Center

Macrocyclic polymers have attracted considerable attention because of their special physical properties which result from the specific topological constraints of their circular architecture.[1–3] Synthesis of well-defined macrocyclic polymers has been reported by several research groups. This synthesis typically involves aromatic anion initiators that form α,ω-polymeric dianions and difunctional electrophiles to link two living ends of the same polymer together.[4–10] Quirk et al. have reported a novel approach of using 1,3-bis(1-phenylethylenyl)benzene (DDPE) as a living coupling agent for the preparation of four-heteroatom star-shaped block copolymers.[11,12] This methodology has been successfully extended to include macrocycle synthesis. In the living-coupling-agent method, a difunctional living polymer reacts with a DDPE molecule to produce a living macrocyclic product. The latter can then undergo a functionalization reaction to yield a cyclic polymer with two attached functional groups. Macrocyclic polystyrenes and polybutadienes with attached hydroxyl groups have been prepared in this way.[13–15]

Extensive studies have also been conducted on the dilithium initiator formed by the addition reaction of secbutyllithium and DDPE in the presence of preformed lithium alkoxides.[13,16] Research has shown that high 1,4-microstructure and monomodal narrow molecular weight distribution can be achieved in polybutadiene and poly(styrene-*b*-butadiene-*b*-styrene) (SBS) triblock copolymers that are prepared using this dilithium initiation system.[13,16] This article describes the further extension of using the living-coupling-agent methodology for the synthesis of macrocyclic block copolymers with well-defined molecular structure: macrocyclic poly(styrene-*b*-butadiene) (SB) block copolymers formed by coupling a dilithium poly(styrene-*b*-butadiene-*b*-styrene) block copolymer and a cyclic polybutadiene with two attached polystyrene branches formed by coupling a dilithium heteroarm star-shaped block copolymer.

CONCLUSION

Macrocyclic block copolymers can be prepared using traditional anionic methodology in which a difunctional electrophile is used to link a difunctional living block copolymer, as reported by Gan et al.[17] This research together with the previous work has clearly demonstrated a novel concept of using living coupling agents for macrocyclic block copolymer synthesis.[13-16,18] In this new method, DDPE or its derivatives can be used as both a difunctional initiator and a living coupling agent. The living-coupling-agent methodology allows one to prepare A-B-A linear triblock copolymers and macrocyclic AB block copolymers, as well as various macromolecules with controlled molecular structures.

REFERENCES

1. Rempp, P.; Strazielle, C.; Lutz, P. In *Encyclopedia of Polymer Science and Technology;* Kroschwitz, J. Ed.; John Wiley & Sons: New York, 1987; Vol. 9, p 193.
2. Semlyen, J. A. In *Cyclic Polymers;* Semlyen, J. A. Ed.; Elsevier: New York, 1986; p 1.
3. Semlyen, J. A. *Adv. Polym. Sci.* **1976**, *21*, 43.
4. Hild, G.; Kohler, A.; Rempp, P. *Eur. Polym. J.* **1980**, *16*, 525.
5. Hild, G.; Strazielle, C.; Rempp, P. *Eur. Polym. J.* **1983**, *19*, 721.
6. Vollmert, B.; Huang, J. X. *Makromol. Chem. Rapid Commun.* **1980**, *1*, 332.
7. Geiser, D.; Hocker, H. *Polym. Bull. (Berlin)* **1980**, *2*, 591.
8. Roovers, J.; Toporowski, P. M. *Macromolecules* **1983**, *16*, 843.
9. Roovers, J.; Toporowski, P. M. *J. Polym. Sci. Polym. Phys. Ed.* **1988**, *26*, 1251.
10. Yin, R.; Hogen-Esch, T. E. *Polym. Prepr., Am. Chem. Soc. Div. Polym. Chem.* **1992**, *33*, 239.
11. Quirk, R. P.; Ignatz-Hoover, F. In *Recent Advances in Anionic Polymerization;* Hogen-Esch, T. E.; Smid, J. Eds., Elsevier: New York, 1987; p 393.
12. Quirk, R. P.; Shock, L. E.; Lee, B. *Polym. Prepr. Am. Chem. Soc. Div. Polym. Chem.* **1989**, *30*, 113.
13. Ma, J.-J. Ph.D. Thesis, University of Akron, 1991.
14. Quirk, R. P.; Ma, J.-J. *Polym. Prepr., Am. Chem. Soc. Div. Polym. Chem.* **1988**, *29*(2), 10.
15. Quirk, R. P.; Ma, J.-J. *Polym. Prepr., Am. Chem. Soc. Div. Polym. Chem.* **1992**, *33*(1), 976.
16. Quirk, R. P.; Ma, J.-J. *Polym. Int.* **1991**, *24*, 197.
17. Gan, Y.; Zoller, J.; Hogen-Esch, T. E. *Polym. Prepr. Am. Chem. Soc. Div. Polym. Chem.* **1993**, *34*(1), 69.
18. Ma, J.-J. *Polym. Prepr. Am. Chem. Soc. Div. Polym. Chem.* **1993**, *34*, 626.

ACKNOWLEDGMENT

The author is grateful to Dr. R.P. Quirk (Institute of Polymer Science, the University of Akron, OH) for his consistent support and valuable contributions.

MACROCYCLIC POLYMERS (with Controlled Dimensions)

Alain Deffieux
EWSCPB
Laboratoire de Chimie des Polymères Organiques
Université Bordeaux-l -France

A specific interest in macrocyclic polymers of higher molecular weights has grown rapidly in the last decade for various reasons.[1] These involve the need of experimental verification of theoretical predictions concerning the solution and bulk behavior of cyclic chains and the observation of original physical and mechanical characteristics associated to the cyclic architecture.[2–9] The investigation of the specific properties of monomacrocyclic and plurimacrocyclic polymers remained restricted, however, because of the extremely limited availability of the corresponding pure compounds.

Two different strategies have been used so far for the synthesis of ring polymers. The first is based on the concurrent formation of linear and cyclic macromolecules in equilibrium, in systems containing reactive functions in their backbones. The second involves the end-to-end closure of ditelechelic linear polymers in highly diluted conditions.

MACROCYCLES OBTAINED FROM LINEAR-RING CHAIN EQUILIBRIA

The prerequisite to this cyclization procedure is the presence of labile linkages in the backbone of the macromolecules such as polycondensates (e.g., polyesters, polycarbonates) or polymers formed by ring-opening polymerization of heterocycles and cyclic olefins; polyalkenamers belong to this category.

In these macromolecular systems, rings are formed through a unimolecular process involving the intramolecular attack of one of the reactive functions of the polymer backbone by the active chain end, as illustrated in **Scheme I**. The reaction can take place either in the course of the propagation reaction, or through rearrangement of linear polymer chains, in the presence of an appropriate catalyst. As can be predicted from the unimolecular and bimolecular reaction mechanisms involved, the cyclic/linear ratio is strongly dependent on the polymer concentration. These equilibria have been studied theoretically.[2,10] The size of the macrocycles and their relative proportion is mainly determined by entropic factors (chain flexibility and conformation).

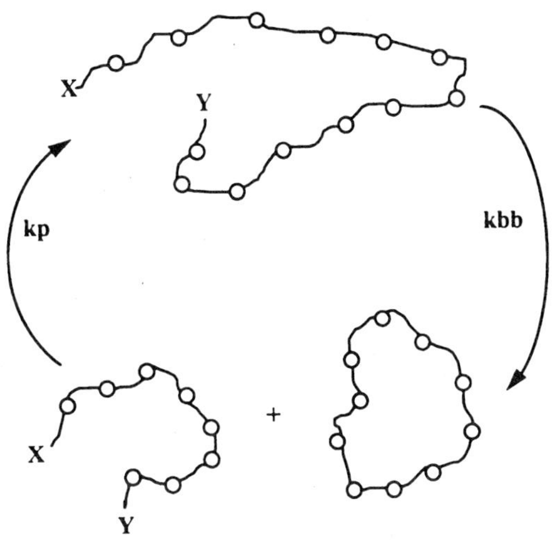

SCHEME I. Formation of macrocycles in systems exhibiting ring-chain equilibria.

MACROCYCLIC POLYMER SYNTHESIS BY END-TO-END RING CLOSURE TECHNIQUES

Although this method of cyclization does not exclude macromolecules having labile functions in the main chain, studies have been conducted mainly on polymers with nonreactive linkages, such as carbon–carbon backbone polymers. The strategy is based on an entirely different principle and involves the end-to-end ring closure of a linear polymer carrying reactive functions at both chain ends. In order to favor the intramolecular coupling yielding cyclization against chain extension, the ring-closure reaction is performed in highly diluted conditions. One interesting advantage of this method deals with the possible control of the size and distribution of the macrocyclic polymers formed because the dimensional characteristics of the macrocyclic polymers are directly related to their linear precursors. The macrocyclic polymers have usually been synthesized by living polymerization.

Plurimacrocyclic Macrocycles

Because the synthesis of well-defined monomacrocyclic polymers was not totally under control until recently, the preparation of polymers with a pluricyclic architecture remained a limited research domain. Recent progress in end-to-end polymer cyclization procedures have afforded interesting possibilities in this area.

Bicyclic "Eight-Shaped" Polymers

The first direct attempt to synthesize polymers with a bicyclic "eight-shaped" chain architecture was described by Antonietti et al. in 1988.[39] The strategy applied is derived from the one used in the synthesis of monomacrocycles by bimolecular end-to-end coupling. The use of a tetrafunctional compound, $SiCl_4$, instead of a difunctional molecule, allowed the reaction of two bifunctional polystyrene chains on one coupling molecule.

Very recently, using a different strategy, Schappacher et al. employed a unimolecular ring-closure process developed for monomacrocycles to synthesize well-defined bicyclic eight-shaped poly(chloroethyl vinyl ether)s.[12] The general approach consists of the synthesis of a linear polymer precursor bearing two series of dual functions that can be directly coupled under appropriate conditions.

Macrotricyclic Poly(vinyl ethers)

The possibility for the preparation of macromolecules of controlled size with other types of pluricyclic architectures, using the same unimolecular cyclization procedure, has also been explored.[13]

Though covering a very broad range of possible structures and architectures, macrocyclic polymers may be considered a special class of macromolecules, as recognized by star polymers and dendrimers, with the common features associated with the cyclic-chain architecture. In particular these include the properties resulting from the absence of chain ends and from the reduction of the volume occupied by cyclic chains, as compared to linear macromolecules of the same molar mass. Some of the solution and bulk properties relevant to these characteristics have already been identified, such as lower chain mobility, a higher glass transition temperature of oligomers, lower viscosity, and so on. Much more work has to be done to identify new polymer domains, however, in which the specific properties of cyclic or pluricyclic polymers may bring about special interest for such architectures, as illustrated by some reported examples. The recent progress in the procedures and techniques available for the selective synthesis of polymers with various macrocyclic architectures has opened up investigation into a broad new area of polymer science.

REFERENCES

1. *Cyclic Polymers*; Semlyen, J. A., Ed.; Elsevier Applied Science: London, 1986.
2. Jacobson, H.; Stockmayer, W. H. *J. Chem. Phys.* **1950**, *18*, 1600.
3. De Gennes, P. G. *J. Chem. Phys.* **1971**, *55*, 572.
4. Di Marzio, E. A.; Guttman, C. M. *Macromolecules* **1987**, *20*, 1403.
5. Yang, A. J. M.; Di Marzio, E. A. *Macromolecules* **1991**, *24*, 6012.
6. Semlyen, J. A. *Pure Applied Chem.* **1987**, *53*, 1797.

7. McKenna, G. B.; Hadziioannou, G.; Lutz, P.; Hild, G.; Strazielle, C.; Straupe, C.; Rempp, P.; Kovacs, A. J. *Macromolecules* **1987**, *20*, 1851.
8. McKenna, G. B.; Hostetter, B. J.; Hadjichristidis, N.; Fetters, L. J.; Plazk, D. J. *Macromolecules* **1989**, *22*, 1834.
9. Horbach, A.; Vernaleken, H.; Weirauch, K. *Macromol. Chem.* **1980**, *181*, 111.
10. Suter, U. W. *Comprehensive Polymer Science;* Pergamon: Oxford, 1989; Vol. 5, 91.
11. Antonietti, M.; Fölsch, K. J. *Makromol. Chem., Rapid Comm.* **1988**, 9, 423.
12. Schappacher, M.; Deffieux, A. submitted for publication in *Macromolecules*, **1995**, *28*, 2629.
13. Schappacher, M.; Deffieux, A. *Macromolecules* **1992**, *25*, 6744.

Macrocrosslinkers

See: Macromonomeric Initiators (Macroinimers)

Macrogels

See: Microgels, Macrogels (by Photocrosslinking)

Macroinimers

See: Macromonomeric Initiators (Macroinimers)

Macromolecular Engineering

See: Tailor-Made Polymers

MACROMOLECULAR RECOGNITION (New Supramolecular Structures)

Akira Harada
Department of Macromolecular Science
Faculty of Science
Osaka University

Our research indicates that cyclodextrins form complexes with various polymers that produce crystalline compound with high selectivity.[1] Cyclodextrins have been found to recognize structures and molecular weights of polymers. We have succeeded in preparing polyrotaxanes in which many cyclodextrins are threaded on a polymer chain.[2–4] In addition, we have prepared polymers starting from the polyrotaxanes.[15]

PREPARATION AND PROPERTIES

Polyrotaxane

A classic example of a supermolecule is provided by the rotaxanes, in which a molecular rotor is threaded to a linear axile. We have succeeded in preparing compounds in which many cyclodextrins are threaded on a single poly(ethylene glycol) chain and are trapped by capping the chain with bulky end groups.

The product is insoluble in water and dimethylformamide although the components: α-cyclodextrin, poly(ethylene glycol) bisamine, bis(2,4-dinitrophenyl)–poly(ethylene glycol), and even α-cyclodextrin–poly(ethylene glycol) bisamine complexes are water soluble. The product is, however, soluble in dimethyl sulfoxide and 0.1 N NaOH. The hydroxyl groups of α-cyclodextrin (pK_a = 12) can be ionized at this pH, so that α-cyclodextrin becomes soluble in an aqueous medium. Neutralizing this solution by adding 0.1 M HCl instantly produced a precipitate. This reaction is reversible.

We have not yet determined how adjacent cyclodextrins are oriented. But the fact that the product is insoluble in water, but soluble in 0.1 N NaOH, shows that strong hydrogen bonds exist between α-cyclodextrins, suggesting that they alternate as shown in **Scheme I**.

Molecular Tube: Synthesis of a Tubular Polymer from Threaded Cyclodextrins

Recently, much attention has been focused on design and fabrication of nanoscale structures. Carbon nanotubes (4–30 nm), for example, have been constructed by an arc-discharge evaporation method similar to that of fullerene synthesis. It is more convenient to construct smaller subnano tubes chemically. Cyclodextrins are one of the most promising candidates for the components of a molecular tube; they contain cylindrical cavities about 0.7 nm deep, with diameters of 0.45 nm, 0.7 nm, and 0.85 nm for α-cyclodextrin, β-cyclodextrin, and γ-cyclodextrin, respectively. We have prepared rotaxane supermolecules in which many cyclodextrins are threaded on a polymer chain, and have reported polyrotaxanes with many threaded cyclodextrins. By removing the bulky ends of the polymer thread, the tube can be unthreaded and can act as a host for reversible binding of small molecules.

Double-Stranded Inclusion Complexes of Cyclodextrin Threaded on Poly(ethylene glycol)

In the course of the experiments on the preparation of inclusion complexes of cyclodextrins with poly(ethylene glycol), we found that γ-cyclodextrin formed a trace amount of complexes with poly(ethylene glycol). The amount of the complexes formed is so small that we could not characterize them. However, we found that some poly(ethylene glycol) derivatives, such as bis(3,5-dinitrobenzoyl)–poly(ethylene glycol) ($PEG–DNB_2$ and bis(2,4-dinitrophenyl-amino)–poly(ethylene glycol) ($PEG–DNP_2$), formed complexes with γ-cyclodextrin to give crystalline compounds in high yields, although α-cyclodextrin did not form complexes with these poly(ethylene glycol) derivatives because the substituents at the end groups are too large to penetrate α-cyclodextrin cavities. We found that these complexes are composed of double chains of PEGS and γ-cyclodextrins by using poly(ethylene glycol) with fluorescent probes at the chain ends.[6]

The diameter of γ-cyclodextrin cavity is 8.5–9 Å, which is twice as large as that of α-cyclodextrin (4.5 Å). But the depth of the cavity of γ-cyclodextrin is the same as that of α-cyclodextrin and β-cyclodextrin (7Å), which corresponds to the length of 2 ethylene glycol units. Molecular model studies indicate that a γ-cyclodextrin cavity is large enough to accommodate a double chain of poly(ethylene glycol), although α-cyclodextrin cavities are too small to fit two chains of poly(ethylene glycol). This double-stranded inclusion phenomenon is reminiscent of inclusion of the double helix of deoxyribonucleic acid (DNA) by DNA polymerases.

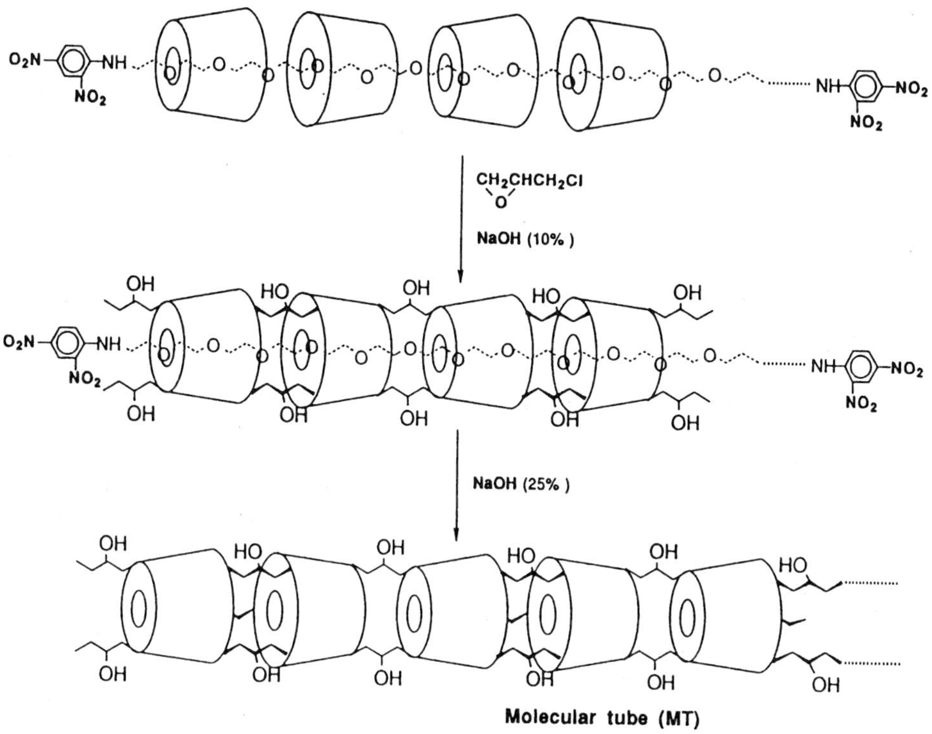

I

REFERENCES

1. Harada, A. *Polym. News* **1993**, *18*, 358.
2. Harada, A.; Li. J.; Kamachi, M. *Nature* **1992**, *356*, 325.
3. Harada, A.; Li. J.; Nakamitsu, T.; Kamachi, M. *J. Org. Chem.* 1993, *58*, 7524.
4. Harada, A.; Li, J.; Kamachi, M. *J. Am. Chem. Soc.* **1994**, *116*, 3192.
5. Harada, A; Li, J.; Kamachi, M *Nature* **1993**, *364*, 516.
6. Harada, A.; Li, J.; Kamachi, M. *Nature* **1994**, *370*, 126.

MACROMONOMER THERMOPLASTIC ELASTOMERS

Mostafa A. H. Talukder
Chemistry and Materials Branch
Research and Technology Division
China Lake Naval Air Warfare Center
Weapons Division

The concept of thermoplastic elastomers (TPEs) is unique[1–7] in terms of design, development, and applications. TPEs possess both glass transition temperature (T_g) and melting point (T_m). Thus, unlike a physical blend, a TPE is equivalent to a chemical blend of a thermoplastic and an elastomer without having any problem of compatibility. In short, TPEs are materials that combine the processing characteristics of thermoplastics with the physical properties of vulcanized rubbers.

Macromonomer TPEs are a new class of TPE based on the macromonomer concept. The comb-shaped polymers are designed in such a way that the polymer backbone forms the elastomeric phase, and the branches the hard phases, giving polymers with TPE properties without the usual $(AB)_n$ or ABA structure. The new design and approach are expected to improve primarily the processability of the polymer by decreasing the melt viscosity of the polymer due to the effects of polymer branching on the viscoelastic properties of the polymers.

SYNTHESIS OF MACROMONOMERS

Macromonomers are low molecular weight polymers or oligomers with polymerizable end groups (an unsaturation or a cycle able to undergo ring-opening polymerization). Their molecular weights usually range from 1,000 to 30,000, and they provide easy access to graft copolymers upon copolymerization with acrylic or vinylic comonomers. Via the polymerizable group, macromonomers are conveniently subjected to copolymerization with a conventional monomer to produce a family of graft copolymers possessing the pendant chain of the macromer. Each macromonomer that is incorporated into the chain results in a graft. This method allows the control of the lengths of the grafts, which is an important feature for many applications.

Although most macromonomers are synthesized via anionic or cationic polymerizations under living conditions, other methods, such as free radical polymerization or group transfer polymerization, are also used to prepare macromonomers. Recently, a unique class of macromonomers has been reported[8] in which the macromolecular part is a highly branched dendritic macromolecule.

POLYMERIZATION OF MACROMONOMERS

The ability of the macromonomers to polymerize and copolymerize with other monomers has been studied by many researchers using different mechanisms. As a general rule, vinyl- or acrylic-terminated macromonomers can be polymerized by a free radical mechanism. The methacrylate-terminated macromonomer can be polymerized by free radical as well as anionic polymerization. The cyclic ether-terminated macromonomer can be polymerized by anionic and cationic polymerization mechanisms. The main aspect to consider in this respect is the low molar concentration of macromonomers. Anionic polymerization[9] is unequivocally the most effective way to polymerize macromonomers for well-defined structure. Although the growth is slow, the lifetime of the active sites is practically unlimited. Similarly, cationic polymerization is carried out for cyclic macromonomers to prepare graft copolymers.

MACROMONOMER TPEs

The polymerization of macromonomers by different techniques, such as anionic polymerization technology, has led to the development of many well-defined graft copolymers or side-chain polymers.[10–13,15] The interest in graft copolymers arises from the strong repulsions between chemically different sequences, yielding intramolecular phase separation in solution as well as in bulk, where surface accumulation tends to occur. This behavior has led to a number of specific applications of graft copolymers as coatings, surface modifiers, adhesion factors, surface tension modifiers, etc. The special case of amphiphilic graft copolymers in which hydrophilic grafts are linked to a hydrophobic backbone or vice versa has attracted much attention in recent years, and the use of macromonomers for such synthesis has been quite successful. The potential applications include coatings, adhesives, compatibilizers, emulsifiers, biomaterials, synthetic membranes, etc.

Feinberg[16] copolymerized methacrylate-terminated polystyrene macromonomer with other acrylic monomers by free radical polymerization to form a phase-separated graft copolymer with a rubbery acrylic backbone and a glassy polystyrene side chain. The macromonomer-based TPEs are useful as pottants for photovoltaic cells.

The polymers currently being used as pottants are predominantly plasticized poly(vinyl butyral) and crosslinked ethylene-vinyl acetate (EVA) copolymer. These materials have inherent disadvantages related to chemical resistance and the need for crosslinking the EVA systems. In comparison, the macromonomer TPEs are thermally reversible crosslinkings of the polystyrene domains and improve the photo and oxidative stability of the acrylate backbone.

Mancinelli et al.[14] developed an acrylic hot melt pressure sensitive adhesive (HMPSA) using the macromonomer approach. Acrylic HMPSA has economic and environmental advantages over the current solvent- and emulsion-based acrylic polymers. The materials also undergo chemical crosslinking on drying.

This author[17–24] designed and developed a new TPE useful as a binder for explosives and propellants, based on the macromonomer technique. The primary criteria of these macromonomer TPEs are such that the polymers have low melt viscosity and have improved compatibility with a nitramine oxidizer in the formulation. The design is based on the macromonomer structure and the incorporation of a comonomer with similar structure to nitramine oxidizer, in the macromonomer TPEs. The first criterion of low melt viscosity is achieved by the effect of the side chains of the macromonomers, and the second criterion of compatibility is achieved by the structural similarity of a binder with other ingredients, such as cyclic nitramine oxidizers in the explosive or propellant formulation.

CONCLUSION

Side-chain polymers, in general, have been studied extensively, but the study of the macromonomer TPEs, in particular, is very limited, not to speak of the systematic study. The present work along with the limited literature cited above shows the tremendous potentiality of macromonomer TPEs for different applications. The structural architecture of macromonomer polymers, in general, including macromonomer TPEs is able to modify a polymer according to need and desire.

FUTURE WORK

The future studies of macromonomer TPEs will include the effects of structure and length of the side chains on the viscoelastic properties of the macromonomer TPEs. A semi-empirical relationship is derived for the effects of the branching on the melt viscosity of macromonomer TPEs.

ACKNOWLEDGMENTS

The work was funded by the Naval Air Warfare Center Weapons Division Independent Research program and the Office of Naval Technology (ONT) Explosive Block Program. LNA was prepared by Dr. William Norris. Dr. Thomas Stephens measured the viscoelastic properties and Dr. Rena Yee ran the compatibility tests.

REFERENCES

1. Legge, N. R.; Holden, G.; Schroeder, H. E. *Thermoplastic Elastomers, A Comprehensive Review;* Hansler: New York, 1987.
2. *Block and Graft Polymerization* Ceresa, R. J. Ed.; John Wiley & Sons: New York, 1972.
3. Schroeder, H. E. *Kautschak Gummi Kunststoffe* **1982**, *35*, 661.
4. *Handbook of Thermoplastic Elastomers* Walker, B. M. Ed., Van Nostrand Reinhold: New York, 1979.
5. Melville, H. W. *J. Chem. Soc.* **1946**, 414.
6. Szwarc, M. *Nature* **1956**, *178*, 1168.
7. Szwarc, M.; Levy, M.; Milkovich, R. *J. Am. Chem. Soc.* **1956**, *78*, 2656.
8. Hawker, C. J.; Frechet, J. M. J. *J. Am. Chem. Soc.* **1990**, *112*, 7638.
9. Masson, P.; Beinert, G.; Franta, E.; Rempp, P. *Polym. Bull.* **1982**, *7*, 17.
10. Revillon, A; Amaide, T. H. *Polym. Bull.* **1982**, *6*, 235.
11. Kennedy, J. P et al. *Polym. Bull.* **1982,** *8*, 551, 563, 571.
12. Schulz, G. O.; Milkovich. R. *J. Appl. Polym. Sci.* **1982**, *27*, 4773.
13. Schulz, G. O.; Milkovich, R. *J. Polym. Sci., (Polym. Chem. Ed.)* **1984,** *22*, 1633.
14. Mancinelli, P. A. et al. *Polym. Prepr.* **1985,** *26*(1).
15. Rempp, P.; Franta, E. *Polymer Prepr.* **1986,** *27*(1), 181.
16. Feinberg, S. C. *Polymer Preprints, ACS* **1985,** *26*(2), 296.
17. Talukder, M. A. H.; Norris, W. P.; Lindsay, G. A. *Synthesis and Polymerization of Linear Nitramine Acrylate* Second Pacific Polymer Conference, Otsu, Japan, November 26–29, 1991.
18. Talukder, M. A. H.; Norris. W. P.; Stephens, T. S.; Yee, R. Y.; Nadler, M. P.; Nissan, R. A.; Lindsay, G. A. *New Macromer Thermoplastic Elastomers* IUPAC International Symposiums on New Polymers, Kyoto, Japan, November 29–December 1, 1991.
19. Talukder, M. A. H. *Macromer Thermoplastic Elastomers* Third Pacific Polymer Conference, Gold Coast, Australia. Preprints, **1993,** *3*, 785.

20. Talukder, M. A. H.; Norris, W. P.; Merwin, L. H.; Stephens, T S.; Yee; R. Y. *Macromer Thermoplastic Elastomers* Tri-Service Propellant Formulators Meeting, NASA Kennedy Space Center, Florida, April 12, 1994.
21. Talukder, M. A. H.; Norris, W. P.; Hoover, J. M.; Stephens, T. S.; Nissan, R. A.; Nadler, M. P.; Yee, R. Y. *Macromer Thermoplastic Elastomers as Binders for Explosives and Propellants* Propellant Development and Characterization Meeting, JANNAF, NASA Kennedy Space Center, Florida, April 13–15, 1994.
22. Talukder, M. A. H.; Norris, W. P.; Lindsay, G. A.; Hoover, J. M. *Energetic Macromer Thermoplastic Elastomers As Binders For Explosives And Propellants* Navy Case No. 075,863, October 18, 1993.
23. Talukder, M. A. H. *Macromer Thermoplastic Elastomers* Navy Case No. 075.864, October 18, 1993.
24. Talukder, M. A. H. *Preparation Of Polyethylene Methacrylate Macromer In High Yield,* Patent disclosure (June 22, 1994).

MACROMONOMERIC INITIATORS (Macroinimers)

Baki Hazer
Department of Chemistry
TÜBITAK-Marmara Research Center
and
Department of Chemistry
Karadeniz Technical University

Both block and graft copolymers can form two-phase morphologies. In contrast to two-phase physical blends, the two-phase block and graft copolymer systems have covalent bonds between the phases, which considerably improves their mechanical strengths owing to domains-induced phase segregation serving as physical crosslinking and reinforcement sites. Several macrointermediates were used to obtain this kind of copolymer. One type, macromonomeric initiators, or "macroinimers," have the properties of macromonomers, macrointermediates, macrocrosslinkers, and macroinitiators.

MACROCROSSLINKERS

Macrocrosslinkers are macromonomers having two vinyl ends. Macrocrosslinkers are usually poly(ethylene glycol) dimethacrylate macromers and can be used as crosslinkers in the free radical polymerization.[2–4]

MACROINITIATORS

Macroinitiators are macromolecules having peroxygen or/and azo groups and can thermally initiate a vinyl polymerization to obtain block copolymers in one step. Macroazoinitiators can be synthesized by the reaction of azobis(cyano-pentanol) with poly(ethylene glycol) having isocyanate terminal groups.[5,6] Laverty and Gardlund reacted poly(ethylene glycol) with 4,4′-azobiscyanopentanoyl chloride to prepare polyazoesters.[7]

Macroazo or peroxy initiators can initiate a vinyl polymerization, and block copolymers can be obtained in one step.

By combining macromonomers, macrocrosslinkers, and macroinitiators in a macrostructure, macroinimers were prepared to obtain crosslinked or branched block copolymers.[8] Macroinimers contain both vinyl group and azo group attached poly(ethylene glycol) blocks, which have high hydrophilicity, flexibility, and ion absorbability.[9] They can thermally homopolymerize by themselves or copolymerize with a vinyl monomer. In both cases, crosslinked or branched copolymers occur while macroinitiators give only linear block copolymers.[10] The vinyl end groups are effective to obtain crosslinked or branched block copolymers. When macroinimers having only one vinyl group instead of two were used in a vinyl polymerization, soluble branched copolymers can be obtained.[8] Block copolymers by macroinimers were formed without homopolymers while macroinitiators or macromers via free radical mechanism give some related homopolymers.[11]

APPLICATIONS

Macroinimers lead to the crosslinked block copolymers. Thus they can be used in the preparation of the block copolymer resins. Sulfonation of the styrene-PEG crosslinked block copolymers was used on ion exchange resins.[12] Hazer et al.[13] used the mixture of styrene and urethane type macroinitiator in the impregnation of wood to gain improvements in water repellency and mechanical properties. By the thermal polymerization of the mixture into wood, the impregnation and chemical modification of hydroxyl end groups of wood were brought about. Therefore water repellency and mechanical properties of wood increase. Grafting reactions of polybutadiene with macroinimer was also carried out very recently. Hydrophobic polybutadiene matrix gains hydrophilicity with PEG units of macroinimers.[14]

REFERENCES

1. Ito, K.; Hashimura, K.; Itsuno, S.; Yamada, E. *Macromolecules* **1991**, *24*, 3977, Sawhney, A. S.; Pathak, C. P.; Hubbell, J. A.; *Macromolecules* **1993**, *26*, 581–587.
2. Matsumoto, A.; Yonezawa, S.; Oiwa, M. *Makromol. Chem., Rap. Commun.* **1990**, *11*, 25.
3. Matsumoto, A.; Yonezawa, S.; Oiwa, M. *Eur. Polym. J.* **1988**, *24*, 703.
4. Shah, A. C.; Parsons, I. W.; Haward, R. N. *Polymer* **1980**, *21*, 825.
5. Furukawa, J.; Takamori, S.; Yamashita, S. *Angew. Makromol. Chem.* **1967**, *1*, 92.
6. Yürük, H.; Özdemir, A. B.; Baysal, B. M. *J. Appl. Polym. Sci.* **1986**, *31*, 2171.
7. Laverty, L. J.; Gardlund, Z. G. *J. Polym. Sci., Polym. Chem. Ed.* **1977**, *15*, 2001.
8. Hazer, B. *Macromol. Rep.* **1991**, *A28*, 47.
9. Hazer, B. *Makromol. Chem.* **1992**, *193*, 1081.
10. Hazer, B.; Erdem, B.; Lenz, R. W. *J. Polym. Sci. A: Poly. Chem.* **1994**.
11. Hazer, B. *Handbook of Polymer Science and Technology;* Cheremisinoff, N. P., Ed.; Marcel Dekker: New York, 1989; Chapter 4.
12. Savaşkan, S.; Beşirili, N.; Hazer, B. *J. Appl. Polym. Sci.* **1996**, *59*, 1515-1524.
13. Hazer, B.; Örs, Y.; Alma, M. H. *J. Appl. Polym. Sci.* **1993**, *47*, 1097.
14. Hazer, B. *Macromol. Chem. Phys.* **1995**, *196*, 1945-1952.

Macromonomers

See: *Branched Polymers*
Comb Polymers (Poly(ethylene oxide) Side Chains)
Functional Polymers (Telechelics and Macromonomers)
Functionalized Polymers (Telechelics and Macromonomers, via Cationic Polymerization)
Ion-Chelating Polymers (Medical Applications)

MACROMONOMERS (Preparation, Polymerizability, and Applications)

Yasuhisa Tsukahara
Department of Materials Science
Kyoto Institute of Technology

Macromonomer is an abbreviation of *macromolecular monomer*, defined as a reactive oligomer of polymer in which a polymerizable functional group is connected to a chain end. Macromonomers consist of two parts: one part is a polymer chain having some kinds of physical properties and the other part is the end-functional group having some extent of chemical reactivity for polymerization reactions.

Macromonomers are important precursors for preparations of various kinds of functional graft copolymers of well-defined structure.[1–6] In these cases, macromonomers are mono-functional, i.e., there is one polymerizable end group per molecule. However, more recently, telechelic of bi-functional and multi-functional macromonomers are being used for the formation of functional polymer networks or gels.[7,8] Macromonomers are also members of the end-functional reactive oligomers.[9] One can utilize the unique properties of oligomers differently from those of small monomers and polymers such as nonvolatility, high solubility, etc., by using macromonomers.

PREPARATION

Macromonomers are now prepared via various routes that are usually classified according to either the type of polymerization in the preparation[5,10] or the polymerizable end group[11,12] or the polymer chain.[13] Among these, there are several common preparation processes, irrespective of the type of polymerization and the structure of macromonomers, which are the end capping (deactivation) method of living polymers, initiation method, utilization of chain transfer reaction, polyaddition reaction, and transformation of the end group of preformed end-functional polymers.

POLYMERIZATION

Polymerizations of macromonomers are carried out by either the chain polymerization or the condensation polymerization depending on the type of the polymerizable end group. Homopolymerizations of macromonomers produce the specific multibranched polymers, while statistical copolymerizations of macromonomers with small monomers produce graft copolymers of wide variety. Compared with the conventional small monomers, polymerizations of macromonomers are accompanied by the following characteristic features because of the influence of the connected polymer or oligomer chain: distribution of reaction species in the polymerization media often becomes heterogeneous owing to the interaction between polymer chains or polymer and solvent, high viscosity of the polymerization media from the beginning of the polymerization reaction, low monomer concentration because the polymerizable end groups are always diluted by their own polymer chain, propagation steps are repeats of a polymer–polymer reaction but not polymer–small molecule reactions, and high segment density around the propagating active site due to the specific multibranched structure. These could make the polymerization behavior of macromonomers very different from those of the small monomers corresponding to the polymerizable end group, although the intrinsic reactivity of the end group always plays an essential role in the polymerizability of the macromonomer.

Homopolymerization

The attainable degree of polymerization (DP) of the polymerization product, i.e., polymacromonomer, was at first considered to be very low owing to the steric effect or the excluded volume effect associated with the specific multibranched structure of the growing chain in the chain polymerization. However, recent studies in radical polymerizations have revealed that the high segment density around the propagating radical does not necessary limit the growing to a large DP.[14,15]

Polymerizations of amphiphilic macromonomers in micelles produce polymacromonomers of large DP owing to the increased local concentration and orientation of the polymerizable end group in micelles.[16]

Copolymerization

There are basically three major methods for the preparation of graft copolymers, that is: "grafting onto" the backbone polymer by coupling reactions, "grafting from" the backbone by polymerization, and "grafting through" the polymerizable functional group by copolymerization with small comonomers. The "macromonomer method" corresponds to the third method and has been widely employed during this decade.[3–5]

PROPERTIES AND APPLICATIONS

Macromonomers in the oligomer region are used directly as a reactive reagent in the reaction injection molding and as a reactive compatibilizer in polymer blends and alloys.[4] However, in most cases, they are used as the precursor of the building block for multibranched polymers, graft copolymers, and network polymers.

Multibranched Polymers

Polymerizations of macromonomers are a good way to produce regular, multibranched polymers with very high branch-density. Such polymacromonomers have regular branch length, regular branch number, and regular branching period. Thus, polymacromonomers are considered to be interesting model-branched polymers for the study of the effect of branching architecture on both molecular and bulk properties.[17–91,21]

The specific multibranched structure of polymacromonomers makes their solution properties very different from those of linear polymers. The polymacromonomers behave like rigid spheres in the low-Mw region. As the branch number increases to a great extent, the chain length ratio of the backbone to the branch becomes large. Thus, the molecular shape changes gradually from that of the spherical star-branched polymer to the anisotropic comb-branched polymer, in which the central backbone chain has large stiffness owing to the axisymmetric tethered polystyrene chains around it.[19,20]

The bulk properties of polymacromonomers are also different from those of the linear homologs. The glass transition temperature (T_g) is an example.[17] T_g of polymacromonomers decreases as the branch length decreases but remains almost constant against the change in the branch number.

Graft Copolymers

The well-defined structure of the graft copolymers and wide choices in both macromonomers and comonomers in the "macromonomer method" make it possible to design the molecular structure of the graft copolymer suitable for each application in the area of amphiphilic polymers and conventional block and graft copolymers. Actually, the graft copolymers are being used as a polymeric surface modifier, compatibilizer, colloidal stabilizer, adsorbents for chromatographic separation, soap-free microspheres, thermoplastic elastomers or gels utilizing their amphiphilic, surface active properties, and the microphase-separated structure.

Network Polymers

Telechelic macromonomers are used as polymeric crosslinkers. Copolymerizations of the telechelic macromonomers with the conventional small comonomer produce various kinds of multi-component polymer networks or amphiphilic copolymer gels. This provides an alternative method of the interpenetrating networks (IPNs) for the preparation of the multi-component networks. Water-swellable amphiphilic networks are used for drug slow-release materials.[7] Amphiphilic networks are also useful for other pharmaceutical, detergent, cosmetic, coating, adhesive, and agricultural fields.

REFERENCES

1. Tsukahara, Y. In *"Macromolecular Design: Concept and Practice"*, ed. by Mishra, M. K. Polymer Frontiers International: New York 1994; Chapter 5; *J. Soc. Rubber Industry Japan*, **1992**, *66*, 635.
2. Milkovich, R. J.; Chiang, M. T. U.S. Patents 3 786 116, 3 842 050, 3 842 057 3 842 058, 3 842 059, 1974; Milkovich, R. J. *ACS Polym. Prepr.* **1980**, *21*, 40.
3. Yamashita, Y.; Tsukahara, Y. In "Modification of Polymers," Carraher, C.; Moore, J., Eds.; Plenum: New York, 1983; p 131.
4. Yamashita, Y., Ed., *"Chemistry and Industry of Macromonomers;"* IPC: Tokyo, 1989; Huthig & Wepf Verlag: Basel, 1993.
5. Rempp, P.; Franta, E. *Adv. Polym. Sci.* **1984**, *58*, 1.
6. Asami, R.; Kondo, Y.; Takaki, M. In *"Recent Advances in Anionic Polymerization;"* Hogen-Esch, T. E.; Smid, J., Eds.; Elsevier: Amsterdam 1986; p 381.
7. Kennedy, J. P.; Ivan, B. *"Carbocationic Macromolecular Engineering: Theory and Practice;"* Hanser: Munich, 1991.
8. De Clercq, R.; Goethals, E. J. *Macromolecules* **1992**, *25*, 1109.
9. Mishra, M. K. Ed., *"Macromolecular Design: Concept and Practice;"* Polymer Frontiers International: New York, 1994.
10. Chujo, Y.; Yamashita, Y. In *"Telechelic Polymers: Synthesis and Applications;"* Goethals, E. J., Ed.; CRC: Boca Raton, 1989, Chapter 8.
11. Meijs, G. F.; Rizzardo, E. *J. Macromol. Sci., Chem. Phys.* **1990**, *C30*, 305.
12. Kobayashi, S.; Uyama, H., Chapter 1 of Reference 10.
13. Ito, K. *J. Adhesion Soc. Japan* **1988**, *24*, 374.
14. Tsukahara, Y.; Mizuno, K.; Segawa, A.; Yamashita, Y. *Macromolecules* **1989**, *22*, 2869.
15. Tsukahara, Y.; Tsutsumi, K.; Yamashita, Y.; Shimada, S. *Macromolecules* **1990**, *23*, 5201.
16. Ito, K.; Tanaka, K.; Tanaka, H.; Imai, G.; Kawaguchi, S.; Itsuno, S. *Macromolecules* **1991**, *24*, 2348.
17. Tsukahara, Y.; Tsutsumi, K.; Okamoto, Y. *Makromol. Chem., Rapid Commun.* **1992**, *13*, 409.
18. Tsukahara, Y.; Kohiya, S.; Tsutsumi, K.; Okamoto, Y. *Macromolecules* **1994**, *27*, 1662.
19. Wintermantel, M.; Schmidt, M.; Tsukahara, Y.; Kajiwara, K.; Kohiya, S.; *Macromol. Rapid Commun.* **1994**, *15*, 279.
20. Halperin, A.; Tirrell, M.; Lodge, T. P. *Adv. Polym. Sci.* **1991**, *100*, 31.
21. Tsukahara, Y. *J. Macromol. Sci., Pure and Appl. Chem., Macromolecular Report* **1995**, *A32*(Suppl.5&6), 821.

MACROMONOMERS (Soap-Free Emulsion and Dispersion Polymerization)

Koichi Ito
Department of Materials Science
Toyohashi University of Technology

Among a variety of methods for preparing polymeric microspheres, the macromonomer technique is unique in that the macromonomers themselves act as reactive emulsifiers or dispersants to be covalently bound, i.e., graft copolymerized, on the surface, so that the resulting soap-free emulsions or dispersions are very effectively sterically stabilized against flocculation.[1–5] Furthermore, the microspheres obtained are usually more or less monodispersed around the size of submicron (emulsion) and micron (dispersion) orders.

The first example of the presently well-known "macromonomer technique" was indeed the nonaqueous dispersion (NAD) system developed by ICI group.[6,7] The counterpart of NAD is the emulsion system in water or the dispersion system in an alcoholic medium using a hydrophilic macromonomer.

The mechanism of particle formation and growth is of great importance for particle design but appears to have been established only in selected cases. We have recently proposed a simple mechanism for a dispersion system that could successfully account for the particle size, and it will be described here.[12–14]

DISPERSION SYSTEM

This system starts with a clear solution including monomer, macromonomer, and initiator. Almost immediately upon polymerization and copolymerization, insoluble polymers emerge from the solution to give a dispersion. It has been generally found that, after a very early stage, the particle size increases with conversion while the total particle number remains constant, i.e., the average volume of each particle increases in proportion to the conversion.[8,9,12–14]

In the present model, macromonomer as a reactive dispersant copolymerizes with the substrate monomer, which is present in a very large molar excess, to produce graft copolymers, the trunk chains of which serve as the anchor into the insoluble polymer core of the same constitution, while the graft chains, or the macromonomer segments, extend into the continuous medium to function as the steric stabilizer.

In conclusion, the simple model for sterically stabilized particles appears to be fairly satisfactory to explain the dispersion macromonomers. The final particle size is thus mainly controlled by the polymerizability and copolymerizability of the substrate monomer and the macromonomer together with the conformational expansion of the macromonomer chain in the medium.

EMULSION SYSTEM

In this aqueous system, water-soluble macromonomers are used to emulsify the insoluble (immiscible) substrate monomer together with a radical initiator that is also usually water-soluble. Therefore, polymerization proceeds without significant change in appearance to give final polymeric emulsions of submicron order in general. The compositions of the isolated polymers have often been found to be very close to those in feed, indicating the azeotropic conversion of the substrate monomer and the macromonomer.[10,11,15] It appears therefore that polymerization and copolymerization proceed within and on the surface of the emulsified droplets, which includes the water-insoluble substrate monomer and the water-soluble macromonomer in nearly the same composition as in the feed.

Apart from detailed mechanism of the emulsion formation, nucleation, and growth of the macromonomer copolymerization system, the resulting emulsions should be of the essentially same structure as the core-shell type as discussed in the dispersion system, and they should be sterically stabilized against coagulation by the soluble macromonomer chains copolymerized.

In conclusion, the application of the macromonomer technique to the dispersion and emulsion systems appears to be promising in view of a variety of potential uses including coatings, medicals, polymeric catalysts, etc. Also important, this technique generally provides simple, well-defined, polymeric microspheres, which will be a definite advantage in controlling the microsphere production and in elucidating the mechanism of particle formation and growth.

REFERENCES

1. Napper, D. H.; Gilbert, R. G. *Compreh. Polym. Sci.* **1989**, *4*, 171.
2. Candau, F. *Compreh. Polym. Sci.* **1989**, *4*, 225.
3. Dawkins, J. V. *Compreh, Polym. Sci.* **1989**, *4*, 231.
4. Walbridge, D. J. *Compreh. Polym. Sci.* **1989**, *4*, 243.
5. Kobayashi, S.; Uyama, H. *Kobunshi Ronbunshu (Jpn. J. Polym. Sci. Techn.)* **1993**, *50*, 209.
6. Barrett, K. E. J. *Dispersion Polymerization in Organic Media*; Wiley-Interscience: London, 1975.
7. Waite, F. A. *J. Oil Col. Chem. Assoc.* **1971**, *54*, 342.
8. Kobayashi, S.; Uyama, H.; Choi, J. H.; Matsumoto, Y. *Proc. Jpn. Acad. Ser. B* **1991**, *67*, 140.
9. Kobayashi, S.; Uyama, H.; Lee, S. W.; Yamamoto, Y. *J. Polym. Sci.: Part A: Polym. Chem.* **1993**, *31*, 3133.
10. Kobayashi, S.; Uyama, H.; Yamamoto, I. *Makromol. Chem.* **1990**, *191*, 3115.
11. Uyama, H.; Honda, Y.; Kobayashi, S. *J. Polym. Sci.: Part A: Polym. Chem.* **1993**, *31*, 123.
12. Nugroho, M.; Chao, D.; Kawaguchi, S.; Ito, K. *Polym. Prepr., Jpn.* **1993**, *42*, 2220.
13. Nugroho, M.; Kawaguchi, S.; Ito, K. *Macromolecular Reports* **1995**, *A32*, 593.
14. Kawaguchi, S.; Winnik, M. A.; Ito, K. Gordon Conf. Polymer Coll., June–July 1993, Tilton, NH; *Macromolecules* **1995**, *28*, 1159.
15. Chao, D.; Ito, K. 34th IUPAC Symp. Macromol., Prague, 1992, Prepr. 1P-19.

MACROMONOMERS (Synthesis, Polymerization, and Utilization)

Yves Gnanou
LCPO, ENSCPB, CNRS
Université Bordeaux I

The quest for new multiphase systems and also the lack of methods of synthesis have both contributed to the present surge of research in the field of macromonomers. Described some fifteen years ago by Milkovich the macromonomer technique has rapidly emerged as one of the most powerful methods of synthesis of two phase systems.[1,2] Fully aware of the prospects opened by the macromonomer method, Milkovich coined the terms "Macromonomer" and "Macromer" (abbreviation of macromolecular monomer) and registered the latter acronym as a trademark.

MACROMONOMERS VIA ANIONIC POLYMERIZATION

The success of the macromonomer technique as one of the most versatile methods of macromolecular engineering owes much to anionic polymerization. Although sporadic attempts at fitting polymer chains with unsaturations have been made as early as 1960, it appears that the domain of macromonomers truly upped and developed only after Milkovich reported his anionically-based strategy of synthesis of macromonomers. He indeed end-capped living polystyrene carbanions with ethylene oxide and acryloyl chloride to derive PS macromonomers of unambiguous functionality.[2]

Since that pioneering work, almost all categories of monómers polymerizing anionically have been subjected to macromonomer synthesis.[3–6] The deactivation of living anionic sites by an unsaturated electrophile turns out to be the privileged method of synthesis. The selection of the appropriate electrophilic unsaturated reactant should be made with care so as to avoid reactions between the unsaturation introduced and the "living" sites to be deactivated. Four types of electrophilic function, which are alkyl halides, benzyl halides, chlorosilanes, and acyl chlorides, borne by either styryl, methacryloyl, or alkyl unsaturations have been used for this purpose. In the category of monomers polymerizing anionically by ring-opening, oxirane, hexamethylcyclotrisiloxane, and ε-caprolactone are the three cyclic monomers that have been subjected to macromonomer synthesis. Advantage has been taken of the "living" character of each polymerization to endow the growing chains with various unsaturations.

In some cases, macromonomers have been obtained from an unsaturated anionic initiator. The difficulty to find compounds endowed with both polymerizable group and a function able to trigger an anionic process is the main reason for the small number of macromonomers prepared by this method. Monomers exhibiting a high electroaffinity such as oxirane, lactones, siloxane, and lactam rings are well suited for the approach involving an unsaturated initiator.

MACROMONOMERS BY CATIONIC POLYMERIZATION

In contrast to anionic polymerization, examples of monomers polymerizing cationically and under "living" conditions were restricted, until recently, to a smaller number.

In the category of heterocycles, there are only three monomers: THF, *N*-*t*-butylaziridine, and alkyloxazoline, which are known to give rise to a "living" or a pseudo-living process. The sole well-defined macromonomers, obtained via cationic processes in this category were thus from these three monomers.[3–6] Both initiating and deactivation methods have been used for the synthesis of macromonomers of polyTHF and polyalkyloxazolines. As to macromonomers of poly(*t*-butylaziridine)s, they have been prepared by deactivation of living *t*-butylaziridine with methacrylic acid.

The development of macromonomers from vinyl monomers and via cationic polymerization has been thwarted for a long time, mainly because of the lack of "livingness" of these cationic processes. The only example of macromonomers from this kind of monomers is drawn from Kennedy's work, who made use of his "inifer" method to synthesize ω-styryl-polyisobutylene macromonomers.

Macromonomers of poly(vinyl ether)s have been generated by deactivation of living sites, using an unsaturated alcohol such as 2-hydroxyethyl methacrylate, and also by the unsaturated initiator pathway.

MACROMONOMERS BY GROUP TRANSFER POLYMERIZATION (GTP)

The group transfer polymerization of methacrylic monomers is now acknowledged by many as a mere anionic polymerization in which most of the active species are in the dormant form. Although GTP requires both specific catalysts, the end-capping agent that was used to derive macromonomers of PMMA was exactly the same as that employed in anionic polymerization, that is, vinyl benzylbromide.[7] Macromonomers of PMMA have been also prepared upon using a silyl ketene acetal derivative carrying a styrenic moiety.[7]

MACROMONOMERS BY RADICAL POLYMERIZATION

Because many vinyl monomers do not lend themselves to a "living" polymerization, there is no other way of preparing macromonomers but to resort to a free radical process. Two steps are generally required to gain access to macromonomers via radical polymerization. The first step is the synthesis of ω-functional polymers, obtained through the use of an appropriate transfer agent or from a functional initiator. The second step consists of the introduction of the unsaturation upon using the terminal function. This can be achieved by reaction with an unsaturated compound carrying an antagonist function.

The procedure described above has been applied to the synthesis of several macromonomers from monomers such as vinylpyrrolidone, acrylamide, vinyl chloride, and avinylidene fluorides.[3–6] Even monomers amenable to a "living" polymerization such as MMA or styrene have been subjected to the above procedure because the ionic "living" processes involve very stringent conditions.

MACROMONOMERS FROM ω-FUNCTIONAL POLYMERS

One of the easiest ways to gain access to macromonomers is to take advantage of the presence of reactive functions in the ω-position to carry out coupling reactions and introduce the chosen unsaturation. Examples of this pathway, which are based on the chemical modification of ω-functional polymers into ω-unsaturated ones, are numerous.[6] Gnanou et al. started from mere commercial ω-OH-PEO to end-cap the latter polymer with either styryl or methacryloyl unsaturations.[8]

Macromonomers have been derived from lignin, various proteins, polypeptides, ω-carboxypolystyrene, ω-hydroxy-polymethacrylates, epoxy resins, poly(alkylene oxide) etc., upon treating the latter polymers with reagents such as methacryloyl chloride and glycidyl methacrylate.

MACROMONOMERS BY STEP-GROWTH PROCESSES

Polycondensates that result from AA + BB type step-growth processes are fitted with antagonist functions at each of their extremities when the reaction is conducted in stoichiometric conditions. Provided one of the initial reactants is a divinylic compound, ω-unsaturated polycondensates are formed. This strategy has been applied in several instances to produce macromonomers.

REACTIVITY OF MACROMONOMERS

The macromonomer method offers undisputed advantages over other techniques of macromolecular engineering for the synthesis of well-defined, branched architectures. Graft copolymers of infinite variety are now accessible, just because the choice of macromonomers is endless. The field of highly compact polymers is also concerned with the macromonomer technique. Homopolymerization of macromonomers can indeed afford branched polymers of well-defined structure.

HOMOPOLYMERIZATION OF MACROMONOMERS

Although polymers with particularly high compacity have been recently obtained by polymerization of macromonomers, this method of preparation of highly compact polymers is generally deemed less attractive than other synthetic pathways.

It is only recently that Tsukahara et al. reported that polymacromonomers with high degrees of polymerization could be obtained by mere increase of the overall concentration of macromonomer in the reaction medium.[10] The authors of this study argued that high concentrations induce slower termination reactions, which allows the radical to grow further and build longer chains.

Ito showed, for instance, that the polymerization of PEO macromonomers is considerably faster in water than in benzene, because in the former medium macromonomers tend to organize in micelles with the hydrophobic polymerizable groups entrapped inside.[10a]

COPOLYMERIZATION OF MACROMONOMERS

Determination of the Macromonomer Reactivity Ratio

The composition of the graft copolymer that results from the copolymerization of a macromonomer with a comonomer is seldom that expected because the behavior of a macromonomer in a given reaction medium can hardly be anticipated.

When dealing with macromonomers, the classical method employed for determining the reactivity ratio is not workable because the experiments with a high concentration of macromonomer cannot be carried out in practice. However, the copolymerization involving macromonomers exactly fulfills the condition defined by Jaacks in his single experiment method.[11]

FACTORS INFLUENCING THE COPOLYMERIZATION OF MACROMONOMERS

The majority of studies that have addressed the question of the macromonomer behavior during copolymerization reaction brought its lack of reactivity to light and revealed a dependence of the reactivity on the molar mass.[6,9,12]

APPLICATIONS DERIVED FROM MACROMONOMER-BASED GRAFT COPOLYMERS

Several macromonomers are now industrially produced and marketed for use as precursors to graft copolymers.[14,15] The number of applications that resort to the macromonomer technique keeps increasing and ever diversifying.

Among the fields that have much benefited from the emergence of macromonomers, that concerning the surface modification of polymers is particularly noteworthy as shown by several recent technological breakthroughs.[16,17] The advantage provided by copolymers with respect to the surface accumulation has been extensively exploited; fluorine- or silicone-containing graft copolymers derived from the copolymerization of appropriate macromonomers are currently added to paints and coatings to improve the water repellent or lubricating properties of their surfaces. Other applications based on the surface accumulation phenomenon include adhesives with polar groups on their surface and materials with antistatic surface.[13]

The domain of dispersion stabilizers also provides several examples of applications that have been made possible thanks to the macromonomer method.[20]

The surface activity of graft copolymers can also serve to compatibilize two polymers that would undergo macrophase separation if they were blended.[15]

In contrast to graft copolymers, the only known example of polymacromonomers that have reached the level of industrial production and utilization were produced by the DuPont company.[21] The branched polymacromonomers that result from the homopolymerization of PMMA macromonomers are indeed used in paints to reduce their viscosity and improve their processability.

To fully exploit the potentiality of the macromonomer technique and to establish it as one of the best methods of engineering of biphasic systems, two series of problems need to be first solved. Well-defined macromonomers are still costly, despite the progress made in their synthesis. Macromonomers are, for instance, not accessible by emulsion polymerization. Also, finding the experimental conditions that are favorable to the incorporation of the macromonomer in the copolymers is cumbersome.

Once these drawbacks are overcome, one might then envisage the application of the macromonomer technique in further fields of industry.

REFERENCES

1. Milkovich, R. U.S. Patent 3 786 166, 1974.
2. Schulz, G.; Milkovich, R. *J. Appl. Polym. Sci.* **1992**, *27*, 4773; *J. Polym. Sci., Chem. Ed.* **1984**, *22*, 3795; **1984**, *22*, 1633.
3. Rempp, P.; Franta, E. *Adv. Polym. Sci.* **1984**, *1*, 58.
4. Percec, V.; Pugh, C.; Nuyken, O.; Pask, S. *Comprehensive Polymer Science*; Allen, G.; Bevington, J. C. Eds.; Pergamon: New York, 1989; Vol. 6, 281.
5. Chujo, Y. *Chemistry and Industry of Macromonomers*; Hüthig & Wepf: Heidelberg, 1993; Chapter II.
6. Gnanou, Y. *Ind. J. Technol.* **1993**, *31*, 317.
7. Asami, R.; Kondo, Y.; Takaki, M. *Polym. Prepr., Am. Chem. Soc., Div. Polym. Chem.* **1986**, *27*(1), 186.
8. Gnanou, Y.; Rempp, P. *Makromol. Chem.* **1987**, *188*, 2111.
9. Tsukahara, Y. "Chemistry and Industry of Macromonomers," Hüthig & Wepf: Heidelberg, 1993; Chapter IV.
10. Tsukahara, Y.; Tsutsumi, K.; Yamashita, Y.; Shimada, S. *Macromolecules* **1990**, *23*, 5201.
10a. Ito, K.; Kobayaski, H. *Polym. J.* **1992**, *24*(2), 199.
11. Jaacks, V. *Makrom. Chem.* **1972**, *161*, 161.
12. Meijs, G.; Rizzardo, E. *J. Macromol. Sci., Rev. Macromol. Chem. Phys.* **1990**, *C30*, 305.
13. Tsukahara, Y.; Tanaka, M.; Yamashita, Y. *Polym. J.* **1987**, *19*, 1121.
14. Yamashita, Y.; Kato, H.; Kojima, S. "Chemistry and Industry of Macromonomers"; Yamashita, Y. Ed.; Hüthig & Wepf: Heidelberg, 1993; Chapter IX.
15. Tsuda, T.; Nakanishi, K. "Chemistry and Industry of Macromonomers"; Yamashita, Y. Ed.; Hüthig & Wepf: Heidelberg, 1993; Chapter X.
16. Nakazaki, T.; Chujo, Y.; Tsukahara, Y.; Yamashita, Y.; Shiga, T. *Polym. Prepr. Jpn.* **1984**, *33*, 2103.
17. Tsukahara, Y.; Tsuruta, Y.; Kohno, K.; Yamashita, Y. *Kobunshi Ronbunshu* **1990**, *47*, 361.
18. Yamashita, Y. *Polym. Bull.* **1982**, *7*, 289.
19. Jpn. Kokai Tokkyo Koho JP58/154766, Toagosei Chemical Industry.
20. Yamashita, Y. *Dyestuff and Chemicals* **1984**, *31*(3), 59.
21. Simms, J. U.S. Patent 3 716 506, Du Pont de Nemours.

MACROMONOMERS (Through Isomerization Polymerization)

Yukio Nagasaki
Department of Materials Science and Technology
Science University of Tokyo

The polymerization systems involving isomerization reactions are continuously providing novel polymers with new repeating units and new end groups by exploring new monomers and new polymerization systems by means of new analytical methods.

Lithium alkylamide is known as a useful agent not only for the metalation reaction of active hydrogen compounds, but also for the anionic additions and polymerization reactions of conjugated olefins such as dienes and styrenes.[1–4] The reactivity of lithium alkylamide is strongly affected by its alkyl substituents. By controlling the reactivity of the alkylamide, we have found a new metalation reaction of 4-methylstyrene derivatives without any side reactions toward the vinyl groups of the compounds.

Lithium diisopropylamide (LDA) was found to induce the metalation reaction of 4-methylstyrene (MST) to form metalated MST, 4-vinylbenzyllithium (VBL), without any side reactions such as LDA-initiated polymerizations.[5] Using VBL as the initiator, we found a new isomerization polymerization system of 4-methylstyrene derivatives.

ISOMERIZATION POLYMERIZATION OF 4-(TRIMETHYLSILYLMETHYL)STYRENE (SMS)[6,7]

To make the monomer SMS undergo an anionic polymerization in the equilibrium system containing LDA and SMSLi, the reaction was carried out using a relatively high concentration of SMS. The idea was to synthesize novel SMS-macro-monomers initiated with SMSLi in the equilibrated state. The obtained oligomers have a UV absorption at 296 nm, which suggests that SMSLi has induced the polymerization reaction of SMS.

Another key consideration in the anionic polymerization of SMS is the mode of enchainment of the repeating units. Two possible modes may be envisioned. The first one is the classical mode in which SMSLi initiated the vinyl polymerization of SMS to form polyvinyl type macromonomers.

The second mode is a novel one in which a one-to-one addition takes place between SMS and SMSLi followed by a transmetalation reaction by DPA. Upon repeating this addition cycle, a polyaddition type of macromonomer was obtained. We defined the polymerization modes including the polyaddition reactions as "LDA Induced Anionic Isomerization Polymerization of 4-Methylstyrenes."

We concluded that the polymerizations of SMS in the presence of LDA proceeded as follows: the addition of SMSLi to the double bond of SMS proceeded predominantly during the initial stage. Rapid transmetalations by DPA then took place, especially at increased temperature and/or higher concentrations of DPA, to form low molecular weight (LMW) oligomers (mostly dimer and trimer). In the high monomer conversion stage, preformed LMW oligomers themselves were metalated by LDA, then the self-addition reactions between the oligomers took place.

LDA induced isomerization polymerization of 4-methylstyrenes derivatives provided a new polymerization route for macromonomers with certain functional groups with predictable extent. The generalized polymerization scheme is summarized in **Scheme I**.

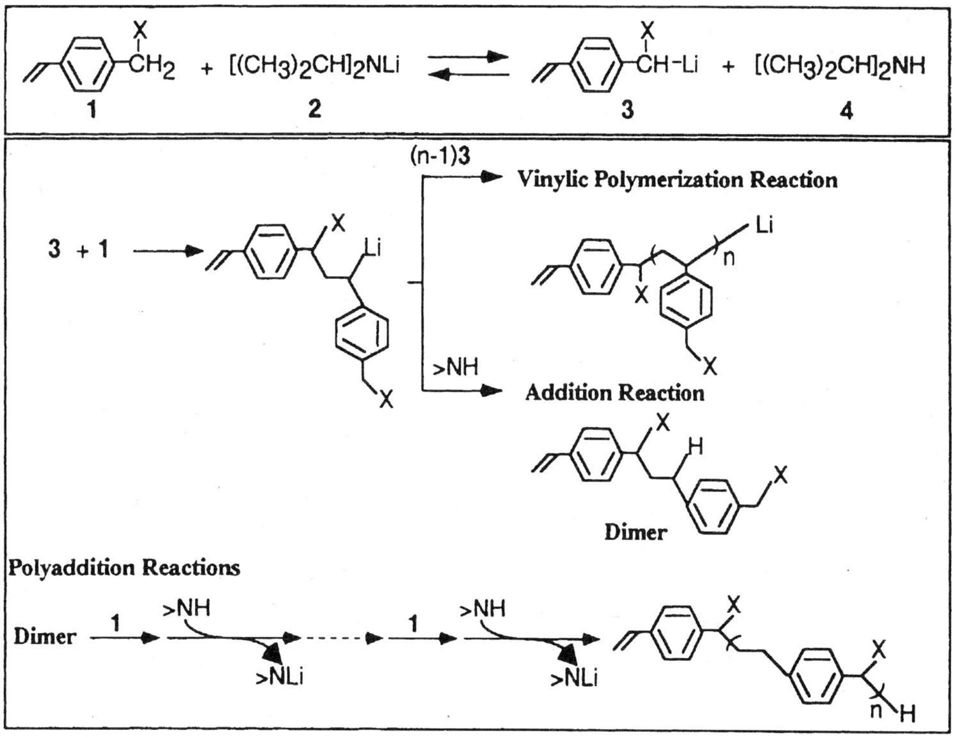

SCHEME I. LDA-Induced isomerization polymerizations of 4-methylstyrenes.

REFERENCES

1. Wakefield, B. J. *Organolithium Methods*; Academic: London, 1988.
2. Tsuruta, T. In *Polymeric Amines and Ammonium Salts*; Goethals, E. J., Ed.; Pergamon: Oxford, 1979; pp 163–172.
3. Cheng, T. C. In *Anionic Polymerization, Kinetics, Mechanism, and Synthesis*; McGrath, J. E., Ed.; American Chemical Society: Washington, DC, 1981; Vol. 166, pp 513–528.
4. Angood, A. C.; Hurley, S. A.; Teit, P. J. T. *J. Polym. Sci., Polym. Chem. Ed.* **1973**, *11*, 2777.
5. Nagasaki, Y.; Tsuruta, T. *Makromol. Chem.* **1986**, *187*, 1583.
6. Nagasaki, Y.; Tsuruta, T. *Makromol. Chem., Rapid Commun.* **1989**, *10*, 403.
7. Nagasaki, Y.; Tsuruta, T. *Macromol. Sci. Phys.* **1995**, *196*, 513.

Macronets

See: Crosslinked Networks, Macronets

MACROPOROUS POLYMERS (Polymeric Porogens to Control Pore Size)

Daniel Horák* and Milan J. Beněs
Institute of Macromolecular Chemistry
Academy of Sciences of The Czech Republic

When a monovinylic monomer (e.g., styrene–ST) is copolymerized with a multivinylic monomer (crosslinking agent, e.g., divinylbenzene—DVB) by suspension polymerization in the presence of an inert diluent (porogen) soluble in both comonomers, resin beads are produced that have permanent porosity in the dry state and are called macroporous (macroreticular). This contrasts with the suspension polymerization in the absence of a diluent when the beads have porosity only in the swollen state (gel porosity) and no permanent porosity in the dry state. The diluents used can be either solvating (e.g., toluene or dichloroethane for ST-DVB resins) or nonsolvating (e.g., aliphatic hydrocarbons or organic alcohols for ST-DVB resins), and either low molecular weight or macromolecular compounds. They control pore size, pore size distribution and total pore volume.[1]

While the low molecular weight diluents used as porogens are common and have frequently been described in the literature,[2–5] few studies have been devoted to polymeric porogens.[1,6,7] In addition, most of the papers describe copolymers of ST with DVB.[1,6–8] This chapter reports on macroporous matrices in which polymers or their solutions in solvating or nonsolvating diluents are used as porogens.

FORMATION OF MACROPOROUS STRUCTURE–MECHANISM

The formation of porous structure by phase separation during polymerization carried out in the presence of diluents has been well described in several reports.[1,6,9,10] The same mechanism is effective for polymerization in the presence of a polymeric porogen of sufficiently high molecular weight and concentration in the polymerizing mixture.

In the early stage, the polymeric porogen is soluble in the reaction medium because the monomers (possibly also a low-molecular-weight diluent) act as a solvent. As the monomers are gradually converted into copolymers, the newly formed crosslinked molecules (nuclei) become insoluble. The monomer solution of a polymeric porogen (and a low-molecular-weight diluent, if added) acts as a monomer reservoir for the growing nuclei. The phase containing diluents strongly decreases in volume by the loss of solvating monomer and, at a certain stage of network formation, phase separation occurs between the copolymer and polymeric porogen. In porogenic domains, depending on the molecular weight of the polymeric porogen, entanglement of polymeric porogen with newly formed chains of the network takes place. This explains an extreme sensitivity of the pore size distribution to the molecular weight of the polymeric porogen. The entanglement phenomenon also may explain the dependence of the pore size distribution on the concentration of polymeric porogen.[9]

MORPHOLOGY OF MACROPOROUS RESINS

The morphology (texture) of macroporous resins is complex. They have two levels of bead substructure, with the bead consisting of randomly packed microspheres that are packed with nuclei (elementary particles).[3,9] The pores are voids between these particles. The voids are irregularly shaped. Since there is more than one family of particles, more than one family of pores (micropores, mesopores, macropores) are formed in some instances.

PROPERTIES OF MACROPOROUS RESINS PREPARED WITH NEAT POLYMER AS A POROGEN

Macroporous resins with polymeric porogens can be prepared only under specific conditions determined by the type of polymeric porogen, its amount and molecular weight, and by the degree of crosslinking (percent of crosslinking agent in the monomeric mixture).[11] Polymeric porogens appear to be more effective than conventional solvating or nonsolvating low-molecular-weight diluents at low concentrations of both the crosslinking agent and porogen.[6] This is caused by the high molar volume of polymeric porogens. Increase of molar volume of polymeric porogens shifts the beginning of phase separation toward lower conversion of monomers during polymerization.

Polystyrene is the most studied polymeric porogen in preparation of widely used ST-DVB copolymers.[10,12–15] It is a chemically similar agent to the chains of the three-dimensional network (its interaction parameter is close to zero).

TYPES OF POLYMERIC POROGEN

Addition of polymers other than polystyrene to the polymerizing mixture of ST and DVB has been studied only to a limited extent. Nevertheless, the effect of the type of polymeric porogen on pore size is great. The greater the difference in the chemical composition of the polymeric porogen and the polymeric product prepared (incompatibility), the sooner the two polymeric phases in the polymerizing mixture separate, the larger are the pores formed. High porosities can thus be achieved even with low concentration of highly incompatible polymeric porogen.

*Author to whom correspondence should be addressed.

The following polymers were used to prepare macroporous ST-DVB resins: poly(vinyl acetate),[10,16,17] poly(methyl methacrylate),[3,10,17] poly(methyl vinyl ether), or an ethylene oxide-propylene oxide copolymer,[18] copolymer of maleic acid ester with vinyl acetate or acrylonitrile.[16]

PROPERTIES OF MACROPOROUS RESINS PREPARED WITH POLYMER AND A NONSOLVATING DILUENT AS POROGENS

Another combination of porogens used in preparation of macroporous resins consists of a polymer and nonsolvating diluent. Such resins differ from those prepared in the presence of a polymeric porogen–solvating diluent mixture. Whereas increase in the amount of solvating diluent increases the pore volume within certain limits without significantly changing the pore size distribution, increase in concentration of the nonsolvating diluent increases both the pore volume and mean pore size.

ADVANTAGES AND DISADVANTAGES OF POLYMERIC POROGENS

An integral part of the production of macroporous resins is removal of porogens by solvent extraction after polymerization. In the ST-DVB resins, extraction is carried out with a solvent, such as dichloroethane or toluene, which strongly interacts with the ST-DVB copolymer chains.[1] In contrast to the low-molecular-weight porogens, polymeric porogens are difficult to remove from the final product.

Complicated removal of polymeric porogens is the major obstacle to their widespread application in preparation of macroporous resins. However, the advantage of macroporous resins prepared with addition of polymeric porogens is that the obtained pore size distributions are relatively narrow. Moreover, it is possible to obtain permanent porosity with a relatively low concentration of polymeric porogen. By combining the variable concentration and molecular weight of the polymeric porogen and, if needed, the concentration and type of low-molecular-weight porogen, it is possible to prepare copolymers with predetermined properties.

APPLICATIONS

Macroporous resins have been widely used as polymeric adsorbents, ion exchangers, supports for catalysts and reagents, and packings of chromatography columns. However, polymer-supported reagents and catalysts for organic synthesis based on conventional macroporous or gel matrices often lack high activity.[8] To overcome the limited accessibility of active sites in supported reagents and catalysts porous polymer particles with large pores are required.[3,7,19] Such structures can be achieved with polymeric porogens.[9]

REFERENCES

1. Sederel, W. L.; de Jong, G. J. *J. Appl. Polym. Sci.* **1973**, *17*, 2835.
2. Millar, J. R.; Smith, D. G.; Marr, W. E.; Kressman, T. R. E. *J. Chem. Soc.* **1963**, 218.
3. Guyot, A.; Bartholin, M. *Progr. Polym. Sci.* **1982**, *8*, 277.
4. Švec, F.; Hradil, J.; Čoupek, J.; Kálal, J. *Angew. Makromol. Chem.* **1975**, *48*, 135.
5. Horák, D.; Švec, F.; Ilavský, M. et al. *Angew. Makromol. Chem.* **1981**, *95*, 117.
6. Seidl, J.; Malinský, J.; Dušek, K.; Heitz, W. *Adv. Polym. Sci.* **1967**, *5*, 113.
7. Bacquet, M.; Lemaguer, D.; Caze, C.; *Eur. Polym. J.* **1988**, *24*, 533.
8. Revillion, A.; Guyot, A.; Yuan, Q.; da Prato, P. *React. Polym., Ion Exch., Sorbents* **1989**, *10*, 11.
9. Guyot, A. *Synthesis and Separations Using Functional Polymers*, Wiley: Chichester, 1988; Chapter I.
10. Seidl, J. *Dissertation Ion Exchange Skeletons*, Institute of Macromolecular Chemistry, Czechoslovak Academy of Sciences, Prague, Czechoslovakia, 1964.
11. Seidl, J.; Malinský, J.; Rahm, J. *Chem. Listy* **1963**, *58*, 651.
12. Seidl, J.; Malinský, J. *Chem. Prum.* **1963**, *13*, 100.
13. Abrams, I. M. *Ind. Eng. Chem.* **1956**, *48*, 1469.
14. Seidl, J.; Malinský, J.; Krejcar, E. *Chem. Prum.* **1965**, *15*, 414.
15. Dušek, K.; Seidl, J.; Malinský, J. *Chem. Prum.* **1963**, *13*, 662.
16. Resindion, S. *Pat. F.* **1962**, 1,295,537.
17. Tai, H.; Zhou, C. *Hebei Gongxueyuan Xuebao* **1991**, *20*, 95; CA 117: **1992**, *112*, 626.
18. Dale, J. A. *Eur. Pat. Appl.* **1985**, 135,292.
19. Brunelet, T.; Gelbard, G.; Guyot, A. *Polym. Bull.* **1981**, *5*, 145.

MAGNETIC FIELD POLYMERIZATION

Cristofor I. Simionescu, Aurica P. Chiriac, and Mihai V. Chiriac
"P. Poni" Institute of Macromolecular Chemistry

Magnetic effects can be of technological interest because they offer a new way of performing radical processes.[1] This is due to the fact that through very weak perturbations of the magnetic field one can control chemical kinetics and thus the course and the rate of reactions that normally require much higher chemical energies. The fundamental articles of Steiner and co-workers[2,3] specify the main requirement of the magnetically sensitive reactions, namely the existence of intermediates with one unpaired electron spin. The polymerization process that contains such species, and whose presence determines the subsequent evolution of reactions, consequently will be influenced by the existence of an external continuous magnetic field and will exhibit magnetokinetic effects.

The literature covering the field, with an impressive evolution in the last two decades, justifies the magnetic field influence in chemical processes through radical pairs mechanism.[4–19] The registered magnetokinetic effects are attributed to changes in spin multiplicity of the radical pairs under the influence of the field. For polymerization processes, the effects are mainly evidenced during the initiation step,[20–31] and their origins have been well treated in the reviews of Steiner.[2,3] Radical pairs, generated through well-known methods, present a spin correlation according to the multiplicity of their direct precursors. The spin conservation rule in chemical reactions is also succeeded from energetic reasons in subsequent processes of the radical pairs. Thus, singlet radical pairs will recombine only to singlet products, whereas triplet radicals will merely react to triplets, if enough energy is available.

The spin motion, respectively, the singlet–triplet transitions in the radical pairs' states, is magnetic field dependent. As a result of changes in the radical pairs' multiplicity, there is a modification of the ratio of cage radicals to escape radicals. In this way the initiation and also the kinetics of the reactions are affected.

CONDITIONS AND METHODS OF RADICAL POLYMERIZATION IN A MAGNETIC FIELD

Interpretation and justification of the changes in the polymerization processes developed in a magnetic field require the treatment of all elements that can be potentially affected by the presence of the magnetic field. The reaction conditions are the classical ones, but the polymerization ampules are placed between the poles of an electromagnet. Appropriate geometries of poles provide a uniform field for different forms of the vessels.

Initiation is the main step for both the polymerization process and the evaluation of appeared magnetokinetic effects.

Phenomenologically, the basic mechanism of processes developed in a magnetic field may become obvious in gas, liquid, and solid states in micellar or biological systems. However, in liquid state the influence of the magnetic field is more easily detected. In polymerization processes, only by following the conversion can one distinguish between magnetic-field-dependent and -independent reactions. Investigation of photosensitive molecules in magnetic fields evidenced a reproducibility of the kinetic curves in many photoreactions.[32,33]

Monomers with polar, ionizable groups are affected by a magnetic field, which can deform the molecules by increasing distance interactions and by modifying angles between chemical bonds.[34]

The interdependence between a series of methacrylic esters having different magnetic susceptibilities and their bulk polymerization in a magnetic field was evidenced.[35]

The magnetic field effect concretized in reducing reaction's induction period and increasing both syntheses' rates and conversions is directly dependent on the magnetic susceptibility of the methacrylic monomer.

Classic methods of polymerization can be performed in a continuous external magnetic field, function of monomer and polymer properties, respectively, on end uses of the synthetized product. Bulk and solution polymerization (in polar–$CHCl_3$ and nonpolar–C_6H_6 solvents) as well as emulsion polymerization were studied.[36] The most striking magnetokinetical modifications are registered in emulsion polymerization reactions.

Generally, the reaction conditions affect the development of polymerization processes and also the evolution of the magnetic-field-dependent reactions.

MAGNETIC FIELD INFLUENCE ON POLYMER PROPERTIES

In addition to the modifications brought to the evolution of the polymerization process, the magnetic field influences the properties of the resulting polymers. This acts as a supplementary argument justifying interest in the field.

The observed changes in the properties of polymers are attributed to the catalytic effect of the field on the molecules, which can be deformed through growing distance interactions and modification of angles between bonds.[37]

The macromolecular compounds obtained in a magnetic field present higher molecular weights as compared to their homologues synthesized in the absence of the field.[12,13,27,30,35,36,38,39] This is attributed to the action of the magnetic field on the macromolecular chains, termination by recombination being favored to the detriment of disproportionation.

Some studies mentioned the magnetic field action on the modification of the dielectric and mechanical properties of polymers as well as on the conductibility of polymers with included paramagnetic centers.[40–43]

Taking into account the special effects generated by a magnetic field, both on the kinetics of polymerization processes and on the modification of polymer characteristics, the interest in the development of this new technique of polymerization with emphasis on theoretical aspects is fully justified.

REFERENCES

1. Salikhov, K. M.; Molin, Y. N.; Sagdeev, R. Z.; Buchachenko, A. L. *Spin Polarization and Magnetic Effects in Radical Reactions;* Elsevier: Amsterdam, 1984.
2. Steiner, U. E.; Ulrich. T. *Chem. Rev.* **1989**, *89*, 51.
3. Steiner, U. E.; Wolff, H. J. *Photochemistry and Photophysics* CRC: Boca Raton, Boston, 1991.
4. Figueras Roca, F. *Ann. Chim.* **1967**, *2*, 255.
5. Sokolik, I. A.; Frankevich, E. L. *Usp. Fiz. Nauk* **1973**, *111*, 261.
6. Atkins, P. W.; Lambert, T. P. *Annu. Rep. Prog. Chem.* **1975**, *72A*, 67.
7. Buchachenko, A. L. *Russ. Chem. Rev. (Engl. Transl.)* **1976**, 45, 761.
8. Sagdeev. R. Z.; Salikhov, K. M.; Molin, Y. N. *Russ. Chem. Rev.* **1977**, *46*, 569.
9. Geacintov, N. E.; Swenberg, C. E. *Luminescences Spectroscopy*; Academic: London, 1978.
10. Lin, S. H.; Fujimura, Y. *Excited States;* Academic: New York, 1979.
11. Molin, Y. N.; Sagdeev, R. Z.; Salikhov, K. M. *Sov. Sci. Rev. Sect. B, Chem. Rev.* **1979**, *1*, 67.
12. Turro, N. J.; Kräutler, B. *Acc. Chem. Res.* **1980**, *13*, 369.
13. Turro, N. J. *Pure Appl. Chem.* **1981**, *53*, 259.
14. Turro, N. J.; Weed, G. C. *J. Am. Chem. Soc.* **1983**, *105*, 1861.
15. Turro, N. J.; Kräutler, B. *Isotopes Org. Chem.* **1984**, *6*, 107.
16. Hoff, A. J. *Q. Rev. Biophys.* **1981**, *14*, *599*.
17. McLauchlan, K. A. *Sci. Prog.* **1981**, *67*, 509.
18. Nagakura, S.; Hayashi, H. H. *Radiat. Phys. Chem.* **1983**, *21*, 91.
19. Gould, I. R.; Turro, N. J.; Zimmt, M. B. *Adv. Phys. Org. Chem.* **1984**, *20*, *1*.
20. Tanimoto, Y.; Nishino, M.; Itoh, M. *Bull. Chem. Soc. Jpn.* **1985**, *58*, 3365.
21. Molin, Y. L.; Sagdeev, R. Z.; Leshina, T. V. *Magnetic Resonance and Related Phenomena;* Springer: Berlin, 1979.
22. Samorskaya, T. G.; Skrunts, L. K.; Kiprianova, L. A.; Levit, I. P.; Gragerov, I. P. *Dokl. Phys. Chem.* **1985**, *283*, 697.
23. Perito, P.; Corden, B. B. *J. Am. Chem. Soc.* **1987**, *109*, 4418.
24. Turro, N. J.; Chow, M. F.; Rigandy, J. *J. Am. Chem. Soc.* **1979**, *101*, 1302.
25. Turro, N. J.; Chow, M. F. *J. Am. Chem. Soc.* **1980**, *102*, 1190.
26. Turro, N. J.; Chow, M. F.; Rigandy, J. *J. Am. Chem. Soc.* **1981**, *103*, 7218.
27. Imoto, M. *Makromol. Chem., Rapid Commun.* **1981**, *2*, 703.

28. Imoto, M. *J. Macromol. Chem. Sci.* 1987, *A24*, 111.
29. Vebeneev, A. A. *Izv. Akad. Nauk* **1989**, *5*, 1163.
30. Golubkova, N. A. *Doklad. Akad. Nauk* **1988**, *300*, 147.
31. Simionescu, C. I.; Chiriac, A.; Neamtu, I.; Rusan, V. *Makromol. Chem. Rapid Commun.* **1989**, *10*, 601.
32. Jonglie, Chen; Anders, H.; Ranby, B. *Chim. J. Polym. Sci.* **1993**, *11*, 1.
33. Huang Junlian; Hu Jouqian; Song Qinhua *Polymer* **1994**, *35*, 1105.
34. Buchachenko, A. L.; Sagdeev, R. E.; Salikhov, K. M. *Magnetic and Spin Effects in Chemical Reactions;* Nauka: Moscow, 1978.
35. Simionescu, C. I.; Chiriac, A. P.; Chiriac, M. V. *Polymer* **1993**, 18, 3917.
36. Simionescu, C. I.; Chiriac, A. P. *Coll. & Polym. Sci.* **1992**, *270*, 753.
37. Roosevelt, F. *Rev. Quim. Ind.* **1971**, *40*, 474.
38. Hanabusa, K. *Kobunshi Ronbunshu* **1989**, *46*, 269.
39. Klimciuk, E. S. *Izv. Akad. Nauk* **1988**, 2736.
40. Casagrainde, C. *Am. Chem. Soc. Polym. Prep.* **1983**, *24*, 273.
41. Frankevich, E. L. *Himia Visokih Energhii* **1985**, *14*, 283.
42. Pankov, S. P. *Vysokomol. Soedin.* **1982**, *24*, 1701.
43. Sokolik, I. A. *Vysokomol. Soedin.* **1987**, *3*, 203.

MAGNETIC FIELD PROCESSING

Raimond Liepins
Materials Science and Technology Division
Los Alamos National Laboratory

Certain property enhancements may be obtained in polymeric materials if they are processed in a magnetic field. Just as electric fields are used in the processing of materials on an industrial scale (e.g., piezoelectric polymers), so will magnetic fields be used once they have been explored more in these applications.

MAGNETIC FIELDS

Magnetic fields are ubiquitous. In the universe, the geomagnetic field reaches as high as 10^8 tesla (T) in neutron stars.[1] Continuous man-made magnetic fields as high as 30 T are available. Because of severe power and tensile strength limitations of the coil materials at the present state of technology, the continuous fields will be limited to 40–50 T. Pulsed magnetic fields are less restricted by power and materials limitations and thus fields up to ~100 T are possible. Explosive flux compression systems provide fields up to 1000 T for about 10 μs.[2]

MATERIAL RESPONSE TO MAGNETIC FIELDS

All materials respond to a magnetic field. In general, the response may be repulsion (negative magnetic susceptibility) representing diamagnetism, or attraction (positive magnetic susceptibility) representing para- and ferromagnetism. However, the intensity of the response (induced magnetization) in diamagnetic materials is more than 6 orders of magnitude smaller than the magnetization developed in ferromagnetic materials.

Most organic and inorganic polymeric materials are diamagnetic. In less than 1% of presently known compounds does the ferromagnetism mask the universal property of diamagnetism of materials.[3,4] Our discussion of magnetic field processing of polymers (and composites) is limited to diamagnetic polymers.

POLYMERS STUDIED

The following is a list of most of the polymers that have been used in these studies: poly(γ-benzyl) glutamate),[5–7] aromatic polyesters,[8–19] polysiloxanes,[20,21] polyamide,[22] epoxies,[23,24] organometallic polymers,[25,26] poly(1,4-phenylene-2,6-benzobisthiazole),[27] and other liquid crystalline polymers.[28–35] In all of this work, correlations with mechanical properties or processing/fabrication have not been of interest. The availability of this fundamental information, however, is of great value to one undertaking magnetic field processing studies.

SUMMARY

The use of magnetic field in the processing of materials is the most economical and convenient means for imposing a high-energy density on a material. The application of a magnetic field in processing is completely free from any kind of contamination. It is uniform throughout the mass of the material to be processed, and it is dissipationless. By its very nature it acts on the microstructure of the material to be processed.

EARLY MAGNETIC FIELD PROCESSING STUDIES OF POLYMERIC MATERIALS

Except in metallurgy, magnetic field processing of materials has not been widely practiced in the West. However, there is ample literature on polymeric materials processing, predominantly from Russia, Latvia, Uzbekistan, and Japan. Almost all of this work has been reported in journals that are not easily accessible and are primarily available only in Russian. These references are listed in *Polymeric Materials Encyclopedia*, Vol. 6, pp 3973–3974.

CURRENT MAGNETIC FIELD PROCESSING STUDIES OF POLYMERIC MATERIALS

The polymeric materials studied have been:

- hydroxypropyl cellulose
- various polysilanes
- poly(vinylidene fluoride) and its copolymers
- polyphenylquinoxaline
- polyaniline
- poly(*p*-phenylene terephthalamide) (KEVLAR™) with pendant toly-oxy-4′-azobenzene-4″-azobenzene groups (pendant mesogen)

The neat polymeric material processing effort was extended also to some typical composite reinforcing agents: carbon and graphite fibrils and silicon carbide whiskers.[36] The objective was to examine the type and extent of the reinforcing agent response to a low (~1 T) magnetic field.

Two types of commercially available epoxy prepregs, nylon net and Kevlar fiber reinforced, have also been processed in a magnetic field as has an inorganic polymeric matrix material LUDOX AS-40 (a silicate).

PROPERTIES

By using magnetic fields it is possible to orient the backbone, the side chains, or both, in a polymer. The alignment can be very significant in liquid crystalline polymers at low magnetic fields because of the inherent cooperativity of molecules in liquid crystalline phases and the anisotropic susceptibilities of polymer chains in general. The physical properties of polymeric, ceramic, and metal matrix composite materials are also dependent on the arrangement of the reinforcing phases, which are often of a particulate or fibrillar nature. Alignment of the reinforcing phase in general leads to strengthening of the composite. The added benefit of particle alignment is the minimization of the free volume and hence the shrinkage that takes place during drying and sintering of the ceramic.

APPLICATIONS

Property Improvements

It has been claimed that processing epoxies neat or in composites leads to increased mechanical strength, better adhesion, and reduced rate of water absorption. Similar results have also been claimed for phenolics, phenol-formaldehyde, and furfural and urea–formaldehyde resins. Claims for such other polymers as polyethylene and polystyrene have not been confirmed.[37]

Composite Processing

Many of the claimed magnetic field effects in composite processing are complicated by the thermal and extremely high magnetic field gradients also existing in the techniques used. The improvements are there, but their true origin is not clear.

Z-Direction Reinforcement

In the composite area, Z-direction reinforcement by means of magnetic fields is possible. This technology can be used not only for polymer matrix but also for carbon–carbon, ceramic, and metal–matrix composites. Furthermore, magnetic field orienting in the Z-direction lends itself easily to continuous fabrication technologies.

Fiber Spinning

Highly oriented polymer microds (fibers) have been produced by fast cooling of samples in the presence of a magnetic field.[11] The implications of this are clear for the use of magnetic field processing in large-scale fiber spinning.

Ceramic Processing

The preliminary work indicates great potential for magnetic fields in particulate alignment in the sol–gel processes for low-temperature ceramic fabrication.

ACKNOWLEDGMENTS

The author would like to thank Frank D. Gac for encouragement and for providing the facilities for completion of this work; John M. Ziegler for providing the polysilane samples; Wayne L. Bongianni for providing the polyvinylidene fluoride and its copolymer samples; Collin P. Sadler for providing the prepreg and graphite samples; Debra A. Wrobleski for providing the polyaniline samples; and Randy K. Jahn for all engineering help. Most of this work was supported by the Office of Industrial Processes, U.S. Department of Energy.

REFERENCES

1. Muira, N. *Physical Phenomena at High Magnetic Fields;* Manousakis, E.; Schlottmann. P.; Kumar, P.; Bedell, K. S.; Mueller, F. M. Eds.; Addison-Wesley: New York, 1992, p 589.
2. Manousakis, E.; Schlottmann, P.; Kumar, P.; Bedell, K. S.; Mueller, F. M. Eds. *Physical Phenomena at High Magnetic Fields* Addison-Wesley: New York, 1992.
3. Selwood, P. W. *Magnetochemistry*, 2nd ed.; Interscience: New York, 1965.
4. Haberditzl, W. *Magnetochemie*; Akademie: Berlin, 1968.
5. Wilkes, G. L. *Polymer Lett.* **1972**, *10*, 935.
6. Sridhar, C. G.; Hines, W. A.; Samulski, E. T. *J. Chem. Phys.* **1974**, *61*, 947.
7. Kishi, R.; Sisdo, M.; Tazuke, S. *Macromolecules* **1990**, *23*, 3868.
8. Moore, R. C.; Denn, M. M. Lawrence Berkeley Laboratory Report LBL-21921, 1986.
9. Hardouin, F.; Achard, M. F.; Gasparoux, H.; Liebert, L.; Strzelecki, L. *J. Poly. Sci., Polym. Phys. Ed.* **1982**, *20*, 975.
10. Liebert, L.; Strzlecki, L.; VanLuyen, D.; Levelut, A. M. *Eur. Poly J.* **1981**, *17*, 71.
11. Maret, G.; Blumstein, A. *Mol. Cryst. Liq. Cryst.* **1982**, *88*, 295.
12. Blumstein, A.; Vilasagar, S.; Ponrathnam, S.; Clough, S. B.; Blumstein, R. B. *J. Polym. Sci., Polymer Phys. Ed.* **1982**, *20*, 877.
13. Anwar, A.; Windle, A. H. *Polymer* **1993**, *34*, 3347.
14. Sigaud, G.; Yoon, D. Y.; Griffin, A. C. *Macromolecules* **1983**, *16*, 875.
15. Stuppand, S. I.; Moore, J. S. *Polym. Mater. Sci. Eng.* **1986**, *54*, 136.
16. Moore, J. S.; Stupp, S. I. *Macromolecules* **1987**, *20*, 282.
17. Talroze, R. U.; Shibayew, V. P.; Plate, N. A. *Polym Sci. USSR* **1983**, *25*, 2863.
18. Stupp, S. I. *Chem. Eng. Prog.* **1987**, *83*, 17.
19. Hudson, S. D.; Thomas, E. L. *Polym. Prep.* **1990**, *31*, 379.
20. Achard, M. F.; Sigaud, G.; Hardouin, F.; Weill, C.; Finkelmann, H. *Mol. Cryst. Liq. Cryst.* **1983**, *92*, 111.
21. Casgrande, C.; Veyssie, M.; Weill, C.; Finkelmann, H. *Mol. Cryst. Liq. Cryst.* **1983**, *92*, 49.
22. Panar, M.; Beste, L. F. *Macromolecules* **1977**, *10*, 1401.
23. Barclay, G. G.; McNamee, S. G.; Ober, C. K. *Polym. Mater. Sci. Eng.* **1990**, *63*, 387.
24. Ober, C. K.; Barclay, G. G.; Papathomas, K. I.; Wang, D. W. *Mater. Res. Soc. Symp. Proc.* **1991**, *203*, 265.
25. Takahasi, S.; Takai, Y.; Morimoto, H.; Sonogashira, K.; Hagihara, N. *Mol. Cryst. Liq. Cryst.* **1982**, *82*, 139.
26. Takahasi, S.; Takoi, Y.; Morimoto, H.; Sonogashira, K. *Chem. Comm.* **1984**, *1*, 3.
27. Srinivasarao, M.; Berry, G. C. *Mol. Cryst. Liq. Cryst.* **1992**, *223*, 99.
28. Nole, C.; Monnerie, L.; Achard, M. F.; Hardouin, F.; Sigaud, G.; Gasparoux, H. *Polymer* **1981**, *22*, 578.
29. Martins, A. F.; Ferreira, J. B.; Volino, F.; Blumstein, A.; Blumstein, R. B. *Macromolecules* **1983**, *16*, 279.
30. Huser, B.; Spiess, H. W. *Macromol. Chem., Rapid Commun.* **1988**, *9*, 337.
31. Maret, G.; Blumstein, A.; Vilasagar, S. *Polym. Prepr.* **1981**, *22*, 246.
32. Stupp, S. I.; Lin, H. C.; Wake, D. R. *Chem. Mater.* **1992**, *4*, 947.
33. Volino, F.; Allonnean, J. M.; Giroud-Godquin, A. M.; Blumstein, R. B.; Stickles, E. M.; Blumstein, A. *Mol. Cryst. Liq. Cryst. Letters* **1984**, *102*, 21.
34. Roth, H. *Acta Polymerica* **1989**, *40*(9), 564.
35. Srinivasarao, M. *Macromol. Chem.* **1992**, *3*, 149.

36. Liepins, R.; McFarlan, J.; Jorgensen, B.; Benicewicz, B.; Jahn, R.; Cash, D.; Milewski, J. V. *Polymer Bull.* **1988**, *19*, 903.
37. Maret, G.; Schickfus, M. V.; Wendorff, J. H. *Physique Sous Champs Magnetiques Intense. Colloques Int. CNRS 242*, Paris, 1975; p 71.

Magnetic Materials

See: *Conjugated Polyradicals*
Dendritic Polyradicals
Polyradicals (Synthesis and Magnetic Properties)

Maleation

See: *Maleic Anhydride Grafted Polypropylene (Solid Phase)*
Maleic Anhydride, Grafting
Maleic Anhydride-Polybutadiene, Ene Reaction

Maleic Anhydride

See: *Anticancer Polymeric Prodrugs, Targetable*
Ethylene-Propylene Rubber Modification (Maleic Anhydride and Derivatives)
Maleic Anhydride Alternating Copolymers (Configurations of Cyclic Units)
Maleic Anhydride Grafted Polypropylene (Solid Phase)
Maleic Anhydride, Grafting
Maleic Anhydride-Polybutadiene, Ene Reaction
Personal Care Application Polymers (Acetylene-Derived)
Poly(styrene-co-N-alkylmaleimide)s (by Reactive Extrusion; Miscibility in Blends)
Reactive Processing, Thermoplastics
Styrene-Maleic Anhydride Copolymer

MALEIC ANHYDRIDE ALTERNATING COPOLYMERS (Configurations of Cyclic Units)

Kiyohisa Fujimori* and Paul G. Brown
Department of Chemistry
University of New England

The free-radical copolymerization of the electron-accepting monomer, maleic anhydride (MA), with various electron-donating comonomers has long been of interest because of its ability to produce rigidly alternating copolymers. While the mechanism of this type of copolymerization has been continuously debated and extensively investigated by means of copolymer compositions and comonomer unit sequence distributions, comparatively little attention has been given to the stereochemistry of the resulting copolymers. The configuration of the cyclic anhydride units in these copolymers, in particular, may be of use in elucidating the mechanism of copolymerization in alternating MA copolymers as a consequence of the ability of MA to form both *cis* and *trans* isomers during addition reactions.

*Author to whom correspondence should be addressed.

Under normal polymerization conditions, free-radical addition to MA during propagation can be considered to be under thermodynamic control, resulting in a predominance of the more favorable *trans* form of the anhydride unit being produced. However, if electron donor–acceptor (EDA) complexes participate in the polymerization process, as suggested by Bartlett and Nozaki,[1] such copolymers may be produced via a propagating mechanism not necessarily under thermodynamic control. Arcus and Bose[2–4] suggested that if EDA complexes, formed between the electron-deficient MA monomers and the electron-rich comonomers, were responsible for the alternating nature of MA copolymers, then such interactions might have specific steric requirements that could lead to stereoregular copolymers being produced. They, therefore, proposed that the opening of the double bond on the MA units should occur in either all *trans* or all *cis* fashions under these conditions. Unfortunately, the early search for stereoregularity in alternating MA copolymers was hampered by the lack of an adequate means by which to determine the configuration of comonomer units in copolymers.

COPOLYMER OF MA WITH STYRENE AND SUBSTITUTED-STYRENE DERIVATIVES

Most recently, the free-radical copolymerization of MA with styrene (ST) and substituted-ST derivatives has been studied by Brown and Fujimori[7–9] and Brown, Fujimori, and Tucker[10] with particular attention being paid to the relationship between the configuration of the cyclic anhydride units and the alternating tendency of the comonomer units in these copolymers. The configuration of cyclic anhydride units was determined directly from resonances in the ^{13}C NMR spectra corresponding to methine carbons of the MA units. The assignment of these resonances to the *cis* and *trans* forms of the anhydride units was achieved through the use of the model compound 2,3-dimethylsuccinic anhydride[10] and the assignments presented in previous work on related MA copolymers.[5,6]

The assignments have allowed the mechanism of alternating MA copolymerization to be investigated on a stereochemical basis by observing the ratio of *cis* to *trans* configurations of MA units and the alternating tendency of the respective comonomer units in these copolymers, over a range of copolymer compositions.

MA/*p*-Methoxystyrene Copolymers

Copolymers of MA with *p*-methoxystyrene (*p*-MST) were prepared in methyl ethyl ketone (MEK) at 1.000 mol/L and 50°C over a wide range of comonomer feed compositions,[10] and their compositions, sequence distributions, and the configurations of MA units within the copolymers determined via ^{13}C NMR spectroscopy.

It was found that the copolymers became more alternating as the fraction of MA in the comonomer feed (f_0) increased, reaching complete alternation at $f_0 \approx 0.30$. Interestingly, it was also found that the ratio of *cis* to *trans* configurations of MA units increased as the copolymers became more alternating, reaching an almost constant value of 1.3 ± 0.2 once the fraction of MA in the copolymers (F_0) reached 0.50 and the copolymers became completely alternating, indicating a link between the alternating tendency of the comonomer units and the stereochemistry of the MA units in these copolymers.

A similar study was undertaken in which MA/*p*-MST copolymers were prepared in $CHCl_3$ but under otherwise identical conditions.[9] This was done to see if the link between the alternating tendency of the comonomer units and the stereochemistry of the MA units observed in the copolymers prepared in MEK could be reproduced in a less polar solvent. Once again the alternating tendency of the copolymers was found to increase as f_0 increased, reaching complete alternation at around $f_0 \approx 0.30$. In addition, the *cis/trans* ratio for MA units also increased in the same manner as before, again reaching an almost constant value of 1.3 ± 0.1 once the copolymers became completely alternating.

MA/Styrene Copolymers

The same trends were observed in this work as were found with the MA/*p*-MST copolymers prepared in MEK and $CHCl_3$, that is, the copolymers prepared in the less polar solvent, $CHCl_3$, were more strongly alternating overall than those prepared in MEK and had a higher *cis/trans* ratio before complete alternation of the comonomer units was attained. Also, the *cis/trans* ratio of copolymers prepared in each solvent again converged to a comparable and constant value once complete alternation occurred, with *cis* MA units being in predominance.

Interpretation of Trends

The same trends in stereochemistry and alternating tendency observed in the above MA copolymerization systems have also been consistently found in copolymerization systems where MA has been replaced by the related cyclic comonomer citraconic (α-methylmaleic) anhydride.[11–13]

The formation of alternating sequences largely through EDA complex addition during propagation would explain why the *cis/trans* ratio converges to an almost constant value in each solvent when the copolymers become completely alternating if complex addition resulted in both *cis* and *trans* MA units being added to the growing radical chain in a fixed proportion.

Hence, if both free and complexed comonomers add to form quasialternating copolymers, with the majority of alternating sequences being formed via the above mechanism, this would explain why the *cis* content increases along with the tendency of the comonomer units to alternate in each of the systems studied. Complex addition, in the manner described above, would therefore, also explain why copolymers prepared in a less polar solvent, where complexation is favored, have a stronger tendency to alternate coupled with a higher overall *cis/trans* ratio where the copolymers were not completely alternating compared with those prepared in a more polar solvent, where complexation is disfavored.

While this does not conclusively prove that EDA complexes participate in alternating MA copolymerization, it does provide plausible explanations for the observed trends by assuming that complex participation occurs under strictly defined conditions.

REFERENCES

1. Bartlett, P. D.; Nozaki, K. *J. Am. Chem. Soc.* **1946**, *68*, 1495.
2. Arcus, C. L. *J. Chem. Soc.* **1955**, *51*, 2801.
3. Arcus. C. L. *J. Chem. Soc.* **1957**, *53*, 1189.
4. Arcus, C. L.; Bose, A. *Chem. Ind. (London)* April 4, 1959; p 456.
5. Ratzsch, M.; Zschoche, S.; Steinert, V.; Schlothauer, K. *Makromol. Chem.* **1986**, *187*, 1669.
6. Koenig, K. E. *Macromolecules* **1983**, *16*, 99.
7. Brown, P. G.; Fujimori, K. *Polym. Bull. (Berlin)* **1992**, *29*, 85.
8. Brown, P. G.; Fujimori, K. *Macromol. Rapid Commun.* **1994**, *15*, 61.
9. Brown, P. G.; Fujimori, K. *Polymer* **1995**, *36*, 1053.
10. Brown, P. G.; Fujimori, K.; Tucker, D. J. *Polym. Bull. (Berlin)* **1992**, *27*, 543.
11. Brown, P. G.; Fujimori, K. *Polym. Bull. (Berlin)* **1992**, *29*, 213.
12. Brown, P. G.; Fujimori, K. *Polym. Bull. (Berlin)* **1993**, *30*, 641.
13. Brown, P. G.; Fujimori, K. *Eur. Polym. J.* **1994**, *30*, 1097.

MALEIC ANHYDRIDE GRAFTED POLYPROPYLENE (Solid Phase)

Viera Khunova*
Faculty of Chemical Technology
Department of Plastics and Rubber
Slovak Technical University

Z. Zamorsky
Faculty of Technology
Technical University Brno

Effective application of polymer composites is determined by the interfacial interactions between polymer and filler. In polymer composites based on polyolefin and inorganic fillers, the role of interaction is more critical because of the nonpolar hydrophobic nature of the polymer phase and the hydrophilic character of inorganic fillers.

The polarity of the polymer matrix can be increased by several methods. One of the most common processes is to use a functional monomer with a pendant reactive polar group. The most widely used reactive functionalities are unsaturated acids and their derivatives. The maleic-anhydride-grafted polyethylene or polypropylene is at the present time a classical example of the reactive functionality.

Considerable effort has been devoted to developing methods for grafting maleic anhydride (MAH) to the backbones of a variety of saturated and unsaturated polymeric materials. In general, MAH grafting has been continuously explored as a technique to improve the properties of polymers by providing polarity to promote hydrophilicity and adhesion, to give functionality for crosslinking and other chemical modifications, and to promote compatibility with other materials.[1]

Recent works indicate that the grafting of maleic anhydride onto polypropylene is of particular interest for purposeful improving of the interphase adhesion and the mechanical properties of polymer composites with inorganic fillers.[2–7]

The grafting of MAH on polypropylene has been studied under a variety of conditions by free radical, ionic, mechanochemical, and free-radical initiators, or radiation-initiation techniques in both the solution and the melt phases.[1]

Solution-phase modification has been realized in various solvents such as benzene, xylene, toluene, *n*-heptane, and carbon tetrachloride.[8–10] The level of maleic anhydride grafting

*Author to whom correspondence should be addressed.

depends on the used free-radical initiator and the solvent. The economic feasibility and the application potential of a solution-phase grafting is in question because of the low grafting efficiency[8] and laborious procedure. Moreover, the process does not seem to be prospective at the industrial large-scale because it is not acceptable from an environmental point of view.

The second and more frequently used mode is *melt-phase modification* accomplished by blending melted polymer with maleic anhydride and peroxide.[11–12] The reaction of maleic anhydride with molten polyolefins in the presence of a peroxide initiator is processed as a heterogeneous reaction accompanied by crosslinking and/or degradation reaction.[13–16] However, these undesirable side reactions are difficult to suppress.

The reactive processing in an inert atmosphere has proved to be very effective in minimizing the extent of the side reactions that frequently accompanied the modification of polymers in the melt-phase.[6] The graft efficiency depends on the processing conditions and the type of organic peroxide used.[12] The purification of melt-phase maleated polyolefins is possible; however, the use of solvents is necessary.[17] The maleic anhydride not grafted is usually removed by dissolving the MAP-PP in hot xylene followed by precipitation and vacuum drying.[12]

Thus, ecological as well as economic requirements resulted in research activities focused at a development of an effective and ecologically friendly way of polypropylene maleation. Proposed mode of *solid-phase modification* is carried out directly by mixing the maleic anhydride and the peroxide initiator with powder polymer in a batch reactor at temperature below the melt temperature of polypropylene.

APPLICATIONS

The major part of the research was to investigate the influence of solid-phase maleated polypropylene on the properties of polymer composites with inorganic fillers.

Besides the positive influence of MAH-PP on mechanical properties, the increase of creep resistance and improved processability of particulate composites were observed.[20] A significant influence of maleated polypropylene on the properties of polymer composites with short glass fibers has also been observed. The addition of only 2 wt % MAH-PP is sufficient to cause the tensile strength to increase by more than 100% in both the polyethylene and the polypropylene composites.

The positive effects of solid-phase modified MAH-PP on compatibility, processing, structure, and mechanical properties of polypropylene-polyamide fibers have also been demonstrated.[21,22]

Summaries of the most important factors that determine the performance of solid-phase maleated polypropylene follow:

Solid-phase maleated polypropylene is feasible to use in polymer composites with both particulate and fiber fillers for purposeful improvement of mechanical properties.

The polymer matrix need not be modified as a whole; an addition of a small amount of maleated polypropylene is sufficient for a significant improvement of compatibility, processing, and mechanical properties of polymer composites and polymer fibers.

By application of solid-phase maleated polypropylene, a high-stress composite can be prepared even if the filler is the dominant phase in composites.

REFERENCES

1. Trivedi, B. C.; Culberston, B. M. *Maleic Anhydride;* Plenum: New York, 1982; Chapter 11.
2. Jančář, J.; Kummer, M.; Kolařík, J. *Interfaces in Polymer, Ceramic, and Metal Matrix Composites;* Ishida, H., Ed.; Elsevier Science: New York, 1988; 705.
3. Pukánsky, B.; Tudos, E; Jančář, J.; Kolařík, J. *J. Mat. Sci. Letters* **1989**, *8*, 1040.
4. Jančář, J.; Kuera, J. *Polym. Eng. Sci.* **1990**, *30*, 714.
5. Takase, S.; Shiraishi, T. *J. Appl. Polym. Sci.* **1989**, *37*, 645.
6. Olsen, D. J. Proc. 50th Annual Tech. Conf., ANTEC, 1986, 1991.
7. Wendt, V. *Acta Polymerica* **1988**, *39*, 260.
8. Vijayalakshmi, N. S.; Murthy, R. A. N. *J. Appl. Poly Sci.* **1992**, *44*, 1377.
9. Mitsui Toatsu Chemicals, Jap. Patent 23, 134 1980.
10. Minoura, Y.; Ueda, M.; Mizunuma, S.; Oba, M. *J. Appl. Polym. Sci.* **1969**, *13*, 1625.
11. Coran. A. I.; Patel, R. P. *Rubber Chem. Technol.* **1983**, *56*, 1045.
12. Callais, P. A.; Kazmierczak, R. T. Proc. 47th Annual Tech. Conf. ANTEC, 1368, 1989 and 1990.
13. Bräuer, M.; Jähnichen, K.; Muller, U.; Zeppenfeld, G. *Plaste und Kautschak* **1988**, *35*, 42.
14. Gaylord, N. G.; Mehta, M. *J. Polym. Sci. Polym. Lett. Ed.* **1982**, *20*, 481.
15. Gaylord, N. G.; Mishra, M. K. *J. Polym. Sci. Polym. Lett. Ed.* **1983**, *21*, 23.
16. Gaylord, N. G.; Mehta, R.; Mohan, D. R.; Kumar, V. *J. Appl. Polym. Sci.* **1992**, *44*, 1941.
17. Inoue, Takyuki, Hattori, Masafumi, Hayama, Kazuhide, Maruta and Riichiro, Eur. Patent 86-303835, 1986.
18. Rengarajan, R.; Vicic, M.; Lee, S. *Polymer* **1989**, *30*, 933.
19. Borsig, E.; Hrčková, L. *J. Macromol. Sci.-Pure Appl. Chem.* **1994**, *A31*, 1447.
20. Muras, J.; Zámorský, Z. *Plasty a kaučuk* **1991**, *28*, 289.
21. Jambrich, M.; Ďuračková, O. *Fibres Textil. in East. Eur.* **1993**, *2*, 34.
22. Gróf, I.; Ďuračková, O.; Jambrich, M. *Colloid Polymer Sci.* **1992**, *270*, 22.

MALEIC ANHYDRIDE, GRAFTING

Norman G. Gaylord
Charles A. Dana Research Institute for Scientists Emeriti
Drew University

The presence of pendant carboxylic acid functionality on a polymer chain provides opportunities for promoting adhesion to substrates and interaction with other polymers to generate covalent or hydrogen bonds leading to compatibilization, which is generally necessary in the preparation of polyblends and polymer alloys from normally incompatible polymers.

Copolymerization of monomers, capable of undergoing radical polymerization with maleic anhydride (MAH) results in the formation of carboxyl-containing copolymers. However, polymers that require metal-containing or organometallic catalysts for their synthesis, such as high-density polyethylene (HDPE), linear low-density polyethylene (LLDPE), polypropylene (PP), and ethylene–propylene copolymer rubber (EPR, EPDM) cannot be prepared in the presence of carboxylic acid monomers since the latter deactivate the catalyst or otherwise interfere in the polymerization. Although acrylic acid (AA) may be used as the carboxyl-containing monomer in lieu of MAH, it also interferes with metal-containing catalysts.

PREPARATION

The reaction of MAH with a preformed polymer, in the melt, results in the appendage or grafting of MAH as individual succinic anhydride or MAH units on the polymer. The use of AA in lieu of MAH in the reaction with the polymer results in the grafting of AA as oligomeric branches.

Although MAH cannot be homopolymerized in the presence of a radical catalyst under normal conditions, homopolymerization can be induced when the reaction is conducted at a temperature where the catalyst has a short half-life, that is, where it is undergoing rapid decomposition.[1] When a preformed polymer is present under these conditions, the MAH is grafted onto the molten polymer to produce pendant anhydride functionality.

Although the presence of a peroxide is necessary to permit the reaction of MAH with a saturated polymer, no peroxide is necessary when the polymer is unsaturated. The uncatalyzed "ene" reaction of MAH with an unsaturated polymer results in appendage of a succinic anhydride moiety to the polymer chain and a shift of the double bond.

This reaction has been used in the preparation of an EPDM-MAH adduct by the reaction of EPDM with MAH in a screw extruder at 260°C.[2]

APPLICATIONS

Carboxyl-containing polymers have improved adhesion to substrates and fillers. The presence of clay or other fillers or reinforcing fibers such as glass fibers during the MAH-peroxide–polymer reaction or the addition of such fibers to an MAH–peroxide–polymer reaction product results in coupling and compatibilization to yield a composite with improved tensile strength, impact resistance, and higher modulus.[3]

A polymer containing grafted MAH has anhydride and carboxyl groups that interact with functional groups on another polymer, which are capable of forming covalent or hydrogen bonds therewith. Thus, laminates or coatings based on MAH-grafted polymers have good adhesion to other polymers in the form of fabricated shapes or films.

In lieu of mixing incompatible polymers and a compatibilizing block or graft copolymer, the latter may be generated *in situ* by reaction of suitable functionality on the polymers. Carboxyl-containing polymers participate in the *in situ* generation of a compatibilizing agent, for example, from a mixture with nylon, through covalent and/or hydrogen bonding.

High-impact, "supertough" nylon is obtained from a blend of MAH-containing EPDM and nylon 6,6.[4]

An oil-resistant thermoplastic elastomer, Geolast,™ is prepared by melt blending PP containing about 10% PP-g-MAH and nitrile rubber (NBR) containing 0.2–3% amine-terminated liquid NBR.[5]

The properties of polyblends containing polar thermoplastic polymers and a S-Bd-S triblock copolymer are greatly improved by the presence of as little as 2% MAH grafted on the latter.

A rapidly growing use of the polymer-MAH-peroxide reaction is in the reclamation of polymer scrap, which contains numerous incompatible polymers.

REFERENCES

1. Gaylord, N. G. *J. Macromol. Sci.-Rev. Macromol. Chem.* **1975**, *C13*, 235.
2. Caywood, S. W., Jr. U.S. Patent 3 884 882, 1975.
3. Gaylord, N. G.; Ender, H.; Davis, L.; Takahashi, A. In *Modification of Polymers* Caraher, C. E.; Tsuda, M., Eds.; ACS Symposium Series 121; American Chemical Society: Washington, DC, 1980; p 469.
4. Epstein, B. N. U.S. Patent 4 174 358, 1979.
5. Coran, A. Y.; Patel, R. *Rubber Chem. Technol.* **1983**, *56*, 1045.

MALEIC ANHYDRIDE–POLYBUTADIENE, ENE REACTION

Martha Albores-Velasco and F. J. Alfan
Facultad de Quimica
UNAM

The insertion of maleic anhydride (MA) into polymers has enormous technological importance, since maleic anhydride allows polymers to be mixed with other polymers in order to produce blends with improved properties. In the case of polybutadiene (PB), a dynamic vulcanization that has been defined as the crosslinking of one polymer during its molten state, mixing with another polymer or with other polymers and this can be used to obtain new elastomeric, thermoset compositions.[1]

Alder, LeBras and Compagnon, Dalalande, and Pinnazi confirmed the addition reaction of maleic anhydride with rubber in the solid phase and in solution.[2–5] The nature of this reaction is complex: it generally occurs with indiscriminate side reactions, including branching, crosslinking, and oligomerization of maleic anhydride; however, it has been used (for example) to form an alkenyl succinic functionality in polymers such as EPDM (ethylene-propylene-diene monomer) rubber.[6]

The ene reaction is a thermal reaction defined by Alder as the indirect substituting addition of a compound with an electron-deficient double bond (enophile) to an alkene possessing an allylic hydrogen atom (ene) with migration of the ene double bond and 1,5-hydrogen shift.[2] A typical enophyle is maleic anhydride, whose reaction with olefins can be formulated as in **Structure 1**.

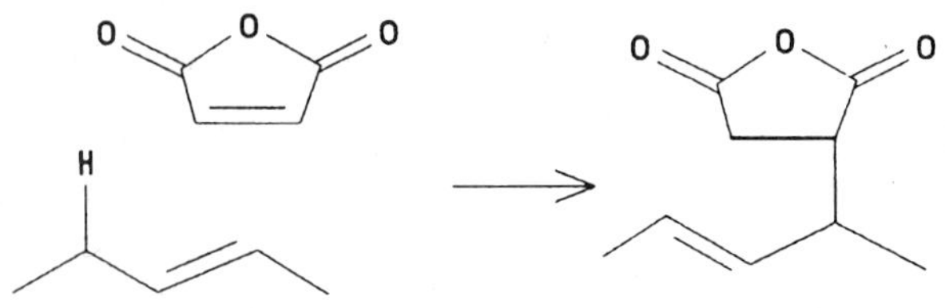

1

Ideally, the ene reaction of maleic anhydride with polybutadiene provides the polymer with a succinyl group that permits the curing of the polymer with ethylene diamine or magnesium oxide.

The aim of the present work was to investigate ene reactions of two polybutadienes of different molecular weights with maleic anhydride and maleimides, to establish experimental parameters for the reaction and to look for a relationship between structure and reactivity of maleic anhydride derivatives. A low temperature was selected to carry out the reaction to avoid crosslinking of the polybutadiene.

We conclude that ene reactions at low temperatures (90°C) are slow and are still complicated by crosslinking of the polybutadiene. This crosslinking decreases when low molecular weight polymers or partially saturated polybutadienes are used. Hydrochlorination is an alternative to hydrogenation for saturation of the double bonds.

REFERENCES

1. Coran, A. Y. *Rubber Chem. Technol.* **1991,** *63,* 599.
2. Alder, K.; Pascher, F.; Schmitz, A. *Chem. Ber.* **1943,** *76B,* 27.
3. Le Bras, P.; Compagnon, *Compt. Rend.* **1941,** *212,* 616.
4. Dalande, A. *Rubber Chem. Technol.* **1948,** *21,* 344.
5. Pinazzi, L. C.; Danjard, J. C.; Pautrat, R. *Rubber Chem. Technol.* **1963,** *36,* 282.
6. Tessier, M.; Marechal, E. *J. Polym. Sci. Polym. Chem. Ed.* **1988,** *26,* 2785.

MALEIMIDE COPOLYMERS (N-Substituted)

Vitan Bonev Konsulov
Chemistry Faculty
University "Konstantin Preslavski"

Polymaleimides are a new class of polymers obtained on the basis of N-substituted maleimides (RMI) or bismaleimides (BMI). RMIs are derivatives of maleic anhydride (MA). They are represented in a limited way in Trivedi and Culbertson's monograph and in Rzaev's monograph.[1,2] RMI homopolymerization and copolymerization, the maleimide copolymer properties, and applied fields are described in my survey.[3]

The RMIs have a higher polymerization capacity than MA. The polymerization[4–28] is conducted in radicals by azobisisobutyronitrile (AIBN), benzoic peroxide (BPO), UV, or Co initiating irradiation;[29] anionic (n-C_4H_9Li; *tert*-C_4H_9Li; *tert*-C_4H_9OK; C_2H_5MgBr);[28,30–36] or zwitterionic mechanisms,[37–41] or by retro-Diels-Alder reaction,[32,42] and Michael's addition polymerization.[39,40]

N-Arylmaleimides are crystal substances with a high melting point.[43] Many *N*-alkylmaleimides are oily liquid products. RMIs are used as comonomers for the maleimide copolymers. Free-radical copolymerization generally uses vinyl monomers such as styrene, α-methyl styrene, vinyl acetate, methyl metacrylate, olefins, vinyl chloride, etc. The maleimide copolymers are characterized by high glass transition temperature (T_g) and they are thermostable in high temperatures.

COPOLYMERIZATION AND COPOLYMERS IN N-SUBSTITUTED MALEIMIDES WITH STYRENE AND α-METHYLSTYRENE

Several alternating copolymers are derived by the radical copolymerization of RMI with styrene and α-methylstyrene (MCT).[24,29,44-80] The monomer reactivity ratios r_1 and r_2 for all monomer pairs are smaller than one and incline to zero. They prove the copolymerization alternating character. Researchers suppose that the formation of alternating copolymers results from the charge-transfer complex (CTC).

Many researchers offer explanations of the alternating copolymerization character with the CTC between the donor monomer ST and the acceptor RMI.[29,47–51,55,56,58–61]

Turner et al. suggest an original method for obtaining new reactive copolymers by an alternating copolymerization of *N*-allylmaleimide (AMI) with vinylbenzyl chloride (VBC) or other substituted styrene.[64]

Many papers report maleimide (MI) copolymer syntheses with MA or other N-substituted maleimides.[70,71]

Takeishi et al. use a new method for RMI radical copolymerization with ST.[40] They initiate it with CTC, formed between thiophenol and *N*-PMI.

COPOLYMERIZATION AND COPOLYMERS OF N-SUBSTITUTED MALEIMIDES WITH ACRYLIC AND METHACRYLIC MONOMERS

Several functional maleimide copolymers with acrylic and methacrylic monomers are obtained by radical polymerization.[40,44–47,60,62,63,77–93] PMI copolymerizations with methyl methacrylate (MMA) in benzene solution, dioxane, THF, and tetrachloromethane are the most fully studied. Regardless of the solution the methacrylic radicals are more reactive than the maleimide ones.

My colleague and I report the synthesis of a new functional copolymer poly(*N*-(4-bromophenyl) maleimide-*co*-methyl methacrylate) are shown in **Structure 1**.[60]

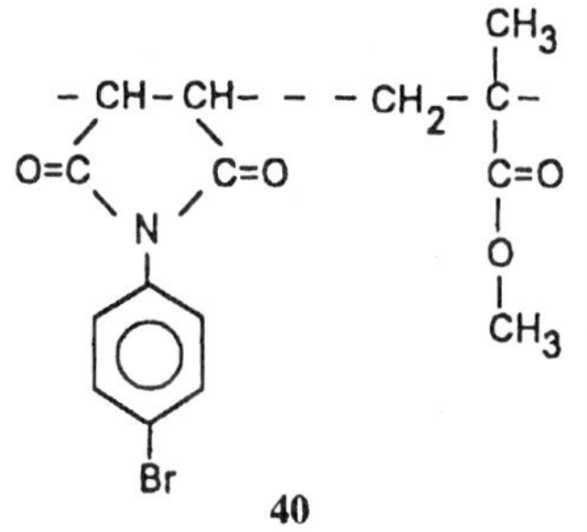

1

Our comparative study of PMI and BPMI copolymerizations with ST and MMA reveals that the copolymerizations flow statistically in contrast to the alternating maleimide copolymerization with ST. The substituent of the p-position (Br-atom) influences the reactive capacity of BPMI-radical in comparison with the phenylmaleimide radical.

COPOLYMERIZATION AND COPOLYMERS OF N-SUBSTITUTED MALEIMIDES WITH OTHER MONOMERS

Maleimide copolymers can be obtained by the radical copolymerization of N-substituted maleimides with vinyl chloride (VC).[94–97] The copolymerizations of *N*-cyclohexylmaleimide (CHMI), *N*-phenylmaleimide (PMI), *N*-(4-methylphenyl) maleimide (MPMI), and *N*-(2,4-dimethylphenyl) maleimide (DMPMI) are carried out in solution (DMF, THF, $CHCl_3$), in bulk or suspension at 50°C. It is determined that RMIs form more reactive radicals and greater contents in the copolymers regardless of polymerization method. The maleimide reactivity decreases at homogeneous copolymerization in solution in the following order: CHMI, DMPMI, MPMI. The copolymerization reactivity ratios for the studied RMIs are higher in solution copolymerization than bulk copolymerization.

Thermostable maleimide copolymers with olefins are obtainable.[93,97] CHMI copolymerizations with isobutene (IB), isooctene (IO) and *cis-* and *trans*-2-butene (CB,TB) are held in benzene at 60°C.[97] The yield, the molecular weight, and the olefin unit contents depend on the olefin structure (alkyl residuum) and they decrease in the following order: IB, IO, CTB. An alternating copolymer is obtained at PMI copolymerization with ethylene.

The obtained olefin maleimide copolymers have a high thermostability—up to 335–340°C. The CHMI copolymer glass transition temperature with IB is 180°C; 157°C for the PMI copolymer with ethylene.[98]

REFERENCES

1. Trivedi, B. C.; Culbertson, B. M. *Maleic Anhydride*; Plenum: New York, NY, 1982.
2. Rzaev, Z. M. *Polymers and Copolymers of Maleic Anhydride*: Elm: Bacu, 1984.
3. Konsulov, V. B.; Konsulova, L. V. *Chemia Industrya* **1990**, *61*, 21.
3a. Akijama, M.; Shimizu, K.; Narita, M. *Tetrahedron Lett.* **1976**, *13*, 1015.
4. Jousset, P. Fr. Patent 1248070, 1962.
5. Joshi, R. M. *Makromol. Chem.* **1963**, 62, 140; *J. Polym. Sci.* **1962**, *60*, 56.
6. Cubbon, R. C. P. *Polymer* **1965**, *6*, 419; *J. Polym. Sci., Polym. Symp.* **1967**, *16*, 387.
7. Kojima, K.; Yoda, M.; Marvel, C. S. *J. Polym. Sci.* **1966**, *A1*(4), 1121.
8. Kogiya, T.; Izu, M.; Kawai, S.; Fukui, K. *J. Polym. Sci.* **1966**, *B4*(6), 387.
9. Nakayama, Y.; Smets, G. *J. Polym. Sci.* **1967**, *A4*(5), 1619.
10. Barrales-Rienda, J. M.; Ramos, J. G. *J. Polym. Sci., Polym. Symp.* **1973**, *42*, 1249.
11. Barrales-Rienda, J. M.; Ramos, J. G.; Chaves, N. S. *Eur. Polym. J.* **1977**, *13*, 129.
12. Barrales-Rienda, J. M.; Ramos, J. G.; Chaves, M. S. *Brit. Polym. J.* **1977**, *9*, 6.
13. Elsabee, M. Z.; Mokhtar, S. *Eur Polym. J.* **1983**, *19*, 451.
14. Simionescu, C. I.; Grigoras, M. *J. Polym. Sci., Polym. Lett. Ed.* **1990**, *28*, 39.
15. Kurmanliev, O. S.; Omasheva, A. V.; Shaihutdinov, E. M.; Sutralin, M. A. *Vysokomol. Soedin.* **1986**, *B28*, 849.
16. Otsu, T.; Tatsumi, A.; Matsumoto, A. *J. Polym. Sci., Polym. Lett. Ed.* **1986**, *24*, 113.
17. Sato, T.; Arimoto, K.; Tanaka, H.; Ota, T.; Doichi, K. *Macromolecules* **1989**, *22*, 2219.
18. Barrales-Rienda, J. M.; Fernandez-Martin, F.; Galicia, C. R.; Chaves, M. S. *Makromol. Chem.* **1983**, *184*, 2643.
19. Sandreczki, T. C.; Brown, I. M. *Macromolecules* **1990**, *23*, 1979.
20. Van Paesschen, G.; Timermann, D. *Makromol. Chem.* **1964**, *78*, 112.
21. Ivanov, V. S.; Smyrnova, V. K.; Sidorova, T. I. *Vysokomol. Soedin.* **1969**, *B11*, 372.
22. Mikhailov, A. I.; Muginova, I. I.; Ivanov, V. S.; Barkalov, I. M.; Goldanskii, V. I. *Vysokomol. Soedin.* **1976**, *A18*(6), 1226.
23. Ivanov, V. S. *Radiacionnaja Chemia Polymerov;* Chemia, L., 1988.
24. Fujita, T.; Okuda, Y.; Yoshihara. M.; Maeshima, T. *J. Macromol. Sci., Chem.* **1988**, *A25*, 327.
25. Cole, T.; Heller, H. C. *J. Chem. Phys.* **1965**, *42*, 1668.
26. Zott, H.; Hensinger, H. *Eur. Polym. J.* **1978**, *14*, 89.
27. Aida, H.; Takase, V.; Nozi, T. *Makromol. Chem.* **1989**, *190*, 2821.
28. Matsumoto, A.; Oki, Y.; Otsu, T. *Polym. J.* **1991**, *23*, 201.
29. Butler, G. B.; Joyce, K. C. *J. Polym. Sci.* **1968**, *C22*, 45.
30. Okamoto, Y.; Nakano, T.; Kabayashi, H.; Hatada, K. *Polym. Bull.* **1991**, *25*, 5.
31. Sheremeteva, T. V.; Larina, G. N.; Tsvetkov, V. N.; Shtennikova, I. N. *J. Polym. Sci.* **1968**, *C22*, 185.
32. Wipfelder, E.; Heusinger, H. *J. Polym. Sci.* **1978**, *16*, 1779.
33. Pyriadi, T. M.; Kaleefa, H. *J. Polym. Sci., Polym. Chem. Ed.* **1984**, *22*, 129; Pyriadi, T. M.; Harwood, H. J. *Polym. Prepr.* **1970**, *11*, 60.
34. Hagiwara T.; Sato, J.; Hamana, H.; Narita, T. *Makromol. Chem.* **1987**, *188*, 1825.
35. Hodge, P.; Khoshdel, E.; Naim, A. A. *Polym. Commun.* **1986**, *27*, 322.
36. Lin, Y.; Huang, S. J. *Polym. Prepr.* **1989**, *30*(1), 248.
37. Sahu, U. S.; Bhadani, S. N. *Makromol. Chem., Rapid Commun.* **1982**, *3*, 103
38. Ivanov, A. A.; Primele'ys, E. *Vysokomol. Soedin.* **1984**, *A26*, 1300.
39. Jakovlev, S. A.; Heryng, Z.; Wagner, T.; Jar, V. *Vysokomol Soedin.* **1984**, *B26*, 30.
40. Takeishi, M.; Arimori, S.; Iwasaki, N.; Sekiya, K.; Sato, R. *Polym. J.* **1992**, *24*, 365.
41. Rivas, B. L.; Pizarro, G. C. *Eur. Polym. J.* **1989**, *25*(3), 231.
42. Aponte, M. A.; Butler, G. B. *Polym. Prepr.* 1983, *24*(2), 362.
43. Ratzsch, M.; Steinert, V.; Giese, B.; Farshchi, H. *Makromol. Chem., Rapid Commun.* **1989**, *10*, 195.
44. Tawney, P. O. Ger. Patent 1000288, 1959.
45. Barb, D. A.; Rose, J. B. G.B. Patent 976976, 1964.
46. Nield, E. U.S. Patent 3 676 404, 1972.
47. Barrales-Rienda, J. M.; Gonzalez de la Campa, J. J.; Gonzalez Ramos, J. *J. Macromol. Sci., Chem.* **1977**, *A11*, 267.
48. Seiner, J. A.; Litt, M. *Macromolecules* **1971**, *4*, 308.
49. Coleman, L. E.; Conrady, J. A. *J. Polym. Sci.* **1959**, *38*, 241.
50. Rzaev, Z. M.; Dzhafarov, R. V. *Azerb. Khim. Zh.* **1983**, *6*, 89.
51. Abayasekara, D. R.; Ottenbrite, R. M. *Polym. Prepr.* **1986**, *27*(1), 462.
52. Narita, M.; Teremoto, T.; Okawara, M. *Bull. Chem. Soc. Jpn.* **1972**, *45*, 3149.
53. Urishido, K.; Koike, K.; Kitano, H.; Kobayashi, M. *Kobunshi Ronbunshu* **1990**, *47*, 79.
54. Yashihara, M.; Asakura, J.; Takahashi, H.; Maeshima, T. *J. Macromol. Sci., Chem.* **1983**, *A20*, 123.
55. Mohamed, A. A.; Jebrael, F. H.; Elsabee, M. Z. *Macromolecules* **1986**, *19*, 32.
56. Sato, T.; Masaki, K.; Seno, H.; Tanaka, H. *Makromol. Chem.* **1993**, *194*, 849.
57. Younes, U. E. U.S. Patent 4 508 883, 1985.
58. Konsulov, V. B.; Grozeva, Z. S.; IUPAC 32nd Int. Symp. Macromol., Kyoto, 1988, Prepr. MACRO '88, 1988, 678.
59. Konsulov, V. B.; Grozeva, Z. S. *Biotechnol. Chem.* **1989**, *1*, 19.
60. Konsulov, V. B.; Grozeva, Z. S. *Chemia Industrya* **1991**, *62*(6), 23.
61. Konsulov, V. B.; Kirilov, V. *Chemia Industrya* **1991**, *62*(2), 22.
62. Oishi, T.; Moriwaki, M.; Momoi, M.; Fujimoto, M. *J. Macromol. Sci., Chem.* **1989**, *A26*(6), 861.
63. Oishi, T.; Saeki, K.; Fujimoto, M. *J. Polym. Sci., Polym. Chem. Ed.* **1989**, *27*, 1429.

64. Turner, S. R.; Anderson, C. C.; Kolterman, K. M. *J. Polym. Sci., Polym. Lett. Ed.* **1989**, *27*, 253.
65. Hagiwara, T.; Suzuki, J.; Takenchi, K.; Hamana, H.; Narita, T. *Macromolecules* **1991**, *26*, 6856.
66. Nayakawa, K.; Yamakita, H.; Kawase, K. *J. Polym. Sci.* **1972**, *Al*(10), 1363.
67. Abayasekara, D. R.; Ottenbrite, R. M. *Polym. Prepr.* **1984**, *25*, 164.
68. Konsulov, V. B. *Modifikacia Polymerov, Zbornik prispevkov, Smolenice* **1987**, 39.
69. Konsulov, V. B.; Pyruleva, J. G.; Grozeva, Z. S. 31st IUPAC Macromol. Symp., Abst. Modif. Polym. 1987, 153.
70. Florjanczyk, Z.; Krawiec, W. *Makromol. Chem.* **1989**, *190*, 2144.
71. Florjanczyk, Z.; Krawiec, W.; Such, K. *J. Polym. Sci., Polym. Chem. Ed.* **1990**, *28*, 795.
72. Semchikov, J. D.; Scherbakov, V. I.; Hvatova, N. L.; Stolyarova, N. E.; Dyachkovskaya, O. S.; Razuvaev, G. A. *Vysokomol. Soedin.* **1982**, *B24*, 516.
73. Mamedova, S. G.; Rzaev, R. M.; Medyakova, L. V.; Rustamov, F. B.; Askerova, N. M. *Vysokomol. Søedin.* **1987**, *A29*, 1922.
74. Fles, D. D.; Vucovic, R.; Ranogajec, F.; Zuanic, M.; Fles, D. *Polymeri* **1989**, *10*, 143.
75. Fles, D. D.; Vucovic, R.; Ranogajec, F. *J. Polym. Sci., Polym. Chem. Ed.* **1989**, *27*, 3227.
76. Turner, S. R.; Wardle, R.; Thaler, W. A. *J. Polym. Sci., Polym. Chem. Ed.* **1984**, *22*, 2281.
77. Rittel, A.; Kucharski, M. *Acta Polym.* **1989**, *40*, 62.
78. Rittel, A. *Acta Polym.* **1989**, *40*, 268.
79. Oishi, T.; Okamoto, N.; Fujimoto, M. *J. Polym. Sci., Polym. Chem. Ed.* **1986**, *24*, 1185.
80. Oishi, T.; Yoshida, M.; Momoi, M.; Fujimoto, M. *Kobunshi Ronbunshu* **1989**, *46*, 763.
81. Yamada, M.; Takase, I.; Mishima, T. *Kobunshi Kagaku* **1967**, *24*, 326.
82. Kumo, V.; Koike, K.; Kitano, H.; Kabayashi, M. *Kobunshi Ronbunshu* **1990**, *47*, 8(1), 79.
83. Yamaguchi, H.; Minoura, Y. *J. Polym. Sci.* **1970**, *A-1*, 8, 1467.
84. Aida, H.; Kimura, M.; Fukuoka, A.; Hirode, T. *Kobunshi Kagaku* **1971**, *28*, 354.
85. Kumar, A. *J. Macromol. Sci.* **1987**, *A24*(6), 711.
86. Borbelly, J.; Kelen, T.; Konsulov, V.; Grozeva, Z. *Polym. Bull.* **1991**, *26*, 253.
87. Konsulov, V. B.; Grozeva, Z. S. Book of Abstrakts (34th IUPAC Int. Symp. on Macromolecules, Prague), 1992, 110.
88. Finter, J.; Widmer, E.; Zweifel, H. *Angew. Makromol. Chem.* **1984**, *128*, 71.
89. Gong, B.-M.; Chien, J. W. *J. Polym. Sci., Polym. Chem. Ed.* **1989**, *27*, 1149.
90. Pradny, M. *Makromol. Chem.* **1989**, *190*, 2229.
91. Simionescu, C. I.; Grigorash, M.; Onofrei, G. *Makromol. Chem.* **1985**, *186*, 1121.
92. Dean Barri, D. U.S. Patent 4 514 543, 1985.
93. Matsumoto, A.; Kubota, T.; Ito, H.; Otsu, T. *Mem. Fac. Eng. Osaka City Univ.* **1990**, *31*, 47.
94. Otsuka, M.; Matsuoka, K.; Takemoto, K.; Imoto, M. *Kogyo Kagaku Zasshi* **1970**, *73*, 1062.
95. Horing, S.; Jakowlew, S.; Ulbricht, J. *Plaste und Kautschuk* **1982**, *29*, 622.
96. Horing, S.; Jakowlew, S.; Ulbricht, J. *Plaste und Kautschuk* **1983**, *30*, 305.
97. Otsu, T.; Matsumoto, A.; Ito, H. *Chem. Express* **1990**, *5*, 901.
98. Schmidt-Naake, C.; Schmidt, H.; Zitanski, B.; Berger, W. *Makromol. Chem.* **1990**, *191*, 3033.

Maleimides

See: Functional Polymaleimides, N-Protected
Maleic Anhydride-Polybutadiene, Ene Reaction
Maleimide Copolymers (N-Substituted)
Maleimides, Fluorine-Containing
Poly(styrene-co-N-alkylmaleimide)s (by Reactive Extrusion; Miscibility in Blends)

MALEIMIDES, FLUORINE-CONTAINING

Akira Nagai* and Akio Takahashi
Hitachi Research Laboratory
Hitachi Ltd.

Maleimides, and in particular bismaleimides, are interesting materials for electronics industry applications because of their network structure formed in the thermosetting process. The cured network structures provide good heat resistance and the bismaleimides have a number of other excellent properties. They contain unsaturated double bonds, which undergo thermal polymerization without formation of volatile by-products. They lend, therefore, no possibility for void formation in the molding process. On this point, they differ favorably from other condensation-type polyimides. Consequently, bismaleimide resins are used as multilayer circuit boards in the electronics industry and as fiber-reinforced plastics in the aerospace industry.[1–3]

Several physical properties are required for high-performance electronic materials, especially packaging materials such as multilayer circuit boards and encapsulating molding compounds. These are a low dielectric constant, good thermal stability, and excellent flame retardancy. The introduction of fluorine-containing groups into the molecular structure is expected to improve all these properties simultaneously. Therefore, fluorine-containing maleimide compounds have been synthesized and the relationship between their properties and chemical structure has been studied with a view toward achieving high-performance electronic applications.

PREPARATION

In general, maleimide compounds are prepared by chemical dehydration (cyclization) with dehydration agents such as acetic anhydride or phosphorus pentaoxide, following Searle's procedure.[6] This method gives maleimides within a relatively short reaction time and at a low reaction temperature. It has been frequently used for industrial syntheses. The fluorine-containing maleimides are also synthesized by reaction of the corresponding fluorine-containing amines with maleic anhydride, with substantial modifications for the original reaction conditions of Searle.[7,8] However, yields are low for high fluorine content maleimides due to lowering of the nucleophilic reactivity of the amino groups. Moreover, purification of products is relatively difficult because of contamination from several catalysts such as tertiary amine or metal acetate salts.

Recently, a more effective process for high fluorine content compounds has been reported by Misra and Tesoro.[9] This process requires a relatively high temperature and longer reaction

*Author to whom correspondence should be addressed.

time to get thermal dehydration as compared with Searle's method. Moreover, an excess amount of maleic anhydride is used.[10] This process, however, seems effective for amines that have amino groups of low reactivity and high fluorine content.

This one-step method gives a cleaner product and higher yield compared to the Searle's conventional procedure. The lower product yield in the conventional method may have resulted from the low nucleophilic reactivity of the amino groups in the presence of ortho-substituted by fluorine. This effect of fluorine compounds is in agreement with behavior in the synthesis of fluorine-containing phenylmaleimides.[11]

CHEMICAL STRUCTURES, PROPERTIES, AND APPLICATIONS

Many fluorine-containing maleimide compounds can be obtained from the corresponding fluorine-containing amines by the thermal dehydration process described above. The earliest fluorine-containing maleimides synthesized were phenylmaleimide derivatives.[7]

These maleimides are finding applications as detergents, lubricating oil additives, crosslinking agents of polymers and rubbers, and so on. These maleimides should be easy to polymerize, and give thermally stable polymers. The polymerization reactivities of these compounds were investigated with free-radical, anionic catalyst, and irradiation processes.[12] The fluorine substituent seems to lower thermal polymerization reactivity of double bonds in the maleimide ring due to its electron-withdrawing effect.[13–15]

Other phenylmaleimide derivatives have been synthesized, mainly because of their thermal stability as polymers and for biochemical applications such as fungicidal activity.[17–19] The thermal stability of polymers obtained with other halogenated monomers has been investigated, and the fluorine polymer has better thermal stability than chlorine- and bromine-containing compounds.[20]

The bismaleimide compound that has flexible ether groups and a long phenoxy chain (**Structure 1**) has been used as heat-resistant material in place of the conventional bismaleimide.[21] This new type of bismaleimide has a high molecular weight, which makes it possible to reduce the crosslinking density of the cured products. Consequently, these bismaleimides have good mechanical properties and excellent toughness. The introduction of flexible and long chains into structures is also effective for getting good solubility.

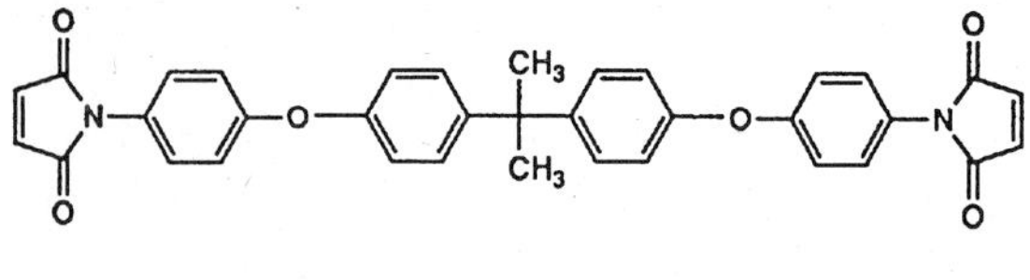

1

The introduction of fluorine-containing groups within this new bismaleimide structure was carried out to get good thermal stability and flame retardancy.[16,25] The cured products also have excellent electrical properties, such as low dielectric constant and dispersion. They are expected to be used in the electronics industry as packaging materials.

The effect of fluorine-containing groups on melting point has been discussed.[16] The melting point tends to be lowered by introducing fluorine groups. In general, the melting point tends to be high with increasing molecular weight because the cohesive energy between molecules in the crystal state is large.

The fluorine-containing maleimides are also applied for enhancement of thermal stability in transparent polymers. They are expected to be used as optical materials.

Fluorine-containing maleimides are expected to find application as packing materials in high-performance liquid chromatography (HPLC) because their polymers have excellent insolubility resistance for HPLC.

FLUORINE-CONTAINING MALEIMIDES IN PATENTS

In Japanese industry, fluorine-containing maleimides have been extensively developed for applications in electronics materials. They are used for insulating materials in packaging technology or optical devices. They feature excellent heat resistance, a low dielectric constant, low dispersion, and good flame retardancy.

Because of the relationship between chemical structure and physical properties (particularly the dielectric constant, thermal stability, and flame retardancy) of maleimide compounds with fluorine-containing aromatic structures, they are the preferred choice for high-performance electronics materials. Numerous novel fluorine-containing maleimide compounds will no doubt be synthesized for such applications in the future.

REFERENCES

1. Takahashi, A. et al. *Proc. Printed Circuit World Conv. III (IPC)* **1984**, *W.C. III-14*.
2. Takahashi, A. et al. *IEEE Trans. CHMT* **1990**, *13*, 1115.
3. Takahashi, A. et al. *IEEE Trans. CHMT* **1992**, *15*, 418.
4. Mijovic, J.; Pearce, E. M.; Foun, C-C. *Adv. Chem. Ser.* **1984**, *208*, 293.
5. Stenzenberger, H. D.; Konig, P. *High Perf. Polym.* **1989**, *1*, 239.
6. Searle, N. E. U.S. Patent 2 444 536, 1948.
7. Barrales-Rienda, J. M.; Ramos, J. G.; Chaves. M. S. *J. Fluorine Chem.* **1977**, *9*, 293.
8. Sakai T. T.; Dallas, J. L. *FEBS Lett.* **1978**, *93*, 43.
9. Misra A. C.; Tesoro, G. *Polymer* **1992**, *33*, 1083.
10. Vygodskii, Ya. S.; Adigezalov, V. A.; Askadskii, A. A.; Slonimskii, G. L.; Bagirov, Sh. T.; Korshak, V. V.; Vinogradova, S. V.; Nagiev, Z. M. *Polymer Sci U.S.S.R.* **1980**, *21*, 2951.
11. Nagai A.; Takahashi, A. *Polymer J.* **1994**, *26*, 357.
12. Barrales-Rienda, J. M.; Ramos, J. G.; Chaves, M. S. *J. Polym. Sci.: Polym. Chem. Ed.* **1979**, *17*, 81.
13. Nishimura, S.; Nagai, A.; Takahashi, A.; Narita, T.; Hagiwara, T.; Hamana, H. *Polymer J.* **1990**, *22*, 171.
14. Matsumoto, A.; Kubota, T.; Otsu, T. *Macromolecules* **1990**, *23*, 4508.
15. Mirsa, A. C.; Tesoro, G. C.; Hougham, G.; Pendharker, S. *Polymer* **1992**, *33*, 1078.
16. Nagai, A.; Takahashi, A.; Suzuki, M.; Mukoh, A. *J. Appl. Polym. Sci.* **1992**, *44*, 159.
17. Duncan, R. J. S.; Wilkinson, S.; Wrigglesworth, R. *J. Immunol. Methods* **1985**, *80*, 137.
18. Tsuda, M.; Nakajima, T.; Kasugai, H.; Kawada, S.; Yamaguchi, I.; Misato, T. *Nippon Noyaku Gakkaishi* **1976**, *1*, 101.
19. Watanabe, S.; Igarashi, Y.; Yagami, K.; Imai, R. *Pestic. Sci.* **1991**, *31*, 45.
20. Patel, R. D.; Patel, M. R.; Bhardwaj, I. S. *Thermochimica Acta* **1981**, *51*, 373.

21. Nagai, A.; Takahashi, A.; Suzuki, M.; Katagiri, J.; Mukoh, A. *J. Appl. Polym. Sci.* **1990**, *41*, 2241.
22. Maglio, G.; Palumbo, R.; Serpe, L. *New Polymeric Mater.* **1992**, *3*, 103.

MARINE ADHESIVE PROTEINS, SYNTHETIC

Hiroyuki Yamamoto
Institute of High Polymer Research
Faculty of Textile Science and Technology
Shinshu University

Nature's powerful adhesives, secreted by marine mollusks such as mussels, oysters, and barnacles, which must routinely cope with the force of surf and tides, are simple proteins. Some of them have been identified as l-β-3,4-dihydroxyphenyl-α-alanine (Dopa) containing proteins. They are called marine adhesive proteins and their adhesive properties have been investigated for use as adhesives for biotechnological (medical and dental) purposes.[1,2] Two approaches (polymer chemical and gene technology) provide competitive strategies for preparing these marine adhesive proteins. The author's group first synthesized a polyphenolic decapeptide (Ala-Lys-Pro-Ser-Tyr-Hyp-Hyp-Thr-Dopa-Lys)$_n$ containing two Dopa residues from the blue mussel *Mytilus edulis* by polycondensation,[3] and two other groups prepared a precursor form Ala-Lys-Pro-Ser-Tyr-Pro-Pro-Thr-Tyr-Lys containing two Tyr residues using genetic engineering technology.[4] The latter genetic product was converted to the final adhesive protein by a modification reaction using an oxidase tyrosinase.

More recently, among these marine adhesive proteins whose amino acid compositions and sequences have already been determined,[5,6] the barnacle adhesive arthropodin proteins with random sequences,[7] the adhesive proteins (Ala-Gly-Dopa-Gly-Gly-X-Lys) (X, hydrophobic amino acids) of a Chilean mussel *Aulacomya ater* Molina,[8] the cuticle collagens of the Polychaete (*Nereis japonica*),[9] and, although it is not of marine animal origin, the adhesive protein (Gly-Gly-Gly-Tyr-Gly-Gly-Tyr-Gly-Lys)$_n$ of the vitellaria of the liver fluke (*Fasciola hepatica*) have also been synthesized.[10]

PROPERTIES

Marine adhesive proteins insolubilize and adhere to the surfaces of a variety of substrates, such as rocks (granite), glass, and plastics, in a watery environment. During the course of our polymer chemical studies on marine adhesive proteins as bioadhesives, the role of Lys and Tyr residues in adhesive proteins was found to be more important to adhesion than previously thought.[11,12]

APPLICATIONS

The bonding strengths of a variety of synthetic polypeptides in water or in organic solvent systems have been reported.[13] As a next step, biological adhesion on glass surfaces precoated with synthetic adhesive proteins has also been examined.

In conclusion, we propose the following requirements for a biological adhesion:

- High molecular weight PLL is effective for wettability and adhesion.
- Cationic Lys residues are more effective than anionic Glu residues.
- Lys and Gly residues in adhesive proteins are effective for adhesion.
- The primary structures of the adhesive proteins have significance.
- In fish sperm cell adhesion, the Lys residues are important.

REFERENCES

1. Waite, J. H. U.S. Patent 4 585 585, 1986.
2. Benedict, C.; Picciano, P. *Adhesives from Renewable Resources;* American Chemical Society: Washington, DC, 1989; pp 465–483.
3. Yamamoto, H. *J. Chem. Soc. 1* **1987a**, 613.
4. Strausberg, R.; Link, R. *Trends in Biotechnology* **1990**, *8*, 53.
5. Waite, J. H. *Comp. Biochem. Physiol.* **1990**, *97B*, 19.
6. Naldrett, M. J. *J. Marine Biol. Ass. U.* K **1993**, *73*, 689.
7. Yamamoto, H.; Nagai, A. *Marine Chem.* **1992**, *37*, 131.
8. Yamamoto, H.; Yamauchi, S.; Ohara, S. *Biomimetics* **1993**, *1*, 219.
9. Yamamoto, H.; Takimoto, T. *J. Mater. Chem.* **1991**, *1*, 947.
10. Yamamoto, H.; Ohkawa, K. *Amino Acids* **1993**, *5*, 71.
11. Yamamoto, H.; Ohara, S.; Tanisho, H.; Ohkawa, K.; Nishida, A. *J. Colloid Interface Sci.* **1993**, *156*, 515.
12. Ohara, S.; Ohkawa, K.; Yamamoto, H. *J. Adhesion Soc. Japan* **1993**, *29*, 345.
13. Yamamoto, H. *J. Adhesion Sci. Technol.* **1987b**, *1*, 177.

MARINE ANTIFOULING COATING MATERIALS

Yukio Imanishi
Department of Material Chemistry
Kyoto University

Organisms that live by attaching themselves to marine rocks and constructions, ships' hulls, fishing nets, and cooling water pipes in electric power stations—such as acorn barnacles, mussels, and algae of various kinds—are called adhesion organisms. Those that cause economic damage (including functional decline in the apparatus or the spoiling of scenery) are called marine-fouling organisms.

The conventional coating materials fall into four groups. First is a dispersion of organotin compound in the usual coating materials. The repellent activity is based on the dissolution of the organotin compound; hence it is biohazardous. Second are polyacrylates of organotin compounds. The ester linkage connecting the organotin compound to the polymer backbone is slowly hydrolyzed and the organotin compound is released into seawater. This is biohazardous, also. Third is polyvinylphenol (Maruka linker PHMC, Maruzen Petrochemical Company). The repellent activity of polyvinylphenol should be based on that of polyphenols and is not very high. Fourth are silicone copolymers. These polymers might be repellent by a different mechanism from the other three types of polymers. Recently, a class of water-based nonstick coatings have been prepared by self-assembly and immobilization of reactive polymeric surfactants.[4] These polymer surfactants contain pendant perfluoroalkyl groups that become oriented so as to yield surfaces with very low energy. The authors claim that coatings with these properties might be used to prevent adhesion of biological foulants.

It has been reported that the repellent activity of alkylphenols is strongly dependent on the nature of the alkyl group.[3] The repellent activity of alkylphenols against the blue mussel is highest when the number (n) of carbon atoms in the alkyl group is 8 or 9.

The repellent activity of alkylphenols relates to the effect of alkylphenols on the fluidity and packing of the lipid membrane. Synthesis of polymers possessing branched-alkylphenol-type structures in the main chain and in the side chains has been suggested.

POLYMERS CARRYING BRANCHED-ALKYLPHENOL-TYPE STRUCTURES

Polymers possessing branched-alkylphenol-type structures in the main chain and in the side chains were synthesized by Sasano et al.[5] *p-t*-Butoxystyrene (TBST) was polymerized with azobisisobutyronitrile (AIBN) as an initiator in a toulene solution under nitrogen atmosphere at 65–75°C to yield poly(TBST) (PTBST).

PTBST was treated with trifluoromethanesulfonic acid (TFMSA) in trifluoroethanol under nitrogen atmosphere at –5°C for 15 min. Different amounts of TFMSA used resulted in different degrees of deblocking reaction to yield *p*-vinylphenol/TBST copolymer of different compositions: P(VP-*co*-TBST).

TBST was copolymerized with 2,3-dimethyl-1,3-butadiene (DB) under conditions similar to those of the homopolymerization.

P(TBST-*co*-DB) was deblocked under conditions similar to the deblocking of PTBST, obtaining P(VP-*co*-DB).

POLYMERS CARRYING SIDE CHAINS POSSESSING QUASISOLUBLE ALKYLBENZENE STRUCTURE

Preliminary experiments with polymers carrying branched-alkylphenol-type structures revealed that alkyl-phenol components should exist in the aqueous phase to exhibit repellent activities. However, solubilization of repellent substances should be strictly forbidden to avoid a biohazard. To reconcile the former condition with the latter, polymers carrying quasisoluble alkylphenol structure were considered, in which repellent-active alkylphenol groups are connected to the insoluble polymer matrix through water-soluble spacer chains (e.g., polyoxyethylene chain). In this way, poly(*p*-nonylphenyl polyoxyethylene ether methacrylate), P(NPM-*n*E), was synthesized.

These polymers do not bear alkylphenol groups because of difficulties in synthesis. Since alkylbenzene groups exist, these polymers may be useful as the reference for alkylphenol-carrying polymers.

REPELLENT, ANTIBACTERIAL, AND ANTIFOULING ACTIVITIES OF POLYMERS

With regard to the antibacterial activity, polymers with protected phenolic groups—PTBST, P(TBST-*co*-DB)—are inactive, but polymers carrying phenolic groups—P(VP-*co*-DB), PVP—are moderately active. In the case of P(VP-*co*-TBST), the activity was low, but copolymers having a high content of VP were slightly active. These observations indicate the contribution of alkylphenol group to antibacterial activity.

The phenolic group appears to make an important contribution to the antifouling activity against attachment of slime and algae. The alkylphenol group should inhibit attachment and growth of bacteria and algae. Polymers carrying the quasisoluble alkylbenzene group are also antifouling.

REFERENCES

1. Ina, K.; Etoh, H. *Kagaku to Seibutsu* **1990**, *28*, 132.
2. Ina, K. *Kagaku to Kogyo* **1991**, *44*, 666.
3. Takasawa, R.; Etoh, H.; Yagi, A.; Sakata, K.; Ina, K. *Agric. Biol. Chem.* **1990**, *54*, 1607.
4. Schmidt, D. L.; Coburn, C. E.; DeKoven, B. M.; Potter, G. E.; Meyers, G. F.; Fischer, D. A. *Nature* **1994**, *368*, 39.
5. Sasano, T.; Takeuchi, T.; Kimura, S.; Imanishi, Y. *Polym. Prepr. Jpn.* **1994**, *43*, 1010.

Matrix Polymerization

See: Template Polymerization

MATRIX POLYMERIZATION (Overview)

Ivan M. Papisov
Moscow State Automobile and Road Technical University

Matrix polymerization, also termed template polymerization, replica polymerization, or matrix polyreaction, is the polymerization of monomers or oligomers with formation of macromolecules whose structure and rate of formation are determined by the information recorded in the structure of other macromolecules: matrices or templates.

In common synthetic polymer chemistry, matrix polymerization dealing with ordinary monomers and polymers may be regarded as a simplest model of matrix synthesis. Several reviews have considered in detail the concept and different aspects of matrix polymerization.[1–6]

According to the conventional scheme of matrix polymerization, mimicking matrix synthesis of nucleic acids, the daughter macromolecule grows along the macromolecule of the matrix due to attachment of these macromolecules to one another by appropriate bonds between their structural (monomer) units. In general, the bonds as well as those binding monomer molecules with structural units of a matrix are of a noncovalent nature (electrostatic, H-bonds, etc.); thus, the matrix may be regarded as unidimensionally adsorbent. **Scheme I** corresponds to chain (a) and condensation (b) matrix polymerization.

"Adsorption" of monomers (comprising oligomers in condensation matrix polymerization) on the matrix may be relatively strong (providing formation of stable matrix–monomer

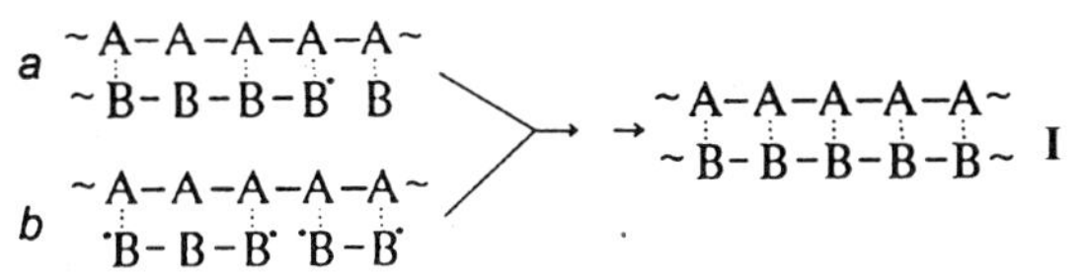

complexes), reversible, or even practically negligible, depending on the energy of the bonds.

The products of matrix polymerization are polycomplexes formed by matrix and daughter macromolecules (see Scheme I, right). Polycomplexes (also called interpolymer or polymer–polymer complexes) may be and usually are prepared by mixing their preformed macromolecular components in a common solvent. Being, as a rule, of a definite composition, polycomplexes possess their own chemical and other properties, which differ from those of their components. Cooperation of interchain bonds systems results in high stability of polycomplexes to dissociation even in very dilute solutions. They can be separated into free components only under special conditions. Formation of a polycomplex is an important sign of matrix polymerization and, consequently, in it each macromolecule of the matrix "works" only one and "dies", being included in the product of the reaction.

STRUCTURAL AND KINETIC EFFECTS IN MATRIX POLYMERIZATION

Within the structure of a daughter macromolecule, a matrix can control: the content of isomers of monomer units and their sequence, composition and sequence of monomer units (in copolymerization), degree of polymerization, and chain conformation. The perfection of the final products can be controlled as well.

The overall rate of matrix polymerization can also be affected by change in the local concentration of monomers inside molecular coils of a matrix as well as by change in the rates of initiation and termination.

It should be mentioned that the solubility of a polycomplex forming in the course of matrix polymerization, as a rule, is less than that of free matrices.

Kinetic Effects

The first trustworthy data concerning matrix polymerization taking place in accordance with Scheme I were published by Kargin, Kabanov, and others. They investigated spontaneous polymerization of 4-vinylpyridine (4Vpy) in presence of polyacids such as poly(styrene sulfonic acid) and poly(acrylic acid) (PAA) taken as the matrix. Essential acceleration of matrix polymerization was observed in comparison with polymerization in the presence of low molecular analogs of the matrix used. First it was assumed that daughter chains represent polycations of usual structures grown via 1-2 addition,[7-10] but Salamone et al.[11-13] proved and then Kabanov et al.[14] confirmed a stepwise 1–6 addition mechanism resulting in formation of polyionene, namely poly(1,4-pyridiniumdiylethylene).

In the majority of systems investigated, intervention of a matrix was shown to increase polymerization rate. However, as soon as a growing daughter chain is attached to the matrix, the latter exerts its own influence on the rates of propagation and termination steps that, in principle, may result in an increase, decrease, or invariability of overall polymerization rate depending on how the rates of the steps change in presence of the matrix.

The matrix macromolecule as microreactor creates its own microsurrounding medium in which the acts of growing and termination are forced to take place, making matrix polymerization in some cases much less sensitive to the nature of a solvent than the blank process.

Polymerization Degree of Daughter Polymer

The most spectacular effects may be expected to appear in condensation matrix polymerization, namely, formation of high molecular fraction of daughter polymer due to concentration on the matrix of all growing chains that are longer than the "critical" one. This effect may be considerably amplified due to the tendency of interacting oligomer–polymer systems to self-organization manifesting in nonequiprobably ("all or nothing" in the limit case) distribution of oligomers between polymer chains.[1]

The Role of Molecular Recognition

The recognition seems to be the necessary step of matrix polymerization, except in the limiting cases dealing with formation of stable complex or covalent bonds between monomer molecules and the matrix. Then, relatively short ("precritical") chains grow independently and only after the recognition does the matrix begin its proper work.

Several essential features of matrix polymerization arise as consequences of the molecular recognition: First, no matter how rigidly a matrix controls the structure of a daughter chain, the latter, in general, contains a fragment or fragments formed independently of the matrix. The fraction of the fragments changes with change in the conditions of polymerization affecting the stability of the polycomplex forming.

Second, if two or more matrices of different structures are simultaneously present in the reaction system, then the growth of daughter chains is controlled only by the strongest matrix, that is, by that giving the most stable polycomplex with the growing chains.

Third, in principle it is possible to create systems where the liberation of a matrix takes place in the process of growth of a daughter chain.

THE PRODUCTS OF MATRIX POLYMERIZATION

The products of conventional matrix polymerization represent polycomplexes formed by matrix and daughter polymers or composites comprising the polycomplex and free matrix or the excess of polymer formed ("polycomplex composites") depending on whether the matrix or the monomer(s) was (were) taken in the excess.

Formation of a perfect double-stranded structure seems to be essential for further modification of ladder-like polycomplexes into ladder polymers using appropriate interchain reactions.[15]

OTHER MATRIX AND PSEUDO-MATRIX PROCESSES

Not only can linear macromolecules be used as matrices; different surfaces of various species such as solid particles, micelles, and films also may be used. Control of propagation in polymerization both on uni- and two-dimensional adsorbents is based on the same principles. For example, oriented films of alternating copolymer covering oriented fibers of polyamide-6,6 were obtained in the copolymerization of acrylic acid and vinylidene chloride in presence of the fibers.[16] The matrix in this case not only controls elementary acts of the process but predetermines the orientation of forming daughter chains.

One of the most fruitful extensions of matrix polymerization seems to be the use of macromolecules to control the processes of low molecular weight compound condensation with formation of a new (e.g., inorganic) phase, if the surfaces of the new phase particles are capable of interacting with the macromolecules, with formation of cooperative systems of noncovalent bonds. The nucleus of the new phase, being much smaller in size than the macromolecule, plays the role of the daughter chain, and the macromolecule, that of the matrix (or the role of the "oligomer" and the "polymer," respectively. Then, the stability of the matrix-particle complex must depend on the size of the particle.

Much published data concerning the formation of a new phase in the presence of macromolecules seem to be dealing with such processes. If these processes are considered as matrix or pseudo-matrix condensation, effective methods of controlling them may be developed based on the principles of matrix polymerization.

REFERENCES

1. Kabanov, V. A.; Papisov, I. M. *Vysokomol. Soed.* **1979**, *A21*, 243.
2. Challa, G.; Tan, Y. Y. *Pure Appl. Chem.* **1981**, *53*, 627.
3. Bamford, C. H. *Chem. Australia* **1982**, *49*, N 9, 341.
4. Tsuchida, E.; Abe, K. *Adv. Polym. Sci.* **1982**, *45*, 1.
5. Papisov, I. M.; Litmanovich, A. A. *Adv. Polym. Sci.* **1989**, *90*, 140.
6. Kabanov, V. A. *Polymerization in Organized Media;* Gordon and Breach Scientific: Philadelphia, PA, 1992; Chapter 7.
7. Kargin, V. A.; Kabanov, V. A.; Kargina, O. V. *Dokl. Akad. Nauk SSSR* **1965**, *161*, 1131.
8. Kabanov, V. A. et al. *J. Polym. Sci.* **1967**, *C16*, 1079.
9. Kabanov, V. A. et al. *J. Polym. Sci.* **1968**, *C23*, 357.
10. Petrovskaya, V. A.; Kabanov, V. A.; Kargin, V. A. *Vysokomol. Soed* **1970**, *A12*, 1645.
11. Salamone, J. S.; Snider, B.; Fitch, W. L. *Polym. Prepr.* **1970**, *2*, 652.
12. Salamone, J. S.; Snider, B.; Fitch, W. L. *Macromolecules* **1970**, *3*, 707.
13. Salamone, J. S.; Snider, B.; Fitch, W. L. *Polymer Preprints* **1972**, *13*, 276.
14. Kabanov, V. A.; Kargina, O. V.; Petrovskaya, V. A. *Vysokomol. Soed* **1971**, *A13*, 348.
15. Litmanovich, A. A.; Markov, S. V.; Papisov, I. M. *Dokl. Akad Nauk SSSR* **1984**, *278*, 676.
16. Tsetlin, B. L.; Golubev, V. N. *Dokl. Akad. Nauk SSSR* **1971**, *201*, 881.

Mechanical Stress

See: Stress-Induced Chemiluminescence Imaging

Mechano–Chemical Initiation

See: Block Copolymers (From Macroinitiators)
Mechano–Chemical Polymerization
Mechano Ions (Macro Ionic Products by Mechanical Fracture)

MECHANO–CHEMICAL POLYMERIZATION

Masato Sakaguchi
Ichimura Gakuen College

Various mechanical operations such as grinding, machining, kneading, mixing, and vibration milling degrade polymeric materials and form free radicals in bulk or on the fresh surface produced by the mechanical operations. The mechanical operations cause homolytic scission of carbon–carbon bonds of polymer main chain and produce so-called mechano radicals, which lead to mechano-chemical phenomena in or on the polymeric materials.

One of these phenomena is mechano–chemical polymerization. Numerous studies of this polymerization in solid state have been carried out based on mechanical actions such as grinding or mixing, mostly in a vibration mill with polymer blends, polymer–monomer blends, or polymerfilled systems. Extensive reviews on this topic have been presented by Watson,[1] Ceresa,[2] Sohma and Sakaguchi,[3] and Casale and Porter.[4]

The main advantages of these polymerizations include many combinations of monomers and polymers without solvent or catalyst, and in some cases the polymerization makes progress at 77 K.[5,6] The disadvantages include the relatively high energy consumption and low concentration of the mechano radical. It is also hard to recognize a quantitative reaction mechanism because of the very complicated phenomena.

RADICAL POLYMERIZATIONS INITIATED BY PTFE MECHANO RADICALS

Methyl Methacrylate (MMA) Monomer

Solid PTFE is mechanically fractured by a vibration glass-ball mill in vacuum at 77 K and produces a powder with fresh surface.

The formation of mechano-radicals is a characteristic feature of polymeric materials. After the milling, MMA monomer was contacted with the PTFE mechano radicals.

The radical polymerization of MMA from the fresh surface of PTFE is initiated by the PTFE mechano radical at 77 K and produces block copolymer of PTFE-PMMA anchored on the PTFE surface.

CATIONIC POLYMERIZATION INITIATED BY PVDF MECHANO CATIONS

The ESR spectrum from milled PVDG with isobutyl vinyl ether (IBVE) under vacuum at 77 K in the dark shows two types of PVDF mechano radicals and cation radicals. The cation radicals of IBVE are induced by electron transfer from IBVE to PVDF mechano cations. The cation radicals of IBVE initiate cationic polymerization of IBVE and produce propagating cation radicals of PIBVE. PVDF mechano cations initiate cation polymerization of IBVE and produced poly(isobutyl vinyl ether) (PIBVE) at 77 K under vacuum without any catalyst, solvent, or counter ion. The molecular weights of the PIBVE produced by the cationic polymerization were $\overline{Mw} = 6.2 \times 10^4$, $\overline{Mn} = 2.3 \times 10^4$.

REFERENCES

1. Watson, W. F. *Chemical Reactions of Polymers*; Fetters, E. M. Ed.; Wiley: New York, 1964; p 1085.
2. Ceresa, R. J. *Block and Graft Copolymerization*; Wiley: New York, 1973 and 1974; Vols. 1 and 2.

3. Sohma, J.; Sakaguchi, M. *Adv. Polym. Sci.* **1976**, *20*, 109.
4. Casale, A.; Porter, R. S. *Polymer Stress Reactions*; Academic: New York, 1978; Vols. 1 and 2.
5. Sakaguchi, M.; Sohma, J. *J. Appl. Polym. Sci.* **1978**, *22*, 2915.
6. Sakaguchi, M. *Fuel* **1981**, *60*, 136.

MECHANO IONS (Macro Ionic Products by Mechanical Fracture)

Masato Sakaguchi
Ichimura Gakuen College

Submitting polymeric materials to mechanical action such as grinding and vibration-milling makes them change to small pieces or powder. The carbon–carbon bond composing the main chain of the polymeric materials is subject to two types of carbon–carbon bond scission.

The first is a homolytic carbon–carbon bond scission that produces macro-free radicals, so-called mechano radicals.

The second is a heterolytic scission of the carbon–carbon bond that produces macro ionic products, so-called mechano anions and mechano cations. The evidence for these products is reported by Sakaguchi et al.[1–5]

REFERENCES

1. Sakaguchi, M. et al. *Polymer* **1984**, *25*, 944.
2. Sakaguchi, M. et al. *Polym. Commun.* **1985**, *26*, 142.
3. Sakaguchi, M. et al. *J. Polym. Sci. Polym. Phys. Ed.* **1987**, *25*, 1431.
4. Sakaguchi, M. et al. *J. Polym. Sci. Polym. Phys. Ed.* **1988**, *26*, 1307.
5. Sakaguchi, M. et al. *Macromolecules* **1989**, *22*, 1277.

Melamine Resin

See: Melamine Resin (Pyrolysis)
Melamine Resins (Overview)
Powder Coatings (Overview)

MELAMINE RESIN (Pyrolysis)

Toshimi Hirata* and Akio Inoue
Forestry and Forest Products Research Institute

Great quantities of melamine–formaldehyde resin are used in the production of adhesives, moldings, and coatings. The bonding strength of melamine resin adhesives is reduced by heat. In addition, melamine resin seems to have a high potential for combustion toxicity by producing an acute gas toxicant, hydrogen cyanide, in the pyrolysis. However, only a few studies have been carried out on this pyrolysis.[1,2]

The chemical structure of the melamine–formaldehyde resin network varies considerably depending on synthesis conditions. A typical structure is given in **Figure 1**. It contains linkages with different bond energies, which makes thermal degradation of melamine–formaldehyde resin complex because bonds are irregularly severed and recombined upon heating.

*Author to whom correspondence should be addressed.

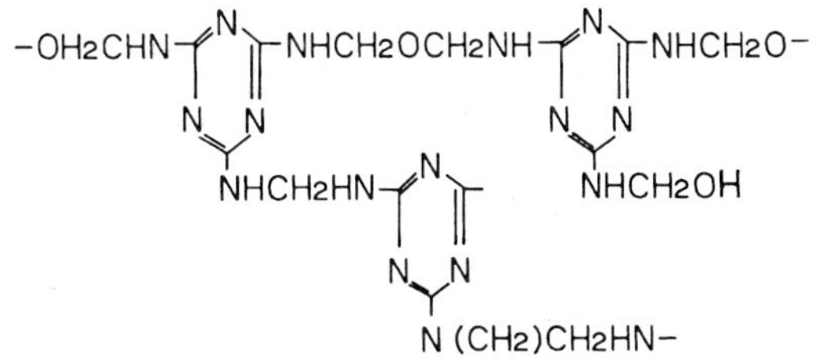

FIGURE 1. Schematic structure of melamine–formaldehyde resin.

We may conclude that the thermal stability of melamine–formaldehyde resin varies depending on curing time and that the pyrolysis progresses through several different steps. The first significant reaction of pyrolysis is the elimination of free side-groups such as methylol groups, which causes slow mass loss. The second and third pyrolysis reactions are the elimination of unstable and stabilized triazine rings, respectively, which give the highest and second highest rates of mass loss, respectively. Finally, char residue, which for the most part consists of carbon and oxygen, is produced through decomposition of hydrogen- and nitrogen-containing structures derived primarily from melamine. The activation energy for mass loss is highest for the third reaction and lowest for the first reaction.

REFERENCES

1. Sekine, Y. *J. Chem. Soc. Jpn., Ind. Chem. Sec.* **1960**, *63*, 1657.
2. Hirata, T. et al. *J. Appl. Polym. Sci.* **1991**, *42*, 3147.

MELAMINE RESINS (Overview)

Mitsuo Higuchi
Kyushu University

Research on the chemistry and application of melamine resin (1,3,5-triazine-2,4,6-triamine polymer with formaldehyde, CAS Registry Number: 9003-08-1) started in the late 1930s when Swiss CIBA developed the industrial production of melamine using dicyandiamide.[1] A variety of melamine-based resins are now widely used as adhesives for wood, resins for decorative laminates, varnish, moldings, and for improving the properties of paper and cellulosic textiles.

The reactions and preparation of the melamine resins in connection with the applications are briefly described in the following section, and newly obtained information on the polymeric structures of melamine-based composite resins is described in the final section.

REACTIONS AND PREPARATION OF THE MELAMINE RESINS

Melamine resin alongside urea resin, is classified as one of the amino resins of importance. The two are similar in their reactions, production, and applications and are often used in blended forms. The melamine resin has higher durability against heat and moisture but is higher in price.

Melamine reacts with formaldehyde in weakly alkaline-aqueous media to form methylol compounds. As the melamine is hexa-functional, mono- to hexa-methylol monomers can be formed. Condensation reactions take place under neutral and acidic conditions forming methylene or dimethylene ether bonds and the hardening of the resin occurs forming a three-dimensionally crosslinked network.

Polymeric Structures of Melamine-Based Composite Adhesives

Melamine resins blended with urea resins or so-called melamine–urea (cocondensed) resins have been used as adhesives for wood since the melamine resin began to take part in the production of bonded wood products. No successful NMR techniques have been reported for determining the amounts of methylene and dimethylene ether bonds that might be formed between melamine and the other components in the hardened resins. However, recently the curing behavior of these resins has been investigated in detail by use of a technique that isolates the three-dimensionally crosslinked parts from the resins in the course of curing, and analyzes their composition. And the polymeric structures of the melamine-based composite adhesives have been deduced from the information obtained by the investigations of the curing behavior of the adhesives and by the analysis of the behavior of a cured resin toward acid hydrolysis.[2]

REFERENCES

1. Swiss Patent 189 406, 1937; 199 784, 1938; 200 244, 1938.
2. Higuchi, M.; Roh, J-K.; Tajima, S.; Irita, H.; Honda, T.; Sakata, I. *Abstracts of Papers, Adhesives & Bonded Wood Symposium*, Seattle, WA 1991; Forest Products Society: Madison, WI, 1994 p 429.

MELANIN (Structure and Chemico-Physical Properties)

Yasuhiro Miyake
Department of Polymer Science
Faculty of Science
Hokkaido University

Yoshinobu Izumi
Macromolecular Research Laboratory
Faculty of Engineering
Yamagata University

Ryusuke Kona
Department of Applied Physics
The National Defense Academy

Melanin is a biopolymer that occurs widely in animals and vegetables. The term usually refers to black or brown organic pigments. According to the reviews,[1–4] melanins have been divided into three classes according to their original sources and colors: eu-melanin (ευ=good), phaeo-melanin (πηαεο=dusky) and allo-melanin (αλλο=other).

Eu-melanin is black and is found in hair, melanoma, the midbrain, the inner ear, the ink sacs of cephalopods, and so on. It is composed of C, H, O, and N, or synthesized with tyrosine or 3,4-dihydroxyphenylalanine (dopa). Phaeomelanin varies from brown to yellow and is found in blond hair, red fox fur, bird feathers, and so on. It is composed of C, H, O, N, and S or synthesized with tyrosine and cysteine. Allo-melanin is a black substrate found in seeds, mushrooms, bacteria, ebony, and so on, and is composed of C, H, and O or synthesized with catechol.

Most research has been done on eu-melanins. The study of melanin structures has encountered great difficulties because of melanin's insolubility in organic solvents except for alkaline solution, making their purification difficult. Because of this insolubility, melanin has been thought to be highly crosslinked. Moreover, melanin is thermally stable to > 200°C.

STRUCTURAL STUDIES

Eu-melanin is an irregular copolymer composed mainly of indole rings, but a theory that dopamelanin is a homopolymer of indolequinone type has also been proposed.[3] On the other hand, different molecular models have been presented by X-ray diffraction analysis of tyrosine–melanin.[4,6] According to X-ray analysis, melanin is composed of four or eight basic planar indolequinone molecules as a monomer unit; planar monomer units tend to form a lamellar structure and a graphite-like stacking with planar layers. In this model, melanin is treated as a homopolymer, connected with four planar indolequinones as a monomer unit or constituted as a large disk-like molecule with more than four planar indolequinones as a base unit. Melanin can be dissolved in a group of aqueous alkaline solutions that are good solvents for cellulose.[7] The shapes of these melanins are elliptic and cylinderlike and obtained in various dimensions.[8,9] It has not been proved experimentally that natural and synthetic melanins are identical.

CHEMICO-PHYSICAL PROPERTIES

Mechanical Behavior[10,11]

Anomalous absorption and dispersion of sound waves in 20% hydrated synthetic diethylamine (DEA) dopamine–melanin were measured at temperatures ranging from −80°C to 50°C. The longitudinal attenuations per wavelength at frequencies between 1 Mhz and 5 Mhz showed an enormous amount of absorption above room temperature, attributed to hydration of the sample.

Sound absorption is largely affected by time and annealing temperature. The relaxation process depends on preparation methods, molecular weight, and pH of samples. It is reported that both resonance and relaxation phenomena have been found. Their origin has not yet been explained clearly.

Magnetic Properties

Melanin has many radicals that are considered to be semiquinone type.[1,2,5] The ESR spectra of various melanins were quite similar, regardless of natural or synthetic origins.[2,3] Each spectrum, except that of phaeo-melanin, is a single broad, slightly asymmetric absorption line with a range of 2.003 – 2.004 g value, which is close to the 2,0023 value of a free radical. The values implied an electron delocalization over one or two aromatic rings. Phaeo-melanin showed ESR spectra with hyperfine splitting.[12,13]

Specific Heat Measurement[14]

Specific heats have been measured on 1% hydrated melanin, 20% DEA-doped melanin, and melanosomes from human malignant melanoma over a low temperature range. A transition near 1.9 K and an unusually high linear contribution to temperature dependency have been bound. The former may be associated with a magnetic transition, possibly from paramagnetism to antiferromagnetism, which is naturally related to unpaired electrons associated with free radicals. The linear contribution may be associated with a special magnetic origin rather than with the amorphous structure. These anomalies may arise as a result of electron–photon coupling and high densities of unpaired spins. The chemico-physical properties of melanins are strongly influenced by circumstances, hydration, doped metal ions, doped organic solvents, pH, and so on.

REFERENCES

1. Nicolaus, R. A. *Melanins;* Hermann: Paris, 1968.
2. Blois, M. S. Jr. *Biology of Normal and Abnormal Melanocytes* Kawamura, T. et al., Eds.; University of Tokyo: Tokyo, 1971.
3. Swan, G. A. *Fortsch. Chem. Org. Naturst.* **1974**, *31*, 521.
4. Shiu-Shin, C. Ph.D. Thesis University of Houston: 1977.
5. Blois, M. S. Jr. et al. *Biophys. J.* **1964**, *4*, 471.
6. Zajac, G. W. et al. *Biochim. Biophys. Acta.* **1994**, *1199*, 271.
7. Henley, D. *Arkiv Kemi* **1961**, *18*, 327.
8. Miyake, Y. et al. *Structure and Function of Melanin* Jimbow, K. Ed. Sapporo Medical College: Sapporo, Japan, 1987; Vol. 4.
9. Miyake, Y. *Proceedings of the 23rd Meeting of Hokkaido Branch, The Society of Polymer Science, Japan* **1989**, *23*, 25.
10. Kono, R. et al. *J. Appl. Phys.* **1979**, *50*, 1236.
11. Kono, R. et al. *J. Chem. Phys.* **1981**, *75*, 4654.
12. Sealy, R. C. et al. *Science* **1982**, *217*, 545.
13. Sealy, R. C. et al. *Proc. Natl. Acad. Sci. U.S.A.* **1982**, *79*, 2885.
14. Mizutani, U. et al. *Nature* **1976**, *259*, 505.

Melt Spinning

See: Engineering Plastics, Melt Spinning
Textile Fibers (Structure and Properties)

MELT SPINNING (Polymer Formation Under High Stress Conditions)

Roland Beyreuther, Harald Brünig, and Roland Vogel
Institute of Polymer Research

MELT SPINNING AS A SPECIAL PROCESS OF POLYMER PROCESSING

Synthetic fibers made of organic high polymers are spun according to two fundamentally different technologies:

- Spinning from a 10–20% polymer solution by continuous precipitation in a so-called spin bath (wet spinning). Typical fibers spun this way are: polyacrylonitrile fibers (PAN), and cellulosic fibers spun from the viscous spin solution in an intermediate step (VIS).
- Spinning from a polymer melt by continuous takeup of separate thin melt streams (melt spinning). Typical fibers spun this way are: polyamide 6 fibers (PA 6, trade names Perlon, Caprolan, Celon, Enkalon, Grilon, etc.); polyamide 6.6 fibers (PA 6.6, trade names Nylon, Antron, Edlon, etc.); and polyester fibers (PET, trade names Dacron, Trevira, Diolen, Terylene, Grilene, etc.)

In wet spinning only the dissolved polymer (the clearly smaller part) in the solution is changed into the fiber form. In melt spinning 100% of the polymer melt is changed to the fiber form. The advantages of melt spinning are not limited to its simple technology, lower energy consumption, and lower use of chemicals. Melt spinning can be carried out with higher productivity and fiber formation velocities that are several orders of magnitude higher.

These advantages are one reason that 75% of the 18.4 million tons of man-made fibers produced per year are melt-spun.[1] But the basic condition for melt spinning is that the polymer must be thermoplastic, that is, must be meltable.

Melt spinning is a special process of thermoplastic polymer formation in both technological and polymer physical aspects. More detailed information on melt-spin technology and melt-spin polymers is given in publications by Falkai,[2] Rogovin,[3] Ziabicki,[4] and—especially on high-speed spinning technologies—by Ziabicki and Kawai.[5]

Peculiarities of Melt Spinning

A number of structural and rheological conditions must be fulfilled for a thermoplastic to be spinnable. These conditions have not yet been completely established for closed-fiber formation. Researchers are trying to develop a model to describe fiber spinnability. This requires establishing the cause–effect relationship between the polymer structure and its (high speed) formation properties. Researchers also must understand why only some thermoplastics are spinnable and what techniques are needed to change nonspinnable thermoplastics into spinnable types.

RHEOLOGICAL ASPECTS OF MELT SPINNABLE POLYMERS

Melt spinning is a process with high demands on the mechanical properties of the processed melt. The melt is carried by a pressure drop through the die. Because all polymer melts show viscoelastic behavior, elastic energy is stored during the flow in the die. Immediately after the outflow of the melt, the stored elastic energy relaxes. This causes the typical extrudate to swell. Application of gravity, inertia, and air drag extends the melt uniaxially in the flow direction while it is cooled down simultaneously. These forces act against the rheological force, which consists of an elastic and a viscous part of strain. The viscosity must be sufficiently large to resist the applied forces. However, it must not be so large as to prevent the necessary deformation before the start of solidification. The elastic part of deformation must be small because an elastic deformation happens more rapidly than a viscous one. Therefore a considerable elastic deformation leads, because the high forces applied, to a cohesive fracture of the fiber in molten state.

In contrast to a polymer solid, the rheological properties of a polymer melt point include low stresses and long relaxation times. The long relaxation times suggest that only large-scale molecular motions are involved. The details of the chemical structure of the polymers that are averaged out can be ignored. According to Marrucci,[8] only the following structural features are retained in the rheological properties of a polymer melt of homopolymers or statistical copolymers:

- The chain flexibility, as measured by the persistence length. It is determined with the structure of the monomer units.
- The averaged molecular weight.
- The molecular weight distribution.
- The long-chain branch structure, if any.

Outlook

A great deal of research must still be done to improve the description of the fiber formation process. Some of the topical fields for modeling are:

- the multifilament spinning process, with different air friction and cooling conditions for each fiber within the bundle;[7–9]
- new polymers high-performance polymers, and blends;[10,11]
- rheological behavior, crystallization effects, and fiber structure;[12–14]
- necking and radial differentiation of the fiber;[15,16]
- spinning stability and draw resonance;[17] and
- special manufacturing processes, including a melt-blown, nonwoven spunbonded process.[18–20]

REFERENCES

1. *Chemiefasern/Textilindustrie* **1994**, *44/96*, 500.
2. Falkai, B. *Synthesefasern: Grundlagen, Technologie, Verarbeitung und Anwendung*; Verlag Chemie: Weinheim, Germany, 1981.
3. Rogovin, Z. A. *Chemiefasern, Chemie-Technologie*; Georg Thieme Verlag: Stuttgart, New York, 1982.
4. Ziabicki, A. *Fundamentals of Fiber Formation*; Wiley: London, 1976.
5. Ziabicki, A.; Kawai, H. *High Speed Fiber Spinning-Science and Engineering Aspects*; John Wiley & Sons: New York, 1985.
6. Marrucci, G. *Makromol. Chem. Macromol. Symp.* **1993**, *69*, 181.
7. Matsuo, T. et al. *Preprints Int. Symp. Chemiefaserstoffe*; Kalinin: All Union Chemical Society: USSR, 1977; p 206.
8. Dutta, A. *Polym. Eng. Sci.* **1987**, *27*, 1050.
9. Schöne. A.; Brünig, H. *Arch. Mech.* **1990**, *42*, 571.
10. Jiang, H. et al. *Mater. Sci. Technol.* **1993**, *12*, 597.
11. Kudo, K. et al. *J. Appl. Polym. Sci.* **1994**, *52*, 861.
12. Dumazet. Ph. et al. *J. Polym.* **1994**, *35*, 2823.
13. Chan, T. W.; Isayew, A. I. *Polym. Eng. Sci.* **1994**, *34*, 461.
14. Parravicini, L. et al. *J. Appl. Polym. Sci.* **1994**, *52*, 875.
15. Ziabicki, A.; Tian, J. J. *Non-Newtonian Fluid Mech.* **1993**, *47*, 57.
16. Zahorski, S. *Non-Newtonian Fluid Mech.* **1993**, *50*, 65.
17. Lee, W. S.; Park, C. W. *Annu. Tech. Conf.-Soc. Plast. Eng.* **1992**, *50*, 2183.
18. Chen, G. Y. et al. *J. Appl. Polym. Sci.* **1992**, *44*, 447.
19. Haji, N. et al. *INDA J. Nonwoven Res.* **1991**, *4*, 16.
20. Milligan, M. W. et al. *J. Appl. Polym. Sci.* **1992**, *44*, 279.

Melting

See: Polymer-Polymer Mixtures (under High Pressure; Physical Properties)

Membrane Oxygenators

See: Lung, Artificial

Membranes

See: Cell Entrapment [Poly(carbamoyl sulfonate) Hydrogels]
Cellulose (From Fly Cotton Mill Waste)
Cellulose (Overview)
Functional Fibers
Gas Separation Membranes
Intelligent Membranes
Ion-Exchange Membranes
Ion-Selective Electrodes and Biosensors
Ion-Selective Microchemical Sensors (Cardiology)
Ionomeric Membranes, Perfluorinated
Ionomers (Overview)
Lipid Membranes (High Stability, from Archaebacterial Extermophiles)
Membranes (Overview)
Membranes, Porous (Synthesized from Microemulsions)
Nonthrombogenic Materials
Perfluorinated Ionomers (Overview)
Pervaporation Membranes (Overview)
Pervaporation Membranes (for Separating Organic Solvents)
Pervaporation Membranes (Materials)
Polybenzimidazoles (Overview)
Poly(imide siloxane)s
Polyoxadiazoles and Polytriazoles
Polypyrroles (From Isoporous Membranes)
Reactive Oligomers (Overview)
Stimuli-Responsible Microcapsules

MEMBRANES (Overview)

D. Paul and K-V. Peinemann
GKSS Research Center

A membrane is an intervening phase that acts as a barrier to the transport of matter between phases on either side.[1] Man-made membranes are thin polymer films whose chemical and physical properties and morphology allow the separation of chemical mixtures without altering the components. Membrane development begins at the point where polymers with specific molecular and supramolecular structures have been developed for the separation of specific substances, or when they are already available.

Membrane rejection is caused by pores in the membrane (substances with geometric dimensions larger than the pore diameter cannot permeate; this is the so-called sieving mechanism) or by different solubilities and diffusivities of components in dense membranes (solute-diffusion mechanism).

The driving force for the transport of a component across the membrane[2] is the difference of concentration, pressure, electrical potential, or temperature. Whereas the porosity of a membrane (important for the pressure-driven process of microfiltration and ultrafiltration of liquids) is determined by the formation of the membrane from a polymer, the permeability of gases or vapors is dependent on their solubility and diffusion by a difference of concentration in the supramolecular polymer structure. Polymers for membranes in electrodialysis (electrical potential as driving force) need a high equivalent concentration of ionic groups to separate charged components of a solution. There are many different demands on the properties of polymers for membranes.

MATERIALS

The principal demands on the properties of membranes are high flux or permeability and high selectivity or rejection. Other desirable properties include pH and temperature stability, mechanical stability for handling, long lifetime without fouling, low compressibility and, in medicine, high biocompatibility. Low material cost is also important.

Cellulose derivates were the first polymers used to produce synthetic membranes for technical applications. Today, cellulose acetates (CA) are still widely used to produce reverse osmosis membranes for the desalination of sea and brackish water. Although CA membranes exhibit lower water fluxes than thin-film composite membranes (e.g., aromatic polyamides), they are still competitive because of their much better resistance to oxidizing agents such as chlorine or ozone. Polyolefins (PE, PP) are important for the production of microfiltration membranes. The polyolefins, especially isotactic polypropylene, are excellent solvent-resistant polymers, which are also stable in environments with extreme pH values.

Silicones, or polyorganosiloxanes, are rubbery amorphous polymers. Polydimethylsiloxane (PDMSi) is an especially important membrane material, because it shows a significantly higher permeability to gases and organic vapors than all other technical polymers. Another important class of membrane polymers are polysulfones (Psu) and poly(ether sulfone)s (PESu). They are widely used to produce ultrafiltration membranes. The polysulfones can be fabricated into ultrafiltration membranes quite easily by phase inversion. Polysulfone membranes are also widely used as a support for the fabrication of thin-film composite membranes. The majority of today's reverse osmosis membranes consist of a polysulfone support covered by a thin selective layer of aromatic polyamide (PARA).

The polyimides (PI) have excellent thermal stability and high chemical stability. Despite good water fluxes and rejections, the market for polyimide ultrafiltration membranes is limited because of insufficient long-term stability at high or low pH values.

Polyphosphazene (PPN), poly(ether ether ketone) (PEEK), modified polyacrylonitrile (PAN mod), and polyelectrolyte complexes (PEL) are characterized by enormous variability as membrane polymers: PPN can be derivated in a wide range. PEEK is normally not soluble, but by introduction of cardiogroups solubility in chlorinated hydrocarbons is reached. Acrylonitrile is the basic material for different copolymers; for example, PAN mod may be a reactive copolymer with carboxylic groups. PEL, formed by ionomers, consists of cationic and anionic natural or synthetic polymers.

APPLICATION

For application the membranes have to be fixed in a technical device, a so-called module.[17]

Medicine

In recent years ~ 50% of the membrane market has been in medical therapy and diagnostics. For extracorporeal blood detoxification by hemodialysis or hemofiltration, > 400 types of dialyzers and filters with membranes of different polymers are on the market.[18]

Some 80% of hemodialyzers are made from cellulose membranes. Most of them are hollow fibers. In order to obtain higher blood compatibility, cellulose hollow fiber membranes were modified by additives (e.g., diethylaminethylcellulose), or high flux membranes from polysulfone or polycarbonate are applied.

Membranes are increasingly used to influence transport for the drug delivery systems. Research projects include liver assist systems with liver cells fixed onto the surface of membranes. Encapsulation of Langerhan's islets in membrane capsules from polyelectrolytes marks a possible way to bioartificial organs.

Technical Applications

Microfiltration now accounts for nearly 50% of the total membrane systems market (medical applications excluded).

A major breakthrough in membrane technology was the development of integral-asymmetric cellulose acetate membranes by Loeb and Sourirajan.[6] The water flux through this new membrane type was so high that membrane technology could compete with distillation for desalination of sea and brackish water. Asymmetric cellulose acetate membranes sill hold a significant portion of the water desalination market because they are inexpensive and resistant against such sterilizing agents as chlorine. For most reverse osmosis applications, however, thin-film composite membranes are used.[3]

A significant application of membrane technology is the production of ultrapure water for the microelectronic industry.

Membranes find a broad application in the treatment of wastewater for pollution control or recovery of high-value products. Examples are pain recovery and pollution abatement for electropainting operations and recovery of sizing chemicals in the textile industry. Another well-established application concerns oil separation from waste oil/water emulsion. The largest applications of membrane technology in the food and dairy industry are concentrations of cheese whey and milk and clarification of fruit juice and alcoholic beverages. Microfiltration, ultrafiltration, and reverse osmosis all use porous membranes. Typical pore sizes are 0.05–10 μm for microfiltration, 1–100 nm for ultrafiltration, and < 2 nm for reverse osmosis. Mainly pore-free membranes are being used for pervaporation and gas separation. Pervaporation has two main applications: removal of small amounts of water from organic solvents and removal of small quantities of organics from water.

The most important applications of gas separation by membranes are air separation for enriched nitrogen or oxygen production, hydrogen/nitrogen separation in ammonia plants, hydrogen/carbon monoxide separation for adjusting syngas composition, and carbon dioxide/methane separation in natural gas purification. Commercial polymers such as cellulose acetate, polysulfone, and polymethylpentene are used to make gas separation membranes, but "tailor-made" polymers with high selectivities are becoming more and more important.[7] The removal and recovery of organic vapors from contaminated gas streams by membrane separation are finding increased acceptance.

TRENDS

Separation processes with polymer membranes are an attractive alternative for product separation. A general advantage of membrane processes is their simplicity. They can easily be combined with other separation processes, leading to optimized technologies.

New or modified polymers with high transport rates for permeants, combining separation and reaction with catalytic active membranes, and membranes as support for reactive groups (e.g., fixed carrier[8] will be developed. Temperature-stable polymer membranes as reactive separating systems will also be available in the future.

REFERENCES

1. Gekas, V. *Desalination* **1988**, *68*, 77.
2. Pusch, W. *Desalination* **1986**, *59*, 105.
3. Petersen, R. J. *J. Membr. Sci.* **1993**, *83*, 81.
4. Rautenbach, R.; Albrecht, R. *Membrane Processes*; John Wiley & Sons: New York, 1989.
5. Sigdell, J. E. *Artif. Organs* **1986**, *10*, 156.
6. Loeb, S.; Sourirajan, S. ACS Advances in Chemistry, Series *38*; Washington, DC, 1963; p 117.
7. Koros, W. J.; Fleming, G. K. *J. Membr. Sci.* **1993**, *83*, 1.
8. Way, J. D.; Noble, R. D. *Membrane Handbook*; Van Nostrand Reinhold; New York, **1992**; Chapter Facilitated Transport, 44, p 833.

MEMBRANES, POROUS (Synthesized from Microemulsions)

H. Michael Cheung and W. R. Palani Raj
Department of Chemical Engineering
The University of Akron

The polymerization of monomer-containing microemulsions as a potential route for synthesizing polymer with specific morphological and compositional characteristics has recently received considerable attention.[1] Microemulsions are thermodynamically stable, microstructured systems containing hydrophilic or hydrophobic components that are stabilized using surface active species. We have examined the polymerization of monomers incorporated in a particular class of microemulsions known as Winsor-IV systems.[2] These are macroscopically monophasic, transparent, isotropic systems that exhibit organization at a microstructural level.[3,4] The labile structural organization of these systems is dependent on composition and consists of aqueous droplets dispersed in oil at low aqueous contents, which change droplets of oil dispersed in the aqueous phase at high aqueous content. At intermediate aqueous content bicontinuous microemulsions are formed. In bicontinuous microemulsions the oil and aqueous domains form an interconnected structure.[5,6]

Work on polymerizing monomer containing Winsor-IV microemulsions indicates the possibility of forming polymeric materials with controlled morphological features. The morphology of the polymer obtained has significant resemblance to the structural characteristics of the microemulsion precursor. The polymerization of monomer present in the dispersed phase in microemulsions exhibiting droplet structure results in the formation of stable microlatex particles.[1] Closed-cell porous polymer is obtained by polymerization of monomer present in the continuous phase of microemulsions with droplet structure.[7–10] Recent research on the polymerization of monomer-containing microemulsions with bicontinuous structure indicates that polymerization of microemulsions exhibiting bicontinuous characteristics is a potential route for synthesizing open-cell porous polymeric materials.[7–20]

Our earlier research on polymerization of microemulsions of methyl methacrylate (MMA), acrylic acid (AA), ethyleneglycol dimethacrylate (GDMA), sodium dodecylsulfate (SDS), and water had indicated the formation of open-cell porous solids from microemulsions showing inferential evidence of a bicontinuous structure.[11–14] It was also observed that the morphology of the porous structure in the polymer could be varied by varying the structure of the microemulsion precursor through compositional changes. There have recently been efforts to utilize polymerization in microemulsions to form porous polymeric membranes.[21–24]

PROPERTIES

Microemulsion Characterization

Conductivity studies were carried out using microemulsion samples in which the ratio of MMA to AA was maintained constant and the amount of aqueous surfactant solution was varied. The conductivity of the microemulsion shows a small increase up to 20% aqueous content. A marked change in the conductivity behavior of microemulsions is observed at aqueous contents above 20%, accompanied by a sharp increase in conductivity up to 80% aqueous content. At aqueous contents above 80% the conductivity does not change significantly with aqueous content.

The viscosity of the microemulsion and the intensity of scattered light were also measured as a function of aqueous content, and the trend was qualitatively similar to the variation in conductivity. These results also point to variation in microemulsion structure with aqueous content.

Polymerization Studies

Detailed investigations on the polymerization of microemulsions with different structures were performed and the polymer characterized using scanning electron microscopy (SEM), thermogravimetric analysis, BET adsorption studies, and differential scanning calorimetry. Thermogravimetric analysis was used to confirm continuity in the pore structure by studying the drying rate curve obtained on drying the porous materials. BET adsorption studies were used to determine the surface area of the porous materials. The pore size distribution of the water-saturated open-cell materials was determined by freezing-point depression measurements using a differential scanning calorimeter. These studies[7–10] conclusively indicated that polymer from water-in-oil droplet microemulsions had a closed-cell porous structure and bicontinuous microemulsions could be polymerized to form open-cell porous materials.

Permeability Measurements

The permeability of the membranes obtained from precursor microemulsions containing AA and MMA in the ratio 4:1 was measured using an aqueous acrylic acid solution of conductivity 2750 μS/cm. The permeability of the membrane is directly related to the increase in conductivity of distilled water in the permeability cell. The permeability of membranes formed from microemulsions containing aqueous contents < 20% is low compared to membranes from microemulsions of higher aqueous content. A sharp increase in permeability of the membranes to acrylic acid is observed at aqueous contents of the precursor microemulsion exceeding 20%.

The results of permeability measurements of the membranes are consistent with the porous characteristics observed from the earlier studies of the polymer morphology[7,10] and the results of the swelling studies. As the water content of the precursor microemulsion is increased, the porosity of the membrane obtained increases. In addition, the extent of swelling of membranes increases with water content of the precursor microemulsion. These effects result in an increase in the number and size of pathways through which the diffusion of acrylic acid could occur.

Tensile Properties

The results of tensile measurements on membranes synthesized from microemulsions were determined as a function of the aqueous content of the microemulsions. The results indicate the membranes to be weak and soft. The tensile properties show a decreasing trend in the elongation and tensile strength at yield of the membranes with increasing water content of the precursor microemulsion. The elongation and tensile strength at yield are inversely related to the water content of the precursor microemulsion. This could be due to the increasing porosity of the membrane with increasing water content of the precursor and consequent decrease in the polymer content of the system, which is the stress-bearing component. The elongation at yield and the tensile strength at yield are rather low to permit the application of these membranes directly in separations, requiring significant differential pressures to be applied across the membrane. This indicates that reinforcement is needed to make the membranes suitable for such applications.

APPLICATIONS

We have demonstrated feasibility of forming porous polymeric membranes by polymerizing bicontinuous microemulsions. The porosity of the membrane and hence its permeability can be altered by selecting microemulsions of suitable composition. This technique offers the interesting possibility of developing membranes with modified surface characteristics by functionalizing the surface using polymerizable surfactants or comonomers during the formation of the microemulsion. We have also shown the development of an asymmetric pore morphology in the membrane depending on the conditions of application. This technique of membrane synthesis provides the opportunity to design membranes of specific pore morphology and permeability by selecting a microemulsion precursor of suitable structure.

REFERENCES

1. Candau, F. In *Polymerization in Organized Media*; Paleos, C. M., Ed.; Gordon and Breach: Philadelphia, 1992; p 215.
2. Winsor, P. A. *Trans. Faraday Soc.* **1948**, *44*, 376.
3. Chen, S. J. et al. *J. Phys. Chem.* **1984**, *88*, 1631.
4. Loic, A. et al. In *Microemulsion Systems*; Rosano, H. L.; Clausse, M., Eds.; Marcel Dekker: New York, 1987; p 225.
5. Clausse, M. et al. *J. Colloid Interface Sci.* **1982**, *87*, 584.
6. Warr, G. et al. *J. Phys. Chem.* **1988**, *92*, 774.
7. Palani Raj, W. R. et al. *Langmuir* **1991**, *7*, 2586.
8. Sasthav, M. et al. *J. Colloid Interface Sci.* **1992**, *152*, 376.
9. Palani Raj, W. R. et al. *Langmuir* **1992**, *8*, 1931.
10. Palani Raj. W. R. et al. *J. Appl. Polym. Sci.* **1993**, *47*, 499.
11. Stoffer, J. O.; Bone, T. *J. Dispersion Sci. Tech.* **1980**, *1*, 37.
12. Stoffer, J. O.; Bone, T. *J. Dispersion Sci. Tech.* **1980**, *1*, 393.
13. Stoffer, J. O.; Bone, T. *J. Polym. Sci.: Polym. Chem. Ed.* **1980**, *18*, 2641.
14. Gan, L. M.; Chew, C. H. *J. Dispersion Sci. Tech.* **1984**, *5*, 179.
15. Chew, C. H.; Gan, L. M. *J. Polym. Sci.: Polym. Chem. Ed.* **1985**, *23*, 2225.
16. Haque, E.; Qutubuddin, S. *J. Polym. Sci., Part C: Polym. Lett.* **1988**, *26*, 429.
17. Qutubuddin, S. et al. In *Polymer Association Structures: Microemulsions and Liquid Crystals*; El-Nokaly, M. A. Ed.; ACS Symposium Series 384; American Chemical Society: Washington, DC, 1989; p 64.
18. Menger, F. M. et al. *J. Am. Chem. Soc.* **1990**, *112*, 1263.
19. Menger, F. M. et al. *J. Am. Chem. Soc.* **1990**, *112*, 6723.
20. Sasthav, M.; Cheung, H. M. et al. *Langmuir* **1991**, *7*, 1378.
21. Palani Raj, W. R. et al. *Polymer* **1993**, *34*(15), 3305.
22. Pashley, R. M. et al. *Patent Cooperation Treaty Int. Appl.*, WO 9 201 506.
23. Gupta, B.; Eicke, H. F. *In Polymer Science: Symposium Proceedings of Polymer '91*, Vol. 2; Sivaram, S., Ed.; Tata McGraw-Hill: New Delhi, 1991; p 681.
24. Palani Raj, W. R. Ph.D. Dissertation, University of Akron, 1994; p 21.

Mesophase Pitch

See: Carbon Fibers (Overview)

Metal-Chelating Polymers

See: Adsorptive Resins
Chitin (Metal Ion Chelation and Enzyme Immobilization)
Dental Polymers (Glass-Ionomers Cements)
Dental Polymers (Hydrogels)
Ion-Chelating Polymers (Medical Applications)
Metal Complexation Polymers
Metal Complexes, Macromolecular
Metal-Containing Monomers (Polymerization)
Metal Ion Binding Resins (Synthesis and Analytical Properties)
Metal-Polymer Complexes
Molecular Recognition (Peptide-Based Systems)

Metal Clusters

See: *Metal Clusters (in Polymers, Synthesis and Characterization)*
Metal Complexes, Macromolecular
Metal Nanoclusters (within Block Copolymer Domains)

METAL CLUSTERS (in Polymers, Synthesis and Characterization)

Galo Cárdenas-Trivino
Departmento de Polímeros
Facultad de Ciencias Químicas
Universidad de Concepción

Kenneth J. Klabunde
Department of Chemistry
Kansas State University

We have created a synthetic approach to preparing metal clusters trapped in solid organic polymers on the basis of our experience in the synthesis of colloidal metals in nonaqueous solvents.[1] Wright[2] has reported interesting approaches to the synthesis of polymer-trapped metal clusters based on atom agglomeration in organic solvents. This method, which involves deposition of metal vapor with low-temperature organic monomers followed by controlled atom accretion, is wide in scope and can be employed with a variety of metals and solvents. It allows fairly large-scale preparation of such colloidal sols, which are of great interest in colloid chemistry and other areas.

METAL MONOMER-COLLOIDS

The first time that a relatively nonpolar organic solvent, styrene, was found to allow formation and stabilization of metal colloids was reported earlier.[3] This must be due to the ligating action of the unsaturated bonds in styrene. The importance of this special ligating/solvating effect was further appreciated when we realized that metal particles in liquid styrene did not exhibit an electrophoretic mobility, as they do in the more polar solvents. Thus, the particles are not charged. This should be because of the ligating action of the unsaturated bonds in styrene.

Methyl methacrylate and other monomers behave similarly. The unsaturated bonds in MMA probably coordinate as shown in **Structure 1**:

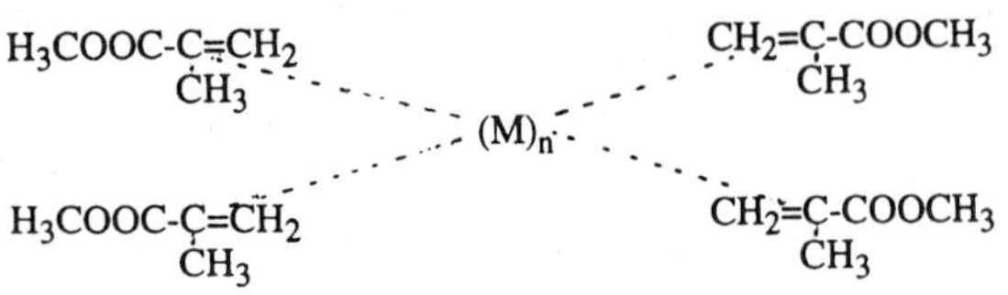

1

During polymerization the metal clusters tend to weakly agglomerate until solidification eventually traps them.

In several cases the stability of the colloids can be related to the inherent stability of the metal clusters; thus Au, Ag, and Pd in vinyl acetate show stability over 3 months, probably because of the oxidative stability of the metals. On the other hand, metals that oxidize more readily, such as Cd, Zn, In, Sn, and Ge, do not form stable colloids.

METHOD

A metal atom reactor has been used to prepare metal clusters.

Bulk Radical Polymerization

AIBN and benzoyl peroxide (BPO) radical initiators have been used. The AIBN was recrystallized from chloroform and BPO from ethanol. Cationic and anionic polymerization were attempted without any success, probably because of complexation with metal atoms.

Thermal Studies and Decomposition Kinetics

The synthesis of polymers with metal incorporated from colloids and with metal dispersed in methyl,[4] ethyl, and butyl methacrylates[7] has been reported. The polymers prepared have a wide range of molecular weight, stability, morphology, and colors according to the metal. The amount of metal incorporated was very low, ranging from 0.05 to 2.0%. However, this small amount of metal is enough to change some physical properties of the polymers.

The average molecular weight ($\overline{M}_v$) of these copolymers ranges from 1.0 to 7.9×10^5.[8] These copolymers exhibit a low amount of metal incorporated, ranging from 0.84 to 2.0% (w/w). The presence of these metal clusters is responsible for the differences in thermal decomposition temperatures and those in decomposition activation energy values.[6]

These copolymers degrade in a single step, losing most of the weight ~ 300°C. The decomposition temperatures of the copolymers are higher than those for acrylonitriles with the same metal doped.[5,9]

Potential Applications

Nowadays industry is trying to find new thermostable polymers to be used in aircraft, electronics, and car parts. The polymer industry is also interested in finding new materials with better mechanical and physical properties. There have been no reports concerning polymers with incorporated metals prepared by chemical liquid deposition.

REFERENCES

1. Cárdenas-Triviño, G. et al. *Langmuir* **1987**, *3*, 986.
2. Wright, R. Presented at 195th National Meeting of the American Chemical Society, Los Angeles, September 1988; I and EC, 32.
3. Cárdenas-Triviño, G. et al. *Bol. Soc. Chil. Quím.* **1990**, *35*, 223; Klabunde, K. J. *J. App. Polymer Sc. Appl. Polymer Symp.* **1991**, *49*, 15.
4. Cárdenas-Triviño, G. et al. *Thermochim. Acta* **1991**, *176*, 233.
5. Cárdenas-Triviño, G. et al. *Thermochim. Acta* **1992**, *188*, 221.
6. Cárdenas-Triviño, G. et al. *Thermochim. Acta* **1993**, *230*, 259.
8. Cárdenas-Triviño, G. et al. *Intern. J. Polymeric Mater.* **1994**, *26*, 199.
9. Cárdenas-Triviño, G.; Salgado, E. et al. *Thermochim. Acta* **1992**, *198*, 123.

METAL COMPLEX CATALYSTS, POLYMERIC

Nobukatsu Nemoto*
Sagami Chemical Research Center

Nobuo Takamiya
Advanced Research Center for Science and Engineering
Waseda University

Application of polymer–metal complexes as polymeric catalysts could induce polymeric effects such as protection of active sites, concentration effects, environmental effects (formation of hydrophobic or electrostatic fields), multifunctionalization, and so on.[1,2] Polymer–metal complex catalysts often accelerate redox reactions because their redox potential differs from that of the corresponding monomolecular metal complexes.

Our research group has taken notice of certain redox reactions catalyzed by two polymer–metal complex catalysts. One catalyst is chemically and thermally stable metallophthalocyanine attached to polymer backbone; the other is a polymer-bound pyridine–Cu complex. Both polymeric catalysts have been used as models for metalloenzymic reactions. For instance, Shirai[3–6] et al. revealed the catalase-like activity of Fe-phthalocyanine-containing polymers for the decomposition of hydrogen peroxide. This research group has also reported the catalase-like activity of Mn-phthalocyanine-containing polymers.[7,8] Additionally, autoxidation reactions of various thiols catalyzed by metallophthalocyanines attached to polymer backbones have been reported by several research groups,[9–28] and the reaction mechanisms were studied in detail.[13,15,16]

On the other hand, Cu(II) ions complexed with polymeric ligands catalyze the oxidation of various organic substrates, which is treated as a biomimetic reaction catalyzed by an enzyme composed of metalloprotein. Since Pecht et al. showed that a Cu(II) complex with poly(l-histidine) ligands is effective in the oxidation of hydroquinone, ascorbic acid, and other aromatic alcohols,[29,30] many studies on Cu(II) complex catalysts with polymeric ligands have been performed.[31–39] In addition, there have been many reports that Cu(II) ions complexed with polymeric ligands are used as catalysts for the oxidative coupling polymerization of phenol derivatives.[2,40–45] These reactions provide poly(*p*-phenylene oxide) derivatives, which are available as engineering plastics. These catalytic polymerization reactions can also be treated as a kinetically biomimetic reaction.[46–49]

On the other hand, linear polyorganosiloxanes (POSs) are known to show desirable characteristics, such as flexibility of the main chain, low glass transition temperature, hydrophobicity based on alkyl side chain, thermostability, stability against atomic oxygen, and physiological inertness.[50,51] Recently, many studies on the preparation of POSs with various pendant metal complexes[32,52–64] have been performed. We believe that the use of linear POSs as polymeric catalyst is of interest, because the flexibility and hydrophobicity of the main chain would have an effect on the formation of a hydrophobic microenvironment, the inclusion of substrates, and the change in polymer conformation required for the inclusion of substrates or for the activation process.

*Author to whom correspondence should be addressed.

We will discuss our recent studies of two polymer–metal complex catalysts composed of siloxane backbone. One catalytic reaction is the oxygenation of 3-methylindole catalyzed by phthalocyaninatocobalt(II) covalently bound to a POS in organic solvents. It was revealed that the change in the conformation of the polymeric catalyst plays an important role in the oxygenation owing to the flexibility of the siloxane backbone. Another catalytic reaction is hydroquinone oxidation catalyzed by pyridine-Cu(II) complexes attached to POSs. The microenvironment around the active sites was controlled by variation in the structure of polymeric ligands (e.g., functional groups in the side chain, the number of side-chain methylene groups, or copolymerization with dimethylsiloxane units). Both hydrophobicity based on alkyl side chains and flexibility of the siloxane backbone contribute to the acceleration of the rate of hydroquinone oxidation.

REFERENCES

1. Kaneko, M.; Tsuchida, E. *J. Polym. Sci., Macromol. Rev.* **1981**, *16*, 397.
2. Tsuchida. E.; Nishide, H. *Adv. Polym. Sci.* **1977**, *24*, 1.
3. Shirai, H. et al. *Makromol. Chem.* **1980**, *181*, 565.
4. Shirai, H. et al. *Makromol. Chem.* **1980**, *181*, 575.
5. Shirai, H. et al. *J. Polym. Sci., Polym. Lett. Ed.* **1983**, *21*, 157.
6. Shirai, H. et al. *J. Polym. Sci., Polym. Chem. Ed.* **1984**, *22*, 1309.
7. Kimura, M. et al. *Macromol. Chem. Phys.* **1994**, *195*, 2423.
8. Kimura. M. et al. *Macromol. Chem. Phys.* **1994**, *195*, 3499.
9. Brouwer, W. M. et al. *Polym. Bull.* **1982**, *8*, 245.
10. Brouwer, W. M. et al. *J. Mol. Catal.* **1984**, *22*, 297.
11. Brouwer, W. M. et al. *Makromol. Chem.* **1984**, *185*, 363.
12. Brouwer, W. M. et al. *J. Mol. Catal.* **1985**, *29*, 335.
13. Brouwer, W. M. et al. *J. Mol. Catal.* **1985**, *29*, 347.
14. Brouwer, W. M. et al. *J. Mol. Catal.* **1985**, *31*, 169.
15. Buck, T. et al. *J. Mol. Catal.* **1991**, *70*, 259.
16. Shirai, H. et al. *J. Phys. Chem.* **1991**, *95*, 417.
17. Schutten, J. H. et al. *Makromol. Chem.* **1979**, *180*, 2341.
18. Schutten, J. H.; German, A. L. *J. Mol. Catal.* **1979**, *5*, 109.
19. Schutten, J. H.; Beelen, T. P. M. *J. Mol. Catal.* **1981**, *10*, 85.
20. van Herk, A. M. et al. *J. Mol. Catal.* **1988**, *44*, 269.
21. van Herk, A. M. et al. *Br. Polym. J.* **1989**, *21*, 125.
22. van Streun, K. H. et al. *Eur. Polym. J.* **1987**, *23*, 941.
23. van Streun, K. H. et al. *Makromol. Chem.* **1990**, *191*, 2181.
24. van Welzen, J. et al. *Makromol. Chem.* **1987**, *188*, 1923.
25. van Welzen, J. et al. *Makromol. Chem.* 1988, *189*, 587.
26. van Welzen, J. et al. *Makromol. Chem.* **1989**, *190*, 2477.
27. Zwart, J. et al. *J. Mol. Catal.* **1977/78**, *3*, 151.
28. Zwart, J.; van Wolput, J. H. M. C. *J. Mol. Catal.* **1979**, *5*, 51.
29. Pecht, I. et al. *Nature* **1965**, *207*, 1386.
30. Pecht. I. et al. *J. Am. Chem. Soc.* **1967**, *89*, 1587.
31. Frey, J. W. et al. *Makromol. Chem.* **1987**, *188*, 821.
32. Koyama, N. et al. *Polymer* **1986**, *27*, 293.
33. Sato, M. et al. *J. Polym. Sci., Polym. Chem. Ed.* **1977**, *15*, 2059.
34. Sato, M. et al. *Makromol. Chem.* **1979**, *179*, 601.
35. Sato, M. et al. *J. Polym. Sci., Polym. Chem. Ed.* **1979**, *17*, 2729.
36. Sato, M. et al. *J. Polym. Sci., Polym. Chem. Ed.* **1980**, *18*, 101.

37. Yamashita, K. et al. *Makromol. Chem., Rapid Commun.* **1988**, *9*, 705.
38. Yamashita, K. et al. *J. Macromol. Sci., Chem.* **1990**, *A27*, 897.
39. Yamashita, K. et al. *Polymer* **1993**, *34*, 2638.
40. Challa, G.; Meinders, H. C. *J. Mol. Catal.* **1977**, *3*, 185.
41. Meinders, H. C. et al. *Makromol. Chem.* **1977**, *179*, 1019.
42. Nishikawa, H.; Tsuchida, E. *Eur. Polym. J.* **1977**, *13*, 269.
43. Tsuchida, E. et al. *Makromol. Chem.* **1973**, *164*, 203.
44. Tsuchida, E. et al. *Makromol. Chem.* **1974**, *175*, 3047.
45. Tsuchida, E.; Nishide, H. *Modification of Polymers;* Carraher, C. E. Jr.; Tsuda, M. Eds.; ACS Symposium Series, American Chemical Society: Washington, DC, 1980; Vol. 121, Chapter 11.
46. Chen, W.; Challa, G. *Polymer* **1990**, *31*, 2171.
47. Chen, W. et al. *Macromolecules* **1991**, *24*, 3982.
48. Koning, C. E. et al. *Polymer* **1987**, *28*, 2310.
49. Viersen, F. J. et al. *J. Polym. Sci., Polym. Chem. Ed.* **1992**, *30*, 901.
50. Yilgör, I.; McGrath, J. E. *Adv. Polym. Sci.* **1988**, *86*, 1.
51. Mark, J. E. *Silicon-Based Polymer Science;* Ziegler, J. M.; Gordon, F. W., Eds.; Advances in Chemistry Series, American Chemical Society: Washington, DC, 1990; Vol. 224, Chapter 11.
52. Awl, R. A. et al. *J. Polym. Sci., Polym. Chem. Ed.* **1980**, *18*, 2663.
53. Duczmal, W. et al. *J. Organomet. Chem.* **1986**, *317*, 85.
54. Ejike, E. N. et al. *J. Appl. Polym. Sci.* **1989**, *38*, 271.
55. Inagaki, T. et al. *J. Chem. Soc., Chem. Commun.* **1989**, 1181.
56. Kavan, V.; Capka, M. *Coll. Czechoslov. Chem. Commun.* **1980**, *45*, 2100.
57. Macurek, M. L. et al. *Polymer* **1980**, *21*, 369.
58. Nagai, K. et al. *Makromol. Chem., Macromol. Symp.* **1992**, *59*, 257.
59. Nemoto, N. et al. *Makromol. Chem.* **1989**, *190*, 2303.
60. Nemoto, N. et al. *J. Inorg. Organomet. Polym.* **1991**, *1*, 211.
61. Nemoto, N. et al. *J. Mol. Catal.* **1991**, *70*, 151.
62. Nemoto, N. et al. *Makromol. Chem.* **1992**, *193*, 59.
63. Nemoto, N. et al. *J. Polym. Sci., Polym. Chem. Ed.* **1994**, *32*, 2457.
64. Pittman, C. U. Jr. et al. *J. Polym. Sci., Polym. Chem. Ed.* **1975**, *13*, 39.

Metal-Complexation Materials

See: Cobalt Carbonyl Complexes (With Main Chain Acetylenic Groups)
Crown Ether Ion-Exchange Resins
Crown Ethers, Polymeric (Catalytic Activity)
Metal Complexation Polymers
Polyelectrolyte Complexes (In Immunology)

METAL COMPLEXATION POLYMERS

Kurt E. Geckeler
Institute of Organic Chemistry
University of Tuebingen

Metal ions are not only valuable intermediates in metal extraction, but also important raw materials for technical applications. Accordingly, complexation of metal ions is an important technique for winning metals from various sources (hydrometallurgy) and for the removal of metal ions from solutions in municipal and industrial wastes.

Generally, metal ion complexation has to take into consideration several phenomena: (1) the nature of the specific metal ion, its microenvironment, its interaction with other metal ions or species present in solution, and its redox reactions and speciation behavior; (2) the type of polymeric ligand, its chemistry, selectivity to metal ions, stability, and its swelling characteristics, if crosslinked resins are used; and (3) physicochemical factors, such as kinetics, that are relevant under complexation conditions. Usually the term metal-complexing (or chelating) polymers refers to polymers that bind metal ions by coordinating interactions; however, ionic interactions cannot always be excluded.

The development of polymeric materials for metal ion complexation and removal has dramatically accelerated.[1–6]

PREPARATION

In general, polymers for metal ion complexation can be prepared either by derivatization of a basis polymer (precursor) with the desired ligand or by polymerization of the corresponding monomeric ligand derivative. Ligand polymerization embraces all classical methods of polyreactions.[2,5,8,9] There are two criteria for developing functional polymers for the complexation of metal ions. First, the polymer backbone must be compatible with the complexation process, and second, selective ligands must be found for certain metals and then bound to the polymer. Examples of synthetic procedures of chelating resins for the complexation of metals have been extensively described.[2–4]

PROPERTIES

Polymers containing ligand moieties can perform several functions in contact with metal ion solutions, including separation, removal, concentration, and fractionation. All of these processes, resulting from metal ion complexation by polymer ligands, are influenced by ligand selectivity and matrix structure among others. Several functional hydrophilic polymers with chelating ligands used for metal ion complexation have been investigated for complexation in the homogeneous phase. These hydrophilic, noncrosslinked, functional polymers that are able to complex metal ions have been termed polychelatogens.[7,11]

Among the hydrophilic polymers reported, poly(ethylenimine) (PEI) has been used frequently because it is commercially available and contains primary, secondary, and tertiary amino groups, which explain its basic properties and provide access to various derivatives. Functionalizing reagents for PEI include aldehydes, ketones, alkyl halides, isocyanates and thioisocyanates, epoxides, cyanamides, guanides, ureas, acids, and anhydrides.[12] Metal complexation of PEI and its derivatives has been described in detail.[7,12–17]

Ligands and Selectivity

A convenient approach for the synthesis of metal-complexing polymers is the attachment of appropriate and selective ligands to a precursor polymer, thus transferring the complex-forming properties of the ligand to the desired polymer backbone. The ligand donor atoms are mainly from the nonmetallic elements of groups V and VI, among which nitrogen, oxygen, and sulfur are the most common.[18] Polymeric ligands with metal-complexing moieties such as hydroxyl, carboxylic acid, ketone, phosphonic acid, nitrogens, and thiol functions are suitable for metal ion complexation.[7,10,13] These groups are introduced into the polymer by covalent attachment to the polymer backbone.

The ligand attached to the polymer backbone also influences the binding capacity of the polymer.

APPLICATIONS

Complexation of metal ions is an important process in analytical chemistry, hydrometallurgical extraction, and water-treatment processes. Therefore, metal complexation polymers have found such application as removal of metal contaminants and extraction of metals from waters. Recovery of metals from diluted solutions is one classical application for polymeric ligands. Trace metal analysis is a major analytical application. The procedure is based on the preconcentration of metal ions on soluble polymers followed by direct analysis or after elution, thus avoiding laborious separation steps.

Preconcentration of trace metals is important in analytical chemistry. In many cases metal ions must be separated from other constituents in the solution that could interfere with the ionic species.[19,20] Therefore, selective preconcentration depends on the nature of the sample, the type and concentration of metal ions, and the method to be used for determination.

For practical analytical applications it is important to combine preconcentration and separation methods based on metal complexing polymers, especially in the presence of interfering components in the analyte solution. In this way, preconcentration by functional polymers was successfully applied to the analysis of heavy metals in drinking and river water in conjunction with flame atomic absorption spectroscopy.[7,21]

Analysis in geology, biology, and environmental science is a more recent application.[22–25] As the determination of radionuclides, especially plutonium, has received increasing interest because of widespread radioactive pollution, efforts have been made to apply functional polymers to their separation and enrichment. A series of actinides was also investigated for separation using a polymeric phosphonic acid for complexation.[23]

Other applications in metal-ion research, such as host–guest chemistry, bioligands, affinity and exchange chromatography, and membrane transport can be envisaged. The scope of applications of metal complexation polymers should increase considerably in the future.

REFERENCES

1. Tsuchida, E.; Nishide, H. *Adv. Polym. Sci.* **1977**, *24*, 1.
2. Warshawsky, A. *Angew. Makromol. Chem.* **1982**, *109/110*, 171.
3. Warshawsky, A.; Kahana, N. *Polym. Sci. Technol.* **1982**, *16*, 227.
4. Sahni, S. K; Reedijk, J. *J. Coord. Chem. Rev.* **1984**, *591*.
5. Hodgkin, J. H. In *Encyclopedia of Polymer Science,* 2nd ed. Kroschwitz, J. I. Ed.; Wiley: New York, 1985; Vol. 3, p 363.
6. Albright, R. L.; Yarnell, P. A. In *Encyclopedia of Polymer Science,* 2nd ed.; Kroschwitz, J. I. Ed.; Wiley: New York, 1987; Vol. 8, p 341.
7. Geckeler, K. E. et al. *Sep. Purif. Methods* **1988**, *17*, 105.
8. Geckeler, K. E.; Stirn, J. *Naturwissenschaften* **1993**, *80*, 487.
9. Geckeler, K. E.; Zhou, R. *Naturwissenschaften* **1993**, *80*, 270.
10. Goethals, E. J. Ed., *Polymeric Amines and Ammonium Salts;* Pergamon: Elmsford, NY, 1980.
11. Spivakov, B. Ya. et al. *Nature* **1985**, *315*, 313.
12. Geckeler, K. E. et al. *Pure Appl. Chem. (IUPAC)* **1980**, *52*, 1883.
13. Rivas, B. L.; Geckeler, K. E. *Adv. Polym. Sci.* **1992**, *102*, 171.
14. Geckeler, K. E. et al. *Angew. Makromol. Chem.* **1991**, *193*, 195.
15. Geckeler, K. E. et al. *Angew. Makromol. Chem.* **1992**, *197*, 107.
16. Shkinev, V. M. et al. *Sep. Sci. Technol.* **1987**, *22*, 2165.
17. Rivas, B. L.; Geckeler, K. E. *Bol. Soc. Chil: Quim.* **1994**, *39*, 107.
18. Bell, F. *Metal Chelation, Principles and Applications;* Oxford University Press: Oxford, UK, 1977.
19. Prasolova, O. D. et al. *J. Anal. Chem.* **1993**, *48*, 85.
20. Spivakov, B. Ya. et al. *Pure Appl. Chem.* **1994**, *66*, 631.
21. Geckeler, K. E. et al. *Fresenius Z. Anal. Chem.* **1989**, *333*, 763.
22. Geckeler, K. E. et al. *Naturwissenschaften* **1993**, *80*, 556.
23. Novikov, A. et al. *Chem. Technol.* **1993**, *45*, 464.
24. Palmer, V. et al. *Angew. Makromol. Chem.* **1994**, *215*, 175.
25. Zhou, R. et al. *Water Res.* **1994**, *28*, 1257.

METAL COMPLEXES, MACROMOLECULAR

Dieter Wöhrle and Günter Schnurpfeil
Institut für Organische und Makromolekulare Chemie
Universität Bremen

Macromolecular metal complexes (MMC) are well known for their active and selective functions in biological matter. Dioxygen transport is realized in hemoglobin and myoglobin containing Fe(II)-porphyrin. Metalloenzymes like carboxypeptidase A, different cytochromes, nitrogenase, reduced nicotinamide adenine dinucleotide (NADH)-Q-reductase, and others catalyze various reactions and electron transfer processes. The complex system of photosynthesis consisting of metal clusters like Mn and electron transfer metal complexes is responsible for water splitting into dioxygen and activated hydrogen. Various other examples are known and fulfill different functions *in vivo*.

Detailed analysis of the primary, secondary, tertiary, and quarternary structures shows in most cases a complicated arrangement of metals with their surrounding. Either metal ions are surrounded by a specific ligand which is embedded or combined in a polymer or structured matrix with all parts (metal ion, ligand, polymer) essential for the specific function, or a metal cluster of defined size is included in a polymer or structured matrix. Metal ions like Ca, Mg, Na, K, and Mn are going for more ionic or coordinative interactions with ligand atoms or groups whereas with Pt, Hg, Cd, and Pb the interactions are more covalent, and with Ni, Cu, and Zn they are not so uniform. Monographs and reviews describe in detail biological macromolecular metal complexes.[1–5]

From these considerations a MMCA consists of a metal ion with a ligand in a high molecular environment. The other possibility is a metal cluster in a high molecular compound. The increasing knowledge about natural macromolecular metal complexes stimulated chemists to synthesize artificial systems. The combinations of different materials are interesting because such so-called composites exhibit new material properties which may be important for different applications. MMCs are classified now in three types:

- In type I MMC a metal complex (chelate) or metal salt is bound in the side chain of a linear or crosslinked organic or inorganic polymer. The other possibility considers the binding on the surface of an organic or inorganic high molecular weight compound.

- Type II MMCs are those where a bifunctional or higher functional metal complex (chelate) is part of a polymer chain or network.
- For type III MMCs we consider the situation where a metal complex (chelate) or metal cluster is incorporated by physical interaction into an organic or inorganic high molecular weight compound.

The activities in the field of macromolecular metal complexes are summarized in some monographs, several reviews,[1,5–33] and are documented in a series of international conferences.[34–37]

TYPE I MMC: METAL COMPLEXES BOUND TO ORGANIC AND INORGANIC HIGH MOLECULAR SUPPORTS

Linear (soluble), crosslinked (gel-type, macroporous, and macroreticular), and surfaces (nonporous) of organic or inorganic high molecular compounds containing suitable groups can bind metal salts, complexes, or chelates. The supports are natural or synthetic. The interactions for the binding of the metal salt, complex, or chelate are according to the nature of the reagents: coordinative, ionic covalent, or π-binding. The preparation is realized as follows:

- Support → support functionalization → binding of the metal salt/complex/chelate → purification from not bound reagents → analytical characterization → investigation of properties (**Equation 1**).

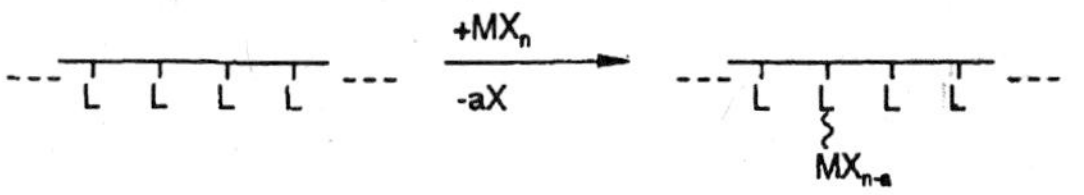

1

- Preparation of a vinyl-substituted ligand or metal complex/chelate → home- or copolymerization → analytical characterization → metalation in the case of an employed ligand → analytical characterization → investigation of properties (**Equation 2**).

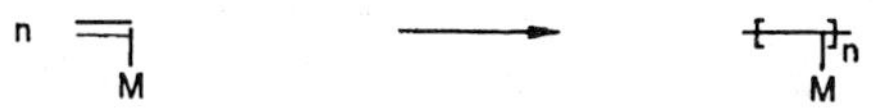

2

Type I MMCs are interesting mainly as ion exchanger, selective metal ion binding, electron or photoelectron transfer, catalyst or photocatalyst, selective gas binding or separation, sensor, drug carrier in pharmacology, and luminescence probe.[1] We mention a few examples by classification in natural polymers, organic polymers, and inorganic polymers as supports.

Macromolecular Metal Complexes at Natural Polymers

Natural polymers containing N, P, O, and S donor groups can coordinate with metal ions or complexes. The multifunctionality makes it difficult to determine the interacting groups.[39]

Examples for metal salt binding are gelatin,[40,41] cyclodextrines,[42,43] chitin and chitosan,[44,45] and humic and fulvic acids.[46]

Macromolecular Metal Complexes at Synthetic Organic Polymers

Binding of Metal Salts, Complexes, or Chelates at Reactive Polymers

A linear or crosslinked polymer with ligand groups containing oxygen, nitrogen, sulfur, phosphorus, or arsen donor atoms interacts ionically or coordinatively with metal salts, complexes, or chelates from solution. The other possibilities use covalent or π-binding of metal complexes or chelates at a polymer backbone or its functional groups.

Alcohols, Ethers, Carboxylic Acids, Diketones

Hojo and several others have studied the reaction of polyvinylalcohol with transition metal ions in aqueous and nonaqueous medium.[47–50] Values of complex binding constants and of thermodynamic parameters increase as follows: $Co^{2+} < Ni^{2+} < Zn^{2+} < Cu^{2+}$.

Linear or crosslinked poly(oxyethylene)–metal salt (Li^+, Na^+, K^+, Mg^{2+}, Ba^{2+}) complexes obtained by direct interaction of both components are important as solid polymer electrolytes.[1,7,33,51] Polymer crown-ethers, macrocyclic ketones, and cryptandes as chelating ligands were intensively studied for alkali- and earth-alkali-metal ion binding but not so much for transition metal ion binding.[1]

Pomogailo and others used mainly homo- and copolymers of acrylic, methacrylic, sulfonic, and phosphorus acids as macroligands.[37,38,55] Crosslinked polyacids are of great importance as sorbents, cationites, and ampholytes. Crosslinking influences composition, structure, and strength of bonds.

Examples of poly(β-diketonates) for metal ion binding are polymethacryloylacetone and pivaloylacetone.[56–60]

Amines

Linear and crosslinked polyethyleneimine shows high ability for the binding of metal ions and high capacity.[60–63]

Sulfur and Phosphorus-Containing Ligands

Due to the lower electronegativity, sulfur-containing ligands are weaker electron donors than oxygen-containing ligands.

Schiff'bases

For Schiff'base chelates of N_2O_2- or N_3O_2-type binding over the diamino-bridge or the annealated benzene rings are realized.[64,65]

Pyridyls, Bipyridyls, and Other Heterocycles

Homo- and copolymers of vinylpyridines are known for demineralization of organic solvents or liquids and desalinization (as membranes) of water. In addition to polymers with pyridyl units, N-containing heterocycles like polymers with bipyridyl, imidazoles, or benzimidazoles possess excellent donor properties for transition metal ion binding (Co^{2+}, Pd^{2+}, Ru^{2+}, Re^{2+}, etc.), which is due to the unshared electron pair at the nitrogen in combination with the aromatic π-system leading to σ-donor and π-acceptor behavior.[1,7,33,66–69]

Macrocyclic Metal Chelates

Wöhrle and several others have reviewed covalent binding of different porphyrins and phthalocyanines.[21,26,27,30,70] Binding of porphyrins at the end group of methoxypolyoxyethylene is carried out to use these MCCs as sensitizers in photodynamic therapy of cancer.[71]

Polymeric π-Complexes of Transition Metal Ions

MMCs of π-type are formed by interaction of aromatic groups (for example polystyrene, polyvinylanthracene, polyvinylnaphthalene, etc.) or unsaturated fragments (for example dienes) of a polymer chain acting as π-donor with transition metal halides ($AlCl_3$, $TiCl_4$, $VOCl_3$, etc.).[1,72]

Polymerization of Vinyl Ligands or Metal Complexes

The principal preparation for polymerization of vinyl ligands or metal complexes was shown in Equation 10.[1,7,21,34,52,53] Compared to the polymer binding of complexes or chelates, the preparation of suitable vinyl ligands or complexes and chelates followed by homo- or copolymerization involves more difficulties. But the incorporation of the complex or chelate is run by more controlled polyreactions. If a vinyl ligand or a metal complex or chelate is employed, difficulties during the polymerization are related to the solubility of the compounds used, and resulting polymers and possible chain (electron) transfer are initiated by transition metals during the polymerization.

Macromolecular Metal Complexes of Inorganic High Molecular Weight Compounds

We investigated metal salts, complexes, and chelates on inorganic supports as heterogeneous catalysts because their reactions are less limited by diffusion. The surface of inorganic supports like silica gel can be modified by amines, nitriles, pyridine, and other nitrogen heterocycles, or by phosphorous-containing ligands for coordinative binding of transition metal ions, complexes, or chelates.

Another possibility is the preparation of inorganic–organic composites. The support surface is activated by mechanical, chemical, or irradiated-chemical treatment, followed by graft polymerization of a suitable monomer (vinylpyridine, acrylic acid, etc.)—also gas phase grafting—with ligand groups.[1,7]

TYPE II MMCS: POLYMERS WITH METAL COMPLEXES OR METALS IN THE BACKBONE

In addition to C, N, O, S, P, and Si, which are well-known as chain-forming or participating elements, most other elements of the periodic system (except hydrogen, noble gases, and twelve elements of Group I A and VII B) are capable of being part of a polymer chain. These elements include many metals or semimetals. We describe a few characteristic examples. Wöhrle and others review them.[21,23,26,27,74] Many polymers are less soluble and therefore, difficult to characterize.

Homochain Metal Polymers

Wöhrle and others summarize the preparation and properties of these polymers.[26,75,76] One general synthetic method is the reaction between dihalides in the presence of sodium. Due to the weak bonds in the chain, these metal polymers are interesting in microlithography, as precursors for ceramics, and as photoconductors due to overlapping of metal–metal–orbitals.

Heterochain Metal Polymers

A metal M and another element X are part of a polymer chain forming polar bonds.[10–12,26,77,78] These polymers are often obtained by the polycondensation of bifunctional metal halides (M = B, Si, Ge, Sn, Pb, As, Sb, Ni, Pd, Pt, Ti, Zr, Hf) with a bifunctional Lewis base like diols, diamines, dihydrazines, dihydrazides, dioximes, diamideoximes, dithiols, and diacetylenes or by polyaddition of a bifunctional metal hydride with bifunctional alkenes.

Polymers with transition metals (stabilized by CO, PR_3, bulky organic rests) σ-bonded to carbon atoms are gaining interest.

Metal and Ligand as Part of a Polymer Chain or Network

AuI, AuCN, $PdCl_2$, and others form polymer coordination compounds in the solid state.[8]

Ligand of a Metal Complex as Part of a Polymer Chain or Network

A ligand is part of a polymer chain or network and includes metal ions in the ligand donor cavities. These polychelates are prepared by either the reaction of a bifunctional- or higher functional-substituted ligand or metal complex with another bifunctional or higher functional reagent, or the reaction of a bis-ligand or ligand precursor with a multivalent metal ion.[1,21,26]

Cofacial Stacked Polymeric Macrocyclic Metal Complexes

Stacked arrangements in a face-to-face orientation of macrocyclic metal complexes (porphyrins, phthalocyanines, naphthalocyanines, and others) are realized by connecting the metal ions in the core with different bivalent atoms or groups. This subject has been intensively reviewed.[23,26,27,79–85] The interest in the materials is connected with their properties: electrical conduction, photoconduction, nonlinear optical, and electroluminescent behavior.

Metallocenes as Part of a Polymer Chain

Disubstituted π-electron-rich charged aromatics connected by π-bonds like ferrocene or ruthenocenylenes can be incorporated into a polymer chain directly connected via their 1,1′-positions or in the polyesters, polyamides, polyalkenes, and polysilanes employing disubstituted derivatives in normal polycondensation or addition reactions.[26,86–91]

TYPE III MMCS: COMPLEXES, ZERO-VALENT METALS OR METAL CLUSTERS INCORPORATED IN INORGANIC OR ORGANIC HIGH MOLECULAR COMPOUNDS

The "physical" incorporation of metal complexes and metal clusters in high molecular organic and inorganic "host" system has been studied in detail. Multiple and dynamic interactions by different secondary binding forces influence the properties of these composite materials. Metal clusters stabilized in a polymer environment or after incorporation in the framework of cages and cavities are interesting as catalysts, photocatalysts, nonlinear optics, hole burning materials, and for other optoelectronics applications.

In principle it is possible to include every metal complex or chelate, either monomolecular or aggregated, into an organic polymer, depending on the solubility and the concentration of the complex.

REFERENCES

1. *Macromolecule Metal Complexes;* Ciardelli, F.; Tsuchida E.; Wöhrle, D. Eds.; Springer: Heidelberg, Germany, 1996.
2. *Bioinorganic Catalysis;* Reedijk, J., Ed.; Marcel Dekker: New York, NY, 1993.
3. *Metal Ions in Biological Systems;* Sigel, H., Ed.; Marcel Dekker: New York, NY.
4. *Metal Ions in Biology;* Spiro, T. G., Ed.; John Wiley & Sons: New York, NY, 1985.
5. Tsuchida, H.; Masuda, H. In *Macromolecular Complexes. Dynamic Interactions and Electronic Processes;* Tsuchida, E., Ed.; VCH: New York, NY, 1991.
6. Pomogailo, A. D. *Polymeric Immobilized Metallocomplex Catalysts;* Nauka: Moscow, Russia, 1988.
7. Pomogailo, A. D.; Uflyand, I. E. *Macromolecular Metal Chelates;* Nauka: Moscow, Russia, 1991.
8. Ray, N. H. *Inorganic Polymers;* Academic: New York, NY, 1978.
9. *Inorganic Polymers;* Stone, F. G. A.; Graham, W. A. G., Eds.; Academic: New York, NY, 1962.
10. *Organometallic Polymers;* Carraher, C. E.; Sheats, J. E.; Pittmann, C. U., Eds.; Academic: New York, NY, 1978.
11. *Advances in Organometallic and Inorganic Polymer Science;* Carraher, C. E.; Sheats, J. E.; Pittmann, C. U., Eds.; Marcel Dekker: New York, NY, 1982.
12. *Metal-Containing Polymeric Systems;* Sheats, J. E.; Carraher, C. E.; Pittmann, C. U., Eds.; Plenum: New York, NY, 1985.
13. Scientific Journal: *J. Inorg. Organomet. Polym.* Plenum: New York, NY.
14. Kepler, B. K. *Metal Complexes in Cancer Chemotherapy;* VCH: New York, NY, 1993.
15. Hartley, F. R. *Supported Metal Complexes. A New Generation of Catalysts;* Reidel: Dordrecht, The Netherlands, 1985.
16. Yermakov, Y. I.; Kuznetsov, B. N.; Zakharov, V. A. *Catalysis by Supported Complexes;* Elsevier: Amsterdam, The Netherlands, 1981.
17. Yermakov, Y. I.; Likholobov, V. *Homogeneous and Heterogeneous Catalysis;* VNU Science: Utrecht, The Netherlands, 1987.
18. *Zeolite Chemistry and Catalysis, Studies in Surface Science and Catalysis;* Jacobs, P. A. et al., Eds.; Elsevier: Amsterdam, The Netherlands, 1991; Vol. 69.
19. *Zeolites as Catalysts, Sorbents and Detergent Builders,* Studies in *Surface Science and Catalysis;* Karge, H.; Weitkamp, J., Eds.; Elsevier: Amsterdam, The Netherlands, 1989; Vol. 46.
20. Photochemistry in Organized and Constrained Media; Ramamurthy, V. Ed.; VCH: New York, NY, 1991.
21. Wörhle, D. *Adv. Polym. Sci.* **1983**, *50*, 45.
22. Korshak, V. V.; Kozyreva N. M. *Russ. Chem. Rev.* **1983**, *54*, 1091.
23. Hanack, M.; Lang, M. *Ad. Mater.* **1994**, *6*, 819.
24. Ozin, G. A.; Gil, C. *Chem. Rev.* **1989**, *89*, 1749.
25. Sherrington, D. C. *Pure Appl. Chem.* **1988**, *60*, 401.
26. Wörhle, D. *Polymers with Metals in the Backbone;* In *Handbook of Polymer Synthesis;* Kricheldorf, H. Ed.; Marcel Dekker: New York, NY, 1992; Vol. B, p 1133.
27. Wörhle, D. *Phthalocyanines in Polymer Phases;* In *Phthalocyanines, Properties and Applications*; Leznoff, C. C.; Lever, A. P. B., Eds.; VCH: New York, NY, 1989; Vol. 1, p 55.
28. Pomogailo, A. D. *Russ. Chem. Rev.* **1992**, *61*, 133.
29. Pomogailo, A. D.; Uflyand, I. E. *Adv. Polym. Sci.* **1990**, *97*, 61.
30. Tsuchida, E.; Nishide, H. *Adv. Polym. Sci.* **1977**, *24*, 1.
31. Kaneko, M.; Wörhle, D. *Adv. Polym. Sci.* **1988**, *84*, 141.
32. Hanack, M.; Deger, S.; Lange, A. *Coord. Chem. Rev.* **1988**, *83*, 115.
33. Biswas, M.; Mukherjee, A. *Adv. Polym. Sci.* **1994**, *115*, 89.
34. MMC II, *J. Macromol. Sci.-Chem.* **1988**, *A25*, Vols. 10 and 11.
35. MMC III, *J. Macromol. Sci.-Chem.* **1990**, *A26*, Vols. 2 and 3; *A27*, Vols. 9–11.
36. MMC IV, *Macromol. Chem. Macromol. Symp.* **1992**, *59*.
37. Efendiev, A. A. et al. MMC V, *Macromol. Symp.* **1994**, *80*.
38. Pomogailo, A. D.; Wörhle, D. In *Macromolecule Metal Complexes;* Springer: Heidelberg, Germany, 1996, p 11.
39. Reedijk, J. In *Macromolecule Metal Complexes;* Ciardelli, F. et al., Eds.; Springer: New York, NY, 1996; p 131; and MMC V. Macromol. Symp. 1994, 80.
40. Tanaka, K. *J. Nat. Chem. Lab. Ind.* **1975**, *70*, 203.
41. Kostromina, N. A. et al. *Teor. Eksp. Khim.* **1975**, *15*, 297.
42. Okamoto, Y. *J. Macromol. Sci.* **1987**, *A24*, 455.
43. Naoshima, Y. et al. *J. Macromol. Sci.* **1986**, *A23*, 861.
44. Millish, F.; Hellmuth, E. W.; Huang, S. Y. *J. Polym. Sci., Polym. Chem. Ed.* **1975**, *13*, 2143.
45. Lopez de Alba, P. L. et al. *J. Radionanal. and Nucl. Chem. Lett.* **1987**, 118, 99.
46. Zhorobekova, S. Z. *Makroligandnye svoistva guminovykh kislot Frunze* **1987**.
47. Hojo, N.; Shirai, H.; Hajashi, S. *J. Polym. Sci. Polym. Symp.* **1974**, *47*, 299.
48. Hojo, N.; Shirai, H.; Hajashi, S. J. *J. Chem. Soc. Jap. Chem. and Chem. Ind.* **1972**.
49. Suzuki, T. et al. *Polym. J.* **1983**, *15*, 409.
50. Pomogailo, A. D.; Lisitskaya, D. A.; Kritskaya, D. A. *Kompleksnye metalloorganicheskie katalizatory polimerizatsii olefinov;* Chernoglovka, Institut Khimicheskoi Fiziki Acad. *Nauk SSSR*, 1983.
51. Takeoka, S. et al. *Macromolecules* **1991**, *24*, 2003.
52. Chiang, G. T. et al. *Macromolecules* **1985**, *18*, 825.
53. Giles, R. M. J.; Greenhall, P. M. *Polym. Commun.* **1986**, *27*, 360.
54. Sinta, R.; Lamb, B.; Smid, J. *Macromolecules* **1983**, *16*, 1383.
55. Ohkamoto, Y. In *Macromolecakar Complexes, Dynamic Interactions and Electronic Processes;* Tsuchida, E., Ed.; VCH: New York, NY, 1991.
56. Davydova, S. L.; Plate, N. A. *Coord Chem. Rev.* **1975**, *16*, 195.
57. Teyssie, M. P.; Teyssie, P. *J. Polym. Sci.* **1961**, *50*, 253.
58. Teyssie, M. P.; Teyssie, P. *Makromol. Chem.* **1963**, 66, 133.
59. Patel, M. N.; Patel, J. R.; Patel, S. H. *J. Marcromol. Sci.* **1988**, *A25*, 211.
60. Geckeler, K. et al. *Pure and Appl. Chem.* **1980**, *52*, 1883.
61. Bauer, E.; Geckeler, K.; Weingartner, K. *Makromol. Chem.* **1980**, *181*, 585.
62. Antonelli, M. L. et al. *J. Polym. Sci., Polym. Lett. Ed.* **1980**, *18*, 179.
63. Harris. C. S.; Shriver, D. E; Ratner, M. A. *Macromolecules* **1986**, *19*, 987.
64. Wöhrle, D.; Buttner, P. *Poly. Bull.* **1985**, *13*, 57.
65. Aeissen, H.; Wöhrle, D. *Makromol. Chem.* **1981**, *182*, 2961.
66. Sherrington, D. C.; Rusanov, A. L. *MMC V, Macromol. Symp.* **1994**, *80*.
67. Keneko, M.; Yamada, A. *Adv. Polym. Sci.* **1984**, *55*, 1.

68. Kaneko, M.; Yamada, A. In *Metal-Containing Polymeric Systems;* Sheats, J. E.; Carraher, C. E.; Pittman, C. U., Eds.; Plenum: New York, NY, 1985; p 249.
69. Sasaki, T.; Matsunaga, F. *Bull. Chem. Soc. Jpn.* **1986**, *41*, 2440.
70. Hanabusa, K.; Shirai, H. In *Phthalocyanines, Properties and Applications;* Leznoff, C. C.; Lever, A. P. B., Eds.; VCH: New York, NY, 1989; Vol. 2, p 197.
71. Wöhrle, D. et al. *Makromol. Chem., Macromol. Symp.* **1992**, *59*, 17.
72. Hirai, H. *J. Macromol. Sci.* **1990**, *A27*, 1293.
73. Pomogailo, A. D. et al. *Kinetics and Catalysis* **1985**, *26*, 1104.
74. Dey, A. K. *J. Indian Chem. Soc.* **1986**, *63*, 357.
75. Miller, R. D. *Angew. Chem. Adv. Mater.* **1989**, *10*, 1773.
76. West, R. J. *Organomet. Chem.* **1986**, *300*, 327.
77. Carraher, C. E. *J. Chem. Ed.* **1981**, *58*, 921.
78. Carraher, C. E.; Scott, W. J.; Schroeder, J. A. *J. Macromol. Sci.-Chem.* **1981**, *A15*(4), 625.
79. Davidson, P. J.; Lappert, M. F.; Pearce, R. *Chem. Rev.* **1976**, *2*, 219.
80. Schrock, R. R.; Parshall, G. W. *Chem. Rev.* **1976**, *2*, 243.
81, Schrock, R. R.; Parshall, G. W. *Methoden der Organischen Chemie* (Houben-Weyl) Thieme-Verlag: Stuttgart, Germany, 1987; Vol. E20.
82. Lang, H. *Angew. Chem.* **1994**, *106*, 569.
83. Dirk, C. W. et al. In *Advances in Organometallic and Inorganic Polymer Science;* Carraher, C. E. et al., Eds.; Marcel Dekker: New York, NY, 1986; p 275.
84. Schultz, H. et al. *Structure and Bonding* **1991**, *74*, 41.
85. Marks, T. J. *Angew Chem.* **1990**, *102*, 886.
86. Neuse, E. W.; Rosenberg, H. *Metallocene Polymers;* Marcel Dekker: New York, NY, 1970.
87. Neuse, E. W. *J. Macromol. Sci.-Chem.* **1981**, *A16*, 3.
88. Manners, I. *Adv. Mater.* **1994**, *6*, 68.
89. Manners, I. *Adv. Mater.* **1994**, *6*, 564.
90. Nuyken, O. et al. *Makromol. Chem., Macromol. Symp.* **1991**, *44*, 195.
91. Rosenblum, M. *Adv. Mater.* **1994**, *6*, 159.

METAL-CONTAINING MONOMERS (Polymerization)

Anatolii D. Pomogailo
Institute of Chemical Physics
USSR Academy of Sciences

We use the term metal-containing polymers to mean the products of homo- and copolymerization of metal-containing monomers (MCM). The latter includes both multiple bond, capable of polymerization transformations, and the chemically bound equivalent of metal (both transition and nontransition metals).[1–5]

MCM CLASSIFICATION

From our point of view, the most effective classification is based on the type of bond between the metal and organic parts of the monomer molecule. By this principle, MCM can be classed into the following major types: monomers with a covalent (σ), ionic, donor-acceptor (nv), chelate, and π-bound metal.[6,7]

Multiple bonds can also be presented, not only by double bond but also triple, allene, and diene bonds as well as their combinations.

The correlation between the electronic structure and properties of MCM as well as their capacity for polymerization transformations is an important problem in the chemistry of MCM. In the MCM of donor-acceptor (nv) type, the functional groups (including the heteroatoms with an unshared pair of electrons) take the part of n-donors, and unsaturated hydrocarbons are π-donors. The former present unshared pairs for the formation of coordination bonds, and the latter give the π-electronic system. Such MCMs are more often formed in the case of transition metal compounds. As a rule, π-MCMs are characteristic for the transition metals of VIA, VIIA, and VIII groups of the periodic table of the elements. MCM of ionic type are more characteristic for nontransition metals; MCM-true organometallic compounds (with metal-carbon bond) occupy only a small place in the MCM chemistry. The most characteristic representatives of each of the above-mentioned types of MCM are presented in **Table 1**. It can be stated that the useful methods of preparation of practically all types of MCM, with the possible exceptions of organometallic, cluster, and polynuclear MCMs, have been developed.

HOMOPOLYMERIZATION OF METAL-CONTAINING MONOMERS

Although homopolymerization of MCM is carried out, as a rule, according to traditional schemes, the basic states, constituting the polymerization process, are characterized by their own features. This is related to the reactions of initiation and chain termination.

The interval of reactivity of MCM is extremely large, i.e., from polymerizable monomers already on the stage of synthesis to monomers incapable of polymerization even under sufficiently hard conditions.

One of the difficulties of the radical polymerization of MCM consists in selection of the appropriate solvent. The polymer formed is often already precipitated at small transformation degrees.

The metal-containing group as specific substituent affects all elementary stages (especially during copolymerization). Thus, in the case of MCM-pure organometallic compounds of transition metals, the most important problem is to avoid the elimination of metal hydride during polymerization.[8]

Such transformations are not practically observed for MCM with polymerizable groups of other types (styrene, (meth)acrylate, etc.).

The choice of initiator for polymerization is important. MCMs can affect the initiator dissociation and initiation rate; sometimes the chemical interaction of MCM with the initiator occurs. The nature of the polymerized system as well as of the metal ion is important to the procedure.

MCM COPOLYMERIZATION

MCM copolymerization is the most widespread means of preparation of metal-containing polymers, because it is often used for modification of properties of traditional polymers. However, joint polymerization adds to the possibilities for study of statistical processes and factors, their influence on the reactivity of multiple bonds of monomers, and for exposure of latent effects, intrinsic for MCM. We pay special attention to the analysis of copolymerization constants, characterizing the relative

TABLE 1. The Main Representatives of MCM

MCM σ-type	MCM ionic type	MCM nv-type	MCM π type	MCM chelate type	MCM cluster type
a) the natural organometalic compounds $(CH_2=CH)_mMCl_{n-m}$; 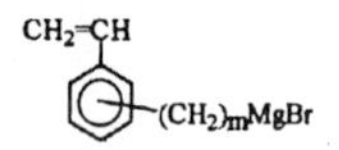 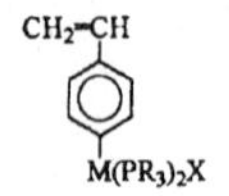 M=Pt, Pd: R=Et, Bu $Et_3Sn-CH=CH-C\equiv CR$; $RC\equiv C-M(PBu_3)_2-C\equiv CR$ (M = Pt, Pd, Ni; R = Ph, $-C\equiv CR$) b)organoelemental compounds $M(OR)_{n-m}(OR'CH=CH_2)_m$, $M(OR)_{n-m}(OR'C(R'')=CH_2)_m$ (M = Ti, V, Zr, Sn)	$M(OCOC(R)=CH_2)_n$ (R = H, CH_3); $(C_5H_5)_2M(OCOC(R)=CH_2)_2$ (M = Ti, V, Zr) 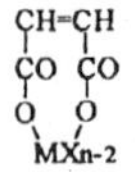 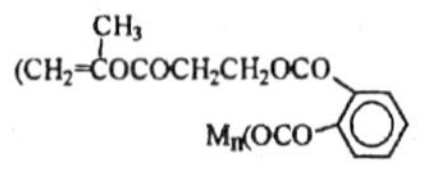	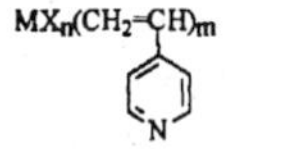 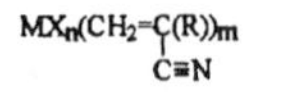 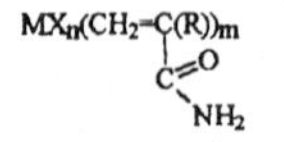 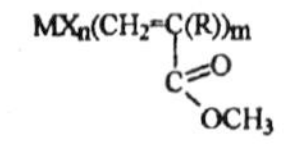 R = H, CH_3 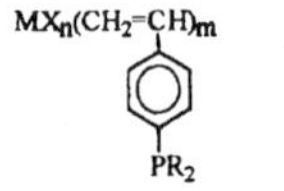	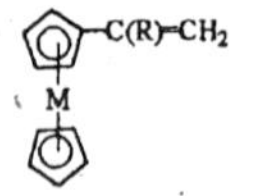 M = Fe, Os, Ru R = H, CH_3 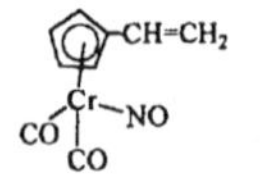 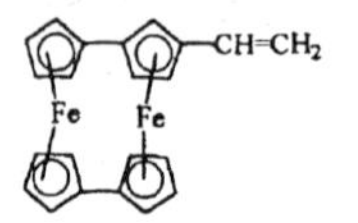 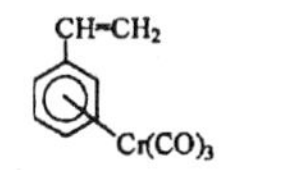 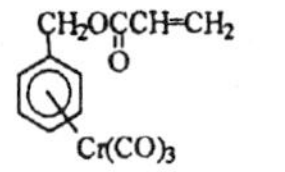 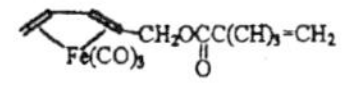	a) intramolecular metallochelates 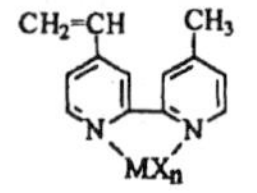 b) intracomplex compounds 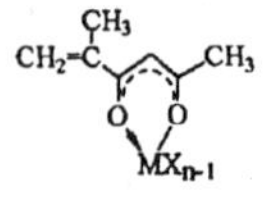 c) macrocyclic complexes 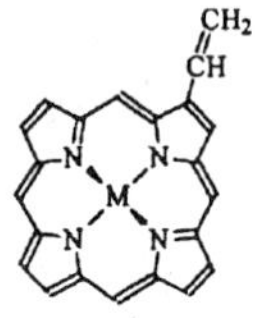	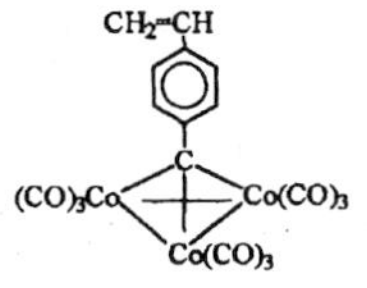 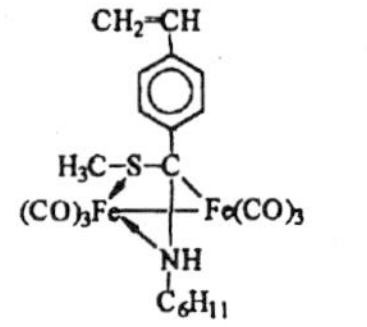 $Rh_6(CO)_{15}(4\text{-}VPy)$ $Rh_6(CO)_{14}(4\text{-}VPy)_2$ 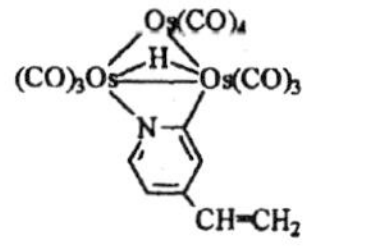 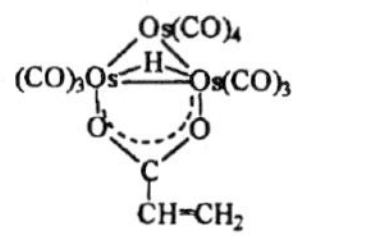 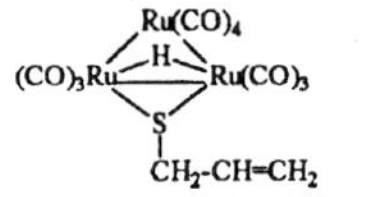

activity of monomers during addition to their "own" and "strange" radicals, to Q (it characterizes resonance stabilization of the monomer during copolymerization) and e (factor reflecting the measure of polar effect of substituent at multiple bond) parameters, to distribution of units along the chain, etc.

The MCM state (in dissociated or undissociated form) has paramount importance during copolymerization of salts of alkaline and alkaline-rare metals. We note the substantial influence of reaction medium, in particular, of ionic force and polarity of solvent.

Generally speaking, the contribution of complex formation into MCM copolymerization is more substantial than into homopolymerization.

MCM copolymerization in aqueous solutions proceeds in a more complex manner; the character of monomer mixture composition versus copolymer composition curves, at different pH of medium, is substantially different. This is conditioned by electrostatic interactions between the growing radical and monomer (ionized or un-ionized). The types of comonomer, counterion, and steric restrictions play an important part.

The distribution of MCM units in polymeric chain is an important characteristic of copolymerization products that defines many of their properties. In such systems the tendency to alternation is more often pronounced.

Two principally different cases are possible during copolymerization of nv-type MCM, stemming from metal ion's ability for additional coordination (extracoordination) with the comonomer. In the first case the complex–radical copolymerization is brought about.[9]

The copolymerization of π-type MCMs has been studied as extensively as their homopolymerization. More monomers can participate in copolymerization than in homopolymerization, because those MCMs incapable of homopolymerization can enter into copolymerization with traditional monomers. There are known cases where MCMs do not polymerize according to any definite mechanism, but enter into copolymerization according to a given mechanism.

Nonradical initiation of copolymerization, as well as the formation of block-copolymers by the synthesis of "living" polymers from MCMs which are capable of initiating comonomer polymerization, has been approved to higher degree for these monomers than for other monomers. It is important that steric hindrances engendered by the presence of metallocycles in such molecules, which are substantial during homopolymerization, take less part in copolymerization.

BASIC APPLICATIONS OF METAL-CONTAINING POLYMERS

Wide interest in metallopolymers is explained by their use as catalysts for different processes (so-called immobilized metallocomplexes or catalysts on polymeric carriers).[10] Substantial attention is being paid to the study of biological activity of metal-containing copolymers: inhibition of different tumor cells, bactericidal properties, fungicidal, insecticidal, and antivirus activity, etc. This field of applications is probably connected with the presence of metal in such products, which also determines their application as polymeric electrolytes: they are magneto- and electroactive, as well as stable to radiation materials and selective sorbents of metal ions. Another field involves the use of metallomonomers and copolymers to improve the properties of traditional polymers as additives and ingredients of polymeric compositions.

REFERENCES

1. Sheats, J. E. *J. Macromol. Sci.-Chem.* **1981,** *A15.* 1173.
2. Sheats, J. E. In *Kirk-Othmer Encyclopedia of Chemical Technology;* Wiley: New York, NY, 1982; Vol. 15, 184.
3. *Metal-containing Polymer Systems;* Carraher, C. E.; Sheats, J. E.; Pittman, C. U., Eds. Plenum: New York, NY, 1985.
4. Kochkin, D. A.; Azerbaev, I. N. *Olovo- i svinetsorganicheskie monomery i polimery;* Nauka: Alma-Ata., Russia, 1968.
5. Pomogailo, A. D.; Savost'yanov, V. S. *Metallosoderzhashchie monomery i polymery na ikh osnove;* Khimiya: Moscow, Russia, 1988.
6. Pomogailo, A. D.; Savost'yanov, V. S. *Russ. Chem. Rev.* **1983,** *52,* 973.
7. Pomogailo, A. D.; Savost'yanov. V. S. *Russ. Chem. Rev.* **1991,** *60,* 762.
8. Pittman, C. U., Jr. *Organometallic Reactions,* Marcel Dekker: New York, NY; 6, 1.
9. Kabanov, V. A.; Zubov, V. P.; Semchikov, Y. D. *Complex – Radical Polymerization;* Khimiya, Moscow, Russia, 1987.
10. Pomogailo, A. D.; Savost'yanov, V. S. *Synthesis and Polymerization of Metal-Containing Monomers;* CRC: Boca Raton, FL, 1994.

Metal-Containing Polymers

See: Cobalt Carbonyl Complexes (With Main Chain Acetylenic Groups)
Colloidal Metal
Conducting Polymers (Nonconjugated)
Dendrimers, Luminescent and Redox-Active
Ferrocene-Backbone Polymers
Germanium-Containing Polymers (Overview)
Metal Clusters (in Polymers, Synthesis and Characterization)
Metal Complex Catalysts, Polymeric
Metal Complexation Polymers
Metal Complexes, Macromolecular
Metal-Containing Monomers (Polymerization)
Metal-Containing Silicon-Based Polymers
Metal Ion Binding Resins (Synthesis and Analytical Properties)
Metal Nanoclusters (within Block Copolymer Domains)
Metallized Polyimides
Metal-Polymer Complexes
Metallocenophanes, Strained (Thermal Ring-Opening Polymerization)
Metalloenzymes, Artificial (Polyethylenimine-Based)
Metallomesogen Cu(II) Polymer
Metallophthalocyanines, Polymeric (Overview)
Metallophthalocyanines, Polymeric (Types of Structures)
Metallosupramolecular Helicates
Molecular Recognition (Peptide-Based Systems)
Organo-Tin Polymers
Organometallic Polymers, Cobalt-Containing
Organometallic Polymers (Transition Metals in Main Chain)
Organometallic Polymers, Fluoroarylene Bridged (Rigid Rods to Segmented Chains)
Phosphinated Polystyrene-Bound Metal Complexes (Pd and Rh)
Photoredox Langmuir-Blodgett Films (for Light Energy Conversion)
Phthalocyanines, Polymeric
Poly(1,1′-Ferrocene-Arylene)s (Pd-Catalyzed)
Poly(phenylene vinylene)s (by the Heck Coupling Reaction)
Polysilanes (Ring-Opening Polymerization of Cyclotetrasilanes)
Radiopaque Polymers
Stacked Transition Metal Macrocycles (Semiconductive Properties)
Supported Polythioethers and Polythiacrownethers
Supramolecular Structures (Bipyridine Block Copolymers with Cu(I) Ions)
Tautomer Polymerization

METAL-CONTAINING SILICON-BASED POLYMERS

Keith H. Pannell and Vyacheslav V. Dementiev
Department of Chemistry
University of Texas at El Paso

Arthur F. Diaz
IBM Almaden Research Center

The addition of transition metal groups to silicon-containing polymers can be expected to modify several properties of the materials, including thermal stability, redox, magnetic, and optical behavior. We overview metal-substituted polysilanes and the recently developed polyferrocenylenesilanes to illustrate the potential for this type of material.

METAL-SUBSTITUTED POLYSILANES

High molecular weight polysilanes containing ferrocenyl groups, $(\eta^5\text{-}C_5H_5)Fe(\eta^5\text{-}C_5H_4)$ (Fc), can be synthesized, as copolymers, via the Wurtz-type coupling reaction.[1,2]

Attempts to produce high molecular weight materials with only $Fc(Me)SiCl_2$ as the monomer resulted in low molecular weight oligomers. The copolymers exhibited the characteristic photochemical depolymerization noted for polysilanes which is the source of their potential as photoresist materials. However, the presence of the Fc substituents resulted in a significant retardation of this depolymerization, presumably due to the ability of ferrocene to quench the triplet states responsible for the polysilane photochemistry.[2]

Alkyl- and aryl-substituted polysilane homopolymers are electroactive and can be electrooxidized, but they decompose rather quickly.[3,4]

Related transition metal-substituted polysilanes containing $Cr(CO)_3$ and $(\eta^5\text{-}C_5H_5)Fe(CO)_2$ groups have been reported. Incorporation of the metal was both at the monomer stage prior to polymerization, and by reaction of the polymers with the metal systems.[5–7]

Tungsten and molybdenum analogs have also been reported and used as preceramic materials.[8–10]

FERROCENYLENESILANE POLYMERS

Ferrocenophanes with a single atom bridge between the two cyclopentadienyl rings possess nonparallel rings.[11–13] Consequently, [1]-silyl-ferrocenophanes exhibit thermally induced ring opening to form high molecular weight materials in which both the transition metal ferrocene groups and silicon atoms are present in the backbone, whereas the corresponding [2]-silyl-ferrocenophanes do not exhibit such polymerization. Manners and Pannell have studied the thermal ring opening of [1]-silyl-ferrocenophanes to yield high molecular weight materials, **Equation 1**.[12–18]

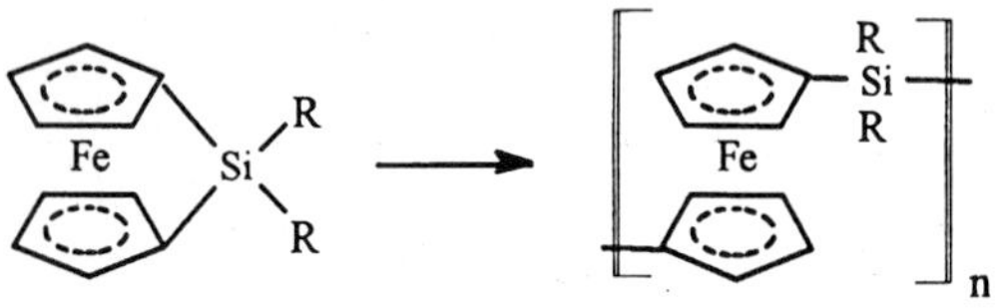

1

The yields of these polymers are generally high, 40–60%, and the molecular weights (mono-modal) range from 30,000 to in excess of 10^6. In certain cases the molecular weights can be dramatically enhanced by performing the thermal ring opening polymerization in solution rather than in the melt.[13]

Manners has published several short reviews of ferrocenylenesilane polymers.[19,20]

REFERENCES

1. Zuaodung, D.; Xiaoyao, W.; Jing, L. *J. Shandong University* **1987**, *22*, 115.
2. Pannell, K. H.; Rozell, J. M.; Zeigler, J. M. *Macromolecular* **1988**, *21*, 276.
3. Diaz, A. F.; Miller, R. D. *J. Electrochem. Soc.* **1985**, *132*, 834.
4. Diaz, A. F. et al. *J. Electrochem. Soc.* **1991**, *138*, 742.
5. Pannell, K. H.; Rozell, J. R.; Vincenti, S. P. *Silicon-Based Polymer Science: A Comprehensive Resource;* Zeigler, J. M.; Fearon, F. W. G. Eds.; Advances in Chemistry Series No. 224; American Chemical Society: Washington, DC, 1990; Chapter 20.
6. Tilley, T. D.; Woo, H-G. *Am. Chem. Soc., Polym. Div.* **1990**, *31*, 228.
7. Vincenti, S. P.; Soto, L.; Pannell, K. H. *46th SW Regional Mtg., Am. Chem. Soc.* San Antonio, October 1991, Abstract 347.
8. Chandra, G.; Zank, G. A. U.S. Patent 4 762 895, 1988.
9. Burns, G. T.; Zank, G. A. U.S. Patent 4 762 895, 1990.
10. Burns, G. T.; Zank, G. A. U.S. Patent 4 945 072, 1990.
11. Stoeckli-Evans, H.; Osborne, A. G.; Whiteley, R. H. *Helv. Khim. Acta* **1976**, *59*, 2402.
12. Finckh, W. et al. *Organometallics* **1993**, *12*, 823.
13. Pannell, K. H. et al. *Organometallics* **1994**, *13*, 3644.
14. Foucher, D. A.; Tang, B. Z.; Manners, I. J. *Am. Chem. Soc.* **1992**, *114*, 6246.
15. Nguyen, M. Y. et al. *SPIE Proc.* **1993**, *1910*, 230.
16. Foucher, D. A. et al. *Macromolecules* **1993**, *26*, 2878.
17. Nguyen, M. T. et al. *Chem. Mater.* **1993**, *5*, 1389.
18. Foucher, D. A. et al. *Macromolecules* **1994**, *27*, 3992.
19. Manners, I. *Adv. Mater.* **1994**, *33*, 989.
20. Manners, I. *J. Inorg. Organomet. Polym.* **1993**, *3*, 185.

Metal-Coordinating Polymers

See: *Coordination Polymers, Dithiooxamides*
Ion-Selective Reagents (Dual Mechanism Bifunctional Polymers)
Polybenzimidazoles (Overview)
Stacked Transition Metal Macrocycles (Semiconductive Properties)
Supported Polythioethers and Polythiacrownethers

Metal Deactivators

See: *Antioxidants*

METAL ION BINDING RESINS (Synthesis and Analytical Properties)

Bernabé L. Rivas
Departamento de Polímeros
Facultad de Ciencias Químicas
Universidad de Concepción

Coordinating and ion exchange polymers for metal ions are of great interest in the different branches of chemistry, in water treatment, environmental protection, chemical analysis, separation of metals, etc.[1–10] The rising world prices of metals coupled with the introduction of strict regulations against the pollution of the environment have focused attention on the recovery of metals from lean ores and wastes. These insoluble (water) polymers with a crosslinked matrix contain the ligand function, which is only a small part of the three-dimensional macromolecular matrix. The base polymers may be natural such as cellulose, chitin, and chitosan, or synthetic such as polyamines, polyacrylamides and poly(acrylic acid). The most common coordinating atoms present at the main or side chain are nitrogen, oxygen, phosphorous, and sulfur.

The complexation or ion exchange process is governed by many different structural-environment characteristics of the polymer support; that is, the nature of the polymer backbone, type and extension of crosslinking agent, swelling capacity, separation of the coordinating or ion exchange group from the polymer backbone, and steric factors.[11,12]

Two main types of processes play lead roles in the extraction of metal ions: liquid–liquid contact processes, and liquid–solid contact processes in which the metal ion is adsorbed on a solid matrix or precipitated from an aqueous solution.

The functional group used in organic exchangers vary widely. In resins like polystyrene copolymerized with divinylbenzene, mainly sulfonic acid groups are introduced as acidic groups or amino groups as basic groups. They have many advantages and can be easily produced. The two main disadvantages are low selectivity and relatively slow attainment of equilibrium. Another versatile polymer is polyethylenimine which may be modified by polymer-analogous crosslinking; alkylation, or the grafting of polyethylenimine onto poly(*p*-chlorostyrene); or by spontaneous copolymerization of ethyleneimine derivatives.

Since such a wide range of materials and methods of synthesis are possible, it is not surprising that the exchanger's physical form may vary from rock-hard material to soft gel. The desirable properties of chelating exchangers are a high capacity for the metal(s) of interest, high selectivity, fast kinetics, and high mechanical strength and toughness of the exchanger particles.[13] Unfortunately, the last two properties are competitive. One of the most versatile polymers is polyethylenimine (PEI) which by itself is a complexing polymer but may be modified to get an insoluble (water) resin.[9,10,14,15]

ACKNOWLEDGMENTS

The author thanks FONDECYT and Dirección de Investigación, Universidad de Concepción, for financial support.

REFERENCES

1. Kaneko, M., Tsuchida, E. *Macromol. Rev.* **1981**, *16*, 397.
2. Lindsay. D. et al. *React. Polym.* **1990,** *12*, 59.
3. Wegscheider, W.; Knapp, G. *Crit. Rev. Anal. Chem.* **1981**, *11*, 79.
4. Hodgkin Jonathan, H. *Encycl. Polym. Sci. Eng.* **1985**, *3*, 363.
5. Shmuckler, G. *Encycl. Polym. Sci. and Tech.* Wiley-Interscience: New York, NY, 1976.
6. Blasius, E., Brozio, R. In *Chelates in Analytical Chemistry* Flashka, H. A.; Barnard, J. A. Eds.; Marcel Dekker: New York, NY, 1967; Vol. 1, pp 50.
7. Pittman, C. U. Jr. In *Polymer Supported Reactions in Organic Synthesis;* Hodge, P.; Sherrington, D. C. Eds., Wiley: New York, NY, 1980.
8. Sherrington, D. C.; Hodge, P. *Synthesis and Separations Using Functional Polymers;* Wiley: New York, NY, 1988; Chapter 10.
9. Rivas B. L.; Geckeler, K. E. *Adv. Polym. Sci.* **1993**, *102*, 171.
10. Rivas B. L.; Bartulin, J. *Bol. Soc. Chil. Quim.* **1986**, *31*, 37.
11. Drago, R. S.; Gaul, J. H. *Inorg. Chem.* **1979**, *18*, 2019.
12. Nishide, H.; Shimidzu, N.; Tsuchida, E. *J. Appl. Polym. Sci.* **1982**, *27*, 4161.
13. Vernon, F. *Hydrometallurgy* **1979**, *4*, 147.
14. Shepherd, E. J.; Kitchener, J. A. *J. Chem. Soc.* **1957**, 86.
15. Saegusa, T.; Kobayashi, S.; Yamada, A. *Macromolecules* **1975**, *8*, 390.

METAL NANOCLUSTERS (within Block Copolymer Domains)

R. E. Cohen,* R. T. Clay, J. F. Ciebien, and B. H. Sohn
Department of Chemical Engineering
Massachusetts Institute of Technology

METAL NANOCLUSTERS–GENERAL

Metal nanoclusters are aggregates of metal atoms with as few as ten and up to several thousand atoms in a single cluster. Typical dimensions[1,2] are from about 10 to 100 Å in diameter with corresponding surface area to volume ratios as high as 0.3 $Å^{-1}$. the properties of metal nanoclusters are interesting because they are different from both the bulk metal and the individual atom.[2,3] Applications are being explored in the areas of advanced materials and heterogeneous catalysis.

*Author to whom correspondence should be addressed.

Quantum size effects arise in metal nanoclusters as a result of their incompletely developed band structure.[2,4] The nanoclusters exhibit absorption band widths which are larger than the bulk material.[5] Nonlinear optical phenomena[4–6] which can accompany this band gap modification have potential application in telecommunications and computing. The high conductivity and small size of metal nanoclusters make them good candidates for the conductive domains in high dielectric constant composites.[7,8]

In the field of heterogeneous catalysis,[9] a wide variety of metal and metal alloy nanoclusters are employed on various inorganic supports. In other cases, clusters of catalytic metals have been stabilized by polymers or surfactants,[10–15] through the use of polymeric gels,[16] or embedded in polymer films.[17–19]

Metal nanoclusters can be prepared through a variety of techniques including metal acid or salt reduction, metal evaporation, and sputtering.

POLYMER-SUSPENDED METAL NANOCLUSTERS

There are few feasible techniques for the synthesis of uniformly sized and dispersed nanoclusters in polymer films. One promising approach involves the alkylidene complex initiated,[20] ring opening metathesis polymerization (ROMP) of block copolymers, in which one of the blocks contains either coordinated metal atoms or coordinating groups capable of sequestering metal atoms during post polymerization processing.[21] The block copolymers can exhibit various equilibrium and kinetically locked microphase-separated morphologies,[22] and the reduction of the metal complexes within the spatially confined regions of these morphologies produces metal clusters in the polymer film.

CONCLUSIONS

Polymers show great promise as nanocluster supports since they can be transparent, permeable, and easily processed. Metal nanoclusters in block copolymers are particularly attractive because of the possibility of obtaining precisely tailored composites, consisting of spherical, cylindrical, or lamellar domain confined nanoclusters, surrounded by a glassy or rubbery matrix. These materials may eventually have applications in high selectivity catalysis, optical switches, and photoluminescent devices.

The feasibility of synthesizing metal nanoclusters *in situ* within copolymer microdomains has been demonstrated. Cluster synthesis inside the nanoscale morphologies is made possible using organometallic monomers in one block, or through chemical treatment of purely organic block copolymers containing metal sequestering groups in one block.

REFERENCES

1. Burton, J. J. *Sintering and Catalysis*, 1st ed.; Kuczynski, Ed.; Plenum: New York, NY, 1975; p 17.
2. Henglein, A. *Chem. Rev.* **1989**, *89*, 1861.
3. Bradley, J. S. et al. *Chem. Mater.* **1992**, *4*, 1234.
4. Stucky, G. D. *Naval Research Reviews* **1991**, *3*, 28.
5. Hache, F.; Ricard, D.; Flytzanis, C. *J. Opt. Soc. Am. B.* **1986**, *3*, 1647.
6. Stucky, G. D., Mac Dougall, J. E. *Science* **1990**, *247*, 669.
7. Laurent, C.; Kay, E. *J. Appl. Phys.* **1988**, *64*, 336.
8. Canet, P. et al. *J. Appl. Phys.* **1992**, *72*, 2423.

9. Satterfield, C. N. *Heterogeneous Catalysis In Industrial Practice*, 2nd ed.; McGraw-Hill: New York, NY, 1991.
10. Hirai, H. *J. Macromol. Sci.-Chem.* **1979**, *A13*, 633.
11. Hirai, H.; Chawanya, H.; Toshima, N. *Bull. Chem. Soc. Jpn.* **1985**, *58*, 682.
12. Bradley, J. S. et al. *J. Mol. Catal.* **1987**, *41*, 59.
13. Andrews, M. P.; Ozin, G. A. *Chem. Mater.* **1989**, *1*, 174.
14. Toshima, N.; Takahashi, T. *Bull. Chem. Soc. Jpn.* **1992**, *65*, 400.
15. Touroude, R. et al. *Colloids Surfaces* **1992**, *67*, 9.
16. Ohtaki, M. et al. *Bull. Chem. Soc. Jpn.* **1990**, *63*, 1433.
17. Park, K. M.; Shim, I. W. *J. Appl. Polym. Sci.* **1991**, *42*, 1361.
18. Hwang, S. T.; Shim, I. W. *J. Appl. Polym. Sci.* **1992**, *46*, 603.
19. Chu, J. W.; Shim, I. W. *J. Mol. Catal.* **1993**, *78*, 189.
20. Schrock, R. R. et al. *J. Am. Chem. Soc.* **1990**, *112*, 3875.
21. Rempp, P.; Merrill, E. W. *Polymer Synthesis*, 2nd ed.; Huthig & Wepf Verlag Basel: New York, NY, 1991.
22. Cohen, R. E.; Bates, F. S. *J. Polym. Sci.: Polym. Phys. Ed* **1980**, *18*, 2143.

METAL-POLYMER COMPLEXES

Yoshimi Kurimura and Masao Kaneko
Faculty of Science
Ibaraki University

Polymer-metal complexes are composed of polymeric ligand and metal ions. In solution, polymer-metal complexes form microheterogeneous regions occupied by polymer-backbone where physicochemical properties differ from those of the bulk solution. Most significant reaction patterns of polymer-metal complexes are due to the characteristic nature of these microheterogeneous regions. Polymer-metal complexes show unique characteristics in absorption spectrum, coordination structure, stability, redox reactions, catalytic activities, electrochemical reactions, and other areas compared to those of corresponding low molecular metal complexes. They often show a specific catalytic activity based on the large ligand comparable to metal enzymes; they are considered as a synthetic enzyme analogue. Complexation of polymeric ligand with metal ions as well as ligand substitution reaction of polymer-metal complexes are utilized to separate metal ions and/or small molecules.[1–5]

The conformation effect of polymer chain is also important in regulating the reactivity of polymer-metal complexes, as observed in enzymes.

PROPERTIES AND APPLICATIONS

Redox Reactivity

Redox reactivity of a polymer-metal complex depends on four effects: (a) electrostatic, (b) hydrophobic, (d) neighboring group, and (d) conformational, of the surrounding polymer chain. A distorted structure or, on the other hand, a highly ordered structure brought about by the surrounding polymeric chain can drastically change the redox chemistry of the central metal ion, as is often seen in metallobenzymes.

Separation of Ions or Molecules

A polymeric ligand is used to selectively bind a specific metal ion from a mixture, to isolate important metals from sea or wastewater, or to remove toxic metal ions from wastewater. This is based on different formation constants between the polymeric ligand and various metal ions or molecules. Recognition of a molecule or ions takes place by this difference in complex formation. A polymeric ligand is usually used in an insoluble resin form to separate a specific metal ion from liquid containing metal ions. A chelate resin containing iminodiacetic acid groups is a good polymeric ligand for this purpose.

Polymer-metal complex can be used for separating organic compounds, such as amino acid and nitrogen-containing compounds from their mixture by utilizing their ability to coordinate with metal complex. Optical resolution of racemic compound is also possible when the polymer-metal complex is surrounded by asymmetric structures.

Catalytic Activity

Polymer-metal complex is used as a catalyst for various reactions. It often shows higher selectivity and activity for substrate than corresponding low molecular weight metal complex, and high stability with repeated use. When the metal complex is immobilized on an insoluble resin, the catalyst is easy to reuse by simply separating it after the catalysis. Neighboring functional groups of a polymeric ligand can form a specific catalyst site with the metal complex center. Metal ions in a polymer complex are concentrated locally near the polymer ligand, which often enhances the complex reaction rate. On the other hand, confining of metal complexes to a polymer chain can prohibit their interaction due to the steric factor of the polymer, which can stabilize an active complex structure.

Electric Conductivity

Polymer-metal complex shows electric conductivity or semiconductivity depending on the doped or nondoped state. π-Conjugated polymers, such as polyacetylene, polyphenylene, polyaniline, and polypyrrole, become π-conductive when doped with transition or alkaline metal cation.

Polymeric metal-phthalocyanine formed by cofacial stacking of metal-phthalocyanine by bridging ligand is electroconductive without doping because of the conjugated π-electron structure.

Photochemical Reactions

Photocurrent is induced by a polymer membrane containing $Ru(bpy)_3^{2+}$ coated on an electrode in the presence of some acceptor molecule, such as methylviologen, which is present either in a solution contacting the membrane or on the Ru complex membrane as a second layer. When utilizing quenching by dioxygen of a photoluminescence from $Ru(bpy)_3^{2+}$ confined in a polymer membrane, the oxygen concentration can be determined by measuring the luminescence intensity of the polymer membrane.

REFERENCES

1. Kurimura, Y. *Adv. Polym. Sci.* **1989**, *90*, 105.
2. Kaneko, M.; Tsuchida, E. *J. Polym. Sci., Macromol. Rev.* **1981**, *16*, 397.
3. *Macromolecular Complexes - Dynamic Interaction and Electronic Processes*; Tsuchida E. Ed., VCH: New York, 1991.
4. Kaneko, M.; Woehrle, D. *Adv. Polym. Sci.* **1987**, *84*, 141–228.
5. The references cited in the reviews 1 to 4.

METALLIZED POLYIMIDES

Eugene Khor
Department of Chemistry
National University of Singapore

The impetus for the incorporation of metallic species into polyimides is to obtain new materials with unique properties. The most desirable property is the enhancement of electrical conductivity while retaining the inherently high temperature stability of polyimides. Traditionally, metal flakes, powder, or graphite is used to render polymers electrically conductive. High loading-levels are necessary, with the attendant reduction in polymer stability. In producing metallized polyimides, the high thermal stability of the polymer is exploited. Soluble metallic species that undergo chemical decomposition during the thermal curing (300°C) of polymer are employed to give semiconductive and conductive polyimides at a very low dopant loading, in most instances preserving the polymer's properties. Such electrically conductive polyimides could aid in dissipating charge buildup in advanced polymers intended for aerospace applications. Furthermore, these materials could be useful as conductive tapes, photosensitive materials, and flexible heating tapes. Finally, a knowledge of the behavior of metal-polyimides interaction is vital for the semiconductor industry where the insulator used is typically polyimide.

The first attempt at interaction dates back to 1959, when Endrey added silver acetate to poly(amide acid) (polyimide precursor). Subsequent casting of the mixture onto glass plates and curing gave polyimide films that exhibited a metallic luster, and were tough and flexible.[1] At about the same time Angelo used copper compounds to produce polyimide films that had four orders of magnitude reduction in electrical resistivity.[2] This was followed by the use of acetylene black to produce elastic polyimide films that had an electrical resistivity of 33.7Ω.[3] Thereafter, many reports using various metal sources such as titanium oxide, palladium and cobalt chloride, and organo-tin compounds were mentioned.[4] This approach, where any metallic species that came to mind was tried, endured until the late 1970s. The results from this exploratory period sparked the ensuing formal and more comprehensive study of metallized polyimides. In late 1978, Professor Larry Taylor at Virginia Tech began what can be called the concerted effort in the study of metallized polyimides.

SUMMARY

The study of metallized polyimides has progressed from a random use of metal compounds to a mature stage where an understanding of the interactions of metal compound–polyimide is available. The various metal additives that have been used in metallized polyimides display a myriad of properties. The type of metallized polyimide obtained depends on the manner in which the metal compound decomposes under the imidization temperature and the metal's interaction with its environment, i.e., water of imidization, solvent, and atmosphere. How this interaction occurs is determined by the metal's chemical behavior broadly representative of its position in the periodic table. We know the means to produce either a surface rich in metal or one which disperses the metal uniformly throughout the polymer bulk. The manipulation of this knowledge into practical applications represents the future of metallized polyimides.

REFERENCES

1. Endrey, A. L.; E. I. Du Pont de Nemours & Co., U.S. Patent 3 073 784, 1963.
2. Angelo, R. J.; E. I. Du Pont de Nemours & Co., U.S. Patent 3 073 785, 1963.
3. Berr, C. E. et al. Belg. Patent 630 749, 1963.
4. Khor, E. Ph.D. Dissertation, Virginia Polytechnic Institute and State University, 1983.

Metallocalixarenes

See: Calixarenes (as Supramolecular Cyclic Oligomers)

METALLOCENE/ALUMINOXANE CATALYSTS (Olefin Polymerization)

Incoronata Tritto,* San Xi Li, Maria Carmela Sacchi, and Paolo Locatelli
Istituto di Chimica delle Macromolecole

Alkylaluminoxanes, obtained by controlled hydrolysis of aluminum alkyls, are oligomeric compounds of general formula $[-Al(R)O-]_n$, that contain at least one bridging oxo group between two aluminum atoms. They have been studied since the 1960s as catalysts for polymerization of monomers like epoxides, aldehydes and olefins.[1–5] Kaminsky's discovery that aluminoxanes are formed by adventitious addition of water to Cp_2MX_2/AlR_nX_{3-n} (Cp = C_5H_5; M = Ti, Zr; X = halide; R = hydrocarbyl) and that they are extremely effective cocatalysts for olefin polymerization has made them a potentially important class of industrial compounds.[6–8] Indeed, this discovery has led to the development of new metallocene/alkylaluminoxane catalytic systems that make possible the synthesis of entirely new families of polymers. Since the synthesis of the metallocene that gives isotactic polypropylene, i.e., Brintzinger's and Ewen's bridged rac-ethylene-bis(indenyl)$TiCl_2$,[9,10] hundreds of new metallocenes have been obtained by changing either the bridge between the cyclopentadienyl ligands or the Cp ligands and their substituents in addition to the metal (Ti,Zr,Hf). Additionally, copolymers with uniform comonomer incorporation and narrow molecular weight distribution; i.e., with improved product consistency or clarity, or with elastomeric properties, can be obtained.[9–12] Kaminsky and Möhring and Coville have reviewed some of the early literature in this field. Industrial development of these systems has been slowed down mainly because they require much expensive methylaluminoxane and heterogenization.[7,8]

STRUCTURAL CHARACTERIZATION OF ALUMINOXANES

Methylaluminoxane $[-Al(CH_3)O-]_n$, which was the first aluminoxane and the most frequently used as a cocatalyst for metallocenes, is a mixture of oligomers with an average molecular

*Author to whom correspondence should be addressed.

mass between 1000 and 1500 g/mol, measured in benzene by cryoscopy. Sinn has reviewed methylaluminoxane (MAO) synthesis and characterization methods.[13] Ethylaluminoxanes and isobutylaluminoxanes, which are less expensive and produce solutions more stable than MAO, can be obtained free of the starting alkylaluminum.[14] However, their catalytic activity does not seem suitable for industrial use. Modified methylaluminoxanes, which contain isobutyl groups, have been developed as alternatives to methylaluminoxane for some applications, due to their solubility in non-aromatic hydrocarbons and to the higher stability of their solutions.[15]

MODEL METALLOCENE CATIONIC COMPLEXES

The primary role of MAO is believed to be the alkylation of the transition metal and the production of cation-like d_0 14-electron alkyl complexes of the type Cp_2MR^+ as catalytically active species. Growing indirect evidence that MAO generates metallocene cations (Cp_2MR^+) in these systems has been furnished by works of model intermediate trapping, and model synthesis.[16–19]

CONCLUSIONS

In conclusion, the three-part method that we first used; that is, the combination of ^{13}C NMR analysis of equilibria involved in metallocene/aluminoxane systems; the ^{13}C NMR study of polymerization *in situ* of ^{13}C enriched ethylene; and the determination of the polymerization catalytic activity, allowed us to find evidence of the function of MAO and $AlMe_3$ contained in it.

Indeed, we have obtained direct evidence of MAO ability:

- To alkylate the transition metal
- To produce and stabilize cation-like catalytically active species **4** Cp_2TiMe^+Cl-$[AlMeO]_n^-$ and species **8** Cp_2TiMe^+Me-$[AlMeO]_n^-$ in titanocene-MAO systems. They can be aptly described as loosely associated ion pairs.

Concerning the role of $AlMe_3$ contained in MAO solutions, it has been shown that:

- $AlMe_3$ is bound mainly to MAO in agreement with[14]
- Some "free" $AlMe_3$ that may exist in solution is not the cocatalyst in the metallocene-MAO based catalytic systems
- The $AlMe_3$ content influences the formation of either active or inactive species

Therefore, there should be an amount of $AlMe_3$ which is crucial for obtaining high activity.

What remains unclear is how the ratio between active and inactive species changes according to the aging and type of MAO solution, as we have sometimes obtained different results by varying MAO solutions. Direct characterization of MAO structures, moreover, is not possible and our and other authors' findings cannot furnish direct evidence either for or against hypothesized or new structures.

REFERENCES

1. Colclough, R. O.; Gee, G.; Jagger, A. H. *J. Polym. Sci.* **1960**, *48*, 273.
2. Vanderberg, E. J. *J. Polym. Sci.* **1960**, *47*, 489.
3. Ishida, S. I. *J. Polym. Sci.* **1962**, *62*, 1.
4. Longiave, C.; Castelli, R. *J. Polym. Sci.* **1963**, *4C*, 387.
5. Sakharovskaya, G. B. *Zh. Obsch. Chim.* **1969**, *39*, 788.
6. Sinn, H.; Kaminsky, W. *Adv. Organomet. Chem.* **1980**, *18*, 99, and references therein.
7. Kaminsky, W. *Catalysis Today* **1994**, *20*, 257.
8. Möhring, P. C.; Coville, J. N. *J. Organom. Chem.* **1994**, *479*, 1.
9. Fink, G.; Mülhaupt, R.; Brintzinger, H. H. Eds.; *Ziegler Catalysts* Springer: Berlin 1995, in press.
10. MetCon '93, May 26–29, 1993 Houston, TX, (Proceedings).
11. Soga, K.; Terano, M. Eds.; *Catalyst Design for Tailor-Made Polyolefins* Elsevier: Amsterdam, 1994.
12. Stepol '94 Volume of *Macromolecular Symposia* lectures presented at "Synthetic, Structural and Industrial Aspects of Stereospecific Polymerization" (Stepol '94) June 6–11, 1994 Milan, Italy, in press.
13. Sinn, H. 1988.
14. Howie, M. S. Proceedings of MetCon '93, May 26–29, 1993 Houston, TX, p 253.
15. Malpass, D. B. Third International Business Forum on Specialty Polyolefins (SPO '93) organized by Schotland Business Research Inc., Houston, TX, September 21–23, 1993.
16. Eisch, J. J.; Piotrowski, A. M.; Brownstein, S. K.; Gabe, E. J.; Lee, F. L. *J. Am. Chem. Soc.* **1985**, *107*, 7219.
17. Eisch, J. J.; Caldwell, K. R. in Moser, W. R.; Slocum, D. W. Eds. *Homogeneous Transition Metal Catalyzed Reactions* Advances in Chemistry Series, No. 230; American Chemical Society: Washington, DC, 1992, p 575 and references therein.
18. Jordan, R. F.; Bajgur, C. S.; Willett, R.; Scott, B. *J. Am. Chem. Soc.* **1986**, *108*, 7410.
19. Jordan, R. F. *Adv. Organomet. Chem.* **1991**, *32*, 325.

METALLOCENE CATALYSTS

Peter J. T. Tait
University of Manchester Institute of Science and Technology (UMIST)

While the development of highly active metallocene polymerization catalyst systems has been one of the major achievements in polymerization catalysis during the past fifteen years or so, success in this area has been achieved mostly by discoveries of suitable co-catalyst activation systems. The use of metallocenes as polymerization catalysts goes back to 1957 when Breslow and Newburg and Natta and co-workers discovered that homogeneous catalyst systems derived from cyclopentadienyl titanium dichloride (Cp_2TiCl_2) and diethyl-aluminum chloride (Et_2AlCl) could catalyse the polymerization of ethene under similar conditions to those required for heterogeneous Ziegler-Natta catalysts.[1,2] These catalyst systems, however, showed very low activities for ethene polymerization, and were inactive for propene polymerization. Subsequent investigations by Reichert and Meyer established surprisingly that the addition of small amounts of water could activate catalyst systems of the type $Cp_2TiEtCl$-$AlEtCl_2$.[3] However, it was when studying halogen-free catalyst systems such as Cp_2ZrMe_2-$AlMe_3$ that Sinn, Kaminsky and co-workers noticed that the addition of water to the co-catalyst in this system produced a highly active catalyst system for ethene polymerization.[4] Attention was then focused on the co-catalyst activation system, $AlMe_3$-H_2O, and it was established that partial hydrolysis of the $AlMe_3$ took place with the formation of methyl aluminoxane (MAO). Additionally Sinn, Kaminsky and co-workers observed that MAO-activated homogeneous metallocene catalysts could polymerize propene and higher α-olefins.[4]

Subsequent discoveries during 1984 and 1985 by Ewen and by Kaminsky, Brintzinger, and co-workers that chiral C_2-symmetry metallocenes such as rac-ethylene bis(tetrahydroindenyl)zirconium dichloride could produce isotactic polypropene, followed by Ewen's discoveries of a catalyst for the preparation of syndiotactic polypropene, and the development of aluminoxane-free metallocene catalysts have led to intensive interest in this area of polymerization catalysis.[5–8]

INDUSTRIAL USAGE

In spite of the high promise of being able to tailor make polymers for specific applications the industrial exploitation of metallocene catalysts has been very slow to get started, and it is only within the past four to five years that significant progress has been made.

LLDPE Copolymers

The first commercial process to use these new metallocene catalysts was that devised by Exxon who commissioned a 15,000-ton plant at Baton Rouge, Louisiana, U.S.A., in 1991. An achiral single-site metallocene catalyst system is believed to be used. A wide range of "Exxact" copolymers of ethene with butene-1 and hexene-1 comonomers can be produced, and already more than 40 grades of resin are promised.

In a further development, Exxon and the Mitsui Petrochemical Industries (Tokyo) have announced a strategic alliance to extend single-site catalyst technology to gas-phase polymerization processes, which are widely understood to have a very great industrial potential.[11,12]

In 1993, Dow began the operation of a LLDPE plant using a cyclopentadienyl amide titanium catalyst. The use of a short dialkylsilyl bridging group produces an open and highly reactive metal atom, hence the term "Constrained Geometry Catalysts" has been applied to this class of catalysts.

Isotactic Polypropene

An industrial process using metallocene-based catalysts for the production of isotactic polypropene was reported in 1990 by the Japanese company, Chisso Corporation (Tokyo).

Syndiotactic Polypropene

Fina and Mitsui Toatsu have developed a pilot plant for the production of syndiotactic polypropene in 1992 using a zirconium catalyst system along with methyl aluminoxane as co-catalyst. A bridge molecule between the catalyst's cyclopentadiene and fluoroenyl groups is claimed to be the key to obtaining syndiotacitity.

Syndiotactic Polystyrene

Syndiotactic polystyrene may be prepared using a family of catalysts containing a single cyclopentadienyl ligand, $CpTiX_3$, and a suitable co-catalyst Idemitsu Kosan and Dow are presently involved in a joint program to commercialize the production of this polymer.

Ethylene-Propylene Rubbers

In a recent and highly significant development of the production of EP-type rubbers using single-site catalysts is promised by DuPont and Dow. The material produced is directly equivalent to EPDM.[14]

The advent of highly efficient metallocene catalyst systems will bring considerable changes to the polymer industry. The use of these catalyst systems allows unprecedented control over catalyst activity and performance as well as over comonomer incorporation. For the first time in the history of polymerization catalysis it is possible to prepare concisely defined highly active complexes of a single type, and so control the regiospecificity and stereospecificity of polymerization processes. Metallocene catalyst systems, additionally, are better than conventional Ziegler-Natta catalysts, at polymerizing monomers of varying sizes and types, which leads to improved comonomer incorporation and copolymers with a more homogeneous composition. As a result tailor-made polymers for specialty end-uses can be produced, and for this reason many groups such as BP, BASF, DSM, Fina, Himont, Hoechst, Mitsubishi, Mitsui Toatsu, Mobil, Phillips, Sumitomo, Union Carbide, Quantum and many others are currently developing potential commercial processes for the production of polyethene, polypropene, and ethene-propene copolymers.

For more detailed reviews, the reader should consult those of Horton, Gupta et al., Soares and Hamielec, and Brintzinger et al.[9,10,15,16]

REFERENCES

1. Breslow, D. S.; Newburg, N. R. *J. Am. Chem. Soc.* **1957**, *79*, 5072; *ibid.* **1959**, *81*, 81.
2. Natta, G.; Pino, P.; Mazzanti, G.; Lanzo, R. *Chim. Ind.* (*Milan*) **1957**, *39*, 1032.
3. Reichert, K. H.; Meyer, K. R. *Makromol. Chem.* **1973**, *169*, 163.
4. Sinn, H.; Kaminsky, W.; Vollmer, H. J.; Woldt, R. *Angew. Chem.* **1980**, *92*, 396.
5. Ewen, J. A. *J. Am. Chem. Soc.* **1984**, *106*, 6355.
6. Kaminsky, W.; Külper, K.; Brintzinger, H. H.; Wild, F.R.W.P. *Angew. Chem. Int. Ed. Engl.* **1985**, *24*, 507.
7. Ewen, J. A.; Jones, R. L.; Razavi, A.; Ferrara, J. D. *J. Am. Chem. Soc.* **1988**, *110*, 6255.
8. Hlatky, G. G.; Turner, H. W.; Eckman, R. R. *J. Am. Chem. Soc.* **1989**, *111*, 2728.
9. Brintzinger, H. H.; Fisher, D.; Mülhaupt, R.; Rieger, B.; Waymouth, R. M. *Angew. Chem.* Int. Ed. Engl. **1995**, *34*, 1143.
10. Horton, A. D. TRIP, **1994**, 2(5), 158.
11. Wood, A.; Chynoweth, E. *Chem. Week* **1992**, *May*, 53.
12. Wood, A. *Chemical Week* **1992**, *April*, 12.
13. Chowdhury, J.; Moore, S. *Chem. Eng.* **1993**, *April*, 39.
14. Shaw, D. *Eur. Rubber J.* **1995**, *July/August*, 40.
15. Gupta, V. K.; Satish, S.; Bhardwaj, I. S. *Rev. Macromol. Chem. Phys.* **1994**, *C34*(3), 439.
16. Soares, J. B. P.; Hamielec, A. E. *Polymer Reaction Engineering* **1995**, *3*(2), 131.

METALLOCENE CATALYSTS (Cationic Group 4 Metal Alkyl Complexes, Olefin Polymerization)

Manfred Bochmann
School of Chemical Sciences
University of East Anglia

The ability of Group 4 metallocenes, notably bis(cyclo-pentadienyl)titanium dichloride, to act as homogeneous catalysts for the polymerization of ethylene was discovered shortly after the invention of the heterogeneous Ziegler–Natta catalysts.[1,2] They were, however, prone to reduction and could not compete in activity with the heterogeneous systems, particularly in the polymerization of higher 1-alkenes, such as propene. Important improvements were brought about by the use of methylaluminoxane (MAO) as the activator.[3] Although excessive amounts of MAO must be used to be effective, catalysts based on Cp_2ZrCl_2/MAO mixtures are beginning to be introduced in many industrial processes, either replacing existing technology or opening the way for new ranges of polymer and copolymer products. These developments have received a further boost by the discovery that zirconocenes with rigidly connected cyclopentadienyl ligands (*ansa*-zirconocenes) of the type first developed by Brintzinger et al. catalyze the polymerization of propene with a remarkable degree of stereoselectivity.[4–9] However, the polymerization mechanism and nature of the catalytically active species remained obscure. The following account summarizes the events that led from the identification of the active species to its synthesis and the development of highly productive aluminum-free catalysts.

MECHANISTIC ASPECTS

In principle, the relative simplicity of homogeneous catalyst systems should make it easier to understand the reaction mechanism. In the case of metallocene-based polymerization catalysts, however, the apparent high reactivity of the catalytically active center, the presence of aluminum alkyls or alkyl halides in excess, and the existence of numerous side reactions and deactivation processes prevent the unequivocal identification of the reaction steps of the catalytic cycle.[3]

The concept that coordinatively unsaturated cationic metal alkyl species of the type $[Cp_2TiR]^+$ might be involved in polymerization goes back to a proposal by Shilov et al. in the 1960s but received little attention and could at the time not be supported by preparative studies.[10,11] The suggestion was revived in 1985 when Eisch et al. reported the isolation and structural characterization of a cationic titanium vinyl complex, derived apparently from the insertion of an acetylene into the Ti–Me bond of a $[Cp_2TiMi]^+$ intermediate, and it was suggested that olefin polymerization might involve mechanistically related steps.[12] At around the same time, Fink et al. proposed a mechanistic scheme based on the polymerization of ^{13}C-labeled ethylene by a $Cp_2TiCl_2/AlEtCl_2$ catalyst and suggested that adducts reacted with further aluminum alkyl to give a spectroscopically unobservable activated Complex C*, which was able to react with the monomer in a fast chain-growth reaction and could be stabilized to give rise to a series of observable temporarily dormant higher titanium alkyls.[13,14]

LIGAND-STABILIZED CATIONIC ALKYL COMPLEXES $[CP_2MR(L)]^+$

Fink's findings could suggest a mechanism involving a cationic electronically unsaturated intermediate as the catalytically active species. The existence of such species was supported further in 1986 by the synthesis of the first isolable cationic titanium alkyl complexes.[15]

Whereas cationic titanium complexes with donor ligands such as nitriles $[Cp_2TiMi(NCR)]^+$ (R = Me, Pr, Ph) do not undergo ligand exchange with weaker ligands, such as tetrahydrofuran (THF) and for the same reason do not react with olefins or acetylenes, Jordan et al. demonstrated that the zirconium THF complex $[Cp_2ZrMe(THF)]^+$, prepared by an oxidation route, was sufficiently labile to catalyze the polymerization of ethylene.[15–17]

CONCLUSION

Over the last 10 to 15 years the chemistry of metallocene-based polymerization catalysts has developed on two fronts. The main features of the polymerization mechanisms have been ascertained through detailed mechanistic studies and rational syntheses of postulated intermediates. At the same time, significant advances in ligand synthesis have led to catalysts with excellent productivity to give poly(1-alkene)s, particularly polypropene, with much improved stereoregularity and molecular weight. Soluble and supported metallocene-based catalysts are now beginning to be used on an industrial scale, in part replacing existing heterogeneous catalysts. These second-generation catalysts use MAO as the activator. There is every prospect that with improved synthetic methods and enhanced catalyst lifetimes these will in time be replaced by low-aluminum or aluminum-free cationic metal alkyl catalysts.

REFERENCES

1. Natta, G.; Pino, P.; Mazzanti, G.; Lanzo, R. *Chim. Ind.* **1957**, *39*, 1032b.
2. Breslow, D. S.; Newburg, N. R. *J. Am. Chem. Soc.* **1957**, *79*, 5073.
3. Sinn, H. J.; Kaminsky, W. *Adv. Organonomet. Chem.* **1980**, *18*, 99.
4. Smith, J. A.; Seyerl, J. V.; Huttner, G.; Brintzinger, H. H. *J. Organomet. Chem.* **1979**, *173*, 175.
5. Wild, F. R. P.; Zsolnai, L.; Huttner, G.; Brintzinger, H. H. *J. Organomet. Chem.* **1982**, *232*, 233.
6. Ewen, J. A. *J. Am. Chem. Soc.* **1984**, *106*, 6355.
7. Ewen, J. A.; Haspeslagh, L.; Atwood, J. L.; Zhang, H. *J. Am. Chem. Soc.* **1987**, *109*, 6544.
8. Ewen, J. A.; Jones, R. L.; Razavi, A.; Ferrara, J. D. *J. Am. Chem. Soc.* **1988**, *110*, 6255.
9. Kaminsky, W.; Külper, K.; Brintzinger, H. H.; Wild, F. R. W. P. *Angew. Chem. Int. Ed. Engl.* **1985**, *24*, 507.
10. Zephirova, A. K.; Shilov, A. E. *Dokl. Akad. Nauk. SSSR* **1961**, *136*, 599.
11. Dyachkovskii, F. S. *Coordination Polymerization*; Chien, J. C. W., Ed.; Academic: New York, 1975; p 199.
12. Eisch, J. J.; Piotrovski, A. M.; Brownstein, S. K.; Gabe, E. J.; Lee, F. L. *J. Am. Chem. Soc.* **1985**, *107*, 7219.
13. Fink, G.; Fenzl, W.; Mynott, R. *Z. Naturforsch. B* **1985,** *40b*, 158.
14. Fink, G.; Mynott, R.; Fenzl, W. *Angew. Makromol. Chem.* **1987**, 154, 1.

15. Bochmann, M.; Wilson, L. M. *J. Chem. Soc., Chem. Commun.* **1986**, 1610.
16. Bochmann, M.; Wilson, L. M.; Hursthouse, M. B.; Short, R. L. *Organometallics* **1987**, *6*, 2556.
17. Jordan, R. F.; Bajgur, C. S.; Willet, R.; Scott, B. *J. Am. Chem. Soc.* **1986**, *108*, 7410.

METALLOCENE CATALYSTS (Group 4 Elements, New Polymeric Materials)

Baotong Huang and Jun Tian
Changchun Institute of Applied Chemistry
Chinese Academy of Sciences

The emergence and evolution of soluble metallocene catalysts brought forth a renaissance in polymeric materials from Ziegler-Natta catalysis.[1] Some existing polyolefins have been bestowed new structures and properties, and polymeric materials heretofore unattainable have been given birth. Syntheses of isotactic polypropylene (*i*-PP) and other stereoregular Pps illustrate the extent to which ingenious catalyst design can reach.

POLYETHYLENE, ETHYLENE/α-OLEFIN COPOLYMERS

Unlike traditional Ziegler-Natta catalysts, homogeneous catalyst systems of group 4 metallocene/methylaluminoxane (MAO) are characterized by their single-sites in polymerization-active centers and narrow polymer MWD (M_w/M_n <3). Polyethylenes obtained have 0.9≠1.2 methyl, 1.1–1.8 vinyl and 0.2 *trans* ethenylene groups in 1000 carbon atoms. MW of PE is controlled by polymerization temperature, amount of metallocene used, concentration of ethylene, and addition of hydrogen. In ethylene homopolymerization activities of typical homogeneous catalyst systems, zirconocene is most active, but its activity drops sharply on substituting MAO with ethyl, *i*-propyl, or *i*-butyl aluminoxane.[2] Ethylene polymerization behavior is heavily influenced by size and electronic properties of substituents on the ligand groups. Electron-releasing groups favor higher activity, while electron-withdrawing groups and spatially hindering bulky substituents lower activity. Aging of the catalyst, solvent and additives also affects ethylene homopolymerization.

Copolymerization of ethylene with propylene, 1-butene, 1-hexene, 4-methyl-1-pentene, 1-octene or other α-olefins to yield LLDPE are realized with these soluble catalysts. Generally speaking, copolymerizability is higher in bridged metallocenes than unabridged, with monomer composition in the copolymers approximating that of the monomer feeds.[3] Bridged hafnocenes give better copolymerizability and higher MW of the copolymers. Among bridged metallocenes, unsymmetric ziconocenes (*Cs*-symmetry) show the strongest copolymerizability, enabling the content of α-olefin in the copolymer to reach as high as 40 mol%. This revolutionary, precise control of molecular architecture, unattainable via existing conventional multisite catalysts, makes possible today's production of various kinds of LLDPE with improved physical properties.

EDPM elastomers with no crystallinity from ethylene/propylene/ethylidene norbornene or 1,4-hexadiene may also be prepared with metallocene catalyst systems in higher activity and narrow MWD than with vanadium catalysts.

STEREOSELECTIVE POLYMERIZATION OF PROPYLENE AND OTHER α-OLEFINS

Polypropylene with various stereospecificity may be prepared by metallocene catalyst systems depending on variations in the stereoarrangements of ligands in the metallocenes. The mode of chain growth in propylene polymerization also varies with the catalyst used. There is currently no other catalyst that can polymerize propylene solely into respective stereospecific varieties of random, isotactic, syndiotactic, hemiisotactic or isotactic block.

Atatic Polypropylene (*a*-PP)

While traditional heterogeneous catalysts do not give completely amorphous PP, achiral C_{2v}-symmetric metallocene/MAO yields, with high activity, completely random atactic PP, a product that has found wide-ranging commercial application. The MW of *a*-PP depends heavily on the substituents on the ligands and the angle between the two ligands.

Isotactic Polypropylene (*i*-PP)

The two variously substituted ring ligands of *ansa*-metallocenes may, through chelating bridges ($-CH_2CH_2-$, *i*-Pr, $(Me)_2Si$, etc.), assume rigid conformation of C_2-symmetry with specific chiral stereoselectivity. Changes in ligand structure may change spatial symmetry of the metallocenes, and change in the chelating bridge also modifies the spatial coordination mode of the ligands. All these subtle changes in catalyst structure profoundly affect the behavior of stereospecific polymerization of α-olefins.

Using Metallocenes with an Ethylene Bridge ($-CH_2CH_2-$)

Ewen first noticed that the homogeneous catalyst system composed of sterically rigid $Et(Ind)_2TiCl_2$(Ind:indenyl) (mixture of racemic and meso forms) and MAO could polymerize propylene into 63% isotactic and 37% atactic PP.[3] He confirmed that isotactic PP resulted from catalytic polymerization by the chiral racemic isomers, in which chain growth conformed to growth model of enantiomorphic-site control.

$-CH_2CH_2-$ bridging beside the central zirconium atom exerts strong conformational rigidity, thus confining ligands on the metal and achieving spatial chirality.

MW of the *i*-PP obviously depends on polymerization temperature, decreasing abruptly from 300,000 to 12,000 following a rise in polymerization temperature from –20°C to 60°C. High MW polypropylene is obtained using analogous hafnocene complex *rac*-$Et(THInd)_2HfCl_2$, though lower in polymerization activity.

Using Metallocenes with a Dimethylsilyl Bridge ($-(me)_2Si-$)

A C_2-symmetric metallocene with a Si-atom bridging the two ligands, due to higher stereorigidity and favorable electronic characteristics, makes propylene polymerize with higher MW and isotacticity. The high MW is attributed mainly to the electron-donating effect of the alkyl substituents, which effectively lowers Lewis acidity of the central metal atom to alleviate β-H chain-transfer, while the high isotacticity benefits mainly from rigidity of the ligand framework, which can be strengthened further by alkyl substituents on the most favorable positions.

Aromatic substitution on appropriate positions of the zirconocene ligand frame gives much higher activity, stereospecificity and PP MW than those of any previously described metallocene systems and comparable to those obtained with commercial heterogeneous catalysis.[4] The substituent must be in the "right" position to be effective.

Syndiotactic Polypropylene (*s*-PP)

s-PP has been prepared in low yields at low temperatures (< –50°C) with homogeneous vanadium Ziegler-Natta catalysts. The chain growth is chain-end controlled.

Ewen and co-workers reported highly active syndiospecific polymerization of propylene at higher temperatures (70°C) with stereorigid *ansa*-*i*-Pr(Cp)(Flu)Zr (Hf)Cl_2/MAO.

Isotactic Block and Hemiisotactic Polypropylenes

Polymerization of propylene with Cp_2TiPh_2/MAO at low temperatures gives products characterized by isostereoblock structure, with adjacent isotactic blocks in opposite stereo-orientation (see Figure 2).[3] Monomer insertion is realized by *cis*-1,2 scheme under chain-end control. With (neomenthyl-Cp)$_2ZrCl_2$/MAO, PP with similar structural characteristics can also be prepared. Block length in these polymers is related to polymerization temperatures, the longer the length, the lower the temperature. Polymers with shorter blocks have lower melting temperatures but good flexibility.

As the polymerization process is affected by both enthalpy and entropy, the stereospecificity varies from isotactic, random to syndiotactic as polymerization temperature rises. The formation of isotactic and syndiotactic PP is governed by combination of chain-end and enantiomorphic-site controls when unsymmetrical titanocene with a bulky substituent on one of the cyclopentadienyl rings is used.[5,6]

With MAO, *i*-Pr(3-MeCp)(Flu)$ZrCl_2$, a metallocene with no element of symmetry in the molecule, gives PP with hemiisotactic structure as one of the two interchangeable diastereomers favors high stereoselectivity in propylene insertion, while the other has almost no stereocontrol because substituents are present at β-position of the Cp ring.[8]

Thermoplastic Elastomers

Thermoplastic elastomers of narrow MWD are formed in stereospecific propylene polymerization with *rac*-[*anti*-CH_3CH(Cp*)(INd)]TiX_2 (X=Cl, Me; C*: tetramethylcyclopentadienyl)/MAO catalyst system.[9] Polymer chains are composed of alternating stereoregular crystalline syndiotactic and stereoirregular noncrystallizable segments, in structure of [(am-PP)$_n$(cry-PP)$_m$]$_p$ (T_m 67–71°C).

POLYMERIZATION OF CYCLIC OLEFINS

While ring-opening reactions occur in polymerization of cyclic olefins, such as cyclopentene, norbornene and 1,4,5,8–dimethylene-1,2,3,4,4a,5,8,8a-octahydronaphthalene (DMON), no ring-opening occurs when homogeneous metallocene catalysts are used.[10] With *ansa*-zirconocenes, the activity of homopolymerization is rather high, and the non-melting, highly crystalline polymers have good heat-resistance. Even copolymers of different ring monomers could be crystalline.

ansa-Metallocenes have higher polymerization activities than unbridged metallocenes. They also have superior copolymerizability in norbornene/ethylene copolymerization.

HOMOPOLYMERIZATION OF STYRENE AND STYRENE/ETHYLENE COPOLYMERIZATION

Homopolymerization of Styrene and Conjugated Diene

Highly syndiotactic polystyrene (*s*-PS) was first obtained by Ishihara and co-workers, using soluble $CpTiCl_3$/MAO catalyst.[11] It is now realized that almost all systems composed of Ti(III) or Ti(IV) compound/MAO soluble in aromatics could polymerize styrene into highly syndiotactic. Even δ-$TiCl_3$ or AA$TiCl_3$, when activated with MAO, could yield a mixture of isotactic and syndiotactic polystyrenes. Monocyclopentadienyl titanium derivatives are the most active, possessing high syndiotactic polymerization activity for styrene, alkyl-substituted styrenes, and styrenes with electron-drawing substituents.

Styrene copolymerizes with conjugated dienes in random, with a higher activity of the dienes than styrene. $CpTiCl_3$/MAO catalyst could polymerize conjugated dienes into *cis*-1,4- and 1,2-structure or mixtures of various structures, e.g., *cis*-1,4-*iso*-poly[(E)-2-methyl-1,3-pentadiene],*cis*-1,4-poly(2,3-dimethyl-1,3-butadiene), 1,2-*syndio*-poly(4-methyl-1,3-pentadiene) and 1,2-*cis*-*iso*-poly[(Z)-1,3-pentadiene].

Copolymerization of Styrene and Ethylene

Although group 4 metallocenes polymerize ethylene and α-olefins in high activity, their polymerization activity towards styrene and conjugated dienes is very low. Monocyclopentadienyl titanium compounds, however, not only are effective in homo- or copolymerization of styrene or conjugated dienes, but also polymerize ethylene and α-olefins.

POLYMERIZATION OF POLAR MONOMERS

Polymerizations involving polar groups have not been feasible with traditional Ziegler-Natta catalysts because of their sensitivity to polar groups and the consequent poisoning and deactivation. Polymerization active centers of group 4 metallocenes are cationic in nature, being able to polymerize olefins even in solvents with Lewis base character. Waymouth realized Ziegler-Natta polymerization of α-olefins containing functional groups like (*i*-Pr)$_2$N, Ph_2P and (*t*-Bu)(CH_3)$_2$SiO, and also cyclopolymerization of 4-trimethylsiloxyl-1,6-heptadiene. Polyelectrolytes could be prepared therefrom by acidolysis.[12]

REFERENCES

1. Sinn, H.; Kaminsky, W.; Vollmer, H. J.; Woldt, R. *Angew. Chem., Int. Ed. Engl.* **1980**, *19*, 390.
2. Gianetti, E., Nicoletti, G. M.; Mazzocchi, R. *J. Polym. Sci., Polym. Chem.* **1985**, *23*, 2117.
3. Ewen, J. A. *J. Am. Chem. Soc.* **1984**, *106*, 6355.
4. Spaleck, W.; Küber, F.; Winter, A.; Rohrmann, J.; Bachmann, B.; Antberg, M.; Dolle, V.; Paulus, E. F. *Organometallics* **1994**, *13*, 954 and references therein.
5. Erker, G. *Pure Appl. Chem.* **1992**, *64*, 393.
6. Ewen, J. A.; Jones, R. L.; Razavi, A.; Ferrara, J. D. *J. Am. Chem. Soc.* **1988**, *110*, 6255.

7. Tian, J.; Hu, L.; Shen, Q.; Huang, B. T. *Chinese Sci. Bull. (Engl.)* **1993**, *39*, 703.
8. Ewen, J. A.; Elder, J.; Jones, R. L.; Haspeslagh, L.; Atwood, J. L., Bott, S. G.; Robinson, K. *Makromol. Chem., Macromol. Symp.* **1991**, *48/49*, 253.
9. Mallin, D. T.; Rausch, M. D.; Lin, Y.-G.; Dong, S.-H.; Chien, J. C. W. *J. Am. Chem. Soc.* **1990**, *112*, 2030.
10. Kaminsky, W.; Spiehl, R. *Makromol. Chem.* **1989**, *190*, 515.
11. Kaminsky, W.; Renner, F. *Makromol. Chem., Rapid Commun.* **1993**, *14*, 239.
11. Ishihara, N.; Seimiya, T.; Kuramoto, M.; Uoi, M. *Macromolecules* **1986**, *19*, 2465.
12. Coates, G. W.; Waymouth, R. M. *J. Am. Chem. Soc.* **1993**, *116*, 91.

Metallocenes

See: *Ferrocene-Backbone Polymers*
Metal Complexes, Macromolecular
Metallocene/Aluminoxane Catalysts (Olefin Polymerization)
Metallocene Catalysts
Metallocene Catalysts (Cationic Group 4 Metal Alkyl Complexes, Olefin Polymerization)
Metallocene Catalysts (Group 4 Elements, New Polymeric Materials)
Olefin Polymerization Catalysts (Silica-Supported Titanium Tetrachloride)
Olefin Polymerization Catalyses, Soluble (Group 4, Cationic Active Centers)
Organometallic Polymers (Transition Metals in Main Chain)
Polyethylene (Commercial)
Polyethylene (High Density Preparation)
Poly(1,1′-Ferrocene-Arylene)s (Pd-Catalyzed)
Polyolefins (Overview)
Polypropylene (Commercial)
Polypropylene, Atactic (High Molecular Weight)
Polypropylene, Hemiisotactic
Styrene, Stereospecific Polymerization
Supported Ziegler-Natta Catalysts (on Silica, for Olefin Polymerization)

METALLOCENOPHANES, STRAINED (Thermal Ring-Opening Polymerization)

Ian Manners, John K. Pudelski, and Daniel A. Foucher
Department of Chemistry
University of Toronto

Interest in preparing soluble, well-defined, and high molecular weight transition metal-containing polymers stems from the possibility of accessing processable materials with novel electronic, optical, magnetic and preceramic properties.[1–19] Development of transition metal-containing polymers has generally been limited by the lack of straight-forward synthetic routes to these materials. It is not surprising, therefore, that the resulting transition metal-containing polymeric products are often of low molecular weight.[7] Ring-opening polymerization routes generally proceed via chain growth processes and therefore offer a distinct advantage over condensation routes in the preparation of soluble, high molecular weight materials that can be fully characterized and structurally well defined. Until recently, however, the possibility of preparing processable, high molecular weight transition metal-containing polymers via ring-opening routes had been virtually unexplored.[20,21]

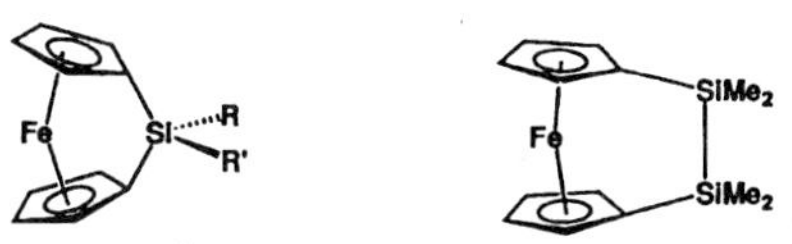

Figure 1. Silicon-bridged [1]ferrocenophane (**1**), and disilane-bridged [2]ferrocenophane (**3**).

In 1992 our group reported the preparation of soluble, high molecular weight polyferrocenylsilanes via the thermal ring-opening polymerization (TROP) of silicon-bridged [1]ferrocenophanes **1** (**Figure 1**).[22–24] These polyferrocenylsilane materials have been thoroughly characterized and exhibit a wide variety of interesting structural, electronic and preceramic properties.[22–26] A critical feature of the TROP reactions is the dependence of polymerizability on strain in the ferrocenophane monomers. Conversely, an analogous disilane-bridged [2]ferrocenophane **3** (Figure 1) failed to undergo TROP even under forcing conditions. Analysis of this species provided evidence of significantly lower strain relative to the analogous[1]ferrocenophanes.[27]

Insight linking monomer strain to polymerizability has greatly expanded the scope of the metallocenophane TROP reaction. Polyferrocenylgermane and polyferrocenylphosphine[28–33] materials have been prepared via TROP reactions of germanium-bridged [1]ferrocenophanes and phosphorus-bridged [1]ferrocenophanes, respectively. Studies of the latter species indicate that they, like their [1]ferrocenophane analogs, are highly strained. We found that hydrocarbon-bridged [2]ferrocenophanes, which in contrast to their disilane-bridged analogs are known to be highly strained, readily undergo TROP affording polyferrocenylethylene materials.[34] Most recently, hydrocarbon-bridged [2]ruthenocenophanes have been prepared and these strained species undergo TROP to afford polyruthenocenylethylenes.[35] Like their polyferrocenylsilane analogs all of these new transition metal-containing polymeric materials have been extensively characterized and exhibit a variety of interesting properties.

PREPARATION AND TROP STUDIES OF DISILANE-BRIDGED [2]RUTHENOCENOPHANES

Disilane-bridged [2]ruthenocenophanes and *bis*(disilane)-bridged [2]ruthenocenophanes (**Figure 2**) may also be prepared in a straightforward manner.[36] Thus the reaction of $Li_2[C_5H_4SiMe_2]_2$ with *cis*-$RuCl_2(DMSO)_4$ gives disilane-bridged **11** in 50% yield as a pale yellow crystalline material which may be purified via vacuum sublimation. Similarly, reaction of $Li_2[C_5H_3(SiMe_2)_2]_2$ with *cis*-$RuCl_2(DMSO)_4$ affords *bis*(disilane)-bridged **12** in 48% yield, also as a pale yellow crystalline solid which may be purified via vacuum sublimation.

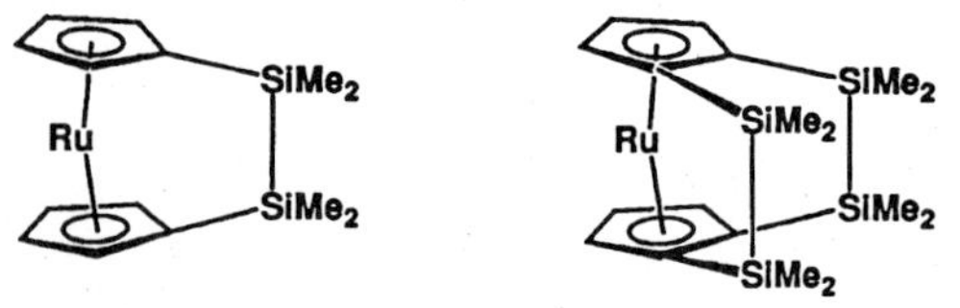

11 12

Figure 2. Disilane- and *bis*(disilane)-bridged [2]ruthenocenophanes.

CONCLUSIONS

Thermal ring-opening polymerization of strained metallocenophanes provides a general route to high molecular weight metal-containing polymeric materials.[7] Since the polymers prepared via TROP routes are generally soluble these materials have been fully characterized and are structurally well-defined. Strained monomer structures are critical for TROP reactivity since unstrained metallocenophanes have been found to be unreactive. Strain in metallocenophane monomers can be probed effectively by X-ray crystallographic analysis where findings such as Cp ring tilting and deviation of the exocyclic bonds to the bridging groups from the Cp ring planes are indicative of strain. Metallocenophane strain can also be probed by other techniques including ^{13}C NMR and UV/Vis spectroscopy and DSC.

Future efforts in this field are targeted at gaining a complete understanding of the mechanism(s) of the TROP reactions of strained metallocenophanes. Mechanistic insight will facilitate TROP of new, more complex metallocenophane monomers and should allow for complete control of the TROP of established metallocenophane monomers. Control of metallocenophane TROP should ultimately result in control of the properties of the resulting metal-containing polymeric products.

REFERENCES

1. Mark, J. E.; Allcock, H. R.; West, R. *Inorganic Polymers*, Prentice Hall: Englewood Cliffs, NJ, 1992.
2. Pittman, C. U.; Carraher, C. E.; Reynolds, J. R. In *Encyclopedia of Polymer Science and Engineering;* Mark, H. E; Bikales, N. M.; Overberger, C. G.; Menges, G. Eds.; Wiley: New York, 1989; Vol. 10.
3. Sheats, J. E.; Carraher, C. E.; Pittman, C. U. *Metal Containing Polymer Systems*, Plenum: New York, 1985.
4. *Inorganic and Metal-Containing Polymeric Materials*, Sheats, J. E.; Carraher, C. E.; Pittman, C. U.; Zeldin, M.; Currell, B. Eds.; Plenum: New York, 1989.
5. Zeldin, M.; Wynne, K.; Allcock, H. R. Eds., *Inorganic and Organometallic Polymers*, ACS: Washington, DC, 1988.
6. Young, R. J.; Lovell, P. A. *Introduction to Polymers*, 2nd ed.; Chapman & Hall: New York, 1992.
7. Manners, I. *Adv. Organomet. Chem.* **1995**, *37*, 131.
8. Patterson, W. J.; Marus, S.; Pittman, C. H. *J. Polym. Sci., Part A-1* **1974**, *12*, 837.
9. Fyfe, H. B.; Mlekuz, M.; Zagarian, D.; Taylor, N. J.; Marder, T. B. *J. Chem. Soc., Chem. Commun.* **1991**, 188.
10. Hagihara, N.; Sonogashira, S.; Takahashi, S. *Adv. Polym. Sci.* **1981**, *41*, 149.
11. Tenhaeff, S. C.; Tyler, D. R. *Organometallics* **1992**, *11*, 1466.
12. Brandt, P. F.; Rauchfuss, T. B. *J. Am. Chem. Soc.* **1992**, *114*, 1926.
13. Davies, S. J.; Johnson, B. F. G.; Khan, M. S.; Lewis, J. *J. Chem. Soc., Chem. Commun.* **1991**, 187.
14. Lewis, J.; Khan, M. S.; Kakkar, A. K.; Johnson, B. F. G.; Marder, T. B.; Fyfe, H. B.; Whittmann, F.; Friend, R. H.; Dray, A. E. *J. Organomet. Chem.* **1992**, *425*, 165.
15. Gonsalves, K.; Zhanru, L.; Rausch, M. D. *J. Am. Chem. Soc.* **1984**, *106*, 3862.
16. Nugent, H. M.; Rosenblum, M.; Klemarczyk, P. J. *Am. Chem. Soc.* **1993**, *115*, 3848.
17. Wright, M. E.; Sigman, M. S. *Macromolecules* **1992**, *25*, 6055.
18. Sturge, K. C.; Hunter, A. D.; McDonald, R.; Santarsiero, B. D. *Organometallics* **1992**, *11*, 3056.
19. Dembek, A. A.; Fagan, P. J.; Marsi, M. *Macromolecules* **1993**, *26*, 2992.
20. Roesky, H. W.; Lücke, M. *Angew. Chem. Int. Ed. Engl.* **1989**, *28*, 493.
21. Roesky, H. W.; Lücke, M. *J. Chem. Soc., Chem. Commun.* **1989**, 748.
22. Foucher, D. A.; Tang, B.-Z.; Manners, I. *J. Am. Chem. Soc.* **1992**, *114*, 6246-6248.
23. Foucher, D. A.; Ziembinski, R.; Tang, B.-Z.; Macdonald, P. M.; Massey, J.; Jaeger, C. R.; Vancso, G. J.; Manners, I. *Macromolecules* **1993**, *26*, 2878–2884.
24. Foucher, D.; Ziembinski, R.; Petersen, R.; Pudelski, J.; Edwards, M.; Ni, Y.; Massey, J.; Jaeger, C. R.; Vancso, G. J.; Manners, I. *Macromolecules* **1994**, *27*, 3992–3999.
25. a) Foucher, D. A.; Honeyman, C. H.; Nelson, J. M.; Tang, B.-Z., Manners, I. *Angew. Chem. Int. Ed. Engl.* **1993**, *32*, 1709. b) Nguyen, M. T.; Diaz, A. F.; Dement'ev, V. V.; Pannell, K. H. *Chem. Mater.* **1993**, *5*, 1389.
26. Tang, B.-Z.; Petersen, R.; Foucher, D. A.; Lough, A. J.; Coombs, N.; Sodhi, R.; Manners, I. *J. Chem. Soc., Chem. Commun.* **1993**, 523–525.
27. Finckh, W.; Tang, B.-Z.; Foucher, D. A.; Zamble, D. B.; Ziembienski, R.; Lough, A.; Manners, I. *Organometallics* **1993**, *12*, 823–829.
28. Foucher, D. A.; Manners, I. *Makromol. Chem., Rapid Commun.* **1993**, *14*, 63–66.
29. Foucher, D. A.; Edwards, M.; Burrow, R. A.; Lough, A. J.; Manners, I. *Organometallics* **1994**, *13*, 4959.
30. Foucher, D. A.; Ziembinski, R.; Rulkens, R.; Nelson, J.; Manners, I. In *Inorganic and Organometallic Polymers II*; Wisian-Neilson, P.; Allcock, H. R.; Wynne, K. J. Eds.; American Chemical Society: Washington, DC, 1994; Vol. ACS Symposium Series 572; p 449.
31. Honeyman, C. H.; Foucher, D. A.; Dahmen, F. Y.; Rulkens, R.; Lough, A. J.; Manners, I. *Organometallics* **1995**, *14*, 5503.
32. Ziembinski, R.; Honeyman, C. H.; Mourad, O.; Foucher, D.; Rulkens, R.; Liang, M.; Ni, Y.; Manners, I. *Phosphorus, Sulfur and Silicon* **1993**, *76*, 219–222.
33. Manners, I. *J. Inorg. Organomet. Polym.* **1993**, *3*, 185–196.
34. Nelson, J. M.; Rengel, H.; Manners, I. *J. Am. Chem. Soc.* **1993**, *115*, 7035–7036.
35. Nelson, J. M.; Lough, A. J.; Manners, I. *Angew. Chem. Int. Ed. Engl.* **1994**, *1333*, 989–991.
36. Nelson, J. M.; Lough, A. J.; Manners, I. *Organometallics* **1994**, *13*, 3703.

METALLOENZYMES, ARTIFICIAL (Polyethylenimine-Based)

Junghun Suh
Department of Chemistry
Seoul National University

Artificial enzymes as effective as natural enzymes are not expected anytime soon. However, even artificial enzymes with limited efficiency may be useful for many practical applications;

some primitive forms of artificial enzymes have been commercialized. Since many catalytic roles are played by metal ions in various organic reactions,[1,2] useful artificial enzymes exploiting metal ions as catalytic groups are more likely than those consisting of only organic catalytic functional groups.

Catalytic principles exploited by enzymes must be incorporated in relatively small spaces on synthetic molecules in devising artificial enzymes. Furthermore, to achieve cooperative catalytic participation of several functional groups in the chemical transformation of the bound substrate, convergent catalytic groups must be finely aligned. Creating such active sites on molecular skeletons provided by relatively small synthetic hosts would not be easy. For this reason, polyethylenimine (PEI) has been exploited as a molecular backbone in the design of artificial enzymes.[3]

PROPERTIES AND APPLICATIONS

Creation of Macrocyclic Metal Centers on PEI

Molecular architecture with very tightly bound metal ions must be developed to design artificial metalloenzymes. Tight binding of metal ions may be achieved by employing multiaza macrocyclic complexes as part of the catalytic units. In addition, multiaza macrocyclic metal complexes themselves manifest the ability to catalyze many organic reactions.[2] Many multiaza macrocyclic metal complexes are prepared by the condensation of carbonyl compounds with multiamines in the presence of metal ions.[6] In this regard, PEI can be used as a synthon of macrocyclic complexes as well as the backbone of polymeric macrocycles.

The macrocycle-containing PEIs possess fixed metal centers which are not removed by repetitive dialysis. Creation of tight metal centers, recognition of anions, and acceleration of deacylation of the bound anionic ester are achieved by the macrocycle-containing PEI derivatives.

Selective Molecular Recognition of Transition State by a PEI-Based Metal Center

Selective recognition of transition state is the major aim in the design of artificial enzymes.

Artificial Metallophosphoesterases Built on PEI

Currently, there is much interest in catalytic hydrolysis of phosphate esters in connection with the biological functions of transphosphorylation reactions, such as information storage and processing, energy storage, regulation, signal transduction, and cofactor chemisty.[7] Several metalloenzymes are known to catalyze hydrolysis of phosphate diesters and phosphate monoesters. For example, Zn^{II} plays essential roles in alkaline phosphatase, 5′-nucleotidase, fructose-1,6-biphosphatase, phosphodiesterase, cyclic nucleotide phosphodiesterase, and nuclease S1. Phosphate diesters are very stable against hydrolysis. The best synthetic catalysts designed so far for hydrolysis of phosphate monoesters and diesters are metal complexes, and the catalytic efficiency of the metal complexes depends heavily on the nature of the metal ion and the structure of the ligands.

Major characteristics of actions of metalloenzymes reproduced by the artificial metallophosphoesterases developed in this investigation include essential involvement of metal ions, complex formation with the substrates, and fast catalytic transformation within the resultant supramolecular complexes. Moreover, the metal ions of the artificial metalloenzymes play several catalytic roles both in the complexation process and the catalytic transformation step.

Cooperation of Macrocyclic Metal Centers with Organic Functional Groups in Artificial Metalloesterases Built on PEI

For the design of effective artificial metalloenzymes, it is essential to accomplish cooperation of metal ions with organic functional groups located within the molecular framework of the catalysts. Although some successful results have been reported for catalytic cooperation of metal ions with organic groups, the organic catalytic groups were attached to the substrate molecules as neighboring groups in most of the studies conducted to date.[1,2] As a next step in the elaboration of the PEI-based artificial metalloenzymes, therefore, we tried to achieve cooperation of the metal centers with organic functional groups incorporated to the catalysts.

Additional catalytic elements such as 4-*N*,*N*-dialkyl-aminopyridine (DAP) and lauryl group have been incorporated to PEI together with macrocyclic metal centers.[5] 4-*N*,*N*-Dialkylaminopyridines are among the best catalytic groups for ester hydrolysis due to strong nucleophilicity of the pyridyl group and easy breakdown of the resulting acylpyridine intermediates. Lauryl groups are known to create hydrophobic microdomains on PEI, which would affect the catalytic outcome of the metal centers and dialkylaminopyridines.

PEI-Based Artificial Metalloenzymes Equipped with β-Cyclodextrin as Specific Binding Site

As for binding sites for certain hydrophobic moieties, we have attached β-cyclodextrin (CD) to PEI.[4] CD is a cyclic compound composed of seven molecules of D-glucose with the shape of a pail without a bottom. One of the hydroxyl groups of CD was converted into *p*-toluenesulfonate ester, which was reacted with PEI to obtain a PEI derivative (CD-PEI) containing CD. CD-PEI manifested high affinity towards *t*-butylphenyl compounds and accelerated deacylation of esters containing *t*-butylphenyl moieties. To design effective artificial enzymes based on PEI equipped with CD as the binding site, introduction of catalytic groups to the polymer backbone in the vicinity of the CD cavity has been attempted.

In order to obtain artificial metalloesterases equipped with CD cavities as binding sites, CD-PEI derivatives containing macrocyclic metal (Zn^{II}, Ni^{II}, or Co^{II}) centers were prepared through metal-template condensation of CD-PEI with glyoxal or butanedione and were subsequently acetylated with acetic anhydride to block nucleophilic amines of PEI.[9]

The metal centers of the PEI-based artificial metalloesterases may play specific catalytic roles such as activation of electrophiles or provision of metal-bound hydroxide nucleophiles.[1,2]

In the ester hydrolysis by the PEI derivatives containing both CD moieties and the macrocyclic metal centers, the CD cavities provide binding sites for the substrates, recognizing the *t*-butylphenyl moieties. On the other hand, the macrocyclic metal centers act as catalytic sites for the ester hydrolysis. In this regard, the PEI derivatives mimic metalloesterases.

Perspectives

To design effective artificial metalloenzymes, both formation of the substrate-catalyst complex and catalytic turnover rate of the complex must be enhanced. Complex formation of the PEI-based artificial metalloenzymes with substrates may be facilitated by introducing a multiple number of interactions in the molecular recognition of substrates. Strategies to increase the catalytic turnover rate for the artificial metalloenzymes built on PEI include introduction of reactive catalytic functional groups in close proximity to the bound substrate and cooperation among several catalytic groups. In addition, it is also desirable to reduce the nonproductive binding through introduction of catalytic groups in planned positions close to the binding sites. This may be accomplished by the site-directed construction of metal centers and organic functional groups in the vicinity of the CD cavities of CD-containing PEI.

Future studies will also be aimed at achieving regioselectivity and stereoselectivity with the PEI-based metalloenzymes. In addition, extension of the PEI-based artificial enzymes toward the area of oxidation-reduction reactions is also desirable since metal ions participate not only as Lewis acid catalysts but also as redox catalysts in biological systems.

REFERENCES

1. Suh, J., *Acc. Chem. Res.* **1992**, *25*, 273.
2. Suh, J., *Perspective on Bioinorganic Chemistry*, Hay, R. W.; Dilworth, J. R.; Nolan, K. B., Eds.; JAI: Greenwich, CT, 1996; Vol. 3, in press.
3. Klotz, I. M., *Enzyme Mechanisms*, Page, M. I. and Williams, A., Eds., The Royal Society of Chemistry: Cambridge, UK, 1987, Chapter 2.
4. Suh, J.; Lee, J. S. H.; Zoh, K. D., *J. Am. Chem. Soc.* **1992**, *114*, 7961.
5. Lee, K. H.; Suh, J., *Bioorg. Chem.* **1994**, *22*, 95.
6. Curtis, N. F., *Comprehensive Coordination Chemistry*, G. Wilkinson, Ed.; Pergamon: Oxford, 1987; Vol. 2, Chapter 21.2.
7. Dugas, H., *Bioorganic Chemistry, A Chemical Approach to Enzyme Action*, 2nd ed.; Springer-Verlag: New York, 1989.
8. Dietrich, M.; Münstermann, D.; Suerbaum, H.; Witzel, H., *Eur. J. Biochem.* **1991**, *199*, 105.
9. Zoh, K. D.; Lee, S. H.; Suh, J., *Bioorg. Chem.* **1994**, *22*, 242.

METALLOMESOGEN (Cu(II) POLYMER

Fernando R. Díaz,* N. Valdebenito, and M. Marcos
Facultad de Química
Pontificia Universidad Católica de Chile

Organometallic compounds which exhibit liquid crystalline behavior are referred to as metallomesogens. Although the first metal containing liquid crystals were reported a long time ago, they attained prominence in the last 10 years. These compounds offer new possibilities for liquid crystals, because they introduce d-block metals that give colored complexes or exhibit paramagnetism. This, in turn, increases their potential applications.

Metal bonding also allows the formation of a liquid crystal compound from non-mesogenic ligands.

Little work has been done about metallomesogen polymers.[1–4] In addition to the properties introduced by the transition metals, the polymeric structure helps to stabilize the mesophase.

*Author to whom correspondence should be addressed.

PROPERTIES

A polarizing optical microscope, an Olympus BH-2 equipped with a Lynkam hot stage, showed the texture of the mesophase. The complex exhibited enantiotropic liquid-crystalline behavior forming a smectic mesophase between 157 and 199°C after the sample was annealed for 1 hour.

REFERENCES

1. Carfagna, C.; Caruso, U.; Roviello, A.; Sirigu, A. *Makromol. Chem. Rapid. Commun.* **1987**, *8*, 345.
2. Caruso, U.; Roviello, A.; Sirigu, A. *Macromolecules* **1991**, *24*, 2606.
3. Marcos, M.; Oriol, L.; Serrano, J. L.; Alonso, P. J.; Puertolas, J. A. *Macromolecules* **1990**, *23*, 5187.
4. Marcos, M.; Serrano, J. L. *Adv. Mater.* **1991**, *30*, 256.

Metallophthalocyanines

See: Metallophthalocyanines, Polymeric (Overview)
Metallophthalocyanines, Polymeric (Types of Structures)

METALLOPHTHALOCYANINES, POLYMERIC (Overview)

S. Venkatachalam, Veena Vijayanathan, and
V. N. Krishnamurthy
Vikram Sarabhai Space Center

Metallophthalocyanines or phthalocyaninato metals and their polymers possess extended conjugated structures and show interesting optical, electrical, and electrochemical properties. They have excellent stability against heat, light, moisture, and air and hence have received considerable attention in the quest for environmentally stable electrically conducting polymers.[1–14] Their structural similarity to chlorophyll has made them useful for applications in artificial solar cells.[5] Their unique electrical and optical properties are mainly due to the large π electron system that characterizes these compounds. The phthalocyanine ring consists of a 42 π electron conjugated system delocalized over the 40 carbon atoms and 8 nitrogen atoms forming the planar macrocycle. Excitation of the mobile π electron by thermal or optical means from the valence band containing the highest occupied molecular orbital (HOMO) to the conduction band containing lowest unoccupied molecular orbital (LUMO) results in useful properties.[15–18]

Besides their traditional uses as dyes, pigments, and catalysts, phthalocyanines are of particular interest in many fields of basic and applied research as molecular units to build up electroresponsive materials in electronic, optoelectronic, and molecular electronic applications.[14,19,20]

POLYMERIC PHTHALOCYANINES

Polymeric phthalocyanines have been well known for many years, and they are prepared from aromatic tetranitriles such as tetracyanobenzene (TCNB) (1,2,4,5-benzenecarbonitrile) or aromatic dianhydrides such as pyromellitic dianhydride (PMDA).[21–27] When these compounds are heated with metal or metal salts, urea and catalyst polymeric phthalocyanines are formed as shown in **Equation 1**.

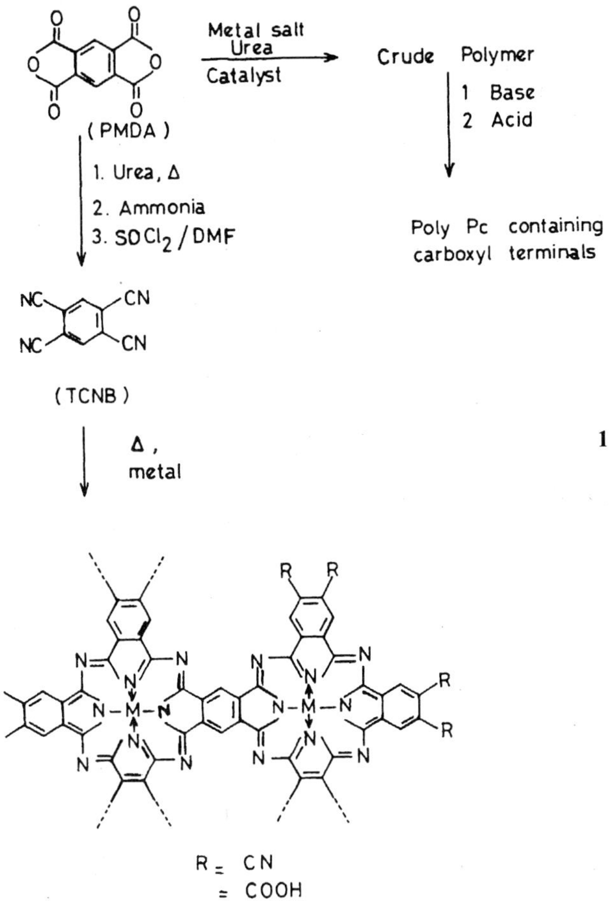

1

ELECTRICAL CONDUCTIVITY

Polymeric phthalocyanines exhibit larger electrical conductivity than the unsubstituted monomeric phthalocyanines.[8,9,28–32] This is because of the effective intramolecular charge transfer occurring through the formation of extended conjugated structures. More effective interactions are obtained in polymers derived from linearly fused phthalocyanine macrocycles such as naphthalocyanine compounds, which exhibit lower oxidation potential, narrow band gap, and better electrical conductivity without doping.[33–37]

MODIFICATION OF PHTHALOCYANINE POLYMER TO IMPROVE PROCESSIBILITY

Due to the presence of rigid structures, phthalocyanine polymers are not soluble in common organic solvents and they are not meltable. Hence formation of thin films or coating of phthalocyanine polymer becomes difficult. Moreover, samples prepared from solution cast techniques from polymers dissolved in dipolar aprotic solvents result in brittle films. To make use of these materials for specific applications, refinement of the structure of the polymer by chemical modifications, copolymerization, or making blends of these polymers with other conducting and nonconducting materials is resorted to.[38–43]

CATALYTIC AND ELECTROCATALYTIC PROPERTIES OF PHTHALOCYANINES

Electrocatalysis is very important for many industrial applications. Redox-active metallophthalocyanine-coated electrodes are capable of providing active sites for effecting selective and efficient chemical reactions at moderate potentials. Hence they have received considerable attention for many photochemical, electrochemical, and semiconductor applications.[5,8,9,44–50] They adsorb oxygen, and the adsorbed oxygen acts as a dopant, which makes it a p-type semiconductor. The resulting polymer shows various layers of electrical conductivity depending on the extent of partial oxidation.[5,8,9,29,30,45–47,51–56] The redox reactions of phthalocyanines also bring about chemical reactions and electrocatalysis.

Metallophthalocyanines and their polymers are capable of effecting photoreduction of water to hydrogen by using visible light excitation.[57]

CHARGE STORAGE DEVICES

The specific functions of phthalocyanine polymers could be exploited by incorporating them into suitable conducting polymer matrices. For example, incorporation of phthalocyanine polymers into conducting polymers is carried out by electrochemical polymerization of the monomers such as pyrrole, aniline, etc. in the presence of the salts of phthalocyanines. These polymers show sufficient stability toward air and electrolyte in addition to ease of processing and control over ionization potential, and hence, are suitable for use as electrodes in aqueous secondary batteries.[42,43] When these modified electrodes are cycled in aqueous electrolytes the polymer undergoes reversible redox reactions.[40,43] Upon reduction of the film, the polymeric phthalocyanine carboxylate anion, being a large one, remains in the matrix and the electrolyte counter-cation migrates into the matrix to conserve electroneutrality of the film. Following oxidation the cation gets released from the matrix. This is analogous to the behavior of polypyrrole doped with large anions.[59–64] This kind of pseudo-cathodic incorporation of cations into the polymers makes them suitable candidates for use as electrode materials in rechargeable batteries.[9,65–69] The specific weight is lower than that of ordinary inorganic materials, and hence cells of high energy density can be obtained. Most of the conducting polymer-based batteries use Li or Li/Al alloy as the counter-electrode. The polymer is used either as an anode or cathode.[70]

Incorporation of phthalocyanine polymers into acetylene black or graphite results in electroactive materials which are useful as rechargeable battery electrodes. Discharge and charge characteristics of various phthalocyanine cathodes coupled with Li anode have been studied.[66–69]

Thin films of polypyrrole doped with polymeric phthalocyanines were found to exhibit stable conductivity and reversible electro-chemical behavior. The oxidation-reduction process in the resulting composite film was largely influenced by the nature of the incorporated anion.

NLO PROPERTIES OF PHTHALOCYANINE POLYMERS

Phthalocyanines and their polymers possessing -CN and -COOH terminals have suitable extended structures that facilitate solitonic, polaronic, and bipolaronic conformational deformation,[31,32,71,72] and hence are expected to show large ultrafast NLO properties. The combination of σ and π electron systems in polymeric phthalocyanine structures can support an excited electronic state, which allows these polymers to undergo redox processes in a wide range of potentials.[8,9,29,30,41,42,44,55,59,74] These properties make them useful as semiconductors, electrochemical catalysts, and materials for NLO. The excitations of chains of polymers are strongly coupled to local bond order, and therefore they show a strong nonlinear response. These polymers can be easily incorporated electrochemically into other conducting polymer matrices[38–43,71] so that polymeric thin films possessing both processibility and electrochemical properties can be realized.

REFERENCES

1. Moser. F. H.; Thomas, A. L. *The Phthalocyanines;* CRC: Boca Raton, FL, 1983; Vols. 1 and 2.
2. Moser, F. H.; Thomas, A. L. *Phthalocyanines;* ACS Reinhold: New York, 1963.
3. Lever, A. B. P. *Adv. Inorg. Chem., Radio Chem.* **1965,** *7,* 27.
4. Leznoff, C. C.; Lever, A. B. P. Eds.; *Phthalocyanines Properties and Applications;* VCH: New York, 1989.
5. Simon, J. S.; Andre, J. J. *Molecular Semiconductors;* Springer: Berlin, 1985; Chapter III.
6. Katon, J. E. Ed.; *Organic Semiconducting Polymers;* Marcel Dekker: New York, 1968.
7. Guttmann, F.; Lyons, L. E. *Organic Semiconductors;* Wiley: New York, 1967.
8. Wohrle, D. *Adv. Polym. Sci.* **1983,** *50,* 45.
9. Kaneko, M.; Wohrle, D. *Adv. Polym. Sci.* **1988,** *84,* 141.
10. Dirk, C. W.; Mintz, E. A.; Schoch, K. F.; Marks, T. J. In *Advances in Organometallic and Inorganic Polymer Science*; Carraher, C. E.; et al., Eds.; Marcel Dekker: New York, 1981; p 275.
11. Kuznosoff, P. M.; Nohr, R. S.; Wynne, K. J.; Kenney, M. E. In *Advances in Organometallic and Inorganic Polymer Science;* Carraher, C. E. et al., Eds.; Marcel Dekker: New York, 1981; p 301.
12. Marks, T. J. *Angew. Chem. Int. Ed. Eng.* **1990,** *28,* 857.
13. Marks, T. J. *Science* **1985,** *227,* 881.
14. Venkatachalam, S.; Krishnamurthy, V. N. *Ind. J. Chem.* **1994,** *33A,* 506.
15. Eley, D. D.; Parfitt, G. D. *Trans. Faraday Soc.* **1955,** *51,* 1529.
16. Elvidge, J. A.; Linstead, R. P. *J. Chem. Soc.* **1955,** 3536.
17. Orti, E.; Bredas, J. L.; Clarisse, C. *J. Chem. Phys.* **1990,** *92*(2), 1228.
18. Orti, E.; Bredas, J. L. *J. Chem. Phys.* **1986,** *89,* 1009.
19. Lever, A. B. P.; Hempstead, M. R.; Leznoff, C. C.; Lin, W.; Melnik, M.; Nevin, W. A.; Seymour, P. *Pure Appl. Chem.* **1986,** *58,* 1467.
20. Collins, R. A.; Mohammad, K. A. *J. Phys. S.* **1988,** *2,* 154.
21. Meltzer, Y. L. *Phthalocyanine Technology: Chemical Process Review;* Noyce Data: Park Ridge, NJ, 1970; p 19.
22. Marvel, C. S.; Rassweller, J. W. *J. Am. Chem. Soc.* **1958,** 80, 1197.
23. Epstein, A.; Wildi, B. S. *J. Chem. Phys.* **1964,** *32,* 324.
24. Wildi, B. S.; Katon, J. E. *J. Polym. Sci.* **1964,** *A2,* 4709.
25. Berlin, A.; Sherlie, A. I. *Inorganic Macromol. Rev.* **1971,** *1,* 235.
26. Bannehr, R.; Meyer, G.; Wohrle, D. *Polym. Bull.* **1980,** *2,* 841.
27. Wohrle, D. *Adv. Polym. Sci.* **1972,** *10,* 35.
28. Venkatachalam, S.; Rao, K. V. C.; Manoharan, P. T. *Polymer* **1989,** *30,* 1633.
29. Bannehr, R.; Jaeger, N.; Schumann, B.; Wohrle, D. *J. Mol. Catal.* **1983,** *21,* 255.
30. Wohrle, D.; Schumann, B.; Schmidt, V.; Jaeger, N. *Makromol. Chem. Macromol. Symp.* **1987,** *8,* 195.
31. Venkatachalam, S.; Rao, K. C.; Manoharan, P. T. J. *Polym. Sci Polym. Phys. Ed.* **1994,** *32(B),* 37.
32. Venkatachalam, S.; Rao, K. V. C.; Manoharan, P. T. *Synth. Met.* **1988,** *26,* 237.
33. Blohm, M.; VanDort, P. C.; Pickett, J. E. *Chem. Tech.* **1992,** *105.*
34. Orti, E.; Crespo, R.; Piqueras, M. C.; Virula, P. M.; Thomas, F. *Synth. Met.* **1993,** *55–57,* 4513.
35. Orti, E.; Piqueras, M. C.; Crespo, R.; Thomas, F. *Mol. Cryst. Liq. Cryst.* **1993,** *234,* 241.
36. Deger, S.; Hanack, M. *Synth. Met.* **1986,** *13,* 319.
37. Tsuchida, E.; Onhashi, Y.; Ohno, H.; Matsuda, H.; Nakanishi, H.; Kato, M.; Deger, S.; Behnisch, R.; Hanack, M. *Synth. Met.* **1989,** *33,* 37.
38. Venkatachalam, S.; Vijayan, T. M.; Packirisamy, S. *Makromol. Chem. Rapid. Commun.* **1993,** *14,* 703.
39. Venkatachalam, S.; Prabhakaran, P. V. *Eur. Polym. J.* **1993,** *29,* 711.
40. Skotheim, T.; Rosenthal, M. V.; Linkous, C. A. *J. Chem. Soc. Chem. Commun.* **1985,** 612.
41. Bull, R. A.; Fan, F. R.; Bard, A. J. *J. Electrochem. Soc.* **1984,** *131,* 687.
42. Veena, V.; Venkatachalam, S.; Krishnamurthy, V. N. *Polymer* **1993,** *34,* 1095.
43. Veena, V.; Venkatachalam, S.; Krishnamurthy, V. N. *Polymer* (communicated) to be published.
44. Wohrle, D.; Kaune, H.; Schumann, B. *Makromol. Chem.* **1986,** *187,* 2947.
45. Kaneko, M.; Manecke, G. *Makromol. Chem.* **1974,** *175,* 2795.
46. Kaneko, M.; Manecke, G. *Makromol. Chem.* **1974,** *175,* 2811.
47. Kaneko, M.; Manecke, G. *Makromol. Chem.* **1974,** *175,* 3401.
48. Loutfy, R. O.; Sharp, J. H.; Hsiao, C. K.; Ho, R. J. *Appl. Phys.* **1981,** *52,* 5218.
49. Loutfy, R. O.; McIntyre, L. F. *Can. J. Chem.* **1983,** *61,* 72.
50. Maitrot, M.; Guilland, G.; Boudjema, B.; Andre, J. J. Strzelecka, H.; Simon, J.; Even, R. *Chem. Phys. Lett.* **1987,** *133,* 59.
51. Inoue, H.; Kida, Y.; Imito, E. *Bull. Chem. Soc. Japan.* **1967,** *40,* 184.
52. Inoue, H.; Kida, Y.; Imito, E. *Bull. Chem. Soc. Japan.* **1968,** *41,* 684.
53. Inoue, M.; Kida, Y.; Imito, E. *Bull. Chem. Soc. Japan.* **1968,** *41,* 692.
54. Baker, D. J.; Boston, D. R.; Bailar, J. C. *Inorg. Nucl. Chem.* **1973,** *35,* 153.
55. Schumann, B.; Wohrle, D.; Jaeger, N. *J. Electrochem. Soc.* **1985,** *132,* 2144.
56. Jacobs, R. C. M.; Janssen, L. J. J.; Barendrecht, E. *Electrochim. Acta* **1985,** *30,* 1313.
57. Darwent, J. R.; Douglas, P.; Harriman, A.; Porter, G.; Richoux, M. C. *Coord. Chem. Rev.* **1982,** *44,* 83.
58. Rieke, P. C.; Armstrong, N. R. *J. Am. Chem. Soc.* **1984,** *47,* 106.
59. Lever, A. B. P.; Wilshire, J. P. *Can. J. Chem. 54,* 2514.
60. Iyoda, T.; Ohtani, A.; Shimidzu, T.; Honda, K. *Synth. Met.* **1987,** *18,* 725.
61. Schimidzu, T.; Ohtani, A.; Iyoda, T.; Honda, K. *J. Chem. Soc. Chem. Commun.* **1986,** *1414,* 1416.
62. Schimidzu, T.; Ohtani, A.; Iyoda, T.; Honda, K. *J. Chem. Soc. Chem. Commun.* **1987,** 327.
63. Schimidzu, T.; Ohtani, A.; Iyoda, T.; Honda, K. *J. Electroanal. Chem.* **1987,** 224.
64. De Paoli, M. A.; Panero, S.; Paserini, S.; Scrosati, B. *Adv. Mater.* **1990,** *2,* 480.
65. Green, J. M.; Faulkner, L. R. *J. Am. Chem. Soc.* **1983,** *105,* 2950.
66. Wohrle, D.; Kirschemann, M.; Jaeger, N. I. *J. Electrochem. Soc.* **1985,** *132,* 1150.
67. Yamaki, J.; Yamaji, A. *J. Electrochem. Soc.* **1982,** *129,* 5.

68. Japan Kokai Tokkyo Koho, Jp 62 202 464; *Chem. Abstr.* **1988**, *108*, 59451 z.
69. Okhada, S.; Yaki, J. I. *J. Electrochem. Soc.* **1989**, *136*, 2437.
70. Genies, E. M. In *Science and Applications of Conducting Polymers;* Salaneck, W. R., Ed.; Adam Hilger: New York, 1991; p 99, and reference therein.
71. Veena, V.; Venkatachalam, S.; Krishnamurthy, V. N., *Synth. Met.* (in press).
72. Venkatachalam, S.; Rao, K. V. C.; Manoharan, P. T. In *Frontiers In Polymer Research*; Prasad, P. N.; Nigam, J. K., Eds.; Plenum: New York, 1991; p 431.
73. Venkatachalam, S.; Rao, K. V. C.; Manoharan, P. T. In *Polymer Science—Contemporary Topics*; Sivaram, S., Ed.; Tata Elsevier: New Delhi, 1991; p 727.
74. Takeshita, K.; Aoyama, Y.; Ashida, M. *Bull. Chem. Soc. Jpn.* **1991**, *64*, 1167.

METALLOPHTHALOCYANINES, POLYMERIC (Types of Structures)

Hirofusa Shirai and Mutsumi Kimura
Department of Functional Polymer Science
Faculty of Textile Science and Technology
Shinshu University

Metallophthalocyanines (MPc) are compounds that contain a very stable pigment called macrocyclic (MPc) tetraazaporphyrin. A metal ion of MPc is present in a specific conjugated π electron environment, which is similar to hemes *in vivo* (**Structure 1**). MPc have properties that endow them with many varied uses, they can be catalysts, sensitizers, or electrical conductors, as well as used in electrical switching between conducting states, and in electrochromism. This versatility has brought MPc much attention.

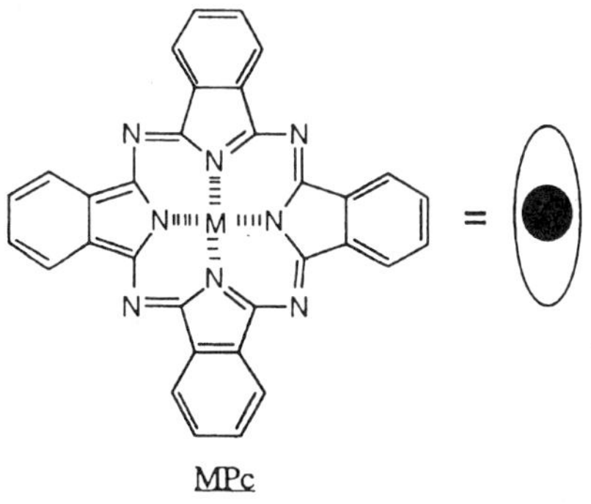

1

The poor solubility of MPc results in major problems in utilizing these functions however. Improvements in the solubility of MPc are achieved through the introduction of functional groups and attachment to polymers.

POLYMERIC MPc IN A POLYMER NETWORK

The preparation of this type of polymer can be divided as follows. Polymerization of monomers with two reactive groups, such as 1,2,4,5-benzenetetracarboxylic acid dianhydride and 1,2,4,5-tetracyanobenzene (see **Structure 2**).[1–5]

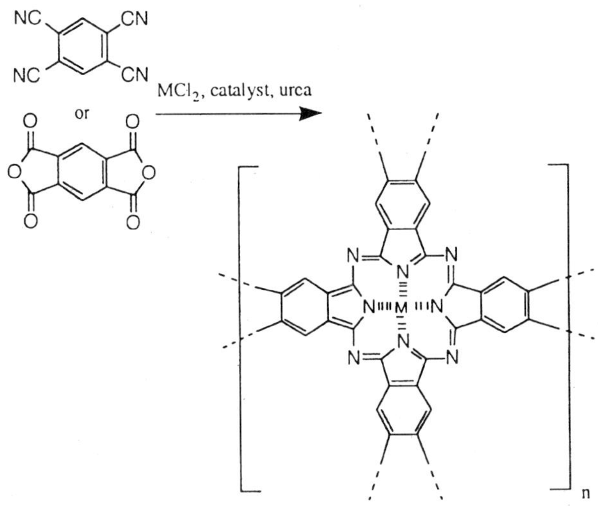

2

MPc-CONTAINING POLYMERS IN SIDE CHAINS

Polymerization of MPc Monomer

Monomers containing MPc can be polymerized to produce polymeric MPc. Two fundamentally different polymerization methods that result in polymeric MPc have been reported in the literature. One of these involves polymerization of vinyl monomer-containing MPc.

Vinyl Polymerization or Copolymerization of Vinyl Monomer-Containing MPc

We have previously synthesized vinyl monomer-containing phthalocyanine and polymerized this vinyl monomer with several other kinds of monomers.[6] As comonomers, the acrylamide derivatives help to dissolve MPc in water.

Polycondensation or Copolycondensation of MPc-Containing Free Functional Groups

Polycondensation or copolycondensation of MPc is the second polymerization method resulting in polymeric MPc found in the literature.[7]

Polymeric Reaction Between MPc and Polymers

MPc can be immobilized to several types of polymers through a covalent or coordinative bond. The properties of polymers containing MPc such as solubility and thermal stability, are changed by the combination. MPc can be linked by the covalent bond to several kinds of polymers. We synthesized poly(2- and 4-vinylpyridine-*co*-styrene) with covalently bonded Fe(III)-Pc. The immobilization reaction of Fe(III)-2,9,16,23-tetracarboxyphthalocyanine tetraacid chloride on the copolymer was carried out through the Friedel–Crafts reaction.[8–10]

POLYMERIC MPc THROUGH THE LINKAGE OF CENTRAL METALS

The linear polymers of MPc can be achieved by stacking or bridging the central metals of MPc. The type of formation of these polymers depends on the kind of central metal in the MPc.

Polymeric MPc Through Covalent Bonds

This type of polymeric MPc can be obtained by the treatment of MPc with tetravalent metals and a bifunctional axial ligand such as $Si(Pc)(OH)_2$, $Ge(Pc)(OH)_2$, and $Sn(Pc)(OH)_2$.[14–16]

Polymeric MPc Through Coordinative Bonds

A low molecular weight MPc and a bifunctional ligand yields a linear polymer complex through the coordination bonds. The cofacial linkage of MPc, which has a transition metal as a central metal of MPc, such as Co(II), Fe(II), and Ru(II), is formed by the addition of bifunctional ligands.[14–16] The cofacial stacking of MPc results in high electrical conductivity of the polymers.

Polymeric MPc Through Covalent and Coordinative Bonds

These types of polymers are formed through one σ-bond and one coordinative bond on MPc. Fluoroaluminum and fluorogallium phthalocyanine form a linear polymer through the electrostatic interaction and stacking of planar MPc rings.[17,18] MPc with Mn(III), Cr(III), Fe(III), Co(III), Rh(III), or Ru(III) as a central metal can form polymers linked with cyanide.[19,20]

REFERENCES

1. Wöhrle, D.; Marose, U.; Knoop, R. *Makromol. Chem.* **1984**, *186*, 2229.
2. Cherkashina, L. G.; Berlin, A. A. *Vysokromol. Soedin.* **1966**, *8*, 627.
3. Snow, A. W.; Griffith, J. R.; Marullo, N. P. *Macromolecule* **1984**, *17*, 1614.
4. Keller, T. M.; Price, T. R.; Griffith, J. R. *Org. Coat. Plast. Chem.* **1980**, *43*, 804.
5. Krejia, L.; Plewka, A. *Electrochim. Acta* **1980**, *25*, 1283.
6. Shirai, H.; Yagi, S.; Suzuki, A.; Hojo, H. *Macromol. Chem.* **1979**, *178*, 1889.
6. Kimura, M.; Dakeno, T.; Adachi, E.; Koyama, T.; Hanabusa, K.; Shirai, H. *Macromol. Chem. Phys.* **1994**, *195*, 2423.
7. Shirai, H.; Takemae, Y.; Kobayashi, K.; Kondo, Y.; Hirabaru, O.; Hojo, N. *Makromol. Chem.* **1984**, *185*, 1359.
8. Shirai, H.; Tsuiki, H.; Masuda, E.; Koyama, T.; Hanabusa, K. *J. Phys. Chem.* **1991**, *95*, 417.
9. Shirai, H.; Maruyama, A.; Hatano, J.; Kobayashi; K., Hojo, N. *Makromol. Chem.* **1980**, *181*, 565.
10. Shirai, H.; Higaki, S.; Hanabusa, K.; Kondo, Y.; Hojo, N. *J. Polym. Sci.* **1984**, *22*, 1309.
11. Dirk, C. W.; Inabe, T.; Schoch, K. F.; Marks, T. J. *J. Am. Chem. Soc.* **1983**, *105*, 1539.
12. Orthmann, E. A.; Wegner, G. *Makromol Chem., Rapid Commun.* **1986**, *4*, 687.
13. Zhou, X.; Marks, T. J.; Carr, S. H. *Mol. Cryst. Liq. Cryst.* **1985**, *118*, 357.
14. Mertz, J.; Hanack, M. *Chem. Ber.* **1987**, *120*, 1307.
15. Keppeler, U.; Hanack, M. *Chem. Ber.* **1986**, *119*, 3363.
16. Kobel, W.; Hanack, M. *Inorg. Chem.* **1986**, *25*, 103.
17. Linsky, J. P.; Paul, T. R.; Nohr, R. S.; Kenney, M. E. *Inorg. Chem.* **1980**, *19*, 3131.
18. Klofta, T. J.; Rieke, P. C.; Linkous, C. A.; Buttner, W. J.; Nanthakumar, A.; Mewborn, T. D.; Armstrong, N. R. *J. Electrochem. Soc.* **1985**, *32*, 2134.
19. Hanack, M.; Hedtmann-Rein. C.; Datz, A.; Keppeler, U.; Münz, Z. *Synyh. Met.* **1987**, *19*, 787.
20. Hanack, M.; Hedtmann-Rein, C. *Z Naturforsch.* **1985**, *40b*, 1087.

Metalloporphyrins

See: Immortal Polymerization

METALLOSUPRAMOLECULAR HELICATES

Edwin C. Constable and Diane R. Smith
Institut für Anorganische Chemie der Universität Basel

Supramolecular chemistry is "the chemistry of the intermolecular bond, covering the structures and functions of the entities formed by the association of two or more chemical species."[1] Self-assembly is an important feature of supramolecular chemistry. *Metallosupramolecular chemistry* focuses on interactions between (transition) metal ions and polydenate ligands.

Transition metal ions have preferred coordination numbers (the number of donor atoms bonded to the metal) and coordination geometries (the spatial arrangements of the ligands and donor atoms). Metallosupramolecular chemistry matches these characteristics with the bonding properties of polydenate ligands (the number, type, and spatial distribution of donor atoms) and uses this molecular coding to control the assembly of supramolecular complexes with highly specific shapes and properties. The use of metallosupramolecular principles for the synthesis of metal-containing polymers containing multiple (double or triple) helical topologies is discussed.[2]

PRINCIPLES OF ASSEMBLY

Double-helical structures result from inter-twisting molecular threads. The incorporation of metal-binding domains into the molecular threads allows metal ions to be utilized in the assembly of the helical species (**Figure 1**).

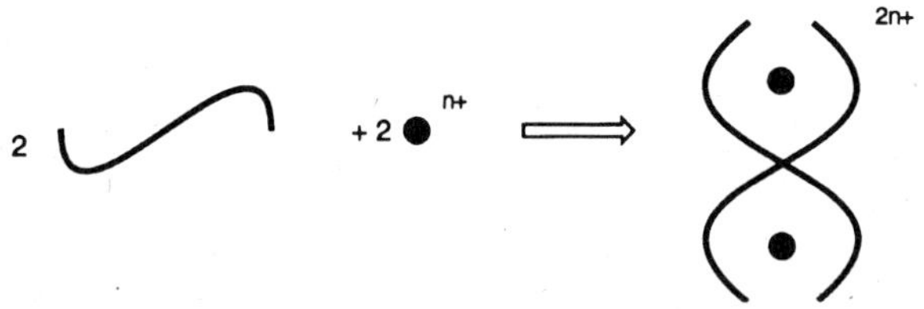

FIGURE 1. The assembly of a dinuclear double helix from the interaction of molecular threads containing two metal-binding domains and appropriate metal ions (l).

Helices are inherently chiral and may be right-handed (Δ) or left-handed (λ). When achiral ligands and metal ions react together, they form helical complexes, and the product consists of a racemic mixture of both enantiomers.

2,2′:6′,2″:6″,2‴-Quaterpyridines

Linking two 2,2′-bipyridine units through the 6-position yields the ligand 2,2′:6′,2″:6″,2‴-quaterpyridine (qtpy). This ligand has four nitrogen donor atoms, and a number of metal-binding modes might be envisaged.

The first example of the designed preparation of a dinuclear double-helical complex was described by Lehn and co-workers, who used the steric interactions between the substituents on a tetramethyl-substituted qtpy (5,5′,3″,5‴-Me_4 qtpy) to force a twist in the inter-annular bond between the two central pyridine rings and divide the ligand into two didentate domains.[3,4]

Six Donor Atoms Per Thread - the Sexipyridines

The molecular thread may be extended by linking three or more didentate bpy domains together. Directly linking them through the 6-position of the ring creates the potentially hexadenate ligand 2,2′:6′,2″:6″,2‴:6‴,2⁗:6⁗,2′′′′′-sexipyridine (spy).

Spacing Six Donor Atoms

The incorporation of spacer groups between the didentate or tridentate domains removes the ambiguity of the bonding modes and codes specifically for dinuclear or trinuclear double helicates. An example of an "extended spy" ligand that is partitioned into two tridentate type domains and forms a double-helical diruthenium(II) complex has been described.[5]

TRIPLE HELICATES

Molecular threads composed of separated didentate bpy metal-binding domains are ideally suited to the formation of double helical structures with metal ions that have a preference for tetrahedral geometry. In the same way that two molecular threads containing didentate domains bind to tetrahedral metal centers to build double-helical complexes, three threads, in principle, will bind to octahedral centers to give triple-helical complexes.

THE FUTURE—A NEW METHODOLOGY?

We have not ventured to discuss the helication behavior of ligands which contain an odd number of donor atoms in the ligand strands. These systems have been intensively investigated, and the partitioning of the donor sets follows the metallosupramolecular principles we developed. Ligands such as 2,2′:6′,2″:6″,2‴:2‴,6⁗-quinquepyridine (qpy) form double-helical complexes with a wide range of metal ions. The total of ten nitrogen donor atoms present in a double-helical array of two qpy ligands can be partitioned between six-coordinate and four-coordinate metal ions (a [6+4] helicate), two six-coordinate metal ions with two ancillary ligands ([6+4+2]) or two five coordinate metal ions [5+5]).[4] Recently, some preliminary studies upon the coordination behavior of the higher oligopyridines have been reported.[6]

We believe that the application of these principles will allow the design and preparation of a new series of metals containing polymers in which a central metallic core is surrounded by a helically chiral ligand coating. Still, the design of ever larger multi-domain ligands presents synthetic and solubility problems. We have now shown that it is not necessary to build, for example, oligopyridine ligands with excessively large numbers of pyridine rings. Double-helical decanuclear systems can be based upon ten octahedral metal centers with two ligand threads, each with ten tridentate domains. However, this is not essential. If the ligand threads can be slipped with respect to each other, it is possible to build polymer with smaller building blocks. Consider a combination of spy and tpy ligands. If the first metal center is coordinated to a tpy ligand, it can only address a tridentate domain of the spy ligand. This leaves a second tridentate domain available for coordination to a second metal center, although the second metal center can only use three of the six donor atoms of the next spy ligand to be presented. In effect, this is a divergent step-wide strategy for the synthesis of metallosupramolecular polymers. We are currently addressing these question.[7,8]

The chirality of the helical polymers has already been addressed by Lehn. In the future, new generations of chiral ligands are to be expected.

Finally, it should be possible to introduce variety in the metallic core of the helical polymers. It already has been shown that the use of simple metallosupramolecular principles allows the designed preparation of specific heteronuclear systems.[9–11]

We look forward to future developments in this topical, exciting, and enjoyable area. We thank all of our collaborators in the past and the funding agencies that allow us to continue working in this area.

REFERENCES

1. Lehn, J. M. *Angew. Chem. Int. Ed. Engl.* **1988**, *27*, 89.
2. Constable, E. C. *Chem. Ind.* **1994**, 56.
3. Lehn, J. M.; Sauvage, J. P.; Simon, J.; Ziessel, R.; Piccinni-Leopardi, C.; Germain, G.; Declerq, J.-P.; Van Meerssche, M. *Nouv. J. Chim.* **1983**, *7*, 413.
4. Gisselbrecht, J. P.; Gross, M.; Lehn, J. M.; Sauvage, J. P.; Ziessel, R.; Piccinni-Leopardi, C.; Arrirta, J. M.; Germain, G.; Van Meerssche, M. *Nouv. J. Chim.* **1984**, *8*, 661.
5. Crane, J. D.; Sauvage, J. P. *New. J. Chem.* **1992**, *16*, 649.
6. Potts, K. T.; Keshavarzk, M.; Tham, F. S.; Raiford, K. A. G.; Arana, C.; Abruna, H. D. *Inorg. Chem.* **1993**, *32*, 5477.
7. Cathey, C. J.; Constable, E. C.; Hannon, M. J.; Tocher, D. A.; Ward, M. D. *J. Chem. Soc., Chem. Commun.* **1990**, 621.
8. Constable, E. C.; Hannon, M. J.; Cargill Thompson, A. M. W.; Tocher, D. A.; Walker, J. V. *Supramol. Chem.* **1993**, *2*, 243.
9. Constable, E. C.; Walker, J. V. *J. Chem. Soc., Chem. Commun.* **1992**, 884.
10. Constable, E. C.; Edwards, A. J.; Raithby, P. R.; Walker, J. V. *Angew. Chem.* **1993**, *32*, 1465.
11. Piguet, C.; Hopfgartner, G.; Bocquet, B.; Schaad, 0.; Williams, A. F. *J. Am. Chem. Soc.* **1994**, *116*, 9092.

METATHESIS DEGRADATION, UNSATURATED POLYMERS

Klaus Hummel* and Robert Saf
Institute for Chemical Technology of Organic Materials
Technical University Graz

Metathesis of unsaturated polymers (MUP) belongs to the same group of reactions as metathesis of low-molecular-weight olefins, ring opening metathesis polymerization (ROMP), or

*Author to whom correspondence should be addressed.

metathesis condensation polymerization of acyclic dienes (ADMET).

The characteristic reaction in MUP degradation is the scission of double bonds in the polymer backbone followed by a recombination of alkylidene groups. As a result, the units of the polymer are considered to come from one double bond in the backbone to the next. For example 1,4-polybutadiene (**Equation 1**):

$$\ldots\!-\!\underbrace{(CH_2\text{-}CH{=}CH\text{-}CH_2)_n}_{\text{polymer}}\!-\!\ldots \qquad \underbrace{{=}CH\text{-}CH_2\text{-}CH_2\text{-}CH{=}}_{\text{unit}} \qquad (1)$$

As **Equation 2** shows with the example of the unit $=CH-(CH_2)_n-CH=$, the formal result of MUP degradation can be the insertion of units between alkylidene groups $R_1CH=$ of a low molecular-weight olefin $R_1-CH=CH-R_1$ and, if possible, a cyclization of the units.

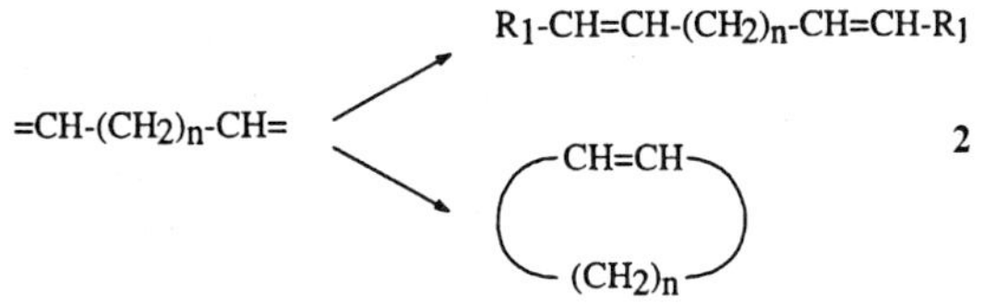

2

MUP degradation dates from the late sixties and the early seventies.[1–3] Progress is documented in several surveys.[4–7]

GENERAL

For MUP degradation, a high activity of the metathesis catalyst is important. For an uncomplicated MUP degradation to low-molecular-weight products, the catalyst action should be sufficiently free of side reactions. Three generations of catalysts are used for MUP degradation. In the beginning tungsten catalysts (especially Wcl_6 and derivatives) and cocatalysts such as $C_2H_5AlCl_2$ were applied. Later, the organoaluminums were substituted by tetraalkyltins (e.g., $(CH_3)_4Sn$). Well-defined carbene catalysts of the Schrock or Grubbs type are preferred now.[8,9]

In principle, all low-molecular-weight olefins can be used for MUP degradation. However, symmetric olefins such as ethylene, 2-butene, 3-hexene, 4-octene, or 6-dodecene are preferred for subsequent analysis of the product mixture using GC–MS.

MUP degradation succeeds with swollen crosslinked polymers (gels) as well as homogeneous polymer solutions.

The degradation of polymers with substituted double bonds (e.g., 1,4-polyisoprene) also is possible in principle.

POSSIBLE APPLICATIONS IN THE INDUSTRY

The recovery of the monomers is possible with polymers such as polypentenamer where ring-chain equilibria exist; see **Equation 3**.

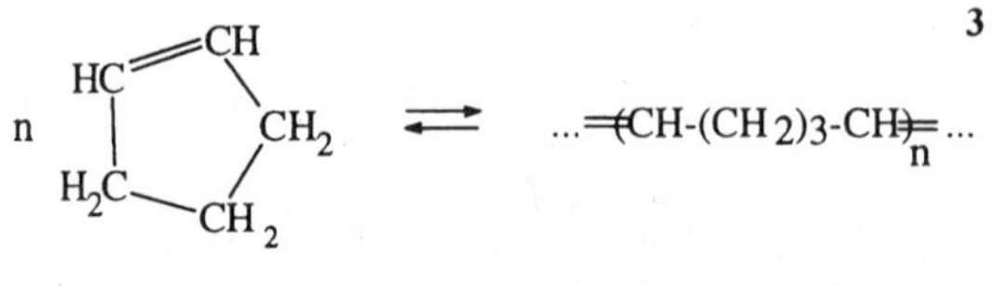

3

INVESTIGATIONS WITHOUT IDENTIFICATION OF LOW-MOLECULAR-WEIGHT DEGRADATION PRODUCTS

For some applications it is sufficient to degrade a crosslinked polymer to a soluble one. A degradation of the surface is used in etching. The total degradation can be used for filler and blend component determination. For other miscellaneous applications of this type, see the published surveys.[4–7]

CONCLUSION

MUP degradation is a valuable tool for special laboratory investigations of unsaturated polymers. In combination with GC-MS analysis, it is suitable for the elucidation of the structure of units present in a polymer even in fractions between 1.0 and 0.01 mol%. A characteristic field of application is the degradation of crosslinked polymers for various analytical purposes.

REFERENCES

1. Hummel, K.; Streck, R.; Weber, H., *Naturwissenschaften* **1970**, *57*, 194.
2. Ast, W.; Hummel, K., *Naturwissenschaften* **1970**, *57*, 545.
3. Michajlov, L.; Harwood, H. J.; Craver, C. D., *Polymer Characterization, Interdisciplinary Approaches, Ed.;* Plenum: New York, 1971.
4. Hummel, K., *Pure Appl. Chem.* **1982**, *54*, 351.
5. Hummel, K., *Olefin Metathesis and Polymerization Catalysts,* Imamoğlu, Y.; Zümreoğlu-Karan, B.; Amass, A. J., Eds., Kluwer Academic: Dordrecht, Boston, London, 1990; p. 209.
6. Hummel, K., *J. Macromol. Sci. - Pure Appl. Chem.* **1993**, *A30*, 621.
7. Ivin, K. J., *Olefin Metathesis*, Academic: London, 1983.
8. Wagener, K. B.; Puts, R. D.; Smith, D. W., *Makromol. Chem., Rapid Commun.* **1991**, *12*, 419.
9. Thorn-Csányi, E., Communication, ISOM 10 Symposium, Tihany (Hungary), 1993, 27.6.-2.7.

Metathesis Polymerization

See: *Liquid Crystalline Polyacetylene Derivatives*
Metathesis Polymerization, Cycloolefins
Molecular Recognition (Macromolecular Ionophore)
Phenylacetylene (Stereospecific Living Polymerization, Rh Complexes)
p-Phenylenevinylene Oligomers, Home-, and Copolymers (Metathesis Preparation)
Poly(Cyclopentadienylene Vinylene)
Polymerization Separation (Ring-Opening Metathesis Polymerization)
Polyphenylacetylene-Based Permselective Membranes
Polysilylpropynes (Steric Effects on Material Properties)
Ring-Opening Metathesis Polymerization (Formation of Cyclic Butadiene Oligomers)
Silacyclobutanes and Related Compounds (Ring-Opening Polymerization)
Uniform Polymers

METATHESIS POLYMERIZATION, CYCLOOLEFINS

Yuri V. Korshak
The Mendeleev University of Chemical Technology

Metathesis polymerization of cycloolefins and the metathesis reaction of linear olefins both proceed through the complete scission of carbon–carbon double bonds and their successive recombination to give unsaturated polymer of the same chemical composition as the initial monomer (**Equation 1**)

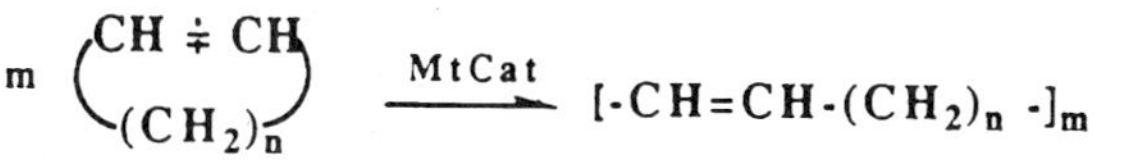

1

where: **MtCat**-metathesis catalyst, and **n** ≥ **2**.

The metathesis polymerization permitted to prepare polymers from cycloolefins, which previously were not considered productive monomers. Among them one can find monocycloolefins, such as cyclobutene, cyclopentene, cycloheptene, cyclooctene, cyclodecene, and cyclododecene, as well as bicyclic compounds, such as norbornene and its derivatives, norbornadiene and dicyclopentadiene.

The unique feature of metathesis polymerization is that the overall amount of carbon–carbon double bonds in the polymer remains the same as in the monomer. This means that double bonds do not disappear in the course of metathesis polymerization (on the contrary to all other polymerization reactions).

As typical catalytic reactions, metathesis polymerization and metathesis reaction of linear olefins proceed under the same type of homogeneous or heterogeneous catalysts mostly based on W, Mo, and Re compounds.

Cycloolefins in the absence of linear olefins produce high molecular weight compounds with carbon–carbon double bonds in the backbones called polyalkenamers or polyalkenylenes.[1,2]

However, even without linear olefins, carbon–carbon double bonds of polymeric macromolecules take part in intra– and intermolecular metathesis reactions, and that has a strong effect on MWD and the values of molecular weights.

In the presence of linear olefins, cross–metathesis reaction between polymer chains and linear olefins occurs, resulting in chemical cutting of polyalkenylenes molecules and the formation of short–chain linear species.

CHAIN POLYMERIZATION MECHANISM VS. STEP REACTION

Chauvin and Herisson, on the basis of their data on cross–metathesis of cyclic and linear olefins, concluded in 1971 that the stepwise mechanism for metathesis polymerization of cycloolefins could not explain many features of the kinetic behavior and suggested chainwise growth through metal carbene complexes.[8] According to the chain-wise mechanism, the metathesis polymerization of cycloolefins occurs as ring-opening polymerization via complete scission of carbon–carbon double bonds through the reaction with metal carbene precursors to give new active carbene species (**Equation 2**).[8]
where: X– is a ligand attached to the metal atom.

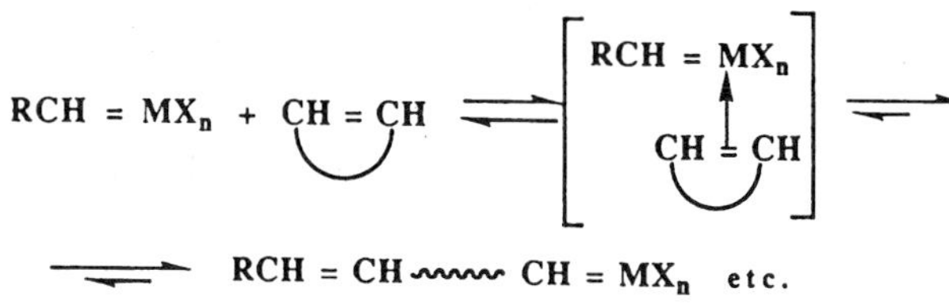

2

Transalkyledenation reaction proceeds at an extremely high rate and the equilibrium state during the metathesis transformation of linear olefins may be achieved in a few minutes. The metathesis polymerization of cycloolefins is also a very fast reaction and occurs at rates similar to those of ionic polymerization.

METAL CARBENE SPECIES IN CYCLOOLEFIN POLYMERIZATION

Catalysts for Metathesis Polymerization

Most catalysts used for cycloolefin polymerization are homogeneous systems, but there are also a few heterogeneous catalytic systems that are very effective in the metathesis polymerization of cycloolefins.

However, heterogeneous catalytic systems did not attain much application in metathesis polymerization of cycloolefins, whereas homogeneous catalysts are very important. Among transition metal compounds one can find those based mainly on W, Mo, and Re, through Ru, Os, Ir, Nb, Ta, Ti, and Cr were shown to possess some metathetical activity. The most active homogeneous Ziegler–type metathesis catalysts are composed of tungsten and molybdenum halides or mixed halide complexes in combination with metalalkyls of I-IV groups metals: e.g., $Wcl_6/Al(C_2H_5)Cl_2$ or $WCl_2{\cdot}2P(C_6H_5)_3(NO)_2/Al_2(CH_3)_3Cl_3$.[9]

BACK-BITING FORMATION OF MACROCYCLES

The monomer–polymer equilibrium in metathesis polymerization appears as a ring-opening or ring-closure reaction; i.e., it is an intermolecular/intramolecular process. These back-biting reactions are important in metathesis polymerization, and the polymerization–depolymerization equilibrium is just a unique example (special case) of the general phenomena, which includes equilibrium between linear chains and cyclic species formation.[10]

The formation of cyclic oligomers occurs in metathesis polymerization of virtually all cycloolefins[3,5,6,11–13] and has a tendency to reach the equilibrium state.

LINEAR OLEFINS AS MOLECULAR WEIGHT REGULATORS

Cross-metathesis reactions between cyclic and linear olefins is a chain transfer reaction to linear olefins and provides the effective molecular weight regulation of polyalkenylenes.

METATHESIS POLYCONDENSATION OF LINEAR α,ω–DIOLEFINS

This unusual reaction can occur in the course of metathesis transformation with linear hydrocarbons possessing at least two terminal olefinic groups.

The peculiarity of this reaction is that it proceeds as a typical equilibrium polycondensation process to give the whole spectrum of stable low molecular weight linear oligomers, but it governs by the chain reaction mechanism with metal carbene active species.

METATHESIS DEGRADATION OF UNSATURATED CARBON–CARBON POLYMERS

Polymer degradation of unsaturated carbon–chain polymers occurs as a cross–metathesis reaction between polymer molecules and linear olefins.

This reaction was applied to chemical analysis of various kinds of polydienes. In most cases polymers degraded to low molecular weight substances which could be identified by various methods.

PROPERTIES AND APPLICATION OF CYCLOOLEFINS POLYMERS

Rubbers and Plastics from Cycloolefins

Polyalkenamers prepared by ring-opening metathesis polymerization of monocycloolefins are usually crystalline substances with rather low melting temperatures depending on configuration of their double bonds.

The highest melting temperature, 145°C, makes *trans*-poly(1-butenylene) (1.4-*trans*-polybutadiene) a thermoplastic material, whereas all other polyalkenylenes are elastomers rather than plastics. Low glass transition temperatures up to −140°C for *cis*-poly(1-pentenylene) and −108°C for *cis*-poly(1–octenylene) are of the same order as for synthetic 1.4-*cis*-polybutadiene rubbers (−100°C ÷ −115°C) that is important for their practical application.

Poly(1-pentenylene) or polypentenamer is a good general purpose rubber whose vulcanizates are comparable with those of polybutadiene or polyisoprene.[4] However, it has no special or unusual properties to compensate for its more expensive production. This is why polypentenylene did not become a commercial product until recently.

Poly(1-octenylen) or polyoctenamer is a commercial product with a tradename VESTSNAMER® developed by Hüls AG.

Poly(1-dodecenylene) or polydodecenamer is a thermoplastic material with a melting point between 61 and 74°C (*trans*-70%) for partially crystalline samples, which is too low to compete with polyethylene.[14,15]

Polymers of bicycloolefins or bicyclodienes are high molded rigid substances.

Polynorbornene with high *trans*-double bonds content and molecular weights more than 10^6 is a tough thermoplastic material melting in 170–205°C range, but becomes elastomeric after absorption oil up to 500% of the polymer weight. It is used mostly in noise and vibration damping applications.[14]

Polydicyclopentadiene is produced by reaction injection molding technique under trade names "Metton" or "Telene." These polymers have the unique property balance of impact and stiffness comparable to that of engineered thermoplastics.[14]

Metathesis Approaches to Polyacetylene Sustnesis

Metathesis polymerization of unsaturated carbocycles was used to prepare polyacetylene from its polymeric precursors.[16,17]

Another new and more direct route to polyacetylene is a metathesis ring-opening polymerization of 1.3.5.7–cyclooctatetraene, first proposed by Korshak et al. in 1985 (**Equation 3**).[18]

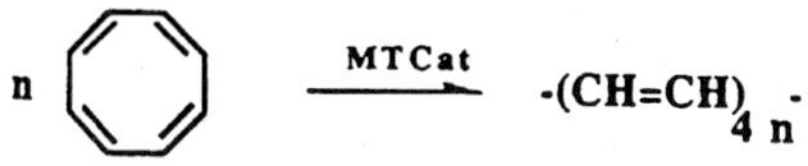

3

The polyacetylene films were similar to Shirakawa and Feast polyacetylene films, with a high degree of crystallinity and stereotacticity.[19,20] However, the reversible, photo-induced *cis/trans* isomerization was observed for the polyacetylenes; this can be a consequence of macrocyclic polyene species formed during metathesis exchange reactions.[20]

REFERENCES

1. Huggins, M. L. *J. Polym. Sci.* **1952**, *8*, 257.
2. *Pure Appl. Chem.* **1976**, *48*, 375.
3. Kuteinikov, V. M.; Korshak, Yu. V.; Dolgoplosk, B. A. *Trans. Mosk. Khim. Technol. Inst.* **1975**, *86*, 117.
4. Haas, F.; Nützel, K.; Parnpus, G.; Theisen, D. *Rubber Chem. Technol.* **1970**, *43*, 1116.
5. Korshak, Yu. V.; Vadanyan, L. M.; Dolgoplosk, B. A. *Proc. Acad. Sci. USSR, Phys. Chem. Sec.* **1973**, *208, 148.*
6. Tlenkopachev, M. A.; Kop'eva, I., A.; Bychkova, N. A.; Korshak, Yu. V.; Timofeeva, G. I.; Tinyakova, E. I.; Dolgoplosk, B. A. *Proc. Acad. Sci. USSR. Chem. Sec.* **1976**, *227*, 279.
7. Natta, G.; Dall'Asta, G.; Bassi, J. W.; Carella, G. *Makromol. Chem.* **1966**, *91*, 87.
8. Herisson, J. L.; Chauvin, Y. *Makromol. Chem.* **1971**, *141*, 161.
9. Zuech, E. A. *Chem. Commun.* **1968**, 1182.
10. Jacobson, H.; Stockmayer, W. H. *J. Chem. Phys.* **1950**, *18*, 1600.
11. Scott, K. W.; Calderon, N.; Ofstead, E. A.; Judy, W. A.; Ward, J. P *Am. Chem. Soc., Adv. Chem. Ser.* **1969**, *91*, 399.
12. Höcker, H.; Reimann, W.; Reif, L.; Riebel, K. *J. Mol. Catal.* **1980**, *8*, 191.
13. Reif, L.; Höcker, H. *Makromol. Chem., Rapid Commun.* **1981**, *2*, 183.
14. Streck, R. In *Olefin Metathesis and Polymerization Catalysts;* Imamoglu, Y.; Zümreoglu-Karan, B.; Amass, A. J., Eds.; Kluwer Academic: Dordrecht/Boston/London, NATO ASI Ser., SerC: Math. Phys. Sci., 1990; 326, Part 3, 439.
15. Vardanyan, L. M.; Korshak, Yu. V.; Teterina, M. P.; Dolgoplosk, B. A. *Proc. Acad. Sci. USSR, Chem. Sec.* **1972**, *207*, 859.
16. Edwards, J. H.; Feast, W. J. *Polymer* **1980**, *21*, 595.
17. Edwards, J. H.; Feast, W. J.; Bott, D. C. *Polymer* **1984**, *25*, 395.
18. Korshak, Yu. V.; Korshak, V. V.; Kanischka, G.; Höcker, H. *Makromol. Chem., Rapid Commun.* **1985**, *6*, 685.
19. Tlenkopachev, M. A.; Korshak, Yu. V.; Orlov, A. V.; Korshak, V. V. *Proc. Acad. Sci., Phys. Chem. Sec.* **1986**, *291*, 1036.
20. Berdyugin, V. V.; Burshtein, K. Ya.; Shorygin, P. P.; Korshak, Yu. V.; Orlov, A. V.; Tlenkopachev, M. A. *Proc. Acad. Sci. USSR, Phys. Chem. Sec.* **1990**, *312*, 410.

METHACRYLIC RESINS (Improved Heat Resistance)

Leonardo Canova* and L. Fiore
EniChem-Istituto G. Donegani

The methacrylic resins, usually methacrylic esters such as poly(methyl methacrylate) (PMMA) [9011–14–7], have some

*Author to whom correspondence should be addressed.

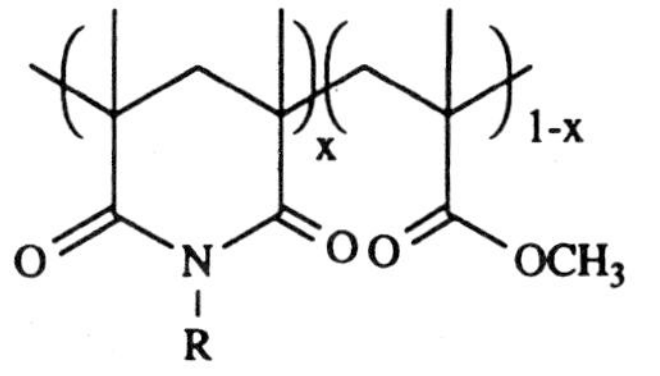

FIGURE 1. PMMA-I-R, where R=H, alkyl or aryl group (for example: CH_3 or C_6H_5).

major advantages, such as excellent transparency, weatherability, and mechanical properties. But they also have some disadvantages including brittleness and low service temperature.[1–4] The durability of PMMA compositions has been improved by adding special acrylic rubbers, thus obtaining materials having the same optical properties of PMMA. Thermal-property improvements have required substantially modifying the chemical structure of PMMA, leading to the stiffening of polymeric chains. This has been done by inserting appropriate units along the polymeric chains, both by copolymerizing and chemically modifying preformed polymethacrylic resins. Interesting results have been obtained by copolymerizing for instance, methyl methacrylate (MMA) with maleic anhydride or imide, or with methacrylates having a bulky ester group, such as tricyclodecyl methacrylate.[4] Better results have been achieved in post-reactions,[4–7] where methacrylic units of the polymer react with particular functional groups already present in the macromolecule, or with other molecules such as amines, making different kinds of six-membered cycles along the polymeric chains. The modified poly(methyl methacrylate) so obtained is named PMMA-I-R (when it contains glutarimidic cycles) or PMMA-GA (if it has glutaric anhydride cycles).

These units give PMMA high thermal properties allowing it to achieve T_g values higher than 165°C.[4,5]

The general structure of the polymers designed with PMMA-I-R, is reported in **Figure 1**. These polymers can also contain the following units: methacrylic acids and glutaric anhydride, as shown in **Figure 2**.

APPLICATIONS

Controlling the level of imidization permits the preparation of PMMA-I-CH_3 having properties tailored to requirements of the market. The combination of higher Vicat temperature, modulus, high thermal stability, chemical resistance, glass clarity, weather stability, scratch resistance, and resistance to UV discoloration, suggest the use of PMMA-I-CH_3 for example, in such applications as automobile headlamp lenses and reflectors, heat-resistant lighting fixture covers, optical waveguides,[11,15] appliances, and medical equipment.[14] Their oxygen permeability 2.5–3.0 $cm^3 \times ml/100\ in^2 \times day \times atm$, as against 3.0–6.0 for PET, suggest their use in packaging.[11,16,17]

These polymers have to be processed at temperatures of about 40°C higher than PMMA, to obtain a similar melt viscosity.[44]

PMMA–I–CH_3 are also very interesting engineering materials, particularly when used with glass-fiber or carbon-fiber reinforcements.[18–21] Moreover, there is intense commercial interest in preparing polymer blends to combine their attractive features and to improve thermal and mechanical properties of a

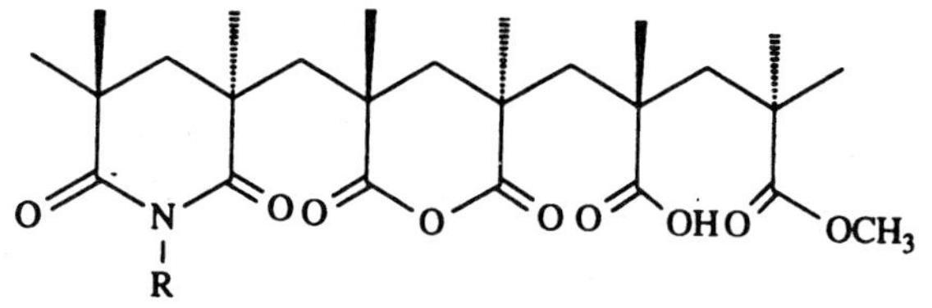

FIGURE 2. PMMA-I-CH_3 having acid and anhydride units.

particular material. There are many patents claiming blends with SAN, AES, MABS, and ABS to increase the service temperature of AN polymers or ABS or for use in optical information storage disks, optical fibers, lenses, and instrument panels with SAN, with polyamide, with PVC, with PET or PBT, with polyoxymethylenes, with PVOH as compatibilizer for polymer blends of aromatic polycarbonate with SAN, with LC co-polyester and/or benzimidazole polymers or with polyoxyphenylene.[19,22–48]

Polymer blends with SAN copolymers have been studied with poly(vinyl chloride) and polyamides.[8,10,12,13,33,49–55]

REFERENCES

1. Anonymous, *Modern Plastics Special Buyers' Guide Issue and Encyclopedia '94;* Specification/Materials; McGraw Hill: New York, 1993; 108.
2. Kline, B. B.; Novak, R. W. In *Encyclopedia of Polymer Science and Engineering;* Bikales, N. M. Ed.; John Wiley & Sons: New York, 1981; p 1, 234–299.
3. Kline, B. B.; Novak, R. W. In Kirk-Othmer *Encyclopedia of Polymer Science and Engineering*, 3rd ed.; John Wiley & Sons: New York, 1981; p 15, 377–398.
4. Stickler, M.; Rhein, T. *Ullman's Encyclopedia of Industrial Chemistry;* VCH: Weinheim, 1992; Vol. A 21, 473–86.
5. Hallden-Abberton, M. P. *Ullman's Encyclopedia of Industrial Chemistry;* VCH: Weinheim, 1992; Vol. A 21, Chapter 5, 267–272.
6. Tampellini, E.; Canova, L.; Cinquina, P.; Gianotti, G. *Polymer* **1994**, *35*, 367.
7. Abis, L.; Canova, L. *Macromolecular Reports* **1994**, *A31* (Suppls. 6 and 7), 1229.
8. Patterson, J. R.; Cinoman, D. S.; Dunkelberger, D. L. *Angew. Makromol. Chem.* **1989**, *171*, 175.
9. Bortnick, N. M.; Cohen, L. A.; Freed, W. T.; Fromuth, H. C.; Hallden-Abbenon, M. P. EP Patent 216 505, 1987.
10. Hallden-Abberton, M. P.; Maxson, P.; Cohen, L. A.; Harvey, N. G. *Res. Discl.* **1991**, *321*, 68.
11. Anonymous, *Modern Plastics Int.* March 33, 1990.
12. Pacielio, G.; Saccani, A.; Sandrolini, F.; Frezzotti, D.; Callaioli, A.; Ravanetti, G. P. *J. Appl. Polym. Sci.* **1994**, *52*(12), 1765.
13. Majumdar, B.; Keskkula, H.; Paul, D. R. *Polym. Prepr. (Am. Chem. Soc., Div. Polym. Chem.)* **1993**, *34*(32), 844.
14. Rohm and Haas, KAMAX Resins Technical Notes, 1990.
15. Takashi, Y.; Tsuruyoshi, M.; Katsuhiko, S.; Yoshihiro, U.; Ryuji, M. EP Patent 256 765, 1988.
16. Anonymous, *Modern Plastics Int.* April 22, 1989.
17. Anonymous, *Plastics Technology* February 23, 1988.
18. Katsuhiko, H.; Kyoaki, N. JP Patent 04 145 161, 1992.

19. Bright, T. A.; McKee, G. E.; Deckers, A.; Knoll, M. DE Patent 4 138 572, 1993.
20. Isao, S.; Kozi, N.; Masaru, M.; Hisao, A.; Hideaki, M. U.S. Patent 4 908 402, 1990.
21. Isao, S.; Kozi, N.; Masaru, M.; Hisao, A.; Hideaki, M. U.S. Patent 62 045 642, 1987.
22. Canova, L.; Albizzati, E.; Giannotta, G. EP Patent 382 167, 1989.
23. Cohen, L. A.; Freed, W. T. EP Patent 570 135, 1993.
24. Steas, W. H. U.S. Patent 4 436 871, 1994.
25. Weese, R. H.; Yarnell, T. M. U.S. Patent 4 217 424, 1980.
26. Staas, W. H. EP Patent 95 274, 1982.
27. Johnson, P. B. EP Patent 515 095, 1992.
28. Besecke, S.; Deckers, A.; Guentherberg, N.; Seitz, F. DE Patent 4 225 875, 1994.
29. Teruhisa, K.; Shinji, D. EP Patent 464 561, 1992.
30. Inskip, H. K.; Waggoner, M. G. EP Patent 94 215, 1983.
31. Naoki, Y.; Akira, N.; Atsunori, K. EP Patent 502 480, 1992.
32. Maxson, P.; Hallden-Abberton, M. P. *Res. Discl.* **1990**, *320*, 978.
33. Dintinger, W. N.; Maxson, P.; Goldman, T. D. *Res. Discl.* **1990**, *320*, 956.
34. Gia, H. B. WO Patent 9 215 618, 1992.
35. Fromuth, H. C.; Cohen, L. A.; Freed, W. T.; Cinoman, D. S. EP Patent 494 542, 1992.
36. Doak, K. W. U.S. Patent 4 595 727, 1986.
37. Kopchik, R. M. U.S. Patent 4 255 322, 1981.
38. Teruhisa, K.; Shinji, D. EP Patent 464 560, 1992.
39. Tetsuo, K.; Masahiro, K. DE Patent 4 317 919, 1993.
40. Kozi, N. EP Patent 352 624, 1990.
41. McKee, G. E.; Deckers, A.; Andreas, Kielhorn, S.; Zeiner, H. DE Patent 4 132 638, 1993.
42. Hideaki, O.; Satoshi, H.; Toshiaki, S.; Takashi, O.; Shiro, O.; Yoshifumi, M. JP Patent 05 214 186, 1993.
43. Freed, W. T.; Work, W. J.; Amici, R. M.; La Fleur, E. E.; Carson, W. G. EP Patent 373 911, 1990.
44. Bortnick, N. M.; Queenan, R. B.; Harvey, N. G.; Cinoman, D. S.; Hallden-Abberton, M. P.; Cohen, L. A.; Goldman, T. D. EP Patent 438 239, 1991.
45. Chen-Tsai, C. H.; Sanchez, I. C.; Isaac, Burton, W. L. U.S. Patent 5 017 411, 1991.
46. Meller, D. S. U.S. Patent 4 254 232, 1978.
47. Bortnick, N. M., Hallden-Abberton, M. P.; Work, W. J. JP Patent 06 145 460, 1992.
48. Isao, S.; Hisao, A.; Koji, N.; Hideaki, M.; Masaru, M. JP Patent 62 048 753, 1987.
49. Fowler, M. E.; Paul, D. R.; Cohen, L. A.; Freed, W. T. *J. Appl. Polym. Sci.* **1989**, *37*(2), 513.
50. Kotnis, M. A.; Spruiell, J. E. *Polym. Eng. Sci.* **1989**, *29*, 1528
51. Patterson, J. *Mod. Plast.* **1991**, *68*(12), 71.
52. Triacca, V. J.; Ziaee, S.; Barlow, J. W.; Keskkula, H.; Paul, D. R. *Polymer* **1991**, *32*(8), 1401.
53. Majumdar, B.; Keskkula, H.; Paul, D. R. *J. Appl. Polym. Sci.* **1994**, *54*(3), 339.
54. Majumdar, B.; Keskkula, H.; Paul, D. R. *J. Appl. Polym. Sci., Part B: Polym. Phys.* **1994**, *32*(12), 2127–33.
55. Majumdar, B.; Keskkula, H.; Paul, D. R. *Polymer* **1994**, *35*(15), 3164.
56. Staas, W. H. U.S. Patent 4 415 706, 1982.

METHACRYLOYL ISOCYANATE AND METHACRYLOYLCARBAMATE

Satoshi Urano
Research Center
Nippon Paint Company, Ltd.

Methacryloyl isocyanate (MAI) is a unique and versatile bifunctional monomer, combining in one molecule a reactive acylisocyanate group and a methacrylic polymerizable double bond. Each group can be used independently, leaving the other available for subsequent reactions.

Methacryloylcarbamate (MAC), which corresponds to masked MAI, has a acylcarbamate group and a methyacryl group in one molecule (**Structure 1**).

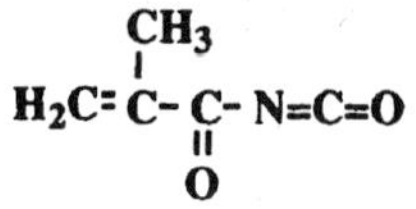

1

C.A.S. Number	4474-60-6
Appearance	Clear, colorless liquid
Molecular weight	111.1
Percent Isocyanate (functionality), weight	37.8% (theoretical)
Boiling point	121.5 - 122.5 °C/760mmHg(45 °C/45mmHg)
Freezing point	-48 °C
Flash point	26 °C
Density, 21 °C	1.0731
50 °C	1.0544
Refractive index	1.451
Solubility	Miscible with hexane, benzene, toluene, xylene, acetone, ethyl acetate, chlorobenzene, methylene chloride, and ethoxyethyl acetate

Note: The above data come from our current knowledge and are only for information purposes. They do not imply any liability on our part and must not be considered as specifications.

METHACRYLOYL ISOCYANATE (MAI)

Characteristics of MAI

Reactivity of Acylisocyanate Group

Alcohols, Amines and Mercaptan

MAI reacts with hydroxyl, amino and mercapto compounds to produce urethane, urea and thiourethane compounds (**Equation 1**).[1] This addition reaction is normally instantaneous because the isocyanate group is activated by the electrowithdrawing carbonyl group.

Amides

MAI reacts with primary amides and cyclic amides (lactams), but the reaction is very slow at ambient temperature.[1]

Hydrolysis

MAI reacts with H_2O momentarily to product methacrylamide below ambient temperature. 1,3–Dimethacryloyl urea is also formed under high concentration of MAI as the hydrolysis is an exothermic reaction.

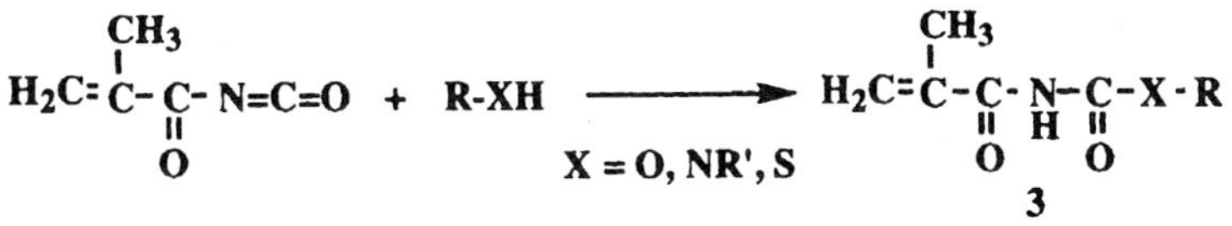

1

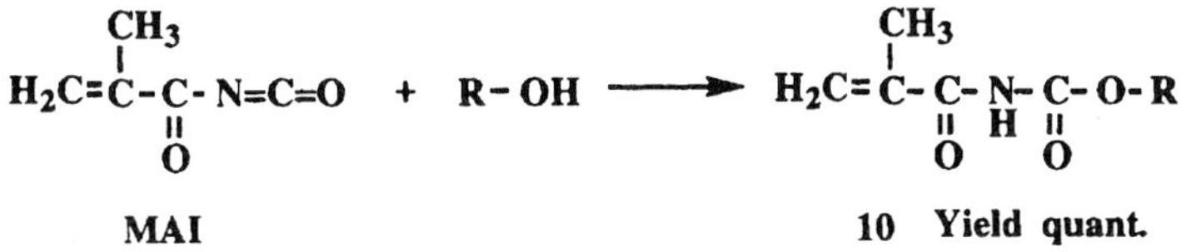

2

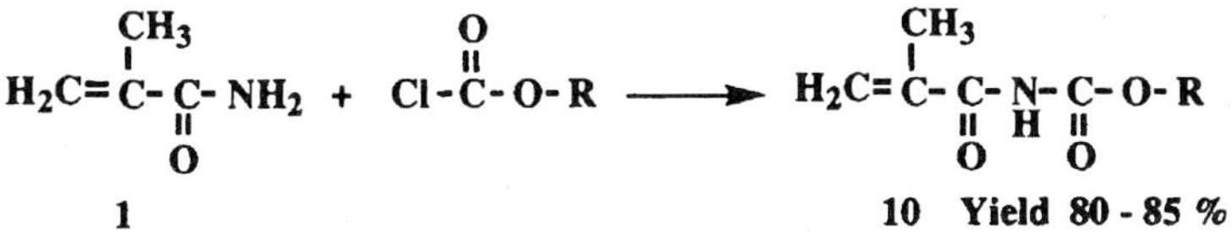

3

Other Active Hydrogen Compounds

MAI reacts with carboxylic acids to form a mixture of complex compounds at room temperature. Active hydrogen compounds such as cyclohexane–1,3–dione produce 1:1 adduct in the presence of a base catalyst.

Reactivity of Methacryloyl Group

Radical Polymerization

MAI can be polymerized with a variety of acrylic or styrenic monomers.

Reactivity of Acylisocyanate Group in MAI Polymer[2]

When MAI is homopolymerized or copolymerized with acrylic or styrenic monomers, the acylisocyanate group maintains completely and the reactivity of pendant acylisocyanates remains high.

Application of MAI

MAI is expected to be a key compound, providing many kinds of derivatives and applications. MAI derivatives are grouped into the following three categories:

- Vinyl-functionalized derivatives
- Polyacylisocyanate resins
- Vinyl-functionalized resins[4]

Vinyl-Functionalized Derivatives

Vinyl-functionalized derivatives can be produced easily by reacting MAI with functional compounds that have an active hydrogen. They contain essentially three groups: the vinyl group, the acylurethane or urea group, and a group that originates from a functional compound.

These functional monomers are used in oxygen-permeable contact lenses, photochromic materials and high-refractive-index, good adhesive polymer, nonbleeding UV absorbing agent,[5] nonlinear optical polymer materials,[6] and coatings.

Polyacylisocyanate Resins[3]

Homopolymer and copolymers of MAI, as stated above, maintain the reactivity of acylisocyanate groups. Hence, MAI polymers can be used for not only intermediate product of functional polymers but also coatings as a crosslinker.

Vinyl-Functionalized Resins[7]

MAI can be used to graft a methacryl functionality onto various polyfunctional materials of high and low molecular weight.

Modifications of epoxy resins, fluorine-containing resins, silicone resins and polymer particles are reported. These vinyl-functionalized resins are applied for crosslinking agents that can be cured by a radical initiator, irradiation or addition reactions of polyamines, graft polymers and enhancing agents of physical properties.

METHACRYLOYLCARBAMATE (MAC)

MAC is a generic term of N-methacryloylcarbamates. There are many MACs having various kinds of ester substituents.

Preparation

There are two pathways for getting MAC. One is the reaction of MAI with an alcohol and the other is the reaction of methacrylamide with a chloroformate (**Equations 2** and **3**).[1,8]

Application of MAC Polymer

Besides the crosslinking reaction, MAC monomer and polymer can be used to mediate for functional monomers and polymers. High reactive acylcarbamate group makes it possible to graft a functional group to a polymer chain. Easy handling may open up new avenues of use for MAC as a masked MAI.

Application for Coatings[9]

The remarkable feature of MAC polymer as a crosslinker is low temperature curing based on the high reactivity of acylcarbamate group.

REFERENCES

1. Urano, S.; Mizuguchi, R.; Tsuboniwa, N.; Aoki, K.; Suzuki, Y.; Itoh, T., Nippon Paint Co. U.S. Patent 4 935 413, 1986.
2. Aoki, K.; Urano, S.; Umemoto, H.; Mizuguchi, R. *Polym. Prepr. (Am. Chem. Soc., Div. Polym. Chem.)* **1988**, *29*, 423; *Chem. Abstract* **1988**, *109*, 211520b.
3. Aoki, K.; Urano, S.; Tsuboniwa, N.; Mizuguchi, R. U.S. Patent 4 788 256, 1988.
4. Urano, S.; Aoki, K.; Tsuboniwa, N.; Mizuguchi, R.; Tsuge, O. *Prog. Org. Coat.* **1992**, *20*, 471.
5. Kumagai, S.; Kashiwai, K.; Suga, A.,Ipposha Oil Industries Co., Ltd. Jpn. Patent 90 180 909, 1990; *Chem. Abstract* **1990**, *113*, 232242k.
6. Urano, S.; Tsuboniwa, N.; Kawakami, T.; Wakita, Nippon Paint Co., Ltd. and Matsushita Electric Industrial Co., Ltd. European Patent 478 268, 1992; *Chem. Abstract* **1992**, *117*, 36192n.
7. Aoki, K.; Kanda, K.; Urano, S.; Mizuguchi, R., Nippon Paint Co., Ltd. U.S. Patent 4 902 727, 1990; *Chem. Abstract* **1987**, *106*, 178218x.
8. Tsuboniwa, N.; Urano, S.; Yamanaka, E. U.S. Patent 5 187 306, 1992.
9. Tobinaga, K.; Sakamoto, H.; Shirakawa, S.; Urano, S.; Tsuchiya, Y., Nippon Paint Co., Ltd. European Patent 431 554, 1991; *Chem. Abstract* **1991**, *115*, 21034m.

Methacryloylcarbamate

See: Methacryloyl Isocyanate and Methacryloylcarbamate

Methyl Methacrylate Polymerization

See: Poly(methyl methacrylate)

α-Methyl Styrene Polymerization

See: Living Carbocationic Polymerization, Olefins

Micellar Enzymology

See: Reverse Micelles (Microcontainers for Functional Polymers)

MICELLE-FORMING MONOMERS, PHOTOPOLYMERIZATION

F. Candau* and R. Zana
Institut C. Sadron

Much work has been performed in the field of polymerization of vesicles, monolayers and multilayers.[1,2] In contrast, there have been only a few reports on the polymerization of micelle-forming monomers in the micellar state. This is probably because vesicles are characterized by lifetimes that are much longer than micelles (weeks or months as opposed to 10^{-3} to 1 s), which permits topochemical polymerization, i.e., with preservation of the initial vesicle structure. IN micelles, the question regarding whether the micellar structure is retained, altered, or destroyed upon polymerization is still a matter of debate.

Several recent reviews have been issued on polymerizable surfactants and their use in emulsion polymerizaiton.[3–5] This overview is devoted to the most recent results of studies dealing with photoinitiated (UV irradiation) homopolymerization of micelle-forming monomers, above their critical micellar concentration; i.e., in the micellar state. This report also covers the possibility of a topochemical polymerization and presents a general mechanism that takes into account the various parameters involved in this type of polymerization, particularly the micelle lifetime.

Until now, the most thoroughly investigated polymerizable surfactants have been the salts of 10-undecenoic acid and the 4-vinylbenzylalkyldimethylammonium chlorides (referred to as Cn-STY, n being the carbon number of the alkyl chain).

SYNTHESIS OF POLYMERIZABLE SURFACTANTS

Many polymerizable surfactants have been synthesized[6–21] with all possible combinations between the various types: ionic, nonionic, and zwitterionic, and of classical polymerizable groups: acrylate, methacrylate, acrylamide, vinyl, allyl, diallyl, etc.

CHARACTERIZATION OF THE MICELLAR SYSTEMS BEFORE POLYMERIZATION

The two main numbers characterizing a micellar surfactant solution are the critical micelle concentration (CMC), which is the concentration above which all the surfactant added to the solution is solubilized under the form of micelles; and the micelle aggregation number, N, which is the number of surfactants forming a micelle.

As a rule, the measured CMCs are close to those of the non-polymerizable homologues.[9]

The micelle aggregation numbers of the Cn-STY surfactants, where the polymerizable group is on the polar head, do not differ much from those of their non-polymerizable analogues.[9]

POLYMERIZATION KINETICS MICELLAR EFFECTS

The few reported kinetic studies on micellar polymerization permit some general conclusions that probably apply to various types of micellar systems and of methods of micellar polymerization.

i) Studies of polymerization of surfactants of low reactivity (sodium 10-undecenoate[4,8,11,22,23] or allyldimethyldodecylammonium bromide[24]) have led to contradictory results. Nevertheless, they show that polymerization does not take place at very low concentration, but can occur slightly below and above CMC.[8,11,13]
ii) Surfactants with polymerizable groups of sufficient reactivity (acrylate, vinyl, acrylamido, etc.) can polymerize at concentrations both above their CMC (micellar state) and below this concentration[9,10] or in apolar solvent[25–27] (molecularly dispersed state). However, the rate of polymerization is much greater above CMC than in the molecularly dispersed state, owing to the high local monomer concentration in the micelles, which is close to that of the liquid or molten state (about 10 M). Also a

*Author to whom correspondence should be addressed.

favorable relative orientation of the monomers in the micelles may further increase the polymerization rate.[4]

iii) The rate of polymerization, R_p, increases with chain length of the surfactant at concentrations above CMC (micellar state),[9,10,26,27] but depends only a little on this quantity in isotropic solution (dispersed state).

MOLECULAR WEIGHT OF THE POLYMERS

The determination of the molecular weight of polymers obtained through polymerization of surfactant micellar solutions is crucial as it immediately reveals whether the initial micellar structure is preserved or altered upon polymerization.

The results define two groups of polymers:

- polymers with high molecular weights ($M_w > 10^5$);
- polymers with low molecular weight ($M_w < 10^5$).

Surfactants containing rather reactive groups, that is most of the systems investigated, usually fall in the first class.[6,7,9,19,25–29]

POLYMERIZED SYSTEMS: STRUCTURE AND PROPERTIES

Reports on structures and properties of polymerized micellar systems are rather scarce because the systems often phase–separate or precipitate upon polymerization. Such is often the case for systems based on highly reactive monomers.[16,25–27] It must be emphasized that the transparency of the polymerized systems is not in relation with the molecular weight of the formed polymers.

In several instances it was shown that the interior of the polymerized particles is more rigid or compact than that of the unpolymerized micelles.

The most comprehensive structural studies of polymerized micellar systems have been performed on the Cr-STY surfactants, with n = 8, 12 and 16 by means of light scattering, quasi-elastic light scattering, viscosity, spectrofluorometry, time-resolved fluorescence quenching, and transmission electron microscopy at cryogenic temperature.[9,30] The results show that the polymerized C8-STY and C12-STY systems behave like polyelectrolyte solutions, showing in particular a maximum in the viscosity vs. concentration plot.

MECHANISM OF MICELLAR POLYMERIZATION

The mechanisms presented below have been used for the interpretation of the results for Cn-STY surfactants, using a rather oil-soluble initiator but they are likely to apply to a variety of systems. The following mechanisms have been postulated depending on the CMC of the surfactant.[10]

Low CMC Surfactants (Case of C16-STY; CMC = 0.29 m*M*)

Under the usual experimental conditions (surfactant concentration around 100 m*M*), before polymerization, most of the surfactant is in the micellar state, and most of the initiator radicals are in the micelles. However, only a small fraction of micelles contain one radical, and the initiation most likely starts within these micelles.

If the monomer has a low reactivity, the oligomeric radical formed during the micelle lifetime is very short and does not prevent the micelle breakdown that brings the radical in the aqueous phase. Several events can then occur, similar to the case of high CMC surfactants (see below), which all lead to low molecular weight polymers.[10]

High CMC Surfactants (C8-STY; CMC ≈ 80 m*M*)

In contrast to the preceding case, at the surfactant concentrations used (0.1–0.3 *M*), the fractions of free and micellized surfactant are comparable and an important part of the initiator is in the aqueous phase.[10] As a result, homogeneous nucleation, i.e., in the intermicellar phase, can effectively compete with micellar nucleation. Nevertheless, owing to the high local surfactant concentration in the micelles, both initiation and polymerization proceed at much higher rates in the micellar than in the intermicellar phase.

In micellar solutions at concentrations only slightly above CMC (or below CMC) the polymerization in the intermicellar phase becomes predominant, accounting for the large decrease of polymerization rate when the concentration is very close to or below CMC.

In conclusion, whatever the reactivity of the surfactant monomer and its alkyl chain length, which determines the CMC and the micelle lifetime, the polymerization cannot be topochemical.[10,17]

USES OF POLYMERIZED SURFACTANT SYSTEMS

This method offers an easy path to the preparation of polymers from monomers of low reactivity. Such polymers are difficult to obtain by conventional methods, in isotropic solutions. The oligomeric aggregates formed have a zero CMC, do not exhibit monomer exchange, and are of interest for potential applications, such as charge separation in photoredox reactions, conversion and storage of solar energy, or antistatic treatment of fabrics.[3]

For more reactive monomers, high molecular weight polymers can be prepared at high reaction rates. Both characteristics are rarely obtained in homogeneous free radical polymerization.

The catalytic activity of the polymers formed is frequently enhanced compared to that of the initial micelles.[31] The polysoap structure, which consists of covalently linked hydrophobic microdomains, is quite attractive for many types of medical and pharmaceutical applications. Studies cover drug carriers and controlled release, adjuvants in immunology to fix or stabilize enzymes, antibodies, and lower toxicity of polysoap-bound drugs.[32] The very small or zero CMC of aggregates of oligomers and polysoaps provides solubilization capacity even at a very low content of polymerized surfactant (microencapsulation).

Polymerizable surfactants are also used for latex stabilization to prevent desorption-causing flocculation.[5]

REFERENCES

1. Fendler, J. H. *Science* **1984**, *223*, 888, and references therein.
2. Ringsdorf, H.; Schlarb, B.; Venzmer, J. *Angew. Chem., Int. Ed.* **1988**, *27*, 113, and references therein.
3. Paleos, C. M.; Malliaris, A. J. *Macromol. Sci.-Rev. Macromol. Chem. Phys.* **1988**, *C28*, 403.
4. Paleos, C. M. *Polymerization in Organic Media;* Paleos, C. M., Ed.; Gordon and Breach Sci.: Philadelphia, 1992; Chapter V.
5. Guyot, A.; Tauer, C. *Adv. Polym. Sci.* **1994**, *111*, 45.

6. Voortmans, G.; Verbeeck, A.; Jackers, C.; De Schryver, F. C. *Macromolecules* **1988**, *21*, 1977.
7. Voortmans, G.; Jackers, C.; De Schryver, F. C. *Brit. Polym. J.* **1989**, *21*, 161.
8. Denton, J. M.; Duecker, D. C.; Sprague, E. D. *J. Phys. Chem.* **1993**, *97*, 756.
9. Cochin, D.; Candau, F.; Zana, R. *Macromolecules* **1993**, *26*, 5755.
10. Cochin, D.; Zana, R.; Candau, F. *Macromolecules* **1993**, *26*, 5765.
11. Arai, K.; Sugita. J.; Ogiwara, Y. *Makromol. Chem.* **1987**, *188*, 2511.
12. Arai, K.; Maseki, Y.; Ogiwara, Y. *Makromol. Chem. Rapid Commun.* **1987**, *8*, 563.
13. Arai, K.; Miyahara, S. *Makromol. Chem.* **1990**, *191*, 2647.
14. Arai, K.; Miyahara, S.; Okabe, T. *Makromol. Chem.* **1991**, *192*, 2183.
15. Arai, K. *Makromol. Chem.* **1993**, *194*, 1975.
16. Hamid, S. M.; Sherrington, D. C. *Polymer* **1987**, *28*, 325 and 332.
17. Hamid, S. M.; Sherrington, D. C. *J. Chem. Soc., Chem. Comm.* **1986**, 936.
18. Nagai, K.; Satoh, H.; Kuramoto, N. *Polymer* **1992**, *33*, 5303.
19. Lerebours, B.; Perly, B.; Pileni, M. P. *Chem. Phys. Lett.* **1988**, *147*, 503.
20. Pucci, B.; Polidori, A., Rakotomanomana, N.; Chorro, M.; Pavia, A. A. *Tetrahedron Lett.* **1993**, *34*, 4185.
21. Anton, P.; Laschewsky, A. *Makromol. Chem.* **1993**, *194*, 601.
22. Larrabee, C. E.; Sprague, E. D. *J. Polym. Sci., Polym. Lett. Ed.* **1979**, *17*, 749.
23. Paleos, C. M.; Stassinopoulou, C.; Malliaris, A. *J. Phys. Chem.* **1983**, *87*, 251.
24. Paleos, C. M.; Dais, P.; Malliaris, A. *J. Polym. Sci., Polym. Chem. Ed.* **1984**, *22*, 3383.
25. Ito, K.; Tanaka, K.; Tanaka, H.; Imai, G.; Kawagachi, S.; Itsumo, S. *Macromolecules* **1991**, *24*, 2348.
26. Nagai, K.; Ohishi, Y.; Inaba, H.; Kudo, S. *J. Polym. Sci., Polym. Chem. Ed.* **1985**, *23*, 1221.
27. Nagai, K.; Ohishi, Y. *J. Polym. Sci., Polym. Chem. Ed.* **1987**, *25*, 1.
28. Yeoh, K. W.; Chew, C.; Gan, L.; Koh, L. *J. Macromol. Sci. Chem.* **1989**, *A26*, 663.
29. Gan, L. M.; Chew, C. *Organized Solutions;* Friberg, S. E.; Lindman, B., Eds.; Surfactant Science Series, Dekker: NY. 1992, *44*, 327.
30. Cochin, D.; Candau, F.; Zana, R.; Talmon, Y. *Macromolecules* **1992**, *25*, 4220.
31. Boyer, B.; Lamatie, G.; Moussamou-Missima, J. M.; Pavia, A. A.; Pucci, B.; Roques, J. P. *Tetrahedron Lett.* **1991**, *32*, 1191.
32. Laschewsky, A. *Advances in Polymer Science;* **1995**, *124*, 1.

MICELLE-POLYMER COMPLEXES

Paul L. Dubin,* Yingjie Li, and Pratap Bahadur
Department of Chemistry
Indiana University-Purdue University at Indianapolis

Complex formation between surfactant and polymers (including polyelectrolytes) has been the subject of intense fundamental and technologic research.[1–12] From a practical point of view, knowledge acquired from the study of polymer-surfactant interactions can be applied to important industrial and biological processes. Industrial situations in which polymers and surfactants are used conjointly include enhanced oil recovery by polymer-micelle flooding,[13] rheology control,[14] drug release from pharmaceutical tablets[15] and various applications in cosmetic formulations.[16] Polymer-micelle association also can be viewed as a type of polymer-colloid interaction. In particular, the interaction of polyelectrolytes with oppositely charged micelles may prove to be a useful model for polyion-colloid systems. Such systems are represented in a wide range of situations, including water purification, precipitation of bacterial cells with polycations,[17] and stabilization of preceramic suspensions.[18] In the biological realm, they are central to immobilization of enzymes in polyelectrolyte complexes[19] and purification of proteins by selective precipitation and coacervation.[20–23] In fact, the fundamental interactions that govern the nonspecific association of DNA with basic proteins[24,25] also must be identical to the ones that control binding charged colloids to oppositely charged polymers.

For convenience of presentation, research in this field will be divided into the following topics: nonionic polymer-anionic surfactant; nonionic polymer-cationic surfactant; nonionic polymer-nonionic surfactant; polyelectrolyte-oppositely charged mixed micelle; and polyanion (or polycation)-anionic (or cationic) surfactant. The article also will discuss the unique properties of polymer-micelle systems and their application.

NONIONIC POLYMER-ANIONIC SURFACTANT COMPLEX

Nonionic polymer and anionic surfactant constitute the most investigated polymer-surfactant complex system. The association between nonionic polymers, such as poly(ethylene oxide) (PEO) and polyvinylpyrrolidone (PVP), and sodium dodecyl sulfate (SDS) has been verified by a variety of techniques.[26–36] With increasing surfactant concentration, no interaction between polymers and surfactants is detected until a critical aggregation concentration, known as CAC**, is reached.[10,26,30]

The CAC is usually lower than the normal CMC in the absence of polymers.[10,26,30] This is a clear indication of interaction between polymer and surfactant. Above the CAC, surfactants form aggregates bound to polymer. The polymer chains become saturated with surfactant aggregates at a concentration C_2, above which polymer-surfactant complexes coexist with free micelles.[26,30]

In recent years, stead-state and time-solved fluorescence have been used to investigate the microenvironment in the polymer-micelle system and to determine the aggregation number of micelles bound to polymers (N). In several systems N is smaller than that in free micelles (N_0) because polymer stabilizes the micelles.

NONIONIC POLYMER-CATIONIC SURFACTANT COMPLEX

In general, nonionic polymers interact strongly with anionic surfactants, but weakly, if at all, with cationic surfactants. Why

*Author to whom correspondence should be addressed.

**The term CAC has been used frequently in the literature. We should point out that CAC may not necessarily correspond to a surfactant concentration at which *surfactant aggregates* start to bind to polymers. It is possible that at this surfactant concentration, *monomeric surfactant* starts to bind to the polymer chain. Therefore, CAC should be understood as a surfactant concentration at which interaction between polymers and surfactants (either surfactant aggregates or single surfactant molecules) takes place and complex starts to form. CAC values from different techniques could be different.

this is so is not clear. For example, this behavior has been explained as due to bulkiness of the cationic head group,[37–39] electrostatic repulsion between polymer and surfactant due to the possible positive charge of polymer upon protonation,[32] and possible difference in interaction of anions and cations with the hydration shell of the polymer which favors interaction with anionic surfactants.[41]

Recently, interactions between nonionic polymers and cationic surfactants have been observed when hydrophobic polymers are used. In this case, the association of such hydrophobic polymers with surfactants may stabilize the polymers in aqueous solution.

NONIONIC POLYMER-NONIONIC SURFACTANT COMPLEX

The interaction between nonionic polymers and nonionic surfactants is usually very weak. However, since the driving force for polymer-micelle interaction is the reduction in Gibbs energy of the total system, interaction between nonionic polymer and nonionic surfactant could occur if a sufficiently hydrophobic polymer is used.

POLYELECTROLYTE-OPPOSITELY CHARGED MIXED MICELLE COMPLEX

In contrast to the interaction between nonionic polymers and ionic surfactants, the interaction between polyelectrolytes and oppositely charged surfactants is dominated by electrostatic forces, although hydrophobic interactions may play a secondary role.[5,8,10] It is useful, at the outset, to distinguish between strong polyelectrolytes, such as polystyrenesulfonate (PSS), which are always 100% ionized, and weak polyelectrolytes, such as poly(acrylic acid) (PAA), whose charge depends upon pH for example. In general, the interaction between polyelectrolytes and oppositely charged surfactants starts at a very low surfactant concentration (CAC), usually a few orders of magnitude lower than the CMC of the free surfactant. Unlike the nonionic polymer-ionic surfactant system, the complex usually cannot coexist with free micelles because precipitation is observed as addition of ionic surfactant brings the polyelectrolyte close to charge neutralization. In some cases, however, further addition of ionic surfactant may resolubilize the precipitate.[5] Phase separation effects result in restriction of most studies of strong polyelectrolytes with oppositely charged surfactants to surfactant concentrations well below CMC. There have been a few reviews on the interaction between polyelectrolytes and oppositely charged surfactants.[1–12]

POLYANION-ANIONIC SURFACTANT COMPLEX

Binana-Limbelé and Zana[41] investigated the effects of a polyanion, NaPAA, on CMC and N of SDS, using conductivity and time-resolved fluorescence. While no complex formation was observed, NaPAA decreases CMC and increases N. Therefore, NaPAA behaves like a small electrolyte. However, if the polymer is hydrophobically modified, some interactions between polyanion and anionic surfactant could be observed.

Although polymer-surfactant complexes have been the subject of extensive research, a clear picture of their structure and energetics still has not emerged. At least to some extent, basic questions remain. The first concerns the thermodynamics of complex formation, that is, what is the driving force for the complex formation for different polymers and different surfactants? A related question is the nature of the interaction. For example, why cationic surfactants interact weakly with nonionic polymers is still unexplained. The second question is about the mechanism of complex formation. For example, it is still not known whether surfactants aggregate first and then bind to the polymers, or aggregate around one or more *already* bound surfactant molecules, or whether these processes take place simultaneously. The third question regards the structure of the complex. For example, it is not know if polymer chains reside inside the surfactant aggregates or wrap around the surface. It is also not clear if the surfactant aggregates restructure upon binding and how the conformation of polymer changes upon binding. The fourth question regards the dependence of the interaction, the binding mechanism and the structure of the complex on variables such as the chemical structure of polymers and surfactants, ionic strength, and temperature. The dependence of the complex structure on length changes, size, and concentration of polymers and surfactants also is important.

PHASE BEHAVIOR OF POLYMER-MICELLE COMPLEXES

Phase diagrams of binary systems viz surfactant-water and polymer-water have long been utilized to study solute-solvent and solute-solute interactions and to provide useful information about aggregation (for surfactant-water systems) and conformation (for polymer-water systems).

RHEOLOGICAL BEHAVIOR OF POLYMER-MICELLE COMPLEXES

The presence of surfactant can substantially increase the viscosity and elasticity of polymer solutions. Since the use of ultrahigh molecular weight polymers to achieve high viscosity is complicated by mechanical shear degradation (chain backbone scission), polymer-surfactant complexes seem to offer a good route to rheological control. An increase in the viscosity of PEO solutions in the presence of SDS was observed long ago.[4]

Novel gels were obtained from very dilute solution of a cationic cellulose derivative in the presence of small amounts of anionic surfactants by Goddard et al.[42,43]

APPLICATION OF POLYMER-MICELLE COMPLEXES

An important role may be played by polymer-micelle interactions in surface conditioning and detergency; the antiredeposition agents, used to improve detergency by avoiding soil redeposition during the rinsing cycle of the washing process, are water-soluble polymers.[44,45]

Polyelectrolyte-surfactant complexes have been used to separate and fractionate polyelectrolytes.[46] Colloid-enhanced ultrafiltration (CEUF) methods are novel separation processes used to remove organic and inorganic species from ultrafiltration. Use of polyelectrolyte-surfactant complexes in CEUF has been reported by Uchiyama et al.[47] Thermoreversible gels may provide a suitable liquid carrier for the release of water-soluble drugs.[15,48] More details above the application of polymer-micelle complexes are provided by Goddard.[49]

ACKNOWLEDGMENT

The support of grant DMR9311433 from the National Science Foundation is gratefully acknowledged.

REFERENCES

1. Breuer, M. M.; Robb, I. D. *Chem. Ind.* **1972**, 530.
2. Robb, I. D. In *Anionic Surfactants, Physical Chemistry of Surfactant Action* Lucassen-Reynders, E. H., Ed.; Marcel Dekker: New York, 1981; p 109.
3. Tsuchida, E.; Abe, K. *Adv. Polym. Sci.* **1982**, 45.
4. Goddard, E. D. *Colloids Surf.* **1986**, *19*, 255.
5. Goddard. E. D. *Colloids Surf.* **1986**, *19*, 301.
6. Saito, S. In *Nonionic Surfactant, Physical Chemistry;* Schick, M. J. Ed.; Surfactant Science Series 23; Marcel Dekker: New York, 1987; Chapter 15.
7. Smid, J.; Fish. D. *Encyclopedia of Polymer Science and Technology* Wiley-Interscience: New York, 1988; Vol. 11, p 720.
8. Hayakawa, K.; Kwak, J. C. T. In *Cationic Surfactants, Physical Chemistry;* Rubingh, D. N., Holland, P. M., Eds.; Marcel Dekker: New York, 1991; Chapter 5, p 189.
9. Piculell, B.; Lindman, B. *Adv. Colloid Interface Sci.* **1992**, *41*, 149.
10. Lindman, B.; Thalberg, K. In *Interactions of Surfactants with Polymers and Proteins* Goddard, E. D., Ananthapadmanabhan, K. P. Eds.; CRC: Boca Raton, FL, 1993; Chapter 5.
11. Lindman, B.; Khan, A. et al. *Pure Appl. Chem.* **1993**, *65*, 953.
12. Brackman, J. C.; Engberts, J. B. F N. *Chem. Soc. Rev.* **1993**, *22*, 85.
13. Desai, N. N; Shah, D. O. *Polym. Prepr.* **1981**, *22*(2), 39.
14. Brackman, J. C. *Langmuir* **1991**, *7*, 469.
15. Alli, D.; Bolton, S; Gaylord, N. S. *J. Appl. Polym. Sci.* **1991**, *42*, 947.
16. Goddard, E. D. *J. Soc. Cosmet. Chem.* **1990**, *41*, 23.
17. Kawabata, N.; Hayashi, T; Nishikawa, M. *Bull. Chem. Soc. Jpn.* **1986**, *59*, 2861.
18. Cesarano, J., III; Aksay, I. A. *J. Am. Ceram. Soc.* **1988**, *71*, 1062.
19. Margolin, A.; Sherstyuk, S. F. et al *Eur. J. Biochem.* 1985, *146*, 625
20. Clark, K. M; Glatz, C. E. *Biotechnol. Prog.* 1987, 3, 241.
21. Fisher, R. P.; Glatz, C. E. *Biotechnol. Bioeng.* **1988**, 32, 777.
22. Bozzano, A. G.; Andrea, G.; Glatz, C. E. *J. Membr. Sci.* **1991**, *55*, 181.
23. Dubin, P. L.; Strege, M. A; West, J. In *Large Scale Protein Purification;* Ladish, M., Ed.; American Chemical Society: Washington, DC, 1990; Chapter 5.
24. Shaner, S. L.; Melancon, P. et al. *Cold Spring Harbor Symp. Quant. Biol.* **1983**, *47*, 463.
25. von Hippel, P. H.; Bear, D. G. et al. *Annu. Rev. Biochem.* **1984**, *53*, 389.
26. Jones, M. N. *J. Colloid Interface Sci.* **1967**, *23*, 36.
27. Cabane, B. *J. Phys. Chem.* **1977**, *81*, 1639.
28. Cabane, B; Duplessix, R. *J. Phys. (Paris)* **1982**, *43*, 1529.
29. Wan-Badhi, W. A.; Wan-Yunus et al. *J. Chem. Soc. Faraday Trans.* **1993**, *8*, 2737.
30. Fishman, M. L.; Eirich, F. R. *J. Phys. Chem.* **1971**, *75*, 3135.
31. Gilányi, T; Wolfram, E. *Colloids Surf.* **1981**, *3*, 181.
32. Moroi, Y.; Akisida, H. et al. *Colloid Interface Sci.* **1977**, *61*, 233.
33. Shirahama, K; Tohdo, M.; Murahashi, M. *J. Colloid Interface Sci.* **1982**, *86*, 282.
34. Turro, N. J., Baretz, B. H; Kuo, P. L. *Macromolecules* **1984**, *17*, 1321.
35. Schwuger, M. J. *J. Colloid Interface Sci.* **1973**, *43*, 491.
36. François, J.; Dayantis, J; Sabbadin, J. *Eur. Polym. J.* **1985**, *43*, 491.
37. Nagarajan, R. *Colloids Surf.* **1985**, *13*, 1.
38. Nagarajan, R. *J. Chem. Phys.* **1989**, *90*, 1980.
39. Ruckenstein, E.; Huber, G.; Hoffmann, H. *Langmuir* **1987**, *3*, 382.
40. Witte, F. M., Engberts, J. B. F. N. *Colloids Surf.* **1989**, *36*, 417.
41. Binana-Limbelé, W.; Zana, R. *Colloids Surf.* **1986**, *21*, 483.
42. Goddard, E. D.; Leung, P. S.; Padmanabham, K. P. A. *J. Soc. Cosmet. Chem.* **1991**, *42*, 19.
43. Leung, P. S.; Goddard, E. D. *Langmuir* **1991**, *7*, 608.
44. Cahn, A.; Lynn, J. In *Encyclopedia of Chemical Technology;* Wiley-Interscience: New York, 1983; p 332.
45. Vogel, F. *Chem. Uns.-Zeit.* **1986**, *20*, 156.
46. Laurent, T. C.; Scott, J. E. *Nature* **1964**, *202*, 661.
47. Uchiyama, H.; Christian, S. D. et al. *J. Colloid Interface Sci.* **1994**, *163*, 493.
48. Lindman, B.; Carlsson, A. et al. *L'actualitee Chimique* May–June 1991, p 181.
49. Goddard, E. D. In *Interactions of Surfactants with Polymers and Proteins* Goddard, E. D.; Ananthapadmanabhan, K. P., Eds.; CRC: Boca Raton, FL, 1993; Chapter 10.

Micelles

See: Block Copolymer Micelles (Overview)
Block Copolymer Micelles (For Drug Targeting)
Block Copolymer Micelles (Microcontainers for Drug Targeting)
Block Copolymer Micelles (1. Structure and Properties)
Block Copolymer Micelles (2. Fluorometric Studies and Computer Modeling)
Block Copolymers (Micellization in Solution)
Micelle-Forming Monomers, Photopolymerization
Micelle-Polymer Complexes
Molecular Assemblies (Polymerization in Aqueous Solution)
Poly(ethylene oxide) Macromonomers (Micellar Polymerization)
Reverse Micelles (Microcontainers for Functional Polymers)
Supramolecular Formation and Structure (PEO-PPO-PEO Block Copolymers)
Surface Active Monomers
Surfactants, Polymeric (Overview)
Surfactants, Polymeric (Compatibilizers for Blends)

MICROBIAL POLYHYDROXYALKANOATES (Poly[(*R*)-3-hydroxybutyrate-*co*-3-hydroxypropionate])

Yoshiharu Doi,* H. Abe, E. Shimamura, M. Hiramitsu, and S. Nakamura
Polymer Chemistry Laboratory
The Institute of Physical and Chemical Research (RIKEN)

Microbial polyhydroxyalkanoates (PHA) are a biodegradable thermoplastic with a wide range of physical properties.[1–3]

*Author to whom correspondence should be addressed.

A wide variety of bacteria synthesize optically active homopolymers and copolymers of (*R*)-3-hydroxy-alkanoic acids (3HA) ranging from 4 to 14 carbon atoms as an intracellular storage material of carbon and energy.[4–6] Saturated, unsaturated, halogenated, branched, and aromatic side chains in 3HA monomeric units have been found in the sequence of microbial PHA.[7–16] In addition, some bacteria have been found to produce copolymers containing hydroxyalkanoate monomeric units without side chains such as 3-hydroxypropionate, 4-hydroxybutyrate, and 5-hydroxyvalerate.[17–22] These microbial copolyesters have attracted much attention as environmentally degradable thermoplastics for a wide range of agricultural, marine, and medical applications.[23] The microbial polyesters are degradable in soil, sludge, or sea water. Many microorganisms, such as bacteria and fungi, secrete extracellular PHA depolymerases to degrade environmental microbial polyesters and utilize the decomposed compounds as nutrients.

This paper surveys the microbial synthesis of copolymers of (*R*)-3-hydroxybutyrate and 3-hydroxypropionate, P(3HB-*co*-3HP) by *Alcaligenes eutrophus* and *Alcaligenes latus* from various carbon substrates, and discusses the physical properties and biodegradabilities of P(3HB-*co*-3HP) copolymers.

PREPARATION

A. eutrophus (ATCC 17699) and *A. latus* (ATCC 29713) were used as bacteria for the microbial synthesis of P(3HB-*co*-3HP). Polyester synthesis was carried out by a two-stage fermentation of *A. eutrophus* and by a single-stage fermentation of *A. latus* (**Scheme I**).[17,18,24]

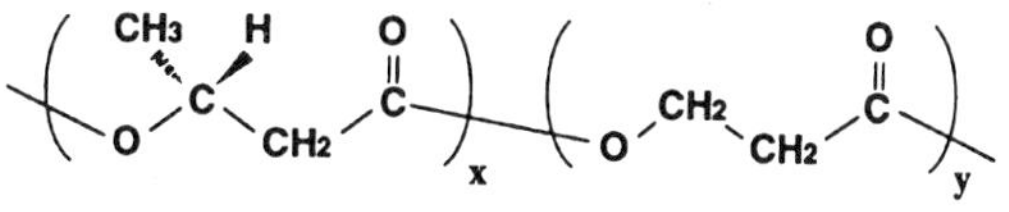

I

(*R*)-3HB **3HP**

Many bacteria, including *A. eutrophus*, accumulate P(3HB) within cells when growth is limited by the depletion of an essential nutrient but the cells have an excess of carbon source. However, *A. latus* accumulates P(3HB) from sucrose during the logarithmic growth phase without limitation of nutrients. Therefore, *A. latus* was used as a bacterium for industrial production of P(3HB).[25] Recently, *A. latus* was found to produce P(3HB-*co*-3HP) copolymers.[24]

PROPERTIES

The glass-transition temperature (T_g) decreased from 4 to −19°C as the 3HP fraction was increased from 0 to 100 mol%. The melting temperature (T_m) decreased from 177 to 44°C with the 3HP fraction and then increased to 77°C as 3HP fraction was increased from 67 to 100 mol%. The enthalpy of fusion (Δh_m) showed a trend similar to that of the T_m data.

The enzymatic degradations of six samples of P(3HB-*co*-3HP) containing 0.7, 11, 20, 37, and 43 mol% 3HP were carried out on the solution-cast films for 2 h at 37°C in an aqueous solution of PHB depolymerase from *Alcaligenes faecalis*.[24] The rate of enzymatic degradation was accelerated by the incorporation of 3HP units in a P(3HB) sequence.

A marked increase in the rate of enzymatic degradation was observed on the films of poly[(*R*)-3-hydroxybutyrate-*co*-4-hydroxybutyrate] (P(3HB-*co*-4HP) and poly[(*R*)-3-hydroxybutyrate-*co*-(*R*)-3-hydroxyhexanoate] (P(3HB-*co*-3HH)).[21,26]

REFERENCES

1. Marchessault, R. H.; Bluhm, T. L.; Deslands, Y.; Hamer, G. K.; Orts, W. J.; Sundararajan, P. R.; Taylor, M. G.; Bloembergen, S.; Holden, D. A., *Makromol. Chem., Macromol. Symp.* **1988**, *19, 235.*
2. Holmes, P. A., *Developments in Crystalline Polymer–2*, Bassett, D. C., Ed.; Elsevier: London, 1988.
3. Doi, Y., *Microbial Polyesters*, VCH: New York, 1990.
4. Anderson, A. J.; Dawes, E. A., *Microbiol. Rev.* **1990**, *54, 450.*
5. Steinbüchel, A.; *Biomaterials*, Byrom, D., Ed.; Macmillan: Basingstoke, 1991.
6. Müller, H. M.; Seebach, D., *Angew. Chem. Int. Ed. Engl.* **1993**, *32, 477.*
7. De Smet, M. J.; Eggink, G.; Witholt, B.; Kingma, J.; Wynberg, H., *J. Bacteriol.* **1983**, *154, 870.*
8. Doi, Y.; Tamaki, A.; Kunioka. M.; Soga, K., *Appl. Microbiol. Biotechnol.* **1988a**, *28, 330.*
9. Brandl, H.; Gross, R. A.; Lenz, R. W.; Fuller, R. C., *Appl. Environ. Microbiol.* **1988**, *54*, 1977.
10. Lageveen, R. G.; Huisman, G. W.; Preusting, H.; Ketelaar, P.; Eggink, G.; Witholt, B., *Appl. Environ. Microbiol.* **1988**, *54*, 2924.
11. Fritzsche, K.; Lenz, R. W.; Fuller, R. C., *Int. J. Biol. Macromol.* **1990a**, *12*, 85.
12. Doi, Y.; Abe, C., *Macromolecules* **1990**, *23*, 3705.
13. Abe, C.; Taima, Y.; Nakamura, Y.; Doi, Y., *Polym. Commun.* **1990**, *31*, 404.
14. Kim, Y. B.; Lenz, R. W.; Fuller, R. C., *Macromolecules* **1992**, *25*, 1852.
15. Fritzsche, K.; Lenz, R. W.; Fuller, R. C., *Int. J. Biol. Macromol.* **1990b**, *12*, 92.
16. Kim, Y. B.; Lenz, R. W.; Fuller, R. C., *Macromolecules* **1991**, *24*, 5256.
17. Nakamura, S.; Kunioka, M.; Doi, Y., *Macromol. Rep. (A)* **1991**, *28*, 15.
18. Hiramitsu, M.; Doi, Y., *Polymer* **1993**, *34*, 4782.
19. Shimamura, E.; Scandola, M.; Doi, Y., *Macromolecules* **1994a**, *27*, 4429.
20. Doi, Y.; Kunioka, M.; Nakamura, Y.; Soga, K., *Macromolecules* **1988b**, *21*, 2722.
21. Nakamura, S.; Doi, Y.; Scandola, M., *Macromolecules* **1992**, *25*, 4237.
22. Doi, Y.; Tamaki, A.; Kunioka, M.; Soga, K., *Makromol. Chem., Rapid Commun.* **1987**, *8*, 631.
23. Holmes, P. A., *Phys. Technol.* **1985**, *16, 32.*
24. Shimamura, E.; Kasuya, K.; Kobayashi, G.; Shiotani, T.; Shima, Y.; Doi, Y., *Macromolecules* **1994b**, *27*, 878.
25. Hänggi, U. J., *Novel Biodegradable Microbial Polymers,* Dawes, E. A., Ed.; Kluwer Academic: Dordrecht, *1990.*
26. Doi, Y.; Kanesawa, Y.; Kunioka, M.; Saito, T., *Macromolecules* **1990**, *23*, 26.

Microbial Polymers

See: Bacterial Polyesters (Containing Unusual Substituent Groups)
Microbial Polyhydroxalkanoates (Poly[(R)-3-hydroxybutyrate-co-3-hydroxypropionate])

Poly(β-hydroxyalkanoates) (Microbial Synthesis and Chemical Modifications)
Protein-Based Polymeric Materials (Synthesis and Properties)
Recycling, Thermodynamic (by Ring-Closing Depolymerization)
Welan Microbial Polysaccharide

Microcapsules

See: *Microcapsules (From Two-Dimensional Networks)*
Microcapsules, Molecular Complexes (Dissolution of Drugs by Cyclodextrin)
Microencapsulated Particles
Microencapsulation (Artificial Cells)
Microencapsulation (Drug Release)
Stimuli-Responsive Microcapsules
Thermodegradable Polymers (Azo Groups in Main Chain)

MICROCAPSULES (From Two-Dimensional Networks)

Heinz Rehage*
Institute für Umweltanalytik
Universität-Gesamthochschule Essen

A. Burger
Hydrotec AG

The tensioactive properties of acrylic or methacrylic diesters can be used to synthesize two-dimensional model networks at the interface between oil and water. We have systematically studied the adsorption process of the amphiphilic ester molecules and the photogelation reactions which finally lead to the formation of ultrathin crosslinked membranes.

Biological membranes are among the most important components of living organisms.[1,2] An impressive example of the importance of membranes is flow-induced deformation in thin vessels. Red blood cells, which have typical diameters of 7–8 μm, are able to flow though constrictions or narrow vessels with diameters of about 2 μm. The high deformation results from the viscoelastic properties of the membrane, which consists of a fluid-like lipid bilayer connected to a rubber-elastic, quasi-two-dimensional spectrin network.

In view of its multifunctional character, the structure of a biological membrane is complex. Consequently, elucidating the exact correlation between microscopic membrane properties and fundamental deformation processes is difficult.

The advantage of model systems lies in their simple structure and the possibility of changing all parameters that might be useful for detailed investigations.

Based on these ideas we have synthesized two-dimensional networks by interfacial polymerization of tensioactive methacrylate diesters.

*Author to whom correspondence should be addressed.

MONOMERS

Preparation of Diesters

To investigate networks with a wide variety of elastic properties, we synthesized a series of diesters with different chain lengths and different surface active properties (**Structure 1**).

- 1,n-alkanedioldimethacrylate:

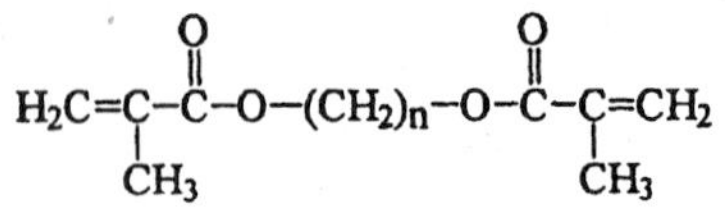
1

with n = 4, 6, 8, 10, 12 respectively.

- 3,16-dioxa-1,18-octadecanedioldimethacrylate (**Structure 2**)

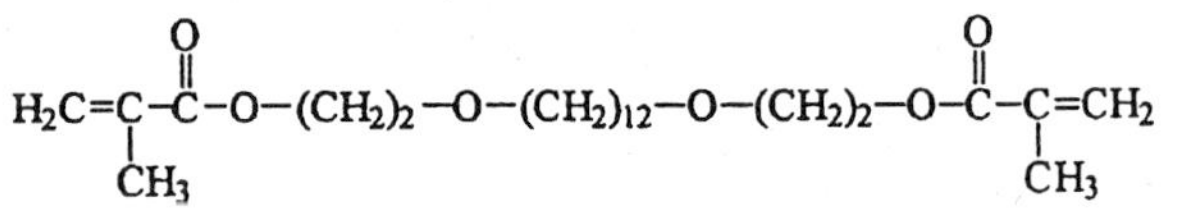
2

- 3,6,19,22-tetraoxa-1,24-tetracosanedioldimethacrylate (**Structure 3**).

TWO-DIMENSIONAL NETWORKS

The polymerization of the monomers was induced by UV irradiation with a low pressure mercury lamp.

The Two-Dimensional Sol–Gel Transition

Linear chains are formed in the first part of the reaction. After some time crosslinking sets in and finally reaches a critical yield value where the low viscous sol state is transformed into a continuous two-dimensional gel.

The Kinetics of Gelation

To get more information on the gelation process, the evolution of the two-dimensional storage modulus μ' and the two-dimensional loss modulus μ'' was recorded as a function of the reaction time by oscillating surface viscosimeters. If polymerization is restricted to the interface, the storage modulus must have a limiting value. This is in excellent agreement with the experimental data.

The formation of crosslinked membranes depends on several parameters, such as the surface concentration of the diesters, the molar mass, and adsorption potential of the monomers. The presence of inhibitors, photosensibilisators, and traces of surface active compounds influence the kinetics of gelation and the elastic properties of the network.[3] Under defined experimental conditions, however, the elastic properties of the networks can easily be controlled.

The Elastic Properties of Two-Dimensional Networks

Summarizing the experimental results, one can conclude that the two-dimensional networks exhibit typical features of rubber-elastic structures. By adjusting the light intensity surface concentration or chemical structure of the reactive diesters, it

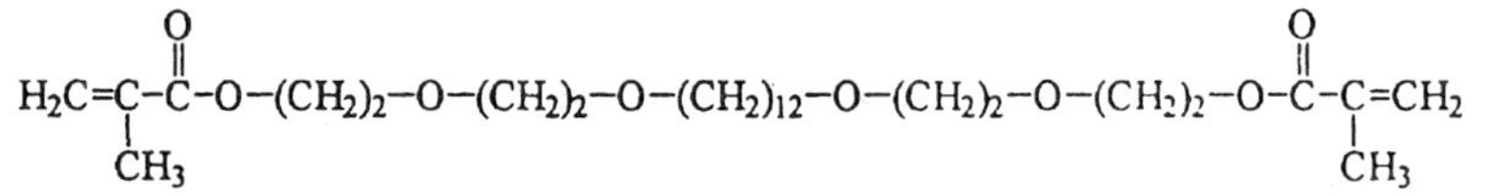

$$H_2C=\underset{\underset{CH_3}{|}}{C}-\overset{\overset{O}{\|}}{C}-O-(CH_2)_2-O-(CH_2)_2-O-(CH_2)_{12}-O-(CH_2)_2-O-(CH_2)_2-O-\overset{\overset{O}{\|}}{C}-\underset{\underset{CH_3}{|}}{C}=CH_2$$

3

is possible to change the physical properties of these structures to desired values. The two-dimensional networks are, therefore, excellent model systems for the fundamental studies of rheological research.

MICROCAPSULES

The photoinduced polymerization and encapsulation of freely suspended oil droplets in water was carried out in a quartz vessel. A control experiment in a surface rheometer was carried out under the same experimental conditions to determine the rheological properties of the flat membranes. In this way, it is possible to prepare microcapsules with well-known elastic properties.

Deformation of Microcapsules Theory

In recent years many new theories have been developed to describe the shear-induced deformation of microcapsules. The basic features and the interesting rheological process of these systems have been thoroughly reviewed in one article of Barthès-Biesel.[4] According to theoretical models, the deformation depends on several parameters, such as the viscosity ratio of the internal and external fluid phases, the capillary number where the influence of the interfacial tension is now replaced by the elastic modulus of the surrounding membrane, and the Deborah-number β, which describes the ratio of the externally imposed shear time and the intrinsic response time of the crosslinked membrane.

Experimental Results

At very low values of shear rate, a deformed microcapsule can still be described by an ellipsoidal geometry. In elevated flow fields, however, one observes striking deviations. In comparison to liquid oil droplets, this occurs at rather low values of the velocity gradient. This principal difference is due to the fluid-like interface of emulsion droplets, whereas the viscoelastic membrane of microcapsules takes up more energy which cannot be dissipated into the internal fluid. It is worthwhile to mention that at equal values of the capillary number, microcapsules tend to show much larger deformations. And even the breaking mechanism is very different.

Encapsulated particles show a rather complicated behavior. It is often observed that the membrane tears at different positions, and there are probably small mechanical defects in the chemical structure of the surrounding network that are responsible for this behavior. Consequently, microcapsules tend to break at a random position.

CONCLUSION

Ultrathin polymer networks are of great interest for many applications in science and industry, and they can even be used to form microcapsules and artificial cells. Since the elastic properties of the enclosing membranes can be varied over many orders of magnitude, such encapsulated oil droplets provide simple model systems for fundamental studies of complex biological flow processes. In former investigations, we have systematically studied the diffusion process of molecules through these membranes.[5] The pore structure of the crosslinked membranes is rather homogeneous and it can easily be adjusted chemically so that sieving and filtering processes become possible on a molecular scale. This might be useful for controlled-release of certain drugs, or for processes where special molecules must be removed from a large amount of similar compounds. Properly adjusting the network elasticity breaks the capsules at well-defined external forces. Such processes are interesting for technical applications where toxic substances, such as herbicides or fungicides, must be released under defined circumstances. In the quiescent state, these harmful compounds are encapsulated so that they cannot directly contact human skin. At elevated shear rates, however, the capsules break and this can be used to release these compounds in technical applications, such as spraying procedures. These and other methods are possible only if the physical properties of the ultrathin membranes are precisely adjusted to the desired process. Systematic studies of these phenomena offer new insights into basic features of living cells and may also open interesting new applications in science and industry.

REFERENCES

1. Burchard, W. *Chem. Unserer Zeit* **1989**, *23*, 7.
2. Burchard, W. *Chem. Unserer Zeit* **1989**, *23*, 69.
3. Burger, A. *Ultradünne Polymernetzwerke und Mikrokapseln als Modellsysteme;* Thesis: Bayreuth, 1994.
4. Barthès-Biesel, D. *Physica A* **1991**, *172*, 103–124.
5. Burger, A.; Leonhard, H.; Rehage, H.; Wagner, R.; Schwoerer, M. *Macromol. Chem. Phys.* **1995**, *196*, 1–46.

MICROCAPSULES, MOLECULAR COMPLEXES (Dissolution of Drugs by Cyclodextrin)

Yasuo Nozawa
School of Pharmaceutical Sciences
University of Shizuoka

Microcapsules, small containers of 5–300 μm in size, are spherical-shaped particles having a wall membrane of polymeric materials. A number of books and reviews concerning microcapsules or microencapsulations have been presented by several investigators.[1–6] In a broad sense microcapsules, liposomes, and molecular complexes are recognized to be a homologous series of microcapsules. The former, liposomes, having a spherical shape of 0.5 μm, were first found by Stockenius in a water suspension of phospholipid.[7] Various inclusion compounds were reviewed by Atwood, in which cyclodextrin (CD) complexes were noted.[8]

The techniques for CD complexing are divided into two procedures practiced in wetted and dried processes. The former contains several ways as preparing in aqueous solution, in kneading, and in heating.[9–11] In the latter, there are many ways such as mix-grinding by various mills.

Efficient preparations of the complexes are often found in such mechanical procedures as solid phase mixing. The roll-mixing procedures possess some favorable features as a simple and continuous operation without organic solvents. Therefore, dissolution of practically insoluble drugs and properties of medicinal molecular capsules are examined in various guest compounds, and mechano–chemical complexing is pursued with some water insoluble medicinal drugs.

PREPARATION AND PROPERTIES

Dissolution and Complexing of Drugs

Dissolution of some practically insoluble drugs is improved in β CD aqueous solution owing to complex forming in aqueous media. Dissolution modes of pharmaceuticals exhibited characteristic patterns of interaction between guest compounds and β CD in its aqueous solution.[13]

At least two water molecules were necessary to form the β CD complexes.[14] In practice, water required for complexing can be supplied from the atmosphere during the roll-mixing procedures, because the β CD molecules are likely to absorb moisture, and it can be seen they are clearly capable of providing a stable crystalline form of β CD 12 hydrates.[15]

To confirm the complex formation, the solid phase roll-mixing will be further examined below in systems of β CD with drugs such as IP, TB, and IM.

SOLID PHASE COMPLEXING

Ibuprofen-β CD System

The roll-mixing products of β CD with various guest compounds were transformed readily to an amorphous solid form according to X-ray diffraction measurements.[12] The roll-mixed system of IP with β CD exhibited amorphous halo patterns having three hillocks around 12°, 18°, and 24° in 2θ. The halo pattern coincided to that of the roll-mixed inclusion complexes, whereas the halo pattern of the roll-treated β CD alone exhibited an inner peak deviating a little at around 13°.

Tolbutamide-β CD System

Guest compound TB is considered to complex with β CD through the results obtained in the dissolution test described above. The complex ratio of the inclusion compound prepared from aqueous solution exhibited a stoichiometrically definite value of around 2.5 in the mixing ratio from 1.0 to 3.0, whereas the complex ratio of the roll-mixtures reached unity at roughly both mixing ratio 1.0 and 0.5. These results suggest that a characteristic equimolar complex was formed in the powder state of the roll mixtures.

Indomethacin-β CD Systems

IM-β-CD complexing in aqueous solution was studied by solubility, ultraviolet, circular dichroism, spectroscopies, H^1-NMR, and kinetic method.[16] A certain IM complex is also detected in the powder roll-mixture. The complex ratio increased up to 1.0 with increasing the roll-mixing periods.

The results so far obtained consequently suggest the presence of novel high-energy complexes of drugs with β CD, which is not able to be obtained in the wetted system. Mechanical powder roll-mixing is a favorable tool for easily preparing molecular capsules with guest compounds capable of providing inclusion complexes.

It can be expected that techniques for molecular complexing are widely applicable to insoluble and labile medicinal drugs for pharmaceutics, to inorganic compounds, to liquid crystals for chemical technology, to nutrition and flavor enhancers for food processing, to perfume and coloring agents for cosmetics manufacturing, and to fertilizer and feeds for agricultural promotion and environmental preservation.

REFERENCES

1. Kondo, T.; Koishi, M. *Microcapsules—Preparation, Properties and Applications*; Sankyo Shuppan: Tokyo, 1977.
2. Kondo, T. *Microcapsules—Their Preparation and Properties* In "*Surface and Colloid Science*" Matijevic, E., Ed.; Plenum: New York, 1978; Vol. 10.
3. Kondo, A. *Microcapsule Process and Technology* Marcel Dekker: New York, 1979.
4. Lim, F. *Biomedical Applications of Microencapsulation* CRC: Boca Raton, FL, 1984.
5. Deasy, P. B. *Microencapsulation and Related Drug Processes* Marcel Dekker: New York, 1984.
6. Gucho, M. H. *Capsule Technology and Microencapsulation* Noyes Data: Park Ridge, NJ, 1972; *Microcapsules and Microencapsulation Techniques* Noyes Data: Park Ridge, NJ, 1976; *Microcapsules and Other* Capsules Noyes Data: Park Ridge, NJ, 1979.
7. Stockenius, W. *J. Biophys. Biochem. Cytol.* **1959**, *5*, 491.
8. Atwood, J. L.; Davies, J. E. D.; MacNicol, D. D. *Inclusion Compounds*, Academic: London, 1984; Vols. 1–3.
9. Cramer, F.; Henglein, F. M. *Chem. Ber.* **1957**, *90*, 2561.
10. Torricelli, C.; Martini, A.; Muggetti, L.; Eli, M.; De Ponti, R. *Int. J. Pharm.* **1991**, *75*, 147.
11. Nakai, Y.; Yamamoto, K.; Terada, K.; Watanabe, D. *Chem. Pharm. Bull.* **1987**, *35*, 4609.
12. Nozawa, Y.; Yamamoto, A. *Pharm. Acta Helv.* **1989**, *64*, 24.
13. Pauli, W. A.; Lach, J. L. *J. Pharm. Sci.* **1965**, *54*, 1745.
14. Manor, P. C.; Saenger, W. *J. Am. Chem. Soc.* **1974**, *96*, 3630.
15. Lindner, K.; Saenger, W. *Carbohyd. Res.* **1982**, *99*, 103.
16. Hoshino, T.; Tagawa, Y.; Hirayama, F.; Otagiri, M.; Uekama, K. *Yakagaku Zasshi* **1982**, *102*, 1184.

Microcontainers

See: Reverse Micelles (Microcontainers for Functional Polymers)

Microelectronics Packaging Materials

See: Poly(imide siloxane)s

Microemulsion Polymerization

See: Flocculants (Organic, Overview)
Membranes, Porous (Synthesized from Microemulsions)

MICROEMULSION POLYMERIZATION (Overview)

Leon Ming Gan* and C. H. Chew
Department of Chemistry
National University of Singapore

Unlike emulsion polymerization, which has been extensively studied over many decades,[1,2] microemulsion polymerization became known only about 1980 as a result of studies on microemulsions following the 1974 oil crisis. The goals of microemulsion polymerization are to produce stable microlatexes (d<100 nm) of uniform sizes and microporous polymeric materials.[3,4] In addition, the thermodynamic stability and the optical transparency of microemulsions are advantageous for photochemical or other reactions.

Microemulsions are transparent liquid systems consisting of water, oil and surfactant or cosurfactant, which is usually a medium-chain-length alcohol, such as pentanol. Cosurfactants are commonly used in the formation of microemulsions that are stabilized by single-chain anionic surfactants (such as sodium dodecylsulfate). However, the presence of a cosurfactant may not be necessary for microemulsions using nonionic surfactants; some cationic surfactants; or double-chain surfactants. A typical example of a double-chain surfactant is sodium 1,4-bis(2-ethylhexyl)sulfosuccinate which is commonly known as Aerosol OT (AOT). When appropriate amounts of oil, water and surfactant with or without a cosurfactant are mixed, thermodynamically stable microemulsions form spontaneously.[5,6]

Depending on the proportion of components and the HLB (hydrophile-lipophile balance) value of surfactant used, the formation of microemulsion droplets can be swollen with oil and dispersed in water (normal or o/w microemulsions), or swollen with water and dispersed in oil (inverse or w/o microemulsions). In the intermediate regions between o/w and w/o microemulsions, there may exist bicontinuous microemulsions whose aqueous and oily domains are interconnected randomly to form sponge-like structures.

POLYMERIZATION

Water-in-Oil (Inverse) Microemulsions

Basic w/o microemulsions are formed from at least four components, even in the absence of cosurfactant; i.e., water, monomer, oil and surfactant. Water–soluble monomers, such as acrylamide (AM), acrylic acid (AA), 2-hydroxyethyl methacrylate (HEMA) and urea-formaldehyde (UF), can be polymerized in water-swollen droplets of w/o microemulsions. These water-soluble monomers also play the role of cosurfactants, thus leading to an increase of micellar solubilization capacity of a w/o microemulsion.

*Author to whom correspondence should be addressed.

Microemulsions are commonly stabilized with a double-chain anionic surfactant AOT. The most investigated monomer in w/o microemulsions is AM because of wide applications of the corresponding polymer.[12,13] The AOT inverse microemulsions require a high ratio (>2.5) of surfactant to monomer for producing clear or slightly bluish microlatexes.

The w/o microemulsions can also be formulated with a polymerizable surfactant (SAAS or PUD), cosurfactant (AA or HEMA) and oil (MMA in the continuous medium). With the water content between 10 wt% and 20 wt%, these w/o microemulsions can be copolymerized to form transparent polymeric solids with microporous structures (close-cell type). In addition, the systems can also be incorporated with a crosslinker (EGDMA) to enhance the stability of the microstructures.

Oil-in-Water Microemulsions

With a suitable three-component systems, the polymerization study of monomer can be carried out in both emulsions and microemulsions using three identical components: monomer, surfactant and water. The ternary system can be continuously changed from turbid emulsions to transparent microemulsions by only increasing the surfactant concentration at a fixed weight ratio of monomer to water. The first report on such a study is the comparison of MMA polymerization in emulsions and microemulsions which were stabilized by CTAB.[7]

Stable PMMA latexes of 16 nm-30 nm (in R_h) were prepared from the above-mentioned emulsion and microemulsion polymerization initiated by KPS. Molecular weights of PMMA were in the range of $5\text{-}7 \times 10^6$.

Winsor I-Like Systems

In a recent study, a new system associated with an o/w microemulsion has been successfully used for polymerizing styrene up to 15 wt% using only about 1 wt% surfactant.[9]

This new system is rather similar to a Winsor I system; i.e., an organic phase containing small portions of water and surfactant is in equilibrium with an o/w microemulsion. The slight difference between two systems is that the new system consists of only a pure styrene phase (upper phase) which is placed on the top of a ternary o/w microemulsion (lower phase). Strictly speaking, this new system is not identical to the Winsor I system. Hence, we refer to this new polymerization system as "Winsor I-like system." It can be prepared simply by topping off a ternary microemulsion with a certain amount of styrene without disturbing both phases.

Polystyrene particles formed in the microemulsion phase are viewed as "seeds" for further growth of polymer particles by continuously recruiting monomer from the styrene phase.

PROPERTIES

Dimensions of Latex Particles

The size of latex particles has usually been determined by quasi-elastic light scattering (QELS) and transmission electron microscope (TEM).

The latex particles in all cases are larger than those of the initial microemulsion droplets (r_h<5 nm). They are around 10–20 nm for latexes prepared in w/o and o/w microemulsions.

However, they are bigger (25–50 nm) for those prepared in bicontinuous microemulsions and in Winsor I-like systems.

The preparation of monodisperse polymer particles by the conventional microemulsion polymerization technique seems to be less encouraging due to the continuous nucleation of polymer particles throughout the whole course of polymerization.[8,11,16] But the new "Winsor I-like" process[9] has potential for preparing near monodisperse polymer particles.

Molecular Weights

The molecular weights of various polymers prepared in microemulsions are generally above one million because of a low radical concentration in each microlatex particle ($n<0.5$). A decrease in polymer molecular weight is usually observed upon increasing the initiator concentration, raising the polymerization temperature or adding an alcohol cosurfactant.[15] Higher molecular weights of polymers are obtained for larger monomer/surfactant ratios,[14,17] a similar trend for particle size.[14,17]

Porosity of Polymeric Materials

The polymerization of a microemulsion containing a polymerizable oil continuous phase, a polymerizable surfactant and/or cosurfactant can produce porous solid polymers of micron or submicron sizes. The morphology and porosity of microstructures are affected by the type and the concentration of surfactant and/or cosurfactant.

The water content in the precursor microemulsions affects the morphology of the resulting polymeric materials. The open-cell structures of polymeric solids can be obtained from those precursor bicontinuous microemulsions with water contents higher than 20 wt% up to about 50 wt%. When the water contents in the similar systems are about 12–20 wt%, it is likely to obtain polymeric materials with close-cell structures.

APPLICATIONS

Polymerization in microemulsions provides a novel technique for the production of low viscosity, optical transparency and high stability of microlatexes of high molecular weight polymers. The drawback of w/o and o/w microemulsion polymerization is the need for a high surfactant concentration to stabilize a low polymer content. The problem has been overcome recently by polymerizing monomers of much higher contents in bicontinuous microemulsions or in Winsor I-like systems using much lower amounts of surfactant. The ultimate goal is to use a polymerizable surfactant that can be copolymerized with the monomer to form an inherent stable latex of uniform size or a solid polymeric material with microstructures.

Microlatexes

Thermodynamically stable microlatexes in the nano size range (20–30 nm) seem to hold promising applications in the field of microencapsulation and targeted delivery of drugs.[18–21]

Polymers

Most of the applications for high molecular weight polymers prepared in emulsions can, in principle, be extended to the microemulsion polymerization process.[22,23] For the polymers with core-shell type structures, they may be better prepared by the polymerization in Winsor I-like systems.

The synthesis of porous polymeric materials (membranes) by the polymerization in bicontinuous microemulsions could have a significant technological importance.

REFERENCES

1. *Emulsion Polymerization;* Piirma, I., Ed.; Academic: New York, 1982.
2. El-Aasser, M. S. *Scientific Methods for the Study of Polymer Colloids and Their Applications;* Candau, F.; Ottewill, R. H., Eds.; Kluwer: London, 1990; Chapter 1; Napper, D. H.; Gilbert, R. G. *Scientific Methods for the Study of Polymer Colloids and Their Applications;* Candau, F.; Ottewill, R. H., Eds.; Kluwer: London, 1990; Chapter 1.
3. Caudau, F. *Encyclopedia of Polymer Science and Engineering*, 2nd Ed.; Mark, H.; Bikales, N. M.; Overberger, C. G.; Menges, G., Eds.; John Wiley & Sons: New York, 1987; Vol. 9, pp 718–724.
4. Candau, *E Polymerization in Organized Media;* Gordon & Breach: Paris, 1992; Chapter 4.
5. *Microemulsions: Structure and Dynamics;* Friberg, S. E.; Bothorel, P., Eds.; CRC: New York, 1986.
6. *Microemulsions and Related Systems;* Bourrel, M.; Schechter, R. S., Eds.; Dekker: New York, 1988.
7. Gan, L. M.; Chew, C. H.; Ng, S. C.; Loh, S. E. *Langmuir* **1993**, *9*, 2799.
8. Gan, L. M.; Chew, C. H.; Lee, K. C.; Ng, S. C. *Polymer* **1994**, *35*, 2659.
9. Gan, L. M.; Lian, N.; Chew, C. H.; Li, G. Z. *Langmuir* **1994**, *10*, 2197.
10. Gan, L. M.; Chieng, T. H.; Chew, C. H.; Ng, S. C. *Langmuir* **1994**, *10*, 4022.
11. Candau, E; Yong, Y. S.; Fitch, R. M. *J. Polym. Sci., Polym. Chem. Ed.* **1985**, *23*, 193.
12. Candau, F. *Scientific Methods for the Study of Polymer Colloids and Their Applications;* Candau, F.; Ottewill, R. H., Eds.; Kluwer: London, 1990; Chapter 3.
13. Thomas, W. M.; Wang, D. W. *Encyclopedia of Polymer Science and Engineering*, 2nd ed.; Mark, H.; Bikales, N. M.; Overberger, C. G.; Menges, G., Eds.; John Wiley & Sons: New York, 1985; Vol. 1, pp 169–211.
14. Gan, L. M.; Chew, C. H.; Lye, I. *Makromol. Chem.* **1992**, *193*, 1249.
15. Gan, L. M.; Chew, C. H.; Lye, I.; Ma, L.; Li, G. *Polymer* **1993**, *34*, 3860.
16. Guo, L. M.; Sudol, E. D.; Vanderhoff, J. W.; El-Aasser, M. S. *J. Polym. Sci., Polym. Chem. Ed.* **1992**, *30*, 691, 703.
17. Candau, F.; Buchert, P. *Colloids Surf.* **1990**, *48*, 107.
18. Rembaum, A.; Dreyer, W. J. *Science* **1980**, *208*, 364.
19. Gasco, M. R.; Trotta, M. *Int. J. Pharm.* **1986**, *29*, 267.
20. Cadic, Ch.; Dupuy, B.; Baquez, Ch.; Ducassou, D. *Innov. Techn. Biol. Med.* **1990**, *11*, 412.
21. Vauthier-Holtzscherer, C.; Benabbou, S.; Spenlehauer, G.; Veillard, M.; Couvreur, P. *S.T.P. Pharm. Sci.* **1991**, *1*, 109.
22. Kozakiewicz, J. J.; Lipp, D. W. U.S. Patent 285 938, 1988.
23. Candau, F.; Buchert, P.; Esch, M. French Patent 8 817 306, 1988.

MICROEMULSION POLYMERIZATION (Functionalization of Microlatexes)

Markus Antonietti
Max Planck Institut für Kolloid und Grenzflächenforschung

WHICH SURFACTANTS ALLOW FORMATION OF A POLYMERIZABLE MICROEMULSION?

Most microemulsions described in the literature are based on the fragile SDS/pentanol or the cationic DTAB or CTAB system. It is obvious that there is great interest in finding new surfactant systems which also allow the formation of polymerizable microemulsions, such as surfactants which allow the synthesis of even smaller latexes or surfactants which work more effectively for interface stabilization.

Functionalization of the Particle Surface

It has been shown that polymerization in microemulsions allows the synthesis of ultrafine polymer lattices in a size range between 5 nm < R < 50 nm. These lattices are well-defined with respect to their size and their chemical composition. The very large interface of such lattices of up to 300 Hm^2/g is easily modified by simple copolymerization with functional comonomers, thus ending in surfaces with high densities of functional groups.[3,4] This convenient, one-step surface functionalization results from the fact that the more polar functional monomers are enriched at the oil/water interface and even act as a cosurfactant.

For special target functionalities, the surfactant system for microemulsification, as well as the type of comonomer wearing the appropriate functional group, must be mutually chosen. This was recently demonstrated for the synthesis of bipyridine-functionalized latex particles.[5] An advantage to the functional copolymerization in microemulsions is that the functionalities may be polymers themselves, such as block copolymers[4] or proteins,[2] a technique which was called "modular functionalization"[1] and which results in hierarchical polymer superstructures.

For modular functionalization, the incorporated building blocks must again be sufficiently surface active or amphilic. It has been shown that polystyrene/poly(4-vinyl-pyridine)-block copolymers are willingly incorporated into the standard system styrene and crosslinker/CTAB/water,[3,4] when a majority of the pyridine units is protonated.

Such a latex particle possesses polyvinylpyridinium tentacles (ca. 3000) protruding from the surface; these chains are anchored by the polystyrene blocks that are embedded in the glassy polystyrene core.

These special microlattices, when neutralized, effectively bind ions of transition- and heavy metals at the interface via complexation (due to neighborhood effects much stronger than a low molecular weight complexer); standard cations such as sodium or calcium remain unconfined.

Synthesis of Specially Functionalized Polymers in Microemulsions

The polymerization in the microenvironments of the parental microdroplets and the presence of the large interface between two phases gives rise to effects that allow the synthesis of special functional polymer systems. For instance, new copolymers, such as systems with nonstandard copolymerization parameters, hydrophilic-hydrophobic copolymers and polyampholytes with very high charge density, can be synthesized.

REFERENCES

1. Antonietti, M.; Basten, R.; Lohmann, S., *Macromol. Chem. Phys.*, accepted.
2. Antonietti, M.; Nestl, T., *Macromol. Rapid Comm.* **1994**, *15*, 111.
3. Antonietti, M.; Lohmann, S.; van Niel, C., *Macromolecules* **1992**, *25*, 1139.
4. Antonietti, M.; Bremser, W.; Lohmann, S., *Prog. Coll. Polym. Sci.* **1992**, *89*, 62.
5. Antonietti, M.; Lohmann, S.; Eisenbach, C., *Macromol. Chem. Phys. Rapid Comm.*, accepted.

MICROEMULSION POLYMERIZATION (Oil-in-Water)

Jorge E. Puig
Departamento de Ingeniería Química
Universidad de Guadalajara

Emulsion, miniemulsion and, more recently, microemulsion polymerization are processes that can produce polymeric particles of large molecular weights (10^5 to 10^7) and small sizes (10–1000 nm) dispersed in aqueous or nonaqueous media.[1–6] In these processes fast reaction rates and large molecular weights can be achieved simultaneously because free radicals that initiate and propagate the reaction are isolated in small loci.[1,4]

Polymerization in microemulsions has been achieved with monomers located in either the continuous[7,8] or the dispersed domains[9–22] and in the bicontinuous region.[23] Polymerization in oil-in-water (o/w) microemulsions has some advantages over emulsion polymerization. Emulsions are turbid and opaque, but microemulsions are usually transparent or translucent and so are particularly suitable for photochemical reactions. Microemulsion polymerization also allows the preparation of stable, monodisperse microlatexes containing particles an order of magnitude smaller than those produced with classical emulsion polymerization. A drawback of microemulsion polymerization, however, is the large amount of surfactant required for microemulsion formation.

KINETICS OF MICROEMULSION POLYMERIZATION

Although both emulsion and microemulsion polymerization processes can produce polymeric particles with fast reaction rates and large molecular weights (> 10^6) dispersed in an aqueous media, their kinetics differ.

In microemulsion polymerization, only two rate intervals are observed.[11,16,17,20,21,24] In interval 1 the polymerization rate increases with time because the number of propagation sites increases. During this interval the diffusive transport of monomer from uninitiated droplets maintains a constant concentration of monomer in the growing particles. The nucleation stage ends when all the microemulsion droplets have disappeared either by becoming dead polymer particles or by diffusion of monomer to the monomer-swollen reacting particles. Once the uninitiated droplets vanish, reaction rate slows down steadily (interval 2) since the concentration of monomer within each particle decreases as the reaction proceeds. No apparent constant rate period nor gel effect are observed in microemulsion polymerization, in contrast to conventional emulsion polymerization.[25] Other researchers favor the mechanism of continuous nucleation.[26,27]

Effect of Monomer Concentration

Most of the monomer in emulsion polymerization is in the emulsified droplets, which do not play any role other than as a source of monomer. Hence, the initial monomer concentration has little or no influence on reaction rate in this process.[1] In microemulsion polymerization, both reaction rate and conversion increase with increasing monomer concentration.[8,9,12,16,24] The increase in the rate of polymerization is due to the increasing size of the microemulsion droplets with increasing monomer concentration.

Effect of Temperature

The effect of temperature on microemulsion polymerization rate has been examined for styrene[11,16] and for MMA[24] using water-soluble or water-insoluble initiators. As expected, overall reaction rates are faster with increasing temperature for both types of initiators because the rate of initiator decomposition grows rapidly with temperature. Final conversion also augments with increasing temperatures because the mobility of the macromolecules increases with temperature, especially when the reaction temperature exceeds the glass transition of the polymer-monomer mixture in the reacting loci.

Effect of Initiator

The polymerization rate in microemulsion systems is very sensitive to the type and concentration of initiator and surfactant. Both polymerization rate and overall conversion increase with increasing initiator concentration.[11,16,21,24] This is the result of the larger flux of free radicals into the swollen micelles.

Effect of Electrolytes

The effect of inorganic electrolytes in emulsion polymerization is complex and depends on several factors such as nature and concentration of electrolyte, type and concentration of surfactant, type of initiation, etc.[30–32] Usually, electrolytes are added to increase the ionic strength of the aqueous solution (or to adjust the pH), and thereby to increase the average particle size and to broaden the particle size distribution in the final latex. However, addition of electrolytes may cause emulsion instability during the reaction.

The effect of adding electrolytes on kinetics of microemulsion polymerization has been examined.[28,29,33] Addition of electrolytes significantly alters the phase behavior of water-oil-surfactant systems et al.[34,35]

Effect of Alcohol

Alcohols dramatically modify the phase behavior of oil-water-surfactant systems.[35]

CONCLUSION

Microemulsion polymerization is a new, challenging and rapidly developing field with potential applications in paints, adhesives, microencapsulation, hydrogels, etc. One of the main problems to overcome in microemulsion polymerization is to find better formulations to reduce the large ratio of surfactant/monomer(s) required by this process.

REFERENCES

1. Piirma, I. *Emulsion Polymerization;* Academic: New York, 1976.
2. Chamberlain, B. J.; Napper, D. H.; Gilbert, R. G. *J. Chem. Soc. Faraday Trans. 1* **1982**, *78*, 591.
3. Choi, Y. T.; El-Aasser, M. S.; Sudol, E. D.; Vanderhoff, J. W. *J. Polym. Sci. Polym. Chem. Ed.* **1985**, *23*, 2973.
4. Candau, F. In *Encyclopedia of Polymer Science and Engineering;* Mark, H. E; Bikales, N. M.; Overberger, C. G.; Menges, G. Eds.; John Wiley & Sons: New York, 1987; Vol. 9.
5. Candau, F. In *Polymerization in Organized Media;* Paleos, C. M. Ed.; Gordon & Breach Sci.: Philadelphia, 1992.
6. Dunn, A. S. In *Comprehensive Polymer Science;* Eastwood, G. C.; Ledwith, A.; Sigwalt, P., Eds.; Pergamon: New York, 1988.
7. Stoffer, J. O.; Bone, T. *J. Dispersion Sci. Technol.* **1980**, *1*, 37.
8. Stoffer, J. O.; Bone, T. *J. Polym. Sci. Polym. Chem. Ed.* **1980**, *18*, 2641.
9. Candau, F.; Leong, Y. S.; Pouyet, G.; Candau, S. *J. Colloid Interface Sci.* **1984**, *101*, 167.
10. Candau, F.; Leong, Y. S.; Fitch, R. *J. Polym. Sci. Polym. Chem. Ed.* **1985**, *23*, 193.
11. Guo, J. S.; El-Aasser, M. S.; Vanderhoff, J. W. *J. Polym. Sci. Polym. Chem. Ed.* **1989**, *27*, 691.
12. Pérez-Luna, V. H.; Puig, J. E.; Castaño, V. M.; Rodriguez, B. E.; Murthy, A. K.; Kaler, E. W. *Langmuir* **1990**, *6*, 1040.
13. Feng, L.; Ng, K. Y. S. *Macromolecules* **1990**, *23*, 1048.
14. Feng, L.; Ng, K. Y. S. *Colloids Surfaces* **1991**, *53*, 349.
15. Gan, L. M.; Chew, C. H., Lye, I.; Imae, T. *Polym. Bull.* **1991**, *25*, 193.
16. Gan, L. M.; Chew, C. H.; Lye, I. *Makromol. Chem.* **1992**, *193*, 1249.
17. Gan, L. M.; Chew, C. H.; Lee, K. C.; Ng, S. C. *Polymer* **1993**, *34*, 3064.
18. Gan, L. M.; Chew, C. H.; Lye, I.; Ma, L.; Li, G. *Polymers* **1993**, *34*, 3860.
19. Texter, J.; Oppenheimer, L.; Minter, J. R. *Polym. Bull.* **1992**, *27*, 487.
20. Full, A. P.; Puig, J. E.; Gron, L. U.; Kaler, E. W.; Minter, J. R.; Mourey, T. H.; Texter, J. *Macromolecules* **1992**, *25*, 5157.
21. Puig, J. E.; Pérez-Luna, V. H.; Pérez-González, M.; Macías, E. R.; Rodríguez, B. E.; Kaler, E. W. *Colloid Polym. Sci.* **1993**, *271*, 114.
22. Rodríguez-Guadarrama, L. A. M.S. Thesis, Universidad de Guadalajara, Mexico, 1992.
23. Haque, E.; Qutubuddin, S. *J. Polym. Sci. Polym. Lett. Ed.* **1988**, *26*, 429.
24. Rodríguez-Guadarrama, L. A.; Mendizabal, E.; Puig, J. E.; Kaler, E. W. *J. Applied Polym. Sci.* **1993**, *48*, 775.
25. Odian, G. *Principles of* Polymerization; John Wiley & Sons: New York, 1981.
26. Guo, J. S.; Sudol, E. D.; Vanderhoff, J. W.; El-Aasser, M. S. *J. Polym. Sci. Polym. Chem. Ed.* **1992**, *30*, 691.
27. Bléger, F.; Murthy, A. K.; Pla, F.; Kaler, E. W. *Macromolecules* **1994**, *27*, 2559.
28. Full, A. P.; Kaler, E. W.; Ceja, J.; Puig, J. E. *Macromolecules* **1996**, accepted.
29. Arellano, J. M.S. Thesis, University of Guadalajara, Mexico, 1994.
30. Mateo, J. L.; Cohen, I. *J. Polym. Sci. Polym. Chem. Ed.* **1964**, *2*, 711.
31. Dunn, A. S.; Al-Shahib, W. A. G. R. *Br. Polym.* **1978**, *10*, 137.
32. Dunn, A. S.; Said, Z. F. M. *Polymer* **1982**, *23*, 1172.
33. Puig, J. E.; Full, A. P.; Kaler, E. W. *Soc. Platics Engs. Techn. Papers* **1992**, *38*, 2587.
34. Kahlweit, M.; Lesner, E.; Strey, R. *J. Phys. Chem.* **1983**, *87*, 5032.
35. Kahlweit, M.; Lesner, E.; Strey, R. *J. Phys. Chem.* **1984**, *88*, 937.

Microemulsions

See: Globular Polybases, Microemulsion-like

MICROENCAPSULATED PARTICLES

Yasuo Hatate* and Hidekazu Yoshizawa
Department of Applied Chemistry and Chemical Engineering
Faculty of Engineering
Kagoshima University

Encapsulated particles are used in cosmetics, paints, drugs and other products.[1–3] Polymeric materials play various important roles through encapsulation. For example, they isolate containing particles from the reactive factors in the ambiente for the protection, preservation and/or easy treatment; they control the release rate of the particles (e.g., drugs) by using the various films and fix the particles (e.g., pigments) to the paper by melting.[4–11] This chapter focuses on the solid state materials (=particles), which dominate the key characteristics of the polymeric particles. A few such polymer particles from recent publications will be introduced as illustrations of particle–encapsulated polymeric spheres.

CARBON BLACK ENCAPSULATED POLYMERIC PARTICLES

Toner powder (copying ink), which usually consists of particles 5–15 μm thick containing carbon black, is important in plain-paper copiers because precise copies can be obtained only by using good toner with a uniform particle size of 5–10 μm and reasonable electrostatic charge of 15–40 $\mu C \cdot g^{-1}$.

STEARIC ACID ENCAPSULATED POLYMERIC PARTICLES

Many studies of sustained release of active agent from polymeric particles have been performed since microcapsules and microspheres for medical applications were introduced by Chang.[12] The sustained release rate has been known to depend heavily on preparation conditions of polymeric particles.[13,14]

We prepared stearic-acid-encapsulated polymeric particles by *in situ* polymerization of styrene and divinylbenzene.

CISPLATIN-ENCAPSULATED POLYMERIC PARTICLES

Recently, many investigations on microencapsulation with various biodegradable polymers have been proposed to overcome the serious side effects of antitumor agents. The objectives of microencapsulation are to keep an optimum drug concentration in tissue and to localize the drug delivery to malignancy cells.[15–18] Cisplatin (CDDP) is one of the most potent antitumor agents, and physiological saline-dissolving CDDP is in clinical use today. Whereas serious side effects are also known to be caused by CDDP, microencapsulation of cisplatin is a useful technique to reduce the drug's side effects. However, current CDDP encapsulated polymeric particles have a significant initial burst.[19–21] The preparation of CDDP encapsulated polymeric particles whose initial burst is diminished is expected in clinical use. We investigated the mechanism of the initial burst in CDDP encapsulated polylactide particles and prepared CDDP encapsulated polymeric particles without an initial burst.[22]

*Author to whom correspondence should be addressed.

REFERENCES

1. Kondo, T.; Koishi, M. *Shinpan Maikurokapuseru-Sono-Seiho-Seishitsu Oyo*, 2nd ed.; Sankyo Shuppan Co.: Tokyo, 1989.
2. Vandegaer, J. E. *Microencapsulation Processes and Applications;* Plenum: New York, NY, 1974.
3. Yoshizawa, H.; Hatate, Y. *Chemical Engineering* **1993**, *38*, 405.
4. Cremers, H. F. M.; Kwon, G.; Bae, Y. H.; Kim, S. W.; Verrijk, R.; Noteborn, H. P. J. M.; Feijen, J. *Biomaterials* **1994**, *15*, 38.
5. Iso, M.; Shirahase, T.; Hanamura, S.; Urushiyama, S.; Omi, S. *J. Microencapsulation* **1989**, *6*, 165.
6. Kim, Y. H.; Bae, Y. H.; Kim, S. W. *J. Controlled Release* **1994**, *28*, 143.
7. Kokufuta, E.; Shimizu, N.; Nakamura, I. *Biotech. Bioeng.* **1988**, *32*, 289.
8. Kwon, G. S.; Bae, Y. H.; Cremers, H.; Feijen, J.; Kim, S. W. *J. Controlled Release* **1992**, *22*, 83.
9. Longo, W. E.; Iwata, H.; Lindheimer, T. A.; Goldberg, E. P. *Polymer Preprints* **1983**, *24*, 56.
10. Watarai, H.; Hatakeyama, S. *Analy. Sci.* **1991**, *7*, 487.
11. Yoshizawa, H.; Uemura, Y.; Kawano, Y.; Hatate, Y. *J. Chem. Eng. Japan* **1993b**, *26*, 692.
12. Chang, T. M. S. *Science* **1964**, *146*, 524.
13. Hatate, Y.; Hamada, H.; Nagata, H.; Imafuku, T. *Kagaku Kogaku* **1987**, *51*, 519.
14. Hatate, Y.; Hamada, H.; Nagata, H.; Sakaki, T.; Kasamatsu, K.; Ikari, A.; Nakashio, F. *Reports of the Asahi Glass Foundation for Industrial Technology* **1988**, *52*, 247.
15. Ohya, Y.; Takei, T.; Kobayashi, H.; Ouchi, T. *J. Microencapsulation* **1993**, *10*, 1.
16. Shimoda, M.; Aikou, T.; Natsugoe, S.; Shimazu, H.; Nakamura, K. *Anti-Cancer Drug Design* **1993**, *8*, 127.
17. Tsai, D. C.; Howard, S. A.; Hogan, T. F.; Malanga, C. J.; Kandzari, S. J.; Ma, J. K. H. *J. Microencapsulation* **1986**, *3*, 181.
18. Wada, S. Doctoral Thesis, Kyoto University, Kyoto, Japan 1991.
19. Hagiwara, A.; Takahashi, T.; Sasabe, T.; Itoh, M.; Yoneyama, C.; Shimotsuma, M.; Iwamoto, A.; Wada, R.; Hyon, S.-H.; Ikada, Y.; Kusanoi, Y.; Muranishi, S. *Igaku No Ayumi* **1990**, *152*, 613.
20. Ike, O.; Shimizu, Y.; Wada, R.; Hyon, S.-H.; Ikada, Y. *Biomaterials* **1992**, *13*, 230.
21. Sakakura, C.; Takahashi, T.; Hagiwara, A.; Itoh, M.; Sasabe, T.; Lee, M.; Shibayashi, S. *J. Controlled Release* **1992**, *22*, 69.
22. Yoshizawa, H.; Nagai, H.; Uemura, Y.; Tokuda, K.; Kumanohoso, T.; Natsugoe, S.; Aikou, T.; Hatate, Y. *Proc. Taipei-Kyushu Symp. Chem. Eng.* **1994**, 225.

MICROENCAPSULATION (Artificial Cells)

Thomas Ming Swi Chang
Artificial Cells and Organs Research Centre
Faculty of Medicine
McGill University

Artificial cells prepared from microencapsulation of biologically active materials were first reported in 1957 and 1964.[1–3] They were first prepared as ultrathin polymer membranes of cellular dimensions microencapsulating the proteins and enzymes extracted from biological cells.[23] This was followed by the encapsulation of biological cells, adsorbents, magnetic materials, drugs, vaccines, hormones, and many other biologically active materials for applications in biotechnology and medicine.[3–7]

Like biological cells, artificial cells contain biologically active materials. However, the content of artificial cells can be more varied than biological cells. The membranes of artificial cells also can be extensively varied using synthetic or biological materials.

METHODS OF PREPARATION

Principles of Methods of Preparations

Many methods are available to prepare artificial cells. The most commonly used approaches are based on the following principles. Small artificial cells in the micron dimensions are prepared by emulsification procedures that are usually modifications of the original basic procedures.[2–6] Here materials for microencapsulation are dissolved or suspended in an aqueous solution. An emulsion is formed, and membranes are formed on the surface of each microdroplet. The microcapsules are resuspended in an aqueous medium. Smaller artificial cells of nanometer dimensions are formed based on the same principles except the initial emulsions formed are much smaller. Larger artificial cells, especially those in the millimeter dimensions, are prepared based on modifications of the original drop method.[2,5–7]

These are generally used to microencapsulate cells or microorganisms in tissue engineering. A spray technique can also be used to encapsulate particulate matter.[2]

Membranes Used for Encapsulation

Different types of synthetic polymers are used, and variations in configuration are possible.[10] A single, ultrathin polymer membrane is the most common one. The unlimited variations in polymers available allow for possible variations in permeability, biocompatibility, and other characteristics.[2–11] Artificial cells can also be made to contain smaller "intracellular compartments."[5–7]

MICROENCAPSULATION OF BIOACTIVE SORBENTS

Bioactive sorbents represent the simplest form of artificial cells already used in routine clinical applications for humans. Sorbents such as activated charcoal, resins, and immunosorbents could not be used in direct blood perfusion because particulate embolism and blood cells were removed. However, sorbents such as activated charcoal inside artificial cells no longer cause particulate embolism and blood cells removal.[4,8,9] This application was developed and used successfully in patients.[9,12,16]

Clinical Uses

Microencapsulation has become a routine treatment for adult and pediatric patients who have been poisoned.[16,17] This applied to the many cases in which the toxin can be adsorbed by activated charcoal.

Microencapsulation of Immunosorbents

Immunosorbents, like other sorbents described above, also have problems when they contact blood directly. The same ultrathin coating used in sorbents has been applied to immunosorbents to prevent these problems.[18] This has been tested clinically in patients.[10]

MICROENCAPSULATION OF CELLS OR MICROORGANISMS

The first encapsulation of biological cells, which was based on a drop method, was reported in 1965.[5] It was proposed that "...protected from...immunological process...encapsulated endocrine cells might survive and maintain an effective supply of hormone. For organ deficiency...cultures of liver cells...in artificial cells".[5] This original drop method for cell encapsulation involves chemically crosslinking the surface of aqueous droplets that contain cells.[5,6] This was modified into the following drop technique that uses milder physical crosslinking, which resulted in alginate–polylsine–alginate (APA) microcapsules containing cells.[19,20]

Microencapsulated Islets for Diabetes Mellitus

The research showed that islets inside artificial cells are prevented from immunorejection after implantation into animals. Islets can remain viable and continued to secrete insulin to control the glucose levels of diabetic rats.

Microencapsulated Hepatocytes for Liver Failure

We found that artificial cells containing hepatocytes increased the survival time of fulminant hepatic failure in rats.[21]

Microencapsulation of Genetically Engineered Microorganisms

We studied microencapsulated genetically engineered microorganisms by using *Escherichia coli* DH5 cells with the *Klebsiella aerogens* urease gene.[22,23] Overall, urea removal efficiency of microencapsulated genetically engineered bacteria is 10–30 times higher than the best available urea removal systems available.

Microencapsulation of Cholesterol-Removing Microorganisms

We selected *Pseudomonas pictorum* (ATCC #23328) as another model system because it can degrade cholesterol.[24,25]

MICROENCAPSULATION OF ENZYMES AND MULTIENZYME SYSTEMS

Artificial cells protect the enclosed enzyme from immunological rejection or tryptic enzymes.[3–8]

Urea Removal

We showed that artificial cells containing urease can convert urea to ammonia, which is then removed by ammonia adsorbent.[3–8]

Enzyme Therapy

Artificial cells have been used in hereditary enzyme defects, including our earliest use of a replacement for catalase in acatalasemic mice.[26] This also has been studied for asparagine removal in the treatment of leukemia in animals.[27]

Multienzyme System

We have prepared artificial cells that contain multienzyme systems with cofactor recycling.[28] This approach can convert metabolic wastes such as urea and ammonia into essential amino acids such as leucine, isoleucine, and valine, which are required by the body.[29]

RED BLOOD CELL SUBSTITUTES

Two major approaches exist: modified hemoglobin and perfluorochemicals. Detailed reviews in the field are available.[30,31]

BIODEGRADABLE ARTIFICIAL CELLS

Biodegradable artificial cells also are used for drug delivery. We have used crosslinked protein and biodegradable polylactide artificial cells.[3,5–8,13] Many groups are extending these approaches for drug delivery use (e.g., medications, hormones, peptides, and proteins).[10,14,15]

FUTURE PERSPECTIVES

The author wrote in his 1972 book on artificial cells: "'Artificial cell' is a concept; the examples described...are but physical examples for demonstrating this idea. In addition to extending and modifying the present physical examples, completely different systems could be made available to further demonstrate the clinical implications of the idea of 'artificial cells.'"[8] An entirely new horizon is waiting impatiently to be explored. This future perspective is even more valid now.

REFERENCES

1. Chang, T. M. S. In *Encyclopedia of Human Biology*; Dulbecco R., Ed.; Academic: San Diego, 1991; Chapter 1, p 377.
2. Chang, T. M. S. Research Report for honours physiology, Medical Library, McGill University, 1957.
3. Chang, T. M. S. *Science* **1964**, *146*, 524.
4. Chang, T. M. S. *Trans. Am. Soc. Artif. Intern. Organs* **1966**, *12*, 13.
5. Chang, T. M. S., Ph.D. Thesis, McGill University, 1965.
6. Chang, T. M. S.; MacIntosh, F. C.; Mason, S. G. *Can. J. Physiol. Pharmacol.* **1966**, *44*, 115.
7. Chang, T. M. S.; MacIntosh, F. C.; Mason, S. G. Canadian Patent 873 815, 1971.
8. Chang, T. M. S. *Artificial Cells* C.C. Thomas: Springfield, IL, 1972.
9. Chang, T. M. S. *Artificial Cells, Blood Substitutes and Immobilization Biotechnology* **1994**, *22*, vii.
10. Chang, T. M. S. In *Microcapsules and Nanoparticles in Medicine and Pharmacology*; Donbrow, M., Ed.; CRC: Boca Raton, FL, 1992; pp 323–339.
11. Chang, T. M. S. *Biotechnology Annual Review* **1995**, *1*, 267–296.
12. Chang, T. M. S. *Kidney Int.* **1975**, *7*, 5387.
13. Chang, T. M. S. *J. Bioeng.* **1976**, *1*, 25.
14. Jalil, R. Nixon, J. R. *J. Microencaps.* **1990**, *7*, 297.
15. Mathiowitz, E.; Langer, R. In *Microcapsules and Nanoparticles in Medicine and Pharmacology*; Donbrow, M., Ed.; CRC: Boca Raton, FL, 1992; pp 99–123.
16. Chang, T. M. S.; Coffey, J. F.; Barre, P.; Gonda, A.; Dirks, J. H.; Levy, M.; Lister, C. *Can. Med. Assoc. J.* **1973**, *108*, 429.
17. Winchester, J. F. *Replacement of Renal Function by Dialysis*; Maher, J. F., Ed.; Kluwer Academic: Boston, MA 1988; pp 439–459.
18. Chang, T. M. S. *Trans. Am. Soc. Artif. Intern. Organs* **1980**, 26, 546.
19. Lim, F.; Sun, A. M. *Science* **1980**, *210*, 908.
20. Goosen, M. F. A.; O'Shea, G. M.; Gharapetian, H. M.; Chou, S.; Sun, A. M. *J. Biotechnol. Bioeng.* **1985**, *27*, 146.
21. Wong, H.; Chang, T. M. S. *Int. J. Artif. Organs* **1986**, *9*, 335.
22. Prakash, S.; Chang, T. M. S. *Biomaterials, Artificial Cells and Immobilization Biotechnology* **1993**, *21*, 629.
23. Praskan, S.; Chang, T. M. S. *J. Biotech. Bioeng.* **1995**, *46* 621.
24. Garofalo, F.; Chang, T. M. S. *Biomat. Artif. Cells and Artif. Organs* **1989**, *17*, 271.
25. Garofalo, F.; Chang, T. M. S. *Applied Biochemistry and Biotechnology* **1991**, *27*, 75.
26. Chang, T. M. S.; Poznansky, M. J. *Nature* **1968**, *218(5138)*, 242.
27. Chang, T. M. S. *Nature* **1971**, *229(528)*, 117.
28. Chang, T. M. S. *Methods in Enzymology* **1987**, *136*, 67.
29. Gu, K. F.; Chang, T. M. S. *Applied Biochemistry and Biotechnology* **1990**, *26*, 263.
30. Chang, T. M. S., Ed.; *Blood Substitutes and Oxygen Carriers*; Marcel Dekker: New York, 1992.
31. Chang, T. M. S. *Artificial Cells, Blood Substitutes and Immbolization Biotechnology* **1994**, *22*, 123.

MICROENCAPSULATION (Drug Release)

Shinzo Omi
Graduate School of Bio-Applications and Systems Engineering
Tokyo University of Agriculture and Technology

Microcapsules (MC) are small containers of polymer film (membrane, shell) in which core materials such as drugs, chemicals and flavors are encapsulated. The core materials can be in gaseous, liquid or solid state as well as in emulsion or suspension, leading to the abundance of MC goods commercially available.[1] Illustration of two typical microcapsules are shown in **Figure 1**.

Desired functions to be exerted from MC are

- protection of core materials from the environment
- controlled exchange of substances with the environment
- apparent transformation of the state of core materials

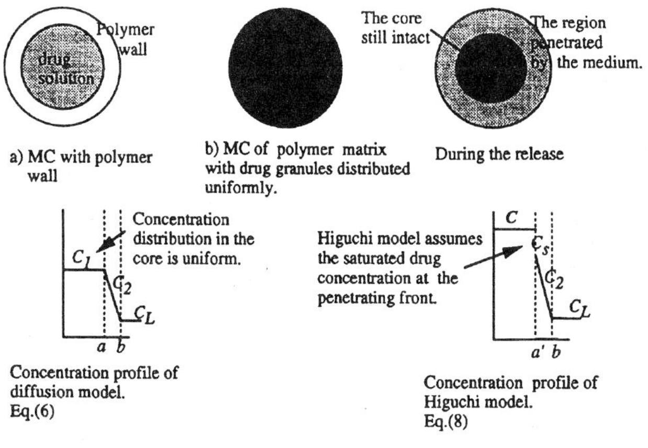

FIGURE 1. Two typical microcapsules (MC) with drug release models.

Liquid crystals and radioactive wastes belong to the first category because we use encapsulation to effect complete isolation. Controlled drug release is a typical example of the second category, and the time-span of the release is widely distributed.

CONTROLLED RELEASE OR CORE MATERIALS

The release profile of core materials from microcapsules can be distinguished in two extreme modes, step-wise or controlled.

Step-Wise (Instantaneous) Release

Step function mode requires an instantaneous release that is triggered by a particular stimulation from outside.

Controlled (Sustained) Release

Diffusivity or permeability of the wall polymer, thickness of the wall, surface area (average size), and loading of the core material become important parameters of the mass transfer that governs the controlled release.[2]

REFERENCES

1. Deasy, P. B. *Microencapsulation and Related Drug Processes* Marcel Dekker: New York and Basel, 1984.
2. Bruck, S. D. Ed.; *Controlled Drug Delivery* CRC: Boca Raton, FL, 1983; Vols. 1 and 2.

Microgels

See: *Colloidal Dispersions, Nanogels*
Microgels (Intramolecularly Crosslinked Macromolecules)
Microgels, Macrogels (by Photocrosslinking)
Network Polymers (Vinyl-Type)
Network Structure Formation (Radical Polymerization)

MICROGELS (Intramolecularly Crosslinked Macromolecules)

Werner Funke
Institute of Technical Chemistry
University of Stuttgart

The number of publications on microgels over the last 40 years shows the interest in this species of macromolecules increased significantly since the beginning of 1970.

Microgels originally were a nuisance to the science of polymers. They interfered with the characterization of macromolecules by scattering light, and they blocked pipes and valves in the equipment used to produce some polymers.

Microgels are also of interest in understanding and imitating the structure of crosslinked polymers. Results of the polymerization of crosslinked polyester resins indicated that such polymers have an inhomogeneous network structure consisting of strongly crosslinked cores embedded in a weakly crosslinked matrix.[1–4] Inhomogeneously crosslinked polymers are not a common subject at this time, because vulcanized rubber was the standard polymer network for studies on the network formation. As a matter of fact these networks are rather homogeneously crosslinked. However, it is well known that many, if not most, crosslinked polymers have an inhomogeneous structure.[5–8]

This knowledge has stimulated the synthesis of reactive microgels.[9] They were used as cores to prepare networks of a defined inhomogeneity by copolymerizing them with other monomers or oligomers, such as styrene or unsaturated polyesters. It was expected that these polymers should have interesting mechanical properties.

DEFINITIONS

The term microgel is only appropriate in the swollen state when such polymers are dissolved. Another name should be used when microgels have been transferred to the solvent-free, solid state microgels. The most convincing name for microgels in the solid, non-swollen state would be "intramolecularly crosslinked macromolecules," because they form colloidal or true solutions that are similar to those of linear or branched macromolecules. However, allowing for the common usage and for sake of simplicity the name "microgel" is used in this paper unless this causes confusion.

CHARACTERISTIC PROPERTIES OF MICROGELS

The most significant characteristics of microgels are as follows:

- Three-dimensional network structure.
- Intramolecular crosslinks.
- Constant surface and a spherical, ellipsoidal or irregular shape.
- Solubility in solvents to form colloidal or normal solutions.

The microgels' size of order is approximately < 100 nm (5–70 nm) and their molar mass is frequently > 1,000,000 g/mol.

NEWER EXPERIMENTAL RESULTS

It is interesting to determine which relationship exists between intrinsic viscosity and molecular mass in case of microgels. As expected, the intrinsic viscosity of weaker crosslinked microgels was higher than that of stronger-crosslinked ones. According to the Einstein law of coil densities, these microgels do not change with molecular mass but they decrease with increasing crosslink density. Moreover, it could be shown that the intrinsic viscosity was larger the better the solvent has been.[10] This kind of microgel does not behave like rigid spheres; they have about the same density from their center out to their periphery. Therefore, they must have a very homogeneous structure.

APPLICATIONS OF MICROGELS

Intramolecularly crosslinked macromolecules, which in solution are microgels, may be or already are used in some cases for scientific and technical purposes. These include:

- Preparation of inhomogeneous polymer networks with a defined structure.[11,12]
- Preparation of interpenetrating networks.[13,14]
- Preparation of core-shell polymers.[15–20]

- Models for studying topochemical organic reactions at polymer surfaces.[21–25]
- Models for studying physico-chemical properties at the transition of macromolecules to polymer particles.
- Relation between crosslink density and intrinsic viscosity.[10,25]
- Use as a binder component of high-solid coating materials.[26]
- Improvement of flow properties of water-borne paints.[33–38]
- Improvement of pigment orientation in automotive coatings.[27,28]
- Preparation of shock-resistant organic coatings.[29]
- Reinforcement of plastic materials.[30,31]
- Biomedical and diagnostic application as carriers for enzymes.[32,33]

A special technical importance that microgels have attained is as a binder component of organic coating materials. In this area they allow the preparation of high-solid coating systems with favorable rheological properties and improved orientation of flake pigments (e.g., in automotive coatings).[9a,34–38]

REFERENCES

1. Gallagher, L.; Bettelheim, F. A. *J. Polym. Sci.* **1962**, *58*, 697.
2. Funke, W. *Kolloid Ztsch.-Ztsch. fuer Polymere* **1964**, *197*, 71.
3. Funke, W. *Adv. Polymer Sci.* **1965**, *4/2*, 157.
4. Yang, Y. S.; Lee, L. J. *Polymer* **1988**, *29*, 1793.
5. Funke, W. *J. Polym. Sci.* **1967**, *C16*, 1497.
6. Dusek, K.; Prins, W. *Fortschrittsber. Hochpolymerenforschung* **1969**, *6*, 1.
7. Racich, J. L.; Koutsky, J. A. *J. Appl. Polym. Sci.* **1976**, *20*, 2111.
8. Okasha, R.; Hild, G.; Rempp, P. *Europ. Polym. J.* **1979**, *15*, 975.
9. Funke, W. *Chimia* **1968**, *22*, 11.

9a. Backhouse, A. J. *J. Coatings Technol.* **1982**, *54, No. 747*, 83.

10. Bolle, Th. Thesis, University Stuttgart, 1993.
11. Funke, W. *Brit. Polym. J.* **1989**, *21*, 107.
12. Antonietti, M.; Rosenauer, Ch. *Macromolecules* **1991**, *24*, 3434.
13. Straehle, W.; Funke, W. *Angew. Makromolek. Chem.* **1979**, *76/77*, 259.
14. Chen, H.; Chen, J. M. *J. Appl. Polym. Sci.* **1993**, *50*, 495.
15. Straehle, W.; Funke, W. *Makromolek. Chem.* 1977, *60/61*, 111.
16. Park, M. H.; Saito, R.; Ishizu, K.; Fukutomi, T. *Polym. Commun.* **1988**, 29, 230.
17. Park, M. H.; Ishizu, K.; Fukutomi, T. *Polymer* **1989**, *30*, 202.
18. Cha, X.; Yin, R.; Zhang, X.; Chen, J. *Macromolecules* **1991**, *24*, 4985.
19. Saito, R.; Kotsubo, H.; Ishizu, K. *Europ. Polym. J.* **1991**, *27*, 1153.
20. Kim, K. S.; Cho, S. H.; Kim, Y. J. *Polym. J.* **1993**, *25*, 847.
21. Obrecht, W.; Seitz, U.; Funke, W. *Macromolek. Chem.* **1976**, *177*, 2235.
22. Seitz, U. *Makromolek. Chem.* **1977**, *178*, 168.
23. Huang, Y.; Seitz, U.; Funke, W. *Makromolek. Chem.* **1985**, *186*, 273.
24. Kleiner, B.; Joos-Müller, B.; Funke, W. *Makromolec. Chem. Rapid Comm.* **1989**, *10*, 345.
25. Kjellqvist, K. Progr. Org. Coatings **1994**, *24*, 209.
26. Patent CO8F 8-30 - DE 3 302 738, Nippon Paint Co., Osaka, Japan.
27. Yabuta, M.; Sasaki, Y. *Proc. ACS-PMSE* **1990**, *63*, 772.
28. Yabuta, M. *J. Japan Soc. Col. Materials (Shikizai Kyokaishi)* **1990**, *63*, 209.
29. Antonietti, M. *Angewandte Chem.* **1988**, *100*, 1813.
30. Zhu, Z.; Xue, R.; Yu, Y. *Angew. Makromolek. Chem.* **1989**, *171*, 65.
31. Ishikura, S.; Ishii, K. *J. Japan Soc. Col. Materials (Shikizai Kyokaishi)* **1990**, *63*, 143.
32. Seitz, U.; Pauly, H. E. *Angew. Makromolek. Chem.* **1979**, *76/77*, 319.
33. Hosaka, S.; Murao, Y.; Tamaki, H.; Masuko, S.; Miura, K.; Kawabata, Y. *Polymer International* **1993**, *30*, 505.
34. Funke, W. *J. Oil Col. Chem. Assoc.* **1977**, *60*, 438.
35. Chatta, M. S.; Cassata, J. C. *Proc. ACS-PMSE* **1985**, *52*, 326.
36. Kashihara, A.; Ishii, K.; Ishikura, K.; Mizuguchi, R. *Proc. ACS-PMSE* **1985**, *189*, 86.
37. Ishikura, S.; Ishii, K.; Mizuguchi, R. *Progr. Org. Coatings* **1988**, *15*, 373.
38. Yagi, T.; Saito, K.; Ishikura, S. *Progr. Org. Coatings* **1992**, *21*, 25.

MICROGELS, MACROGELS (by Photocrosslinking)

Yoichi Shindo and Masatoshi Hasegawa
Department of Chemistry
Faculty of Science
Toho University

There have been many studies on the conditions and mechanisms of crosslinking emulsion polymerization and photopolymerization for the formation of a three-dimensional network of polymer molecules.[1–5] Technological advances have made photopolymers increasingly important for application in fields such as printing, coating, microlithography, and electronic device manufacturing. The gel formation due to photogenerated crosslinks in solid state has been investigated.[6–8] The gel consists mainly of intermolecular crosslinks, and intramolecular crosslinking is negligible for photoirradiation.[7] However, when polymer molecules are crosslinked in solution, both intra- and intermolecular crosslinks are formed.[12–15] According to the Ziegler dilution law, diluting the systems increases the probability of intramolecular crosslinking. This can be forced up to a threshold where no more macroscopic gelation occurs and only microgels are formed.[9–11] Microgels may be considered as species intermediates between large macromolecules and small particles, having some properties characteristic of each. Compared with the procedure of emulsion polymerization or photopolymerization, the photodimerization process is advantageous for the formation of microgels or macrogels. This paper describes the preparation and properties of microgels and macrogels by the photodimerization of polymers with pendent cinnamoyl groups or styrylpyridinium groups in solution.[16–18]

APPLICATIONS

Microgels or macrogels consisting of the densely intramolecular-crosslinked structure are easily prepared by cyclodimerization caused by u.v. light irradiation polymers with pendent cinnamoyl groups in organic solvents such as THF or dichloromethane. The film-forming property of the gels is sufficient by themselves, and provides improved coating and adhesion properties. In particular, microgels have photosensitive surface-activated characteristics, and thus are suitable vehicle components for high-solid systems in the technical and industrial areas.

The entrapping method of the biocatalysts in polymer gels due to photocrosslinking technique has merits owing to the mild network formation of polymer gels where no change in pH or temperature is necessary. The immobilization requires no covalent bond formation, between the matrix and enzyme molecules and thus maintains the native properties of the enzymes.

Photosensitive poly(vinyl alcohols) are useful for entrapping enzymes because of their excellent water solubility and high photosensitivity.

REFERENCES

1. Smets, G.; Aerts, A.; Van Erm, J. *Polym. J.* **1980**, *12*, 539.
2. Antonietti, M. *Angew. Chem. Int. Ed. Engl.* **1988**, *27*, 1743.
3. Funke, W. E. *J. Coatings Technol.* **1988**, *60*, 68.
4. Shindo, Y.; Sugimura, T.; Horie, K.; Mita, I. *Eur. Polym. J.* **1990**, *26*, 683.
5. Sasa, N., Yamaoka, T. *Adv. Mater.* **1993**, *6*, 417.
6. Reither, A.; Egerton, P. L. *Macromolecules* **1979**, *12*, 670.
7. Egerton, P. L.; Pitts, E.; Reither, A. *Macromolecules* **1981**, *14*, 95.
8. Matsumoto, A.; Fukazawa, A.; Oiwa, M. *J. Appl. Polym. Sci.* **1983**, *28*, 11.
9. Arbogast, W., Horvath, A.; Vollmert, B. *Macromol. Chem.* **1980**, *181*, 1513.
10. Burchard, W., Schmidt, M. *Macromolecules* **1981**, *14*, 370.
11. Antonietti, M.; Ehlich, D.; Folsch, K. J.; Sillescu, H.; Schmidt, M., Lindner, P. *Macromolecules* **1988**, *21*, 736.
12. Soper, B.; Harward, R. N.; White, E. F. T. *J. Polym. Sci.* **1972**, A-1, 2545.
13. Horie, K.; Otagawa, A.; Muraoka, M.; Mita, I. *J. Polym. Sci., Polym. Chem. Ed.* **1975**, *13*, 445.
14. Matsumoto, A.; Yokoyama, S.; Khono, T.; Oiwa, M. *J. Polym. Sci. Phys. Ed.* **1977**, *15*, 127.
15. Dusek, H.; Gordon, M.; Ross-Murphy, S. B. *Macromolecules* **1978**, *11*, 236.
16. Shindo, Y.; Sato, H.; Sugimura, T.; Horie, K.; Mita, I. *Eur. Polym. J.* **1989**, *25*, 1033.
17. Shindo, Y.; Sugimura, T.; Horie, K.; Mita, I. *Eur. Polym. J.* **1986**, *22*, 859.
18. Shindo, Y.; Hasegawa, M.; Sugimura, T.; Ebisuno, T.; Mitsuda, M. *Joint Symposium on Polymer Gels and Networks Prepr., Jpn.* 1993; p 88.

Micromechanical Properties

See: *Deformation and Fracture (Micromechanical Mechanisms)*

Microspheres

See: *Catalyses (by Polyelectrolytes and Colloidial Particles)*
High Solids Coatings (Microspheres from Nonaqueous Polymer Dispersions)
Macromonomers (Soap Free Emulsion and Dispersion Polymerization)
Microgels (Intramolecularly Crosslinked Macromolecules)
Microspheres, Core-Shell Type (from Block Copolymers)
Monodisperse Polymer Particles (Nonmagnetic and Magnetic)
Precipitation Polymerization
Spheres, Poly(butylene terephthalate)

MICROSPHERES, CORE-SHELL TYPE (from Block Copolymers)

Koji Ishizu
Department of Polymer Science
Tokyo Institute of Technology

Block and graft copolymers composed of incompatible block segments form a microphase-separated solid-state structure. We established a novel synthesis method of the core-shell type microspheres by crosslinking the segregated chains in spherical microdomains (see **Figure 1**).[1–5] These core-shell type microspheres also stabilized in good solvent by highly branched arms.

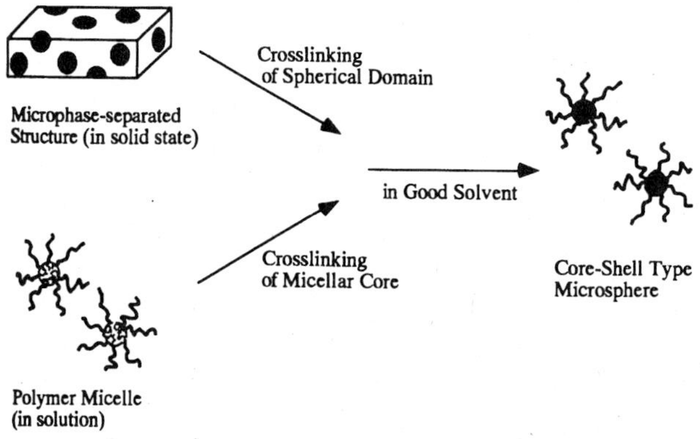

FIGURE 1. Schematic representation of synthesis routes on core-shell type microspheres.

Such chain conformation of core-shell microspheres is similar to star-shaped polymers or $(AB)_n$ star-block polymers. According to theoretical results,[6] the central cores of star polymers do not interpenetrate each other beyond the overlap concentration (C*). Stars with many arms (the critical number of arms is estimated to be of order 10^2) are expected to form a crystalline array near the C* concentration.

In a previous work, the author studied the morphological behavior of core-shell type microspheres from casting solution onto a carbon substrate.[7] In recent reports, we prepared the core-shell type microspheres using relatively small-diameter particles. Film formation of these microspheres was investigated through SAXS and TEM observations.[7]

PREPARATION

A well-defined poly[styrene(S)-b-4-vinylpyridine(4VO)] diblock copolymer SV was prepared by the sequential anionic addition polymerization using *n*-butyllithium (*n*-BuLi) as an initiator in tetrahydrofuran (THF) at –78°C.

APPLICATIONS

The core-shell type microspheres can be considered as the smallest units in the microphase-separated structure. However, the properties of the microspheres in the solvent are governed by the shell chains in a good solvent. Therefore, the core-shell microspheres can be used as composite materials in blends that allow the introduction of spherical microdomains into a matrix.

This study investigated the morphologies of blends of well-ordered core-shell microspheres with lamellar AB-type block copolymers.[8]

It was suggested that microspheres separation with three phases could be obtained by blending the microsphere and block copolymer.

REFERENCES

1. Ishizu, K.; Fukutomi, T. *J. Polym. Sci., Polym. Lett. Edn.* **1988**, *26*, 281.
2. Ishizu, K. *Polymer* **1989**, *30*, 793.
3. Ishizu, K. *Polym. Commun.* **1989**, *30*, 209.
4. Ishizu, K.; Önen, A. *J. Polym. Sci., Polym. Chem. Edn.* **1989**, *27*, 3721.
5. Ishizu, K.; Saito, R. *Polym. Plast. Technol. Eng.* **1992**, *31*, 607.
6. Witten, T. A.; Pincus, P. A.; Catea, M. E. *Europhys. Lett.* **1986**, *2*, 137.
7. Ishizu, K. *J. Colloid Interface Sci.* **1993**, *156*, 299.
8. Saito, R.; Kotsubo, H.; Ishizu, K. *Polymer* **1994**, *35*, 1580.
9. Ishizu, K.; Sugita, M.; Kotsubo, H.; Saito, R. *J. Colloid Interface Sci.* **1995**, *169*, 456.

MICROWAVE ABSORBING MATERIALS

Thomas J. Fabish
Alcoa Laboratories

Controlling the power absorbed in the interaction of a microwave with condensed phase materials is basic to the application of microwave energy in science and technology.

The material factors governing energy absorption and wave propagation are of particular interest in this report.

Materials occupy an important role in achieving reduced radar signatures for aircraft and ships, in managing returns from structures affecting radar navigation on the waterways, and in the construction of many devices vital to microwave communication. Perhaps the most familiar commercial application of moderately high-power microwave technology is the preparation of foods utilizing consumer or industrial size microwave ovens. While food preparation can usually proceed through direct interaction between the microwave energy and the dielectric food product, a number of advantages to taste, appearance, consistency, and preparation time can be gained through proper design of the food package.[1,2] Moderate- to high-power microwave materials processing is another application gaining greater attention as scientists look at anticipated trends in primary energy sources and to the effects of processing on final properties.[3,4] Of more fundamental interest, it may be possible to alter reaction pathways in biological systems using microwave excitation and in some chemical processes of scientific and commercial importance.[5] Advantages in chemical synthesis and end-use properties of microwave processed materials relative to thermal processing apparently are linked to elemental excitation mechanisms that are accessible in the microwave region of the electromagnetic spectrum and to the substantially different energy gradients that are produced within the material in the two energy regimes.

This report focuses on the development of design processes for consumer microwave products, notably packaging for microwave food preparation.

A central objective is the development, verification, and application of a predictive engineering model for microwave-solid interaction. This model relates important performance parameters such as medium wave propagation factors and heat generation rate to measurable materials parameters and sample dimensions in a computationally simple and physically transparent manner. The simplifying assumptions and analytical model are presented, and two kinds of laminates are considered in detail for their analytical simplicity and their versatility in simulating important aspects of microwave food packaging.

ANALYTICAL MODEL

Criteria Governing Choice of Materials

The present development effort aims to create overlayers for aluminum foil that will render the laminate active in microwave packaging applications. In addition to acceptable power absorption, the efficiency of thermal transport through the absorber to a load or heat sink must be adequate to control safety and performance related to device temperature.

Several kinds of magnetic powders were evaluated for use as fillers to enhance power absorption. Candidate magnetic powders were chosen mainly on the basis of their magnetic response in the microwave region. In particular, the highest relative magnetic loss was sought. Two dielectric particulate fillers were chosen for extensive evaluation for the wide range they offered in dielectric response. Both the real and imaginary terms of the dielectric response function are important to absorber design for their influence on reflection coefficient and wave propagation factors. The polymer functions mainly as a binder for the particulate fillers, placing few demands on polymer performance.

Materials

Magnetic powders investigated included Grade E iron manufactured by GAF Corporation, a nickel-zinc ferrite in two different size mixes and a lithium-nickel-zinc-manganese ferrite (Ferrite 50) from Trans-Tech, Inc., Gaithersburg, Maryland, and a manganese-zinc ferrite from Titan Corp., Valparaiso, Indiana. Of the four magnetic particles tested, the Grade E iron provided superior magnetic loss at 2.45 Ghz so this material was used in nearly all of the development tests and will be the only magnetic powder discussed in detail.

Composite test films of various thickness were fabricated using three different methods. In the first, the particulate and polymer were cosprayed directly onto aluminum foil. The second and third fabrication methods both started with mechanically blended particulate and polymer that was heat pressed onto foil or extruded into sheet and laminated to foil. Two polymer matrices were employed, polyerethane (Miles, Inc. Desmothen 670A80) for the sprayed films and polypropylene for the mechanically blended films. Iron-loaded film was made by all three methods.

SUMMARY

Consideration of elementary aspects of the microwave/solid-interaction leads to a simplified model of the role of materials in managing the distribution of incident power between reflected, absorbed, and transmitted components that proves rather robust in interpreting the results of power absorption experiments performed in a multimode microwave capacity operated at moderate power levels. The model is predictive in that the input consists of device dimensions and effective medium magnetic and dielectric response functions that are independently measurable. Materials were shown to be key to successful absorber design for applications where device shape is not the dominant factor. Three particulate materials used as fillers in a polymer matrix and applied in two basic laminate forms were shown to provide broad latitude in optical property design opportunities to meet criteria for power absorption and transmission that are relevant to diverse applications.

ACKNOWLEDGMENTS

The author recognizes valuable contributions to this work from colleagues at Alcoa Technical Center, including A. Bension and J. D. Davidson for materials and experiments, W. W. Hyland and R. J. Perry for the optical constants measurements, and D. A. Smith, J. E. Stillwagon, T. L. Levendusky, G. H. Armstrong, and B. O. Hall for guidance, and C. L. Rohrer for a critical reading of the manuscript. Also, L. N. Boretzky and D. C. McClurg of Miles, Inc., Pittsburgh, PA, supplied sprayed test panels and materials, and S. Nadel of Airco Solar Products, Concord, CA, provided TiN_x-coated samples.

REFERENCES

1. Harrison, P. *Packaging Tech. and Sci.* **1989**, *2*, 5.
2. Mermelstein, N. H. Ed., *Food Tech.* **1989**, *January*, 117.
3. Chen, M.; Siochi, E. J.; Ward, T. C.; McGrath, J. E. *Polym. Engr. Sci.* **1993**, *33*(17), 1092.
4. Agrawal, Raj K.; Drzal, L. T. *J. Adhesion* **1989**, *29*, 63.
5. Julien, H. *Proceedings PMSE* **1992**, *66*, 378.

Microwave Processing

See: Epoxy Resins (Microwave Processing)

MIKTOARM STAR POLYMERS

Nikos Hadjichristidis, Hermis Iatrou, and Yiannis Tselikas*
Department of Chemistry
University of Athens, Greece

Jimmy W. Mays
Department of Chemistry
University of Alabama at Birmingham

Star polymers are branched polymers with several linear chains ("arms") attached to a single branch point. Miktoarm (mikto from the Greek word μικτός meaning a group of different elements) star polymers are star polymers with chemically different arms (see **Figure 1**).

*Author to whom correspondence should be addressed.

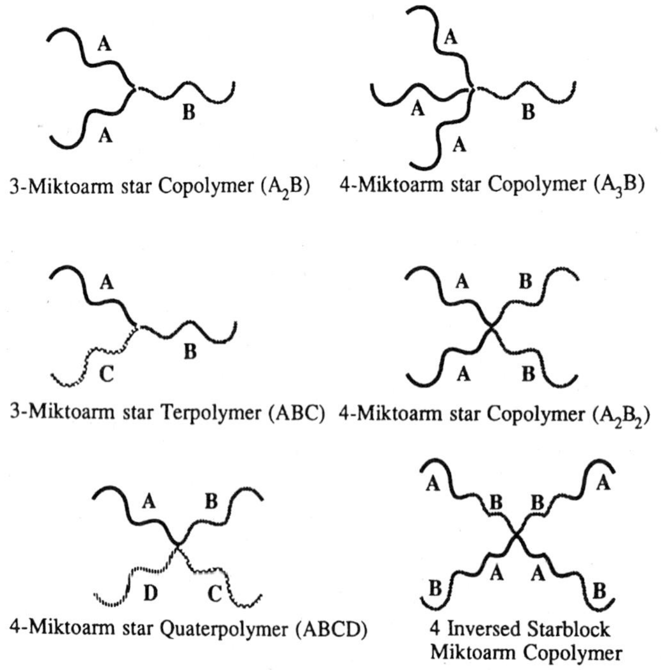

FIGURE 1. Some examples of miktoarm stars.

Model miktoarm star polymers have been prepared only recently (almost entirely by anionic polymerization techniques). Therefore, this review focuses on the synthesis of these molecules. Only limited results exist on properties and potential applications of these materials.

STRATEGIES AND METHODS OF PREPARATION

Three general methods based on anionic polymerization techniques exist for preparing miktoarm stars.

Divinylbenzene Approach

In this approach, the polymerization of a small amount of divinylbenzene (DVB) is initiated by the living ends of the polymeric chains. This leads to the formation of star molecules, comprised of a crosslinked poly(divinylbenzene) (PDVB) core connected to the polymeric chains. The core should contain the same number of living anions as the star's arms, because there should be no termination of anions during this linking reaction. These "living" star polymers may then be used to initiate the polymerization of another monomer, creating new branches that grow out from the core.

This method suffers from architecture limitations since it is only possible to prepare miktoarm stars of the A_NB_N structure.

Miktoarm stars prepared by this method will not exhibit high molecular and compositional homogeneity. Also, it is difficult to control the precise number of arms, and the weight of the PDVB core can constitute a large percentage of the total weight of the star, influencing the properties of the material.

1,1-Diphenylethylene Derivatives Approach

This approach pioneered by Yamagishi et al., involves the reaction of living polymeric chains with a 1,1-diphenylethylene (DPE) derivative that serves as a coupling agent.[1] A second monomer is added to the living coupled product, and a miktoarm star of the A_2B_2 type is formed.

A weak point of this method is that different rate constants exist for the reaction of the first and second living polymeric chain with the linking agent in hydrocarbon solvents. This leads to the formation of the dianion and a monoanion during the addition of the living polymeric chains.

The Chlorosilane Approach

In 1988 Pennisi and Fetters prepared three-arm asymmetric PS and PB stars of the A_2B type.[2] These homopolymer stars had two arms of equal length and a longer or shorter third arm. The synthetic approach involves the reaction of living arm B with an excess of methyltrichlorosilane followed, after removal of the excess linking agent, by the addition of a slight excess of living arm A.

Mays employed the same approach to prepare a near-monodisperse 3-miktoarm star copolymer of the A_2B type, where A is PI and B is PS.[3] This method was later applied by Hadjichristidis and co-workers to the synthesis of the various combinations of A_2B-type stars incorporating PI, PS, and PBD.[12] Samples having molecular weights well over 100,000 have been prepared.[4] The same procedures were employed by Tselikas and Hadjichristidis for the synthesis of 4-miktoarm star copolymers of the A_3B type, except the linking agent used was $SiCl_4$.[5]

Similar procedures have been employed by Iatrou and Hadjichristidis for the synthesis 3-miktoarm star terpolymers of the ABC-type and of A_2B_2 and ABCD co- and quaterpolymers.[6,7]

The chlorosilane approach is very powerful, and it can be used for the synthesis of other miktoarm stars (e.g., $(AB)_2A_2$, $(AB)_2B_2$, $(AB)_2BC$, AB_5, etc.), as well as for the synthesis of more complicated architectures such as the A_3BA_3 "super-H" polymer.[8]

PROPERTIES AND APPLICATIONS OF MIKTOARM STARS

Miktoarm stars have been synthesized only recently, and relatively little is known about their properties and potential uses.

Dilute Solution Properties

Linear diblock copolymers show little evidence of substantial segregation of the blocks.[9] The two chains in a diblock are statistically already separated in space, and only a few heterocontacts are available for further expansion of the molecule. In the case of A_2B and A_2B_2 miktoarm stars, analysis of intrinsic viscosity and hydrodynamic radius data suggests that a small expansion of the copolymers occurs in selective and in nonselective solvents.[4] This is in contrast to observations for linear diblock copolymers and is due to the increased occurrence of heterocontacts in the miktoarm star architecture, i.e., "crowding" of the different segments together near the star's center.[9]

Morphology and Phase Behavior in Bulk

Model diblock, triblock, and starblock copolymers have been investigated in the past in terms of their bulk morphologies in the weak and strong segregation regimes. These studies revealed that block copolymers form well-defined microphase-separated morphologies with various symmetries.[10–12] In the case of branched block copolymers, theoretical work is beginning to appear, and experimental investigations are at an early stage.[13-16] Bright field transmission electron microscopy (TEM) has been applied to A_2B and A_3B miktoarm star copolymers.[17,18]

The morphologies observed for the A_2B and A_3B miktoarm star copolymers seem to be in fairly good agreement with these predicted theoretically by Milner, at least on a qualitative level of comparison.[16] However, a spherical structure is not observed in the phase diagram of the A_3B material at high PS content as predicted. The triconnected cubic structure is observed at around 50% PS, as expected.[16]

APPLICATIONS

These new materials may find application as compatibilizers of two, three, or even four different homopolymers. Already, the A_2B miktoarm stars have been observed to be more effective than simple diblocks at compatibilizing A/B blends.[19]

A substantial impact of the development and study of miktoarm stars will be enhancement of our understanding of graft copolymers, of which the A_2B material is the simplest model structure. A systematic study of the effects of branching on morphology and phase behavior has been initiated recently.[20] Balazs and collaborators have demonstrated that grafts offer substantial advantages over blocks in compatibilizing immiscible blends, and they have suggested some optimal architectures.[15,21]

ACKNOWLEDGMENT

Hermis Iatro, Yiannis Tselikas and Nikos Hadjichristidis are grateful to the Greek General Secretariat for Research and Technology and Jimmy W. Mays is grateful to the Army Research Office for support (33198-CH-DPS).

REFERENCES

1. Yamagishi, A.; Szwarc, M.; Tung, L.; Lo, G. Y.-S. *Macromolecules* **1978**, *11*, 607.
2. Pennisi, R. W.; Fetters, L. J. *Macromolecules* **1988**, *21*, 1094.
3. Mays, J. W. *Polym. Bull.* **1990**, *23*, 247.
4. Iatrou, H.; Siakali-Kioulafa, E.; Hadjichristidis, N.; Roovers, J.; Mays, J. W. *J. Polym. Sci., Polym. Phys. Ed.* **1995**, *33*, 1925.
5. Tselikas, Y.; Hadjichristidis, N., to be published.
6. Iatrou, H.; Hadjichristidis, N. *Macromolecules* **1992**, *25*, 4649.
7. Iatrou, H.; Hadjichristidis, N. *Macromolecules* **1993**, *26*, 2479.
8. Iatrou, H.; Avgeropoulos, A.; Hadjichristidis, N. *Macromolecules* **1994**, *27*, 6232.
9. Prud'homme, J.; Roovers, J.; Bywater, S. *Eur. Polym. J.* **1972**, *8*, 901.
10. Molau, G. E. *Block Copolymers;* Agarwal, S. L., Ed.; Plenum: New York, 1970.
11. Thomas, E. L.; Anderson, D. M.; Henkee, C. S.; Hoffman, P. *Nature* **1988**, *334*, 598.
12. Bates, F. S.; Fredrickson, G. H. *Annu. Rev. Phys. Chem.* **1990**, *41*, 525.
13. Benoit, H.; Hadziioannou, G. *Macromolecules* **1988**, *21*, 1449.
14. Olivera de la Cruz, M.; Sanchez, I. C. *Macromolecules* **1986**, 19, 2501.
15. Shinozaki, A.; Jasnow, D.; Balazs, A. C. *Macromolecules* **1994**, *27*, 2496.
16. Milner, S. T. *Macromolecules* **1994**, *27*, 2333.
17. Hadjichristidis, N. et al. *Macromolecules* **1993**, *26*, 5812.
18. Tselikas, Y.; Iatrou, H.; Hadjichristidis, N., to be submitted.
19. Hadjichristidis, N., unpublished results.
20. Mays, J. W.; Gido, S. P.; Russell, T. P. et al., work in progress.
21. Gersappe, D. et al. *Science* **1994**, *265*, 1072.

Miniemulsion Polymerization

See: Vinyl Chloride (Minisuspension Polymerization)

Minisuspension Polymerization

See: Vinyl Chloride (Minisuspension Polymerization)

MODEL SILICONE NETWORKS

Enrique M. Vallés* and Marcelo A. Villar
Planta Piloto de Ingenieria Quimica
-UNS-CONICET

Silicone is the popular term used to describe a family of organo-silicon compounds. These materials' characteristics include their outstanding resistance to aging and weathering, their ability to maintain good physical properties over a wide temperature range, their low surface tension, and their incompatibility with water and most organic materials.[1–4]

Silicone networks also have been the subject of extensive scientific research. Model silicone networks (those prepared by end-linking functionally terminated polymer chains), have been widely used to test molecular theories of rubber elasticity.[5–8] In the last twenty years, the development of this theory has made significant progress, and the experimental contribution of silicone networks has been closely related to this progress. The current theory allows us to predict the elastic modulus from the molecular structure.

NETWORK PREPARATION

Polymer networks initially were prepared by introducing random chemical crosslinks on high molecular weight molecules. The necessity of well-characterized networks has focused the attention on some alternative synthetic procedures. In particular, those that link long chains of a given linear polymer by their ends have been quite successful in obtaining model networks.[8] In this type of reaction, the end-functional groups at the extremes of a linear prepolymer are reacted with crosslinking agents having a functionality of 3 or more. If every polymer chain has two terminal reactive groups and each group effectively reacts with different crosslinker molecules, an ideal network is obtained. The average chain length between crosslinking points is equal to the length of the bifunctional molecules used to prepare the model network. Their molecular structure may be determined by any of the numerous standard characterization techniques available to study linear chains in solution prior to crosslinking. Furthermore, thanks to modern anionic polymerization procedures, it is possible to synthesize telequelic monodisperse chains. These can be used to build model networks with chains of narrow molecular weight distribution.[10–12] This has been of great importance in the evaluation of molecular theories, and silicone networks have played a significant role in this area.[13-15]

Model silicone networks have been synthesized using bifunctional prepolymers, principally the Pt catalyzed hydrosilation reaction between silane groups from the crosslinker molecules and vinyl groups on the chain ends of the prepolymer molecules.[6,7] Model silicone networks have also been prepared from hydroxyl-terminated polydimethylsiloxane chains using crosslinkers containing ethoxy groups and a stannous catalyst.[5,16,17]

RUBBER ELASTICITY

Shear Elastic Modulus

Rubber elasticity theory relates the elastic modulus of a rubber network with the concentration of network chains, ν.

In the last years, a considerable number of studies have been done on model silicone networks.[5,7,8,17,18–27]

Ultimate Properties

The ultimate properties of model PDMS networks are considerably better than those of regular networks.[28–31] Their ultimate strength and maximum extensibility can significantly exceed the values obtained for regular PDMS networks

*Author to whom correspondence should be addressed.

crosslinked by radiation curing. The improvement is attributed to the fact that model networks possess few dangling end irregularities.

VISCOELASTIC PROPERTIES

The nonequilibrium mechanical properties of rubber materials also are of great technological importance. However, very little work has been done to explain the influence of network defects on non-equilibrium properties.[32] In particular pendant chains have a strong influence on viscoelastic properties, increasing the relaxation times considerably.[33–35] The origin of this was explained by de Gennes: pendant chains relax by a different reptation mechanism than free linear chains, with movements from the free end to the end that is attached to the network structure. Due to the additional restrictions imposed by this mode of reptation, relaxation times of networks with pendant chains increase exponentially with the number of entanglements.

Bibbó and Vallés found that pendant chains have a strong influence on the loss modulus of PDMS networks.[36]

CONCLUSION

Preparing and studying model networks with controlled structure can provide a great deal of valuable information on rubber-like elasticity and other related problems. Particularly, networks prepared with a known amount of defects, such as pendant chains, trapped molecules, and loops are necessary to understand the influence of these kinds of defects on equilibrium and dynamic mechanical properties.

REFERENCES

1. Hays, W. R.; Kehrer, G. P.; Monroe, C. M. *112th Meeting of the Rubber Division, American Chemical Society* Cleveland, OH, 1977.
2. Smith, A. L. *Analysis of Silicones, Chemical Analysis*; John Wiley & Sons: New York, NY, 1974; Vol. 41.
3. Lynch, W. *Handbook of Silicone Rubber Fabrication*; Van Nostrand Reinhold: New York, NY, 1978.
4. Meals, R. *Silicones* Encyclopedia of Chemical Technology, 2nd ed.; John Wiley & Sons: New York, NY, 1969; Vol. 18.
5. Mark, J. E.; Sullivan, J. L. *J. Chem. Phys.* **1977**, *66*, 1006.
6. Vallés, E. M.; Macosko, C. W. *Rubber Chem. Technol.* **1976**, *49*, 1232.
7. Vallés, E. M.; Macosko, C. W. *Macromolecules* **1979**, *12*, 673.
8. Mark, J. E. *Rubber Chem. Technol.* **1981**, *54*, 809.
9. Flory, P. J. *Principles of Polymer Chemistry* Cornell University: Ithaca, NY, 1953.
10. Morton, M.; Fetters, L. J.; Inomata, J. Rubio, D. C.; Young, R. N. *Rubber Chem. Technol.* **1976**, *49*, 303.
11. Morton, M. *Anionic Polymerization: Principles and Practice*; Academic: New York, NY, 1983.
12. Rempp, P.; Herz, J.; Hild, G.; Picot, C. *Pure Appl. Chem.* **1975**, *43*, 77.
13. Heinrich, G.; Straube, E.; Helmis, G. *Adv. Polym. Sci.* **1988**, *85*, 33.
14. Mark, J. E. *Adv. Polym. Sci.* **1982**, *44*, 1.
15. Mark, J. E. *Frontiers in Materials Technology*; Meyers, M. A.; Inal, O. T., Ed.; Elsevier Science: The Netherlands, 1985; Chapter XI.
16. Llorente, M. A.; Mark, J. E. *J. Chem. Phys.* **1979**, *71*, 682.
17. Yark, J. E.; Llorente, M. A.; *J. Am. Chem. Soc.* **1980**, *102*, 632.
18. Falender, J. R.; Yeh, G. S. Y.; Mark. J. E. *J. Chem. Phys.* **1979**, *70*, 5324.
19. Gottlieb, M.; Macosko, C. W.; Lepsch, T. C. *J. Polym. Sci., Polym. Phys. Ed.* **1981**, *19*, 1603.
20. Gottlieb, M.; Macosko, C. W.; Benjamin, G. S.; Meyers, K. O.; Merrill, E. W. *Macromolecules* **1981**, *14*, 1039.
21. Llorent, M. A.; Mark, J. E. *Macromolecules* **1980**, *13*, 681.
22. Llorente, M. A.; Andrady, A. L.; Mark, J. E. *J. Polym. Sci., Polym. Phys. Ed.* **1980**, *18*, 2663.
23. Mark, J. E.; Rahalkar, R. R.; Sullivan, J. L. *J. Chem. Phys.* **1979**, *70*, 1794.
24. Macosko, C. W.; Benjamin, G. S. *Pure Appl. Chem.* **1981**, *53*, 1505.
25. Meyers, K. O.; Bye, M. L.; Merrill, E. W. *Macromolecules* **1980**, *13*, 1045.
26. Oppermann, W.; Rehage, G. *Coll. Polym. Sci.* **1981**, *259*, 1177.
27. Vallés, E. M.; Rost, E. J.; Macosko, C. W. *Rubber Chem. Technol.* **1984**, *57*, 55.
28. Andrady, A. L.; Llorente, M. A.; Mark, J. E. *J. Chem. Phys.* **1980**, *72*, 2282.
29. Andrady, A. L.; Llorente, M. A.; Mark, J. E. *J. Chem. Phys.* **1980**, *73*, 1439.
30. Andrady, A. L.; Llorente, M. A.; Sharaf, M. A.; Rahalkar, R. R.; Mark, J. M. *J. Appl. Polym. Sci.* **1981**, *26*, 1829.
31. Llorente, M. A.; Andrady, A. L.; Mark, J. E. *J. Polym. Sci., Polym. Phys. Ed.* **1981**, *19*, 621.
32. Kramer, O. *Br. Polym. J.* **1985**, *17*, 129.
33. Curro, J. G.; Pincus, P. *Macromolecules* **1983**, *16*, 559.
34. de Gennes, P. G. *Scaling Concepts in Polymer Physics* Cornell University: Ithaca, NY, 1979.
35. Tsenoglou, C. *Macromolecules* **1989**, *22*, 284.
36. Bibbó, M. A.; Vallés, E. M. *Macromolecules* **1984**, *17*, 360.

MOLECULAR ASSEMBLIES (Polymerization in Aqueous Solution)

Kenichiro Arai
Faculty of Engineering
Gunma University

Molecular assemblies in aqueous solution include plain monolayer and multilayer membranes, vesicles, micelles, and others in extended forms. Amphiphilic molecules form bilayer assemblies, including monolayer and multilayer membranes and vesicles, such as naturally occurring phospholipids generally composed of a hydrophilic head group and two long hydrophobic tails although some synthetic amphiphiles with one or three hydrophobic tails are also found to form bilayer assemblies.[1,2] The bilayer assemblies are two-dimensional fluid, and the structures are dynamic but relatively stable and highly ordered.

Polymerization of the amphiphiles in the molecular assemblies has been studied to enhance the chemical and colloidal stability of the assembly structures and to use the polymerized assemblies for various applications. Polymerizations of the bilayer assemblies and applications of the polymerized assemblies have been reviewed extensively.[3–8]

POLYMERIZATION OF BILAYER ASSEMBLIES

Amphiphiles such as phospholipids (e.g., phosphatidylcholine (lecithin) can form assemblies in the form of bimolecular plain layers, vesicles, monomolecular layers, and multilayers. Polymerization of these assemblies is not difficult, because of its rather stable and highly ordered structure. However, in this case, it is important to avoid altering the innate characteristics of these assemblies, including the fluidity of the membrane and orderedness of the assembly structure.

Polymerization of Polymerizable Amphiphiles

Polymerizable amphiphiles have been prepared by introducing a reactive group into the amphiphilic molecules. In many cases, the reactive groups are double bonded and the polymerization may be initiated by usual free-radical initiators or by photoirradiation.

Other Methods for Preparing Polymerized Bilayer Assemblies

Polymerized bilayer assemblies also may be prepared by polymerizing amphiphiles in isotropic solution and the subsequent formation of the assemblies from the prepolymerized amphiphiles. In this method, the main problem is accommodating three-dimensional polymer molecules in two-dimensional assemblies.

If the head group of the amphiphilic molecules are ionic, the polymerized assemblies may be prepared by the electrostatic association of preformed assemblies with ionic polymerizable monomers as counter ions and the subsequent polymerization of the monomers in the assemblies.

Polymers with ionic groups may be used to form the polymerized assemblies instead of the monomers.

In another method, hydrophobic interaction between the hydrophobic groups in the polymers and the hydrophobic tails of the amphiphilic molecules in the assemblies is used to bind the polymer to the preformed assemblies.

Characteristics of Polymerized Assemblies

Assembly characteristics are altered more or less by the polymerization of the assemblies. Polymerization of reactive groups attached to hydrophobic tails significantly decreases the mobility of the molecular chains in the assemblies. Polymerization of the reactive groups at the hydrophilic head group applied fewer influences, but it alters the surface structure of the assemblies.

Applications

Applications of polymerized bilayer assemblies include biocompatible materials for medical uses, catalysts for specific biochemical and chemical reactions, and drug delivery systems for medicine and agricultural chemicals.

Polymerization of Micelles

Polymerization of amphiphilic molecules in micelle form has also been investigated. The polymerization, however, has inherent difficulties that contrast with amphiphiles in bilayer assemblies, because micelles are in dynamic equilibrium with their monomers in which the exchange rate of the monomers for the micelles is higher. A typical example is polymerization of sodium undecenoate, which is an aliphatic carboxylate with a double bond at the hydrophobic chain end.

Polymerization of 1-O-3-(4-vinylphenyl)propyl-β-d-glucose in micelle form by free-radical initiators was also examined.[9]

In another example, the polymerization of allyldimethyldodecylammonium bromide created an identity of the DP of the polymerized micelle with the association number of the monomer micelle.[10]

Besides direct observation of micelle structures, researchers have investigated other applications of polymerized micelles.[11–13]

REFERENCES

1. Hargreaves, W. R.; Deamer, D. W. *Biochemistry* **1978**, *17*, 3759.
2. Kunitake, T.; Kimizuka, N.; Higashi, N.; Nakashima, N. *J. Am. Chem. Soc.* **1984**, *106*, 1978.
3. Fendler, J. H.; Tundo, P. *Acc. Chem. Res.* **1984**, *17*, 3.
4. Fendler, J. H. *Isr. J. Chem.* **1985**, *25*, 3.
5. O'Brien, D. F.; Kingbiel, R. T.; Specht, D. P.; Tyminsky, P. N. *Ann. N. Y Acad. Sci.* **1985**, *446*, 282.
6. Bader, H.; Dorn, K.; Hupfer, B.; Ringsdorf, H. *Adv. Polym. Sci.* **1985**, *64*, 1.
7. Ringsdorf, H.; Schlarb, B.; Venzmer, J. *Angew. Chem. Int., Ed. Engl.* **1988**, *27*, 113.
8. Paleos, C. N. *Rev. Macromol. Chem. Phys.* **1990**, *C30*, 379.
9. Nagai, K.; Elias, H. G. *Makromol. Chem.* **1987**, *188*, 1095.
10. Paleos, C. M.; Davis, P.; Malliaris, A. *J. Polym. Sci., Polym. Chem. Ed.* **1984**, *22*, 3383.
11. Arai, K.; Miyahara, S.; Okabe, T. *Makromol. Chem.* **1991**, *192*, 2183.
12. Toshima, N.; Takahashi, T.; Hirai, H. *Chem. Lett.* **1986**, *35*.
13. Toshima, N.; Takahashi, T.; Hirai, H. *Chem. Lett.* **1987**, 1031.

Molecular Complexes

See: *Inclusion Complexes (Overview)*
Microcapsules, Molecular Complexes (Dissolution of Drugs by Cyclodextrin)
Molecular Complexes [Poly(ethylene oxide) and Urea]

MOLECULAR COMPLEXES [Poly(ethylene oxide) and Urea]

Bogdan Georgiev Bogdanov
Central Research Laboratory
Bourgas Technological University

Urea is well known to form crystalline molecular complexes (inclusion compounds, adducts) with various low molecular weight straight chain compounds such as n-alkanes.[1–14] Complexes with polymers were first found with poly(ethylene oxide) (PEO).[15,16] More recently, molecular complexes between urea and other linear homopolymers such as polyethers, polybutadiene, polyethylene, polypropylene, polyesters, poly(L-lactic acid) and block copolymers of PEO were prepared and characterized.[18–34]

It has been reported that PEO and other linear polymers also form complexes with thiourea.[16,33,35–38]

Crystallization of urea-hydrocarbon systems proceeds with the formation of hexagonal urea lattices in which guest molecules are enclosed by *trans* (zig-zag) conformation.[39] The urea channel structure is stable only when the channels are packed densely with guest molecules. Removal of the guest molecules results in a collapse of the channel structure with the urea recrystallizing in the tetragonal phase of "pure" urea.[6,40–42]

For polyethers -[-$(CH_2)_m$-O-]-$_n$ with $m > 3$, the urea adducts are similar to those obtained with n-alkanes.[43] However the crystal structures of complexes between PEO ($m = 2$) (including its oligomer and higher molecular weight PEO) and urea are different from those of urea-n-alkane complexes.[13,16,37,44–46]

A molecular complex (MC) of PEO contains about 2 moles of urea per monomeric ethylene oxide unit.[36,43] The crystalline structure of the MC differs from the structures of PEO and urea themselves. The melting temperature of MC determined calorimetrically is 145°C, while the PEO melting point is 64°C, and urea melts at 134°C. Infrared spectroscopic (IR) studies show that hydrogen bonds form between the oxygen atoms of the ethylene oxide units and the amino groups of urea. These bonds are susceptible to partial destruction as the MC melts.[47] It was assumed that the conformation of PEO macromolecules in the crystalline state of the MC is similar to that of the initial bulk PEO (i.e., tgg in a 7_2 helix).[46,48,49]

Low-molecular-weight PEO with a number-average molecular weight from 400 to 2000 was found to crystallize in two modifications of the MC with urea (Form I and II). The metastable crystalline molecular complex (Form II) has a melting temperature T_m = 107°C that is lower than Form I molecular complex, which is stable at room temperature.[49]

This chapter deals with properties of molecular complexes between urea and polyethers and their dependencies of various factors.

PREPARATION

Several methods for PEO-urea molecular complexes (MC) preparation are known.

The first mixes a solution of PEO with solid urea.[7,16,17,35–37,48,50]

A second method uses coprecipitation of benzene solution of PEO with urea solution in methanol.[36,37,51]

Another method immerses a bulk PEO (thin oriented films) in methanolic solution of urea at room temperature.[34,46]

A molecular complex of urea and polyethers with $m > 3$ is prepared like the PEO-urea complex by adding a methanolic solution of urea to benzene solution of polyethers or by immersing polyethers directly in a methanolic or an ethanolic solution of urea.[18]

STRUCTURE AND PROPERTIES

Crystal Structure (Polymorphism) and Structural Transformations at Heating and Cooling

X-Ray Analysis

The results of X-ray analysis show that all urea-n-alkane adducts have hexagonal structure.[37,57]

The Form II crystal modification is found for the complexes of urea with ethylene oxide oligomers ($\overline{M}_n$ from 400 to 1000). The Form II complex has a tetragonal unit cell (a = 7.30 Å, c = 19.51 Å) containing eight molecules of urea and eight ethylene oxide units.

The Form II modification is metastable. For molecular complex of urea and PEO with $\overline{M}_n$ = 1000 a solid-solid transition to the conventional Form I crystal occurs when heating up to a temperature between the melting points of both forms (from 117 to 125°C).[34]

PROPERTIES AND PARAMETERS

Complexes of urea with ethylene oxide oligomeric mixtures ($\overline{M}_n$ from 200 to 20000) were investigated by IR-absorption.[49] The features of the bands mainly due to urea and PEG, respectively, were different in both modifications (Forms I and II) of MC. In the region of 1400 to 800 cm^{-1} many bands of the molecular complex were split and the spectrum of Form I resembled the spectrum of PEO more than that of Form II of MC.[49]

On the basis of IR spectra it has been considered that PEO molecules in the urea complexes assume a helical conformation that is similar to the 7_2 helix in the ordinary PEO crystal.[37,46]

The results from our study of IR absorption confirm an earlier suggestion of the existence of a high-temperature metastable crystalline molecular complex Form II between high-molecular weight PEO and urea.[51,53,55] As mentioned above the Form II crystal modification of urea-PEO complex is obtained from Suehiro and Nagano using low-molecular-weight PEO.[49]

The complexation of urea and high-molecular-weight PEO ($\overline{M}_v$ = 2.7.10^6) is studied in a temperature interval from −130 to 150°C by IR spectroscopy.[54,56]

Thermodynamic Parameters of Physical Transformation (T_m, T_c, Δh_m, T_g)

The melting temperature of PEO increases from 51 to 64°C when the molecular weight ranges from 1500 to 300000.[37] The melting temperature of urea-PEO complex is higher than the melting temperature of urea (132.7°C) and ranges from 138 to 145°C depending on the molecular weight of PEO.[58,59] A steep increase of the melting temperature of Form I with molecular weight is observed for PEOs up to 2000 and then the increase becomes significantly less.[49]

Solubility

The molecular complexes between urea and PEO are soluble in the common strong solvents of both components (PEO and urea): water, tetrahydrofuran, dimethylformamide. They are not soluble in ether, cyclohexane or heptane.[37]

Thermal Stability

The thermal decomposition of the high-molecular weight ($\overline{M}_v$ = 2.7 × 10^6) PEO-urea complex is investigated by means of a derivatograph (DTA).[52] It is established that PEO-urea MC decomposes on heating with an absorption of heat, like pure urea. The DTA thermal curves indicate two basic steps of decomposition of both the urea and PEO-urea complex, with endotermic peaks at 245 and 385°C. The characteristic decomposition temperatures of the mass losses of PEO-urea MC are determined as T_o (beginning) = 145 to 150°C and T_{10} (10% decomposition) = 185 to 190°C.

APPLICATIONS

There is no information about particular application of PEO-urea molecular complex itself. A limited number of references consider the use of PEO-urea systems in compositions for water-soluble paints with an improved adhesion, in pharmaceuticals for external application, in pesticidal compounds, in cleansing compositions and for improvements in strength and crack resistance of PEO-based materials.[60–64] The complexation between PEO and urea can be use for fractionation or separation of PEOs and their copolymers as well.[19,32,37,44]

REFERENCES

1. Bengen, M. F. German Patent 869070, 1940.
2. Bengen, M. F.; Schlenk, Jr., W. *Experimentia* **1949**, *5*, 200.
3. Angla, B., *Comptes Rend.* **1949**, *565*, 204.
4. Zimmerschied, W. J.; Dinerstein, R. A.; Weitkamp, A. W.; Marschner, R. F. *Ind. Eng. Chem.* **1950**, *42*, 1300.
5. Takemoto, K. *Hosetsu Kagobutsu no Kagaku (Chemistry of Inclusion Compounds) (Modern Chemistry Series 40)*, Tokyo Kagaku Dojin: Tokyo, 1969.
6. McAdie, H. G.; Frost, G. B. *Can. J. Chem.* **1958**, *36*, 635.
7. Shimada, S.; Tanigawa, T.; Kashiwabara, H. *Polymer* **1980**, *21*, 10, 1116.
8. Patrilyak, K. I. *Zh. Fiz Khim.* **1980**, *549*, 2207.
9. Tolmachev, V. V.; Martirocov, R. A.; Gaile, A. A.; Simenov, L. V. *Gh. Phys. Chim.* **1984**, *58*, 2, 319.
10. Izumikawa, S.; Yoko, K. *Calorimetry and Therm. Anal.* **1985**, *12*, 57.
11. Matishev, V. A. *Izv. Vuzov "Neft i gaz"* **1981**, *2*, 45.
12. Lee, K J.; Mattice, W. L.; Snyder, R. G. *J. Chem. Phys.* **1992**, *96*(12), 9138.
13. Harris, K. D. M. *J. Solid State Chem.* **1993**, *106*, 83.
14. Shannon, I. J.; Harris, K. D. M.; Rennie, A. J. O.; Webster, M. B. *J. Chem. Soc. Fara.* **1993**, *89*(12), 2023.
15. Barker, G. E.; Ranauto, H. J. *J. Am. Oil. Chem. Soc.* **1955**, *32*, 49.
16. Parrod, J.; Kohler, A. *Compt. Rend. Hebd. Seances Acad. Sci.* **1958**, *246*, 1046.
17. Harkema, S.; Van Hummel, G. J.; Daasvant, K.; Reinhoudt, D. N. *J. C. S. Chem. Comm.* **1981**, 368.
18. Suehiro, K.; Urabe, A.; Yoshitake, Y; Nagano, Y. *Makromol. Chem.* **1984**, *185*, 11, 2467.
19. Schmidt, G.; Enkelmann, V.; Westphal, U.; Droescher, M.; Wegner, G. *Colloid Polym. Sci.* **1985**, *263*, 120.
20. Chenite, A.; Brisse, F. *Macromolecules* **1992**, *25*, 776.
21. Suehiro, K.; Kuramori, M. *J. Macromol. Sci. Phys.* **1994**, *B33*, 1.
22. Tonelli, A. E. *Polym. Prepr. (Am. Chem. Soc., Div. Polym. Chem.)* **1992**, *33*(1), 551.
23. Tonelli, A. E. *Polymer* **1994**, *35*, 573.
24. Yokoyama. F.; Monobe, K. *Polymer* **1980**, *21*, 968.
25. Hort, Y.; Tanigawa, T.; Shimada, S.; Kashiwabara, H. *Polym. J.* **1981**, *13*, 293.
26. Yokoyama, F.; Monobe, K. *J. Polym. Sci., Polym. Lett. Esd.* **1981**, *19*, 91.
27. Yokoyama, F.; Monobe, K. *Polymer* **1983**, *24*, 149.
28. Yokoyama, F: Monobe, K. *Nippon Kagaki Kaishi* **1983**, *2*, 259.
29. Yokoyama, F.; Monobe, K. *Mem. Sch. Eng., Okayama Univ.* **1985**, *19*(2), 1.
30. Mayer, R. *Bull. Sos. Chim. Fr.* **1966**, *9*, 2998.
31. Chenite, A.; Brisse. F. *Macromolecules* **1993**, *26*, 3055.
32. Hild, G. *Bull. Sos. Chim. Fr.* **1969**, *12*, 4531.
33. Hild, G. *Rev. Gen. Caout. Plast.* **1969**, *46*, 771.
34. Suehiro, K.; Nagano., Y. *Makromol. Chem., Rapid Commun.* **1983**, *4*, 137.
35. Parrod, J.; Kohler, A. *J. Polym. Sci.* **1960**, *48*, 457.
36. Bailey, Jr., F. E.; France, H. G. *J. Polym. Sci.* **1961**, *49*, 397.
37. Hild, G. *Bull. Sos. Chim. Fr.* **1969**, *8*, 2840.
38. Yan, X.; Zhang, B.; Yao, S.; Xu, X.; Wang, D.; Qian, B. *Gaofenzi Cailiao Kexue Yu Gongcheng* **1992**, *8*, 56.
39. Smith, A. E. *Acta Cryst.* **1952**, *5*, 224.
40. McAdie, H. G. *Can. J. Chem.* **1962**, *40*, 2195.
41. Harris, K. D. M. *J. Phys. Chem. Solids* **1992**, *53*, 529.
42. Wyckoff, R. W. J.; Corey, R. B. *Z. Kristallogr.* **1934**, *89*, 462.
43. Chenite, A.; Brisse, F. *Macromolecules* **1991**, *24*, 2221.
44. Parrod, J.; Kohler, A. *Int. Symp. Macromol. Chem., Mosscau*, June 14–19 **1960**, *Abstract, III*, 54.
45. Hill, F. N.; Bailey, Jr., F. E.; Fitzpatrick, J. T. *Ind. Eng. Chem.* **1958**, *50*, 5.
46. Todokoro, H.; Yoshihara, T.; Chatani, Y.; Murahashi, S. *J. Polym. Sci.* **1964**, *B2*, 363.
47. Tarnoruckij, M. M.; Shkoljnikova, L. S.; Loreij, A. K.; Dindoin, V. I.; Zykova, L. I. *Vysokomolekul. Soed.* **1975**, *B17*, 817.
48. Shimada, Sh.; Fujiwara, A.; Kashiwabara, H. *Polym. J.* **1981**, 13, 8, 769.
49. Suehiro, K.; Nagano, Y. *Makromol. Chem.* **1983**, *184*, 669.
50. Shimada, Sh.; Fujimara, A.; Kashiwabara, H. *Polym. J.* **1981**, *13*, 769.
51. Bogdanov, B.; Michailov, M.; Uzov, Ch.; Gavrailova, G. *Makromol. Chem. Phys.* **1994**, *195*, 2227.
52. Gurova, K.; Michailov, M.; Bogdanov, B.; Uzov, Chr. *J. Therm. Anal.* **1994**, *41*(1), 173.
53. Bogdanov, B.; Michailov, M.; Uzov, Ch. *J. Polym. Mater.* **1990**, *7*(2), 145.
54. Bogdanov, B.; Michailov, M.; Uzov, Kh.; Gavrailova, G. *J. Polym. Sci., Part B: Polym. Phys.* **1994**, *32*(2), 387.
55. Bogdanov, B.; Michailov, M.; Gavrailova, G.; Uzov, Chr. *10th national conference on molecular spectroscopy with international participation*, Bulgaria, Blagoevgrad, August 29th-September 3s, 1988, Abstracts, F 0 37/A, p. 59.
56. Bogdanov, B.; Michailov, M.; Gavrailova, G.; Uzov, Chr. *Analytical Laboratory*, (Sofia) **1992**, *1*, 1.
57. Ghoihman, A. Sh.; Solomko, V. P. *Polymer inclusion compounds* (Russ) (*Vysokomolecularnie coedinenja vklutschenia*), Nauk. Dumka. Kiev, 1982; pp 192.
58. Knight, H. B.; Witnauer, L. P.; Coleman, J. E.; Noble, N. R.; Swern, D. *Anal. Chem.* **1952**, *24*, 1331.
59. Miguel, R.; Perie, J. J.; Klaene, A. *Bull. Soc. Chim.* **1964**, 2275.
60. U.S. Patent 4 173 554 (Cl. 260/292), 1978.
61. JP Patent 82 142 259 (Cl. A61L15/06), 1982.
62. JP Patent 04 193 804 (Cl. A01N37/36), 1992.
63. JP Patent 05 310 542 (Cl. A61K7/075), 1993.
54. Kashikar, S. P.; Goethals, E. J. *New Polymeric Mater.* **1993**, *4*, 1.

Molecular Composites

See: *Anoxic Polymer Materials*
Bisphenol-A-Polycarbonate/Polyester Blends
Inclusion Polymerization
Molecular Composites (High Modulus Polyimide/Polyimide Films)

Molecular Composites (Liquid Crystalline Polyamide and Amorphous Polyimide)
Molecular Composites (Polyimide/Polyimide, Laminate Processability)
Molecular Composites (Third Generation Polymers)
Polybenzimidazoles (Overview)
Poly[N,N′-(sulfo-p-phenylene)terephthalamide]

MOLECULAR COMPOSITES (High Modulus Polyimide/Polyimide Films)

Rikio Yokota
Institute of Space and Astronautical Science
Ministry of Education

A molecular composite (MC) is defined as a polymer blend consisting of two or multi-components dispersed at the molecular level. MC can reinforce a ductile matrix polymer with high modulus, rigid-rod polymer molecules, and overcome problems such as interfacial adhesion, thermally induced stresses, and voids in macroscopic composite.

In polyimides, even the rigid-rod polyimide(PI) precursor, polyamic acid (PAA), is freely soluble in polar solvents. It gives good processability to the fibers and films for polyimides. Yokota developed high modulus and high strength polyimide films from thermal imidization after cold drawing of polyamic acid films.[1,2] Furthermore, high strength and high modulus polyimide/polyimide(PI/PI) MC films, in which rigid-rod polyimides embedded in a ductile matrix of flexible polyimides were prepared using the blend of PAA stage in solution.[3,4] This article presents the preparation and properties of high strength and high modulus PI/PI MC films.

PREPARATIONS

Rigid polybiphenyltetracarboxydiimide with *p*-phenylene PI(BPDA/PDA) and flexible oxydianiline (ODA) containing polyimide PI(BPDA/ODA) have been used for the successful system of the PI/PI Mcs. **Figure 1** shows the chemical structure and symbol of polyimides.

PROPERTIES OF HIGH MODULUS POLYIMIDE MC FILMS

The glass transition temperature systematically moves to the higher temperature with an increasing content of rigid PI(BPDA/PDA). This indicates the rigid (PI(BPDA/PDA) molecules are well mixed in a matrix of flexible PI(BPDA/ODA) molecules. Temperature dependence of thermal expansion coefficients below T_g also becomes lower in proportion to rigid PI content, and the curves for the elongation in the stretching direction for the 60% drawn film has a small negative temperature dependence.[4] These results show that the rigid PI molecules work as reinforcements and molecularly disperse in a matrix PI(BPDA/ODA) molecules.

The temperature dependence of the loss modulus for the polyimide blend film containing 70% rigid PI(BPDA/PDA) with and without cold drawning indicates a single absorption peak around 340°C due to T_g of MC. The storage modulus of the drawn film is remarkably high in comparison to that of the undrawn films, and the fall of the storage modulus E′ over the T_g region is small. These data provide evidence of the polyimide molecular composite two polyimides dispersed at molecular level.

The tensile modulus of the 60% drawn PI/PI MC films increases in proportion to the reinforcement of PI(BPDA/PDA) content. After annealing at 330°C for 2 hours, the modulus becomes higher by about 1.5 times in comparison to the nonannealed samples. The ultimate tensile strength also increases as the PI(BPDA/PDA) content increases.

REFERENCES

1. Kochi, M.; Uruji, T.; Iizuka, T.; Mita, I.; Yokota, R. *J. Polymer Sci., Polym. Lett.* **1987**, *25*, 441.
2. Kochi, M.; Yokota, R.; Iizuka, T.; Mita, I. *J. Polymer Sci., Part B: Polymer Phys.* **1990**, *28*, 2463.
3. Yokota, R.; Horiuchi, R.; Kochi, M.; Soma, H.; Mita, I. *J. Polymer Sci., Part C: Polymer Lett.* **1988**, *26*, 215.
4. Yokota, R.; Horiuchi, R.; Kochi, M.; Takahashi, C.; Soma, H.; Mita, I. *Polyimides: Material, Chemistry and Characterization* Feger, C., Ed.; Elsevier Sci.: New York, NY, 1989; 13.

MOLECULAR COMPOSITES (Liquid Crystalline Polyamide and Amorphous Polyimide)

Yu-Der Lee* and Ken-Yuan Chang
Department of Chemical Engineering
National Tsing Hua University

"Molecular Composite" (MC) is a newly developed concept recently introduced in composite materials and polymer science. The concept consists of combining two components with a dissimilar characteristic. For example, a stiff and strong rigid-rod polymer can be dispersed in a matrix of flexible polymer

*Author to whom correspondence should be addressed.

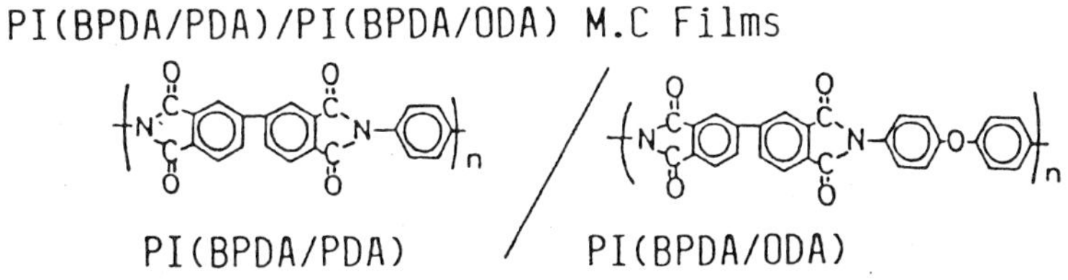

FIGURE 1. Structures and symbols of polyimides.

at a molecular level. Having a high aspect ratio, this rigid-rod polymer serves as a reinforcing element similar to the fiber employed in traditional fiber-filled composites.

Polyimides (PI) are one of the most important polymers with a high thermal stability despite low solubility in common organic solvents. In this article, block copolymers of liquid-crystalline poly(p-benzamide) (PBA) and amorphous PI are synthesized. For copolymerization to occur, however, organic soluble polyimides are the only choice for the observed system of PBA/PI molecular composites. Some major approaches toward synthesizing organic soluble polyimides were suggested by Harris et al.[1] According to previous studies concerning organic soluble polyimides in the observed laboratory, polyimides prepared from 3,3′,4,4′-benzophenonetetracarboxylic dianhydride (BTDA)/2,3,5,6,-tetramethyl-*p*-phenylene diamine (TMPD) are selected for block copolymerization with PBA.[2] The product obtained is confirmed to be a block copolymer.

PREPARATION

Preparation of Acid-Terminated PBA Prepolymers

Acid-terminated PBA prepolymer was prepared using Yamazaki's phosphorylation reaction.[3] A bifunctional monomer, terephthalic acid (TA), was applied to control molecular weight and end-group functionality.

Preparation of Amine-Terminated Polyimide Prepolymer

Amine-terminated polyimide prepolymer was synthesized from BTDA and excess TMPD using a conventional low-temperature solution polycondensation reaction. The imidization was restrained to a special approach.

Preparation of PBA/PI Block Copolymers

PI prepolymer solution was diluted with NMP, pyridine, and LiCl and mixed with the aforementioned PBA prepolymer solution. A further phosphorylation reaction between these two prepolymers proceeded to produce block copolymers.

PROPERTIES

PBA/PI Block Copolymers

Inherent viscosities of copolymers were determined by a Ubbelhode viscometer at 0.5 g/dL polymer solution in DMAc containing 4% LiCl at 30°C. The inherent viscosities of block copolymers are markedly larger than those of corresponding prepolymers. The significant increase in inherent viscosity of the final product after extraction verifies that the final product is a block copolymer with high molecular weight.

Infrared spectra analysis provided more evidence and confirmed that a successful copolymerization was achieved under such reaction conditions.

Solubility and Lyotropic Behavior

Copolymer solutions with increasing concentrations in NMP-LiCl have been prepared to be characterized for their lyotropic behavior. Measurements taken by a Brookfield viscometer revealed that the bulk viscosity of copolymer solution initially increased with concentration, reached a maximum with the onset of liquid crystalline order, and then decreased with a further increase of concentration. This behavior is a unique character of liquid-crystalline polymers and a powerful recognition of mesomorphic state. The concentration at which the peak of maximum viscosity appears is referred to as a critical concentration (C_{cr}).

Direct observation of liquid crystalline texture in a microscope also can identify a lyotropic state. A liquid-crystalline polymer will yield an optically anisotropic phase and exhibit colored domains when a thin layer of solution is constructed and magnified between crossed polarizers. The PBA/PI block copolymers could form anisotropic (liquid crystalline) solutions at room temperature in DMAc-LiCl, NMP-LiCl, and sulfuric acid, and they revealed a thread-like texture whenever concentrations were higher than C_{cr}

The block copolymer, consisting of rigid PBA and flexible PI, could be appropriately processed into a MC. This new material might be applied in highly oriented fibers with excellent mechanical properties and thermal stability.

ACKNOWLEDGMENT

A portion of this article has been published in the *Journal of Polymer Science, Polymer Chemistry*, Vol. 31, pp 2775–2784.

REFERENCES

1. Harris, F. W; Feld, W. A.; Lanier, L. H. *Polym. Prepr.* **1976**, *17*(2), 353.
2. Lee, H-R; Lee, Y-D. *J. Polym. Sci. Polym. Chem.* **1989**, *27*, 1486.
3. Yamasaki, N.; Matsumoto, M.; Higashi, F. *J. of Polym. Sci., Polym. Chem. Ed.* **1975**, *13*, 1373.

MOLECULAR COMPOSITES (Polyimide/Polyimide, Laminate Processability)

Tsutomu Takeichi
School of Materials Science
Toyohashi University of Technology

Molecular composite (MC), a kind of polymeric material, disperses rigid polymer in flexible polymer in the molecular order. High performance is expected if the rigid polymer is well dispersed in the flexible polymer, because this improves adhesion between the rigid polymer and the flexible polymer, increases aspect ratio, and decreases the defects of the rigid polymer.

Recently, a new type of molecular composite was proposed utilizing rigid polyimide as reinforcement and flexible polyimide as the matrix.[1] Researchers obtained high modulus and high strength MC films by drawing the films at the stage of poly(amide acid)s. The polyimide/polyimide MC (PI/PI MC) is superior to the previously developed MC systems.

Processability remains a problem with the MC systems. The thermoplastic polymer tends to induce segregation when heated for laminate processing. Use of thermoset polymers has not been successful either.[2] However, PI/PI MC that is processable as laminate resin should be possible using thermoset-type polyimide if two components are properly designed. During our study on the crosslink sites for polyimides and polyamides, we found the acetylene unit introduced into the polymer backbone played an excellent crosslink site.[3–6] Thus we designed a PI/PI MC system consisting of rigid polyimide and flexible polyimide having acetylene units in the backbone, which was found to give MC with excellent properties and laminate processability.[7]

REFERENCES

1. Yokota. R.; Horiuchi, R.; Kochi, M.; Soma, H.; Mita, I. *J. Polym. Sci. Part C: Polym. Lett.* **1989**, *26*, 215.
2. Chuah, H. H.; Tan, L.-S.; Arnold, F. E. *Polym. Eng. Sci.* **1989**, *29*, 107.
3. Takeichi, T.; Stille, J. K. *Macromolecules* **1986**, *19*, 2093, 2103, 2108.
4. Takeichi, T.; Date, H.; Takayama. Y. *J. Polym. Sci., Part A: Polym. Chem.* **1990**, *28*, 1989, 3377.
5. Takeichi, T.; Ogura, S.; Takayama, Y. *J. Polym. Sci., Part A: Polym. Chem.* **1994**, *32*, 579.
6. Takeichi, T.; Kobayashi, A.; Takayama, Y. *J. Polym. Sci., Part A: Polym. Chem.* **1992**, *30*, 2645.
7. Takeichi, T.; Takahashi, N.; Yokota, R. *J. Polym. Sci., Part A: Polym. Chem.* **1994**, *32*, 167

MOLECULAR COMPOSITES (Third Generation Polymers)

Sanjay Palsule
Department of Chemistry
Heriot-Watt University

Polymer materials have been classified into three generations. Homopolymeric and copolymeric, thermoplastics, thermosets, and elastomers have been classified as the first generation polymers. Particulate and fiber reinforced polymer matrix composites and polymer blends have been classified as the second generation polymers. Molecular composites based on rigid and flexible polymers have been classified as the third generation polymers.[1–3]

MISCIBILITY AND STRUCTURE OF A MOLECULAR COMPOSITE

Structure of a molecular composite is governed by the miscibility between the rigid and the flexible polymer because phase separation adversely affects the structure and properties of a molecular composite. It is difficult to achieve a miscible molecular composite owing to the low combinatorial entropy of mixing of a rigid and a flexible polymer and high tendency of self alignment for the rigid polymer.[9]

However, recent investigations indicate that there are four ways in which a miscible molecular composite could be obtained:

(1) By introducing a favorable mixing entropy by attaching a flexible side chain to the rigid rod.[11,12]

(2) By incorporating complimentary functional groups on flexible coils and on side chains attached to rigid rods that are capable of forming strong specific interactions between the rod and the coil.[13–15]

(3) By blending a flexible and a rigid polymer capable of forming intermolecular specific interactions (hydrogen bonds) that result in an exothermic heat of mixing.[16,17]

(4) By enhancing combinatorial entropy of mixing by solubility parameter approach. That is by blending a flexible polymer with a rigid polymer made up of groups chemically similar to groups in the flexible polymer but imparting rod-like conformation to the rigid polymer.[2,6–8,10,18,19]

PHASE SEPARATION

Most of the molecular composites exhibit limited miscibility and undergo thermally induced phase separation.

MORPHOLOGY

Morphology of a molecular composite is governed by miscibility.

DYNAMIC MECHANICAL THERMAL PROPERTIES

In addition to indicating miscibility in a molecular composite, dynamic mechanical thermal measurements also indicate dynamic modulus and damping of the molecular composite. Damping indicates imperfections in elasticity. Maximum damping occurs in the glass transition region and tan-δ (α) peak passes through a maximum. In this region much of the energy used to deform the material is dissipated as heat and is manifested in transition of the material from glassy to the rubbery state. In the region beyond the temperature of maximum α damping the material behaves like a leather.

VISCOELASTIC BEHAVIOR AND DYNAMIC MODULUS

The modulus of a polymer as a function of temperature indicates the viscoelastic properties of the material. In the glassy region, the modulus of molecular composite increases with increasing amounts of rigid polymer in the molecular composite.

MECHANICAL PROPERTIES

In a molecular composite, the ultimate reinforcement is a single extended rigid macromolecule.

Stress–Strain Relationship

The stress–strain behavior is at best only a rough guide to behavior of a molecular composite in an engineered object owing to its viscoelastic behavior. Palsule studied the stress–strain behavior of anoxic molecular composites of PAI/PSI.[2,6]

Young's Modulus

The Young's modulus of a molecular composite is higher than the modulus of the constituent flexible polymer and increases with increasing amounts of the rigid polymer. The modulus in the glassy state is determined by the aspect ratio of the reinforcing rigid polymer and strength of the intermolecular forces.[1,2,4,20] An increase in the amount of the rigid polymer in a molecular composite increases the strength of the intermolecular forces and thereby the modulus of the molecular composite.[2]

Tensile Strength

The tensile strength of several miscible molecular composites is known to increase with increasing amounts of the rigid polymer. Palsule states that tensile strength of a molecular composite is governed by the strength of the covalent bonds, intramolecular forces of the rigid polymer and miscibility between the rigid and the flexible polymer.[2] An increase in the amount of rigid polymer in the miscible molecular composite increases the amount (number) of covalent bonds carrying the applied tensile loads and thereby increases the tensile strength.

Yield Strength

Yield strength of a miscible molecular composite is known to increase with increasing amounts of reinforcing rigid macromolecules in the molecular composite. However, the yield strength of anoxic molecular composites decreases with increasing amounts of reinforcing rigid polymer.[2]

With the increase in the amount of rigid polymer in the molecular composite, the rigid macromolecules preferentially align owing to inter- and intramolecular interactions.[9,21] Thus less energy is dissipated in orienting the macromolecules and more energy is available to stretch the molecules during plastic deformations, which increases the yield strength of molecular composites.

Elongation at Break

Elongation at break is governed by the energy required to orient and stretch the molecules in the direction of applied load. With the increasing amounts, the rigid macromolecules in a molecular composite are preferentially aligned. Thus less energy is required to reorient molecules in a direction of elongation and more energy is available to stretch the molecules before breaking which increase the elongation at break.[2]

Toughness

Toughness is related to impact strength and indicates the energy that a material can absorb before breaking. Toughness is directly proportional to elongation at break.

An increase in the amount of rigid polymer in a molecular composite increases the amounts of energy-absorbing rigid molecule that buckle but do not break while absorbing energy and increase toughness.

A NEW THERMODYNAMIC AND MECHANICAL APPROACH TO MOLECULAR COMPOSITES

Owing to its miscible nature, a molecular composite is homogeneous and its properties are governed by properties of the constituent polymers, in particular the flexible polymer. Because of its miscible homogeneous nature, there is no segregated phase that could support loads acting on a molecular composite. Thus the miscible homogeneous nature, absence of load bearing segregated phase, and limited properties of the constituent polymer characterize a molecular composite as a material with properties better than those of the constituent flexible polymer, but not as good as those of a fiber reinforced composite.

Thus mechanics and thermodynamics govern that a molecular composite should be based on a flexible polymer of poor mechanical properties but special molecular properties (e.g., optical, electrical, or surface properties). For example, the anoxic molecular composites of low surface energy organosiloxane copolymer reinforced molecularly by a rigid polymer, exhibiting properties better than those of the constituent organo-siloxane polymer, but not as good as those of a fiber reinforced composite, are emerging as advanced materials for low earth orbit spacecrafts.[5,22]

ACKNOWLEDGMENTS

Support of the work by the Rotary Foundation of Rotary International (1990-91) and Materials & Processes Division of European Space Agency (1991-1996) is gratefully acknowledge.

REFERENCES

1. Palsule, S. *ESA Journal* **1993**, *17*, 133.
2. Palsule, S. Ph.D. Thesis, Heriot-Watt University, Edinburgh (UK) 1994.
3. Palsule, S. *Polym. Eng. Sci.*, submitted 1996a.
4. Hwang, W. F.; Wiff, D. R.; Benner, C. L.; Helminiak, T. E. *J. Macromol. Sci. Phys.* **1983**, *B22*, 231.
5. Palsule, S. European Patent Application, (To ESA), EP632100A1, 1995.
6. Palsule, S. *Polym. Eng. Sci.*, submitted 1996b.
7. Palsule, S. *Polym. Eng. Sci.*, submitted 1996c.
8. Palsule, S. *Polym. Eng. Sci.*, submitted 1996d.
9. Flory, P. J. *Macromolecules* **1978**, *11*, 1138.
10. Palsule, S.; Cowie, J. M. G. *Polym. Bull.* **1994**, *33*, 241.
11. Ballauff, M. *Mol. Cryst. Liq. Cryst.* **1986**, *136*, 175.
12. Ballauff, M. *J. Polym. Sci. Part B, Polym. Phys.* **1987**, *25*, 739.
13. Painter, P. C.; Tang, W. L.; Graf, J. F.; Thomson, B.; Colman, M. M. *Macromolecules* **1991**, *24*, 3929.
14. Stein, R. S.; Setumadhavan, M.; Gaudiana, R. A.; Adams, F.; Guarrera, D.; Roy, S. K. *Pure and Appl. Chem.* **1992**, *29*, 517.
15. Tang, W.-L.; Colman, M. M.; Painter, P. C. *Macromol. Symp.* **1994**, *84*, 315.
16. Kyu, T.; Chen, T. I.; Park, H. S.; White, J. L. *J. Appl. Polym. Sci.* **1989**, *37*, 201.
17. Park, H. S.; Kyu, T. *Polym. Compos.* **1989**, *10*, 429.
18. Fukai, T.; Yang, J. C.; Kyu, T.; Cheng, S. Z. D.; Lee, S. K.; Hsu, S. L. C.; Harris, F. W. *Polymer* **1992**, *33*, 3621.
19. Palsule, S. *Polym. Eng. Sci.*, submitted 1996e.
20. Takayanagi, M. *Pure and Appl. Chem.* **1983**, *55*, 819.
21. Flory, P. J. *Macromolecules* **1978**, *11*, 1141.
22. Palsule, S. *Aerospace Polymers and Composites* 1st ed., Praxis: Chichester, U.K., 1996.

Molecular Recognition

See: *Antibody-Polymer (Porphyrin) Complexes*
Calixarenes (as Supramolecular Cyclic Oligomers)
Cyclodextrins (Thermodynamic Functions of Molecular Recognition)
Enzyme Mimics (Cyclodextrins)
Host-Guest Chemistry [Using α-Helical Poly(L-lysine)]
Macromolecular Recognition (New Supramolecular Structures)
Matrix Polymerization (Overview)
Metal-Polymer Complexes
Metalloenzymes, Artificial (Polyethylenimine-Based)
Molecular Recognition (Interpolymer Interactions)
Molecular Recognition (Macromolecular Ionophore)

Molecular Recognition (Peptide-Based Systems)
Molecular Shape Recognition
Poly(α-amino acid) Spherical Particles

MOLECULAR RECOGNITION (Interpolymer Interactions)

Ivan M. Papisov and A. A. Litmanovich
Moscow State Automobile and Road Technical University

Molecular recognition in interpolymer interactions is defined as selective noncovalent binding of a given macromolecule—the "recognizer"—with a certain macromolecule(s) in a mixture of macromolecules differing from one another in their structure or length, or both, each of which taken separately can bind with the recognizer. This phenomenon has been reviewed in detail.[1–5] Molecular recognition is a consequence of cooperativity of systems of intermolecular noncovalent bonds between structural (monomer) units of interacting macromolecules; a fragment of the bond systems may be schematically represented as shown in **Scheme I**, where A and B denote the units of respective macromolecules having affinity for one another. Examples of intermolecular A...B bonds are interactions between poly(methacrylic acid) (PMAA) and poly(ethylene

-A-A-A-A-A-
 ⋮ ⋮ ⋮ ⋮ ⋮ **1**
-B-B-B-B-B-

oxide) (PEO), and between poly(acrylic acid) (PAA) and polyvinylpyridine (PVPy).

The products of interpolymer interactions are termed polycomplexes or alternatively interpolymer complexes of polymer–polymer complexes; particular names such as interpolyelectrolyte complexes and stereocomplexes dealing with polycomplexes formed by, respectively, oppositely charged and stereoisomeric macromolecules, are in use as well. At mutual saturation of macromolecular components each polycomplex is usually characterized by definite composition; in the majority of cases, the stoichiometry of A and B units is 1:1.

Table 1 summarizes the available literature data concerning systems in which molecular recognition comprising interpolymer substitution reactions was studied.

APPLICATIONS

Molecular recognition may be effectively used in fractionation of nonuniform polymers and mixtures of polymers, separation of free macromolecules from their polycomplexes with other macromolecules or species, etc. High sensitivity of a macromolecule-recognizer in molecular recognition to very minor differences in molecular structure of macromolecules and species offers the unique possibility of precise addressing of the recognizer to a certain molecular structure ("target"). This property of macromolecular reagents is a very attractive possibility for creation of biologically active macromolecular species intending for recognition of some definite structures in living organisms.[8a]

Molecular recognition plays a very important role in *Matrix Polymerization (Overview)* (see article). Recognition of growing particles in the course of new phase formation in the presence of macromolecules, in principle, offers the possibility to create nano-structure systems.[9,10]

REFERENCES

1. Kabanov, V. A.; Papisov, I. M. *Vysokomol. Soed.* **1979**, *A21*, 243.
2. Tsuchida, E.; Abe, K. *Adv. Polym. Sci.* **1982**, *45*, 1.
3. Papisov, I. M.; Litmanovich, A. A. *Adv. Polym. Sci.* **1989**, *90*, 140.
4. Papisov, I. M.; Litmanovich, A. A. In *Macromolecular Reactions*; Plate, N. A., Ed.; John Wiley & Sons: New York, NY, 1995; Chapter 7.
5. Kabanov, V. A. *Vysokomol. Soed.* **1994**, *A36*, 183.
6. Papisov, I. M.; Nedyalkova, Ts. I.; Avramtchuk, N. K.; Kabanov, V. A. *Vysokomol. Soed.* **1973**, *A15*, 2003.
7. Abe, K.; Koide, M.; Tsuchida, E. *Macromolecules* **1977**, *10*, 1259.
8. Kokufuta, E.; Nakamura, I.; Yokota, A. *Polymer* **1983**, *24*, 1031.
8a. Kabanov, V. A.; Kabanov, A. V. *Vysokomol. Soed.* **1994**, *A36*, 198.
9. Papisov, I. M.; Osada, Y.; Okudzaki, H.; Ivabushi, T. *Polym. Sci.* **1993**, *35*, 105.
10. Papisov, I. M.; Yablokov, Yu. S.; Prokof'ev, A. I. *Polym. Sci.* **1994**, *36*, 291.
11. Papisov, I. M.; Nekrasova, N. A.; Pautov, V. D.; Kabanov, V. A. *Dokl. Akad. Nauk SSSR* **1974**, *214*, 861.
12. Litmanovich, A. A. *Vestnik Moskovskogo Universiteta, seriya khimiya* **1978**, *19*, 617.
13. Korugic-Perkovic, L.; Ferguson, I. *Polymeri (SFRY)* **1983**, *4*, 301.
14. Litmanovich, A. A.; Anufrieva, E. V.; Papisov, I. M.; Kabanov, V. A. *Dokl. AN SSSR* **1979**, *246*, 923.
15. Litmanovich, A. A.; Papisov, I. M.; Kabanov, V. A. *Eur. Polym. J.* **1981**, *17*, 981.
16. Litmanovich, A. A.; Papisov, I. M.; Kabanov, V. A. *Vysokomol. Soed.* **1980**, *A22*, 1180.
17. Kikuchi, Y.; Kubota, N. *Makromol. Chem. Rapid. Commun.* **1985**, *6*, 387.
18. Izumrudov, V. A.; Bronich, T. K.; Zezin, A. B.; Kabanov, V. A. *Dokl. AN SSSR* 1984, 278, 404.
19. Baranovsky, V. Y.; Gnatko, N. N.; Litmanovich, A. A.; Papisov, I. M. *Vysokomol. Soed.* **1988**, *A31*, 984.
20. Korobko, T. A.; Izumrudov, V. A.; Zezin, A. B. *Vysokomol. Soed.* **1993**, *A35*, 87.
21. Korobko, T. A.; Izumrudov, V. A.; Zezin, A. B.; Kabanov, V. A. *Vysokomol. Soed.* **1994**, *A36*, 223.
22. Korobko, T. A.; Izumrudov, V. A.; Zezin, A. B.; Kabanov, V. A. *Doklady RAN* **1994**, *338*, 1.
23. Izumrudov, V. A.; Platonova, O. A.; Korobko, T. A. *Vestnik Moskovskogo Universiteta, Seriya 2, Khimiya* **1994**, *35*, 4.
24. Litmanovich, A. A.; Kirsh, Yu. E.; Papisov. I. M. *Vysokomol. Soed.* **1978**, *B20*, 83.

MOLECULAR RECOGNITION (Macromolecular Ionophore)

Kazuaki Yokota and Toshifumi Satoh
Division of Molecular Chemistry
Graduate School of Engineering
Hokkaido University

Toyoji Kakuchi
Division of Bio Science
Graduate School of Environmental Earth
Hokkaido University

TABLE 1. Investigated Systems

No.	Recognizer P_r	Competitors ΣP_i or P	Type of reaction and details	References
1	PMAA (polymer)	PVPd + PEO PVPd + PVA PEO + PVA	MR, substitution MR in matrix polymerization mechanism of substitution (proposal)	**6**
2	PEO (oligomer)	PMAA + PAA	MR, substitution; substitution in matrix polymerization	**11**
3	PMAA (polymer)	PVPd + PEO, PVPd + PAAm, PVPd + PVA, PEO + PAAm PEO + PVA, PAAm + PVA	MR, substitution; Inversion of MR caused by change in DP of PEO	**7**
4	PMAA (polymer)	PVPd + PEO	MR; inversion of MR caused by change in DP of competitors	**12**
5	PAA (polymer)	PVPd + PEO	MR	**13**
6	PVPd or PEO (oligomers)	PMAA (i:h:s = 6:41:53) + PMAA (i:h:s = 14:32:54)	MR; dependence on nature of P_r and DP of P_r; evaluation of $\Delta\Delta G°$	**14,15**
7	PVPd (oligomer)	PMAA + copoly(MAA AA)	MR; dependence on DP of P_r and composition of copolymer; evaluation of $\Delta\Delta G°$	**15,16**
8	PMAA (polymer)	PVPd + P2VP PVPd + PEI	MR; inversion of MR caused by change in pH	**7**
9	PEAD	PVSA + PGA	MR; inversion of MR caused by change in pH	**17**
10	P4VP (intermediate DP)	PMAA (polymer) + PPh (oligomer)	Substitution; inversion of MR caused by change in DP of PPh and in nature and concentration of small ions	**18**
11	PMAA (polymer)	PVPd + PEO-ML	MR in matrix polymerization	**19**
12	P4VP-Alk	PMAA + $C_nH_{2n+1}COONa$	Substitution; inversion of MR caused by change in n, temperature, length and structure of Alk, nature of water/alcohol solvent, pH, and degree of alkylation and nature and concentration of small ions, DP of P_r and PMAA	**20–23**
13	PAA (polymer)	PVPd (polydisperse)	Fractionation over DP	**24**
14	PAA (polymer) PAA (oligomer) PEO (oligomer)	PEO (polydisperse oligomer) PEO (polydisperse polymer) PAA (polydisperse polymer)	Fractionation over DP; dependence on width of initial MWD and temperature	**8**
15	PAA (polymer)	P4VP-49 (nonuniform)	Fractionation over composition	**15,16**

MR, molecular recognition; DP, degree of polymerization; PAA, poly(acrylic acid); PAAm polyacrylamide; PEAD, poly(diethylamino-dextran); PEI, polyethyleniminepoly; PEO, poly(ethylene oxide); PEO-ML, poly(ethylene oxide) monolaurate; PGA, poly(glutamic acid); PMAA, poly(methacrylic acid); PPh, sodium polyphosphate; PVA, poly(vinyl alcohol); PVPd, poly(*N*-vinylpyrrolidone); PVSA, poly(vinylsulfuric acid); P2VP, poly(2-vinylpyridine); P4VP, poly(4-vinylpyridine); P4VP-Alk, poly(*N*-alkyl-4-vinylpyridine); P4VP-49, poly(4-vinylpyridine) quaternized with ethyl bromide to 49 mol %; i:h:s, ratio of iso-, hetero-, and sindio-triads.

Ionophores are receptors that form stable, lipophilic complexes with charged hydrophilic species containing metal ions and organic molecules, and thus transport them into lipophilic phases across natural or artificial membranes.[1] The metal ion is entrapped in a cavity-like structure formed by the ionophore consisting of the cyclic or open-chain molecule. An example of the cyclic ionophore is given of valinomycin that consists of 12 subunits connecting with alternate peptide and ester bonds. The ligand cavity fills K+ of optimum size and thereby the affinity to K^+ is 10^4 times greater than that to Na^+. Monesin, as a remarkable example of the open-chain ionophores, is a polyether antibiotic with a terminal carboxy group. In the complex, the polyether antibiotics are wrapped around the metal ion, but the ion selectivities are somewhat lower than those of the cyclic ionophores according to the structural features.

Crown ether is one artificial ionophore capable of mimicking effectively natural, cyclic ionophore. Polymeric crown ethers also act as ionophores. The interest in macromolecular ionophores has been limited to the synthesis and use of polymeric crown ethers and little attention has been given to the synthesis of the macromolecular ionophores consisting of tetrahydrofuran and/or tetrahydropyran subunits. The ionophore belonging to this type can be found in three papers. The first macromolecular open-chain ionophore is threo-α,ω-poly(2,6-tetrahydrofurandiyl), which was synthesized through the ring expansion of the oxirane derived from the polymer of butadiene.[2] The second is poly(7-ocanorbornene) resulting from metathesis polymerization, which binds various cations containing methylene blue and rhodamine 6G.[3] Unlike the crown ethers, these macromolecular ionophores form helical conformers capable of varying their pitch and cavity size to optimize coordination with a given cation. The third is poly[(1→6)-2,5-anhydro-3,4-di*O*-ethyl-D-glucitol] (**6b**) which was derived from the cyclopolymerization of diepoxide, namely, 1,2:5,6-dianhydro-3,4-di-*O*-ethyl-D-mannitol (**5b**), as shown in **Scheme I**.[4] Polymer **6b** is characterized as a new macromolecular ionophore that has the chiral recognition ability toward racemic α-amino acids in addition to the binding property for various sizes of cations, such as alkali metal ions, methylene blue, and rhodamine 6G. Detailed descriptions are given for the new macromolecular ionophore.[4,5]

I

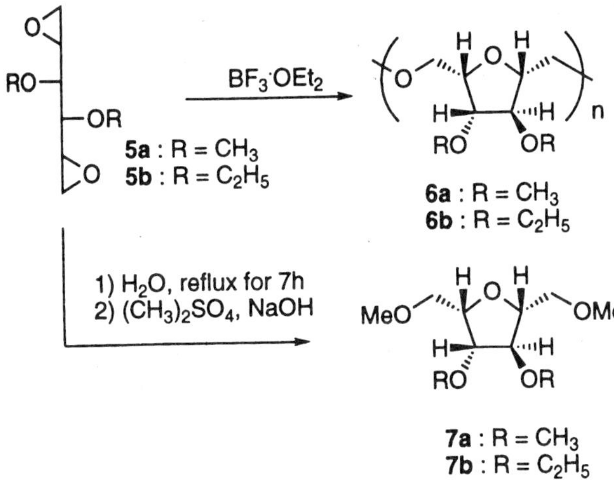

PROPERTIES OF CYCLIC STRUCTURE OF POLYMERS 6A AND 6B

To elucidate the cyclic structure of polymer **6b**, diepoxide **5b** was hydrolyzed to form a cyclic unimer and then its hydrolyzed product was treated with dimethyl sulfate.

REFERENCES

1. Hilgenfeld, R.; Saenger W. In *Topics in Current Chemistry 98, Host Guest Complex Chemistry Vol. II*, F. Vögtle, Ed.; Springer-Verlag: Berlin, 1982; Chapter 1.
2. Schultz, W. J.; Etter, M. C.; Pocius, A. V.; Smith, S. *J. Am. Chem. Soc.* **1980**, *102*, 7981.
3. Novak, B. M.; Grubbs, R. H. *J. Am. Chem. Soc.* **1988**, *110*, 960.
4. Hashimoto, H.; Kakuchi, T.; Yokota, K. *J. Org. Chem.* **1991**, *56*, 6470.
5. Kakuchi, T.; Harada, Y.; Satoh, T.; Yokota, K.; Hashimoto, H. *Polymer* **1994**, *35*, 204.

MOLECULAR RECOGNITION (Peptide-Based Systems)

Takatoshi Kinoshita
Department of Materials Science and Engineering
Nagoya Institute of Technology

There has been intense interest in recent years in the modeling of native proteins such as enzymes, hormones, and membrane proteins. As part of the protein mimetic systems we have shown two approaches for mimicking intelligent functions of membrane proteins: molecular design of membrane protein structural modules along with functional analysis of the designed elements incorporated in bilayer membrane systems; and functional modeling of biological membranes, such as stimulus-response coupling, information transfer, energy transformation, molecular recognition, and others.[1]

Here we report our recent development in molecular recognition with polypeptide-based systems, with emphasis on the amino acid recognition on the periphery of the α-helix rod using ternary complex species, α-helical polypeptide ligand-metal ion-guest amino acid species, as an additional example of the protein mimetic systems.

It has been recognized that metal ion-containing proteins, such as metalloenzymes, cleverly make the higher order structure participating in the specific reactions in a controlled manner.[2,3]

Many studies of metalloenzyme models using amino acid, oligo- and polypeptides have been carried out hoping to match or exceed the catalytic activity of metalloenzymes.[4–6]

Meanwhile, the complexes in metalloproteins serve as part of the substrate recognition site besides being post of the catalytic reaction center. We considered, therefore, that metal-ligand complexes with an unsaturated coordination site may be available for molecular recognition using their ability to form ternary complexes, i.e., ligand-metal ion-guest species.

A number of studies are concerned with the complexation behavior of polypeptides; however, most have been carried out with natural coordination amino acids. The pyridyl ligand-Cu^{2+} pair is known to form four complex species that can be distinguished spectroscopically. Therefore we selected poly(N^{ω}-2-pyridylmethyl-l-glutamine) (P2PG) as a rigid polymer carrying well-characterized ligands in the side chains.[7–9] The conformation of P2PG and its complex formation behavior with Cu^{2+} in 2,2,2-trifluoroethanol (TFE) were investigated by CD and absorption spectroscopy. Furthermore, we tried to determine whether the Cu^{2+}–P2PG complexes are able to serve as a molecular recognition system for the enantiometric indole ring derivatives.

REFERENCES

1. Kinoshita, T. *Prog. Polym. Sci.* **1995**, *20*, 527.
2. Colman, P. M.; Jansonius, J. N.; Matthews, B. W. *J. Mol. Biol.* **1972**, *70*, 701.
3. Williams, R. J. P. *Mol. Phys.* **1989**, *68*, 1.
4. Sigel, H.; Martin, R. B. *Chem. Rev.*, **1982**, *82*, 385.
5. Handel, T.; DeGrado, W. F. *J. Am. Chem. Soc.* **1990**, *112*, 6710.
6. Karlin, K. D. *Science* **1993**, *261*, 701.
7. Nagata, Y.; Kanuka, S.; Kinoshita, T.; Takizawa, A.; Tsujita, Y.; Yoshimizu, H. *Bull. Chem. Soc. Jpn.* **1993**, *66*, 2972.

8. Nagata, Y.; Kanuka, S.; Kinoshita, T.; Takizawa, A.; Tsujita, Y.; Yoshimizu, H. *Biopolymers* **1994**, *34*, 701.
9. Nagata, Y.; Kinoshita, T.; Takizawa, A.; Tsujita, Y.; Yoshimizu, H. *Bull. Chem. Soc. Jpn.* **1994**, *67*, 773.

MOLECULAR SHAPE RECOGNITION

Yoshihiro Ito
Division of Material Chemistry
Faculty of Engineering
Kyoto University

The design and synthesis of molecular pairs that interact complementarily are important to developing new drugs, chemical sieves, sensors, and even catalysts. Two main approaches were carried out to experimentally design the molecular shape by researchers. Onc is by molecular imprinting, another by selection from a molecular library.

PREPARATIONS AND PROPERTIES

Molecular Imprinting

The idea of using a specific imprint molecule to coordinate the assembly of monomers or polymers around a molecule of interest, to thereby create a specific host, was raised by Wulff in the early 1970s.

Dabulis and Klibanov[1] reported molecular imprinting of proteins and other macromolecules. They used a nature of protein ligand memory in organic solvents.

Using monomers, they are polymerized in the presence of a print molecule with the functionality found in the monomer units. Two different approaches have been developed: covalent and noncovalent orchestration.[2]

The covalent approach uses reversible covalent binding of the imprint molecule to the monomer to define the imprint-molecule-monomer interaction, whereas the noncovalent approach allows a cocktail of functionalized monomers to 'prearrange' around the imprint molecule by noncovalent interactions (i.e., ionic, hydrophobic, hydrogen bonding, etc.) After completion of polymerization, the imprint molecule is removed from the polymer, leaving a polymer with recognition sites complementary to the imprint species in both shape and function and which has a macroporous structure allowing imprint molecule diffusion into and out of the polymer matrix.

Using vinyl polymerization, Heilman and Maier[3] reported amorphous silicon dioxide imprinted with a transition state analogue by a modified sol-gel process.

Researchers have succeeded in orchestrating the binding of functional monomers to only two or three sites on each template. The resulting imprints thus recognize two or three sites on their target molecules; antibodies, in contrast, bind their targets in about 10 places. This scarcity of binding sites can limit the specificity of some artificial imprints.[4]

Selection from Library

Recent developments in biological science providcd a new method to find complementary molecular shape pairs. The method selects desired molecules from a combinatorial molecules group, and is now used in pharmaceutical drug design.[5–7]

In making a combinatorial library of dipeptide composed of 20 natural amino acids, one must create a mixture of $20,^2$ or 400, dipeptide. Repeated seven times, this procedure would result in a mixture of $20,^8$ or about 26 billion, octapeptides. This mixture is useless by itself. However, researchers are developing ingenious methods for fishing useful compounds out of this complex mix and for labeling the compounds as they are being synthesized so that they can later be identified if they prove to have useful biological activity. The molecular library is usually chemical and biological.

Chemical Method

Chemical libraries are generally classified as either ordered or disordered (random).

Ordered Library

In an ordered library researchers know the place of each compound.

Random Library

The random library is generated by the "divide and pool" method. In this process, peptide-synthesis beads are segregated into individual reaction vessels for the coupling of specific acids, the beads are then combined, mixed to homogeneity, and redivided into separate reaction vessels for subsequent coupling steps.

There are two approaches to find a target molecule in this library. The first is to pan library compounds freed from support into solution, and the second is to pan library molecules linking to the supports.

Chemical Library for Molecular Diversity

One advantage of using chemical methods to prepare a library is incorporating non-natural components in the molecular diversities. Therefore, in addition to biopolymers (such as oligopeptides and oligonucleotides) libraries of non-natural polymers were also constructed.[9–12]

Biological Method

The biological method usually provides libraries of peptides by splicing a random mixture of synthetic DNA molecules encoding the peptides of interest to the gene encoding a readily expressed protein.[13–15] This DNA construct is introduced into an appropriate expression system where, upon translation, the resulting peptide is synthesized as a fusion protein. One of the most common expression systems fuses the random sequence to the gene II or gene VIII coat protein of filamentous phage particles.

APPLICATIONS

These techniques are mainly applicable to the development of drugs, separation material for chromatography, catalysts for organic synthesis, and biosensors. The molecular imprinting technique was used for chromatographic or sensoring materials, and also was used to construct artificial enzymes.[16] Molecular evolution engineering was applied for synthesis of RNA catalysts.[17,18] On the other hand, a new application is being developed. The photolithographic techniques were used to facilitate sequency analysis by generating miniaturized arrays of densely packed oligonucleotide probes.[19] This is considered a powerful tool for rapid investigations in human genetics and diagnostics, pathogen detection, and DNA molecular recognition.

REFERENCES

1. Dabulis, K.; Klibanov, A. M. *Biotech. Bioeng.* **1992**, *39*, 176.
2. Mosbach, K. *TIBS* **1994**, *19*, 9.
3. Heilmann, J.; Maier, W. F. *Angew. Chem. Int. Ed. Engl.* **1994**, *33*, 471.
4. Flam, F. *Science* **1994**, *263*, 1221.
5. Alper, J. *Science* **1994**, *264*, 1339.
6. Baum, R. M. *Chem. Eng. News* **1994**, *Feb.* 7, 20.
7. Mitchison, T. J. *Chem. Biol.* **1994**, *1*, 3.
8. Jung, G.; Beck-Sickinger, A. G. *Angew. Chem. Int. Ed. Engl.* **1992**, *31*, 367.
9. Liskamp, R. M. J. *Angew. Chem. Int. Ed. Engl.* **1994**, *33*, 633.
10. Campbell, D. A.; Bermak, J. C. *J. Am. Chem. Soc.* **1994**, *116*, 6039.
11. Kessler, H. *Angew: Chem. Int. Ed. Engl.* **1993**, *32*, 543.
12. Wyatt, J. R.; Vickers, T. A.; Roberson, J. L.; Buckheit, Jr., R. W.; Klimkait, T.; DeBaets, E.; Davis, P. W.; Rayner, B.; Imbach, J. L.; Eckers, D. J. *Proc. Natl. Acad. Sci., USA* **1994**, *91*, 1356.
13. Chiswell, D. J.; McCafferty, J. *TIBTECH* **1992**, *10*, 80.
14. Geisow, M. *TIBTECH* **1992**, *10*, 299.
15. Little, M.; Fuchs, P.; Breitling, F.; Dubel, S. *TIBTECH* **1993**, *11*, 3.
16. Beach, J. V.; Shea, K. J. *J. Am. Chem. Soc.* **1994**, *116*, 379.
17. Prudent, J. R.; Uno, T.; Schultz, P. G. *Science* **1994**, *264*, 1924.
18. Lorsch, J. R.; Szostak, J. W. *Nature* **1994**, *371*, 31.
19. Pease, A. C.; Solas, D.; Sullivan, E. J.; Cronin, M. T.; Holmes, C. P.; Fodor, S. P. A. *Proc. Natl. Acad. Sci., USA* **1994**, *91*, 5022.

MOLECULAR WEIGHT DISTRIBUTION (Ionic Polymerization)

Deyue Yan
Department of Applied Chemistry
Shanghai Jiao Tong University

The statistical nature of polymerization reactions makes it impossible to characterize a polymer by a single molecular weight and only a molecular weight distribution (MWD) can completely specify the chain lengths. Polymers with the identical monomeric unit may differ in MWD, with consequent differences in physical and mechanical properties. The MWD of a polymer is dependent on the polymerization mechanism and reaction conditions. In order to control and regulate the MWD of a polymer during polymerization, it is necessary to study the relationship between MWD and the reaction conditions of the polymerization with a certain mechanism. The reaction scheme of an anionic or a cationic polymerization is rather simple because of the absence of both combination termination and disproportionation termination. Therefore the MWD functions of the polymers generated from ionic polymerization can be rigorously derived by way of kinetic analysis.

LIVING POLYMERIZATION

Heretofore a lot of living polymerizations involving anions, cations, coordination complexes and group-transfer mechanisms have been reported. Taking advantage of living polymerization, a number of polymers with novel architectures (such as telechelic polymers and star-shaped polymers) were synthesized.

Poisson Distribution

It is well known that the product of a fast initiated living polymerization has a Poisson distribution.[1]

Gold Distribution

A living polymerization with slow initiation results in the polymer with Gold distribution.[2]

Höcker Distribution

The ionic polymerization initiated by a bifunctional initiator is usually applied to synthesize the telechelic polymer. If the two function groups of the bifunctional initiator have the same reaction activity, the resultant polymer possesses Höcker distribution.[3]

YLJ Distribution

The ionic polymerization initiated instantaneously by a multifunctional initiator produces a star-shaped polymer with poisson distribution. The star-branched polymer generated from a multifunctional ionic polymerization with slow initiation falls in with YLJ distribution.[4] It is evident that both Gold distribution and Höcker distribution are the special examples of YLJ distribution.

INFLUENCE OF CHAIN TRANSFER AND TERMINATION ON MWD

It is known that chain transfer and termination play an important role in some cationic polymerization systems.

Transfer to Monomer

A number of authors[5–8] have theoretically studied the reaction of chain transfer to monomer.

Spontaneous Transfer

For the polymer formed from the cationic polymerization with instantaneous initiation and spontaneous transfer, the MWD function is expressed in Reference 9. The effect of both spontaneous transfer and monomer transfer on MWD has been also studied.[10]

Spontaneous Termination

Spontaneous termination presents in some cationic polymerization systems.[11]

Impurity Termination

The impurity termination is important for both anionic and cationic polymerization.[12]

Monomer Termination

Anionic polymerization of methyl methacrylate (MMA) in nonpolar solvents is more complex than in polar solvents. A number of experimental data reveal that both the vinyl double bond and carbonyl group of MMA monomer are active to organometallic reagents. Glusker et al.[13] deemed the initiation of the anionic polymerization of MMA in nonpolar solvent to be a competitive reaction, and the reaction to occur by both a Michael addition across the carbon-carbon double bond of the monomer which results in the active species to propagate, and the carbonyl addition which leads to the formation of the by-products inactive in the polymerization reaction. Bywater et al.[14] considered that there might be a wide variation in reactivity of the species with different chain length in the first few minutes of polymerization to produce the wide distribution. Supposing both propagation and termination rate constants change with the degree of polymerization for the anionic polymerization of polar monomer in nonpolar solvent, the molecular parameters of the resulting polymer were derived by Feng and Yan.[15]

The effect of monomer transfer or monomer termination on the MWD of the polymer generated from a bifunctional initiator has been also reported.[4,16]

Up to now the effects of most of elementary reactions in ionic polymerizations on the MWD of resulting polymers have been illuminated. One can evaluate the molecular parameters of the polymers formed in ionic polymerization in terms of the pertinent reaction conditions. The theoretical formulae given above are convinced to be helpful for the macromolecule design, control and regulation of ionic polymerizations.

REFERENCES

1. Flory, P. J. *J. Am. Chem. Soc.* **1940**, *62*, 1561.
2. Gold, L. *J. Chem. Phys.* **1958**, *28*, 91.
3. Höcker, H. *Makromol. Chem.* **1972**, *157*, 187.
4. Wang, Z.; Yan, D. *J. Polym. Sci.-Phys.* **1987**, *25*, 1491.
5. Kyner, W. T.; Radok, J. R. M.; Wales, M. *Chem. Phys.* **1959**, *30*, 363.
6. Nanda, V. S. *Trans. Faraday Soc.* **1964**, *60*, 949; Jain, S. C.; Nanda, V. S. *Indian J. Chem.* **1975**, *13*, 614; ibid. *Eur. Polym. J.* **1977**, *13*, 137; ibid. *J. Polym. Sci.* **1970**, *B8*, 843.
7. Guyot, A. *J. Polym. Sci.* **1968**, *B6*, 123.
8. Peebles, L. H. *J. Polym. Sci.* **1969**, *B7*, 75.
9. Yuan, C.; Yan, D. *Makromol. Chem.* **1987**, *188*, 341.
10. Yuan, C.; Yan. D. *Polymer* **1988**, *29*, 924.
11. Yan, D.; Yuan, C. *J. Macromol. Sci.-Chem.* **1986**, *A23*, 769.
12. Yan, D.; Yuan, C. *J. Macromol. Sci.-Chem.* **1986**, *A23*, 781.
13. Glusker, D. L. *J. Polym. Sci.* **1961**, *49*, 297.
14. Cottam, B. J.; Wiles, D. M.; Bywater, S. *Can. J. Chem.* **1963**, *41*, 1905.
15. Feng, J.; Yan, D. *Makromol. Chem.*; T&S Ed.; **1993**, *2*, 129.
16. Feng, J.; Yan, D. *Polym. Int.* **1991**, *26*, 81.

Molecularly Doped Polymers

See: Charge Transport Polymers (for Organic Electroluminescent Devices)

Monoclonal Antibodies

See: Antibody-Polymer (Porphyrin) Complexes

Monodisperse Networks

See: Polydimethylsiloxane Networks, Monodisperse

Monodisperse Particles

See: Colloidal Crystals, Polymeric
Dispersion Polymerization (Overview)
Monodisperse Polymer Particles (Methods of Preparation)
Monodisperse Polymer Particles (Nonmagnetic and Magnetic)
Monodisperse Polymer Particles (Preparation and Applications)

MONODISPERSE POLYMER PARTICLES (Methods of Preparation)

Shiro Kobayashi and Hiroshi Uyama
Department of Molecular Chemistry and Engineering
Faculty of Engineering
Tohoku University

Monodisperse (uniform) polymer particles receive much attention because of their various potential uses in technical and biomedical fields. Until now, particles with narrow size distribution in the size range of 10 nm to 300 μm have been prepared by various polymerization methods. Soap-free emulsion polymerization techniques are often used to prepare monodisperse particles in less than 1 μm diameter. Dispersion polymerization in polar media and seeded polymerization generally afford polymer particles of more than 1 μm diameter. Here we review preparation of monodisperse polymer particles and polymerization techniques. We also briefly describe some applications on monodisperse particles.

PREPARATION

Emulsion Polymerization

Soap-free emulsion polymerization is a very attractive and convenient method to prepare monodisperse polymer particles containing no emulsifiers.[1] Resulting particles are stabilized by ionic species derived from an initiator, and their diameter is normally more than 0.4 μm. Monodisperse polystyrene (Pst), poly(methyl methacrylate) (PMMA), and poly(vinyl acetate) particles were obtained through this technique.[2] Particle size was controlled by the initiator type and concentration.

Monodisperse copolymer particles in the sub-micron range were also obtained by soap-free emulsion copolymerization.[3]

Soap-free emulsion copolymerization of styrene with *N*-substituted acrylamide produced monodisperse particles with thermo-sensitive properties.[4,5]

Monodisperse polymer particles bearing a sugar derivative on the surface were prepared using surface-anchored phenylboronic acid group.[6]

Polymer particles with hydrophilic polymer chains on the surface were prepared by soap-free emulsion copolymerization of styrene with a macromonomer of the hydrophilic polymer. The copolymerization with poly(2-alkyl-2-oxazoline) (PROZO) macromonomer bearing a polymerizable styryl group at the chain end afforded monodisperse Pst particles in the submicron range.[7]

Emulsion copolymerizing styrene with a block-type amphiphilic PROZO macromonomer with a polymerizable styryl group was performed.[7a]

Phospholipids on the latex surface were immobilized by emulsion copolymerizing styrene with polymerizable and nonpolymerizable phospholipids.[8]

Monodisperse crosslinked Pst particles from 200 nm to 800 nm were prepared by soap-free emulsion copolymerization of styrene with DVB.[8a] Relatively monodisperse fine Pst microgel ranging from 10 nm to 60 nm was obtained using microemulsion technique.[9,10]

Monodisperse, multi-hollow particles were obtained by two-stage, emulsion terpolymerization of styrene, butyl acrylate, and methacrylic acid, followed by treatment of the resulting polymer latex with alkaline and acid.[11,12]

Seeded Polymerization

Several methodologies of seeded polymerizations have been reported, and among them, seeded emulsion polymerization and molecular diffusion process are often used to prepare large, uniform particles.

Porous monodisperse particles in the range of 10 μm were prepared by seeded emulsion copolymerization of styrene and DVB in the presence of micron-size Pst seed particles and/or a diluent solvent.[13]

In the molecular diffusion process, seed particles dispersed in the media swell by the addition of a monomer emulsion (low molecular weight compound), which diffuses from the emulsion droplets into the seed particles.[14] The subsequent polymerization of the swollen particles produces larger particles.

Macroporous micron-size uniform particles were prepared by a multi-step swelling and polymerization method in the presence of uniform seed particles.[15]

Dispersion Polymerization

There have been many studies of dispersion polymerization of acrylic esters in non-polar media to develop nonaqueous, dispersion-coating technology.[16] Recently, dispersion polymerization of styrene in polar media has been extensively investigated because it conveniently affords micron-size polymer particles in a single step.

Micron-size uniform particles of highly hydrophilic polymer were obtained by dispersion polymerization technique.[17,18] *N*-Vinylformamide (NVF) dissolves in water and various organic solvents except hydrocarbons, but its polymer dissolves only in water. Dispersion polymerization of NVF was performed in methanol using PEtOZO stabilizer to give monodisperse particles in the micron range. The copolymerization with a crosslinking monomer produced monodisperse hydrogel microsphere.

The dispersion polymerization generally involves the use of polymeric stabilizer. The amount of the stabilizer is not low because only a small amount of its acts to stabilize the particles and the rest exists in solution. Kobayashi has used macromonomers as a stabilizer in the dispersion polymerization to produce particles with stabilizer polymer chains chemically bound to the surface.[19] The macromonomer acts as a stabilizer as well as a comonomer.

PROZO macromonomer was effective for the preparation of monodisperse Pst and PNVF particles.[17,18,19a]

Other Methodologies

Large, monodisperse particles up to 30 μm were prepared in the space shuttle under a microgravity environment where the gravitational effects of creaming and settling can be eliminated.[20] On earth, using a rotary-cylinder reactor eliminated such effects.[21]

Micron-size monodisperse glycidyl methacrylate copolymer particles were prepared by precipitation copolymerization with functional monomers such as methacrylic acid and 2-hydroxyethyl methacrylate in organic media.[22–24]

Large monodisperse particles of poly(2-hydroxyethyl methacrylate) and poly(glycerol monomethacrylate) in the 50–400 μm in diameter range was prepared by Colvin.[25]

APPLICATIONS

Monodisperse particles have been applied in various fields. Monodisperse latex has served as the standards for size in measurement of electron microscopes for years. Strict monodispersity and sphericity are necessary for this purpose.

Monodisperse latex has also been applied to a carrier of antibody for latex diagnosis. The antibody immobilized on the particle surface is specifically reacted with antigen to lead to the aggregation of the latex, which can then be visually observed.

Uniform particles in the micron range are used to support HPLC column. Macroporous monodisperse poly(styrene-*co*-DVB) particles with uniform pore size were tested as a support of size-exclusion chromatography and resulted in high efficiency and range of separation.[26,27]

Monosized hydrophilic methacrylate particles (diameter of 20 μm) bearing a dye affinity ligand were tested as a support of high performance affinity chromatography and exhibited high capacities and low back pressure.[28]

Preparation and applications of monosized magnetic particles have been developed by Ugelstad.[29,30]

REFERENCES

1. Goodall, A. R.; Wilkinson, M. C.; Hearn, J. *J. Polym. Sci., Polym. Chem. Ed.* **1977**, *15*, 2193.
2. Tamai, H.; Fujii, A.; Suzawa, T. *J. Colloid Interface Sci.* **1987**, *116*, 37.
3. Shirahama, H.; Suzawa, T. *J. Appl. Polym. Sci.* **1984**, *29*, 365.
4. Kawaguchi, H.; Hoshino, F.; Ohtsuka, Y. *Makromol. Chem., Rapid Commun.* **1986**, *7*, 109.
5. Hoshino, F.; Fujimoto, T.; Kawaguchi, H.; Ohtsuka, Y. *Polym. J.* **1987**, *19*, 241.
6. Tsukagoshi, K.; Kawasaki, R.; Maeda, M.; Takagi, M. *Chem. Lett.* **1994**, 681.
7. Kobayashi, S.; Uyama, H.; Yamamoto, I. *Makromol. Chem.* **1990**, *191*, 3115.

7a. Uyama, H.; Matsumoto, Y.; Kobayashi, S. *Chem. Lett.* **1992**, 2401.

8. Yamaguchi, K.; Watanabe, S.; Nakahama, S. *Makromol. Chem., Rapid Commun.* **1986**, *10*, 1195.

8a. Zou, D.; Derlilch, V.; Gandhi, K.; Park, M.; Sun, L.; Kriz, D.; Lee, Y. D.; Kim, G.; Aklonis, J. J.; Salovey, R. *J. Polym. Sci., Polym. Chem. Ed.* **1990**, *28*, 1909.

9. Antonietti, M.; Bremser, W.; Müschenborn, D.; Rosenauer, C.; Schupp, B.; Schmidt, M. *Macromolecules* **1991**, *24*, 6636.
10. Antonietti, M.; Lohmann, S.; Niel, C. V. *Macromolecules* **1992**, *25*, 1139.
11. Okubo, M.; Kanaida, K.; Fujimura, M. *Chem. Express* **1990**, *5*, 797.
12. Okubo, M.; Ichikawa, K.; Fujimura, M. *Colloid Polym. Sci.* **1991**, *269*, 1257.
13. Cheng, C. M.; Micale, F. J.; Vanderhoff, J. W.; El-Aasser, M. S. *J. Polym. Sci., Polym. Chem. Ed.* **1992**, *30*, 235.
14. Jansson, L. H.; Wellons, M. C.; Poehlein, G. W. *J. Polym. Sci., Polym. Lett. Ed.* **1983**, *211*, 937.
15. Hosoya, K.; Fréchet, J. M. J. *J. Polym. Sci., Polym. Chem. Ed.* **1993**, *31*, 2129.
16. Barrett, K. E. J. *Dispersion Polymerization in Organic Media* Wiley: London, 1975.
17. Uyama, H.; Kato, H.; Kobayashi, S. *Chem. Lett.* **1993**, 261.
18. Uyama. H.; Kato, H.; Kobayashi, S. *Polym. J.* **1994**, *26*, 858.
19. Kobayashi, S.; Uyama, H. *Kobunshi Ronbunshu* **1993**, *50*, 209.

19a. Kobayashi, S.; Uyama, H.; Lee, S. W.; Matsumoto, Y. *J. Polym. Sci., Polym. Chem. Ed.* **1993**, *31*, 3133.

20. Vanderhoff, J. W.; El-Aasser, M. S.; Micale, F. J.; Sudol, E. D.; Tseng, C. M.; Silwanowicz, A.; Kornfeld, D. M.; Vicente, F. A. *J. Disp. Sci. Tech.* **1984**, *5*, 231.

21. Kim, J. H.; Sudol, E. D.; El-Aasser, M. S.; Vanderhoff, J. W.; Kornfeld, D. M. *Chem. Engi. Sci.* **1988**, *43*, 2025.
22. Hosaka, S.; Murao, Y.; Masuko, S.; Miura, K. *Immunol. Commun.* **1983**, *12*, 509.
23. Hosaka, S.; Miura, K.; Masuko, S.; Murao, Y. *Kobunshi Ronbunshu* **1993**, *50*, 295.
24. Hosaka, S.; Murao, Y.; Tamaki, H.; Masuko, S.; Miura, K.; Kawabata, Y. *Polym. International* **1993**, *30*, 505.
25. Colvin, M.; Chung, S. K.; Hyson, M. T.; Chang, M.; Rhim, W. K. *J. Polym. Sci., Polym. Chem. Ed.* **1990**, *28*, 2085.
26. Kulin, L. I.; Flodin, P.; Ellingsen, T.; Ugelstad, J. *J. Chromatogr.* **1990**, *514*, 1.
27. Fréchet, J. M. J. *Makromol. Chem., Macromol. Symp.* **1993**, *70/71*, 289.
28. Clonis, Y. *J. Chromatogr.* **1987**, *407*, 179.
29. Ugelstad, J.; Berge, A.; Ellingsen, T.; Aune, O.; Kilaas, L.; Nilsen, T.-N.; Schmid, R.; Stenstad, P.; Funderud, S.; Kvalheim, G.; Nustad, K.; Lea, T.; Vartdal, F.; Danielsen, H *Makromol. Chem., Macromol. Symp.* **1988**, *17*, 177.
30. Ugelstad, J.; Mork, P. C.; Schmid, R.; Ellingsen, T.; Berge. A. *Polym. International* **1993**, *30*, 157.

MONODISPERSE POLYMER PARTICLES (Nonmagnetic and Magnetic)

John Ugelstad,* Ruth Schmid, Oddvar Aune, Jon Bjørgum, Lars Kilaas, and Per Stenstad
SINTEF
Applied Chemistry

Arvid Berge
Department of Industrial Chemistry
University of Trondheim

Arne Skjeltorp
Institute for Energy Technology

Tore Lindmo
Department of Physics
University of Trondheim

Lars Kornes
DYNAL A/S

The present contribution deals with monodisperse particles prepared by the method of "activated swelling." This method allows for the preparation of highly monodisperse polymer particles in a size range of 1 to 200 µm and provides unparalleled freedom of choice as regards the type of vinyl monomers to be used, the degree of crosslinking, and the option of producing macroporous structures.

The first part of the article treats nonmagnetic monodisperse polymer particles, and the second part deals with magnetic monodisperse particles.

NONMAGNETIC PARTICLES

Procedures for preparation of monodisperse polymer particles by "activated swelling" and the basic thermodynamic principles behind this method have been treated at length in a number of papers.[1–3]

*Author to whom correspondence should be addressed.

In order for this activation to take place, the monodisperse particles used as seed must contain a considerable amount of a relatively low molecular weight, highly water insoluble compound, here called a "Y compound." When dispersed in water, particles containing Y compounds can absorb 10 to 1000 times more slightly water soluble compounds, here called "Z compounds," than can particles containing only polymer.

The preparation of polymer particles by dispersion polymerization in organic media has recently been subject to intense research; it has been found that, under very controlled reaction conditions, one can directly produce monodisperse particles in sizes up to about 10 µm from a number of monomers.[4,5] One may use monodisperse particles from dispersion polymerization as seed for the preparation of much larger monodisperse particles through activated swelling.

Preparation of Monodisperse Particles

The first step in the production of monodisperse particles by activated swelling is the preparation of monodisperse seed particles that are partly or totally made up of oligomeric substances.

Normally, after the first step, we have particles that contain 50–90% by volume of an oligomeric substance and therefore are able to absorb 10 to 1000 times more slightly water soluble Z compounds than can particles of the same size consisting of pure polymer. For preparation of polymer particles by the activated swelling method, the Z compounds normally are vinyl compounds; they may include divinyl or trivinyl compounds in order to achieve crosslinking and inert compounds (porogens) in order to produce macroporous structures. Addition of the Z compounds in the form of a finely divided dispersion in water facilitates swelling.

The essential feature of the activated swelling method is that we first may prepare monodisperse droplets that contain all the ingredients necessary for production of the particles, that is, the monomer mixture including the crosslinking components and optionally the porogens.

A special advantage of the activated swelling method is that the original seed particles make up only a small fraction of the final particles less than 1%—allowing a high degree freedom in the choice of monomers and of inert liquids for production of macroporous particles. This freedom leads to very good control of pore size and pore size distribution.

Applications

Chromatography

The preparation of real monodisperse polymer particles by the method of activated swelling was a breakthrough for chromatography.[8] Highly monodisperse macroporous systems with particles of exactly the same size, ranging from 1 to more than 200 µm in diameter, can be tailor-made with respect to pore volume, pore size, and pore size distribution. Because the preparation method is highly reproducible and easy to scale up, one also achieves excellent column-to-column reproducibility.

Particle-Based Immunoassays

Polymer particles are used extensively as solid-phase material in immunoassays of the so-called sandwich or immunometric type, in which molecules of the analyte to be measured are captured by antibodies immobilized on a solid-phase surface, and the amount of bound analyte is determined by adding a second, labeled indicator antibody that binds to another epitope of the analyte. Immobilizing the antibody on a solid phase facilitates the washing off of unbound labeling reagent prior to measuring the amount of bound analyte-label complex. A solid-phase material in the form of microscopic particles is advantageous in that the solid phase is distributed throughout the reaction volume during the incubation, giving minimal diffusion distance for the analyte to reach a binding partner.[7] Although these particles have many advantages as a solid phase, the isolation and washing procedures involved in the application are still a serious disadvantage. These procedures are avoided in the flow cytometric method.

Models for Studies of Physical Processes

Much of our understanding of nature can be expressed in terms of a hierarchy of "particles" and interactions between them. The organization of small particles to form extended structures is an important process in many areas of science and technology. The availability of monosized polymer microspheres has allowed the construction of experimental models for a variety of equilibrium and nonequilibrium processes. These include studies of crystal growth, deposition, fracture processes, and complex dynamics of interacting particles.[8]

The experimental model systems were essentially two dimensional. The microspheres were dispersed in a liquid (water or organic liquid) and confined to a monolayer between two plane parallel glass plates. The separation between the plates could be spaced evenly by using a small fraction of spheres that were slightly larger than the particles used in the experiments. The interaction between the particles could be controlled via the electrolyte concentration in the surrounding fluid. The microspheres were also dispersed in magnetic fluids or ferrofluids in order to produce magnetic holes. The interaction could then be varied by use of external fields.

In 1983 it was discovered that it was possible, by dispersing monosized nonmagnetic spheres/particles in a ferrofluid, to produce a system of interacting magnetic dipoles that have been called "magnetic holes".[9] The concept is very simple and corresponds to the "magnetic analog" of Archimedes's principle.

MAGNETIC MONODISPERSE POLYMER PARTICLES

Magnetizable particles have found widespread use in various fields of biochemistry, biology, and medicine. Different types of particles and methods of preparation have been reviewed.[10,11] Most of the commercially available particles are relatively small and polydisperse. Frequently they are prepared by the precipitation of magnetic compounds in the presence of polymers of synthetic or natural origin, or they are formed by the treatment of flakes of magnetite.

The successful application of monodisperse magnetic particles for separation and isolation procedures in biology and medicine is based on several basic principles. Compared to alternative separation methods, monodisperse magnetic particles have the following advantages:

- The separation time is short, typically from 10 seconds to a couple of minutes.
- The target material for the magnetic beads with affinity ligands can be anything from small haptens or oligonucleotides up to whole cells or even multicellular organisms.
- The target material can be isolated magnetically from quite crude liquids such as blood, tissue homogenates, and soil samples.
- The reaction kinetics of the attachment of particles to target molecules or cells is very fast.
- Magnetic solid phase handling of biochemical reactants simplifies reaction steps such as enzymatic reactions, change of buffer conditions, and washing steps.
- The magnetic particle-based solid phase is dispersed and can be handled like a liquid; hence, reaction steps can be fully automated.

Preparation of Monodisperse Magnetic Particles

The most common types of magnetizable particles, which are used in cell separation, microbiology, and molecular biology, are all prepared from monodisperse porous polymer particles with very fine pores. The magnetite is formed inside the pores of the particles, which have been pretreated so that they carry oxidizing groups covalently bound to the surface of the pores. Treatment of these particles with Fe^{2+} salts under mildly alkaline conditions causes magnetic iron oxides to precipitate as fine grains distributed evenly throughout the volume of the particles. This procedure ensures that the particles become superparamagnetic; they have no remanent magnetism after being subjected to magnetic fields. Finally, most of the particles are treated with polymerizable substances that fill up the pores to give a smooth surface. This treatment also introduces surface groups that may be activated to bind affinity ligands such as antibodies, other proteins, lectins, and nucleic acids. The ligands are covalently attached to the surface in such a way that their affinity functions are retained.

Currently, two types of monosized magnetic polymer particles are commercially available under the trade name Dynabeads®. Dynabeads M-280 (2.8 μm) and Dynabeads M-450 (4.5 μm) (Dynal A/S, Oslo, Norway) both contain about 25% magnetic oxides by weight.

Applications

The use of Dynabeads for magnetic separation in biochemistry, biology, and medicine has been cited in about 1500 scientific papers since 1986, when the beads became commercially available. Most often the method is called "immunomagnetic separation" (IMS).

Selective Separation of Eukaryotic Cells

In the selective separation of cells by magnetic beads, several methods have been applied for the attachment of the particles to the target cells by use of monoclonal antibodies (mAbs).

Removal of cancer cells from bone marrow. The use of monodisperse magnetic beads for selective isolation of cells was initiated by Kemshead, Rembaum, and Ugelstad, who used magnetic beads for removal of neuroblasts from bone marrow in connection with autologous bone marrow transplantation.[12]

Isolation of subsets of cells. Monodisperse magnetic beads have been used extensively in selective separation of a wide variety of cells, both singular cell types and groups of cells. Dynabeads are used extensively for isolating of cells from blood to be used in tissue typing.[13]

Isolation of subsets of cells free of magnetic beads. In some instances it is desirable to release the cells from the beads after isolation. The methods for isolation of pure subsets of single cell types or classes of cells that are free from magnetic beads may be divided into negative selection and positive selection.

- *Negative selection of cells* means purification of the target cells by immunomagnetic depletion of the other cell types present in the sample.
- *Positive cell separation* means isolation of the target cells from the cell suspension by the appropriate combination of mAbs and beads.

Isolation of subcellular compartments. A special type of particle with an extremely smooth surface is used in the isolation of organelles. These particles are used in the "free flow" immunoisolation system without pelleting and resuspending.[14,15]

Microbiology

Selective separation of procariotic cells. Immunomagnetic separation has many advantages compared to conventional methods in microbiology and seems to have great potential in the detection and identification of bacteria.

Molecular biology. New applications of magnetic beads are multiplying in the field of molecular biology. Dynabeads, most often the M-280 beads, have proven to be a valuable tool for the isolation, identification, manipulation, and genetic analysis of specific nucleic acid sequences, both DNA and RNA.

Immunoprecipitation, Immunoassays

Dynabeads M-280 or M-450 coated with specific monoclonal or polyclonal antibodies have been applied to isolate proteins both for analytical purposes and for purification. In both cases, close to 100% of target protein in a sample is captures within a few minutes.

REFERENCES

1. Ugelstad, J.; Mørk, P. C.; Kaggerud, K. H.; Ellingsen, T.; Berge, A. *Adv. Colloid Interface Sci.* **1980**, *13*, 101.
2. Ugelstad, J.; Mørk, P. C.; Mfutakamba, H. R.; Soleymani, E.; Nordhuus, I.; Schmid, R.; Berge, A.; Ellingsen, T.; Aune, O.; Nustad, K. *Science and Technology of Polymer Colloids;* NATO ASI Series E67; M. Nijhoff: Boston, 1983; p 51.
3. Ugelstad, J.; Berge, A.; Ellingsen, T.; Schmid, R.; Nilsen, T. N.; Mørk, P. C.; Stenstad, P.; Hornes, E.; Olsvik, Ø. *Prog. Polym. Sci.* **1992**, *17*, 87.
4. Barret, K. E. *Dispersion Polymerization in Organic Media;* John Wiley & Sons: London, 1975.
5. Wang, Q.; Fu, S.; Yu, T. *Prog. Polym. Sci.* **1994**, *19*, 703.
6. Ugelstad, J.; Söderberg, L.; Berge, A.; Bergström, J. *Nature* **1983**, *303*, 95.
7. Millán, J. L.; Nustad, K.; Nøgaard- Pedersen, B. *Clin. Chem.* **1985**, *31*, 54.
8. Meakin, P.; Skjeltorp, A. T. *Adv. Physics* **1993**, *42*, 1.
9. Skjeltorp, A. T.; *Phys. Rev. Lett.* **1983**, *51*, 2306.
10. Platsoucas, C. D. *Future Directions in Polymer Colloids;* NATO ASI Series E138; M. Nijhoff: Dordrecht, The Netherlands, 1987; p 321.
11. Pieters, B. R.; Williams, R. A.; Webb, C. *Colloid and Surface Engineering, Applications in the Process Industries;* Butterworth Heineman: Oxford, 1990; p 248.
12. Treleaven, J. G.; Gibson, F. M.; Ugelstad, J.; Rembaum, A.; Phillip, T.; Caine, C. D.; Kemshead, J. T. *Lancet* **1984**, *14*, 70.
13. Vartdal, F.; Bratlie, A.; Thorsby, E. *Transpl. Proc.* **1988**, *20*, 384.
14. Howell, K. E.; Schmid, R.; Ugelstad, J.; Grünberg, J. *Methods in Cell Biology;* Academic: New York, 1993; p 265.
15. Howell, K. E.; Crosby, J. R.; Ladinsky, M. S.; Jones, S. M.; Schmid, R.; Ugelstad, J. *Advances in Biomagnetic Separation;* Eaton: Natick, MA, 1994; p 195.

MONODISPERSE POLYMER PARTICLES (Preparation and Applications)

Harald D. H. Stöver
Department of Chemistry
McMaster University

Kai Li
Corporate Research and Development
3 M Canada Incorporated

Monodisperse polymer particles are batches of particles having nearly identical diameters, ranging from a few nanometers up to hundreds of micrometers. They are usually prepared by polymerization of one or more monomers in aqueous emulsion or organic solution, and they are widely used due in part, to their narrow size distribution. High-volume applications include latex paint and adhesives prepared by aqueous emulsion polymerization techniques, where the uniform particle size facilitates film formation.

This article concentrates on monodisperse polymer particles in the 1–20 micron diameter region, excluding most emulsion polymers, except those prepared by swelling techniques.

Processes to form monodisperse particles may be based on emulsion, dispersion, or precipitation polymerization processes, but share three essential features: (1) Particle initiation is rapid and usually complete within the first few minutes of polymerization. (2) The growing particles become stabilized against homo-coagulation at this point. (3) From this point onward, no new particles are formed.[1] Instead, all subsequently formed polymer is generated within the existing particles or rapidly adsorbed by them.

This article outlines some of the newly emerging methods and applications of monodisperse polymer microspheres.

SEEDED POLYMERIZATIONS

In seeded polymerizations a highly monodisperse polymer latex is formed first by emulsion or dispersion polymerization. In a second step, this seed latex is swollen with additional monomer and other additives that either facilitate monomer uptake by the seed particles (activators, i.e., low molecular weight hydrocarbons) or generate porosity in the final particles (porogens, i.e., nonsolvents or linear polymers). In these processes, the total particle number is determined by the concentration of the seed latex at the start of the second stage. This so-called two-step activated swelling method was pioneered by Ugelstad and coworkers. The key to the method is the use of the low molecular weight activator, such as heptane, decane, or ether, that allows a seed latex to be swollen by up to 200 times its original volume with monomer, while still remaining colloidally stable.[2–7] The resulting stable monomer emulsion is then polymerized, with the swollen monomer droplets being the locus of polymerization.[2]

The seeded swelling technique also is used to prepare both soluble and crosslinked microspheres containing different functional groups.

UNUSUAL MORPHOLOGIES

Copolymerization of different monomers in such systems often leads to phase separation and interesting particle morphologies that are visible by scanning or transmission electron microscopy. Other classical shapes from seeded swelling systems include particles with raspberry-like surfaces, and dumbbell-shaped particles that are prepared using lightly crosslinked seed particles.[8,9]

Monodisperse polymer particles, due to their even size and controlled properties, are becoming valuable platforms for a large variety of chemical and separation processes.

REFERENCES

1. Arshady, R. *Colloid Polym. Sci.* **1992**, *270*, 717–732.
2. Ugelstad, J.; Kaggerud, K. H.; Hansen, F. K.; Berge, A. *Makromol. Chem.* **1979**, *180*, 737.
3. Ugelstad, J.; Mork, P. C.; Kageud, K. H.; Ellipses, T.; Berge, A. *Adv. Colloid. Interface Sci.* **1980**, *13*, 101.
4. Ugelstad, J.; Mork, P. C.; Mfutakamba, H. R.; Soleimany, E.; Nordhuns, I.; Schmid, R.; Berger, A.; Ellingsen, T.; Aune, O.; Nustad, K. *Science and Technology of Polymer Colloids* Poehlein, G. W.; Ottewill, R. H.; Goodwin, J. W., Eds.; M. Nijhoff: Boston, 1983; pp 88–90.
5. Ugelstad, J.; Mfutakamba, H. M.; Mork, P. C.; Ellingsen, T.; Berge, A.; Schmid, R.; Holm, L.; Jorgedal, A.; Hansen, F. K.; Nustad, K. *J. Polym. Sci., Polym. Symp.* **1985**, *72*, 225.
6. Ugelstad, J.; Berge, A.; Ellingsen, T.; Aune, O.; Kilaas, L.; Nilsen, T. N.; Schmid, R.; Stenstad, P.; Funderud, S.; Kvalheim, G.; Nustad, K.; Lea, T.; Fartdal, F. *Makromol. Chem. / Makromol. Symp.* **1988**, *17*, 1776.
7. Ellingsen, T.; Aune, O.; Ugelstad, J.; Hagen, S. *J. Chromatogr.* **1990**, *535*, 147–161.
8. Okubo, M.; Katsuta, Y.; Matsumoto, T. *J. Polym. Sci. Polym. Chem. Ed.* **1981**, *18*, 481.
9. Sheu, H. R.; El-Aasser, M. S.; Vanderhoff, J. W. *J. Polym. Sci.: Part A: Polym. Chem.* **1990**, *28*, 629–651.

MONODISPERSE POLYMERS

Yusuke Kawakami and Atsunori Mori
Graduate School of Materials Science
Japan Advanced Institute of Science and Technology

Monodisperse polymers with narrow molecular weight distributions have been obtained by living polymerization. The first living polymerization, reported in 1956 by M. Szwarc, was the anionic polymerization of styrene, which yielded a polystyrene with a narrow molecular weight distribution.[1] Synthetic polymer chemists developed living polymerization extensively. A number of living polymerizations have been found: not only anionic but also cationic, ring-opening, metathesis, coordination polymerizations that yield a wide variety of monodisperse polymeric materials. **Table 1** lists monodisperse polymeric

TABLE 1. Representative Monodisperse Polymers

Monomer (Polymer)	Initiator	*Mw*/*Mn*	References
Styrene	*n*-BuLi	1.03	2, 3
	Sodium naphthalide	1.04$\sim$1.10	1, 5
	Cumylpotassium	1.06	4
	α-Methylstyrene tetramer dianion	1.05$\sim$1.12	5, 65
	$SnCl_4$/n-Bu_4NCl	$\sim$1.2	
	BPO - TEMPO	1.19	10
4-(t-Butyldimethylsiloxystyrene)	Lithium naphthalide	$\sim$1.1	6
3-(t-Butyldimethylsiloxystyrene)	BuLi, etc.	1.04$\sim$1.08	6
4-Cyanostyrene (Structure 7)	1,1-Diphenylhexyllithium	1.04$\sim$1.06	8, 9
(Structure 3)	Cumylpotassium	1.08	66
(Structure 5)	Potassium naphthalide, etc.	1.04$\sim$1.22	67
4-Methoxystyrene	HI-ZnI_2/n-Bu_4NCl	1.08	9
α-Methylstyrene	*n*-BuLi	1.05	68
	$SnBr_4/CH_2Cl_2$	1.14	69
1,3-Butadiene	*n*-BuLi	$\sim$1.1	5
Isoprene	*n*-BuLi	$\sim$1.1	5
2-Triisopropoxysilyl-1,3-butadiene	1,1-Diphenylhexyllithium-LiCl	1.02	11
Propylene	$V(acac)_3$ (Structure 13) - Et_2AlCl	1.14	12
Isobutene	BCl_3/*t*-$ROOCCH_3$ or *t*-$ROCH_3$	—	13
Trimethylsilylpropyne	$NbCl_5$	$\sim$1.07	14
2-Butylacetylene	$MoOCl_4Sn$ - EtOH	1.12	15
Norbornene	Titanacyclobutane complex	1.08$\sim$1.14	16
1,2-Disubstituted norbornadiene	Molybdenum complex	$\sim$1.05	17
Methyl methacrylate	Sodium biphenyl	$\sim$1.1	5
	1,1-Diphenylhexyl-lithium-LiCl	1.03	18
	Lanthanoid complex	1.03	25, 26
	Ketene silyl acetal/Fluoride anion etc.	$\sim$1.1	20-24
	Aluminum porphyrin complex	1.19	27-30
	SmI_2–HMPA	1.06	31
	t-BuMgBr-$MgBr_2$ (Toluene)	1.1$\sim$1.2	32
	t-BuLi-AlR_3 (THF)	1.13$\sim$1.18	32
	t-BuLi–(Structure 20) (Toluene)	1.08	33
Alkyl acrylates	Sodium biphenyl	1.04	34
	Lanthanoid complex	1.03$\sim$1.16	69
	Aluminum porphyrin complex	1.11$\sim$1.43	70
	Cobalt porphyrin complex	1.1$\sim$1.2	36

TABLE 1. Representative Monodisperse Polymers (continued)

Monomer (Polymer)	Initiator	*Mw/Mn*	References
t-Butyl crotonate	*s*-BuLi	1.01	35
Isobutyl vinyl ether	$HI-I_2$	1.04	37
	$HI-ZnI_2$	1.06	37
	$CF_3COOH/ZnCl_2$/ether	1.07	38
t-Butyldimethylsilyl vinyl ether	$ZnCl_2$	~1.2	71, 72
Vinyl acetate	$Al(iBu)_3$/2,2′-bipyridyl /TEMPO	1.1~1.2	39
1,2-Diisocyano arenes	Palladium complex	1.08	40
α-Methylbenzyl isocyanide	π-Allylnickel complex	1.1~1.6	41
n-Hexylisocyanate	$TiCl_3(OCH_2CF_3)$	1.1~1.3	42
Ethylene oxide	KOH	—	46, 47
	Aluminum porphyrin complex	1.1~1.2	48
Propylene oxide	Aluminum porphyrin complex	1.1~1.2	48, 49
Oxetane	ROH/Spirosilane (Structure 22)	—	73
	Aluminum porphyrin complex	1.11	50
Tetrahydrofuran	$EtOSO_2CF_3$	—	57-59
ε-Caprolactone	$(i\text{-}PrO(_2AlOZnOAl(Oi\text{-}Pr)_2$	1.3~1.5	43
	Et_2AlOMe	1.03~1.13	44
	Aluminum porphyrin complex	1.19	52
	Lanthanoid complex	1.06	55
δ-Valerolactone	Aluminum porphyrin complex	1.12	53
	Lanthanoid complex	1.07	56
β-Lactones	Aluminum porphyrin complex	1.09	51
	RCOONa	—	45
Propylene sulfide	Zinc porphyrin complex	1.04~1.10	54
Oxazolines	$EtOSO_2CF_3$	1.1~1.3	60
N-t-Butylaziridine	Et_3OBF_4	—	61
α-Amino acids N-carboxylic acid anhydride [Poly(α-amino acids)]	RNH_2	—	62
(Structure 21) [Poly(dimethylsilylene)]	*n*-BuLi	1.3~1.5	63
Cyclic dimethylsiloxanes [Poly(dimethylsiloxane)]	$LiOSiMe_3$	~1.1	64

compounds classified by the type of monomer structures and the methods of polymerization. Their polydispersity values (*Mw/Mn*) and references are also given.

Since monodisperse polymers synthesized by living polymerization possess living ends, these living polymers can initiate further living polymerization using different monomers to yield well-defined monodisperse block copolymers. Living polymers synthesized by bifunctional initiators possess living ends at both terminals and produce ABA-type block copolymers.[74,75] The coupling of an anionic living polymer with a cationic living polymer also produces the corresponding block copolymer.[76] A functional group transformation of living end has also been studied; it yields block copolymers by the transformation of an anionic living end to a different type of anionic or cationic initiator.[77,78]

The transformation of the living end affects not only the initiator but also several monomeric functional groups, yielding a variety of macromonomers. Thus, living polymerization of the macromonomer will lead to well-defined monodisperse graft copolymers.

As the method of living polymerization develops further, a greater variety of monodisperse polymers, including block and graft copolymers, will be produced. The methodology must be available to design materials of well-defined structure in processed form in order to develop the desired properties to the maximum extent.

REFERENCES

1. Szwarc, M.; Levy, M.; Milkovich, R. *J. Am. Chem. Soc.* **1956**, *78*, 2656.
2. Fujimoto, T.; Narukawa, H.; Nagasawa, M. *Macromolecules* **1970**, *3*, 57.
3. Morton, M.; Rembaum, A. A.; Hall, J. L. *J. Polym. Sci. A.* **1963**, *1*, 461.
4. Yen, S. P. S. *Makromol. Chem.* **1965**, *81*, 152.
5. Fetters, L. J. *J. Polym. Sci.* **1969**, *26*, 1.
6. Hirao, A.; Kitaura, K.; Nakahama, S. *Macromolecules* **1993**, *26*, 4995
7. Ishizone, T.; Hirao, A.; Nakahama, S. *Macromolecules* **1991**, *24*, 625.
8. Ishizone, T.; Sugiyama, K.; Hirao, A.; Nakahama, S. *Macromolecules* **1993**, *26*, 3009.
9. Higashimura, T.; Kojima, K.; Sawamoto, M. *Polym. Bull.* **1988**, *19*, 7.
10. Georges, M. K.; Veregin, R. P. N.; Kazmater, P. M.; Hamer, G. K. *Macromolecules* **1993**, *26*, 2987.
11. Ozaki, H.; Hirao, A.; Nakahama, S. *Macromolecules* **1992**, *25*, 96.
12. Doi, Y.; Ueki, S.; Keii, T. *Macromolecules* **1979**, *12*, 814; Doi, Y.; Suzuki, S.; Soga, K. *Macromolecules* **1986**, *19*, 2896.
13. Faust, R.; Kennedy, J. P. *Polym. Bull.* **1986**, *15*, 317.
14. Fujimori, J.; Masuda, T.; Higashimura, M. *Polym. Bull.* **1988**, *20*, 1.
15. Nakano, M.; Masuda, T.; Higashimura, *T. Macromolecules* **1994**, *27*, 1835.
16. Gilliom, L. R.; Grubbs, R. H. *J. Am. Chem. Soc.* **1986**, *108*, 733.
17. Sunaga, T.; Ivin, K. J.; Hofmeister, G. E.; Oskam, J. H.; Schrock, R. R. *Macromolecules* **1994**, *27*, 4043.

18. Varshney, S. K.; Hautekeer, J. P.; Fayt, R.; Jérome, R.; Tessyié, Ph. *Macromolecules* **1990**, *23*, 2618.
19. Wang, J.-S.; Jérome, R.; Bayard, Ph.; Tessyié, Ph. *Macromolecules* **1994**, *27*, 4913 and references therein.
20. Webster, O. W.; Hertler, W. R.; Sogah, D. Y.; Fahnhan, W. B.; Rajan-Babu, T. V. *J. Am. Chem. Soc.* **1983**, *105*, 5706.
21. Hertler, W. R.; Sogah, D. Y.; Webster, O. W.; Trost, B. M. *Macromolecules* **1984**, *17*, 1415.
22. Rajan-Babu, T. V. *J. Org. Chem.* **1984**, *49*, 1083.
23. Webster, O. W. *Polym. Prepr.* **1986**, *27*, 161.
24. Concerning the mechanistic studies on group-transfer polymerization: Müller, A. H. E. *Makromol. Chem., Makromol. Symp.* **1990**, *32*, 87.
25. Yasuda, H.; Yamamoto, H.; Yokota, K.; Miyake, S.; Nakamura, A. *J. Am. Chem. Soc.* **1992**, *114*, 4908.
26. Yasuda, H.; Yamamoto, H.; Yamashita, M.; Yokota, K.; Nakamura, A.; Miyake, S.; Kai, Y.; Kanehisa, N. *Macromolecules* **1993**, *26*, 7134.
27. Kuroki, M.; Aida, T.; Inoue, S. *J. Am. Chem. Soc.* **1987**, *109*, 4737.
28. Kuroki, M.; Watanabe, T.; Aida, T.; Inoue, S. *J. Am. Chem. Soc.* **1991**, *113*, 5903.
29. Sugimoto, H.; Kuroki, M.; Watanabe, T.; Kawamura, C.; Aida, T.; Inoue, S. *Macromolecules* **1993**, *26*, 3403.
30. Adachi, T.; Sugimoto, H.; Aida, T.; Inoue, S. *Macromolecules* **1992**, *25*, 2880.
31. Nomura, R.; Toneri, T.; Endo, T. *Polym. Prepr. Jpn.* **1994**, *43*, 158.
32. Hatada, K.; Ute, K.; Tanaka, K.; Okamoto, Y.; Kitayama, T. *Polym. J.* **1986**, *18*, 1037.
33. Kitayama, T.; Zhang, Y.; Hatada, K. *Polym. J.* **1994**, *26*, 868.
34. Fayt, R.; Forte, R.; Jacobs, C.; Jérome, R.; Ouhadi, T.; Teyssié, Ph.; Varshney, S. K. *Macromolecules* **1987**, *20*, 1442.
35. Kitano, T.; Fujimoto, T.; Nagasawa, M. *Macromolecules* **1974**, *7*, 719.
36. Wayland, B. B.; Poszmik, G.; Mukerjee, S. L.; Fryd, M. *J. Am. Chem. Soc.* **1994**, *116*, 7943.
37. Miyamoto, M.; Sawamoto, M.; Higashimura, T. *Macromolecules* **1985**, *18*, 123.
38. Aoshima, S.; Higashimura, T. *Macromolecules* **1989**, *22*, 1009.
39. Mardare, D.; Matyjaszewski, K. *Macromolecules* **1994**, *27*, 645.
40. Ito, Y.; Ihara, E.; Murakami, M.; Shiro, M. *J. Am. Chem. Soc.* **1990**, *112*, 6446.
41. Deming, T. J.; Novak, B. M. *Macromolecules* **1991**, *24*, 6043.
42. Patten, T. E.; Novak, B. M. *J. Am. Chem. Soc.* **1991**, *113*, 5065.
43. Hamitou, A.; Jérome, R.; Hubert, A. J.; Teyssié, Ph. *Macromolecules* **1973**, *6*, 651.
44. Duda, A.; Florijanczyk, Z.; Hofman, A.; Slomkowski, S.; Penczek, S. *Macromolecules* **1990**, *23*, 1640.
45. Slomkowski, S.; Penczek, S. *Macromolecules* **1976**, *9*, 367; **1980**, *13*, 229.
46. *Ring-opening Polymerization;* Ivin, K. J.; Saegusa, T., Eds.; Elsevier: London, 1984; Vols. 1–3.
47. Tsuruta, T.; Kawakami, Y. In *Comprehensive Polymer Science;* Allen, G.; Bevington, J. C., Eds.; Pergamon: Oxford, New York, 1989; Vol. 3, p 457.
48. Aida, T.; Inoue, S. *Macromolecules* **1981**, *14*, 1166.
49. Kuroki, M.; Aida, T.; Inoue, S. *Makromol. Chem.* **1988**, *189*, 1305.
50. Takeuchi, D.; Watanabe, Y.; Aida, T.; Inoue, S. *Polym. Prepr. Jpn.* **1994**, *43*, 171.
51. Yasuda, T.; Aida, T.; Inoue, S. *Macromolecules* **1983**, *16*, 1792.
52. Shimasaki, K.; Aida, T.; Inoue, S. *Macromolecules* **1987**, *20*, 3076.
53. Endo, M.; Aida, T.; Inoue, S. *Macromolecules* **1987**, *20*, 2982.
54. Aida, T.; Kawaguchi, K.; Inoue, S. *Macromolecules* **1990**, *23*, 2612.
55. Takemoto, Y.; Yamamoto, H.; Yasuda, H. *Polym. Prepr. Jpn.* **1991**, *40,* 1933.
56. Yamashita, M.; Yamamoto, H.; Yasuda, H. *Polym. Prepr. Jpn.* **1991**, *40,* 1936.
57. Kobayashi, S.; Danda, H.; Saegusa, T. *Macromolecules* **1974**, *7*, 415.
58. Penczek, S.; Matyjaszewski, K. *J. Polym. Sci., Polym. Symp.* **1976**, *56,* 255.
59. Takaki, M.; Asami, R.; Kuwabara, T. *Polym. Bull.* **1982**, *7*, 521.
60. Kobayashi, S.; Igarashi, T.; Moriuchi, Y.; Saegusa, T. *Macromolecules* **1986**, *19*, 535.
61. Goethals, E. J. *J. Polym. Sci., Polym. Symp.* **1976**, *56,* 255.
62. Sisido, M.; Imanishi, Y.; Higashimura, T. *Makromol. Chem.* **1977**, *178*, 3107.
63. Saskamoto, K.; Obata, K.; Hirata, H.; Nakajima, M.; Sakurai, H. *J. Am. Chem. Soc.* **1989**, *111*, 7641.
64. Kawakami, Y.; Miki, Y.; Tsuda, T.; Murthy, R. A. N.; Yamashita Y. *Polym. J.* **1982**, *14*, 913.
65. Ishizone, T.; Kato, R.; Ishino, Y.; Hirao, A.; Nakahama, S. *Macromolecules* **1991**, *24*, 1449.
66. Tsuda, K.; Ishizone, T.; Hirao, A.; Nakahama, S.; Kakuchi, T.; Yokota, K. *Macromolecules* **1993**, *26*, 6985.
67. Fujimoto, T.; Ozaki, N.; Nagasawa, M. *J. Polym. Sci. A* **1965**, *1*, 461.
68. Higashimura, T.; Kamigaito, M.; Kato, M.; Hasebe, T.; Sawamoto, M. *Macromolecules* **1993**, *26*, 2670.
69. Morimoto, M.; Ihara, E.; Yasuda, H. *Polym. Prepr. Jpn.* **1994**, *43*, 159.
70. Hosokawa, Y.; Kuroki, M.; Aida, T.; Inoue, S. *Macromolecules* **1991**, *24*, 824.
71. Sogah, D: Y.; Webster, O. W. *Macromolecules* **1986**, *19*, 1775.
72. Kawakami, Y.; Aoki, T.; Yamashita, Y. *Polym. Bull.* **1987**, *18*, 473.
73. Sogah, D. Y. *Polym. Prepr.* **1991**, *32*, 307.
74. Guyot, P.; Favier, J. C.; Uytterhoeven, H.; Fontanille, M.; Sigwalt, P. *Polymer* **1981**, *22*, 1724.
75. Guyot, P.; Favier, J. C.; Fontanille, M.; Sigwalt, P. *Polymer* **1982**, *23*, 73.
76. Yamashita, Y.; Nobutoki, K.; Nakamura, Y.; Hirota, M. *Macromolecules* **1971**, *4*, 548.
77. Quirk, R. P.; Chem, W.-C. *Macromolecules* **1986**, *19*, 1291.
78. Hurley, J.; Richards, D. H.; Stewart, M. J. *Br. Polym. J.* **1986**, *18*, 181.
79. Richards, D. H. *Br. Polym. J.* **1980**, *12*, 89.

MONOLAYER AND LANGMUIR-BLODGETT FILMS (of Protein)

Yoshiyuki Kondo* and Yoshiro Ogoma
Department of Functional Polymer Science
Faculty of Textile Science and Technology
Shinshu University

Many studies on the biological functions of proteins have been using Langmuir-Blodgett (LB) films.[1–4] It is difficult to prepare a monolayer at the air–water surface for LB films of protein because it seems to be impossible for proteins to form their films without surface denaturation.

*Author to whom correspondence should be addressed.

Calmodulin (CaM), a heat-stable and multifunctional Ca^{2+}-binding protein responsible for the Ca^{2+} stimulation of cyclic nucleotide phosphodiesterase (PDE), is known to play a central role in the Ca^{2+}-dependent regulation of eukaryotic cells.[5] CaM is a single 148-amino-acid polypeptide (MW 17,000) containing four Ca^{2+}-binding sites. It is heat-stable and resistive for surface denaturation. Therefore, the use of CaM as a biosensor, particularly as a Ca^{2+} sensor, seems feasible.

BIOLOGICAL ACTIVITY OF CAM LB FILM

Retaining the activity of CaM in LB film is very important for the practical use of this film as a biosensor.

From the results discussed above, we conclude that CaM, which is heat-stable, resistive to surface denaturation, and a receptor for Ca^{2+}, can retain its secondary structure and its activity even in LB film, and that the use of this film as a Ca^{2+} sensor will be feasible in the future. For this purpose, it is necessary to study the state of CaM molecules in LB film. However, it is difficult to realize the state of protein in monolayer and LB film. Recently, direct observations of surface monolayer at the air–water interface were made by use of fluorescence microscopy[16,17] and Brewster angle microscopy.[18–20] Therefore, further details of the properties of LB films of protein will be presented in the near future.

REFERENCES

1. Langmuir, I.; Schaefer, V. J.; Wrinch, D. W. *Science (Washington, D.C.)* **1937**, *85*, 76.
2. Cheesman, D. F.; Davies, J. T. *Adv. Protein Chem.* **1954**, *9*, 439.
3. Yamashita, T.; Bull, H. B. *J. Colloid Interface Sci.* **1967**, *24*, 310.
4. Ishii, T.; Muramatsu, M. *Bull. Chem. Soc. Jpn.* **1970**, *43*, 2364.
5. Kakiuchi, S.; Hidaka, H.; Means, A. R. *Calmodulin and Intracellular Ca^{2+} Receptors;* Plenum: New York, 1982.
6. Lösche, M.; Sackmann, E.; Möhwald, H. *Ber. Bunsenges. Phys. Chem.* **1983**, *87*, 848.
7. Shimomura, M.; Fujii, K.; Karg, P.; Frey, W.; Sackrnann, E.; Meller, P.; Ringsdolf, H. *Jpn. J. Appl. Phys.* **1988**, *27*, L1761.
8. Höing, D.; Möbius, D. *J. Phys. Chem.* **1991**, *95*, 4590.
9. Höing, D.; Möbius, D. *Thin Solid Films* **1992**, *210/211*, 64.
10. Hénon, S.; Meunier, *J. Thin Solid Films* **1992**, *210/211*, 121.

MONOMER-ISOMERIZATION POLYMERIZATION (2-Butene with Ziegler-Natta Catalysts)

Kiyoshi Endo
Department of Applied Chemistry
Faculty of Engineering
Osaka City University

Polymerizations of vinyl monomer usually proceed through double bond opening without any rearrangement of chemical structure. However, isomerization reactions sometimes occur in such polymerizations. In the case of monomer-isomerization polymerization, such isomerization takes place prior to the polymerization.

Three research groups reported that a high-molecular-weight polymer was obtained by polymerization of 2-butene with Ziegler-Natta catalysts.[1–3]

All experimental results gathered in the study of polymerization of 2-butene with Ziegler-Natta catalysts led to the following conclusion: The 2-butene isomerized to polymerizable 1-butene prior to the propagation, and then the 1-butene homopolymerized to give poly(1-butene).

Monomer-isomerization polymerization thus can be defined as polymerization in which the charging monomers first isomerize to polymerizable monomers, which then homopolymerize to yield high polymers consisting of the isomerized units exclusively. In other words, monomer-isomerization polymerization consists of two distinct and independent-reactions. One is the isomerization reaction of charged monomer to yield polymerizable monomers. The other is the polymerization of the isomerized monomers. Both reactions proceed simultaneously and independently in the system.

Otsu called this procedure "monomer-isomerization polymerization" and developed it as a new route for the synthesis of polymers from various internal olefins.[4,5]

The difference between monomer-isomerization polymerization of internal olefins with Ziegler-Natta catalysts and isomerization polymerization of branched 1-olefins with cationic catalysts is that, in the former process, isomerization to polymerizable 1-olefins takes place before polymerization; in the latter, the propagating cationic species itself isomerizes to the most stable cationic species during propagation.

PREPARATION AND PROPERTIES

Internal olefins such as 2-butene have not been known to polymerize with Ziegler-Natta catalysts due to steric hindrance of the alkyl substituents,[6] and a high-molecular-weight homopolymer of 2-butene has not been synthesized. However, butene can isomerize with transition metals, including Ziegler-Natta catalysts, to yield an equilibrium mixture of 1-butene, *cis*- and *trans*-2-butene.[7] Among these butenes, only 1-butene is a polymerizable 1-olefin with Ziegler-Natta catalysts. If catalysts can catalyze both reactions in the same system, polymers will be obtained from the polymerization of 2-butene. In fact, when *cis*- or *trans*-2-butene is polymerized in the presence of Ziegler-Natta catalysts such as $TiCl_3$-$Al(C_2H_5)_3$ catalyst at a relatively high temperature, the process yields high-molecular-weight polymers.[8] Product analysis using NMR and IR spectroscopy confirmed that the polymer formed was poly-1-butene.[2,9]

Monomer-isomerization polymerization of 2-butene with Ziegler-Natta catalysts can be summarized as follows: This polymerization reaction occurs at two different catalyst sites, and the concentration of 1-butene produced by isomerization in the system is limited by thermodynamic factors. However, the isomerization rate is dependent on the reaction conditions and on the nature of the transition metal component. In the second step, 1-butene polymerizes and is momentarily removed from the dynamic prevailing at the catalyst site. The process is a coordination polymerization. The resulting polymer is poly(1 butene); it contains considerable amounts of isotactic fraction.

REFERENCES

1. Symcox, R. O. *J. Polym. Sci., Polym Lett. Ed.* **1964**, *2*, 947.
2. Shimizu, A.; Otsu T.; Imoto, M. *J. Polym. Sci., Polym. Lett. Ed.* **1965**, *3*, 449.

3. Iwamoto, A.; Yuguchi, S. *Bull. Chem. Soc. Jpn.* **1967**, *40*, 159.
4. Kennedy J. P.; Otsu, T. *Adv. Polym. Sci.* **1970**, *7*, 369.
5. Otsu, T.; Shimizu, A.; Itakura, K.; Imoto, M. *Makromol. Chem.* **1968**, *123*, 284.
6. Natta, G.; Dall'Asta, G.; Mazzanti, G.; Pasquon, I.; Valassiri, A.; Zambelli, A. *J. Am. Chem. Soc.* **1961**, *83*, 3343.
7. Cramm, R.; Lindsey, R. V., Jr. *J. Am. Chem. Soc.* **1966**, *88*, 1535.
8. Shimizu, A.; Otsu T.; Imoto, M. *J. Polym. Sci.* A-1 **1966**, *4*, 1579.
9. Endo, K.; Ueda, R.; Otsu, T. *J. Polym. Sci., Polym. Chem. Ed.* **1991**, *29*, 807.

Mortar

See: Concrete-Polymer Composites

Multicomponent Polymer Networks

See: Graft and Network Copolymers (by Macromolecular Ion Coupling Reactions)

Multicomponent Resins

See: Polyethylene, Multicomponent

MULTIHOLLOW PARTICLES

Masayoshi Okubo
Chemical Science and Engineering
Faculty of Engineering
Kobe University

Polymer particles with one or more hollows on the inside have received much attention since thermoplastic styrene/acrylic polymer particles containing one hollow at the center have been produced by alkali swelling of carboxylated polymer particles with core-shell structures.[1] The hollow is filled with water in polymer emulsion, and with air in a dried state. Such particles, under the name Ropaque Particle®, are commercially supplied as hiding or opacifying agents in coating and molding compositions by Rohm & Haas Co.

Similar hollow particles have been produced by seeded emulsion copolymerization of methyl methacrylate, divinylbenzene, and methacrylic acid in the presence of polystyrene particles.[2] Production of these particles is based not on alkali swelling but on shrinkage of volume with polymerization.

The author and coworkers have been investigating the control of carboxyl groups within carboxylated polymer particles prepared by emulsion copolymerization with an unsaturated acid monomer such as methacrylic acid.[3] In the process, polymer particles containing many hollows on the inside were produced.

MULTIHOLLOW POLYMER PARTICLES BY THE STEPWISE ALKALI/ACID METHOD

Polymer particles containing many hollows in the inside were produced by treating styrene-butyl acrylate-methacrylic acid terpolymer particles (74.3/17.0/8.7, molar ratio) stepwise with alkali and acid.[4] The particles and method were named "multihollow particles" and "stepwise alkali/acid method," respectively.

The formation of multihollow structures can be controllable in the acid treatment process as well as in the alkali treatment process. The formation of multihollow structures seems to proceed as follows: In the early stage of the acid treatment process, the polymer wall (i.e., shell) is formed quickly at the surface of the alkali-swollen particle, because the "soluble" polymer segments containing ionized carboxyl groups are precipitated by their deionization. The shell prevents shrinkage to the original state. The fixation of polymer molecules proceeds gradually in the inside with the diffusion of acid through the shell, resulting in the multihollow structure. This formation mechanism is similar to that of an asymmetrical polymer membrane consisting of skin and sponge layers formed by the wet cast method.[5]

REFERENCES

1. Kowalski, A. *USP* **1984**, *4*, 427, 836.
2. Sakurai, F.; Kasai, K.; Hattori, M.; Kondo, M. (Japan Synthetic Rubber Co.) *JSR Technical Review* **1991**, *98*, 53.
3. Okubo, M.; Xu, D. H.; Kanaida, K.; Matsumoto, T. *Coll. Polym. Sci.* **1987**, *265*, 246.
4. Okubo, M.; Kanaida, K.; Fujimura, M. *Chemistry Express* **1990**, *5*, 797.
5. Matsumoto, T.; Nakamae, K.; Ochiumi, T.; Horie, S. *J. Membrane Sci.* **1981**, *9*, 109.

Multilayer Assemblies

See: Photoreactive Langmuir-Blodgett-Kuhn Assemblies (Functionalized Liquid-Crystalline Side Chain Polymers)

MULTILAYER FILMS (Polyelectrolytes)

Gero Decher
C.N.R.S.
Institut Charles Sadron (CRM)

The methods most commonly used for the preparation of ultrathin multilayer films have been the Langmuir-Blodgett (LB) technique and molecular self-assembly based on chemisorption.[1–12] In recent years, my coworkers and I have established a self-assembly technique that is based on physisorption from solution and utilizes the electrostatic attraction between opposite charges.[13–19]

PREPARATION

The buildup of heterofunctional multilayers is straightforward and is easily achieved because it is a directed assembly process in which the layer sequence is exactly controlled. Every adsorption step is self-regulating, leading to individual layers of a defined thickness. Depending on the application, the multilayers may be fabricated on planar surfaces (for basic research or for optical or sensing applications), on latex particles (e.g., for catalysis), or on custom surfaces (e.g., for antistatic coatings).

The top of **Figure 1** shows how one might envision the consecutive adsorption of polyanions and polycations; the bottom shows the actual procedure as it is being carried out.

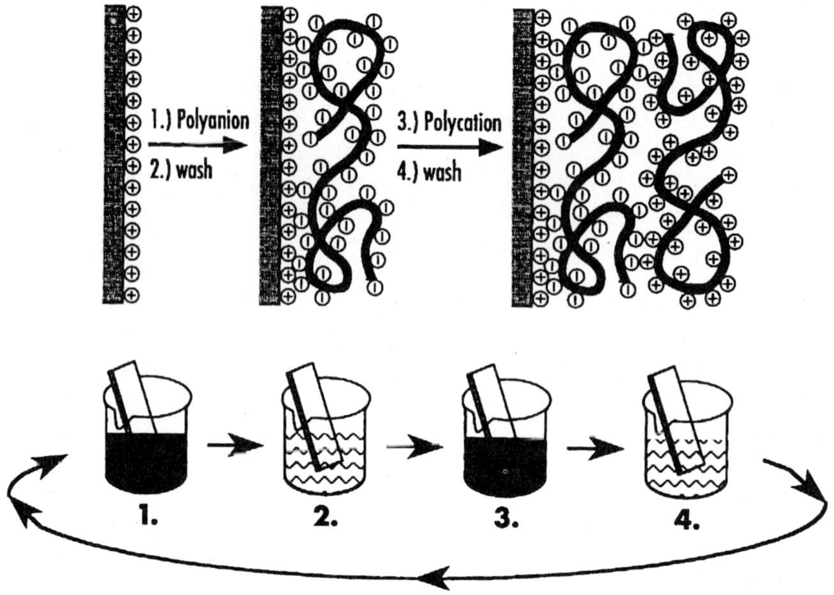

FIGURE 1. *Top*: Schematic for the buildup of multilayer assemblies by consecutive adsorption of anionic and cationic polyelectrolytes (cyclic repetition of steps 1 to 4). It is not implied that the symbols used for the polyelectrolytes represent their actual structure in solution or after the adsorption. For reasons of clarity, the counterions were omitted in this drawing. *Bottom*: Schematic of the deposition procedure, showing adsorption on a glass slide in regular beakers.

The key to successful multilayer deposition is surface refunctionalization in each adsorption step.[13–15,20] In the case of polyions, this means that the adsorption conditions must lead to a consecutive change of the sign of the film surface charge after deposition of a given polyanion or polycation in any deposition sequence. For the simplest case of a $(AB)_n$ two-component film in which A is a polyanion, B is a polycation, and n is the number of repetitions of this sequence, linear film growth is observed over several tens to several hundreds of layers when adsorption times are in the minute range.

PROPERTIES

The conformation of polyelectrolytes in solution depends strongly on the concentration of added electrolyte. At low salt concentrations, the polyelectrolytes are rather extended because of the repulsion of charges along the polymer backbone. At high salt concentrations, the polyions coil up because the added salt screens this repulsion. We have shown that added electrolyte also has an influence on the conformation within the adsorbed layer and thus on the architecture of the resulting film. The possibility for control of the average incremental film growth is remarkable.[16]

APPLICATIONS

Although the technique of consecutive electrostatically driven adsorption of polyanions and polycations is presently not yet well enough developed for practical applications, the prospects for future use are quite promising. Indeed, several patents have been filed.[17] A tremendous advantage over other molecular deposition techniques is that no dedicated and sensitive equipment (e.g., ultrahigh vacuum apparatus of Langmuir troughs) is needed. Moreover, the adsorption is carried out from aqueous solutions, which makes the technique environmentally benign.

The system with which we have done most of our work is polystyrene sulfonate/polyallylamine, but other polyelectrolytes and functional polymers are easily incorporated into the films as well.

REFERENCES

1. Ulman, A. *An Introduction to Ultrathin Organic Films: From Langmuir-Blodgett to Self-Assembly;* Academic: Boston, 1991.
2. Blodgett, K. B.; Langmuir, I. *Phys. Rev.* **1937**, *51*, 964.
3. Gaines, G. L. *Insoluble Monolayers at Liquid-Gas Interfaces;* Interscience: New York, 1966.
4. Kuhn, H.; Möbius, D.; Bücher, H. In *Physical Methods of Chemistry, Part 3B;* Weissberger, A.; Rossiter, P., Eds.; Wiley-Interscience: New York, 1972; p 577.
5. Merrifield, R. B. *Science (Washington, DC)* **1965**, *150*, 178.
6. Netzer, L.; Iscovici, R.; Sagiv, J. *Thin Solid Films* **1983**, *99*, 235.
7. Maoz, R.; Netzer, L.; Gun, J.; Sagiv, J. *J. Chim. Phys.* **1988**, *85*, 1059.
8. Lee, H.; Kepley, L. J.; Hong, H.-G.; Mallouk, T. E. *J. Am. Chem. Soc.* **1987**, *110*, 618.
9. Cao, G.; Hong, H.-G.; Mallouk, T. E. *Acc. Chem. Res.* **1992**, *25*, 420.
10. Tomalia, D. A.; Naylor, A. M.; Goddard, W. A., III, *Angew. Chem., Int. Ed. Engl.* **1990**, *29*, 138.
11. Tomalia, D. A.; Durst, H. D. *Supramolecular Chemistry I--Directed Synthesis and Molecular Recognition;* Springer: Berlin, 1993; p 193.
12. Fodor, S. P. A.; Read, J. L.; Pirrung, M. C.; Stryer, L.; Lu, A. T.; Solas, D. *Science (Washington, DC)* **1991**, *251*, 767.
13. Decher, G.; Hong, J.-D. *Makromol. Chem., Macromol. Symp.* **1991**, *46*, 321.
14. Decher, G.; Hong, J.-D. *Ber. Bunsenges. Phys. Chem.* **1991**, *95*, 1430.
15. Decher, G.; Hong, J.-D.; Schmitt, J. *Thin Solid Films* **1992**, *210/211*, 831
16. Decher, G.; Schmitt, J. *Progr. Colloid Polym. Sci.* **1992**, *89*, 160.
17. Decher, G.; Hong, J.-D. Eur. Patent 0 472 990 A2, 1992.

18. Lvov, Y.; Decher, G.; Möhwald, H. *Langmuir* **1993**, *9*, 481.
19. Lvov, Y.; Haas, H.; Decher, G.; Möhwald, H.; Kalachev, A. *J. Phys. Chem.* **1993**, *97*, 12835.
20. Decher, G.; Hong, J.-D.; Lowack, K.; Lvov, Y.; Schmitt, J. In *Self-Reproduction of Supramolecular Structures;* Colonna, S. et al., Eds.; NATO ASI Series C: Mathematical and Physical Sciences; Kluwer Academic: Dordrecht, The Netherlands, 1994; Vol. 446, 267.

Multilayer Structures

See: Compatibilizers, Polymeric (Recycling of Multilayer Structures)

Myoglobin

See: Hemoproteins

N

Nanocomposites

See: *Electrically Conducting Composites*
Inorganic Nanostructured Materials
Organic/Inorganic Composites

Nanogels

See: *Colloidal Dispersions, Nanogels*

Nanostructured Materials

See: *Inorganic Nanostructured Materials*

Nanotubes

See: *Peptide-Based Nanotubes (New Class of Functional Biomaterials)*

Natural Rubber

See: *Epoxidized Natural Rubber*
Interpenetrating Polymer Networks, Rubber-Based
Natural Rubber (Chemical Modification)
Natural Rubber (Hevea brasiliensis)
Natural Rubber Latex (Polymeric Flocculant for Tin Tailings Slurries)
Tack, Elastomers
Vulcanization (Overview)

NATURAL RUBBER (Chemical Modification)

D. J. Hourston and J. O. Table
IPTME
Loughborough University

CONSTITUTION OF NATURAL RUBBER

Freshly tapped natural rubber latex is a whitish fluid with a density of between 0.975 and 0.980 g/ml and a pH of 6.5 to 7.0. Because latex is a natural product, the composition varies widely, depending on its source.

The rubber hydrocarbon in fresh hevea latex is predominantly *cis*-1,4-polyisoprene. The soluble fraction has, typically, a number-average molecular weight of about 300,000.

Natural rubber latex coagulates within a few hours of leaving the tree. Putrefaction then sets in at a later stage. Thus, preservation is necessary. Many preservatives have been used with NR latex, but the most common and efficient is ammonia or systems containing ammonia.

THE CHEMICAL MODIFICATION OF NATURAL RUBBER

Natural rubber comes from the tree with a predetermined structure from which it derives its excellent physical properties for serving as a general purpose elastomer. There is, however, an ever-present demand for improvement in areas such as vulcanization, ageing, bonding, processability, and abrasion resistance and adhesion. Also, properties such as gas permeability, oil and flame resistance have to be adjusted to fit certain applications.

Improvements based on compounding alone are not always sufficient to meet these targets. Therefore, other methods such as modification of the natural rubber molecule itself have to be pursued.

Modification of natural rubber can be achieved by one of the following means:

- Changing the structure, or geometry, of the natural rubber molecule without introducing new material.
- Attaching to the NR molecule of groups having specific physical characteristics or chemical reactivities.
- Grafting short or long chains of a different polymer type on to the natural rubber hydrocarbon.

Ene Reactions

These belong to a group of simple chemical reactions that do not proceed via reactive intermediates, but by direct reaction between the parent molecules.

The ene reaction has found two important applications: in rubber-bound antioxidants and urethane vulcanization.[1] Azodicarboxylates have been examined in some detail as ene reagents, as it is known that the ene reaction, like that of most olefins, is insensitive to radical initiators or scavengers and to solvent type.[1a]

The ene reaction has been used to modify natural rubber. The C-nitroso and the activated azo compounds are useful ene reagents.

Several authors have found experimentally that graft copolymers with an elastomeric backbone and hard side chains can behave as thermoplastic rubbers.[2–4]

Azo modification can be used to graft many polymers to natural rubber. This can be achieved by introducing a hydroxy group to the synthetic polymer of desired molecular weight by anionic polymerization. Campbell et al. reported a high yield on reacting such a functionalized polymer with an azo chloride.[4a]

MODIFICATION OF NATURAL RUBBER WITH PERACIDS

Most of the recent papers on chemical modification of natural rubber are about epoxidation reactions.[5,6] The material is of interest in that, although capable of crystallizing like NR and, thus, giving "non-black" stocks a high strength, it also exhibits a good level of oil resistance and low air permeability similar to that of nitrile rubber.[7]

The epoxidation reaction with peroxy acids is stereospecific, and, thus, epoxidized natural rubber (ENR) retains the stereoregular all-*cis*-1,4-configuration of natural rubber.

Epoxidation of NR results in a systematic increase in polarity. This will enhance the hydrocarbon oil resistance and also raise the glass transition temperature. In turn, the T_g will influence such properties as abrasion resistance, resilience and road grip. An increase in the level of epoxidation also decreases the gas permeability.

POLYMERIZATION OF VINYL MONOMERS IN NATURAL RUBBER LATEX

The literature contains several examples of cases where various vinyl monomers have been polymerized in natural rubber.[8–11] The polymerization can be achieved in the dry state, in solution, or in the latex. The modification of NR in the latex form is clearly more challenging.

Polymerization in the latex involves blending monomer with a stabilized latex or adding an emulsified monomer to the latex, and then introducing an initiator that brings about polymerization of the added monomer under convenient conditions.

PREVULCANIZED LATEX

A prevulcanized latex is one which is chemically modified by the addition of vulcanizing ingredients, so that on drying it gives a vulcanized film. The prevulcanized latex is particularly useful in the dipping industry, and also in applications in other fields, such as textiles and adhesives. Prevulcanization can be achieved using sulfur, sulfur-donors, peroxides or irradiation.

REFERENCES

1. Baker, C. S. L. *Kautschuk und Gummi* **1983**, *36*, 677.
1a. Hoffman, H. M. R. "The ene reaction" *Angew. Chem. Int. Ed.* **1969**, *8(8)*, 556.
2. Falk, J. C.; Schlott, R. J.; Hoeg, D. F.; Pendleton, J. F. *Rubb. Chem. Technol.* **1973**, *46*, 1044.
3. Sigwalt, P.; Polton, A.; Miskovic, M. *J. Polym. Symp.* **1976**, *56*, 13.
4. Foss, R. P.; Jacobson, H. W.; Cripps, H. N.; Sharkey, W. H. *Macromolecules* **1976**, *9*, 373.
4a. Campbell, D. S.; Mente, P. G.; Tinker, A. J. U.S. Patent Appln. No. 374866, 1982.
5. Hashim, A. S.; Kohjiya, S. *Kautsh. Gummi Kunstst.* **1993**, *46*, 208.
6. Roy, S.; Namboodri, C. S. S.; Maiti, B. R.; Gupta, B. R. *Polym. Eng. Sci.* **1993**, *33*(2), 96.
7. Baker, C. S. L.; Gelling, I. R.; Azemi, S. B. *J. Nat. Rubb. Res.* **1986**, *1*, 135.
8. Hourston, D. J.; Romaine, J. *J. Appl. Polym. Sci.* **1991**, *43*, 2207.
9. Hourston, D. J.; Romaine, J. *J. Appl. Polym. Sci.* **1990**, *39*, 1587.
10. Erbil, H. Y. *Nat. Rubb. Res.* **1986**, *1*, 2345.
11. Burfield, D. R.; Ng, S. C. *Eur. Polym. J.* **1978**, *14*, 799.

NATURAL RUBBER (*Hevea brasiliensis*)

W. Stephen Fulton and W. M. H. Thorpe
Tun Abdul Razak Laboratory
The Malaysian Rubber Products' Research Association

Most of the dry natural rubber used today is obtained from the *Hevea brasiliensis* tree as latex, which is a colloidal dispersion consisting of nonrubber substances and rubber particles in an aqueous serum phase. The rubber hydrocarbon constitutes between 30% and 45% of the whole latex, while the nonrubber substances account for between 3% and 5%. Rubber particles vary in size from 0.15 μm to 3 μm and contain 90–95% natural rubber (*cis*-1,4-polyisoprene) with a molecular weight distribution of $10^5 - 10^7$ g · mol^{-1}.

Malaysia, Indonesia, and Thailand are the three main rubber-producing countries in the world, and in 1994 they contributed 74% to the world's total rubber production, which was some 5 million tons.[1] In recent years the relative consumption of natural rubber has remained fairly constant despite competition from synthetic rubber.

On average, rubber trees are productive for 25 years, coming to maturity after 5 years, and peaking in production after 12 years.

PRODUCTION METHODS OF DRY RUBBER

Traditional Grades

Classification of traditional grades, such as ribbed smoked sheet (RSS), is based upon the source of raw materials and processing methods. Visual grading and comparison with international samples depends upon factors such as color, cleanliness, and uniformity of appearance as described in the "Green Book," which lists all 35 grades.

Heveaplus MG

Heveaplus MG is the trade name for materials prepared by the graft polymerization of methyl methacrylate in natural rubber latex. The majority of the poly(methyl methacrylate) is grafted onto the rubber molecules as long side chains but a percentage (~20%) is present as free poly(methyl methacrylate). The two main uses for Heveaplus MG rubber are as adhesives and as a means of producing hard vulcanizates without the disadvantage of poor processing associated with high filler levels.[2]

APPLICATIONS

Natural rubber is used predominantly in the manufacture of tires where its excellent tack, low heat build-up, low rolling resistance, and good low-temperature performance make it the preferred elastomer.

REFERENCES

1. International Rubber Study Group, *Rubber Statistical Bulletin* **1994**, *49*, 3.
2. Wheelans, M. A. *N.R. Technology* **1977**, *8*, 69.

NATURAL RUBBER LATEX (Polymeric Flocculant for Tin Tailings Slurries)

C. C. Ho* and K. C. Lee
Department of Chemistry
University of Malaya

E. B. Yeap
Department of Geology
University of Malaya

Dewatering is the science of separating water from solids. In any dewatering process, two aspects are important: the total

*Author to whom correspondence should be addressed.

amount of removable water and speed. The dewatering of tin tailings, slurry, confined to so-called ex-mining ponds, has never been satisfactory. Like most other mining slurries, the main difficulty lies not in clarifying the supernatant water but in removing water from the soft, settled clays.

Natural rubber (NR) latex is obtained from the rubber tree *Hevea brasiliensis.*

RESULTS AND DISCUSSION

Electrophoretic mobility data indicate that both the latex and clay particles are negatively charged at pH 7.[1] Thus, both the latex and slurry dispersions were stable in their original state. Under normal conditions of stable dispersion, the NR latex particles would cream slowly and the clay particles settle very slowly under gravity.

Information on the interaction between the latex and clay particles of differing sizes is obtained through studies on the adsorption of smaller particles on larger ones. Thus it was found that the adsorption of latex particles on the clay surface was almost spontaneous and the extent of adsorption depended strongly on the clay mineral composition, the pore fluid chemistry of the slurry, and also the composition of the latex. The effect of slurry and latex concentrations, pH of the mixed dispersion, inorganic salt type, and its concentration on the effectiveness of solid-liquid separation was investigated.[2]

APPLICATIONS

The aim is to find an acceptable reclamation option for tin tailings slurries. This option must be technically sound, economical, and permanent (problem not transferred to another site). The reclaimed slime preferably can be used as foundation material. Based on the known properties of slurry studied in this work, the option is to increase the solid concentration of slurry with chemical reagents. The flocculated slurry is then allowed to settle to the bottom of the pond followed by surcharging, thus transforming the consolidated slurry to foundation grade material.

CONCLUSION

In contrast to the water-soluble synthetic polymeric flocculants used in flocculation of colloidal dispersions, the present system of mixed dispersion of latex-slurry operates via a heterocoagulation process in the presence of metallic ions acting as electrostatic bridges linking the dissimilar particles together. Since both the latex and the clay particles were polydisperse but comparable in size, small, negatively charged latex particles can adsorb on large negatively charged clay particles or vice versa through these electrostatic bridges. Intrinsic to this behavior is the anisotropic character of the clay particles: the surface charges on the basal surface and the edge of the plate-like clay particles are different and strongly pH-dependent. Although both the latex and clay particles are polydisperse in size and the dispersions complex in chemical compositions, a total heterocoagulation of the latex and clay particles, with complete removal of disperse particles from the liquid phase, could be achieved at optimum dosage of latex and electrolyte at the natural pH of the slurry (near neutral). Underdosing of latex results in incomplete clay particles removal, whereas overdosing latex leaves behind excess free latex particles. The presence of sufficient metallic ions such as $CaCl_2$ is essential as electrostatic bridges especially at alkaline pH. Since only a very small amount of NR latex (0.1% w/v) is required (to heterocoagulate completely at 8% w/v slurry) and is effective at neutral pH, the system is extremely attractive as a cost-effective and environment friendly method of flocculating slurries (mainly clay particles) from mining sources. On-site evaluation of consolidated slurry is still in progress.

ACKNOWLEDGMENT

This project was supported by the International Development Research Center (IDRC), Canada. A studentship to K. C. Lee by the University of Malaya is gratefully acknowledged. We would also like to thank R. N. Yong (McGill University, Montreal, Canada) and B. K. Tan for useful discussions.

REFERENCES

1. Ho, C. C.; Lee, K. C.; Yeap, E. B. In *Colloid Polymer Particles*; Goodwyn, J. W.; Buscall, R., Eds.; Academic: London, 1995; p 277–306.
2. Ho, C. C.; Lee, K. C.; Yeap, E. B. In *Progress in Pacific Polymer Science 3*; Ghiggino, K. P., Ed.; Springer-Verlag: Berlin, 1994; p 337–350.

Nerve Regeneration

See: Regeneration Templates, Artificial Skin and Nerves

NETWORK POLYMERS (Chemorheology)

Saburo Tamura
Miyagi Polytechnic College

Chemorheology can be defined as the study, using rheological methods, of structural changes in polymers due to chemical, mechanical or mechanochemical causes. Chemorheology of rubber-like network polymers was theoretically introduced by Tobolsky,[1] who has further developed the area with his co-workers.[2] The topic has also been explored by Berry, Watson, and co-workers,[3,4] Murakami and Temura and co-workers,[5,6] and others.[7] It has been applied mainly in investigations of degradation of rubber vulcanizates. Network polymers have been used under different conditions and then subjected to various types of structural changes, that is, degradations, depending on circumstances. Although the quantitative analysis of degradation of network polymers has been difficult, it has progressed, using chemorheology.

PRINCIPLE OF CHEMORHEOLOGY

Chemorheology of rubber-like network polymers has usually been investigated by measuring the structures decay of a stretched sample held at constant extension: this is called a continuous chemical stress relaxation (CSR). The CSR-method can continuously give information on degradation of rubber networks occurring with observation time. The basis of chemorheology is due to classical rubber-like elasticity.[1,5,6]

Mechanisms of Chemical Stress Relaxation of Network Polymers

The structure of crosslinked rubbers is generally very complicated. Besides having chemical crosslinkages, the network contains defects and entanglements which act as physical crosslinks; therefore, physical stress relaxation due to flow of polymer chains, not accompanying chemical reaction of networks, has usually been observed. Also, actual network has a random distribution of chain length. Degradation of network polymers is divided into two different types: scission and crosslinking. The scission is classified into four modes:[1,5,6]

- random scission along main chains
- scission at crosslinkages
- selective scission of main chain near crosslinkage
- exchange reaction at crosslinkages

APPLICATION OF CHEMORHEOLOGY

Chemorheology is useful for investigating degradation of network polymers at initial stage. This method is simple to measure, and continuously gives information on degradation process with time.

Chemorheology of Crosslinked EPDM and NR

EPDM (ethylene-propylene-diene rubber) is thermally more stable than diene rubbers. Up to an appropriate high temperature of around 423 K, EPDM is little cleaved even in the presence of oxygen. Therefore, CSR of crosslinked EPDMs[6] having different crosslinking structures at high temperature can give information on stability and scission mechanisms of crosslinkages.

Since NR has many carbon–carbon double bonds in the main chain, chemorheological behavior of the vulcanizates is more complicated.[8,9,11,12] In inert gas and in an appropriate temperature range, the stress relaxation shows the characteristic behavior depends only on the crosslinking structures. In the presence of enough oxygen, however, the chemical stress relaxation of NR vulcanizates is largely caused by the scission of main chain.

Crosslinking Reaction in Network Polymers

In the degradation of diene rubber networks, especially, the crosslinking reaction sometimes occurs together with the cleavage of networks. The crosslinking reaction depends on the surrounding condition; for example, under photo-irradiation, the number of crosslinkages newly formed in sulfur-cured NR gradually increases with an increase in oxygen content in an atmosphere.

If a network polymer is extended below the glass transition temperature (T_g), it will be cleaved mechanically to generate polymer radicals. The reactivity of the polymer radicals can be followed by chemorheology.

Dynamic Chemorheological Methods

Dynamic rheological properties are also available in the investigation of degradation of network polymers. The storage modulus at a given frequency is closely related to the number of effective chains.[13]

In conclusion, the chemorheology is widely applicable to the study of degradation of network polymers.

REFERENCES

1. Tobolsky, A. V. *Properties and Structure of Polymers;* John Wiley & Sons: New York, 1963.
2. Tobolsky, A. V.; Mercurio, A. *J. Polym. Sci.* **1959**, *36*, 467.
3. Berry, J. P.; Watson, W. F. *ibid.* **1955**, *18*, 201.
4. Dunn, J. R.; Scanlan, J.; Watson, W. F. *Trans. Faraday Soc.* **1958**, *27*, 730.
5. Murakami, K.; Tamura, S.; Oikawa, H. *Degradation and Stabilization of Polymers;* Jellinek, H. H. G., Ed.; Elsevier: New York, 1983; Chapter 9.
6. Murakami, K.; Ono, K. *Chemorheology of Polymers;* Elsevier: New York, 1980.
7. Osthoff, R. C.; Bueche, A. M.; Grubb, W. T. *J. Am. Chem. Soc.* **1954**, *76*, 4659.
8. Tamura, S.; Murakami, K. *Polymer* **1973**, *14*, 569.
9. Tamura, S.; Murakami, K. *Rubber Chem. Technol.* **1975**, *48*, 141.
10. Murakami, K.; Tamura. S. *Polym. J.* **1971**, *2*, 328.
11. Tamura, S.; Murakami, K. *Polymer* **1976**, *17*, 325.
12. Tamura. S.; Murakami, K. *ibid.* **1978**, *19*, 801.
13. Yano, S. *Rubber Chem. Technol.* **1981**, *54*, 1.

NETWORK POLYMERS (Vinyl-Type)

Akira Matsumoto
Department of Applied Chemistry
Faculty of Engineering
Kansai University

Network polymers—or simply, networks—are commonly referred to as crosslinked polymers. Such polymers are also called thermosetting polymers or thermosets as infusible, cured materials in which the term curing often replaces crosslinking. Whereas linear and branched polymers can usually be dissolved in some solvent and in most instances they will melt and flow, they are said to be thermoplastic.

Although the insoluble, infusible properties result in the difficulty of a full understanding of network polymers as the lack of suitable analytical methods, crosslinking leads to the formation of networks: it is accomplished during the polymerization of multifunctional monomers instead of difunctional ones producing linear polymers or by various chemical reactions to link together the preformed polymer chains. Thus, the phenolic polymers and aminoplastics via the addition-condensation reaction of formaldehyde with phenol and urea or melamine, alkyd resins and polyimide resins as polycondensates, polyurethanes, epoxide resins, and styrene-divinylbenzene crosslinked copolymers and unsaturated polyesters as addition polymers are typical examples of the network polymers formed through the polymerizations and/or copolymerizations of multifunctional monomers.

Crosslinking not only leads to a wide range of commercially important network polymers, but also is accompanied by a number of extreme changes of the physical properties of polymers which depend on the level of crosslinking. Thus, network polymers with a low crosslinking density lose their flow properties, but still undergo a reversible deformation; the polymers will exhibit elastic properties as elastomers. The higher the crosslinking density, the more rigid the polymer. Rigid plastics characterized by high chain rigidity and high resistance to deformation, quite different from the flexible plastics, are produced by extensive crosslinking.

Recently, the network polymers have become interested not only in high-performance materials, but also in high-functional polymer gels. Here a gel is defined as a crosslinked polymer network swollen in a liquid medium. Some gels are chemically crosslinked by covalent bonds, whereas others are crosslinked physically by weak forces such as hydrogen bonds, van der Waals forces, or hydrophobic and ionic interaction. Gels are characterized by their equilibrium and dynamic and kinetic properties that depend on the gel state represented by osmotic pressure, temperature, solvent composition, and degree of swelling.

The term gel, as the opposite of sol, is often used synonymously with networks or insoluble materials. Sol and gel are separated using an appropriate extracting reagent in which the sol is soluble. If the gel particles are very small, they are called microgels. Microgels behave as tightly packed spheres suspended in solvents; they have attracted considerable interest in recent years with the development of solid-phase synthesis and techniques for immobilizing catalysts. An interesting network polymer is the interpenetrating polymer network (IPN). An IPN is obtained by carrying out a polymerization with crosslinking in the presence of another, already crosslinked polymer or by simultaneously polymerizing two different types of monomers through the different processes such as chain-growth and step-growth polymerizations.

MICROGELATION

The crosslinkage between primary chains may accelerate the intermolecular interaction between polymer chains as a reflection of decreased interaction of the polymer segment with solvent and/or unreacted monomer around the crosslinked unit. This may be accompanied by an enhanced occurrence of intramolecular crosslinking. The intensified intramolecular crosslinking induces a local microsyneresis leading to the formation of microgel. Numerous reports have been published on the microgel formation in the homopolymerization of multivinyl monomers and their copolymerizations with monovinyl monomers.[2–7]

SOLVENT EFFECT AS REFLECTION OF THERMODYNAMIC EXCLUDED VOLUME EFFECT

Four factors, including cyclization, reduced reactivity of prepolymer, intramolecular crosslinking, and microgelation, are not a primary factor for greatly delayed gelation in the radical polymerization of multivinyl monomers. These four factors have been discussed often, but we have to look for other reasons for delayed gelation from ideal network formation, especially at an early stage of polymerization, i.e., at least up to the theoretical gel point. Thus, the significance of the thermodynamic excluded volume effect on the intermolecular crosslinking between high-molecular-weight prepolymers was proposed.[9]

In conclusion, the thermodynamic excluded volume effect is a primary factor in the great deviation of actual gel point from the theoretical one early in polymerization, especially up to the theoretical gel point. Its significance should be reduced with conversion as a result of increased prepolymer concentration and concurrently, at a stage of polymerization beyond the theoretical gel point, the intramolecular crosslinking becomes more important with the progress of polymerization as a secondary factor. The latter leads to the restriction of segmental motion of prepolymer and, moreover, imposes the steric hindrance, inducing the significance of the reduced reactivity of prepolymer as a tertiary factor.

The thermodynamic excluded volume effect should be reflected also as a solvent effect on gelation. That is, a delayed gelation is expected in a good solvent compared to a poor solvent, as opposed to the earlier report.[2] Although the gelation in the same copolymerization systems was rechecked in detail, it was clearly delayed in a good solvent as a reflection of the thermodynamic excluded volume effect.[9] A similar solvent effect was also observed in the gelation behavior of the copolymerization of MMA with oligoglycol dimethacrylate and with trimethylolpropane trimethacrylate.[10]

REFERENCES

1. Walling, C. *J. Am. Chem. Soc.* **1945**, *67*, 441.
2. Horie, K.; Otagawa, A.; Muraoka, M.; Mita, I. *J. Polym. Sci. Polym. Chem. Ed.* **1975**, *13*, 445.
3. Galina, H.; Dusek, K.; Tuzar, Z.; Bohdanecky, M.; Sokr, J. *Eur. Polym. J.* **1980**, *16*, 1043.
4. Spevacek, J.; Dusek, K. *J. Polym. Sci. Polym. Phys. Ed.* **1980**, *18*, 2027.
5. Shah, A. C.; Parsons, I. W.; Haward, R. N. *Polymer* **1980**, *21*, 825.
6. Leicht, R.; Fuhrmann, J. *Polym. Bull.* **1981**, *4*, 141.
7. Staudinger, H.; Husemann, E. *Chem. Ber.* **1935**, *68*, 1618.
8. Matsumoto, A.; Ohtsuka, S.; Ohara, Y.; Oiwa, M. *ACS Polym. Prep.* **1979**, *20*(1), 921.
9. Matsumoto, A.; Oiwa, M. *3rd Japan-China Symposium on Radical Polymerization;* Osaka, 1984; p 29.
10. Matsumoto, A.; Matsuo, H.; Ando, H.; Oiwa, M. *Eur. Polym. J.* **1989**, *25*, 237; *Kobunshi Ronbunshu* **1989**, *46*, 583.

NETWORK STRUCTURE FORMATION (Radical Polymerization)

V. P. Roshchupkin* and S. V. Kurmaz
Institute of Chemical Physics
Russian Academy of Sciences

Originally radical polymerization of multifunctional monomers and oligomers was considered a readily available avenue for producing highly crosslinked polymer networks with desired structure and properties.[1] It was suggested that the only problem was to choose the proper size and structure of the initial oligomer. However, extensive studies have shown that the polymerization is not limited to chemical crosslinking and that the relationship between initial oligomer structure and the resulting polymer structure and properties is not so straightforward as it might appear. In many instances this process is accompanied by dramatic physical and structural changes, such as micro-phase separation, vitrification of reaction medium, and heterogeneous

*Author to whom correspondence should be addressed.

polymer network formation.[2–4] These physical and structural changes in turn often significantly affect both polymerization kinetics and the structure and properties of the resulting polymers.[5,6] Because of the strong interrelationship between chemical and physical phenomena characteristic of crosslinking polymerization it may be considered as a reaction system self-organization process.

STRUCTURAL DEVELOPMENT DURING RADICAL POLYMERIZATION OF MULTIFUNCTIONAL MONOMERS

A distinguishing feature of multifunctional monomer polymerization is formation of polymer gel even at conversion about 1 mol%. This prevents traditional sample removal techniques to examine reaction kinetics and polymer structure formation except where primary polymer products can be isolated and studied as in polymerization in dilute solution or in the presence of catalytic chain transfer.[6,7]

As a consequence of the multifunctional monomer radical polymerization, the glassy network polymers obtained in bulk polymerization consist of highly crosslinked microgel particles embedded in a much less crosslinked matrix.

EFFECT OF STRUCTURE DEVELOPMENT ON KINETICS AND THERMODYNAMICS OF THREE-DIMENSIONAL RADICAL POLYMERIZATION

The structure development during three-dimensional radical polymerization produces the spatial kinetic non-uniformity of elementary reactions of initiation, propagation, and termination because of variation in the local molecular packing density and in the molecular mobility.[3,10] As a result, processes in the different structural regions are distinguished by their rates.

PECULIARITIES OF STRUCTURE DEVELOPMENT DURING RADICAL CROSSLINKING COPOLYMERIZATION

The reactivity ratios of monovinyl monomers determined from linear copolymerization cannot be put to use in three-dimensional crosslinking copolymerization of multifunctional monomers for non-uniformity of elementary reactions throughout the inhomogeneous reaction medium. IR-spectroscopy monitoring, in real time the conversion of each of the comonomers, revealed that the kinetics of their copolymerization depend critically on the crosslinker content.[8]

REFERENCES

1. Berlin, A. A.; Matveyeva, N. G. *Macromol. Rev.* **1977**, *12*, 1.
2. Roshchupkin, V. P.; Ozerkovskii, B. V.; Kalmikov, Ju. B.; Korolev, G. V. *Visokomolek. Soed.* **1977**, *A19*, 699, 7020.
3. Kochervinskii, V. V.; Kerapetjan, Z. A.; Roshchupkin, V. P.; Smirnov, B. R.; Korolev, G. V. *Visokomolek Soed.* **1975**, *A17*, 2425.
4. Roshchupkin, V. P.; Ozerkovskik, B. V. *Karbozepnii Polimeri* Nauka: Moskva, 1977; p 132.
5. Roshchupkin, V. P.; Ozerkovskii, B. V.; Karapetjan, Z. A. *Visokomolek. Soed.* **1977**, *A19*, 2239.
6. Ozerkovskii, B. V.; Plotnikov, V. D.; Roshchupkin, V. P. *Visokomolek. Soed.* **1983**, *A25*, 1816.
7. Kast, H.; Funke, W. *Macromol. Chem.* **1979**, *180*, 1335.
8. Kurmaz, S. V.; Roshchupkin, V. P. *Visokomolek. Soed.* **1994**, *36*, 2121, 2123, 2247.
9. Berlin, A. A.; Matveyeva, N. G. *Macromol. Rev.* **1980**, *15*, 107.
10. Roshchupkin, V. P.; Gafurov, U. G. *Dokl. Akad. Nauk SSSR.* **1977**, *235*, 869.

Networks

See: *Crosslinked Networks, Macronets*
Graft and Network Copolymers (By Macromolecular Ion Coupling Reactions)
Interpenetrating Polymer Networks (Overview)
Interpenetrating Polymer Networks, Rubber-Based
Macromonomers (Preparation, Polymerizability, and Applications)
Microcapsules (From Two Dimensional Networks)
Model Silicone Networks
Network Polymers (Chemorheology)
Network Polymers (Vinyl-Type)
Network Structure Formation (Radical Polymerization)
Ordered Epoxy Networks
Organic/Inorganic Composites
Organic/Inorganic Composites
Polydimethylsiloxane Networks, Monodisperse

Nitrile Rubber

See: *Blends, Interchain Crosslinking*
Reactive Liquid Rubbers (Epoxy Toughening Agents)

NITROCELLULOSE

Jun-ichi Kimura
Propellants and Explosives Laboratory (PEL)
The First Research Center
Technical Research and Development Institute
Japan Defense Agency

Nitrocellulose is the nitrate ester of cellulose, a natural polymer composed of β-1,4-anhydroglucose units shown below. It was the first chemical derivative of cellulose, and the first practical synthesis of nitrocellulose was performed in 1846 (**Structure 1**).

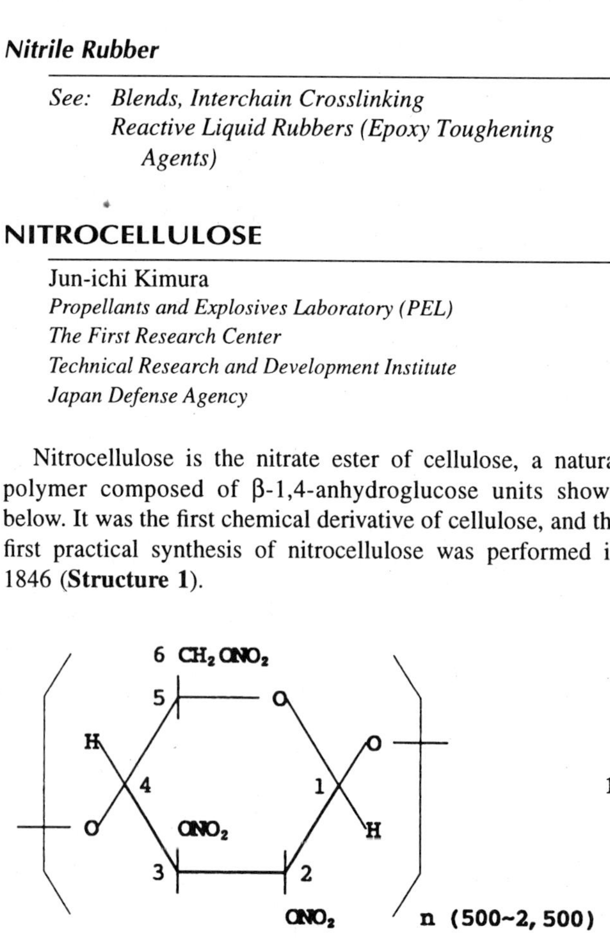

1

The name nitrocellulose is a misnomer. It implies a functional group NO_2 attached directly to the carbon. The term nitrocellulose, however, is so heavily engraved in the literature, it will continue to be used in the future.

A problem with nitrocellulose is its lack of stability, and exothermic decomposition of nitrocellulose has occasionally caused the material to autoignite during storage.

PREPARATION

The first synthesis of nitrocelluloses using a mixture of nitric acid and sulfuric acid was performed by F. Schonbein in 1846. Variations of his synthetic method are still used to produce commercial nitrocellulose. Detailed reviews of the synthetic method of nitrocellulose can be seen in a number of excellent articles.[1-4]

APPLICATION

The largest use of nitrocellulose is in protective and decorative coatings for automobile and wood furniture.

The oldest and presently second largest application of nitrocellulose is for military use as explosives and propellants.

An important invention for the practical use of nitrocellulose was brought about in 1869 by Nobel, who discovered that amine compounds, such as diphenylamine, act as stabilizers of nitrocellulose and nitrocellulose-based propellants. Amine stabilizers are believed to act as absorbers of nitrogen oxides, which retard the autocatalytic decomposition of nitrocellulose. Very recently, the author found experimental facts that diphenylamine actually acts as a chain-breaking antioxidant that intercepts peroxy radicals to form hydroperoxides.[5]

DEGRADATION

Thermal Decomposition

Although the thermal decomposition of nitrocellulose has been well studied, recent experimental findings remain unexplained, based on generally accepted kinetics and mechanism of the thermal and oxidative decomposition. The underlying mechanisms, therefore, seem more complex than the currently accepted theory.

Within the last ten years, advances in chemical instrumentation provided significant information on the decomposition of nitrocellulose below 100°C. the highly sensitive NOx chemiluminescence analyzer revealed the following facts:[6] (1) activation energy measured 43 kcal/mol (180 kJ/mol) above 90°C and 25.6 kcal/mol (107 kJ/mol) below 90°C; (2) the ratio of NO/NO_2 increased with increasing pressure and most of the NOx evolved at atmospheric pressure was NO gas. In contrast, nitrate esters such as nitroglycerin (NG) and pentaerythrytoltetranitrate (PETN) evolve mostly NO_2 gas when they are heated.

The author reported a new finding. Nitrocellulose emits ultra weak light when it is heated in nitrogen atmosphere.[7]

A possible precursor of these light-emitting species are hydroperoxides and peroxy radicals.

Oxidative Decomposition

Little attention has been paid to the autoxidation of nitrocellulose probably because it was presumed that oxidation by nitrogen oxides would predominate the autoxidation. The author has conducted a systematic research on the effect of oxygen to the decomposition of nitrocellulose by chemiluminescence technique. These experimental facts imply that some unstable intermediates are accumulated during storage, probably due to autoxidation.

Photodegradation

The degradation of nitrocellulose upon exposure to sunlight prompted numerous studies aimed at improving and predicting its weatherability. When exposed to ultraviolet (UV) light, nitrocellulose films changed in molecular weight, evolution of gaseous products, yellowing due to the formation of carbonyl groups and the loss of mechanical properties.

Metal ions accelerated the photodegradation in the following order: $Fe^{2+} > Fe^{3+}j >> Ni^{2+} > Cr^{3+}$. Negative ions such as NO_2^-, NO_3^-, and SO_4^{2-} did not alter the photodegradation.[8]

REFERENCES

1. Brewer, R. J.; Bogam, R. T. *Encyclopedia of Polymer Science and Engineering*, 2nd ed.; John Wiley & Sons: New York, 1985; Vol. 3, pp 142–147.
2. Bogam, R. T.; Kuo, C. M.; Brewer, R. J. *Encyclopedia of Chemical Technology*, 3rd ed.; John Wiley & Sons: New York, 1979; Vol. 3.
3. Urbansky, T. *Chemistry and Technology of Explosives*; Pergamon: New York, 1965; Vol. 3.
4. Miles. F. D. *Cellulose Nitrate*; Oliver and Boyd: London, 1955.
5. Kimura, J., unpublished, 1995.
6. Volltrauer, H. N.; Fontijin, A. *Combustion Flame* **1981**, *41*, 313.
7. Kimura, J. *Prop. Explos. Pyrotech.* **1989**, *14*, 89.
8. Osada, H.; Hara, Y. *J. Japan Explos. Soc.* **1971**, *32*, 148.

No-Catalyst Copolymerization

See: Spontaneous Copolymerization

Nonaqueous Dispersions

See: Dispersion Polymerization (Overview)
High Solids Coatings (Microspheres from Nonaqueous Polymer Dispersions)

Nonconjugated Polymers

See: Conducting Polymers (Nonconjugated)

NONLINEAR OPTICAL MATERIALS

Sukant Tripathy,* Jeng-I Chen, Sutiyao Marturunkakul, and Jayant Kumar
University of Massachusetts, Lowell

Nonlinear optics is a growing field of science and technology that is expected to play an important role in photonics. Photonics, an analog of electronics, is a technology in which photons instead of electrons carry information.[1,2] The significance of photonics arises from its advantages over electronics in that photonics provides larger bandwidths, faster response times,

*Author to whom correspondence should be addressed.

and less noise from extraneous electromagnetic fields. Photonics holds application potential in many areas of present and future information and image processing technologies.[3–6] The role of nonlinear optical (NLO) process is to provide key functions necessary for the photonic technology. Abilities to change properties of light, such as its frequency, phase, amplitude, or transmission characteristics when the light passes through an NLO-active medium are examples of phenomena that are potentially useful in photonics. Thus, advancements in the field of nonlinear optics will have a direct effect on the progress of photonic technology.

Although NLO materials have been investigated for more than three decades, considerable effort has gone into the development and investigation of polymeric NLO materials only in the last decade. A significant progress in material development efforts toward practical applications has been observed in recent years. Extensive reviews of researches in the field of nonlinear optics are available.[7–11]

PREPARATION AND PROPERTIES

Third-Order Materials

The organic materials are superior to inorganics both in the speed of response and in the magnitude of the third-order effect. The largest third-order susceptibilities have been observed in macromolecules with π-conjugation along an extended backbone.[4,12] The π electrons, distributed along the backbone, react quickly when the electrical environment is changed by other molecules, an electric field, or light. This quick response as well as the extensive spreading of the π cloud are the source of a large, fast, third-order response in these quasi one-dimensional systems. The third-order index, n_2 of polydiacetylenes, is one of the largest indices demonstrated to hold within the range of the transparent frequencies of a material. A variety of other types of conjugated polymers, for example, polyacetylenes, polythiophenes, and poly(phenylene) vinylene)s, have also been investigated for their third-order nonlinearities.[2,13–16] Although the optical nonlinearities of materials have been continuously improved over the past decade, no material has yet been identified for commercial application.

Second-Order Materials

These molecules possess donor-acceptor groups attached to an aromatic ring system that increases charge transfer through π-electron delocalization.[17] Such dye molecules are characterized by intramolecular charge transfers that give rise to large ground and excited state dipole moments, and second-order molecular hyperpolarizability, β. The β can be experimentally determined by electric-field-induced second harmonic generation (EFISH) and solvatochromic techniques.[18–21]

NLO-active chromophores act as vital and active components in organic/polymeric second-order NLO materials. These materials are prepared with chromophores integrated in various formats. To prepare polymeric NLO materials, for example, chromophores are dispersed in polymer matrices, covalently bonded to either side chains or main chains of polymers, or introduced into polymer networks. The development of NLO materials requires simultaneous optimization of properties, such as thermal stability of materials, temporal stability of nonlinearities, optical properties, and processability.

To summarize some of the materials that have been developed during the past decades, it is convenient to subdivide the organic second-order NLO materials into crystalline, liquid crystalline, and noncrystalline materials.

Organic Crystalline and Liquid Crystalline Materials

The growth of noncentrosymmetrical crystals provides a high concentration of nonlinear groups and a high polar order. Quite a few organic crystals are characterized by their large second-order nonlinearities. A fairly comprehensive list of them is provided in the literature.[5,23,24] Materials such as 2-methyl-4-nitroaniline (MNA), 4-dimethylamino-4′-nitrostilbene (DANS), 2-N-cyclooctylamino-5-nitropyridine (COANP) have very large nonlinearities with potential applications for SHG and EO modulation. However, there has not been much success so far due to difficulties in crystal growth and processing.

Liquid crystalline materials have already been used in commercially available products such as flat panel displays, liquid crystal light valves and spatial light modulators. Recently, ferroelectric liquid crystals have been developed to exhibit second-order NLO responses.[22,25] The optical nonlinearities of these materials are relatively small.

Noncrystalline Materials

Noncrystalline second-order NLO materials can be classified based on the preparation methods as follows: (1) Langmuir–Blodgett and self assembled films and (2) electrically poled polymeric films.

Langmuir–Blodgett and Self-Assembled Films. An LB film with alternating layers, where one or both layers possess a nonlinear moiety needs to be assembled in such a way that little cancellation of the nonlinearity occurs in order to exhibit second-order NLO properties.[26–31] Unfortunately, these films, particularly the alternating layer type, involve a painfully tedious method of preparation, they tend to be fragile and the nonlinear component is quite diluted by the noncontributing aliphatic chains.

Self-assembly technique offers an alternative route to prepare monolayer or multilayer thin films of highly ordered multifunctional compounds in a more effective and timesaving manner. This technique presents some advantages over the LB technique in that each layer in the self-assembled films can be covalently bonded. Moreover, no excess deposition can take place as it is limited by the reactive sites on the layer surface.[32] A number of self-assembled films exhibiting second-order NLO effects have been prepared and studied.[32–34]

Recently, a soluble unsymmetrically substituted polydiacetylene has been reported to exhibit SHG signal from spin-coated or cast films without recourse to poling. It is suggested that the spontaneous orientation of the polymer chains during the film formation is responsible for the acentric alignment of the polymer. LB films prepared from such polymer also exhibit SHG signal.[35,36]

Electrically Poled Polymeric Films. Polymeric films incorporating NLO moieties have attracted a great deal of attention and are being seriously investigated for practical applications, especially for EO modulation.

Among various formats of the second-order NLO polymers, no particular polymer or system has yet been identified as a clear-cut winner. Most of the research has been focused on enhancing the optical nonlinearity and its temporal stability at elevated temperatures. With the rapid research progress in recent years, polymeric materials exhibiting large and stable NLO properties for real applications may be anticipated in the near future. Other properties of materials, such as optical loss, optical power handling, processability, reproducibility, are also considered to be equally important.[37] Moreover, other process and fabrication problems may also be encountered during device fabrication. A successful development of a practical device for NLO application will require optimization of properties in all aspects.

APPLICATIONS

At the present time, third-order NLO materials have not yet found commercial applications in photonics as the third-order NLO effects are still too small. A lot of progress still needs to be made and perhaps some new material breakthroughs have to be realized before a realistic application can be found.

There has been no commercially available device made of second-order NLO polymers, while some organic materials are already being used in frequency doubling lasers and EO devices. Prototype devices made of polymeric materials for frequency doubling and EO devices are already being prepared and studied, while properties of NLO polymers are continuously improved. The rapid progress in material development thus far is very optimistic and it is expected that properties of NLO polymers will reach the desired goals in the future. Before NLO polymeric materials can find widespread use in realistic applications, it has to be proven that they can be made economically, are compatible with production lines already established in the microelectronic industry, and are reproducible in their properties.

REFERENCES

1. Stucky, G. D.; Phillips, M. L. F.; Gier, T. E. *Chem. Mater.* **1989**, *1*, 492.
2. Prasad, P. N.; Williams, D. J. *Introduction to Nonlinear Optical Effects in Molecules and Polymers;* John Wiley & Sons: New York, 1991.
3. Thakur, M.; Tripathy, S. *Encyclopedia of Polymer Science and Engineering*, 2nd ed.; Wiley: New York, 1985; Vol. 5, pp 756–771.
4. Tripathy, S.; Cavicchi, E.; Kumar, J.; Kumar, R. S. *Chemtech* **1989**, *19*, 620; Tripathy, S.; Cavicchi, E.; Kumar, J.; Kumar, R. S. *Chemtech* **1989**, *19*, 747.
5. Eaton, D. F. *Science* **1991**, *253*, 281.
6. Eaton, D. F. *Chemtech* **1992**, *22*, 308.
7. Chemla, D. S.; Zyss, J. *Nonlinear Optical Properties of Organic Molecules and Crystals;* Academic: Orlando, FL, 1987; Vols. 1 and 2.
8. Cavicchi. J.; Kumar, J.; Tripathy, S. *Optical Techniques to Characterize Polymer Systems;* Elsevier: New York, 1989; pp 395–391.
9. Miyata, S. *Nonlinear Optics: Fundamentals, Materials and Devices;* Elsevier Science: Holland, 1992.
10. Burland, D. M.; Miller, R. D.; Reiser, O.; Twieg, R. J.; Walsh, C. A. *J. Appl. Phys.* **1992**, *71*, 410.
11. Zyss, J. *Molecular Nonlinear Optics: Materials, Physics, and Devices;* Academic: San Diego, CA, 1994.
12. Chang, T. Y. *Optical Engineering* **1981**, *20*, 220.
13. Kajzar, F.; Etemad, S.; Baker, G. L.; Messier, J. *Synth. Met.* **1987**, *17*, 563.
14. Torruellas, W. E.; Neher, D.; Zanoni, R.; Stegeman, G. I.; Kajzar, F.; Leclerc, M. *Chem. Phys. Lett.* **1990**, 175, 11.
15. Halliday, D. A.; Bum, P. L.; Bradley, D. D. C.; Friend, R. H.; Gelsen, O. M.; Holmes, A. B.; Kraft, A.; Martens, J. H. F.; Pichler, K. *Adv. Mater.* **1993**, *5*, 40.
16. Brédas, J. L.; Adant, C.; Tackx, P.; Persoons, A.; Pierce, B. M. *Chem. Rev.* **1994**, *94*, 243.
17. Lalama, S. J.; Garito, A. F. *Phys. Rev. A* **1979**, *20*, 1179.
18. Levine, B. F.; Bethea, C. G. *J. Chem. Phys.* **1975**, *63*, 2666.
19. Singer, K. D.; Garito, A. F. *J. Chem. Phys.* **1981**, *75*, 3572.
20. Teng, C. C.; Garito, A. F. *Phys. Rev. B.* **1983**, *28*, 6766.
21. Oudar, J. L.; Zyss, J. *Phys. Rev. A* **1982**, *26*, 2016.
22. Zyss, J.; Ledoux, I. *Chem. Rev.* **1994**, *94*, 77.
23. Morichere, D.; Dentan, V.; Kajzar, F.; Robin, P.; Levy, Y.; Dumont, M. *Optics Commun.* **1989**, *74*, 68.
24. Twieg, R. J.; Jain, K. *Nonlinear Optical Properties of Organic and Polymeric Materials;* ACS Symposium Series 233; American Chemical Society: Washington, DC, 1982; Chapter 3.
25. Walba, D. M.; Ros, M. B.; Clark, N. A.; Shao, R.; Robinson, M. G.; Liu, J-Y.; Johnson, K. M.; Doroski, D. *J. Am. Chem. Soc.* **1991**, *113*, 547.
26. Girling, I. R.; Cade, N. A.; Kolinsky, P. V.; Earls, J. D.; Gross, G. H.; Peterson, I. R. *Thin Solid Films* **1988**, *132*, 101.
27. Anderson, B. L.; Hoover, J. M.; Lindsay, G.; Higgins, B. G.; Stroeve, P.; Kowel, S. T. *Thin Solid Films* **1989**, *179*, 413.
28. Takahashi. T.; Miller, P.; Chen, Y. M.; Samuelson, L.; Galotti, D.; Mandal, B. K.; Kumar, J.; Tripathy, S. K. *J. Polym. Sci., Polym. Phys.* **1993**, *31*, 165.
29. Hodge, P.; Ali-Adib, Z.; West, D.; King, T. *Macromolecules* **1993**, *26*, 1789.
30. Ashwell, G. J.; Gongda, Y.; Lochun, D.; Jackson, P. D. *Polym. Prepr., Am. Chem. Soc. Div. Polym. Chem.* **1994**, *35*, 185.
31. Bosshard, Ch.; Otomo, A.; Stegeman, G. I.; Küpfer, M.; Flörsheimer, M.; Günter, P. *Appl. Phys. Lett.* **1994**, *64*, 2076.
32. Li, D.; Swanson, B. I.; Robinson, J. M.; Hoffbauer, M. A. *J. Am. Chem. Soc.* **1993**, *115*, 6975.
33. Li, D.; Ratner, M. A.; Marks, T. B.; Zhang, C. H.; Yang, J.; Wong, G. K. *J. Am. Chem. Soc.* **1990**, *112*, 7389.
34. Lundquist, P. M.; Yitzchaik, S.; Zhang, T.; Kanis, D. R.; Ratner, M. A.; Marks, T. J.; Wong, G. K. *Appl. Phys. Lett.* **1994**, *64*, 2194.
35. Tripathy, S. K.; Kim, W. H.; Bihari, B.; Cheong, D. W.; Kumar, J. *Mat. Res. Soc. Symp. Proc.* **1994**, *328*, 433.
36. Kim, W. H.; Bihari, B.; Moody, R. A.; Kumar, J.; Tripathy, S. K. *Macromolecules* **1995**, *28*(1).
37. Mortazavi, M. A.; Knoesen, A.; Kowel, S. T.; Higgins, B. G.; Dienes, A. *J. Opt. Soc. Am. B* **1989**, *6*, 733.

Nonlinear Optical Polymers

See: *Conjugated Ladder Polymers*
Conjugated Polymers (Insulating and Conducting Forms)
Conjugated Polymers (Prepared by Organometallic Processes)
Enzyme-Catalyzed Polymerization
Langmuir–Blodgett Films (Organic Polymers for Photonics Applications)
Liquid Crystalline Elastomers

Liquid Crystalline Polymers, Side Chain
Metallophthalocyanines, Polymeric (Overview)
Nonlinear Optical Materials
Nonlinear Optical Polymers [Poly(arylene ethynylene) Derivatives]
Nonlinear Optical Polymers, Thermoset
Photochromic Films, Azo Dyes (Photoinduced Anisotropy and Photoassisted Poling)
Poly(arylene vinylene)s (Mechanistic Control of a Soluble Precursor Method)
Polyazines (Third-Order Nonlinear Optical Properties)
Polydiacetylenes, Unconventional
Polyphenylacetylene
Poly(phenylene ethynylene) (Synthesis and Optical Properties)
Poly(p-phenylene vinylene)s (Method of Preparation and Properties)
Polysilanes (Overview)
Polysilanes (Bearing Hydroxy Groups and Derivatives)
Polythiophenes, Substituted (Regioselectivity, Electrical and Optical Properties)
Polyureas (Second-Order Nonlinear Optical Properties)
Rigid-Rod Polybenzoxazoles and Polybenzothiazoles (PBZT and PBO)
Rigid-Rod Poly(p-phenylene benzobisthiazole)
Small-Bandgap Polymers (Special Class of Organic Conducting Compounds)
Thiophene Copolymers (Electrical and 3rd-Order NLO Properties)
Topochemical Polymerization (Diacetylenes)

NONLINEAR OPTICAL POLYMERS [Poly(arlene ethynylene) Derivatives]

J. Le Moigne
Institut de Physique et Chimie des Matériaux de Strasbourg

While electronic conduction needs an electron delocalization associated with a high electron mobility between molecules, nonlinear optical properties only need a large electron delocalization and an optical transparency. In polymers, the former is often restricted to polyacetylene, polythiophene, polyphenylene or polyanilines and their derivatives, while for the latter a large class of conjugated main chain or side chain polymers is open.[1] One of the most studied π-conjugated polymer, the poly(*p*-phenylene vinylene) (PPV) and its family, have attracted special attention for their high value of the third order susceptibility and their high electronic conductivity. In contrast, the acetylene analogue of PPVA has received much less attention, mainly due to the low solubility of the compound and the low molecular weight of the polymer chain.

This article provides an overview of the investigations on poly(arylene ethynylene) derivatives in the domain of nonlinear optics, emphasizing the importance of synthetic methods on the properties of such materials. The main aim is to outline and discuss the principal results concerning the development of a novel class of a processable polymer for third order nonlinear applications.

PREPARATION

General Methods

The ideas of using the alternating aryl and ethynyl group in the main chain of a polymer was introduced ten years ago.[2] This polyester chain combines the properties of an aryl and acetylenic functions to create a new family of linear or bidimensional polymers.[3–6] Recently, higher molecular weights of the π-conjugated compounds, which have been developed for nonlinear optics, were obtained by improving the solubility. The main chains having alkyl side chains was introduced in order to increase the solubility.

The phenylenylene polymers were synthesized by a condensation method known as Heck coupling.[7–9] A substituted diethynyl arene reacts with a disubstituted halogeno-aryl compound in the presence of a catalytic amount of a palladium complex. This general route is described in the general scheme on **Figure 1**, where R_1, R_2, R_3, R_4 could be the same or different (H, NO_2, alkyl-ether, alkyl-thioether or alkyl-ester…). As aryl units homo or copolymers with phenyl, thienyl, anthryl, or stilbene groups could be obtained.[10]

Oligomers Synthesis

Oligomers attempt to obtain the nonlinear optical properties of high polymers with lower defect in the chain and higher solubility. The grafting of shorter side chains than in polymers increases the nonlinear active part in the macroscopic solid.

Step by Step Synthesis

A new synthetic, stepwise and repetitive approach to oligomer building was reported first by Zhang and more recently applied by Schumm and coworkers using linear phenylacetylene sequences.[12,13] This technique doubles the molecular length at each iteration. The synthetic scheme follows a multistep reaction using a double protective unit based on the triazene phenyl trimethyl silyl acetylene.

Oligomers with Random Distribution

A series of oligomers of the poly(phenylene ethynylene) with alkyl ether and alkyl esters side groups were synthesized by the palladium coupling method.

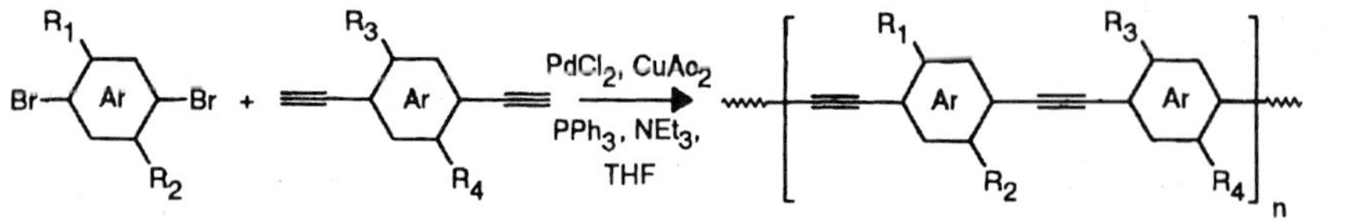

FIGURE 1. Heck coupling reaction of aryl dihalide with diethynyl compound with the complex catalysts Pd-copper triphenyl phosphine.

CHARACTERIZATION

Polymers and Oligomers in Solution

By using NMR ^1H and ^{13}C, Ir and Uv-Visible spectroscopies, molecular weights determination by size exclusion chromatography (SEC) the fundamental properties of these conjugated molecules were investigated. The stiffness of polymer chains was estimated using viscosity investigations.

Molecular Weight and Stiffness

The molecular weights are determined by a single light scattering method on an ethynyl series that includes pyridine, alkylthiophene and selenophene in homopolymers and copolymers.[11] $\overline{M}_w$ values as high than 9.6×10^2 were found by these authors. Also, large refractive indices in the range of 0.29 and 0.34 $m^3\ kg^{-1}$ are observed for these polymers. The large refractive indices are confirmed in other series (phenyl-alkyloxy, etc.) that are consistent with the high polarizability of the phenylethynyl units.

Solid State Characterization

Thermal Properties

The stability and thermal behaviors were performed for only two homopolymers, the first with one ester substituent on each phenyl ($pPYCOOC_{12}$) and the second with disubstituted ether side chains ($pPYOC_{12}$).

The solid films of those polymers, obtained by casting the concentrated solutions on glass plates, are observed in polarized light. Only a homeotropic orientation was detected. Nevertheless, these films exhibited a strong birefringence after rubbing the surface or shearing the melt. This birefringence could be induced at room temperature. By heating over 70°C, the polymers became more fluid, the samples remained highly birefringent until 150°C.

Orientation

The liquid crystal properties enabled the orientation of the polymers by shearing or methods such as thermal gradient.

CONCLUSION

The poly(arylene ethynylene) derivatives, mainly 1,4-substitutions, that maintain a linear conformation of the main chain are well suited for nonlinear optics. The optical properties as well as other physical properties are greatly dependent of the synthetic methods. A new synthetic route, using a step-by-step iteration, leads to well-defined oligomers, while the polycondensation using a Heck palladium catalyst successfully gives high molecular weight polymers. These linear rigid-rod conjugated polymers, in which solubility depends on the nature and the length of the side chains, could give good quality films by casting in chloroform or toluene. Such polymers exhibit fairly strong nonresonant $\chi^{(3)}$ values in the range of 10^{-9} to 10^{-10} esu for the most efficient of them, as revealed by THG and DFWM techniques.

REFERENCES

1. Garito, A.; Jen, A. K-Y.; Lee, C. Y-C.; Dalton, L. R. *Electrical, Optical and Magnetic Properties of Organic Solid State Materials,* Proc. Mat. Res. Soc. **1993**, *328*, Part 11.
2. Lakmikantham, M. V.; Vartikar, J.; Kwan, Yu Jen; Cava, M. P.; Huang, W. S.; MacDiarmid, A. *Polym. Prep.* **1983**, *24*, 75.
3. Sonogashira, K.; Tohda, Y.; Hagihara, N. *Tetrahedron Lett.* **1975**, 4467.
4. Tohda, Y.; Sonogashira, K.; Hagihara, N. *Synthesis* **1977**, 777.
5. Sasabe, H.; Wada, T.; Hosoda, H.; Ohkawa, H.; Yamada, A.; Garito, A. F. *Mol. Cryst. Liq. Cryst.* **1990**, *189*, 155.
6. Rutherford, D.; Stille, J. K.; Elliott, C. K.; Reichert, V. R. *Macromolecules* **1992**, *25*, 2294.
7. Dieck, H. A.; Heck, R. F. *J. Organomet. Chem.* **1975**, *93*, 259.
8. Austin, W. B.; Bilow, N.; Kelleghan, W. J.; Lau, K. S. Y. *J. Org. Chem.* **1981**, *46*, 2280.
9. Heck, R. F. *Palladium Reagents in Organic Syntheses;* Academic: 1990; p 299.
10. Moroni, M.; Le Moigne, J.; Luzzati, S. *Macromolecules* **1994**, *27*, 562.
11. Yamamoto, T.; Tagaki, M.; Kizu, K.; Kubota, K.; Kabara, H.; Kirihara, T.; Kaino, T. *J. Chem. Soc. Chem. Comm.* **1993**, 797.
12. Zhang, J.; Moore, J.; Xu, Z.; Aguire, R. *J. Am. Chem. Soc.* **1992**, *114*, 2273.
13. Schumm, J. S.; Pearson, D.; Tour, J. S. *Angew. Chem. Int. Ed. Engl.* **1994**, *33*, 1360.

NONLINEAR OPTICAL POLYMERS, THERMOSET

P. J. Shannon,* W. M. Gibbons, and S. T. Sun
Wilmington Research Center
Alliant Techsystems Incorporated

Organic polymers comprising dyes with intramolecular charge transfer between a donor and acceptor in a π conjugated system have been of interest as nonlinear optical (NLO) media.[1–6]

We review our work on a novel thermoset for NLO materials based on the hydrosilation reaction of tetramethylcyclotetrasiloxane (TMCTS) with olefin containing NLO chromophores.[7,8] This new thermoset resin offers many advantages including high loading of chromophores, high thermal stability and high optical clarity.

DISCUSSION

Dye Synthesis and Reactivity

The dyes investigated in the thermosetting hydrosilation reactions are listed in **Table 1**. Dyes **1** through **6** contained highly reactive allyl groups as hydrosilation active moieties. The methacrylates **8** and **9** and the cyclopentyl analog **7** were designed with two features in mind. Each possessed carbon–carbon double bonds with different reactivities toward hydrosilation. The double bonds were covalently attached to two different sites on the NLO π system. Multiple reactivities of the double bonds would allow the prepolymer to form in a low temperature cure and still allow a high degree of ordering during the poling process. The high temperature crosslinking reaction would lock in the chromophore alignment with multisite covalent bonding of the chromophore. This feature would

*Author to whom correspondence should be addressed.

TABLE 1. Dyes Useful in Preparing Crosslinked Organosilicon Polymers

1	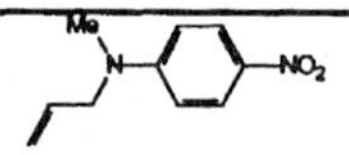
2	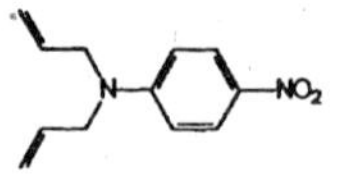
3	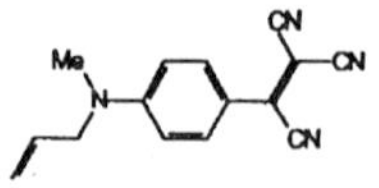
4	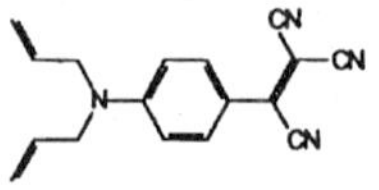
5	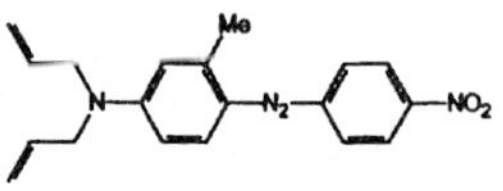
6	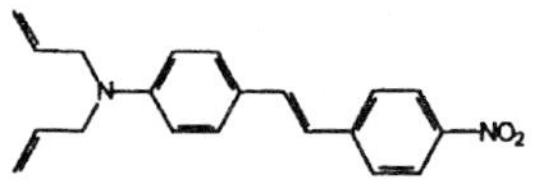
7	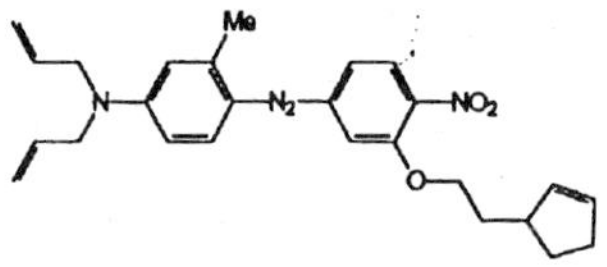
8	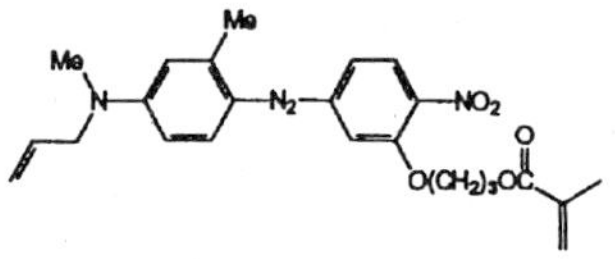
9	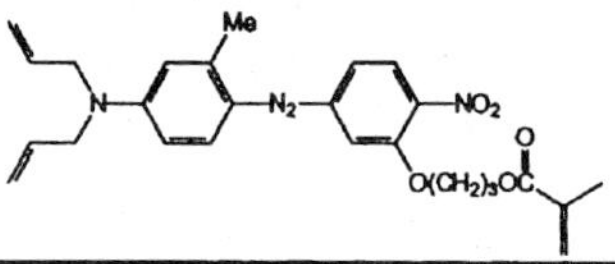

significantly reduce the mobility of the chromophore and result in poled polymer films with high thermal stability of the SHG signal.

In general the dyes showed high reactivity toward hydrosilation.

Prepolymer and Film Formation

We prepared crosslinkable prepolymers by hydrosilation of the desired chromophores with TMCTS in toluene.

Thin films of the prepolymers were spun onto soda lime glass substrates and subjected to a precure heat cycle at 130 to 190°C.

SHG Measurements

Second-harmonic generation experiments were used for the quick screening of the polymer thin films.

SUMMARY

We have combined the excellent properties of the hydrosilation thermosetting matrix with chromophores that possess high second-order polarizabilities to produce novel thermosetting NLO materials. Formulations have been illustrated that feature high nonlinearities and reasonable thermal stability. In the cases where the NLO chromophore possessed the capability for multisite covalent bonding, which locked in the orientation of the chromophore, the SHG thermal stability has been exceptional. Optimization of the chemistry, curing, and poling schedule should lead to systems that exhibit excellent nonlinear optical and thermal properties.

ACKNOWLEDGMENTS

The authors thank Scott Beard and Sadhana Mital for technical support.

REFERENCES

1. Willard, C. S.; Williams, D. J. *J. Ber. Bunsenges. Phys. Chem.* **1987,** *91,* 1304.
2. Urlich, D. R. *Mol. Cryst. Liq. Cryst.* **1990,** *189,* 3.
3. Singer, K. D.; Holland, W. R.; Kuzyk, M. G.; Wolk, G. L. *Mol. Cryst. Liq. Cryst.* **1990,** *189,* 123.
4. Mohlmann, G. R. *Synth. Met.* **1990,** *37,* 207.
5. Eich, M.; Bjorklund, G. C.; Yoon, D. Y. *Polym. Advan. Tech.* **1990,** *1,* 189.
6. Burland, D. M.; Miller, R. D.; Walsh, C. A. *Chem. Rev.* **1994,** *94,* 31.
7. Gibbons, W. M.; Grasso, R. P.; O'Brien, M. K.; Shannon, P. J.; Sun, S. T. *Macromolecules* **1994,** *27,* 771.
8. Gibbons, W. M.; Grasso, R. P.; O'Brien, M. K.; Shannon, P. J.; Sun, S. T. *Appl. Phys. Lett.* **1994,** *64,* 2628.

NONTHROMBOGENIC MATERIALS

Shoji Nagaoka
Department of Industrial Chemistry
Faculty of Engineering
Tokyo Metropolitan University

The problem of thrombosis associated with the use of synthetic materials implanted in the blood stream continues to be a serious one. "Nonthrombogenic Materials" prevent thrombus formation by the physicochemical properties of their surfaces. They are separate from "Antithrombogenic Materials" that immobilized various antithrombotic agents such as albumin, heparin, prostaglandin, urokinase, and selected conjugates.[1]

Thrombogenesis on the surface of artificial materials is triggered by the absorption of plasma proteins or platelets and their deterioration. Therefore, to prepare nonthrombogenic material without using bioactive substances, it is desirable to reduce the interaction between blood constituents and the surface of the material. Abuchowski has reported that poly(ethylene oxide) (PEO) chains covalently bound to bovine serum albumin significantly reduce the immunogenicity in rabbits, since the flexible and hydrophilic PEO shell with bound water surrounding the albumin covers the antigenic determinants.[2]

We prepared hydrogels with long poly(ethylene oxide) chains and investigated their interactions with blood components *in vitro* and *in vivo* evaluations.

MATERIALS

Polymers with PEO chains were synthesized by the random copolymerization of methoxy poly(ethylene glycol) monomethacrylate (MnG; n, chain length of PEO) with methyl methacrylate (MMA) and acrylonitrile (AN) using AIBN as an initiator. Photo-induced graft copolymerization of MnG to poly(vinyl chloride) (PVC) with photo sensitive dithiocarbamate groups also was carried out. Details of this polymerization were reported elsewhere.[3] The obtained polymers were referred to as P(MMA-co-MnG), P(AN-co-MnG) and P(VC-g-MnG) respectively and "PEO-hydrogels" is a general term for these polymers (**Equation 1**).

$$\text{MnG: } CH_2 = C(CH_3)CO\text{-}(OCH_2CH_2)_nOCH_3 \qquad (1)$$
$$n = 4, 9, 15, 23, 50, 100$$

BASIC PROPERTIES OF PEO HYDROGELS

The adsorption of blood constituents onto polymer surfaces with similar water contents and surface chemical states determined by ESCA, but with different PEO chain lengths (n), has been studied using rabbit blood. It was found that plasma protein adsorption and platelet adhesion onto these polymers significantly decreased with increased PEO chain length, and very few platelets were observed when PEO chain length approached 100.

We expect the rapid movement of long hydrated PEO chains to influence the microhemodynamics at the blood-material interface. The microstream of water combined with flexible long PEO chains prevents stagnation, consequent adhesion and denaturation of plasma proteins which causes platelet adhesion, more effectively than short PEO chains or other hydrophilic polar groups.

As this PEO chain movement on the surface of the material resembles that of the cilia of some micro-organism, we call these long flexible PEO chains on the surface "molecular cilia".[4]

IN VIVO AND *EX VIVO* EVALUATIONS OF PEO-HYDROGELS

PEO chains were found to be more flexible in the graft copolymers than in the random copolymers, owing to the microdomain structure of the graft copolymers. We have carried out *in vivo* and *ex vivo* evaluations of PEO-hydrogels using the graft copolymer with the longest (n = 100) PEO chains, P(VC-g-M100G). Following the implantation of a catheter made of P(VC-g-M100G) into the femoral vein of mongrel dogs, neither platelet adhesion nor fibrin deposition was observed on its surface for up to 72 days.[3]

CONCLUSION

The surface dynamics hypothesis for nonthrombogenicity has evolved considerably over the past years, and can now be modeled.[4] There is direct evidence for the repulsive force produced by steric exclusion. Some quantitative measure of the actual motion or relaxation times is now available through the studies reported herein, and *in vitro* and *in vivo* blood studies confirmed the PEO-hydrogel surfaces merit in clinical use.

REFERENCES

1. Kim, S. W. *ASAIO* **1987**, *10, 82.*
2. Abuchowski, A.; van Es, T.; Palczuk, N. C.; Davis, F. F. *J. Biol. Chem.* **1977**, *252*, 3578.
3. Mori, Y.; Nagaoka, S. *Trans. Am. Soc. Artif. Intern. Organs* **1982**, *28*, 459.
4. Nagaoka, S. *ASAIO* **1987**, *10*, 76.

Nonthrombogenic Polymers

See: *Nonthrombogenic Materials*
Nonthrombogenic Polymers (Multiblock Copolymers of Polyether and Polyamide)

NONTHROMBOGENIC POLYMERS (Multiblock Copolymers of Polyether and Polyamide)

Nobuhiko Yui
Japan Advanced Institute of Science and Technology

The development of blood-contacting devices has required designing nonthrombogenic polymers with high reliability. Thrombogenesis has been clarified to be initiated via activation pathways of both coagulation factors and platelets. In our previous studies, it was demonstrated that the crystallinity of polyamides is closely connected with platelet adhesion *in vitro*. The crystallinity of polyamides is an important factor in designing nonthrombogenic polymers. Semicrystalline polymers undergo a very limited crystallization, which results in the formation of a microstructure composed of ordered crystalline and disordered amorphous phases. Thus, this type of microstructure could affect the fate of platelets on these polymer surfaces. From this perspective, synthesis, characterization, and blood-contacting properties of multiblock copolymers consisting of amorphous polyether and semicrystalline polyamide segments have been thoroughly investigated in the last decade. Representative of such nonthrombogenic polymers is a series of multiblock copolymers consisting of poly(propylene oxide) (PPO) and nylon 6,10 segments.

Our systematic studies have demonstrated that platelet adhesion and activation *in vitro* and nonthrombogenicity *in vivo* on these multiblock copolymer surfaces are greatly reduced, mediated by their surface microdomain structures. Therefore, it has been proposed that prevention of contact-induced activation of platelets on crystalline–amorphous microdomain-structured polymer surfaces is a critical step in the design of improved nonthrombogenic polymers.[1,2]

It is recognized that platelet activation in biological circumstances is mediated by an elevation of cytoplasmic free calcium levels ($[Ca^{2+}]i$).[3,4] Thus, the mechanistic aspect on platelet activation in relation to intracellular calcium levels forms an important investigative basis for progress in nonthrombogenic polymers.[5] We demonstrated that hydrophobic polymer surfaces, such as polystyrene (Pst), stimulate a drastic increase in $[Ca^{2+}]i$ in platelets, indicating the importance of intracellular calcium concentration as a secondary messenger involved in polymer–platelet interaction.[6] Furthermore, we have suggested the possibility of platelet activation occurring on Pst surfaces via a different pathway from thrombin–stimulated activation.[7,8] It is thus proposed that a change in the membrane glycoprotein–cytoskeleton network may be important in initiating the activation of platelets on polymer surfaces.

Interaction of platelets with microdomain–structured copolymer surfaces has been estimated in terms of evaluating intracellular calcium change in platelets. Recently, we reported that increases in $[Ca^{2+}]i$ in platelets contacting our designed nonthrombogenic multiblock copolymer surfaces are strongly reduced by both modifying the surface microstructure of the polymer itself and adsorbing plasma proteins onto the polymer surface. This result suggests that platelets contacting a particularly microstructured surface are prevented from contact-induced activation in terms of a cytoplasmic calcium-mediated activation pathway.

PREPARATION

Multiblock copolymerization of poly(propylene oxide) (PPO) and nylon 610 is carried out by an interfacial polycondensation reaction.

CONCLUSION AND APPLICATION

Multiblock copolymers consisting of polyether and polyamide segments form a microstructure composed of crystalline and amorphous phases, the surfaces of which is responsible for excellent nonthrombogenic properties. The surface microstructure is controlled by the crystallization of polyamide segments in copolymers. Regulated adsorption layer of plasma proteins on these copolymer surfaces is achieved via transient exchanges of adsorbed proteins with plasma. The copolymer surfaces exhibits few effects on changes in both $[Ca^{2+}]i$ and membrane fluidity of platelets. We concluded that these copolymer surfaces do not initiate the cytoplasmic calcium–mediated activation processes of platelets.

Multiblock copolymers consisting of polyether and polyamide segments offer not just good nonthrombogenicity but also the ability to be fabricated by injection molding, compression molding, or solvent casting. Thus, these materials have been considered for use in a number of disposable, blood-contacting devices, such as catheters, blood bags, and extracorporeal devices, some of which are now being developed and commercialized in Terumo Corporation, Japan.

REFERENCES

1. Okano, T.; Uruno, M.; Sugiyama, N.; Shimada, M.; Shinohara, I.; Kataoka, K.; Sakurai, Y. *J. Biomed. Mater. Res.* **1986**, *820*, 1035.
2. Yui, N.; Kataoka, K.; Sakurai, Y.; Aoki, T.; Sanui, K.; Ogata, N. *Biomaterials* **1988**, *9*, 225.
3. Feinstein, M. B.; Egan, J. J.; Shaafi, R. I.; White, J. *Biochem. Biophys. Res. Commun.* **1983**, *113*, 598.
4. Ware, J. A.; Johnson, P. C.; Smith, M.; Salzman, E. W. *J. Clin. Invest.* **1986**, *77*, 878.
5. Yui, N. *Hyomen (Surface in Japanese)* **1991**, *29*, 1021.
6. Yui, N.; Kataoka, K.; Sakurai, Y.; Fujishima, Y.; Aoki, T.; Maruyama, A.; Sanui, K.; Ogata, N. *Biomaterials* **1989**, *10*, 309
7. Yui, N.; Suzuki, K.; Okano, T.; Sakurai, Y. *Jpn. J. Artif. Organs* **1992**, *21*, 222.
8. Yui, N.; Suzuki, K.; Okano, T.; Sakurai, Y.; Ishikawa, C.; Fujimoto, K.; Kawaguchi, H. *J. Biomater. Sci. Polym. Ed.* **1993**, *4*, 199.

Nonwovens

See: Aramid Fibers and Nonwovens

NORBORNADIENE POLYMERS

Takashi Iizawa
Department of Chemical Engineering
Faculty of Engineering
Hiroshima University

Tadatomi Nishikubo
Department of Applied Chemistry
Faculty of Engineering
Kanagawa University

There are many ways to transform solar energy besides solar batteries. One of them is a solar energy exchange-storage system based on isomerization between norbornadiene (NBD) and quadricyclane (QC) (**Scheme I**). The photochemical valence isomerization of NBD to QC can store solar energy as strain energy in a QC molecule and the QC molecule can then release about 96 kJ/mol of thermal energy upon contact with a certain catalyst.

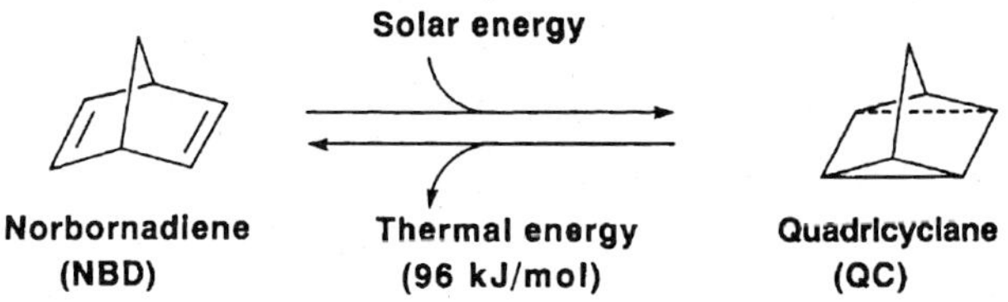

Scheme I. Solar energy exchange-storage system based on isomerization between NBD and QC.

Polymer–containing NBD moiety (NBD polymer) is another conventional solar energy exchange–storage system because of the storage of solar energy and exchange into thermal energy without any deformation and destruction of the polymer shape. Many synthetic methods have been developed to prepare NBD polymers. The most useful and easiest method is to react polymers with NBD derivatives. The substitution reaction of poly(4-chloromethylstyrene) (PCMS) with potassium salt of carboxylic acid or phenol derivatives of NBD proceeds quantitatively using phase–transfer–catalyst (PTC) in an aprotic polar solvent under mild conditions when slightly excess amounts of the salt were used.[1–4] It is possible to synthesize NBD polymers from polymers with chloromethyl groups such as poly(2-chloroethyl vinyl ether) and polymer obtained from polyaddition of diepoxide with diacyl chloride using the PTC method.[2,5] Furthermore, NBD polymers are synthesized from poly(2,6-dimethyl-*p*-phenylene oxide) (PPO) after bromination of pendant methyl groups in PPO.[6]

The reaction of poly(glycidyl methacrylate) and its copolymers (PGMA) of methyl methacrylate with carboxylic acid derivatives of NBD given NBD polymers with a high degree of functionality (DF) using tetrabutylammonium bromide (TBAB) as a catalyst in DMF.

The synthesis of polymers from radical polymerization of the corresponding NBD monomers has also been reported.[8]

The properties of the obtained NBD polymers, such as the photoreactivity and the capacity of heat storage, have also been reported. Furthermore, it might be available as a new photoresponsible polymer in optoelectronics switch and device technology and other fields of imaging technology. This chapter describes the synthesis of NBD polymers and their properties.

PROPERTY

Photoreactivity of NBD Polymer

The photochemical valence isomerization of pendant NBD to the QC moiety proceeds smoothly in the film state similar to the corresponding low molecular weight of NBD compound and the polymer solution upon irradiation with sunlight of high-pressure mercury lamp.

Capacity of Heat Storage of NBD Polymer

Each NBD polymer has a much smaller capacity of heat storage per weight than the low molecular weight NBD. This means that introducing NBD into the polymer and chromophore causes the capacity of heat storage per weight to decrease greatly.

APPLICATIONS

Although many studies of NBD polymers have been reported, there are many problems to overcome its practical use. NBD polymer has advantages such as high photoreactivity, high storage stability, and resistance to side reactions. If both isomerizations between NBD and QC are controlled, the NBD polymer system can be designed with a wide variety of properties between exchanger and storage material of solar energy. This system is of interest in a new type of micro–heater which not only exchanges solar energy but also generates heat gradually after the irradiation. It can be utilized in new materials, such as fiber, film, seat, clothes, and paint in the fields of textile, building, and agriculture. Further applications of polymers containing NBD or QC moiety to other uses, such as optoelectronic[10] and energetic binder for solid rocket propellants,[9] have recently been investigated.

REFERENCES

1. Nishikubo, T.; Sahara, A.; Shimokawa, T. *Polym. J.* **1987**, *19*, 991.
2. Nishikubo, T.; Shimokawa, T.; Sahara, A. *Macromolecules* **1989**, *22*, 8.
3. Nishikubo, T.; Hijikata, C.; Iizawa, T. *J. Polym. Sci., Part A: Polym. Chem.* **1991**, *29*, 671.
4. Iizawa, T.; Hijikata, C.; Nishikubo, T. *Macromolecules* **1992**, *25*, 21.
5. Kameyama, A.; Watanabe, S.; Kobayashi, E.; Nishikubo, T. *Macromolecules* **1992**, *25*, 2307.
6. Iizawa, T.; Sueyoshi, T.; Hijikata, C.; Nishikubo, T. *J. Polym. Sci., Part A: Polym. Chem.* **1994**, *32*, 3091.
7. Nishikubo, T.; Kawashima, T.; Watanabe, S. *J. Polym. Sci., Part A: Polym. Chem.* **1993**, *31*, 1659.
8. Kamogawa, H.; Yamada, M. *Bull. Chem. Soc. Jpn.* **1986**, *59*, 1501.
9. Wright, M. E.; Allred, G. D. *J. Org. Chem.* **1993**, *58*, 4122.
10. Morino, S.; Watanabe, T.; Magaya, Y.; Yamashita, T.; Horie, K.; Nishikubo, T. *J. Photopolym. Sci. Technol.* **1994**, *7*, 121.

Novolak Resins

See: *Novolak Resins (For Microlithography)*
Novolak Type Epoxy Resins
Phenolic Resins (Improvement in Heat Resistance)

NOVOLAK RESINS (for Microlithography)

Leonard E. Bogan, Jr.
Rohm and Haas Company

There are historical and functional reasons for using novolak resins as the primary solid components of most photoresists used today in integrated circuits. As early as 1949, novolak resins were used to formulate positive photoresists for the manufacture of printing plates.[1] When high-resolution lithography (microlithography) became necessary for the fabrication of microelectronic devices, the existing technology for macrolithography including novolak resins was borrowed. Novolak resins are still used because they have the right blend of properties for this demanding task: they form defect-free films, they dissolve in high-pH, aqueous solutions at controllable rates, they react or interact with other materials in ways that affect the dissolution rates, of the novolak resins, they resist etches commonly used in the fabrication of integrated circuits, and they perform reliably. An appreciation of the properties needed in a photoresist binder, and for the good performance of novolak resins in this capacity, requires some understanding of microlithography and integrated circuit (IC) manufacture.

Figure 1 shows the steps used to apply a pattern of silicon dioxide to the surface of a silicon wafer.

In device manufacture, a similar series of steps is followed for each layer of pattern used, and complex devices may require hundreds of layers. A defect in any layer can render the device inoperative, so it is imperative that the yield for each layer be extremely high. Photoresists based on novolak resins have a proven record of such reliability.

In a positive photoresist, the unexposed portion of the coating is not removed by the developer. In a conventional positive resist, novolak resin is formulated with a naphthoquinone diazide, which inhibits dissolution. Upon exposure to the proper dose and type of radiation, the naphthoquinone diazide rearranges to form an indenecarboxylic acid, which accelerates dissolution of the exposed portion of the coating.

RESIN DISSOLUTION RATE

Transfer of an image from a mask to the photoresist coating depends ultimately on a difference in the dissolution rates of exposed and unexposed resist (Figure 1, step 5). In order to faithfully reproduce features with dimensions of less than 0.5 µm, controlling resist dissolution rate is essential. Although photoactive compounds and crosslinkers will affect it, the dissolution rate of the resist is controlled by the choice of novolak resin.

The primary factor determining the dissolution rate of a novolak resin is its composition. The dissolution rate of a novolak polymer decreases with increasing molecular weight, as expected,[1,2–4] but this effect is small compared to the effect of resin composition.[5,6] In general, the incorporation of *p*-cresol or dimethylphenols slows resin dissolution rate, while the incorporation of *m*-cresol and especially resorcinols speeds dissolution rate.[3–5] Theoretically, resin dissolution rate could be easily controlled by adjusting the ratio of *m*-cresol to *p*-cresol in a copolymer, but variations in the synthesis make fine control impossible.

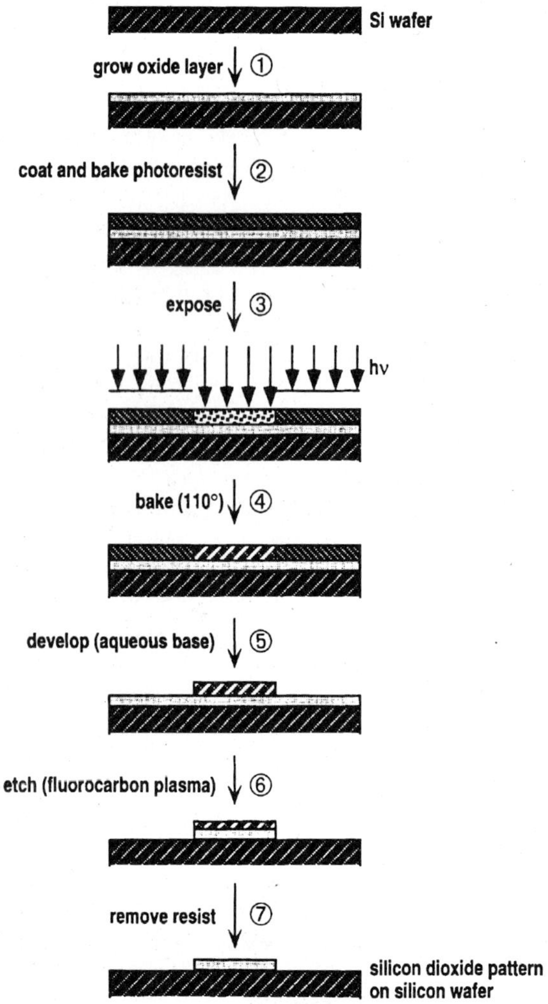

FIGURE 1. Fabrication of a silicon dioxide pattern on the surface of a silicon wafer using microlithography.

RESIN PREPARATION

These resins can be prepared by the conventional reaction of the cresols with formaldehyde in water with an acid catalyst, but several modifications of this method have been reported. Many workers have reported the use of organic solvents in this reaction.[3,4,7-11] This is particularly important for the preparation of resins with softening points above 100°C. It has been reported that the use of more polar solvents yields resins with increased molecular weight, and that solvents with more hydrogen–bonding capability yield resins with slower dissolution rates.[7]

RESIN CHARACTERIZATION

Novolak resins for use in photoresists are usually characterized by composition, dissolution rate, and either glass transition temperature (T_g) or thermal flow temperature (T_{flow}). Other characteristics sometimes reported are degree of *o,o′* bonding, degree of C-4 substitution, branch density, and optical density.

REFERENCES

1. Hanabata, M.; Furuta, A. *Kobunshi Ronbunshu* **1989**, *46*, 15–19.
2. Zampini, A.; Turci, P.; Cernigliaro, G. J.; Sandford, H. F.; Swanson, G. J.; Meister, C. C.; Sinta, R. *Proc. SPIE—Int. Soc. Opt. Eng.* **1990**, *1262*, 501–12.
3. Blakeney, A. J.; Jeffries, A. T.; Wehrle, J. J.; Gardner, W. M. *Proc. SPIE—Int. Soc. Opt. Eng.* **1988**, *920*, 339–348.
4. Khanna, D. N.; Durham, D. L.; Seyedi, F.; Lu, P. H.; Perera, T. *Polym. Eng. Sci.* **1992**, *32*, 1500–8.
5. Pampalone, T. R. *Solid State Technol.* **1984**, 115–120.
6. EP 435502 to Rohm and Haas, July 3, 1991.
7. Sobodacha, C. J.; Lynch, T. J.; Durham. D. L.; Paradis, V. R. *Proc. SPIE—Int. Soc. Opt. Eng.* **1993**, *1925*, 582–92.
8. Miloshev, S.; Novakov, P.; Dimitrov, V.; Gitsov, I. *Polymer* **1991**, *32*, 3067–70.
9. Toukhy, M. A.; Beauchemin, B. T. *Polym. Microelectron. Proc. Int. Symp.* **1990**, 1990, 363–74.
10. JP 5017548 to Sumitomo, January 26, 1993.
11. Bogan, L. E. *Macromolecules* **1991**, *24*, 4807–12.

NOVOLAK-TYPE EPOXY RESINS

Kiichi Hasegawa
Plastics Department
Osaka Municipal Technical Research Institute

Novolak-type epoxy resins are useful and important epoxy resins. There are two main kinds of novolak-type epoxy resin: phenol and *o*-cresol. Recently, however, naphthol novolak-type epoxy resin and biphenol novolak-type epoxy resin have also appeared on the market. Synthesizing these resins is similar to that of the generally used bisphenol-A type epoxy resin; the resins are produced by reacting novolaks with epichlorohydrin.

Because of their high functionality, cured novolak-type epoxy resins have a high degree of crosslinking. Due to good heat resistance these resins are suitable for use in microelectronics, especially as encapsulation materials for semiconductors.

In general, novolaks have many isomers and molecular distribution. This article describes the results of a study of the relationship between properties and structures of epoxy resins prepared from novolaks in which the position of methylene linkage, the number, and kind of substituents were different.

STRUCTURE AND VISCOELASTIC PROPERTIES OF EPOXY RESINS PREPARED FROM TWO-NUCLEI PHENOLIC COMPOUNDS[1,2]

Two-nuclei phenolic compounds are named bisphenol-F. They are prepared with a large excess of phenol to formaldehyde. Commercial bisphenol-F is the mixture of *p,p′* (30%–35%), *o,p′* (50%–55%) and *o,o′* (10%–15%). In general, bisphenol-F type epoxy resin is prepared from the mixed isomers. To clarify the relationship between structure and properties, the isomers were isolated and epoxidized. The conclusion obtained are as follows:

1. A bulky and rigid substituent such as t-butyl or phenyl group effectively improved T_g of the cured resin, whereas a flexible substituent such as nonyl group reduced T_g

2. The introduction of substituents to positions where the mobility of the main chain was restricted resulted in a considerably higher T_g

STRUCTURE AND VISCOELASTIC PROPERTIES OF EPOXY RESINS PREPARED FROM MULTI-NUCLEI PHENOLIC COMPOUNDS[3–5]

In microelectronics, *o*-cresol novolak-type epoxy resins that consist of multi-nuclei compounds, such as 3-nuclei compounds, are generally used. It was clear that T_g of cured resins were closely related to the number of functional groups of epoxy resins. That is, the more functional groups, the higher T_g given by the cured resins, hexa functional epoxy resins especially did not give a clear T_g up to 300°C. In addition, the higher linearity in the main chain of epoxy resins gave a cured resin with a higher T_g.

REFERENCES

1. Hasegawa, K.; Fukuda, A.; Tonogai, S.; Horiuchi, H. *Jpn. J. Polym. Sci. Tech.* **1983**, *40*, 321.
2. Hasegawa, K.; Fukuda, A.; Tonogai, S.; Horiuchi, H. *Jpn. J. Polym. Sci. Tech.* **1984**, *41*, 575.
3. Hasegawa, K.; Fukuda, A.; Tonogai, S. *Jpn. J. Polym. Sci. Tech.* **1986**, *43*, 529.
4. Hasegawa, K.; Fukuda, A.; Tonogai, S. *J. Appl. Polym. Sci.* **1989**, *37*, 3423.
5. Hasegawa, K.; Fukuda, A.; Tonogai, S.; Uede, K. *J. Appl. Polym. Sci.* **1989**, *38*, 1581.

Nucleating Agents

See: Additives (Property and Processing Modifiers)
Poly(aryl ether ether ketone) [PEEK]
(Conformational Order Studied Using FTIR)
Polypropylene (Nucleating Agents)

NUCLEIC ACID ANALOGS

Takehiko Wada and Yoshiaki Inaki*
Faculty of Engineering
Osaka University

Kiichi Takemoto
Faculty of Science and Technology
Ryukoku University

Nucleic acids are one of the most important biopolymers and play an important role in replication and transcription of genetic code and protein synthesis. These functions are realized by specific hydrogen bonding between nucleic acid bases, such as adenine–thymine and cytosine–guanine. To deeply understand the functions of nucleic acids, synthesis of nucleic acid analogs, which specifically bind to DNA or RNA and regulate gene expression, has received much study. Synthetic nucleic acids, however, easily decompose and decrease the function by enymolysis and hydrolysis. To avoid these problems, we proposed using nucleic acid analogs, which exhibit more stable chemical bonding between nucleic acid moieties. A number of syntheses of stable nucleic acid analogs have been reported and are found to form polymer complexes in organic solvents by complementary hydrogen bonding between nucleic acid bases. Applications of these nucleic acid analogs for polymeric drugs, HPLC resins for specific separation of nucleic acid fragments, photo-resists, photo-reversible recording systems, and others, were also reported. These nucleic acid analogs are containing nucleic acid bases. Concerning the advances in related fields, reviews have been given,[1–3] and the reader is referred to those in connection with this article.

The nucleic acid bases, such as adenine and thymine, are relatively hydrophobic and nucleic acid analogs are soluble in organic solvents but insoluble in water. However, nucleic acids are generally only soluble in water and denatured in organic solvents. Thus, it is necessary to synthesize water-soluble nucleic acid analogs to develop a new material target for the natural nucleic acids. For this purpose, we synthesized water-soluble nucleic acid analogs.

This article first describes the syntheses and properties of nucleic acid analogs and then next explains the syntheses and properties of water-soluble nucleic acid analogs.

PREPARATION

Whereas nucleic acids are polyphosphodiester, the backbone polymers of these analogs are synthetic and nucleic acid bases are present in the side chains.

POLY(*N*-VINYL) DERIVATIVES

We prepared poly(1-vinyluracil) and poly(9-vinyladenine) by free radical polymerization of the corresponding monomer.[4,5] We prepared poly(1-vinylcytosine) and poly(9-vinylhypoxanthine) by polymer reaction of poly(4-ethoxy-1-vinyl-2-pyrimidone) and poly(9-chloro-9-vinylpurine), respectively.[6,7] We studied the formation of the polymer complex by the specific base pair between *N*-vinyl polymers, and with polynucleotides.[8–10]

POLYMETHACRYLATE DERIVATIVES

Polymethacrylate derivatives, such as poly[1-(2-methacryloyloxyethyl)uracil] (polyMAOU) and poly[9-2-methacryloyloxyethyl)adenine] (polyMAOA) were prepared by the free radical polymerization of the corresponding monomers.[11,12] The stereoregular polymethacrylate derivatives were prepared by the polymer reactions of the nucleic acid bases with the stereoregular poly(2-bromoethyl methacrylate)s which were obtained by anionic polymerizations.[13] The polymethacrylates containing pyrimidine bases, such as polyMAOU, can be readily obtained from the reaction of the cyclic compound with polyacrylic acid and polymethacrylic acids.[14,15]

POLYMETHACRYLAMIDE DERIVATIVES

We prepared the polymethacrylamide and polyacrylamide derivatives of nucleic acid bases by free-radical polymerizations of corresponding monomers.[16]

*Author to whom correspondence should be addressed.

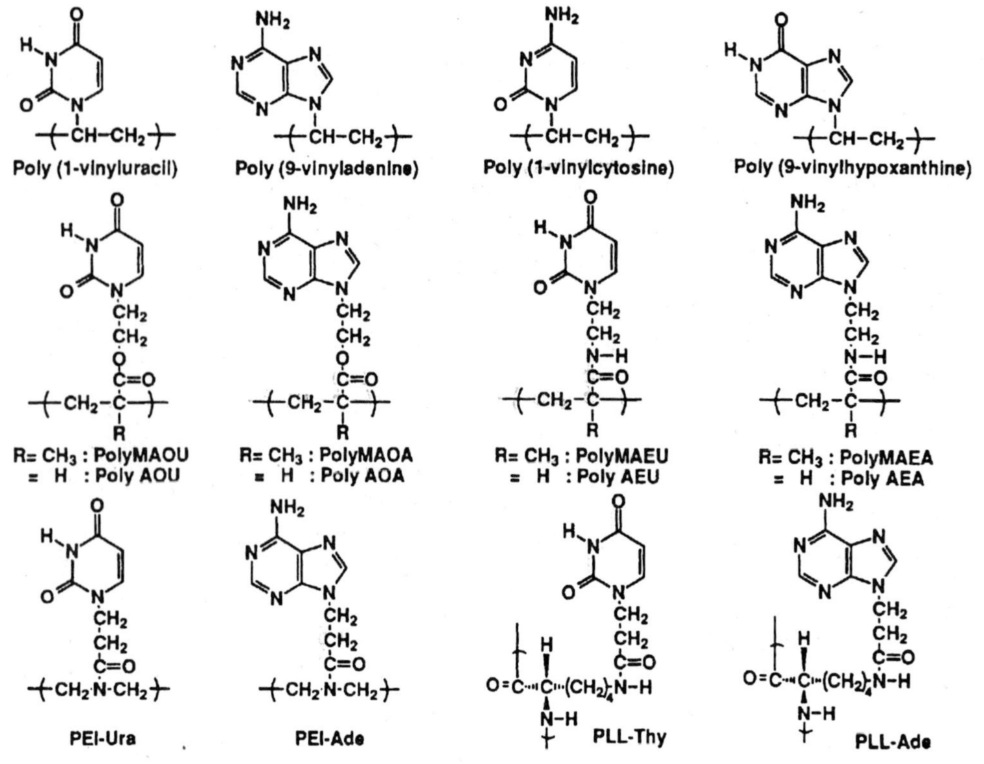

FIGURE 1. Example of synthetic nucleic acid analogs.

POLYETHYLENIMINE DERIVATIVES

Polyethylenimine is highly effective against ascite tumors. The high toxicity of this polymer is eliminated, however, without significantly reducing its efficacy, by grafting it to monomeric serine and/or histidine residues. Recently, new nucleic acid analogs with a polyethylenimine backbone and nucleic acid base derivatives as pending side chains have prepared.[17–42]

WATER-SOLUBLE NUCLEIC ACID ANALOGS

It is well known that introducing polar groups to the polymer side chain is effective for synthesizing water-soluble polymer derivatives. The polar group we used[26–31] was a nonionic hydroxyl group, not anionic or cationic groups. The synthesis of water-soluble nucleic acid analogs was achieved by introducing a hydroxyl group, such as homoserine or serine, to the polymer side chain.

PROPERTIES

Interaction of Synthetic Polymers with Polynucleotides

The formation and stoichiometry of the polymer complex between nucleic acid analogs are affected by several factors. One is the property of the polymer backbone: its flexibility, steric regularity, electric charge, branching, and molecular weight. These can reflect the compatibility and penetration ability of the polymer and the stability of the polymer complex. A polynucleotide has a flexible, sterically regular, and negatively charged polymer backbone, whereas the backbone of the synthetic analogs is probably less flexible, sterically inhomogeneous, and neutral. Other important factors for the specific polymer complex are temperature and solvent conditions for the interaction. Further, the pH and concentration of salt are important for the complex formation in an aqueous solution.[32,33]

APPLICATIONS

Polymeric Drugs

A polymeric drug is a polymer that contains a drug unit as part of the polymer backbone, as a terminal unit, or as a pendant unit. In some cases the polymer itself functions as the therapeutic agent, where the low molecular weight analogs are inactive. Many useful drugs are known in the derivatives of the nucleic acid moieties. Therefore, much attention is being directed to the synthetic nucleic acid analogs as polymeric drugs. Many polymers are reported to be under development as anticancer polymeric drugs, which contain the derivatives of nucleic acid base and nucleoside: 5-fluorouracil, arabino cytidine, and 6-mercaptopurine in the side chain. These polymers are designed to release the low molecular weight drug in a biological system containing hydrolyzable bonds in the side chain or in the main chain.[35–37]

We conducted a fundamental study relating to the polymeric drugs for the polymers containing uridine in the side chain and the main chain.[38]

Another interesting subject of the synthetic nucleic acid analogs as polymeric drugs is the interferon inducer. The most effective synthetic inducers of β-interferon are found among nucleic acids and polynucleotides, such as double-stranded helical poly(C) · poly(I) complex. Activity of poly(C) · poly(I) complex is, however, rapidly decreased by enzymolysis. Synthetic nucleic acid analogs are incapable of inducing interferon in human fibroblast cell in culture. However, the complex of water-soluble polyethylenimine derivatives containing cytosine and homoserine as a spacer (PEI-Hse-Cyt) with poly(I) is a very effective inducer of β-interferon in fibroblast human cell. A reason for this effectiveness may be the stability of PEI-Hse-Cyt · poly(I) complex and the high uptake of this complex by cells.[2]

High Performance Liquid Chromatography (HPLC)

The base pairing between nucleic acid bases is a complementary and specific interaction, but the stacking interaction is not specific. The specific interaction of nucleic acids can be applied to high performance liquid chromatography (HPLC) for specifically separating nucleic acid fragments.

In our previous studies, nucleic acid bases were immobilized on silica gel and applied to HPLC for separation of nucleic acid fragments. The complementary separations were observed in the methanol mobile phase, but the separation in the aqueous mobile phase was explained by a hydrophobic interaction.[39–49] To separate the nucleic acid fragments, however, a water or water–methanol mobile phase should be used because the materials are water-soluble.

REFERENCES

1. Takemoto, K.; Inaki, Y. *Adv. Polym. Sci.* **1981**, *41*, 1.
2. Pitha, J. *Adv. Polym. Sci.* **1983**, *50*, 1.
3. *Functional Monomers and Polymers;* Takemoto, K.; Inaki, Y.; Ottenbrite, R., Eds., Marcel Dekker: New York, 1987.
4. Kondo, K.; Iwasaki, H.; Nakatani, K.; Ueda, N.; Takemoto, K.; Imoto, M. *Makromol. Chem.* **1969**, *125*, 42.
5. Kaye, H. *J. Polym. Sci.* **1969**, *B7*, 1.
6. Pitha, J.; Pitha, J. M.; Stuart, E. *Biochem.* **1971**, *10*, 4596.
7. Pitha, P. M.; Michelson, A. M. *Biochim. Biophys. Acta* **1970**, *204*, 381.
8. Pitha, P. M.; Pitha, J. *J. Biopolym.* **1970**, 9, 965.
9. Kaye, H. *J. Am. Chem. Soc.* **1970**, *92*, 5777.
10. Kawakubo, F.; Kondo, K.; Takemoto, K. *Makromol. Chem.* **1973**, *169*, 37.
11. Kondo, K.; Iwasaki, H.; Ueda, N.; Takemoto, K.; Imoto, M. *Makromol. Chem.* **1968**, *120*, 21.
12. Kita, Y.; Inaki, Y.; Takemoto, K. *J. Polym. Sci. Polym. Chem. Ed.* **1980**, *18*, 427.
13. Inaki, Y.; Takada, H.; Kondo, K.; Takemoto, K. *Makromol. Chem.* **1977**, 178, 365.
14. Inaki, Y.; Futagawa, H.; Takemoto, K. *Org. Prep. Proc. Int.* **1980**, *12*, 275.
15. Kita, Y.; Futagawa, H.; Inaki, Y.; Takemoto, K. *Polym. Bull.* **1980**, *2*, 25.
16. Inaki, Y.; Sugita, S.; Takahara, T.; Takemoto, K. *J. Polym. Sci.; Polym. Chem. Ed.* **1986**, *24*, 3201.
17. Overberger, C. G.; Inaki, Y. *J. Polym. Sci. Polym. Chem. Ed.* **1979**, *17*, 1739.
18. Morishima, Y.; Overberger, C. G. *J. Macromol. Sci.* **1979**, *A13*, 573.
19. Overberger, C. G.; Morishima, Y. *J. Polym. Sci. Polym. Chem. Ed.* **1980**, *18*, 1247.
20. Overberger, C. G.; Morishima, Y. *J. Polym. Sci. Polym. Chem. Ed.* **1980**, *18*, 1267.
21. Overberger, C. G.; Morishima, Y. *J. Polym. Sci. Polym. Chem. Ed.* **1980**, *18*, 1433.
22. Overberger, C. G.; Ludwick, A. G. *J. Polym. Sci. Polym. Chem. Ed.* **1982**, *20*, 2123.
23. Overberger, C. G.; Ludwick, A. G. *J. Polym. Sci. Polym. Chem. Ed.* **1982**, *20*, 2139.
24. Smith, Jr., W. T.; Brahme, N. A. *J. Polym. Sci. Polym. Chem. Ed.* **1985**, *23*, 879.
25. Inaki, Y.; Futagawa, H.; Takemoto, K. *J. Polym. Sci. Polym. Chem. Ed.* **1980**, *18*, 2959.
26. Wada, T.; Inaki, Y.; Takemoto, K. *Polym. J.* **1988**, *20*, 1059.
27. Wada, T.; Inaki, Y.; Takemoto, K. *Polym. J.* **1989**, *21*, 11.
28. Wada, T.; Inaki, Y.; Takemoto, K. *J. Bioactive Compatible Polym.* **1989**, *4*, 25.
29. Wada, T.; Inaki, Y.; Takemoto, K. *J. Bioactive Compatible Polym.* **1992**, *7*, 25.
30. Takemoto, K.; Wada, T.; Mochizuki, E.; Inaki, Y. *Polym. Mater. Sci. Eng.* **1990**, *62*, 558.
31. Takemoto, K.; Wada, T.; Mochizuki, E.; Inaki, Y. *Biotechnology and Polymers* **1991**, 31.
32. Thiele, D.; Guschlbauer, W.; Faver, A. *Biochim. Biophys. Acta* **1972**, *272*, 22.
33. Fujioka, K.; Baba, Y.; Kagemoto, A.; Fujishiro, R. *Polym. J.* **1980**, *12*, 843.
34. Overberger, C. G.; Inaki, Y.; Nambu, Y. *J. Polym. Sci., Polym. Chem. Ed.* **1979**, *17*, 1759.
35. Gebelein, C. G. *ACS Symp. Ser.* **1982**, *186*, 193.
36. Kato, Y.; Saito, M.; Fukushima, H.; Takeda, Y.; Hara, T. *Cancer Res.* **1984**, *44*, 25.
37. Seita, T.; Kinoshita, M.; Imoto, M. *J. Macromol. Sci. Chem.* **1973**, *A7*, 1297.
38. Mochizuki, E.; Inaki, Y.; Takemoto, K. *Polym. Prep. Jpn.* **1983**, *32*, 1923.
39. Rata, Y.; Seita, T.; Hashimoto, T.; Shimizu, A. *J. Chromatogr.* **1977**, *134*, 204.
40. Inaki. Y.; Nagae, S.; Miyamoto, T.; Sugiura, Y.; Takemoto, K. *Poly. Mater. Sci. Eng.* **1987**, *57*, 286.
41. Takemoto, K.; Inaki, Y. *Functional Monomers and Polymers;* Takemoto, K.; Inaki, Y.; Ottenbrite, R. M., Eds.; Marcel Dekker: New York, 1987; p 149.
42. Inaki, Y.; Takemoto, K. *Current Topics in Polymer Science;* Ottenbrite, R. M., et al., Eds.; Hanser: 1987; p 79.
43. Inaki. Y.; Takemoto, K. *Nucleic Acids Res., Sym. Ser.* **1988**, *19*, 45.
44. Inaki, Y.; Sugiura, Y.; Miyamoto, T.; Hojho, H.; Nagae, S.; Takemoto, K. *Nucleic Acids Res., Symp. Ser.* **1988**, *20*, 59.
45. Nagae, S.; Miyamoto, T.; Inaki, Y.; Takemoto, K. *Anal. Sci.* **1988**, *4*, 575.
46. Inaki, Y.; Nagae, S.; Miyamoto, T.; Sugiura, Y.; Takemoto, K. *Applied Bioactive Polymeric System;* Gebelein, C. G.; Carraher, Jr., C. E.; Foster, V. R., Eds.; Plenum: New York, 1988; p 185.
47. Nagae, S.; Miyamoto, T.; Inaki, Y.; Takemoto, K. *Polym. J.* **1989**, *21*, 19.
48. Nagae. S.; Suda, Y.; Inaki, Y.; Takemoto, K. *J. Polym. Sci. Part A. Polym. Chem.* **1989**, *27*, 2593.
49. Nagae, S.; Inaki, Y.; Takemoto, K. *Polym. J.* **1989**, *21*, 425.

NUCLEIC ACID INTERACTIONS (with *cis*-Platin)

Akihiro Kagemoto,* Satoshi Kimura, Makoto Satou, and Yoshihiro Baba
Laboratory of Chemistry
Department of General Education
Osaka Institute of Technology

It is well known that deoxyribonucleic acid (DNA) in living cells is the functional macromolecule with a double-stranded helical structure by the complementary interaction between purine and pyrimidine bases of DNA. It is also very important for understanding the development of function *in vivo* to inform about the physiological function from the mechanism of genetic information controlled by the DNA molecule, and to obtain information about the stability of a double-stranded helical structure of DNA and/or the interactions between DNA and a relatively large molecule such as protein[1,2] or relatively small one such as metallic ions.[3,4] It is well known that various platinum complexes are used as antitumor agents in living cells. *cis*-Diamine-dichloroplatinum(II), *cis*-platin, currently is the most widely used anticancer agents except for the *trans*-isomer.

Recently, considerable attention focused on *cis*-platin and the interaction between it and nucleic acids as a genetic substance in living cells. The relationship has been studied by many investigators using various methods, such as ^{13}C NMR,[5,6] X-ray diffraction,[7] atomic absorption,[8] electrophoresis,[9] circular dichroism,[10] UV,[11] and molecular dynamics simulation.[12,13]

In our previous paper,[14] the thermal behaviors of *cis*-platin binding to DNA in dilute solutions have been thermodynamically investigated by flow microcalorimeters and a differential scanning calorimeter. We reported that the thermodynamic quantities for binding *cis*-platin to DNA have been determined from the heat of mixing. Consequently, *cis*-platin has bound predominantly to the d(GpG) sequence rather than d(ApA) sequence of DNA and its binding energy has been estimated at about –106 kJ per mole of *cis*-platin.

There have been no studies on the binding position and/or an interaction mode of *cis*-platin with the base sequences of DNA in the cancer cells.

To obtain information for this article about the thermodynamic properties of DNA-I extracted from cancer cells, the interactions between DNA-I and *cis*-platin as an antitumor agent were studied using a microcalorimeter, a spectrophotometer, and a spectropolarimeter. Furthermore, to obtain further information about *cis*-platin's binding position to base sequences in DNA-I, the thermal properties of DNA solutions containing 26% (DNA-II), 42% (DNA-III), and 72% (DNA-IV) of guanine–cytosine (GC) base pairs of DNA in dilute solutions with various concentrations of *cis*-platins were also investigated with a differential scanning calorimeter (DSC).

CONCLUSION

To obtain information about DNA modes of binding of *cis*-platin, the thermal behaviors of DNA-I (GC:44%), DNA-II (GC:26%), DNA-III (GC:42%), and DNA-IV (72%) with various guanine–cytosine base pairs (GC) of DNA for *cis*-platin have been studied by means of differential scanning calorimetry (DSC).

We have concluded that the *cis*-platin molecule binds predominantly to d(GpG) or d(ApG) sequences rather than to d(ApA), and its corresponding enthalpy changes are estimated to be about 26.0 kJ mol^{-1} (mol is mole of base pairs) for $\Delta H_{GC/GC}$, and 5.55 kJ mol for $\Delta H_{AT/GC}$, under the assumption that $\Delta H_{AT/AT}$ is nearly zero, similar to ΔH_{AU} for duplex-I, which is independent of r. From these results, we have concluded that *cis*-platin binds predominantly to d(GpG) sequence more than to the d(ApG) sequence, and DNA forms the DNA–*cis*-platin complex accompanying interaction of *cis*-platin with d(GpG) sequence, and the mode of binding of DNA with *cis*-platin in the d(GpG)–*cis*-platin complex is the chelation of *cis*-platin to N_7–N_7 of two adjacent guanines as pointed out from the X-ray diffraction studies;[7] consequently, the one of three hydrogen bonds formed between GC base pairs by complementary interaction in DNA is weak and may transform into the sequence, such as duplex-I with a double-stranded helical structure with increasing the concentration of *cis*-platin.

*Author to whom correspondence should be addressed.

REFERENCES

1. Fujioka, K.; Baba, Y.; Kagemoto, A. *Polym. J.* **1979**, *11*, 509.
2. Kagemoto, A.; Irie, H.; Aramata, M.; Baba, Y. *Thermochim. Acta* **1991**, *181*, 155.
3. Matuoka, Y.; Nomura, A.; Tanaka, S.; Baba, Y.; Kagemoto, A. *Thermochim. Acta* **1990**, *163*, 147.
4. Baba, Y.; Kagemoto, A. *Biopolymers* **1974**, *13*, 339.
5. Caradonna, J. P.; Lippard, S. J. *J. Am. Chem. Soc.* **1982**, *104*, 5793.
6. den Hartog, J. H. J.; Altona, C.; van Boom, J. H.; van der Marel, G. A.; Haasnoot, C. A. G.; Reedijk, J. *J. Biomol. Struct. Dyn.* **1985**, *6*, 1137.
7. Sherman, S. E.; Wang, D. G. A. H. J.; Lippard, S. *J. Science* **1985**, *230*, 412.
8. Fichtinger-Schepman, A. M. J.; van der Veer, J. L.; den Hartog, J. H. J.; Lohman, P. H. M.; Reedijk, J. *Biochem.* **1985**, *24*, 707.
9. Rice, J. A.; Crothers, D. M.; Pinto, A. L.; Lippard, S. J. *Proc. Natl. Acad. Sci. USA.* **1988**, *85*, 4158.
10. Srivastava, R. C.; Froehlich, J.; Eichhorn, G. L. *Biochim.* **1978**, *60*, 879.
11. Horacek, P.; Drobnik, J. *Biochim. Biophys. Acta* **1971**, *254*, 341.
12. Kozelka, J.; Chottard, J. C. *J. Am. Chem. Soc.* **1985**, *107*, 4079.
13. Kozelka, J.; Petsko, G. A.; Lippard, S. *J. Biophys. Chem.* **1990**, *35*, 155.
14. Kagemoto, A.; Takagi, H.; Naruse, K.; Baba, Y. *Thermochim. Acta* **1991**, *190*, 191.

Nucleic Acids

See: Biopolymers (Overview)
DNA (Bending and Twisting Elasticity)
DNA (Intercalation with Drugs)
DNA Degradation (Atomic Target Method Using Synchrotron Radiation)
Nucleic Acid Analogs
Nucleic Acid Interactions (with cis-platin)
Nucleohistone Complexes (Protein-Nucleic Acid Interactions)

Polyelectrolyte Complexes (Targeting of Nucleic Acids)
Polynucleotide Analogues

NUCLEOHISTONE COMPLEXES (Protein-Nucleic Acid Interactions)

Maria Sakarellos-Daitsiotis
Department of Chemistry
University of Ioannina

Deoxyribonucleic Acid (DNA) is closely associated with a wide variety of different DNA-binding proteins. Interactions between proteins and nucleic acids govern packaging, replication, recombination, restriction, and transcription, and they are intimately linked to gene expression, cell division, and differentiation. Many proteins, particularly those serving as repressors, transcriptional activators, and restriction endonucleases, have extremely high specificity. For instance, they are able to find a single site of 10–20 base pairs in a background of approximately 10^6–10^9 base pairs. Therefore, it is particularly important to determine the molecular basis of specificity, which requires characterizing the conformational properties of the protein, the DNA target site, and the changes of the specific complex.[1–4]

Eucaryotic DNA, however, is tightly complexed with specialized proteins—histones—that "package" it and help to regulate its activity. The histones constitute a well-defined class of structural proteins with low sequence specificity and their association with DNA leads to the formation of nucleosomes, the unit particles of chromatin. These proteins play a central role in DNA folding, which is critical for packing the DNA in an orderly way. The exact manner in which a region of the genome is folded in a particular cell can determine the activity of the genes in that region.[5–7]

Here we address the contribution of histones into chromosomes, the DNA polymorphism induced by environmental effects, and the interactions between DNA and sequential polypeptides applied as histone models.

ROLE OF HISTONES IN THE COMPACTION OF DNA

DNA is an enormously long, unbranched, and linear polymer containing many millions of nucleotides arranged in an irregular but nonrandom sequence. Genetic information, dependent on the exact order of the nucleotides, is directly related to the problem of storing a large amount of genetic material in a small amount of space.

The DNA-binding proteins in eucaryotes are divided into two general classes: the histones and the nonhistone chromosomal proteins. The latter designation includes hundreds or thousands of different proteins with many different functions and specificities. Histones occur in enormous quantity (about 60 million copies of each type) with a total mass nearly equal to that of the DNA cell. The nucleohistone complexes, known as chromatin, can be dissociated after treatment with salt or dilute acid. Five types of histones, H1, H2A, H2B, H3, and H4, are obtained with fractionation by ion-exchange chromatography with molecular weights ranging from about 11 to 21 KD.

Histones are among the most highly conserved proteins known. A major advance in our present understanding of chromatin structure came in 1974 from Roger Kornberg, who proposes that chromatin is made up of repeating unit particles now known as nucleosomes. Each nucleosome contains a set of eight histone molecules—two copies each of H2A, H2B, H3, and H4—forming a protein core around which the double-stranded DNA is wound. The remainder of the DNA, called the linker, joins the adjacent nucleosomes and contributes to the flexibility of the chromatin fiber.

INDUCED DNA POLYMORPHISM

Contrary to RNA, which is found only in the A form, DNA can adopt several conformations depending on environmental conditions, such as counterions, relative humidity, polarity, and the presence of macromolecules like poly(ethylene oxide)s, polyacrylates, polyamines, and basic polypeptides.[8,9] The four principal structures of the traditional polymorphs of DNA are the A, B, C, and D types.

DNA is wound around globular histone octamers to form nucleosome cores. However, DNA does not necessarily need histones to form ordered, compact structures. It has the intrinsic ability to do so on its own under certain conditions. A spontaneous condensation due to lateral (side-by-side) aggregation takes place and even higher order organization is achieved and DNA arranges ultimately into lamellar microcrystals. This kind of self-organization is relevant to the packing of DNA in phage heads and therefore is particularly interesting.[10,11]

MODEL SYSTEMS COMBINING NUCLEIC ACIDS AND SYNTHETIC SEQUENTIAL POLYPEPTIDES: A CIRCULAR DICHROISM APPLICATION

Model studies using synthetic simplified polypeptides, i.e., homopolymers, random copolymers, or sequential polypeptides, have been widely used to understand the role of peptide primary sequence and secondary structure on the mode of binding with DNA. Sequential polypeptides have been used as models for lysine or arginine-rich histones, their conformation and interactions with DNA were investigated and compared with data obtained from reconstituted nucleohistones. The use of simplified model polypeptides proved quite reasonable since histones are conformationally flexible molecules and their structure in solution is strongly dependent upon the environment.[12–15]

Polypeptides containing arginine have not been reported frequently as models in DNA–protein interactions, due to synthetic problems, although arginine is found in direct contact with nucleic acids. Moreover, evidence has been provided for the formation of two hydrogen bonds between the guanidinium group of arginine and the guanine/cytosine bases, as an element of a selective amino acid-nucleotide recognition. To evaluate the role of varying lengths of the hydrocarbon side-chains and bulk of the Xaa residue, the L–Ala, Val, Leu, Ile, Nva, and Nle amino acids were introduced in the $(\text{L–Arg–Xaa–Gly})_n$ synthetic polypeptides. Using circular dichroism spectroscopy, we found that these polypeptides have conformational characteristics similar to those of arginine-rich histones. the technique of continuous-flow salt gradient linear dialysis was applied for mixing DNA with polypeptides at various ratios (r).[16–22]

Investigation of the effect of the ionic strength on the DNA–(L–Arg–Xaa–Gly–)$_n$ (Xaa = L–Ile, Nle, Nva) points out that in the salt range from 0.14 M NaCl (physiological conditions) to 0 in which neither the polypeptides nor the free DNA show any structural modifications, the DNA–polypeptide complexes adopt the 10.2 B-DNA form. When salt content falls below a critical value (<0.3 NaCl) the ellipticity value of the DNA–polypeptide decreases from $[\theta]_{275} = 8.4 \times 10^{-3}$ for DNA alone to $[\theta]_{275} = 2.5 \times 10^{-3}$ depending on the polypeptide sequence. These experimental findings indicate the importance of the ionic strength as a critical parameter in the process of DNA–protein interactions, as it has been also found for chromatin.[23–25]

REFERENCES

1. Freemont, P. S.; Lane, A. N.; Sanderson, M. R. *Biochem. J.* **1991**, *278*, 1.
2. Harrison, S. C.; Aggarwal, A. K. *Annu. Rev. Biochem.* **1990**, *59*, 933.
3. Record, Jr., M. T.; Ha, J.-H.; Fisher, M. A. *Methods in Enzymology;* Academic: New York, 1991; 208, 291.
4. Lohman, T M. *Crit. Rev. Biochem.* **1986**, *19*, 191.
5. Revzin, A. In *Nonspecific DNA-Protein Interactions;* Revzin, A., Ed.; CRC: Boca Raton, FL, 1990; p 5.
6. Kowalczykowski, S. C.; Bear, D. G.; von Hippel, P. H. In *The Enzymes;* Boyer, P. D., Ed.; Academic: New York, 1981; vol. 14, p 373.
7. Behe, M. J. In *Nonspecific DNA-Protein Interactions;* Revzin, A., Ed.; CRC: Boca Raton, FL, 1990; p 229.
8. Simpson, R. T.; Stafford, R. W. *Proc. Natl. Acad. Sci. USA* **1983**, *80*, 51.
9. Pennings, S.; Muyldermans, S.; Wyns, L. *Biochemistry* **1986**, *25*, 5043.
10. North, A. C. T.; Rich, A. *Nature* **1961**, *191*, 1242.
11. Seeman, N. C.; Rosenberg, J. M.; Rich, A. *Proc. Nat. Acad. Sci. USA* **1976**, *73*, 804.
12. Hélène, C.; Maurizot, J.-C. *CRC Crit. Rev. Biochem.* **1981**, *10*, 213.
13. Schwartz, A. M.; Fasman, G. D. *Biochemistry* **1977**, *16*, 2287.
14. Barbier, B.; Caille, A.; Brack, A. *Biopolymers* **1984**, *23*, 2299.
15. Mantel, R.; Fasman, G. D. *Biochemistry* **1976**, *15*, 3122.
16. Bokma, T. J.; Johnson, W. C. Jr.; Blok, J. *Biopolymers* **1987**, *26*, 893.
17. Lancelot, G.; Mayer, R.; Hélène, C. *J. Amer. Chem. Soc.* **1979**, *101*, 1569.
18. Seeman, N. C.; Rosenberg, J. M.; Rich, A. *Proc. Natl. Acad. Sci. USA* **1976**, *73*, 804.
19. Panou-Pomonis, E.; Sakarellos, C.; Sakarellos-Daitsiotis, M. *Biopolymers* **1986**, *25*, 655.
20. Tsikaris, V.; Sakarellos-Daitsiotis, M.; Sakarellos, C.; Marraud, M. *Biopolymers* **1988**, *27*, 213.
21. Tsikaris, V.; Sakarellos-Daitsiotis, M.; Panou-Pomonis, E.; Sakarellos, C.; Marraud, M. *Int. J. Peptide Protein Res.* **1989**, *33*, 195.
22. Carrol, D. *Anal. Biochem.* **1971**, *44*, 496.
23. Panou-Pomonis, E.; Sakarellos, C.; Sakarellos-Daitsiotis, M. In *Proceedings of the Ninth American Peptide Symposium;* Deber Chemical: Rockford, Illinois, 1985; p 197.
24. Panou-Pomonis, E.; Sakarellos, C.; Sakarellos-Daitsiotis, M. *Eur J. Biochem.* **1986**, *161*, 185.
25. Tsikaris, V.; Panou-Pomonis, E.; Sakarellos, C.; Sakarellos-Daitsiotis, M.; Marraud, M. *Int. J. Biol. Macromol.* **1991**, *13*, 349.

NYLON 3

Junzo Masamoto
Polymer Development Laboratory
Asahi Chemical Industry Company, Ltd.

Breslow et al.[1] reported that nylon 3 [sometimes described as nylon 3 or poly(β–alanine)] is obtained using hydrogen transfer polymerization of acrylamide in the presence of an anionic catalyst.

Nylon 3, $(CH_2CH_2CONH)_n$ is a highly crystalline polymer with a high density of amide groups and is considered useful as a textile.

Some properties of nylon 3 are thought to be different from those of other polyamides. Its glass transition temperature (T_g) is believed to be much higher than that of ordinary polyamides. Because the high density of the amide groups leads to a high absorption of water, the nylon 3 may be expected to be similar to silk or cotton. Considering the chemical structure, Young's modulus of nylon 3 should be comparable to that of silk.

After Brewlow's report,[1] various studies were completed.[2–9] Currently, nylon 3 is produced commercially by Asahi Chemical as an excellent stabilizer for polyoxymethylene.

METHODS FOR THE PREPARATION OF NYLON 3[10]

The following methods were reported for the preparation of nylon 3: anionic ring-opening polymerization of β-lactam,[11] ring-opening polymerization of 8-ring dilactam (1.5-diazacyclooctane-2,6-dion);[12] thermal polymerization of ethylene cyanhydrine, and alternative copolymerization of carbon monoxide and ethylene imine.[13]

The polymerization method of α-amino acid to nylon 3 poly(β-alanine) was also studied. For example, nylon 3 was synthesized from β-alanine *N*-carboxyanhydride(NCA), *N*-dithiocarbonyl ethoxycarbonyl-β-alanine and *N*-carbothiophenyl-β-alanine.[14]

The use of various active β-alanine esters for the preparation of poly(β-alanine) was proposed by Takemoto and his coworkers.[15]

The direct condensation of β-alanine to obtain nylon 3 was reported by Konishi et al.[16] In some cases a high molecular weight nylon 3 was reported.

NYLON 3 FIBER[17–20]

Nylon 3 melts at about 340°C with considerable decomposition. Therefore, melt-spinning cannot be used but wet or dry spinning may be chosen for fiber preparation.

SOLUBILITY OF NYLON 3[17]

Nylon 3 is insoluble in water and most organic solvents, as well as in aqueous basic solutions and dilute acid solutions. However, nylon 3 is soluble in water above 140°C–170°C, strong inorganic acids, organic acids, some aqueous and methanolic inorganic salts, hot aqueous phenol, and aqueous chloral hydrate. As for inorganic salts, $Ca(SCN)_2$, LiCl, $CaBr_2$, $ZnCl_2$, and $SbCl_3$ are effective solvents. Among these solvents, only formic acid is available as a useful spinning dope for the production of the nylon 3 fiber.

WET SPINNING OF NYLON 3[19,20]

Wet spinning of nylon 3 is only possible when the spinning dope uses formic acid solution. Masamoto et al.[20] studied various solvents and established these conclusions. The available reduced viscosity of the polymer ranged from 0.9 to 3.0, and the preferable polymer concentration was from 38 to 40 wt%. In wet spinning, the combination of the solvent and coagulant is important. The coagulants that exhibit fiber forming abilities are organic solvents, such as propanol, i-propanol, butanol, hexanol, diisopropylether, dioxane, tetrahydrofuran, methyl acetate, ethyl acetate, i-propyl acetate, butyl acetate. Water and water-containing formic acid were not useful coagulants.

DRY SPINNING OF NYLON 3[18]

Dry spinning of nylon 3 using formic acid needs a higher molecular weight polymer than does wet spinning. With the polymer having an average molecular weight of 90,000 (reduced viscosity: 0.9) the filaments could not be wound continuously. However, a polymer with an average molecular weight of 240,000 (reduced viscosity: 1.8) could be spun dry and filaments could be wound continuously. In the dry spinning of nylon 3, the spinnability of the spinning dope is an important factor in fiber-forming ability.

DRAWING OF NYLON 3 FIBERS[20]

In such media as silicone oil, Wood's metal, and hot air, which are thought to have no interaction with polyamides, the nylon 3 fibers could be stretched above their glass transition temperature (170°–180°C).

CRYSTAL STRUCTURE OF NYLON 3[21-24]

The crystal structure of nylon 3 was determined by Masamoto et al.[21] using the X-ray diffraction pattern. The four crystal structures of nylon 3 have the designated modifications: I, II, III, and IV.

Nylon 3 has a monoclinic unit cell unlike other ordinary odd-numbered nylons, such as nylons 7, 9 or 11, which have a triclinic or pseudohexagonal unit cell.

PROPERTIES OF NYLON 3[10,20]

Although drawn nylon 3 fibers are brittle, a wet-heat treatment improves the tensile elongation properties. The mechanical properties were as follows: tensile strength, 2–3 g/d; tensile elongation, 10%–20%; Young's modulus, 800–1200 kg/mm^2. Young's modulus was quite high, comparable to that of silk.

Because nylon 3 had a high amide concentration, a high moisture regain was expected. The moisture regain curve was also quite similar to that of silk.

APPLICATION

Although the nylon 3 fibers exhibited properties quite similar to those of silk, the nylon 3 fibers have not yet been commercialized. Currently, various applications of nylon 3 are being investigated, but the only commercially viable application is as a stabilizer for polyoxymethylene.

As a stabilizer for polyoxymethylene, the copolyamide (nylon 6-66-10) is known to be a formaldehyde scavenger. Recently, the Asahi Chemical researchers determined that nylon 3 was an excellent stabilizer for polyoxymethylene.[25]

We completed several experiments using nylon 3 as a molecular composite.[26,27]

REFERENCES

1. Breslow, D. S.; Hulse, G. E.; Matlack, A. S. *J. Am. Chem. Soc.* **1957**, *79*, 3760; Matlack, A. S. U.S. Patent 2 672 480, 1954, assigned to Hercules Powder Co.
2. Ogata, N. *Bull. Chem. Soc. Jpn.* **1960**, *33*, 906; *J. Polym.* Sci. **1960**; *46*, 271; *Makromol. Chem.* **1960**, *40*, 55.
3. Tani, H.; Oguni, N.; Araki, T. *Makromol. Chem.* **1964**, *76*, 82.
4. Trossarelli, L.; Guaita, M.; Camino, G. *ibid.* **1967**, *105*, 285.
5. Masamoto, J.; Yamaguchi, K.; Kobayashi, H. *Kobunshi Kagaku* **1969**, *26*, 631.
6. Gamino, G.; Guaita, M.; Trassarelli, L. *Makromol. Chem.* **1970**, *136*, 155; *J. Polym. Sci., Lett. Ed.* **1977**, *15*, 417; Gamino, G.; Lim., S. L.; Trossarelli, L. *Eur. Polym. J.* **1977**, *13*, 473: Gamino, G.; Costa, M.; Trassarelli, L. *J. Polym. Sci., Polym. Chem. Ed.* **1980**, *18*, 377.
7. Nakayama, H.; Higashimura, T.; Okamura, S. *Kobunshi Kagaku* **1966**, *23*, 433; **1966**, *23*, 439; **1966**, *23*, 537; **1967**, *24*, 427.
8. Masamoto, J.; Ohizumi, C.; Kobayashi, H. *Ibid.* **1969**, *26*, 638.
9. Glickson, J.; Applequist, Y. *Macromolecules* **1969**, *2*, 628.
10. Masamoto, J. Poly-β-alanine Fiber, "Polyaminoacids-Application and Survey"; Fujimoto, Y. Kodansha, Ed.; 1974; p 234; Kobunshi **1975**, 24, 187.
11. Bestian, R. *Angew. Chem.* **1968**, *80*, 304; Kodaira, T.; Miyake, H.; Hayashi, K.; Okamura, S. *Bull. Chem. Soc. Jpn.* **1965**, *38*, 1788.
12. Lautenschlanger, H. U.S. Patent 3 126 353, 1964, assigned to BASF.
13. Kagiya, T.; Narisawa, S.; Manabe, K.; Fukui, K. *J. Polym. Sci., B* **1965**, *3*, 617; Kagiya, T.; Narisawa, S.; Ichida, T.; Fukui, K.; Kondo, M. *J. Polym. Sci., A-1* **1966**, *4*, 293.
14. Birkhofer, L.; Modic, R. *Liebigs Ann. Chem.* **1959**, *628*, 162; Noguchi, J.; Hayakawa, T. *J. Am. Chem. Soc.* **1954**, *76*, 2846; Higashimura, T.; Katoh, H.; Suzuoki, K.; Okamura, S. *Makromol. Chem.* **1966**, *90*, 243; Kricherdorf, H. R. *Makromol. Chem.* **1973**, *173*, 13; Kricheldof, H. R.; Muelhaupt, R. *ibid.* **1979**, *180*, 1419; *J. Macromol. Sci., Chem.* **1980**, *A14*, 349.
15. Hannabusa, K.; Kondo, K.; Takemot, K. *Makromol. Chem.* **1979**, *180*, 307; Hanabusa, K.; Shrai, H.; Hojo, N.; Kondo, K.; Takemoto, K. *ibid.* **1982**, *183*, 1101; Kondo, K.; Miwa, Y.; Takemoto, K. *ibid.* **1983**, *184*, 1171; Hanabusa, K.; Kato, K.; Shirai, H.; Hojo, N. *J. Polym. Sci., Part C, Polym. Lett.* **1987**, *24*, 311; Hanabusa, K.; Tutsumi, H.; Kurose, A.; Shirai, H.; Hayakawa, H.; Hojo, N. *J. Polym. Sci., Part* A. *Polym. Chem.* **1989**, *27*, 1665.
16. Higashi, F.; Sano, K.; Murakami, T. *Sen-i Gakkaishi* **1981**, *37*, T481; Sakabe, H.; Nakamura, H.; Konishi, T. *ibid.* **1989**, *45*, 493; Kimura, H.; Sakabe, H.; Itoh, T.; Konishi, H. *ibid.* **1991**, *47*, 447.
17. Masamoto, J.; Kaneko, Y.; Sasaguri, K.; Kobayashi, H. *Sen-i Gakkaishi* **1969**, *25*, 525; Masamoto, J.; Sasaguri, K.; Ohizumi, C.; Kobayashi, H. *ibid.* **1970**, *26*, 239.
18. Masamoto, J.; Sasaguri, K.; Kobayashi, H. *ibid.* **1970**, *26*, 246.
19. Masamoto, J.; Yamaguchi, K.; Kobayashi. H. *ibid* **1969**, *25*, 533; Masamoto, J.; Ohizumi, C.; Kobayashi, H. *ibid.* **1970**, *26*, 16; Masamoto, J.; Miyake, A.; Ohizumi, C.; Kobayashi, H. *ibid.* **1970**, *26*, 102, Masamoto, J.; Ohizumi, C.; Kobayashi, H. *ibid.* **1970**, *26*, 138.
20. Masamoto, J.; Sasaguri, K.; Ohizumi, C.; Yamaguchi, K.; Kobayashi, H. *J. Appl. Polym. Sci.* **1970**, *14*, 667.
21. Masamoto, J.; Sasaguri, K.; Ohizumi, C.; Kobayashi, H. *J. Polym. Sci., A-2* **1970**, *8*, 1703.

22. Munoz-Guerra, S.; Fernandez, J. M.; Rodriguez-Galan, A.; Subirana, J. *J. Polym. Sci., Polym. Phys. Ed.* **1985**, *23*, 733.
23. Masamoto, J.; Kobayashi, H. *Kobunshi Kagaku* **1970**, *27*, 220.
24. Masamoto, J.; Kaneko, Y.; Kobayashi, H. *ibid.* **1970**, *27*, 301.
25. Yamamoto, G.; Misumi, T. U.S. Patent 4 855 365, 1989; U.S. Patent 5 015 70 7, 1991, assigned to Asahi Chemical.
26. Mppre, D. R.; Mathias, L. J. *Polym. Material Sci. Eng.* **1985**, *53*, 693; *J. Appl. Polym. Sci.* **1986**, *32*, 6299.
27. Yasu, M.; Rikukawa, M.; Sanui, K.; Ogata, N. *Polym. Prepr. Jpn.* **1994**, *43*, 3153.

NYLON 6 (Depolymerization)

Ricardo Vera-Graziano* and Francisco Diaz-Camacho
Instituto de Investigaciones en Materiales
UNAM

Nylon 6, [poly(ε-caprolactam), CAS 25038-54-4], is an important commercial thermoplastic,[1–3] heavily reclaimed for recycling by both chemical and mechanical techniques.[4–7] Its mechanical recycling usually renders products of lower value than the originals, so several studies have focused on efficient chemical techniques for recovering the monomer: ε-caprolactam (ε-caprolactam).[8–18]

Acid-catalyzed depolymerization of nylon 6 is the most common process in industry; orthophosphoric acid has usually been used as a catalyst but a large number of other acids have been found effective.[14–18,20] The ε-caprolactam is obtained in aqueous solution and must be made alkaline before recovering the monomer by distillation.

Alkali-catalyzed depolymerization of nylon 6 has also been reported, though less often.[11–13] These reports show inconsistencies in their results: Takashi et al.,[11] showed that for any reaction time, monomer recovery is minimum at an NaOH–catalyst concentration of 1.5 wt%; Mukherjee and Goel,[12] report an ε-caprolactam maximum recovery at 1.0 wt% NaOH–catalyst concentration; and Luederwald and Aguilera[13] show that the monomer yield increases initially with catalyst concentration and levels off at higher concentrations. How differences in nylon 6 water content, molecular weight, and reaction temperature lead the role of the alkaline-catalysts is not yet completely understood.[19,20]

Based on those studies, a technique for alkaline depolymerization of melted nylon 6 (ADP/N–6) has been developed and is presented here. By this technique the ε-caprolactam is directly obtained in high yields as a crystalline solid ready to be used for repolymerization.

CHARACTERIZATION

During depolymerization, dimers, trimers, and bigger cyclic caprolactam-like compounds may be formed. These and other volatile additives or impurities may flow with the ε-caprolactam to the special trap where the monomer is recovered. Thus, it is important to determine the structure and purity of the recovered ε-caprolactam samples.

*Author to whom correspondence should be addressed.

POLYMERIZATION OF RECOVERED CAPROLACTAM

To further test the usefulness of the ε-caprolactam samples obtained from reclaimed nylon 6, materials were polymerized by the ring-opening reaction of ε-caprolactam, described elsewhere.[24] The reagent-grade monomer was also polymerized of the obtained polymers.

The IR spectra of all polymers correspond to the nylon 6 standard spectrum,[22] and their melting points also correspond to nylon 6 (220 ± 2°C).[23] It should be mentioned that the ε-caprolactam obtained from reclaimed nylon 6/polyurethane copolymer with additives was successfully repolymerized without previous purification.

REFERENCES

1. *Modern Plastics Encyclopedia;* Graff, G.; Kreiser, K., Eds.; Modern Plastics International, McGraw-Hill: New York, Editions, 1990–1994.
2. *Encyclopedia of Polymer Science and Engineering*, 2nd ed.; Mark, H. F.; Bikales, N. M. et al., Eds.; John Wiley & Sons: New York, 1988; 18, p 425.
3. *Anuario Estadistico de la asociacion Nacional de la Industria Quimica*; ANIQ Eds., Mexico, 1992.
4. Mikula, F.; Petru, K. *Intern. Chem. Eng.* **1969**, *9(2)*, 288.
5. Nelson, W. E. *Nylon Plastics Technology*, 1st ed.; Butterworth: Boston, 1976.
6. Reimschussel, H. K. *J. Polym. Sci.* **1977**, *65*, 12.
7. Sidney, P E. *Plastics Extrusion Technology Handbook*, 1st ed.; Industrial: New York, 1981; 1–53.
8. Vainiota, S. *Am. Ind. Hyg. Asoc.* **1989**, *50*(8), 396.
9. Langhamm, M. *Makcromoleculare Chemie-Macromolec. Chem. Phys.* **1986**, *187*(4), 829.
10. Gupta, A.; Gandhi, S. K. *J. Appl. Polym. Sci.* **1982**, *27*, 1099.
11. Takashi, O.; Yasushi, H.; Yukihiro, Y. *Nippon Kagaku Kaishi* **1977**, *7*, 1041.
12. Mukherjee, A. K.; Goel, D. K. *J. Appl. Polym. Sci.* **1978,** *22*(2), 361.
13. Luederwald, I.; Aguilera, C. *Makromol. Chem. Rapid Commun.* **1982**, *3*, 343.
14. Dmitrieva, L. A.; Bychkov, N.; Kharitonov, M. USSR Patent 374 305, March 20, 1973.
15. Konomi, T. *J. Polym. Sci. Part A-1. Polym. Chem.* **1970**, *8*(5), 1261.
16. Petru, K. Czech. Patent 143 502, November 15, 1971.
17. Mikula, F.; Petru, K. *Chem. Prum.* **1967**, *17*(3), 1329.
18. Smith, S. *J. Polym. Sci.* **1958**, *30*, 459.
19. Ogale, A. A. *J. Appl. Polym. Sci.* **1984**, *29*, 3947.
20. Bonfield, J. H.; Richard, C. H. Fr. Patent 1 389 100, February 12, 1965.
21. Vera-Graziano, R.; Diaz-Camacho, F. *International Symposium on Polymers;* Cancún, Mexico, Preprints, 211, November 3–7, 1993.
22. *Standard Spectra Collection;* Sadtler Research Laboratories, Div. of Bio-rad Lab.: PA, 1980.
23. Brandrup, J.; Immergut, E. H. *Polymer Handbook*, 3rd ed.; John Wiley & Sons: New York, 1989.
24. Mathias, L. J.; Vaidya, R. A.; Canterberry, J. B. *J. Chem. Educ.* 1984, *61*(9), 805.

Nylons

See: *Bulk Polymerization*
Engineering Thermoplastics (Survey of Industrial Polymers)
Ferroelectric Polymers (Structural Phase Transitions)
Fibers (Stainproofing)
Melt Spinning (Polymer Formation Under High Stress Conditions)
Nylon 3
Nylon 6 (Depolymerization)
Nylons (By High Pressure Solid-State Polycondensation)
Nylons 3, Helical [Poly(α-alkyl-β-L-aspartate)s]
Polyamide Fibers, Aliphatic (Nylon 66 and Nylon 6)
Polyamides (Impact Toughness and Toughening)
Polyamides, Alicyclic and Aromatic (Ferroelectric Hysteresis and Pyroelectricity)
Polyamides, Glycine-Containing
Polyamides, Lactam-Based
Poly(ether amide) Triblock and Star Block Copolymers
Poly(ether-block-amide) Copolymers (Characterization of Structure and Polymorphism)
Polyphthalamides
Powder Coatings (Overview)
Pyroelectricity
Reaction Injection Molding (RIM) Materials
Solid-State Polyamidation
Textile Fibers (Structure and Properties)
Ultrasonically Assisted Polymer Synthesis

NYLONS (by High Pressure Solid-State Polycondensation)

Tokimitsu Ikawa
Department of Chemistry
Faculty of Science
Okayama University of Science

Thermally induced, solid-state polycondensation (SSP), utilizing the molecular arrangement of nylon monomer crystals is considered effective in preparing large nylon single crystals, though SSP must be done under reduced pressure to remove the H_2O produced in polycondensation.[1] SSP as a crystallization method was first applied to 6-amino-*n*-caproic acid (6ACA), and then to 11-amino-*n*-undecanoic acid (11AUA).[2–6]

Single crystals of nylon 6 and 11 prepared by SSP were observed under electron microscopy. However, the resulting nylon crystals were macroscopically composed of randomly oriented crystallites. This may be due to the destruction of crystals caused by explosive dehydration of the evolved H_2O. SSP of various ω-amino acids and nylon salts from diamines–dicarboxylic acids have since been studied.[7–11]

We conducted high pressure, solid-state polycondensation (HPSSP) of 11AUA and 12-amino-*n*-dodecanoic acid (12ADA) single crystals, for HPSSP is sure to provide topotactic reaction field as compared with SSP. However, HPSSP is unable to remove the evolved H_2O from the reaction field in P. In addition, raising the polymerization pressure is unfavorable for polycondensation, as represented by $(\delta \ln K/\delta P) = -\Delta V/RT$. Despite these disadvantages, we obtained well oriented nylon 11 and 12 crystals.[12,13] Recently, various polyimides, aramides and polybenzoxazole have been prepared by HPSSP.[14,15] This paper investigates the effects of monomer crystallite sizes, CH_2 numbers in monomer, polymerization temperature and pressure on HPSSP of various ω-amino acids and aliphatic nylon salts, and discusses the mechanism of HPSSP.

PREPARATION METHOD OF NYLON CRYSTALS BY HPSSP

Monomer Crystals

We used glycine, β-alanine, 4-amino-*n*-butyric acid (4ABA), 6ACA, 8-amino-*n*-octanoic acid (8AOA), 11AUA, and 12ADA to prepare nylon 2, 3, 4, 6, 8, 11, and 12, respectively. Nylon mm salts (m = 4, 6, 7, 8, 9, 10, and 12) were also used to prepare nylon mm.

Preparation of Nylon Crystals by HPSSP

Monomer single crystals (SC) sealed in a teflon tube with silicone oil, which is carefully dehydrated, were put in a high-pressure autoclave and polymerized at various temperatures and pressures ranging from 196 Mpa to 490 Mpa. Polymerization temperatures under high pressure, T_p(P), were chosen to avoid melting monomer crystals.

CHARACTERIZATION OF NYLON CRYSTALS AND MECHANISM OF HPSSP

Nylons from ω-Amino Acids

In contrast to SSP, glycine, β-alanine, and 4ABA were polymerized by HPSSP, but other ω-amino acids were easily polymerized to nylon 6, 8, 11, and 12, respectively.

Mechanism of HPSSP of ω-Amino Acids

A scanning electron micrograph of 12ADA and the resulting nylon 12 crystals are shown in **Figures 1a** and **1b**, respectively. The original shape was maintained in nylon 12 like a single crystal after polymerization.

In general, the T_ms of nylons produced by HPSSP are higher by 20° ~ 30°C than those of nylons produced by conventional methods. These high T_ms and large crystallite sizes suggest that the one-dimensional chain growth in a monocrystallite form longer chain crystals.

HPSSP of Nylon Salts

The mechanism of HPSSP of nylon salt crystals is essentially the same with ω-amino acid crystals. The melting point of nylon mm is by 10° – 25°C higher than those of nylon mm prepared by conventional methods.

APPLICATION

Nylons obtained by HPSSP method have a higher melting point due to the development of longer chain crystals, but are very brittle because of high crystallinity without chain entanglement. Accordingly, successive processing accompanied by

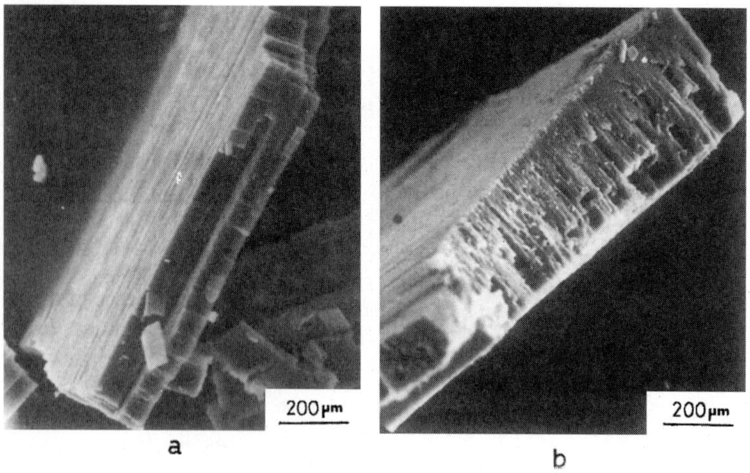

FIGURE 1. SEM of (a) 12ADA, and (b) the resulting nylon 12 crystal.

splintering among monocrystallites or crystals under high pressure after HPSSP may give moldings high dimensional stability and high thermal resistance. Further, high modulus and high tenacity nylons may be prepared by applying the HPSSP method to drawn nylon fibers or films.

REFERENCES

1. Shulz, G. W. *Z. Phisik. Chem. (A)* **1938**, *182*, 127.
2. Morosoff, M.; Lim, D.; Morawetz, H. *J. Am. Chem. Sci.* **1954**, *86*, 3167.
3. Macchi, E.; Morosoff, M.; Morawetz, H. *J. Polym. Sci.* **1968**, *6*, A-l, 2033.
4. Macchi, E. *J. Polym. Sci.* **1972**, *6*, 45.
5. Macchi, E. *Makromol. Chem.* **1979**, *180*, 1603.
6. Macchi, E. *Makromol. Chem., Rapid Commun.* **1980**, *1*, 563.
7. Oya, S.; Tomioka, M.; Araki, T. *Kobunshi Kagaku* **1966**, *23*, 415.
8. Kiyotukuri, T.; Otsuki, F. *Kobunshi Kagaku* **1972**, *29*, 159.
9. Yamazaki, T. Thesis Kyoto Univ., 1983.
10. Kimura, H.; Sakabe, H.; Itoh, T.; Konishi, T. *Sen-i Gakkaishi* **1992**, *48*, 246; *ibid.* **1992**, *48*, 376.
11. Kimura, H.; Itoh, T.; Fujii, A.; Sakabe, H.; Konishi, T. *Sen-i Gakkaishi* **1994**, *50*, 395.
12. Ikawa, T. et al. *Sen-i Gakkaishi* **1986**, *42*, T403.
13. Ikawa, T. et al. *Sen-i Gakkaishi* **1988**, *44*, 385.
14. Itoya, K.; Kumagai, Y.; Kakimoto, M.; Imai, Y. *Polym. Prep. Jpn.* **1992**, *41*, 355.
15. For example, Imai, Y. *Polym. Prepri.* **1994**, *35*, No. 1, 395.

NYLONS 3, HELICAL [Poly(α-Alkyl-β-l-Aspartate)s]

S. Muñoz-Guerra, F. López-Carrasquero,
J. M. Fernández-Santin, and J. A. Subirana
Department d'Enginyeria Química
Universitat Politècnica de Catalunya

Nylon 3 is a noncommercial synthetic polymer whose structure and properties fit well in the pattern of behavior characteristics of nylons.[1,2] However, the high concentration of amido groups present in the chain of this nylon places it close to polypeptides. Since earlier times, the chemical modification of nylon 3 has attracted interest as a promising route of access to materials with properties close to natural silk. Earlier investigations were addressed to the development of C-alkyl and C-aryl substituted nylons 3 with the expectation that they would be more soluble than the parent polymer.[3] Certain types of helical conformations unusual among polyamides, have been claimed for some of these modified nylons 3.[4,5]

The chemical design of polyamides able to adopt regular helical conformations is certainly appealing because novel properties without precedents in this family of polymers could be brought out by these means. Since carboxylate side groups are recognized to be highly efficient promoters of α-helical conformations in proteins, the insertion of such groups on nylon 3 looked to be a convenient method to mimic polypeptide structures and properties in this nylon. Some years ago we found that poly(α-isobutyl-β-l-aspartate), which is a stereoregular derivative of nylon 3 carrying an isobutoxycarbonyl group attached to the β-carbon of the repeating unit, was able to assume helical conformations of α-helix type.[6] These findings were the starting point of a sustained investigation of these alkoxycarbonyl substituted nylons 3, first limited to the isobutyl derivative and later extended to the whole family of [poly(α-alkyl-β-l-aspartate)s].[7,8]

PREPARATION

The reaction pathway leading to the preparation of [poly(α-alkyl-β-l-aspartate)s] from l-aspartic acid is depicted in **Scheme I**.

PROPERTIES

General

[Poly(α-alkyl-β-l-aspartate)s] are nonhygroscopic, while solid polymers with an appearance that changes from powdery to fibrous as the length of the alkyl side chain increases. Their solubility also depends on their constitution, the higher members being readily soluble in halogenated solvents as chloroform and dichloromethane; common solvents to the whole series are

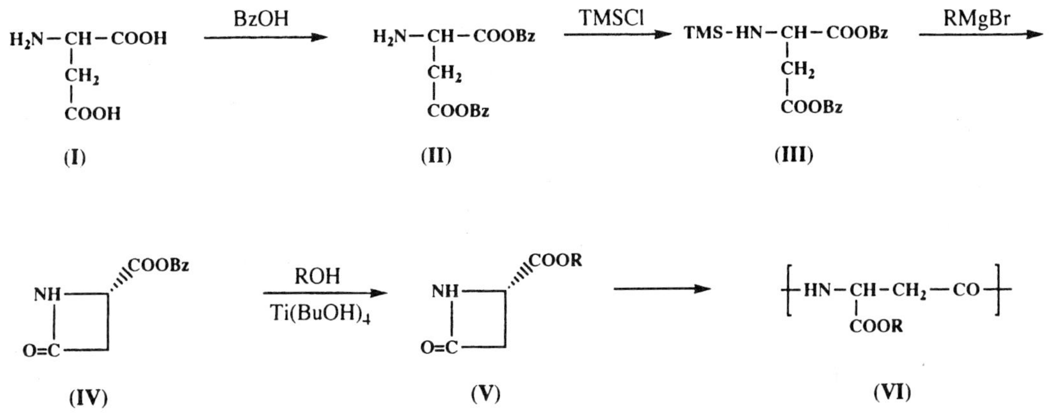

I

trifluoroacetic acid and trifluoroethanol. These polymers readily form films by casting that may be easily oriented in fibers by stretching under moderate heating. The density of these polymers ranges from 1.36 to 1.01 g mL^{-1} as the number of carbons contained in the alkyl side chain increases from 1 to 22.

Polymer Structure

Although all these polymers seem to adopt a helical conformation in the solid state, the level of crystallinity attained in each case largely depends on the nature of the alkyl side chain.[9,17] Members with alkyl side chains containing four or fewer carbon atoms crystallize in well-ordered, three-dimensional structures, usually displaying crystal polymorphism. At the other end, the highest members of the series (n = 18 and 22) adopt biphasic structures with the main chains and the alkyl side chains crystallizing separately in the manner that stereoregular comb-like polymers usually do. Intermediate members ($6 \leq n \leq 12$) form structures composed of rigid helical backbones with attached side chains remaining in a liquid disordered state, although in certain cases (n = 8) crystallization may be induced by thermal treatment.

[Poly(α-alkyl-β-l-aspartate)s] tend to be arranged in helical conformation even in solution when solvents of well recognized helicogenic character, such as chloroform or trifluoroethanol, are used. The transition from helix to random-coil may be promoted by adding small amounts of strong organic acids or by heating. These helices are significantly less stable than the α-helix in polypeptides.[9,18]

Thermal Properties

Nylon 3 melts at a temperature near 340°C with considerable decomposition. When side groups are attached to the main chain, the melting point of this nylon changes to an extent that depends on both the degree of substitution and the precise nature of the substituents. Melting decomposition temperatures ranging from 268°C to 410°C are reported for C-methyl substituted nylons 3.[4,19] [Poly(α-alkyl-β-l-aspartate)s] melt with decomposition within the temperature range of 250–350°C.

Properties Related to the Helical Structure

Properties directly related to the rigid rod nature of the helical chains of these polymers have been observed in the particular case of poly(α-isobutyl-β-l-aspartate), which is the member of the series that has been most extensively studied to date.[20]

Solutions of this polymer of M_w = 76,000 in dichloromethane at concentrations above 20% form highly birefringent liquid crystal phases displaying regular extinction lines characteristic of a cholesteric structure. The cholesteric structure tends to be retained in the solid films that result from slow evaporation of such solutions.[21]

ACKNOWLEDGMENTS

This research has been supported by the Spanish Ministerio de Educatión y Ciencia with grants PR-84-0161, PA-86-0128, PB-90-0597, and PB-93-0960 and by the Spanish chemical firm, ERT.

REFERENCES

1. Kohan, M. I. *Nylon Plastics* John Wiley & Sons: New York, 1973.
2. Muñoz-Guerra, S.; Fernández-Santin, J. M.; Rodríguez-Galán, A.; Subirana, J. A. *J. Polym. Sci.* **1985**, *23*, 733.
3. Bestian, H. *Angew. Chem., Int. Ed. Engl.* **1968**, *7*, 4278.
4. Eisenbach, C. D.; Lenz, R. W.; Duval, M.; Marchessault, R. H. *Makromol. Chem.* **1979**, *180*, 429.
5. Prieto, A.; Iribarren, I.; Muñoz-Guerra. S.; Bui, C.; Sekiguchi, H. *Crystallization of Polymers,* Dosíere, M. Ed.; Kluwer: Dordrecht-Boston-London, 1992.
6. Fernández-Santin, J. M.; Aymami, J.; Rodríguez-Galán, A.; Muñoz-Guerra, S.; Subirana' J. A. *Nature (London)* **1984**, *311*, 53.
7. Muñoz-Guerra, S. *Makromol. Chem., Macromol. Symp.* **1991**, *48/49*, 71.
8. *Trends in Macromolecular Research,* 1994, *1*, 181.
9. Fernández-Santin, J. M.; Muñoz-Guerra. S.; Rodríguez-Galán, A.; Aymamí, J.; Lloveras, J.; Subirana, J. A.; Giralt, E.; Ptak, M. *Macromolecules* **1987**, *20*, 62.
10. Graf, R.; Lohaus, G.; Börner, K.; Schmidt, E.; Bestian, H. *Angew. Chem.* **1962**, *74*, 15523.
11. Reimschuessel, H. K. *Polym. Prep.* **1977**, *18*, 91.
12. Lenz, R.; Guerrin, Ph. *Polym. Sci. Technol.* **1983**, *23*, 219.
13. Rodríguez-Galán, A.; Muñoz-Guerra, S.; Subirana, J. A.; Chuong, B.; Sekiguchi, H. *Makromol. Chem. Macromol. Symp.* **1986**, *6*, 277.
14. Saltzmann, T. N.; Ratcliffe, R. W.; Christensen, B. G.; Bouffard, F. A. *J. Am. Chem. Soc.* **1980**, *102*, 6163.
15. García-Alvarez, M.; López-Carrasquero, F.; Tort, E.; Rodríguez-Galán, A.; Muñoz-Guerra, S. *Synthet. Commun.* **1994**, *24*, 745.

16. López-Carrasquero, F.; Martinez de Ilarduya, A.; Muñoz-Guerra, S. *Polym. J.* **1994a**, *26*, 694.
17. López-Carrasquero, F.; García-Alvarez, M.; Muñoz-Guerra, S. *Polymer* **1994**, *35*, 4502.
18. Macromol Chem. Phys. **1995**, *196*, 253.
19. Schmidt, E. *Angew. Makromol. Chem.* **1970**, *14*, 185.
20. Montserrat, J. M.; Muñoz-Guerra, S.; Prieto, A.; Rodríguez-Galán, A.; Subirana, J. A.; Vives, J. Pat. ES 2 005 185; CA 112 140 105u, 1990.
21. Montserrat, J. M.; Muñoz-Guerra, S.; Subirana, J. *Makromol. Chem., Macromol. Symp.* **1988**, *20/21*, 319.

O

Odorants

See: *Additives (Property and Processing Modifiers)*

Oil Recovery Application Polymers

See: *Water-Soluble Polymers (Oil Recovery Applications)*

Olefin-Carbon Monoxide Copolymers

See: *Olefin-Carbon Monoxide Copolymers (Overview)*
Olefin-Carbon Monoxide Copolymers, Alternating

OLEFIN-CARBON MONOXIDE COPOLYMERS (Overview)

Fabio Garbassi and A. Sommazzi
Enichem S.p.A.
Istituto G. Donegani

Carbon monoxide can be copolymerized with various hydrocarbon monomers to give polyketones. These polymers are known, but only recently have they generated interest both in academic and industrial circles, because of the low cost of carbon monoxide feedstock and the potential use as starting materials for the preparation of other types of functionalized polymers.

It is necessary to distinguish two main families of olefin/CO copolymers, depending on their composition:

- copolymers containing less than 50 mol% of CO (**Scheme I**):

$$\text{-CH}_2\text{-CH}_2\text{-CO-(CH}_2\text{-CH}_2)_n\text{-CO-(CH}_2\text{-CH}_2)_m\text{-CO-CH}_2\text{-CH}_2\text{-} \qquad \textbf{I}$$

- copolymers displaying a perfectly alternating structure along the chain, thus having 50 mol% of CO and olefin, respectively (**Scheme II**).

$$\text{-CH}_2\text{-CH}_2\text{-CO-(CH}_2\text{-CH}_2\text{-CO-)}_n\text{-CH}_2\text{-CH}_2\text{-CO-} \qquad \textbf{II}$$

NONALTERNATING COPOLYMERS

Random nonalternating copolymers with a CO content less than 50 mol% have been obtained mainly via free-radical mechanism of γ-ray irradiation.

Random copolymers with a molecular weight up to 8000 are obtained, depending on reaction conditions. Variables such as pressure, composition of reaction mixture, temperature and solvent greatly influence the composition, yield, molecular weight and other properties of the polyketones formed.

The above processes give rise to polymers that are photodegradable, due to the presence of the photoactive CO moiety. At room temperature, chain cleavage is induced by UV radiation, mainly following the Norrish-II type process, with the formation of a methyl and vinyl end group, respectively.[7–9]

Although many studies have been developed on the free-radical initiated copolymerization of CO and ethylene, it is possible to replace ethylene in part or completely with other olefins or functionalized vinyl compounds. However, lower yields and molecular weight values are generally obtained. Propylene, butadiene, allyl esters, vinyl acetate, vinyl chloride, acrylonitrile and tetrafluoroethylene have been used to form copolymers with CO or terpolymers with CO and C_2H_4.[1,2,4] The copolymerization of CO with cyclic olefins such as cyclohexene, cyclopentadiene, and with vinyl monomers has been also reported.[10–21]

PROPERTIES OF NONALTERNATING COPOLYMERS

Polymers obtained by the copolymerization of CO and olefins range from liquids of low molecular weight to crystalline solids. Their properties are strongly dependent on the CO content, and that can be determined by elemental analysis. IR and NMR spectroscopies and X-ray diffraction have been used for structural characterization.[3,5,6,22–26]

The melting point of the copolymers depends on the CO content.[27] Melting temperature slowly increases up to 35–40% CO, then increases fast, reaching a value of 240 to 260°C at 50 mol%. The heat and entropy of fusion have been found lower than polyethylene values. Mechanical properties, like torsional modulus, exhibit a similar behavior depending on CO content.[6,6a]

ALTERNATING COPOLYMERS

Perfectly alternating copolymers can be obtained using catalysts based on metal transition compounds.

A new class of catalysts was patented by Shell, constituting a technological breakthrough for what concerns molecular weights and productivity.[31,32] The active form of such catalysts are bidentate Pd complexes with P or N ligands.[31–37] They often require the presence of a protonic acid coactivator.[31–34] The use of chelating bidentate ligands, in particular 1,3-bis(diphenylphosphino)-propane, brings an increase in copolymerization rate and catalyst lifetime.[38]

After the discovery of the Shell-type catalyst, the field showed an increasing interest, as demonstrated by the great number of recent papers and patents.[30,36–68] Such interest was caused by the low cost of the monomers and, more likely, by the unusual properties of the materials obtained.

The metal-catalyzed alternating copolymerization of carbon monoxide with α-olefins is much less known than with ethylene. Most studies are reported in the patent literature.[39,40,51–61] Only recently some scientific papers have been published.[39,40,56–69] The alternating copolymerization of CO with styrene or styrene derivates has been studied.[39–61] The insertion of styrene is regiospecific and stereospecific.[42–45,48,50,55] Preliminary investigations indicate that perfectly alternating copolymers can be prepared with IB Group metal catalysts.[70,71] An advantage of these catalyst systems is their low cost with respect to palladium systems.

Heterogeneous catalytic systems have been claimed in some patents. These catalysts are prepared by anchoring palladium complexes or bidentate ligands to an oxide carrier, like silica.[72–74] These heterogeneous catalysts reduce the reactor fouling.

CHARACTERIZATION OF ALTERNATING COPOLYMERS

The perfectly alternating ethylene-carbon monoxide copolymer is a white powder of high crystallinity (35 to 50% as determined by X-ray diffraction). Its alternating structure has been confirmed by elemental analysis. IR and NMR spectra.[22,30,38,41] Average molecular weights near 40,000 are easily obtained with molecular distribution ranging from 3 to 8. The polymer is insoluble in many common organic solvents, probably because of its high crystallinity, but it is soluble in solvents like *m*-cresol, *o*-Cl-phenol, and hexafluoroisopropanol that have high polarity and acidity.

By thermogravimetric analysis, the polymer appears stable up to 250°C. The DSC curve exhibits a T_g transition near room temperature (25 to 35°C) and the T_m peak at 250 to 260°C. Due to the narrow temperature interval between melting and degradation temperature, a third monomer (i.e., propylene) is commonly introduced in the chain structure in an amount of 5–10%. Its presence decreases T_m to 200°C without substantially affecting the degradation temperature.

PROPERTIES OF ALTERNATING COPOLYMERS

Et/CO copolymer has no practical relevance. On the contrary, ethylene/propylene/CO terpolymers are going to be commercialized by Shell, after ten years of intensive research and development work, which has produced more than 200 patents, approximately equally divided between process and applications issues. Shell has announced the decision to design a commercial plant, for a capacity of 10,000 m.t./y, that is expected to be on-stream in the mid-1990s.[75] Some polymer grades named Carlton® have been recently put on the market for experimental production.[76]

Fibers prepared by solution spinning exhibit a modulus up to 27 Gpa and a tensile strength near 2 Gpa.[81–84] Films also have been manufactured.[77–79] They are important in perspective for the interesting permeability to gases exhibited by polyketones.[79,80,85,86] In comparison with other polymers, polyketones are situated at the level of PET, a barrier polymer used in the food packaging industry to manufacture bottles for mineral water and carbonated drinks. The presence of the carbonyl chromofore can make the alternating polyketone more photodegradable than the corresponding polyolefins (even if it is less photodegradable than random low-CO containing copolymers). Another interesting aspect connected to the CO moiety is it can be chemically modified to obtain functional polymers.

Alternating polyketones are new, interesting materials that have the potential of finding their space among existing polymers. Their properties are at the border between commodity polymers (like PE and PVC) and structural, medium performance polymers like polyamides and polyesters. As a newcomer in a crowded field with much interpolymer competition, they have to find their way by the identification of specific niche applications, where cost/performance ratio could constitute a plus with respect to existing materials.

REFERENCES

1. Ballauf, F.; Bayer, O.; Leichmann, L. (to Faberfabriken Bayer), G. Patent 863 711, 1941.
2. Brubaker, M. M. (to Du Pont) U.S. Patent 2 495 286. 1950.
3. Brubaker, M. M.; Coffman, D. D.; Hoehn, H. H. *J. Am. Chem. Soc.* **1952**, *74*, 1509.
4. Barb, W. G. *J. Am. Chem. Soc.* **1953**, *75*, 224.
5. Colombo, P.; Kukacka, I. E.; Fontana, J.; Chapman, R. N.; Steinberg, M. J. *J. Polym. Sci. [A-1]* **1966**, *4*, 29.
6. Colombo, P.; Steinberg, M.; Fontana, J. *J. Polym. Sci. [B]* **1963**, *1*, 447.

6a. Hudgin, D. E. U.S. Patent 3 948 873, 1976.

7. Hartley, G. H.; Guillet, J. E. *Macromolecules* **1986**, *1*, 165.
8. Guillet, J. E.; Dhanraj, J.; Golemba, F. J.; Hartley, G. H. *Adv. Chem. Ser.* **1968**, *85*, 272.
9. Heskins, M.; Guillet, J. E. *Macromolecules* **1970**, *3*, 224.
10. Kagiya, T.; Kondo, M.; Fukyi, K. *J. Polym. Sci. [A-1]* **1969**, *7*, 2793.
11. Weintraub, L.; Hofmann, J.; Manson, J. *Chem. Ind.* **1965**, 1976.
12. Weintraub, L. U.S. Patent 3 790 460, 1974.
13. Otsuka, M.; Yasuhara, Y; Takemoto, K.; Imoto, M. *Makromol. Chem.* **1967**, *103*, 29.
14. Kawai, W.; Ichihashi, T. *J. Polym. Sci. [A-1]* **1972**, *10*, 1709.
15. Kawai, W. *J. Polym. Sci. Pol. Chem.* **1977**, *15*, 1479.
16. Kawai, W.; Ichikashi, T. *J. Polym. Sci. Polym. Chem.* **1974**, *12*, 1041.
17. Kawai, W.; Ichikashi, T. *J. Makromol. Sci. Chem. [A]* **1977**, *11*, 1097.
18. Kawai, W. *Eur. Polym. J.* **1974**, *10*, 805.
19. Ratti, L.; Visani, F.; Ragazzini, M. *Eur Polym. J.* **1973**, *9*, 429.
20. Ratti, L.; Visani, F. U.S. Patent 3 823 116, 1974.
21. Braun, D.; Sonderhof, D. *Eur. Polym. J.* **1982**, *18*, 141.
22. Wu, T. K.; Ovenall, D. W.; Hoen, H. H. *Applications of Polymer Spectroscopy*, 1st Ed., Brame, E. G., Ed., Academic: New York, 1978; p 19.
23. Bruch, H. D.; Payne, W. G. *Macromolecules* **1986**, *19*, 2712.
24. Chatani, Y.; Takizawa, T.; Murahashi, S.; Sakata, Y.; Nishimura. Y *J. Polym. Sci.* **1961**, *55*, 811.
25. Chatani, Y.; Takizawa, T.; Murahashi, S. *J. Polym. Sci.* **1962**, *62*, S 27.
26. Alfonso, G. C.; Fiorina, L.; Martuscelli, E.; Pedemonte, E.; Russo, S. *Polymer* **1973**, *14*, 373.
27. Garbassi, F.; Sommazzi, A. *Polymer News* **1995**, *20*, 201.
28. Starkweather, H. W. *J. Polym. Sci. Polym. Phys.* **1977**, *15*, 247.
29. Starkweather, H. W. *Encyclopedia of Polymer Science and Engineering*, 2nd ed.; John Wiley & Sons: New York, 1987; Vol. X, p 369.
30. Zhao, A. X.; Chien, J. C. W. *J. Polym. Sci. Polym. Chem.* **1992**, *30*, 2735.
31. Drent, E. E. P. Appl. 121 965, 1984.
32. Drent, E. E. P. Appl. 181 014, 1986.
33. van Broekhoven, J. A. M.; Drent, E.; Klei, E. E. P. Appl. 213 671, 1987.
34. van Broekhoven, J. A. M.; Drent, E. E. P. Appl. 235 865, 1987.
35. Sen, A. *Chemtech* **1986**, 48.
36. Sommazzi, A.; Garbassi, F.; Milani, B. *8th Int. Symp. Homogeneous Catalysis*, Amsterdam, 1992.
37. Sommazzi, A.; Garbassi, G.; Mestroni, G.; Milani, B. U.S. Patent 5 310 871, 1994.
38. Drent, E.; van Broekhoven, J. A. M.; Doyle, M. J. *J. Organom. Chem.* 1991, 4/7, 235.
39. Drent, E. P. Appl. 229 408, 1987.
40. Drent, E. U.S. Patent 4 778 279, 1988.

41. Lai, T. W.; Sen, A. *Organometallics* **1984**, *3*, 866
42. Brookhart, M.; Rix, F. C.; De Simone, J. M.; Barborak, J. C. *J. Am. Chem. Soc.* **1992**, *114*, 5894.
43. Brookhart, M.; Wagner, M. I.; Balavoine, G. G. A.; Haddon, H. A. *J. Am. Chem. Soc.* **1994**, *116*, 3641.
44. Barsacchi, M.; Consiglio, G.; Medici, L.; Petrucci, G.; Suter, U. W. *Angew Chem. Int. Ed. Engl.* **1991**, *30*, 989.
45. Corradini, P.; De Rosa, C.; Panunzi, A.; Petrucci, G.; Pino, P. *Chimia* **1990**, *44*, 52.
46. Batistini, A.; Consiglio, G. *Organometallics* **1992**, *11*, 1766.
47. Pisano, C.; Nefkens, S. C. A.; Consiglio, G. *Organometallics* **1992**, *11*, 1975.
48. Barsacchi, M.; Batistini, A.; Consiglio, G.; Suter, W. E. *Macromolecules* **1992**, *25*, 3604.
49. Sen, A. *Acc. Chem. Res.* **1993**, *26*, 303.
50. Busico, V.; Corradini, P.; Landriani, L.; Trifuoggi, M. *Makroml. Chem., Rapid Commun.* **1993**, *14*, 261.
51. Pisano, C.; Mezzetti, A.; Consiglio, G. *Organometallics* **1992**, *11*, 20.
52. Pisano, C.; Consiglio, G.; Sironi, A.; Moret, M. *J. Chem. Soc. Chem. Commun.* **1991**, 421.
53. Milani, B.; Alessio, E.; Mestroni, G.; Sommazzi, A.; Garbassi, F.; Zangrando, E.; Bresciani-Pahor, N.; Randaccio, L. *J. Chem. Soc. Dalton Trans.* **1994**, *13*, 1903.
54. Jiang, Z.; Adams, S. E.; Sen, A. *Macromolecules* **1994**, *27*, 2694.
55. Jiang, Z.; Adams, S. E.; Sen, A. *Polym. Preprints* **1994**, 676.
56. Drent, E.; Wife, R. L. U.S. Patent 4 970 294, 1990.
57. van Leewen, P. W. N.; Roobeck, C. F.; Wong, P. K. E. P. Appl. 393 790, 1990.
58. Wong, P. K. E., P. Appl. 384 517, 1990.
59. van Deursen, J. H.; van Doorn, J. A.; Drent, E. E. P. Appl. 390 237, 1990.
60. Drent, E. E. P. Appl. 390 292, 1990.
61. van Doorn, J. A.; Wong, P. K.; Sudmeier, O. E. P. Appl. 376 364, 1989.
62. Batistini, A.; Consiglio, G.; Suter, U. W. *Angew Chem. Int. Ed. Engl.* **1992**, *31*, 303.
63. Batistini, A.; Consiglio, G.; Suter, U. W. *Angew. Chem. Int. Ed. Engl.* **1992**, *104*, 306.
64. Chien, J. C. W.; Zhao, A. X.; Xu, F. *Polym. Bull.* **1992**, *28*, 315.
65. Wong, P. K.; van Coorn, J. A.; Drent, E.; Sudmeyer, O.; Still, H. A. *Ind. Eng. Chem. Res.* **1993**, *32*, 986.
66. Jiang, Z.; Dahlen, G. M.; Houseknecht, K.; Sen, A. *Macromolecules* **1992**, *25*, 2999.
67. Jiang. Z.; Dahlen, G. M.; Houseknecht, K.; Sen, A *Polym. Prepr. Am. Chem. Soc., Div. Polym. Chem.* **1992**, *33*(1), 1233.
68. Xu, F. Y.; Zhao, A. X.; Chien, J. C. W. *Makroml. Chem.* **1993**, *194*, 2579.
69. Sen, A.; Brumbaugh, J. S.; Lin, M. *J. Mol. Catal.* **1992**, *73*, 297,
70. Sommazzi, A.; Lugli, G.; Garbassi, F.; Calderazzo, F. E. P. Appl. 560 455, 1993.
71. Sommazzi, A.; Lugli, G.; Garbassi, F.; Calderazzo, F. E. P. Appl. 560 456, 1993.
72. Sommazzi, A.; Lugli, G.; Garbassi, F.; Calderazzo, F.; Belli Dell'Amico, D. E. P. Appl. 559 288, 1993.
73. Sommazzi, A.; Lugli, G.; Garbassi, F.; Calderazzo, F.; Belli Dell'Amico, D. E. P. Appl. 559 289, 1993.
74. Wong, P. K. E. P. Appl. 404 228, 1990.
75. *Chemical Week*, **1994**, June 22, 15.
76. Shell, *Carilon® thermoplastic polymers*, Information Sheet, 1994.
77. Gerlowski, L. E. et al., U.S. Patent 4 892 697, 1990.
78. Klingensmith, G. B. et al., U.S. Patent 5 073 327, 1991.
79. Gerlowski, L. E. et al., U.S. Patent 5 077 385, 1991.
80. Gerlowski, L. E. et al., E. P. Appl. 445 865, 1991.
81. Beyen, J. M. et al., E. P. Appl. 310 171, 1989.
82. van Breen, A. W. et al., E. P. Appl. 360 358, 1990.
83. Lommerts, B. J. et al., E. P. Appl. 456 306, 1991.
84. Rutten, H. J. J. *PCT WO* 90/14453, 1990.
85. Meusitieri, G.; Del Nobile, M. A.; Sommazzi, A.; Nicolais, L. *J. Polym. Sci.* Part B **1995**, *33*, 1365.
86. Del Nobile, M. A.; Mensitieri, G.; Nicolais, L.; Sommazzi, A.; Garbassi, F. *J. Appl. Polym. Sci.* **1993**, *50*, 1261.

OLEFIN–CARBON MONOXIDE COPOLYMERS, ALTERNATING

Gennadii P. Belov
Institute of Chemical Physics
Russian Academy of Sciences

The copolymers of carbon monoxide with olefins are of considerable interest from the following standpoints:

- Carbon monoxide is a comparatively cheap starting material of virtually unlimited availability.
- Reactive carbonyl groups in the backbone should confer greater photo- and biodegradability to the copolymers when compared to corresponding polyolefins.
- The copolymer adheres well to the surface of inorganic materials and metals; it has good fiber and film-forming properties.
- The copolymer can be used as an excellent starting material for preparing a wide range of other functional polymers because the presence of a carbonyl group in its polymeric chain facilitates chemical modifications of this copolymer.
- Olefin savings may reach 50%.

METHODS OF PRODUCING COPOLYMERS

Three synthetic methods have been developed for copolymerization of carbon monoxide with olefins. Copolymerizations may be initiated by free radicals or induced by γ-rays, obtaining random copolymers. In addition, a number of transitional metal compounds are effective catalysts for alternating copolymerization of carbon monoxide with olefins.

STRUCTURE AND PROPERTIES

The crystal structures of CO-C_2H_4 copolymers with a different C_2H_4/CO ratio have been determined.[10–12] In 1961, Chatani et al.[10] reported on the crystal structure of an alternating copolymer of carbon monoxide and ethylene (POK) from oriented film samples. To avoid confusion with poly(aryl ketone)s like PEEK and PEK, Lommerts abbreviated this polymer as polyolefin ketone–POK. The polymer was prepared by γ-radiation-initiated copolymerization with a melting point of 175–185°C. Lommerts et al.[12] designated this structure as a polyolefin(C_2)–ketone-β (POK-β).

More recently, a catalytic polymerization method was developed by Drent et al.[5,6] and Belov et al.,[7–9] resulting in a perfectly alternating copolymer with a melting point 250–260°C, which is designated polyolefin(C_2)–ketone-α (POK-α).

The calculated crystalline density of the POK-β structure is 1.297 g/cm³, whereas the density of the POK-α structure is 1.383 g/cm³.

DEGRADATION

Guillet and Gooden carried out extensive studies on the photochemistry of CO-C_2H_4 copolymers.[14–18]

The degradation mechanism for these materials is similar to that of ketone carbonyl copolymers except that as the carbonyl group is now directly located in the polymer backbone. Norrish types I and II reactions give chain scission directly. The pathway therefore is similar, involving rapid embrittlement and fragmentation followed by eventual slow assimilation by microorganisms.

An interesting feature of this perfectly alternating copolymer is that Norrish-II reactions are inhibited because of absence of CH_2-groups at the γ-position of the carbonyl groups, resulting in an improved UV stability. It has also been shown by Gooden et al.[18] that for extended (crystalline) material, the contribution of the Norrish-II type of chain scission to the total photodegradation is greatly decreased.

DERIVATIZATION REACTIONS

Because of the ease with which the carbonyl group can be chemically modified, the copolymers of carbon monoxide and olefins should be excellent starting materials for the synthesis of other classes of functionalized polymers.[19–28]

APPLICATIONS

POKs with a low carbon monoxide content are used as photodegradable plastics, for example as a six-pack connector.[29,30] Manufacturers of the material claim it has good clarity and processing properties, good storage stability, and degrades to safe by-products.

As with ketone carbonyl photodegradable material, the photosensitive group is chemically bonded to the polymer, so there is no chance of it leaching out into food if used for packing. Rate of degradation can be controlled by adjusting the percentage of carbon monoxide in the copolymer. Some C_2H_4-CO products are reported to be brittle enough to disintegrate upon handling after only a week of UV exposure.

Lommerts[13] demonstrated the potential of perfectly alternating POK for industrial and technical fiber applications. The excellent high-temperature strength (up to 3.8 Gpa), creep resistance, and dimensional stability of these materials are of paramount importance for various advanced industrial fiber end uses.

REFERENCES

1. Iwashita, Y.; Sakurada, M. *Tetrahedron Lett.* **1971**, *26*, 2409.
2. Sen, A.; Brumbaugh, J. S. *J. Organometal. Chem.* **1985**, *279*, C5.
3. Sen, A.; Brumbaugh, J. S.; Lin, M. *J. Molec. Catal.* **1992**, *73*, 297.
4. Consiglio. G.; Studer, B. et al. *J. Molec. Catal.* **1990**, *58*, L9. S
5. Drent, E. Europ. Patent 121 965, 1984.
6. Drent, E.; von Broekhoven, J. A. M.; Doyle, M. J. *J. Organometal. Chem.* **1991**, *417*, 235.
7. Chepaikin, E. G.; Bezruchenko, A. P.; Belov, G. P. Rus. Patent 1 636 417,1987.
8. Chepaikin, E. G.; Bezruchenko, A. P., et al. *Vysokomolek. Soed. (Rus.)* ser. B. **1990**, *32*, 593.
9. Belov, G. P.; Chepaikin, E. G., et al. *Polym. Sci.* **1993**, *35*, 1323.
10. Chatani, Y.; Takizawa, T., et al. *J. Polym. Sci.* **1961**, *55*, 811.
11. Chatani, Y.; Takizawa, T.; Murahashi. S. *J. Polym. Sci.* **1962**, *62*, S27.
12. Lommerts, B. J.; Klop, E. A.; Aerts, I. *J. Polym. Sci., Polym. Phys. Ed.* **1991**, *31*, 1319.
13. Lommerts, B. I. "Structure development in polyketone and polyalcohol fibers", Dissertation, Rijksuniversiteit Groningen, 1994.
14. Hartley, G. H.; Guillet, J. E. *Macromolecules* **1968**, *1*, 165.
15. Guillet, J. E.; Dhanraj, J., et al. *Adv. Chem. Ser.* **1968**, *85*, 272.
16. Heskins, M.; Guillet, J. E. *Macromolecules* **1970**, *3*, 224.
17. Gooden, R.; Hellman, M. Y., et al. *Macromolecules* **1984**, *17*, 2830.
18. Gooden, R.; Davis, D. D., et al. *Macromolecules* **1988**, *21*, 1212.
19. Brubaker, M. M.; Coffman, D. D.; Hoehn, H. H. *J. Am. Chem. Soc.* **1952**, *74*, 1509.
20. Hoehn, H. H., U.S. Patent 2 495 255, 1950.
21. Coffman, D. D.; Hoehn, H. H.; Maynard, J. T. *J. Am. Chem. Soc.* **1954**, *76*, 6394.
22. Kleiner, H.; Wilms, H. U.S. Patent 2 846 406, 1958.
23. Johnson, J. R. U.S. Patent 2 527 821, 1950.
24. Scott, S. L. U.S. Patent 2 495 293, 1950.
25. Mortenson, C. W. U.S. Patent 2 396 963, 1946.
26. Upson, R. W. U.S. Patent 2 599 501, 1952.
27. Michel, R. H.; Murphy, W. A. *J. Polym. Sci.* **1955**, *55*, 741.
28. Schreiber, R. R. U.S. Patent 2 542 782, 1951.
29. Bremer, W. P. *Polym. Plast. Technol. Rng.* **1982**, *18*, 137.
30. Leaversuch, R. *Mod. Plast. Int.* **1987**, *17*, n. 10, 94.

Olefin Polymerization

See: *Captodative Compounds, Polymerization*
Captodative Olefins
Living Carbocationic Polymerization, Olefins
Metallocene Catalysts (Cationic Group 4 Metal Alkyl Complexes, Olefin Polymerization)
Olefin Polymerization by 2,ω-Linkage (Migratory Nickel-Phosphorane Catalysts)
Olefin Polymerization Catalysts (Experimental and Theoretical Characterization)
Olefin Polymerization Catalysts (Silica-Supported Titanium Tetrachloride)
Olefin Polymerization Catalysts, Soluble (Group 4, Cationic Active Centers)
Rare Earth Polymerization Initiators
Supported Catalysts (Lewis Acid and Ziegler-Natta)
Supported Ziegler-Natta Catalysts (on Silica, for Olefin Polymerization)
Supported Ziegler-Natta Catalysts [Titanium (IV) Chloride on Reduced n-type Oxides]

OLEFIN POLYMERIZATION BY 2,ω-LINKAGE (Migratory Nickel-Phosphorane Catalyst)

Gerhard Fink,* V. M. Möhring, A. Heinrichs, C. Denger, R. H. Schubbe, and P. H. Mühlenbrock
Max-Planck-Institut für Kohlenforschung

Polymerization of α-olefins with various catalysts (e.g., with Ziegler catalysts) leads to 1,2-linked, comb-like polymers. Using nickel-catalysts, built from zero-valent nickel-compounds like $Ni(cod)_2$ and bis(trimethylsilyl)aminobis(trimethylsilyl)iminophosphorane in equimolar ratio, a new type of polymer was obtained.[1,2] In contrast to the 1,2-linked polymers, this polymer contains only methyl groups in well-defined distances in polymer chain, depending on the length of the α-olefin (**Figure 1**).

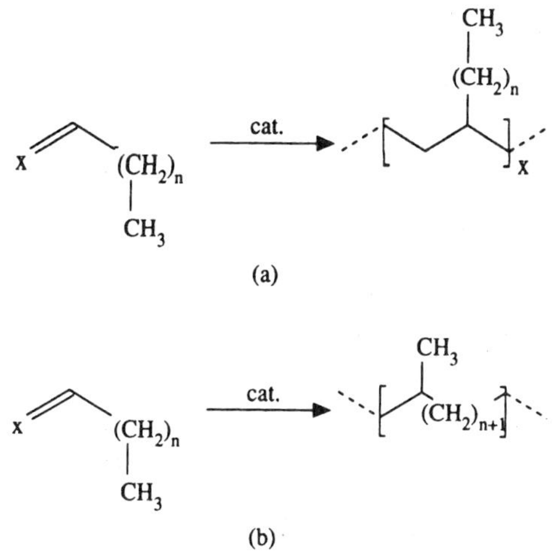

FIGURE 1. Polymerization of α-olefins under 1,2- and 2,ω-linkage.

The insertion of the α-olefin occurs regioselectively, only 2,ω-coupling is observed. We decided to call this α-olefin polymerization 2,ω-polymerization.[1-3]

Figures 2 and **3** show more details of the resulting polymers, include the polymerization of 1-pentene or 4-methyl-1-pentene.

The polymers contain only methyl branches, regularly spaced along the chain with a separation corresponding to the chain length of the monomer. Thus, in the polymer of a linear α-olefin with n ($-CH_2-$) groups, the distance between two methyl branches is (n+1) ($-CH_2-$) groups. According to this mechanism we can predict the structure or the resulting polymer.

Taking into account all these results, a scheme was developed which explains the origin of the special structure of the poly(α-olefin)s. Its main points are

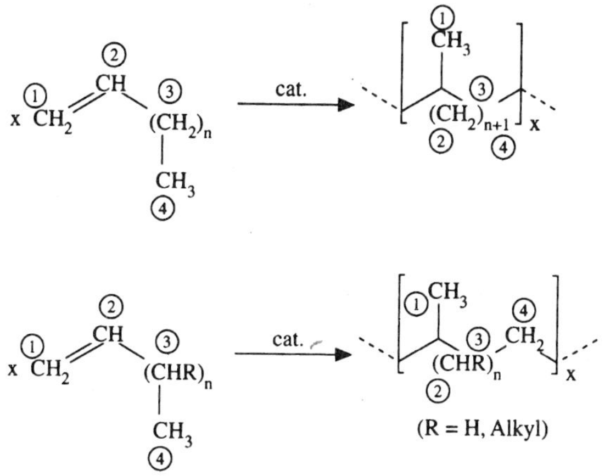

Example : Poly(1 - pentene)
Poly[2,5 - (1 - pentene)]

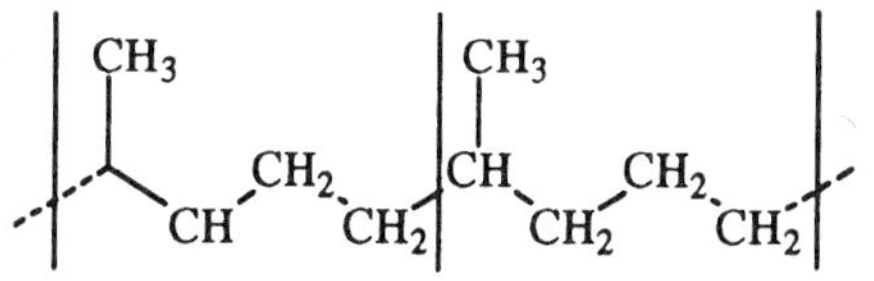

Strongly alternating ethylene/propylene copolymer

Example : Poly(4 - methyl - 1 - pentene)
Poly[2,5 - (4 - methyl - 1 - pentene)]

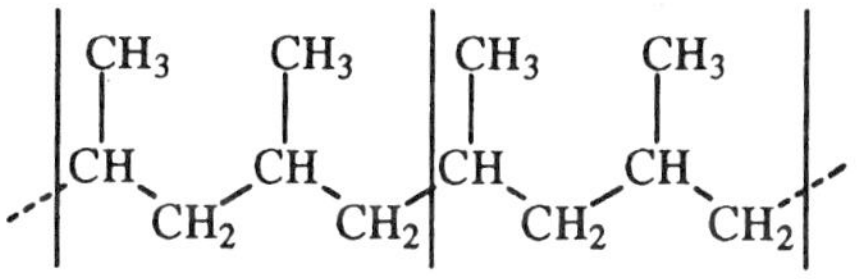

FIGURES 2 and 3. Examples for the 2-ω-linkage.

- The monomer can only insert into a primary nickel-alkyl bond at the end of the growing polymer chain.
- The insertion is regioselective, only C_ω-C_2-coupling of the growing chain with the next monomer takes place.
- The nickel catalyst "migrates" along the polymer chain between two insertions. During this migration transfer reactions can occur, but there are no insertions.

REFERENCES

1. Möhring, V. M.; Fink, G. *Angew. Chem.* **1985**, *97*, 982.
2. Möhring, V. M. Dissertation, H.-H.-University Düsseldorf, 1985.
3. Fink, G. In *"Recent Advances in Mechanistic and Synthetic Aspects of Polymerization,"* Fontanille, M.; Guyot, A. Eds., NATO ASI, Series C: Math. Phys. Sci., 215; Reidel: Dordrecht, 515, 1987.

*Author to whom correspondence should be addressed.

OLEFIN POLYMERIZATION CATALYSTS (Experimental and Theoretical Characterization)

Daniel E. Damiani,* Alfredo Juan, and Maria L. Ferreira
Planta Piloto de Ingenieria Quimíca
Universidad Nacional del Sur-CONICET

Since the excellent properties of $MgCl_2$-supported Ziegler-Natta catalysts for polypropylene production were discovered, a large number of papers have been published. Due to the complex nature of these catalysts, some important aspects, including type and structure of active sites, the role that they support, the electron donor and the monomer play in the active site formation, and the mechanisms of deactivation are still under discussion.[1–10]

The combination of a solid (i.e., $MgCl_2$) acting as a support, an electron donor (internal donor) and $TiCl_4$ constitutes a precatalyst for pro-chiral olefin polymerization. The catalyst is completed by the addition of an aluminum alkyl and another electron donor (external donor). The role of these donors is more or less complex. However, the internal donor must impose some restriction to the coordination capacity of titanium. This should result in a template effect that induces the insertion of the incoming monomer to the growing chain in such a way that only one form of polymer out of all possible forms is mainly obtained. Although this is not the only role of the internal donor, it is required to confer selectivity to the catalyst.

We selected a $MgCl_2$/internal donor/$TiCl_4$ system which was characterized by means of conventional experimental techniques during the preparation procedure. The experimental information allowed us to speculate on the structure of the surface compounds formed. By means of a semiempirical molecular orbital calculation we investigated theoretical models that represent the proposed structures to determine the possibility of their occurrence. The combination of experimental and theoretical results could better define the convenience of further experimental work.

A literature survey led us to conclude that besides aromatic esters, silyl-ethers and sterically hindered secondary amines were among the most used catalysts modifiers.[11–15] In most cases these compounds were used as external rather than internal electron donors, and although some experimental results are available, data on the characterization of a $MgCl_2$/secondary amine/$TiCl_4$ system is scarce.[11,12,14] Consequently we decided to investigate the precatalyst formed by $MgCl_2$/2,2,6,6-tetramethylpiperidine (TMPIP)/$TiCl_4$. Therefore, we will report theoretical results related to the role of TMPIP as internal donor, compare them to experimental results and, based on the combination of both, predict catalytic behavior.

DISCUSSION

The combination of experimental and theoretical characterization predicts that TMPIP is not an adequate internal donor, because it is placed in an undesirable way and will produce less stereospecific sites on both planes and one predominant form of site that is undesirable. With these results in mind, we could say that a precatalyst where the internal donor is TMPIP would produce a nonstereospecific catalyst when it is activated (with $AlEt_3$, for example), and probably with low productivity. Experimental results confirm this proposition.

CONCLUSIONS

Based on the combination of experimental and theoretical tools, we characterized a Ziegler-Natta precatalyst. According to the results it could not be an attractive material since sites precursors formed on it could lead not only to stereospecific sites but also to nonselective and unstable reaction sites that may conduce to a less active, less selective catalyst than the well studied one using ethylbenzoate as internal donor. Besides the physical characterization indicates a lower surface area which may result in an inefficient use of the supported Ti and in a poor particle shape control. This attempt proved to be useful in helping to guide the experimental work.

ACKNOWLEDGMENTS

The authors thank the Universidad Nacional del Sur and The Consejo Nacional de Investigaciones Cientificas y Ténicas for their financial support.

REFERENCES

1. Dumas, C.; Hsu, C. C. *Rev. Macromol. Chem. Phys.* **1984**, *C24*(3), 355.
2. Chien, J. C. W.; Wu, J. C.; Kuo, C. *J. Polym. Sci. Polym. Chem. Ed.* **1982**, *20*, 2019.
3. Chien, J. C. W.; Wu, J. C. *J. Polym. Sci. Polym. Chem. Ed.* **1982**, *20*, 2445.
4. Chien, J. C. W.; Wu, J. C.; Kuo, C. *J. Polym. Sci. Polym. Chem. Ed.* **1983**, *21*, 725.
5. Chien, J. C. W.; Wu, J. C.; Kuo, C. *J. Polym. Sci. Polym. Chem. Ed.* **1983**, *21*, 737.
6. Chien, J. C. W.; Kuo. C. *J. Polym. Sci. Polym. Chem. Ed.* **1985**, *23*, 731.
7. Chien. J. C. W.; Hu, Y. *J. Polym. Sci. Polym. Chem. Ed.* **1988**, *26*, 2003.
8. Keszler, B.; Bodor, G.; Simon, A. *Polymer* **1982**, *23*, 916.
9. Kashiwa, N.; Kawasaki, M.; Yoshitake, J. *Studies in Surface Science and Catalysis;* Elsevier: Amsterdam, 1986; Vol. 56.
10. Chien. J. C. W.; Weber, S.; Hu, Y. *J. Polym. Sci., Part A* **1989**, *27*, 1489.
11. Langer, A.; Burkhardt, T.; Steger, J. *Polym. Sci. Technol.* **1983**, *19*, 225.
12. Dumas, C.; Hsu. C. *J. Appl. Polym. Sci.* **1989**, *37*, 1605; 1625.
13. Soga, K.; Shiono, T.; Doi, Y. *Makromol. Chem.* **1988**, *189*, 31.
14. Harkonen, M.; Seppala, J. V. *Makromol. Chem.* **1991**, *192*, 721.
15. Proto, A.; Oliva, L.; Pellechia, C.; Sivak, A.; Cullo, L. *Macromolecules* **1990**, *23*, 2904.

*Author to whom correspondence should be addressed.

OLEFIN POLYMERIZATION CATALYSTS (Silica-Supported Titanium Tetrachloride)

Dong-ho Lee
Department of Polymer Science
Engineering College
Kyungpook National University

Amorphous silica is employed as a support for the chromium-based Phillips catalyst used worldwide for the preparation of high density polyethylene (HDPE) in both gas phase and slurry processes.[1] However, the Phillips catalyst is not suitable for the preparation of all HDPE and linear low density polyethylene (LLDPE) products because the chromium oxide catalyst is capable of incorporating less comonomer than the titanium-based Ziegler-Natta catalyst.[2]

Ziegler-Natta catalysts are used for the preparation of PE with a relatively narrow molecular weight distribution (i.e., M_w/M_n values of 3–7)[3] while the Phillips catalyst is used for PE markets that required a relatively broad molecular weight distribution (i.e., M_w/M_n values of 8–30).[4]

The supported catalyst is a milestone in the development of the olefin polymerization industry. During the past two decades, many efforts have been devoted to the enhancement of the catalyst efficiency of Ziegler-Natta catalysts for the simpler and more economical processes.[5–11] In preparation of the supported Ziegler-Natta catalysts for olefin polymerization, various supports of silica, alumina, and magnesium compounds such as $MgCl_2$,[6–9] and $Mg(OR)_2$[12,13] had been used to increase the catalyst activity.

Recently, soluble metallocene catalysts had been also supported on silica to initiate the olefin polymerization with not only methylalumoxane but ordinary alkylaluminum such as trimethylaluminum and triethylaluminum, and to increase the molecular weight of polymer.[14–16] In addition to the above inorganic supports, it has been announced that an organic material, cyclodextrin, also can be used as a support.[17]

Since the heterogeneous supported catalysts are able to replicate their morphology into the morphology of the produced polymer particles, the polymer can be made to be spherical with a controlled diameter, particle size distribution, and compactness by using the spherical silica as a support of transition metal catalyst as well as metallocene compounds.[18] Therefore it is still desirable to support Ziegler-Natta catalysts on silica for a gas phase or slurry process of olefin polymerization although it is not necessary.

This article includes the preparation, properties and application of silica-supported $TiCl_4$ catalysts for the olefin polymerization.

PREPARATION AND PROPERTIES

Silica-Supported Catalysts

For the control of polymer morphology, $TiCl_4$ can be supported on spherical silica by treating the calcinated silica with $TiCl_4$ directly.

The obtained SiO_2/ $TiCl_4$ catalyst can be used for the polymerizations of ethylene and propylene. For ethylene polymerization, the catalyst activity is low but the crystallinity of the resulting PE is high.[19] In addition, the catalyst shows low activity and yields PP with low isospecificity for propylene polymerization.[20]

During polymerization, owing to the alkylation and reduction of titanium compound by alkylaluminum cocatalyst, nearly all supported -O-Ti anchor bonds are broken and highly dispersed $TiCl_3$ crystals are formed.[21]

Modified Silica-Supported Catalysts

Since the simple silica-supported Ti catalysts show a low catalyst activity for ethylene as well as propylene polymerization and a large production of atactic PP (low isospecificity) as described previously, many preparation methods have been suggested for the improvement of catalyst performances.

For the improvement of catalyst activity, the silica was reacted with a magnesium compound such as $MgCl_2$ or alkylmagnesium (MgR_2) before supporting $TiCl_4$ to obtain the so-called bimetallic supported catalysts used in ethylene polymerization. These kinds of $SiO_2/MgCl_2/TiCl_4$ bisupported catalysts can be prepared by different methods.[22–25]

In $MgCl_2$ supported catalyst for propylene polymerization, the internal donor (ID) such as monoester or diester is necessary to be added in catalyst for the improvement of isospecificity of polypropylene (PP).

APPLICATIONS

As mentioned above, the various silica-supported $TiCl_4$ catalysts can be used for the manufacture of HDPE, LLDPE, iPP, and other α-olefin homopolymers and copolymers.

REFERENCES

1. Daniel, M. P. *Ind. Eng. Chem. Res.* **1988**, *27*, 1559.
2. Finogenova, L. T.; Zakharov, V. A.; Buniyat-Zade, A. A.; Bukatov, G. D.; Plaksunov, T. K. *Polym. Sci. USSR* **1980**, *22*, 448.
3. Hoff, R. E.; Pullukat, T. J.; Dombro, R. A. In *Advances in Polyolefins;* Seymour, R. B.; Cheng, T., Eds.; Plenum: New York, 1987; 241
4. Nowlin, T. E. *Prog. Polym. Sci.* **1985**, *11*, 29.
5. Quirk, R. P.; Hsieh, H. L.; Tait, P. J. T., Eds., *"Transition Metal Catalyzed Polymerizations; Alkenes and Dienes"* Part A and B; Harwood Academic: New York, 1983.
6. Kissin, Y. V. *Isospecific Polymerization of Olefins with Heterogeneous Ziegler-Natta Catalysts;* Springer-Verlag: New York, 1985.
7. Quirk, R. P.; Hoff, R. E.; Klingensmith, G. B.; Tait, P. J. T.; Goodall, B. L. *Transition Metal Catalyzed Polymerizations, Ziegler-Natta and Methathesis Polymerizations;* Cambridge University: New York, 1988.
8. Kaminsky, W.; Sinn, H. Eds., *Transition Metala and Organometallics as Catalysts for Olefin Polymerization;* Springer: Berlin, 1988.
9. Keii, T.; Soga, K., Eds., *Catalytic Olefin Polymerization;* Kodansha: Tokyo, 1990.
10. Guyot, A.; Spitz, R. Eds., *Makromol. Chem., Macromol. Symp.* **1993**, 66.
11. Soga, K.; Terano, M. Eds., *Catalyst Design for Tailor-made Polyolefins;* Kodansha: Tokyo, 1994.
12. Lee, D. H.; Jeong, Y. T.; Soga, K.; Shiono, T. *J. App. Polym. Sci.* **1993**, *47*, 1449.
13. Lee, D. H.; Jeong, Y. T. *Eur. Polym. J.* **1993**, *29*, 883.
14. Chein, J. C. W.; He, D. *J. Polym. Sci., Polym. Chem. Ed.* **1991**, *29*, 1603.

15. Soga, K.; Kaminaka, M. *Makromol. Chem. Rapid Commun.* **1992**, *13*, 221.
16. Kaminsky, W.; Renner, F. *Makromol. Chem. Rapid Commun.* **1993**, *14*, 239.
17. Lee, D. H.; Yoon, K. B. *Makromol. Chem. Rapid Commun.* **1994a**, 15, 841.
18. Boor, Jr., J. *Ziegler-Natta Catalysts and Polymerization;* Academic: New York, 1979; p 7.
19. Munoz-Escalona, A. In *Transition Metal Catalyzed Polymerizations; Alkenes and Dienes*; Quirk, R. P.; Hsieh, H. L.; Klingensmith, G. B.; Tait, P. J. T., Eds.; Harwood Academic: New York, 1983; 323.
20. Yermakov, Y. I.; Kuznetsov, B. N.; Zakharov, V. A. In *Catalysis by Supported Complexes;* Elsevier: Amsterdam, 1981; Chapter 5.
21. Ris, T.; Dahl, I. M.; Ellestad, O. H. *J. Mol. Chem.* **1983**, *18*, 203.
22. E Patent 55 605. 1981, Union Carbide.
23. Soga, K. In *History of Polyolefins;* Seymour, R. B.; Cheng, T., Eds.; D. Reidel: New York, 1986; p 243.
24. Spitz, R.; Pasquent, V.; Guyot, A. In *Transition Metala and Organometallics as Catalysts for Olefin Polymerization;* Kaminsky, W.; Sinn, H., Eds.; Springer: Berlin, 1988; p 406.
25. Pasquent, V.; Spitz, R. *Makromol. Chem.* **1990**, *191*, 3087.

OLEFIN POLYMERIZATION CATALYSTS, SOLUBLE (Group 4, Cationic Active Centers)

Jun Tian and Baotong Huang*
Changchun Institute of Applied Chemistry
Chinese Academy of Sciences

Group 4 metallocene catalyst system Cp_2TiX_2/R_nAlCl_{3-n}** as a soluble catalyst for ethylene (but not for propylene) polymerization, which had incidentally served as a much simpler and more efficient model system for mechanistic studies of Ziegler-Natta catalysis, gained its thrust only after its union with methyaluminoxane (MAO) as discovered by Kaminsky.[1-3]

This renaissance in Ziegler-Natta chemistry makes possible not only the ingenious control of polyolefin microstructures by choice of appropriate ligands in homogeneous transition metal catalyst precursors but also considerable advances in direct understanding of the nature of the catalytic center and the mechanism of olefin polymerization hitherto unavailable with conventional heterogeneous Ziegler-Natta catalysts. The next significant move was the discovery by Jordan of one-component metallocenes, riding off the aluminum activator.[4,5] There are now even nonmetallocene one-cyclopentadienyl (mono-Cp) and Cp-free soluble catalysts for olefin polymerization.

Various lines of evidence from spectroscopic studies of reactions and investigation of model complexes have provided a cationic working rationale for the active components of the potent and highly stereoselective homogeneous and even heterogeneous olefin polymerization catalysts. The key steric and electronic properties of 14-electron group 4 metallocene $Cp_2M(R)^+$ catalyst species are: the d° metal electron configuration, the highly unsaturated metal center, and the availability of vacant coordination sites *cis* to alkyl ligands. The active species of group 4 mono-cp and Cp-free olefin polymerization catalysts are now also proven cationic analogous to these involved in metallocene-based catalytic systems.

*Author to whom correspondence should be addressed.
**In this article Cp: cyclopentadienyl.

ALUMINUM-BASED METALLOCENES

Recognition of the Cation-Like Character of Metallocene Catalysts

Participation of a highly electrophilic, cationic, metallocene alkyl complex Cp_2TiEt^+ as the active species in ethylene polymerization in methylene chloride with homogeneous metallocene system Cp_2TiCl_2/Et_2AlCl was first suggested from ingenious electrodialysis studies.[6]

Recent ^1H and ^{13}C NMR studies on the adducts formed between $Cp_2Ti(Me)Cl$ or $Cp_2Ti(Me)_2$ and MAO or Me_3Al at Al/Ti ratio of 1–40 show that MAO is a better alkylating agent, having a greater capacity for providing and stabilizing the "cation-like" complexes than Me_3Al, giving $Cp_2TiMe^+.^-MAO$ complex.[7]

Supported Catalysts

It was soon found that low catalytic activity of ZrR_4 (R=alkyl, allyl, benzyl), in the absence of aluminum cocatalysts, was greatly increased on silica. Support of a soluble group 4 organometallic catalyst on silica or alumina has been commercial practice in many instances.

AL-FREE, ONE-COMPONENT METALLOCENE CATALYSTS

Base-Coordinated

Recognition of the "cation-like" nature of soluble metallocene catalysts saw the gradual but rapid evolution in synthesizing aluminum-free complexes for olefin polymerization. Complexes of this type are generally available from neutral Cp_2MR_2 precursors via protonlysis (by HNR^+), one-electron oxidation (by Cp_2Fe^+, Ag^+, etc.) or R-abstraction (by $B(C_6F_5)_3$, Ph_3C^+, etc.). The first Al-free, one-component discrete metallocene catalyst for olefin polymerization, $[Cp_2ZrMeTHF)^+][^-BPh_4]$, as reported by Jordan, was prepared by reacting equimolar Cp_2Zre_2 and Ag $[Bph_4]$ in acetonitrile followed by crystallizing the product in THF.[4,5]

Base-Free

The ensuing decrease in polymerization activity and requirement for solvents of higher dielectric constant, owing to the relativity strong coordination of the existing Lewis base such as THF and thus the difficulty in its displacement by the incoming olefin molecules, was soon circumvented by synthesizing base-free cations on modifying the ligands or the counterion.

MECHANISTIC CONSIDERATIONS

Model Analysis

Model of interaction between the growing chain-end and the cationic active center and the fixed ligands determines propagation and termination of the polymer chain through probable β-H or β-CH_3 elimination, and thus the microstructure, MW and the chain-end structure of the polymer product. Among them, agostic metal-hydrogen (M···H-C) interactions are potentially important structural features of unsaturated electrophilic d° alkyl organometallics that are key intermediates in olefin polymerization which are believed to play an important role, both in the stabilization of the catalytic species and in the stereospecificity of the alkene insertion reaction.

Base-coordinated "cationic-like" metallocene $[Cp_2MR(L)^+][^-BPh_4]$ (L-labile ligands), though low or even not at all, in olefin polymerization activity, because of their ability to form quantitative olefin insertion products of relatively high stability, properly serve as model complexes for fine structure analysis and thus make possible insight into the physical picture of fundamental processes in olefin polymerization, as described in the following work.

Studies on Theoretical Calculations

Much sophisticated theoretical calculations not only support the rationality of the cationic nature of the catalytic centers in olefin polymerization but also afford precise further understanding of the intricate chemical changes in the behavior of olefin in the cationic species.

NONMETALLOCENE—MONO-Cp AND Cp-FREE CATALYSTS

Mono-Cp Type

Monocyclopentadienyl compounds of group 4 metals have been used in producing high MW polyethylene and stereoirregular and poorly regioregular polypropylene as well as syndiotactic polymerization of styrene.

Cp-Free Type

Cp-free $M(CH_2Ph)_{4-n}X_n$(M=Ti, Zr; X=halogen; n=0.12) complexes in the presence of MAO polymerize ethylene in high activity, propylene partially isospecifically, styrene into *syndio*-PS, and butadiene in *cis*-1,4-manner.

REFERENCES

1. Natta, G.; Pino, P.; Mazzanti, G.; Lanzo, R. *Chim. Ind. (Milan)* **1957**, *39*, 1032.
2. Breslow, D. S.; Newberg, N. R. *J. Am. Chem. Soc.* **1957**, *79*, 5073; **1959**, *81*, 81.
3. Sinn, H.; Kaminsky, W.; Vollmer, H. J.; Woldt, R. *Angew. Chem.* **1980**, *92*, 396; *Angew. Chem., Int. Ed. Engl.* **1980**, *19*, 390.
4. Jordan, R. F.; Bajgur, C. S.; Willett, R.; Scott, B. *J. Am. Chem.* Soc. **1986**, *109*, 7410.
5. Jordan, R. F.; Dasher, W. E.; Echols, S. F. *J. Am. Chem. Soc.* **1986**, *108*, 1718.
6. Dyachkovski, F. S.; Shilova, A. K.; Shilov, A. E. *J. Polym. Sci., Part C* **1967**, *16*, 2333.
7. Tritto, I.; Li, S.; Sacchi, M. C.; Zannoni, G. *Macromolecules* **1994**, *26*, 711.

Oligomers

See: *Block Copolymers (By Step-Growth Polymerization)*
Cyclic Oligomers of Engineering Thermoplastics
End Functionalized Oligomers (by Chain-Transfer Technique)
Fluoroacrylic Acid Oligomers
Fluorosilicon Oligomers
Ionene Oligomers
Nonlinear Optical Polymers [Poly(arylene ethynylene) Derivatives]
Perfluorocyclobutane Aromatic Ether Polymers
p-Phenylene Vinylene Oligomers, Homo- and Copolymers (Metathesis Preparation)
Phosphorus-Containing Polymers (From Dicarboxylic Acids with C-P Bonds)
Polyamides, Glycine-Containing
Polypeptide Restricted Backbone Conformation (Cα,α-Symmetrically Disubstituted Glycines)
Vegetal Biomass (2. Oligomers and their Polymerization)

Olypiadane

See: *Polyrotaxanes (Overview)*

Optical Brighteners

See: *Additives (Property and Processing Modifiers)*

Optical Fibers

See: *Plastic Optical Fibers*
Poly(fluoroalkyl-α-fluoroacrylate)s

OPTICALLY ACTIVE POLYAMIDES, COUMARIN DIMER (Chiral Stationary Phases for HPLC)

Kazuhiko Saigo
Department of Chemistry and Biotechnology
Graduate School of Engineering
The University of Tokyo

Coumarin (2-chromenone, 2*H*-benzopyran-2-one), a naturally occurring aromatic flavor, is found in various plants, particularly in some kinds of fruits. In 1902, Ciamician and Silber found that coumarin dimerizes upon photoirradiation in ethanol.[1] There are four types of dimers for coumarin: *anti* head-to-head, *syn* head-to-head, *anti* head-to-tail, and *syn* head-to-tail types. In our laboratory, racemic *anti* head-to-head coumarin dimer was found to react very smoothly with aliphatic and aromatic diamines to give the corresponding racemic polyamides.[2,3]

Since *anti* head-to-head coumarin dimer is a chiral compound of C_2 symmetry, its optically active form can be potent as a chiral component in optically active polymer synthesis. Optically active polyamides, derived from optically active *anti* head-to-head coumarin dimer, may be suitable synthetic polymers to mimic the sophisticated functions displayed by natural polymers, since the polyamides can be expected to form an ordered conformation resulting from the rigid, propeller-like structure of the coumarin dimer component. These considerations attracted us to resolve *anti* head-to-head coumarin dimer and to synthesize optically active polyamides from the optically active form.

PREPARATION OF OPTICALLY ACTIVE POLYAMIDES

The ring-opening polyaddition reaction of (−)-*anti* head-to-head coumarin dimer ((−)-**1**) proceeds very smoothly in DMAc with diamines, such as 1,6-hexanediamine, piperazine, 1,4-phenylenediamine, 1,3-xylylenediamine, 4,4′-oxybisaniline, and *N,N*-dimethyl-1,6-hexanediamine, to give the corresponding optically active polyamides **3** having a C_2-symmetric component in the main chain (**Scheme I**).[4,5]

Optically active polyamide **3b**, derived from (−)-**1** and piperazine, is obtained without any precipitation during the polyaddition reaction, whereas the corresponding racemic polyamide deposits at the early stage of polymerization. Moreover, all of **3** are more soluble in aprotic polar solvents than are the corresponding racemic polyamides.

Optically active *O,O′*-dimethylated polyamides are easily prepared by an interfacial polycondensation reaction of optically active bis(acid chloride) with diamines in a chloroform/1.0 M NaOH mixture at room temperature by using an equimolar amount of benzyltriethylammonium chloride as a phase-transfer catalyst.[6]

Optically active *O,O′*-dicarbamoylated polyamides are obtained from **3** by a polymer reaction with propyl and phenyl isocyanates in DMAc at 50–70°C in the presence of a catalytic amount of triethylamine.[7]

APPLICATIONS

In order to quantitatively evaluate the chiral-recognition ability of polyamides, silica gel columns for HPLC, coated with these optically active polyamides as a chiral stationary phase, were prepared and used for the optical resolution of racemates having aromatic group(s) as well as a hydrogen accepting and/or donating group, which can interact with the functional groups in the polyamides, such as the phenolic hydroxyl (methoxyl, carbamoyloxy) groups, aromatic groups, and amide linkages nearby asymmetric centers. The chiral-recognition ability of polyamides **3** is diverse and strongly depends on the structure of the diamine component in the main chain.[8,9]

REFERENCES

1. Ciamician, G.; Silber, P. *Chem. Ber.* **1902**, *35*, 4128.
2. Hasegawa, M.; Yonezawa, N.; Kanoe, T.; Ikebe, Y. *J. Polym. Sci., Polym. Lett. Ed.* **1982**, *20*, 309.
3. Hasewaga, M.; Saigo, K.; Katsuki, H.; Yonezawa, N.; Kanoe, T. *J. Polym. Sci., Polym. Chem. Ed.* **1983**, *21*, 2345.
4. Yonezawa, N.; Kanoe, T.; Saigo, K.; Chen, Y.; Tachibana, K.; Hasegawa, M. *J. Polym. Sci., Polym. Lett. Ed.* **1985**, *23*, 617.
5. Saigo, K.; Chen, Y.; Yonezawa, N.; Kanoe, T.; Tachibana, K.; Hasegawa, M. *Macromolecules* **1986**, *19*, 1552.
6. Chen, Y.; Saigo, K.; Yonezawa, N.; Hasegawa, M. *Bull. Chem. Soc. Jpn.* **1987**, *60*, 1895.
7. Saigo, K.; Nakamura, M.; Adegawa, Y.; Hasegawa, M. *Chem. Lett.* **1989**, 337.
8. Saigo, K.; Chen, Y.; Yonezawa, N.; Tachibana, K.; Kanoe, T.; Hasegawa, M. *Chem. Lett.* **1985**, 1891.
9. Chen, Y.; Saigo, K.; Yonezawa, N.; Tachibana, K.; Hasegawa, M. *Bull. Chem. Soc. Jpn.* **1987**, *60*, 3341.

Optically Active Polymers

See: *Asymmetric Polymerization (Overview)*
Asymmetric Synthesis (Using Chiral Polymers From (+)-Camphor)

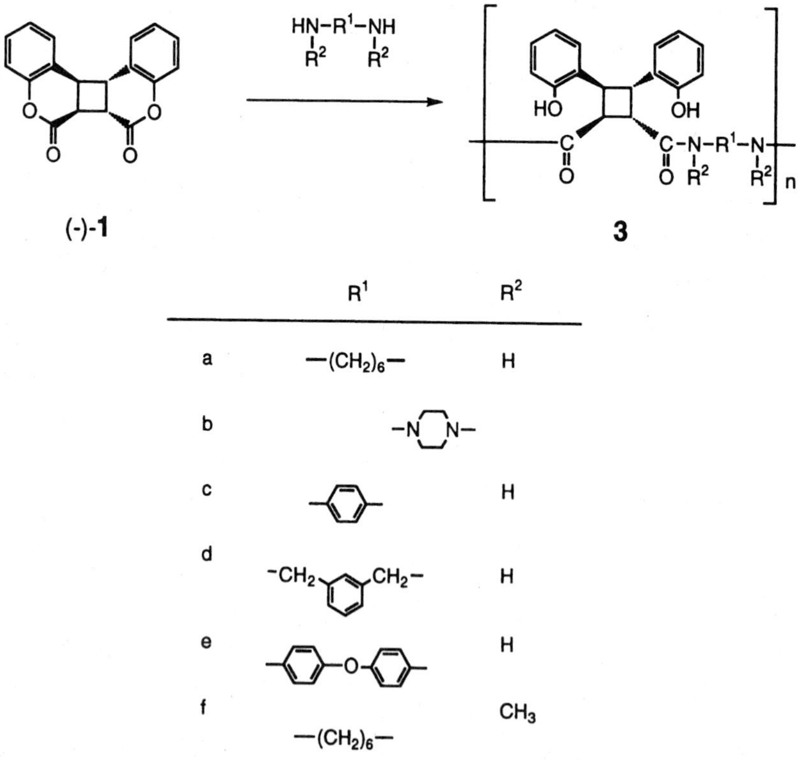

I

Biodegradable Optically Active Polyamides (Stereoregular Polyamides from L-Tartaric Acid)
Chiral Polyanilines
Cyclopolymerization (Cyclization Efficiency and Configuration Control)
Cyclopolymerization, Enantioselective Cationic
Liquid Crystalline Polymers, Side Chain
Metallocene Catalysts (Group 4 Elements, New Polymeric Materials)
Optically Active Polyamides, Coumarin Dimer (Chiral Stationary Phases for HPLC)
Optically Active Polymers (Overview)
Photochromic Optically Active Polymers
Polyglycidols, Polythioglycidols
Polyisocyanides
Polyisopeptides
Polynucleotide Analogues
Stereoregular Polymethacrylates

OPTICALLY ACTIVE POLYMERS (Overview)

Francesco Ciardelli* and Mauro Aglietto
Dipartimento di Chimica e Chimica Industriale
University of Pisa

Because this Encyclopedia contains several contributions devoted to optically active polymers having particular structural features or specific functions, the present chapter is mainly devoted to introduce the reader to the fundamental aspects responsible for the optical activity of synthetic macromolecules.

Though naturally occurring macromolecules are generally formed starting with a single enantiomer of chiral monomers, most synthetic polymers are not optically active even if often chiral. The observation of optical activity in these last polymers is connected to particular requisites normally deriving from the use of optically active monomers or asymmetric polymerization reactions of racemic or prochiral monomers. In few cases optically active polymers were obtained by separation of enantiomeric macromolecules from a racemate.

According to its main objective, this article reports first on the molecular requisites necessary for a synthetic macromolecular system to show optical activity, and successively on the influence on chiroptical properties (optical rotation and circular dichroism) of configurational and conformational contributions.

MOLECULAR REQUISITES FOR OPTICAL ACTIVITY

In case of both low molecular weight molecules and macromolecules, optical activity can be observed only in a chiral molecule, which can be identified by looking at their symmetry properties. A molecule is chiral if all its allowed conformations lack reflection symmetry elements.

In this connection macromolecules differ from conventional low molecular weight molecules because they possess a substantially linear structure along the chain backbone. Accordingly, analysis of the symmetry properties has been carried out on the basis of three different models: the infinite length chain, the finite length chain with equal end groups, and the finite length chain with different end groups.

OPTICAL ACTIVITY AND STRUCTURE

Configurational Optical Activity

The presence in the macromolecules of stereogenic centers with a predominant absolute configuration makes the polymeric system optically active.

Polymerization of Optically Active Monomers

This approach is certainly the most widely used for its relative simplicity and enormous potentiality. The only limits are the availability of optically active monomers with adequate enantiomeric purity and a polymerization process not producing appreciable racemization.

Vinyl Monomers. Several optically active 1-olefins were polymerized in the presence of isotactic specific Ziegler-Natta catalysts to polymers whose specific rotatory power was increasing in absolute value with isotacticity degree and with reducing the distance between side chain chiral center and double bond, thus indicating a macromolecular effect connected with cooperative interactions between chiral units.

More recently the new metallocene catalysts have allowed the preparation of syndiotactic polymers from (S)-4-methyl-1-hexene (I, n = 1, $R^1 = C_2H_5$, $R^2 = CH_3$). This polymer shows optical rotation of the same order of magnitude as the isotactic polymer thus showing that the enantiomeric purity of the side chains prevails on the stereoregularity along the backbone according to the model previously proposed.[8,9]

In case of vinyl ethers,[11] vinyl ketones, acrylates or methacrylates,[3] and acryl or methacrylamides monomers of high enantiomeric purity can be easily prepared starting with synthetic or naturally occurring chiral alcohols or α-amino acids.[1,11–26] The corresponding polymers with different stereoregularity degrees can be obtained by various chain polymerization reactions.

Chiral poly(crown ether)s with high enantiomeric purity have been described as obtained through cyclopolymerization of divinyl ethers.

Cyclic Monomers. Properly substituted chiral cyclic monomers, such as propylene oxide, can provide polymers with asymmetric centers in the backbone. Starting with a single enantiomer and using a nonracemizing process, isotactic optically active polymers are obtained. The systems more extensively investigated are those obtained by ring-opening polymerization of optically active *N*-carboxy-anhydrides, epoxides,[2] episulfides, aziridines, lactides, and lactones.

New optically active polymers have been recently reported as obtained by polymerization of optically active 2,4-disubstituted-2-oxazolines.[27]

Optically Active Polymers by Stepwise Polymerization. Optically active monomers giving polycondensation or polyaddition step-grown polymerization can be used to prepare optically active polymers with the chiral centers either in the backbone or in the side chains. Several examples of polyesters, polyamides, and polyurethanes were reported obtained starting with optically active diacids, glycols, or diamines, but the number of possible structures is really very large.

*Author to whom correspondence should be addressed.

Asymmetric Polymerization of Prochiral Monomers

Starting with prochiral monomers, optically active polymers can be obtained only if the catalyst or initiator is able to induce one prevailing absolute configuration or chiral conformation in the formed macromolecules (asymmetric polymerization). This process was reviewed exhaustively in a recent paper and the reader is referred to this excellent contribution for detailed information.[8] Here the general aspects are summarized with reference to systems where asymmetric control of configuration is observed in the macromolecule backbone. In particular it is important to consider that even if the capacity of the catalyst or initiator to chirally discriminate the enantiofaces of the prochiral monomer is a necessary condition, this may be not sufficient unless the macromolecules formed in this process have the molecular requisites for showing measurable chiroptical properties as discussed earlier.

Alternating copolymerization has been also used for producing polymers with stereogenic centers in the backbone primarily by using maleic anhydrides, maleinimides, and their derivatives in combination with a comonomer with an electron-rich double bond.

Stereoselective Polymerization of Racemic Monomers

This process consists of the preferential polymerization of a single enantiomer from a racemic monomer mixture obtained by an optically active polymerization agent (catalyst or initiator).[1]

Template Copolymerization of Vinyl Monomers to Achieve Asymmetrically Controlled Sequences

In copolymers of two vinyl monomers, the absolute configuration of the backbone stereogenic centers depends on sequence distribution.[4] Thus these copolymers can display measurable chiroptical properties, provided sequence distribution is controlled to yield situations favoring a specific configuration. This has actually been achieved by copolymerizing several vinyl monomers with an optically active template having attached to it two monomeric units in a fixed geometry.

Conformational Optical Activity

Conformational optical activity referring to macromolecular compounds is used to indicate the presence of secondary structures involving the macromolecule as a whole or a substantial fraction of it, with a predominant handness.

Polymerization of Achiral Monomers

Polymers from achiral isonitriles obtained in the presence of Ni-catalysts have been separated into fractions with opposite optical rotation due to the one-sense helical conformation of the backbone.[5,28]

Polymerization of Prochiral Monomers with Bulky Side Groups

In the case of tritylmethacrylate the polymerization with an optically active catalyst gives polymers with chiroptical properties related to the one screw sense helical conformation.[6,78]

Copolymerization of an Achiral Monomer with a Chiral Comonomer

The inclusion of the two monomers in the same chain induces the structural units of the achiral monomer to assume a chiral conformation. Thus, if only one enantiomer of the chiral monomer is used, the units of the achiral comonomer can display induced optical activity mainly of conformation origin.

Examples of this method are offered by the co-isotactic copolymers of optically active α-olefins with achiral comonomers as 4-methyl-1-pentene, styrene, and vinylnaphthalenes obtained by Ziegler-Natta catalyst and anionic copolymers of TrMA with (S)-α-methylbenzylmethacrylate.[30–34]

A similar approach has been more recently extended to free radical copolymerization of several achiral chromophoric monomers with (-)-methylacrylate or methacrylate.[35]

APPLICATIONS OF OPTICALLY ACTIVE POLYMERS

Though optically active polymers can be the bases of materials with unique mechanical and functional properties, their use is still limited. Some applications have already been reported in a previous paper.[7]

Recent examples are in the same line and refer mainly to asymmetric synthesis where the optically active polymer acts as chiral reagents, chiral membranes, and chiral supports for enantioselective molecular recognition and separation of enantiomers.[35–40]

REFERENCES

1. Pino, P. *Adv. Polym. Sci.* **1965**, *4*, 393.
2. Osgan, M.; Price, C. C. *J. Polym. Sci.* **1959**, *34*, 153.
3. Beredjick, N.; Schuerch, C. *J. Chem. Soc.* **1956**, *78*, 2646.
4. Wulff, G.; Zabrocki, K.; Hohn, I. *Angew. Chem., Int. Ed.* **1978**, *17*, 535.
5. Nolte, R. J. M.; van Beijnen, A. J. M.; Drenth, W. *J. Ant. Chem. Soc.* **1974**, *96*, 5932.
6. Okamoto, Y.; Suzuki, K.; Hota, K.; Hatada, K.; Yuki, H. *J. Am. Chem. Soc.* **1979**, *101*, 473.
7. Ciardelli, F. In *Encyclopedia of Polymer Science and Engineering;* 2nd ed.; John Wiley & Sons: New York, 1987; Vol. 8, p 463.
8. Okamoto, Y.; Nakano, T. *Chem. Rev.* **1994**, *94*, 349.
9. Pino, P.; Ciardelli, F.; Zandomeneghi, M. *Annu. Rev. Phys. Chem.* **1970**, *21*, 561.
10. Zambelli, A.; Grassi, A.; Galimberti, M.; Perego, G. *Makromol. Chem., Rapid Commun.* **1992**, *13*, 407.
11. Lorenzi, G. P.; Benedetti, E.; Chiellini, E. *Chim. Ind. Milan* **1964**, *46*, 1474.
12. Liquori, A. M.; Pispisa, B. *J. Polym. Sci. Part* B **1967**, *5*, 375.
13. Vukovic, R.; Fles, D. *J. Polym. Sci., Part A-1* **1975**, *13*, 49.
14. Pieroni, O.; Ciardelli, F.; Botteghi, C.; Larducci, L.; Salvadori, P.; Pino, P. *J. Polym. Sci., Part C* **1969**, 22, 993.
15. Allio, A.; Pino, P. *Helv. Chim. Acta* **1974**, *57*, 616.
16. Arcus, C. L.; West, D. W. *J. Chem. Soc.* **1959**, 2699.
17. Schulz, R. C.; Hilpert, H. *Makromol. Chem.* **1962**, *55*, 132.
18. Schulz, R. C. *Z. Naturforsh.* **1964**, *19b*, 387.
19. Klabunovskii, E. I.; Petrov, Yu I.; Shvartsman, M. I. *Vysokomol. Soed.* **1964**, 6, 1487.
20. Liu, J.-H.; Kondo, K.; Takemoto, K. *Makromol. Chem. Rapid Commun.* **1982**, *3*, 215.
21. Nguyễn-tâ-Thiê, Suter, U. W.; Pino, P. *Macromol. Chem.* **1983**, *184*, 2335.
22. Kulkarni, R. K.; Morawetz, H. *J. Polym. Chem.* **1961**, *54*, 491.

23. Whistler, R. L.; Panzer, H. P.; Roberts, H. J. *J. Org. Chem.* **1961**, *26*, 1583.
24. Kaiser, E.; Schulz, R. C. *Makromol. Chem.* **1968**, *81*, 273.
25. Braud, C.; Vert, M.; Sélégny, E. *Makromol. Chem.* **1974**, *175*, 775.
26. Camail, M.; Maesano, J. C.; Margaillan, A.; Pautasso, J. P.; Vernet, J. L. *Eur. Polym. J.* **1994**, *30*, 485.
27. Guo, X. Q.; Schulz, R. C. *Polym. Int.* **1994**, *34*, 229.
28. Drenth, W.; Nolte, R. J. M. *Acc. Chem. Res.* **1979**, *12*, 30.
29. Okamoto, Y.; Shohi, H.; Yuki, H. *J. Polym. Sci. Polym. Lett. Ed.* **1983**, *21*, 601.
30. Carlini, C.; Ciardelli, F.; Pino, P. *Makromol. Chem.* **1968**, *119*, 244.
31. Ciardelli, F.; Salvadori, P.; Carlini, C.; Chiellini, E. *J. Am. Chem. Soc.* **1972**, *94*, 6536.
32. Ciardelli, F.; Righini, C.; Zandomeneghi, M.; Hug, W. *J. Phys. Chem.* **1977**, *81*, 1948.
33. Yuki, H.; Ohta, K.; Okamoto, Y.; Hatada, K. J. *Polym. Sci. Polym. Lett. Ed.* **1977**, *15*, 589.
34. Okamoto, Y.; Suzuki, K.; Yuki, H. *J. Polym. Sci. Polym. Chem. Ed.* **1980**, *18*, 3043.
35. Adjidjonou, K.; Caze, C. *Eur. Polym. J.* **1994**, *30*, 395.
36. Liu, J.-H.; Lin, S.-R.; Kuo, J.-C. *J. Polym. Sci., Part A* **1987**, *25*, 2521.
37. Yashima, E.; Noguchi, J.; Okamoto, Y. *J. Appl. Polym. Sci.* **1994**, *54*, 1087.
38. Sellergren, B.; Andersson, L. *J. Org. Chem.* **1990**. *55*, 3381.
39. Lepisto, M.; Sellergren, B. *J. Org. Chem.* **1989**, *54*, 6010.
40. Calmes, M.; Daunis, J.; Ismaili, H.; Jacquier, R.; Koudou, J.; Nkusi, G.; Zouanate, A. *Tetrahedron* **1990**, *46*, 6021.

Optoelectronic Polymers

See: *Electroluminescent Polymers (Overview)*
Photochromic Films, Azo Dyes (Photoinduced Anisotropy and Photoassisted Poling)
Photoluminescent Polymers
Photopolymer Materials (Development of Holographic Gratings)
Polyaniline Network Electrodes (Enhanced Performance of Polymer LEDs)
Polyphenylacetylene
Poly(p-phenylene vinylene)s (Methods of Preparation and Properties)

Ordered Bicontinuous Double-Diamond Morphology

See: *Block Copolymers (Ordered Bicontinuous Double-Diamond Morphology)*

ORDERED EPOXY NETWORKS

S. Jahromi
Department of Polymer Technology
Faculty of Chemical Engineering and Materials Science
Delft University of Technology

Recently, ordered networks have become the subject of intensive research in the field of liquid crystalline (LC) polymeric materials. This can partly be attributed to the fact that, although main-chain LC polymers have excellent properties, like a high modulus in the direction of the orientation, the properties transverse to the direction of the orientation are rather poor. These problems can be overcome by the introduction of crosslinks between the chains, which improves the dimensional stability of these ordered systems.

Ordered networks can be classified in several categories as described below.

LC THERMOSETS

These are ordered networks that, because of the high degree of crosslink density, display no thermal transitions like smectic to nematic or LC to isotropic.

LC ELASTOMERS

These are lightly crosslinked either side- or main-chain LC polymers.[1–5] LC elastomers are quite interesting systems owing to the fact that the macroscopic orientation can be achieved by applying mechanical fields.[6–10] As a consequence, fascinating opto-electronic properties can be realized by mechanical deformation and orientation.[11]

ANISOTROPIC GELS

These are ordered networks swollen either in an isotropic or anisotropic solvent. Ordered networks swollen in low molecular weight (LMW) LC solvents form another interesting class of LC materials because of, for example, their intriguing electro-optical properties.

NETWORKS WITH NONLINEAR OPTICAL (NLO) PROPERTIES

These are materials where NLO active molecules are chemically bonded to the network.[12–21] The presence of crosslinks ensures the orientational stability of these systems in contrast with the linear NLO polymeric materials where the degree of polar orientation of the NLO molecules usually decreases as a function of time.

This article deals with the first category, LC thermosets. These materials are interesting because of their excellent optical and mechanical properties.[22,23] The basic idea in the production of highly crosslinked and highly ordered networks is to orient LMW LC monomers uniaxially and subsequently freeze in the orientation by polymerization and crosslinking.

In this paper we will present, briefly, part of the activities carried out at the polymer department of the Delft University of Technology regarding preparation and properties of ordered epoxy networks.[24–27] Our research started with synthesizing a series of LMW LC diepoxides.[24]

The polymerization was carried out in the LC phase with the aid of a cationic initiator under the influence of ultraviolet light. In this way, highly ordered networks could be produced that maintained their LC-like order up to the degradation temperature.

Although networks prepared by photoinitiated chain polymerization are very well suited for application as thin layers, there are certain drawbacks, especially when preparing bulk materials.

As an alternative, we studied the copolymerization reaction of the LC diepoxide I with a series of aromatic diamines.[25] LC epoxy-amine systems are excellently suitable to investigate network formation in an ordered state.

Recently, several publications have appeared concerning the copolymerization of LC diepoxides with aromatic diamines.[28–33] These papers have actually been published simultaneously with our studies. As stated earlier, conventional epoxy networks have excellent mechanical and thermal properties. The objective of nearly all these investigations has been to improve these properties by incorporating LC-like structures into these systems.

Here, we will demonstrate that it is indeed possible to produce highly macroscopically ordered epoxy-amine networks based on LMW LC monomers.

RESULTS AND DISCUSSION

Preparation

In **Scheme I**, the synthetic route for the preparation of the LC diepoxide I is shown. Other LC-diepoxide monomers are prepared by slight modifications of Scheme I. A more detailed synthetic procedure has been reported earlier.[24]

The LC diepoxide I is chosen for copolymerization with aromatic diamines because it has a broad nematic range and a good reactivity toward amines.

As mentioned earlier, our main objective was to study in detail the physical properties of ordered networks. In this regard, there are two topics that are of primary importance. In

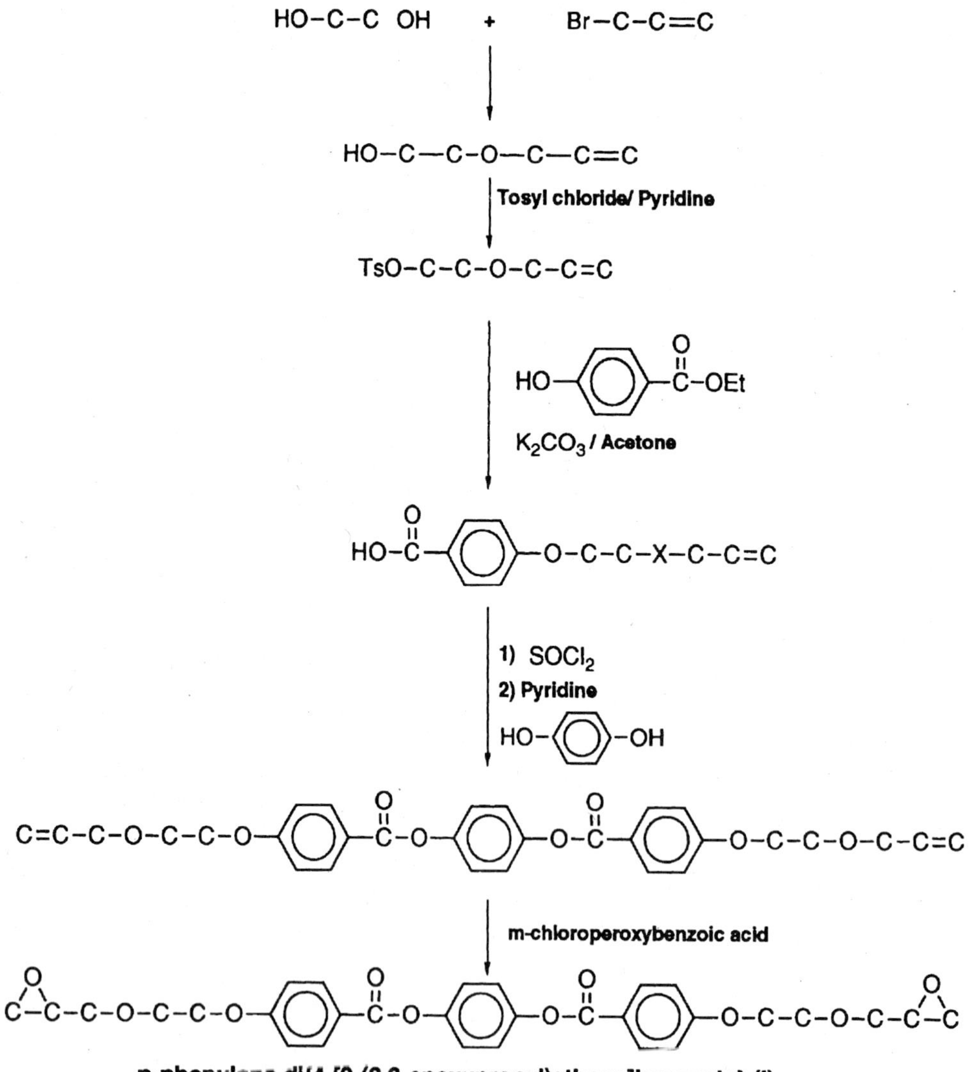

p-phenylene di{4-[2-(2,3-epoxypropyl)ethoxy]benzoate} (I)

Crystal $\xrightarrow{124^\circ C}$ Nematic $\xrightarrow{186^\circ C}$ Isotropic

SCHEME I. Synthetic route to monomer I.

the first place, of course, the state of order of these systems should be investigated. The question is how the orientational order changes as a function of the degree of polymerization and, especially, what the effect is of the crosslinking reaction. The second point of interest is the elastic properties of the ordered networks. The rigid character of the monomers and the ordered structure are the two main factors that distinguish the present networks from conventional crosslinked polymers.

Time-Resolved Measurements of the Degree of Order During the Polymerization Reaction

It can be concluded that the orientational order increases during the chain extension process up to the point where it becomes fixed as a result of the crosslinking reaction.

Dynamic Mechanical and Thermal Properties of Macroscopically Ordered Epoxy-Amine Networks

The viscoelastic response of the ordered epoxy-amine networks, especially in the rubber state, is quite unique.

In the present case, the distances between the crosslink junctions are too short and especially too rigid to behave as Gaussian chains.

The second factor that distinguishes the present networks from conventional systems is the existence of the long range orientational order. Earlier investigations pointed out that it is probably very difficult to make a direct distinction between the influence of both effects and therefore to determine which factor is playing a more important role.[27]

The next interesting point that should be considered is the behavior of the rubber modulus of the ordered epoxy networks as a function of the temperature. The rubber modulus is, in contradiction with the theoretical prediction, decreasing as a function of the temperature. Similar behavior was observed for crosslinked LC main chain polymers.[34,35]

The thermal expansion of the macroscopically ordered epoxy(I)-amine(II) networks is highly anisotropic. The coefficient of thermal expansion is, in the direction of macroscopic order, almost zero below T_g, and even negative above T_g. This anisotropic behavior is characteristic of highly oriented polymers and has also been observed for networks prepared from LC diacrylates.[36–38]

ACKNOWLEDGMENTS

The author would like to thank J. Lub and B. Norder for their contributions. The financial support of the Dutch Ministry of Economic Affairs (IOP-IC technology program) is gratefully acknowledged.

REFERENCES

1. Finkelmann, H.-J.; Rehage, G. *Makromol. Chem., Rapid. Commun.* **1981**, *2*, 317.
2. Zentel, R.; Reckert, G. *Makromol. Chem.* **1986**, *187*, 1915.
3. Bualek, S.; Zentel, R.; *Makromol. Chem.* **1988**, *189*, 791.
4. Zentel, R.; *Angew. Chem. Adv. Mater.* **1989**, *101*, 1437.
5. Zentel, R.; Reckert, G.; Bualek, S.; Kapitza, H. *Makromol. Chem.* **1989**, *190*, 2869.
6. Finkelmann, H.-J.; Koch, H.-J.; Gleim, W.; Rehage, G. *Makromol. Chem. Rapid Commun.* **1984**, *5*, 287.
7. Mitchell, G. R.; Davis, F. J.; Ashman, A. *Polymer* **1987**, *28*, 639.
8. Mitchell, G. R.; Davis, F. J.; Cuo, W.; Cywinski, R. *Polymer* **1991**, *32*, 1347.
9. Degert, C.; Davidson, P.; Megtert, S.; Petermann, D.; Mavzac, M. *Liq. Cryst.* **1992**, *12*, 779.
10. Legge, C. H.; Davis, F. J.; Mitchell, G. R. *J. Phys. II (France)* **1991**, 1, 1253.
11. Meier, W.; Finkelmann, H. *Makromol. Chem. Rapid Commun.* **1990**, *11*, 1253.
12. Zentel, R.; Jungbauer, D.; Twieg, R. J.; Yoon, D. Y.; Willson, C. G. *Makromol. Chem.* **1993**, *194*, 859.
13. Jeng, J. R.; Chen, Y. M.; Kumar, J.; Tripathy, S. *K.J.M.S.— Pure Appl. Chem.* **1992**, *A29*, 1115.
14. Tweig, R.; Ebert, *M.;* Jungbauer, D.; Lux, M.; Reck, B.; Swalen, J.; Teraoka, I.; Willson, C. G.; Yoon, D. Y.; Zentel, R. *Mol. Cryst. Liq. Cryst.* **1992**, *217*, 19.
15. Xu, C.; Wu, B.; Dalton, L. R.; Shi, Y.; Ranon, P. M.; Steier, W. H. *Macromolecules* **1992**, *25*, 6714.
16. Yu, L.; Chan, W.; Bao, Z. *Macromolecules* **1992**, *25*, 5609.
17. Chen, M.; Dalton, L. R.; Yu, L. P.; Shi, Y. Q.; Steier, W. H. *Macromolecules* **1992**, *25*, 4032.
18. Hashidate, S.; Nagasaki, Y.; Kato, M.; Okada, S.; Matsuda, H.; Minami, N.; Nakanisi, H. *Polym. Adv. Tech.* **1992**, *3*, 145.
19. Chen, M.; Yu, L.; Dalton, L. R.; Shi, Y.; Steier, W. H. *Macromolecules* **1991**, *24*, 5421.
20. Mandal, B. K.; Lee, J. Y.; Zhu, X. F.; Chem, Y. M.; Parkeenavincha, E.; Kumar, J. Tripathy, S. K. *Synth. Met.* **1991**, *41-43*, 3143.
21. Eich, M.; Reck, B.; Yoon, D. Y.; Willson, G.; Bjorklund, G. C. *J. Appl. Phys.* **1989**, *66*, 3241.
22. Hikmet, R. A. M.; Lub, J.; Broer, D. J. *Adv. Mater.* **1991**, *3*, 392.
23. Hikmet, R. A. M.; Broer, D. J. *Polymer* **1991**, *32*, 1627.
24. Jahromi, S.; Lub, J.; Mol, G. N. *Polymer* **1994**, *35*, 622.
25. Jahromi, S.; Mijs, W. J. *Mol. Cryst. Liq. Cryst.* **1994**, *250*, 209.
26. Jahromi, S. *Macromolecules* **1994**, *27*, 2804.
27. Jahromi, S.; Kuipers, W. A. G.; Norder, B.; Mijs, W. J. *Macromolecules* **1995**, *28*, 2201.
28. Su, W.-F. A. *J. Polym. Sci.: Part A: Polym. Chem.* **1993**, *31*, 3251.
29. Carfagna, C.; Amendola, E.; Giamberini, M. *Macromol. Chem. Phys.* **1994**, *195*, 279.
30. Mallon, J. J.; Adams, P. M. *J. Polym., Sci., Part A: Polym. Chem.* **1993**, *31*, 2249.
31. Rozenberg, B. A.; Gur'eva, L. L. *Proc. Am. Chem. Soc., Div. Polym. Mat.: Sci. Eng.* **1992**, *66*, 162.
32. Chein, L.-C.; Lin, C.; David, S. F.; McCargar, J. W. *Macromolecules* **1992**, *25*, 133.
33. Barclay, G. G.; McNamee, S. G.; Ober, C. K.; Papathomas, K. I.; Wang, D. W. **1992**, *30*, 1831, 1845.
34. Hanus, K.-H. Pechhold, W., Soergel, F., Stoll, B., Zentel, R. *Colloid Polym. Sci.* **1990**, *268*, 222.
35. Pakula, T.; Zentel, R. *Makromol. Chem.* **1991**, *192*, 2401.
36. Choy, C. L. In *Development in Oriented Polymers—I* Ward, I. M., Ed.; Applied Science: London, 1982.
37. Clough, S. B.; Blumstein, A.; Hsu, E. C. *Macromolecules* **1976**, *9*, 123.
38. Broer, D. J.; Mol, G. N. *Polym. Eng. Sci.* **1991**, *31*, 625.

ORGANIC/INORGANIC COMPOSITE MATERIALS

James M. O'Reilly and Bradley K. Coltrain
Eastman Kodak Company

Organic-inorganic composites, or OIC, represent a relatively new class of materials with properties intermediate between organic polymers and inorganic materials. The composites are intended to combine the advantages of organic polymers, such as flexibility, toughness, and coatability, with those of ceramics or glasses, such as hardness, durability, and high refractive index. Ideally, the organic and inorganic components would be reacted or blended to produce single-phase composites. Unfortunately, organic polymers and inorganic glasses tend to be immiscible, so some degree of phase separation occurs. Much of the work on OICs has centered around controlling the degree of phase separation to produce homogeneous, though not single phase, composites. Typically, this has involved the use of sol-gel chemistry, or the hydrolysis and condensation of inorganic monomers, to produce the inorganic component. Sol-gel chemistry allows the formation of an inorganic network at low temperatures at which the organic polymers can survive. This OIC approach is in contrast to typical filled systems in which inorganic particles or fibers are added to polymers for reinforcement. The homogeneity of composites produced by this later approach is limited by the particle size of the filler, as well as the ability of the polymer to disperse it.

OIC research has evoked considerable interest in the last few years. The present article intends to provide an understanding of OIC materials by reviewing some of this previous work.

Some of the original work in this area was by Helmut Schmidt and co-workers, who adopted the perspective of organic modified glasses or ORMOSILS.[1–10] These were produced via a sol-gel route using principally silicon alkoxides containing nonhydrolyzable organic substituents [$R'Si(OR)_3$].

Another approach to organic modification of inorganic glasses was taken independently by the groups of Mackenzie and Klein.[11,12] In these cases, inorganic xerogels were first produced via sol-gel chemistry and then were imbided with methyl methacrylate monomer that was polymerized *in situ* to produce a co-continuous composite.

Saegusa has produced silicates with controlled porosity by using linear organic polymers, generally oxazolines or polymers containing a carboxylic amide group capable of hydrogen bonding to a silicate network.[13,14]

The largest area of OIC research activity has been inorganic modification of organic polymers, that is, the organic polymer is the dominant phase. Two approaches are normally utilized. In the first case, a "blend" is produced simply by reacting inorganic monomers *in situ* in linear organic polymers. Homogeneous, transparent OICs with highly dispersed, inorganic oxide phases can be produced, if appropriate polymer backbones are used.[15–29] For example, transparent composites possessing high values for the rubbery plateau modulus above the polymer glass transition temperature were obtained by reacting tetraethoxysilane (TEOS) in poly(vinyl acetate), poly(methyl methacrylate), and poly[bis(methoxy) (ethoxyethoxy)phosphazene].[15–21]

The second approach to predominantly organic OICs, pioneered by Wilkes and Mark and co-workers, utilized organic polymers modified with trialkoxysilane moieties to facilitate crosslinking between the polymer and the growing inorganic oxide network.[32–57] Cross-reaction retards phase separation, frequently producing homogeneous hybrid materials. Typically, low-molecular weight organic oligomers were end-capped with trialkoxysilanes and then co-reacted with monomers, such as TEOS.

Homogeneous OIC materials can be produced either as "blends" or as crosslinked networks. Unfortunately, only a limited number of polymers interact strongly enough with the growing inorganic network to retard phase separation. Further, pendant trialkoxysilane allows polymer crosslinking to compete with the reaction between the organic polymer and the inorganic network, which can cause phase separation. It has been shown that in the case of polymers capable of forming homogeneous composites, either with or without trialkoxysilane functionalization, increases in modulus with increasing inorganic are observed in either case in roughly comparable amounts.[61] The polymer T_g does not shift in the "blended" systems, but does shift higher in the systems with trialkoxysilane functionality. The key result is that very similar composite properties can be obtained with or without trialkoxysilane functionality as long as the composite is highly homogeneous.[61]

One final approach to homogeneous OIC materials involves simultaneous polymerization of both inorganic and organic monomers to produce semi-interpenetrating networks.[62–68]

CHEMISTRY

The "sol-gel" chemistry employed in the formation of OIC networks will be described. The typical monomers are generally metal alkoxides, with tetraethoxysilane (TEOS) being one of the more widely studied. Several reviews describe this chemistry in detail, particularly for silicon-based systems.[69–71]

PHYSICAL PROPERTIES

One of the main objectives of OIC research is to prepare materials with the desirable properties of inorganic constituents and the favorable processability of the organic constituents. There are many examples where silica can be incorporated into OIC to increase the modulus and the strength of composites. Titania can be used similarly to produce high-refractive index optical films or elements. Alumina can be incorporated into composites to increase the thermal conductivity.

Glass Transition

The glass transition temperature of an OIC is an important property of OIC and can be controlled by composition, molecular weight, and crosslinking.

Mechanical Properties

The mechanical properties of OIC are perhaps one of the most important aspects of these materials. Mechanical properties of OIC are not dramatically different from polymers with fillers of the same composition and loading.

Refractive Index

OIC materials offer the opportunity to fine-tune the refractive index. The high-refractive index of TiO_2, Ta_2O_5, and GeO_2 offer the potential of producing OIC with an index of refraction greater than the 1.7–1.8 value obtained in the most refractive polymers.

APPLICATIONS OF ORGANIC-INORGANIC COMPOSITES

The principal goal of OIC research is to replace other composites or materials and to achieve enhanced properties with better processability and lower cost. OICs have been designed for structural elements, optical components, coatings, matrices or hosts for solutes, electrical and thermal conductors, and many others.

To date, OIC materials have not proven useful as structural composites owing to the complexity of producing defect-free composites in the high shrinkage process. However, the use of sol-gel and OIC to increase the compressive strength of composites is a good application of this technology.[72]

The opportunities for the use of OICs as optical components are myriad, and although there are many practical uses, none are commercially significant. Thin-film applications are more promising and many sol-gel and OIC examples exist.

The most promising applications of OIC materials are as thin-film protective coatings. The chemistry allows for readily coatible solutions, which is difficult for ceramics. Very abrasion-resistant coatings have been produced for protection of a variety of plastics.[7–21,51]

ACKNOWLEDGMENTS

We thank C. J. T. Landry for comments on this review. We gratefully acknowledge the contributions of our colleagues to this research, W. Ferrar, L. Kelts, C. J. T. Landry, M. Landry, V. K. Long, D. Perchak, J. Pochan, G. Rakes, J. Sedita, S. Tunney, S. R. Turner, and N. Zumbulyadis.

REFERENCES

1. Schmidt, H. *J. Non-Cryst. Solids* **1985**, *73*, 681.
2. Schmidt, H.; Scholze, H.; Tunker, G. *J. Non-Cryst. Solids* **1986**, *80*, 557.
3. Schmidt, H. *Mater. Res. Soc. Symp. Proc.* **1984**, *32*, 327.
4. Philipp, G.; Schmidt, H. *J. Non-Cryst. Solids* **1986**, *82*, 31.
5. Schmidt, H. J. *Non-Cryst. Solids* **1989**, *112*, 419.
6. Gautier-Luneau. I.; Mosset, A.; Galy, J.; Schmidt, H. *J. Mater. Sci.* **1990**, *25*, 3739.
7. Schmidt, H.; Kaiser, A. U.S. Patent 4 440 745, 1984.
8. Schmidt, H.; Tunker, G.; Scholze, H. U.S. Patent 4 374 696, 1983.
9. Scholze, H.; Schmidt, H. U.S. Patent 4 238 590, 1980.
10. Scholze, H.; Schmidt, H. U.S. Patent 4 243 692, 1981.
11. Pope, E. J. A.; Asami, M.; Mackenzie, J. D. *J. Mater. Res.* **1989**, *4*, 1018.
12. Klein, L. C.; Abramoff, B. *Polym. Prepr.* **1991**, *32*(3), 519.
13. Saegusa, T. *Polym. Prepr.* **1993**, *34*(1), 804.
14. Saegusa, T.; Chujo, Y. *Makromol. Chem. Macromol. Symp.* **1992**, *64*, 1.
15. Fitzgerald, J. J.; Landry, C. J. T.; Pochan, J. M. *Macromolecules* **1992**, *25*(14), 3715.
16. Fitzgerald, J. J.; Landry, C. J. T.; Schillace, R. V.; Pochan, J. M. *Polym. Prepr.* **1991**, *32*(3), 532.
17. Landry, C. J. T.; Coltrain, B. K.; Brady, B. K. *Polymer* **1992**, *33*(7), 1486.
18. Landry, C. J. T.; Coltrain, B. K.; Wesson, J. A.; Lippert, J. L.; Zumbulyadis, N. *Polymer* **1992**, *33*(7), 1496.
19. Landry, C. J. T.; Coltrain, B. K.; Landry, M. R.; Fitzgerald, J. J.; Long, V. K. *Macromolecules* **1993**, *26*, 3702.
20. Coltrain, B. K.; Fertar, W. T.; Landry, C. J. T.; Molaire, T. R.; Zumbulyadis, N. *Chem. Mater.* **1992**, *4*, 358.
21. Ferrar, W. T.; Coltrain, B. K.; Landry, C. J. T.; Long, V. K.; Molaire, T. R.; Schildkraut, D. E.; ACS Symp. Ser. 572, Inorganic and Organometallic Polymers II, ACS, Washington, DC, 1993; p 258.
22. Mauritz, K. A.; Warren, R. M. *Macromolecules* **1989**, *22*, 1730.
23. Mauritz, K. A.; Storey, R. F.; Jones, C. K. *ACS Symp. Ser.* **1989**, *395*, 401.
24. Mauritz, K. A.; Jones, C. K. *J. Appl. Polym. Sci.* 1990. *40*. 1401.
25. Davis, S. V.; Mauritz, K. A. *Polym. Prepr.* **1992**, *33*(2), 363.
26. David, I. A.; Scherer, G. W. *Polym. Prepr.* **1991**, *32*(3), 530.
27. Wung, C. J.; Pang, Y.; Prasad, P. N.; Karasz, F. E. *Polymer* **1992**, *32*, 605.
28. Ravaine, D.; Seminel, A.; Charbouillot, Y.; Vincens, M. *J. Non-Cryst. Solids* **1986**, *82*, 210.
29. Morikawa, A.; Iyoku, Y.; Kakimoto, M.; Imai, Y. *Polym. J.* **1992**, *21*, 107.
30. Nandi, M.; Conklin, J. A.; Salvati, Jr., L.; Sen, A. *Chem. Mater.* **1991**, *3*, 201.
31. Nandi, M.; Conklin, J. A.; Salvati, Jr., L.; Sen, A. *Chem. Mater.* **1990**, *2*, 772.
32. Mark, J. E. *Br. Polym. J.* **1985**, *17*, 144.
33. Mark, J. E.; Ning, Y.-P. *Polym. Bull.* **1984**, *12*, 413.
34. Mark, J. E.; Sur, G. S. *Polym. Bull.* **1985**, *14*, 325.
35. Mark, J. E.; Sun, C.-C. *Polym. Bull.* **1987**, *18*, 259.
36. Sur, G. S.; Mark, J. E. *Eur. Polym. J.* **1985**, *21*, 1051.
37. Mark, J. E.; Wang, S.-B. *Polym. Bull.* **1988**, *20*, 443.
38. Clarson, S. J.; Mark, J. E. *Polym. Comm.* **1987**, *28*, 249.
39. Sun, C.-C.; Mark, J. E. *Polymer* **1989**, *30*, 104.
40. Glaser, R. H.; Wilkes, G. L. *Polym. Bull.* **1988**, *19*, 51.
41. Huang, H. H.; Orler, B.; Wilkes, G. L. *Macromolecules* **1987**, *20*, 1322.
42. Huang, H. H.; Orler, B.; Wilkes, G. L. *Polym. Bull.* **1985**, *14*, 557.
43. Wilkes, G. L.; Orler, B.; Huang, H. H. *Polym. Prepr.* **1985**, *26*(2), 300.
44. Huang, H. H.; Glaser, R. H.; Wilkes, G. L. *ACS Symp. Ser.* **1988**, 360, 354.
45. Glaser, R. H.; Wilkes, G. L. *Polym. Bull.* **1989**, *22*, 527.
46. Glaser, R. H.; Wilkes, G. L. *J. Non-Cryst. Solids* **1989**, *113*, 73.
47. Noell, J. L. W.; Wilkes, G. L.; Mohanty, D. K.; McGrath, J. E. *J. App. Polym. Sci.* **1990**, *40*, 1177.
48. Rodrigues, D. E.; Brennan, A. B.; Betrabet, C.; Wang, B.; Wilkes, G. L. *Chem. Mater.* **1992**, *4*, 1437.
49. Huang, H. H.; Wilkes, G. L.; Carlson, J. G. *Polymer* **1989**, *30*, 2001.
50. Huang, H. H.; Wilkes, G. L. *Polym. Bull.* **1987**, *18*, 455.
51. Betrabet, C.; Wilkes, G. L. *Polym. Prepr.* **1992**, *33*, 286.
52. Huang, H.-H.; Wilkes, G. L. *Polym. Bull.* **1987**, *18*, 455.
53. Huang, H.-H.; Wilkes, G. L.; Carlson, J. G. *Polymer* **1989**, *30*, 2001.
54. Mourey, T. H.; Miller, S. M.; Wesson, J. A.; Long, T. E.; Kelts, L. W. *Macromolecules* **1992**, *25*, 45.

55. Wei, Y.; Bakthavatchalam, R.; Whitecar, C. K. *Chem. Mater.* **1990**, *2*, 337.
56. Kohjiya, S.; Ochiai, K.; Yamashita, S. *J. Non-Cryst. Solids* **1990**, *119*, 132.
57. Coltrain, B. K.; O'Reilly, J. M.; Turner, S. R.; Sedita, J. S.; Smith, V. K.; Rakes, G. A.; Landry, M. R. "Proceedings of Fifth Annual International Conference on Crosslinked Polymers", Luzerne, Switzerland, 1991; p 11.
58. Coltrain, B. K.; Rakes, G. A.; Smith, V. K. U.S. Patent 5 019 607, 1991.
59. Wang, S.; Ahmad, Z.; Mark, J. E. *Chem. Mater.* **1994**, *6*, 943.
60. Wang, S.; Ahmad, Z.; Mark, J. E. *Polym. Bull.* **1993**, *31*, 323.
61. Coltrain, B. K.; Landry, C. J. T.; O'Reilly, J. M.; Chamberlain, A. M.; Rakes, G. A.; Sedita, J. S.; Kelts, L. W.; Landry, M. R.; Long, V. K. *Chem. Mater.* **1993**, *5*, 1445.
62. Ellsworth, M. W.; Novak, B. M. *Polym. Prepr.* **1992**, *33*(1), 1088.
63. Ellsworth, M. W.; Novak, B. M. *Chem. Mater.* **1993**, *5*, 839.
64. Ellsworth, M. W.; Novak, B. M. *J. Am. Chem. Soc.* **1991**, *113*, 2756.
65. Novak, B. M.; Davies, C. *Macromolecules* **1991**, *24*, 5481.
66. Novak, B. M.; Ellsworth, M.; Wallow, T.; Davies, C. *Polym. Prepr.* **1990**, *31*(9), 698.
67. Novak, B. M.; Ellsworth, M. W. *Mater. Sci. Eng.* **1993**, *A162*, 257.
68. Novak, B. M.; Auerbach, D.; Verner, C. *Chem. Mater.* **1994**, *6*, 289.
69. Brinker, C. J. *J. Non-Cryst. Solids* **1988**, *100*, 31.
70. Coltrain, B. K.; Kelts, L. W. *ACS Adv. Chem. Series* **1994**, *234*, 403.
71. Brinker, C. J.; Scherer, G. S. *Sol-Gel Science*, Academic: San Diego, CA. 1990.
72. Kovar, R. F.; Lusignea, R. W.; Haghighat, R. R.; Pantano, C.; Thomas, E. L. *Mater. Res. Soc. Symp. Proc.* **1990**, *175*, 193.

ORGANIC/INORGANIC COMPOSITES

Garth L. Wilkes* and Jianye Wen
Department of Chemical Engineering and Polymer Materials and Interfaces Laboratory
Virginia Polytechnic Institute and State University

A new class of composite materials has been developed in recent years through the combination of inorganic metal alkoxides and low molecular weight organics and polymers (often functionalized) by the sol-gel approach. Unlike the traditional composite materials, which have macroscale domain size of micron and even millimeter scale, many of these new inorganic/organic composites are nanoscopic with the physical constraint of several nanometers as the minimum size of the components or phases. The term "nanocomposite" is also used by some researchers. As a result of this nanoscale domain size, these inorganic/organic composite materials may possess properties that the traditional composite materials do not have, such as optical transparency. Through the combinations of different inorganic and organic components, different types of primary and secondary bonding, and different processing methods, materials with new properties can be produced for electrical, optical, structural, or related applications. The inorganic/organic composite materials made in this way, which have also been termed "ceramers" by Wilkes et al.[1] and "ormosils" by Schmidt,[2] are of considerable interest because of their potential for providing unique combinations of properties that cannot be achieved by conventional methods. For example, the brittleness, optical, and nonlinear optical characteristics of glasses can be greatly modified by incorporating organic materials into the inorganic network. This text will briefly review the synthesis, structure–property response, and applications of some of these new composite materials.

*Author to whom correspondence should be addressed.

GENERAL CONSIDERATIONS OF THE SOL-GEL PROCESS

Sol-gel reactions are also the key reactions in the preparation of inorganic/organic composite materials. They can be divided into two generalized steps: hydrolysis of metal alkoxides to produce metal hydroxides; this is followed by polycondensation of the hydroxyl groups to form a three dimensional network—often SiO_2.

CLASSIFICATION OF INORGANIC/ORGANIC COMPOSITES

One major advantage of the sol-gel process is the low temperature processing of ceramics to final materials at temperatures that are substantially lower than conventional ceramic processing conditions. On the basis of the molecular structure of inorganic and organic components, the connection between inorganic and organic phases, and the preparation technique, the inorganic/organic composite materials prepared by the sol-gel process can be divided into several groups.

Group I: Incorporation of Oligomeric/Polymeric Materials into the Inorganic Network with or without Covalent Bonds between Two Phases

Incorporation through Covalent Bonding

As mentioned above, the sol-gel process for metal alkoxides can be used to produce ceramics and glasses at relatively low temperatures. In the presence of preformed oligomers or polymers, generally functionalized with hydroxyl or trialkoxysilyl groups, inorganic/organic composite materials can be formed via the co-condensation of functionalized oligomers or polymers with metal alkoxides. An early example was that of the polydimethylsiloxane(PDMS)-TEOS system first studied by Wilkes and co-workers[2] and later by other researchers.[6–11]

Some other examples of functionalized oligomers and polymers that have been studied are polystyrene,[17] polyoxazolines (POZO),[18] polyimides,[19–21] polyamide,[21] poly(ether ketone) (PEK),[22] poly(ethylene oxide),[23] polybutadiene,[8,24] and poly(methyl methacrylate) (PMMA).[25,26]

Incorporation without Chemical Bonds between the Inorganic and Organic Phases

Inorganic/organic composite materials can also be formed through the sol-gel process without forming chemical bonds between two phases by mixing polymers with metal alkoxides in a common solvent or by impregnating them inside the porous inorganic network commonly found in oxide xerogels. In the first approach, polymers can be trapped within the oxide gel network if the hydrolysis and condensation of metal alkoxide are carried out in the presence of preformed polymers. Optically transparent composite materials can be obtained if there is no phase separation during both the gel-forming and the drying process. A number of organic polymers have been incorporated into inorganic network through this approach, such as PMMA,[28] polycarbonate,[28] poly(vinyl alcohol),[29,30] poly(methyl oxazoline),[31] poly(ethyl oxazoline),[32] poly(vinyl acetate),[28] poly(acylic acid),[33] poly(2-vinylpyridine),[34] polyacrylonitrile,[34] and cellulosics.[35,36] In the second approach, the hybrid materials can be obtained through the impregnation of porous oxide gels with polymerizable organic monomers. The impregnation of porous oxide gels with organics is followed by an *in situ* polymerization initiated by thermal or irradiation methods. For example, the transparent silica gel-PMMA composite can be prepared through the impregnation of porous silica gel with methyl methacrylate (MMA).[37,38]

Group II: Incorporation of Small Organic Molecules

Organics Embedded within Inorganic Networks

The basic idea in the synthesis of these composite materials is very similar to those discussed above except that small molecular weight organic molecules are used instead of oligomers and polymers. The doping or organic dyes into inorganic glasses is a well-studied area. The main driving force behind the study is the development of new optical materials in the applications of photophysical, electrical, and nonlinear optical (NLO) devices.[39–42]

Organoalkoxysilanes as the Precursors of Composite Networks

The ≡Si–C– bond opens the possibility to introduce organic groups R into an inorganic network by the sol-gel process. This approach has been extensively studied by Schmidt and coworkers since the early 1980s.[2,5,43–50] Specifically, organic molecules have been chemically bonded into the inorganic networks by using bifunctional, trifunctional, or polyfunctional alkoxysilanes ($R_nSi\ (OR')_{4-n}$, n = 1–3, R′ = alkyl) as the only or one of the precursors. Bifunctional alkoxysilanes have to be used in the presence of higher functionality precursors in order to form a three-dimensional network. Trifunctional alkoxysilanes, $RSi(OR')_3$, are the most common precursors to introduce organic groups within an inorganic network because a variety of such silanes is commercially available.

Group III: *In Situ* Formation of Inorganic Species within a Polymeric Network

Mark et al. also developed another kind of composite material by the sol-gel approach in the 1980s.[7,58–66] In his approach, the hydrolysis and condensation reactions of the sol-gel process are carried out within a polymeric matrix to generate *in situ* ceramic particles such as silica or titania within the polymer matrix. Of primary interest was the reinforcement provided by these fillers in elastomeric matrices. PDMS has been most studied in this regard because of its great miscibility with TEOS.

Group IV: Interpenetrating Networks and Simultaneous Formation of Inorganic and Organic Phases

A significant disadvantage of the sol-gel process is the removal of solvent and byproducts such as excess water and alcohol. Moreover, many polymers have poor solubility in the water/alcohol medium of the sol-gel solution. As a result, the approach of forming a composite material by the incorporation of preformed polymers discussed above is limited by these factors. Efforts to solve these problems are focused on forming composite materials by interpenetrating networks and simultaneous formation of inorganic and organic phases. By using triethyoxysilane as the precursor with the R group being a polymerizable epoxy group, Schmidt has synthesized SiO_2- or SiO_2/TiO_2-epoxide composite materials.[46–48]

Group V: Other Systems

Group V is designated to include all composite materials that do not belong to any of the above categories. A typical example of these materials is the so-called "nonshrinking" sol-gel composite materials developed by Novak and Ellsworth.[67–72]

Usuki and co-workers developed a new strategy for synthesizing inorganic/organic nanocomposite materials by exploiting the unique intercalation and self-assembling characteristics of layered ceramics.[70–72]

In a related approach, Giannelis also synthesized new inorganic/organic nanocomposite materials by exploiting the unique intercalation and self-assembling characteristics of layered ceramics.[73–75] In this approach, the intercalation of single polymer chains can be achieved either by *in situ* polymerization or by direct intercalation.

STRUCTURE-PROPERTY FEATURES

The systematic study of the general structure–property behavior for group I composites has received considerable attention because of the work of Wilkes and co-workers—particularly where the oligomeric/polymeric species were the dominant network material.

Microwave curing processing has also been utilized to promote the sol-gel reaction. The use of microwave processing is a quick and efficient method for preparing the ceramers into high-strength materials. The morphology for the ceramers made by microwave processing and thermal curing is nearly identical in that they display an average correlation length of ca. 10 nm.

APPLICATIONS

The incorporation of organic/oligomeric/polymeric materials into the inorganic networks by the sol-gel process offers a great opportunity to optimize specific properties independently, which have led to the development of a whole array of systems that have new applications or will have potential applications. So far, the number of commercial sol-gel products is still comparatively small but the promise of new technological uses remains. Some applications and potential applications of these composite materials are summarized as follows.

Scratch- and Abrasive-Resistant Hard Coatings and Special Coatings

The sol-gel process has been extensively applied in the preparation of coating materials because the shrinkage problem in the preparation of monolithic materials by the sol-gel process can be avoided. The introduction of organic/polymeric components into the inorganic network reduces the extent of shrinkage because less byproducts are produced. It also brings flexibility to the brittle inorganic network and, most importantly, increases the adhesion between the coating and polymer substrate.

Schmidt has developed a series of scratch- and abrasive-resistant coating materials that are based on Al_2O_3, ZrO_2, TiO_2, or SiO_2 as network formers and epoxy or methacrylate groups bound to Si via a ≡Si–C bond.[12-16,46–48] They can be thermally cured at low temperature and thus can be applied to organic polymers. Another series of abrasive-resistant hard coating materials that are based on organoalkoxysilanes (such as triethocysilane functionalized DETA or glycerol) and metal alkoxides as the precursors of composite networks have also been developed by Wilkes and co-workers (G. L. Wilkes, unpublished results).[51–57] They have been applied on PMMA, polycarbonate, and CR-39 lenses. The abrasion resistance of these coated polymers is improved substantially. High refractive index optical coatings can also be obtained.[26a,27]

NLO Materials

Organic materials play an important role in the development of optical systems such as fiber optics, wave guides, optical sensors, and NLO devices because they have higher optical response than many inorganic compounds. However, most organic compounds and polymers in their pure state are not good photonic media because of high optical losses. However, many inorganic glasses are excellent photonic media because of their low optical losses and good mechanical properties. The sol-gel technique offers a strong opportunity to produce useful optical materials by incorporating organic dyes or π conjugated polymer into the inorganic networks. Inorganic/organic composite materials made by sol-gel process exhibit exciting prospects for the applications in these areas because they combine the optical response of the organic structure with good optical quality, good mechanical property, and environmental stability of inorganic glasses. Depending on the type of organic materials used, second-order NLO materials, third-order NLO materials, and other optical materials can be obtained.[3,4,27,39–42,76,77]

Adhesives, Contact Lenses

Based on organoalkoxysilanes as one of the precursors of composite networks, hot melt adhesives for glass containers have been developed.

Hard contact lens materials were also developed by the hydrolysis and condensation of an epoxide-substituted alkoxysilane and titanium alkoxides. Glycol groups formed by the epoxide radicals offer good wettabilities. High oxygen permeabilities result from the silicone-like structure elements.[46–48]

Reinforcement of Elastomers and Plastics

It is very important in the area of polymers (especially for elastomers) to improve their mechanical properties because they often lack certain necessary mechanical properties. Reinforcement of elastomers is normally achieved by blending fillers (e.g., carbon black) into the polymers. It is difficult to control the morphological structure of the resulting material, particularly the degree of particle dispersion, resulting in inhomogeneous distributions of agglomerated silica particles.[78,79] *In situ* generation or inorganic filler, typically silica, through the sol-gel process provides an opportunity to overcome these problems.

Catalyst and Porous Supports, Adsorbents

Another important application of these inorganic/organic composite materials is to produce highly porous metal oxides having nanoscale pores. This can be readily done by calcination of the organic/inorganic composite at a temperature below the fusion temperature of metal alkoxide, normally 400–500°C. The resulting materials may find applications such as gas and liquid adsorbents, membranes, carrier for catalysts and enzymes, sensors, or ion exchangers.[5,31,43,44,82–85]

CONCLUSIONS AND FUTURE RESEARCH

The sol-gel process opens the possibility for the preparation of a whole new array of composite materials that display unique properties through the incorporation of organic/oligomeric/polymeric components with appropriate metal alkoxide or related species to produce organic/inorganic hybrid materials—many of them of network character. Depending on the chemical structure of metal alkoxides and organic components and the reaction conditions, the resulting composite can vary from soft and flexible to brittle and hard materials. It is possible to shift the morphology of these composite materials from that of a continuous glassy inorganic phase with separated organic domains to a continuous flexible organic domain with separated inorganic phases. While commercial applications of these materials are still relatively small, it will undoubtedly grow. Although the synthesis and application of these new composite materials have attracted interest from polymer scientist, ceramist, inorganic and organic chemist, and physicist, the cooperation between them is very important to understand the basic chemistry in conjuction with structure in order to speed up the applications of these new materials.

ACKNOWLEDGMENT

The authors acknowledge the financial support of the Air Force Office of Research.

REFERENCES

1. Wilkes, G. L.; Orler, B.; Huang, H. *Polym. Prep.* **1985**, *26*, 300.
2. Schmidt, H. *J. Non-Cryst. Solids* **1985**, *73*, 681.
3. Dislich, H. *Angew. Chem.* **1971**, *83*, 428.
4. Klein, L. C. *Sol-Gel Technology for Thin Films, Fiber, Preforms, Electronics and Specialty Shapes;* Noyes: Park Ridge, NJ, 1988.
5. Schmidt, H. *Mat. Res. Soc. Symp. Proc.* **1984**, *39*, 327
6. Parkhurst, C. S.; Doyle, L. A.; Silverman, L. A.; Singh, S.; Anderson, M. P.; McClurg, D.; Wnek, G. E.; Uhlmann, D. R. *Mat. Res. Soc. Symp. Proc.* **1986**, *73*, 769.
7. Mark, J. E.; Sun, C. C. *Polym. Bull.* **1987**, *18*, 259
8. Kohjiya, S.; Ochiai, K.; Yamashita, S. *J. Non-Cryst. Solids* **1990**, *119*, 132.
9. Chung, Y. J.; Ting, S. J.; Mackenzie, J. D. *Mat. Res. Soc. Symp. Proc.* **1990**, *180*, 981.
10. Morita, K.; Hu, Y.; Mackenzie, J. D. *Mat. Res. Soc. Symp. Proc.* **1992**, 271, 693.
11. Wen, J.; Mark, J. E. *Polym. J.* **1995**, *27*, 492.
12. Huang, H.; Wilkes, G. L. *Polym. Bull.* **1987**, *18*, 455.
13. Huang, H.; Orler, B.; Wilkes, G. L. *Macromolecules* **1987**, *20*, 1322.

14. Huang, H.; Glaser, R. H.; Wilkes, G. L. *Inorganic & Organometallic Polymers:* Zeldin, M.; Wynne, K. J.; Allcock, H. R., Eds; ACS Symposium Series 360; American Chemical Society: Washington, DC, 1987; p 354.
15. Huang, H.; Willies, G. L. *Polym. Prep.* **1987**, *28*, 244.
16. Huang, H.; Glaser, R. H.; Wilkes, G. L. *Polym. Prep.* **1987**, *28*, 434.
17. Chujo, Y.; Ihara, E.; Kure, S.; Suzuki, K.; Saegusa, T. *Makromol. Chem. Macromol. Symp.* **1991**, *42/43*, 303.
18. Morikawa, A.; Iyoku, Y.; Kakimoto, M.; Imai, Y. *J. Mat. Chem.* **1992**, *2*, 679.
19. Nandi, M.; Conklin, J. A.; Salvati, Jr., L.; Sen. A. *Chem. Mat.* **1991**, *3*, 201.
20. Wang, S.; Ahmad, Z.; Mark, J. E. *Chem. Mater.* **1994**, *6*, 943.
21. Ahmad, Z.; Wang, S.; Mark, J. E. In *Hybrid Organic-Inorganic Composites;* Mark, J. E., Ed.; ACS Series 585. Washington, DC, 1995; p 291.
22. Noell, J. L. W.; Wilkes, G. L.; Mohanty, D. K.; McCrath, J. E. *J. Appl. Polym. Sci.* **1990**, *40*, 1177.
23. Boulton, J. M.; Thompson, J.; Fox, H. H.; Gorodisher, I.; Teowee, G.; Calvert, P. D.; Uhlmann, D. R. *Mat. Res. Soc. Symp. Proc.* **1990**, *180*, 987.
24. Surivet, F.; Lam, T. M.; Pascault, J. P.; Pham, Q. T. *Macromolecules* **1992**, *25*, 4309.
25. Coltrain, B. K.; Landry, C. J. T.; O'Reilly, J. M.; Chamberlain, A. M.; Rakes, G. A.; Sedita, J. S. S.; Kelts, L. W.; Landry, M. R.; Long, V. K. *Chem. Mater.* **1993**, *5*, 1445.
26. Wei, Y.; Bakthavatchalam, R.; Whitecar, C. K. *Chem. Mater.* **1990**, *2*, 337.

26a. Wang, B.; Wilkes, G. L.; Smith, C. D.; McCrath. J. E. *Polym. Com.* **1991**, *32*, 400.

27. Wang, B.; Wilkes, G. L.; Hedrick, J. C.; Liptak, S. C.; McCrath, J. E. *Macromolecules* **1991**, *24*, 3449.
28. Landry, C. J. T.; Coltrain, B. K. *Polym. Prep.* **1990**, *32*, 514.
29. Messermith, P. B.; Stupp, S. I. *Polym. Prep.* **1991**, *32*, 536.
30. Suzuki, F.; Onozato, K. *J. Appl. Polym. Sci.* **1990**, *39*, 371.
31. Saegusa, T. *J. Macromol. Sci., Chem.* **1991**, *A28*, 817.
32. David, I. A.; Scherer, G. W. *Polym. Prep.* **1991**, *37*, 530.
33. Nakanishi, K.; Soga, N. *J. Non-Cryst. Solids* **1992**, *139*, 1.
34. Novak, B. M.; Auerbach, D.; Verrier, C. *Chem. Mater.* **1994**, *6*, 282.
35. Lu, S.; Melo, M. M.; Zhao, J.; Pearce, E. M.; Kwei, T. K. *Macromolecules* **1995**, *28*, 4908.
36. Novak, B. M.; Ellsworth, N.; Davies, C. *Extended Abstracts*, Japan-U.S. Joint Seminar on Inorganic and Organometallic Polymers, 1991; p 160.
37. Pope, E. J. A.; Asami, M.; MacKenzie, J. D. *J. Mat. Res.* **1989**, *4*, 1017; *J. Non-Cryst. Solids* **1988**, *101*, 198.
38. Abramoff, B.; Klein, L. C. In *Sol-Gel Optics I;* Mackenzie, J. D.; Ulrich, D. R., Eds.; Proc. SPIE 1328: San Diego, CA, 1990; p 241.
39. Avnir, D.; Levy, D.; Reisfeld, R. *J. Phys. Chem.* **1984**, *88*, 5956.
40. Avnir, D.; Braun, S.; Ovadia, L.; Ottologhim, M. *Chem. Mater.* **1994**, *6*, 1605.
41. Tanaka, H.; Takahashi, J.; Tsuchiya, J. *J. Non-Cryst. Solids* **1989**, *109*, 164.
42. Pang, Y.; Samoc, M.; Prasad, P. N. *J. Chem. Phys.* **1991**, *94*, 5382.
43. Schmidt, H.; Philipp, G. *J. Non-Cryst. Solids* **1984**, *63*, 983.
44. Schmidt, H.; Scholze, H.; Kaiser, A. *J. Non-Cryst. Solids* **1984**, *63*, 1.

44a. Schmidt, H.; Scholze, H.; Tunker, G. *J. Non-Cryst. Solids* **1986**, *80*, 557.

45. Schmidt, H. *J. Non-Cryst. Solids* **1989**, *112*, 419.

45a. Schmidt, H. In *Ultrastructure Processing of Ceramics, Glasses, and Composite*; Ulhmann, D. R.; Ulrich, D. R., Eds.; Tuscon, 1989.

46. Schmidt, H.; Wolter, H. *J. Non-Cryst. Solids* **1990**, *121*, 428.
47. Schmidt. H. *J. Sol-Gel Sci. & Tech.* **1995**, *5*, 115.
48. Schmidt, H. *Mat. Res. Soc. Symp. Proc.* **1990**, *180*, 961.
49. Luneau, I. G.; Mosset, A.; Galy, J.; Schmidt, H. *J. Mat. Sci.* **1990**, *89*, 3739.
50. Nab, R.; Schmidt, H. In *Sol-Gel Optics I*; Mackenzie, J. D.; Ulrich, D. R., Eds.; Proc. SPIE 1378: San Diego, CA, 1990; p 958.
51. Wang, B.; Wilkes, G. L. *J. Macromol. Sci., Pure Appl. Chem.* **1994**, *A31*, 349.
52. Betrabet, C.; Wilkes, G. L. *Polym. Prep.* **1992**, *33*(2), 286.
53. Tamami, B.; Betrabet, C.; Wilkes, G. L. *Polym. Bull.* **1993**, *30*, 393.
54. Tamami, B.; Betrabet, C.; Wilkes, G. L. *Polym. Bull.* **1993**, *30*, 39.
55. Wen, J.; Vasudevan, V. J.; Wilkes, G. L. *J. Sol-Gel Sci. Tech.* **1995**, *5*, 115.
56. Wen, J.; Wilkes, G. L. *J. Inorg. Organomet. Polym.* **1995**, *5*, 343.
57. Wen, J.; Jordens, K.; Wilkes, G. L. *Mat. Res. Soc. Symp. Proc.* **1996**, in press.
58. Mark, J. E.; Pan, S. J. *Makromol. Chem. Rapid Comm.* **1982**, *3*, 681.
59. Mark, J. E. *Frontiers of Macromolecular Science;* Saegusa, T.; Higashimura, T.; Abe, A., Eds.; Blackwell Scientific: Oxford, 1989.
60. Mark, J. E. *J. Appl. Polym. Sci., Appl. Polym. Symp.* **1992**, *50*, 273.
61. Mark, J. E.; Wang, S.; Xu, P.; Wen, J. *Mat. Res. Soc. Symp. Proc.* **1992**. *274*, 77.
62. Ning, Y. P.; Tang, M. Y.; Jiang, C. Y.; Mark, J. E.; Roth, W. C. *J. Appl. Polym. Sci.* **1984**, *29*, 3209.
63. Sohoni, G. B.; Mark, J. E. *J. Appl. Polym. Sci.* **1992**, *45*, 1763.
64. Wen, J.; Mark, J. E. *Macromol. Rep.* **1994**, *A31*, 429.
65. Mark, J. E.; Jiang, C.; Tang, M. *Macromolecules* **1984**, *17*, 2613.
66. Tang, M.; Mark, J. E. *Polym. Eng. Sci.* **1985**, *25*, 29.
67. Novak, B. M.; Davis, C. *Macromolecules* **1991**, *24*, 2481.
68. Ellsworth, M. W.; Novak, B. M. *J. Am. Chem. Soc.* **1991**, *113*, 2756.
69. Ellsworth, M. W.; Novak, B. M. *Chem. Mater.* **1993**, *5*, 839.
70. Usuki, A.; Kujima, Y.; Kawasumi, M.; Okada, A.; Fukushima, Y.; Kurauchi, T.; Kamigaito, O. *J. Mat. Res.* **1993**, *8*, 1179.
71. Kojima, Y.; Usuki, A.; Kawasumi, M.; Okada, A.; Kurauchi, T.; Kamigaito, O. *J. Polym. Sci., Polym. Chem.* **1993**, *31*, 983.
72. Kojima, Y.; Usuki, A.; Kawasumi, M.; Okada, A.; Fukushima, Y.; Kurauchi, T.; Kamigaito, O. *J. Mat. Res.* **1993**, *8*, 1185.
73. Mehrotra, V.; Giannelis, E. P. *Solid State Commun.* **1991**, *77*, 155.
74. Giannelis, E. P.; Mehrotra, V.; Tse, O.; Vaia, R. A.; Sung, T. C. *Synthesis and Processing of Ceramics: Scientific Issues;* Rhine, W. E. et al., Eds.; MRS: Pittsburgh, PA, 1992.
75. Giannelis, E. P. *J. Min. Met. Mater. Soc.* **1992**, *44*, 28.
76. Prasad, P. N. *Polymer* **1991**, *32*, 1746.
77. Prasad, P. N.; Bright, F. V.; Narang, U.; Wang, R.; Dunbar, R. A. *Proc. Am. Chem. Soc.* **1994**, *70*, 349.
78. Polmanteer, K. E.; Lentz, C. W. *Rubber Chem. Technol.* **1975**, *48*, 795.
79. Hamilton, J. R. *Silicone Technology;* Interscience: New York, 1970.
80. Wang, C. J.; Pang, Y.; Prasad, P. N. *Polymer* **1991**, *32*, 605.
81. Mauriu, K. A.; Warren, R. M. *Macromolecules* **1989**, *22*, 1730.
82. Saegusa, T.; Chujo, Y. *J. Macromol. Sci., Chem.* **1990**, *A27*, 1603.
83. Saegusa, T. *Proc. Am. Chem. Soc.* **1994**, *70*, 371.
84. Shea, K. J.; Loy, D. A. *Chem. Mat.* **1989**, *1*, 572.
85. Hardy, A. B.; Rhine, W. E.; Bowen, H. K. *Mat. Res. Soc. Symp. Proc.* **1990**, *180*, 1009.

Organic/Inorganic Hybrid Polymers

See: *Bridged Polysilsesquioxanes (Highly Porous Hybrid Organic-Inorganic Materials)*
Inorganic Nanostructured Materials
Inorganic/Organic Hybrid Polymers (High Temperature, Oxidatively Stable)
Organic/Inorganic Composite Materials
Organic/Inorganic Composites
Organic/Inorganic Polymer Hybrids

ORGANIC/INORGANIC POLYMER HYBRIDS

Yoshiki Chujo
Division of Polymer Chemistry
Kyoto University

Recently, composite materials of inorganic materials and organic polymers have attracted attention in the development of high-performance, highly functional polymeric materials, for example, addition of inorganic materials as a filler in organic polymers and coating organic polymers on the surface of metals.

This chapter describes organic-inorganic polymer hybrids prepared by the sol-gel procedure. In these hybrids, the organic polymer segments are dispersed in the silica gel matrix at the molecular level.

ORGANIC-INORGANIC POLYMER HYBRIDS THROUGH HYDROGEN BONDING

Preparations of polymer hybrids was carried out by means of a sol-gel reaction of silicates in the presence of organic polymers as illustrated in **Equation 1**. Typical examples of organic polymers are poly(2-methyl-2-oxazoline), poly(*N*-vinylpyrrolidone), and poly(*N,N*-dimethylacrylamide).[1,2] One organic polymer was dissolved in ethanol with tetraethoxysilane. The resulting mixture was subjected to an acid-catalyzed sol-gel reaction. After drying for several days, a homogeneous and transparent glass material was produced. These colorless, transparent homogeneous hybrids were obtained in a wide range of compositions, which demonstrated the properties of organic polymers.

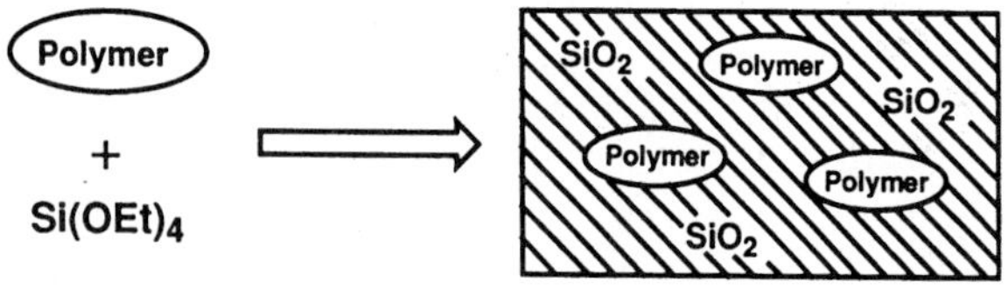

1

The homogeneity of the present hybrid suggests that the organic polymer segments and the inorganic one were blended at the molecular level. This may be due to strong hydrogen bonding between amide and silanol groups.

Organic polymer segments are dispersed in the hybrid matrix without aggregation. Accordingly, the properties of organic polymers, as an assembly such as thermal stability or glass transition temperature, differ from those of organic polymers themselves.

POLYMER HYBRIDS STARTING FROM NATURAL ORGANIC RESOURCES

Various functional groups are know as acceptors of hydrogen bonding in addition to amide groups, for example carbamate, carbonate, or ester groups. This means that introduction of these functional groups to organic materials makes it possible to form homogeneous polymer hybrids. For example, sucrose,[3] lignin,[4] and chitosan[5] gave only turbid materials because of their aggregation during sol-gel reaction of silicates. The hydroxyl groups of these materials were reacted with isocyanate to form carbamate groups. Urethane-modified materials (sucrose, lignin, and chitosan) were found to produce homogeneous, transparent polymer hybrids with silica gel by the sol-gel procedure.

COMPATIBILIZER FOR POLYMER HYBRIDS

Poly(2-methyl-2-oxazoline) has high compatibility with polar organic polymers such as poly(vinyl chloride) or polyamides.[6,7] On the other hand, poly(vinyl chloride) or poly(methyl mechacrylate) forms only heterogeneous materials with silica gel by sol-gel procedure. In the combination of poly(vinyl chloride) or poly(methyl methacrylate) with silicate, however, addition of poly(2-methyl-2-oxazoline) produced homogeneous ternary polymer hybrids.[8] In other words, poly(2-methyl-2-oxazoline) plays a role of compatibilizer between organic polymers and inorganic materials.

POROUS SILICA BY PYROLYSIS OF POLYMER HYBRIDS

Blending at molecular-level dispersion between silica gel and organic polymers also has been supported by the number, size, and surface area of the pores of porous silica gel prepared after burning the above hybrids at 600°C.[9] No fusion of silica gel takes place at this temperature.

Porous silica also was produced from polymer hybrids by solvent extraction. Soxhlet extraction of polymer hybrids with protic solvent, which cleaves the hydrogen bonding in the hybrid matrix, eliminated organic polymers completely, producing porous silica.

POLYMER HYBRIDS CONTAINING TRANSITION METAL SALTS

In the system of polymer hybrid between poly(2-methyl-2-oxazoline) and silica gel, transition metal salts, such as cupric chloride, cupric acetate, or nickel chloride, were added to form homogeneous colored polymer hybrids. In these materials, transition metal salts are dispersed in the matrix.[10] The polymer hybrids thus obtained can be used as an effective transition metal catalyst-supported inorganic material (ceramics).

Pyrolysis of these polymer hybrids containing metal salts produced porous silica with metals after elimination of organic polymer components. These materials can be used as a highly effective catalyst.

CONCLUSIONS

Molecular hybrids between organic polymers and silica gel are expected to have many possibilities as new composites. Hybrids may show intermediate properties between plastics and glass (ceramics). In addition, the composition of these hybrids can be widely varied. In other words, the hybrids can be used to modify organic polymer or inorganic glassy materials. Porous silica with controlled pore size also is important as a catalyst for size-selective organic reactions.

REFERENCES

1. Saegusa, T.; Chujo, Y. *J. Macromol. Sci. Chem.* **1990,** *A27,* 1603.
2. Chujo, Y.; Ihara, E. et al. *Makromol. Chem., Macromol. Symp.* **1991,** *42/43,* 303.
3. Chujo, Y.; Kure, S. et al. *Polym. Prep. Jpn.* **1992,** *41,* 733.
4. Chujo, Y.; Fujita, T.; Tamaki, R. *Polym. Prep. Jpn.* **1994,** *43,* 1960.
5. Tamaki, R.; Chujo, Y. *Polym. Prep. Jpn.* **1994,** *43,* 1960.
6. Saegusa, T.; Chujo, Y. *Makromol. Chem., Macromol. Symp.* **1990,** *33,* 31.
7. Saegusa, T.; Chujo, Y. *Makromol. Chem., Macromol. Symp.* **1991,** *51,* 1.
8. Chujo, Y.; Matsuki, H.; Tamaki, R. *Polym. Prep. Jpn.* **1993,** *42,* 2973.
9. Chujo, Y.; Matsuki, H.; Yazawa, T. *Polym. Prep. Jpn.* **1993,** *42,* 3545.
10. Chujo, Y.; Kure, S. et al. *Proc. Jpn. Acad.* **1994,** *70*(B), 138.

ORGANOBORON MAIN CHAIN POLYMERS

Yoshiki Chujo
Division of Polymer Chemistry
Graduate School of Engineering
Kyoto University

In organic synthesis, organoboron compounds are known as versatile reagents or reaction intermediates for preparation of a wide variety of functional compounds.[1–4] Hydroboration reaction (addition of B-H to C=C) is a well known tool for preparation of various alkylborane compounds in organic synthesis, which takes place under mild conditions almost quantitatively. However, direct use of this reaction in polymer synthesis has been limited. This chapter describes the synthesis of organoboron main chain polymers and their use as a novel type of reactive polymers.

ORGANOBORON POLYMERS PREPARED BY HYDROBORATION POLYMERIZATION

We explored a polyaddition between diene and monoalkylborane and termed this "hydroboration polymerization."[5] The organoboron polymers obtained are effectively converted to polymers with various functional groups. In other words, polymers with organoboron units in the main chains are a new type of reactive polymers. In this section, the scope of these hydroboration polymerizations is described as novel methods for preparation of organoboron polymers.

Thexylborane was used as a monoalkylborane (bifunctional hydroborane) component in hydroboration polymerization because thexylborane is known to be stable without disproportionation. As a diene monomer, relatively longer chain dienes were used to avoid a competitive cyclization reaction, shown in **Scheme I**, which proceeds in THF at 0°C under nitrogen without catalyst.[5]

I

Hydroboration polymerization can be applied to various combinations of dienes such as 1,9-decadiene, *p*-divinylbenzene, *p*-diallylbenzene, bis(allyl ether)s of ethylene glycol, tetramethylene glycol, triethylene glycol, hydroquinone, and bisphenol-A with thexylborane to produce corresponding organoboron polymers, which were stable against protic solvents such as water and alcohol under nitrogen. Under air, these polymers were unstable enough to decompose as usual for organoboron compounds.

VERSATILE REACTIONS OF ORGANOBORON POLYMERS

Organoboron polymers prepared by hydroboration polymerization have a new structure consisting of C-B bonds in the main chain. Generally, they can be regarded as a polymer homologue of trialkylborane, which is a versatile compound in organic synthesis. In other words, the polymers obtained by hydroboration polymerization are new reactive polymers.

Versatile reactions of organoboron main chain polymers prepared by hydroboration polymerization are demonstrated in **Scheme II**. These conversions demonstrate the useful and versatile characteristics of organoboron main chain polymers as polymeric precursors to give functional polymers.[6]

POLY(ORGANOBORON HALIDE)S

We reported a polyaddition between boron tribromide and terminal diynes as a novel methodology for preparation of poly(organoboron halide)s,[7] which we called "haloboration polymerization."

POLYCYCLODIBORAZANES

Hydroboration reaction of cyano groups gives iminoborane species, which dimerize to form a boron-nitrogen four-membered ring (cyclodiborazane) in the case of appropriate borane used.[8] When this reaction is used for bifunctional monomers, formation of polymeric materials consisting of cyclodiborazane units can be expected.

REFERENCES

1. Brown, H. C. Hydroboration; W. A. Benjamin: New York, 1962.
2. Brown, H. C. *Organic Synthesis via Boranes;* Wiley-Interscience: New York, 1975.
3. Pelter, A.; Smith, K.; Brown, H. C. *Borane Reagents*; Academic: London, 1988.
4. *Current Topics in the Chemistry of Boron;* Kabalka, G. W., Ed.; The Royal Society of Chemistry: Cambridge, 1994.
5. Chujo, Y.; Tomita, I. et al. *Macromolecules* **1991,** *24,* 345.
6. Chujo, Y. *J. Macromol. Sci., Pure Appl. Chem.* **1994,** *A31,* 1647.
7. Chujo, Y.; Tomita, I.; Saegusa, T. *Macromolecules* **1990,** 23, 687.
8. Hawthorne, M. F. *Tetrahedron* **1962,** *17,* 112.

Organometallic Polymers

See: *Conducting Polymers (Nonconjugated)*
Organometallic Polymers, Cobalt-Containing
Organometallic Polymers (Transition Metals in Main Chain)
Organometallic Polymers, Fluoroarylene Bridged (Rigid Rods to Segmented Chains)

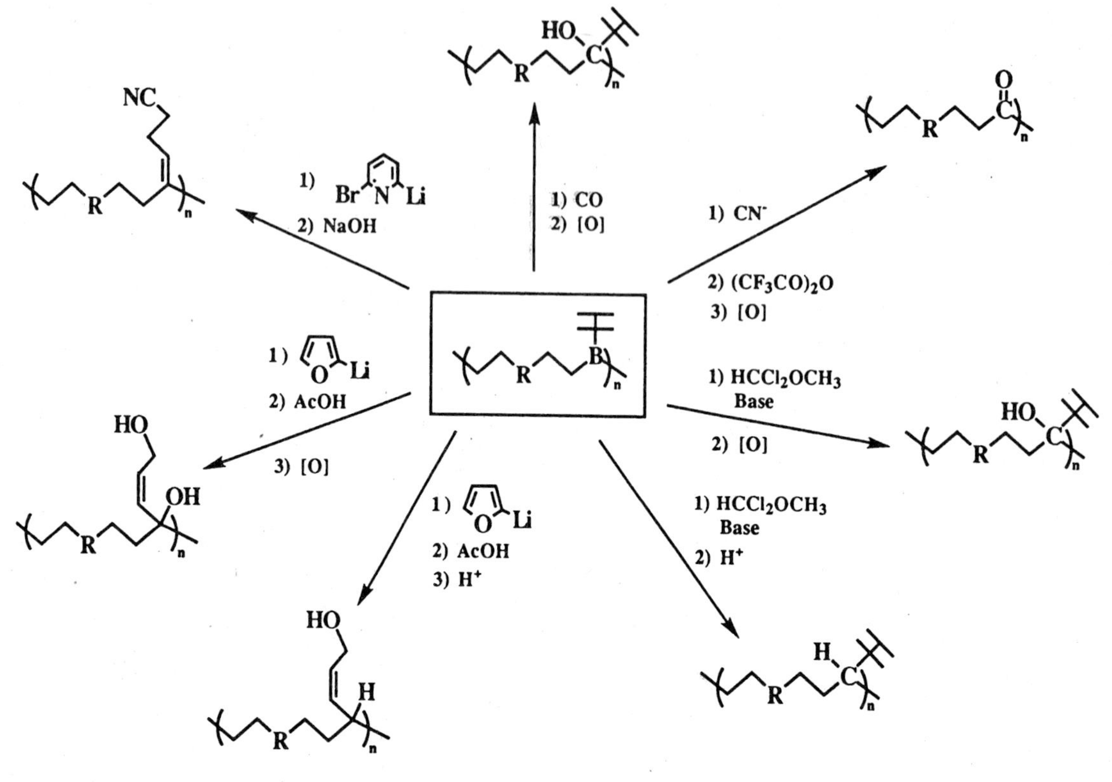

II

ORGANOMETALLIC POLYMERS (Transition Metals in Main Chain)

Shigetoshi Takahashi
The Institute of Scientific and Industrial Research
Osaka University

Kenkichi Sonogashira
Faculty of Engineering
Fului University of Technology

An organometallic polymer is defined as one in which the metal-containing portion is either incorporated as an integral part of the polymer or bound to the polymer structure by a covalent bond.

Organometallic polymers containing transition metals in the main chain are attracting interest because of their potential as specialty materials. They offer unusual electrical, magnetic, and optical properties and might also be useful as synthetic pyrolytic precursors to new ceramics. We will describe two typical classes, metallocene polymers and M-C(sp) σ-bound polymers, of organometallic polymers containing transition metals in the main chain.

Synthesis and study of polymers and oligomers are extremely active aspects of organometallic chemistry because the resulting materials may be expected to possess many novel thermal, electronic, and optical properties.[2]

ORGANOMETALLIC POLYMERS CONTAINING METALLOCENES IN THE MAIN CHAIN

Ferrocene-containing polymers prepared by vinyl polymerization or polycondensation have been studied for the past 40 years.[3] Organometallic polymers containing metallocene in the main chain or side chain are especially interesting because such polymers have been used in modified electrodes, electrochemical sensors, and nonlinear optical devices.[4]

ORGANOMETALLIC POLYMERS CONTAINING M-C (sp) σ-BONDS IN THE MAIN CHAIN

Polymer structures constructed by a σ-bonding between metal and carbon atoms seem to be the most primitive organometallic polymers containing metal atoms in the main chain. Transition metals are unlikely to form stable polymers, though nontransition metal polymers like -$[SiR_2(CR_2)_m]_n$- are known, because in general a sigma bond between transition metal and sp^3-carbon atoms is thermodynamically and kinetically unstable, and cleavage of the bond leading to the decomposition of polymer structures may easily occur. However, M-C(*sp*-hybrid) sigma bonding is relatively stable because of stabilization by dπ-pπ interaction and the ionic character between transition metal and *sp*-carbon atoms. Both stable σ-alkynyl transition metal complexes and organometallic polymers called "transition metal-poly-yne polymers" are known.[1]

Polymer Synthesis of Metal-Poly-Yne Polymers

The first synthesis[5] of metal-poly-yne polymers was reported by Hagihara et al. in 1977. The polymers consist of platinum atoms and butadiynediyl units, and were made by a novel method[6] that uses a copper(I) catalyst and an amine for condensation polymerization between platinum halides and terminal acetylenes. This method gives platinum-poly-yne polymers having a variety of polymer sequences with butadiyne and diethynylbenzene bridging ligands.[1,5] Palladium- and nickel-poly-yne polymers[7,8] were also synthesized by the same or a modified method.

ORGANOMETALLIC POLYMERS CONTAINING M-C(sp^2) σ-BONDS IN THE MAIN CHAIN

In special cases, a σ-bond between a transition metal and *sp*2-hybrid carbon is stable enough to form isolable polymeric molecules. The first example of an arene-bridged organometallic polymer was reported by Hunter et al.[9]

ORGANOMETALLIC DENDRIMERS

Several reports describe the successful routes to organic[10] and organometallic[11] dendrimers. Organometallic complexes may be useful as building blocks with structural variability and can also provide a new strategy for the construction of dendrimers having novel topologies.

A new methodology for the construction of three-dimensional network macromolecules was developed for the synthesis of ferrocene containing dendrimers.[12]

REFERENCES

1. Hagihara, N., et al. *Adv. Polym. Sci.* 1981, *41*, 149; Chisholm, M. H. *Angew. Chem., Int. Ed. Engl.* **1991**, *30*, 673.
2. Nalwa, H. S. *Appl. Organometal. Chem.* **1991**, *5*, 349.
3. Neuse, E. W. *Adv. Macromol. Chem.* **1968**, *1*, 1; Lorkowski, H. J. *Top. Curr. Chem.* **1967**, *9*, 207; Pittman, C. U. Jr. *J. Paint Technol.* **1967**, *39*, 585; Carraher, C. E. Jr. et al., Eds., *Organometallic Polymers;* Academic: New York, 1978.
4. Murray, R. W. *Electroanal. Chem.* **1983**, *3*, 246; Hale, P. D. et al. *J. Am. Chem. Soc.* **1989**, *111*, 3482; Wright, M. E.; Svejda, S. A. *Materials for Nonlinear Optics Chemical Perspectives* Marker, S. R. et al., Eds.; ACS Symposium Series 455; American Chemical Society: Washington, DC, 1991; p 603.
5. Hagihara, N. et al *Macromolecules* **1977**, *10*, 879.
6. Sonogashira, K. et al. *J. Organomet. Chem.* **1978**, *145*, 101; Takahashi, S. et al., *Macromolecules* **1978**, *11*, 1063.
7. Takahashi, S. et al. *J. Polym. Sci., Polym. Chem. Ed.* **1982**, *20*, 565.
8. Sonogashira, K. et al. *J. Organometal. Chem.* **1980**, *188*, 237.
9. Sturge, K. C. et al. *Organometallics* **1992**, *11*, 3056; McDonald, R. et al. *Organometallics* **1992**, *11*, 893.
10. Tomalia, D. A. et al. *Angew. Chem., Int. Ed. Engl.* **1990**, *29*, 138; Hodge, P. *Nature* **1993**, *362*, 18.
11. Bunz, U. H. F.; Enkelmann, V. *Organometallics* **1994**, *13*, 3823, and references cited therein.
12. Galloway, C. P.; Rauchfuss, T. B. *Angew. Chem., Int. Ed. Engl.* **1993**, *32*, 1319.

ORGANOMETALLIC POLYMERS, FLUOROARYLENE BRIDGED (Rigid Rods to Segmented Chains)

Allen D. Hunter
Department of Chemistry
Youngstown State University

Andrew X. Guo
Department of Chemistry
University of Alberta

Over the past 20 years, there has been extensive interest in the synthesis of organic polymers having delocalized π systems. The most widely studied materials are *trans*-polyacetylene; various polyaromatic systems, such as poly(*p*-phenylene), polythiophene, and polypyrrole, and their derivatives.

Transition metal-containing polymers also have the potential for conjugation down their backbones. In addition, work on organometallic catalysts[1] has demonstrated that the electronic properties of such discrete species can generally be tuned independently of steric factors, and presumably of processing characteristics such as melting point and solubility.

Very little work has been reported in which the metal atoms of an organometallic polymer are in the polymer backbone and are attached to the bridging ligands by metal–carbon covalent bonds. One class of such materials is compounds in which metal centers are bridged by diisocyanoarenes.[1a–3,5,8–14]

Another class includes acetylide-bridged complexes that have been reported for Cu, Ni, Pd and Pt.[4,6,7,15–24] These main-chain organometallic polymers display interesting conjugation, stability, and processability characteristics and demonstrate the potential of such materials for polymer applications.

Our work on potentially conjugated main-chain organometallic polymers has focused on two classes of such materials. The first class has square-planar metal-phosphine centers bridged by electron deficient aromatic ligands (**Structure 1**):

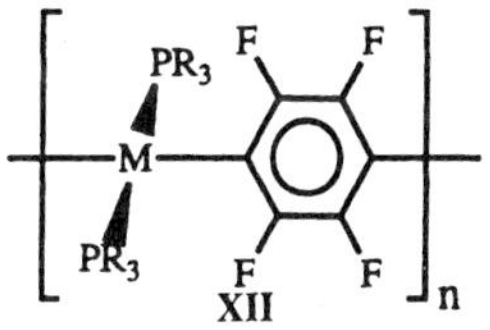

XII

1

The second class has octahedral metal-phosphine centers bridged by bifunctional isonitrile ligands (**Structure 2**).

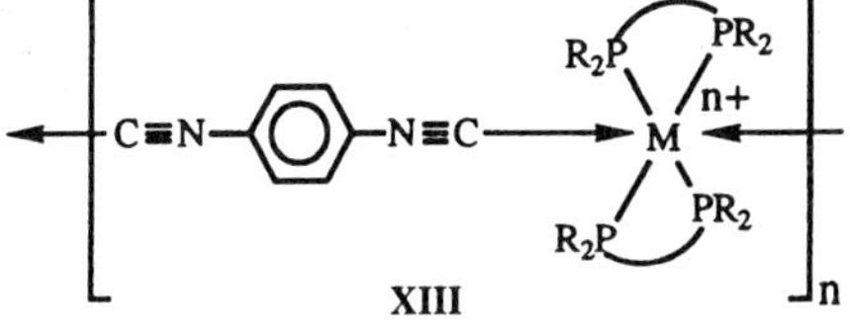

XIII

2

APPLICATIONS

In general, these studies demonstrate that rationally designed organometallic polymers and alternating copolymers containing nickel–phosphine units or organosilicon units alternating with highly fluorinated arenes in their backbones can be readily prepared in moderate to high molecular weights. These new polymers display excellent thermal, oxidative, and hydrolytic stabilities superior to those of many conjugated organic polymers. As originally hoped, we have demonstrated that those characteristics most closely related to polymer processing properties (i.e., solubility, melting point, glass transition temperature) can be rationally tuned to give materials that are readily characterizable and easily processable. We are attempting to extend these studies and improve upon the electronic, nonlinear optical, and mechanical properties of these materials.

REFERENCES

1. Tolman, C. A. *Chem. Rev.* **1977**, *77*, 313.
1a. Collman, J. P.; McDevitt, J. T.; Leidner, C. R.; Yee, G. T.; Torrance, J. B.; Little, W. A. *J. Am. Chem. Soc.* **1987**, *109*, 4606.
2. Feinstein-Jaffe, I.; Maisuls, S. E. *J. Organomet. Chem.* **1987**, *326*, C97.
3. Feinstein-Jaffe, I.; Frowlow, F.; Wackerle, L.; Goldman, A.; Efraty, A. *J. Chem. Soc., Dalton Trans.* **1988**, 469.
4. Fyfe, H. B.; Mlekuz, M.; Zargarian, D.; Taylor, N. J.; Marder, T B. *J. Chem. Soc., Chem. Commun.* **1991**, 188.
5. Kobel, W.; Hanack, M. *Inorg. Chem.* **1986**, *25*, 103.
6. Ogawa, H.; Onitsuka, K.; Joh, T.; Takahashi, S.; Yamamoto, Y; Yamazaki, H. *Organometallics* **1988**, *7*, 2257.
7. Keaton, E. *Organic Semiconducting Polymers;* Marcel Dekker: New York, 1968.
8. Bennett, D. W. et al. *Inorg. Chem.* **1988**, *27*, 2945.
9. Chatt, J.; Elson, C. M.; Pombeiro, A. J. L.; Richards, R. L.; Royston, G. H. D. *J. Chem. Soc., Dalton Trans.* **1978**, 165.
10. Fischer. H.; Seitz, F.; Muller, G. *Chem. Ber.* **1987**, *120*, 811.
11. Hieber, W.; von Pigenot, D. *Chem. Ber.* **1956**, *89*, 193.
12. Hoskins, B. F.; Robson, R. *J. Am. Chem. Soc.* **1990**, *112*, 1546.
13. Singleton, E.; Oosthuizen, H. E. *Adv. Organometal. Chem.* **1983**, *22*, 209.
14. Yamamoto, Y. *Coord. Chem. Rev.* **1980**, *32*, 193.
15. Amer, A. et al. *J. Polym. Sci. Polym. Lett.* **1984**, *22*, 77.
16. Davies, S. J.; Johnson, B. F. G.; Khan, M. S.; Lewis, J. *J. Chem. Soc., Chem. Commun.* **1991b**, 187.
17. Guenther, M. D.; Bezoari, M. D.; Kovacic, P. *J. Polym. Sci., Polym. Lett.* **1984**, *22*, 65.
18. Sonogashira, K.; Ohga, K.; Takahashi, S.; Hagihara, N. *J. Organomet. Chem.* **1980**, *188*, 237.
19. Sonogashira, K.; Kataoka, S.; Takahashi, S.; Hagihara, N. *J. Organomet. Chem.* **1978**, *319*, 160.
20. Takahashi, S.; Murata, E.; Sonogashira, K.; Hagihara, N. *J. Polym. Sci.*, Part A: *Polym. Chem.* **1980**, *18*, 661.
21. Takahashi, S.; Morimoto, H.; Murata, E.; Kataoka, S.; Sonogashira, K.; Hagihara, N. *J. Polym. Sci.*, Part A: *Polym. Chem.* **1982b**, *20*, 565.
22. Takahashi, S.; Kariya, M.; Yatake, T.; Sonogashira, K.; Hagihara, N. *Macromolecules* **1978**, *11*, 1063.
23. Takahashi, S.; Kariya, M.; Yatake, T.; Kataoka, S.; Sonogashira, K.; Hagihara, N. *J. Polym Chem.* **1982a**, *20*, 565.
24. Ogawa, H.,; Joh, T.; Takahashi, S., Sunugashira, K. *J. Chem. Soc., Chem. Commun.* **1985**, 1220.

ORGANOMETALLIC POLYMERS, COBALT-CONTAINING

Takeshi Endo* and Ikuyoshi Tomita
Research Laboratory of Resources Utilization
Tokyo Institute of Technology

Polymers containing carbon-metal bonds in the main chain connected with appropriate organic systems (like a conjugated one) may have potential functions, such as electron conductivity, optical nonlinearity, and liquid crystallinity. Further, based on the reactivity and the catalytic activity of the organometallics, organometallic polymers (i.e., the polymer-homolog of organometallics) may serve as interesting reactive materials and polymeric catalysts.

It is well known that some transition metals react with two equivalents of acetylenes to give metallacyclopentadienes that sometimes show good air stability. Among them, derivatives of cobaltacyclopentadiene[1] are easy to prepare, stable under air, and have catalytic activities[2] as well as moderate reactivities[3] to be converted to cyclic compounds. Therefore, polymers containing cobaltacyclopentadiene moieties as repeating units may show interesting properties and reactivities.

PREPARATION AND PROPERTIES OF ORGANOMETALLIC POLYMERS

Organocobalt polymers (**3a–c**) can be prepared by the reaction of $CpCo(PPh_3)_2$ (**1**) and the corresponding diynes (**2a–c**) in toluene at 50–60°C for 1–3 days (**Scheme I**).[4] The molecular weights of the resulting polymers reached up to 2.0×10^5 when the monomers were purified enough.[5,6] The polymers thus obtained are quite stable under air and can be kept at least for several months without any change in molecular weight and structure.

APPLICATIONS

The macromolecular design of organocobalt polymers can be made easily by using designed diynes. These polymers have unique structures connected with metallacyclopentadiene rings. By using the same synthetic methodology, the metal included in the main chain might be varied. Characteristics such as electron conductivity, nonlinear optical properties, and liquid crystallinity may be revealed, by the appropriate molecular design. Partial decomposition of the organocobalt polymers by the pyrolysis may give an interesting catalyst system bounded on the organic polymers.

Organocobalt polymers were converted to organic polymers with 2-pyridone, dithiolactone, and pyridine moieties in the main chain by the reactions with isocyanates, carbon disulfide, and nitriles. Polymers with other ring systems such as thiophene, pyrrole, and benzene may be prepared starting from organocobalt polymers. In those cases, metallacycles in the main chain were rearranged to other cyclic systems by the polymer reactions. The polymer reactions accompanying the reconstruction of the main chain were quite rare; that may be, however, served as novel polymer synthetic methods in the near

*Author to whom correspondence should be addressed.

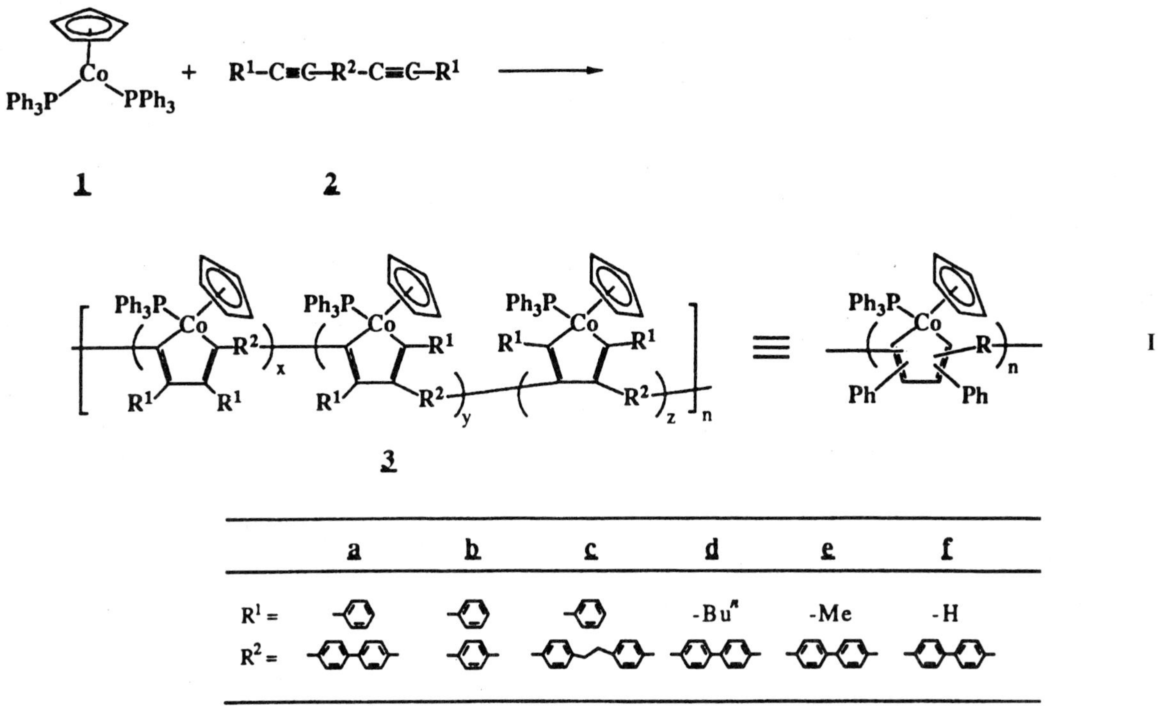

future. In this sense, organocobalt polymers may be applied as polymer precursors to various kinds of functional polymers.

REFERENCES

1. Yamazaki, H.; Wakatsuki, Y. *J. Organomet. Chem.* **1977**, *139*, 157.
2. Wakatsuki, Y.; Yamazaki, H. *J. Chem. Soc. Dalton Trans.* **1982**, 1923.
3. Hong, P.; Yamazaki, H. *Synthesis* **1977**, 50.
4. Tomita, I.; Nishio, A.; Igarashi, T.; Endo, T. *Polym. Bull.* **1993**, *30*, 179.
5. Tomita, I.; Nishio, A.; Igarashi, T.; Endo, T. *Polym. Prepr. Jpn.* **1993**, *42*, 3575.
6. Tomita, I.; Nishio, A.; Endo, T. *Pacific Polym. Prepr.* **1993**, *3*, 661.

ORGANOTIN POLYMERS

Rongbao Wei, Liang Ya, Wu Jinguo, and Xue Qifeng
Department of Chemical Engineering
Tianjin Institute of Technology

The first organotin compound was prepared by Frankland in 1849. In the past decade or so, organotin compounds have received great attention because of their characteristic properties.[1]

Carraher et al.[2–6] have performed much work on organotin polymers. Through interfacial synthesis, they reacted bisalkyltin dichlorides (R_2SnCl_2) with diacids, diols (diphenos), and diamines to produce a series of organotin polymers.

One important use of these compounds is in the plastics industry as PVC heat stabilizers, in film for food packings, and in PVC plastic articles. They have excellent transparency. Other applications include antiseptic and antifouling compounds in industry and agriculture. Simple organotin compounds such as $R_2Sn(SCH_2COOC_8H_{17}\text{-}i)_2$ And $R_2Sn(OOCC_{11}H_{23}\text{-}n)_2$, in which *R* is methyl, *n*-butyl, or *n*-octyl, are extensively used. On the other hand, organotin polymers are seldom used.

We have prepared a series of new organotin polyesters, polyethers, polyamines, polythioethers, polyaminoesters, and polyetheresters through dialkyl or bisalkoxycarbonylalkyl dichlorides reacting with organic diacids, diols or diphenols, diamines, dithiols, amino acids, and hydroxy acids (**Structure 1**).[7–11]

The polymers synthesized were determined by TG/DTG, IR, $\overline{M}_n$, etc., and tested for their effect on the heat stabilizers of PVC resins.

CONCLUSIONS

A number of new organotin polyesters, polyethers, polyamines, poly(amine ester)s, poly(ether ester)s, and polythioethers have been prepared by interfacial polycondensation. Molecular weights ($\overline{M}_n$) are about 10^3. They are soluble in $CHCl_3$, DMSO, DMF, DMAC and methylpyridine; they can inhibit PVC decomposition at temperatures as high as 190–269°C. Therefore, these polymers have a potential of application stabilizers on PVC products.

REFERENCES

1. Davies, A. G.; Smith, P. J. *Adv. Inorg. Chem. Radiochem.* **1980**, *23*, 1.
2. Carraher, C. E., Jr.; Dammeier, R. *Makromol. Chem.* **1971**, *141*, 245–251.
3. Carraher, C. E., Jr.; Dammeier, R. *Makromol. Chem.* **1970**, *135*, 153.
4. Carraher, C. E., Jr.; Scherubel, G. A. *Makromol. Chem.* **1972**, *152*, 61–66.
5. Carraher, C. E., Jr.; Winter, D. O. *Makromol. Chem.* **1971**, *141*, 250–264.

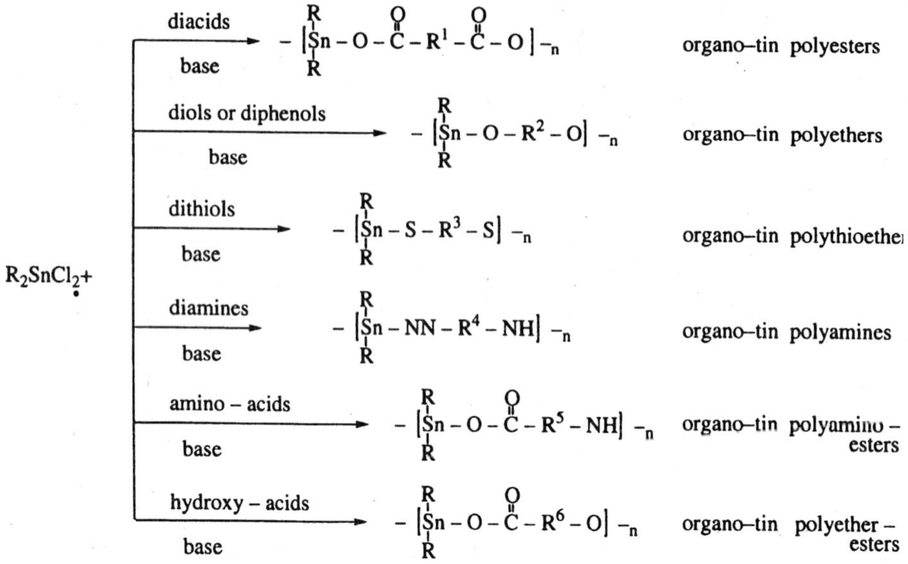

1

R = n – Bu, Me, $CH_3OCOCH_2CH_2$–, $C_2H_5OCOCH_2CH_2$–, $CH_3OCOCH(CH_3)CH_2$–, $CH_3CH_2CH_2CH_2OCOCH_2CH_2$–,

R^1 = $-(CH_2)-_n$ n=0 – 12, o, m, p–C_6H_4–, –C ≡ C–

R^2 = $-C_6H_4-$, $-C_6H_4-SO_2-C_6H_4-$, $-C_6H_4-CMe-C_6H_4-$, $-(CH_2)-_n$ n = 2, 4, 6

R^3 $-(CH_2)-_n$ n = 2, 3, 6

R^4 $-(CH_2)-_n$ n = 2 – 8, 10, $-C_6H_4-CH_2-C_6H_4-$, $-C_6H_4-CH_2-CH_2-C_6H_4-$, $-C_6H_4-SO_2-C_6H_4-$, $-C_6H_4-O-C_6H_4-$

R^5 $-(CH_2)-n$ n = 1, 3, 5

R^6 p – phenylene

6. Carraher, C. E., Jr.; Winter, D. O. *Makromol. Chem.* **1972**, *152*, 55–59.
7. Rongbao, Wei. et al. *Acta Polym. Sinica* **1992**, *4*, 403.
8. Rongbao, Wei. et al. *Tianjin Technol.* **1989**, *1*, 1.
9. Rongbao, Wei. et al. *Tianjin Technol.* **1989**, *2*, 1.
10. Rongbao, Wei. et al. *Chin. React. Polym.* **1993**, 2(1), 23.
11. Rongbao, Wei. et al. *Tinjin Hua Gong* **1991**, *1*, 4.

Organs, Artificial

See: Hybrid Artificial Organs

ORIENTAL LACQUER

Ju Kumanotani
Institute for Industrial Science
The University of Tokyo

Natural oriental lacquer is made from the sap of lacquer trees and has been used in China for 5000 years.[1] The lacquer is curable at ambient temperature or by heating and is a hard high-gloss polymeric material with high resistance to water and acid and is very durable. It can be pigmented red with HgS or black with iron. In addition, the lacquer exhibits good adhesion on wood, paper, textiles, iron, and clay.

In Asia, three kinds of lacquer trees are grown: *Rhus vernicifera* (in China, Korea, and Japan), *R. succedanea* (in Taiwan and Vietnam), and *Merranhohoea usitate* (in Burma and Thailand). the sap from *R. vernicifera* is the most highly prized.

COLLECTION OF SAP AND ITS BLEND

Oriental lacquer is a 100% solid, biodegradable material that is curable at room temperature. Its use as a renewable resource in the future may be beneficial environmentally compared with the use of synthetic coatings. However, the wide use of this lacquer has some disadvantages: the low yield or low productivity of sap from lacquer trees: the heavy labor involved in sap collection; the low drying rate because it is controlled by diffusion of laccase or oxygen, preventing drying at a higher temperature, and the effect of causing an allergic reaction in those who come in contact with the lacquer.

PROPERTIES OF CONSTITUENTS OF SAP (*R. vernicifera,* CHINA AND JAPAN)

Urushiol

Urushiol is a mixture of 3-substituted catechol derivatives with 15 carbon chains with 0 to 3 olefins. It is sometimes contaminated with other phenolic lipids such as those having

17 carbon chains instead of 15. The triene side chain of urushiol makes up 55–70 of the urushiol in sap from *R. vernicifera*.[2–6]

Laccase, Stellacyanin, and Peroxidase

Laccase, a reductase–oxidation copper glycoprotein is most important for the polymerization of urushiol. Stellacyanin, a copper glycoprotein, has a molecular weight of 20,000 with a ratio of sugar to protein of 4060, although its role in sap is not yet elucidated.[7]

Lacquer peroxidase is found in *R. vernicifera* and *R. succedanea*, with a molecular weight of 50,000 and 45,000, respectively.[8,9] It may participate in the decomposition and subsequent polymerization of the hydroperoxides of the unsaturated side-chain moiety of urushiol.

Plant Gum

Polysaccharide isolated from the sap of *R. vericifera* (China) by ion-exchange chromatography (IEC) is highly branched and composed of two fractions with weight-average molecular weights of 84,000 and 27,700 determined by aqueous-phase gel permeation chromatography (GPC).

Lacquer Glycoprotein

This glycoprotein has not yet been clarified. Its ratio of sugar to protein is 1:9 (w/w), and its molecular weight is expressed in terms of 7800_n-SS-, where *n* is given a value from 1 to 10.

POLYMERIZATION OF URUSHIOL IN DRYING OF THE LACQUER

Theoretical arguments for the reactivity of the possible reaction species have been given, and studies of model reactions have been done.[10–21]

DURABILITY OF THE LACQUER

In contrast to sap films, the lacquer film, which is composed of densely packed grains (<0.1 mm in diameter), is highly durable for aerobic oxidation.[22–24] It is postulated that the wall of each grain is composed of polysaccharides and that the inside is polymerized urushiol. Each grain is held together with polymerized urushiol and glycoproteins, and the inside of the grain is prevented from aerobic oxidation because polysaccharides act as barriers to oxygen.

SYNTHESIS OF ORIENTAL LACQUER

Oriental lacquer is a complex bioproduct. Except for the probable bio- or chemical synthesis of each constituent in the future, the synthesis of the lacquer may be tentatively classified into three groups. The first is the lacquer-like coatings independent of constituents such as cardanolformaldehyde resin or blends of synthetic resins into oriental lacquer. The second is the synthesis of phenolic lipids, such as urushiol, in the sap of lacquer trees that may be cured in the presence of laccase or polyphenol oxidase. The third is morphological homologs, which can make a densely packed grain structure and is degradable like the oriental lacquer. However, laccase (pycnoporus coccineus, IFO 4923) or polyphenol oxidase (from basidiomycetes) are now available commercially, and the polysaccharides and glycoproteins can be supplied from the sap of lacquer trees.[25–27] Blends of the different kinds of oriental lacquer or prepared urushiol or laccase added to the blends will make a new oriental lacquer. The substantial property and structure correlation of the oriental lacquer is revealed.[28] Oriental lacquers should be synthesized in this way.

REFERENCES

1. Jiago, Hu. In *International Symposium on Conservation and Restoration of Cultural Property— Conservation of Far Eastern Art Objects*; Tokyo National Research Institute of Cultural Properties: Tokyo, Japan, 1980; pp 89–112.
2. Majima R. *Ber.* **1922**, *55*, 172.
3. Hashimoto, O.; Tajima, T. *Mokuzai Kagakukaishi* **1975**, *21*, 675.
4. Smyes, W. F.; Dawson, C. R. *J. Am. Chem. Soc.* **1954**, *76*, 2959.
5. Yamauchi, Y.; Oshima R.; Kumanotani, J. *J. Chromtogr.* **1982**, *243*, 71
6. Tyman, J. H. P.; Mathews, A. J. *J. Chromatogr.* **1982**, *235*, 149.
7. Peisach, J.; Levine, W. G.; Blumberg, W. E. *J. Biol. Chem.* **1967**, *242*, 2847.
8. Suzuki, S.; Yoshimura, T.; Sakurai, T. *J. Inorg. Biochem.* **1991**, *44*, 267.
9. Sakurai, T. *J. Inorg. Biochem.* **1992**, *48*, 299.
10. Kato, T.; Kumanotani, J. *Bull. Chem. Soc. Jpn.* **1969**, *42*, 2378.
11. Kumanotani, J. Presented in *Fatipec XIII*; Cannes, France, 1976.
12. Kumanotani, J. *Makromol. Chem.* **1979**, *179*, 471.
13. Oshima, R.; Yamauchi, Y.; Watanabe, C.; Kumanotani, J. *J. Org. Chem.* **1985**, *50*, 2613.
14. Oshima R.; Kumanotani, J. *Rev. Lahnoamer Quim* **1988**, *19/3*, 121.
15. Kumanotani, J. *Chemistry and Chemical Industry* (*Jpn*) **1984**, *3*, 151.
16. Kumanotani, J. *Chem. Educ.* (*Jpn*) **1988**, *36*, 233.
17. Kumanotani, J. *Urushi*; J. Paul Getty Museum: Los Angeles, CA, 1988; p 243.
18. Kumanotani, J.; Kato, T.; Hikosaka, A. *J. Polym. Sci., Part C* **1969**, *23*, 519.
19. Kato, T.; Kumanotani, J. *J. Polym. Sci.* **1969**, *A-1*, 1455.
20. Kato, T.; Yokoo, Y.; Taniai, T.; Kumanotani, J. *Can. J. Chem.* **1969**, *47*, 2106.
21. Takada, M.; Yamauchi, Y.; Oshima, R.; Kumanotani, J.; Seno, M. *J. Org. Chem.* **1988**, *53*, 3078.
22. Kumanotani, J. *PMSE* **1988**, *57*, 278.
23. Kumanotani, J. *Progress in Organic Coatings* **1995**, *26*, 163.
24. Lalyanaraman, B.; Felix, C. C.; Sealy, R. C. *Environmental Health Perspectives* **1985**, *64*, 185.
25. Kumanotani, J.; Inoue, K.; Achiwa M.; Chen, L. W. *PMSE* **1985**, *13*, 163.
26. Kumanotani, J. In *Renewable Resource Materials*; Carraher, C. E., Jr.; Sperling, L. H., Eds.; Plenum: New York, 1986; p 171.
27. Oda, Y.; Adachi, K.; Iita, I.; Ito, M.; Aso, Y.; Igarashi, H. *Agric. Biol. Chem.* **1991**, *55*, 1393.
28. Iwatsuki, N. The Univ. of Tokyo, unpublished results.

Ormosils

See: Organic/Inorganic Composite Materials
Organic/Inorganic Composites

Orthoesters

See: Bicyclic and Spirocyclic Monomers, Oxygen-Containing (Ring-Opening Polymerization)

Orthopedic Polymers

See: *Biodegradable Polymers (Orthopedic Applications)*
Bioerodible Polymers A. Controlled Drug Release
B. Fracture Fixation
Bone Cements, Bioresorbable
Fracture Immobilization Polymers

Oxazoline Polymerization

See: *Cyclic Imino Ethers (Ring-Opening Polymerization)*
Cyclic Imino Ethers (Ring-Opening Polymerization of 1,3-Oxazo Monomers)
Double Isomerization Polymerization
2-Oxazolines, Polymerization
Poly(N-acylethylenimine)s
Polyalkylenimines
Zwitterionic Polymerization (Aziridine, Oxazoline Derivatives and Anhydrides)

2-OXAZOLINES, POLYMERIZATION

Andrzej Dworak, Barabar Trzebicka, and Wojciech Walach
Institute of Coal Chemistry
Polish Academy of Sciences

The ring-opening polymerization of 2-oxazolines was discovered in the mid-1960s by Seeliger, Tomalia, Bassiri and Litt, and Kagiya.[1–4] It is initiated by Lewis acids, protonic acids, alkyl and alkene halogenides, sulfonate esters, and chloroformates among others.

Since its discovery, the mechanism of this cationic (or electrophilic) polymerization has been studied in great detail. It is probably one of the best understood mechanisms in polymer chemistry, and is set forth in **Structure 1**.

COPOLYMERIZATION

Zwitterionic Copolymerization

2-Oxazolines undergo "spontaneous" copolymerization with some electrophilic comonomers without added initiator. Alternating copolymerization of 2-methyl-2-oxazoline with β-propiolactone was the first observed example of such a process.[6] In this reaction, a zwitterion is first formed from 2-oxazoline (the nucleophilic monomer N) and an electrophilic monomer E.[7,8] This step is followed by the reaction of the zwitterions, as shown in **Structure 1** and **Equation 1**.

Alternating copolymers are formed by this mechanism. Macrozwitterions may also react with each other to form macrocycles.[7]

STAR POLYMERS

Trifunctional initiators lead to star polymers when initiation is fast enough. For this purpose initiators containing bromomethyl groups were used: 1,3,5-tris(bromomethyl)-benzene or cyclotriphosphazene containing CH_2Br groups. Chloroformates of triols were also applied.[9]

BLOCK COPOLYMERIZATION

The living character of active species makes possible the "one pot" method, in which one of the monomers is polymerized first, then the second monomer is added to the living system.[10]

This method was applied to the synthesis of diblock ("one pot, two stage") and triblock ("one pot, three stage") copolymers. In most cases, these were copolymers of 2-oxazolines with alkyl substituents of different lengths.[10a]

Block Copolymers through Macroinitiators

Macroinitiators are generally prepared by the introduction into the polymer chain of a group that can initiate the polymerization of oxazolines. For the synthesis of polyether–oxazoline block copolymers, polyether alcohols or glycols were tosylated, nosylated, or converted into trifluoromethanesulfonates or chloroformates, which initiate the polymerization of oxazolines, although in some cases the process is slow.[5,9,11–17]

GRAFT POLYMERS

Graft polymers of oxazolines are obtained either by grafting 2-oxazoline monomers onto chains containing suitable initiator groups or by homo- or copolymerization of oxazoline-containing macromonomers.

Initiator Groups

A large variety of polymers contain groups that either initiate oxazoline polymerization or can be converted into initiating groups.

Macromonomers

Macromonomers that have an oxazoline ring and a polymerizable double bond lead to a great variety of graft polymers.

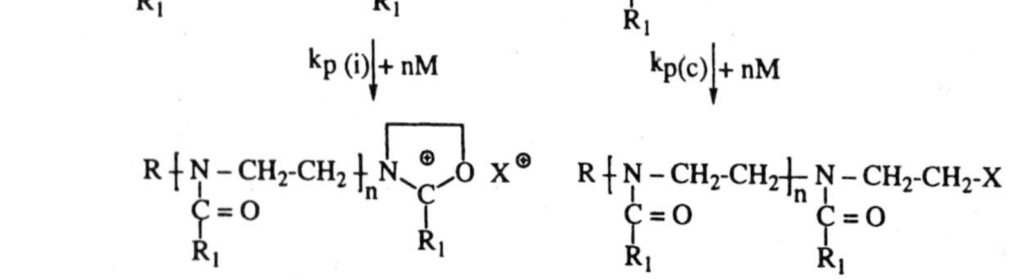

1

2

N + E ⟶ $^{+}NE^{-}$

RING OPENING ADDITION POLYMERIZATION OF 2-OXAZOLINES

Carboxylic acids, thiols, epoxides, and isocyanates react with 2-oxazolines.[19–24] These reactions were used for polymer modifications and polymerizations.[25]

The reaction of bis-oxazoline with dithiols and bisphenols takes a similar course.[22,26]

POLYMERS CONTAINING OXAZOLINE RINGS

2-Oxazolines that contain a substituent with a double bond may be polymerized through this bond.

2-Isoprenyl-2-oxazoline provided the first example of such polymerization.[19,27] Later, this and other 2-oxazolines with vinyl substituent were polymerized or copolymerized with vinyl monomers.[28–31] Such polymers may be modified or crosslinked by the ring-opening reactions of oxazolines.[20,21,28]

CROSSLINKING OF POLYOXAZOLINES

Several methods for the crosslinking of polyoxazolines have been described. In most cases, crosslinking was used to stabilize polyoxazoline gels.

POLYMER PROPERTIES AND APPLICATIONS

Polyoxazolines with lower alkyl substituents (methyl, ethyl) are highly soluble in water.

The hydrophilic character of poly(2-methyl-2-oxazoline) and poly(2-ethyl-2-oxazoline) can be combined with the hydrophobic (lipophilic) properties of polymers of 2-oxazolines with higher alkyl, perfluoroalkyl, or phenyl substituents. Gels containing chain units from both these monomers are amphiphilic; i.e., they swell both in water and in organic solvents.[32,33]

Surfactant properties result when hydrophilic oxazolines are connected with hydrophobic elements to polymer chains.

Due to their amphiphilic properties, oxazoline polymers have also been used as emulsifiers for styrene/water, butyl acrylate/water, and molten salts/polybutadiene, and in the synthesis of polymeric microspheres.[34–38] These properties also make it possible to use polyoxazolines as phase transfer catalysts.[18]

The miscibility of polyoxazolines with many commodity polymers was investigated. Polyoxazolines are miscible, within wide limits, with polymers containing proton-donating groups: acrylic acid home- and copolymers, poly(vinyl) phenol), and phenol–formaldehyde resins.[39] Blends of polyalkyloxazolines (methyl-, ethyl-, and propyl-) are miscible with poly(vinyl chloride), poly(vinyl fluoride), poly(vinylidene fluoride), and polystyrene.[40] These properties have led to the application of polyoxazolines as compatibilizers and their use for reactive blending.[41–43]

A significant application of polyoxazolines is their use to obtain linear, high-molecular-weight polyethylenimine; the polymerization of aziridines yields low-molecular-weight, branched products.

Other reported applications include adhesives, materials approved for contact with food, antistatica, flocculants, quenchants, and cosmetics.

Reviews on the polymerization of 2-oxazolines have been written by Kobayashi and Saegusa.[44,45]

REFERENCES

1. Seeliger, W.; Aufderhaar, E.; Diepers, W.; Feinauer, R.; Nehring, R.; Thier, W.; Hellmann, H. *Angew. Chem.* **1966**, *78*, 613.
2. Tomalia, D. A.; Sheets, D. P.; *J. Polym. Sci., Polym. Chem. Ed.* **1966**, *4*(9), 2253.
3. Bassiri, T. G.; Levy, A.; Litt, M. H. *J. Polym. Sci., Polym. Lett. Ed.* **1967**, *5*, 871.
4. Kagyia, T.; Narisawa, S.; Maeda T.; Fukai, K. *J. Polym. Sci., Polym. Lett., Ed.* **1966**, *4*, 441.
5. Kobayashi, S.; Morikawa K.; Shimizu, N.; Saegusa, T. *Polym. Bull. (Berlin)* **1984**, *11*(3), 253.
6. Saegusa, T.; Ikeda, H.; Fujii, H. *Macromolecules* **1972**, *5*(4), 354.
7. Saegusa, T.; Kobayashi, S.; Kimura, Y.; Ikeda H. *J. Macromol. Sci., Chem.* **1975**, *A9*(5), 641.
8. Lee, K I.; Lee, M. H. *Polymer* **1993**, *34*(3), 650.
9. Dworak, A.; Schulz, R. C.; *Makromol. Chem.* **1991**, *192*(2), 437.
10. Kobayashi, S.; Saegusa, T. *Makromol. Chem., Suppl.* **1985**, *12*, 11.

10a. Kobayashi, S.; Igarashi, T.; Moriuchi, Y; Saegusa, T. *Macromolecules* **1986**, *19*(3), 535.

11. Simionescu, C. I.; Rabia, I. *Polym. Bull. (Berlin)* **1983**, *10*(7–8), 311.
12. Percec, V. *Polym. Bull. (Berlin)* **1981**, *5*(11–12), 643.
13. Miyamoto, M.; Sano, S.; Saegusa, T.; Kobayashi, S. *Eur. Polym. J.* **1983**, *19*(10_11), 953.
14. Miyamoto, M.; Aoi, K.; Yamanaku, H.; Saegusa, T. *Polym. J. (Tokyo)* **1992**, *24*(4), 405.
15. Schulz, R. C.; Dworak, A. *Makromol. Chem., Macromol. Symp.* **1994**, *85*, 203.
16. Kobayashi, S.; Uyama, H.; Narita, Y. *Macromolecules* **1990**, *23*(1), 353.
17. Simionescu, C. I.; David, G.; Grigoras, M. *Eur. Polym. J.* **1988**, *24*(9), 849.
18. Kahovec, J.; Jelinkova, M.; Janout, V. *Polym. Bull. (Berlin)* **1986**, *15*(6), 485.
19. Kagiya, T.; Matsuda, T.; Makato, M.; Hirata, R. *J. Macromol. Sci., Chem.* **1972**, *6*(7), 1354.
20. Nishikubo, T.; Iizawa, T.; Tokairin, A. *Makromol. Chem., Rapid Commun.* **1981**, *2*(1), .91
21. Nishikubo, T.; Tokairin, A.; Torikai, S.; Iizawa, T. *Makromol. Chem.* **1985**, *186*(4), 675

22. Nishikubo, T.; Iizawa, T.; Tokutabe, N.; Sugiyama, S. *J. Polym. Sci., Polym. Lett., Ed.* **1980**, *18*, 1761.
23. Feinauer, R.; Seeliger, W. *Liebigs Ann. Chem.* **1966**, *698*, 174.
24. Nehring, R.; Seeliger, W. *Liebigs Ann. Chem.* **1966**, *698*, 67.
25. McManus, S.; Patterson, W.; Pittman, C. *J. Polym. Sci., Polym. Chem., Ed.* **1975**, *13*(7), 1721.
26. Nishikubo, T.; Iizawa, T.; Tokairin, A. *Makromol. Chem.* **1981**, *185*, 1307.
27. Kagiya, T.; Matsuda, T. *Polym. J.* (Tokyo) **1977**, *3*(3), 307.
28. Nishikubo, T.; Tokairin, A.; Takahashi, M.; Nosaka, W.; Iizawa, T. *J. Polym. Sci., Polym. Chem., Ed.* **1985**, *23*(6), 1805.
29. Nishikubo, T; Iizawa. T.; Takahashi, M.; Tokairin, A. *J. Polym. Sci., Polym. Chem., Ed.* **1987**, *25*(11), 2931.
30. Dibona, M.; Fibiger, R. F.; Gurnee, E.; Shuetz, J. *J. Appl. Polym. Sci.* **1986**, *31*(5), 1509.
31. Miyamoto, M.; Sano, Y.; Kimura, Y.; Saegusa, T. *Makromol. Chem.* **1986**, *187*(8), 1807.
32. Chujo, Y.; Sada, K; Matsumoto, Y.; Saegusa, T. *Polym. Bull.* (*Berlin*) **1989**, *21*(4), 353.
33. Chujo, Y.; Sada, K.; Saegusa, T. *Macromolecules* **1990**, *23*(5), 1234. 2/,1880.
34. Hsieh, B. R.; Litt, M. H. *Macromolecules* **1986**, *19*, 516.
35. Litt, M. H.; Hsieh, B. R.; Krieger, I. M.; Chen, T. T.; Lu, M. L. *J. Colloid Interface Sci.* **1987**, *115*(2), 312.
36. Litt, M. H.; Chen, T. T.; Hsieh, *J. Polym. Sci., Polym. Chem., Ed.* **1986**, *24*, 3407.
37. Litt, M. H.; Lin, C. H.; Krieger, I. M. *J. Polym. Sci., Polym. Chem., Ed.* **1990**, *28*(10), 2777.
38. Uyama, H.; Seto, M.; Matsumoto, Y.; Kobayashi, S. *Bull. Chem. Soc. Jpn.* **1993**, *66*(10), 3124.
39. Lin, P.; Clash, C.; Pearce, E. M.; Kwei, T. W.; Aponte, M. A. *J. Polym. Sci., Polym. Phys., Ed.* **1988**, *26*(3), 603.
40. Kobayashi, S.; Kaku, M.; Saegusa, T. *Macromolecules* **1988**, *21*(2), 334.
41. Liu, N. C.; Baker, W. E.; Russell, K. E. *J. Appl. Polym. Sci.* **1990**, *41*(9–10), 2285.
42. Lee, K. I.; Lee, M. H. *Polym. Prepr., Am. Chem. Soc. Div. Polym. Chem.* **1990**, *31*(2), 466.
43. Mulhaupt, R.; Muller, M.; Worner, H. *Polym. Bull.* (*Berlin*) 1995.
44. Kobayashi, S.; Saegusa, T. In *Ring Opening Polymerization;* Ivin, K. J.; Saegusa, T., Eds., Elsevier Applied Science: Essex, UK 1984; Vol. 2, p 761.
45. Kobayashi, S. *Prog. Polym. Sci.* **1990**, *15*, 751

OXIDATIVE COUPLING

Herbert Naarmann
BASF Plastics Research Laboratory

Oxidative coupling is a condensation process in which aromatic or heteroaromatic compounds eliminate hydrogen to form new C→C or C→heteroatom bonds. Depending on the starting materials, the process results in dimers, oligomers, or polymers. **Figure 1** is a general scheme for oxidative coupling reactions.

As can be seen from Figure 1, hydrogen is eliminated and condensation products are formed. Oxidizing agents (chemical or electrochemical) serve as catalysts. If necessary, complexing agents, activators, or Lewis acids can be added.

The potential afforded by the oxidant/Friedel-Crafts complexing agent reaction pair was considerably extended. In principle, all Friedel-Crafts catalysts and the usual dehydrogenation agents can be used for oxidative coupling.

STARTING MATERIALS

The various types of oxidative coupling reactions are presented in Figure 1. Type I reactions use aromatics and heterocycles as starting materials, Type II reactions use alkyl-substituted aromatics, and Type III reactions use phenols or aromatic amines.

Type I Reactions: Aromatics and Heterocycles

Benzene

The stepwise synthesis of polyphenylenes starts with a cationic radical that leads to rodlike molecules, which branch and crosslink to produce layer lattices (like graphite).

All the condensation products are crosslinked to some extent and completely insoluble.

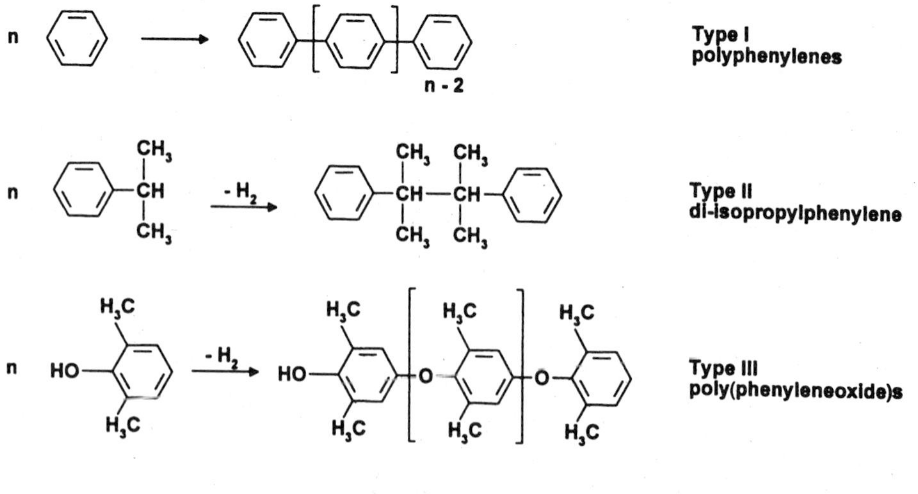

FIGURE 1. Oxidative coupling reactions.

Condensed Aromatic Hydrocarbons

Condensed aromatic hydrocarbons include diphenyl, terphenyl, napthaquin, anthracene, pyrene, phenanthrene, and chrysene.

Coupling two different aromatic compounds has the advantage that the resulting polycondensates are meltable and sometimes soluble, and have higher Dps (e.g., benzene/terphenyl, naphthalene/benzene).[5,5a,6]

Heterocycles

Several reports have been made on the oxidative coupling of heterocycles, including pyrrole, thiophene, indole, carbazole, quinoline, quinaldine, imidazole, pyridine, dithienyl, furan, and substituted derivatives and condensed systems.[1–5,6,8] In each case the products are dark-colored and insoluble.

Halogenated Aromatics

Under standard conditions ($AlCl_3/CuCl_2$, room temperature), chlorobenzene, 1,2,4-triclorobenzene, and fluorobenzene produce oligomers bonded at the ortho position.[1–4,9,10]

Type II Reactions: Alkyl-Substituted Aromatics

Alkyl-substituted aromatics, such as toluene, isopropylbenzene, or acrylbiscyano acetate, are starting materials for Type II reactions.[1–5,11,12] The use of toluene leads to oligomers with a high proportion of ortho bonds. Isopropylbenzene results in dimers and oligomers.

Type III Reactions: Phenols or Aromatic Amines

Phenols

The oxidative coupling of phenols in Type III reactions yields polycondensates with O as the binding link, as shown in **Structure 1**.[14]

Structure 1 shows the condensation of 2,6-dimethylphenol to a specific polymer—PPE or poly[oxy(2,6-dimethyl-1,4-phenylene)]. The dehydrogenation is carried out by use of copper salts that have been complexed with amines in the presence of oxygen or air.

In these reactions the phenolic oxygen is incorporated as a binding link in the polymer chain. The polymers are soluble and have mole weights lower than 50,000.

Aromatic Amines

The oxidative coupling of aromatic amines in Type III reactions yields polycondensates with-NH-or-N-as the binding link. **Structure 2** below summarizes the condensation process. AB stands for aniline black.

Recent studies of polyaniline suggest that the polymer can exist in a wide range of structures.[15] These can be regarded as copolymers of reduced (amine) and oxidized (imine) units, of the form shown in **Structure 3**.

CATALYSTS

Oxidation to dimers, oligomers, or polymers can be carried out not only chemically but also electrochemically (e.g., by anodic oxidation). Electrochemical (anodic) oxidation is a powerful tool.

PROPERTIES AND APPLICATIONS

The properties and applications of materials are always linked. Polyphenylenes and polythiophenes are no exception. Such products have been known for a very long time—some of them for a century—but the recent discovery of new properties such as electrical conductivity, photovoltaic effect, and thermoelectric power has opened up completely new applications in the high-technology sector.[1–4,16–21]

Type I Reaction Products

Type I reaction products include polyphenylenes, polyaromatics, and polyheterocycles.[1–4,16–19] In terms of its properties, polyaniline also belongs to these materials. Compounds of this type are deeply colored and insoluble. They show thermostabilities greater than 200°C. Polyphenylenes are even stable up to about 500°C.

Most striking, however, are the electronic properties of these materials.[20] The conductivity of the oxidatively coupled products lies between 10^{-4} and $5 \cdot 10^{-1}$ S/cm.[1–4] Even higher conductivities are attained by doping.

Additional applications are as follows:[22]

- Drug release (e.g., heparin, penicillin)
- Deposition of conducting materials (e.g., pyrrole on nylon, glass fiber fleeces, C fibers)

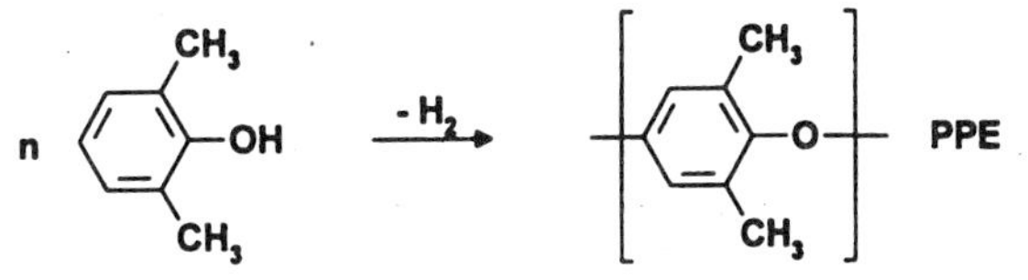

1

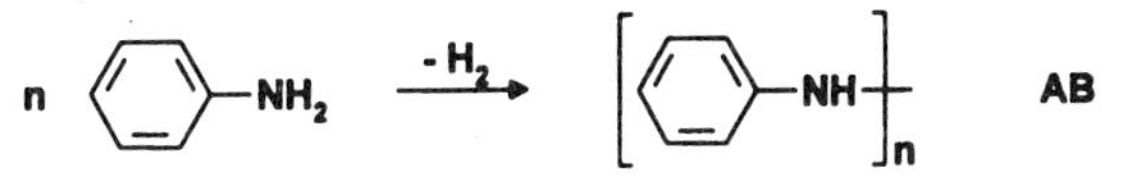

2

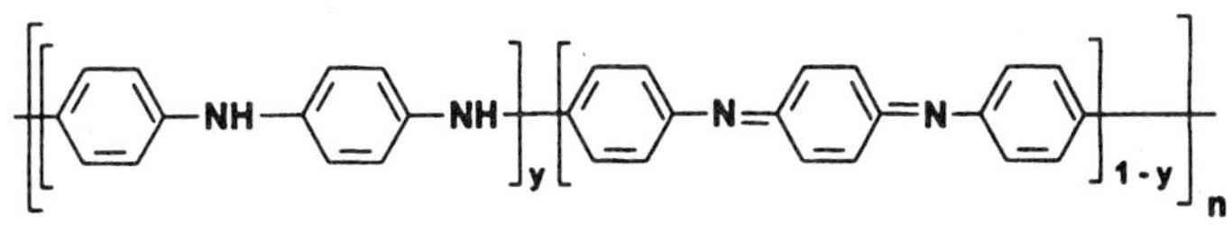

3

- Deposition of metals (Cu) also on deposited electrically conducting polymers (printed circuits)
- Deposition of structured electrically conducting polymers by irradiation
- Elastic polypyrroles (elongation > 100%)
- Polymeric shish kebab structures
- Piezoceramic membranes
- Ion-exchange membranes
- Self-dopant systems with ionic side groups or CT-complexing in the main chain
- Optical data storage
- A new generation of "intelligent" materials
- Electrochromic displays
- Light-emitting diodes, and
- Field transistors[18,19,23,24]

Type II Reaction Products

Type II reaction products are light-colored, soluble compounds with molecular weights up to about 5000. They are meltable up to 200°C, but they sometimes decompose. They do not exhibit the unusual electrical properties of Type I reaction products.

Compounds of type II are thermolabile and decompose into radicals at temperatures between 40 and 300°C, depending on the substituents. They can therefore be used as radical generators (e.g., as C-C-labile radical starters in polymerization reactions) or as radical catchers (e.g., as scavengers in flameproofing to suppress radical chain reactions).[25]

Type III Reaction Products

Type III reaction products with ether links (as illustrated in Structure 1) have high molecular weights greater than 50,000.[13] An unusual example is poly(2,6-dimethyl phenyl ether), which is thermostable, has a glass transition temperature of 250°C, and is miscible with polystyrene in all proportions, permitting the manufacture of polymer blends.

The versatility of poly(phenylene ether) blends is opening up many new fields of application in the automotive, appliance, and electronics industries. Blends based on poly(phenylene ether)s have new property profiles, including flexural modulus values of 2500–9000 N/mm^2, heat distortion temperatures of 85–200°C, and notched impact strength (Izod) of 100–400 J/m.

REFERENCES

1. Naarmann, H.; Beck, F. Presented at the meeting of the Gesellschaft Deutchfer Chemiker, Munich, 12 October 1964, 1964.
2. Naarmann, H.; Beck, F.; Kastring, E. G. DE Patent 117 529 (BASF), 1964.
3. Naarmann, H.; Beck, F.; Kastring, E. G. GB Patent 109 213 7 (BASF), 1964.
4. Kauffmann, Th. et al. *Phys. Chem. Ber.* **1981**, *114*, 3667–3673.
5. Kricheldorf, H. B.; Schwarz, G. *Handbook of Polymer Synthesis:* Kricheldorf, H. B., Ed.; Marcel Dekker: New York, 1991, Part B, pp 1647–1668.

5a. Miller, L. I.; Bilow, N. U.S. Patent 3 677 976, 1972.

6. Miller, L. I.; Bilow, N. U.S. Patent 3 578 611, 1971.

6a. Lamb, S. B.; Kovacic, P. *J. Polym. Sci., Polym. Chem. Ed.* **1980**, *18*, 2423.

7. *Synth. Met.* **1991**, *41*(1–2).
8. *Synth. Met.* **1989**, *28*(1–3).
9. Hsing, C. F.; Jones, M. B.; Kovacic, P. *J. Polym. Sci. Polym. Chem. Ed.* **1981**, *19*, 973.
10. Naarmann, H.; Lübcke, E. DE Patent 1 256 421 (BASF), 1964.
11. de Jongh, H. A. P. et al. *J. Polym. Sci. Polym. Chem. Ed.* **1973**, *11*, 345.
12. de Jongh, H. A. P. et al. *J. Org. Chem.* **1971**, *36*, 3160.
13. Viehe, G. et al. *Polym. Bull.* **1980**, 2, 363, 417, 689.
14. Brandt, H. In *Houben-Weyl, Melhoden der organizchen Chemie*, Vol. E 20/2; Georgthieme Verlag, 1987; pp 1380–1389.
15. MacDiarmid, A. G. et al. *Synth. Met.* **1987**, *18*, 85.
16. Elsenbaumer, R. L.; Shacklett, L. W. In *Handbook of Conducting Polymers;* Skotheim, T. A., Ed.; Marcel Dekker: New York, 1986; Vol. 1, p 216.
17. Kossmehl, G. A. In *Handbook of Conducting Polymers,* Skotheim, T. A., Ed.; Marcel Dekker: New York, 1986; Vol. 1, p 351.
18. Street, G. B. In *Handbook of Conducting Polymers;* Skotheim, T. A., Ed.; Marcel Dekker: New York, 1986; Vol. 1, p 265.
19. Tourillon, G. In *Handbook of Conducting Polymers;* Skotheim. T. A., Ed.; Marcel Dekker: New York, 1986; Vol. 1, p 293.
20. Ellis, I. R. In *Handbook of Conducting Polymers;* Skotheim, T. A., Ed.; Marcel Dekker: New York, 1986; Vol. 1, p 489.
21. Gazard, M. In *Handbook of Conducting Polymers;* Skotheim. T. A., Ed.; Marcel Dekker: New York, 1986; Vol. 1, p 673.
22. Naarmann, H. *J. Polym. Sci. Polym. Symp.* **1994**, *75*, 53.
23. Schwoerer, M. *Phys. Bl.* **1994**, *49*(1), 52.
24. Garnier, F. *Adv. Mat.* **1989**, *1*, 117.
25. Burger, H. et al., al. DE Patent 1 255 302 (BASF), 1965.

Oxidative Degradation

See: Antioxidants and Stabilizers
Degradation (Weatherability)
Poly(vinyl chloride) (Mechanisms of Stabilization)

Oxidative Polymerization

See: Captodative Compounds, Polymerization
Enzyme-Catalyzed Oxidative Polymerization (Aromatic Compounds)
Oxidative Coupling
Polypyrrole (Processable Dispersions)
Uniform Polymers

Oxiranes

See: Ring-Opening Coordination Polymerization (by Soluble Multinuclear Aldoxides)

Oxygen Copolymerization

See: Vinyl Polyperoxides (Oxygen Copolymerizations)

P

Pacemakers

See: *Blood Compatible Polymers*

Packaging

See: *Compatibilizers, Polymeric (Recycling of Multilayer Structures)*
Ethylene-Vinyl Alcohol Copolymer (Overview)
Microwave Absorbing Materials

Paint

See: *Alkyd Emulsions*
Conducting Polymer Colloids
Electroconductive Paints
Resins and Paints (Analytical Pyrolysis)

Paired Polymers

See: *Interpolymers*

PALLADIUM-CATALYZED SYNTHESIS (Monomers and Polymers)

Walter Heitz and Andreas Greiner
Fb Physikalische Chemie
Polymere und Zentrum für Materialwissenschaften
Philipps-Universität Marburg

PD-CATALYZED CHAIN-GROWTH REACTIONS

The Pd-catalyzed alternating copolymerization of CO and ethylene is of significant interest. Several Pd catalysts for this copolymerization have been described in the patent literature but without any mechanistic information.[1] Sen and Lai described a novel Pd catalyst with the general formula $[Pd(PPh_3)_n(CH_3CN)_{4-n}](BF_4)_2$ (n = 1, 2, 3) for the rapid dimerization of ethene and for the alternating copolymerization of CO and ethene (**Structure 1**).

$$H_2C=CH_2 + CO \xrightarrow{[Pd]} \left[CH_2-CH_2-\overset{O}{\overset{\|}{C}} \right]_n$$

1

Mechanistic Aspects

$[Pd(PPh_3)_n(CH_3CN)_{4-n}](BF_4)_2$ (n = 1, 2, 3) catalyzes the rapid dimerization of ethylene in the absence of CO and the formation of C_4H_g, indicating β-hydrogen abstraction.[2] If CO is present, an alternating insertion of ethylene and CO most likely is for the copolymer formation. The competing β-hydrogen abstraction is suppressed by the faster insertion of CO into the Pd-alkyl bond.

The Pd-catalyzed alternating copolymerization of α-olefins gives interesting results with regard to the regio- and stereoselectivity of the reaction.

Styrene and CO were copolymerized using phenanthroline-modified Pd catalysts.[4] The tacticity of the copolymers depended strongly on the structure of the ligands.

The Pd-catalyzed homopolymerization of selected olefins is possible if β-hydrogen elimination can be avoided. Risse and Mehler obtained soluble polynorbonene with $[Pd(CH_3CH)_4\text{-}[BF]_2$ as catalyst.[5,6]

PD-CATALYZED STEP-GROWTH REACTIONS

Pd-catalyzed step-growth reactions involve a change of the oxidation state of Pd species. A wide variety of products can be obtained by the initial reaction of haloarenes or analogous arenes with nucleophiles in the presence of a Pd(0) catalyst and a base.

The Pd-catalyst coupling reaction of haloarenes and olefins is known as the Heck-reaction.[7,8] In many cases Pd(II) salts are used as the initial Pd species and are reduced *in situ* by a base such as a trialkylamine. Control of the regioselectivity of Heck-type polyreactions is crucial for rod-like poly(arylene vinylene)s.

Poly(arylene vinylene)s have been prepared by coupling of bromostyrene or of dibromoarenes with ethylene divinylarenes respectively (**Structure 2**).[18–21]

2

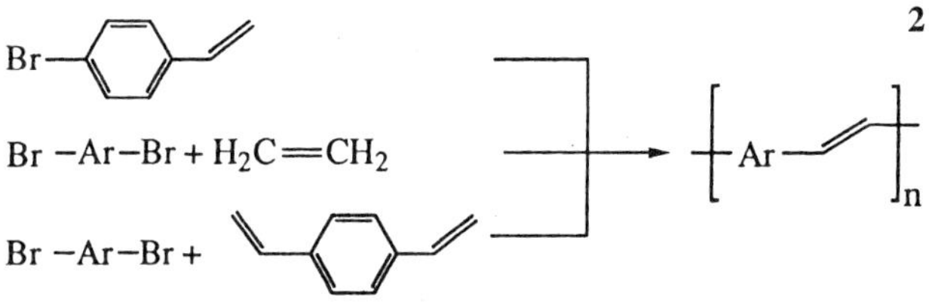

A wide variety of poly(arylene acetylene)s were prepared by Pd-catalyzed coupling of dihaloarenes and diarylene acetylenes or haloarylene acetylenes in the presence of catalytic amounts of Pd(0) complexes, Pd(II) complexes, or Pd(II) salts, catalytic amounts of copper salts, and a trialkylamine (**Structure 3**).[13–18]

$$Hal-Ar-Hal + HC\equiv C-Ar-C\equiv CH \xrightarrow{[Pd]}$$
$$\left[-Ar-C\equiv C- \right]$$

3

Bis(tributylstannyl)acetylenes and dialkoxy-substituted dibromoarenes have been successfully used for the Pd-catalyzed synthesis of poly(arylene acetylene)s.[19]

The Pd(0)-catalyzed coupling of aromatic boronic acids and haloarenes, known as Suzuki coupling, was introduced for the synthesis of hyperbranched polyphenylenes by Kim and Webster.[20,21]

Schlüter et al. extended this synthetic route to linear polyphenylenes.[22] The feasibility of this reaction for the synthesis of a wide variety of linear polyphenylenes was demonstrated extensively in the literature.[23–29]

Aryl-aryl coupling reactions of haloarenes and metallated arenes in the presence of a Pd catalyst have been widely used for the synthesis of polyarylenes. Yamamoto et al. have prepared poly(2,5-thienylene) by the coupling of 2,5-dibromothiophene in the presence of $PdCl_2$, NEt_4, PPh_3, and an excess of zinc.[30]

REFERENCES

1. Sen. A. *Adv. Polym. Sci.* **1986**, *73/76*, 125.
2. Sen, A.; Lai, T.-W. *J. Am. Chem. Soc.* **1981**, *103*, 4627.
3. Sen, A.; Lai, T.-W. *J. Am. Chem. Soc.* **1982**, *104*, 3520.
4. Sen, A.; Jiang, Z. *Macromolecules* **1993**, *26*, 911.
5. Mehler, C.; Risse, W. *Makromol. Chem., Rapid Commun.* **1991**, *12*, 255.
6. Mehler, C.; Risse, W. *Macromolecules* **1992**, *25*, 4226.
7. Heck, R. F. *Organic React.* **1981**, *27*, 345.
8. Heck, R. F. *Pure Appl. Chem.* **1978**, *50*, 691.
9. Heitz, W.; Brügging, W.; Freund, L.; Gailberger, M.; Greiner, A.; Jung, H.; Kampschulte, U.; Niessner, N.; Osan, F.; Schmidt, H.-W.; Wicker, M. *Makromol. Chem.* **1988**, *189*, 119.
10. Greiner, A.; Heitz, W. *Makromol. Chem. Rapid Commun.* **1988**, *9*, 581.
11. Martelock, H.; Greiner, A.; Heitz, W. *Makromol. Chem.* **1991**, *192*, 967.
12. Ashai Glass Co. Ltd. Jpn. Patent 57 207 618, 1981; *Chem. Abstract* **1983**, *99*, 23139k.
13. Sanechika, K.; Yamamoto, T.; Yamamoto, A. *Bull Chem. Soc. Jpn.* **1984**, *57*, 752.
14. Trumbo, D. L.; Marvel, C. S. *J. Polym. Sci., Polym. Chem. Ed.* **1987**, *25*, 839.
15. Kondo, K.; Okuda, M.; Fujitani, T. *Macromolecules* **1993**, *26*, 7382.
16. Moroni, M.; Le Moigne, J. *Macromolecules* **1994**, *27*, 562.
17. Beginn, C.; Grazulevicius, J. V.; Strohriegel, P. *Macromol. Chem. Phys.* **1994**, *165*, 2353.
18. Varlemann, Y. Ph.D. Thesis, University of Marburg, 1992.
19. Giesa, R.; Schulz, R. C. *Macromol. Chem.* **1990**, *191*, 857.
20. Miguara, N.; Yanagi, T.; Suzuki, A. *Synth. Commun.* **1981**, *11*, 513.
21. Kim, Y. H.; Webster, O. W. *Polym. Prepr.* **1988**, *29*(2), 3102.
22. Rehahn, M.; Schlüter, A.-D.; Wegner, G.; Feast, W. J. *Polymer* **1989**, *30*, 1060.
23. Fahnenstich, U.; Koch, K.-H.; Müllen, K. *Makromol. Chem. Rapid Commun.* **1989**, *10*, 563.
24. Rehahn, M.; Schlüter, A.-D.; Wegner, G. *Makromol. Chem. Commun.* **1990**, *11*, 535.
25. Rehahn, M.; Schlüter, A.-D.; Wegner, G. *Makromol. Chem.* **1990**, *191*, 1991.
26. Knapp, R.; Rehahn, M. *Makromol. Chem., Rapid Commun.* **1993**, *14*, 451.
27. Rau, I. U.; Rehahn, M. *Makromol. Chem.* **1993**, *194*, 2225.
28. Schmitz, L.; Rehahn, M.; Ballauff, M. *Polymer* **1993**, *34*, 646.
29. Rau, I. U.; Rehahn, M. *Acta Polym.* **1994**, *45*, 3.
30. Yamamoto, T.; Osakada, K.; Wakabayashi, T.; Yamamoto, A. *Makromol. Chem., Rapid Commun.* **1985**, *6*, 671.

Panel Products

See: Cellulose-Filled Composites

Paper

See: Alginate Fibers (High Performance Papers)
Cellulose (Overview)
High Performance Paper
Paper Coatings

PAPER COATINGS

Pierre Lepoutre
Department of Chemical Engineering
University of Maine

Printing papers and paperboard are often coated with pigmented systems for aesthetic reasons.[1] For good print quality, paper must have a smooth surface with a fine capillary structure.

Paper coating is basically a cheap paint consisting of a pigment and a small amount of polymeric binder. Unlike most surfaces to which paints are applied, printing papers generally have a very limited use-life and moderate performance requirements. Hence, paper coatings are formulated at high pigment volume concentration with inexpensive low-refractive-index pigments such as platelike clay and ground rhombic natural calcium carbonate. Titanium dioxide is used only parsimoniously in some formulations requiring very high opacity. Other pigments, mineral or organic, may be added for special properties. The binder may be a starch or a latex, or a combination thereof. The latexes are usually copolymers of styrene, butadiene, acrylic, and vinyl acetate monomers. In addition to pigment and binder, other additives may be present as processing aids. The pigmented coating is applied to paper as an aqueous dispersion with the highest possible proportion of solids in order to minimize drying costs and an undesirable roughening of the basestock by water.

COATING STRUCTURE

The final, dry coating layer thus consists of pigment particles bonded together by the polymeric binder. The amount of binder, the more expensive component of the coating, is usually kept small (5–15 parts per hundred parts by weight of pigment or 12–40 parts by volume) and is much less than the amount required to fill the interstices between the pigment particles.

OPTICAL PROPERTIES

Gloss, particularly print gloss, is often sought after because it enhances the print quality. Gloss is a function of the refractive index of the surface and of its topography. The first step in achieving high gloss is to produce a smooth surface. Smoothness is improved when the surface voids, which constitute the sheet roughness, are filled by the coating.

MECHANICAL PERFORMANCE REQUIREMENTS

The main objective in applying pigmented coatings on printing papers is to raise the print quality, that is, to ensure the uniform transfer of the ink film, and its rapid and uniform setting at the surface of the sheet.[2]

ROLE OF THE POLYMERIC BINDER

The main function of the binder is to provide cohesion to the coating and adhesion of the coating to the base paper. A basic requirement is the formation of an interface between binder and pigment–or binder and fiber–through which stresses can be transmitted.

ROLE OF POLYMERIC ADDITIVES

Polymers are also used for other functional properties. Low-molecular-mass anionic polyelectrolytes such as polyacrylate sodium adsorb on pigment surfaces and stabilize the suspension. Carboxymethyl cellulose, starch derivatives, some natural gums and other hydrocolloids, and hydrophobically modified cellulosic polymers are used to slow down the dewatering of the coating suspension on paper, preventing solids from building up too rapidly; otherwise, the blade scraping action would crease scratches and other runnability problems.

CONCLUSION

Polymers play an important role in paper coating. They help control the rheology of the coating suspension during high-speed application and provide cohesion to the porous network of pigment particles in the finished coated paper. Incorporation of desirable functional groups into the polymer composition, or at the surface of the latex particles, enhances interactions with pigment and fiber surfaces and improves adhesion. Increasingly, latexes are tailor-made for a particular grade at a particular mill. By controlling the bulk composition of the polymer and its surface energy and chemistry, formulators can obtain a range of coating structures to suit particular printing and converting requirements.

REFERENCES

1. Hagemeyer, R. W. In *Pulp and Paper Chemistry and Technology* 3rd ed.; Casey, J. P., Ed.; Wiley-Interscience: New York, 1983; Vol. 4, p 2013ff.
2. Aspler, J.; Lepoutre, P. *Prog. In Organic Coatings* **1991**, *19*, 333.

PARACRYSTALS

Hans Bradaczek
Institut für Kristallographie
Freie Universität Berlin

Rolf Hosemann's intention in developing the theory of the paracrystal was to create to link between the crystalline and the amorphous state of matter.[1,2]

If a probability distribution of the position of a neighboring particle (atom) is used to describe a structure, a crystal can be described by a pointlike distribution, and a nearly amorphous structure by a very broad distribution. Between these extremes a wide variety of more or less distorted structures can be considered. Polymers are a favored object of the theory of the paracrystal.

One of the major advantages of this theory is the (principally) simple mathematical formulation in Fourier space, which represents the profile of the diffraction pattern. The degree of the paracrystalline distortion can be derived from the measured profile of the reflections.

THE PARACRYSTALLINITY OF POLYMERS

Bear and Rugo's detection of crystal structures in diffraction patterns of seagull quill ceratine in 1951 came as a great surprise because at that time knowledge about partly crystalline structures, especially in natural products, was very poor.[3] Hosemann used to take this early example to prove the universality of the paracrystalline theory.

In the following years, several X-ray investigations of biological macromolecules were carried out. Almost all of them revealed paracrystalline distortions. One of the most significant examples was the detection of paracrystalline distortions in the collagen of rat's tail sinew.[4]

Paracrystalline distortions could be detected even in bacterial multilayers.[5]

Although a gap exists between the *a priori* condition and the real existence of microparacrystals, **Figure 1** tries to explain this phenomenon at least in part.

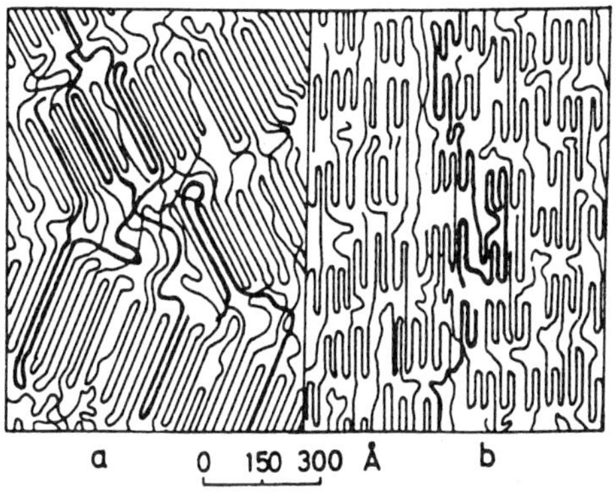

FIGURE 1. Schematics of a rubber-like network demonstrating the existence of microparacrystals. *Source:* Hoseman, R. R., *Progr. Colloid. Polym. Sci.,* **1988**, *77*, 15-25.

ACKNOWLEDGMENT

The author thanks Prof. Dr. Gerhard Hildebrandt for revising the manuscript and offering many suggestions, and Dr. Wilhelm Uebach for calculations.

Although the theory of the paracrystal was Rolf Hosemann's favorite research subject, he was unable to finish his work. One of the aims of this article is to encourage young colleagues to continue his research.

REFERENCES

1. Hosemann, R. *Z. Phys.* **1950**, *128*, 1–35.
2. Hosemann, R. *Z. Phys.* **1950**, *128*, 465–492.
3. Bear, R. S.; Rugo, H. J. *Ann. N.Y. Acad. Sci.* **1951**, *53*, 627.
4. Newesely, H.; Hosemann, R.; Uther, B. Z. *Naturforsch., C* **1980**, *35*, 177–187.
5. Labischinski, H.; Bradaczek, H. *Kohlrausch: Praktische Physik*; B. G. Teubner: Stuttgart, 1985; Chapter 10, p 759.

PARACYANOGEN*

Leon Maya
Chemical and Analytical Sciences Division
Oak Ridge National Laboratory

Paracyanogen is a polymeric material that in all probability never will be a high-volume article of commerce. It merits attention because of historical and theoretical considerations. This material is one of the first synthetic polymers ever made. It was observed by Gay-Lussac in 1816 during experiments on the pyrolysis of mercuric cyanide and has attracted the attention of many chemists up to the present.[1]

PREPARATION

The fact that paracyanogen forms under a wide variety of conditions does not clarify the more difficult questions regarding the properties of this material and how these relate to a molecular structure. In this connection, the question arises whether the polymeric materials observed through the years since the initial discovery by Gay-Lussac are the same regardless of the preparation procedure. Labes, on the basis of his own work and a critical review of the literature, concluded that there are actually two distinct polymeric materials formed from cyanogen.[3,4] One is called polycyanogen; it is a linear polymer with a backbone made of the repeating –C=N– unit with pendant CN moieties on carbon. The other is called paracyanogen; its structure is unknown, but it is often thought to be a ladder polymer, as described in **Structure 1**.[2]

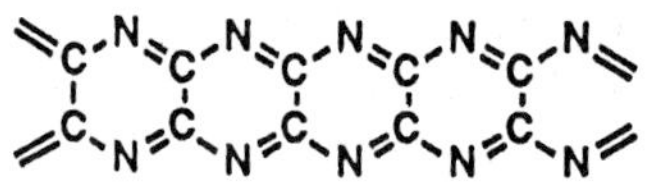

1

STRUCTURE

An open disordered structure, first suggested by Sidwick, and represented in **Structure 2**, appears to be a more rational representation of paracyanogen.[8] This structure appears to be consistent with the spectroscopic observations, including ^{13}C NMR, which shows most of the carbon as sp^2 with a relatively small shoulder due to sp^1 carbon of the terminal nitrile moieties in the polymer.[5] Structure 6 is also consistent with observations regarding electrical conductivity that characterize paracyanogen as an insulator that nevertheless conducts at higher fields, evidently after overcoming a relatively high activation energy.[4,6,7]

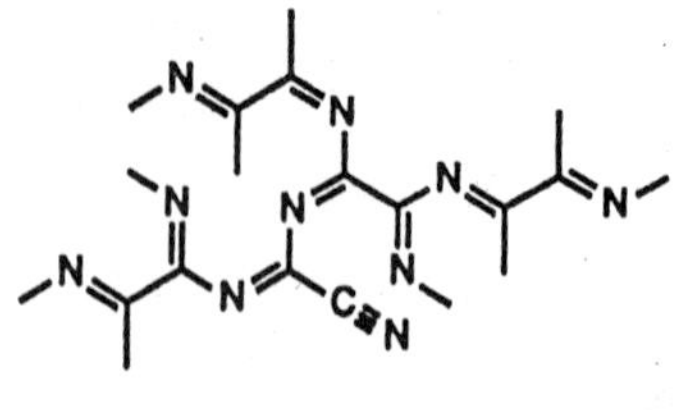

2

THEORETICAL CONSIDERATIONS

The connection of paracyanogen with carbon nitride is interesting. The polymer is a potential product of sputtering or chemical vapor deposition involving carbon and nitrogen species. Such conditions were recently reported in what appears to be the first successful preparation of carbon nitride films.[9]

In a more speculative vein, paracyanogen has been linked with prebiotic organic matter on earth. It has been suggested that "intractable heteropolymers" present in carbonaceous chondrites and organics in cometary ice might have contributed to the development of life on earth.[10]

ACKNOWLEDGMENT

This work was supported by the Division of Materials Sciences, Office of Basic Energy Sciences, U.S. Department of Energy, under Contract No. DE-AC05-84OR21400 with Martin Marietta Energy Systems, Inc.

REFERENCES

1. Gay-Lussac, J. L. *Ann. Chem.* **1816**, *53*, 139.
2. Bircumshaw, L. L.; Taylor, F. M.; Whiffen, D. H. *J. Chem. Soc.* **1954**, 931.
3. Labes, M. M. *Mol. Cryst. Liq. Cryst.* **1989**, *171*, 243.
4. Chen, J. H.; Labes, M. M. *J. Polym. Sci., Polym. Chem. Ed.* **1985**, *23*, 517.
5. Maya, L. *J. Polym. Sci., Polym. Chem. Ed.* **1993**, *31*, 2595.
6. Cuomo, J. J.; Leary, P. A.; Yu, D.; Reuter, W.; Frisch, M. *J. Vac. Sci. Technol.* **1979**, *16*, 299.
7. Fabian, M. *J. Mater. Sci.* **1967**, *2*, 424.
8. Sidwick, N. V. *The Organic Chemistry of Nitrogen* Clarendon: Oxford, **1937**; p 299.
9. Niu, C.; Lu, Y. Z.; Lieber, C. M. *Science (Washington, DC)* **1993**, *261*, 334.
10. Chyba, C. F.; Thomas, P. J.; Brookshaw, L.; Sagan, C. *Science (Washington, DC)* **1990**, *249*, 366.

Paracyclophane Polymerization

See: Vapor Deposition Polymerization ([2.2]Paracyclophane and Analogues)

Particleboard

See: Carbon Composites

Particles

See: Deformation and Fracture (Micromechanical Mechanisms)
Precipitation Polymerization (Overview)
Precipitation Polymerization
Vesicles, Polymerization in

Pearl Polymerization

See: *Suspension Polymerization*

Pectins

See: *Gums (Overview)*

PEEK

See: *Poly(aryl ether ketone)s*

PEKK

See: *Poly(aryl ether ketone)s*

PEPTIDE-BASED NANOTUBES (New Class of Functional Biomaterials)

M. Reza Ghadin* and Jeffrey D. Hartgerink
Departments of Chemistry and Molecular Biology
Scripps Research Institute

Many forms of nanostructured materials have thus far been prepared; however, one particular architecture, nanotubular objects, has received much attention lately.[1] Nanotubular objects include polymerized cyclodextrins, all carbon graphite nanotubes, boron-nitride nanotubes, silicon–cast cylindrical micelles, inorganic zeolites, gold nanotube membranes, and self-assembled peptide nanotubes.[2–8] The reason for the recent surge of activity in nanotube design lies in their fascinating structure and their potentially wide range of applications. Potential uses of nanotubes include molecules wires, cell membrane channels, catalysts, size-separation devices, and novel means for drug delivery. In designing a nanotubular object, four important variables need to be controlled: internal and external surface chemical properties, internal pore diameter, and length. Of the above mentioned methods for preparing nanotubes, self-assembling peptide nanotubes offer complete control over the external surface characteristics and internal pore diameter. The ability to easily tailor two of the most important properties of nanotubes, makes the self-assembling peptide nanotubes particularly well-suited for a number of potential applications.

DESIGN PRINCIPLES

Cyclic Nature of the Peptide Subunits

Cyclic peptides constitute the building blocks of self-assembling peptide nanotubes (**Figure 1**). These building blocks offer three key advantages: easy synthesis, highly variable pore diameter, and variable outer surface chemistry. The peptides can be prepared using standard solid-phase peptide synthesis[10] followed either by solid-phase[11] or solution-phase[12] cyclization protocols, depending on the peptide sequence. These methods allow simple and rapid preparation of cyclic peptides. The upper limit to ring size has not yet been explored. Thus far, eight-, ten-, and twelve-residue cyclic peptides have been successfully used in the design and construction of nanotubes–these tubular constructs have internal van der Waals diameters of 7.5, 10, and 13 Å, respectively.

*Author to whom correspondence should be addressed.

Alternating D/L Stereo Chemistry

The fundamental requirement for cyclic peptides to form nanotubes is the alternating D- and L-amino acid configuration, which constrains the peptide to a flat ring-shaped geometry with the amide backbone functionalities oriented perpendicular to the plane of the ring structure.

β-Sheet Hydrogen Bonding

Because of the predisposition of the backbone carbonyls and NHs perpendicular to the plane of the cyclic peptide structure, peptide subunits can stack to form complementary β-sheetlike intermolecular hydrogen bonding networks. In this configuration, all of the backbone amide groups can participate in hydrogen bonding. This provides the primary driving force for the self-assembly process.

Ring-Stacking Arrangements

A cyclic peptide with *N*-amino acids can stack onto another subunit of the same ring size in N/2 possible relative registers. This is because only homochiral amino acids may form cross-strands near neighbor pairs.

SELF-ASSEMBLY OF SOLID-STATE NANOTUBULAR ARRAYS

Six different cyclic peptide sequences have been designed and employed in the construction of solid-state nanotubes in order to study the scope and limitation of the approach. These include five eight-membered peptide subunits: *cyclo*-[(L-Gln-D-Ala-L-Glu-D-Ala)$_2$] *cyclo*-[(L-Gln-D-Ala)$_4$], *cyclo*-[(L-Gln-D-Val)$_4$], *cyclo*-[(L-Gln-D-Leu)$_4$], and *cyclo*-[(L-Gln-D-Phe)$_4$], as well as a larger twelve-membered ring structure: *cyclo*-[(L-Gln-D-Ala-L-Glu-D-Ala)$_3$].

The peptide subunit *cyclo*-[(L-Gln-D-Ala-L-Glu-D-Ala)$_2$] was used to prepare the first nanotubular array system.[9]

CYLINDRICAL PEPTIDE DIMERS

To gain a detailed understanding of the structural aspects of the nanotubes and thermodynamics of the self-assembly process, a "minimal" peptide cylinder based on the assembly of two peptide subunits was designed. Methylation of alternate amide backbone functionalities (at all L- or all D-residues) yielded peptides in which one face of the cyclic peptide structure had no hydrogen bond donor potential. In addition, the steric interactions of the methyl groups block the approach of cyclic peptides attempting to hydrogen bond from the *N*-methylated face of the ring structure. Therefore, a solution of a cyclic peptide in which all D-amino acids are *N*-methylated can only form dimeric hydrogen bonded cylindrical structures. Furthermore, because of the imposed facial selectivity in the intermolecular hydrogen bonding interactions between subunits, only the antiparallel ring stacked arrangements can be formed.

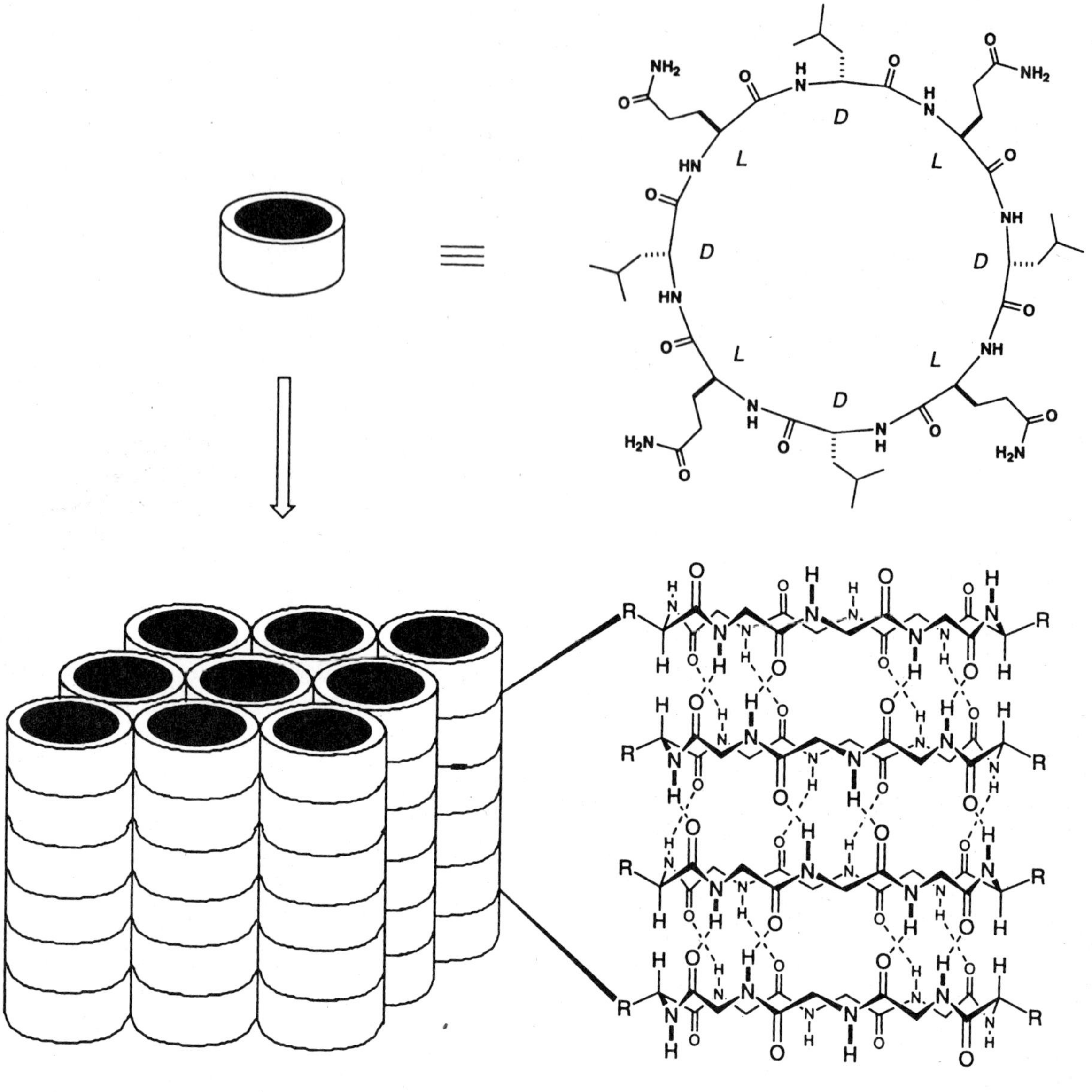

FIGURE 1. Schematic representation of the structure and assembly mechanism of peptide nanotubes. Top left and right show a single cyclic peptide which can self-assemble into nanotubes (lower left). These nanotubes are held together by an antiparallel hydrogen bonding network (lower right).

DESIGN OF TRANSMEMBRANE ION CHANNELS AND PORE STRUCTURES

Various cylindrical architectures have been proposed for use as cell membrane ion channels and pore structures.

The first sets of experiments were aimed at determining the efficiency of the channel-forming peptides in promoting proton transport in large unilaminar vesicles.[13] Liposomes prepared for such studies contained 5(6)-carboxyfluoroscein, as the fluorescence pH indicator. A plot of the change in fluorescence vs. time shows that the peptide nanotube is as effective or even better than the natural channel-forming counterparts.

To determine the ability of a peptide nanotube to transport small molecules across lipid bilayers, in this case glucose, we prepared an enzyme-coupled detection systems.[14] Transport experiments demonstrate the ability of peptide nanotubes to function as highly efficient membrane ion channels and size-selective pore structures. These characteristics now provide the basis for the future design of biologically active cytotoxic agents.

CONCLUSION

We have described a new functional nanostructured tubular biopolymeric system. The ease with which peptide nanotubes can be constructed, design flexibility in varying the chemical and structural properties of the cyclic peptide building blocks, and the tunable inner pore diameter of such ensembles, represent some of the most desirable characteristics of this new material. Electron microscopy has indicated formation of extensive, highly ordered, tubular structures. Studies with *N*-methylated cyclic peptides with the aid of solution–phase ^{1}H NMR and X-ray structure analyses have unequivocally documented the basic repeating motif and the mode of intermolecular hydrogen bonding interactions in such self-assembled tubular constructs. Membrane channel studies have shown that the nanotubes can function as highly efficient ion channels and pore structures with potential applications to biological settings.

REFERENCES

1. Ghadiri, M. R. *Adv. Mater.* **1995**, *7*, 675.
2. Harada, A.; Li, J.; Kamachi, M. *Nature* **1993**, *364*, 516.
3. Iijima, S. *Nature* **1991**, *354*, 56.
4. Chopra, N. G.; Luyken, R. J.; Cherry, K.; Crespi, V. H.; Cohen, M. L.; Louie, S. G.; Zettl, A. *Science* **1995**, *269*, 966.
5. Monnier, A.; Schuth, F.; Huo, Q.; Kumar, D.; Margolese, D.; Maxwell, R. S.; Stucky, G. D.; Krishnamurty, M.; Petroff, P.; Firouzi, A.; Janicke, M.; Chmelka, B. F. *Science* **1993**, *261*, 1299.
6. Kresge, C. T.; Leonowicz, M. E.; Roth, W. J.; Vartuli, J. C.; Beck, J. S.; *Nature* **1992**, *359*, 710.
7. Davis, M. E.; Saldarriaga, C.; Montes, C.; Garces, J.; Crowder, C. *Nature* **1988**, *331*, 689.
8. Nishizawa, M.; Menon, V. P.; Martin, C. R. *Science* **1995**, *268*, 700.
9. Ghadiri, M. R.; Granja, J. R.; Milligan, R. A.; McRee, D. E.; Khazanovich, N. *Nature* **1993**, *366*, 324.
10. Schnolzer, M.; Alewood, P.; Jones, A.; Alewood, D.; Kent, S. B. H. *Int. J. Peptide Protein Res.* **1992**, *40*, 180.
11. Rovero, P.; Quartara, L., Fabbri, G. *Tet. Lett.* **1991**, *32*, 2639.
12. Ghadiri, M. R.; Kobayashi, K.; Granja, J. R.; Chadha, R. K.; McRee, D. E. *Angew. Chem. Int. E. Engl.* **1995**, *34*, 93.
13. Ghadiri, M. R.; Granja, J. R.; Buehler, L. K. *Nature* **1994**, *369*, 301.
14. Granja, J. R.; Ghandir, M. R. *J. Am. Chem. Soc.* **1994**, *116*, 10785.

Perfluorinated Ion-Exchange Membranes

See: Ion-Exchange Membranes

Perfluorinated Ionomers

See: Ionomeric Membranes, Perfluorinated Ionomers (Overview)
Perfluorinated Ionomers (Overview)
Perfluorinated Ionomers (Anisotropic Ionic Conductivity)

PERFLUORINATED IONOMERS (Overview)

Michel Pineri, Pierre Aldebert, and Gérard Gebel
CEA-Département de Recherche Fondamentale sur la Matière Condensée
SESAM/Laboratoire de Physico-Chimie Moléculaire

Perfluorinated ionomers (PFI) are obtained by copolymerization of tetrafluoroethylene and perfluorovinyl ether monomers whose general formula is **Equation 1**:

$$CF_2 = CF - \left(O - \underset{\displaystyle CF_3}{\underset{|}{CF}} - CF_2 \right)_m - O - (CF_2)_n - X \qquad \mathbf{1}$$

where X is either SO_2F or $COOCH_3$.

The thermoplastic resin is laminated to produce a film whose thickness is between 50 and 200 μm. Hydrolysis is carried out by soaking at room temperature in alkaline hydroxide solutions to produce the ionic form; further exchange by soaking in the appropriate aqueous electrolyte solution results in the desired neutralized or acidic form. The ion exchange capacity ranging from 2–0.5 meq/g (equivalent weight from EW = 500–2000 g/eq) depends on the degree of copolymerization.

The first commercially available PFI membrane was Nafion® by E. I. DuPont de Nemours Co. This sulfonated PFI corresponds to values m = 1 and n = 2 in the general formula. The most studied membrane is Nafion® 117 (EW: 1100, 75 μm thick).

Two different approaches were followed for preparing carboxylic membranes: either chemical modification of the starting sulfonic material or direct synthesis from different monomers.[1]

SRUCTURE AND PROPERTIES

A major property of these ionomers is the ability to absorb large amounts of polar solvents. The ion clustering, which is specific of ionomers,[2–4] is modified upon swelling and leads to a continuous ionic phase above the percolation threshold; the resulting ionic conductivity and anion versus cation selectivity are the basic properties for applications of these membranes.

Membranes

Swelling Properties

Water and aqueous electrolyte swelling have been widely reported in literature.[1,5,6]

Solvent-Polymer Molecular Interactions

Preferential sorbtion sites for polar solvent are expected to be the ionic groups.[7] Evidence of water–ion interactions was obtained from infrared measurements by frequency shifts of the bands which were attributed to the symmetric stretching vibrations of the SO_3^-.[1,8,9] Various states of hydrate associations have been shown, for example, in simple electrolyte solutions.

Phase Segregation at High Solvent Content

Because of the hydrophobicity of the perfluorinated matrix, an ion-solvent phase is expected to be formed within the polymer. Small-angle scattering (SAS) and electron microscopy techniques are the most suitable techniques to characterize such reverse micelles.

Ionic Conductivity

Ionic conductivity depends on both the number of cationic species and on their mobility. Mobility is related to the ion diffusion coefficient and to the water diffusion because of the strong association between ions and water molecules. A continuous phase of ionic domains is necessary to get ion conductivity throughout the membrane. Such a condition is fulfilled by coalescence of the ionic clusters.

SOLUTIONS

Preparation

While low equivalent weight (EW < 900) Nafion® membranes are easily soluble in many polar solvents, high EW (> 1000) membranes are difficult to solubilize. In the early 1980s, an efficient method to obtain solutions using 1:1 water/alcohol mixtures at 250°C under pressure in an autoclave was proposed.[10,11]

Structure and Properties

Structural studies of long pendant chain perfluorinated ionomer solutions were mainly conducted using small angle neutron and X-ray scattering techniques. The structure of these solutions is well described by a rod-like model. A phenomenological model was recently proposed for describing the structural evolution from the dry membrane to solutions.[12]

APPLICATIONS

Nafion® membranes were first synthesized by DuPont in 1962; their ability to work as electroactive separators in chlor-alkali cells was demonstrated in 1964.[13] Further improvements, such as fabric reinforced Nafion® (1970) for mechanical properties and the introduction of a carboxylic layer (1971) to improve OH^- rejection, have led these membranes to be attractive alternatives to mercury or asbestos chlor-alkali cells.

Another promising application of these membranes is the relatively low temperature acidic fuel cells fed for instance by hydrogen or methanol.[14]

Other applications, even prospective, of PFI are far less important, especially from an industrial point of view. In the field of electrolytic uses, the regeneration of spent chromic acid plating baths, the electrowinning of certain metals, and the electrosynthesis of both inorganic and organic chemicals are relevant.[15] The nonelectrolyte applications, in which the electric potential is not used as a driving force, mainly concern preconcentration devices, in which properties are used in organic synthesis and are related to the superacidic properties of the protonic perfluorosulfonic ionomers.

REFERENCES

1. *Perfluorinated ionomer membranes*; Eisenberg, A.; Yeager, H. L., Eds.; ACS Symposium series 180, American Chemical Society: Washington, DC, 1982.
2. Wilson, F. C.; Longworth, R.; Vaughan, D. J. *Polym. Prep.* **1968**, *9*, 505.
3. Eisenberg, A. *Macromolecules* **1970**, *3*, 147.
4. Gierke, T. D., presented at Electrochemical Society Fall Meeting Atlanta, GA, USA, October 1977.
5. Morris, D. R.; Sun, X. *J. Appl. Polym. Sci.* **1993**, *50*, 1445.
6. Puspa, K. K.; Nandan, D.; Iyer, R. M. *J. Chem. Soc., Faraday Trans 1* **1988**, *84*, 2047.
7. Mauritz, K. A.; Hora, C. J.; Hopfinger, A. J. In *Ions in Polymers*; Eisenberg, A., Ed.; Adv. Chem. Ser. 187, ACS: Washington, DC, 1980; p 123.
8. Heitner-Wirguin, C. *Polymer* **1979**, *20*, 371.
9. Kujanski, W.; Nguyen, Q. T.; Néel, J. *J. Appl. Polym. Sci.* **1992**, *44*, 951.
10. Grot, W. G.; Chadds, C. European Patent 0 066 369, 1982.
11. Martin, C. R.; Rhoades, T. A.; Ferguson, J. A. *Anal. Chem.* **1982**, *54*, 1641.
12. Loppinet, B. Ph.D. Thesis, Université J. Fourier, Grenoble, 1994.
13. Grot, W. G. *Proc. of the Electrochem. Soc.* **1986**, *86*(13), 1.
14. Dhar, H. P. *J. Electroanal. Chem.* **1993**, *357*, 237.
15. Yeager, H. L. *Polym. Prep.* **1988**, *29*(2), 440.

PERFLUORINATED IONOMERS (Anisotropic Ionic Conductivity)

Robert B. Moore,* Kevin M. Cable, and Kenneth A. Mauritz
Department of Polymer Science
University of Southern Mississippi

The desirable properties of perfluorosulfonate ionomers (PFSIs) stem from their complex chemical and supermolecular structures.

As expected, the structure–property relationships of PFSIs are strongly influenced by the presence of backbone crystallinity and ionic aggregates. Because the PFSIs are semicrystalline and contain ionic aggregates, and these supermolecular structures act as barriers to flow, as-received (i.e., commercially available) PFSIs are not melt processable. However, we have recently shown that by eliminating or weakening these barriers to flow, a melt-processable form of PFSIs is produced.[1] Through this melt-processable material we are able to deform or orient PFSIs in order to impart new solid-state properties.

By neutralizing PFSIs with alkylammonium counterions [e.g., tetrabutylammonium (TBA^+)], the influence of electrostatic crosslinks via ionic aggregation can be significantly weakened.

In addition to the reduction of electrostatic effects, the barriers to flow caused by PFSI crystallinity may be eliminated by evaporating solutions of the polymer at sufficiently low temperatures. Previous morphological investigations of solution-cast PFSIs have shown that low-temperature evaporation yields a noncrystalline material, termed "recast" PFSI.[2]

The dramatic change in melt-flow behavior for the TBA^+-form PFSIs allows for permanent deformation (i.e., chain flow) when the materials are subjected to uniaxial stress. In contrast, the strong electrostatic crosslinks in PFSIs containing small inorganic ions yield an elastic behavior that would tend to restore the isotropic morphology after removal of the extensional stress. Previous studies of ionomer deformation have shown that uniaxial extension alters the spatial distribution of ion pairs and aggregates and/or affects the general shape of the ionic domains.[3–7] Thus, in light of these fundamental results, it is reasonable to expect that changes in the bulk properties of PFSIs may be realized if uniaxial extension has a consequential effect on the nanoscale organization of the ionic domains. In this article we report a procedure for altering the morphology of PFSI membranes that yields a material exhibiting anisotropic ionic conductivity.

*Author to whom correspondence should be addressed.

CONCLUSIONS

Uniaxial extension of TBA+-form PFSI membranes at elevated temperatures results in a material with an oriented morphology that persists after the samples are removed from the extensional stress. This oriented morphology is proposed to consist of elongated ionic domains and oriented polymer chains. Surface conductivity measurements from the oriented Na+-form membranes show that ionic conductivity in the direction parallel to the stretching direction was 40% greater than the conductivity in the perpendicular direction. This anisotropic behavior suggests that intra- and interaggregate ionic transport is enhanced in the parallel direction relative to transport in the perpendicular direction. Thus, in agreement with previous morphology and transport models for PFSIs, the potential energy barrier for ion transport in these uniaxially extended membranes is envisioned as being lowest when traversing the elongated ionic domains and minimized when interaggregate transport occurs along a direction of the preferential chain orientation.

ACKNOWLEDGMENT

We gratefully acknowledge funding for this research by the National Science Foundation/Electric Power Research Institute (Grant No. DMR-9211963) and the Air Force Office of Scientific Research (Grant No. F49620-93-0189DEF).

REFERENCES

1. Moore, R. B.; Cable, K. M.; Croley, T. L. *J. Membr. Sci.* **1992**, *75*, 7.
2. Moore, R. B.; Martin, C. R. *Macromolecules* **1988**, *21*, 1334.
3. Gierke, T. D.; Munn, G. E.; Wilson, F. C. *J. Poly. Sci., Polym. Phys. Ed.* **1981**, *19*, 1687.
4. Fujimura, M.; Hashimoto, T.; Hiromichi, K. *Macromolecules* **1981**, *14*, 1309.
5. Fujimura, M.; Hashimoto, T.; Hiromichi, K. *Macromolecules* **1982**, *15*, 136.
6. Roche, E. J.; Stein, R. S.; Russell, T. P.; MacKnight, W. J. *J. Polym. Sci., Polym. Phys. Ed.* **1980**, *18*, 1497.
7. Visser, S. A.; Cooper, S. L. *Macromolecules* **1992**, *25*, 2230.

Perfluorinated Polyethers

See: Fluoropolymer Coatings (New Developments)

Perfluoroalkoxyvinyl Copolymers

See: Engineering Thermoplastics (Survey of Industrial Polymers)

PERFLUOROCYCLOBUTANE AROMATIC ETHER POLYMERS

David A. Babb, Katherine S. Clement, and W. Frank Richey
Central Research and Development
Organic Products Research
Dow Chemical Company

Bob R. Ezzel
Central Research and Development
Advanced Polymeric Systems Lab

The thermal [$2\pi + 2\pi$] cyclodimerization of fluorinated carbon–carbon double bonds is a chemical reaction which has been well documented.[1–3] The adaptation of this reaction to the preparation of high molecular weight polymers, however, has been a relatively recent development.[4,5] This article outlines the development of the chemistry for the preparation of high molecular weight polymers and the characterization of the resulting thermoplastic and thermoset products.

The thermal cyclodimerization of fluorinated olefins is obviously not a concerned pericyclic reaction (disallowed via Woodward-Hoffman rules), but has been shown to be a stepwise radical reaction that starts with predominantly head-to-head bond formation, creating the most thermodynamically stable diradical intermediate, and resulting in the formation of a 1,2-substituted hexafluorocyclobutane ring.[6] The reaction is highly selective toward cyclodimerization and proceeds at relatively mild temperatures (140°C–200°C).

Our work has involved the adaptation of this chemistry to the formation of high molecular weight polymers. The aromatic trifluorovinyl ether functionality was chosen over other possible options for several reasons. The synthetic methods used are straight forward and provide a variety of useful monomers from common phenolic starting materials using a single synthetic scheme. The aromatic trifluorovinyl ethers are more stable to spontaneous oxidation than the perfluoroalkyl ether analogs, and undergo cyclodimerization under milder conditions. Finally, the presence of an aromatic ring provides a handle for chemical derivations that allow incorporation of the trifluorovinyl ether group into a variety of polymeric systems.

The chemistry was developed along three different avenues aimed at incorporation the hexafluorocyclobutane functionality into polymers. This first approach involved the preparation of high molecular weight polymers from trifluorovinyl ether-containing monomers. Monomers that have two dimerizable perfluorocyclobutane rings result in thermoplastic polymers, whereas polymers with three or more dimerizable trifluorovinyl ether groups result in thermoset polymers (**Figure 1**).

The second approach to incorporating the hexafluorocyclobutane group into polymers has been to end-cap low molecular weight thermoplastic oligomers with aromatic trifluorovinyl ether end groups. Thermally curing these oligomers than causes chain extension via cyclodimerization, and results in high molecular weight polymers with hexafluorocyclobutane rings dispersed more widely along the backbone.

The third method developed for the incorporation of the hexafluorocyclobutane ring into polymers has involved the preparation of 1,2-bis(aryl ether)hexafluorocyclobutane monomers with reactive groups attached to the aromatic rings. These monomers are then polymerized via conventional polymerization reactions.

APPLICATIONS

The combination of properties and processability that are accessible with polymers of these types make the resulting materials useful in a variety of applications. Prepolymers of the thermoset products are, in general, compatible with a wide range of organic solvents, making possible a number of processing methods. In general the polymers provide processability with the ability to tailor the viscosity and the percent solids of a prepolymer

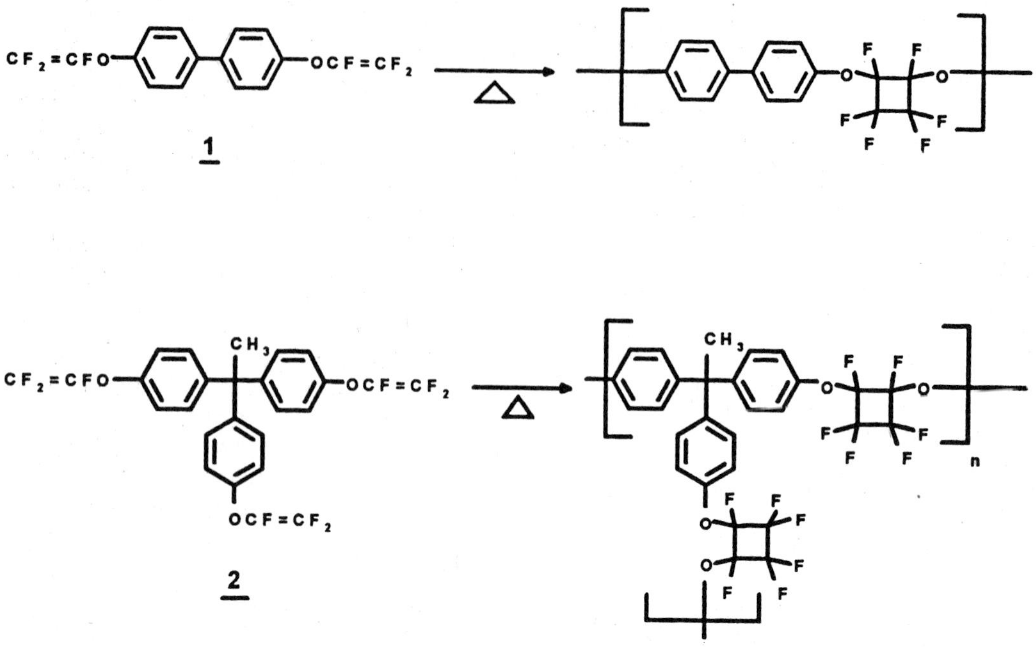

FIGURE 1. Bifunctional and trifunctional monomers, thermally cured to produce thermoplastics and thermosets, respectively.

solution. The high degree of selectivity of the cyclodimerization reaction allows the macromolecular morphology of the matrix to be designed. A specified degree of crosslinking may be incorporated into a substantially linear polymer simply by adding a trifunctional monomer. Thermoplastic polymers may be extruded, solvent cast, or compression molded. The resulting polymers, in general, exhibit excellent electrical properties and may also have excellent optical transparency depending on the nature of the aromatic group chosen.

REFERENCES

1. Perry, D. R. *Fluorine Chemistry Reviews* **1967**, *1*, 2.
2. Sharkey, W. H. *Flourine Chem. Rev.* **1968**, *2*, 1–53.
3. Chambers, R. D. *Flourine in Organic Chemistry*; John Wiley & Sons: New York, 1973.
4. Babb, David A.; Ezzell, Bob R.; Clement, Katherine S.; Frank Richey, W.; Kennedy, Alvin P. *J. Polym. Sci., Part A: Polym. Chem.* **1993**, *31*, 3465.
5. Yamomoto, M.; Swenson, D. C.; Burton, D. J. *Macromol. Symp.* **1994**, *82*, 125.
6. Bartlett, P. D.; Montgomery, L. K.; Seidel, B. *J. Am. Chem. Soc.* **1964**, *86*, 616.

Perfluorodioxole Polymers

See: Fluorinated Plastics, Amorphous

Perfluoropolymers

See: Engineering Thermoplastics (Survey of Industrial Polymers)
Fluorinated Plastics, Amorphous
Fluoropolymer Coatings (New Developments)
Perfluorinated Ionomers (Overview)
Tetrafluoroethylene Copolymers (Overview)

Periodic Copolymers

See: Alternating Ethylene-Vinyl Alcohol Copolymer
Zwitterionic Polymerization (Overview)

PERIODIC POLYMERS

Shiro Kobayashi,* Satoru Iwata, and Hiroshi Uyama
Department of Molecular Chemistry and Engineering
Faculty of Engineering
Tohoku University

Many studies have been carried out on the synthesis and properties of alternating copolymers in which equimolar quantities of two monomers are sequenced in an alternating fashion as given by $(AB)_n$.[1] In addition to these alternating copolymers, other types of sequence-regulated polymers are known where two, three, or more monomeric units appear in an ordered sequence. These periodically sequence-regulated polymers are to be called periodic copolymers, periodic terpolymers, periodic quarterpolymers, etc. depending on how many kinds of monomer are involved. Examples are:

-AABAABAABAAB- or $(AAB)_n$
-AABBAABBAABB- or $(AABB)_n$
-ABACABACABAC- or $(ABAC)_n$
-ABCDABCDABCD- or $(ABCD)_n$

*Author to whom correspondence should be addressed.

There are named periodic polymers and given as

poly(A-*per*-A-*per*-B)
poly(A-*per*-A-*per*-B-*per*-B)
poly(A-*per*-B-*per*-A-*per*-C)
poly(A-*per*-B-*per*-C-*per*-D), respectively.

A complete regulation of polymer sequence composed of three and more monomeric units has not yet been attained by the polymerization of vinyl monomers. However, a few periodic polymers have been synthesized by unique combinations of two or more monomers based on the following methodologies: zwitterionic polymerizations, polymer modifications, biradical copolymerizations, biotechnological synthesis, etc.

PREPARATIONS

AB_2 or A_2B-Type Periodic Copolymer

AB_2 or A_2B-type copolymers are obtained by the spontaneous copolymerization via zwitterion intermediate. The most important factor to control the sequences of monomeric unit in polymer chain is the combination choice of a nucleophilic monomer M_N and an electrophilic monomer M_E.

The combined use of salicylyl phenyl phosphonite (SPO) as M_N and benzaldehyde (BA) as M_E gives the AB_2-type copolymer [poly(SPO-*per*-BA-*per*-BA)] comprising phosphinate and ester groups in the main chain (**Scheme I**).[2]

Cyclic imino ethers, 2-oxazoline (OZO) and 5,6-dihydro-4H-1,3-oxazine (OZI) have served as M_N, and copolymerize with succinic anhydride (SAn) and glutaric anhydride (GAn) to afford 2:1 sequence-regulated copolymers in which one OZO or OZI unit is incorporated in a ring opened form and the other unit is formed via reaction of the carbon-nitrogen double bond of OZO or OZI.[3]

$A_{1/2}B$ Type Copolymer

2,4-Bis(4-methoxyphenyl)-1,3,2,4-dithiadiphosphetane-2,4-disulfide (Lawesson's reagent, LR) reacts with cyclic ethers, ethylene oxide (EO) and oxetane (OX) to yield the copolymers [poly(1/2LR-*alt*-EO) and poly(1/2LR-*alt*-OX), respectively] having 1/2:1 composition of LR and cyclic ether.[6]

ABC-Type Terpolymer

ABC-type periodic terpolymer is obtained by the combination of ethylene phenyl phosphonite (EPO) as M_N with two kinds of M_E, a vinyl compound having an electron withdrawing group [acrylonitrile (AN) or methyl acrylate (MA)], and carbon dioxide.[4]

Periodic Poly(Amino Acid)s

Recently, periodically sequence-regulated polypeptides $[(AlaGly)_x ZGly]_n$ (Z = Ala, Asn, Asp, Glu, Leu, Met, Phe, Thr, Tyr, and Val) have been biologically synthesized.[11] The resulting poly(amino acid)s possessed exact structure and uniformity of chain length.

PROPERTIES

Most of the periodic copolymers have low molecular weight or have unusual repeating units. Hence, the systematic examination of the properties for these periodic copolymers has not yet been completed. The properties of the virtually periodic copolymers have been investigated in comparison with those of homopolymers, alternating copolymers, and statistic copolymers.

REFERENCES

1. Cowie, J. M. G. *Alternating Copolymers*; Plenum: New York, 1985.
2. Saegusa, T.; Kobayashi, T.; Chow, T. Y.; Kobayashi, S. *Macromolecules* **1979**, *12*, 533.
3. Kobayashi, S.; Isobe, M.; Saegusa, T. *Macromolecules* **1982**, *15*, 703.
4. Saegusa, T.; Kobayashi, S.; Kimura, Y. *Macromolecules* **1977**, *10*, 68.
5. Deguchi, Y.; Fournier, M. J.; Mason, T. L.; Tirrell, D. A. *Pure Appl. Chem.* **1994**, *A31*, 1691.
6. Yokota, K.; Hirabayashi, T. *Macromolecules* **1981**, *14*, 1613.

Permselective Membranes

See: *Polyphenylacetylene-Based Permselective Membranes*

Peroxide Initiators

See: *Block Copolymers (From Macroinitiators)*
Fluoropolymers (Preparation by Aromatic Fluoroalkylation)
Fluorosilicon Oligomers
Peroxide Initiators (Overview)
Peroxide Initiators (for High Solids Polyacrylate Resins)
Peroxide Initiators (for Water-Based Acrylate Emulsion Polymerization)

PEROXIDE INITIATORS (Overview)

Jose Sanchez and Terry N. Myers
Buffalo Research and Development Center
Elf Atochem North America, Incorporated

Peroxide initiators (peroxides) are chemical substances that possess one or more oxygen-oxygen bonds and have the general structures $ROOR^1$ or ROOH where R and R^1 are inorganic or organic groups. Under the influence of heat or light, the weak

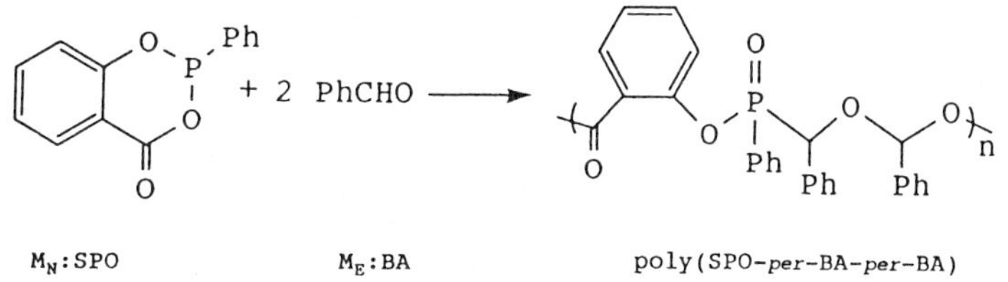

I

oxygen-oxygen bond cleaves to produce a pair of free radicals (radicals) (**Equation 1**):

$$ROOR^1 \xrightarrow{\Delta\ or\ h\nu} RO + OR^1 \ (\textit{or other radicals}) \qquad 1$$

The radicals produced are highly reactive species that possess a free electron (unbounded or unpaired) and have very short half-life times (10^{-3} second or less).[1] Oxygen-centered radicals such as alkoxy radicals (e.g., *tert*-amyloxy) or acyloxy radicals can further decompose to produce lower energy alkyl radicals and a ketone or carbon dioxide by a β-scission process.

Peroxides can also undergo activated or electron-transfer (redox) processes to produce radicals, e.g., **Equation 2**:

$$ROOH\ (\text{or}\ ROOR^1) + M^{+n} \rightarrow RO^{\cdot} + M^{+(n+1)} + HO^{-}\ (\text{or}\ R^1O^{-}) \qquad 2$$

PEROXIDE HALF-LIFE CHARACTERISTICS

Thermal decomposition rates for peroxides are affected by many factors including pressure, viscosity, peroxide physical state, peroxide concentration, solvent polarity, peroxide structure, and temperature. Increases in pressure and viscosity retard decomposition rates when the increases are relatively large.[2,3] Typically, solid peroxides are much more stable in the solid state than in the solution state. Dibenzoyl peroxide (BPO; mp, 106–107°C) [94-36-0] is commercially attractive in large part because of this phenomenon.

Organic Peroxides

There are approximately 100 different organic peroxides, in well over 300 formulations (e.g., neat liquids and solids, pastes, powders, solutions, dispersions, and emulsions) that are commercially produced throughout the world, primarily for the polymer and resin industries.[4–9]

There are seven classes of organic peroxides that are commercially produced for the polymer and resin industries. These classes include diacyl peroxides, *tert*-alkyl peroxyesters (including *OO-tert*-alkyl *O*-alkyl monoperoxycarbonates), dialkyl peroxydicarbonates, di(*tert*-alkylperoxy)ketals, di-*tert*-alkyl peroxides, *tert*-alkyl hydroperoxides, and ketone peroxides. The 10-hr, 1-hr, 1-min, and 1-sec half-life temperatures (HLTs) of many commercial organic peroxides are listed in **Table 1**. The 10-hr HLTs of the commercial organic peroxides range from ~ 20°C to ~130°C.

Inorganic Peroxides

World consumption of inorganic peroxides for the polymer and resin industries is small compared to the consumption of organic peroxides. The major inorganic peroxides that are used in these industries are hydrogen peroxide {*7722-84-1*] ammonium peroxydisulfate {$(NH_4)_2S_2O_8$} [*7727-54-0*], potassium peroxydisulfate ($K_2S_2O_8$) [*7727-21-1*], and sodium peroxydisulfate ($Na_2S_2O_8$) [*7775-27-1*]. These initiators are normally used at ambient temperature together with an activator in redox polymerization systems.

Safe Handling of Peroxides

Care must be exercised when storing, handling, and using peroxides. In their pure or highly concentrated states, they can decompose (some violently) when exposed to excessive temperatures. Contamination, particularly of hydroperoxides and ketone peroxides, can also lead to violent decomposition. Because some peroxides are shock or friction sensitive in the pure state, they are generally desensitized through formulation into solutions, pastes, or powders with inert diluents.

Applications of Peroxides

Because they produce highly reactive radicals, peroxides are used in many commercial radical reactions. In the various polymer industries they are used as radical sources for polymerizing and copolymerizing vinyl monomers, curing unsaturated polyester resins and elastomers, crosslinking olefin polymers and copolymers, grafting vinyl monomers onto polymers, modifying polypropylene, and compatibilizing polymer blends and alloys.

The effectiveness of a peroxide in a given application depends on its thermal decomposition rate and the efficiency of the radicals generated in carrying out the desired reaction.

High Pressure Ethylene Polymerization

Combinations of organic peroxides covering a broad half-life temperature range are used in high pressure ethylene polymerizations and copolymerizations.[10]

Vinyl Chloroide (VCl) Polymerizations

In general, VCl suspension polymerizations are carried out in the 45–80°C temperature range over periods of 3–10 hr. The peroxides used in polymerization of VCl monomer have 1-hr HLTs of ~ 50°C to ~ 85°C.[12]

Acrylic Monomer Polymerization

The classes of organic peroxides having broad half-life characteristics are usually used in the commercial polymerization and copolymerizations of acrylic monomers. Applications range from acrylic lattices to acrylic rods and sheets to high solids acrylic solution coatings (HSCs).

Styrene Polymerizations

Peroxides are used for the production of styrene homo- and copolymers and impact styrene polymers such as high-impact polystyrene (HIPS) and acrylonitrile-butadiene-styrene (ABS) polymers.

Curing of Allyl Diglycol Carbonate (ADC)

ADC monomer is cured with peroxides over a broad temperature range (35–120°C) and time (up to 12 hr) to produce shatter-resistant and optically clear sheets and lenses.

Curing of Unsaturated Polyester Resins

Approximately 30% of peroxide-cured unsaturated polyester resins are cured at elevated temperatures in molding applications.

TABLE 1. Half-life Temperatures for Commercial Peroxides

Chemical Name	Solvent[a]	E_a[b] kJ/mole	ln(A)	Half-life Temperature, °C 10-hr	1-hr	1-min	1-sec
Class: Diacyl Peroxides							
Di(2,4-dichlorobenzoyl) peroxide [*133–14–2*]	Benzene	117.2	32.3	54	73	111	160
Diisononanoyl peroxide [*58499–37–9*]	TCE	129.7	35.9	61	78	114	158
Dibenzoyl peroxide [*94–36–0*]	Benzene	128.9	33.9	73	92	131	179
Di(decanoyl) peroxide [*762–12–9*]	TCE	128.4	34.9	65	83	120	165
Dilauroyl peroxide [*105–74–8*]	TCE	130.1	35.6	64	81	117	162
Succinic acid peroxide [*123–23–9*]	Acetone	97.5	23.7	66	90	143	214
Class: Dialkyl Peroxides							
1,3(4)-Di(1-(*tert*-butylperoxyl)-1-methylethyl)benzene [*25155–25–3*]	Dodecane	154.0	36.4	119	139	181	231
2,5-Di(*tert*-butylperoxy)-2,5-dimethyl-3-hexyne [*1068–27–5*]	Dodecane	159.4	36.6	131	152	194	246
2,5-Di(*tert*-butylperoxy)-2,5-dimethylhexane [*78–63–7*]	Dodecane	155.6	36.7	120	140	181	232
Di-α-cumyl peroxide [*80–43–3*]	Decane	154.0	36.6	117	137	178	228
Di-*tert*-amyl peroxide [*10508–09–5*]	Dodecane	159.4	37.5	123	143	184	233
Di-*tert*-butyl peroxide [*110–05–4*]	Decane	164.8	38.4	129	149	189	238
tert-Butyl α-cumyl peroxide [*3457–61–2*]	Dodecane	158.2	37.0	124	144	185	235
Class: Diperoxyketals							
1,1-Di(*tert*-amylperoxy)cyclohexane [*15667–10–4*]	Dodecane	144.8	36.6	93	112	150	197
1,1-Di(*tert*-butylperoxy)-3,3,5-trimethylcyclohexane [*6731–36–8*]	Dodecane	148.5	37.5	96	115	153	199
1,1-Di(*tert*-butylperoxy)cyclohexane [*3006–86–8*]	Dodecane	144.8	36.2	97	116	155	203
2,2-Di(*tert*-amylperoxy)propane [*3052–70–8*]	Dodecane	144.3	34.8	108	128	170	222
2,2-Di(*tert*-butylperoxy)butane [*2167–23–9*]	Dodecane	143.5	34.5	107	127	169	221
Ethyl 3,3-di(*tert*-amylperoxy)butyrate [*67567–23–1*]	Dodecane	148.5	35.6	112	132	173	224
Ethyl 3,3-di(*tert*-butylperoxy)butyrate [*55794–20–2*]	Dodecane	151.5	36.2	114	134	175	225
n-Butyl 4,4-di(*tert*-butylperoxy)valerate [*995–33–5*]	Dodecane	147.7	35.6	109	129	170	220
Class: Peroxydicarbonates							
Di(2-ethylhexyl) peroxydicarbonate [*16111–62–9*]	TCE	127.6	36.7	49	66	99	140
Di(2-phenoxyethyl) peroxydicarbonate [*41935–39–1*]	TCE	128.0	36.8	50	67	101	142
Dicyclohexyl peroxydicarbonate [*1561–49–5*]	TCE	123.8	35.3	50	67	102	145
Di(hexadecyl) peroxydicarbonate [*26322–14–5*]	TCE	125.1	35.8	50	67	101	143
Diisopropyl peroxydicarbonate [*105–64–6*]	TCE	123.8	35.2	50	67	102	146
Di-*n*-propyl peroxydicarbonate [*16066–38–9*]	TCE	128.4	37.1	50	66	99	140
Di-*sec*-butyl peroxydicarbonate [*19910–65–7*]	TCE	116.7	32.5	51	69	107	154
Class: Peroxyesters							
2,5-Di(2-ethylhexanoylperoxy)-2,5-dimethylhexane [*13052–09–0*]	Decane	131.4	34.8	73	91	129	176
2,5-Di(benzoylperoxy)-2,5-dimethylhexane [*2618–77–1*]	Benzene	152.3	38.2	100	118	156	202
3-Hydroxy-1,1-dimethylbutyl 2-ethylperoxyhexanoate [*95732–35–7*]	AMS	118.0	31.2	65	84	125	177
3-Hydroxy-1,1-dimethylbutyl peroxyneodecanoate [*95718–76–0*]	TCE	111.3	32.3	37	54	91	136
3-Hydroxy-1,1-dimethylbutyl peroxyneoheptanoate [*110972–57–1*]	AMS	115.1	33.2	41	58	94	139
α-Cumyl peroxyneodecanoate [*26748–47–0*]	TCE	111.3	32.0	38	56	93	139
α-Cumyl peroxyneoheptanoate [*104852–44–0*]	TCE	115.9	33.4	43	60	96	141
Di-*tert*-butyl diperoxyphthalate [*2155–71–7*]	Benzene	158.2	39.6	104	122	159	203
OO-*tert*-amyl O-(2-ethylhexyl) monoperoxycarbonate [*70833–40–8*]	Dodecane	150.6	37.8	99	117	155	201
OO-*tert*-butyl O-(2-ethylhexyl) monoperoxycarbonate [*34443–12–4*]	Dodecane	131.8	31.7	100	121	166	222
OO-*tert*-butyl O-isopropyl monoperoxycarbonate [*2372–21–6*]	Benzene	142.3	35.2	99	118	159	208
tert-Amyl 2-ethylperoxyhexanoate [*686–31–7*]	TCE	141.8	38.4	73	90	125	167
tert-Amyl peroxyacetate [*690–83–5*]	Dodecane	139.3	34.1	100	120	162	214
tert-Amyl peroxybenzoate [*4511–39–1*]	Dodecane	138.5	33.9	100	120	162	214
tert-Amyl peroxyneodecanoate [*68299–16–1*]	TCE	120.5	34.5	46	64	99	143
tert-Amyl peroxypivalate [*29240–17–3*]	TCE	117.6	32.2	55	74	112	161
tert-Butyl 2-ethylperoxyhexanoate [*3006–82–4*]	Dodecane	142.7	38.1	77	95	130	173
tert-Butyl 3,5,5-trimethylperoxyhexanoate [*27836–52–8*]	Benzene	138.9	33.7	101	122	164	217
tert-Butyl peroxymaleate [*1931–62–0*]	Acetone	113.0	26.9	87	111	161	226
tert-Butyl peroxyacetate [*107–71–1*]	Decane	137.7	33.3	102	123	166	219
tert-Butyl peroxybenzoate [*614–45–9*]	Dodecane	132.2	31.4	104	125	171	228
tert-Butyl peroxyisobutyrate [*109–13–7*]	Decane	123.8	31.1	82	102	146	200
tert-Butyl peroxyneodecanoate [*26748–41–4*]	TCE	119.2	33.7	48	66	102	147
tert-Butyl peroxyneoheptanoate [*26748–38–9*]	TCE	113.4	31.0	53	72	112	162
tert-Butyl peroxypivalate [*927–07–1*]	TCE	117.6	32.0	58	76	116	165

[a]TCE = trichloroethylene, AMS = α-methylstyrene.
[b]To change from kilojoules to kilocalories, divide by 4.184.

Curing of Elastomers, Crosslinking of Polyolefins, and Controlled-Rheology (Vis-Breaking) of Polypropylene

In these applications, organic peroxides are usually blended into molten polymers at relatively high temperatures before decomposition of the peroxide.

Curing of Elastomers

Curing of elastomers, such as ethylene-propylene rubber (EPR), ethylene-propylene-diene rubber (EPDM), nitrile rubber, silicone rubbers, and fluoroelastomers, is conducted at 95°C to 215°C.

Crosslinking of Polyolefins

Polyolefins such as low-density polyethylene (LDPE), high-density polyethylene (HDPE), and linear low-density polyethylene (LLDPE) are exclusively crosslinked with dialkyl peroxides.

Controlled Rheology (Vis-Breaking) of Polypropylene (PP)

A low concentration (typically 400–700 ppm) of dialkyl peroxide is used to extruder-modify PP to improve the processing and molding characteristics of virgin PP.

Reactive Extrusion

Reactive extrusion is a growing application area for chemically modifying polymers with peroxides. It is used for preparing polymeric blends and alloys and for grafting of monomers onto polymers, for example, grafting of maleic anhydride onto polyolefins such as polypropylene.[12]

REFERENCES

1. Griller, D.; Ingold, K. U. *Acc. Chem. Res.* **1976**, *9*(1), 13.
2. Halle, R. *Plast. Compound* **1992**, *3*(2), 73–77.
3. Pryor, W. A.; Morkved, E. H.; Bickley, H. T. *J. Org. Chem.* **1972**, *37*, 1999–2005.
4. Immergut, E. H., Eds; John Wiley & Sons: New York 1989. Davies. A. G. *Organic Peroxides*, Butterworths: London 1961.
5. Hawkins, E. G. E. *Organic Peroxides*; Spon, E. and F. F. London, 1961.
6. Tobolsky, A. V.; Mesrobian, R. B. *Organic Peroxides*, Interscience: New York, 1954.
7. Swern, D. Ed., *Organic Peroxides*, Wiley-Interscience: New York, 1970; Vol. 1.
8. Swern, D. Ed., *Organic Peroxides*, Wiley-Interscience: New York, 1971; Vol. 2.
9. Swern, D. Ed., *Organic Peroxides*, Wiley-Interscience: New York, 1972; Vol. 3.
10. Beach, D. L.; Kissin, Y. V. In *Encyclopedia of Polymer Science and Engineering* Kroschwitz, J. I., Ed.; Wiley-Interscience: New York, 1986; Vol. 6, pp 383–490.
11. Smallwood, P. V. In *Encyclopedia of Polymer Science and Engineering*, Kroschwitz, J. I., Ed.; Wiley-Interscience: New York, 1989; Vol. 17, pp 295–329.
12. Callais, P. A.; Kazmierczak, R. T. *Proc. 47th ANTECH* 1989, pp 1368–1370; Callais, P. A.; Kazmierczak, R. T. *Proc. 48th ANTECH* 1990; pp 1921–1923; Callais, P. A.; Palys, L. H.; Kazmierczak, R. T. *Proc. 51st ANTECH* 1993; pp 2539–2543.

PEROXIDE INITIATORS (for High Solids Polyacrylate Resins)

Lea L. Anderson and Ginger G. Myers
Akzo Nobel Chemicals Incorporated

Over the past several years, increasingly stringent environmental regulations have exerted pressure to reduce levels of organic solvents in coatings applications. One solution is higher solids acrylate resins. These resins are most commonly used in automotive refinishing and coating applications. Resins are typically high in polymer content with the lowest possible solvent content to give reasonable flow properties. Initiator properties contribute to the final resin molecular weight, viscosity, and polydispersity.

Because initiators can be a significant expense in resin manufacture, and can affect batch cycle times and conversion, we have evaluated a large number of organic peroxides for high solids resins. We have chosen two procedures for evaluating organic peroxides: a high solids model system and the standard ampoule technique.

ACKNOWLEDGMENTS

Kai-Yun Wang and Ruth Gallagher of Akzo Nobel Chemicals Inc. for their contributions to the experimental portion of this paper.

PEROXIDE INITIATORS (for Water-Based Acrylate Emulsion Polymerizations)

Lea L. Anderson and Wilfried M. Brouwer
Akzo Nobel Chemicals Incorporated

PART I: REDUCTION OF RESIDUAL MONOMER IN EMULSION-BASED ACRYLIC RESINS

Complete conversion of monomers into resinous polymer is a desire of every resin producer, as residual monomers may give a pungent odor to the end product. Toxicity may also be an important issue. Various means exist to decrease residual monomers in (co)polymer resins, such as:

- Temperature increase.
- Addition of reactive comonomer.
- Use of the proper organic peroxides.

Emulsion-Based Polymerization

In an emulsion polymerization, irrespective of using a water or oil soluble initiator, part of the initiator will be present in the water phase and therefore efficiency is only somewhat affected by the amount of polymer in the particle phase.[1]

Conclusions

Organic (hydro)peroxides, such as *tert*-butyl hydroperoxide and various peroxyesters, are more effective than persulfates and hydrogen peroxide for reducing residual monomer in a latex when a redox system is used. The advantages of such a system are that no reactor capacity is consumed as the system is charged batchwise to a latex storage tank at ambient temperature, and that there will be less increase in the ionic strength, which gives all the additional advantages (less corrosion, lower water sensitivity in application, etc.).

PART II: THE USE OF HYDROPEROXIDES AS INITIATORS IN EMULSION POLYMERIZATIONS OF ACRYLATES

Currently, manufacturers of emulsion-based polyacrylate resins primarily use persulfates as initiators. However, there is a high degree of interest in using organic peroxides in this application. One obvious advantage of peroxides over persulfates is their much less corrosive nature.

Conclusions

Organic hydroperoxides can be effective initiators in the polymerization of acrylates in emulsion systems. Three hydroperoxides were studied and all were found to function as suitable initiators. Comparison of potassium persulfate to the hydroperoxide *Trigonox* A-W70 showed that the hydroperoxide gave a resin with a much lower molecular weight and a much more desirable pH.

REFERENCES

1. Ballard, M. J. et al. *Macromolecules* **1986**, *19*, 1303.

PERSONAL CARE APPLICATION POLYMERS (Acetylene-Derived)

Robert B. Login and Eugene S. Barabas
International Specialty Products Corporation

A key point in designing polymers for personal care applications such as hair fixatives and conditioners is to start from monomers that afford water-soluble or removable polymers. In addition, such monomers must contribute to the polymer adhesive and film-forming properties and the resulting polymers must exhibit a low order of toxicity. Both vinylpyrrolidone (VP) and methyl vinyl ether (MVE) are premier examples of such monomers.

POLY(VINYL ETHERS)

Poly(vinyl ether)s have been used successfully in a number of applications. The cosmetics industry, however, has yet to find use for these attractive polymers.

VINYL ETHER BASED POLYMERS FOR PERSONAL CARE APPLICATIONS

Although a wide variety of vinyl ether monomers are possible, only the methyl derivative, MVE, has achieved significance in polymers designed for personal care applications. The most outstanding feature of MVE is its ability to copolymerize by free-radical initiation to form alternating copolymers with electron-deficient monomers, such as maleic anhydride (MAn).[1–3]

POLYVINYLPYRROLIDONE

In the field of cosmetics, polyvinylpyrrolidone (PVP) has the historic significance of being the first synthetic polymer to replace natural polymers like shellac in cosmetic formulations.[4]

Because of its chemical structure, and particularly because of the presence of the lactam group, it exhibits substantivity to hair, and instead of welding the hairs together it coats them with smooth and lustrous films. The film is clear and nongreasy but rewettable and easy to remove. Because of these advantageous properties, a whole series of PVP-based products (e.g., wave sets, shampoos, lotions, hair tints, and hair grooming gels) appeared on the market. PVP contributed to the field of hair care cosmetics a reliable noncorrosive film former with solubility in propellants used in pressurized products and other properties the industry had been searching for.[5]

The polymer was found to be just as useful in skin care applications. PVP increases the lubricity and emolliency of skin care preparations and ensures their moisture balance. PVP also improves the performance of skin and eye make-ups, face rouges, and lipsticks. In these applications PVP acts as a complexing agent and dispersing aid.[6] However, in suntan lotions and deodorants PVP works mostly as a film former and fixative. But PVP performs also as a protective colloid in stabilizing creams, emulsions, lotions, and foam products.[7]

POLY(VINYLPYRROLIDONE-*co*-VINYLACETATE) COPOLYMERS

When the hydrophilicity of PVP is too high in certain applications, it can be controlled by introducing less water-sensitive monomers into the polymer chain. Vinyl acetate (VA) is particularly suitable for this purpose because it copolymerizes readily with VP. The reactivity ratios are 3.4:0.195 (VP:VA).[8]

P(VP/VA) copolymers form clear films and because of the effect of the less hydrophilic VA units have less tendency for tackiness under high humidity conditions. The copolymers can be prepared by free-radical polymerization both in solution and emulsion or by irradiation.[9–12]

CATIONIC COPOLYMERS OF VINYLPYRROLIDONE

Quaternary Derivatives

The propensity of monomeric fatty alkyl and polymeric quaternary ammonium derivatives to be substantive to the hair and skin is well known.[13–15] This observation stimulated the design of cationic PVP derivatives. After all, PVP is strongly adhesive to the skin and hair. Copolymers with high ratios of VP would also be expected to give clear, hard, nonflaking, lustrous, and readily water-soluble film-forming polymers. The first candidates to appear commercially were based on quaternized copolymers of dimethylaminoethyl methacrylate (DMAEMA) and VP (polyquaternium-11).[16,17] They were followed by copolymers of methyl vinylimidazolium chloride (polyquaternium-16) and most recently, methacrylamidopropyltrimethyl-ammonium chloride (MAPTAC) (polyquaternium-28).[18,19] Replacement of the ester in DMAEMA with the amide analog yields copolymers with superior hydrolytic resistance and, hence, stable performance in alkaline formulations such as those designed to bleach or perm the hair.

THREE-COMPONENT COPOLYMERS

A further refinement to the above idea is to replace some or all of the VP in the copolymer with vinyl caprolactam. This change gives a terpolymer that exhibits enhanced moisture resistance without the loss of the interaction of the lactam with hair, enhancing the hydrophobic amine/hair interactions.

Alkylated Polyvinylpyrrolidone Derivatives

Alkylated PVP resins are also surface active, readily depositing at the interfaces, and like PVP, they retain the lactam's ability to complex through hydrogen and electrostatic bonding. Both of these attributes can be combined in sunscreen formulations where alkylated PVPs function as water-proofing film formers.[20]

REFERENCES

1. Biswas, M.; Mazumdar, A.; Mitra, P. In *Encyclopedia of Polymer Science and Engineering*; John Wiley & Sons: New York; 1989; Vol. 17, p 446.
2. Cowie, J. M. G. In *Comprehensive Polymer Science*; Allen, G.; Bevington, J., Eds.; Pergamon: 1989.
3. Trivedi, B. C.; Culbertson, B. M. *Maleic Anhydride*; Premium: New York, 1982.
4. Corbett, J. F. *The Chemistry of Hair Care Products*; ISDC: August 1976; p 285.
5. Lochead, R. Y. *Cosmetics and Toiletries* **1988**, *103*.
6. Prescott, F. J.; Hahnel, E.; Day, D. *Drug. Cos. Ind.* **1963**, *93*(4–5), 443.
7. *Kollidon Grades*; Bulletin MEF 129e; BASF: Ludwigshafen, Germany, 1986; p 14.
8. Greenley, R. Z. In *Free Radical Reactivity Ratios in Polymer Handbook*; Brandrup, J.; Immerut, E. H., Eds.; John Wiley & Sons: New York, 1989.
9. Straub, F.; Sporr, H.; Schenk, H. U.; Schwartz, W. Ger. Offen. 2 730 019, 1979.
10. Barabas, E. S.; Cho, J. R. U.S. Patent 4 554 311, 1985.
11. Barabas, E. S.; Fein, M. R. U.S. Patent 3 691 625, 1972.
12. Tashmukhamedov, S. A. *Uzb. Khim. Zh.* **1972**, *16*(3), 54.
13. Hunting, A. L. L. *Cosmetics & Toiletries* **1984**, *99*, 99.
14. Jachowicz, J.; Berthiaume, M.; Garcia, M. *Coll. Polym. Sci.* **1985**, *263*, 847.
15. Kamath, Y. K.; Dansizer, C. J.; Weigmann, H-D. *J. Appl. Polym. Sci.* **1984**, *29*, 1011.
16. Barabas, E. S.; Fein, M. M. U.S. Patent 3 910 862, 1975.
17. Valan, K. U.S. Patent 3 914 403, 1975.
18. *Gafquat 734, 755N and HS-100: Cationic Conditioning Copolymers*; Bulletin Q0394; International Society Products: Wayne, New Jersey, 1994.
19. *Gafquat HS-100 Resin*; Bulletin; International Specialty Products: Wayne, NJ, 1989/90.
20. Oteri, R.; Johnson, S.; Dastis, S. *A New Waterproofing Agent for Sunscreen Products, Cosmetics and Toiletries* **1987**, *102*, 102.

Personal Care Polymers

See: Hair and Skin Care Biomaterials (Cosmetology and Dermatology)
Hair and Skin Cosmetics
Personal Care Application Polymers (Acetylene-Derived)
Silicone Polymers, Organo-Modified (Application in Personal Care Products)
Water-Swellable Polymers (Carbomer Resins)

Pervaporation Membranes

See: Pervaporation Membranes (Overview)
Pervaporation Membranes (for Separating Organic Solvents)
Pervaporation Membranes (Materials)

PERVAPORATION MEMBRANES (Overview)

Takashi Iwatsubo,* A. Yamasaki, and K. Mizoguchi
National Institute of Materials and Chemical Research

Pervaporation (PV) is a separation process of liquid mixtures that uses synthetic membranes in which the liquid mixture is directly contacted with one side of the membrane (i.e., the feed or upstream side) and vacuum or inert gas flow is applied on the other side (i.e., the permeate or downstream side). The mixture permeates, according to the gradient of the chemical potential, across the membrane and is taken out as vapor on the permeate side. The vapor mixture is cooled and liquefied. Separation takes place because of the difference in the permeation rates of the permeants.

PV has advantages for the separation of the ethanol/water mixture, especially for dehydration over distillation in the energy consumption as well as the capability of the separation of an azeotropic mixture (96 wt % ethanol).

THEORY OF PV TRANSPORT

Performance of PV

There are two aspects of the performance of the PV process: the permeation flux and the separation factor. Permeation flux is defined as the amount permeated per unit time and unit membrane area under a steady-state condition.

It is well known that PV performance depends on operational conditions, such as feed liquid composition, temperature, pressure (i.e., upstream and downstream), and membrane thickness.

Transport Mechanism of PV

Transport of PV is explained by the solution (sorption)–diffusion theory. According to the theory, transport takes place by the following three steps: dissolution of the liquid mixture into the membrane on the feed side, diffusion through the membrane due to the chemical potential gradient, and evaporation at the downstream side.[1] The second step is considered to be the rate-determining step. Sorption equilibrium is assumed on both sides of the membrane.

MEMBRANES FOR PV

Selection of Membranes

The selection of proper membranes is a key to successful membrane separation not only for PV. There are several aspects of membrane selection, of which selection of membrane materials is the most important. Material selection means the selection of the chemical structure of polymers, and the key factor is affinity of the materials to the components. For example,

*Author to whom correspondence should be addressed.

hydrophilic materials such as PVA and PAN are used for dehydration membranes, whereas hydrophobic materials such as polydimethylsiloxane (PDMS) are selected for the removal of organics from the water stream.

Higher Order Structure

It is well known that the membrane structure can be altered by membrane formation conditions such as the species of solvent in the casting solution and solvent evaporation time or temperature, and the structural (morphological) changes can reflect the membrane performance.

Modification of the Membranes

Many kinds of modification have been tested to improve membrane performance. Crosslinking is effective in reducing the swelling of the membrane to increase the selectivity. Surface treatment using low temperature plasma is also an effective method to control the hydrophobicity and hydrophilicity in plasma polymerization and in plasma graft.[2–4] The introduction of functional materials to membranes is also an effective way to improve the performance.

Examples of PV Membranes

Alcohol/Water

Many studies have been done on the PV separation of the ethanol/water mixture. Especially for dehydration, extraordinarily high separation factors have been reported in the literature. All of the dehydration membranes are hydrophilic and applied to a higher ethanol concentration range. These membranes are also effective for the dehydration of other aqueous systems such as acetic acid/water and low-chain alcohols/water.

Removal of Trace VOCs From Water

Recently, contamination of underground water by organic solvents such as chloroform has become a serious environmental problem. PV has been applied to the removal of these components from contaminated water, where hydrophobic membranes such as elastomer are used.

Separation of Organic–Organic Mixture

It is very difficult to obtain highly selective membranes for a given organic–organic mixture.

CONCLUSION

PV is now commercialized for the dehydration of an alcohol/water mixture. Intensive studies will be continued in the fields of VOC removal in the water stream and organic–organic mixtures of near boiling points isomers. The mechanism of PV has not been resolved yet, partly because experimental measurements are always necessary for the estimation of the PV performance for a given system. Because PV has definite advantages over other separation processes, more research should be performed in the future.

ACKNOWLEDGMENT

We are very grateful to Professors Lichtenthaler and Thompson.

REFERENCES

1. Binning, R. C.; Lee, R. J.; Jennings, J. F.; Martin, E. C. *Ind. Eng. Chem.* **1961**, *53*, 45.
2. Masuoka, T.; Iwatsubo, T.; Mizoguchi, K. *J. Membrane Sci.* **1992**, *69*, 109.
3. Masuoka, T.; Iwatsubo, T.; Mizoguchi, K. *J. Appl. Polym. Sci.* **1992**, *46*, 311.
4. Hirotsu, T. In *Pervaporation Membrane Separation Processes*; Huang, R. Y. M., Ed.; Elsevier: Amsterdam, The Netherlands, 1991; Chapter 11.

PERVAPORATION MEMBRANES (for Separating Organic Solvents)

Yu Nagase* and Kiyohide Matsui
Sagami Chemical Research Center

Membrane separation processes have been used in industrial applications, such as microfiltration, ultrafiltration, reverse osmosis, desalination of brine, salt manufacturing from sea water, and oxygen enrichment of air. Separation through membranes is one of the most promising achievements of energy-saving technology. The separation of organic liquid mixtures by use of the pervaporation technique through a synthetic membrane was first suggested by Binning and Lee.[1,2] The term "pervaporation" was proposed as far back as 1917 but was rarely used until recently.[3] This technique is a liquid-phase permeation system, in which the liquid feed mixture is maintained in contact with membranes and the permeate is removed from the opposite side of the membrane as a vapor under reduced pressure.

Recently, the separation of an alcohol/water mixture by using the pervaporation technique with a polymer membrane has been given much attention. The process is expected to provide a continuous and economical method for concentrating alcohol produced by, for example, the fermentation of biomass such as starch, sugar, and cellulose.

To achieve the selective permeation of alcohol, it is very important to enhance the solubility of alcohol over water in a polymer membrane because of the higher diffusivity of water compared with alcohol. A crosslinked polydimethylsiloxane (PDMS) membrane, called a "silicone membrane," has been known to show a selective permeation of alcohol and high permeability at the pervaporation of an aqueous alcohol solution.[3]

Masuda et al. have found that many kinds of substituted acetylenes are polymerizable by using group V and VI transition metals as catalysts.[4] One poly(substituted-acetylene), poly(1-trimethylsilyl-1-propyne) (PTMSP), shows the greatest permeability coefficients of several gases among the polymer membranes and selective permeation of alcohol at pervaporation of an aqueous alcohol solution.[5–7] However, PTMSP membranes suffer from relatively low selectivity in the separation of gases and the alcohol/water mixture.

This article deals with our recent studies on two kinds of membrane materials that show an organic-permselective property

*Author to whom correspondence should be addressed.

at the pervaporation of an aqueous organic solution. One of them is a PTMSP-based membrane containing PDMS or a fluoroalkyl group in the side chain, which attains higher permselectivity of alcohol compared with a PTMSP homopolymer membrane. The other is a PDMS-grafted polyimide membrane, which is durable to any organic solvent owing to the toughness of its main-chain component, in contrast to most of the polymer membranes containing a PDMS segment. If membrane materials are insoluble in various organic solvents, pervaporation can be extended to separate many kinds of organic liquid mixtures and becomes a more promising separation process.

SEPARATION PROPERTIES

Chemically Modified PTMSP Membranes

Preferential permeation of ethanol was observed for all the modified PTMSP membranes prepared, at every composition in the feed solution.

A part of the copolymers exhibits improved selectivity of ethanol compared with the PTMSP homopolymer. In the case of PTMSP/PDMS graft copolymers, the content of ethanol in the vapor permeated through the membranes is higher than that through the PTMSP homopolymer membrane.

PI/PDMS Graft Copolymer Membranes

The gas permeability coefficients of the PI/PDMS graft copolymers, whose PDMS contents are more than 50 wt %, are much higher than those of PI homopolymer membranes of the same order of magnitude as that of the PDMS membrane. It is suggested that a continuous phase of PDMS exists in the copolymer membrane. This allows us to expect that the copolymers exhibit the selective permeation of organic molecules with high durability as separating membranes.

REFERENCES

1. Binning, R. C.; Lee, R. J. U.S. Patent 2 953 502, 1960.
2. Binning, R. C.; Lee, R. J.; Jennings, J. F.; Martin, E. C. *Ind. Eng. Chem.* **1961**, *53*, 45.
3. Kimura, S.; Nomura, T. *Maku* **1983**, *8*, 177.
4. Masuda, T. Higashimura, T. *Advances in Polym. Sci.* **1987**, *81*, 121.
5. Masuda, T.; Isobe, E.; Higashimura, T. *J. Am. Chem. Soc.* **1983**, *105*, 7473.
6. Masuda, T.; Tang, B.; Higashimura, T. *Polym. J.* **1986**, *18*, 565.
7. Ishihara, K.; Nagase, Y.; Matsui, K. *Makromol. Chem., Rapid Commun.* **1986**, *7*, 43.

PERVAPORATION MEMBRANES (Materials)

Rüdiger N. Lichtenthaler* and Claudia Staudt-Bickel
Angewandte Thermodynamik
Physikalisch-Chemisches Institut
Ruprecht-Karls-Universität Heidelberg

In this review, we discuss the development and state of the art of pervaporation membranes. This is because in some areas pervaporation is already established industrially (e.g., dehydration of organic solvents), whereas in other areas further research and development need to be done (e.g., separation of organic mixtures). In this respect, the development of new membranes is most important, which means the development and modification of polymers because almost all pervaporation membranes are nonporous polymeric membranes. We will discuss how macroscopic and microscopic properties of a membrane polymer affect its separation characteristics and the methods and tools for preparing the most suitable membrane for a given separation.

*Author to whom correspondence should be addressed.

PRINCIPLES OF PERVAPORATION

Pervaporation is the transfer of mass from a liquid phase to a vapor phase through nonporous polymeric membranes.[1] **Figure 1** shows the principles of pervaporation and the general flow scheme of such a membrane separation unit. The liquid feed mixture to be separated (mass flux J^F) flows along one side of the membrane while the various feed components are permeating into and through the membrane at different rates. Therefore, the liquid retentate (mass flux J^R) leaving the unit on the same side of the membrane as the feed enters is depleted in the components permeating preferentially. Consequently, the vaporous permeate (mass flux J^P) collected on the other side of the membrane is enriched in these preferentially permeating components.

The driving force for the mass transport of each component, i, through the nonporous membrane is the difference, $\Delta\mu_I$ (T,p, $x_{i=m\ldots n}$), of the chemical potential between feed side and permeate side, depending on temperature, T, pressure, p, and the mole fractions, x_i, of all n components present in the mixture. In pervaporation, $\Delta\mu_i$ mainly is achieved by keeping the permeate pressure, p^P, much lower that the feed pressure, p^F, and the retentate pressure, p^R.

SELECTION AND MODIFICATION OF MEMBRANE POLYMERS

The selection of a membrane usually is made on the basis of solubility and diffusivity data of the various components of a mixture in the membrane polymer.

Solubility and diffusivity of low molecular weight solutes in polymers strongly depend on the molecular size and shape of the solute, the polymer–solute interactions, and the chemical and physical structure of the polymer.

Crosslinking

A widely used way to modify membrane polymers is crosslinking. This can be done chemically by adding suitable crosslinking agents and physically by curing the membranes at high temperatures or by radiation.

Polymer Blends

Many efforts have been made to improve membrane separation behavior by mixing polymers (i.e., no covalent bonds exist between the different types of polymers). Two kinds of blends can be distinguished: homogeneous blends, in which the different polymers are miscible in the whole composition range, and heterogeneous blends, in which the different polymers are not completely miscible. In the latter case, domains of any one polymer in the matrix of the other polymers can be observed.

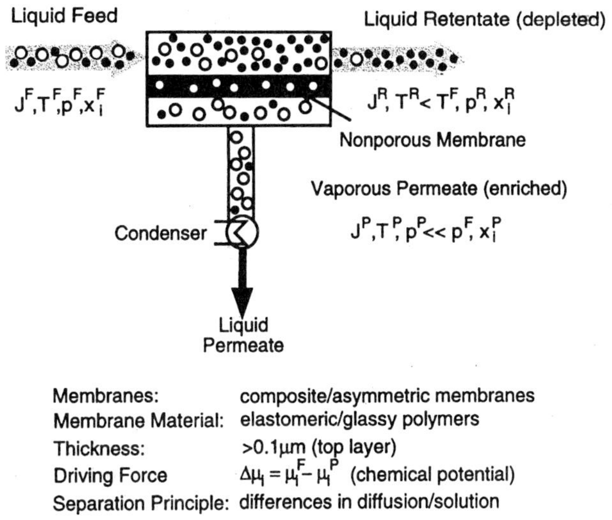

FIGURE 1. Principles of pervaporation.

Therefore, only homogeneous blends are considered to be potential membrane materials, because heterogeneous blends will not have enough mechanical strength as a thin membrane. Blend membranes have been used to separate water/ethanol mixtures.[1] The blend of the two copolymers poly(acrylonitrile-*co*-acrylic acid) (PAN-co-AA) and poly(acrylonitrile-*co*-vinylpyridine) (PAN-co-VP) was found to form more selective membranes than either of the copolymers.

A poly(vinyl alcohol)-chitosan (PVA-CS) blended composite membrane has been prepared for the dehydration of alcohols.[2]

The separation of propanol isomers has been investigated using a membrane of poly(vinyl alcohol) blended with cyclodextrin (PVA-CD).[3]

Remarkable separation of aromatic/aliphatic hydrocarbon feed mixtures (i.e., benzene/cyclohexane and cyclohexene/cyclohexane) has been obtained by blend membranes of cellulose acetate and various types of polyphosphonate (CA-PPN).[4,5]

Copolymers

Various elastomeric polyurethane (PU) polymers have been used for membranes for the separation of aromatic and aliphatic hydrocarbons (e.g., toluene/cyclohexane mixtures).[6]

Membranes prepared from poly(*N*-vinylpyrrolidone-*co*-acrylonitrile) (PNVP-co-AN) selectivity separate cyclohexanone and cyclohexanol form a mixture of these two components with cyclohexane.

Grafting

Grafting is an excellent technique for modifying membrane polymers. The irradiation technique can be applied directly to a membrane already formed if the starting trunk polymer is partially crystalline. The strongly interacting groups in the grafted polymer will then consist of trunk polymer crystallites and graft copolymer amorphous regions. The grafted side chains of the amorphous part cause selective sorption and diffusion, while the crystallites of the trunk material prevent excessive swelling of the membrane that might lead to the loss of selectivity.

Ion-exchange membranes are most suitable for the dehydration of organic liquids because these membranes are hydrophilic. The choice of counterion is very important because it has an affect on the interaction between the ionic groups and water molecules. The stronger the interaction of the counterion with the water molecules, the more selectivity a polymer will absorb and diffuse water with respect to the other (less polar) components.

Polyion complex hollow-fiber membranes used for the separation of water/ethanol mixtures have been prepared by partially hydrolizing PAN to introduce carboxylic groups, which were converted to polyion complexes.[8]

Plasma Polymers

Plasma-polymerized films with perfluorocarbon monomers were deposited onto a porous membrane substrate of polysulfone.[9] Hydrophobic membranes are obtained with the separation characteristics for ethanol/water mixtures, depending on when plasma polymerization occurs.

Methanol-selective plasma-polymerized membranes have been used successfully to remove methanol from various organic mixtures.[10,11]

The plasma technique also has been used to prepare plasma-grafted pervaporation membranes.[7]

If the grafted layer is not mainly formed on the surface of the substrate but in its pores, the method is often called plasma-graft filling polymerization.[12] In this way pervaporation membranes have been prepared for the separation of organic liquid mixtures.[13–15]

Nonvolatile Additives

Various investigations have been performed to enhance membrane selectivity and/or flux by incorporating nonvolatile additives (e.g., zeolites or salts) into the membrane polymer. In a series of papers covering zeolite-filled elastomeric membranes, various aspects of these composite membranes have been investigated.[16–20]

Inorganic salts have also been used as fillers. In this way homogeneous cellulose films were modified, which feature high selectivity and low fluxes for the dehydration of alcohol/water mixtures by vapor permeation.[21]

CONCLUSION

Pervaporation is regarded as the membrane process most suitable to separate volatile components present in low concentration from liquid mixtures. For the dehydration of organics, pervaporation is used industrially, and for the separation of volatile organics from aqueous solution, it is already used in some technical applications. The separation of organic mixtures on an industrial scale will only be possible if significant improvements are made in membrane selectivity and permeability. Furthermore, the membranes have to have long-term stability at operating conditions in the chemical environment they are exposed to. Design and development of suitable membrane polymers are a challenge for polymer chemists. It is very likely that by using the methods and techniques discussed in this review, together with new concepts for membrane and process design, this challenge will be met.

REFERENCES

1. Jo, W. H.; Kang, Y. S.; Kim, H. J. *J. Membr. Sci.* **1993**, *85*, 81–88.
2. Wu- L-G.; Zhu, C. L.; Liu, M. In *Proceedings of the 6th International Conference on Pervaporation Processes in the Chemical Industry*; Bakish Materials Corporation: Englewood, NJ, 1992; pp 438–444.
3. Miyata, T.; Iwamoto, T.; Uragami, T. *J. Appl. Polym. Sci.* **1994**, *51*, 2007.
4. Cabasso, I.; Jagur-Grodinski, J.; Vofsi, D. *J. Appl. Polym. Sci.* **1974**, *18*, 2137.
5. Cabasso, I. *Ind. Eng. Chem. Prod. Res. Dev.* **1983**, *22*, 313.
6. Ohst, H.; Hildenbrand, K.; Dhein, R. In *Proceedings of the 5th International Conference on Pervaporation Processes in the Chemical Industry*; Bakish Materials Corporation: Englewood, NJ, 1991; pp 7–21.
7. Hirotsu, T. In *Pervaporation Membrane Separation Processes*; Huang, R. Y. M., Ed.; Elsevier Science: Amsterdam, The Netherlands, 1991; Chapter 11, 461–490.
8. Tsuyumoto, M.; Karakane, H.; Maeda, Y.; Tsugaya, H. In *Proceedings of the 4th International Conference on Pervaporation Processes in the Chemical Industry*; Bakish Materials Corporation: Englewood, NJ, 1989; pp 157–168.
9. Masuoka, T.; Mizoguchi, K.; Hirasa, O.; Yamauchi, A. In *Proceedings of the 3rd International Conference on Pervaporation Processes in the Chemical Industry*; Bakish Materials Corporation: Englewood, NJ, 1988; pp 143–149.
10. Brüschke, H. E. A.; Schneider, W. H.; Scholz, H.; Steinhauser, H. In *Proceedings of the 6th International Conference on Pervaporation Processes in the Chemical Industry*; Bakish Materials Corporation: Englewood, NJ, 1992; pp 423–429.
11. Brüschke, H. E. A. In *Proceedings of the Aachener Membran Kolloquium*; GVG VDI Verlag: Düsseldorf, Germany, 1993; 263–272.
12. Yamaguchi, T.; Nakao, S.; Kimura, S. *Macromolecules* **1991**, *24*, 5522–5527.
13. Yamaguchi, T.; Nakao, S.; Kimura, S. *Ind. Eng. Chem. Res.* **1991**, *31*, 1914–1919.
14. Yamaguchi, T.; Nakao, S.; Kimura, S. *Ind. Eng. Chem. Res.* **1993**, *32*, 848–853.
15. Yamaguchi, T.; Yamahara, S.; Nakao, S.; Kimura, S. *J. Membr. Sci.* **1994**, *95*, 39–49.
16. Hennepe, H. J. C.; Bargeman, D.; Mulder, M. H. V.; Smolders, C. A. *J. Membr. Sci.* **1987**, *35*, 39–55.
17. Hennepe, H. J. C.; Boswerger, W. B. F.; Bargeman, D.; Mulder, M. H. V.; Smolders, C. A. *J. Membr. Sci.* **1994**, *89*, 185–196.
18. Jia, M.; Peinemann, K. V.; Behling, R. D. *J. Membr. Sci.* **1991**, *57*, 289–296.
19. Duval, J-M.; Folkers, B.; Mulder, M. H. V.; Desgandchamps, G.; Smolders, C. A. *J. Membr. Sci.* **1993**, *80*, 189–198.
20. Bartels-Caspers, C.; Tusel-Langer, E.; Lichtenthaler, R. N. *J. Membr. Sci.* **1992**, *70*, 75–83.
21. Jansen, A. E.; Versteeg, W. F.; van Engelenburg, B.; Hanemaaijer, J. H. In *Gas Separation Technology*; Vasant, E. F.; Dewolfs, R., Eds.; Elsevier Science: Amsterdam, The Netherlands, 1990; pp 413–427.

PERYLENE POLYMERS

Klaus Müllen,* Heribert Quante, and Nicholas Benfaremo
Max-Planck-Institute for Polymer Research

The development of the chemical industry has always been closely associated with dyestuffs. Dyestuffs originally were used mostly as textile and vat dyes. However, with the development of new fabrics and materials, new chromophores have been needed to meet changing demands.[1,2] A particularly interesting class of these compounds is high-performance dyes and pigments. Special qualities of these materials are light weight, weather and high-temperature stability, and resistance to acids and bases. These qualities are advantageous in the use of pigments and dyes in auto coatings; lacquers; exterior paints and coatings; and colored, exterior-use plastics. These stringent demands are also necessary because of drastic processing temperatures that range from 150 to 350°C. Acid- or base-catalyzed polymerizations and acidic polymers (e.g., poly(vinyl chloride) require acid- and base-resistant chromophores.[3]

Two of the most prominent representatives of this class of high-performance dyes and pigments are the perylenes and perylene-3,4:9,10-tetracarboxylic bisimides. Because of their extreme insolubility, these compounds were originally used exclusively as pigments. It has only been since the successful preparation of soluble analogs in the 1970s, however, that perylene has also been used in dyes.[4–7]

Currently, another area of color chemistry, functional dyes, is experiencing great development.[8–15] Applications range from laser dyes to tracers used in high-sensitivity analytical investigations and in fluorescence immune tests.[16–21] As photoefficient materials, perylenetetracarboxylic bisimides have become established in photovoltaic and photocopying processes and in the development of optical switches.[22–32] Perylenetetracarboxylic bisimide films, used as coverings in greenhouses, promote plant growth as they convert a portion of the sunlight into a wavelength suitable for plants.[33]

PERYLENE POLYMERS

High solubility and a low tendency to migrate–often mutually exclusive properties–are required for dye applications. An

*Author to whom correspondence should be addressed.

attractive alternative to monomeric chromophores is soluble, polymeric materials that can be easily mixed into the worked material and that simultaneously show no migration. In principle, the structure of polymeric perylene species offers many options to bring about miscibility.

Perylene as a Component in Side-Chain Polymers

Perylene can be grafted onto the main chain of a polymer after being suitably functionalized. To this end, *p*-vinylbenzal-3-aminoperylene,3-acrylamidoperylene, methacrylamidoperylene,3-(α-acryloyloxy)ethylperylene, and 3-vinylperylene have been used.[34–36]

In contrast to their homopolymerization, monofunctionalized perylenes are well suited for copolymerization; therefore, commodity polymers such as poly(methyl methacrylate) and polystyrene can be easily colored.

Perylene as a Component in Main-Chain Polymers

Because of the poor solubility of most perylene derivatives, there are only a few examples known of perylene main-chain polymers.

A basic premise in approaches toward the successful preparation of high molecular weight, soluble, perylene main-chain polymers is to use the most highly soluble perylene building blocks possible.

Perylenetetracarboxylic Bisimides as Components in Side-Chain Polymers

Due to the limited solubility of perylenetetracarboxylic bisimides, only side-chain polymers with tetraphenoxy-substituted species are known. Proceeding from the readily available *N,N'*-dibutyl-1,6,7,12-tetraphenoxyperylene-3,4:9,10-tetracarboxylic bisimide many monofunctionalized perylenetetracarboxylic bisimides can be made accessible by the base-induced transimidization (alkylation) with bromoalkanes.

The attachment of the chromophores onto liquid crystalline main-chain polymers has made application areas for liquid crystalline chromophores accessible.

Liquid crystalline, dye-containing polymers obtained through copolymerization are ideal candidates for use in thermal writing displays and, when combined with low molecular weight liquid crystals or mixtures of liquid crystals, electrooptical displays of the guest-host type.[37,38]

Perylenetetracarboxylic Bisimides as Components in Main-Chain Polymers

Perylenetetracarboxylic bisanhydrides can undergo polycondensation with a variety of alkyl- and aryldiamines as, for example, 1,4-diamino-2,5-di-*t*-butylbenzene.[39–43]

An additional possibility for the synthesis of the polyimides is the generation of a bifunctional perylenetetracarboxylic bisimide, followed by subsequent polymerization with suitable bifunctional components.

Only in the last two years, since the development of soluble perylenetetracarboxylic bisanhydrides [**Figure 1** (44)], has the field of polymeric perylenetetracarboxylic bisimides been accessible. Their polycondensation with diamino components yields easily soluble polyimides [Figure 1 (46)] with molecular weights as high as 35,000.[44] Characteristic of these materials are their outstanding film-forming properties that allow for the manufacture of free-standing films. The special advantage of these main-chain polymers lies in their ability to achieve a high optical density in comparison with perylenetetracarboxylic side-chain polymers.

Although most of the synthetic perylene polymers synthesized to date have no significant industrial applications, because of their poor solubility, the development of soluble monomeric derivatives has led to a renaissance in the area. The synthetic possibilities now opened for the construction of perylene polymers will occupy a significant position in the development of future technologies.

REFERENCES

1. Zollinger, H. *Color Chemistry*; Verlag, Chemie: Weinheim, Germany, 1987.
2. Gordon, P. F.; Gregory, P.; *Organic Chemistry in Colour* Springer Verlag: Berlin, Germany, 1987.
3. Christie, R. M. *Polym. Int.* **1994**, *34*, 351.
4. Graser, F. D.O.S. 2 139 688, 1973.
5. Iden, R.; Seybold, G. D.O.S. 3 434 059.
6. Iden, R.; Stange, A.; Eilingfeld, H. *BMFT* Research Report; T84-164.
7. Seybold, G.; Stange, A. D.O.S. 3 545 004.
8. Fabian, J.; Zahradnik, R. *Angew. Chem.* **1989**, *101*, 693.
9. Fabian, J.; Nakazumi, H.; Matsuoka, M. *Chem. Rev.* **1992**, *92*, 1197.
10. Gupta, M. C. *Appl. Opt.* **1984**, *23*, 3950.
11. Griffiths, J. *Chem. Ber.* **1986**, *22*, 997.
12. Takagi, K.; Kawabe, M.; Matsuoka, M.; Kitao, T. *Dyes Pigm.* **1985**, *6*, 177.
13. Maeda, M. *Laser Dyes*; Academic: New York, 1984.
14. Harker, J. M.; Santana, G. R.; Taft, S. G. *IBM J. Res. Dev.* **1981**, *25*, 677.
15. Loutfy, R. O.; Hor, A-M.; Hsiao, C-K.; DiPaola-Baranyi, G.; Kazmaier, P. M. *Pure Appl. Chem.* **1988**, *60*, 1047.
16. Langhals, H. Z. *Anal. Chem.* **1985**, *320*, 361.
17. Langhals, H. *Chem. Ind.* **1985**, *37*, 470.
18. Langhals, H.; Schott, H.; Schwendener, R. A. D.O.S. 39 352 579, 1989.
19. Reisfeld, R.; Seybold, G. *Chimia* **1990**, *44*, 295.
20. Löhmannsröben, H. G.; Langhals, H. *Appl. Phys. B* **1989**, *48*, 449.
21. Drexhage, K. H. *Topics Appl. Phys.* **1973**, *1*, 142.
22. Loutfy, R. O.; Hor, A. M.; Katzmeier, P.; Tam, M. *J. Imag. Sci.* **1989**, *33*, 151.
23. Loutfy, R. O.; Hor, A. M. *Can. J. Chem.* **1983**, *61*, 901.
24. Law, K.-Y. *Chem. Rev.* **1993**, *93*, 449.
25. Schlettwein, D.; Wöhrle, D.; Karmann, E.; Melville, U. *Chem. Mater.* **1994**, *6*, 3.
26. Schlosser, E. G. *J. Appl. Photogr.* **1978**, *4*, 118.
27. Wiedemann, W. EP 0 040 402, 1981.
28. Tang, C. W. EP 0 000 829A, 1980.
29. Neumann, P.; Etzbach, K-H.; Hoffmann, G. D.O.S. 3 339 540 A1, 1985.
30. Nakazawa, T.; Fushida, A.; Kamezaki, Y. E.P. 0 100 581, 1984.
31. O'Neil, M. P.; Niemczyk, M. P.; Svec, W. A.; Gosztola, D.; Gaines III, G. L.; Wasielewski, M. R. *Science* **1992**, *257*, 63.
32. Panayotatos, P.; Bird, G.; Sauers, R.; Piechowski, A.; Husaln, S. *Solar Cell* **1987**, *21*, 301.
33. Ito, N.; Aiga, H. *Jpn. Kokai Tokkyo Koko* JP 62 132 693.

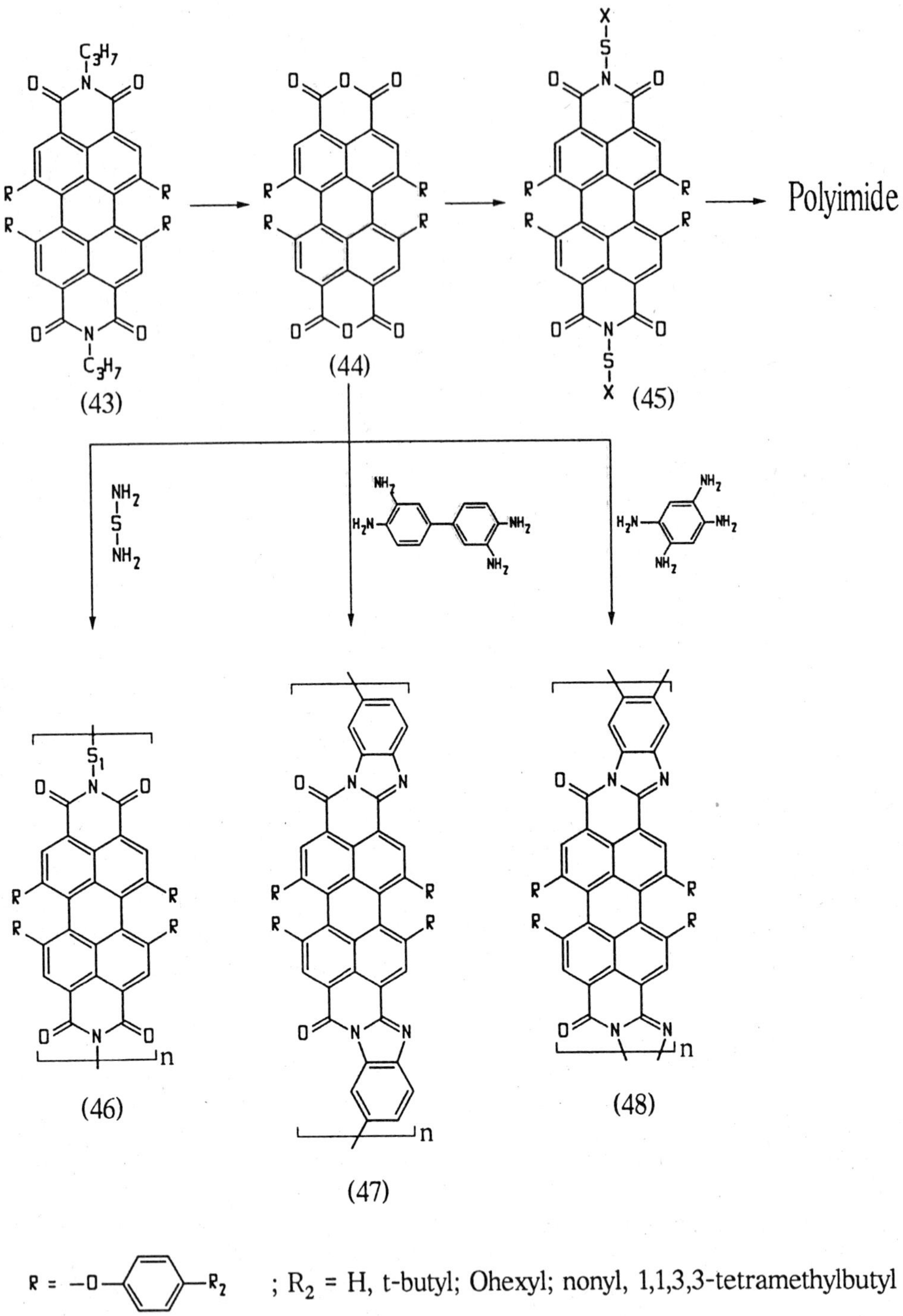

46: S_1 = $(CH_2)_8$; $C_6H_4OC_6H_4$; C_6H_4-CH_2-C_6H_4

45: S = C_6H_4 X = Br; S = C_6H_4 X = NH_2

FIGURE 1. Soluble main-chain polyperylenetetracarboxylic bisimides.

34. Kamogawa, *J. Polym. Sci., Part A-1* **1972**, *10*, 1345.
35. Jeon, I. R.; Noma, N.; Shirota, Y. *Mol. Cryst. Liq. Cryst.* **1990**, *190*, 1.
36. Loer, B.; Kerrmann, H. D.O.S. 2 137 288, 1973.
37. Beck, K. H.; Etzbach, K-H.; Schmidt, H-W. EP 0 422 535A1, 1990.
38. Etzbach, K-H.; Beck, K. H.; Wagenblast, G. EP 0 422 538A1, 1990.
39. Ono, Y.; Yokoi, M.; Yamazaki, K.; Hotta, H.; Kobayashi, K.; Yamada, T.; Kojima, H. JP 04 264 451.
40. Langhals, H. *Chem. Phys. Lett.* **1986**, *150*, 321.
41. Karayannidis, G. P.; Sideridou-Karayannidou, I. *J. Macromol. Sci., Chem.* **1986**, A*23*.
42. Lyapunov, V. V.; Lyakh, E. N.; Solomin, V. A.; Zhubanov, B. A. *Dokl. Akad. Nauk.* **1992**, *326*(1), 106.
43. Ananiichuk, N. A.; Kutna, L. V.; Rozhanchuk, V. N.; Tret'yakov, Y. P. *Khim. Volokna* **1980**, *3*, 8.
44. Dotcheva, D.; Klapper, M.; Müllen, K. *Macromol. Chem.* **1994**, *195*, 1905.

Phase-Transfer Catalysts

See: *Catalysts, Polymeric (Higher Activity than Monomeric Analogs)*
Phase-Transfer Catalysts, Polymeric (Overview)
Phase-Transfer Catalysts, Polymeric (Structure and Activity)
Phase-Transfer Catalyzed Polymer Synthesis
Step Polymerization Catalysts
Supported Chiral Catalysts

PHASE TRANSFER CATALYSTS, POLYMERIC (Overview)

Rayna Stamenova,* T. Tsanov, and C. Tsvetanov
Institute of Polymers
Bulgarian Academy of Sciences

Polymeric phase transfer catalysts (PTCs) are insoluble polymers having covalently bound functional groups active as catalysts for reactions between anions and neutral organic substrates. Reactions are conducted in a three-phase system consisting of mutually insoluble aqueous and organic layers. Ionic reagents (i.e., salts, bases, and acids) are dissolved in the aqueous phase, and the substrate is dissolved in the organic phase (liquid–liquid PTC). Alternatively, ionic reagents can be used in the solid state as a suspension in the organic medium (solid–liquid PTC). The catalyst is a third insoluble phase. At the end of the reaction, it can be isolated by simple filtration and recycled. The active functional groups may be quaternary ammonium or phosphonium ions, crown ethers, cryptands, grafted poly(ethylene glycol)s (PEGs), or analogues of dipolar aprotic solvents.

SYNTHESIS OF CATALYSTS

Polymeric PTCs are classified into three groups: supported onium salts, supported crown ethers and cryptands, and supported solvents and co-solvents. Some examples of polymeric PTCs are shown in **Figure 1.**

*Author to whom correspondence should be addressed.

ANALYSIS OF CATALYSTS

The great difficulty of analyzing insoluble materials is a reason for inadequate characterization of polymeric PTCs. Elemental analysis of nitrogen, phosphorus, or halogen is commonly used. The accuracy of elemental analysis is very poor for catalysts with a lower degree of functionalization. Therefore, elemental analysis and weight changes should be considered only qualitative. IR spectra and ^{13}C NMR spectra are also only qualitatively useful.

GENERAL PRINCIPLES AND MECHANISM OF CATALYSIS

In the absence of PTC, the reactions proceed at a very low rate or not at all. In the simplest case of nucleophilic substitution reactions, Starks offered a now classic diagram of the phase transfer catalytic cycle (**Equation 1**):[3]

$$\begin{array}{llll} \text{QNu} + \text{R-X} & \longrightarrow & \text{R-Nu} + \text{QX} & \text{Organic phase} \\ \uparrow\downarrow & & \downarrow\uparrow & \\ \text{QNu} + \text{MX} & \rightleftharpoons & \text{M-Nu} + \text{QX} & \text{Aqueous phase} \end{array} \quad (1)$$

The lipophilic cation, Q^+, or some equivalent cation solvator is soluble in both aqueous and organic phases and when in contact with the aqueous reservoir of salt, exchanges anions with the excess of anions in the salt solution. Once the nucleophile is in the nonpolar organic media, the displacement can take place with product formation. The agent, Q^+, having a strong affinity for the organic solvent, is typically a tetraalkylonium ion, NR_3^+, PR_3^+, or a complexing agent such as crown ether, cryptand, or linear polyether. Generally, the polymer-supported PTCs are regarded as heterogeneous catalysts. Nevertheless, because the polymer support is usually an inert matrix, the catalysis may be described as homogeneous catalysis in a liquidlike phase within the swelled polymer. The most important rate-limiting factors, which determine the catalytic activity of PTCs are intrinsic reactivity, mass transfer, and intraparticle diffusion.

The structure of the active site of a polymeric PTC plays a vital role for the catalytic activity. The hydrophilic–lipophilic balance of the polymer support is highly important for the thermodynamic and kinetic aspects of intrinsic activity.

An interesting example of a polymeric PTC is the gel from crosslinked PEO (cr-PEO) where the immobilization is realized without participation of an inert polymer support.[1,2]

PEGs supported to crosslinked polystyrene are remarkably active and stable PTCs for the alkylation of nitriles, ketones, and alcohols.[4]

APPLICATIONS

Polymeric PTCs can be used for any organic reaction that is catalyzed by a soluble PTC. Some significant examples of applications of polymer-supported PTCs are displacement reactions; alkylation eliminations; base-promoted reactions such as isomerizations, nucleophilic additions, and condensations; oxidations; and asymmetric catalysis.[5] The mechanisms of action of polymer PTC are now well known enough to allow the design of large-scale processes and the execution of laboratory-scale

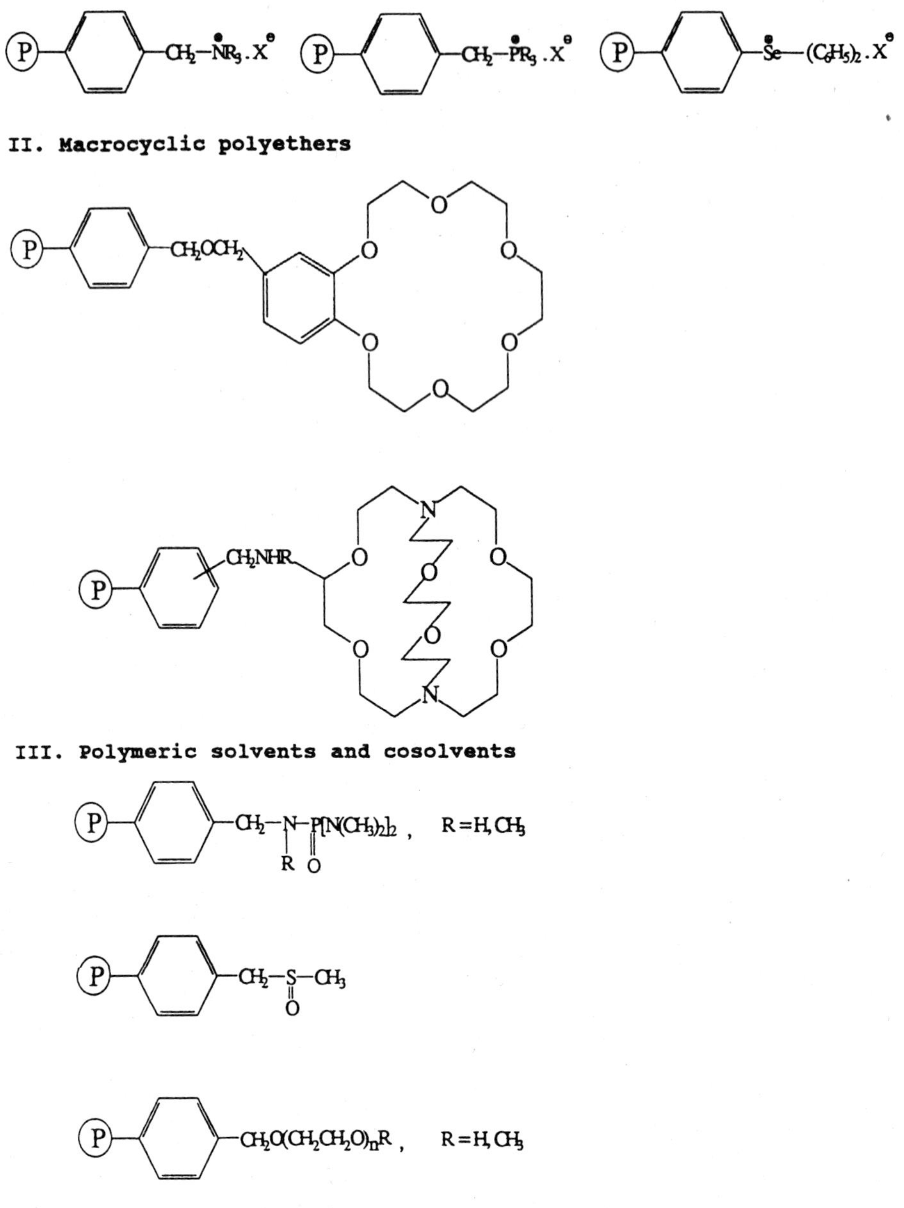

FIGURE 1. Polymer-supported phase transfer catalysts.

synthesis. Although polystyrene is most used as the support for PTC, networks based on cr-PEO are also worthy because of their stability in alkali media, sufficient mechanical strength, and lower price. They have the greatest synthetic utility for alkylation in the presence of alkaline hydroxides.

REFERENCES

1. Tsanov, T.; Stamenova, R.; Tsvetanov, C. *Polymer* **1993**, *34*, 616.
2. Tsanov, T.; Stamenova, R.; Tsvetanov, C. *Polym. J.* **1993**, *25*, 853.
3. Starks, C. M. *J. Am. Chem. Soc.* **1971**, *93*, 195.
4. Kimura, W.; Kirszensztejn, P.; Regen, S. L. *J. Org. Chem.* **1983**, *48*, 385.
5. Tomoi, M.; Ford, W. T. *Synthesis and Separations Using Functional Polymers*; Sherrington, D. C.; Hodge, P., Eds.; John Wiley & Sons: 1988; Chapter 5, p 182.

PHASE TRANSFER CATALYSTS, POLYMERIC (Structure and Activity)

Kazuichi Tsuda
Department of Applied Chemistry
Nagoya Institute of Technology

In 1975–76, insoluble polymer-supported onium ions and polyethers were found to be the phase transfer catalysts by three groups.[1-3] Regen called his reaction system a three-phase reaction, because the reaction mixtures consist of a swollen polymer gel, an aqueous salt solution, and an organic solution. Polymeric phase transfer catalysts offer big advantages in organic synthesis: ease of separation of the polymer catalyst from reaction mixtures by filtration and recycling of the catalysts. However, the disadvantages are expensive synthetic cost and lower activity or slower rates compared with low molecular weight phase transfer catalysts.

Regen, Sherrington, Tomoi, and many authors reviewed polymer-supported phase transfer reaction.[10–18] Here, I review polymeric phase transfer catalysis from the standpoint of the relationship between structure and activity of newly synthesized phase transfer catalysts such as sulfonium salt, sulfoxide, and acetoamide-type catalysts.

CATALYST TYPE

There are three types of polymeric phase transfer catalysts from the standpoint of the structure of the functional group: onium salts, cosolvent, and crown ether and polyether. Gel-type backbone polymers are prepared by the ordinary radical copolymerization of *p*-chloromethylstyrene with styrene or by chloromethylation of polystyrene.

Hydrophilic polymers, polyethylenimine, polyacrylamide, and poly(ethylene glycol), are used for the swollen or homogeneous reaction system and aqueous media. Recently, poly(α-amino acid)s such as polyglutamate and polyalanine have been used as backbones.[19–22]

Onium Salts

The onium salts type of phase transfer catalyst is the most common. Ammonium salts, phosphonium salts, and sulfonium salts are known as the polymer-supported onium salt catalysts. These onium salts catalysts are suitable for nucleophilic displacement reactions.

Cosolvent and Solvent

Dipolar aprotic solvents such as hexamethylphosphoramide (HMPA), dimethyl sulfoxide (DMSO), *N,N*-dimethylformamide (DMF), and sulfones are also efficient functional groups of polymeric phase transfer catalysts.[4–9,24,25] These polymers are regarded as containing active sites of typical dipolar aprotic solvents.

Crown Ether, Cryptand, and Polyether

There are two kinds of polymer-supported phase transfer catalyst of this type. One is the crown ether structure introduced onto the polymer backbone, using the condensation polymerization for the preparation. The other is the pendant crown ether polymer catalyst. The cryptand-type of polymeric-phase transfer catalysts is also prepared by a method similar to that used for the crown ether polymer.

Polymers containing pendant poly(ethylene glycol)s (PEG) and macromers containing PEG units are obtained by the polymerization of corresponding monomers or by polymer modification of *p*-chloromethylstyrene with PEG or its monoalkylethers.[26–32]

Phase Transfer Catalysts on Inorganic Supports

Silica gel or alumina are used as the supports, and the coupling reaction of a catalytic moiety and an inorganic support is carried out by silane coupling reagents.[33]

RELATION BETWEEN THE STRUCTURE AND THE ACTIVITY

Polymer Structure

Any polymers that are inert to the catalytic reaction could be used as catalyst supports. Many kinds of polymers, inorganic compounds, and some of the natural products are used as the supports, and most of them are lipophilic in nature.

Spacer

Spacer chain length between the active site and the polymer backbone remarkably affects the catalytic activity. Active sites separated by spacers from the polymer backbone are more active than small organic ones.

Distribution of the Active Group

The distribution of the active group and functionalization is more easily controlled by the copolymerization method compared with the chemical modification method.

Degree of Crosslinking

The degree of crosslinking affects the swelling of polymeric catalysts, and part of the swelling ability comes from the porosity of the polymer. The highly crosslinking polymer swells less, and the diffusion of the reactant to the active site becomes difficult.

Microenvironment

The microenvironment is one of the most important factors for reactivity. When the reaction anions in the phase transfer catalyst are in a lipophilic region, high activity is attained.

Solvent and Substrate

Generally, highly swollen polymer supports in a lipophilic solvent are usually more active than less swollen catalysts, and substrates which much more interact with the lipophilic polymer phase than organic layer extremely access the reaction. Highly swollen polymer supports in a lipophilic solvent are usually more active than less swollen catalysts.

MECHANISM OF CATALYSIS

Tomoi and Ford proposed that catalytic activity depends on three factors–intrinsic reactivity, mass transfer, and intraparticle diffusion–because the polymer-supported phase transfer catalysts are regarded as heterogeneous catalysts.[15] However, we made many soluble, homogeneous polymeric catalysts; the third factor was not essential in these cases.

MECHANISTIC PROBLEM

Other factors that affect the activity of polymeric phase transfer catalysts are catalyst particle size, crosslinking, and stirring speed.

Details of New Polymeric Phase Transfer Reaction

As previously mentioned, many kinds of polymeric phase transfer catalyses, such as the onium salts type, the solvent and cosolvent type, and the crown ether, cryptand, and polyether type have been reviewed.[10–16] Therefore, our work deals mainly with new types of polymeric phase transfer catalysis, such as sulfonium salt, sulfoxide, and amide polymer catalyst.

Sulfonium Salt

Sulfonium salts work as highly efficient catalysts even in the presence of strong bases. However, separation of the products and recovery of the catalysts were not easy in these catalytic reaction systems. Therefore, to develop the availability of these catalysts, sulfonium salt moieties are incorporated into crosslinked polystyrene.[23]

Synthesis of Polymeric Sulfonium Salt Catalyst

The polymeric sulfonium salt catalyst is synthesized from the reaction of polystyrene, diphenyl sulfoxide, and anhydrous aluminum chloride.

These sulfonium salt polymers work as effective phase transfer catalysts for the substitution reactions with other nucleophilic reagents.

Sulfoxide Type

The catalytic activity of polymeric sulfoxides can be attributed to the initial coordination of the metal cation to the sulfinyl oxygen atom at the interface and results in the transfer of the cation and the anion from the aqueous phase to the organic phase.[7,9] Sulfones behave similarly to the sulfoxide, but their activities are lower than sulfoxides.

Phosphoric Triamide Type

Tomoi found that polymers carrying hexamethylphosphoric triamide (HMPA) have a catalytic activity similar to onium salts and crown ethers.[4]

Amide Type

N-Methyl-*N*-(*p*-vinylbenzyl) acetamide (MBAA) was synthesized by the reaction of *p*-chloromethylstyrene with *N*-methylacetamide in the presence of sodium hydride. This monomer was readily polymerized or copolymerized with styrene by AIBN. The resulting polymers served as effective phase transfer catalysts for several nucleophile substitution reactions under liquid–liquid biphase conditions.

Alkylated Nylon 66 as Phase Transfer Catalysts

In the case of polyacrylamides, the catalytic activity is increased by substitution of an amino hydrogen with a methyl group.[34] This means that polymeric amides containing active sites along the backbone, such as alkylated nylon 66, may work as phase transfer catalysts.

DISADVANTAGES

Polymeric phase transfer catalysts have many advantages compared with organic low molecular weight phase transfer catalysts. However, they have disadvantages such as high cost, a decrease in activity by recycling, and catalytic activity is lower than organic catalysts, in many cases.

CONCLUSION

These polymeric catalysts serve as effective phase transfer catalysts for several reactions under two-phase or three-phase conditions. The catalytic activity of the copolymers containing styrene units is generally higher than that of homopolymers. These polymers extract alkali metal cations more effectively than their monomeric analogues, and this extraction ability increases with increasing density of active sites, suggesting their cooperative effect.

New applications of polymer-supported catalysts are desirable. Because polymeric phase transfer catalysts are expensive, the reactions that result in valuable products are more suitable. When the catalyst can be recycled and preserve high activity, polymer supported phase transfer catalysts will be used in many reactions.

For the synthesis of excellent catalysis, the design of the polymer structure (backbone structure) and the control of conformation are essential. Because synthetic polymers have a limit for these purposes, a polypeptide backbone catalyst should be considered. Polypeptide has many excellent characteristics, such as easy control of conformation and microenvironment and hydrophilic–lipophilic balance. Polypeptide-supported catalysts have a promising future.

REFERENCES

1. Regen, S. L. *J. Am. Chem. Soc.* **1975**, *97*, 5956.
2. Brown, J. M.; Jenkins, J. *J. Chem. Soc., Chem. Commun.* **1976**, 458.
3. Cinquini, M.; Colonna, S.; Molinari, H.; Montanari, F.; Tundo, P. *J. Chem. Soc., Chem. Commun.* **1976**, 394.
4. Tomoi, M.; Takubo, T.; Ikeda, M.; Kakiuchi, H. *Chem. Lett.* **1976**, 473.
5. Regen, S. L.; Nigam, A.; Besse, J. J. *Tetrahedron Lett.* **1978**, 2757.
6. Janout, V.; Cefelin, D. *Collectt Czech. Chem. Commun.* **1982**, *47*, 1818.
7. Kondo, S.; Ohta, K.; Tsuda, K. *Makromol. Chem., Rapid Commun.* **1983**, *4*, 145.
8. Janout, V.; Kahovec, J.; Hrudkova, H.; Svec, F. *Polym. Bull.* **1984**, *11*, 215.
9. Kondo, S.; Ohta, K.; Ojika, R.; Yasui, H.; Tsuda, K. *Makromol. Chem.* **1985**, *186*, 1.
10. Regen, S. L. *Angew. Chem., Intern. Ed. Engl.* **1979**, *18*, 421.
11. Sherrington, D. C. *Polymer-Supported Reactions in Organic Synthesis*; Hodge, P.; Sherrington, D. C., Eds.; Wiley: Chester, U.K. 1980.
12. Akelah, A.; Sherrington, D. C. *Chem. Rev.* **1981**, *81*, 557.
13. Akelah, A.; Sherrington, D. C. *Polymer* **1983**, *24*, 1369.
14. Ford, W. T.; Tomoi, M. *Adv. Polym. Sci.* **1984**, *55*, 49.
15. Tomoi, M.; Ford, W. T. *Polymeric Phase Transfer Catalysts: Synthesis and Separations Using Functional Polymers*; John Wiley & Sons: New York, 1988.
16. Mathur, N. K.; Narang, C. K.; Williams, R. E. *Polymers as Aid in Organic Chemistry*, Academic: New York, 1980.
17. Dehmlow, E. V.; Dehmlow, S. S. *Phase Transfer Catalysis*, 2nd ed.; Verlag Chemie: Weinheim, Germany, 1983.
18. Keller, E. *Phase-Transfer Reactions*; George Theme Verlag: Stuttgart, Germany, 1986–78; vols. 1 and 2.
19. Inoue, S.; Kawano, Y. *Makromol. Chem.* **1979**, *180*, 1405.
20. Anzai, J.; Ueno, A.; Suzuki, Y.; Osa, T. *Makromol. Chem., Rapid Commun.* **1982**, *3*, 55.
21. Julia, S.; Guixer, J.; Masana, J.; Rocas, J.; Colonna, S.; Annuziata, R.; Molinari, H. *J. Chem. Soc., Perkin Trans. 1* **1982**, 1317.
22. Colonna, S.; Molinari, H.; Banfi, S.; Julia, S.; Masana, J.; Alvarez, A. *Tetrahedron* **1983**, *39*, 1635.
23. Kondo, S.; Murayama, T.; Takeda, Y.; Tsuda, K. *Makromol. Chem., Rapid Commun.* **1988**, *9*, 625.

24. Kondo, S.; Inagaki, Y.; Tsuda, K. *J. Polym. Sci., Polym. Lett. Ed.* **1984**, *22*, 249.
25. Kondo, S.; Yasui, H.; Ohta, K.; Tsuda, K. *J. Chem. Soc., Chem. Commun.* **1985**, 400.
26. Hamaide, T.; Mariaggi, N.; Foureys, J. L.; Perchec, P. I.; Guyot, A. *J. Polym. Sci., Polym. Chem. Ed.* **1984**, *22*, 3091.
27. Ito, K.; Tsuchida, H.; Hayashi, A.; Kitano, T.; Yamada, E.; Matsumoto, T. *Polym. J.* **1985**, *17*, 827.
28. Itsuno, S.; Yamazaki, K.; Arakawa, F.; Kitano, T.; Ito, K.; Yamada, E.; Matsumoto, T. *Kobunshi Ronbunshu* **1986**, *43*, 91.
29. Wakui, T.; Xu, W. Y.; Chen, C. S.; Smid, J. *Makromol. Chem.* **1986**, *187*, 533.
30. Regen, S. L.; Dulak, L. *J. Am. Chem. Soc.* **1977**, *99*, 623.
31. McKenzie, W. M.; Sherrington, D. C. *J. Chem. Soc., Chem. Commun.* **1978**, 541.
32. Yanagida, S.; Takahashi, K.; Okahara, M. *Yukugaku* **1979**, *28*, 14.
33. Tundo, P.; Venturello, P. *J. Am. Chem. Soc.* **1979**, *101*, 6606.
34. Kimura, Y.; Regen, S. L. *J. Org. Chem.* **1983**, *48*, 1533.

PHASE-TRANSFER CATALYZED POLYMER SYNTHESIS

Louis M. Leung
Department of Chemistry
Hong Kong Baptist University

A wide range of polymers has been prepared successfully using the phase-transfer catalysis (PTC) method.[3,3a] They include homopolymers and copolymers that are normally prepared from a polycondensation method, with functional groups such as amide, ether, thioether, ester, thioester, carbonate, thiocarbonate, and sulfonate.[10–27] Specialized polymers such as polyene and block copolymers have also resulted using the same PTC process.[2,4–9] In addition to polymer synthesis, PTC has been used in the modification and functionalization of polymers for applications, including ion-exchange resins, polymer-supported chemical reactions, and ion-selective electrodes.[1,28–30] The polymer prepared from the PTC method is reported to have a high molecular weight (MW) and narrow polydispersity (index <1.4), with specific end groups, and are linear chains that have no branching.[1–3,11,12] As an example, the PTC synthesis of polythiocarbonate and polythiocarbamate, using carbon disulfide as the nucleophile, is discussed in the next section.[31–33]

SULFUR-CONTAINING POLYMERS

Traditionally, sulfur-containing polymers have been synthesized through the use of toxic or expensive chemicals such as thiophosgene dithiols, and bischloroformate in a multistep reaction.[22–26] Recent studies on the synthesis of dithiocarbonate and trithiocarbonate have found that carbon disulfide is a low cost and effective sulfur-containing nucleophile.[29,30] Under strong base conditions, CS_2 becomes trithiocarbonate dianion (CS_3^{2-}) and will undergo a PTC-aided nucleophilic substitution reaction with an alkyl dihalide to produce polytrithiocarbonate.[31] In the presence of a deprotonated glycol or diamine (primary or secondary), the intermediates produced are dithiocarbonate dianions (or dixanthate anions) or dithiocarbamate dianions. These two dianions will undergo the same nucleophilic substitution reaction with alkyl dihalide to produce poly(*S*-dithiocarbonates) and polydithiocarbamates.[32,33]

PROPERTIES

All the polymers were colored, ranging from yellowish to brown. All the sulfur-based polymers have excellent resistance to chemicals and solvents.

REFERENCES

1. *Crown Ethers and Phase Transfer Catalysts in Polymer Science*; Methias, L. J.; Carraher, C. E., Jr., Eds.; Plenum: New York, 1984.
2. *Phase-Transfer Catalysis: New Chemistry, Catalysis and Applications*; Starks, C. M., Ed.; American Chemical Society: Washington, DC, 1987.
3. Boileau, S. In *New Methods for Polymer Synthesis*; Migs, W. J., Ed.; Plenum, New York, 1992.

3a. *Phase-Transfer Reactions. A Fluka Compendium*; Keller, W. E., Ed.; Georg Thieme Verlag: New York, 1986 and 1987; Vol. 1 and 2.

4. Leung, L.; Yam, C. M. *Polym. Bull.* **1993**, *30*, 629.
5. Petit, G. J. *J. Polym. Sci., Chem. Ed.* **1980**, *18*, 345.
6. Swatos, W. J.; Gordon, B., III. *Polym. Prepr.* **1990**, *31*(1), 505.
7. Gilch, H. G.; Wheelwright, W. L. *J. Polym. Sci., Part A1* **1966**, *4*, 1337.
8. Yamazaki, N.; Imai, Y. *Polym. J.* **1983**, *15*(12), 905.
9. Leung, L.; Chik, G. *Polymer* **1993**, *34*, 5174.
10. Banthia, A. K.; Lundeford, D.; Webster, D. C.; McGrath, J. E. *J. Macromol. Sci., Chem. Ed.* **1981**, *A15*, 943.
11. Gerbi, D. J.; Dimotsis, G.; Morgan, J. L.; Williams, R. F.; Kellman, R. *J. Polym. Sci., Polym. Lett. Ed.* **1985**, *23*, 551.
12. Yamazaki, N.; Imai, Y. *Polym. J.* **1983**, *15*, 603.
13. Sandler, S. R.; Karo, W. *Polymer Synthesis*; Academic: New York, 1980; Vol. 3.
14. Ueda, M.; Oishi, Y.; Sakai, N.; Imai, Y. *Macromolecules* **1982**, *15*, 248.
15. Shaffer, T. D.; Percec, V. *J. Polym. Sci., Chem. Ed.* **1986**, *24*, 451.
16. Shaffer, T. D.; Kramer, M. C. *Makromol. Chem.* **1990**, *191*, 71.
17. Imai, Y.; Tassavori, S. *J. Polym. Sci., Chem. Ed.* **1984**, *22*, 1319.
18. Podkoscielny, W.; Szubinska, S. *J. Appl. Polym. Sci.* **1988**, *35*, 1853.
19. Soga, K.; Hosoda, S.; Ikeda, S. *J. Polym. Sci., Chem. Ed.* **1979**, *17*, 517.
20. Rokicki, G.; Kuran, W.; Kielkiewicz *J. Polym. Sci., Chem. Ed.* **1982**, *20*, 967.
21. Rokicki, G.; Pawlicki, J.; Kuran, W. *Polym. J.* **1985**, *17*, 509.
22. Tagle, L. H.; Diaz, F. R.; Salas, P. *J. Macromol. Sci., Chem.* **1989**, *A26*(9), 1321.
23. Berti, C.; Marianucci, E.; Pilati, F. *Polym. Bull.* **1985**, *14*, 85.
24. Montaudo, G.; Puglisi, C.; Berti, C.; Marianucci, E.; Pilati, F. *J. Polym. Sci., Part A* **1989**, *27*, 2657.
25. Montaudo, G.; Puglisi, C.; Berti, C.; Marianucci, E.; Pilati, F. *J. Polym. Sci., Part A* **1989**, *27*, 2277.
26. Soga, K.; Imamura, H.; Sato, M.; Ikeda, S. *J. Polym. Sci., Chem. Ed.* **1975**, *14*, 677.
27. Sato, M.; Yokoyama, M. *Makromol. Chem.* **1984**, *185*, 629.
28. Chan, W. H.; Lee, A. W. M.; Wong, C. S.; Choi, W. K.; Fung, K. K. *Anal. Lett.* **1990**, *23*, 659.
29. Lee, A. W. M.; Chan, W. H.; Wong, H. C. *Synth. Commun.* **1988**, *18*(13), 1531.
30. Lee, A. W. M.; Chan, W. H.; Wong, H. C.; Wong, M. S. *Synth. Commun.* **1989**, *19*(3, 4), 547.

31. Leung, L.; Chan, W. H.; Leung, S. K. *Polym. Sci., Part A* **1993**, *31*, 1799.
32. Leung, L.; Chan, W. H.; Fung, S. M.; Leung, S. K. *J. Macromol. Sci., Pure Appl. Chem.* **1994**, *A3*(4), 495.
33. Leung, L.; Wong, T. M. Manuscript in preparation.

Phase Transitions

See: Crystals, Phase Transitions

PHENOLIC COMPOSITES

Phillip A. Waitkus
Plastics Engineering Company

The principles governing the structure of natural composites, such as that of a tree, are of considerable interest, because these natural materials combine several functions to achieve very synergistic, highly engineered results. The presence of a synergy of properties, and the property of adaptability, characterizes composites in the natural world. It seems reasonable then to use this abstract definition as a model for artificial composites. The purpose of this article is to describe how currently produced phenolic composites conform in many ways to this model.

Most of the work with phenolic composites to date has been done with resoles. Unless otherwise indicated, this discussion is limited to this type of resin, cured without acids.

EFFECTS OF HIGH-ENERGY RADIATION

The consensus regarding the resistance of phenolics to the effects of γ- and neutron radiation places the damage threshold at about 10^8 Gy. Ivanoff concluded that a safe γ-radiation design limit for phenolics (type unspecified) is ~ 10^7 Gy.[2] Most workers using inorganic reinforcements, especially graphite or glass fibers, observed a gradual increase in physical properties up to the radiation damage threshold. It was postulated that the increase in physical properties with irradiation is a result of a free-radical reaction that grafts the phenolic polymer to the reinforcement, thereby enhancing the composite's properties.[1]

EFFECT OF ACTINIC RADIATION

The use of phenolic materials–resins and monomers–as UV and oxidative stabilizers for thermoplastics is well known. There are, however, few reports on the effect of this form of radiation on the physical properties of phenolic composites.

EFFECT OF HEAT

Although phenolics have a long history of use in high-temperature applications, only recently have the physical chemical phenomena, accompanying exposure to heat, been understood. Korb, Landi, Morrison, and Sepe have shown that the material's glass transition temperature (T_g) increases in a predictable way with increasing length of exposure and increasing exposure temperature.[3–7]

CORROSION RESISTANCE

Murphy submerged pultruded phenolic resole specimens in numerous potentially corrosive agents for one year to evaluate their potential loss in flexural strength. Of the agents tested, hydrocarbons, jet fuel, and ethylene glycol showed slight increases in flexural strength. Dilute acids, bases, salts, and methanol showed only slight to moderate degradation, whereas strong acids and bases showed substantial strength degradation. Oxidizing media such as sodium hypochlorite degrade the surface finish of phenolic moldings.

FLAME RESISTANCE

Among the commercial polymers, phenolics exhibit low flammability and are one of the lowest smoke-generating polymers known.[8,9] Because the polymer lacks a ceiling temperature for depolymerization to its constituent monomers, thermal degradation proceeds to form relatively high yields of structural carbon char. This char is predominantly vitreous or glassy carbon.[10]

CONCLUSION

The usefulness of phenolic composites stems from a fortunate combination of strength with good resistance to flame, chemical agents, and energy across a broad range of the electromagnetic spectrum. If these materials are caught on fire, there is ample evidence that they will neither produce large amounts of toxic fumes or smoke nor easily lose stiffness from the heat. These are the synergy of properties inherent in phenolic composites.

An interesting aspect of phenolic composites is their ability to adapt to stress. This has been observed on exposure to moderate to high doses of ionizing radiation, heat, and certain chemicals. The combination of properties and desirable phenomena listed in this discussion, together with the adaptability of this material to many externally applied stresses, completes the analogy of phenolic composites to naturally occurring ones.

REFERENCES

1. Ambardanishvili, T. S.; Dunua, V. Y.; Knadze, G. I.; Kolomiytsev, M. A. *Bull. Acad. Sci. of Georgian SSR* **1973**, *69*(3), 569–572.
2. Ivanov, V. S. In *New Concepts in Polymer Science*; de Jonge, C. R. H. I., Ed.; VSP: Utrecht, The Netherlands, 1992, p 199.
3. Korb, L. *Mod. Plast.* **1976**, 90.
4. Landi, V. R.; Mersereau, M.; Dorman, S. E. "The Effects of Molding Time and Temperature on the Modulus and Glass Transition of Phenolics;" Soc. Plast. Eng. Regional Technical Conference: Philadelphia, PA, 1985.
5. Landi, V. R.; Mersereau, M. "The Glass Transition in Novolac Phenolic Molding Compounds and the Kinetics of Its Development During Cure and Postcure–Part 1"; Soc. Plast. Eng. Annual Technical Conference: Philadelphia, PA, 1986; p 1369.
6. Morrison, T. N.; Waitkus, P. A. "Studies of the Post Bake Process of Phenolic Resins"; Phenolic Molding Division; Soc. Plast. Ind.: Cincinnati, OH, 1987.
7. Sepe, M. P.; 37th Annual Thermoset Molding Conference, 1991, Sponsored by the Department of Engineering Professional Development, University of Wisconsin-Madison, William Wuerger, Program Director.
8. Loche, H. W.; Strass, E. L.; Conley, R. L. *J. Appl. Polym. Sci.* **1965**, *9*, 2799.
9. Hilado, C. S.; Machado, A. M. *J. Fire and Flammability* **1979**, *9*, 367.
10. Knop, A.; Scheib, W. *Chemistry and Application of Phenolic Resins*; Springer-Verlag: New York, 1979; p 246.

PHENOLIC RESINS (Improvement in Heat Resistance)

Akinori Fukuda
Osaka Municipal Technical Research Institute

There are two approaches to improve the heat resistance of phenolic resins. One is for widely used commercial resin by means of changing either the formulation of the molding compound or the molding condition. The other is to create a new phenolic resin system (e.g., using a chemically modified phenolic moiety or a new kind of curing agent). The former consists of an increase in crosslinking density of cured resin by either increasing the amount of the curing agent or by increasing postcuring (i.e., after curing) moldings. In the latter case, some methods can increase the functionality of a novolac molecule (e.g., the use of high molecular weight novolac) or a triazine or phosphazene ring or metal can be introduced into the resin.

COMMERCIAL TWO-STAGE PHENOLIC RESINS

Improvement of heat resistance of phenolic resins is attributed to an increase of crosslink density by increasing the amount of curing agent and increasing aftercure of molded parts. As curing agents, methylol groups to resol and hexamethylene tetramine (hexamine) to novolac are used. This paper focuses on hexamine.

Heat treatment is a useful method for improving the performance of phenolics in industry because the process of the treatment is simple and the improvement of heat resistance is remarkable.

CHANGING THE STRUCTURE AND PROPERTIES OF PHENOLIC RESIN

Increase in Molecular Weight or Functionality of Novolac

In phenolic resin, an increase of molecular weight of novolac means an increase in functionality of the novolac molecule. Therefore, a good way to increase the heat resistance of the resin to increase the molecular weight of novolac.

Polyvinylphenol

A kind of high molecular weight novolac was manufactured by the polymerization of vinylphenol prepared by dehydration of ethylphenol by the Maruzen Chemical Co.

Poly(hydroxyphenyl maleimide)

A series of homopolymers of hydroxyphenyl maleimide and its copolymer with another vinyl monomer was prepared by radical polymerization. These polymers provided phenolics with both heat resistance and toughness by modification of the resin.

Decrease in Concentration of the Phenolic Hydroxyl Group

Aiming to increase thermal stability (antipyrolysis) and toughness and to decrease water absorption, many studies were done to decrease the concentration of the hydroxyl group.

PHENOLIC RESINS (Improvement of Toughness)

Akihiro Matsumoto
Osaka Municipal Technical Research Institute

Phenolic resin has desirable characteristics such as superior mechanical strength, heat resistance and dimensional stability, as well as high resistance against various solvents, acids and water. Its ability to maintain good mechanical strength at high temperatures is its best characteristic. In comparison with other resins, it is also flame resistant and emits smaller quantities of smoke with relatively low toxicity upon incineration.

Phenolic resin is brittle and not very tough, as are many other thermosetting resins. In order to improve the toughness of phenolic resin, various methods have been tested and adapted. These include:

- Use of reduced amount of curing agent
- Introduction of butadiene bond, alkylene ether chain or urethane bond between the adjacent phenol nuclei
- Modification with drying oil
- Use of plasticizing additive such as natural rubber or butadiene acrylonitrile rubber (NBR)
- Addition of reinforcing material or filler such as cloth pieces and grass fibers

IMPROVING TOUGHNESS OF PHENOLIC RESIN

Thermosetting resins, such as phenolic resin, may be brittle because of the crosslinking structure that prevents plastic deformation induced by slippage of molecular chains. Taking this into account, the potential toughening mechanisms for phenolic resin could be classified as toughening by homogeneous and miscible structure; microdomain dispersed phase structure; or cocontinuous two-phase structure.

Even though there are not as many reports published on toughening mechanisms for phenolic resin as for epoxy resin, the above classification is based on the author's analysis of recently published reports on phenolic resin.

Phenolics

See: *Antioxidants*
Benzoxazine Monomers and Polymers (New Phenolic Resins by Ring-Opening Polymerization)
Novolak Resins (For Microlithography)
Novolak-type Epoxy Resins
Phenolic Composites
Phenolic Resins (Improvements in Heat Resistance)
Phenolic Resins (Improvement of Toughness)
Phenolics (Linear and Cyclic Oligomers)
Polyacenic Semiconductor Materials (Application to Rechargeable Batteries)
Powder Coatings (Overview)
Resins (Applications in Coatings Industry)
Step Polymerization Catalysts

PHENOLICS (Linear and Cyclic Oligomers)

Yoshiaki Nakamoto,* Tada-Aki Yamagishi,
and Shin-ichiro Ishida
Department of Chemistry and Chemical Engineering
Faculty of Technology
Kanazawa University

Phenol-formaldehyde resin is one of the oldest synthetic plastics and is widely used by taking advantage of its heat resistance and electric insulation.[1] In 1944, formation of a cyclic oligomer[2] was obtained in the condensation of *p*-substituted phenols with formaldehyde under alkaline conditions. In the 1980s, Gutsche[3] established the synthesis of cyclic oligomers and named them Calixarenes (**Scheme I**). Recently, the chemical and physical properties of calixarene have attracted the interest of host-guest chemists because of the existence of a cavity in the molecule and the ring arrangement of phenolic hydroxyl groups. Calixarenes are chemically modified and applied to many fields, e.g., novel hosts, ionophore and chromophore, phase transition catalyst, ion-selective electrode, monolayer membrane, and liquid crystal.[4,5]

Phenolics have the variety of methylene linkage (*ortho–ortho',* *ortho–para',* and *para–para'*) and the degree of branch. Linear phenolic oligomers with definite structure are desirable as a model compound to elucidate the structure-properties relationship of phenolics. In this chapter, we will limit the discussion of phenolic oligomers with definite structures, including cyclic and linear forms, and will focus on their characteristics, as well as inclusion and assembly properties.

PREPARATION

Calixarenes are prepared from *p*-substituted phenols and formaldehydes by basic catalysts.

*Author to whom correspondence should be addressed.

PROPERTIES

Calixarenes are characterized by high melting points above 250°C, particularly those with hydroxyl groups. For example, *p-tert*-butylcalix[4]arene melts at 342–344°C, *p-tert*-butylcalix[6]arene at 380–381°C, and *p-tert*-butylcalix[8]arene at 411–412°C. On the other hand, the melting points of linear oligomers are lower than those of calixarenes. In both calixarenes and linear oligomers, the ester and ether derivatives melt at lower temperatures than parent compounds.

Calixarenes have low solubility in both aqueous solutions and organic solvents. However, calixarenes have sufficient solubility (a few wt %) in dichloromethane, chloroform, tetrahydrofuran (THF), pyridine, and N,N-dimethylformamide (DMF). Linear oligomers have good solubility in organic solvents such as benzene, acetone, methanol, THF, pyridine, DMF, and dimethyl sulfoxide (DMSO).

APPLICATIONS

Inclusion of Organic Molecules

Calixarenes have a cavity large enough to allow substrate inclusion in the molecule. The application of calixarenes for host-guest chemistry has been widely investigated. For example, *p-tert*-butylcalix[4]arene forms a complex with toluene (Host:Guest = 1:1)[6] and anisole (H:G = 2:1).[7]

Linear oligomers of *p*-substituted phenols also have been reported to form a complex with organic molecules, determined by ^{1}H-NMR and TG/DTA measurements.[8]

Ionophore

It is well known that *p-tert*-butylcalix[4]arene has four conformation isomers: cone-, partial cone-, 1,2-alternate-, and 1,3-alternate-conformer (**Scheme II**). The cone type shows good selectivity and the highest extraction value for Na^+. The partial cone type shows the selectivity for K^+. The 1,3-alternate type shows little selectivity, but the highest extraction toward Rb^+ and Cs^+. The cavity in the calixarenes becomes looser and larger

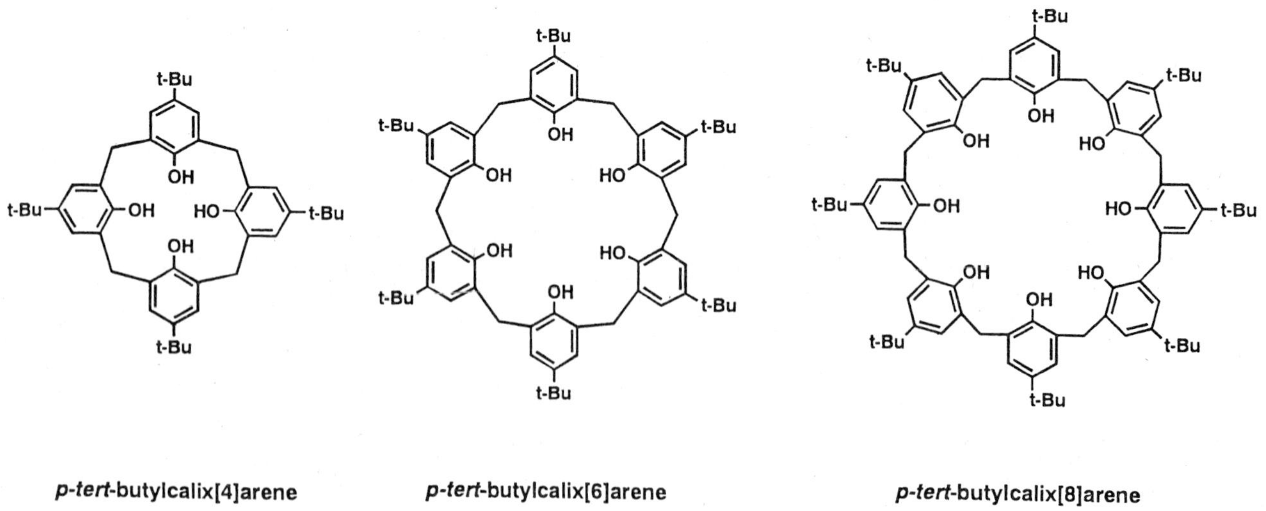

SCHEME I. *p-tert*-Butylcalix[n]arenes (n = 4, 6, 8).

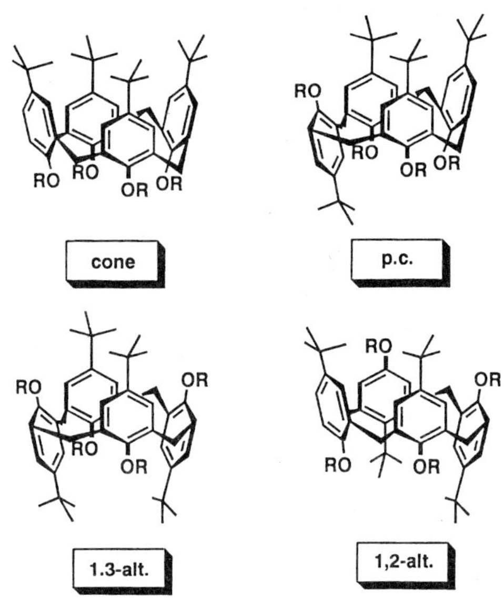

SCHEME II. Conformational isomers of *p-tert*-butylcalix[4]arene.

as the conformation changes from cone to partial cone, and to 1,3-alternate.

Molecular Assembly

A cone-shaped calixarene is analogous to a disc, so it is expected to form a discotic-like liquid crystal. The azomethine-type calixarenes are reported to form a columnar liquid crystalline phase when heated.[9] The tungsten-oxo azophenylcalix[4]arenes complex also forms discotic mesophases in a wide range of temperatures.[10] These mesophases consist of columnars, in which cone cores exhibit a head-to-tail arrangement. These results will lead to new host-guest interactions for a calixarene in a liquid crystalline state.

REFERENCES

1. Knop, A.; Pilato, L. A. *Phenolic Resins*; Springer-Verlag: New York, 1985.
2. Zinke, A.; Ziegler, E. *Ber.* **1944**, *77*, 264.
3. Gutsche, C. D. *Acc. Chem. Res.* **1983**, *16*, 161.
4. Gutsche, C. D. *Calixarenes*; The Royal Society of Chemistry: Cambridge, 1989.
5. *Calixarenes: A Versatile Class of Macrocyclic Compounds*; Vicens, J.; Böhmer, V., Eds.; Kluwer Academic: Dordrecht, 1991.
6. Andreetti, G. D.; Ungaro, R.; Pochini, A. *J. Chem. Soc., Chem. Commun.* **1979**, 1005.
7. Ungaro, R.; Pochini, A.; Andreetti, G. D.; Sangermano, U. *J. Chem. Soc., Perkin Trans.* **1984**, *2*, 1979.
8. a) Sone, T.; Ohba, Y.; Yamazaki, H. *Bull. Chem. Soc. Jpn.* **1989**, *62*, 1111; b) Ohba, Y.; Moriya, K.; Sone, T. *Bull. Chem. Soc. Jpn.* **1991**, *64*, 576.
9. a) Komori, T.; Shinkai, S. *Chem. Lett.* **1992**, 901; b) Komori, T.; Shinkai, S. *Chem. Lett.* **1993**, 1455.
10. Xu, B.; Swager, T. M. *J. Am. Chem. Soc.* **1993**, *115*, 1159.

PHENYLACETYLENE (Stereospecific Living Polymerization, Rh Complexes)

Yasuhisa Kishimoto, Ryoji Noyori,* Peter Eckerle,
Tatsuya Miyatake, and Takao Ikariya
ERATO Molecular Catalysis Project
Research Development Corporation of Japan

Polyacetylenes[1] with alternating olefinic bonds along the main chain are of significant interest because of their unique physical properties such as photoconductivity,[2] optical nonlinear susceptibility[3] and magnetic susceptibility.[4] The presence of aryl or alkyl substituents is crucial for obtaining sufficient stability, appropriate chain conformation and stiffness, and high solubility in organic solvents.

Although stereo-controlled, living polymerization[5] is an indispensable tool for designing highly functional materials based on compositionally pure monodisperse polymers, efficient methods for acetylene derivatives have remained rare. The only example is the metathesis polymerization of alkylated acetylenes, particularly *tert*-butylacetylene with 88–97% *cis*-stereoselectivity achieved by group 6 metal catalysts.[6]

Polymerization of phenylacetylene (PA) has been extensively studied to obtain stereo-controlled polymers that might lead to new synthetic functional materials. $MoCl_5$- or $MoO\text{-}Cl_4$-based three component systems[7] and well-characterized Mo alkylidene complexes,[8] are known to be effective for living polymerization of *o*-substituted PAs, but the stereochemistry of the main chain of the polymer is, in many cases, a mixture of cis and trans configurations.

Many Rh(I) complexes promote various reactions of aromatic substituted acetylenes such as dimerization,[9] linear oligomerization,[10] and polymerization.[11] The polymers obtained with these Rh-based initiators have *cis*-stereochemistry with respect to the C=C double bond in the main chain, but efficient control of molecular weight and molecular weight distribution of the polymers has not been achieved. A well-defined Rh complex is required to attain stereospecific living polymerization of phenylacetylenes.

Desirable stereospecific living polymerization of PA and its *p*-substituted derivatives has been accomplished by using a newly developed organo-rhodium(I) complex, $Rh(C{\equiv}CC_6H_5)(nbd)[P(C_6H_5)_3]_2$ (**1**, nbd = 2,5-norbornadiene), as an initiator.[12] This chapter describes its characteristic features and performance, **Scheme I**.

LIVING NATURE OF POLYMERIZATION

The living nature of this polymerization is demonstrated by the following observations. The M_n value of the products increases proportionally to the conversion of the parent and substituted PAs in both ether and THF, while the molecular-weight distribution remains within a narrow range throughout the polymerization. A higher M_n value, up to ca. 2×10^5, can be attained by increasing the monomer/initiator ratio or by decreasing initiator concentration.

*Author to whom correspondence should be addressed.

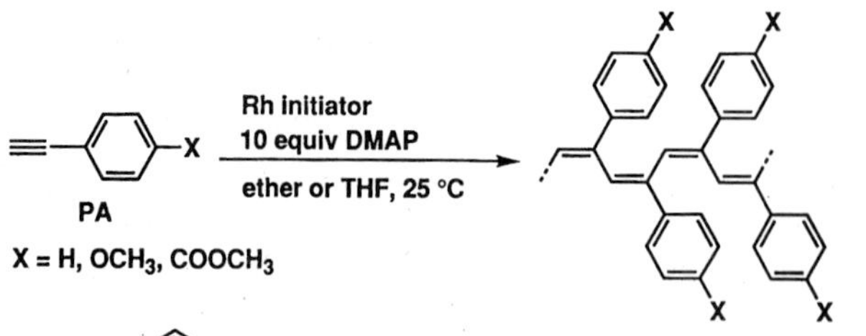

I

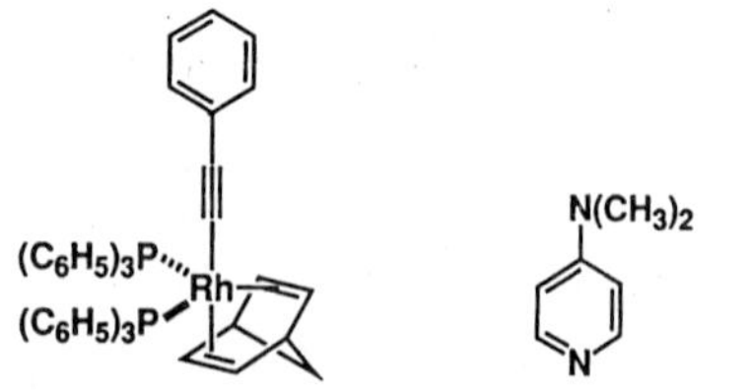

Rh initiator DMAP

Direct evidence that this polymerization is living is provided by isolation of a polymer with an active end from the reaction mixture without treatment with acetic acid.[13]

The living nature of the Rh-initiated reaction allows synthesis of an AB-type block copolymer from different PAs.

PROPERTIES AND STRUCTURE OF POLYPHENYLACETYLENES

The polyphenylacetylene obtained by this method is a yellow to red-brown color compound and is readily soluble in most common aprotic solvents such as toluene, dichloromethane, and THF, but only slightly soluble in ether.

With regard to the structure of poly(phenylacetylene)s, four stereoisomers are possible in terms of the configuration of the C=C double bond and the conformation of the C–C single bond of the polymer main chain.

NMR studies of the polymers indicate that the soluble polymers have a *cis* configuration at the C=C double bond of the polymer chain.

REFERENCES

1. Costa, G. *Comprehensive Polymer Science*; Allen, G.; Bevington, J. C., Eds.; Pergamon: Oxford, U.K., 1989; Vol. 4, Chapter 9.
2. Kang, E. T.; Neoh, K. G.; Masuda, T. et al. *Polymer* **1989**, *30*, 1328.
3. (a) Neher, D.; Wolf, A.; Bubeck, C.; Wegner, G. *Chem. Phys. Lett.* **1989**, *163*, 116. (b) Le Moigne, J.; Hilberer, A.; Strazielle, C. *Macromolecules* **1992**, *25*, 6705.
4. (a) Ovchinnikov, A. A. *Theor. Chim. Acta* **1978**, *47*, 297. (b) Rossitto, F. C.; Lahti, P. M. *Macromolecules* **1993**, *26*, 6308.
5. Examples of stereo-controlled living polymerization: (propene) (a) Doi, Y.; Suzuki, S.; Soga, K. *Macromolecules* **1986**, *19*, 2896. (2-vinylpyridine) (b) Soum, A.; Fontanille, M. *Makromol. Chem.* **1980**, *181*, 799. (cycloalkene) (c) McConville, D. H.; Wolf, J. R.; Schrock, R. R. *J. Am. Chem. Soc.* **1993**, *115*, 4413. (d) Oskam, J. H.; Schrock, R. R. *J. Am. Chem. Soc.* **1993**, *115*, 11831. (methyl methacrylate) (e) Hatada, K.; Ute, K.; Tanaka, K et al. *Polym. J.* **1986**, *18*, 1037. (f) Kitayama, T.; Shinozaki, T.; Sakamoto, T. et al. *Makromol. Chem., Suppl.* **1989**, *15*, 167. (g) Yasuda, H.; Yamamoto, H.; Yokota, K. et al. *J. Am. Chem. Soc.* **1992**, *114*, 4908. (1,2-diisocyanoarenes) (h) Ito, Y.; Ihara, E.; Murakami, M.; Shiro, M. *J. Am. Chem. Soc.* **1990**, *112*, 6446.
6. (a) Nakano, M.; Masuda, T.; Higashimura, T. *Macromolecules* **1994**, *27*, 1344. (b) Nakayama, Y.; Mashima, K.; Nakamura, A. *Macromolecules* **1993**, *26*, 6267.
7. (a) Masuda, T.; Yoshimura, T.; Fujimori, J.; Higashimura, T. *J. Chem. Soc., Chem. Commun.* **1987**, 1805. (b) Masuda, T.; Mishima, K.; Fujimori, J. et al. *Macromolecules* **1992**, *25*, 1401.
8. Schrock, R. R.; Luo, S.; Zanetti, N. C.; Fox, H. H. *Organometallics* **1994**, *13*, 3396.
9. (a) Yoshikawa, S.; Kiji, J.; Furukawa, J. *Makromol. Chem.* **1977**, *178*, 1077. (b) Carlton, L.; Read, G. *J. Chem. Soc., Perkin Trans. 1* **1978**, 1631. (c) Boese, W. T.; Goldman, A. S. *Organometallics* **1991**, *10*, 782.
10. (a) Singer, H.; Wilkinson, G. *J. Chem. Soc. (A)* **1968**, 849. (b) Kern, R. J. *Chem. Commun.* **1968**, 706.
11. Examples of stereoregular polymerization, though not living in nature, are: (a) Furlani, A.; Napoletano, C.; Russo, M. V.; Feast, W. J. *Polym. Bull.* **1986**, *16*, 311. (b) Furlani, A.; Licoccia, S.; Russo, M. V. et al. *J. Polym. Sci., Part A, Polym. Chem.* **1986**, *24*, 991. (c) Furlani, A.; Napoletano, C.; Russo, M. V. et al. *J. Polym. Sci., Part A, Polym. Chem.* **1989**, *27*, 75. (d) Tabata, M.; Yang, W.; Yokota, K. *Polym. J.* **1990**, *22*, 1105. (e) Haupt, H.-J.; Ortmann, U. Z. *Anorg. Allg. Chem.* **1993**, *619*, 1209. (f) Goldberg, Y.; Alper, H. *J. Chem. Soc., Chem. Commun.* **1994**, 1209.
12. (a) Kishimoto, Y.; Miyatake, T.; Eckerle, P. T. et al. *Abstract of 35th IUPAC International Symposium on Macromolecules* **1994**, 108. (b) Kishimoto, Y.; Eckerle, P.; Miyatake, T. et al. *J. Am. Chem. Soc.* **1994**, *116*, 12131.
13. Schäfer, M.; Mahr, N.; Wolf, J.; Werner, H. *Angew. Chem., Int. Ed. Engl.* **1993**, *32*, 1315.

p-PHENYLENE VINYLENE OLIGOMERS, HOMO- AND COPOLYMERS (Metathesis Preparation)

Emma Thorn-Csányi* and Peter Kraxner
Institut für Technische und Makromolekulare Chemie
Universität Hamburg

Conjugated polymers have received great interest because of properties such as electroconductivity, nonlinear optical response, and electro- and photoluminescence.[1]

*Author to whom correspondence should be addressed.

Such polymers, especially poly(*p*-phenylene vinylene) (PPV), [*96638-49-2*], [*87092-55-5*], and [*26009-24-5*] garnered renewed attention since serving as an acting element in light-emitting devices.[2] In the past, its electroconductivity[3–6] was a major field of investigation; now, optical properties are the major interest of several research groups, around the world.[7–10]

Several methods of preparation have been used in the last 35 years: poly-Wittig,[11] poly-Horner,[12] poly-Heck,[13] and the sulphonium salt route,[14] now the most common method.[2,7,15–18]

This paper deals with a new synthetic route based on olefin metathesis.

For preparation of polymers, two metathetic routes are possible: ring opening metathesis polymerization (ROMP) or acyclic diene metathesis (ADMET) polycondensation.

p-PHENYLENE VINYLENE OLIGOMERS AND POLYMERS

Depending on the method of polymer formation, two types of substrates are suitable for metathesis–forming PPV: cyclic olefins as well as acyclic dienes, which have to contain *p*-phenylene vinylene units. The first direct metathesis synthesis of PPV succeeded using [2.2]-paracyclophane-1,9-diene (PCPDE) as substrate[19] (**Figure 1**).

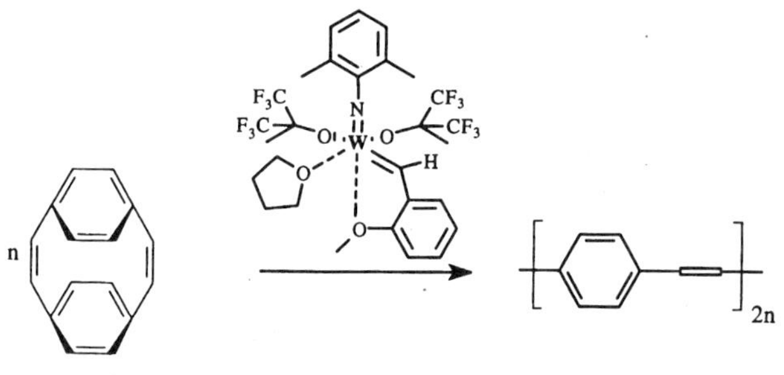

FIGURE 1. ROMP of PCPDE.

In addition to polymers, cyclic oligomers were detected by mass spectroscopy.[20] The formation of rings with 4, 5, and 6 *p*-phenylene vinylene units proves the existence of a backbiting reaction during polymerization. 1,4-Divinylbenzene (DVB) is suitable for ADMET polycondensation with stable metal carbene complexes as catalysts (**Figure 2**).

PV-COPOLYMERS

Although metathesis polycondensation of PCPDE and DVB yielded PPV, products obtained still lack processibility. To circumvent this problem, several methods have been used such as introducing substituents[21] (s. substituted PPVs) or synthesizing PPV via soluble precursor polymers.[14] Metathetic precursor routes were performed for polyacetylene[22] and PPV.[23]

SUBSTITUTED PPVS

In summary, metathesis polycondensation opened a way for preparation of products with longer PV sequences. By increasing chain length, processibility and solubility decrease because of physical crosslinking. To overcome this problem, preparation of substituted divinylbenzenes (DRDVB) was performed. A variety of 2,5-alkyl-substituted DVBs can be easily prepared.[24] Diheptyl substituted DVB (DHepDVB) is given[25] as an example. In analogy to the conversion of unsubstituted DVB the formational of all-trans-poly(2,5-diheptyl-1,4-phenylene vinylene) (DHepPPV) was expected.

Metathesis reaction of 1,4-divinylbenzene is a versatile means of synthesizing *p*-phenylene vinylene oligomers of different chain length. The reaction is stereoselective, forming an all-trans-product. Lower oligomers can be isolated and used as telomers for copolymerization. By selecting suitable reaction conditions, chain length of the PV-oligomers is preserved. By using different "comonomers" (also in the form of polymers) as well as by introduction of substituents in DVB, our method of preparing copolymers offers the possibility of obtaining a whole range of new products with valuable properties. In particular, substituted DVBs enlarge the applicability of PV-oligomers and -polymers not only by improving solubility of higher oligomers, but by providing important new properties such as liquid crystallinity.

REFERENCES

1. Brédas, J. L.; Silbey, R. *Conjugated Polymers*; Kluwer, Boston, 1991.
2. Friend, R. H.; Burroughes, J. H. et al. *Nature* **1990**, *347*, 539.
3. Karasz, F. E.; Wnek, G. E. et al. *Polymer* **1979**, *20*, 1441.
4. Murase, I.; Ohnishi, T. et al. *Polymer Commun.* **1984**, *25*, 327.
5. Hörhold, H.-H.; Helbig, M. *Makromol. Chem., Macromol. Symp.* **1987**, *12*, 229.
6. Brédas, J. L.; Beljonne, D. et al. *Synth. Metals* **1991**, *43*, 3743.
7. Schwoerer, M.; Gmeiner, J. et al. *Acta Polymer* **1993**, *44*, 201.
8. Brédas, J. L.; Shuai, Z.; Su, W. P. *Chem. Phys. Lett.* **1994**, *228*, 301.
9. Heeger, A. J.; Pakbaz, K. et al. *Synth. Metals* **1994**, *64*, 295.
10. Friend, R. H.; Cacialli, F. et al. *Synth. Metals* **1994**, *67*, 157.
11. McDonald, R. N.; Campbell, T. W. *J. Am. Chem. Soc.* **1960**, *82*, 4669.
12. Drehfahl, G.; Hörhold, H.-H.; Wildner, H. *Nachr. Chem. Techn.* **1965**, *13*, 451.
13. Heitz, W.; Greiner, M. *Makromol. Chem., Rapid Commun.* **1988**, *9*, 581.
14. Wessling, R. A.; Zimmerman, R. G. U.S. Patent 3 401 152, 1968.
15. Karasz, F. E.; Lenz, R. W. et al. *J. Polym. Sci., Polym. Chem.* **1988**, *26*, 3241.

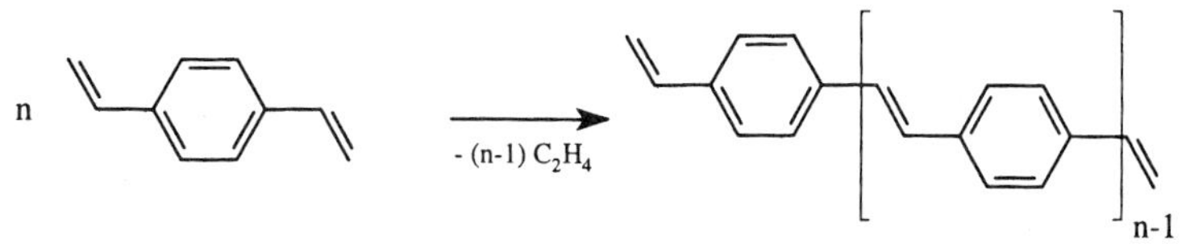

FIGURE 2. ADMET of DVB.

16. Saito, S.; Momii, T. et al. *Chem. Lett.* **1988**, 1201.
17. Müllen, K.; Garay, R. O. et al. *Adv. Mater.* **1993**, *5*, 561.
18. Jin, J.-I.; Lee, Y.-H.; et al. *Macromolecules* **1994**, *27*, 5239.
19. Thorn-Csányi, E.; Höhnk, H.-D. *ISOM 9* (21-26.07. 1991), Collegeville, PA, *J. Mol. Catal.* **1992**, *76*, 101.
20. Kraxner, P. *Thesis*, Untersuchungen zur ringöffnenden metathetischen Polymerisation von [2.2]Paradyclophan-1,9-dien und dessen Copolymerisation mit Cycloocten, University of Hamburg, *Germany*, 1993.
21. Hörhold, H.-H.; Helbig, M. et al. *Z. Chem.* **1987**, *27*, 126.
22. Feast, W. J.; Edwards, J. H. *Polymer* **1980**, *21*, 595.
23. Grubbs, R. H.; Conticello, V. P.; Gin, D. L. *J. Am. Chem. Soc.* **1992**, *114*, 9708.
24. Kraxner, P. Doctoral thesis, Metathetische Polykondensation von substituierten 1,4-Divinylbenzolen: Ein neuer Weg zu löslichen Poly(*p*-phenylen vinylen)-Derivaten, University of Hamburg, Germany, 1996.
25. Thorn-Csányi, E. Kraxner, P. *Macromol. Rapid Commun.* **1995**, *16*, 147.

Phillips Catalysts

See: Supported Chromium Polymerization Catalysts (on Silica, Phillips Catalysts)

Phosphazene Polymers

See: Polyphosphazenes

PHOSPHINATED POLYSTYRENE-BOUND METAL COMPLEXES (Pd and Rh)

Kiyotomi Kaneda
Department of Chemical Engineering
Faculty of Engineering Science
Osaka University

Use of homogeneous metal catalysts on an industrial scale still has practical problems, such as, difficulties associated with corrosion, plating out on the reactor wall and catalyst or ligand recovery from the reaction mixture. The issue of metal separation from products is of interest because of the need to reuse catalysts and the importance of excluding metal and ligand contaminants from products made of high performance materials.

Chemically bonding metal complexes to an insoluble solid is designed to overcome these difficulties for homogeneous catalysts. Catalysts prepared this way have attributes of those usually classified as both heterogeneous and homogeneous, and can be regarded as a new class of hybrid catalysts.[1]

I describe studies on the preparation of phosphinated polystyrene-bound metal complexes (metal:Pd, Rh), and on their use in catalysis of various organic reactions.[2]

APPLICATIONS

The polymer-bound Pd and Rh complexes were applicable to catalysts for various organic reactions.

Catalysis of Polymer-Bound Pd(II) Complex

Hydrogenation[5,6]

The phosphinated polystyrene-bound palladium(II) complexes are extremely efficient catalysts for hydrogenation of conjugated olefins and allylic compounds at ambient temperature and pressure of 1 atm or below. Conjugated dienes selectively reduced to monoenes; cyclopentadiene gave exclusively cyclopentene. The polymer-bound palladium complex is divalent before and after being used for the hydrogenation. It can be reused without loss of the catalytic activity.

The phosphinated polystyrene-bound palladium(II) complex is also an effective catalyst for hydrogenation of acetylenes.

Codimerization[7] *and Cooligomerization*[8]

Selective codimerization of various acetylenes and allylic halides proceeds smoothly with homogeneous palladium(II) catalysts, which provides a useful 1,4-diene synthesis without isomerization to 1,3-dienes.[9]

The polymeric palladium(II) complex catalyzes cotrimerization of styrene and ethylene (1 atm) in the presence of $AgBF_4$.

Catalysis of Polymer-Bound Pd(0) Complex

Carbon-Carbon Bond Formation[3]

The polymer-bound palladium(0) complex has high catalytic activities for carbon-carbon bond formation reactions involving the oxidative addition of organic halides to the metal: vinylic hydrogen substitution with aryl halides; acetylenic hydrogen substitution with aryl halides; and vinylic halogen substitution with Grignard reagents.[10]

Telomerization of Butadiene with Active Hydrogen Compounds[11]

The polymer-bound Pd(0) complex also had high catalytic activities for telomerization of butadiene with various active hydrogen compounds (alcohols, amines, carboxylic acids, phenols, and water) to give 2:1 adducts.

Catalysis of Polymer-Bound Rh(II) Complex[4]

The polymer-bound Rh(II) complex is an efficient catalyst for hydrogenation of monoenes and dienes, but it displayed no activity for acetylene hydrogenation at room temperature under 1 atm or below.

REFERENCES

1. Grubbs, R. H. *CHEMTECH* **1977**, *7*, 512.
2. Heterogenization of Pd complexes using phosphinated polystyrenes also has been studied by Bailor,[(a)] Trost,[(b)] and Pittman,[(c)] independently. (a) Bruner, H. Bailor, J. C. *Inorg. Chem.* **1976**, *12*, 1465. (b) Trost, B. M.; Keinan, E. J. *J. Am. Chem. Soc.* **1978**, *100*, 7779. (c) Pittman, C. U.; Ng, Q. *J. Organomet. Chem.* **1978**, *153*, 85.
3. Terasawa, M.; Kaneda, K. et al. *J. Organomet. Chem.* **1978**, *162*, 403.
4. Kaneda, K.; Terasawa, M. et al. *Chem. Lett.* **1976**, 995.
5. Terasawa, M.; Kaneda, K. et al. *J. Catal.* **1978**, *51*, 406.
6. Terasawa, M.; Kaneda, K. et al. *J. Catal.* **1979**, *57*, 315.
7. Kaneda, K.; Uchiyama, T. et al. *Chem. Lett.* **1976**, 449.
8. Keneda, K ; Terasawa, M. ct al. *Tetrahedron Lett.* **1977**, *34*, 2957.
9. Kaneda, K.; Uchiyama, T. et al. *J. Org. Chem.* **1979**, *44*, 55
10. a) Tsuji, J. *Organic Synthesis with Palladium Compounds*; Springer-Verleg: Berlin, 1980. b) Heck, R. F. *Palladium Reagents in Organic Syntheses*, Academic: London, 1985.
11. Kaneda, K.; Kurosaki, H. et al. *J. Org. Chem.* **1981**, *46*, 2356.

PHOSPHOLIPID BILAYERS

Mario Suwalsky
Faculty of Chemical Sciences
University of Concepción

Phospholipids are large natural molecules that present several interesting properties. With proteins, they constitute the major component of biological membranes. A cell membrane separates two aqueous compartments (plasma and cytoplasma) from each other, where entirely different processes occur. It is not surprising, therefore, that the phospholipids are arranged in asymmetric bilayers.

The lipid bilayers present different physicochemical properties in their internal and external moieties. The apolar interior is a relatively homogeneous region of hydrocarbon chains, where hydrophobic interactions and van der Waals attractive and steric repulsive forces predominate. On the other hand, the hydrophilic head group region contains charged groups in contact with water.

In contact with water, phospholipids spontaneously form into higher molecular aggregates if their concentration surpasses a critical concentration called the "critical micellar concentration."[1] The type of aggregates that will be formed depends on several internal and external factors such as the chemical structure of the phospholipid, its shape, surface charge, concentration, temperature, pH, and ionic environment. Some different types of molecular aggregates that phospholipids can form are shown in **Figures 1** and **2**.

The phase diagrams of phospholipids in water as a function of concentration and temperature are complex.[2,3]

Two types of idealized phase transitions can be described for phase changes that phospholipids can undergo as a function of concentration, temperature, and degree of hydration. The first, thermotropic, occurs mainly for phospholipids that exhibit limited hydration.

The second type of phase transitions are lyotropic.

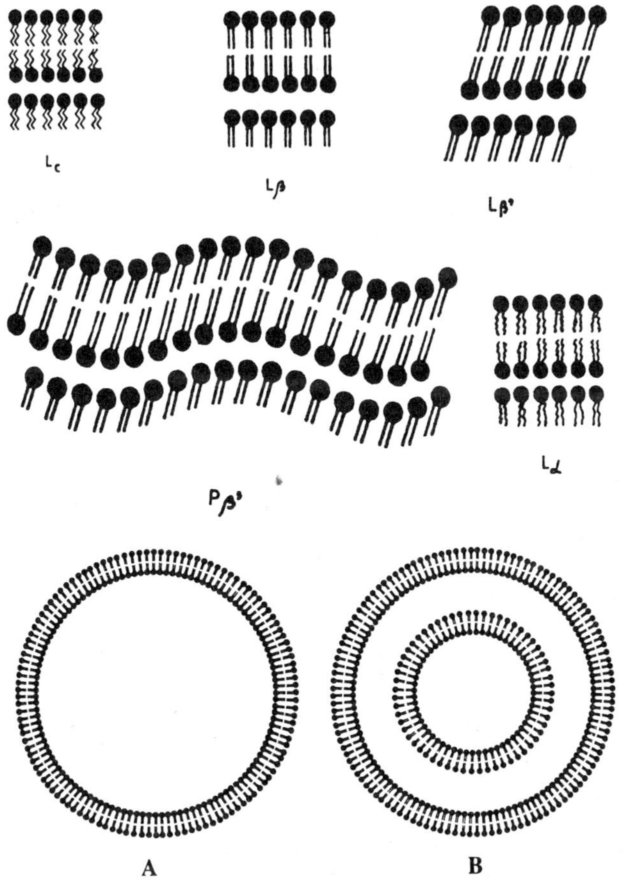

FIGURE 1. Diagrammatic representation of different phospholipid bilayer phases; L_c, crystalline; L_β, gel with untilted chains; L_β', gel with tilted chains; P_β', gel, rippled; L_α lamellar with disordered chains; A, unilamellar vesicle; B, multilamellar vesicle.

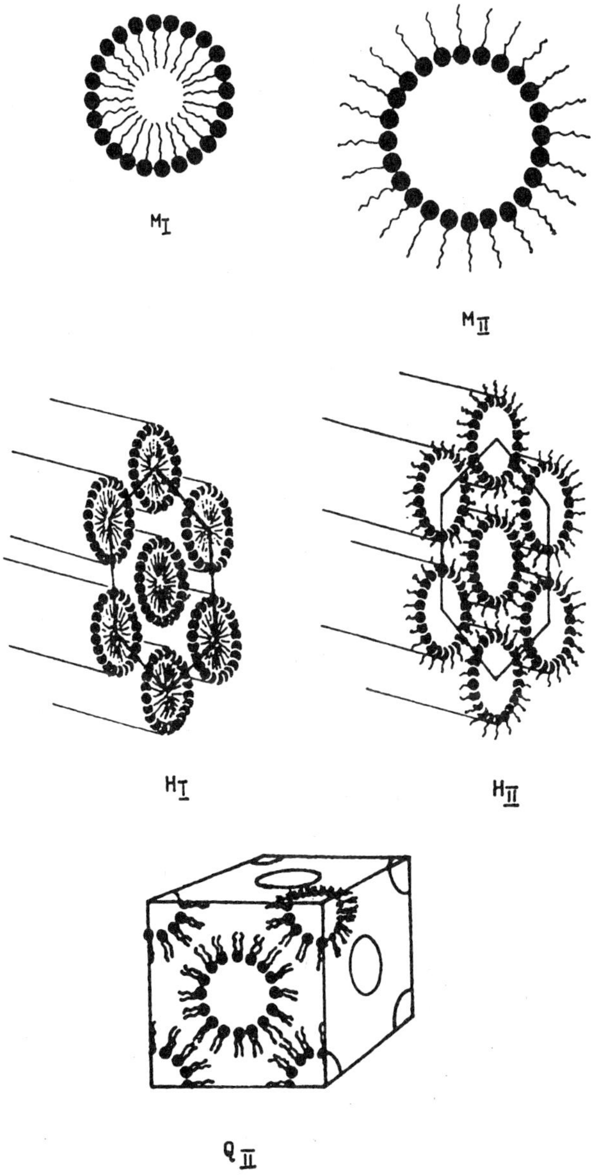

FIGURE 2. Diagrammatic representation of some nonbilayer phases: M_I, normal micelle; M_{II}, inverted micelle: H_I, normal hexagonal; H_{II}, inverted hexagonal; Q_{II}, inverted cubic.

STRUCTURAL CHARACTERISTICS OF PHOSPHOLIPID BILAYERS

Of all the phases described above, the most relevant are phospholipid bilayers. In fact, they are highly related to structural properties and functions of cell membranes, which are highly complex entities.

The phospholipid molecules have the following common features: the head group and the two hydrocarbon chains lie in the same plane; the head group is oriented perpendicularly to the main direction of the two hydrocarbon chains; and, while one of the hydrocarbon chains is completely extended, the second makes an almost right-angle turn in its initial part.

APPLICATIONS

The main applications of phospholipid bilayers are in the pharmaceutical industry and in basic research. In fact, under the form of liposomes they are used as multipurpose vehicles for delivery of drugs, genetic material, and cosmetics. Instead of being diluted in blood, destroyed during absorption, or the source of toxic effects, these compounds can reach target sites in concentrated doses.[4–6]

Moreover, phospholipid bilayers are useful models for studying the mechanism of action of therapeutic drugs,[7] their toxic effects on cell membranes,[8] as well as those other compounds of biological relevance such as pesticides[27] and metallic cations.[9]

ACKNOWLEDGMENTS

Support from research grants 1930504 of FONDECYT and 92.13.94-1 of the Research Council of the University of Concepci6n are gratefully acknowledged.

REFERENCES

1. Sackmann, E. *Biophysics* Springer-Verlag: Berlin **1983**; Chapter 12.2.
2. Seddon, J. M. *Biochim. Biophys. Acta* **1990**, *1031*, 1.
3. Lindblom, G.; Rilfors, L. *Biochim. Biophys. Acta* **1989**, *988*, 221.
4. Ostro, M. J. *Sci. Am.* **1987**, *256*, 102.
5. Weiner, N.; Martin, F.; Riaz, M. *Drug Dev. Ind. Pharm.* **1989**, *15*, 1523.
6. Lasic, D. *Am. Scient.* **1992**, *80*, 20.
7. Suwalsky, M.; Sánchez, I. et al. *Biochim. Biophys. Acta* **1994**, *1195*, 189.
8. Suwalsky, M.; Frías, J. Z. *Naturforsch.* **1993**, *48c*, 632.
9. Suwalsky, M.; Ungerer, B. et al. *Bol. Soc. Chil. Quim.* **1994**, *39*, 315.

PHOSPHOLIPID-IMMOBILIZED SURFACES

Kazuo Yamaguchi*
Department of Materials Science
Faculty of Science
Kanagawa University

Masa-aki Wakita
Kurita Central Laboratories
Kurita Water Industries Ltd.

Shinji Watanabe
Department of Materials Science
Faculty of Engineering
Kitami Institute of Technology

Immobilization of phospholipids on the surface of an insoluble support is achieved by two different approaches. One is to synthesize polymeric latex for a solid support and simultaneously immobilize phospholipids on the surface on the basis of emulsion polymerization, using polymerizable phospholipids as an emulsifier.[1,2] Another approach is to immobilize phospholipids bearing carboxyl and amino groups on the surface of silica gel with a reactive group through covalent linkage.[3,4]

Methods for phospholipids immobilized on the surface and the application of immobilized phospholipids on silica gel are described as follows.

PREPARATION AND PROPERTIES

Polymerizable and Functional Phospholipids

There are two types of polymerizable and functional phospholipids: natural phosphatidylcholines containing two alkyl chains, and unnatural phosphocholine and phosphate bearing a single chain. Styryl group, methacrylate, 1,3-diene and 1,3-diyne moieties, and isonitriles are introduced into phospholipids as polymerizable groups.[6] Functional groups other than polymerizable ones include carboxyl and amino groups,[3,5] azo group for a radical initiator,[7] sulfur compounds such as thiol, sulfide and disulfide chemically adsorbed on gold,[8,9] and chloro- and alkoxysilyl groups reactive to OH group on inorganic solid surface.[10] Among these compounds, the phospholipids described below, as well as others, are shown in **Scheme I**.

Emulsion Polymerization Using Polymerizable Phospholipids

Emulsion polymerization using polymerizable phospholipids with single and double alkyl chains (**1–3**) as an emulsifier has been reported. The phospholipids are used to immobilize them on the resulting latex surface; they also function as stabilizer of latex–particle dispersion into water.

More recently, the unnatural phospholipid bearing methacrylate (**4**) was reported as a comonomer for emulsion polymerization of methyl methacrylate or styrene.[11,12] A phosphocholine with an azo group (**5**) was attempted as a radical initiator for emulsion polymerization.[11]

Polymerizable isothiuronium salts (**6–8**) were used as a comonomer for emulsion polymerization of styrene, followed by alkaline hydrolysis of immobilized isothiuronium salts into thiols.

Immobilization of Functional Phospholipids on Silica Gel Surface

Another approach to immobilization of surface phospholipids is to use phospholipids with carboxyl and amino groups (**9** and **10**) to react with amino and carboxyl groups on silica gel.

Phospholipids immobilized on silica gel have been obtained by coupling of carboxyl groups at the terminal alkyl chain of phospholipid (**9**) with an amino group on silica gel.[3]

APPLICATIONS

Phospholipids immobilized on silica gel were applied to chromatographic purification of proteins and peptides.

REFERENCES

1. Yamaguchi, K.; Watanabe, S.; Nakahama, S. *Makromol. Chem.* **1989**, *190*, 1195.

*Author to whom correspondence should be addressed.

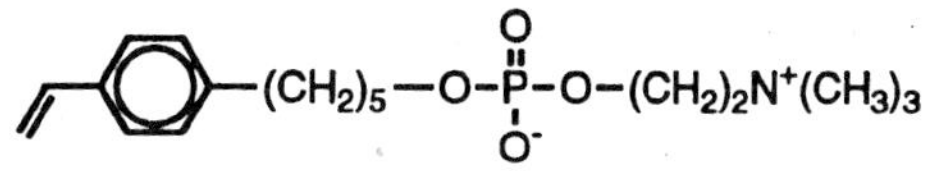

1

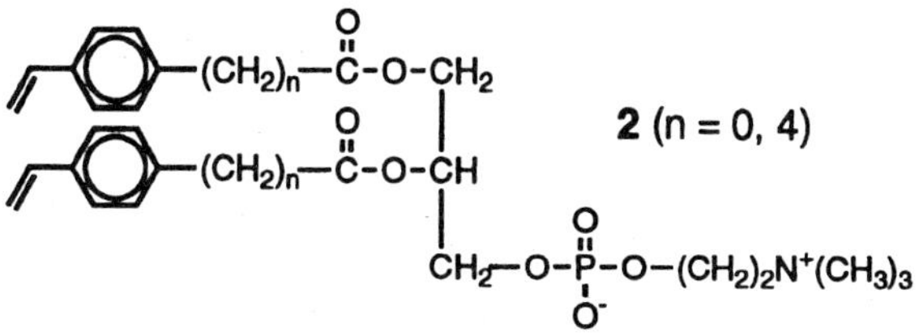

2 (n = 0, 4)

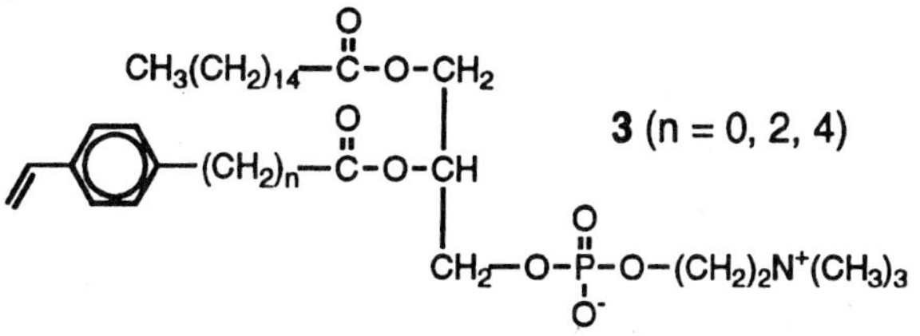

3 (n = 0, 2, 4)

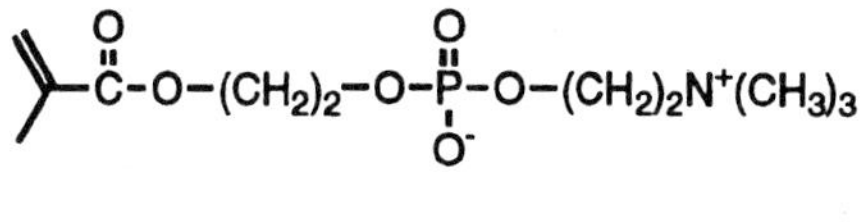

4

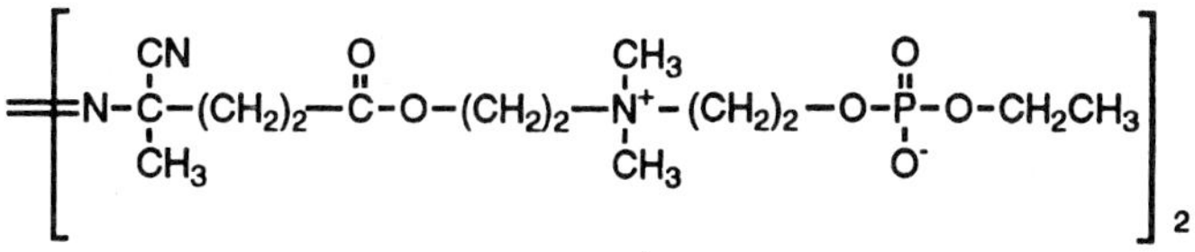

5

I

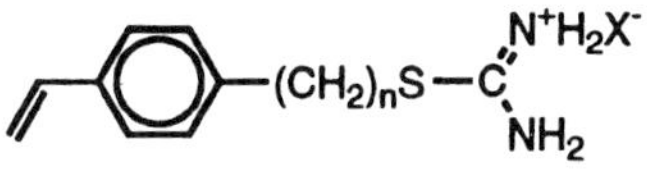

6; n=4, X=Br
7; n=1, X=Cl

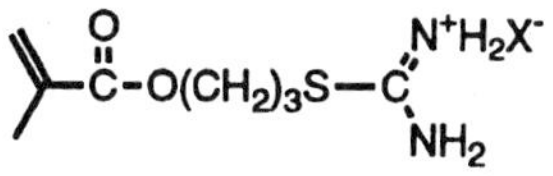

8

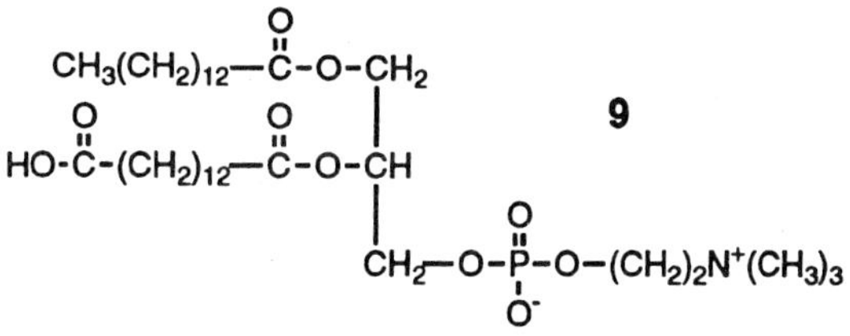

9

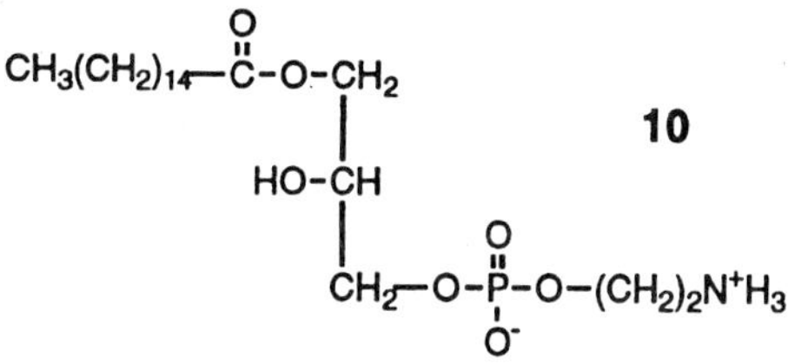

10

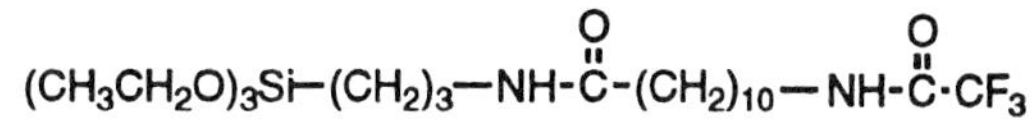

11

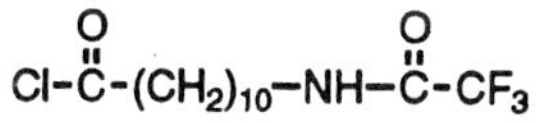

12

2. Watanabe, S.; Ozaki, H. et al. *Makromol. Chem.* **1992**, *193*, 2781.
3. Pidgeon, C.; Venkataram, U. V. *Anal. Biochem.* **1989**, *176*, 36.
4. Markovich, R. J.; Stevens, J. M.; Pidgeon, C. *Anal. Biochem.* **1989**, *182*, 237.
5. Kallury, K. M. R.; Lee, W. E.; Thompson, M. *Anal. Biochem.* **1992**, *64*, 1062.
6. Singh, A.; Schnur, J. M. *Phospholipids Handbook* Marcel Dekker: New York, 1993; Chapter 7.
7. Sugiyama, K.; Ohga, K.; Kikukawa, K. *Makromol. Chem. Phys.* **1994**, *195*, 1341.
8. Fabianowski, W.; Coyle, L. C. et al. *Langmuir* **1989**, *5*, 35.
9. Lang, H.; Duschl, C.; Vogel, H. *Langmuir* **1994**, *10*, 197.
10. Durrani, A. A.; Hayward, J. A.; Chapman, D. *Biomaterials* **1986**, *7*, 121.
11. Sugiyama, K.; Shiraishi, K. et al. *Polym. J.* **1993**, *25*, 521.
12. Sugiyama, K.; Aoki, H. *Polym. J.* **1994**, *26*, 561.

Phospholipids

See: *Phospholipid Bilayers*
Phospholipid-Immobilized Surfaces

Phosphorus-Containing Polymers

See: *Cyclolinear Organophosphazene Polymers*
Cyclomatrix Phosphazene Polymers
Epoxy-Imide, Phosphorylated Epoxy-Imide Polymers
Ferrocene-Backbone Polymers
Flame Retardant Finishing
Flame Retardants (Overview)
Phosphinated Polystyrene-Bound Metal Complexes (Pd and Rh)

Phosphorus-Containing Polymers (Overview)
Phosphorus-Containing Polymers (From Dicarboxylic Acids with C-P Bonds)
Poly(alkylene phosphate)s
Polyphosphazene Block Copolymers
Polyphosphazenes (Solution Properties)
Polyphosphonates and Polyphosphates (Low Temperature Solution Polycondensation)

PHOSPHORUS-CONTAINING POLYMERS (Overview)

Choichiro Shimasaki and Hiromi Kitano
Department of Chemical and Biochemical Engineering
Toyama University

To obtain polymeric materials with various functionalities, syntheses of many kinds of phosphorus-containing polymers have been examined (**Scheme I**). Phosphorus-containing polymers are in general heat resistant, fire resistant, and adhesive. The polymers have been used as fire-retardant materials, ion exchange resins, dental composites, and medical materials. As for phosphorus-containing polymers, their fire-resistant properties are affected not only by the content of phosphorus but also by the molecular structure. Though phosphorus-containing polymers are expensive and labile to hydrolysis, their commercial potential is promising.

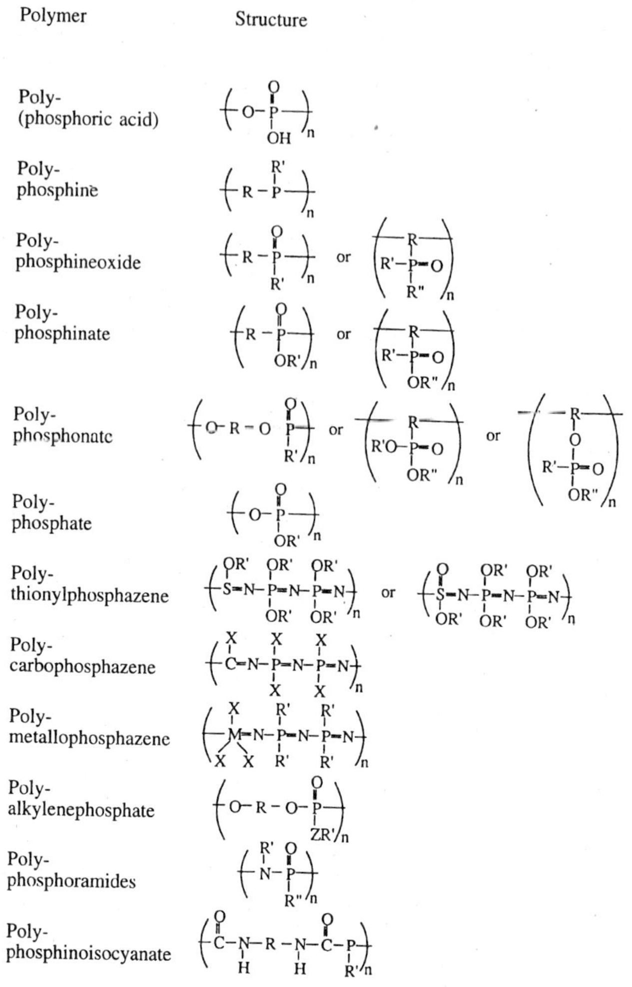
Polymer
Structure
Poly-(phosphoric acid)
Poly-phosphine
Poly-phosphineoxide
Poly-phosphinate
Poly-phosphonate
Poly-phosphate
Poly-thionylphosphazene
Poly-carbophosphazene
Poly-metallophosphazene
Poly-alkylenephosphate
Poly-phosphoramides
Poly-phosphinoisocyanate

SCHEME I. Chemical structure of phosphorus-containing polymers.

POLY(PHOSPHORIC ACID)

Poly(phosphoric acid) was introduced to a compound that could be derived from ordinary (ortho)phosphoric acid by the abstraction of water in a gold or platinum crucible. The polymer has a structure of anhydrous phosphate and can be written as $HO[P(O)(OH)O]_nH$ in general.

Ammonium polyphosphate can be obtained by the direct action of ammonia on poly(phosphoric acid), or by the heating of orthophosphoric acid with urea. When sodium dihydrogen orthophosphate is heated above 240°C, the intermediate pyrophosphate produced below 240°C may be changed to a long-chain material known as "Maddrell's salt" or sodium trimetaphosphate.

Another chain variety is "Kurrol's salt," which is made by slow cooling of a metaphosphate melt with $Na_2O/P_2O_5 = 1$ and seeding under the correct conditions. Kurrol's and Maddrell's salts are both high molecular weight polyphosphates that exhibit the properties of high polymers. Potassium polyphosphate can be made by the same method as sodium salts.

POLYPHOSPHINE

Polymers with polyphosphine structure in the main chain can be obtained by the polyaddition of primary phosphine to non-conjugated dienes.

Polyphosphineoxide

Polyphosphineoxide is a stable polymer and can be used as fire-resistant materials. This polymer can be obtained by the radical polymerization of dichloroalkylphosphine with olefins and subsequent hydrolysis, cationic ring-opening polymerization of cyclic phosphinite, or the Arbuzov reaction of bromophenylphosphinite.

Polyphosphonate

Polyphosphonate homopolymers are obtained by heating phenyl(vinyl)phosphinic acid or its esters. Salts with many metals can be highly polymeric and insoluble in most solvents. A typical formula for these compounds is $[M(R_2PO_2)_n]$, where *M* is a divalent metal such as Mo, Zn, Pb, Be, or Cu, and R is an alkyl or aryl group.

Polyphosphate

Syntheses of polyphosphate are classified into polycondensation and ring-opening polymerization. Pentavalent phosphorus compound reacts with aliphatic alcohol to give polyphosphate. Resinous poly(aryl phosphate)s are prepared by condensation of a mixture of monohydroxyl and dihydroxyl benzene derivatives with phosphorus oxychloride, or by first condensing the monohydroxyl benzene and then the dihydroxyl benzene with phosphorus oxychloride.

A ring-opening polymerization of cyclic phosphate ester, 2-alkoxy-2-oxo-1,3,2-dioxaphosphorane, gives a solid polyphosphate with a large molecular weight in the presence of anionic catalyst (such as n-C_4H_9Li or t-C_4H_9OK) if the purity of starting monomer is sufficiently high.

Polyphosphazenes

The products obtained by reacting PCl_5 with NH_4Cl include various chains of composition $(PNCl_2)_n{\cdot}PCl_4$. Some of the products, such as poly(dichlorophosphazene), can be highly polymeric with n > 15,000 or more and can approach the limiting composition $(PNCl_2)_n$. This product can be made directly by heating cyclic trimer, $(PNCl_2)_3$, in the region of 250–300°C. The polymer is a soft colorless transparent soluble elastomer consisting of simple linear chains with a rather broad molecular weight distribution.

By the substitution of chlorine atoms with suitable moiety in polydichlorophosphazene, which is soluble in organic solvents, moisture-stable polyphosphazenes can be prepared.

Both polydibromophosphazene and polydifluorophosphazene are pale amber or colorless rubbery solids, that can be prepared by heating their respective cyclic trimers or tetramers under suitable conditions.

Polythionylphosphazene

To obtain polythionylphosphazenes, cyclothionylphosphazenes are polymerized, and the active chlorine atoms in the polymer obtained are reacted with nucleophiles in a way similar to that used for polyphosphazenes.

APPLICATIONS

Food and Dentifrice Applications

A number of different chain phosphates, known as condensed phosphoric acids, have been employed to a considerable extent in the manufacture of sausage products. Insoluble sodium metaphosphate, Maddrell's salt, has been used as toothpaste for many years.

Detergent-Building Applications

When straight soaps are used in hard water, insoluble soaps form as flocculant precipitates until the calcium and magnesium ions are used up. When a soap-regenerating agent, condensed phosphates, is added to either a flocculated or dispersed lime soap, the forming ability is reestablished.

Metal Surface Cleaning and Water Treatment Applications

More work has been done on metal cleaning than other materials, because metals must be cleaned during manufacture as well as in their finished form. Long-chain phosphate esters have been utilized in some patented cleaning compositions.

Successful removal of calcium and magnesium ions by precipitation, so-called chemical softening, was the first method used to soften water. All of the chain phosphate anions are strong complexing agents that tie up the cations, such as Ca^{2+} and Mg^{2+}, of hard water so that the usual precipitates attributable to hardness cannot form.

Glasses Application

Phosphate glasses have relatively specialized applications and are used in much more limited quantities. Although simple two-component Na_2O/P_2O_5 glasses are frequently used as water softeners and sequestering agents in detergents, various soluble multicomponent glasses have been patented for the same purpose, such as $MgO/Na_2O/K_2O/P_2O_5$ glasses.

Slowly soluble phosphate glasses are used as micronutrient carriers in agriculture.

Fluorophosphate glasses based on $Al(PO_3)_3$ and LiF appear to have good water resistance and they possess special optical properties.

Fire-Retardant Applications

In current practice, polymeric phosphate esters are used almost exclusively as additives to modify the properties of established organic polymers. In addition to phosphate esters, many organophosphorus compounds, such as polymerized phosphonates, polymerized phosphites, polymerized phosphine oxides, and polymerized phosphinates based on P–C linkages are available for fireproofing and as fire retardants.

Biopolymers Applications

Among biopolymers consisting of polysaccharides, proteins, and nucleic acids, all nucleic acids are phosphate esters. Only a few proteins and saccharides are found in phosphorylated form.

PHOSPHORUS-CONTAINING POLYMERS (from Dicarboxylic Acids with C–P Bonds)

Gueorgui Borissov
Institute of Polymers
Bulgarian Academy of Science

Since 1950, research has been timed at synthesizing phosphorus-containing monomers (PCMs) and introducing them into the polymer molecule. It is reasonable to prepare and apply PCMs similar to those already used. In this respect, phosphorus-containing dicarboxylic acids [PDAs] deserve special attention, because they could participate in common reactions for polymer synthesis.

The preparation of appropriate PCMs is important for synthesizing phosphorus-containing polymers [PCP] with higher fire resistance without impairing their performance characteristics.

PHOSPHORUS-CONTAINING MONOMERS (PCMs)

PDAs with C–P bonds are relatively stable to hydrolysis and can be easily purified. PDAs can be divided into two groups. One includes acids that have two carboxylic groups bonded to the P atom separately, that is, the P atom is in the main chain (**Structure 1**). The outer group includes PDAs in which the two carboxylic groups are bonded with the same organic fragment (i.e., the P atom enters as a substituent in the side chain) (**Structure 2**).

HOOC-R-P(O)(R^{1}RCOOH), **1**

HOOC-R-COOH
|
$R^1P(O)R^2$ **2**

Recent patent data show that PDAs are applied in many fields other than synthesis of polymers. Some of these suggestions include synergistic dispersing agents for pigments and fillers in water,[4] herbicides,[5] sequestrants in cleaning solutions,[6,7] corrosion inhibitors,[8–10] chelating agents for metal ions, softeners for PVC,[11] catalysts,[12] light stabilizers for PVC,[13] intermediates for agrochemicals and polymeric additives,[14,15] finishing agents for textiles,[16] industrial cooling water treatment,[17] and treatment of water or aqueous systems to hinder the deposition of solid material.[18]

PHOSPHORUS-CONTAINING OLIGOMERS (PCOs) AND PHOSPHORUS-CONTAINING POLYURETHANES (PCPus)

There is relatively scant information in the literature about the synthesis of PCOs with terminal hydroxyl groups. They can be divided into two groups: those with phosphorus in the main chain and those with P as a substituent (in the side chain).

PCOs are usually low-melting resinous compounds or light-colored viscous liquids. PCOs with phosphorus in the main chain are highly soluble in chloroform, *m*-cresol, and dimethylformamide.[20] The category also includes those PCOs with P in the side chain in chloroform, dichlorethane, dioxane, and dimethylformamide.[22] Almost all PCOs have been used in the synthesis of phosphorus-containing polyurethanes and copolyurethanes (PCPus).[19–22] PCOs also find application in compositions for solid polyurethane.[21]

PHOSPHORUS-CONTAINING POLYMERS ON THE BASIS OF PHOSPHORUS-CONTAINING DICARBOXYLIC ACIDS WITH C–P BONDS

A considerable number of PDAs with C–P bonds, as well as their derivatives, have been synthesized. PCPs were obtained mainly from PDA with carboxyl groups bonded to phosphorus via aromatic rings or containing aromatic rings.

PHOSPHORUS-CONTAINING POLYESTERS (PCPes) WITH PHOSPHORUS IN THE MAIN CHAIN

There are relatively few phosphorus-containing polyesters (PCPes) prepared from PDAs with P-bonded carboxylic groups via aliphatic fragments.

Phosphorus-containing copolyesters are of considerable interest and data is much more abundant. Most of the research is aimed at synthesis of phosphorus-containing poly(ethylene terephthalate) (PET).[3,24,26,27,29]

PHOSPHORUS-CONTAINING POLYAMIDES (PCPas) WITH PHOSPHORUS IN THE MAIN CHAIN

The synthesis of polyamides on the basis of PDAs and diamines has been well described.[1,2,23,25,28,30,31–39]

The synthesis of polyamidoesters from PDA, diamines, and diols has been reported.[26,40,41] It has been found that PCPas melt at lower temperature, but possess higher solubility and better adhesion compared with similar phosphorus-free polyamides.[36] Almost all PCPas exert improved resistance to combustion.

PHOSPHORUS-CONTAINING POLYMERS WITH PHOSPHORUS IN THE SIDE CHAIN

Data about PCPs based on PDAs with phosphorus in the side chain are scant. Some experiments on the synthesis of PCPes from the diester of 2,3-dialkoxycarbonylpropanephosphinic acid $[(RO)_2P(O)CH_2(COOR^1)]$-CH_2-$COOR^1$ and diols have been carried out.[21] This acid has four ester groups (two alkoxycarbonyl and two alkoxyphosphoryl), which can be transesterificated.

PHOSPHORUS-CONTAINING POLYBENZIMIDAZOLES (PCbis)

PCPbis have been synthesized by the procedures used to prepare phosphorus-free polybenzimidazoles. They are usually yellow-brownish amorphous powders[43] or yellowish nonmelting powders.[44] They possess high thermal stability similar to that of the phosphorus-free derivatives. They decompose at temperatures 500°C[42] or > 400°C.[45] All PCPbis are distinguished by their fire resistance. The PCPbis are characterized by better solubility than that of the phosphorus-free analogues. Poly(-1,3,4-oxadiazole)s (PCPods) are similar to PCPbis. Synthesized PCPods are amorphous substances of relatively good thermal stability and improved fire resistance. A synergistic effect between phosphorus and chlorine has been observed.[46]

CONCLUSION

The most intense research on phosphorus-containing polymers (PCPs) was carried out in the 1960s. The data show that the synthesis of PCP is similar to that used for the preparation of phosphorus-free polymers. The introduction of phosphorus into the polymer molecule does not significantly change its characteristics. However, the improvement of fire resistance is often accompanied by deterioration of other important properties. Light stability, adhesion, and solubility of polymers are also improved by the introduction of phosphorus.

REFERENCES

1. Bride, M. H. et al. *J. Appl. Chem.* **1961**, *11*(9), 352.
2. Petrov, K. A. et al. *Sb., Khim. And Prim. Fosfoorgan. Soedin.*; Izd. Acad. Nauk SSSR: Moskva, 1962; p 285.
3. Sivriev, Hr. et al. *Eur. Polym. J.* **1979**, *15*, 771.
4. Mozanek, Y. et al. Eur. Pat 340 583, 1989, *Chem. Abstract* **1990**, *112*, 200713e.
5. Tokahashi, H. et al. Jpn. Kokai 1976, 7688636; *Chem. Abstract* **1977**, *84*, p 51576a.
6. Auel, I. Ger. Offen 2 333 151, 1975; *Chem. Abstract* **1975**, *82*, 171202p.
7. Maier, L. Ger. Offen 2 617 906, 1976; *Chem. Abstract* **1977**, *86*, 72870g.
8. Block, H. D.; Kollfoss, H. Ger. Offen 2 621 604, 1977; *Chem. Abstract* **1978**, *88*, 89844c.
9. U.S. Patent 4 138 413, 1979; *Chem. Abstract* **1979**, *91*, 62552u.
10. Greaves, B.; Marschall, A. Ger. Offen DE 3 314 008, 1983; *Chem. Abstract* **1984**, *100*, 11436f.
11. Thomm, H. P. et al. Ger. Offen 2 647 042, 1978; *Chem. Abstract* **1978**, *89*, 189946d.

12. Haeussle, P. Ger. Offen 2 902 202, 1980; *Chem. Abstract* **1981**, *94*, 103555e.
13. Ried, W. Y. et al. Eur. Pat. 152 097, 1985; *Chem. Abstract* **1986**, *104*, 130841b.
14. Imomura, K. et al. JP 6222792, 1987; *Chem. Abstract* **1987**, *107*, p 217840m.
15. Imomura, K. et al. JP 6222791, 1987; *Chem. Abstract* **1987**, *107*, p 217841.
16. Wilhelm, D.; Fietier, Y. EP 484 196, 1992; *Chem. Abstract* **1992**, *117*, 152736s.
17. Fabrini, R. M. BR 9101514, 1992; *Chem. Abstract* **1993**, *119*, 119981m.
18. Greaves, B.; Yugham, P. Ger. Offen 3 039 356, 1981; *Chem. Abstract* **1981**, *95*, 156314c.
19. Laszkievicz, B. *Appl. Polym. Sci.* **1967**, *11*, 2295.
20. Borissov, G.; Sivriev, Hr. *Commun. Depart. Chem., Bulg. Acad. Sci.* **1970**, *3*, 465.
21. Borissov, G. et al. *Eur. Polym. J.* **1988**, *24*, 741.
22. Borissov, G.; Kraicheva, I. *Commun. Depart. Chem., Bulg. Acad. Sci.* **1979**, *12*, 599.
23. Petrov, K. A. et al. *Zh. Obshch. Khim.* **1960**, *30*, 3000.
24. Borissov, G. et al. *Visokomol. Soedin.* **1970**, *12*, 620.
25. Petrov, K. A.; Parshina, V. A. *Zh. Obshch. Khim.* **1960**, *30*, 1342.
26. Korshak, V. V. et al. *Izv. Akad. Nauk SSSR, Ser. Khim.* **1964**, 1292.
27. Borissov, G.; Sivriev, Hr. *Macromol. Chemie.* **1972**, *158*, 215.
28. Korshak, V. V. et al. *Plast. Massi.* **1963**, *11*, 11.
29. Borissov, G.; Devedjiev, I. *Commun. Depart. Chem., Bulg. Acad. Sci.* **1972**, *5*, 553.
30. Yokoyama, M.; Konyo, S. *Kogyo Kogaku Zasshi* **1967**, *70*, 1836 *Chem. Abstract* **1962**.
31. Frunze, T. M. et al. *Izv. Akad. Nauk SSSR, Ser. Khim.* **1958**, 783.
32. Korshak, V.V. et al. *Sb. Khimiya and Prim. Fosfoorgan. Soedin*; Izd. Akad. Nauk SSSR: Moskva, 1962; p 247.
33. Medved, T. Ya. et al. *Visokomol. Soed.* **1963**, *5*, 1309.
34. Korshak, V. V. et al. *Visokomol. Soed.* **1963**, *5*, 1309.
35. Yokoyama, M.; Konyo, S. *Kogyo Kogaku Zasshi* **1967**, *70* 1455 *Chem. Abstract* **1968**, 69437u.
36. Pellon, J.; Carpenter, W. G. *J. Polym. Sci.* **1969**, *A1*, 863.
37. Vinokurova, G. M.; Aleksandrova, I. A. *Izv. Akad. Nauk SSSR, Ser. Khim.* **1967**, 362.
38. Vinokurova, G. M. et al. *Izv. Akad. Nauk SSSR, Ser. Khim.* **1967**.
39. Borissov, G.; Sivriev, Hr. *Eur. Polym. J.* **1973**, *9*, 717.
40. Korshak, V. V. et al. *Visokomol. Soed.* **1963**, *5*, 969.
41. Korshak, V. V. et al. *Izv. Akad. Nauk SSSR, Ser. Khim.* **1964**, 897.
42. Korshak, V. V. et al. *Visokomol. Soed.* **1964**, *6*, 1251.
43. Korshak, V. V. et al. *Visokomol. Soed.* **1966**, *8*, 777.
44. Hsu, Y-W. et al. *Ko Fen Tsu T'ung Hsun* **1965**, *7*(4), 252.
45. Sivriev, Hr.; Borissov, G. *Eur. Polym. J.* **1977**, *13*, 25.
46. Borissov, G. et al. *Faserforsch. U. Textiltechnik Z. Polymerforsch.* **1977**, *28*, 601.

PHOTOOXIDATION, POLYOLEFINS

Graeme A. George
School of Chemistry
Queensland University of Technology

The chemical process of photooxidation has been studied in detail[1,2,5] and has been shown to be a free-radical chain reaction. Differences in the rate of photooxidation of polyolefins may be rationalized through an understanding of the rates of the elementary steps in the chain reaction of initiation, propagation, branching, and termination dependent on the polymer structure.[6] By either accelerating or inhibiting these elementary reactions through the use of additives in the polymer, the rate of photooxidation may be controlled. Changes in the mechanical properties of the polymer on environmental exposure result from changes in the length of the polymer chain (molecular weight), the forces between the polymer chains, and the crystallinity and morphology of the polymer due to these chemical reactions. Photooxidative degradation of polyolefins is thus an interplay between free-radical chemistry and physical processes such as oxygen diffusion, free volume changes, and recrystallization.[4]

ELEMENTARY CHEMICAL PROCESSES IN PHOTOOXIDATION

The current understanding of the free-radical oxidation sequence in polymers was developed from studies of liquid hydrocarbons,[7] but more recently the heterogeneous nature of oxidation in the solid state has been shown to control the environmental durability of solid polymers.[8,9]

PHYSICAL EFFECTS IN PHOTOOXIDATION

The solid polyolefin shows many different features of photooxidation that are not related to simple hydrocarbon free-radical chemistry. These include thickness, residual and applied mechanical stress, processing history, and orientation. The low diffusion coefficient of molecules in the solid polymer means that many reaction processes are diffusion controlled, resulting in an oxidation product profile that is highest at the surface and, beyond a characteristic surface layer thickness (dependent on the sample and exposure conditions), declines towards the center.

Although the slow diffusion of oxygen into the bulk of the polymer will result in a heterogeneous oxidation of the solid polyolefin normal to the surface, there is also evidence that the surface is itself very nonuniform in oxidation.[10]

One of the suggested reasons for localized oxidation is the presence of metallic impurities, particularly titanium polymerization catalyst residues.

An applied tensile stress results in increased chain scission on outdoor exposure of polypropylene[11,12] but the effects on tensile strength of thicker samples are minimal. The normal approach to controlling the photooxidative lifetime of polymers has been to incorporate additives that either inhibit or enhance the photochemical free-radical chain reaction.

CONTROLLED PHOTOOXIDATION OF OLYOLEFINS

The major effort in controlling the rate of photooxidation of polyolefins by additives has been in stabilization of the polymer in order to increase the outdoor lifetime in engineering applications. The additives are either pigments, including carbon black, which must be dispersed in the polymer during processing (generally from a masterbatch), or organic additives, which are synthesized with the appropriate pendant alkyl groups to achieve optimum solubility.[2] The stabilizers either lower the rate of initiation of the oxidation, or interrupt the sequence of chemical reactions that lead to chain scission and oxidation product formation.

Recently, there has been considerable research in the accelerated photooxidation of polyolefins, principally for use as agricultural film which will photodegrade to a mulch after the growing season.[3]

Stabilization

The stabilizers present in a commercial polyolefin article will include processing antioxidants (designed to stabilize the polymer in the melt), thermal antioxidants for inhibiting oxidation in the solid state during service, and photostabilizers designed for inhibiting the photooxidative reaction sequence described above. No single compound can meet all of these requirements, and the technology of polyolefin photostabilization involves optimization of stabilizer selection to ensure maximum efficiency (generally through synergism) and elimination of any antagonistic effects that different types of stabilizers may have.

Prooxidants

The most effective prooxidants for controlled-lifetime applications appear to be dithiocarbamate iron(III) complexes. On UV irradiation in oxygen they generate the superoxide anion ($O_2^{-\cdot}$), which rapidly attacks the polyolefin as well as a thiyl radical that abstracts a hydrogen atom from the polyolefin to generate an alkyl radical and so commence the oxidation sequence.

REFERENCES

1. Hawkins, W. L. *Polymer Degradation and Stabilization*; Springer-Verlag: Berlin, 1984.
2. Rabek, J. F. *Photostabilization of Polymers–Principles and Applications*; Elsevier Applied Science: London, 1990.
3. Gilead, D.; Scott, G. *Developments in Polymer Stabilization*; Scott, G., Ed.; Applied Science: London, 1982; Chapter 4.
4. Vink, P. *Degradation and Stabilization of Polyolefins*; Allen, N. S., Ed.; Applied Science: London, 1983; Chapter 5.
5. Al-Malaika, S. *Comprehensive Polymer Science*, Allen, G.; Bevington, J., Eds.; Pergamon: Oxford, 1989; Vol. 6; Chapter 19.
6. MacCallum, J. R. *Comprehensive Polymer Science*, Allen, G.; Bevington, J., Eds.; Pergamon: Oxford, 1989; Vol. 6; Chapter 18.
7. Howard, J. A.; Ingold, K. W. *Can. J. Chem.* **1968**, *46*, 2655 and 2661.
8. Roginsky, V. A. *Developments in Polymer Degradation-5*; Grassie, N., Ed.; Applied Science: London, 1982, Chapter 6.
9. George, G. A.; Ghaemy, M. *Polym. Deg. Stab.* **1991**, *34*, 37.
10. Knight, J. B. et al. *Polymer* **1985**, *26*, 1713.
11. O'Donnell, B.; White, J. R. *Polym. Deg. Stab.* **1994**, *44*, 211.
12. Baumhardt-Neto, R.; De Paoli, M.-A. *Polym. Deg. Stab.* **1993**, *40*, 53, 59.

Photoacid Generators

See: *Chemically Amplified Resists (for Deep-UV Lithography)*
Chemically Amplified Resists (Introduction and Recent Developments)
Chemically Amplified Resists (New Generations)

Photoactive Polymers

See: *Light Harvesting Polymers*
Signal Transduction Composites (Biomaterials with Electroactive Polymers)

Photoassisted Poling

See: *Photochromic Films, Azo Dyes (Photoinduced Anisotropy and Photoassisted Poling)*

Photochemical Switching

See: *Ionic Conductivity Switching (Photochromic Crown Compounds)*

PHOTOCHROMIC FILMS, AZO DYES (Photoinduced Anisotropy and Photoassisted Poling)

Michel Dumont
Institut d'Optique Théorique et Appliquée (Unité Associée au CNRS n °14)

Research on photochromic thin films is aimed at many kinds of applications in optoelectronics and can be divided into two principal categories. The first is information and image processing, in which the index of refraction or absorption is modified by a writing beam and monitored by a reading beam. This class includes optical memories, which need long-term storage capability; spatial light modulators (for incoherent–coherent image conversion); and real-time holography, which needs a fast response (for image correlations, deformation analyses, etc.).

Besides simple photochromism or more or less reversible bleaching (which can be called "intensity" or "population" effects), all these write-and-read properties can be achieved by polarization effects in which a polarized writing beam modifies the angular distribution of anisotropic molecules and produces a macroscopic anisotropy. The reading beam then probes the dichroism or the birefringence of the film. Photoinduced anisotropy (PIA) is known from the beginning of this century under the name of "Weigert effect," but new interest has recently been focused on it, in view of potential applications.

The second class of applications is guided optics: Light is guided in a sandwich planar structure on which an optical circuitry is drawn. Optical bleaching is one of the most interesting ways to draw desired patterns. Organic guiding films are studied for their nonlinear optical properties, which lead to second-harmonic generation and electrooptic effects. Electrooptic effects are very promising in telecommunication networks, for modulating and multiplexing optical signals, in combination with laser diodes and waveguide amplifiers.

PHOTOISOMERIZATION OF AZOBENZENES

Although they are not the only class of photoisomerizable molecules exhibiting orientational properties, we will restrict our discussion to the derivatives of azobenzene. This type of molecules was widely studied in the 1960s and the 1970s, mainly by Russian authors[1] who recognized that PIA of azo dyes in viscous solutions is a consequence of the *cis-trans* photoisomerization of the N=N double bond, which induces rotation of molecules.

Photoisomerization of azobenzenes is discussed in great detail by Rau,[2] who gives a complete bibliography. The simplified scheme is presented in **Figure 1**, which shows DR1 (Disperse Red 1), the most commonly used derivative of azobenzene because of its large β and μ_0 values, which are due to its donor and acceptor groups. The *trans* form of azo dyes presents a strong visible absorption band (peak at 490 nm for DR1). The absorption of one photon raises molecules to electronically excited states; a nonradiative decay brings them back to the ground state, either in *trans* form or in the *cis* form (Figure 1b). At room temperature, the *cis* form spontaneously relaxes toward the *trans* form, with a time constant that depends on the end radicals and on the environment (i.e., the nature of the physical or chemical links with the host medium). The reverse photoisomerization (*cis* to *trans*) also exists, but with a smaller absorption cross-section in the visible range.

GENERAL DISCUSSION

The origin of photoinduced rotations is clearly related to photoisomerization of molecules. Other azobenzene derivatives have shown the same properties.

The mechanism is still under discussion. Some people think that the conformation change pushes the neighbor polymer chains and increases the free volume for a short time, after which the relaxation of the polymer again clamps the dye molecule. The ultimate efficiency of photoassisted poling is not yet established.

RELATED TOPICS

Photoinduced anisotropy is widely studied in connection with liquid crystals by mixing an azo dye with a liquid crystal, by linking azo-dye radicals to liquid crystal polymer chains,[4,5] or by covering the walls of a conventional liquid crystal cell with Langmuir-Blodgett films containing azo dyes:[6,7] the light induced alignment of dye molecules is transferred to the whole liquid crystal sample.

CONCLUSION

Optical ordering of chromophores in polymer films is very attractive for applications, but efficiency and time constants have to be considered carefully in each case. One has to keep in mind that these processes are rather slow compared to pure optical processes. Hole burning, whose time constant depends on the pumping intensity (milliseconds), is fast enough for applications such as holographic correlators. On the opposite, reorientation, which is a random diffusion mechanism, is basically slow (minutes), but the resulting order can be permanent (but erasable and rewritable) in well designed polymers. This process is particularly suitable for information storage and for gratings or waveguides drawing.

Optical poling deserves special attention. Only the illuminated areas of poled, and it is possible to make successive poling with different orientations of the field, because the dark areas do not lose the orientation from the previous poling; these operations are not possible with thermal poling. In particular, for harmonic generation in wave guides, it is possible to pole alternately reversed zones in order to achieve quasiphase matching. This can be done through masks or by moving the sample under a laser spot while modulating the field.[3] This last method is also possible through thermal poling by heating with a strong laser, but thermal diffusion limits the resolution. The main difficulty of this second-harmonic generation method is the absorption of the harmonic, in the useful spectral ranges, by most of the dyes.

Optical mechanisms are well understood and are promising, but the chemical tailoring of polymers, in order to adjust the angular mobility of chromophores to the right value for each application, is the clue to success.

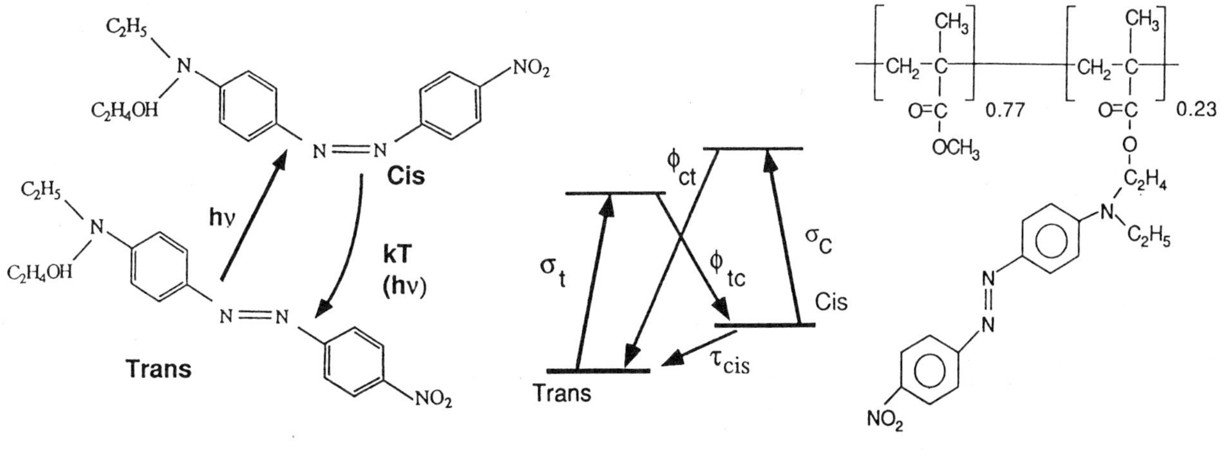

FIGURE 1. *Trans-cis* photoisomerization of DR1 (disperse red 1), with a simplified diagram of energy levels. A copolymer of DR1 and PMMA, poly(methyl methacrylate), is shown on the right-hand side.

REFERENCES

1. Neporent, B. S.; Stolbova, O. V. *Opt. Spektrosk* **1961**, *10*, 287 [*Opt. Spectrosc.* **1961**, *10*, 146]; *Opt. Spektrosk* **1963**, *14*, 624 [*Opt. Spectrosc.* **1963**, *14*, 331] Makushenako, A. M. et al. *Opt. Spektrosk* **1971**, *31*, 557 [*Opt. Spectrosc.* **1971**, *31*, 295]; *Opt. Spektrosk* **1971**, *31*, 741 [*Opt. Spectrosc.* **1971**, *31*, 397].
2. Rau, H. In *Photochemistry and Photophysics*; Rabek, J. F., Ed., CRC: Boca Raton, FL, 1990; Vol. II, Chapter 4, p 119.
3. Yilmaz, S. et al. *Appl. Phys. Lett.* **1994**, *64*, 2770.
4. Petri, A. et al. *Liq. Cryst.* **1993**, *15*, 113.
5. Wiesner, U. et al. *Makromol. Chem.* **1990**, *191*, 2133; Wiesner, U. et al. *Makromol. Chem. Rapid Comm.* **1991**, *12*, 457; *Liq. Cryst.* **1992**, *11*, 251.
6. Sekkat, Z. et al. *Opt. Commun.* **1994**, *111*, 324.
7. Palto, S. P. et al. *J. de Physique II* **1995**, *5*, 133.

Photochromic Materials

See: Ionic Conductivity Switching (Photochromic Crown Compounds)
Photochromic Films, Azo Dyes (Photoinduced Anisotropy and Photoassisted Poling)
Photochromic Optically Active Polymers
Photochromic Poly(α-aminoacid)s
Photochromic Reactions, Polarization-Induced (in Polymer Solids)

PHOTOCHROMIC OPTICALLY ACTIVE POLYMERS

Carlo Carlini, Luigi Angiolini, and Daniele Caretti
Dipartimento di Chimica Industriale e dei Materiali
University of Bologna

Photochromic compounds are able to change their absorption spectra when exposed to light or dark conditions. This process is reversible when the photochromic moieties exist, as usually occurs, in two different forms, whose relative concentration depends on the wavelength of the incident light.[1]

Polymers bearing side-chain photochromic groups, in addition to the color change, have recently attracted a great deal of interest because the photoisomerization of the chromophores can induce reversible variations of their macromolecular structure and hence their physical properties.[2,3] Therefore, photochromic polymeric materials have been considered hopeful candidates for chemical photoreceptors, optical data storage systems in microelectronics,[1,4–8] and devices converting light into mechanical energy.[9,10]

Most biological polymers and photoresponsive systems that are able to convert light into physiological response are optically active.

Few examples of synthetic optically active polymers with photochromic side-chain are known, however, and their potential applications have not yet been exploited. They have mainly been investigated for their photoresponsive properties. Photoresponsive optically active polymers, in this context, are not simply optically active macromolecules with pendant photochromic moieties, but integrated systems where, on one side, the above chromophores show induced dichroic bands and, on the other side, the macromolecules as a whole respond to the photoisomerization by undergoing reversible conformational changes. Our discussion is therefore devoted to the synthesis, structural characterization, and properties of optically active polymers having both photochromic groups and chiral moieties in the side chains.

PREPARATION AND PROPERTIES

The most widely used synthetic approach for obtaining optically active polymers with side-chain photochromic and chiral moieties is based on the copolymerization of an optically active unsaturated monomer with an achiral comonomer containing the proper photochromic moiety. Because of the presence of optically active co-units, the resulting copolymeric macromolecules may therefore exist, at least for sections, in a dissymmetric conformation with a prevailing handedness.[11] As a consequence of cooperative effects, the achiral co-units can in fact be forced to assume a molecular arrangement of one single chirality, thus allowing a chiral perturbation of the electronic transitions of the photochromic groups to be observed. The amount of chain sections having an ordered secondary structure with a prevailing screw sense is usually dependent on main-chain stereoregularity, comonomer structure, copolymer composition, and sequence distributions of the repeating co-units.[11]

The first example is offered by the copolymers of (–)menthyl acrylate (MtA) with *trans*-4-acryloxystilbene (AS)[12] or *trans*-4-vinylstilbene (VS),[13] prepared by free-radical initiation, [poly(MtA-*co*-AS) [*84059-29-0*] and poly(MtA-*co*-VS) [*90168-61-9*], respectively].

Optically active acrylic and methacrylic copolymers bearing side-chain photochromic *trans*-azobenzene chromophores have also been described. In particularly, copolymers of MtA and (–)-menthyl methacrylate with *trans*-4-acryloxyazobenzene as well as with *trans*-4-methacryloxyazobenzene [poly(MtA-*co*-AAB) [*90168-63-1*] and poly(MtMA-*co*-MAB), respectively] have been prepared by free-radical initiation and their photochromic behavior and chiroptical properties at different *trans/cis* azobenzene ratios investigated.[14,15]

trans→cis photoisomerization of the azobenzene moieties produces substantial changes in chiroptical properties of all the copolymer systems.

Optically active methacrylic polymers with well-defined stereoregularity and bearing L-lactoyl (MLH) as well as *trans*-4-phenylazobenzoyl groups (MABH) anchored to the main chain of the same macromolecule through acyl hydrazide moieties, have also been recently described.[16]

These compounds represent the first example of optically active polymers with a hydrocarbon backbone, containing chromophores absorbing in two different UV spectral regions, to be used as conformational probe and photochromic moiety, respectively.

Optically active polymers bearing side-chain phenylazonaphthalene moieties have been prepared by free-radical copolymerization of MtA with either *trans*-4-(1-naphthylazo)phenyl acrylate or *trans*-4-(phenylazo)-1-naphthyl acrylate [poly(MtA-*co*-NAPA) [*154753-50-1*] and poly(MtA-*co*-PANA) [*154753-48-7*], respectively].[17]

Optically active polyisocyanates having side-chain azobenzene chromophore have been synthesized by anionic polymerization of (+)(R)-2,6-dimethylheptylisocyanate with hexyl isocyanate and either 4-phenylazobenzyl isocyanate or 3-(4-phenylazo phenoxy)propyl isocyanate [poly(DMHI-*co*-HI-*co*-PABI) [*146915-03-9*] and poly(DMHI-*co*-HI-*co*-PAPI) [*146915-05-1*], respectively].[18]

Optical rotatory dispersion and CD measurements support the helical polymer structure with one predominant twist sense attributable to the presence of the chiral co-units. The polymeric systems give rise to completely reversible photochemical *trans*→*cis* isomerization of azobenzene chromophores with significant change of CD spectra in the $n \rightarrow \pi^*$ and $\pi \rightarrow \pi^*$ regions of the photochromic group, but without any CD variation in the 200–300 nm region, where the main-chain chromophore is located.

POTENTIAL APPLICATIONS

Optically active photochromic polymers, like other photochromic compounds, may be applied in different technological fields, such as optical information storage, waveguides, chemical photoreceptors, and special devices based on molecular switches and photoresponsive materials. In particular, the photoresponse may be used in several applications, including photomodulation of viscosity, photovariation of NLO properties, photocontrol of drug delivery, and membrane permeability. However, for most of these applications photochromic nonoptically active polymers are the best candidates. Optically active photochromic polymers must therefore be used for special devices sensitive to the reversible change of chiroptical properties of the macromolecules, induced by the photoisomerization of the side-chain photochromic groups. Indeed, photochromic optically active polymers have been proposed[19,20] in optical information storage for designing "chiroptical molecular switches" consisting of a reading system able to reveal a record memorized by light irradiation, by monitoring the optical rotation in the place of absorbance, thus avoiding possible memory erase during the reading process. Finally, optically active photochromic polymers with chiral smectic C* phases may be considered interesting materials for applications implying the photomodulation of their ferroelectric properties.

In conclusion, although no practical applications of synthetic photochromic optically active polymeric materials are presently known, they cannot be excluded in the near future if one takes into account the rapidly increasing sophistication of advanced materials, particularly in microelectronics and optical devices.

REFERENCES

1. Dürr, H.; Bouas-Laurent, H., Eds.; *Photochromism. Molecules and Systems*; Elsevier: Amsterdam, 1990.
2. Kumar, G. S.; Neckers, D. C. *Chem. Revs.* **1989**, *89*, 1915.
3. McArdle, C. B., Ed.; *Applied Photochromic Polymer Systems*; Blackie: Glasgow, 1991.
4. Tazuke, S., Ed.; *Photoresponsive Materials, Proc. MRS Internat. Meeting on Advanced Materials;* Materials Research Society: Pittsburgh, 1989; Vol. 12.
5. Kaempf, G. *Polymer. J.* **1987**, *19*, 257.
6. Ichimura, H. et al. *Makromol. Chem., Rapid Commun.* **1989**, *10*, 5.
7. Schmidt, H. W. *Angew. Chem. Adv. Mater.* **1989**, *101*, 964.
8. Heller, H. G. *IIE Proc.* **1983**, *130*, 209.
9. Zimmermann, E. K.; Stille, J. K. *Macromolecules* **1985**, *18*, 321.
10. Smets, G. J. *Molecular Models of Photoresponsiveness*; Montagnoli, G.; Erlanger, B. F., Eds.; Plenum: New York, 1983; p 281.
11. Ciardelli, F. et al. *Pure Appl. Chem.* **1982**, *54*, 521.
12. Altomare, A. et al. *Polymer* **1982**, *23*, 1355.
13. Altomare, A. et al. *Macromolecules* **1984**, *17*, 2207.
14. Altomare, A. et al. *J. Polym. Sci. Polym. Chem. Ed.* **1984**, *22*, 1267.
15. Altomare, A. et al. *Polym. J.* **1988**, *20*, 801.
16. Angiolini, L. et al. *Makromol. Chem., Rapid Commun.* **1993**, *14*, 771.
17. Angiolini, L. et al. *J. Polym. Sci. Part A: Polym. Chem.* **1994**, *32*, 1159.
18. Müller, M.; Zentel, R. *Makromol. Chem.* **1993**, *194*, 101.
19. Suzuki, Y. et al. *Polym. Bull.* **1987**, *17*, 285.
20. Feringa, B. L. et al. *Tetrahedron* **1993**, *49*, 8267.

PHOTOCHROMIC POLY(α-AMINOACID)S

Osvaldo Pieroni*
CNR-Institute of Biophysics
and
Department of Chemistry and Industrial Chemistry
University of Pisa

Adriano Fissi
CNR-Institute of Biophysics
Galina Popova
Mendeleyev University of Chemical Technology of Russia

Photochromic polymers are able to respond to light by giving reversible variations of their structure and conformation which, in turn, are accompanied by variations of their physical and chemical properties. Photochromic polymers, therefore, may be highly promising materials in devices that can be photoregulated. Poly(α-aminoacid)s and polypeptides are special polymers because they can exist in disordered or regularly folded structures typical of those existing in proteins, such as α-helix and β-structures. When photochromic molecules such as azobenzene and spiropyran compounds are attached to poly(α-aminoacid)s, the photoreactions of the photosensitive molecules can induce order–disorder structural changes in the attached macromolecules.

Because the photostimulated order–disorder structural changes take place as cooperative transitions, these macromolecular systems actually behave as amplifiers and transducers of the primary photochemical event occurring in the photosensitive side chains. Moreover, the various polypeptide structures are characterized by typical values of optical rotatory power and standard circular dichroism (CD) spectra, so the photochromic behavior involves not only variations of the absorption spectra, but is also accompanied by large and reversible changes of optical activity (optical rotatory power and circular dichroism spectra). For these reasons, poly(α-aminoacid)s with photochromic side chains are of growing interest in photooptical technology and are good candidates for optical molecular switches and photomodulable devices.

*Author to whom correspondence should be addressed.

PROPERTIES

Photoregulation of the Macromolecular Structure

Photochromic poly(α-aminoacid)s were first described in 1967 by Goodman et al.,[1] who prepared *p*-phenylazophenylalanine polymers by synthesizing the azo-modified amino acid and then polymerizing the corresponding *N*-carboxyanhydride. The *trans-cis* photoisomerization of the azobenzene units was found to induce optical activity changes. Such changes, however, were not the consequence of photoinduced variations of the macromolecular secondary structure. Photoinduced conformational changes were first reported by Ueno et al.[2–6] in 1977 in copolymers of *p*-phenylazo-benzyl-L-aspartate and benzyl-L-aspartate. Irradiation with UV light and the consequence *trans-cis* isomerization of the photochromic side chains produced an inversion of the sign of the CD spectra, indicating a conformational variation from left-handed helix to right-handed helix.

Photochromic poly(α-aminoacid)s containing spiropyran units in the side chains have been obtained by reacting poly(L-glutamic acid) and poly(L-lysine) with spiropyran reagents.[9–13] All the modified polymers, having various contents of photochromic units, resulted in soluble hexafluoro-2-propanol, where they exhibited an intense reverse photochromism, a photochromic behavior opposite to that usually observed in organic solvents.

Photostimulated "Aggregation-Disaggregation" Processes and Photocontrol of Solubility

Azo-modified polymers of L-glutamic acid undergo aggregation phenomena on aging in trimethylphosphate/water solution, as indicated by the variation of light-scattering intensity and CD spectra during the time.[7,8] The time dependence of CD spectra is characterized by the gradual appearance of an intense CD couplet attributable to the interactions of the azo side chains, together with a progressive distortion of the α-helix pattern. Such distortions are typical of those produced by light-scattering effects as a result of formation of aggregates among helical polypeptide chains.

Photoresponse in Monolayer and LB Systems

Poly(L-lysine) containing 43 mol% azobenzene units has been reported to form a stable monolayer at the water/air interface.[14] When the polypeptide monolayer was kept at constant area, irradiation with 365 nm and 450 nm light, alternately, produced reversible changes of the monolayer's surface pressure.

A detailed investigation of Langmuir–Blodgett films of photochromic polypeptides has been carried out by Menzel et al.,[15] who obtained LB films from poly(L-glutamate)s bearing azobenzene moieties in the side chains coupled to the main chain with spacers of different lengths. The polymers have a "hairy rod" structure, that is, a stiff polymer backbone with flexible side chains.

Photoresponsive Membranes

Reversible photoinduced changes of membrane functions have been observed in polypeptide membranes prepared from poly(L-glutamic acid) containing 12–14 mol% azobenzene groups in the side chains.[16] The water content of the membrane was found to increase by irradiation at 350 nm (*trans*-to-*cis* isomerization) as a result of the different polarity and hydrophobicity between the *trans* and the *cis* isomers. As a consequence, the membrane potential and conductivity were found to change on irradiation.

POSSIBLE APPLICATIONS

As with other photochromic compounds, photochromic poly(α-aminoacid)s can find applications in various fields of science and engineering, including optical data storage, integrated optics, optical processing, and molecular switches. In particular, because the light signal produces large variations of the macromolecular structure, which are accompanied by reversible variations of the physical properties of materials, photochromic poly(α-aminoacid)s are good candidates for photoresponsive materials. Indeed, several photoresponse effects have been reported, including light-induced variations of viscosity and solubility, photocontrol of membrane permeability, photoregulation of binding and releasing of drugs, and photomechanical effects. The inverse temperature transition exhibited by elastin polypeptides was photomodulated in an azo-modified elastinlike polypeptide.[17] Moreover, photochromic poly(α-amino acid)s are optically active polymers, so the photoisomerization reaction of the photochromic groups also produces large and reversible changes of the optical rotatory power and CD spectra. They have been proposed, therefore, as "chiroptical molecular switches" for optical recording, in which a record written by irradiation can be read by monitoring the optical rotation instead of the absorbance. This procedure should prevent the record from being erased by the read-out process.[18,19]

However, although the studies to date are elegant and impressive, photochromic poly(α-aminoacid)s, like other photochromic polymers, have not yet been used in practical systems. They probably require further research work, but in our opinion they will become important materials in the future.

REFERENCES

1. Goodman, M.; Falxa, M. L. *J. Am. Chem. Soc.* **1967**, *89*, 3863.
2. Ueno, A. et al. *J. Polym. Sci., Polym. Lett.* **1977**, *15*, 407.
3. Ueno, A. et al. *Bull. Chem. Soc. Jpn.* **1979**, *52*, 549.
4. Ueno, A. et al. *Macromolecules* **1980**, *13*, 459.
5. Ueno, A. et al. *Makromol. Chem.* **1981**, *182*, 693; Ueno, A. et al. *Makromol. Chem., Rapid Commun.* **1984**, *5*, 639.
6. Ueno, A. et al. *J. Am. Chem. Soc.* **1981**, *103*, 6410.
7. Pieroni, O. et al. *J. Am. Chem. Soc.* **1985**, *107*, 2990.
8. Ciardelli, F. et al. *J. Chem. Soc.; Chem. Commun.* **1986**, 264; Fissi, A.; Pieroni, O. *Macromolecules* **1989**, 22, 1115.
9. Ciardelli, F. et al. *J. Am. Chem. Soc.* **1989**, *111*, 3470.
10. Pieroni, O. et al. *J. Am. Chem. Soc.* **1992**, *114*, 2734.
11. Fissi, A. et al. *Biopolymers* **1993**, *33*, 1505.
12. Cooper, T. M. et al. *Photochem. Photobiol.* **1992**, *55*, 1.
13. Pachter, R. et al. *Biopolymers* **1992**, *32*, 1129; Cooper, T. M. et al. *Appl. Opt.* **1993**, *32*, 674.
14. Malcolm, B. R.; Pieroni, O. *Biopolymers* **1990**, *29*, 1121.
15. Menzel, H. et al. *Thin Solid Films* **1991**, *223*, 181; Menzel, H. et al. *Macromolecules* **1993**, *26*, 3644; Menzel, H. *Macromolecules* **1993**, *26*, 6226.

16. Kinoshita, T. et al. *J. Chem. Soc.; Chem. Commun.* **1984**, 929; Kinoshita, T. et al. *Macromolecules* **1986**, *19*, 51; Sato, M. et al. *Polym. J.* **1988**, *20*, 761.
17. Strzegowski, L. A. et al. *J. Am. Chem. Soc.* **1994**, *116*, 813.
18. Suzuki, Y. et al. *Polym. Bull.* **1987**, *17*, 285.
19. Feringa, B. L. et al. *Tetrahedron* **1993**, *49*, 8267.

PHOTOCHROMIC REACTIONS, POLARIZATION-INDUCED (in Polymer Solids)

Qui Tran-Cong
Department of Polymer Science and Engineering
Kyoto Institute of Technology

Photochromic reactions have been of continuing interest to chemists as well as material scientists in the past several decades because of their potential applicability. A photochromic reaction is defined as a reversible chemical reaction that proceeds forward by irradiation with light and backward by heat.[1] Because of the reversible changes in absorption spectra and thus in refractive index, photochromic reactions have been used for optical information storage[2] as well as wave-guiding materials.[3] From the viewpoint of practical application, polymers doped with appropriate photochromic molecules are potentially suitable candidates among organic materials. For this reason, many studies on photochromic reactions in polymer solids have been performed in an attempt to clarify the correlations between the structures of glassy polymers and the kinetics of photochromic reactions.[4–5]

We will summarize recent efforts to create and control the spatial distribution of refractive index of amorphous polymers by taking advantage of polarization-induced photochromic reactions.

POLARIZATION-INDUCED PHOTOCHROMIC REACTIONS IN BULK POLYMERS

Characteristics of Samples

Samples used in this work are so-called jaw-photochromic molecules.[6] These are bis (9-anthrylmethyl) ether (BAME) derivatives,[7,8] 9-(hydroxymethyl-10-[(naphthyl-methoxy) methyl] anthracene (HNMA),[9] and tetraethyl {3,3} (1,4) naphthaleno (9,10) anthracenophane-2,2,15,15-tetracarboxylate (abbreviated hereafter as cyclophane).[10–12] The chemical structures of these photochromic molecules are illustrated in **Figure 1**. The polymer used in this work is poly(methyl methacrylate) (PMMA, $M_w = 1.4 \times 10^5$, $M_w/M_n = 1.7$).

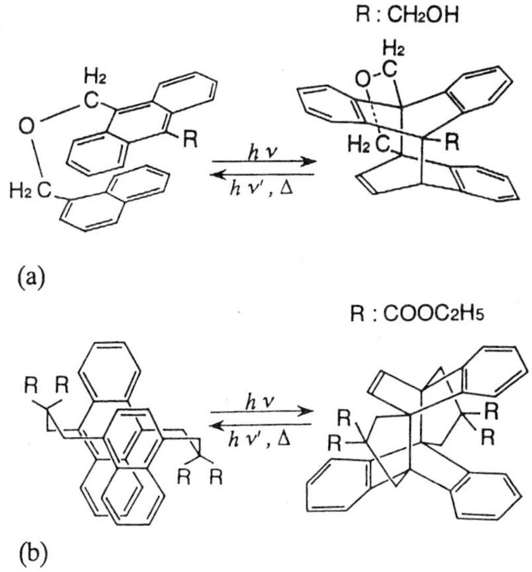

FIGURE 1. (a) 9-(hydroxymethyl)-10-[(naphthylmethoxy) methyl] anthracene (HNMA); (b) tetraethyl {3,3} (1,4) naphthaleno (9,10) anthracenophane-2,2,15,15-tetracarboxylate (cyclophane).

CONCLUDING REMARKS AND PERSPECTIVE

We have shown that a spatial distribution of refractive index of amorphous polymers can be generated and controlled by taking advantage of photochromic reactions induced by linearly polarized light.

The magnitude of the refractive-index distribution can be improved by using photochromic molecules undergoing reactions without requiring intramolecular conformational transitions. Among the well-known organic molecules exhibiting photochromism, such as azobenzene, spiropyrans, indigos, bichromophoric molecules such as BAME and HNMA, or fulgides,[13] naphthaleno anthracenophane is potentially one of the best candidates for this purpose because of its specific cage structure, the large normalized refractive-index difference[2] and the relatively high thermal stability of the photodimers (~ 130°C in PMMA films). The large changes in refractive index of this cyclophane in a number of polymers were also revealed by holographic relaxation (the so-called forced Rayleigh scattering) experiments.[14]

So far, only linearly polarized light has been used to induce the formation of a refractive-index distribution in polymer solids. The combination of circularly polarized light with photochromic reactions in polymer solids has the potential to provide polymers with highly ordered distribution of refractive index. Such polymeric materials would be important in optoelectronics because their optical properties are comparable to those of liquid crystals.

Annealing after irradiation with linearly polarized light can provide a simple method to measure the extremely slow reorientational relaxation of photochromic dopants in glassy polymers. This technique is simple and complementary to the more sophisticated methods such as monitoring the temporal behavior of the second-harmonic generation signals[15,16] or the modified fluorescence depolarization method.[17] Furthermore, the relations between the reorientation dynamics of photochromic molecules with different sizes and various local motions of polymers in the glassy state might provide some helpful guiding principles for controlling the stability of host/guest polymer materials using nonlinear optical dyes.

ACKNOWLEDGMENT

This work is financially supported by the Ministry of Education, Science and Culture, Japan (Grant-in-Aid, Nos. 63044083, 01044081 and 06239238) and the Ogasawara Foundation for the Promotion of Science and Engineering, Japan

(1993). Chemical synthesis of the photochromic cyclophane used in this work was performed in the group of Professor Duy H. Hua (Department of Chemistry, Kansas State University, Manhattan, Kansas).

REFERENCES

1. Brown, G. H., Ed. *Photochromism*; Techniques of Chemistry, Vol. III; Wiley-Interscience: New York, 1971.
2. Tomlinson, W. J.; Chandross, E. A. In *Advances in Photochemistry*; Pitts, J. N. Jr.; Hammond, G. S., Eds.; John Wiley & Sons: New York, 1980; Vol. 12, p 201.
3. Bennion, I. et al. *Radio. Electron Eng.* **1983**, *53*, 313.
4. Smets, G. *Adv. Polym. Sci.* **1983**, *50*, 17.
5. Horie, K.; Mita, I. *Adv. Polym. Sci.* **1989**, *88*, 77.
6. Bouas-Laurent, H. et al. *Pure Appl. Chem.* **1980**, *52*, 2633.
7. Tran-Cong, Q. et al. *Polymer* **1986**, *29*, 2261.
8. Tran-Cong, Q.; Han, C. C. U.S. Patent 4 772 745, 1988.
9. Tran-Cong, Q. et al. *Macromolecules* **1990**, *23*, 3002.
10. Shinmyozu, T. et al. *Chem. Lett.* **1978**, 405.
11. Tazuke, S.; Watanabe, H. *Tetrahedron Lett.* **1982**, *23*, 197.
12. Hua, D. H. et al. *Acta Cryst. C* **1994**, *50*, 1090.
13. Heller, H. G.; Oliver, S. *J. Chem. Soc. Perkin Trans. II* **1981**, 197.
14. Kim, H. et al. *Polym. Commun.* **1991**, *32*, 108.
15. Eich, M. et al. *J. Appl. Phys.* **1989**, *66*, 3241.
16. Hampsch, H. L. et al. *Polym. Commun.* **1989**, *30*, 40.
17. Cicerone, M. T.; Ediger, M. D. *J. Phys. Chem.* **1993**, *97*, 10489.

Photoconducting Polymers

See: Light Harvesting Polymers
Photoconductive Phthalocyanine Polymers
Polysilanes (Overview)
Poly(N-vinylcarbazole)

PHOTOCONDUCTIVE PHTHALOCYANINE POLYMERS

Hong-Zheng Chen,* Mang Wang, and Shi-Lin Yang
Department of Polymer Science and Engineering
Zhejiang University

Photoconductivity is a process in which the charge carrier may be generated and transported under illumination. Photoconductive materials have been widely used in photocopiers, laser printers, and electrophotographic printing plates. Phthalocyanine (Pc) compounds attract much attention as organic photoconductive materials because of their low price, stability, low toxicity and especially broad IR and visible region response.[1–5] However, their insolubility and difficulty in film forming prevent further detailed study, and their photosensitivity and lifetime in the visible region are unfavorable compared with inorganic photoconductive materials. One way to solve these problems is to synthesize phthalocyanine polymers (PPcs). Works in this area have evolved along three principal structural themes (**Figure 1**):

*Author to whom correspondence should be addressed.

- Plane phthalocyanine polymers,[6–17] including the bridging of Pc units by exocyclic groups attaching to the benzene rings (Figure 1a, top) and a single benzene ring being fused into two Pc cycles (Figure 1a, bottom).
- Linear phthalocyanine polymers,[18–23] in which the central metal ions are bridged by ligands to form the polymeric linkages (Figure 1b).
- Polymers containing Pc units,[24–35] in which the Pc units are covalently bonded or inserted into the polymer chains (Figure 1c).

PHOTOCONDUCTIVITY OF PHTHALOCYANINE POLYMERS

Plane Phthalocyanine Polymers

Plane PPcs were the first PPcs to be studied on their synthesis and physical, electrical, and catalytic properties. Their photoconductivity was first reported by Meier in 1985.[36]

Linear Phthalocyanine Polymers

Linear PPcs exhibit photoconductivities because their one-dimensional structures are favorable to single migration of the charge carriers.

Polymers Containing Pc Units

Polymers containing Pc units are neither planar π-conjugated structures like plane PPcs, whose huge π conjugation is favorable to the photoconductive property, nor one-dimensional structures like linear PPcs, whose one-dimensional linear models help the transfer of charge carriers along the chain. But they also exhibit excellent photoconductivity.

We think that there may be three ways to design photoconductive polymer molecules with high photosensitivity.

- Preparing polymers with huge π conjugation systems, which may help the charge carriers move and hence increase photoconductivity.
- Synthesizing polymers of one-dimensional linear structure, which is favorable to single migration of the charge carriers.
- Using polymers with charge transfer, which can enhance photoconductivity rapidly.

REFERENCES

1. Moser, F. H.; Thomas, A. L. *The Phthalocyanines*; CRC: Boca Raton, FL, 1983.
2. Loutfy, R. O. et al. *J. Imag. Sci.* **1985**, *29*, 116.
3. Wagner, H. J. et al. *J. Mater. Sci.* **1982**, *17*, 2781.
4. Wang, M. et al. *Electrophotography* **1989**, *28*, 134.
5. Wang, M. et al. *J. Photochem. Photobiol., A: Chem.* **1990**, *53*, 431.
6. Inoue, H. et al. *Bull. Chem. Soc. Jpn.* **1965**, *38*, 2214.
7. Inoue, H. et al. **1967**, *40*, 184.
8. Inoue, H. et al. **1968**, *41*, 684.
9. Inoue, H. et al. **1968**, *41*, 692.
10. Drinkard, W. C.; Bailar, J. C. Jr. *J. Am. Chem. Soc.* **1959**, *81*, 4795.
11. Boston, D. R.; Bailar, J. C. Jr. *Inorg. Chem.* **1972**, *11*(7), 1578.
12. Wöhrle, D.; Meyer, G. *Makromol. Chem.* **1980**, *181*, 2127.
13. Wöhrle, D.; Preußner, E. **1985**, *186*, 2189.

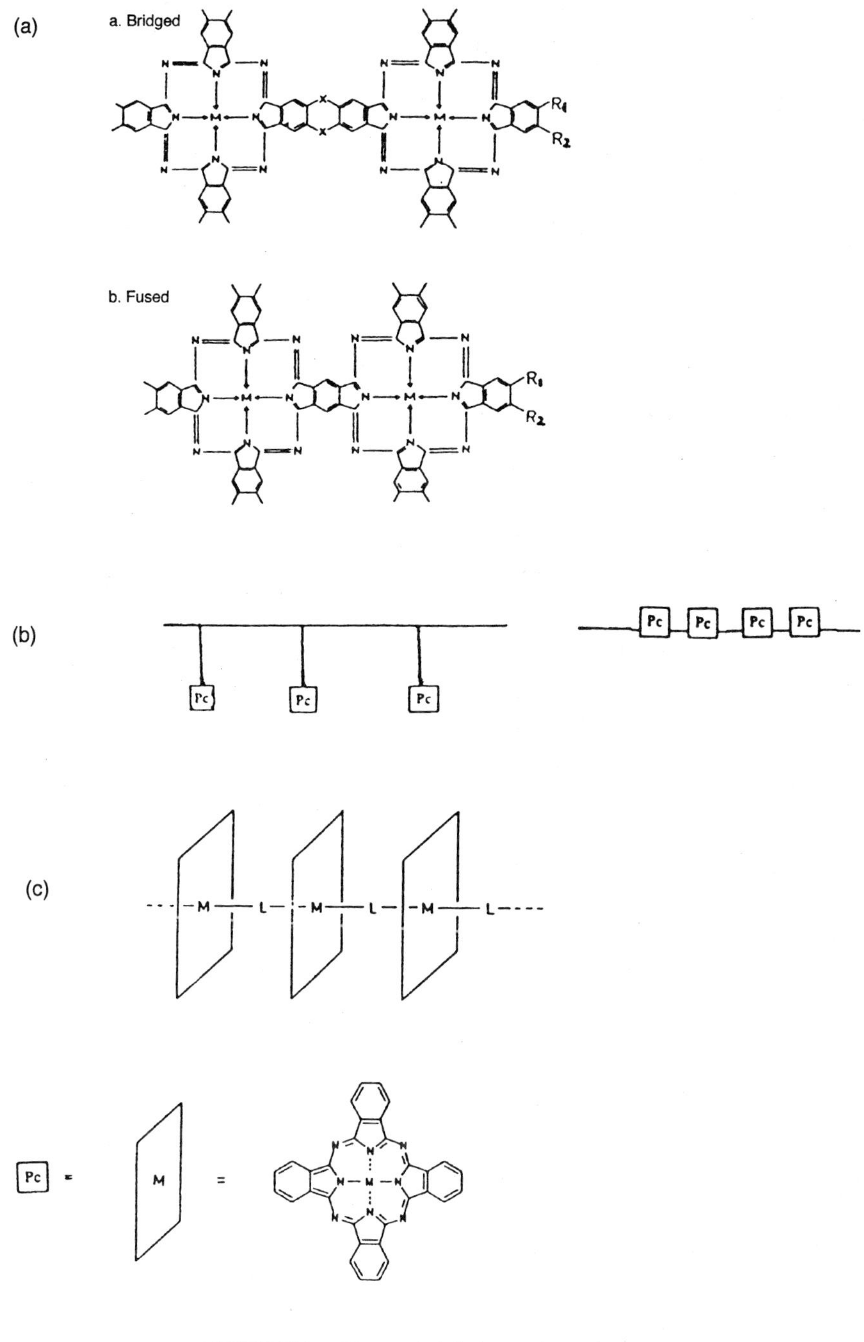

FIGURE 1. Types of phthalocyanine polymers.

14. Wöhrle, D. et al. **1985**, *186*, 2209.
15. Wöhrle, D.; Schulte, B. **1985**, *186*, 2229.
16. Wöhrle, D.; Frawczyk, G. **1986**, *187*, 2535.
17. Wöhrle, D.; et al. **1986**, *187*, 2947.
18. Schneider, O.; Hanack, M. *Angew. Chem. Int. Ed. Engl.* **1980**, *19*(5), 392.
19. Fischer, K.; Hanack, M. *Chem. Ber.* **1983**, *116*, 1860.
20. Skotheim, T. A. *Handbook of Conducting Polymers*; Marcel Dekker: New York, 1986.
21. Metz, J. et al. Naturforsch. Z. **1983**, *38b*, 378.
22. Orthmann, E.; Wegner, G. *Angew. Chem. Int. Ed. Engl.* **1986**, *25*(12), 1105.
23. Keppeler, U. et al. *Angew. Chem.* **1987**, *99*, 349.
24. Kupchan, S. M. et al. *J. Chem. Soc., Chem. Commun.* **1976**, 86.
25. Schutten, J. H.; Zwart, J. *J. Mol. Catal.* **1979**, *5*, 109.
26. Shirai, H. et al. *J. Polym. Sci., Polym. Lett. Ed.* **1979**, *17*, 661.
27. Gebler, M. *J. Inorg. Nucl. Chem.* **1981**, *43*, 2759.
28. Wöhrle, D.; Krawczyk, G. *Polym. Bull.* **1986**, *15*, 193.
29. *Jpn. Kokai Tokkyl Koho*, JP60 201 345 (85 201 345).
30. Higaki, S. et al. *Makromol. Chem.* **1980**, *181*, 565.
31. Chen, H. Z. et al. *J. Appl. Polym. Sci.* **1992**, *46*, 1033.

32. Chen, H. Z. et al. *J. Polym. Sci.* **1993**, *31*, 1165.
33. Chen, H. Z. et al. *J. Appl. Polym. Sci.* **1993**, *49*, 679.
34. Chen, H. Z. et al. *J. Photochem. Photobiol., A: Chem.* **1993**, *70*, 179.
35. Shen, J. L. Master Thesis, Zhejiang University, 1993.
36. Meier, H. et al. *Polym. Bull.* **1985**, *13*, 43.

PHOTOCROSSLINKABLE PHOTOPOLYMERS (Effect of Cinnamate Group Structure)

Jerzy Paczkowski
Department of Chemistry and Chemical Engineering
Technical and Agricultural University

Photopolymers are polymers that undergo phototransformation under direct UV or Vis irradiation and form an insoluble product. Most photocrosslinkable polymers have been prepared by reacting a functionally reactive photosensitive moiety with a polymer backbone.[1–6]

The classic example of crosslinking by an excited state of chromophores occurs in poly(vinyl cinnamate).

The photochemistry of the crosslinking process involves the well-known organic solid-state reaction of the dimerization of cinnamic acid,[7] which forms three different types of crystal. In the crystalline phase the efficiency of the cyclodimerization is determined entirely by the dimensions of the crystal lattice. The Schmidt's[8] photopochemical principle of the [2+2] photodimerization of cinnamic acid correlates the symmetry of the monomer packing with the stereochemistry of the dimer (**Structure 1**).

Crosslinks are formed by photoaddition between the excited cinnamoyl group from one chain and a group-state cinnamoyl group from another chain (**Structure 2**).

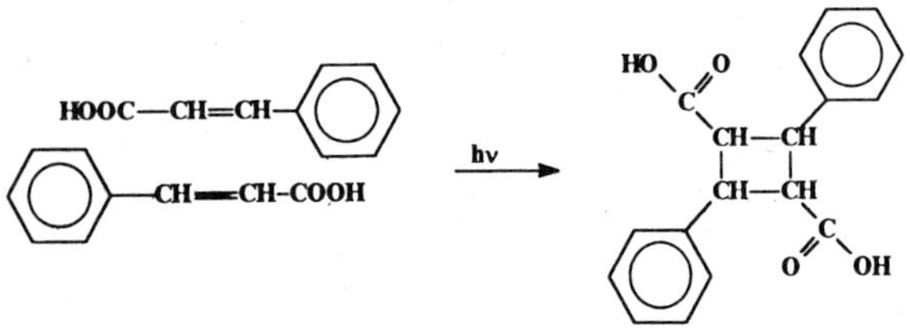

1

Cinnamic acid Truxilic acid

The crosslinking-generating photoreaction, from a practical point of view, is reasonably efficient. The density of crosslinking leading to the insoluble product is very low, and in order to reach the critical hardening point it is sufficient to transform ~ 0.02% of the cinnamate double bonds into crosslinking bonds.[9,10] The combination of critical density of crosslinking and reasonably high quantum yield of the crosslink formation (Φ=0.5±0.05 according to Kirsh,[9] Φ=0.33 according to Tanaka,[11] and Φ=0.14 according to Reiser and Egerton[12]), gives a high photochemical amplification effect. However, the photographic performance of poly(vinyl cinnamate) is poor when it is exposed to a medium-pressure mercury lamp (because of the mismatch between the absorption spectrum of the cinnamoyl group and the spectral emission of the mercury lamp). The improvement of the efficiency of photocrosslinking can be obtained by the following methods: employing a suitable sensitizer,[1,13] introducing as many light-sensitive groups as possible to the polymeric chain, preparing the proper mixture of two polymeric systems,[14] and modifying the photofunctional group[6,15] or the replacement of the cinnamoyl group with a chromophore that absorbs at longer wavelength.[16]

The efficiency of [2+2] cycloaddition for cinnamates may be strongly affected by the structure of the cinnamate pendant group introduced to the polymeric chain. Therefore, it is important to describe the effect of the substitution in the phenyl ring of cinnamic acid, as well as the nature of photochemical and photophysical processes leading to a change in sensitivity of photopolymers containing more complex light-scattering cinnamate groups.

EFFECT OF SUBSTITUTION IN PHENOL RINGS ON PHOTOCHEMICAL PROPERTIES OF PHOTOPOLYMERS WITH CINNAMATE PENDANT LIGHT-SENSITIVE GROUP

General and Spectral Sensitivity

The general structure of the photopolymers with pendant cinnamate derivative group can be described as shown in **Structure 3** where: R is –H, –CH_3, –OCH_3, –Cl, –Br, –I, –NO_2, –OH in ortho, meta and para positions.

Based on data of spectroscopic and photochemical properties, one can conclude that:

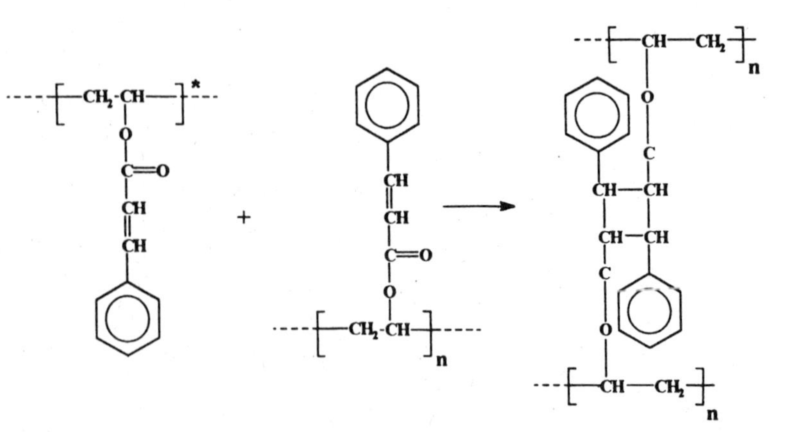

2

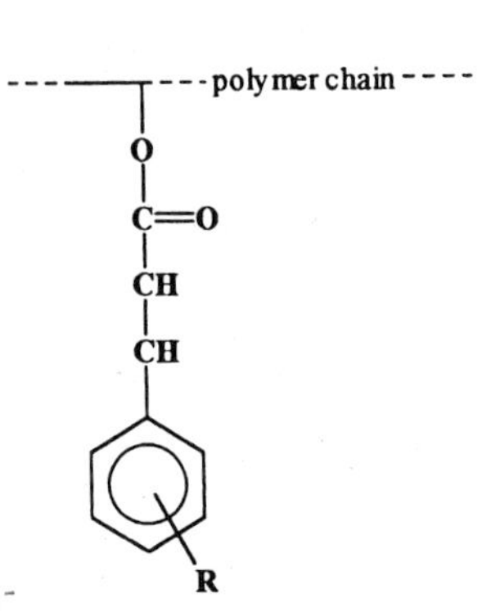

3

- Substituents in the *ortho* position cause a reduction in general light sensitivity. High general sensitivity of 2-iodocinnamate is probably caused by secondary reactions taking place in the irradiated layer.
- Substituents in the *para* position essentially enhances the average reactivity, with the exception of a strong electron-withdrawing substituent NO_2 that indicates the opposite effect.
- There is no general rule describing the effect of substituent in *meta* position on the average sensitivity of the photopolymer.

ACKNOWLEDGMENT

This work supported by the State Committee for Scientific Research (KBN), Grant No. 7 7709 92 03.

REFERENCES

1. Minsk, L. M. et al. *J. Appl. Poly. Sci.* **1959**, *2*, 302; Minsk, L. M.; Van Densen, W. P. U.S. Patent 2 690 966, 1948.
2. Nishikubo, T. et al. *J. Polym. Sci., Part A: Polym. Chem. Ed.* **1983**, *21*, 2035.
3. Iizawa, T. et al. *Macromol. Chem.* **1983**, *186*, 2297.
4. Nishikubo, T. et al. *Macromol. Chem.* **1985**, *186*, 1555.
5. Nishikubo, T. et al. *J. Polym. Sci., Part A: Polym. Chem. Ed.* **1983**, *21*, 2291.
6. Sierocka, M. et al. *Polym. Photochem.* **1984**, *4*, 207.
7. Stobbe, H. *Ber* **1912**, *45*, 3396; Stobbe, H.; Lehfeldt, A. *Ber. B*, **1925**, *58*, 2415.
8. Cohen, M. D. et al. *J. Chem. Soc.* **1964**, 2000; Schmidt, G. M. J. *J. Chem. Soc.* **1964**, 2014.
9. Kirsh, Y. et al. *Zh. Fiz. Khim.* **1964**, *39*, 1886.
10. Nakamura, K. et al. *Bull. Chem. Soc. Jpn.* **1968**, *41*, 1765.
11. Tanaka, H. et al. *J. Polym. Sci., Part A-1* **1972**, *10*, 1729.
12. Reiser, A.; Egerton, P. L. *Photogr. Sci. Eng.* **1979**, *23*, 144.
13. Robertson, E. M. et al. *J. Appl. Polym. Sci.* **1959**, *2*, 308.
14. Watanabe, S.; Ichimura, K. *J. Polym. Sci., Polym. Chem. Ed.* **1982**, *20*, 3267.
15. Sierocka, M. et al. *Vysokomol. Soed.* **1984**, *A26*, 250.
16. Reiser, A. *Photoreactive Polymers. The Science and Technology of Resists*; John Wiley & Sons: New York, 1989.

Photocrosslinking

See: *Microgels, Macrogels (by Photocrosslinking)*
Photocrosslinkable Photopolymers (Effect of Cinnamate Group Structure)
Photocrosslinking (Overview)
Photoinitiators (for Photocuring)
Polyimides (Negative-type Photosensitive Precursors)

PHOTOCROSSLINKING (Overview)

Bengt Rånby*
Department of Polymer Technology
Royal Institute of Technology

*Author to whom correspondence should be addressed.

Baojun Qu
Structure Research Laboratory
University of Science and Technology of China

Wenfang Shi
Department of Applied Chemistry
University of Science and Technology of China

The field of photocrosslinking of monomers, oligomers, and polymers has grown into an important branch of polymer science. It constitutes the basis of a considerable number of commercial applications, not only in conventional areas of thin layer materials, such as coatings, inks, photoresists, adhesives, photoimaging, and photolithography, but also in new domains using photocrosslinked polymeric materials in thick layers, such as insulating materials on wire and cable, hot water pipe, shrinkage tube and hose, and foams.

The term *photocrosslinking of polymer* in this chapter is defined as the process whereby light (UV, visible, or laser light) is used to induce the crosslinking of preexisting high polymers. This means that the light irradiation of polymers carrying more than two reactive groups per chain (or of blends with a photoinitiator or a sensitizer) initiates crosslinking to a three-dimensional network structure. The reaction may also occur with an added crosslinking reagent which carries two or more functional groups per molecule. This article is mainly focused on the crosslinking and applications of highly viscous prepolymers, polymer melts, and solid polymer matrices.

BASIC PRINCIPLE AND PROCESSES

Photocrosslinking requires light absorption in the UV-visible spectral region (typically from 250 to 780 nm) by functional groups in monomers, oligomers, polymers or additives to yield electronically excited singlet and triplet states, which lead to crosslinking directly, or by energy transfer to form reactive intermediates, such as free radicals, reactive cations, and so forth. The reactive species formed subsequently initiate polymerization or undergo crosslinking reactions. Photocrosslinking processes can be induced by the use of photoinitiated crosslinking systems containing photoinitiators, photocrosslinkers, and photocrosslinkable polymers, leading to the crosslinking of macromolecular chains.

PHOTOCROSSLINKING OF MAJOR POLYMER MATERIALS

Polyethylene

Polyethylene (PE) is the thermoplastic material produced in the largest quantities in the world. There has been considerable interest in converting PE to a thermosetting material by crosslinking methods, in order to combine its low cost, easy processing and good properties with dimensional stability. It is well known that a modest crosslinking of PE can considerably improve its thermal stability and resistance to electrical discharge, to solvents, to creep, and to environmental stress-cracking.

The photoinitiated crosslinking of polyethylenes [HDPE, linear LDPE (LLDPE) and LDPE] and subsequent chemical analysis and determination of physical properties have been extensively investigated. The processes involve the use of photoinitiator, mainly aromatic ketones such as benzophenone (BP), anthraquinone (AQ), and their derivatives.[1–19,27]

Polypropylene

The photocrosslinking of polypropylene (PP) is less successful than that of PE due to the higher rate of main-chain scission in PP during UV-irradiation. There are only few reports on the crosslinking of PP at present.

Polystyrene

Photocrosslinking of polystyrene and its copolymers can be carried out by addition of a photoinitiator or incorporated photoactive groups into the polymer.

Poly(vinyl chloride)

Poly(vinyl chloride) (PVC) is highly sensitive to photodegradation because dehydrochlorination and formation of polyene. There are only a few papers on the photocrosslinking of PVC.

Elastomers

Elastomers can be classified into natural rubber (polyisoprene) and synthetic polymers, such as polybutadiene (BR) and styrene-butadiene copolymers (SBR); polyisobutylene and its copolymers; ethylene-propylene rubbers and terpolymers (EPDM); polychloroprene (CR); and fluoroelastomers. Crosslinking of natural rubber and synthetic rubbers is of great industrial importance in obtaining a variety of elastomers.[20]

To date, crosslinking with sulfur or peroxide is the most widely used technique. Commercial interest in photocrosslinkable elastomers started around 1980 and resulted in a number of patents, but it is limited to photocrosslinking of thin films of commercial elastomers.

Polysiloxanes

Siloxanes containing reactive groups, for example (meth) acryloxy groups, are crosslinked by UV radiation (in the presence of a sensitizer) due to the double bonds.[21]

Photocrosslinking rates of silicone elastomers can be substantially increased by incorporating photocrosslinkable pendent vinyl groups to the polysiloxane chain.

Polyesters

Polyesters can be classified as saturated polyesters, alkyd resins, and unsaturated polyesters (UPE). The UPEs are distinguished from alkyds in that their source of unsaturation is other than drying oils. The photocrosslinking of polyesters contains unsaturated groups, which are usually based on maleic anhydride, and diols such as ethylene glycol, propylene glycol, diethyleneglycol, and 1,3-butanediol. The photocrosslinking of UPE has been studied for many years. In 1967, UV-sensitive polyester compositions based on the benzoin ethers of secondary alcohols is photoinitiators were introduced.[22] These systems combined excellent storage stability with high photoreactivity. With technical improvements of UV curing lamps and ancillary equipment, UV-curable UPE resins have been extensively used in various industrial applications.

Poly(meth)acrylates

Photocrosslinking of polyacrylates (PA) and polymethacrylates (PMA), and poly(methyl methacrylate)s (PMMA) has been extensively studied.

Polyurethanes

Photocrosslinking of polyurethanes (PUR) can be carried out by photoactive groups incorporated into PUR.

Poly(vinyl alcohol) and Poly(vinyl acetate)

Photocrosslinking of Poly(vinyl alcohol) (PVA) has been extensively studied. Modified photocrosslinkable PVA with pendent, highly reactive groups, such as stilbazolium groups, styrylpyridinium or styrylquinolinium groups exhibit high photosensitivity, even if the content of photosensitive groups is very low.[23–26]

GENERAL ADVANTAGES AND LIMITATIONS OF PHOTOCROSSLINKING

Advantages and limitations of photocrosslinking depend mainly on the particular applications and the specific requirements. The UV curing of coatings has rapidly advanced new resin formulations, and new applications have been developed because of its unique advantages. Some important features of general validity are as follows:

Advantages

- Rapid crosslinking in a one-step process forms the basis for on-line production units;
- Crosslinking can be carried out at room temperature for thin coatings and for thicker samples in the melt;
- Environmental pollution problems are absent because the use of organic solvents can often be avoided;
- The equipment is inexpensive and simple, UV light sources are readily available, and operation and maintenance are easy; and
- Energy consumption is low.

Limitations

- Limited thickness of the layers which can be photocrosslinked;
- Problems with pigmented layers which scatter and absorb light; and
- Oxygen inhibition or rate decrease may occur with radical initiating systems.

REFERENCES

1. Chen, Y. L.; Rånby, B. *J. Polym. Sci., Part A: Polym. Chem.* **1989**, *27*, 4051.
2. Chen, Y. L.; Rånby, B. *J. Polym. Sci. Part A: Polym. Chem.* **1989**, *27*, 4077.
3. Qu, B. J.; Shi, W. F.; Rånby, B. *J. Photopolym. Sci. Tech.* **1989**, *2*, 269.
4. Fouassier, J. P. *J. Chem. Phys.* **1983**, *80*, 339.
5. Qu, B. J. et al. *Macromolecules* **1992**, *25*, 5215.
6. Qu, B. J. et al. *Macromolecules* **1992**, *25*, 5220.
7. Qu, B. J.; Rånby, B. *J. Appl. Polym. Sci.* **1993**, *48*, 701.
8. Qu, B. J.; Rånby, B. *J. Appl. Polym. Sci.* **1993**, *48*, 711.
9. Yan, Q.; Xu, W. Y.; Rånby, B. *Polym. Eng. Sci.* **1991**, *31*, 1561.
10. Yan, Q.; Xu, W. Y.; Rånby, B. *Polym. Eng. Sci.* **1991**, *31*, 1567.
11. Chen, Y. L.; Rånby, B. *J. Polym. Sci., Part A: Polym. Chem.* **1990**, *28*, 1847.

12. Chen, Y. L.; Rånby, B. *Polym. Adv. Tech.* **1990**, *1*, 103.
13. Yan, Q.; Rånby, B. *Polym. Eng. Sci.* **1992**, *32*, 831.
14. Yan, Q.; Xu, W. Y.; Rånby, B. *Polym. Eng. Sci.* **1994**, *34*(5), 446.
15. Zamotaev, P. V.; Litsov, N. I.; Kachan, A. A. *Vysokomol. Soedin, Ser. B* **1982**, *24*(8), 577.
16. Zamotaev, P. V. et al. *Vysokomol. Soedin, Ser. A* **1985**, *27*(10), 2072.
17. Zamotaev, P. V.; Litsov, N. I.; Kachan, A. A. *Polym. Photochem.* **1986**, *7*(2), 139.
18. Zamotaev, P. V.; Luzgarev, S. V. *Angew. Makromol. Chem.* **1989**, *173*, 47.
19. Zamotaev, P. V.; Chodak, I. *Angew. Makromol. Chem.* **1993**, *210*, 119.
20. Coran, A. Y. In *Science and Technology of Rubber* Eirich, F. R., Ed.; Academic: New York, NY, 1978, Chapter 7.
21. Panda, S. P. *J. Appl. Polym. Sci.* **1974**, *18*(8), 2317.
22. Bayer, Ger. Patent DAS 1 694 149, 1967.
23. Ichimura, K.; Watanabe, S. *J. Polym. Sci., Polym. Lett. Ed.* **1980**, *18*(9), 613.
24. Ichimura, K.; Watanabe, S. *J. Polym. Sci., Polym. Chem. Ed.* **1982**, *20*(6), 1419.
25. Ichimura, K. *J. Polym. Sci., Polym. Chem. Ed.* **1982**, *20*(6), 1411.
26. Ichimura, K. *Makromol. Chem.* **1987**, *188*(12), 2973.
27. Kachan, A. A. *Ukr. Khim. Zh. (Russ. Ed.)* **1990**, *56*(3), 298–306.

PHOTODEGRADABLE PLASTICS

Norman S. Allen
Manchester Metropolitan University

Ricardo Acosta-Ortiz
Centro de Inventigacion en Quimica Aplicada

Photodegradation's effectiveness as a method of combating plastic pollution was recognized in the mid-1960s. Since then numerous patents have been filed on the subject. The photodegradation of plastics can be enhanced by the introduction of chromophoric groups either in the backbone of the polymeric chain or by mixing with polyolefins. The main applications for these types of materials are in packaging, mulching foils in agriculture, and litter bags. Polyethylene is the best degradation option for packaging applications because it has an excellent water barrier and gas permeability.

PHOTO-OXIDATION MECHANISMS

Impurities and inherent defects in plastics make them prone to degradation when they are exposed to ultraviolet light. The rates of photodeterioration of plastics vary with their structure and the nature of the environment. These rates will range from a few months outdoors for unstabilized polypropylene to several years for tetrafluoroethylene. Degradation and oxidation processes are manifested by chemical and physical changes. Examples of physical changes are discoloration, crazing, loss of gloss, erosion cracking, and loss of tensile strength and extensibility. These are associated with chain scission processes although in a number of cases crosslinking may occur. Chemical changes, however, may involve the formation of functional groups specific to a particular type of polymer.

In general, the mechanisms involved in photodegradation of all polymers are free-radical.[1]

Photo-oxidation could be initiated by the excitation of oxygenated products such as hydroperoxides, carbonyl groups, or peroxides developed during processing.[2]

The effect of auto-oxidation on a polymer can be crosslinking or chain splitting. Cleavage of the polymer chain may occur either by photochemical reactions such as Norrish I and Norrish II reactions or by the cleavage of a macroalkoxy radical to render a ketone or aldehyde and a terminal macroradical. Crosslinking results from the reaction between two radicals, thus increasing the molecular weight.

INHERENTLY UNSTABLE POLYMERS

Ethylene Carbon Monoxide Copolymers

One effective method to obtain photodegradable plastics is to copolymerize ethylene monomer with variable amounts of carbon monoxide in the presence of Pd (II) systems to get perfectly alternating copolymers.[3,4]

Vinyl Ketone Copolymers

Another commercial approach to photodegradable plastics is the manufacture of copolymers of vinyl ketones with other suitable monomers like ethylene,[5,6] styrene, methyl methacrylate,[7] and vinyl chloride. The rate of photodegradation depends on the concentration of carbonyl groups. The physical properties of these copolymers are comparable to polyethylene's, making them attractive as packaging material.

Copolymers with Double Bonds

Another approach to developing photosensitive plastics is to reduce the stability of the material by attaching double bonds to the backbone of the polymer. Double bonds are prone to oxidation, thus generating oxygenated compounds such as hydroperoxides and carbonyl groups.

PHOTOSENSITIZERS

Transition Metal Salts

Transition metal ions can effectively catalyze the homolysis of hydroperoxides to give hydroxyl and alkoxyl radicals which are very reactive and can continue with the photooxidation chain reaction.

Transition Metal Complexes

The metallic complex called ferrocene has attracted much interest.[8] This compound is normally quite stable to light but in the presence of organic halides it decomposes to give free radicals.

Among the factors controlling the rate of photo-oxidation of plastics, the behavior of transition metal acetylacetonates[9] and thiolates are interesting.

These antioxidant-photoactivator systems have proved reproducible and reliable in practice and they form the basis of the Scott–Gilead process. In agricultural technology, particularly in hot climates, multicomponent systems provide the increased induction periods required for some crops.

Carbonylic Compounds

Carbonyl compounds were among the first photosensitizers and include aromatic ketones, aldehydes, and quinones. These compounds sensitize through a primary photochemical process of hydrogen atom abstraction from the polymer involving the photoexcited triplet state of the carbonyl group.

CONCLUSIONS

It is evident from the above that several approaches are available for producing photodegradable plastics to solve certain environmental problems. Unlike in biodegradable plastics, effective photosensitized degradation can reduce the molecular weight of the polymer to a form which is acceptable to society. This is especially important in agricultural film usage where the Scott–Gilead process, for example, has proved effective. Time-controlled degradation is all important for adapting to a particular environment.

Finally, one important feature concerning the photoactivity of prodegradant additives relates to their pro-oxidant behavior during thermal processing. This is the ability of the additive to operate as a thermal sensitizer during processing, generating photoactive ketonic and hydroperoxide groups on the polymer backbone. These functional groups may then contribute to the subsequent photosensitized degradation of the polymer material.

REFERENCES

1. McKellar, J. F.; Allen, N. S. *Photochemistry of Man-Made Polymers*; Applied Science: London, England, 1979.
2. Gugumus, F. *Polymer Degradation & Stability* **1993**, *39*, 117.
3. Xu, F. Y.; Chien, J. C. W. *Macromolecules* **1993**, *26*, 3485.
4. Drent, E.; Van Brockhoven, J. A. M.; Doyle, M. J. *J. Organomet. Chem.* **1991**, *417*, 235.
5. Guillet, J. E. U.S. Patents 3 753 962; 3 811 931; 3 853 814; and 3 860 533.
6. Guillet, J. E. *In Polymers and Ecological Problems* Plenum: New York, NY, 1973; p 1.
7. Amerik, Y.; Guillet, J. E. *Macromolecules* **1971**, *4*, 375.
8. Zelenkova, T. P. et al. *Zh. Prikladnoi Khimii* **1978**, *1*, 151.
9. Ogiwara, Y.; Kubota, H.; Kimura, Y. *J. Polym. Sci., Polym. Chem. Ed.* **1978**, *16*, 2865.

Photodegradation

See: *Additives (Types and Applications)*
Environmentally Degradable Polymers
Light Stabilizers (Overview)
Nitrocellulose
Photo-Oxidation, Polyolefins
Photodegradable Plastics
Polysilanes (Overview)
Polystyrene and Derivatives, Photolysis
Poly(vinyl chloride) (Mechanisms of Stabilization)
Ring-Opening Polymerization, Free Radical
Weathering of Polymers (Methodology and Limitations of Accelerated Testing)

PHOTODIMERIZABLE POLY(CROWN ETHER)S

Masamitsu Shirai,* Masahiro Tsunooka, and Makoto Tanaka
Department of Applied Chemistry
College of Engineering
University of Osaka Prefecture

*Author to whom correspondence should be addressed.

Poly(crown ether)s can be expected to show a higher binding ability for large alkali metal cations than monomeric analogs, because of cooperative action of two neighboring crown ether units in the polymer chain.[1] The cooperative action depends on the relative position of the neighboring crown ether moieties attached to the polymer backbone. The cation binding properties of poly(crown ether)s are affected by factors such as the distance between two adjacent crown ether groups in a polymer molecule,[2] the tacticity of polymers,[3] and the chain rigidity and helicity.[4] The fixation of the relative position of the two crown ether groups can be expected to enhance the binding ability of the polymer for large alkali metal cations. Here we report the synthesis and the cation binding properties of polymers which have both crown ether moieties as cation binding sites and cinnamic acid ester moieties as photodimerizable groups.[5] The fixation of the neighboring crown ether units by the phototransformation in the presence of template cations largely enhances the binding ability for large cations.

CATION BINDING PROPERTIES

Table 1 lists the alkali cation binding properties of (**5a-5c**) and (**6a-6c**) (**Scheme I**) at 35°C.

TABLE 1. Extraction of Picrate Salts by 5a-c and 6a-c into a Chloroform Phase[a]

	f (%)					
Cation	**5a**	**5b**	**5c**	**6a**	**6b**	**6c**
Li^+	0.3	0.9	1.9	0.0	0.5	0.6
Na^+	0.9	1.7	3.0	0.4	3.0	3.8
K^+	27.8	44.0	56.9	16.5	28.6	34.9
Rb^+	7.3	35.3	42.9	9.9	26.2	32.8
Cs^+	1.4	7.9	11.6	27.5	70.7	82.4
NH_4^+	0.0	1.6	1.9	0.7	3.1	4.4

[a]At 35 °C. [Picric acid] = 6.55×10^{-5} mol/L, [MOH] = 1×10^{-2} mol/L, [Crown units] = 3×10^{-4} mol/L.

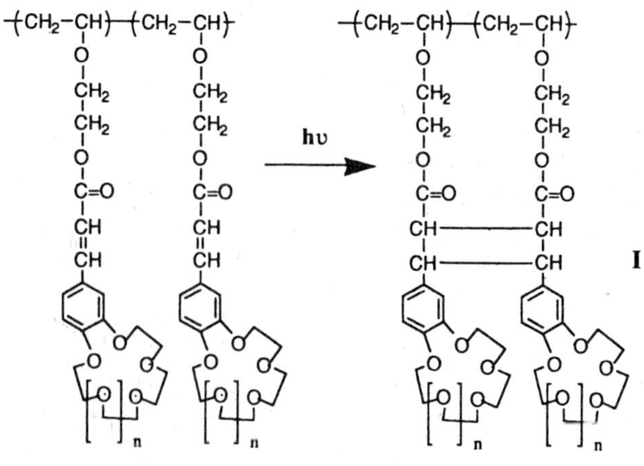

I

5a : n=1
6a : n=2

5b, 5c : n=1
6b, 6c : n=2

REFERENCES

1. Smid, J.; Sinta, R. *Host Guest Complexes Chemistry III*; Springer-Verlag; New York, NY, 1984; Chapter IV.
2. Yagi, K.; Ruiz, J. A.; Sanchez, M. C. *Makromol. Chem., Rapid Commun.* **1980**, *1*, 263.
3. Kimura, K.; Maeda, T.; Shono, T. *Makromol. Chem.* **1981**, *182*, 1579.
4. Anzai, J.; Ueno, A.; Suzuki, Y. et al. *Makromol. Chem., Rapid Commun.* **1982**, *3*, 55.
5. Shirai, M.; Ueda, A.; Tanaka, M. *Makromol. Chem.* **1985**, *186*, 2519.

Photoemulsion Polymerization

See: Poly(vinyl alcohol) (High Molecular Weight)

Photografting

See: Polyethylene Film Gels (Photografting of Hydrophilic Monomers)

Photoinduced Anisotropy

See: Photochromic Films, Azo Dyes (Photoinduced Anisotropy and Photoassisted Poling)

PHOTOINDUCED POLYMERIZATION

Christian Decker
Laboratoire de Photochimie Générale
Université de Haute Alsace

Light-induced polymerization is one of the most efficient processes currently used for producing rapid polymer materials having well-defined characteristics. Essentially all monomers are undergoing polymerization when exposed to UV radiation in the presence of an adequate photoinitiator. With multifunctional monomers, the chain reaction develops extensively within a fraction of a second under intense illumination to generate a highly crosslinked polymer network. By this process, called photocuring, a quasi-instant solidification of a liquid resin can be achieved on order and selectively in the illuminated areas.

Besides its rapidity, the UV-curing technology offers a number of advantages, such as a low energy consumption, no emission of solvent, ambient temperature operations, and the production of polymer materials having tailor-made properties. Photoinitiation is also unique in that it allows a precise control of the duration and rate at which the initiating species are produced. This explains why photoinduced polymerization has attracted so much interest during the last decade and has found a growing number of industrial applications. UV-curable resins are being increasingly employed as fast drying adhesives and protective coatings for all kinds of materials, such as metals, plastics, wood, paper, and optical fibers. In photolithography, high resolution relief images are being produced by patterning of negative working photoresists, either by UV exposure through a mask or by laser direct imaging.

PHOTOINITIATORS FOR LIGHT-INDUCED POLYMERIZATION

The photoinitiator (PI) plays a key role in light-induced polymerization because it governs both the reaction rate and the depth of cure. Great progress has been achieved with the development of ever more efficient photoinitiators for both radical- and cationic-type polymerizable resins.[4,5] Most research efforts in this area are focused on the synthesis of new photoinitiators best suited to the photocuring of pigmented resins and of cationic systems, and to polymerization induced by visible light.

Photoinitiators for Pigmented Systems

Achieving a fast, deep, thorough cure of thick pigmented coatings by UV radiation remains one of the great challenges in photocuring, because the strong absorption and the light-scattering by the pigment particles prevent the penetration of photons in the deep-lying layers.

Visible Photoinitiators Systems

Photopolymerizable systems which can be cured rapidly by visible radiation are needed for some specific end-uses, such as laser direct imaging, holography, or photopolymerization color printing. A large variety of dyes have been tested as initiators or sensitizers, but they generally prove much less efficient than UV photoinitiators.[4,6]

Cationic Photoinitiators

The photoinitiators used in cationic polymerization are mainly diaryliodonium or triarylsulfonium salts which generate Brönsted acids when they are exposed to UV radiation.[5]

UV-CURABLE MONOMERS AND OLIGOMERS

One of the prime objectives in UV-curing chemistry is to create new monomers that will undergo fast and extensive polymerization to yield polymer materials having well-designed properties. Very promising results have been obtained with some novel acrylate, epoxy-silicone, and vinyl ether monomers.

Acrylate Monomers ($R - COO - CH = CH_2$)

Introducing a carbamate or an oxazolidone function in the structural unit of an acrylate monomer was found to greatly increase its reactivity.[7] The light-induced polymerization proceeds substantially faster than with conventional mono- or diacrylate monomers, and almost as fast as with the highly reactive triacrylates.

Epoxy-Silicone Monomers

Photoinitiated cationic polymerization has not experienced as fast a growth in UV-curing applications as free radical polymerization, mainly because of cure speed, product properties, and cost considerations. This situation may change in the future with the appearance of new highly reactive epoxy monomers containing siloxane groups, which were found to polymerize surprisingly fast upon UV irradiation in the presence of an onium salt.[8,9]

Vinyl Ether Monomers ($R - O - CH = CH_2$)

Vinyl ethers are among the most reactive monomers known to polymerize by a cationic mechanism and are therefore, increasingly used in UV-curing applications.[10,11]

LASER-INDUCED POLYMERIZATION

Laser-assisted processing is a very efficient way of achieving ultrafast polymerization of photosensitive resins, because of the distinct advantages of these powerful sources of coherent radiation: very large light intensity, monochromatic emission, high imaging resolution, and great penetration into organic materials.

Multifunctional acrylate monomers have been shown to polymerize after a few-millisecond exposure to an argon ion[1–3] or krypton ion[12,13] laser beam tuned to their continuous emission in the UV or visible range, with formation of a totally insoluble polymer.

Acrylate monomers can also be polymerized quasi-instantly by exposure to pulsed lasers emitting an intense flash of UV radiation during a few nanoseconds.[14,15]

CONCLUSION

Photoinduced polymerization of multifunctional monomers is one of the best methods available to quasi-instantly generate tridimensional polymer networks. Because of its distinct advantages regarding process facility and product quality, the radiation curing technology has found its major openings in the coatings industry and in photolithography. With the advent of laser-curable resins, new applications have appeared in microelectronics, holography, and 3D-prototyping, as well as for the production of fast-drying adhesives and composite materials.

Significant progress has been achieved in UV-curing chemistry, with the recent development of very efficient photoinitiators and highly reactive monomers and functionalized oligomers. The performance of such novel compounds is best assessed by real-time infrared spectroscopy, a technique which permits continuous monitoring of the rapid formation of the polymer, in the millisecond timescale. The important kinetic parameters of different types of light-induced polymerization have thus been determined *in situ* under the same conditions as those used in most industrial applications. Such evaluation, together with mechanistic studies, is necessary for a better understanding and control of the manifold processes involved in ultrafast photocuring reactions.

Considering the many advantages to be gained by using UV- and laser-curable systems, we anticipate that this advanced technology will continue to expand and attract attention in an ever-growing number of industrial sectors.

REFERENCES

1. Decker, C. *Polym. Photochem.* **1983**, *3*, 131.
2. Decker, C. In *Materials for Microlithography-Radiation Sensitive Polymers*; Thompson, L. F.; Willson, C. G.; Fréchet, J. M. J., Eds.; American Chemical Society: Washington, DC, 1984; p 207.
3. Decker, C.; Moussa, K. *Macromolecules* **1989**, *22*, 4455.
4. Deitliker, K. In *Chemistry and Technology for UV and EB Formulation for Coatings, Inks and Paints*; Oldring, P. K. T., Ed.; SITA Technology: London, England, 1991; Vol. 3, p 59.
5. Crivello, J. V.; Deitliker, K. In *Chemistry and Technology for UV and EB Formulation for Coatings, Inks and Paints*; Oldring, P. K. T., Ed.; SITA Technology: London, England, 1991; Vol. 3, p 327.
6. Eaton, D. F. *Adv. Photochem.* **1986**, *13*, 427.
7. Decker, C.; Moussa, K. In *Radiation Curing of Polymeric Materials*; Hoyle, C. E.; Kinstle, J. F., Eds.; ACS Symp. Ser. 417; American Chemical Society: Washington, DC, 1990; p 439; and In *Radiation Curing in Polymer Science and Technology*; Fouassier, J. P.; Rabek, J. F., Eds.; Elsevier: London, England, 1993; p 33.
8. Eckberg, R. P.; Riding, K. D. In *Radiation Curing of Polymeric Materials*; Hoyle, C. E.; Kinstle, J. F., Eds.; ACS Symp. Ser. 417; American Chemical Society: Washington, DC, 1990; p 382.
9. Crivello, J. V.; Lee, J. C. In *Radiation Curing of Polymeric Materials*; Hoyle, C. E.; Kinstle, J. F., Eds.; ACS Symp. Ser 417; American Chemical Society: Washington, DC, 1990; p 398.
10. Lapin, S. C. In *Radiation Curing of Polymeric Materials*; Hoyle, C. E.; Kinstle, J. F., Eds.; ACS Symp. Ser. 417; American Chemical Society: Washington, DC, 1990; p 363.
11. Okamoto, Y.; Klemarczyk, P.; Levandoski, S. *Proc. RadTech '92 Conf.* Boston, MA, 1992; p 559.
12. Decker, C.; Moussa, K. *Makromol. Chem.* **1990**, *191*, 963.
13. Decker, C. *Eur. Polym. Paint Colour J.* **1992**, *182*, 383.
14. Decker, C. *J. Polym. Sci., Polym. Chem. Ed.* **1983**, *21*, 2451.
15. Hoyle, C. E. In *Radiation Curing Science and Technology*; Pappas, S. P., Ed.; Plenum: New York, NY, 1992; p 57.

Photoinitiators

See: *Block Copolymers (From Macroinitiators)*
Dental Polymers (Overview)
Living Radical Polymerization
Photocrosslinking (Overview)
Photoinduced Polymerization
Photoinitiator (for Free Radical Polymerization)
Photoinitiators (for Photocuring)
Photoinitiators, Water-Soluble (for Vinyl Polymerization)
Photopolymer Materials (Development of Holographic Gratings)
Photopolymerization (Initiated by Charge-Transfer Complexes)
Polysilanes (Overview)
Radiation Curing

PHOTOINITIATORS (for Free Radical Polymerization)

H. F. Gruber
Institute of Chemical Technology of Organic Materials
Vienna University of Technology

Photoinitiators (PIs) or photoinitiator systems are molecules or combinations of molecules initiating polymerization reactions when exposed to irradiation. The field of photoinitiated polymerizations is of substantial scientific and industrial interest because of applications such as photoresists, flexographic printing plates, photopolymerizable inks, coatings, and adhesives. These are a few of the more prominent uses that rely primarily on photoinitiated free radical polymerizations. The wide acceptance of UV curing in the polymer industry results not only from the many practical applications but also from increasing economical pressures, calling for the development of solvent-free systems, which are readily transformed at room temperature with low energy requirements.[1–9]

Photopolymerization may be achieved by the utilization of PIs, photocrosslinking agents, and photocrosslinkable polymers. Photocrosslinking agents and photocrosslinkable polymers are used primarily in photoimaging applications, PIs are more broadly used in both photoimaging and UV-curing processes.

PRINCIPLES OF PHOTOINITIATION

A PI can be defined as a molecule which absorbs energy, either directly or indirectly, from a photon and subsequently initiates polymerization. The PI is consumed during the initiation process. PIs absorb light in the UV-visible spectral range (250–450 nm) and convert this light energy into chemical energy in the form of reactive intermediates, such as free radicals ($R\cdot$) or reactive cations (R^+), which subsequently initiate polymerization of monomers and oligomers.

Most applications of UV curing are carried out in air. Oxygen generally acts to inhibit radical photopolymerizations by quenching the triplet excited PI as well as by reaction with radicals.

Another problem in photopolymerization concerns curing of pigmented coatings. Pigments have their own intense absorption which overlaps to a great extent with the absorption of the PI. They thus compete with the PI for the available light and inhibit the efficiency generation of initiating radicals. Pigments also aggravate the light penetration problem by scattering light, thus preventing it from traveling as far into the film as even the Lambert–Beer law would predict.

PHOTOINITIATORS FOR FREE RADICAL UV-CURING

PIs are generally divided into two classes, namely Type I and Type II PIs according to the basic mechanism.

Type I Photoinitiators

Upon irradiation Type I PIs undergo fragmentation to yield initiating radicals. The majority of Type I PIs are aromatic carbonyl compounds containing suitable substituents which facilitate direct photofragmentation.

Type II Photoinitiators

Type II PIs undergo mainly two reaction pathways: hydrogen abstraction by the excited initiator, and photoinduced electron transfer, followed by fragmentation.

TYPE I PHOTOINITIATORS

Benzoin Derivatives

Benzoin derivatives, notably benzoin ethers, were among the earliest patented PIs. They are frequently used as PIs for vinyl polymerization and they comprise the greatest part of the total amount of commercially used free radical PIs.[10–13]

Benzil Derivatives

Benzilketals represent a very important class of PIs including 2,2-dimethoxy-2-phenylacetophenone (DMPA), which is commercially available and one of the most frequently used PIs.[14–18]

α-Hydroxyalkylphenones

The most important commercially available products of this class are 2-hydroxy-2-methyl-1-phenylpropan-1-one (HMPP) and 1-hydroxy-cyclohexyl-phenylketone (HCPK) which are promoted by their manufactures for their high storage stability and nonyellowing characteristic in contrast to benzoin ethers and DMPA.[19–22]

α-Aminoalkylphenones

α-Aminoalkylphenones are highly reactive Type I PIs, the most efficient of which are submitted at the 4-position of the benzoyl moiety by strongly electron donating substituents.[23,24]

Acylphosphinoxides

Acylphosphineoxides and acylphosphonates are a class of PIs with high efficiency, particularly suitable for photocuring of thick films (up to 20 mm), white pigmented formulations based on acrylates and styrene containing unsaturated polyester (UP) resins, as well as glass fiber-reinforced polyester laminates with reduced transparency.[21,25–27]

TYPE II PHOTOINITIATORS

The activity of most bimolecular PIs for UV curing relies on the well-known photoreduction of aromatic ketones.

Benzophenone–Amines

The readily available inexpensive benzophenone–amine systems are the best known Type II PIs used in low price UV-curing applications.[28–30]

Thioxanthone–Amines

Thioxanthone and its derivatives are more efficient PIs than benzophenones with respect to the increased absorption in the range from 380 to 420 nm.

PHOTOINITIATORS FOR VISIBLE LIGHT CURING

There is a wide variety of dyes or colored compounds for use as light-absorbing components in PI systems, permitting the initiation of polymerization by means of visible light; nevertheless, systems of this kind have not yet received widespread commercial utilization.[31–33] Some visible photoinitiating systems suffer from lower storage stability than is expected from a UV-curable system, requiring handling of initiators and formulations under safety light such as yellow or red light. Visible PIs are used for photopolymer printing plates and dental components, for example.

MACROMOLECULAR PHOTOINITIATORS

Macromolecules containing covalently bonded photoinitiating groups have received considerable attention.[34–41] They combine the properties of a PI with the typical properties of a polymer. Compared to low molecular PIs they exhibit various advantages such as low volatility (hence no odor problems and reduced toxicity), low migration tendency of the PI and photoproducts, and therefore reduced yellowing.

WATER SOLUBLE PHOTOINITIATORS

Although most of the conventional light-curable systems contain no solvents, some drawbacks arise from the use of the reactive diluents and monomers: relatively high viscosity of the formulations, irritation and odor problems, and migration of residual monomers. There is a new trend towards the development of water-borne UV-curable formulations, using water-thinnable systems, emulsified oligomers, aqueous dispersions, or aqueous solutions of monomers.[42–44]

CONCLUSIONS

Photoinduced radical polymerization is a technology enjoying increasing commercial interest because of its potential in a wide variety of applications. Applications are emerging in the coating industry (anticorrosion coatings, protective varnishes, highly flexible and thermally resistant coatings on electric sheets, varnishes for glass cloth), in laminates (UV-laminated adhesives, glass laminates), and in the graphic arts (UV-finishing of cosmetic and external food packaging, varnishing, UV marking inks). We expect new applications in leather finishing, pressure-sensitive adhesives, composites, dental sealant materials, coatings for different glass substrates, three-dimensional curing of objects, and modeling. Another promising area is the application of laser-induced processes in monomeric and polymeric materials to photoimaging, microelectronics, three-dimensional machining, holographic optical elements, or information recording and storage.[45,46] Development of new reactive monomers and oligomers and water-borne UV-curable formulations will presumably accompany the market growth in the field of photopolymerization. Many basic studies such as photopolymerizations in direct and reverse micelles and microemulsions, vesicles, multilayers, and crystals in the solid state, topochemical photopolymerization; UV-curing of encapsulated liquid crystals, magnetic field effects on emulsion photopolymerization; photoinitiated charge transfer polymerization; and concurrent cationic and radical photopolymerization and surface photopolymerization of adsorbed or oriented molecules are also expected to become important as high-tech applications.

REFERENCES

1. Dietliker, K. K. In *Chemistry & Technology of UV & EB Formulation for Coatings, Inks & Paints*; SITA Technology: London, UK, 1991; Vol. 3.
2. Gruber, H. F. *Prog. Polym. Sci.* **1992**, *17*, 953.
3. Pappas, S. P. In *Encyclopedia of Polymer Science and Technology*; John Wiley & Sons: New York, NY, 1988; Vol. 11, p 186.
4. Fouassier, J. P. In *Polymerization Science and Technology*; Allen, N. S., ed.; Elsevier: London, UK, 1989.
5. Fouassier, J. P. *Prog. Org. Coat.* **1990**, *18*, 229.
6. Allen, N. A.; Edge, M. *J. Oil Color Chem. Assoc.* **1990**, *73*(11), 438.
7. Bett, S. J.; Dworjanyan, P. A.; Garnett, J. L. *J. Oil Color Chem. Assoc.* **1990**, *73*(11), 446.
8. Kloosterboer, J. K. *Adv. Polym. Sci.* **1988**, *84*, 3.
9. Senich, G. A.; Florin, R. E. *J. Macromol. Sci. Rev. Macromol. Chem. Phys.* **1984**, *C24*(2), 239.
10. Hutchinson, J.; Ledwith, A. *Polymer* **1973**, *14*, 405.
11. Kuhlmann, R.; Schnabel, W. *Polymer* **1977**, *18*, 1163.
12. Lissi, E. A.; Garrido, J. *J. Polym. Lett. Ed.* **1984**, *22*, 391.
13. Hageman, H. J.; Jansen, L. G. *Makromol. Chem.* **1988**, *189*, 2781.
14. Kirchmayr, R.; Berner, G.; Husler, R. et al. *Farbe u. Lack* **1982**, *88*, 910.
15. Lougnot, D. J.; Fouassier, J. P. *Makromol. Chem. Rapid Commun.* **1983**, *4*, 11.
16. Kurreck, H.; Kristie, B.; Lubitz, W. *Angew. Chem. Int. Ed.* **1984**, *23*, 173.
17. Phan, X. T. P. *J. Radiat. Curing* **1986**, *13*, 18.
18. Jaegermann, P.; Lendzian, F.; Rist, G. et al. *Chem. Phys. Lett.* **1987**, *140*, 615.
19. Herz, C. P.; Eichler, J. *Farbe u. Lack* **1979**, *85*, 983.
20. Ohngemach, J.; Neisius, K. H.; Eichler, J. et al. *Kontakte* **1980**, *3/80*, 15.
21. Baxter, J. B.; Davidson, R. S.; Hageman, H. J. et al. *Polymer* **1988**, *29*, 1575.
22. Böumer, W. *Kontakte* **1989**, *3/89*, 42.
23. Dietliker, K. K. In *Chemistry & Technology of UV & EB Formulation for Coatings, Inks & Paints*; SITA Technology: London, UK, 1991; Vol. 3, p 151.
24. Desobry, V.; Dietliker, K. K.; Huesler, R. et al. (Ciba Geigy AG) Eur. Patent EP 284 561, 1988.
25. Sumiyoshi, T.; Schnabel, W.; Henne, H. *J. Photochem.* **1986**, *32*, 119.
26. Baxter, J. E.; Davidson, R. S.; Hageman, H. J. *Polymer* **1988**, *29*, 1569.
27. Baxter, J. E.; Davidson, R. S.; Hageman, H. J. *Eur. Polym. J.* **1988**, *24*, 419 and 551.
28. Dietliker, K. K. In *Chemistry & Technology of UV & EB Formulation for Coatings, Inks & Paints*; SITA Technology: London, UK, 1991; Vol. 3, p 189.
29. Li, T. *Polym. Bull.* **1990**, *24*, 397.
30. Mateo, J. L.; Bosch, P.; Catalina, F. et al. *J. Polym. Sci. Chem.* **1991**, *29*, 1955.
31. Dietliker, K. K. In *Chemistry & Technology of UV & EB Formulation for Coatings, Inks & Paints*; SITA Technology: London, UK, 1991; Vol. 3, p 228.
32. Timpe, H. J.; Neuenfeld, S. *Kontakte* **1990**, *2*, 28.
33. Eaton, D. F. *Adv. Photochem.* **1986**, *13*, 427.
34. Carlini, C.; Gurzoni, F. *Polymer* **1983**, *24*, 599.
35. Carlini, C. *Br. Polym. J.* **1986**, *18*, 236.
36. Ahn, K. D.; Ihn, J. J.; Kwon, I. C. *J. Macromol. Sci. Chem.* **1986**, *A23*, 355.
37. Köhler, M.; Ohngemach, J. *Polym. Paint Color J.* **1988**, *178*, 203.
38. LiBassi, G.; Nicora, C.; Cadona, L. et al. (Fratelli Lamberti) Eur. Patent Appl. 161 463, 1985.
39. Chiang, W.; Chan, S. *Angew. Makromol. Chem.* **1990**, *179*, 57.
40. Önen, A.; Yagci, Y. *J. Macromol. Sci. Chem.* **1990**, *A27*, 743.
41. Gruber, H. F.; Klos, R.; Greber, G. *J. Macromol. Sci. Chem.* **1991**, *A28*(9), 925.
42. Lougnot, D. J.; Fouassier, J. P. *J. Polym. Sci. Chem.* **1988**, *26*, 1021.
43. Arnoldus, R. *Polym. Paint Color J.* **1990**, *180*, 434.
44. Fouassier, J. P.; Burr, D.; Weider, F. *J. Polym. Sci.* **1991**, *29*, 1319.
45. Fouassier, J. P. *J. Photochem. Photobiol.* **1990**, *A51*, 67.
46. *Lasers in Polymer Science and Technology: Applications*; Fouassier, J. P.; Rabek, J. F., Eds.; CRC: Boca Raton, FL, 1989.

PHOTOINITIATORS (for Photocuring)

Norman S. Allen
Chemistry Department
Manchester Metropolitan University

During the last twenty years the field of radiation curing science and technology has grown from a subject of esoteric research specialities into major industrial developments and is now a field of central importance in polymer science and technology. Applicàtions include resists, barrier coatings, paper, board and metal coatings, dentistry and imaging science, and printing. Inherent in these technologies is the use of a photoactivator system which is capable of absorbing the incident UV and visible radiation wavelengths used for converting a monomer or prepolymer system into a crosslinked network.[1-5] Two areas of importance are the development of monomeric and polymeric photoinitiators with reactive functionalities capable of co-reaction and the extension of their absorption spectrum, thus imparting greater spectral sensitivity to the visible region of the spectrum. The latter is now of utmost importance particularly for laser-induced polymerizations and holography.

THE COATING

Several types of UV-light curable monomers and prepolymers have been developed. In the industry there are two basic types of systems. The first are called diluent monomers and are divided into what may be called first and second generation types. First generation types are typical simple mono- and difunctional structures such as 2-ethylhexyl acrylate, methyl methacrylate, and styrene. The second generation type are usually di- and multifunctional acrylates such a glycerol propoxylate triacrylate.

PHOTOINITIATORS AND MECHANISMS

To induce the photopolymerization or photocrosslinking of an acrylated system two types of initiators are available. The first induces a free-radical chain process in which low molecular weight monomers and prepolymers are converted by absorption of UV-visible light into highly crosslinked, solvent, and chemically resistant films. The second type of initiator polymerizes via an ionic mechanism (usually cationic) and thus has the advantage of being oxygen insensitive. It is widely used for ring opening reactions, for example, epoxy resins.

Radical Addition

The basic mechanism for any photocurable free-radical system involves the formation of free-radical species through the absorption of light by the photoinitiator. The photophysical and photochemical properties of the photoinitiators, therefore are extremely important in controlling their reactivity and should possess the following properties:

- high absorptivity in the region of activation which will depend upon the application and light source used
- high quantum yield for free radical formation
- adequate solubility in the resin system used
- high storage stability
- odorless and nonyellowing qualities
- nontoxicity, low cost, ease of handling

There are two basic categories of photoinitiators which meet these requirements. The first group are now called type I photoinitiators.[1-3,6,7] Examples of such structures are illustrated in **Figure 1** which provides a reasonable coverage of the different structures available. The second group are called type II photoinitiators and examples of these structures are shown in **Figure 2**.[1-3,6,7]

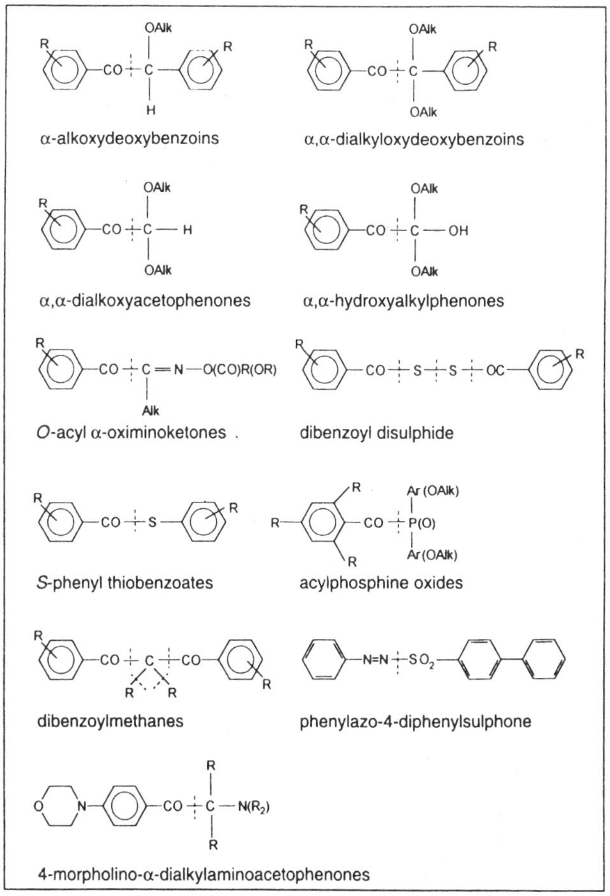

FIGURE 1. Some typical photofragmenting Type I photoinitiators.

FUTURE DEVELOPMENTS

Photoinitiator chemistry is both an exciting and rapidly expanding field with many new developments. Future directives include the following:

- the development of cheap co-reactive initiators with no toxicity, low migration, and good solubility
- improved water-soluble photoinitiators
- improved cost-effective visible sensitisers which bleach
- less toxic and cheap cationic initiators
- more durable UV-curable packages
- more effective initiators for epoxidized rubbers
- more effective initiators for crosslinking thermoplastics for improved property requirements such as flammability and physical properties

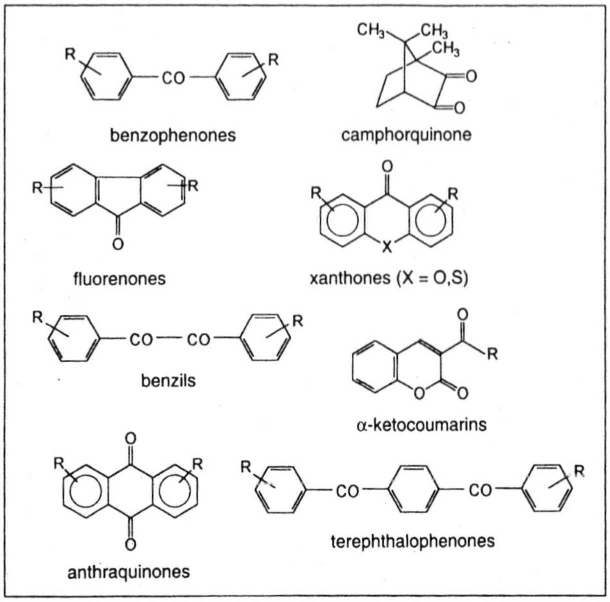

FIGURE 2. Some typical type II hydrogen atom abstracting photoinitiators.

REFERENCES

1. *Photopolymerisation and Photoimaging Science and Technology*; Allen, N. S., Ed.; Elsevier Applied Science: Barking, UK, 1989.
2. Allen, N. S.; Edge, M. *J. Oil. Col. Chem. Assoc.* **1990**, *11*, 438.
3. *Radiation Curing of Polymers II*; Randell, D. R., Ed.; Royal Society of Chemistry; Cambridge, UK, 1991.
4. Davidson, R. S. *J. Photochem. and Photobiol. A: Chem. Ed.* **1993**, *69*, 263.
5. Davidson, R. S. *J. Photochem. and Photobiol. A: Chem. Ed.* **1993**, *73*, 81.
6. Bottcher, H. *Technical Applications of Photochemistry*; Deutsche Verlag fur Grandstoffindustrie: Leipzig, Germany, 1991.
7. Oldring, P. K. T. *Chemistry and Technology of UV and EB Formulations for Coatings, Inks and Paints*; SITA Technology: London, UK, 1991; Vols. I–IV.

PHOTOINITIATORS, WATER-SOLUBLE (for Vinyl Polymerization)

Maria Victoria Encinas* and E. A. Lissi
Departamento de Quimica
Facultad de Ciencias
Universidad de Santiago de Chile

In spite of their practical interest, few systems that act as photoinitiators of the free radical polymerization of vinyl monomers in aqueous media are not well characterized, and very few photoinitiation quantum yields have been reported over a significant range of experimental conditions. We have classified photoinitiation systems as either unimolecular or bimolecular processes. Photoinitiators that bear similar chromophore groups are discussed together.

UNIMOLECULAR PHOTOINITIATORS

Peroxides

Peroxides show absorption in the 200–300 nm region of the nσ* type. This absorption leads to repulsive states.[2] Because the excited state is repulsive, excited molecules cleave the O–O bond with nearly unitary efficiency, almost independent of the nature of the medium.[3] Hydrogen peroxide, peroxydisulphate, and peroxydiphosphate have been used as photoinitiators of polymerization in aqueous solution.[4–6]

Azo Compounds

Alkylazo compounds present at nπ* band of relatively low molar absorptivity in the 350 nm region. These compounds are readily dissociated by light to give free radicals. Water-soluble azo compounds have been extensively studied as thermal and photochemical sources of free radicals at relatively low temperatures, and their decomposition mechanism has been thoroughly investigated.[7,8]

Azoisobutyramide has been described as a water-soluble photoinitiator.[9] Encinas and Lissi reported that α,α'-azobis(2-amidino)propane hydrochloride (ABAP) and 4,4'-azobis(4-cyanovaleric) (ABCV) are efficient photoinitiators of vinylpyrrolidone and 2-hydroxyethyl methacrylate polymerization in aqueous solution.[10,11]

Carbonyl Compounds

Alkylphenones undergo α-cleavage from the excited triplet state and generate a benzoyl and an alkyl radical. A series of water-soluble photoinitiators based on 4-substituted (2-hydroxy-2-propyl)phenyl ketones and on benzoylmethyl thiosulphates has been developed.[12,13] On the basis of their curing speed, it has been proposed that these compounds would be suitable photoinitiators in water-based curing systems or in aqueous solution of monomers.

α-Dicarbonyl Compounds

In protic solvents (methanol or water), 2,3-butanedione is solvated to give an α-substituted alkanone.[15,16] Irradiation of the solvated species at 300 nm in the presence of vinyl acetate or vinylpyrrolidone produces a significant amount of polymer.

Phosphines

Acylphosphine oxides and acylphosphonates undergo α-scission upon irradiation, giving radicals that are capable of initiating the polymerization of vinyl monomers.[17] Majima et al. have reported that lithium- and magnesium phenyl-2,4,6-trimethylbenzoylphosphinates are effective water-soluble photoinitiators for the polymerization of acrylamide and methacrylamide.[18]

Copper(II) Complexes

Copper(II)-bis(aminoacid) chelates with several amino acids (glutamic acid, serine, or valine) irradiated at 365 nm initiate the polymerization of acrylamide in aqueous solutions, at pH ≥ 7.[19,20]

*Author to whom correspondence should be addressed.

BIMOLECULAR PHOTOINITIATION SYSTEM

Coordination Complexes

Many metal complexes have been tested as photoinitiators.[21] Iron(III), cobalt(III), and chromium(VI) complexes are among the most studied photoinitiators for vinyl polymerization in aqueous media.[22–27] The process involves a charge transfer reaction, which gives rise to a radical from the co-substrate and the reduced form of the metal.

Aqueous polymerization of vinyl monomers photoinitiated by tri-nuclear transition metal complexes has been reported.[28–30]

Carbonyl Compounds

Most of the polymerizations in aqueous solution photoinitiated by aromatic carbonyl compounds comprise benzophenone or thioxanthone derivatives in the presence of aliphatic amines as hydrogen donors.[1,31,32] The sensitizers have been made water soluble by attaching to their molecules ionic groups such as trimethylammonium or sulphonates.

Aromatic Hydrocarbons

Encinas and Lissi have shown that vinyl monomers sensitized by pyrene in the presence of amines polymerize only in polar media.[33,34] Recently, this system has been extended to the polymerization of vinylpyrrolidone in aqueous media by use of pyrene ionic derivatives.[14]

Dye-Sensitized Systems

Water-soluble dyes appear to be ideal photoinitiators because of their high absorption in the visible region. However, our understanding of the mechanism involved in dye-sensitized photopolymerization is limited. Usually, such photoinitiator systems consist of a dye and a reducing agent such as an amine, sulphinate, phosphine, enolate, α-aminosulfone, or carboxylate.

ACKNOWLEDGMENT

The authors acknowledge FONDECYT, Grant #1941068, for supporting this work.

REFERENCES

1. Lissi, E. A.; Encinas, M. V. *Photochemistry and Photophysics* CRC: Boca Raton, FL, 1991; Vol. 4, Chapter 4.
2. Evleth, E. M. *J. Am. Chem. Soc.* **1976**, *98*, 1637.
3. Lissi, E. A. *Can. J. Chem.* **1974**, *52*, 2491.
4. Dainton, F. S.; Sisley, W. D. *Trans. Faraday Soc.* **1963**, *59*, 1369.
5. Roffey, C. G. *J. Oil Col. Chem. Assoc.* **1985**, *68*, 116.
6. Lenka, S.; Nayak, P. L. *J. Photochem.* **1987**, *36*, 365.
7. Nuyken, O.; Kerber, R. *Makromol. Chem.* **1978**, *179*, 2845.
8. Nuyken, O.; Knepper, T.; Voit, B. *Makromol. Chem.* **1989**, *190*, 1015.
9. Bianchi, J. P.; Price, F. P.; Zimm, B. H. *J. Polym. Sci.* **1957**, *25*, 27.
10. Encinas, M. V.; Lissi, E. A. *Eur. Polym. J.* **1992**, *28*, 471.
11. Encinas, M. V.; Lissi, E. A.; Martinez, C. *Eur. Polym. J.*, in press.
12. Koehler, M.; Ohngemach, J. *Polym. Paint Colour J.* **1988**, *178*, 403.
13. Bassi, G. L.; Broggi, F.; Revelli, A. *Polym. Paint Colour J.* **1989**, *179*, 684.
14. Encinas, M. V.; Lissi, E. A.; Majmud, C.; Cosa, J. J. *Macromolecules* **1993**, *21*, 6284.
15. Encinas, M. V.; Garrido, J.; Lissi, E. A. *J. Polym. Sci., Polym. Chem. Ed.* **1985**, *23*, 2481.
16. Miyata, K.; Nakashima, K.; Koyanagi, M. *Bull. Chem. Soc. Jpn.* **1989**, *62*, 367.
17. Schnabel, W. *J. Radiat. Curing* **1986**, *13*, 26.
18. Majima, T.; Schnabel, W.; Weber, W. *Makromol. Chem.* **1991**, *192*, 2307.
19. Namasivayam, C.; Natarajan, P. *J. Polym. Sci., Polym. Chem. Ed.* **1983**, *21*, 1371.
20. Namasivayam, C.; Natarajan, P. *J. Polym. Sci., Polym. Chem. Ed.* **1983**, *21*, 1385.
21. Yang, D. B.; Kutal, C. In *Radiation Curing: Science and Technology*; Pappas, S. P., Ed.; Plenum: New York, 1992; p 21.
22. Okimoto, T.; Inake, Y.; Takemoto, K. *J. Macromol. Sci., Chem.* **1973**, *7*, 1537.
23. Abbas, A. *J. Photochem.* **1986**, *35*, 87.
24. Bhaduri, R. *Makromol. Chem.* **1977**, *178*, 1385.
25. Aslam, M.; Anwaruddin, Q.; Natarajan, L. V. *Polym. Photochem.* **1984**, *5*, 41.
26. Robert, B.; Bolte, M.; Lemaire, J. *J. Chim. Phys., Phys.-Chim. Biol.* **1985**, *82*, 361.
27. Galcera, T.; Jouan, X.; Bolte, M. *J. Photochem. Photobiol. A: Chem. Ed.* **1988**, *45*, 249.
28. Muralidharan, G.; Anwaruddin, Q.; Natarajan, L. V. *J. Macromol. Sci., Chem.* **1983**, *A19*, 501.
29. Iwai, K.; Uesugi, M.; Takemura, F. *Polym. J.* **1985**, *17*, 1005.
30. Iwai, K.; Uesugi, M.; Sakabe, T.; Hazama, C.; Takemura, F. *Polym. J.* **1991**, *23*, 757.
31. Catalina, F.; Peinado, C.; Corrales, T. *Rev. Plast. Mod.* **1992**, Vol. 63, 561.
32. Davidson, R. S. *J. Photochem. Photobiol. A: Chem. Ed.* **1993**, *73*, 81.
33. Encinas, M. V.; Majmud, C.; Garrido, J.; Lissi, E. A. *Macromolecules* **1989**, *22*, 563.
34. Encinas, M. V.; Majmud, C.; Lissi, E. A.; Scaiano, J. C. *Macromolecules* **1991**, *24*, 2111.

Photoisomerization

See: *Ionic Conductivity Switching (Photochromic Crown Compounds)*
Norbornadiene Polymers
Photochromic Films, Azo Dyes (Photoinduced Anisotropy and Photoassisted Poling)
Photochromic Optically Active Polymers
Photochromic Poly(α-aminoacid)s
Photoreactive Langmuir-Blodgett-Kuhn Assemblies (Functionalized Liquid Crystalline Side Chain Polymers)

Photoluminescence

See: *Photoluminescent Polymers*
Polyaddition, Radical

PHOTOLUMINESCENT POLYMERS

Steven Holdcroft and Mohamed S. A. Abdou
Department of Chemistry
Simon Fraser University

In this paper, the photoluminescence of polymers is described. Attention is given to the two types of photoluminescent polymers, their photophysical properties, and their application.

Photoluminescent polymers fall into two distinct classes: lumiphore-bearing polymers and π-conjugated polymers. Other polymers that do not clearly fall into these categories exist; these are dealt with under the heading "*Miscellaneous Polymers*". It should be recognized that overlap exists between the various class types.

LUMIPHORE-BEARING POLYMERS

Polymers bearing aromatic functionality are usually photoluminescent. Many common monomers bear such groups. Thus, conventional ionic or free radical polymerization can be used to yield the desired photoluminescent polymer. **Table 1** shows the structure of common photoluminescent vinyl polymers.

TABLE 1. Common Photoluminescent Polymers

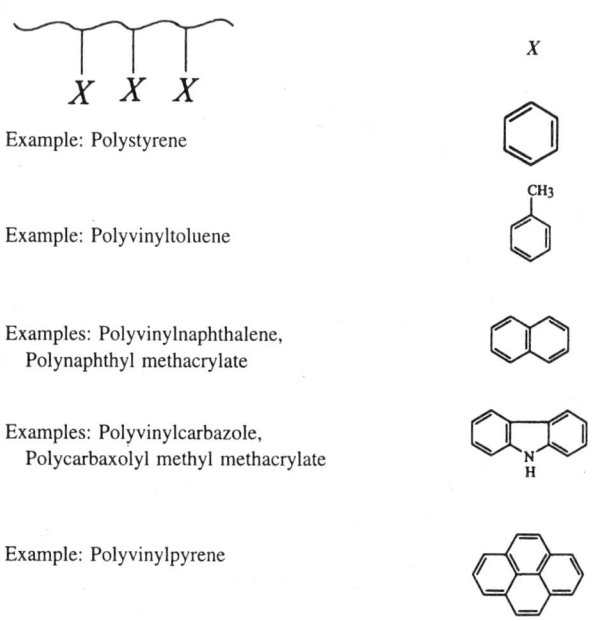

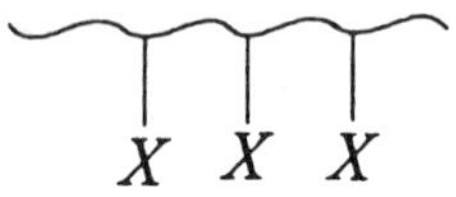	*X*
Example: Polystyrene	
Example: Polyvinyltoluene	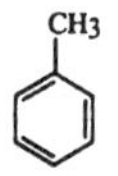
Examples: Polyvinylnaphthalene, Polynaphthyl methacrylate	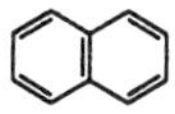
Examples: Polyvinylcarbazole, Polycarbaxolyl methyl methacrylate	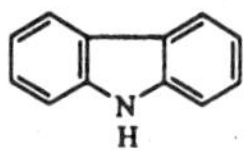
Example: Polyvinylpyrene	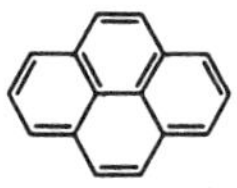

An alternative route is to react the lumiphore with a functionalized polymer. The reactions of lumiphoric diazo compounds and alcohols with polycarboxylic acids serve as examples.[1]

Anionic polymerization is useful for incorporating lumiphores into the chain in a more controlled manner. By suitable choice of initiator and terminator, appropriate lumiphores can be attached to the chain termini.

Studies of Polymer Dynamics and Molecular Structure

Photoluminescence is a nondestructive technique that provides conformational and dynamical information on polymer chains. It enables investigations of polymers in the solid state and in solution, in a wide variety of solvents and electrolytes, in dilute and concentrated polymer solutions, and at various temperatures. If the polymer of interest is not inherently photoluminescent, then it can be made so by incorporation of photoluminescent labels. Because photoluminescence, is an example sensitive technique, the extent of labeling need only be very small. Luminescent probes can be incorporated into a wide variety of polymers with molecular precision, and with little or no modification of the physical properties of the host polymers. Information on the polymer is transmitted photonically, and the lifetimes and emission characteristics of the probes are often medium-sensitive.

Polymers with Terminal Lumiphores

Perhaps the simplest yet most elegant type of photophysical probe uses polymers with terminal lumiphores, for which emission is somehow modified when the two ends come into close proximity.

Polymers with Pendant Lumiphores

Polymerization of vinyl monomers juxtaposes substituents along the polymer chain. When the substituents are aromatic, excimer formation can occur for segmental conformations that bring substituents into a face-to-face arrangement within a distance of ca. 4 Å.

Compatibility of Polymers and Polymer Morphology

Polymers that are chemically dissimilar are usually incompatible and tend to phase-separate. Photoluminescence spectroscopy, and in particular nonradiative energy transfer, is useful for gaining detailed information at the molecular level because the distance chromophores must approach one another for energy transfer is on the order of nanometers. Thus, the efficiency of energy transfer in blends of two polymers, one labeled with an energy donor and the other with an acceptor, provides information on the polymers' compatibility.

Studies of Energy Transfer and Light Harvesting

The natural photosynthetic process is made possible by the efficiency with which a multichromophoric arrangement absorbs photonic energy and directs it to a reaction center. Similarly, efficient energy transfer in synthetic materials is a central issue. Energy migration and energy transfer processes in polymers can be studied by time-dependent and steady-state luminescence spectroscopy.[2,3] The depolarization of fluorescence upon excitation with polarized light provides a means by which energy transfer along the chain can be determined. Alternatively, an appropriately located energy quencher can be used to capture the migrating energy.

Transport Properties

Luminescence is an important tool in studying the transport of small molecules through polymers.[4] For this purpose, lumiphores are used as probes, except that here the measurable parameter is the quenching of the luminescence of the probe by diffusants and penetrants.

π-CONJUGATED POLYMERS

Photoluminescence

Unsaturated polymers containing alternate single and double bonds are an emerging class of luminescent polymer.

In contrast to polymers possessing fluorescent labels, π-conjugated polymers derive their luminescence from the inherent nature of the polymer backbones. Thus, photophysical studies provide fundamental information on the electronic structure of these polymers as well as their environment. Their photophysical properties have also attracted significant technical interest because of potential applications in opto-electronics and electro-optics. The phenomenon of electroluminescence from π-conjugated polymers for display technologies, for example, illustrates the need to understand their photophysical and photochemical processes in intricate detail.[5] Such polymers luminesce because of the delocalized π-conjugated system.

The photoluminescence of several conjugated polymers–polyacetylenes, poly(*p*-phenylene vinylenes), and polythiophenes–has been extensively studied.

Electroluminescence

The search for efficient electroluminescent materials for flat panel displays has driven efforts to understand the photophysical phenomenon of conjugated polymers. Organic polymers offer a number of potential advantages over conventional inorganic thin film displays, including tunable emission and the cost advantage of fabricating large area displays. Interest in the electroluminescence of conjugated organic polymers were sparked by a report that poly(*p*-phenylene vinylene) emitted visible light when a potential was applied across a thin film of the polymer sandwiched between conductive electrodes.[6,7]

Electroluminescence spectra are similar to the corresponding photoluminescence spectra, which indicate a common excited state.

A number of other polymers have been investigated with a view to either improving efficiency or tuning the color of emission. These include poly(*p*-phenylene), polythiophene, and poly(alkyl fluorene)s and their derivatives.

MISCELLANEOUS POLYMERS

Interest in new luminescent materials has developed largely as a result of the rapidly growing photonics and optoelectronic technologies. Emerging photoluminescent polymers, include polysilanes, polyimides, polyrotaxanes, polyporphyrins, and supramolecular systems.

REFERENCES

1. Anufrieva, E. V.; Gotlib, Y. Y. *Adv. Polym. Sci.* **1981**, 40.
2. Fox, M. A.; Jones, W. E. Jr.; Watkins, D. M. *Chem. Eng. News,* March 15, **1993**, p. 38.
3. Webber, S. E. *Chem. Rev.* **1990**, *90*, 1469.
4. MacCallum, J. R.; Rudkin, A. L. *Eur. Polym. J.* **1979**, *14*, 655.
5. Friend, R. H.; Bradley, D. D. C.; Townsend, P. D. *J. Phys. D: Appl. Phys.* **1987**, *20*, 1367.
6. Burroughes, J. H.; Bradley, D. D. C.; Brown, A. R.; Marks, R. N.; Mackay, K.; Friend, R. H.; Burns, P. L.; Holmes, A. B. *Nature (London)* **1990**, *347*, 539.
7. Bradley, D. C. *Adv. Mater.* **1992**, *4*, 756.

PHOTOMODIFICATION (Pendant Acyloxyimino Groups to Amino Groups)

Masahiro Tsunooka
Department of Applied Chemistry
College of Engineering
Osaka Prefecture University

Photochemical reactions are very important for the development of new photoreactive (photofunctional) polymers. Photoreactive polymers are very widely used as photosensitive polymers in the printing industry and as photoresists in the electronics industry.

An important property of polymers is their solubility, which is related to their polarity. In the photolysis of pendant acyloxyimino (AOI) groups in a polymer matrix, we have found that AOI groups can be transformed effectively into amino groups.[1,2] The resulting amino groups can easily be changed to ammonio groups by the reaction of the resulting polymers with acids.[3] These processes can increase the polarity of the original polymers remarkably, leading to the formation of a hydrophilic polymer surface and changes in solubility.[3–5]

Furthermore, the pendant amino groups resulting from the photolysis of pendant AOI groups are effective for the crosslinking of polymers bearing pendant epoxy groups.[6]

This paper describes the photochemical reactions of polymers bearing AOI groups and their applications.[11]

PREPARATION[1,2]

O-Acryloyl acetophenone oxime (AAPO) copolymers with styrene (St) or methyl methacrylate (MMA) were prepared by solution polymerization in benzene under vacuum at 60°C.

PROPERTIES

Photochemical Reactions of Copolymers Bearing Pendant Acyloxyimino (AOI) Groups in the Solid Phase[1,2]

Formation of Amino Groups and Double Bonds

An outline of the photoreactions of pendant AOI groups is shown in **Scheme I**.

Surface Properties of AAPO Copolymers

Changes of Contact Angle of Water on AAPO Copolymers by UV Irradiation[3]

It is expected that the transformation of amino groups into ammonio groups will result in modification of the polarity of polymers.

Photoinitiated Crosslinking of AAPO–Glycidyl Methacrylate (GMA) Copolymers[6]

Because amino groups can react with epoxy groups, AAPO (12.6)–GMA copolymer films could be crosslinked by UV irradiation, and the degree in insoluble fraction increased with an increase in the content of amino groups formed by hydrolysis of imino groups.[6]

Similarly, O-acyl oximes were found to be very good photobase generators.[7–10]

APPLICATIONS

Color Filters

Irradiated parts of AAPO copolymer films can be dyed by acid dyes. This result suggests that they can be used as the films for color filters that separate natural light into three light components (blue, green, and red) in a camera tube.

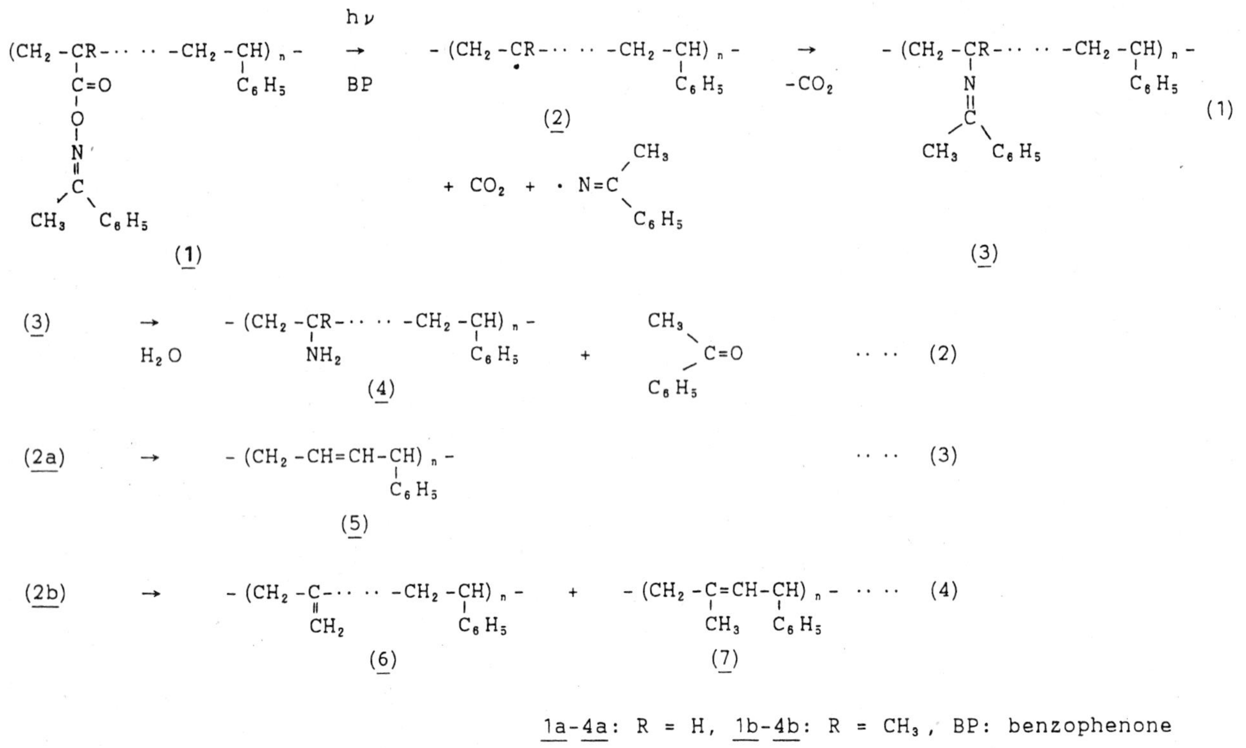

SCHEME I. Photochemical reactions of AAPO-St copolymers.

Dual-Tone Photoresists

A very promising application for AAPO copolymers that change their dissolution properties after irradiation is their use in dual-tone photoresists, which fabricate fine patterns without swelling at development.

Crosslinkers in UV Curing

O-Acyl oximes and AAPO polymers are photobase generators, and the amines resulting from the photolysis of pendant AOI groups can be used as crosslinkers for epoxides and isocyanates.

REFERENCES

1. Song, K. H.; Tsunooka, M.; Tanaka, M. *J. Photochem. Photobiol., A: Chem.* **1988**, *44*, 197.
2. Song, K. H.; Tsunooka, M.; Tanaka, M. *Makromol. Chem., Rapid Commun.* **1988**, *9*, 519.
3. Song, K. H.; Iwamoto, A.; Tsunooka, M.; Tanaka, M. *J. Appl. Polym. Sci.* **1990**, *39*, 1769.
4. Song, K. H.; Tonogai, S.; Tsunooka, M.; Tanaka, M. *J. Photochem. Photobiol., A: Chem.* **1989**, *49*, 269.
5. Tsunooka, M.; Tanaka, M. *Polymers for Microelectronics–Science and Technology* Kodansha: Tokyo, 1990; p 103.
6. Song, K. H.; Urano, A.; Tsunooka, M.; Tanaka, M. *J. Polym. Sci., Polym. Lett. Ed.* **1987**, *5*, 417.
7. Ito, K.; Nishimura, M.; Sashio, M.; Tanaka, M. *Chem. Lett.* **1992**, 1153.
8. Ito, K.; Shigeru, Y.; Tsunooka, M. *J. Photopolym. Sci., Technol.* **1994**, *7*, 75.
9. Ito, K.; Nishimura, M.; Sashio, M.; Tsunooka, M. *J. Polym. Sci., Polym. Chem. Ed.* **1994**, *32*, 1793.
10. Ito, K.; Nishimura, M.; Sashio, M.; Tsunooka, M. *J. Polym. Sci., Polym. Chem. Ed.* **1994**, *32*, 2177.
11. Tsunooka, M. *Polym. News* **1994**, *19*, 73.

PHOTOPOLYMER MATERIALS (Development of Holographic Gratings)

L. V. Natarajan,* R. L. Sutherland, V. Tondiglia, and T. J. Bunning
Science Applications International Corporation

W. W. Adams
Materials Directorate Wright Laboratory
Wright Patterson Air Force Base

In recent years, much attention has been focused on holographic processes and on the development of novel materials for high-performance holograms. This interest is largely due to the need for expanding the information storage technology presently available. Holographic storage has long held promise for large digital storage capacity because the information packing densities can be considerably increased by using three-dimensional (3-D) storage techniques in the form of interference patterns. Holographic storage also will enable fast data transfer

*Author to whom correspondence should be addressed.

rates because it permits the reading and writing of data to proceed simultaneously as parallel processes. Recent development in materials, spatial light modulators (SLMs), and charge coupled device (CCD) arrays have brought the promise close to reality. In addition to their use for information storage, holograms have demonstrated applications in the field of electrooptics, in devices such as wide-angle beam deflectors, tunable reflection filters, high-speed beam steering, variable-power lens systems, and electrooptical diffraction optics.

For many years, the most widely used holographic materials have been silver halide photographic emulsion and dichromated gelatin. However, these materials have some drawbacks, such as high-grain noise, the need for wet processing, and environmental sensitivity. During the past two decades, synthetic polymers have emerged as alternatives to the dichromated gelatin and silver halide emulsions.

PHOTOPOLYMER RECORDING MATERIALS

The photopolymerization process in the recording of holograms generally starts with liquid prepolymer compositions and involves free radical polymerization of vinyl-related monomers. Cationic photopolymerization reactions are also used for recording holograms. In writing holograms, photopolymerizable materials are either used as a liquid syrup placed between glass slides or as dry films coated on glass or plastic sheets.

Another class of photosensitive materials for recording holograms is the photocrosslinkable polymers. Crosslinkable polymers are usually doped with metal ions or dyes for photosensitization.

FLUID COMPOSITIONS

Acrylate-based monomers have been the popular choice for many of the liquid compositions for photopolymerization.

Recently, Fimia et al. demonstrated that a combination of 4,5-diiodosuccinyl fluorescein and N,N-dimethyl aniline serves as an efficient photoinitiating system for hologram formation in a mixture of 2-hydroxymethyl methacrylate (HEMA) and ethylene glycol dimethacrylate (EGDMA).[1,2]

Kawabata et al. developed a novel photopolymer system consisting of a radically polymerizing monomer (RPM)–for example, a difunctional acrylate type–and a cationically polymerizable monomer (CPM) for writing reflection holograms.[3]

DRY FILM PHOTOPOLYMER MATERIALS

Most of the dry film formulations tested so far contain acrylate-based monomers. In the past 20 years, Du Pont chemists have successfully developed high diffraction efficiency holograms based on acrylates containing photopolymer dry films.[4,5] Acrylamide-based dry films have also been used to produce high-efficiency holograms.

LIQUID CRYSTAL COMPOSITE PHOTOPOLYMER MATERIALS

Some interesting recent studies have dealt with holographic gratings written on polymer containing low-molecular-weight liquid crystals or liquid crystal polymers. The combination of liquid crystals, which are electrooptical materials, with photopolymerizable monomers offers a unique single-system approach to the economical fabrication of electrically switchable holograms. Switchable holographic gratings are desirable for a wide range of applications in diffractive optics. Liquid crystal materials are useful for these devices because of their large field-induced birefringence.

Recent reports have described the recording of switchable holograms by photopolymerization of liquid crystalline monomers.[6,7]

CONCLUSIONS

To a large extent, the choice and development of photopolymer materials will depend on the type of application demanded. In general, however, materials that have the advantage of self-development *in situ*, good shelf life, broad spectral sensitivity, low noise, good photosensitivity, and high diffraction efficiency are desirable. For practical devices in imaging, information processing, and diffractive optics, an economical and switchable holographic material is desired that allows high diffraction efficiency, low switching fields, fast response and decay speeds, and ease of processing. Recent progress in this respect is impressive. Still, there is a growing need to produce polymer materials that can offer as much storage capability as dichromated gelatin and as great sensitivity as silver halide. Polymer materials have definite advantages over photographic plates and dichromated gelatin in ease of processing and environmental stability.

REFERENCES

1. Fimia, A.; Lopez, N.; Mateos, F.; Sastre, R.; Pineda, J.; Amat-Guerri, F. *Appl. Opt.* **1993**, *20*, 3706.
2. Mallavia, R.; Amat-Guerri, F.; Fimia, A.; Shastre, R. *Macromolecules* **1994**, *27*, 2643.
3. Kawabata, M.; Sato, A.; Sumiyoshi, I.; Kubota, T. *Proc. SPIE* **1993**, *1914*, 66.
4. Monroe, B. M.; Smothers, W. K.; Keys, D. E.; Krebs, R. R.; Mickish, D. J.; Harrington, A. F.; Armstrong, S. R.; Chan, D. M. T.; Weathers, C. I. *J. Image Sci.* **1991**, *35*, 19.
5. Weber, A. M.; Smothers, W. K.; Trout, T. J.; Mickish, D. J. *Proc. SPIE* **1990**, *1212*, 30.
6. Zhang, J.; Carlen, C. R.; Palmer, S.; Sponsler, M. B. *Proc. SPIE* **1993**, *2042*, p 235.
7. Zhang, J.; Sponsler, M. B. *J. Am. Chem. Soc.* **1992**, *114*, 1506.

PHOTOPOLYMER MATERIALS (for Computer-Generated Optics)

Yuri B. Boiko
School of Physics
The University of Melbourne

V. S. Soloviev
Image Processing Systems Institute of the Samara Aerospace University

Computer-generated optics are optical elements whose diffractive structure is calculated numerically and used for wavefront transformation of optical waves. Photopolymer materials are suitable for the recording of computer-generated optical elements with a relief representation of diffractive structure, in which thickness of the polymer layer at each point is proportional to the phase shift of the incident input wave. Photopolymerizable layers have been successful in recording relief structures of laser focusing elements for the far-IR region (including CO_2 laser radiation with wavelength 10.6 μm)[1–3] and fly-eye arrays (microlens arrays) for visible light.[4]

Although photopolymerizable materials provide relief patterning through wet processing after exposure to the image, the technological approach described here is wet-processing-free and relies on the self-developing nature of the photopolymerization process.

In general, photopolymer-based technologies for computer-generated optics relief fabrication include the following stages (after the element structure is calculated and reproduced on the amplitude mask as transparency [or optical density] levels in multiple gray scale):

- preparation of the photopolymerizable recording layer
- photopolymer imaging through the amplitude mask
- self-development of the recorded relief structure (performed in the dark)
- fixing of the recording by overall exposure or by temperature processing.

PREPARATION

Two approaches employing photopolymers for computer-generated optic recording have been investigated: layers of liquid photopolymerizable compositions (LPPC) and dry photopolymer (DP) films.[2,3]

PROPERTIES

Relief formation is a self-developing process. It takes place after exposure when the exposed photopolymer layer is isolated from the initiating radiation (i.e., it is placed into the dark). The properties of the relief formation are quite similar for both LLPC and DP films.

APPLICATIONS

The resolution of relief hologram recording on photopolymerizable layers is limited to low spatial frequencies, normally ~10–20 mm^{-1}, quite sufficient for recording diffractive optical elements in the far-IR region (including those for CO_2 laser radiation with 10.6 μ wavelength). Additionally, because the procedure does not involve wet processing, the optical quality of the element surface allows fabrication of microoptical elements for visible light, such as fly-eye arrays.

Layers of LPPC appear to be more efficient than DP films at present in terms of resolving power, as well as in relief depth values for each given thickness of the recording layer.

Another important feature of relief recording is its linearity in transforming the amplitude function of the photomask into the relief structure of the recorded kinoform.

Linearity of relief recording in photopolymers seems to depend on photochemical transformations of the sensitizer and the initiator during exposure.

REFERENCES

1. Boiko, Yu. B.; Soloviev, V. S. et al. *Proc. SPIE* **1990**, *1238*, 253.
2. Boiko, Yu. B.; Soloviev, V. S. et al. *Proc. SPIE* **1993**, *2042*, 271.
3. Boiko, Yu. B.; Soloviev, V. S. et al. *Appl. Opt.* **1994**, *33*, 787.
4. Soloviev, V. S. et al. *Proc. SPIE* **1993**, *2042*, 248.

Photopolymerization

See: *Liquid Crystalline State Polymerization*
Micelle-Forming Monomers, Photopolymerization
Photoinitiators (for Free Radical Polymerization)
Photoinitiators, Water-Soluble (for Vinyl Polymerization)
Photopolymerization (Initiated by Charge-Transfer Complexes)
Self-Assembled Polymers (Photopolymerizable Bolaform Amphiphiles)
Topochemical Photopolymerization (Types of Topochemical Behavior)

PHOTOPOLYMERIZATION (Initiated by Charge-Transfer Complexes)

Subasini Lenka* and P. L. Nayak
Department of Chemistry
Ravenshaw College

During the past two decades, the photochemistry of polymers has been of central importance. Research on photoinitiated polymerization has given rise to entirely new applications in microelectronics, such as in resist barrier coatings, encapsulants, and printed circuit board technologies.

Photopolymerization initiated by charge-transfer complexes has attracted attention because of its versatility for initiating polymerization in areas such as copolymerization and graft copolymerization. We have published a review of photoinduced graft copolymerization onto selected fibers[1] and have outlined the use of some charge-transfer complexes.

Photoinitiated polymerizations can be classified into four major categories:[2–28]

- Photoinitiated radical polymerization
- Photoinitiated charge-transfer polymerization
- Photoinitiated cation-radical polymerization
- Photoinitiated simultaneous radical and cation-radical polymerization

PHOTOPOLYMERIZATION INITIATED BY CHARGE-TRANSFER COMPLEXES

Photoinitiated radical polymerization by charge-transfer complexes may be represented as follows (**Equation 1**):

$$I + h\nu \rightarrow I^{*} \qquad (1)$$

*Author to whom correspondence should be addressed.

Photoinitiators in the presence of a co-initiator form charge-transfer complex (CT-complex) and subsequently dissociate into radicals, initiating free-radical polymerization (**Equation 2**).

$$\underset{\text{CT}-\text{Complex}}{I^{*} + AH \longrightarrow [I \ldots . AH]} \longrightarrow \dot{I}H + \dot{A} \qquad (2)$$

Ghosh and Chakraborty[28] reported that photopolymerization of methyl methacrylate and some other vinyl monomers with chlorine as the photoinitiator. Low concentrations of chlorine easily induced photopolymerization of MMA at 40°C. The kinetic data indicate that polymerization follows a radical mechanism involving the formation of a charge-transfer complex between the monomer and chlorine, which decomposes resulting in the formation of a radical that initiates the polymerization.

Lenka and Nayak[29] reported the photopolymerization of methyl methacrylate using α-Picoline-Br_2 charge-transfer complex as the photoinitiator.

Lenka and Nayak[30] have reported the photopolymerization of methyl methacrylate using isoquinoline–bromine charge-transfer complex as the photoinitiator.

Photopolymerization of vinyl monomers using chlorine and chlorine-charge transfer complexes has been investigated by a number of workers. Lenka and Nayak[31] have reported the photopolymerization of methyl methacrylate using isoquinoline–chlorine charge-transfer complex as the initiator.

Lenka and Nayak[32,33] have reported the photopolymerization of MMA using α-picoline-SO_2 and isoquinoline-sulfur dioxide charge-transfer complexes as the photoinitiators.

CONCLUSION

Photopolymerization initiated by charge-transfer complexes is extremely useful for initiating polymerization of vinyl monomers. This field is attractive because it does not require sophisticated apparatus and hence the method could be taken up by industry for production of numerous vinyl polymers. This method could also be used for grafting vinyl monomers to natural and synthetic fibers for better commercial value.

REFERENCES

1. Nayak, P. L.; Lenka, S. *J. Macromol. Sci. Rev. Macromol. Chem. Phys.* **1991**, *31*(1), 91.
2. Taylor, H. S.; Bates, J. R. *J. Am. Chem. Soc.* **1927**, *49*, 2438.
3. Staudinger, H. *Die hochomolekularen Organischem Verbcndungen*; Julius Springer: 1932; p 151.
4. Whitmore, F. C. *Ind. Eng. Chem.* **1934**, *26*, 94.
5. Price, C. C. *Ann. N.Y. Acad. Sci.* **1934**, *44*, 35.
6. Allen, N. S.; Rabak, J. F. *New Trends in the Photochemistry of Polymers*; Elsevier: London, 1985.
7. Block, H. et al. *Polymer* **1971**, *12*, 271.
8. Eleinger, C. P. *Adv. Macromol. Chem.* **1968**, *1*, 169.
9. Fouassier, J. P. *J. Chim. Phys.* **1983**, *80*, 339.
10. Fouassier, J. P. et al. *J. Radiat. Curing* **1983**, *10*, 9.
11. Fouassier, J. P. et al. *Polym. Photochem.* **1984**, *5*, 57.
12. Heine, H. G. et al. *Angew. Chem. Ind. Ed. Engl.* **1972**, *11*, 974.
13. Hutchinson, J.; Leelwith, A. *Adv. Polym. Sci.* **1974**, *14*, 49.
14. Kato, M.; Yoneshige, Y. *Makromol. Chem.* **1973**, *164*, 159.
15. Kinstle, J. F.; Watson, S. L. Jr. *J. Radiat. Curing* **1974**, *1*, 2.
16. Labana, S. S. *J. Macromol. Sci. Rev. Macromol. Chem. C* **1974**, *11*, 299.
17. Ledwith, A. *J. Oil Colour Chem. Assoc.* **1976**, *59*, 157.
18. Ledwith, A. *Pure Appl. Chem.* **1977**, *49*, 431.
19. Ledwith, A. et al. *Macromolecules* **1975**, *8*,
20. McGinniss, V. S. *J. Radiat. Curing* **1975**, *2*, 3.
21. Osborn, C. L. *J. Radiat. Curing* **1975**, *3*, 2.
22. Oster, G.; Yang, N. *Chem. Revs.* **1968**, *68*, 125.
23. Pappas, S. P. *Prog. Org. Coat.* **1974**, *2*, 333.
24. Pappas, S. P.; McGinniss, V. D. Technology Vol. I; Pappas, Ed., Stanford Technology Marketing: Norwalk, CT, 1978; p 1.
25. Rober, J. R. *Photochem. Photobiol.* **1968**, *7*, 5.
26. Rabek, J. F.; Rabek, T. I. *J. Appl. Polym. Sci.* **1963**, *7*, 537.
27. Ranaweera, R. P. R.; Scott, G. *Eur. Polym. J.* **1976**, *12*, 591.
28. Ghosh, P.; Chakraborty, S. *J. Polym. Sci., Polym. Chem. Ed.* **1975**, *13*, 1531.
29. Lenka, S. et al. *Polym. Photochem.* **1984**, *4*, 83.
30. Lenka, S. et al. *Polym. Photochem.* **1984**, *4*, 167.
31. Lenka, S. et al. *J. Polym. Sci., Polym. Chem. Ed.* **1984**, *22*, 429.
32. Nayak, P. L. et al. *Polym. Photochem.* **1985**, *6*, 163.
33. Lenka, S. et al. *J. Polym. Sci., Polym. Chem. Ed.* **1987**, *25*, 703.

PHOTOREACTIVE LANGMUIR-BLODGETT-KUHN ASSEMBLIES (Functionalized Liquid Crystalline Side Chain Polymers)

Wolfgang Knoll
Max-Planck-Institute für Polymerforschung
and
Frontier Research Program
The Institute of Physical and Chemical Research (RIKEN)

Functional polymers are considered to be potential candidates for the materials needed in future devices for electro-optical and photonic applications.[1–4] One of the very promising concepts to derive from the required functionalization of the matrix polymer involves the covalent coupling of the purposefully tailored chromophores to a polymeric backbone, which not only results in an enhanced stability and good processability of the material, but in many cases leads to a liquid crystalline phase behavior that offers the great potential to link the self-ordering tendency of these systems with an optical control of their mesogenic character.[5]

A major technological breakthrough, however, is hampered by the fact–along with other reasons–that a detailed understanding of the electronic and steric coupling of the photo-isomerizing chromophores to their liquid-crystalline environment is not yet derived and that many basic questions about the matrix-dependent photophysical and photochemical processes in these complex systems still need to be answered experimentally as well as theoretically.

A particularly useful model structure for studying these problems can be prepared by the Langmuir-Blodgett-Kuhn (LBK) technique, provided the chemical nature of the system of interest is sufficiently amphiphilic so that its intrinsic tendency to self-organize can be used to prepare stable monolayers

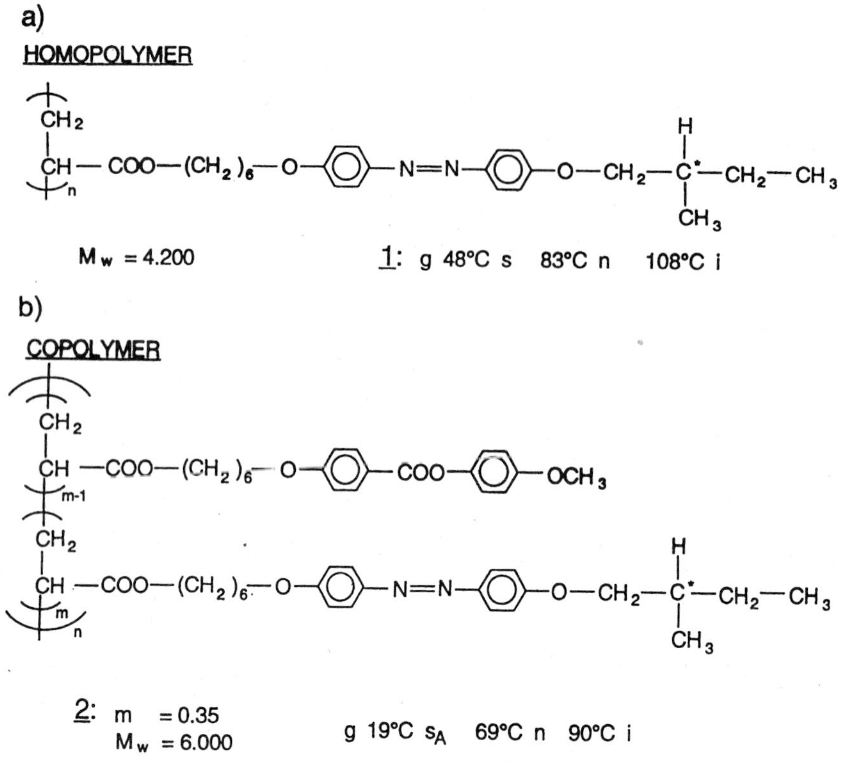

FIGURE 1. Structural formula of the liquid crystalline side chain homopolymer (a) and the copolymers (b) used in these studies.

at the air/water interface.[6] Their transfer onto solid substrates then allows smectic multilamellar assemblies to be built up with an unmatched control of order, orientation, and organization of the individual units.[7]

PREPARATION AND PROPERTIES

Bulk Liquid Crystalline Polymer Properties

One attractive feature of liquid crystalline side-group polymers as materials for electro-optical and information storage device applications is the possibility to introduce functional groups in addition to other sidechains that control the mesophase types and their temperature ranges.

In order to study the influence of chiral groups on the mesophase behavior of a liquid crystalline azobenzene containing polymer, side-group copolymers that combine chiral azobenzene moieties with achiral mesogenic units were prepared together with the corresponding homopolymers.[8] The structural formulas are given in **Figure 1**.

Characterization of Langmuir Monolayers

All polymeric systems were characterized at the air/water interface by π-A isotherms on pure water at different temperatures.

The most stable monolayers were obtained with the homopolymer (Figure 1a). The area per repeat unit in the range of A = 0.25 nm^2 fits to a structural picture where the chromophore mesogens (in the *trans* state) behave like the hydrophobic tails of classical low molecular weight amphiphiles, i.e., sticking out from the water away from the polymeric backbone.

Langmuir–Blodgett–Kuhn Multilayer Assemblies

Structural Analysis by X-Ray Reflectivity Measurements

Structural characterization of the prepared multilayer assemblies was obtained by X-ray reflection measurements from samples consisting of various layers, transferred on to hydrophobic (by treatment with hexamethyldisilazane (HMDS) in $CHCl_3$) float glass substrates.

Surface Plasmon (SP) Spectroscopy

If a p-polarized laser beam is totally internally reflected from the Au- or Ag-coated (d ≈ 50 nm) base of a prism, a sharp dip in the reflected intensity is observed when probed as a function of the angle of incidence, θ, indicating the excitation of a surface electromagnetic mode called a *surface plasmon.* In particular, any thin film coating like an LBK mono- or multilayer will cause an angular shift of the SP resonance, from which the optical thickness of the film can be derived.

LBK multilayers of this photoreactive liquid crystalline polymer can be well described by multiples of (optically) identical double layers. Also, no systematic changes of the index of refraction for increasing layer thicknesses is found.

Vibrational Spectroscopy

IR dichroism measurements are another powerful means to derive information about orientational properties, this time with respect to vibrational band transition moments. In particular, if IR spectra can be taken in transmission as well as in reflection geometry, anisotropic orientation of IR bands can be analyzed.

In our case this could be achieved, however, only for homopolymer 1.

Structural Model of Homopolymer 1 and Copolymer 2

The X-ray data of the homopolymer samples clearly indicated the formation of highly ordered double-layer structures corresponding to the expectations for a Y-type LBK deposition process. Together with the surface plasmon, and in particular, with the waveguide optical experiments, one can conclude that each dipping cycle adds another double layer to the multilayer assembly, more or less identical in its properties to the layers already deposited.

CONCLUSIONS

A prerequisite for multilayer preparation according to the LBK technique is the formation of stable monolayers at the water/air interface. For classical low molecular mass amphiphiles this typically requires the compression of the monolayer to a highly condensed, usually solid-analogue state. Photo-isomerization reactions, however, need for the trans-*cis* conversion substantial degree of disorder, with free volume in the immediate environment of the azobenzene chromophores. These two, mutually exclusive principles require structural concepts that compromise both needs in order to allow for the build-up of photoreactive LBK films. It has been shown that copolymers with only a certain percentage of azobenzene chromophore-bearing side chains meet these conditions.

ACKNOWLEDGMENTS

I would like to thank the many colleagues that contributed to this work. In particular, I am grateful to M. Sawodny, A. Schmidt, and M. Stamm, C. Urban, and H. Ringsdorf who kindly provided the materials.

REFERENCES

1. Sasaki, A. *Mol. Cryst. Liq. Cryst.* **1986**, *139*, 103.
2. Attard, G.; Williams, G. *Nature* **1987**, *326*, 544.
3. Nakamura, T.; Ueno, T.; Tani, C. *Mol. Cryst. Liq. Cryst.* **1989**, *169*, 167.
4. Ruebner, J.; Ruhmann, R.; Rodekirch, G. *Plaste Kautsch.* **1989**, *36*, 253.
5. Schmidt, H. W. Ph.D. Thesis, Universität Mainz, 1984.
6. Kuhn, H.; Möbius, D.; Bücher, H. in *Physical Methods of Chemistry*; Weissberger, A.; Rossiter, B. W.; Eds.; John Wiley & Sons: New York, 1972, Part III B, Chap VII.; Kuhn, H.; Möbius, D. in *Investigations of Surfaces and Interfaces–Part B* Rossiter, B. W.; Baetzold, R. C., Ed. Physical Methods of Chemistry Series, 2nd ed., Vol. IXB; Wiley: New York, 1993.
7. Gaines, G. L. *Insoluble Monolayers at Liquid-Gas Interfaces*; Wiley-Interscience: New York, 1966.
8. Ringsdorf, H.; Urban, C.; Knoll, W.; Sawodny, M. *Makromol. Chem.* **1992**, *193*, 1235.

PHOTOREDOX LANGMUIR–BLODGETT FILMS (for Light Energy Conversion)

Tokuji Miyashita
Institute for Chemical Reaction Science
Tohoku University

Photoinduced electron transfer reactions play an important role in many areas of science and in photofunctional devices.[1] Photoredox processes are based on photoinduced electron transfer reaction, and an efficient photoredox cycle strongly depends on control of vectorial photoinduced electron transfer in systems. Various photoredox couples incorporated into organized molecular assemblies such as micelles, microemulsion, and lipid bilayers have been investigated from the viewpoint of solar energy conversion and artificial photosynthesis.[2] Tris(2,2'-bipyridine)ruthenium complex, $Ru(bpy)_3^{2+}$, is one of the best known redox-active sensitizers. The charge transfer excited state of $Ru(bpy)_3^{2+}$ is quenched by both electron acceptor and donor, yielding a powerful oxidant and reductant, respectively. $Ru(bpy)_3^{3+}$ and $Ru(bpy)_3^{+}$ are thermodynamically capable of oxidizing water to O_2 and of reducing to H_2, respectively.[3,4] Therefore, there are many studies on photochemistry of ruthenium complexes incorporated into various molecular assemblies to use as a photoredox center in artificial photosynthesis or photogalvanic cell.[5–7] Recently, as the most elaborate organized molecular assembly, Langmuir–Blodgett (LB) films have attracted much attention.[8] The monolayers are excellent molecular environments for the photoredox process since reactants can be placed at the known distance within nanometer thickness, and a desired molecular array can be fabricated by choosing monolayers deposited in LB systems.

We have continued to investigate the preparation of polymer LB films.[9] *N*-alkylacrylamide structure has an excellent ability to form a stable monolayer and LB multilayers. Recently, moreover, we attempted to incorporate various photofunctional groups into polymer LB multilayer assemblies in the form of copolymers of the *N*-alkylacrylamide monomer.[10–13] We describe the preparation and some properties of polymer LB films containing ruthenium complex (**Scheme I**) as a novel photoredox film.

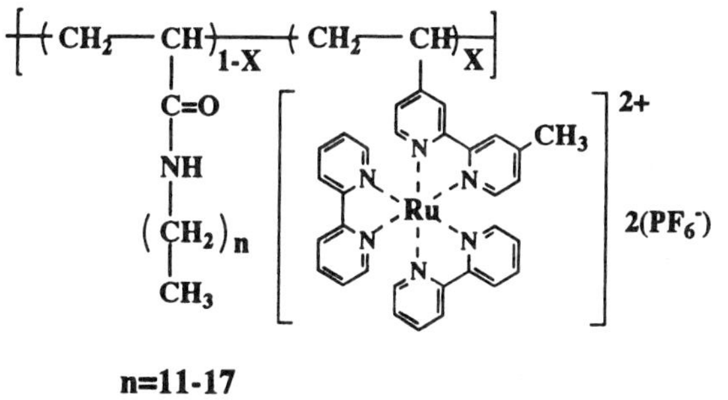

SCHEME I. Copolymers of *N*-alkylacrylamides and ruthenium complex.

REFERENCES

1. Kavarnos, G. J. *Fundamentals of Photoinduced Electron Transfer*; VCH: New York, 1993.
2. Fendler, J. H. *Membrane Mimetic Chemistry*; Wiley-Interscience: New York, 1982.
3. Bock, C. R.; Meyer, T. J.; Whitten, D. G. *J. Am. Chem. Soc.* **1975**, *97*, 2909.
4. Tokel-Tavvoryan, N. E.; Hemingway, R. E.; Bard, A. J. *J. Am. Chem. Soc.* **1973**, *95*, 6582.

5. Brugger, P. A. *J. Am. Chem. Soc.* **1980**, *102*, 2461.
6. Eltorki, F. M.; Schmel, R. H.; Reed, W. F. *J. Chem. Soc., Faraday Trans.* 1 **1989**, *85*, 349.
7. Miyashita, T.; Matsuda, M. *J. Phys. Chem.* **1981**, *85*, 3122.
8. Kuhn, H.; Möbius, D.; Bücher, H. *Physical Methods of Chemistry*; Wiley-Interscience: New York, 1972.
9. Miyashita, T. *Prog. Polym. Sci.* **1993**, *18*, 263.
10. Miyashita, T.; Mizuta, Y.; Matsuda, M. *Br. Polym. J.* **1990**, *22*, 327.
11. Murakata, T.; Miyashita, T.; Matsuda, M. *Macromolecules* **1989**, *22*, 2706.
12. Miyashita, T.; Yatsue, T. et al. *Thin Solid Films* **1989**, *179*, 439.
13. Miyashita, T.; Yatsue, T.; Matsuda, M. *J. Phys. Chem.* **1991**, *95*, 2448.

Photosensitive Polymers

See: *Cardo Polyesters (Heat Resistant and Photosensitive)*
Functionalized Polymers (Comb, Rotaxanic, and Dendrimeric Structures)
Ionic Conductivity Switching (Photochromic Crown Compounds)
Microgels, Macrogels (by Photocrosslinking)
Norbornadiene Polymers
Photochromic Optically Active Polymers
Photochromic Poly(α-aminoacid)s
Photocrosslinkable Photopolymers (Effect of Cinnamate Group Structure)
Photodimerizable Poly(crown ether)s
Photomodification (Pendant Acyloxyimino Groups to Amino Groups)
Photopolymer Materials (Development of Holographic Gratings)
Photopolymer Materials (for Computer-Generated Optics)
Polys[(disilanylene) oligophenylene]s
Polyimides (Negative-type Photosensitive Precursors)
Poly(N-vinylcarbazole)
Reversibly Crosslinked Gels
Stimuli-Responsive Polymers (Thermodynamics in Drug Delivery Technology)

Photosensitizers

See: *Photodegradable Plastics*

Photostabilizers

See: *Additives (Agents for Sustaining Properties)*
Additives (Property and Processing Modifiers)
Hindered Amine Light Stabilizers, Monomeric
Photo-oxidation, Polyolefins
Polyethylene (Stabilization and Compounding)

Phthalocyanine-Containing Polymers

See: *Metal Complex Catalysts, Polymeric*
Metallophthalocyanines, Polymeric (Overview)
Metallophthalocyanines, Polymeric (Types of Structures)
Photoconductive Phthalocyanine Polymers
Phthalocyanines, Polymeric
Stacked Transition Metal Macrocycles (Semiconductive Properties)

PHTHALOCYANINES, POLYMERIC

Padma Vasudevan* and Neelam Phougat
Center for Rural Development and Technology
Indian Institute of Technology

H. S. Nalwa
Hitachi Ltd.

Phthalocyanine (H_2Pc) monomer is a highly conjugate-daromatic macromolecule having four isoindolene units linked by aza-bridges to form a cyclic system.

Based on their high chemical and thermal stability and highly conjugated aromatic structure H_2Pc and its derivatives find a variety of applications such as dyes, catalysts, electrocatalyst, gas detectors, high temperature lubricants, rectifying barriers, and photosensitizers.[1–10]

Polymerization leads to a higher degree of conjugation, augmenting the electron delocalization. This results in the modification of its properties depending on the bonding arrangement between the monomer units.[11–17]

The following are the possible modes of bonding the monomer into oligomers and polymers:

- Polymerization by bonding through ligand.
- Polymerization by bonding through metal to form stacks.
- Multinuclear phthalocyanine.
- Polymer bound phthalocyanine.

SYNTHESIS AND PROPERTIES

Polymerization by Bonding Through Ligand

In these polymers, generally adjacent phthalocyanine rings are bonded together by a shared benzene ring, forming a two-dimensional structure.

Polymeric phthalocyanines are generally purified by heating at ~400°C in vacuum.[20] Pure polymeric metal phthalocyanines are greenish blue, blue, or purple in color. In comparison to monomers, polymers have a lower solubility or are insoluble in organic solvents and concentrated H_2SO_4. In general, pMPc have less thermal and chemical stability than the monomer.

Under inert and oxidative conditions polymeric metal phthalocyanines are stable up to 500°C and 340°–360°C, respectively.

Electrical Conductivity

Polymerization increases the degree of conjugation leading to increased conductivity by as much as 7-fold.[18,21–37,67,68,70–72,98] Electrical conductivity of pMPc is reported to be in the range of 10^{-1} Ω^{-1} cm^{-1} to 10^{-13} Ω^{-1} cm^{-1}.

*Author to whom correspondence should be addressed.

Effect of Doping on Conductivity

After doping with electron acceptors conductivity is reported to increase.[38]

Effect of Heat Treatment on Conductivity

Heat treatment of phthalocyanine has been shown to produce a positive effect in terms of its increased electrical conductivity.

Dielectric Properties and Pyroelectric Effect

Polymeric CuPc, pCoPc, and pFePc exhibit a pyroelectric effect.[24,39]

Superconductivity

Some oligo and polyphthalocyanines are reported to show high Tc-superconductivity.

Bridged Phthalocyanine (Bonding Through Metal)

This is a class of 1-dimensional metal phthalocyanine polymer where central metal ion of phthalocyanine is axially linked with bifunctional ligand acting as the bridging ligand L. Solubility of polymer in organic solvent can be increased by introducing bulky substituents in the phthalocyanine ring.[40,41] Bridged phthalocyanine can be classified into three groups:

- Type A bridged phthalocyanine: Having two coordinate bonds between ligand and metal.
- Type B bridged phthalocyanine: Having one covalent and one coordinate bond between ligand and metal.
- Type C bridged phthalocyanine: Having two covalent bonds between ligand and metal.[42,43]

Mixed valence compounds such as $PyPcFeCNPcFe(NH_3)$ are also reported.[44]

Properties of Bridged Phthalocyanine Complexes

Bridged one-dimensional metal phthalocyanine complexes $[PcML]_n$ exhibit good semiconducting properties, even without central oxidative doping.[45–59] $[PcML]_n$ with a transition-metal atom. Fe, Ry, Co, Cr, Mn, Os, and Rh in their various oxidation states as the central metal atom M, and bidentate ligand L, CN^- pyz, SCH^-, tz, p-dib, bpy, pyrimidazene, azide(N_3^-), 1,4-diazabicyclo [2.2.2] octane (dabco), pyridazine as the bridging ligand, show room temperature powder conductivities of 10^{-10}–10^{1} Ω^{-1} cm^{-1} without doping.[45–49,60–67,69,74,99] They exhibit conductivities 10^5–10^6 times higher than the corresponding monomers $PcML_2$, which have higher conductivity than PcM.[50]

Phthalocyanine Bound to Polymer

Polymer-bound phthalocyanine has the combined property of phthalocyanine and polymer (solubility, thermal, and mechanical behavior). The polymer chain can also bond covalently to central metal ion of phthalocyanine. A covalent bond can also be formed between the polymer backbone and the ligand of the metal chelate.

Polynuclear Phthalocyanine

Binuclear, trinuclear, tetranuclear, and pentanuclear phthalocyanines are reported.[75–78,94,97] The benzo group of two or more phthalocyanine molecules is attached to the bridging atom or molecule.

APPLICATIONS

Based on their characteristic properties such as color, high chemical and thermal stability, and semiconductivity, polymeric phthalocyanines find wide-ranging applications. They are used in electrochromic devices, in trays for automated processing of Si wafers, in electrographic photoreceptors, in electrically conductive compositions, in fabrication of photovoltaic cells, in making electrically conductive circuit substrates, in switching devices, as a deodorant, and for screening out UV light.[19,73,79–96]

Phthalocyanine polymers exhibit properties that are generally comparable to epoxy or polyimide resins.

Polymeric metal phthalocyanines are also used in optical recording material, in semiconductor devices, as catalysts, and electrocatalysts.

CONCLUSION

The above discussions indicate that polymeric phthalocyanines with versatile properties have a great potential in wide-ranging applications. They have already entered frontier areas such as nonlinear optics and superconductivity. More concerted research and development in these areas would offer new opportunities.

REFERENCES

1. Ting, R. Y.; Cottington, R. L. *Natl. SAMPE Tech. Conf.* **1980**, *12*, 714.
2. Anon *Res. Discl.* **1985**, *254*, 304.
3. Kamimura, H.; Murakami, K.; Shimada, M. et al. *Chem. Abstract 111*, P235268t.
4. Bogaty, H.; Brown, K. C.; Loveless, N. P. et al. *Chem. Abstract 102*, P209134r.
5. Tokyo Ohka Kogyo Co. Ltd. *Chem. Abstract 98*, P55774j.
6. So, F. F.; Forrest, S. R. *J. Appl. Phys.* **1988**, *63*, 442.
7. Arishima, K.; Hiratsuka, H.; Tate, A. et al. *Appl. Phys. Lett.* **1985**, *46*, 279.
8. Kato, M.; Nishioka, Y.; Kaifu, K. et al. *Appl. Phys. Lett.* **1985**, *46*, 196.
9. Milgrom, L. R. PCT Int. Appl. WO 92 15 099 (Cl. H01B1/12), 03 Sep 1992, GB Appl. 91/3 991, 26 February 1991; pp 11.
10. Anderson, T. L.; Komplin, G. C.; Pietro, W. J. *J. Phys. Chem.* **1993**, *97*, 6577.
11. Willis, M. R. *Mol. Cryst. Liq. Cryst.* **1988**, *171*, 217.
12. Allcock, H. R. *Chem. Eng. News* **1985**, *63*, 22.
13. Snow, A.; Griffith, J. *Encycl. Polym. Sci. Eng.* **1987**, *11*, 212.
14. Berezin, B. D.; Lomova, T. *Izv. Vyssh. Uchebn. Zaved, Khim. Khim. Tekhnol.* **1987**, *30*, 3.
15. Shirai, H. *Hyomen* **1987**, *25*, 422.
16. Hanack, M.; Datz, A.; Fay, R. et al. In *Handbook of Conducting Polymers*; Skotheim, T., Ed.; Marcel Dekker: New York, 1986.
17. Wohrle, D.; Meyer, G. *Kontakte (Darmstodt)* **1985**, 38.
18. Venkatachalam, S.; Prabhakaran, P. V. *Eur. Polym. J.* **1993**, *29*, 711.
19. Marvuglio, P.; Sharrock, R. F.; Kennedy, R. J. *J. Oil Colour Chem. Assoc.* **1978**, *61*, 79.
20. Dainippon Ink and Chemicals, Inc. Kawamura Physical and Chemical Research Institute *Chem. Abstract 103*, P90526y.

21. Meier, H.; Albrecht, W.; Tschirwitz, U. et al. *Ber. Bunsenges. Physik, Chem.* **1973**, *77*, 843.
22. Drinkard, W. C.; Bailer, J. C. Jr. *J. Am. Chem. Soc.* **1959**, *81*, 4795.
23. Nalwa, H. S.; Sinha, J. M.; Vasudevan, P. *Makromol. Chem.* **1981**, *182*, 811.
24. Bijwe, J.; Pandey, P. S.; Vasudevan, P. et al. *Eur. Polym. J.* **1987**, *23*, 167.
25. Anasuya, M. V.; Natarajan, T. S.; Rangarajan, G. et al. *Solid State Commun.* **1991**, *77*, 661.
26. Dirk, C. W.; Inabe, T.; Lyding, J. W. et al. *J. Polym. Sci. Polym. Symp.* **1983**, *70*, 1.
27. Teng, G.; Dalal, N. S. *Synth. Met.* **1989**, *32*, 63.
28. Inoue, H.; Kida, Y.; Imoto, E. *Bull. Chem. Soc. Japan* **1968**, *41*, 684.
29. Dulov, A. A.; Sherle, A. I.; Abramova, L. A. et al. *Vysokomol Soedin., Ser. A* **1991**, *33*, 380.
30. Nalwa, H. S. *Appl. Organomet. Chem.* **1991**, *5*, 203.
31. Hanack, M.; Datz, A.; Fischer, K. et al. *Polym. Prepr. (Am. Chem. Soc., Div. Polym. Chem.)* **1982**, *23*, 126.
32. Shirai, H.; Kobayashi, K.; Takemae, Y. et al. *Polym. Prepr. (Am. Chem. Soc., Div. Polym. Chem.)* **1979**, *20*, 482.
33. Gould, R. D. *J. Appl. Phys.* **1982**, *53*, 3353.
34. Dulov, A. A.; Abramova, L. A.; Sherle, A. I. et al. *Elektron. Org. Mater.* **1985**, 242.
35. Venkatachalam, S.; Vijayan, T. M.; Packirisamy, S. *Makromol. Chem.; Rapid Commun.* **1993**, *14*, 703.
36. Shirai, H.; Kobayashi, K.; Takemae, Y. et al. *Makromol. Chem.* **1979**, *180*, 2073.
37. Venkatachalam, S.; Prabhakaran, P. V. *Eur. Polym. J.* **1993**, *29*, 711.
38. Nalwa, H. S. *Polym. Commun.* **1990**, *31*, 232.
39. Tyagi, S.; Santosh, S.; Mathur, C.; Vasudevan, P. *J. Mater. Sci. Lett.* **1989**, *8*, 893.
40. Kuznesoff, P. M.; Nohr, R. S.; Wynne, K. J. et al. *J. Makromol. Sci. Chem.* **1981**, *A16*, 299.
41. Schoch, K. F. Jr.; Kundalkar, B. R.; Marks, T. J. *Org. Coat. Plast. Chem.* **1979**, *41*, 127.
42. Wu, J.; Pan, Z.; Du, X. et al. *J. Polym. Sci., Part C: Polym. Lett.* **1988**, *26*, 433.
43. Hanack, M.; Metz, J.; Pawlowski, G. *Chem. Ber.* **1982**, *115*, 2836.
44. Hanack, M.; Hirsch, A. *Synth. Met* **1988**, *29*, F9.
45. Metz, J.; Hanack, M. *J. Am. Chem. Soc.* **1983**, *105*, 828.
46. Hanack, M.; Muntz, X. *Synth. Met.* **1985**, *10*, 357.
47. Diel, B. N.; Inabe, T.; Jaggi, N. K. et al. *J. Am. Chem. Soc.* **1984**, *106*, 3207.
48. Schneider, O.; Hanack, M. *Angew. Chem.* **1983**, *95*, 804.
49. Kobel, W.; Hanack, M. *Inorg. Chem.* **1986**, *25*, 103.
50. Zhao, S.; Zheng, Q.; Xie, J. *Huaxue Xuebao* **1989**, *47*, 842.
51. Hedtmann-Rein, C.; Keppeler, U.; Muenz, X. et al. *Mol. Cryst. Liq. Cryst.* **1985**, *118*, 361.
52. Toscano, P. J.; Marks, T. J. *Mol. Cryst. Liq. Cryst.* **1985**, *118*, 337.
53. Van der Pol, J. F.; Neeleman, E.; Miltenburg, J. C. et al. *Macromolecules* **1990**, *23*, 155.
54. Hanack, M.; Deger, S.; Keppeler, U. et al. *Conduct. Polym., Proc. Workshop 1986* **1987**, 173-8.
55. Hanack, M.; Lange, A.; Rein, M. et al. *Synth. Met.* **1989**, *29*, F1.
56. Hanack, M.; Deger, S.; Lange, A. et al. *Synth. Met.* **1986**, *15*, 207.
57. Dirk, C. W.; Mintz, E. A.; Schoch, K. F. Jr. et al. *J. Macromol. Sci., Chem.* **1981**, *A16*, 275.
58. Linsky, J. P.; Paul, T. R.; Nohr, R. S. et al. Report 1980, TR-6; Order No. AD-A088159, 23 pp. Avail. NTIS. From Gov. Rep. Announce. Index (U.S.) 1980, 80(26), 5074.
59. Ziegler, C.; Bufler, J.; Martin, D. et al. *Synth. Met.* **1993**, *55*, 91.
60. Hanack, M. *Israel J. Chem.* **1985**, *25*, 205.
61. Hussein, N. A. (Univ. Nottingham, Nottingham, UK). 1990. 332 pp. Avail Univ. Microfilms Int., Order No. BRD-91051. From Diss Abstract Int. B, **1991**, *51*, 4422.
62. Hayashida, S. *Kobunshi Kako* **1992**, *41*, 343.
63. Deger, S.; Hanack, M. *Synth. Met.* **1986**, *13*, 319.
64. Hanack, M. *Chimia* **1983**, *37*, 238.
65. Hanack, M.; Hayashida, S.; Grosshans, R. *Mol. Cryst. Liq. Cryst. Sci. Technol., Sec. A* **1992**, *217*, 197.
66. Hanack, M. *GIT Fachz Lab.* **1987**, *31*, 75.
67. Pohl, H. A.; Rembaum, A.; Henry, A. *J. Am. Chem. Soc.* **1962**, *84*, 2699.
68. Berlin, Y. A.; Danielyan, N. G.; Beshenko, S. I. et al. *Dokl. Akad. Nauk USSR* **1985**, *282*, 347.
69. Djurando, D.; Hamwi, A.; Fabre, C. et al. *Synth. Met.* **1986**, *16*, 227.
70. Nalwa, H. S.; Vasudevan, P. *J. Mat. Sci. Lett.* **1983**, *2*, 77.
71. Assour, J. M.; Harrisen, S. E. *J. Chem. Phys.* **1964**, *40*, 365.
72. Epstein, A.; Bildi, B. S. *Symp. on Electrical Cond. Inorg. Solids*; Interscience: New York, 1961.
73. Tachibana, K. *Chem. Abstract 106*, P206276k.
74. Hanack, M.; Datz, A.; Fischer, K. et al. *Polym. Prepr. (Am. Chem. Soc., Div. Polym. Chem.)* **1982**, *23*, 126.
75. Lam, H.; Marcuccio, S. M.; Svirskaya, P. I. et al. Report 1989, TR-24; Order No. AD-A205868, 50 pp. Avail NTIS. From Gov. Rep. Announce. Index (U.S.) 1989, 89, Abstract No. 937,238.
76. Hempstead, M. R. Avail NLC. From Diss. Abstract Int. B 48, Pt. 1, 1988, 3564.
77. Nevin, W. A.; Hampstead, M. R.; Liu, W. et al. Report 1986, Tr-10; Order No. AD-A174289/9/GAR, 47 pp. Avail NTIS. From Gov. Rep. Announce. Index (U.S.) 1987, 87, Abstract No. 707,280.
78. Dodsworth, E. S.; Lever, A. B. P.; Seymour, P. et al. Report 1985, TR-3; Order No. Ad-A160784/5/GAR, 44 pp. Avail NTIS. From Gov. Rep. Announce. Index (U.S.) 1986, 86, Abstract No. 605,017.
79. Hotomi, H. *Chem. Abstract 109*, P30275e.
80. Li, H.; Guarr, T. F. *Electroanal. Chem.* **1991**, *297*, 169.
81. Yamana, M.; Kanda, K.; Kashiwazaki, N. et al. *Jpn. J. Appl. Phys., Part 2* **1989**, *28*, L1592.
82. Kashiwazaki, N.; Yamana, M.; Nakano, T. et al. *Kino Zairyo* **1989**, *9*, 46.
83. Osada, Y.; Mizumoto, A.; Tsuruta, H. *J. Macromol. Sci., Chem.* **1987**, *A24*, 403.
84. Li, H.; Gaur, T. F. *J. Electroanal. Chem.* **1991**, *297*, 169.
85. Kunimura, S.; Nakayama, S. *Chem. Abstract 114*, P44545h.
86. To, Y.; Nishioka, Y. *Chem. Abstract 110*, P182939h.
87. Kajiura, S.; Maeda, M.; Mizushima, K. et al. *Chem. Abstract 105*, P15211e.
88. Yamada, A.; Shigehara, J.; Miyazaki, S. et al. *Chem. Abstract 111*, P235055w.
89. Hotomi, H.; Iino, S. *Chem. Abstract 111*, P67883u.
90. Brokken-Zijp, J. C. M.; Gerards, L. E. H.; Hanack, M. *Chem. Abstract 109*, P74859c.
91. Yamada, A.; Osada, Y. *Chem. Abstract 106*, P167292r.
92. Yamada, A., Osada, Y. *Chem. Abstract 106*, P94546q.
93. Loutfy, R. O.; Hsiao, C. K. *Polym. Prepr. (Am. Chem. Soc., Div. Polym. Chem.)* **1982**, *23*, 237.
94. Leznoff, C. C.; Ham, H.; Marcuccio, S. M. et al. Report 1987, TR-12; Order No. AD-A177316/7/GAR, 19 pp. Avail NTIS. From Gov. Rep. Announce. Index (U.S.) 1987, 87, Abstract No. 723,714.

95. Taomoto, A.; Asakawa, S.; Machida, I. et al. *Chem. Abstract 113*, P69717h.
96. Shirai, H. *kagaku to Kogyo (Osaka)* **1989**, *63*, 142.
97. Lam, H. C. Y. 1989 Avail NLC. From Diss Abstract Int. B **1990**, 3995.
98. Shulga, Y. M.; Bulatov, A. V.; Hanack, M. *Synth. Met.* **1993**, *55*, 62.
99. Wei, H. H.; Shyu, H. L. *Polyhedron* **1985**, *4*, 979.

Phthalonitrile Resins

See: *High Temperature Phthalonitrile Resins*

Piezoelectric Materials

See: *Amphiphilic Gas Sensors (Based on Polymer-Coated Piezoelectric Crystals)*
Liquid Crystalline Elastomers
Poly(vinylidene fluoride) (Overview)
Vinylidene Fluoride-Based Thermoplastics (Applications)

Pigments

See: *Alkyd Emulsions*
Carbon Black
Carbon Black (Graft Copolymers)
Colorants (Overview)
Paper Coatings
Polyethylene (Stabilization and Compounding)
Powder Coatings (Overview)

Pitch

See: *Coal Tar Pitch*

Plasma Expanders

See: *Dextran*

Plasma Polymerization

See: *Carbon Fibers (Plasma Surface Treatments)*
Hydrophilic Surfaces (Plasma Treatment and Plasma Polymerization)
Pervaporation Membranes (Materials)
Polyester Fabrics (Surface Modification by Low Temperature Plasma)
Polyethylene (Surface Functionalization)
Ultrathin Films (by Plasma Deposition)

PLASTIC OPTICAL FIBERS

Toshikuni Kaino
NTT Opto-electronics Laboratories

Plastic optical fibers, POFs for short, have many advantages over their glass counterparts. They are easier to handle because of their good ductility and light weight, easier to splice to each other and to light sources because of their large core diameter and high numerical aperture (NA), insensitivity to vibrational stress, and high flexibility notwithstanding their larger fiber diameters. Also these fibers are low in manufacturing cost. Therefore they are expected to be applied as a short distance optical signal transmission medium for certain kinds of data links.

There are three types of optical fibers depending on the core-cladding structure: step index (SI) fibers, graded index (GI) fibers, and single mode fibers. Poly(methyl methacrylate) (PMMA) and polystyrene (PS) are normally used as a core because these polymers are easily purified at the monomer level and no condensation reaction is involved in obtaining transparent polymer.

As a cladding material, the refractive index should be less than 2 to 5% of that of the core material. So PMMA and fluorinated polymers are used as claddings onto PS and PMMA core, respectively.

Because polymers are available over a wide range of refractive indices, the high NA of POFs is easily attainable, which brings about the ease of connection characteristics.

FABRICATION OF POF

To minimize the POF optical loss, the selection of polymerization techniques is important to limit the choice of polymer. High purity polymers can be obtained efficiently by purification at the monomer stage. Light scattering caused by the presence of a particle of contaminants within the core is a major potential source of optical fiber loss.

POFs are manufactured by a melt spinning process, a SI fiber with core-cladding structure being obtained by extrusion.

GI-POFs

In a graded index (GI) fiber, the refractive index varies radially and hence increases the signal bandwidth. So the GI fiber is expected to be a high-speed, data-link, optical-signal transmission medium.

Concerning GI plastic rods with the required quadratic refractive index profile, a two-step copolymerization and a photo-copolymerization technique have been reported.[1,2]

High Thermal Resistance POFs

Higher thermal resistance is required for POF data link systems, especially in the automotive field. POFs are widely used as optical signal transmission media and light guides in automotive applications. In such a system, POFs are exposed to various temperature conditions.[3] Preferably, the temperature at the high end should be higher than 125°C. Copolymerization of MMA with higher thermal resistant cycloaromatic ester acrylate comonomers, such as adamantylmethacrylate-MMA copolymer with thermal resistance temperature of about 125°C, was fabricated.

APPLICATION OF POFs

Application areas for POFs include data-transmission systems in automobiles, aircraft, LANs, and inter- and intra-office network systems. Therefore these areas would provide a good market for POFs. The small diameter, less than 50 μm, of glass optical fibers does create handling problems in some applications.

POF can offer significant advantages over silica glass fiber in short-distance data transmission systems such as optical data links and those found in aircraft and automobiles, where ease of handling and installation are more important than optical attenuation.[4]

REFERENCES

1. Ohtsuka, Y.; Nihei, E.; Koike, Y. *Appl. Phys. Lett.* **1990**, *57*, 120.
2. Koike, Y.; Nihei, E.; Tanio, N.; Ohtsuka, Y. *Appl. Opt.* **1990**, *29*, 2686.
3. Aoyagi, T. *Proc. SPIE* **1987**, *840*, 10.
4. Glen, R. M. *Chemtronics* **1986**, *1*, 98.

Plasticization

See: *Ionomers (Overview)*

Plasticizers

See: *Additives (Property and Processing Modifiers)*
Additives (Types and Applications)
Plasticizers (Polymer-Plasticizer Interactions)
Poly(vinyl chloride) (Structural Order)
Silicone Sealants

PLASTICIZERS (Polymer-Plasticizer Interactions)

Maria Pizzoli and Mariastella Scandola
Dipartimento di Chimica "G. Ciamician" della
Universita di Bologna
Centro di Studio per la Fisica delle Macromolecole del C.N.R.

Plasticizers are substances (usually low molecular weight diluents) that are incorporated into plastic materials to improve their workability and increase flexibility. Polymer and plasticizer form a homogeneous single–phase mixture; intimate mixing of the two components leads to substantial physical property modifications with changing plasticizer content.[1]

In addition to depressing the characteristic temperature (T_g) of the glass-to-rubber transition (where the polymer chain undergoes cooperative segmental motions), the plasticizer also may affect the so–called secondary transitions (that involve local motions of specific molecular groups). Moreover, in partially crystalline polymers (where a 3-dimensionally ordered crystalline phase coexists with disordered amorphous domains) the presence of a plasticizer influences both crystallization and melting.

GLASS TRANSITION OF PLASTICIZED POLYMERS

Plasticization increases polymer workability essentially through a decrease of the polymer T_g. A wide range of studies have analyzed the dependence of T_g depression on the type and amount of plasticizer, and a number of interpretations about the molecular origin of the phenomenon have been suggested. An intuitive picture represents the thoroughly dispersed plasticizer, which is well above its own T_g, as contributing a large amount of free volume to the polymer–plasticizer mixture and promoting the onset of cooperative macromolecular motions at lower temperatures.

EFFECT OF PLASTICIZER ON POLYMER SECONDARY RELAXATIONS

Secondary relaxations in glassy polymers can originate from local motions of a great variety of molecular groups, according to the polymer–specific chemical structure.[2,3] The different molecular origin of secondary relaxation phenomena in polymers makes it impossible to give a univocal description of the effects produced by addition of plasticizers. However, due to the local character of the motions underlying secondary relaxations, changes are likely to occur only if the plasticizer-polymer interaction directly involves the molecular group responsible for the relaxation.

CRYSTALLIZATION AND MELTING OF PLASTICIZED POLYMERS

Crystallinity in polymers can develop only in the range between glass transition and melting temperatures.[4,5] Addition of plasticizer to a crystallizable polymer induces not only a strong T_g depression but also a more moderate decrease of T_m. As a consequence, the crystallization window (the T_m-T_g range) broadens and shifts to lower temperatures, affecting the crystallization behavior of the polymer.

POLYMERIC PLASTICIZERS

Polymeric plasticizers are low T_g polymers that form homogeneous mixtures when blended with a second polymeric component with higher T_g.

Polymers plasticized by polymers are miscible binary polymer blends, a subject extensively investigated in recent years.

REFERENCES

1. Sears, J. K.; Touchette, N. W. *Encyclopedia Polym. Sci. Eng.*: Wiley: New York, 1989; Supplement.
2. McCrum, N. G.; Read, B. E.; Williams, G. *Anelastic and Dielectric Effects in Polymeric Solids*; Wiley: New York, 1967.
3. Heijboer, J. *Secondary Loss Peaks in Glassy Amorphous Polymers in Molecular Basis of Transitions and Relaxations*; Meier, D. J., Eds., Gordon and Breach Science: New York, 1978; p 75.
4. Wunderlich, B. *Macromolecular Physics*; Academic: New York, 1976; Vol. 2.
5. Mandelkern, L. *Crystallization and Melting in Comprehensive Polymer Science*; Pergamon: New York, 1989; Vol. 2, Chapter 11.

Pleistomers

See: *Ultrahigh Molecular Weight Polymers*

Plywood

See: *Wood Composites (High Performance)*

Polyacenaphthylene

See: *Acenaphthylene, Polymerization*

POLYACENIC SEMICONDUCTOR MATERIALS (Application to Rechargeable Batteries)

Kazuyoshi Tanaka* and Tokio Yamabe
Division of Molecular Engineering
Graduate School of Engineering
Kyoto University

Shizukuni Yata
Battery Business Promotion
Kanebo Ltd.

The science and technology of carbon allotropy has recently boomed with the inclusion of such new members as fullerenes and carbon nanotubes. Such "pure" carbon allotrope is interesting in its own right, but marginal members of the allotrope are also of interest because they are expected to show a variety of properties lacked by pure carbon materials. A simplified diagram of such an "extended" carbon family, including the conventional polymers such as polyacetylene and polyethylene, shows a rather ambiguous region called amorphous carbon (a-C) near the one-dimensional graphite (1D-G) line connecting polyacene and graphite.[1,2] The a-C consists of many fractions of condensed aromatic rings incompletely grown and is often respected as defective graphite or even "garbage". Coal, coke, charcoal, pitch, and others are classified into a-C and have seldom been considered as electronic materials that can be compared to graphite.

The polyacenic semiconductor material (PAS) obtained by the pyrolytic treatment of phenol–formaldehyde resin is one of the members of this "garbage" group with only exception that it is used as versatile electronic material that can be doped and undoped like other electrically conductive polymers. It can also accommodate bulky and small dopants in a more stable manner compared with graphite and most of the conductive polymers. Here we summarize our research on the structures and properties of PAS materials and application to the electrode material of rechargeable batteries.

PREPARATION

Normal PAS

Normal PAS samples are prepared by the pyrolysis of phenol–formaldehyde resin molded beforehand.[3,4] The degree of carbonization can be controlled by the pyrolytic temperature (T_p). The PAS material thus prepared has proved that it can be doped with both an electron acceptor (iodine or bromine) and electron donor (lithium or sodium).

Porous PAS

For the sake of doping more bulky dopants, such as AsF_5 or ClO_4, a porous PAS has been developed by pyrolysis of phenol–formaldehyde resin molded with $ZnCl_2$. $ZnCl_2$ is removed in a later step by thorough washing with HCl and water. It has been proved that the porous PAS material thus prepared can be doped with AsF_5 in vapor phase and with $FeCl_3$, $NbCl_5$, $MoCl_5$, $TaCl_5$, or WCl_6 in liquid phase.[6,7] The porous PAS material is used for the electrode in rechargeable batteries as mentioned later.

ELECTRONIC PROPERTIES

Electrical Transport

Measurements of DC electrical transport properties such as DC conductivity, thermoelectric power and magnetoresistance have been performed for the pristine and the heavily doped normal PAS.[8] All of the pristine PAS samples are p-type on the basis of the sign of thermoelectric powers. This signifies that there are some intrinsic levels in all of the pristine PAS samples probably concerning their structures.

APPLICATIONS

Electrochemical Properties

Electrochemical measurements on the porous PAS materials were studied to examine their applicability to the electrodes in secondary batteries.[5,9]

PAS Batteries

Based on the examination of the test–type Li/PAS battery showed a high energy density in the range of 150–160 Wh/l under the combination of, for instance, the 2% p-type and 6% n-type dopings.[10] This is superior characteristics to conventional Ni–Cd or lead–acid batteries. Furthermore, the coin-type Li/PAS battery with 20 mm diameter and 2 mm thickness was found to flow the amount of charge of ca. 12 mAh in the discharging process from 4.0–1.5 V.

Deeply Li-Doped PAS beyond C_6Li Stage

It is of interest to examine how much Li can be doped in the PAS material since the utilization of deeply Li-doped PAS materials as the electrode will further improve the capacity of batteries.[11]

Moreover, in the present PAS electrode, the most deeply doped state up to the 1000 mAh/g corresponding to [C]/{Li} = 2.2 has been achieved. Based on the XPS measurement, the doped Li in such a deep level is not in the metallic but in the partially ionic (δ+ where δ<1) states. It is further possible that the PAS sample doped up to the C_2Li stage could show as high an energy density *per volume* as lithium metal.

*Author to whom correspondence should be addressed.

REFERENCES

1. Yamabe, T.; Tanaka, K.; Ohzeki, K.; Yata, S. *Solid State Commun.* **1982**, *44*, 823.
2. Tanaka, K.; Yamashita, S.; Yamabe, H.; Yamabe, T. *Synth. Met.* **1987a**, *17*, 143.
3. Yamabe, T.; Tanaka, K.; Ohzeki, K.; Yata, S. *J. Phys. (Paris) Colloq. Suppl. No. 6* **1983**, *44*, C3–545.
4. Tanaka, K.; Ohzeki, K.; Yamabe, T.; Yata, S. *Synth. Met.* **1984a**, *9*, 41.
5. Yata, S.; Hato, Y.; Sakurai, K. et al. *Synth. Met.* **1987**, *18*, 645.
6. Tanaka, K.; Koike, T.; Yamabe, T.; Yata, S. *Synth. Met.* **1989a**, *29*, E489.
7. Tanaka, K.; Takeda, S.; Koike, T. et al. *Synth. Met.* **1993a**, *58*, 123.
8. Tanaka, K.; Yamanaka, S.; Koike, T. et al. *Phys. Rev.* **1985**, *B32*, 6675.
9. Yata, S.; Osaki, T.; Hato, Y. et al. *Synth. Met.* **1990b**, *38*, 177.

10. Yata, S.; Sakurai, K.; Osaki, T. et al. *Synth. Met.* **1990c**, *38*, 185.
11. Yata, S.; Kinoshita, H.; Komori, M. et al. *Synth. Met.* **1993**, *55–57*, 388.

Polyacetals

See: *Acetals Resins*
Acetal Resins (Homopolymers Copolymers)
Bicyclic Acetals, Ring-Opening Polymerization

POLYACETYLENE FILMS

Kazuo Akagi* and Hideki Shirakawa
Institute of Materials Science
University of Tsukuba

Ordinary polyacetylene film is composed of fibrils that are bundles of polyene chains. Since the fibrils are randomly oriented, the inherent electrical conductivity of polyacetylene chain is depressed owing to fibril contact resistance. This makes it difficult for polyacetylene to become a complete one-dimensional conductor at the macroscopic level. Nowadays, the primary concern is how to align the fibrils of polyacetylene film in order to achieve higher electrical conductivity.

In the present study, we have developed two kinds of solvent-free polymerization methods that are called *solvent evacuation* (*SE*)[1] and *intrinsic non-solvent* (*INS*)[2,3] methods. The polyacetylene films prepared by the *SE* method showed extremely high mechanical strengths characterized by both Young's modulus of 100 GPa and tensile strength of 0.9 GPa, in addition to high bulk density above 1.0 g/cm^3 and electrical conductivity over 2.2×10^4 S/cm enhanced by mechanical stretching of 6–8 times. Besides, the *INS* method produced high quality polyacetylene films, whose mechanical and electrical properties (i.e., stretchability of 6–9 times and conductivity of $2.2–4.3 \times 10^4$ S/cm) are comparable with and/or superior to those of the films prepared by the *SE* method.

We have confirmed that these methods are promising approaches to synthesize highly conducting polyacetylene films with high stretchability that should be available for new electrical and electronic materials.

EXPERIMENTAL

Solvent Evacuation (SE) Method

Figure 1 shows a schematic representation of the catalyst preparation. Polymerization was carried out at –78°C by introducing acetylene gas onto the catalyst. The present films, unlike usual films, were found not shiny but dull and/or almost black, and also hard and tough irrespective of thickness. *Cis* contents of the present films were 80–90%.

Intrinsic Non-Solvent (INS) Method

Schematic representation of this method is also shown in Figure 1. The polymerization was carried out for 0.5–1 hour. The polyacetylene film prepared was washed several times by toluene cooled at –78°C and dried through vacuum pumping. We emphasize here that all procedures from the catalyst preparation to the end of polymerization are free from solvent.

*Author to whom correspondence should be addressed.

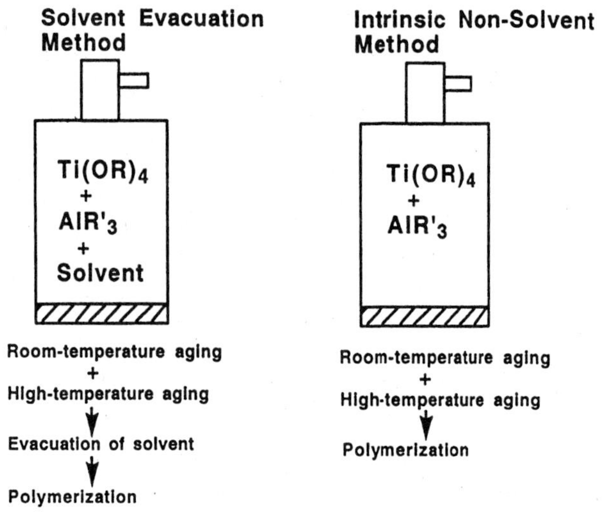

FIGURE 1. Schematic representation of two kinds of solvent-free polymerization methods.

RESULTS AND DISCUSSION

For simplicity, we will first report results from the solvent evacuation (SE) method in the section, and then report those from the *intrinsic non-solvent* (*INS*) method in the last section.

Polymerization Behavior

We have performed two preliminary experiments to deepen our knowledge about the present catalyst system and the catalysis of acetylene polymerization. (i) Effect of thermal treatment of the catalyst upon the catalytic activation was examined by focusing on production yield of polyacetylene gels. (ii) Effect of *SE* on polymerization behavior was examined through formation of polyacetylene films.

Next, we synthesized polyacetylene films by means of *SE* method linked with the thermal treatment of catalyst.

MECHANICAL PROPERTIES

The stretchability of the films was more enhanced as the catalyst was aged at higher temperature and for longer time within the regions of aging temperature and time concerned here. Actually, the maximum draw ratio of 8.6 was obtained with the aging at 150°C for 5 h. Polyacetylene films with thickness of several to about 100 μm were then subjected for stress–strain tests. Pronounced increase of mechanical strength is recognized in stretched films, especially *trans* stretched one.

Young's modulus and tensile strength of *trans* films stretched by 7–8 times were 100 GPa and 0.9 GPa, respectively. Those of *cis* films with the same draw ratio were 25–30 GPa and 0.6 GPa, respectively. Here it is worthwhile to emphasize that mechanical strengths of the present *trans* films are comparable to those of well known engineering plastics, such as Kevlar, poly(*p*-phenylene terephthalamide), whose Young's modulus and tensile strength are 132 GPa and 3.9 GPa, respectively.

X-RAY DIFFRACTION AND DSC ANALYSES

As the mechanical stretching proceeds, the half width of the X-ray diffraction peak ($\Delta 2\theta = 2.0°$) decreased drastically to the value of 1.5 (draw ratio of 6). This indicates that the stretching increases the order of the crystal structure.

A characteristic feature of the present film is that the DSC main peak shifts to higher temperature (155°C) and the half width of X-ray diffraction peak decreases from 2.0–1.5 in $\Delta 2\theta$ with increase of the draw ratio. This means that the less ordered crystalline structure of as-grown film changes to more ordered one in accompany with the stretching.

RELATION OF ORIENTATION ANGLE AND DRAW RATIO

We evaluated an alignment of polyene chains of film using an orientation angle (θ), which is defined as a statistically averaged value between the polyene chain and the stretching direction of film assuming that the polyene chain is cylindrically symmetric.[4,5] The film stretched by 7 times shows an orientation angle less than 10°, demonstrating that the highly stretched film is composed of highly aligned polyene chains, irrespective of *cis* and *trans* form.

MORPHOLOGY

The morphology of the thick film was different between the surface layer part and the internal part.

The stretchability and the properties of the films depend on how uniformly and how highly the granular structure is drawn as a whole. The stretchability of the granular structure is affected by its morphology and/or its higher order structure.

ELECTRICAL CONDUCTIVITY

Doping was carried out by exposing polyacetylene films to vapor of iodine (I_2). The most favorable results were obtained aging at 150°C for five hours, where parallel and perpendicular conductivities were respectively 2.2×10^4 and 3.2×10^2 S/cm and electrical anisotropy about 70. These values are evidently due to high stretchability of the film with draw ratio of ca. 7–8. At the same time, it must be stressed that in the case of the aging at 110°C for 3–5 hours, the unstretched film having such a relatively high conductivity as 1.0×10^3 S/cm can also exhibit high conductivity over 2.0×10^4 S/cm after stretching even though the draw ratio was 4–5. This result indicates that inherently high conductivity and high stretchability of unstretched as-grown films, both of which are respectively correlated with conjugation length and mechanical strength, are indispensable factors for achievement of much higher electrical conductivity in the stretched film.

INTRINSIC NONSOLVENT (INS) POLYMERIZATION

Polyacetylene films obtained here are black and dull color with no metallic luster, as observed in films prepared by the *SE* method. Such an appearance presents a striking contrast to typical S-type films with metallic luster that are synthesized using a room-temperature aged catalyst with solvent.

In the present case using a high-temperature aged catalyst without solvent, the optimal Al–Ti ratio is found to be 4–5, especially for synthesis of highly stretchable polyacetylene film.

The as-grown film has a high bulk density of 1.0 ~ 1.15 g/cm^3, very close to a true density (1.20 g/cm^3) of polyacetylene,[6] a *cis* content of 85 ~ 95%, and a maximum stretchability of 6 ~ 9 times. The electrical conductivity of iodine–doped stretched film is 2.2×10^4 ~ 4.3×10^4 S/cm. The *cis* film stretched up to the maximum draw ratio showed mechanical strengths of 28 GPa in Young's modulus and 800 MPa in tensile strength. The *trans* film, prepared by thermal heating of the *cis* film at 150°C for 30 min, showed Young's modulus of 40 GPa and tensile strength of 2100 MPa. It is worth emphasizing that the tensile strength (2100 MPa) of the *trans* film is the highest value among those reported so far and even comparable to that (3900 MPa) of a typical engineering plastic called Kevlar, poly(*p*-phenylene terephthalamide).

It is of keen interest to compare the present results with those by the *SE* method. The Young's modulus of the *trans* film (40 GPa) by the *INS* method is about one third smaller than that (100 GPa) by the *SE* method, but the tensile strength (2100 MPa) by the former is about 2 times higher than that (900 MPa) by the latter. This means that the polyacetylene film synthesized with the *INS* method is more easily elongated but less breakable than the film with the *SE* method.

The mechanical stretchability (l/l_0) of the film by the *INS* method is somewhat superior to that by the *SE* method, giving rise to a higher conductivity of stretched film after iodine doping.

CONCLUSION

We have developed two kinds of solvent-free acetylene polymerization methods. Polyacetylene films prepared by the *SE* and *INS* methods have been found to exhibit extremely high mechanical strengths characterized by Young's modulus of 100 GPa or tensile strength of 2.1 GPa, in addition to high bulk density more than 1.0 g/cm^3 and high electrical conductivity of 2.0 ~ 4.0×10^4 S/cm enhanced by uniaxial alignment of polyene chains associated with mechanical stretching of 8 ~ 9 times.

We are now prepared to say that high mechanical strength is a prerequisite for achievement of high electrical conductivity.

REFERENCES

1. Sinclair, M.; Moses, D.; Akagi, K.; Heeger, A. J. *Nonlinear Optical Properties of Polymers, Materials Research Society Symposium Proceedings* Heeger, A. J.; Orenstein, J.; Ulrich, D. R., eds.; Materials Research Society, Pittsburgh, 1988, Vol. 109, p 205.
2. Akagi, K.; Sakamaki, K.; Shirakawa, H. *Macromolecules* **1992**, *25*, 6725 *Synth. Met.* **1993**, *55*, 779.
3. Akagi, K.; Sakamaki, K.; Shirakawa, H.; Kyotani, H. *Synth. Met.* **1995**, *69*, 29.
4. Zbinden, R. *Infrared Spectroscopy of High Polymers*; Academic: New York, 1964; p 166.
5. Galtier, M.; Charbonnel, M.; Mountanen, A.; Ribet, J. L. *Polymer* **1984**, *25*, 1253.
6. Abadie, M. J. M.; Hacene, S. M. B.; Cadene, M.; Rolland, M. *Polymer* **1986**, *27*, 2003.

Polyacetylenes

See: *Acetylene Polymers*
Contact Lenses, Gas Permeable
Electrically Conducting Composites
Liquid Crystalline Polyacetylene Derivatives
Living Coordination Polymerization
Polyacetylene Films
Poly(cyclopentadienylene vinylene)
Polyphenylacetylene
Polyphenylacetylene-Based Permselective Membranes
Poly(phenylene ethynylene) (Synthesis and Optical Properties)
Polysilylpropynes (Steric Effects on Material Properties)

Polyacrolein

See: *Acrolein (Polymerization, Utilization)*

Polyacrylamide

See: *Acrylamide (Polymerization and Application)*
Acrylamide, Inverse Emulsion Polymerization (in Supercritical Carbon Dioxide)
Drag-Reducing Polymers
Inverse-Emulsion/Suspension Polymerization
Water-Soluble Polymers (Oil Recovery Applications)
Water-Soluble Polymers (Rheological Properties)

Polyacrylates

See: *Acrylic Elastomer Curing*
Acrylic Rubber
Peroxide Initiators (for High Solids Polyacrylate Resins)
Water-Borne Coatings (Urethane/Acrylic Hybrid Polymers)

Poly(acrylic acid)

See: *Dental Polymers (Hydrogels)*
Drag-Reducing Polymers
Water-Swellable Polymers (Carbomer Resins)

Polyacrylonitrile

See: *Carbon/Carbon Composites (with Surface Treated Carbon Fibers)*
Carbon Fibers (Overview)
Polyacrylonitrile Gel Fibers
Precipitation Polymerization (Overview)
Thermoreversible Gelation (Polyacrylonitrile Solutions)

POLYACRYLONITRILE GEL FIBERS

Derek M. Ölé Kiminta
Department of Chemical Engineering and Chemical Technology
Imperial College of Science Technology and Medicine
University of London

In recent years, certain high-strength fibers have found their way outside the classical textile industry and into various applications, such as reinforcement of thermoplastics and duraplastics. Indeed, they are being used in buildings, boats, spacecrafts and automobiles. At initial stages of processing, polyacrylonitrile is taken in a bundle of filaments and annealed in air at about 230°C at a constant low tension. The fibers are then boiled in a NaOH aqueous solution for 45 min and then cleaned and flashed with distilled water to remove traces of loose polymers and impurities. They are then dried in low heat in a vacuum. In this process the initial stages are systematic; that is, bulk PAN was converted into fiber, which was conditioned to suit the processing needed and alignment of the molecular chains along the fiber structure. It was then heated further by stabilizing it in oxygen at 200–300°C. The stabilized fiber was then carbonized at 920°C (typically above 1000°C) in an inert environment (N_2) for 10 hours. In a recent study,[1,2] ammonia accelerated the stabilization of certain acrylic precursors in the production of carbon fibers. The fibers are composed of fibrils that are aligned parallel to the fiber axis during manufacturing (drawing). The slow oxidation is essential during the conversion of polyacrylonitrile into carbon fibers where stabilization and orientation must be maintained to avoid fusion during the high temperature treatment. PAN precursor fiber has fine filaments and is considered more suitable because its oxidation time is shorter.

GENERAL DISCUSSION ON PAN FIBER PRODUCTION PROCESSES

During carbon fiber production, a high degree of orientation along the fiber axis is needed. To this end, a stretching treatment on the precursor fibers should be done during the pyrolysis process (**Figure 1**).

Three advantages of PAN-based fibers are a high degree of molecular orientation, a high melting point and a much greater yield of the carbon fibers.

The only disadvantage of PAN fibers is shrinking, which usually disturbs the orientation that affect the mechanical properties of the carbon fibers.

Shrinkage can be prevented by applying tension to the gel fibers or slightly stretching the precursors during treatment.

Interestingly, the network of PAN gel fibers contains cationic groups and ionic groups, respectively, making the PAN gel fibers amphoteric. The nitrile group is known to be highly polar. Because of this, dipole–dipole forces may result and usually act as crosslinks. This makes the polymer soluble only in highly ionized solvents.

When PAN precursor fibres are being converted into carbon fibers, first they are oxidized within an atmosphere selected, at temperatures ranging from 200–300°C and must be placed under low tension at this stage for better orientation and to prevent fiber shrinkage. Once the thermoplastic PAN fibers are

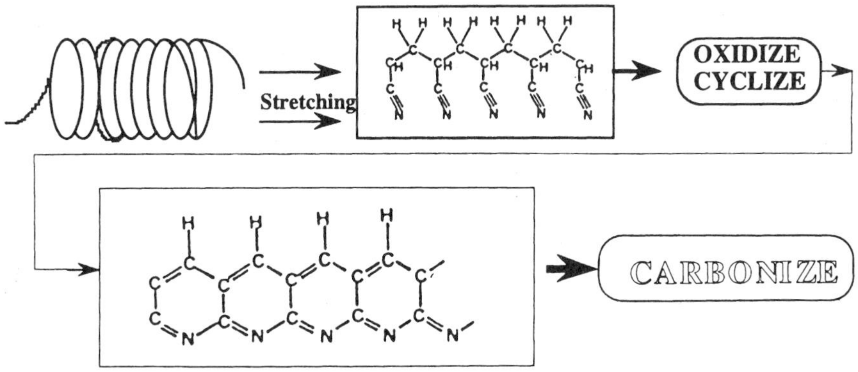

FIGURE 1. A schematic illustration of stages through which PAN gel fibers must go through to acquire necessary mechanical properties and preferred orientation. *Source*: Reference 3.

converted into a nonplastic cyclic compound (ladder), heat treatment processing is continued at much higher temperatures. The fibers are then carbonized at temperatures of 1000°C and higher.

CONCLUSION

This study suggests that an important change takes place during the stabilization of acrylic fibers where active sites form atoms into the polymer, hence initiating polymerization of nitrile groups that form a condensed heterocyclic ring structure. The samples that were heated in the tube–oven with oxygen-free nitrogen exhibited an extreme sensitivity toward oxygen when below 100°C. On the synchroton scattering and time resolved studies of the structural changes taking place during the stabilization reactions, we strongly believe that by using about 5% ammonia with air in the stabilization environment, the time required to stabilize can be reduced in half. Also, we assume that the conditions and method of arranging and rearranging the fiber structure, which is possibly only during stabilization, can possibly design, custom tailor, or influence the forming microstructures of the fiber for a suitable specific use of the product (carbon fibers).

REFERENCES

1. Watt, W. *Int. Conf. Carbon*, London, 1970.
2. Kiminta, D. M. Ö.; Bhat, G. *ORNL Annual Report (summer '91) (ORNL/SURA)*, 1991, Paper presented at Oak Ridge National Laboratory (ORNL), October 4, 1991.
3. Ross, J. H. *Appl. Polymer. Symp.* **1976**, *29*, 151.

POLY(N-ACYLETHYLENIMINE)S

Cristofor I. Simionescu and Geta David
Department of Macromolecules
"Gh. Asachi" Technical University

Poly(N-acylethylenimine)s are ethylenimine polymers resulting from ethylenimine 1- and 2-oxazoline derivatives.

The growing interest in their study, having a fundamental or applicative character, is related either to the synthesis chemistry or to the special properties that make them useful materials in various areas.

SYNTHESIS

Monomers

The most convenient and versatile method to prepare poly(N-acylethylenimine)s is the cationic ring-opening polymerization of 2-substituted-2-oxazolines, developed in 1960s.

HOMOPOLYMERS

Polymerization Mechanism

The polymerization of 2-oxazolines is characterized by initiation with an electrophilic reagent,[2–6] including Lewis acids and their salts, protonic acids, alkyl esters such as chloroformates, sulfonate esters (i.e., methyl tosylate methyl triflate) and sulfonic anhydrides, alkyl halides, allylic halides, propargylic halides, organic electron acceptors.

Polymerization proceeds by ring opening via cationic propagating species.

Usually the initiator interacts with the nitrogen lone pair of electrons withdrawing the electron density from the ring, allowing the generation of the N-substituted-2-oxazolinium cation.

When organic electron acceptors are used as electrophile reagents, the initiation takes place via an intermediate n–π charge transfer.[6]

Saegusa[7] evidenced that in the polymerization of cyclic iminoethers, the propagation can proceed by ionic **11** or covalent **12** species, or concurrently from both, as a consequence of counterion nature (especially its nucleophilicity) (**Scheme I**).

The propagation exhibits living characteristics, the onium or covalently bonded species being stable when favorable conditions (high purity of the components, low monomer–initiator ratio, polar solvent with no protic hydrogens) are realized.

Branched Poly(N-acylethylenimine)s

In the last decade many papers were concerned with obtaining star-shaped poly(N-acylethylenimine)s. The general method is to use multifunctional initiators, that is,[2,4,8] chloroformates of

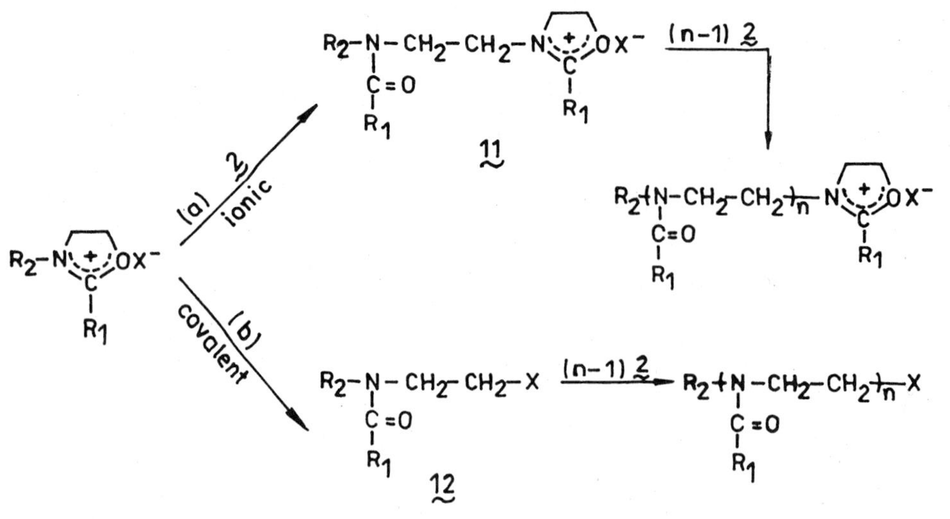

I

2,2-*bis*(hydroxymethyl)-1-butanol, 1,3,5-*tris*(bromomethyl)benzene, allylic tetrahalide.

COPOLYMERS

Statistical and Alternating Copolymers Zwitterionic Copolymerization

The most recent data on statistical or alternating copolymers of 2-alkyl-2-oxazolines are related to their spontaneous copolymerization via zwitterion intermediates.

Evidenced by Saegusa and Kobayashi, the zwitterionic copolymerization, considered also by some authors as a spontaneous alternating copolyaddition, is still an interesting intensively studied subject.[9,10] It allows the obtaining of copolymers from pairs of nucleophilic and electrophilic monomers not usually able to copolymerize by radical or ionic mechanism.

Graft Copolymers

Recently, a number of poly(N-acylethylenimine) macromonomers with styryl, (meth)acryl, glycol, vinyl ester and butadiene type polymerizable groups have been developed by the research group of Kobayashi with the aim of designing a variety of well-defined graft copolymers, usually possessing surfactant properties.[11–15] Hyperbranched polyethylenimine-poly(N-propionylethylenimine) copolymers were synthesized by Tomalia by a "graft on graft" procedure.[16]

Block Copolymers

The synthetic procedures yielding block copolymers containing poly(N-acylethylenimine) sequences are preferentially based on the living nature of cationic ring-opening polymerization of 2-substituted-2-oxazolines.

The use of polyfunctional initiators yields star block copolymers.[1,2]

PROPERTIES AND APPLICATIONS

Morphological Features

As a consequence of their highly symmetric backbones with a 2_1 helical conformation most poly(N-acylethylenimine) homopolymers, block copolymers and even random copolymers from 2-substituted-2-oxazoline comonomers can easily crystallize in a similar manner, with the lateral chains alternating on each side of the main chain.

Polymer Modification

The reaction used most often to modify poly(N-acylethylenimine) homo- and copolymers is hydrolysis to polyethylenimine under acidic or alkaline conditions.[1]

Poly(N-Acylethylenimine)s-Polymers for Advanced Technologies

Many possible applications of poly(N-acylethylenimine)s are available based on their properties; i.e., hydrophilicity or hydrophobicity, depending on side–chain nature; miscibility with commodity polymers; biocompatibility or no acute toxicity; high surface activity; adhesivity; gel and chelating properties; and ability to be easily modified. Nonionic surfactants, exchange and chelating resins, catalysts, coatings, adhesives, nonionic gels, engineering plastics, solid polymer electrolytes and biomedicine include these polymeric materials in the great family of polymers for advanced technologies.[1]

Considering the data presented here, poly(N-acylethylenimine)s still remain a subject of major interest more than 20 years after their discovery.

REFERENCES

1. Kobayashi, S. *Progr. Polym. Sci.* **1990a**, *15*, 751.
2. Dworak, A.; Schulz, R. C. *Makromol. Chem.* **1991**, *192*, 437.
3. Kobayashi, S.; Uyama, H.; Narita, Y. *Macromolecules* **1990b**, *23*, 353.
4. Kobayashi, S.; Uyama, H.; Narita, Y.; Ishiyama, Y. *Macromolecules* **1992a**, *25*, 3232.
5. Kobayashi, S.; Uyama, H.; Mori, T.; Narita, Y. *Chem. Lett.* **1991a**, 1771.
6. Simionescu, C. I.; David, G.; Grigoras, M. *Eur. Polym. J.* **1987a**, *23*, 689.
7. Saegusa, T.; Kobayashi, S. *Makromol. Chem., Macromol. Symp.* **1986**, *1*, 23 and references therein.
8. Cai, G.; Litt, M. H. *J. Polym. Sci.; Part A: Poly. Chem.* **1989**, *27*, 3603.

9. Saegusa, T.; Kobayashi, S.; Kimura, Y. *Pure Appl. Chem.* **1976**, 48I, 307.
10. Simionescu, C. I.; Negulescu, I. I. *Tratat de chimia compusilor macromoleculari*; Romanian Academy Ed., Bucuresti, 1993; Vol. 4.
11. Uyama, H.; Kobayashi, S. *Macromolecules* **1991**, *24*, 614.
12. Kobayashi, S.; Kaku, M.; Sawada, S.; Saegusa, T. *Polym. Bull.* **1985**, *13*, 447.
13. Kobayashi, S.; Masuda, E. Shoda, S.; Shimano, Y. *Macromolecules* **1989**, *22*, 2878.
14. Kobayashi, S.; Uyama, H.; Shirasaku, H. *Makromol. Chem., Rapid Commun.* **1990d**, *11*, 11.
15. Kobayashi, S.; Uyama, H.; Narita, Y. *Makromol. Chem., Rapid Commun.* **1992b**, *13*, 337.
16. Tomalia, D. A.; Hedstrand, D. M.; Ferrito, M. S. *Macromolecules* **1991**, *24*, 1435.

Polyacylhydrazines

See: Carbonylation Polymerization (Palladium-Catalyzed)

Polyacylisocyanate

See: Methacryloyl Isocyanate and Methacryloylcarbamate

Poly[(acyloxy)aloxane]s

See: Alumina Fibers (from Poly[(acyloxy)aloxane]s

Polyaddition

See: Bisallenes, Polyadditions
Electron Transfer Reactions (Novel Polymerization by Divalent Samarium)
Poly(alkylene phosphate)s
Polyesters [Bis(cyclic ether) and Diacyl Chloride Polyaddition]

POLYADDITION, RADICAL

Eiichi Kobayashi
Department of Industrial Chemistry
Science University of Tokyo

Addition of -OH, -COOH, -SH, -COSH, >P-H, >NH and Si-H to carbon–carbon double or triple bonds may take place by electrophilic, nucleophilic, or radical addition in two-step processes. Sometime, simultaneous addition takes place. For instance, hydrogen sulfide and mercaptans add to olefins and acetylenes by an electrophilic mechanism in the absence of initiator but the reaction is usually very slow. However, in the presence of free radical initiator, these compounds containing an -SH group add to olefins and acetylenes to give anti-Markownikoff orientation products. Recently, the geometric structure of addition products and transition state of addition reactions have been investigated to clarify polyaddition mechanisms and polymer structures.

TRANSITION STATE

Radical addition takes place in two steps, radical addition to olefins or acetylenes, and hydrogen transfer between the intermediate carbon radical and -SH group. The reaction order, substituent effect on the relative reactivity, and quantum chemical calculation of molecular orbital are effective diagnostic tool to estimate transition state.

POLYADDITION

Few efforts have been made for the polyaddition of dithiol to diolefins, except for hexamethylenedithiol to 1,5-hexadiene to give a crystalline polyalkylene sulfide.[1–5] Telechelic monodispersed dithiols were synthesized from 2-mercaptoethyl ether with 1,5-hexadiene or 1,9-decadiene by slow dropwise addition of the diene into an excess of the dithiols.[6] Telechelic polymers also synthesized from 1,3-benzenedithiol with 1,4-divinylbenzene.[7] Heat-resistant polysulfide was obtained by addition of 1,3-benzenedithiol in a strained and unsaturated ring compound such as norbornadiene.[8] The author and co-workers have extensively studied polymer syntheses in solution and in solid-state.[9–11] Solid-state polymerization of 1,4-benzenedithiol (BDT) to 1,4-diethynylbenzene (DEB) gave a thin polymer film onto glass plate.[10] The polymer film is constructed by a layer structure. The phase transition of the layer structure is reversibly effected by photon mode.[12,13]

REFERENCES

1. Marvel, C. S.; Chambers, R. R. *J. Am. Chem. Soc.* **1948**, *70*, 993.
2. Marvel, C. S.; Aldrich, P. H. *J. Am. Chem. Soc.* **1950**, *72*, 1978.
3. Marvel, C. S.; Hill, Jr., H. W. *J. Am. Chem. Soc.* **1951**, *73*, 481.
4. Marvel, C. S.; Markhart, Jr., A. H. *J. Am. Chem. Soc.* **1951**, *73*, 1064.
5. Marvel, C. S.; Kotch, A. *J. Am. Chem. Soc.* **1951**, *73*, 1100.
6. Ameduri, B.; Berrada, K.; Boutevin, B.; Bourden, R. D.; Pereira, L. *Polymer Bull.* **1991**, *26*, 377.
7. Nuyken, O.; Reuschel, G.; Siebzehnübl, F. *Makromol. Chem., Macromol. Symp.* **1989**, *26*, 313.
8. Nuyken, O.; Völkel, T.; Pöhlmann, T. *Makromol. Chem.* **1991**, *192*, 1959.
9. Kobayashi, E.; Ohashi, T.; Furukawa, J. *Makromol. Chem.* **1986**, *187*, 2525.
10. Kobayashi, E.; Ohashi, T.; Furukawa, J. *J. Polymer Sci. Part A. Polymer Chem.* **1987**, *25*, 2077.
11. Kobayashi, E. *High Polymers, Japan* **1989**, *38*, 836.
12. Kobayashi, E.; Ohashi, T.; Furukawa, J. *Polymer Adv. Tech.* **1991**, *2*, 301.
13. Kobayashi, E.; Ohashi, T.; Furukawa, J. *Polymer Adv. Tech.* **1992**, *3*, 81.

Poly(β-alanine)

See: Nylon - 3

Polyalkenamers

See: Metathesis Polymerization, Cycloolefins

Polyalkenylenes

See: Metathesis Polymerization, Cycloolefins

POLY(n-ALKYL L-GLUTAMATE) (Structure and Molecular Mobility in Solid and Liquid Crystalline State)

Takeshi Yamanobe
Department of Chemistry
Gunma University

Poly(γ-n-alkyl L-glutamate) (PALG) has the structure shown in **Figure 1**. It is well known that the poly(L-glutamate) derivatives adopt a rapid α-helical main chain conformation. However, the n-alkyl side chain is flexible. Therefore, PALG has both rigid and flexible parts. Depending on the ratio of rigid and flexible parts, PALGs show interesting features.[1] The polymers with n-alkyl side chains longer than a n-decyl group form a side chain crystalline phase composed of paraffin-like crystallites. The crystallization forces the α-helical main chain to pack into a characteristic layer structure and the crystallites are located between the layers of the main chain.

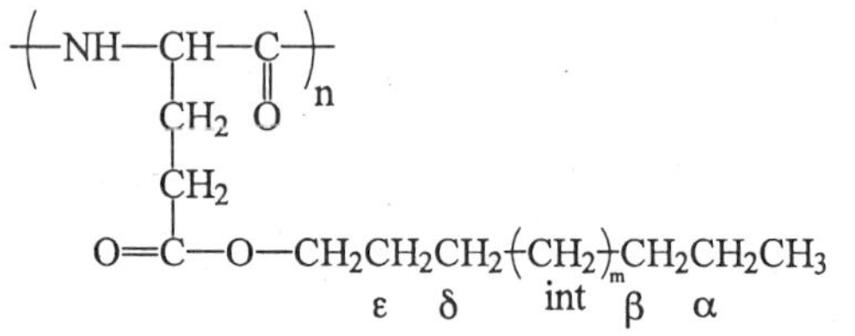

FIGURE 1. Structure of poly(γ-n-alkyl L-glutamate).

As temperature increases, the crystallites of the n-alkyl side chain melt and the polymers form a thermotropic cholesteric liquid crystal. The configurational freedom of the n-alkyl side chain within the melt has increased and the molecular motion of the side chains is free from the restraint of the α-helical main chain. In other words, the side chain within the melt behaves as a solvent, which makes the rigid main chain easy to rearrange. By further elevation of the temperature above 200°C, the cholesteric liquid crystal changes to a columnar hexagonal phase.[2] Thus, the mobility of the side chain plays an important role in the formation of the liquid crystalline state.

MOLECULAR MOTION OF THE MAIN CHAIN

The mobility of the main chain is much slower than that of the side chain for all PG–ns. The molecular motion rate can be deduced from the ^{13}C NMR spectrum. As the temperature is increased, a progressive broadening of the CO(amide) and C_α peaks is observed.[3] As suggested by Lyerla et al.,[4] the broadening phenomenon can be explained on the basis that the reorientation rate of the main chain in the liquid crystalline state is close to the decoupling frequency for the dipolar interaction with the protons. This reduces the efficiency of the decoupling and causes a maximum line width of the carbons when the molecular motion occurs at the frequency that corresponds to the amplitude of the proton decoupling field (about 60 kHz). Therefore, it can be said that the main chain is undergoing reorientation at an approximate frequency of 60 kHz.

A mode of molecular motion can be obtained by measuring the ^2H NMR spectrum. To get information only about the main chain of PG–n, the ^1H of the NH group in the main chain is replaced with ^2H. At low temperature, the spectrum shows the typical powdered pattern represented by three different principal values. As the temperature increases, the two principal values that are perpendicular to the helical axis become averaged and the principal value that is parallel to the helical axis does not change. This means that the main chain is undergoing rotation about the helical axis at an approximate frequency of 60 kHz.

At low temperatures the side chain crystallite is formed and the main chain takes an α-helical conformation. The side chain crystallite transforms into a rotator phase below the melting point. Above melting point of the side-chain crystallite, the n-alkyl side chain moves like the liquid paraffins in the liquid crystalline state. Even in the liquid crystalline state, the main chain takes an α-helix conformation with rotations about the α-helix at 60 kHz.

REFERENCES

1. Watamabe, J.; Ono, H.; Uematsu, I.; Abe, A. *Macromolecules* **1985**, *18*, 2141.
2. Watanabe, J.; Takashina, Y. *Macromolecules* **1991**, *24*, 3423.
3. Yamanobe, T.; Tsukahara, M.; Komoto, T.; Watanabe, J.; Ando, I.; Uematsu, I.; Deguchi, K.; Fujito, T.; Imanari, M. *Macromolecules* **1988**, *21*, 48.
4. Lyerla, J. R.; Yannoni, C. S.; Fyfe, C. A. *Acc. Chem. Res.* **1982**, *15*, 208.

POLY(ALKYLENE NAPHTHALATE)

Chun-Shan Wang and Yih-Min Sun
Department of Chemical Engineering
National Cheng Kung University

Although the demand of poly(ethylene terephthalate) (PET) and poly(butylene terephthalate) (PBT) is great, better thermal and mechanical properties are desired for some applications. Whereas polyimides are superior in these properties, their high price and processing problems remain obstacles. The newly developed high performance polyalkylene naphthalate seems to fill the gap between the two. Polyalkylene naphthalate with a naphthalene ring and flexible aliphatic diol structure is easy to process and exhibits superior physical and mechanical properties than widely-used polyesters. Of poly(alkylene naphthalate)s, poly(ethylene 2,6-naphthalate) (PEN) and poly(butylene 2,6-naphthalate) (PBN) have attracted worldwide attention because of their superior thermal, mechanical, barrier and chemical resistance properties. Much research is under way to develop commercial applications for these new high performance polyesters. In essence, poly(alkylene naphthalate)s extend the limits of polyesters, improving performance of existing applications while making polyesters desirable in entirely new applications.

PREPARATION

Poly(alkylene naphthalate)s can be produced[1–3] either via direct esterification (**Equation 1**) or ester exchange process (**Equation 2**).

The attempted preparation of high molecular weight polyalkylene naphthalates via the interfacial process has not been successful owing to the very low solubilities of polyalkylene

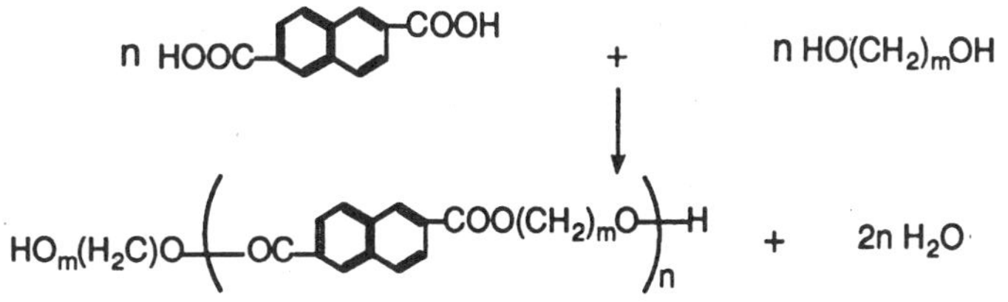

1

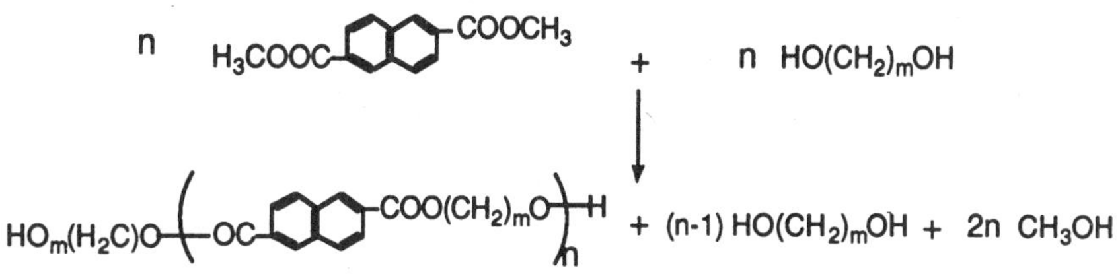

2

naphthalates in the reaction solvents, such as methylene choride dimethylformamide etc.

The transesterification process is generally the preferred process for the manufacture[4–14] of polyalkylene naphthalate.

Studies on the formation of polyethylene naphthalate were reported in detail.[15,16] The transesterification of 2,6-dimethyl naphthalate with ethylene glycol was kinetically investigated in the presence of various catalyst at 185°C.[15]

PROPERTIES

PEN structure is shown in **Structure 1**.

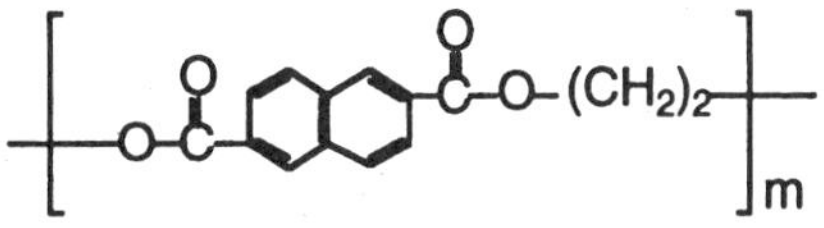

1

Although PET film has well balanced mechanical, electrical and chemical properties and has been widely applied in various fields. The trend toward miniaturization, better properties and safety consideration has demanded higher performance film and PEN film has fulfilled these requirements. PEN with its rigid and flat naphthalene ring structure provides superior mechanical, electrical and chemical properties than that of PET. PEN has specific gravity slightly lower, crystallization rate about seven times slower and melt viscosity about ten times larger than that of PET.

Biaxually stretched PEN film can be produced in the same manner as PET film and therefore, the same equipments and the film making technology for PET can be applied in the manufacturing of PEN film. During processing, the higher melting point, T_g and melt viscosity of PEN than that of PET has to be considered.

Mechanical Properties

PEN film has higher tensile strength and Young's modulus, lower elongation than that of PET.

Thermal Properties

Dimensional Stability

The T_g of PEN is about 45°C higher than that of PET. Usually, higher modulus means higher thermal shrinkage for thermoplastics, however, PEN is not the case. PEN film has higher modulus than that of PET film but has lower thermal shrinkage.

Oligomers

PET is notorious for its migration of oligomers to surface upon heating (surfacing of oligomers is observed at 110°C and cloudiness of PET film is noticeable after 1 minute at 180°C). No surfacing of oligomer is observed on PEN film at 160°C, the migration of oligomers at 180°C is only one tenth that of PET.

Optical Properties

Both PEN and PET are transparent in the visible light region. In the UV light region, PET transmits light up to 310 nm while PEN film blocks light below 380 nm wave length.

APPLICATIONS

Poly(ethylene naphthalate) (PEN)

PEN has thermal, mechanical, barrier and chemical resistance properties which are superior to those of PET. In essence, PEN extends the limits of polyesters, improving performance of existing applications while making polyesters desirable in entirely new applications.

Recording Tape[17–21]

Using PEN film as the substrate for videotape will help develop LP (long playing time), and HD (high density) tapes.

Other Films[22–26]

Dimensional stability also makes PEN an excellent choice for membrane touch switches such as those found on appliances, computers, and other equipment. PEN is also a leading candidate for electrical films, such as electrical insulations for motors, transformers, generators, wire, and cable.

Packaging[27–33]

PEN's benefits really shine in food and medical packaging, where it could bring the advantage of plastics to products ordinarily packaged in other materials. PEN bottles had excellent barrier properties against water vapor, CO_2 and O_2 as much as five times greater than PET, and expected to be used for returnable/refillable and hot-fill applications. PEN also provides protection from UV light degradation.

Fiber[34–36]

PEN's modulus, tenacity and thermal characteristics, coupled with its low thermal shrinkage, make it an attractive candidate for reinforcing. With improved chemical resistance as well, PEN fibers are excellent choice for hosing and belts.

POLY(BUTYLENE NAPHTHALATE) (PBN)

PBN with excellent mechanical strength, heat resistance, dimensional stability, resistance to chemical, acid and alkaline and impermeability to gasoline and gasohol is particularly suited for automobile parts that contact fuel, oil, and combustion gas.[37] Fiber-reinforced PBN for high temperature and humidity circumstances[38] and tough elastomeric fibers[39] have been reported.

REFERENCES

1. Wang, S. C.; Tseng, T. S., Proceedings of 15th Polymer Conference, Taiwan, R. O. C., 1992; p 597.
2. Wang, C. S.; Sun, Y. M., Proceedings of 15th Polymer Conference, Taiwan, R. O. C., 1993; p 67.
3. Yamamoto, N. Jpn. Kokay 94–16797, 1994.
4. Duling, I. N. U.S. Patent 3 436 376, 1969.
5. Browne, A. A., McIntyre, J. E. U.S. Patent 3 842 040, 1974.
6. Shima, T.; Yamashiro, S.; Aoki, H.; Shimoma, M. Jpn. Kokay 73–40918, 1973.
7. Sakamoto, S.; Watanabe, T.; Sato, Y. Jpn. Kokay 87–250027, 1987.
8. Shima, T. and coworkers U.S. Patent 3 819 584, 1974.
9. Kanai, T. and coworkers U.S. Patent 3 935 166, 1976.
10. Sano, K. and coworkers U.S. Patent 3 951 905, 1976.
11. Kuratsuji, T. and coworkers U.S. Patent 3 961 008, 1976.
12. Shimotsuma, S. and coworkers U.S. Patent 3 963 844, 1976.
13. Aoyama, M. Jpn. Kokay 94–56975, 1994.
14. Lee, K. Y.; Cho, B. H.; Kim, Y. W. U.S. Patent 5 294 695, 1994.
15. Wang, C. S.; Sun, Y. M. *J. Polym. Sci. Part A: Polym. Chem.* **1994**, *32*, 1295.
16. Wang, C. S.; Sun, Y. M. *J. Polym. Sci. Part A: Polym. Chem.* **1994**, *32*, 1305.
17. Hatayama, T. and coworkers EP 5391662A2, 1993.
18. Tahoda, T. and coworkers Jpn. Kokay 91–224,722, 1991.
19. Chuujou, T. and coworkers WO 93/09166, 1993.
20. Hamano, H. and coworkers EP 542511A2, 1993.
21. Sasaki, S. and coworkers Jpn. Kokay 92–10,919, 1992.
22. Sakamoto, S. and coworkers Jpn. Kokay 92–198,328, 1992.
23. Sato, H. and coworkers Jpn. Kokay 92–345,617, 1992.
24. Ishibarada, M.; Hayashi, S.; Saito, S. *Polymer* **1986**, *27*, 349.
25. Ichikawa, P. and coworkers Jpn. Kokay 91–183,185, 1991.
26. Tsukao, K. and coworkers Jpn. Kokay 92–53,739, 1992.
27. Onda, T. and coworkers Jpn. Kokay 92–100,914, 1992.
28. Nagai, A. and coworkers Jpn. Kokay 92–153,312, 1992.
29. Yamamoto, H. and coworkers Jpn. Kokay 90–217,222, 1990.
30. Yamamoto, H. and coworkers Jpn. Kokay 91–39,250, 1991.
31. Yamamoto, H. and coworkers Jpn. Kokay 92–39,024, 1992.
32. Yamamoto, H. and coworkers Jpn. Kokay 92–39,025, 1992.
33. Yamamoto, H. and coworkers Jpn. Kokay 92–52,124, 1990.
34. Yamamoto, H. and coworkers Jpn. Kokay 81–49,014, 1981.
35. Yamamoto, H. and coworkers Jpn. Kokay 83–98,426, 1983.
36. Nagai, A. and coworkers Jpn. Kokay 93–163,612, 1993.
37. Hatayama, T. and coworkers EP 539166A2, 1993.
38. Hatayama, T. and coworkers Jpn. Kokay 92–20,560, 1992.
39. Yamada, Y. and coworkers Jpn. Kokay 93–140,818, 1993.

Poly(alkylene phosphate)

See: Phosphorus-Containing Polymers (Overview)

POLY(ALKYLENE PHOSPHATE)S

Stanislaw Penczek and Grzegorz Lapienis
Center of Molecular and Macromolecular Studies
Polish Academy of Sciences

There is a number of comprehensive reviews devoted to the synthesis and properties of polymers containing phosphorus atoms in the main and/or side chains.[1-7] Among various structural types, poly(alkylene phosphate)s constitute only a small group of polymers containing phosphorus atoms in the main chain.

There are three ways for the synthesis of poly(alkylene phosphate)s:

1. polymerization of cyclic monomers
2. polycondensation
3. polyaddition

POLYMERIZATION OF CYCLIC MONOMERS

Polymerization of cyclic monomers recently has been used in the preparation of polymers bearing resemblance to nucleic and teichoic acids as well as bioanalogous polymers: nontoxic, water-soluble, and biodegradable.[4,5,8–13]

Cyclic Phosphates

Polymerization of the 5-membered cyclic monomers, if properly conducted, gave high molecular weight polymers. In the case of the much less strained 6-membered ones, it was impossible to prepare higher polymers, because of an extensive chain transfer.[14,15]

Anionic Polymerization

Anionic polymerization of cyclic phosphate esters has been extensively studied for 2-alkoxy-2-oxo-1,3,2-dioxaphospholanes (**1**) (Structure 1 in **Equation 1**).[16,19,20,22–24]

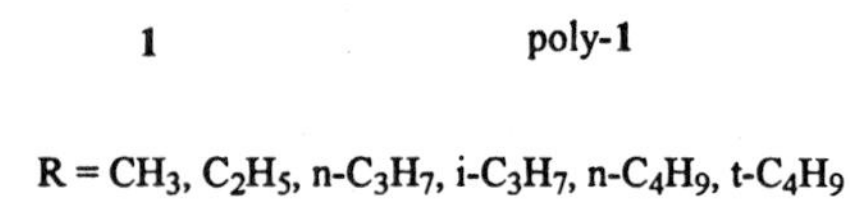

1 **poly-1** (1)

R = CH_3, C_2H_5, n-C_3H_7, i-C_3H_7, n-C_4H_9, t-C_4H_9

Metal alkyls, and particularly aluminium trialkyls, are the preferred initiators to obtain high-molecular-weight polymers.

The ability of **1** to form high-molecular-weight polymers may be attributed to the high strain of the 5-membered rings

(~29 kJ·mol^{-1}: data from the heat of hydrolysis,[25,26] 15 ± 2 kJ·mol^{-1}: our data from the heat of polymerization),[23] which presumably lowers the probability of side reactions.

The presence of substituents in the 5-membered rings usually decreases the enthalpy of polymerization[27] (cf. lactams and acetals)[29] and leads to the lowering of DP_n of the resulting polymers.[16–19,21,24,29]

Cationic Polymerization

Cationic polymerization of the organic phosphorus containing monomers is less promising than the anionic one.

Cyclic Phosphites and Phosphoramidites

The reversible anionic polymerization of 2-hydro-2-oxo-1,3,2-dioxaphosphorinane (**9**) in bulk or in CH_2Cl_2 solution, initiated with C_2H_5ONa, t-C_4H_9OK, n-C_4H_9Li, or (i-$C_4H_9)_3$ Al at 25–45°C leads to the high molecular weight linear polymer with ($\overline{M}_n$ up to 10^5.[31] Aluminum alkyls again seem to be the initiators of preference (**Equation 2**).

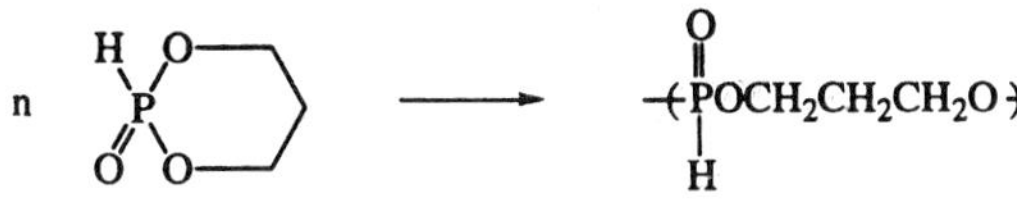

2

9 poly-**9**

For trivalent phosphorus cyclic compounds only the anionic polymerization of 2-diethylamino-1,3,2-dioxa-phosphorinane has been studied.[32,33]

Polycondensation

The usual method of the synthesis of polyesters, like the condensation of dicarboxylic acids with diols, cannot be applied for the phosphoric acids, because H_3PO_4 acts as a powerful dehydrating agent.

The usual method of preparation of poly(alkylene phosphate)s is based on the esterification of aliphatic diols with equimolar amounts of dichlorophosphates (**Equation 3**).

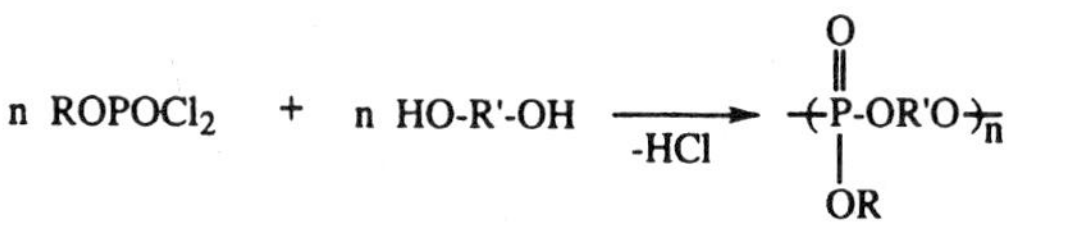

3

16 **17** **18**

With aliphatic diols polycondensation proceeds faster than with bisphenols, but some side reactions are difficult to eliminate and lead to the low-molecular-weight products formation ($\overline{M}_n$ in the range of a few thousands).[30,34,35]

Poly(hydrogen phosphonate)s

Condensation of diols with dialkyl (mostly dimethyl and diethyl) or diaryl (mostly diphenyl) phosphonate gives linear poly(hydrogen phosphonate)s. Hexamethylene diol,[36–38] 1,12-dodecanediol,[39] *trans*-1,4-cyclohexanedimethanol,[40] glycerol and pentaerythritol[41] as aliphatic diols have been used (**Equation 4**).

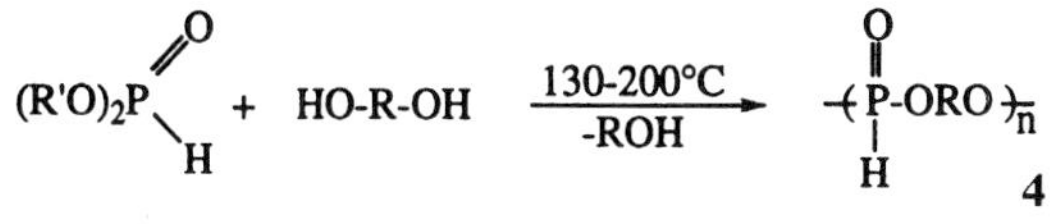

4

26 **27**

R'=$CH_3C_2H_5$

Polyaddition

An interesting novel family of water-soluble bioanalogous polymers was prepared by direct addition of phosphoric acid to diepoxides.[43] Under certain conditions, polymers, which are insoluble but high-swelling in water, are obtained (the degree of swelling approaches 10^3%).

Some Properties and Applications of Poly(alkylene phosphate)s

High-molecular-weight poly(alkylene phosphate)s were obtained as models of biopolymers and bioanalogous polymers.[10–13]

Amphiphilic polyphosphates containing in the main chain poly(ethylene glycol) molecules and in the side chains long-chain aliphatic alcohols, form spontaneous emulsions consisting of empty spheres. These spherical particles have a narrow size distribution and their water emulsion is stable for several months.[42]

Simple poly(alkylene phosphate)s were used for studies of cation binding; studies by conductivity gave the binding sequence similar to that reported for DNA, namely: $Ca^{+2} > Mg^{+2} > K^{+} > Na^{+}$.[44]

REFERENCES

1. Allcock, H. R. *Phosphorus-Nitrogen Compounds*; Academic: New York, 1972.
2. Gefter, E. L. *Organophosphorus Monomers and Polymers*; Pergamon: Oxford, 1962.
3. Kobayashi, S.; Saegusa, T. *Alternating Copolymers*; Cowie, J. M. G., Ed.; Plenum: New York, 1985; Chapter V.
4. Lapienis, G.; Penczek, S. *Ring-Opening Polymerization*; Ivin, K. J.; Saegusa, T., Eds.; Elsevier: New York, 1984; Vol. 2, Chapter XIII.
5. Penczek, S.; Lapienis, G. *Handbook of Polymer Synthesis* Kricheldorf, H. R., Ed.; Marcel Dekker: 1992; Chapter XVII.
6. Sander, M.; Steininger, E. *J. Macromol. Sci. Revs. 1967*, C1, 1; *1968*, c2, 1.
7. Sandler, S. R.; Karo, W. *Polymer Synthesis*; Academic: New York, 1974; Vol. 1, Chapter XIII.
8. Penczek, S.; Lapienis, G.; Klosinski, P. *Pure Appl. Chem.* **1984**, *56*, 1309; *Phosphorus Sulfur* **1986**, *27*, 153.
9. Penczek, S.; Biela, T.; Klosinski, P.; Lapienis, G. *Makromol. Chem., Macromol. Symp.* **1986**, *6*, 123.
10. Penczek, S.; Klosinski, P. *Models of Biopolymers by Ring-Opening Polymerization*; Penczek, S., Ed.; CRC: Boca Raton, Florida, 1990; Chapter IV.
11. Penczek, S.; Klosinski, P. *Progress in Biomedical Polymers*; Gebelein, C. G.; Dunn, R. L., Eds.; Plenum: New York, 1990; p 291.

12. Penczek, S.; Klosinski, P. *Biomimetic Polymers*; Gebelein, C. G., Ed.; Plenum: New York, 1990; p 243.
13. Penczek, S.; Klosinski, P. *Biomimetic Polymers*; Gebelein, C. G.; Ed.; Plenum: New York, 1990; p 223.
14. Lapienis, G.; Penczek, S. *Macromolecules* **1977**, *10*, 1301.
15. Lapienis, G.; Penczek, S. *J. Polym. Sci., Polym. Chem. Ed.* **1977**, *15*, 371.
16. Libiszowski, J.; Kaluzynski, K.; Penczek, S. *J. Polym. Sci., Polym. Chem. Ed.* **1978**, *16*, 1275.
17. Yasuda, H.; Sumitani, M.; Lee, K.; Araki, T.; Nakamura, A. *Macromolecules* **1982**, *15*, 1231.
18. Fontaine, L.; Derouet, D.; Brosse, J. C. *Eur. Polym. J.* **1990**, *26*, 865.
19. Vandenberg, E. J. U.S. Patent 3 520 849, 1970; C.A., **1970**, *73*, 67314.
20. Vogt, W.; Pflüger, R. *Makromol. Chem., Suppl.* **1975**, *1*, 97.
21. Klosinski, P.; Penczek, S. *Macromolecules* **1983**, *16*, 316.
22. Penczek, S.; Libiszowski, J. *Makromol. Chem.* **1988**, *189*, 1765.
23. Sosnowski, S.; Libiszowski, J.; Slomkowski, S.; Penczek, S. *Makromol. Chem., Rapid Commun.* **1984**, *5*, 239.
24. Yasuda, H.; Sumitani, M.; Nakamura, A. *Macromolecules* **1981**, *14*, 458.
25. Cox, J. R., Jr.; Wall, R. E.; Westheimer, F. H. *Chem. Ind.* **1959**, 929.
26. Westheimer, F. H. *Acc. Chem. Res.* **1968**, *1*, 70.
27. Ivin, K. J. *Reactivity Mechanisms and Structure in Polymer Chemistry*; Jenkins, A. D.; Ledwith, A., Eds.; Wiley: New York, 1974; Chapter XVI.
28. Penczek, S.; Kubisa, P.; Matyjaszewski, K. *Adv. Polym. Sci.* **1980**, *37*, 1.
29. Lapienis, G.; Penczek, S. *J. Polym. Sci., Polym. Chem. Ed.* **1990**, *28*, 1519.
30. Munoz, A.; Navech, J.; Vives, J.-P. *Bull. Soc. Chim. France* **1966**, 2350.
31. Kaluzynski, K.; Libiszowski, J.; Penczek, S. *Makromol. Chem.* **1977**, *178*, 2943.
32. Kaluzynski, K.; Penczek, S. *Makromol. Chem.* **1987**, *188*, 1713.
33. Pretula, J.; Kaluzynski, K.; Pencrek, S. *J. Polym. Sci. Polymer Chem. Ed.* **1984**, *22*, 1251.
34. Derouet, D.; Piatti, T.; Brosse, J. C. *Eur. Polym. J.* **1987**, *23*, 657.
35. Yvernault, T.; Guilloton, O. *C. R. Hebd. Seances Acad. Sci., Ser. C.* **1963**, *257*, 1923.
36. Borisov, G.; Tcholakov, G. *Commun. Dep. Chem. Bulg. Akad. Sci.* **1970**, *3*, 13.
37. Petrov, K. A.; Nifanteev, E. E.; Goltsova, R. G. *Vysokomol. Soedin.* **1964**, *6*, 1545.
38. Petrov, K. A.; Nifanteev, E. E.; Goltsova, R. G.; Korneev, S. M. *Vysokomol. Soedin.* **1964**, *6*, 68.
39. Gray, G. M.; Branham, K. S.; Ho, L.-H.; Mays, J. W.; Bharara, F. C.; Hajipetrou, A.; Beal, J. B. *Inorganic and Organometallic Oligomers and Polymers*; Harrod, J. F.; Lane, R. M., Eds.; Kluwer Academic: 1991; p 249.
40. Coover, H. W., Jr.; McConnell, R. L. U.S. Patent 3 271 329, 1966; C.A. **1966**, *65*, 20238a.
41. Petrov, K. A.; Nifanteev, E. E.; Goltsova, R. G.; Solntseva, L. M. *Vysokomol. Soedin.* **1962**, *4*, 1219.
42. Penczek, S.; Pretula, J. *Macromolecules* **1993b**, *26*, 2228.
43. Nyk, A.; Klosinski, P.; Penczek, S. *Makromol. Chem.* **1991**, *192*, 833.
44. Ostrowska-Czubenko, J.; Wodzki, R. *Colloid & Polymer Sci.* **1988**, *266*, 35.

POLYALKYLENIMINES

Keigo Aoi and Masahiko Okada
Department of Applied Biological Sciences
Faculty of Agricultural Science
Nagoya University

Polyalkylenimines are polyamines whose structure is classified into linear and branched types as shown below (**Equation 1**).

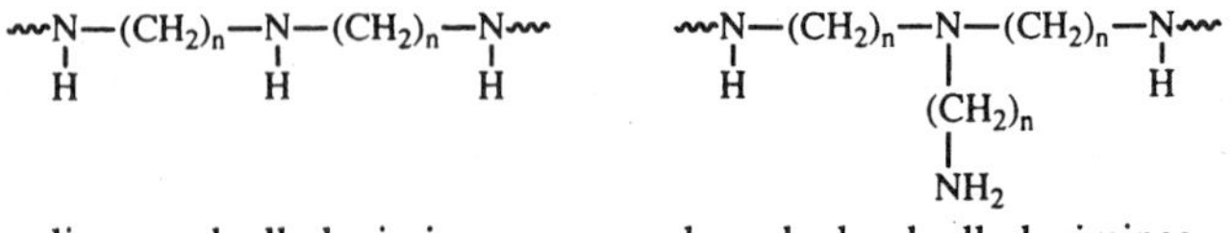

linear polyalkylenimines — branched polyalkylenimines

where n = 2, polyethylenimines
n = 3, polytrimethylenimines **1**

Linear polyalkylenimines have amino groups only in the main chain, and branched polyalkylenimines in both the main and side chains.

PREPARATION OF POLYALKYLENIMINES

Polyalkylenimines are prepared by ring-opening polymerization of cyclic amines **1–4, 7, 8** amides **5,6,** or imino ethers (**Structures 1–8**).[1] Five- and six-membered cyclic imines do not have polymerizability due to decreased ring strain.

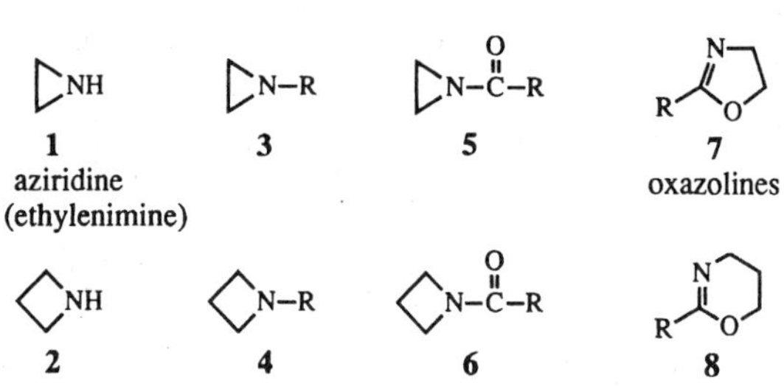

1 aziridine (ethylenimine); **3**; **5**; **7** oxazolines; **2** azetidine; **4**; **6**; **8** oxazines

Linear Polyalkylenimines

Linear polyethylenimine was first prepared by the alkaline hydrolysis of poly(*N*-formylethylenimine), which is derived by the isomerization polymerization of unsubstituted 2-oxazoline.[2]

Linear polyethylenimine was also prepared by the cationic ring-opening polymerization of *N*-(2-tetrahydropyranyl)aziridine, followed by hydrolysis in 0.1 N aqueous hydrogen chloride at room temperature.[5]

Linear polytrimethylenimine was prepared by the alkaline hydrolysis of poly(*N*-formyltrimethylenimine), which is the ring-opened polymer of unsubstituted 5,6-dihydro-4*H*-1,3-oxazine.

N-alkyl-substituted Polyalkylenimines

N-Alkyl-substituted polyalkylenimines are generally prepared by ring-opening polymerization of *N*-alkyl-substituted cyclic imines with various kinds of cationic initiators (e.g., Lewis acids, protic acids, and alkyl halides).[7]

Branched Polyalkylenimines

Branched polyethylenimine is the most important polyamine from the viewpoint of industrial applications. Branched polyethylenimine is produced through a cationic polymerization of ethylenimine with chain-growth mechanism. Ethylenimine is highly toxic and must be handled with extreme caution. Explosive polymerization occurs with Lewis and Brϕnsted acids.[8] Generally, polyethylenimine is believed to be highly branched, possessing primary, secondary, and tertiary amine groups in a ratio of approximately 1:2:1.[4,5]

Azetidine is also polymerized with cationic initiators to give branched polytrimethylenimine. Polytrimethylenimine thus obtained contains 20% tertiary, 60% secondary, and 20% primary amino functions.[9]

Properties and Applications

Linear polyethylenimine is insoluble in benzene, diethyl ether, acetone, and water at room temperature, and soluble in hot water. The melting point (T_m) and the glass transition temperature (T_g) are 59°–60°C and –23.5°C, respectively.[2,3] Anhydrous linear polyethylenimine forms a double-stranded helical structure, which is the first example of double-stranded helices stabilized by intermolecular hydrogen bonds among synthetic polymers.[10] This polymer shows highly hygroscopic properties.

Linear polytrimethylenimine is soluble in methanol and ethanol, and insoluble in water, whereas it is very hygroscopic. Its melting temperature depends on the content of water; 74°–76°C for the polymer containing water of 9.3 wt% and 81°–84°C for that of 17 wt%.[6]

Branched polyethylenimines are generally water-soluble, that is, hydrophilic. They are weak bases and show cationic character depending upon the extent of protonation, thus offering a wide variety of physicochemical properties of industrial importance.

Preparation of optically active polypropylenimine was achieved by the ring-opening polymerization of optically active propylenimine with acid catalysis.[11]

Polyethylenimines with mesogenic side chains were prepared by partial or complete alkylation of a linear polyethylenimine with mesogenic groups, followed by quaternization with bromohydric acid or dimethyl sulfate. Quarternized polymers exhibit thermotropic smectic A phases.[12]

Besides these applications, alkylenimine polymers have been used in a wide variety of applications, such as adhesives,[13–16] separations,[17–22] paper,[23–27] plastics,[29,30] and textiles.[30–34]

REFERENCES

1. Tomalia, D. A.; Killat, G. R. *Encyclopedia of Polymer Science and Engineering*; John Wiley & Sons: New York, 1985; Vol. 1, p 680.
2. Saegusa, T.; Ikeda, H.; Fujii, H. *Macromolecules* **1972**, *5*, 108.
3. Tanaka, R.; Ueoka, I.; Takaki, Y.; Kataoka, K.; Saito, S. *Macromolecules* **1983**, *16*, 849.
4. Dick, C. R.; Ham, G. E. *J. Macromol. Sci., Chem.* **1970**, *4*, 1301.
5. Weyts, K. F.; Goethals, E. J. *Polym. Bull.* **1988**, *19*, 13.
6. Saegusa, T.; Nagura, Y.; Kobayashi, S. *Macromolecules* **1973**, *6*, 495.
7. Goethals, E. J. *Ring-Opening Polymerization*; Elsevier Applied Science: Essex, U.K. 1984; Vol. 2, Chapter 10.
8. Weightman, J.; Hoyle, J. P. *J. Am. Med. Assoc.* **1964**, *189*, 543.
9. Schacht, E. H.; Goethals, E. J. *Makromol. Chem.* **1974**, *175*, 3447.
10. Chatani, Y.; Kobatake, T.; Tadokoro, H.; Tanaka, R. *Macromolecules* **1982**, *15*, 170.
11. Minoura, Y.; Takebayashi, M.; Price, C. C. *J. Am. Chem. Soc.* **1959**, *81*, 4689.
12. Masson, P.; Heinrich, B.; Frere, Y.; Gramain, P. *Macromol. Chem. Phys.* **1994**, *195*, 1199.
13. Lacy, G. H.; Chervenak, R. R. U.S. Patent 3 140 196, July 7, 1964.
14. Lindsey, W. B. U.S. Patent 3 361 586, January 2, 1968.
15. Brit. Patent 1 128 085, September 25, 1968.
16. Shirona, K.; Yoshitake, T. Jpn. Patent 74 29 886, August 8, 1974.
17. Okhara, O.; Shirai, S.; Iihashi, M.; Umemoto, S. Jpn. Kokai 75 78 583, June 26, 1975.
18. Seko, M.; Miyake, T.; Takeda, K.; Imamura, K. Jpn. Kokai 76 37 087, March 29, 1976.
19. Avny, Y.; Porath, D. *J. Macromol. Sci. Chem.* **1976**, *10*, 1187.
20. Saegusa, T.; Kobayashi, S.; Keiza, H. *Polym. J.* **1978**, *10*, 403.
21. Bresak, A. F.; Sokol, P. E.; Tolgyesi, E. Can. Patent 1 112 774, November 17, 1981.
22. Bartulen, J.; Maturana, H. A.; Perich, I. M.; Rivas, L. B. *Bol. Soc. Chil. Quim.* **1982**, *27*, 180.
23. Miyabe, Y. Ger. Offen. 2,362,077, June 27, 1974.
24. Roberts, E. J.; Rowland, S. P. U.S. Patent 3 885 069, May 20, 1975.
25. Leffler, C. T.; Wagner, F. L. S. Afr. Patent 7 701 063, November 25, 1977.
26. Philipp, B.; Luu, H. T.; Lins, K. J.; Daroydoff, W.; Arnold, K. *Acta Polym.* **1980**, *31*, 654.
27. Valendo, P. F.; Tsmyg, N. G.; Gembitskii, P. A. USSR Patent 796 290, January 15, 1981.
28. Neth. Appl. 6 601 400, August 4, 1966.
29. Lupinski, J. H.; Koppel, D. K.; Hertz, J. J. *J. Polym. Sci., Part C* **1967**, *16*, 1561.
30. Roberts, E. J. U.S. Patent Appl. 279 919, 1972.
31. Miller, S. L.; Dickert, Y. J. U.S. Patent 4 113 674, September 12, 1978.
32. Bertoniere, R. N.; Rowland, S. P. *Text Res. J.* **1978**, *48*, 385.
33. Coiner, M. H.; Haylock, J. C.; Newland, J. H.; Schwartz, J. L. U.S. Patent 4 126 415, November 2, 1978.
34. Evans, E. G. Eur. Patent Appl. 30 786, June 24, 1981.

Polyallenes

See: Allene Polymerization

Poly(amic acid)s

See: Polyimides (Introduction and Overview)
Polyisoimides

Poly(amide imide)s

See: Engineering Thermoplastics (Survey of Industrial Polymers)
Polyamideimides (with Improved Melt Processability)
Polyimides (from N-Silylated Aromatic Diamines)

POLY(AMIDE IMIDE)S (with Improved Melt Processability)

Kil-Yeong Choi, Mi Hie Yi, Moon, and Young Jin
Advanced Polymer Division
Korea Research Institute of Chemical Technology

Aromatic polyimides are well recognized as a class of high performance materials because of their remarkable thermal and oxidative stabilities and excellent electrical and mechanical properties for long time periods of operation.[1] It is well known that these properties are attributed to rigidity of chemical structure and strong interaction between polyimide chains.[2] Unfortunately, this strong interaction and the rigidity of the polymer chains make the polymer intractable. Poor thermoplastic fluidity and bad solubility are major problems for wide application of polyimides. Thus, to overcome these processing problems various approaches have been carried out by incorporating flexible units such as –NHCO–, –O–, and $-SO_{2}-$, some of which are commercialized.[3]

Among them, PAI [poly(amide imide)] is the most successful material for melt processing. It is injection moldable and has excellent thermal, physical, and mechanical properties. However, PAI's drawbacks are high melt viscosity, high fabrication temperature and long post-curing time.[4–6] Thus, the molecular weight of commercialized PAI, Torlon®, is intentionally reduced by introducing post-curable trimellitic anhydride groups at ends of the polymer.[7]

It has been reported that the hydrogen atoms of amide groups in aramid were replaced by alkyl groups to reduce the strong intermolecular hydrogen bonding.[8] The modified aromatic polyamide showed, surprisingly, thermotropic liquid crystal behavior and increased solubility. This is one example to improve processability by reduction of chain–chain interactions.

Therefore, we have undertaken research on amorphous poly(amide imide)s to reduce the chain–chain interactions by introducing of N-alkyl group[9,10] or cyclic aliphatic diamines in PAI for good processability.[11,12]

RESULTS AND DISCUSSION

N-Alkylated PAI

To reduce chain–chain interactions of PAI, hydrogen atoms of the amide group in PAI were replaced with various alkyl groups such as the methyl, ethyl, and n-butyl groups.

The inherent viscosities (η_{inh}) of polymers obtained ranged from 0.51 ~ 0.53 dL/g, which is comparable to that of Torlon® (η_{inh} = 0.54 dL/g) used. This implies that the amide bond and imide bond were not cleaved during *N*-alkylation. The degree of substitution of *N*-alkylated PAI, calculated from ^{1}H-NMR spectra, was controlled by varying a ratio of alkyl halide to amide unit in PAI (Torlon®), and it was strongly dependent on the chain length of a substituent.

The solubility of the polymers was similar to that of Torlon®; they dissolved in aprotic polar solvents, such as NMP, DMAc, and DMF.

Thermal stability of the modified PAIs was analyzed by TGA and the results were similar to that of Torlon®. All the polymers prepared in this work were amorphous with a glass transition temperature (T_g) ranging from 238°C to ~276°C.

The *N*-alkylated PAIs showed significant lower melt viscosities than Torlon®, which is close to that of Ultem®.

New Aliphatic-Aromatic Copoly(amide imide)s

It is well known that the aliphatic moiety as comonomer in aromatic copolymer reduces the rigidity of chain. The shortcomings of using aliphatic diamines are deterioration of thermal properties of the copolymers; for example, T_g and T_m are decreased. However, alicyclic compounds have better thermal properties since they have a more rigid chemical structure than linear aliphatic compounds. There are several candidates of cyclic compounds for processable PAI, such as cyclohexane diamines and isophorone diamine, which are commercially available.

New PAIs Containing Isophorone Diamine Moiety

To study PAIs containing isophorone diamine moiety, we applied two different synthetic routes, as shown in **Scheme I**.

CONCLUSION

We have shown that reduction of chain–chain interaction of polymer backbone by *N*-alkylation or by introducing isophorone diamine as a comonomer effectively improves melt processability of poly(amide imide)s. *N*-alkylated PAIs showed melt viscosity of 10^2–10^4 poise, which are much lower than that of the conventional PAI, Torlon®. And they exhibited good thermal stabilities and slightly lower glass transition temperatures which ranged from 238 to 276°C, depending on substituents and degree of substitution.

Preliminary results of copolymerization with various alicyclic diamines suggested that IPDA is more preferable as a commoner for PAI.

The present PAIs are considered to be new candidates for melt processable high-performance polymer materials.

REFERENCES

1. Mooney, P. J. *High Temperature Plastics* Business Communications 1991, 39.
2. Wilson, D.; Stenzenberger, H. D.; Hergenrother, P. M. *Polyimide*; Blackie: Glasgow UK, 1991; 58.
3. Wilson, D.; Stenzenberger, H. D.; Hergenrother, P. M. *Polyimide*; Glasgow, Blackie: UK, 1991; 58. Reference 2, p 260.
4. Otani, R. *Plastics Age* **1993**, 126.
5. Koga, K. *Plastics* **1990**, *41*, 3, 50.
6. Bolon, D. A.; Gorczyca, T. B. *Proceedings of Second International Conference on Polyimides* **1985**, 306.
7. Hanson, R. B. U.S. Patent 4 136 085.
8. Takayanagi, M.; Katayose, T. *J. of Polym. Sci: Polym. Chem. Ed.* **1981**, *19*, 1133.
9. Choi, K-Y. *Proceedings of Workshop on Thermally Stable Polymers*, 1992; 45.
10. Choi, K-Y.; Yi, M. H. *Die Angewandte Makromolekulare Chemie*, **1984**, *222*, 103.
11. Yi, M. H.; Jin, M. Y. and Choi, K.-Y. *Die Angewandte Makromolekulare Chemie*, **1995**, *233*, 89.
12. Choi, K.-Y.; Jin, M. Y.; Yi, M. H. *Proceedings of 35th IUPAC International Symposium on Macromolecules*, 1994, 246.

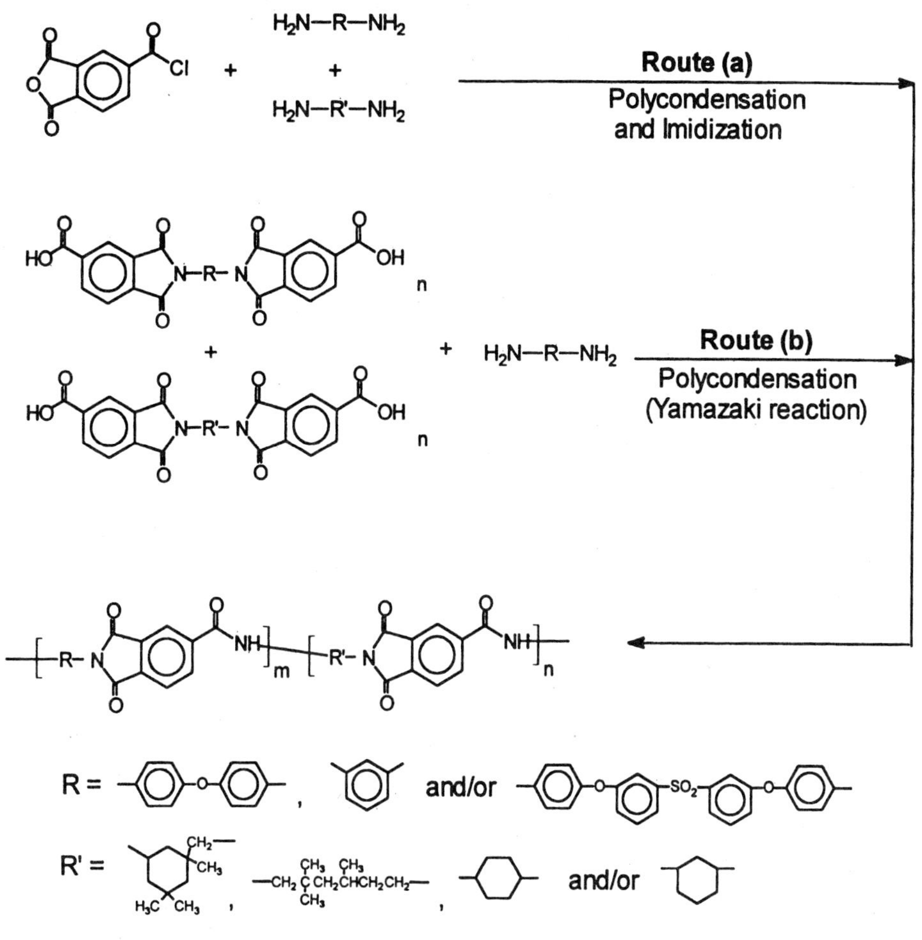

SCHEME I. Synthetic route of new PAIs.

POLYAMIDE FIBERS, ALIPHATIC (Nylon 66 and Nylon 6)

Kunihiko Okajima, Chihiro Yamane, and Fumiaki Ise
Fundamental Research Laboratory of Natural and Synthetic Polymers
Asahi Chemical Industry Company, Ltd.

Polyamide fiber is classified largely into two types depending on spinning technology and tenacity level relating to polymer chain rigidity. One is aliphatic polyamide fiber represented by nylon 66 and nylon 6, produced by the melt-spinning method. Their tenacity level is at most 10 g/d. The other is aromatic polyamide (aramid) fiber with rigid and symmetrical structure, mainly produced by the air-gap/wet-spinning method on so-called lyotropic mesophase dope. Its tenacity level is usually more than 20 g/d. For nylon 66 and nylon 6 fibers, their fiber-forming process and final usage are essentially the same, differing only in source polymer and properties. Thus, they are regarded as a family fiber. Therefore, it is wise to describe their synthesis, properties, and applications in comparative manner. The rheology of their melt, their structural formation during melt-spinning, their spinning-line analysis and the effect of water on their structural formation all determine fiber properties.

SYNTHESIS OF SOURCE POLYMERS

Source polymers suitable for nylon 66 and nylon 6 fiber formation were first synthesized based on polycondensation reaction of hexamethylenediamine and adipic acid by Carothers in 1932,[1] and ring-opening polymerization of caprolactam by P. Schlack[2] in 1938, respectively, as shown in **Equations 1** and **2**.

$$n\,H_2N(CH_2)_6NH_2 + n\,HOOC(CH_2)_4COOH \rightarrow HN\text{-}((CH_2)_6NHCO(CH)_4CO)n\text{-} + H_2O \quad (1)$$

$$n(CH_2)_5CONH \rightarrow \text{-}((CH_2)_5CONH)n\text{-} \quad (2)$$

These polymerizations are usually carried out under phosphor compound catalysts.

PRODUCTION OF NYLON 66 AND NYLON 6 FIBERS

Carothers was also the first to propose the melt-spinning method for nylon 66 in 1937 within Du Pont.[3] Commercial production of a long-fiber named nylon 66 started in 1939. Nylon 6 fiber was commercialized by Toray in 1951 with introduction of spinning technology for nylon 66. Melt-spinning is constituted of the melting and mixing of polymer, the extrusion of the melt, the cooling of the extruded strand, the drawing or annealing and the winding. All these processes influence the final yarn properties. At present, melt-spinning is essentially categorized into three types depending on drawing technology and spinning speed.

PROPERTIES

Fundamental properties of these two polymers are listed in **Table 1**. Most physical constants are in the same range. Typical differences are seen in crystallization rate, crystal structure, and molecular relaxation factors, all of which determine fiber properties.

USAGE AND APPLICATION

Nylon fiber is generally used as multifilament or monofilament. Advent of polyamide fiber first threatened the rayon market and then constituted an outstanding position as yarn for tirecord and yarn for fiber-reinforcement for industrial materials such as rubber belt, car hood conveyor bag, and air-bag for motor vehicles which require high tenacity, high elastic recovery and good adhesive ability. Carpet, computer ribbon, fishery net, fishing gut and tennis gut are also main usage owing to the above characteristic properties of nylon. Coverage ability for inorganic material and pigments insures the usage as mat, artificial lawn, and fibrous grinder. It is recognized as most suitable yarn for pantyhose and sportswear because of its high toughness and heat resistance.

The melt blend fiber from nylon 66 and nylon 6 has been commercialized for leg material (material for pantyhose, etc.) for the first time by Asahi Chemical Industry Co. Ltd. The blend fiber is said to have better transparency, higher tenacity (especially burst strength) and easier finishing nature than nylon 66 homofiber.

REFERENCES

1. Carothers, W. H.; Hill, R. *J. Am. Chem. Soc.* **1932**, *54*, 1579.
2. Schlack, P., German Patent 748 253, 1938; U.S. Patent 2 241 321, 1941.
3. Carothers, W. H., U.S. Patent 2 071 253, U.S. Patent 2 130 948, 1938.

POLYAMIDE THERMAL STABILITY

Linda J. Broadbelt*
Department of Chemical Engineering
Northwestern University

Michael T. Klein, Barry D. Dean, and Stephen M. Andrews
Amoco Performance Products, Incorporated
Department of Chemical Engineering
University of Delaware

The polymers in the polyamide family are used for their combination of elasticity, toughness, and abrasion resistance. Aromatic containing polyamides are used as specialty fibers that have unusual heat resistance and outstanding strength to weight ratios.[1] When exposed to elevated melt-processing temperatures and harsh processing environments, thermal degradation reactions occur according to a complex network of parallel and series steps. Increased understanding of this network is a reasonable first step in the development of strategies for minimizing thermal degradation or its deleterious effects.

Analyses of polyamide thermal degradation provided insight into product identification and several possible degradation mechanisms. This motivated the current work into the fundamentals underlying aliphatic–aromatic polyamide thermal degradation using quantitative model compound kinetic studies. This should allow clearer focus on reaction mechanisms and, further, permit resolution of pathways and rate constants. N,N'-dihexyisophthalamide (DHI), was chosen as a model of a polyamide synthesized from an aromatic acid and an aliphatic amine.

The physical properties of high performance aromatic polyamides are a function of the relative positions of the ring substituents. Poly(*p*-phenylene terephthalamide), more commonly known by its trade name, Kevlar®, is an aromatic polyamide synthesized from terephthalic acid and *p*-phenylene diamine. It is used extensively for its high strength-to-weight ratio in applications such as bulletproof vests. Nomex®, the all-meta analogue of Kevlar®, is synthesized from isophthalic acid and *m*-phenylene diamine. It is virtually nonflammable and is used as a flame retardant material and in other industrial applications where unusual heat resistance is required. Kevlar® is readily crystallized because of the para configuration that allows the individual linear polymer chains to pack tightly together with interchain hydrogen bonding. Its density is quite high, having a value of 1.44 g/cm^3. In contrast, the metal polymer, with its less linear structure, does not pack as tightly and has a lower density of 1.38 g/cm^3. A copolymer or polymer blend composed of varying proportions of these two different monomeric configurations would accommodate to some degree the two homopolymers' properties. Alteration of the relative amounts of the homopolymers allows more flexible design of polymer properties.

These practical issues motivated comparative investigation of DHI and its para counterpart, *N,N'*-dihexylterephthalamide (DHT). Particular attention paid was to yields and selectivities of DHI and DHT product analogs afforded under the same reaction conditions. These comparisons provided insight into the probable behavior of a copolymer or polymer blend under harsh thermal processing environments.

ELUCIDATION OF REACTION PATHWAYS

Analysis of the disappearance rates and the changes in reaction product selectivities from thermolysis of DHI provided valuable insight into the reaction pathways, kinetics and mechanisms underlying the thermal reactions of aliphatic-aromatic polyamides.

REFERENCE

1. Billmeyer, F. W. *Textbook of Polymer Science*; Wiley: New York, 1984.

*Author to whom correspondence should be addressed.

Polyamides

See: *Aramid Fibers and Nonwovens*
Aramids (Processable Precursors)
Aramids and Aromatic Polyimides, Soluble (Bulky Substituents)

TABLE 1. Properties of Polyamides

		Nylon 66	
Melting point (°C)[a]		264	[illegible]
Glass-transition temperatures (°C)[a]			
	oven-dry	78	76
	50% RH	35	20
	100% RH	−15	−22
Heat of melting (cal/mol)[b]		75	45
Crystallization rate (sec^{-1})		1.66[c]	0.14[d]
Max. temp. of crystallization rate (°C)		150[c]	145.6[d]
Modulus of crystal ($dyne/cm^2$)[e]		α-crystal 175×10^4	α-crystal 168×10^4 γ-crystal 28×10^4
Molecular cohesive energy (cal/mol)		3.4	—
Specific heat(cal/g/°C)[f]			
	at melting point	0.75	0.66
	at 20 °C	0.40~0.55	0.4~0.46
Thermal conductivity (kcal/mhr/°C)		0.22	0.23
Linear expansion coefficient[g] (cm/cm/°C)		7.0×10^{-5}	6.5×10^{-5}
Crystal structure[g]		α: Triclinic β: Triclinic Pseudo hexagonal	α: Monoclinic β: Monoclinic γ: Monoclinic
Density (g/cm^3) at 25 °C[g]		α-crystal 1.24 amorphous 1.09	α-crystal 1.23 γ-crystal 1.17 amorphous 1.11
Electric resistance (Ω.cm)[h]		4.5×10^{13}	7×10^{14}
Dielectric constant (at 60Hz)[h]		4.0	4.1
Dynamic viscoelasticity[i] (at 0.1 Hz–110 Hz)	log tan δ	−1.3~−0.9	−0.9~−0.8
	T_{max} (°C)	98~140	80~140
	E' ($dyne/cm^2$)	$5\times10^9\sim1\times10^{11}$	$4\times10^9\sim3\times10^{10}$
Relaxation time(T_1/sec by ^{13}CNMR)[j]			
	amorphous phase	~1.37	~0.72
	meso phase	~9.2	~5.53
	crystalline phase	~82.5	32.3~58.5
Permeability coefficient (cm^3-NPT/m^2/mil/24hr/atm)[k]	O_2	80	40
	CO_2	140	75
	N_2	5	14
Evolved gases by pyrolysis		Ammonia, cyanogen, CO_2	
Dyeability		Acid, dispersible (atmospheric pressure)	Acid, dispersible (atmospheric pressure)
Biodegradation[l]		Degradable by wood-rotting Basidiomycetes	
Moisture regain(%) (at 20 °C×65%)		4.5	4.5
Cryoscopic nature[m]		Dropped by water Mp =178.5 °C(N66/water=1/1.96)	

Sources:
[a]Kohan, M. I., et al. *Nylon Plastics*; John Wiley & Sons: New York, 1973.
[b]Dole, M.; Wunderlich, B. *Macro. Chem.* **1959**, *34*, 29
[c]Magill, J. H. *Polymer* **1962**, *2*, 221.
[d]Magill, J. H. *Polymer* **1962**, *3*, 655.
[e]Tadokoro, H. *Kobunshi* **1978**, *27*(10), 729.
[f]Vieweg, R.; Muller, A., et al; Kunststoff Handbuch VI Polyamide; Carl Hanser Verlag: 1966.
[g]Muller, A.; Pfuger, R. *Plastics* **1959**, *24*(263), 350.
[h]Fukumoto, O. *Plastic zairyoukouza*, Polyamide resin, Nikkan kougyou, 1970; Vol. 16.
[i]Okajima, K.; Tomokiyo, M.; Koizumi, T., unpublished results.
[j]Okajima, K.; Koizumi, T., unpublished results.
[k]Gotou, K. *Koubunnshi Kaishitu gijyutu*; Kagakukougyousya, 1972; Vols. 1 and 2.
[l]Japan Open Patent H5-230273.
[m]Saitoh, M.; Hattori, M., unpublished results.

Bicyclic Lactams, Ring-Opening Polymerization
Biodegradable Optically Active Polyamides (Stereoregular Polyamides from L-Tartaric Acid)
Condensation Polymers (Synthesis from Silyated Diamines)
Cyclolinear Organophosphazene Polymers
Diphenimides (Ring-Opening Polymerization)
Enzymatic Polymerization
Heteroaromatic Polymers (Five-Membered Rings in Main Chain)
Nonthrombogenic Polymers (Multiblock Copolymers of Polyether and Polyamide)
Nylon 3
Nylon 6 (Depolymerization)
Nylons (By High Pressure Solid-State Polycondensation)
Nylon 3, Helical [Poly(α-alkyl-β L-aspartate)s]
Optically Active Polyamides, Coumarin Dimer (Chiral Stationary Phases for HPLC)
Phosphorus-Containing Polymers (From Dicarboxylic Acids with C-P Bonds)
Polyamide Fibers, Aliphatic (Nylon 66 and Nylon 6)
Polyamide Thermal Stability
Polyamideimides (with Improved Melt Processability)
Polyamides (Impact Toughness and Toughening)
Polyamides, Alicyclic and Aromatic (Ferroelectric Hysteresis and Pyroelectricity)
Polyamides, Glycine-Containing
Polyamides, Lactam-Based
Poly(ether amide) Triblock and Star Block Copolymers
Poly(ether-block-amide) Copolymers (Characterization of Structure and Polymorphism)
Polyimidazolylamides
Polyisophthalamides (Thermal Stability Enhancement by Pendent Groups)
Polyisophthalamides (Water-Uptake Enhancement)
Polyphthalamides
Poly[N,N'-(sulfo-p-phenylene)terephthalamide]
Powder Coatings (Overview)
Solid-State Polyamidation
Thermodegradable Polymers (Azo Groups in Main Chain)

POLYAMIDES (Impact Toughness and Toughening)

Dongming Li* and H.-C. Wang
Baytown Polymers Center
Exxon Chemical Company

A. F. Yee
Department of Materials Science and Engineering
University of Michigan

*Author to whom correspondence should be addressed.

Polyamides are polymers that contain recurring amide groups as integral parts in their main chains. Although polyamides from numerous combinations of diacids, diamines, amino acids and their copolymerization are possible, the most important commercial ones are polyamide 6 (PA 6 or nylon 6) and polyamide 6,6 (PA-6,6 or nylon 6,6). This article is devoted mainly to these two polyamides and covers the impact toughness and fracture mechanism of polyamides, blending as a toughening approach, the toughening mechanisms of polyamide blend systems and the use of plastic/elastomer core/shell toughener to improve the stiffness/toughness balance of toughened polyamide blends.

IMPACT TOUGHNESS AND FRACTURE MECHANISM OF POLYAMIDES

PA-6 and PA-6,6 have excellent resistance to chemicals, fatigue and abrasion, and high mechanical strength. They are tough engineering resins with high tensile strength and elongation, even at impact rates, and they exhibit strong resistance to macroscopic fracture in the falling weight and unnotched Izod tests.[1,2] The weakness of PA-6 and PA-6,6 is their notch sensitivity.

The high toughness of polyamides in the absence of notches, in contrast to their low notched Izod impact strength, is generally viewed as a reflection of polyamide's good resistance to crack initiation but low resistance to crack propagation.

Polyamides, although highly deformable in shear and uniaxial tensile tests, are capable of only very limited volume dilatation under a dilatational stress. When the dilatational strain reaches a critical value, polyamides undergo internal crazing. Narisawa and Ishikawa demonstrated the existence of such internal crazes in a low speed plane strain bending test of notched PA-6 bars.[3]

The internal craze grows along a plane normal to the major tensile stress, and quickly changes into a crack that leads to the failure of the bar.

An understanding of the fracture mechanism of unmodified polyamides is important because it shows that the way to effectively toughen polyamides under the more severe plane strain condition is to prevent, postpone or terminate the crazing–cracking process.

BLENDING AS A TOUGHENING APPROACH

Blending polyamides with other polymers, primarily elastomers and soft plastics, has been studied extensively as a remedy for the low notched impact strength of polyamides. However, due to the immiscible nature of polyamides with most other polymers, simple physical blending results almost exclusively in blends with poor mechanical properties due to the coarse dispersed phase morphology and poor interfacial adhesion.[4–6] Compatibilization is needed to convert such immiscible blends into commercially useful products.

A compatibilizer is usually in the form of a block or graft copolymer with one portion identical to or miscible with the polyamide phase and the other identical to or miscible with the dispersed phase.

APPROACHES TO ACHIEVING HIGH TOUGHNESS

Many factors affect the notched Izod impact strength of polyamide blends, such as modifier concentration, size, interparticle distance, mechanical properties of the modifiers and adhesion between the modifier and the matrix. With the modifier concentration held at a sufficient and constant value and with good adhesion, particle size is the most influential factor.

Compatibilization combined with intensive mixing is the most common approach to reduce the size of a poorly dispersed modifier into the optimum size range to achieve successful toughening.

It was recently discovered that if a polymer forms dispersed phases too fine to toughen polyamides, another polymer less miscible with polyamide can be added to bring the particle size up to the optimum range to achieve successful toughening.

Processing conditions can also affect the phase morphology and the impact toughness of polyamide blends.

IMAGING TOUGHENING MECHANISMS

The toughening mechanism of polymer blends has been under study for many years. Some possible functions of a modifier in the toughening of polymers could be (a) stretching and tearing; (b) to cause multiple crazing; (c) to create stress concentration and induce the shear yielding of matrix and (d) cavitation and inducing shear yielding and plastic dilatation.

CORE–SHELL TOUGHENER AS A NEW APPROACH TO BALANCED STIFFNESS/TOUGHNESS

Polymer blending is a successful approach to toughened polyamides. However, the impact toughness increase of elastomer toughened polyamides usually comes at the expense of reduced stiffness, as measured by modulus. The use of plastics as tougheners for polyamides, although leading to less stiffness loss, does not always lead to sufficient toughness, especially at low temperatures. It is highly desirable to design a toughener that will achieve effective toughening with less elastomer and less stiffness drop. To this end, a toughener of core–shell morphology with Exxpro elastomer as the shell has been created in a melt-blending process and found to effectively toughening PA-6 with 10 to 50% less Exxpro elastomer.[7]

Creating core–shell type tougheners has been an interesting area in toughening research.

CONCLUSION

Polyamides are versatile materials with good overall properties, but they are notch sensitive and have low notched impact strength. The cause of their low impact strength and notch sensitivity is the ease with which fracture occurs through a crazing–cracking mechanism under a dilatation stress associated with notches. Blending polyamides with other polymers, either elastomers or plastics, provides remedies to this problem. High toughness polyamides blends can be made by blending. The modifiers are found to effectively toughen polyamides by a cavitation mechanism. To minimize the stiffness loss usually accompanying elastomer toughening, a core–shell type toughener with plastics as the core, a reactive elastomer as the shell has been designed to reduce the elastomer needed to achieve the required toughness increase. With this special core–shell toughener design, the dispersed phase size is not affected because the reactive elastomer is the shell. As much as 40% of the elastomer can be replaced by a plastic core, which means only about half the regular amount of elastomer is needed. This translates into reduced loss in stiffness and enhancement of other properties, such as heat distortion temperature and creep resistance.

REFERENCES

1. Heater, J. B.; Lacey, E. M. *Mod. Plast.* **1964**, *41*, 123.
2. Flexman, Jr., E. A. *Polym. Eng. Sci.* **1979**, *19*, 564.
3. Narisawa, I.; Ishikawa, M. In *Advances in Polymer Science 91/92*; Kaush, H. H., Ed.; Springer-Verlag: Berlin, 1990; p 353.
4. Cimmino, S.; D'Orazio, L.; Greco, R.; Maglio, G.; Malinconico, M.; Mancarelia, C.; Martuscelli, E.; Palumbo, R.; Ragosta, G. *Polym. Eng. Sci.* **1984**, *24*, 48.
5. Ide, F.; Hisegawa, A. *J. Appl. Polym. Sci.* **1974**, *18*, 963.
6. La Mantia, F. P.; Valenza, A. *Eur. Poly. J.* **1989**, *24*, 553.
7. Li, D.; Ban, L. L.; Wang, H.-C.; Yu, T. C. *Exxon Company Report* ET.6CL.94, 1994.

POLYAMIDES, ALICYCLIC AND AROMATIC (Ferroelectric Hysteresis and Pyroelectricity)

Yukinobu Murata
Osaka Prefectural College of Technology

K. Tsunashima
Toray Industries Incorporated

N. Koizumi
Institute for Chemical Research
Kyoto University

Piezoelectric properties were extensively studied for aliphatic polyamides (nylon) with odd-numbered carbon atom sequence such as nylon 9 and nylon 11.[5,11,13] Polarization reversal and switching phenomenon were observed for nylon 7 and nylon 11.[4] Ferroelectricity of these nylons was attributed to the oriented δ' phase with the hydrogen bond weakened by quenching.[1,2] In connection with hydrogen bonding in these nylons, ferroelectric polarization was examined for polyureas and polyurethanes and was attributed to polar crystals.[14,15] Alicyclic polyamides prepared from 1,3-bis(aminomethyl)cyclohexane (BAC) and aromatic polyamides from *m*-xylylenediamine (MXD) are poorly crystalline or amorphous in quenched samples, and aromatic polyamide obtained from isophthalic acid (I) and copolyamide from 70 mol% I and 30 mol% terephthalic acid are amorphous. The ferroelectric hysteresis and pyroelectricity[6–10] are described for these alicyclic and aromatic polyamides in this chapter.

PREPARATION

Alicyclic and aromatic polyamides were prepared by melt-polycondensation from diamines and dicarboxylic acids. Aromatic polyamides were obtained from *m*-xylylenediamine (MXD) and aliphatic dicarboxylic acids (ADC) of the carbon number n = 6 to 11, and 13 and were designated as nylon MXDn.[3] Alicyclic polyamides were made from 1,3-bis(aminomethyl)cyclohexane (BAC) and ADC of the carbon number = 6, 7, and 10 and were abbreviated to nylon BACn.

PROPERTIES

The *D–E* hysteresis loop was observed in quenched nylons MXDn regardless whether n was odd or even.[6,7] Since the number of amide groups per unit volume decreases with increasing n, the origin of the *D–E* hysteresis loop of nylon MXDn is related to the amide group. Yoda and Matsubara reported a polar crystal structure for nylon MXD6 where the chain segment between the aromatic nuclei was twisted and two amide groups in one repeating unit were oriented perpendicular to each other.[16] However, the crystal density of the structure proposed by Yoda and Matsubara was smaller than that of the quenched specimen.[12] Thermal behavior indicates that nylons MXDn are poorly crystalline or amorphous in quenched samples. The origin of ferroelectricity in MXDn may be attributable to the polar structure of Yoda and Matsubara or to amide groups in amorphous regions.[6]

Quenched nylons BAC6, BAC7, and BAC10 showed the *D–E* hysteresis curves with remanent polarizations of about 30 mCm^{-2} although the measurement was made at a elevated temperature of 323K for nylon BAC6.[8,9] Nylons BACn exhibited the thermal behavior similar to nylons MXDn and therefore were amorphous or poorly crystalline in quenched samples.

REFERENCES

1. Balizer, E.; Fedderly, J.; Haught, D.; Dickens, B.; DeReggi, A. S. *CEIDP Annual Report* **1991**, *IEEE 91 CH3055-1*, 193.
2. Balizer, E.; Fedderly, J.; Haught, D.; Dickens, B.; DeReggi, A. S. *J. Polym. Sci. Polym. Phys. Ed. Part B* **1994**, *32*, 365.
3. Carlston, E. F.; Lum, F. G. *Ind. Eng. Chem.* **1957**, *49*, 1239.
4. Lee, J. W.; Takase, Y.; Newman, B. A.; Scheinbeim, J. I. *J. Poly. Sci. Polym. Phys. Ed. Part B* **1991**, *29*, 279.
5. Mathur, S. C.; Scheinbeim, J. I.; Newman, B. A. *J. Appl. Phys.* **1984**, *56*, 2419.
6. Murata, Y.; Tsunashima, K.; Koizumi, N.; Ogami, K.; Hosokawa, F.; Yokoyama, K. *Jpn. J. Appl. Phys.* **1993**, *32*, L849.
7. Murata, Y.; Tsunashima, K.; Koizumi, N. *CEIDP Annual Report* **1993**, *IEEE 93CH3269-8*, 269.
8. Murata, Y.; Tsunashima, K.; Koizumi, N. *Jpn. J. Appl. Phys.* **1994**, *33*, L354.
9. Murata, Y.; Tsunashima, K.; Koizumi, N. *Proc. 8th International Symposium on Electrets*, Paris, 1994b; 709.
10. Murata, Y.; Tsunashima, K.; Umemura, J.; Koizumi, N. *CEIDP Annual Report* **1994**, *IEEE 94CH3456-1*, 779.
11. Newman, B. A.; Chen, P.; Pae, K. D.; Scheinbeim, J. I. *J. Appl. Phys.* **1980**, *51*, 5161.
12. Ota,–T.; Yamashita, M.; Yoshizaki, O.; Nagai, E. *J. Polym. Sci. A-2* **1996**, *4*, 959.
13. Scheinbeim, J. I. *J. Appl. Phys.* **1981**, *52*, 5939.
14. Tasaka, S.; Shouko, T.; Inagaki, N. *Jpn. J. Appl. Phys.* **1992**, *31*, L1086.
15. Tasaka, S.; Shouko, T.; Asami, K.; Inagaki, N. *Jpn. J. Appl. Phys* **1994**, *33*, 1376.
16. Yoda, N.; Matsubara, I. *J. Polym. Sci. Part A* **1964**, *2*, 253.

POLYAMIDES, GLYCINE-CONTAINING

Juan A. Subirana, Sebastián Muñoz-Guerra, and Jordi Puiggali

Departament d'Enginyeria Química Escola Tècnica Superior d'Enginyers Industrials de Barcelona Universitat Politècnica de Catalunya

Glycine is the simplest amino acid present in proteins. Because it lacks a side chain, its conformational features are different from those found in the other α-substituted amino acids.

As a polyamide, polyglycine is equivalent to nylon 2, because it has two carbon atoms in each residue. However, polyglycine itself has not found any commercial applications because, owing to the high density of hydrogen bonds, it is infusible and insoluble in most solvents. It can be solubilized only at about 100° in formic acid and other aggressive solvents.

Owing to its unique conformational properties, glycine residues have been incorporated in various ways into conventional polyamides. The structure and properties of such copolymers are the subject of this review. The fact that glycine is a natural amino acid might increase the biodegradability of commercial nylons, a feature of considerable interest for the applications of these polymers either as disposable or biomedical materials.

GLYCINE CONTAINING POLYPEPTIDES

Before analyzing polyamides, it is wise to consider the structure of polyglycine and glycine rich proteins. Polyglycine itself has been found to occur in two forms, I and II. Form I has extended chains, somewhat different from a polypeptide β sheet, since molecules with two mirror image conformations are found in the crystal.[8] Form II is a unique structure that forms a threefold helix, stabilized by hydrogen bonding with six neighbors in the crystal.[5]

Copolymers of glycine with other amino acids have been studied in detail in the past, not only as models for silk and collagen, but also as models of other natural proteins and, in general, as polypeptides that might have unique structures.[9,14]

SEQUENTIAL COPOLYMERS OF GLYCINE AND ω-AMINO ACIDS

Sequential copolymers of glycine and ω-amino acids have been synthesized either by the active ester polymerization technique or by the azide method.[1,4,6,10–13,15]

Structural studies on nylons 2/3, 2/5, 2/6, 2/11, 2/12, and 2/3/3 show that two crystalline forms named form I and form II (by analogy with the modifications of polyglycine to which they are closely related), are stabilized depending on the crystallization conditions.[4,6,10,13,15] Form II appears to be the most stable structure because it is always the predominant one.

BIODEGRADATION TESTS

The degradation of nylon 2/6 by both fungi and bacteria was studied by Bailey et al. by measuring the amount and rate of carbon dioxide liberated.[2,3] The fungus *Aspergillus niger* completely degraded the copolymer in about three weeks and the bacteria *Flavobacterium*, *Flavobacterium spp.*, and *Alcaligenes* all degraded the material but did not give a quantitative yield of carbon dioxide.

COPOLYMERS OF GLYCINE AND COMMERCIAL NYLONS BASED IN ALIPHATIC DIAMINES AND DIACIDS

Nylon 2/6,6 has been synthesized by interfacial polymerization of N-glycyl hexanediamine with adipoyl chloride.[7] A number average molecular weight around 8400 was determined by

GPC in m-cresol at 130°. Thermal analysis and X-ray diffraction data indicate that the polymer is semicrystalline. A characteristic intense reflection around 4,15 Å is found as in nylon 2/n. Thermogravimetric analysis indicates a decomposition temperature around 340°C, much higher than its melting temperature. Thus the polymer may be spun into fibers at 240°C and tensile properties have been measured. Biodegradability of nylon 2/6,6 was also studied and compared with nylon 2/6. Scanning electron pictures showed that films of nylon 2/6,6 were more readily degradable than nylon 2/6 under attack by fungi.

CONCLUSION

Sequential copolymers of glycine show unique conformational features owing to the tendency of the glycine residue to adopt a conformation similar to that found in helical polyglycine II. As a result the polyamides from this family do not usually form sheet structures, but instead they form a network of hydrogen bonds with their six neighboring chains in the crystal.

However, conventional polyamides in which glycine residues are incorporated at random do maintain the typical sheet structure of the parent polyamide.

ACKNOWLEDGMENTS

The work carried out in the Department of the authors in Barcelona was possible thanks to the enthusiastic work of many graduate students and collaborators who are included in the references. It was supported by grants PA86-0218, PB91-0588, and PB93-1067 from the DGICYT, by the CEE (BRITE-EURAM, Project BE-3106-89) and by grants MAT89-1136-CE (CICYT) and QFN91-4204 (CIRIT-CICYT).

REFERENCES

1. Bailey, W. J.; Okamoto, Y. *Polym. Prepr.* **1971**, *12*, 177.
2. Bailey, W. J.; Okamoto, Y.; Kuo, W.; Narita, T. *Proc. 3rd Int. Biodegradation Symp.*; Applied Science: Barking, 1976.
3. Bailey, W. J.; Gapud, B. *Ann. N.Y. Acad. Sci.* **1985**, *42*, 446.
4. Bermúdez, M.; Puiggalí, J.; Muñoz-Guerra, S. *Macromolecules* **1994**, *27*, 6325.
5. Crick, F. H. C.; Rich, A. *Nature* **1955**, *176*, 780.
6. Gonsalves, K. E.; Chen, X. *Polym. Commun.* **1990**, *31*, 312.
7. Gonsalves, K. E.; Chen, X.; Wong, T. K. *J. Mater. Chem.* **1991**, *1*, 643.
8. Lotz, B. *J. Mol. Biol.* **1974**, *87*, 169.
9. Lotz, B.; Brack, A. In *Applied Fibre Science, Vol. 3*, Happey F., Ed.; Academic: London 1979, 371.
10. Puiggalí, J.; Muñoz-Guerra, S.; Lotz, B. *Macromolecules* **1986**, *19*, 119.
11. Puiggalí, J.; Muñoz-Guerra, S. *J. Polym. Sci., Polym. Phys. Edn.* **1987**, *25*, 513.
12. Puiggalí, J.; Muñoz-Guerra, S.; Subirana, J. A. *Polymer* **1987**, *28*, 205.
13. Puiggalí, J.; Muñoz-Guerra, S. *J. Polym. Sci. Phys. Edn.* **1989a**, *27*, 1563.
14. Traub, W. In *First Cleveland Symposium on Macromolecules*; Elsevier: Amsterdam, 1977; 23.
15. Vidal, X. Ph.D. Thesis, Universitat Politècnica de Catalunya, Barcelona, 1993.

POLYAMIDES, LACTAM-BASED

Rudolf Puffr and Jaroslav Stehlíček
Institute of Macromolecular Chemistry
Academy of Sciences of the Czech Republic

Linear aliphatic polyamides represented by the formula X-$[NH(CH_2)_{n-1}CO]_x$-Y, where n is the number of carbon atoms of aminoacid, x is degree of polymerization, X = H, and Y = OH (or X and Y are the residues of a catalyst) are frequently referred to as **polyamide n** (abbrev. **PAn**) or **nylons n**. The preparation consists in the condensation of ω-aminoacids or in the ring-opening polymerization of lactams $CONH(CH_2)_{n-1}$ or their derivatives. The nomenclature of polyamides is either monomer-based, such as polyglycine, poly(ε-caprolactam) for PA2 and PA6, or derived from the repeating constitutional unit, such as poly[imino(1-oxo-1,12-dodecanediyl)] for PA12.

Other type is formed by condensation of aliphatic diamines with dicarboxylic acids and is marked as **polyamide m,n** (abbrev. **PAm,n**) or **nylon m,n** X-$[NH(CH_2)_mNHCO$-$(CH_2)_{n-2}CO]_x$-Y.

The most important members of both above series are the twins PA6 and PA6,6 which rank among the first produced and still very important polymers with the worldwide production of 5 Mt/year in 1994.

Much endeavor has been devoted to investigation and testing of poly(4,4-dimethyl-3-propanelactam) and PA4 as fiber-forming polymers with properties approaching natural silk.[1,2,5,6]

PREPARATION

The ring-opening polymerization of lactams consists of reversible transamidation reactions in which cyclic amides are converted into linear ones forming an equilibrium mixture: linear polymer–cyclic oligomers–monomer. Cleavage of the amide bond in lactams proceeds by a nucleophilic or electrophilic attack of added or formed species in the solvolytic (namely, hydrolytic -h, acidolytic -ac, or aminolytic -am), cationic (-c) or anionic (-a) initiation and propagation.

The growth centers are neutral, non-ionic even in the ionic polymerizations. The scheme also suggests that several parallel propagation reactions occurs in most mechanisms, of both chain-growth type and step-growth type.

Consequently, molecular weight (MW) is not a simple function of initiator concentration. Polyamides with a narrow MW distribution were prepared only by the anionic polymerization of 4-membered lactams where only one propagation reaction is kinetically significant, namely the acylation of strongest nucleophile with the strongest electrophile.[6] Equilibrium lactam-polyamide is governed by thermodynamics of the ring-opening reaction in liquid phase. The equilibrium lactam concentration increases with increasing temperature. If the polymerization temperature is low enough to allow polymer crystallization, the equilibrium is established only in the amorphous phase and the yield of polymer increases.

Bulk hydrolytic polymerization initiated with water at 240–300°C is the most common production method for PA6 and PA12. The mechanism and kinetics are well established.[3,7]

Anionic polymerization is the fastest polymerization of lactams characterized by the lowest activation energy. Any base able to produce N-anion of the lactam may act as a catalyst.

Growth centers of the N-acyllactam type are formed either spontaneously by a slow initiation reaction of the lactam anion with lactam or, more often, N-acyllactam or its precursor is added into so called *activated* (or assisted) *polymerization.*[3,7] This is the only method allowing polymerization of stable rings, such as are 2-pyrrolidone and 2-piperidone[3], suspension polymerizations of lactams,[15,16] and reaction processing technologies.[2–17]

Monomer casting consists of the anionic polymerization (mostly of 6-hexanelactam) carried out in simple inexpensive molds. It is suitable for small-series production of moldings of almost unlimited range of shapes and weight up to several hundreds kg.

The higher molecular weight and crystallinity of cast PA6 results in higher strengths, moduli, better wear, and fatigue resistance and lower creep as compared to molded PA6.[18]

Reaction injection molding (RIM) of 6-hexanelactam, which has been commercialized under trade name NyRIM®, consists of the anionic polymerization proceeding on the oligooxypropylene initiator carrying N-acyllactam endgroups. The product is a multiblock copolymer PA6-polyether.[19-20]

Continuous anionic polymerization of 6-hexanelactam in twin screw extruders has technically and economically interesting applications.

PROPERTIES

Basic Types of Polyamides

In the homologous series of PAn and PAm,n, the increasing concentration of amide groups in the chain results in the increasing melting temperature T_m and glass-transition temperature T_g, density, water sorption, strengths, and moduli and the decreasing elongation and toughness. Similar effect on properties has degree of crystallinity. An exception is water sorption, which increases with the increasing fraction of noncrystalline phase.

Modified Polyamides

Formation of composites or chemical modification of existing basic polyamides may improve some properties that are insufficient for certain applications, or provide new properties in a more efficient and economical way than the development of new polymer types.[17,21,22]

Filled Polyamides

Particulate and platelike fillers (minerals, synthetic inorganic compounds, or metal powders) increase melt viscosity, material rigidity, creep resistance, and heat-deflection temperature (HDT) and reduce molding shrinkage, water sorption, and thermal expansion. Filling may produce special grades, such as self-extinguishing, non-inflammable, antistatic, semiconducting polyamides, and polyamides with improved sliding properties.

Reinforcing fillers also increase strength and, in some cases, notched impact resistance.

In *molecular composites*, reinforcement of polyamide matrix by stiff (mostly aromatic) polymers that can aggregate to liquid crystals is expected to occur at molecular and/or supermolecular levels.

Polyamide Blends

The preparation of polymer blends in an efficient extruder or batch mixer is the most often used method to extend the assortment of polymer materials. The aim is to improve some polyamide properties, such as to enhance mechanical properties (PA6/polyester), extend the temperature region of application (PA6/polycarbonate), reduce water sorption and permeability of gases, and increase chemical resistance.[23–24]

Copolymers

Melting temperature, T_g, crystallinity and mechanical moduli of polyamides can be decreased and toughness, water sorption, solubility, dyeing ability, and adhesivity (such as, in hot-melt adhesives) increased in *random copolyamides.*

Grafting of polyamides with unsaturated monomers is usually carried out heterogeneously and has been utilized for the modification of polyamide fibers, textiles, and membranes. The main aim is to enhance hydrophilicity, dyeing, and antistatic properties.

Block copolymers based on polyamides have been broadly investigated and are manufactured as materials for hydrophilic fibers, elastic fibers, tire cords reducing flat spotting of tires, and new engineering plastics.

APPLICATIONS

The main use of polyamide production is for synthetic fibers, but the largest growth can be seen in the application of polyamides as engineering plastics as a consequence of an ever-enlarging assortment with a broad region of properties.

Polyamide fibers will keep their important position with respect to other chemical fibers, depending on the utilization of progressive modifications of the materials and development of new textile technologies.

The majority of polyamide plastics (69%) are injection-molded parts, components, and final products. Extruded films, sheets, pipes, and other profiles represent 23%, monomer-cast molding 4%, and products of other technologies, including sintered powders, 4%. Exemplary applications in mechanical engineering are covers and mechanical parts of machines, such as gears, chains, bearings, joints, and couplings; in electronics sockets, plugs, switches, and whole series of component parts for TV sets, videorecorders, acoustic instruments, electric tools, and household machines; sport outfits, such as bicycle rims, skiing boots, helmets; medical instruments, and surgical implants. Two-layered PA12/polyethylene films with excellent barrier properties are used as tubular packing materials of meat products; PA12 hoses are suitable for applications where resistance to hydraulic fluids, fuels, and oils is required.

REFERENCES

1. Bestian, H. *Angew. Chem.* **1968**, *80*, 304.
2. Barnes, C. E. *Lenzinger Ber.* **1987**, *62*, 62.
3. Sekiguchi, H. *Ring-Opening Polymerization*, Evin, K. J.; Seagusa, T. Eds.; Elsevier: New York, NY 1984; Vol. 2, Chapter 12.
4. *Lactam-Based Polyamides*; Puffr, R.; Kubánek, V. Eds., CRC: Boca Raton, FL, 1991; Vol. 1.
5. *Lactam-Based Polyamides*; Puffr, R.; Kubánek, V. Eds., CRC: Boca Raton, FL, 1991; Vol. 2.

6. Hashimoto, K.; Hotta, K.; Okada, M.; Nagata, S. *J. Polym. Sci. Part A: Polym. Chem.* **1995**, *33*, 1995.
7. Šebenda, J. see Reference 4, Chapter 2.
8. Tai, K.; Arai, Y.; Tagawa, T. *J. Appl. Polym. Sci.* **1983**, *28*, 2527.
9. Srivastava, D.; Gupta, S. K. *Polym. Eng. Sci.* **1991**, *31*, 596.
10. Ramesh, G. M.; Gupta, S. K. *Polymer* **1993**, *34*, 1716.
11. Gaymans, R. J.; Amirtharaj, J.; Kamp, H. *J. Appl. Polym. Sci.* **1982**, *27*, 2513.
12. Mizerovskii, L. N.; Siganov, D. L. *Vysokomol. Soedin.* Ser. B **1991**, *33*, 845.
13. Mizerovskii, L. N.; Siganov, D. L.; Silant'eva, V. G.; Artsis, E. S. *Vysokomol. Soedin. Ser. A* **1991**, *33*, 967.
14. Kulkarni, M. R.; Gupta, S. K. *J. Appl. Polym. Sci.* **1994**, *53*, 85.
15. Biernacki, P.; Wlodarczyk, M. *Eur. Polym. J.* **1980**, *16*, 843.
16. Russo, S.; Biagini, E.; Bontà, G. *Makromol. Chem. Macromol. Symp.* **1991**, *48/49*, 31.
17. Neuhäusl, E. see Reference 5, Chapter 11–13.
18. Carlion, G. see Reference 27, Chapter 13.
19. Hedrick, R. M.; Gabbert, J. D. *Reaction Injection Molding, ACS Symp. Ser.* **1985**, *270*, 135.
20. Yeh, J. L.; Kuo, J. F.; Chen, C. Y. *J. Appl. Polym. Sci.* **1993**, *50*, 1671.
21. *Nylon Plastics*; Kohan, M. I., Ed.; John Wiley & Sons: New York, 1973.
22. Stehlíček, see Reference 5, Chapter 9.
23. Fakirov, S.; Evstatiev, M.; Petrovich, S. *Macromolecules* **1993**, *26*, 5219.
24. Gattiglia, E.; Turturro, A.; Lamantia, F. P.; Valenza, A *J. Appl. Polym. Sci.* **1992**, *46*, 1887.

Polyamidoamines

See: *Dendritic Polymers, Divergent Synthesis (Starburst Polyamidoamine Dendrimers)*
High Pressure Polymerization Processing (High Temperature Thermosetting Polymers)
Ion-Chelating Polymers (Medical Applications)

Polyamine Hardners

See: *Polyurethanes, Blocked Copolymer (Reactive Modifiers for Epoxy Resins)*

Poly(amine quinone)s

See: *Quinone-Amine Polyurethanes*
Quinone-Diamine Polymers

Polyamines

See: *Condensation Polymers (Synthesis from Silylated Diamines)*
Cyclic Imino Ethers (Ring-Opening Polymerization of 1,3-Oxazo Monomers)
Polyalkylenimines
Polyamines (Ru-Catalyzed Polycondensation of Diamines and Diols)
Polyvinylamine (Overview)
Polyvinylamine
Powder Coatings (Overview)

POLYAMINES (Ru-Catalyzed Polycondensation of Diamines and Diols)

Masaru Yoneyama
Department of Biological and Chemical Engineering
Faculty of Engineering
Gunma University

Generally, polyamines have been synthesized by cationic ring-opening polymerization of ethyleneimines (aziridines), trimethyleneimines (azetidines), and 2-oxazolines, while there are relatively few reports on the synthesis of condensation-type polyamines.[1–3]

In the field of organic synthesis, it has been known that, using ruthenium catalyst, alcohols were readily activated and reacted with both aromatic and aliphatic amines to give unsymmetrical secondary or tertiary amines.[4–7] Moreover, N-alkylation is regulated at the N-mono- or N,N-dialkylation stage by controlling a molar ratio of alcohol to amine. It was anticipated that this reaction could be utilized in a polycondensation process for linear polyamines. Therefore, a detailed investigation of ruthenium-catalyzed polycondensation of diamines with diols was undertaken (**Equation 1**). The following discussion is the recent results for a novel synthesis and characterization of poly(secondary amine)s from aromatic di(primary amine)s and bis(benzyl alcohol)s using ruthenium complexes.

$$H_2N\text{-}R\text{-}NH_2 \;+\; HO\text{-}R'\text{-}OH \xrightarrow[\text{Ru cat.}]{-H_2O} \left[NH\text{-}R\text{-}NH\text{-}R' \right]_n \qquad (1)$$

PREPARATION

Detailed reaction condition for the ruthenium-catalyzed polycondensation of 3-aminophenyl 4'-aminophenyl ether (**1a**) with p-xylylene glycol (**2a**) has been examined in order to determine optimal condition (**Equation 2**).

$$\underset{1}{H_2N\text{-}R\text{-}NH_2} \;+\; \underset{2}{HOCH_2\text{-}Ar\text{-}CH_2OH} \xrightarrow[\text{Solvent}]{\text{Ru cat.}} \underset{3}{\left[NH\text{-}R\text{-}NH\text{-}CH_2\text{-}Ar\text{-}CH_2 \right]_n} \qquad (2)$$

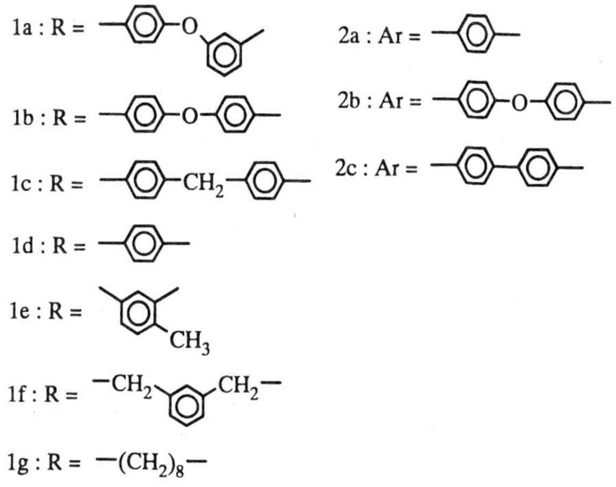

A variety of poly(secondary amine)s **3aa–3ec** were prepared from the combination of aromatic diamines **1a–e** and bis(benzyl alcohol) **2a–c** (Equation 2). Bis(benzyl alcohol) **2b** and **2c** were prepared by the reduction of corresponding diacids with lithium aluminum hydride.[8]

PROPERTIES

Solubility

These polyamines except for polyamine **3ec** are readily dissolved in *N,N*-dimethylacetamide, formic acid, and *o*-chlorophenol.

Thermal Behavior

The thermal stability of the polymer was examined by thermogravimetry. All polymers were stable up to 210°C and decomposed rapidly around 420°C in air, while this rapid degradation was not determined in nitrogen. The glass transition temperatures of polyamines were determined by differential thermal analysis to be around 110°C. All polyamines did not show any melting temperature, which was supported by the observation that they exhibited no distinct DTA endotherms below 200°C. Thus, the polyamines obtained are considered to be the new engineering plastics.

APPLICATION

Condensation-type polyamines were mainly investigated as thermally stable polymers.

REFERENCES

1. Kobayashi, S. *Prog. Polym. Sci.* **1990**, *15*, 751.
2. Goethals, E. J. *Ring-Opening Polymerization*, Ivin, K. J.; Saegusa, T., Eds.; Elsevier: New York, NY, 1984; Vol. 2, p 715.
3. Kobayashi, S.; Saegusa, T. *Ring-Opening Polymerization*, Vol. 2, Ivin, K. J.; Saegusa, T., Eds.; Elsevier: New York, NY, 1984, p 765.
4. Murahashi, S.; Kondo, K., Hakata, T. *Tetrahedron Lett.* **1982**, *23*, 229.
5. Watanabe, Y.; Tsuji, Y.; Ohsugi, Y. *Tetrahedron Lett.* **1981**, *22*, 2667.
6. Watanabe, Y.; Tsuji, Y.; Ige, H.; Ohsugi, Y.; Ohta, T. *J. Org. Chem.* **1984**, *49*, 3359.
7. Huh, K.; Tsuji, Y.; Kobayashi, M.; Okuda, F.; Watanabe, Y. *Chem. Lett.* **1988**, 449.
8. Baltensperger, U.; Gunter, J. G.; Kagi, S.; Kahr, G.; Marty, W. *Organometallics* **1983**, *2*, 571.

Polyaminimides

See: Aminimides, Monomers and Polymers

POLY(α-AMINO ACID) SPHERICAL PARTICLES

Chuichi Hirayama, Hirotaka Ihara, Shoji Nagaoka, and Masayo Sakata
Department of Applied Chemistry
Faculty of Engineering
Kumamoto University

It is very attractive to produce spherical particles from poly(α-amino acid)s, especially in the application field of packing materials for separation and carrier particles. In general, the suspension polymerization is a good technique for obtaining spherical particles from polymerizable materials. In the case of poly(α-amino acid)s, unfortunately, it is difficult to get spherical particles using this technique because poly(α-amino acid)s with high molecular weight can be obtained only by polymerization of *N*-carboxyanhydrides of α-amino acids (NCAs), which are too reactive and unstable in aqueous or protic solvent solutions. Therefore, it is necessary to develop a new approach for sphering poly(α-amino acid)s.

In this article, we describe an unique sphering technique for poly(α-amino acid)s, and chromatographic applications using the particles.

FORMING SPHERES AND CREATING POROSITY

We have newly developed the "suspension evaporation" method in order to obtain spherical particles directly from poly(α-amino acid)s.[1–5] The method is based on the fact that poly(α-amino acid)s become insoluble through partially produced β-structures when poly(α-amino acid)s are reprecipitated from the solution: the method includes the preparation of suspension particles and gradual removal of solvent from them. **Figures 1a** and **1b** show electron micrographs of typical examples prepared from poly(γ-methyl L-glutamate) (PMLG) using this method. The particles are clearly spherical.

The spherical PMLG particles can be readily macroreticulated by adding a diluent in the process of sphering and then removing the diluent with a proper solvent.[4,5]

APPLICATION TO SIZE-EXCLUSION CHROMATOGRAPHY

We have applied a size-exclusion chromatography (SEC) technique to estimate the pore-size distribution of PMLG particles.

In this study, various diluents have been used for macroreticulation of PMLG particles.

As it is desirable that aqueous SEC is applicable at high flow rate, it is very important to check the flow rate resistant in column chromatography. A remarkable higher flow rate was obtained with PMLG without any crosslinking treatment, compared with other aqueous SEC packings. The PMLG particles show minimal swelling. These results indicate that hard gels may be produced by selection of the appropriate material with hydrogen bonding ability, and it is not necessary to depend on special crosslinking such as with conventional SEC packings.

MOLECULAR-SHAPE RECOGNITION IN MICROPOROUS PMLG PARTICLES.[6]

Hydrophilic compounds such as D_2O, methanol, ethylene glycol, and polysaccharides were eluted in order of their molecular sizes in the column packed with porous PMLG particles. This shows that these solutes were eluted in SEC separation mode and that PMLG particles do not interact with these hydrophilic solutes. However, these columns provided higher retention capacity[7] for hydrophobic compounds (such as propyl alcohol, butyl alcohol and pentyl alcohol) than for hydrophilic compounds. These substances were eluted in order of their alkyl lengths, which were the same as those of reversed-phase liquid

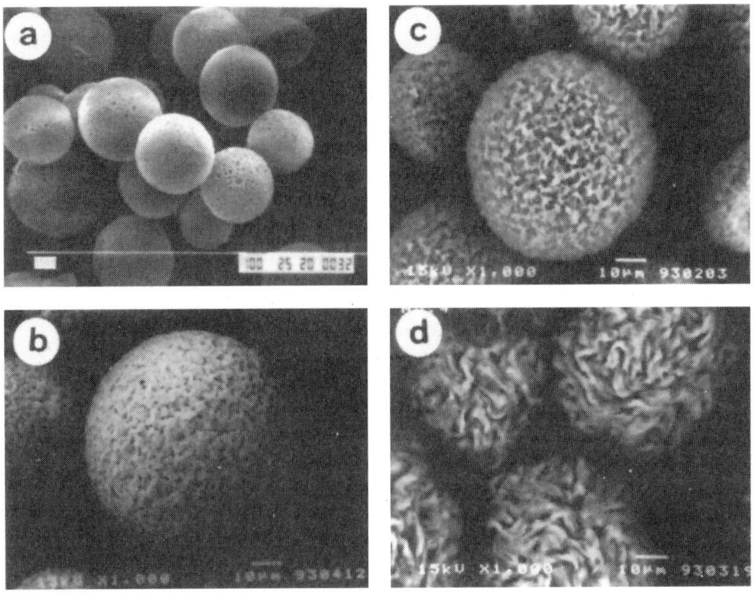

FIGURE 1. Typical scanning electron micrographs of porous PMLG particles. Diluents: (a), decahydronaphthalene; (b)*, di(2-ethylhexyl)phthalate; (c)*, heated at 220°C for 70 min in methanol; (d)*, exposed to formic acid vapor at 30°C for 24 h. Scale bars shows 10 μm. The samples marked "*" were observed under 0.6 torr. without drying. The other samples were observed after freeze-drying.

chromatography (RPLC) packings such as butylated silica gels (C_4-silica).[8]

CHEMICAL MODIFICATION OF PMLG PARTICLES

PMLG particles are extremely suitable for applications as carriers for affinity chromatography. First, because the particles themselves are extremely heard, and the residual amino acid (methoxycarbonyl) groups hardly ever show hydrolyzation in a pH range of 2–10, but a variety of ligands can be introduced using aminolysis reaction and transesterification through catalyzation by strong acid.

CONCLUSION

In the past, development of organic polymer packings for use in aqueous SEC has centered mainly around polysaccharides with few unusual adsorption characteristics, especially dextran, agarose, and cellulose. However, with the marked increase in demand for liquid chromatography, desirability for high-speed processing and separation of large amounts has increased, and aqueous SEC must be able to respond to these needs. The authors believe that accurate control of hardness and pore size characteristics of the packings will alleviate these problems, and they are searching for packing materials that can be used in place of polysaccharides. The results of this search clearly showed that hydrophilic polymers showing strong intermolecular hydrogen bonding, such as poly(amino acid)s, form extremely hard spherical particles, and, depending on the method used for creating porosity, may produce a variety of packings for use with aqueous SEC, having exclusion molecular weight (10^3–10^7). We know that these new hard particles will be applicable as packing materials for various affinity chromatographies and carrier particles.

REFERENCES

1. Ihara, H.; Yoshinaga, T.; Motozato, Y.; Hirayama, C. *Polym. J.* **1985**, *17*, 1301.
2. Hirayama, C.; Ihara, H. *J. Chromatogr.* **1985**, *347*, 357.
3. Ihara, H.; Yoshinaga, T.; Hirayama, C. *J. Chromatogr.* **1986**, *362*, 197.
4. Hirayama, C.; Ihara, H.; Nagaoka, S.; Furusawa, H.; Tsuruta, S. *Polym. J.* **1986**, *22*, 614.
5. Ihara, H.; Furusawa, H.; Li, X.; Hirayama, C. *J. App. Polym. Sci.* **1991**, *42*, 347.
6. Hirayama, C.; Ihara, H.; Nagaoka, S.; Syono, T. *Chem. Lett.* **1992**, 971.
7. The retention capacity was determined by $100 \times V_e/V_t$ (%), where V_e and V_t are the elution volume of solutes and the total column volume, respectively. V_{30} shows retention capacity at 30°C.
8. C_4-silica (YMC-Packed Column A-802) was purchased from Yamamura Chemical Laboratories Co., Ltd.

Poly(amino acid)s

See: *Fluorine-Containing Poly(amino acid)s (Conformation and Gas Permeability)*
Photochromic Poly(α-aminoacids)
Poly(α-amino acid) Spherical Particles
Poly(α-amino acids) (Biodegradation, Medical Applications)
Pseudo-Poly(amino acid)s (Overview)

POLY(α-AMINO ACIDS) (Biodegradation, Medical Applications)

Toshio Hayashi
Research Institute for Advanced Science and Technology
Osaka Prefecture University

There are two different mechanisms for hydrolysis. Polymers that are decomposed by enzyme-specific reactions are called "enzymatically degradable polymers," and polymers that are decomposed by contact with water or serum are "nonenzymatically degradable polymers."[1,2]

Natural polymers such as polypeptides, polysaccharides, polynucleotides, and bacterial polyesters are generally degraded in biological systems by hydrolysis followed by oxidation. Most of these biopolymers are enzymatically biodegradable, except for highly crystalline celluloses and "hard" proteins like keratin and silk fibroin. However, most nonenzymatically degradable polymers are synthetic polymers, including aliphatic polyesters and polycarbonates prepared from fatty acids and diols.

It might seem that natural polymers would be well suited for use as biomedical materials because of their structural similarity to components in host tissues. However, there are some significant disadvantages to natural polymers. These include strong antigenestic and physiological activities and the potential for rejection, a difficulty in evaluating degradation rates *in vivo* because of differences in enzyme concentrations in different parts of living tissues, and the mechanical strength of natural polymers is generally insufficient. For these reasons, their application as biomedical materials has been limited to only a few specific areas.

Polypeptides for biomedical uses are classified into two categories: natural and synthetic polypeptides. A brief overview of some important polypeptides is provided below.

POLYPEPTIDES OF NATURAL ORIGIN

Collagen

Collagen is the major component of all mammalian tissues including skin, bone, cartilage, tendons, and ligaments. Collageneous materials have been extensively studied and used as biodegradable sutures, artificial skin, wound dressing, tendons, and for vessel replacement.[3–7] Various crosslinking treatments such as glutaraldehyde, formaldehyde, and carbodiimide have been used for decreasing the biodegradation rate.[8,9] Major drawbacks of solid crosslinked collagen implants include a lack of deformability and flexibility and low tensile strength. To remedy these problems, a new collagen prosthesis was developed as a dural substitute.[10]

Gelatin

Gelatin is a water soluble, biodegradable polymer with extensive industrial, pharmaceutical, and biomedical uses and has been commonly used for coating and microencapsulating various drugs as well as preparing of biodegradable hydrogels.[11,12,14,15] Gelatin is a variant of collagen, that is, a denatured collagen.

Other Proteins

An elastin-fibrin biomaterial that acts like a connective tissue matrix has recently been developed.[16] The protein that is most widely used as a microsphere material is serum albumin from bovine, human, or other appropriate species.[17] Albumin was also used as a crosslinking agent to prepare biodegradable hydrogels.[18]

SYNTHETIC POLYPEPTIDES

Synthetic polypeptides have several potential advantages as biomaterials. Many types of synthetic polypeptides have been prepared for biomedical applications, such as sutures, artificial skin substrates, and drug delivery systems.[20–22] Side chains offer sites for the attachment of drugs, crosslinking agents, or pendant groups that can be used to modify the physico-chemical properties of the polymer. Because these polymers release naturally occurring amino acids as the primary products of polymer backbone cleavage, their degradation products may be expected to show a low level of systemic toxicity.

Water-Soluble Synthetic Polypeptides

Among the potentials of biodegradable medical applications, water-soluble polypeptides may be very useful for protein conjugates and drug delivery systems. They are typical biodegradable polymers and thus these controlled release systems offer the distinct advantage that no residual polymer remains following drug release of polymer biodegradation. Various types of water-soluble polypeptides were prepared and they investigated the relation between their molecular structure and enzymatic hydrolysis behavior *in vitro* to simulate *in vivo* degradation.[24–27]

Synthetic Polypeptide Hydrogels

Hydrophilic polypeptide membranes may offer potential for biodegradable medical applications, such as temporary artificial skin substitutes in burn therapy, temporary barriers to prevent adhesion between tissue planes damaged by either accident or surgery, polymer carriers for conjugates coupled to proteins for therapeutic use, and drug delivery systems.[19] Until now, many types of polypeptide hydrogels have been prepared and their properties and enzymatic degradable behaviors *in vitro* and *in vivo* have been investigated from the viewpoints of medical applications.[28–36]

Synthetic Polypeptide Fibers

Synthetic polypeptide fibers have been also investigated for possible uses of biomedical applications, such as a surgical suture. Catgut is still the only available suture as an enzymatically degradable fiber, but does not meet well the requirement that suture materials need to retain a sufficient strength during the early phase of wound healing and must be completely absorbed with minimal tissue reactions.[3] One of the most essential properties required for biodegradable materials is the retention of mechanical strength until biotissue repair is completed.

Hayashi et al. investigated the biodegradation behaviors of hydrophilic copolypeptide fibers *in vitro*, including enzymatic degradation and mechanical properties, to simulate *in vivo* fiber degradation.[37,38]

REFERENCES

1. Kimura, Y. *Biomedical Applications of Polymeric Materials*, CRC Press: Boca Raton, 1993; p. 163.
2. Hayashi, T. *Prog. Polym. Sci. 18* **1994**, *19*, 663.

3. Okada, T.; Hayashi, T.; Ikada, Y. *Biomaterials* **1992**, *13*, 448.
4. Yannas, I. V.; Burke, J. F. *J. Biomed. Mater. Res.* **1980**, *14*, 65.
5. Doillon, C. J. *J. Biomater. Appl.* **1988**, *2*, 562.
6. Kato, Y. P.; Christiansen, D.; Hahn, R. A. *Biomaterials* **1989**, *10*, 38.
7. Chvapil, M.; Moore, W. S.; Noishiki, Y. *Angio. Arch.* **1985**, *9*, 7.
8. Cote, M. F.; Doillon, C. J. *Biomaterials* **1992**, *13*, 612.
9. Visset, C. E. *Biomaterials* **1992**, *13*, 34.
10. Pietrucha, K. *Biomaterials* **1991**, *12*, 320.
11. Tabata, Y.; Uno, K.; Ikada, Y.; Muramatsu, S. *J. Pharm. Pharmacol.* **1993**, *45*, 303.
12. Tabata, Y.; Ikada, Y. *J. Controlled Release* **1993**, *27*, 79.
13. Ziegler, G. R. *Biotechnol. Progr.* **1991**, *7*, 283.
14. Shinde, B. G.; Nithianandam, V. S.; Kaleem, K.; Erhan, S. *Bio-Medical Mater. Eng.* **1992**, *2*, 123.
15. Shinde, B. G.; Erhan, S. *Bio-Medical Mater. Eng.* **1992**, *2*, 127.
16. Collet, D.; Lefebvre, F.; Quentin, C.; Rabaud, M. *Biomaterials* **1991**, *12*, 763.
17. Yan, C.; Kitano, H.; Ise, N. *Biotechnol. Appl. Biochem.* **1988**, *10*, 13.
18. Shalaby, W. S. W.; Chen, M.; Park, K. *J. Bioact. Compat. Polym.* **1992**, *7*, 257–264.
19. Anderson, J. M.; Spilizewski, K. L.; Hiltner, A. *Biocompatibility of Tissue Analogs*; CRC: Boca Raton, 1985; p 67.
20. Spira, M.; Fissette, J.; Hall, C. W.; Hardy, S. B.; Gerow, F. J. *J. Biomed. Mater. Res.* **1969**, *3*, 213.
21. Aiba, S.; Minoura, N.; Fujiwara, Y.; Yamada, S.; Nakagawa, T. *Biomaterials* **1985**, *6*, 290.
22. McCormick-Thomson, L. A.; Duncan, R. *J. Bioact. Biocompat. Polym.* **1989**, *4*, 242.
23. Salthouse, T. N. *J. Biomed. Mater. Res.* **1976**, *10*, 197.
24. Hayashi, T.; Iizuka, Y.; Oya, M.; Iwatsuki, M. *J. Appl. Polym. Sci.* **1991**, *43*, 2223.
25. Hayashi, T.; Tabata, Y.; Nakajima, A. *Polymer J.* **1985**, *17*, 463.
26. Hayashi, T.; Iwatsuki, M. *Biopolymers* **1990**, *29*, 549.
27. Iizuka, Y.; Oya, M.; Iwatsuki, M.; Hayashi, T. *Polymer J.* **1993**, *25*, 285.
28. Nakanishi, E.; Shimoizu, Y.; Ogura, K.; Hibi, S.; Hayashi, T. *Polymer J.* **1991**, *23*, 1061.
29. Nakanishi, E.; Sugiyama, E.; Shimizu, Y.; Hibi, S.; Maeda, M.; Hayashi, T. *Polymer J.* **1991**, *23*, 983.
30. Nakanishi, E.; Hamada, K.; Sugiyama, E.; Hibi, S.; Hayashi, T. *Polymer J.* **1991**, *23*, 1053.
31. Hayashi, T.; Takeshima, K.; Tabata, Y.; Nakajima, A. *Polymer J.* **1985**, *17*, 1149.
32. Hayashi, T.; Takeshima, K.; Nakajima, A. *Polymer J.* **1985**, *17*, 1273.
33. Hayashi, T.; Nakanishi, E.; Nakajima, A. *Polymer J.* **1987**, *19*, 1025.
34. Hayashi, T.; Iizuka, Y.; Oya, M.; Iwatsuki, M. *Biomaterials* **1993**, *14*, 497.
35. Hayashi, T.; Nakanishi, E.; Iizuka, Y.; Oya, M.; Iwatsuki, M. *European Polymer J.* **1995**, *31*, 453.
36. Yoda, R.; Komatsuzaki, S.; Nakanishi, E.; Hayashi, T.; Kawaguchi, H. *Biomaterials* **1995**, *16*, 1203.
37. Hayashi, T.; Iwatsuki, M. *Sen-i Gakkaishi* **1988**, *44*, 19.
38. Hayashi, T.; Ikada, Y. *Biomaterials* **1990**, *11*, 409.

Poly(p-aminostyrene)

See: Polystyrene and Derivatives, Photolysis

Polyampholytes

See: Amphoteric Latex
Flocculants (Organic, Overview)
Polyampholytes (Overview)
Polyampholytes (Properties in Aqueous Solution)
Polyzwitterions (Overview)
Zwitterionic Polymers (Dilute Solution and Rheological Properties)
Zwitterionic Polymers (Ionic-Conductivity)

POLYAMPHOLYTES (Overview)

Charles L. McCormick and Erich E. Kathmann
Department of Polymer Science
University of Southern Mississippi

Polymers possessing ionic groups pendent to or on the backbone are perhaps the most important class of macromolecules, ranging from biopolymers such as polynucleotides and proteins to technologically important viscosifiers and polysoaps. These ion-containing polymers may be divided into two groups: polyelectrolytes and polyampholytes. The former have *either* anionic or cationic groups along the chain while the latter have *both* anionic and cationic groups.

In contrast to polyelectrolytes, structure property relationships of polyampholytes are governed by coulombic attractions between anionic and cationic mer units. When an anionic or cationic species is in sufficient excess (≥ 10–15 mol%), charge repulsions induce an extended conformation of the chain resulting in rheological behavior typical of polyelectrolytes. As the molar ratio of anionic to cationic species approaches one, coulombic interactions lead to globule-like conformations and, in many cases, insolubility in deionized water. These attractive interactions may be screened by the addition of electrolytes or change in pH, which induces a transition to a random coil conformation, often facilitating solubility. This behavior, commonly referred to as the "antipolyelectrolyte effect," leads to an enhancement in viscosity and higher values of second virial coefficients in the presence of electrolytes. Several theories have been advanced to explain this behavior.[1–3,15]

Polyampholytes in which both the anionic and cationic groups are pH responsive were the first systems to be studied extensively.[2–26]

Polyampholytes in which the anionic species is based on a sulfonate group have been investigated.[27–46] The sulfonate group, a much weaker base than the carboxylate group, will remain ionized in aqueous media over the useful pH range. Studies of these systems have focused on rheological behavior in the presence of electrolytes. Copolymers in which the cationic/anionic charge ratio approximates one are, in some cases, insoluble in deionized water and require the addition of a critical concentration of salt to achieve solubility.[30,33,44] Upon further addition of salt, these systems display "antipolyelectrolyte" behavior as evidenced by an increase in viscosity. Polyampholytes of this type that also incorporate a nonionic hydrophilic termonomer exhibit a tendency to intermolecularly associate in deionized water.[36,38,46] These associations can be disrupted by the addition of electrolytes.

Polyampholytes have been prepared from zwitterionic monomers in which the anionic and cationic groups are incorporated into a single mer unit.[47–92] The anionic group is typically a sulfonate,[47–77] carboxylate,[87–92] or phosphate[78–83] moiety, while the cationic moiety is usually an ammonium species. Polyzwitterions containing the sulfonate moiety (sulfobetaines) have been the most thoroughly investigated. In comparison with copolymers containing sulfonate and ammonium groups on different mer units, the majority of the poly-(sulfobetaines) systems are insoluble in deionized water with a few exceptions.[47,55,57] The addition of a critical concentration of salt is again required to achieve solubility and to enhance viscosity.

Although polyampholytes have not been extensively used in commercial applications, their unusual solution properties present unique opportunities for formulation in the presence of electrolytes. These include areas such as personal care, enhanced oil recovery, and flocculation. Polyampholytes have also shown potential for use as drag-reducing agents. Intramolecular associations may also be utilized in drug delivery or controlled release applications because these systems can be designed to act as polyelectrolytes or polyampholytes when pH sensitive functionalities are employed in the original polymer. The reader is referred to two recent review articles for other potential applications of polyampholytes.[93,94]

In continuing efforts to understand structure/property relationships of polyampholytes, we have initiated a systematic study investigating macroscopic properties as affected by the nature, distribution, and concentration of the charged groups incorporated into the polymer. To elucidate these effects, three types of polyampholytes have been synthesized. The first type, termed a high charge density polyampholyte, consists of a copolymer containing varying molar amounts of cationic and anionic comonomers. The second type, a low charge density polyampholyte, incorporates a significant amount of a nonionic, hydrophilic termonomer along the polymer chain. Finally, the third type, termed polybetaine or polyzwitterion, couples the anionic and cationic units in the same mer unit. Types 1 and 2 have been prepared with various ratios of cationic-to-anionic monomers while Type 3 inherently contains a charge balance.

CONCLUSIONS

In this report we have illustrated the effects of molecular architecture of polyampholytes on solution behavior under specified conditions of concentration, pH, and added electrolytes. The nature of the charged groups as well as the amounts incorporated onto the polymer chain induce different solubility and rheological behavior. For high charge density systems, as the molar ratio of anionic to cationic species approaches one, polyampholytic behavior is observed as evidenced by an increase in viscosity upon the addition of electrolytes. The same behavior is observed for low charge density systems but on occasion complicated by the presence of intermolecular associations. Polyampholytes containing ionic species sensitive to changes in pH demonstrate the ability to behave as polyelectrolytes or polyampholytes dependent on the pH of the aqueous environment.

REFERENCES

1. Higgs, P. G.; Joanny, J. F. *J. Chem. Phys.* **1991**, *94*(2), 1543.
2. Katchalsky, A.; Miller, I. R. *J. Polym. Sci.* **1954**, *13*, 57.
3. Merle, Y. *J. Phys. Chem.* **1987**, *91*, 3093.
4. Alfrey, T.; Morawetz, H. *J. Am. Chem. Soc.* **1952**, *74*, 436.
5. Alfrey, T.; Fuoss, R.; Morawetz, H.; Pinner, H. *J. Am. Chem. Soc.* **1952**, *74*, 438.
6. Ehrlich, G.; Doty, P. *J. Am. Chem. Soc.* **1954**, *76*, 3764.
7. Marvel, C. S.; Moyer, W. W. *J. Am. Chem. Soc.* **1957**, *79*, 4990.
8. Marvel, C. S.; DeTommaso, G. L. *J. Org. Chem.* **1960**, *25*, 2207.
9. Furukawa, J.; Kobayashi, E.; Toshiki, D. *J. Polym. Sci. Polym. Chem. Ed.* **1979**, *17*, 255.
10. Merle, Y.; Merle-Aubry, L.; Selegny, E. In *Polymeric Amines and Ammonium Salts (IUPAC)*; Goethals, E., Ed.; Pergamon: Elmsford, NY, 1980; p 113.
11. Fukutomi, T.; Horikoshi, T.; Ishizu, K. *Polym. J.* **1984**, *16*(8), 619.
12. Chang, C.; Muccio, D. D.; St. Pierre, T. *J. Polym. Sci. Polym. Symp.* **1986**, *74*, 17.
13. Vrancken, M.; Smets, G. *J. Polym. Sci.* **1954**, *14*, 521.
14. Alfrey, T.; Pinner, S. H. *J. Polym. Sci.* **1957**, *23*, 533.
15. Mazur, J.; Silberberg, A.; Katchalsky, A. *J. Polym. Sci.* **1959**, *35*, 43.
16. Jordan, D. O.; Kuruscev, T. *Polymer* **1960**, *1*, 185.
17. Ascoli, F.; Botre, C. *J. Polym. Sci.* **1962**, *62*, S56.
18. Allison, J. P.; Marvel, C. S. *J. Polym. Sci. Part A* **1965**, *3*, 137.
19. Panzik, H.; Mulvaney, J. E. *J. Polym. Sci. Polym. Chem. Ed.* **1972**, *10*, 3469.
20. van der Does, L.; Hofman, J.; van Utteren, T. E. C. *J. Polym. Sci. Polym. Chem. Ed.* **1973**, *11*, 169.
21. Bekturov, E. A.; Kudaibergenov, S. E.; Sigitov, V. B. *Polymer* **1986**, *27*, 1269.
22. Roy-Chowdhury, P.; Dewhare, A. R. *J. Appl. Polym. Sci.* **1974**, *18*, 2471.
23. Dewhare, A. R.; Roy-Chowdhury, P. *J. Appl. Polym. Sci.* **1976**, *20*, 1673.
24. McCormick, C. L.; Johnson, C. B. *J. Macromol. Sci. Chem. Part A* **1990**, *27*, 539.
25. McCormick, C. L.; Salazar, L. C. *J. Appl. Polym. Sci.* **1993**, *48*, 1115.
26. Nonaka, T.; Egawa, H. *Bull. Chem. Soc. Jpn.* **1980**, *53*, 1632.
27. Salamone, J. C.; Watterson, A. C.; Hsu, T. D.; Tsai, C. C.; Mahmud, M. U. *J. Polym. Sci. Polym. Lett. Ed.* **1977**, *15*, 487.
28. Salamone, J. C.; Watterson, A. C.; Hsu, T. D.; Tsai, C. C.; Mahmud, M. U.; Wisniewski, A. W.; Israel, S. C. *J. Polym. Sci. Polym. Symp.* **1978**, *64*, 229.
29. Salamone, J. C.; Tsai, C. C.; Watterson, A. C. *J. Macromol. Sci, Part A* **1979**, *13*(5), 665.
30. Salamone, J. C.; Quach, L.; Watterson, A. C.; Krauser, S.; Mahmud, M. U. *J. Macromol. Sci. Part A* **1985**, *22*, 653.
31. Salamone, J. C.; Tsai, C. C.; Olson, A. P.; Watterson *Ions in Polymers, Adv. Chem. Ser. 187*; Eisenberg, A., Ed. American Chemical Society: Washington, DC, 1980; p 337.
32. Corpart, J.; Selb, J.; Candau, F. *Makromol. Chem., Makromol. Symp.* **1992**, *53*, 253.
33. Corpart, J.; Candau, F. *Macromolecules* **1993**, *26*, 1333.
34. Corpart, J.; Selb, J.; Candau, F. *Polymer* **1993**, *34*, 3873.
35. Skouri, M.; Munch, J. P.; Candau, S. J.; Neyret, S.; Candau, F. *Macromolecules* **1994**, *27*, 69.
36. Salamone, J. C.; Ahmed, I.; Rodriguez, E. L.; Quach, L.; Watterson, A. C. *J. Macromol. Sci. Chem. Part A* **1988**, *25*, 811.
37. Salamone, J. C.; Rodriguez, E. L.; Lin, K. C.; Quach, L.; Watterson, A. C.; Ahmed, I. *Polymer* **1985**, *26*, 1234.
38. Peiffer, D. G.; Lundberg, R. D. *Polymer* **1985**, *26*, 1058.

39. Peiffer, D. G.; Lundberg, R. D.; Duvdevani, I. *Polymer* **1986**, *27*, 1453.
40. Salamone, J. C.; Tsai, C. C.; Watterson, A. C.; Olson, A. P. In *Polymeric Amines and Ammonium Salts (IUPAC)*; Goethals, E., Ed.; Pergamon: Elmsford, N.Y., 1980; p 105.
41. Salamone, J. C.; Mahmud, N. A.; Mahmud, M. U.; Nagabhushanam, T.; Watterson, A. C. *Polymer* **1982**, *23*, 843.
42. McCormick, C. L.; Johnson, C. B. *Macromolecules* **1988**, *21*, 687.
43. McCormick, C. L.; Johnson, C. B. *Macromolecules* **1988**, *21*, 694.
44. McCormick, C. L.; Salazar, L. C. *Macromolecules* **1992**, *25*, 1896.
45. McCormick, C. L.; Johnson, C. B. *Polymer* **1990**, *31*, 1100.
46. McCormick, C. L.; Salazar, L. C. *Polymer* **1992**, *33*, 4384.
47. Hart, R.; Timmerman, D. *J. Polym. Sci.* **1958**, *28*, 638.
48. Schulz, D. N.; Peiffer, D. G.; Agarwal, P. K.; Larabee, J.; Kaladas, J. J.; Soni, L.; Handwerker, B.; Garner, R. T. *Polymer* **1986**, *27*, 1734.
49. Huglin, M. B.; Radwan, M. A. *Polymer International* **1991**, *26*, 97.
50. Schulz, D. N.; Kitano, K.; Danik, J. A.; Kaladas, J. J. *Polym. Mat. Sci. Eng.* **1987**, *147*, 149.
51. Liaw, D.; Lee, W.; Whung, Y.; Lin, M. *J. Appl. Polym. Sci.* **1987**, *34*, 999.
52. Liaw, D.; Lee, W.; Lin, M. *J. Macromol. Sci. Pure and Applied* **1993**, *A30*, 51.
53. Salamone, J. C.; Volksen, W.; Israel, S. C.; Olson, A. P.; Raia, D. C. *Polymer* **1977**, *18*, 1058.
54. Salamone, J. C.; Volksen, W.; Olson, A. P.; Israel, S. C. *Polymer* **1978**, *19*, 1157.
55. Wielema, T. A.; Engberts, J. B. F. N. *Eur. Polymer J.* **1987**, *23*, 947.
56. Monroy Soto, V. M.; Galin, J. C. *Polymer* **1984**, *25*, 121.
57. Monroy Soto, V. M.; Galin, J. C. *Polymer* **1984**, *25*, 254.
58. Galin, M.; Marchal, E.; Mathis, A.; Meurer, B.; Monroy, Y. M.; Galin, J. C. *Polymer* **1987**, *28*, 1937.
59. Zheng, Y. L.; Knoesel, R.; Galin, J. C. *Polymer* **1987**, *28*, 2297.
60. Knoesel, R.; Ehrmann, M.; Galin, J. C. *Polymer* **1993**, *34*, 1925.
61. Konack, C.; Rathi, R. C.; Kopeckova, P.; Kopecek, J. *Polymer* **1993**, *34*, 4767.
62. Konack, C.; Rathi, R. C.; Kopeckova, P.; Kopecek, J. *Macromolecules* **1994**, *27*, 1992.
63. Lee, W.; Tsai, C. C. *Polymer* **1994**, *35*, 2211.
64. Itoh, Y.; Abe, K.; Senoh, S. *Makromol. Chem.* **1986**, *187*, 1691.
65. Koberle, P.; Laschewsky, A.; Lomax, T. D. *Makromol. Chem. Rapid Commun.* **1991**, *12*, 427.
66. Laschewsky, A.; Zerbe, I. *Polymer* **1991**, *32*, 2070.
67. Laschewsky, A.; Zerbe, I. *Polymer* **1991**, *32*, 2081.
68. McCormick, C. L.; Salazar, L. C. *Polymer* **1992**, *33*, 4617.
69. Kathmann, E. E.; Davis, D. D.; McCormick, C. L. *Macromolecules* **1994**, *27*, 3156.
70. Zheng, Y.; Galin, M.; Galin, J. C. *Polymer* **1988**, *29*, 724.
71. Bazuin, C. G.; Zheng, Y.; Muller, R.; Galin, J. C. *Polymer* **1989**, *30*, 654.
72. Ehrmann, M.; Mathis, A.; Meurer, B.; Scheer, M.; Galin, J. C. *Macromolecules* **1992**, *25*, 2253.
73. Mathis, A.; Zheng, Y.; Galin, J. C. *Makromol. Chem. Rapid Commun.* **1986**, *7*, 333.
74. Davidson, N. S.; Fetters, L. J.; Funk, W. G.; Graessley, W. W.; Hadjichristidis, N. *Macromolecules* **1988**, *21*, 112.
75. Fetters, L. J.; Graessley, W. W.; Hadjichristidis, N.; Kiss, A. D.; Pearson, D. S.; Younghouse, L. B. *Macromolecules* **1988**, *21*, 1644.
76. Bredas, J. L.; Chance, R. R.; Silbey, R. *Macromolecules* **1988**, *21*, 1633.
77. Rego, J. M.; Huglin, M. B. *Polym. J.* **1991**, *23*(12), 1425.
78. Yasuzawa, M.; Nakaya, T.; Imoto, M. *J. Macromol. Sci. Chem. Part A* **1986**, *23*(8), 963.
79. Nakaya, T.; Toyoda, H.; Imoto, M. *Polym. J.* **1986**, *18*(11), 881.
80. Umeda, T.; Nakaya, T.; Imoto, M. *Makromol. Chem. Rapid Commun.* **1982**, *3*, 457.
81. Nakaya, T.; Yasuzawa, M.; Imoto, M. *Macromolecules* **1989**, *22*, 3180.
82. Ueda, T.; Oshida, H.; Kurita, K.; Ishihara, K.; Nakabayashi, N. *Polym. J.* **1992**, *24*(11), 1259.
83. Hamaide, T.; Germanaud, L.; Le Perchec, P. *Makromol. Chem.* **1986**, *187*, 1097.
84. Pujol-Fortin, M. L.; Galin, J. C. *Macromolecules* **1991**, *24*, 4523.
85. Pujol-Fortin, M. L.; Galin, M.; Galin, J. C. *Macromolecules* **1991**, *24*, 6443.
86. Pujol-Fortin, M. L.; Galin, J. C.; Morawetz, H. *J. C. Polymer* **1994**, *35*, 1462.
87. Ladenheim, H.; Morawetz, H. *J. Polym. Sci.* **1957**, *26*, 251.
88. Rosenheck, K.; Katchalsky, A. *J. Polym. Sci.* **1958**, *32*, 511.
89. Topchiev, D. A.; Mkrtchyan, L. A.; Simonyan, R. A.; Lachinov, M. B.; Kabanov, R. A. *Polym. Sci. U.S.S.R. (Eng. Transl.)* **1977**, *A19*, 580.
90. Wielema, T. A.; Engberts, J. B. F. N. *Eur. Polymer J.* **1988**, *24*, 647.
91. Hsu, Y. G.; Hsu, M. J.; Chen, K. M. *Makromol. Chem.* **1991**, *192*, 999.
92. Kathmann, E. E.; McCormick, C. L. *Polym. Prepr. Am. Chem. Soc. Div. Polym. Chem.* **1994**, *35*(2), 641.
93. Bekturov, E. A.; Kudaibergenov, Rafikov, S. R. *J. Macromol. Sci.-Rev. Macromol. Chem. Phys.* **1990**, *C30*(2), 233.
94. Salamone, J. C.; Rice, W. C. *Encyclopedia of Polymer Science and Engineering, 2nd ed.*: Wiley-Interscience: New York, 1988; Vol. 11, p 514.

POLYAMPHOLYTES (Properties in Aqueous Solution)

F. Candau and J. F. Joanny
Institut Charles Sadron (CRM-EAHP)
CNRS-ULP Strasbourg, France

Polyampholytes (PA) represent a special class of polyions that contain both positively and negatively charged units on the same polymer chain. The introduction of ionic groups of opposite charges along the backbone leads to a complex solution behavior that is essentially controlled by electrostatic interactions.

A renewed interest for these compounds appeared in the 1980s when several groups reported novel methods of synthesis based on the polymerization of zwitterionic monomers, ion-pair comonomers.[30–33,35–41,44–46] or charged anionic and cationic monomers.[17–25,28a,41] These studies were also stimulated by recent achievements obtained both theoretically and experimentally in the field of polyelectrolytes.

Several reviews have appeared on amphoteric polymers.[3–6,40] In particular, the complexation ability of PA with transition metal ions, dyes, surfactants, and proteins has been thoroughly investigated by Bekturov.[6] The present overview is essentially devoted to synthetic linear polyampholytes derived by free radical polymerization techniques and their properties in aqueous solution.

VARIOUS TYPES OF POLYAMPHOLYTES

Annealed Polyampholytes

Annealed PA are chains formed with strong or weak acids mixed with strong or weak bases in various combinations. The ionic character of these amphoteric materials depends on the pH of the medium; they behave as polycations in acid solution and as polyanions in alkaline solution. The pH value at which a polymer chain is electroneutral corresponds to the isoelectrical point (IP). Most studies dealt essentially with copolymers of (meth)acrylic acid with vinylpyridines or dialkylaminoethyl(meth)acrylates.

Quenched Polyampholytes

The studies reported on quenched PA are far more recent than those on annealed PA. In a quenched polyampholyte, the number of positively and negatively charged monomers and their distribution along the chain are imposed by the chemistry. In this case, the monomers retain their respective charges over a wide range of pH (strong polyampholytes) in contrast with annealed PA where the net charge and the charge distribution are monitored by the pH of the solution (weak PA).

Two routes have been envisioned for the synthesis of quenched PA. First, copolymerization in aqueous solution[15,17–25] or in a microemulsion[8–11,42] of charged anionic and cationic monomers mixed in varying proportions in the presence of their nonpolymerizable inorganic counterions. The net charge is therefore directly related to the monomer feed composition.

The second method that involves a two-step procedure was devised and extensively studied by Salamone and his group.[30–33,35–41,44–46]

In this method, each monomer is a copolymerizable gegenion to its oppositively charged counterpart and nonpolymerizable ions are absent.

The subsequent (co)polymerization by emulsion or solution techniques of these ion-pair monomers results in a copolymer containing equimolar amounts of cationic and anionic charges along the polymer chains with no inorganic counterions.

Microstructure

The few experimental data available indicate that the microstructure depends not only on the type of monomer pairs used but also on the experimental conditions (solvent, range of pH) and on the method of synthesis.

Annealed Polyampholytes

In this type of PA, the charge distribution is an annealed variable since it is adjusted by tuning the pH while the position of the monomers along the chain is a quenched variable.

Usually radical polymerization of acidic and basic monomers results in statistical copolymers whose composition is determined by their corresponding monomer reactivity ratios. A classic example is that of the copolymerization of 2-vinylpyridine (weak base) and methacrylic acid (weak acid), which leads to statistical copolymers (r_{2vp} = 1.55 and r_{MA} = 0.58).[1]

The synthesis of PA with a well-defined arrangement of the monomers, as an alternating structure, is more difficult to achieve and requires most often specific methods like polymerization of betaines,[34,47] charge-transfer polymerization,[32] or special modification reactions.[6,13,36] Nevertheless, the free-radical copolymerization of acid and base units was reported to lead to alternated PA in the following cases:

- By copolymerization in aqueous solution of maleic acid (MA) and allyl or diallyl amine derivatives.[22] Perfectly alternated PA resulted, which is not surprising since it is well known that α, β disubstituted olefin cannot homopolymerize.
- By copolymerization of acrylic acid and 2-ethyl-2-oxazoline.[29] The system polymerizes spontaneously via a complex charge transfer mechanism through zwitterion intermediates.
- By copolymerization of a strong base (DEAEM) and a weak acid (methacrylic acid).[2] At high pH (pH = 7.2), the two monomers are completely ionized and copolymers with a strong tendency to alteration were obtained. At this pH value, the microstructure is controlled by the self-neutralization of acid (base) units by adjacent base (acid) units.

Quenched PA

The microstructure of ion-pair copolymers was found to depend on the nature of the monomers used. If both ionic components have similar structures (case of 3-methacrylamidopropyltrimethylammonium2-acrylamide2-methylpropanesulfonate), random copolymers are produced. If both ionic moieties are of opposite polarity (case of 4-vinyl-piridinium-p-styrenesulfonate), spontaneous alternating copolymerization occurs most likely by a charge transfer mechanism.[35,36]

In general, fully charged polyampholytes prepared in solution from charged cationic and anionic monomers exhibit a tendency for alternation, as a result of the strong electrostatic attractive forces acting between the opposite charges.[11,16–18,22] However, microemulsion polymerization leads to almost random polyampholytes, that is, to copolymers more homogeneous in composition than those prepared in solution. This result was accounted for by the marked differences in mechanism and microenvironment between the two processes.[11]

The molecular weights of polyampholytes prepared by the various methods cover a wide range of values: Mw ~ 2×10^5 for annealed polyampholytes,[12] Mw ~ 10^6–2×10^7 for quenched polyampholytes. The high values found for the latter are mainly due to the high reactivities of the monomers involved and/or to the process. For example, microemulsion polymerization leads to ultra high molecular weights because of the high local monomer concentration within the particles (~50%) and the fact that the propagating radicals grow in relative isolation.[7]

PROPERTIES IN AQUEOUS SOLUTION: EXPERIMENTAL RESULTS

Solubility and Conformation

The solution behavior of polyampholytes is essentially controlled by the competition between repulsive (polyelectrolyte effect) and attractive (polyampholyte effect) electrostatic interactions. These are directly related to the strength of the acid or base groups and the value of the pH (annealed PA) or to the copolymer composition (quenched PA).

The addition of salt screens the electrostatic interactions and weakens the attractions.

The amount of salt required to dissolve the polymer depends on the nature of the salt with data following the Hofmeister series.[6,9] the net charge,[9] and the charge density (the lower the charged density, the less added salt).[19,27,40,41]

APPLICATIONS

The dual-charge nature of polyampholytes described in this review results in some unique physico-chemical properties that can be taken advantage of in a number of applications. The antipolyelectrolyte behavior reflected by the rise in viscosity upon increasing the ionic strength favors their use for all applications where thickeners are required in brine solution, as for example in flocculation and oil recovery processes. Note that the ultra high molecular weight PA (> 10^7) produced by a microemulsion polymerization route has great potential for these purposes.

Although this is not yet commercially exploited, some applications should be quite sensitive to the effect of microstructure. Random PA were shown to be suitable desalinating agents, in contrast with alternated PA.[14a]

Several recent studies have illustrated the utility of ampholytic polymers. Water absorbency studies conducted on polyampholytes grafted onto polysaccharides[35,45] and on NaAMB-AMPDAC-AM terpolymers[29] showed good swelling properties and high absorbency both in deionized water and in salt solution. Mumick et al. have compared the drag reduction behavior of several well-characterized polyampholytes and observed a dependence of drag reduction effectiveness on copolymer structure, composition, and solvation.[25]

Because of their ability to bind to low molecular weight substances (metal ions, surfactants, dyes, drugs, etc.) synthetic polyampholytes can be used as selective chelating agents, as pigment-retention aids, and in paper manufacturing.[6,40]

Polyampholytes may also interact with high molecular weight compounds and this can be employed for protein separation.[27b] The precipitation of the PA around its isoelectrical point facilitates polymer recycling after protein separation. Other polyampholyte-mediated-protein separation methods are ion-exchange displacement chromatography and aqueous, two-phase partitioning.[28] In the same area, another promising application could be in controlled drug delivery, based on the fact that some PA are able to bind to the above compounds at definite pH values and to release them at the isoelectrical point.[6]

Recent experiments have also shown that polyampholytes can adsorb on charged latex particles even if their net charge has the same sign as that of the latex.[14b,27a]

REFERENCES

1. Alfrey, Jr. T.; Morawetz, H. *J. Am. Chem. Soc.* **1952**, *74*, 436.
2. Alfrey, Jr. T.; Overberger, C. G.; Pinner, S. H. *J. Am. Chem. Soc.* **1953**, *75*, 4221.
3. Bekturov, E. A.; Bakanova, Z. K. *Synthetic Water-Soluble Polymers in Solution*; Hüthig & Wepf; Basel, 1986; Chapter 4.
4. Bekturov, E. A.; Kudaibergenov, S. E. *J. Macromol. Sci.-Phys.* **1986**, *B25*, 133.
5. Bekturov, E. A.; Kudaibergenov, S. E. *Makromol. Chem., Macromol. Symp.* **1989**, *26*, 281.
6. Bekturov, E. A.; Kudaibergenov, S. E.; Rafikov, S. R. *Rev. Macromol. Chem. Phys.* **1990**, *C30*, 233.
7. Candau, F. In *Polymerization in Organized Media*; Paleos, C. M., Ed.; Gordon and Breach Sci.: Philadelphia, 1992, pp 215–282.
8. Corpart, J. M.; Selb, J.; Candau, F. *Makromol. Chem., Macromol. Symp.* **1992**, *53*, 253.
9. Corpart, J. M.; Candau, F. *Macromolecules* **1993**, *26*, 1333.
10. Corpart, J. M.; Candau, F. *Colloid Polym. Sci.* **1993**, *271*, 1055.
11. Corpart, J. M.; Selb, J.; Candau, F. *Polymer* **1993**, *34*, 3873.
12. Ehrlich, G.; Doty, P. *J. Am. Chem. Soc.* **1954**, *76*, 3764.
13. Furukawa, J.; Kobayashi, E.; Doi, T. *J. Polym. Sci., Polym. Chem. Ed.* **1979**, *17*, 255.
14. Hahn, M.; Jaeger, W.; Schmolke, R.; Behnisch, J. *Acta Polymerica* **1990**, *41*, 107.

14a. Jeffrey, E. A.; Hodgkin, J. H.; Solomon, D. H. *J. Macromol. Sci. Chem.* **1976**, *A10*, 943.

14b. Joanny, J. F. *J. Phys. II. (France)* **1994**, *4*, 1281.

15. Johnson, C. B.; McCormick, C. L. *Polym. Mater. Sci. Eng.* **1987**, *57*, 154.
16. Kathmann, E. E.; Salazar, L. C.; McCormick, C. L. *Polym. Prepr.* **1991**, *32*, 98.
17. McCormick, C. L.; Johnson, C. B. *Macromolecules* **1988**, *21*, 686.
18. McCormick, C. L.; Johnson, C. B. *Macromolecules* **1988**, *21*, 694.
19. McCormick, C. L.; Salazar, L. C. *Am. Chem. Soc. Polym. Prepr.* **1989**, *30*, 344.
20. McCormick, C. L.; Johnson, C. B. *Polymer* **1990**, *31*, 1100.
21. McCormick, C. L.; Johnson, C. B. *J. Macromol. Sci.-Chem.* **1990**, *A27*, 539.
22. McCormick, C. L.; Salazar, L. C. *Macromolecules* **1992**, *25*, 1896.
23. McCormick, C. L.; Salazar, L. C. *Polymer* **1992**, *33*, 4384.
24. McCormick, C. L.; Salazar, L. C. *J. Appl. Polym. Sci.* **1993**, *48*, 1115.
25. Mumick, P. S.; Welch, P. M.; Salazar, L. C.; McCormick, C. L. *Macromolecules* **1994**, *27*, 323.
26. Neufeld, C. H.; Marvel, C. S. *J. Polym. Sci.: Part A-1* **1967**, *5*, 537.
27. Neyret, S. *Thesis of University L. Pasteur*, Strasbourg, 1995.

27a. Neyret, S.; Ouali, L.; Candau, F.; Pefferkorn, E. *176*, 86, 1995.

27b. Patrickios, C. S.; Jang, C. J.; Hertler, W. R.; Hatton, T. A. *Polym. Prepr., Am. Chem. Soc. Div. Polym. Chem.* **1993**, *34*, 1.

28. Patrickios, C. S.; Hertler, W. R.; Abbott, N. L.; Hatton, T. A. *Macromolecules* **1994**, *27*, 930.

28a. Peiffer, D. G.; Lundberg, R. D. *Polymer* **1989**, *26*, 1058.

29. Rivas, B. L.; Canessa, G. S.; Pooley, S. A. *Polym. J.* **1989**, *25*, 225.
30. Salamone, J. C.; Watterson, A. C.; Hsu, T. D.; Tsai, C. C.; Mahmud, M. U. *J. Polym. Sci. Polym. Lett. Ed.* **1977**, *15*, 487.
31. Salamone, J. C.; Tsai, C. C.; Olson, A. P.; Watterson, A. C. *Polym. Prepr., Am. Chem. Soc. Div. Polym. Chem.* **1978**, *19*, 261.
32. Salamone, J. C.; Watterson, A. C.; Hsu, T. D.; Tsai, C. C.; Mahmud, M. U.; Wisniewski, A. W.; Israel, S. C. *J. Polym. Sci., Polym. Symp.* **1978**, *64*, 229.
33. Salamone, J. C.; Tsai, C. C.; Watterson, A. C. *J. Macromol. Sci.-Chem.* **1979**, *A13*, 665.
34. Salamone, J. C.; Tsai, C. C.; Olson, A. P.; Watterson, A. C. *J. Polym. Sci. Polym. Chem. Ed.* **1980**, *18*, 2983.
35. Salamone, J. C.; Watterson, A. C.; Quach, L.; Raheja, M. K. *Polym. Prepr., Am. Chem. Soc. Div. Polym. Chem.* **1985**, *26*, 196.
36. Salamone, J. C.; Quach, L.; Watterson, A. C.; Krauser, S.; Mahmud, M. U. *J. Macromol. Sci.-Chem.* **1985**, *A22*, 653.
37. Salamone, J. C.; Rodriguez, E. L.; Lin, K. C.; Quach, L.; Watterson, A. C.; Ahmed, I. *Polymer* **1985**, *26*, 1234.
38. Salamone, J. C.; Raheja, M. K.; Anwaruddin, Q.; Watterson, A. C. *J. Polym. Sci., Polym. Lett. Ed.* **1985**, *23*, 655.
39. Salamone, J. C.; Ahmed, I.; Elayaperumal, P.; Raheja, M. K.; Watterson, A. C.; Olson, A. P. *Polym. Mater. Sci. Eng.* **1986**, *55*, 269.
40. Salamone, J. C.; Rice, W. C. *Encyclopedia of Polymer Science and Engineering*; Mark, H. F.; Bikales, N. M.; Overberger, C. G.; Menges, G. Wiley: New York, 1987; Vol. 11, p 514.

41. Salamone, J. C.; Ahmed, I.; Rodriguez, E. L.; Quach, L.; Watterson, A. C. *J. Macromol. Sci.-Chem.* **1988**, *A25*, 811.
42. Skouri, M.; Munch, J. P.; Candau, S. J.; Neyret, S.; Candau, F. *Macromolecules* **1994**, *27*, 69.
43. Vrancken, M.; Smets, G. *J. Polym. Sci.* **1954**, *14*, 521.
44. Watterson, A. C.; Parkeenvincha, E.; Salamone, J. C. *Polym. Prepr.* **1990**, *31*, 492.
45. Watterson, A. C.; Murata, N.; Salamone, J. C. *Polym. Prepr.* **1990**, *31*, 494.
46. Watterson, A. C.; Liang, C. H.; Salamone, J. C. *Polym. Prepr.* **1990**, *31*, 497.
47. Wielema, T. A.; Engberts, J. B. *Eur. Polym. J.* **1987**, *23*, 947.

Polyanhydrides

See: Controlled Drug Delivery Systems
Poly(terephthalic acid anhydride) (as a Latent Monomer)

POLYANILINE (Electrochemical Synthesis)

Ljerka Duić
Faculty of Chemical Engineering and Technology
University of Zagreb

The early works on polyaniline (PANI), from the first electrochemical oxidation of aniline carried out by Letheby until the late 1960s,[3] has been reviewed extensively by Nelson.[5,7] In recent times a widespread interest in the electrochemical synthesis of PANI arose with the work of Diaz and Logan, when high quality films of conductive PANI and platinum substrate were prepared.[1]

Over the years a number of studies, concerned with the products of aniline oxidation, proposed various structural forms.

MacDiarmid et al. designated four different forms of PANI subunits, two of which 1A and 2A, are free amine forms, whereas two others, 1S and 2S, are ammonium salt forms, as shown in **Figure 1**.[6]

If one takes a tetramer as a repetitive unit in PANI chain, the emeraldine salt results, consisting of alternating 1S and 2S units, which Paul and co-workers proved to be a more conductive form of PANI, the protonation of which depends on pH of

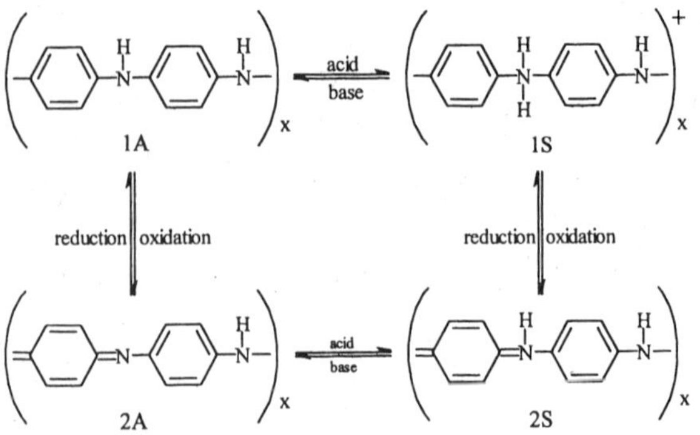

FIGURE 1. The four idealized forms of PANI that can be reversibly interconverted by chemical or electrochemical oxidation or reduction.

the solution and amounts to about 50% at pH 0.[2,6,8] This special quality, which has been recognized only at the beginning of the 1980s, has been a major reason for the great attention drawn recently to PANI, especially by electrochemists.

At the same time, as the new era of conducting polymers (a subgroup of "synthetic metals") emerged, it triggered not only scientific but also practical interest. The research on new devices, employing PANI as electronic conductor of special qualities, is receiving great attention, particularly in special areas such as variety of sensors, electroactive material in batteries, electrochromic devices, and electromagnetic shielding.

ELECTROCHEMICAL SYNTHESIS

The usually applied techniques in anodic oxidation of aniline, in the electrochemical synthesis of PANI, are constant current (CC), constant potential (CP) technique, and cyclic voltammetry (CV), in which the potential is linearly scanned up to a switching potential and then reversed to its initial value at a programmed rate.

The electrochemical polymerization of aniline is usually carried out at about ambient temperature (20°–25°C), although it has been noted that the rate of the reaction is temperature dependent in the range of 0° to 60°C.

The Influence of a Counter-Ion on the Reaction Rate and on the Morphology of PANI

It is well known that PANI is in a conducting state only when protonated or rather when its charge, owing to the protonation, is neutralized by a counter-ion. During the electrochemical synthesis the counter-ion is easily administered from the supporting electrolyte, usually an inorganic acid, and gets incorporated into PANI film creating corresponding PANI.

PROPERTIES

Polyaniline is generally amorphous owing to the weakly polar covalent bonding between monomer units, although Wessling and Volk succeeded in turning PANI into partly crystalic material.[9]

PANI films turn from transparent yellow to green to dark blue to black when the redox potential is varied from –0.2 to 1.0 V (versus SCE). The yellow to green transition was found to be reversible, and the repetition of 10^6 times was achieved.[4]

The conductive form of PANI is associated with intermediate oxidation states, that is, emeraldine, which features an equal number of imine and amino nitrogens. In the neutralized free-base form PANI is an insulator. The conductive salt form is obtained upon treatment with acids.

The conductivity of PANI can vary from 10^{-13} Scm^{-1} to 350 Scm^{-1}, and it depends not only on oxidation level, but also on pH of the solution in which PANI is equilibrated, as well as on the degree of PANI fibers orientation (stretching). The transition from conductive to nonconductive PANI is achieved by increasing pH above 4, and the highest conductivity is obtained in the potential range of 0.3–0.5 V versus SCE.

PANI synthetized with usual small anions is insoluble in water and in most organic solvents. It is slightly soluble in pyridine and dimethylformamide and in concentrated sulfuric acid. Solubility of PANI can be significantly increased if synthetized using large, organic anions.

In a reduced state PANI has one acid-based equilibrium that corresponds to the protonation of amino nitrogen at pK = 3. In an oxidized state there is an additional protonation of the imino nitrogen at pK = 6. Dissociation constants depend on the type of anion incorporated by PANI, which is influenced by electrostatic interactions between the positive and the negative charge of anion. Protonation of PANI causes changes in the geometry of PANI chain.

APPLICATIONS

PANI-coated electrodes have been studied toward applications in a number of areas. One of the areas is electrochromic display devices based on PANI electrochromic properties. Studies are also carried out to utilize the so-called "memory effect." The electrodeposition of PANI on the surface of semiconductor photoanodes has been shown to decrease photocorrosion. The electrocatalytic behavior of PANI electrodes has been established on several inorganic and organic systems. PANI is also a good material for immobilization of enzymes, which results in specific PANI enzyme electrodes used as sensors to specific reactants.

Investigations are also carried out in producing a "solid-state" PANI-based transistor. However, the use of PANI as an electrode in a rechargeable solid-state battery is one of the earliest applications of PANI. Both aqueous and nonaqueous cells have been extensively investigated and in 1988 the Bridgeston Corporation produced Li-PANI battery for the market. It is realistic to expect in the future that PANI, owing to its unique and useful properties, will find wide application in specific areas.

REFERENCES

1. Diaz, A. F.; Logan, J. A. *J. Electroanal. Chem.* **1980**, *111*, 111.
2. Chiang, J. C.; MacDiarmid, A. G. *Synth. Met.* **1986**, *13*, 193.
3. Duić, L.; Mandić, Z.; Kovaćićek, F. *J. Polym. Sci., Part A* **1994**, *32*, 105.
4. Kobayashi, T.; Yoneyama, H.; Tamura, H. *J. Electroanal. Chem.* **1984**, *161*, 419.
5. Letheby, H. *J. Chem. Soc.* **1862**, *15*, 161.
6. MacDiarmid, A. J.; Chiang, J. C.; Halpern, M.; Huang, W. S.; Mu, S. L.; Somasiri, N. D. L.; Wu, W.; Yaniger, S. I. *Mol. Cryst. Liq. Cryst.* **1985**, *121*, 173.
7. Nelson, R. F. *Technique of Electroorganic Synthesis*; Weinberg, N. L., Ed.; John Wiley & Sons: New York, 1974; Chapter V.
8. Paul, E. W.; Ricco, J.; Wrighton, W. S. *J. Phys. Chem.* **1985**, *89*, 1441.
9. Wessling, B.; Volk, H. *Synth. Met.* **1986**, *16*, 127.

POLYANILINE (Gold Recovery)

E. T. Kang,* Y. P. Ting, and K. G. Neoh
Department of Chemical Engineering
National University of Singapore

K. L. Tan*
Department of Physics
National University of Singapore

*Author to whom correspondence should be addressed.

Gold recovery from primary and secondary sources, such as ores, leach solutions, electronic scraps, and waste electroplating solutions, has emerged as an important technology.[1] Its recovery by less energy intensive processes, such as reduction and accumulation by activated carbon,[2] biomass and biomaterials,[3,4] polymeric adsorbents,[5] and electroless plating,[6] has also been actively investigated. Gold extraction from chloride solutions has risen to prominence during the last two decades because this recovery route does not have the adverse environmental effects of cyanidation.[7] As the demand for gold increases, extraction of the metal from acid solutions needs to be accomplished with greater efficiency.

In a parallel development, electroactive (conductive) polymers have emerged as a unique class of electronic materials since the mid-1970's.[8] The N-containing conductive polymers, such as polyaniline (PAN) and polypyrrole (PPY), are of particular interest owing to their high electrical conductivity, environmental stability, and interesting redox properties associated with the chain nitrogens.[9] The aniline polymers have the general formula $[-B-NH-B-NH)_y\ (-B-N{=}Q{=}N-)_{1-y}]_x$, in which B and Q denote the C_6H_4 rings in the benzenoid and quinonoid forms, respectively.[10–12] Thus, the aniline polymers are basically poly(*p*-phenyleneimineamine)s, in which the intrinsic redox states can range from that of the fully reduced leucoemeraldine (LM, y = 1), through that of the 50% oxidized emeraldine (EM, y = 0.5), to that of the fully oxidized pernigraniline (PNA, y = 0). The polymer can achieve its highly conductive state either through the protonation of the imine nitrogens (=N–) in its EM oxidation state, or through the oxidation of the amine nitrogens (–NH–) in the fully reduced LM state. Thus, by coupling the increase in the intrinsic oxidation state of the polymer, and the subsequent reprotonation and reduction of the intrinsically oxidized polymer in acid media, spontaneous and sustained reduction of precious metals, such as gold, to their elemental form can be achieved.[13–16]

REFERENCES

1. Arbiter, N.; Han, K. N., Eds., *Gold–Advances in Precious Metals Recovery*; Gordon and Breach Sci.: NY, 1990.
2. McDougall, G. J.; Hancock, R. D. *Minerals Sci. Engrg.* **1980**, *12*, 85.
3. Hosea, M.; Greene, B.; McPherson, R.; Henzl, M.; Alexander, M. D.; Darnall, D. W. *Inorg. Chim. Acta* **1986**, *123*, 161.
4. Kuyucak, N.; Volesky, B. *Biorecovery* **1989**, *1*, 189.
5. Akser, M.; Wan, R. Y.; Miller, J. D. *Solvent Extraction and Ion Exchange* **1986**, *4*, 531.
6. Okinaka, Y. *Gold Plating Technology*, Electrochem.: NY, 1974; Chapter 11.
7. Palmer, B. R. *Mineral Processing and Extractive Metallurgy Review* **1990**, *6*, 127.
8. Billingham, N. C.; Calvert, P. D. *Adv. Polym. Sci.* **1989**, *90*, 1.
9. Tan, K. L.; Tan, B. T. G.; Kang, E. T.; Neoh, K. G. *J. Chem. Phys.* **1991**, *94*, 5382.
10. Green, A. G.; Woodhead, A. E. *J. Chem. Soc.* **1910**, 1117.
11. MacDiarmid, A. G.; Chiang, J. C.; Richter, A. F.; Epstein, A. J. *Synth. Met.* **1987**, *18*, 285.
12. MacDiarmid, A. G.; Yang, L. S.; Huang, W. S.; Humphrey, B. D. *Synth. Met.* **1987**, *18*, 393.
13. Kang, E. T.; Ting, Y. P.; Neoh, K. G.; Tan, K. L. *Polymer* **1993**, *34*, 4994.

14. Kang, E. T.; Neoh, K. G.; Tan, K. L. *Adv. Polym. Sci.* **1993**, *106*, 135.
15. Kang, E. T.; Neoh, K. G.; Tan, K. L. *Surf. Interf. Anal.* **1993**, *20*, 833.
16. Ting, Y. P.; Neoh, K. G.; Kang, E. T.; Tan, K. L. *J. Chem. Tech. Biotechnol.* **1994**, *59*, 31.

POLYANILINE NETWORK ELECTRODES (Enhanced Performance of Polymer Light-Emitting Diodes)

A. J. Heeger,* Y. Yang, E. Westerweele, C. Zhang, Y. Cao, and P. Smith
UNIAX Corporation

Since the π-π* band gap of luminescent polymers can be controlled through changes in the molecular structure, emission over the entire visible spectrum is available.[1-10] Moreover, as a result of the significant processing advantages of semiconducting (conjugated) polymers, uniform and flexible thin films can be fabricated from solution in relatively large areas. Because of these features, there is growing interest in polymer light-emitting diodes (LEDs) for use in display applications.

Flexible "plastic" LEDs were demonstrated by using low surface resistance films of the metallic form of polyaniline (PANI) as the flexible, transparent, hole-injecting electrode.[9] PANI has advantages over indium/tin-oxide (ITO) as a transparent anode in addition to mechanical flexibility. PANI/ITO bilayer electrodes have been demonstrated to yield improved performance over either PANI or ITO separately as transparent anodes for use in polymer LEDs.[11] The PANI/ITO bilayer electrode combines the advantages of the higher work function of PANI with the lower surface resistance of ITO.

We describe here the *in situ* generation of high surface area network electrodes using blends of PANI protonated with camphor-sulfonic acid (CSA) in a low molecular weight polyester resin (PES) on ITO as the anode. Since PANI-CSA is insoluble in xylene, only the polyester in the PANI:PES blend is removed by the xylene during application of the semiconducting luminescent polymer layer, that is, during spin-casting from xylene solution. Consequently, this procedure leads to the formulation of a rough and porous PANI-CSA network on the ITO and to a significant increase in the contact area to the active MEH-PPV layer.

CONCLUSION

In summary, high surface area, transparent, network electrodes made of polyaniline in its conducting form can be fabricated from dilute blends of PANI-CSA in polyester or PMMA. The PANI-network electrode improves charge carrier injection into the polymer LED, thereby leading to lower voltage operation, improved efficiency and an increase in brightness at low operating voltages. For 10% PANI in the initial blends used to form the electrode, the brightness at 4 volts is increased by nearly two orders of magnitude over that obtained from devices using the simple PANI/ITO bilayer as the anode. Polymer LEDs fabricated with the PANI-CSA network as anode and Al:Li (0.2% Li) as the cathode turn on at 1.7 volts, and achieve a brightness in excess of 400 cd/m^2 at 3 volts with a measured external quantum efficiency of 2.23% photons/electron.

*Author to whom correspondence should be addressed.

ACKNOWLEDGMENTS

We thank Dr. F. L. Klavetter for providing materials and for advice and assistance with materials processing, Dr. I. D. Parker of UNIAX Corp. for important discussions of the polymer LED as a tunnel diode, and Dr. C. Y. Yang of the Institute for Polymers and Organic Solids, University of California at Santa Barbara, for the TEM images of the PANI network which stimulated this work.

REFERENCES

1. a) Burn, P. L.; Holmes, A. B.; Kraft, A.; Bradley, D. D. C.; Brown, A. R.; Friend, R. H. *J. Chem. Soc., Chem. Commun.* **1992**, *1*, 32; b) Burn, P. L.; Holmes, A. B.; Kraft, A.; Bradley, D. D. C.; Brown, A. R.; Friend, R. H.; Gymer, R. W. *Nature* **1992**, *356*, 47; c) Burroughes, J. H.; Bradley, D. D. C.; Brown, A. R.; Marks, R. N.; Mackay, K.; Friend, R. H.; Burns, P. L.; Holmes, A. B. *Nature* **1990**, *347*, 539.
2. Brown, A. R.; Bradley, D. D. C.; Burn, P. L.; Burroughes, J. H.; Friend, R. H.; Greenham, N.; Holmes, A. B.; Kraft, A. *Appl. Phys. Lett.* **1992**, *61*, 2793.
3. Grem, G.; Leditzky, G.; Ullrich, B.; Leising, G. *Adv. Mater.* **1992**, *4*, 36.
4. a) Ohmori, Y.; Uchida, M.; Muro, K.; Yoshino, K. *Jpn. J. Appl. Phys.* **1991**, *30*, L1941; b) Ohmori, Y.; Uchida, M.; Muro, K.; Yoshino, K. *Solid State Commun.* **1991**, *80*, 605.
5. Ohmori, Y.; Morishima, C.; Uchida, M.; Yoshino, K. *Jpn. J. Appl. Phys.* **1992**, *31*, L568.
6. Braun, D.; Gustafsson, G.; McBranch, D.; Heeger, A. J. *J. Appl. Phys.* **1992**, *72*, 564.
7. Braun, D.; Heeger, A. *J. Appl. Phys. Lett.* **1991**, *58*, 1982.
8. Vestweber, H.; Greiner, A.; Lemmer, U.; Mahrt, R. F.; Richert, R.; Heitz, W.; Bassler, H. *Adv. Mater.* **1992**, *4*, 661.
9. Gustafsson, G.; Cao, Y.; Treacy, G. M.; Klavetter, F.; Colaneri, N.; Heeger, A. *J. Nature* **1992**, *357*, 477.
10. Zhang, C.; Höger, S.; Pakbaz, K.; Wudl, F.; Heeger, A. J. *J. Elect. Materials* **1993**, *22*, 413.
11. Yang, Y.; Heeger, A. *J. Appl. Phys. Lett.* **1994**, *64*, 1245.

Polyanilines

See: *Chiral Polyanilines*
Conducting Polymer Colloids
Conducting Polymers (For Langmuir–Blodgett Film Fabrication)
Conducting Polymers (Self-Acid-Doped, Conjugated)
Conductive Elastomeric Blends
Electrically Conducting Composites
Electrochemical Polymerization
Electroconductive Composites (Emulsion Pathways)
Enzyme-Catalyzed Polymerization
Intelligent Membranes
Magnetic Field Processing
Oxidative Coupling
Polyaniline (Electrochemical Synthesis)
Polyaniline (Gold Recovery)
Polyaniline Network Electrodes (Enhanced Performance of Polymer LED's)
Polyanilines, Oxidation States

POLYANILINES, OXIDATION STATES

Luiz H. C. Mattoso
Empresa Brasileira de Pesquira Agropecuária

Alan G. MacDiarmid
Chemistry Department
University of Pennsylvania

The term "polyaniline" (PAni) as commonly employed, refers to a class of polymers of which the base form has the generalized composition and consists of alternating reduced (y) and oxidized (1-y), repeat units (**Structure 1**).[2,8] In principle y can be varied continuously from one, the completely reduced polymer, to zero to give the completely oxidized polymer (**Structures 2** and **3**). The terms leucoemeraldine, protoemeraldine, emeraldine, nigraniline, and pernigraniline, used in the following discussion will refer to the different oxidation states of the polymer where y = 1, 0.75, 0.5, 0.25, and 0, respectively. These names apply to polymers either in the base form, such as emeraldine base or in the protonated salt form, such as emeraldine hydrochloride.[2,8] Although, in principle, the oxidation states can vary continuously, it will be shown that PAnis can exist in only three discrete oxidation states: leucoemeraldine, emeraldine, and pernigraniline, the other oxidation states being a mixture of these.

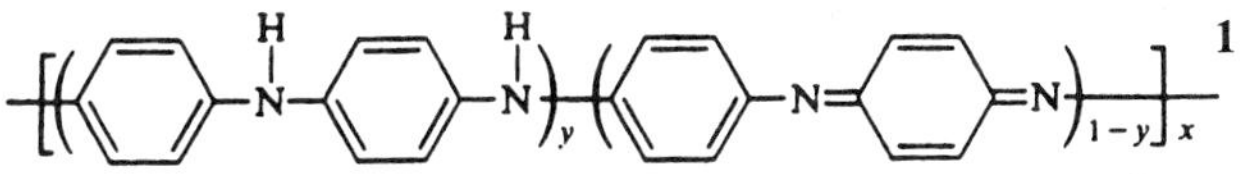

1

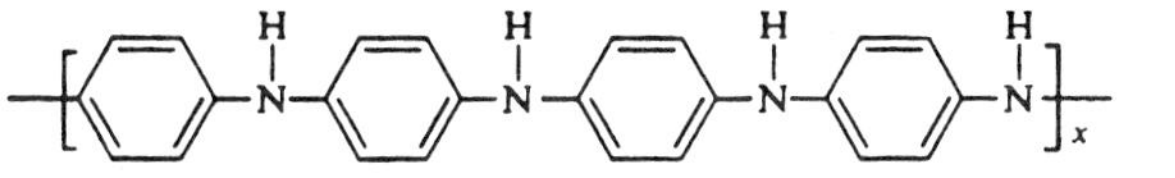

2

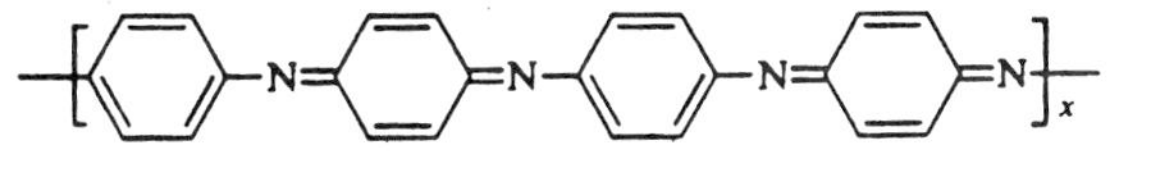

3

SYNTHESIS OF POLYANILINE IN DIFFERENT OXIDATION STATES

Emeraldine Oxidation State

The partly protonated emeraldine salt can be synthesized easily by the oxidative polymerization of aniline in aqueous acidic media using a variety of oxidizing agents, the most commonly used being ammonium peroxydisulphate, $(NH_4)_2S_2O_8$, in aqueous 1.0 M HCl.[2,9] Such emeraldine salt form has been widely investigated since it is the most conductive form of polyaniline, in addition to being very stable.

Synthesis of Leucoemeraldine

Leucoemeraldine was first synthesized in 1910.[4] It can be conveniently prepared as an analytically pure, off-white powder by the reduction of emeraldine base, the most commonly used reducing agents being phenyl hydrazine or hydrazine.[4,5,12]

Synthesis of Pernigraniline

Pure pernigraniline has recently been synthesized for the first time.[14] Synthesis of the analytically pure, dark-purple, partially crystalline powder has been accomplished by the controlled oxidation of emeraldine base with m-$Cl(C_6H_4)C(O)OOH/N(C_2H_5)_3$ in NMP containing a trace of $CrCl_3$.

DOPING OF THE POLYANILINES

Protonic Acid Doping

The emeraldine base form of PAni is the first well established example of the doping of an organic polymer to a highly conducting regime by a process in which the number of electrons associated with the polymer remains unchanged during the doping process.[2,8] This is accomplished by treating emeraldine base with aqueous protonic acids and is accompanied by a 10 order of magnitude increase in conductivity (to 1–5 S/cm, compressed powder pellet) reaching a maximum in 1.0 M aqueous HCl.

Oxidative Doping

Leucoemeraldine base can be controllably and oxidatively doped to the highly conducting regime analogously to $(CH)_x$. This can be accomplished either electrochemically or chemically.[10]

Although the electrochemical doping procedure is a convenient method for small amounts of polymer, the chemical oxidation is suited for the preparation of large quantities.

APPLICATIONS

There has been great excitement about the technological applications of conducting polymers owing to their unique possibility of combination of properties from polymers (processability, chemical stability, and flexibility) and metals (electrical conductivity and optical and magnetic properties). The latest progress on the fabrication of polyaniline and its blends has attracted an enormous industrial interest.[1] In particular, this is true when the polymer is processed in the form of films and fibers readily obtained in the conductive doped state by the conventional processing techniques commonly used to commercial polymers.[7]

Rechargeable batteries of polyaniline, for example, have been sold since 1987.[3] Thin coating of doped polyanilines on nylon and polyester fibers have been fabricated for textiles.[6] Also, composites and blends of polyaniline with several conventional plastics for use in electromagnetic shielding, microwave absorption, aerospace and related purposes have been produced and/or displayed by various industries worldwide.[11] Moreover, light emitting diodes, chemical sensors, electrochromic displays, anodic passivation, corrosion prevention of metals, and electromechanical devices using polyanilines and/or other conducting polymers have also been demonstrated.[13] Some of these applications are already being developed at the industrial level for commercialization. The future of the applications of conducting polymers seems highly promising, especially when one considers the immense progress that may still be achieved in areas that are at an embrionic stage, as is the case of molecular electronics.

REFERENCES

1. Cao, Y.; Smith, P.; Heeger, A. J. *Synth. Met.* **1994**, *55–57*, 3514.
2. Chiang, J.-C.; MacDiarmid, A. G. *Synth. Met.* **1986**, *13*, 193.
3. Enomoto, T.; Allen, D. P. *Bridgestone News Release*, September 9, 1987; *Chem. Week*, October 14, 1987, p 40.
4. Green, A. G.; Woodhead, A. E. *J. Chem. Soc. Trans.* **1910**, *97*, 2388.
5. Green, A. G.; Woodhead, A. E. *J. Chem. Soc. Trans.* **1912**, *101*, 1117.
6. Gregory, R. V.; Kimbrell, W. C.; Kuhn, R. H. *Synth. Met.* **1993**, *28*, C823.
7. Ikkala, O. T.; Laakso, J.; Vakiparta, K.; Virtanen, E.; Ruohonen, H.; Jarvinen, H.; Taka, T.; Passiniemi, P.; Osterholm, J. E.; Cao, Y.; Andreatta, A.; Smith, P.; Heeger, A. J. *Proceed. of the International Conference on Science and Technology of Synthetic Metals*, published in Synth. Met., **1995**, *69*, 97.
8. MacDiarmid, A. G.; Chiang, J.-C.; Richter, A. F.; Epstein, A. J. *Synth. Met.* **1987a**, *18*, 285.
9. MacDiarmid, A. G.; Chiang, J.-C.; Richter, A. F.; Somasiri, N. L. D.; Epstein, J. A. In *Conducting Polymers*, Alcacer, L., ed.; D. Reidel: Dordrecht; 1987b; pp 105–120.
10. MacDiarmid, A. G.; Epstein, A. J. *Faraday Discuss. Chem. Soc.* **1989**, *88*, 317.
11. MacDiarmid, A. G.; Epstein, A. J. *II Brazilian Polymer Conference*, São Paulo, Brazil, October 1993; p 544 and references therein.
12. Ray, A. Ph.D. Dissertation, University of Pennsylvania, 1989a.
13. Roth, S.; Graupner, W. *Synth. Met.* **1993**, *55–57*, 3623; references therein.
14. Sun, Y.; MacDiarmid, A. G.; Epstein, A. J. *J. Chem. Soc. Chem. Commun.* **1990**, 529.

Polyanions

See: Biologically Active Polyanions
Flocculants (Organic, Overview)

Polyaromatics

See: Aromatic Hydrocarbon-Based Polymers
Enzymatic Polymerization

Poly(aryl ether benzimidazole)s

See: Polybenzimidazoles (Overview)

POLY(ARYL ETHER ETHER KETONE)[PEEK] (Bulk Crystallization Kinetics)

F. J. Medellin-Rodriguez
CIEP-Fac. de C. Quimicas
UASLP, Mexico

P. J. Phillips
Department of Materials Science and Engineering
University of Tennessee

Attempts at understanding the bulk crystallization kinetics of semicrystalline polymers have paralleled the study of the molecular process of crystallization. Practical commercial reasons involving simulations have justified the continual study of bulk behavior, the molecular approach being justified for scientific reasons in addition. Over the last few years, there has been an increasing interest in understanding the bulk crystallization process of synthetic PEEK because of its outstanding chemical, thermal, and mechanical properties.[1] The chemical structure of PEEK is based on an aromatic backbone in a sequence consisting of two ether groups and one carbonyl group each separated by benzene rings. PEEK crystallizes into an orthorhombic crystalline structure, single crystals having been grown from α-chloronaphthalene and benzophenone solutions.[5,8,9,16] Some studies of the morphology as well as the molecular kinetics and the molecular conformation as a function of thermal properties have also been reported.[2,3,6,10–15] However, interpretation of the bulk growth kinetics of the polymer has remained a subject of controversy.[4,7]

In summary, the overall process of crystallization of PEEK is a complex combination of parallel mechanisms of nucleation and growth. Therefore, simple models must be discarded for simulation purposes. Additional work is needed on the bases of the mechanisms enunciated here before single representational equations can be developed for simulation purposes.

REFERENCES

1. Atwood, T. E.; Dawson, P. C.; Freeman, J. L.; J. Hoy, L. R.; Rose, J. B.; Staniland, P. *Polymer* **1981**, *22*, 1096.
2. Bassett, D. C.; Olley, R. H.; Al Raheil, I. A. M. *Polymer* **1988**, *29*, 1745.
3. Blundell, D. J.; Osborn, B. N. *Polymer* **1983**, *24*, 953.
4. Cebe, P. *Polymer Eng. and Sci.* **1988**, *28*, 1192.
5. Dawson, P. C.; Blundell, D. J. *Polymer* **1980**, *21*, 577.
6. Deslandes, Y.; Sabir, F-N.; Roovers, J. *Polymer* **1991**, *32*, 1267.
7. Jonas, A.; Legras, R. *Polymer* **1991**, *32*, 2691.
8. Lovinger, A. J.; Davis, D. D. *J. Appl. Phys.* **1985**, *58*, 2843.
9. Lovinger, A. J.; Davis, D. D. *Macromolecules* **1986**, *19*, 1861.
10. Medellin-Rodriguez, F. J.; Phillips, P. J., ANTEC Technical Papers **1991**, *37*, 883.
11. Medellin-Rodriguez, F. J.; Phillips, P. J. *Macromolecules* **1995**, *28*, 7744.
12. Medellin-Rodriguez, F. J.; Phillips, P. J., to be published.
13. Nguyen, H. X.; Ishida, H. *A. C. S. Div. Polym. Chem. Polym. Preprints* **1985**, *26*, 273.
14. Nguyen, H. X.; Ishida, H. *J. of Polym. Science: B: Polymer Physics* **1986a**, *24*, 1079.
15. Olley, R. H.; Bassett, D. C.; Blundell, D. J. *Polymer* **1986**, *27*, 344.
16. Rueda, D. R.; Ania, F.; Richardson, A.; Ward, I. M.; Balta Calleja, F. J. *Polym. Commun.* **1983**, *24*, 258.

POLY(ARYL ETHER ETHER KETONE)[PEEK] (Conformational Order Studied Using FTIR)

F. J. Medellin-Rodriguez
CIEP-Fac. de C. Quimicas-UASLP, Mexico

P. J. Phillips
Department of Materials Science and Engineering
University of Tennessee

In the recent past, aromatic polymers have attracted attention for high engineering applications. Among these outstanding chemical properties (radiation resistance, low flammability, and toxic gas emission), very good mechanical properties, and very high thermal stability have made synthetic PEEK one of the most studied members of the poly(aryl ether ketone) family.[1] Such properties together with the desired melt processability associated with semicrystalline polymers has increased interest in using PEEK as the matrix in reinforced composites, in aerospace applications, in electrical coatings, and in the replacement of automotive components.[3]

Understanding the mechanisms through which PEEK first nucleates and then crystallizes has been the main motivation in some previous research.[2,5] Nguyen and Ishida first characterized the conformational changes of PEEK as a result of thermal treatments using infrared spectroscopy–FTIR.[6,7] From analyses made on KBr powder and through IR spectra subtraction it was concluded that PEEK is able to develop "local-order" through the diphenyl ether moieties of the macromolecule. The technique indicated motivating results, but a systematic analysis of nucleation-property local-order-content relations was not made. The main purpose of the present work is therefore to give a review of "local-order," how it varies with time in solution and in the melt, and how it influences nucleation and growth.

PROPERTIES

PEEK resins are expected to show different properties as a result of the treatments even if the only difference between them is the physical form of the specimen. This is because additives play a relatively important role in such behavior.

Nucleating Agents and Impurities

In practice, resins with added nucleating agents will nucleate first then crystallize faster, when they are either cooled from the melt or crystallized from the amorphous state, than those without them. The effect on linear cooling is seen as a displacement of the overall crystallization peak to higher temperatures (of the order of 5°C) when crystallization is carried out dynamically.

Solvent Effects

IR spectroscopy can be used to analyze the possibilities of chemical reaction and/or thermal degradation of the original macromolecules in the refluxing solution as long as the effects are measurable in the FTIR. It has to be also considered at this point that dissolution of PEEK in α-chloronaphthalene at high temperature can remove antioxidants leaving a polymer susceptible to oxidation then degradation through branching and even crosslinking.[2]

In summary, we have concluded that, as reported by Nguyen and Ishida, conformational changes are the only detectable changes in the IR and are to be correlated with the DSC traces shown earlier.[6,7]

LOCAL ORDER

Nguyen and Ishida first proposed "local-order" in the molten state of PEEK.[6,7]

A comparison between as-received and low-reflux-time-reprocessed PEEK was made and there was an indication that melts of as-received PEEK have a higher content of local-order.[4] As-received resins crystallize in about 15 seconds from the melt at the mentioned temperature. However, solvent treated samples had longer crystallization induction periods, lower nucleation density and even allowed crystallization rates to be measured over the entire crystallization temperature range.[4]

It seems to be a reasonable hypothesis that the "local order" observed in the FTIR corresponds to an ordering of molecular sections in a direction toward the conformations found in the crystalline state. As such it would explain the dependence of both the primary nucleation behavior and the radical growth rates on "local order."

REFERENCES

1. Atwood, T. E.; Dawson, P. C.; Freeman, J. L.; Hoy, L. R. J.; Rose, J. B.; Staniland, P. *Polymer* **1981**, *22*, 1096.
2. Jonas, A.; Legras, R. *Polymer* **1991**, *32*, 2691.
3. Kemish, D. J. *Modern Plastics Encyclopedia*; McGraw-Hill, Ed.; 1991.
4. Medellin-Rodriguez, F. J.; Phillips, P. J., to be published.
5. Mercier, J. P. *Polym. Eng. and Science* **1990**, *30*, 270.
6. Nguyen, H. X.; Ishida, H. *A.C.S. Div. Polym. Chem. Polym. Preprints* **1985**, *26*, 273.
7. Nguyen, H. X.; Ishida, H. *J. of Polym. Science: B: Polymer Physics* **1986a**, *24*, 1079.

POLY(ARYL ETHER KETONE) (Functionalization)

Jacques Roovers
Institute for Environmental Research and Technology

Fei Wang
EIC Laboratories, Incorporated

All-aromatic poly(ether ketone)s are high performance polymers used as matrix components in structural composites with glass fibers or carbon fibers. The semi-crystalline polymers show excellent mechanical properties, a high heat deflection temperature, excellent solvent resistance, and oxygen and radiation resistance. It is desirable to take advantage of the same good properties in functional applications of these polymers, such as coatings, membranes, and absorbents.

ALKYL SUBSTITUTED POLY(ARYL ETHER KETONE)S

Amorphous *tert*-butyl-substituted poly(aryl ether ether ketone) (PEEK) is prepared by nucleophilic substitution of *tert*-butylhydroquinone with difluorobenzophenone under mild conditions in N-methylpyrrolidinone (NMP) at 170°C.[1–4]

Methyl substituted PEEK has been prepared by nucleophilic substitution in NMP at 180°C[6] or alternatively in the melt with the bis-trimethylsilyl derivative of methylhydroquinone (**Scheme I**).[5]

Soluble methyl-substituted PEEK is the starting material for extensive functionalization by bromination. About 40% of methyl groups can be converted to bromomethyl before an appreciable amount (5%) of dibromomethyl substitution is observed.[6]

Long-chain alkyl substituents result in comb type polymers with aliphatic-aromatic character. An interesting approach based on the naphthalene unit has been described recently.[7]

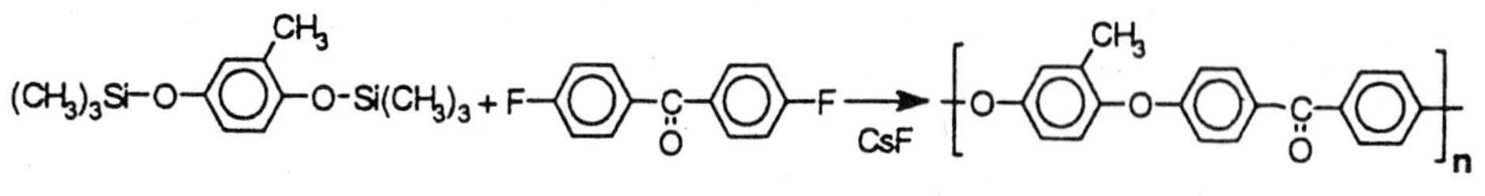

I

SULFONATION

For convenient sulfonation the poly(ether ketone) should contain the ether-phenylene-ether link as in PEEK. In that case, sulfonation occurs during dissolution of the dry semicrystalline polymer in concentrated H_2SO_4 of more than 94.9%, preferably 97.4%.[8,9] Monosulfonation proceeds to completion (**Structure 1**):

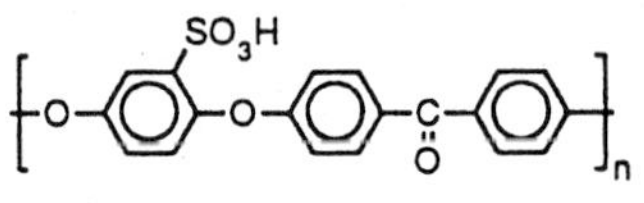

1

CARBONYL FUNCTIONALIZATION

Besides *tert*-butyl substituted PEEK,[1–4] two routes to amorphous prepolymers of poly(ether ketone)s have been actively pursued. One involves the phenylketimine derivative of difluorobenzophenone[11] that can be used to prepare ketimine modified PEEK at 180°C in NMP. The soluble prepolymer can be hydrolysed with dilute acid to yield very fine powders of PEEK.[10–13]

Biphenol–based ketimine modified poly(ether ether ketone)s also have been prepared.[13,14] The second method involves the ethyleneglycolacetal of 4,4'-dihydroxybenzophenone to prepare amorphous poly(ether ketone).[15]

CROSSLINKING

Methods have been sought to use the carbonyl group for crosslinking, for example, by formation of ketimine bridges. This requires incorporation of primary (aromatic) amine groups, either endstanding[11] or along the polymer chain.[16]

A newer method for crosslinking involves incorporation of a fraction of 2,5-bis(benzoyl) benzocyclobutene groups into the polymer.

END-FUNCTIONALIZATION

PEEK polymers prepared by the industrially used nucleophilic aromatic substitution have either hydroxyl or fluorine endgroups.

Amino-terminated poly(ether ketones) have been prepared by a one-step[2,11,17,18] or two–step process.[2] In each case, *m*-aminophenol has been used as the end-capping compound. The amino-terminated polymers were chain extended with benzophenone tetracarboxyl acid anhydride to form imide linkages.

IRRADIATION

In comparison with other polymers, PEEK is resistant to radiation.[19–21] Radiation effects are noted only at very high doses. Electron beam irradiation of amorphous PEEK in air leads to increased glass transition,[22] reduced crystallinity,[22,23] lower melting temperature,[23] and reduced elongation at break.[19] All these effects can be ascribed to the formation of branched and crosslinked materials. However, chain scission also must occur to some extent as inferred from increased relaxation processes in a more disordered sample after irradiation.[22]

REFERENCES

1. Mohanty, D. K.; Lin, T. S. et al., *31st International SAMPE Symposium* **1986**, *31*, 945.
2. Bennett, G. S.; Farris, R. J. *J. Polym. Sci.: Part A: Polym. Chem.* **1994**, *32*, 73.
3. Risse, W.; Sogah, D. Y. *Macromolecules* **1990**, *23*, 4029.
4. Risse, W.; Sogah, D. Y.; Boettcher, F. P. *Makromol. Chem., Macromol. Symp.* **1991**, *44*, 185.
5. Kricheldorf, H. R.; Delius, U.; Tönnes, K. F. *New Polym. Mater.* **1988**, *1*, 127.
6. Wang, F.; Roovers, J. *J. Polym. Sci.: Part A: Polym. Chem.* **1994**, *32*, 2413.
7. Ritter, H.; Thorwith, R. *Makromol. Chem.* **1993**, *194*, 1469.
8. Bishop, M. T.; Karasz, F. E. et al. *Macromolecules* **1985**, *18*, 86.
9. Jin, X.; Bishop, M. T. et al. *Brit. Polym. J.* **1985**, *17*, 4.
10. Roovers, J.; Cooney, J. D.; Toporowski, P. M. *Macromolecules* **1990**, *23*, 1611.
11. Mohanty, D. K.; Lowery, R. C. et al. *32nd International SAMPE Symp.* **1987**, *32*, 408.
12. Brink, A. E.; Gutzeit, S. et al. *Polymer* **1993**, *34*, 825.
13. Lindfors, B. E.; Mami, R. S. et al. *Makromol. Chem., Rapid Commun.* **1991**, *12*, 337.
14. Mohanty, D. K.; Senger, J. S. et al. *33rd International SAMPE Symp.* **1988**, *33*, 970.
15. Kelsey, D. R.; Robeson, L. M. et al. *Macromolecules* **1987**, *20*, 1204.
16. Mohanty, D. K.; Wu, S. D.; McGrath, J. E. *Polymer Preprints* **1988**, *29*, 352.
17. Lyle, G. D. et al. *33rd SAMPE Symp.* **1988**, *33*, 1080.
18. Corfield, G. C.; Wheatley, G. W.; Parker, B. G. *J. Polym. Sci.: Part A: Polym. Chem.* **1992**, *30*, 845.
19. Sasuga, T.; Hayakawa, N. et al. *Polymer* **1985**, *26*, 1039.
20. Hegazy, El-S.A.; Susaga, T. et al. *Polymer* **1992**, *33*, 2897.
21. Hegazy, El-S.A.; Sasuga, T. et al. *Polymer* **1992**, *33*, 2904.
22. Sasuga, T.; Hagiwara, M. *Polymer* **1985**, *26*, 501.
23. Yoda, O. *Polymer* **1984**, *25*, 238.

POLY(ARYL ETHER KETONE KETONE)S AND COPOLYMERS (Synthesis, Structure, Morphology, and Properties)

Stephen Z. D. Cheng* and Rong-Ming Ho
Department of Polymer Science
Maurice Morton Institute
The University of Akron

Benjamin S. Hsiao
Central Research and Development Department
Experimental Station
E. I. du Pont de Nemours and Company

*Author to whom correspondence should be addressed.

Poly(aryl ether ketone)s and their copolymers are the most common members of a new class of high-temperature, high-performance engineering thermoplastics with potential applications as matrix resins in composites. Among the poly(aryl ether ketone)s, poly(ether ether ketone) (PEEK) was the first member to be commercialized in this family, which was developed by ICI.

PEKK is typically prepared by way of a Friedel Crafts condensation polymerization of diphenyl ether (DPE), terephthaloyl chloride (TCl), and/or isophthaloyl chloride (ICl). As they are prepared, these polymers can be thought of as consisting of "phthalate diads" containing linkages such as -DPE-T-DPE-T or -DPE-I-DPE-I-. For poly(ether ketone ketone) with all-*para* linkages in phenylene groups [PEKK (T/I = 100/0)], its chemical structure is (**Structure 1**):

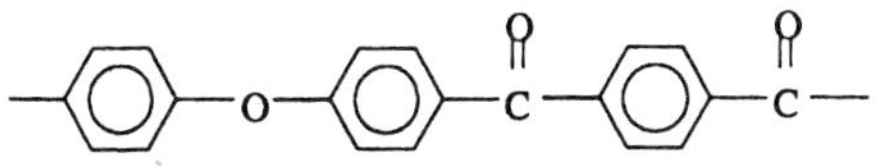

1

Poly(ether ketone ketone) with one *meta*-linked phenylene in the backbone [PEKK(T/I = 0/100)] has the chemical structure of (**Structure 2**):

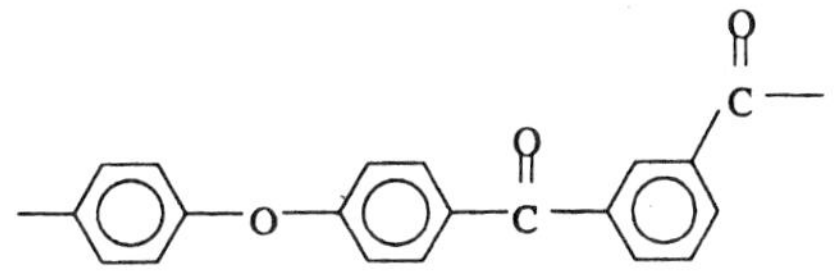

2

While a PEKK containing perfect alternated T and I, PEKK(T/I = 50/50), can be viewed as a homopolymer and the chemical compositions of both T and I are 50%. The chemical structure is (**Structure 3**).

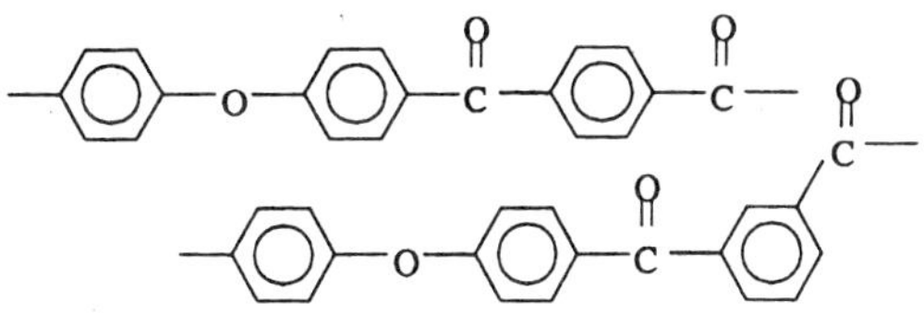

3

CRYSTAL STRUCTURE

The PEEK crystal structure is very similar to the crystal structure of poly(*p*-phenylene oxide) (PPO).[2] Chains in crystals adopt an extended conformation with the phenylene rings alternating at angles of ca. ± 37° to the plane of the zigzag backbone.

PEKK(T/I = 100/0)

For PEKK with all *para*-linked phenylene groups, PEKK(T/I = 100/0), based on wide-angle X-ray diffraction (WAXD) experiments, the crystal possesses a two-chain orthorhombic unit cell with dimensions a = 0.769 nm, b = 0.606 nm, and c = 1.016 nm (form I).[1] Recently, μm-sized single lamellar crystals have been successfully grown and provided an opportunity for the investigation of electron diffraction (ED).

Because the glass transition temperature of PEKK(T/I = 100/0) has been reported to be ca. 165°C and its equilibrium melting temperature is ca. 410°C, the crystallization window is thus over 200°C.[4] AT 265°C the maximum crystallization rate was found.

PEKK(T/I = 0/100)

The crystal structure of [PEKK(T/I = 0/100)] containing one *meta*-linkage in the repeat unit (three phenylene groups) has also been studied.[5] Only one crystal form has been found based on WAXD and ED experiments.

PEKK(T/I = 50/50)

Crystal structure and morphology of [PEKK(T/I = 50/50)] have been investigated by transmission electron microscopy (TEM), ED, WAXD, and differential scanning calorimetry (DSC) experiments.[6]

CRYSTAL MORPHOLOGY

The morphological study of PEEK crystallized from the melt was first reported by Lovinger et al. for thin film samples and Bassett et al. for replicas permanganic etching.[7–9] They showed that PEEK spherulites consists of narrow lamellae having the b-axis of the unit cell in the radial direction. Recently, a second morphological form of PEEK in the thin films has been identified at high crystallization temperatures. This consists of large (μm-sized), faceted single crystals.

PEKK(T/I = 100/0)

In the early study of the crystal morphology of PEKK, TEM images of the replica have also been briefly reported by Gardner et al.[4] Form I can always be isolated when PEKK(T/I = 100/0) film is crystallized from the melt.

PEKK(T/I = 0/100)

It is interesting to find that in contrast to both cases of PEEK and PEKK(T/I = 100/0), isolated PEKK(T/I = 0/100) single lamellar crystals in μm-size cannot be formed.[3,5,7] Despite our extensive efforts toward isothermal crystallization at different temperatures and using sample preparation procedures, large single crystals have not been observed.

PEKK(T/I = 50/50)

Different from PEEK and PEKK(T/I = 50/50), isolated μm-size single lamellar crystals in PEKK)T/I = 50/50) are not formed despite extensive efforts on isothermal crystallization at different temperatures, times, and various sample preparation procedures. However, relatively regular shaped, more or less flat-on single crystals crystallized at high crystallization temperatures can be obtained in PEKK(T/I = 50/50).

CRYSTALLIZATION, MELTING, AND GLASS TRANSITION

PEEK exhibits a lower degree of crystallinity (less than ca. 40%) than polymers such as polyethylene or polyoxymethylene, possibly owing to the high rigidity of the chain.[10] The addition of keto linkages inevitably increases the stiffness of the molecule, as reflected by the increase of glass transition temperature, crystallization rate, and the decrease of crystallinity.[4,11] An increase of keto linkages also enhances the packing efficiency of the unit cell structure, thereby producing a higher melting temperature.[4] The incorporation of *meta*-linked phenylene content is observed to increase the chain flexibility and, thus, decrease the glass transition temperature.

MECHANICAL PROPERTIES

PEKK polymer is being developed as a potential high-performance thermoplastic matrix system for advanced composites.[12–14] The PEKK with all-*para* linkages in phenylene groups, PEKK(T/I = 100/0), has a high glass transition temperature (T_g = 165°C), but melts too close to its thermal degradation temperature to be of practical use. By incorporating *meta*-linkages into PEKK structure, the melting temperature can be lowered into a usable range while retaining a high glass transition temperature. Both PEKK(T/I = 70/30) and PEEK have excellent flame resistance properties with a heat release rate lower than 65/65 as required by FAA for commercial aircraft interior. The structure, property, and processing relationship of using PEKK as high performance thermoplastics can be easily tailored by changing the chemical compositions. This certainly broadens another dimension of using PEKK polymers.

REFERENCES

1. Attwood, T. E.; Dawson, P. C.; Freeman, J. L.; Hoy, L. R. J.; Rose, J. B.; Staniland, P. A. *Polymer* **1981**, *22*, 1096.
2. Boon, J.; Magre, E. P. *Makromol. Chem.* **1969**, *126*, 130.
3. Ho, R. M.; Cheng, S. Z. D.; Hsiao, B. S.; Gardner, K. H. *Macromolecules* **1994**, *27*, 2136.
4. Gardner, K. H.; Hsiao, B. S.; Matheson, R. R.; Wood, B. A. *Polymer* **1992**, *33*, 2484.
5. Ho, R. M.; Cheng, S. Z. D.; Fisher, H. P.; Eby, R. K.; Hsiao, B. S.; Gardner, K. H. *Macromolecules*, in press.
6. Ho, R. M.; Cheng, S. Z. D.; Hsiao, B. S.; Gardner, K. H. *Macromolecules*, submitted.
7. Lovinger, A. J.; Davis, D. D. *J. Appl. Phys.* **1985**, *58*, 2843.
8. Lovinger, A. J.; Hudson, S. D.; Davis, D. D. *Macromolecules* **1992**, *25*, 1752.
9. Olley, R. H.; Bassett, D. C.; Blundell, D. J. *Polymer* **1986**, *27*, 344.
10. Cheng, S. Z. D.; Cao, M. Y.; Wunderlich, B. *Macromolecules* **1986**, *19*, 1868.
11. Jog, J. P.; Nadkarni, V. M. *J. Appl. Polym. Sci.* **1986**, *32*, 3317.
12. Chang, I. Y. *SAMPE Q.* **1988**, *19*(4), 29.
13. Chang, I. Y. *Composites Science and Technology* **1985**, *24*, 61.
14. Pratte, J. F.; Krueger, W. H.; Chang, I. Y. *SAMPE International Symp.* **1988**, *34*, 2229.

Poly(aryl ether ketone)s

See: *Condensation Polymers (Using Phosphorus Pentoxide/Methanesulfonic Acid)*
Crosslinking Agents, Phenyltriazenes (for Polyimides and Other Aromatic Polymers)
Cyclic Oligomers of Engineering Thermoplastics
Dendrimers, Poly(ether ketone)
Engineering Plastics, Melt Spinning
Engineering Thermoplastics (Survey of Industrial Polymers)
Fluoropolymers (Trifluoromethylated PES and PEEK)
Poly(aryl ether ether ketone) [PEEK] {Bulk Crystallization Kinetics)
Poly(aryl ether ether ketone) [PEEK] (Conformational Order Studied Using FTIR)
Poly(aryl ether ketone) (Functionalization)
Poly(aryl ether ketone ketone)s and Copolymers (Synthesis, Structure, Morphology, and Properties)
Poly(aryl ether ketone)s (Overview)
Poly(aryl ether ketone)s
Poly(aryl ether ketone)s, Amine-Terminated
Poly(arylene ether nitrile)s
Poly(ether ether ketone) (Dynamic Relaxation Characterization)
Poly(ether sulfone)s and Poly(ether ketone)s (Methyl-Substituted Aromatic)
Polyimide Blends
Reissert Polymers
Step Polymerization Catalysts

POLY(ARYL ETHER KETONE)S (Overview)

Peter Wolf
BASF AG

The rapid development of high-performance plastics has been triggered by the dynamic and often spectacular growth in the last few decades of high-technology industries such as microelectronics and aerospace travel. One of the most important polymer families in this class are the highly promising poly(aryl ether ketone)s (PAEK), which display a wide range of outstanding properties.

The class of poly(aryl ether ketone)s consists of several structurally different polymers. Because each member of this family is composed of identical building blocks, mostly benzene groups connected by ether and ketone links, they are frequently identified by the number of bridging groups within a repeating unit, characterizing the molecular structure.

The ratio and the sequence of ether and carbonyl groups significantly affects the physical properties of a PAEK polymer. Higher numbers of carbonyl groups result in more rigid polymer chains and thus, higher glass transition and melting temperatures in these semi-crystalline materials.

PREPARATION

The chemical structure of PAEKs suggests that two polymers forming synthetic approaches can be used. The first route involves the formation of ether linkages in the polymerization step, but an alternative route is the formation of carbonyl bridges during polymerization. Both approaches have been realized on a commercial scale and are called nucleophilic and electrophilic routes, respectively.

During the nucleophilic process, suitable bisphenols react with activated dihalogen-substituted aromatics in the presence of a base (**Scheme I**).[1] Victrex PEEK®, for instance, is produced by this process.

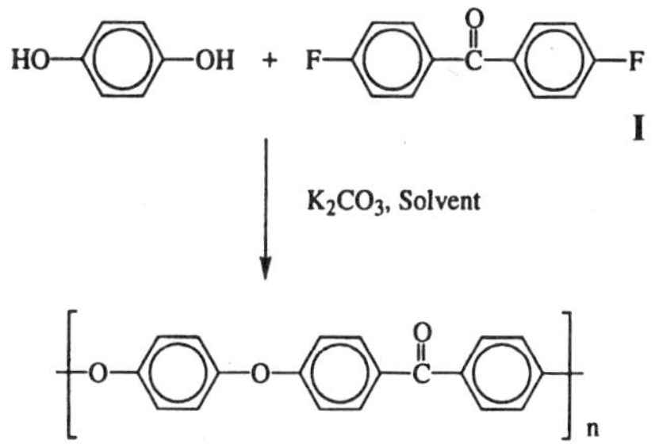

I

In the electrophilic process, which is used to produce Ultrapek®, aromatic diacid dichlorides are converted with monomers that are acylable using Friedel–Crafts chemistry (**Scheme II**).[2,3]

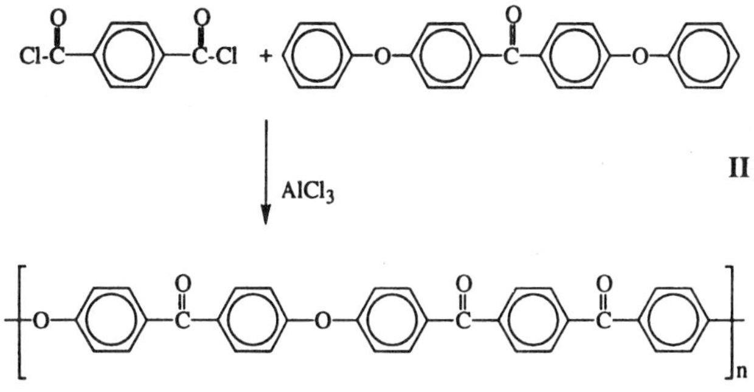

II

Other synthetic routes to PAEKs have been developed but so far have not been used on a commercial scale.

PROPERTIES

The chemical structure of PAEKs is responsible for a unique property spectrum rarely found in other polymers and that include the following:

- high sustained service temperatures up to 280°C,
- heat deflection temperatures up to 350°C,
- good mechanical properties at high temperatures,
- excellent creep resistance,
- excellent tribological behavior,
- good chemical and hydrolytic resistance,
- low flammability even without flame retardants, and
- high resistance against nuclear radiation.

The term high-temperature-resistant thermoplastics, frequently used to described PAEKs, does not describe this polymer class adequately because it highlights only one particular PAEK feature. Instead PAEKs should be classified as high-performance polymers.

PROCESSING

PAEKs are melt processed according to standard methods used for conventional thermoplastics. Most important are injection molding, extrusion, compression molding, and transfer molding techniques.

APPLICATIONS

PAEK Market

PAEK parts are primarily used in the following markets: general engineering, automotive, electronics, medicine, appliances, and the aerospace industry. In these applications, PAEKs are used increasingly to replace metals, thermosets, or ceramics. For complicated part geometries, PAEKs frequently offer economic advantages over other materials because they are easy to process, a significant cost saving for nonthermoplastic materials.

Outlook

Introducing PAEKs to the marketplace advanced the properties of thermoplastics and in the future may have an impact comparable to the development of fluoroplastics in the 1950s. As their use becomes more widespread and better known, PAEKs will find new applications. With poly(aryl ether ketone)s, an upper limit in the thermal stability of thermoplastic organic polymers has been reached because of the principle stability of organic matter. Further developments, therefore, will not be in new structures but in reducing PAEK's high prices, which may be achieved by optimizing manufacturing techniques. To this end, the electrophilic synthetic route shows promise as does the development of new polymer blends that maintain PAEK properties while significantly reducing PAEK prices. Although PAEKs will continue to establish themselves as important engineering plastics with above-average growth rates, the success of these developments will determine the ultimate potential for this class of materials.

REFERENCES

1. Attwood, T. E.; Dawson, P. C.; Freeman, J. L. et al. *Polymer* **1981**, *22*, 1096.
2. Dahl, K. British Patent 1 387 303, 1975.
3. Jansons, V.; Dahl, K. *Makromol. Chem., Macromol. Symp.* **1991**, *51*, 87.

POLY(ARYL ETHER KETONE)S (Including Thioether Ketones)

Haruyuki Yoneda* and Isaburo Fukawa
Asahi Chemical Industry Company

Poly(aryl ether ketone)s are thermoplastic, semi-crystalline polymers with good processability, excellent thermal stability, and mechanical properties at high temperatures.[1,2,3] These polymers are composed mostly of 1,4-phenylene, and keto and ether groups.[4] Values for T_m and T_g depend on the polymer structure (the repeating units), and the higher the ratio of keto to ether is in the repeating unit, the higher the polymer's T_m and T_g.

Poly(arylether thioether ketone)s and poly(arylthioether ketone)s, in which parts of ether groups are substituted by

*Author to whom correspondence should be addressed.

thioether groups, also exist. These polymers are also semi-crystalline and non-flammable because they contain an S atom in their polymer structure.

PREPARATION

Preparation methods for poly(aryl ether ketone)s are generally by nucleophilic or electrophilic substitution reactions.

CHEMICAL PROPERTIES

Chemical Resistance

There are no common room-temperature solvents for poly(aryl ether ketone)s. At higher temperatures, poly(aryl ether ketone)s dissolve in solvents such as hydrofluoric acid, trifluoromethanesulfonic acid, dichlorotetrafluoroacetonemonohydorate, phenol-1,2,4-trichlorobenzene (TCB), benzophenone, and diphenylsulfone.

Hydrolysis Resistance

Poly(aryl ether ketone)s resist hydrolysis even at elevated temperatures, and after exposure to water for 3000 h at 280°C and 18MPa. No significant degradation of flexural or tensile properties occurs.

Sulfonation

The sulfonation of poly(aryl ether ketone)s depends on their polymer structure. Poly(aryl ether ketone)s containing phenylene groups deactivated by electron-withdrawing keto groups, such as PEK and PEKK, are difficult to sulfonate. However, phenylene groups linked with two ether groups (–O–O–O–) such as PEEK, PEEKK, and PEEEK are easily sulfonated in sulfonic acid.

Radiation Resistance

Poly(aryl ether ketone)s shows excellent resistance to hard radiation.

APPLICATIONS

Poly(aryl ether ketone)s have the following features: high continuous working temperature, good chemical resistance, good hydrolysis resistance, and low flammability. These polymers are used in heavy duty applications: as injection mouldings (electrical connecters, automobile engine components, valve components, and hot-water meters), wire and cable coatings (computer cables, oil and gas wells, and aircraft), composites (aircraft) films (flexible printed circuit boards and wire wrapping), and others (filament coating, rotational moulding, and bearings).

REFERENCES

1. Staniland, P. A. *Comprehensive Polymer Science*; Pergamon: Oxford, 1989; Vol. 5.
2. Crritchley, J. P.; Knight, G. J.; Wright, W. W. *Heat-Resistant Polymer*; Plenum: New York, 1983.
3. May, R. *High Performance Polymers and Composites*; John Wiley & Sons: New York, 1991.
4. T. E. Attwood, P. C. Dawson, J. L. Freeman, L. R. J. Hoy, J. B. Rose, and P. A. Starit and *Polymer*, 22, 1103 (1981).

POLY(ARYL ETHER KETONE)S, AMINE-TERMINATED

Greggory S. Bennett
3M Center

The aerospace and electronics industries, where precision performance is demanded, have created a diverse market for specialized polymers. Materials possessing a combination of high strength-to-weight ratio, thermal and chemical stability, low dielectric constants, high glass transition temperatures, toughness, and processability are often required. Polymers that have many of the properties described above are known as high-performance polymers. Two general classes of high-performance polymers exist: thermosets and thermoplastics. Each system has inherent advantages and disadvantages associated with it for high-performance applications.

THERMOPLASTIC/THERMOSET HYBRID MATERIALS

A wide variety of functionalized rubbers increase the toughness of epoxy-based thermosets.[1–8]

Although rubber toughening of epoxy systems is an effective method to increase the fracture energy, the incorporation of rubber into the matrix results in reduced bulk properties.

Another drawback associated with rubber toughening is that the relative improvements in the rubber-toughened neat resins do not translate to comparably tougher composites and adhesives.[9–11]

In an attempt to address the deficiencies associated with rubber toughening, the effect of incorporating engineering thermoplastics as toughening agents has also been evaluated. The incorporation of tough, high glass transition temperature, solvent-resistant materials, having high moduli, into epoxy/amine systems alleviates the softening and T_g-lowering phenomena associated with adding a rubbery second phase.

Reactive aromatic thermoplastic oligomers have been used to incorporate the advantageous properties of aromatic thermoplastics into thermosetting systems. The molecular weight, reactive end group, and chemical composition of the aromatic thermoplastic can be tailored to incorporate desired properties.

AMINE-TERMINATED POLY(ARYL ETHER KETONE)S

Amine-terminated poly(aryl ether ketone)s are a relatively new class of materials that have been specifically designed to impart their advantageous properties into thermoset systems generally lacking the toughness and chemical resistance inherent to poly(aryl ether ketone)s.

The synthesis of amine-terminated poly(aryl ether ketone)s of controlled molecular weight takes advantage of step-growth polymerization, a reaction of (A–A) monomers with (B–B) monomers, where A and B refer to the reactive functional groups of each step-growth monomer.

An excess of (A–A) or (B–B) molecules is added to the system after the reaction has gone to completion to effectively endcap the poly(aryl ether ketone). Terminal amine functionality is best incorporated in a second synthetic step.

A one-step synthetic method can also be achieved through step-growth polymerization.

The properties of amine-terminated poly(aryl ether ketone)s are, not surprisingly, very similar to their nonfunctional analogs. Their T_g is a function of molecular weight and the nature of the backbone. Values of T_g ranging from 120 to 177°C are known. Decomposition temperatures are generally between 400 and 500°C. Common mechanical properties include a Young's modulus of ~ 2.5 GPa, an ultimate stress of 65 MPa, a breaking stress of ~ 60 MPa, and a breaking strain of 3 to 6%.[15]

The primary use of these materials is as modifiers for epoxy thermosets. The amine-terminated poly(aryl ether ketone)s can be designed to be miscible with commercial epoxy resins. Upon curing, two-phase separation into an epoxy-rich phase and a thermoplastic-rich phase results. Four distinct morphologies, a thermoset-rich continuous phase with thermoplastic-rich inclusions, a thermoplastic-rich and thermoset-rich mixed continuous phase, a thermoplastic-rich continuous phase with thermoset-rich inclusions, and a miscible system, have been observed.

The final thermal and mechanical properties are a function of the resulting morphology, which is a function of the loading level of the reactive thermoplastic oligomer, its molecular weight, its chemical structure, and the processing conditions. Increases in toughness of three- to sevenfold have been observed without a significant drop in the modulus. The low viscosity of the thermoset can be maintained when the molecular weight of the reactive thermoplastic oligomer is controlled. The ductility and chemical resistance of the continuous phase appears to be the primary factor controlling the resulting mechanical and chemical properties.[12–14]

REFERENCES

1. McGarry, F. J.; Willner, A. M. "Research Report R68-8" Massachusetts Institute of Technology: Cambridge, MA, 1968.
2. Sultan, J. N.; McGarry, F. J. *Polym. Eng. Sci.* **1973**, *13*(1), 29.
3. *Rubber Modified Thermosets*; Riew, C. K., Ed.; ACS Symposium Series; American Chemical Society: Washington, DC 1984.
4. Kirshenbaum, S. L.; Gazit, S.; Bell, J. P. *Rubber Modified Epoxies*; Am. Chem. Soc. Ser., 1984; 163.
5. Dusek, K.; Lednicky, F.; Lunak, S.; Mach, M.; Duskova, D. In *Rubber Modified Epoxies*; Am. Chem. Soc. Ser., 1984; 27.
6. Yorkgitis, E. M.; Tran, C.; Eiss, N. S.; Hu, T. Y.; Yilgor, I.; Wilkes, G. L.; McGrath, J. E. *Rubber Modified Epoxies*; Am. Chem. Soc. Ser., 1984; 137.
7. Verchere, D.; Sautereau, H.; Pascault, J. P. *Polymer* **1989**, *30*, 108.
8. Bascom, W. D.; Ting, R. Y. *J. Mater. Sci.* **1981**, *16*, 2657.
9. Hunston, D. L.; Bascom, W. D. In *Rubber Modified Epoxies*; Am. Chem. Soc. Ser., 1984; 83.
10. Gilwee, W. J.; Nir, Z. In *Rubber Modified Epoxies*; Am. Chem. Soc. Series 321; 1984.
11. Sanjana, Z. N.; Testa, J. H. *30th National SAMPE Symposium* **1985**, 1221.
12. Thompson, S. A.; Farris, R. J. Ph.D. Thesis, University of Massachusetts, Amherst, 1987.
13. Bennett, G. S.; Farris, R. J.; Thompson, S. A. *Polymer* **1991**, *32*, 1633.
14. Bennett, G. S.; Farris, R. J. Ph.D. Dissertation, University of Massachusetts, Amherst, 1992.
15. Bennett, G. S.; Farris, R. J. *Polym. Eng. Sci.* **1994**, *32*, 73.

Poly(aryl ether nitrile)s

See: Poly(arylene ether nitrile)s

Poly(aryl ether sulfone)s

See: Condensation Polymers (Using Phosphorus Pentoxide/Methanesulfonic Acid)
Crosslinking Agents, Phenyltriazenes (for Polyimides and Other Aromatic Polymers)
Cyclic Oligomers of Engineering Thermoplastics
Engineering Thermoplastics (Survey of Industrial Polymers)
Fluoropolymers (Trifluoromethylated PES and PEEK)
Poly(arylene ether sulfone)s (Semicrystalline and Amine Functionalized)
Poly(ether sulfone)s
Poly(ether sulfone)s (Properties, Applications, and Trends)
Poly(ether sulfone)s and Poly(ether ketone)s (Methyl-Substituted Aromatic)
Poly(p-phenylene sulfone)
Solubility Parameters (A Synthetic Approach)
Step Polymerization Catalysts
Sulfur Dioxide Copolymerization

POLY(ARYL ETHER SOLUFONE)S

Robert A. Clendinning and B. L. Dickinson
Amoco Polymers Business Group

SYNTHESIS

The discovery, development, and commercialization of poly(aryl ether sulfone)s by Union Carbide Corporation, Inc. and by Imperial Chemical Industries, Ltd. in the 1960s was called a technological breakthrough. UDEL® polysulfone and VICTREX® PES opened the way to a large number of successful commercial polymers including RADEL® R poly(phenyl sulfone), RADEL® A poly(ether sulfone), VICTREX® HTA poly(ether sulfone), VICTREX® PEEK poly(ether ketone), KADEL® poly(ether ketone), and ULTEM® poly(ether imide). The outstanding properties of these polymers and the chemistry by which they are made has spurred a high level of research interest in this field of polymer chemistry.

All of the above polymers are made commercially by aromatic nucleophilic substitution reactions, which will be the main focus of this article. Among other methods that have been used to prepare these polymers are electrophilic substitution, the Ullman reaction, nickel coupling reactions, transetherification reactions, and decarboxylation reactions.

PROPERTIES

Three polysulfone-based polymers have been developed to reach commercial importance. They are bisphenol A polysulfone, polyether sulfone, and polyphenyl sulfone. **Structures 1–3**, of these resins follow.

Bisphenol A Polysulfone (CAS 25154-01-2)

1

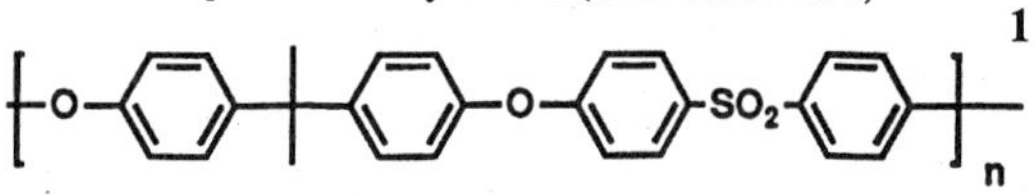

Poly(ether sulfone) (CAS 113569-14-5)

2

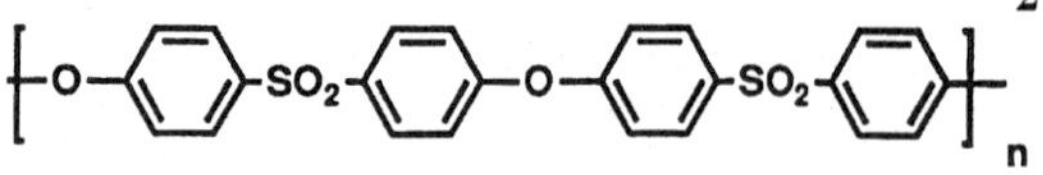

Poly(phenyl sulfone) (CAS 25608-64-4)

3

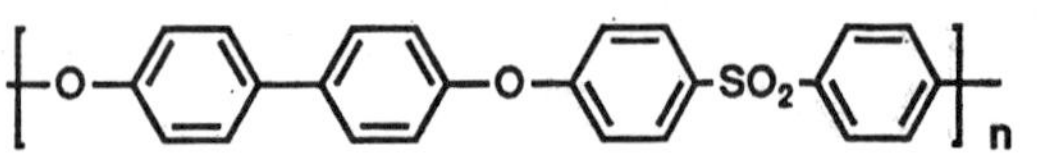

Sulfone-Based Engineering Polymers

All are tough, rigid, amorphous thermoplastics with high tensile strength, low creep, and excellent hydrolytic stability. Chain rigidity is derived from the relatively inflexible and immobile phenyl and SO_2 groups, and toughness from the connecting ether oxygens.

MECHANICAL PROPERTIES

All of the polysulfones are characterized by toughness and excellent practical impact strength.

THERMAL PROPERTIES

The heat distortion temperatures of polyphenyl and polyether sulfone are similar and are higher than bisphenol A polysulfone because of their higher glass-transition temperatures.

FLAMMABILITY

Both poly(ether sulfone) and poly(phenyl sulfone) exhibit excellent resistance to burning and require no flame retardant additives for most applications. Both polymers also show very low smoke emission.

ELECTRICAL PROPERTIES

Low dielectric constants, and low loss factors are seen even in the microwave range for the sulfone polymers. Dielectric strengths and resistivities are high.

CHEMICAL AND SOLVENT RESISTANCE

Polysulfones and poly(aryl ethers) in general have excellent hydrolytic stability compared to polycarbonates, polyesters, and polyetherimides.

APPLICATIONS

Polysulfone resins find greatest use in areas where property requirements such as heat resistance and hydrolytic stability exceed those of lower cost engineering plastics.

Poly(aryl ether thioether ketone)s

See: Poly(aryl ether ketone)s

POLY(ARYL PREHNITIMIDE)S

Zhi Yuan Wang* and Yu Qi
Ottawa-Carleton Chemistry Institute
Department of Chemistry
Carleton University

Aromatic polypyromellitimides are one of the most important high-performance polyimides derived from pyromellitic dianhydride (PMDA) and aromatic diamines, such as Kapton® poly(*N,N'*-bis-phenoxyphenylpyromellitimide) made from PMDA and oxydianiline (ODA). They generally have stiff, planar chemical structures that lend themselves to efficient packing, which in part allows the formation of long-range order. Due to their high crystallinity and strong interchain interactions, poly(pyromellitimide)s generally exhibit good solvent resistance or poor solubility.

Many new polyimides have been synthesized with an emphasis on structural modifications of the diamine and dianhydride monomers. A wide variety of flexible bridging units can be incorporated into the diamine monomers that impart mobility to the otherwise rigid polymer backbone.[1]

Being the smallest and rigid monomer for commercial polyimides, PMDA has recently been subjected to structural modifications by introducing pendant groups, aiming at increasing the solubility of poly(pyromellitimide)s without substantially decreasing the rigidity. Conceivably, substitutions at the 3,6-positions of PMDA would produce a series of new monomers. Indeed, PMDA has been modified by introducing phenyl, trifluoromethyl, and trifluoromethoxy at the 3,6-positions (**Figure 1**).[2–6]

It is generally recognized that a branched, asymmetric molecule tends to be more soluble than its linear, symmetric isomer.

Benzene-1,2,3,4-tetracarboxylic dianhydride or so-called prehnitic dianhydride (**2a**) is a structural isomer of PMDA (Prehnitic acid: 1,2,3,4-benzenetetracarboxylic acid). Aromatic polyimides based on prehnitic dianhydride are still unknown, except for an ill-defined, water-soluble copolymer derived from (**1a** and **2a**) with 3,3'-diaminodiphenylether described in a patent.[7] In light of the effect of the catenation pattern on the solubility of polyimide, we have recently designed and synthesized phenylated isomeric PMDA, diphenylprehnitic dianhydride.[8,9] Two anhydride groups in (**2b**) are locked on one benzene ring

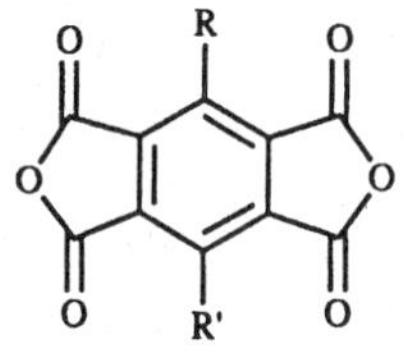

1a, R = R' = H
b, R = Ph, R' = H
c, R = R' = Ph
d, R = R' = CF_3

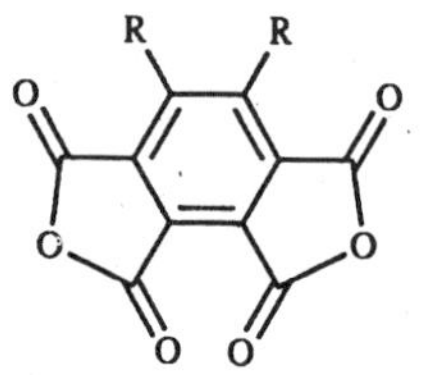

2a, R = H
b, R = Ph

FIGURE 1. Pyromellitic dianhydride, prehnitic dianhydride, and their derivatives.

*Author to whom correspondence should be addressed.

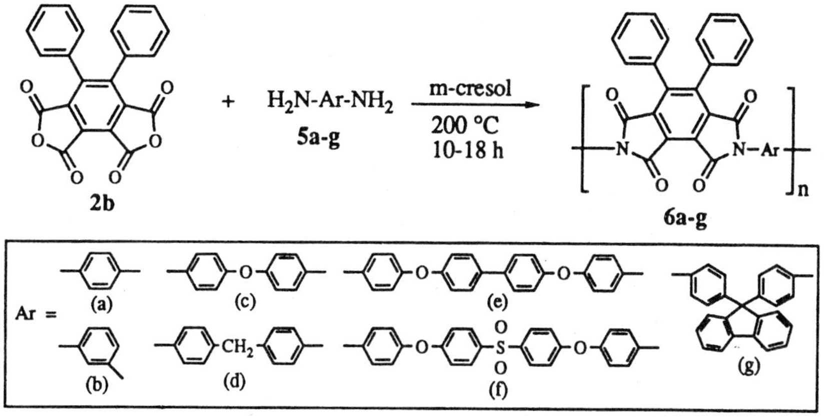

I

at the 1,2,3,4-positions, which means that the resulting polyimides should have a rigid but bent diimide unit. Of particular interest is the effect of the bent but still rigid dianhydride functionality on the solubility and other properties of these polyimides.

CHARACTERIZATION

One-step polymerizations with various aromatic diamines (**5a–5g**) were best carried out in *m*-cresol in the presence of a catalytic amount of isoquinoline at 200°C (**Scheme I**). Although dianhydride (**2b**) is less reactive than PMDA by comparing the chemical shifts of their carbonyl carbons, high molecular weight polyimides could be readily obtained with aromatic diamines in ~18 h, except for less reactive diamines such as 4,4'-diaminophenyl sulfone.[10]

The structures of the synthesized polyimides were characterized by FTIR and high-field ^{1}H and ^{13}C NMR spectrometries. By comparing IR spectra of the monomer (**2b**), model diimide, and polyimides synthesized, the complete imidization in polymerization was confirmed.

PROPERTIES

Solubility

Polyimides derived from rigid dianhydrides without a flexible bridging unit such as PMDA are practically insoluble and have to be applied as poly(amic acid) solutions in amide-type solvents, which are subsequently cured at elevated temperatures to effect imidization. In comparison, poly(aryl prehnitimide)s (**6**) have much improved solubilities than do poly(aryl pyromellitimide)s. Except for (**6e**), they are readily soluble at ambient temperatures in many common organic solvents such as TCE, DMAc, NMP, dimethylsulfoxide (DMSO), and *N,N*-dimethylformamide (DMF), *m*-cresol, and even pyridine.

Enhanced solubility due to the bent prehnitimide is also evident in the case of copolyimides derived from (**1a** and **2b**) with linear, rigid *p*-phenylenediamine.

Viscosity

Intrinsic viscosities of all polyimides (**6a–6g**) ranged from 0.45 to 1.81 dL/g. The formation of transparent, creasable films from polymers **6a–6f** indicates that these polyimides have reasonably high molecular weights and some degree of toughness.

Thermal Properties

In general, with increasing chain stiffness, the detection of glass transitions by DSC becomes more difficult. Polyimides **6b**, **6c**, **6e**, and **6f** showed distinct high T_gs ranging from 362 to 437°C. The T_gs of polyimides **6a** and **6g** were not detectable by DSC.

Like many aromatic polyimides and poly(aryl pyromellitimide)s, all poly(aryl prehnitimide)s displayed excellent thermal stabilities.

These poly(aryl prehnitimide)s appear to have better tensile moduli than poly(4,4'-oxydiphenylene pyromellitimide), and the tensile modulus of the latter nonoriented thin film has been reported to be 2.1 GPa.[11]

CONCLUSIONS

A series of new soluble polyimides with high molecular weights has been synthesized from diphenylprehnitic dianhydride. The results presented here demonstrate that changing the substitution pattern of the anhydride groups on a core benzene ring can greatly affect the solubility of resulting polyimides. The unique bent prehnitimide structure greatly enhances the solubility of polyimide in common organic solvents. All poly(arylprehnitimide)s synthesized have very high glass transition temperatures, excellent thermal stabilities, and good mechanical properties. These thermal and thermomechanical properties, along with exceptionally good solubility in water-miscible solvents such as THF, DMF, and DMSO, make some of these polyimides potentially excellent candidates for use as asymmetric membranes in high-temperature separation. Work is in progress on the syntheses and evaluations of copolyimides from diphenylprehnitic dianhydride and other commercially available dianhydrides.

ACKNOWLEDGMENT

We gratefully acknowledge financial support from the Natural Sciences and Engineering Research Council of Canada.

REFERENCES

1. Wilson, D.; Stenzenberger, H. D.; Hergenrother, P. M. *Polyimides*; Blackie and Son: Glasgow, 1990.
2. Harris, F. W.; Hsu, S. L. C. *High Perf. Polym.* **1989**, *1*, 3.

3. Giesa, R.; Keller, U.; Eiselt, P.; Schmidt, H-W. *J. Polym. Sci., Part A, Polym. Chem.* **1993**, *31*, 141.
4. Lukmanov, V. G.; Alkeseevaand, L. A.; Yagupolskii, L. M. *Zh. Org. Khim.* **1974**, *10*, 2000.
5. Matsuura, T.; Ishizawa, M.; Hasuda, Y.; Nishi, S. *Macromolecules* **1992**, *25*, 3540.
6. Auman, B. C. *Polym. Prepr., Am. Chem. Soc. Div. Polym. Chem.* **1994**, *35*(2), 749.
7. Suzuki, S.; Kaneda, I.; Takahashi, M.; Nagai, H. Jpn Patent 71 16 906, May 10, 1971; *Chem. Abstract* **1971**, *75*, 141,367k.
8. Wang, Z. Y.; Qi, Y. *Macromolecules* **1994**, *27*, 625.
9. Wang, Z. Y.; Qi, Y. *Macromolecules* **1995**, *28*, 4207.
10. Ando, S.; Matsuura, T.; Sasaki, S. *J. Polym. Sci., Part A, Polym. Chem.* **1992**, *30*, 2285.
11. Mita, I.; Kochi, K.; Hasegawa, M.; Iizaka, T.; Soma, H.; Yokota, Y.; Horiuchi, in *Polyimides: Materials, Chemistry and Characterization*, Feger, C.; Khojasteh, M.; McGrath, J., Eds.; Elsevier: Amsterdam, 1989; p 1.

Poly(aryl sulfide)s

See: Poly(p-phenylene sulfide) (Synthesis by p-Dichlorobenzene and Sodium Sulfide)
Poly(phenylene sulfide) and Copolymers (Melt Preparation via a New Process)
Polysulfides (Prepared from Sulfur Dioxide)

Poly(aryl thioether ketone)s

See: Poly(aryl ether ketone)s
Poly(thioether ketone and sulfone)s

Poly(aryl thioether sulfone)s

See: Poly(thioether ketone and sulfone)s

Polyarylates

See: Cyclic Oligomers of Engineering Thermoplastics
Polyarylates, Dicyanovinyl-Containing (Thermal Properties)
Polyesters, Aromatic (Amorphous and Liquid Crystalline)
Pseudo-Poly(amino acid)s (Overview)

POLYARYLATES, DICYANOVINYL-CONTAINING (Thermal Properties)

Myoung-Seon Gong*
Department of Chemistry
Dankook University

Byung-Gu Kim and Chulhee Kim
Department of Polymer Science and Engineering
Dankook University

Polyarylates are aromatic polyesters based on diphenols and aromatic dicarboxylic acids. As a result of their processability, weatherability, high softening temperatures, and excellent physical and mechanical properties, polyarylates have been under intensive industrial development.[1] The dicyanovinylidene group ($C = C(CN)_2$) can be considered structurally equivalent to the carbonyl group in various chemical reactions.[2] Thus, polyenaryloxynitriles can be considered analogous to polyesters. Recently, it was reported that polyenaminonitriles and polyenaryloxynitriles, which contained the dicyanovinyl group, are thermally curable and thus show good thermal stabilities.[3–10] Novel polyarylates, which undergo heat curing as a result of inserting the dicyanovinyl group in the main chain, are of a considerable interest and their preparation and thermal properties have been investigated.

PROPERTIES

The polymerization of Bisphenol A with different contents of **1** and terephthalate units was attempted to obtain a controlled amount of enaryloxynitrile units in the main chain of aromatic polyesters. Copolyarylates containing dicyanovinyl groups were also obtained from 4,4'-sulfonyldiphenol, 4,4'-thiodiphenol, and phenolphthalein (**Scheme I**).

*Author to whom correspondence should be addressed.

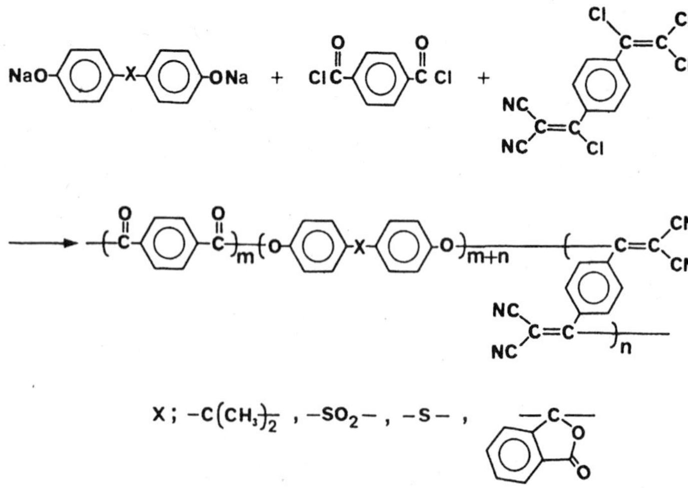

I

The polyarylates containing dicyanovinyl groups cured thermally without release of volatile byproduct. The polymers were no longer soluble in a solvent for the untreated polymers. This observation may be attributable to further chemical change of dicyanovinyl groups. The thermal decomposition of these polymers may begin at 380°C. All the polymers show excellent thermal stabilities despite their low molecular weights.

It is reported that among several possible curing reactions, poly(enaminonitriles) undergoes intramolecular cyclization to amino-cyano quinoline units as well as crosslinking.[3–5] In the case of polyarylates containing dicyanovinyl groups, the curing reaction is assumed to proceed by intermolecular and intramolecular addition of dicyanovinyl groups, but we have been as yet unable to assign a particular pathway for the chemical curing process.

REFERENCES

1. Dickinson, B. L. "*Polyarylate, Modern Plastics Encyclopedia*", McGraw-Hill: New York, 1982; 59(10A).
2. Wallenfels, K.; Friedrich, K.; Rieser, J. *Angew. Chem. Int., Ed. Engl.* **1976**, *15*, 261.
3. Moore, J. A.; Robello, D. R. *Polym. Prepr. (Am. Chem. Soc., Div. Polym. Chem.)* **1986**, *27*(2), 127.
4. Moore, J. A.; Robello, D. R. *Macromolecules* **1986**, *19*, 2667.
5. Moore, J. A.; Robello, D. R. *Macromolecules* **1989**, *22*, 1084.
6. Moore, J. A.; Mehta, P. G. *Polym. Mater. Sci. Eng.* **1990**, *63*, 351.
7. Moore, J. A.; Mehta, P. G. *Polym. Mater. Sci. Eng.* **1991**, *64*, 835.
8. Kim, S. T.; Lee, J. O.; Kim, Y. S.; Gong, M. S. *Polymer (Korea)* **1991**, *15*, 95.
9. Moon, H. S.; Kim, S. T.; Gong, M. S. *Makromol. Chem., Rapid Commun.* **1991**, *12*, 591.
10. Moon, H. S.; Kim, J. S.; Gong, M. S. *Polymer J.* **1993**, *25*, 193.

POLY(ARYLENE ETHER NITRILE)S

Shigeru Matsuo
Central Research Laboratory
Idemitsu Kosan

Many kinds of poly(arylene ether)s (CAS Registry No. 111867-27-7) have been prepared via nucleophilic aromatic substitution polymerization.[1–3] Especially semicrystalline poly(arylene ether ether ketone)s and related materials exhibit excellent hydrolytic, thermal, and dimensional stabilities, and they have good mechanical properties.[4] Poly(arylene ether nitrile) (PEN™) is also a semicrystalline polymer, which has the structure of poly(*m*-phenylene ether) with pendent nitrile groups.[5–7] Pendent nitrile groups are expected to have some advantages; for instance, they may promote adhesion with various materials and they present a potential crosslinking site.[8–10] This report is concerned with the synthesis and characterization of the semicrystalline poly(arylene ether nitrile) and related polymers.

PREPARATION

Poly(arylene ether nitrile)s are prepared by nucleophilic aromatic substitution polymerization of 2,6-dihalobenzonitrile with various bisphenols using *N*-methyl-2-pyrrolidone (NMP) as a solvent in the presence of sodium carbonate or potassium carbonate (**Scheme I**).

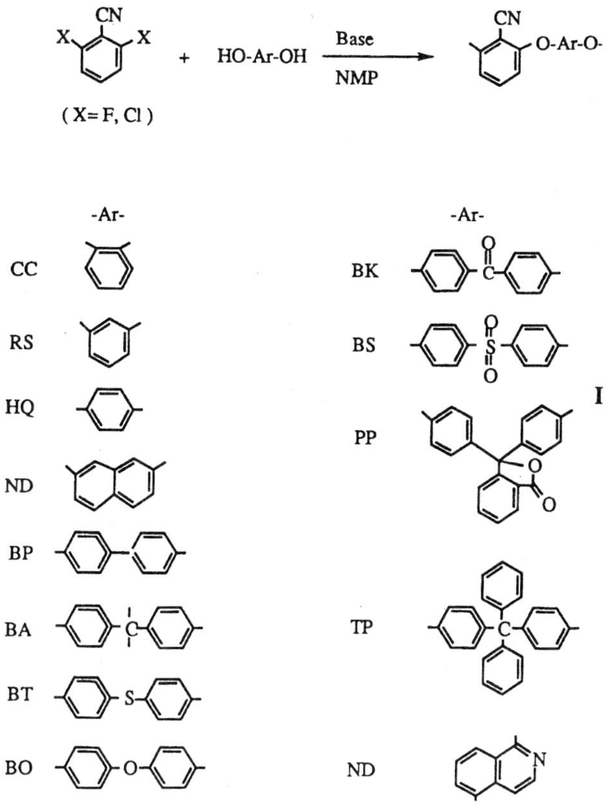

I

PROPERTIES

The thermal decomposition temperatures of the polymers varied widely with the nature of bisphenol components. The polymers derived from bisphenols with electron-withdrawing groups such as ketone and sulfone exhibited lower decomposition temperatures than the other bisphenol-based polymers.

The resorcinol-based poly(arylene ether nitrile) is now available commercially from Idemitsu Material Ltd. under the trade name PEN. PEN can be molded by ordinary injection or extrusion machinery. The moldability of this polymer at 350°C is better than that of PEEK at 390°C. The poly(arylene ether nitrile) had a low heat distortion temperature (HDT), just above its T_g, which was very similar to PEEK, but the glass-fiber-filled material was 330°C due to its high melting temperature and its high crystallinity.

Tensile and flexural strengths of the poly(arylene ether nitrile) were better than those of PEEK. The scanning electron microscopy photograph of the glass-fiber-filled specimen after the tensile test indicates that the poly(arylene ether nitrile) adheres to the glass surface fairly well. The pendent nitrile group may give dipole–dipole interaction forces and good adhesion to glass fibers; hence, excellent mechanical properties result. Other key properties include creep resistance, wear resistance, and chemical resistance.

APPLICATIONS

PEN is used for engineering and electrical parts, such as bearings, washers, rollers, industrial belts, compressor components, integrated circuit carriers, connectors, and pipes for water purification.

REFERENCES

1. Clendinning, R. A.; Farnham, A. A.; Hall, W. F.; Johnson, R. N.; Merriam, C. N. *J. Polym. Sci., Part A1* **1967**, *5*, 2375.
2. Attwood, T. E.; Dawson, P. C.; Freeman, J. L.; Hoy, L. R.; Rose, J. B.; Staniland, P. A. *Polymer* **1981**, *22*, 1096.
3. Cludby, M. E. A.; Feasey, R. G.; Jennings, B. E.; Jones, M. E. B.; Rose, J. B. *Polymer* **1965**, *6*, 589.
4. Critchely, J. P.; Knight, G. J.; Wright, W. W. *Heat-Resistant Polymers*; Plenum: New York, 1983.
5. Matsuo, S.; Murakami, T.; Takasawa, R. *J. Polym. Sci., Polym. Chem. Ed.* **1993**, *31*, 3439.
6. Matsuo, S.; Murakami, T. Eur. Patent 192 177, 1986.
7. Matsuo, S.; Murakami, T.; Nagatoshi, K.; Bandou, T. Eur Patent 243 000, 1987; *Chem. Abstract* **1987**, *108*, 132788W.

POLY(ARYLENE ETHER SULFONE)S (Semicrystalline and Amine Functionalized)

A. Recca,* F. A. Bottino, G. Di Pasquale, and A. Pollicino
Istituto Chimico
Facoltà di Ingegneria
Università di Catania

P. T. McGrail and P. A. Staniland
Imperial Chemical Industries
Wilton Materials Research Centre

Poly(ether sulfone) (PES) and poly(ether ether sulfone) (PEES) are engineering polymers that have been commercialized for specialized markets, and have potential uses in areas such as matrix materials for thermoplastic based composites.[1] Although their continuous-use temperature is limited to ~ 180°C, their use in environmental highly sensitive structural part seems unlikely because they are not melt-crystallizable. To overcome these problems there is a considerable amount of research being carried out to (a) develop new semicrystalline polysulfones with improved solvent resistance, and (b) make existing systems crosslinkable by minor chemical modification.

AMINE FUNCTIONALIZED POLYSULFONES

Several studies have been reported in literature on the use of thermoplastic polymers, and in particular, of PES as a toughening agent for epoxy systems.[2–8]

The basic approach of the research in this field is to synthesize functionalized poly(ether sulfone), and then to react or blend them with the more common thermosetting component to arrive finally at composite materials. One functionalization that allows to crosslink the PES with resins like epoxy is the introduction of amino groups. Two approaches have emerged on this topic in the last few years. The first is concerned with the preparation of amino terminated oligomers, the second with the introduction of amino functional groups in the PES backbone.

REFERENCES

1. Kennedy, P. E. *Thermally Stable Polymers. Synthesis and Properties*; Marcel Dekker: New York, 1980.
2. Hedrick, J. L.; Yilgor, I.; Wilkes, G. L.; McGrath, J. E. *Polym. Bull.* **1985**, *13*, 201.
3. Bucknall, C. B.; Partridge, I. K. *Polym. Eng. Sci.* **1986**, *26*, 54.
4. Bucknall, C. B.; Partridge, I. K. *Brit. Polym. J.* **1983**, *15*, 71.
5. Bucknall, C. B.; Partridge, I. K. *Polymer* **1983**, *24*, 639.
6. Raghava, R. S. *Natl. SAMPE Symp. Exhib.* **1983**, *28*.
7. Ibrahim, A. M.; Quinlivan, T. J.; Seferis, J. C. *Polym. Prepr.* **1985**, *26*(1), 277.
8. Yamanaka, K.; Ionue, T. *Polymer* **1989**, *30*, 662.

*Author to whom correspondence should be addressed.

Poly(arylene ethynylene)s

See: Nonlinear Optical Polymers [Poly(arylene ethynylene) Derivatives]

POLY(ARYLENE-1,3,4-OXADIAZOLE)S

Burkhard Schulz** and Ludwig Brehmer
Research Group Thin Organic Films
Institute of Solid State Physics
University of Potsdam

Aromatic poly(1,3,4-oxadiazole)s are known to have a combination of valuable properties, such as thermal and hydrolytic stability, strength, fatigue resistance, and a relatively low density.[1] They form fibers, films, and membranes.[2–4] However, these polymers often exhibit poor solubility in many solvents, and they are frequently processed from sulfuric acid solutions.[5]

Although poly(arylene-1,3,4-oxadiazole)s have a combination of excellent properties, their commercial application is limited. Nevertheless, the discovery of new organic solvents for fully aromatic poly(1,3,4-oxadiazole)s, a better understanding of their solid-state structures, and systematic investigations of properties such as electrochemical behavior offer new chances to get special materials for applications in microelectronics or sensories.[6]

PROPERTIES

Poly(arylene-1,3,4-oxadiazole)s are soluble only in strong acids such as sulfuric acid, PPA, and triflouromethane sulfonic acid; they are insoluble in organic solvents. However, the incorporation of flexible bonds or voluminous substituents make poly(1,3,4-oxadiazole)s soluble in amidic organic solvents such as dimethylformamide (DMF), dimethylacetamide (DMAc), and *N*-methylpyrrolidinone (NMP) (**Structure 1**).[7,8,27]

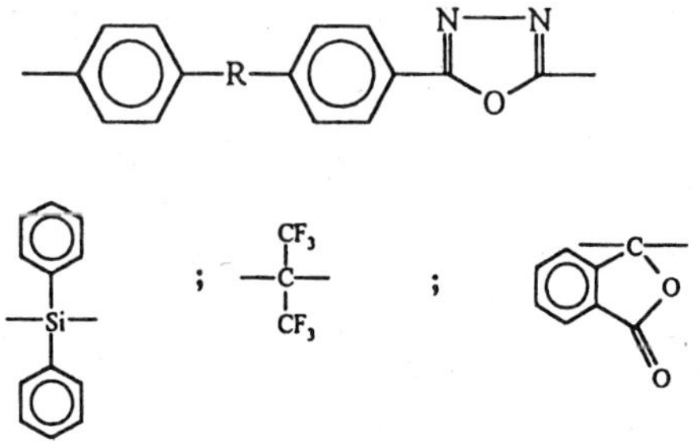

1

**Author to whom correspondence should be addressed.

The structures and properties of polyoxadiazoles in solution are investigated in detail for poly(*p*-phenylene-1,3,4-oxadiazole) (POD).[9,10] Contrary to other polyheterocyclics, no liquid-crystalline phases in POD solutions were observed. But copolyesters containing both 1,3,4-oxadiazole rings and naphthylene units show lyotropic phases.[11–13]

Recently it was shown that poly(arylene-1,3,4-oxadiazole)s are soluble in an aprotic solvent system consisting of a nitroalkane and a Lewis acid.[6]

Solid poly(arylene-1,3,4-oxadiazole)s have densities in the 1.2–1.4 g/cm^3 range. They are thermally stable up to 450°C and have good stability toward hydrolytic attack.[14–17]

Little is known about the electrical properties of poly(arylene-1,3,4-oxadiazole)s. Like other aromatic or conjugated polymers, they are insulators and show electrical conductivity only after doping with iodine or Lewis acids.[18]

APPLICATIONS

Applications of poly(arylene-1,3,4-oxadiazole)s are known only for POD. This polymer is mainly used as a fiber.[19] POD fibers have a combination of good properties (i.e., strength, stiffness, resistance, and low density).[20–22] Wet-spun POD fibers can be pyrolyzed at 1300–2800°C to form carbon fibers.[23] They show higher electric conductivity than other carbon fibers derived from other heat-resistant polymers.[24]

Poly(ether imide 1,3,4-oxadiazole)s, poly(1,3,4-oxadiazole amide imide)s, and poly(ether ketone 1,3,4-oxadiazole)s yielded flexible, creasable films showing excellent thermal stability. The high T_g values, good mechanical behavior, and dielectric properties of these polymers make them useful as coatings in microelectronic fabrication or as matrix resin in fiber-reinforced composite applications.[25,26] The combination of their thermostability, fatigue resistance, and electronic properties offers the possibility to construct new special materials on the basis of poly(arylene-1,3,4-oxadiazole)s.

REFERENCES

1. Nanjan, M. J. In *Encyclopedia of Polymer Science and Engineering*; Mark, H. F., et al., Ed.; John Wiley & Sons: New York, 1988; Vol. 12.
2. Bühler, K.-U. *Spezialplaste*; Akademie-Verlag: Berlin, Germany, 1978.
3. Korshak, V. V.; Berestneva, G. L.; Vinogradova, S. V.; Gergaja, M. S.; Tur, D. R. *Chem. Gheterocycl. Soed.* **1971**, *11*, 1457.
4. Cassidy, P. E.; Fawcett, N. C. *Macromol. Sci., Rev. Macromol. Chem.* **1979**, *C17*, 209.
5. Jain, S. R.; Sridhora, K.; Thangamathesvoran, P. M. *Progr. Polym. Sci.* **1993**, *18*, 997.
6. Schulz, B.; Leibnitz, E. *Acta Polymerica* **1992**, *43*, 343.
7. Kovacs, H. N.; Delman, A. D.; Simms, B. B. *J. Polym. Sci., Part A1* **1970**, *8*, 869.
8. Hensema, E. R.; Sena, M. E. R.; Mulder, M. H. V.; Smolders, C. A. *J. Polym. Sci., Part A, Polym. Chem.* **1994**, *32*, 513.
9. Schelle, H.; Hofmann, D.; Weigel, P. *Acta Polymerica* **1989**, *40*, 450.
10. Hofmann, D.; Weigel, P.; Ganster, J.; Fink, H. P. *Polymer* **1991**, *32*, 284.
11. Guseva, M. A.; Raskina, A. B.; Volochina, A. W.; Kudravzev, G. I.; Kwitko, U. I.; Rudaja, C. U. *Chem. Volokna* **1986**, 15.
12. Brydon, D. L.; Fisher, I. S.; Emans, J.; Smith, D. M.; MacDonald, W. A. *Polymer* **1989**, *30*, 619.
13. Brydon, D. L.; Fisher, I. S.; Emans, J.; Smith, D. M.; Bowen, T.; Morvey, I. W. *Polymer* **1993**, *34*, 4481.
14. Frazer, A. H.; Sarasohn, J. M. *J. Polym. Sci., Part A1* **1966**, *4*, 1649.
15. Burja, A. J.; Okromshedlidze, N. P.; Dubkova, V. I. *Dokl. Akad. Nauk BSSR* **1989**, *33*, 233.
16. Lavrenko, P. N.; Okatova, O. V.; Melnikov, A. B. *Vysokomol. Soed.* **1981**, *A23*, 532.
17. Grebenkin, A. N.; Antonov, N. G.; Kolzov, A. J. *Vysokomol. Soed.* **1989**, *B31*, 203.
18. Tsutsui, T.; Tukuta, Y.; Hora, T.; Sato, S. *Polym. J.* **1987**, *19*, 719.
19. Yang, H. H. *Aromatic High-Strength Fibers*; John Wiley & Sons: New York, 1989.
20. Jones, R. S.; Soehngen, J. W. *J. Appl. Polym. Sci.* **1980**, *25*, 315.
21. Bunajeva, A. S.; Martimova, G. A.; Pedshenko, D. S. *Khim. Volokna* **1984**, 37.
22. Taeger, E.; Mieck, K-P.; Banke, K. H. *Chemiefaser/Textilindustrie* **1991**, *93*, T28.
23. Shiotani, K.; Ishii, M.; Shioja, M. *Sen-i-Gakkaishi* **1991**, *47*, 234.
24. Shiotani, K.; Shioja, M.; Kikutnai, T.; Takadu, A. *Sen-i-Gakkaishi* **1992**, *48*, 379.
25. Mercer, F. W. *High Performance Polymers* **1992**, *4*, 73.
26. Bruma, M.; Schulz, B.; Mercer, F. W. *J. Macromol. Sci. A-Pure and Applied Chem.* **1995**, *A32*, 259.
27. Vinogradova, S. V.; Korshak, J. V.; Tur, D. R. *Vysokomol. Soed.* **1972**, *A14*, 915.

POLY(ARYLENE VINYLENE)S (Mechanistic Control of a Soluble Precursor Method)

Paul M. Lahti
Department of Chemistry
University of Massachusetts

Among highly conjugated polymers, the poly(arylene vinylene) (PAV) family has attracted considerable attention. While the PAVs have been studied for quite some time, recent increased interest has been based on a variety of strategies to make processible PAVs as electrooptic materials.

In the late 1960s and early 1970s, Wessling and coworkers at Dow Chemical Company, developed a soluble precursor route to PAVs, specifically to poly(1,4-phenylene vinylene)s, PPVs.[1,2] During the 1970s, Hörhold's group in Germany further investigated this route, in addition to their work on PAV synthesis by a related dehydrohalogenation polymerization.[3–5] The basic idea of soluble precursor routes to conjugated polymers is to make a nonconjugated, conformationally flexible precursor polymer that can be processed and/or oriented. Upon appropriate treatment–typically thermolytic–the final conjugated material is produced as films, fibers, bubbles, or other forms allowed by the processing of the precursor. The Wessling methodology is summarized in **Equation 1**.

We have been interested in being able to synthesize as broad a variety of PPV and PAV polymers as possible using the Wessling methodology. In order to maximize the D.P. and yield of high polymer, we found it of use to understand and control the mechanism of this fascinating process.

Overall, studies of mechanistic control in the Wessling process have proven to be very productive in terms of improving the molecular weight and yield of polyelectrolyte precursors to PAVs and PPVs. While some drawbacks exist for PAV synthesis

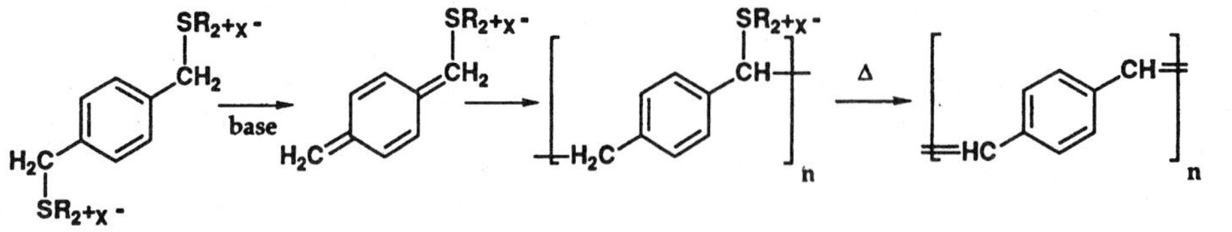

1

BIS-SALT (1) XYLYLENE (2) POLYELECTROLYTE (3) PPV (4)

by this methodology the ease of PAV processibility and high degrees of conjugation in the final eliminated products make this one of the most desirable methods for making highly conjugated polymeric systems with potentially interesting electrooptic properties.

APPLICATIONS

PAVs are presently being tested for a number of applications, mostly electrooptic. The semiconducting capability of PAV films and fibers upon doping with electron acceptors–and, to a lesser extent, donors–has been well established.

The ability to process the soluble precursor form of a PAV and then to "lock" its morphology in conjugated, full eliminated form has also drawn attention to possible use of PAVs as nonlinear optical (NLO), photoconductive, or electroluminescent (EL) materials.[6–8]

ACKNOWLEDGMENTS

This work would not have been possible without the efforts of the members of my research group, who are acknowledged in various citations herein. A particularly superb review of the PAV literature up to 1990 is included in the thesis of my former colleague, Dr. F. R. Denton III.[44] I am indebted to my colleagues from the Department of Polymer Science and Engineering, Profs. F. E. Karasz and R. W. Lenz, for their collaborating during the course of this work. Finally, I am grateful for financial support from the Air Force Office of Scientific Research (University Research Initiative), the University of Massachusetts Materials Research Laboratory, and the National Science Foundation.

REFERENCES

1. Wessling, R. A.; Zimmerman, R. G. U.S. Patent 3 402 152, 1968.
2. Wessling, R. A. *J. Polym. Sci., Polym.* Sympos. **1985**, *72*, 55.
3. Hörhold, H. H.; Palme, H.-J.; Bergmann, R. *Faserforsch. Textiltech. Z. Polymerforschung* **1978**, *29*, 299.
4. Hörhold, H. H.; Ozegowski, J.-H.; Bergmann, R. *J. Prakt. Chem.* **1977**, *319*, 622.
5. Hörhold, H. H.; Raabe, D. *Acta Polymerica* **1979**, *30*, 86.
6. Obrzut, J.; Obrzut, M. J.; Karasz, F. E. *Synth. Met.* **1989**, *29*, E103.
7. Rentsch, S.; Yang, J. P.; Li, H. L.; Lenzner, M.; Bergner, H. *Synth. Met.* **1991**, *41–43*, 1369.
8. Frankevich, E. L.; Lymarev, A. A.; Sokolik, I.; Karasz, F. E.; Blumstengel, S.; Baughman, R. H.; Hörhold, H. H. *Phys. Rev. B* **1992**, *46*, 9320.

Polyaspartates

See: Nylons 3, Helical [Poly(α-alkyl-β]-L-aspartate)s]

POLYAZINES (Third-Order Nonlinear Optical Properties)

Hari Singh Nalwa
Hitachi Research Laboratory
Hitachi Ltd., Japan

Organic π-conjugated polymers have emerged as one of the key materials for third-order nonlinear optics. Both experimental and theoretical studies have found that the third-order optical nonlinearities increase as the length of the π-electron conjugation increases.[1–3] Polyazine is isoelectronic with polyacetylene because it consists of pairs of nitrogen atoms substituted for pairs of carbon atoms in the polyacetylene chain. The peculiar arrangement of carbon and nitrogen atoms in the polyazine backbone offers not only environmental stability but also optical transparency when compared with polyacetylene. Theoretical and experimental results on the third-order nonlinear optical properties of well-defined chain-length oligoazines and polyazines are discussed.

THIRD-ORDER NONLINEAR OPTICAL PROPERTIES

Polyazine is a simple linear chain of atoms with alternating single and double bonds, but with pairs of nitrogen atoms substituted for pairs of carbon atoms in a polyacetylene chain. Therefore, polyazine can be viewed as isoelectronic with polyacetylene. The π-band structure of polyazine is different from that of polyacetylene because polyazine has four π bonds–lower two filled and upper two empty–whereas polyacetylene has only two bands.[4] The significant difference between third-order nonlinear optical properties of polyacetylene and polyazine derivatives occurs because of the presence of nitrogen atoms in the π-electron backbone, which causes the bonding pattern among two conjugated polymers to differ. Polyacetylene structure supports soliton-like defects, whereas polyazine leads to a polaron or bipolaron-type deformation.[5] In polyazine, defects are located at a carbon or at a nitrogen atom; in polyacetylene, defect centers are carbon atoms. Therefore, the origin of nonlinear optical responses in these two conjugated polymers should be quite different. Polyacetylene has a solitonic mechanism for optical nonlinearity, whereas the species that contribute to the optical nonlinearity in a polyazine π-conjugated backbone could be photoinduced excitons, polarons, and/or bipolaron.[6] The photoinduced excitons are coupled with the pairs of nitrogen and carbon atoms in the polyazine π-conjugated chain. Polyazines seem to be very promising π-conjugated systems for third-order nonlinear optics, and their potential needs to be further explored.

REFERENCES

1. *Nonlinear Optical Properties of Organic Molecules and Crystals*; Chemla, D. S.; Zyss, J., Eds.; Academic: New York, 1987.
2. Zyss, J. *Molecular Nonlinear Optics*; Academic: New York, 1993.
3. Nalwa, H. S. *Adv. Mater.* **1993**, *5*, 341.
4. Euler, W. B.; Hauer, C. R. *Solid State Commun.* **1984**, *51*, 473.
5. Su, W. P.; Schrieffer, J. R.; Heeger, A. J. *Phys. Rev. B* **1980**, *22*, 2099.
6. Nalwa, H. A.; Hamada, T.; Kakuta, A.; Mukoh, A. *Nonlinear Optics* **1993**, *7*, 155.

Polyazoles

See: Poly(benzobisthiazole-imide)s

Polyazomethines

See: Condensation Polymers (Synthesis from Silylated Diamines)

Polybenzidine

See: Enzyme-Catalyzed Polymerization

Polybenzimidazobenzophenathrolines

See: Conjugated Ladder Polymers

POLYBENZIMIDOLES (Overview)

Eui-Won Choe*
Robert L. Mitchell Technical Center
Hoechst Celanese Corporation

David D. Choe
New Jersey Institute of Technology

Polybenzimidazoles are a class of linear heterocyclic polymers that contain benzimidazole nuclei as a repeating unit. Benzimidazole groups exhibit excellent inherent thermal stability, resistance to chemicals, acid and basic hydrolysis, and temperatures, but lack a thermo-oxidative stability.

Wholly aromatic PBI having excellent thermal and chemical stabilities considerably higher than aliphatic series was reported by Vogel and Marvel in 1960.[1,2] The excellent thermal and inherent nonflammability of the wholly aromatic PBI led to tremendous advances in the high temperature resistant polymer field. One of the polybenzimidazoles, poly[2,2'-(m-phenylene)-5,5'-bibenzimidazole], was commercialized by Celanese Corp., in 1983, and the fiber made from this polymer was trade-marked as "PBI®," the name that will be used throughout this article to abbreviate the polymer. The CAS registry numbers for PBI are 25928-81-8 and 25734-65-0.

This article describes the synthesis of monomers and polymers, properties, and applications with an emphasis on the development and current trends of polybenzimidazoles since 1985. The current trends in polybenzimidazole research and development include molecular composites, catalyst support, metal sorbents, PBI blends with polyimides, polyarylates, fluoro-containing polyamideimide blends, fluoro-containing polyimides, poly(4-vinylpyridine), or poly(N-vinyl-2-pyrrolidone).

POLYMER SYNTHESIS AND PROPERTIES

Many aromatic polybenzimidazoles, characterized by a high thermal stability, have been reported and reviewed.[3–14] A series of wholly aromatic polybenzimidazoles with a structure repeat unit of **Structures 1** and **2** showed excellent thermal and chemical stabilities. They were prepared by many polycondensation methods, melt/solid-state polycondensation, solution polymerization, and suspension, utilizing bis(o-diamine)s and difunctional acids or derivatives, or nucleophilic substitution reaction of AB monomers.

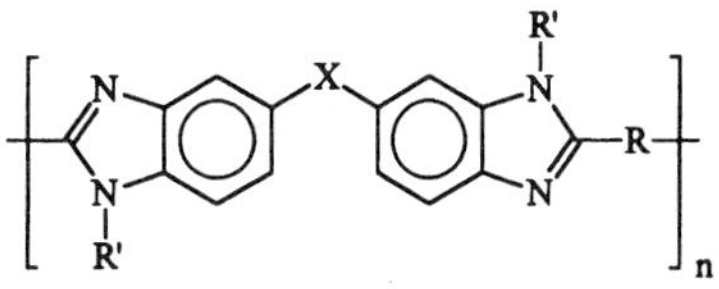

1

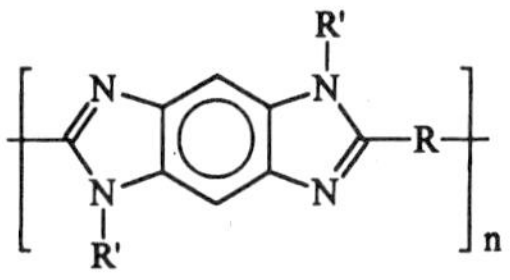

2

Stabilization of PBI with Sulfuric Acid

The mechanical properties of PBI are in the conventional textile range rather than in the range of reinforcement fibers. The particularly desirable attributes of PBI are its high temperature performance, high moisture regain, chemical resistance, and nonburning characteristics. Burning tests revealed that PBI fabric would not burn in air but would begin to shrink when exposed to a naked flame. To minimize shrinkage at elevated temperatures, the drawn yarn is subjected to acid treatment. The fiber is treated with aqueous sulfuric acid to form a salt with an imidazole ring structure. This salt rearranges under thermal treatment to ring sulfonated PBI.

Notable is the significant improvement of dimensional stability after stabilization treatment, which reduces flame shrinkage from 50 to 6%. Results of other flame tests support the outstanding performance of the polymer as a fabric.[15]

Stabilization of PBI with Phosphoric Acid

The treatment of PBI film with phosphorus oxychloride was found by Dynatech Company and FRL Albany International to significantly increase its resistance to burning. Subsequently, the use of phosphoric acid was found to be equally effective.

APPLICATIONS

PBI has an inherent nonflammability that is attributed to the imidazole structure and has enormous utility in a wide variety of applications.

As Catalyst Support

Polybenzimidazole was used as support for the preparation of polymer-supported catalysts for both reduction and oxidation.

*Author to whom correspondence should be addressed.

The high thermal stability of PBI support at an elevated temperature is an extremely valuable feature for this application.

Metal Sorbents

Chemically modified PBIs with chelating ligands via epoxidation have been used for hydrometallurgical applications and are excellent metal-selective sorbents.

PBI Paper

Sulfonated PBI was utilized to make high temperature-resistant synthetic nonflammable paper with a LOI of 41%.[16]

PBI Performance Parts

PBI is a unique nonflammable and non-smoking material due to the inherent property of benzimidazole moiety. PBI Performance Parts are marketed under the trade name of Celazole® by Hoechst Celanese Corporation.

PBI Matrix Resin

PBI has been used as a matrix resin for composite preparation. PBI resin is a 45% solution of low IV PBI (Mw 12,000–20,000) in DMAc having a solution viscosity of 3–6000 poise. PBI prepregs using graphite, quartz, and ceramic in fabric, tow, and tape forms can be fabricated by several processing techniques well known in industries such as autoclave processing, compression molding, or filament winding.

CONCLUSIONS

This article describes the synthesis of monomers and polymers, properties, and applications with an emphasis on the development and current trends of polybenzimidazoles since 1985. The current trends in polybenzimidazole research and development include molecular composites, catalyst support, metal sorbents, and PBI blends with various high performance polymers.

REFERENCES

1. Polymer Resists Heat, *C & E News*, November 28, 1960.
2. Vogel, H. A.; Marvel, C. S. *J. Polym. Sci.* **1961**, *50*, 511.
3. Lee, Stoffey, Neville *New Linear Polymers*; McGraw-Hill: New York, 1967; Chapter 9.
4. Jones, J. I. *Macromol. Sci.* **1968**, *C2*, 303.
5. Frazer, A. H. *High Temperature Resistant Polymers*; Interscience: New York, 1968, pp 138.
6. Levine, H. H. *Encycl. Polym. Sci. Technol.* **1969**, *11*, 188.
7. Korshak, V. V. *Heat Resistance Polymers*; I Zdatel'stvo "Nauka,": Moscow, 1969, Israel translation, Keter: Jerusalem, 1971; pp 244–248.
8. Koton, M. M. In *Advances in Macromolecular Chemistry*; Pasika, W. M., Ed.; Academic: 1970; 2, 175.
9. Critchley, J. P. *Progress in Polymer Science*; Pergamon: 1970; Vol. 2, p 58.
10. Korshak, V. V.; Teplyakov, M. M. *J. Macromol. Sci. Revs. Macromol. Chem.* **1971**, *C5*(2), 409.
11. Cassidy, P. E. *Thermally Stable Polymers*; Marcel Dekker: NY and Basel, 1980; p 163.
12. Neuse, E. W. *Advances in Polym. Sci.* **1982**, *47*, 1.
13. Buckley, A.; Stuetz, D.; Serad, G. A. *Encyclo. Polym. Sci. Eng.*; Kroschwitz, J. I., Ed.; New York, NY 1987; 11, 572–601.
14. Preston, J.; Black, W. B. *J. Polym. Sci.* **1965**, *B3*(10), 845; *Am. Chem., Soc. Div. Polym. Chem., Preprints* **1965**, *6*(2), 757.
15. Sousa, J. A. et al. *Natick Tech. Rpt. TR-80/013*, 1980.
16. Ramirez, J. E.; Dwigging, C. F. *Converting and Packaging* **1985**, *1*(2), 12–14.

Polybenzimidazoles

See: *Condensation Polymers (Using Phosphorus Pentoxide/Methanesulfonic Acid)*
Phosphorus-Containing Polymers (From Dicarboxylic Acids with C-P Bonds)
Polybenzimidazoles (Overview)
Polyimide Blends
Poly(terephthalic acid anhydride) (as a Latent Monomer)

Polybenzoazoles

See: *Heteroaromatic Polymers (Five-Membered Rings in Main Chain)*

POLY(BENZOBISTHIAZOLE-IMIDE)S

Kunio Kimura
Faculty of Environmental Science and Technology
Okayama University

Aromatic heterocyclic rodlike polymers, such as polybenzothiazoles and polybenzoxazoles, called polyazoles, have been of interest because of their excellent mechanical properties and their relatively good solubilities. Polyazoles can be processed from strong acid solutions such as methanesulfonic acid, 95% sulfuric acid, or polyphosphoric acid in the fully cyclized forms into highly ordered fibers. The heterocyclic units enhance the solubilities of stiff polymers by protonation.[1,2]

Combining the imide and azole units could be of interest in producing new processable high-performance polymers possibly with improved solubilities compared with polyimides and perhaps excellent mechanical properties and thermal stabilities.

Here, the synthesis and characterization of poly[(benzol-[1,2-d;4,5-d']bisthiazole-imide)]s containing a direct attachment of the imide unit and the thiazole unit are described, giving particular attention to the preparation method, solubilities, and thermal properties.

PREPARATION OF POLY(BENZOBISTHIAZOLE-IMIDE)S

Poly[(benzo[1,2-d;4,5-d']bisthiazole-imide)]s [(**Figure 1 (IV)**)] are prepared by the two-step polymerization procedure involving ring-opening polyaddition and subsequent cyclodehydration from 2,6-diaminobenzo[1,2-d;4,5-d']bisthiazole [Figure 1 (**I**)] and aromatic tetracarboxylic dianhydrides [Figure 1 (**II**)]. The isolable intermediates, poly-[(benzo[1,2-d;4,5-d']bisthiazole-amic acid)]s [Figure 1 (**III**)], are obtained in *N*-methyl-2-pyrrolidone (NMP) at 25°C and converted into poly[(benzo[1,2-d;4,5-d']bisthiazoleimide)]s by heat treatment at 250°C under reduced pressure with no side reactions.

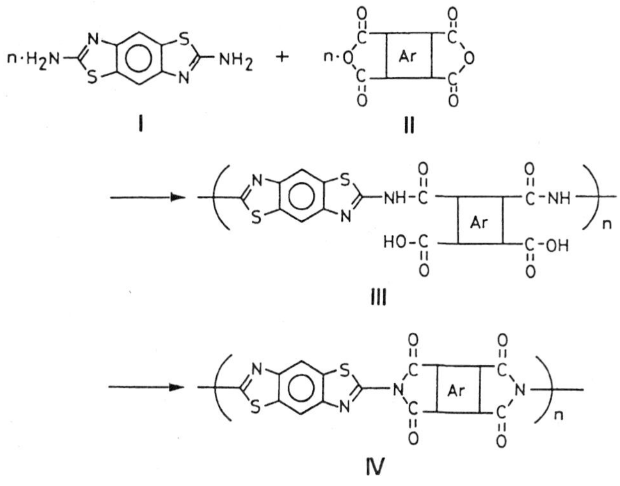

FIGURE 1. Two-step polymerization procedure for obtaining poly(benzobisthiazole-imide)s (IV).

PROPERTIES OF POLY(BENZOBISTHIAZOLE-IMIDE)S

The results indicate that Structure **IV** in Figure 1 has excellent thermal stability, that is, 5% loss of weight is found from 453°C to 516°C in air and from 468°C to 519°C in nitrogen. Furthermore, all poly[(benzo[1,2-d;4,5-d']bisthiazole-imide)]s have high char yields at 550°C in both air and nitrogen. It is concluded that these polymers are a new class having excellent thermal stability and solubility.

REFERENCES

1. Cassidy, P. E. *Thermal Stable Polymers*; Marcel Dekker: New York, 1980.
2. Wolfe, J. F.; Loo, B. H.; Arnold, F. E. *Macromolecules* **1981**, *14*, 915.

Polybenzothiazoles

See: *Condensation Polymers (Synthesis from Silylated Diamines)*
Condensation Polymers (Using Phosphorus Pentoxide/Methanesulfonic Acid)
Heteroaromatic Polymers (Five-Membered Rings in Main Chain)
Poly(benzobisthiazole-imide)s
Poly(terephthalic acid anhydride) (as a Latent Monomer)
Rigid-Rod Polybenzoxazoles and Polybenzothiazoles (PBZT and PBO)

Polybenzoxazinone

See: *Condensation Polymers (Using Phosphorus Pentoxide/Methanesulfonic Acid)*

Polybenzoxazoles

See: *Carbonylation Polymerization (Palladium-Catalyzed)*
Condensation Polymers (Synthesis from Silylated Diamines)
Condensation Polymers (Using Phosphorus Pentoxide/Methanesulfonic Acid)
Heteroaromatic Polymers (Five-Membered Rings in Main Chain)
High Pressure Polycondensation (for High Temperature Polymers)
Poly(terephthalic acid anhydride) (as a Latent Monomer)
Rigid-Rod Polybenzoxazoles and Polybenzothiazoles (PBZT and PBO)

Poly(γ-benzyl L-glutamate)

See: *Liquid Crystalline Polymers (Polypeptides)*

Polybenzyls

See: *Aromatic Hydrocarbon-Based Polymers*

Polybetaines

See: *Polyampholytes (Overview)*
Polyzwitterions (Overview)

Polybibenzoates

See: *Liquid Crystalline Polymers (Polyesters of Bibenzoic Acid)*

POLYBITHIOPHENES (Advanced Derivatives)

Mario Leclerc* and Karim Faïd
Départment of Chimie
Université de Montréal

The multiple potential technological applications of conducting polymers have been and remain the main driving force in the development of this class of materials, with great promises in various fields including rechargeable batteries, molecular electronics, and linear or nonlinear optics. Polythiophene and its derivatives are among the most investigated materials in this research area with some of this work summarized in different reviews.[1-5]

Unsubstituted polythiophene is insoluble and infusible making its study and utilization quite difficult. Various derivatives have been produced by the straightforward modification of the starting thiophene monomer by covalently grafting functional groups at the β-positions.

One approach to the problem of tailoring the physical properties of the polythiophene derivatives was the design of novel structures with an adequate choice of the nature and positions of the substituents, various symmetrically and asymmetrically disubstituted bithiophenes and terthiophenes.[6-8] For instance, the design of "quantum dots" along the main polymeric chains by an insertion of adequate potential barriers can be of great interest for the control of the electronic properties of the resultant materials.

*Author to whom correspondence should be addressed.

In this paper, we want to focus on advanced polythiophene derivatives produced mainly by the chemical polymerization of new bithiophene derivatives, describing both their syntheses and properties. Novel developments in the charge carrier generation and transport in these materials will be presented and discussed.[9,10]

PREPARATION

Monomer Syntheses

Symmetrically and asymmetrically disubstituted bithiophene (such as 4,4'-dibutoxy-2,2'-bithiophene, 4-butoxy-4'-decyl-2,2'-bithiophene, 4-bromo-4'-octyloxy-2,2'-bithiophene, 3,3'-dibutoxy-2,2'-bithiophene, and 3-butoxy-3'-decyl-2,2'-bithiophene) have been synthesized through the selective modification of 3,3'-dibromo-2,2'-bithiophene and 4,4'-dibromo-2,2'-bithiophene according to **Scheme I**.[7]

Polymerization

All polymers were obtained by chemical oxidation of the corresponding dimers using iron trichloride according to a procedure similar to that reported in the literature.[11]

PROPERTIES

It appears that the careful molecular design of the starting bithiophene dimers, through the precise control of the nature and the position of the substituents, allows a fine tuning of the physical properties of the polythiophene derivatives. This synthetic approach is extremely versatile and in addition to being useful in the design of advanced polythiophenes can be also used to produce a whole new category of materials where the charge transport properties can be adjusted at will through the confinement and localization of radical-cations by designing "*quantum discontinuities*" (quantum wells or dots) along the backbone by alternating electrodonating and electrowithdrawing substituents.

These localization phenomena may also be very helpful for the development of polymeric light-emitting diodes or magnetic polymeric materials.[12–15] The enhancement of the polaron population available for the radiative decay process through the molecular design of quantum dots along the polymer backbone is highly desirable.

Similarly, the development of magnetic organic materials is based mainly on the utilization of two types of building blocks, a spin-containing unit, and a ferromagnetic coupling unit to form a polaronic ferromagnet. In these materials, the stabilization of radical cations is crucial to ensure the formation of high-spin structures by disfavoring the bipolaron formation. The design of materials which allows the confinement and localization of polarons, and thus the reduction of the recombination processes, may lead to higher spin concentration and improved magnetic properties.

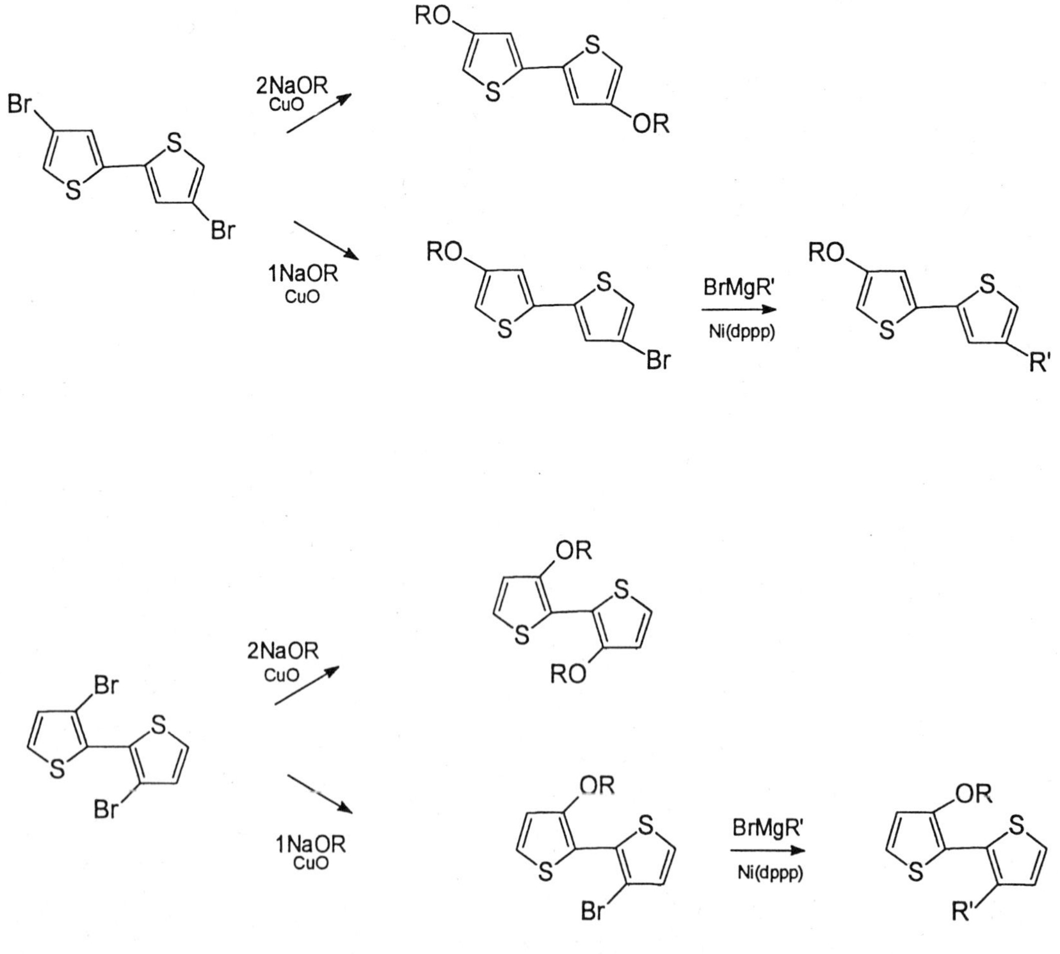

I

REFERENCES

1. Gustafsson, G.; Inganäs, O.; Salaneck, W. R.; Laakso, J.; Loponen, M.; Taka, T.; Osterholm, J. E.; Stubb, H.; Hjertberg, T. In *Conjugated Polymers*; Brédas, J. L.; Silbey, R.; Eds.; Kluwer Academic: Dordrecht, The Netherlands, 1991; pp 315–363.
2. Reynolds, J. R.; Baker, C. K.; Jolly, C. A.; Poropatic, P. A.; Jose, J. P. *Conductive Polymers and Plastics*; Margolis, J. M., Ed.; Chapman and Hall; 1989; Chapter 1.
3. Naarmann, H.; Theophilou, N. *Electroresponsive Molecular and Polymeric Systems*; Skotheim, T. A., Ed.; Marcel Dekker: New York, 1988; Chapter 1.
4. Tourillon, G. *Handbook of Conducting Polymers*; Skotheim, T. A., Ed.; Marcel Dekker; New York, 1986.
5. Roncali, J. *Chem. Rev.* **1992**, *92*, 711.
6. Leclerc, M.; Daoust, G. *J. Chem. Soc., Chem. Commun.* **1990**, 273.
7. Faïd, K.; Cloutier, R.; Leclerc, M. *Macromolecules* **1993**, *26*, 2501.
8. Faïd, K.; Leclerc, M. *J. Chem. Soc., Chem. Commun.* **1993**, 962.
9. Faïd, K.; Leclerc, M. *Chem. Mater.* **1994**, *6*, 107.
10. Faïd, K.; Leclerc, M.; Nguyen, M.; Diaz, A. *Macromolecules* **1995**, *28*, 284.
11. Sugimoto, R.; Takeda, S.; Gu, H. B.; Yoshino, K. *Chem. Express* **1986**, *1*, 635.
12. Meyers, F.; Heeger, A. J.; Brédas, J. L. *J. Phys. Chem.* **1992**, *97*, 2750.
13. Burn, P. L.; Kraft, A.; Baigent, D. R.; Bradley, D. D. C.; Brown, A. R.; Friend, R. H.; Gymer, R. W.; Holmes, A. B.; Jackson, R. W. *J. Amer. Chem. Soc.* **1993**, *115*, 10117.
14. Swanson, L. S.; Shinar, J.; Brown, A. R.; Bradley, D. D. C.; Friend, R. H.; Burn, P. L.; Kraft, A.; Holmes, A. B. *Phys. Rev. B* **1992**, *23*, 15072.
15. Murray, M. M.; Kaszynski, P.; Kaisaki, D. A.; Chang, W.; Dougherty, D. A. *J. Am. Chem. Soc.* **1994**, *116*, 8152.

POLYBIURETS

Keisuke Kurita
Department of Industrial Chemistry
Faculty of Engineering
Seikei University

Polyureas are most easily synthesized by polyaddition of diisocyanates with diamines. They have useful properties, but solubility is generally poor because of strong intermolecular hydrogen bonding between the urea linkages. As a urea-relating connecting block, the biuret linkage is interesting in view of the development of new types of polymeric materials with advanced functions.

PROPERTIES

Polybiurets prepared from chloroformyl isocyanate and *N,N'*-bis(trimethylsilyl)diamines were semicrystalline materials with a low degree of crystallinity (< 30%) even after annealing. Aliphatic polybiurets showed glass transition temperatures of 60–100°C in a level similar to polyamides and polyureas. The melting temperatures were, however, 120–190°C, much lower than those of the corresponding polyamides and polyureas.

Polyaddition of benzyloxyamine and diisocyanates produced white fibrous materials, which were readily soluble in common organic solvents such as tetrahydrofuran, dioxane and pyridine, in addition to DMAc and DMSO. Debenzylated derivatives, poly(*N*-hydroxybiuret)s, showed similar high solubility. This is in sharp contrast to the poor solubility of polyureas.

APPLICATIONS

Two procedures for the synthesis of polybiurets have been developed recently, and the interesting properties of polybiurets are being disclosed. Unlike ordinary polyureas, they are characterized by low melting temperatures and excellent solubility in organic solvents and will find various practical applications because of the easily processable polar structures.

POLYBOROSILOXANES

Yoshimoto Abe* and Takahiro Gunji
Department of Industrial Chemistry
Faculty of Science and Technology
Science University of Tokyo

Much attention has been paid to inorganic polymers during the scientific and technologic development of nuclear energy and space research. The objective is to obtain more thermally stable polymers than polysiloxanes in order to withstand ultimate conditions.

Polyborosiloxanes have been an exploration target because of their thermodynamically stable backbone linkage, despite their high susceptibility to hydrolysis. They cannot be used themselves as high-temperature polymers but show interesting properties and can be used as precursors for ceramic binders, additives and coatings. They are also investigated in sol-gel chemistry as precursor sols and/or borosilicate glasses.

The reaction of dimethyl- or diphenyldichlorosilane[1] with boric acid in the molar ratio of Si/B = 3/2 produces a polymer (**Equation 1**). Polyborodiphenylsiloxane also is prepared by a reaction of diphenylsilanediol with boric acid, where dibutyl ether and toluene are used as a solvent to remove the water formed by azeotropic distillation to give white solids (Equation 1).

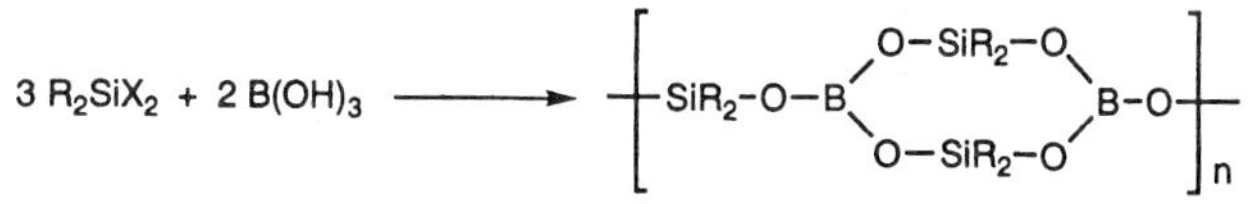

R = Me (X = Cl) and Ph (X = Cl or OH) **1**

Polyborosiloxanes are used as precursors for ceramics and glasses, which are prepared by the reaction of diphenylsilanediol with boric acid or by a sol-gel process.

PROPERTIES

Usually polyborosiloxanes obtained by polycondensation of difunctional silicon compounds and boric acid or borates are relatively low-molecular-weight polymers of M_n (from several hundred to a thousand). They are liquids, highly viscous liquids or solids, which are soluble in alcohols, acetone, THF, dioxane, toluene and pyrrolidones.

---

*Author to whom correspondence should be addressed.

As is suggested from the ionic character of borosiloxane linkage, it easily undergoes nucleophilic attack by water to form boric acid. This creates a problem because polyborodimethylsiloxanes are not practically used by themselves in contact with air moisture.

An outstanding physical property of polyborosiloxanes is their behavior as plastics or elastomers and/or rheology ("bouncing putty"), depending on the molar ratio Si/B (less than 100) and molecular weight.[3–6] "Bouncing putty" demonstrates contradictory behavior: it flows to change its shape as a plastic and also bounces like a golf ball when force is applied. Properties are characterized as polymers prepared by heterofunctional polycondensation of linear low molecular weight ω-ω'-hydroxy–or acetoxypolysiloxanes and boric acid or trialkoxyborane.

APPLICATIONS

High susceptibility to hydrolysis of polyborosiloxanes makes it difficult to use them as industrial materials, but they can be made more stable elastomers by vulcanization of the composition Si/B higher than 100 and a molecular weight around $3–5 \times 10^5$. Since the elastomers have excellent electrical properties, they can be used like polysiloxane rubbers in the electrical industry.

Polyborosiloxanes have been regarded as precursors for borosilicates and ceramics rather than organic polymers because the precursor method, as well as the sol-gel process, provides the materials for bulk bodies, powders, thin films and fibers with high purity and molecular homogeneity under mild conditions.

Polyborosiloxanes can be used as binders, coatings and heat-resistant paints.[2] When polyborodiphenylsiloxane is heated to 1500–1700°C under N_2 or Ar stream, β-SiC and B_4C are formed,[1] which means they are a precursor for ceramics and are used as binders to prepare SiC and Si_3N_4 sintered bodies.[7,8]

Polyborosiloxanes have a potential for highly functional coatings such as heat-resistant paints for farinfrared ray irradiators, or exhaust of cars and electrical wires.[9,10]

REFERENCES

1. Yajima, S.; Hayashi, J.; Okamura, K. *Nature* 1977, *266*, 521.
2. Yajima, S. *Ceramic Bull.* 1983, *62*, 893.
3. Wick, M. *Kunststoffe* 1960, *50*, 433.
4. McGregor, R. R.; Warrick, E. L. U.S. Patent 2 431 878, 1947.
5. McGregor, R. R. *Silicones and Their Uses*; McGraw-Hill: New York, 1954; p 186.
6. Martin, F. S. U.S. Patent 2 609 201, 1952.
7. Yajima, S.; Shishido, T.; Hamada, M. *Nature* 1977, *266*, 522.
8. Yajima, S.; Okamura, K.; Hasegawa, Y. *Ger. Offen.* 2 846 529.
9. Tomida, K.; Miyashita, N.; Hashimoto, H. *Showa Electric Wire & Cable LTD's Review* 1989, *39*, 321.
10. Kiuchi, Y.; Sukegawa, N. et al. *ibid.*, 1991, *41*, 79.

Polybutadiene

See: *Acrylonitrile-Butadiene-Styrene*
Butadiene Polymerization (Supported Lanthanide Catalysts)
Butadiene Polymerization, Neodymium-Catalyzed
Conducting Polymers (Nonconjugated)
Interpenetrating Polymer Networks, Rubber-Based
Maleic Anhydride-Polybutadiene, Ene Reaction
Polybutadienes
Rare Earth-Initiated Polymerization (of Dienes)
Supported Catalysts (Lewis Acid and Ziegler-Natta)

Poly(butadiene-acrylonitrile)

See: *Unsaturated Polyester Resins (Toughened with Liquid Rubber)*

Poly(butadiene-co-isoprene)

See: *Rare-Earth Initiated Polymerization (of Dienes)*

POLYBUTADIENES

Patricia Dreyfuss

Polybutadienes [9003-17-2] are obtained primarily from the interaction of 1,3-butadiene with a variety of initiators and catalysts.[1–3] Currently available polybutadiene materials range from general purpose noncrystallizing rubbers,[5–8] to highly crystalline plastics,[1,3,9] to liquid, castable resins.[10–14]

MICROSTRUCTURE

When 1,3-butadiene is polymerized, the resulting polymer contains double bonds, which may exist as any one of three isomers: *cis*-1,4 [40022-03-5], *trans*-1,4 [40022-02-4] and vinyl [26160-94-5] (**Scheme I**).

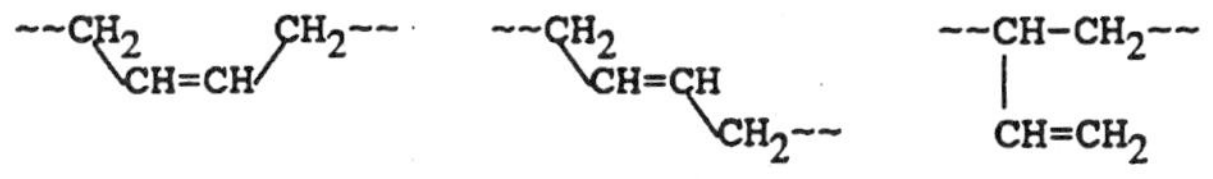

I

cis-1,4 *trans*-1,4 vinyl (1,2)

SYNTHESIS

Polybutadienes are prepared using free radical, anionic and Ziegler–Natta initiating systems. Emulsion or solution methods of polymerization are used. The method determines the microstructure and molecular weight of the product formed.

Free radical polymerization of butadiene in solution gives polymers of low molecular weight, 1000–4000 g/mol [25038-44-2].[1,11]

Free radical polymerization of butadiene in emulsion gives amorphous polymers [25038-44-2] and copolymers with high *trans*-1,4 content and high molecular weight. The microstructure produced depends on temperature.

PROPERTIES AND APPLICATIONS

Properties of polybutadiene vary greatly with microstructure.[4,9] With the exception of silicone rubbers, 1,4-polybutadienes have the lowest glass transition temperatures of all commercial elastomers and offer unusually good performance at temperatures at least as low as –80°C. Polymers of randomly mixed microstructure (40–85% *cis*) are amorphous and have intermediate properties. Polymers of uniform microstructure (98–100% *cis* or *trans*) are crystalline and the melting temperature is a function of microstructure. Applications depend on properties.

High *cis*-1,4-Polybutadiene Rubber

High *cis*-1,4-polybutadiene has useful physical properties. It crystallizes upon being stretched and thus has high tensile strength. It has excellent flexibility at low temperatures. It possesses low hysteresis, excellent dynamic properties, and outstanding resilience, tear strength and abrasion resistance. But compared to natural rubber it has low tack and green strength.

High *trans*-1,4-Polybutadiene

High *trans*-1,4-polybutadiene is a tough crystalline elastomer with good hardness, thermoplasticity and poor hysteresis. At the highest *trans* content, the polymer is difficult to stabilize and dissolve. It has two melting temperatures that depend on the % of *trans* moiety, which in turn is determined by the catalyst system used.

High Vinyl Polybutadienes

These 1,2 addition products consists of three main groups: syndiotactic, isotactic and atactic. High molecular weight solids and low molecular weight liquid resins are known. Each structure produces polymers with unique physical, mechanical and rheological properties.

Syndiotactic polybutadiene is a unique material that combines the properties of plastic and rubber.[1,15]

High (99%) isotactic polybutadiene has been known since 1965, when Natta reported crystallographic data and other properties of the crystalline polymer.[16] Its properties have not excited sufficient interest to justify major studies or commercial development.

Atactic (amorphous) high 1,2-polybutadiene has excited considerable interest in recent years.[1] Its low hysteresis and good wet grip/rolling resistance characteristics make this polymer a potential candidate for tire tread formulations with lower tire energy loss and better fuel economy.

Medium Vinyl Polybutadiene

Medium vinyl polybutadienes have approximately 35–55% vinyl moieties.[3] Their T_gs are in the range –50 to –70°C and they have wear, wet-skid, heat buildup and rolling resistance properties, which are similar to those for SBR prepared by emulsion polymerization.[3]

Liquid Polybutadienes

High vinyl (>90%) liquid polybutadienes have been used to make thermosetting resins with excellent electrical properties, high thermal distortion temperatures, low water absorption and ready cure with peroxide catalysts. Solvent resistance to aromatic and aliphatic hydrocarbons is poor, but solvent resistance to oxygenated organic solvents is good.

Liquid polybutadienes with functional endgroups undergo chemical reactions typical of the functional group. For example, polybutadienes α,ω-hydroxyl endgroups can be fabricated in polyurethanes by reaction of the diols with isocyanates or into polyesters by reaction with some acids, acid chlorides or acid anhydrides.

Other Applications

A polybutadiene with 60% *trans*-1,4, 20% *cis*1-4, and 20% vinyl configuration is used in the manufacture of ABS polymers.[4] Polybutadiene also is used in the manufacture of impact-modified polystyrene.

REFERENCES

1. Halasa, A. F.; Massie, J. M. in *Kirk-Othmer Encyclopedia of Chem. Technol.*, 4th ed.; John Wiley & Sons: New York, VCH: New York; 1993; 8, 1031.
2. Dreyfuss, P. *International Encyclopedia of Composites* **1991**, 4, 136.
3. Haynes, A. C. *Developments in Diene-Based Rubbers, in Handbook of Elastomers, New Developments and Technology;* Bhowmich, A. K.; Stephens, H. L., Eds.; Marcel Dekker: New York, 1988; Chapter 26, 761–774.
4. Ulrich, H. *Introduction to Industrial Polymers,* 2nd ed.; Hanser: Munich, 1993; 48, 63–64, 180.
5. Whitby, G. S., Ed.; *Synthetic Rubber,* Wiley: New York, 1954.
6. Kuzma, L. J. in *Rubber Technology,* 3rd ed.; Morton, M., Ed.; Van Nostrand Rheinhold: New York, 1987; 235–259.
7. Hammer, R. S.; Railsback, H. E. in *Rubber Technology,* 2nd ed.; Morton, M., Ed.; Van Nostrand Rheinhold: New York, 1973; 199–219.
8. Svedlik, J. F. in *The Vanderbilt Rubber Handbook;* Winspear, G., Ed., Vanderbilt, R. T.; Norwalk, CT, 1968; 89–98.
9. Stempel, G. H. in *Polymer Handbook,* 2nd ed.; Brandrup, J.; Immergut, E. H., Eds.; Wiley: New York, 1975; V-1–V-5.
10. Stander, M. in *Handbook of Composites;* Lubin G., ed.; Van Nostrand Reinhold: New York, 1982; 50–56.
11. Edwards, D. C. Chapter 4, *Liquid Rubber,* in *Handbook of Elastomers, New Developments and Technology;* Bhowmich, A. K.; Stephens, H. L., Eds.; Marcel Dekker: New York, 1988; 101–140.
12. ARCO Chemical Company, "Poly BS Resins," General Bulletin, New York, January 1978.
13. ARCO Chemical Company, "Poly BD Liquid Resins in Urethane Elastomers," Product Bulletin BD-3, New York, October 1974.
14. BF Goodrich Chemical Company, "Hycar Elastomers," Product Description, Cleveland.
15. Bhowmick, A. K.; Stephens, H. L. *Additional Types of Thermoplastic Elastomers,* in *Handbook of Elastomers, New Developments and Technology,* Chapter 12; Bhowmich, A. K.; Stephens, H. L., Eds.; Marcel Dekker: New York, 1988; 411–442.
16. Natta, G. *Science* **1965**, *147*, 269.

Poly(1-butene)

See: Monomer-Isomerization Polymerization (2-Butene with Ziegler–Natta Catalysis)
Nylon 6 (Depolymerization)
Polyolefins (Overview)

Poly(butylene naphthalate)

See: Poly(alkylene naphthalate)

Poly(butylene terephthalate)

See: *Poly(butylene terephthalate) (Annealing Behavior, Thermal Analysis)*
Poly(butylene terephthalate)/Polycarbonate Blends
Spheres, Poly(butylene terephthalate)

POLY(BUTYLENE TEREPHTHALATE) (Annealing Behavior, Thermal Analysis)

Richard E. Robertson* and Junkyung Kim
Macromolecular Science and Engineering Center and Department of Materials Science and Engineering
The University of Michigan

Mark E. Nichols
Ford Research Laboratory

Scanning calorimetry has been used extensively to examine changes in the thermal behavior of polymers as a result of annealing below the melting temperature.[1] A typical thermogram, obtained from well below the annealing temperature to above the melting point, is often perplexing because of the occurrence of multiple melting peaks.

Poly(butylene terephthalate) (PBT) is a polymer that often displays multiple peaks.[2–5] Depending on its thermal history, PBT can exhibit up to three apparent melting peaks during thermal analysis. In addition, two different types of spherulites can be present: one having the usual extinction pattern when viewed between crossed polarizers, with extinction parallel with the axes of the polarizers, and the other having the extinction pattern rotated by 45°C.[3,4] Because the latter unusual spherulites can be transformed to usual ones by annealing near the melting temperature, an association of these spherulites with specific melting peaks seems plausible. But the association cannot be more than partially true, because Yeh and Runt[2] found, from combined light scattering and thermal analysis studies, that the usual spherulites contribute to each endothermic peak resulting from annealing. On the other hand, an explanation for the multiplicity of endothermic peaks in terms of distinct structures is unnecessary because as many as three melting peaks can occur during thermal analysis of polymer specimens containing only one type of spherulite.[6] Such behavior can occur if the crystals melt over a range of temperatures, as usually happens. For example, simultaneous melting and recrystallization of a distribution of melting points explained the occurrence of up to three melting endotherms during thermal analysis of a series of PBT specimens that had been cooled at various constant rates from the melt.[7]

A feature of DSC thermograms more difficult to understand is the occurrence after annealing of a second peak below the original peak that appears to grow with annealing at the expense of the original peak. This behavior was found with PBT when annealed for 24 h at temperatures from 130° to 198°C and with poly(phenylene sulfide) (PPS) when annealed for times up to 3996 min at 250°C.[9] Growth of a lower melting peak at the expense of an upper (original) peak during annealing is problematic. In the absence of a chemical change, such behavior is generally proscribed by thermodynamics. When annealed near but below the melting temperature, a crystal generally cannot transform from a higher melting state to a lower one.

Two possible explanations for the apparent growth of a lower melting peak at the expense of a higher one during the anneal are that it is not real but arises as a misinterpretation of the DSC thermogram, or that it is real and is coupled to crystallization of previously noncrystalline material.[8] To the extent that the latter is correct, annealing results in a "deperfection" of some of the crystals.

*Author to whom correspondence should be addressed.

REFERENCES

1. Wunderlich, B. *Macromolecular Physics*; Academic: New York, 1976; Vol. 2.
2. Yeh, J. T.; Runt, J. *J. Polym. Sci., Part B, Polym. Phys.* **1989**, *27*, 1543.
3. Stein, R. S.; Misra, A. *J. Polym. Sci., Polym. Phys. Ed.* **1980**, *18*, 327:
4. Ludwig, H. J.; Eyerer, P. *Polym. Eng. Sci.* **1988**, *28*, 143.
5. Hobbs, S. Y.; Pratt, C. F. *Polymer* **1975**, *16*, 462.
6. Nichols, M. E.; Robertson, R. E. *J. Polym. Sci., Part B, Polym. Phys. Ed.* **1992**, *30*, 305.
7. Nichols, M. E.; Robertson, R. E. *J. Polym. Sci., Part B, Polym. Phys. Ed.* **1992**, *30*, 755.
8. Kim, J. K.; Nichols, M. E.; Robertson, R. E. *J. Polym. Sci., Part B, Polym. Phys.* **1994**, *32*, 1607.
9. Chung, J. S.; Cebe, P. *J. Polym. Sci., Part B, Polym. Phys.* **1992**, *30*, 163.

POLY(BUTYLENE TEREPHTHALATE)/ POLYCARBONATE BLENDS

Masami Okamoto** and Tomoko Okuyama
Toyobo Research Center

Takashi Inoue
Department of Organic and Polymeric Materials
Tokyo Institute of Technology

The poly(butylene terephthalate) (PBT)/bisphenol-A-polycarbonate (PC) blend was developed by General Electric Co. in 1979. The present consumption of the PBT/PC blend in the world is about 20,000 tons per year, and the application is mostly automobile exterior parts for which high heat resistance, excellent impact strength, and chemical resistance are required. It is said that, for the design of an excellent blend the key is careful control of the exchange reaction, i.e., ester-exchange between PC and PBT should go on during melt processing at an optimum level.[1] According to our recent studies,[2] the exchange reaction is not so important, but the mixing and demixing process under shear field in the extruder is primarily important for structure development.

STRUCTURE DEVELOPMENT IN PBT/PC BLEND

There are many experimental studies on the two-phase morphology of PBT/PC blend.[3–7] Among them, Hobbs et al. reported that a highly interconnected two-phase structure

**Author to whom correspondence should be addressed.
Present affiliation: Advanced Polymeric Materials Engineering, Toyota Technological Institute.

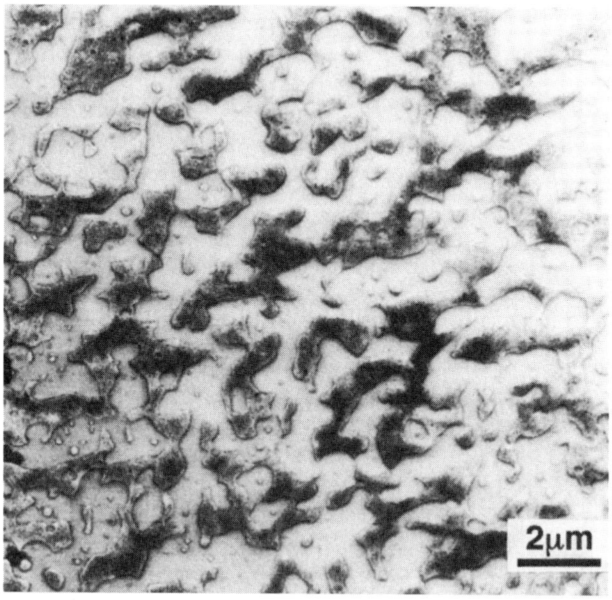

FIGURE 1. Transmission electron micrograph (RuO_4) of a melt compounded and quenched 50/50 PBT/cPC blend.

developed in melt-mixed blends.[4] It suggested to us a possibility of structure formation via spinodal decomposition.

LCST Phase Diagram and Demixing Mechanism

Figure 1 is transmission electron microscopy (TEM) photograph of a 50/50 PBT ($M_n = 47 \times 10^3$; Toyobo Plastic Division)/PC-copolymer (cPC; $M_n = 24 \times 10^3$, $T_g = 192°C$, APEC HTKU-9360, Bayer product) blend. One sees a high connectivity of both phases and regular spacing of the phases. That is, the results by Hobbs et al.[6] is nicely reproduced here. The two-phase structure with unique periodicity and phase connectivity is one of the hallmarks of spinodal decomposition.

Thus, both PBT/PC and PBT/cPC systems exhibit LCST phase behavior and a plausible scenario is presented for the development of bicontinuous structure in the melt-extruded blends in terms of the spinodal decomposition.

APPLICATIONS

Listed in **Table 1** are typical properties of two PBT/PC/rubber alloys; one has a PBT matrix and the other has a PC matrix. Both alloys have excellent toughness, especially at very low temperature, high heat resistance, and resistance to gasoline and other hydrocarbons. Superior flowability and processability make it possible to mold very large parts. The potential applications are for car bumpers, instrument housings, cable connectors, protective helmets, ski boot parts, and castors for furniture.

REFERENCES

1. Bertolucci, M. D.; DeLaney, D. E. *Soc. Plant. Eng. NATEC* September 20–27, 1983.
2. Okamoto, M.; Inoue, T. *Polymer* **1994**, *35*, 257.
3. Delimoy, D.; Bailly, C.; Devaux, J.; Legras, R. *Polym. Eng. Sci.* **1988**, *28*, 104.
4. Hobbs, S. Y.; Dekkers, J. M. E.; Watkins, V. H. *J. Mater. Sci.* **1988**, *23*, 1219.
5. Hobbs, S. Y.; Watkins, V. H.; Bendler, J. T. *Polymer* **1990**, *31*, 1663.
6. Wahrmund, D. C.; Paul, D. R.; Barlow, J. W. *J. Appl. Polym. Sci.* **1987**, *22*, 2155.
7. Devaux, J.; Godard, P.; Mercier, J. P. *Polym. Eng. Sci.* **1982**, *22*, 229.

POLY(*tert*-BUTYLSTYRENE)

Jimmy W. Mays,* Jian Zhou, and William K. Nonidez
Department of Chemistry
University of Alabama at Birmingham

Nikos Hadjichristidis
Department of Chemistry
University of Athens

The synthesis of poly(*m-tert*-butylstyrene) was first reported in 1946.[1] Eight years later polymers of *p-tert*-butylstyrene were reported.[2] Much of the early interest in these polymers was due to their higher heat distortion temperatures as compared with

*Author to whom correspondence should be addressed.

TABLE 1. Typical Properties of PBT/PC/Rubber Alloys

Properties	Units	ASTM method	A–101 (PBTmatrix)	A–102 (PCmatrix)
Tensile strength	MPa	D–638	52	57
Ultimate elongation	%	D–638	19	17
Flexural strength	MPa	D–790	74	83
Flexural modulus	GPa	D–790	2.1	2.1
Impact strength +23 °C	J/m	D–256	530	600
notched −40 °C	J/m		120	180
Heat distortion 1.81 MPa	°C	D–648	83	127
temperature 0.45 MPa	°C		151	146
Falling dart +23 °C	J	D–3029	45	37
impact −40 °C	J		42	35
Specific gravity		D–792	1.20	1.18
Mold shrinkage				
flow	mm/mm		0.01	0.008
cross flow	mm/mm		0.013	0.01

polystyrene.[2,3] Today's commercial poly(*tert*-butylstyrene) (PtBS) products are mixed (95% *p*, 5% *m*) isomer materials. In addition to their high glass transition temperatures (T_g) of about 130–140°C,[4,5] other advantages include low vapor pressure and high flash point of the monomer and low polymerization shrinkage (7% versus 17% for polystyrene).[6] Applications include use as tackifiers, as viscosity-index improvers in motor oils, and as components in thermoplastic elastomers and associative thickeners.

SYNTHESIS

tert-Butylstyrene (tBS) is an extremely versatile monomer in that addition polymerizations may be conducted thermally,[2] by UV radiation,[1] and by free-radical initiators,[2] Ziegler–Natta catalysts,[7,8] anionic initiators,[9] and cationic initiators.[10,11] Emulsion polymerization of tBS has also been reported.[12,13] Thus, tBS displays a versatility in polymerizations that, as might be expected, rivals that of styrene.

Some of the most interesting synthetic work to date with PtBS has involved block copolymers, which were first synthesized by Fetters and co-workers.[9]

PROPERTIES

PtBS, as produced by nonstereospecific polymerization mechanisms, is amorphous and stereoirregular (55% syndiotactic, 15% isotactic, and 30% heterotactic diads).[14] These polymers have a density of about 0.95 g/cm[3,8,14,15] and reported T_g values range from about 126°C to 148°C.[4,5,11,15–19] The more recent data on high molecular weight materials suggest a limiting T_g value of approximately 145°C. The isotactic polymer has a melting temperature of about 300°C,[8] but even highly crystalline materials are reported to be soluble in common solvents at room temperature.[19]

The dilute solution properties of PtBS indicate that this chain behaves as a typical random coil in solution, with chain flexibility slightly less than that for polystyrene due to the presence of the bulky *tert*-butyl substituent, which hinders rotation about main-chain bonds.

APPLICATIONS

Initial interest in PtBS arose due to the high heat distortion temperature of this polymer compared with PS,[2] and patents related to this issue have appeared.[3,20] Although PtBS offers this and other advantages (lower monomer volatility, higher flash point, less shrinkage on polymerization) relative to PS, its higher cost has limited commercialization of PtBS molding resins.

REFERENCES

1. Marvel, C. S.; Allen, R. E.; Overberger, C. G. *J. Am. Chem. Soc.* **1946**, *68*, 1088.
2. Ferstandig, L. L.; Butler, J. C.; Straus, A. E. *J. Am. Chem. Soc.* **1954**, *76*, 5779.
3. Levine, I. E.; Elwell, W. E. U.S. Patent 2 723 261, 1955.
4. *Polymer Handbook*, 3rd ed.; Brandrup, J.; Immergut, E. H., Eds.; Wiley-Interscience: New York, 1989.
5. Malhotra, S. L.; Lessard, P.; Blanchard, L. P. *J. Macromol. Sci., Chem.* **1981**, *A15*, 121.
6. Brighton, C. A.; Pritchard, G.; Skinner, G. A. *Styrene Polymers: Technology and Environmental Aspects*; Applied Science: London, 1979; pp 262–266.
7. Overberger, C. G.; Nozakura, S. *J. Polym. Sci., A* **1963**, *1*, 1439.
8. Saunders, F. L. *J. Polym. Sci., A-1* **1967**, *5*, 2187.
9. Fetters, L. J.; Firer, E. M.; Defauti, M. *Macromolecules* **1977**, *10*, 1200.
10. Weir, N. A.; Milkie, T. H.; Nicholas, D. *J. Appl. Polym. Sci.* **1979**, *23*, 609.
11. Kennedy, J. P.; Meguriya, N.; Kezler, B. *Macromolecules* **1991**, *24*, 6572.
12. Westby, M. J. *Coll. Polym. Sci.* **1988**, *266*, 46.
13. Satpathy, U. S.; Dunn, A. S. *Brit. Polym. J.* **1988**, *20*, 521.
14. Kucukyavuz, Z.; Kucukyavuz, S. *Eur. Polym. J.* **1978**, *14*, 867.
15. Puleo, A.; Muruganandam, N.; Paul, D. R. *J. Polym. Sci., Polym. Phys. Ed.* **1989**, *27*, 2385.
16. Hseih, D. T.; Peiffer, D. G. *J. Appl. Polym. Sci.* **1993**, *47*, 1469.
17. Rappon, M.; Chuenarm, A.; Duggal, A. J.; Gill, H.; Bhaovibul, O.; Syvitski, R. T. *Eur. Polym. J.* **1991**, *27*, 365.
18. Gustafsson, A.; Wiberg, G.; Gedde, U. W. *Polym. Sci. Eng.* **1993**, *33*, 549.
19. Rinaldi, P. L.; Tokles, M.; Hatvany, G. S.; Harwood, H. J *J. Am. Chem. Soc.* **1992**, *114*, 10651.
20. Saunders, F. L.; Richardson, E. H. U.S. Patent 3 414 647, 1968.

Polycaprolactam

See: *Nylon 6 (Depolymerization)*
Polyamide Fibers, Aliphatic (Nylon 66 and Nylon 6)
Polyamides, Lactam-Based

Polycaprolactone

See: *Controlled Drug Delivery Systems*
Poly(ε-caprolactone) (Overview)

POLY(ε-CAPROLACTONE) (Overview)

Joseph V. Koleske
Consolidated Research, Incorporated

Poly(ε-caprolactone) is the polymer of a cyclic ester, ε-caprolactone, which can be prepared by the Baeyer–Villiger reaction for the oxidation of cyclic ketones to lactones.[1] The colorless monomer readily absorbs water and, because water can act as an initiator leading to an α-hydroxy-, ω-carboxyl-terminated polymer, it should be dried before polymerization to ensure a low acid number in the final product.

In the presence of an active hydrogen initiator, preferably from hydroxyl groups, and a small amount of a suitable catalyst, ε-caprolactone will readily polymerize at temperatures of 110°C to 200°C. Suitable catalysts include dibutyltin dilaurate, stannous octanoate, and zinc octanoate. The polymerization can be depicted in a single sense as (**Equation 1**).

COMMERCIAL PRODUCTS AND END USES

Poly(ε-caprolactone) is commercially available in a variety of molecular weights that vary from ~200 to ~100,000. Molecular weight is controlled by use of dry monomer (water would act as an initiator) and addition of a specific amount of active-hydrogen initiator. Initiators are usually diols, such as 1,4-butanediol, ethylene glycol, diethylene glycol, neopentyl glycol, and similar compounds, when polycaprolactone diols are

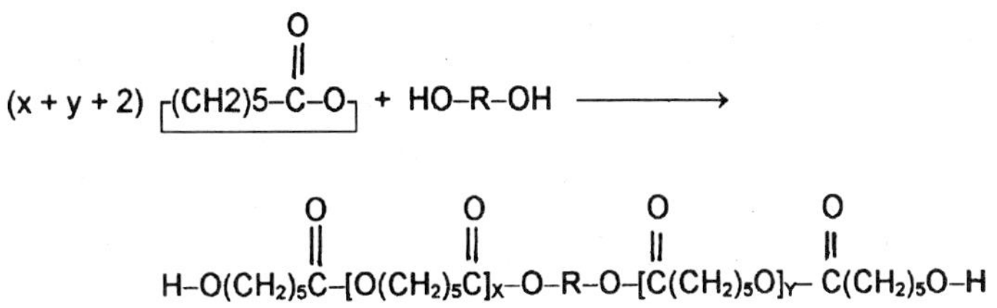

1

prepared; or triols, such as trimethylolpropane, when polycaprolactone triols are prepared.

Polycaprolactone polyols are used mainly in the manufacture of polyurethane coatings, adhesives, and shaped articles.

Polycaprolactone polyol-based polyurethanes have high tensile strength, good hydrolytic stability, good resiliency, excellent tear and cut-through resistance, low temperature flexibility, excellent abrasion resistance, and very little color when prepared from aliphatic isocyanates.

Although low molecular weight polycaprolactones range from liquids to hard waxes, the high molecular weight polymer is a strong, ductile polymer with excellent mechanical characteristics. It has a low melting point and a low glass transition temperature.[2]

Poly(ε-caprolactone) has the unique characteristic of being miscible with almost all other polymers.[3,4]

High molecular weight poly(ε-caprolactone) is usually used as an additive to other polymers to obtain special effects, but it is used as the major ingredient in many formulations. Among the various end uses, one can mention hot-melt or other adhesives,[5] orthopedic casts and splints,[6,7] low-shrink glass-reinforced styrene/polyester molding additive, as a polymeric plasticizer,[4] art and craft materials (Friendly Plastics), molding materials in toy manufacture, mold releases in phenolic compositions,[8] box-toe and heel counters in shoe manufacture, time-release drug[9] and herbicide[10,11] agents, biodegradable plastics,[12,13] and other end uses.

REFERENCES

1. Starcher, P. S.; Phillips, B. *J. Am. Chem. Soc.* **1958**, *80*, 4079.
2. Koleski, J. V.; Lundberg, R. D. *J. Polym. Part A-2* **1969**, *7*, 795.
3. Brode, G. L.; Koleske, J. V. *J. Macromol. Sci.-Chem.* **1972**, *A6*(6), 1109.
4. Koleske, J. V. Ch. 22 in *Polymer Blends*, Vol. 2; Paul, D. R.; Newman, S., Eds.; Academic: New York, 1978.
5. Lundberg, R. D. et al. U.S. Patent 3 641 204, 1972.
6. Phillips, B. et al. U.S. Patent 3 692 023, 1972.
7. *Polymer News* **1980**, *3*(1), 28.
8. Soldatos, A. C. U.S. Patent 3 629 364, 1971.
9. Schindler, A. et al. *J. Polym. Sci., Polym. Chem. Ed.* **1982**, *20*, 319.
10. "Herbicide Fibers Being Field Tested," In *Southern Research Bulletin*, Birmingham, AL, Fall-Winter 1983–84.
11. *Chem. Week*, May 23, 1984; p 32.
12. Potts, J. E. et al. SPE Tech. Paper 21 February 25, 1975; p 567.
13. Fields, R. D. et al. *J. Appl. Polym. Sci.* **1974**, *18*(2), 3571.

Poly(carbamoyl sulfonate)

See: *Cell Entrapment [Poly(carbamoyl sulfonate) Hydrogels]*

Polycarbodiimides

See: *High Pressure Polymerization Processing (High Temperature Thermosetting Polymers)*

Poly(carbon disulfide)

See: *Charge Transport Polymers (for Organic Electroluminescent Devices)*

POLYCARBONATE/ACRYLIC BLENDS (Phase Morphology)

Chi-Cheng Hung, W. G. Carson, and S. P. Bohan
Bristol Analytical Research
Rohm and Haas Company

Several studies have suggested that most polycarbonate/poly(methyl methacrylate)(PC/PMMA) blends made by solution-casting film are not thermodynamically stable with a lower critical solution temperature (LCST).[1–3] In addition, cloud-point estimates suggest that either the temperature span between the LCST and T_g is too small to obtain miscible blends by melt processing[1–3] or that the LCST is below T_g.[4] Although most of the polycarbonate and acrylic polymer blends that have been studied previously were prepared by solution-casting films, two were prepared by melt processing, which is more commonly used commercially. It has been shown by dynamic mechanical analysis (DMA) that melt-processed blends have two distinct phases.[1]

We will investigate the morphology of melt-blended PC/poly(methyl methacrylate-*co*-cyclohexyl methacrylate) (PMCHM) and PC/PMMA by NMR, differential scanning calorimetry (DSC), and transmission electron microscopy (TEM).

We also have studied the effect of copolymerizing a second monomer, cyclohexyl methacrylate (CHM), with methyl methacrylate (MMA) and have observed the morphology of blends of this acrylic copolymer with polycarbonate by NMR, DSC, and TEM. It is known that a thermodynamic LCST of PC/PMMA blends can be elevated by comonomerization, which allows specific polymer–polymer interactions that are unavailable in the blends of the homopolymers.[5] The effect may be ascribed to either a strong interchain attraction mechanism (such as hydrogen bonding, if appropriate comonomers are used) or by a strong intrachain repulsion of the added comonomer with the original monomer units.

We have investigated the miscibility of two melt-processed acrylic polymer and polycarbonate blends by solid-state NMR, DSC, and TEM. All three methods agree well on the morphology of PC/PMCHM and PC/PMMA blends. We observe the comonomerization effect on PC/PMCHM blend, the miscibility of which can be explained by a strong repulsion within the acrylic copolymer. The PC domains in PC/PMCHM and PC/PMMA blends were analyzed by imaging analysis. Though the PC/PMCHM blends are miscible, we are able to observe a small amount of PC in domains with an approximate size of 500 Å by TEM. The upper limit of the domain size in this miscible blend as measured by NMR is ~40 Å in radius.

ACKNOWLEDGMENTS

We are greatly thankful to Dr. Phelps Johnson for preparing the materials and to Joan Cronin and Martin Mastovich for TEM measurements. We would also like to thank Edward Chiemingo and Edwin Berner for DSC measurements.

REFERENCES

1. Chiou, J. S. et al. *J. Polym. Sci., Polym. Phys.* **1987**, *25*, 1459.
2. Kyu, T.; Saldanha, J. M. *Macromolecules* **1988**, *21*, 1021.
3. Gardlund, Z. G. *ACS Symp. Series* **1984**, Vol. 206, p 129.
4. Nishimoto, M. et al. *Polymer* **1991**, *32*(7), 1274.
5. Paul, D. R.; Newman, S., Eds.; *Polymer Blends*; Academic: New York, 1978.

Polycarbonates

See: *Bisphenol-A-Polycarbonate/Polyester Blends*
Bulk Polymerization
Cyclic Carbonates (Ring-Opening Polymerization)
Cyclic Oligomers of Engineering Thermoplastics
Engineering Thermoplastics (Survey of Industrial Polymers)
Enzymatic Polymerization
Fluorine-Containing Polymers (Polycarbonates and Polyformals)
Poly(butylene terephthalate)/Polycarbonate Blends
Polycarbonate/Acrylic Blends (Phase Morphology)
Polycarbonates (Overview)
Poly(propylene carbonate)
Pseudo-Poly(amino acid)s (Overview)
Recycling Thermodynamic (by Ring-Closing Depolymerization)
Ring-Opening Polymerization, Anionic (with Expansion in Volume)
Ring-Opening Polymerization, Cationic (with Expansion in Volume)
Step Polymerization Catalysts

POLYCARBONATES (Overview)

Kalyan Sehanobish,* H. T. Pham, and C. P. Bosnyak
The Dow Chemical Company

Polycarbonate (PC) resins can be divided into two classes: aliphatic and aromatic PCs. Aliphatic PCs are not widely used as thermoplastics but can be used as a comonomer or a cocondensate with other polycondensates.[1] For example, aliphatic PC can be used to modify an aromatic PC bisphenol A PC as a high–heat elastomer.[2] Other uses of aliphatic carbonates are as plasticizers for poly(vinyl chloride) (PVC)[3] and as stabilizers for PVC.[4]

The most common aromatic PC, poly(bisphenol A carbonate) (structural), known as PC, is an engineering thermoplastic material well known for its high–impact toughness, glass transition temperature (150°C), and high optical clarity.[5,6] PCs have wide applications ranging from consumer products such as beer glasses to instrument panels in automobiles.

SYNTHESIS OF POLYCARBONATES

Poly(bisphenol A carbonate) is a condensation polymer usually synthesized from bisphenol A, a monophenol chain terminator to control molecular weight, and phosgene. Phosgene, however, is not used in melt polymerization. Addition of polyfunctional monomer can yield branched or crosslinked systems.

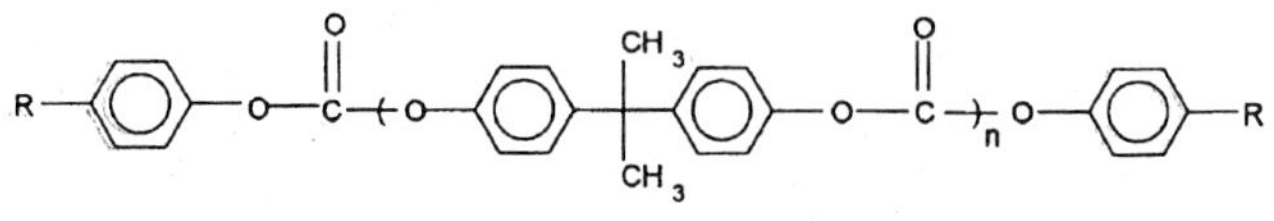

STRUCTURE 1. Bisphenol A polycarbonate (PC).

IGNITION RESISTANCE

PC has an oxygen index from 26% to 28% (ASTM D2863) and is rated as V2 at 1.6 mm, according to UL94.

STABILITY OF POLYCARBONATES

PC is usually fabricated at high–melt temperatures (about 300°C and above) where thermal degradation can occur. Yellowing is usually the first observation of thermal degradation,[8] followed by molecular weight degradation, particularly in the presence of basic or iron impurities.

PC itself exhibits remarkable resistance to photo-oxidation.[9] However, with prolonged exposure to UV radiation, PC yellows.

PC has an equilibrium moisture content of 0.2 at 25°C and 60% RH and 0.9% at 100°C, but it does not noticeably affect the mechanical integrity of the PC performance.

RHEOLOGY

Shear viscosity as a function of shear rates is different compared to other thermoplastics such as polystyrene or polyethylene. PC has a larger range of shear rates at which the shear viscosity of PC is Newtonian.

Most aspects of viscoelastic behavior of PC have been reported.[7] The elastically recoverable strain after cessation of steady-shear flow for PC is in the order of 1.0 to 2.5 depending on molecular weight, rate of shear deformation and temperature.[10]

PROCESSING

PCs are usually made into articles by profile extrusion, injection molding, and blow molding. Other processes, such as solvent cast film of high molecular weight PCs in liquid crystal display screens and thermoforming, also are used. Some PCs can be processed using single-screw extruders.

Injection molding grade PC resins contain mold release agents such as long–chain carboxylic acid esters at amounts less than 1% by weight.

*Author to whom correspondence should be addressed.

PC does not have broad resistance to aromatic, chlorinated, and polar solvents. For example, cracking can be induced at low tensile stresses in liquids such as acetone, and PC can be dissolved in toluene and methylene chloride. However, PC is resistive to non-polar liquids such as aliphatic hydrocarbons.

APPLICATIONS

Construction

PCs are used in windows, bathroom stalls, laminated walls, and skylight roofing, all of which require transparency and high toughness. For outdoors, special UV-stabilized grade PCs are used to maintain transparency and physical properties. Scratch-resistant coatings are used in sheeting products for heavy traffic areas.

Business Equipment

PCs are commonly used in business equipment housing such as portable computers, projectors, camera housings, and telecommunication equipment.

Automotive

PCs are used in a variety of places such as headlamp covers, tool boxes, instrument panels, and automotive body panels.

Optical Lenses

PC has a refractive index of 1.586 (ASTM D-542)—similar to glass. The advantages of PC lenses are their light weight and toughness, which makes it well suited to safety optical spectacle lenses, sunglasses, and goggles. Lenses made from PC have a scratch resistant coating.

Optical Medium

PC is the main component in compact discs and optical discs.

Medical

Application of PCs include syringes, medical tubings such as kidney dialysis tubings, and pacemaker components.

Consumer Products

General grade PCs also can be used in consumer products such as transparent or colored dinnerware or glassware, which are generally injection molded. Other use of PCs are water bottles (blow molded), milk bottles, containers and toys.

Appliances

PCs also are used in appliance parts such as crisper trays in refrigerators and housing appliances that must be high impact and heat resistant.

POLYCARBONATE BLENDS AND COPOLYMERS

There are two major trends for modification of homopolymer bisphenol A PC properties: blending with another polymer such as acrylonitrile butadiene styrene copolymer, or with polyesters, and copolymerization.

REFERENCES

1. Freitag, D.; Grigo, U. et al. *Encyclopedia of Polymer Science and Engineering*, 2nd ed.; Vol. 11; Mark, J. et al., Eds.: Wiley-Interscience: New York, 1987; Vol. 11.
2. Schreckenberg, M.; Konig, K. et al. Ger. Patent (DOS), 2 837 325 (Bayer), 1978.
3. Hostettler, F.; Cox, E. F. U.S. Patent 3 305 605 (Union Carbide), 1964.
4. Minnagawa, M. and coworkers, U.S. Patent 4 255 301 (Argus Chemical Corp.), 1978.
5. Odian, G. *Principles of Polymerization* 2nd ed.; Wiley-Interscience, New York, 1981.
6. Pakull, R.; Grigo, U.; Freitag, D. *Polycarbonates. RAPRA Review Reports* **1991**, *42*.
7. Schnell, H. *Chemistry and Physics of Polycarbonates*; Wiley-Interscience: New York, 1964.
8. Lee, L.-H. *J. Polym. Sci., Polym. Chem. Ed.* **1964**, *2*, 2859.
9. Davis, A.; Golden, J. H. *J. Macromol. Sci., Rev. Macromol. Chem. Phys.* **1969**, *C3*, 49.
10. Pham, H. T.; Bosnyak, C. P. et al. *J. Appl. Polym. Sci.* **1993**, *48*, 1425.

Polycarbophosphazene

See: Phosphorus-Containing Polymers (Overview)

Poly(carborane-siloxane-acetylene)s

See: Inorganic/Organic Hybrid Polymers (High Temperature, Oxidatively Stable)

Polycarbosilanes

See: Polysilanes (Overview)
Preceramic Polymers

Polycatenanes

See: Polyrotaxanes (Overview)

Polycations

See: Flocculants (Organic, Overview)
Flocculants, Cationic
Polyelectrolytes, Cationic

Polychloroprene

See: Chloroprene Rubber

Polychlorotrifluoroethylene

See: Engineering Thermoplastics (Survey of Industrial Polymers)
Tetrafluoroethylene Copolymers (Overview)

Polycondensation

See: Carbenoid Polycondensation
Condensation Polymers (Synthesis from Silylated Diamines)
Condensation Polymers (Using Phosphorus Pentoxide/Methanesulfonic Acid)
High Pressure Polycondensation (for High Temperature Polymers)
Optically Active Polyamides, Coumarin Dimer (Chiral Stationary Phases for HPLC)
Poly(alkylene phosphate)s
Polyamines (Ru-Catalyzed Polycondensation of Diamines and Diols)

Poly(ethylene terephthalate) (Chemistry and Preparation)
Polyimides, Soluble Rigid-Rods (via Pd-Catalyzed Polycondensation)
Polyphosphonates and Polyphosphates (Low Temperature Solution Polycondensation)
Solid-State Polyamidation
Step Polymerization Catalysts

Polycyanoacrylates

See: *Cyanoacrylates*
Poly(2-cyanoacrylate)

POLY(2-CYANOACRYLATE)S

Michael Stein and Werner Weitschies
Schering AG

The discovery of 2-cyanoacrylates in 1950 provided a basis for a radically new type of adhesive. Conventional adhesives require heat, pressure, catalysts, or evaporation of a solvent. The 2-cyanoacrylates, however, polymerize at room temperature without solvent or catalyst. They are, therefore, widely used as "instant glues" in industry,[1] as biodegradable tissue adhesives in surgery,[2,3] or—after polymerization—as drug carriers in the form of nanospheres in experimental drug formulations.[4]

CHEMICAL STRUCTURE

The chemical structure of poly(2-cyanoacrylate)s is derived from ethane, which is disubstituted at C-1 (**Figure 1**). A variety of chains is available as –R. In the technical field methyl, ethyl, or isopropyl 2-cyanoacrylate are generally used; in medical areas the butyl ester is used; and for research purposes isobutyl, hexyl, or ethoxyethyl esters are synthesized.

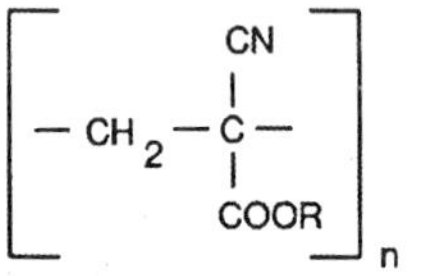

FIGURE 1 Chemical structure of monomeric 2-cyanoacrylates.

POLYMERIZATION

Polymerization usually follows a base-catalyzed anionic mechanism. However, radically or photochemically initiated polymerization is also possible. In any case, the reaction is highly exothermic as can be shown and evaluated by DSC.[5] Anionic polymerization can be initiated at low or ambient temperatures even by such weak bases as water, alcohols or amines.

The bonding action observed when a 2-cyanoacrylate is placed between two adherents, such as glass or metal, is the result of this anionic polymerization, which is obviously catalyzed by trace amounts of water or other weak bases present on the adherent surfaces. The bonding effect is maximized when the monomer is spread into a very thin film. It is apparent from the polymerization mechanism that acidic substances exert an inhibitory effect on the reaction. Thus, they can be used as reaction modifiers.[6] The speed of polymerization in a given compound is a function of the side chain.[7,8] 2-Cyanoacrylates with side chains of 4 or 6 carbon atoms polymerize within seconds.

PROPERTIES

Monomer

Pure cyanoacrylate monomers are colorless, clear liquids of low viscosity at room temperature. They exhibit pungent, penetrating odors which may be mildly lacrimatory. This irritant effect is less pronounced in higher esters.

Monomeric 2-cyanoacrylates should not be stored or shipped in containers made of glass, aluminum, or PVC, because of the acrylates' reduced stability. The most suitable container material is polyethylene.[10]

Polymer

Generally, 2-cyanoacrylate polymers are brittle substances of little flexibility, which sometimes limits their practical usefulness in bonding soft materials. They are X-ray amorphous and do not show any recrystallization on storage.[11] The polymers are soluble only in polar solvents such as acetone, methylene chloride, or chloroform.

APPLICATIONS

2-Cyanoacrylate-based adhesive formulations are used in the electrical and electronics industries as well as in engineering and automobile, ship, and aircraft construction. However, because of their susceptibility to heat, most 2-cyanoacrylic esters should not be used in applications where they will be subjected to temperatures > 70°C for a long time.[9]

In addition to technical applications, 2-cyanoacrylates are often helpful in medicine as tissue adhesives. In this area, the butyl ester is mainly used because of its good compromise between biological tolerance and degradability.[12,13] Esters with shorter side chain are irritating to skin and will degrade rapidly, sometimes creating inflammation reactions. Higher esters are well tolerated by tissue, but exhibit relatively long degradation times.

2-Cyanoacrylates also are used in interventional neuroradiology as embolic material for endovascular therapy of arteriovenous malformations.[14] They are also applied for the infarction of tumors.[15] Drug-loaded nanospheres based on 2-cyanoacrylates are prepared in pharmaceutical research as controlled-release formulations or as drug carriers for targeting.[4]

REFERENCES

1. Coover, Jr., H. W. et al. *SPE J.* **1959**, *15*, 413.
2. Matsumoto, T. et al. *Arch. Surg.* **1967**, *94*, 153.
3. Refojo, M. F. et al. *Surv. Ophthalmol.* **1971**, *15*, 217.
4. Couvreur, P. *CRC Crit. Rev. Therap. Drug Carrier Sys.* **1988**, *5*, 1.
5. Stein, M.; Stangl, R. *Acta Pharm. Technol.* **1990**, *36*, 95.
6. Stein, M. *J. Appl. Polym. Sci., Appl. Polym. Symp.* **1991**, *48*, 441.
7. Collins, J. A. et al. *Arch. Surg.* **1966**, *93*, 428.
8. Leonard, F. et al. *J. Appl. Polym. Sci.* **1966**, *10*, 1617.
9. Coover, H. M.; McIntire, J. M. In *Encyclopedia of Polymer Science and Engineering*, John Wiley & Sons: New York, 1985; p 299.
10. Stricker, H. *Arch. Pharm.* **1967**, *300*, 19.
11. Stein, M. *8th Eur. Conf. Biomat.*, Heidelberg, 1989; p 18.
12. Harper, M. C.; Ralston, M. *J. Biomed. Mater. Res.* **1983**, *17*, 167.
13. Galil, K. A. et al. *J. Biomed. Mater. Res.* **1984**, *18*, 601.
14. Brothers, M. F. et al. *Am. J. Neuroradiol.* **1989**, *10*, 777.
15. Papo, J. et al. *Am. J. Radiol.* **1981**, *137*, 781.

Polycyanogen

See: Paracyanogen

Polycyanurates

See: Cyanate-Based Resins (Polycyanurates)
Cyanate Ester Resins
High Pressure Polymerization Processing (High Temperature Thermosetting Polymers)

Polycyclodiborazanes

See: Organoboron Main Chain Polymers

POLY(CYCLOPENTADIENYLENE VINYLENE)

Franz Stelzer* and Michael Schimetta
Christian Doppler Laboratorium für Katalytische Polymerisation
Institut für Chemische Technologie Organischer Stoffe

Günther Leising
Institut für Festkörperphysik
Technische Universität Graz

Because of the increased interest in conjugated polymers in the past decade, many different types have been investigated. One of the polymers most often investigated is polyacetylene, which has the highest conductivities after doping with iodine. Our polymer, poly(1,3-cyclopentadiene-1,4-ylene vinylene) (PCPDV), (**Structure 1**) can be considered a special form of polyacetylene in which at least each first and second double bond is *trans*, *cisoid*. Other important precursor routes to conjugated polymers include synthesis of poly(*p*-phenylene) via a precursor polymer synthesized from the biotechnologically produced monomer *cis*-cyclohexa-1,3-dien-5,6-diol[1] and of poly(*p*-phenylene vinylene) (PPV) via polymeric sulfonium salts as developed by Wessling.[2] These polymers have three characteristics in common with PCPDV and the synthetic route described in this chapter.

- The conjugated rings as are connected by a vinylic double bond.
- the polymer is easily synthesized via a precursor route similar to the synthesis of PPP if starting with a diester of norbornenediol, and
- with respect to their electronic structure they are similar to PPV as the simplest poly(arylene vinylene) after *n*-doping the cyclopentadienylene units.

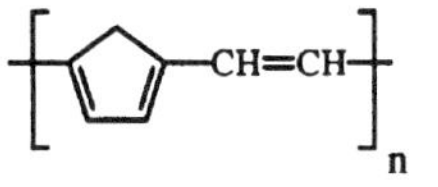

1

In addition, theoretical considerations of PPV promote the synthesis of PCPDV because it is the carbocyclic, nonaromatic analogue of poly(thienylene vinylene), (PTV) and poly(furylene vinylene) (PFV).

*Author to whom correspondence should be addressed.

Well known precursor routes such as the Wessling route to PFV PPV,[2] or a similar one to PTV,[3] cannot be applied successfully because of the nonaromatic character of cyclopentadiene. Coupling reactions of the Wittig or Heck type, or condensation polymerization, are also not applicable and have not been reported until now.

SYNTHETIC ROUTE VIA RING-OPENING METATHESIS POLYMERIZATION

Asrar[4] of Monsanto Company patented the precursor route to PCPDV by ring-opening metathesis polymerization (ROMP) of norbornenediol derivatives—mainly diacetates—followed by elimination in alkoxide media at elevated temperature. This provided a way to get a thermoplastic bulk polymer that could be fabricated via simple thermoplastic form-giving processes.

In recent years several investigations have been made on the stereochemistry of the polymerization of norbornenediol diesters,[5] the thermal elimination of different functional groups, and the characterization of the resulting PCPDV.[6,7] A generalized route is shown in **Figure 1**.

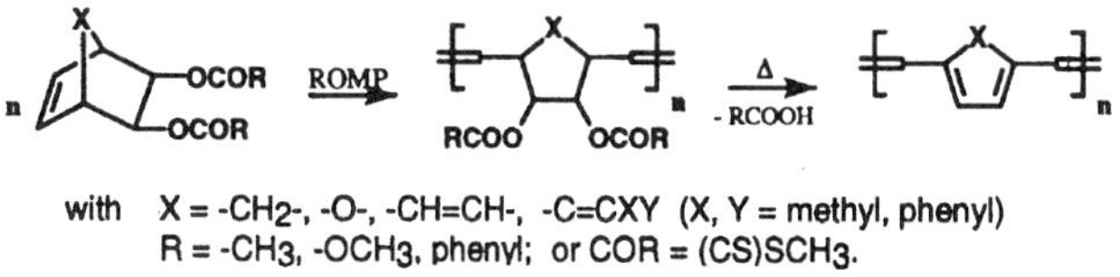

FIGURE 1. Synthesis of poly(cyclopentadienylene vinylene) and similar polymers.

APPLICATIONS

Almost all applications of these polymers make use of conjugated double-bond systems of the final polymer. As polyacetylene itself is highly unstable, it is hoped that other less labile polymers show similar electro/photoactive properties and are much less sensitive to oxygen and humidity. In addition, it is necessary to keep the conjugated double bonds in plane. It is not possible to name special applications for PCPDV alone, but only for polyacetylenes and poly(arylene vinylene)s in general. This is easily understood if one considers PCPDV a bridged polyacetylene. PCPDV is a very interesting material in the spectrum of conjugated polymers. Improvement of the purity and reproducibility in the synthetic method is highly desirable because a number of possible applications are feasible for this electroactive polymer in the fields of electronics, optoelectronics, and photonics. So one can think of applications in electroluminescence devices such as light-emitting diodes (LEDs), which show different colors than PPV (yellow green),[8] precursor PPP obtained in a way similar to the route to PCPDV in this paper (blue)[9] or with quantum well PPP (blue).[10]

An important field for the near future should be applications in photovoltaic devices. The possible use of polyacetylene has already been shown; therefore a protected PA should be useful. More detailed information about such devices can be found in overviews by Kanicki[11] or by new solar cells with PPV as active material as shown by Marks.[12] Polymers of this type have also been used for capacitors, gas sensors, and electrodes for batteries.[13]

Another application of PCPDVs could be in the field of photoresists if they are combined with the use of photoacid generators. Thus the precursor polymer can be used for direct imaging for black/white contrast, as a standard photoresist or, finally, as a photofabricated conducting structure.

ACKNOWLEDGMENTS

We want to acknowledge the financial support from the Austrian Science Foundation, Vienna (project P9093-CHE), and from the Christian Doppler Society (Vienna).

REFERENCES

1. Ballard, D. G. H. et al. *J. Chem. Soc., Chem. Commun.* **1983**, 954; Ballard, D. G. H. et al. *Macromolecules* **1988**, *21*, 294.
2. Wessling, R. A.; Zimmermann, R. G. U.S. Patent 3 401 152, 1968.
3. Murase, I. et al. *Polym. Commun.* **1987**, *28*, 229–31.
4. Asrar, J. U.S. Patent 4 816 521, 1989.
5. Schimetta, M.; Stelzer, F. *Macromolecules* **1994**, *27*, 3769.
6. Zenkl, E. et al. *Polymers for Microelectronics*; ACS Symposium Series 537 Ed. Thompson, L. F.; Willson, C. G.; Tagawa S., American Chemical Society: Washington, DC, 1994.
7. Schimetta, M. Ph.D. Thesis, Graz University of Technology, 1994.
8. Burroughes, J. H. et al. *Nature* **1990**, *347*, 539.
9. Grem, G. et al. *Adv. Mat.* **1992**, *4*, 36.
10. Grem, G. et al. *Chem. Mater.* **1995**, *7*, 2.
11. Kanicki, J. *Handbook of Conducting Polymers*; Skotheim, T. A., Eds.; Marcel Dekker: New York, 1986.
12. Marks, R. N. *J. Phys.: Condens. Mat.* **1994**, *6*, 1379.
13. Zeibt, G. (edit.), Poly-conjugated Materials, E-MRS, North-Holland, Amsterdam, 1992.

Polydepsipeptides

See: Biodegradable Polydepsipeptides (Synthesis, Characterization, and Degradation)

Polydiacetylenes

See: Conjugated Polymers (Insulating and Conducting Forms)
Langmuir-Blodgett Film Polymerization
Liquid Crystallization State Polymerization
Polydiacetylenes, Unconventional
Topochemical Polymerization (Diacetylenes)

POLYDIACETYLENES, UNCONVENTIONAL

Takeshi Ogawa
Instituto de Investigaciones en Materiales
Universidad Nacional Autónoma de México

Polydiacetylenes, PDAs, are polymers obtained by the polymerization of 1,3-butadiynes, so called diacetylenes (DAs). Some PDAs undergo solid-state polymerization by irradiation, heat, or pressure, giving completely crystalline, highly conjugated 1,4-PDAs. This type of polymerization had been known for some time, but it was clarified in 1969 by Wegner[1] and has been named "topochemical polymerization," as shown in **Figure 1**. The polymerization of DAs depends on the substituent groups, **R** and **R'** on the 1 and 4 carbons, and not every DA undergoes topochemical polymerization. It is said that the crystal packing favorable for topochemical polymerization is such that the distance between the adjacent DAs is <4 Å and the angle between the DA rod and the stacking axis of PDA is 45°C.[2]

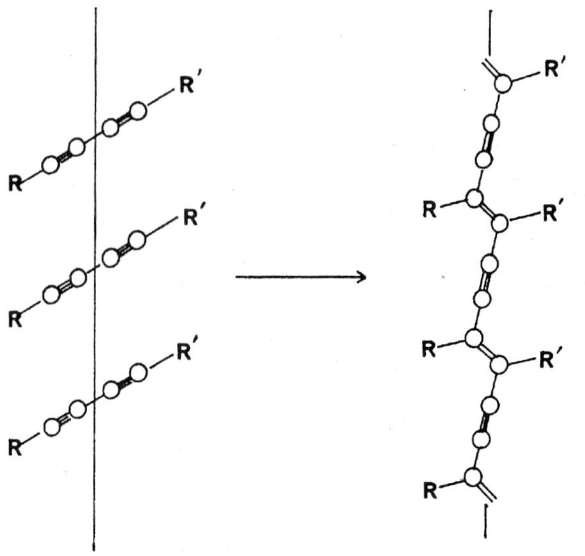

FIGURE 1. Schematic representation of topochemical polymerization of diacetylenes.

PDA systems were extensively investigated from 1970 to 1985, with principal interest in their NLO applications. However, because of the poor processibility of such PDAs, no practical applications have so far been found. Interesting materials for applications in optics, optoelectronics, electronics, and so on could be obtained if PDAs could be made into films, sheets, or filaments of any desired dimensions with excellent optical qualities. For this purpose the following methods are being developed and are thought to be promising for practical applications of PDAs:

- The preparation of processible polymers with DA groups in their main or side chains, which after processing into desired forms are subjected to heat, irradiation, or both simultaneously to cross-polymerize the DA groups and develop PDA networks in the materials.
- The polymerization of DAs in the molten state.
- The dispersion of PDA microcrystals in host polymers.

PREPARATION

DA-Containing Polymers

Polymers with DA groups in their main chains can be cross-polymerized by irradiation or heating, and PDA networks are formed. If the polymers are soluble in solvents or can be melted, they can be processed into desired shapes by conventional methods such as casting and molding before cross-polymerization. These polymers can be synthesized by the polycondensation of difunctional monomers, or by the oxidative coupling polymerization of bisacetylenic monomers using a cuprous salt and amine base.

Polymers with PDA networks in their side chains can be obtained by the polymerization of DA-containing vinyl monomers, followed by heat or radiation treatment.

Polymerization of DAs in the Molten State

Some DAs undergo polymerization in the molten state by heating or irradiation, giving red to brown, highly transparent materials. It should be noted that not all DAs undergo such amorphous-state polymerization. Some give a deep red, highly transparent glass, whereas some undergo topochemical polymerization only in the crystalline state.

Dispersion of Topochemically Polymerizable DAs in Host Polymers

If DA microcrystals are dispersed in a transparent polymer and the crystal size is small enough relative to the light wavelengths, the composite materials become transparent and will have properties of topochemically polymerized PDAs.

PROPERTIES AND APPLICATIONS

Diacetylene-Containing Polymers

Not all DA-containing polymers are topochemically cross-polymerizable. Some undergo cross-polymerization only in the molten state, and some hardly cross-polymerize on heating or by irradiation. A favorable crystal packing of the DA moiety is required for topochemical cross-polymerization to take place.

Nonlinear Optical Properties of PDA-Containing Films

Because of the PDA network, these films exhibit third-order NLO properties. The polymers with topochemically polymerizable DA groups are crystalline, and therefore the transparency of films seems to be unsatisfactory for applications. Those with DA groups that undergo amorphous-state polymerization give films with excellent transparency, and they seem to be more promising from a practical point of view.

AMORPHOUS PDAS

Amorphous PDAs are interesting materials for optical applications because of their excellent optical transparencies and highly conjugated systems. However, there have been very few studies about them. Not all DAs undergo polymerization in the molten state below their thermal decomposition temperatures. Many photosensitive DAs that undergo topochemical polymerization do not polymerize in the amorphous state, and some do not polymerize appreciably under any conditions.

If extremely high crosslinking is achieved, extremely hard materials may be obtained, and DA-containing divinyl compounds can be considered to be model compounds for obtaining such systems.

REFERENCES

1. Wegner, G. *Naturforsch* **1969**, *24B*, 824.
2. Baughman, R. H.; Yee, K. C. *Macromol. Rev.* **1978**, *13*, 219.

Poly(diallyldimethylammonium chloride)

See: Cyclopolymerization (Overview)

Polydiazetidinediones

See: High Pressure Polymerization Processing (High Temperature Thermosetting Polymers)

POLY(DICHLOROPHENYLENE OXIDE)S (by Pyridine/Copper Complex)

Duygu Kısakürek
Department of Chemistry
Middle East Technical University

Poly(dichlorophenylene oxide) can be synthesized by polymerization of 2,4,6-trichlorophenol (TCP) or 4-bromo- or 4-iodo-2,6-dichlorophenol((4Br-DCP) and (4I-DCP), respectively).

Hay and his group[1] used copper (I) and pyridine instead of silver for polymerization of a 2,6-disubstituted phenols at room temperature. However, no appreciable reaction occurred when 2,6-dichlorophenol was oxidized, even at 100°C. At the same time, Blanchard et al.[2] described preparation and characterization of di-2,4,6-trichlorophenolate)bis(pyridine) copper(II) complex as well as their decomposition under a variety of conditions yielding poly(chlorophenylene ether)s, which are highly branched as evidenced by low intrinsic viscosity values.

Stamatoff reported that polymerization by free-radical initiated decomposition of sodium 2,6-dichloro-4-bromophenolate with benzoylperoxide under a variety of conditions gave linear poly (2,6-dichlorophenylene oxide) with bromine content *ca.* 1% with high molecular weights.[3] Subsequently, Harrod carried out studies on phenoxo complexes of copper (II) by preparing series of (2,4,6-trichlorophenoxo) copper(II) complexes containing various amine ligands.[4]

Thermal decomposition of bis(2,4,6-trichlorophenoxo)-bis(pyridine) copper(II) ($Py_2Cu(TCP)_2$) complex in solution is induced by oxygen on free-radical initiators. Preceding an induction period, radical formation was observed after the complete disappearance of the complex.

^{1}H-NMR studies have shown that a bromine atom in the para position of 2,4,6-trihalophenoxides results in polymers with higher linearity,[7,8,12] independent of the method of initiation and the type of ligand substituted.[5–9,12]

PREPARATION

The easiest and the cheapest way to synthesize poly(dichlorophenylene oxide) is by decomposition of trihalophenoxo metal complexes.

PROPERTIES AND APPLICATION

Characterization of polymers synthesized by any method was achieved by using ^{1}H NMR. The spectra indicate that polymers synthesized from $Py_2Cu(TCP)_2$ are highly branched, whereas those from *bis*(4-bromo-2,6-dichlorophenoxo)-bis(pyridine) ($Py_2Cu(4Br\text{-}DCP)_2$) are mainly linear structures independent of the method of initiation.

Fire resistant poly(dichlorophenylene oxide)s from the decomposition of $Py_2Cu(TCP)_2$, using any method of initiation, are highly branched and high Tg (> 170°C) with molecular weights (M_n = 2000 to 3000) by electroinitiation, (M_w = 20000–30000) in solid state, and (M_w = 50000–150000) in solution.[10,11,13,14] However, high values in molecular weight were observed from decomposition of $Py_2Cu(4Br\text{-}DCP)_2$ complexes.[6,12–16]

REFERENCES

1. Hay, A. S. *J. Polym. Sci.* **1962**, *58*, 581.
2. Blanchard, H. S.; Finkbeiner, H. L.; Russell, G. H. *J. Polym. Sci.* **1962**, *58*, 469.
3. Stamatoff, G. S. U.S. Patent 3 257 388, 1966.
4. Harrod, J. F. *Can. J. Chem.* **1969**, *47*, 637.
5. Harrod, J. F.; van Gheluwe, P.; Kısakürek, D.; Shaver, A. *Macromolecules* **1981**, *14*, 565.
6. Kısakürek, Binboğa, N.; Harrod, J. F. *Polymer* **1987**, *28*, 1767.
7. Saçak, M.; Akbulut, U.; Kısakürek, D.; Toppare, L. *Polymer* **1989**, *30*, 928.
8. Kısakürek, D.; Yiğit, S. *Eur. Polym. J.* **1991**, *27*, No. 9, 955.
9. Şanli, O.; Kısakürek, D. *Makromol. Chem.* **1992**, *193*, 619.
10. Şen, Ş.; Kısakürek, D. *Polymer* **1993**, *34*(19), 4146.
11. Baştürkmen, M.; Kısakürek, D. *Polymer* **1993**, *34*(3), 625.
12. Aras, L.; Şen, Ş.; Kısakürek, D. *Polymer* 1994, in press.
13. Binboğa, N. *The Solution Properties of Dihalophenylene Oxide Polymers*, M.Sc. Thesis, METU, Ankara, Turkey, 1984.
14. Baştürkmen, M. Synthesis and Characterization of bis(Pyridine) bis(Trichlorophenoxo) Cu(II) and bis(Pyridine) bis(Trichlorophenoxo) Co(II) Complexes in Solid State, M.Sc. Thesis, METU, Ankara, Turkey, 1991.
15. Bilir, N. *Polymerization of bis(Pyridine) bis(Trihalophenoxo) Cu(II) Complexes with Various Halogens in Solution*, M.Sc. Thesis, METU, Ankara, Turkey, 1992.
16. Akbaş, M. *Polymerization of bis(Pyridine) bis(Trihalophenoxo) Cu(II) Complexes with Different Halogens in Para Position in Solid State*, M.Sc. Thesis, METU, Ankara, Turkey, 1992.

Polydicyclopentadiene

See: *Metathesis Polymerization, Cycloolefins*
Reaction Injection Molding (RIM) Materials

Poly(1,5-dihydroxynaphthalene)

See: *Enzyme-Catalyzed Oxidative Polymerization (Aromatic Compounds)*

Polydimethylsiloxane

See: *Controlled Drug Delivery Systems*
Polydimethylsiloxane Networks, Monodisperse
Silicone Elastomers
Silicone Release Coatings
Silicone Rubber Latex
Silicone Sealants

POLYDIMETHYLSILOXANE NEWORKS, MONODISPERSE

Judith Stein
General Electric Corporate Research and Development

In recent years, many researchers have investigated the viscoelastic properties of model silicone networks.[1]

Networks with controlled and well-defined distances between crosslinks can be prepared by using "monodisperse" polymers. The term "monodisperse" will refer to siloxanes with narrow molecular weight distributions ($M_w/M_n < 1.3$). The distances between crosslinks are precisely predetermined by the molecular weight of the siloxane polymer. The development of these networks has allowed the study of the effect of defect sites and unincorporated chains on the mechanical properties.

PREPARATION OF MONODISPERSE POLYMERS

There are five methods for the production of monodisperse siloxane polymers: anionic polymerization, fractional precipitation, supercritical extraction, polycondensation, and linear homologation.

FORMATION OF MONODISPERSE NETWORKS

Monodisperse networks can be prepared by either condensation of α, ω siloxanediol polymers with alkoxysilanes[7] or hydrosilylation of vinyl-terminated siloxanes with hydrosilane crosslinkers.[4] In either case the stoichiometry must be carefully controlled. The amount of crosslinker required can be calculated using both GPC and IR and NMR spectroscopies (**Schemes I** and **II**).

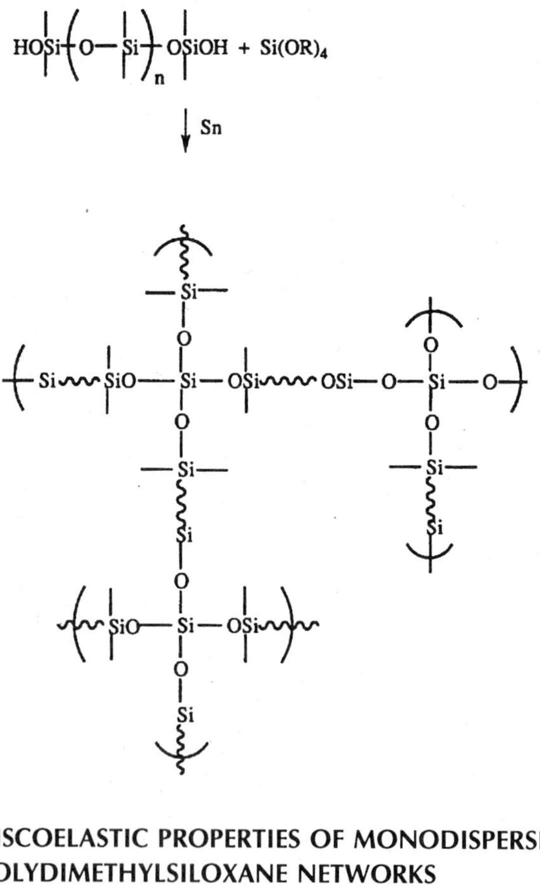

I

VISCOELASTIC PROPERTIES OF MONODISPERSE POLYDIMETHYLSILOXANE NETWORKS

Networks prepared using monodisperse polymers are useful in characterizing the effect of average chain length and chain length distribution on the mechanical properties of networks. Mark[5,8] has described the results of equilibrium tensile measurements on networks prepared from fractionated polydimethylsiloxane.

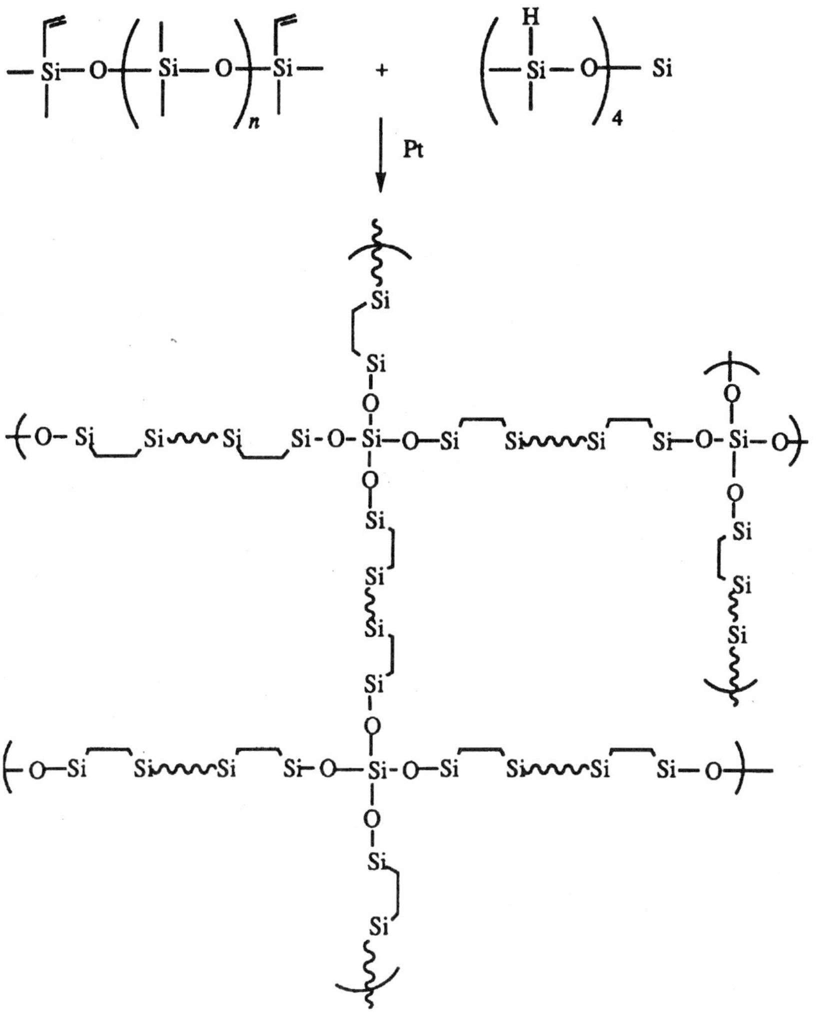

II

In 1980, Kosfeld[2] and co-workers reported dynamic mechanical measurements of networks prepared by telechelic condensation of monodisperse polydimethylsiloxane with tetraethylorthosilicate.

Meyers[3] determined the tensile and swelling behavior of polydimethymethylsiloxane networks of high junction functionality prepared by hydrosilylation.

Rennar[6] has examined the mechanical behavior and swelling properties of networks prepared from endlinking fractionated vinyl terminated polydimethylsiloxanes with pentamethylcyclopentasiloxane. The frequency dependence of the storage modulus G' for polydimethylsiloxane networks of different M_n of the chains was measured.

Bontems[4] has also found a good correlation between theoretical predictions and experimental value of the tensile modulus for polydimethylsiloxane networks prepared from monodisperse vinyl-terminated siloxanes in intermediate molecular weight regions.

APPLICATIONS

Monodisperse polydimethylsiloxane networks have been employed primarily to validate various aspects of the theories of rubber elasticity. Villar[9] has used monodisperse monofunctional polydimethylsiloxane pendant chains to determine the influence of defect sites on viscoelastic 5 properties. Additionally, some researchers have suggested that molecular weight distributions may affect other properties such as adhesion.[4] Garrido[10] has examined the diffusion of monodisperse polydimethylsiloxane through polydisperse model networks by pulsed field gradient NMR spectroscopy, and although he did not find a dependency on the molecular weight distribution, it would be interesting nonetheless to examine diffusion through monodisperse networks. Although there has been no commercial utilization of monodisperse networks to date, several potential applications such as size-selective membranes and sensors can be envisioned.

REFERENCES

1. Mark, J. E. *Advances in Polymer Science, 44*; Springer-Verlag: Berlin, 1982.
2. Kosfield, R. et al. *Polym. Bull.* **1980**, *3*, 603.
3. Meyers, K. O. et al. *Macromolecules* **1980**, *13*, 1045.
4. Bontems, S. L. et al. *J. Poly. Sci., Part A* **1993**, *31*, 2697.
5. Mark, J. E.; Sullivan, J. L. *J. Chem. Phys.* **1977**, *66*, 1006.
6. Rennar, N. *Kaust. + Gummi.* **1989**, *6*, 480.
7. Mark J. E. et al. *J. Chem. Phys.* **1979**, *70*, 1794.
8. Mark, J. E. *Makromol. Chem. Suppl.* **1979**, *2*, 87.

9. Villar, M. A. et al. *J. Appl. Polym. Sci.: Appl. Polym. Sym.* **1991**, *49*, 115.
10. Garrido, L. et al. *J. Polym. Sci., Part B: Polym. Phys.* **1988**, *26*, 2367.

Polydioxanone

See: Bioerodible Polymers A. Controlled Drug Release B. Fracture Fixation

Polydioxines

See: Ladder Polymers (Methods of Preparation)

POLY(1,3-DIOXOLANE)

Jeanne François and Emile Franta
Institut Charles Sadron/ULP

Poly(1,3-dioxolane) (PDXL) is a perfectly alternating copolymer constituted of the same units as poly(methylene oxide) (PMO) and poly(ethylene oxide) (PEO): PDXL: $[CH_2 - CH_2 - O - CH_2 - O]_n$. This structure can be expected to endow the chains with intermediate properties between those of PMO and PEO, particularly in water. PEO is very soluble in water, whereas PMO is soluble only at high temperature under pressure. Only recently was it possible to prepare model macromolecules whose molecular weight and end groups could be controlled. This prompted us to study the properties of PDXL in water and to compare them to those of PEO.

SYNTHESIS OF POLY(1,3-DIOXOLANE)

Chwialkowska[1] used fluorosulfonic esters, triflic esters, and triflic anhydride as initiators. These strong alkylating agents give rise to a living polymerization as no transfer to monomer occurs; termination with sodium methanolate endows the chains with $-OCH_3$ end groups.

Franta et al.[2] used the activated monomer mechanism developed by Penczek[3] for polymerization of substituted oxiranes.

PHYSICAL PROPERTIES

Crystallinity and Morphology

Crystallization kinetics, morphologies, and melting temperatures of PDXL samples of molecular weight higher than 8000 were thoroughly studied by Geil, Alamo, Azchambault and Neron.[5–8] Their results agree and lead to a description of several types of morphologies: classical Maltese cross spherulites, spherulites exhibiting wide uniformly oriented plates, and disordered structures.

Melia[9] and more recently Benkhira et al.[10] measured glass transition temperature, T_g, melting temperature, T_m, and the heat of fusion, ΔH_m. The first authors found: a PDXL of $M_n = 2800$ (end groups unknown) for $T_g = 209$ K, $T_m = 325$ K and estimated the heat of fusion of the 100% crystalline polymer: $\Delta H_m = 225 \pm 4$ J g^{-1}. For a polymer ($M_w = 30000$) carrying a OCH_3 type group at both ends, Benkhira et al.[11] obtained $T_g = 211$ K and $T_m = 328$ K.

PROPERTIES IN AQUEOUS SOLUTION

Poly(1,3-dioxolane) is water soluble at room temperature. Its structure, between that of PMO (insoluble in water) and PEO (highly water soluble), makes it a good candidate to determine the influence of structural factors on water solubility. No previous reports have been published before those of Benkhira et al.[9,11] who investigated this system in the entire range of concentration. In particularly, they established the phase diagram in water, measured the partial specific volume, determined the conformation of PDXL in dilute and semidilute aqueous solutions, and studied its interactions with sodium dodecyl sulfate.

CONCLUSION

Recent work has allowed synthesis of well defined poly(1,3-dioxolane) samples whose end groups can be controlled. This made possible study of the properties of this polymer in water solution. Its chemical composition should cause it to be more soluble than polyethylene oxide, but the opposite is true, indicating the importance of structural parameters in the solubility of PEO: Presence of an acetal unit in the repeating structure of PDXL lowers interaction with water. As a consequence, the solubility domain is much more limited than that of PEO, but the thermodynamic and conformational properties in dilute and semidilute regimes at room temperature correspond to good solvent conditions, however with a lower second virial coefficient than that of PEO. Interactions of PDXL with sodium dodecyl sulfate also exemplify the role played by ethylene oxide type units.

These results have been established on polymers carrying OCH_3 end-standing groups, it is expected that their replacement by two hydroxyl functions will modify solubility in water.

REFERENCES

1. Chwialkowska, W.; Kubisa, P.; Penczek, S. *Makromol. Chem.* **1982**, *183*, 753.
2. Franta, E.; Kubisa, P. et al. *Makromol. Chem.* (Macromol. Symposium) **1988**, *13/14*, 127.
3. Penczek, S. *ACS Polym. Preprints* **1988**, *29*(2), 38.
4. Sasaki, S.; Takahashi, Y.; Tadokoro, H. *J. Polym. Sci., Polym. Phys. Ed.* **1972**, *10*, 2363.
5. Geil, P. H. Polymer Reviews, Intersciences: New York *Polymer Single Crystals.*
6. Alamo, R.; Fatou, J. G.; Guzman, J. *Polymer* **1982**, *23*, 374.
7. Archambault, P.; Prud'homme, R. E. *J. Polym. Sci.* **1980**, *18*, 35.
8. Nèzon, M.; Tardif, A.; Prud'homme, R. E. *Eur. Polym. J.* **1976**, *12*, 605.
9. MeCa, T. P. *J. Appl. Chem.* **1964**, *14*, 461.
10. Benkhira, A.; Franta, E.; François, J. *Macromolecules* **1992**, *25*, 5697.
11. Benkhira, A.; Franta, E.; François, J. *Macromolecules* **1994**, *27*, 3963.

POLY[(DISILANYLENE) OLIGOPHENYLENE]S

Mitsuo Ishikawa* and Joji Ohshita
Department of Applied Chemistry
Faculty of Engineering
Hiroshima University

*Author to whom correspondence should be addressed.

Organosilicon polymers have become the subject of intense research in numerous laboratories.[1] Of these, the alternating polymers composed of a disilanylene unit and π-electron system show many unusual properties, because of the delocalization of σ-electrons in the silicon–silicon bonds through π-system in the polymer chain.[2–15] This unique behavior can be applied to the production of functionality materials. Indeed, these polymers are photoactive, and some of these can be used as a top imaging layer of the double-layer photoresist system. It has also been reported that when the alternating disilanylene polymers are doped with oxidizing agents such as ferric chloride, antimony pentafluoride, and iodine, they become conducting.

To date, many papers dealing with the synthesis and applications of this type of polymers have been published. In this article we report the synthesis and photochemical and conducting properties of poly[(disilanylene)phenylenes] and poly[(disilanylene)oligophenylenes] ($[SiR_2SiR_2(C_6H_4)_m]_n$, where R = Me, Et, Ph; and m = 1–4).

PREPARATION

Poly[(disilanylene)phenylenes][16,17]

The reaction of *p*-bis(chloromethylphenylsilyl)benzene with sodium dispersion in toluene at 80–90°C, followed by reprecipitation of the resulting products from benzeneethanol and then from benzene-2-propanol gave poly[*p*-(1,2-dimethyl-1,2-diphenyl disilanylene)phenylene (**1a**) in 45% yield. *P*-Bis(chloroethylmethylsilyl)benzene and *p*-bis(chlorodiethylsilyl)benzene readily react with sodium metal under the same conditions to afford the respective poly[*p*-(disilanylene)phenylenes] (**1b** and **1c**) in good yields. Poly[*o*- and *m*-(disilanylene)phenylenes] (**1d–1g**) were also prepared by similar condensation of the corresponding bis(chlorosilyl)benzenes with sodium dispersion in refluxing toluene (**Equation 1**).

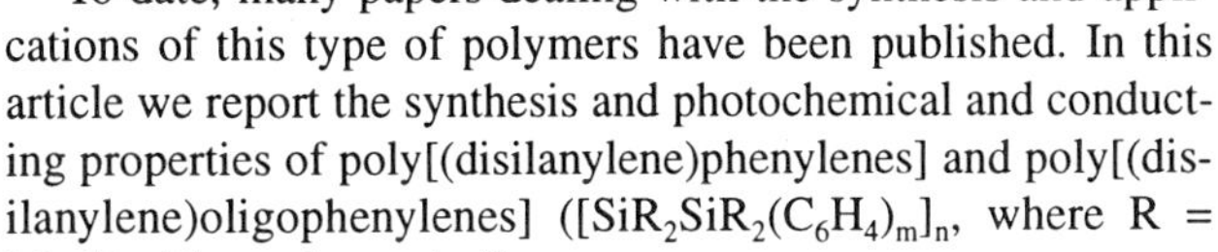

1a *p*-, R^1 = Ph, R^2 = Me
1b *p*-, R^1 = Et, R^2 = Me
1c *p*-, R^1 = R^2 = Et
1d *o*-, R^1 = Ph, R^2 = Me
1e *o*-, R^1 = R^2 = Me
1f *m*-, R^1 = Ph, R^2 = Me
1g *m*-, R^1 = R^2 = Me

Poly[(disilanylene)oligophenylenes][18]

Poly[4,4'-(disilanylene)biphenylenes] were readily obtained by the sodium condensation of 4,4'-bis(chlorosilyl) biphenyls in refluxing toluene, while poly[(disilanylene)terphenylenes] were synthesized by the nickel-catalyzed cross-coupling reactions of dibromobenzene with the corresponding di-Grignard reagents prepared from bis(4-bromophenyl) disilanes.

PROPERTIES

Like other polymers having a regular alternating arrangement of a disilanylene unit and π-electron system, the present polymers show a characteristic absorption band in the UV region. They are significantly red-shifted due to delocalization of σ-electrons in the silicon–silicon bonds through the π-system. Poly[(disilanylene)phenylenes] (**1a–1g**) are highly photoactive.

In marked contrast to poly[(disilanylene)phenylenes], poly[(disilanylene)oligophenylenes] were found to be less photoactive in solution.

Poly[(disilanylene)oligophenylenes] are insulators like other organosilicon polymers, but they become conducting by treatment of the thin solid films of the polymers with a dopant.

APPLICATIONS[19,20]

We have studied lithographic applications of these polymers and demonstrated that polymer (**1a**) can be used as the top imaging layer for the double-layer positive photoresist systems.

ACKNOWLEDGMENT

This work was supported by a Grant-in Aid for Scientific Research from the Ministry of Education, Science and Culture, Japan. We thank Shin-Etsu Chemical Co. Ltd., Nitto Electric Industrial Co. Ltd., Dow Corning Asia Ltd., Toshiba Silicone Co. Ltd., Kaneka Corporation, Sumitomo Electric Industries, and Japan High Polymer Center for financial support.

REFERENCES

1. Zeldin, M.; Wynne, K. J.; Allcock, H. R. *Inorganic and Organometallic Polymers*; ACS Symposium Series 360; American Chemical Society: Washington, DC, 1988.
2. Ohshita, J.; Kanaya, D.; Ishikawa, M.; Yamanaka, T. *J. Organomet. Chem.* **1989**, *369*, C18.
3. Ishikawa, M.; Hasegawa, Y.; Kunai, A.; Yamanaka, T. *Organometallics* **1989**, *8*, 2741.
4. Iwahara, T.; Hayase, S.; West, R. *Macromolecules* **1990**, *23*, 1298.
5. Ishikawa, M.; Hasegawa, Y.; Kunai, A.; Yamanaka, T. *J. Organomet. Chem.* **1989**, *381*, C57.
6. Ijadi-Maghsoodi, S.; Barton, T. J. *Macromolecules* **1990**, *23*, 4485.
7. Ishikawa, M.; Hatano, T.; Hasegawa, Y. et al. *Organometallics* **1992**, *11*, 1604.
8. Ishikawa, M.; Horio, T.; Hatano, T.; Kunai, A. *Organometallics* **1993**, *12*, 2078.
9. Ohshita, J.; Matsuguchi, A.; Furumori, K. et al. *Macromolecules* **1992**, *25*, 2134.
10. Brefort, J. L.; Corriu, R. J. P.; Gerbier, P. et al. *Organometallics* **1992**, *11*, 2500.
11. Hong, H. H.; Weber, W. P. *Polym. Bull.* **1989**, *22*, 363.
12. Ohshita, J.; Kanaya, D.; Ishikawa, M.; Yamanaka, T. *Chem. Express* **1990**, *5*, 489.
13. Hu, S.; Weber, W. P. *Polym. Bull.* **1989**, *21*, 133.
14. Ohshita, J.; Kanaya, D.; Ishikawa, M. et al. *Macromolecules* **1991**, *24*, 2106.
15. Chicart, P.; Corriu, R. J. P.; Moreau, J. J. E. *Chem. Mater.* **1991**, *3*, 8.
16. Nate, K.; Ishikawa, M.; Ni, H. et al. *Organometallics* **1987**, *6*, 1673.
17. Ishikawa, M.; Sakamoto, H.; Ishii, M.; Ohshita, J. *J. Polym. Sci., Part A: Polym. Sci.* **1993**, *31*, 3281.
18. Ohshita, J.; Watanabe, T.; Kanaya, D. et al. *Organometallics* **1994**, *13*, 5002.
19. Nate, K.; Inoue, T.; Sugiyama, H.; Ishikawa, M. *J. Appl. Polym. Sci.* **1987**, *24*, 2445.
20. Ishikawa, M.; Nate, K. *Inorganic and Organometallic Polymers*; ACS Symposium Series 360; American Chemical Society: Washington, DC, 1988; Chapter 16.

Polydithiocarbamate

See: *Phase-Transfer Catalyzed Polymer Synthesis*

Polydithiocarbonate

See: *Phase-Transfer Catalyzed Polymer Synthesis*

Polydithiones

See: *Ladder Polymers (Methods of Preparation)*

Polydodecenamer

See: *Metathesis Polymerization Cycloolefins*

Polyelectrolyte Complexes

See: *Enzyme Immobilization (Polyion Complexes)*
Polyelectrolyte Complexes (Overview)
Polyelectrolyte Complexes (In Immunology)
Polyelectrolyte Complexes (Targeting of Nucleic Acids)
Poly(vinylpyridinium salts)
Stimuli-Responsive Microcapsules
Superabsorptive Polymers (from Natural Polysaccharides and Polypeptides)

POLYELECTROLYTE COMPLEXES (Overview)

Joachim Kötz
Universität Potsdam

Polyelectrolyte complexes, called symplexes, are formed when oppositely charged macromolecules interact in solution via predominantly coulombic forces.

Depending on the reaction conditions and the nature of the components used, the complex formation process and the properties of the complexes as polymer materials can vary from highly swollen hydrogels to compact precipitates. Potential areas of application of polyelectrolyte complexes are membranes for dialysis and ultrafiltration, coatings on films and fibers, microcapsules, and implants for medical use. Symplex formation as a process has relevance to flocculation phenomena in water processing and paper manufacture, or as a general method for analysis, binding, and separation of polyelectrolytes. The leading role of medical applications derives from the similarities of polyelectrolyte complexes to protein structures: ionic charge characteristics, hydrogel structure, and permeabilities to water and to solutes existing in body fluids.

PREPARATION

Generally, three principal routes to symplexes are known: a polyacid–polybase reaction, a polysalt–polysalt reaction, and a matrix polymerization. Only in the first case does the neutralization result in an exothermic reaction (reaction heats in the order of 2 kcal/mol),[1] but in general the binding between two oppositely charge polyelectrolytes is nearly athermal. The process of complex formation is entropy driven because of the release of counterions, which are no longer restricted by association or binding to the polymer backbone-chain. Tsuchida demonstrated the exponential increase of the complex formation constant by increasing the chain length of the components up to six monomer units.[2] Above this critical chain length no significant influence of the molecular weight on the complex formation constant was found.

Considering the numerous and often contradictory phenomena of symplex formation arising from the state of dispersity of the system, depending on the polymer concentration of the reacting components in water, we have to discern between homogeneous one-phase systems and phenomena of phase separation. Macroscopic homogeneous systems can be obtained three ways: at very low polymer concentrations because of the formation of very small symplex aggregates, at an excess of one component because of the formation of sequential soluble complexes, or at high polymer concentrations from the disaggregation of the complexes. Phenomena of phase separation include turbid symplex dispersions, symplex flocs, and coacervation into a polymer-rich and a polymer-poor phase. The subdivision into macroscopic homogeneous systems and turbid systems is often rather unsharp because it is an arbitrarily nonphysical one, depending on the solution power of the human eye.

PROPERTIES

Colloidal Stability in Solution

We turn to the colloidal stability of the resulting complex aggregates. The phase behavior at the titration endpoint can be thus classified: stable dispersions, film-forming systems, and flocculating systems. The last one can be further subdivided into large voluminous and small sticky flocculent systems.[3]

Note the following general trends:

- Flocculation occurs when strong-acid, fully ionized polyanions and strong basic polycations of high charge density are used.
- When we use weak polyelectrolyte components of low molecular mass, dispersions predominate. By increasing the molecular mass we increase the flocculation tendency.
- Films form favorably when the two components vary significantly in molecular mass.
- The hydrophoby of the components influences the floc behavior; that means an increase of the hydrophobic portion of the polyelectrolytes results in the formation of "sticky" systems.

Symplex Precipitates

The precipitated polyelectrolyte complex hydrogels can be isolated, washed, and characterized. The general properties of solid symplexes are quite similar to those of other hydrogels: hard and brittle solids and leathery or rubbery when saturated with moisture.

Stoichiometry

The stoichiometry of precipitated and purified symplexes is normally determined by elemental analysis. With many systems of linear structure, 1:1 symplexes formed even at a rather large difference in charge density.[5]

Swelling Behavior

The swelling behavior of symplexes is important for many applications. Normally, the swelling minimum occurs at the equivalence point of cationic to anionic groups.[6]

Supermolecular Structure and Morphology

On the supermolecular level, the complex aggregates can be described by two borderline models: a ladder model with well-defined electrostatic bonds between the charge centers and the Michaels "scrambled-egg" model with a more statistical charge compensation.[4]

APPLICATION

Generally, we consider two areas of application: process-oriented or product-oriented.

Symplex Formation as a Process

Potential areas of symplex formation as a process are as follows:

- binding of pharmaceutical products
- immobilization of enzymes and biological substances
- flocculation in water processing
- analyzing polyelectrolytes in laboratory analyses

Symplexes as Polymeric Materials

Symplexes may be used as the following polymer materials: membranes, coating films, and fibers, microcapsules, and implants. Their properties such as their high degree of hydrophilicity, biocompatibility, mechanical strength in the swollen state, and permeability impart special advantages for medical use.

Bixler and Michaels successfully applied polyelectrolyte complex membranes consisting of polyelectrolyte components with strong acid and strong basic functional groups as ultrafiltration membranes.[7] Antithrombogenic properties and high dialysis coefficient for blood toxins make symplex membranes important for artificial kidneys.[7]

Symplexes can be applied, generally, as prosthetic materials for body repair, coatings and components of heart valves and artificial hearts, or as contact lenses because of their properties in the swollen state and good biocompatibility. Other symplex products are of minor importance compared to symplex membranes.

REFERENCES

1. Falb, R. D.; Grode, G. A.; Takahashi, M. T. et al. *Characteristics of Heparinized Surfaces* Advances in Chemistry 87; American Chemical Society: Washington, DC, 1968.
2. Tsuchida, E.; Osada, Y. *Makromol. Chem.* **1974**, *175*, 593.
3. Kötz, J.; Köpke, H.; Schmidt-Naake, G, et al., submitted for publication in *Polymer.*
4. Michaels, A. S.; Miekka, G. *J. Phys. Chem.* **1961**, *65*, 1765.
5. Philipp, B.; Linow, K-J.; Dautzenberg, H. *Acta Chim. Hung.* **1984**, *117*(1), 67.
6. Vogel, M. K.; Cross, R. A.; Bixler, H. J. et al. *J. Macromol. Sci. Chem.* **1970**, *4*, 675.
7. Bixler, H. J.; Michaels, A. S. *Encyc. Polym. Sci. Tech.* **1969**, *10*, 765.

POLYELECTROLYTE COMPLEXES (In Immunology)

Mamed I. Mustafaev
Department of Chemistry
Baku State University

A. Sezai Saraç
Department of Chemistry
Istanbul Technical University

POLYELECTROLYTES

In this article we attempt to interpret the growing literature dealing with immunologically active synthetic polymers, specifically the important physico-chemical characteristics of the macromolecules. There has been a steady increase in the number of publications about immunologically active synthetic high molecular weight compounds of polyelectrolyte (PE) origin: macromolecules bearing positive and negative charges or capable of being charged in physiological conditions have been shown to have immunoadjuvant activity in several model systems.

POLYELECTROLYTE COMPLEXES

Mixing of PE with soluble bifunctional low molecular weight complex-forming ligands allowed us to obtain soluble polyelectrolyte complex (PEC) and to control the hydrophobic-hydrophilic balance of their macromolecules, their conformation and, correspondingly, their supramolecular structure by using very simple experimental designs. Thus, we established a correlation between the structure of PEC and their immunomodulating activity.

Polyelectrolyte–SAS Complexes

One of the approaches to compactization of PE macromolecules in aqueous systems is the loading of polyionic chains with lateral nonpolar radicals or surfactants (Surface active substance—SAS). This can be achieved either through covalent binding of SAS to functionally important groups of PEs or by their binding, that is, the formation of mobile electrostatic bonds between SAS and PE.[4,6,14,32–41] At a definite level of equilibrium between the free energy of PE and its hydrophobicity, hydrophobization of the macromolecule leads to marked conformational changes, such as compactization of PE chains.

We performed studies on mice immunized with SAS or PE–SAS complexes at varying concentrations of SAS (PE concentration was 0.8–1.0 mg per animal; the sRBC dose was 10^7). The formation of complexes at low values of the [SAS]/[PE] ratio was accompanied by the augmentation of the immunostimulating activity of PE. The PE–SAS complexes having a more folded structure of their constituent components carrying adhesive hydrophobic groups of SAS in the polyionic molecule enabled the adjuvant effect of PE.

Polyelectrolyte-Metal Complexes (PMC)

The increasing interest in PMCs is primarily due to the crucial role of metal ions in biological processes and also to the unique capabilities of the PMC proper, whose physicochemical characteristics differ drastically from those of the original components, polymer and metal (Me).[35–41]

Data from physico-chemical analyses of the PE behavior in aqueous solutions containing low molecular weight complex-forming additives (SAS, transient metal ions) as well as immunological data provide compelling evidence that in all the systems under study the mechanisms underlying the immunomodulating effects of PEC are universal, coupled with the structural and chemical transitions and conformational state of PEC.

Polyelectrolyte-Hapten

The first artificial antigen based on a synthetic PE is the electrostatic complex of trinitrophenol (TNP) with poly(2-methyl-5-vinylpyridine) and poly(4-vinylpyridine) with molecular mass 10^5.[42]

Polyelectrolyte-Protein (Polysaccharide)

Another group of artificial immunogens based on nonnatural PE carriers includes polyelectrolyte complexes (or conjugates) of proteins, polypeptides, and polysaccharides antigens.[1–3,5,7–31] It has been shown that the binding of weakly immunogenic antigens of different origin to synthetic PE intensifies the immune response of organisms to the test antigens.

REFERENCES

1. Kabanov, V. A.; Mustafaev, M. I.; Nekrasov, A. V. et al. *Dokl. Akad. Nauk. SSSR* **1984**, *274*, 998.
2. Kabanov, V. A.; Mustafaev, M. I.; Petrov, R. V. et al. *Proc. Third Joint Sov.-Swed. Symp. On Phys.-Chem. Bio.* **1983**, 23.
3. Petrov, R. V.; Mustafaev, M. I.; Norimov, A. S. *Sov. Med. Rev., D. Immunol.* Harwood Acad; Publ. GmbH U.K. **1992**, 1–113.
4. Mustafaev, M. I.; Goncharov, V. V.; Evdakov, V. P. et al. *Dokl. Akad. Nauk SSSR* **1975**, *225*, 721–724.
5. Kabanov, V. A.; Mustafaev, M. I.; Norimov, A. S. et al. *Dokl. Akad. Nauk. SSSR* **1978**, *243*, 1130–1133.
6. Kabanov, V. A.; Mustafaev, M. I.; Goncharov, V. V. *Dokl. Akad. Nauk SSSR* **1979**, *244*, 1261–1265.
7. Kabanov, V. A.; Mustafaev, M. I.; Goncharov, V. V. et al. *Dokl. Akad. Nauk. SSSR* **1980**, *250*, 1504–1507.
8. Khaitov, R. M.; Mustafaev, M. I.; Norimov, A. S. et al. *Bull. Exp. Bio. and Med.* **1985**, *11*, 597–600.
9. Khaitov, R. M.; Mustafaev, M. I.; Norimov, A. S. *Immunologiya* **1986**, *2*, 22–24.
10. Mustafaev, M. I.; Norimov, A. S. *Immunologiya* **1988**, *6*, 30–35.
11. Mustafaev, M. I. *Proc. Sec. All-Union Symp. on Interpolymeric Complexes* **1989**, 326–330.
12. Mustafaev, M. I.; Norimov, A. S. Zavgorodny, S. G. et al. *Immunologiya* **1989**, *4*, 88–89.
13. Mustafaev, M. I.; Babakhnin, A. A.; Popov, A. N. et al. *Molek. Bio.* **1990**, *24*, 358–369.
14. Mustafaev, M. I.; Man'ko, V. M.; Sokolova, E. A. et al. *Immunologiya* **1990**, *6*, 48–51.
15. Mustafaev, M. I.; Norimov, A. S. *Biomed. Sci.* **1990**, *1*, 274–278.
16. Norimov, A. S.; Mustafaev, M. I.; Romanova, R. Y. et al. *Bull. Exp. Bio. and Med.* **1990**, *12*, 78–82.
17. Petrov, R. V.; Khaitov, R. M.; Norimov, A. S. et al. *Dokl. Akad. Nauk SSSR* **1979**, *249*, 249–252.
18. Petrov, R. V.; Kabanov, V. A.; Khaitov, R. M. et al. *Zh. Mikrobio.* **1981**, *2*, 58–63.
19. Petrov, R. V.; Khaitov, R. M.; Norimov, A. S. et al. *Immunologiya* **1982**, *6*, 52–54.
20. Petrov, R. V.; Kabanov, V. A.; Khaitov, R. M. et al. *Dokl. Akad. Nauk. SSSR* **1985**, *283*, 744–748.
21. Abramenko, T. V.; Vinogradov, I. V.; Mustafaev, M. I. et al. *Zh. Mikrobio.* **1983**, *11*, 86.
22. Goncharov, V. V.; Mustafaev, M. I.; Frolov, T. V. *Twelfth All-Union Conf. on Electron Microscopy* 1982; 27.
23. Mustafaev, M. I.; Norimov, A. S. *First All-Union Cong. on Immunol.* 1989; 160.
24. Petrov, R. V.; Mustafaev, M. I.; Norimov, A. S. et al. *Dokl. Akad. Nauk. SSSR* **1990**, *312*, 505.
25. Popov, A. N.; Mustafaev, M. I.; Voitenko, V. G. et al. *Second All-Union Symp. on Interpolymeric Complexes* 1989; 327.
26. Vinogradov, I. V.; Kabanov, V. A.; Mustafaev, M. I. et al. *Dokl. Akad. Nauk. SSSR* **1981**, *25*, 1504.
27. Khaitov, R. M.; Mustafaev, M. I.; Norimov, A. S. et al. *Bull. Exp. Bio. and Med.* **1985**, *11*, 597.
28. Petrov, R. V.; Kabanov, V. A.; Mustafaev, M. I. et al. *Molek. genet., mikrobio. i virusol.* **1986**, *6*, 30.
29. Zemskov, V. M.; Khramsov, A. V.; Mustafaev, M. I. et al. *Immunologiya* **1989**, *5*, 49.
30. Petrov, R. V.; Mustafaev, M. I.; Norimov, A. S. *Biomed. Sci.* **1991**, *1*, 250.
31. Mustafaev, M. I.; Norimov, A. S. *Biomed. Sci.* **1990**, *1*, 274.
32. Schwartz, A.; Perry, J.; Burch, J. *Surfaktanty i detergenty*; Inostrannaya Literatura: Moscow, USSR, 1960; p 145.
33. Anufrieva, E. V.; Panarin, E. P.; Pautov, V. D. *Vysokomol. Soedin.* **1981**, *6*, 1222.
34. Bekturov, E. A.; Legkunets, R. E. *Assotsiatsiya polymerov s melkimi molekulami*; Nauka: Alma-Ata, USSR (Kazakhstan), 1983; p 155.
35. Man'ko, V. M.; Sokolova, E. A.; Gadzhiev, R. I. et al. *Immunologiya* **1991**, *1*, 75.
36. Carraher, C. E. In *Biomedical and Dental Applications of Polymers*; Genelein, C. G.; Koblitz, F. F., Eds.; Plenum: New York, NY, 1980, p 215.
37. Rothenbacher, H.; Schermenn, A. R. *J. Nutrition* **1980**, *110*, 1648.
38. Prohaska, I. R.; Zukasewyez, O. A. *Science* **1983**, *213*, 559.
39. Hart, D. A. *Cell Immunol.* **1982**, *71*, 159.
40. Hart, D. A. *Cell Immunol.* **1982**, *71*, 169.
41. Fracex, P. I.; Haas, S. M. *J. Nutrition* **1977**, *107*, 1889.
42. Petrov, R. V.; Evdakov, V. P.; Khaitov, R. M. et al. *Dokl. Akad. Nauk. SSSR* **1977**, *236*, 1260.

POLYELECTROLYTE COMPLEXES (Targeting of Nucleic Acid)

Alexander V. Kabanov
Department of Pharmaceutical Sciences
College of Pharmacy
University of Nebraska Medical Center

The delivery of nucleic acids into intact cells is a vital step in many biotechnological methods as well as in rapidly developing gene therapies.[1–7] This chapter reviews the formation and behavior of the polyelectrolyte complexes of nucleic acids and discusses some key results on their use in the delivery of genetic material into cells.

POLYELECTROLYTE COMPLEXES OF NUCLEIC ACIDS

Formation and Solution Behavior of the Complexes

The formation and behavior of the polyelectrolyte complexes of the nucleic acids are very similar to those of complexes formed by synthetic polyions.[8–11] These complexes are spontaneously produced during the mixing of nucleic acid and polycation aqueous solutions.[13,14]

Properties of Nucleic Acids in the Complexes

Incorporation of the nucleic acid in the polyelectrolyte complexes leads to significant changes in its properties, specifically, in its compaction, recharging, and stabilization against digestion by enzymes. These property changes have very important implications on the targeting of the nucleic acids. The binding of polycations induces DNA compaction because of the compensation of electrostatic charges of DNA and hydrophobic interactions of the complexes sites. As a result of compaction, the penetration of DNA in tissues and its incorporation in endocytic vesicles that provide for biopolymer transport into a cell is facilitated.

CONCLUSION

The polyelectrolyte complex is one of the most promising tools for gene delivery. Until now, significant efforts have been made in the development of DNA carriers on the basis of polylysine conjugates capable of specific interactions with target cell receptors. However, the practical significance of this approach for *in vivo* gene delivery and transfection may be hindered by the problem of immunoreactivity of such conjugates, an area which has not yet been considered. The immune response of an organism to the haptene is usually significantly amplified after its attachment to a polycation.[15,16] Such conjugates may cause immunological reactions and related side effects. Therefore we believe that much effort must be made to develop delivery systems which do not contain biospecific components and which are not recognized by the immune system.[12] One promising example of such systems is a cationic block copolymer comprising poly(ethylene oxide) and polyspermine segments.[17] This copolymer forms a micelle-like complex with DNA in which mutually neutralized polycation and DNA chains comprise a hydrophobic core and poly(ethylene oxide) chains provide for a nontoxic and nonimmunogenic shell. Similar systems are currently being developed in several laboratories worldwide.

REFERENCES

1. *DNA Cloning: A Practical Approach*; Glover, D. M., Ed.; IRL: Oxford, U.K., 1985; Vols. 1 and 2.
2. Karlsson, S. *Blood* **1991**, *78*, 2481.
3. Einerhand, M. P.; Valerio, D. *Curr. Top. Microbiol. Immunol.* **1992**, *177*, 217.
4. Morsy, M. A.; Mitani, K.; Clemens, P. et al. *JAMA* **1993**, *270*, 2338.
5. Dorudi, S.; Nothover, J. M.; Yile, R. G. *Br. J. Surg.* **1993**, *80*, 566.
6. Walsh, C. E.; Liu, J. M.; Miller, J. L. et al. *Proc. Soc. Exp. Biol. Med.* **1993**, *204*, 289.
7. *Gene Therapy for Neoplastic Diseases*; Huber, B. E.; Lazo, J. S., Eds.; New York Academy of Sciences: New York, NY, 1994.
8. Kabanov, V. A. *Pure Appl. Chem., Macromol. Chem.* **1973**, *8*, 121.
9. Lysaght, M. J. In *Polyelectrolytes*; Frisch, K. C.; Klempner, A. V.; Patsis, A. V., Eds.; Technomic: Westport, CT, 1976; p 34.
10. Tsuchida, E.; Abe, K. *Adv. Polym. Sci.* **1982**, *45*, 1.
11. Kabanov, V. A.; Zezin, A. B. *Macromol. Chem. Suppl.* **1984**, *6*, 259.
12. Kabanov, A. V.; Alakhov, V. Y.; Chekhonin, V. P. In *Sov. Science Review D. Physiochemical Biology*; Skulachev, V. P., Ed.; Harwood: Glasgow, Scotland, 1992; Vol. 11, Part 2, p 1.
13. Miller, I. R.; Bach, D. *Biopolymers* **1968**, *6*, 169.
14. Kabanov, V. A.; Kabanov, A. V.; Astafieva, I. V. *Polym. Prepr.* **1991**, *32*, 592.
15. Kabanov, V. A.; Petrov, R. V.; Khaitov, R. M. In *Sov. Science Review D. Physiochemical Biology*; Skulachev, V. P., Ed.; Harwood: Glasgow, Scotland, 1984; Vol. 5, p 277.
16. Kabanov, V. A. *Macromol. Chem., Macromol. Symp.* **1986**, *1*, 101.
17. Kabanov, A. V.; Vinograder, S. V.; Suzdaltseva, Yu, G. et al. *Bioconjugate Chem.* **1995**, *6*, 639–643.

Polyelectrolyte-Lipid Complexes

See: *Polyelectrolyte-Surfactant Complexes (Solid State Morphology)*

Polyelectrolyte-Surfactant Complexes

See: *Micelle-Polymer Complexes*
Polyelectrolyte Complexes (In Immunology)
Polyelectrolyte-Surfactant Complexes (Binding Isotherms)
Polyelectrolyte-Surfactant Complexes (Solid State Morphology)

POLYELECTROLYTE-SURFACTANT COMPLEXES (Binding Isotherms)

Toshio Shimizu
Department of Information Science
Faculty of Science
Hirosaki University

Surfactant ions interact strongly with oppositely charged polyions, and the binding has been recognized as site specific. The binding isotherms of surfactant ions by polyions of opposite charge are usually very steep, indicating a cooperative nature of binding similar to micelle formation. The electrostatic interaction is the primary factor for determining this strong and site-specific character of the binding. The highly cooperative binding is ascribed to the contribution of hydrophobic interaction between alkyl chains of neighboring bound surfactants.[1–25]

In this article we report binding isotherms for the binding of cationic surfactants [dodecylpyridinium ions (DP^-)] by alternating copolymers of maleic acid and ethylene (MAE), ethyl vinyl ether (MAEVE), styrene (MASt), and indene (MAIn), in the presence of NaCl, based on the use of a surfactant-ion-selective, solid-state membrane electrode.

TWO-STEP BINDING ISOTHERMS

In the binding isotherms for MAEVE, MASt, and MAIn, a two-step binding behavior can be seen: the first binding mode shifts to the lower concentrations region with increasing size of the hydrophobic side chain on the copolymers, and the second step is in almost the same position as the one-step isotherm observed for MAE.

From these observations of the behavior of the two-step binding isotherms, it seems reasonable to conclude that in the first binding mode the surfactants are bound not only by an electrostatic but also a hydrophobic interaction.

There is only one charged carboxyl available to combine with the side chain in the monomeric unit, in the case of $\alpha = 0.5$. It is reasonable to assume that a number of ionized groups will remain isolated from the hydrophobic side chains. These unoccupied charged carboxyls would contribute independently to the binding of the surfactants, similar to the case of $\alpha = 1.0$.

SALT CONCENTRATION DEPENDENCE

A distinctive feature of the salt concentration dependence of the binding isotherms for MASt is that the two-step behavior is more obvious at higher salt concentration. The binding site of the first mode may be a combination of a charged site with a hydrophobic site, whereas the second binding mode may be ascribed to an isolated charge.

THE PH CHANGE: UPTAKE AND RELEASE OF H^+ IONS

Values for pH increase, accompanied by the sharp rise in the binding isotherms, and then decrease in the leveling-off stage of the binding process.

As the cooperative binding proceeds, bound surfactants lying close to each other form themselves into clusters on the polyion. The clusters of the bound surfactants would make the environment more apolar in their vicinity; as a result the dielectric constant in the neighborhood of the polyion decreases to a great (or small) extent.

ACKNOWLEDGMENT

The author is grateful to Dr. J. C. T. Kwak of Dalhousie University for his continued collaboration and encouragement, and also to Dr. K. Hayakawa of Kagoshima University for his continued help.

REFERENCES

1. Goddard, E. D.; Hanmnan, R. B. *J. Coll. and Interface Sci.* **1976**, *55*, 73.
2. Satake, I.; Yang, J. T. *Biopolymers* **1976**, *15*, 2263.
3. Hayakawa, K.; Kwak, J. C. T. *J. Phys. Chem.* **1982**, *86*, 3866.
4. Hayakawa, K.; Santerre, J. P.; Kwak, J. C. T. *J. Phys. Chem.* **1983**, *87*, 506.
5. Malovikova, A.; Hayakawa, K.; Kwak, J. C. T. *J. Phys. Chem.* **1984**, *88*, 1930.
6. Malovikova, A.; Hayakawa, K.; Kwak, J. C. T. In *Relation Between Structure and Performance of Surfactants*; Rosen, M. J., Ed.; ACS Symposium Series: American Chemical Society: Washington, DC, 1984; p 226.
7. Shirahama, K.; Tashiro, M. *Bull. Chem. Soc. Jpn.* **1984**, *54*, 375.
8. Shimizu, T.; Seki, M.; Kwak, J. C. T. *Colloids and Surfaces* **1986**, *20*, 289.
9. Shimizu, T.; Seki, M. *Rep. Prog. Polym. Phys. Jpn.* **1985**, *28*, 21.
10. Skerjanc, J.; Kogej, K.; Vesnaver, G. *J. Phys. Chem.* **1988**, *92*, 6382.
11. Thalbery, K.; Lindman, B. *J. Phys. Chem.* **1989**, *93*, 1478.
12. Thalbery, K.; Lindman, B.; Bergfeldt, K. *Langmuir* **1991**, *7*, 2893.
13. Hayakawa, K.; Kwak, J. C. T. In *Cationic Surfactants: Physical Chemistry, Surfactant Science Series*; Rubingh, D. N.; Holland, P. M., Eds.; Marcel Dekker: New York, NY, 1991; p 189.
14. Lindman, B.; Thalberg, K. In *Interactions of Surfactants with Polymers and Proteins*; Goddard, E. D.; Ananthapadmanabhan, K. P., Eds.; CRC: Boca Raton, FL, 1993; Chapter 5.
15. Shimizu, T.; Minakata, A.; Kwak, J. C. T. *J. Polym. Sci. Phys. Ed.* **1985**, *23*, 1139.
16. Schwarz, G. *Eur. J. Biochem.* **1970**, *12*, 442.
17. Chu, D.; Thomas, J. K. *J. Am. Chem. Soc.* **1986**, *108*, 6270.
18. Chandar, P.; Somasundaran, P.; Turro, N. J. *Macromolecules* **1988**, *21*, 950.
19. Limbela, W. B.; Zana, R. *Colloids and Surfaces* **1986**, *21*, 483.
20. Gao, Z.; Wasylishen, R.; Kwak, J. C. T. *Macromolecules* **1989**, *22*, 2544.
21. Shimizu, T.; Kwak, J. C. T. *Colloids and Surfaces* **1994**, *82*, 163.
22. Shimizu, T.; Seki, M. *Rep. Prog. Polym. Phys. Jpn.* **1987**, *30*, 31.
23. Shimizu, T. *Colloids and Surfaces* **1994**, *84*, 239.
24. Shimizu, T.; Suzuki, K.; Sugiyama, N. et al. *Rep. Prog. Polym. Phys. Jpn.* **1985**, *28*, 25.
25. Shimizu, T. *Colloids and Surfaces* **1995**, *94*, 115.

POLYELECTROLYTE-SURFACTANT COMPLEXES (Solid State Morphology)

Markus Antonietti* and Christian Burger
Max-Planck-Institut für Kolloid-und Grenzflächenforschung

In contrast with the solution properties of polyelectrolyte–surfactant complexes (PE–Surfs), relatively little is known about the solid state structure and the material properties of these molecules. This is somewhat hard to understand since PE–Surfs easily redissolve in organic solvents and can be casted to films of high homogeneity and optical quality.

PHASE MORPHOLOGIES

A typical example for the order phenomena in solid PE–Surfs is given by the complex between polyacrylic acid and dodecyltrimethylammonium-counterions (PAA–C_{12}). The purified complex is a rubbery, highly deformable material which is highly birefringent in the oriented state.[1]

Although it is amorphous on a local length scale (as seen by wide angle X-ray scattering), we observe a set of extremely narrow small-angle scattering peaks which can be indexed according to a face centered cubic (FCC) lattice. Since we know from the mechanical as well as optical properties that the polymer phase should exhibit cylindrical domains, we propose the structure model shown in **Figure 1**.

POLYELECTROLYTE-LIPID COMPLEXES

Not only surfactant phases, but lipid phases too can be stabilized by complexation with polymeric counterions. First

*Author to whom correspondence should be addressed.

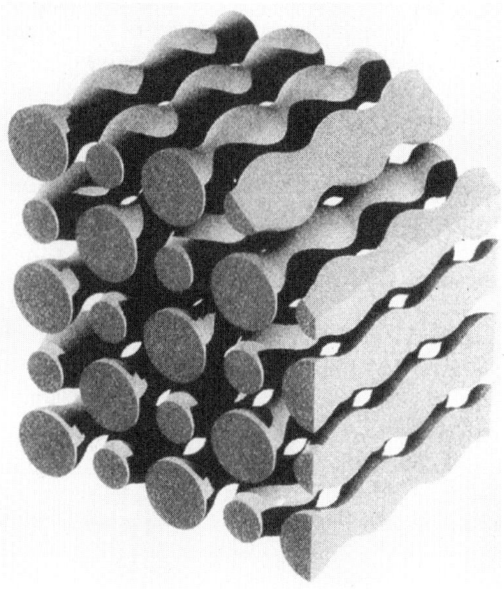

FIGURE 1. Model of the amphotropic phase structure of PAA–C_{12}.

experiments in this direction were described by Kunitake[2] and Okahata[3,4] who complexed model lipids with synthetic as well as normal polyelectrolytes. They found that the resulting materials can be dissolved in organic solvents; the casted films essentially maintain a stack-of-bilayers morphology as shown by X-ray analysis.

PE–SURFS AND PE–LIPIDS AS FUNCTIONAL MATERIALS

The high degree of order of PE–Surfs and PE–Lipids makes them likely as a matrix for functional groups which might be inserted as cosurfactants or even simply mixed in. Some special properties also arise from the functionality of the PE–Surfs themselves.

SUMMARY

The principle of polyelectrolyte–surfactant complexation allows the design and simple synthesis of a variety of highly ordered materials that are far from being thoroughly explored. The variation of polyelectrolyte properties (backbone, tacticity, distance between charges, hydrophobicity, etc.) and of surfactant properties (length and number of tails, disk- or platelet geometry, fluorinated surfactants, etc.) enables the combination of highly ordered, self-organizing materials with special properties of functional polymer systems.

REFERENCES

1. Antonietti, M.; Conrad, J. *Angew. Chem. Int. Ed.* **1994**, *33*, 1869.
2. Shimomura, M.; Kunitake, T. *Polym. J.* **1984**, *16*, 187.
3. Okahata, Y.; Taguchi, K.; Seki, T. *J. Am. Chem. Soc.* **1985**, *107*, 5300.
4. Okahata, Y.; Enna, G. *J. Phys. Chem.* **1992**, *92*, 4246.

Polyelectrolytes

See: *Amphiphilic Polymers (Binding Properties for Small Molecules)*
Amphiphilic Polymers (Fluorescence)
Bioartificial Materials
Biologically Active Polyanions
Block Copolymer Micelles (1. Structure and Properties)
Catalyses (by Polyelectrolytes and Colloidal Particles)
Cellulosic Ethers, Cationic
Conducting Polymers (Self-Acid-Doped, Conjugated)
Flocculants (Organic, Overview)
Flocculants, Cationic
Globular Polybases, Microemulsion-like
Hair and Skin Care Biomaterials (Cosmetology and Dermatology)
Hair and Skin Cosmetics
Hydrophobic Polyelectrolytes (Conformational Transitions)
Ion-Chelating Polymers (Medical Applications)
Ionene Oligomers
Ionomers (Solution Behavior)
Micelle-Polymer Complexes
Multilayer Films (Polyelectrolytes)
Personal Care Application Polymers (Acetylene-Derived)
Poly(arylene vinylene)s (Mechanistic Control of a Soluble Precursor Method)
Polyelectrolyte Complexes (Overview)
Polyelectrolyte Complexes (In Immunology)
Polyelectrolyte-Surfactant Complexes (Binding Isotherms)
Polyelectrolyte-Surfactant Complexes (Solid State Morphology)
Polyelectrolytes (Association of Hydrophobic Counterions)
Polyelectrolytes (Chain Models of Polymers)
Polyelectrolytes (Counterion Binding and Hydration)
Polyelectrolytes (Special Properties in Solution)
Polyelectrolytes (Solution Properties)
Polyelectrolytes, Cationic
Polyelectrolytes, Rigid-Rod [Poly(p-phenylene)-based]
Polynucleotide Analogues
Polysoaps
Poly(vinylpyridinium salts)
Polyvinyltetrazoles
Styrenesulfonic-Acid and its Salts (Polymerization and Applications)
Supported Thiazolium Salt Catalysts (for Acyloin Condensation)
Xanthan Gum (Conformation and Solution Properties)

POLYELECTROLYTES (Association of Hydrophobic Counterions)

Hiroshi Ochiai* and Tomoyuki Itaya
Department of Chemistry
Faculty of Education
Hiroshima University

Polyelectrolytes exhibit various interesting phenomena because of their dual character as highly charged electrolytes and flexible chain molecules.[1] One of the most characteristic properties in aqueous media is the counterion binding by the polyion. As the result of the counterion binding, a layer of highly concentrated counterions is formed around the polyion surface. Therefore, the solution properties of polyelectrolytes should be dependent on the nature of this ionic layer. If the counterion has a hydrophobic moiety, the counterion binding will be influenced by its hydrophobicity, chemical structure, bulkiness, and so on.[2,3] In addition, the hydrophobic counterions bound to the polyion can associate with each other through hydrophobic or van der Waals interactions, or both, around the polyion.[4] Thus, the interaction between the polyion and the hydrophobic counterions should result from a combination of two types of processes: one is the electrostatic interaction between a polyion and counterions, and the other is the hydrophobic interaction between the hydrophobic counterions themselves.

On the interaction between polyions and hydrophobic counterions, Duxbury has carried out several studies of the solution properties of polyelectrolyte-hydrophobic dye systems.[5] It is well known that the dyes are electrostatically bound to ionic sites of the polyelectrolyte and undergo effective aggregation, resulting from either hydrophobic interaction or some dye–dye interaction. Nevertheless, details of these interactions between dyes and the effects of polyelectrolytes on the solution properties are not clear-cut, because of their complicated structure. To elucidate the binding of hydrophobic counterions to the polyion and the interaction between hydrophobic counterions themselves around it, it is helpful to use structurally simple aromatic counterions such as alkylbenzenesulfonate, naphthalenesulfonate, and anthraquinone sulfonate ions.[6–10] In this article, the spectroscopically explored association behaviors of hydrophobic counterions bound to a polycation are described in relation to the abnormal viscosity and solubility behaviors of poly(allylammonium chloride) (PAAHCl) observed in their sodium salt solutions.

MATERIALS

Poly(allylammonium chloride) (PAAHCl) was purified by reprecipitation with acidic methanol (0.5 *M* HCl). All the hydrophobic counterions were used as the sodium salts: benzenesulfonate (BS), *p*-methylbenzenesulfonate (MBS), and *p*-ethylbenzenesulfonate (EBS).

ASSOCIATION OF COUNTERIONS BY HYDROPHOBIC INTERACTION[6,7,11]

In viscosity plots of PAAHCl in both NaBS and NaMBS solutions, we observe the typical viscosity behavior of polyelectrolytes. However, in NaEBS solutions the reduced viscosity η_{sp}/c rapidly decreases at lower polymer concentration when the solution is diluted with salt solutions of 6×10^{-3} *M* or higher. Moreover, the polymer precipitates readily above a salt concentration of 1×10^{-2} *M*. Judging from the fact that PAAHCl does not precipitate in aqueous 3 *M* NaCl solution, the EBS ion precipitates the polymer more readily than the simple hydrophilic ion, Cl^-.[12] In the PAAHCl–NaEBS system, the hydrophilic Cl^- around $PAAH^+$ is exchanged with a second hydrophobic counterion EBS upon dilution with NaEBS solutions. Consequently, the hydration state around $PAAH^+$ changes to a more hydrophobic character, which facilitates effective contraction and precipitation of the polymer coils.

ASSOCIATION OF COUNTERIONS BY CHARGE-TRANSFER INTERACTION[9,10]

We note that hydrophobic, stacking, and charge-transfer interactions between the counter ions are considerably influenced by the charge density of the polyion and the addition of another simple electrolyte such as NaCl, since the reaction field of interaction is offered essentially by the high electrostatic potential of the polyion.[7–10]

The importance of interactions between aromatic rings in aqueous media has been recognized in host–guest chemistry and protein stability. Generally, we cannot experimentally investigate hydrophobic or stacking interaction between aromatic compounds such as benzene, naphthalene, and anthracene in aqueous solution because of their poor solubility. However, the present polyion–hydrophobic(aromatic) counterion system is very suitable for clarifying the interaction between aromatic rings in aqueous media. In the future, the investigation of the interaction between aromatic counterions bound to the polyion will explicate the interaction between aromatic rings in detail. Simultaneously, that would be very advantageous for understanding the properties and functions of polyelectrolytes.

*Author to whom correspondence should be addressed.

REFERENCES

1. Oosawa, F. *Polyelectrolytes* Marcel Dekker: New York, NY, 1971.
2. Satoh, M.; Yoda, E.; Hayashi, T.; Komiyama, J. *Macromolecules* **1989**, *22*, 1808.
3. Satoh, M.; Yoda, E.; Komiyama, J. *Macromolecules* **1991**, *24*, 1123.
4. Herkstoreter, W. G.; Matic, P. A.; Hartman, S. E. et al. *J. Polym. Sci., Polym. Chem. Ed.* **1983**, *21*, 2473.
5. Duxbury, D. F. *Chem. Rev.* **1993**, *93*, 381.
6. Itaya, T.; Ueda, K.; Ochiai, H. *Polymer J.* **1992**, *24*, 539.
7. Itaya, T.; Ueda, K.; Ochiai, H.; Imamura, A. *Polymer J.* **1993**, *25*, 545.
8. Itaya, T.; Ochiai, H.; Ueda, K. et al. *Macromolecules* **1993**, *26*, 6021.
9. Itaya, T.; Ochiai, H.; Kawabata, Y. et al. *Bull. Chem. Soc. Jpn.* **1994**, *67*, 1538.
10. Itaya, T.; Kawabata, Y.; Ochiai, H. et al. *Bull. Chem. Soc. Jpn.* **1994**, *67*, 2047.
11. Itaya, T.; Ochiai, H.; Aoyama, T. et al. *J. Polym. Sci., Polym. Phys. Ed.* **1994**, *32*, 171.
12. Ochiai, H.; Handa, M.; Matsumoto, H. et al. *Makromol. Chem.* **1985**, *186*, 2547.

POLYELECTROLYTES (Chain Models of Polyions)

Esen A. Bekturov, L. A. Bimendina, and S. E. Kudaibergenov
Institute of Chemical Sciences
Kazakh Academy of Sciences

Polyelectrolytes are divided into polyacids, polybases, and polyampholytes. Depending on the charge density in the chain, polyelectrolytes are divided into weak and strong. The charge of weak polyelectrolytes is determined by dissociation constants of ionic groups and depends essentially on the pH of the solution. Strong polyelectrolytes in water solutions are totally ionized independent of the solution's pH.

Typical weak polyacids are poly(acrylic acid) and poly(methacrylic acid).

Strong polyacids are poly(ethylenesulfonic acid), poly(styrenesulfonic acid), and poly(phosphoric acid).

Weak polybases include poly(4-vinylpyridine), polyethyleneimine, and polyvinylamine.

Strong polybases can be obtained by complete alkylation of nitrogen (sulfur, phosphorus) atoms of weak polybases.

Polymer soaps are also strong polyelectrolytes.[1] Their molecules contain both ionic groups and long nonpolar side graftings.

In water media, macromolecules of polymer soaps form the intramolecular micelles stabilized by hydrophobic interactions of long nonpolar side chains.

Polyampholytes contain both acidic and basic groups which can possess different charges. The charge of polyampholyte changes depending on the pH of the solution.

Polymeric betaines are polyampholytes too. They have positive and negative charges in one link of their side chain, in which the counterions are connected with polyions by a chemical bond.

PECULIARITIES OF POLYELECTROLYTE SOLUTIONS

Polyelectrolytes are considerably distinguished from both nonionic polymers and low molecular weight electrolytes. Peculiarities of polyelectrolyte behavior are conditioned by the charges of the macromolecule.[2–5] Strongly diluted polyelectrolyte solution with flexible chains consists of small parts in which the polyions with a high density of fixed charges create high local electrostatic potentials that cause high concentration of counterions (ionic atmosphere). One can consider the ionic atmosphere as the spherical region about the central ion where opposite charges predominate. Dilution of the solution results in increased volume where counterions are distributed. Therefore the screening of fixed charges of polyions is decreased, their reciprocal repulsion is increased, and the polyions swell ("polyelectrolyte effect").

THE THEORY OF POLYELECTROLYTE SOLUTIONS

The theory of polyelectrolytes solution must describe the changes in distribution of fixed charges leading to the changes of polyion sizes.[8,9] However, the determination of polyion electrostatic potentional is a difficult task. One simple method is to calculate the potential of the electrostatic field for the polyion at a given conformation and a given disposition of fixed charges. One usually uses spherical models of polyions or models displaying their chain structure for calculating the electrostatic potential of the polyion.

SPECIFIC BINDING OF COUNTERIONS

The binding of counterions caused by coulombic interactions of charges is called the binding of ionic atmosphere and is predicted by general electrostatic theories.[6,7] Specific binding of counterions by polyions is determined by the size of the polyion charged group, the radius of hydrated and dehydrated counterions, and the energy of solvability of the counterions. Additional interactions can play an essential role in the binding of ions; hydrophobic interactions are an example.

The solubility of polyelectrolytes and proteins in salt solutions is influenced by the binding of small ions by polyions. As a rule the interaction of low molecular weight electrolytes in water solutions of polyelectrolytes decreases their solubility (salting-out effect). The salting-out effect is very useful for the purification of proteins.

APPLICATION OF POLYELECTROLYTES

Polyelectrolytes attract the attention of investigators because of their original properties such as solubility in water, ability to interact with charged particles and surfaces, and chemical and biological stability. They are widely used in various fields of industry, agriculture, medicine, biotechnology, and electronics, as flocculants, coagulants, structurization agents, prolongers of drugs, biocatalysts, sensors, and so forth. With the help of gels, membranes, and films of polyelectrolytes it is possible to regulate the water regime in the ground, to purify waste-water, to disinfect ground and water from radionuclides, and to create artificial nutrient media and muscles, as well as chemomechanical devices.

REFERENCES

1. Armstrong, R. W.; Strauss, U. In *Encyclopedia of Polymer Science and Technology*; John Wiley & Sons: New York, NY, 1969; Vol. 10, p 781.
2. Zezin, A. B. *Polyelectrolytes in Encyclopedia Polymers*; Sovetskaya Entsiklopedia: Moscow, 1977; Vol. 3, p 89.
3. Bekturov, E. A.; Bakauova, Z. K. *Synthetic Water-Soluble Polymers in Solutions*; Huttig & Wepf: New York, NY, 1986; p 242.
4. *Developments in Ionic Polymers* 2nd ed.: Wilson A, D.; Prosser, H. I., Eds.; Elsevier: London, U.K., 1986.
5. Eagland, D. *Solution Properties of Water-Soluble Polymers* In *Water Science Reviews*; Franks, F., Ed.; Cambridge Univ: Cambridge, U.K., 1985; Vol. 4, p 40.
6. Tanford, C. *Phisicheskaya Chimiya Polimerovs*: Chimiya: Moscow, 1965; p 574.
7. Moravetz, H. *Makromolekules in Rastvorakh*, Mir: Moscow, 1967.
8. Rice, S. A.; Nagasawa, M. *Polyelectrolyte Solutions*; Academic: New York, NY, 1961.
9. Oosava, F. *Polyelectrolytes*; Dekker: New York, NY, 1971.

POLYELECTROLYTES (Counterion Binding and Hydration)

Mitsuru Satoh* and Jiro Komiyama
Department of Polymer Chemistry
Tokyo Institute of Technology

The counterion binding phenomena of polyelectrolytes have been investigated in many areas because of essential participation in biological systems (e.g., conformational transition of DNA, polypeptides, and polysaccharides,[1] in ion-exchange membranes (recovery and separation of precious and heavy metals and ion selectivity), and in polyion gels (controlled release of medicines and phase transition). Theoretical elucidation, however, is behind experimental studies except for the Manning Theory.[2]

Counterion binding has been categorized into the following two binding modes: counterion condensation and site binding. (The Manning Theory is well known as a typical model for the former). Excess binding energies due to hydration change of counterions, which may explain ion specificities often observed for the counterion binding, has been incorporated into the Manning Theory.[3]

COUNTERION CONDENSATION

Theoretical formulation for the counterion condensation mode was given by Oosawa and Manning.[2,4,5] They used somewhat difficult polymer models and theoretical approaches to attain the same theoretical prediction: counterions are "condensed" onto polyions above a certain charge density.

Application of Manning Theory

The Manning Theory has been successfully tested for predictability on various solution properties of polyelectrolytes. For example, Kowblansky et al.[6] compared ion activity coefficients of sodium halides in aqueous polyphosphate (PP) and carrageenan solutions with the prediction by The Manning Theory. Agreement was satisfactory especially for the former polymer. Similar examinations have been made for counterion diffusion coefficients.[7,8]

A significant success of the Manning Theory may be analysis of helix–coil transition of DNA.

SPECIFIC BINDING AND DEHYDRATION

In specific counterion binding, the binding degrees are dependent by definition on the counterion species. Specificity should come from differential free energy changes for ion–pair formations in which no or one solvent molecules intervene between a counterion and a polymer charge. Therefore, some hydration (solvation) changes should accompany specific binding[9] just as in the site binding mode. Actually, the degrees of dehydration have been used as a measure of the specificity of binding.[10] Another and more direct measure of specificity may be a deviation from theoretical predictions, typically by the Manning Theory. If deviations from the prediction on, for example, counterion activity depend on the counterion species, the extent of the deviation may be correlated with differential free energy changes for the dehydration.

Structural Hydration Interaction (SHI) Model by Desnoyers[11] and Its Application to Polyion Systems

According to the model, ion specificity is explained in terms of structure maker and structure breaker. Small ions such as Li^- and F^- are structure makers because they construct a hydration structure due to their strong electrostatic field, and large ions such as Cs^+ and I^- are structure breakers because they destroy water structure around them due to their weak field.

It seems worthwhile to apply the SHI model to polyelectrolyte systems. According to the SHI model, specific counterion bindings are expected for a couple of polymer charged groups and counterions both of which have a like hydration structure, namely two structure makers or two breakers.

Counterion Binding and Hydrophobic Hydration

Counterion binding phenomena are complicated because of the coexistence of different (incompatible) hydration structures in a polyelectrolyte as well as mutual perturbation. This seems to prevent theoretical treatment for specific binding systems from being developed. Combined use of condensation theories such as the Manning Theory and the SHI model may be most effective for semi-quantitative understanding of specific counterion binding. However, extensive studies with computer simulation may be promising for prediction of counterion specificities.

REFERENCES

1. Nilsson, S.; Piculell, L. *Macromolecules* **1991**, *24*, 3804.
2. Manning, G. S. *J. Chem. Phys.* **1969**, *51*, 924.
3. Manning, G. S. *Acc. Chem. Res.* **1979**, *12*, 443.
4. Onishi, T.; Imai, N.; Oosawa, F. *J. Phys. Soc., Japan* **1960**, *15*, 896.
5. Oosawa, F. *Polyelectrolytes* Dekker: New York, 1971.
6. Kowblanskey, M.; Tomasula, M.; Ander, P. *J. Phys. Chem.* **1978**, *82*, 1491.
7. Lubas, W.; Ander, P. *Macromolecules* **1980**, *13*, 318.
8. Henningson, C. T.; Karluk, D.; Ander, P. *Macromolecules* **1987**, *20*, 1286.
9. Koda, S.; Nomura, H.; Nagasawa, M. *Biophys. Chem.* **1985**, *23*, 147.
10. Strauss, U. P.; Leung, Y. P. *J. Am. Chem. Soc.* **1965**, *87*, 1476.
11. Desnoyers, J. E.; Arel, M. et al. *J. Phys. Chem.* **1969**, *73*, 3346.

POLYELECTROLYTES (Special Properties in Solution)

Eckhard Nordmeier

COUNTERION BINDING

We can distinguish three different modes of counterion binding: territorial binding, site binding, and hydrophobic binding (adsorption).

Under purely long-range electrostatic interactions, counterions tend to be nonspecifically bound with polyions. They are accumulated around the polyion skeleton within a territory of

*Author to whom correspondence should be addressed.

volume V_p. At one extreme a territorial bound counterion is dehydrated and localized near a polyion group, but it is not covalent or ionic associated with the group. At the other extreme the counterion is delocalized. It retains the fully hydrated state while moving in an unrestricted and random way along the axis of the polyion chain.

If site-specific interactions are dominant, the counterions may bind on specific polyion sites. Then, they are fixed to a certain site and cannot move. Of course, the type of site binding can be different. It can be covalent, ionic, or non-ionic. Tight, localized binding is also expected when the nearest-neighbor charge spacing on the polyion is sufficiently close so that two or more groups can cooperate in holding the counterion.

Hydrophobic binding takes place when the counterions are organic. For instance, the binding of aromatic counterions by polyions containing the aromatic ring is particularly strong. However, for the biologist, territorial and site binding of inorganic counterions is of major practical interest.

DIFFUSION PROCESSES

Small ions play a dual role in polyion diffusion. First, they form clouds surrounding each polyion. The clouds reduce the polyion surface charge and thereby they reduce the magnitude of the polyion–polyion interactions. Second, the polyions move through neighbor clouds nearby. Since the small ions have their own dynamic there is a significant coupling of the small ions diffusion with the polyion diffusion.

Two different groups of theories on diffusion are known. The first group are the polyelectrolytes theories developed by Schurr et al. and Magdelenat et al. describing polyion spheres.[1,2] The second group are the nonpolyelectrolyte theories of Maeda et al. and Fujime et al. handling semiflexible rods.[3–5]

REFERENCES

1. Schurr, J. M. *CRC Crit. Rev. Biochem.* **1977**, *4*, 371.
2. Magdelenat, H.; Turq, M.; Chemia, M. *Biopolymers* **1976**, *15*, 175.
3. Maeda, T.; Fujime, S. *Macromolecules* **1984**, *17*, 1157.
4. Maeda, T.; Fujime, S. *Macromolecules* **1984**, *17*, 2381.
5. Fujime, S.; Maeda, T. *Macromolecules* **1985**, *18*, 191.

POLYELECTROLYTES (Solution Properties)

Ichiro Noda* and Yoshio Muroga
Department of Applied Chemistry
Nagoya University

The solution properties of chainlike polyelectrolytes can be roughly divided into electrochemical and polymeric properties. The electrochemical properties, such as osmotic coefficient, activity coefficient, and transference number, depend on charge density but not on molecular weight. The polymeric properties, such as intrinsic viscosity, sedimentation and diffusion coefficients, and viscoelastic properties, depend on the molecular weight as well as the charge density. For the study of polymeric properties, in particular, it is essential to use polyelectrolytes with narrow molecular weight distributions (NMWD).

*Author to whom correspondence should be addressed.

SOLUTION PROPERTIES

Electrochemical Properties

Even in aqueous solutions of strong polyelectrolytes, such as sodium polystyrenesulfonate (NaPSS), the electrochemical properties (e.g., the activity coefficient of the counterion) are extremely different from simple electrolyte solutions and are practically equal to the sum of the contributions from polyelectrolyte in salt-free solution and from simple salt solution when simple salt is added. These remarkable features of the electrochemical properties can be interpreted by assuming that most of the counterions are bound to a polyion so that the charges of polyions are almost cancelled. Thus, we call this phenomenon "ion-binding." However, this term does not literally mean binding; rather, the counterions are strongly attracted around a polyion skeleton by a high electrostatic field due to many charges fixed on a polyion.[2]

Polymeric Properties

The polymeric properties of polyelectrolytes are primarily related to both chain conformations and solution structures. The conformation of polyelectrolytes at the infinite dilution in the presence of abundant added salt can be obtained in a manner similar to that for non-ionic polymer solutions, but its determination is difficult at low added-salt concentration C_s, particularly in salt-free solutions, because it changes greatly with polymer concentration C.

Chain Conformation

In general, the conformation of polymer chains is determined by short- and long-range effects. Short-range effects are the interaction between neighboring monomers that gives rise to the chain stiffness. The chain stiffness is expressed in terms of a statistical segment length b or a persistence length p, which is equal to b/2 for a Gaussian chain. Long-range effects are the interaction between segments along the chain far apart; they are called excluded volume effects.

APPLICATIONS

Polyelectrolytes have been attracting much attention as functional polymers in applications because they have various characteristic abilities such as increasing the viscosity of aqueous fluids, dispersing or aggregating finely divided solids in aqueous mediums, and absorbing large amounts of aqueous solutions to achieve the Donnan equilibrium and hold them in polyion networks.[1] One or more of these properties are utilized when polyelectrolytes are used as additives to injection fluids for oil recovery, as flocculants to improve wastewater, or as polymeric absorbents on disposable diapers or sanitary napkins. Ion-exchange resins and ionic polymer membranes composed of crosslinked polyelectrolytes are used to produce pure water or NaCl from saline water and NaOH from aqueous NaCl solution. Moreover, polyelectrolytes have attracted much interest in biology and medicine.[3]

REFERENCES

1. *Water Soluble Polymers*; Bikales, N. M., Ed.; Plenum: New York, 1973.
2. Nagasawa, M. *Polyelectrolytes*; Sélegny, E. et al., Eds.; D. Reidel: Dordrecht, The Netherlands, 1974.
3. Hutchinson, F. G. In *Development in Ionic Polymers-2*; Wilson, A. D.; Prosser, H. J., Eds.; Elsevier: London, 1983; p. 303.

POLYELECTROLYTES, CATIONIC

Claude Loucheux
Université des Sciences et Technologies de Lille
Laboratoire de Chimie Macromoléculaire URA CNRS 351

Jeanne François
Institut Charles Sadron CNRS-ULP

Due to their easy synthesis and their large domain of application, synthetic cationic polyelectrolytes were developed long ago, in both academic and industrial laboratories. The present paper is voluntarily limited to nitrogen-containing cationic derivatives.

SYNTHESIS OF POLYBASES

In this section we review some important polybases, which are at the origin of cationic polyelectrolytes by neutralization, alkylation, and N-oxidation reactions.

Polyethyleneimine (PEI) and Derivatives

Commercial PEI is a highly branched material.[1] It is usually obtained by ring-opening polymerization of ethyleneimine with cationic initiators. Primary, secondary, and tertiary amino groups are present simultaneously: typically 25% primary amino groups, 50% secondary, and 25% tertiary.

Polyvinylpyridines

In the important family of polyvinylpyridines, i.e., poly(4-vinylpyridine) (P4VP), poly(2-vinylpyridine) (P2VP) and poly(2-methyl-5-vinylpyridine) (P2M5VP), the radical homo- and copolymerization of the corresponding monomers is well known.

Polyvinylimidazoles

The vinylimidazole group is very often introduced in polymeric systems of biological interest. Poly(*N*-vinylimidazole) and poly[4(5-)vinylimidazole] are obtained by conventional free radical polymerization of the corresponding monomers.

Polyvinylamidoamines

The polyaddition of primary monoamines or bis (secondary) amines to bis-acrylamides gives the large family of poly(amidoamine)s.[2]

Poly(aminoalkyl acrylate)s and Poly(aminoalkyl methacrylate)s

This family of polymers is especially interesting industrially. Poly(dimethylaminoethyl acrylate) is the main polymer in this series. The corresponding monomer results from a transesterification reaction of methyl acrylate by dimethylaminoethanol in the presence of a catalyst.

Polyvinylamine

Polyvinylamine is the simplest primary polyamine but, of course, it cannot be synthesized from a monomer. Laboratory synthesis allows obtaining pure well-known polymers. Hydrazinolysis of polyvinylphthalimide[3,4] or poly(*N*-vinylsuccinimide)[5] is a convenient method.

Polyallylamines

Polyallylamines are commercial products.[6] They can be obtained by free radical polymerization of the monomer.

Polysaccharide Derivatives

Chitosan, which is obtained by deacetylation of chitin, is interesting because chitin is the second most abundant source of natural polysaccharides (after cellulose). Chitosan bears a primary amine function and behaves like a classical polyelectrolyte.[7]

Another polysaccharide cationic polyelectrolyte is obtained from guar gum by reaction of primary alcohol functions of the heterocycles with 2,3-epoxypropyltrimethyl ammonium chloride.[8] All these polymers are produced for industrial purposes.

As a general remark, all amine monomers are more or less toxic and special care has to be taken for their handling.

REACTIONS OF POLYBASES FOR OBTAINING POLYELECTROLYTES

Acidic neutralization and alkylation of polyamines are direct ways of obtaining cationic polyelectrolytes. Both reactions are accompanied by conformational changes.

GENERAL PROPERTIES AND USES

A considerable amount of work has been devoted to the properties of polybases and their cationic polyelectrolyte derivatives in solution. The main interest in such polymers is related to their partial or complete solubility in water. In aqueous solution, their conformation results mainly from an equilibrium between the electrostatic repulsions and the hydrophobic or dipolar (hydrogen bonds) attractions and depends on various intrinsic and extrinsic parameters. The intrinsic parameters are the nature of the basic group (primary, second, or tertiary amine) and its basic strength, which varies with its position inside the main chain (PEI) or in lateral groups (P2VP and P4VP), the nature of the hydrophobic substituents, and the tacticity and the molecular weight of the polymer. The extrinsic parameters usually considered are the pH, the average ionic strength of the medium, the nature of the added salt, and the temperature.

Conformational Properties

In **Figure 1** we have schematized some typical behaviors of cationic polyelectrolytes in aqueous solution, which are expected when the electrostatic interactions dominate all the other types of interactions, and in the absence of specific ion–polyion interactions:

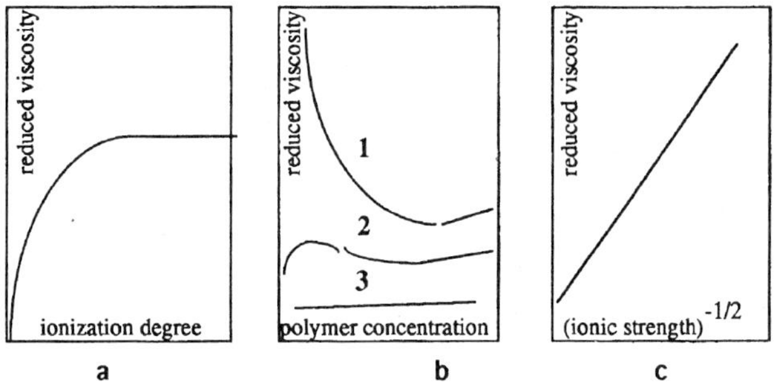

FIGURE 1. Typical viscosimetric behaviors of a polyelectrolyte in aqueous solution.

- The decrease of pH which induces the creation of charges is expected to provoke a chain expansion (represented here by the reduced viscosity η_{red} and up to a critical charge parameter x. At lower pH, the counterion condensation takes place and the chain dimensions do not vary (Figure 1a).
- The screening of the electrostatic interactions can be achieved either by increasing the polymer concentration itself or by increasing the concentration of a neutral salt. The typical curves of the polymer concentration dependence of η_{red} are shown in Figure 1b. Curve 1 schematizes the behavior observed in salt-free solutions. Curve 3 identical to the curves obtained for uncharged polymer and corresponds to a salt concentration large enough to completely screen out the electrostatic interactions. Curve 2 reflects the behavior expected for an intermediate salt concentration. Figure 1c represents the variation of η_{red} versus $I^{-1/2}$, the extrapolation at infinite ionic strength giving the dimensions of the chains without electrostatic interactions.

Industrial Uses

Cationic polyelectrolytes have industrial importance. Quaternized poly(dimethylaminoethyl acrylate)s and poly(diallyldimethylammonium chloride) are widely used in the treatment of urban or industrial waste waters. Copolymers with acrylamide of these monomers are used for the same purpose.

Many different cationic polyelectrolytes are or have been used in hair care [for instance, poly(diallyldimethylammonium chloride) as preconditioner and 2-hydroxypropyltrimethylammonium chloride ethers of hydroxyethylcellulose in shampoo formulations, etc.]. A number of these cationic polyelectrolytes are known commercially as "polyquaterniums".[9,10] Cationic guars are used industrially as thickeners, as retention agents for fibers (especially in the paper industry), as sedimentation and flocculation agents, and as conditioning agents for cosmetics and hair care products. Due to their positive charges, cationic polyelectrolytes induced the flocculation of colloidal particles, humic substances, etc. in water.[11,12]

In the presence of negatively charged polyelectrolytes, inter-polyelectrolyte complexes can be obtained.[13] Especially, nonstoichiometric inter-polyelectrolyte complexes[14] are a new family of enzyme carriers for α-chymotrypsin, urease, and penicillin amidase. These complexes give the possibility to use them as reversible immobilized enzymes. They combine the advantages of homogeneous catalysis and easy separation of products. Moreover, multienzyme systems are possible. The interaction of cationic polyelectrolytes with negatively charged biological macromolecules is well known (nucleic acids, proteins, etc.).

REFERENCES

1. Gaylord, N. G. In *Encyclopedia of Polymer Science and Technology* 1st ed.; Wiley: New York; 1964; Vol. 1, p 734.
2. Ferruti, P.; Barbucci, R. *Adv. Polym. Sci.* **1984**, *58*, 55.
3. Reynolds, D. D.; Kenyon, W. O. *J. Amer. Chem. Soc.* **1947**, *69*, 911.
4. Katchalsky, A.; Masur, J.; Spitnik, P. *J. Polym. Sci.* **1957**, *23*, 513.
5. Bayer, E.; Gekler, K.; Wingard, R. E., Jr. *Makromol. Chem.* **1980**, *181*, 585.
6. Boseki, Nitto *Technical Bulletin of PAA* **1989**, p 1–22.
7. Demarger-André, S.; Domard, A. *Carbohydrate Polymers* **1993**, *22*, 117.
8. Goddard, E. D. *J. Soc. Cosmet.* **1990**, *41*, 23.
9. *International Cosmetic Ingredients Dictionary*, Monographs p 571.
10. Lockhead, R. Y. *Cosmetics and Toiletries* **1988**, *103*, 23.
11. Rebhun, M. *Water Sci. Tecnol.* **1993**, *27*(11), 1.
12. Hirtzel, C. S.; Rajagopalan, R. *Colloidal phenomena* Noyes: Park Ridge, NJ, USA, 1985; p 50.
13. Bekturov, E. A.; Bakanova, Z. Kh. *Synthetic Water-Soluble Polymers in Solution*; Hüthig and Wepf: Basel, Switzerland; 1986.
14. Zezin, A. B.; Izumrudov, V. A.; Kabanov, V. A. *Makromol. Chem. Macromol. Symposium* **1989**, *26*, 249.

POLYELECTROLYTES, RIGID-ROD [Poly(*p*-phenylene)-Based]

Matthias Rehahn*
Polymer Institute
Karlsruhe University

Iris U. Rau
Kunststofflaboratorium, BASF AG

Synthetic rodlike polyelectrolytes are of growing importance in both fundamental and applied science.[1-3] In contrast to the more common flexible-chain polyelectrolytes, their persistence length is assumed to be independent of ionic strength. They therefore represent useful probes for exploring polyelectrolyte behavior in the absence of conformational effects. Moreover, rod-like polyelectrolytes attract attention because of their unusual rheological behavior and ability to form lyotropic mesophases.

Novel procedures have been developed for the preparation of poly(*p*-phenylene)-(PPP-)based polyelectrolytes. These successes were rendered possible by the excellent thermal and chemical stability of the rod-like PPP backbone[4-7] and by the

*Author to whom correspondence should be addressed.

highly efficient Pd-catalyzed polycondensation methods available for PPP synthesis.[8–12] These transition-metal-catalyzed coupling reactions tolerate a variety of functional groups in monomers and are therefore ideal for the preparation of PPP polyelectrolytes.

Pd-CATALYZED DIRECT SYNTHESIS OF PPP POLYELECTROLYTES

Wallow and Novak[13,14] first applied Pd-catalyzed polycondensation of bromoaromatics and benzeneboronic acid derivatives to prepare carboxylated PPP derivatives in aqueous solution (**Scheme I**).

Child and Reynolds[15] recently described the *in aqua* Pd-catalyzed synthesis of sulfonatoalkoxy-substituted PPP **6** (**Scheme II**).

PPP POLYELECTROLYTES VIA Pd-CATALYZED PRECURSOR SYNTHESIS

Rulkens, Schulze, and Wegner[16] recently succeeded in preparing a sulfonated PPP derivative **10** via its 4-methyl-phenylsulfonate-substituted precursor PPP **9** (**Scheme III**).

PROPERTIES

A conclusive picture of the properties of rigid-rod PPP polyelectrolytes is not yet available. Only some isolated data are published. Consequently, a short summary is provided here.

The carboxylated PPP polyelectrolyte **3**[17,18] dissolves in dilute aqueous base as its sodium, potassium, or triethylammonium salt. The free acid is completely insoluble in all common organic solvents. Concentrated solutions of **3** (>5% w/v) can be prepared in basic aqueous DMF (75:25 water/DMF). These solutions tend to form extremely viscous gels.

Liquid crystallinity is not observed in solutions of **3**; however, thin films cast from concentrated solutions of **3** are highly birefringent, confirming the polymer's anisotropic structure. Thermogravimetric analysis (TGA; under N_2) shows the onset of thermal decomposition above 520°C, while differential scanning calorimetry (DSC) shows no apparent softening temperature. Both methods indicate that **3** undergoes dehydration at 300–330°C to form a polyanhydride structure; this process is confirmed by IR spectroscopy.

For PPP polyelectrolyte **6**,[15] TGA measurements (under N_2) show an onset of decomposition at ~250°C. The polymer loses 30% of its mass between 300 and 350°C and retains 60% of its mass at 800°C.

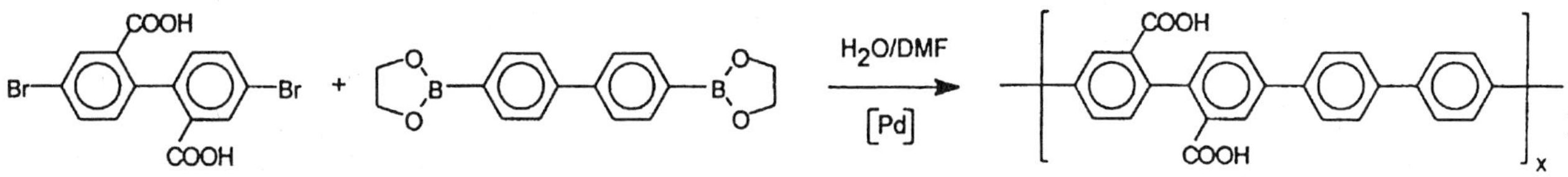

1 **2** **3**

I

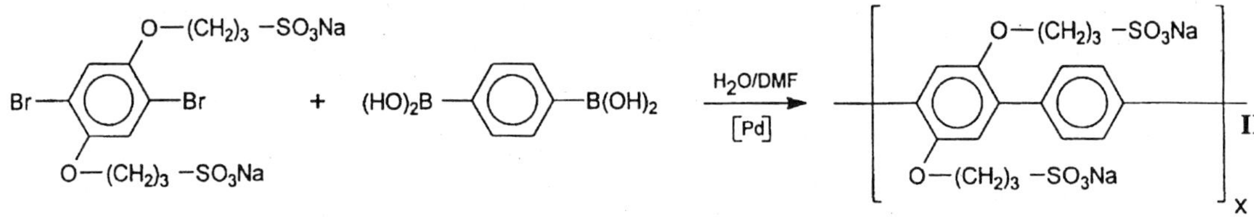

4 **5** **6**

II

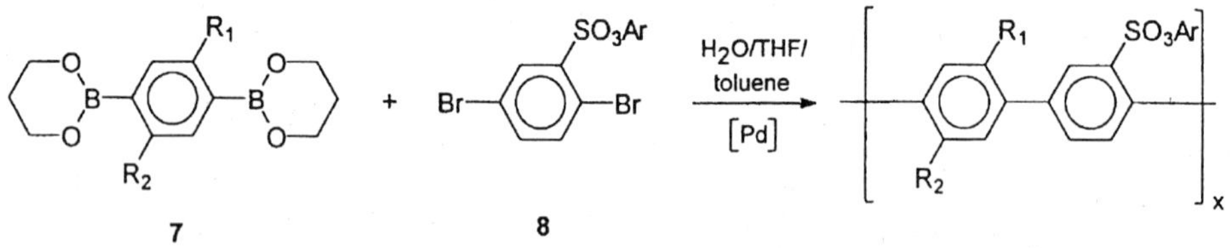

7 **8** **9**

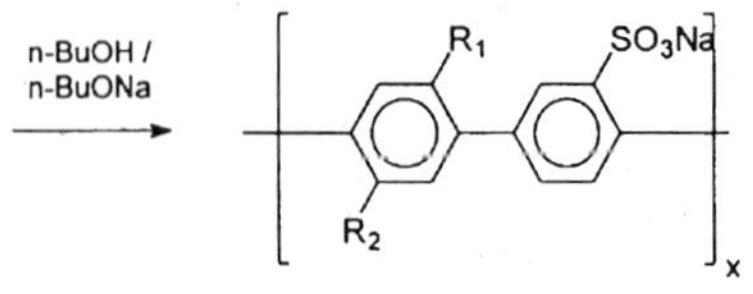

10

III

R1 = methyl
R2 = dodecyl
Ar = 4-methylphenyl

Solutions of PPP polyelectrolyte **10** in DMSO show birefringence above a concentration of ~5% (w/v).[16] Moreover, according to TGA measurements, sulfonated PPP polyelectrolyte **10** is more stable than its ester **9**, having two stages of weight loss: from 360°C to 500°C, there is a weight loss of 21%. Up to 600°C, **10** loses 84% of its original weight. DSC measurements show a small endothermic peak (20°C), indicative of the melting of the alkyl side chains, and a T_g at 130°C ($\Delta C_p = 0.2\ J{\cdot}g^{-1} \cdot K^{-1}$).

REFERENCES

1. Gray, F. M. *Solid Polymer Electrolytes*; VCH: New York, 1991.
2. MacCallum, J. R.; Vincent, C. A. *Polymer Electrolyte Reviews*; Elsevier: London, 1987.
3. Kesting, R. E. *Synthetic Polymeric Membranes* 2nd ed.; Wiley: New York, 1985.
4. Tour, J. M. *Adv. Mater.* **1994**, *6*, 190.
5. Gorman, C. B.; Grubbs, R. H. In *Conjugated Polymers* Bredas, J. L.; Silbey, R., Eds.; Kluwer Academic: Dordrecht, The Netherlands, 1991; p 1.
6. Naarmann, R. In *Conjugated Polymeric Materials: Opportunities in Electronics, Optoelectronics, and Molecular Electronics* Bredas, J. L.; Chance, R. R., Eds., Kluwer Academic: Dordrecht, The Netherlands, 1990; Vol. 182, p 11.
7. Elsenbaumer, R. L.; Shacklette, L. W. In *Handbook of Conducting Polymers* Skotheim, T. A., Ed.; Dekker: New York, 1986; Vol. 1, p 213.
8. Wallow, T. I.; Novak, B. M. *J. Org. Chem.* **1994**, *59*, 5034.
9. Schlüter, A-D.; Wegner, G. *Acta Polymer.* **1993**, *44*, 59.
10. Rehahn, M. et al. *Polymer* **1989**, *30*, 1060.
11. Miller, R. B.; Dugar, S. *Organometallics* **1984**, *3*, 1261.
12. Miyaura, N. et al. *Synth. Commun.* **1981**, *11*, 513.
13. Wallow, T. I.; Novak, B. M. *Polym. Prepr. (Am. Chem. Soc., Div. Polym. Chem.)* **1991**, *32*, 191.
14. Wallow, T. I.; Novak, B. M. *J. Am. Chem. Soc.* **1991**, *113*, 7411.
15. Child, A. D.; Reynolds, J. R. *Macromolecules* **1994**, *27*, 1975.
16. Rulkens, R. et al. *Macromol. Rapid Commun.* **1994**, *15*, 669.
17. Wallow, T. I.; Novak, B. M. *Polym. Prepr. (Am. Chem. Soc., Div. Polym. Chem.)* **1992**, *33*, 908.
18. Wallow, T. I.; Novak, B. M. *Polym. Prepr. (Am. Chem. Soc., Div. Polym. Chem.)* **1992**, *33*, 1218.

Polyemeraldine Salts

See: *Chiral Polyanilines*
Polyanilines, Oxidation States

Poly(enol ester)s

See: *Enol Ester Polymers*

POLY(ESTER AMIDE) AND POLY(ESTER IMIDE) RESINS (Binders for Air-Drying Protective Coatings)

P. H. Gedam* and V. S. Kishanprasad
Indian Institute of Chemical Technology

*Author to whom correspondence should be addressed.

Early development of poly(ester amide)s and poly(ester imide)s resulted from structural modification of polyamides and polyimides to easily processable products for insulating and high-temperature coatings, adhesives, laminates, and composites, but not for air-drying protective coatings.

A few disclosures have been made, mostly in patent literature, of air-drying poly(ester imide) resins.[1,2–4] They were prepared by reacting the fatty acids obtained from soybean and linseed oils, aromatic polycarboxylic acids, glycerol or other polyols, and aromatic diamines.

We will describe our work on the synthesis, characterization, and evaluation of air-drying oil-modified poly(ester amide)s[5–7] and poly(ester imide)s.[8,9] The resins having structural variations with respect to the unsaturated hydrocarbon chains and ester-amide/ester-imide linkages were designed by adopting the concept of oil length, which is commonly used in formulating alkyd resins having a wide range of film properties.

POLY(ESTER AMIDE) RESINS

Preparation

The compositions of linseed oil, diethanolamine, and phthalic anhydride were used as components of the poly(ester amide) resins of 55, 60, and 65% oil length.

Resin Characteristics

The data on the characteristic properties of the resins, such as the amine values, acid values, hydroxy values, densities, and molecular weight values, increase with decreasing oil length of the resin. The data on drying properties reveal that poly(ester amide) resins possessed good drying properties, the lower oil length resin drying faster than the longer oil length one.

Coating Properties

The data indicate that the poly(ester amide) resins and their iron-oxide paints possess excellent coating properties, including adhesion to mild steel substrate, tensile strength, and toughness of their free films.

POLY(ESTER IMIDE) RESINS

The air-drying poly(ester imide) resins of 60, 65, and 70% oil lengths were prepared from *N*-(carboxymethyl)trimellitimide (CMT), linseed oil, and glycerol.

Resin Characteristics

The results indicate that the properties of poly(ester imide) resins vary with their oil lengths. The acid values, hydroxy values, densities, and molecular weights increase with decreasing oil length of the resins.

Coating Properties

All the poly(ester imide) resins reported here have exhibited fast drying properties. The values of the adhesion, scratch hardness, tensile strength, and toughness of the zero-PVC paints, i.e., the 60, 65, and 70% oil length resin films, decrease with increasing oil length of the poly(ester imide) resins. This indicates that the lower oil length resin possesses superior coating properties.

CONCLUSION

The oil-modified poly(ester amide) and poly(ester imide) resins discussed here have excellent film-forming properties and can be used as binders in formulating air-drying protective coatings.

REFERENCES

1. Wright, H. J.; Dupris, R. N. *Ind. Eng. Chem.* **1946**, *38*, 1303.
2. Dr. Beck and Co. GmbH. France Patent 91 327, 1968.
3. Dr. Beck and Co. GmbH. Brit. Patent 996 649, 1965.
4. Miroslav, S.; Jaroslan, K. Czech. Patent 303 444, 1978.
5. Praveenkumar, T. K. et al. *J. Coat. Technol.* **1986**, *58*(737), 61.
6. Praveenkumar, T. K. et al. *Paint and Resin* **1988**, *10*, 13.
7. Praveenkumar, T. K. et al. *Farbe and Lacke* **1988**, *94*, 259.
8. Praveenkumar, T. K. Ph.D. Thesis, Osmania University, 1987.
9. Kishanprasad, V. S.; Gedam, P. H. *J. Appl. Polym. Sci.* **1991**, *43*, 1709.

Poly(ester amide)s

See: *Poly(ester amide) and Poly(ester imide) Resins (Binders for Air-Drying Protective Coatings)*
Poly(ester amides) (Overview)
Poly(ester imide) Compatibilizer (in Thermotropic Liquid Crystalline Polymer Blends)

POLY(ESTER AMIDE)S (Overview)

M. Srinivasan,* H. Sankarasubramanian, and P. S. Suresh
Dr. Bharat Ram Research and Development Centre
SRF Ltd.

The class poly(ester amide)s can be either a group of copolymers, blends, and alloys, or a combination of both. In this area the main interest has been on ordered poly(ester amide)s that have regular enchainment of ester and amide functionality and are semicrystalline with excellent fiber forming properties.[1–3] A number of reports on block poly(ester amide)s designed to obtain high tenacity yarns have also appeared.[5,6] Unsaturated poly(ester amide)s, block and random, have been prepared to improve the properties of the unsaturated polyester resins after crosslinking with styrene.[7,8] The present review focuses on the copoly(ester amide)s, which should be classed as random copolymers. They have shown excellent potential as sandwich adhesives for leather/nylon/leather and also as compatibilizers in polyester-polyamide blending. Surprisingly, these copoly(ester amide)s were functioning effectively at 5 to 10 phr level, creating excellent miscibility between the amides and esters with no phase separation in nylon 6 PEt blends.

The poly(ester amide)s[9] were prepared from a preformed polyesterdiacid. The well characterized poly(ester diacid) is evaluated for acid number and reacted with 10% excess of an appropriate diamine to get the polyester amine salt (PEtAS). The PEtAS is the subjected to copolymerization with two other monomers such as caprolactam(2) and N–6,6 salt in varying percentages of PEtASA.

The relative viscosities of poly(ester amide)s prepared from PEtAS remained low in the range of 1.06–1.09, but the terpolyesteramide prepared with varying weight ratios of PEtAS to N–6,6 salt and 40% caprolactam had relative viscosities in the range of 1.43–2.43.

The poly(ester amide)s have high melting points compared to the terpoly(ester amide).

All the terpoly(ester amide)s were found to be soluble in concentrated sulphuric acid, *m*-cresol, methonol, chloroform, and *N*-methylpyrollidone.

The thermal stability of the materials was studied using thermogravimetric analysis (TGA). The onset of decomposition was obtained between 240°C and 280°C.

Ay-Ling Liang et al. reported[10] the thermotropic liquid crystalline behavior of a series of aliphatic-aromatic poly(ether ester amide)s, which were prepared by direct polycondensation. All of the poly(ether ester amide)s exhibit thermotropic liquid crystalline behavior as observed by optical polarizing microscopy.

REFERENCES

1. Cartaldo, L.; De Candia, F.; Magolic, G.; Palumbo, R.; Strazza, G. *J. Appl. Polym. Sci.* **1982**, *27*, 1809.
2. Kricheldoy, H. R.; Koshig, J. *Eu. Polym. J.* **1978**, *14*, 923.
3. Katayame, S.; Mumkami, T.; Takahashi, Y.; Serita, H.; Buchi, Y.; Ito, T. *J. Appl. Polym. Sci.* **1976**, *20*, 975.
4. Preston, J. *J. Polym. Sci.* A-1, **1972**, *8*, 3135.
5. Akya, K. *Angew. Makromol. Chem.* **1969**, *7*, 79.
6. Huet, J. M.; Marechal, E. *Eu. Polym. J.* **1974**, *10*, 757.
7. Huezowshi, P.; Kapko, J. *Makromol. Chem.* **1978**, *179*, 571.
8. Malascot, G.; Jasse, B. *Eu. Polym. J.* **1975**, *11*, 669.
9. Srinivasan, M. SRF Ltd., Madras, India, unpublished results.
10. Liang, Ay-Ling; Hsu, Keh-Ying; Li, Chien-Hui; Chang, Teh-Cou *Polym.* **1992**, *33*, 2217.

POLYESTER FABRICS (Surface Modification by Low Temperature Plasma)

Masayuki Yamagami and Hirofumi Miyoshi
Research Institute for Advanced Science and Technology
University of Osaka Prefecture

Although synthetic fibers have disadvantages, such as electrostatic charging and poor water absorption, they have advantages over natural fibers in being tough, durable, and mass producible. As a solution to these disadvantages, however, low temperature plasma has recently drawn attention. Low temperature plasma is suitable for surface modification of fibers and fabrics without affecting their strength, because high energy particles, including electrons, react at very shallow positions, that is, within several nanometers, below the material's surface.[1]

There are three approaches to the surface modification of fibers or fabrics by low temperature plasma technology, as follows:

*Author to whom correspondence should be addressed.

SIMPLE PLASMA TREATMENT

This fundamental method is based on the fact that plasma treatment of fabrics with an inorganic gas, such as gaseous oxygen, nitrogen, or argon, results in the formation of active situs, etching, and hydrophilicity on the fiber surface. These phenomena have been studied for application to weight-loss processing of various synthetic fibers,[2] shrink-resistant finishing of wool,[3] hyperchromic processing of polyester fabrics (PET fabrics),[4–7] and sophisticated functionalization of cotton products.[8,9]

PLASMA POLYMERIZATION

Plasma treatment of various fabrics with a polymerizing gas, such as gaseous acetone (CH_3COCH_3)[10] or tetrafluoromethane (CF_4),[11] easily achieves surface modification as a result of adhesion of the resulting polymer to the fabric surface. However, this treatment poses a durability problem in the case of apparel products (which require washing fastness), because washing can release the adhering matter from the fabric surface.

PLASMA GRAFTING

Two methods are available for this treatment: simultaneous plasma discharge graft polymerization, in which plasma is discharged to a fabric impregnated with a monomer, typically a vinyl monomer; and plasma predischarge graft polymerization, in which a fabric is activated by plasma discharge in the absence of monomers, and immediately brought into contact with a monomer.[12] In the simultaneous discharge method, hydrophilicity is conferred by impregnating PET fabrics with polyethylene glycol, followed by plasma discharge;[13] plasma discharge results in homopolymer production simultaneously with monomer graft polymerization on the fiber. The resulting homopolymer can be lost upon washing, leading to deterioration of the surface-modifying effect. In the predischarge method, hydrophilicity is conferred by discharging oxygen plasma to PET fabrics and then bringing the fabric into contact with acrylic acid.[17] In this method, the monomer undergoes graft polymerization on the plasma-activated fiber, because the monomer is not directly exposed to the plasma. The modified fabric surface layer is therefore, dominated by chemically bound graft polymers and the functionalized fabric possesses improved durability.

PLASMA GRAFT POLYMERIZATION

The present study[15] is summarized as follows:

- Simple plasma treatment of PET fabrics for water absorbing processing led to the attachment of hydrophilic groups to the fabric surface, resulting in markedly improved water absorption. However, this water-absorbing effect decreased significantly 2–3 days later.
- Plasma polymerizing treatment of PET fabrics using ethylene trifluoride resulted in high water repellency as a result of polymer adhesion to the fabric surface. However, polymer adhering to the fabric surface dropped off upon washing, indicating lack of durability. Although plasma polymerizing treatment is one of the most common practices in thin-film formation and coating technologies, its application to apparel materials appears to be inappropriate.
- Investigation of plasma grafting treatment has been limited to a few studies, because of the high cost of the two-stage reaction system, or the use of vinyl monomers. In the present study, plasma grafting treatment using hydrophilic monomers conferred a high antistatic property upon PET fabrics. Also, plasma grafting treatment using water-repelling monomers conferred sophisticated function upon waterproofed fabrics. Both showed good washing fastness in the washing test.

REFERENCES

1. Rakowski, W.; Okoniewski, M.; Bartos, K.; Zawadzki, J. *Melliand Textilberichte* **1982–4**, *63*, 307.
2. Yasuda, T. *J. Appl. Polym. Sci., Appl. Polym. Symp.* **1984**, *38*, 201.
3. Wakita, T. *Dyeing Industry* **1987**, *35*, 306.
4. Hashizume, S. *Kakogijitsu* **1984**, *19*, 381.
5. Nishikawa, A.; Ejiri, K. Hayashi, F.; Miyazaki, K. *Dyeing & Finishing* **1990**, *42*, 409.
6. Nishikawa, A.; Miyazaki, K.; Ueno, M.; Ejiri, K. *Dyeing & Finishing* **1989**, *41*, 501.
7. Koide, K. Seki, M. *Dyeing Industry* **1989**, *37*, 64.
8. Kubota, S. *Dyeing Industry* **1988**, *36*, 66.
9. Miyahara, S.; Ohishi, K. *Dyeing & Finishing* **1993**, *45*, 401.
10. Wakida, T.; Niu, S.; Kawamura, H. *Textile Res. J.* **1993**, *63*, 433.
11. Goto, T.; Wakida, T. *Chem. Express* **1991**, *6*, 711.
12. Hirotsu, T. *Sen-i Gakkaishi* **1985**, *41*, 388.
13. Goto, T.; Wakita, T. *Journal of the Textile Society of Japan* **1990**, *46*, 192.
14. Pei, J. H.; Zhou, B. G. *Mater. Chem. Phys.* **1988**, *20*, 99.
15. Yamagami, M.; Taniguchi, T.; Hayashi, F.; Harigai, S. *Dyeing Industry* **1989**, *37*, 613.

POLY(ESTER IMIDE) COMPATIBILIZER (in Thermotropic Liquid Crystalline Polymer Blends)

Yongsok Seo,* Sang Mook Lee, Soon Man Hong, Seung Sang Hwang, Tae Suk Park, and Kwang Ung Kim
Polymer Processing Laboratory
Korea Institute of Science and Technology

For the past two decades, there have been intensive studies for the binary or ternary blends of flexible chain polymers and thermotropic liquid crystalline polymers (TLCPs) because TLCP has the potential to generate *in situ* reinforced composites of high mechanical performance and has a variety of processing options.[1–10] Most TLCPs are immiscible with general thermoplastics. Even though the mechanical properties of *in situ* composites can be enhanced by the formation of fine TLCP fibrils with a large aspect ratio in flow direction, the adhesion of composites at the phase boundary is weak due to immiscibility. How to improve the poor mechanical properties in the direction

*Author to whom correspondence should be addressed.

normal to flow while keeping high performance in the other direction has been the main concern of recent studies.

There are two general ways to improve the miscibility and hence the weak adhesion at the phase boundary. First is the synthesis of miscible TLCPs with blending matrix polymers. The miscibility of TLCP should be good to uniformly disperse TLCP in the matrix polymer, and the dispersed TLCP phase should be easily deformed to long fibril shape. Miscibility of commercially available TLCP has been investigated, but no break-through result has been obtained yet.[8,9] The second method is to provide compatibility (rather than miscibility) by the addition of a third component as a compatibilizer. This is a well-known technique in the polymer blending and compounding industries. However, it has not been applied to *in situ* composites yet. In this study, we present our recent results about the compatibilized *in situ* composite.

Poly(ether imide) (PEI), Ultem 1000 of General Electric (United States), was used as a matrix polymer. It has attracted a lot of attention for its ability to form *in situ* composites with various TLCPs due to its high thermal resistance and processibility in conventional processing machines.[4,10] The chemical structure of the repeating unit of Ultem 1000 is shown in **Structure 1**.

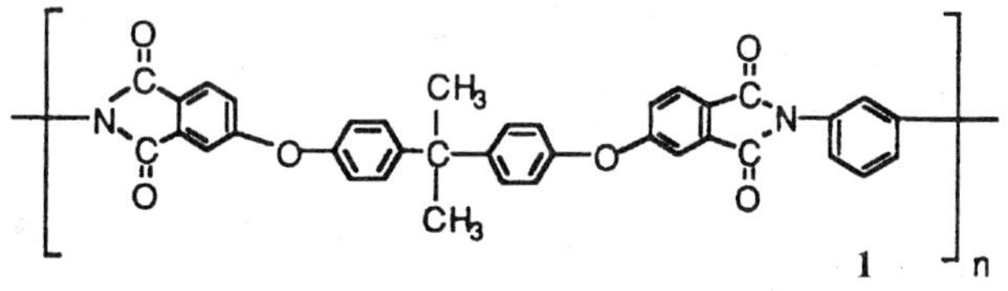

1

Thermotropic liquid-crystalline poly(ester amide) (PEA), Vectra B 950 of Hoechst Celanese (United States), was a chosen TLCP. Its chemical structure is shown in **Structure 2**.

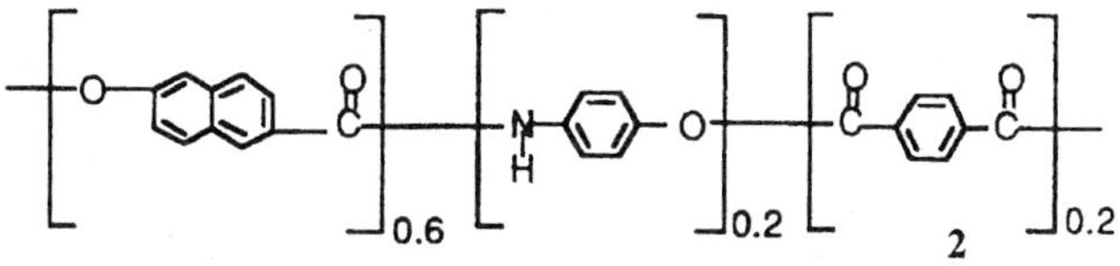

2

The compatibilizer that we synthesized for compatibilization of PEI and TLCP is a poly(ester imide) (PEsI) having the chemical structure shown in **Structure 3**.

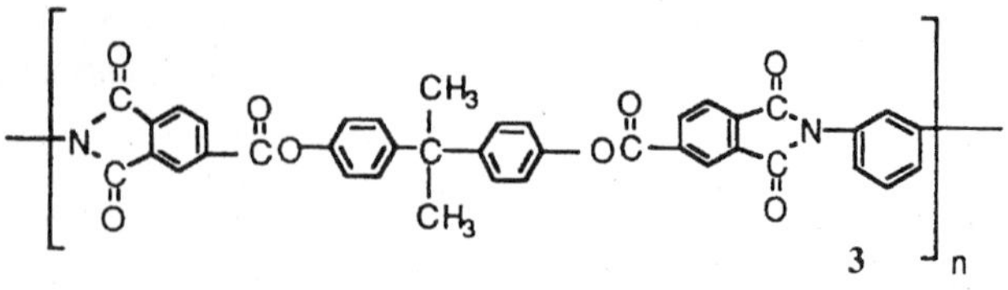

3

Morphological evidence demonstrated that the addition of the proper amount of PEsI reduces the TLCP particle size and induces fine distribution. However, the optimum amount of compatibilizer for the best mechanical properties and dispersion of TLCP phase was observed. An excess amount of PEsI coalesces the TLCP particles. The effectiveness of the compatibilizer depends on the absolute content of the compatibilizer. This is not peculiar to PEsI; it also happens to other compatibilized *in situ* composite systems and compatibilized polymer blends.[11]

With the goal in mind that compatibilizer can be applied to *in situ* composite, this study could be expanded to explore the development of *in situ* coupling between thermally stable thermotropic fibers and incompatible thermoplastic matrices by using compatibilizers in much the same fashion as in polymer blends. This study would stimulate more interest in the development of the compatibilized *in situ* composite.

REFERENCES

1. *Thermotropic Liquid Crystal Polymer Blends*; LaMantia, F. A., Ed.; Technomic: Lancaster, PA, 1994.
2. Seo, Y.; Hong, S. M.; Hwang, S. S.; Park, T. S.; Kim, K. U.; Lee, S.; Lee, J. *Polymer* **1995**, *36*, 515.
3. Seo, Y.; Hong, S. M.; Hwang, S. S.; Park, T. S.; Kim, K. U.; Lee, S.; Lee, J. *Polymer* **1995**, *36*, (P525).
4. Bafna, S. S.; Sun, T.; deSouza, J. P., Baird, D. G. *Polymer* **1995**, *36*, 259.
5. Ryu, C.; Seo, Y.; Hwang, S. S.; Hong, S. S.; Park, T. S.; Kim, K. U. *Int. Polym. Proc.* **1994**, *9*, 266.
6. Lee, W. C.; DiBenedetto, T. *Polymer* **1993**, *34*, 684.
7. Bretas, R. E. S.; Baird, D. G. *Polymer* **1992**, *24*, 5233.
8. Shin, B. Y.; Chung, I. J. *Polym. Eng. Sci.* **1990**, *30*, 13.
9. Datta, A.; Chen, H. H.; Baird, D. G. *Polymer* **1993**, *34*, 759.
10. Lee, S.; Hong, S. M.; Seo, Y.; Park, T. S.; Kim, K. U.; Lee, J. W. *Polymer* **1994**, *35*, 519.
11. Seo, Y.; Hong, S. M.; Hwang, S. S.; Kim, B. C. ACS Symposium series, in press, 1996.

Poly(ester imide)s

See: Poly(ester amide) and Poly(ester imide) Resins (Binders for Air-Drying Protective Coatings)
Poly(ester imide) Compatibilizer (in Thermotropic Liquid Crystalline Polymer Blends)

Polyester Resins

See: Dimer Acid-Based Polyesters
Resins (Applications in Coatings Industry)
Resins and Paints (Analytical Pyrolysis)
Unsaturated Polyester Resins (Overview)
Unsaturated Polyester Resins (Polyester Composites)
Unsaturated Polyester Resins (Toughening with Elastomers)
Unsaturated Polyester Resins (Toughening with Liquid Rubber)

Poly(ester urethane)s

See: Diisocyanates (In Situ Generation)

Polyesters

See: *Bacterial Polyesters (Containing Unusual Substituents Groups)*
Bicyclic Lactones, Ring-Opening Polymerization
Biodegradable Polyesters (Chemically-Induced Surface Degradation)
Biodegradable Polyesters (Theoretical Modeling of Degradation)
Biodegradable Polyesters, Cyclic Ethers
Bisphenol-A-Polycarbonate/Polyester Blends
Bulk Polymerization
Carbenoid Polycondensation
Carbonylation Polymerization (Palladium-Catalyzed)
Cardo Polyesters (Heat Resistance and Photosensitive)
Dimer Acid-Based Polyesters
Engineering Thermoplastics (Survey of Industrial Polymers)
Enzymatic Polymerization
Heteroaromatic Polymers (Five-Membered Rings in Main Chain)
Hyperbranched Aliphatic Polyesters
Hyperbranched Polyesters
Liquid Crystalline Aromatic Polyesters
Liquid Crystalline Polymers (Polyesters)
Liquid Crystalline Polymers (Polyesters of Bibenzoic Acid)
Liquid Crystalline Polymers (Thermotropic Polyesters)
Liquid Crystalline Polymers (Wholly Aromatic Polyesters)
Melt Spinning (Polymer Formation Under High Stress Conditions)
Microbial Polyhydroxyalkanoates (Poly[(R)-3-Hydroxybutyrate-co-3-Hydroxypropionate])
Perfluorocyclobutane Aromatic Ether Polymers
Phosphorus-Containing Polymers (From Dicarboxylic Acids with C-P Bonds)
Photocrosslinking (Overview)
Poly(alkylene naphthalate)
Polyacrylates, Dicyanovinyl-Containing (Thermal Properties)
Poly(butylene terephthalate (Annealing Behavior, Thermal Analysis)
Poly(butylene terephthalate)/Polycarbonate Blends
Poly(ε-caprolactone) (Overview)
Polyester Fabrics (Surface Modification by Low Temperature Plasma)
Polyesters [Bis(Cyclic Ether) and Diacyl Chloride Polyaddition]
Polyesters (Derived from Renewable Resources)
Polyesters (Synthesis from Lactones)
Polyesters, Aromatic (Amorphous and Liquid Crystalline)
Poly(ethylene terephthalate) (Chemistry and Preparation)
Poly(ethylene terephthalate) (Dielectric Properties)
Poly(ethylene terephthalate) (Heat Setting and Thermal Stabilization)
Poly(ethylene terephthalate) (Modified with Ionic Groups)
Poly(ethylene terephthalate) (Ultra-High Molecular Weight)
Poly(ethylene terephthalate) Copolymers
Poly(ethylene terephthalate) Degradation (Chemistry and Kinetics) Poly(β-hydroxyalkanoates) (Microbial Synthesis and Chemical Modifications)
Powder Coatings (Overview)
Reactive Liquid Rubbers (Epoxy Toughening Agents)
Recycling, Thermodynamic (by Ring-Closing Depolymerization)
Ring-Opening Coordination Polymerization (by Soluble Multinuclear Aldoxides)
Step Polymerization Catalysts
Textile Fibers (Shingosen)
Unsaturated Polyester Resins (Overview)
Unsaturated Polyester Resins (Polyester Composites)
Whiskers of Rigid-Rod Polymers

POLYESTERS [Bis(cyclic ether) and Diacyl Chloride Polyaddition]

Tadatomi Nishikubo* and Atsushi Kameyama
Department of Applied Chemistry
Faculty of Engineering
Kanagawa University

Oxetanes are also a class of cyclic ethers that react toward various nucleophiles, but the utility of oxetanes as starting materials in synthetic organic chemistry has not been sufficiently enhanced. On the polymer synthesis, it is predominantly cationic ring-opening polymerization of oxetanes that has been investigated. We found that oxetanes react with carboxylic acid derivatives, as well as oxiranes, in the presence of quaternary onium salts. The reaction of oxetanes with active esters[1] and acyl chlorides[2] give high yields with the corresponding products. The advantage of these reactions with the "neutral catalysts" is selective transformation of the cyclic ethers under mild conditions without any side reactions such as polymerizations of the used cyclic ethers.

The class of ester-formation reactions can be extended to synthesis of polyesters by polyaddition of bis(cyclic ether)s containing oxirane and oxetane moieties with bis(aryl ester)s[3]

*Author to whom correspondence should be addressed.

and diacyl chlorides.[4] The polyaddition of bis(cyclic ether)s with diacyl chlorides give polyesters with reactive chloromethyl groups at their side chains. The polyaddition is a unique and successful method for the synthesis of reactive polymers, compared with the conventional method of reactive polymer synthesis. This is because reactive polymers have been prepared predominantly by selective polymerization of the monomers that have polymerizable and reactive groups, such as chloromethyl groups. Here, novel polyester synthesis by the polyaddition of bisoxiranes (**Scheme I**) and bisoxetanes with diacyl chlorides using quaternary onium salts is demonstrated. Polymer reactions and functionalization of the polyesters as applications are also described.

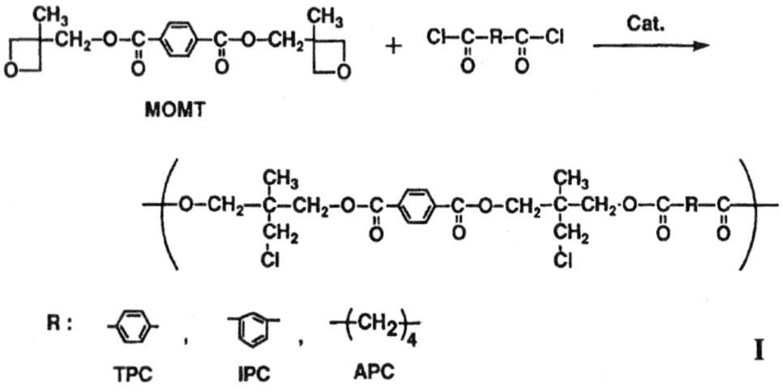

I

REFERENCES

1. Nishikubo, T.; Sato, K. *Chem. Lett.* **1991**, 697.
2. Sato, K.; Kameyama, A.; Nishikobo, T. *Macromolecules* **1992**, *25*, 1198.
3. Kameyama, A.; Watanabe, S.; Kobayashi, E.; Nishikubo, T. *Macromolecules* **1992**, *25*, 2307.
4. Kameyama, A.; Yamamoto, Y.; Nishikubo, T. J. *Poly. Sci., Part A: Chem. Ed.* **1993**, *31*, 1639.

POLYESTERS (Derived from Renewable Resources)

Matthias Ballauff and R. Storbeck
Polymer-Institut
Universität (T.H.) Karlsruhe

The synthesis of well-defined and stable polymers directly from polysaccharides, is rendered a difficult task because of the great number of hydroxy-functionalities along the chain.

In recent years it had been demonstrated that the dianhydrohexitols 1,4:3,6-dianhydro-D-sorbitol (DAS), 1,4:3,6-dianhydro-L-iditol (DAL) as well as 1,4:3,6-dianhydro-D-mannitol (DAM) deriving from starch may be used for the synthesis of well-defined polyesters,[1–5] polyamides,[6] and polyurethanes.[7] Thus, polyesters containing 1,4:3,6-dianhydro-D-sorbitol could be shown[4,5] to be thermally stable up to elevated temperatures. Of particular interest is the high glass transition temperature T_g of polyesters deriving from DAS and terephthalic acid, which is located around 200°C if molecular weights are high enough.[4] This work surveys the synthesis and the properties of polyesters made of well-defined diols and dicarboxylic acids deriving from starch. It becomes evident that these materials exhibit interesting properties and should be able to compete with diols and diacids deriving from an oil-based chemistry.

POLYESTERS FROM DIANHYDROHEXITOLS

Thiem and Lüders[1,2] were the first to explore the potential of the dianhydrohexitols DAS, DAM, and DAI for the synthesis of polyesters in a systematic fashion. The resulting materials were found to exhibit rather low molecular weights and a low solubility. To improve the low solubility of the materials, the unsubstituted terephthalic acid dichloride **1a** was replaced by the 2,5-diethoxyterephthalic acid dichloride **1b** to yield polyesters **2b–4b** (**Figure 1**).

Polycondensation of the dianhydrohexitols in solution proceeds without difficulties if the weak base pyridine is used. Using the procedure given recently,[4] polyester with degree of polymerization around 100 can be reached without difficulty.

An important prerequisite for further investigations by thermal analysis and possible applications is the sufficient thermal stability of the polyesters. It could be demonstrated[4] that polyesters **2a–4a** exhibit excellent thermal stability up to ~ 350°C under an atmosphere of nitrogen. Thermogravimetry under an ambient atmosphere leads to decomposition at ~ 300–320°C.

Since the two hydroxy functionalities form an angle different from 180° for all three dianhydrohexitols, polyesters **2–4** are random coils in solution with Mark-Houwink exponents between 0.5 and 0.7.[8] Thus, polyester **2b** which can be dissolved easily in common organic solvent and studied by static light scattering has an exponent of 0.52 in dimethylformamide at 30°C (K = 0.16dl/g) whereas a value of 0.63 results for chloroform at 25°C (K = 0.064dl/g). The results suggest that incorporation of the stiff DAS-units along the polyester chain will strongly increase the glass transition temperature.

ACKNOWLEDGMENT

Financial support by the Bundesministerium für Forschung und Technologie, Schwerpunkt "Nachwachsende Rohstoffe," and by the Cerestar GmbH is gratefully acknowledged.

REFERENCES

1. Thiem, J.; Lüders, H. *Starch/Stärke* **1984**, *36*, 170.
2. Thiem, J.; Lüders, H. *Polymer Bull.* **1984**, *11*, 365.
3. Braun, D.; Bergmann, M. *J. Prakt. Chem.* **1992**, *334*, 298.
4. Storbeck, R.; Rehahn, M.; Ballauff, M. *Makromol. Chem.* **1993**, *194*, 53.
5. Storbeck, R.; Ballauff, M. *Polymer* **1993**, *34*, 5003.
6. Thiem, J.; Bachmann, F. *Makromol. Chem.* **1991**, *192*, 2163.
7. Thiem, J.; Lüders, H. *Makromol. Chem.* **1986**, *187*, 2775.
8. Storbeck, R. *Synthese und Charakterisierung von Polyestern auf Basis nachwachsender Rohstoffe*, Dissertation, Karlsruhe, 1994.

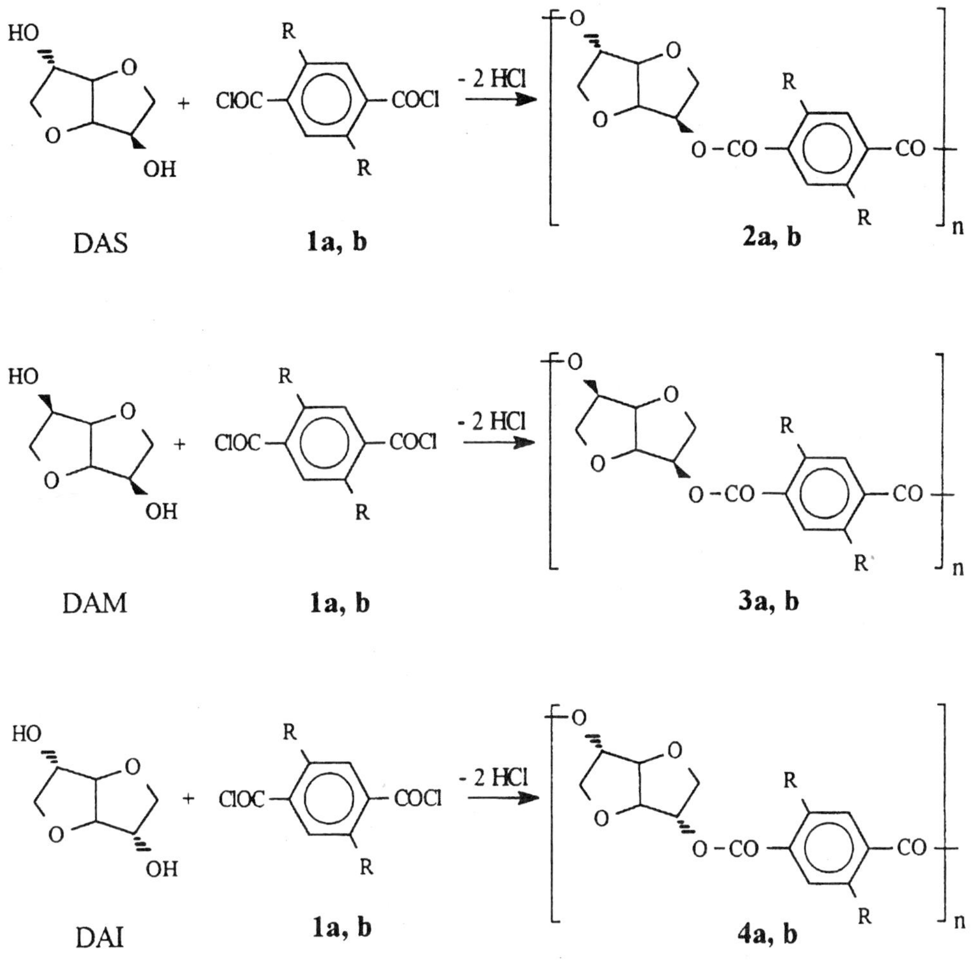

FIGURE 1. Synthesis of polyesters from DAS, DAM, and DAI and terephthalic acid. *Source*: Reference 14. With permission.

POLYESTERS (Synthesis from Lactones)

Zbigniew Jedliński
Institute of Polymer Chemistry
Polish Academy of Sciences

Ring-opening polymerization of lactones and related cyclic monomers (e.g., glycolide, lactide) has become a very important method for preparation of aliphatic main-chain polyester. The polymerization of lactones enables the proper design of synthesis and preparation of "tailor made" polymers exhibiting high molecular weight, unimodal molecular weight distribution, and other useful properties, such as mechanical strength, biodegradability, bioresorbability, and nonimmunogenity, which are important for their applications in medicine and pharmacology. The following general factors influence the course of lactones polymerization: (1) size of a lactone monomer ring, (2) position and kind of substituents at the ring, (3) possible reaction mechanism, stereochemistry, and side reactions, (4) certain reaction variables, that is, type of initiator, solvent, monomer concentration, temperature, etc.

SYNTHESIS OF POLYESTERS FROM LACTONES

Anionic Polymerization

Anionic polymerization has an essential role in the polyester synthesis from lactones because it enables proper design of polymer molecular weight, structure, and properties. Initially β-lactone polymerization with organic bases, such as pyridine, tertiary amines, and phosphines was studied.

Important progress was recently observed in the anionic polymerization of β-lactones initiated by strong bases like alkali metal alcoholates.

As novel initiators for anionic β-lactones polymerization, alkali metal supramolecular organic complexes of potassium and sodium with suitable ligand, for example, 18-crown-6, have been used. Polymerization of β-propiolactone, β-butyrolactone, and other lactones, as well as preparation of various block copolymers using this novel initiator system is possible.[1]

The polymerization of many racemic and optically active substituted β-lactones, and structure of resulting polymers has been studied using various initiators.[2]

The polymerization is an equilibrium reaction yielding chain macromolecules and cyclic oligomers due to "back-biting". The polymerization of lactones with anionic initiators is of importance, because living polymers and block copolymers with well defined structures and narrow molecular weight distribution can be synthesized. Using this approach many block copolymers having ester and ether blocks have been synthesized.[3]

Insertion Polymerization

Insertion (coordinative) polymerization proceeds with initiators capable of monomer complexation. The most common coordinative initiators are: aluminum, zinc, magnesium, titanium, tin and other heavy metal alkoxides, halides, or alkyls, and also bimetallic alkoxides. Addition of water or alcohols is said to activate some of these initiators. Polymerizability of lactones in the insertion polymerization is similar to that found for anionic polymerization and depends on the size of the monomer ring. The mechanism of this kind of polymerization involves coordination of a monomer on the metal site of the initiator and, after the acyl-oxygen bond cleavage insertion of the monomer units into the growing polymer chain occurs.

Recently much attention has been given to β-butyrolactone polymerization using insertion polymerization technique.[4] This monomer is of significant importance, because its R-polymers are analogous to biodegradable polymer, poly(β-hydroxybutyrate), produced in nature by some bacteria and fungi.

Cationic Polymerization of Lactones

The polymerization of β-lactones has been studied using many cationic initiators: protic acids, Lewis acids, alkylating agents (e.g., stable carbenium salts), acylating agents (e.g., acylium ions).

POLYESTERS FROM GLYCOLIDES AND LACTIDES

Glycolides and lactides are cyclic dimers of glycolic and lactic acids, respectively. The polymerization of these monomers was studied using aluminum, titanium, zirconium, and tin alkoxides, as well as some cationic and anionic initiators.[5] Various block copolymers of lactide and other monomers have been studied, because they are expected to find applications in medicine as biodegradable materials.[6,7]

REFERENCES

1. Jedliński, Z. *Makromol. Chem., Macromol. Symp.* **1992**, *60*, 235.
2. Grenier, D.; Prud'homme, R. E.; Leborgne, A.; Spassky, N. *J. Polym. Sci., Chem.* **1981**, *19*, 1781.
3. Kurcok, P.; Penczek, J.; Franek, J.; Jedliński, Z. *Macromolecules* **1992**, *25*, 2285.
4. Gross, R. A.; Zhang, Y.; Konrad, G.; Lenz, R. W. *Macromolecules* **1988**, *21*, 2657.
5. Jedliński, Z. *Handbook of Polymer Synthesis*; Marcel Dekker: New York, Basel, Hong Kong, 1992, Chapter 11.
6. Jedliński, Z.; Walach, W.; Kurcok, P.; Adamus, G. *Makromol. Chem.* **1991**, *192*, 2051.
7. Jedliński, Z.; Kurcok, P.; Walach, W.; Janezek, H.; Radecka, I. *Makromol. Chem.* **1993**, *194*, 1681.

POLYESTERS, AROMATIC (Amorphous and Liquid Crystalline)

Barry D. Dean
Amoco Polymers Business Group

Aromatic polyesters are typically classed as either amorphous (also referred to as polyarylate) or liquid crystalline (LCP).

Although amorphous polyarylates were initially viewed as a low-cost material to fill a performance void between polycarbonate and high-end engineering thermoplastics such as poly(ether sulfone) and poly(ether imide), the cost performance capability has not materialized and only very limited commercial success has been realized. However, liquid crystalline polyesters continue to receive increasing research and development attention, as evidenced by the continuum of increasing patent activity and patent assignees.[1]

SYNTHETIC PROCESSES

There are five synthetic routes generally accepted as capable of producing high molecular weight polyarylate or LCP: acid chloride, diacetate, diphenyl ester, phenol silyl ether, and promoted esterification. Direct condensation of an aromatic diacid with a diphenol generally results in only low molecular weight oligomer and, even when catalyzed, requires long reaction times and is limited to wholly aromatic monomer structures; thus, the use of lower cost but less stable alkylidene diphenols such as bisphenol A is precluded. The catalyzed direct condensations exhibit limitations for production of useful molecular weight for commercial applications.

COMMERCIAL FEASIBILITY

Of the process chemistry routes discussed for making aromatic polyesters, the diacetate route clearly has evolved as the preferred commercial method for making either amorphous, aromatic polyester, or liquid crystalline polyesters based on process patents issued to known commercial suppliers.

STRUCTURE-PROPERTY RELATIONSHIPS

Amorphous Polyarylate

The amorphous polyarylates, aside from mechanical properties similar to polycarbonate, exhibit markedly higher glass transition temperature (T_g) and higher practical temperature utility.

In addition to good thermal and UV resistance, polyarylates offer excellent dimensional stability, electrical properties, and inherent flame retardancy, based on rapid char formation. Over all, the property performance is very well balanced for consideration for design in a variety of applications such as coatings for enhanced weatherability, automotive lamp housings, and connectors (electrical and non-electrical) requiring snap-fit capability.

Liquid Crystalline Polyester

Thermotropic liquid crystalline polyesters generally have rigid, rod-like structures, for example, aromatic rings (benzene) attached in para positions via esters linkages. The homopolyester of *p*-hydroxy benzoic acid (ultimate rigid rod structure) exhibits a melting point of ~ 610°C and from a practical viewpoint is not readily melt processable. Modification of the rigid, rod-like structure to decrease the melting temperature to below 400°C, rendering the polyesters melt processable, can be addressed by a variety of structural modifications: introduction of flexible spacer units into the rigid rod structure, substitution of the aromatic rings, modification with rod-like comonomers, or introduction of rigid kinks into the rigid rod structure.

Comparison of both neat and filled thermotropic LCP with amorphous polyarylate highlights the exceptional mechanical properties, dimensional stability, and electrical properties reflective of liquid crystalline polyesters. In addition, as crystalline (highly ordered) materials, thermotropic LCPs exhibit good chemical resistance, excellent barrier properties, and inherent flame retardancy.

REFERENCE

1. Kirsch, M. A.; Williams, D. J. *CHEMTECH* **1994**, *April.*

Poly(ether amide)s

See: *Poly(ether amide) Triblock and Star Block Copolymers*
Poly(ether-block-amide) Copolymers (Characterization of Structure and Polymorphism)

POLY(ETHER AMIDE) TRIBLOCK AND STAR BLOCK COPOLYMERS

Richard J. Clark* and C. S. Henkee
Huntsman Corporation

Nylon 6 and nylon 6,6 have been modified with polyethers to produce poly(ether amide) segmented block copolymers.[1] These poly(ether amide) block copolymers are typically referred to as thermoplastic elastomers (TPE) or copolyamides (COPA). Impact performance, tensile properties, and flexural moduli of COPA can be tailored by adjusting the polyamide to polyether ratio.[2]

A new approach has been undertaken to synthesize polyether modified polyamides: amine terminated polyethers are used as the initiators for the polymerization of caprolactam to produce triblock and star block poly(ether amide) copolymers.

CHARACTERIZATION OF BLOCK COPOLYMERS

Initially, the thermal behavior of the triblock and three arm, star block copolymers were examined. As the amount of amine terminated polyether is increased from 0–40% in the block copolymer, the crystalline melting point of the polymer remains constant. At the same time, work was underway with poly(ether amide) segmented block copolymers. As the amount of polyether is increased in the segmented block copolymer, the crystalline melting point of the block copolymer is reduced to lower temperatures. It became apparent that the poly(ether amide) triblock and star block copolymers may have a potential performance advantage over the traditional segmented block copolymers.

As the amount of amine terminated polyether is increased in the block copolymer, an increase in the flexural modulus at 5% incorporation followed by a reduction in the flexural modulus/strength is observed. Tensile elongation remains relatively constant until 20% polyether is incorporated into the block copolymer. At 20%, the tensile elongation drops off dramatically. The notched izod performance greatly increases with the incorporation of polyether, but at 20%, the notched izod performance begins to decrease. Low temperature (–30°C) impact performance is also improved by the incorporation of the amine terminated polyether.

Significant physical property differences are observed when the midblock of the block copolymer is composed of poly(ethylene glycol) rather than poly(propylene glycol).

The tensile elongation is significantly increased by the incorporation of poly(ethylene glycol) midblock. It is believed that the improved elongation is due to molecular structure of the poly(ethylene glycol).

As with most polyamides, the question of moisture absorption needs to be addressed. Star block copolymers having 5% and 10% poly(propylene glycol) midblocks have a lower level and a lower rate of moisture absorption than polycaprolactam, whereas star blocks with 15% poly(propylene glycol) midblocks absorb just slightly more water than polycaprolactam. In the case of star block copolymers containing poly(ethylene glycol) midblocks, as the ratio of polyether to polyamide increases, the level and the rate of moisture absorption increases.

Poly(ether amide) triblock and star block copolymers have been synthesized. These block copolymers have excellent impact performance and a higher modulus at low levels of polyether amine incorporation than polycaprolactam. The effect of increasing the flexural modulus/strength when incorporating low levels (5–10%) of amine terminated poly(propylene glycol) as well as increasing the notched izod impact performance is unusual and potentially a very useful response.

The use of poly(propylene glycol) and poly(ethylene glycol) midblocks produces polymers with significantly different levels of moisture absorption. This characteristic of the polyether midblock allows these block copolymers to be tailored for a wide range of uses from engineering to fabric/textile applications.

ACKNOWLEDGMENTS

The authors would like to thank C. A. Armstrong, E. R. Villarreal, and E. D. Riffe for their help in obtaining the data. The authors would also like to thank G. P. Speranza and R. J. Dominguez for their concepts and suggestions, and Robert Zimmerman and Jeff Vipond for their help in preparing experimental polyether amines.

REFERENCES

1. Legge, N. R. *Elastmeries* **1991**, *September*, 14.
2. U.S. Patent 4 717 763, 1988.

*Author to whom correspondence should be addressed.

POLY(ETHER-*block*-AMIDE) COPOLYMERS (CHARACTERIZATION OF STRUCTURE AND POLYMORPHISM)

Galen R. Hatfield*
Washington Research Center
W. R. Grace and Company

Marvin Havens
Cryovac Division
W. R. Grace and Company

Poly(ether-*block*-amide) (PEBA) copolymers (**1**) provide a wide range of combined thermoplastic and rubber properties as a result of many composition variables that may be incorporated into the polymer structure.[1] It is possible to produce materials with Shore Hardness ranging from 75D to 60A. Vicat softening points ranging from 60°C to 191°C are also possible. It is even possible to tailor electrical properties ranging from good insulators to surface resistivities of 10^8 Ω-cm²/cm without conductive additives such as carbon black (**Structure 1**).

$$\mathrm{HO\text{-}[\text{-}\overset{\displaystyle O}{\overset{\|}{C}}\text{-}PA\text{-}\overset{\displaystyle O}{\overset{\|}{C}}\text{-}O\text{-}PE\text{-}O\text{-}]\text{-}H} \qquad \mathbf{1}$$

Given the wide range of physical properties that can be designed into these polymers, it is not surprising that they are suitable for a number of commercial applications. For example, athletic footwear and power transmission belts utilize the resiliency (low hysteresis), flexibility over wide temperature ranges, and abrasion resistance properties of PEBA copolymers. Coiled telephone and electrical cords manufactured from PEBA copolymers have good thermoformability, which is a benefit for manufacturing (processability) and end use (holding the coil) purposes. Medical materials fabricated from PEBA copolymers are also used for their good but flexible strength as well as for their high moisture vapor transmission rates (MVTRs).

Nylons are historically and commercially important polymers for a variety of desired properties such as strength, chemical resistance, etc. However, one drawback of nylons is their stiffness, which is difficult to modify with plasticizers because of the high processing temperatures and high solid phase crystallinity that tend to vaporize or phase separate most plasticizers. One answer has been block copolymerization, which can incorporate distinct physical and/or chemical modifications into a unique polymeric molecule that avoids problems with mixing, migration, vaporization, etc. With poly(ether-*block*-amide) copolymers, the "hard" or stiff blocks of nylon are interspersed with "soft" or flexible blocks of polyethers, which are coupled together by ester linkages. Each block, either the nylon or polyether, may vary in length as well as composition within its respective chemical families. The relative concentration of blocks may also vary. All these options offer a wide array of ultimate physical and chemical properties.

For the hard blocks, commercial PEBA copolymers start with the common lactams for either nylon 6 (caprolactam) or nylon 12 (lauryl lactam) or a blend of the two. The initial reaction converts monomers to a nylon prepolymer. A small amount of diacid is included to limit the molecular weight of the polyamide block and to provide acid termination. For the soft blocks, commercial PEBA copolymers use one of three polyethers: poly(ethylene oxide), poly(propylene oxide), or poly(tetramethylene oxide). Physically and chemically, these are highly flexible polyether chains with alcohol terminations and are available commercially with different molecular weights. These polyether blocks are then added to the acid terminated polyamide blocks which are incorporated into the final polymer by ester linkages. The range of recipe variations is the key to the range of physical properties.

*Author to whom correspondence should be addressed.

APPLICATIONS

PEBA copolymers are readily adaptable to a variety of traditional thermoplastic processes including injection molding, extrusion, blow molding, thermoforming, and rotational molding. Because PEBA copolymers are related to polyamides, they may be extruded in a similar manner, although they may require more attention to drying (<0.10% moisture). This is related to the polyether block which, depending on type and relative amount, can be hygroscopic. For harder grades with more nylon, injection molding and extrusion temperatures may be in the range of that for polyamides, that is, up to 280°C. For softer grades with more polyether, temperatures may be as low as 160°C. Depending upon the application, PEBA copolymers may be assembled or fabricated by traditional methods, including insert molding, bonding, and welding. Articles made with PEBA copolymers take advantage of nylon-like properties such as chemical resistance, particularly to oils and greases, heat resistance, and abrasion resistance but with a range of hardnesses and vicat softening points. For these reactions, PEBA copolymers can be found in articles such as athletic footwear to medical films.

ACKNOWLEDGMENTS

We wish to thank Jerry Sakowski (Oneida Research Services, Inc.) and Mike Jones for their technical contributions to this work. We would also like to thank J. Onuferko for her continued support of this work.

REFERENCE

1. Koch, R. B. *Adv. Polym. Tech.* **1982**, 2, 160.

Poly(ether carbonate)s

See: Poly(propylene carbonate)

POLY(ETHER ETHER KETONE) (Dynamic Relaxation Characteristics)

Douglass Kalika** and R. K. Krishnaswamy
Department of Chemical and Materials Engineering
University of Kentucky

**Author to whom correspondence should be addressed.

The development of certain classes of engineering thermoplastics based on wholly-aromatic backbone structures and encompassing exceptional thermal and mechanical performance properties, has led to the application of these melt-processable materials in a variety of areas where such resins have not traditionally been employed. In the area of fiber-reinforced composites, poly(ether ether ketone) (PEEK) has been explored as a replacement for thermosetting matrices, with the PEEK thermoplastic offering potential advantages in terms of processing and ultimate material toughness while overcoming deficiencies in thermal and chemical resistance typically attributed to thermoplastics in general. The successful application of PEEK to these systems requires an understanding as to how the processing conditions and history of the material influence the resultant semicrystalline morphology and how this morphology impacts macroscopic composite properties.

In this article, we will describe the dynamic relaxation characteristics of poly(ether ether ketone), paying particular attention to the influence of semicrystalline morphology on the relaxation behavior of the material. Because of its relatively stiff backbone, PEEK behaves as a "low crystallinity" polymer in that it can be quenched from the melt to a wholly-amorphous glass, thus affording an opportunity to investigate a wide range of crystalline morphologies as established over both cold-crystallization and melt-crystallization histories.[1] The thermal, structural, and morphological properties of PEEK have been extensively investigated over the course of the past decade.[2–17]

CONCLUSIONS

We have described the dynamic relaxation characteristics of amorphous and semicrystalline PEEK for samples with a wide range of thermal and chemical exposure history. The presence of crystallinity has a significant impact on the observed relaxations, because of the constraining influence of the crystalline phase on the motions inherent in both the glass-rubber (α) and sub-glass (β) relaxations. For the α relaxation, the presence of cold-crystallinity leads to a broadening of the relaxation and a positive offset in the measured isochronal relaxation temperatures; the extent of the offset decreases somewhat with increasing crystallization temperature (i.e., increasing amorphous interlayer thickness). For solvent-crystallized samples, this offset is even more pronounced and suggests the development of a tighter crystalline morphology in the solvent-crystallized case. Comparison of the α relaxation intensity for the amorphous and cold-crystallized samples indicates the presence of a finite rigid amorphous phase fraction: a progressive mobilization of the rigid amorphous phase is evident in the temperature range between T_g and T_m. The β relaxation also displays a sensitivity to the presence of crystallinity which stands in contrast to the sub-glass relaxation characteristics reported for more flexible semicrystalline polymers such as PET. Specifically, a bimodal relaxation behavior is observed which encompasses essentially non-cooperative motions which originate in the bulk amorphous phase and motions of a somewhat more cooperative nature which originate in the vicinity of the crystal-amorphous interphase. The influence of the crystalline phase on the β relaxation is thus manifested in an overall positive shift in the measured relaxation temperature upon the introduction of crystallinity and a disproportionate reduction in the observed relaxation intensity.

REFERENCES

1. Boyd, R. H. *Polymer* **1985**, *26*, 323.
2. Blundell, D. J.; Osborn, B. N. *Polymer* **1983**, *24*, 953.
3. Cheng, S. Z. D.; Cao, M. Y.; Wunderlich, B. *Macromolecules* **1986**, *19*, 1868.
4. Cebe, P.; Hong, S. *Polymer* **1986**, *27*, 1183.
5. Velisaris, C. N.; Seferis, J. C. *Polym. Eng. Sci.* **1986**, *26*, 1574.
6. Lee, Y.; Porter, R. S. *Macromolecules* **1987**, *20*, 1336.
7. Lee, Y.; Porter, R. S. *Macromolecules* **1988**, *21*, 2770.
8. Lee, Y.; Porter, R. S.; Lin, J. S. *Macromolecules* **1989**, *22*, 1756.
9. Jonas, A.; Legras, R.; Issi, J.-P. *Polymer* **1991**, *32*, 3364.
10. Dawson, P. C.; Blundell, D. J. *Polymer* **1980**, *21*, 577.
11. Wakelyn, N. T. *J. Polym. Sci., Polym. Lett.* **1987**, *25*, 25.
12. Hay, J. N.; Langford, J. I.; Lloyd, J. R. *Polymer* **1989**, *30*, 489.
13. Bassett, D. C.; Olley, R. H.; Al Raheil, I. A. M. *Polymer* **1988**, *29*, 1745.
14. Lovinger, A. J.; Davis, D. D. *Macromolecules* **1986**, *19*, 1861.
15. Lovinger, A. J.; Hudson, S.; Davis, D. D. *Macromolecules* **1992**, *25*, 1752.
16. Marand, H.; Prasad, A. *Macromolecules* **1992**, *25*, 1731.
17. Medellin-Rodriguez, F. J.; Phillips, P. J. *Polym. Eng. Sci.* **1990**, *30*, 860.

Poly(ether ether ketone)s

See: Poly(aryl ether ketone)s

Poly(ether imide)s

See: Cyclic Oligomers of Engineering Thermoplastics
Cyclolinear Organophosphazene Polymers
Poly(ester imide) Compatibilizer (in Thermotropic Liquid Crystalline Polymer Blends)
Polyimide Blends
Polyimides (Introduction and Overview)

Poly(ether ketone amide)s

See: Macrocyclic Aramids

Poly(ether ketone)s

See: Poly(aryl ether ketone)s

Polyether-Polysiloxane Copolymers

See: Silicone Polymers, Organo-Modified (Application in Personal Care Products)

Poly(ether sulfone)s

See: Poly(aryl ether sulfone)s

POLY(ETHER SULFONE)S (Properties, Applications, and Trends)

Anton Ticktin and Karin Elbl-Weiser
BASF AG

DEFINITION

Among the amorphous high performance thermoplastics polysulfones (PSU) and poly(ether sulfone)s (PES) have the longest history and still have a promising future with consumption growth rates predicted at 10% a year.[1] From the chemical point of view both of these polymers are characterized by aryl groups linked by sulfo (-SO_{-2}) and ether (O) groups. Polysulfone, in contrast to poly(ether sulfone) has additional isopropylidene (-C($CH_{-3})_2$) links which alter the physical properties of the material, especially the glass transition point, T_g. A third type of sulfone polymer, a poly(aryl sulfone) (PAS) was introduced in the market in 1983 and is derived from hydroquinone. A fourth type, the poly(phenyl sulfone) (PPSU), was introduced in 1976 and features, in addition to the sulfo- and ether- bridges, directly linked aryl groups.

SYNTHETIC ROUTES TO POLY(ETHER SULFONE)S

Electrophilic Aromatic Substitution

Polycondensation via sulfonylation is possible with two different monomer types as educts. One option is to start with monosulfonylchlorides, which yield the desired polymers in a self-condensation. This provides the advantage of being independent of the stoichiometric balance and provides an all-*para* structure which is not active for further sulfonylation to yield branched polymers.[4] The other option is a condensation of AA- and BB-monomers, that is the reaction of bis(sulfonyl chloride)s with aromatic ethers.

Nucleophilic Aromatic Substitution

Synthesis of poly(ether sulfone)s by polyetherification is possible when alkali metal phenoxides react with activated aromatic halides.

POLYMER STRUCTURE AND PROPERTIES

Aromatic rings linked only by sulfone groups yield polymers which are crystalline. Poly(phenyl sulfone) (**Structure 1**) melts and decomposes above 500°C.[3] For poly(biphenyl sulfone) (**Structure 2**) a melting range of 410–430°C has been reported.[4] These polymers do not show thermoplastic processability. Introduction of ether linkages makes the polymers more flexible and destroys the crystallinity. The C-SO_2-C bond angle in aryl sulfones is 105°C, whereas the COC angle is 124°C.[5] This mismatch reduces the chain symmetry and the polymers are amorphous.

The thermal stability of amorphous PES is determined by the glass temperature. Structural elements with dipolar character or large sterical requirements increase the glass temperatures of the polymers.

Poly(ether sulfone) shows good chemical resistance for an amorphous thermoplastic, but certain organic chemicals do initiate environmental stress cracking. As the molecular mass of PES increases, the resistance to chemicals improves, and the likelihood of stress cracking is reduced. Even at elevated temperatures PES is resistant to water and aqueous solutions, aqueous inorganic acids, organic acids, alkalis, aliphatic hydrocarbons (such as gasoline and kerosene), alcohols, amines, most detergents and sterilizing agents, oils, and grease. Moldings can withstand brief exposure to aromatic solvents such as benzene, xylene, and toluene. Amides and chlorinated hydrocarbons dissolve PES.

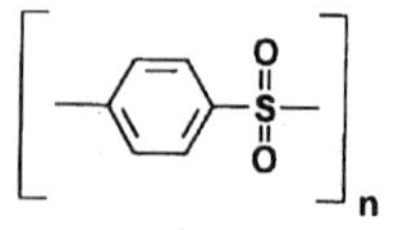

1

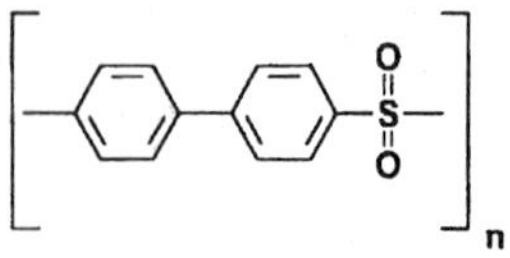

2

APPLICATIONS

Poly(ether sulfone) has important material properties, of which the following are of particular commercial interest:

- high temperature stability
- good mechanical properties such as rigidity, tensile strength, and toughness
- chemical resistance
- hydrolysis resistance
- inherent fire resistant
- good dielectric properties

However, these properties are commercially advantageous only because the processing of PES is similar to that of any other thermoplastic and it can be easily injection molded, extruded, blow molded, or machined.

FUTURE TRENDS

The world consumption of poly(ether sulfone), which in 1994 was about 3000 tons, is expected to grow strongly in the next 5 years.[1] The market will be split among polysulfone, poly(ether sulfone), and the partly transparent poly(ether imide). Considerable research is already being devoted to the possibility of using polymer blends either incorporating the above plastics as a filler or as the base matrix to further enhance the property profile of high performance materials.

The impetus for future applications will come predominantly from two areas. First, metals, thermosets, ceramics, and glass will continue to be replaced by PES plastics as the cost, weight, and processing advantages become more widely known. Especially in the area of metal or thermosetting plastics, where often the main criterion is temperature stability, high-performance plastics will continue to make inroads. There are already on the market PES products with optimized frictional coefficients designed specifically to compete with metals.[6] The cost effectiveness of injection molding as opposed to machining metal will certainly also be a relevant factor in changing from metals to plastic. Moreover, glass will increasingly be replaced by lighter materials capable of being reused.

Second, in the high technology sector more demands will be placed on plastic materials able to withstand higher temperatures under strict tolerance limits. In the electronics sector, plastics must not only be thermally and dimensionally stable and flame retardant but also must be resistant to cleaning agents such as acidic solutions for fat removal. The same arguments apply to the automobile parts that must be lighter but resistant to higher temperatures. The high-performance plastics will thus be found in an even greater number of applications and as electronics are incorporated into more equipment, machines, instruments, and household appliances, the demand for poly(ether sulfone) will continue to grow.

REFERENCES

1. Haley, M. J.; Jakobi, R.; Tashiro, M. "High Performance Thermoplastics" SRI Report, 1993; Stanford Research Institute: Stanford, CA.
2. Rose, J. B. *Chimia* **1974**, *28*, 561.
3. Rose, J. B. *Polymer* **1974**, *15*, 456.
4. 3M Company, U.S. Patent 3 321 449, 1967.
5. Attwood, T. E.; Cinderey, M. B.; Rose, J. B. *Polymer* **1993**, *34*, 2155.
6. BASF AG, Product Information, Ultrason KR4113.

POLY(ETHER SULFONE)S AND POLY(ETHER KETONE)S (Methyl-Substituted Aromatic)

Yoshio Imai
Department of Organic and Polymeric Materials
Tokyo Institute of Technology

Aromatic poly(ether sulfone)s and poly(ether ketone)s—including polysulfone (PSF), poly(ether sulfone) (PES), and poly(ether ether ketone) (PEEK) are a class of high performance engineering plastic materials used widely in a variety of industrial fields for many years.[1–5]

Most of the aromatic polyethers reported on are composed of unsubstituted phenylene rings. Less is known about the aromatic polymers that have methyl substituents on the phenylene rings.[6–8] The presence of methyl groups renders these polymers interesting with respect to physical and chemical properties.

SYNTHESIS AND PROPERTIES OF METHYL-SUBSTITUTED AROMATIC POLYETHERS

Aromatic poly(ether sulfone)s and poly(ether ketone)s were obtained readily by the aromatic nucleophilic substitution polycondensation of methyl-substituted biphenyl-4,4'-diols with activated aromatic dihalides (4,4'-dichlorodiphenyl sulfone and 4,4'-difluorobenzophenone). This occurred in a mixed solvent system of *N*-methyl-2-pyrrolidone (NMP) and toluene in the presence of anhydrous potassium carbonate (**Scheme I**), according to a conventional procedure for polyether synthesis.[10,11]

The introduction of methyl substituents to the phenylene rings greatly improved the solubility of these polymers.

All of the soluble aromatic polyethers afforded readily transparent and tough films by casting from the chloroform solutions.

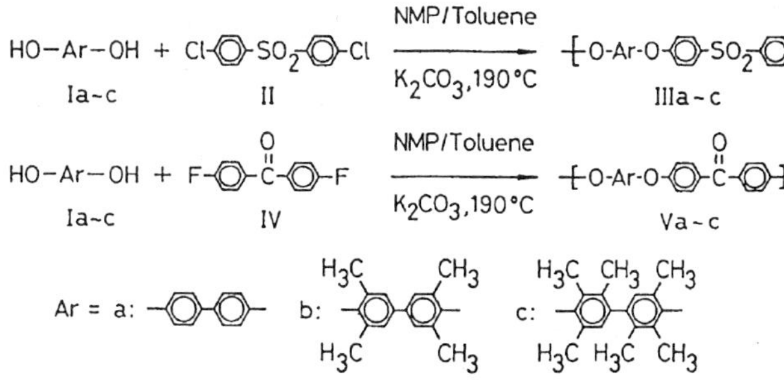

I

The tensile strength and tensile modulus of the poly(ether sulfone)s increased as the number of methyl substituents on the phenylene rings rose while the elongation at break and fracture energy decreased.

EFFECT OF METHYL SUBSTITUTION ON THE T_g OF AROMATIC POLYETHERS

In the case of the poly(ether sulfone) series, the T_gs of the two methyl-substituted polymers, TMBP–PES and HMBP–PES, were 60–70°C higher than those of the unsubstituted polymer BP–PES. A similar relationship was observed in the bisphenol–A based poly(ether sulfone)s (BIS–PES); the T_g of the tetramethyl-substituted polymer TMBIS-PES was about 60°C higher than that of the unsubstituted polymer BIS-PES.

PHOTOCHEMICAL CROSSLINKING BEHAVIOR OF AROMATIC POLY(ETHER KETONE)S

Films made from the methyl-substituted aromatic poly(ether ketone)s, TMBP–PEK and JMBP-PEK, were subjected to UV radiation at room temperature. Both poly(ether ketone)s became insoluble after the radiation. The gelation time—that is, the time for becoming insoluble—of the HMBP–PEK was only 2 minutes and was much shorter than that of the TMBP–PEK.[9]

REFERENCES

1. Johnson, R. N. *Encycl. Polym. Sci. Technol.* **1969**, *11*, 447.
2. *High Performance Polymers: Their Origin and Development* Seymour, R. B.; Kirshenbaum, G. S., Eds.; Elsevier: New York, 1986.
3. Johnson, R. N.; Farnham, A. G.; Clendinning, R. A.; Hale, W. F.; Merriam, C. N. *J. Polym. Sci., A-1* **1967**, *5*, 2375.
4. Attwood, T. E.; Barr, D. A.; King, T.; Newton, A. B.; Rose, J. B. *Polymer* **1977**, *18*, 359.
5. Attwood, T. E.; Dawson, P. C.; Freeman, J. L.; Hoy, L. R.; Rose, J. B.; Staniland, P. A. *Polymer* **1981**, *22*, 1096.
6. Mohanty, D. K.; Sachdera, Y.; Hedrick, J. L.; Wolfe, J. F.; McGrath, J. E. *Am. Chem. Soc., Polym. Prepr.* **1974**, *25*(2), 19.
7. Mohanty, D. K.; McGrath, J. E. *Polym. Sci. Technol.* **1985**, *31*, 113.
8. Aitken, C. L.; McHattie, J. S.; Paul, D. R. *Macromolecules* **1992**, *25*, 2910.
9. Keitoku, F.; Kakimoto, M.; Imai, Y. *J. Polym. Sci., Part A, Polym. Chem.* **1994**, *32*, 317.
10. Viswanathan, R.; Johnson, B. C.; McGrath, J. E. *Polymer* **1984**, *25*, 1827.
11. Mohanty, D. K.; Hedrick, J. L.; Gobetz, K.; Johnson, B. C.; Yilgor, I.; Yilgor, E.; Yang, R.; McGrath, J. E. *Am. Chem. Soc., Polym. Prepr.* **1982**, *23*(1), 284.

Poly(ether urethane)s

See: *Diisocyanates (In Situ Generation)*
Polyurethane Elastomers, Segmented (Non-Hydrogen Bonding Systems)
Unsaturated Polyester Resins (Toughening with Elastomers)

Polyethers

See: *Bicyclic Ethers, Ring-Opening Polymerization*
Contact Lenses, Gas Permeable
Cyclic Oligomers of Engineering Thermoplastics
Cyclolinear Organophosphazene Polymers
Dendritic Polyradicals
Heteroaromatic Polymers (Five-Membered Rings in Main Chain)
Nonthrombogenic Polymers (Multiblock Copolymers of Polyether and Polyamide)
Perfluorocyclobutane Aromatic Ether Polymers
Poly(dichlorophenylene oxide)s (by Pyridene/Copper Complex)
Poly(ether amide) Triblock and Star Block Copolymers
Poly(ether-block-amide) Copolymers (Characterization of Structure and Polymorphism)
Polyethers (1,3,4-Thiadiazole-Containing)
Polyglycidols, Polythioglycidols
Poly(propylene oxide)
Reactive Liquid Rubbers (Epoxy Toughening Agents)
Ring-Opening Coordination Polymerization (by Soluble Multinuclear Aldoxides)
Solid Polymer Electrolytes (Polyether Blends and Composites)
Zinc Catalysts (Polymerization of Styrene and Methyl Methacrylate)

POLYETHERS (1,3,4-Thiadiazole-Containing)

Yasuo Saegusa,* Takeshi Iwasaki, and Shigeo Nakamura
Department of Applied Chemistry
Faculty of Engineering
Kanagawa University

The method most frequently utilized for the preparation of aromatic polyethers is aromatic nucleophilic substitution polymerization, in which aromatic dihalides activated by strongly electron-attracting groups, most commonly a carbonyl or sulfone group (e.g., 4,4'-difluorobenzophenone and bis(4-chlorophenyl) sulfone) are used as partners of bisphenols. Dihalides containing heterocyclic nuclei, such as 1,3,4-oxadiazole, phenylquinoxaline, benzoxazole, pyridine 1,2,4-triazole, and benzothiazole are also used as activated monomers in this type of reaction.[1,3–15] In these halides, the heterocycles not only serve as activating groups but also stabilize the negative charge of a Meisenheimer complex formed through the polymer-forming nucleophilic-substitution reaction and, thereby, greatly promote polyetherification. Moreover, the resulting heterocycle-containing polyethers also have some of the favorable properties of heterocyclic polymers such as extremely high glass transition temperature (T_g) and thermal stability.

PREPARATION

We report here the synthesis and properties of a series of 1,3,4-thiadiazole-containing poly(ethers) (designated in **Scheme I** as **III**) by the fluorine displacement polymerization of novel activated difluoride: 2,5-bis(4-fluorophenyl)1,3,4-thiadiazole (**I**), which contain 1,3,4-thiadiazole (the sulfur analog of 1,3,4-oxadiazole) as an activating group and seven kinds of aromatic diols (**II**) having a variety of ring structures (Scheme I).

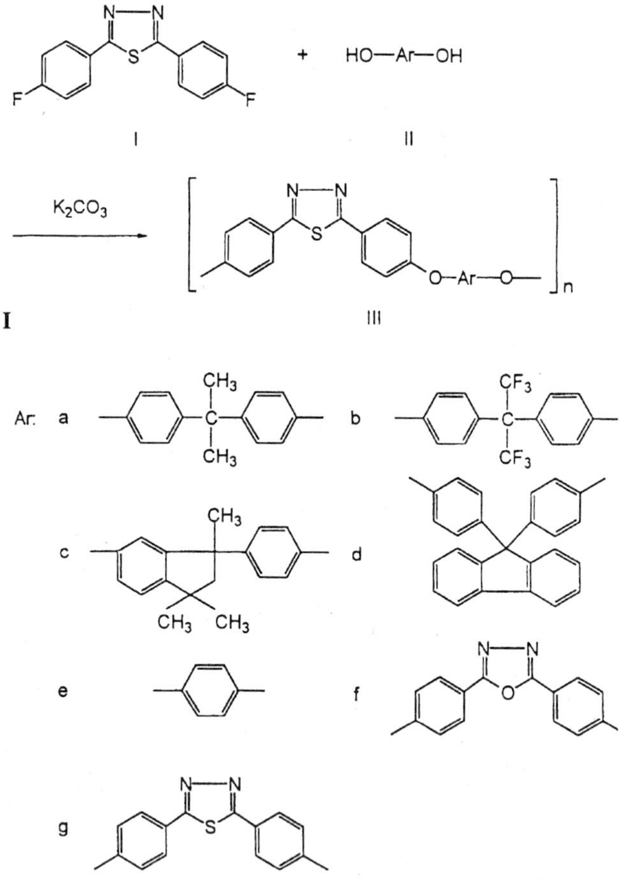

I

PROPERTIES

Polyethers **IIIe**, **IIIf**, and **IIIg** containing highly rigid *p*-phenylene, diphenyloxadiazole, and diphenylthiadiazole units, respectively, in the polymer backbone were highly crystalline, judging from their X-ray diffraction diagrams, and were soluble only in concentrated sulfuric acid and/or *o*-chlorophenol and a mixture of phenol and *sym*-tetrachloroethane (3:2 v/v).[16–18] In contrast, polyethers **IIIa**, **IIIb**, **IIIc**, and **IIId**, which contain bulky and distorted and/or unsymmetrical diphenylisopropylidene, diphenylhexafluoroisopropylidene, phenylindane,

*Author to whom correspondence should be addressed.

and diphenylfluorene units, respectively, were amorphous and also dissolved easily in a variety of organic solvents such as NMP, *m*-cresol, chloroform, and *sym*-tetrachloroethane, in addition to the above mentioned solvents, at room temperature. The high crystalline characteristics and limited solubility behavior of polyethers **IIIe–IIIg** are attributed to their rigid main-chain structures.

Transparent and tough films that are colorless to slightly yellow could be cast from the NMP solutions of amorphous and soluble polyethers **IIIa–IIId**. The mechanical properties of the films were excellent, their tensile strength, elongation at break, and tensile moduli being recorded in the ranges of 48–72 MPa, 5–7%, and 1.3–1.9 GPa, respectively.

The T_gs of amorphous polyethers **IIIa–IIId** determined by differential scanning calorimetry ranged from 204 to 299°C, depending on the bisphenols used. No melting temperature was detected for the highly crystalline polyethers **IIIe–IIIg** below their initial decomposition temperatures of about 400°C. All the polyethers showed a similar pattern of decomposition with no weight loss below 400°C in air; the observed temperatures at 10% weight loss were between 464 and 513°C. Thus, the high T_gs and/or thermal stability of these 1,3,4-thiadiazole-containing polyethers are almost comparable to those of the heterocycle-containing polyethers and copolyethers already reported.[2–4,6–10,13,15,19–21]

APPLICATIONS

Most of the 1,3,4-thiadiazole-containing polyethers obtained are characterized by high solubility in organic solvents and excellent mechanical properties and thermal stability, and, hence, considered to be one of the promising high-temperature film and plastic materials.

REFERENCES

1. Johnson, R. N.; Farnham, A. G.; Clendinning, R. A.; Hale, W. F.; Merriam, C. N. *J. Polym. Sci., A-1* **1967**, *5*, 2375.
2. Percec, V.; Wang, J. H.; Okita, S. *J. Polym. Sci., Part A: Polym. Chem.* **1991**, *29*, 1789.
3. Carter, K. R.; Jonsson, H.; Twieg, R.; Miller, R. D.; Hedrick, J. L. *Polym. Prepr. Div. Polym. Chem. Am. Chem. Soc.* **1992**, *33*(1), 388.
4. Hedrick, J. L. *Polym. Bull.* **1991**, *25*, 543.
5. Harris, F. W.; Korleski, J. E. *ACS Div. Polym. Matls.: Sci. Eng.* **1989**, *61*, 870.
6. Hedrick, J. L.; Labadie, J. W. *Macromolecules* **1988**, *21*, 1883.
7. Hedrick, J. L.; Labadie, J. W. *ACS Div. Polym. Matls.: Sci. Eng.* **1988**, *59*, 42.
8. Hedrick, J. L.; Labadie, J. W. *Macromolecules* **1990**, *23*, 1561.
9. Hedrick, J. L.; Hilborn, J. G.; Labadie, J. W.; Volksen, W. *Polym. Bull.* **1989**, *22*, 47.
10. Hedrick, J. L.; Hilborn, J. G.; Oalmer, T. D.; Labadie, J. W.; Volksen, W. *J. Polym. Sci., Part A: Polym. Chem.* **1990**, *28*, 2255.
11. Hedrick, J. L.; Russell, T. P.; Labadie, J. W.; Hilborn, J. G.; Palmer, T. D. *Polymer* **1990**, *31*, 2384.
12. Hilborn, J. G.; Labadie, J. W.; Hedrick, J. L. *ACS Div. Polym. Matls.: Sci. Eng.* **1989**, *60*, 522.
13. Hilborn, J. G.; Labadie, J. W.; Hedrick, J. L. *Macromolecules* **1990**, *23*, 2854.
14. Kricheldorf, H. R.; Schwarz, G.; Erxleben, J. *Makromol. Chem.* **1988**, *189*, 2255.
15. Hedrick, J. L. *Polym. Bull.* **1990**, *24*, 371.
16. Saegusa, Y.; Sekiba, K.; Nakamura, S. *J. Polym. Sci., Part A: Polym. Chem.* **1990**, *28*, 3647.
17. Saegusa, Y.; Koshikawa, T.; Nakamura, S. *J. Polym. Sci. Part A: Polym. Chem.* **1992**, *30*, 1369.
18. Saegusa, Y.; Takashima, T.; Nakamura, S. *J. Polym. Sci., Part A: Polym. Chem.* **1992**, *30*, 1375.
19. Bass, R. G.; Srinivasan, K. R.; Smith, J. G. *Polym. Prepr. Div. Polym. Chem. Am. Chem. Soc.* **1991**, *32*(2), 160.
20. Connell, J. W.; Hergenrother, P. M.; Wolf, P. *ACS Div. Polym. Matls.: Sci. Eng.* **1990**, *63*, 366.
21. Connell, J. W.; Croall, C. I. *Polym. Prepr. Div. Polym. Chem. Am. Chem. Soc.* **1991**, *32*(2), 162.

Polyethylene

See: *Chlorinated Polyethylene*
Chlorosulfonated Polyethylene
Living Coordination Polymerization
Metallocene Catalysts (Group 4 Elements, New Polymeric Materials)
Photo-oxidation, Polyolefins
Photocrosslinking (Overview)
Polyethylene (Commercial)
Polyethylene (High Density Preparation)
Polyethylene (Low Density Preparation)
Polyethylene (New Aspects of Morphology)
Polyethylene (Surface Functionalization)
Polyethylene Film Gels (Photografting of Hydrophilic Monomers)
Polyethylene Foams
Polyethylene, High Performance (Gel-Like Spherulite Press Method and Ultradrawing)
Polyethylene, Multicomponent
Polyethylene, Recycled
Polyolefins (Overview)
Polyolefins (Solid State Nuclear Magnetic Resonance)
Powder Coatings (Overview)
Supported Catalysts (Lewis Acid and Ziegler-Natta)
Supported Chromium Polymerization Catalysts (on Silica, Phillips Catalysts)
Surface Modification (by Low Energy Electron Beam)
Ultra-Strong Polyethylene Fibers

POLYETHYLENE (Commercial)

James F. Ross and John L. MacAdams
Polyolefins Research Center
Quantum Chemical Company

Traditionally, commercial polyethylenes have been classified into three major groups based on both the manufacturing process and the polymer properties:

- Low Density, or LDPE;
- High Density, or HDPE;
- Linear Low Density, or LLDPE.

Sometimes, polyethylene with densities below 0.91 are considered to be a fourth group, the very- or ultra-low density polyethylenes (VLDPE or ULDPE).

Today, according to the Society of the Plastics Industry, commercial production of polyethylene in the United States alone is approximately ten million metric tons per year, of which about 30% is LDPE, 45% is HDPE, and 25% is LLDPE.

PROCESSES

Low Density Polyethylene (LDPE)

LDPE is produced by a free-radical catalyzed reaction using oxygen or other free radical initiators such as organic peroxides or azo compounds. Synthesis conditions are usually 250–300°C outlet temperature, 120–280 MPa (pressure). Nominal reactor residence times are about 10–50 seconds.

High Density Polyethylene (HDPE)

These polymers are produced as a slurry in an inert hydrocarbon diluent or in gas phase fluidized beds. To produce copolymers, a very small amount of alpha olefin comonomer, usually butene-1 or hexene-1, is used to maintain the density in the 0.945–0.965 range.

Transition metal catalysts are used to manufacture HDPE. Chromium catalysts tend to produce broader molecular weight distribution (MWD) than Ziegler types. Recently developed metallocene catalysts can be employed to give radically different polymers having either a very narrow, or even bimodal, MWD.

Linear Low Density Polyethylene (LLDPE)

Linear low density polyethylenes are obtained by incorporating sufficient alpha olefin comonomers to yield polymers with densities in the 0.910–0.940 density range. Butene-1 is the usual comonomer, but either hexene-1, octene-1, or 4 methylpenetene-1 is employed to give enhanced physical and optical properties, albeit at a higher production cost.

Synthesis conditions, as well as equipment, are similar to those employed for HDPE and, in fact, many commercial fluid bed installations have been designed to switch back and forth between HDPE and LLDPE as market conditions dictate.

PROPERTIES OF COMMERCIAL POLYMERS[1,2]

In today's competitive marketplace, prime grade commercial polyethylenes must be both processable and uniform.

The main characteristics or properties which determine a resin's processability are molecular structure, uniformity, and additive content.

Film Grades

Almost all polyethylene film is fabricated as either cast film or blown film. In general, cast films have a better appearance and gauge thickness is more readily controlled. Blown films are more evenly oriented in machine and transverse directions, thus providing greater toughness.

Clarity and Haze

Slight haziness is a characteristic of all polyethylene resin. Surface roughness diffuses light passing through the film and contributes to its haziness. Surface roughness is a function both of the extrusion conditions and the fundamental structure of the polymer. The partly crystalline, partly amorphous structure of the resin also contributes to haziness.

Gloss

Generally, polyethylene items with high gloss sell better than products with a dull appearance. Glossier items, however, show fingerprints more clearly than less glossy products. Therefore, the importance of high gloss depends on the application.

Extrusion Coating Grades[3]

In extrusion coating, a thin film of molten polyethylene is pressed into or onto a substrate. Polyethylene is now being coated commercially on many substrates, including paper, paperboard, cellophane, polyester and other plastic films, metal foil, cloth, and glass fiber.

A polyethylene coating provides a heat sealable surface, increased tear and crease resistance, and an excellent barrier against moisture, grease, and oil. Other properties of polyethylene coated materials are: flexibility, even at refrigeration temperatures; printability (with surface treatment); resistance to chemicals; absence of taste and odor; nontoxicity; good scuff and puncture resistance, (especially important for puncture-prone substrates, such as aluminium foil); good appearance, especially gloss; and low material costs, especially for thinner gauges.

Injection Molding Grades[4]

Injection molding is essentially a three-step process. In the first, the polymer is heated, melted, and mixed into a homogeneous viscous liquid. In the second step, this molten polymer is forced, or injected, into a relatively cool mold. The third step is the cooling and subsequent ejection of the finished part from the mold.

The three steps mentioned above give rise to three sets of desired physical properties. In the first and second steps, the low-melt viscosity dictates the need for a high melt index. In the third step of injection molding, it is often important that the polymer have consistent shrinkage on cooling and solidification.

Blow Molding Grades[5]

Blow molding is a widely used method of producing polyethylene bottles and other hollow shapes. Blow-molding grades must possess a combination of a high-melt strength and good molding properties, as well as environmental stress-crack resistance and toughness. Most blow-molding grades on the market today are made from HDPE, but some LDPE or LLDPE is employed for flexible (squeeze) bottles.

Wire and Cable Grades

In general, polyethylene's electrical properties yield a tough, flexible, and chemically inert insulation that does not absorb water and remains flexible at very low temperatures. The dielectric strength of polyethylene varies very little with frequency or higher ambient temperatures. However, it is markedly affected by the quality of the coating, which must adhere tightly to the wire conductor without voids or air spaces.

High-density polyethylenes are preferable to lower density resins for some wire and cable coating applications. They have higher abrasion resistance and can better withstand mechanical abuse. Their tear resistance is greater than low density's as is their tensile and tear strength.

Rotomolding Polymers

For rotomolding large hollow items, such as barrels, special powdered grades of polyethylene are offered.

Metallocene Catalyzed Polyethylenes

Recently, limited quantities of polyethylenes produced by means of metallocene catalysts have appeared on the market. These resins are unique because metallocene catalysts are homogeneous, unlike heterogeneous transition metal catalysts used in conventional low-pressure processes. Nor are they like free-radical catalysts which produce polymers with a wide range of chain lengths and chain spacing. Because they are homogeneous (single site), metallocene catalysts produce polymers with markedly narrower molecular weight distributions than heretofore available. The copolymers possess very narrow compositional distributions as well.

These resins have been used as seal layers in multilayer films for packaging applications. These films have enhanced dart impact strength and greater clarity. Single layer films, however, are tacky. Their narrow molecular weight distribution makes them more difficult to process.

Ethylene–octene copolymers or proprietary terpolymers have been produced which are more easily processed into film. Metallocene catalyzed resins also appear promising as impact modifiers. At present, however, the market for metallocene catalyzed polyethylenes has not been adequately defined.

REFERENCES

1. *Petrothene Polyolefins. . .a Processing Guide* 5th ed.; USI Chem. Div., Nat'l Distillers and Chem.; Quantum Chemical: Cincinnati, OH, 1986.
2. DuBois, H.; John, F. W. *Plastics*, 6th ed.; Van Nostrand; 1981.
3. Fenner, R. T. *Principles of Polymer Processing*; MacMillan: 1979.
4. Weir, C. I. *Introduction to Injection Molding* Soc. Plastics Engineering: 1975.
5. *Molding of Plastics*; Bikales, N. M., Ed.; Wiley-Interscience: 1971.

POLYETHYLENE (High Density Preparation)

John Scheirs*
ExcelPlas Australia

George Evens
DSM Research

The technologies that have emerged for the production of high-density polyethylene (HDPE) are solution, slurry, and gas-phase polymerization. Important considerations in the manufacture of HDPE are ease of process control, degree of automation, absence of reactor fouling, minimal formation of waxes, high rate of ethylene conversion, product flexibility, space-time yield, and on-stream time (i.e., reliability). The fundamental differences in the various polymerization processes reflect the different approaches that have been devised to remove the substantial heat of polymerization of ethylene.

*Author to whom correspondence should be addressed.

SOLUTION POLYMERIZATION

Solution processes for the preparation of HDPE allow high rates of production with a relatively compact reactor. Solution-based processes enable production of HDPE over a broad range of densities but are limited in the range of melt indices they can elicit.

High-density polyethylene has a melting point of about 135°C. At ambient temperatures it does not dissolve in any solvent. HDPE does, however, dissolve in hydrocarbons of sufficient molecular weight at the solubility point, which is close to its melting point. This establishes the first condition for a solution process, that is, the reactor temperature must be high enough to prevent the polymer from crystallizing. In general, temperatures between 150°C and 250°C are used for solution polymerization.

In solution polymerization, HDPE dissolves as it is formed in a hydrocarbon solvent (commonly, cyclohexane). A good example of the solution polymerization method of ethylene is that used by DuPont and called Sclairtech®.

Catalyst for the Solution Process

The catalysts used in the solution-polymerization of HDPE vary depending on the type of process being used. Two groups of catalyst families can be distinguished: catalysts using a titanium/magnesium combination and catalysts based on a titanium/vanadium combination. The way the catalyst is prepared can make a tremendous difference in activity and selectivity.

Although solution processes are best for the production of very low density polymers, there is a restriction in the lower limit of the density due to the catalyst used.

The discovery of highly active single-site catalysts was a breakthrough in polyethylene catalyst research. These catalysts are a further development of the metallocene/methylaluminoxane catalysts invented by Professor Kaminsky.[1] These new catalysts have only one active site.

The advantages of single-site catalysts are: their very high activity (also at high polymerization temperatures), their high activity towards α-olefins, and their ability to copolymerize a wide range of comonomers. An additional feature of these catalysts is the production of polymer chains ending with a vinyl group. Due to the high reactivity of the single-site catalyst for α-olefins, the vinyl chain ends can be copolymerized. Herein lies a major advantage of the solution process. Because the process can operate at very high ethylene conversions, the ratio of ethylene to α-olefin in the reaction medium is reduced to low values. Using this relatively high concentration of unsaturated chain ends, long chain branching is introduced while MWD is kept very narrow.

SLURRY POLYMERIZATION

Slurry polymerization is the oldest and a still widely used method for producing PE because of its flexibility and process engineering maturity. It is capable of producing a very wide range of MWs of HDPE from a very low MW to an ultra-high molecular weight polyethylene (UHMWPE).[2]

In slurry polymerization of HDPE, the reaction occurs below the melting point of the polymer in a hydrocarbon diluent. As soon as the polymer is formed, it crystallizes and solid particles begin to grow. In the slurry process, the polymerization temperature must be below the melting point of the polymer; that is, as the density decreases, the polymerization temperature must decrease.

Different diluents have different solvency abilities. As a rule of thumb, the lower the MW of a linear alkane, the worse the solvency for PE.

Among the advantages of the slurry process are: the mild operating conditions, easy heat removal, high monomer conversions, the high solids content, and the ability to make a broad range of MWs.

Catalysts Used in Slurry Polymerization

The catalyst in a slurry process is not only important for the resulting polymer structure but it determines the morphology of the polymer powder that is produced.

The first generation Ziegler and Phillips catalysts were not very active and had to be removed from the polymer through a complex extraction process. The second generation catalysts make use of a support. The Ziegler catalysts are predominantly supported on an activated $MgCl_2$. Other modifiers can also be added, such as Lewis bases and silicium compounds. An aluminum alkyl is generally used as an activator.

High-efficiency catalyst systems for slurry polymerization are generally based on transition-metal compounds supported on a high-surface area silica substrate.

GAS-PHASE POLYMERIZATION

Gas-phase polymerization is the newest technology introduced for the manufacture of HDPE. It is characterized by its low capital investment costs, low operating costs, and by the absence of hydrocarbon solvents or diluents in the manufacturing process. Gas-phase technology was pioneered by Union Carbide and patented under the Unipol process.

In the gas-phase process, polymerization takes place in a fluidized-bed reactor at a temperature below the melting point of the polymer (the reactor is usually maintained at 100°C). Since polymerization and crystallization occur almost simultaneously, the growing polymer crystallizes under a shear-field.[3,4]

Unipol Catalysts

The Unipol process is generally based on three distinct catalyst systems. The polydispersity of the HDPE produced in the Unipol process is largely determined by the catalyst type. For instance, catalysts based on $TiCl_4/R_3Al$, chromocene, and silyl chromate produce polymers of narrow, medium, and broad MWDs, respectively.[5–8]

Another class of organochromium catalysts used in the Unipol process is based on bis(triphenyl silyl) chromate which is combined with the surface silanol groups on silica.[9–14] Unlike the original Phillips chromium oxide catalyst, this system needs to be reduced to an active polymerization state using an aluminum alkyl (i.e., aluminum ethoxide). It has the advantage of being responsive to hydrogen, thus allowing the MW of the polymer to be regulated as in the case of titanium-based catalysts. This catalyst produces HDPE of a broad MWD as a result of the chemical heterogeneity on the catalyst surface, which yields sites of differing polymerization activity and differing rates of chain propagation. A variation of the Unipol silyl chromate system is achieved through the addition of titanium alkoxide compounds. Titanium alkoxide effectively reduces silyl chromate structures to chromium (III) titanates, which then form surface titanyl chromates by oxidation with dry air.

HDPE produced by Unipol silyl chromate (and Phillips-type catalysts) are linear polymers with one vinyl end group per chain).

The effects of residual chromium catalyst residues on the stability of HDPE has been studied in detail by Scheirs et al.[3,4] It has been found that although the new high-efficiency polymerization catalysts leave smaller concentrations of residues in the final polymer, this does not necessarily result in higher polymer stability. It was found that chromium residues as low as 1 ppm can have a quite a deterimental effect on the thermooxidative stability of HDPE. This is because the chromium residues, in the form of chromium (III) oxide, are efficient prooxidants which when distributed on high-surface-area silica fragments increase their effective concentration and impact on polymer stability.

REFERENCES

1. Sinn, H.; Kaminsky, W. *Adv. Organomet. Chem.* **1980**, *18*, 99.
2. Beach, D. L.; Kissin, Y. V. "Ethylene Polymers." In *Encyclopedia of Polymer Science and Engineering* 2nd ed.; Mark, H. F.; Kroschwitz, J. I.; Eds.; John Wiley & Sons: New York, 1989; Vol. 6.
3. Scheirs, J.; Bigger, S. W.; Billingham, N. C. *J. Polym. Sci., Polym. Chem. Ed.* **1992**, *30*, 1873.
4. Scheirs, J.; Bigger, S. W.; Billingham, N. C. *Poly. Deg. and Stab.* **1992**, *38*, 139.
5. Karol, F. J.; Johnson, R. N. *J. Polym. Sci., Polym. Chem. Ed.* **1975**, *13*, 1607.
6. Karol, F. J. In *Encyclopedia of Polymer Science and Technology-Supplement*; John Wiley & Sons: New York, **1976**; Vol. 1, p 120.
7. Wagner, B. E.; Goeke, G. L.; Karol, F. J. Eur. Patent Appl. 55 605, 1981.
8. Wagner, B. E.; Goeke, G. L.; Karol, F. J. U.S. Patent 4 303 771, 1981. *Chem. Abstract* **1981**, *96*, p 21841o2.
9. Goeke, G. L.; Hamer, A. D.; Karol, F. J. Eur. Patent Appl. 3 836, 1979. *Chem. Abstract* **1979**, *92*, 23177b.
10. Goeke, G. L.; Wagner, B. E.; Karol, F. J. U.S. Patent 4 302 565, 1981.
11. Karol, F. J.; Goeke, G. L.; Wagner, B. E.; Fraser, W. A.; Jorgensen, R. J.; Friis, N. U.S. Patent 4 302 566, 1981.
12. Karol, F. J.; Munn, W. L.; Goeke, G. L.; Wagner, B. E.; Maraschin, N. J. *J. Polym. Sci., Polym. Chem. Ed.* **1978**, *16*, 771.
13. Karol, F. J.; Wagner, B. E.; Levine, I. J.; Goeke, G. L.; Noshay, A. In *Advances in Polyolefins* Plenum: New York, 1987, p 337.
14. Karol, F. J.; Wu, C. *J. Polym. Sci., Polym. Chem. Ed.* **1974**, *12*, 1549.

POLYETHYLENE (Low Density Manufacture)

John Scheirs
ExcelPlas Australia

Low-density polyethylene (LDPE) is commercially manufactured by the high-pressure (HP) polymerization of ethylene by a free-radical reaction in either a tubular or autoclave reactor. LDPE polymers contain alkyl branches (2 to 8 carbon atoms long) as well as long-chain branches (LCBs). Since these alkyl substituents cannot be accommodated in the crystalline lattice of polyethylene, the crystallinity and density of LDPE are reduced relative those of a linear poly(ethylene) (PE). LDPEs have densities in the range of 0.915 to 0.930 g/cm^3. Commercial grades of LDPE are approximately 45–55% crystalline and have a melting point in the region of 110–115°C (in contrast to HDPE, which has a melting point of 132°C).

Over the years, the production of LDPE has increased steadily and it has become the most widely used plastic in the production of such items as packaging film, shrink/stretch film, trash bags, meat/food wrap film, cable sheathing, greenhouse and mulch film, flexible piping, buckets, and stackable boxes, Tupperware™ containers. It is also used extensively in laminates with cartonboard/aluminium foil for packaging of liquids.

There are two commercial routes to the production of LDPE, namely, the tubular process and the autoclave process. The relative usage of tubular and autoclave processes is about equal.

MECHANISM OF HIGH-PRESSURE POLYMERIZATION

The first step in the polymerization process is that of initiation. The peroxide initiator decomposes by homolytic cleavage of the peroxidic bond to yield free radicals which add to the ethylene to initiate the growth of the polymer chain. In the propagation step, chain growth proceeds by the continued addition of ethylene units to the macroalkyl radical. Finally, chain growth is completed by the mutual termination of polymer radicals, chain transfer with ethylene, beta-scission of tertiary polymer radicals, or chain-transfer with a solvent (chain-transfer agent). Other radical reactions that can occur during polymerization are intramolecular chain transfer and intermolecular chain transfer. In intramolecular chain transfer, short-chain alkyl branches (usually butyl) form as a result of back-biting reactions. In intermolecular chain transfer, LCBs form by the extractions of hydrogen from a completed polymer chain by a macroalkyl radical, followed by the addition of another macroalkyl radical to the newly formed radical.

Peroxide Initiators

High-pressure polymerization of ethylene is usually achieved by using a mixture of organic peroxides (either alkyl or acyl peroxides), each of which has a specific temperature range over which decomposition occurs. Typical peroxides are *t*-butyl hydroperoxide, *t*-butyl perneodeconate, *t*-butyl peroxypivalate, and t-butyl peroxybenzoate.

Chain Transfer Agents

Chain transfer agents (CTA) are used to control the MW of LDPE by stopping the growth of one macroalkyl radical and beginning the growth of another. Chain transfer agents are usually solvent molecules which are added to the reactor in small amounts, for example, methyl ethyl ketone (MEK) or isopropanol are commonly used chain transfer agents.

POTENTIAL PROCESS PROBLEMS

The use of highly flammable ethylene combined with high temperatures and ultrahigh pressures makes LDPE polymerization somewhat difficult to control at times.

CONTROL OF POLYMER PROPERTIES BY ADJUSTING REACTOR VARIABLES

By careful manipulation of the reactor conditions, the properties of the LDPE product can be tailored to suit specific application areas. For example, high MW products are usually made by decreasing temperature and increasing pressure. The density of the resin, which is a function of short-chain branching, can be increased by decreasing the reaction temperature or increasing the reaction pressure. Conversely, the LCB is increased by raising the peak reaction temperature and decreasing the pressure.

SPECIALTY GRADES OF LDPE

Until the mid-1980s the basic structure and properties of LDPE remained essentially unchanged. In recent years, however, three new types of LDPE have emerged. These are very low-density PE (VLDPE), ultra-low density PE (ULDPE), and high MW, low-density PE (HMW-LDPE). Cd Chemie is producing a VLDPE (known as Lotrene™ 1001) with a melting point of 104°C for special packaging/sealing applications. The single site, metallocene catalysts recently developed by Exxon and Mitsui can produce VLDPE grades with a density as low as 0.86 g/cm^3. Dow has recently developed a range of ULDPEs targeted at medical plastic applications. Known as Attane™ 4601, 4602, 4603, and 4701, these resins have twice the puncture resistance of commonly used medical packaging materials, such as ethylene-vinyl acetate (EVA) copolymers, and are resistant to gamma sterilization.[1] Quantum has developed an HMW-LDPE (known as Petrothene™ NA 355) that combines the strength of a fractional melt index PE and the clarity of a 2 melt index resin. Because of its high melt strength, it can be processed to produce the higher clarity and tensile properties that come from orientation. Where both strength and clarity are needed, NA 355 can be a cost-competitive alternative to low EVA copolymers.[2]

REFERENCES

1. Dow Chemical article. In *Plastics World* **1991**, *August*, p 92.
2. Quantum article. In *Plastics World* **1990**, *May*, p 33.

POLYETHYLENE (New Aspects of Morphology)

Genevieve Delmas-Patterson and H. Phuong Nguyen
Chemistry Department
University of Quebec at Montreal

The properties of polyethylene (PE, CAS 9002-88-4) were investigated for several decades using various techniques.[1,2] At room temperature, PE is described as a semi-crystalline solid whose relative fractions of crystalline and amorphous phases can be obtained using, for instance, the sample density, the density of the pure orthorhombic crystal (1.0), and that of the pure amorphous PE (0.86 g cm^{-3}). A sample's crystallinity can also be obtained by comparing its enthalpy of melting (or of dissolution), ΔH_m, to that of a perfect crystal, $\Delta H_{m,o}$ (293 J/g). For linear PE, the calorimetric crystallinity rarely exceeds 0.7 for melt-crystallized samples and 0.8 for solution-grown crystals.

The mechanical properties are a rough average of the two phases. The amorphous phase made of disordered chains contributes slightly to the overall mechanical properties. Surprisingly, a solid with a fraction equal to 0.30 of disordered chains has the strength observed on PE samples. This strength is particularly high for linear PE with a high molecular weight and a low crystallinity as its density (0.94 g cm^{-3}) and its low ΔH_m value (180–190 J/g) indicate. Indeed, PE seems to be endowed with an unmelting order or cohesion undetected by differential scanning calorimetry (DSC). It is probably a short-range order that does not contribute to wide-angle X-ray scattering.

Other observations indicate unmelted order in PE melts or solutions. In PE melts, there are persistent bi-refringence[1] and two relaxation times,[3] and in PE solutions, gels and memory exist. Thermoreversible gels should not occur in a polymer that crystallizes easily.[4,5] With the gel route, highly drawable films are prepared that lead, after drawing, to high modulus fibers otherwise difficult to obtain.

This paper gives an example of a trace in a slow temperature ramp (T ramp) showing second melting and crystallization. We propose that this behavior occurs because of a physical network in the melt, and we will show how the network accounts for the ambiguities described.

CONCLUSION

Research in our group has shown that network formation in PE melts and solutions is ordered and melts under certain conditions. Because of the strain's effect on T_m and on the kinetics of melting, fast DSC experiments do not meet these conditions, and incomplete melting typically occurs. The strained phase has an order with too short a range to be detected by X-ray diffraction. This new phase is sensitive to the sample history and aging. The network fraction in the sample and the strain in the ordered phase can be measured by the enthalpy of fusion and by the range of temperatures for the phase change. Both can be altered by appropriate treatment. Other techniques sensitive to the strain between the chains should be developed to quantify the strained phase without melting it. The analysis of the rocking vibrations of the methylene segments seems to support the phase analysis from slow calorimetry.[6,7] Results from slow calorimetry obtained on other polymers suggest that these new morphological features for PE apply with some variation to other chain molecules.

ACKNOWLEDGMENT

The authors are grateful to the National Science and Engineering Research Council of Canada for supporting this research.

REFERENCES

1. Wunderlich, B. *Macromolecular Physics* Academic: New York, 1980; Vol. 3.
2. Wunderlich, B. *Macromolecular Physics* Academic: New York, 1976, Vol. 2.
3. De Gennes, P. G. *Macromolecules* **1984**, *17*, 703.
4. Narth, K. A.; Barham, P. J.; Keller, A. *Macromolecules* **1982**, *15*, 464.
5. Zwijnenburg, A.; Pennings, A. J. *J. Colloid Polym. Sci.* **1975**, *253*, 452.
6. Bernazzani, P.; Delmas, G. Université du Québec à Montréal, unpublished results.
7. Bernazzani, P. M.Sc. Thesis, Université du Québec à Montréal, 1994.

POLYETHYLENE (Stabilization and Compounding)

John Scheirs
ExcelPlas Australia

Compounding is the general term for the operation converting the polyethylene (PE) produced in a polymerization reactor into a pelletized form suitable for use by a downstream processor. In the compounding process, the polymer is physically and chemically modified by incorporating various additives. The selection of additives and of compounding conditions depends on the end application of the compounded product.

In any compounding operation, the first step is stabilizing and incorporating other additives. Stabilizing the polymer is important to prevent thermal oxidation and discoloration during melt processing and to impart long-term heat and light stability. Stabilizers are the most widely used additives in the PE industry. Other additives commonly added to high-density polyethylene (HDPE) are stearates (e.g., calcium and zinc stearates to neutralize acidic catalyst residues and to impart lubricity), pigments (e.g., carbon black, cadmium yellow, and titanium dioxide), metal deactivators (e.g., oxalyl bishydrazide for PE cable insulation), processing aids (e.g., Dynamar®, Ucarsil®, Viton® A, to improve flow of the molten polymer), and optical brightners (e.g., Uvitex® OB to enhance appearance). Additives commonly added to low-density polyethylene (LDPE) include antioxidants (hindered phenolics and phosphates for process stabilization), antiblock compounds (e.g., silica, talc, and kaolin to reduce sticking of adjacent film layers), and slip agents (e.g., erucamide and oleamide to reduce friction).[3]

STABILIZERS AND OTHER ADDITIVES

Antioxidants and Processing Stabilizers

The most widely used and reported antioxidant for PE is Irganox® 1076. This antioxidant, also known as octadecyl 3-(3,5-di-*tert*-butyl-4-hydroxyphenyl)propionate, has gained widespread acceptance in the industry since the early 1980s, when it began to replace the more volatile butylated hydroxytoluene (BHT or 2,6-di-*tert*-butyl-4-methyl phenol).

Irganox 1076 belongs to a class of antioxidants known as hindered phenols that act as radical scavengers (also known as radical traps or chain-breaking antioxidants).[1]

UV Stabilizers

There are two classes of UV stabilizers used in PEs: UV absorbers and hindered-amine light stabilizers (HALS). A common UV absorber is Chimassorb® 81 (or Cyasorb® 531), which is chemically known as 2-hydroxy-4-*n*-octyloxybenzophenone. These additives absorb UV radiation in the range of 300–360 nm, the spectral region of sunlight most damaging to polymers.

Hindered-amine light stabilizers (HALS) have become increasingly important in the UV stabilization of PE and have generally displaced UV absorbers.[2] HALS inhibit photooxidation in PE extremely well. The mechanism by which HALS function involves their oxidation to a nitroxyl radical that scavenges alkyl radicals. This reaction then produces hydroxylamines that can react with peroxyl radicals and, in turn, regenerate nitroxyl radicals. This cycling of nitroxyl radicals accounts, in part, for the high effectiveness of HALS, even at low concentrations.

Pigments

Carbon black is probably the most common pigment and UV screener used in PE. Usually added in concentrations between 2 and 3 wt %, carbon black protects against UV light. When compounded into LDPE, carbon black can inhibit thermal oxidation for ~30 h at 140°C.[4]

Common Additives Used in LDPE

Silica is commonly used as an antiblocking agent for LDPE film-grade resins. Erucamide, chemically known as *cis*-13-docosenamide, is a fatty acid amide CH_3–$(CH_2)_7$–CH=CH–$(CH_2)_{11}$–CO–NH_2 commonly used in LDPE film to provide "slip." Erucamide, which is incompatible with PE, migrates to the surface of the LDPE film over ~24 h. It is essentially a surface-active molecular lubricant used to lower the film's coefficient of friction (COF), allowing adjacent surfaces to slide smoothly over one another and bags to slide smoothly in automatic operations such as form, fill, and seal.

COMPOUNDING PRINCIPLES

Compounding primarily involves the melt plasticization of a PE reactor product, generally accompanied by additives. Two types of processes are necessary for compounding PE and additives: distributive mixing and dispersive mixing. Distributive mixing is essential for homogenizing the polymer and for distributing the additive properly. Dispersive mixing is essential for carbon black, which tends to form agglomerates.

PRODUCT PROBLEMS RELATED TO COMPOUNDING

Gel Contamination

Gels are the most common problem in compounding PE and the number one source of complaints for LDPE. Gels are small regions of loosely crosslinked polymer that generally form in an extruder during compounding.

Black Speck Contamination

Common sources of contamination in HDPE are black specks, local burnt regions within pellets. These are small areas of highly degraded polymer or degraded additive that have been carbonized because of excessive residence time, excessive shearing, or the presence of hot spots (localized high-temperature regions) in an extruder.

Extraneous Contamination

Typical contaminants in commercial PE are degraded polymer, metal, sand, paper, and fiber. Degraded polymer contamination (known in the industry as burnt resin) is usually a translucent brown color and shatters under pressure.

Odor

Odor in PE is generally associated with the polymer's oxidation, the additives, or both.

Pellet Geometry

Poor pellet geometry is the term that describes any deviation in pellet shape from that which is standard or desired. The shape of pellets can influence their behavior in extruders and injection molding machines. There are many variations of poor pellet geometry, including match-sticks, jackstraws, bullet-heads, shorts, tails, streamers, fines, and angel-hair.

Color

During PE compounding, the natural pellets are frequently discolored, usually yellowing, although pink and grey hues can also develop. In nearly all cases, the discoloration is related to the additives present.

The most well-known origin of discoloration in PE is the yellowing imparted by stilbene quinone compounds that are the rearrangement products (usually dimers) of hindered phenolic antioxidants.[5]

PE discoloration can be masked with optical brightners such as Uvitex OB, which mask yellowing by emitting into the blue spectrum.

REFERENCES

1. Al-Malaika, S.; Scott, G. In *Degradation and Stabilization of Polyolefins*; Allen, N., Ed.; Applied Science: London, 1983; p 247.
2. Bigger, S. W.; Delatycki, O. *J. Polym. Sci., Polym. Chem. Ed.* **1989**, *27*, 63.
3. Breuer, T. E.; Martin, N. B.; Tomlinson, H. H. et al. *ANTEC '84*, 1984; 141.
4. Hawkins, W. L. In *Polymer Stabilization*; Hawkins, W. L., Ed.; Wiley-Interscience: New York, 1972; Chapter 2.
5. Henman, T. J. *Developments in Polymer Degradation*; Elsevier: London, 1985; Vol. 6, p 107.

POLYETHYLENE (Surface Functionalization)

Ging-Ho Hsiue
Department of Chemical Engineering
National Tsing Hua University

Polyethylene (PE) was first produced on a commercial scale in 1939. PE's most attractive features, apart from its low price, are excellent electrical insulation properties over many different frequencies; very good chemical resistance; good processability, toughness, and flexibility; and in thin films of certain grades, transparency.[1] However, PE's hydrophobicity limits its usefulness. To solve this problem, the tuning of PE's surface properties by surface functionalization became imperative. Various surface modifications of thin PE films were performed to achieve functionalization and good mechanical properties. The techniques included high-energy-radiation grafting, the plasma technique, chemical modification, photochemical grafting, macroinitiator grafting, and anionic grafting.[2–13] As a result,

PE's surface properties have been exploited successfully to produce materials with many applications such as separation membranes, biomaterials, biosensors, polymeric catalysts, flame-retardant materials, and materials for recovering Uranium.[14–23]

High-energy-radiation-induced grafting provides a convenient way to functionalize the surfaces of PE thin films. High-energy radiation has high penetration power and emits no pollution. Free radicals were brought about upon radiation, and the grafting was then completed via the free radicals. The surface functionalization of PE thin films via high-energy radiation resulted in a few applications such as PE-g-acrylic acid (PE-g-AA) for separation membranes, sulfonated PE-g-styrene for electric dialysis, and PE-g-vinyl pyridine for ion exchange membranes.[2,14,24–26] Moreover, PE-g-AA could be modified further for use in biomedical applications. For instance, PE-g-AA-HEMA (2-hydroxy ethyl methacrylate) was used as an anti-blood-clotting material and a biosensor, respectively.[19,20,27–29] Hsiue et al. also reported that PE-g-AA was used to immobilize glucose oxidase and urease.[27–29] Humidity sensors and flame-retardant materials prepared via radiation-induced surface functionalization of PE have also been reported.[15,16]

The plasma technique is another important tool that permits hydrophilic functional groups such as $-NH_2$, -COOH, and -OH to be functionalized onto PE surfaces. This technique is of great interest because it is simple; offers diversity in functionalization, especially for polar functional groups; proceeds quickly; allows the extent of functionalization to be controlled; and produces a uniform functionalized surface and unaltered bulk properties. The applications resulting from the plasma technique resemble those from high-energy-radiation grafting.

Ion implantation is also used for the surface functionalization of organic materials.[12] Oxidized structures and double bonds in PE are created by implanting F^+ ions. Because the thickness of the surface layer affected by ion implantation does not exceed several tenths of a micrometer, the mechanical properties are virtually unaltered. The electric conductivity of polymers can be increased by 14 orders of magnitude. The wettability and the sticking properties also improve.

Postpolymerization for polymer surface functionalization is a broadly used synthetic scheme to alter the polymers' chemical and physical properties.[10] This approach depends on physically entrapping relatively short, terminally functionalized oligomers into a host polymer film. Various surface-grafting chemistry and surface-grafted surfaces can be obtained as a result.

Photografting is a useful technique for introducing various functional groups into polymer films.[6–8] Generally, this technique uses UV light and an initiator to achieve surface functionalization. Light in the near UV is advantageous because it is selectively absorbed by the initiator without affecting the polymer. The most commonly used initiators for grafting are benzophenone or its derivatives. These initiators create grafting sites when UV radiation removes hydrogen from the polymer films. This technique has been applied for the surface modification of polymers, the immobilization of enzymes, the photostabilization of polypropylene, the syntheses of insoluble polymeric ligand for adsorption of metallic ions, and for polymer used in humidity sensors.

CONCLUSION

The merits of polyethylene are its low price, good processability, excellent electrical insulation properties, good chemical resistance, toughness, and flexibility. However, the hydrophobicity of PE hinders its usefulness considerably. PE thin films were surface-modified in many ways to achieve functionalization and good mechanical properties. Here, the techniques included high-energy-radiation grafting and the plasma technique. Applications such as gas separation membranes, ion exchange membranes, biomaterials, biosensors, and polymeric catalysts were obtained as a result.

REFERENCES

1. Brydson, J. A. *Plastics Materials*, 4th ed.; Butterworth Scientific: London, 1982; Chapter X.
2. Gupta, B. D.; Chapiro, A. *Eur. Polym. J.* **1989**, *25*, 1137.
3. Mercx, F. P. M. *Ind. Eng. Chem. Res.* **1993**, *32*, 848.
4. Iwata, H.; Ikada, Y.; Hata, Y. et al. *J. Polym. Sci., Polym. Chem.* **1988**, *26*, 3309.
5. Suzuki, M.; Ikada, Y.; Iwada, H. et al. *Macromolecules* **1986**, *19*, 1804.
6. Edge, S.; Walker, S.; Feast, W. J. et al. *J. Appl. Polym. Sci.* **1993**, *47*, 1075.
7. Kubota, H. *Eur. Polym. J.* **1992**, *28*, 267.
8. Kubota, H.; Hata, Y. *J. Appl. Polym. Sci.* **1990**, *41*, 689.
9. Morgan, P. W.; Corelli, J. C. *J. Electrochem. Sci.* **1982**, *129*, C329.
10. Bergbreiter, D. E.; Srinivas, B.; Gray, H. N. *Macromolecules* **1993**, *26*, 3245.
11. Gray, H. N.; Bergbreiter, D. E. *ACS Polym. Mater. Sci. Eng.* **1993**, *206*, 178.
12. Svorcik, V.; Rybka, V.; Endrst, R. et al. *J. Appl. Polym. Sci.* **1993**, *49*, 1939.
13. Chtourou, H.; Riedl, B.; Kokta, B. V. et al. *J. Appl. Polym. Sci.* **1993**, *49*, 361.
14. Guthrie, J. T.; Kotov, S. *J. Appl. Polym. Sci.* **1989**, *37*, 39.
15. Sakai, Y.; Sadaoka, Y.; Rao, V. L. *J. Mater. Sci.* **1989**, *24*, 101.
16. Kaji, K.; Yoshizawa, I.; Kohara, K. et al. *J. Appl. Polym. Sci.* **1994**, *51*, 841.
17. Kabay, N.; Katakai, A.; Sugo, T. et al. *J. Appl. Polym. Sci.* **1993**, *49*, 599.
18. Garnett, J. L.; Kenyon, R. S.; Levot, R. et al. *J. Macromol. Sci. Chem.* **1980**, *A14*, 87.
19. Mullerschulte, D.; Horster, F. A. *Polym. Bull.* **1982**, *7*, 77.
20. Mullerschulte, D.; Horster, F. A. *Polym. Bull.* **1982**, *7*, 395.
21. Yamaguchi, T.; Nakao, S.; Kimura, S. *Ind. Eng. Chem. Res.* **1993**, *32*, 848.
22. Kubota, H. *Eur. Polym. J.* **1993**, *29–4*, 551.
23. Weaver, J. K.; Derkash, R. S.; Greenwald, A. S. *Clin. Orthopaed. Relat. Res.* **1993**, *290*, 224.
24. Chen, W. K. W.; Mesrobian, R. B. *J. Polym. Sci.* **1957**, *18*, 903.
25. Hsiue, G. H.; Huang, W. K. *J. Appl. Polym. Sci.* **1985**, *31*, 1119.
26. Hsiue, G. H.; Wu, J. H. *J. Chin. Inst. Chem. Eng.* **1987**, *18*, 33.
27. Hsiue, G. H.; Wang, C. C. *J. Appl. Polym. Sci.* **1990**, *40*, 235.
28. Hsiue, G. H.; Wang, C. C. *Biotechnol. Bioeng.* **1990**, *36*, 811.
29. Hsiue, G. H.; Wang, C. C.; Chen, C. Y. et al. *Angew. Makromol. Chem.* **1990**, *197*, 149.

POLYETHYLENE FILM GELS (Photografting of Hydrophilic Monomers)

Mitsuo Hirata
Department of Industrial Chemistry
Nihon University

We found in a series of experiments that photografting hydrophilic monomers onto hydrophobic substrates such as polyethylene (PE) and polytetrafluoroethylene (PTFE) enhanced their wettability and adhesion. Grafted layers of the surfaces absorbed a significant amount of water, which is characteristic of hydrogels.[1] Hydrophilic monomers with a hydrophobic part in the structure grafted better onto these surfaces. For example, methacrylic monomers were more effective than acrylic monomers in modifying hydrophobic substrates. Hydrophilic monomers could not be photografted directly onto PTFE surfaces, and PTFE gels were difficult to obtain when PTFE surfaces were photografted by a pretreatment of oxygen plasma.[2]

We focused our efforts on preparing strong gels by grafting hydrophilic monomers throughout PE films. We used the following monomers: methacrylic acid (MAA, acidic), methacrylamide (MAAm, neutral), 2-(dimethylamino)ethyl methacrylate (DMA, basic), and *N*-isopropylacrylamide (NIPAAm, thermoresponsive).

SUMMARY

Results of Grafting MAA, MAAm, and DMA on PE Film Gels

- MAAm and DMA monomers promote the surface and the internal graftings of PE films, respectively;
- the grafted PE films can accommodate a significant amount of water, and the amount of absorbed water increases in the order of DMA > MAA > MAAm. In particular, the DMA-g-PE films have the most distinguished water absorption; and
- the electrical conductivity of each grafted PE film sharply increases around the Qc, the grafted amount at which the grafting front reaches the film's center.

Conclusions on Functionalities on Stimuli-Responsive PE Film Gels Prepared by Grafting MAA, DMA, and NIPAAm Monomers

- permeabilities of MAA-g-PE and DMA-g-PE films vary with a change in the pH of the medium;
- on-off regulations of permeation for NIPAAm-g-PE and DMA-g-PE films are repeated in response to the alternating temperature change below and above their respective Ttrs; and
- the grafted PE films have adequate mechanical strength in the swollen state to be useful as functional membranes.

ACKNOWLEDGMENT

The author wishes to thank Dr. K. Matsuda for his comments and help.

REFERENCES

1. Yamada, K.; Tsutaya, H.; Tatekawa, S. et al. *J. Appl. Polym. Sci.* **1992**, *46*, 1065.
2. Yamada, K.; Hayashi, K.; Sakasegawa, K. et al. *Nippon Kagakukaishi* (in Japanese) **1994**, 427.

POLYETHYLENE FOAMS

Guillermina Burillo
Instituto de Ciencias Nucleares UNAM

Esbaide Adem
Instituto de Física UNAM

Expanded polyolefins (foams) offer many unique properties to the cellular plastic industry. Foams of polyethylene (PE) are tough, flexible, chemical and abrasion resistant, and offer excellent electrical and thermal insulation properties; the mechanical properties are intermediate between rigid and highly flexible foams. The density ranges from 0.032 g/c^3 to that of the solid polymer. The cell structure is usually closed-cell. Highly expanded PE foams are potentially the most inexpensive of cellular plastics, but the need for expensive processing techniques has caused the cost per unit volume to exceed that of low-density polystyrene (PS) and polyurethane foam.

Polyethylene can be crosslinked chemically using certain peroxides[1–3] or with high-energy radiation.[4–6]

PREPARATION

Expanded plastic or plastic foam consists of two main phases: a gaseous phase (blowing agent) and a solid phase (plastic). The chemical nature of the plastic is the dominant factor in determining foam properties and the composition of the gaseous phase has an important effect on some properties such as thermal insulation.

The blowing or foaming agents for the polyolefins are normally gaseous materials or decomposable solids. About 1000 different compounds have been proposed as chemical blowing agents; of these, only about a dozen are commercially significant,[7] and only three of them have obtained commercial value as foaming agents for polyolefins: 4,4'-oxybis(benzenesulfonyl hydrazide), azodicarbonamide, and *N,N'*-dinitroso pentamethylenetetramine.

Physical blowing agents should be gaseous under normal conditions to avoid condensation in the foam cells, but if easily liquefied under process conditions many difficulties are avoided.

Chemically Crosslinked PE Foams

To promote the formation of small, uniform, stable cells in the product, gas release must occur after the formation of the crosslinking bonds has taken place. Therefore, the decomposition temperature of the crosslinking agent must be less than the decomposition temperature of the blowing agent. Dicumyl peroxide in LDPE or 2,5-dimethyl-2,5-di(tertiary butylperoxy) hexyne in HDPE are usually selected as chemical crosslinking agents when azodicarbonamide is used as the blowing agent.

Radiation Crosslinked PE Foams

The advantage of using radiation for crosslinking comes in freeing the temperatures needed for gas release from the relatively narrow range of temperatures required to begin peroxide crosslinking. AZD is the blowing agent usually chosen because of its relatively high gas yield (230 c^3/g of nitrogen, carbon monoxide, and carbon dioxide) and low cost plus 66–68% of residues.

Post-Crosslinking Process

Post-crosslinking of low-expansion foamed PE also can be carried out easily in-line by electron beam irradiation and on continuous sheets or profiles for other uses to improve solvent resistance, thermoforming properties, and crack resistance.

Flame-Retardant Foams

In radiation-crosslinking foaming processes, additives such as dyes, antioxidants, and especially flame-retardants are often used, thus modifying the irradiation process.

PROPERTIES

Polyethylene foam possesses desirable properties obtained from polyethylene resin, namely resilience, good elasticity, and toughness. It also has the attributes of a closed-cell foam: low water absorption, good energy absorption, and low thermal conductivity. But generally, the surface is rough and the sheet gauge is not precise. One of the main deficiencies is its poor resistance to creep.

Use of crosslinked polyethylene allows exercise of more control over the formation of cells. Foams derive their properties in part from the polymer material of the cell walls, and in part from the foam density; cell size and cell structure are also important variables in a foam.

Polyethylene has one inherent advantage over other materials: its physical properties can be varied over wide limits, from a very flexible rubberlike foam to a rigid foam.

Properties of Post-Crosslinked Foams

Post-crosslinked foams have increased solvent resistance for bottle closures, improved thermoforming properties for structured foam sheets, and increased crack resistance for reinforcing components of luggage and shoes.

APPLICATIONS OF EXPANDED POLYETHYLENE

Applications of expanded polyethylene range across many fields:

- Industrial uses include pipe insulation, insulation of air conditioners, adhesive tape, core cushion of adhesive tape, gaskets, packing, hospital bedding, coated tapes, replacements for metal fasteners.
- Among construction uses are insulation of corrugated iron roofing, backing for floor tiles, joint filling and sealing, wall coverings, furniture, warehouse pallets, and paintbrush handles.
- In the automobile industry they are used for carpet padding, trunk mats, wheel housing covers, vinyl roof padding, headliners, molded door panels, seat trim padding, truck interior linings, die cut gaskets, and mounting tapes.
- Sports applications include protective athletic equipment, life jackets, inner liner of athletic footwear, toys for fun and learning, bathroom mats, floats for water-ski belts, and buoyant jackets.
- Agricultural uses include insulation of tunnel culture livestock, expansion joints on roofs, walls, and concrete slabs.
- They also find applications in other miscellaneous products, such as orthopedic support material in splints and shoes, helmet liners, bra-cups, and jewelry case liners.

Low density polymer foams in the 0.032–0.160 g/cm^3 density range are used widely for package cushioning because the energy absorption under continued impact provides protection for delicate electronic parts, for polished surfaces, and for heavy metal assemblies. The automotive industry is using LDPE foams for many gasketing and sealing applications (such as body sealing gaskets) to keep out wind and water.

High-density PE foams in the density range of 0.16–0.64 g/cm^3 are moving into areas previously occupied by products of solid polymers or natural cellular materials such as cork or wood.

REFERENCES

1. Precopio, F. M.; Gilbert, A. R. (to General Electric) U.S. Pat 2 888 424, 1959.
2. Benning, C. J. *J. of Cell Plast.* February 1967, *3* No. 2, 62.
3. Kemmotsu, T.; Okada, M.; Ono, T. *Radiat. Phys. Chem.* **1993**, *42*(1–3), 97.
4. Brasch, A. (to Electronized Chemicals), U.S. Patent 2 981 668, 1961.
5. Trageser, D. A. *Radiat. Phys. Chem.* **1977**, *9*, 261.
6. López, M. A.; Burillo, G.; Charlesby, A. *Radiat. Phys. Chem.* **1994**, *43*(3), 227.
7. Lasman, H. R. *SPE Journal* **1962**, *18*, 1184.

Poly(ethylene glycol)s

See: Anticancer Polymeric Prodrugs, Targetable
Cyclodextrin-Polymer Inclusion Complexes
Poly(ethylene oxide) (Overview)
Poly(ethylene oxide) (Applications in Drug Delivery)

POLYETHYLENE, HIGH PERFORMANCE (Gel-Like Spherulite Press Method and Ultradrawing)

Toshihiko Ohta
Faculty of Human Life Science
Osaka City University

The gel-like spherulite (GSP) method is an effective way to produce high-performance organic materials, especially polyolefin, which is used for materials that have a large cross-sectional area, such as rods and thicker tapes.[1]

Even in the gel-spinning method, strength and modulus of filaments rapidly decrease while diameters of a single filament are increased up to 0.1 mm.[1] High-performance plastic rods are required as a replacement for steel wire, because the latter has some essential problems such as electromagnetic interference, corrosion, and weight. Hence, several kinds of plastic composite rods reinforced by glass fiber or aramid fiber were used. Development of polyoxymethylene (POM) rod was also in progress,[2] and researchers found that high-strength (2 GPa) and high-modulus (150 GPa) polyethylene (PE) rods having a diameter of 1.0 mm could be continuously produced by this method. Such high-performance organic materials were unknown before this development.

THE GSP METHOD

This method is characterized by ultradrawing of a pressed sheet formed by compression of an accumulated material of gel-like spherulites obtained from a semidilute solution of ultra-high molecular weight PE (UHMW-PE).[1] The gel-like spherulites formed from 2.0 wt % decalin solution of UHMW-PE ($M_w = 2 \times 10^6$, Hizex 240M) were generally used, because the upper limit of solution concentration, where gel-like spherulites of UHMW-PE were formed separately, was 2.0–2.5 wt %.

Each spherulite shows gel-like behavior because it contains a large amount of the solvent (about 90 wt %).

CONCLUSION

From research we draw the following conclusions:

- The increase in strength and modulus by ultradrawing of the GSP sheets is independent of the initial solution concentration, mainly due to the small difference of chain entanglement density between both GSP sheets, which was caused by disentanglement between polymer chains during formation of a gel-like spherulite.
- The slope of Griffith plot of the drawn tapes from GSP sheets is in the same range as those of the gel-spinning method and of the fibrillar-crystal-growing method.
- The slopes of strength and modulus versus draw ratio were considerably lower in the GSP sheets than in the gel-casting sheet, mainly due to the lower chain entanglement density in the GSP sheets than in the gel-casting sheet, resulting from the difference in the solution cooling rate between the GSP sheets and the gel-casting sheet.
- The maximum draw ratio (λ_{max}) could be predicted from the drawing stress at 100°C of a predawn tape in the GSP method and the gel-casting method.

We concluded that the influence of spherulites' size and of no (of fewer) chain connections between spherulites on the ultradrawing behavior in the GSP method could be ignored, and the ultradrawing in this method could be explained by the deformation of three-dimensional networks composed of chain entanglement points, the same as that in the gel-casting/spinning method. Furthermore, we conclude that in the specimen having a higher chain entanglement density, the drawing stress was more effectively transferred to each polymer chain through the chain-entanglement points, and consequently the increase in strength and modulus induced by chain extension was promoted more in the ultradrawing process.

REFERENCES

1. Ohta, T. et al. *Polymer* **1989**, *30*, 2170.
2. Ishida, S. *Sen-i Gakkaishi* **1987**, *43*, 143.

POLYETHYLENE, MULTICOMPONENT

Ted M. Pettijohn
Witco Corporation

Multicomponent polyethylene [9002-88-4] is a term used to describe blends of two or more distinctly different polyethylene component resins. The component resins are selected in such a way as to give the resulting blend enhanced physical properties.[1] The components vary in molecular weight, molecular weight distribution, and both short-chain (comonomer content) and long-chain branching content.

Blends of two components are the most common multicomponent system and are often referred to as bimodal resins. Bimodal describes the appearance of the molecular weight distribution as determined by size exclusion chromatography.

PREPARATION

Four methods of preparing multicomponent resins exist: multiple reactor processes, extruder (or melt) blending, multiple catalysts in one reactor, and multi-site catalyst in one reactor. These resins are produced commercially using multiple-reactor technology.[2] There are two major configurations: reactors-in-series and reactors-in-parallel.

PROPERTIES

In single-component of unimodal resins, there exists a distinct balance between a resin's stiffness and its environmental stress crack resistance (ESCR).

Processability, toughness, and ESCR can all be enhanced by blending two or more polymers to produce a multicomponent resin.

The key to achieving those enhanced properties is through the proper selection of the component resins. If those components are not used in the correct ratios, the properties of the resulting blends can be and often are worse than the unimodal resins.

APPLICATIONS

The primary applications of multicomponent polyethylene resins are in film, pipe, and blow-molding markets.[3] High-density, high-molecular-weight multicomponent polyethylene film resins are used to produce pond liners and trash and merchandise bags. The multicomponent resins are stronger than conventional unimodal linear, low-density resins.

Improved ESCR is the driving factor behind multicomponent resin use in pipe and blow-molding applications. Pipe produced with multicomponent resins is stronger and has a longer useful life. Drums and household and industrial containers are the primary markets of the blow-molding resins. Examples of these products include oil, antifreeze, detergent, and cleanser bottles.[4]

REFERENCES

1. Bailey, F. W.; Whitte, W. M. (to Phillips Petroleum Company) U.S. Patent 4 461 873, 1984.
2. Lundeen, A. J.; Feig, J. E. (to Occidental Chemical Corp.) U.S. Patent 5 236 998, 1993 and references therein.
3. Leaversuch, R. D. *Modern Plast.* **1990**, *August*, 52.
4. Schut, J. H. *Plastics Tech.* **1990**, *November*, 25.

Poly(ethylene naphthalate)

See: Poly(alkylene naphthalate)

Poly(ethylene oxide)

See: Comb Polymers (Poly(ethylene oxide) Side Chains)
Molecular Complexes [Poly(Ethylene Oxide) and Urea]
Nonthrombogenic Materials
Poly(ethylene oxide) (Overview)
Poly(ethylene oxide) (Applications in Drug Delivery)
Poly(ethylene oxide) Macromonomers (Micellar Polymerization)
Poly(ethylene oxide) Side-Chain Acrylates (Synthesis and Polymerization)
Solid Polymer Electrolytes
Solid Polymer Electrolytes (Polyether Blends and Composites)
Supramolecular Formation and Structure (PEO-PPO-PEO Block Copolymers)
Surface-Modified Polymers
Uniform Polymers

POLY(ETHYLENE OXIDE) (Overview)

Joseph V. Koleske
Consolidated Research Incorporated

Ethylene oxide (EO) is one monomer that can be polymerized to form a product having a very wide range of molecular weights that often have a narrow molecular weight distribution. Molecular size varies from the dimer, diethylene glycol, to polymers having a molecular weight of several million, all of which are commercially available. Polymers of ethylene oxide generally are known by two distinguishing terms: poly(ethylene glycol) for those compounds having hydroxyl end groups and a molecular weight of about 200–20,000, and poly(ethylene oxide) or polyoxyethylene for those products with a molecular weight >20,000.

PREPARATION

Ethylene oxide is the simplest member of the 1,2- or vicinal epoxides and has a molecular weight of 44.05.

It is a colorless gas with a boiling point of 10.4°C at 1 atmosphere. At safe exposure levels, ethylene oxide is odorless and cannot be detected. When it can be detected by smell through its etherlike odor, at about 500 ppm, safe-exposure limits have been exceeded.

This 3-member ring epoxide is polymerized by a ring-opening reaction that is initiated by acid, base, or ionic coordinate mechanisms.

COPOLYMERS

A wide variety of copolymers containing ethylene oxide are known. Anionic initiators are usually used when copolymers are prepared.

With anionic initiation, the relative reaction rates of the oxirane monomers are much nearer to each other and random copolymers can be formed. If the monomers are fed in different ratios, block copolymers are formed, with the nature of the blocks depending on monomer feed composition.

Researchers have prepared block copolymers of ethylene oxide with styrene.[2–4] Star-shaped block copolymers[5] and multifunctional branched or graft copolymers of ethylene oxide with polystyrene have been prepared from chlormethylstyrene.[6]

Ethylene oxide has been grafted to a variety of substrates such as silica,[7] gelatin and collagen,[8] and acrylic acid.[9]

PHYSICAL CHARACTERISTICS

Poly(ethylene oxide)s are considered to be soluble in water in all proportions. However, a recent study indicates that in very dilute solution (0.02–0.06%) an unknown fibrillar phase with a crystal-like behavior exists.[10] It is also soluble in organic solvents such as acetonitrile, chloroform, ethyl acetate, ethoxyethanol, dichloroethylene, isopropanol (91%), methyl *i*-butyl ketone, tetrahydrofuran, toluene, and similar compounds.[11,12] The polymer has a solubility parameter[13] of 10.3 $(cal/cm^3)^{0.5}$ The Mark–Houwink expression relating intrinsic viscosity and molecular weight in water at 30°C for the molecular weight range of 10^4 to 10^7 is given by (**Equation 1**):

$$[\eta] = 1.25 \times 10^{-4} M^{0.78} \qquad (1)$$

When crystallized from the bulk, polyoxyethylene has a lamellar structure.[14,15]

High molecular weight polymers have a glass transition temperature of −52°C.[16] The polymer has good mechanical properties. It can be characterized as a strong, tough material with an elongation of 700–1200% and an engineering tensile strength at break of 1800–2400 psi (12.4 to 16.5 Mpa).[17]

ASSOCIATION COMPLEXES

Homopolymers of ethylene oxide will form association or interpolymer complexes with a number of other polymers.[1,18,19] The complexes are formed because of the strong hydrogen-bonding affinity of the ether oxygen atoms in the molecular chains. These complexes include those with poly(acrylic acid) and poly(methacrylic acid),[20–23] phenolics,[24] *p*-hydroxybenzoic acid–formaldehyde copolymers,[25,26] polysaccharides,[27,28] and the like.

A wide variety of electrolytes are known to associate with polyoxyethylene.

A number of salts such as potassium carbonate, zinc sulfate, potassium fluoride, sodium chloride, potassium bromide, and others will decrease the lower consolute temperature of polyoxyethylene and cause it to precipitate from aqueous solution.[1,29–32]

END USES

Polymers of ethylene oxide are used in a number of industries and in a variety of end uses.[1] The polymers often are used as additives, and small amounts impart special characteristics to a product.

The end uses include binders for ceramic green ware, wood products and artifact preservation, sizings, adhesives, surfactants, chromatography, water treatment and flocculation, hydrodynamic drag-reduction agents; functional fluids such as metal-working lubricants, hydraulic fluids, drilling fluids, metal-cleaning formulations, and quenchants; flocculants, thermal sportswear, personal care products such as lotions, lipsticks, chapsticks, ointments, shaving creams, adhesives, and skin lubricants; pharmaceutical products such as suppositories, enzyme modifiers, emollients, tablet and specialty coatings, and binders; textile-manufacture products such as dye assistants, antistatic agent additives, fiber lubricants, wastewater flocculants; and many others.

REFERENCES

1. Bailey, F. E. Jr.; Koleske, V. *Alkylene Oxides and Their Polymers*; Marcel Dekker: New York, NY, 1991.
2. Hill, F. N.; Fitzpatrick, J. T. U.S. Patent 2 866 761, 1961.
3. Richards, D. H.; Szwarc, M. *Trans. Faraday Soc.* **1959**, *56*, 1644.
4. Quirk, R. A.; Ma, J. J. *J. Polym. Sci., Part A: Polym. Chem.* **1988**, *26*, 2031.
5. Xie, Hongquan; Xia, Juan *Makromol. Chemie* **1987**, *188*, 2543.
6. Alteres, T. *J. Polym. Sci.* **1985**, *13*, 4131.
7. Tajouri, T.; Facchini, L.; Legrand, A. P.; Balard, P.; Papirer, E. *Chim. Phys. Phys.-Chem.-Biol.* **1987**, *84*(2), 243; *Chem. Abstract* **1987**, *107*, 59750.
8. Gantar, A. et al. *Polymer* **1987**, *28*, 1403.
9. Sun, F. et al. *Xuexiao Huaxue Xuebao* **1987**, *8*(7), 658; *Chem. Abstract* **1988**, *108*, 38551.
10. Klenin, V. I.; Kolnibolochuk, N. K.; Solonina, N. A. *Vysokomol. Soedin. Ser. A.* **1988**, *30*(10), 2076; *Chem. Abstract* **1989**, *110*, 58539.
11. Bailey, F. E., Jr.; Koleske, J. V. *Poly(ethylene oxide)*; Academic: New York, NY, 1976.
12. Bailey, F. E., Jr.; Koleske, J. V. In *Nonionic Surfactants, Physical Chemistry*; Schick, M. J., Ed.; Marcel Dekker: New York, NY, 1987; p 927.
13. Galin, P. M. *Polymer* **1983**, *24*, 323.
14. Geil, P. H. *Polymer Single Crystals*; Wiley-Interscience: New York, NY, 1963.
15. Keller, A. In *Growth and Perfection of Crystals*; Doremus, R. H. et al., Eds.; John Wiley & Sons: New York, NY, 1958; p 1.
16. Faucher, J. A.; Koleske, J. V.; Santee, E. R.; Stratta, J. J.; Wilson, C. W. *J. Appl. Phys.* **1966**, *37*; 3962.
17. Johari, G. P.; Hallbrucker, A.; Mayer, E. *J. Polymer Sci. Part B: Polym. Phys.* **1988**, *26*, 1923.
18. Bailey, F. E.; Lundberg, R. D.; Callard, R. W. *J. Polym. Sci. Part A* **1966**, *4*, 1563.
19. Labouc, A.; Rabadeux, J. C. *Eur. Polym. J.* **1988**, *24*, 603.
20. Smith, K. L.; Winslow, A. E.; Peterson, D. E. *Ind. Eng. Chem.* **1959**, *51*, 1361.
21. Elias, H. G. *Angew Chem.* **1961**, *73*, 209.
22. Chatterjee, S. K.; Chatterjee, N.; Riess, G. *Makromol. Chemie* **1982**, *183*, 481.
23. Chen, H. L.; Morawetz, H. *Eur. Polym. J.* **1983**, *19*, 923.
24. Chatterjee, S. K.; Sethi, K. R. *J. Polymer Sci., Chem. Ed.* **1983**, *21*, 1045.
25. Chatterjee, S. K.; Sethi, K. R. *Polymer Comm.* **1983**, *24*, 253.
26. Chatterjee, S. K.; Sethi, K. R. *J. Macromol. Sci. - Chem.* **1984**, *A-21*, 765.
27. Ohno, H.; Takinishi, Y.; Tsuchida, E. *E. Makromol. Chem. Rapid Comm.* **1981**, *2*, 511.
28. Sjoberg, A.; Karlstrom, G. *Macromolecules* **1989**, *22*, 1325.
29. Bailey, F. E., Jr.; Callard, R. W. *J. Appl. Polym. Sci.* **1963**, *1*, 56.
30. Amu, T. C. *Polymer* **1982**, *23*, 1775.
31. Stockmayer, W. H.; Fixman, M. J. *Polym. Sci.* **1963**, *C1*, 137.
32. Ananthapadmanabahn, K. P.; Goddard, E. D. *Langmuir* **1987**, *3*, 25.

POLY(ETHYLENE OXIDE) (Applications in Drug Delivery)

Neil B. Graham and Marion E. McNeill
Department of Pure and Applied Chemistry
University of Strathclyde

Key properties that make poly(ethylene oxide) (PEO) attractive as a biomaterial are biocompatibility, hydrophilicity, and versatility. The simple water-soluble linear polymer can be modified by chemical interaction to water-insoluble but water-swellable hydrogels retaining the desirable properties associated with the ethylene oxide part of the structure.

Poly(ethylene glycol)s (PEG) first appeared in the U.S. Pharmacopoeia in 1950; since then they have been used increasingly for a variety of pharmaceutical applications.[1]

PEO HYDROGELS

Methods of Forming PEO Hydrogels[4]

1. Radiation or radical crosslinking of $\overline{M}_w$ high PEO.
2. Entanglement crosslinking of high $\overline{M}_w$ PEO.
3. Crosslinking by reaction of the hydroxyl groups of PEG.
4. Building a block copolymer along with a hydrophobic component.
5. Complexation between high $\overline{M}_w$ PEO and hydrogen-bonding polymers such as poly(acrylic acid).

Properties

Glass Transition Temperature, T_g

The temperature at which the glassy amorphous domains of PEG become rubbery, T_g, is typically –60°C increasing to approximately –45°C in the xerogels. Both linear and crosslinked PEOs are therefore rubbery, semicrystalline polymers at working temperatures. Addition of 1 mol water per ether group of PEG lowered the T_g to –73°C.[6] Further addition of water then increased the T_g to a constant value of –65°C.

Interaction of PEO with Solvents

The degree of swelling and diffusion of solutes through water-swollen hydrogels is of primary interest but the ability of non-aqueous solvents to swell a crosslinked PEO hydrogel or dissolve a PEO copolymer provides a means of incorporating pharmaceutically active compounds, which are less soluble in water, into a hydrogel matrix.

Water in hydrogels may exist in different states. Jhon and Andrade labelled three types of water: bound, interfacial, and free (or bulk) water.[8] McNeill and Graham also identified these three types of water in hydrated PEO hydrogels using a differential scanning calorimetry, DSC.[7]

Assuming water exists in hydrogels in different states as argued above, diffusion of water-soluble molecules would be expected to be different through the free water and bound water.

APPLICATIONS

Where prolonged medication is necessary, repeated doses providing instant delivery cause fluctuations of the drug concentration in the plasma; initially peaks may reach a toxic level dropping to troughs at an ineffective level until the next dose is administered. One of the goals of controlled delivery is obtaining therapeutically optimum drug concentrations for a prolonged period. In hydrophobic polymers water-soluble drugs have low flux rates. Migration of drugs take place along the matrix continuum, sometimes called "partition" diffusion. The diffusion coefficients of drugs in these polymers will be of the order 10^{-8}–10^{-11} cm^2s^{-1}. In hydrogels with more than 50% water, transport of water-soluble drugs is primarily through aqueous microchannels and the mechanism is termed "pore" diffusion. The diffusion coefficient will be 10^{-6}–10^{-7} cm^2s^{-1}. Diffusion coefficients of drugs with molecular weights <500 are 10^{-5} cm^2s^{-1} order in water.

Poly(ethylene oxide)-based hydrogels are particularly biocompatible and exhibit low levels of blood complement activation and thrombogenicity.[2,3] PEO hydrogels containing medication can be conceived in many different shapes and sizes and inserted into accessible body cavities such as the conjunctival sac of the eye, the ear, mouth, rectum, uterus, and vagina, and then retrieved afterwards. They can be swallowed to release the active additive in the stomach or, after a delay, in the intestines. When hydrated they may contain several times their dry weight in water without feeling wet. Being non-toxic they can be placed over an inflamed wound to cool, protect from airborne bacteria, and release a therapeutic agent. They can also be used as transdermal patches.

Crosslinked Hydrogels

Solution of Drug in a Fully Hydrated Hydrogel

Most nonbiodegradable polymer-based drug delivery systems are capable only of continuously releasing useful amounts of low molecular weight compounds.[9] Hydrated polymers containing more than 50% water can provide high fluxes of water-soluble agents and moderate fluxes of macromolecules.[8,10,11]

Dispersion of Drug in an Initially Dry Hydrogel

Unlike release of solute from a solution in a fully swollen hydrogel where controlling variables such as D_p and monolith size are constant and measurable, the mechanisms which control the rate of release of dispersed water-soluble drugs in an initially dry hydrogel are complex and the result of a combination of several interrelated controlling factors that are changing with time.

Microgels. Poly(ethylene glycol)/urethane crosslinked microgels have been prepared as either water-soluble or water-insoluble forms.[12,13] The water-soluble forms are envisaged as solubilizing aids for some low-solubility drugs. The water-insoluble forms crystallize to spherical granules that can be admixed with a drug and formed into tablets by melting or compression. On swelling in water they provide essentially constant drug release without an initial surge.

Biodegradable. In designing a polymeric implant for sustained drug delivery it may be advantageous to use a biodegradable polymer that will be bioabsorbed, avoiding the necessity to remove the implant when it is exhausted.

Block Copolymers

A useful property of poloxamer gels is their thermal reversibility. Aqueous solutions of Pluronic F127 at concentrations > 20% are liquid at < 25°C and gel at higher temperatures.[5] Therapeutic agents in Pluronic F127 have been applied to burns.

Because of their surface-active characteristics and hydrophile/lipophile balance, poloxamers are useful for solubilizing drugs. Suppositories of poloxamer gels have been found to compare favorably with other suppository materials, giving more sustained and uniform release of the drug from the formulation.

PEO Complexes

Peppas and Klier reviewed the structure and swelling behavior of poly(ethylene glycol)/poly(methacrylic acid) complexes.[14] Polymer discs were swollen in solutions of guaiacol glyceryl ether at pH 4 and pH 8 then transformed to water to monitor the release of the drug. The diffusion coefficient of the solute in the highly swollen uncomplexed disc was 2.56 × $10^{-6}cm^2s^{-1}$ but only 3 × $10^{-9}cm^2s^{-1}$ from the complexed disc. Poly(methacrylic acid-g-ethylene glycol) can be prepared in the form of microspheres approximately 25 μm in diameter by means of a free-radical suspension polymerization on which streptokinase can be immobilized resulting in enhanced thrombus dissolution.[15]

A range of PEG-ethyl cellulose polymer films containing caffeine and salicylic acid was prepared and studied by Samuelov et al.[16]

REFERENCES

1. Heochst *Brochure on Polyethylene Glycols*, 1984.
2. Gaylord, N. G. *Polyethers*; Interscience: New York, 1963.
3. Union Carbide, *Carbowax Polyethylene Glycols*, 1986.
4. Graham, N. B. In *Hydrogels in Medicine and Pharmacy* Peppas, N., Ed.; 1987; Vol. II, 95.
5. Garcia Sagrado, F.; Guzman, M.; Molpeceres, J.; Aberturas, M. R. *Pharm. Tech. Eur.* **1994**, *6*(5), 46.
6. Graham, N. B.; Zulfiqar, M.; Nwachuku, N. E.; Rashid, A. *Polymer* **1989**, *30*, 528.
7. McNeill, M. E.; Graham, N. B. *J. Biomater. Sci. Polymer*, Ed. **1993**, *4*(3), 305.

8. Jhon, M. S.; Andrade, J. D. *J. Biomed. Mater. Res.* **1973**, *7*, 509.
9. Langer, R.; Hsieh, D. S. T.; Rhine, W.; Folkman, J. *J. Membr. Sci.* **1980**, *7*, 334.
10. Graham, N. B. In *Poly(ethylene glycol) Chemistry: Biotechnical and Biomedical Applications*; Milton Harris, J., Ed.; Plenum: New York, 1992; p 263.
11. Sato, S.; Kim, S. W. *Int. J. Pharm.* **1984**, *22*, 229.
12. Graham, N. B. Polymerization process, U.K. Patent G.B. 290264.
13. Graham, N. B.; Mao, J. Microgels, U.K. Patent Application No. 940-5264.4.
14. Klier, J.; Peppas, N. A. In *Absorbent Polymer Technology: Studies in Polymer Science, No. 8*; Brannon-Peppas, L.; Harland, R., Eds.; Elsevier: Amsterdam, 1990; p 147.
15. Drummond, R. K.; Klier, J.; Alameda, J. A.; Peppas, N. A. *Macromolecules* **1989**, *22*, 3816.
16. Samuelov, Y.; Donbrow, M.; Friedman, M. *J. Pharm. Sci.* **1979**, *68*, 325.

POLY(ETHYLENE OXIDE) MACROMONOMERS (Micellar Polymerization)

Koichi Ito
Department of Materials Science
Toyohashi University of Technology

Poly(ethylene oxide) (PEO) macromonomers have a general structure given by **Structure 1**:

$$RO\!-\!\!\left[CH_2CH_2O\right]_n\!\!-\!V \qquad \mathbf{1}$$

where V is the α-end group that can polymerize and R is the ω-end group that may be usually alkyl or hydrogen. They are amphiphilic and soluble in a variety of solvents including water, alcohols, benzene, and even petroleum, depending on the nature of the end groups and the PEO chain length (n).

MICELLAR POLYMERIZATION

Most characteristic of the micellar polymerization of the PEO macromonomers in water is the unusually high rate of conversion together with the very high degree of polymerization attained.[1–4]

Considering the structure of the macromonomers, the hydrophilic PEO chains will clearly constitute the surrounding shell of the micelle, while the hydrophobic polymerizing α-end groups will be concentrated or organized in the core of the micelle, and the ω-end groups will either extend out of the shell or enter into the core depending on their hydrophilic or hydrophobic nature.

REFERENCES

1. Ito, K.; Tanaka, K.; Tanaka, H.; Imai, G.; Kawaguchi, S.; Itsuno, S. *Macromolecules* **1991**, *24*, 2348.
2. Ito, K.; Hashimura, K.; Itsuno, S.; Yamada, E. *Macromolecules* **1991**, *24*, 3977.
3. Chao, D.; Itsuno, S.; Ito, K. *Polym. J.* **1991**, *23*, 1045.
4. Ito, K.; Tomi, Y.; Kawaguchi, S. *Macromolecules* **1992**, *25*, 1534.

POLY(ETHYLENE OXIDE) SIDE-CHAIN ACRYLATE (Synthesis and Polymerization)

Bunichiro Yamada
Material Chemistry Laboratory
Faculty of Engineering
Osaka City University

To obtain high polymer from vinyl monomer, a chain reaction involving propagating radical as chain carrier has to proceed with sufficiently long kinetic chain length. A large number of monomers including monosubstituted ethylenes, 1,1-disubstituted ethylenes, and 1,2-disubstituted ethylenes have been polymerized. Substituted acrylic esters have offered a large number of polymers bearing a wide variety of groups among the series of substituted styrene, acrylonitrile, and acrylic ester, whereas the styrene and acrylonitrile analogs involve monomers without polymerizability.[1] Therefore, a functional group or segment exhibiting strong characteristics is often bound to the acryloyl moiety to yield polymer with the desired functionality. Poly(ethylene oxide) segment ($-(CH_2CH_2O)_n-$: PEO) and cyclic oligoethylene oxide, and crown ether, that exhibit unique characteristics can be bound to the acryloyl moiety in several manners. PEO is usually expected to provide hydrophilic character or water solubility. Besides this, PEO has been used in combination with alkaline metal salt to compose ionic conductive material. Crown ethers have been known to show strong tendency to complex with metallic cation and some other cations depending on the size of the cyclic structure. The polymer-bearing PEO or crown ether can be obtained from the acryloyl monomer consisting of the corresponding components when the monomer fulfills the requirements for polymerization.

PREPARATION

Polymer and Copolymer

A variety of acrylic and methacrylic PEO esters including the divinyl monomers of which the polymerization is readily initiated with common initiator such as 2,2'-azobisisobutyronitrile in bulk or solution are commercially available.

Because of amphiphilic nature, the monomer consisting of methacryloyl moiety and PEO (**3**, $CH_2 = C(CH_3)\,CO(OCH_2CH_2)_nOR^1$) have been polymerized in water forming micelle and in homogeneous benzene solution.[2] The polymerization of the macromonomers is faster than the low molecular weight methacrylic ester whereas the polymerization is carried out at a low macromonomer concentration such as of the order of 10^{-2} mol/L. The amphiphilic macromonomers result in micelles that concentrate and orientate the hydrophobic methacrylate groups in water. Furthermore, the alkyl end group consisting of a certain number of carbons or above render to organize smaller micelles involving looplike structures of PEO. The fast polymerization of the macromonomers in benzene is ascribable to slow diffusion of the propagating radical.

Ito et al. ran the copolymerization of styrene with methacrylate with PEO chain as esters alkyl group and the reactivities of the methacrylates toward polystyrene radical were determined.[6]

The reactivities of the methacrylates with short PEO chain ($n < 4$) are quite similar to that of methyl methacrylate indicating that the steric and polar effects of the ester alkyl group on the reactivity are not significant. The reactivity apparently decreases with an increase in the chain length of PEO regardless of the terminal group in comparison with the methacrylate with short PEO.

PROPERTIES OF POLYMER

Introduction of PEO as ester alkyl group or α-substituent of an acrylate brings about considerable drop of T_g of polymer. Hydrophilicity of the polymer increases with an increase in PEO portion in the monomer unit.

The polymers from the monomers bearing PEO exhibit unique characteristics other than hydrophilic or amphiphilic character. Alkaline metal salt such as lithium perchlorate is miscible with PEO, and ion conductive character may be given at temperatures above the melting point of the crystalline PEO. Restricted mobility of the cation below the melting point may diminish the conductivity. Polymers from PEO acrylate (**3**, $CH_2 = CHCO(OCH_2CH_2)nOCH_3$) and PEO methacrylate (**3**, $CH_2 = C(CH_3)CO(OCH_2CH_2)_nOCH_3$) have been chosen instead of PEO alone to use the rubber state of these polymers with suitable crosslinking.[7]

Poly(PEO itaconate)s were also employed for the same purpose in the hope of obtaining the complex with weak tendency to crystallize because of low T_g of the itaconates.[8]

Acrylic and methacrylic esters bearing 12-crown-4,15-crown-5, and 18-crown-6, which can be polymerized, as well as alkyl acrylate and methacrylate and their polymers, exhibited much larger extractability of alkali metal salts from aqueous solution than the corresponding crown ethers or bis(crown ethers)s. Furthermore, much higher selectivity of Na^+ from K^+ was observed for the polymers of methacrylic ester bearing 15-crown-5 and 18-crown-6-moieties than for the corresponding crown ethers.[3–6]

REFERENCES

1. Yamada, B.; Kobatake, S. *Prog. Polym. Sci.* **1994**, *19*, 1089.
2. Ito, K.; Tanaka, K.; Tanaka, H.; Imai, G.; Kawaguchi, S.; Itsuno, S. *Macromolecules* **1991**, *24*, 2348.
3. Kimura, K.; Tamura, H.; Maeda, T.; Shono, T. *Polym. Bull.* **1979**, *1*, 403.
4. Varma, A. J.; Majewicz, T.; Smid, J. *J. Polym. Sci.: Polym. Chem. Ed.* **1979**, *17*, 1573.
5. Maeda, T.; Ouchi, M.; Kimura, K.; Shono, T. *Chem. Lett.* **1981**, *1981*, 1573.
6. Ito, K.; Tsuchida, H.; Hayashi, A.; Kitano, T.; Yamada, E.; Matsumoto, T. *Polym. J.* **1985**, *17*, 827.
7. Bannister, D. J.; Davies, G. R.; Ward, I. M. *Polymer* **1984**, *25*, 1600.
8. Cowie, J. M. G.; Ferguson, R. *J. Polym. Sci. Polym. Phys. Ed.* **1985**, *23*, 2181.

POLYETHYLENE, RECYCLED

John Scheirs
Excel Plas Australia

Environmental pressures during the past decade have caused recycled plastics to become a predominant subject in today's plastics industry. Plastic packaging is a high volume market owing to the many advantages of plastics over other traditional materials; however, it is also the most visible in the waste stream and has received a great deal of public criticism. Recycling of plastics is faced with many barriers. Achieving a clean recycled resin continues to be a challenge. Unfortunately, there are an increasing number of products on the market that contain two or more plastic materials and these materials are not always easily separated or even recognized by the consumer. One approach to this problem is to utilize the post-consumer waste in a commingled (mixed) form.[1] However, the properties of mixed polymer recycle blends are inferior to that of sorted material. One of the most recognizable post-consumer plastic containers is the high-density polyethylene (HDPE) milk bottle, and these yield a recycle stream that is highly homogeneous and consistent. Their high volume usage particularly in the U.S. and Australia results in large quantities of recycled HDPE material. Presently about 60% of milk bottles in the U.S. are recycled and about 50% in Australia.

In the last five years, the recycling of HDPE from post-consumer milk bottles has gone from the domain of backyard scrap merchants to the technologically advanced recovery facilities of large PE firms. Next to PET drink bottles, HDPE milk bottles or "jugs" (as they are known in the U.S.) are undoubtedly the largest single source of recycled plastic. Furthermore, this HDPE recycle is most in demand because it is unpigmented and is the most consistent stream of recycled plastic available. In the U.S. and Australia the majority of milk, water, and juice bottles are made from HDPE. In Europe, though milk is more commonly packaged in cartonboard cartons and glass bottles, HDPE is still used widely for detergent and shampoo bottles. For instance, more than 60% of household packaging waste in Germany is composed of polyolefins. This mainly consists of HDPE bottles and LDPE film.[2]

Recycled PE can also originate from post-industrial PE waste, which includes in-house scrap and off-cuts.

LDPE recycling is also widespread, although not to the same extent as HDPE recycling. The majority of LDPE that is recycled originates from post-industrial waste such as bundle-shrinkwrap used to stabilize loads on pallets as well as greenhouse films and mulch films. There is only a limited proportion of recycled LDPE that can be classified as post-consumer recycle (PCR).

PROPERTIES OF RECYCLED HDPE

The advantages of recycling HDPE from milk and juice bottles are: these bottles are easily recognizable and are in abundant supply; the resultant recycled product has consistent melt flow index (MFI) and density from lot-to-lot over time; because dairy resin is a tightly specified blow-molding resin, recycling of HDPE milk containers can potentially give a clean, natural-colored product; and studies have shown that the resultant recycle has essentially the same rheological properties as the virgin resin and has not undergone any noticeable thermal degradation.

APPLICATIONS FOR MODIFIED HDPE RECYCLE

There are numerous current commercial applications for modified recycled HDPE and these are summarized in **Table 1**.

TABLE 1. Current Commercial Applications for Recycled HDPE

Application	Modifier	Comments
Mobile garbage bins	50% LLDPE	Structural foam process
Forklift pallets	50% LLDPE	Structural foam process
Milk crates	70% LLDPE	Injection molded
Garden trellis	50% LLDPE	Structural foam process
Wheelbarrow body	50% LLDPE	Structural foam process (nitrogen in melt)
Compost bins	20% LLDPE	thick sections, UV stabilized with carbon black
Spirally-wound pipe	50% butene HDPE	Ribloc™ pipe for sewer replacement
Artificial lumber	50% sawdust	For decking, patios e.g., Trex™ by Mobil
Trailer boards	Carbon black	Tongue and groove boards (Eaglebrook Plastics)
Oil bottles	Carbon black	Dark colored bottles with no critical ESCR
Detergent bottles	50% bimodal HDPE	Alathon™ L5040 imparts ESCR requirements
Grocery bags	50% HMW-HDPE	Bimodal Alathon™ 5005 imparts high tenacity
Shopping bags	75% LLDPE	Blend matches performance of Dowlex™ 2101
Overwrap film	50% LDPE	Low tear strength advantageous for application
Curbside boxes	50% LLDPE	For curbside collection and recycling
Pails	50% LLDPE	Made by Tucker Household Goods (US)
Highway posts	Glass microspheres	Hollow spheres lighten and cheapen posts
Structural beams	Glass fiber	For marine piers (Plastics Lumber Co.)
Storm water grates	20% barium sulfate	High specific gravity filler ensures heavy part
Garden edging	30% SB rubber	Highly flexible strip (e.g. Flexiedge™)
Inspection pits	5% carbon black	Carbon black reinforces and stabilizes molding
Soaker hose	70% rubber crumb	Incompatibility b/w rubber and PE gives porosity
Antigraffiti panels	70% Tirecycle™ 801	Exudation of oil extender prevents marking
Roofing shingles	60% Tirecycle™	High UV and thermal stability, rigidity
Castors, bin wheels	30% Tirecycle™	Good abrasion resistance
Sand sausages	50% rubber crumb	Soil stabilization in coastal areas
Rubber mats	50% Tirecycle™ 801	Mats for tray trucks

CONTAMINATION

Contamination in recycled PE can arise from a number of sources: from multi-component systems that use dissimilar polymers such as PP closures, adhesive-backed paper labels, and even through the incorporation of additives such as pigments; during use (e.g., by the contents of the packaging); during collection (e.g., owing to consumers' mixing plastic types); by the environment (e.g., soil in LDPE mulch film); and by reprocessing (e.g., gels and black specks).

ACKNOWLEDGMENTS

The author would like to acknowledge the following people for their contributions to this paper in the form of correspondence and valuable advice: Philip Blatz, DuPont Central Research and Development, Wilmington, Delaware; Bruce Kuiken, Quantum Chemical Co., Heath, Ohio; Susan Selke, Associate Professor, Michigan State University; Claude White, Lyondell, Alathon Polymers Division, Texas; Francis White, Graham Engineering Co., York, Pennsylvania; Alexander Zahavich, Uniplast Industries, Inc., Orillia, Ontario, Canada.

REFERENCES

1. Umstadter, L. W. "Recycling Commingled Plastics" Florida Recycling Fair, February, 1991. Proceedings available from Prof. Charles Beatty, University of Florida, Gainesville, Florida, USA.
2. Brandrup, J. *Die Makromol. Chem., Macromol. Symp.*: W. Heitz, Ed. **1992**, *57*, 57.

Poly(ethylene terephthalate)

See: *Poly(ethylene terephthalate) (Chemistry and Preparation)*
Poly(ethylene terephthalate) (Dielectric Properties)
Poly(ethylene terephthalate) (Heat Settings and Thermal Stabilization)
Poly(ethylene terephthalate) (Modified with Ionic Groups)
Poly(ethylene terephthalate) (Ultra-High Molecular Weight)
Poly(ethylene terephthalate) Copolymers
Poly(ethylene terephthalate) Degradation (Chemistry and Kinetics)

POLY(ETHYLENE TEREPHTHALATE) (Chemistry and Preparation)

Saleh A. Jabarin
Polymer Institute
The University of Toledo

Poly(ethylene terephthalate) (PET) is a polycondensation polymer that is most commonly produced from a reaction of ethylene glycol (EG) with either purified terephthalic acid (PTA) or dimethyl terephthalate (DMT), using a continuous melt-phase polymerization process. In many cases, melt-phase polymerization is followed by solid-state polymerization.

Melt-phase polycondensation is used to prepare fiber-grade PET or a precursor resin which is then solid-state polymerized to achieve higher molecular weight or intrinsic viscosity. Melt polymerization is usually carried out at around 285°C. Due to increased rate of thermal degradation of PET by further increase in temperature or time of polymerization, final intrinsic viscosity (I.V.) is usually kept to below 0.6.

The process of solid-state polycondensation is used for preparation of PET for container and other applications requiring an I.V. greater than 0.6. This process minimizes degradation and side reaction product formation that would occur during continued melt-phase polymerization. Solid-state polymerization is carried out by heating the solid low I.V. resin below its crystalline melting point (~225°C) but above its glass transition temperature, Tg (~80°C). Usually the process is carried out at temperature ranges between 180 and 240°C, at which the polymer end groups are sufficiently mobilized for reaction to take place. Reaction byproducts are removed by a flow of inert gas or by maintaining reduced pressure.

MELT-PHASE POLYCONDENSATION

PET can be prepared by direct esterification of terephthalic acid and ethylene glycol or transesterification of dimethyl terephthalate with ethylene glycol.

Side Reactions During Synthesis

Commercial synthesis of PET does not lead entirely to a pure linear poly(ethylene terephthalate) structure. The bulk polymer made by the melt-phase process contains small amounts of cyclic ethylene terephthalate such as trimers, tetramers, and pentamers.[1,2] In addition, it contains small amounts of ether linkages.

The polycondensation step during melt-phase polymerization also can produce gaseous degradation products such as acetaldehyde.

Catalyst systems used in esterification and polycondensation reactions during commercial synthesis of PET are of prime importance. The catalyst system affects the reaction rate, color, and side reactions. Among systems most frequently used are the oxides and the acetates of antimony, lithium, calcium, magnesium, manganese, zinc, germanium, and their mixtures.

The standard process used for production of PET from EG and PTA or DMT also has been modified by addition of a small amount of a comonomer to create PET-copolymer resins. These copolymers often exhibit improved processing characteristics and physical properties. Such comonomer modifiers include purified isophthalic acid (IPA), cyclohexanedimethanol (CHDM), diethylene glycol (DEG), and dimethyl-2,6-naphthalenedicarboxylate (NDC).

SOLID-STATE POLYMERIZATION

Chemical and Physical Processes of Solid-State Polymerization

At the completion of melt-phase polymerization, at least three kinds of PET chains are important for solid-state polymerization: chains terminated with hydroxyl end groups, chains terminated with carboxyl end groups, and chains containing a vinyl ester end group.

During the solid-state polymerization of PET, various reactions can occur. The main polycondensation reaction takes place when two chains having hydroxyl end groups interact to produce a larger molecular weight chain and ethylene glycol as a byproduct.

Esterification can produce higher molecular weight PET and water as side products. This reaction occurs between PET chains with hydroxyl and carboxyl end groups.

The reaction of vinyl ester end groups with hydroxyl end groups produces a higher molecular weight PET and acetaldehyde product.

REFERENCES

1. Heffelfinger, C. J.; Knox, K. L. in *The Science and Technology of Polymer Films*, Sweeting, O. J., Ed.; Wiley-Interscience: New York, 1971; Vol. II.

2. Goodman, I. Polyesters, in *Encyclopedia of Chemical Technology* Othmer, K., Ed.; Kirk-Othmer copublishers, Interscience: New York, NY, 1968; 2nd ed., Vol. 16.

POLY(ETHYLENE TEREPHTHALATE) (Dielectric Properties)

Gu Xu* and Craig Zaluski
McMaster University

Poly(ethylene terephthalate), or PET, is a typical member of the polyester family composed of repeated units of (-CH_2CH_2-OOC-C_6H_4-COO-) containing a phenyl group (C_6H_4). Whereas PET fibers (with trade names of Terylene, Dacron, etc.) are now largely used for the textile apparel and carpet market, PET films (branded as Mylar, Melinex, etc.) offer a huge variety of applications in the electrical and power industries as well as in the electronic industry. This is because of their excellent electrical insulating and good thermal–mechanical properties.

As PET finds favor as an electrical insulator, its dielectric properties have become a critical concern in optimization of applications. For example, to ensure efficient transmittance of an electrical signal via a cable, the dielectric loss (DL) of the cable sheathing must be kept to a minimum. In other cases, the dielectric breakdown strength (DBS) becomes crucial when a high voltage or high electric field is involved.

Dielectric loss studies of PET were first carried out by Reddish[1] using PET fiber melts. Subsequent studies were provided by Huff and Muller,[2] Krum and Muller,[3] Saito and Nakajima,[4] Yamafuji,[5] Hellwege and Langebin,[6] Berestneva et al.,[7] Mikhailov et al.,[8] and Saito.[9] The DBS measurement of PET was initially given by Inuishi and Powers in 1957,[18] where only a dc voltage was applied. Later, Park, Hara et al.[13–15] published a series of results for the effects of mechanical stresses, heat treatment, and temperature on the DBS.

PREPARATION AND DIELECTRIC PROPERTY MEASUREMENT

PET is made by condensation from equimolar amounts of ethylene glycol ($HOCH_2CH_2OH$) and terephthalic acid (*p*-$HOOCC_6H_4COOH$). To get PET films with excellent mechanical properties, high thermal stability, and relatively high dielectric strength, the polymer must undergo extrusion casting followed by biaxial stretching and heat treating. The extruded polymer is quenched so that the film remains completely amorphous and no spherulitic crystallization occurs prior to the film heat and drawing treatments.

DIELECTRIC PROPERTIES

DL and permittivity describe the efficiency and capacity of an insulator to store electric charges representing energy or information. The DBS provides the maximum voltage applicable across the material under specified conditions.

*Author to whom correspondence should be addressed.

Dielectric Loss and Permittivity of PET

Little difference is observed between the permittivity of the PET films and that of the melt fibers. For frequencies higher (or lower) than 1 kHz, the DL maxima appeared at higher (lower) temperatures, following approximately the Williams-Landel-Ferry or (Vogel-Tammann-Fulcher) equation and the Arrhenius equation for α and β relaxation peaks, respectively.[1,10,11]

Moisture sorption was, in general, found to increase both the DL and the permittivity.[16,17]

DBS of PET

In cases of dc voltage, the most probable process leading to room-temperature breakdown is failure at a defect.[18] In the presence of moisture, the breakdown strength becomes time dependent for long durations of the applied field, and the prebreakdown current increases with time at high voltages. This indicates segregation of the moisture to form regions of higher conductivity and lower breakdown strength. Prebreakdown current measurements also revealed the presence of highly localized field emission current, which has an exponential field dependence. It is believed that dielectric breakdown occurs as a result of local heating from field emission current.[18]

APPLICATIONS

PET films have found widespread use in many applications from parcel strapping and recording tapes to photographic and photoresist films for the manufacture of printed microcircuits. Other applications include dielectrics in capacitors, motor and generator insulation, membrane touch switches, liquid crystal display laminations, and wire and cable insulation.

The concerns of dielectric properties in the selection of insulating materials have received much attention. Park et al.[13] have investigated the potential of PET for insulation of superconducting magnet coils in fission reactors, where the insulator must withstand electrical stresses under high mechanical loads, extreme temperatures, and intense radiation. Chen et al.[12] investigated the microwave processing of polymers and revealed the importance of dielectric properties on the power absorption of the polymer, where the power absorption is directly proportional to the dielectric loss of the polymer.

REFERENCES

1. Reddish, W. *Trans. Faraday Soc.* **1950**, *46*, 459.
2. Huff, K.; Muller, F. H. *Kolloid Z.* **1957**, *153*, 5.
3. Krum, F.; Muller, F. H. *Kolloid Z.* **1959**, *164*, 8.
4. Saito, S.; Nakajima, T. *J. Polym. Sci.* **1959**, *36*, 533.
5. Yamafuji, K. *J. Phys. Soc. Jpn.* **1960**, *15*, 2295.
6. Hellwege, K. H.; Langebin, G. *Kolloid Z.* **1960**, *172*, 44.
7. Berestneva, G. L. et al. *High Mol. Wt. Compds.* **1960**, *2*, 1739.
8. Mikhailov, G. P.; Borissova, T. I. *Polym. Sci. USSR* **1961**, *2*, 387.
9. Saito, S. *Kolloid Z.* **1963**, *189*, 116.
10. O'Reilly, J. M.; Sedita, J. S. *J. Non-Cryst. Solids* **1991**, *131*, 1140.
11. Lightfoot, S.; Xu, G. *Polym.-Plast. Tech. Eng.* **1993**, *32*, 21.
12. Chen, M. et al. *Polym. Eng. Sci.* **1993**, *33*, 1092.
13. Park, C. H. et al. *IEEE Trans. Elect. Insul.* **1982**, *EI-17*, 234.

14. Park, C. H. et al. *IEEE Trans. Elect. Insul.* **1982**, *EI-17*, 546.
15. Park, C. H. et al. *IEEE Trans. Elect. Insul.* **1985**, *EI-20*, 567.
16. Rowland, S. P. *Water in Polymers* American Chemical Society: Washington, DC, 1980.
17. Xu, G. et al. *J. Appl. Phys.* **1989**, *66*, 5290.
18. Inuishi, Y.; Powers, A. D. *J. Appl. Phys.* **1957**, *28*, 1017.

POLY(ETHYLENE TEREPHTHALATE) (Heat Setting and Thermal Stabilization)

Saleh A. Jabarin
Polymer Institute
The University of Toledo

Poly(ethylene terephthalate) (PET) is a crystallizable polymer whose morphology can vary widely depending on the fabrication process. The polymer can be obtained as a "glassy" or "amorphous" transparent solid by rapidly quenching the melt below the glass transition temperature, T_g. Amorphous PET is of little commercial significance because it has low mechanical properties, high gas permeation rates, low dimensional stability, and high extensibility. When the PET is heated above its T_g, it crystallizes rapidly, forming an opaque material exhibiting spherulitic superstructures. This morphology can also be obtained by slow cooling of the polymer melt.

ORIENTATION AND STRAIN-INDUCED CRYSTALLIZATION

Orientation can change the internal morphological structure of PET, and proper orientation can greatly improve its mechanical, thermal, transport, and optical properties. The mechanism of molecular orientation and the structure of oriented PET have been studied.[1–3] The effects of draw ratio, temperature, stretch rate, and molecular weight on levels of crystallinity, and physical and mechanical properties have also been documented.[4–8]

PET shows strain-induced crystallization upon substantial orientation.[1,9–11] The orientation temperature range for PET is between the glass transition temperature (T_g ~ 80°C) and the onset of thermal crystallization (T_c ~ 120°C). Heating the oriented PET relieves stress, and the strained chains relax to a more random conformation, which causes shrinkage to occur.

An important commercial application of the strain-induced crystallization of PET is demonstrated by stretch blow-molded beverage containers.

BASIC PRINCIPLES OF HEAT SETTING

The annealing of oriented PET causes the polymer to crystallize. The amount of crystallinity and the rate of crystallization are dependent on annealing or crystallization temperature and the amount of orientation in the amorphous phase.

Higher degrees of crystallinity are obtained by heat setting at higher temperatures. In addition, the time of crystallization (heat-setting time) is very short, on the order of a few seconds. The results for thermal crystallization without orientation indicate that the half-time of crystallization at 150°C, for example, is 2 min.

APPLICATIONS

A review of technical patent and conference literatures indicates frequent use of terms such as heat stabilization, heat setting, thermal stabilization, and annealing. Often the term "heat setting" is interchangeably used with the term "annealing," defined in Webster's Dictionary as "to subject (glass or metal) to heat and slow cooling so as to toughen and reduce brittleness or to temper." In the context of PET container technology, one may assume that heat stabilization or thermal stabilization is achieved by heat setting. In other words, the molecular structure produced by orientation is further stabilized by exposing the oriented structure, in a constrained manner, to temperatures higher than the orientation temperature for a short period of "annealing" time. As the polymer is oriented and annealed, the crystallites may increase in number and size. As the crystallites grow, they encompass and lock in place many polymer chains that are already in an aligned position. Therefore, these chains are not free to shrink when exposed to higher temperatures experienced during storage and hot-filling.

The degree of crystallinity, the degree of orientation, and the morphology of oriented PET determine barrier, mechanical, and shrinkage behavior as well as other physical properties.

REFERENCES

1. Thompson, A. B. *J. Polym. Sci.* **1959**, *34*, 741.
2. Heffelfinger, C. J.; Schmidt, P. G. *J. Appl. Polym. Sci.* **1965**, *9*, 2661.
3. Bumbleton, J. M.; Bowles, B. B. *J. Polym. Sci., A-2* **1966**, *4*, 951.
4. Spruiell, J. E. et al. *Trans. Soc. Rheol.* **1972**, *16*, 535.
5. DeVries, A. J.; Bonnebat, C. *Polym. Eng. Sci.* **1976**, *16*, 93.
6. Jabarin, S. A. *Polym. Eng. Sci.* **1984**, *24*, 376.
7. Cakmak, M. et al. *Polym. Eng. Sci.* **1986**, *16*, 292.
8. Chandran, P.; Jabarin, S. *ANTEC* 1991; Vol. xxxvii, p. 880, *ANTEC* 1992; Vol. xxxviii, p 653.
9. Kawaguchi, T. *J. Appl. Polym. Sci.* **1961**, *16*, 482.
10. Misra, A.; Stein, R. S. *J. Polym. Sci., Polym. Phys. Ed.* **1979**, *17*, 235.
11. Jabarin, S. A. *Polym. Eng. Sci.* **1992**, *32*, 1341.

POLY(ETHYLENE TEREPHTHALATE) (Modified with Ionic Groups)

Riccardo Po'
Enichem S.p.A.
Istituto Guido Donegani

Incorporation of a small amount of ionic units within a polymer chain may significantly change physical properties from those of its corresponding neutral homologue. In particular, the interaction and aggregation of ionic groups affect the polymer structure and thus the rheological properties, mechanical properties, and attitude to miscibility with other polymers. Usually polymers containing a level of ionic modification up to about 15% are referred to as "ionomers."

Although physical properties can be changed over a broad range by changing the level of ionic modification, relatively little work has been accomplished on ionomeric poly(ethylene terephthalate) (PET) compared with other polymer systems.[1–5]

PREPARATION AND PROPERTIES

PET Ionomers with Sulfur-Containing Groups

From an industrial point of view, the most important ion-modified PET contains 1–5 mol % of 5-sodiosulfoisophthalate (Na-SIP) units (**Structure 1**).

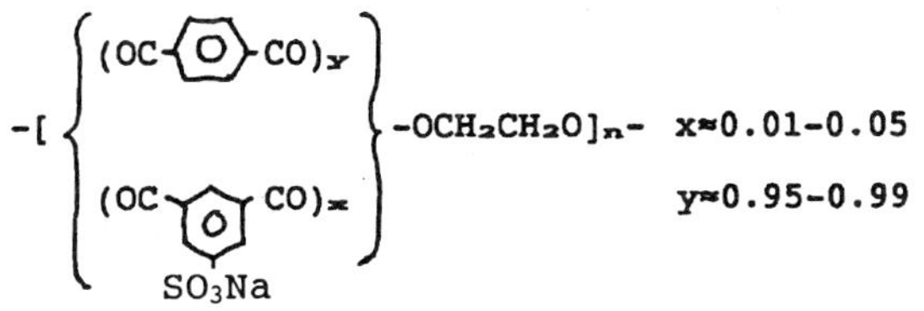

1

Greener et al.[6] interpreted the rheological and dynamic–mechanical behavior of Na-SIP modified PET in terms of the restricted mobility model proposed by Eisenberg et al.[7] This model assumes that ionic moieties at high ionic concentration aggregate into "multiplets" typically consisting of 2–8 ion pairs. Neighboring multiplets aggregates themselves into clusters, which create a contiguous phase of restricted mobility. When the ion concentration is low, a distinct phase cannot exist, but nonetheless a physical cross-link can occur. Indeed, increases in melt viscosity and flow activation energy were observed with increasing sulfonation.

Increases in melt viscosity[8] and decreases in crystallinity and fiber orientation[8,9] in PET containing 1.5–4.5 mol% of Na-SIP were observed by several other authors; mechanical properties of fibers were studied by Bhattacharya.[10]

PET Ionomers with Phosphorus-Containing Groups

Troev et al. published several papers on PET modified with phosphite[45,46] and phosphonate[47–51] salts. The obtained materials show good flame-retardant properties because of the presence of phosphorus atoms, satisfactory thermal resistance, and poor hydrolytic resistance.

PET Ionomers Having Carboxylate Groups

Metal carboxylate groups have been introduced in PET through three ways: copolymerization, reactive modification, and grafting.

APPLICATIONS

Sulfonated PET is widely used to make fibers dyeable with basic dyes and fibers with antipilling and antistatic properties. 5-Sodiosulfoisophthalic acid is the commercially used monomer, but none of the other sulfonic comonomers[11–46] has ever found commercial application.

In one case the high elongational melt viscosity of PET ionomers has been suggested as an advantage in the manufacture of PET bottles through the extrusion-blow-molding process.[39]

Phosphorus-containing PET was proposed as flame retardant material, but this property is imparted by the phosphorus, not by the ionic structure.

Finally, Simionescu has suggested that the use of his polychelates as semiconductor materials.[52]

REFERENCES

1. Eisenberg, A.; King, M. *Ion-Containing Polymers: Physical Properties and Structure*; Academic: New York, 1977.
2. MacKnight, W. J.; Earnest, Jr. T. R. *J. Polym. Sci., Macromol. Rev.* **1981**, *16*, 41.
3. Tant, M. R.; Wilkes, G. L. *J. Macromol. Sci., Rev. Macromol. Chem. Phys.* **1988**, *C28*, 1.
4. Fitzgerald, J. J.; Weiss, R. A. *J. Macromol. Sci., Rev. Macromol. Chem. Phys.* **1988**, *C28*, 99.
5. Hara, M.; Sauer, J. A. *J. Macromol. Sci., Rev. Macromol. Chem. Phys.* **1994**, *C34*, 325.
6. Greener, J. et al. *Macromolecules* **1993**, *26*, 6416.
7. Eisenberg, A. et al. *Macromolecules* **1990**, *23*, 4098.
8. Hong, Z. et al. *J. Appl. Polym. Sci.* **1987**, *34*, 1353.
9. Timm, D. A.; Hsieh, Y. L. *J. Polym. Sci., Polym. Phys.* **1993**, *31*, 1873.
10. Bhattacharya, S. D.; Jawale, C. J. *Man-Made Text. India* **1991**, *34*, 358.
11. BP 868 496, 1961, DuPont; *Chem. Abstract* **1961**, *55*, 24035i.
12. Sakurai, K. et al. JP 62 11443, 1962, Teijin; *Chem. Abstract* **1963**, *59*, 799f.
13. Shima, T.; Funahashi, K. JP 63 15098, 1963, Teijin; *Chem. Abstract* **1963**, *59*, 15431f.
14. Horn, C. F. U.S. Patent 3 184 434, 1965, Union Carbide Co.
15. Horn, C. F.; Kincaid, H. S. U.S. Patent 3 256 317, 1966, Union Carbide Co.
16. Horn, C. F.; Vineyard, H. U.S. Patent 3 280 169, 1966; Union Carbide Co.
17. Pich, R. FR 1 492 279, 1967, Societe Rhodiaceta; *Chem. Abstract* **1968**, *68*, 96749.
18. Tanaka, T. et al. U.S. Patent 3 541 050, 1969, Toyo Rayon Co.
19. BP 1 151 596, 1969, Celanese Co.; *Chem. Abstract* **1969**, *71*, 40158.
20. Stackman, R. W.; Sargent, D. E. U.S. Patent 3 507 835, 1970, Celanese Co.
21. Tanaka, T. et al. DE 1 806 884, 1970, Toyo Rayon Co.; *Chem. Abstract* **1970**, *73*, 16194.
22. Lappin, G. R. et al. U.S. Patent 3 528 947, 1970, Eastman Kodak Co.
23. Tanaka, T. et al. BP 1 214 087, 1970; Toray; *Chem. Abstract* **1971**, *74*, 32650.
24. FR 1 595 743, 1970, Toyo Rayon Co.; *Chem. Abstract* **1971**, *74*, 88592.
25. Trapasso, L. E.; Stackman, R. W. U.S. Patent 3 592 796, 1971, Celanese Co.
26. Price, J. A.; Stewart, M. J. U.S. Patent 3 624 034, 1971, FMC Co.
27. Keck, M. H. U.S. Patent 3 554 975, 1971, Goodyear Tire and Rubber Co.
28. Caldwell, J. R. U.S. Patent 3 657 193, 1972, Eastman Kodak Co.
29. Moebius, H. et al. U.S. Patent 3 663 508, 1972, Hoechst.
30. Stewart, M. J. U.S. Patent 3 766 143, 1973, FMC Co.
31. Radlamann, E. et al. DE 2 224 255, 1973, Bayer AG; *Chem. Abstract* **1974**, *80*, 109668.
32. Radlamann, E.; Nischk, G. U.S. Patent 3 821 281, 1974, Bayer AG.
33. Weedon, G. C. U.S. Patent 3 812 178, 1974, Allied Chem. Co.
34. Juelke, C. V.; Cook, N. K. U.S. Patent 3 816 520, 1974, Celanese Co.
35. Hahn, M.; Okuzumi, Y. U.S. Patent 3 817 934, 1974, Goodyear Tire and Rubber Co.
36. Henry, C. L.; McNeely, G. W. U.S. Patent 3 856 753, 1974, Akzona.
37. Tryon, S. U.S. Patent 3 860 527, 1975, FMC Co.
38. McNeely, G. W. U.S. Patent 4 029 638, 1977, Akzona.

39. Sinker, S. M. et al. U.S. Patent 4 554 328, 1985, Celanese Co.
40. Fock, J. U.S. Patent 4 598 141, 1986. Th. Goldschmidt AG.
41. O'Neill, G. U.S. Patent 4 579 936, 1986, Eastman Kodak Co.
42. Suzuki, T. et al. U.S. Patent 4 622 381, 1986, Teijin.
43. Kawaguchi, K. et al. EP 517 511, 1992, Polyplastic Co.
44. Po, R. et al. EP 581 341, 1993, Enichem.
45. Troev, K. et al. *Eur. Polym. J.* **1979**, *15*, 437.
46. Troev, K. et al. *Eur. Polym. J.* **1979**, *15*, 1143.
47. Troev, K. et al. *Eur. Polym. J.* **1981**, *17*, 27.
48. Troev, K. et al. *Eur. Polym. J.* **1981**, *17*, 31.
49. Troev, K. et al. *Eur. Polym. J.* **1993**, *29*, 1205.
50. Troev, K. et al. *Eur. Polym. J.* **1993**, *29*, 1211.
51. Troev, K. et al. *Eur. Polym. J.* **1993**, *29*, 1499.
52. Simionescu, C.; Vasiliu-Oprea, C. *Adv. Chem. Ser.* **1973**, *128*, 68.

POLY(ETHYLENE TEREPHTHALATE) (Ultra-High Molecular Weight)

Susumu Tate
Toyobo Research Institute
Toyobo Company, Ltd.

Poly(ethylene terephthalate) (PET) is widely used in fibers, films, bottles for beverages, etc. Some authors claim that high molecular weight is essential for high-tenacity and high-modules fibers using PET.[1,2] However, PET with 0.6–1.0 dL/g of intrinsic viscosity ($M_n = 2–4 \times 10^4$) is obtained from conventional polymerization methods.

Several methods for obtaining ultrahigh molecular weight PET, (UHMW-PET) with > 2 dL/g of intrinsic viscosity (Mn > 1.0×10^5) have been proposed.

- **Melt-phase polymerization**. UHMW-PET with an intrinsic viscosity (IV) of 2.3 dL/g was obtained from melt-phase polymerization using an addition of long-chain aliphatic acid with glycol.[3] UHMW-PET with IV ≥ 2 dL/g from thin film-state polymerization has been patented.[4]
- **Solid-state polymerization**. Li-Chen Hsu[5,6] obtained UHMW-PET with IV ≈ 2 dL/g by special solid-state polymerization. Cohn[7,8] reported that solid-state polymerization using a porous and fibrous structure of PET prepolymer obtained by precipitating from PET solution produced UHMW-PET with IV = 2–5 dL/g (Mn = $1.1–4.2 \times 10^5$).
- **Swollen-state polymerization**. A new polymerization technique for UHMW-PET has been reported. UHMW-PET has been obtained by polymerization at the swollen state in specific solvents under bubbling nitrogen gas at atmospheric pressure (swollen-state polymerization).[9–11] Suitable solvents could swell PET but did not dissolve it. Solvents with solubility similar to PET can attain higher ultimate intrinsic viscosity at the same polymerization temperature. The rate of swollen-state polymerization was related to the degree of swell; that is, to the contents of impregnated solvent in swollen PET. When a hydrogenated terphenyl was used as the solvent, UHMW-PET with an intrinsic viscosity 2–3 dL/g was obtained and a honeycomb-like structure was observed in the UHMW-PET particles.

APPLICATIONS

UHMW-PET has not yet been produced commercially. M. Ito and others[1,2] prepared high tensile strength fiber from UHMW-PET with IV = 3.6 dL/g. Fiber having 17 g/d of tensile strength was obtained from solution spinning of hexafluoro-2-propanol/dichloromethane mixture.

Some patents[12,13] disclosed high-tenacity and high-modulus fibers (15–18 g/d and 230–250 g/d respectively from special spinning of UHMW-PET under reduced viscosity. UHMW-PET made by swollen-state polymerization has a honeycomb-like structure; easy-to-impregnate solvent reduces melt viscosity. UHMW-PET impregnated with solvent, for example methylnaphtharene, was convenient for spinning to reduce melt viscosity. Fiber from reduced viscosity spinning (RVS) apparently has higher tenacity for conventional industrial fibers (8–9 g/d tensile strength and 100–120 g/d initial modulus), but does not exceed 20 g/d of tensile strength believed to be the minimum for superhigh tenacity fibers. Further approaches for application of UHMW-PET are expected.

REFERENCES

1. Ito, M. et al. *Sen-i Gakkaishi* **1992**, *48*, 569.
2. EP-A-0359692, 1989.
3. EP-A-0181498, 1985.
4. EP-A-0182352, 1985.
5. Hsu, L.-C. *J. Macromol. Sci.* 1967, *B1*(4), 801.
6. Hsu, L.-C. *Cryogenic Properties of Polymers*; Marcel Dekker: New York, 249.
7. U.S. Patent 4 917 845, 1990.
8. Cohn, G. ACS Prepr. 1989, *30*, 160.
9. U.S. Patent 4 613 664, 1986.
10. U.S. Patent 4 742 151, 1988.
11. Tate, S. et al. *Polymer* 1993, *34*, 4974.
12. JP 03-294539, 1991.
13. JP 06-17313, 1994.

POLY(ETHYLENE TEREPHTHALATE) COPOLYMERS

Naoto Tsutsumi*
Kyoto Institute of Technology

Minoru Nagata
Kyoto Prefectural University

There is great industrial motivation for incorporating a third component into poly(ethylene terephthalate) (PET) to overcome the undesirable properties (poor dyeability, pilling, stiff "hand", low moisture regain, and generation of static charges) encountered in developing PET fibers for clothing materials. Morimoto has reviewed the extensive improvement studies.[1] These modifications keep the third component small to change only the objective properties of PET fiber.

We focus here on the enhancement of dyeability, water absorption, alkali hydrolysis, flame retardancy, thermal shrinkage, and biodegradability in PET copolymers with a third component.

*Author to whom correspondence should be addressed.

PROPERTIES

Thermal and Physical

Gordon and Mera prepared PET copolymers with 2,3-dialkyl-1,4-butanediols by melt polycondensation, where the dialkyl group is longer than the butyl group.

Kiyotsukuri et al. used another approach for copolymerization of alkyl disubstituted aliphatic diols.[2] They used 2,2-dialkyl-1,3-propanediols with the alkyl side group shorter than the butyl group as a third component.

Dyeability and Water Absorption

The incorporation of comonomer with alkyl side chain significantly enhances the dye uptake, whereas the incorporation of linear comonomer does not change the dyeability.

Another modification approach utilizes grafting of vinyl monomer onto PET fibers. Photoinitiated grafting[4] and benzoyl-peroxide-initiated grafting[5] of acrylic acid onto PET fibers improved their dyeability for methylene blue dye.

Thermal Shrinkage

Kiyotsukuri et al. studied the thermal shrinkage of the drawn (4×) PET copolymer with 2,2-dialkyl-1,3-propanediols.[2] Incorporation of third comonomer increases thermal shrinkage and excellent 100% thermal shrinkage was achieved when 20 mol % of propanediol with ethyl and butyl side groups and propanediol with ethyl side groups comonomers were incorporated in the copolymer.

Alkali Hydrolysis

Kiyotsukuri et al. obtained interesting results for alkali hydrolysis for PET copolyester incorporating PEG or PBG as a third component.[3] PBG enhanced resistance to alkali hydrolysis, whereas copolyester film with PEG quickly and completely dissolved in alkali solution. This may be ascribed to higher hydrophilicity of the PEG unit. Incorporation of 2,2-dialkyl-1,3-propanediols enhances resistance to alkali hydrolysis.[2] Bulky alkyl side groups adjacent to the ester linkage cause steric hindrance which may suppress the alkali hydrolysis.

Flame Retardancy

Improving flame retardancy by incorporating a third component has been studied for PET copolymers with 2,5-dichloro- and 2,5-dibromoterephthalic acids,[6] with bis(β-hydroxyethyl)2,5-dichlorohydroquinone, bis (β-hydroxyethyl)2,5-dibromohydroquinone, bis(β-hydroxyethyl)-tetrachlorobisphenol A and bis(β-hydroxyethyl)-tetrabromobisphenol A,[7] and with bisphenols.[8] Flame retardancy was enhanced by the incorporation of halogenated comonomers.

Biodegradability

Plastics wastes motivate the study of biodegradable PET. Reed and Gilding reported the preparation and biodegradability of PET copolymers with incorporated PEG.[9] They found that degradation occurred by hydrolysis of ester linkage accelerated by the addition of enzymes. By modification of PET with L-lactic acid or oligooxyethylenes, Niekraszewicz obtained biodegradable, cheap, and environmentally safe polymeric materials.[10]

APPLICATIONS

PET has been used widely as fibers, plastics, and molding materials. Since a small amount of incorporated third component improves only the objective properties and leaves the rest of PET's properties unchanged, the application field of the copolymer is the same as PET polymer's. And environmental concerns and changes of lifestyle are giving them new applications.

PET copolymers with alkyl side chain comonomers show promise for use as heat-shrinkable film or wrap.[2] Large amounts of PEG with various molecular weights incorporated to PET produce biodegradable elastomers applicable to medicine, especially surgery.[9]

Environmental problems demand the development of biodegradable polymeric materials to replace our current nonbiodegradable plastic packaging. Biodegradable polymeric films and sheets also have agricultural uses. Improving biodegradability of PET by copolymerization will intensify in the future.

REFERENCES

1. Morimoto, S. *Man-Made Fibers, Science and Technology*; Mark, H. F.; Atlas, S. M.; Cernia, E., Eds.; Wiley-Interscience: New York, NY, 1968; Vol. 3.
2. Kiyotsukuri, T.; Masuda, T.; Tsutsumi, N. *Polymer* **1994**, *35*, 1274.
3. Kiyotsukuri, T.; Masuda, T.; Tsutsumi, N. et al. *Polymer* **1995**, *36*, 262.
4. Yao, Z.; Rånby, B. *J. Appl. Polym. Sci.* **1990**, *39*, 1459.
5. Saçak, M.; Oflaz, F. *J. Appl. Polym. Sci.* **1993**, *42*, 1909.
6. Kiyotsukuri, T.; Hashimoto, Y.; Nagata, M. *Sen-i Gakkaishi* **1981**, *37*, T-97.
7. Nagata, M.; Kiyotsukuri, T.; Uchino, N. *Sen-i Gakkaishi* **1981**, *37*, T-369.
8. Bajaj, P.; Khanna, D. N. *Textile. Res. J.* **1981**, *51*, 696.
9. Reed, A. M.; Gilding, D. K. *Polymer* **1981**, *22*, 499.
10. Niekraszewicz, A. *Polimery (Warsaw)* **1993**, *38*, 399.

POLY(ETHYLENE TEREPHTHALATE) DEGRADATION (CHEMISTRY AND KINETICS)

Saleh A. Jabarin
Polymer Institute
The University of Toledo

Poly(ethylene terephthalate) (PET) like other polyesters can experience various degradation processes, which may be classified as follows:[1–3]

- thermal degradation under the influence of heat alone,
- oxidative degradation upon heating in the presence of atmospheric oxygen,
- hydrolytic degradation in the presence of moisture,
- photo-oxidative degradation under the influence of light and oxygen,
- radiochemical degradation under the influence of ionizing radiation, and
- chemical degradation in the presence of various reagents, including alcohols, amines, and oxidizing acids.

It is generally known that degradation of the PET chain may accompany the conversion of PET pellets into fabricated products such as films and sheets by extrusion or injection molding.

THERMAL DEGRADATION

Thermal degradation results from the interaction of heat with PET chains to cause random chain scission and the formation of vinyl esters and carboxyl end groups as initial products. In addition it may also lead to:[4–16]

- reduction in molecular weight or intrinsic viscosity (I.V.);
- formation of additional carboxyl end groups;
- discoloration;
- formation of volatile products, mainly acetaldehyde but also including smaller amounts of carbon dioxide and carbon monoxide, plus traces of ethylene, water, methane, benzene, and 2-methyldioxolane;
- formation of anhydride groups; and
- formation of low molecular weight nongaseous products such as terephthalic acid and oligomers of PET.

THERMAL-OXIDATIVE DEGRADATION

Thermal-oxidative degradation, which is the reaction of oxygen with PET at elevated temperatures can lead to:[7,15,17–22]

- reduction of molecular weight;
- production of acetaldehyde and other gaseous products;
- formation of branched chains and, in extreme cases, gelation; and
- discoloration.

FORMATION OF CYCLIC OLIGOMERS

In PET, as in many other condensation polymers, ring formation competes with polymerization and provides a possible mechanism for polymer degradation.[23]

The concentration and proportions of cyclic oligomers in different melt-polymerized samples of PET are remarkably similar; this observation suggests that oligomer formation is an equilibrium-controlled process.

HYDROLYTIC DEGRADATION

The chemical reaction of PET with water at elevated temperatures leads to a reduction in molecular weight and the formation of carboxyl and hydroxyl end groups.[24]

SUMMARY

The degradation processes of PET are controlled by two classes of factors:

Physical factors such as:

- processing temperatures,
- residence time in the melt,
- drying temperature and time,
- melt and drying environments, and
- moisture content.

and chemical factors which affect the detailed molecular structure of the polymers:

- molecular weight or intrinsic viscosity,
- catalyst system,
- methods of polymerization: melt phase versus solid state,
- diethylene glycol content,
- copolymerization: the kind and amount of comonomer modifier,
- monomer: DMT (dimethyl terephthalate) versus TPA (terephthalate acid),
- polymerization conditions, and
- stabilizer system.

The degradation behavior of PET is not just a function of the initial I.V. or molecular weight, but also of the chemistry of the PET resin.

REFERENCES

1. Buxbaum, L. H. *Angew. Chem., Int. Ed.* **1968**, *7*, 182–190.
2. *Chemical Reactions of Polymers*; Fettes, E. M., Ed.; Interscience: New York, NY, 1964; pp 501–514, 553–561, 608–610.
3. Conley, R. T.; Gaudiana, R. A. In *Thermal Stability of Polymers*; Conley, R. T., Ed.; Marcel Dekker: New York, NY, Abstr. **1970**; Vol. 1, pp 347–456.
4. Goodings, E. P. *Society of Chemical Industry (S.C.I.) Monograph No. 13* **1961**, 211–228.
5. Ritchie, P. D. In *High Temperature Resistance and Thermal Degradation of Polymers; Society of Chemical Industry (S.C.I.) Monograph No. 13* **1961**, 107–131.
6. Pohl, H. A. *J. Amer. Chem. Soc.* **1951**, *73*, 5660.
7. Marshall, I.; Todd, A. *J. Chem. Soc., Faraday Trans.* **1953**, *49*, 67.
8. Madorsky, S. L. *Thermal Degradation of Organic Polymers*; Interscience: New York, NY, 1964; pp 272–276.
9. Zimmerman, H. *Faserforsch. Textiltech.* **1962**, *11*, 481.
10. Zimmerman, H.; Leibnitz, E. *Faserforsch. Textiltech.* **1965**, *16*, 282; *Chem. Abstr.* **1965**, *63*, 8532a.
11. Zimmerman, H.; Chu, D. D. *Faserforsch. Textiltech.* **1973**, *24*, 445; *Chem. Abstract* **1975**, *82*, 8678m.
12. Schaaf, E.; Zimmerman, H. *Faserforsch. Textiltech.* **1974**, *25*, 434; *Chem. Abstr.* **1975**, *82*, 98679u.
13. Hergenrother, W. L. *J. Polym. Sci., Part A-1* **1974**, *12*, 875.
14. Newkirk, A. E. *Anal. Chem.* **1960**, *32*, 1558.
15. Jabarin, S. A.; Lofgren, E. A. *J. Polym. Eng. Sci.* **1984**, *24*, 1056.
16. Zimmerman, H.; Kim, N. T. *J. Polym. Eng. Sci.* **1980**, *20*, 680.
17. Spanninger, P. A. *J. Polym. Sci. Part A-1* **1974**, *12*, 709.
18. Nealy, D. L.; Adams, L. J. *J. Polym. Sci., Part A-1* **1971**, *9*, 2063.
19. Yoda, K. et al. *J. Appl. Poly. Sci.* **1970**, *14*, 2357.
20. Buxbaum, L. H. *Polym. Prepr.* **1967**, *8*, 552.
21. Zimmerman, H.; Schaaf, E.; Seganowa, A. *Faserforsch. Textiltech.* **1971**, *22*, 255; *Chem. Abstract* **1971**, *75*, 36923r.
22. Zimmerman, H.; Becker, D. *Faserforsch. Textiltech.* **1973**, *24*, 479; *Chem. Abstract* **1974**, *81*, 38036b.
23. Davies, T. In *Chemical Reactions of Polymers*; Fettes, E. M., Ed.; Interscience: New York, NY, 1964; pp 501–514.
24. Ravens, D. A. S.; Sisley, J. E. In *Chemical Reactions of Polymers*; Fettes, E. M., Ed.; Interscience: New York, NY, 1964; pp 551–564.

Polyethylenimine

See: *Metal Complexation Polymers*
Metal Ion Binding Resins (Synthesis and Analytical Properties)
Metalloenzymes, Artificial (Polyethylenimine-Based)
Polyalkylenimines
Polyethylenimine (Overview)
Synzymes

POLYETHYLENIMINE (Overview)

Ichimoto Akasaki
Nippon Shokubai Company, Ltd.

Polyethylenimine (PEI) is a water-soluble polymer which is rich in reactivity and very high in cationization density. With its unique characteristics fully exploited, PEI is now in widespread use as a water treating agent, chelating agent, adhesive, fiber treating agent, enzyme fixing agent, and so on, in many important industrial fields.

PREPARATION

PEI is prepared by ring-opening polymerization of ethyleneimine monomer using acid catalyst. PEI is not a perfectly linear polymer but partly contains branched chains of primary, secondary, and tertiary amino nitrogen as shown in **Structure 1**.

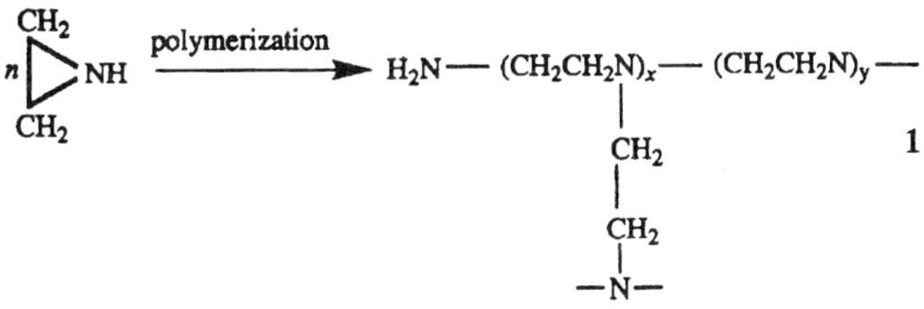

1

Branching is caused by the high reactivity of the activated hydrogen coupled to nitrogen atoms, which makes it substantially difficult to suppress the branching perfectly in the course of the ethyleneimine polymerization process. PEI is only produced by Nippon Shokubai Co., Ltd. (Japan) and BASF-AG (Germany).

PEI has a branched chain structure but when produced from ethyloxazoline it can have a linear chain structure.

When polyethyloxazoline is hydrolyzed the product is crystalline polymer and insoluble in water.

PROPERTIES

The major features of PEI are

- it has the highest cation density among existing materials,
- is very rich in reactivity,
- has built-in amino groups in the main chain,
- is water-soluble polymer at any desired ratio, and
- has branched structure.

Viscosity

Viscosity of PEI shows a great dependence on its molecular weight, concentration, and temperature.

Solubility

Water is the optimal solvent for PEI, and both can be blended at any desired ratio. PEI can also dissolve in lower alcohol such as methanol, ethanol, and ethylene glycol.

Safety

Ethyleneimine, the raw material of PEI, is known as a highly reactive and strongly toxic chemical substance. But PEI produced by its polymerization does not require particular handling.

CHEMICAL REACTIVITY[1–7]

Not only does PEI have a variety of applications with its structure intact, but its rich chemical reactivity and the resultant chemical modification offer widespread utilization in fine chemical and other industries.

INDUSTRIAL USE

- Paper[8–15]
- Adhesives[16–24]
- Ink and Photography[25–27]
- Textiles[28–32]
- Flocculants, Clarificants, and Floatating Agents[33–41]
- Chelating Resins, Ion-Exchanging Resins, Separation Membranes, and Adsorbents[42–51]
- Cosmetics Toiletries[52–56]
- Lubricants, Rust Inhibitors, and Dispersants[57–63]
- Plating Agents[64–69]
- Biological, Medical and Enzymatic Applications[70–81]
- Cement Additives, and Agents for Oil Excavation[82–84]

PEI is an intermediate for polymer reactions because it has a number of amino groups in its molecular structure. It will be widely used in the future as a reactive intermediate for new functional materials.

ACKNOWLEDGMENT

Part of this contribution was published by CMC Co., Ltd. (Japan) *The Preparations and Applications of Reactive Polymer.*

REFERENCES

1. A. C. C. U.S. Patent 2 887 405; Bayer A. G. Jpn. Kokai 67489, 1986; Fuji Electric Jpn. Kokai 221088, 1985.
2. I G. U.S. Patent 2 223 930; General Aniline and Film; U.S. Patent 2 272 489; Denki Kagaku Kogyo: Jpn. Kokai 198190, 1984.
3. General Aniline and Film: U.S. Patent 2 222 208; Nitto Denko: Jpn. Kokai 271791, 1987.
4. Motecatini Edison: U.S. Patent 3 037 835; Miyoshi Oil and Fat: Jpn. Kokai 185322, 1982; 120879, 1983; 173378, 1984.

5. Bayer A. G. Belg. Patent 622716; Nippon Soda: Jpn. Kokai 162682, 1983; Miyoshi Oil and Fat: Jpn. Kokai 185322, 1982; Nippon Shokubai: Jpn. Kokai 135585, 1985.
6. Bayer A. G. U.S. Patent 2 912 296; I.C.I. U.S. Patent 3 049 468; Br. Patent 755 478.
7. N.K.K., Yushiro Chemical Ind. Jpn. Kokai 260695, 1985; Tokuyama Soda: Jpn. Kokai 78906, 1985.
8. Poshmann, F. J. *Pulp Paper Mag. Can.* **1968**, *69*(8), T210.
9. Jujo Paper: Jpn. Kokai 120486, 1982; Canon: Jpn. Kokai 198186, 1984; Nippon Shokubai: Jpn. Kokai 76386, 1985.
10. Asahi Glass: Jpn. Kokai 41996, 1983.
11. Nippon Shokubai: Jpn. Kokai 71796, 1985.
12. Ciba Geigy A. G. Eur. Patents 176479; 172 139.
13. Dow Chemical Company: U.S. Patent 3 662 528.
14. Sumitomo Chemical: Jpn. Kokai 29508, 1976.
15. Lion: Jpn. Kokai 137587, 1984.
16. Araki, M., Ed.; Converting Technical Institute: Tokyo Saishin-Kako.
17. Nippon Shokubai: Jpn. Kokai 80015, 1982; 80472, 1982.
18. Nippon Soda: Jpn. Kokai 162682, 1983.
19. Nitto Denko: Jpn. Kokai 118775, 1985.
20. Monsanto: Jpn. Kokai 246, 1986.
21. Denki Kagaku Kogyo: GB 2 101 141.
22. Hoechst A. G. Jpn. Kokai 21171, 1986.
23. H. B. Fuller: Jpn. Kokai 287978, 1986.
24. Sekisui Chemical: Jpn. Kokai 270135, 1987.
25. Konica: Jpn. Kokai 255944, 1987; 255946, 1987; 257155, 1987.
26. Konica: Jpn. Kokai 286038, 1987.
27. Xerox: Jpn. Kokai 119280, 1987; 119279, 1987.
28. Unitika: Jpn. Kokai 47350, 1982.
29. Miyoshi Oil and Fat: Jpn. Kokai 120879, 1983; 173378, 1984.
30. Teijin: Jpn. Kokai 9995, 1985.
31. Nitibi: Jpn. Kokai 184113, 1987.
32. U.S. Dept. of Agriculture: U.S. Patent (Appli) 4 328 01, 1974.
33. Joergensen, S. E. Jpn. Kokai 127453, 1982.
34. Diafloc: Jpn. Kokai 87000, 1986.
35. Kurita Ind. Jpn. Kokai 82911, 1984.
36. Mitsubishi Rayon: Jpn. Kokai 175543, 1985.
37. Hakuto Kagaku: Jpn. Kokai 74607, 1986;Ipposha Oil Ind. Jpn. Kokai 68506, 1987.
38. Shinnippon Seitetsu: Jpn. Kokai 158111, 1987.
39. Mather, G. G. J. *Environ. Sci.* **1983**, *26*(3), 29–34.
40. Miyoshi Oil and Fat: Jpn. Kokai 207591, 1982.
41. Hoechst, A. G. DE 3 404 538.
42. Asahi Chemical Ind. Jpn. Kokai 37087, 1976.
43. Unitika Jpn. Kokai: 20224, 1983; 150432, 1983.
44. Miyoshi Oil and Fat: Jpn. Kokai 189127, 1984.
45. Kuraray Chemical: Jpn. Kokai 11230, 1985; JPN Min. Agency Ind. Sci. and Tech. Jpn. Kokai 210532, 1985.
46. Vysoka Skol. Chem. Tech. Jpn. Kokai 264 325, 1985.
47. Chia, E. S. K. *J. Appl. Polym. Sci.* **1975**, *19*, 1347; Nitto Denko: Jpn. Kokai 27101, 1982; Toray Ind. Jpn. Kokai 197 105, 1987; JPN Min. Agency Ind. Sci. and Tech. Jpn. Kokai 55304, 1984; 55305, 1984; 28803, 1985; Teijin: Jpn. Kokai 179 188, 1984; 38603, 1986.
48. Tsai, M. F.; Levy, M. *Ion-Exch. Technol.* **1984**, 533–541.
49. Olympus Kogaku Kogyo: Jpn. Kokai 128 342, 1985.
50. Kawasaki Jukogyo: Jpn. Kokai 227 821, 1986; Mitsubishi Jukogyo: Jpn. Kokai 112 662, 1987; 112 663, 1987.
51. U.O.P. Inc. U.S. Patent 4 511 654.
52. Okumura, T. *Fragrance J.* **1979**, *37*, 82–87.
53. Procter and Gamble: Jpn. Kokai 195 097, 1987.
54. Colgate Palmolive: Br Patent 1 277 357; Gillette: U.S. Patent 3 862 310; Reckitt and Colmann: Jpn. Kokai 65106, 1976.
55. Shiseido: Jpn. Kokai 56117, 1986.
56. Procter and Gamble: Eur. Patent 111 984.
57. N.K.K., Yushiro Chemical Ind. Jpn. Kokai 184595, 1985; 260695, 1985; Nihon Parkerizing: Jpn. Kokai 253128, 1986.
58. Nippon Oil: Jpn. Kokai 112895, 1985; Kao, Kawasaki Seitetsu: Jpn. Kokai 246293, 1986.
59. Lion: Jpn. Kokai 136 429, 1986; 146 331, 1986; 288 153, 1987; 223 060, 1987; 288 154, 1987; Dai-Ichi Kogyo Seiyaku: Jpn. Kokai 263 627, 1986.
60. Hitachi Maxell Ind. Jpn. Kokai 139 923, 1986.
61. Sanyo Chemical Ind. Jpn. Kokai 225 297, 1986.
62. Dai-Ichi Kogyo Seiyaku, Nippon COM: Jpn. Kokai 20389, 1988; Dai-Ichi Kogyo Seiyaku: Jpn. Kokai 20390, 1988; 20391, 1988; 20392, 1988.
63. Nippon Shokubai: Jpn. Kokai 65940, 1988.
64. Seiko Instr. And Elec. Jpn. Kokai 101 687, 1982; 67889, 1983; 20488, 1984; Kawasaki Seitetsu: Jpn. Kokai 36495, 1983.
65. Arakawa Chemical Ind. Jpn. Kokai 174 485, 1982.
66. Hooker Chem. and Plas., Occidental Chem. Jpn. Kokai 188 693, 1982.
67. Nihon Kagaku Sagyo: Jpn. Kokai 9593, 1986.
68. OMI INT: Jpn. Kokai 41787, 1986.
69. Okuno Pharm. Ind. Jpn. Kokai 272 387, 1986.
70. Miles: Jpn. Kokai 118 793, 1982; 110 190, 1982; 58072, 1985; Ceskoslovenska Akad. Ved, Vojtisek V. Jpn. Kokai 155 092, 1983; Genex: Jpn. Kokai 190 393, 1983; Nippon Shokubai: Jpn. Kokai 160 885, 1985.
71. Mitsui Sugar: Jpn. Kokai 39794, 1982.
72. Miles: Jpn. Kokai 78588, 1983.
73. Fuji Electric: Jpn. Kokai 221 088, 1985; 176 587, 1985; 66982, 1985.
74. Bayer A. G. Jpn. Kokai 67489, 1986.
75. Miles: Jpn. Kokai 166 888, 1987.
76. Kobayashi, Kuratomi: Jpn. Kokai 152 478, 1983; 2684, 1985.
77. Block Eng. Jpn. Kokai 155 760, 1984; 131 165, 1984.
78. Toray Ind. Jpn. Kokai 39297, 1984.
79. Baker J. T. Chem. Jpn. Kokai 260 593, 1985; 173 000, 1984.
80. Mitsubishi Rayon: Jpn. Kokai 251 770, 1986; 36323, 1987; 38171, 1987; 38173, 1987.
81. Nitto Denko: Jpn. Kokai 83885, 1987; 296 876, 1987; NOK Jpn. Kokai 294 084, 1987; Ajinomoto: Jpn. Kokai 44885, 1988.
82. Nisso Master Builders: Jpn. Kokai 156 563, 1983.
83. Halliburton: U.S. Patent 4 742 094; Hughes Tool: U.S. Patent 4 469 518, Dow: JP 44008, 1987.
84. Fujisawa Pharmaceutical: Jpn. Kokai 83663, 1986.

Poly(p-ethylphenol)

See: Enzyme-Catalyzed Polymerization

POLY(1,1'-FERROCENE-ARYLENE)S (Pd-Catalyzed)

Matthias Rehahn* and Ralf Knapp
Polymer Institute
Karlsruhe University

The discovery of ferrocene (dicyclopentadienyliron) in 1951 has provided access to a fascinating class of organometallic compounds.[1,2] The challenging synthetic and structural possibilities of these metallocenes have spurred considerable research.[3,4] Compared to the widely explored chemical of nonpolymeric metallocenes, however, the scope of research work performed on macromolecular metallocene compounds has been limited. The reason for this deficit may be found in the multitude of competing side reactions to which ferrocene and other metallocenes are prone, efficiently antagonizing clean and homologous propagation steps. This assessment is particularly valid for polymetallocenes which contain the metallocene moiety as the integral part of their polymer backbone, such as poly(1,1'-ferrocene-arylene)s A (**Scheme I**).[5–7] But these main-chain polymetallocenes offer considerable potential for both fundamental research and technical applications. The electric, magnetic, and optical properties of these polymers provide a focus of increasing interest for both chemists and physicists.[5,6,8–16]

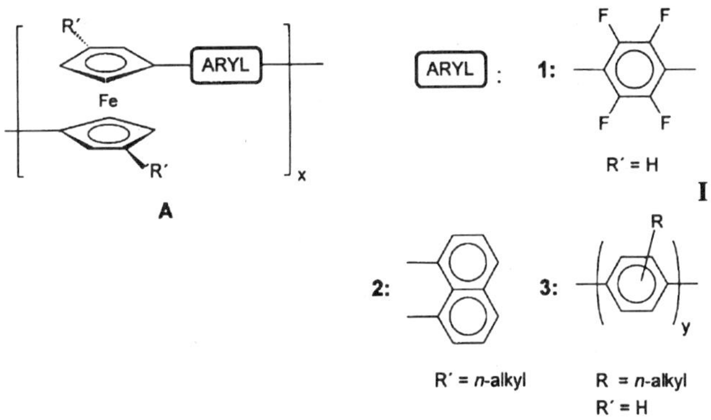

I

In the 1960s Rosenberg described the preparation of a polymer **A1** (Scheme I) via polycondensation of 1,1'-dilithiumferrocene and halogen-containing comonomers such as hexafluorobenzene or 1,1'-bis(pentafluorophenyl)ferrocene.[6,17–19] Supported by the concept of solubilizing sidechains,[20] two methods were published in 1993, both of which allow the preparation of constitutionally well-defined, high molecular weight poly(1,1'-ferrocene-arylene)s **A2** and **A3**, respectively. Naphthalene-bridged polymers **A2** were prepared by polymerization of the 1,8-bis(cyclopentadienyl)naphthalene dianion through its reaction with an iron salt.[21,22] Simultaneously with the development of polymers **A2**, a synthesis for poly(1,1'-ferrocene arylene)s **A3** was developed.[23] This latter method involves a Pd-catalyzed polycondensation reaction of bromoaromatics and arylboronic acid derivatives,[24–27] which is remarkable for its regioselectivity, high yields, and exceptional tolerance toward many functional groups that may be present in the starting materials.[28,29] However, despite these attributes only poor results are generally obtained when this reaction is applied to couple electron-rich aromatics such as ferrocene. In order to profit from this reaction for the synthesis of poly(1,1'-ferrocene-arylene)s such as **A3**, an alternative strategy must be chosen which provides the opportunity to circumvent the carbon–carbon bond formation between ferrocenylene and phenylene moieties as the polymer propagation process. Based on these considerations, the concept outlined in **Scheme II** was developed. Here, poly(1,1'-ferrocene-arylene) **1** is prepared via Pd-catalyzed polycondensation of 1,1'-bis(4-bromophenyl)ferrocene **2** and 2,5-dihexylbenzene-1,4-diboronic acid **3**. Now, despite the presence of the electron-rich ferrocene moieties in monomer **2**, advantage can be taken of the favorable reactivity of the bromine functionalities bonded at phenyl groups.

*Author to whom correspondence should be addressed.

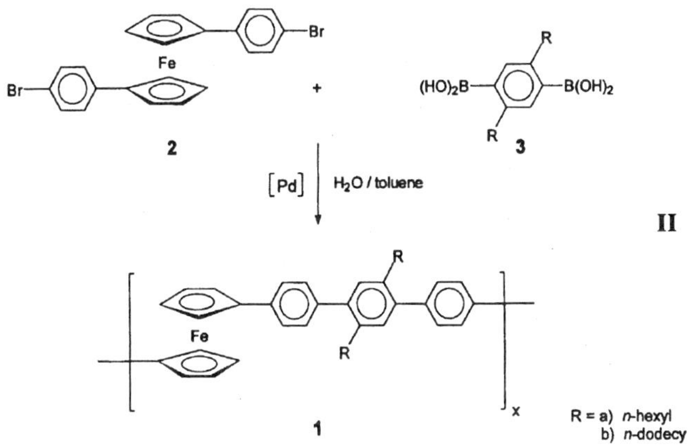

II

PROPERTIES

At this writing, we are carrying out a profound analysis of the properties of the poly(1,1'-ferrocene arylene)s; therefore, only a short summary of the available information is provided.[30]

According to differential scanning calorimetry (DSC), poly(1,1'-ferrocene-*p*-terphenylene)s **1** exhibit a glass transition at 80°C (**1a**, *R* = *n*-hexyl) and 20°C (**1b**, *R* = *n*-dodecyl). No semicrystallinity or formation of mesophases was observed for polymers **1** by DSC, wide-angle X-ray scattering (WAXS), or polarization microscopy.

REFERENCES

1. Kealy, T. J.; Pauson, P. L. *Nature* **1951**, *168*, 1039.
2. Miller, S. A.; Tebboth, J. A.; Tremaine, J. F. *J. Chem. Soc.* **1952**, 632.
3. Ferrocenes: Togni, A.; Hayashi, T., Eds.; VCH: Weinheim, Germany, 1995.
4. Rosenblum, M. *Chemistry of the Iron Group Metallocenes*; Wiley-Interscience: New York, NY, 1965; Part I.
5. Neuse, E. W. *J. Macromol. Sci.* **1981**, *A16*, 3.
6. Neuse, E. W.; Rösenberg, H. *Rev. Macromol. Chem.* **1970**, *5*, Part 1.
7. Paushkin, Y. M.; Polak, L. S. et al. *J. Polym. Sci.* **1964**, *C4*, 1481.
8. Nalwa, H. S. *Appl. Organomet. Chem.* **1990**, *4*, 91.
9. Hmyene, M.; Yassar, A. et al. *Adv. Mater.* **1994**, *6*, 564.

10. Kollmar, C.; Couty, M.; Kahn, O. *J. Am. Chem. Soc.* **1991**, *113*, 7994.
11. Chi, K. M.; Calbrese, J. C. et al. *Organometallics* **1991**, *10*, 668.
12. Wright, M. E.; Toplikar, E. G. et al. *Macromolecules* **1994**, *27*, 3016.
13. Wright, M. E.; Toplikar, E. G. et al. *Macromolecules* **1992**, *25*, 1838.
14. Nalwa, H. S. *Appl. Organomet. Chem.* **1991**, *5*, 349.
15. Niishikata, Y.; Morikawa, A. et al. *J. Chem. Soc., Chem. Commun.* **1989**, 1772.
16. Wrighton, M. S. *Acc. Chem. Res.* **1979**, *12*, 303.
17. Rosenberg, H. U.S. Patent 3 442 130, 1969.
18. Rosenberg, H.; Hedberg, F. L. 3rd Int. Symposium Organomet. Chem.; Munich, Germany, 1967; Abstract p 108.
19. Rosenberg, H.; Barton, J. H.; Hollander, M. M. 2nd Int. Symposium Organomet. Chem.; Madison, WI, 1965; Abstract p 42.
20. Ballauff, M. *Angew. Chem., Int. Ed. Engl.* **1989**, *28*, 253.
21. Rosenblum, M. *Adv. Mater.* **1994**, *6*, 159.
22. Nugent, H. M.; Rosenblum, M.; Klemarczyk, P. *J. Am. Chem. Soc.* **1993**, *115*, 3848.
23. Knapp, R.; Rehahn, M. *Makromol. Chem., Rapid Commun.* **1993**, *14*, 451.
24. Rehahn, M.; Schlüter, A.-D.; Wegner, G. *Makromol. Chem.* **1990**, *191*, 1991.
25. Rehahn, M.; Schlüter, A.-D.; Wegner, G. et al. *Polymer* **1989**, *30*, 1060.
26. Miller, R. B.; Dugar, S. A. *Organometallics* **1984**, *3*, 1261.
27. Mijaura, N.; Yanagi, T.; Suzuki, A. *Synth. Commun.* **1981**, *11*, 513.
28. Schlüter, A.-D.; Wegner, G. *Acta Polymer* **1993**, *44*, 59.
29. Kalinin, V. N. *Synthesis* **1992**, 413.
30. Knapp, R.; Rehahn, M., unpublished results.

Polyferrocenes

See: Ferrocene-Containing Polymers

POLYFLUOROACRYLATES (Effects of Side-Chain H/F Variations)

Allen D. Hunter
Department of Chemistry
Youngstown State University

X. Andrew Guo
Department of Chemistry
University of Alberta

The broad resistance to physical and chemical attack imparted by fluorination of polymers suggested that the properties of acrylics could be enhanced by the introduction into the structures of substantial amounts of fluorine, provided such an addition does not compromise other characteristic acrylic properties. Fluorinated polymers also possess a range of unusual surface chemical properties which could lead to greater versatility in the acrylics.[3,4] For these reasons, various polyfluoroacrylates have been studied.[1–7]

Using an emulsion polymerization technique, Gilbert prepared some fluorinated polyacrylates and polymethacrylates including poly(1,1,1,3,3,3-hexafluoro-2-pentafluorophenylisopropyl acrylate), which his patents claimed as useful textile impregnants exhibiting excellent water-repellency.[7] Later Griffith and Roitman studied a series of polyfluoroacrylates including poly(1,1,1,3,3,3-hexafluoro-2-phenylisopropyl acrylate), prepared by solution polymerization in fluorocarbon solvents, to evaluate the unique surface (wetting) properties of these materials.[3,4,6] Most recently, Snow has also examined the chemical–vapor absorption property of this polymer, along with some other polyfluoroacrylates.[5] However, no undertakings have been made to systematically investigate the structure–property relationships as a function of the fluorine content in the side chains of these species, and information regarding the thermal behavior of these materials is completely lacking. We have therefore begun to investigate polyfluoroacrylates.[1,2]

EXPERIMENT

Monomers

We have described elsewhere[1,2] the general procedures for the syntheses and characterizations of the five monomers discussed in this work: monomer (**1**), 2-2',3',5',6'-tetrafluorophenylisopropylacrylate; monomer (**2**), 2-pentafluorophenylisopropyl acrylate; monomer (**3**), 1,1,1,3,3,3-hexafluoro-2-pentafluorophenylisopropyl acrylate;[7] monomer (**4**), 1,1,1,3,3,3-hexafluoro-2-phenylisopropyl acrylate;[3–6] monomer (**5**), 2-phenylisopropyl acrylate.[8]

RESULTS AND DISCUSSION

We specifically chose the monomers to prepare polymers having different fluorine contents [ranging from 0%, polymer (**5**), to ca. 54%, polymer (**3**), by weight] and distributions (i.e., substitutions on the arenes or on the CX_3 groups) in their side chains. This enabled us to explore the effect of these differences on the physical properties of these materials.

We chose AIBN as the initiator since its initiation can be readily effected under mild conditions (40–70°C). As a result, five addition polymers (polyacrylates) were conveniently and successfully prepared using solution techniques as **Equation 1** shows.

Freestanding transparent films of these polymers can be easily cast from these solutions. On the other hand, transparent colorless objects (such as rods) of these polymers can be made through bulk polymerization. These objects resemble Plexiglas® in both appearance and in hardness, probably due to their structural similarity, although quantitative measurements have not yet been made.

Studies on the morphologies of these polyacrylates by transmission electrons/scanning electron micrography revealed that they are all amorphous materials.[1,2,9] A general trend in T_g is that the fluorinated polymers possess a higher T_g than the nonfluorinated ones (by 20–30°C), regardless of whether the substitution of fluorine is to the CX_3 groups or to the arenes. These polymers soften and flow at temperatures above their T_gs, which indicates that they are readily processable thermoplastics.

Polymer (**3**) shows the highest thermal stability of all five polyacrylates. Specifically, under N_2 it does not start to decompose until 380°C (360°C in air), and loses 50% of its weight at 510°C (470°C in air). This thermal stability, incorporating its solvent-resistance, flame-retardance, and processability (T_g = 83°C), indicates that this material may be of interest in the area of specialty engineering plastics.

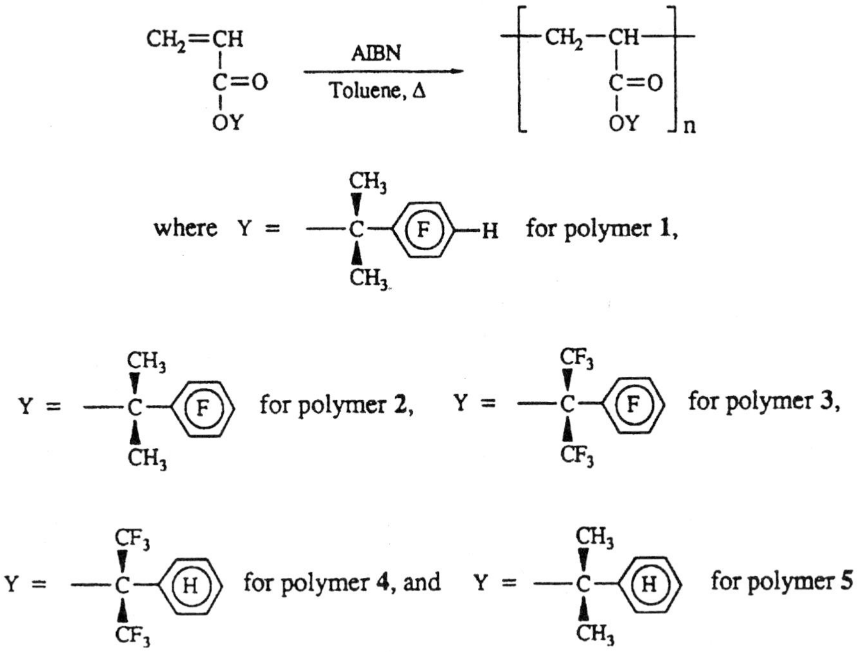

1

REFERENCES

1. Guo, X. A. Ph.D. Thesis, University of Alberta, Edmonton, Canada, 1994.
2. Guo, X. A.; Hunter, A. D.; Chen, J. *J. Polym. Sci., Part A: Polym. Chem.* **1994**, *32*, 47.
3. Griffith, J. R.; O'Reer, J. G. *Polym. Sci. Technol.* **1981**, *14* (*Biomed. Dent. Appl. Polym.*), 373.
4. Griffith, J. R.; O'Reer, J. G. *Org. Coat. Plast. Chem.* **1980**, *42*, 204.
5. Snow, A. W.; Sprague, L. G. et al. *J. Appl. Polym. Sci.* **1991**, *43*, 1659.
6. Roitman, J. N.; Pittman, A. G. *J. Polym. Sci., Polym. Chem. Ed.* **1974**, *12*, 1421.
7. Gilbert, E. E.; Farah, B. S. U.S. Patent 3 544 535, 1970.
8. Kishida, K.; Hasegawa, A. et al. Japan Kokai 78 34 853, 1978.
9. Guo, X. A.; Oak, N.; Hunter, A. D., unpublished observations.

POLY(FLUOROALKYL α-FLUOROACRYLATE)S

Tetsuo Shimizu
Research and Development Department
Daikin Industries Ltd.

Several methacrylate and acrylate polymers having fluoroalkyl side chains have been industrially used as oil and water repellent agents, moisture-proof coatings, clad materials for polymeric optical fibers, and so on. Meanwhile, poly(fluoroalkyl α-fluoroacrylate)s, which exhibit high glass transition and degradation temperatures compared with the corresponding polymethacrylates–polyacrylates, have been applied to optical materials with high heat resistance and low attenuation loss.[1,2]

Comparatively easy methods have been developed for the synthesis of the α-fluoroacrylate monomers.[3] They readily undergo radical polymerization using conventional initiators to produce corresponding homopolymers with high molecular weights.

Our structural studies pointed out that radically polymerized poly(fluoroalkyl α-fluoroacrylate)s show crystalline properties even in the homologues with short side chains whose carbon number is 2–5, whereas the corresponding short fluoroalkyl polymethacrylates–polyacrylates are amorphous.[4,5] So far, we have found crystalline properties only in the polymethacrylates–polyacrylates homologues with crystallizable longer fluoroalkyl side chains. We are preparing further structural studies on poly(fluoroalkyl α-fluoroacrylate)s having various side chains.[6,7]

PREPARATION

Monomer Synthesis

Fluoroalkyl α-fluoroacrylate monomers are generally prepared by reactions of α-fluoroacryloyl halides or α-fluoroacrylic acid with the corresponding fluoroalcohols.

Polymerization and Polymer Characterization

Fluoroalkyl as well as alkyl α-fluoroacrylate monomers readily undergo radical polymerization. In addition, fluoroalkyl α-fluoroacrylates can be easily copolymerized with other vinyl monomers, such as alkyl α-fluoroacrylates or alkyl methacrylates.[8,9]

Thermal Properties

The T_gs of poly(fluoroalkyl α-fluoroacrylate)s are a little lower than those of the poly(α-chloroacrylate) analogues, but generally 1–30 degrees higher than those of the polymethacrylate analogues. The increase in T_g of poly(α-fluoroacrylate)s compared with that of the corresponding polymethacrylates is ascribed to the enhancement of the intermolecular interaction of the main chains.

Poly(fluoroalkyl α-fluoroacrylate)s possess outstanding thermal stability. Thermal degradation temperatures of the polymers are the highest among four poly(α-substituted acrylate)s, and ca. 70°C higher on average than the corresponding polymethacrylates according to the thermogravimetry in air.

Optical Properties

The introduction of C—F bonding into the hydrocarbon polymer in general decreases the refractive index by an amount proportional to the density of fluorine, since a fluorine atom or a C—F bonding has small polarizability. The refractive indices of poly(fluoroalkyl α-fluoroacrylate)s are lower than those of the corresponding polymethacrylates.

Surface Properties

Poly(fluoroalkyl α-fluoroacrylate)s exhibit lower critical surface tension than the corresponding polymethacrylates.

APPLICATIONS

We use poly(fluoroalkyl α-fluoroacrylate)s as cladding materials for polymeric optical fibers with poly(methyl methacrylate) cores, taking advantage of their improved thermal properties and toughness as compared to poly(fluoroalkyl methacrylate)s. In the preparation of the cladding materials, however, we often copolymerize fluoroalkyl α-fluoroacrylates to adjust certain properties. The patent literature claims a range of comonomers. For example, methyl α-fluoroacrylate[8] or acrylic acid[9] decreases the crystallinity of the polymers or improves the interfacial adhesion between the core and the cladding.

To extend the use of polymeric optical fibers for longer distance optical communications, the intrinsic attenuation losses of the core polymer must be reduced. These losses are dominated by CH overtone absorptions in a spectral region between 600 and 900 nm. α-Fluoroacrylate polymers are well-suited for core materials, offering a reasonable compromise between polymerizability under radical conditions and physical properties such as attenuation losses, thermal resistence, and hydrophobicity.[1]

REFERENCES

1. Ishiwari, K.; Ohmori, A., Koizumi, S. *Nippon Kagaku Kaishi* **1985**, *10*, 1924.
2. Theis, J.; Groh, W.; Shütze, G.; Wieners, G. *Ann. Tech. Conf. (ANTEC), Soc. of Plast. Eng.* **1990**, 893.
3. Wakselman, C. *Macromol. Symp.* **1994**, *82*, 77; Boguslavskaya, L. S.; Chuvatkin, N. N. *Macromol. Symp.* **1994**, *82*, 51.
4. Shimizu, T.; Tanaka, Y.; Kutsumizu, S.; Yano, S. *Macromolecules* **1993**, *26*, 24, 6694.
5. Shimizu, T. et al. *Polym. Prepr. Jpn.* **1994**, *43*, 4, 1557.
6. Koizumi, S.; Tadano, K. et al. *Macromolecules* **1992**, *25*, 6563.
7. Shimizu, T.; Tanaka, Y.; Kutsumizu, S.; Yano, S. *Macromol. Symp.* **1994**, *82*, 173.
8. Ohmori, A. et al. (Daikin Kogyo) U.S. Patent 4 720 166, 1988.
9. Yamamoto, T.; Nishida, K. (Mitsubishi Rayon) Japan Kokai 61103107, 1986.

Poly(fluoroethylene-vinyl ether)

See: Fluoropolymer Coatings (New Developments)

Polyformaldehyde

See: Polyoxymethylenes

Polyformals

See: Fluorine-Containing Polymers (Polycarbonates and Polyformals)

Polygalacturonans

See: Hemicelluloses

Polygermanes

See: Germanium-Containing Polymers (Overview)

Polyglycidol

See: Polyglycidols, Polythioglycidols

POLYGLYCIDOLS, POLYTHIOGLYCIDOLS

Alain Le Borgne*
Laboratoire de Physicochimie des Biopolymères
CNRS-Thiais

Nicolas Spassky and Daniel Taton
Laboratoire de Chimie Macromoléculaire
Université Pierre et Marie Curie-Paris

Polyglycidol is a member of the wide family of poly- and oligoethers derived from oxiranes. These polymers have been used as surfactants, plasticizers, adhesives, and coatings, as well as prepolymers for preparing polyurethane elastomers and foams[1–4] . . . due to their broad physical properties. Commercial polyoxiranes are, however, based on a limited number of monomers: ethylene oxide, propylene oxide, butylene oxide and epichlorohydrin. Surprisingly, the polymerization of glycidol has received little attention (**Equation 1**).

1

Polythioglycidol, the polythiirane analog structurally related to polyglycidol, is also examined in this review.

POLYGLYCIDOL

The physical and chemical properties of glycidol are the subject of a detailed monograph that can be consulted by the interested reader.[5]

SYNTHESIS OF POLYGLYCIDOLS

In the case of oxiranes, there are three main mechanisms involving different modes of ring-opening of the cycle: acid-initiated (with Bronsted or Lewis acids), base-initiated (with

*Author to whom correspondence should be addressed.

alkali hydroxides or alkoxides) and coordinate-initiated (with oxides or alkoxides of zinc, aluminium...).

The polymerization of glycidol was carried out both with and without protection of the hydroxyl group with a suitable protective group in order to suppress secondary reactions during the polymerization that could lead to polymers (or oligomers) with irregular structures.

Polymerization of Unprotected Glycidol

The direct polymerization of glycidol, either by anionic or by cationic species leads to highly branched polymers. Studies of the polymerization may lead to better understanding of the conditions required to prepare polymers with controlled architecture such as dendritic polymers,[7] a field growing interest.

Polymerization of Protected Glycidol

To obtain polyglycidol with a regular structure, the hydroxyl group of the monomer must be protected before its polymerization. The protective group must lead to a monomer easy to purify, particularly by distillation; be stable during the polymerization process; and be easy to remove without alterating the polymer backbone.

The following protective groups have been used leading to the corresponding monomers: trimethylsilyl glycidylether, terbutyl glycidylether, 1-ethoxy-1-ethyl glycidylether.

PROPERTIES OF POLYGLYCIDOL

Chemical Properties

The chemical reactivity of the polymer arises from the lateral primary hydroxyl group. Polyglycidol can be derivatized with isocyanates (ethyl isocyanatoacetate, chloroethyl isocyanate) leading to the corresponding polyurethanes,[11] esterified with formic acid producing the corresponding poly(glycidyl formate)[9] or with more complex carboxylic acids like 3-(5-fluorouracil-1-yl)propanoic acid.[10]

Other potential modifications of hydroxyl groups (such as oxidation, etherification...) have not been examined. There are no reports concerning the stability of polyglycidol. The polymer probably behaves like other polyethers and is degraded thermally or chemically (with acids, ozone...).[4] The biodegradability of polyglycidol has not yet been investigated.

Solubility

Polyglycidol is soluble in water, methanol (in the case of low molecular weight polymers), pyridine, polar solvents (such as dimethylformamide and dimethyl sulfoxide) but insoluble in acetone, acetonitrile, dioxane, chloroform and apolar solvents (toluene, hexane, etc.).

Thermal Properties

Polyglycidol is a highly hydrophilic polymer. Small amounts of residual water (6–12%) in the polymer markedly affect its glass transition temperature, lowering it to –30° to –40°C from the usual –8° to –12°C in the dry state, based on DSC studies.[8] In contrast to polypropylene oxide, polyglycidol is reluctant to crystallize, even in the isotactic form. Poly(R)glycidol does, however, crystallize on stretching, relaxing, restretching and then holding.

POTENTIAL APPLICATIONS

Contrary to the heavily produced polyethers already cited, which are used in a wide range of domains (surfactants, lubricants, pharmaceuticals, cosmetics and others...[4]), polyglycidol has not found thus far many applications.

Polyglycidol has several characteristics that can offer opportunities for specific application: it belongs to the limited family of synthetic, water-soluble polymers and it can be considered a combination of polyvinyl alcohol and polyethylene oxide, and viewed as a simplified model of polysaccharide.

It can be also prepared in both chiral forms since (R) or (S) glycidols are easily available via the asymmetric epoxidation reaction[6] and are now being produced on a pilot scale by ARCO Chemical Co.

Finally, the presence of a functionalizable group in repeating unit allows chemical modification or fixation of specific groups (bioactive, liquid crystalline, with non linear optics properties...) to prepare functional polymers.

POLYTHIOGLYCIDOLS

The synthesis of polythioglycidols has received very little attention, but the polymerization of thioglycidol has been studied by Bonnans-Plaisance et al. using quaternary ammonium salts of dithiocarboxylic acids as initiators (such as tetramethylammonium dithiobenzoate).[12,13] Racemic homopolymers and copolymers (with methyl thiirane) have been prepared without protection of the hydroxyl group.

Polythioglycidols are soluble only in highly polar hydrogen-bonding solvents such as dimethylsulfoxide or dimethylformamide but are insoluble in water (contrary to polyglycidols) as well in tetrahydrofuran, acetone and chlorinate solvents. These polymers have not received further characterization.

REFERENCES

1. Gagnon, S. D. *Encyclopedia of Polymer Science and Engineering*, 2nd ed.; John Wiley & Sons: New York, 1986; Chapter "1,2 epoxide Polymers", Vol. 6, p 273.
2. Boileau, S. *Comprehensive Polymer Science* **1989**, Pergamon: Oxford; 2.3, Chapter "Anionic Ring-Opening Polymerization: Epoxides and Episulfides", p 467.
3. Inoue, S.; Aida, T. *Handbook of Polymer Synthesis* (Part A), Dekker, M.; New York, 1992; Chapter 8.
4. Bailey, F. E. Jr., Koleske, J. V. *Alkylene Oxides and their Polymers*; Dekker: New York, 1991.
5. Kleeman, A.; Wagner, R. *Glycidol, Properties, Reactions, Applications*; Alfred Hüthig, Verlag: Heidelberg, 1981.
6. Gao, Y.; Hanson, R. M.; Klunder, J. M.; Ko, S. Y.; Masamune, H.; Sharpless, K. B. *J. Am. Chem. Soc.* **1987**, *109*, 5765.
7. Tomalia, D. A.; Naylor, A. M.; Goddard, W. A. *Angew. Chem. Int. Ed. Engl.* **1990**, *29*, 138.
8. Vandenberg, E. J. In *Coordination Polymerization* in "Polym Science and Technology", Price, C. C.; Vandenberg, E. J., Eds.; Plenum: New York, 1983; 19, p 11.
9. Taton, D.; Le Borgne, A.; Sepulchre, M.; Spassky, N. *Macromol. Chem. Phys.* **1994**, *195*, 139.
10. Ouchi, T.; Yuyama, H.; Inui, T.; Murakami, H.; Fujie, H.; Vogl, O. *Eur. Polym. J.* **1986**, *22*, 537.
11. Cohen, N. L. *J. Polym. Sci. Pol. Chem. Ed.* **1975**, *13*, 1993.

12. Bonnans-Plaisance, C.; Levesque, G. *Macromolecules* **1989**, *22*, 2020.
13. Bonnans-Plaisance, C.; Levesque, G. *Polymer* **1991**, *32*, 1318.

Polyglycolides

See: *Biodegradable Polyesters (Chemically-Induced Surface Degradation)*
Biodegradable Polyesters (Theoretical Modelling of Degradation)
Bioerodible Polymers A. Controlled Drug Release B. Fracture Fixation
Controlled Drug Delivery Systems

Polyheterocycles

See: *Heteroaromatic Polymers (Five-Membered Rings in Main Chain)*
Polyethers (1,3,4-Thiadiazole-Containing)

POLY(HEXAFLUORO-1,3-BUTADIENE)

Tadashi Narita
Department of Environmental Engineering
Saitama Institute of Technology

The polymerization of hexafluoro-1,3-butadiene (CF_2=CF CF=CF_2) (HFBD) has been difficult because HFBD is a nonconjugated monomer. In contrast to the trans planar structure of 1,3-butadiene (CH_2=CH-CH=CH_2), HFBD is in a nonplanar cisoid conformation and the dihedral angle between C(1)=C(2)- and -C(3)=C(4) is 47.4 ± 2.4° determined by gas phase electron diffraction.[1] The infrared spectra and Raman spectra of gaseous and solid HFBD show it to be consistent with a gauche (C_2) structure and yields a barrier to planarity at the trans conformation of 986 ± 150 cm^{-1} (2.85 kcal/mol).[2] The results of ^{19}F NMR measurements support the twisted structure of HFBD.[3]

PREPARATION

Anionic Polymerization

Anionic polymerization of HFBD with cesium fluoride as initiator in toluene or tetrahydrofuran (THF) produces high yields of poly(HFBD) at 60°C after seven days.[4] No polymer is obtained without cesium fluoride at 60°C in toluene and THF. Rubidium fluoride also shows some polymerization of HFBD in THF.

The polymer obtained is white and solid. The polymer is insoluble in ordinary organic solvents and slightly soluble in hexafluorobenzene. The polymer is, therefore, precipitated as the polymerization reaction proceeds and gel permeation chromatogram is not recorded. However, the materials should be of high molecular weight, since the intrinsic viscosities of hexafluorobenzene-soluble parts of poly-(HFBD) produced with cesium *tert*-butoxide in toluene and in THF at 60°C for three hours were 0.85 and 0.15, respectively, and films were obtained by treating the samples with a roller.

Polymerization of HFBD failed when using typical anionic polymerization initiators.

Radical Polymerization

The thermal reactions of HFBD are investigated at temperatures between 150° to 180°C, and hexafluorocyclobutene, dimers, which are formed in largest amount, and trimers are isolated as the principal products. The yields of radical polymerization of HFBD are from 3% to 5% by bis(trifluoromethyl)nitroxide and 15% to 18% with benzoyl peroxide at 60°C for one week. The melting point of the hexafluorobenzene-soluble fraction is 138 ~ 141°C, [η] is 0.08 in hexafluorobenzene at 30.1°C and molecular weight is 3530 (Rast method). The bulk polymerizations of HFBD using di-*tert*-butyl peroxide as initiator at 115°C for two weeks give elastomeric gum with 64% conversion.

Radiation-induced polymerization of HFBD using γ-rays and high-energy electrons is studied. The monomer undergoes 1,2- and 1,4-polymerizations to yield a solid polymer of molecular weight ~5000.

PROPERTIES

The thermogravimetric analysis (TGA) of poly(HFBD) obtained by anionic polymerization shows that poly(HFBD) is highly stable against heat. The polymer obtained in toluene begins to decompose at 360°C and that yielded in THF at 200°C. The results of differential scanning calorimetric (DSC) analyses indicate that poly(HFBD) shows no peak below the decomposition temperature. In the case of poly(tetrafluoroethylene) (PTFE), an endotherm appears at 327°C, which is the melting point of the polymer. The melting point of poly(HFBD) might be higher than the decomposition temperature. The softening point of the poly(HFBD) produced in toluene is 339°C. The softening point of the polymer obtained in toluene is higher than that of PTFE.

REFERENCES

1. Chang, C. H.; Andreassen, A. L.; Bauer, S. H. *J. Org. Chem.* **1971**, *36*, 920.
2. Wurrey, C. J.; Bucy, W. E.; During, J. R. *J. Chem. Phys.* **1977**, *67*, 2765.
3. Manatt, S. L.; Bowers, M. T. *J. Am. Chem. Soc.* **1969**, *91*, 4381.
4. Narita, T.; Hagiwara, T.; Hamana, H.; Sezaki, M.; Nagai, A.; Nishimura, S.; Takahashi, A. *Macromolecules* **1989**, *22*, 1989.

Poly(1-hexene)

See: *Living Coordination Polymerization*

Polyhydrazides

See: *Polyoxadiazoles and Polytriazoles*

Polyhydroxyalkanoates

See: *Bacterial Polyesters (Containing Unusual Substituent Groups)*
Microbial Polyhydroxyalkanoates (Poly[ER-3-hydroxybutyrate-CO-3-hydroxypropionate])
Poly(β-hydroxyalkanoates) (Microbial Synthesis and Chemical Modifications)

POLY(β-HYDROXYALKANOATES) (Microbial Synthesis and Chemical Modifications)

Jorge Romero Garcia
Centro de Investigación en Quimica Aplicada

Many microorganisms produce energy storage materials that are generally described as poly(β-hydroxyalkanoate) (PHA) polyesters.[1] However, these macromolecules are often copolymers containing different alkyl groups at the β-position[2] that possess the general formula shown in **Figure 1**.

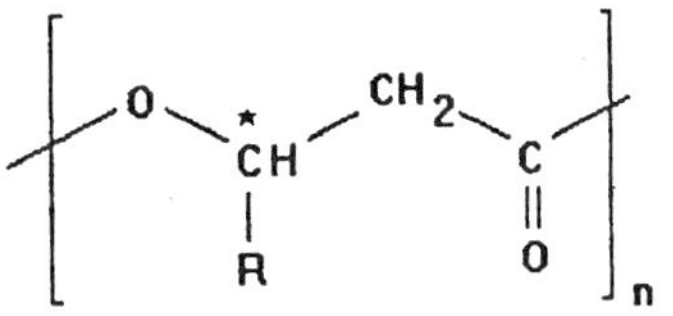

FIGURE 1. R = n-alkyl side chain of variable length.

These optically active biopolymers are synthesized and deposited intracellularly as inclusion bodies and can amount to 80% (dry weight) if the bacteria are cultivated in the presence of excess carbon and if one nutrient limits growth such as N, P, S, Mg, or O_2.[3] The synthesis of the most well-known polyester, poly(β-hydroxybutyrate) (PHB), has been studied extensively in *Alcaligenes euthrophus, Zoogloea ramigera*, and *Azotobacter beijerinki*.[4–13] These organisms can metabolize carbon sources such as glucose, fructose, alcohol, or organic acids to PHA via acetyl coenzyme A.[14] *Alcaligenes euthrophus* can also synthesize a random copolymer of 3-hydroxybutyrate and 3-hydroxyvalarate P(3HB-*co*-3HV), from propionate or valerate and glucose.[15,16] Similarly, a random copolyester containing HB and 4-hydroxybutyrate (4HB) monomer units can be produced when the microorganism is cultivated in the presence of 4-hydroxybutyrate, 4-chlorobutyrate, or γ-butyrolactone as a single carbon source.[17–19]

Fluorescent pseudomonads belonging to the rRNA homology group, such as *Pseudomonas oleovorans* can produce medium side-chain PHA copolymers from alkanes, alkenes, and alkanoates when the microorganism is grown under nutrient-limiting conditions.[20–24] Those PHAs have R groups varying between C_3 and C_9 in length. *P. oleovorans* is also capable of producing PHA with olefinic, ester, and slightly branched terminal side chains.[2,25–27] A variety of other PHA homopolymers and copolymers bearing functional groups in the side chain are also synthesized by *Pseudomonas oleovorans* if organic compounds containing functional groups are included in the growth medium. PHA with halogen atoms, cyano groups, unsubstituted phenyl, and phenoxy groups has been reported.[28–31,32]

Pseudomonas oleovorans is a versatile microorganism able to produce functionalized PHA from various substrates. However, sometimes long periods of time are needed to get good yields of PHA. Moreover, that bacteria is unable to use certain substrates to produce PHA with specific functional groups on the side chains of the polyester. Chemical modification of unsaturated pendant groups, provides an alternative functionalized PHA polymers so far not reported by biosynthetic methods. Chemical modifications could have some limitations, for example, low conversion yields of C = C, to functionality and the purification of the final products. Nevertheless, advantages including good PHA yields in short periods of time produced by *P. oleovorans* grown in alkanes and alkenes, and regulation of the mole percent of olefinic groups,[22] make chemical modification a good method for making functional PHA derivatives with different physical and thermal characteristics from their precursors.

REFERENCES

1. Steinbüchel, A. *Biomaterials. Novel Materials from Biological Sources*; Macmillan: Basingstoke, U.K. 1991; Chapter III.
2. Hazer, B. et al. *Macromolecules* **1994**, *27*, 45.
3. Dawes, E. A.; Senior, P. J. *J. Adv. Microb. Physiol.* **1973**, *10*, 135.
4. Schindler, J. S.; Schlegel, H. G. *Biochemistry* **1963**, *154*, 339.
5. Haywood, G. W. et al. *FEMS Microbiol. Lett.* **1988**, *52*, 259.
6. Haywood, G. W. et al. *FEMS Microbiol. Lett.* **1989**, *57*, 1.
7. Anderson, J. A. et al. *Int. J. Macromol.* **1990**, *12*, 102.
8. Shuto, H. et al. *Eur. J. Biochem.* **1981**, *118*, 53.
9. Nakada, H. et al. *J. Biochem.* **1981**, *89*, 625.
10. Fukui, T. et al. *Eur. J. Biochem.* **1982**, *127*, 423.
11. Senior, P. J.; Dawes, E. A. *Biochem. J.* **1973**, *134*, 225.
12. Tal, S. et al. *J. Gen. Microbiol.* **1990**, *136*, 645.
13. Tal, S. et al. *J. Gen. Microbiol.* **1990**, *136*, 1191.
14. Haywood, G. W. et al. *Biotechn. Lett.* **1989**, *11*, 471.
15. Holmes, P. A.; Collins, S. H. *Japan Kokai No.* 150393, 1982.
16. Bloembergen, S. et al. *Macromolecules* **1986**, *19*, 2865.
17. Doi, Y. *Microbial Polyesters*; VCH: Weinheim, 1990.
18. Kunioka, M. et al. *Polym. Commun.* **1988**, *29*, 174.
19. Doi, Y. et al. *Macromolecules* **1988**, *21*, 2727.
20. Smet, M. J. et al. *J. Bacteriol.* **1983**, *154*, 870.
21. Brandl, H. et al. *Applied Env. Microbiol.* **1988**, *54*, 1977.
22. Lageveen, R. G. et al. *Appli. Env. Microbiol.* **1988**, *54*, 2924.
23. Gross, R. A. et al. *Macromolecules* **1989**, *22*, 1105.
24. Huisman, G. W. et al. *Appli. Env. Microbiol.* **1989**, *55*, 1949.
25. Fritzsch, K. et al. *Int. J. Biol. Macromol.* **1990**, *12*, 85.
26. Scholz, C. et al. *Macromol. Chem. Phys.* **1994**, *195*, 1405.
27. Scholz, C. et al. *Macromolecules* **1994**, *27*, 6358.
28. Doi, Y.; Abe, C. *Macromolecules* **1990**, *23*, 3705.
29. Kim, Y. B. et al. *Macromolecules* **1991**, *24*, 5256; **1992**, *25*, 1852.
30. Lenz, R. W. et al. *J. Bioactive Compat. Polymers* **1991**, *6*, 382.
31. Hori, K. et al. *Biotechnol. Lett.* **1994**, *16*, 501.
32. Ritter, H.; V-S., A. *Macromol. Chem. Phys.* **1994**, *195*, 1665.

Poly(2-hydroxyethyl methacrylate)

See: *Contact Lenses, Gas Permeable*
Cornea, Artificial (Hydrophilic Polymeric Sponges)
Hydrogel Biomaterials (HEMA-Based)

Poly(N-hydroxypropylmethacrylamide)

See: *Anticancer, Polymeric Prodrugs, Targetable*

Poly(p-hydroxystyrene)

See: *Polystyrene and Derivatives, Photolysis*

POLYIMIDAZOLYLAMIDES

Paul G. Rasmussen* and Ramachandran P. Subrayan
Departments of Chemistry and Macromolecular Science and Engineering
The University of Michigan

Recently, imidazole-based polyamides, particularly those derived from 4,5-dicyanoimidazoles, have been synthesized, including both AB- and AABB-type polyamides. The high nitrogen and low hydrogen content of dicyanoimidazoles suggests that the polymers synthesized from them might be useful in applications requiring thermal stability and low flammability. In addition, the electron withdrawing effect of cyano groups should increase the oxidative stability of the polyamides by lowering the lowest occupied molecular orbit (LUMO) energy levels.[1]

Moderate molecular-weight polyamides have been synthesized from 2-amino-4-cyano-1-methyl-5-imidazolecarboxylic acid (**4a**) and its regioisomer, 2-amino-5-cyano-1-methyl-4-imidazolecarboxylic acid (**4b**) (see **Scheme I**).[2,3]

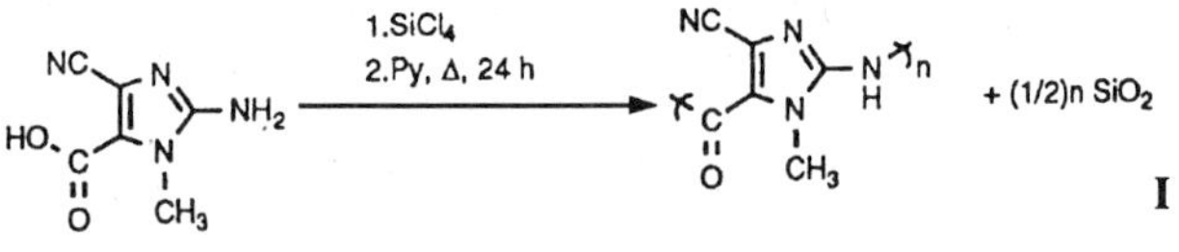

4a
4b (regioisomer)

15a (80%)
15b (50%, regioisomer)

I

SYNTHESES AND PROPERTIES OF AB POLYIMIDAZOLYLAMIDES FROM DICYANOIMIDAZOLES

Monomer Synthesis

The amino acids **4a** and **4b** were prepared by a multistep synthesis, starting from 2-amino-4,5-dicyanoimidazole.

Synthesis of Polyimidazolylamides and Model Compounds

Polymerization of the amino acid (**4a** or **4b**) was accomplished using silicon tetrachloride (Scheme I).

Polymerization of the amino acid chlorides (**12a** and **12b**) was also attempted in a variety of solvents with pyridine and DMAP. The greatest solubility was obtained with hexamethylphosphoramide (HMPA).[5] For example, polyimidazolylamide **15a** was prepared by dissolving the amine hydrochloride acid chloride **12a** in HMPA with pyridine and DMAP. Polyamides with inherent viscosities ranging from 0.10–0.22 dl/g were obtained by this method. The addition of lithium or calcium salts did not increase the solubility of the polyimidazolylamide in contrast to what has been observed for aromatic polyamide systems (**Scheme II**).

Characterization of Polyimidazolylamides

The polyimidazolylamides were light brown to red-orange in color, depending on regioisomeric structure. Polymer **15a** was red to red-orange, while **15b** was yellow to tan.

*Author to whom correspondence should be addressed.

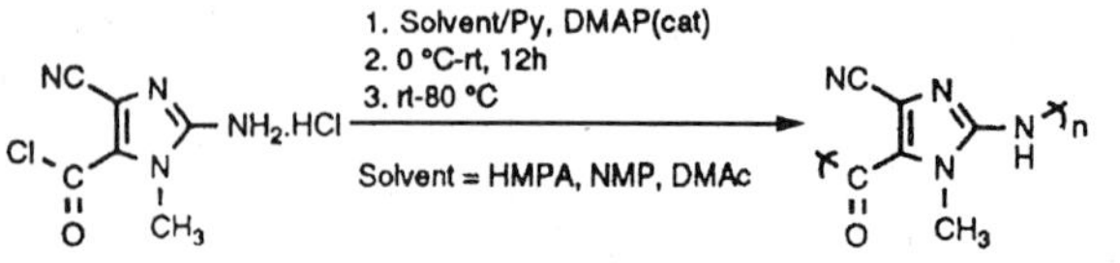

II

12a
12b (regioisomer)

15a (40-75%)
15b (50-70%, regioisomer)

The polyimidazolylamides were isolated as fine powders that appeared nonmicrocrystalline by low-power optical microscopy. Powder X-ray diffraction studies (Guinier) indicated little to no crystallinity.

Viscosity-average molecular weights (M_v) of approximately 1800–4000 were obtained.

Approximate number-average molecular weights (M_n) were found by end group analysis of ^{1}H NMRs. Values ranged from 1500–3000 for the polyimidazolylamides.

Conclusions

Polyimidazolylamides, a new class of heteroaromatic polyamides, were prepared from the polymerization of amino acids and amino acid chlorides to yield low molecular-weight polymers. In air and nitrogen atmospheres, these materials were thermally stable in excess of 300°C.

SYNTHESIS AND PROPERTIES OF AABB POLYIMIDAZOLYLAMIDES

Monomer Synthesis

The AABB polyimidazolylamides have been synthesized from 1-methyl-4,5-imidazoledicarboxylic acid. The diacid is converted to the diacid chloride **20**, in good yield, using oxalyl chloride ($(COCl)_2$) and a catalytic amount of dimethylformamide (DMF) at room temperature.

Polyamide Synthesis

The monomer 1-methyl-4,5-imidazoledicarboxylic acid is an AA monomer that has been the subject of earlier studies toward the syntheses of aliphatic-aromatic polyamides.[7,8] The syntheses and characterization of polyamides derived from the diacid and hexamethylenediamine (HMDA), neopentyldiamine (NPDA), and ethylenediamine (EDA) will be discussed here.

The polymerization of **20** with HMDA gave the polyamide designated as 6-IM. The monomer **20** polymerized with NPDA, to give the polymer designated NEO-IM and EDA, giving the polyamide 2-IM. Solution polymerization was found to be more suitable than interfacial for these polyamides. Polymerizations were conducted in methylene chloride (CH_2Cl_2) or N-methylpyrrolidinone (NMP) at 0°C (**Scheme III**). The isolated yields of the polymerizations ranged from 30–60%.

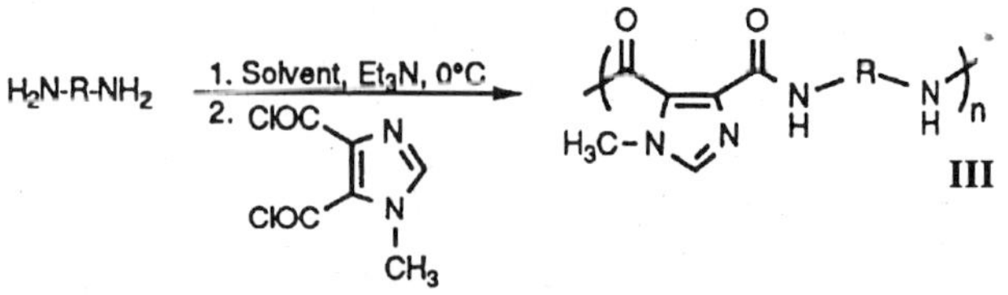

20

III

Polymer Characterization

The number average molecular weight (M_n) of the polymers, as determined by gel permeation chromatography (GPC), was found to be 6500 for 6-IM and 4500 for NEO-IM. The molecular-weight distributions were quite broad.

The T_g of the polyamides ranged from 86–189°C. The 6-IM polymer has the lowest T_g at 86°C, due to the relatively long and flexible diamine portion. The shorter aliphatic chain in 2-IM reduces the polymer chain flexibility, producing a T_g of 189°C.

The polyamides form transparent but brittle films upon solvent casting. The polymers are thermally stable to temperatures equal to or in excess of 419°C. The polymers showed at least partial solubility in a wide range of solvents, from chlorinated ones such as CH_2Cl_2 and chloroform, to polar aprotic solvents such as DMF and NMP.

Conclusions

The molecular weight of the polyamide synthesized from 1-methyl-4,5-imidazoledicarbonyl chloride and HMDA (6-IM), first reported by Takahashi and coworkers, has been improved, and two new polyamides have been synthesized from **20** and NPDA (NEO-IM) and EDA (2-IM). The molecular weights of the polyamides were moderate. Model studies did show the evidence of imidization, which is a contributing factor to reducing the molecular weights. The polymers possess moderate T_gs, are amorphous, thermally stable, and soluble in a wide range of solvents.

REFERENCES

1. Allan, D. S.; Bergstrom, D. F.; Rasmussen, P. G. *Synth. Met.* **1988,** *25,* 139.
2. Thurber, E. L.; Rasmussen, P. G. *J. Polym. Sci., Polym. Chem. Ed.* **1993,** *31,* 351.
3. Thurber, E. L.; Subrayan, R. P.; Rasmussen, P. G. *Contemporary Topics in Polymer Science*; Plenum: New York, 1992; Vol. 7, pp 95–102.
4. Bouck, K. J.; Rasmussen, P. G. *Macromolecules* **1993,** *26,* 2077.
5. HMPA is a carcinogen and extreme caution should be observed in its handling.
6. Panor, M.; Beste, L. F. *Macromolecules* **1977,** *10,* 1396.
7. MacDonald, R. N.; Cairncross, A.; Sieja, J. B.; Sharkey, W. H. *J. Polym. Sci., Polym. Chem. Ed.* **1974,** *12,* 663.
8. Takahashi, K.; Suzuki, K.; Ishiguro, H.; Zaima, T.; Mitsuhashi, K. *Nippon Kagaku Kaishi* **1974,** 801.

Poly(imide amide)s

See: Carbonylation Polymerizations (Palladium-Catalyzed)

POLYIMIDE BLENDS

Zhiliu Feng* and Hao Tang
Changchun Institute of Applied Chemistry
Chinese Academy of Sciences

Our experience has also been that polyimides (PI) with different properties can be obtained by changing the structures of the dianhydride or diamine, especially the dianhydride's length. The shorter the length of the dianhydride, the more rigid the PI's molecular chain is, hence, the higher the T_g and tensile modulus are. Simultaneously, the length of annealing time also has an obvious effect on PI's structure and properties, as the degree of molecular order and the strength of its emission and excitation fluorescence spectra all increase with an increase in the annealing time.[1] These results imply that PIs obtained by general processing are in a meta-stable state, and with annealing time, they gradually become stable.

However, the general rule is that if a PI exhibits specific properties or some improvement in one property, it is achieved at the expense of other properties. Therefore, new methods for producing PI materials with well-balanced or improved properties are needed to meet the demand of high-tech applications. Recently, blending PIs with PIs or with other high-performance polymers proved to be an easier, practical way to obtain specific properties for some applications.

THERMOPLASTIC BLENDS

Thermoplastic PI blends can be prepared by solution blending or melt blending two PIs with different structures or by blending their precursors with successive imidization. Because the components' compatibility (or miscibility) is the main factor affecting the properties of a polymer blend, considerable research has been directed at studying this relationship.

Although polyimides are generally immiscible with each other for thermodynamic reasons, the quasi-compatible, interpenetrating mixtures of two different polyimides will exhibit the characteristics of a compatible mixture.[3] Because polyimide applications are at temperatures below the glass transition, phase separation, even if occurs, will do so at an exceedingly slow rate.

One of the most common features of PI-PI blends is that they exhibit a glass transition temperature well below those of the polyimides alone.[2] This phenomenon has also been found in polyimide and poly(ether sulfone) (PES) blends and in thermoplastic polyimide–thermosetting polymer systems.[4–6]

THERMOPLASTIC AND THERMOSETTING BLENDS

The purpose of blending thermosetting PI with thermoplastic PI is to overcome the former's brittleness and to improve the blend's processability and thermal stability. In this case, during blending at least one PI is synthesized or crosslinked in the presence of the other. Therefore, the thermoplastic PI is linear while the thermosetting PI is crosslinked, and a semi-interpenetrating polymer network results. Much research has been done on this type of PI blend to obtain a strong, heat-resistant plastic material for practical applications.[7]

BLENDS WITH OTHER HIGH-PERFORMANCE POLYMERS

Poly(aryl ether ketone)

Poly(aryl ether ketone) (PAEK), especially poly(ether ether ketone) (PEEK) is a crystalline polymer with a high melting point of 250–400°C depending on the molecular composition

*Author to whom correspondence should be addressed.

and is well-known for its excellent mechanical properties, environmental resistance, and continuous use at high temperatures. Used for engineering applications, PAEK is especially suitable as a matrix material for thermoplastic composites because of its good adhesion to glass and carbon fibers, which is due to the formation of the transcrystalline region at the fiber surface.[8–11] By contrast, the polyimides that are used to blend with PAEK are amorphous polymers with a high T_g, and their resistance to chemicals is lower than that for PEEK. Therefore, blending PEEK with PI is an interesting way to combine the complementary properties of both polymers.

Polybenzimidazole

Aromatic polybenzimidazoles (PBI) and aromatic polyimides may be miscible at a range of compositions and structural variations.[12]

PEI-PEK-C and PEI-PES Blends

Wang et al. used a viscometry method to study the miscibility of a PEI and phenolphthalein poly(ether ether ketone) (PEK-C) blend.[13] The results they obtained indicated that the blends' Huggins coefficients exhibited a positive deviation from simple additivity, while the inherent viscosities of the blends exhibited a negative deviation. The values for the viscometry, calculated from these parameters, were all >0. Obviously, the blends were miscible throughout the composition range.

The blends of PEI and poly(ether sulfone) (PES) appeared to be miscible throughout the composition range when the blends were cast as films or precipitated from a solution of solvents such as CH_2Cl_2 and DMAc.[4] After annealing above their T_gs, the blends were phase separated irreversibly, indicating that the miscibility of PI-PES blends prepared from solution blending was unstable and possibly caused by the relatively rapid removal of solvent or by the solvent residue. Either a linear dependence or a negative T_g deviation from the linearity was observed.

Other Blends

Jo et al. studied the miscibility of blends of poly(ether imide) (PEI) and poly(ethylene terephthalate) (PET).[14] Single and composition-dependent T_gs were observed throughout the composition range, indicating that the blends were miscible in the amorphous region.

Recently, research was conducted using thermoplastic polymers, especially PI, to toughen epoxy resins.[6,15] This thermoplastic PI modification offers three advantages over the conventional rubber-toughening process: it operates in thermosetting resins that are too tightly crosslinked to absorb significant amounts of energy by yielding; it avoids the loss of modulus that results from adding rubber particles; and it can enhance the blends' softening temperature.

Recently, PIs blended with liquid-crystalline polymers have become a popular way to improve PI's processability and to obtain molecular composites with good mechanical properties, especially the tensile modulus and strength.[16,17] Preliminary work at our institute has showed that adding liquid-crystalline polymers to PI achieves these results.

REFERENCES

1. Luo, H.; Feng, H.; Tang, H. et al. *Polymer*, submitted.
2. Han, B. J.; Park, J. M.; Gryte, C. C. *Polym. Eng. Sci.* **1993**, *33*, 901.
3. Ree, M.; Yoon, D. Y.; Volksen, W. *Polymeric Mater. Sci. Eng. Prepr.* **1989**, *60*, 179.
4. Liang, K.; Grebowicz, J.; Valles, E. et al. *J. Polym. Sci., Polym. Phys. Ed.* **1992**, *30*, 465.
5. Pascal, T.; Mercier, R.; Sillion, B. *Polymer* **1990**, *31*, 78.
6. Hourston, D. J.; Lane, J. M.; MacBeath, N. A. *Polym. Int.* **1991**, *26*, 17.
7. Burks, H. D.; St. Clair, T. L. *SAMPLE Quarterly* **1987**, *October*, 1.
8. Jones, D. P.; Leach, D. C.; Moore, D. R. *Polymer* **1985**, *26*, 1385.
9. Seferis, J. C. *Polym. Comput.* **1986**, *7*, 158.
10. Lee, Y.; Porter, R. S. *Polym. Eng. Sci.* **1986**, *26*, 633.
11. Hsiao, B. S.; Chen, E. J. H. *Proc. Mater. Res. Soc. Symp.* **1990**, 170.
12. Leung, L.; Williams, D. J.; Karasz, F. E. et al. *Polym. Bull.* **1986**, *16*, 457.
13. Wang, W.; Sun, Z.; Zhang, P. et al. *Acta Polymerica Sinica* **1992**, *6*, 748.
14. Jo, W. H.; Lee, M. R.; Min, B. G. et al. *Polym. Bull.* **1994**, *33*, 113.
15. Bucknall, C. B.; Gilbert, A. H. *Polymer* **1989**, *30*, 213.
16. Sun, T.; Baird, D. G.; Huang, H. H. et al. *J. Composite Mater.* **1991**, *25*, 788.
17. Baird, D. G.; Bafna, S. S.; Souza, J. P. D. *Polym. Composites* **1993**, *14*, 214.

POLYIMIDE, NEW-TPI (Semicrystalline Thermoplastic)

Peggy Cebe*
Department of Physics and Astronomy
Tufts University

Benjamin S. Hsiao
Central Research and Development Experimental Station
E. I. DuPont de Nemours, Incorporated

NEW-TPI is a novel, crystallizable, thermoplastic polyimide produced by Mitsui Toatsu Chemical Company (Tokyo, Japan). NEW-TPI offers a processing advantage over other polyimides. It is fully imidized in the solid state, and can be melt-processed by normal methods, such as injection molding and extrusion, used to process other high-temperature thermoplastics.

The thermoplasticity and crystallizability of NEW-TPI result from the structure of the chemical repeat unit. Incorporation of meta linkages, and phenyl-ether moieties in the polymer backbone result in increased chain flexibility.

Since NEW-TPI is both crystallizable and thermoplastic, some of its most important properties relate to thermal behavior and crystallization kinetics. Thermal properties have been studied by many groups.[1,3,11] They include the high glass transition (250°C) and melting point (388°C). The heat of fusion, infinite crystal melting point, thermal expansion and degradation, and

*Author to whom correspondence should be addressed.

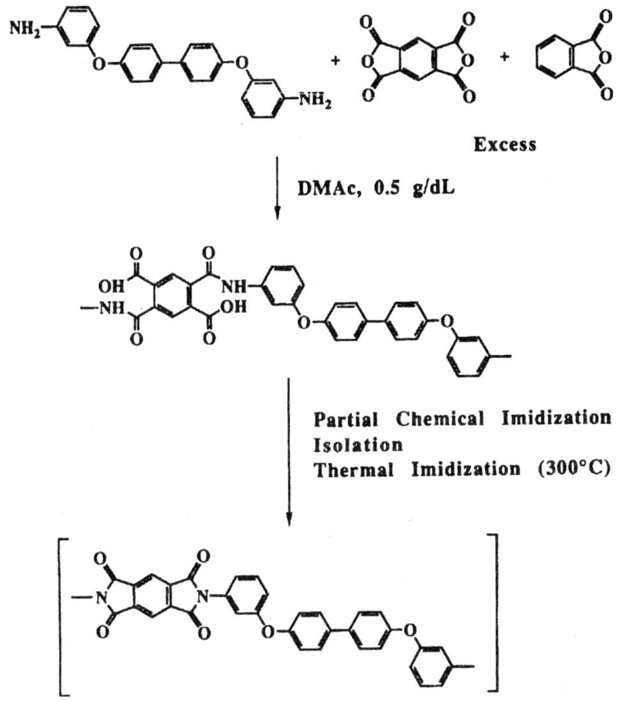

FIGURE 1. Scheme of NEW-TPI polymerization.

effects of annealing treatments on properties are described. Crystallization kinetics studies indicate that NEW-TPI crystallizes relatively slowly and can develop a modest mass fraction crystallinity from 0.20 up to 0.45.[4,12,13]

PREPARATION AND PROCESSING

NEW-TPI represents the first commercially available crystallizable thermoplastic member of the polyimide class. Its chemical structure, shown above, results from the polymerization of a dianhydride and a diamine. The synthesis of the diamine component, 4,4'-bis(3-aminophenoxy)-biphenyl, is described in a U.S. Patent 5 077 436 issued to Mitsui Toatsu (MTC) in December 1991.[2]

The polymerization scheme is shown in **Figure 1** and is described in several U.S. Patents awarded to MTC.[2]

As is typical of the general class of polyimides, solvent resistance of NEW-TPI is excellent. The resistance of NEW-TPI to various acids, bases and organic solvents, has been reported by Mitsui Toatsu.[18] Only concentrated nitric acid and potassium hydroxide affect both the amorphous and semicrystalline films; the amorphous NEW-TPI swells while the crystalline film crazes slightly. Amorphous films swell slightly in concentrated sulfuric acid and exhibit slight crazing in dichloromethane and chloroform. The long exposures used in these tests, and the minimal effects seen, certainly allow us to conclude that NEW-TPI is extremely resistant to attack by acids, bases, and organics. And, the semicrystalline film is more resistant than the amorphous film.

PROPERTIES

Thermal Properties

In amorphous NEW-TPI T_g is about 250°C, while in semicrystalline film T_g can be near 256°C. The heat distortion temperature (HDT) occurs below T_g, at 238°C, in amorphous film. However, semicrystalline film benefits from the reinforcing effect of crystals, and has higher HDT of 256°C.

Crystallization Kinetics

NEW-TPI has relatively slow kinetics compared with other high-performance thermoplastic polymers such as PET, PPS, PEEK, or PEKK.[16,19–28] The half-time for crystallization of NEW-TPI is on the order of minutes, while for the other polymers it is on the order of seconds. NEW-TPI also has a small crystallization temperature window, defined as the range between the glass transition and infinite crystal melting point. This window is 150°C for NEW-TPI, much smaller than for PET (200°C), PPS (220°C), or PEEK (250°C).

Structure

Using WAXS on films "drawn four times at 290°C then annealed at 290°C for 30 min with fixed ends" Okuyama et al. investigated the crystal structure of NEW-TPI.[15] They concluded that this polymer has a unit cell with two monomer units per cell, one at the center of four equivalent units. The unit cell parameters are: a = 7.89 Å, b = 6.29 Å, and c (fiber axis) = 25.11 Å, $\alpha = \beta = \gamma = 90°$. The fiber axis length is very close to the length of the extended monomer unit, indicating that NEW-TPI crystals have very straight chains.

Mechanical Properties

Viscoelastic properties at high temperature have been reported by Hou and Reddy.[6] These authors studied the effects of melt holding time and temperature on the dynamic storage modulus, G', and loss modulus G".

Aihara and Cebe studied the effects of orientation on NEW-TPI properties at elevated temperature.[14]

Effects of electron irradiation on mechanical properties has been studied by Sasuga, et al.[5,11]

Electrical Properties

High-temperature dielectric properties of NEW-TPI have been reported by our group as a function of electric field frequency.[13,17]

REFERENCES

1. Hergenrother, P. M. *SPE Conference on High Temperature Polymers and Their Uses*; Case Western Reserve University, October 2–4, 1989.
2. U.S. Patents 4 937 316, 1990; 5 013 817, 1991; 5 043 419, 1991; 5 077 436, 1991.
3. Brillhart, M. V.; Cebe, P. *J. Polym. Sci.: Polym. Phys.* **1995**, *33*, 927.
4. Friler, J. B.; Cebe, P. *Polym. Eng. & Sci.* **1993**, *33*, 587.
5. Hirade, T.; Hama, Y.; Sasuga, T.; Seguchi, T. *Polymer* **1991**, *32*, 2499.
6. Hou, T. H.; Reddy, R. M. *SAMPE Quarterly* **1991**, *22*(2), 38.

7. Hsiao, B.; Sauer, B.; Biswas, A. *J. Polym. Sci.: Poly. Phys. Ed.* **1993**, *32*, 737.
8. Huo, P. P.; Friler, J. B.; Cebe, P. *Polymer* **1993**, *34*, 4387.
9. Lu, S. X.; Cebe, P.; Capel, M. *J. Appl. Polym. Sci.* **1995**, *57*, 1359.
10. Mitsui Toatsu Chem., Inc. Tokyo, Japan, Technical Data Sheet: A/00 "Typical Properties of NEW-TPI Molded Parts" 1987.
11. Sasuga, T. *Polymer* **1991**, *32*, 1539.
12. Hsiao, B. S.; Sauer, B. B. *J. Polym. Sci., Polym. Phys. Ed.* **1993**, *31*(8), 901.
13. Huo, P. P.; Cebe, P. *Polymer* **1993**, *34*, 696.
14. Aihara, Y.; Cebe, P. *Polym. Eng. Sci.* **1994**, *34*, 1275.
15. Okuyama, K.; Sakaitani, H.; Arikawa, H. *Macromol.* **1991**, *25*, 7261.
16. Sauer, B.; Hsiao, B. *Polymer* **1993**, *34*(15), 3315.
17. Cebe, P.; Huo, P. *Thermochimica Acta* **1994**, *238*, 229.
18. Mitsui Toatsu Chem., Inc. Tokyo, Japan, Technical Data Sheet: E/00 "Chemical Resistance of NEW-TPI", 1987.
19. Blundell, D. J.; Osborn, B. N. *Polymer* **1983**, *24*, 953.
20. Cebe, P.; Hong, S.-D. *Polymer* **1986**, *27*, 1183.
21. Chung, J. S.; Cebe, P. *Polym. Sci., Polym. Phys.* **1992**, *30*, 163.
22. Gardner, K. H.; Hsiao, B. S.; Matheson, R. R.; Wood, B. A. *Polymer* **1992**, *33*(12), 2483.
23. Groeninckx, G.; Reynaers, H.; Berghmans, H.; Smets, G. *J. Polym. Sci. Polym. Phys. Ed.* **1980**, *18*, 1311.
24. Lee, Y.; Porter, R. S. *Macromol.* **1988**, *21*, 2770.
25. Lopez, L. C.; Wilkes, G. L. *Polymer* **1988**, *29*, 106.
26. Lovinger, A. J.; Davis, D. D.; Padden, Jr., F. J. *Polymer* **1985**, *26*, 1595.
27. Palys, L.; Phillips, P. J. *J. Polym. Sci. Polym. Phys. Ed.* **1980**, *18*, 829.
28. Vilanova, P. C.; Ribas, S. M.; Guzman, G. M. *Polymer* **1985**, *26*, 423.

POLY(IMIDE SILOXANE)S

Jerald Rosenfeld, H. R. Acharya, and J.-O. Choi
Occidental Chemical Corporation

Toshio Suzuki
Sumitomo Bakelite Company, Ltd.

Poly(imide siloxane)s refer to a class of polyimides that have polysiloxane segments incorporated into the polymer backbone. This siloxane incorporation imparts interesting properties that are distinct from the original polyimides or polysiloxanes. The most predominant applications have been in the area of microelectronics packaging; the other applications have been membrane separations, wire and cable insulation, high temperature polymer electrolytes, and others.

The purpose of this paper is to describe the structure-property relationships of polyimidesiloxanes and their relevance in development of applications, mainly in microelectronic packaging.

In an approach initiated by General Electric (G.E.) and continued at M&T Chemical, siloxane diamines were used as one of the diamines to make condensation polyimides, and they eventually led to the development of soluble, fully imidized, high molecular weight poly(imide siloxane)s. Improved formulations with increased solubility and high glass transition temperatures were developed at Occidental Chemical Corporation in conjunction with Sumitomo Bakelite in the 1980s.[1] Also, much research in this area was carried out by J.E. McGrath and coworkers at Virginia Polytechnic Institute.[2]

The main use of this type of polymer has been in microelectronics. The trend in microelectronics has been toward continued miniaturization of integrated circuits, which allows fabrication of smaller, faster, and more powerful computers. In earlier ICs or chips, ceramics were predominantly used as intermetallic dielectrics that separate the layers of conductors and for encapsulation, which protects the chip from physical and environmental damage. More recently, polymers have gained acceptance in these applications.[3]

Polyimides, in particular, have found wide use as intermetallic dielectrics.

Poly(imide siloxane)s which are copolymers of polyimide and polydimethylsiloxane, have improved properties compared to polyimides. By incorporation of siloxane, the polymers show improved adhesion, form stable solutions in the fully imidized form in various organic solvents, and, therefore, do not have to be stored cold like the polyimide precursors. Also, fully imidized poly(imide siloxane)s can be processed at much lower temperatures than polyimides because only the solvent needs to be removed; or 150 to 250°C versus 300 to 400° for polyimides. Also, no water is evolved in the process. In some applications, the polyimidesiloxanes are used as "hot melt" adhesives where there is no solvent removal, and a strong bond can be formed in one second or less.

Poly(imide siloxane)s generally absorb less water and have somewhat lower dielectric constants than conventional polyimides. In addition, they adhere to many substrates very well and do so without the need of an adhesion promoter. Typically, their modulus is lower than that of polyimides, and they tend to cause less stress buildup. Thermal stability is reduced somewhat compared with polyimides but is adequate for most microelectronic applications.[5] What makes poly(imide siloxane)s so different from polyimide homopolymers is their thermoplasticity and adhesive characteristics. They can act like a "hot melt" adhesive yet maintain strength at high temperatures (> 200°) owing to the high glass transition temperature (T_g) of the imide block. For example, they can be used to bond an Integrated Circuit (IC) chip in place and hold it at elevated temperatures during wire bonding. Also, unlike epoxies, they are reworkable owing to their thermoplasticity. A faulty chip can be removed with heat or solvent.

PREPARATION OF POLY(IMIDE SILOXANE)S

Polyimides are generally prepared in a polar solvent, such as NMP, at room temperature or by addition of a dianhydride to a solution of a diamine and then allowed to stir for several hours, all under moisture-free conditions.[4]

Poly(imide siloxane) copolymers are prepared in a similar manner. The dianhydride(s) can be added to the diamine(s) or the addition can be reversed (**Figure 1**).

However, because of their solubility, the poly(imide siloxane)s can be imidized in solution.

Photosensitive polyimides have been the subject of intense research in the past two decades. Photosensitive polyimides and poly(imide siloxane)s can be made by incorporating various photosensitive groups into the polymer backbone or sidegroups.

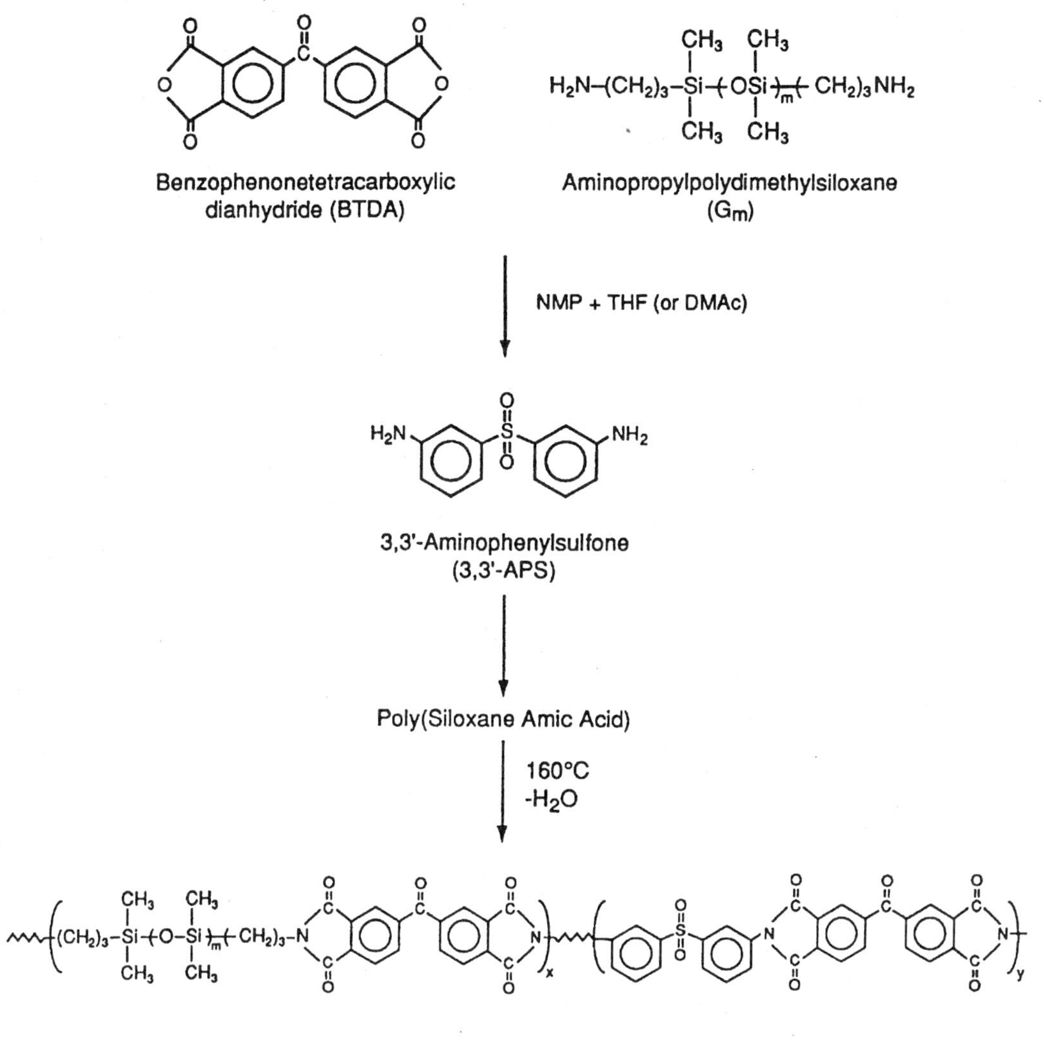

FIGURE 1. Poly(imide siloxane) formation. *Source*: Reference 2.

STRUCTURE VS. PROPERTIES

The poly(imide siloxane)s structure consists of the "rigid" or "glassy" imide segments and the "soft" or "rubbery" siloxane segments. The properties of the polyimidesiloxane depend on the monomer constituents that make up the polymer.

Morphology and Thermal Properties

Polyimidesiloxane films are generally clear to the eye. Therefore, detection of the multi-phase structure must be done by thermal methods.

T_g and Solubility

T_g and solubility of poly(imide siloxane)s are determined primarily by the dianhydride(s) and diamine(s) that form the polyimide blocks in the polymer. As the level of the siloxane in the polymer is increased, the polyimide blocks are shortened. As the imide blocks are shortened, imide T_g drops and solubility increases.

Physical Properties

At low siloxane content, poly(imide siloxane)s are rigid and more like polyimides. As siloxane content is increased, flexibility also increases. At levels higher than 40% siloxane, siloxane is the continuous phase and the materials are thermoplastic elastomers.

Electrical Properties

Dielectric constants of about 2.8 to 3.2 are reported for poly(imide siloxane) compositions compared to 3 to 3.5 for polyimide homopolymers. High dielectric breakdown strength and resistivity values complete an overall favorable electrical property picture.[5]

Water Absorption

High water absorption tends to reduce electrical properties and can promote corrosion (especially in the presence of ions) at the metal surface, leading to device failure. Water absorption values of 0.8% and as low as 0.2% are reported for poly(imide siloxane)s compared to as high as 3.5% for polyimides.[5]

Adhesion

Lupinski and Policastro point out that good adhesion is essential to protection from corrosion.[7] They also note that polyimides and polysiloxanes by themselves have poor adhesion.

Introduction of the siloxane blocks into the polyimide backbone flexibilizes the chain and increases mobility and allows the contact with the surface necessary to form a good adhesive bond. Lupinski and Policastro, in fact, report much higher peel strengths for poly(imide siloxane)s compared with polyimides.[7]

MICROELECTRIC PACKAGING APPLICATIONS

Poly(imide siloxane)s have many inherent properties attractive for packaging applications in the microelectronic industry. These properties include low water-absorption, low stress, high purity (low ionic impurities), excellent dielectric properties, high thermal stability, and room-temperature storage. A significant feature of this family is primer-less adhesion to substrate and leadframe materials commonly used in microelectronic packaging, such as Si, metals (Cu, Alloy 42, Al, Ag, Au, and others), inorganics (Al_2O_3, AlN, BN) and organic polymers (polyimides).

Besides the above-mentioned properties, poly(imide siloxane)s possess two important attributes: they are thermoplastic in nature, or they soften when heated to a sufficient temperature above the glass transition (T_g); hence, they make excellent candidates for "hot melt" adhesives; and second, they are soluble in different solvents with a wide range of boiling points.

Poly(imide siloxane) applications are broadly classified as: adhesives (films and pastes) and coatings.

Coatings

The coatings are usually solutions of poly(imide siloxane)s in suitable solvents formulated with appropriate additives.

The coatings are used in interlayer dielectrics, wafer-level passivation, photoresists, die-coat, and encapsulants.

Miscellaneous Applications

It is well known that when siloxane-containing polymers are exposed to aggressive oxygen environments, the surface siloxane segments convert to a ceramic-like silicate (SiO_2), which provides a protective overlayer for the bulk material. This transformation is of extreme interest to both the electronics and aerospace industries for the enhancement of oxygen plasma etch resistance and for the *in situ* formation of protective coatings on organic polymeric materials.[6]

Tian et al. have investigated poly(imide siloxane)s as high-temperature electrolytes and found that incorporation of the siloxane segment into the backbone imparts flexibility to the backbone and enhances the ionic conductivity.[8] This conductivity was about 100 times higher than for a polyimide prepared with no siloxane in it.

ACKNOWLEDGMENTS

The authors are grateful for the kind support and encouragement from our colleagues at Occidental Chemical Corporation, especially David Tang, Sergio Rojstaczer, Timothy Meterko, and George Kosanovich. We also appreciate the assistance from K. Tanaka, of Sumitomo Bakelite, Japan.

REFERENCES

1. U.S. Patents 3 325 450 and 4 973 645, Kuckertz, H. *Makromol. Chem.*, 1966, 98, 101; U.S. Patent 4 395 527.
2. Bott, R. H.; Summers, J. D.; Arnold, C. A.; Taylor, L. T.; Ward, T. C.; McGrath, J. E. *J. Adhesion* **1987**, *23*, 67; Summers, J. D.; Arnold, C. A.; Bott, R. H.; Taylor, L. T.; Ward, T. C.; McGrath, J. E. *A.C.S. Polymer Preprints* August, 403, 1990.
3. Soane, D. S.; Martynenko, Z. *Polymers in Microelectronics*; Elsevier: New York, 1989.
4. Lai, J. H., Ed., *Polymers for Electronic Applications*; CRC: Boca Raton, Florida, 1989.
5. Lee, C. J. *J. of Electronic Materials* **1989**, *18*, 313.
6. Arnold, C. A.; Summers, J. D.; McGrath, J. E. *Polym. Eng. and Sci.* **1989**, *29*, No. 20, 1413.
7. Lupinski, J. H.; Policastro, P. P. *Polymer News* **1990**, *15*, 71.
8. Tian, S. B.; Pak, Y. S.; Xu, G. "Polyimide-Polysiloxane-Segmented Copolymers as High-Temperature Polymer Electrolytes," *J. Polym. Sci.: Part B: Polymer Physics* **1994**, *32*, 2019.

Polyimides

See: *Anoxic Polymer Materials*
Aramids and Aromatic Polyimides, Soluble (Bulky Substituents)
Carbonylation Polymerization (Palladium-Catalyzed)
Condensation Polymers (Synthesis from Silylated Diamines)
Conducting Polymers (for Langmuir-Blodgett Film Fabrication)
Contact Lenses, Gas Permeable
Crosslinking Agents, Phenyltriazenes (for Polyimides and Other Aromatic Polymers)
Cyclolinear Organophosphazene Polymers
Engineering Thermoplastics (Survey of Industrial Polymers)
Fluorinated High Performance Polymers (Imide and Hexafluoroisopropylidene Groups)
Heteroaromatic Polymers (Five-Membered Rings in Main Chain)
High Pressure Polycondensation (for High-Temperature Polymers)
Metallized Polyimides
Molecular Composites (High Modulus Polyimide/Polyimide Films)
Molecular Composites (Liquid Crystalline Polyamide and Amorphous Polyimide)
Molecular Composites (Polyimide/Polyimide, Laminate Processability)
Pervaporation Membranes (for Separating Organic Solvents)
Poly(aryl prehnitimide)s
Poly(benzobisthiazole-imide)s
Polyimide Blends
Polyimide, NEW-TPI (Semicrystalline Thermoplastic)
Poly(imide siloxane)s

Polyimides (Arylene Ether Diamines having Cyclic Side Groups)
Polyimides (Containing Ether Linkages: Adhesive Properties)
Polyimides (from Cyclic Unsaturated Imides)
Polyimides (from N-Silylated Aromatic Diamines)
Polyimides (from Tricyclic Heteroaromatic Dyes)
Polyimides (Introduction and Overview)
Polyimides (Negative-type Photosensitive Precursors)
Polyimides (via Interfacial Polyfunctional Condensation)
Polyimides, High Performance Films
Polyimides, Precursors (Dye-Containing Side-Group)
Polyimides, Processable (Modified with other Rigid Groups)
Polyimides, Soluble Rigid-Rods (via Pd-Catalyzed Polycondensation)
Polyisoimides

POLYIMIDES (Arylene Ether Diamines Having Cyclic Side Groups)

Chin-Ping Yang* and Jiun-Hung Lin
Department of Chemical Engineering
Tatung Institute of Technology

Incorporation of any arylene ether linkage into a polymer macrochain is known to impart processability to the polymer with little reduction in thermal stability.[1,2] On the other hand,, a polymer chain with a cyclic side cardo group[3,4] (for example, a fluorene, phthalide or phthalimidine group) generally can possess a favorable combination of properties such as good thermo-oxidative stability, relatively high glass transition temperature and good mechanical properties. Hence, combining aryl ether and cardo units with the polymer chain may increase solubility of polymers without extreme loss of thermal stability.

PREPARATION

Monomer Synthesis

The syntheses of diamines **II** containing a preformed cyclic side group of aryl ether linkages were achieved by condensation of p-chloronitrobenene with four cardo ring-containing bisphenols, individually, producing corresponding dinitro compounds **I**, followed by catalytic reduction (**Equations 1** and **2**).

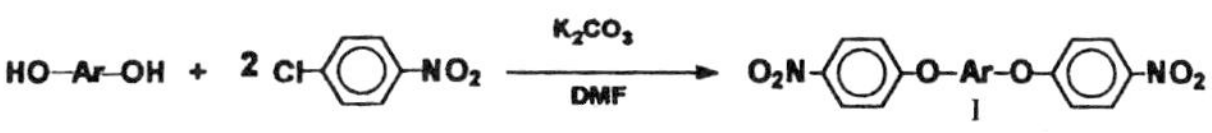
1

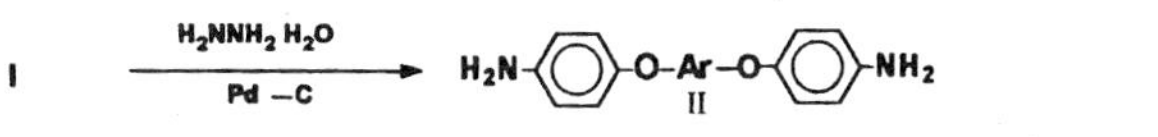
2

*Author to whom correspondence should be addressed.

PREPARATION OF POLYIMIDES

The polyimides **C-Va-e** and **T-Va-e** were synthesized by the two-step polymerization method using diamines **II** and aromatic tetracarboxylic dianhydrides, which involved ring-opening polyaddition and subsequent cyclodehydration (**Equations 3** and **4**).[5–7] The inherent viscosities of the resulting poly(amic acid)s were in the range of 0.60–2.07 dL/g. Flexible and transparent polyimide films could be obtained by thermal treatment. Alternatively, chemical treatment of poly(amic acid)s with a mixture of acetic anhydride and pyridine also is effective in obtaining polyimides.[8] Using this procedure, polyimides had good solubility in NMP with inherent viscosities of 0.53–1.39 dL/g.

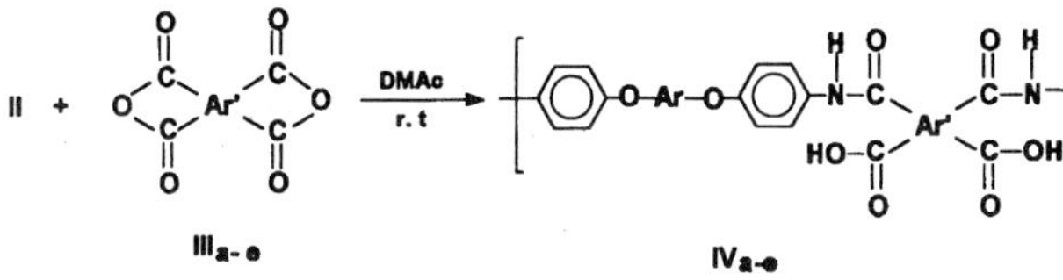
3

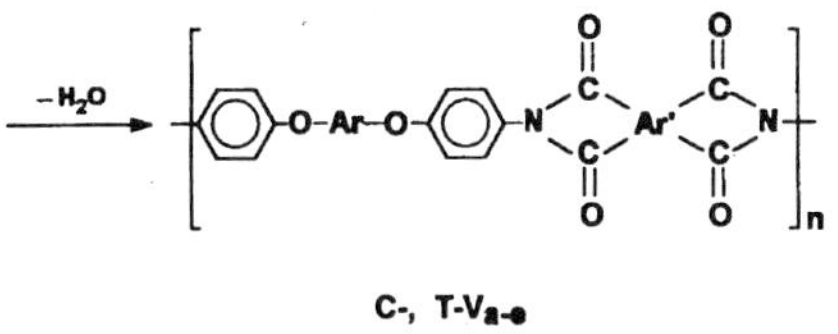
4

PROPERTIES OF POLYIMIDES

Except for polyimides **C-Va**, most of resultant polyimides by chemical cyclodehydration were soluble in organic solvents such as DMAc, NMP and *o*-chlorophenol. However, polymers **T-Va-e** prepared by thermal cyclodehydration had limited solubility. When the chemically cyclodehydrated polyimides were heated at 250°C for another 30 min in an air oven to achieve complete imidization, solubility did not change and was the same as the corresponding polyimides **C-Va-e**.

Thermal behavior of these polymers was evaluated by means of DSC and TG. In nitrogen or air atmosphere, the 10% weight loss temperature is up to 581°C. Tetracarboxylic dianhydrides had significant influence on thermal and thermooxidative stability of the polyimides.

DSC measurements were conducted with a heating rate of 20°C/min. The polyimides had T_gs in the range of 260–351°C. This order was comparable to the decreasing order of stiffness and polarity of the polymer backbones.

REFERENCES

1. Hale, W. F.; Farnham, A. G. et al. *J. Polym. Sci., Polym. Chem. Ed.* **1967**, *5*, 2399.
2. Malichenko, B. F.; Sherikova, V. V. et al. *Vysokomol. Soedin., Ser. B* **1972**, *14*, 423.
3. Vinogradova, S. V.; Ya, S. V. *Russ. Chem. Rev.* **1973**, *42*, 551.
4. Vinogradova, S. V.; Korshak, V. V. *J. Macromol. Sci. Chem.* **1974**, *11*, 45.
5. Dine-Hart, R. A.; Wright, W. W. *J. Appl. Polym. Sci.* **1967**, *11*, 609.
6. Sroog, C. E.; Endrey, A. L. et al. *J. Polym. Sci., Part A.* **1965**, *3*, 1373.

7. Sroog, C. E. *J. Polym. Sci., Part C.* **1967**, *16*, 191.
8. Sroog, C. E. *Macromolecular Synthesis*; Moore, J. A., Ed.; John Wiley & Sons: New York, 1977; Coll. Vol. 1, 295.

POLYIMIDES (Containing Ether Linkages; Adhesive Properties)

Katsuya Asao and Motonori Mitoh
Osaka Prefectural Industrial Technology Research Institute

Polyimides are known for their superior thermal, thermooxidative, and chemical resistance. Therefore, they are used as adhesives, films, composite matrices, coatings, membranes, and so on in severe environments.[1]

In this study, aromatic polyimides containing ether linkages were prepared. These polyimides were thermoplastic and indicated excellent thermal stability. For instance, a polymer designated the PET-R/BTDA polyimide in **Table 1** exhibited interesting thermal and distinguished adhesive properties. These polymers were prepared from commercially available 3,3',4,4'-benzophenonetetracarboxylic dianhydride (BTDA) and aromatic diamines containing ether linkages by the usual procedure. In this article, we describe thermal and adhesive properties of prepared polyimides.[2]

PROPERTIES

Thermal Properties of Polyimides

The polyimides obtained have good thermal stability as suggested by thermogravimetric analysis (TGA) where the temperatures of 5% weight loss (T_d) in air and in nitrogen were the range of 517°C from 544°C.

Glass transition temperature (T_g) and crystalline transition temperature (T_m) as determined by differential scanning calorimetry (DSC) on the polyimide films are listed in Table 1.

Adhesive Properties of Polyimides

The TPE-R/BTDA and ASD/BTDA adhesives exhibited superior strengths. All failure modes were primary cohesive.

CONCLUSIONS AND APPLICATIONS

Several kinds of polyimides were prepared by the reaction of BTDA with commercially available diamines containing ether linkages. These polyimides exhibited excellent thermal and thermooxidative stability. TPE-Q/BTDA, TPE-R/BTDA, BAPB/BTDA, and DPE/BTDA were semi-crystalline with T_g from 241°C to 357°C and T_ms between 349°C and 485°C. Especially the TPE-R/BTDA showed interesting thermal properties and high adhesive properties.

These polyimides would be applied to films, composite matrices, membranes, carbon materials, engineering materials, and microparticles otherwise adhesive.

REFERENCES

1. Wilson, D.; Stenzenberger, H. D.; Hergenrother, P. M. *Polyimides*; Blackie & Son Ltd.: London, 1990.
2. Asao, K.; Hirohata, T.; Nishimura, H. *J. Adhesion Soc. Japan* **1993**, *29*, 395.

TABLE 1. Polyimides Containing Ether Linkages

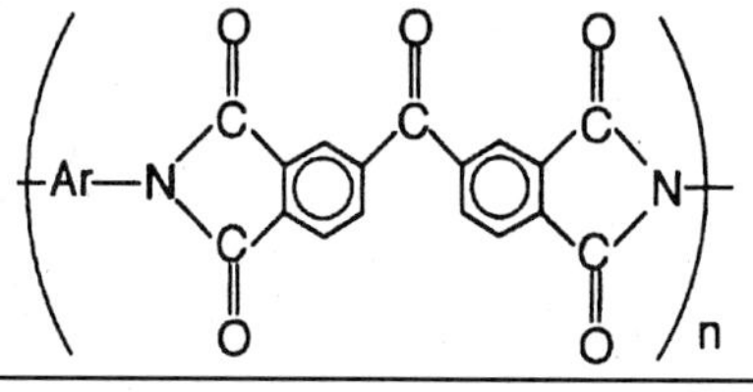

Polyimide designation	Ar	Polyamic acid ηinh (dL/g)	Td(C°) In nitrogen	Td(C°) In air	Polyimide Tg1 (°C)	Polyimide Tg2 (°C)	Polyimide Tm1 (°C)	Polyimide Tm2 (°C)
TPE-Q/BTDA	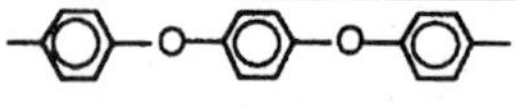	0.80	530	517	ND	271	453	ND
TPE-R/BTDA	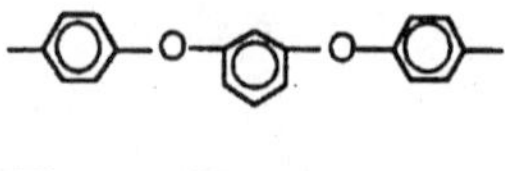	0.56	537	522	357	241	426	349
BAPB/BTDA	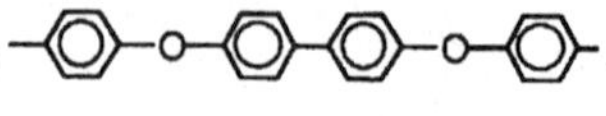	0.87	541	524	ND	272	470	ND
DPE/BTDA	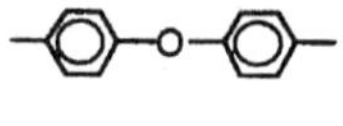	0.71	544	542	ND	293	485	ND
ASD/BDTA	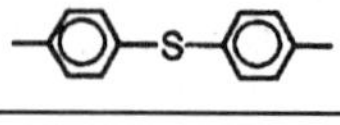	0.42	535	532	291	298	ND	ND

ND = not detected.

POLYIMIDES (from Cyclic Unsaturated Imides)

Thanun M. Pyriadi
Department of Chemistry
College of Science
University of Baghdad

HOMOPOLYMERIZATION

Homopolymerization of cyclic imides containing vinyls are considered, such as N-substituted maleimides, itaconimides, or citraconimides.[2,3] Numerous types of such monomers have been prepared in our laboratories, and most of them polymerized free radically or anionically resulting in low molecular weight polyimides. This is in contrast to the step growth polycondensation of primary diamines with dianhydrides, which yields high molecular weight polyimides.[6]

It is interesting to notice that maleimide and N-substituted maleimides readily polymerize free radically and anionically while maleic anhydride does not.

Citraconimide and N-substituted citraconimides did not homopolymerize, but could enter cyclopolymerization reactions owing to pi-pi interactions of two vinyls.[1] N-substituted itaconimides, however, readily polymerized radically and anionically.[5]

AMIC ACID POLYMERIZATION

N-substituted maleamic acids could not be homopolymerized free radically or ionically via the vinyl functions.[2] This was expected since the vinyls are 1,2-substituted. However, N-substituted amic acids may enter polycondensation polymerization through their caroxyl groups provided that there is present either another carboxyl group or hydroxyl group on the same monomer.[3,4]

MODIFICATION OF POLYIMIDES TO ION-EXCHANGE RESINS

A new successful attempts have been made to convert or modify some specific N-substituted imides to serve as ion-exchange resins.[7,8] For example, crosslinked poly(*N*-phenylmaleimide), and poly(*N,N*-hexamethylene-bis-maleimide) were prepared by free radical polymerization of the corresponding imides in benzene. The precipitated polyimides were partially hydrolyzed to obtain some pendant carboxylic groups on the polymeric chains. The last compounds were found to be useful cationic exchange resins of good capacities.

REFERENCES

1. Pyriadi, T. M.; Mutar, E. H. "Synthesis and free radical polymerization of N-substituted citraconimides," *J. Polym. Sci., Polym. Chem.* **1980**, *18*, 2535.
2. Pyriadi, T. M. "Chemistry of imides and isoimides," Ph.D. Thesis, The University of Akron, Akron, Ohio 1970.
3. Pyriadi, T. M.; Smoka, N. "Homopolymerization of four new hydroxy-amic acids and polycondensation of glycerol with dibasic amic acids," *J. Macromol. Sci., Chem. A* **1987**, *24*(7), 829.
4. Pyriadi, T. M.; Smoka, N. "Polycondensation of some dibasic amic acids with diols," Proceedings of Fourth Congress of the Research Council, Baghdad, Iraq; October 23, 1986.
5. Pyriadi, T. M.; Fraih, M. "Synthesis and polymerization of N-arylitaconimides free radically and anionically," *J. Macromol. Sci., Chem. A* **1982**, *18*(2), 159.
6. Stevens, M. P. in "Polymer Chemistry," Addison-Wesley: Reading, MA, 1975.
7. Pyriadi, T. M.; Farhan, M. "Preparation of ion-exchange resins from poly(N-phenyl), and N,N'-hexamethylene-maleimides," Iraqi Patent 2003. Presented at the IUAPC symposium, Genoa, Italy, May 1987. Proceedings; paper no. 10.5, p 276.
8. Pyriadi, T. M.; Ahmad, A. "Preparation of new ion-exchange resion from grafting N-allyl diphenamic acid on poly(vinyl alcohol) followed by sulfonation of the phenyl rings," Iraqi Patent 2507, 1993.

POLYIMIDES (from *N*-Silylated Aromatic Diamines)

Yoshio Imai*
Department of Organic and Polymeric Materials
Tokyo Institute of Technology

Yoshiyuki Oishi
Department of Applied Chemistry and Molecular Science
Iwate University

Recently, we have investigated in detail the synthesis of polyimides and poly(amide-imide)s by the *N*-silylated diamine method and have obtained interesting results compared with the conventional method. We utilized this method further for the synthesis of a variety of polyimides such as random and block copolyimides, tetraphenylthiophene-containing polyimides, hydroxyl-containing polyimides, and polydimethylsiloxane-poly(imide) block copolymers. In addition, this method was successfully applied to the preparation of ultrathin films of aromatic polyimides and poly(amide-imide)s through the vapor deposition polymerization.

PREPARATION OF POLYIMIDES

N-trimethylsilylated amines react with various electrophiles such as acid anhydrides and chlorides.[3] The reaction of an *N*-silylated amine with a cyclic anhydride afforded the intermediate ring-opened adduct, silylated amic acid, which eliminated trimethylsilanol at an elevated temperature giving a cyclic imide.

The application of this reaction to the synthesis of aromatic polyimides was reported by Klebe in the patent literature.[1] The ring-opening polyaddition of *N*-silylated aromatic diamines to aromatic tetracarboxylic dianhydrides afforded poly(amic acid trimethylsilyl esters), which in turn were converted thermally to aromatic polyimides with the elimination of trimethylsilanol.

We have reinvestigated the polyimide-forming reaction starting from *N,N'*-bis(trimethylsilyl)-substituted aromatic diamines and aromatic tetracarboxylic dianhydrides (**Equation 1**).[4]

The presence of trimethylsilyl groups provides solubility for the silylated diamines and silylated poly(amic acid)s in nonpolar solvents. Colorless, transparent, and flexible films of the silylated precursor polymers were prepared by casting directly from the reaction solutions. Desilylation of the silylated polymers with an alcohol such as methanol proceeded smoothly, affording the corresponding poly(amic acid)s.

*Author to whom correspondence should be addressed.

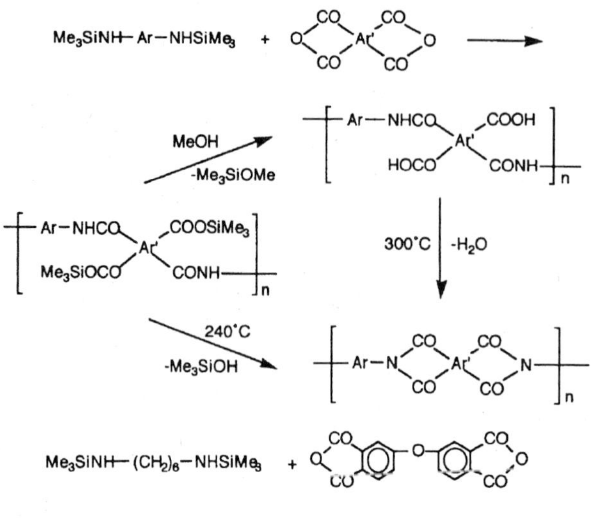

1

The films of the silylated precursor polymers were converted by subsequent thermal treatment at 240°C for 1 hour to yellow, transparent, and tough films of the aromatic polyimides without the formation of side products such as isoimide.

The *N*-silylated diamine method starting from *N,N'*-bis(trimethylsilyl)-*p*-phenylenediamine and pyromellitic dianhydride gave a transparent film of poly[*N,N'*-(*p*-phenylene)pyromellitimide] with lower crystallinity, compared with the film from the conventional diamine method. Bulkiness of the trimethylsilyl group of the silylated precursor polymer may reduce close-packing ability of the precursor polymer chains, and hence the loose-packing may remain mostly unchanged in the poly(imide) chains after the thermal cyclization. The other characteristics such as chemical resistance, thermal stability, and tensile properties of the polyimide films prepared by the *N*-silylated diamine method were almost the same as those of the polyimide films obtained by the conventional method.

Korshak and his group have reported that polyimides containing aliphatic groups, which were difficult to prepare in high molecular weight by the conventional diamine method, could be obtained satisfactorily starting from *N*-silylated aliphatic diamines and aromatic dianhydrides.[2]

It is very difficult to prepare polydimethylsiloxane-polyimide block copolymers having high molecular weights, because most of the polymerizations could not proceed in homogeneous solutions owing to the large difference of solubility between polydimethylsiloxane (PDMS) and polyimide. The *N*-silylated diamine method was successfully utilized for the preparation of these block copolymers starting from telechelic amine-terminated PDMS oligomer (PDMS-diamine).[5]

The polyimide synthesis via the silylation method is applicable to vapor deposition polymerization, which means a wholly dry process for the preparation of ultrathin films of aromatic polyimides on appropriate substrates.

PREPARATION OF POLY(AMIDE IMIDE)S

The *N*-silylated diamine method for the preparation of poly(amide imide)s has several advantages over the conventional method: (1) High-purity *N*-silylated diamines can be obtained simply by distillation. (2) *N*-silylated diamines are readily soluble in organic solvents, and hence a variety of solvents can be used as reaction media. (3) *N*-silylated aromatic diamines are more reactive than the parent diamines toward aromatic acid chlorides in polar solvents, and consequently higher molecular weight poly(amide-amic acids) can be obtained. (4) The polymerization proceeds under neutral reaction conditions with the elimination of trimethylsilyl chloride, and therefore the polymerization solution, which does not contain any hydrochlorides, can be used directly for the solution casting to prepare polymer films. (5) The trimethylsilyl chloride recovered from the polymerization system can be recycled as a silylating agent for aromatic diamines.

In the case of preparation of an ultrathin film of aromatic poly(amide-imide) by the silylation method through vapor deposition polymerization, the evaporation of *N,N'*-bis-(trimethylsilyl)-*p*-phenylenediamine together with 4-(chloroformyl)phthalic anhydride under vacuum afforded an ultrathin film of the poly(amide amic acid trimethylsilyl ester) on a substrate, which in turn was converted to the poly(amide imide) by thermal treatment.[6] This *N*-silylated diamine method had the advantage of causing no corrosion of a metal substrate by the trimethylsilyl chloride by-product, unlike the hydrochloric acid generated by the conventional method.

REFERENCES

1. Boldebuck, E. M.; Klebe, J. F. U.S. Patent 3 303 157, 1967; *Chem. Abstract* **1967**, *66*, 96125f.
2. Korshak, V. V.; Vinogradova, S. V.; Vygodskii, Y. S.; Nagiev, Z. M.; Urman, Y. G.; Alekseeva, S. G.; Slonium, I. Y. *Makromol. Chem.* **1983**, *184*, 235.
3. Klebe, J. F. *Adv. Org. Chem.* **1972**, *8*, 97.
4. Oishi, Y.; Kakimoto, M.; Imai, Y. *Macromolecules* **1991**, *24*, 3475.
5. Kajiyama, M., Oishi, Y.; Kakimoto, M.; Imai, Y. *Kobunshi Ronbunshu* **1992**, *49*, 235.
6. Takahashi, Y.; Iijima, M.; Oishi, Y.; Kakimoto, M.; Imai, Y. *Macromolecules* **1991**, *24*, 3543.

POLYIMIDES (from Tricyclic Heteroaromatic Dyes)

Himansu M. Gajiwala*
Department of Chemistry
Tuskegee University

Robert Zand
The Macromolecular Research Center
The University of Michigan

Studies of the synthesis and properties of polyimides that incorporate tricyclic heteroaromatic compounds is a major research effort in our laboratories.[1–4] Although polyimides represent an extensively studied class of thermally stable polymers, only a few polyimides that incorporate tricyclic heterocycles have been investigated.[5–12] The formation of a ladder structure that utilizes aromatic or heterocyclic rings in the main chain frequently confers enhanced stability to the polymer, since such a structure is inherently more stable than a single stranded chain.[13,14] On this rationale, we have synthesized several new polyimides that incorporate different tricyclic heteroaromatic compounds. As all the selected compounds are commercially available dyes, the resultant polyimides are expected to show good optical properties as well as exhibit good thermal properties.

Computer simulation work has shown that tricyclic heteroaromatic diamines like thionine, proflavine, and acridine yellow are planar systems because of the resonance throughout the molecule. As a result, for these polymers, the electron delocalization should result in a higher heat stability and good mechanical properties as well as better solubility characteristics. Thionine is a charge diamine, proflavine is a diamine containing a nitrogen heteroatom, and the diamine acridine yellow is similar to proflavine but possesses two additional methyl groups ortho to the amine linkages. The heterocyclic systems are expected to give better processing properties to the resultant polyimides. The charged system of thionine and the methyl groups of acridine yellow are expected to give enhanced solubility to the resultant polyimides by expanding the polyimide chains.

PROPERTIES

Reaction

Polyimides by the Two-Step Low-Temperature Reaction

The rate of polyamic acid formation from thionine and pyromellitic dianhydride (PMDA) was found to increase when the concentration of the reactants was increased. However, the rate of the polymerization reaction was found to increase with a rise in temperature for thionine. This type of exception to the general trend is possible when two different, very unreactive monomers do not condense exothermically with each other at ambient temperature. In these cases, the molecular weight of the polyamic acid increases as the temperature is increased.[15–17] This result demonstrates that tricyclic diamines thionine is a relatively unreactive diamine. However, thionine was the only diamine found to react with PMDA at ambient temperature, unlike proflavine and acridine yellow. This may be due to the fact that thionine, being a charged diamine, is slightly more reactive compared to the uncharged tricyclic heteroaromatic diamines, resulting in reaction with highly reactive dianhydride (like PMDA) at ambient temperature.

The polyamic acid obtained from thionine and PMDA can be converted to polyimide either by thermal or by chemical means. Imide linkage formation, by the chemical cyclization using acetic anhydride system, replaces the chloride counter ion by an acetate counter ion, as proven by the model compound characterization using NMR and mass spectral data.

Polyimides by a Single-Step High-Temperature Reactions

All polyimides were synthesized by reacting one of the heteroaromatic diamines with one of the indicated dianhydrides at the appropriate elevated temperature.

Structure-Property Relationship of Different Polyimides

An assessment of the polyimides made with thionine, (with the chloride counter ion), proflavine, or acridine yellow as diamines and different dianhydride moieties, reveals that the polyimide of each diamines has its unique solubility characteristics but nearly the same thermal stability.

Results indicate that the polyimide obtained from proflavine and PMDA shows better solubility than the polyimide obtained from acridine yellow and PMDA. This descrepancy may be due to the fact that the introduction of the methyl group, ortho to the amino linkage of the diamine, hinders the free rotation of the imide nitrogen of the polyimide in the case of acridine yellow making the polymer more rigid, and less soluble.

Polymer chain rigidity, interchain separation, and charges on the polymer backbone all play an important role in determining the solubility of the given polymeric chain.

Liquid Crystalline Property

Examination of the polymers for liquid crystalline behavior at dilute polymer concentrations led to the finding that the polyimide obtained from benzophenone tetracarboxylic dianhydride and proflavine exhibited behavior that we interpret as lyotropic liquid crystalline behavior.

APPLICATIONS

These polyimides can be utilized in several applications, and some of them are mentioned below. All these polyimides, being thermally stable and having relatively low dielectric properties, can be used in high-temperature applications in the aerospace and electronics industries. Most of these polyimides have a tendency to form coils and are resistant to common chemicals. These properties can be utilized in semipermeable membrane applications. In addition, as all these polyimides were synthesized from dyes, they can be used in the non-linear optics and electro-optics area.

*Author to whom correspondence should be addressed.

REFERENCES

1. Gajiwala, H. M. Synthesis and Characterization of Thermally Stable and/or Conductive Polymers, Ph.D. Thesis, University of Michigan, 1992.
2. Gajiwala, H. M. *Macromolecules* **1993**, *26*, 5976.
3. Patil, R. D. Synthesis and Characterization of Polyimides from Acridine Yellow, M.S. Thesis, Tuskegee University, 1994.
4. Patil, R. D.; Jackson, J. L.; Gajiwala, H. M. *Polymer Preprints* **1994**, *35*(1), 507.
5. Cassidy, P. E. *Thermally Stable Polymers, Synthesis and Properties*; Marcel Dekker: New York, 1980.
6. Critchley, J. P.; Knight, G. J.; Wright, W. W. *Heat-Resistant Polymers—Technologically Useful Materials*; Plenum: New York, 1983.
7. Mital, K., Ed., *Polyimides: Synthesis, Characterization and Applications*; Plenum: New York, 1984; Vols. 1 and 2.
8. Bessonov, M. I.; Koton, M. M.; Kudryavtsev, V. V.; Laius, L. A. *Polyimides—Thermally Stable Polymers*; Plenum: New York, 1987.
9. Wilson, D.; Stenzenberger, H. D.; *Polyimides*: Chapman and Hall: New York, 1990.
10. Nium, K.; Nakamichi, K.; Takatura, R.; Toda, F.; Uno, K.; Iwakura, Y. *J. Polym. Sci.—Polym. Chem. Ed.* **1979**, *17*, 2371.
11. Nium, K.; Nakamichi, K.; Toda, F.; Uno, K.; Hasegava, M.; Iwakura, Y. *J. Polym. Sci.—Polym. Chem. Ed.* **1980**, *18*, 377.
12. Nium, K.; Hirohashi, R.; Toda, F.; Hasegawa, M.; Iwakura, Y. *Polymer* **1981**, *22*, 649.
13. Overberger, C. F.; Moore, J. A. *Adv. Poly. Sci.* **1970**, *7*, 113.
14. Bell, V. L.; Stump, B. L.; Gager, H. *J. Polym. Sci.; Polym. Chem. Ed.* **1976**, *14*, 2275.
15. Wilson, D.; Stenzenberger, H. D., Eds., *Polyimides*; Chapman and Hall: New York, 1990; Chapter 1.
16. Young, P. R.; Chang, A. C. *SAMPE Proc.* **1983**, *28*, 824.
17. Harris, F. W.; Hsu, S. L.-C. *High Performance Polymers* **1989**, *1*, 3.

POLYIMIDES (Introduction and Overview)*

Cyrus E. Sroog
Polymer Consultants Incorporated

IN GENERAL

Achievement of a successful route to high-molecular-weight, aromatic polyimides was described first in Endrey in 1962 and in successive DuPont patents.[1,2] In this method, the synthesis was conducted in two stages, first to prepare a soluble poly(amic acid) that was then converted to the desired polyimide. Subsequent publications in the U.S. and Great Britain followed.[3–5]

The preparation of poly(amic acid) proceeds by reaction of a dianhydride and a diamine at ambient temperatures in polar, aprotic solvents such as NMP, DMAc, or DMF. The poly(amic acid) is then dehydrated to the polyimide as illustrated in the following equations (**Scheme I**):

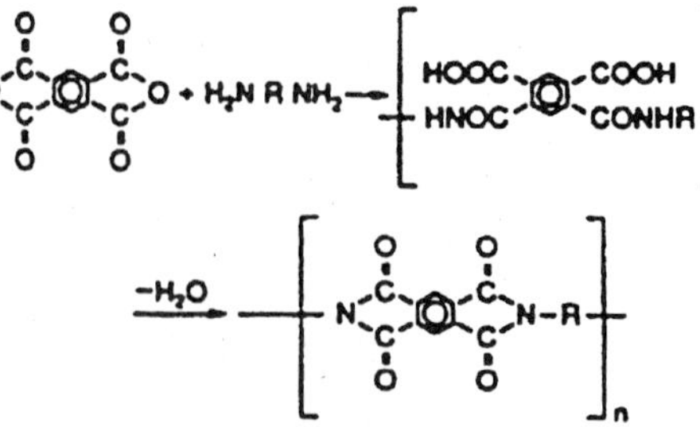

I

The two-step reaction initiated the enormous research and industrial activity with polyimides. It made possible the shaping of the soluble, tractable poly(amic acid) and then its conversion to the insoluble, intractable polyimide.

*Significant portions of this article appeared earlier in *Progress in Polymer Science*, Vol. 16, pp 561–694, 1991. Permission to utilize this material is gratefully acknowledged.

End-capping of the amic acid chain can be accomplished, for example, by incorporation of phthalic anhydride in the polymerization recipe. It is also possible to end-cap with such compounds as maleic anhydride or nadic anhydride to provide a site for subsequent crosslinking.

A recent important paper by Dr. Echigo of Unitika Ltd.[6] has described the use of a mixed solvent of THF/Methanol. This results in easier removal of solvent from cast amic-acid and lower viscosity of poly(amic acid) solution compared with conventional solvents.

PROPERTIES OF AROMATIC POLYIMIDES

Hydrolytic Stability

The wholly aromatic polyimides are known widely for the unique combination of thermooxidative stability with mechanical and electrical properties useful over an extremely wide temperature range. Hydrolytic stability is also very important; poor hydrolytic stability could render the remaining properties essentially useless or of limited significance.

Radiation Resistance

The radiation resistance of aromatic polyimides is outstanding.[5]

Crystalline and Amorphous Transitions of Polyimides

Crystalline Transitions

Polyimides as a polymer class should be very crystalline, possessing planar ring structures, extensive opportunity for charge transfer interaction, and strong capacity for general chain-to-chain polar interaction. Nevertheless, polyimides are derived from highly disordered soluble precursors, and conversion to imide takes place in a formed structure (for example, film, fiber, etc.) with high and increasing viscosity (during conversion). Aromatic polyimides as a group tend therefore to be amorphous or to have very low crystallinity.[7]

Amorphous Transitions

The generalizations regarding crystalline melting point apply also to T_g. A single aromatic ring (for example, PMDA) will produce a polyimide with a higher T_g than that from a multi-ring dianhydride or those from dianhydrides bridged by non-ring connecting groups. In the case of diamines, the same relationship prevails.

Polyimide Solubility

Aromatic polyimides as a class are inherently insoluble in common solvents. Alkyl group substitution enhances solubility.

Introduction of siloxane structures contributes to enhanced solubility.

Polyimide Color

Aromatic polyimides have been recognized from the earliest days of their invention as highly colored polymers with colors ranging from yellow to deep red. Achievement of reduced color requires first rigorous purification of starting materials, but beyond monomer purification, there are significant structural considerations related to color reduction and elimination. These are inclusion of bulky electron withdrawing groups such as SO_2 or $C(CF_3)_2$, and groups such as oxygen or sulfur.[7,8]

POLYIMIDE APPLICATIONS

Films

Films of aromatic polyimides represent the largest end-use area for the polyimides. Polyimide films are prepared by casting solutions of poly(amic acid) precursor onto a substrate and then converting to the polyimide film.

Properties of Polyimide Films

Polyimides have been recognized from the earliest days of their invention for their outstanding thermal stability both in air and anaerobically. In addition to thermal durability, the early references to polyimide films described and emphasized their breadth of useful properties over an extremely broad temperature range.

These properties in addition to thermal durability, included excellent mechanical properties from 4°K to 573°K, outstanding electrical properties with stability over wide ranges of relative humidity, insensitivity to solvents, and excellent radiation resistance.

Polyimide Fibers

Early reference to polyimide fibers appears in work by Irwin.[9]

Separation Applications

A significant property of polyimides is their utility as separation membranes to permit selective permeation of gases and water vapor.

ELECTRONIC APPLICATIONS

Circuit Preparation

Background

Polyimides have played an increasing role in electronics for a variety of applications owing to key polyimide properties. These include thermal stability, capability of solution coating of soluble poly(amic acid) precursors, toughness of polyimide film after final curing, excellent dielectric properties, and excellent dimensional stability of the final converted polyimide. Use includes base sheet film for flexible circuitry, planarization coatings for computer chips, interlayer dielectrics, and alpha particle barriers. Crosslinked polyimides have found significant utility in circuit boards. An important response to the increased complexity of circuits and the cost burden of multi-step processing has been to use photosensitive polyimide precursors directly.

Yoda and Miramoto[10] have pointed out that a useful photosensitive system has three key elements: (a) the polyimide precursor should be soluble in organic solvents and processable for microcoating on a substrate, (b) a sharp contrast of polymer solubility should exist between the photosensitive precursor and the crosslinked portion and (c) high resolution should result after photoexposure.

Polyimide Polymer Properties in Electronic Applications

In terms of polyimide properties, reliability translates to need for reduced coefficients of expansion, reduced sensitivity to moisture, reduced dielectric constant of the polyimide, and increased purity; these are the focus of principal developing demands placed on polyimide materials. There are, in addition, several further property needs that, while less critical, are still important. These include: increased thermal durability; adherability to copper, ceramics, and other materials, as well as self-adhesion, and enhanced chemical resistant to solvents, copper etchants and a variety of other materials.

Polyimide Blends

Molecular Composites of Polyimides

There has been increasing activity in recent years in preparation of molecular composites of polyimides of considerably varying structure. The work has strong product implications in widely differing fields such as electronics and reinforced aerospace composites. In each case, a "soft component" is needed to provide adhesion to, for example, a silicon wafer in electronics or to a reinforcing fiber in an aerospace composite. Also required is a "hard component" to provide certain mechanical properties, such as low TCE, in electronics and high modulus/tensile strength for aerospace composites.

Polymer Blends

Jaffe, in a study of polyimide based blends,[11] described polyimide/benzimidazole (PBI) blends and polyimide/polyimide blends. The results reported indicated possibilities for miscible blends over a wide composition range.

Reinforced Composites

For high performance composites, an area of great importance for polyimides, composites consist of fiber dispersed in a polymer matrix.

Advanced composites offer a unique strength to weight ratio, making them unusually attractive to the aircraft industry. The potential for weight reduction ranges from the present level of 10% to projected reduction of 50%. Polyimides useful in advanced composites include capped polyimides (maleimide, nadimide, and acetylene capped) as well as conventional polyimides.

Polyimide Adhesives

The broad utility of polyimides in many cases requires the use of adhesives. Polyimide use can be limited if the available adhesives are inadequate in properties to match the outstanding property combinations available in the polyimide substrates.

To serve as an adhesive, a polyimide-based material must retain a high level of thermal and thermo-oxidative stability, have resistance to solvents, and exhibit very low moisture absorption. The adhesives, ideally, will also not be the source of deformation during end-use. In effect, the polyimide adhesive system must be capable of flow under conditions required for development of adhesion, but in most cases should have very limited flow under conditions of use.

Polyimide Foams

Product areas for foams are those that require the combination of polyimide thermal durability and high-temperature mechanical strength with foam-type properties. Application areas include aircraft/aerospace, where foam's light weight is of value in wall and ceiling panels, sections of bulkhead, acoustical and thermal insulation, and vibration insulation. Other product areas include electronic instrument protection, marine uses, and applications such as cryogenic blankets.

Polyimides with Liquid Crystal (LC) Characteristics

There has long been speculation that polyimides could in some structures exhibit LC properties; an increase of polyimide research in this area has developed in recent years.

Recent research by Kricheldorf et al. has resulted in an LC poly(imide/ester)[12] in which all components are aromatic. The polymer retains nematic properties to 500°C, with T_g 187°C.

REFERENCES

1. Endrey, A. L. Can. Patent 645,073, E. I. DuPont & Co. 1962; Can. Patent 659 328, 1963.
2. Endrey, A. L. U.S. Patent 3 179 630, 1965; U.S. Patent 3 179 631, 1965.
3. Jones, J. L.; Ochynski, F. W.; Rackley, F. A. Chem. Ind. 1686 1962.
4. Bower, G. M.; Frost, L. W. *J. Polym. Sci.* **1963**, *A-1*, 3135.
5. Sroog, C. E.; Endrey, A. L.; Abramo, S. V. et al. *Polym. Sci.* **1965**, *A-3*, 1373.
6. Echigo, Y. et al. Proceedings 5th International Conference on Polyimides, Ellenville, NY, November 25, 1994; p 93.
7. Sroog, C. E. *Prog. Polym. Sci.* **1991**, *16*, 561–694.
8. St. Clair, A. K.; St. Clair, T. L. U.S. Patent 4 595 548, NASA, 1986.
9. Irwin, R. S. U.S. Patent 3 415 782, E. I. DuPont & Co., 1968.
10. Yoda, N.; Miramoto, H. *J. Macromolek Sci. Chem.* **1984**, *A21* (13 and 14), 1641.
11. Jaffe, M. Symp. Recent Advances in High Performance Polymers, San Diego, CA, January 1990.
12. Kricheldorf, H. et al. *J. Polym. Sci., Pt. A, Polym. Chem.* **1993**, *31*, 279–282.

POLYIMIDES (Negative-Type Photosensitive Precursors)

J. Tanaka, F. Shoji, and Fumio Kataoka*
Production Engineering Research Laboratory
Hitachi Ltd.

M. Kojima
Yamazaki Works
Hitachi Chemical Company Ltd.

In this article, the properties of negative-type, photosensitive, polyimide precursors using aromatic azide photocrosslinkers are described. As shown in **Scheme I** azide derivatives (**2**) used in the present precursors have a dialkylamino group at one terminal end of the molecule for incorporating the photosensitive azido group to polyamic acids (**1**), which are precursors of polyimides. The azide group is incorporated into the polyamic acids by a salt formation between the carboxyl group of the polymers and the amino group of the photocrosslinkers giving the photosensitive precursors (**3**).

I

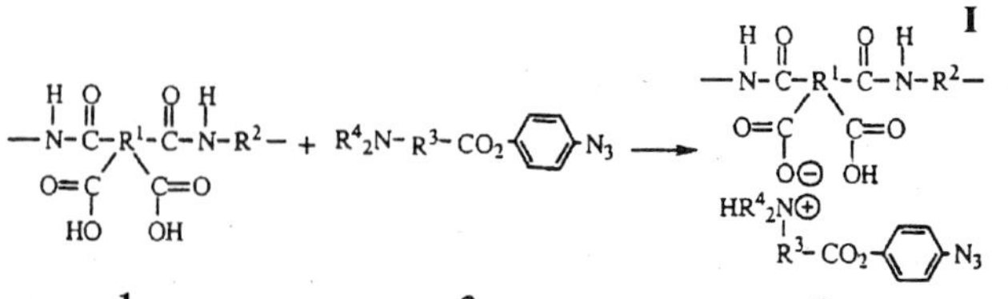

1 2 3

The precursor film formed on the substrate is soluble in solvent and is converted to an insoluble material by photocrosslinking after UV exposure. The unexposed areas of the film are eliminated in the development step, resulting in patterns of the precursor. The resultant pattern-formed layer is then heated to remove the photosensitive group and then imidized to give patterns of polyimides.

PREPARATION

Azide Photocrosslinkers

Azide photocrosslinkers used in the photosensitive precursors are easily prepared through a reaction between acid chlorides of aromatic azides with dialkylaminoalcohols followed by treatment with sodium hydroxide. Aromatic azide derivatives containing both a *p*-azidobenzoyl group and a *p*-azidocinnamoyl group were synthesized by this method.

Photosensitive Precursors

The synthesis of polyamic acid prepared from 3,3', 4,4'-biphenyltetracarboxylic acid dianhydride and 4,4'-diaminodiphenylether was used as a base polymer of the photosensitive precursor.

PROPERTIES

Process for Patterning

A photosensitive precursor solution deposited on a substrate is dried at the prebaking step using a batch convection oven or a hot plate resulting in the formation of a photosensitive film.

Two kinds of exposure systems, deep-UV and mid-UV lithography, can be used for the present materials depending on the composition of the photosensitizers.

The photocrosslinked precursor resulting from UV irradiation is converted to polyimide by thermal curing.

Sensitization at Longer Wavelengths

These polyimide precursors are not limited to use with Deep-UV light sources. It is possible to sensitize at longer wavelengths by adding a photosensitizer.

Resulting Polyimides

Mechanical properties and heat resistivity of polyimides derived from their photosensitive precursors are usually affected

*Author to whom correspondence should be addressed.

by photosensitive additives or photoreactive side chains. Effects on mechanical properties such as flexibility and tensile strength are very small and this indicates that the polyimide structure is preserved intact during the curing process. These characteristics are adequate for use as an insulator in various applications.

POLYIMIDES (Via Interfacial Polyfunctional Condensation)

Leo Wang Chen*
Institute of Materials Engineering
National Taiwan University

Y. T. Chern
Institute of Chemical Engineering
National Taiwan Institute of Technology

In some studies,[1] from 3,7-bis(2,2-dicyanomethyliden)-pyromellitide and diamine under ambient temperature imide structure can form directly. Quite a lot of research has been done concerning modification of the heat resistance and the processing properties of polyimide. Yet, so far, synthesis of polyimide by interfacial polymerization was rare. Also, the method of interfacial polymerization can be applied to preparing composite membranes, which provides higher selectivity without severely reducing membrane flux.[2–6] This method is attempted for preparing polyimide membranes in this study.

Interfacial polycondensation has been reported in the literature. Most of these studies[7–9] characterize the interfacial reaction of difunctional reagents and give no report on the reaction of polyfunctional reagents. In this research,[10–13] the interfacial polyfunctional of 1,2,4,5-benzenetetracyl chloride (BTAC) dissolved in dichloromethane and the diamine dissolved in water was studied. By altering various factors, such as adding a phase transfer agent, changing the temperature, adjusting pH value, applying agitation, and changing the composition, the chemical structure of the films is controlled.

Also, we set different reaction conditions and employed ATR for analyzing the chemical structures on both sides of the film during the reaction. Heat-resistant polyimide films were prepared therefrom. Composite membranes were prepared by the interfacial polycondensation of a water-soluble diamine, (4,4'-methylene dianiline (MDA), or ethylene diamine (EDA) with an organic-solvent (hexane)-soluble BTAC on top of a porous polysulfone support. The composite membranes performed well for gas separation application, for example, CO_2 and CH_4 gas mixtures.

CURING REACTION OF POLY(BTAC-MDA) FILM

It is found that the heat resistance of the film that had been through the first thermal treatment stage, as well as the thermal cracking temperature of the film, is lower than 350°C. The thermal cracking temperature is taken by measuring the temperature at 5% weight loss. Because the film contains amide groups, the heat resistance of the film is poor. When the film has been through the secondary stage of thermal treatment, its thermal cracking temperature is nearly 500°C. Summarizing the

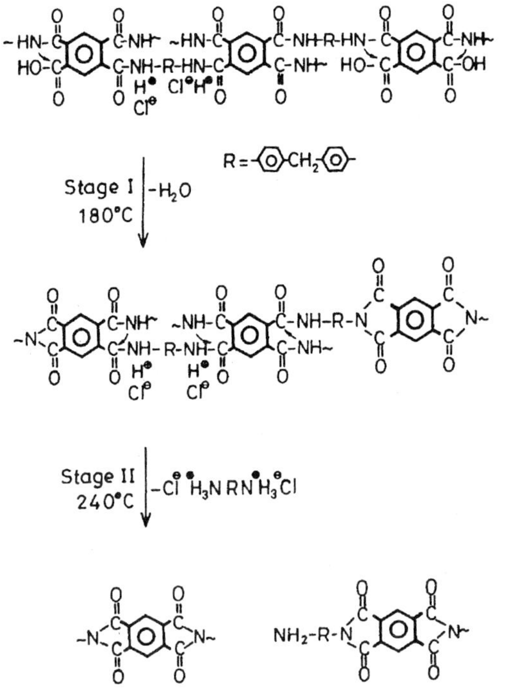

I

above results, the heat absorption of the band at 220–300°C is the imidization of the adjacent carbon on the benzene ring linked with functional groups of amide. The typical reaction is shown in **Scheme I**.

THE GAS PERMEABILITY OF THE HEAT-TREATED POLY(BTAC-MDA) COMPOSITE FILMS

Stiffer polymers generally have a higher mobility (diffusivity) selectivity, because they behave more like "molecular sieves."[14] Such polymers are able to discriminate between penetrant molecules of different sizes and shapes. This is evident from the fact that the permeability of the Poly(BTAC-MDA) composite film, treated at 135°C in a nitrogen atmosphere for 4 hours, versus different gases decreases in the order $P(CO_2) > P(O_2) > P(N_2) > P(CH_4)$.

It is also known that increasing "kinetic" diameters of the penetrant molecules will decrease the diffusion coefficients of these gases in glassy polymers. Hence, the overall permselectivity of the Poly(BTAC-MDA) composite film treated at 135°C in a nitrogen atmosphere for 4 hours is probably controlled by the mobility selectivity. The increase of the $\alpha^*(O_2/N_2)$ value of the films was less than that of the $\alpha^*(CO_2/CH_4)$ value. This may be due to the fact that the difference between the molecular kinetic diameter between CO_2 (3.3) and CH_4 (3.8) is greater, but is smaller between O_2 (3.46) and N_2 (3.64). It is reasonable considering that the large differences in the size of the penetrant molecules and the greater molecular-sieving effect.

THE GAS PERMEABILITY OF THE HEAT-TREATED POLY(BTAC-EDA) COMPOSITE FILMS

The permselectivity ($\alpha^*(CO_2/CH_4)$ or $\alpha^*(O_2/N_2)$) of the Poly(BTAC-EDA) composite films increased after heat treatment. The reason for the increase in permselectivity was similar to the reason mentioned in the above Poly(BTAC-MDA)

*Author to whom correspondence should be addressed.

composite film system. But the permselectivity ($\alpha^*(CO_2/CH_4)$ and $\alpha^*(O_2/N_2)$) was not high. This is probably due to the Poly(BTAC-EDA) composite film still containing considerable amounts of soft, aliphatic amide chains present after thermal treatment.[10,11] Softer polymers generally have a lower diffusivity selectivity. After heat treatment, the order of the permeation coefficient of this composite film is $P(CO_2)>P(O_2)>P(CH_4)>P(N_2)$. And, the composite films being more permeable to CH_4 than to N_2, this is probably due to the presence of a considerable quantity of aliphatic chains ($-CH_2-CH_2-$) in the Poly(BTAC-EDA) composite film caused by particularly excellent solubility for methane.

REFERENCES

1. Kim, J. H.; Moore, J. A. *Macromolecules* **1993**, *26*, 3510.
2. Lonsdale, H. K. *Polym. Eng. Sci.* **1985**, *25*, 1074.
3. Bartels, C. R. *J. Membrane Sci.* **1989**, *45*, 225.
4. Sugawara, S.; Lonno, M.; Saito, S. *J. Membrane Sci.* **1989**, *43*, 313.
5. Bartels, C. R.; Kreuz, K. L.; Wachtel, A. *J. Membrane Sci.* **1987**, *32*, 291.
6. Cadotte, J. E.; King, R. S.; Majerle, R. J. et al. *Macromol. Sci. Chem.* **1981**, *A15*(5), 727.
7. Tsai, H. B.; Lee, Y. D. *J. Polym. Sci. Polym. Chem.* **1987**, *25*, 3405.
8. Oishi, Y.; Kakimoto, M. A.; Iami, Y. *Macromolecules* **1988**, *21*, 547.
9. Podkoscielny, W.; Wdowicka, D. *J. Appl. Polym. Sci.* **1988**, *35*, 1779.
10. Chern, Y. T.; Chen, L. W. *J. Appl. Polym. Sci.* **1991**, *42*, 2543.
11. Chern, Y. T.; Chen, L. W. *J. Macromol. Sci. Chem.* **1991**, *A28*(1), 105.
12. Chern, Y. T.; Chen, L. W. *J. Appl. Polym. Sci.* **1991**, *42*, 2535.
13. Chern, Y. T.; Chen, L. W. *J. Appl. Polym. Sci.* **1992**, *44*, 1087.
14. Stern, S. A.; Shah, V. M.; Hardy, B. J. *J. Polym. Sci. Polym. Phys.* **1987**, *25*, 1263.

POLYIMIDES, HIGH PERFORMANCE FILMS

Dmitri Likhatchev and Ricardo Vera-Graziano
Instituto de Investigaciones en Materiales

Polyimides have achieved wide acceptance in the modern industry because of their excellent thermal and oxidative stability, resistance to practically all kinds of radiation and organic chemicals, and good dielectric and mechanical properties. These polymers make up approximately 40% of the commercially available, heat-resistant, nonmetallic materials where prolonged use at temperatures up to 300°C is required. In general, they are marked as high-performance films, molded parts, wire-coating enamels, adhesives matrix resins for advanced composites, foams, and fibers.[1]

SYNTHESIS OF LINEAR AROMATIC POLYIMIDES

The tremendous progress made in this field since the late 1950s has been due to the development of a two-stage polycondensation via soluble, fabricable, intermediate prepolymer, commonly called poly(amic acid).[4–9] A one-step, high-temperature polycondensation method for the synthesis of aromatic polyimides, soluble in phenolic solvents, was developed in the 1970–80s.[10–12] Nowadays, these are the only two principal methods for producing commercial high-performance polyimide films.

PROPERTIES OF LINEAR AROMATIC POLYIMIDES

The properties of linear aromatic polyimides vary dramatically depending on the presence and position of the so-called bridging group such as —O—, —CO—, —CH_2—, hexafluoropropane, etc., between the aromatic rings in the diamine and/or dianhydride moieties.

The macromolecules of the polyimides with no bridging group in their backbone have the only possible rod-like conformation. The correspondent polymer films are considered to be very tough and rigid, but brittle. They exhibit neither T_g, nor T_m up to temperatures of intensive thermal decomposition.

The polyimides with one, or several, bridging groups in the dianhydride moiety exhibit just slightly different levels of properties since free rotation on such a group is hindered by strong interaction of imide-aromatic units. Thus, such a backbone is still too rigid. However, when a bridging group is placed in the diamine moiety the rotational barrier for the attached segments of the chain was found to be no more than 5–7 kcal/mol.[14,15] Such macromolecules can have a set of possible conformations, and exist in the form of thermodynamic coils typical for conventional flexible-chain polymers (practically the only difference is the length of stiff units, and this is usually higher for the polyimides). In practice, the polyimides having these structures possess an optimum balance of thermal and mechanical properties. For this reason they have found a great number of practical applications.

Thermo-Oxidative Stability

Outstanding thermal and thermo-oxidative stability is common for the whole class of linear aromatic polyimides. For most of them the temperatures of 5% weight loss in air are in the range of 430–575°C.

In practice, the very high thermal and thermo-oxidative stability of aromatic polyimides allows them to retain a reasonable level of physical and mechanical properties after prolonged use at elevated temperatures.

Radiation, Hydrolytic, and Chemical Stability

Aromatic polyimides are extremely stable to high-energy radiation. They are superior to practically all commodity polymers with respect to the stability under irradiation by electron, proton, or neutron beams, γ-rays and UV light.

The hydrolytic stability of aromatic polyimides was studied by Sroog et al.[6] by the retention of film flexibility in boiling water.

All aromatic polyimides can be dissolved in strong mineral acids such as fuming nitric acid or concentrated sulfuric acid, especially after heating.[1–3] The resulting solutions are unstable and their viscosity decreases notably with time. Dilute acids do not affect aromatic polyimides.

Mechanical and Electrical Properties

Generally, the tensile modulus and tensile strength of non-oriented polyimide films decrease, and the elasticity increases, with increasing chain flexibility, for example, presence, position, and number of bridging groups.

It is important to note that the level of mechanical properties depends dramatically on the synthetic method and the conditions of the film preparation.

The films made from polymers with more rigid backbones demonstrate significantly higher tensile modulus, tensile strength, and T_g, but lower elasticity than their more flexible counterparts.

Polyimide Color

Aromatic polyimides range in color from amber-yellow to deep red, depending on the structure. The most probable origin of the color may be chromophoric units,[16] charge-transfer complexing,[17] presence of isoimide defect sites,[13] or impurities of starting materials. St. Clair et al. have shown that the inclusion of bulky electron withdrawing groups such as SO_2 or $C(CF_3)_2$ generally afford polyimides with less color or practically no color.

Health and Safety

Fully imidizated polyimides can be used in contact with food and in medical applications, usually with FDA approval. However, most of the starting materials used for their preparation require prudent handling. Dianhydride monomers are skin and eye irritants and several aromatic diamines (for instance, benzidine) are suspected carcinogens.

APPLICATIONS OF LINEAR AROMATIC POLYIMIDES

Films of aromatic polyimides represent the largest end-use area for these polymers. The unique combination of outstanding mechanical, electrical, and chemical properties and ability to retain these properties over a wide range of temperatures where other engineering materials do not function are the reasons for their wide acceptance in modern industry.

The electrical properties of polyimides are well suited for applications in the electrical and electronic industries, where they are used as wire and cable insulators; motor and generator armature slot liners; transformer and capacitor insulators; substrates for printed circuits; and magnetic recording and pressure-sensitive tapes and tubings. High-heat connectors, switches, housings, and controls also use polyimides instead of glass and ceramics.

Because of the high strength and thermal stability of high-performance polyimide films, the thickness of the correspondent electrical insulation can be 2- to 3-fold lower than that of conventional polymers. This leads to 25–50% savings in weight and dimensions for electrical equipment and this is especially important for aerospace applications. The chemical resistance of polyimides toward hot oils and greases, fuels, and lubricants allows their use in the petrochemical industry and transportation.

Polyimide films withstand better (than other polymers) soldering, etching, and other fabricating procedures used for production of microelectronic devices. Colorless polyimides are very promising for optoelectronic applications.

Melt-processible polyimides are increasingly used for load-bearing applications such as struts, chassis, and brackets in automotive and aircraft industries. Other applications include advanced composite structures, adhesives, coatings, and fibers for protective clothing.

REFERENCES

1. Bessonov, M. I.; Koton, M. M.; Kudryavtsev, V. V. et al. *Polyimides—Thermally Stable Polymers*; USSR, Plenum: USA, 1987.
2. Sroog, C. E. *Prog. Polym. Sci.* **1991**, *16*, 561.
3. Wilson, D.; Stenzenberger, H. D.; Herenrother, P. M. *Polyimides*, 1st ed.; Blackie & Son Ltd.: London, GB, 1990.
4. E. I. Du Pont de Nemours & Co., French Pat. 1 239 491, 1960.
5. Bower, G. M.; Frost, L. W. *J. Polym. Sci.* **1963**, *A-1*, 3135.
6. Sroog, C. E.; Endrey, A. L.; Abramo, S. V. et al. *J. Polym. Sci.* **1965**, *A-3*, 1373.
7. Endrey, A. L. Can. Patent 645 073, E. I. Du Pont & Co., 1962; Can. Patent 659 328, E. I. Du Pont & Co. 1963; Endrey, A. L. U.S. Patent 3 179 630, 1965; U.S. Patent 3 179 631, 1965; U.S. Patent 3 179 633, 1965; U.S. Patent 3 179 635, E. I. Du Pont & Co., 1965.
8. Edvwards, W. M.; U.S. Patent 3 179 614, 1965; U.S. Patent 3 179 634, E. I. Du Pont & Co., 1965.
9. Adrova, N. A.; Koton, M. M.; Moskwina, E. M. *Dokl. Akad. Nauk SSSR*, (Engl. Transl.), **1965**, *165*(5), 1171.
10. Vinogradova, S. V.; Slonimskii, G. K.; Vygodskii, Ya S. et al. *Polym. Sci. USSR*, **1969**, 11(12), 3098. Vinogradova, S. V.; S. Vygodskii, Ya S. et al. *Polym. Sci. USSR*, **1969**, *12*(9), 2254.
11. Sasaki, Y. et al. U.S. Patent 4 290 936, 1981; U.S. Patent 4 247 442, 1981; U.S. Patent 4 473 523, 1984, Ube Industries.
12. Harris, F. W.; Feld, W. A.; Lanier, L. H. In *Applied Polymer Symposium* Platzer, N., Ed.; Wiley: New York, 1975; 26, 421.
13. Nurmukhametov, R. N.; Likhatchev, D. Yu. et al. *Polym. Sci. USSR*, **1989**, *31*(2), 434.
14. Zhang, R.; Mattice, W. L. *Macromolecules* **1993**, *26*, 6100.
15. Birshtein, T. M.; Gorunov, A. N. *Vysokomol. Soed. (Polym. Sci. USSR)*, **1979**, *A21*(9), 1990.
16. St. Clair, A. K.; St. Clair, T. L. U.S. Patents 4 595 548 and 4 603 601, NASA 1986.
17. Koton, B. V.; Gordina, T. A.; Voishev, V. S. et al. *Polym. Sci. USSR*, **1977**, *19*(3), 711; Bikson, B. R.; Freimanis, Ya. F. *Polym. Sci. USSR*, **1970**, *12*(1), 81.

POLYIMIDES, PRECURSORS (Dye-Containing Side-Groups)

Matthias Rehahn* and Ludwig Schmitz
Polymer Institute
Karlsruhe University

In the last few years considerable attention has also been directed to polyimides that additionally contain dye functionalities, either incorporated in, or attached to, the polymer main-chain.[4,6,7,9,13,14] This interest stems mainly from the expectation that these materials are photoactive and/or photoconductive and can thus be used for optical devices or as nonlinear-optical materials. Synthesis of dye-containing polyimides, however, proved to be difficult. Methods that already start from dye-containing monomers have serious limitations that are associated with the widespread low chemical and thermal stability, particularly of many azo compounds.[12,15] Only a limited number of largely mainchain pigmentary polyimides are available via direct syntheses. Thus, as an alternative, precursor strategies

*Author to whom correspondence should be addressed.

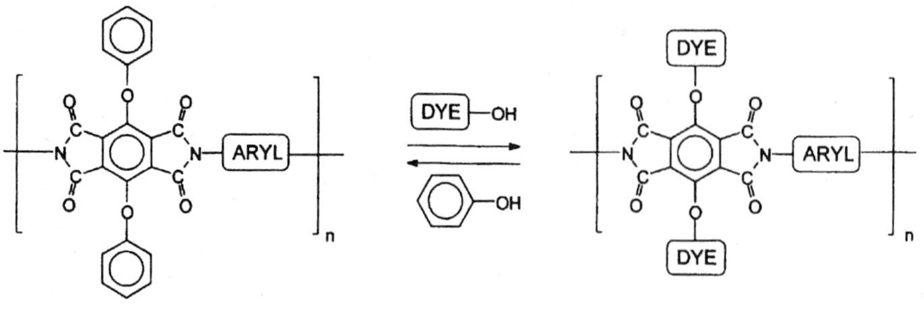

I

are required for the synthesis of pigmentary polyimides, especially for those that bear the dye subunit as a lateral substituent.

STRATEGY

The precursor strategy under consideration involves the Pd-catalyzed synthesis of soluble, rodlike precursor polyimides such as 1, having 3,6-diphenoxy substituted pyromellitic diimide units.[2,3] Subsequent nucleophilic substitution of the phenoxy substituents of 1 allows the introduction of the desired lateral dye substituents, leading to the pigmentary polyimides 2 (**Scheme I**).[3] As many dye molecules (as well as many other desired substituents) are stable under the conditions of the employed nucleophilic substitution, this strategy should open up general access to well-defined functionalized polyimides 2 of high molecular weight. The choice of the dye component to be introduced into the precursor polyimide 1 is restricted in only two respects: On one hand, appropriate dyes must have just one phenolic hydroxy group, the *ortho*-positions, which should be unsubstituted because of the sterical requirements of the *Meisenheimer* transition complex involved in the substitution reaction.[11] However, a minimum thermostability is required for the dyes to be introduced (*ca.* 120°C).

The development of the present precursor strategy has basically been related to the development of a novel Pd-catalyzed synthesis of aromatic polyimides, which is described in more detail under Polyimides (Soluble, Rigid Rods, Via Pd-Catalyzed Polycondensation) M. Reahn, L. Schwitz. In contrast to the classical polyimide synthesis that is, in general, strongly affected by bulky substituents of the pyromellitic diimide units,[8,10] the Pd-catalyzed method allows the preparation of precursor polyimides 1 of high molecular weight. This is because the critical imide formation step is already performed in the monomer synthesis, where the low yields of imidizations involving 3,6-disubstituted pyromellitic dianhydrides may be accepted.

PROPERTIES

A profound analysis of the properties of these novel pigmentary polyimides is presently being carried out. Moreover, it is intended to widen the scope of the nucleophilic substitution reaction with the aim of preparing further attractively functionalized polyimides.

REFERENCES

1. Giesa, R.; Keller, U.; Eiselt, P. et al. *J. Polym. Sci: Part A: Polym. Chem.* **1993**, *31*, 141.
2. Schmitz, L.; Rehahn, M. *Macromolecules* **1993**, *26*, 4413.
3. Schmitz, L.; Rehahn, M.; Ballauff, M. *Polymer* **1993**, *34*, 646.
4. Combellas, C.; Petit, M. A.; Thiebault, A. *Makromol. Chem.* **1992**, *193*, 2445.
5. Giesa, R.; Keller, U.; Schmidt, H.-W. *Polym. Prepr.* **1992**, *33*, 396.
6. Groh, W.; Lupo, D.; Sixl, H. *Angew. Chem. Int. Ed. Engl. (Adv. Mater.)*, **1989**, *28*, 1548.
7. Hall, H. K.; Kuo, T.; Leslie, T. M. *Macromolecules* **1989**, *22*, 3525.
8. Harris, F.; Hsu, S. L. C. *High Perform. Polym.* **1989**, *1*, 3.
9. Messier, J.; Kayar, F.; Prasad, P. et al. Eds.; *Nonlinear Optical Effects in Organic Polymers*, Academic: Kluver, Dortrecht, 1989.
10. Brandelik, D.; Field, W. A.; Arnold, F. E. *Polym. Prepr.* **1987**, *28*(1), 88.
11. Maiti, S.; Mandall, B. K. *Prog. Polym. Sci.* **1986**, *12*, 11.
12. Guthrie, J. T. *In: Encyclopedia of Polymer Science and Engineering*, John Wiley & Sons: New York, 1986; Vol. 5, 277.
13. Williams, D. J. *Angew. Chem. Int. Ed. Engl.* **1984**, *23*, 690.
14. Williams, D. J. Ed.; *Nonlinear Optical Properties of Organic and Polymeric Materials*, ACS Symposium Series 233, American Chemical Society, Washington, DC, 1983.
15. Dawson, D. J. *Aldrichimia Acta* **1981**, *14*(2), 23.

POLYIMIDES, PROCESSABLE (Modified with Other Rigid Groups)

Jin Chul Jung*
Department of Materials Science and Engineering
Pohang University of Science and Technology

Polyimides have been known to be one of the most thermally stable polymers. However, all of the polyimides synthesized for the purpose of high-temperature properties have more or less difficulties in processing[1,2] owing to their generally low solubilities and even lower fusibilities, and as is well known, they are normally processed into final products by way of their precursors such as poly(amic acid)s or poly(ester amide)s.

To balance the thermal stability and the processability, numerous methods to obtain polyimides with chemically modified chain structures have been introduced[1,2] and some of them have reached successful commercialization.[3,4] These methods include incorporation of hinge atoms or kink units or flexible spacer units either into dianhydride fragment or diamine fragment or both.

Incorporation into the main chain of other rigid groups, whose rigidity is not far lower or higher than that of the imide group, is another important method for preparing polyimides

*Author to whom correspondence should be addressed.

with improved processability and balanced thermal resistance. The processability of these rigid polyimides is improved by distortion of the main chain regularity, and the thermal stability is maintained by retained rigidity of the main chain.

SYNTHESIS AND PROPERTIES

Although amide bond is a bit less rigid than imide bond, poly(amide imide)s are certainly the most important type of modified polyimide[1] because a series of the poly(amide imide)s[1] has already been commercialized by Amoco under the trade name of Torlon®.

Hydrazide group is more rigid than amide group owing to a stronger hydrogen-bonding capability. A polyhydrazide containing pyromellitimide unit has been prepared[5] by condensation of isophthaldihydrazide with pyromellitic dianhydride. In a likewise manner, polyhydrazideimides with a wide range of different structures[8] such as poly(ester hydrazide imide)s and poly(amide hydrazideimide)s[9] have been prepared.

Some polyimides with spiroladder structure units have been prepared and their solubilities and fusibilities were investigated. Such an example[6] is prepared from 1,4-diamino-1,4-diaminomethylcyclohexane and 1,4,5,8-naphthalenetetracarboxylic dianhydride.

A polyimide whose benzimide ring is fused to an oxoisoindolobenzothiadiazine dioxide ring has been prepared[7] by condensation of diaminobenzene sulfonamide with pyromellitic dianhydride followed by two consecutive thermal cyclodehydration reactions.

Polyimides with good solubilities in spite of multifused ring systems in both diamine and dianhydride fragments are derived from dianhydride and diamine monomers containing one or two hetero atoms in their fused ring systems.[9,10]

Benzimidazole group is another rigid structure whose rigidity is as high as to be compared to benzene structure in the same chemical environment. A wide variety of poly(benzimidazole imide)s have been prepared.[11,12]

Incorporation of pyrrole units is achieved by condensation polymerization of dianhydride or diamine monomers containing pyrrole units with other imidization counterparts.

Poly(benzthiazole imide)s are an important class of modified polyimides having superior mechanical and high-temperature properties and a wide variety of poly(benzthiazole imide)s have been prepared and characterized and their physical properties investigated.[8,11–13] They are normally prepared by condensation of dianhydrides with diamines containing benzthiazole fragments that are commonly synthesized by condensation of *o,o*'-diaminodithiophenols with aminocarboxylic acids.

Recently, homo-[8,11,15] and copoly[(1,3,4-oxadiazole) imide]s[14] have drawn interest, because the copolymers with proper compositions are soluble in DMAc and processable to films and fibers.

Poly(benzoxazole imide)s also possess very high thermal resistance owing to the heterocyclic aromaticity of the benzoxazole ring. A number of poly(benzoxazole imide)s have been prepared[12,13,16] exclusively by polymerization of dianhydrides with benzoxazole-containing diamines.

Poly(quinoxaline imide)s are perhaps the most important modified polyimides since they have not only extremely high thermal stability but also excellent mechanical properties. This materials behavior results from the ladder-like structure of the quinoxaline rings. Most of the poly(quinoxaline imide)s have been prepared by polyimidization of diamine monomers containing quinoxaline fragments,[12,17–21] but pyrazine-2,3,5,6-tetracarboxylic dianhydride has also been polymerized with 2,5-diaminothiadiazole[22] to obtain a poly(pyrazine imide).

APPLICATIONS

Of the many polyimides discussed above only the poly(amide imide)s have been commercialized and marketed by Amoco under the trade name of Torlon. It has various unfilled, filled, and reinforced grades depending on the final applications. They are used for high-temperature electrical connectors and other electrical, electronic, and aerospace applications.[4]

Applications of poly(amide imide)s in automobile industry is noticeable. Poly(amide imide)s filled with polyterafluroethylene and some inorganic fillers are used for making various automotive parts owing to their thermal stability and lubricity.[4]

REFERENCES

1. Volksen, W. *Adv. Polym. Sci.* **1994**, Vol. *117*, p 111.
2. Bessonov, M. I.; Koton, M. M.; Kudryavtsev, V. V. et al. *Polyimides Thermally Stable Polymers* Translated from Russian by Backingowsky, L. V.; Chlenov, M. A. Translation Edited by Wright, W. W. Consultants Bureau: New York and London, 1987.
3. Critchley, J. P.; Knight, G. J.; Wright, W. W. *Heat-Resistant Polymers*; Plenum: New York and London, 1984; pp 186.
4. Seymour, R. B. *Polymers for Engineering Applications*; ASM International: New York, Brisbane, Toronto, Singapore, 1987; p 131.
5. Loncrinin, D. F.; Walton, W. L.; Hughes, R. B. *J. Polym. Sci. A-1* **1966**, *4*, 440.
6. Heller, J.; Hodgkin, J. H.; Martinelli, F. J. *J. Polym. Sci. Polym. Lett* **1968**, *6*, 153.
7. Imai, Y.; Koga, H. *J. Polym. Sci. Polym. Chem.* **1973**, *11*, 2623.
8. Frost, L. W.; Bower, G. M.; Freeman, J. H. et al. *J. Polym. Sci. A-1* **1968**, *6*, 215.
9. Subramanian, P.; Srinivasan, M. *J. Polym. Sci. Polym. Chem.* **1988**, *26*, 1553.
10. Mitra, P.; Biswas, M. *J. Polym. Sci. Polym. Chem.* **1990**, *28*, 3 795.
11. Preston, J.; Black, W. B. *J. Polym. Sci. A-1* **1967**, *5*, 2429.
12. Preston, J.; Dewinter, W. F.; Black, W. B. et al. *J. Polym. Sci. A-1* **1969**, *7*, 3027.
13. Preston, J.; Dewinter, W. F.; Black, W. B. *J. Polym. Sci.-1* **1969**, *7*, 283.
14. Hedrick, J. L. *Polymer* **1992**, *33*, 3375.
15. Thaemlitz, C. J.; Weikel, W. J.; Cassidy, P. E. *Polymer* **1992**, *33*, 3278.
16. Sakaguchi, Y.; Kato, Y. *J. Polym. Sci. Polym. Chem.* **1993**, *31*, 1029.
17. Overberger, C. G.; Moore, J. A. *Adv. Polym. Sci.* **1971**, Vol. 7, p 113.
18. Korshak, V. V.; Berestmeva, G. L.; Aksenova, T. S. et al. *Polym. Sci. U.S.S.R.* **1987**, *29*, 1767.
19. Akutsu, F.; Kuze, S.; Mtsuo, K. et al. *Makromol. Chem. Rapid Comm.* **1990**, *11*, 673.
20. Hedrick, J. L.; Labadie, J. W.; Russell, T. P. in Feyer, C.; Khojasteh, M. M.; McGrath, J. E. *Polyimides: Materials, Chemistry and Characterization*; Elsevier: Amsterdam, 1989; p 61.
21. Hedrick, J. L.; Labadie, J. W.; Russell, T. P. et al. *Polymer* **1991**, *32*, 950.
22. Hirsch, S. S. *J. Polym. Sci. A-1* **1969**, *7*, 15.

POLYIMIDES, SOLUBLE RIGID-RODS (Via-Pd-Catalyzed Polycondensation)

Matthias Rehahn*
Polymer Institute
Karlsruhe University

and

Lüdwig Schmitz
Hoechst AG

STRATEGY

An alternative strategy for the synthesis of rigid-rod polyimides **1a**(**A**) of high molecular weight was developed recently with the aid of a Pd-catalyzed polycondensation reaction (**Scheme I**, method A).[5,6] In contrast to the classic polyimide synthesis,[1–4,7,8] polycondensation of a dibromo compound such as **2** and a benzene-1,4-diboronic acid derivative **3** allows the imide structure to be preformed during the synthesis of monomer **2**.

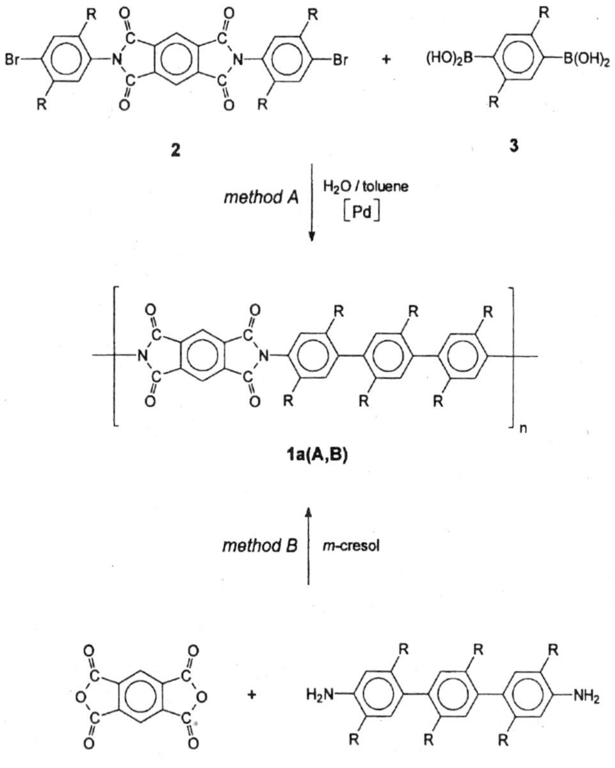

In order to compare this new approach to the classic one, pyromellitic dianhydride **4** was reacted with terphenyldiamine **5** to give a constitutionally identical polymer **1a**(**B**) (Scheme I, method B). Thus, it was possible to compare the efficiency of the two methods with regard to the structural regularity and number-average degree of polymerization (P_n) of the products **1a**. Dodecyl side chains ensure sufficient solubility of the polymers **1a**(**A**,**B**) to perform a full molecular characterization.

PREPARATION

For the synthesis of polyimide **1a**(**A**) (Scheme I; method A), equimolar amounts of monomer **2** and 2,5-didodecyl-1,4-benzenediboronic acid **3** are treated in the presence of catalytic amounts of $Pd(Ph_3)_4$ in a heterogeneous system of water (Na_2CO_3)/toluene. Polymer **1a**(**A**) is recovered as a slightly yellowish material in virtually 100% yield. Polymerizations according to method B are carried out by reacting the diaminoterphenyl derivative **5** with equimolar amounts of pyromellitic dianhydride **4** in *m*-cresol. The water formed in the course of the cyclization step is removed azeotropically. Work-up here results in a slightly yellow, fibrous material **1a**(**B**). The solubility of both polymers **1a**(**A**,**B**) in common organic solvents is high enough to record high-resolution NMR spectra. The NMR spectra of both polymers proved to be very similar.

GPC points towards almost identical molecular weights for the two polymers **1a**(**A**) and **1a**(**B**). Membrane osmometry in *o*-dichlorobenzene finally leads to the conclusion that a similar P_n was achieved for polyimides **1a**(**A**) and **1a**(**B**) (M_n = 20,000 and 22,000, respectively). These values correspond to P_n = 14 and 15, respectively, and prove unambiguously that both methods are of comparable efficiency for a polyimide synthesis.

PROPERTIES

Aliphatic side chains attached to an aromatic polymer backbone generally limit the thermal stability of these materials. In differential thermal analysis measurements, polymers **1** show a pronounced loss of weight at temperatures above 380°C. Stability for several minutes is given up to ~280°C. Differential scanning calorimetry curves of low molecular weight polyimides **1a** (P_n = 7) show an endotherm at ~70°, and a second endotherm at 230°C. The first is associated with a so-called sidechain melting process; the second probably represents the transition into the isotropic melt.[6]

REFERENCES

1. Takekoshi, T. *Adv. Polym. Sci.* **1990**, *94*, 1.
2. Sillon, B. In *Comprehensive Polymer Science*; Allen G.; Berington, J. C., Eds.; Pergamon: Oxford, 1989; Vol. 5, p 499.
3. Bessonov, M. I. et al. *Polyimides: Thermally Stable Polymers*; Consultants Bureau: New York, 1986.
4. Mittal, K. L. *Polyimides—Characterization and Applications*; Plenum: New York, 1986.
5. Schmitz, L. et al. *Polymer* **1993**, *34*, 646.
6. Helmer-Metzmann, F. et al. *Makromol. Chem.* **1992**, *193*, 1847.
7. Cassidy, P. E.; Aminabhavi, T. M. *Polym. News* **1990**, *14*, 362.
8. Buys, H. C. W. M. et al. *J. Appl. Polym. Sci.* **1990**, *41*, 1261.

Polyiminocarbonates

See: Pseudo-Poly(amino acid)s (Overview)

Polyiodide Ion

See: Poly(vinyl alcohol)-Iodine Interactions

*Author to whom correspondence should be addressed.

Polyisobutylene

See: *Head-to-Head Polymers*
Isobutylene Copolymers (Commercial)
Polyolefins (Overview)
Supported Catalysts (Lewis Acid and Ziegler-Natta)
Supported Lewis Acid Catalysts (on Polypropylene; Recoverable and Reusable)

Poly(isobutylene-co-isoprene)

See: *Butyl and Halobutyl Rubbers*
Butyl Rubber (for Chemical Protective Clothing)
Isobutylene Copolymers (Commercial)

Polyisocyanates

See: *Powder Coatings (Overview)*

POLYISOCYANIDES

Frank Millich
University of Missouri–Kansas City

Polyisocyanides were first discovered and extensively characterized at the University of Missouri–Kansas City.[1–6] Three catalytic polymerization systems were identified and developed.[3–5,7] Because of the relatively recent history of this polymer class, commercial uses have not yet been developed. However, the structure of polyisocyanides has drawn profound academic interest in the physical and chemical aspects.

Polyisocyanides are poly(vicinal imines) and are formed by cationically initiated α,α-addition of monoisocyanides (**Equation 1**):

$$[R-\overset{+}{N}\equiv\overset{-}{C}: \leftrightarrow R-N=\ddot{C}:] \xrightarrow{H^+} -(\underset{\underset{N-R}{\|}}{C})_n- \qquad \mathbf{1}$$

The backbone chain element is a single carbon atom. Consequently, even modest-sized R-groups of the Schiff base pendant substituent sterically force a helical conformation of the backbone chain. The term "isocyanides" is now preferred, but until the mid-1960s *Chemical Abstracts* referred to the monomers as "isonitriles," and the first polymers were named accordingly.

Polyisocyanides have been previously reviewed or updated in the chemical literature at least every five years since 1971 by Millich and Nolte.[6,8–14] Accordingly, this review has given special attention to a representative selection of publications that appeared between 1987–1993.

PREPARATIONS

The Millich group at first observed that polyisocyanides precipitated spontaneously from stored, neat liquid monomers in which the polymers had low solubility.[3] They next observed that α-phenylethyl isocyanide polymerized on the ground-glass surfaces of glassware joints during distillation of monomer. When coating glass surfaces with initiators of various types, only cationic coating led to high polymer.[4] With certain limitations, cationic initiation is generally a successful procedure.

The polymerization rates with secondary isocyanides, and tert-butyl isocyanide especially, are slow, requiring several days, because the placement of monomer onto the growing chain end is a sterically difficult problem. The polymerization suggests a "living polymer," since extension of reaction-time yields increased molecular weight and yield, notwithstanding the possibility of some chain termination, which nearly all cationic systems tend to show at room temperature and above.[9]

In a study of isocyanide complexes of metal ions, Nolte observed the formation of polymers.[19,20] Among active divalent complex ions of Co, Pd, Cu, Pt, and Ni, the last was found to be the best, as chlorides or acetylacetonides. As with protonic initiation, the polymerizations are general for most isocyanides, conducted in the presence of a nonsolvent, progress at room temperature, are cationically propagated, and can show a sensitivity to the presence of oxygen.[16,21–23]

Deming and Novak recently discovered what may be the premier catalyst system for polyisocyanides. They found that $[(\eta^3\text{-}C_3H_5)Ni(OC(O)CF_3)]_2$ is capable of catalyzing a "living" polymerization of isocyanides, and they also achieved a homogeneous catalyst.[21,24] Some virtues of this type of catalyst are even more rapid polymerization rates, quantitative yields, very narrow molecular-weight polydispersities, and molecular-weight control.

A large number of isocyanides have been polymerized and copolymerized to date, chiefly by the use of nickel complexes, which afford rapid syntheses, and by protonic acid, which lead to higher molecular weights at room temperature and below.[8,14,25,26] Aryl isocyanides are more inclined to polymerize; tert-alkyl isocyanides catenate very slowly, for obvious steric reasons; primary-alkyl, especially below R = butyl, suffers in yield from competing cationic reactions, although β-phenyl isocyanide and some others present little difficulty; sec-alkyl isocyanides succeed fairly consistently; and vinyl-conjugated isocyanides yield variously some soluble and some insoluble polymers.[25] Some polyisocyanides show low ceiling temperatures.[9] Copolymers also have been formed by isocyanides with other isocyanides, and with other types of monomers, for example, diazomethane.[15,17,18]

The "grandfather" model polyisocyanide, the first synthesized high polymer of this class and thus extensively characterized, is poly(α-phenylethyl isocyanide). Most of the historic molecular characterizations were carried out on this polymer, partly due to its convenient solubility in common solvents.

The first polyisocyanides synthesized were achieved with bulky N-substituents, that is, 2°-alkyl groups. The resultant tight helical conformation common to *sec*- and tert-alkyl polyisocyanides leads to rigid rod overall conformation. The rigidity of such macromolecules severely lowers solubilities in a wide range of solvents, except for acids, such as di- and trichloroacetic acid. Polyisocyanides that are only moderately rigid, for example, poly(α-phenylethyl isocyanide), are soluble in a wider but limited number of solvents, especially those high in polarity.[10] A handy index of rigidity is the magnitude of the Mark-Houwink constant, **a**, which occurs in the range 1–2 for non-N-primarily-alkyl polyisocyanides.[27] Polyisocyanides afford a substantial class of rigid-rod polymers for the study of physical phenomena. Three areas in particular have drawn scientific investigation: liquid crystals, viscosity theory, and helix conformation.

APPLICATIONS

The novel structure of polyisocyanides offers many academic points of interest. However, the full development of applications for these polymers and copolymers lies ahead. Among the areas that are rich in possibilities for scientific investigation and exploitation are the large number of N-atoms and their basicity, the large amount of unsaturation and the reactivity of imine functional groups, and the tight helical conformations and the rigid rods that are formed by many members of this family.

A copious number of potential applications have been studied by Nolte and Drenth and coworkers.[14] In one reported application, closed vesicles formed by the dispersal of surfactant amphiphiles in water were stabilized by imbibing with polymerizing oligoisocyanide. Vesicles formed from polyisocyanides are the subject of great interest as models of biological membranes with ion-transport channels as carriers of drugs and as microreactors.[29–32]

The magnetic properties of rigid high-spin poly(N-oxypiperidyl isocyanide) were investigated and found to be paramagnetic with an estimated spin density of 10^{21} spins/gram.

In a remarkable experiment, the resolution of poly(tert-butyl isocyanide) into enriched fractions of enantiomeric helices were achieved by column chromatography.[28,33] The resolving column packing used, poly[(s)-(+)-sec-butyl isocyanide], polymerized on porous beads, which indicates another potential application of some polyisocyanides. Since it is naturally insoluble in all nonacidic solvents, it is not even necessary to crosslink the polymer for use as a column packing if it is of sufficiently high molecular weight.

Thus, polyisocyanides have already demonstrated the potential for valuable applications.

REFERENCES

1. Millich, F. *Polymerization of Isonitriles* research proposal submitted to Petroleum Research Fund, 1962; Grant A-1777, 1963.
2. Millich, F.; Sinclair, R. G. *Abstracts of Papers* Abstracts 58 16th Midwest Conference, American Chemical Society, Kansas City, MO; November 19, 1964; ibid; idem, 1st Midwest Regional Conference, American Chemical Society; Abstract 408, November 4, 1965, Kansas City, MO.
3. Millich, F.; Sinclair, R. G. *Polym. Prepr., Am. Chem. Soc., Div. Polym. Chem.*, 6, 736. Presented at the 150th National Meeting of the American Chemical Society: Atlantic City, NJ, 1965.
4. Millich, F.; Sinclair, R. G. *J. Polym. Sci.* **1968**, *22*, 23.
5. Millich, F.; Sinclair, R. G. *J. Polym. Sci. A-1* **1968**, *6*, 1417.
6. Millich, F. *J. Polym. Sci. Macromol. Rev.* **1980**, *15*, 207.
7. Millich, F.; Wang, G-M. *Abstracts of Papers* Abstract 617, 6th Midwest Chemistry Conference of the American Chemical Society: Lincoln, NE, 1970.
8. Millich, F. In *Encyclopedia of Polymer Science and Technology*; Mark, H. F.; Gaylord, N. G.; Biklaes, N. M., Eds.; Wiley-Interscience: New York, 1971; Suppl. Vol. 15, p 395.
9. Millich, F. *Chem. Rev.* **1972**, *72*, 101.
10. Millich, F. *Adv. Polym. Sci.* **1975**, *19*, 117.
11. Millich, F. In *Preparation and Properties of Stereoregular Polymers*; Lenz, R. W.; Ciardelli, Eds.; D. Reidel: Dordrecht, Holland, 1980; p 195.
12. Millich, F. In *Encyclopedia of Polymer Science and Engineering*; Mark, H. F.; Bikales, N. M.; Overberger, C. G.; Menges, F. M., Eds.; John Wiley & Sons: 1988, p 383.
13. Drenth, W.; Nolte, R. J. M. *Acc. Chem. Res.* **1979**, *12*, 30.
14. Nolte, R. J. M.; Drenth, W. In *New Methods for Polymer Synthesis*; Mijs, W. J., Ed.; Plenum: New York, 1992; Chapter 9.
15. Chenvanij, J. Ph.D. Dissertation, University of Missouri, Kansas City, 1973.
16. Millich, F.; Baker, G. K. *Macromolecules* **1969**, *2*, 122.
17. Yamamoto, Y.; Hagihara, N. *Nippon Kagaku Zasshi* **1968**, *89*, 898.
18. Iwatsuki, S.; Ito, K.; Yamashita, Y. *Kogyo Kagaku Zasshi* **1967**, *70*, 1822.
19. Nolte, R. J. M.; Drenth, W. *Recl. Trav. Chim. Pays-Bas* **1973**, *92*, 788.
20. Nolte, R. J. M.; Stephany, R. W.; Drenth, W. *Recl. Trav. Chim. Pays-Bas* **1973**, *92*, 83.
21. Deming, T. J.; Novak, B. M. *Macromolecules* **1991**, *24*, 6043.
22. Deming, T. J.; Novak, B. M. *J. Amer. Chem. Soc.* **1993**, *115*, 9101.
23. Deming, T. J.; Novak, B. M. *Macromolecules* **1993**, *26*, 7092.
24. Deming, T. J.; Novak, B. M. *Macromolecules* **1991**, *24*, 326.
25. King, R. B.; Borodinsky, L.; Greene, M. J. *J. Polym. Sci., Polym. Chem. Ed.* **1987**, *25*, 2165.
26. King, R. B.; Green, M. L. *J. Polym. Sci., Polym. Chem. Ed.* **1987**, *25*, 907.
27. Millich, F.; Hellmuth, E. W.; Huang, S. Y. *J. Polym. Sci., Polym. Chem. Ed.* **1975**, *13*, 2143.
28. Nolte, R. J. M.; van Beijnen, A. J. M.; Drenth, W. *J. Am. Chem. Soc.* **1974**, *46*, 5932.
29. van Beijnen, A. J. M.; Nolte, R. J. M.; Zwikker, J. W.; Drenth, W. *Recl. Trav. Chim. Pays-Bas* **1982**, *101*, 409.
30. van der Linden, J. H.; Schoonman, J.; Nolte, R. J. M.; Drenth, W. *Recl. Trav. Chim. Pays-Bas* **1984**, *103*, 260.
31. Roks, M. F. M. Ph.D. Thesis, University of Utrecht, Utrecht, The Netherlands, 1987.
32. van Esch, J.; Roks, M. F. M.; Nolte, R. J. M. *J. Amer. Chem. Soc.* **1986**, *108*, 6093.
33. van Beijnan, A. J. M.; Nolte, R. J. M.; Drenth, W. *Recl. Trav. Chim. Pays-Bas* **1980**, *99*, 121.

Polyisocyanurates

See: Step Polymerization Catalysts

POLYISOIMIDES

Keisuke Kurita
Department of Industrial Chemistry
Seikei University

Polyisoimides should have great potential as precursors for polyimides because they could be transformed into polyimide without forming water to produce void-free materials (**Equation 1**). Furthermore, they would be highly soluble with low glass transition temperatures because of their bent molecular structures, resulting in insoluble and rigid rodlike polyimides on isomerization. These properties may be important for developing new composite materials from polyimides and their high-performance films.

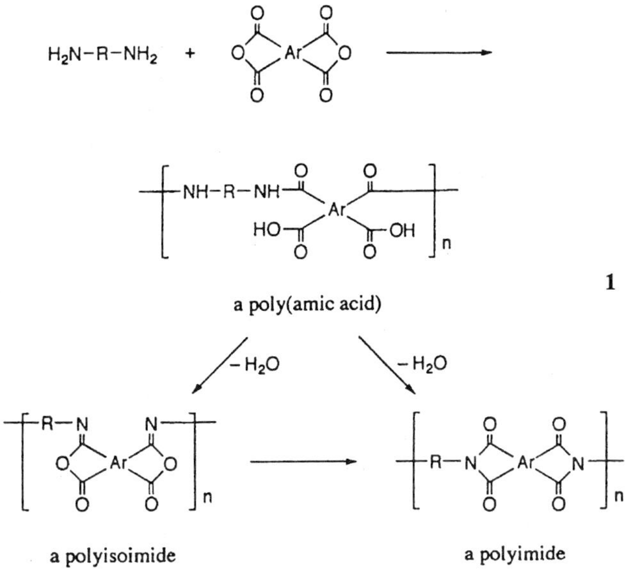

1

PREPARATION

Isoimide compounds are synthesized by dehydration cyclization of amic acids prepared from amines and cyclic dicarboxylic anhydrides such as phthalic and maleic anhydrides. Common dehydrating agents such as acetic anhydride, however, usually afford imide rings instead of isoimide rings, and trifluoroacetic anhydride[1,2] or *N,N'*-dicyclohexylcarbodiimide (DCC)[3] is necessary for achieving the preferential formation of isoimide rings. Moreover, isoimide rings are susceptible to hydrolysis and to other nucleophilic attacks.[4–6] The synthetic difficulty and inherent instability of isoimide structures have restricted the quantitative studies of polyisoimides.

PROPERTIES

Copoly(imide isoimide) films derived from 4,4'-diaminodiphenyl ether and pyromellitic dianhydride were transformed into polyimide films slowly at 350°C in air.[7] The initial isoimide content of 94% decreased to 16%, 12%, and 3% after 25 min, 90 min, and 22 h, respectively.

Unlike the rigid, rodlike polyimides, polyisoimides may be tractable because of their unique molecular structure.

APPLICATIONS

Polyisoimides are characterized by their high solubility, low glass transition temperatures, and their conversion to polyimides without producing harmful water. Although the stability of the polyisoimides depends on their molecular structures and the process described here requires an additional step compared with the conventional direct imidization of poly(amic acid)s, the characteristics of polyisoimides would be significant in various fields. Acetylene-terminated imide and isoimide oligomers are commercially available. They can be cured and are desirable as matrix resins for composites with high mechanical properties and thermooxidative stability.

REFERENCES

1. Tsou, K. C.; Barrnett, R. J.; Seligman, A. M. *J. Am. Chem. Soc.* **1955**, *77*, 4613.
2. Roderic, W. R.; Bhatia, P. L. *J. Org. Chem.* **1963**, *28*, 2018.
3. Cotter, R. C.; Sauers, C. K.; Whelan, J. M. *J. Org. Chem.* **1961**, *26*, 10.
4. Hedaya, E.; Hinman, R. L.; Theodoropulos, S. *J. Org. Chem.* **1966**, *31*, 1311.
5. Wallace, J. S.; Tan, L.-S.; Arnold, F. E. *Polymer* **1990**, *31*, 2411.
6. Kurita, K.; Suzuki, Y.; Enari, T. et al. *J. Polym. Sci., Part A: Polym. Chem.* **1994**, *32*, 393.

POLYISOPEPTIDES

Hikasu Sekiguchi and B. Coutin
Laboratoire de Chimie Macromoléculaire
Université Pierre et Marie Curie

A polyisopeptide is a nonpeptidic polymer of α-amino acid, or more exactly, of a tri- or higher multifunctional amino acid with one of the α-functional (amino or carboxyl) group of each repeating unit bound to the third functional group of the vicinal monomer unit. For example, when the third functional group is an amino or carboxylic group, the main chain has a polyamide structure (**Structures 1** and **2**).

Polyisopeptides are characterized by their peptide-like (*N*-acyl α-amino acid, or α-amino acid amide) repeating groups, half of each one making part of the main chain and the other half being in the side chain. They are therefore polymers of α-amino acids in the same way as polypeptides, of which they are isomers and from which they differ only in the disposition of a part of the peptide group. This is why we have proposed the terminology "polyisopeptides" to this family of polymers.

The third functional group of the α-amino acid, used in the chain formation, may be other than an amino or carboxyl group, in which case the main chain of the resulting polymer may have various chemical structures.

It is expected that polyisopeptides possess the physical properties (melting, solubility, transformability, etc.) of synthetic

$$NH_2-\underset{\displaystyle CO_2R}{\underset{|}{CH}}-X-CO_2H \longrightarrow \left[NH-\underset{\displaystyle CO_2R}{\underset{|}{CH}}-X-CO\right] \qquad 1$$

$$HOCO-\underset{\displaystyle NHQ}{\underset{|}{CH}}-Y-NH_2 \longrightarrow \left[CO-\underset{\displaystyle NHQ}{\underset{|}{CH}}-Y-NH\right] \qquad 2$$

(with R=H pr or alkyl, Q = H or acyl)

linear polymers without losing completely the biochemical properties of peptide groups (susceptibility to enzymatic hydrolysis, to biodegradation, etc.). The final degradation products of polyisopeptides are α-amino acids, biocompatible and bioresorbable. All these advantages allow us to consider that polyisopeptides have potential for biomedical applications.

BIOLOGICAL SYNTHESIS

ε-Poly(L-lysine) can be produced biologically by *Streptomyces albulus subsp. lysinopolymerus.*[2] No other polyisopeptide has so far been prepared by biological means.

CHEMICAL SYNTHESIS

Activation of Carboxyl Groups

This is probably the most universal method of polyisopeptide synthesis. Particularly for the synthesis of diamine-diacid-type polyisopeptides, the title method is the only one available. In certain cases, the method is combined with the activation of an amino group, namely by trimethyl silylation.

In practice, a few means of carboxyl activation are available, principally the active ester method, the acid chloride method, and the imide method (others also exist).

ω-Poly(amino acid) Polymers

β-Poly(α-benzyl L-aspartate) was prepared by polycondensation of α-benzyl β-pentachlorophenyl L-aspartate, followed by catalytic hydrogenation.[4]

ε-Poly(L-lysine) and δ-poly(L-ornithine) were obtained by the polycondensation of pentachlorophenyl ester of their trimers, respectively.[7–9]

Diamine-Diacid Polymers

Probably because the synthesis is much easier, diamine-diacid type polyamides on the basis of a multifunctional α-amino acid as one of the monomers are more numerous than the ω-polyamide type ones. A characteristic feature of polyamides of this type is the existence of structural isomers for a given couple of monomers.

STRUCTURE

Polyisopeptides are nonpeptidic polymers of multifunctional α-amino acids, with their main chain containing the structural elements of polyamides, polypeptides, and other aliphatic polymers. When they are derived from optically active monomers, they are stereoregular polymers. When they contain the monomer units in an ordered way, they are structurally regular.

Many polyisopeptides and copolyisopeptides are expected to be crystalline, with β-zigzag chain arrangement. Exceptions are β-poly(L-aspartic α-isobutyl ester) and γ-poly-(L-glutamic α-benzyl ester), which were found to be helical, somewhat similar to the corresponding polypeptides.[5,6]

CHEMICAL MODIFICATIONS

All polyisopeptides bear at least one kind of functional group, protected or not. These are amino and/or carboxylic groups. They are part of the peptidelike groups that connect the repeating units (or the comonomer units), and they endow the main chain amide groups with peptide-like properties.

Thus, it is expected that the polymers are biodegradable and biocompatible (i.e., they are potential polymers for medical uses). For example, it would be possible to fix a drug on the backbone, with or without a spacer.

PHYSICAL PROPERTIES

It is expected that polyisopeptides with polyamide chains exhibit the well-known thermal behavior represented by the zigzag curve of melting points depending on the pair-impair number of carbon atoms of the monomer units.

The conformations of several polyisopeptides have been studied. β-Poly(α-isobutyl L-aspartate) can exist in two types of helical structure in the solid state: a hexagonal conformation and a tetragonal one, due to the presence of the isobutyl side group.[3,5] This polymer was also shown to be highly birefringent in solvents like chloroform and able to form lyotropic liquid crystal.[10] Films with cholesteric structure could be obtained by simple casting.

APPLICATIONS

Many of the polyisopeptides are potential biomaterials. By the modification of their functional side groups they can be transformed into water-soluble materials and used for immobilizing drugs or herbicides. Some attempts at degradation and drug delivery have already succeeded.

α-β-Poly(*N*-2-hydroxyethyl-D,L-asparagine) was described as a new plasma expander and used as a carrier for acetylsalicyclic acid and acetamidobenzoic acid.[11,12]

REFERENCES

1. Bechaouch, S.; Coutin, B.; Sekiguchi, H. *Macromol. Rapid Commun.* **1994**, *15*, 125.
2. Sakai, H.; Shima, S. Jpn Kokai Tokkyo Koho, JP 53 72896, 1978; *Chem. Abstract* **1979**, *91*, p 173372e.
3. Muñoz-Guerra, S. *Makromol. Chem. Macromol. Symp.* **1991**, *48/49*, 71.
4. Kovacs, J.; Ballina, R.; Rodin, R. L. et al. *J. Am. Chem. Soc.* **1965**, *87*, 119.
5. Fernández-Santin, J. M.; Muñoz-Guerra, S.; Rodríguez-Galán, A. et al. *Macromolecules* **1987**, *20*, 62.
6. Puiggalí, J.; Muñoz-Guerra, S.; Rodríguez-Galán, A.; Alegre, C.; Subirana, J. A. *Makromol. Chem., Macromol. Symp.* **1988**, *20/21*, 167.
7. Kushwaha, D. R. S.; Mathur, K. B.; Balasubramanian, D. *Biopolymers* **1980**, *19*, 219.
8. Szókán, G.; Gyenes, M.; Tyihák, E.; Szende, B. *Peptides 1982*; de Gruyter, W., Ed.; Berlin, 1983; p 203.
9. Mathur, K. B.; Pandey, R. K.; Jagannadham, M. V.; Balasubramanian, D. *Int. J. Peptide Protein Res.* **1981**, *17*, 189.
10. Montserrat, J. M.; Muñoz-Guerra, S.; Subirana, J. A. *Makromol. Chem., Macromol. Symp.* **1988**, *20/21*, 319.
11. Neri, P.; Antoni, G.; Benvenuti, F. et al. *J. Med. Chem.* **1973**, *16*, 893.
12. Giammona, G.; Charlisi, B.; Palazzo, S.; *J. Polym. Sci.: Part A: Polym. Chem. 25*, 2813 (1987).

Polyisophthalamides

See: Polyisophthalamides (Enhancement of water-uptake)
Polyisophthalamides (Thermal Stability Enhancement by Pendent Groups)

POLYISOPHTHALAMIDES (Enhancement of Water-Uptake)

Javier de Abajo,* José G. de la Campa, and
Angel E. Lozano
Instituto de Ciencia y Tecnologia de Polimeros, CSIC

Jack Preston
Department of Polymer Science
Camille Dreyfus Laboratory
Research Triangle Institute

Polyisophthalamides belong to the class of high-performance polymers that, in the form of fibers, are termed aramids. The polymers on which aramids are based, that is, wholly aromatic polyamides, offer an outstanding balance of thermal and mechanical properties, but their inherent molecular rigidity is also the origin of a remarkable intractability. Indeed, such polymers have melting temperatures beyond their decomposition temperatures, so that they cannot be processed in the molten state. Furthermore, wholly aromatic polyisophthalamides (PIPAs) are insoluble in all common organic solvents and can be dissolved only by highly polar aprotic solvents containing salts. Thus, PIPAs can be processed from solution into fibers, films, coatings, and more recently, into semipermeable membranes, in the form of hollow fibers or flat membranes.[2–4]

WATER UPTAKE

The ability of a polymeric material to absorb water is a crucial property for many applications. For instance, water absorbed can be an effective plasticizer, lowering the glass transition temperature, particularly in polar polymers. The amount of water uptake is also very important for electrical applications because electrical properties, such as dielectric constant or resistivity, can dramatically change as a function of the water absorbed. The hydrophilicity of polymers is also of major importance in the field of semipermeable membranes, particularly for membranes used for the filtration of aqueous solutions and for water purification.[5–7]

PIPAs are polar polymers, with a high density of polar -CO-NH- linkages and, consequently, they can absorb comparatively high amounts of water. Depending on the chemical structure and the degree of crystallinity, the water uptake of PIPAs can range between 2–3% and 15% or more. The most representative member of this family of polymers is poly(*m*-phenyleneisophthalamide), MPD-I, which can be prepared by *m*-phenylenediamine and isophthaloyl chloride by conventional means. Based on this polymer, the DuPont Company marketed the heat-resistant fiber, Nomex®, in the 1960s. It can be considered the first man-made, high-performance organic fiber, with a glass transition temperature of about 275°C and with excellent thermooxidative and fire resistance.[1,2] MPD-I in the form of a fiber (for example, Nomex®) is probably in the highest degree of crystallization attainable and it still, while in that state, absorbs about 6.5% water (w/w) at room temperature and approximately 65% relative humidity (r.h.). MPD-I prepared in the laboratory and isolated as small porous pearls, with rather low molecular order and essentially amorphous, can take up to 9% water.[8]

*Author to whom correspondence should be addressed.

Chemistry offers two main alternatives to enhance the hydrophilicity of MPD-I: either by treating the polymer with suitable reagents to modify its original chemical structure or by synthesizing new polyisophthalamides using novel monomers to obtain modified PIPAs. Considering the fact that MPD-I and related PIPAs are nonmelting polyamides, which hardly dissolve in organic media, the approach of using new monomers to obtain modified PIPAs appears as the most suitable route. From the start, it was realized that the introduction of polar substituents or side groups with an affinity for water should result in an enhancement of the water absorption. Thus, PIPAs have been prepared with carboxylic, carboxamide, sulfonic and sulfonate side groups.[7,9–11]

In recent years we have carried out a systematic study on the effect of several substituents on the properties of PIPAs and particularly on the amount of the water uptake. A set of 5-substituted isophthalic acid derivatives were synthesized and purified for step-growth polymerization, most of them as diacid chlorides.[12–16] They were combined with aromatic diamines to yield a rather large family of modified PIPAs, covering a wide range of structures and properties.

It can be concluded that there are many possibilities to modify the hydrophilicity of polyisophthalamides by using monomers bearing groups of different chemical structure. A conscious choice of the side groups, either on the isophthaloyl unit or on the diamine unit, can provide substantial enhancement of water uptake. Apart from polarity, which is a major factor, other factors playing a significant structure/property role are the size of the substituents, the degree of polymer crystallinity, the effect of chain separation, the limitation of interchain hydrogen bonding, and the lowering of cohesive energy density.

REFERENCES

1. Lee, H.; Stoffey, D.; Neville, K. *New Linear Polymers*; McGraw-Hill: New York, 1967; p 129–170.
2. Scott, K. Water desalination, in *Handbook of Industrial Membranes*, Elsevier: Oxford, 1995.
3. Preston, J. *Aromatic Polyamides*, in *Encyclopedia of Polymer Science and Engineering*; Wiley-Interscience: New York, 1988; Vol. 11, p 381.
4. Yang, H. H. *Aromatic High-Strength Fibers*; Wiley-Interscience: New York, 1988; p 66–289.
5. Sedlácek, B.; Jahovec, J., Eds. *Synthetic Polymeric Membranes*; deGruyter: Berlin, 1987.
6. Staude, E. *Makromol. Chem.* **1982**, *Supp Vol. 3*, 139.
7. Sumimoto, H.; Hashimoto, K. *Adv. Polym. Sci.* **1985**, *64*, 63.
8. de la Campa, J. G.; Guijarro, E.; Serna, F. J.; de Abajo, J. *Eur. Polym. J.* **1985**, *21*, 1013.
9. Bogdanov, M. N.; Kharkov, S. N.; Spirina, I. A.; Leshiner, A. V.; Plyashkevich, L. A. *Polym. Sci. USSR* **1966**, *7*, 898.
10. Hinderer, H. E.; Smith, R. W.; Preston, J. *Appl. Polym. Symp.* **1973**, *21*, 1.
11. Walch, A.; Lukas, H.; Klimmek, A.; Pusch, W. *Polym. Lett.* **1974**, *12*, 697.
12. Lozano, A. E.; de la Campa, J. G.; de Abajo, J. *Makromol. Chem. Rapid Commun.* **1990**, *11*, 471.
13. Lozano, A. E.; de Abajo, J.; de la Campa, J. G. *J. Polym. Sci. A. Polym. Chem.* **1993**, *31*, 1203.
14. Lozano, A. E.; de la Campa, J. G.; de Abajo, J. *J. Polym. Sci. A, Polym. Chem.* **1993**, *30*, 1327; *ibid.* **1993**, *31*, 1383.
15. Lozano, A. E.; de la Campa, J. G.; de Abajo, J. *Polymer* **1994**, *35*, 872; *ibid*, **1994**, *35*, 1317.
16. Meléndez, A.; de la Campa, J. G.; de Abajo, J. *Polymer* **1988**, *29*, 1142.

POLYISOPHTHALAMIDES (Thermal Stability Enhancement by Pendent Groups)

Javier de Abajo, José G. de la Campa, and Angel E. Lozano
Instituto de Ciencia y Tecnologia de Polimeros
CSIC

Jack Preston
Department of Polymer Science
Camille Dreyfus Laboratory
Research Triangle Institute

Some recent examples of PIPAs prepared from isophthalic acid or isophthaloyl chloride and novel diamines have demonstrated a real improvement in both processability and thermal resistance. They are actually soluble in polar organic solvents at relatively high concentration and show glass transition temperatures (T_gs) over 300°C and thermal resistance (estimated by TGA) comparable to that of MPD-I.

In this report, a comparative study of the effect of various bulky pendent groups anchored to the 5-position of the isophthaloyl moiety of polyisophthalamides is made for the first time. For the preparation of the polymers, suitable monomers have been chosen, based on isophthalic acid and isophthaloyl chloride, and they have been combined with *m*-phenylenediamine to get a set of poly(*m*-phenyleneisophthalamide)s great enough to establish reliable structure–properties relationships.

THERMAL PROPERTIES OF PIPAS

The structures of the polymers used for this study are shown in **Table 1**, along with the values of inherent viscosities, thermal transitions, and heat resistance as measured by thermogravimetric analysis.

Polymer **1** is poly(*m*-phenyleneisophthalamide) (MPD-I), the simplest, wholly aromatic polyisophthalamide, marketed under the trade name Nomex® and Fenilon®. It was prepared from *m*-phenylenediamine and isophthaloyl chloride with η_{inh} 0.91 dl/g, and was taken as the reference or control for the rest of the PIPAs, what can be considered as derived from it, that is, as substituted or modified PIPAs. The substituents can either increase intermolecular attractions through changes in molecular dipole moment and/or polarizability, or separate the chains through steric interactions. These factors depend, obviously, on the size and type of substituent.

The thermal behavior was evaluated by differential scanning calorimetry and thermogravimetric analysis. Neither crystallization exotherms nor melting endotherms were observed by DSC in the range 50–400°C, and only the inflection associated to the T_g could be detected on the calorimetric curves. The values of T_gs listed in Table 1 show that the introduction of pendent groups caused a change in the T_g that greatly depended on its chemical nature.

While PIPAs with phenoxy, benzamide, benzoyl, or benzoate pendent groups showed a lower decomposition temperature (T_d) than the unsubstituted polymer **1**, those PIPAs containing heterocycle pendent groups, showed higher T_ds, by 20–30°.

It can be said that the introduction of bulky pendent groups is a very effective approach to modify the thermal properties of aromatic polyamides. By means of relatively easy and inexpensive chemical means, monomers can be prepared that provide the structural characteristics needed for the improvement of thermal transitions and thermal resistance without impairing other properties. Furthermore, the solubility in polar organic solvents was enhanced by the incorporation of pendent groups in all cases, and properties such as mechanical strength and chemical resistance were not essentially changed in any case.

TABLE 1. Thermal Properties of Modified Poly(*m*-phenylene Isophthalamide)s

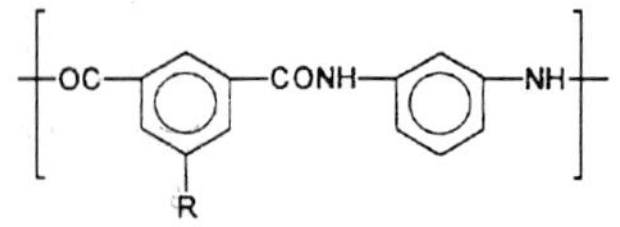

Polymer	R	η_{inh}, dl/g	Tg, °C	Td, °C
1	H—	0.91	276	450
2	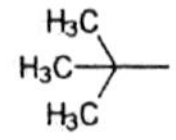	0.68	295	455
3	O_2N—	0.80$_a$	309	390
4	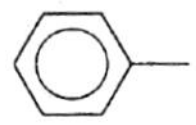	0.66	290	450
5	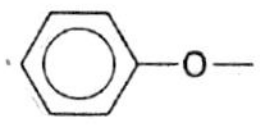	0.74	245	430
6	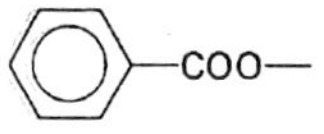	1.12	275	365
7	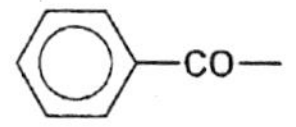	0.80	255	405
8	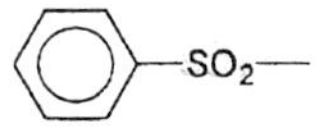	0.31	277	430
9	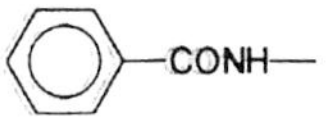	0.72	276	415
10	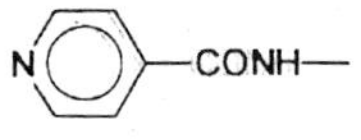	0.66	320	390
11[b]	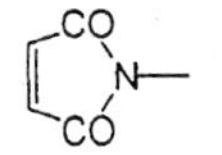	0.83	287	450
12[b]	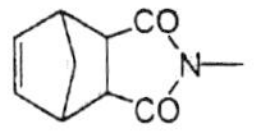	1.04	—	240(450)
13	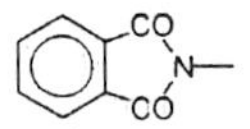	0.65	315	470
14	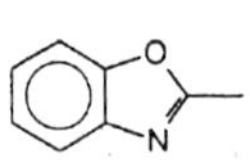	0.63	315	475
15	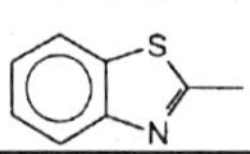	0.58	305	480

[a]Measured in H_2SO_4.
[b]Diamine unit was DDE instead of MPD.

Polyisoprene

See: *Conducting Polymers (Nonconjugated)*
Diene Rubbers, Conventional (Ozone Degradation and Stabilization)
Eucommia Ulmoide Gum
Natural Rubber (Chemical Modification)
Natural Rubber (Hevea brasiliensis)
3,4-Polyisoprene, Crystalline
Rare Earth-Initiated Polymerization (of Dienes)
Supported Catalysts (Lewis Acid and Ziegler-Natta)

3,4-POLYISOPRENE, CRYSTALLINE

Adel F. Halasa and W. L. Hsu
Tire Materials and Compound Research Corporate Research
The Goodyear Tire and Rubber Company

Amorphous 3,4-polyisoprene made by organolithium catalysts in the presence of polar modifiers or made by transition metal catalysts has been reported in the literature.[1–4] The first report on the preparation of crystalline 3,4-polyisoprene was made by Chinese workers using $Fe(AcAc)_3$ in the presence of aluminum alkyls and a nitrogen bearing donor ligand; but its tacticity was not defined.[5,6]

Sun and Wang utilized a polymerization catalyst based on $Fe(AcAc)_3$/donor ligand/$Al(iBu)_3$ in aromatic solvents.[5] The mode of addition was pointed out in their paper to be very critical in obtaining high conversion with a maximum degree of crystallinity. Hsu and Halasa later showed that the addition of a protic additive such as water or methanol enhanced the rate of polymerization and produced gel-free polymer with a high degree of crystallinity in high yield.[7]

POLYMER CHARACTERIZATION

Crystalline 3,4-polyisoprenes discussed in this paper were prepared using an *o*-phenanthroline chelated $Fe(AcAc)_3$ that was reduced with a partially hydrolyzed trialkyl aluminum.[7] The polyisoprenes produced were analyzed by ^{1}H NMR/^{13}C NMR. They contained 70–81% 3,4-polyisoprene and 30–19% *cis* 1,4-polyisoprene. The DSC shows the Tg of these polymers to be about +8°C, and to have several melting points (T_m) in the range of 40–90°C, depending on thermal history. Wide Angle X-ray Scattering showed a high degree of crystallinity for stretched samples (40–45%) and 20–30% for unstretched samples. This shows that this polymer can be strain crystallized like other stereoregular polymers.

CONCLUSION

The preparation of what is believed to be syndiotactic 3,4-polyisoprene was made by $Fe(AcAc)_3$ reduced by a partially hydrolyzed trialkyl aluminum in the presence of the chelating, nitrogen-bearing *o*-phenanthroline. The degree of crystallization was enhanced by the addition of water to the catalyst. The ^{13}C NMR spectra was complex but did reveal the 3,4 units are in "head to tail" sequences.

REFERENCES

1. Halasa, A. F.; Hall, J. E.; Lohr, D. F. *J. Polym. Sci. Poly. Chem. Ed.* **1981**, *19*, 1937.
2. Düch, M. W.; Grant, D. M. *Makromolekulare Chem.* **1970**, *3*, 165.
3. Natta, G.; Porri, L.; Carbonaro, A. *Makromolekulare Chem.* **1964**, *77*, 126.
4. Zeigler, K. *Angew. Chem.* **1936**, *49*, 499.
5. Sun, Q.; Wang, F. *Acta Polymerica Sinica* **1988**, *2*, 145.
6. Qiu, Z. W.; Chen, X.; Sun, B.; Zhoul, Z.; Wang, F. *J. Macromol. Sci. Chem.* **1988**, *A25*(2), 127.
7. Hsu, W. L.; Halasa, A. F. paper presented at the 144th ACS Rubber Division Meeting, Orlando, Florida, October 1993, to be published in *Rubber Chem. Technol., 1994.*

Poly(N-isopropylacrylamide)

See: *Colloidal Dispersions, Nanogels*
Enzyme-Poly(N-isopropylacrylamide) Conjugates (Thermoreversibly Soluble/Insoluble)
N-Isopropylacrylamide Copolymers (Drug Delivery)
Poly(N-isopropylacrylamide) Hydrogels

POLY(*N*-ISOPROPYLACRYLAMIDE) HYDROGELS

Agneza Safranj and Masaru Yoshida
Department of Material Development
Takasaki Radiation Chemistry Research Establishment
Japan Atomic Energy Research Institute

Ryoichi Katakai
Department of Chemistry
Faculty of Engineering
Gumma University

Poly(*N*-isopropylacrylamide) (PNIPAAm) shows a typical thermal reversibility in aqueous solutions, precipitating from solution above a critical temperature called the lower critical solution temperature (LCST) and redissolving below this temperature. The LCST of PNIPAAm is around 32°C. Crosslinked gels collapse and deswell above the LCST, and reswell and expand below this critical temperature.[1,2] PNIPAAm has been used as a model system to demonstrate the validity of theories describing the coil-globule transition and folding and unfolding of biopolymers, and volume-phase transitions of gels.[3,5]

PREPARATION

The PNIPAAm hydrogel can be synthesized at room temperature from an aqueous solution of the monomer by using a redox initiator, ammonium persulfate and *N,N,N′,N′*-tetramethylethylenediamine in the presence of *N,N′*-methylenebisacrylamide, as a crosslinker.[1,2,5,6] Since the LCST of PNIPAAm is around 32°C, the polymerization at room temperature proceeds in a homogeneous solution. PNIPAAm hydrogels can also be synthesized by starting the polymerization below the LCST, then elevating the temperature above it. In this way macroporous gels with fast temperature response were obtained.[8,9]

We applied the method of radiation-induced simultaneous polymerization and crosslinking for gel preparation.[10] This method has several advantages: it is a simple and additive-free process at all temperatures, the degree of crosslinking can be easily controlled by irradiation conditions, and the economy is competitive when compared with conventional methods. It is also possible to combine into one step the synthesis and sterilization of the product.[11]

PROPERTIES

The most important property of PNIPAAm hydrogels is their temperature-induced inverse volume phase transition in aqueous solutions. The volume phase transition of the gels is similar to the coil-globule transition of the polymer chains and the thermoshrinking behavior is caused by a critical balance of hydrophobic and hydrophilic groups at the polymer chain.[4,12] Binkert et al. measured the radius of gyration for polymers of molecular weight of 7 million and found it to be about 135 nm.[13] In hydrogel, polymer chains keep this extended conformation although the chains are held together at their crosslink points, as manifested by a highly swollen state. With increasing temperature the hydrogen bonding weakens, leading to a reduction in the structuring of water around the hydrophobic groups. As this structured water is released, the interactions between hydrophobic side groups of the polymer increase. Above the LCST, which is around 32°C for PNIPAAm, these hydrophobic interactions become dominant, meaning that the entropic term becomes dominant and the free energy of mixing takes a positive value. This leads to a collapse of the PNIPAAm chains from the expanded coil to a compact globule conformation (radius of gyration becomes only 16 nm, as measured by Binkert et al.).[13] In the case of the crosslinked PNIPAAm, this refolding of the polymer causes the shrinking of the gel.

APPLICATIONS

Thermally responsive gels have been proposed for different applications in drug delivery devices and separation processes. When applied in controlled drug delivery, the gels respond either to the stimulus by shrinking and pumping out the drug, or they can function as membranes whose permeability can be turned on and off. Hoffman and coworkers studied extensively the use of PNIPAAm gels and its copolymers as temperature and pH-sensitive drug-delivery systems.[6,7,14,15]

The application of PNIPAAm in biotechnology was also investigated Park and Hoffman prepared PNIPAAm gel beads with immobilized enzyme.[16,17] The shrinking of the gel upon heating squeezed out the product and remaining substrate from inside the gel, while the swelling, upon cooling, drew in the substrate. Besides the use of immobilized biocatalysts in chemical conversion, Kokufuta attempted to directly convert the energy of immobilized enzyme reactions into mechanical work by immobilizing urease into NIPAAm-acrylic gels, and Concanavalin A into NIPAAm gels.[18] Our research on PNIPAAm gels has been carried out with applications such as drug delivery and bioseparation in mind. However, we have not yet focused on a development of any specific application, except to use NIPAAm as a comonomer in synthesis of detectors with high sensitivity to low LET particles.[19] For biomedical applications, to date, we have used crosslinked poly(acryloyl-L-proline alkyl ester) gels, as discussed in another chapter of this Encyclopedia (Acryloyl-L-Proline Alkyl Ester Hydrogels).

REFERENCES

1. Hirokawa, Y.; Tanaka, T. *J. Chem. Phys.* **1984**, *81*, 6379.
2. Otake, K.; Inomata, H.; Konno, M., Saito, S. *Macromolecules* **1990**, *23*, 283.
3. Matsuyama, A.; Tanaka, F. *J. Chem. Phys.* **1991**, *94*, 781.
4. Otake, K.; Inomata, H.; Konno, M.; Saito, S. *J. Chem. Phys.* **1989**, *91*, 1345.
5. Shibayama, M.; Tanaka, T.; Han, C. C. *J. Chem. Phys.* **1992**, *97*, 6829.
6. Hoffman, A. S.; Afrassiabi, A.; Dong, L. C. *J. Controlled Release* **1986**, *4*, 213.
7. Hoffman, A. S. *J. Controlled Release* **1987**, *6*, 297.
8. Kabra, B. G.; Gehrke, S. H. *Polymer Communications* **1991**, *32*, 322.
9. Wu, X. S.; Hoffman, A. S.; Yager, P. *J. Polym. Sci.: Part A: Polym. Chem.* **1992**, *30*, 2121.
10. Nagaoka, N.; Safranj, A.; Yoshida, M.; Omichi, H.; Kubota, H.; Katakai, R. *Macromolecules* **1993**, *26*, 7386.
11. Rosiak, J. M.; Olejniczak, J. *Radiat. Phys. Chem.* **1993**, *42*, 903.
12. Taylor, L. D.; Cerankowski, L. D. *J. Polym. Sci., Polym. Chem. Ed.* **1975**, *13*, 2551.
13. Binkert, T.; Oberreich, J.; Meewes, M.; Nyffenegger, R.; Ricka, J. *Macromolecules* **1991**, *24*, 5806.
14. Dong, L. C.; Hoffman, A. S. *J. Controlled Release* **1990**, *13*, 21.
15. Dong, L. C.; Hoffman, A. S. *J. Controlled Release* **1992**, *19*, 171.
16. Park, T. G.; Hoffman, A. S. *Appl. Biochem. Biotechnol.* **1988**, *19*, 1.
17. Park, T. G.; Hoffman, A. S. *J. Biomed. Mater. Res.* **1990**, *24*, 21.
18. Kokufuta, E. *Advances in Polymer Science* **1993**, *110*, 157.
19. Ogura, K.; Hattori, T.; Hirata, M.; Asano, M.; Yoshida, M.; Omichi, H.; Nagaoka, N.; Kubota, H.; Katakai, R. *Radiat. Meas.* **1995**, *25*, 159.

POLYITACONATES (Monosubstituted Synthesis and Solution Properties)

Deodata Radiéc and Ligia Gargallo
Depto. Química Física Facultad de Química
Pontificia Universidad Católica de Chile

Poly(monoitaconates) may be considered as typical comb-like-polymers depending on the length of the side chain.[2,3] The effect of the length of the side chain and the presence of the carboxylic group have been taken into account to explain the particular conformational behavior of some poly(monoitaconates).[2,3]

MONOMERS AND POLYMERS PREPARATION

Esterification of Itaconic Acid

Monoitaconates (2-methylenesuccinates) are prepared by esterification of itaconic acid (2-methylene succinic acid) or itaconic anhydridre (β-methyl hydrogen itaconate) with the corresponding alcohols according to the procedure described by Baker et al.[9]

Polymerization

Radical polymerization of monoitaconates is carried out in bulk at the melting point temperature, under N_2 atmosphere in the presence of 2,2'-azoisobutyonitrile (AIBN) as the initiator. Polymerization time is about 48 h and conversion of monomer to polymer is 45%. It is known that polymerizations of diphenyl and ditoluyl itaconates can be carried out in solid state.[1,10]

SOLUTION PROPERTIES

General Solution Behavior

Poly(monoitaconates) have been studied from different points of view. Viscometric behavior of these polymers show very interesting phenomena. The first members of the series that is, poly(monomethyl itaconate) and poly(monoethyl itaconate), behave as polyelectrolytes in polar solvents such as methanol or water.[8,11] Similar behavior is observed for fractions of poly(monobenzyl itaconate), poly(monoethylphenyl itaconate and poly(monopropylphenyl itaconate).[4,7]

On the other hand, polymonoalkyl itaconates containing long side chains may be considered as typical comblike polymers. Polymonoitaconates containing octyl, decyl, dodecyl, and tetradecyl groups as side chains show a particular conformational behavior in solution and in general depict a deviation from expected properties for flexible linear polymers, mainly owing to high rigidity, aggregation, and chain-thickness effect.[2,3]

Deviation from Classical Solution Behavior

Because of the high rigidity shown by poly(*n*-alkyl itaconates) with long side chains, Kuhn-Mark-Houwink-Sakurada relation deviates from the linearity usually found in this kind of plot for most polymeric systems.[2] This behavior is explained by assuming that the viscosity dependence at high molecular weights corresponds to that of a nondraining coil slightly perturbed by the excluded volume effect. However, large values for the second viral coefficient A_2 are found. These results are interpreted in terms that the viscosity of a dilute solution may be affected not only by hydrodynamic interaction of chains but also by the interaction of the side chains, in many respects similar to the short-range forces excerted by hydrogen bonds in polypeptides, and they can compensate the effects.[2] The high A_2 values observed are a reflection of the excluded volume effect.[2] The second viral coefficient A_2 decreases with increasing molecular weight and the molecular weight dependence of the experimental [η] reflects the influence of chain stiffness and on excluded volume in a combined way. For these polymers unperturbed dimensions cannot be obtained by the classical semiempirical methods such as Stockmayer-Fixman.[12]

Photobehavior of Aromatic Polymonoitaconates

Photobehavior of dilute solutions of aromatic polymonoitaconates and small model molecules has also been studied together with copolymers containing monoitaconate units.[13]

Interpolymer Complex Formation

Interpolymer complex formation in solution has been reported in mixtures containing polymonoitaconates, specifically poly(monobenzyl itaconate) and poly(monomethyl itaconate) with poly(*N*-vinyl-2-pyrrolidone) (PVP).[5,6,14,15]

ACKNOWLEDGMENTS

We express our thanks to DIUC, Ponticicia Universidad de Chile Católica de Chile and FONDECYT for partial financial support.

REFERENCES

1. Velicković, J.; Plavsić, M. *Eur. Polym. J.* **1983**, *19*, 1171.
2. León, A.; Gargallo, L.; Horta, A.; Radić, D. *J. Polym. Sci., Polym. Phys. Ed.* **1989**, *27*, 2337.
3. León, A.; Gargallo, L.; Radić, D.; Horta, A. *Polymer* **1992**, *32*, 761.
4. Gargallo, L.; Radić, D.; Bruce, D. *Polymer* **1993**, *34*, 4773.
5. Pérez-Dorado, A.; Fernández-Piérola, I.; Baselga, J.; Gargallo, L.; Radić, D. *Makromol. Chem.* **1989**, *190*, 2975.
6. Pérez-Dorado, A.; Fernández-Piérola, I.; Baselga, J.; Gargallo, L.; Radić, D. *Makromol. Chem.* **1990**, *191*, 2905.
7. Yazdani-Pedram, M.; Gargallo, L.; Radić, D. *Eur. Polym. J.* **1985**, *21*, 707.
8. Gargallo, L.; Yazdani-Pedram, M.; Radić, D.; Horta, A. *Eur. Polym. J.* **1989**, *25*, 1059.
9. Baker, B. R.; Schaub, R. E.; Williams, J. H. *J. Org. Chem.* **1952**, *17*, 122.
10. Velicković, J.; Plavsić, M. *Eur. Polym. J.* **1976**, *12*, 151.
11. Gargallo, L.; Yazdani-Pedram, M.; Radić, D.; Horta, A.; Bravo, J. *Eur. Polym. J.* **1993**, *29*, 609.
12. Stockmayer, W. H.; Fixman, M. *J. Polym. Sci.* **1963**, *C-1*, 137.
13. Abuin, E.; Lissi, E.; Gargallo, L.; Radić, D. *Makromol. Chem., Macromol. Symp.* **1986**, *2*, 155.
14. Bimendina, L. A.; Bekturov, E. A.; Tleubaeva, G. S.; Frolova, V. A. *J. Polym. Sci., Polym. Symp.* **1979**, *66*, 9.
15. Radić, D.; Opazo, A.; Guerrero, X.; Gargallo, L. *Inter. J. Polymeric Mater.* **1972**, *18*, 19.

Polyketene

See: Poly(β-ketone)

Poly(ketene acetal)

See: Poly(β-ketone)

Polyketide

See: Poly(β-ketone)

POLY(β-KETONE)

Shiro Kobayashi and Shin-ichiro Shoda
Department of Materials Chemistry
Graduate School of Engineering
Tohoku University

A poly(β-ketone) (poly(ketene)) is a polymer having a repeating 1,3-diketone unit in the main chain. Since a keto function exists as equilibrium with an enol form, the structure of poly(β-ketone) can be expressed as a mixture of a keto unit and an enol unit (**Figure 1**).

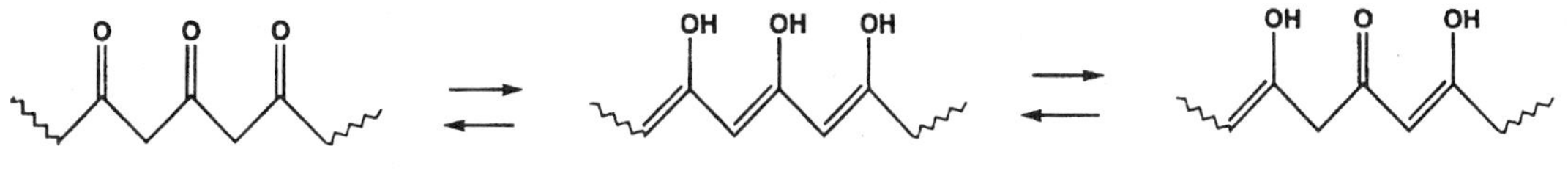

FIGURE 1. Structure of poly(β-ketone), which exists in a ketoenol equilibrium.

PREPARATION

A polyketide skeleton has also been constructed via a biogenetically modeled route, that is, by using a stepwise C–C bond-forming reaction (for example, Claisen condensation) starting from various β-keto compounds.[1–4]

Another efficient route for 1,3-diketone unit formation can be achieved by the homopolymerization of a ketene derivative, which contains the C=C and C=O group, both of which are polymerizable in the presence of anionic or a cationic initiator. The regioselective polyaddition of C=C bond unit in ketene affords the polyketone structure, whereas the incorporation of the C=O bond forms an ester moiety as undesired unit in the main chain. The reaction mode of ketone polymerization changes greatly depending upon the class of initiators employed. The poly(β-ketone) moiety can be produced preferentially when a cationic initiator is used.[5]

The ring-opening polymerization of diketene also affords polymers having an ester moiety as well as a keto moiety, and the ratio of these two moieties changes depending upon the source of initiator. When $AlCl_3$ or BF_3-OEt_2 is used as a catalyst, poly(β-ketone) type polymer can be obtained predominantly.

Recently, based on findings concerning the ketene formation from an acyl cation by a deprotonation process, a new synthetic route for preparation of poly(β-ketone)s has been developed.[6] Acetyl chloride and substituted enolizable acetyl halides containing at least one α-hydrogen atom undergo Friedel-Crafts dehydrohalogenative polymerization to poly(β-ketone)s.

The polymerization mechanism has been explained by assuming an initial ionization of acetyl chloride-aluminum trichloride complex to the acetyl cation followed by slow deprotonation, leading to a ketene formation.

In addition to the direct conversion of ketene or diketene derivative to poly(β-ketone)s, several indirect methodologies have recently been developed.

In spite of its unique structure derived from the 1,3-diketone unit structure in the main chain, poly(β-ketone) has not yet been applied to industrial purpose. However, this polymer has a potential utility as a high-performance polymer having an electronic or optical properties in the field of materials science.

REFERENCES

1. Harris, T. M.; Harris, C. M. *Tetrahedron* **1977**, *33*, 2159.
2. Bringmann, G. *Angew. Chem. Int. Ed. Engl.* **1982**, *21*, 200.
3. Mahalingam, S.; Kuzuma, P. C.; Lee, J. Y.-C.; Harris, T. M. *J. Am. Chem. Soc.* **1985**, *107*, 7760.
4. Yamaguchi, M.; Hasebe, K.; Minami, T. *Tetrahedron Lett.* **1986**, *27*, 2401.
5. Oda, R.; Munemiya, N.; Okano, M. *Makromol. Chem.* **1961**, *43*, 149.
6. Olah, G. A.; Zodok, E.; Edler, R.; Adamson, D. H.; Kasha, W.; Prakash, G. K. S. *J. Am. Chem. Soc.* **1989**, *111*, 9123.

Polyketones

See: *Olefin-Carbon Monoxide Copolymers (Overview)*
Ring-Opening Polymerization, Radical (With no Shrinkage in Volume)

Polylactides

See: *Biodegradable Copolymers [comparing poly(L-lactic acid)]*
Bioerodible Polymers A. Controlled Drug Release B. Fracture Fixation
Controlled Drug Delivery Systems
Stereo-Block Copolymers (of L- and D-Lactides)

Polylactones

See: *Poly(ε-caprolactone) (Overview)*

Poly(L-lysine)

See: *Host-Guest Chemistry [using α-Helical Poly(L-Lysine)]*

Polymaleimides

See: *Maleimides*

Polymer Brushes

See: *Crosslinked Polymer Brushes*

Polymer-Ceramic Nanocomposites

See: *Inorganic Nanostructured Materials*

POLYMER DISPERSED LIQUID CRYSTAL DISPLAY (Driving Circuit for HDTV)

Shoou-Jinn Chang
Department of Electrical Engineering
National Cheng Kung University

The arrival of the high-definition television (HDTV) in the next few years is going to create a demand for large screen, high-resolution color displays. Liquid crystal display (LCD) projector TV has shown great advantages over the cathode ray tube (CRT) projection system for its compactness and lightness.[1] However, the twist nematic liquid crystal (TNLC) requires two polarizers. These polarizers limit the transmittance of light intensity. The electrooptical material consists of dispersions of liquid crystal droplets in polymer film, referred to by the term Nematic Curvilinear Aligned Phase (NCAP) or Polymer Dispersed Liquid Crystal (PDLC), is an alternate material that can also be used for projection HDTV application. The droplets in PDLC have positive dielectric anisotropy and their radius sizes can be preselected between 0.1 μm and 10 μm.[2] A PDLC light valve is constructed by sandwiching a 10 to 25 μ layer of PDLC material between two transparent conducting electrodes. Already, full-color TV displays using the PDLC materials have been demonstrated. In the projection TV developed by Kunigita, the PDLC system can achieve a contrast ration greater than 200.[3] For such a PDLC system, the interface between PDLC material and its addressing active matrix is an important issue. PDLC displays require a driving transistor array with higher driving voltages, more current, and less leakage.

ADVANTAGES AND DISADVANTAGES OF PDLC

PDLCs are light scattering materials that operate on the principle of electrically modulating the refractive index of a

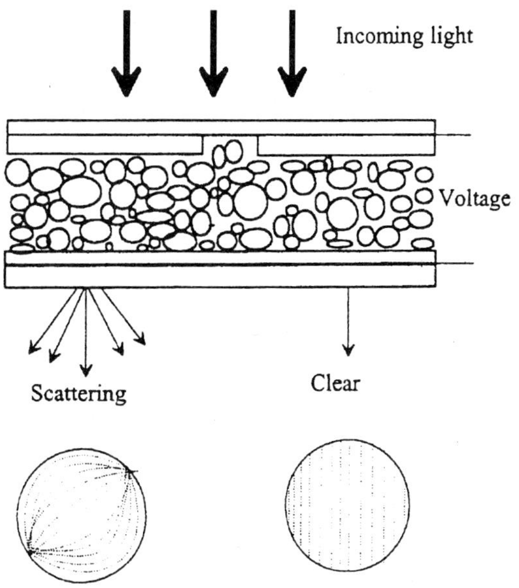

FIGURE 1. Basic device structure of a PDLC light valve.

liquid crystal to match or mismatch the refractive index of an optically isotropic, transparent solid.[4] The advantages of PDLC materials are that they provide higher optical efficiency because they do not require the polarizers, they block light by scattering, producing less heat; and the switching and relaxation times are shorter than those of TNLC and STNLC. The disadvantages of PDLC are low resistivity, typically about 10^{10} Ω-cm; and small dielectric constant, about 10.

OPERATION PRINCIPLE OF PDLC

A PDLC light valve is shown in **Figure 1**. The principle of operation of the valve is to control the light-scattering properties of the liquid crystal droplets by applying a voltage between the electrodes. In the off state (no electric field applied), the film is opaque because of the light-scattering properties of randomly oriented, liquid crystal molecules in the droplets. The application of an electric field across the film causes the liquid crystal molecules to align in the direction parallel to the electric field. If the refractive index of the polymer material matches that of the aligned liquid crystal, the incident light can travel through the film without being reflected at the interface of liquid crystal and polymer; therefore, the film becomes transparent.

REFERENCES

1. Morozumi, S. *Mol. Cryst. Liq. Cryst.* **1991**, *199*, 47.
2. Doane, J. W.; Vaz, N. A.; Wu, B. G.; Zumer, S. *Appl. Phys. Lett.* **1986**, *48*, 269.
3. Kunigita, M.; Hirai, Y.; Ooi, Y.; Niiyama, S.; Asakawa, T.; Masumo, K.; Kumai, H.; Yuki, M.; Gunjima, T. *SID 90 Digest*; p 227.
4. Craighead, H. G.; Chang, J.; Hackwood, S. *Appl. Phys. Lett.* **1982**, *40*, 22.

Polymer Electrolytes

See: Solid Polymer Electrolytes

Polymer-Induced Epitaxy

See: Epitaxial Films (Epitaxy During Polymerization)
Ultrathin Films (Self-Assembled)

Polymer-Micelle Complexes

See: Micelle-Polymer Complexes

Polymer-Polymer Complexes

See: Matrix Polymerization
Molecular Recognition (Interpolymer Interactions)
Polyitaconates (Monosubstituted Synthesis and Solution Properties)
Polymer-Polymer Interactions (Fluorescence Studies)
Polypeptide-Polypeptide Complexes
Poly(N-vinyl-2-pyrrolidone) Interactions
Stereo-Block Copolymers (of L- and D- Lactides)
Stereocomplexes (between Enantiomeric or Diastereomeric Polymers)
Template Polymerization (Overview)
Template Polymerization (Kinetics)

Polymer-Polymer Interactions

See: Polysaccharide-Polysaccharide Interactions

POLYMER-POLYMER INTERACTIONS (Fluorescence Studies)

Inés Fernandez de Piérola*
Departamento de Química Física
Universidad Nacional de Educación a Distancia (UNED)

Teresa Dib Zambon Atvars
Instituto de Química
Universidad Estadual de Campinas

Catalina Salom and Margarita Gonzalez Prolongo
Departamento de Materiales y Producción Aerospacial
ETSI Aeronauticos
Universidad Politécnica

Fluorescence methods are now being considered very useful in studying many different polymer properties both in solution or in bulk. Polymer compatibility and, in general, polymer–polymer interactions have also been studied by means of fluorescence techniques, including fluorescence quenching, fluorescence depolarization, excimer or exciplex fluorescence, and non-radiative energy transfer process.[1,2]

Several techniques based on fluorescence measurements have been proposed. Morawetz was the first one in to classify them in four groups:[3,4] fluorescence quenching, fluorescence depolarization, excimer or exciplex fluorescence, and non-radiative energy transfer.

*Author to whom correspondence should be addressed.

POLYMER BLENDS AND POLYCOMPLEXES

Polymer–polymer interactions can be either of dispersive or dipolar types and specific, for example, hydrogen bonds. The first type of interactions are weak and, since changes of the configurational entropy are small, polymer blends characterized by non-specific interactions are usually incompatible.[5] If the two mixing polymers are able to form hydrogen bonds or if they have complementary ionic groups, the enthalpy of mixing is enough to stabilize a macrocomplex, polycomplex or polysalt, respectively.

Fluorescence spectroscopy was utilized to study the driving forces producing miscibility in polymer blends. One of the most studied systems is composed of poly(acrylic acid) (PAA) and poly(ethylene glycol) (PEO), because these polymers have long been known to form complexes in aqueous solutions.[6] Morawetz et al., using dansyl-labeled PAA, observed that the dansyl moiety fluoresces much more intensely on complexation with POE than in water, reflecting exclusion of the water molecules of the PAA surface and suggesting a strong interaction between these two polymers.[7]

Although the cooperative nature of the association process between PAA and PEO has been well established,[8] only the results of *excimer fluorescence* of pyrene attached to monodisperse PEO at both chain ends demonstrated that the hydrophobic driven forces are responsible for the complex formation.[10]

Other studies using spectroscopy of these polymer blends have employed the fluorescence depolarization technique. This method is based on the idea that the degree of depolarization of the fluorescence will reflect the mobility of the polymer segments attached to the fluorophore.[9] Heyward's group applied the *fluorescence depolarization* technique to study the complexation of acenaphthylene labeled PAA with PEO.[9]

In other cases the relative intensity of the *excimer monomer fluorescence* can also be used as a probe for studies of polymer-blend miscibility in solid state, as in the case of polystyrene (PS) with poly(vinyl methyl ether) (PVME), poly(2-vinylnaphthalene) (P2VN) with either poly(cyclohexyl methacrylate) (PCMA) or polystyrene[12] and polystyrene with polybutadiene.[13]

Another way to study miscibility of polymer blends is based on the *fluorescence quenching* processes. In this case the emission intensity of a chromophore decreases in the presence of certain molecules or chemical groups called quenchers.

In changing the fluorophore-labeled polymer, one produces different photophysical systems, which allows the measurement of different properties, such as monomer fluorescence intensity showing its dependence on the rigidity of the media; solvatochromism of the label showing the dependence of the fluorescence of some molecules on the dye-polymer interactions; relative intensity of excimer/monomer forms showing its dependence on the conformation of the macromolecular chains; fluorescence depolarization showing the dependence of the anisotropy on the rigidity of the polymer chain; and miscibility of polymers by non-radiative energy transfer process producing decrease of the fluorescence intensity.

FLUORESCENCE OPTICAL MICROSCOPY

Fluorescence optical microscopy (FOM) is a technique that combines the fluorescence properties of some molecules with optical microscopy. This technique has initially been applied in a number of fields outside polymer science, principally cell biology. Applications of this technique in polymer science include determination of additives in semicrystalline polymers;[16] studies of distribution of oxidative degradation in polymers,[16] studies of morphology of polymer blends;[17–19] studies of diffusion in polymers;[20] analysis and distribution of finishes on fibers, yarns and fabrics;[21,22] and *in situ* chemical analysis of polymers in blends.[23]

POLYMER INTRINSIC FLUORESCENCE

Very often, fluorescence techniques have been applied on labeled polymers but there are also some studies on polymer–polymer miscibility based on the intrinsic fluorescence of one of the mixing polymers. Polymer compatibility studies based on the intrinsic excimer emission of some polymers have already been mentioned.[11–15]

SOLVATOCHROMIC CHROMOPHORE

Fluorescence spectra of some chromophores like dansyl (DNS, N,N-dimethyl-1-amine-5-naphthalenesulfonic acid) and prodan (N,N-dimethyl-2-amine-6-propionylnaphthalene) depend on the polarity of the medium. For example, in aqueous solution, the maximum emission of DNS is observed at 500 nm, whereas in solvents of lower polarity, its spectrum is blue shifted, peaking at 460 nm in THF. The solvatochromic effect in the fluorescence of DNS was found to be very useful in studies of microheterogeneous or biphasic polymer systems.[24–26]

Fluorescence techniques provide very useful information on several aspects of polymer–polymer interactions, in particular on the distribution of polymers in microdomains in nanoscales that cannot be detected by other macroscopic techniques.

ACKNOWLEDGMENTS

Financial support by CICYT (Spain) under grant MAT93/0167 and by FAPESP, PADCT/CNPq and FINEP (Brazil), is gratefully acknowledged.

REFERENCES

1. Winnik, M. A., Ed. *Photophysical and Photochemical Tools in Polymer Science*; NATO ASI Series; D. Riedel: New York 1986; Vol. 182.
2. Nishihima, Y. J. *Polym. Sci. Part C* **1970**, *31*, 353.
3. Morawetz, H. *Science* **1979**, *203*, 405.
4. Morawetz, H. *Science* **1988**, *240*, 172.
5. Paul, D. R.; Newman, S., Eds., *Polymer Blends*; Academic: New York, 1978.
6. Osaka, Y. *J. Polymer. Sci. Polym. Chem. Ed.* **1979**, *17*, 3485.
7. Morawetz, H.; Piérola, I. F.; Jachowicz, J.; Chen, H. L. *Polym. Prepr.* **1982**, *23*, 12.
8. Bednar, B.; Morawetz, H.; Shafer, J. A. *Macromolecules* **1984**, *17*, 1634.
9. Heyward, J. J.; Ghiggino, K. P. *Macromolecules* **1989**, *22*, 1159.
10. Oyama, H. T.; Tang, W. T.; Frank, C. W. *Macromolecules* **1987**, *20*, 474, 1839.
11. Oyama, H. T.; Tang, W. T.; Franck, C. W. *Macromolecules* **1989**, *22*, 1255.
12. Franck, C. W.; Zin, W.-C. In *Photophysics of Polymers*; ACS Symposium Series 358, Hoyle, C. E.; Torkelson, J. M., Eds.; Washington, DC 1987; p 18.
13. Tsai, F-J.; Torkelson, J. M. *Macromolecules* **1988**, *21*, 1026.

14. Gelles, R.; Franck, C. W. *Macromolecules* **1982**, *15*, 1486.
15. Larbi, F. B. C.; Malone, M. F.; Winter, H. H.; Halary, J. L.; Leviet, M. H.; Monnerie, L. *Macromolecules* **1988**, *21*, 3532.
16. Billigham, N. C.; Calvertin, P. D. In *Developments in Polymer Characterization*; Dawkins, J. V., Ed.; Applied Science: Ripple Road (England), 1982; Vol. 3, p 229.
17. Martins-Franchetti, S. M.; Atvars, T. D. Z. *Eur. Polym. J.* **1995**, *31*, 467.
18. Dibbern-Brunelli, D.; Atvars, T. D. Z. *Appl. Polym. Sci.* **1995**, *55*, 889.
19. Lin, L.; Sosnowski, S.; Chaffey, C. E.; Balke, S.; Winnik, M. A. *Langmuir* **1994**, *10*, 2496.
20. Billigham, N. C.; Calvertin, P. D.; Uzuner, A. *Polymer* **1990**, *31*, 258.
21. Heilweil, H. G. *Notes on Research* 463, 1992.
22. Biguelin, A. C.; Nunes, S. P.; Atvars, T. D. Z. *Technical Report*; Rhodia, S. A. 1993.
23. Dibbern-Brunelli, D.; Atvars, T. D. Z. *J. Appl. Polym. Sci.* **1995**, *58*, 779.
24. Chen, H. L.; Morawetz, H. *Macromolecules* **1982**, *15*, 1445.
25. Parreño, J.; Piérola, I. F. *Polymer* **1990**, *31*, 1768.
26. Parreño, J.; Piérola, I. F. *Polym. Net. Blends* **1991**, *1*, 81.

POLYMER-POLYMER MIXTURES (Under High Pressure; Physical Properties)

Chitoshi Nakafuku
Faculty of Education
Kochi University

The effect of pressure on the melting and crystallization behavior of polymers has been studied by many investigators. Phase diagrams of certain crystalline polymers were determined, and books on this research area have been published.[1–3] Pressure dependence of T_m of many crystalline polymers is determined by differential thermal analysis (DTA) under high pressure or high-pressure dilatometry. Optical microscopic observations of the melting and crystallization process of polymers under high pressure have been performed by using diamond anvil cell. X-ray diffraction measurements for crystalline polymers under high pressure were also performed using diamond anvil cell or high-pressure and cell equipped, spindle-shaped beryllium window.

The phase behavior or melting and crystallization behavior of polymer–polymer mixtures under high-pressure is a very important subject to study from scientific and engineering points of view. High pressure has considerable effects on these behaviors of polymers.

PRESSURE CHANGE OF THE PHASE DIAGRAM OF UCST AND LCST

McMaster has theoretically investigated the miscibility of the polymer–polymer mixture using Prigogine-Flory theory.[4] He also discussed the pressure change of the LCST behavior of the polymer–polymer system. The case of increase of the critical temperature with pressure and the case of decreasing the critical temperature at lower pressure and then increasing with pressure were predicted. Patterson and Robard have extensively used the same Prigogine-Flory theory and discussed the thermodynamics of compatible polymer mixtures.[5] According to them, the pressure dependence of the LCST of polymer mixtures is about 2 K MPa^{-1}.

Pressure change of the phase diagram of polymer–polymer or polymer–oligomer mixtures was investigated by Rostami and Walsh.[6–8] They derived an equation of the difference of chemical potentials as functions of many thermodynamic quantities in the mixture for each component using the combinatorial entropy of mixing by the Flory-Huggins theory and Flory and co-workers theory. They obtained binodal and spinodal curve by differentiating the equations under pressure and comparing with the experimental data for the above mixtures. The pressure change of the UCST behavior of the oligomer mixtures was determined.

MELTING OF POLYMERS IN THE BINARY MIXTURE UNDER HIGH PRESSURE

The melting temperature of crystalline polymers increases with pressure. The value of the pressure dependence of T_m (dT_m/dP) of polymers at 0.1 MPa ranges from about 0.12 K MPa^{-1} or 0.15 K MPa^{-1} of poly(ethylene oxide) (PEO) to 1.52 K MPa^{-1} or 0.97 K MPa^{-1} of PTFE.[9–12]

Melting and crystallization process of PE and isotactic poly(propylene) (iPP) mixture was studied up to 500 MPa by DTA.[13]

The pressure change of T_m of PEO in the miscible mixture with poly(methyl methacrylate) (PMMA) was studied by high pressure DTA.[14]

Melting behaviors of poly(ε-caprolactone) (PCL) in the mixture with copolymer of styrene-acrylonitrile was studied under high pressure up to 500 MPa. These two polymers are miscible at 0.1 MPa and T_m depression of PCL was observed. The depression rate of T_m was almost the same as that of 0.1 MPa up to 500 MPa.

CRYSTALLIZATION OF POLYMERS IN THE BINARY MIXTURE UNDER HIGH PRESSURE

When crystallization of PE from the melt is performed under high pressure, so-called extended chain crystal (ECC) is formed. T_m of ECC of PE is higher than 140°C and the band structure is recognized in the observation by electron microscopy. The effect of mixing iPP on the crystallization of PE under high pressure was studied by the author.[13] Basically, mixing iPP had no serious effect on the formation of ECC of PE in the crystallization under high pressure. T_m of the high-pressure crystallized PE in the binary mixture was about 140°C at 0.1 MPa.

In a semi-miscible polymer mixture including a crystalline polymer, the crystallization of the crystalline polymer is interfered with by increasing pressure.

REFERENCES

1. Wunderlich, B. *Macromolecular Physics*; Academic: New York, 1980; Vol. 3.
2. Bassett, D. C. *Developments in Crystalline Polymers-1*; Applied Science: Essex, England, 1982; Chapter III.
3. Kovarskii, A. L. *High Pressure Chemistry and Physics of Polymers*; CRC: Boca Raton, FL, 1994.
4. McMaster, L. P. *Macromolecules* **1973**, *6*, 760.
5. Patterson, D.; Robard, A. *Macromolecules* **1978**, *11*, 690.
6. Walsh, D. J.; Rostami, S. *Macromolecules* **1985**, *18*, 216.

7. Rostami, S.; Walsh, D. J. *Macromolecules* **1985**, *18*, 1228.
8. Rostami, S.; Walsh, D. J. *J. Polym. Eng. Sci.* **1987**, *27*, 315.
9. Li, W.; Rodosz, M. *Macromolecules* **1993**, *26*, 1417.
10. Tsujita, Y.; Nose, T.; Hata, T. *Polym. J.* **1974**, *6*, 51.
11. McGeer, P. L.; Duus, H. C. *J. Chem. Phys.* **1952**, *20*, 1813.
12. Starkweather, H. W. Jr.; Zoller, P.; Jones, G. A.; Vega, A. J. *J. Polym. Sci., Polym. Phys. Ed.* **1982**, *20*, 751.
13. Nakafuku, C. *Polym. J.* **1983**, *15*, 641.
14. Nakafuku, C.; Toyonaga, N. *Polymer* **1992**, *33*, 2370.

Polymer-Polymer Reactions

See: Interpolymers

Polymer Reactions

See: Chlorinated Polyethylene
Fluoropolymers (Preparation by Fluorination of Polymers)
Polyvinyltetrazoles
Reactive Processing, Thermoplastics
Reissert Polymers
Silk Fibers (Chemical Modification)
Supported Catalysts (Lewis Acid and Ziegler-Natta)
Wood (Chemical Modification)
Wood, Chemically Modified (Thermoplasticization and Solubilization)

Polymer-Small Molecule Interactions

See: Amphiphilic Polymers (Binding Properties for Small Molecules)
Poly(vinyl alcohol)-Iodine Interactions
Poly(vinyl alcohol)-Ion Complex Gels
Poly(N-vinyl-2-pyrrolidone) Interactions

Polymer-Surfactant Interactions

See: Poly(N-vinyl-2-pyrrolidone) Interactions

POLYMERIC DRUGS

Masayuki Kuzuya* and Shin-ichi Kondo
Laboratory of Pharmaceutical Physical Chemistry
Gifu Pharmaceutical University

Polymeric drugs can be divided into two types. One is that polymers act themselves as genuine drugs. For example, it is known that poly(acrylic acid) acts as an anticancer agent. The other is that the parent drugs are attached to a polymeric carrier, and the drugs can be released from the polymeric carrier by hydrolytic, enzymatic, or oxidative cleavages. This type of polymeric drug can be regarded as polymeric prodrugs. The majority of polymeric drugs thus far investigated are polymeric prodrugs.

A prodrug is a pharmacologically inactive derivative of a parent drug molecule that requires spontaneous or enzymatic transformation within the body in order to release the active drug, and that has improved delivery properties over the parent drug molecule. Thus, the prodrug approach has been used frequently to solve pharmaceutical formulation problems, such as stability and solubility.

*Author to whom correspondence should be addressed.

PREPARATION OF POLYMERIC PRODRUG[1–3]

In principle, the polymeric prodrugs can be prepared in two different ways. One approach is to substitute reactive groups of the polymers with functional groups of biologically active compounds. It is often observed, however, that some undesired side reactions occur in such reactions, and the substitution reactions of this type sometimes do not proceed to completion because of severe steric hindrance by neighboring side-chain groups. However, such problems can be removed by spacing the reactive group of several carbon atoms from the polymer main chain. The other approach is the polymerization of drug molecules with polymerizable functional groups such as vinyl groups. This approach has the advantage that the monomer can be highly purified and the polymerized or compolymerized with any number of desirable comonomers.

RECENT ADVANCES IN POLYMERIC PRODRUGS

Passive Targeting

The physicochemical properties of polymer, such as the hydrophilic-lipophilic balance, the electric charge, the molecular weight, and so on, are of essential significance with respect to their interaction with biological structures. Passive targeting is the method that the polymeric prodrug could be selectively collected in specific organs, tissues, or cells by adjusting its physicochemical property.

Active Targeting

Selective destruction of cells (cancer, etc.) is required for the treatment of various diseases. A homing device should provide selective binding of the molecule to its target. The antibodies, hormone, lectin, or sugar chain can be utilized as homing device. The polymeric prodrugs possessing these functions can be actively targeted to specific organs, tissues, or cells.

Lysosomotropic Polymeric Prodrug

The term "lysosomotropic" designates all substances that are taken up selectively into lysosomes, irrespective of their chemical nature or mechanism of uptake. The materials that can be taken up into lysosomes by endocytosis include a variety of small molecules, all major groups of macromolecules, insoluble particles, and so on.[4]

The lysosomotropic polymeric prodrug, which must be endocytizable, should be degradable by the lysosomes, although this may not be an absolute requirement. As a rule, the linkage chosen should render the drug pharmacologically inactive, be stable in the environments to which the complex is to be exposed before reaching the lysosomes, and be broken within the lysosomes.

Hybrid Polymeric Prodrug

The hybrid polymeric prodrug, which possesses two or more kinds of drug as a side chain, is expected to have improved drug efficacy because of the sequential and/or simultaneous action of some kinds of drugs in the body.

Novel Syntheses of Polymeric Prodrugs by Mechanochemical Polymerization

Kuzuya et al. (1991)[5] recently reported the detailed electron spin resonance (ESR) study on mechanochemical solid state polymerization of acrylamide and presented the first example of polymeric prodrugs prepared by mechanochemical solid state polymerizations of several methacryloyl vinyl monomers derived from bioactive compounds.[6]

Kondo et al.[7] prepared hybrid polymeric prodrugs for chemoembolization. This hybrid polymeric prodrug was synthesized by the mechanochemical copolymerization of the methacryloyl derivatives of 5-fluorouracil (5FU) and 2,4,6-triiodophenol (contrast medium) with acrylamide. This hybrid polymeric prodrug released a significant amount of 5FU in rat serum at 37°C. It was also shown that the embolization formed by this polymeric prodrug was directly observed by skiagraphy. These results suggest that the hybrid polymeric prodrug containing anticancer agent and contrast medium is useful for chemoembolization.

REFERENCES

1. Joshi, H. N. *Pharm. Technol.* **1988**, *12*, 118.
2. Scheler, W. *Makromol. Chem., Macromol. Symp.* **1987**, 12, 1.
3. Kalal, J. *Makromol. Chem., Macromol. Symp.* **1987**, 1, 259.
4. DE Duve, C.; DE Barsy, T.; Poole, B.; Trouet, A.; Tulkens, P.; Van Hoof, F. *Biochem. Pharmc.* **1974**, 23, 2495.
5. Kuzuya, M.; Kondo, S.; Noguchi, A.; Noda, N. J. *Polym. Sci., Polym. Chem.* **1991**, 29, 489.
6. Kuzuya, M.; Kondo, S.; Noguchi, A. *Macromolecules* **1991**, 24, 4047.
7. Kondo, S.; Kuzuya, M. *DDS* **1994**, 9, 315.

Polymeric Initiators

See: Initiators, Polymeric

Polymeric Scale Inhibitors

See: Scale Inhibitors, Polymeric

2,ω-Polymerization

See: Olefin Polymerization by 2,ω-Linkage (Migratory Nickel-Phosphorane Catalyst)

POLYMERIZATION SEPARATION (Ring-Opening Metathesis Polymerization)

James G. Hamilton
School of Chemistry
The Queen's University of Belfast

In spite of many excellent chromatographic methods, the separation of isomeric compounds can be time consuming and can lead to low yields and impure products. In some cases these problems can be overcome, at least in part, by using a polymerization separation technique that is based on the simple principle of rendering involatile or insoluble one isomer of a pair by its polymerization. The unreacted isomer can then be recovered either by treating the reaction mixture with a solvent, in which case the polymerized isomer precipitates, leaving the unreacted isomer in solution, or more conveniently, by subjecting the reaction mixture to high vacuum and collecting the remaining isomer in a cold trap.

The method has been applied so far to the separation of linear olefins using Ziegler-Natta polymerization and of cyclic olefins using Ring-Opening Metathesis Polymerization (ROMP) and pure isomers have been recovered from pairs that, owing to very similar physical and chromatographic properties, would have been difficult to separate by more conventional means.[1–4]

APPLICATIONS

In the area of ROMP (and polymerization in general), it is frequently the case that ease of polymerization is strongly affected by variations in the position of substituent groups on potential monomers. Therefore, this separation method may be found to work where a suitable polymerization mechanism exists, and where such structural variations occur between a pair of isomers.

The recent development of well-defined ROMP initiators has expanded the range of potential monomers and has increased the degree of control over the polymerization reaction.[5,6] These developments should allow many other potential applications of the polymerization-separation method to be explored.

REFERENCES

1. Endo, K.; Tsujinoto, M.; Otsu, T. *J. Polym. Sci., Part A: Polym. Chem.* **1993**, *31*, 2177.
2. Hamilton, J. G.; Rooney, J. J.; Snowden, D. G. *J. Polym. Sci., Part A: Polym. Chem.* **1994**, *32*, 993.
3. Ivin, K. J. *Olefin Metathesis*; Academic: London, 1983.
4. Breslow, D. S. *Prog. Polym. Sci.* **1993**, *18*, 1141.
5. Gagné, M. R.; Grubbs, R. H.; Feldman, J.; Ziller, J. W. *Organometallics* **1992**, *11*, 3933.
6. Schrock, R. R. *Pure & Appl. Chem.* **1994**, *66*, 1447.

Polymethacrylamides

See: Charge Transport Polymers (for Organic Electroluminescent Devices)

Poly(p-methoxystyrene)

See: Polystyrene and Derivatives, Photolysis

Poly(methyl methacrylate)

See: Poly(methyl methacrylate) (Overview)
Stereoregular Polymethacrylate Macromonomer
Stereoregular Polymethacrylates
Uniform Polymers
Vinylidene Fluoride-Based Thermoplastics (Blends with Other Polymers)
Zinc Catalysts (Polymerization of Styrene and Methyl Methacrylates)

POLY(METHYL METHACRYLATE) (Overview)

Michael S. Cholod and Hsing-Yeh Parker
Rohm and Haas Company

Otto Rohm and Otto Haas pioneered the commercial application of acrylate and methacrylate polymers.[1,2] Their first commercial (meth)acrylate product was a polyacrylic interlayer for safety glass in 1928. This was shortly followed by cast polymethyl methacrylate [9011-14-7] sheet initially used in both North America and Europe for aircraft glazing during the Second World War. The sheets were cast between two glass plates and then heated and vacuum-formed to the shape of aircraft canopies. Shortly after the war, new markets for poly(methyl methacrylate) sheets in glazing and lighting fixtures arose. At about that time molding pellets were introduced into many automotive applications, once heat and light stable colorants were found. In the 1960s processes for continuous casting of sheet and continuous bulk polymerization to yield pellets were commercialized.[3,4] During the early 1980s continuous melt-calendered sheet was developed and replaced cell cast sheet in many applications where lower cost was a factor.

PREPARATION

Poly(methyl methacrylate) is made commercially by four different polymerization methods: bulk casting, suspension polymerization, bulk polymerization, and emulsion polymerization. Each of these methods utilizes free radical chemistry. Other forms of polymerization (for example, anionic, group transfer, and other types of "living" polymerizations) are known but are not commercially important today.

Commercially available methyl methacrylate (MMA) polymers have molecular weights ranging from about 70,000 to about 200,000 for those polymers produced for molding or extrusion applications. Cast sheets and polymers made by emulsion polymerization can have molecular weights greater than 1,000,000. MMA polymerization is accompanied by the liberation of the heat of polymerization, (13.8 kcal/mol) which must be controlled to allow for safe operation and to prevent unwanted side-reactions. Volumetric shrinkage during polymerization is about 21%.

PROPERTIES

The glass transition temperature of poly(methyl methacrylate) depends on the chain tacticity and can range from about 45°C for isotactic poly(methyl methacrylate) to greater than 130°C for syndiotactic poly(methyl methacrylate). Tacticity of poly(methyl methacrylate) depends on the method and temperature of polymerization.[5] Tacticity of anionically polymerized poly(methyl methacrylate) is strongly influenced by the solvent and cationic species used and can be adjusted over a wide range. The tacticity of emulsion-made poly(methyl methacrylate) is usually 50–70% syndiotactic, ~ 30% atactic and ~ 10% or less isotactic. Commercial bulk polymerized poly(methyl methacrylate) has a T_g in the 110–115°C range.

Poly(methyl methacrylate) made by emulsion polymerization can reach several million in molecular weight.

Optical Properties

Poly(methyl methacrylate) polymers occupy a unique position among commercial polymers in their superior optical properties, ease of formation, and stability to light[3] and heat. Poly(methyl methacrylate) is available with almost perfect transmission in the visible range. Poly(methyl methacrylate) transmits well into the ultraviolet range, making it a frequently used material for medical applications in which UV spectroscopic analytical techniques are increasingly utilized. Some commercial grades of poly(methyl methacrylate) are supplied with ultraviolet absorbers, such as benzotriazoles, that protect the polymer against long-term degradation by UV light, especially at those wavelengths present in sunlight.

APPLICATIONS

Polymers made from poly(methyl methacrylate) and copolymers that are largely poly(methyl methacrylate) are used primarily in plastics applications. These poly(methyl methacrylate) products have a unique combination of properties, including crystal clarity, resistance to light and weathering, breakage resistance, and mechanical strength, making them unique among plastics. Perhaps the best known product is Plexiglas® sold by AtoHaas, N.A., Inc. The hardness of poly(methyl methacrylate) allows it to be substituted for glass, metals, and wood in many applications.

Today the principal injection molding uses for poly(methyl methacrylate) are in automotive taillight lenses, numerous medical applications, and glazing (both in cell cast and extrusion melt-calendered sheet). The high optical clarity and low birefrigence of poly(methyl methacrylate) make it attractive for use in laser-disk applications, although it has not been widely used for compact disks owing to its relatively high co-efficient of expansion when exposed to water vapor, a critical property for one-sided disks. Methyl methacrylate polymers are also used extensively for extrusion applications[6] such as sheet, film, profiles, and rods.

Emulsion polymers of methacrylates and acrylates are usually used as specialty polymers, such as processing aids and impact modifiers for thermoplastics, adhesives, caulks, and coatings or binders in floor polishes, paints, and paper applications.

REFERENCES

1. Hochheiser, S. *Rohm & Haas*; University of Pennsylvania: Philadelphia, 1986.
2. Cholod, M. S.; Miller, G. W. *Polymers, Polymethylmethacrylate*; In Encyclopedia of Chemical Processing and Design McKetta, J. J., Ed.; Marcel Dekker: New York, 1992; pp 288–346.
3. Opel, C. J.; Bottoms, P. H. U.S. Patent 2 376 371, to Swedlow: April 2, 1968.
4. Salkind, M.; Riddle, E. H.; Keefer, R. W. *Ind. Eng. Chem.* **1958**, *51*, 1232, 1928.
5. Thompson, E. V. *J. Polymer Science, Part A-2* **1966**, *4*, 199–208.
6. Rohm & Haas Co., *Plexiglas V-Series Acrylic Plastic Extrusion Resins*; Bulletin PL-132G, April 1988.

Poly(3-methyl-1-pentene)

See: Polyolefins (Solid State Nuclear Magnetic Resonance)

Poly(4-methyl-1-pentene)

See: Polyolefins (Overview)

Polymethylenes

See: Captodative Compounds, Polymerization
Uniform Polymers

POLYMETHYLPHENYLSILOXANE (Photophysics, Conformation, and Dynamics)

Arturo Horta and Inés Fernandez de Piérola
Departamento de Fisicoquímica (CTFQ)
Facultad de Ciencias
Universidad a Distancia (UNED)

Antonio L. Macanita
Instituto de Tecnología Química e Biológica

Silicones are quite well-known materials with application in fields as different as biomedical devices, waterproofing agents, seals, hoses, O-rings, or as the stationary phase in chromatography.[1] These exists an evident connection between the applicability of these materials and the conformational and dynamic properties of their siloxane chain structure.

CONFORMATIONAL CHARACTERISTICS

When the chromophores responsible for the photophysical behavior are side groups rigidly attached to the polymer backbone, as is the case in polymethylphenylsiloxane (PMPS), then the existence of an excimer-forming site (EFS) and the formation of excimers are dependent on the conformational characteristics of the chain. Ideally, if the excited state lifetime is shorter than the average time required for a transition between conformational states, then excimers would form only at the EFS existent at the time of the excitation, and excimer emission would reflect the relative abundance of such EFS in the instantaneous (equilibrium) distribution of conformational states in the macromolecule. On the other hand, if the excited state lifetime is longer, transitions between conformations have time to occur, and the excimer emission would reflect the dynamics of such transitions.[2]

The analysis of EFS, or performed dimers in the ground state, is thus relevant for photophysics.

The published conformational studies of PMPS are based on the work of Mark and Ko.[3] These authors obtained conformational energies for PMPS combining known interaction energies for the related chain of polydimethylsiloxane (PDMS) with new results from semiempirical interatomic potentials. They found that, contrary to what happens in the vinyl chains, where steric repulsions due to atom crowding are the dominant interactions among side groups, in the case of PMPS the interactions between pendant groups are attractive, particularly when two phenyl groups are involved, owing to the relatively large intergroup separations dictated by the long Si—O bond and the wide Si—O—Si angle that build the siloxane backbone.

The characteristic ratio, ($C_\infty = \langle r^2\rangle_0/nl^2$) of PMPS with $W_m = 0.51$ has been obtained by Salom et al. from intrinsic viscosity and quasi-elastic light scattering.[5] The value determined by quasi-elastic light scattering ($C_\infty = 8.05$), is in good agreement with a previous experimental result for an unspecified tacticity.[6] The agreement with the value predicted by the rotational isomeric state model, according to the calculations of Mark and Ko, for this tacticity ($C_\infty = 5$–6), is rather modest.[3]

The temperature coefficient of dimensions, $d \ln \langle r^2\rangle_0/dT$, was determined by Llorente et al. from force-temperature thermoelastic measurements on elastomeric networks obtained by crosslinking PMPS chains.[4] Their results ($10^3 . d \ln \langle r^2\rangle_0/dT = 0.64 \pm 0.13\ K^{-1}$ for $W_m = 0.51$) are in agreement with previous experiments on samples of unspecified tacticity ($0.61 \pm 0.09\ K^{-1}$).[7]

Electrical birefringence of PMPS solutions was also measured to determine the molar Kerr constant, $\langle_m K\rangle/x$ of the polymer.[8] The value determined is small: $10^{25} . \langle_m K\rangle/x = 1.9\ m^5V^{-2}mol^{-1}$, which can be explained, in part, because the dipole moments of consecutive Si—O and O—Si bonds along the chain backbone suffer a substantial cancellation owing to the wide Si—O—Si angle.

Another property also dependent upon the anisotropic part of the polarizability tensor, $\hat{\alpha}$, is the effective optical anisotropy, $\langle\gamma^2\rangle/x$. This was determined from depolarized Rayleigh scattering experiments on CCl_4 solutions of PMPS samples of different tacticities ($W_m = 0.45$; 0.57; 0.67; 0.75).[9]

The dipole moment has been determined both for the homopolymer PMPS and for copolymers with dimethylsiloxane units.[10,11] The dipole moment ratio, $\langle\mu^2\rangle/xm^2$ (with $m = 0.60$ D), of the homopolymer: $\langle\mu^2\rangle/xm^2 = 0.30$, is similar to that of other siloxane polymers (0.37 for PDMS).[9]

GROUND STATE PREFORMED DIMERS

All the conformational studies referred to up to now take as starting point the energetic parameters first calculated for PMPS by Mark and Ko.[3] Also, they use fixed geometry for the molecule, with the sole exception of allowing a limited fluctuation of the rotational angles around given positions. However, recent studies by Mattice et al. on PDMS and other polydialkyl-siloxanes have shown the need for a more realistic description of siloxane chains.[12–15] These authors modified the starting energetic parameters for alkyl siloxanes to make them compatible with the results determined from *ab initio* quantum computations and the more recent conventional molecular mechanics considerations.[12] Also, they allowed for relaxation of all the atoms by introducing a potential energy consisting of terms for bond bending, bond stretching, interatomic van der Waals interactions, and torsional rotations. Hence, the structure is flexible, in all its internal degrees of freedom and can relax. With such allowances, they found that the stable conformers of PDMS are displaced with respect to the standard positions, for example, the *trans* state is 20° away from the planar conformation, the gauche states are split into two minima, and there are stable states around the *cis* conformation.[12] Also as a result of the new energetic parameters and relaxation, they found for PDMS that the states g^+g^- or g^-g^+ for the pair of bonds Si—O—Si are rather stable, in sharp contrast with the previous model, which assigned a negligible probability to such states.[12] In general, the rather uniform distribution of rotational isomeric states in PDMS makes the adoption of three equally spaced rotational states justifiable only as a convenient mathematical device.[13]

CONCLUSIONS

The fast relaxations involving rotations through backbone bonds of phenylsiloxanes dissolves in common solvents at r.t. take place in hundreds of picoseconds. Within this range, they are faster in dimer and oligomers, slower in cycles, and intermediate in copolymers of methylphenylsiloxane with dimethylsiloxane. Phenylsiloxanes have a large fraction of ground-state, preformed dimers, but these have a negligible influence in the photophysics of these systems, whereas isolated chromophores in copolymers should be always taken into account to extract dynamic information from fluorescent measurements.

The adoption of three equally spaced rotational states makes sense in the case of phenylsiloxanes only as a convenient mathematical device, but probably it does not represent the conformational reality of these systems.

ACKNOWLEDGMENTS

Financial supports from ITQB (Oeiras) and the Instituto Superior (Lisboa) in Portugal, from DGICYT (Spain) under grant N. MAT93-0167 and from the Acción Concertada Spain-Portugal N. HP16B, are gratefully acknowledged.

REFERENCES

1. Allcock, H. R.; Lampe, F. W. *Contemporary Polymer Chemistry*; Prentice Hall: Englewood Cliffs, NJ, USA, 1981.
2. Bahar, I.; Mattice, W. L. *J. Chem. Phys.* **1989**, *90*, 6783.
3. Mark, J. E.; Ko, J. H. *J. Polym. Sci.: Polym. Phys. Ed.* **1975**, *13*, 2221.
4. Llorente, M. A.; Piérola, I. F.; Saiz, E. *Macromolecules* **1985**, *18*, 2663.
5. Salom, C.; Freire, J. J.; Hernández-Fuentes, I. *Polymer* **1989a**, *30*, 615.
6. Buch, R. R.; Klimisch, H. N.; Johannson, O. K. *J. Polym. Sci.* **1970**, *A2*, *8*, 541.
7. de Candia, F.; Turturro, A. *J. Macromol. Sci.-Chem.* **1972**, *A6*, 1417.
8. Mendicuti, F.; Tarazona, M. P.; Saiz, E. *Polym. Bull.* **1985**, *13*, 263.
9. Floudas, G.; Fytas, G.; Momper, B.; Saiz, E. *Macromolecules* **1990**, *23*, 498.
10. Salom, C.; Freire, J. J.; Hernández-Fuentes, I. *Polym. J.* **1988**, *20*, 1109.
11. Salom, C.; Hernández-Fuentes, I. *Eur. Polym. J.* **1989b**, *25*, 203.
12. Bahar, I.; Zúñiga, I.; Dodge, R.; Mattice, W. L. *Macromolecules* **1991**, *24*, 2986.
13. Bahar, I.; Zúñiga, I.; Dodge, R.; Mattice, W. L. *Macromolecules* **1991a**, *24*, 2993.
14. Bahar, I.; Neuburger, N.; Mattice, W. L. *Macromolecules* **1992**, *25*, 4619.
15. Neuburger, N.; Bahar, I.; Mattice, W. L. *Macromolecules* **1992**, *25*, 2447.

Poly(N-methylsilazane)s

See: Polysilazanes (Through Ring-Opening Polymerization)

Poly(p-nitrostyrene)

See: Polystyrene and Derivatives, Photolysis

Polynorbornene

See: Metathesis Polymerization, Cycloolefins

Polynortricyclene

See: Transannular Polymerization

POLYNUCLEOTIDE ANALOGUES

Man Jung Han, Tae Joon Cho, and Ji Young Chang
Department of Applied Chemistry
Ajou University

Polynucleotides play a most essential function in the living organism: the storage and transmission of genetic information by which the nucleic acids are replicated or transcripted, and the proteins are synthesized. It is therefore very interesting to synthesize polynucleotide analogues (PNAs) resembling very closely the natural polymers as model compounds for the investigation of the polynucleotides. Moreover, PNAs can be also used as chemotherapeutic agents, such as interferon inducers or antisense compounds. Comprehensive reviews of the preparation of PNAs[1] and their physiological activities[2] have appeared in the literature.

Several preparation methods of PNA had been developed: Many vinyl monomers containing nucleic acid bases (NABs) were polymerized.[3–9] NABs were also attached to the functional polymers such as poly(α-amino acids),[10,11] poly(vinyl alcohol)[12–15] and polyethylenimine.[16–21] Other methods involved condensation polymerization of ω-hydroxy carboxylic acid[22] or α-amino acid derivatives[23,24] containing the NABs.

Compared with the natural polynucleotides, the synthesized PNAs showed several drawbacks. Most of the reported PNAs exhibited neither good solubilities in water owing to the lack of hydrophilic groups, nor optical activities owing to the absence of sugar moieties on the polymer chain. The alternating sequences between nucleosides and phosphate, observed in natural polynucleotides, were rarely realized in synthetic PNAs. Most of the synthesized PNAs could not be classified as analogous to DNA or RNA owing to the lack of sugar moieties. Consequently the synthesized PNAs hardly showed the physicochemical properties and biological functions similar to those of natural polymers. In order to overcome the drawbacks of the PNAs mentioned above, several new monomers, nucleoside analogues, were synthesized. All of the monomers contained furanose or pyranose rings and vinyl groups that would facilitate radical polymerizations. Copolymerization of them with maleic anhydride or vinylene carbonate gave the alternating copolymers as shown in **Scheme I**.

COPOLYMERIZATION AND HYDROLYSIS OF THE POLYMERS

The copolymerization of the monomers with maleic anhydride or vinylene carbonate were carried out in bulk or in solution in the presence of radical initiators (AIBN or BPO). The anhydride groups of the polymers (**15, 17, 19,** and **21**) were hydrolyzed in water at room temperature, whereas hydrolysis

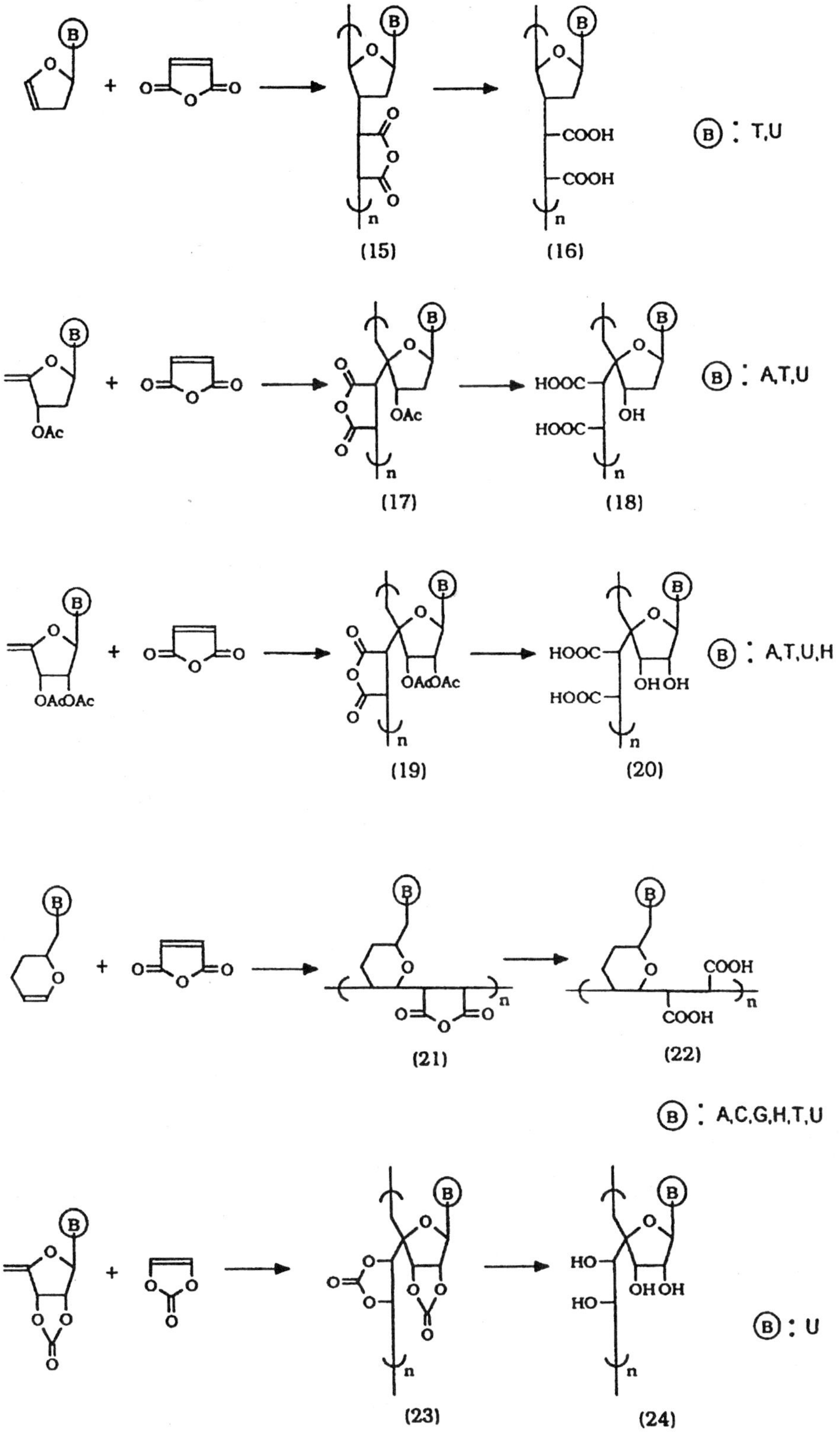

I

of blocking or carbonate groups was accomplished at higher temperatures (80–100°C) with the aid of hydroxide catalyst.

Polymers **16, 18, 20, 22,** and **24** have the alternating sequences between nucleoside analogues and 1,2-dicarboxyethylene, 1,2-dicarboxypropylene, or 1,2-dihydroxypropylene similar to the structures of polynucleotides, which contain the alternating sequences between nucleosides and phosphate. Polymers **16** and **18** are analogous to DNA whereas polymer

20 and **24** with hydroxyl groups on the 2'-position of the furanose rings are RNA analogues.

PROPERTIES AND BIOLOGICAL FUNCTIONS

Most of the synthesized polymers showed physicochemical properties similar to those of the natural polymers: they were soluble in water and optically active owing to the retention of the chiralities of the relevant starting monomers at C1', C2', and C3' during their copolymerization.

REFERENCES

1. Takemoto, K.; Inaki, Y. *Functional Monomers and Polymers*; Marcel Dekker: New York and Basel, 1987; Chapter IV.
2. Pitha, J. *Advance in Polymer Science*; Springer-Verlag: Berlin and Heidelberg, 1983; Vol. 50, p 1.
3. Kaye, H. *Polym. Letter* **1969**, *7*, 1.
4. Kondo, K.; Iwasaki, H.; Nakatani, K.; Ueda, N.; Takemoto, K.; Imoto, M. *Makromol. Chem.* **1969**, *125*, 42.
5. Kaye, H. *J. Polym. Sci.* **1969**, *B7*, 1.
6. Kondo, K.; Iwasaki, H.; Ueda, N.; Takemoto, K.; Imoto, M. *Makromol. Chem.* **1968**, *120*, 21.
7. Kita, Y.; Inaki, Y.; Takemoto, K. *J. Polym. Sci. Chem. Ed.* **1980**, *18*, 427.
8. Kondo, K.; Hisaoka, Y.; Takemoto, K. *Chem. Lett.* **1973**, 125.
9. Kondo, K.; Takemoto, K. *Macromol. Chem. Rapid Commun.* **1980**, *2*, 303.
10. Ishikawa, T.; Inski, K.; Takemoto, K. *Polym. Bull.* **1978**, *1*, 85.
11. Anand, N.; Murthy, N. S. R. K.; Naider, F.; Goodman, M. *Macromolecules* **1971**, *4*, 564.
12. Seitz, T.; Yamauchi, K.; Kinoshita, M.; Imoto, M. *Makromol. Chem.* **1972**, *154*, 263.
13. *Ibid.* **1973**, *163*, 15.
14. Overberger, C. G.; Lu, C. X. *J. Polym. Sci. Chem. Ed.* **1987**, *25*, 1523.
15. Overberger, C. G.; Chang, J. Y. *J. Polym. Sci. Chem. Ed.* **1989**, *27*, 3589.
16. Overberger, C. G.; Inaki, Y. *J. Polym. Sci. Chem. Ed.* **1979**, *17*, 1739.
17. Overberger, C. G.; Morishima, Y. *J. Polym. Sci. Chem. Ed.* **1980**, *18*, 1247.
18. Ludwiek, A.; Overberger, C. G. *J. Polym. Sci. Chem. Ed.* **1982**, *20*, 123.
19. Overberger, C. G.; Lu, C. X. *J. Polym. Sci. Chem. Ed.* **1986**, *24*, 243.
20. Overberger, C. G.; Chang, Y.; Gunn, V. E. *J. Polym. Sci. Chem. Ed.* **1989**, *27*, 99.
21. Overberger, C. G.; Chang, J. Y. *J. Polym. Sci. Chem. Ed.* **1989**, *27*, 4013.
22. Halford, M. H.; Jones, A. S. *J. Chem. Soc.* **1968**, 2667.
23. Buttrey, J. D.; Jones, A. S.; Walker, T. R. *Tetrahedron* **1979**, *31*, 73.
24. Ishirawa, T.; Inaki, Y.; Takemoto, K. *Polym. Bull.* **1978**, *1*, 215.

Polynucleotides

See: Nucleic Acids

Polyoctenamer

See: Metathesis Polymerization, Cycloolefins

POLYOLEFIN GRAFT COPOLYMERS (Prepared by Borane Approach)

T. C. Chung
Department of Materials Science and Engineering
The Pennsylvania State University

Due to low surface energy, lack of chemical functionalities and crystallinity, polyolefins interact poorly with other materials. Polyolefins exhibit inadequate compatibility with other synthetic polymers and virtually no adhesion to metal or glass. If these problems could be overcome the market for polyolefin applications would expand dramatically. Accordingly, chemical modification of polyolefins, especially polypropylene has been an area of intense interest as a way to improve these commodity polymers.

An established technique for improving the interfacial interaction between polymers and other materials is the use of graft and block copolymers as compatibilizers.[1–3] It is desirable to prepare polyolefin graft copolymers containing functional polymers, such as PMMA and PVA, at side chains, which can dramatically increase the interaction of polyolefin with a broad range of polymers containing functional groups. Unfortunately, the chemistry to prepare polyolefin graft copolymers and functionalized polyolefins is limited due mainly to the inert nature of polymers, as well as catalyst poison by functional groups.

We have developed a new method to prepare polyolefin graft copolymers that generally includes most polyolefin backbones and a wide selection of free radically polymerized and ring-opening polymerized grafts. **Equation 1** illustrates the general synthetic schemes.

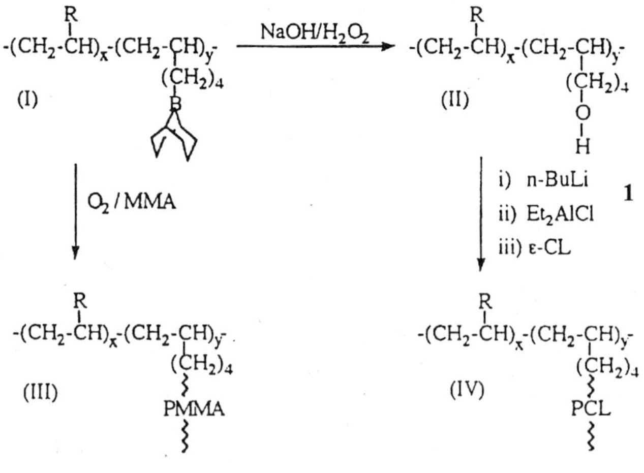

1

R : H, CH_3, C_2H_5, C_6H_{13} etc.

The borane-containing polyolefin (I) can be obtained from both direct and post-polymerization processes.[4–8] The borane groups in polyolefin (I) act as free radical sources under certain oxidative conditions to initiate the graft-from polymerization, such as methyl methacrylate (MMA) and vinyl acetate (VA). The free radical polymerized functional polymers are therefore chemically bonded to the side chains of polyolefin (III). The other type of containing condensation polymer segments (polyester or nylon), also can be prepared by the subsequent functional polyolefins obtained from borane-containing polymers. For example, the hydroxylated polyolefin (II) can initiate ring-opening polymerization of ε-caprolactone (ε-CL). The anionic ring-opening reaction took place at room temperature by insertion and ring opening of ε-CL into the Al-O bond to form polyolefin graft caprolactone copolymers (IV).

APPLICATIONS

Polyolefin graft copolymers are effective compatibilizers to improve interactions between the corresponding polyolefins and other materials, including functional polymers, engineering plastics and substrates.

PP/PMMA Blends[9,11]

Optical microscopy was used to evaluate graft copolymers' ability to act as a phase compatibilizer for blends of PP and PMMA homopolymers. Blend A is 70/30 weight % mixture of PP and PMMA homopolymers that were blended in solution.

Two distinct phases are visible: crystalline PP phase and an amorphous PMMA phase. Within the PP domain, the spherulite size varies greatly with a few extremely large crystallites and predominantly small spherulites. Blend B is a sample of blend A to which 10 weight % of PP-g-PMMA has been added in solution. The added graft copolymer was also 30 weight % PMMA so as not to change the overall composition. The graft copolymer, behaving as a polymeric emulsifier, increases the interfacial interaction between the PP amorphous region and the PMMA to reduce the domain sizes. The most noticeable change in the micrographs is the disappearance of visibly distinct PMMA domains. The large phase-separated PMMA domains are now dispersed into the inter-spherulite regions and cannot be resolved by resolution of the optical microscope. The mode of nucleation within the polypropylene crystallite phase has changed as evidenced by the now relatively homogeneous sperulite size.

PP/PCL Blends[9,12]

SEM operating with secondary electron imaging was used to observe the surface topography of cold fractured film edges.

In the homopolymer blend, the polymers are grossly phase separated as can be seen by the minor component PCL, which exhibits nonuniform, poorly dispersed domains and voids at the fracture surface. The compatibilized blend no longer displays the distinct PCL globules. Rather, a flat mesa-like fracture surface is observed similar to that seen on fractured PP or PP-g-PCL surfaces. Both PP and PCL phases are mechanically tied and the interface is bridged by the graft copolymer extending into both phases tying them by a covalent bond.

PP/PC Blends[12]

It would be extremely advantageous and inexpensive if commodity PP could be effectively compatibilized in blends with the more costly engineering resin polycarbonate. Since PCL and polycarbonate (PC) form a miscible blend, the graft copolymer PP-g-PCL should behave as an emulsifier for PP and PC blends.

A blend containing 70/30/10 of PP/PC/PP-g-PCL, where the graft contains 57 weight % PCL, was used to evaluate the effect of the polymeric compatibilizer on the blend's morphology. Only small distorted spherulites are observed and only a few very small distinct PC phases can be found. The PP-g-PCL is clearly proven to be an effective compatibilizer for PP and PC blends. Films of the compatibilized blends formed in the melt press were optically clear. This is unlike pure i-PP which forms hazy, translucent films. The lack of large spherulites in both the blend and the graft must minimize scattering.

ACKNOWLEDGMENT

The authors would like to thank the Polymer Program of the National Science Foundation for financial support.

REFERENCES

1. Riess, G.; Periard, J.; Bonderet, A. *Colloidal and Morphological Behavior of Block and Graft Copolymers*; Plenum: NY, 1971.
2. Epstein, B. U.S. Patent 4 174 358, 1979.
3. Lohse, D.; Datta, S.; Kresge, E. *Macromolecules* **1991**, *24*, 561.
4. Chung, T. C. U.S. Patents 4 734 472 and 4 751 276, 1988.
5. Chung, T. C. *Macromolecules* **1988**, *21*, 865.
6. Ramakrishnan, S.; Berluche, E.; Chung, T. C. *Macromolecules* **1990**, *23*, 378.
7. Chung, T. C.; Rhubright, D. *Macromolecules* **1993**, *26*, 3019.
8. Chung, T. C.; Rhubright, D. *J. of Polymer Sci. Polymer. Chem. Ed.* **1993**, *31*, 2759.
9. Chung, T. C. U.S. Patent 5 286 800, 1994.
10. Chung, T. C.; Jiang, G. J. *Macromolecules* **1992**, *25*, 4816.
11. Chung, T. C.; Rhubright, D.; Jiang, G. J. *Macromolecules* **1993**, *26*, 3467.
12. Chung, T. C.; Rhubright, D. *Macromolecules* **1994**, *27*, 1313.

Polyolefins

See: Crosslinking (of Polyolefins)
Epitaxial Crystallization (of Linear Polymers)
Foamed Plastics
Functionalized Polyolefins (Using Borane Monomers and Transition Metal Catalysts)
Photo-oxidation, Polyolefins
Polyolefin Graft Copolymers (Prepared by Borane Approach)
Polyolefins (Overview)
Polyolefins (Solid State Nuclear Magnetic Resonance)

POLYOLEFINS (Overview)

Frank M. McMillan
Calsec Consultants

Polyolefins are the world's most important class of polymers in terms of volume usage and the simplest in terms of composition (empirical formula, C_nH_{2n}). They are long-chain, aliphatic hydrocarbons consisting of polymers of alpha-olefins, with molecular weights ranging from a few thousand to a few million. They differ from one another in molecular weight, in the degree of chain branching, in the frequency of (short) side chains, and in the regularity of the chain structure. These differences result from variations in reaction conditions, the type of catalyst used, and the particular olefin used as monomer.

POLYMER TYPES AND POLYMERIZATION PROCESSES

Polyethylene—Low-density (LDPE) (High-pressure, Branched) was discovered accidentally in the laboratories of Imperial Chemical Industries, in England in the 1930s.[1] It is made by free-radical (anionic) polymerization of ethylene under high pressure (15,000–40,000 psig) and high temperature (150–300°C) in the presence of trace amounts of oxygen or peroxides as initiator.

LDPE has a moderate degree of crystallinity (ca. 50%), because the molecular chain has numerous short-chain branches. It is a non-brittle plastic with easy processability, moderate hardness, and low softening temperature. Major uses of LDPE are in packaging film and laminates, paper and foil coating, blow-molded containers, wire insulation, and injection-molded articles.

Polyethylene–high density (HDPE) (Low-pressure, Linear) is made by polymerization processes developed at Karl Ziegler's laboratory in Germany.[10] and by Phillips Petroleum Co. and Standard Oil Co. of Indiana in the U.S. using ionic catalysts.[1,3] The Ziegler type (typically, titanium trichloride and aluminum alkyl) and Phillips type (chromium oxide on silica) catalysts are active at relatively low temperatures (50–120°C) and pressures (100–600 psig).[7]

Because it has little or no branching or side chains, this type of polyethylene has higher crystallinity than low-density polyethylene, and, consequently, higher stiffness, strength, and melting point.

The most important applications of high-density polyethylene are blow-molded containers, extruded pipe, sheet and wire coating, packaging film, and injection-molded articles such as containers, toys, and housewares.

Polyethylene–linear low-density (LLDPE) has a degree of crystallinity and properties generally intermediate between conventional high- and low-density polyethylenes.

Because LLDPE is less expensive to produce (capital costs are lower and it can be made in existing PE plants with some modifications) and has an attractive balance of properties, its consumption has grown rapidly. Major applications are in film, injection molded containers, wire and cable insulation, and extruded pipe and tubing.

Polypropylene (PP), discovered by Natta in Italy and made also by Ziegler in Germany and by Phillips and others in the U.S., is relatively hard, strong, and high melting.[2,4,5]

Polypropylene can be produced in several forms that differ in their stereo configuration. By far the most important commercially is the isotactic form. The syndiotactic form has only recently (1994) been produced on a semi-commercial scale and has not yet found large-scale applications. The atactic form, in which the placement of the methyl groups is random, is a soft, amorphous material that has found limited applications in adhesives and laminates.

Isotactic polypropylene has a high degree of crystallinity owing to excellent fitting between polymer chains, which confers relatively high stiffness, strength, and melting point. This combination of properties enables the polymer to be oriented by drawing or stretching, making it useful for fibers and oriented films. Major applications for polypropylene are in fibers and filaments, molded and extruded articles, and film.

Poly(1-butene)

The isotactic polymer, made with a Ziegler-Natta stereospecific catalyst, is not particularly high melting but is strong and tough.[4,6,8] Polybutene resins, as either homopolymers, copolymers, or blends, are used mainly in pipe, heavy-duty bags and film.

Poly(4-methyl-1-pentene)

The isotactic polymer is a hard, high melting, transparent, and chemically resistant plastic.[6] It is used in applications that demand high clarity but only moderate strength, particularly laboratory and medical "glassware," and, when reinforced with fiberglass, in lightweight moldings.

Copolymers

Small amounts of higher olefins are often added in polyolefin manufacturing processes as co-monomers to moderate crystallinity and improve certain polymer properties, particularly impact resistance, particularly at low temperatures. Most copolymers are random, but block copolymers, including terminal block copolymers, are also produced.

A recently introduced novel copolymer of propylene and 1-octene is characterized by an unusual linear structure with long-chain branching. Produced by use of a homogeneous transition metal catalyst (in contrast to Ziegler/Natta type catalysts, which are heterogeneous), this copolymer displays elastomeric properties, combined with easy processing characteristics. Target applications include packaging, consumer and industrial goods, medical products, wire coatings, and automotive components.

PROCESS AND PRODUCT DEVELOPMENTS

A steady stream of process improvements, particularly the introduction of gas phase processes and catalysts of greater specificity and higher activity, has led to a succession of products of greater variety, lower cost, and better quality. A development of outstanding impact was the introduction in the 1990s of the metallocene catalysts—coordination compounds of transition metals such as zirconium, titanium, and hafnium with cyclopentadiene or polycyclic hydrocarbons containing a cyclopentadiene ring.

They offer the first practical route to a number of polymers of interest, such as syndiotactic polypropylene and a variety of copolymers with valuable combinations of properties. They may thus be said to have opened a new era in polyolefin technology.[9,10]

Blends

For applications that require special balances of properties but do not represent large-volume demands, producers frequently turn to physical blends of polyolefins with other types of polymers. The most widely used blends are those with other polyolefins, particularly ethylene/propylene or ethylene/propylene/diene elastomeric copolymers, which impart improved impact resistance.

Stabilization

The polyolefins are inherently susceptible to degradation by light, oxygen, or heat. Where these conditions will be encountered in service, appropriate stabilizers must be added to the polymer.

Processing (Fabrication, Manufacturing)

The polyolefins are thermoplastics with exceptionally easy processing characteristics; hence, they may be shaped by virtually any of the conventional processes.

REFERENCES

1. Clark, A.; Hogan, J. P.; Banks, R. O. Papers presented Amer. Chem. Soc., Petr. Chem. Div., April 1956; 211.
2. Fawcett, E. W.; Gibson, R. O. *J. Amer. Chem. Soc.* **1934**, 386.
3. McMillan, F. M. *The Chain Straighteners*; London: 1979; *Polyolefins*; Applied Polymer Science, 2nd ed.; Tess and Phoehlein, Eds., ACS Symposium Series 285, Washington, DC, 1985.
4. Natta, G. et al. *J. Amer. Chem. Soc.* **1955**, *77*, 708; *J. Polym. Sci.* **1955**, *16*, 143.
5. Natta, G.; Danusso, F. *J. Polym. Sci.* **1959**, *34*, 3.
6. Natta, G.; Danusso, F., Eds. *Stereoregular Polymers and Stereospecific Polymerization*; Pergamon: New York, 1967.
7. Sailors, H. R.; Hogan, J. P. *Macromolecular Sci.-Chem.* **1981**, *A15*(7), 1377.
8. Schildknecht C. *Vinyl and Related Polymers*; Wiley: New York, 1952.
9. Sinclair, K. B. *New Polyolefins from Emerging Catalyst Technologies*; Plastics Eng. Polyolefins VIII International Conference, Houston, TX, February 21–24, 1993.
10. Ziegler, K.; Martin, H. *Angew. Chem.* **1955**, *67*(19/20), 541.

POLYOLEFINS (Solid State Nuclear Magnetic Resonance)

Piero Sozzani and Roberto Simonutti
Dipartimento di Chimica Organica e Industriale
Università di Milano

Nuclear magnetic resonance spectroscopy (NMR) can be performed on polyolefins both in solution and in the solid state to provide an accurate and often unique characterization of the material. The constitutional and molecular aspects, the structure of the defects, and the sequence of stereochemical relationships among the groups substituted along the polymer chain have been extensively addressed by solution NMR.[1–4] The sequences of the monomer units in copolymers can be also determined.[5,6] In most cases the analysis is performed at high temperature in high boiling solvents, but sometimes this is not feasible owing to the lack of complete solubility or swellability of the polymer, especially at high molecular masses and in crosslinked materials. Today, solid state NMR provides a valuable means for avoiding the solubility problem. Although this implies a sensitive loss of resolution, the materials can be characterized in the bulk, adding considerable information on the single-chain arrangements and collective properties of the macromolecules in the crystalline and in the amorphous phase.[7,8]

The main stream of research to date has addressed the most simple and representative elements of this class of polymers, namely, polyethylene (PE) and polypropylene (PP) in their manifold forms.

Solid state NMR techniques have acquired a wider scope with the introduction of a variety of high resolution experiments, but some low resolution experiments can still add useful data to the description of the system. They can be divided into essentially three different classes with a different emphasis on the typical NMR parameters, such as chemical shifts, relaxation times, and quadrupolar interactions: (i) experiments in high resolution Cross Polarization Magic Angle Spinning (CP-MAS);[7,9,10] (ii) wide-line relaxometry;[11] (iii) deuterium line shape;[12] and (iv) anisotropic information retaining experiments.[13,14] Multidimensional spectroscopy, which can be applied to the above classes, offers an additional means for enriching the available information.[15] The choice of experiments is ample, but implies, of course, considerable expertise and instrumentation.

BRANCHING AND CROSSLINKING

The distribution of the defect in the material may grossly affect its macroscopic properties. Polyethylene can be synthesized in different forms with variable degrees of branching: essentially, low density polyethylene (LDPE), high density polyethylene (HDPE), and a number of exotic materials presenting a controlled branching in the form of a comonomer, or a highly linear structure, such as polymethylene obtained by diazomethane polymerization. Consequently one of the questions asked spectroscopists has been how the branching is divided between the crystalline and the amorphous phase. High resolution ^{13}C solid state NMR has provided an answer by analyzing minor signals mainly arising from chain-end methyls and methyne at the branching points (11 and 40 ppm, respectively).[16]

CONFORMATION, MOBILITY, AND CRYSTAL PACKING

The chain flexibility of high molecular-mass polyolefins produces a virtually infinite number of accessible conformations, and the local conformation affects in turn the chemical shift. The measurement of the resonance frequencies of the carbon nucleus is thus not only a function of the primary structure and of the connectivity, but of the 3D structure of the macromolecule as well. The crystal texture generates a peculiar sequence of conformations, which is reflected on the MAS NMR spectrum.

Syndiotactic polypropylene presents a variety of crystal modifications, the most energetically stable being the number-eight shape helix.[18] Backbone methylenes experience two different conformational environments alternatively. Two sequences of bond conformations, namely, **tggt** and **gttg** (**g** for **gauche**, and **t** for **trans**), can thus reproduce the chain. The γ-**gauche** effect dictates that the outer conformations act on the chemical shift of the central methylene: consequently, the signals must be two **gauche** effects (10 ppm) apart.[19,20]

When a dumbbell of quenched material is stretched at 0°C, only the extended **tttt** conformation is produced, and the resonances containing **gauche** conformations disappear.[21] The chain symmetry is again high, actually showing three signals for three carbons in the monomer unit. When the stretched sample undergoes thermal treatment to above 50°C, the crystalline-stretched form reverses into the helix form and the corresponding change can be directly followed by variable temperature MAS NMR.

MORPHOLOGY, AMORPHOUS PHASE, AND INTERPHASE

By taking advantage of conformational sensitivity, the classic problem of the lamellar surface arrangement has been addressed.

When the *true* amorphous region, far from the boundaries of crystallites, is the object of characterization, the spectroscopist must cope with an enormous range of polymer motion. Several orders of magnitude are covered from low to high temperature, passing through the glass transition temperature. Motional broadening, due to decoupling modulation, or the dispersion of chemical shifts may even quench the amorphous phase signal. Without discussing these phenomena, we shall just point out that correlation times fast enough to produce narrow signals at usual magnetic-field intensities can be reached at over 60°C above the glass transition.[7] Consequently, the conformation distribution of the amorphous phase of isotactic[23] and syndiotactic[20] polypropylene could be determined at high temperature.

^{1}H relaxation times (in particular, rotating-frame relaxation times, $T_{1\rho}$, and spin-lattice, T_1) are a useful tool for the study of polymer morphologies.[25]

^{13}C spectra are well resolved and more information than proton spectra. The possibility of retaining both the high resolution of carbon spectra and the spatial information of proton relaxation is given by a transfer of magnetization from the proton to the carbon. This family of techniques has been extensively applied to the field of polymer blends and block copolymers,[26,27] and has shown its usefulness in highlighting the relationships and the dimensions of the phases in semicrystalline polyolefins.[24,28] Using these procedures, a contrast has been provided among phases: for example, the bulk amorphous phase of syndiotactic polypropylene has been recorded as depleted by the crystalline phase signals and by the signals due to those carbons close to the crystalline phase.[22] That part of the amorphous phase that was relatively far from the crystalline phase could thus be selected.

POLYMORPHISM OF POLYOLEFINS CONTAINING SIDE CHAINS

Isotactic poly(1-butene) possibly crystallizes in three crystal forms. From I to III, the pitch of helix widens, and the dihedral angle between the CH_2CH_3 group and the main chain becomes larger than 60 degrees. For the γ-gauche interaction a shift downfield is generally obtained when the helices are expanded from the 3_1 helix to the 11_3 or 4_1 helix.[29]

Longer side-chain polyolefins may show a wide range of polymorphic behavior owing to the greater conformational freedom of the packing arrangement in the crystalline state. This may end in an extreme simplification of the spectrum, if rapid motions of the side chains are gained, in spite of the stability of the main chain helices: thus isotactic poly[(S)-3,7-dimethyl-1-octene] shows chemical shifts corresponding to solution shifts.[30]

ACKNOWLEDGMENTS

The authors wish to thank CNR and the Ministry for Public Education, Italy for partial support.

REFERENCES

1. Bovey, F. A. *Chain structure and conformation of Macromolecules*; Academic: New York, 1982.
2. Schilling, F. C.; Tonelli, A. E. *Macromolecules* **1980**, *13*, 275.
3. Di Silvestro, G.; Sozzani, P.; Savarè, B. et al. *Macromolecules* **1985**, *18*, 1983.
4. Sozzani, P.; Oliva, C. *J. Magn. Reson.* **1985**, *63*, 115.
5. Randall, J. C. *NMR and Macromolecules, Sequence Dynamic and Domain Structure*; A.C.S.: Washington, DC, 1984.
6. Tonelli, A. E. *NMR Spectroscopy and Polymer Microstructure: the Conformational Connection*; VCH: Deerfield Beach, FL, 1989.
7. Komoroski, R. A. *High-Resolution NMR Spectroscopy of Synthetic Polymers in Bulk*; VCH: Deerfield Beach, FL, 1986.
8. Dybowski, C. R.; Brandolini, A. J. *NMR Spectroscopy of Solid Polymer Systems In Characterization of Solid Polymers*, Spells, S. J., Ed.; Chapman & Hall: London, 1994.
9. Grimmer, R.; Blumich, B. *NMR Basics, Principles and Progress* **1993**, *30*, 1.
10. Laupetre, F. *NMR Basics, Principles and Progress* **1993**, *30*, 63.
11. McBrierty, V. J.; Packer, K. J. *Nuclear Magnetic Resonance Solid Polymers*; Cambridge University: Cambridge, 1993.
12. Jelinski, L. W. *Ann. Rev. Mater. Sci.* **1985**, *15*, 359.
13. Dabbagh, G.; Weliky, D. P.; Tycko, R. *Macromolecules* **1994**, *27*, 6183.
14. Opella, S. J.; Waugh, J. S. *J. Chem. Phys.* **1977**, *66*, 4919.
15. Spiess, H. W. *Chem. Rev.* **1991**, *91*, 1321.
16. VanderHart, D. L.; Perez, E. *Macromolecules* **1986**, *19*, 1902.
17. No reference cited.
18. Allegra, G.; Ganis, P.; Corradini, P. *Makromol. Chem.* **1963**, *61*, 225.
19. Bunn, A.; Cudby, M. E. A.; Harris, R. K. et al. *J. Chem. Soc., Chem. Commun.* **1981**, 15.
20. Sozzani, P.; Simonutti, R.; Galimberti, M. *Macromolecules* **1993**, *26*, 5782.
21. Sozzani, P.; Galimberti, M.; Balbontin, G. *Makromol. Chem. Rapid Commun.* **1992**, *13*, 305.
22. Sozzani, P.; Simonutti, R.; Comotti, A. *Magn. Reson. Chem.* **1994**, *32*, S45.
23. Saito, S.; Moteki, Y.; Nakagawa, M. et al. *Macromolecules* **1990**, *23*, 3256.
24. Kitamaru, R.; Horii, F.; Murayama, K. *Macromolecules* **1986**, *19*, 636.
25. McBrierty, J.; Douglass, D. C. *J. Polym. Sci: Macro. Rev.* **1981**, *16*, 295.
26. Stejskal, E. O.; Schaefer, J.; Sefcik, M. D. et al. *Macromolecules* **1981**, *14*, 275.
27. Enrichs, P. M.; Tribone, J.; Massa, D. J. et al. *Macromolecules* **1988**, *21*, 1282.
28. Packer, K. J.; Poplett, I. J. F.; Taylor, M. J. et al. *Makromol. Chem., Makromol. Symp.* **1990**, *34*, 161.
29. Belfiore, L.; Schilling, F. C.; Tonelli, A. E. et al. *Macromolecules* **1984**, *17*, 2561.
30. Sacchi, M. C.; Tritto, I.; Locatelli, P. et al. *Macromolecules* **1986**, *19*, 1634.

Poly(organoboron halide)s

See: *Organoboron Main Chain Polymers*

Poly(ortho ester)s

See: *Bioerodible Polymers A. Controlled Drug Release B. Fracture Fixation*

Polyoxadiazoles

See: *Condensation Polymers (Using Phosphorus Pentoxide/Methanesulfonic Acid)*
High Pressure Polymerization Processing (High Temperature Thermosetting Polymers)
Poly(arylene-1,3,4-oxadiazole)s
Polyoxadiazoles (Overview)
Polyoxadiazoles and Polytriazoles

POLYOXADIAZOLES (Overview)

Jin Chul Jung
Department of Materials Science and Engineering
Pohang University of Science and Technology

In the oxadiazole ring there are four kinds of different structures, 1,3,4–, 1,2,4–, 1,2,3–, and 1,2,5–oxadiazole. Of them no polymers containing the latter two ring structures have been prepared yet, while a number of polymers having the former two ring structures on their main chains have been known, and quite detailed investigations on structures and properties of the polymers have been made.

Oxadiazole rings are heterocyclic, aromatic with planar structure, and polymers with main chains making up the ring are known to have excellent thermal and hydrolytic stability and film- and fiber-forming property.

SYNTHESIS

Poly(1,3,4-oxadiazole)s

The most common method for preparing poly(1,3,4-oxadiazole)s is the intramolecular cyclodehydration of polyhydrazides **9** in acidic conditions.[1–9] This can be carried out either in bulk by heating above T_g under vacuum or in solution by using strong dehydrating acids such as concentrated sulfuric acid, dichloro- or trifluoroacetic acid, poly(phosphoric acid) (PPA), or methanesulfonic acid (MSA) (**Scheme I**).

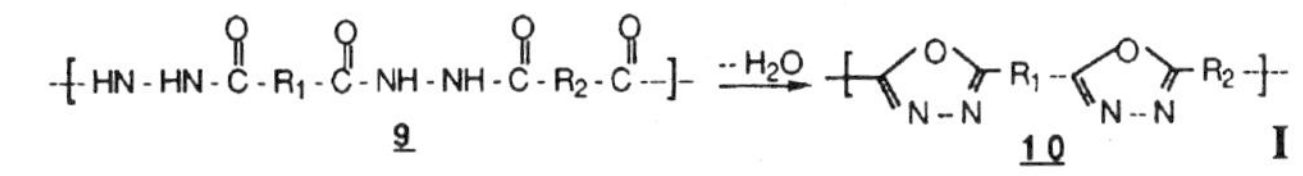

I

Poly(1,2,4-oxadiazole)s

The first poly(1,2,4-oxadiazole)s were prepared by condensation of dicarboxylic acid derivatives **19** with diamideoximes **20** in polar protic solvents at low temperatures in the presence of acid acceptors such as pyridine or triethyl amine, as illustrated in **Scheme II**. In the condensation poly(O-acyl amideoxime)s are

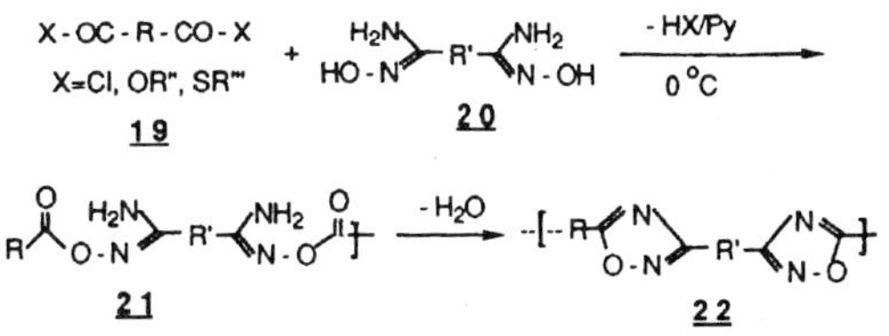

II

obtained as precursors, which are converted to poly(1,2,4-oxadiazole)s by intramolecular cyclodehydration.

The second approach[10,12] to synthesis of poly(1,2,4-oxadiazole)s is 1,3-dipolar addition polymerization of terephthalonitrile oxide with terephthalonitrile in bulk in the presence of small amounts of acidic catalyst like p-toluenesulfonic acid.

PROPERTIES

Both oxadiazole rings with 1,3,4- and 1,2,4-configuration are heterocyclic aromatic, and they exhibit excellent thermal stability and hydrolytic resistance.[13] However, their thermal stability is generally inferior to that of other heterocyclic fused ring systems.

Poly(1,3,4-oxadiazole)s

The physical and solution properties of poly(1,3,4-oxadiazole)s have been described in the previous two review articles.[14,15] Aliphatic poly(1,3,4-oxadiazole)s have densities in the 1.21–1.28 g/cm^3 range[2] and thermal stabilities up to 400–450°C. Aromatic poly(1,3,4-oxadiazole)s are generally only soluble in concentrated sulfuric acid. This means that their main chains have such rigid structure that the interaction between the main chains is very strong. They have thermal resistances up to 400–550°C, depending on the structure of the aromatic spacer units. Although they are very stable against acidic or basic hydrolysis, the oxadiazole ring can be opened.

Recently, electrical and electrochemical properties of both doped and undoped poly(1,3,4-oxadiazole-2,5-diyl-1,4-phenylene)[16] and poly(p-phenylene vinylene)[17] and polyazomethines[18] containing 1,3,4-oxadiazole structure units on their main chains were investigated. It was generally confirmed that the π-electrons of 1,3,4-oxadiazole ring are in resonance with other structure units consisting of the polymers and delocalized along the main chains, although contribution of the 1,3,4-oxadiazole ring is generally smaller than that of 1,4-phenylene unit.

Poly(1,2,4-oxadiazole)s

Physical properties of poly(1,2,4-oxadiazole)s are far less investigated compared to poly(1,3,4-oxadiazole)s. This certainly results from the greater synthetic difficulty and lower molecularity of the poly(1,2,4-oxadiazole)s.

All-aromatic poly(1,2,4-oxadiazole)s are soluble only in strong acidic solvents such as concentrated sulfuric acid, trifluoroacetic acid, or methanesulfonic acid. They have excellent hydrolytic stabilities against both acids and bases and good radiation resistance. They degrade generally in the 300–400°C range.

REFERENCES

1. Frazer, A. H.; Sweeny, W.; Wallenberger, F. T. *J. Polym. Sci. Part A* **1964**, *2*, 1157.
2. Unishi, T.; Hasegawa, M. *J. Polym. Sci. Part A* **1965**, *3*, 3191.

3. Frazer, A. H.; Sarasohn, L. M. *J. Polym. Sci. Part A1* **1966**, *4*, 1649.
4. Iwakura, Y.; Uno, K.; Hara, S. *Makromol. Chem.* **1966**, *95*, 248 and **1967**, *108*, 160.
5. Iwakura, Y.; Uno, K.; Hara, S. *J. Polym. Sci. Part A* **1965**, *3*, 45.
6. Sato, M.; Yokoyama, M. *J. Polym. Sci. Poly. Chem.* **1980**, *18*, 2751.
7. Kaeriyama, K. *J. Appl. Poly. Sci.* **1979**, *24*, 1205.
8. Thaemlitz, C. J.; Weikel, W. J.; Cassidy, P. E. *Polymer* **1992**, *33*, 3278.
9. Holsten, J. R.; Lilyquist, M. R. *J. Polym. Sci. Part A* **1965**, *3*, 3905.
10. Overberger, C. G.; Fujimoto, S. *J. Polym. Sci. B* **1965**, *3*, 735.
11. Imai, Y.; Akiyama, M.; Uno, K. et al. *J. Polym. Sci. B* **1966**, *4*, 305.
12. Itoya, K.; Kakimoto, M.; Imai, Y. et al. *Polym. J.* **1992**, *24*, 979.
13. Tiemann, F.; Krüger, P. *Chem. Ber.* **1884**, *17*, 1685.
14. Hasegawa, M. *Polyhydrazides and Polyoxadiazoles* In *Encycl. Polym. Sci. Technol.* ed.; Vol. 11, pp 169–187.
15. Nanjan, M. J. *Polyhydrazides and Polyoxadiazoles* In *Encycl. Polym. Sci. Eng.* ed.; Vol. 12, pp 332–339.
16. Sinigersky, V.; Wegner, G.; Schopov, I. *Eur. Polym. J.* **1993**, *29*, 617.
17. Janietz, S.; Schulz, B.; Törrönen, M. et al. *Eur. Polym. J.* **1993**, *29*, 545.
18. Saegusa, Y.; Koshikawa, T.; Nakamura, S. *J. Polym. Sci. Polym. Chem.* **1992**, *30*, 1369.

POLYOXADIAZOLES AND POLYTRIAZOLES

Marcel H. V. Mulder
University of Twente

Edwin R. Hensema
Hercules European Research Center

Polymers containing an oxadiazole or a triazole group are known for their excellent thermal properties. In the 1960s, aromatic poly(1,3,4-oxadiazole)s and poly(1,2,4-triazole)s were studied as new high-performance polymers for fiber applications.[1–12] Aromatic poly(1,3,4-oxadiazole)s and poly(1,2,4-triazole)s are chemically resistant and thermally stable polymers since they are only soluble in strong acids and stable up to 450°C. However, despite their specific properties they are not produced commercially. Almost two decades later these polymers were studied as membrane material for gas separation.[13–22] The aromatic poly(1,3,4-oxadiazole)s and poly(1,2,4-triazole)s appeared to be promising new gas separation membrane materials since they combine excellent gas separation properties with a high thermal stability and a good chemical resistance.

The electrical conductive character is another property that has been frequently studied.[23–26] For this application, the polymers are often doped with, for example, Br_2, I_2, or $FeCl_3$. Upon doping, both the conductivity and its solubility increase.

This review focuses on the preparation and properties of aromatic poly(1,3,4-oxadiazole)s and poly(1,2,4-triazole)s, with special attention to the employment as membrane material. For earlier work on these polymers as well as on the work on nonaromatic poly(1,3,4-oxadiazole) and poly(1,2,4-triazole), the reader is referred to the excellent reviews of Cotter and Matzner and Cassidy.[27,28]

SYNTHESIS

Various synthesis routes are known for the preparation of poly(1,3,4-oxadiazole)s and poly(1,2,4-triazole)s.[27,28] Poly(1,3,4-oxadiazole)s either can be made directly by a reaction between a dicarboxylic acid and hydrazine or can be obtained after a cyclodehydration reaction of a polyhydrazide.

Poly(1,2,4-triazole)s can be prepared from polyhydrazides and from poly(1,3,4-oxadiazole)s as precursor polymers (see **Figure 1**).

Polyhydrazides with alternating *para-* and *meta*-phenylene groups in the main chain can be prepared by a polycondensation reaction using terephthaloyl chloride and isophthaloyl dihydrazide as monomers. Conversion of this precursor polymer yields poly(1,3,4-oxadiazole)s and poly(1,2,4-triazole)s with alternating *p-* and *m*-phenylene groups (see reaction route A in Figure 1). Holsten and Lilyquist have extensively studied poly(1,2,4-triazole) synthesis using polyhydrazide as a precursor polymer in combination with aniline.[8]

Employing mixtures of terephthalic and isophthalic acid in different ratios in combination with hydrazine sulfate results in the direct formation of poly(1,3,4-oxadiazole)s where *p-* and *m*-phenylene groups are incorporated randomly (see reaction route B). This reaction route was originally developed by Iwakura et al.[29] By means of a reaction with aniline in poly(phosphoric acid), they can be converted into poly(1,2,4-triazole)s, having the same incorporation of *p-* and *m*-phenylene groups in the main chain as the precursor poly(1,3,4-oxadiazole).

PROPERTIES OF AROMATIC POLY(1,3,4-OXADIAZOLE)S AND POLY(1,2,4-TRIAZOLE)S

Poly(1,3,4-oxadiazole)s

Hensema et al. synthesized and studied a wide variety of poly(1,3,4-oxadiazole)s and their physical properties—e.g., glass transition temperature, degradation temperature, solubility, film-forming properties.[30,31]

For example, poly[(*p*-, *m*-phenylene)-1,3,4-oxadiazole]s with various ratios of *p-* and *m*-phenylene groups in the polymer backbone were prepared according to the reaction scheme in Figure 1B and their physical properties were studied. The thermostability of poly(1,3,4-oxadiazole)s was found to increase with an increasing *p*-phenylene content in the main chain. The thermostability and chemical resistance of the poly(1,3,4-oxadiazole)s are favorable for many applications, but, as a consequence, a more widespread use of these polymers is hampered because of their limited processability. However, units can be incorporated into the polymer backbone to enhance tractability.

Poly(cyclohexane-1,3,4-oxadiazole)

Preparation of alicyclic poly(1,3,4-oxadiazole)s was originally reported by Iwakura et al.[32] The polymer is soluble in formic acid and, despite its aliphatic unit, has a good thermal stability. Its degradation temperature is ~ 100°C lower than that found for poly[(*p*-phenylene)-1,3,4-oxadiazole] (542°C).[31] Incorporation of the flexible cyclohexane unit results in a decrease of the glass transition temperature compared to its aromatic counterpart, poly[(*p*-phenylene)-1,3,4-oxadiazole].

Poly(1,1,3-trimethyl-3-phenylindane oxadiazole)

The 1,1,3-trimethyl-3-phenylindane unit is applied as a solubility-increasing unit for a number of polymers. Hensema[31] and Sena[33] incorporated this unit into poly(1,3,4-oxadiazole) and found that the solubility improved only marginally.

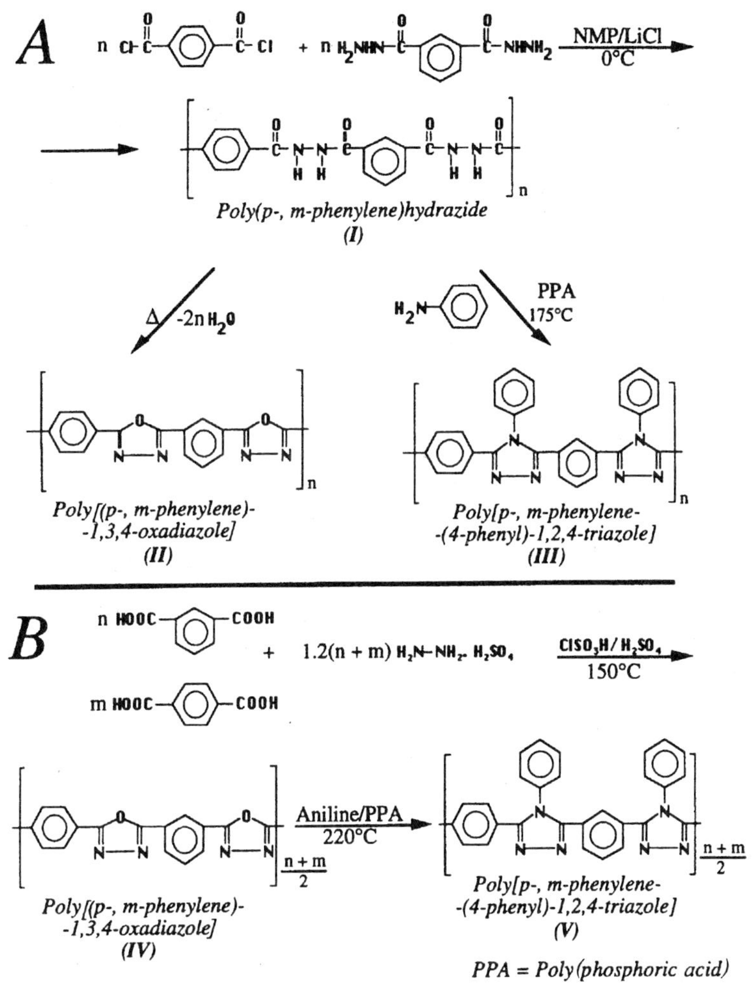

FIGURE 1. Preparation of alternating poly[(*p*-,*m*-phenylene)-1,3,4-oxadiazole] and poly[*p*-,*m*-phenylene-(4-phenyl)-1,2,4-triazole] via poly(*p*-,*m*-phenylene)hydrazide (A) and the direct preparation of random poly[(*p*-,*m*-phenylene)-1,3,4-oxadiazole]s and conversion into poly[*p*-,*m*-phenylene-(4-phenyl)-1,2,4-triazole]s (B).

Poly(5-t-butylisophtalic oxadiazole) (TBI-POD)

The preparation of TBI-POD using oleum as reaction mixture has been reported in a patent by Stephens.[34] The preparation of flexible films from TBI-POD/*m*-cresol solutions is claimed in this patent.

Poly(4,4'-(2,2'-diphenyl)hexafluoropropane oxadiazole)

Incorporation of the polar and bulky 4,4'-(2,2'-diphenyl)hexafluoropropane (HF) moiety into the polymer backbone of, for example, polyimides has been shown to be an effective way of substantially increasing the solubility of these polymers.[35] Hensema[31] and Sena[33] prepared HF-POD directly by the one-step synthesis using 4,4'-(2,2'-diphenyl)-hexafluoropropane dicarboxylic acid and hydrazine as monomers. Flexible films could be cast from a HF-POD/trifluoroacetic acid solution. Solubility tests showed that HF-POD is soluble in THF, chloroform, NMP, *o*-dichlorobenzene, and trifluoroacetic acid. The increased solubility of HF-POD compared to the other poly(1,3,4-oxadiazole)s is believed to be a result of the more loosely packed polymer matrix.

Poly(4,4'-diphenylether oxadiazole)

Hensema[31] and Sena[33,36] incorporated this unit into the poly-1,3,4-oxadiazole backbone and flexible films were obtained from a formic acid solution.

Poly(imide oxadiazole)

Heating of poly(imide hydrazide) in poly(phosphoric acid) gives a poly(imide oxadiazole).[37–39] These polymers are soluble in DMSO, DMAc, and NMP. However, they are highly brittle and difficult to process into a fiber or a film.

Poly(ether ketone oxadiazole)

Oxadiazole groups can also be incorporated into poly(ether ketone).[40,41] In this way a polymer is obtained that is soluble in aprotic solvents.

Poly(ether alkyloxadiazole)s

Polyoxadiazoles have been prepared with aliphatic moieties in the main chain.[42] The glass transition temperature is much lower compared to the aromatic PODs and decreases with increasing number of -CH_2- groups.

Poly(1,2,4-triazole)s

Recently, Hensema et al. have synthesized and studied a large number of different poly(1,2,4-triazole)s in which different units were built *into* and *onto* the polymer backbone. These syntheses were based on the work of Holsten[8] and Korshak[9]. For example, poly[*p*,*m*-phenylene(4-phenyl)-1,2,4-triazole]s with various ratios of *p*-phenylene and *m*-phenylene groups were synthesized starting from the corresponding poly(1,3,4-oxadiazole)s. The glass transition temperature of these polymers was found to increase with increasing *p*-phenylene ratio, which is due to an increased chain stiffness, and upon extrapolation, a theoretical glass transition temperature of 380°C is obtained for poly[*p*-phenylene-(4-phenyl)-1,2,4-triazole].[31]

POLY(1,3,4-OXADIAZOLE)S AND POLY(1,2,4-TRIAZOLE)S AS GAS SEPARATION MEMBRANE MATERIALS

The aromatic poly(1,3,4-oxadiazole)s and poly(1,2,4-triazole)s are promising new gas separation membrane materials since they combine excellent gas separation properties with a high thermal stability and a good chemical resistance.[14–22,43] Possible applications include the separation from air into oxygen and nitrogen and the separation from carbon dioxide from methane.

POLY(1,3,4-OXADIAZOLE)S AS ULTRAFILTRATION MEMBRANE

Recently, the preparation of asymmetric poly(1,3,4-oxadiazole)s ultrafiltration membranes by phase inversion was described.[44] The polymers were prepared according to the procedure of Frazer.[1–7]

REFERENCES

1. Frazer, A. H.; Wallenberger, F. T. *J. Polym. Sci.* **1964**, *A2*, 1137.
2. Frazer, A. H.; Wallenberger, F. T. *J. Polym. Sci.* **1964**, *A2*, 1147.
3. Frazer, A. H.; Sweeny, W.; Wallenberger, F. T. *J. Polym. Sci.* **1964**, *A2*, 1157.
4. Frazer, A. H.; Wallenberger, F. T. *J. Polym. Sci.* **1964**, *A2*, 1171.
5. Frazer, A. H.; Sarasohn, I. M. *J. Polym. Sci.* **1966**, *A1*, *4*, 1649.
6. Frazer, A. H.; Reed, T. A. *J. Polym. Sci.* **1967**, *C19*, 89.
7. Frazer, A. H.; Wilson, D. R. *Appl. Polym. Symp.* **1969**, *9*, 89.
8. Holsten, J. R.; Lilyquist, M. R. *J. Polym. Sci.* **1965**, *A3*, 3905.
9. D'Yachenko, A. V.; Korshak, V. V.; Krongauz, E. S. *Vysokomol. Soedin., Ser. A* **1967**, *9*(10), 2231.
10. Korshak, V. V.; Berestneva, G. L.; Bragina, I. P.; Yeremina, G. V.; Krylova, V. V. *Vysokomol. Soyed.* **1974**, *A16*, 1714.
11. Korshak, V. V.; Krongauz, E. S.; D'Yachenko, A. V. *Doklady Akademii. Nauk SSSR* **1967**, *176*, 338.
12. Korshak, V. V.; D'Yachenko, A. V.; Krongauz, E. S.; Berestneva, G. L. *Vysokomol. Soyed.* **1969**, *A11*, 7.
13. Gebben, B.; Mulder, M. H. V.; Smolders, C. A. *J. Memb. Sci.* **1989**, *46*, 29.
14. Gebben, B. Ph.D. Thesis, University of Twente, The Netherlands, 1988.
15. Hensema, E. R.; Smolders, C. A. In *Gas separation technology, Process Technology Proceedings 8* Vansant, E. F.; Dewolfs, R., Eds.; Elsevier: Amsterdam, 1988; p 373.
16. Hensema, E. R.; Gebben, B.; Mulder, M. H.; Smolders, C. A. *Bull. Soc. Chim. Belg.* **1991**, *100*, 129.
17. Hensema, E. R. Ph.D. Thesis, Enschede, The Netherlands, 1991.
18. Hensema, E. R.; Smolders, C. A. In *Membrane Proceedings Indo-EC Workshop 1991*; Oxford & IBH: New Delhi, India, 1992; p 213.
19. Hensema, E. R.; Mulder, M. H. V.; Smolders, C. A. *J. Appl. Polym. Sci.* **1993**, *49*, 2081.
20. Hensema, E. R.; Sena, M. E. R.; Mulder, M. H. V.; Smolders, C. A. *Gas Sep. Purif.* **1994**, *8*, 149.
21. Hensema, E. R. *Adv. Mater.* **1964**, *6*, 269.
22. Hensema, E. R.; Boom, J. P. *Makromol. Chem., Macromol. Symp.*, in press.
23. Hagiwara, T.; Demura, T.; Iwata, K. *Nippon Kagaku Kaishi* **1986**, *3*, 356; *Chem. Abstract* CA 104:217211v.
24. Janietz, S.; Schulz, B.; Torronen, M.; Sundholm, G. *Eur. Polym. J.* **1993**, *29*, 545.
25. Schulz, B.; Knochenhauer, G.; Brehmer, L.; Janietz, S. *Synth. Met.* **1995**, *69*, 603.
26. Roh, Y. B.; Araki, H.; Yoshino, K.; Takase, M.; Masaaki, B.; Banyjo, T. *Jpn. J. Appl. Phys.* Part 1 **1994**, *33*, 1146.
27. Cotter, R. J.; Matzner, M. In Blomquist, A. T.; Wasserman, H., Eds.; *Organic Chemistry*; Academic: New York, 1972; Vol. 13B, 1.
28. Cassidy, P. E. *Thermally stable polymers*; Marcel Dekker: New York, 1980.
29. Iwakura, Y.; Uno, K.; Hara, S. *J. Polym. Sci.* **1965**, *A3*, 45.
30. Hensema, E. R.; Boom, J. P.; Mulder, M. H. V.; Smolders, C. A. *J. Polym. Sci., Polym. Chem. Ed.* **1994**, *32*, 513.
31. Hensema, E. R.; Sena, M. E. R.; Mulder, M. H. V.; Smolders, C. A. *J. Polym. Sci., Polym. Chem. Ed.* **1994**, *32*, 527.
32. Leibnitz, E. *Angew. Makromol. Chem.* **1993**, *204*, 101.
33. Sena, M. E.; Andrade, C. T. *Polym. Bull.* **1995**, *34*, 439.
34. Stephens, J. R. U.S. Patent 4 487 921, 1984.
35. Harris, F. W.; Lanier, L. H. In *Structure–solubility relationships in polymers*; Harris, F. W.; Seymour, R. B., Eds.; Academic: London, 1977.
36. Sena, M. E.; Andrade, C. T. *Polym. Bull.* **1995**, *34*, 643.
37. Ray, A.; Rao, Y. V.; Bhattacharya, V. K.; Maiti, S. *J. Polym. Sci., Polym. Lett. Ed.* **1983**, *21*, 77.
38. Zhubanov, S. A.; Mashkevich, S. A.; Maimkov, T. P. *Izv. Akad. Nauk SSSR Ser. Khim.* **1973**, *23*, 48.
39. Yang, C. *J. Polym. Sci., Polym. Chem. Ed.* **1979**, *17*, 3255.
40. Mercer, F. W.; McKenzie, M. T.; Merlino, G.; Fone, M. M. *J. Appl. Pol. Sci.* **1995**, *56*, 1397.
41. Mercer, F. W. *High Perform. Polym.* **1993**, *5*, 69.
42. Di Pace, E.; Fichera, A. M.; Laurienzo, P.; Malinconico, M.; Martuscelli, E.; Perenze, N.; Volpe, M. G. *J. Polym. Sci., Polym. Phys.* **1995**, *32*, 1643.
43. Wessling, M.; van den Boomgaard, T.; Mulder, M. H. V.; Smolders, C. A. *Makromol. Chem., Macromol. Symp.* **1992**, *70–71*, 379.
44. Leibnitz, E.; Eisold, C.; Paul, D. *Angew. Makromol. Chemie* **1993**, *210*, 197.

Polyoxazolines

See: Oxazoline Polymerization

Polyoxetanes

See: Telechelic Polyoxetanes

Polyoxometalates

See: Antitumoral and Antiviral Polyoxometalates (Inorganic Discrete Polymers of Metal Oxide)

POLYOXYARYLENESILOXANES

Susan Adams Nye
GE Silicones

Marsha M. Grade
GE Corporate Research and Development Center

In the search for new materials for space applications, Curry and Byrd[1] found that polyoxyarylenediphenylsilanes (**Figure 1**) had excellent stability to high temperatures and ultraviolet radiation. The polymers were made by the reaction of aromatic diols with bis(anilino)-diphenylsilane. Of seven resins evaluated, they found that the polymer synthesized from 4,4'-biphenol had the best thermal characteristics (10%) weight loss (> 600°C). Alternative syntheses have been reported by Krimm and Schnell[2] who employed ring-opening polymerization of cyclic siloxanes to obtain similar materials of high molecular weight. Additionally, MacFarlane and Yankura[3] reported on a polyoxyarylenedimethylsilane (**Figure 2**; X = O, R = H) from the condensation 4,4'-dihydroxydiphenylether with either dichloro- or diethoxydimethylsilanes. Recently, GE researchers have reported that a polyoxyarylenedimethylsiloxane, made from 3,3',5,5'-tetramethyl-4,4'-diphenol (TMBP) (Figure 2; X = nil, R = CH_3) had excellent hydrolytic stability.[4] Apparently the *ortho* methyl substituents provided steric hinderance to furnish this stability. Other attributes of this polymer included its excellent flame properties. It was nonflammable (V-0 by the UL-94 test) and had a high percent char after being exhaustively burned.

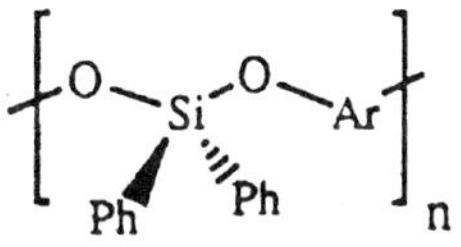

FIGURE 1. Generic polyoxyarylenediphenylsilanes.

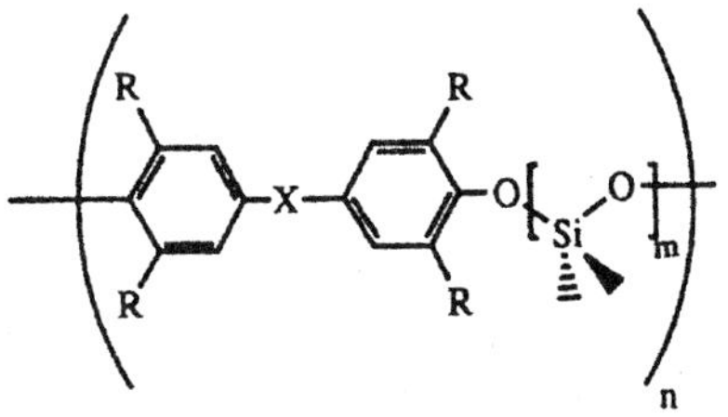

FIGURE 2. Polyoxybiphenylsilanes.

The applications for this type of polymer would include any device, system, or assembly needing thermal stability and/or low flammability.

ACKNOWLEDGMENTS

The authors wish to acknowledge S. S. Kupperblatt, J. L. Webb, K. Sy, J. Carnahan, S. Weissman, N. Marotta, T. Burnell, J. Rich, L. Lewis, J. Cella, P. Anderson, and J. Razzano for their support of this project.

REFERENCES

1. Curry, J. E.; Byrd, J. D. *J. Appl. Polym. Sci.* **1965**, *9*, 295.
2. Krimm, H.; Schnell, H. *Patentschrift 1* (March) **1963**, *136*, 114.
3. MacFarlane, R.; Yankura, E. S. Contract No. *DA-19-020-ORD-5507*, Quarterly Report No. 7, Naugatuck Division of United States Rubber Company, 1963.
4. Webb, J. L.; Nye, S. A.; Grade, M. M. U.S. Patent 5 041 514, August 20, 1991.

Poly(p-oxybenzoate)

See: Liquid Crystalline Aromatic Polyesters
Whiskers of Rigid-Rod Polymers

Polyoxyethylene

See: Poly(ethylene oxide)

POLYOXYMETHYLENE FIBER, SUPERDRAWN

Tamikuni Komatsu
Analytical Research Center
Asahi Chemical Industry Company, Ltd.

The manufacture of high modulus and high tensile strength fibers by molecular orientation of crystalline polymers has been an outstanding event since the invention of nylon fiber in 1938. Polyoxymethylene (POM), the representative crystalline polymer, consists of the repeating unit (-CH_2-O-) and has the theoretical crystalline modulus estimated as 95–150 GPa.[1,2] Also, a POM resin is one of the three biggest engineering thermoplastics, and the preparation of superdrawn POM fibers has long been awaited. In 1983, Nakagawa et al. prepared superdrawn POM fibers having a Young's modulus of 60 GPa at a very progressive strain rate by microwave heating/drawing.[3] The finding of the excellent mechanical properties of these POM fibers encouraged the later development of other high-performance fibers, but the POM fibers had many defects such as conspicuous fibrillation and voids that hindered their industrial applications. In 1991, Komatsu et al. developed a new process for the continuous drawing of polymers in a highly pressurized medium, applied the process to the POM, and solved all the above problems.[4] Today, the super drawn POM fibers are growing as a civil engineering polymer material.

PREPARATION

The superdrawn POM fibers free of voids is prepared by the pressurized drawing method proposed by Komatsu et al. The method consists of continuous two-step drawing under high pressure.[4] The method enables the preparation of superdrawn POM fibers having a dense structure, a draw ratio of 34, and improved mechanical properties.

PROPERTIES

Several research groups have investigated the structures and properties of the POM drawn fibers. Many of the excellent properties of the fibers, which result from the formation of the extended chain crystal, are influenced by the fibrillar structure containing many defects such as voids. The conventional fibers are white and contain many voids appearing as a ladder, connecting fibrils along the fiber axis. The pressurized fibers are transparent and free of voids even at a superdraw ratio.

MECHANICAL PROPERTIES

The modulus of the pressurized fibers increases linearly up to $\lambda = 24$, slowly increases above it, and reaches a maximum 58 GPa at $\lambda = 34$.[5] The modulus of the conventional fibers prepared by drawing in heated silicone oil increases with λ, but is less than that of the pressurized fibers at each λ; a maximum modulus is 50 GPa at $\lambda = 29$. The microwave heating/drawing fibers show a modulus behavior between the pressurized fibers and the conventional fibers, and a maximum modulus is 54 GPa at $\lambda = 34$.

The tensile strength of the pressurized fibers increases linearly to $\lambda = 22$, reaches 2.0 GPa as a maximum, and holds a plateau above it. The conventional fibers increase linearly to $\lambda = 15$, slowly increase in the range $15 \leq \lambda \leq 20$, reach 1.5 GPa as a maximum, and rapidly decrease above 20. The microwave heating/drawing fibers show the same tendency as the conventional fibers except for a maximum 1.7 GPa at $\lambda = 28$.[6]

CREEP BEHAVIORS

The creep moduli and viscosities at a fixed draw ratio are constant relative to the stress. The result is thus analogous to the phenomenon of Hookean elasticity. All parameters increase with the bulk Young's modulus, E, of the fiber, and especially the viscosities increase rapidly at $\lambda \geqq 16$. This is due to an increase in the crystalline component and a decrease in the amorphous component. The rapid changes in the viscosities means that the amorphous chain is transformed from a loose chain into a rigid chain morphology, such as a taut tie molecule that is strongly resistant to shear stress.

CHEMICAL RESISTANCE

A crystalline component is generally resistant to chemicals more than amorphous component. Drawing brings the transformation of the lamella and amorphous component into the oriented crystalline component and an increase in crystallinity. The drawn fibers are therefore improved in chemical resistance.

MISCELLANEOUS PROPERTIES

The annealed superdrawn POM fibers have a linear expansion coefficient as small as 10^{-6} °C^{-1} that is the same as that for quartz optical fibers.[7] A velocity of the second heat shrinkage of the fibers is as small as 10^{-8}–10^{-7} s^{-1} at 100°C.[8] The POM fibers are excellent in impact brittleness compared with glass fibers, and have an especially good low-temperature resistance −200°C. In the sea, the POM drawn fibers are waterproof and almost without adhesion of sea microorganisms.

APPLICATIONS

The superdrawn POM fibers are thermoplastic, weak to bending and abrasion, and large in strain compared with Kevlar. Thus, they can not be used under the environments of dynamic stress and high temperature in the outdoors. The good points of the fibers are high modulus, high strength, toughness, being waterproof, large diameter, and light weight. The fibers are therefore advantageous in the fields of civil engineering and the sea, and are widely used for soil reinforcement.[9]

REFERENCES

1. Sugeta, H.; Miyazawa, T. *Polym. J.* **1970**, *1*, 226.
2. Asahina, M.; Enomoto, S. *J. Polym. Sci.* **1962**, *59*, 101.
3. Nakagawa, K.; Maeda, O.; Yamakawa, S. *J. Polym. Sci., Polym. Lett. Edn.* **1983**, *21*, 933.
4. Komatsu, T.; Enoki, S.; Aoshima, A. *Polymer* **1991**, *32*, 1983.
5. Komatsu, T.; Enoki, S.; Aoshima, A. *Polymer* **1991**, *32*, 1988.
6. Nakagawa, K.; Konaka, T.; Yamakawa, S. *Polymer* **1985**, *26*, 84.
7. Nakagawa, K.; Konaka, T.; Yamakawa, S. Denshi Tsuushin Gakkai Sougou Zenkoku Taikai Prepr. Jpn., *7*, 300, 1983.
8. Komatsu, T.; Enoki, S.; Aoshima, A. *Polymer* **1991**, *32*, 2992.
9. Asahi Chem. Ind. Co. Ltd. and Tokyu Construction Co. Ltd., Geogrid for Soil reinforcement. Technological Authentication No. 0405. The Doboku Kenkyu Center Foundation, Tokyo, Japan 1991.

Polyoxymethylenes

See: *Acetal Resins*
Acetal Resins (Homopolymers, Copolymers, and Block Copolymers)
Engineering Thermoplastics (Survey of Industrial Polymers)
Polyoxymethylene Fiber, Superdrawn
Uniform Polymers

Polyoxypropylene

See: *Poly(propylene oxide)*
Supramolecular Formation and Structure (PEO-PPO-PEO Block Copolymers)

Polypentenamer

See: *Metathesis Degradation, Unsaturated Polymers*
Metathesis Polymerization, Cycloolefins

POLYPEPTIDE-POLYPEPTIDE COMPLEXES

Mu Shik Jhon and Jae Chang Jung
Department of Chemistry
Korea Advanced Institute of Science and Technology

The majority of biological polymers aggregate by secondary binding forces such as electrostatic interactions, hydrogen bonds, van der Waals forces, and charge-transfer interactions.

The aggregation phenomena of more than two different polymer chains in solution caused by secondary binding forces are generally called "polymer–polymer complexes." There have been some reported systems where complexation accompanying conformational changes of polypeptides occurs between polypeptides and different polymers. In complex formation between poly(L-lysine) (PLL) [25104-18-1] and poly(methacrylic acid) (PMAA) [25087-26-7], the effect of tacticity on the complex formation was studied. In this system, the helical structure of PLL is distributed by the complex formation with PMAA.[1] While poly(ethylenimine) [26336-38-9] promotes the helix formation of polypeptide in poly(L-glutamic acid) (PLGA) [25513-46-6]-poly(ethylenimine) complex.[2] Complex formation of PGA with various polycation was studied in previous work.[3] In complex formation by hydrogen bonding between poly(L-proline) (PLP) [25191-13-3] and PMAA, PMAA disturbs the helix formation of PLP when the complex is formed.[4] Very few works about the interpolymer complexation between polypeptides have been studied through synthetic polypeptide–polypeptide complexes. These could provide a useful model system for understanding the factors influencing the structure of protein aggregates and the properties of such aggregates in biological systems. Polypeptide–polypeptide complexes by electrostatic forces in PLGA-PLL were studied.[5–8] Poly(L-arginine) [25212-18-4]-PLGA and poly(L-arginine)-poly(L-aspartic acid) [69307-21-7] complexes were studied.[9] The charge-transfer complexes in block copolymer $(\text{L-tryptophan})_n$-$(\gamma\text{-ethyl-DL-glutamate})_m$ [29531-53-1] with poly(L-histidine hydrochloride) [61857-39-4] were also studied.[10] Complex formation by hydrogen bonding between PLP and PLGA was studied in the authors' work.[11] Our systems are different from the above mentioned works in which we used the two polypeptides having helical conformations, so our systems are a helical–helical type of complex formation between synthetic polypeptides, and the conformational effect on the complex formation can be studied.

PROPERTY

Complex Formation by Electrostatic Interaction

PLGA is α-helix and PLL is random coil at pH 4, and when the two polypeptides mixed, the conformation of PLGA-PLL complex is β-sheet.

At pH 7.0, poly(L-arginine) (PLA) and PLGA have random coil conformation, but when these two polypeptides are mixed, they form a complex containing α-helical conformation.[9] However, as the pH decreases below 2, the intensity of CD spectrum decreases without the serious change of the shape of CD spectrum. This indicates that the carboxyl group of the PLGA's side chain becomes uncharged, and the uncharged carboxyl group cannot interact with guanidinium groups. Thus, the PLA-PLGA complex would be changed to the random coil PLA and the α-helical PLGA.

Complex Formation by Charge Transfer Interaction

Charge transfer complex of polypeptides between poly(L-histidine hydrochloride) as the acceptor and poly(L-tryptophan) [27813-82-7] as the donor was studied.[10] From the CD spectra, it was confirmed that there was no appreciable effect of the secondary structure of the polymeric donor by complex formation.

Complex Formation by Hydrogen Bonding

We studied complex formation by hydrogen bonding between PLGA as the proton donor and poly(L-proline) (PLP) as the proton acceptor.[11] In this system, both component polypeptides can have helical conformation. ΔpH is the difference of pH between the measured value of mixture and the theoretical value calculated from the PLGA and PLP solution before mixing. If the complex is formed through hydrogen bonding, the protons in the solution will be captured by the PLGA-PLP complex and the concentration of protons in the solution will be reduced. In other words, the increasing ΔpH accompanying the complex formation indicates the polymer–polymer complex formation through hydrogen bonding.

As explained previously, in spite of the importance of the polypeptide–polypeptide complex formation, there have been very few works about this subject. And for the known polypeptide–polypeptide complex systems, much research should be done in the future to elucidate the exact conformation of the component polypeptides and structures of complexes.

REFERENCES

1. Shinoda, K.; Sakai, K.; Hayashi, T.; Nakajima, A. *Polymer J.* **1976**, *8*, 208.
2. Sato, H.; Hayashi, T.; Nakajima, A. *Polymer J.* **1976**, *8*, 517.
3. Abe, K.; Tsuchida, E. *Polymer J.* **1977**, *9*, 79.
4. Park, S. M.; Jeon, S. H.; Ree, T. *J. Polym. Sci. Polym. Chem. Ed.* **1989**, *27*, 4109.
5. Hammes, G. G.; Schullery, S. E. *Biochemistry* **1968**, *7*, 3882.
6. Nakajima, A.; Shinoda, K.; Hayashi, T.; Sato, H. *Polymer J.* **1975**, *7*, 550.
7. Domard, A.; Rinaudo, M. *Macromolecules* **1980**, *13*, 898.
8. Domard, A.; Rinaudo, M. *Macromolecules* **1981**, *14*, 620.
9. Mita, K.; Ichimura, S.; Zama, M. *Biopolymers* **1978**, *17*, 2783.
10. Palumbo, M.; Cosani, A.; Terbojevich, M.; Peggion, E. *Biopolymers* **1977**, *16*, 109.
11. Jung, J. C.; Kim, H. D.; Ree, T.; Jhon, M. S. *J. Polym. Sci. Polym. Chem. Ed.* **1993**, *31*, 3377.

POLYPEPTIDE RESTRICTED BACKBONE CONFORMATION ($C^{\alpha,\alpha}$-Symmetrically Disubstituted Glycines)

Ettore Benedetti
Biocrystallography Research Center, C.N.R.
Department of Chemistry
University of Naples

Claudio Toniolo
Biopolymer Research Center, C.N.R.
Department of Organic Chemistry
University of Padova

The recent explosive interest in the study of oligopeptides rich in $C^{\alpha,\alpha}$-symmetrically disubstituted glycines (for review articles see References 1–19) is based on the following factors:

- (2-Aminoisobutyric acid Aib) (**Figure 1**), the prototype of these amino acids, characterizes an important family of natural antibiotics, the peptaibols, which alter the ionic permeability of biological membranes by forming channels.

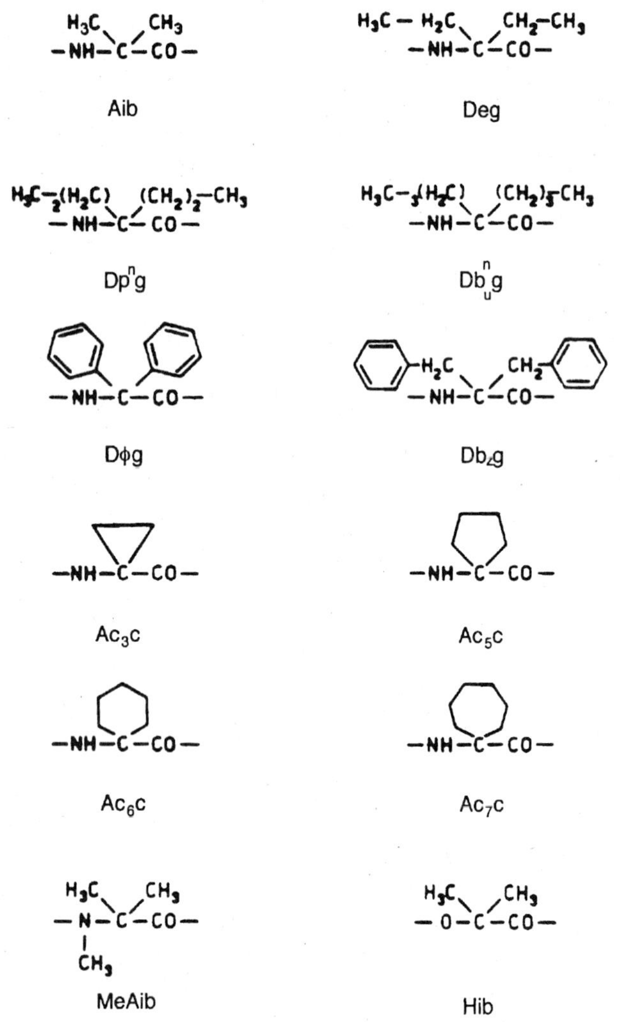

FIGURE 1. The $C^{\alpha,\alpha}$-symmetrically distributed glycines and their isosteres whose preferred structures have been studied to date.

- Tetrasubstitution at the α-carbon in these amino acids results in a severe steric hindrance, the synthesis of the related peptides being greatly complicated by this property.
- Derivatives of these residues show high stability and crystallinity, thus allowing one to perform an X-ray characterization of conformation and electronic structure not only of N- and C-protecting groups, but also of C-activating groups commonly used in peptide synthesis.
- The structural preferences of peptides containing these amino acids are unique, as they possess significant constraints on their conformational freedom. In particular, this point is relevant to (a) the exploitation of these compounds as precise molecular rulers or scaffolding units in the *de novo* design of protein and enzyme mimetics and templates for molecular and chiral recognition study, and (b) the three-dimensional (3D) structure–activity relationships of conformationally constrained enzyme-resistant agonists and antagonist of bioactive peptides.

APPLICATIONS AND FUTURE PROSPECTS

Applications of $C^{\alpha,\alpha}$-symmetrically disubstituted (achiral glycines to chemical and biochemical problems will undoubtedly take advantage of recent improvements in the synthesis of peptides involving severely hindered amino acid residues.

Quite recently, an extremely important and promising synthetic result was reported, namely the incorporation of Aib and Ac$_3$c residues into proteins by site-directed mutagenesis.[22]

In addition to peptaibol antibiotics, Aib is a constituent of the naturally occurring cyclotetrapeptide metabolite chlamydocin. The X-ray diffraction structure of the dehydroderivative of this highly strained compound[23] revealed the Aib residue in the uncommon γ-bend conformation.[20,21] Aib and Ac$_n$c residues have been incorporated in a variety of bioactive peptides. In particular: (*i*) A detailed 3D structure–activity investigation was performed on position 2 analogues of the formylmethionyl tripeptide chemoattractants.[24] This study showed the potential benefit of exploiting the Ac$_n$c series to assess the effect of side-chain hydrophobicity on bioactivity, while keeping the backbone conformation fixed. More recently, the crystal structure of the (Dp$_n$g)^2-analogue was also investigated.[25] (*ii*) The conformational versatility of the N^{α}-acylated tripeptide amide tail of oxytocin was demonstrated by X-ray diffraction analysis of three analogues containing the Ac$_3$c, Ac$_5$c, and Ac$_6$c residues in position 2.[26] Folded and helical Aib-rich peptides were exploited as rigid building blocks in chemical, photochemical, and spectroscopic studies.[27–32] Ac$_3$c is recognized as the biosynthetic precursor of the plant hormone ethylene, the agent responsible for fruit ripening. In addition, short peptides containing this residue were used as latent inhibitors of proteolytic enzymes.[33] Ac$_3$c is also present in a cotoxin, a cyclo-nonadepsipeptide having high pathogenicity in corn leaves.[34] Ac$_5$c, called cycloleucine in the biomedical literature, is a known antitumor agent. Ac$_6$c is a constituent of cycloacillin, an important semisynthetic penicillin. Aib, MeAib, and 2-aminodamantane-2-carboxylic acid or adamantine (Ada) are widely used in amino acid transport studies. Dϕg hydantoin (dilantin or phenytoin) is currently the most widely used anti-epilectic drug.

The electron-deficient 2,2-dimethyl-1*H*-phenalen-1,3-dione-based tetracyclic amino acid was synthesized and proposed to be the chromophore/acceptor residue of choice for light-induced electron transfer peptides.[35]

The first fullerene-based peptide was prepared and characterized. The resulting "ball-and-helix" system was generated by the C_{60} ring moiety and the Aib residue, respectively.[36]

REFERENCES

1. Nagaraj, R.; Balaram, P. *Accts. Chem. Res.* **1981**, *14*, 356.
2. Jung, G.; Bruckner, H.; Schmitt, H. In *Structure and Activity of Natural Peptides*; Voelter, W.; Weitzel, G., Eds.; de Gruyter, Berlin, 1981; p 75.
3. Toniolo, C.; Bonora, G. M.; Bavoso, A.; Benedetti, E.; DiBlasio, B.; Pavone, V.; Pedone, C. *Biopolymers* **1983**, *22*, 205.
4. Spatola, A. In *Chemistry and Biochemistry of Amino Acids: Peptides and Proteins*; Weinstein, B., Ed. Dekker: New York, 1983; Vol. 7, p 267.
5. Mathew, M. K.; Balaram, P. *Mol. Cell. Biochem.* **1983**, *50*, 47.
6. Venkataram Prasad, B. V.; Balaram, P. *CRC Crit. Rev. Biochem.* **1984**, *16*, 307.

7. Benedetti, E.; Di Blasio, B.; Pavone, V.; Pedone, C.; Bavoso, A.; Toniolo, C.; Bonora, G. M.; Leplawy, M. T.; Hardy, P. M. *J. Biosci.* **1985**, *8*, 253.
8. Toniolo, C. *Brit. Polym. J.* **1986**, *18*, 221.
9. Toniolo, C.; Benedetti, E.; Pedone, C. *Gazz. Chim. Ital.* **1986**, *116*, 355.
10. Toniolo, C.; Benedetti, E. *ISI Atlas of Science: Biochemistry* **1988**, *1*, 225.
11. Aubry, A.; Boussard, G.; Cung, M. T.; Marraud, M.; Vitoux, B. *J. Chim. Phys. Phys.-Chim. Biol.* **1988**, *85*, 345.
12. Toniolo, C. *Biopolymers* **1989**, *28*, 247.
13. Uma, K.; Balaram, P. *Indian J. Chem.* **1989**, *28B*, 705.
14. Toniolo, C. *Int. J. Pept. Protein Res.* **1990**, *35*, 287,
15. Karle, I. L.; Balaram, P. *Biochemistry* **1990**, *29*, 6747.
16. Hruby, V. J.; Al-Obeidi, F.; Kazmierki, W. *Biochem. J.* **1990**, *268*, 249.
17. Toniolo, C.; Benedetti, E. *Macromolecules* **1991**, *24*, 4004.
18. Rizo, J.; Gierasch, L. M. *Annu. Rev. Biochem.* **1992**, *61*, 387.
19. Toniolo, C.; Crisma, M.; Formaggio, F.; Valle, G.; Cavicchioni, G.; Précigoux, G.; Aubry, A.; Kamphuis, J. *Biopolymers* **1993**, *33*, 1061.
20. Toniolo, C. *CRC Crit. Rev. Biochem.* **1980**, *9*, 1.
21. Némethy, G.; Printz, M. P. *Macromolecules* **1972**, *5*, 755.
22. Ellman, J. A.; Mendel, D.; Schultz, P. G. *Science* **1992**, *255*, 197.
23. Flippen-Anderson, J. L.; Karle, I. L. *Biopolymers* **1976**, *15*, 1081.
24. Toniolo, C.; Crisma, M.; Valle, G.; Bonora, G. M.; Polinelli, S.; Beker, E. L.; Freer, R. J. Sudhanand, Balaji Rao, R.; Balaram, P.; Sukumar, M. *Pept. Res.* **1989**, *2*, 275.
25. Dentino, A. R.; Raj, P. A.; Bhandary, K. K.; Wilson, M. E.; Levine, M. J. *J. Biol. Chem.* **1991**, *266*, 18460.
26. Fabiano, N.; Valle, G.; Crisma, M.; Toniolo, C.; Saviano, M.; Lombardi, A.; Isernia, C.; Pavone, V.; Di Blasio, B.; Pedone, C.; Benedetti, E. *Int. J. Pep. Prot. Res.* **1993**, *42*, 459.
27. Bortolussi, A.; De Pieri, G.; Signor, A.; Bonora, G. M.; Toniolo, C. *Gazz. Chim. Ital.* **1985**, *115*, 253.
28. Basu, G.; Kubasik, M.; Anglos, D.; Secor, B.; Kuki, A. *J. Am. Chem. Soc.* **1990**, *112*, 9410.
29. Sisdo, M.; Ishikawa, Y.; Harada, M.; Itoh, K. *Macromolecules* **1991**, *24*, 3999.
30. Budt, K. H.; Vatele, J. M.; Kishi, Y. *J. Am. Soc.* **1991**, *108*, 6080.
31. Maruyama, K.; Hashimoto, M.; Tamiaki, H. *Chem. Lett.* **1991**, 1455.
32. Toniolo, C.; Formaggio, F.; Crisma, M.; Valle, G.; Boesten, W. H. J.; Schoemaker, H. E.; Kampuis, J.; Temussi, P. A.; Beker, E. L.; Précigoux, G. *Tetrahedron* **1993**, *49*, 3641.
33. Suckling, C. J. *Angew. Chem. Int. Ed. Engl.* **1988**, *27*, 537.
34. Ueda, K.; Xiao, J. Z.; Doke, N.; Nakatsuka, S. *Tetrahedron Lett.* **1992**, 5377.
35. Kotha, S.; Anglos, D.; Kuki, A. *Tetrahedron Lett.* **1992**, 1569.
36. Prato, M.; Bianco, A.; Maggini, M.; Scorrano, G.; Toniolo, C.; Wuld, F. *J. Org. Chem.* **1993**, *58*, 5578.

Polypeptides

See: *Enzymatic Polymerization*
Enzyme-Degradable Hydrogels
Fluorine-Containing Poly(amino acid)s (Conformation and Gas Permeability)
Glycoproteins (Synthesis by NCA Method)
Host-Guest Chemistry [using α-helical Poly(L-lysine)]
Liquid Crystalline Polymers (Polypeptides)
Marine Adhesive Proteins, Synthetic
Molecular Recognition (Peptide-Based Systems)
Nucleohistone Complexes (Protein-Nucleic Acid Interactions)
Peptide-Based Nanotubes (New Class of Functional Biomaterials)
Photochromic Poly(α-minoacid)s
Poly(n-alkyl L-glutamate) (Structure and Molecular Mobility in Solid Liquid Crystalline State)
Polyamides, Glycine-Containing
Poly(α-amino acid) Spherical Particles
Poly(α-amino acids) (Biodegradation, Medical Applications)
Polyisopeptides
Polypeptide-Polypeptide Complexes
Polypeptide Restricted Backbone Conformation (Cα,α-Symmetrically Disubstituted Glycines)

Polyperoxides

See: *Vinyl Polyperoxides (Oxygen Copolymerization)*

Polyperylene

See: *Perylene Polymers*

Polyphenothiazine

See: *Conjugated Ladder Polymers*

Polyphenoxozine

See: *Conjugated Ladder Polymers*

Polyphenoxazines

See: *Ladder Polymers (Methods of Preparation)*

Poly(phenyl sulfone)

See: *Poly(aryl ether sulfone)s*

POLYPHENYLACETYLENE (Overview)

Anita Furlani* and Maria Vittoria Russo
Department of Chemistry
University "La Sapienza"

Polyphenylacetylene (PPA; *Chemical Abstracts* registry number 25538-69-1) was prepared for the first time many years ago, but the interest in this polymer has increased recently owing to its analogy to polyacetylene.

This review will summarize the researches undertaken in the last 10 years on PPA. Preceeding investigations on PPA are reported in reference.[1]

*Author to whom correspondence should be addressed.

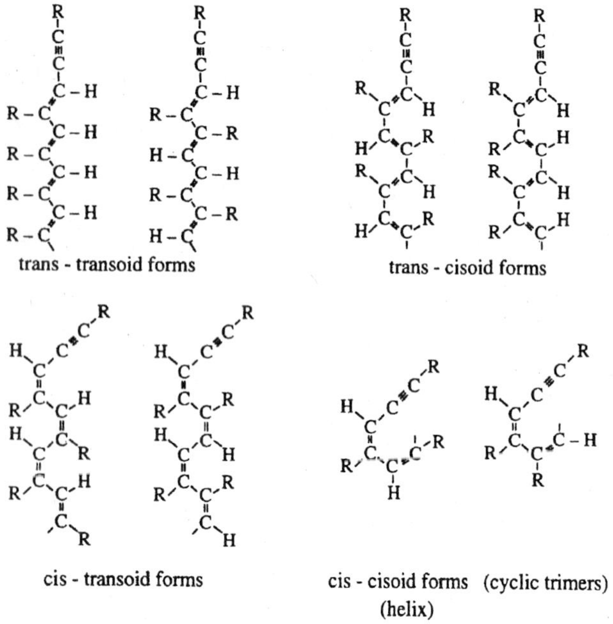

FIGURE 1. Conformers of PPA, depending on the monomer polymerization modes.

Although various catalytic systems such as Ziegler Natta catalysts or transition metal complexes have already been proposed for the polymerization of phenylacetylene (PA),[1] many investigations have been carried out to find new catalysts or new reaction conditions. By polymerization of monosubstituted acetylenes, it is possible to obtain eight different polymeric structures (**Figure 1**), which correspond to possible different regular head-to-tail-tail-to-head or head-to-tail-head-to-tail polymerizations of the monomer molecules. The polymer molecule can also undergo rotations around the single bonds, giving coiled structures or *cis* or *trans* isomerizations.

SYNTHESIS OF PPA

The new studies on the synthesis of PPA can be summarized as follows:

- Investigations on the activity of catalysts
  - W salts or complexes
  - Mo salts or complexes
  - Rh complexes
  - Other transition metal complexes
- Precursor polymers
- Plasma, laser, or photochemical initiators, thermal polymerization.

PLASMA, PHOTOCHEMICAL, AND THERMAL POLYMERIZATION OF PA

Inoue et al.[3–6] investigated the plasma and photochemical polymerization of PA. Films of PPA were obtained. The π electron delocalization of the films, prepared by photochemical polymerization, was more widely extended than that of plasma-polymerized PPA.

Laser-initiated polymerizations of PA were also carried out in the presence of carbonyl catalysts.

Products obtained by thermal polymerization of PA were examined.[7] The low MW of thermal PPA is caused not by inactivation of the active centers due to delocalization of the unpaired electrons on the chain, but by intramolecular cyclization at the end of the chain.

STABILITY AND THERMAL ANALYSIS OF PPA

Several investigations have been carried out on PPA stability. It was found that PPA prepared with the $WOCl_4/Sn(C_6H_5)_4$ catalyst is stable under vacuum in the solid state for at least two years, but it is less stable in THF solution (50 days).[8] Oxygen activates a degradation process leading to a continuous decrease of the MW.[9] The presence of a metathesis catalyst seems to increase the decomposition rate.[10]

STRUCTURAL INVESTIGATIONS

As previously reported, the regular chain formation of the monomer units can give eight different regular isomeric structures of the PPA chains. Irregular distribution of the monomer units in the chain is also possible; therefore, PPA with different structures can be obtained. The PPA obtained with Rh complexes exhibited a high-resolution NMR spectrum,[2] and a *cis*-transoidal head-to-tail structure can be assigned to this polymer.

The PPA prepared with a Nd catalyst was a mixture of *cis-trans* and *cis-cis* conformers.[11] On the basis of IR and high-resolution NMR spectra, it was found that thermally polymerized PPA consists of *trans* and cyclohexadiene units, whereas WCl_6 and Co(acetylacetonate)/$AlEt_3$ catalysts give PPA mainly of linear *cis* and *trans* units.[12]

XPS STUDIES

It is known that PPA is a stable material; however, it undergoes oxidation in oxygen-rich atmosphere.[13] The formation of C-O- and C=O groups is suggested in the C 1*s* XPS spectra by the appearance of components at binding energy (B.E.) = 286.3 and 288.1 eV (C 1*s* main peak position is found at B.E. = 284.6 eV).[14]

CONDUCTIVITY AND PHOTOCONDUCTIVITY OF PPA

The conductivity and photoconductivity of *trans*-PPA were found to be higher than those of the *cis* isomer.[15] The *trans*-PPA, in the form of an amorphous film, is a semiconducting photoconductor.

The conductivities of poly(substituted phenylacetylene)s,[16,17] of PPA, and of some poly(halophenylacetylene)s[18] were investigated. The conductivity increases with increasing temperature and is dependent on the number of conjugated double bonds in the chain and on the stereoregularity of the polymer. The number of double bonds in the chain can be calculated from the position of the UV band of PPA.[18]

The conductivity of PPA increases upon doping with iodine to values in the range 10^{-5}–10^{-3} S·cm^{-1}. The conduction mechanism seems to be ionic.[19] The conductivity of PPA in the presence of various doping agents such as $LiAlH_4$, $AgPF_6$, BF_3, $FeCl_3$, $NaBH_4$, HCl, H_2SO_4, $SnCl_2$, $HClO_4$, NH_3, and HNO_3 was also examined. High conductivity values (10^{-2} S·cm^{-1}) upon doping with iodine (75% wt/wt) were found.[20] Other doping agents were also used to increase the conductivity of PPA.[21] The charge-transfer interaction of PPA with iodine, AsF_5, and 2,3-dichloro-5,6-dicyano-*p*-benzoquinone (DDQ) in solution is associated with the appearance of broad charge-transfer bands extending beyond the absorption edge of the polymer into the near-IR. The metalized films of PPA, pretreated with a dye, can reach high conductivity values.[22]

PHOTOLUMINESCENCE AND PHOTOSENSITIVITY OF PPA

Photoluminescence of PPA and poly(1-chloro-2-phenylacetylene) (ClPPA) at room temperature were observed at 710 nm for PPA and at 480 nm for ClPPA. With decreasing temperature, these peaks shifted to higher energy and their intensity decreased.[23] The photoluminescence of films of *cis*-transoidal and *trans*-cisoidal PPA was examined.

Films of *cis*-PPA obtained by irradiation with ^{60}Co γ-rays were found to be highly photosensitive. The photosensitivity was dependent on the irradiation dose and on the polymer structure. The crucial dose to produce a light response is 5×10^3 Gy.[24]

Electron beams were also used to increase the photosensitivity of PPA prepared with rare earth complexes.[25] The addition of dopants, such as $FeCl_3$ and I_2, significantly enhances the photosensitivity of PPA.

Electrochromic devices, having a poly(substituted phenylacetylene)s color-forming layer and an electrolyte layer between a transparent electrode and a counter electrode, are proposed.[26] Patents for electroactive inks,[27] resist patterns,[28] and photosensitive systems for color printing[29] claim the use of PPA or similar polymers as electroactive or photoactive components.

NONLINEAR OPTICAL (NLO) PROPERTIES

PPA nonlinear effects have been studied by several research groups. One of the pioneering papers was published by Prasad et al.;[30] third-order nonlinear optical interactions of oxidized, doped, and undoped PPA were studied, showing a subpicosecond response, by the degenerate four-wave mixing method.

Investigations concerning thin film design and fabrication, applications of laser Raman optical waveguides, inelastic electron tunneling, picosecond transient gratings, and surface plasmon coupling techniques for the study of the ultrastructure of polymers (among others, PPA) have been also reported.[31] The potential use of PPA as an optically active element in relation to *cis–trans* isomerization has been proposed.[32] An accurate value of the nonlinear susceptibility $\chi^{(3)}$, (7.0 1.0) $\times 10^{-12}$ esu, for PPA has been determined with third-harmonic generation (THG) measurements; the influence of back-reflection effects for different optical configurations is estimated;[33] the effects of the increasing electronic coupling on optical absorption bands related to NLO for PPA derivatives with various substituents at the phenyl rings have been reported more recently,[34,35] as well as scaling laws relating third-order NLO with linear optical properties of PPA in comparison with other polyacetylenes.[36]

An increase of $\chi^{(3)}$ values obtained by THG measurements with increasing λ_{max} of absorption for substituted PPA is also found.[37]

Last but not least, alignment of liquid crystals on rubbed thin films of polystyrene derivatives and PPA in display devices is a new application for polymeric materials.[38] Second- and third-order NLO susceptibilities of polymeric liquid crystals (donor–acceptor-substituted side-chain polymers) have been compared with $\chi^{(3)}$ values of PPA in relation to the polymers structure in a recent paper.[39]

SENSORS AND OTHER MEMBRANE APPLICATIONS

PPA has been tested in several devices for the sensing of gases. PPA is an insulator and its conductivity changes upon absorption of gases; therefore, interdigitated electrode structures covered with a PPA film have been exploited so that even small capacitance changes induced by gases can be detected.

Electrochemical sensors belong to another interesting field of technological applications for PPA. Iodine-doped PPA has been used as the sensor membrane electrode with a quasi-Nernstian response for potentiometric detection of cations (K^+, Ca^{2+}, Cu^{2+}, and La^{3+}).[40]

The influence of HCl doping on the response of a PPA membrane tested as sensor for anions and cations has been reported, in comparison with the behavior of I_2-doped polybenzylpropargylamine membranes.[41]

A side aspect of the use of PPA-based membranes concerns the catalytic activity of I_2-doped PPA in the photodegradation process of organic pollutants in wastewater[42,43] and in the photooxidation of ethoxy- and butoxyethanol.[44]

Finally, the gas permeability of PPA containing CF_3 groups on the benzene rings is noteworthy; the selectivity and permeation behavior of the membranes to oxygen and nitrogen in relationship with the polymer structures have been established.[45] Trimethylsilyl-group-containing PPAs are good materials for the preparation of oxygen and ethanol permselective membranes.[46]

REFERENCES

1. Ehrlich, P.; Anderson, W. A. *Handbook of Conducting Polymers* Dekker: New York, 1986; Vol. 1, p 441.
2. Furlani, A.; Napoletano, C.; Russo, M. V.; Feast, W. J. *Polym. Bull. (Berlin)* **1986**, *16*, 311.
3. Takai, Y.; Inoue, M.; Mizutani, T.; Ieda, M. *J. Phys. D.* **1985**, *18*, 1637.
4. Inoue, M.; Takai, Y.; Mizutani, T.; Ieda, M. *Jpn. J. Appl. Phys. Part 1* **1986**, *25*, 1174.
5. Inoue, M.; Takai, Y.; Mizutani, T.; Ieda, M. *Jpn. J. Appl. Phys. Part 2* **1986**, *25*, L716.
6. Inoue, M.; Fujioka, H.; Sorita, T.; Tanaka, T. *Polym. Prepr. Am. Chem. Soc. Div. Polym. Chem.* **1987**, *28*, 332.
7. Chauser, M. G.; Anisimova, O. S.; Kol'tsova, L. S.; Zaichenko, N. L.; Cherkashin, M. I. *Izv. Akad. Nauk. SSSR, Ser. Khim.* **1988**, 67; *Chem. Abstract* **1988**, *108*, 95232.
8. Vohlidal, J.; Redrova, D.; Pacovska, M.; Sedlacek, J. *Collect. Czech Chem. Commun.* **1993**, *58*, 2651.
9. Sedlacek, J.; Vohidal, J. *Makromol. Chem. Rapid Commun.* **1993**, *14*, 51.
10. Severina, E. N.; Mavrenkova, G. V.; Geiderikh, M. A.; Davydov, B. E. *Vysokomol. Soedin. Ser. A* **1987**, *29*, 739; *Chem. Abstract* **1987**, *107*, 7761.

11. Gruzdeva, V. F.; Bodarenko, G. N.; Prokof'eva, N. I.; Gribov, L. A. *Vysokomol. Soedin. Ser. A* **1992**, *34*, 99; *Chem. Abstract* **1992**, *116*, 195221.
12. Chauser, M. G.; Kol'tsova, L. S.; Vladimirov, L. V.; Urman, Ya. G.; Alekseeva, S. G.; Zaichenko, N. L.; Oleinik, E. F.; Cherkashin, M. I. *Vysokomol. Soedin. Ser. A* **1988**, *30*, 1464; *Chem. Abstract* **1988**, *109*, 74254.
13. Neoh, K. G.; Kang, E. T.; Tan, K. L. *Thermochimica Acta* **1989**, *146*, 251.
14. Kang, E. T.; Neoh, K. G.; Tan, K. L.; Tan, B. T. G. *J. Polym. Sci., Polym. Phys. Ed.* **1989**, *27*, 2061.
15. Hong, H.; Zhou, S.; Jin, X. *Ganguang Kexue Yu Kuang Huaxue* **1992**, *10*, 91; *Chem. Abstract* **1993**, *119*, 9485.
16. Gu, H. B.; Hayashi, S.; Sugimoto, R.; Yoshino, K. *Technol. Rep. Osaka Univ.* **1986**, *36*, 323.
17. Zhou, S.; Hong, H.; He, Y.; Yang, D.; Jin, X.; Qian, R.; Masuda, T.; Higashimura, T. *Polymer* **1992**, *33*, 2189.
18. Janevski, A.; Leben, S.; Sebenik, A.; Osredkar, U. *Polym. Bull. (Berlin)* **1991**, *26*, 473.
19. Leclerc, M.; Prud'homme, R. F. *Macromolecules* **1987**, *20*, 2153.
20. Furlani, A.; Napoletano, C.; Paolesse, R.; Russo, M. V. *Synth. Meth.* **1987**, *21*, 337.
21. Li, Y.; Chen, Z.; Tong, Y. *Gaofenzi Cailiao Kexue Yu Gongcheng* **1990**, *6*, 20; *Chem. Abstract* **1992**, *116*, 141048.
22. Omori, F.; Tanimoto, F.; Kitano, H. *Jpn. Kokai Tokkyo Koho 5 pp.* JP 63303077 A2 881209 Showa - JP 87-137379 870531; *Chem. Abstract* **1989**, *111*, 116476.
23. Hayashi, S.; Yoshino, K. *Technol. Rep. Osaka Univ.* **1986**, *36*, 117.
24. Yang, M.; Zhao, J.; Li, A.; Shen, Z.; Zhang, M.; Lin, S. *J. Polym. Sci., Polym. Chem. Ed.* **1989**, *27*, 3829.
25. Zhao, J.; Yang, M.; Shen, Z. *Polym. J. (Tokyo)* **1991**, *23*, 963.
26. Fujisaka, T.; Inoue, T.; Koremoto, T.; Suezaki, M. *Jpn. Kokai Tokkyo Koho 3 pp.* JP 02102289 A2 900413 Heisei - JP 88-256376 881011; *Chem. Abstract* **1990**, *113*, 88307.
27. Kitano, H.; Tanimoto, F. *Jpn. Kokai Tokkyo Koho 8 pp.* JP 61287971 A2 861218 Showa - JP 85-130304 850614; *Chem. Abstract* **1987**, *107*, 98376.
28. Nakamura, J.; Motoiama, T. *Eur. Patent Appl.* 13 pp. EP 449439 A1 911002 EP 91/301866 910306; *Chem. Abstract* **1992**, *116*, 117251.
29. Robillard, J. J. A. *U.S. Patent 3 pp.* 5053320 A 911001; US 90-909193 900416; *Chem. Abstract* **1992**, *116*, 224736.
30. Prasad, P. N.; Rao, D. N.; Swiatkiewicz, J.; Chopra, P.; Ghoskal, S. K. *Springer Ser. Chem. Phys.* **1986**, *46* (Ultrafast Phenom. 5), 518.
31. Prasad, P. N. *Thin Solid Films* **1987**, *152*, 275.
32. Yoshino, K. *Synth. Meth.* **1989**, *28*, C669.
33. Neher, D.; Wolf, A.; Bubeck, C.; Wegner, G. *Chem. Phys. Lett.* **1989**, *163*, 116.
34. Neher, D.; Kaltbeitzel, A.; Wolf, A.; Bubeck, C.; Wegner, G. *NATO ASI, Ser. E* **1990**, *182*, (Conjugated Polym. Mater.: Oppor. Electron., Optoelectron., Mol. Electron.) 387.
35. Neher, D.; Wolf, A.; Leclerc, M.; Kaltbeitzel, A.; Bubeck, C.; Wegner, G. *Synth. Meth.* **1990**, *37*, 249.
36. Bubeck, C.; Grund, A.; Kaltbeitzel, A.; Neher, D.; Mathy, A.; Wegner, G. *NATO ASI, Ser. E* **1991**, *194*, (Org. Mol. Nonlinear Opt. Photonics) 335.
37. Wada, T.; Masuda, T.; Sasabe, H. *Mol. Cryst. Liq. Cryst. Sci. Technol., Sect. A* **1994**, *247*, 139.
38. Nakajima, K.; Wakemoto, H.; Sato, S.; Yokotani, F.; Ishihara, S.; Matsuo, Y. *Mol. Cryst. Liq. Cryst.* **1990**, *180B*, 223.
39. Smith, D. A. McL.; Coles, H. J. *Liq. Cryst.* **1993**, *14*, 937.
40. Campanella, L.; Salvi, A. M.; Sammartino, M. P.; Tomassetti, M. *Chim. Ind. (Milan)* **1986**, *68*, 71.
41. Campanella, L.; Salvi, A. M.; Morgia, C. *Chim. Oggi* **1988**, *5*, 65.
42. Campanella, L.; Salvi, A. M.; Morgia, C.; Sbardellati, R. *Inquinamento* **1987**, *29*, 32.
43. Campanella, L.; Morgia, C.; Salvi, A. M. *Environ. Technol. Lett.* **1989**, *10*, 241.
44. Minero, C.; Maurino, V.; Campanella, L.; Morgia, C.; Pelizzetti, E. *Environ. Technol. Lett.* **1989**, *10*, 301.
45. Hayakawa, Y.; Nishida, M.; Aoki, T.; Muramatsu, H. *J. Polym. Sci., Polym. Chem. Ed.* **1992**, *30*, 873.
46. Aoki, T.; Nakahara, H.; Hayakawa, Y.; Kokai, M.; Oikawa, E. *J. Polym. Sci., Polym. Chem. Ed.* **1994**, *32*, 849.

POLYPHENYLACETYLENE-BASED PERMSELECTIVE MEMBRANES

Toshiki Aoki* and Eizo Oikawa
Department of Chemistry and Chemical Engineering
Faculty of Engineering
Niigata University

STRUCTURE AND ADVANTAGE

Polyphenylacetylene and its derivatives belong to poly(substituted acetylene)s, which consist of conjugated double bonds as the main chain and phenyl side groups (**Structure 1**).

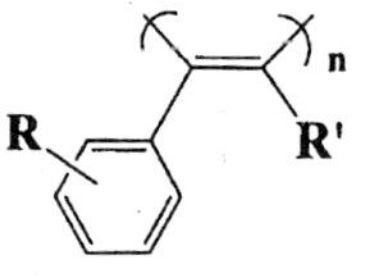

1

R = -H, $-Si(CH_3)_3$, or $-CF_3$

R' = -H, -Cl, $-CH_3$, or $-C_6H_5$

Since the polymers can be easily fabricated into a tough membrane owing to their rigid main chain, they are applied to some permselective membranes. In particular, their performances as oxygen permselective membranes are high. Oxygen-permeation behavior for various polyphenylacetylenes has been investigated.[1–7]

PREPARATION

Monomer Synthesis

Phenylacetylene derivatives are roughly divided into two types: mono-substituted acetylenes and disubstituted acetylenes and in addition, isomers with substituents at different positions on the phenyl ring such as *ortho* or *para* are available.

Polymerization[1–5,8,9]

Catalysts for polymerization of phenylacetylenes are mainly group 5–8 transition metal compounds.

$MoCl_5$ and WCl_6 are effective for a monosubstituted acetylene and a disubstituted acetylene without a bulky substituent, while $NbCl_5$ and $TaCl_5$ can polymerize a disubstituted acetylene with a bulky substituent to yield a high molecular weight polymer.[1,2] The polymerization by these catalysts is thought to be *via* carbene and is called metathesis polymerization.

*Author to whom correspondence should be addressed.

PROPERTIES

Since the polyphenylacetylenes described above are soluble in some solvents such as toluene, chloroform and since tetrahydrofuran and have a very high molecular weight of 10^5 to 10^6, they can be fabricated into a tough membrane by the solvent casting method. Because of the conjugated structure of the polymers, the membranes are yellow, orange, or purple.

Thermal stability is higher owing to the presence of the phenyl group compared with aliphatic polyacetylenes, and thermal decomposition temperature exceeds 300°C. The glass transition temperatures are about 200°C and are much higher than those of the corresponding vinyl polymers.

APPLICATIONS

Oxygen Permeation[1–7]

For an oxygen-permselective membrane, a high oxygen permeability coefficient (Po_2; 1 barrer = 10^{-10} cm^3 (STP) ·cm/cm^2·s·cmHg), a high oxygen separation factor ($\alpha = Po_2/P_{N_2}$), and high stability are required.

Although the Po_2 for polyphenylacetylene (polyPhA) is low, some of the poly(substituted phenylacetylene)s described above show high Po_2 of more than 100. For example, they are poly(*p*-trimethylsilyldiphenylacetylacetylene),[4] and poly[*o,p*-bis(trimethylsilyl) phenylacetylene],[3] all containing bulky substituents.

Since these highly oxygen-permeable polyphenylacetylenes have higher α values than the other polymers with a similar Po_2, they are very good materials for oxygen-permselective membranes.

REFERENCES

1. Masuda, T.; Higashimura, T. *Acc. Chem. Res.* **1984**, *17*, 51.
2. Masuda, T.; Higashimura, T. *Adv. Polym. Sci.* **1986**, *81*, 121.
3. Aoki, T.; Nakahara, H.; Hayakawa, Y.; Kokai, M.; Oikawa, E. *J. Polym. Sci., Polym. Chem. Ed.* **1994**, *32*, 849.
4. Hayakawa, Y.; Nishida, M.; Aoki, T.; Muramatsu, H. *J. Polym. Sci., Polym. Chem. Ed.* **1992**, *30*, 873.
5. Tsuchihara, K.; Masuda, T.; Higashimura, T. *Macromolecules* **1992**, *25*, 5816.
6. Takada, K.; Matsuya, H.; Masuda, T.; Higashimura, T. *J. Appl. Polym. Sci.* **1985**, *30*, 1605.
7. Masuda, T.; Iguchi, Y.; Tang, B.; Higashimura, T. *Polymer* **1988**, *29*, 2041.
8. Aoki, T.; Kokai, M.; Shinohara, K.; Oikawa, E. *Chem. Lett.* **1993**, 2009.
9. Yang, W.; Tabata, M.; Kobayashi, S.; Yokota, K.; Shimizu, A. *Polym. J.* **1991**, *23*, 1135.

Polyphenylacetylenes

See: Polyphenylacetylene (Overview)
Polyphenylacetylene-Based Permselective Membranes
Phenylacetylene (Stereospecific Living Polymerization, Rh Complexes)
Polyradicals (Synthesis and Magnetic Properties)

Polyphenyldiacetylene

See: Polyradicals (Synthesis and Magnetic Properties)

Poly(phenylene ethynylene)

See: Nonlinear Optical Polymers [Polyl(arylene ethynylene) Derivatives]
Polyradicals (Synthesis and Magnetic Properties)

POLY(PHENYLENE ETHYNYLENE) (Synthesis and Optical Properties)

Koichi Kondo
Department of Chemistry
Faculty of Science and Technology
Ritsumeikan University

Polymers constructed with triple bonds in the main chain have attracted a lot of attention because of their unique π-conjugated system, which may be compared with other well-known π-conjugated polymers with double bonds.

Poly(phenylene ethynylene) (**Structure 1**) can be prepared without any difficulty; it has the stable rigid structure consisting of alternating aromatic double bonds and triple bonds, which is different from such other π-conjugated polymers such as polyacetylene and poly(phenylene vinylene).

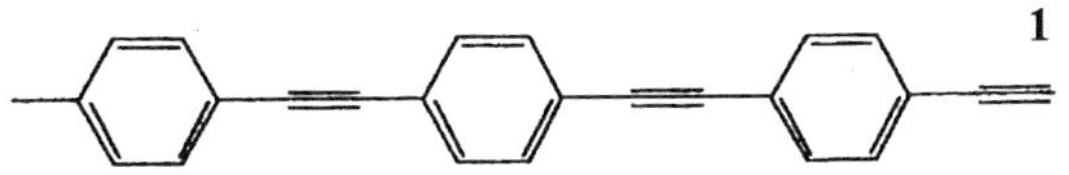

1

Research and development of poly(phenylene ethynylene) has been prevented by its insolubility, although the unique rod-like π-conjugated system is expected to have potential uses in nonlinear optics, molecular electronics, and highly heat-resistant materials.

Synthetic routes to polymer 1 come from the method that is widely used for aromatic acetylenic compound preparation, the palladium-catalyzed coupling of terminal acetylenes and aromatic halides.[1] In principle, the coupling of such bifunctional monomers as 1,4-diethynyl and 1,4-dibromo- or iodobenzene leads to polymer 1. Similarly, 4-bromophenylacetylene can be intramolecularly coupled to afford polymer 1 when the monomer can be obtained.

SYNTHESIS OF POLY(PHENYLENE ETHYNYLENE)

Besides polymer **1**, polymers containing anthryl, pyridyl, and thiophyl moieties were prepared in 70–100% yield. All the polymers that are insoluble in common organic solvents.

The coupling polycondensation demonstrated that most of the polymers containing triple bonds are insoluble in common organic solvents, preventing further research and development.

A lot of effort has been devoted to improving the solubility of polymer 1 in the more than 10 years after the first investigation.

Strategies taken for that purpose are as follows:

- Incorporation of functional groups in the main chain good enough to interact with common organic solvents
- Bringing 1,3-substituted aromatic structure into the *p*-phenylene ethynylene framework to increase flexibility

- Attachment of long alkyl groups to the aromatic ring
- Combinations of the first three strategies

OPTICAL PROPERTIES OF POLY(PHENYLENE ETHYNYLENE)

The improvement in the solubility of poly(phenylene ethynylene) provides a chance to investigate the unique triple-bond π-conjugated system, which confers nonlinear optical properties. The third-order nonlinearity of these polymers is especially intriguing in comparison with the same property of other π-conjugated polymers such as polydiacethylene, polyacethylene, and polyphenylene vinylene), which have been considerably studied. Nevertheless, it is difficult at present to compare the hyperpolarizability ($\chi^{(3)}$ value) of poly(phenylene ethynylene) with that of other π-conjugated polymers and reach any conclusion, since the data are still limited to a few cases, which were determined by different methods.

CONCLUSION

The lack of hydrogen in the main triple bond chain results in an almost carbon-like framework, which leads to a rod-like rigid polymer, decreases in solubility, and gives rise to higher thermal stability. These properties are characteristics of poly(phenylene ethynylene) since the first preparation.

Eventually, the advent of soluble poly(phenylene ethynylene) synthesized by a variety of modifications has made the characterization of it possible and begun to open at last a new era in potential applications.

REFERENCE

1. Takahashi, S.; Kuroyama, Y.; Sonogashira, K.; Hagihara, N. *Synth.* **1980**, 627.

Poly(phenylene oxide)s

See: *Engineering Thermoplastics (Survey of Industrial Polymers)*
Oxidative Coupling

Poly(phenylene sulfide)

See: *Engineering Thermoplastics (Survey of Industrial Polymers)*
Poly(phenylene sulfide) (Elastomer Toughened)
Poly(p-phenylene sulfide) (Synthesis by p-Dichlorobenzene and Sodium Sulfide)
Poly(phenylene and sulfide) and Copolymers (Melt Preparation via a New Process)
Polysulfides (Prepared from Sulfur Dioxide)

POLY(PHENYLENE SULFIDE) (Elastomer Toughened)

Junzo Masamoto
Polymer Development Laboratory
Asahi Chemical Industry Company, Ltd.

Poly(phenylene sulfide) (PPS) features excellent mechanical properties, thermal stability, chemical resistance, flame resistance, and precise moldability. However, PPS is a very brittle material. For use in electrical and electronic parts, automobile parts, and mechanical parts, elastomer-toughened PPS and related compounds are desired.

In development of elastomer-toughened PPS, our goal is to maintain the advantageous thermal, mechanical, and chemical properties of PPS, but to reduce its brittleness by using the polymer alloy method.

PREPARATION OF TOUGHENED PPS

Chemical Treatment of PPS

PPS was chemically treated with diphenylmethane diisocyanate (MDA) by the method described in our patent.[1] PPS was mixed with MDA and extruded in a molten state with a twin extruder.

Elastomer Toughening

A reactive processing method was developed using MDA-treated PPS and an olefinic elastomer with a functional group of maleic anhydride to produce elastomer-toughened PPS. The elastomer alloy of the chemically treated PPS had a notch izod impact strength of about 50 kg·cm/cm, as reported previously,[2] compared with a value of about 1 kg·cm/cm for the original PPS.

PROPERTIES

Elastomer-toughened PPS has applications in two forms: unreinforced and reinforced. Usually, we use our toughened PPS in the reinforced form.

The notched izod impact strength of the general glass-fiber-reinforced PPS is about 7 kg·cm/cm, whereas that of the glass-fiber-reinforced, elastomer-toughened PPS is about 22 kg·cm/cm.

Glass-fiber-reinforced, elastomer-toughened PPS maintains its original thermal properties. Heat deflection temperature is at the same level as the reference, >260°C.

Glass-fiber-reinforced, elastomer-toughened PPS has an impact strength twice that of the linear PPS and 4 times that of the cross-linked PPS.

Since this toughened PPS maintains the advantages of PPS—that is, thermal properties, mechanical properties, chemical properties, and dimensional stability—toughened PPS has been in market development for electrical, electronic, automotive, and mechanical applications, and good results have been obtained.

ACKNOWLEDGMENT

This article was previously published in *Polym. Eng. Sci.* **1996**, *36*(2), 265–270 and appears with permission.

REFERENCES

1. Masmoto, J.; Nakamoto, T.; Kubo, K. U.S. Patent 5 191 020, 1993 (assigned to Asahi Chemical Industry).
2. Masamoto, J.; Kubo, K. *Am. Chem. Soc. Div. Polym. Prepr.* **1991**, *32*(2), 215–216.

POLY(*p*-PHENYLENE SULFIDE) (Synthesis by *p*-Dichlorobenzene and Sodium Sulfide)

Darryl R. Fahey and Jon F. Geibel
Research and Development
Phillips Petroleum Company

Poly(*p*-phenylene sulfide), (PPS), is a perfectly alternating copolymer of sulfur and 1,4-phenylene groups. It possesses a favorable combination of mechanical, electrical, thermal, and flame-resistant properties,[1-4] along with good processability,[5] and has gained considerable market acceptance since its commercial introduction over 20 years ago. Because of its stability at high temperatures, it is classified as an engineering thermoplastic. Isomers and derivatives of PPS have been extensively studied, but they apparently do not possess properties desired in the marketplace and have not been commercialized.

While many synthetic routes to PPS have been patented and published, only procedures based on *p*-dichlorobenzene and sodium sulfide are used commercially (**Equation 1**).

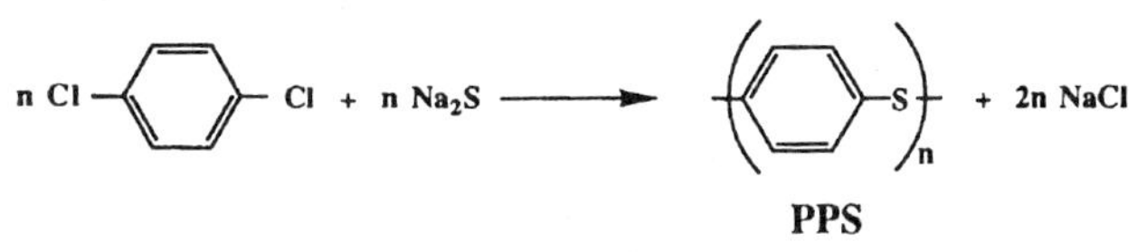

This combination of starting materials produces the necessary 1,4-disubstituted benzene structure, useful (high) molecular weights, and economy of production. Both raw materials are relatively inexpensive commodity chemicals.

EDMONDS AND HILL SYNTHESIS PROCEDURE

The first commercially viable process for the synthesis of PPS was developed by Edmonds and Hill at Phillips Petroleum Company, for which they were granted a U.S. patent in 1967.[6] The process involves the polymerization of *p*-dichlorobenzene and a sodium sulfide source in a polar organic compound at elevated temperature and pressure. A preferred polar organic compound is *n*-methylpyrrolidinone (NMP).

The synthesis of PPS from *p*-dichlorobenzene and sodium sulfide was optimized over the years without the benefit of a detailed understanding of the reaction mechanism. As a result, the process was characterized by puzzling, yet essential, special features.

The polymer produced by the Edmonds and Hill process is low molecular weight in the current product portfolio. It is a linear polymer having approximately 150–200 repeating units. As received from the polymerization process, the polymer is an off-white powder. There are no known solvents for PPS below 200°C,[7] which complicates traditional molecular weight characterization methods.

HIGH MOLECULAR WEIGHT PPS POLYMERIZATION PROCESSES

One application for which the Edmonds and Hill polymer was not well suited was that of fiber spinning. In its uncured state, the Edmonds and Hill polymer did not have sufficient molecular weight to have good fiber properties.

The first recognition that additives could have a substantial and highly desirable effect on the polymerization of linear PPS was made by R. W. Campbell in the laboratories of Phillips Petroleum Company. The Campbell PPS process provided a one-step synthesis of high molecular weight linear PPS directly in the polymerization vessel.[8,9]

The second PPS process for high molecular weight linear PPS practiced commercially was disclosed by Iizuka and coworkers at Kureha Chemical Industry.

Comonomers can be used to create a variety of polymer structures that may impart different, sometimes desirable, properties. For example, even higher molecular weight PPS polymers can be produced by the addition of a small amount of a tri- or tetrafunctional comonomer into either the Edmonds and Hill process or the Campbell process.[10]

Another process for the synthesis of PPS, as well as other poly(arylene sulfide)s and poly(arylene oxide)s, has been published by Dembek et al.[11] Their approach utilized a pentamethylcyclopentadienylruthenium(II) π complex to active *p*-dichlorobenzene toward displacement by a variety of nucleophilic comonomers. Important facets of this approach are the tremendous activation of *p*-dichlorobenzene afforded upon π coordination to the transition-metal group and the improved solubility of the resultant organometallic derivative of PPS. These features allow the polymerization to proceed at very mild conditions.

MECHANISMS

The preparation of PPS is complex, despite the deceivingly simple equation that is usually written for its preparation (Equation 1). Although growth of the polymer may be written as a series of conventional condensation steps, the system displays significant deviations from classical[12,13] A-A + B-B step-growth polymerization behavior. First, at incomplete conversions, polymer (or oligomer) yields and molecular weights are higher than expected.[14-19] Second, perfect 1:1 stoichiometric monomer ratios are not required to achieve high molecular weight polymer. Lastly, the deceptively simple equation typically written for PPS polymerizations (Equation 1) is not, in reality, the conventional A-A + B-B polymerization systems it appears to be. Instead, the NMP proactively participates in the reaction.

Recent experimental studies directly toward understanding the commercial process for PPS synthesis have led to the conclusions that the polymerization occurs by a step-growth process and that the dominant mechanism of each step is an S_NAr displacement of halide by a sulfur nucleophile.[20]

SIDE REACTIONS

The major side reactions occurring during the polymerization are cyclizations. Cyclic oligomers from the tetramer up to the heptadecamer have been detected as components in PPS.[21,23,24]

POSTPOLYMERIZATION TREATMENTS

Curing of PPS

Low molecular weight linear PPS possesses only modest mechanical properties. The polymer, however, displays an important property in that it can be converted to a much tougher material by an oxidative heat treatment. PPS behaves as a true thermoplastic material during normal processing conditions. It can be repeatedly melted and reprocessed with only minor changes in its rheology. However, PPS properties can be altered under more extensive oxidative heat treatment. The change in properties can be beneficial, depending upon the specific end use of the polymer. This behavior has been termed curing.

CONCLUSIONS

What appears to be a simple polymerization of sodium sulfide and *p*-dichlorobenzene has been shown to be quite complex. The molecular weight, molecular weight distribution, polymer architecture, and end-group chemistry are controlled by various factors. These factors include polymerization conditions, recovery techniques, and postpolymerization treatments. PPS characterization methods were presented as needed to shed light on synthetic aspects of the polymerization. A more detailed discussion of characterization methods has recently been published.[22] To the extent that a PPS manufacturer can utilize a wide variety of synthetic methods, it maximizes its probability of being able to customize a polymer for specific applications.

REFERENCES

1. Geibel, J. F.; Campbell, R. W. *Comprehensive Polymer Science, Volume 5; Step Polymerization*; Pergamon: Oxford, 1989; Chapter 32.
2. Lopez, L. C.; Wilkes, G. L. *J. Macromol. Sci., Rev. Macromol. Chem. Phys.* **1989**, *C29*(1), 83.
3. Geibel, J. F.; Campbell, R. W. *Encyclopedia of Chemical Processing and Design, Volume 40;* Marcel Dekker: New York, 1992; pp 94–125.
4. Shue, R. S. *Dev. Plast. Technol.* **1985**, *2*, 259.
5. Gabriele, M. C. *Plastics Technol.* **1992**, *38*, 59.
6. Edmonds, J. T.; Hill, H. W. U.S. Patent 3 354 129 to Phillips Petroleum Company, November 21, 1967.
7. Beck, H. N. *J. Appl. Polym. Sci.* **1992**, *45*, 1361.
8. Hill, H. W. *I & EC Prod. Res. Dev.* **1979**, *18*, 252.
9. Campbell, R. W. U.S. Patent 3 919 177 to Phillips Petroleum Company, November 11, 1975.
10. Scoggins, L. E.; Munro, B. L. U.S. Patent 4 415 729 to Phillips Petroleum Company, November 15, 1983.
11. Dembek, A. A.; Fagan, P. J.; Marsi, M. Macromolecules **1993**, 26, 2992.
12. Stille, J. K. *J. Chem. Ed.* **1981**, *58*, 862.
13. Manaresi, P.; Munari, A. *Comprehensive Polymer Science. Volume 5. Step Polymerization;* Pergamon: Oxford, 1989; Chapter 2.
14. Koch, W.; Heitz, W. *Makromol. Chem.* **1983**, *184*, 779.
15. Koch, W.; Risse, W.; Heitz, W. *Makromol. Chem. Suppl.* **1985**, *12*, 105.
16. Rajan, C. R.; Ponrathnam, S.; Nadkarni, V. M. *J. Appl. Polym. Sci.* **1986**, *32*, 4479.
17. Rajan, C. R.; Nadkarni, V. M.; Ponrathnam, S. *J. Polym. Sci., Part A, Polym. Chem.* **1988**, *26*, 2581.
18. Koschinski, I.; Reichert, K. H. *Makromol. Chem. Rapid Commun.* **1988**, *9*, 291.
19. Park, L. S.; Seo, K. H.; Chang, J. G.; Kwon, Y. H.; Han, S. K.; Cha, I. H. *Polymer (Korea)* **1989**, *13*, 866, also presented at the Pacific Basin Polymer Conference, Maui, Hawaii, 1989.
20. Fahey, D. R.; Ash, C. E. *Macromolecules* **1991**, *24*, 4242.
21. Fahey, D. R.; Hensley, H. D.; Ash, C. E.; Senn, D. R. *ACS PMSE Proceedings* **1992**, *67*, 468.
22. Geibel, J. F.; Leland, J. E. "Polymers Containing Sulfur, Poly(Phenylene Sulfide)," *Kirk-Othmer Encyclopedia of Chemical Technology*, John Wiley & Sons: New York; in press.
23. Reents, W. D. Jr.; Kaplan, M. L. *Polymer* **1982**, *23*, 310.
24. Kaplan, M. L.; Reents, W. D. Jr. *Tetrahedron Lett.* **1982**, *23*, 373.

POLY(PHENYLENE SULFIDE) AND COPOLYMERS (Melt Preparation via a New Process)

David R. Fagerburg
Eastman Chemical Company

Poly(phenylene sulfide) (PPS) has found numerous engineering plastic and some fiber and film applications owing to its excellent combination of chemical resistance and thermal properties.

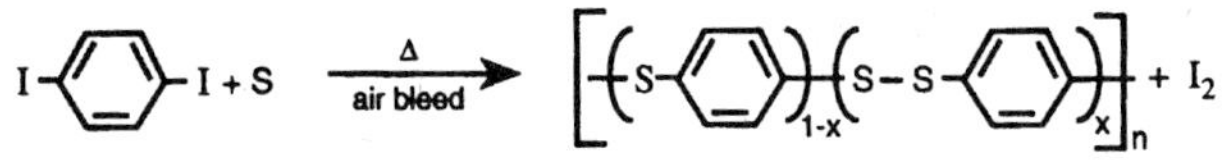

1

We have recently reported[1-3] a new melt-phase polymerization process for the preparation of PPS involving reaction of *p*-diidobenzene and sulfur (**Equation 1**) and some of the characteristics of the resultant polymer. This new melt process has several unique and unusual characteristics. The process can be carried out entirely in the melt using, for example, a final reaction temperature of 300°C (initial reaction temperature is usually 230°C). However, in an unusual twist for PPS chemistry, alternatively the polymer reaction can be stopped at a prepolymer stage by use of a lower final reaction temperature, typically 250°C, much as one would expect to be able to do in the technology normally practiced for production of poly(ethylene terephthalate). This prepolymer can then be solid state polymerized to its final, high molecular weight (e.g., the patent of Fagerburg and Watkins[4]) without an attendant increase in branching. Elemental iodine is the reaction by-product in either the melt-phase-only or the melt-phase–solid-phase reaction sequence, and no ionic by-products are formed.

In addition, unlike other PPS processes, careful control of reactant stoichiometry is not nearly as critical to obtaining high molecular weight. Instead, molecular weight can be controlled by the time-temperature sequence along with reactant stoichiometry. As is also evident from the process in Equation 1 above, by-product iodine can be recovered and used in the synthesis of more monomer (**Equation 2**), thus closing the reaction loop. Thus, our new process effectively makes PPS from benzene and sulfur, basic chemicals.

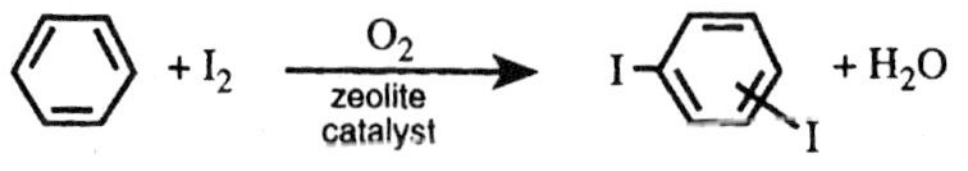

2

The reaction mixture contains all three possible diiodo isomers, with the para isomer predominating by far.

Not surprisingly, it was discovered that the oxyiodination process could be used to prepare diiodo aromatic compounds from a number of aromatic systems such as biphenyl, naphthalene, and diphenyl ether. This leads to a potentially large number of possible poly(arylene sulfide) polymers and copolymers.

Our unique route for PPS synthesis allowed us to investigate several interesting aspects of PPS morphology, only one of which is mentioned here. Because of the presence of heavy iodo end groups in the PPS chains made by this synthetic route, it was possible to use elemental analysis to measure the degree of polymerization (DP) with reasonable accuracy. By using this analysis for a series of oligomers, it was confirmed by hot-stage optical microscopy[5] that spherulitic growth was observed for samples with a DP ≥38, whereas for samples with a DP of 30 the organization showed lamellar sheafs. Chain-extended crystal structures were observed at DP <30. The lamellar thickness of the chain-folded crystallites was determined by small-angle X-ray scattering (SAXS) data to be 10 monomer units in length. Thus, for chain folding to occur with reasonable probability, the average chain must be ≥3 times the length through the lamellae.

Because this process allows the facile synthesis of material from quite low to very high melt viscosity, it was also determined from differential scanning calorimetry (DSC)[6] that the T_g exhibited at least two of the three regions of behavior versus polymerization degree reported by Cowie[7] for amorphous polymers. The region II behavior was linear from a DP of about 20 to a DP of 77, where region I behavior was observed. From that transition one would estimate a T_g at infinite chain length of 108°C, whereas from a plot of T_g versus $1/M_n$, one would calculate 99°C.[8] Our experience with even very high molecular weight materials, as determined from melt viscosity, would support a 99°C value. The transition from region II to region I behavior was far lower than for the amorphous polymers studied in this way.[7]

It is characteristic of PPS from this process that the melt viscosity increases with the time of testing for almost all samples.[2,3] This was found to be due either to continued polymerization or to branching of the polymer chains, with chain extension being the predominate effect.[9] This phenomenon of melt viscosity increase during rheological testing was absent in chain terminated samples.

Disulfide linkages in our polymer are present to varying degrees depending upon preparative conditions and reactant ratios. However, contrary to initial concerns, the presence of disulfide linkages does not seem to adversely affect thermal or oxidative stability as noted from the thermogravimetric analysis.[2]

One side point of interest was that, owing to the disulfide linkages in PPS produced by this new process, crystallization rates may be made slower than for commercial PPSs; however, the rate can also be made equivalent via addition of a nucleation aid.[10]

The heat distortion temperature is, of course, an important reason to use PPS in any application. We have determined (S. Bagrodia et al., unpublished work) that the heat distortion temperature for 30 wt % glass-filled compounds, compounded and injection molded as above, ranged from 257–275°C in the melt viscosity range from 191 to 37,400 poise (at 300°C measurement temperature), with most of the samples tested lying between 266 and 275°C. Likewise, chemical resistance at elevated temperature, 93°C, was also quite good, equivalent to commercial material or somewhat better, with the exception of immersion in 10% NaOH (J. J. Watkins, unpublished work).

ACKNOWLEDGMENTS

I wish to acknowledge the many long hours of indispensable collaborative, co-inventive work of Mark Rule, J. J. Watkins, and P. B. Lawrence, without whom this work would have been far less fun and also considerably more difficult, if even achievable. Also appreciated are the contributions of many analytical personnel making characterization possible, including J. E. Briddell, C. S. Sass, R. L. Zimmerman, D. W. Lowman, and J. D. Cloyd.

REFERENCES

1. Rule, M.; Fagerburg, D. R.; Watkins, J. J.; Fauver, J. S. U.S. Patent 4 786 713, November 22, 1988. *Chem Abstract 110*: 174004j, and subsequent patents.
2. Rule, M.; Fagerburg, D. R.; Watkins, J. J.; Lawrence, P. B. *Makromol. Chem., Rapid Commun.* **1991,** *12,* 221.
3. Rule, M.; Fagerburg, D. R.; Watkins, J. J.; Lawrence, P. B.; Zimmermann, R. L.; Cloyd, J. D. *Makromol. Chem. Symposium* **1992,** *54/55,* 233.
4. Fagerburg, D. R.; Watkins, J. J. U.S. Patent 4 877 862, October 31, 1989.
5. Sass, C. S.; Fagerburg, D. R. *J. Polym. Sci. Part B: Polym. Phys.* **1994,** *32,* 579.
6. Fagerburg, D. R.; Watkins, J. J.; Lawrence, P. B. *Macromolecules* **1993,** *26,* 114.
7. Cowie, J. M. G. *Europ. Polym. J.* **1975,** *11,* 297.
8. Fox, T. G. Jr.; Flory, P. J. *J. Appl. Phys.* **1950,** *21,* 581.
9. Fagerburg, D. R.; Watkins, J. J.; Lawrence, P. B. *J. Appl. Polym. Sci.* **1993,** *50,* 1903.
10. Fagerburg, D. R.; Watkins, J. J.; Lawrence, P. B. U.S. Patent 4 877 851, October 31, 1989.

POLY(*p*-PHENYLENE SULFONE)

Douglas R. Robello*
Imaging Research and Advanced Development
Eastman Kodak Company

Abraham Ulman
Department of Chemistry
Polytechnic Institute of New York

Edward J. Urankar
Department of Chemistry
Baker Laboratory
Cornell University

Polysulfone polymers are widely used in electrical connectors, aircraft components, and other high-performance applications in which toughness, hydrolytic stability, and moderately high heat resistance are important. Although many polymers containing only sulfonyl and aromatic hydrocarbon repeat units have been described in the literature, until now the simplest and most symmetrical derivative, poly(*p*-phenylene sulfone) has not been reported.

In principle, poly(*p*-phenylene sulfone) might be produced by oxidizing poly(*p*-phenylene sulfide). The oxidation of sulfides is

*Author to whom correspondence should be addressed.

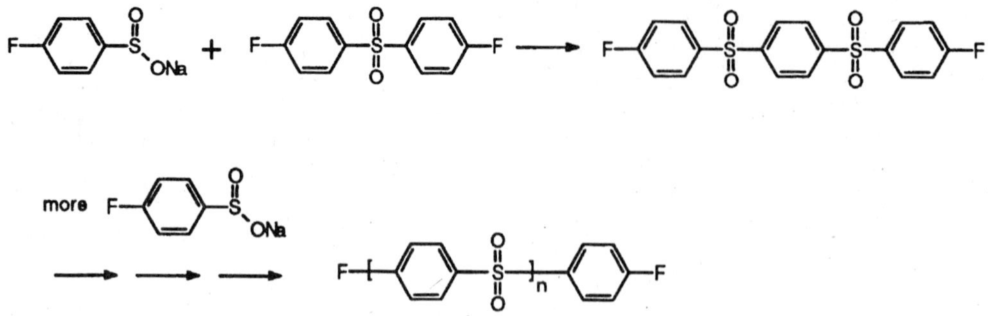

1

I

the most common procedure for the synthesis of low molar mass sulfones. However, one of the strong points of PPS is its chemical resistance. PPS is highly crystalline and therefore insoluble, and chemical reagents cannot easily penetrate the solid. Attempted oxidation of PPS does not produce poly(*p*-phenylene sulfone) but instead slowly crosslinks the polymer.[1]

Lenz et al. and Lovell and Still demonstrated that PPS could be synthesized by the self-condensation of salts of 4-halobenzene thiophenols.[2,3]

In analogy, we proposed that the heretofore unknown poly(*p*-phenylene sulfone) might be synthesized by the self-condensation (via aromatic nucleophilic substitution) of 4-halobenzene-sulfinate salts. We reasoned that compared with the *para*-thioether present in PPS, the *para*-sulfone in poly(*p*-phenylene sulfone) should be much more strongly activating and more easily lead to polymers. All that is required for successful polymerization is a reactive core onto which the monomer can be added. For our purposes, 4,4′-difluorodiphenyl sulfone could act as this core group and initiate the proposed polymerization. In this way, 4,4′-difluorodiphenyl sulfone represents a dimeric unit of poly(*p*-phenylene sulfone) (Structure **1** in **Scheme I**). In this paper, we review the first preparation of poly(*p*-phenylene sulfone), which was synthesized by aromatic nucleophilic substitution, and its physical characterization.

PREPARATION

Poly(*p*-phenylene sulfone) was best prepared by heating freshly dried sodium 4-fluorobenzene sulfinate with a small amount (1-2 mol %) of 4,4′-difluorodiphenyl sulfone as initiator in dimethyl sulfoxide solution at 175°C for 24 h. The polymer was precipitated into water, and the resulting finely divided solid was collected, washed, and dried.

PROPERTIES

Attempts to characterize the polymer were complicated by its insolubility.

The molecular weight of the polymer was estimated from elemental analysis, but the results were confounded by uncertainty in polymer purity and the inability to definitively identify the chain ends. When we assumed that the polymer possessed 4-fluorophenyl end groups, the fluorine analysis of a representative batch of poly(*p*-phenylene sulfone) was consistent with 11 repeat units.

X-ray powder diffraction of poly(*p*-phenylene sulfone) showed 11 peaks between 3 and 50°C. The very small amount of diffuse scattering indicated that the degree of crystallinity in poly(*p*-phenylene sulfone) was high (>85%), a feature that probably is responsible for the poor solubility of the polymer. The melting point of poly(*p*-phenylene sulfone) could not be detected by differential scanning calorimetry up to 500°C, above which temperature decomposition began.

The exceptional thermal stability of poly(*p*-phenylene sulfone) was demonstrated by thermogravimetric analysis, in which decomposition in nitrogen began at 517°C. Remarkably, decomposition in air began at a slightly higher temperature (555°C), a very unusual finding. This effect may be caused by oxidative crosslinking reactions in the polymer in the presence of oxygen. Unfortunately, because of its intractability, films or pellets of poly(*p*-phenylene sulfone) could not be produced.

CONCLUSIONS

The first preparation of poly(*p*-phenylene sulfone) has been achieved by a novel variation on self-condensation, which introduced elements of a chain polymerization mechanism. Although the molecular weight of the title polymer was apparently rather low, the material exhibits unusual thermal stability for an organic polymer. Its high degree of crystallization leads to insolubility and a very high melting point, features that preclude its use in practical applications. However, it may be possible to prepare substituted analogs of poly(*p*-phenylene sulfone) that are more processable yet retain much of its extraordinary stability.

ACKNOWLEDGMENTS

We express our appreciation to P. Mark Henrichs and N. Zumbulyadis for solid-state NMR spectra. We thank the following members of the Eastman Kodak Company Analytical Technology Division: Roger Moody for thermogravimetric and differential scanning calorimetry data, Sandra Finn for Fourier transform-IR spectra, and Thomas Blanton for X-ray diffraction measurements. We acknowledge Scott Tunney for helpful discussions concerning the possible mechanism of the polymerization.

REFERENCES

1. Hawkins, R. T. *Macromolecules* **1976,** *9,* 189 and references therein.
2. Lenz, R. W ; Handlovits, C. E.; Smith, H. A. *J. Polym. Sci.* **1962,** *58,* 351.
3. Lovell, P. A.; Still, R. H. *Makromol. Chem.* **1987,** *188,* 1561.

Poly(p-phenylene-1,4-thiophene)s

See: Liquid Crystalline Polymers (Conjugated Chain)

Poly(phenylene vinylene)s

See: *Aromatic Hydrocarbon-Based Polymers*
Liquid Crystalline Polymers (Conjugated Chain)
p-Phenylenevinylene Oligomers, Homo- and Copolymers (Metathesis Preparation)
Poly(arylene vinylene)s (Mechanistic Control of a Soluble Precursor Method)
Poly(cyclopentadineylene vinylene)
Poly(phenylene vinylene)s (by the Heck Coupling Reaction)
Poly(p-phenylene vinylene)s (Methods of Preparation and Properties)

POLY(PHENYLENE VINYLENE)S (by the Heck Coupling Reaction)

Luping Yu* and Zhenan Bao
Department of Chemistry
University of Chicago

Poly(phenylene vinylene)s (PPVs) are an interesting class of polymers that exhibit high electric conductivity upon doping, high photoconductivity, large third-order nonlinear optical activity, and electroluminescence.[1-8] Different approaches have been developed to synthesize PPV, such as the Wittig reaction and the McMurry condensation.[1,2,9-13] Not all the approaches lead to the formation of polymers with high molecular weights because the unsubstituted PPV is insoluble in common organic solvents.

Recently, the Heck reaction has been explored to prepare PPV and its derivatives.[14,15] Polymers with high molecular weights and well-defined structures can be synthesized. Compared with other methods, this approach is attractive in preparing functionalized PPV owing to mild reaction conditions and the wide availability of different monomers.

THE HECK REACTION

The Heck reaction is a palladium-catalyzed vinylation of organic halides, forming a carbon–carbon bond. Various organic halides, such as aryl, heterocyclic, benzyl, or vinyl bromides or iodides, can be utilized in this reaction.

VARIOUS PPV SYSTEMS FROM THE HECK REACTION AND THEIR PROPERTIES

Heitz et al. were the first to utilize the Heck reaction to synthesize PPV from 4-bromostyrene.[14] The resulting polymer was insoluble with a low molecular weight. Following this earlier work, they conducted research to modify the structure of poly(1,4-phenylenevinylene) to generate soluble, fusible products.[16-18]

Recently, Bao et al. synthesized 2,5-dialkoxyl-substituted PPVs (**Scheme I**).[15] These polymers were soluble in common organic solvents, such as tetrahydrofuran (THF), $CHCl_3$, and toluene. The gel permeation chromatography studies using polystyrene as standards indicated that the number averaged molecular weights of greater than 10,000 Dalton/mole.

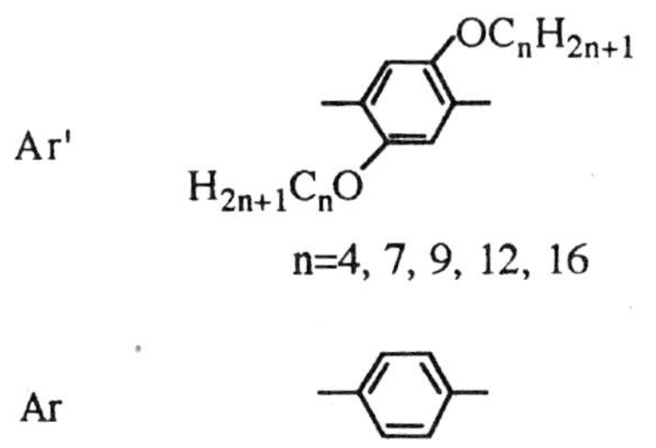

I

In these polymers, nematic textures were observed under a polarizing microscope, and the texture of polymers with side-chains shorter than $n = 16$ was reversible only under a very slow cooling rate owing to the crosslinking. As the side-chain length increased, the melting temperature of the polymer decreased steadily, as did the clearing temperature.

Several physical properties of these polymers have been studied.[15] Polymer system in Structure 4 in Scheme I exhibited third-order optical nonlinearity. All of the polymers were found to be photoconductive upon exposure to a laser beam at wavelength $\lambda = 632$ nm.

The application of the Heck coupling reaction is not limited to the synthesis of PPV and its derivatives; it can also be applied as a useful synthetic tool to prepare various types of conjugated polymers with vinyl linkages. Weitzel et al. have incorporated large, aromatic π-systems such as anthracene into arylenvinylene chains.[19,20]

Owing to the mild reaction conditions, different functional groups can be tolerated in the Heck reaction. The recent interest in photoresponsive polymers led to the synthesis of polymers or oligomers containing porphyrin and metalloporphyrin moieties by Bao et al.[21]

In summary, the Heck reaction has proved to be a very versatile method for synthesizing PPV and its derivatives. By careful design of monomers, new polymers with interesting structures and properties can be synthesized.

REFERENCES

1. Skotheim, T. A., In *Handbook of Conductive Polymers;* Marcel Dekker: Basel, Switzerland, 1989.
2. Reynolds, J. R.; Pomerantz, M. In *Electroresponsive Molecular and Polymeric Systems*; Skotheim, T. A., Ed.; Marcel Dekker: Basel, Switzerland, 1991; Vol. 2, Chapter 4.
3. Lenz, R. B.; Han, C. C.; Lux, M. *Polymer* **1989,** *30,* 1041.
4. Gailberger, M.; Bassler, H. *Phys. Rev. B* **1991,** *44,* 8643.
5. Kaino, T.; Kubodera, K. I.; Tomura, S.; Kurihara, T.; Saito, S.; Tsutsui, T.; Tokito, S. *Electron Lett.* **1987,** *23,* 1095.
6. Burroughes, J. H.; Bradley, D. D.; Brown, A. R.; Marks, R. N.; Mackay, K.; Friend, R. H.; Burns, P. L.; Holmes, A. B. *Nature* **1990,** *340,* 539.
7. Braun, D.; Heeger, A. J. *Appl. Phys. Lett.* **1982,** *58,* 1991.
8. Greenham, N. C.; Moratti, S. C.; Bradley, D. D. C., et al. *Nature* **1993,** *365,* 628.
9. McDonald, R. N.; Campbell, T. N. *J. Am. Chem. Soc.* **1960,** *82,* 4669.
10. Gooding, R.; Lillya, C. P.; Chien, J. C. W. *J. Chem. Soc. Chem. Commun.* **1983,** 151.

*Author to whom correspondence should be addressed.

11. Kossmehl, G.; Wallis, J. *Makromol. Chem.* **1982,** *183,* 331.
12. Rajaraman, L.; Balasurbramanian, M.; Nanjan, M. J. *Curr. Sci.* **1980,** *49,* 101.
13. Feast, W. J.; Millichamp, I. S. *Polym. Commun.* **1983,** *24,* 102.
14. Heitz, W.; Brugging, W.; Freund, L.; Gailberger, M.; Greiner, A.; Jung, H.; Kampschulte, U.; Niebner, N.; Osan, F. *Makromol. Chem.* **1988,** *189,* 119.
15. Bao, Z. N.; Chen, Y. M.; Cai, R. B.; Yu, L. P. *Macromolecules* **1993,** *26,* 5281.
16. Brenda, M.; Greiner, A.; Heitz, W. *Makromol. Chem.* **1990,** *191,* 1083.
17. Greiner, A.; Heitz, W. *Makromol. Chem., Rapid Commun.* **1988,** *9,* 581.
18. Martelock, H.; Greiner, A.; Heitz, W. *Makromol. Chem.* **1991,** *192,* 967.
19. Weitzel, H. P.; Bohnen, A.; Mullen, K. *Makromol. Chem.* **1990,** *191,* 2815.
20. Scherf, U.; Mullen, K. *Synthesis* **1992,** 23.
21. Bao, Z. N.; Chen, Y. M.; Yu, L. P. *Macromolecules* **1994,** *27,* 4629.

POLY(*p*-PHENYLENE VINYLENE)S (Methods of Preparation and Properties)

Bing R. Hsieh
The Wilson Center for Research and Technology
Xerox Corporation

Poly(*p*-phenylene vinylene) (PPV) and related polymers have emerged as the most widely studied *π*-conjugated polymer class, mainly because they show many interesting and potentially useful opto- and photoelectronic properties. Since the discovery of electroluminescence in PPV in **1990,** doping conjugated polymers into their conductive forms is no longer of prime interest. Today, the exploration of natural or pristine conjugated polymers for semiconductor device applications such as photovoltaic cells, field-effect transistors, light-emitting diodes (LEDs), and Schottky diodes has become the major focal point of interest.[1] The PPV family of polymers serves as a prototypical conjugated polymer class for application as well as for fundamental understanding of the electronic processes in conjugated polymers.

PREPARATION

The advances in synthetic methodology for the preparation of thin films and fibers of PPVs enables them for consideration in various applications. The overall methodology can be roughly divided into three categories: precursor approach, side-chain derivatization, and *in situ* polymerization.

The precursor approach relies on the preparation of a soluble precursor polymer that can be cast into thin films and then transformed to the final conjugated polymer films through solid state thermo- or photo conversion. The sulfonium precursor route (SPR) to PPV is particularly well known[2,3] and involves the polymerization of *p*-xylene bis(tetrahydrothiophenium chloride), or one of its analogs or derivatives, in the presence of a base in water or methanol to give the corresponding sulfonium precursor polymer.[2-4]

The side-chain approach involves the polymerization of a highly substituted monomer to a soluble conjugated polymer that can be cast into thin films directly without conversion. The polymerization of *bis*(halomethyl)benzenes in the presence of a large excess of potassium *t*-butoxide to PPVs is referred to as the Gilch route.[8] Recently, the Gilch route has been widely used for the preparation of soluble PPV derivatives[7,9-13] in order to avoid the conversion step and the many problems associated with SPR.

Several elegant *in situ* polymerization processes have been reported for deposition of PPV thin films. Electroreduction of α,α,α,α',α',α'-hexachloro-*p*-xylene.[15] Chemical vapor deposition polymerization of 1,9-dichloro-[2,2]-paracyclophane gave a chlorine precursor polymer thin film which was then thermally converted to PPV.[16]

PROPERTIES

The SPR synthesis has enabled unique processing flexibility to achieve uniaxially or biaxially oriented PPV thin films.[17,18] The mechanical properties of PPV films are anisotropic and dependent upon the degree of molecular orientation. The Young's modulus varies between 2.3 and 37 GPa as a function of draw ratio in the machine direction and between 2.3 and 0.5 GPa in the transverse direction.[17] A typical PPV film has smooth lamellar morphology and about 50% crystallinity.[19]

Structural ordering of PPV also depends on the side groups.[20]

Recent investigation of photoexcitation processes of conjugated polymers has focused on PPVs because of their potential optoelectronic applications.[21] An exciton model[21-25] and a band model[26] have been proposed to interpret the primary photoexcitation in PPVs.

High photoconductivity of PPVs was reported first in 1970,[27] and more recently for the sulfonium PPV.[28,29]

The photoelectronic properties of metal/organic semiconductor interfaces are beginning to receive attention due mainly to the potential semiconductor device applications. We initiated the studies of metal–PPV interface formation in 1991, aiming at understanding the metal–PPV interfacial chemistry and stability.[33] Our results so far suggest that the barrier height at the metal/PPV interface is highly sensitive to the surface preparation and relatively insensitive to the work function of the metals.[34]

Redox reactions (doping) of PPV have been investigated extensively in order to achieve conductive materials. Most of the PPV research prior to the discovery of EL in PPV (1984–1990) dealt with doping PPV to its conductive forms.[21,35] Conductivity as high as 500 S/cm had been obtained for PPV.[36] Unfortunately, the doped PPVs are highly unstable and have very little value for application.

Third-order nonlinear optical properties for PPV with different degrees of conversion were studied; the $\chi^{(3)}$ value increased with the temperature of conversion and reached 7.5×10^{-11} esu for the fully converted PPV.[37] A lower value, of 2.1×10^{-11} esu, was reported for a differently prepared PPV.[6] A high $\chi^{(3)}$ value of 4×10^{-10} esu has been reported for oriented PPV along the draw direction.[38]

APPLICATIONS

Organic photoconductors used in xerographic copiers and printers are the first and only commercial organic semiconductor devices.[39-41] Today, organic photoconductors constitute the largest industrial use of organic semiconductor materials as thin film electronic devices.[39] A typical organic photoconductor is a multilayered device comprising two photoactive layers–namely, the charge-generation layer (CGL) and the charge-transport layer (CTL). Organic pigments such as phthalocyanines and perylenebisimidazoles are the materials used in the CGL,[41] and molecularly doped polymers are being used for the CTL. We excluded PPV for CTL application upon realizing its low charge mobility and charge trapping characteristics.[5,30-32] We demonstrated the possibility of using PPV as a CGL.[30-32] A very high charge generation efficiency, ca. 25%, was observed for PPV in an organic photoconductor-like bilayer device using PPV as the CGL, a molecularly doped polymer as the CTL, and indium tin oxide glass as the substrate. We further elucidated a near-ohmic characteristic between PPV and the molecularly doped polymer layer and an intrinsic photogeneration process for PPV.[30-32]

The discovery of EL in PPV in 1990 by the Cambridge group[42] has rekindled a great deal of interest in organic semiconductor materials and their device applications. Currently, both molecular and polymer-based LEDs are being considered and explored for one of the most challenging and exciting applications of organic LEDs: flat panel display. Organic LEDs, which are being actively pursued in several industrial research laboratories, are likely to be the next commercial organic semiconductor devices.

The concept of multilayered molecular LED device structure originated from Tang et al.[48,49] has been widely adapted for the fabrication of a variety of multilayered PPV EL devices.

Other potential device applications of PPVs include photodiodes, photodetectors, and photovoltaic cells. Optical recording devices based on PPVs,[50] current-controlled electrical switching from a combination of PPV and ferroelectric KH_2PO_4,[51] and metal–insulator–semiconductor field–effect transistors based on DM–PPV[52] are among the novel device applications of PPVs.

CONCLUDING REMARKS

The development of several ingeniously simple synthetic routes toward the formation of fully conjugated PPV thin films is a significant accomplishment. This development is the basis for physical property investigation and commercial applications of PPVs. The overall photoelectrical properties of PPVs resemble closely those of aromatic dyes and pigments, which are the first generation of organic semiconductor materials.[53] It is, therefore, not surprising that PPVs and related conjugated polymers are being seriously considered for semiconductor device applications. One can effectively view PPVs as polymeric dyes and pigments and described the photophysical processes of PPVs by using the exciton model. However, PPVs are much less stable than organic pigments, mainly because the vinylene groups are highly susceptible to photooxidation. Although this instability may ultimately restrict PPVs for commercial device applications, the experience and knowledge gained in the exploration of PPVs will lead to the development of new stable polymeric semiconductor materials in the future.

REFERENCES

1. Schott, M. In *Organic Conductors: Fundamentals and Applications* Farges, J.-P., Ed.; Marcel Dekker: New York, 1994; Chapter 12.
2. Wessling, R. A. *J. Polym. Sci: Polym. Symp.* **1985,** *72,* 55.
3. Hörhold, H. H.; Helbig, M. *Makromol. Chem., Macromol. Symp.* **1987,** *12,* 229.
4. Lenz, R. W.; Han, C. C.; Stenger Smith, J.; Karasz, F. E. *J. Polym. Sci; Part A: Polym. Chem.* **1988,** *26,* 3241.
5. Hsieh, B. R.; Antoniadis, H.; Abkowitz, M. A.; Stolka, M. *Polym. Prepr.* **1992,** *33(2),* 414.
6. Garay, R. O.; Baier, U.; Bubeck, C.; Müllen, K. *Adv. Mater.* **1993,** *5,* 561.
7. Burn, P. L.; Kraft, A.; Baigent, D. R.; Bradley, D. D. C.; Brown, A. R.; Friend, R. H.; Gymer, R. W.; Holmes, A. B.; Jackson, R. W. *J. Am. Chem. Soc.* **1993,** *115,* 10117, and references cited therein.
8. Gilch, H. G.; Wheelwright, W. L. *J. Polym. Sci: A-1* **1966,** *4,* 1337.
9. Wudl, F.; Allemand, P. M.; Srdanov, G.; Ni, Z.; McBranch, D. *ACS Symp. Ser.* **1991,** *455,* 683.
10. Wudl, F.; Hoger, S.; Zhang, C.; Pakbaz, K.; Heeger, A. J. *Polym. Prepr.* **1993,** *34(1),* 197.
11. Zhang, C.; Hoger, S.; Pakbaz, K.; Wudl, F.; Heeger, A. J. *Polym. Prepr.* **1994,** *35(1),* 197.
12. Zhang, C.; von Seggern, H.; Pakbaz, K.; Kraabel, B.; Schmidt, H.-W.; Heeger, A. J. *Synth. Metals* **1994,** *62,* 35.
13. Gettinger, C. L.; Heeger, A. J.; Drake, J. M.; Pine, D. J. *J. Chem. Phys.* **1994,** *101,* 1673.
14. Nishihara, H.; Tateishi, M.; Aramaki, K.; Ohshawa, T.; Kimura, O. *Chem. Lett.* **1987,** 539.
15. Tateishi, M.; Nishihara, H.; Aramaki, K. *Chem. Lett.* **1987,** 1727.
16. Iwatsuki, S.; Kubo, M.; Kumeuchi, T. *Chem. Lett.* **1991,** 1071.
17. Machado, J. M.; Masse, M. A.; Karasz, F. E. *Polymer* **1989,** *30,* 1992.
18. Machado, J. M.; Masse, M. A.; Karasz, F. E.; Burnett, J. M.; Kovar, R. F.; Druy, M. A. *Brit. Polym. J.* **1990,** *22,* 59.
19. Masse, M. A.; Martin, D. C.; Thomas, E. L.; Karasz, F. E. *J. Mater. Sci.* **1990,** *25,* 311.
20. Martens, J. H. F.; Marseglia, E. A.; Bradley, D. D. C.; Friend, R. H.; Burn, P. L.; Holms, A. B. *Synth. Metals* **1993,** 55-57, 449.
21. Friend, R. H.; Bradley, D. D. C.; Townsend, P. D. *J. Phys. D* **1987,** *20,* 1367.
22. Rauscher, U.; Bässler, H.; Bradley, D. D. C.; Hennecke, M. *Phys. Rev. B* **1990,** *42,* 9830.
23. Colaneri, N. F.; Bradley, D. D. C.; Friend, R. H.; Burn, P. L.; Holmes, A. B.; Spangler, C. W. *Phys. Rev. B* **1990,** *42,* 11670.
24. Rothberg, L. J.; Hsu, J. W. P.; Wilson, W. L.; Jedju,T. M.; Yan, M. *Proc. Society of Photo-Optical Instrumentation Engineers* **1993,** 1910, 122.
25. Woo, H. S.; Graham, S. C.; Halliday, D. A.; Bradley, D. D. C.; Friend, R. H.; Burn, P. L.; Holmes, A. B. *Phys. Rev. B.* **1992,** *46,* 7379.
26. Pakbaz, K.; Lee, C. H.; Heeger, A. J.; Hagler, T. W.; McBranch, D. *Synth. Metals* **1994,** *64,* 295.
27. Drefahl, G.; Kuhmstedt, R.; Oswald, H.; Hörhold, H.-H. *Makromol. Chem.* **1970,** *131,* 89.
28. Tokito, S.; Tsutsui, T.; Tanaka, R.; Saito, S. *Jpn. J. Appl. Phys.* **1986,** *25,* L680.
29. Takiguchi, T.; Park, D. H.; Ueno, H.; Yoshino, K.; Sugimoto, R. *Synth. Metals* **1987,** *17,* 657.
30. Lee, C. H.; Yu, G.; Moses, D.; Heeger, A. J. *Phys. Rev. B.* **1994,** *49,* 2396.
31. Antoniadis, H.; Abkowitz, M. A.; Hsieh, B. R. *Appl. Phys. Lett.* **1994,** *65,* 2030.
32. Antoniadis, H.; Abkowitz, M. A.; Hsieh, B. R.; Jenekhe, S. A.; Stolka, M. *Mat. Res. Soc. Symp. Proc.* **1994,** *328,* 377.
33. Gao, Y.; Park, K. T.; Hsieh, B. R. *J. Chem. Phys.* **1992,** *97,* 6991.
34. Gao, Y.; Park, K. T.; Hsieh, B. R. *J. Appl. Phys.* **1993,** *73,* 7894.

35. Jin, J. I.; Park, C. K.; Shim, H. K. *Macromolecules* **1993,** *26,* 1799, and references cited therein.
36. Murase, I.; Ohnishi, T.; Noguchi, T.; Hirooka, M. *Polym. Commun.* **1984,** *25,* 327.
37. Bradley, D. D. C. *Makromol. Chem., Macromol. Symp.* **1990,** *37,* 247.
38. Singh, B. P.; Prasad, P. N.; Karasz, F. E. *Polymer* **1988,** *29,* 1940.
39. Mort. J. *Phys. Today* **1994,** *47(4),* 32.
40. Pai, D.; Springett, B. E. *Rev. Modern Phys.* **1993,** *75,* 163.
41. Law, K. Y. *Chem. Rev.* **1994,** *93,* 449.
42. Burroughes, J. H.; Bradley, D. D. C.; Brown, A. R.; Marks, R. N.; Mackay, K.; Friend, R. H.; Holmes, P. L. *Nature* **1990,** *347,* 539, and references cited therein.
43. Burroughes, J. H.; Jones, C. A.; Friend, R. H. *Nature* **1988,** *335,* 137.
44. Horowitz, G. *Adv. Mater.* **1990,** *2,* 287.
45. Bradley, D. D. C. *Polym. International* **1991,** *26,* 3.
46. Friend, R. H. *Synth. Metals* **1992,** *51,* 357.
47. Bradley, D. D. C. *Adv. Mater.* **1994,** *4,* 756.
48. Tang, C. W.; Van Slyke, S. A. *Appl. Phys. Lett.* **1987,** *51,* 913.
49. Tang,, C. W.; Van Slyke, S. A.; Chen, C. H. *J. Appl. Phys.* **1989,** *65,* 3610.
50. Yoshino, K.; Kuwabara, T.; Iwasa, T.; Kawai, T.; Onoda, M.; *Jpn. J. Appl. Phys.* **1990,** *29,* L1514.
51. Yamamoto, H. *J. Appl. Phys.* **1992,** *71,* 4620.
52. Parker, I. D.; Gymer, R. W.; Harrison, M. G.; Friend, R. H.; Amhed, H. *Appl. Phys. Lett.* **1993,** *62,* 1519.
53. Gutman, F.; Lyons, L. E. *Organic Semiconductors* John Wiley & Sons: New York, 1967.

Polyphenylenes

See: *Aromatic Hydrocarbon-Based Polymers*
Liquid Crystalline Polymers (Conjugated Chain)
Oxidative Coupling
Polyelectrolytes, Rigid-Rod [Poly(p-Phenylene)-Based]
Polyphenylenes (by New Precursor Route)

POLYPHENYLENES (by New Precursor Route)

Kyoji Kaeriyama
Department of Materials Science
Kyoto Institute of Technology

Poly(*p*-phenylene) (ppp) is thermally stable and moderately conducting. Busch et al.[1] reported the preparation of the hexadecamer in 1936. Kovacic and Oziomek[2] subsequently prepared ppp by oxidizing benzene with copper(II) chloride in the presence of Friedel–Crafts catalysts such as aluminum chloride.

One preferable method is to use a soluble precursor route, which allows the precursor film to be processed and pyrolyzed to ppp. Marvel and Hartzel[3] first proposed 1,4-addition polymerization of cyclohexadiene and subsequent dehydrogenation of the resulting polycyclohexadiene to ppp. Ballard et al.[4] subsequently applied this procedure to a soluble precursor route.

A comparatively simple procedure is the electrochemical polymerization of benzene.

Wegner and co-workers[5-7] stressed that coupling of boric acid with bromide according to the Suzuki reaction was a useful route to high-quality ppp, and Wallow and Novak[8] prepared water-soluble ppp containing carboxyl groups in a similar way.

This article describes a new procedure to prepare structurally regular ppp and poly(*m*-phenylene) (pmp) with defined molecular weight from commercially available starting materials. In this procedure, methyl 2,5-dichlorobenzoate (mcba) was polymerized in the presence of zerovalent nickel generated *in situ* by the reduction of nickel(II) bromide with triphenylphosphine and zinc powder. The polymethoxycarbonylphenylenediyl (pmcp) obtained was hydrolyzed to polycarboxyphenylenediyl (pcyp) followed by decarboxylation to produce ppp and pmp.[9,10]

PREPARATION OF PPP

In the present procedure, mcba, purchased from Aldrich Chemical and recrystallized from its methanol solution, was converted into ppp according to **Scheme I**. Pmcp was soluble in chloroform at room temperature and in pyridine and DMF at their refluxing temperatures. It was also possible to make a film of pmcp from its pyridine solution.

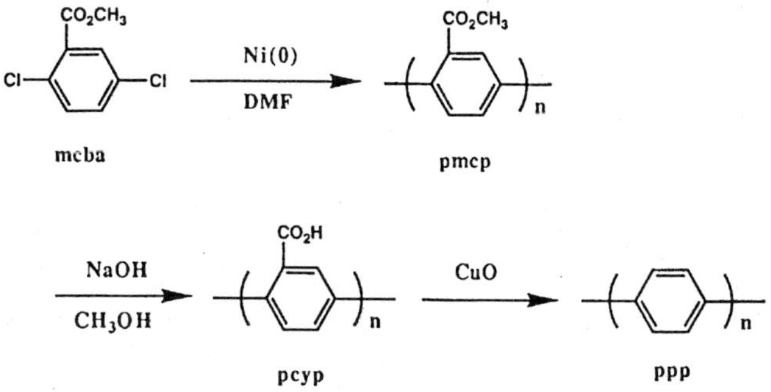

SCHEME I. Route to ppp.

The molecular weight of pmcp was determined by gel permeation chromatography using standard polystyrene samples. The degree of polymerization was 50–100. These values are high for condensation products of dihalobenzenes.

Pmcp was hydrolyzed by refluxing with 10% sodium hydroxide solution in methanol for 1.5 hr. The resulting pcyp was soluble in pyridine, quinoline, and DMF as well as aqueous sodium hydroxide and it was thus possible to cast a film from a solution of these solvents.

The decarboxylation of pcyp was performed by heating to reflux 1.5 g in 24 ml of quinoline in the presence of copper catalysts under argon.

PREPARATION OF PMP

Poly [5-(2-ethyl(hexyloxycarbonyl)phenylene-1,3-diyl] (phcp) was prepared by dehalogenative coupling of 2-ethylhexyl 3,5-dichlorobenzoate (hcba) in a procedure similar to that used to prepare pmcp.

REFERENCES

1. Busch, H.; Weber, W.; Darboven, C.; Renner, W.; Hahn, H. J.; Mathauser, G.; Stärtz, F.; Zitzmann, K.; Engelhardt, H. *J. Prakt. Chem.* **1936,** *146,* 1.

2. Kovacic, P.; Oziomek, J. In *Macromolecular Synthesis* Moore, J. A., Ed.; John Wiley & Sons: New York, 1977; p 109.
3. Marvel, C. S.; Hartzel, G. E. *J. Am. Chem. Soc.* **1959,** *81,* 448.
4. Ballard, D. G. H.; Courtis, A.; Shirby, I. M.; Taylor, S. C. *J. Am. Chem. Soc., Chem. Commun.* **1983,** 954.
5. Schlüter, A. D.; Wegner, G. *Acta Polymer* **1993,** *44,* 59.
6. Vahlenkamp, T.; Wegner, G. *Macromol Chem. Phys.* **1994,** *195,* 1933.
7. Suzuki, A. *Acc. Chem. Res.* **1982,** *15,* 178.
8. Wallow, T. I.; Novak, B. M. *J. Am. Chem. Soc.* **1991,** *113,* 7411.
9. Chaturvedi, V.; Tanaka, S.; Kaeriyama, K. *Macromolecules* **1993,** *26,* 2607.
10. Kaeriyoma, K.; Mehta, M. A.; Chaturuedi, V.; Masuda, H. *Polymer* **1995,** *36,* 3029.

Poly(p-phenylphenol)

See: Enzyme-Catalyzed Polymerization

Polyphenylquinoxaline

See: Magnetic Field Processing

Polyphenylsilsesquioxane

See: Ladder Polyphenylsilsesquioxane

Polyphenylthiazines

See: Ladder Polymers (Methods of Preparation)

Polyphosphates

See: Phosphorus-Containing Polymers (Overview)
Polyphosphonates and Polyphosphates (Low Temperature Solution Polycondensation)

POLYPHOSPHAZENE BLOCK COPOLYMERS

Krzysztof Matyjaszewski*
Department of Chemistry
Carnegie Mellon University

Michael L. White
GE Corporate Research and Development

Polyphosphazenes are an interesting class of inorganic polymers combining high-temperature stability and low-temperature flexibility. The low glass transition temperatures exhibited by most polyphosphazenes result from the unique electronic structure of the backbone.

Polyphosphazenes have been used to produce high-performance seals, gaskets, o-rings, fuel hoses, and vibration shock mounts in military vehicles.[1] Polyphosphazene block copolymers could lead to materials with enhanced physical properties.

*Author to whom correspondence should be addressed.

There are many methods for preparing polyphosphazenes, including the thermal and Lewis acid catalyzed polymerization of hexachlorocyclotriphosphazene, the polycondensation of $Cl_3P = N–P(O)Cl_2$, and the thermal polymerization of phosphoranimines.[3-11] However, polyphosphazene block copolymers have not been prepared using these techniques because of the uncontrolled nature of these processes. In addition, the end groups have not been identified, and relatively harsh synthetic conditions are required (a few days at over 200°C for uncatalyzed processes).

Recently, the synthesis of polyphosphazenes by the anionically initiated polymerization of phosphoranimines was reported.[1,2] Suitable initiators include tetra-*n*-butyl-ammonium fluoride and numerous other alkoxides, aryloxides, amines, and amides.[12-16] Using this technique, it is possible to prepare well-defined polyphosphazenes with relatively controlled molecular weights and low polydispersities.

Although they are a function of the polymerization time, polydispersities are generally between 1.2 and 2.0. Evidence exists for a chain growth process as relatively high molecular weight species are observed at low conversions. A slow macrocondensation process superimposed upon the chain growth mechanism allows the molecular weights to grow after complete monomer conversion. Thus, polydispersities increase with and after the monomer conversion.[13,15,17] The molecular weight can be controlled either by adjusting the polymerization time or by using chain terminators.[18] Additionally, new molecular architectures such as polyphosphazene AB block copolymers that were previously unreported may be accessed.

RANDOM AND GRAFT COPOLYMERS

Polyphosphazene random copolymers have been produced by ring opening polymerization and synthesized from phosphoranimines. Allcock used polydichlorophosphazene as a precursor to random copolymers by using two or more nucleophiles in varying concentrations at the substitution phase.[5] Neilson also reported the formation of random copolymers with various alkyl and aryl groups using the thermal polymerization of phosphoranimines.[10] The anionically initiated polymerization of phosphoranimines has also been used to prepare random copolymers by the simultaneous polymerization of phosphoranimines.[19]

SYNTHESIS

Polyphosphazene block copolymers bearing various combinations of trifluoroethoxy and alkoxyalkoxy groups have been prepared using tetra-*n*-butylammonium fluoride (TBAF) as an initiator by adding a second phosphoranimine after the first converts completely.[14] This method is possible because silyl end groups are present on the polymer chain. *N*-silylated phosphoranimines bearing 2,2,2-trifluoroethoxy groups and various combinations of 2,2,2-trifluoroethoxy, 2-methoxyethoxy, and 2-(2-methoxyethoxy)ethoxy groups can be prepared via the Staudinger reaction between the substituted phosphite and azidotrimethylsilane.[9,17,20] When *N*-*t*-butyl-2,2,2-trifluoroethoxyphosphoranimine is used to end cap the chain before a second monomer is added, the polymerization occurs primarily from the chains N-terminal and little, if any, branching occurs (**Equation 1**).[14]

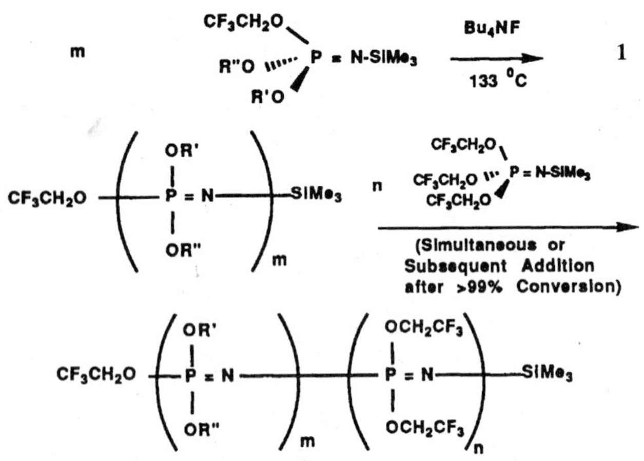

1a. Random R' = $CH_3OCH_2CH_2$-; R" = CF_3CH_2O-
1b. Random R' = $CH_3OCH_2CH_2OCH_2CH_2$-; R" = CF_3CH_2O-
1c. Random R', R" = $CH_3OCH_2CH_2$-
1d. Random R', R" = $CH_3OCH_2CH_2OCH_2CH_2$-
2a. Block R' = $CH_3OCH_2CH_2$-; R" = CF_3CH_2O-
2b. Block R' = $CH_3OCH_2CH_2OCH_2CH_2$-; R" = CF_3CH_2O-
2c. Block R' = $CH_3OCH_2CH_2OCH_2CH_2$-

Considerable evidence suggests that block copolymers form in lieu of two homopolymers when phosphoranimine is added in the presence of TBAF.[14] Size exclusion chromatography (SEC) shows that the molecular weight increases significantly and that molecular weight distributions are always monomodal after polymerization.

The solubility data also supports copolymer formation. The alkoxyalkoxy bearing homopolymers are soluble in $CHCl_3$, but the poly(*bis*-2,2,2,-trifluoroethoxyphosphazene) homopolymer (PBFP) and the random and block copolymers are insoluble in $CHCl_3$. There are also significant differences in the thermal, morphological, and physical properties of the polyphosphazene random and block copolymers.

REFERENCES

1. Singler, R. E.; Schneider, N. S.; Hagnauer, G. L. *Polymer Eng. & Sci.* **1975**, *15*, 321.
2. Allcock, H. R.; Connolly, M. S.; Sisko, J. T. *Macromolecules* **1988**, *21*, 323-334.
3. Allcock, H. R.; Kugel, R. L. *J. Am. Chem. Soc.* **1965**, *87*, 4216.
4. Allcock, H. R. *Chem. Rev.* **1972**, *72*, 315.
5. Allcock, H. R. *Chem. & Eng. News* **1985**, *63*, 22.
6. Allcock, H. R. *J. Inorg. and Organomet. Polym.* **1992**, *2*, 197.
7. Majumdar, A. N.; Young, S. G.; Merker, R. L. *Macromolecules* **1990**, *23*, 14.
8. D'Halluin, G.; De Jaeger, R.; Chambrette, J. P.; Potin, P. *Macromolecules* **1992**, *25*, 1254.
9. Flindt, E.; Rose, H. Z. *Anorg. Allg. Chem.* **1977**, *428*, 204.
10. Neilson, R. H.; Wisian-Neilson, P. *J. Macromol. Sci.-Chem.* **1981**, A16(1), 425.
11. Neilson, R. H.; Wisian-Neilson, P. *Chem. Rev.* **1988**, *88*, 541.
12. Montague, R. A.; Matyjaszewski, K. *J. Am. Chem. Soc.* **1990**, *112*, 6721.
13. Matyjaszewski, K.; Franz, U.; Montague, R. A.; White, M. L. *Polymer* **1994**, *35*, 5005.
14. Matyjaszewski, K.; Moore, M. K.; White, M. L. *Macromolecules* **1993**, *26*, 6741.
15. Matyjaszewski, K.; Cyprk, M.; Dauth, J.; Montague, R.; White, M. *Macromol Symp.* **1992**, *54/55*, 13.
16. White, M. L.; Matyjaszewski, K. M. ACS Symp. Ser. **1995**, *572*, 311.
17. Matyjaszewski, K.; Lindenberg, M. S.; Moore, M. K.; White M. L. *J. Polym. Sci., Polym. Chem. Ed.* **1994**, *32*, 465.
18. Montague, R. A.; Burkus, F. I.; Matyjaszewski, K. *ACS Polym. Prepr.* **1993**, *34(1)*, 316.
19. Matyjaszewski, K.; Lindenberg, M. S.; Moore, M. K.; White, M. L. *J. Inorg. Organomet. Polym.* **1993**, *3*, 317.
20. Staudinger, H.; Meyer, J. *J. Helv. Chim. Acta* **1919**, *2*, 635.

Polyphosphazenes

See: *Contact Lenses, Gas Permeable*
Controlled Drug Delivery Systems
Cyclolinear Organophosphazene Polymers
Cyclomatrix Phosphazene Polymers
Flame Retardants (Overview)
Phosphorus-Containing Polymers (Overview)
Polyphosphazene Block Copolymers
Polyphosphazenes (Solution Properties)
Star-Branched and Crosslinked Polymers (Cyclotriphosphazene Cores)

POLYPHOSPHAZENES (Solution Properties)

María Pilar Tarazona* and Enrique Saiz
Departmento de Química Física
Universidad de Alcalá de Henares

Polyphosphazenes (PPNs) are polymers formed by an inorganic backbone of alternated phosphorus and nitrogen atoms with two lateral substituents over the phosphorous that may be identical or different. The large number of different organic groups that can be attached to the phosphorous atoms make PPNs versatile polymers as several reviews attest.[1,4-6,8,9]

However, most of these reviews focus on the synthesis and applications, while the solution behavior of these polymers is not well understood.

PROPERTIES

In PPNs, the polymeric backbone is formed by conjugated double bonds. Consequently, it would be interesting to determine whether these bonds confer rigidity to the chain. One hint is the chain's glass transition temperature, T_g, which is probably the property measured most for PPNs. Polydichlorophosphazene (PDCP) has a T_g of –65°C, and therefore it behaves as an elastomeric, although hydrolytically unstable material, at room temperature. The T_g values depend strongly on the substituents attached to the phosphorous atoms, but they are, in general, very low especially for those PPNs containing alkoxyde groups.[2,3] At least in the solid state, the PPN chains seem rather flexible.

*Author to whom correspondence should be addressed.

Dimensions

Viscosity measurements provide direct information about the molecular dimensions of polymer chains.

The characteristic ratios of PPNs are larger than those of carbonated chains, which usually range between 4 and 10. This result is not surprising if we consider that P and N atoms form the chain's skeleton. However, large dimensions indicate extended chains that usually mean a polymer with a rigid backbone.

Conformations Adopted in Solution

The analysis of the Mark–Houwink *K* and *a* constants indicates that PPNs behave as flexible polymers in the random-coil state when they are in solution, which concurs with the result obtained in the solid state by examining the low values of T_g that these polymers exhibited. These values indicate that there is a large number of allowed conformations and a relatively easy interconversion among them. However, the study of unperturbed dimensions shows that these polymers must have extended conformations to account for the large dimensions that usually indicate a rigid polymeric backbone.

Theoretical calculations of the conformational energy maps for PDCP provide an easy qualitative explanation for the extended and, at the same time, flexible chains.[7] The energy contours for the N–P–N pair indicate that at least one of these bonds should always be in a trans conformation.

These results indicate that the chain may adopt a large number of conformations in solution or in an amorphous solid state, but because any of those conformations has at least one of every two bonds in a trans orientation, the result is a flexible but extended polymeric chain.

REFERENCES

1. Allcock, H. R. *Chem. Eng. News* **1985,** *63,* 22.
2. Allcock, H. R.; Connolly, M. S.; Sisko, J. T. et al. *Macromolecules* **1988,** *21,* 339.
3. *Polymer Handbook,* 2nd ed.; Brandup, J.; Immergut, E. H., Eds.; Wiley: New York, 1989.
4. Kireev, V. V.; Mitropolskaya, G. I.; Zinovich, Z. K. *Russian Chem. Rev.* **1982,** *51,* 149.
5. Neilson, R. H.; Wisian-Neilson, P. *Chem. Rev.* **1988,** *88,* 541.
6. Potin Ph.; De Jaeger, R. *Eur. Polym. J.* **1994,** *27,* 341.
7. Saiz, E. *J. Polym. Sci., Polym. Phys. Ed.* **1987,** *25,* 1565.
8. Tarazone, M. P. *Polymer* **1994,** *35,* 819.
9. Vinogradova, S. V.; Tur, D. R.; Minosyants, I. I. *Russian Chem. Rev.* **1984,** *53,* 49.

Polyphosphinate

See: Phosphorus-Containing Polymers (Overview)

Polyphosphine

See: Phosphorus-Containing Polymers (Overview)

Poly(phosphine oxide)

See: Phosphorus-Containing Polymers (Overview)

Polyphosphonates

See: Phosphorus-Containing Polymers (Overview)
Poly(alkylene phosphate)s
Polyphosphonates and Polyphosphates (Low Temperature Solution Polycondensation)

POLYPHOSPHONATES AND POLYPHOSPHATES (Low Temperature Solution Polycondensation)

Der-Jang Liaw* and Der-Wun Wang
Department of Chemical Engineering
National Taiwan Institute of Technology

Polyphosphonates and polyphosphates are of commercial interest because of their fine flame retarding characteristics and potential as high-performance plastics.[1,2]

This article describes the low temperature solution polycondensation of phenoxy dichlorophosphate (PDCP) with 4,4′-sulfonyldiphenol (SDP), phenylphosphonic dichloride (PPD) with 2,2-*bis*(4-hydroxyphenyl)hexafluoropropane (bisphenol AF) and its derivatives, and various aryl phosphorodichlorides with bisphenol AF or with 3,3′,5,5′-tetrabromo bisphenol AF (TBPAF).

PROPERTIES

The polyphosphonates and polyphosphates were insoluble in aliphatic hydrocarbons and alcoholic solvents but soluble in chlorinated aliphatic hydrocarbons such as CH_2Cl_2 and $CHCl_3$, and aprotic solvents such as dimethylsulfoxide and dimethylformamide.

We conclude that chlorinated aliphatic hydrocarbons such as methylene chloride or chloroform are the best polymerization solvents. The condensation system requires a moderate base such as triethylamine or tri-*n*-butylamine as an acid acceptor. The PDCP reaction did not proceed in the presence of weaker bases such as pyridine or *N,N*-dimethyl aniline.

The fluorine-containing polyphosphonates had T_gs between 20 and 117°C, and T_ms between 118 and 205°C. Polyphosphonates having flexible ether linkages, which increase free volume, had lower T_gs.[3] The fluorine-containing polyphosphates had T_gs between 81 and 151°C.

Polyphosphonates and polyphosphates, having bromine atoms, had lower thermal degradation temperatures. This is because hydrogen bromide is easily evolved for the bromine-containing polymers during the thermal degradation process.

The polyphosphates and polyphosphonates had higher LOI values and better flame retardation than the bisphenol A- and phenolphthalein-based polyphosphate esters, whose LOI values are 26–44.[8-10] It is tempting to conclude that sulfonyl group and trifluoro methyl groups in the polymer bring about an unusual improvement in flame resistance.[3-7]

ACKNOWLEDGMENT

We thank the National Science Council of the Republic of China for their financial support.

*Author to whom correspondence should be addressed.

REFERENCES

1. Stepniczka, H. E. *J. Fire Flam., Fire Retard. Chem.* **1975,** *2(2),* 30.
2. Imai, Y.; Kamata, H.; Kakimoto, M. *J. Polym. Sci., Chem. Ed.* **1984,** *22,* 1259.
3. Liaw, D. J.; Shen, W. C. *Polymer* **1993,** *34,* 1336.
4. Liaw, D. J.; Chen, P. S. *Polymer* **1995,** *36,* 4491.
5. Liaw, D. J.; Wang, D. W. *Reactive Polymer,* in press.
6. Liaw, D. J.; Wang, D. W., unpublished results.
7. Liaw, D. J.; Chang, P. Submitted for publication in *Polymer.*
8. Annakutty, K. S.; Kishore, K. *Polymer* **1988,** *29,* 762.
9. Kishore, K.; Annakutty, K. S.; Mallick, M. *Polymer* **1988,** *29,* 1273.
10. Kishore, K.; Kannan, P. *J. Polym. Sci., Polym. Chem. Ed.* **1990,** *28,* 3481.

Polyphosphoramides

See: Phosphorus-Containing Polymers (Overview)

Poly(phosphoric acid)

See: Phosphorus-Containing Polymers (Overview)

Polyphthalamides

See: Engineering Thermoplastics (Survey of Industrial Polymers)

POLYPHTHALAMIDES

Robert G. Keske
Amoco Polymers Business Group

Polyamides have found wide applications as engineering resins since their inception.[1] For more than fifty years, in an effort to expand the thermal capabilities of conventional aliphatic polyamides, numerous companies have been investigating the production of polyamides which incorporate substantial portions of aromatic dicarboxylic acids into their structure.[2] Notable among these partially aromatic polyamides are the polyphthalamides (PPA), which are defined as polyamides wherein at least 60 molar percent of the diacid component is derived from terephthalic and/or isophthalic acids.[3] Different processes have been examined during this time, culminating in a number of recent commercial products.

The poly(hexamethylene terephthalamide) (PA 6T) homopolymer, with a crystalline melting point in excess of 360°C, cannot be melt processed without decomposition. Lowering of the melting point is accomplished via copolymer formation in a variety of ways. Most early approaches suffered from economic disadvantages due to the use of monomers available in limited quantities. However, more recent products tend to utilize monomers which are readily available. For example, Ultramid T® from BASF is the copolymer of nylon 6 with PA 6T (PA 6/6T).[4] Zytel® HTN from DuPont replaces some of the hexamethylene diamine in PA 6T with 2-methylpentamethylenediamine.[5] Arlen® products from Mitsui Petrochemical and Amodel® polymers from Amoco are thought to replace portions of the terephthalic acid in PA 6T with either isophthalic or adipic acid, or mixtures thereof. In this manner, products have been obtained with melting points low enough to allow facile melt processing, yet significantly higher than those of conventional nylons (e.g., nylon 6,6). Furthermore, unlike the higher melting nylon 4,6, these polymers have substantially reduced moisture absorption.

DISCUSSION

Several studies have shown that during the production of polyamides from aromatic diacids excessive amounts of side reactions can occur if proper process conditions are not chosen. The primary branching reaction has been attributed to the reaction of an amine endgroup with another amine group (or its derivative) to produce a trifunctional species.

Although early workers concluded that aromatic dicarboxylic acids had essential disadvantages for the preparation of linear polyamides by means of melt polycondensation,[6] later work on to 60–80% terephthalic acid (TA)/ 40–20% isophthalic acid (IA)–HMDA copolymers (PA 6T/6I) showed that acceptable product for fiber applications could be prepared in smaller scale conventional equipment.[7,8]

Although fiber properties were described for the aforementioned 6T/6I/6,6 co- and terpolymers, no commercialization of this type of composition occurred until the discovery of a novel process to greatly reduce the polymerization time required at high melt temperatures.[9] This new process allowed the preparation of compositions having sufficiently high crystallization rates such that the benefits of crystallinity (e.g., high heat deflection temperatures) were realized for injection molded articles.[10]

Another process for the preparation of partially aromatic polyamides involved the extension of a continuous process for the preparation of aliphatic polyamides.[11,12] to those containing substantial portions of aromatic dicarboxylic acids.[13,14]

More recent entrants into the polyphthalamide category are based on 2-methylpentamethylenediamine (2-MPDA) T polymers,[15] which, if fully amorphous, transparent polymers were desired, could optionally contain isophthalic acid or 2-ethyltetramethylenediamine (2-ETDA).[16,17] As these semi-crystalline polymers were somewhat slow to crystallize, applications initially described were limited to fibers[15,16] or amorphous, transparent products.[16,18] This work was extended to semi-crystalline copolymers with PA 6T, wherein a substantial portion of both HMDA and 2-MPDA (and optionally a small amount of 2-ETDA) are included as comonomers.[19]

APPLICATIONS

Polyphthalamides represent a new family of high temperature, high performance thermoplastics. These semi-crystalline resins have excellent physical and mechanical properties, and outstanding dimensional stability. Because of their balance of properties, these resins fit a broad range of applications and uses. Some selected industries where these resins could be used include electrical equipment, sporting goods, telecommunications, automotive, industrial, power equipment, and friction and wear applications. Some of the more common uses in these industries include bearing cages, connectors, housings, lighting components, wiring devices and fasteners; parts in pumps, compressors, generators, electric motors and small engines.

The resins can be combined with a wide variety of types, levels and combinations of fillers and reinforcements. These include a broad range of unreinforced, glass reinforced, mineral and mineral–glass reinforced, impact modified and flame retardant grades. Because of increased thermal capabilities, the resins are able to withstand vapor phase and infrared soldering processing. The lower moisture absorption of polyphthalamides compared to standard nylon 6 or 6,6 polyamides results in lower dimensional and mechanical property changes at all levels of humidity. Finally, these resins demonstrate excellent chemical resistance. They perform well in the presence of most organic chemicals, transportation fluids and aqueous solutions.

REFERENCES

1. Carothers, W. H. U.S. Patent 2 130 523, 1938, to E. I. Du Pont de Nemours and Co.
2. Schlack, P. German Patent 929 151, 1939, to Farbwerke Hoechst AG.
3. *1994 Annual Book of ASTM Standards*; American Society for Testing and Materials: Philadelphia, PA, 1994; Vol. 8.03, p 435. ASTM Designation: D5336-93.
4. Blinne, G.; Baierweck, P.; Gotz, W.; Kopietz, M. *Kunststoffe* **1989,** *79(9),* 814.
5. *Plastics Technol.* **1994,** *40(7),* 65.
6. Kuze, K.; Miwa, S. *Kobunshi Kaguka* **1968,** 25277, 318
7. Chapman, R. D.; Holmer, D. A.; Pickett, Jr. O. A.; Saunders, J. H. U.S. Patent 3 941 755, 1976, to Monsanto Co.
8. Chapman, R. D.; Holmer, D. A.; Pickett, Jr. O. A.; Saunders, J. H. U.S. Patent 4 022 756, 1977, to Monsanto Co.
9. Richardson, J. A.; Poppe, W.; Bolton, B. A.; Paschke, E. E. U.S. Patent 4 603 193, 1986, to Amoco Corp.
10. Poppe, W.; Chen, Y.-T.; Autry, L. W.; Richardson, J. A.; Sinclair, D. P. U.S. Patent 4 603 166, 1986, now Reissue 34,447 to Amoco Corp.
11. Pipper, G.; Schmidt, F.; Theysohn, R.; Riegger, S.; Heil, E.; Fischer, H.; Thoma, R.; Matthies, P. U.S. Patent 4 540 772, 1985, to BASF AG.
12. Reimann, H.; Pipper, G.; Weiss, H-P.; Plachetta, C.; Koch, E. M. U.S. Patent 5 081 222, 1992, to BASF AG.
13. Reimann, H.; Pipper, G.; Weiss, H-P.; Plachetta, C.; Koch, E. M. U.S. Patent 5 218 082, 1993, to BASF AG.
14. Richardson, J. A.; Poppe, W.; Bolton, B. A.; Paschke, E.E. European Patent 0 123 377, 1989, to Amoco Corp.
15. Bier, G.; Blaschke, F.; Funten, H. a.d.; Schade, G. U.S. Patent 4 111 921, 1978, to Dynamit Nobel AG.
16. Schade, G.; Vollkommer, N.; Wemheier, H. U.S. Patent 4 163 101, 1979, to Dynamit Nobel AG.
17. Lahary, P.-Y.; Roy, S. U.S. Patent 5 109 106, 1992, to Rhone-Poulenc Chimie.
18. Lahary, P.-Y.; Coquard, J. U.S. Patent 5 322 923, 1994, to Rhone-Poulenc Chimie.
19. Mok, S. L.; Pagilagan, R. PCT Pub. No. WO 92/10525, 1991, to E. I. Du Pont de Nemours and Co.

Polypropellanes

See: [1.1.1] Propellanes (New Vistas in their Polymer Chemistry)

Polypropylene

See: Bulk Polymerization
Fillers, Polypropylene
Head-to-Head Polymers
Living Coordination Polymerization
Maleic Anhydride Grafted Polypropylene (Solid Phase)
Metallocene Catalysts
Metallocene Catalysts (Group 4 Elements, New Polymeric Materials)
Photo-oxidation, Polyolefins
Photocrosslinking (Overview)
Polyolefin Graft Copolymers (Prepared by Borane Approach)
Polyolefins (Overview)
Polyolefins (Solid State Nuclear Magnetic Resonance)
Polypropylene (Commercial)
Polypropylene (Controlled Degradation)
Polypropylene (Crystallization Kinetics)
Polypropylene (Nucleating Agents)
Polypropylene, Atatic (High Molecular Weight)
Polypropylene Blends and Composites
Polypropylene, Hemiisotactic
Polypropylene, Isotactic (Polymorphism)
Polypropylene, Isotactic (Supermolecular Structure)
Powder Coatings (Overview)
Propylene Polymerization (Stochastic Analysis of the Propagation Mechanism)
Supported Lewis Acid Catalysts (on Polypropylene; Recoverable and Reusable)
Surface Modification (by Low Energy Electron Beam)
Suture Materials (Current Research and Development)

POLYPROPYLENE (Commercial)

Edward P. Moore, Jr.
Montell USA, Incorporated

Polypropylene is a high-volume, commodity polymer possessing characteristics of special materials and has a history of technical innovation. This polymer has numerous grades for specific end uses, and the catalyst is a complex system resulting from major technological achievements. Ironically, the catalyst's sophistication allows the manufacturing process to be simple, economical, flexible, and environmentally friendly. Continuing catalyst developments, such as homogenous catalysts, have great potential for generating new types of polypropylene with expanded applications. New developments are already creating new families of propylene-based materials with properties and potential applications that far surpass conventional polypropylenes.

CURRENT TECHNOLOGY

Current Products

Although technically incorrect, the term *polypropylene* is widely accepted to mean highly isotactic propylene (CAS No. 9003-07-0) homopolymers, propylene-ethylene copolymers, and other co- and terpolymers of propylene that are produced in commercial quantities. There are four types:

- homopolymers–polymers of propylene only;
- random copolymers–polymers of propylene and small quantities of comonomer, usually ethylene, inserted randomly throughout the PP chain;
- impact (or heterophasic) copolymers–matrices of isotactic polymers of propylene with low-to-substantial amounts of olefin copolymer rubber, usually ethylenepropylene rubber, formed in a later stage of the polymerization and dispersed within the matrix; and
- filled and reinforced versions of the above.

Homopolymers

Standard grades of homopolymers are simple in composition, intentionally varying only in molecular weight (melt flow rate). Some differences in isotactic levels occur, but their importance is relatively minor. However, additives and post-reactor treatments can affect the properties.

All polypropylene processes are suitable for making homopolymer, and all are used. Homopolymers are injection molded into small items or where low-temperature impact strength is not critical, such as bottle caps and closures, syringes and disposable medical kits, toys, disposable serviceware, household containers, small automotive and appliance parts, and lawn furniture. However, much is also extruded into unoriented film and oriented into items such as fine fibers, rope, netting, strapping, and biaxially oriented film.[1]

Random Copolymers

Random copolymers contain up to 6% (by weight) of ethylene comonomer inserted at random in the PP chain, providing a lower melting point, improved optics, and lower modulus. Butene is sometimes used as a termonomer with ethylene at low levels to enhance the effect of ethylene or in place of ethylene as the comonomer.[2,3]

Random copolymers are used primarily in areas where improved optical properties or low melting point are important. Unoriented films for apparel packaging and food wrap, clear household containers for food and general storage, medical devices, and heat sealing oriented films are major uses.[1]

Impact Copolymers

Impact copolymers have up to 20% of comonomer (usually ethylene) added at a later stage of polymerization, creating ethylene-propylene rubber (EPR), which forms a separate phase within the homopolymer matrix. The fractions, molecular weights, and comonomer contents of both phases must be carefully selected and controlled to produce effective impact copolymers.

Filled and Reinforced Polypropylene

The most common mineral fillers used in PP are calcium carbonate, talc, and glass fibers. Calcium carbonate and talc are essentially bulk fillers, used up to ~40% by weight, providing increases in stiffness, hardness, and strength with increased density. Glass-reinforced PP requires good adhesion between the glass and the matrix resin and retention of the fiber length for optimum reinforcement.

Atactic PP

Commercial atactic PP (APP), often erroneously called amorphous PP, is low in molecular weight, with high tackiness, making it attractive for adhesive applications. In a relatively sophisticated application, its good resistance and low dielectric constant make it attractive for filling power cable insulation to ensure that it was void free.

NEW DEVELOPMENTS

High Melt Strength PP

Conventional PP with a linear chain suffers from poor melt drawdown characteristics compared with branched polymers such as low-density polyethylene.[4] A radiation method for creating free radicals in PP without the degradation usually encountered was discovered, giving a polymer with high melt strength (HMS). This discovery led to the development of several grades of high melt strength PP for commercialization.

HMS PP can have a beneficial effect on any critical melt phase operations. The most attractive use has been in extrusion coating, in which high drawdown ratios occur. HMS PP has provided uniform coating at significantly higher line speeds than previously possible.[5] Improved uniformity has also been observed in thermoforming and blow molding operations.[6,7] Foamed PP has always been extremely difficult to prepare because of PP's poor melt strength, resulting in the foam cell walls rupturing during expansion. HMS PP has made it possible to prepare excellent quality PP foams.[8]

High Alloy Copolymers

Soft copolymers have been prepared at flexural modulus values from 90–550 MPa (13 to 80K psi) resulting in new fabrication processes and applications. These materials crystallize more slowly to lower levels so that they can be extruded on air-quenched film lines or calendered while maintaining their inherent toughness; conventional PP would give crystalline, brittle films.[10,11] Attractive applications are industrial sacks, diaper backings, and geomembranes. These materials also have sufficient toughness to be used as impact modifiers in polypropylene compounding operations. The low-melting point copolymers provide lower seal initiation temperatures for heat-sealing films. The injection-molding types match the stiffness range provided by externally compounded thermoplastic elastomers and are used in similar applications, such as automotive bumper covers, air dams, and rub strips.

The stiff type, at 2400 MPs (350,000 psi) flexural modulus, is used for injection-molded applications requiring a high heat deflection temperature or great stiffness.

OTHER POLYMERS

Metallocene Catalysts

One of the more exciting areas of research in the last decade has been homogenous catalysts.[12-15] Because the catalyst sites are of only one well-characterized type, polymers are more consistently stereoregular and display narrow molecular weight distributions.

Syndiotactic PP

Syndiotactic PP (sPP, CAS No. 26063-22-9) can be prepared readily with metallocene catalysts. Compared to isotactic PP, sPP exhibits a lower melting point, lower stiffness, slower crystallization, and higher clarity but similar low-temperature impact.

Nonolefin Alloys

With the development of higher porosity PP, it became easier to infuse the PP particle with nonolefin monomers and polymerize it to intimate alloys with polypropylene.[9]

Unsaturated PP

As a first step toward adding functionality to PP, incorporating unsaturation has been studied. One approach was to polymerize with a diolefin comonomer such as butadiene.[16] With metallocene catalysts, unsaturation levels equivalent to 1–6% of butadiene in the copolymer have been achieved.[17] Many other approaches to functionalizing PP have been considered and reviewed.[18]

REFERENCES

1. *Polypropylene 1994: Global Markets, Forces, and Producer Strategies;* Phillip Townsend Associates: Houston, TX, 1994.
2. Galli, P.; Mayfield, J. W.; Brusaferro, A. Presented at the *AIChE Petro Expo '86* New Orleans, LA, April 1986.
3. *Cefor Resins;* Research Product Technical Bulletins SC:2030-94 and SC:1825-93; Shell Chemical: Houston, TX.
4. Cogswell, F. N. *Polymer Melt Rheology;* John Wiley & Sons: New York, 1981.
5. Yoo, H. J.; "Performance of High Melt Strength (HMS) Polypropylene Resins in Extrusion Coating Process," *SPE ANTEC '94*, San Francisco, CA, May 1994.
6. McKugh, K. E.; Ogale, K.; "High Melt Strength Polypropylene for Melt Phase Thermoforming," *SPE ANTEC '90*, Dallas, TX May 1990.
7. Beren, J. R.; Capellman, C. Presented at SAE Int'l Congress and Exposition; Detroit, MI, February/March 1994.
8. Bradley, M.; Phillips, E. M. "Novel Foamable Polypropylene Polymers," *SPE ANTEC '90*, Dallas, TX, May 1990.
9. Cecchin, G. "In-situ Polyolefin Alloys," *Makromol. Symp.* **1994,** *78,* 213-228.
10. Giacobbe, J.; "Evaluation of Air-quenched Blown Film Made with Low Modulus Polyolefins," *SPE ANTEC '92*; Detroit, MI, May 1992.
11. Shah, A. "Low Flexural Modulus Calenderable Polypropylenes," *SPE ANTEC '91*, Montreal, Canada, May 1991.
12. Ewen, J. A.; Elder, M. J.; Jones, R. L. et al. *J. Am. Chem. Soc.* **1984,** *106,* 6355.
13. Corradini, P.; Busico, V.; Guerra, G. *Comprehensive Polymer Science*; Pergamon: Oxford, UK, 1989; Chapter 3, pp 29-50.
14. Ewen, J. A.; Haspeslagh, L.; Atwood, J. L.; Bott, S. G.; Robinson, K. "Metallocene/Polypropylene Structural Relationships," *Makromol. Symp.* **1991,** *48/49,* 253-295.
15. Halterman, R. L. "Synthesis and Applications of Chiral Cyclopentadienyl Metal Complexes," *Chem. Rev.* **1992,** *92,* 965-994.
16. Cecchin, G.; Guglielmi, F.; Zerega, F. U.S. Patent 4 602 077, July 22, 1986.
17. Galimberti, M.; Albizzati, E. "C-NMR Analysis of alpha-olefins Copolymers with 1,3-butadiene," *Makrom. Chem.* **1990,** 192(11), 2591-2601.
18. Simonazzi, T.; DeNicola, A. J.; Aglietto, M. et al. *Comprehensive Polymer Science–First Supplement*; Pergamon: Oxford, UK, 1992; Chapter 7, pp 133-158.

POLYPROPYLENE (Controlled Degradation)

Fernanda M. B. Coutinho* and Marisa C. G. Rocha
Instituto de Macromolécules
Universidade Federal do Rio de Janeiro

Stephen T. Balke
Department of Chemical Engineering and Applied Chemistry
University of Toronto

For the purpose of improving the processability of polypropylene, the molecular weight and the MWD can be modified in a postreactor stage by means of different degradation methods.

Modification of the polypropylene MWD during extrusion can be generally achieved by thermal degradation or by peroxide-promoted degradation.[2,3,5-24] Thermal degradation processes carried out in the presence or absence of oxygen require relatively long processing times (1–24 h) at elevated temperatures (250–450°C) and, therefore, high specific energy inputs.[1,2,6] However, the products obtained showed significant concentrations of low molecular weight and oxygen-containing material, and the achievable reduction in molecular weight is low.[22,24]

The peroxide-promoted degradation of polypropylene during extrusion, also termed vis-cracking[11] or vis-breaking,[25] is now a well-recognized manufacturing process.[11,19,25] The product is termed "controlled-rheology" polypropylene (CR-PP). Controlled rheology polymers, in the broadest sense, are polymers in which the structure or composition has been modified to improve material response during melt processing.[5]

The reactive degradation process provides a straightforward and economical means for producing polypropylene with the molecular weight and MWD required for specific applications.[15] CR-PP resins are typically used in fiber production and fast cycle molding operations.[19]

In recent years, polypropylene with molecular weights of about 40,000 and with very narrow molecular weight distributions have been shown to be valuable materials for the plastics industry.[9] These low molecular weight polypropylene resins (LMWPP) have been currently used as nucleating agents (additives for crystallization), pigment carriers, lubricant agents for vinyl plastics, and flow modifiers in thermoplastic compounds.[28-31] These resins have exceptionally low melt viscosities of 200 poise, independent of the processing shear rate, and 10 times higher melt flow than CR-PP resins, and they retain the high melting point inherent in conventional high molecular weight polypropylene.[26]

The process for preparing the LMWPP is also a peroxide-initiated degradation of a high molecular weight polypropylene by reactive extrusion. One relevant patent is that of Greene and Pieski.[24]

*Author to whom correspondence should be addressed.

PROCESS CONTROL FOR POLYPROPYLENE DEGRADATION

The major effects of the peroxide-promoted polypropylene degradation are the removal of the high molecular weight tail and the corresponding narrowing of the molecular weight distribution.

The decrease of $\overline{M}_w$ and the narrowing of the MWD are associated with the changes in the low properties of polypropylene, such as the reduction in the melt viscosity and the elasticity.

As the viscosity of the polypropylene changes as a result of the degradation, it is possible to monitor the extent of degradation by measuring changes in the viscosity. As a result, control schemes have been developed to continuously produce different grades of degraded polypropylene by using on-line rheometers of capillary or Couette design equipped with data processing systems.[12,19,22]

PROCESSING VARIABLES

The effects of varying process parameters such as initiator concentration, the screw speed, temperature, reaction atmosphere, and method of peroxide addition on the polypropylene degradation processes have been discussed in several papers.[4,6,9,10,13,16,18,21,22,27,29,32-34] These studies provide valuable information on the kinetics of the degradation process.

REFERENCES

1. Kowalski, R. C. *Reactive Extrusion, Principles and Practice;* Hanser: New York 1992; Chapter 1.
2. Kowalski, R. C.; Harrison, J. .; Staton, J. C.; Keller, J. P. U.S. Patent 3 608 001, 1966.
3. Pabedinskas, A.; Cluett, W. R.; Balke, S. T. *Polym. Eng. Sci.* **1991,** *31,* 365.
4. Tzoganakis, C.; Vlachopoulos, J.; Hamielec, A. E.; Shinozaki, D. M. *Polym. Eng. Sci.* **1989,** *29,* 390.
5. Dziemianowicz, T. S.; Cox, W. W. *ANTEC* **1985,** *31,* 549.
6. Xanthos, M. *Reactive Extrusion, Principles and Practices;* Hanser: New York 1992; Chapter 2.
7. Kowalski, R. C.; Harrison, J. W.: Staton, J. C.; Keller, J. P. U.S. Patent 3 551 943, 1971, Exxon.
8. Chen, C.; Nguyen, K. T.; Sanschagrin, B.; Piché, L. *ANTEC* **1994,** *40,* 2041.
9. Rocha, M. C. G.; Coutinho, F. M. B.; Balke, S. T. *Polymer Testing* **1995,** *14,* 369.
10. Rocha, M. C. G.; Coutinho, F. M. B.; Balke, S. T. *Polímeros: Ciência e Technologia* **1994,** Jul/Set, 33.
11. Triacca, V. J.; Gloor, P. E.; Zhur, S.; Hrymak, A. N.; Hamielec, A. E. *Polym. Eng. Sci.* **1993,** *33,* 445.
12. Dumoulin, M. M.; Daigeneault, L. E.; Gendron, R.; Dufour, J. *ANTEC* **1993,** *39,* 1448.
13. Tzoganakis, C.; Tang, Y.; Vlachopoulos, J.; Hamielec, A. E. *Polym. Plast. Technol. Eng.* **1989,** *28,* 319.
14. Lew, R.; Cheung, P.; Balke, S. T. Reactive Extrusion of Polypropylene, In ACS Symposium Series; Provder, T., Ed.; American Chemical Society: Washington, DC, *29,* 1989.
15. Pabedinskas, A.; Cluett, W. R.; Balke, S. T. *Polym. Eng. Sci.* **1989,** *29,* 993.
16. Suwanda, D.; Lew, R.; Balke, S. T. *J. Appl. Polym. Sci.* **1988,** *35,* 1019.
17. Suwanda, D.; Lew, R.; Balke, S. T. *J. Appl. Polym. Sci.* **1988,** *35,* 1033.
18. Tzoganakis, C.; Vlachopoulos, J.; Hamielec, A. E. *Int. Polym. Proc.* **1988,** *3,* 141.
19. Curry, J.; Jackson, S.; Stoehrer, B.; Van der Veen, A. *Chem. Eng. Prog.* **1988,** *84,* 43.
20. Balke, S. T.; Suwanda, D.; Lew, R. *J. Polym. Sci., Polym. Lett.* **1987,** *25,* 313.
21. Tzoganakis, C.; Vlachopoulos, J.; Hamielec, A. E. Presented at the 36th Canadian Chemical Engineering Conference; Sarnia: Ontario, Canada, 1986.
22. Fritz, H. G.; Strohrer, B. *Inter. Polym. Process* **1986,** *1,* 31.
23. Hudec, P.; Obdrzalek, L. *Angew. Macromol. Chem.* **1980,** *89,* 41.
24. Greene, R. E.; Pieski, E. T. U.S. Patent 3 144 436, 1964.
25. Tollefson, N. M. *J. Appl. Polym. Sci.* **1994,** *52,* 905.
26. Bourland, L. *Polym. Prepr.* **1987,** *28,* 388.
27. Suwanda, D. Master of Applied Science Thesis, University of Toronto, 1987.
28. Ryu, S. H.; Gogos, C. G.; Xanthos, M. *ANTEC* **1989,** *35,* 879.
29. Bourland, L. *Plast. Eng.* **1987,** *43,* 39.
30. Bourland, L. *ANTEC* **1988,** *34,* 321.
31. Bourland, L. U.S. Patent 4 501 849, 1985.
32. Ryu, S. H.; Gogos, C. G.; Xanthos, M. *Adv. Polym. Technol.* **1991/1992,** *11,* 121.
33. Tzoganakis, C.; Vlachopoulos, J.; Hamielec, A. E. *Chem. Eng. Progress* **1988,** *84,* 47.
34. Tzoganakis, C.; Vlachopoulos, J.; Hamielec, A. E. *Polym. Eng. Sci.* **1988,** *28,* 170.

POLYPROPYLENE (Crystallization Kinetics)

Robert A. Shanks and Long Yu
Royal Melbourne Institute of Technology

The crystallization of polymers proceeds through a series of stages: melt, nucleation, lamellar growth, spherulite growth, and aggregation growth. The melt crystallizes when the system is cooled below the equilibrium melting temperature (T°_m) to the crystallization temperature ($T_c \leq T_m$). The difference of $\Delta T = T_c - T_m$ is a measure of the supercooling that, in turn, depends on the cooling rate and the nucleation of mechanisms.[1-3]

Methods for evaluating the development of crystallinity in polymers fall into two categories: measurable information, such as density, crystallinity, and thermal energy difference, and direct observation of the light microscopy, the rate of spherulite formation, and the subsequent growth. From this information, the crystallization mechanisms can be deduced.[4]

Techniques for studying the kinetics of pure polymers are still developing, especially for nonisothermal conditions and the methods, models, and techniques for studying pure polymers remain far from perfect. Research of the crystallization of polymer blends and filled polymers is far more recent. However, this research is now attracting widespread interest because polymer blends and filled polymers have grown rapidly in importance, and with current technology, there are fewer dead ends in the development of polymer blends than in that of new resins.

CRYSTALLIZATION OF POLYPROPYLENE (PP) BLENDS AND FILLED PP

Crystallization of Blends

The polymer blends can be miscible or immiscible. The miscible polymer blends are considered a homogenous solution. Adding a second polymer to a semi-crystalline polymer acts as a diluent that can increase crystallinity by decreasing the concentration and the nuclei numbers or increase crystallinity by enhanced nucleation or increased chain mobility.

Most PP blends are immiscible. The discussion of the crystallization of neat polymers directly applies to immiscible polymer blends, in which crystallization occurs within a domain of nearly pure resin.[5] The presence of dispersed particles may strongly affect the overall crystallization growth rates. Adding LLDPE decreases PP's crystallization growth.

Crystallization of Filled PP

Filler can affect PP crystallization significantly. The interface between inorganic filler and PP can provide a nuclei center for PP to enhance the crystallization growth. However, large particles of filler, such as glass fiber, can stop the spherulite growth and a small quantity of filler can produce numerous nuclei centers. PP's crystallization growth usually increases when filler is added.

Nucleation

PP's overall crystallization growth is controlled by two factors: nucleation and spherulite growth. In polymer crystallization, nucleation processes are even more important than in systems of atoms or small molecules. In high molar mass polymers, not only is the initial formation of the crystalline regions dominated by nucleation processes but so is their subsequent development.[3]

When a second component is added, PP nucleation depends on three factors: self-seeding varied with the mixing process; nuclei moving in or out of the PP matrix; and interactions at the interface.[8] LLDPE decreases the nuclei density and poly(styrene-*b*-ethylene-*co*-butene-*b*-styrene) increases it. Furthermore, the nuclei density and the time of the nuclei's appearance is sensitive to the crystallization temperature. At a certain crystallization temperature, the number of spherulites increased in the initial stage and then remained constant after the crystalline fraction reached 10%. The observed number of nuclei decreased as the temperature increased.

Spherulite Growth

For most thermoplastic polymers, their spherulite radical growth rate from the melt is linear and is usually described by its radius.[11]

The effect of another component on the spherulite growth depends on the interfacial energy and particle size. The size of PP spherulites is about 10 μm. Glass fiber can stop PP spherulite growth by simply appearing in the front of the spherulite.[12] However, most immiscible polymer components, such as various PE and compatibilizers, do not affect the growth.[6-10]

REFERENCES

1. Wunderlich, B. *Macromolecular Physics II*; Academic: New York, 1976.
2. Mandelkern, L. *Crystallization of Polymer*; McGraw-Hill: New York, 1964.
3. Price, F. P. In *Nucleation*; Zettlemoyer, A. C., Ed.; Marcel Dekker: New York, 1969.
4. Yu Long, Shanks, R. A.; Stachurski, Z. H. *Polymer in Progress*, in press.
5. Utracki, L. A. *Polymer Alloy and Blends*; Hanser: New York, 1989.
6. Galeski, A.; Pracella, M.; Martuscelli, E. *J. Polym. Sci., Polym. Phys. Ed.* **1984,** *22,* 739.
7. Galeski, A.; Bartczak, Z.; Pracella, M. *Polymer* **1984,** *25,* 1323.
8. Yu Long; Stachurski, Z. H.; Shanks, R. A. *Mater Forum* **1992,** *16,* 173.
9. Cimmino, S.; Martuscelli, E.; Silvestre, C. *J. Polym. Sci., Part B: Polym. Phys.* **1989,** *27,* 1781.
10. Bartczak, Z.; Galeski, A.; Pracella, M. *Polymer* **1986,** *27,* 537.
11. Evans, U. R. *Trans. Faraday Soc.* **1945,** *41,* 365.
12. Mehl, A. N.; Rebenfeld *J. Polym. Sci., Part B: Polym. Phys.* **1993,** *31,* 1677.

POLYPROPYLENE (Nucleating Agents)

Kazuta Mitsuishi
Industrial Technology Center of Okayama Prefecture

Polypropylene (PP) production has responded to market demand with a startling increase, chiefly in industrial and household articles such as injection- or blow-molded products, stretched tapes, and monofilaments, which are characterized by high strength, impact resistance, dimensional stability, chemical resistance, moldability, and recycling adaptability. For molded products and sheets, in particular, characteristics such as stiffness, transparency, and luster are desired. And in the case of block copolymers having impact resistance (now increasing in demand), better transparency is becoming an extremely important goal, for the sake of material design.

With the purpose of imparting these characteristics, researchers have developed the technology of adding crystal nucleating agent to PP in a molten state to raise the rate of crystallization. This process increases crystallinity and hence micronization of spherulites in the course of cooling followed by solidification; and as the result, the PP becomes transparent. Nucleating agents, in addition to improving mechanical properties such as stiffness, help shorten molding cycles because of an increased PP crystallization rate.

TYPES OF NUCLEATING AGENTS

Crystal nucleating agents for PP can be classified largely into organic and inorganic types. Major nucleating agents used for PP are organic agents, such as metal aromatic carboxylates, sorbitol derivatives, and organic phosphates and inorganic agents, such as talc.

POLYMER-TYPE NUCLEATING AGENTS

A polymer-type nucleating agent, because of its polymer characteristics, when added into PP at the time of its polymerization reaction can be expected to better disperse and, hence, show a better nucleating effect contrast to organic agents. Researchers are investigating and adopting polymer-type nucleating agents, polyvinylalcanes (polyvinylcyclohexane),[1] poly-3-methyl-butene,[2] and silane derivatives,[3] for films with high transparency.

Among these, the nucleating effect of polyvinylcyclohexane is excellent, effectively improving optical properties.

OTHER NUCLEATING AGENTS

Carbon, aramid, polyester, cellulose, and glass fibers have been known to work as crystal nucleating agents for PP. The fiber surface itself works to initiate nucleation, and transcrystallinity sometimes develops around the fiber. Chatterjee indicates that, after melting of resin, PP crystal residues remaining on the surface or corners of the fiber work as sites of self-seeded nucleation.[4] Quillin observed growth of transcrystallinity in nonisothermal crystallization of a PP–cellulose system.[5] Among other recent studies are reports on boron nitride,[6] titanium dioxide,[7] rare-earth oxide,[8] and montanates.[9]

SURFACE MODIFICATION OF NUCLEATING AGENT

For inorganic nucleating agents, the powder surface is sometimes chemically modified in order to improve dispersion in the resin. Here the relationship of chemical modification of nucleating agents with T_c and the degree of crystallization is apparent. In all cases, surface modification of the powder suppressed increase of T_c and of the degree of crystallization, somewhat decreasing its effect as a nucleating agent.

To summarize, crystal nucleating agents exert great influence on the crystalline structure and the material properties of polypropylene. Future development of crystal nucleating agents, therefore, will become an elemental technology, to control the higher order structure of polymers, along with the technologies of catalyst preparation and polymerization.

REFERENCES

1. Japan Patent Publication No. Hei 3 74247, 1991, Sumitomo Chemical Co.
2. Japan Patent Application (laid-open) No. Hei 2 272045, 1990, Mitsubishi Petrochemical Co.
3. Japan Patent Application (laid-open) No. Hei 1 262120, 1989, Chisso Co.
4. Chatterjee, A. M.; Price, F. P.; Newman, S. *J. Polym. Sci., Polym. Phys. Ed.* **1975,** *13,* 2385.
5. Quillin, D. T.; Caulfield, D. F.; Koutsky, J. A. *J. Appl. Polym. Sci.* **1993,** *50,* 1187.
6. Morita, Y.; Matsuda, N.; Suga, K.; Takai, T. *Proc. Int. Wire Cable Symp.* **1991,** *40,* 260.
7. Burke, M.; Young, R. J.; Stanford, J. L. *Plast. Rubber. Comp. Process Appl.* **1993,** *20,* 121.
8. Liu, J.; Tang, G.; Qu, G.; Zhou, H.; Guo, Q. *J. Appl. Polym. Sci.* **1993,** *47,* 2111.
9. Piesold, P. *Kunstst. Ger. Plast.* **1992,** *82,* 988.

POLYPROPYLENE, ATACTIC (High Molecular Weight)

Luigi Resconi* and Rosanna Silvestri
Montell Polyolefins
G. Natta Research Center

The steady appearance of patents[1-6] and scientific publications[7-15] on amorphous or largely amorphous polypropylene demonstrates the continuing interest in this polymer.

To be fully amorphous, polypropylene must lack any long-range order in the sequence of chirotopic methynes in the polymer chain (i.e., to be atactic).[16]

Direct consequences of the lack of configurational order on the molecular properties of atactic polypropylene (aPP) are the absence of melting and crystallization transitions, the lack of any crystallinity peak in the X-ray spectrum, a complete solubility in most organic solvents including hydrocarbons, and a high transparency. Other physical properties depend also on the polymer's molecular weight.

Atactic polypropylene so far has been obtained as a by-product in the synthesis of isotactic polypropylene (iPP). This product is actually neither really atactic nor fully amorphous, being a mixture of chains of rather different tacticities and molecular weights.[17-20] Its molecular weights are too low for most practical application, and the lack of a direct, high-yield process able to produce aPP with higher molecular weights has prevented the full investigation of its properties and limited its possible applications. This notwithstanding, atactic polypropylene by-products have found applications as bitumen additive and in hot-melt adhesives. With the advent of more and more stereospecific catalysts for isotactic polypropylene, the availability of aPP-type materials is coming to an end, and as a consequence the interest in these products is rising.

Several catalytic systems have been described for the synthesis of aPP, but none of them display the necessary activity for large-scale aPP production.[1-15]

Group 4 metallocene-based catalysts allow the production of polyolefins in a much wider range of composition and microstructures than any other catalytic systems.[21] In the case of propylene polymerization, both highly isotactic[22-24] and highly syndiotactic[25-27] polymers have been obtained, as well as practically all intermediate microstructures, including ideally atactic polypropylene.[28]

PREPARATION

A suitable catalyst for the high yield synthesis of HMW-aPP is the $Me_2SiFlu_2ZrCl_2$/MAO catalyst (**Figure 1**).

PROPERTIES

HMW-aPP is completely amorphous and soluble in such solvents as diethylether or hexane, and has the additional property of very narrow molecular weight distributions, the latter being typical of metallocene-produced polyolefins.

These materials are characterized by a very low density and a high transparency.

*Author to whom correspondence should be addressed.

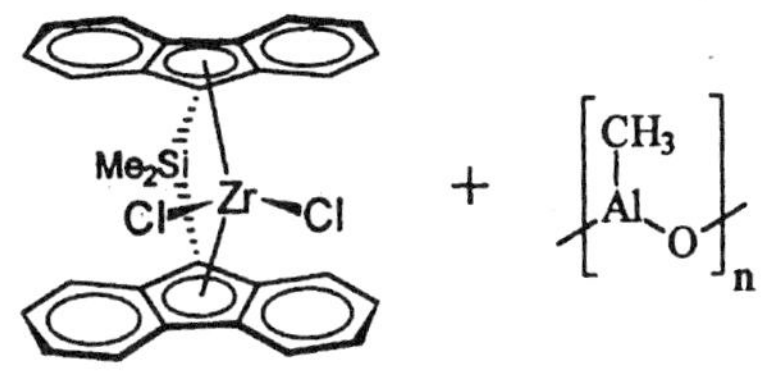

FIGURE 1. The $Me_2Si(9\text{-fluorenyl})_2ZrCl_2$/MAO catalyst.

HMW-aPP shows elastomeric behavior with lack of yielding and high elongation at break.

These samples possess some elastic recovery properties, which increase with molecular weight.

Differential scanning calorimetry analysis of all these samples do not show any melting endotherm, even after storage for one year at room temperature (an obvious consequence of their microstructure), while the commercial atactic sample usually shows a small melting endotherm peak from low molecular weight crystallizable chain segments.

Dynamic-mechanical analysis of these samples shows one single narrow mechanical damping peak, due to the narrow molecular weight distribution. The α transition peak, which is usually found in highly amorphous isotactic samples,[29] has disappeared.

The X-ray diffraction pattern of these samples confirms the total lack of crystallinity of this material.

HMW-aPP can be used in blends with iPP, both homopolymers and copolymers, and also in heterophasic compositions, to improve elasticity, softness and transparency. It shows promise for biomedical applications.

REFERENCES

1. Kakugo, M. et al. (to Sumitomo Chemical) Eur. Patent Application 241 560, 1987.
2. Kakugo, M.; Miyatake, T.; Mizunuma, K.; Yagi, Y. Eur. Patent Application 371 411 to Sumitomo Chem., 1990.
3. Smith, T.; Ames, W.; Holliday, R.; Pearson, N. Eur. Patent Application 232 201 to Eastman Kodak, 1987.
4. Smith, C. Eur. Patent Application 423 786 to Himont, 1991.
5. Pellon, B.; Allen, G. Eur. Patent Application 475 307 to Rexene, 1992.
6. Job, R. U.S. Patent 5 270 410 to Shell Oil, 1993.
7. Schilling, F.; Tonelli, A. *Macromolecules* **1980,** *13,* 270.
8. Suter, U.; Neuenschwander, P. *Macromolecules* **1980,** *14,* 528.
9. Plazek, D. L.; Plazek, D. J. *Macromolecules* **1983,** *16,* 1469.
10. Zhongde, X.; Mays, J.; Xuexin, C.; Hadjichristidis, N.; Schilling, F.; Bair, H.; Pearson, D.; Fetters, L. *Macromolecules* **1985,** *18,* 2560.
11. Soga, K.; Park, J.; Uchino, H.; Uozumi, T.; Shiono, T. *Macromolecules* **1989,** *22,* 3824.
12. Marigo, A.; Marega, C.; Zanetti, E.; Zannetti, R.; Paganetto, G. *Makromol. Chem.* **1991,** *192,* 523.
13. Miyatake, T.; Mizunuma, K.; Kakugo, M. *Makromol. Chem., Macromol. Symp.* **1993,** *66,* 203.
14. Alfonso, G.; Yan, D.; Zhou,Z. *Polymer* **1993,** *34,* 2830.
15. Ittel, S. *ACS Poly. Prepr.* **1994,** *35(1),* 663.
16. Farina, M. *Topics in Stereochemistry* **1987,** *17,* 1.
17. Kawamura, H.; Hayashi, T.; Inoue, Y.; Chûjô, R. *Macromolecules* **1989,** *22,* 2181.
18. Kakugo, M.; Miyatake, T.; Naito, Y.; Mizunuma, K. *Makromol. Chem.* **1989,** *190,* 505.
19. van der Veen, S. *Polypropylene and Other Polyolefins*; Elsevier: Amsterdam, 1990.
20. Busico, V.; Corradini, P.; De Martino, L.; Graziano, F.; Iadicicco, A. *Makromol. Chem.* **1991,** *192,* 49-57.
21. Ewen, J.; Elder, M.; Jones, R.; Haspeslagh, L.; Atwood, J.; Bott, S.; Robinson, K. *Makromol. Chem., Macromol. Symp.* **1991,** *48/49,* 253.
22. Mise, T.; Miya, S.; Yamazaki, H. *Chem. Lett.* **1989,** 1853.
23. Spaleck, W.; Küber, F.; Winter, A.; Rohrmann, J.; Bachmann, B.; Antberg, M.; Dolle, V.; Paulus, E. *Organometallics* **1994,** *13, 954,* and references therein.
24. Stehling, U.; Diebold, J.; Kirsten, R.; Röll, W.; Brintzinger, H. H.; Jüngling, S.; Mülhaupt, R.; Langhauser, F. *Organometallics* **1994,** *13,* 964.
25. Ewen, J. A.; Elder, M.; Jones, R. L.; Curtis, S.; Cheng, H. N. In *Catalytic Olefin Polymerization, Stud. Surf. Sci. Cat.*, Keii, T.; Soga, K., Eds.; Kodansha/Elsevier: 1990; Vol. 56, pp 439-482.
26. Winter, A.; Rohrmann, J.; Antberg, M.; Dolle, V.; Spaleck, W. Eur. Patent Application 387 690 to Hoechst, 1990.
27. Razavi, A.; Atwood, J. *J. Organomet. Chem.* **1993,** *459,* 117.
28. Resconi, L.; Piemontesi, F.; Franciscono, G.; Abis, L.; Fiorani, T. *J. Am. Chem. Soc.* **1992,** *114,* 1025, and references therein.
29. Flocke, H. A. *Kolloid Zeit.* **1962,** *180,* 118.

POLYPROPYLENE BLENDS AND COMPOSITES

Béla Pukánszky
Department of Plastics and Rubber Technology
Technical University of Budapest and Central Research Institute for Chemistry
Hungarian Academy of Sciences

Polypropylene (PP) blends and composites are used in large quantities in many fields. The automotive industry is one of the largest consumers, but these materials are also widely used in the electrical and electronics industry, for household appliances, furniture production, and in many other applications. The success of modified PP lies in its advantageous price, volume, and performance relationship,[1,2] allowing PP blends and composites to successfully penetrate fields traditionally occupied by other materials, such as acrylonitrile-butadiene-styrene (ABS),[3] polyurethanes, or other elastomers. Their application is also being extended to fields in which engineering thermoplastics have been used exclusively up to now.[4]

PP COMPOSITES

Although a great number of materials are used or at least tested as fillers or reinforcements for PP, only three of them have real practical importance: talc, $CaCO_3$, and glass fibers. It is obvious that in spite of the significant effort put into the research and development of glass-fiber-reinforced PP composites, their economic importance and growth rate is smaller than those of particulate-filled materials.[5] Some of the other materials used as filler and reinforcement for PP are mica,[6] glass beads,[7] sepiolite,[8] magnesium hydroxide,[9] wood flour and cellulose,[10] but wollastonite,[11] gypsum,[12] carbon black,[13] clay,[14] metal powders (aluminum, iron, and nickel),[15] steel fibers,[16] silicium carbide,[15] phenolic microspheres,[17] and diverse flame retardants[18] are also mentioned in patents and publications.

Component Properties

The most important particle characteristics are particle size, particle size distribution, specific surface area, and aspect ratio.[19] Although the exact correlation of the aspect ratio of the fillers and reinforcements and the mechanical properties of the composites is a much discussed question, it is clear that the reinforcing effect of the filler increases with increasing aspect ratio.[14] The stiffness of fiber-reinforced composites is much higher than that of the particulate-filled polymers. Anisotropic filler particles orientate during processing, which leads to further reinforcement, or more exactly, to anisotropic properties in the product.[22]

Composition

The modulus of PP composites increases with increasing filler content.[23] Tensile properties (yield stress, tensile strength, elongation) and impact characteristics depend on composition in a more complicated way. These properties are said to decrease with increasing filler content in particulate-filled polymers, while ate least stiffness and strength increase with filling in the case of reinforcements.[24]

Composition dependence of fracture and impact properties of PP composites is more complicated than that of the other mechanical characteristics. Very often a maximum is detected as a function of composition.[26-28]

Interaction

The polymer matrix and the filler particles always interact with each other, but often particle–particle interactions develop in the composites as well. This latter interaction leads to aggregation and inferior mechanical properties–primarily decreased impact resistance.[20] The former interaction results in increased reinforcement.[21]

Structure

The effect of the structure is at least as important in particulate-filled and reinforced PP as in its blends. Three major factors must be considered here. Fillers and reinforcements influence the crystalline structure of the polymer. Talc has a very strong nucleating effect,[30] while on glass fibers a transcrystalline layer forms.[31]

PP BLENDS

The properties of polypropylene blends are determined by the same factors as those of the composites (i.e., component properties, composition, component interaction, and structure). Micromechanical deformations taking place under the effect of external loading are also similar, and composition dependence of properties can be described by the same correlations in blends and composites.[25]

The structure of PP blends varies widely depending on component properties and composition. Due to its crystalline structure, polypropylene forms multiphase blends. Cocrystallization is rarely observed, and its miscibility is limited with most polymers.[32]

The largest commercial importance has been achieved by PP–EPR and PP–EPDM blends. These are used in large quantities, mainly in the automotive industry.

Impact-modified PP has a dispersed-phase structure. Although much attention has been paid to the crystalline morphology of these blends,[35,36] the most important characteristic that determines properties is the size of the dispersed elastomer particles and interfacial adhesion.[33,34] The size of the EPR or EPDM droplets depends mainly on the relative viscosity of the components and on processing conditions.

Thermoplastic elastomers (TPO) based on polypropylene are produced in increasing quantities. TPO is prepared from EPDM block copolymers in which the PP blocks interact with the PP phase of the elastomer, forming relatively large domains acting as physical crosslinks.

PP is blended with most polymers. PP–PE blends have been prepared and intensively studied in the past two decades. The goal of the modification was to improve the low-temperature impact strength and processability of PP.

In the past 10 years interest in blends of PP and engineering thermoplastics has increased significantly. PP–polyamide blends have gained commercial importance already.

MULTICOMPONENT SYSTEMS

Multicomponent systems of PP are prepared in increasing quantities to further improve properties or to combine the benefits of different modifications. Two polymers and an elastomer are combined in PP–EPDM–PE blends,[37] with the goal of further improving impact resistance and processability. Occasionally two fillers, or more exactly a filler and a reinforcement, are added to PP. The reinforcement gives PP the necessary stiffness, while the filler increases dimensional stability, and decreases warping and orientation-dependent shrinkage.

The largest family of multicomponent PP composites contains both an elastomer and a filler.[38] The filler improves stiffness and usually decreases impact resistance, while the elastomer has an inverse effect. The final goal of development in numerous cases is the simultaneous increase of stiffness and toughness; combination of the two additives seems to be the obvious way to achieve it.

CONCLUSIONS

PP blends and composites have great commercial interest and widespread applications. Given the intensity of research and development, this class of materials is expected to grow. Properties of modified PP depend on the same factors for blends and composites–component properties, composition, interaction, and structure. Although much attention is paid to the study of the crystalline morphology of modified PP, dispersed-phase seems to have a greater effect on polymer properties. In development, researchers have focused attention mainly on glass-fiber-reinforced PP and on blends with engineering thermoplastics at the moment.

REFERENCES

1. Aumayr, G. *Kunststoffberater* **1989,** *34,* 63.
2. Grolik, W.; Gübitz, F.; Naumann, H.-G. *Kunststoffe* **1989,** *79,* 1334.
3. Schmidt, H.; Jzquierdo, R. *Kunststoffe* **1988,** *78,* 149.
4. Käufer, H.; Fischer, H. J. *Gummi, Fasern, Kunstst.* **1987,** *40,* 436.

5. Karger-Kocsis, J. In *Polypropylene: Structure, Blends, and Composites*; Karger-Kocsis, J., Ed.; Chapman and Hall: London, United Kingdom, 1995; Vol. 3, p 142.
6. Fenton, M.; Naik, S.; Hawley, G. 39th Ann. Conf., Reinf. Plast. Comp. Inst., SPI **1984,** Sec. 12-C, 1-9.
7. Denault, J.; Vu-Khanh, T. *Polym. Compos.* **1988,** *9,* 360.
8. Acosta, J. L. et al. *J. Appl. Polym. Sci.* **1986,** *32,* 4119.
9. Hornsby, P. R.; Watson, C. L. *Plast. Rubber Process. Appl.* **1989,** *11,* 45.
10. Törmälä, P.; Pääkkonen, E.; Laiho, J. *Kunststoffe* **1985,** *75,* 287.
11. Ramsteiner, F.; Theysohn, R. *Composites* **1984,** *15,* 121.
12. Socha, D. A. Presented at Filplas Conference; Manchester, United Kingdom, April 12-13, 1989.
13. Sumita, M.; Sakata, K.; Asai, S.; Miyasaka, K.; Nakagawa, H. *Polym. Bull.* **1991,** *25,* 265.
14. Riley, A. M.; Paynter, C. D.; McGenity, P. M.; Adams, J. M. *Plast. Rubber Process. Appl.* **1990,** *14,* 85-93.
15. Bigg, D. M. *Polym. Compos.* **1987,** *8,* 115.
16. Bridge, B.; Folkes, M. J.; Jahankhani, H. *J. Mater. Sci.* **1988,** *23,* 1948.
17. Zuchowska, D.; Hlavatá, D. *Eur. Polym. J.* **1991,** *27,* 355.
18. Bajaj, P. et al. *J. Appl. Polym. Sci.* **1987,** *34,* 1785.
19. Pukánszky, B. In *Polypropylene: Structure, Blends and Composites*; Karger-Kocsis, J., Ed.; Chapman and Hall: London, United Kingdom, 1995; Vol. 3, p 1.
20. Svehlova, V.; Poloucek, E. *Angew. Macromol. Chem.* **1987,** *153,* 197.
21. Pukánszky, B.; Turcsányi, B.; Tüdős, F. In *Interfaces in Polymer, Ceramic, and Metal Matrix Composites*; Ishida, H., Ed.; Elsevier: New York, NY, 1988; p 467.
22. Mittal, R. K.; Gupta, V. B.; Sharma, P. *J. Mater. Sci.* **1987,** *22,* 1949.
23. Nielsen, L. E. *Mechanical Properties of Polymers and Composites*; Marcel Dekker: New York, 1974.
24. Vollenberg, P. H. T.; Heikens, D. *J. Mater. Sci.* **1990,** *25,* 3089.
25. Pukánszky, B. *J. Polym. Eng.* **1993,** *12,* 1.
26. Pukánszky, B.; Maurer, F. H. J. *Polymer* **1995,** *36,* 1617.
27. Kučera, J. In *Polymer Composites*; Sedláčtek, B.; de Gruyter, Walter, Eds.; Berlin, 1986; p 544.
28. Kokta, B. V.; Raj, R. G.; Daneault, C. *Polym.-Plast. Technol. Eng.* **1989,** *28,* 247.
29. Trotignon, J. P. et al. In *Polymer Composites*; Sedlátek, B.; de Gruyter, Walter, Eds.; Berlin, Germany, 1986; p 191.
30. Menczel, J.; Varga, J. *J. Thermal. Anal.* **1983,** *28,* 161.
31. Folkes, M. J.; Hardwick, S. T. *J. Mater. Sci. Lett.* **1987,** *6,* 656.
32. Shih, C-K. *Polym. Eng. Sci.* **1987,** *27,* 458.
33. Hayashi, K.; Morioka, T.; Toki, S. *J. Appl. Polym. Sci.* **1993,** *48,* 411.
34. Fortelny, I.; Kamenická, D.; Kovar, J. *Angew. Macromol. Chem.* **1988,** *164,* 125.
35. Choudhary, V.; Varma, H. S.; Varma, I. K. *Polymer* **1991,** *32,* 2534.
36. Greco, R.; Mancarella, C.; Martuscelli, E.; Ragosta, G.; Jinghua, Y. *Polymer* **1987,** *28,* 1929.
37. Flaris, V.; Stachurski, Z. H. *J. Appl. Polym. Sci.* **1992,** *45,* 1789.
38. Kolarik, J.; Jancar, J. *Polymer* **1992,** *33,* 4961.

POLY(PROPYLENE CARBONATE)

Witold Kuran
Department of Chemistry
Warsaw University of Technology

Some important properties of aliphatic polycarbonates [poly(alkylene carbonate)s], such as melting point, resistance to heat and to hydrolysis, and ductility, are too low to fulfill conditions required for plastics that might be useful for engineering thermoplastics. But owing to some of their other properties, aliphatic polycarbonates have found use in a wide range of other applications. Frequently, other functional groups such as ether groups are present in the polycarbonate chain, apart from the carbonate groups. Low molecular weight aliphatic polycarbonates and poly(ethercarbonate)s can be used as the diol components for the synthesis of polyurethanes. High molecular weight aliphatic polycarbonates can be used as binders in powder metallurgy and as binders and lubricants in ceramics.

PREPARATION AND PROPERTIES

Aliphatic polycarbonates may be prepared either by the reaction of phosgene with a diol at low temperature or by ester interchange of a lower dialkyl carbonate, cyclic alkylene carbonate, or diphenyl carbonate with a diol at elevated temperature, preferably in the presence of catalysts such as alkali metal, tin, and titanium compounds. The molecular weights of thus-formed low molecular weight polymers are process-dependent and are in the range of 500 and 5000.[1]

Aliphatic polycarbonates can also be prepared by polymerization of cyclic alkylene carbonates (trimethylene carbonate, neopentylene carbonate) using ionic initiators like electrophilic ones[3,4] or nucleophilic ones[5,6] or coordination catalysts.[7]

Anionic and coordination polymerizations of cyclic aliphatic carbonates with a six-membered ring at moderate temperatures can yield pure, high molecular weight polycarbonates. In contrast, cationic polymerizations result in low-molecular-weight polycarbonates containing a few percent of ether linkages.[7,8]

Five-membered cyclic aliphatic carbonates like ethylene carbonate and propylene carbonate hardly undergo a ring opening. The polymerization of such alkylene 1,2-carbonates has been reported to proceed in the presence of metal alkoxides, metal acetylacetonates, and metal alkyls by partial elimination of carbon dioxide, yielding the respective low molecular weight poly(alkylene ether carbonate)s only at high temperature.[9-12]

Researchers have reported forming low molecular weight poly(propylene carbonate) diols by the copolymerization of propylene oxide and carbon dioxide in the presence of diols using coordination catalysts based on polymer-chelated double metal cyanide complexes.[15]

Another method for the preparation of low molecular weight poly(alkylene carbonate)s is the copolymerization of alkylene oxide and carbon dioxide with aluminum porphyrins as coordination catalysts.

High molecular weight poly(propylene carbonate) was found to be formed for the first time from the alternating copolymerization of propylene oxide and carbon dioxide run with a diethylzinc–water system as a coordination catalyst (**Equation 1**).[17,18]

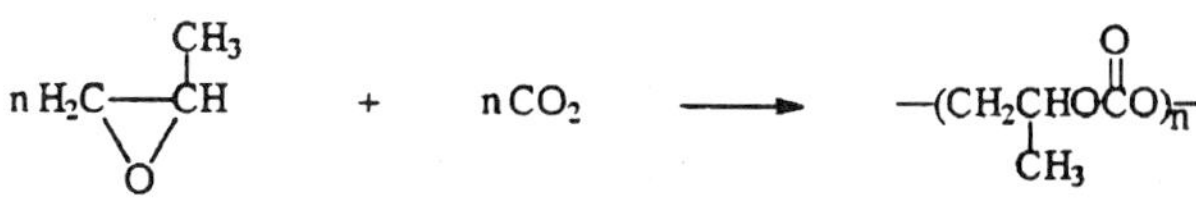

1

According to later studies, binary and ternary catalyst systems containing diprotic (or multiprotic) compound as the partner component with organometallic compound such as diethylzinc (diethylmagnesium or diethylcadmium) formed effective catalysts for the propylene oxide (or other epoxide)–carbon dioxide alternating copolymerization.

Poly(propylene carbonate)s derived from propylene oxide–carbon dioxide copolymerizations catalyzed with zinc-based coordination catalysts are generally polymers of high molecular weight (in the region 50 – 150 $\times 10^3$), but the distribution of their molecular weight is rather broad.[20,21] Polymers yielded by metal acetate and metal carbonate catalysts[24,25] and those yielded by γ-alumina supported catalysts[23] are of lower molecular weights, usually not exceeding 30×10^3.

Properties of aliphatic polycarbonates prepared from the epoxide–carbon dioxide copolymerization are influenced mainly by the properties of the carbonate function of the polymer. Poly(propylene carbonate) contains 47.02 wt % of oxygen. As expected, it shows low combustion heat which is equal to 4,420 kcal/kg.[19] The copolymer dissolves in chlorinated and aromatic hydrocarbons, acetone, tetrahydrofuran, and dioxane, but is insoluble in methanol.

Poly(propylene carbonate) (specific gravity of 1.275 g/cm^3) is characterized by T_g = 30–41°C and displays a glassy-state β-transition at about –85°C. It is an amorphous polymer in nature. It can be molded at 140°C without remarkable decomposition. But it undergoes decomposition by heating at a temperature of 180°C, giving the cyclic carbonate, propylene carbonate, in very high yield.[19,22,26,27]

The thermal stability of poly(propylene carbonate) can be improved to some extent by endcapping. Conversion of the terminal hydroxyl groups of performed propylene oxide–carbon, dioxide alternating copolymers to oxygen–carbon, oxygen–phosphorus and oxygen–sulfur bonds was reported to suppress the thermal decomposition significantly. The endcapping prevents the copolymer depolymerization to the cyclic carbonate via chain unzipping (which involves a nucleophilic hydroxyl group reaction).

It was observed that epoxide–carbon dioxide alternating copolymers are readily hydrolyzed, especially under basic conditions. A possibility of biochemical hydrolysis of such copolymers was also reported.[19,29]

APPLICATIONS

Aliphatic polycarbonates, which can be derived from epoxides and carbon dioxide, especially in the most representative propylene oxide–carbon dioxide alternating copolymer, poly(propylene carbonate), have received considerable attention because of their interesting properties. Their potential applications are wide.

Owing to thermal degradability, high molecular weight poly(propylene carbonate) can be used in powder metallurgy and as a binder and lubricant in ceramics. On temperature degradation, it gives low boiling propylene carbonate, which vaporizes, leaving no sludge, gum, or residue.[19] Some other practical applications such as water and gas barrier adhesives in laminate formation have been attempted.[28] Researchers also describe thin-film plasma polymerized from propylene oxide and carbon dioxide for obtaining permeable membranes.[30] A use of poly(propylene carbonate) in medicine, due to its biodegradability, has also been reported.[19,21]

Use of poly(propylene carbonate) diols and poly(propylene ether carbonate) diols for synthesis of polyurethanes deserves attention.[2,13,14,16,31,32] The reason for this is that thermoplastic poly(carbonate urethane) elastomers exhibit good hydrolytic resistance.

REFERENCES

1. Freitag, D. et al. *Encyclopedia of Polymer Science & Engineering*; 2nd ed.; Wiley-Interscience: New York, NY, 1988; Vol. 11, p 648.
2. Cao, N. et al. *Polymer* **1992**, *33*, 1384.
3. Kricheldorf, H. R.; Dunsing, R.; Serri i Albet, A. *Makromol. Chem.* **1987**, *188*, 2453.
4. Kricheldorf, H. R.; Jenssen, J. *J. Macromol. Sci.-Chem.* **1989**, A26, 631.
5. Kühling, S.; Keul, H.; Höcker, H. *Makromol. Chem., Suppl.* **1988**, *15*, 9.
6. Höcker, H.; Keul, H.; Kuhling, S.; Hovestadt, W.; Muller, A. J. *Makromol. Chem., Macromol. Symp.* **1991**, *44*, 239.
7. Kricheldorf, H. R.; Jenssen, J.; Kreiser-Saunders, I. *Makromol. Chem.* **1991**, *192*, 2391.
8. Brunelle, D. J. *Ring-Opening Polymerization* Hanser: Munich, Germany, 1993; Chapter 11.
9. Soga, K.; Hosoda, S.; Tazuke, Y.; Ikeda, S. *J. Polym. Sci., Polym. Lett. Ed.* **1976**, *14*, 161.
10. Soga, K.; Tazuke, Y.; Hosoda, S.; Ikeda, S. *J. Polym. Sci., Polym. Chem. Ed.* **1977**, *15*, 219.
11. Vogdanis, L.; Heitz, W. *Makromol. Chem., Rapid Commun.* **1986**, *7*, 543.
12. Vogdanis, L.; Martens, B.; Uchtmann, H.; Henzel, F.; Heitz, W. *Makromol. Chem.* **1990**, *191*, 465.
13. Harris, R. F.; McDonald, L. A. *J. Appl. Polym. Sci.* **1989**, *37*, 1491.
14. Kuyper, J.; Lednor, P. W.; Pogany, G. A. (to Shell Internationale Research Maatschappij B. V.) Eur. Patent Appl. 0 222 456, 1987.
15. Chen, L-B., *Makromol. Chem., Macromol. Symp.* **1992**, *59*, 75.
16. Kuran, W.; Listoś, T. Polish Patent 165 607, 1994.
17. Inoue, S.; Koinuma, H.; Tsuruta, T. *J. Polym. Sci., Part B* **1969**, *7*, 287.
18. Inoue, S.; Koinuma, H.; Tsuruta, T. *Polym. J.* **1970**, *2*, 220.
19. Inoue, S. *Chemtech* **1976**, *6*, 588.
20. Rokicki, A.; Kuran, W. *J. Macromol. Sci.-Rev. Macromol. Chem.* **1981**, C21, 135.
21. Inoue, S.; Aida, T. *Handbook of Polymer Synthesis, Pt. A*; Marcel Dekker: New York, 1991; Chapter 8.
22. Rätzsch, M.; Haubold, W. *Faserforsch. Textiltech., Z. Polymerforsch.* **1977**, *28*, 15.
23. Listoś, T.; Kuran, W.; Siwiec, R. *J. Macromol. Sci.-Pure Appl. Chem.* **1995**, A32, 393.
24. Soga, K.; Uenishi, K.; Hosoda, S.; Ikeda, S. *Makromol. Chem.* **1977**, *178*, 893.
25. Soga, K.; Uenishi, K.; Ikeda, S. *J. Polym. Sci., Polym. Chem. Ed.* **1979**, *17*, 415.
26. Udipi, K.; Gillham, J. K. *J. Appl. Polym. Sci.* **1974**, *18*, 1575.
27. Inoue, S.; Tsuruta, T.; Takada, T.; Miyazaki, N.; Kambe, M.; Takaoka, T. *Appl. Polym. Symp.* **1975**, *26*, 257.
28. Dixon, D. D.; Ford, M. E.; Mantell, G. J. *J. Polym. Sci., Polym. Lett. Ed.* **1980**, *18*, 131.
29. Nakano, M. *J. Soc. Synth. Org. Chem. Jpn* **1984**, *42*, 665.
30. Sugiyama, K.; Shiraishi, K.; Ihara, T.; Kiboku, M. *Polym. J.* **1991**, *23*, 1287.
31. Stevens, H. C. (*to Pittsburgh Plate Glass Co.*), U.S. Patent 3 248 416, 1966.
32. Myers, M. O. (*to Down Chemical Co.*), U.S. Patent 4 931 486, 1990.

Poly(propylene glycol)s

See: *Cyclodextrin-Polymer Inclusion Complexes*
Poly(propylene oxide)

POLYPROPYLENE, HEMIISOTACTIC

Piero Sozzani* and Giuseppe Di Silvestro
Dipartimento di Chimica Organica e Industriale
Università di Milano

Hemitactic polymers (from the Greek words: *hemi* meaning half, and *tacticity* meaning order) are characterized by a microstructure containing a series of stereogenic nuclei among the polymer chains, for example, the odd series, that shows a constant and ordered configuration, whereas the configuration of the remaining nuclei is at random. From the symmetry elements defining the relationships among the stereogenic nuclei in the ordered series follows further classifications, in particular hemi*iso*tactic and hemi*syndio*tactic polymers are defined, being derived by an *iso*tactic or a *syndio*tactic configuration of two following carbons in the ordered series.

The concept of hemitacticity was first proposed in 1982 and can be extended to virtually any regular succession of configurations when regularly interleaved to random configurations.[1,2]

PREPARATION

Any synthetic process that is able to produce along the polymer chain two independent series of stereogenic carbon atoms may in principle produce hemitactic structures. In the cases studied up to now, that is, of hemiisotactic polypropylenes (hit-PP), the polymers were synthesized by two routes, one of which is direct (one-step process), and the other one is indirect (two-step process).

Two-Step Process

Hemiisotactic polypropylene (hit-PP) was synthesized for the first time by combining inclusion polymerization and hydrogenation of the product.[1,5] Highly isotactic 1,4-*trans*-poly(2-methyl-1,3-pentadiene) (polymer I) is produced in a first step, and polymer II is obtained by a non-specific reduction of polymer I by diimide in a second step (**Scheme I**).[6,7]

One-Step Process

Soluble metallocene catalysts for olefin polymerization make accessible a direct synthetic way to hit-polypropylene.[8]

The first metallocene catalyst able to produce hit-PP is prepared in such a way that one catalytic site is highly stereospecific and the other one is non-specific.[9-11] The alternate insertion reaction on the two catalytic sites generates the hemiisotactic polymer.

PROPERTIES

The absence of a semicrystalline structure limits possible mechanical applications. No attempt to crystallize was successful, and a typical hit-PP shows similar properties as an atactic polypropylene. However, hit-PPs containing a certain degree of order in the disorder series of carbon atoms may contain stereoblocks, which are able to crystallize, thus inducing thermoplastic properties.

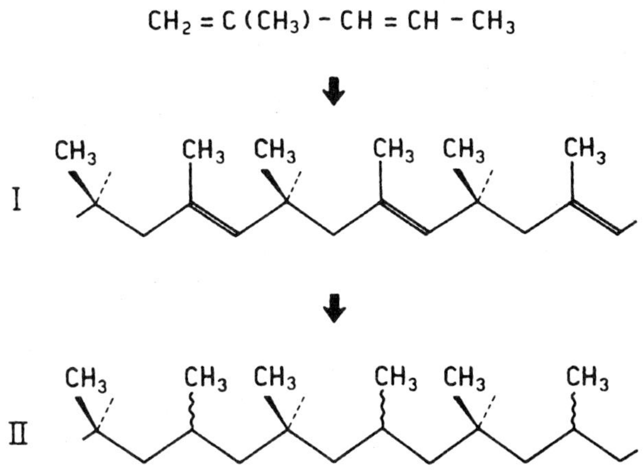

SCHEME I. Two-step process for obtaining hemi*iso*tactic polypropylene.

Nuclear magnetic resonance spectroscopy is a method of choice for a detailed description of polymer chains, being sensitive to stereochemistry distribution of the sequences along polymer chains.[3] The sensitivity is remarkable in polypropylene, and depends on the distribution of conformations in solution; gauche interaction (about 6 ppm upfield in hydrocarbons) is the driving force for the chemical shift and a conformational analysis provides assignments of the resonances with a certain degree of accuracy.[12] The selection rules effective in hit-PP allow the analyst to identify single resonances at a higher level of resolution.[4,13]

ACKNOWLEDGMENTS

This work is dedicated to the memory of the late Professor Farina, who conceived the idea of hemitactic polymers. Thanks are due to the Italian Ministry of Education, which supported the current work.

REFERENCES

1. Farina, M.; Di Silvestro, G.; Sozzani, P. *Macromolecules* **1982,** *15,* 1451.
2. Farina, M.; Di Silvestro, G.; Sozzani, P.; Savarè, B. *Macromolecules* **1985,** *18,* 923.
3. Bovey, F. A. *Chain Structure and Conformation of Macromolecules;* Academic: New York, 1982.
4. Sozzani, P.; Oliva, C. *J. Magn. Reson.* **1985,** *63,* 115.
5. Di Silvestro, G.; Sozzani, P.; Savarè, B.; Farina, M. *Macromolecules* **1985,** *18,* 928.
6. Farina, M. *Inclusion Compounds*; Atwood, J. L. et al., Eds., Pergamon: London, 1984, Vol. II.
7. Di Silvestro, G.; Sozzani, P. Polymerization in clathrates; *Comprehensive in Polymer Science*; Eastmond, G. C., Ed.; Pergamon: London, 1988.
8. Porri, L., Ed.; Papers presented at the International Symposium on Stereochemical Polymerization; *Makromol. Chem., Macromol. Symp.* **1995,** 53.

*Author to whom correspondence should be addressed.

9. Ewen, J. A. *J.A.C.S.* **1984,** *106,* 6355.
10. Ewen, J. A.; Jones, R. L.; Razavi, A. **1988,** *110,* 6255.
11. Ewen, J. A.; Elder, M. J.; Jones, R. L.; Haspelagh, L.; Atwood, J. L.; Bott, S. G.; Robinson, K. *Makromol. Chem., Macromol. Symp.* **1991,** *48/49,* 253.
12. Tonelli, A. E. *NMR Spectroscopy and Polymer Microstructure: The Conformational Connection*; VCH: Deerfield Beach, FL, 1989.
13. Farina, M.; Di Silvestro, G.; Sozzani, P. *Progress in Polymer Sciences* **1991,** *16,* 219.

POLYPROPYLENE, ISOTACTIC (Polymorphism)

Paul J. Phillips and Khaled Mezghani
Department of Materials Science and Engineering
University of Tennessee

Isotactic polypropylene is an important polymer not only commercially, but also scientifically, because of its different morphological behavior.

On the crystal lattice level, iPP exhibits three different morphological forms, α, β. and γ, distinguished by the arrangement of the chains.[2-8] Another form of iPP with a degree of order between crystalline and amorphous phases was first reported by Natta,[1] who named it the "smectic" form, and has been described in some detail by Corradini and co-workers.[9]

This article treats essential results from X-ray and depolarized light microscopy of the supermolecular structure of iPP in four parts: The first three sections concern the crystalline forms, namely α-, β-, and γ-form. The last part of the analysis will be devoted to the mesomorphic form, which is called *smectic*.

α-FORM

The predominant crystal structure of pure (without nucleating agent) isotactic polypropylene, at atmospheric pressure, is the monoclinic α structure. In **1960,** Natta and Corradini[1] calculated the unit cell parameters of the monoclinic α-iPP, which contains four three-fold helical chains, to be: a = 6.65 Å, b = 20.96 Å, and c = 6.50 Å, with α = γ = 90 °, and β = 99.3 °. The space group has been assigned C2/c and Cc for the statistically ordered arrangement of up and down stems in the unit cell.

The spherulitic morphology observed in melt-crystallized iPP was first reported by Padden and Keith who classified the spherulites with respect to their birefringence.[11] The α form of iPP exhibits three types of spherulites: Positive birefringence spherulites (α_I) can be developed at an isothermal crystallization temperature below 132°C. Negative birefringence spherulites (α_{II}) exist at a temperature greater than 138°C. However, the most common spherulites are of mixed birefringence type (α_{III} or α_m); they possess no distinct Maltese cross and are revealed as radiating arrays of intermingled areas of positive and negative birefringence. This type of spherulite has been observed at any temperature up to 150°C. The presence or absence of one type depends not only on the crystallization temperature but also on the film thickness.

Values of $T_m°$ reported in the literature fall into two groups, one around 186°C and the other around 210°C.[12,14-22] Krigbaum, and Miller, have determined $T_m°$ to be 186 ± 2°C, whereas Monnasse and Fatou have extrapolated $T_m°$ to 208°C.[12,16]

It is believed that the higher value, 210°C, is inaccurate since it is caused by lamellae thickening phenomenon.

β-FORM

The β-form of isotactic polypropylene was first identified in 1959 by Keith et al.[23] In the crystallization of conventional iPP grades, a small amount of β-form occurs sporadically at high supercoolings (T_c < 130°C) or in quenched samples. However, under special crystallization conditions, when the temperature gradient method is used[24-26] or when selective β-nucleating agents are present,[4,27-41] a high amount of β-form is produced. For example, quinacridone dyes or triphenodithiazine are β-nucleating agents widely known to produce even 100% of the β-form under the appropriate thermal conditions of crystallization.[38,40]

When crystallized from the melt the β form of iPP displays a strong negative birefringence as commonly observed in polymer spherulites.

Usually, the growth rate of the β form exceeds that of the α form, while its melting point is in the 145–150°C interval, much lower than that of the α form. Similar to the α form, many values of the equilibrium melting point have been reported; it varies between 170 and 200.[26,33,40,42,43] Recently, Mezghani and Phillips[51] reported the value of 174.4°C for $T_m°$ of the β form using the same procedure for determining the most accurate value $T_m°$ of the α form.

γ-FORM

The γ form of polypropylene was first noted during the 1960s and was generated by several methods.[2,3,10,29,44-49] It was largely produced by crystallization at elevated pressures. However, under specific conditions, low molecular weight fractions (M_w < 6000) of iPP can be used to generate γ form or even single crystals.

When the pressure is varied, only the α and γ forms are observed.[44,50] As the crystallization pressure increases the γ form starts to coexist with the α form until it becomes dominant at 2 kbar.[50] Furthermore, it appears from experiment that the lower the supercooling, the higher the amount of γ form produced at a specific pressure.[52,53]

As for the β form, all attempts to obtain oriented fibers of γ phase produced only a transformation into the oriented α crystals.

On the spherulitic level, pure γ form shows negative birefringence.

The melting point of the γ form is mostly reported in the range from 125 to 150°C for low molecular weight samples. In the case of pressure-crystallized samples with high molecular weight iPP, the melting occurs above 150°C. The results of melting at high pressures show that the equilibrium melting point of iPP is linearly proportional to the pressure. The equilibrium melting point of the γ form at atmospheric pressure is the intercept of the line and has a value of 187.6°C.[53]

SMECTIC

The smectic form is an intermediate crystalline order. This mesomorphic form was first named by Natta et al. and was later assigned different names.[1] Miller suggested that the order existing in this form might be of the type described in general terms by Hosemann as "paracrystalline."[54,55] This term was also used by Zannetti et al.[56,57] Wunderlich and Grebowicz proposed the term "condis crystal," which means conformational disordered crystal.[58]

The preparation of this form is very simple: quenching thin sheets of iPP from the melt into ice water. The smectic form is transformable into the α form by annealing at temperatures higher than 70°C.

The density of the smectic form has been reported by Natta et al. to be 0.88 g/cm³, which is higher than that of atactic PP (0.85 g/cm³).

REFERENCES

1. Natta, G.; Corradini, P. *Suppl. Nuovo Cimento* **1960,** *15,* 40.
2. Addink, E. J.; Beintema, J. *Polymer* **1961,** *2,* 185.
3. Turner-Jones, A.; Aizlewood, J. M.; Beckett, D. R. *Makromol. Chem.* **1964,** *75,* 134.
4. Binsbergen, F.L.; Lange, B. G. M. *Polymer* **1968,** *9,* 23.
5. Norton, D. R.; Keller, A. *Polymer* **1985,** *26,* 704.
6. Awaya, H. *Polymer* **1988,** *29,* 591.
7. Brückner, S.; Meille, S. V.; Petraccone, V.; Pirozzi, B. *Progr. Polym. Sci.* **1991,** *16,* 361.
8. Varga, J. *J. Mater. Sci.* **1992,** *27,* 2557.
9. Corradini, P.; Petraccone, C.; Dé Rosa, C.; Guerra, G. *Macromolecules* **1986,** *19,* 2699.
10. Kojima, M. *J. Polym. Sci.* **1968,** A2, *6,* 1255.
11. Padden, F. J.; Keith, H. D. *J. Appl. Phys.* **1959,** *30,* 1479.
12. Monasse, B.; Handin, J. M. *Colloid and Polym. Sci.* **1985,** *263,* 822.
13. Mezghani, K.; Campbell, R. A.; Phillips, P. J. *Macromolecules* **1994,** *27,* 997.
14. Krigbaum, W. R.; Vematsu, I. *J. Polym. Sci., Polym. Phys. Ed.* **1965,** *3,* 767.
15. Miller, R. L.; Seeley, E. G. *J. Polym. Sci., Polym. Phys. Ed.* **1982,** *20,* 2297.
16. Fatou, J. G. *Eur. Polym. J.* **1971,** *7,* 1057.
17. Samuels, R. J. *J. Polym. Sci., Polym. Phys. Ed.* **1975,** *13,* 1417.
18. Martuscelli, E. et al. *Polymer* **1982,** *23,* 229.
19. Martuscelli, E. et al. *Polymer* **1983,** *24,* 693.
20. Bu, H. S.; Cheng, S. Z. D.; Wunderlich, B. *Makromol. Chem.* **1988,** *9,* 76.
21. Cheng, S. Z. D.; Janimak, J. J.; Zhang, A. *Macromolecules* **1990,** *23,* 298.
22. Campbell, R. A. Ph.D. Thesis, University of Tennessee, 1991.
23. Keith, H. D.; Padden, F. J., Jr.; Walter, N. M.; Wyckoff, H. W. *J. Appl. Phys.* **1959,** *30,* 1485.
24. Crissman, J. M. *J. Polym. Sci.* **1969,** A2, 398.
25. Fujiwara, Y. *Colloid Polym. Sci.* **1976,** *253,* 273.
26. Lovinger, A. J.; Chua, J. O.; Gryte, C. C. *J. Polym. Sci., Polym. Phys. Ed.* **1977,** *15,* 641.
27. Leugering, H. J. *Makromol. Chem.* **1967,** *109,* 204.
28. Turner-Jones, A.; Cobbold, A. J. *Polym. Lett.* **1968,** *6,* 539.
29. Morrow, D. R. *J. Macromol. Sci., Phys.* **1969,** B3, 53.
30. Duswalt, A., *Am. Chem. Soc. Org. Coat.* **1970,** *30,* 93.
31. Ullmann, W.; Wendoeff, J. H. *Prog. Coll. Polym. Sci.* **1979,** *66,* 25.
32. Forgacs, P.; Totochko, B. P.; Sheromov, M. A. *Polym. Bull.* **1981,** *6,* 127.
33. Shi, G. Y.; Huang, B.; Zhang, J. Y. *Makromol. Chem., Rapid Commun.* **1984,** *5,* 573.
34. Garbarczyk, J.; Paukszta, D. *Coll. Polym. Sci.* **1985,** *263,* 985.
35. Jacoby, P.; Bersted, H. B.; Kissel, W. J.; Smith, C. E. *J. Polym. Sci.* **1986,** B24, 461.
36. Zhou, G. E.; He, Z. Q.; Yu, J. M.; Han, Z. W. *Makromol. Chem.* **1986,** *187,* 633.
37. Shi, G. Y.; Huang, B.; Coa, Y. H.; He, H. Q.; Han, Z. W. *Polymer* **1986,** *187,* 643.
38. Varga, J. *J. Therm. Anal.* **1986,** *31,* 165.
39. Varga, J.; Schulek-Toth, F. *Makromol. Chem., Macromol. Symp.* **1986,** *5,* 213.
40. Varga, J. *J. Therm. Anal.* **1989,** *35,* 1891.
41. Varga, J. *J. Mater. Sci.* **1992,** *27,* 2557.
42. Samuels, R. J.; Yee, R. J. *J. Polym. Sci., Polym. Phys. Ed.* **1972,** *10,* 385.
43. Dé Rosa, C.; Guerra, G.; Petraccone, V.; Tuzi, A. *Polymer* **1987,** *28,* 143.
44. Kardos, J. L.; Christiansen, J. L.; Baer, E. *J. Polym. Sci.* **1966,** (A2)4, 777.
45. Morrow, D. R.; Newman, B. A. *J. Appl. Phys.* **1968,** *39,* 4944.
46. Pae, K. D.; Morrow, D. R.; Sauer, J. A. *Nature* **1966,** *211,* 514.
47. Awaya, H. *J. Polym. Sci., Polym. Lett.* **1966,** *4,* 127.
48. Pae, K. D. *J. Polym. Sci.* **1968,** *6,* 657.
49. Sauer, J.A.; Pae, K. D. *J. Appl. Phys.* **1968,** *39,* 4959.
50. Campbell, R. A.; Phillips, P. J. *Polymer* **1993,** *34,* 4809.
51. Mezghani, K.; Phillips, P. J. *Polymer* **1995,** *36,* 2407.
52. Mezghani, K.; Phillips, P. J., submitted for publication in *ANTEC.*
53. Mezghani, K.; Phillips, P. J., unpublished results.
54. Miller, R. L. *Polymer* **1960,** *1,* 135.
55. Hosemann, R. *Acta Crystall.* **1951,** *4,* 520.
56. Zannetti, R.; Celotti, G.; Fichera, A.; Francesconi, R. *Makromol. Chem.* **1969,** *128,* 137.
57. Zannetti, R.; Celotti, G.; Armigliato, S. *Eur. Polym. J.* **1970,** *6,* 879.
58. Wunderlich, B.; Grebowicz, J. *J. Adv. Polym. Sci.* **1984,** *60,* 1.

POLYPROPYLENE, ISOTACTIC (Supermolecular Structure)

József Varga
Department of Plastics and Rubber Technology
Technical University of Budapest

Aggregates of chain-folded fibrillar or lamellar primary crystallites with definite geometrical arrangements are designated supermolecular structures (SMS). Several reviews are available on this topic dealing with polymers in general or with isotactic polypropylene in particular.[1-5] During the crystallization of isotactic propylene (iPP), a polymorphic material[6] with several modifications–monoclinic (α), hexagonal (β) and triclinic (γ)–different supermolecular structures may develop, which imply crystallites of the α and β modification.[4] No SMS of the γ modification has been found yet.

The types and structural features of formed SMS are markedly influenced by the thermal conditions of crystallization, by mechanical stresses to the crystallizing system, and by the presence of extraneous materials (heterogeneous ingredients or solvents). In a quiescent melt iPP crystallizes into spherulitic structures, whereas under the mechanical load to the melt, cylindrites are formed. In the presence of extraneous materials with nucleating ability, characteristic transcrystalline structures may develop. Occasional appearance of hedrites of hexagonal shape can be detected during the crystallization of the β modification of iPP (β-iPP) from the melt.[4] The solution crystallization of iPP leads to the formation of quadrites and dendrites.[7-10] SMS of iPP may be readily visualized by polarized light microscopy (PLM).[4,11] The constituents of SMS, the primary crystallites, may be observed by transmission electronmicroscopy (TEM).[3,12,13] Recently, atomic force microscopy (AFM) and scanning electron microscopy (SEM) have been used for the study of morphology of iPP on subspherulitic and molecular levels.[14-17]

REFERENCES

1. Geil, P. H. *Polymer Single Crystals* Interscience: New York, NY, 1963.
2. Wunderlich, B. *Macromolecular Physics*, Academic: New York, NY, 1973; Vol. I.
3. Norton, D. R.; Keller, A. *Polymer* **1985,** *26,* 704.
4. Varga, J. *J. Mater. Sci.* **1992,** *27,* 2557.
5. Varga, J. In *Polypropylene: Structure, Blends and Composites*; Karger-Kocsis, J., Ed.; Chapman and Hall: London, U.K., 1994; Vol. 1, Chapter 3, pp 56-115.
6. Brückner, S.; Meille, S. V. et al. *Progr. Polym. Sci.* **1991,** *16,* 361.
7. Khoury, F. J. *J. Res. Natl. Bur. Stand.* **1966,** *470,* 29.
8. Sauer, J. A.; Morrow, D. R.; Richardson, G. C. *J. Appl. Phys.* **1965,** *30,* 3017.
9. Kargin, V. A.; Gorina, I. I. *Vysokomol. Soedin.* **1965,** *7,* 220.
10. Kojima, M. *J. Polym. Sci. A-2* **1968,** *5,* 597.
11. Haudin, J. M. In *Optical Properties of Polymers;* Meeten, G. H., Ed.; Elsevier: London, U.K., 1986; Chapter 4, p 167.
12. Olley, R. H. *Sci. Prog. Oxford* **1986,** *70,* 17.
13. Vaughan, A. S. *Sci. Prog. Oxford* **1992,** *76,* 1.
14. Lotz, B.; Wittman, J. C. et al. *Polym. Bull.* **1990,** *26,* 209.
15. Schönherr, H. Y.; Snetivy, D.; Vancso, G. J. *Polym. Bull.* **1993,** *30,* 567.
16. Snetivy, D.; Guillet, J. E.; Vancso, G. J. *Polymer* **1993,** *34,* 429.
17. Aboulfaraj, M.; Ulrich, B. et al. *Polymer* **1993,** *34,* 4817.

POLY(PROPYLENE OXIDE)

Witold Kuran
Department of Chemistry
Warsaw University of Technology

Polymerization of propylene oxide, having a highly strained three-membered ring in its molecule, takes place with a ring cleavage involving the C(α)–O or C(β)–O bond (**Equation 1**).

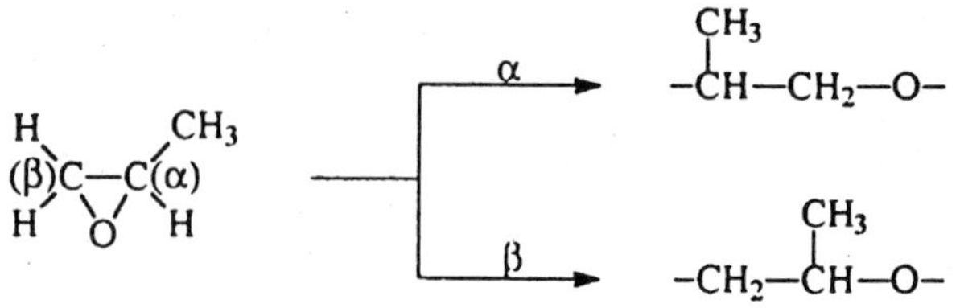

1

Therefore, propylene oxide polymers with head-to-tail, head-to-head and tail-to-tail structures along their chains can be formed. The regioselectivity of the propylene oxide polymerization is strongly dependent on the nature of the initiators.[1] Particularly in anionic and coordination polymerization, the ring-opening predominantly occurs via a C(β)–O bond cleavage, and polymers with mostly regular head-to-tail linkages of monomeric units are formed. But in the cationic polymerization both the C(α)–O and C(β)–O bonds undergo a cleavage, statistically forming irregular polymer chains with head-to-head and tail-to-tail linkages apart from head-to-tail linkages. Regardless of the kind of initiator or catalyst, polymerization of propylene oxide proceeds with inversion of the configuration at the carbon atom of the oxirane ring where cleaved.[2-6]

PREPARATION AND PROPERTIES

Low molecular weight propylene oxide polymers are prepared by polymerizing propylene oxide with active hydrogen compounds as starters using alkali metal alcoholates or alkali metal hydroxides as initiators in most cases. The polymerization of propylene oxide with an alkali applied as initiator in propylene glycol solution or in the presence of water yields a polymer containing hydroxyl end groups, that is, HO-$[CH_2CH(CH_3)O]_n$-H, which is referred to as poly(propylene glycol) {poly[oxy (methyl-1,2-ethanediyl)]}, CAS [25322-694].[8,9] Poly(propylene glycol)s are colorless viscous liquids with the following properties: molecular weight: 150–4000; specific gravity: 0.99–1.02 g/cm^3; refractive index (25°C); 1.443–1.451; viscosity (25°C): 60–1200 mPa·s (milliPascal second) (cP); glass transition temperature: –60 to –75°C. Commercial poly(propylene glycol)s are of rather low molecular weight distribution (M_w/M_n = 1.11 – 1.25). They are soluble in ketones, chlorohydrocarbons, and aromatic hydrocarbons. The solubility in water (which is due to the presence of polar hydroxyl end groups in the polymer) decreases drastically with increasing molecular weight. Poly(propylene glycol)s with molecular weight of about 800–900 are essentially insoluble in water; however, their solubility in hydrocarbons increases with increased molecular weight. When oxypropylated derivatives are obtained by the propylene oxide polymerization with polyfunctional protic compounds as starters, they are commonly named polyether polyols.[7]

Besides from the homopolymers, block copolymers and random copolymers of propylene oxide with ethylene oxide have achieved commercial importance.

Poly(propylene oxide)s are propylene oxide polymers obtained by anionic polymerization without the presence of compounds, containing the hydroxyl group, as additives. They have considerable amounts of unsaturated chain-end groups (allyl and propenyl groups) from chain transfer reactions onto the monomer.

Particular attention has been devoted to living processes. Inoue and others found the homogeneous catalysts like metalloporphyrins of aluminum and zinc, such as (5,10,15,20-tetraphenylporphinato)aluminum chloride [(Tpp)-Al-X] and (5,10,15,20-tetraphenyl-21-methylporphinato)zinc methoxide, to fulfill the desired conditions for control of the molecular weight and narrow molecular weight distribution, and they achieved the living polymerization of propylene oxide.[11-13]

High molecular weight propylene oxide polymers such as polymethyloxirane (CAS 9003-09-2); poly(propylene oxide); and polyoxypropylene are prepared using coordination catalysts with multinuclear metallic species which comprise Group 2 and 3 metals such as aluminum, zinc, and cadmium as well as transition metals such as iron.[1,2,18-30]

Isotactic poly(propylene oxide) (CAS 27082-59-3), crystallizes in orthorhombic cells involving equal number of macromolecules of the opposite configuration.[31] Specific gravity of the crystalline polymer is 1.157 g/cm^3. The viscosity-average molecular weight of poly(propylene oxide) can be calculated according to **Equation 2** (viscosity measured in benzene at 25°C).[32]

$$[\eta] = 1.12 \times 10^{-4} (M)^{0.77} \quad \mathbf{2}$$

Copolymerization of propylene oxide with a small amount (5–35%) of an unsaturated epoxide, such as allyl glycidyl ether, gives rise to poly(propylene oxide-*co*-allyl glycidyl ether), (CAS 25104-27-2), a polyether elastomer with double bonds in pendant substituents. It can be vulcanized with sulfur yielding a specialty elastomer under the registered trademark Parel 58 (Hercules Powder Co.) and Dynagel XP-139 (General Tyre & Rubber Co.).[22,33-35] This polymer, obtained with trialkylaluminum–water–acetylacetone catalyst, is amorphous and contains only ca. 70% head-to-tail linkages. It exhibits excellent low temperature and dynamic properties as well as good ozone and heat resistance similar to that of natural rubber, combined with a better resistance to water and alcohol.

Another commercially important polyether elastomer with excellent low temperature flexibility and low gas permeability is the propylene oxide–epichlorohydrin–allyl glycidyl ether terpolymer [poly(propylene oxide-*co*-epichlorohydrin-*co*-allyl glycidyl ether)] (CAS 25213-15-4). This polyether elastomer is known by the trade name Zeospan.[36]

APPLICATIONS

Polyoxypropylene diols and higher polyols are used for the preparation of polyurethanes and polyurethane elastomers. Most urethane polyols in use are copolymers of propylene oxide and up to 20% ethylene oxide. Important factors in these applications are functionality, concentration, and reactivity of the hydroxyl end groups: In polyether polyols used in the preparation of polyurethanes, a primary hydroxyl group reacts much more rapidly with an isocyanate than a secondary hydroxyl group does.[37]

Block copolymers of propylene oxide and ethylene oxide with a proper balance between hydrophobic oxypropylene units and hydrophilic oxyethylene units are applied as non-ionic surfactants.

Poly(propylene glycol) and random copolymers of propylene oxide and ethylene oxide are used as so-called functional fluids. Their characteristic properties arising from the nature of the polyether backbone suit them to a large variety of applications such as hydraulic fluids, quenchants, lubricants, and surfactants.

Commercially available high molecular weight polymers of propylene oxide, the vulcanizable copolymers with allyl glycidyl ether and the terpolymers with epichlorohydrin and allyl glycidyl ether, although not outstanding in any particular respect, have good properties resembling those of natural rubber. They are applied in the automotive industry as engine mounts.[35]

Polyethers with terminal functionalities, prepared by propylene oxide polymerization (and copolymerization) with coordination catalysts containing mononuclear aluminum species, such as metalloporphyrins,[14,15,38] aluminum Schiff base complexes,[16] and aluminum calixarene derivatives appear to have potential utility.[17]

Crystalline propylene oxide polymers, consisting of isotactic poly(propylene oxide), have no direct utility so far. This is connected with relatively low melting point of such polymers (not higher than 75°C).[39]

REFERENCES

1. Osgan, M.; Price, C. C. *J. Polym. Sci.* **1959,** *34,* 153.
2. Vandenberg, E. J. *J. Polym. Sci., Part A-1* **1969,** *7,* 525.
3. Price, C. C.; Akkapeddi, M. K. et al. *J. Am. Chem. Soc.* **1972,** *94,* 3964.
4. Oguni, N.; Maeda, S.; Tani, H. *Macromolecules* **1973,** *6,* 459.
5. Hasebe, Y.; Tsuruta, T. *Makromol. Chem.* **1987,** *188,* 1403.
6. Watanabe, Y.; Yasuda, T. et al. *Macromolecules* **1992,** *25,* 1396.
7. Newton, R. A. In *Kirk-Othmer Encyclopedia of Chemical Technology*, 3rd ed.; Wiley-Interscience: New York, NY, 1982; Vol. 18, p 633.
8. Weimer, D. R.; Cooper, D. E. *J. Am. Oil Chem. Soc.* **1966,** *43,* 440.
9. Christianson, L. R.; Dheming, M.; Ochoa, E. *J. Cell. Plast.* **1977,** *13,* 111.
10. Biedroń, T.; Kubisa, P. et al. *Polym. Int.* **1995,** *36,* 73.
11. Aida, T.; Inoue, S. *Macromolecules* **1981,** *14,* 1166.
12. Aida, T.; Mizuta, R. et al. *Makromol. Chem.* **1981,** *182,* 1073.
13. Inoue, S. *Catalysis in Polymer Synthesis*; ACS Symposium Series, 496; American Chemical Society: Washington, DC, 1992; Chapter 15.
14. Inoue, S.; Aida, T. *Ring-Opening Polymerization*; Hanser: Munich, Germany, 1993; Chapter 6.
15. Vincens, V.; LeBorgne, A.; Spassky, N. *Makromol. Chem., Rapid Commun.* **1989,** *10,* 623.
16. Vincens, V.; LeBorgne, A.; Spassky, N. *Catalysts in Polymer Synthesis*; ACS Symposium Series 496; American Chemical Society: Washington, DC, 1992; Chapter 16.
17. Kuran, W.; Listoś, T.; Abramczyk, M., unpublished results.
18. Pruitt, M. E.; Baggett, J. M. (Dow Chemical Company) U.S. Patent 2 706 181, 1955.
19. Furukawa, J.; Tsuruta, T. et al. *Makromol. Chem.* **1959,** *32,* 90.
20. Colclough, R. O.; Gee, G.; Jagger, A. H. *J. Polym. Sci.* **1960,** *48,* 273.
21. Vandenberg, E. J. *J. Polym. Sci.* **1960,** *47,* 485.

22. Vandenberg, E. J. *Catalysts in Polymer Synthesis*; ACS Symposium Series 496; American Chemical Society: Washington, DC, 1992; Chapter 1.
23. Machon, J. P.; Sigwalt, P. *Compt. Rend.* **1965,** *260,* 549.
24. Osgan, M.; Teyssié, P. *J. Polym. Sci., Part B* **1967,** *5,* 789.
25. Osgan, M.; Teyssié, P. *J. Polym. Sci., Part B* **1970,** *8,* 319.
26. Tani, H. *Adv. Polym. Sci.* **1973,** *11,* 57.
27. Kuran, W.; Rokicki, A.; Pieńkowski, J. *J. Polym. Sci., Polym. Chem. Ed.* **1979,** *17,* 1235.
28. Inoue, S. *Progr. Polym. Sci.* **1988,** *13,* 63.
29. Tsuruta, T.; Kawakami, Y. *Comprehensive Polymer Science*; Pergamon: Oxford, U.K., 1989; Vol. 3, Chapter 33.
30. Kuran, W.; Listoś, T. *Macromol. Chem. Phys.* **1994,** *195,* 401.
31. Stanley, E.; Litt, M. *J. Polym. Sci.* **1960,** *43,* 453.
32. Allen, G.; Booth, C.; Jones, M. N. *Polymer* **1964,** *5,* 195.
33. Vandenberg, E. J.; In *Kirk-Othmer Encyclopedia of Chemical Technology*, 3rd ed.; Wiley-Interscience; New York, NY, 1979; Vol. 8, p 568.
34. Meissner, B.; Schatz, M.; Brajko, V. *Elastomers and Rubber Compounding Materials*; Elsevier: Amsterdam, The Netherlands, 1989; Chapter 4, p 278.
35. Owens, K.; Kyllingstad, V. L. In *Kirk-Othmer Encyclopedia of Chemical Technology*, 4th ed.; Wiley-Interscience: New York, NY, 1993; Vol. 8, p 1079.
36. Maeda, A.; Inagami, N. *Rubber & Plastics News* **1987,** *42,* 1.
37. Hanna, J. G.; Siggia, S. *J. Polym. Sci.* **1962,** *56,* 297.
38. Jun, C. L.; LeBorgne, A.; Spassky, N. *J. Polym. Sci., Polym. Symp.* **1986,** *74,* 31.
39. Booth, C.; Devoy, C. J.; Gee, G. *Polymer* **1971,** *12,* 327.

POLY(2,6-PYRIDINEDIYLVINYLENE)

M. Judith Percino* and Víctor M. Chapela
Centro de Química
Universidad Autónoma de Puebla

The poly(2,6-pyridinediylvinylene) or poly(pyridine-2,6-diylvinylene) (CAS 26023-00-7) polymer shown on **Figure 1** is a polymer that contains conjugated double bonds in its structure, similar to those in polyacetylene polymers, and also has a nitrogen atom at the pyridine cycle.

Much of the research about this polymer centers on its electrical conductivity properties.[1,2] There are studies where charge-transfer complexes with this polymer show an increment on the polymer electrical conductivity.[3,4] The polymer also shows extra thermal stability. It can be considered a weak base resin because of the presence of pyridine as functional groups. And it can be reacted with hydrochloric acid to form its hydrochloride salt.[5]

*Author to whom correspondence should be addressed.

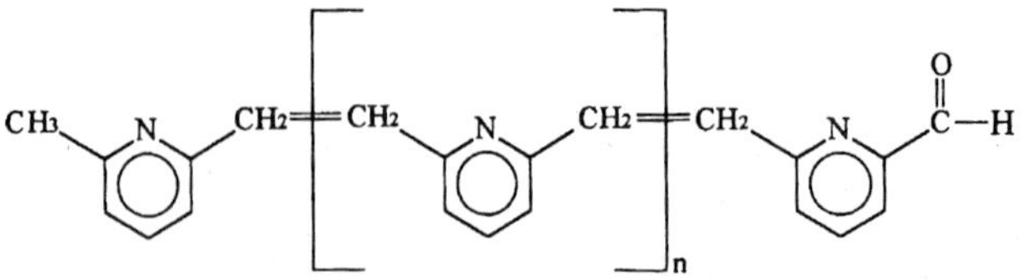

FIGURE 1. Poly(2,6-pyridinediylvinylene).

There have been studies with different polymers such as polystyrylpyridine (PSP) which can be considered similar to each other by virtue of their analogous properties. PSP can be obtained from dimethyl pyridine derivatives and aromatic dialdehydes presenting excellent thermal stability and potential application as ion exchange resins.

PREPARATION

Poly(2,6-pyridinediylvinylene) is synthesized by polycondensation of 6-methyl-2-pyridinecarboxaldehyde which is commercially available, or by oxidation of 2,6-lutidine with either V_2O_5–MoO_3 or V_2O_5–MoO_3–P_2O_5.[6]

PROPERTIES

The polymer is a black powder, insoluble in most solvents (e.g., CCl_4, CCl_3H, CH_3OCH_3, CH_3CH_2OH, Me_2SO, CH_3COOH, HCOOH, CF_3COOH, and concentrated HCl) and with no observable melting point up to 460°C.[5] X-ray spectra show that the polymers are amorphous.

APPLICATIONS

Principal application research efforts for poly(2,6-pyridinediylvinylene) seek to take advantage of its principal properties: the electrical properties for semiconductors, the thermal stability for high temperature applications, and the ion-exchange property to replace less stable resins.

Katon reported on charge-transfer complexes of polymers having pyridine rings with low molecular weight acceptors.[7] Several researchers prepared quaternary salts of the polymer searching for semiconductor behavior.[1-4] Using the same approach, Mizoguchi et al. reported compounds of poly(2-vinylpyridine) or poly(4-vinylpyridine) with tetracyanoquinodimethane (TCNQ), $SbCl_5$, or I_5.[8] Selected changes of resistivities of the polymer and its salts are shown in **Table 1**.

TABLE 1. Selected Changes of Resistivities

	With	Salt name	Resistivity	Reference
Polymer	—	—	10^7–10^9 Ω-cm	4
Polymer	Me_2SO_4	Poly(4-methylpyridinium-2,6-diylvinylene Me sulfate)	1 . 10^5 Ω-cm	4
Polymer -Me_2SO_4	Li-7,7,8,8TCNQ		1 . 10^4 Ω-cm	4

REFERENCES

1. Chumakov, Yu. I.; Stolyanov, Z. E.; Shapovalova, Yu. P. et al. USSR Patent 184 455, 1966; *Chem. Abstract* **1967,** *66,* P29344p.
2. Chiguir, A. N.; Novikova, V. F.; Kalikhman, I. D. et al. *Vysokomol. Soedin. Ser. A* **1969,** *11(8),* 1805-1811.
3. Rembaun, A.; Singer, S. (California Institute of Technology) U.S. 3 538 053, 1970; *Chem. Abstract* **1971,** *74,* P3278h.
4. California Institute of Technology Fr. Patent 2 102 832, 1972; *Chem. Abstract* **1973,** 78 P17003m.
5. Percino, M. J. M. S. Dissertation, Facultad de Química, Universidad Nacional Autonoma de México, 1986.
6. Letis, L.; Skolmeistre, R.; Kalmins, M. *USSR otkrytiya Izobret. Prom. Obraztsy* **1974,** *51(5),* 84; *Chem. Abstract* **1974,** 80.
7. Katon, J. E. *Organic Semiconducting Polymers*; Edward Arnold: London, U.K., 1968.
8. Mizoguchi, A.; Moriga, H.; Shimizu, T.; Amano, Y. *Natl. Tech. Rep.* **1963,** *9,* 407.

Poly(2-pyridone)s

See: *Diyne Cycloaddition Copolymerization (Transition Metal Catalyzed)*
Ladder Polymers (Cycloaddition Copolymerization of Cyclic Diynes)

Polypyromellitimides

See: *Poly(aryl prehnitimide)s*

Poly(2-pyrone)s

See: *Diyne Cycloaddition Copolymerization (Transition Metal Catalyzed)*
Ladder Polymers (Cycloaddition Copolymerization of Cyclic Diynes)

Polypyrrole

See: *Conducting Polymer Colloids*
Conducting Polymers (Self-Acid-Doped, Conjugated)
Conductive Elastomeric Blends
Electrically Conducting Composites
Electrochemical Polymerization
Electroconductive Composites (Emulsion Pathways)
Intelligent Membranes
Oxidative Coupling
Polypyrrole (Processable Dispersions)
Polypyrroles (from Isoporous Membranes)

POLYPYRROLE (Processable Dispersions)

B. M. Mandal, P. Banerjee, and S. N. Bhattacharyya
Polymer Science Unit
Indian Association for the Cultivation of Science

Electrically conducting polypyrrole (PPy) is prepared by electrochemical or chemical oxidative polymerization of pyrrole.[1-8] In chemical oxidative polymerization, ferric chloride is the initiator (oxidant) of choice. It also acts as the dopant.

The polymer is insoluble and infusible and hence difficult to process. Preparing the polymer in dispersion form greatly increases its processability. The dispersion may be used alone or blended with latexes or solutions of other polymers.

SYNTHESIS OF PPY DISPERSIONS

Initiator

PPy dispersions are prepared by the chemical oxidative dispersion polymerization of pyrrole in the presence of a dispersing agent.

Stabilizers and Dispersions Media

Most of the reported dispersion polymerizations of pyrrole have been carried out in aqueous media. Water-soluble polymeric stabilizers such as methyl cellulose (MC),[12] poly(vinyl alcohol-*co*-acetate),[13-16] poly(ethylene oxide),[9,10,15,17] polyvinylpyrrolidone,[13,15,17] poly(2-vinylpyridine),[18] poly(4-vinylpyridine),[19] poly(2-vinylpyridine-*co*-*n*-butyl methacrylate) with 5.4 mol% BM,[15,18] poly(4-vinyl pyridine-*co*-*n*-butyl methacrylate) with 6.1 mol% BM,[15,19,20] poly(vinyl methyl ether) (PVME),[21] ethylhydroxyethyl cellulose (EHEC),[22] and poly(*N,N'*-dimethylaminoethyl methacrylate-*b*-*n*-butyl methacrylate) containing 29 mol% BM[23] are effective. DeArmitt and Armes have also successfully used the low molecular weight surfactant sodium dodecylbenzene sulfonate as stabilizer.[11]

Polymeric stabilizers act by way of adsorption on the PPy particles formed during the polymerization. The adsorption is aided by hydrogen bonding.[9,13,14,18,19] The dispersion is sterically stabilized.[13]

CHARACTERIZATION OF POLYPYRROLE PARTICLES

Stabilizer Content

The maximum stabilizer loading obtained using most of the stabilizers is less than approximately 25%. Exceptions do however exist.

Particle Size

In general, the dispersions contain spherical particles. Exceptions do exist.

Dispersions prepared in aqueous medium have relatively narrow particle size distributions (standard deviation, SD ± 15–20%).[15] Those prepared in aliphatic ester media using poly(vinyl acetate) stabilizer have much higher polydispersity (SD ± 30–40%).[15,24] However, these are results for redispersed particles: our TEM particle size study using the dialyzed original PPy dispersion prepared using either PVME or EHEC stabilizer showed that very small (~20 nm) particles are present in addition to the larger (100–200 nm) particles.[22]

Conductivity

Most of the dispersions with the unadsorbed stabilizers removed do not give free-standing films. The range of the conductivity reported for particles with various stabilizers from the lowest value of 1×10^{-9} S cm^{-1} (film conductivity of MC-stabilized particles with ~90% MC) to as high as 12 (pellet conductivity of PVME-stabilized PPy particles with 6.4% PVME prepared at 2°C).[12,21]

PPy prepared in nonaqueous media like aliphatic esters, acetonitrile–methanol mixture, and ethyl alcohol shows much lower conductivity than that prepared in aqueous medium.[21,24,25]

Redispersibility of Particles

The PPy particles isolated from the original dispersion can be redispersed in solvents with soluble stabilizers. Dried particles are difficult to redisperse, perhaps because drying renders the stabilizer distribution nonuniform.[26]

Conductivity Stability

The conductivity of the stabilizer-coated PPy particles decreases with time like the bare PPy particles prepared without any stabilizer.

PROCESSING OF PPY DISPERSIONS

PPy dispersions can be used to form coatings on various surfaces by conventional processes such as spraying, film casting, spin coating, and so forth. The submicronic PPy particles may also be blended with conventional polymers which can be processed using conventional plastics processing machinery.

POLYPYRROLE-COATED PARTICLES

Electrically conducting PPy coatings have been made on various colloidal particles such as polystyrene (PS) latex,[27] colloidal silica,[28] hematite (both polyhedral and spindle), cerium (IV) oxide, and silica coated with hematite.[29]

Cooper and Vincent prepared PPy-coated beads (~0.05 to ~0.5 mm diameter) of poly(methyl methacrylate) (PMMA) or of a copolymer of methyl acrylate and vinylidene chloride.[16] The beads have a nonconducting polymer core and a conducting shell. The conductivity of the beads glued together with a solution of PMMA in toluene was in the range of 10^{-4} to 10^{-3} S cm^{-1}.

REFERENCES

1. Pron, A.; Kucharski, Z. et al. *J. Chem. Phys.* **1985,** *8,* 5923.
2. Diaz, A. F.; Bargon, J. In *Handbook of Conducting Polymers*; Skotheim, T. A., Ed.; Marcel Dekker: New York, NY, 1986; Vol. 1, Chapter 3.
3. Myers, R. E. *J. Electron. Mater.* **1986,** *2,* 61.
4. Armes, S. P. *Synth. Met.* **1987,** *20,* 367.
5. Rapi, S.; Bocchi, V.; Gardini, G. P. *Synth. Met.* **1988,** *24,* 217.
6. Machida, S.; Miyata, S. *Synth Met.* **1989,** *31,* 311.
7. Kang, E. T.; Neoh, K. G. et al. *Macromolecules* **1991,** *24,* 2822.
8. Whang, Y. E.; Han, J. H. et al. *Synth. Met.* **1991,** *45,* 151.
9. Cawdery, N.; Obey, T. M.; Vincent, B. *J. Chem. Soc., Chem. Commun.* **1988,** 1189.
10. Odegard, R.; Skotheim, T. A.; Lee, H. S. *J. Electrochem. Soc.* **1991,** *138,* 2930.
11. DeArmitt, C.; Armes, S. P. *Langmuir* **1993,** *9,* 652.
12. Bjorklund, R. B.; Liedberg, B. *J. Chem. Soc., Chem. Commun.* **1986,** 1293.
13. Armes, S. P.; Vincent, B. *J. Chem. Soc., Chem. Commun.* **1987,** 288.
14. Armes, S. P.; Miller, J. F.; Vincent, B. *J. Colloid Int. Sci.* **1987,** *118,* 410.
15. Armes, S. P.; Aldissi, M. et al. *J. Colloid Int. Sci.* **1991,** *141,* 119.
16. Cooper, E. C.; Vincent, B. *J. Phys. D., Appl. Phys.* **1989,** *22,* 1580.
17. Markham, G.; Obey, T. M.; Vincent, B. *Colloids Surf.* **1990,** *51,* 239.
18. Armes, S. P.; Aldissi, M. *Polymer* **1990,** *31,* 569.
19. Armes, S. P.; Aldissi, M.; Agnew, S. F. *Synth. Met.* **1989,** *28,* C837.
20. Rawi, Z.; Mykytiak, J.; Armes, S. P. *Colloids Suf.* **1992,** *68,* 215.
21. Digar, M. L.; Bhattacharyya, S. N.; Mandal, B. M. *Polymer* **1994,** *35,* 377.
22. Mandal, T. K.; Mandal, B. M. *Polymer* **1995,** *36,* 1911.
23. Beadle, P.; Rowen, L. et al. *Polymer* **1993,** *34,* 1561.
24. Armes, S. P.; Aldissi, M. *Synth. Met.* **1990,** *37,* 137.
25. French, M. A.; Billingham, N. C.; Armes, S. P. *Synth. Met.* **1993,** *57,* 3556.
26. Armes, S. P.; Aldissi, M. et al. *Langmuir* **1991,** *7,* 1447.
27. Yassar, A.; Roncali, J.; Garnier, F. *Polym. Commun.* **1987,** *28,* 103.
28. Armes, S. P.; Gottesfeld, S. et al. *Polymer* **1991,** *32,* 2325.
29. Partch, R.; Gangolli, S. G. et al. *J. Colloid Int. Sci.* **1991,** *144,* 27.

POLYPYRROLES (from Isoporous Membranes)

R. P. Burford and J. Mansouri
Department of Polymer Science
University of New South Wales

POLYPYRROLE WITH CONTROLLED MORPHOLOGY AND HIGH SURFACE AREA STRUCTURES

For conducting polymers to be switching devices, catalytic supports, and carriers for active ingredients, they must have a high and predictable surface area and controlled morphology. Despite recent attempts to develop soluble precursors, these polymers cannot generally be dissolved, and therefore conventional methods of casting are not applicable.[1,2] Here we discuss alternative approaches.

POLYPYRROLE WITH FIBRILLAR/MICROPOROUS MORPHOLOGY

Highly ordered, high surface area polypyrroles can be made by polymerizing within the pores of an isoporous template membrane. Microporosity can be created by dissolving the template, giving a conducting polymer replicating the template pores. This category of synthesis was first reported by Penner et al. and subsequently by Burford et al.[4,5] This approach has been used to produce nanocylinders, fibers, and hollow tubules and can be achieved either by electrodeposition or by chemical polymerization of pyrrole in the pores of the membrane.[3-13]

PREPARATION METHODS

Host membranes were either polycarbonate or γ-alumina membranes. Polycarbonate membranes (Nucleopore; ca. 10 μm thick) contain cylindrical pores and usually have a homogeneous pore size distribution. These membranes are available in pore diameters ranging from 100 Å to 12 μm, so polypyrrole fibers with a wide range of fibril diameters can be synthesized. γ-Alumina (Anopore) membranes are typically 50 μm thick and have nominal pore sizes of 0.2 μm and gradually tapered honeycomb structures.

If one begins with isoporous templates, one could, with conventional polymers, make replicates by filling pores with the melt or with concentrated solutions. However, since polypyrroles are generally intractable, some form of *in situ* polymerization is necessary. This can be done either chemically, usually by oxidation, or electrochemically.

MORPHOLOGY AND MECHANISM OF GROWTH

Martin et al. have shown that polypyrrole grown through the nucleopore membrane preferentially precipitates on the pore wall to form hollow tubules.[14] Martinez et al. attribute this to the presence of "molecular anchors" (in this case, anionic sites) on the pore walls which bind the nascent polymer to the wall, coating it with a thin film of the heterocyclic polycation.[17]

Fibers or tubes prepared using γ-alumina membrane as template have a dense fibrillar morphology and more uniform cross-sections with similar diameter and geometries which we believe replicates the inherent uniformity of the host membrane.

Electronic Conductivity

When the template method is used to synthesize nanoscopic conductive polymer fibers, electronic conductivities along the axes can be dramatically higher than those for flat films.[3,14,16] Both molecular and supermolecular differences between template-synthesized conductive polymers and conventional forms are responsible for this enhanced conductivity, as the relatively long narrow channels constrain polymerization. Polarized infrared absorption spectroscopy, direct current and optical measurement of conductivity, and X-ray diffraction analysis were used to show that the polymer chains in the narrowest polyheterocyclic fibrils are preferentially oriented parallel to the fibril axes and that the extent of orientation increases as fibril diameter and synthesis temperature decrease.[11,15] Narrowest fibers had the highest conductivities (in siemens, 3200 S cm^{-1}). 30-nm diameter PPy fibrils synthesized at –20°C were almost one order of magnitude more conductive than the best electrochemically synthesized film.

Ionic Transport

Conducting polymers have potential applications as electrode materials in batteries, as electrochromic displays, and as molecular electronic circuit elements.[18-20] Here, these polymers must switch rapidly between the conductive and insulating states.[21,22] This switching process requires that counterions be incorporated into and subsequently expelled from the polymer film. The rate of counterion diffusion is the limiting step in the switching process; it is very slow and decreases with increasing polymer film thickness.[23,24] The slow ion transport in polypyrrole has been attributed to its very dense amorphousness.[16]

The enhancement in ion transport has been explained in terms of a morphological effect: conventional electropolymerized polypyrrole films have a dense amorphous morphology, with poor ion transport characteristics. Here an ion must take a long, tortuous route through the polymer phase (i.e., the thickness of the polymer) and so transport is very slow. In a microporous morphology pore transport is fast but polymer diffusion is still extensive and slow. In fibrillar–microporous morphology (formed by template method) the polymer fibrils are surrounded by solution-filled pores, so an ion will always be as close to the solution as the radius of the polymer fibril.

REFERENCES

1. Aldissi, M. U.S. Patent 4 929 389, 1990.
2. Feast, W. J. Presented at Polymer 91, Melbourne, Australia, February 1991; paper 70.
3. Cai, Z.; Martin, C. R. *J. Am. Chem. Soc.* **1989,** *111,* 4138.
4. Burford, R. P.; Tongtam, T. *J. Mater. Sci.* **1991,** *26,* 3264.
5. Penner, M.; Martin, C. R. *J. Electrochem. Soc.* **1986,** *133,* 2206.
6. Tierney, M. J.; Martin, C. R. *J. Phys. Chem.* **1989,** *93,* 2878.
7. Liang, W.; Martin, C. R. *J. Am. Chem. Soc.* **1990,** *112,* 9666.
8. Mansouri, J.; Burford, R. P. *J. Mater. Sci.* **1994,** *29,* 2500.
9. Martin, C. R.; Van Dyke, L. S. et al. *J. Am. Chem. Soc.* **1990,** *112,* 8976.
10. Van Dyke, L. S.; Martin, C. R. NTIS Report No. AD-216 603/1/HDM, 1990.
11. Cai, Z.; Lei, J. et al. *Chem. Mater.* **1991,** *3,* 960.
12. Martin, C. R. *Adv. Mater.* **1991,** 3(9), 457.
13. Martin, C. R.; Parthasarathy, R.; Menon, V. P. *Synth. Met.* **1993,** *55,* 1165.
14. Martin, C. R.; Liang, W. et al. *Synth. Met.* **1993,** *57,* 3766.
15. Martin, C. R.; Menon, V. P.; Parthasarathy, R. V. *Proc. Am. Chem. Soc.* **1994,** 229.
16. Van Dyke, L. S.; Martin, C. R. *Synth. Met.* **1990,** *36,* 275.
17. Martinez, L.; Hernandez, A.; Tejerina, A. F. *Sep. Sci. Technol.* **1987,** *22,* 85.
18. Kaneto, K.; Maxfield, M. et al. *J. Chem. Soc., Faraday Trans. 1* **1982,** *78,* 3417.
19. Yoneyama, H.; Wakamoto, K.; Tamura, H. *J. Electrochem. Soc.* **1985,** *132,* 2414.
20. Thackeray, J. W.; White, H. S.; Wrighton, M. S. *J. Phys. Chem.* **1985,** *889,* 5133.
21. Naegele, D.; Bitten, R. *Solid State Ionics* **1988,** 28-30, 983.
22. Gazard, M. In *Handbook of Conducting Polymers*; Skotheim, T. A., Ed.; Marcel Dekker: New York, NY, 1986; Vol. 1, Chapter 19.
23. Hug, R.; Farrington, G. C. *J. Electrochem. Soc.* **1985,** *132,* 1132.
24. Hug, R.; Farrington, G. C. *J. Chem. Soc., Faraday Trans 1* **1982,** *78,* 3417.

Polyquincridono

See: Conjugated Ladder Polymers

Polyquinolino

See: Conjugated Ladder Polymers

Polyquinoxalines

See: Conjugated Ladder Polymers
Ladder Polymers (Methods of Preparation)

Polyradicals

See: Conjugated Polyradicals
Dendritic Polyradicals
Ferromagnetic Polymers
Polyradicals (Synthesis and Magnetic Properties)

POLYRADICALS (Synthesis and Magnetic Properties)

Yozo Miura
Department of Applied Chemistry
Faculty of Engineering
Osaka City University

The recent interest in polyradical chemistry is inspired by the expectation that polyradicals might be ferromagnetic materials made from organic elements only.[1-5] The theoretical possibilities and possible models for polymeric ferromagnets have been proposed based on the molecular orbital (MO) and valence bond (VB) theory of the π-electron networks of the alternate hydrocarbons.[6] For the realization of polymeric ferromagnets, much effort has been paid to the syntheses of polyradicals. Herein, molecular design, syntheses, and magnetic characterization of organic magnetic polyradicals are reviewed.

FUNDAMENTAL DIFFERENCES BETWEEN ORGANIC MOLECULE-BASED AND INORGANIC ATOM-BASED MAGNETISM

Whereas inorganic atom-based magnets are comprised of transition metal elements such as Fe, Co, Ni, Nd, Sm, and Eu, organic molecule-based magnets are comprised of nonmetallic light elements such as H, C, N, O, and S. Accordingly, while the unpaired electron spins in organic atom-based magnets occupy the 3d and 4f orbitals, the unpaired electron spins in organic molecule-based magnets occupy the 2s and 2p orbitals (or 3s and 3p).

SYNTHESES AND MAGNETIC CHARACTERIZATION OF ORGANIC POLYRADICALS

The following π-conjugated polyradicals have been synthesized, and their magnetic behaviors have been discussed.

Polyphenylacetylene-Based Polyradicals

Polyphenylacetylene-based polyradicals have been considered to be one of the most promising candidates for polymeric ferromagnets. To date, a variety of polyphenylacetylene-based polyradicals have been prepared by many groups. The first example reported is **Structure 1**, which was prepared by Nishide et al. by the polymerization of 5 with WCl_6 or $MoCl_5$ followed by oxidation with PbO_2 or $K_3Fe(CN)_6$.[7] The number average molecular weights ($\overline{M}_n$) of the polymeric precursor were $0.23 \times 10^4 - 1.8 \times 10^4$, and the spin concentration of the polyradical was 0.38 spin/monomer unit.

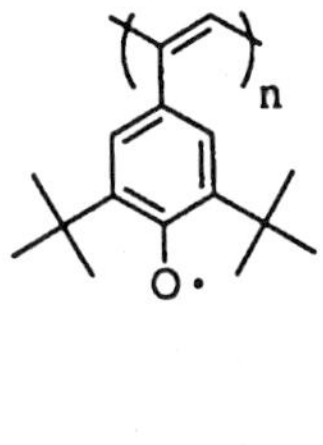

1

Polyphenylacetylene-based nitronyl nitroxide polyradical was prepared by Fujii et al.,[8] Miura et al.,[9] and Saf et al.[10] independently. This radical is quite stable and is not destroyed by oxygen. The spin concentrations of the polyradicals obtained were ~0.5 spin/monomer unit.

Aryl nitroxides are advantageous radicals in the study of polyradical magnetism because they are oxygen-insensitive and the unpaired electron spin is extensively delocalized over the whole of the radicals.

Polyphenylacetylene-based polynitroxides were prepared by Miura et al.[11,12] and Nishide et al.[13]

Polyphenyldiacetylene-Based Polyradicals

It is well known that some 1,3-butadiyne derivatives undergo topochemically controlled solid state polymerization in crystals at high temperature (below their melting points) or by UV-irradiation. If unsymmetrical 1,3-butadiynes having stable free radicals polymerize in a 1,4-fashion, the unpaired electron spins are expected to couple ferromagnetically to give high-spin polyradicals.

Inoue et al. reported the topochemical polymerization of a phenylnitroxide diacetylene and its precursor.[14]

Polytriarylmethyl and Related Polyradicals

π-Conjugated *m*-connected triarylmethyl polyradicals have been suggested to be polymeric ferromagnets polymers. Rajca et al. prepared a variety of *m*-connected π-conjugated triarylmethyl polyradicals.[15,16] Since these polyradicals are oxygen-sensitive, all operations were carried out in an inert gas.

The intermolecular magnetic interactions were antiferromagnetic; bulk ferromagnets were not achieved.

Miura et al. have prepared polyradicals **Structures 2** and **3**. $\overline{M}_n$ values of **2** and **3** were $4.3 \times 10^3 - 4.8 \times 10^3$ and 3.0×10^3, and the spin concentrations were 0.82–0.91 (**2**) and 0.86 spin/monomer unit (**3**), respectively. Therefore, it is obvious that no significant decomposition of the radical moieties took place.

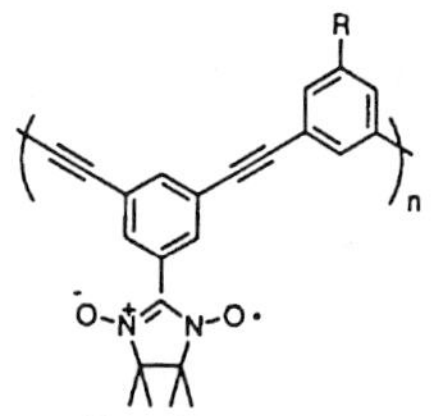

2

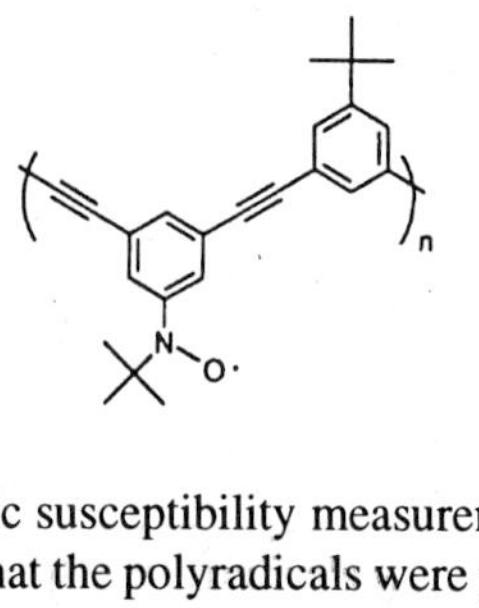

3

The magnetic susceptibility measurements of **2** and **3**, however, revealed that the polyradicals were paramagnetic, showing that the unpaired electron spins to not interact through the 1,3-phenylene diethynylene coupling unit.

Conjugated Polyradical Ions

An early theoretical work by Fukatome et al. suggested that radical cations (polarons) obtained by oxidizing (doping) conjugated organic segments could serve as useful spin carriers of novel magnetic materials.[17]

Kaisaki et al. and Murry et al. prepared one-dimensional polaronic ferromagnets by I_2 or AsF_5 doping of, i.e., **4** and **5**,

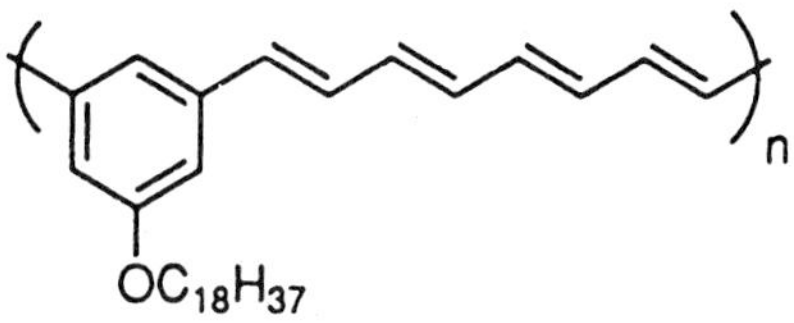

4

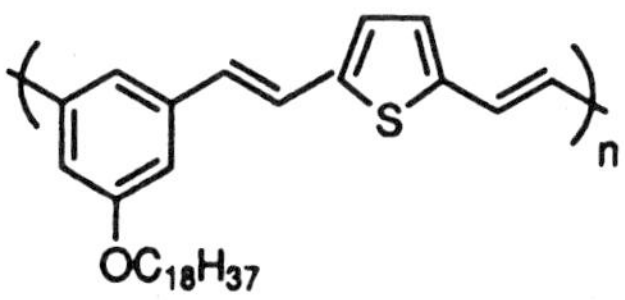

5

prepared by the Wittig or Suzuki reaction.[18,19] Magnetic characterization of the doped polymers by SQUID indicated that ferromagnetic coupling of polarons occurred (S = ~2.6). Heavy doping, however, reduced S to ½, though the spin concentrations were increased.

A further attempt to realize polaronic magnetism is oxidative coupling of 1,3,5-triaminobenzene.[20] Treatment of 1,3,5-triaminobenzene with I_2 in acetic acid gave an insoluble black polymeric material, for which ferromagnetic interaction was claimed.

Unconjugated Polyradicals

The magnetic properties of some unconjugated polyradicals were investigated by several groups.[10,21-25] Since the radical sites are linked to the main chain by the unconjugated σ-bonds, the through-bond interactions between the unpaired electron spins cannot be expected. Accordingly, the magnetic interactions among the unpaired electrons are governed by through-space interactions which, in most cases, will be antiferromagnetic.

REFERENCES

1. Miller, J. S.; Epstein, A. J.; Reiff, W. M. *Chem. Rev.* **1988,** *88,* 201.
2. Dougherty, D. A. *Acc. Chem. Res.* **1991,** *24,* 88.
3. Iwamura, H.; Koga, N. *Acc. Chem. Res.* **1993,** *26,* 346.
4. Iwamura, H.; Miller, J. S. *Mol. Cryst. Liq. Cryst.* **1993,** *232,* 1.
5. Iwamura, H.; Miller, J. S. *Mol. Cryst. Liq. Cryst.* **1993,** *233,* 1.
6. Mataga, N. *Theor. Chim. Acta.* **1968,** *10,* 372.
7. Nishide, H.; Yoshioka, N.; Inagaki, K.; Kaku, T.; Tsuchida, E. *Macromolecules* **1992,** *25,* 569.
8. Fujii, A.; Ishida, T.; Koga, N.; Iwamura, H. *Macromolecules* **1991,** *24,* 1077.
9. Miura, Y.; Inui, K.; Yamaguchi, F.; Inoue, M.; Teki, Y.; Takui, T.; Itoh, K. *J. Polym. Sci., Polym. Chem. Ed.* **1992,** *30,* 959.
10. Saf, R.; Hummel, K.; Gatterer, K.; Fritzer, H. P. *Polym. Bull.* **1992,** *28,* 395.
11. Miura, Y.; Matsumoto, M.; Ushitani, Y. *Macromolecules* **1993,** *26,* 2628.
12. Miura, Y.; Matsumoto, M.; Ushitani, Y.; Teki, Y.; Takui, T.; Itoh, K. *Macromolecules* **1993,** *26,* 6673.
13. Nishide, H.; Kaneko, T.; Yoshioka, N.; Akiyama, H.; Igarashi, M.; Tsuchida, E. *Macromolecules* **1992,** *26,* 4567.
14. Inoue, K.; Koga, N.; Iwamura, H. *J. Am. Chem. Soc.* **1991,** *113,* 9803.
15. Rajca, A.; Utamapanya, S. *J. Am. Chem. Soc.* **1993,** *115,* 2396.
16. Rajca, A.: Utamapanya, S.; Thayunianvan, S. *J. Am. Chem. Soc.* **1992,** *114,* 1884.
17. Fukutome, H.; Takahashi, A.; Ozaki, M. *Chem. Phys. Lett.* **1987,** *133,* 34.
18. Kaisaki, D. A.; Chang, W.; Dougherty, D. A. *J. Am. Chem. Soc.* **1991,** *113,* 2764.
19. Murry, M. M.; Kaszynski, P.; Kaizaki, D. A.; Chang, W.; Dougherty, D. A. *J. Am. Chem. Soc.* **1994,** *116,* 8152.
20. Torrance, J. B.; Oostra, S.; Zazzal, A. *Synth. Met.* **1987,** *19,* 709.
21. Kamachi, M.; Tamaki, M.; Morishima, Y.; Nozakura, S.; Mori, W.; Kishita, M. *Polym. J.* **1982,** *14,* 363.
22. Kamachi, M.; Enomoto, H.; Shibasaki, M.; Mori, W.; Kishita, M. *Polym. J.* **1986,** *18,* 439.
23. Vlietstra, E. J.; Notle, R. J. M.; Zwikker, J. W.; Drenth, W.; Meijer, E. W. *Macromolecules* **1990,** *23,* 946.
24. Dulog, L.; Lutz, S. *Makromol. Chem., Rapid Commun.* **1993,** *14,* 147.
25. Allgaier, J.; Finkelmann, H. *Makromol. Chem., Rapid Commun.* **1993,** *14,* 267.

Polyrotaxanes

See: *Functionalized Polymers (Comb, Rotataxanic and Dendrimeric Structures)*
Ionene Oligomers
Ionene Polymers
Macromolecular Recognition (New Supramolecular Structures)
Polyrotaxanes (Overview)

POLYROTAXANES (Overview)

J. Fraser Stoddart,* Jon A. Preece, and Françisco M. Raymo
School of Chemistry
University of Birmingham

MOLECULAR RECOGNITION AND NANOSCIENCE

Nature has made elegant use of the mutual recognition between molecular entities to construct highly ordered structures and superstructures of nanoscale and larger dimensions,[1] which not only have forms, but also perform functions. If chemists are to vie with nature in the construction of "intelligent structures," they must understand the fundamental basis for the mutual stereoelectric recognition between molecular entities. This understanding is being achieved in small molecular assemblies and supramolecular arrays with the construction of

*Author to whom correspondence should be addressed.

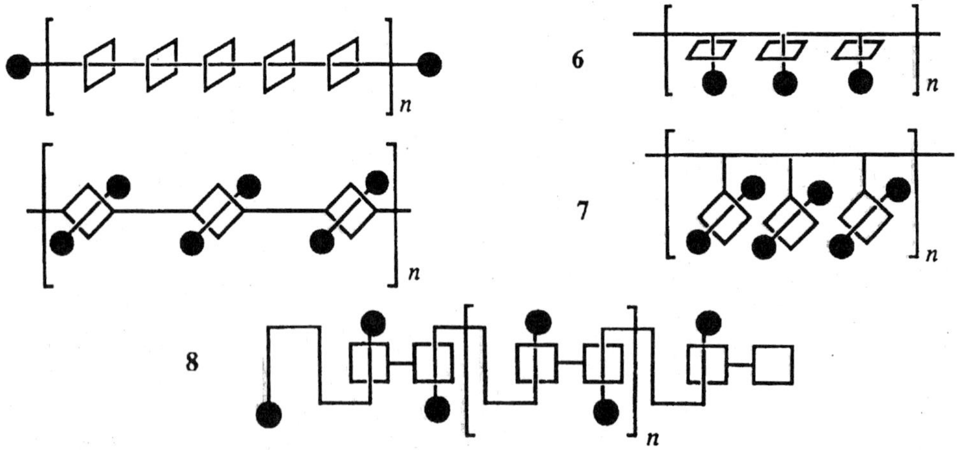

FIGURE 1. Some of the possible polyrotaxane structural types.

molecular[2] and supramolecular[3] systems possessing device-like functions. One of the goals now is to extend this understanding of recognition processes to much larger molecular and supramolecular architectures into the realm of macromolecular systems or polymers.[4] This understanding is necessary if we are to gain insight into the relationship between molecular structures and bulk properties, thus enabling chemists to fabricate, from the microscopic molecular level, designer materials with ever-increasing specifications in a macroscopic sense.

Rotaxanes and Polyrotaxanes

Rotaxanes, and ultimately polyrotaxanes, may be one of the vehicles for such studies of structure-property relationships since the forces that encourage these molecules to form at the molecular level have been shown to live on in their bulk structures afterwards.[5]

Rotaxanes are molecular entities in which a linear molecular component is threaded through the cavity of a macrocyclic molecular component, resulting in a so-called pseudorotaxane architecture, a 1:1 complex in the simplest case. Subsequently, covalent bond formation involving two large "stoppers" at the termini of the linear component, such that the macrocycle cannot pass over the stoppers, leads to a [2]rotaxane–a molecular species made up of two components. The word rotaxane is derived from the Latin *rota* for a wheel and *axis* for an axle.

Polyrotaxane Exemplified

Not surprisingly, the area of polymer chemistry concerned with polyrotaxanes is only just starting to be explored with the advent of self-assembly as a potentially efficient means of constructing them. Indeed, in the last few years, polyrotaxanes belonging to the structural types **4-7**, shown **Figure 1**, have been described in the chemical literature and the reader is directed to a comprehensive review by Gibson et al.[6] for a detailed description of the history, preparation, and properties of these early polyrotaxanes. Several other shorter reviews[7-9] are also available.

In all the approaches that have been devised for the construction of polymers belonging to structure types **4-7**, the polymerization step relies on a classical covalent bond forming reaction of one or more bisfunctionalized entities, albeit in the presence of a macrocyclic compound. In essence, the resulting polymer is formed in a manner that is no different from the preparation of the "naked" polymer without the macrocycle being present.

A Self-Assembly Approach to Constructing Polyrotaxanes

A different approach to the formation of polyrotaxanes being developed by the Birmingham Group makes use of the new concept in chemical synthesis of self-assembly to either

- Thread a macrocycle on to a preformed polymer/oligomer, containing along its backbone recognition sites for the macrocycle. In so doing, it should be possible to construct a very highly ordered polymer at a molecular level and so engineer the bulk properties of the polymer by controlling the number of recognition sites for a macrocycle along the polymer chain
- Form polymers that contain within their repeating units not only covalent bonds but also mechanical links, resulting in molecular architectures of the structure shown in Figure 1, type **8**. This methodology enables the self-assembly step to be incorporated into polymerization reminiscent of the way in which the natural polymer DNA polymerizes.

PROPERTIES AND APPLICATIONS

The properties of polyrotaxanes and polypseudorotaxanes, based on structural types **4-7** shown in Figure 1, have not been described to any great extent in the chemical literature. At present, researchers are largely concerned with synthesis and characterization of these polymers. However, their properties should be markedly different from those of the naked polymers. These changes in properties will be a consequence of the macrocyclic components disguising and shielding parts of the polymer in the solid, fluid, and solution states, and changing the packing of the polymer chains in the solid and fluid state.

It is tempting to envisage that changing the physical properties of a polymer might be translated into applications. For instance, a polymer could be designed to incorporate a prodrug which could then be hydrolyzed to afford an active drug. It

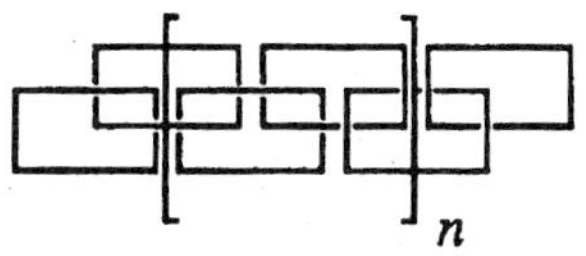

[*n*]Catenane Motifs

28

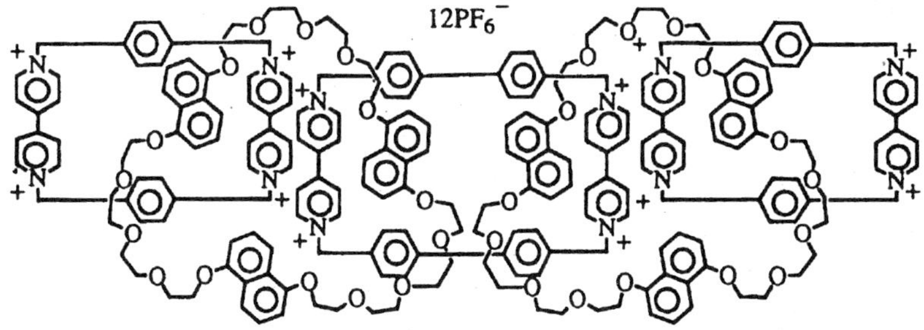

29.12PF$_6$

Olympiadane

FIGURE 2. Polycatenanes and olympiadane–a [5]catenane.

could be imagined that, if the polymer was threaded with macrocycles, such as one of the cyclodextrins[10-12] which then dethread slowly from the polymer *in vivo*, the rate of hydrolysis would be retarded. In principle, the rate of hydrolysis could be controlled by the amount of cyclodextrin threaded on to the polymer. Thus, it might be possible to control the rate of drug release into the body. Also, the mechanical, chemical, electrical, and optical properties of a solid or liquid polymer could be engineered by having varying proportions of a macrocycle threaded on to it. A specific example might be an electrically conducting polymer that could be insulated by being threaded through nonelectrically conducting macrocycles.[13] Such an approach to molecular engineering could be rewarding because electrically active materials are generally chemically reactive. And so, by encapsulating them in an inert macrocycle, the chemical lifetime of the polymer could be increased.

The properties of polymers based on structural type **8** in Figure 1 can only be the subject of conjecture. However, the unusual mechanical linking in polymer backbone ensures that such polymeric architecture will be targeted by chemists, if for no other reason than it is a considerable intellectual challenge. As an aside, other polymeric materials are being designed, synthesized, and characterized, that also have mechanical links in their backbone. These compounds are called polycatenanes **28** (Figure 2); catenane is derived from the Latin word *catena* for a chain. To date, the longest linear catenated compound reported is the [5]catenane–olympiadane **29**.12PF$_6$[14]-containing five interlocked rings. The properties of such compounds again can be a topic for speculation only, but in composition they are the molecular equivalents of the macroscopic chains we use in everyday life. By implication, such polymeric materials also should have novel material properties.

REFERENCES

1. Singer, S. J.; Nicholson, G. L. *Science* **1972,** *15,* 720.
2. Bissell, R. A.; Cordova, E. et al. *Nature* **1994,** *369,* 133.
3. Ballardini, R.; Balzani, V. et al. *Angew. Chem. Int. Ed. Engl.* **1993,** *32,* 1301.
4. Ashton, P. R.; Philp, D. et al. *Makromol. Chem., Macromol. Symp.* **1992,** *54/55,* 441.
5. Ashton, P. R.; Preece, J. A. et al. *Synthesis* **1994,** 1344.
6. Gibson, H. W.; Bheda, M. C.; Engen, P. T. *Progr. Polym. Sci.* **1994,** *19,* 843.
7. Ogino, H. *New J. Chem.* **1993,** *17,* 683.
8. Amabilino, D. B.; Parsons, I. W.; Stoddart, J. F. *TRIP* **1994,** *2,* 146.
9. Wenz, G. *Angew. Chem. Int. Ed. Engl.* **1994,** *33,* 803.
10. Ritter, H. *Macromol. Symp.* **1994,** *77,* 73.
11. Harada, A.; Li, J. et al. *Macromolecules* **1993,** *26,* 5267.
12. Wenz, G.; Keller, B. *Angew. Chem. Int. Ed. Engl.* **1992,** *31,* 197.
13. Dagani, R. *C&E News* **1992,** April 13, 39.
14. Amabilino, D. B.; Ashton, P. R. et al. *Angew. Chem. Int. Ed. Engl.* **1994,** *33,* 1286.

POLYSACCHARIDE-POLYSACCHARIDE INTERACTIONS

Cyrille Rochas
Laboratoire de Spectrométrie Physique
Université Joseph Fourier

François René Taravel
CERMAV Université Joseph Fourier

Polysaccharides offer an exceptional ratio of hydroxyl groups per saccharide residues. Such hydrogen bonding potential has to be taken into account when considering interactions, either in association with neighboring carbohydrate polymers or surrounding water molecules.[1] For polysaccharides forming three-dimensional networks under specific conditions (gel structure), these interactions are hydrogen bonding, dipole and ionic interactions, and solvent partition effects.[2]

In many applications, particularly in food systems, industrial polysaccharides are used in combination with other polymers (including proteins). Thus, the combination of aqueous glycan solutions provides a very effective means of obtaining systems with new and unique properties. Such polymer combinations may either be compatible, leading to synergistic interactions and property enhancements, or incompatible and result in precipitation or phase separation phenomena. However, both compatible and incompatible systems offer advantages that are exploitable in different areas (composite materials, biological processes, and food and nonfood industries).

PROPERTIES

It is of interest to prepare formulations with more than one component, because, although the properties of the individual component are sometimes maintained, generally the properties of the blends are better than or different to those of the individual polymers.[3] It is remarkable that some polymers without gelling properties form strong gels on mixing. This exceptional effect is called "synergism."

Synergistic blends are very attractive because new properties are obtained in comparison to those of individual components, and also it can be of interest to replace part or all of an expensive hydrocolloid with a cheaper one. In addition to the usual applications to food, polysaccharide–polysaccharide interactions play a major role in the architecture of animal or plant cells, and the cell wall can be understood as a two-phase system, a crystalline phase (skeleton) embedded in an amorphous phase (matrix).

Within the cells or in industrial applications the role of these interactions is to preserve a shape, to give consistency, or to retain solvent (generally water). They increase the viscosity of solutions or generate gels.

Polysaccharide–polysaccharide interactions exist not only for systems with two or more components, but also for systems with only one component.

SINGLE COMPONENT SYSTEM

Single polymer based gels are driven by interactions between the macromolecule chains. Every change in the thermodynamic conditions (temperature/concentration) will modify the interactions between the molecules; in addition, the molecular conformation will be changed, as well as the solution state. κ-Carrageenan, a polysaccharide extracted from red seaweeds, is representative of this. A temperature decrease or a concentration decrease can induce a conformational transition and at the same time a sol-gel transition takes place.[4]

DOUBLE COMPONENT SYSTEMS

Depending on the nature of the components and on the gelation mechanism, different structures can be supposed.[5]

The first kind of structure is a network of the first polymer within which the second polymer is entrapped. If the size of the entrapped polymer is smaller than the mesh size of the network, no specific property of the blend is found. This is the usual situation. This kind of gel could have three different advantages. First, before gelation, the viscosity of the nongelling polymer can promote the adjustment of the formulation. Second, the nongelling polymer can move freely in and out of the gel network. This can be used to release more or less rapidly the nongelling polymer from one medium to another one (drug delivery). Third, the nongelling polymer can modify the network of the other polymer.

The second kind of mixed gel is interpenetrating networks. At low temperatures κ-carrageenan-agarose blend was clearly shown to be an interpenetrated network.[6]

The third kind of gel is the result of thermodynamic incompatibility. This incompatibility is driven by the concentration and the molecular weight of the polymers involved. The incompatible amylose-amylopectin blend is the best known system of this type.[7]

The fourth kind of gel can exist if specific interactions between the two polymers exist. Owing to synergism, binding of one polymer to another should result in a new system. Generally, it is not possible to predict the properties of this system.

The fifth kind of gel is not really a new type. Whereas the four previous types of interactions were created by artificial mixing of two different polysaccharides, plant mixtures of interacting polysaccharides can also be extracted *in vivo*. For example, the extract obtained from *Eucheuma cottonii* was proven to be a mixture of 90% κ-carrageenan and of 10% ι-carrageenan.[8] Thus the mechanical properties of the gel prepared are strongly dependent on the ratios of both polysaccharides.

The last kind of gel is a mixed gel with specific and known interactions occurring between two polyelectrolytes bearing opposite electrical charges. Polyanions and polycations interact strongly to form polyanion–polycation complexes whose properties depend on the charge and on the respective concentration of the two polymers. In normal conditions, an insoluble precipitate is formed. Chitosan, a polycationic polymer obtained as a deacetylation product from chitin (from shellfish), can be used with polyanionic substances, such as alginate, carrageenan, or pectin, to produce insoluble materials.[9]

PREPARATION AND APPLICATIONS

Starch–carrageenan associations have been employed for many years in the food industry in order to reduce starch content, thereby reducing the caloric value of preparations.[10] Also guar gum, xanthan gum, arabic gum, and carboxymethylcellulose modify the gelatinization properties of wheat starch, resulting in a decrease in the retrogradation of starch in solution.[11] The synergistic properties of starch-gum combinations are very useful in a wide variety of manufactured food products in which various starches are employed. Different starch-carrageenan combinations are used in several formulations of dairy products to produce mouth-feel and body as well as the desired textural effect for the consumer taste. These range from soups and fluid sauces to full-bodied puddings and pie fillings.

Many different classes of polysaccharides are employed as stabilizers in the ice cream industry. They involve mainly carob and guar gums, carboxymethylcellulose, alginate, and κ-carrageenan. They are used to improve the stability and eating texture of the product.

In the field of electrophoresis the addition of galactomannan to agarose allows new gels to be prepared for electrophoretic separation.[12] It has also been suggested that the interaction involved in alginate-pectin mixed gels could have biomedical applications.[13]

REFERENCES

1. Perez, S. *Water and Biological Macromolecules: Topics and Structural Biology*; Macmillan: 1993; Chapter 10.
2. Morris, E. R. *Carbohydr. Polym.* **1992,** *17,* 65.
3. Morris, E. R. *Foods Gels*; Elsevier Applied Science: London, 1990; Chapter 8.
4. Rochas, C.; Landry, S. *Carbohydr. Polym.* **1987,** *7,* 435.
5. Morris, V. J. *Gums and Stabilizers for the Food Industry 3*; Elsevier Applied Science: London, 1986; Part 1.
6. Zhang, J.; Rochas, C. *Carbohydr. Res.* **1990,** *13,* 257.
7. Kalichevsky, M. T.; Ring, S. G. *Carbohydr. Res.* **1987,** *162,* 323.
8. Rochas, C.; Rinaudo, M.; Landry, S. *Carbohydr. Polym.* **1989,** *10,* 115.
9. Mireles, C.; Martino, M.; Bouzas, J.; Torres, J. A. *Advances in Chitin and Chitosan*; Elsevier Science: 1992.
10. Descamps, O.; Langevin, P.; Combs, D. H. *Food Technol.* **1986,** *40,* 81.
11. Christianson, D. D.; Hodge, J. E.; Osborne, D.; Detroy, R. W. *Cereal Chem.* **1981,** *58,* 513.
12. Cook, R. B.; Witt, H. J. U.S. Patent 4 290 911, 1981.
13. Morris, V. J.; Chilvers, G. R. *J. Sci. Food Agric.* **1984,** *35,* 1370.

Polysaccharides

See: *Acetan*
Agarose (Overview)
Alginate Fibers (High Performance Papers)
Bicyclic Acetals, Ring-Opening Polymerization
Cellulosic Liquid Crystals
Chondroitin Sulfate (Overview)
Dextran
Enzymatic Polymerization
Enzyme-Degradable Hydrogels
Food Polymers
Gelling Agents (Agarose and Carrageenan)
Glycosaminolglycans (Overview)
Guar (Graft Copolymerization)
Gums (Overview)
Hemicelluloses
Heparinoids, Artificial (Biological Activities)
Hyaluronic Acid (Macroionic, Rheological and Lubricity Properties)
Hydrophobized Polysaccharides (Versatile Functions)
Hydroxyethyl Cellulose (Infrared Spectra and Dielectric Properties)
Hydroxyethyl Cellulose (Preparation and Use in Protein Precipitation)
Hydroxyethyl Cellulose (Variables Affecting Hydroxyethylation Reaction)
Lyotropic Polysaccharides
Polyesters (Derived from Renewable Resources)
Polysaccharide-Polysaccharide Interactions
Polysaccharides, Synthetic (Hypoglycemic Activity)
Polysaccharides, Synthetic (Preparation and Biological Activities)
Starch-Based Plastics (Measurement of Biodegradability)
Starch Biodegradation (in Starch-Plastic Blends)
Starch Graft Copolymers
Starch-Polymer Composites
Starch Polymers, Natural and Synthetic
Vegetal Biomass (1. Monomers and their Polymerization)
Vegetal Biomass (3. Polymers, Derivatives and Composites)
Water-Soluble Polymers (Oil Recovery Applications)
Welan Microbial Polysaccharide
Wood, Chemically Modified (Thermoplasticization and Solubilization)
Wood Graft Copolymers
Wood-Plastic Composites
Wood-Polymer Composites
Wound Dressing Materials
Xanthan, Depolymerization
Xanthan Gum (Overview)
Xanthan Gum (Conformation and Solution Properties)

POLYSACCHARIDES, SYNTHETIC (Hypoglycemic Activity)

Kenichi Hatanaka
Department of Biomolecular Engineering
Faculty of Bioscience and Biotechnology
Tokyo Institute of Technology

The preparation of branched polysaccharides, which contain polysaccharide main chain and glycosyl branches, is described. In addition, the hypoglycemic activity of the synthetic (1 → 6)-α-D-glucopyranan having α-glucosidic branches is also reported.

There are three synthetic routes for preparation of branched polysaccharides: glycosylation reaction of natural polysaccharide, polymerization and/or copolymerization of disaccharide derivatives, and glycosylation reaction of synthetic polysaccharide derivatives that have been synthesized by polymerization and/or copolymerization of monosaccharide derivatives.

Of the many kinds of natural polysaccharides, hypoglycemic polysaccharides, Panaxans, have been purified from ginseng roots.[1,2] However, the fine structure of the polysaccharides was not determined; therefore, the relationship between the hypoglycemic activity and the biological effect remained unclear. To determine the structure–function relationship of a compound, the synthetic pure compound is quite effective. For the polysaccharide for which the fine structure is difficult to determine, a synthetic polysaccharide that has the same or similar structure as the natural one becomes quite an important compound in order to better understand the relationship between the biological activity and the chemical structure of the polysaccharide.

PROPERTIES

Hypoglycemic Activity

The synthetic model compound of Panaxan A–i.e., (1 → 6)-α-D-glucopyranan, having only α-D-glycopyranosyl branch at C-3, which is prepared by the chemo-enzymatic method–shows hypoglycemic activity of lowering the blood glucose level on intraperitoneal (*ip*) injection to mice.[3] (1 → 6)-α-D-Glucopyranan having both α-D-glycopyranosyl and β-D-glucopyranosyl branches at C-3, which is not treated with cellulase, also lowers the blood glucose level in mice, indicating that β-D-glucopyranosyl branches at least do not inhibit the hypoglycemic function of the branched polysaccharide. However, the synthetic linear (1 → 6)-α-D-glucopyranan, having no branches, and (1 → 6)-α-D-glucopyranan, having α- D-mannopyranosyl branches, do not show the hypoglycemic activity. Therefore, glucopyranosyl branching units may be essential for the activity. The hypoglycemic activity of synthetic glucose-branched (1 → 6)-α-D-glucopyranans is detected 5 and 10 h after the i.p. injection, and 24 h after the i.p. injection, the blood glucose level is restored to the normal value. The strength of the hypoglycemic effect depends on the polysaccharide structure, such as the degree of branching and the molecular weight. For example, the synthetic branched polysaccharide with a molecular weight of less than 1.0×10^4 does not have hypoglycemic activity.

REFERENCES

1. Konno, C.; Sugiyama, K.; Kano, M.; Takahashi, M.; Hikino, H. *Planta Med.* **1984,** 434.
2. Tomoda, M.; Shimada, K.; Konno, C.; Sugiyama, K.; Hikino, H. *Planta Med.* **1984,** 436.
3. Hatanaka, K.; Song., S.-C.; Maruyama, A.; Kobayashi, A.; Kuzuhara, H.; Akaike, T. *Biochem. Biophys. Res. Commun.* **1992,** *188,* 16.

POLYSACCHARIDES, SYNTHETIC (Preparation and Biological Activities)

Toshiyuki Uryu
Institute of Industrial Science
University of Tokyo

In 1966, cationic ring-opening polymerization of 1,6-anhydro sugars invested by C. Schuerch provided a synthetic dextran, (1 → 6)-α-D-glucopyranan (a suffix *-an* designates *poly*saccharide), with the same structure as natural dextran.[1]

Although there are various kinds of natural polysaccharides containing $NHCOCH_3$, NH_2, SO_3Na, $NHSO_3Na$, COOH, ether (anhydro), and sulfamide groups other than OH group, all of which interfere with ionic ring-opening polymerization, polysaccharides having N_3, NH_2, and $NHSO_3Na$ groups have been prepared by the ring-opening method.[4] In conventional ring-opening polymerization of anhydro sugars, the hydroxyl group is protected by benzyl and *tert*-butyldimethylsilyl groups, which can be removed after polymerization.

PREPARATIONS

(1→6)-α-D-Glycopyranans

A completely α-stereoregular 2,3,4-tri-*O*-benzyl-(1→6)-α-D-glucopyranan is prepared by ring-opening polymerization of 1,6-anhydro-2,3,4,-tri-*O*-benzyl-β-D-glucopyranose according to **Scheme I.** Debenzylation of the polymer is carried out with sodium in liquid ammonia, that is, by Birch reduction, to give OH-free (1→6)-α-D-glucopyranan: synthetic dextran.

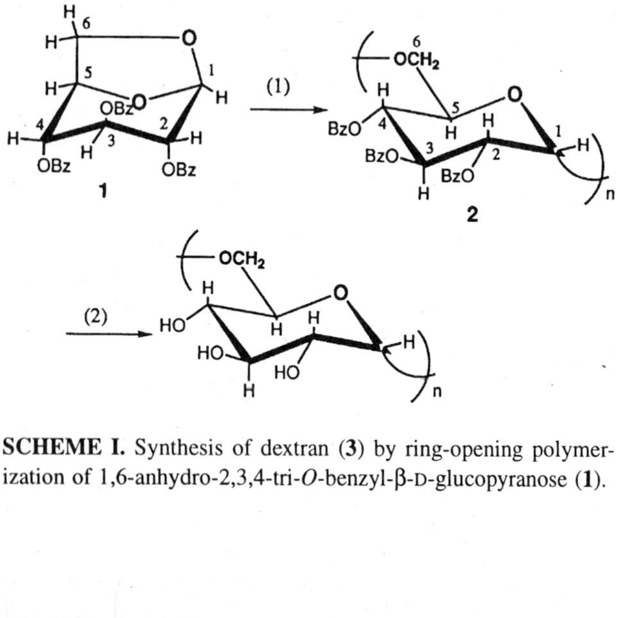

SCHEME I. Synthesis of dextran (**3**) by ring-opening polymerization of 1,6-anhydro-2,3,4-tri-*O*-benzyl-β-D-glucopyranose (**1**).

(1 → 4)-β-D-Ribopyranan and 6-Deoxy-(1 → 4)-α-L-talopyranan: Cellulose-Type Polysaccharides

Because a bicyclic compound 1,4-anhydro-α-D-ribopyranose can be equally regarded as 1,5-anhydro-β-D-ribofuranose, it has two modes of ring-opening polymerizations. When the 1,4-anhydro ring is subjected to scission via oxonium ion mechanism, (1→4)-β-D-ribopyranan, which is called a cellulose-type polysaccharide, (1→4)-β-D-glycopyranan, is formed.[5] On the other hand, when the 1,5-anhydro ring is selectively opened, (1→5)-α-D-ribofuranan is produced. If the sugar is glucose instead of ribose, cellulose will be synthesized by ring-opening polymerization of a 1,4-anhydro-glucose derivative. However, such trials failed to produce cellulose.[6,7]

(1→5)-α-D-Glycofuranans

Selective 1,5-ring-opening polymerization of 1,4-anhydro-2,3-di-*O*-benzyl-α-D-ribopyranose (=1,5-anhydro-2,3-di-*O*-benzyl-β-D-ribofuranose) occurs with Lewis acids such as $BF_3{\cdot}OEt_2$ and PF_5 as catalyst in a short time. The obtained polymer is 2,3-di-*O*-benzyl-(1→5)-α-D-robofuranan.[2] After debenzylation with sodium in liquid ammonia, (1→5)-α-D-robofuranan is obtained.

(1→3)-Glucans and Mannans

Polysaccharide backbones of fungi and mushrooms are composed mainly of 1,3-linked polysaccharides. Cationic ring-opening polymerization of 1,3-anhydro-2,4,6-tri-*O*-(p-bromobenzyl)-β-D-glucopyranose takes place by triflic anhydride or silver triflate catalyst to provide stereoregular (1→3)-α-D-glucopyranans with $\overline{M}_n$ of $11 \times 10^3 - 19 \times 10^3$ and with $[\alpha]_D$ of +259°.[8]

PROPERTIES

Solubility

Synthetic dextran has the same backbone structure as that of natural dextran, (1→6)-α-D-glucopyranan. However, high molecular weight synthetic dextrans have almost no solubility in water because of their completely linear structure. Branched dextrans have higher solubilities than linear ones.[3]

(1→4)-β-D-Ribopyranan and (1→5)-α-D-ribofuranan with $[\alpha]_D$ of −66° and +164°, respectively, are soluble in water. (1→5)-α-D-Lyxofuranan is also soluble in water.

Biological Activities

Synthetic branched dextrans has been used for examining chemical structures producing antibodies *in vivo.*[9] Branched dextrans react with rabbit antibodies to a natural 95%-linear dextran to an increasing extent as the degree of branching increases. Therefore, the antibody has been raised against linear 1,6-α-glucose segments. A synthetic 1,6-α-mannan has been found to induce cutaneous hypersensitivity and humoral antibody formation using guinea pigs and mice.[10]

APPLICATIONS

Synthetic polysaccharides have no practical applications. Since sulfated ribofuranans and ribopyranans exhibit high anti-AIDS (acquired immunodeficiency syndrome) virus activity, their inhibitory effects on virus infection have been examined *in vitro*.[11,12] However, sulfated synthetic polyriboses have high anticoagulant activities which are considered to be side effects against anti-AIDS virus activity.[13] Thus a sulfated curdlan with high anti-AIDS virus activity but low anticoagulant activity has been synthesized.[14] Its phase I/II test for AIDS virus carriers has been carried out.

REFERENCES

1. Ruckel, E. R.; Schuerch, C. *J. Am. Chem. Soc.* **1966,** *88,* 2605.
2. Uryu, T.; Yamanouchi, J. et al. *J. Am. Chem. Soc.* **1983,** *105,* 6865.
3. Ito, H.; Schuerch, C. *J. Am. Chem. Soc.* **1979,** *101,* 5797.
4. Uryu, T.; Hatanaka, K. et al. *Macromolecules* **1983,** *16,* 853.
5. Uryu, T.; Kitano, K. et al. *Macromolecules* **1981,** *14,* 1.
6. Micheel, F.; Brodde, O.-E. *Liebigs Ann. Chem.* **1974,** 702.
7. Uryu, T.; Yamaguchi, C. et al. *Macromolecules* **1985,** *18,* 599.
8. Good, F. J. Jr.; Schuerch, C. *Macromolecules* **1985,** *18,* 595.
9. Torii, M.; Ogawa, S. et al. *J. Biochem.* **1986,** *99,* 263.
10. Grappel, S. F. *Experientia* **1971,** *27,* 329.
11. Yoshida, T.; Katayama, Y. et al. *Macromolecules* **1992,** *25,* 4051.
12. Yoshida, T.; Wu, C. et al. *Macromolecules* **1994,** *27,* 4422.
13. Hatanaka, K.; Yoshida, T. et al. *J. Med. Chem.* **1987,** *30,* 810.
14. Kaneko, Y.; Yoshida, O et al. *Biochem. Pharm.* **1990,** *39,* 793.

Polysilanes

See: *Charge Transport Polymers (for Organic Electroluminescent Devices)*
Magnetic Field Processing
Metal-Containing Silicon-Based Polymers
Polysilanes (Overview)
Polysilanes (Bearing Hydroxy Groups and Derivatives)
Polysilanes (Ring-Opening Polymerization of Cyclotetrasilanes)
Preceramic Polymers
Ultrasonically Assisted Polymer Synthesis

POLYSILANES (Overview)

Samuel P. Sawan* and Sonny A. Ekhorutomwen
Department of Chemistry
Polymer Science/Plastics Engineering Option
University of Massachusetts, Lowell

The polysilanes represent a rather new class of polymers, which have the polymer backbone made up entirely of silicon atoms. These materials differ from other important inorganic polymers, for examples, the siloxanes and the phosphazenes, in which the polymer chains are heteroatomic. The polysilanes are structurally more closely related to the homoatomic organic polymers such as the polyolefins. However, because the units in the main chains are all silicon atoms, the polysilanes exhibit quite unusual properties.

*Author to whom correspondence should be addressed.

The synthesis of the first soluble high molecular weight materials in the late 1970s[1,2] and early 1980s[3-5] has caused an explosive rebirth of interest in polysilanes that continues today.

The cumulated silicon-silicon bonds in the polymer chain allow for extensive electron delocalization to take place. As a consequence, the electronic and photochemical behavior of the polysilanes is very different from that of most other inorganic and organic polymers, in which electron delocalization is much less important. Many of the technical uses, as well as many of the remarkable properties of polysilanes, result from this unusual mobility of the σ electrons.

SYNTHESIS

Wurtz-Type Coupling Reactions

Kipping probably prepared the first substituted polysilanes in the 1920s, by the condensation of dichlorodiphenylsilanes with sodium metal, the so-called Wurtz reductive coupling. It is interesting to note that in spite of considerable efforts to elucidate other useful synthetic schemes, the Wurtz coupling of dichlorosilanes currently remains the most viable general procedure for the preparation of high molecular weight, linear polysilanes. Single dichlorosilanes yield homopolymers, while mixtures of dichlorosilanes lead to copolymers, as shown in **Equations 1** and **2**, respectively.

$$R^1R^2SiCl_2 \xrightarrow[>100°C]{\text{Na, Solvent}} (R^1R^2Si)_n \quad (1)$$

$$R^1R^2SiCl_2 + R^3R^4SiCl_2 \xrightarrow[>100°C]{\text{Na, Solvent}} (R^1R^2Si)_n(R^3R^4Si)_m \quad (2)$$

STRUCTURES OF POLYSILANES

Configurations and Stereochemistry

Polysilanes also exhibit stereoisomerism. Polysilane homopolymers in which all the side groups are identical, such as $(n\text{-}Bu_2Si)_n$, exhibit no stereoisomeric effects, since a plane of symmetry can be drawn through each silicon atom. But in polymers bearing two different substituents on each silicon, for example, $(n\text{-}BuSiMe)_n$, each silicon is a stereogenic center, and the relative configurations of other silicon atoms becomes significant. Thus, polysilanes may exhibit isomerism and thus be found with isotactic, syndiotactic, and heterotactic sequences (or meso and racemic dyads).[11,12]

Ordering in Copolymers

Polysilylene copolymers may have different arrangements of the comonomers: random, blocklike, or ordered, with two kinds of silicon atoms alternating along the polymer chain.

PROPERTIES OF POLYSILANES

Electronic Properties

The polysilanes have attracted much scientific interest due to their unusual electronic characteristics. The σ-σ* delocalization of the Si–Si bonds is crucial for many of the proposed technological applications of polysilanes.

In most polymers, organic or inorganic, delocalization of the σ electrons is not observed due to the high ionization energy of the σ bond. The ionization energies for the Si–Si bonds are much lower than those of π electrons in olefins, and hence polysilanes behave in many ways like conjugated unsaturated carbon polymers. Additionally, the interaction between adjacent silicon orbitals is relatively large. It is convenient to think about this interaction as taking place between neighboring Si–Si bonds, leading to energy splitting by resonance.[13]

There are Si–Si σ antibonding orbitals which are also split by σ* resonance, corresponding to the filled Si–Si bonding orbitals.

Thermochromism and Conformation

Many polysilanes show strong thermochromic behavior.[15,16] For poly(di-*n*-alkylsilylene)s in solution at ordinary temperatures, the limiting value of λ_{max} is 305–315 nm. For polysilanes that contain branched alkyl groups, λ_{max} shifts to longer wavelength, for example, λ_{max} = 326 nm for (cyclohexyl-SiMe)$_n$, probably because more *trans* conformations are present. Many polysilanes show strong reversible thermochromism, with λ_{max} usually moving to longer wavelength as the temperature decreases.[17]

Photochemistry and Photodegradation of Polysilanes

The photochemical properties of the polysilanes are quite unusual and are the major reason for the potential uses of these polymers. Photoscission is usually the main process, especially upon irradiation in air or in solution. Aryl polysilanes, however, show a significant photocrosslinking component. The major reactions in photolysis of polysilanes are homolysis to give silyl radicals, and silylene elimination.[6,7]

APPLICATIONS OF POLYSILANES

Precursors for Silicon Carbide

Interest in polysilanes was aroused in 1975, when Yajima et al.[1,18] found that the permethyl polymer $(Me_2Si)_n$, or its cyclic oligomer $(Me_2Si)_6$, could be transformed into silicon carbide by heating to high temperatures. In this process, either $(Me_2Si)_n$ or the cyclic oligomer $(Me_2Si)_n$ is synthesized from Me_2SiCl_2, which is first converted to the polycarbosilane by heating in air followed by heating in nitrogen to produce the β-SiC.

Polymerization Photoinitiators

In the photolysis of polysilanes, silyl free radicals are produced, which can be used to induce free radical polymerization. Since the silyl radicals can add to carbon-carbon double bonds and begin the formation of polymer chains, polysilanes can be used as radical photoinitiators.[19,20]

Photoresist Applications

The polysilanes are useful as photoresists because they undergo scission or crosslinking when exposed to UV light.[10,14] Both the molar absorptivities and the wavelength of maximum absorption decrease with increasing exposure, so that the polysilanes can be "bleached" photochemically.[21]

At present most microlithography is carried out with visible light. However, there is an increasing need for ever-closer spacing of features on microchips, to increase the speed of operation as well as to permit further miniaturization. At the present maximum resolution of 0.5–1.0 μm, diffraction effects from the light are limiting. For the next increase in resolution, it will be necessary to use light of shorter wavelength. The polysilanes are well suited for this next generation of photoresists, since they are active in the UV region.[22]

Polysilanes also behave as promising photoresists at high energy, in the X-ray region, and can be employed for electron beam imaging without the use of a mask. Furthermore, some polysilanes can be developed without the use of a solvent, using UV laser or deep UV photolysis. These are so called "self-developing" photoresists.[9]

Electrical and Photoconductors

The unusual absorption spectra of the polysilanes suggest potentially interesting conducting and photoconducting characteristics. Although the polysilanes are insulators in the pure state, West and co-workers[23] have demonstrated that the copolymer polysilastyrene can be doped to a semiconducting level with arsenic pentafluoride. The preparation of other electrically conductive polysilanes by doping has been reported by Naarman et al.[24] A number of other workers (e.g., Lee and Sawan)[6] have studied the photoconducting[25,26] and charge-transporting[27-31] characteristics of a variety of polysilane derivatives.

Nonlinear Optical Application

Polysilanes also appear of interest for nonlinear optical (NLO) studies. The first NLO measurement on a polysilane derivative was reported by Kajzar and co-workers.[32] They reported a resonant third-order nonlinear susceptibility ($\chi^{(3)}$) value for poly(phenyl methyl)silane at 1.06 μm of 1.5×10^{-12} esu, as measured by the Maker fringe technique.

REFERENCES

1. Yajima, S.; Hayasi, T.; Omori, M. *Chem. Lett.* **1975,** *9,* 31.
2. Yajima, S.; Okamura, K.; Hayasi, J.; Omori, M. *J. Am. Ceram. Soc.* **1976,** *59,* 324.
3. Hasegawa, Y.; Iimura, M.; Yajima, S. J. *J. Mater. Sci.* **1980,** *15,* 720.
4. Hasegawa, Y.; Okamura, K. *J. Mater. Sci.* **1983,** *18,* 3633.
5. Hasegawa, Y.; Okamura, K. *J. Mater. Sci.* **1985,** *20,* 321.
6. Lee, Y. I.; Hsu, T. M.; Wakin, F. G.; Sawan, S. P. *Polym. Preprint. (Am. Chem. Soc., Div. Polym. Chem.)* **1992,** *33(1),* 1040.
7. Hsu, T. M.; Sawan, S. P. *Polym. Preprint. (Am. Chem. Soc., Div. Polym. Chem.)* **1992,** *33,* 1038.
8. Hsu, T. M.; Sawan, S. P. *Polym. Preprint. (Am. Chem. Soc., Div. Polym. Chem.)* **1992,** *33,* 1036.
9. Zeigler, J. M.; Harrah, L. A.; Johnson, A. W. *SPIE-Advances in Resist Technology and Processing* **1985,** *539,* 166.
10. Gauthier, S.; Worsfold, D. *Macromolecules* **1989,** *22,* 2214.
11. Randall, J. C. *Polymer Sequence Determination*; Academic: New York, NY; 1977; pp 1-40.
12. Bovey, F. A. *Acct. Chem. Res.* **1968,** *1,* 175.
13. Bock, H.; Ensslin, W. *Angew. Chem. Int. Ed. Engl.* **1971,** *10,* 404.
14. Miller, R. D.; Rabolt, J.; Sooriyakumaran, R.; Fleming, W.; Fickes, G. N.; Farmer, B. L.; Kuzmany *In Inorganic and Organometallic Polymers*; Zeldin, M.; Wynne, K. J.; Allcock, H. R., Eds.; A.C.S. Symp. Ser.; ACS: Washington, DC, 1988; Vol. 360, pp 43-60.

15. Trefonas, P.; Damewood, R. Jr.; West, R. *Organometallics* **1985,** *4,* 1314.
16. Harrah, L. A.; Zeigler, J. M. *J. Polym. Sci., Polym. Lett.* **1985,** *23,* 209.
17. Rabolt, J. F.; Hofer, D.; Miller, R. D. *Macromolecules* **1986,** *19,* 611.
18. Yajima, S.; Hayashi, J.; Okamura, K. *Chem. Lett.* **1975,** 1209.
19. West, R.; Wolff, A. R.; Peterson, D. J. *J. Radiat. Curing.* **1986,** *13,* 35.
20. Wolff, R.; West, R. *Appl. Organomet. Chem.* **1987,** *1,* 7.
21. Trefonas, P. et al. *J. Polym. Sci., Polym. Chem. Ed.* **1983,** *22,* 822.
22. Miller, R. D.; Wallraff, G.; Clecak, N.; Sooriyakamaran, R.; Michl, J.; Karatsu, T.; Mckinley, A. J.; Klingensmith, K. A.; Downing, J. *J. Polym. Mater. Sci. Eng.* **1989,** *60,* 49.
23. West, R.; David, L. D.; Djurovich, P. I.; Searley, K. L.; Srinavasan, K. S. V.; Yu, H. *J. Am. Chem. Soc.* **1981,** *103,* 7352.
24. Naarman, H.; Theophilou, N.; Geral, L.; Sledz, J.; Schien, F. German Patent DE 3 634 281; *Chem. Abstr.* **1988,** *109,* 120989u.
25. Kepler, R. G.; Zeigler, J. M.; Harrah, L. A.; Kurtz, S. R. *Phys. Rev. B* **1987,** *35,* 2818.
26. Fujino, M. *Chem. Phys. Lett.* **1987,** *136,* 45.
27. Stolka, M.; Yuh, H. J.; McGrane, K.; Pai, D. M. *J. Polym. Sci., Polym. Chem. Ed.* **1987,** *25,* 823.
28. Abkowitz, M.; Knier, F. E.; Yuh, H. J.; Eeagley, R. J.; Stolka, M. *Solid State Commun.* **1987,** *62,* 547.
29. Abkowitz, M.; Stolka, M. *Phlos. Mag. Lett.* **1988,** *58,* 39.
30. Samuel, L. M.; Sanda, P. N.; Miller, R. D. *Chem. Phys. Lett.* **1989,** *159,* 227.
31. Rice, M. J.; Phillpot, S. R. *Phys. Rev. Lett.* **1987,** *58,* 937.
32. Kajzar, F.; Messier, J.; Rossilio, C. *J. Appl. Phys.* **1986,** *60,* 3040.

POLYSILANES (Bearing Hydroxy Groups and Derivatives)

Shuzi Hayase
Materials and Devices Research Laboratories
Research and Development Center
Toshiba Corporation

One of the most frequently used methods to obtain high molecular weight polysilanes is that using sodium, which eliminates chlorine atoms and puts the Si atoms together (reductive coupling method) in toluene.[1,2] This is a convenient method. However, the structural varieties have been limited to polysilanes bearing alkyl or phenyl groups, because the sodium reacts with functional groups such as hydroxyl, carboxyl, and amino.

Hydroxy substituents appeared attractive, because the OH group is one of the simple and basic functional groups. Once the polysilane is synthesized, various functional groups can be incorporated into the polysilane side chains without damaging the labile silicon chains.

SYNTHESIS OF POLYSILANES BEARING HYDROXYL GROUPS

Polysilanes bearing hydroxyl groups were synthesized by the method described in **Scheme I.**[5-8]

Dichlorosilane monomers whose OH group was protected with silyl ethers were polymerized by the Wurtz coupling reaction. After the polymerization, the silyl ether was removed under mild conditions.

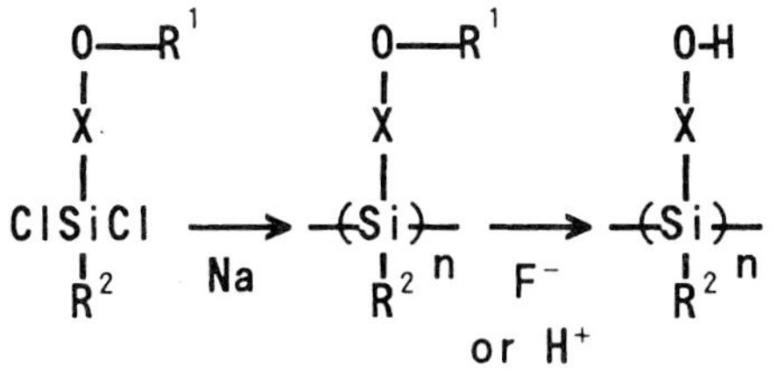

I

INTRODUCTION OF FUNCTIONAL GROUPS

It is easy to introduce various functional groups through the OH group of the polysilanes.

Acid anhydrides and acid chlorides react with polysilanes with hydroxyl groups in the presence of a triethylamine or a *p*-toluenesulfonic acid. Polysilanes bearing carboxylic acids, halogens, and crosslinkable acrylic groups have been synthesized.[8]

PROPERTIES OF POLYSILANES BEARING OH GROUPS

Solubility in Various Solvents

Polysilanes bearing only alkyl or phenyl groups are soluble in hydrophobic solvents such as toluene, xylene, and hexane. On the other hand, polysilanes bearing OH groups are soluble in hydrophilic solvents.[8]

The introduction of carboxylic acids increased the solubility in water containing bases.[8]

Thermal Stability

Thermal stability of polysilanes bearing OH groups is a little less than that of polysilanes bearing only alkyl or phenyl groups, except for the polysilane in which the OH group is substituted at the *para* position of the Si.[8]

APPLICATIONS

Resist Materials for Deep UV Lithography

Polysilane is one of the candidates for deep UV resists.[4,8] Because photodecomposition products of polysilanes bearing only alkyl or phenyl groups are not soluble in basic aqueous solutions, development had to be carried out with organic solvents, such as ketones and alcohols, which is not compatible to currently conducted large scale integration (LSI) lithography, where development uses water containing dilute organic bases. Polysilanes bearing phenolic OH have the potential to be resists that can be developed with basic aqueous solutions.

Polysilane Composites

Polysilanes bearing OH groups are soluble in hydrophilic solvents such as alcohols, making it possible to employ the polysilane groups for reactions in hydrophilic solvents. One representative example is the preparation of a polysilane-ultrafine metal powder composite. We expect unusual optical properties on the basis of quantum effects for these composites.

Highly Oriented Polysilane Thin Films Prepared by the Langmuir-Blodgett Technique

Unusual properties of the polysilanes have been attributed to the extensively delocalized σ electrons of the Si bonds.[3,4] Since the σ electrons are delocalized along the Si main chains, it would be very useful to make oriented polysilane thin films to realize polysilane devices. The Langmuir-Blodgett (LB) technique is an excellent method for preparing ultrafine thin films with a well-defined molecular order.[9-11]

Anisotropy of various physical properties has been observed. For example, the third-order nonlinear optical constant in the direction parallel to the Si–Si main chain was 2.8×10^{11} esu, which is 4 to 5 times as large as that perpendicular to the Si–Si main chain. This suggests that the third-order nonlinear optical properties are strongly dominated by the oscillator strength of the silicon chains. The same conclusion has been drawn by Embs,[12] employing poly(diphenylsilane) derivatives for making oriented LB films.

CONCLUSION

We have succeeded in synthesizing high molecular weight polysilanes bearing OH groups. The synthesis method had to be adjusted, depending on the substituents on Si. Properties for the polysilanes bearing OH groups were different from those for conventional polysilanes bearing only alkyl or phenyl groups. Solubilities, miscibilities, polysilane composites, resist properties, oriented LB films, and the third-order nonlinear optical properties were discussed. The work described here is just the first step in research into polysilanes bearing functional groups, the synthesis of which will open a range of new fields in polysilane design and application.

REFERENCES

1. West, R.; David, L. D.; Djurovich, P. I.; Stearley, K. L.; Srinivasan, K. S. V.; Yu, H. J. *J. Am. Chem. Soc.* **1981,** *103,* 1352.
2. Trujillo, R. E. *J. Organomet. Chem.* **1980,** *198,* C27.
3. West, R. In *The Chemistry of Organic Silicon Compounds*; Patai, S.; Rappoport, Z., Eds.; John Wiley & Sons: New York, 1989; p 1207.
4. Miller, R. D. *Chem. Rev.* **1989,** *89,* 1359.
5. Horiguchi, R.; Ohishi, Y.; Hayase, S. *Macromolecules* **1988,** *21,* 304.
6. Nakano, Y.; Murai, S.; Kani, R.; Hayase, S. *J. Polym. Sci., Polym. Chem. Ed.* **1993,** *31,* 3361.
7. Kani, R.; Nakano, Y.; Majima, Y.; Hayase, S.; Chien-Ha Yuan; West, R. *Macromolecules* **1994,** *27,* 1911.
8. Hayase, S.; Horiguchi, R.; Onishi, Y.; Ushirogouchi, T. *Macromolecules* **1989,** *22,* 2933.
9. Good, R. D.; Stromberg, R. R.; Patrick, R. L. *Technique of Surface and Colloid Chemistry and Physics*; Marcel Dekker: New York, 1972; Vol. 1.
10. Nishikata, Y.; Morikawa, Y.; Takiguchi, Y.; Kanemoto, M.; Kakimoto, M.; Imai, Y. *Jpn. J. Appl. Phys.* **1988,** *27,* L1163.
11. Nishikata, Y.; Morikawa, Y.; Takiguchi, Y.; Kanemoto, M.; Kakimoto, M.; Imai, Y. *Nippon Kagaku Kaishi* **1987,** 2174.
12. Embs, F. W.; Wegner, G.; Neher, D.; Albouy, P.; Miller, R. D.; Willson, C. G.; Schrepp, W. *Macromolecules* **1991,** *24,* 5068.

POLYSILANES (Ring-Opening Polymerization of Cyclotetrasilanes)

Krzysztof Matyjaszewski,* Eric Fossum, and Peter Miller
Department of Chemistry
Carnegie Mellon University

*Author to whom correspondence should be addressed.

Polysilylenes (polysilanes) are an interesting class of polymers that consist of linear chains of silicon atoms bearing two substituents, generally either alkyl or aryl.[1-3]

Polysilylenes have been prepared by several methods, including the reductive coupling of dichlorosilanes,[1-3] dehydrogenative coupling of primary silanes,[4] anionic polymerization of masked disilenes,[5] and ring-opening polymerization (ROP) of cyclotetrasilanes.[6,7] Only the latter two methods provide routes to polysilylenes with controlled structures. Because the electronic properties of polysilylenes depend on the nature of the substituents and their configuration as well as the conformation of the backbone, it is necessary to prepare well-defined polymers with controlled microstructures.[8,9]

ROP of cyclotetrasilanes allows for potential control over the resulting microstructure because monomers with known configurations can be prepared.[10] In addition, ROP affords the possibility of controlling the molecular weight distributions of the polymers, the synthesis of polysilylenes with substituents that would not survive the reductive coupling process, and also the preparation of a variety of copolymers that cannot be prepared by other methods.

CYCLOTETRASILANES WITH METHYL AND PHENYL SUBSTITUENTS

Most of the cyclotetrasilanes used for polymerizations were prepared by dearylation reactions of aryl-containing cyclotetrasilanes using trifluoromethanesulfonic (triflic) acid in methylene chloride, followed by methylation with methylmagnesium bromide in a mixture of benzene/toluene, as depicted in **Scheme I.**

POLYMERIZATION

All of the cyclotetrasilanes described possess sufficient ring strain to be polymerized and have been found to polymerize under anionic conditions employing either *n*-BuLi or silyl cuprates as initiators.

OTHER POLYMERIZABLE CYCLOSILANES

Several other cyclosilanes have been prepared and found to be polymerizable. Suzuki and co-workers have synthesized nonamethylphenylcyclopentasilane, which polymerized anionically at –78°C to afford the corresponding high molecular weight linear polymer.[11] Gauthier and Worsford have prepared polymerizable cyclotetrasilanes utilizing the Wurtz coupling of dichlorosilanes.[12] Both $Me_4Hex_4Si_4$ and $Me_4n\text{-}Prop_4Si_4$ were obtained in reasonable yields. The cyclotetrasilanes were polymerized thermally at temperatures above 100°C giving rise to high molecular weight linear polysilylenes.

SYNTHETIC UTILITY OF ROP OF CYCLOTETRASILANES

Preparation of Block Copolymers

The ROP of cyclotetrasilanes has proven to be a versatile route to a variety of copolymers. A series of block copolymers with both polystyrene and polyisoprene were prepared by initiation of the ROP of **4** with polystyryl lithium and polyisoprenyl lithium, respectively.[13] Molecular weights were in the range M_n=10,000–40,000 with narrow molecular weight distributions, $M_w/M_n < 1.3$.

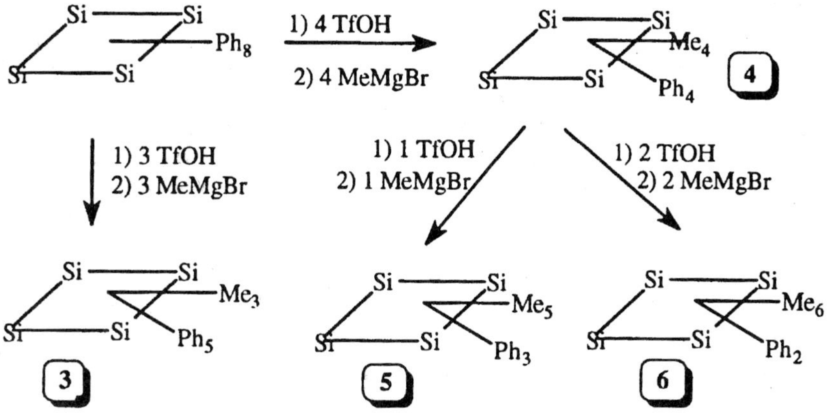

I

Random Copolymers with Ferrocenylsilanes

Cyclotetrasilane **4** was found to undergo thermal ROP at 150°C.[14] High molecular weight (>50,000) random copolymers consisting of poly(ferrocenylsilane)segments and polymethylphenylsilylene segments have been prepared by the simultaneous thermal ROP of **4** with [1]-dimethylsilylferrocenophane.[14]

REFERENCES

1. Michl, J.; Miller, R. D. *Chem. Rev.* **1989,** *89,* 1359.
2. West, R. *J. Organomet. Chem.* **1986,** *300,* 327.
3. Matyjaszewski, K.; Cypryk, M.; Frey, H.; Hrkach, J.; Kim, H. K.; Moeller, M.; Ruehl, K.; White, M. *J. Macromol. Sci. Chem.* **1991,** A28, 1151.
4. Tilley, T. D. *Acc. Chem. Res.* **1993,** *26,* 22.
5. Sakamoto, K.; Obata, K.; Hirata, H.; Nakajima, M.; Sakurai, H. *J. Am. Chem. Soc.* **1989,** *111,* 7641.
6. Cypryk, M.; Gupta, Y.; Matyjaszewski, K. *J. Am. Chem. Soc.* **1991,** *113,* 1046.
7. Chrusciel, J.; Cypryk, M.; Fossum, E.; Matyjaszewski, K. *Makromol. Chem., Macromol. Symp.* **1993,** *73,* 167.
8. Sundararajan, P. R. *Macromolecules* **1988,** *21,* 1256.
9. Welsh, W. J.; Damewood, J. R. Jr.; West, R. C. *Macromolecules* **1989,** *22,* 2947.
10. Fossum, E.: Gordon-Wylie, S. W.; Matyjaszewski, K. *Organometallics* **1994,** *13,* 1695.
11. Suzuki, M.; Kotani, J.; Gyobu, S.; Kaneko, T.; Saegusa, T. *Macromolecules* **1994,** *27,* 2360.
12. Gauthire, S.; Worsfold, D. J. In *Inorganic and Organometallic Oligomers and Polymers*; Harrod, J. F.; Laine, R. M., Eds.; Kluwer: Dordrecht, 1991; p 37.
13. Fossum, E.; Love, J.; Matyjaszewski, K. *J. Organomet. Chem.* **1995,** *499,* 253.
14. Fossum, R.; Matyjaszewski, K.; Rulkens, R.; Manners, I. *Macromolecules* **1995,** *28,* 401.

Polysilazanes

See: *Polysilazanes (Through Ring-Opening Polymerization)*
Preceramic Polymers

POLYSILAZANES (Through Ring-Opening Polymerization)

Alain Soum,* L. Billon, M. Bouquey, S. Bruzard, E. Duguet, and M. Schappacher
Laboratoire de Chimie des Polymères Organiques
Ecole Nationale Supérieure de Chimie et de Physique de Bordeaux

In the field of ceramic precursors, polysilazanes (**Structure 1**) were promising candidates;[1,2] much work has thus been done toward their synthesis through aminolysis and ammonolysis of dichlorosilanes[3,4] or via deamination or redistribution reactions of aminosiliazanes.[5,6]

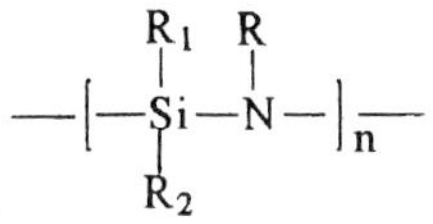

1

Ring-opening polymerization of cyclosilazanes also gives access to polysilazanes, but has been rapidly overlooked because of the difficulties of synthesizing the corresponding monomers.[9,10]

Recently, ring-opening polymerization of cyclosilazanes has been reinvestigated and demonstrated that, under specific conditions, high molar mass polymers with a well defined linear structure can be prepared via anionic or cationic processes. These polysilazanes are semicrystalline and exhibit specific physical properties which make them a new class of polymers.[9,10]

SYNTHESIS OF MONOMERS

Monomers best suited to undergo a ring-opening process, hence to form these new polysilazanes, are cyclosilazanes of general formula **2**. Whatever the polymerization process (anionic or cationic) and the silicon substituents, only *N,N′*-dimethyl 4-membered silazane rings polymerize, whereas 8-members rings oligomerize and 6-membered ones do not sustain ring-opening.

*Author to whom correspondence should be addressed.

Cyclodisilazanes (**Structure 2**) are synthesized according to the method first described by Lienhard et al.[11] and then by Fink.[12]

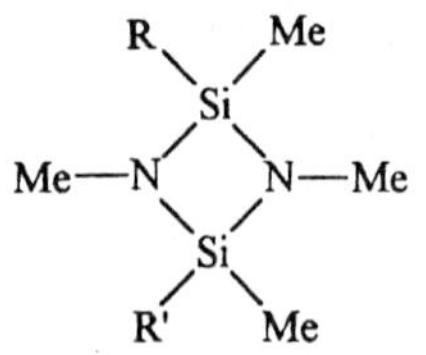

2

R = R' = Methyl

R = R' = Vinyl or Ethyl

R = Methyl, R' = Vinyl, Ethyl or Phenyl

RING-OPENING POLYMERIZATION

Polymerization of cyclic monomers has been attempted through radical, rare earth, or transition metal catalyzed processes, as well as anionic and cationic mechanisms. Only the two latter methods give rise to polymers.

RING-OPENING COPOLYMERIZATION

Random and block copolymers have been prepared through anionic copolymerizations, involving either different cyclodisilazanes[13] or a cyclosilazane and a comonomer such as cyclosiloxanes or vinylics (styrene).

PROPERTIES OF POLY(N-METHYLSILAZANE)

Polysilazanes derived from monomers 2 are white solids (ρ~0.9 g/cm^3) and are soluble in the usual organic solvents. Contrary to what is observed with conventional polysilazanes, certainly due to substitution of the nitrogen atom, those prepared via anionic polymerization are stable under ambient air.

Phase Transitions

DSC analyses show that most of the poly(N-methylsilazane)s synthesized are semicrystalline polymers with a glass transition temperature (T_g) lower than ambient temperature and a melting temperature (T_m) which varies from 130 to 230°C, depending on the structure of the silicon substituent.

Surprisingly, some polysilazanes exhibit a secondary crystalline transition (T_{cc}) below melting temperature.

Another important feature is the low value of the T_g/T_m ratio (T_g/T_m <0.5). This agrees with the high linear structure of polysilazanes and makes the latter polymers similar to polymethylene, polydifluoromethylene, and polyoxymethylene from the phase transition point of view.

The semicrystalline structure of some polysilazanes is demonstrated by X-ray analysis whose diffraction patterns exhibit the presence of an intense peak. The intensity of this peak, which is almost unique, indicates a high degree of orientation of the polymer chains.

Thermal Stability and Mechanical Properties

The [Si-N] bond is known to be more stable than the [Si-O] bond.[14] This is confirmed by TGA analyses, which show that polysilazanes do not present weight loss before 450°C; this is 50°C higher than the corresponding polysiloxanes.

Poly(*N*-methylsilazane)s exhibit interesting mechanical properties which corroborate the DSC analyses. They present a glassy behavior below –50°C and a large viscoelastic domain that ranges from –50 to 200°C depending on the silicon substituents. More important are the values of the storage modulus E′ which are at least ten times higher than those of the corresponding polysiloxanes in glassy and rubbery domains.

APPLICATIONS OF POLY(*N*-METHYLSILAZANE)

Until now, applications of conventional polysilazanes have not been extensively studied except as ceramic precursors (certainly because of poor control of the structure and lack of stability of the oligomers). In the field of precursors to ceramics, poly(*N*-methylsilazane)s are not very useful since the ceramic yield is low (50% at 1000°C) and the carbon free content is high (29% at 1000°C).

Nevertheless, the potential of semicrystalline polysilazanes as materials has not been completely explored. The most promising applications might be in the same area as that of polysiloxanes (hydrophobic behavior, low variations of physical constants with temperature, high chemical and thermal stabilities, and low fire hazard) with the advantages of a semicrystalline thermoplastic polymer (in the process) and a higher mechanical strength. In particular, the higher value of the rubbery modulus of polysilazane networks and its stability in a larger temperature range lead one to expect applications in reinforcement of siloxane elastomers.

REFERENCES

1. Laine, R.; Blum, Y. D. et al. *ACS Symp. Ser.* **1988,** *360,* 142.
2. Mark, J. E.; Allcok, H. R.; West, R. *Inorganic Polymers*; Prentice Hall: Englewood Cliffs, NJ, 1992; Chapter 6.
3. Seyferth, D.; Wiseman, G. H.; Prud'homme, C. *J. Am. Ceram. Soc.* **1983,** *66,* C13.
4. Lebrun, J. J.; Porte, H. Europ. Patent 197863, 1985.
5. Verbeck, W. U.S. Patent 3 853 567, 1974.
6. Gaul, J. H. U.S. Patent 4 340 619, 1982.
7. Küger, C. R.; Rochow, E. G. *J. Polym. Sci.* **1964,** A2, 3179.
8. Seyferth, D.; Wiserman, G. H. *J. Am. Ceram. Soc.* **1984,** *67,* C132.
9. Duguet, E.; Schappacher, M.; Soum, A. French Patent 2674859, 1992.
10. Duguet, E.; Schappacher, M.; Soum, A. *Macromolecules* **1992,** *25,* 4835.
11. Lienhard, K.; Rochow, E. G. *Angew. Chem.* **1963,** *75,* 638.
12. Fink, W. *Helv. Chim. Acta* **1964,** *47,* 498.
13. Bouquey, M.; Soum, A. *Polym. Int.*, in press.
14. Fink, W. *J. of Paint Techn.* **1970,** *42,* 221.

Poly(siloxane-acetylene)s

See: Inorganic/Organic Hybrid Polymers (High Temperature, Oxidatively Stable)

Polysiloxanes

See: Anoxic Polymer Materials
Cyclic Siloxanes (Ring-Opening Polymerization)
Dendrimers, Polysiloxanes

Functional Polysiloxanes
Photocrosslinking (Overview)
Polydimethylsiloxane Networks, Monodisperse
Poly(imide siloxane)s
Polymethylphenylsiloxane (Photophysics, Conformation and Dynamics)
Polyoxyarylenesiloxanes
Recycling, Thermodynamic (by Ring-Closing Depolymerization)
Silicone Elastomers
Silicone Polymers, Organo-Modified (Application in Personal Care Products)
Silicone Release Coatings
Silicone Rubber Latex
Silicone Sealants
Siloxane-Containing Polymers
Siloxane Elastomers
Step Polymerization Catalysts

Polysilsesquioxanes

See: *Bridged Polysilsesquioxanes (Highly Porous Hybrid Organic-Inorganic Materials)*
Silsesquioxane-Based Polymers

Polysilylenes

See: *Polysilanes*

POLYSILYLPROPYNES (Steric Effects on Material Properties)

Chao-fong Tien* and A. C. Savoca
Air Products and Chemicals, Incorporated

Polysilylpropynes (**Structure 1**) are a series of synthetic polymers first developed and investigated by Masuda, Higashimura, and co-workers.[1] These polymers possess a polyacetylene main chain with alternating pendant methyl and substituted silyl groups attached to the polymer's backbone.

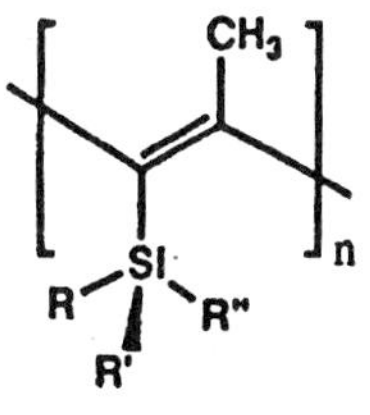

1

Typically, the pendant silicon atom has three organic groups attached, such as methyl, ethyl, phenyl, or higher inert hydrocarbon moieties. Variations in these organic groups impact polymer physical properties. Notable are the effects on gas permeability and glass-transition temperature. The gas permeability of a polysilylpropyne is very sensitive to the substituents on silicon. A three to four order change in oxygen permeability can be achieved by simply replacing one of the organic groups attached to the silicon.[2] On the other hand, irrespective of the organic groups attached to the silicon, the glass-transition temperature of polysilylpropynes are all higher than their decomposition temperature (>300°C).[2,3]

Since it was first reported in 1983 by Masuda, Higashimura, and co-workers, polytrimethylsilylpropyne's (PTMSP) high gas permeability has received considerable attention from both scientific and application point of view.[1] Despite the fact that it is in the glassy state with T_g higher than 300°C, PTMSP exhibits the highest gas permeability known to any polymer, in fact an order of magnitude higher than that of silicon rubber.

Polysilylpropynes can be readily prepared from silylpropyne monomers in high molecular weight in the presence of transition metal halide catalyst, such as niobium or tantalum pentahalide.[1,4]

MAIN CHAIN CONFORMATION

Based on theoretical calculations, the thermodynamically most favorable conformation for polysilylpropyne is a helical chain.[6,7] However, all experimental observations indicate that polysilylpropyne adopts a rigid and an irregular chain conformation instead of the helical chain.[5]

PHYSICAL, MECHANICAL, AND GAS TRANSPORT PROPERTIES

Appearance

Polyacetylene is a deeply colored and crystalline material and is well known for its conducting property. However, polysilylpropynes are colorless, nonconductive, and amorphous materials. Thus, the attachment of methyl and silyl groups to polyacetylene backbone exerts an unexpected effect on the polymer's appearance. As mentioned above, in order to relieve the steric strain between pendant groups, polysilylpropyne exists in a rigid and irregular conformation with main chain π orbitals orthogonal to each other.

The orthogonal conformation will minimize the π orbital overlap. There is virtually no conjugation between a polysilylpropyne's π orbitals. This absence of any extensive conjugation is exemplified by the colorless and non-conductive nature of polysilylpropynes. The rigid and irregular conformation prevents any intimate and orderly packing in the solid state and leads to a polysilylpropyne's amorphous nature.

Environmental and Chemical Stability

Unlike polyacetylene, polysilylpropynes are air stable.[1] Air stability may also be attributed to the lack of π bond conjugation. Electronically, polysilylpropyne's main-chain double bonds are similar to isolated double bonds. But chemically, polysilylpropyne's double bonds are inert to many reactions that are characteristic for isolated double bonds, such as bromination, epoxidation, carbene addition, hydroboration, hydrogenation, hydrosilylation, etc.

Physical and Mechanical Properties

Most polysilylpropynes are soft materials and show relatively large elongation at break. These polymers exhibit initial weight loss at temperature near 300°C in nitrogen. DSC and DMS measurement from −100 to +300°C do not reveal a thermal transition that is characteristic of glass transition temperature.

*Author to whom correspondence should be addressed.

Gas Transport Properties

The most unusual property of polysilylpropyne is gas transport properties. The oxygen permeability of PTMSP is 10 to 15 times higher than that of polydimethylsiloxane. Polydimethylsiloxane had the highest gas permeability known to polymer before 1983 and it is in the rubbery state.[1] Polytrimethylsilylpropyne is a glassy polymer with T_g higher than 300°C. This discovery leads to a new direction in designing polymers with high gas permeability and provides a better understanding of the mechanism of gas transport within polymer matrix.

The oxygen permeability coefficient of polysilylpropynes are inversely proportional to the side chain length. The longer the side chain, the lower the gas permeability coefficient.

APPLICATIONS

One potential application for the highly gas permeable PTMSP is in the area of gas separation membranes. However, high gas permeability is always accompanied by low selectivity. This lack of selectivity severely limits the use of PTMSP as membrane separation barrier when high purity gas is needed. However, there are several novel approaches that significantly expand the potential applications of PTMSP. One is the hybrid approach that combines membrane separation with other separation processes.

Another notable approach is the surface modification of PTMSP to form a composite system.[8] Surface modification is achieved by chemical reaction of PTMSP's surface with some reactive species that generally leads to a very thin skin with better separation characteristics. This thin separation surface is supported by the very permeable PTMSP bulk. In some cases, very high selectivity and permeability can be obtained for the composite system. The high permeable PTMSP also shows promise in areas such as absorbents for organic materials, protective coatings for slow release, contact lens, and pervaporation.

REFERENCES

1. Masuda, T.; Isobe, E.; Higashimura, T.; Tanaka, K. *J. Am. Chem. Soc.* **1983,** *105,* 7473.
2. Savoca, A. C.; Surnamer, A. D.; Tien, C. F. *Macromolecules* **1993,** *23,* 6211.
3. Masuda, T.; Tang, B. Z.; Tanaka, A.; Higashimura, T. *Macromolecules* **1986,** *19,* 1459.
4. Masuda, T.; Isobe, E.; Higashimura, T. *Macromolecules* **1985,** *18,* 641.
5. Masuda, T.; Higashimura, T. *Adv. Polym. Sci.* **1987,** *81,* 121.
6. Clough, S. B.; Sun, X. F.; Tripathy, S. K.; Baker, G. L. *Macromolecules* **1991,** *24,* 4296.
7. Savoca, A. C., unpublished results.
8. Langsam, M. U.S. Patent 4 657 564, 1987.

POLYSOAPS

Ulrich P. Strauss
Department of Chemistry
Rutgers University

Polysoaps are polymers to whose backbone surfactant molecules are attached.[1]

POLYVINYLPYRIDINE-DERIVED POLYSOAPS

A polysoap prepared from poly-2-vinylpyridine by quaternizing approximately one-third of its pyridine groups with *n*-dodecyl bromide showed characteristics typical of all polysoaps so far.[1] Its reduced specific viscosity in aqueous solution was much lower than that of normal polyelectrolytes of comparable molecular weight, indicating a compact molecular structure. The compactness was attributed to intramolecular micelle formation due to hydrophobic interactions between the dodecyl groups. The existence of soap-like micelles was confirmed by the observation that the polysoap solutions solubilized hydrocarbons normally insoluble in water. However, unlike ordinary surfactants, the polysoap required no critical micelle concentration for solubilization to occur.

TRANSITION FROM POLYELECTROLYTE TO POLYSOAP

The striking transition from typical polyelectrolyte to polysoap is illustrated by results found with a series of five poly(4-vinylpyridine) derivatives obtained by quaternizing parts of the pyridine groups with *n*-dodecyl bromide and the remainders with ethyl bromide.[2]

MALEIC ANHYDRIDE-ALKYL VINYL ETHER COPOLYMERS

While the demonstration of the transition between polysoap and polyelectrolyte required a series of various polyvinylpyridine derivatives, certain members of the family of the hydrolyzed copolymers of maleic anhydride and alkyl vinyl ethers allow this transition to be observed with a single compound. Depending on the size of the alkyl group, they fall into three classes of distinct physical behavior. Those members whose alkyl group contains from one to three carbon atoms behave as typical polyelectrolytes at all pH-values; those members whose alkyl groups contain more than nine carbon atoms behave as typical polysoaps.[4-7]

The most interesting behavior is observed with those members with four to eight carbon atoms in their alkyl groups. They behave as polysoaps at low pH where the carboxylate groups are little ionized and as polyelectrolytes at high pH where the carboxylate groups are highly ionized. In an intermediate pH-region they exhibit a transition between the two formations.[4,8]

APPLICATIONS

An early industrial application of polysoaps was based on the combination of their solubilizing ability with their low surface activity, which was utilized to minimize undesirable foaming in the emulsion polymerization of synthetic rubber. Since then, most of the applications that have emerged are based on the high viscosity resulting from the intermolecular association obtainable with polysoaps of the appropriate hydrophobic group content. While simple polymers and polyelectrolytes can produce high viscosities if their molecular weight is large enough, these large macromolecules tend to degrade, with undesirable accompanying viscosity loss. In contrast, hydrophobically modified macromolecules do not have to be of high molecular weight to produce high viscosities through association. Because of the low molecular weight, degradation is insignificant and the viscosity remains high.[9] Furthermore, the viscosity due to simple polyelectrolytes is lowered by added salt, necessarily present in many applications, whereas added salt may raise the viscosity caused by the intermolecular association of polysoaps.[3] The thickening action of polysoaps has been utilized in formulations of coatings, paints, varnishes, cosmetics and other personal care items, in drilling fluids, and in the flooding of oil reservoirs for enhanced oil recovery.[10]

REFERENCES

1. Strauss, U. P.; Jackson, E. G. *J. Polym. Sci.* **1951,** *6,* 649.
2. Strauss, U. P.; Gershfeld, N. L. *J. Phys. Chem.* **1954,** *58,* 747.
3. Strauss, U. P.; Gershfeld, N. L.; Crook, E. H. *J. Phys. Chem.* **1956,** *60,* 577.
4. Dubin, P.; Strauss, U. P. *J. Phys. Chem.* **1967,** *71,* 2757.
5. Martin, P. J.; Morss, L. R.; Strauss, U. P. *J. Phys. Chem.* **1980,** *84,* 577.
6. Ito, K.; Ono, H.; Yamashita, Y. *J. Colloid Sci.* **1964,** *19,* 28.
7. Varoqui, R.; Strauss, U. P. *J. Phys. Chem.* **1968,** *72,* 2507.
8. Martin, P. J.; Strauss, U. P. *Biophys. Chem.* **1980,** *11,* 397.
9. Fernando, R. H.; Lundberg, D. J.; Glass, J. E. *Polymers in Aqueous Media: Performance Through Association; Advances in Chem. Ser. 223* American Chemical Society: Washington, DC, 1989; Chapter XII, Reference 4.
10. McCormick, C. L.; Johnson, C. B. *Polymers in Aqueous Media: Performance Through Association; Advances in Chem. Ser. 223* American Chemical Society: Washington, DC, 1989; Chapter XX, Reference 4.

Polystyrene

See: *Bulk Polymerization*
Degradation (Thermal, Polystyrene and Related Vinyl Polymers)
Foamed Plastics
Gelatin, Physical (Atactic Polystyrene)
Head-to-Head Polymers
Metallocene Catalysts
Photocrosslinking (Overview)
Poly(tert-butylstyrene)
Polystyrene (by Suspension Polymerization)
Polystyrene and Derivatives, Photolysis
Polystyrene, High Impact
Polystyrene Manufacture (Using Bulk Free Radical Polymerization)
Polystyrene, Rubber Toughened
Polystyrene, Stereoregular
Polystyrene, Syndiotactic
Polystyrene, Syndiotactic (Sulfonation and Crystallization)
Styrene, Stereospecific Polymerization
Styrenic Resins
Thermoreversible Gels (Isotactic, Syndiotactic, and Atactic Polystyrene)
Zinc Catalysts (Polymerization of Styrene and Methyl Methacrylate)

POLYSTYRENE (by Suspension Polymerization)

Erol Erbay and Tulin Bilgic
Petkim Petrochemicals Holding Incorporated
Research and Development Centre

The suspension polymerization technique for the production of polystyrene [9003-53-6] was first introduced in the 1940s by the Koppers Chemical Company, then showed rapid development in the 1950s. The suspension process is still in use for the production of polystyrene.[1] However, it has been replaced by more economical techniques such as continuous mass process in the recent years.[2]

The suspension process has several advantages compared with the others. The simplicity of the process, the good control of the polymerization temperature, the lower viscosity of the reaction medium, and the easy removal of the heat of reaction released during the polymerization of styrene are the great advantages of the suspension technology. In addition to general purpose polystyrene, expandable polystyrene and high-impact polystyrene are also produced by this technique.

In the production of polystyrene by the suspension process, styrene, initiators, a suspending agent, and other additives are dispersed in a medium where they are insoluble. Water generally is used as the suspension medium where styrene has very low solubility; however, only 0.062% of styrene is soluble in water at 80°C.[3]

Suspension polymerization is usually carried out as a batchwise process.[4-8] Traditional glass-lined batch reactors have been replaced by larger stainless-steel clad reactors.[9] Dcpending on the hydrodynamic conditions — especially agitation in the reaction kettle — styrene is dispersed as droplets with a diameter of 0.1 to 2 mm. During the polymerization process the diameter of droplet does not change; every droplet behaves as a separate reactor, and a bulk or mass polymerization occurs in each droplet.[10]

Two kinds of suspending agents are generally used. One is the protective colloid, a water-soluble polymer that functions to increase the viscosity of the continuous phase. A wide range of protective colloids are used, such as natural polymers (starch, gelatin), modified natural polymers (methyl and hydroxyethyl cellulose), synthetic polymers such as partially hydrolyzed poly(vinyl acetate), poly(vinyl alcohol), and poly(vinylpyrrolidone), and low-molecular weight agents (poly(glycol ether)s and phthalic acid esters). The other group is water-insoluble inorganic suspending agents, such as hydrolyzed magnesium silicate, bentonite + inorganic salts, titanium dioxide, calcium carbonate + fatty acids, tricalcium phosphate, and zinc oxide. Various suspension stabilizers can be used alone or in combination up to 0.5 wt% concentration. Surfactants such as dodecylbenzene sodium sulfonate and sodium 2-ethylhexyl sulfate are used to modify the surface tension of the medium.

Generally, peroxides and azo compounds are used as polymerization initiators alone or in combination.

It is known that suspension polymerization of styrene is extremely dependent on reactor geometry and agitation, especially from the particle size and distribution points of view. Some investigations in this field have covered the fundamental effect of agitation.[11-16] The type, concentration, and particle size of the suspending agent can also affect the particle size and distribution of the product.[17-19] The bead-size distribution is one of the important quality parameters of the suspension type of polystyrene. This parameter gains particular importance in the case of expandable polystyrene because distributions with mean particle sizes varying between 0.3 and 5 mm, depending on the type, are obtained.[20,21]

APPLICATION

Polystyrene is widely used in toys, decorative and commodity packaging, food packaging, housewares, home appliances, recreational products, electronics, furniture, building and construction (insulation), automotive parts, medical wares, disposables, luggage, and light-duty industrial components.

REFERENCES

1. Meister, B. J.; Malanga, M. T. In: Moore, E. R., Ed.; *Styrene Polymers, Encyc. Polym. Sci. Eng.* 2nd ed.; John Wiley & Sons: New York, 1989; 16, pp 21-62.
2. Maul, J. In: Elvers, B.; Hawkins, S.; Schulz, G., Eds. *Polystyrene, Ullmann's Encyc. Ind. Chem.*; VCH: Weinheim, 1992, A1, pp 615-625.
3. Boundy, R. H.; Boyer, R. F.; Stoesser, S. M. *Styrene: Its Polymers, Copolymers and Derivatives*; Reinhold: New York, 1952.
4. Bishop, R. B. *Practical Polymerization for Polystyrene*; Cahners: Boston, 1971.
5. Albright, L. F. *Processes for Major Addition-Type Plastics and Their Monomers*; McGraw-Hill: New York, 1974.
6. Munzer, M.; Trommsdorff E. In: Schildknecht, C. E.; Skeist, I., Eds.: *Polymerization Processes*; John Wiley & Sons: New York, 1977; pp 106-142.
7. Brighton, C. A.; Pritchard, G.; Skinner, G. A. *Styrene Polymers: Technology and Environmental Aspects*; Applied Science: London, 1979.
8. Svec, P.; Rosik, L.; Horak, Z.; Vecerka, F. *Styrene-Based Plastics and Their Modification*; Ellis Horwood: London, 1989.
9. Simon, R. H. M.; Chappelear, D. C. In: Henderson, J. N.; Boutan, T. C., Eds.; *Polymerization Reactors and Processes; ACS Symposium Series*, 104; American Chemical Society: Washington, DC, 1979; pp 71-112.
10. Hohenstein, W. P.; Vingiello, F.; Mark, H. *India Rubber World* **1994,** *110,* 291.
11. Nomura, M.; Harada, M.; Eguchi, W.; Nagata, S. *J. Appl. Polym. Sci.* **1972,** *16,* 835.
12. Ide, Y.; White, J. L. *J. Appl. Polym. Sci.* **1975,** *19,* 719.
13. Moritz, H. U.; Langner, F.; Reichert, K. H. *Ger. Chem. Eng.* **1979,** *2,* 112.
14. Oldshue, J. Y.; Mechler, D. O.; Grinnell, D. W. *CED* **1982,** *May,* 68.
15. Oldshue, J. Y. *Fluid Mixing Technology*; McGraw-Hill: New York, 1983.
16. Renyun, P.; Qiyun, Z.; Jianging, Z.; Huigen, Y. *Hecheng Xiangjiao Gangye* **1985,** *8,* 305.
17. Beckmann, G. *Chem. Eng. Techn.* **1964,** *36,* 169.
18. Hopff, H. et al. *Chem. Ing. Tech.* **1964,** *36,* 1085.
19. Bilgic, T.; Karali, M.; Savasci, O T. *Angew. Macromol. Chem.* **1993,** *213,* 33.
20. Guenther, F. O. *SPE Trans.* **1962,** *July,* 243.
21. Skinner, J. J.; Eagleton, S. D. *J. Plast. Inst.* **1964,** *32,* 231.

Polystyrene-co-acrylonitrile

See: Styrene-Acrylonitrile Copolymers

POLY(STYRENE-*co*-*N*-ALKYLMALEIMIDE)S (by Reactive Extrusion; Miscibility in Blends)

I. Vermeesch* and Gabriel Groeninckx
ICI Polyurethanes

*Author to whom correspondence should be addressed.

Poly(styrene-*co*-*N*-alkylmaleimide)s (**Figure 1**) are amorphous, transparent copolymers with a high glass transition temperature (T_g). The preparation of these copolymers by a novel technique, reactive extrusion, and the study of the specific interactions within these copolymers and in blends with other polymers are described in this article. Reactive extrusion implies that a reaction such as bulk polymerization, grafting, functionalization, or degradation takes place in the melt in a continuous way by means of an extruder.[1,2]

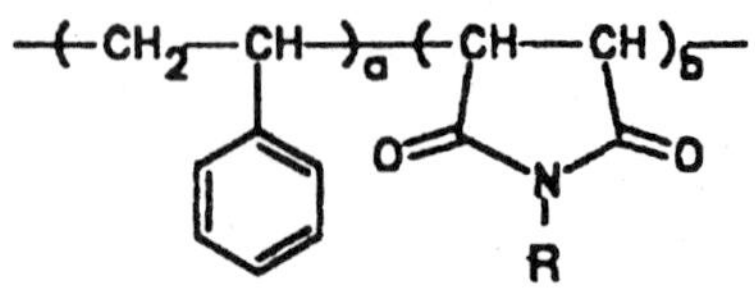

FIGURE 1. Schematic representation of a poly(styrene-*co*-*N*-alkylmaleimide) copolymer.

CONCLUSIONS

Copolymers of *N*-maleimides are readily produced from poly(styrene-co-maleic anhydride) via a one-step imidization with primary *N*-alkyl amines using reactive extrusion. Large amounts of ready-to-use materials may be obtained in small reaction times without catalysts, suspending agents, solvents, processing aids, or stabilization products. Screw-design and devolatilization of the condensation water appear to be primary factors controlling the imidization kinetics.

FTIR characterization of a model amorphous poly(styrene-*co*-*N*-maleimide) copolymer containing 14 wt% maleimide, NMI-14, revealed that self-association occurs through hydrogen bonding between the NH and carbonyl groups of the imides.

Conversion of styrene-*co*-maleic anhydride to styrene-*co*-*N*-maleimide copolymers not only raises the T_g significantly, but also increases the probability of finding miscible or partially miscible polymer blend systems. By DSC and FTIR spectroscopy it has been proven that NMI-14 is miscible on a molecular scale with PMMA and SPVP over the whole blend composition range through the formation of hydrogen bonds.

APPLICATIONS

Poly(styrene-*co*-*N*-alkylmaleimide)s are used industrially in a broad range of applications. In general they are used to increase the heat distortion temperature and the processability of materials. Blends of polyphenylene ether and poly(styrene-*co*-*N*-methylmaleimide)s are used as conductive resins with antistatic properties, excellent heat resistance, and processability.[3] Compounds containing no poly(styrene-*co*-*N*-methylmaleimide) have poor fluidity and are not moldable. Nylon 6,6 is blended with maleic anhydride/*N*-phenylmaleimide/styrene polymers to give a compound with increased dimensional stability.[4] Other examples are blends of chlorinated poly(vinyl chloride) with methyl methacrylate/*N*-phenylmaleimide/styrene polymers, poly(styrene-*co*-*N*-phenylmaleimide) with poly(styrene-*co*-acrylonitrile), and poly(styrene-*co*-*N*-maleimide) with polycarbonate of bisphenol A.[5-7] Glass-fiber reinforced polyamides (PA 6, PA 6,6) blended with poly(styrene-*co*-*N*-maleimide)s are useful for automobile tanks.[8]

ACKNOWLEDGMENTS

I. Vermeesch gratefully acknowledges Monsanto Europe at Louvain-la-Neuve, Belgium, for the use of their twin-screw extruder, Prof. M. M. Coleman for the use of the FTIR and his useful guidelines at Pennsylvania State University, U.S., Prof. G. Maes of the Katholieke Universiteit Leuven, Belgium, for the interesting discussions concerning hydrogen bonding, and the Belgian IWONL and NFWO Institutes for the scholarships during her Ph.D. studies.

REFERENCES

1. Lambla, M. *Comprehensive polymer science: Reactive Processing of thermoplastic polymers*; Leuven: DSM-Chair, 1991.
2. Barth, U. *Papers from a Two-day seminar on Reactive Processing: Practice and Possibilities*; Shawbury: organized by RAPRA Technology Ltd. 1989.
3. European Patent 450483, 1991.
4. Japanese Patent 03207759, 1991.
5. Japanese Patent 03182536, 1991.
6. Dean, B. *J. Appl. Polym. Sci.* **1987,** *33,* 2259.
7. U.S. Patent 4 160 792, 1979.
8. Japanese Patent 03200868, 1989.

POLYSTYRENE AND DERIVATIVES, PHOTOLYSIS

Wolfram Schnabel* and Ivo Reetz
Hahn-Meitner-Institut Berlin GmbH

Polystyrene, **PSt**, and its ring- or chain-substituted derivatives strongly absorb light at wavelengths below 300 nm. Typical absorption spectra possess a broad band between 230 and 290 nm. In the case of **PSt** the wavelength of maximum absorption (λ_{max}) is 262 nm. Substitution of hydrogens at the ring or at the main-chain results in a shift of λ_{max} to higher wavelengths. The monophotonic absorption in this wavelength range reflects the formation of excited singlet states (S_1) of phenyl groups via a $\pi \rightarrow \pi^*$ transition. Deactivation of the excited singlet states occurs mainly by internal conversion, that is, via radiationless $S_1 \rightarrow S_0$ transitions and to a certain extent by luminescence. Moreover, part of the singlet states undergoes intersystem crossing to the triplet manifold or form excimers, that is, dimers of electronically excited and neighboring non-excited phenyl groups.

The extent of chemical deactivation of excited states depends on the physical state of the system: irradiation of the polymers in solution results in much higher quantum yields of bond cleavage than irradiation of the rigid polymer matrix. Principally, chemical processes can originate both from singlet and triplet states, and in some cases triplet states play the major role in this respect. Two kinds of photoproducts can be discriminated: (1) low molecular volatile compounds such as H_2 and alkanes (in the case of alkyl-substituted polystyrenes) or hydrogen halides (in the case of halogen-substituted polystyrenes). (2) Chemical alterations in the residual macromolecules, that is, main-chain rupture, formation of crosslinks, and unsaturations. Since the physical properties of polymers strongly depend on average molar mass, crosslinking and main-chain scission exert a great influence although occurring with low quantum yields. Remarkably, intermolecular crosslinking predominates over main-chain scission as far as **PSt** and some derivatives are concerned. Halogen-substituted polystyrenes are especially prone to crosslinking. Derivatives of **PSt** are applicable as resist materials for photomicrolithographic techniques allowing the generation of microstructures of micrometer, and even submicrometer, dimensions.[1-4]

*Author to whom correspondence should be addressed.

APPLICATIONS: POLYSTYRENE DERIVATIVES AS RESIST MATERIALS

The fabrication of microelectronic devices such as computer chips is based on photomicrolithographic techniques allowing the generation of microstructures of micrometer and submicrometer dimensions. For this purpose etch-resistant materials commonly consisting of synthetic polymers are employed that are coated on the substrate prior to the formation of fine structures and are finally removed from it.[1] The rapidly developing miniaturization continuously asked for new resist materials fulfilling the various technological requirements and, therefore, numerous polymers of different chemical nature were examined. Among those were polystyrene derivatives. It has been pointed out that chloromethylated polystyrenes have attracted some interest because they are readily crosslinked and can, therefore, serve as negative-tone resist materials.[4] Chlorinated and chloromethylated polystyrenes were also proposed as appropriate materials concerning photooxidative imaging at polymer surfaces by deep UV irradiation (λ = 193 nm or 248.4 nm). In this case hydrophobic surface areas are converted into hydrophilic ones by the incorporation of carbonyl and hydroxyl groups. The exposed hydrophilic areas bind water via hydrogen bonding and thus allow the conversion of $TiCl_4$ into TiO_2. The latter protects the underlying organic film from oxidative removal during a subsequent oxygen reactive ion etching step. In this way negative tone images are generated.[2]

Moreover, derivatives of *p*-hydroxystyrene such as poly(acetoxystyrene) have been studied.[5] These polymers undergo a photo-Fries rearrangement.

In this way hydroxyl groups at the phenyl rings are uncovered and the polymer becomes soluble in aqueous alkaline solution and insoluble in common organic solvents. Therefore, it can be employed as an either positively or negatively acting resist, depending on the chemical nature of the developer. Other interesting work concerns the synthesis of copolymers of *p*-(trimethylsilyl)methylstyrene, or *p*-trimethylsilylstyrene and chlorostyrene, or chloromethylstyrene.[6,7]

These copolymers exhibit high plasma resistance and a useful radiation sensitivity. However, the steadily increasing requirements for materials of extremely high radiation sensitivity could not be fulfilled by polymers undergoing single step photoreactions. The problem was solved quite elegantly on the basis of the chemical amplification concept by employing indirectly acting systems in which a catalyst is photochemically generated.[3,8-10] The classical systems, which will only be dealt with here, consist of an onium salt and poly[*p*-(*tert*-butoxycarbonyl)styrene] or poly[*p*-(*tert*-butoxycarbonyloxy)styrene]. Diazonium, iodonium, or sulfonium salts of the structure ArN_2^+

MX_n^-, $Ar_2I^+ MX_n^-$, $Ar_3S^+ MX_n^-$ with MX_n^- being BF_4^-, PF_6^-, AsF_6^-, or SbF_6^- can be used for this purpose.

The photolysis of the onium salts results in the generation of protons that catalyze the removal of the protecting groups from the phenyl rings when the systems are subjected to a post-irradiation thermal treatment at moderate temperatures.

REFERENCES

1. Reiser, A. *Photoreactive Polymers. The Science and Technology of Resists*; Wiley: New York, 1989.
2. Nalamasu, O.; Taylor, G. N. *Proc. SPIE* **1989,** 1086, 186.
3. Ito, H.; Willson, C. G.; Frechet, J. M. J. *Proc. SPIE* **1987,** *771,* 24.
4. Harita, Y.; Kamoshida, Y.; Tsutsumi, K.; Koshiba, M.; Yoshimoto, H.; Harada, K. *SPSE: Unconv. Imag., Sci. Techn. 22nd Symp.* **1982,** 34.
5. Tessier, T. G.; Frechet, J. M. J.; Willson, C. G.; Ito, H. *ACS Symp. Ser.* **1985,** *266,* 269.
6. MacDonald, S. A.; Ito, H.; Willson, C. G. *Microelectron. Eng.* **1983,** *1,* 269.
7. Suzuki, M.; Saigo, K.; Gokan, H.; Ohnishi, Y. *J. Electrochem. Soc.* **1983,** *30,* 1962.
8. Frechet, J. M. J.; Eichler, E.; Ito, H.; Willson, C. G. *Polymer* **1983,** *24,* 995.
9. Frechet, J. M. J.; Bouchard, F.; Houlihan, F. M.; Kryczka, B.; Eichler, E.; Clecak, N.; Willson, C. G. *J. Imaging Sci.* **1985,** *30,* 59.
10. Ito, H.; Ueda, M.; England, W. P. *Macromolecules* **1990,** *23,* 2589.

Polystyrene-b-butadiene-b-styrene

See: Styrene-Butadiene Copolymers

Polystyrene, Carboxylated

See: Compatibilizers, Polymeric

Polystyrene, Halogenated

See: Polystyrene and Derivatives, Photolysis

POLYSTYRENE, HIGH IMPACT

Atsushi Nanasawa
Styrenic Resins Development Department
Asahi Chemical Industry Company Ltd.

Polystyrene is a well-known commodity thermoplastic, and a large quantity of it is produced and consumed for our daily use. There are two types of polystyrene. One is a transparent simple styrene polymer usually called general purpose polystyrene or GPPS in short, and the other is a high-impact polystyrene (HIPS), which is reinforced by a rubber component.

GPPS is a non-crystalline thermoplastic with a glass transition temperature of 100°C. Its advantages are easy processability and high flexional modulus compared to crystal thermoplastics, such as polyethylene and polypropylene, or soft poly(vinyl chloride) with a large amount of plasticizer. Though GPPS is brittle in nature, HIPS is a toughened GPPS, composed by adding rubber, materials without any disadvantages. This invention was able to enlarge the polystyrene market.

PROPERTIES

Considering HIPS as an organic material, (1) processability, (2) toughness, (3) flexional modulus, and (4) appearance of molded products are the important properties.

Processability

For durable use of HIPS, such as TV, VCR, and CRT casings, injection molding is the main procedure. Amorphous HIPS, having a relatively lower melting viscosity than the other rubber reinforced plastics like ABS resin or modified polyphenylene ether (PPE), is a particularly suitable resin for injection molding.

Plasticizer to obtain higher processability is usually chosen from low molecular weight components miscible with polystyrene. It should be noted that residual monomer and oligomer have a similar function.

Toughness

Toughness against a shock is the most expected property for the rubber reinforced polymers. Poor toughness of GPPS can be explained by a polymer nature where molecules are stiff and sparsely entangled.

In order to reinforce GPPS, rubber components are added to avoid brittle fracture by absorbing the energy through craze formation and preventing craze propagation into crack. The rubber component in HIPS is called gel. The functions of gel particles in HIPS are summarized by Bucknall, Haaf, and coworkers, Donald and Kramer, and Kinloch and Young, and explained as follows.[2-4] At room temperature gel particles, whose modulus are much lower than matrix polystyrene, become stress concentration points. When concentrated stress exceeds critical level of craze initiation (σx), craze forms in the equator direction of the gel particles where the stress concentration is highest. These craze formations were observed by transmission electron microscopy.[5] In HIPS a large number of gel particles exist, hence the quantity of these craze formations are enormous and the size is small. During this procedure, the stress is absorbed by a lot of these so-called micro-crazes. As a result, propagation into cracks is avoided. Therefore, HIPS is less likely to fracture than GPPS, which breaks out and cracks more easily. Gel particles are considered to work not only as a micro-craze initiator, but also as a terminator of propagating micro-craze.[6]

There is an optimum gel particle diameter range that maximizes the toughness.[3]

Flexional Modulus

HIgh flexional modulus is one of the advantages of HIPS over other commodity thermoplastics, such as polyethylene, polypropylene, and plasticized poly(vinyl chloride). As mentioned before, HIPS is suitable for injection molding of large-size parts. Large-size parts usually require a hardness that does not bend under a stress. In other words, high flexional modulus is a fundamental property of HIPS.

Modulus is a phenomenon in the range where deformation does not occur. Thus it is estimated by a simple additional rule based on gel phase volume. It is also known that gel with smaller particles gives a slightly higher modulus.[7]

Surface Gloss

Appearance, represented by surface gloss, is an important property sometimes controlling a value of the final product. It is known that appearance widely changes according to molding conditions such as temperature, injection speed, and so on. It is also known that surface gloss becomes better by using a HIPS made of small gel particles.[1]

Surface gloss of rubber reinforced materials relates to the roughness of the surface. It is clear that the HIPS with small and narrowly distributed gel gives excellent surface gloss.

APPLICATION

HIPS is used in various ways. Home electronics casings is one of the largest applications and it is quite easy to find HIPS used for TV and audio sets. Toys or video and audio cassettes are usually made by HIPS. HIPS is one of the principal packaging materials for food, too. The consumption of polystyrene including GPPS and HIPS divided by use can be found in statistics.[8]

REFERENCES

1. Echte, A.; Haaf, F.; Hambecht, J. *Angew. Chem. Int. Ed. Engl.* **1981,** *20,* 344.
2. Haaf, F.; Breuer, H.; Echte, A.; Schmitt, B. J.; Stabenow, J. *J. Scientific and Industrial Research* **1981,** *40,* 659.
3. Donald, A. M.; Kramer, E. J. *J. Appl. Polym. Sci.* **1982,** *27,* 3729.
4. Kinloch, A. J.; Young, R. J. *Fracture Behavior of Polymers*; Elsevier Applied Science: London and New York, 1983.
5. Kambour, R. P.; Russell, R. R. *Polymer* **1971,** *12,* 237.
6. Bucknall, C. B. *Adv. Polym. Sci.* **1978,** *27,* 121.
7. Bucknall, C. B.; Davies, P.; Patridge, I. K. *J. Mater. Sci.* **1987,** *22,* 1341.
8. Taguchi, S.; Matsuzaki, S. *Plastics Age* 1994; 40 (December), 91.

Poly(styrene-co-maleic anhydride)

See: Styrene-Maleic Anhydride Copolymers

POLYSTYRENE MANUFACTURE (Using Bulk Free Radical Polymerization)

Duane B. Priddy
Dow Plastics
The Dow Chemical Company

Polystyrene is a commodity thermoplastic with global sales of >10 billion pounds per year. Polystyrene is a rigid-glass and rather brittle material with a heat distortion temperature ~107°C. It is used to make foam for insulation and packaging, clear film, and is fabricated into many parts. The typical weight average molecular weight (M_w) range of polystyrene for extruded foam board is 200,000 to 230,000. Extruded foam sheet for packaging typically utilizes higher M_w polystyrene (M_w 270,000–330,000). Molding grades of polystyrene typically are in the 250,000–330,000 M_w range, and have a few percent of a plasticizer (e.g., mineral oil) and a mold release agent (e.g., stearic acid) added.

Today, most polystyrene is manufactured via continuous free radical bulk styrene polymerization (CFRBSP), but with the aid of a peroxide initiator. CFRBSP reactors, in general, are of three types: stratified agitated tower, ebullient stirred tank, and recirculated coil.

Since most polystyrene is now produced with the aid of chemical initiators, there has been a significant amount of research aimed at developing new initiator technology for polystyrene. The new polystyrene initiator research activities have focused mainly on increasing the rate at which high molecular weight polystyrene can be manufactured. However, since branching in some polymers is known to impart improvement in certain properties, there has also been some recent literature describing the use of initiators leading to branched polystyrene. This paper will review some of the recent progress in the development of new polystyrene initiators for both polymerization rate enhancement and branching.

CHEMICAL INITIATION OF CFRBSP

Chemical initiators for CFRBSP generally are of three types: azo, tetraphenylethane, or peroxides.[1] Of these types, peroxides are preferred for a variety of reasons: 1) in general peroxides are the least costly; 2) peroxides are more efficient than azo initiators due to fewer "in-cage" side reactions; 3) as a result of in-cage side reactions, azo initiators can leave harmful residues in the polymer; 4) diphenylmethyl radicals formed from tetraphenylethane initiators terminate (by primary radical coupling), as well as initiate resulting in the formation of thermally labile endgroups on the terminal end.

Initiators for CFRBSP Rate Enhancement

A further economically driven objective is to use initiators that increase the rate of CFRBSP to produce high molecular weight polystyrene faster.

In recent years there has been considerable interest in difunctional peroxide initiators. One of the key reasons for the use of difunctional initiators is their theoretical ability to form initiator fragments, which can initiate polymer growth from two different sites wihin the same fragment ultimately leading to "double-ended polystyrene." If double-ended polystyrene chains are produced, higher M_w polystyrene can be produced at faster rates than achievable using monofunctional initiators.[2]

The most common class of difunctional peroxide initiators utilized for continuous bulk S polymerization and copolymerization are the *gem*-bis(*t*-butylperoxy)alkanes also commonly referred to as perketals.

Initiators for Branching

The introduction of long chain branching into polymers is known to improve certain properties (e.g., melt strength), and is practiced commercially for polyethylene and polycarbonate. Branching in polystyrene also leads to changes in properties, but to a lesser extent. The main mode of polymerization used to make branched polystyrene has been batch anionic polymerization.

One of the biggest problems associated with the introduction of branching into free radically produced polystyrene is the formation of gels. Long chain branching is, in fact, naturally present in extremely small amounts within polystyrene produced using CFRBSP.

Efforts aimed at introducing branching during polymerization have included the addition of a small amount of divinyl monomer, vinyl functional initiator, polyfunctional initiator, or vinyl functional chain transfer agent to the polymerization mixture.[3-7] All of these approaches, except for the vinyl functional chain transfer agent, were subjected to CFRBSP in our laboratory, and eventually after a few days of operation, led to gels and reactor fouling.

A very recent approach to the preparation of branched polystyrene, which should eliminate the gel problem during manufacture of CFRBSP, is the introduction of latent functionality into polystyrene.[8] Once the polymer exits the polymerizer, the latent functionality is activated during high temperature devolatilization resulting in the formation of branch points. The functional group used to demonstrate this approach was the benzocyclobutene moiety (BCB) which is stable at temperatures < 180°C, but has a half-life for decomposition at 240°C of 5 min. BCB functional groups were incorporated into polystyrene by initiating polymerization using BCB functional benzoyl peroxide (BCBPO).[9,10]

As the ratio of BCBPO to benzoyl peroxide increases in the initiator mixture, the M_w exponentially increases upon heating at 240°C. The deviation of the M_w measured using GPC versus GPC-LALLS increases with the BCBPO/benzoyl peroxide ratio indicating that the level of branching also increases.

CONCLUSIONS AND FUTURE DIRECTIONS

Most polystyrene is manufactured using CFRBSP because it is the lowest cost process. Economics will continue to force polystyrene manufacture by other more costly processes (e.g., suspension polymerization) to close down. Due to the competitive nature of the polystyrene business, keeping manufacturing cost at a minimum is essential. Thus research will continue to focus mainly upon increasing product yield from existing plants with minimal capital investment. However, a steady level of research aimed at improving product performance will also continue. As with most thermoplastics, improved product performance typically means improving the flow/toughness property balance.

REFERENCES

1. Priddy, D. B. *Adv. Polym. Sci.* **1994,** *114,* 69.
2. Wittmer, P. *Makromol. Chem.* **1988,** *170,* 1.
3. Zhu, S.; Hamielec, A. E. *Macromolecules* **1993,** *26,* 3131.
4. Zhu, S.; Hamielec, A. E. *Makromol. Chem., Macromol. Symp.* **1993,** *69,* 247.
5. Matsubara, T.; Ito, N.; Ishida, Y.; Iwamoto, M. U.S. Patent 4 376 847, 1983.
6. Izumida, K.; Okumura, R. U.S. Patent 5 191 040, 1993.
7. Tung, L. H.; Hu, A. T.; McKinley, S. V.; Paul, A. M. *J. Polym. Sci., Polym. Chem. Ed.* **1981,** *19,* 2027.
8. DeLassus, S. L.; Howell, B. A.; Cummings, C. J.; Dais, V. A.; Nelson, R. M.; Priddy, D. B. *Macromolecules* **1994,** *27,* 1307.
9. Tong, W. H.; Priddy, D. B. U.S. Patent 5 034 485, 1992.
10. Tong, W. H.; Priddy, D. B. U.S. Patent 5 079 322, 1992.

POLYSTYRENE, RUBBER TOUGHENED

Yasushi Okamoto,* H. Miyagi, and H. Kihara

Petrochemicals Research Laboratory

Sumitomo Chemical Company, Ltd.

It is well known that the toughness of glassy polymers can be improved by the dispersion of rubber particles, and that the toughening mechanism of these systems has been studied extensively.[1-7] In high-impact polystyrene (HIPS), the absorption of impact energy is thought to increase mainly by the generation and growth of crazes in the PS matrix.[8] Effects of the rubber content and rubber particle size on the toughening of HIPS have been studied from both scientific and technical aspects. In monomodal rubber particle HIPS (monomodal HIPS), the optimum rubber particle diameter known is 1–2 μm.[9,10]

From practical points of view, HIPS with the dual population of rubber particle sizes (bimodal HIPS) has been studied for its unique impact properties.[11,12] It has been shown that better impact toughness can be achieved in an optimum bimodal HIPS than in a monomodal one with the same rubber content.[13] However, few studies have been concerned with the cause of the increase in impact strength of bimodal HIPS. A model was recently presented to explain the craze growth mechanism in bimodal HIPS by using the finite element method (FEM).[14] This model assumed that small particles initiated crazes, which propagated in the directions of large particles, where they were stopped.

In 1990, a new toughening model was proposed by Gebizlioglu et al.[15] They examined the mechanism in rubber dispersed polystyrene samples that included low M_w polybutadiene (PB) rubber droplets. They concluded that the liquid PB in these pools acts as a plasticizing agent under the prevailing negative pressures of the craze tip and craze borders to result in a greatly increased propensity for crazing at low stresses, thus avoiding early craze fracture from extrinsic flaws. Furthermore, Argon et al. also provided a theoretical model to express the above phenomenon, and craze flow stress based on the model has shown excellent agreement with experimental measurement.[16]

In our work, the crazing process in bimodal HIPS was studied by observation with a transmission electron microscope (TEM).[17] Nonlinear stress analysis around rubber particles was also performed by using FEM.

We recently discovered that cavitation can occur in high impact polystyrene (HIPS) with rubber particles of a rubbery core/rigid polymer shell, and studied the cavitation mechanism in this system.[18] Here, we provided direct verification of the rubber sorption model by microscopic work by using the annealing effect on craze, and providing by FEM analysis that sorption of rubber components to craze fibrils would plasticize the craze.

REFERENCES

1. Bucknall, C. B. *Toughened Plastics*; Applied Science: London, 1977.
2. Kesukkula, H. *Rubber-Toughened Plastics*; American Chemical Society: Washington, DC, 1989; p 289.

*Author to whom correspondence should be addressed.

3. Boyer, R. F.; Kesukkula, H. *Encycl. Polym. Sci. Technol.* **1982,** *13,* 392.
4. Angier, D. J.; Fettes, E. M. *Rubber Chem. Technol.* **1965,** *36,* 1164.
5. Moore, J. D. *Polymer* **1971,** *12,* 478.
6. Donald, A. M.; Kramer, E. J. *J. Appl. Polym. Sci.* **1982,** *27,* 3729.
7. Okamoto, Y.; Miyagi, H.; Uno, T.; Amemiya, Y. *Polym. Eng. Sci.* **1993,** 33(24), 1606.
8. Bucknall, C. B.: Smith, R. R. *Polymer* **1965,** *6,* 437.
9. Bucknall, C. B.; *Polymer Blends*; Paul, D. R.; Newman, S., Eds.; Academic: New York, 1978; Chapter 2, p 99.
10. Donald, A. M.; Kramer, E. J. *J. Mater. Sci.* **1982,** *17,* 2351.
11. BASF, U.S. Patent 4 493 922.
12. Sumitomo Chemical Co., U.S. Patent Appl. 07/223 599.
13. Hobbs, S. Y. *Polym. Eng. Sci.* **1986,** *26,* 74.
14. Wrotecki, C.; de Charentenay, F. X. *Deform. Yield Frac. Polym.* **1988,** *7,* 51.
15. Gebizlioglu, O. S.; Beekham, H. W.; Argon, A. S.; Cohen, R. E.; Brown, H. R. *Macromolecules* **1990,** *23,* 3968.
16. Argon, A. S.; Cohen, R. E.; Gebizlioglu, O. S.; Brown, H. R.; Kramer, K. J. *Macromolecules* **1990,** *23,* 3975.
17. Okamoto, Y.; Miyagi, H.; Kakugo, M.; Takahashi, T. *Macromolecules* **1991,** *24,* 5639.
18. Okamoto, Y.; Miyagi, H.; Mitsui, S. *Macromolecules* **1993,** *26,* 6547.

POLYSTYRENE, STEREOREGULAR

Clara Silvestre* and Sossio Cimmino
Instituto di Ricerca e Tecnologia delle Materie Plastiche
Italian National Council of Research-CNR

Since the successful synthesis of syndiotactic polystyrene, there has been considerable interest in characterizing the material.[1-3] This report updates results on the synthesis, structure, crystallization, thermal behavior, and properties of the stereoregular (isotactic and syndiotactic) polystyrenes.

PREPARATION

Isotactic Polystyrene (iPS)

Most heterogeneous Ziegler–Natta catalysts polymerize styrene to chemically regular isotactic polymers. Catalyst systems most widely used for polymerization include $TiCl_4$, α and δ-$TiCl_3$ in combination with $Al(C_2H_5)_3$, and $VCl_3Al(i\text{-}C_4H_9)_3$.[4-7]

The activation energy of the polymerization reaction with $TiCl_4$-$Al(C_2H_5)_3$ system is 43.9 kJ/mol.[7] It is generally accepted that the mechanism of styrene polymerization with the Ziegler–Natta catalyst system is identical to the mechanism proposed for isospecific alkene polymerization. The reaction is first order in both monomer and titanium.[8-10]

Polystyrene (PS) obtained with the Ziegler–Natta catalyst system is always a mixture of products exhibiting different degrees of isotacticity. For example, PS produced with the α-$TiCl_3$-$Al(C_2H_5)_3$ system contains ca. 70% of a highly isotactic and crystalline product insoluble in boiling toluene with the meso-triad content higher than 95%; ca. 10% of a product of medium isotacticity (insoluble in ketones); and ca. 20% of an amorphous, sterically irregular product soluble in boiling methylethyl ketone.[11]

*Author to whom correspondence should be addressed.

Syndiotactic Polystyrene (sPS)

Syndiotactic PS with high stereoregularity was first obtained by Ishihara et al., followed by Zambelli et al.[1-3] A variety of titanium compounds soluble in aromatic hydrocarbons in the presence of methylalumoxane (MAO) promote syndiotactic polymerization of styrene, with a narrow molecular mass distribution, $M_w/M_n = 2$. Monocyclopentadienyl trialkoxytitanium, monocyclopentadienyl trichlorotitanium, (pentamethylcyclopentadienyl) trichlorotitanium, and (η^5-indenyl) trichlorotitanium are among the most active catalysts.[1-3,12-16]

Some other catalyst systems have been successfully used to obtain syndiotactic-rich polystyrene, like rare earth coordination catalysts, tetrabenzylzirconium activated with MAO, or organometallic derivative of titanium or zirconium activated with tris(pentaflurorphenyl) boron.[17,18]

STRUCTURE

Isotactic Polystyrene

Isotactic polystyrene crystallizes in a trigonal system with unit-cell dimensions of a = 21.9 ± .1A and c = 6.65 ± .05A, where c is the axis of the polymeric chain, as found by Natta and coworkers.[6] It assumes in the crystalline state a 3/1 helical form with a regular repetition of trans and gauche conformation of the skeletal C-C bonds.[5] The number of chains per unit cell is 6.[19] This form is often named α.[20] Another skeletal conformation having a nearly extended form consisting of a repetition of non-staggered trans C-C bonds (with the internal rotation angle twisted by about 20° from the perfectly staggered trans form) has been found in iPS gel formed from solution of bulky non-planar hydrocarbon solvents and by exposure of amorphous iPS films to solvent vapor at room temperature.[20-25] For this form, named β, an orthorhombic cell with a = 21 A, b = 16.4 A, and c = 30 A and with 12 monomers in a repeat distance of 30.6 A was proposed.[21,22]

Syndiotactic Polystyrene

Different crystalline polymorphic forms of sPS were reported. The various forms differ for the conformation of the polymer chains and for the mode of packing of chains having the same conformation. In particular four crystalline modifications of syndiotactic polystyrene were reported. Two of these forms, named α and β, contain planar zig-zag chain conformation with an identiy period of 5.06 A. The two other crystalline forms, described as γ and δ, are characterized by chains in a helix conformation and identiy period of 7.5 A.

MORPHOLOGY

Isotactic Polystyrene

The slow crystallization rate of iPS makes it a convenient model for studies of the morphology, kinetics, and mechanisms of the crystallization from glass, as well as from the melt and from solution. From dilute solution, melt, and glass the iPS crystallizes in a three-fold conformation.

The crystallization from dilute solution produces single crystals. Single crystals were also obtained in samples crystallized in bulk from the melt. The crystallization from glass and the melt produces generally spherulites.

In the case of gel, where the samples are crystallized in the extended conformation, the crystals are of fringed micellar type.[20,25]

Syndiotactic Polystyrene

Different morphologies can be obtained depending on different crystallization conditions. For samples crystallized isothermally from glass at 200°C, lamellae are obtained.[26]

For samples crystallized isothermally from glass at 180°C, conversely, two-dimensional crystals are obtained. Highly oriented structures can be obtained by drawing a molten film. After drawing, shish-kebab structures are observed with a low shish content.

From the melt, syndiotactic PS crystallizes according to a spherulitic morphology.[29,30]

CRYSTALLIZATION BEHAVIOR

Isotactic Polystyrene

The crystallization rate of iPS from melt, glass, and solution is very low, in comparison with polyethylene, isotactic polypropylene, and its syndiotactic isomer.[32-35] The crystallization rate can be increased by the addition of nucleating agent.[36,37]

From the melt and from the glass state similar values of the spherulites growth rate are obtained. The polymer can isothermally crystallize in a very large temperature range (130–210°C).[31,38,39]

From concentrated solution (5% or more), at temperatures below 20°C, or by exposure of amorphous iPS films to solvent at room temperature, gelation sets. The gel is transparent, but it becomes turbid with time over a period of days. The turbidity is attributed to the usual chain folded crystallization within the gel. The gelation is a form of crystallization of fringed micellar type.[24,25]

The very slow crystallization rate allows quenching to the glass state without appreciable crystallization. Quenched isotactic polystyrene is amorphous and optically clear.

Syndiotactic Polystyrene

sPS is a much faster crystallizing polymer than its isotactic isomer.[42] It is possible to note that the spherulite growth rate in sPS is more than one order of magnitude higher than in iPS. Because the two samples present almost the same value of the glass transition temperature, T_g, and both are of high molecular mass, these results indicate that the configuration is one of the factors to take into account in the definition of the crystallization process.

Although the crystallization rate of sPS is very high, it is possible to obtain the completely glassy form by rapid quenching of the melt at 300°C in an acetone/dry ice bath at –80°C.[43] Several solvents induce crystallization of glass sPS, namely dichloromethane, chloroform, and cyclohexane.[43]

In solution sPS can give rise to gels depending on temperature, time, concentration and nature of solvent.[44-47] It was found that 60°C is the critical gelation temperature below which gels do not form. These crystals melt at about 120°C and possess a TTGG chain conformation. In analogy to iPS the gels are transparent and become turbid over a period of time.

THERMAL BEHAVIOR

Isotactic Polystyrene

Isotactic polystyrene is reported as a thermally ideal polymer, because T_g and some thermodynamic properties do not appear to depend on tacticity.[48] For isothermally crystallized iPS (from melt and glass) three melting peaks are always detected on the DSC thermograms.[28,40,49-52]

The value of the equilibrium melting temperature ranges between 238 and 241°C.[28,49,51,52]

Syndiotactic Polystyrene

The melting temperature of sPS is dependent on the crystallization temperature and on the structure of the starting material.

The equilibrium melting point of sPS, measured according to the Hoffman treatment and the melting point depression, occurs between 270 to 275°C, and is much higher than the T_m reported for iPS.[28,42,49,51-55] Higher values of the melting temperature, up to 291.5°C, are also obtained.[56-58]

APPLICATIONS

Isotactic PS is potentially a valuable material because of its high heat resistance and good dielectric properties. Low crystallinity, low rate of crystallization, and high brittleness represent the principal obstacles for commercial application of iPS as a general-purpose plastic.

With respect to iPS, sPS presents enhanced mechanical properties, especially at elevated temperatures, and good solvent resistance. Furthermore, the higher crystallization rate and crystallinity, compared with that of isotactic polystyrene, lead to a potentially commercial application of sPS as a low cost engineering plastic, but no commercial application of this polymer is yet known.

iPS and sPS can be easily distinguished from each other. In fact each polymer has a specific infrared and nmr spectrum, X-ray diffraction pattern, and solvent and thermal stability, and they differ also in melting point, crystallization rate, and crystallinity.[19,27,41,56-68]

REFERENCES

1. Ishihara, N.; Seimiya, T.; Kuramoto, M.; Uoi, M. *Macromolecules* **1986,** *19,* 2464.
2. Ishihara, N.; Kuramoto, M.; Uoi, M. *Macromolecules* **1988,** *21,* 3356.
3. Zambelli, A.; Longo, P.; Pellecchia, C.; Grassi, A. *Macromolecules* **1987,** *20,* 2035.
4. Natta, G.; Pino, P.; Corradini, P.; Danusso, F.; Mantia, E. *J. Am. Chem. Soc.* **1955,** *77,* 1700.
5. Natta, G.; Danussi, F.; Moraglio, G. *Makromol. Chem.* **1958,** *28,* 166.
6. Natta, G.; Corradini, P.; Bassi, I. W. *Nuovo Cimento, Suppl XV* **1960,** 58.
7. Natta, G.; Danusso, F., Eds.; *Stereoregular Polymers and Stereospecific Polymerization* Pergamon: Oxford, UK, 1967; Vols. 1 and 2.
8. Aubert, J. H. *Polymer Prepr. Am. Chem. Soc. Div. Polym. Chem.* **1987,** *28,* 147.
9. Burfield, D. R.: In: Quirk, R. P.; Ed.; *Transition Metal Catalyzed Polymerization: Alkenes and Dienes*; Harwood Academic: New York, 1983; p 171.

10. Kissin, Y.V. In: Quirk, R. P.; Ed.; *Transition Metal Catalyzed Polymerization: Alkenes and Dienes*; Harwood Academic: New York, 1983; p 597.
11. Kissin, Y. V. *Isospecific Polymerization of Olefuns with Heterogeneous Ziegler-Natta Catalysts*; Springer-Verlag: New York, 1985.
12. Chien, J. C. W.; Salajka, Z. *J. Polym. Sci. Phys.* **1991,** *29,* 1253.
13. Chien, J. C. W.; Salajka, Z.; Dong, S. *Macromolecules* **1992,** *25,* 3199.
14. Ready, T. E.; Day, R. O.; Chien, J. C. W.; Rausch, M. D. *Macromolecules* **1993,** *26,* 5822.
15. Zambelli, A.; Pellecchia, C.; Oliva, L.; Longo, P.; Grassi, A. *Makromol. Chem.* **1991,** *192,* 223.
16. Longo, P.; Grassi, A.; Proto, A.; Ammendola, P. *Macromolecules* **1988,** *21,* 24.
17. Yang, M.; Cha, C.; Shen, Z. *Polym. J.* **1990,** *22,* 919.
18. Pellecchia, C.; Longo, P.; Proto, A.; Zambelli, A. *Makromol. Chem., Rapid Commun.* **1992,** *13,* 265.
19. Tadokoro, H. *Structure of Crystalline Polymers* Wiley: New York, 1979.
20. Sundararajan, P. R. *Macromolecules* **1979,** *12,* 573.
21. Sundararajan, P. R.; Tyrer, N. J. *Macromolecules* **1982,** *15,* 1004.
22. Tyrer, N. J.; Bluhm, T. L.; Sundararajan, P. R. *Macromolecules* **1984,** *17,* 2896.
23. Tyrer, N. J.; Sundararajan, P. R. *Macromolecules* **1985,** *18,* 511.
24. Atkins, E. D. T.; Isaac, D. H.; Keller, A.; Mihasaka, K. *J. Polym. Sci., Polym. Phys. Ed.* **1977,** *15,* 211.
25. Atkins, E. D. T.; Hill, M. J.; Jarvis, D. A.; Keller, A.; Sarche, E.; Shapiro, J. S. *Colloid Polym. Sci.* **1984,** *22,* 262.
26. Greis, O.; Xu, Y.; Asano, T.; Petermann, J. *Polymer* **1989,** *30,* 590.
27. De Rosa, C.; Rapacciuolo, M.; Guerra, G.; Petraccone, V.; Corradini, P. *Polymer* **1992,** *33,* 1423.
28. Silvestre, C.; Cimmino, S.; Karasz, F. E.; MacKnight, W. J. *J. Polym. Sci., Polym. Phys. Ed.* **1987,** *25,* 2531.
29. Cimmino, S.; Di Pace, E.; Martuscelli, E.; Silvestre, C. *Polymer Commun.* **1991,** *32,* 251.
30. Koyama, K.; Soeno, T.; Haneda, T.; Yonetake, K.; Ishizuka, O. *Polymer Preprints Japan* **1988,** *37,* 1173.
31. Kenyon, A. S.; Gross, R. C.; Wurstner, A. L. *J. Polym. Sci.* **1959,** *15,* 159.
32. Hoffmann, J. D.; Davis, G. T.; Lauritzen, J. I. *Treatise on Solid State Chemistry* Hannay, N. B., Ed.; Plenum: New York, 1976; Vol. 3.
33. Mandelkern, L. *Crystallization of Polymers*; McGraw-Hill: New York, 1964.
34. Padden F. J.; Keither, H. D. *J. Applied Physics* **1959,** *30,* 1479.
35. Martuscelli, E.; Silvestre, C.; Canetti, M.; De Lalla, C.; Bonfatti, A.; Seves, A. *Makromol. Chem.* **1989,** *190,* 2615.
36. U.S. Patent 3 367 926, February 6, 1968; Voek, J. F. (to Dow Chemical Co.)
37. U.S. Patent 3 886 131, May 27, 1975; Beck H. N. (to Dow Chemical Co.)
38. Amelino, L.; Martuscelli, E.; Sellitti, C.; Silvestre, C. *Polymer* **1990,** *31,* 1051.
39. Martuscelli, E.; Sellitti, C.; Silvestre, C. *Makromol. Chem., Rapid Commun.* **1985,** *6,* 125.
40. Berghmans, H.; Ovemberg, N. *J. Polym. Sci., Polym. Phys. Ed.* **1977,** *15,* 1757.
41. Avrami, M. J. *J. Chem. Phys.* **1939,** *7,* 1103.
42. Cimmino, S.; Di Pace, E.; Martuscelli, E.; Silvestre, C., to be published on *Trends in Polymer Science.*
43. Vittoria, V.; de Candia, F.; Iannelli, P.; Immirzi, A. *Makromol. Chem., Rapid Commun.* **1988,** *9,* 765.
44. Prasad, A. Ph.D. *Dissertation*, The Florida State University, 1989.
45. Prasad, A.; Mandelkern, L. *Macromolecules* **1990,** *23,* 5041.
46. Marand, H. *Chemtracts–Macromolecular Chemistry* **1991,** *2,* 51.
47. Daniel, C.; Dammer, C.; Guenet, J. M. *Polymer* **1994,** *35,* 4243.
48. Karasz, F. E.; Bair, H. E.; O'Reilly, J. M. *J. Phys. Chem.* **1965,** *69,* 2668.
49. Lemstra, P. J.; Kooistra, T.; Challa, G. *J. Polym. Sci., A-2* **1972,** *10,* 823.
50. Ovembergh, N.; Berghmans, H.; Reynaers, H. *J. Polym. Sci., Polym. Phys. Ed.* **1976,** *14,* 1177.
51. Plans, J.; MacKnight, W. J.; Karasz, F. E. *Macromolecules* **1984,** *17,* 810.
52. Pelzbauer, Z.; Manley, R. St. J. *J. Polym. Sci. A-2* **1970,** *8,* 649.
53. Hoffman, J. D.; Weeks, J. J. *J. Chem. Phys.* **1965,** *42,* 4301.
54. Flory, P. J. *Principle of Polymer Chemistry* Cornell University: Ithaca, NY, 1953.
55. Gianotti, G.; Valvassori, A. *Polymer* **1990,** *31,* 473.
56. Arnauts, J.; Berghmans, H. *Polymer Comm.* **1990,** *31,* 343.
57. Gvozdic, N. V.; Mejer, D. J. *Polymer Comm.* **1991,** *32,* 183.
58. Gvozdic, N. V.; Mejer, D. J. *Polymer Comm.* **1991,** *32,* 493.
59. Kobayashi, M.; Akita, K.; Tadokoro, H. *Makromol. Chem.* **1968,** *118,* 324.
60. Kobayashi, M.; Tsmura, K.; Tadokoro, H. *J. Polym. Sci.* **1968,** A-2, *6,* 1493.
61. Kobayashi, M.; Nakaoki, T.; Ishihara, N. *Macromolecules* **1989,** *22,* 4377.
62. Guerra, G.; Musto, P.; Karasz, F. E.; MacKnight, W. J. *Makromol. Chem.* **1990,** *191,* 2111.
63. Reynolds, N. M.; Savage, J. D.; Hsu, S. L. *Macromolecules* **1989,** *22,* 2869.
64. Abis, L.; Albizzati, E.; Conti, G.; Giannini, U.; Resconi, L.; Spera, S. *Makromol. Chem., Rapid Commun.* **1988,** *9,* 209.
65. Gomez, M. A.; Tonelli, A. E. *Macromolecules* **1990,** *23,* 3385.
66. Capitani, D.; De Rosa, C.; Ferrando, A.; Grassi, A.; Segre, A. L. *Macromolecules* **1992,** *25,* 3874.
67. Grassi, A.; Longo, P.; Guerra, G. *Makromol. Chem., Rapid Commun.* **1989,** *10,* 687.
68. Rapacciuolo, M.; De Rosa, C.; Guerra, G.; Mansitieri, G.; Apicella, A.; Del Nobile, M. A. *J. Mater. Sci., lett.* **1991,** *10,* 1084.

Polystyrene, Succinylated

See: Compatibilizers, Polymeric

POLYSTYRENE, SYNDIOTACTIC

Masahiko Kuramoto
Polymer Research Laboratory
Idemitsu Petrochemical Company, Ltd.

Industrial polystyrene, which is a representative general purpose resin, is an amorphous atactic polymer not having a melting point (T_m). It has good stiffness, and optical and electric properties; however, it has weak points for chemical and heat resistance. Atactic polystyrene (APS) is obtained with radical initiator representatively. Since the discovery of Ziegler–Natta catalyst in the 1950s, many efforts to produce stereoregular polymers have been made. In the field of polystyrene, isotactic polystyrene (IPS) was produced in 1955 by Natta. On the other hand, an opposite stereoregular polystyrene, syndiotactic polystyrene (SPS) has been recently synthesized by Ishihara et al., (Idemitsu Kosan Co., Ltd.).[1] SPS is entirely different from conventional polystyrene (GPPS) in structure, physical properties, and synthetic method. The material is a new crystalline thermoplastic. Its melting point (T_m) is 270°C, the crystallization rate is very fast in comparison with that of IPS. And SPS has some superior properties such as heat resistance, chemical resistance, water/steam resistance, and so on. Therefore, it is recognized as a promising crystalline thermoplastic material to be used in engineering applications.

PREPARATION

Catalyst

It is well known that APS is obtained with a catalyst such as radical initiator, and IPS is obtained with a Ziegler–Natta catalyst such as titanium trichloride-trialkylaluminum. Some studies on the catalyst for syndiospecific polymerization of styrene have been reported.[2-9] It was found that SPS is obtained by using a transition metal, especially titanium catalyst, combined with aluminoxane.

For aluminoxane, methylaluminoxane (MAO) is a highly active co-catalyst.[3,7]

PROPERTIES

SPS exhibits several polymorphic crystalline structure. A well-defined X-ray diffraction pattern of SPS is quite different from that of IPS. The identity period measured from the fiber spectrum of this structure is about 5.1Å, which is much smaller than that of crystallized IPS(6.65Å). The identity period is twice as great as that of polyethylene (PE). The result indicates that the crystalline form of SPS has a trans planar-zigzag structure (TT). Helical crystals (TTGG) are formed in the presence of solvent.

The melting point of SPS is about 270°C, which is higher than that of IPS by 40°C. The crystallization rate of SPS is much faster than that of IPS. It was reported that the isothermal spherulitic growth rate of SPS is more than 100 times that of IPS.[10]

SPS has resistance for high temperatures and solvents in addition to properties resembling well-known atactic polystyrene such as low density, high modulus of elasticity, low moisture absorbance, and excellent electricals.

APPLICATIONS

A variety of polymer processing methods, such as injection molding and sheet forming, can be used in the same way as other thermoplastics. Products such as injection mold, film, sheet, and fiber can be produced according to purpose. For structural materials, reinforcing SPS with glass fiber, mineral fillers, or elastomers is one approach to offset the brittleness.

FILM AND PACKAGE APPLICATIONS

Because SPS is a crystalline polymer, mechanical strengths are improved by biaxial orientation, as with other crystalline polymers. Biaxial-stretching is an effective method to improve the mechanical properties without use of reinforcing materials. SPS film has a high tensile strength and a high level of transparency. SPS has resistance to acids, alkalies, hot water, lower humidity expansion, and excellent dielectric properties. Therefore, SPS is considered to be useful as photographic sheet film and electrical insulation film. It is possible to vacuum-form SPS. This thermoformed product is more heat-resistant than CPET and can stand cooking temperatures of up to 260°C when used for ovenware.

SPS is a new polystyrene having the properties of engineering plastics, and it is possible to use ordinary polymer processing. The products are expected to be useful in the fields of automotive parts, electric and electronic parts, industrial machine parts, food packaging, filters, photographic sheet film, and electrical insulation film.

REFERENCES

1. Ishihara, N.; Seimiya, T.; Kuramoto, M.; Uoi, M. *Macromolecules* **1986,** *19,* 2464; U.S. Patent 5 189 125, 525 693; Eur. Patent 0210615, 3675599.
2. Kuramoto, M.; Ishihara, N.; Uoi, M. *Polymer Preprint Japan* **1987,** *36,* 209.
3. Ishihara, N.; Kuramoto, M.; Uoi, M. *Polymer Preprint Japan* **1987,** *36,* 210.
4. Grassi, A.; Pellecchia, C.; Longo, P.; Zambelli, A. *Gazz. Chim. Ital.* **1987,** *117,* 249.
5. Pellecchia, C.; Longo, P.; Grassi, A.; Amendola, P.; Zambelli, A. *Makromol. Chem., Rapid Commun.* **1987,** *8,* 277.
6. Kakugo, M.; Miyatake, T.; Mizunuma, K. *Chemistry Express* **1987,** *12,* 445.
7. Ishihara, N.; Kuramoto, M.; Uoi, M. *Macromolecules* **1988,** *21,* 3356.
8. Zambelli, A.; Oliva, L.; Pellecchia, C. *Macromolecules* **1989,** *22,* 2129.
9. Zambelli, A.; Pellecchia, C.; Oliva, L.; Longo, L. *Makromol. Chem.* **1991,** *192,* 223.
10. Takebe, T.; Funaki, K.; Yamasaki, K. *IPC-92 Preprint* 175 175.

POLYSTYRENE, SYNDIOTACTIC (Sulfonation and Crystallization)

Robert B. Moore,* E. Bruce Orler, and Dorie J. Yontz
Department of Polymer Science
University of Southern Mississippi

Many studies of industrially important ionomers have demonstrated that the desirable chemical and physical properties of these materials are strongly dependent on the size, quantity, and distribution of ionic multiplets and crystalline domains within the polymer.[1]

Previous investigations of semicrystalline ionomers have focused primarily on polyethylene-based systems.[2-5] While the effect of ionic groups on the rate and degree of crystallization has been partially accounted for by the incorporation of noncrystallizable monomer units, the current understanding of the fundamental effects of ionic aggregation on ionomer crystallization remains limited. To address this lack of critical information regarding the complex morphology of semicrystalline

*Author to whom correspondence should be addressed.

ionomers, we have begun an investigation aimed at understanding the link between ionic aggregation and crystallization in model semicrystalline ionomers.

Since ethylene-based ionomer systems possess undesirable microstructural characteristics, and a wealth of experimental and theoretical information regarding ionic aggregation exists for sulfonated, *atactic*, polystyrene (SaPS) ionomers, the optimum choice of a model semicrystalline ionomer for fundamental investigation is lightly sulfonated, stereoregular polystyrene (e.g., sulfonated syndiotactic polystyrene, SsPS).[6] In this study, the influence of ionic interactions on the crystallization kinetics and formation of polymorphic crystalline domains in lightly sulfonated syndiotactic polystyrene ionomers is examined.

CONCLUSIONS

In order to investigate the link between ionic aggregation and crystallization in semicrystalline ionomers, a model system has been prepared by lightly sulfonating syndiotactic polystyrene. The syndiotactic polystyrene (sPS) was lightly sulfonated using acyl sulfate solutions in 1,2,4-trichlorobenzene. From these studies, it is clear that incorporation of small quantities of sulfonate groups onto crystallizable polystyrene chains has a profound effect on the crystallization of syndiotactic polystyrene. In comparison to pure sPS, the rate of crystallization in SsPS ionomers decreases as the level of sulfonation is increased from 0.5 mol% to 2.0 mol%. The rate of crystallization decreases further as the acid-form samples are neutralized to the K^+-form. For the acid-form SsPS materials, the reduced rate of crystallization is primarily attributed to the incorporation of non-crystallizable groups into the crystalline backbone. However, for neutralized SsPS ionomers, strong ionic interactions tend to restrict the mobility of the crystallizable chain segments, and thus reduce the rate of crystallization to a level well below that of H^+-form SsPS (at the same degree of sulfonation). For SsPS containing 2 mol% sulfonation, neutralization clearly slows the rate of crystallization, but does not eliminate the formation of crystalline domains under isothermal crystallization conditions.

With respect to the effects of neutralization on SsPS crystallization, the strength of the ionic interactions, as governed by the choice of neutralizing counterion, is shown to play an important role in the crystallization behavior of SsPS ionomers. At a given crystallization temperature (e.g., $T_c = 210°C$), the strength of the electrostatic crosslinks influences the mobility of crystallizable polymer chains by affecting the rate at which ion-pairs hop from multiplet to multiplet. When SsPS is neutralized with small K^+ ions, the strength of electrostatic crosslinks in the melt are strong compared to those in the same material neutralized with larger Cs^+ counterions. Therefore, the low rate of crystallization observed for the K^+-neutralized SsPS material is attributed to strong ionic interactions, which reduce the chain segment mobility relative to that observed with the Cs^+-neutralized SsPS.

The use of lightly sulfonated syndiotactic polystyrene as an ideal model system for investigations of semicrystalline ionomers is highlighted by the morphological information obtained in this initial study. While the ionic groups in SsPS aggregate into multiples that appear to be quite similar to those found in sulfonated, *atactic* polystyrene, the SsPS chain segments between the ionic groups are still able to crystallize. Moreover, these crystalline domains organize into polymorphic forms and spherulitic superstructures that are comparable to those found in pure sPS. Since the rate and/or degree of crystallization is strongly influenced by the incorporation of ionic groups onto sPS, these studies suggest that the choice of counterion and crystallization conditions may be used to selectively manipulate the composition of the polymeric crystal-structure within SsPS ionomers.

ACKNOWLEDGMENTS

The authors gratefully acknowledge the Mississippi NSF EPSCoR Program (Grant No. EHR-9108767) for financial support and the Dow Chemical Company for supplying the syndiotactic polystyrene.

REFERENCES

1. Eisenberg, A.; Bailey, F. E., Eds.; *Coulombic Interactions in Macromolecular Systems,* ACS Symposium Series 302; American Chemical Society: Washington, DC, 1986.
2. Wilson, F. C.; Longworth, R.; Vaughan, D. J. *Polym. Prepr. (Am. Chem. Soc., Div. Polym. Chem.)* **1968,** 9(1), 505.
3. Marx, C. L.; Cooper, S. L. *J. Macromol. Sci.: Phys. Ed.* **1974,** B9, 19.
4. Otocka, E. P.; Kwei, T. K. *Macromolecules* **1968,** *1,* 401.
5. Tsujita, Y.; Shibayama, K.; Takizawa, A.; Kinoshita, T. *J. Appl. Polym. Sci.* **1987,** *33,* 1307.
6. Fitzgerald, J. J.; Weiss, R. A. *J. Macromol. Sci.: Rev. Macromol. Chem. Phys.* **1988,** C28, 99.

Poly(styrenesulfonic acid)

See: Styrenesulfonic Acid and its Salts (Polymerization and Applications)

Polysulfides

See: Blends, Interchain Crosslinking
Polysulfides (Prepared from Sulfur Dioxide)
Polysulfides (Use as Modifiers in Epoxy Systems)
Polysulfides (Use in Sealants)

POLYSULFIDES (Prepared from Sulfur Dioxide)

Fuminori Akiyama
Institute for Chemical Reaction Science
Tohoku University

Polysulfide polymers are most predominantly used for sealants, adhesives, and coating. The most widely used one may be the mercaptan-terminated polymer, which is merchandized as "Thiokol LP." The polysulfide liquid polymer is usually prepared from bis(2-chloroethyl) formal and sodium polysulfide.[1,2]

Polysulfide polymers can be prepared from various organic or inorganic sulfur compounds, such as alkali metal sulfide or polysulfide, S_8, sulfur chloride, cyclic sulfides, dithiols, and sulfur dioxide (SO_2).

PREPARATION AND PROPERTIES

Two sorts of organic polymer prepared from sulfur dioxide (SO_2) as one of the starting materials have been known. First, polysulfones have been widely investigated since first reported by Solonia.[3] Second, polysulfites have been known to be prepared by the reaction of epoxides with SO_2 using various catalysts since first reported by Schaefer.[4] It has not been reported that polysulfide polymer could be prepared from SO_2 before our study. In this section it will be presented that polysulfide polymers can be prepared from SO_2 with dithiols by using various catalysts; the second is the reaction of cyclopentadiene with unstable sulfur oxides formed by the reduction of SO_2 with benzoin carbanion.

Polysulfide Polymer Prepared from SO_2 and Dithiols

Reaction of 1,2-ethanedithiol, 1,3-propanedithiol, and 1,6-hexanedithiol with SO_2 in the presence of triethylamine at room temperature for a few days was shown to give polysulfide polymers which contain disulfide and trisulfide linkages accompanied by dithiosulfate linkage.[5]

Catalysts other than triethylamine were shown to bring about copolymerization of 1,2-ethanedithiol with SO_2.[6] Reaction of 1,2-ethanedithiol with SO_2 in the presence of manganese dioxide (Mn O_2), lead dioxide (PbO_2), phosphorous pentoxide (P_2O_5), and boron trifluoride etherate (BF_3OEt_2) at room temperature for a few days gave copolymers.

Sulfur content (slightly less than 73%) for the copolymers obtained using P_2O_5 or BF_3OEt_2 indicates that the ratio of trisulfide and disulfide linkages is close to 1, judging from the fact that the expected sulfur contents for $-CH_2CH_2S_2-$ and $-CH_2CH_2S_3-$ are 69.56 and 77.42%, respectively.

Polysulfide polymer, which is prepared from 1,2-ethanedithiol with SO_2 catalyzed by P_2O_5 or BF_3OEt_2 at room temperature, have high trisulfide linkage content and low mp. If reactions of various alkanedithiols with SO_2 are carried out using various catalysts in changing reaction temperature, liquid polysulfide polymers having high trisulfide content can be obtained. Such a liquid polysulfide polymer may be used as sealants if oxidizing reagents or hardening reagents are added. Another use of the polysulfide polymer obtained from dithiol and SO_2 may be a promoter for vulcanization of rubber.

Polysulfide Prepared from the Reaction of Unstable Sulfur Oxides Formed by the Reduction of SO_2 with Benzoin Carbanion

The final reduction product of SO_2 is elemental sulfur. If SO_2 is reduced using a mild reducing reagent, unstable and reactive products $[S_mO_n]$ may be formed. If an organic substance having enough reactivity to react with $[S_mO_n]$ exists in solution when $[S_mO_n]$ is formed, organosulfur compounds may be formed. Benzoin carbanion was used as mild reducing reagent in this study. It was found that SO_2 or carbon dioxide was reduced by benzoin carbanion to give unstable sulfur oxides or carbon oxides.[7] It was revealed that cyclopentadiene could react with $[S_mO_n]$.[8]

Reaction products of cyclopentadiene with $[S_mO_n]$, which were oligomers, were separated by means of column chromatography and TLC.

Elemental analyses of this oligomer indicates that the oligomer contains sulfur and oxygen in the ratio of $SO_{0.14}$. This value means that this oligomer has mainly sulfide structures.

Sulfide structure of the oligomer is considered to be formed by the reaction of cyclopentadiene with $[S_mO_n]$ to give cyclopentadiene-S_mO_n, followed by decomposition of it to give sulfide structure. The strong basicity of the benzoin carbanion solution probably brings about decomposition of cyclopentadiene-S_mO_n to give sulfide structure.

Various vinyl monomers such as styrene, isoprene, phenylacetylene, cyclooctatetraene, 1,3,5,-hexatriene, allene, 1-methylcyclopropene, and cyclopentadiene were tested for the reaction with $[S_mO_n]$. Only cyclopentadiene afforded oligomer having sulfide structure in the reaction with $[S_mO_n]$ formed by the reduction of SO_2 with benzoin carbanion.

REFERENCES

1. Berenbaum, M. B.; Panek, J. B. *Polyethers, Part III Polyalkylene Sulfides and Other Polythioethers* Interscience: 1962; p 43.
2. Bertozzi, E. R. *Rubber Chem. Technol.* **1968,** *41,* 114.
3. Solonina, W. *Zh. Russk, Fizz-khim. Obshch.* 1898, *30,* 826.
4. Schaefer, J.; Katnik, R. J.; Kern, R. J. *Macromolecules* **1968,** *1,* 101.
5. Akiyama, F.; Tatsuta, R.; Matsuda, M. *J. Polym. Sci., Polym. Lett. Ed.* **1979,** *17,* 23.
6. Akiyama, F. *J. Polym. Sci., Polym. Lett. Ed.* **1980,** *18,* 401.
7. Akiyama, F. *Bull. Chem. Soc. Jpn.* **1988,** *61,* 3951.
8. Akiyama, F. *J. Polym. Sci., Polym. Lett. Ed.* **1986,** *24,* 631.

POLYSULFIDES (Use as Modifiers in Epoxy Systems)

Debbie Carpenter
Morton International

There are two methods for formulating liquid polysulfide/epoxy systems. The first method is "flexibilizing" the epoxy system with liquid polysulfide. The addition of liquid polysulfide to an epoxy system improves flexibility without adversely affecting the system's chemical resistance.

The second method focuses on epoxy-terminated liquid polysulfide polymers. Products can be made by co-reacting liquid polysulfide with epichlorohydrin or an epoxy resin to make versatile epoxy formulations without liquid polysulfide's typical mercaptan odor. Formulations based on these products exhibit excellent fuel resistance, good flexibility, and improved adhesion to various substrates.

With more than 40 years of field experience, liquid polysulfides have proven successful. They can easily be compounded into sealants, adhesives, coatings, potting compounds, and flexible molding compositions. Compounds based on these polymers are used in residential and commercial building construction, insulating glass, aerospace, electronic, aviation, and marine applications.

"FLEXIBILIZING" EPOXY SYSTEMS

Epoxy systems have been successfully "flexibilized" by the addition of liquid polysulfide polymers since 1954. The liquid polysulfide improves flexibility without adversely affecting the chemical resistance of the epoxy system.

Liquid polysulfide systems are generally formulated with epoxy resins as two-component room-temperature cure mixes.

This gives good shelf stability and permits easy handling of the formulated product.

One component contains the liquid polymer and curing agent or catalyst. Generally, the formulator uses ½ to 2 times as much liquid polysulfide as epoxy resin. This ratio is varied to get the desired physical properties for the system.

The second component contains the epoxy resin. Most commercially available epoxy resins are suitable for formulating with liquid polysulfide polymers. Liquid resins, such as Bisphenol A or Bisphenol F, tend to be preferred because of their ease of handling. Reactivity of the finished system can even be varied by changing the epoxy resin.

When added to the formulation, the liquid polysulfide acts as a viscosity modifier and reactive diluent. The addition of liquid polysulfide to the formulation permits ease of handling of the mixed material and improved wetting of the substrate. Through proper selection of fillers, surfactants, defoamers, and other additives, the formulator can obtain the desired properties for the specific application.

EPOXY-TERMINATED LIQUID POLYSULFIDE POLYMERS

Although liquid polysulfide polymers are effective modifiers for epoxy systems, the mercaptan odor of these polymers has limited their use in some industrial applications. Epoxy-terminated polysulfide polymers provide all the attributes of polysulfide without the mercaptan odor.

Besides the elimination of the odor, this method has many added benefits. Additional benefits of liquid polysulfide/epoxy systems include: improved chemical resistance, introduction of flexibility, and enhancement of adhesion.

Physical properties of the system can be controlled by varying the molecular weight of the liquid polysulfide used.

POLYSULFIDES (Use in Sealants)

Michael J. Scherrer, David E. Vietti, and
Anthony R. Fiorillo
Morton International

MERCAPTAN-TERMINATED POLYMERS

Chemistry

The polysulfide polymer is formed by the reaction of bischloroethylformal and sodium polysulfide. The polymer is initially made as a high molecular weight aqueous dispersion. It is then split and terminated with mercaptan groups by the addition of NaSH. The resultant polymer is represented in **Equation 1**.

$$HS(C_2H_4OCH_2OC_2H_4SS)_nC_2H_4OCH_2OC_2H_4SH \qquad 1$$

Polysulfide polymer

The formal linkage provides flexibility, and the disulfide linkage imparts excellent chemical and solvent resistance.

Polymer Properties

Polysulfide polymers are available in molecular weights ranging from 1000 to 8000 and with the degree of crosslinking varying between 0 and 2%. Crosslinking is provided by the incorporation of the monomer trichloropropane into the backbone during the polymerization.

Polysulfide polymers have been used as the base polymer in sealants since the early 1950s.

All these polymers are cured through the mercaptan terminals using thiol chemistry involving oxidation and condensation or addition reactions. This chemistry provides the basis of formulating sealants with mercaptan-terminated polymers.

Sealant Applications for Polysulfides

Applications for polysulfide polymers are many and diverse because of their unique properties, such as dimensional stability, flexibility, chemical resistance, low moisture vapor transmission, low gas transmission, and weatherability. This makes them particularly useful in a variety of sealant applications. For example, the outstanding resistance of polysulfides to petroleum products has made them the standard sealant for virtually all aircraft integral fuel tanks and bodies. Another important application is in insulating glass window sealants. Polysulfides also have a long record as construction sealants.

CURING CHEMISTRY

Inorganic and Organic Oxide Curing

Polysulfide polymers are converted to high molecular weight elastomers through the reactive hydrogen of the mercaptan groups.

For a long time, lead IV oxide PbO_2 was the most widely used oxidizing agent for high molecular weight polysulfides ($M_w > 2500$). Since the early 1970s manganese dioxide, MnO_2, has become the predominate oxidizing agent.

The newest curing system is sodium perborate monohydrate ($NaBO_2$, H_2O_2, H_2O). Sodium perborate offers several advantages as a curing agent for building sealants. One is that it is light-colored rather than the dark colors of the manganese or lead oxides. Thus it is nonstaining and can be manufactured in a variety of colors. Sealants cured with sodium perborate have low modulus, excellent elasticity, and outstanding resistance to water, weather, UV light, and mold (even without additives). They have good adhesion to most surfaces and are environmentally friendly.

Additions Reactions

Polysulfide polymers copolymerize with diisocyanates or epoxy resins by addition reactions between the mercaptan terminals and the isocyanate or epoxide groups.

FORMULATING INTO FINISHED SEALANTS

Raw Materials

Polysulfide-based sealants are formulated to obtain the desired properties for a particular application. A typical formulation will contain:

- Polysulfide polymer
- Curing agent
- Cure accelerators (bases) or retarders (acids)
- Fillers

- Plasticizers
- Thixotropes
- Adhesion promoters

HEALTH AND SAFETY

Because they are sulfur-containing polymers, the polysulfides have a characteristic odor. Although the odor is somewhat objectionable, it probably arises from small amounts of lower molecular weight compounds. In fact, the lower molecular weight polymers exhibit the strongest odor. Toxicity tests conducted on a representative liquid polysulfide used in Morton sealant applications indicate that the polymers are not eye irritants and have a low order of oral toxicity (LD_{50} > 5 g/kg). Tests on the lower molecular weight liquid polysulfide products show similar findings. These materials are not eye or skin irritants, do not cause allergic skin reactions, and are not toxic when administered orally (LD_{50} > 3.4 g/kg). Under the criteria set forth under OSHA's Hazard Communication Standard (29 CFR 1910.1200), Morton classifies liquid polysulfide products as nonhazardous. When used in accordance with prescribed procedures, they do not pose a health hazard.

When considering sealants or other formulated products, health and safety considerations relating to other ingredients should be taken into account.

Polysulfites

See: Sulfur Dioxide Copolymerization

POLY[*N,N*′-(SULFO-*p*-PHENYLENE) TEREPHTHALAMIDE]

Yoshiyuki Okamoto and T. K. Kwei
Department of Chemistry
Herman F. Mark Polymer Research Institute
Polytechnic University

Poly(*p*-phenylene terephthalamide) (PPTA) was developed by DuPont under the tradename Kevlar.[1] It is a rigid, rod-like polymer with high tensile strength, stiffness, crystallinity, and thermal stability. This stiff aromatic polyamide is very difficult to process because of its insolubility in common organic solvents.[2,3] To improve the processability of PPTA, a facile derivatization through substitution of the amide hydrogen or sulfonation of the aromatic ring has been investigated.

SYNTHESIS

Attempts at sulfonating PPTA with oleum or SO_3 – DMF complex resulted in sulfonation on the PPTA backbone.[4] Thus, a condensation polymerization of 2,5-diaminobenzene sulfonic acid with terephthaloyl chloride in dimethyl acetamide (DMAc) containing LiCl was performed to obtain poly[*N,N*-(sulfo-*p*-phenylene)terephthalamide] (PPTS) (**Figure 1**).[5-9]

The polymer obtained was isolated by precipitation in methanol; washed thoroughly with water, methanol, and acetone; and dried at 60–70°C *in vacuo* to constant weight. Various molecular weights of PPTS (5000–10,000) were obtained under different temperatures and concentrations of reactants.

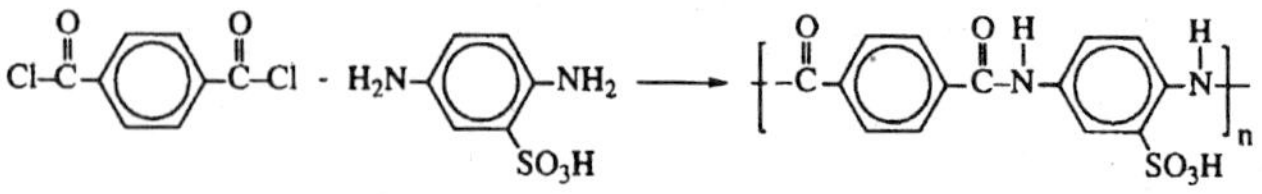

FIGURE 1. Poly[*N,N*′-(sulfo-*p*-phenylene)terephthalamide], (PPTS).

INTRINSIC VISCOSITY AND MOLECULAR WEIGHT

PPTS can be dissolved in a variety of solvents, including DMAc, DMSO, and water. Polymers such as PPTS exhibit strong interactions between themselves or with solvent molecules. Sulfuric acid has the largest dielectric constant, and it prevents interchain hydrogen bonding and minimizes the aggregation of PPTS molecules. DMAc forms very strong complexes with amide groups, thereby reducing but not eliminating intermolecular association. Water, although it has a high dielectric constant, is a poor solvent for PPTS. As a result, PPTS forms gels in aqueous solution at low concentrations, ~ 0.6% g/dL.

T_g AND LIQUID CRYSTALLINE STRUCTURES

The T_g of PPTS is not easily identified from the DSC scan. There appears to be a specific heat increase around 320°C. Thus, the T_g of PPTS is assigned at ~320°C. PPTS forms gel and shows strong birefrigence when dissolved in water or DMAc and shows typical nematic liquid crystalline structure in these solvents. The formation of gels depends on solvent quality; the gels are also thermally reversible. The gelation concentration of PPTS in water is < 1%; in DMAc it is > 5% at room temperature.[10]

ALKALINE METAL SALTS AND ION CONDUCTIVITY

PPTS, though only slightly soluble in water, can be converted to a polyelectrolyte by neutralization with alkali metal hydroxides or carbonates. The salts have much higher solubilities in water than the parent polymer.

These salts can be cast into free-standing films or membranes. The polymeric salts are chemically and thermally stable and can be utilized as high-temperature ion exchangers, gas separation membranes, and alkali ion conductors.

The ionic conductivities of these films were measured at up to 300°C under nitrogen atmosphere. The ionic conductivities of Li and Na salts follow Arrhenius behavior and reach ~ 5 × 10^{-6} S/cm at 250°C.[11]

REFERENCES

1. Bair, T. I.; Morgan, P. W. U.S. Patent 3 673 143, 1974; 3 817 941, 1974.
2. Preston, J. In *Encyclopedia of Polymer Science and Engineering;* Mark, H. F. et al., Eds.; Vol. 11, pp 381–409.
3. Arpin, M.; Strazielle, C. *Polymer* **1977,** *18,* 591.
4. Silver, F. M. *Polym. Prepr.* **1979,** *20(1),* 207; *J. Polym. Sci., Polym. Chem. Ed.* **1979,** *17,* 3519; **1979,** *17,* 3535.
5. Salamone, J. C. et al. *Polym. Prepr.* **1989,** *30,* 281.
6. Salamone, J. C. et al. *Polym. Prepr.* **1989,** *29,* 273.

7. Vandenberg, E. J. et al. *J. Polym. Sci., Part A: Polym. Chem.* **1989,** *27,* 3743.
8. Vandenberg, E. J. et al. *Polym. Mater. Sci. Eng. Proc.* **1987,** *57,* 139.
9. Salamone, J. C. et al. *Polym. Mater. Sci. Eng.* **1987,** *57,* 144.
10. Chu, E. Ph.D. Thesis, Polytechnic University, Brooklyn, NY, 1993.
11. Okamoto, Y. et al. *Solid State Ionics* **1993,** *60,* 131.

Polysulfones

See: Poly(aryl ether sulfone)s

POLYSULFONES

Christos P. Tsonis
Chemistry Department
King Fahd University of Petroleum and Minerals

The free-radically initiated copolymerization of sulfur dioxide with monomers containing a carbon–carbon double or triple bond leading to polysulfones has been the subject of thorough investigations.[1-6] The driving force behind these studies is the potential application of these copolymers as radiation resists, biomaterials, membranes, fire resistants, and polysoaps.[7-12]

SYNTHESIS

Initiators

There are several radical initiator methods employed for the copolymerization of SO_2 with carbon–carbon double or triple bonds. The most common and easy to use are 2,2′-azobiisobutyronitrile (AIBN) and *tert*-butyl hydroperoxide (TBHP).

Alkenes

Terminal and internal monoalkenes copolymerize with sulfur dioxide in the presence of a free-radical generator to produce polysulfones (**Equation 1**).[1-3]

$$CH_2{=}\underset{\displaystyle R}{\underset{|}{C}H} + SO_2 \xrightarrow{\text{Free radical}} \sim CH_2{-}\underset{\displaystyle R}{\underset{|}{C}H}{-}SO_2{\sim} \quad 1$$

The polymer formed is an alternating copolymer regardless of the reaction temperature, radical initiator, and copolymerization feed ratio. Cyclic monomers such as cyclohexane or nonbornene also copolymerize readily with sulfur dioxide to form alternating polysulfone copolymers.[13,15-19]

Both conjugated and nonconjugated dienes react with sulfur dioxide by 1,4-addition in the presence of a free-radical initiator to form alternating polysulfones.

The copolymerization of styrene with sulfur dioxide produces copolymers with variable composition.[20-25]

The styrene derivatives such as *p*-methylstyrene and *p*-bromostyrene copolymerize with sulfur dioxide in the presence of AIBN at temperatures of 0–60°C.[26,27] Although both *p*-methylstyrene and *p*-bromostyrene form polysulfone copolymers with 2 moles of the organic monomer per mole of SO_2, α-methylstyrene does not copolymerize with sulfur dioxide, probably because of steric factors.

Alkynes

Terminal aliphatic acetylenic hydrocarbons, such as 1-heptyne, are also copolymerized with liquid sulfur dioxide in the presence of *tert*-butyl hydroperoxide at low temperature.[14] The resulting copolymers are alternating with 100% trans configuration regardless of the nature of the alkyne, solvent, and reaction temperature. Internal alkynes (e.g., 3-hexyne) are not known to copolymerize with sulfur dioxide.

Vinyl Monomers

Monomers with electron withdrawing groups directly attached to the vinyl carbon usually do not copolymerize with sulfur dioxide at high temperature.[2,28] At relatively low polymerization temperature the composition of the copolymer formed varies and depends on the nature of the organic monomer, feed ratio, and reaction temperature. For example, vinyl chloride gives a 2:1 polysulfone, and tetrafluoroethylene forms a 67:1 copolymer mole ratio, whereas the copolymer composition of an acrylic acid–SO_2 system varies from 7 to 43 acrylic acid units per SO_2 molecule.[22,29-31]

The copolymerization of acrylic monomers, namely acrylonitrile, acrylamide, acrolein, and methacrylate, with liquid sulfur dioxide at low temperature (–50 to –15°C) and high dilution in the presence of *tert*-butyl hydroperoxide, gives polysulfones with relatively high SO_2 incorporation.[32,33]

Kinetics and Mechanism

The free-radically initiated copolymerization of sulfur dioxide with carbon–carbon double or triple bond is a chain growth exothermic process, which is controlled by resonance, inductive, and steric factors possessed by the organic monomer.[28,34-36] To obtain high molecular weight polysulfones the activated monomer and growing polymer chain must be stable and fast enough to allow long copolymer growth.

The ability of a particular organic monomer to copolymerize with sulfur dioxide depends on its ceiling temperature as well.[6]

Studies on the relative reactivities of organic monomers with sulfur dioxide show that the highest copolymerization rates are in reaction systems having an electron donor monomer of low resonance stabilization (e.g., ethylene) and the resulting copolymers have an alternating structure.[35,36] In copolymerization monomer systems containing monomers of high resonance stabilization with electron acceptor properties (e.g., styrene), high degree of homopropagation occurs.

CHARACTERIZATION

Solubility

In general, SO_2 polar units introduced into the hydrocarbon polymer chains tend to decrease their solubility in common solvents because strong polymer–polymer bonds develop. Alkene–SO_2 copolymers with high sulfur dioxide to hydrocarbon ratio (e.g., ethylene–SO_2) are usually soluble in polar solvents such as H_2SO_4 and Me_2SO.[37] Polysulfones such as 1-hexene–SO_2, cyclopentene–SO_2, and styrene–SO_2 copolymers dissolve easily in benzene, dioxane, and tetrahydrofuran, respectively.[38]

The solubility of polysulfones having polar pendant groups, such as acrylamide–SO_2 copolymers, decreases in highly polar solvents (e.g., H_2O) as the amount of SO_2 incorporated into the copolymer increases.[41]

Stability

The stability of polysulfones is relatively moderate to low.[39,40] This is an inherent problem in polysulfones with labile beta-hydrogens.[42-46] Molecular oxygen or a strong base usually abstracts the beta-hydrogen and the polysulfone chain breaks down producing ~CH=CHX and $\sim SO_2$–H fragments.

When polysulfones are exposed to a relatively high temperature or high energy radiation, a random C–S bond cleavage normally takes place instead of crosslinking.[47-50] The decomposition products are mainly sulfur dioxide, olefinic monomer, and other organic compounds.

APPLICATIONS

Sulfur dioxide-based copolymers prepared from olefins, acetylenes, and vinyl monomers have not found applications in the thermoplastic industry mainly because they generally possess low thermal stability and thus fail to undergo molding.[48] They are good coating materials in various minerals (e.g., gypsum) and improve the compatibility of these minerals with polyolefins.[6] They also find applications as antistatic additives for fossil fuels.[6] They are sensitive positive resists and are used in lithographic processes for the electronics industry.[7,8]

Studies are under way to use polysulfones as medical devices in the respiratory system because of their good biocompatibility and high permeability of molecular oxygen and carbon dioxide.[9,10] In addition, polysulfones have flame resistance potential since they are known to be self-extinguishing.[11]

REFERENCES

1. Dainton, F. S.; Ivin, K. J. *Quart. Rev.* **1958,** *12,* 61.
2. Tokura, N. In *Encyclopedia of Polymer Science and Technology*; Wiley-Interscience: New York, NY, 1968; Vol. 9, p 460.
3. Ivin, K. J.; Rose, J. B. *Advances in Macromolecular Chemistry*; Academic: New York, NY, 1968; Vol. 1.
4. Tsuruta, T.; O'Driscoll, K. F. *Structure and Mechanism in Vinyl Polymerization*; Marcel Dekker: New York, NY, 1969.
5. Sandler, S. R.; Karo, W. *Polymer Synthesis*; Academic: New York, NY, 1980; Vol. 3.
6. Fawcett, A. H. *Encyclopedia of Polymer Science and Engineering*; Wiley-Interscience: New York, NY, 1987; Vol. 1, p 408.
7. Lin, B. J. In *Introduction to Microlithography; Theory, Materials and Processing*, ACS Symposium Series 219; American Chemical Society: Washington, DC, 1983; p 287.
8. Tarascon, R. G.; Reichmanis, E. et al. *Polym. Sci. Eng.* **1989,** *29,* 850.
9. Gray, D. N. *Polym. Sci. Technol.* **1981,** *14,* 21.
10. Gray, D. N. *Org. Coat. Plast. Chem.* **1980,** *42,* 616.
11. Tsonis, C. P.; Ali, S. A. et al. *J. Appl. Polym. Sci.* **1989,** *38,* 1899.
12. Anton, P.; Laschewsky, A. *Makromol. Chem., Rapid Commun.* **1991,** *12,* 189.
13. Oster, B.; Lenz, R. W. *J. Polym. Sci., Part A* **1977,** *15,* 2479.
14. Tsonis, C. P.; Ali, S. A.; Wazeer, M. I. *Polymer* **1986,** *27,* 1991.
15. Frederic, D. S.; Cogan, H. D.; Marvel, C. S. *J. Am. Chem. Soc.* **1934,** *56,* 1815.
16. Zutty, N. L.; Wilson, C. W. *Tetrahedron Lett.* **1963,** *30,* 2181.
17. Hill, E. H.; Caldwell, J. R. *J. Polym. Sci.* **1964,** A2, 1251.
18. Zutty, N. L.; Wilson, C. W. et al. *J. Polym. Sci.* **1965,** A3, 2781.
19. Sartori, G.; Lundberg, R. D. *J. Polym. Sci., Polym. Lett. Ed.* **1972,** *10,* 583.
20. Glavis, F. J.; Ryden, L. L.; Marvel, C. S. *J. Am. Chem. Soc.* **1937,** *59,* 707.
21. Walling, C. C. *J. Polym. Sci.* **1955,** *16,* 315.
22. Cais, R. E.; O'Donnell, J. H. *J. Polym. Sci., Polym. Lett. Ed.* **1977,** *15,* 659.
23. Matsuda, M.; Iino, M. *Macromolecules* **1969,** *2,* 216.
24. Matsuda, M.; Iino, M. et al. *Macromolecules* **1972,** *5,* 240.
25. Iino, M.; Thoi, H. H. et al. *Macromolecules* **1979,** *12,* 160.
26. Tokura, N.; Matsuda, M.; Ogawa, Y. *J. Polym. Sci., Part A-1* **1963,** *1,* 2965.
27. Tokura, N.; Matsuda, M.; Arakawa, K. *J. Polym. Sci., Part A-1* **1964,** *2,* 3355.
28. Tokura, N.; Matsuda, M.; Yazaki, F. *Makromol. Chem.* **1960,** *42,* 108.
29. Harman, J.; Joyce, R. M. U.S. Patent 2 411 722, 1946.
30. Tsonis, C. P.; Pasika, W. M. *Polymer* **1980,** *21,* 253.
31. Wazeer, M. I.; Ali, S. A.; Tsonis, C. P. *J. Polym. Sci., Polym. Phys. Ed.* **1988,** *26,* 1539.
32. Tsonis, C. P.; Ali, S. A. *Makromol. Chem., Rapid Commun.* **1989,** *10,* 641.
33. Tsonis, C. P.; Ali, S. A. et al. *Makromol. Chem.* **1992,** *193,* 2175.
34. Ivin, K. J.; Walker, N. A. *J. Polym. Sci., Part A-1* **1971,** *9,* 2371.
35. Florjanczyk, Z.; Florjanczyk, T.; Klopotek, B. *Makromol. Chem.* **1987,** *188,* 2811.
36. Florjanczyk, Z.; Raducha, D.; Zygadlo, E. *Makromol. Chem.* **1989,** *190,* 2149.
37. Ovenall, D. W.; Sudol, R. S.; Cabat, G. A. *J. Polym. Sci., Polym. Chem. Ed.* **1973,** *11,* 233.
38. Ivin, K. J.; Ende, H. A.; Meyerhoff, G. *Polymer* **1962,** *3,* 129.
39. Marvel, C. S.; Weil, E. D. *J. Am. Chem. Soc.* **1964,** *76,* 61.
40. Cook, R. *J. Polym. Sci., Polym. Chem. Ed.* **1978,** *16,* 3001.
41. Firth, W. C.; Palmer, L. E. *Macromolecules* **1971,** *4,* 654.
42. Naylor, M. A.; Anderson, A. W. *J. Am. Chem. Soc.* **1954,** *76,* 3962.
43. Wellish, E.; Gipstein, E.; Sweeting, O. J. *J. Appl. Polym. Sci.* **1964,** *8,* 1623.
44. Gipstein, E.; Wellish, E.; Sweeting, O. J. *J. Org. Chem.* **1969,** *29,* 207.
45. Wellish, E.; Gipstein, E.; Sweeting, O. J. *J. Appl. Polym. Sci.* **1964,** *8,* 1623.
46. Verma, A. K. *Progr. Polym. Sci.* **1986,** *12,* 219.
47. Gipstein, E.; Moreau, W. et al. *J. Appl. Polym. Sci.* **1977,** *21,* 677.
48. Bowmel, J. N.; O'Donnell, J. H. *Polym. Degrad. & Stab.* **1981,** *3,* 87.
49. Kiran, E.; Gillham, J. K.; Gipstein, E. *J. Appl. Polym. Sci.* **1977,** *21,* 1159.
50. Brown, J. R.; O'Donnell, J. H. *Macromolecules* **1972,** *5,* 109.

POLY(TEREPHTHALIC ACID ANHYDRIDE) (as a Latent Monomer)

Ying-Hung So
The Dow Chemical Company

Terephthalic acid (TA) is the monomer for many important polymers including polyesters, polyamides, polybenzothiazoles, and polybenzoxazoles.[1-5] With low chemical reactivity, no melting before decomposition, and limited solubility, except in very polar solvents such as dimethylsulfoxide, dimethylformamide, and 1-methyl-2-pyrrolidinone, TA is often converted to terephthaloyl chloride, diaryl, or dialkyl terephthalates for polymerization reactions.

Carboxylic anhydrides readily undergo nucleophilic reactions; therefore, they are a good reactive form of carboxylic acid.[6] Polycarboxylic anhydrides liberate only one water molecule per repeating unit in polymer formation, which can be a significant advantage in polymerization reactions. The use of poly(terephthalic anhydride) (PTAA) as a latent monomer is illustrated with the preparation of poly(*p*-phenylenebenzobisoxazole) (PBO) and poly(*p*-phenylenebenzobisthiazole) (PBZT) in poly(phosphoric acid) (PPA).[7]

Most polycarboxylic anhydrides melt at lower temperatures than the corresponding diacids.[8] Rabilloud et al. made polybenzimidazles by melt polymerization of poly(isophthalic acid anhydride) and aromatic tetraamines.[9]

PBO AND PBZT FROM PTAA IN PPA

Synthesis of PTAA

Bucher et al. prepared PTAA by heating TA in acid anhydride with a catalytic amount of phosphoric acid.[10]

High molecular weight PTAA was synthesized by the Schotten–Baumann condensation of TA and terephthaloyl chloride (TCl) in the presence of triethylamine with dimethylformamide as the solvent.[11] Number-average repeating units were found to be 43. Melting point was 380°C as determined by DSC (differential scanning calorimetry).

PBO and PBZT from PTAA

PBO and PBZT are polymers with extended rod-like configurations. Fibers from these polymers have excellent tensile strength and tensile modulus, very good chemical resistance, and high thermal stability.[12]

These polymers are prepared from the dihydrochloride salts of 1,3-diamino-4,6-dihydroxybenzene (DADHB), or 1,4-diamino-2,5-dithiobenzene (DADTB) with terephthalic acid (TA); or terephthaloyl chloride in polyphosphoric acid (PPA).

High molecular PBO with intrinsic viscosity of 29 dL/g was produced by the reaction of DADHB dihydrochloride salt with PTAA of particle size 34–78 µm which was prepared by the Schotten–Baumann condensation.

Fiber was extruded from the dope at 165°C out of a 180 µm spinneret with 36 holes vertically downward into a water coagulation bath. As-spun fiber had tensile strength and tensile modulus of 700 ksi [4.83 gigaPascals (Gpa)] and 22 Msi (150 GPa), respectively.

PBZT was produced from 1,4-diamino-3,5-dithiobenzene dihydrochloride and unsieved PTAA prepared by the Schotten–Baumann condensation. Intrinsic viscosity of the polymer was 36 dL/g.

CONCLUSION

Poly(terephthalic acid anhydride) is a convenient latent monomer for PBO and PBZT synthesis. The polyanhydride may directly react with *o*-aminophenol groups to form benzoxazole linkages and it also breaks down in PPA to generate small TA particles. TA generated are activated by PPA through the formation of benzoyl-phosphate anhydride. The melting point of PTAA is 380°C, which may limit these polyanhydrides' use in melt polymerization.

Since polyanhydrides have high reactivity towards nucleophiles, liberate only one water molecule per polymer repeating unit, normally have lower melting points than the corresponding dicarboxylic acids, and are easy to prepare, they are often the monomers of choice in polymer preparation.

ACKNOWLEDGMENT

I thank Miss Beth Kapp for technical assistance, Mr. Duane Krueger for particle size characterization, Dr. Cyrus Crowder for X-ray diffraction studies, and Mr. Jim Sedon and Dr. Don McLemore for helpful discussion.

REFERENCES

1. Jadhav, J. Y.; Kantor, S. W. In *Encyclopedia of Polymer Science and Technology,* 2nd ed.; Mark, H. F.; Kroschmitz, J. I., Eds.; Wiley: New York, NY, 1988; Vol. 12, pp 217-256.
2. Yang, H. H. *Aromatic High-Strength Fibers*; Wiley-Interscience: New York, NY, 1989; Chapter 4.
3. Preston, J. In *Encyclopedia of Polymer Science and Technology,* 2nd ed.; Mark, H. F.; Kroschmitz, J. I., Eds.; Wiley: New York, NY, 1988; Vol. 11, pp 381-409.
4. Yang, H. H. *Aromatic High-Strength Fibers*; Wiley-Interscience: New York, NY, 1989; Chapter 2.
5. Wolfe, J. F. In *Encyclopedia of Polymer Science and Technology,* 2nd ed.; Mark, H. F.; Kroschmitz, J. I., Eds.; Wiley: New York, NY, 1988; Vol. 11, pp 601-635.
6. March, J. *Advanced Organic Chemistry*, 4th ed.; Wiley: New York, NY, 1992; pp 392-393, 418-419.
7. So, Y. H. *J. Polym. Sci., Part A, Polym. Chem.* **1994,** *32,* 1899.
8. Leong, K.; Domb, A. et al. In *Encyclopedia of Polymer Science and Technology,* 2nd ed., suppl.; Mark, H. F.; Kroschmitz, J. I., Eds.; Wiley: New York, NY, 1989; 648-665.
9. Rabilloud, G.; Sillion, S.; Gaudemaris, G. Fr. Patent 1 526 830, 1968; *Chem. Abstract* **1969,** *71,* 4097.
10. Bucher, J. E.; Slade, W. C. *J. Am. Chem. Soc.* **1909,** *31,* 1319.
11. Leong, K. W.; Simonte, V.; Langer, R. *Macromolecules* **1987,** *20,* 705.
12. *The Materials Science and Engineering of Rigid-Rod Polymers*; Adams, W. W.; Eby, R. K.; McLemore, D. E., Eds.; Materials Research Society: Pittsburgh, PA, 1989; Vol. 134.

POLYTERPENE RESINS

Steve Goldschmidt*
Ascona Resins

Joseph J. McBride, Jr.
Formerly: Director of R&D
Arizona Chemical Company

Polyterpene resins are low molecular weight hydrocarbon polymers prepared by cationic polymerization (copolymerization) of monoterpenes (isoprene dimers) such as α- and β-pinene or limonene.

USES

Polyterpene resins are used by the adhesive, sealant, wax coating, and investment casting industries. More specifically, they are generally used in pressure-sensitive adhesives (PSAs), hot-melt adhesives and coatings, nonwoven fabrics, and elastomeric sealants. Some of them are used to manufacture chewing gum.

*Author to whom correspondence should be addressed.

Commerical Products

Commercial products are generally produced by the cationic polymerization of α-pinene, β-pinene, or limonene, the copolymerization of one or more of the preceding raw materials with styrene, C9 hydrocarbons such as vinyl toluene, or C5 hydrocarbons such as piperylene, or the polymerization with phenol or substituted phenols.

For additional reading on polyterpene resins, we recommend the excellent chapter written by Ruckel and Arlt in the Naval Stores Handbook compiled by Zinkel and Russell in 1989.[1] A major contribution to the understanding of polyterpene resins is the paper presented at the 1994 Pulp Chemicals Association meeting by Erwin Ruckel.[2]

Terpene Phenolic Resins

Terpene phenolic resins are manufactured when a higher solubility parameter than the one corresponding to pure polyterpene resins (or hydrocarbon copolymer) is desired for one of the following reasons:

- if the backbone polymer of the adhesive is more polar (e.g., certain ethylene vinyl acetate (EVA)s, neoprene or polyurethanes), and
- if one wants to promote hydrogen bonding across the surface of a polar substrate such as poly(ethylene terephthalate) (PET) or poly(methylmethacrylate).

PROPERTIES

Polyterpene resins are characterized primarily by their:

- *molecular weight*, the most important single property of a terpene resin.[3] The relatively low molecular weights at which the properties of terpene resins plateau (i.e., the minimum molecular weight at which they achieve their desired properties), coupled with their narrow molecular weight distribution and excellent solubility in elastomers, make them uniquely useful for adhesives formulation.
- *molecular weight distribution*, determines (within a defined polyterpene resin type) what the cloud point in a given structural polymer will be.
- *softening point*, is measured by the Ring and Ball method.[4] This value is a complex property that is related to T_g and melt viscosity. It is the characteristic that is typically used to define a resin of this family. Terpene resins are available in softening points from 25°C to 135°C, but most commercial resins have softening points of 85°C, 95°C, 105°C, and 115°C.
- *color*, is read in a 50% heptane solution and defined on the Gardner scale. Most commercial resins have a color of 3–4, but a process for preparing even lower color resins was discovered by Schleunz.[5]
- *"toluene insolubles,"* a measure of the amount of inorganic material (mainly catalyst residue) in the resin.
- *chlorine content*, traces of chlorine, introduced by the catalyst, can be evidenced by the Beilstein test.[6]
- viscosity at compounding temperatures.
- color and thermal stability.
- compatibility with the other components of the adhesive or sealant.

APPLICATIONS

Only three terpenes [α-pinene, β-pinene, and limonene (or dipentene)] have found commercial application in the manufacture of polyterpene resins.

Polyterpene resins are most often used in the formulation of adhesives, coatings, and elastomeric sealants. In adhesives, they are used in hot melts and in PSAs. PSAs might themselves be applied via hot melts, via solvents, or via water emulsions. They include rubber cements. The hot-melt adhesive group includes applications for coatings and investment waxes. The elastomeric sealant group includes sealants, caulks, and can end cements.

REFERENCES

1. Ruckel, E. R.; Arlt, H. G. In *Polyterpene Resins: Naval Stores Production–Chemistry–Utilization*; Pulp Chemicals Association: Panama City, FL, 1989; Chapter 13, p 510.
2. Ruckel, E. R. Presented at the Pulp Chemicals Association Meeting, New Orleans, LA, September 1994.
3. Ruckel, E. R. Presented at the Pulp Chemicals Association Meeting, New Orleans, LA, September 1994.
4. "Ring and Ball Method"; Technical Report No. E28-58T; American Society for Testing and Materials.
5. Schleunz, R. W. U.S. Patent 4 482 688.
6. Feigl, X. In *Spot Tests* Elsevier: 1939; p 261.

Polytetrafluoroethylene

See: *Engineering Thermoplastics (Survey of Industrial Polymers)*
Fluoropolymers
Polytetrafluoroethylene (Overview)
Polytetrafluoroethylene (Effect of γ-Irradiation)
Tetrafluoroethylene Copolymers (Overview)

POLYTETRAFLUOROETHYLENE (Overview)

Dewey L. Kerbow
E. I. DuPont de Nemours and Company, Incorporated

Polytetrafluoroethylene (CAS Registry No. 9002-84-0), or PTFE, exists in commerce today in very much the same form as when it was developed over 50 years ago following its discovery by R. Plunkett.[1]

The unique combination of properties–including chemical resistance, thermal-oxidation resistance, high lubricity, electrical and thermal insulation, low flammability, and excellent weatherability–are derived primarily from two factors: the molecular structure and molecular weight of PTFE.

PREPARATION

Polymerization

The tetrafluoroethylene (TFE) monomer (CAS Registry No. 116-14-3) poses significant handling problems in production, storage, and polymerization. It releases about 44 kjoule/mole upon deflagration (comparable to black powder), liberating CF_4, and carbon. Typically, the monomer is produced and polymerized at the same site and inventories are kept low. Inhibitors such as terpenes or HCl are used to prevent spontaneous polymerization during compression, purification, and storage.[2]

TFE polymerizations are carried out in water anywhere between 10 and 110°C, under pressures from 5 to 30 kg/cm^2. Inorganic initiators such as persulfates are commonly used.

The three major forms of PTFE are known as dispersion, fine powder, and granular. Highly-ordered crystal structures develop because few imperfections can occur and molecular weights can be very high.

An additional product form of PTFE, called micropowder, has been developed for additive uses.[3]

PTFE can incorporate low levels of comonomers and still retain its "homopolymer" status. These monomers are incorporated at levels of up to about 0.5 mole percent to reduce crystallinity, creep, etc. for specific product needs. (When crystallinity is reduced to about 50 percent by higher levels of these same monomers, melt-processable fluoropolymers are obtained.) The common modifiers include chlorotrifluoroethylene, perfluoropropylvinyl ether, perfluorobutyl ethylene, and hexafluoropropylene.

Fabrication Processes

Virgin high molecular weight PTFE melts at 342°C; but its melt-flow rate is extremely low even at 380°C; thus, it cannot be processed by normal thermoplastic techniques. A number of processes have had to be developed or invented to convert the virgin high molecular weight PTFE to useful forms.

PROPERTIES

Physical

The melting point of virgin PTFE is 342–343°C and the remelting temperature is 327°C. Some low molecular weight micropowders melt below 300°C. Extended-chain forms, produced by quenching samples which have been highly elongated slightly below the melting point, can exhibit melting points as high as 380°C.[4]

The specific gravity of the virgin PTFE varies from less than 2.1 to almost 2.3 g/cm^3 depending on several factors. As such, density is an extremely useful parameter for analyzing PTFE and controlling the processing conditions. As polymerized PTFE has the highest density; crystallinity approaches 98%.

PTFE undergoes several transitions between –90°C and the melting point. Density is a useful monitor for the effects of these transitions on physical properties.

The coefficient of friction is about 0.07 for PTFE, making it one of the lowest of any polymer.

Chemical

PTFE is subject to attack by strong nucleophiles, such as molten alkali metals, sodium hydroxide at elevated temperature, sodium/liquid ammonia solutions, and sodium naphthalenide. These can partially defluorinate PTFE surfaces to make them more bondable or printable. Corona discharge and plasma etching perform somewhat the same function.

Extremely strong oxidants, such as H_2O_2/H_2SO4 mixtures, and fluorine and strong fluorinating agents can also degrade PTFE. Mixtures of fine particles of aluminum and magnesium with PTFE powder show high reactivity. Because their reaction is so exothermic these mixtures have applications as military flares.

A limited number of fluorinated solvents provide some solvation of PTFE. True solutions of about 0.5 percent by weight have been produced in perfluorodecaline, fluoroanthracenes, and other similar structures.[5]

Electrical

The high dielectric strength, low dissipation factor, and low dielectric constant are exceptional among plastics. Equally important is the insensitivity of these parameters to moisture and to temperature.

Because PTFE will not char under dielectric breakdown, it will not arc track.

Thermal

PTFE will support combustion only in atmospheres containing more than 95% oxygen. Above 450°C, carbon-chain scission occurs at a measurable rate, with TFE monomer being the primary product.[6] Manufacturers' literature provides warnings to processors of PTFE, including recommendations for venting and handling of tobacco products to prevent contamination.

Between –110 and +327°C, PTFE undergoes 7 first- and second-order transitions. Among these are amorphous transition ranges of ~–110 to –73°C and ~120 to 140°C. Crystalline transitions occur at 19 and 30°C, and over the range of 80 to 110°C. The 19°C transition is important because it occurs within the normal ambient range and because it causes relatively large changes in some properties.

Radiation

The weathering properties of PTFE are excellent since no significant UV absorption occurs. Higher energy radiation, such as electron beams and gamma and X-rays, degrades the carbon chain through ionization, elimination of radicals, and chain scission.

Adsorption, Permeation, Solubility

Adsorption of most chemicals is low because surface energy is low and because solubility is low. Despite the low solubility, the permeation rate can be high because void content, resulting from processing conditions, is a dominant factor in permeation of small molecules such as the permanent gases.

APPLICATIONS

Electrical

The primary use of PTFE is in electrical insulation.

Bearings and Lubricants

Creep properties of molded or machined bushings can cause deformation under high stress, but this same creep contributes to the effectiveness of PTFE additives to lubricants because it allows particles to deform between surfaces to create excellent lubricity.

Coated or Expanded Fabrics

The production of coated cooking utensils is a major consumer application of PTFE. Lubricity and low surface energy contribute value to these items, as well as to industrial applications where moldings, castings, and films can be formed against PTFE sheets for ease of removal and without contaminating the formed object. PTFE can be polymerized directly onto surfaces from monomer vapor or sputtered from a heated polymer source under vacuum to form ultrathin release coatings.

Drip Suppressants and Flow Modifiers

Polycarbonates and other thermoplastics can incorporate PTFE to reduce drips during fire situations. Fibrillation under extrusion shear provides interconnecting links to increase apparent viscosity (melt strength) or modify the flow behavior of thermoplastics.

Other Applications

The recovery and recycling of PTFE is being studied widely. Thermal degradation is a well behaved process used to produce relatively pure monomer.

The chemical industry uses PTFE pipe linings, valve inserts, gaskets, and tank linings to provide protection against corrosive fluids. Smaller quantities are used in the fabrication of lab equipment.

Medical implants such as joint replacements and vascular grafts have been an historically small but important application for PTFE. Most PTFE manufacturers have withdrawn their products from this market because of concerns about unwarranted liability exposure.

Grades of PTFE that can be processed into thinner constructions are being developed. These enable the direct formation of thin tapes and films of thin-wall tubing.

REFERENCES

1. Plunkett, R. J. (to Kinetic Chemicals Inc.) U.S. Patent 2 230 654, 1941.
2. Dietrich, M. A.; Joyce, R. M. (to DuPont) U.S. Patent 2 407 405, 1946.
3. Hendriock, H.-J. *Kunststoffe-Ger. Plas.* **1986,** *76,* 920.
4. Khanna, Y. P. et al. *Macromolecules* **1990,** *23,* 2488.
5. Chu, B.; Wu, C.; Zuo, J. *J. Macromolecules* **1987,** *20,* 700.
6. Siegle, J. C.; Muus, L. T.; Lin, T. P.; Larsen, H. A. *J. Polym. Sci. A* **1964,** *2,* 391.

POLYTETRAFLUOROETHYLENE (Effect of γ-Irradiation)

Mituaki Tutiya
Tokyo Metropolitan Isotope Research Center

Polytetrafluoroethylene (PTFE) is a good example of a polymer[1] with poor radiation resistance. The mechanical properties of PTFE degrade in both tensile strength and elongation upon irradiation. Also, irradiation results in an increase in the degree of crystallinity of PTFE.[2-5] The increased crystallinity is manifested by annealing after irradiation.[5,6] In virgin PTFE at 19°C, the influence of irradiation at doses above 10^8 R on molecular motions of crystalline phases results in a clearly observed lowering of the temperature of the crystalline transition.[2,7-12] Moreover, the melting point of PTFE, T_m, becomes lower at high radiation doses.[3,7,13-14] In the case of irradiation at high temperatures above T_m in vacuum, abnormal irradiation-induced changes have been observed on both the mechanical properties and the degree of crystallinity of PTFE.[9,14,15]

PREPARATION AND PROPERTIES

The PTFE samples used in this experiment were commercial products.[8,14]

Irradiation at Total Dose Below 10^8 R

The tensile strength at the point of breakage of PTFE samples irradiated at room temperature in air decreases abruptly in the dose range from about 10^4 R to 10^5 R.[15] The elongation at the point of breakage increases in the dose range from 10^3 R to 10^5 R and then decreases abruptly.

The increase in NMR crystallinity due only to irradiation was observed for all irradiated samples. The increase in NMR crystallinity due to annealing after irradiation took place abruptly within narrow temperature ranges of 100–150°C and 200–300°C.[6]

The increase in NMR crystallinity of PTFE due only to irradiation at room temperature, and a further increase of crystallinity upon annealing below T_m after irradiation, can be explained by the basic assumptions that the crystallization is attributed to irradiation-induced scission of the main chain and that the rate of crystallization of the amorphous fraction (which contains the intermediate component) is retarded due to its depletion, known as the overlapping effect.[16]

Irradiation at Total Dose Above 10^8 R

When PTFE was irradiated with Co^{60} γ-rays at high doses above 10^8 R in air, the values of mechanical properties in the irradiated PTFE approach zero, and a saturation of increase with a subsequent decrease in NMR crystallinity occurs.

On the other hand, when PTFE is irradiated in vacuum, its mechanical properties do not degrade in the irradiation temperature range from 320°C to 380°C.[9,14,15] An increase in Young's modulus is observed at total doses of 1.1×10^7 R and 1.1×10^8 R. At a total dose of 1.1×10^8 R, the mechanical properties have a tendency to increase in strength and to decrease in elongation.

As radiation dose increases, T_m of the PTFE samples irradiated at 340°C in vacuum is observed to be lowered more markedly than that at room temperature in air.[13,14]

REFERENCES

1. Florin, R. E. *Radiation Chemistry of Fluorocarbon,* in *High Polymers, 25,* Wall, L. A. Ed.; John Wiley & Sons: New York, 1972.
2. Licht, W. R.; Kline, D. E. *J. Polym. Sci., A* **1964,** *2,* 4673.
3. Licht, W. R.; Kline, D. E. *J. Polym. Sci., A2* **1966,** *4,* 313.
4. Peffely, W. M.; Honnold, V. R.; Binder, D. *J. Polym. Sci., A1* **1966,** *4,* 977.
5. Tutiya, M.; Yamamoto, K. *Oyo Buturi* **1965,** *34,* 424.
6. Tutiya, M.; Yamamoto, K. *Jpn. J. Appl. Phys.* **1968,** *7,* 440.
7. Kusy, R. P. *J. Polym. Sci., A1* **1972,** *10,* 1745.
8. Tutiya, M. *Jpn. J. Appl. Phys.* **1970,** *9,* 1204.
9. Tutiya, M. *Jpn. J. Appl. Phys.* **1972,** *11,* 1542.
10. Tutiya, M. *Polym. J.* **1974,** *6,* 39.
11. Rabolt, J. F. *J. Polym. Sci., Phys. Ed.* **1983,** *21,* 1797.
12. Bunn, C. W.; Howell, E. R. *Nature* **1954,** *174,* 549.

13. Tutiya, M.; Sekiguchi, T.; Nonomura, T. *Bulletin of Tokyo Metropolitan Isotope Research Center* **1987,** *4,* 43.
14. Tutiya, M. *Rept. Prog. Polym. Phys. Jpn.* **1993,** *36,* 487.
15. Tutiya, M. *Tokyo Metropolitan Isotope Research Center Annual Report* **1970,** 18.
16. Tutiya, M. *Rept. Prog. Polym. Phys. Jpn.* **1974,** *17,* 515.

Polytetrahydrofuran

See: Telechelic Oligomers (with Cyclic Onium Salt Groups)

Polythiacrownethers

See: Supported Polythioethers and Polythiacrownethers

POLY(2,5-THIENYLENE VINYLENE) AND DERIVATIVES

H. J. Geise,* L. Chen, and J. Briers
Department of Chemistry
University of Antwerp (UIA)

Over the past two decades, the chemistry and physics of organic metals have developed into a rich field of interdisciplinary research with ever-widening perspectives. This development was spurred by the observation that certain polymers, after reaction with oxidizing or reducing agents, are able to conduct electrical current as effectively as copper. Researchers have shown much interest in poly(2,5-thienylene vinylene) (PTV) in recent years. PTV, which has a conjugated system and contains an easily oxidizable heterocycle, can be expected to exhibit a high stability when doped. Because of its lower ionization potential, PTV can be doped using a less strong doping agent, for example, I_2. These advantages make PTV and its derivatives suitable subjects not only for chemists attempting to synthesize polymers wtih good mechanical and electrical properties and well-defined structures, but also for physicists working to clarify the relationship between charge carriers and mobility, and for engineers searching for applications for new materials in the electronics and semiconductor industries.

PREPARATION

In **1960,** McDonald and Campbell showed that the Wittig reaction was suitable for preparing poly(arylene vinylene)s (PAV).[2]

An alternative strategy for improving processability and increasing molecular weight is to adopt a two-step, indirect synthesis in which a soluble, polymeric precursor is synthesized. In **1986,** the precursor route was successfully adopted for the synthesis of PTV.[4] Using this route, researchers can synthesize PTV via pyrolysis of a water-soluble sulfonium polyelectrolyte, derived from 2,5-bis(tetrahydrothiophenium methyl) thiophene chloride by reaction with one equivalent of alkali in an aqueous solution at 0°C under inert conditions.[5] A modification to the reaction was made by polymerizing in a water–methanol mixture. This substitution of the methoxy groups for the sulphonium groups results in a neutral methoxy precursor polymer. The methoxy precursor polymer is stable at room temperature and soluble in organic solvents. A thermal treatment under a vacuum or a nitrogen atmosphere results in the conjugated PTV by elimination of the side groups.

This method of synthesis has several advantags. First, the precursor polymer can be cast into a film, or other desired forms such as fibers, by evaporating the solvent.

Alternative synthetic routes with great promise of versatility have recently been developed. One, proposed in 1991 by Galarini et al. and Bolognesi et al., produced a series of new alkyl-substituted PTVs in a one-step reaction by coupling 2,5-diodothiophene and *trans*-1,2-bis(tri-*n*-butyl-stannyl) ethylene with the help of Pd complexes as catalysts.[6,7] In another proposal, butoxy-substituted PTVs were synthesized using a nickel-catalyzed Grignard coupling of dibutoxythiophene and dichloroethylene.[8]

PROPERTIES

Molecular Orientation and Mechanical Properties

An important feature of the precursor route is its ability to prepare oriented polymer films. Highly oriented PTV was made by uniaxial stretching of the precursor films or fibers during the elimination step at a temperature between 80 and 140°C.

Electrical Properties

The most interesting property of conjugated polymers such as PTV is their electrical conductivity.

PTV is able to conduct electrical current only after reaction with a chemical doping agent. The chemical doping occurs when the polymer reacts with oxidizing agents (p-doping) such as $FeCl_3$, H_2SO_4, AsF_5, and I_2 or with reducing agents (n-doping) such as $NaC_{10}H_8$. However, n-doping of PTV (as well as most other conjugated compounds) does not produce conductive materials with good stability. Therefore, most efforts are directed toward investigations of the more stable p-doped polymers.

At room temperature, pristine PTV shows a dark conductivity of $\sigma = 2.1 \times 10^{-9}$ S/cm. This value is influenced by oxygen absorbed in the film, acting as a weak oxidizing agent.[10] The advantages of PTV include its relatively low bandgap and its low oxidizing potential compared to other conjugated polymers. These make it possible to dope PTV with a relatively weak oxidizing agent such as I_2.

For PTV powder, sythesized via the Wittig route, doped with I_2, and pressed into pellets, a conductivity value of $\sigma = 5 \times 10^{-8}$ S/cm is reported.[3] PTV films synthesized via the precursor route reach conductivities of up to 230 S/cm after doping with I_2. Uniaxial stretching increases the conductivity to 2673 S/cm, measured in the stretching direction.

In general, when the doping level increases, the optical absorption moves through the visible region. Heavily doped PTV films are black with a metallic luster. In the UV-vis spectra, the λ_{max} shifts to higher wavelengths.

APPLICATIONS

Plastic, all-solid-state batteries are under development in several universities and industrial laboratories. Some, based on

*Author to whom correspondence should be addressed.

polypyrrole and polyaniline, have already been commercialized. In this context, one can expect that the high stability of doped PTV will be useful. Because ethoxy-substituted PTV is transparent in the oxidized state, it is of interest to those making electrochromic devices and antistatic layers that are not black.[9] The photoconductivity that was measured after radiating the material with near-IR light opens the possibility of using the material to make opto-electronic devices.[10] Finally, the fabrication of a thin PTV film transistor was reported in 1993.[11] This transistor has a carrier mobility comparable to that of amorphous silicon thin-film transistors. In conclusion, its interesting properties, combined with the many possible applications described in literature, make PTV a promising conducting polymer.

REFERENCES

1. Naarmann, H.; Strohriegel, P. *Handbook of Polymer Synthesis*; Marcel Dekker: New York, 1992; Part B, Chapter 21.
2. McDonald, R. N.; Campbell, T. W. *J. Am. Chem. Soc.* **1960,** *82,* 4669.
3. Kossmehl, G.; Härtel, M.; Manecke, G. *Die Makromol. Chem.* **1970,** *131,* 15.
4. Harper, K.; Watson, W. J. W. European Patent 0 182 548 A3, 1986.
5. Jen, K. Y.; Maxfield, M.; Shacklette, L. W.; Elsenbaumer, R. L. *J. Chem. Soc., Chem. Commun.* **1987,** 309.
6. Galarini, R.; Musco, A.; Pontellini, R.; Bolognesi, A.; Destri, S.; Catellani, M.; Mascherpa, M.; Zhuo, G. *J. Chem. Soc., Chem. Commun.* **1991,** 364.
7. Bolognesi, A.; Catellani, M.; Musco, A.; Pontellini, R. *Synth. Met.* **1993,** 55-57, 1255.
8. Van Dort, P. C.; Pickett, J. E.; Blohm, M. L. *Synth. Met.* **1991,** 41-43, 2305.
9. Kanatzidis, M. G. *Chem. Eng. News* **1990,** 36.
10. Murata, H.; Tokito, S.; Tsutsui, T.; Saito, S. *New Polymeric Mater.* **1990,** 2(1), 11.
11. Fuchigami, H.; Tsumura, A.; Koezuka, H. *Appl. Phys. Lett.* **1993,** 63(10), 1372.

Poly(thienylene vinylene)s

See: Poly(2,5-thienylene vinylene) and Derivatives

Polythienylene

See: Polythiophenes

POLY(THIOETHER KETONE AND SULFONE)S

M. Srinivasan
Dr. Bharat Ram Research and Development Center
SRF Ltd.

Poly(thioether ketone)s are, in a sense, derived from polythioethers through the appropriate choice of monomers. The thermal stability of the poly(thioether)s is probably because their structure involves C–C, C–H and C–S bonds, all of which are thermodynamically stable.

Because these thioethers are largely melt-processable (most of them have poor solubility), their versatility is limited. However, if the polymer backbone flexibility can be enhanced, the nonlinear backbone geometry will allow the polymer to gain higher configurational entropy, thermal stability, and better solubility.[3] One of the ways in which this has been achieved is through the introduction of carbonyl function in the backbone to get poly(thioether ketone)s.

Phosphorus pentoxide-methanesulfonic acid (PPMA) (1:10) has also been used as a powerful dehydrating/cyclizing or condensing agent in the synthesis of poly(thioether ketone and sulfone)s.[7]

Using PPMA in the direct polycondensation, the polymers reported have used thioether sulfones to generate the thioether ketonesulfones (**Scheme I**).[2] In this report, two model compounds were documented using 1,4-bis(p-phenylthio)-benzene (6) (BPTB) and/or bis(p-phenyl (thio) diphenylsulfone) (7) (BPTD) and p-toluic acid (8a). The documented yields of polymers were 89–92%. Using aluminium chloride as Lewis acid in 1,2-dichloroethane as solvent, polycondensation of BPTB and BPTD with various aromatic diacid chlorides has been reported on the synthesis of poly(thioether ketone and sulfone)s.[1] The thermal stability as determined by weight loss in thermogravimetric analysis indicates that the sulfonylthioethers are more stable than the thioether ketones, although both categories had good thermal stability. As per the literature, differential scanning calorimetry (DSC) does not give an endothermic peak, indicating that the T_g is probably well above the decomposition temperature, and X-ray confirms that these polymers were totally amorphous.

Recently, polyamides containing a combination of ether, thioether, sulfone, or carbonyl were also reported obtained by using phosphonylation and solution methods.[4,5]

The poly(thioether amide sulfone)s had higher viscosities than regular polyamides. This may be due to the greater nucleophilicity of the sulfonediamines used. Compared to poly(thioether amide)s, poly(thioether amide sulfone)s showed better thermal stability. Also, DSC, as in the earlier cases, did not indicate glass transition, an observation supported by X-ray that indicated the systems were totally amorphous.

The acid 4,4′-[carbonylbis(*p*-phenylthio)dibenzoic acid, a ketodiacid derived from bis(*p*-phenylthio)-benzophenone, *N,N′*-dimethylcarbonyl chloride, and other sulfone diacids, were also used to prepare polyamides by the phosphorylation method, using triphenyl phosphite.[7] These polymers, which had ketone function, were significantly crystalline, while the sulfones were all amorphous. The thioethersulfones or the sulfide-sulfones (as they are more commonly known) are all much more soluble. This may be due to the flexibilizing effect of the sulfide function. The polymers that had both sulfide and sulfones were the most soluble. Equally important, the thioether ketone amides were thermally more stable than thioethers and thioether sulfones. This observation can be attributed to the oxidative stability, coupled with rigid polymer backbone and hence better packing density. Also, the T_g increased with increasing rigidity of the polymer chain.

The extension of synthetic use of the thioether ketones and sulfone monomers to poly(amide hydrazide)s has been documented through the phosphorylation method.[6,8,9] While viscosities are reasonably good, their thermal stabilities are lower.

Some polyamideimides have been reported from bis(4-trimellitimidophenyl) sulfone.[10] The reported viscosities are moderate and the thermal stability fair.

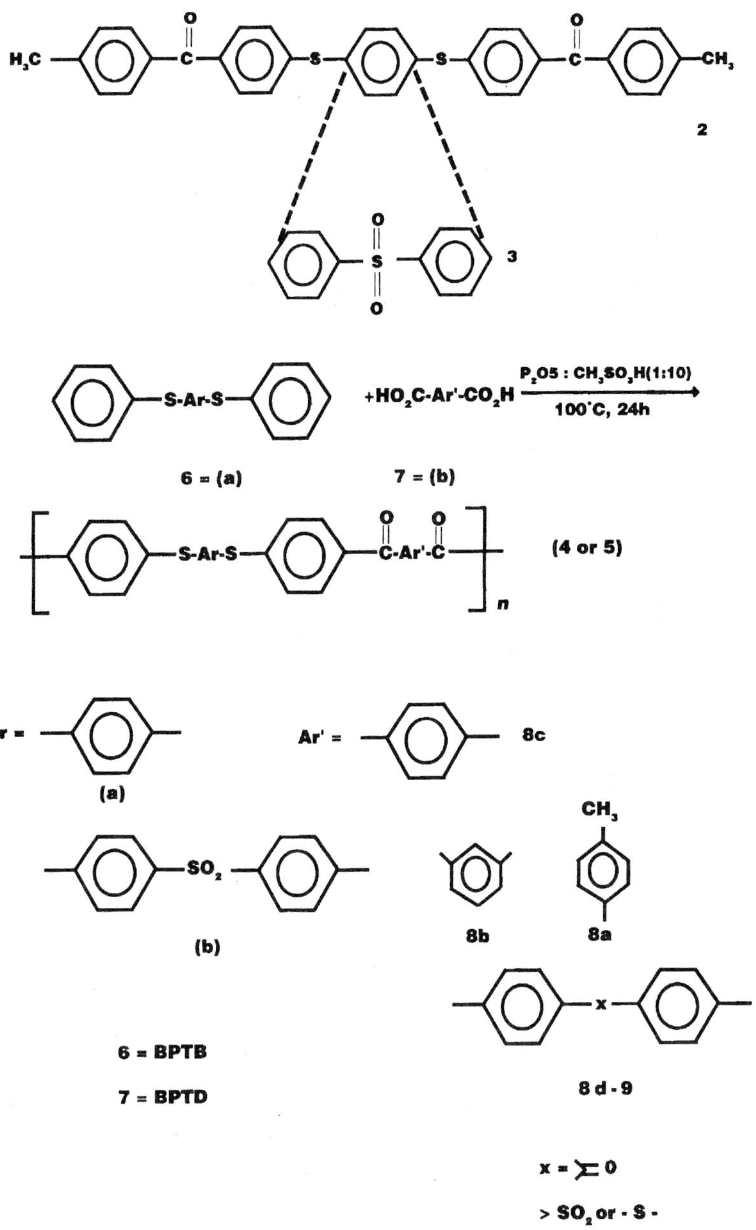

I

To increase processability, solubility, and thermal stability, poly(ester imide)s that incorporate ether and sulfones in the preformed imidedicarboxylic acids have also been documented.[11]

REFERENCES

1. Joseph, K. A.; Srinivasan, M. *Polym. Int.* **1992,** *29,* 121.
2. Joseph, K. A.; Srinivasan, M. *J. Polym. Sci., Polym. Chem. Ed.* **1993,** *31,* 3485.
3. Joseph, K. A. Ph.D. Dissertation, Indian Institute of Technology, 1992.
4. Scariah, K. J.; Krishnamoorthy, V. N.; Rao, K. V. C.; Srinivasan, M. *Makromol. Chem.* **1985,** *186,* 2427.
5. Joseph, K. A.; Srinivasan, M. *Polym. Int.* **1993,** *30,* 257.
6. Joseph, K. A.; Srinivasan, M. *Polym. Int.* **1993,** *30,* 327.
7. Eaton, P. E.; Carlson, G. R. *J. Org. Chem.* **1973,** *38,* 4071.
8. Yamazaki, N.; Higashi, F.; Kawabata, J. *J. Polym. Sci., Polym. Chem. Ed.* **1975,** 13,1343.
9. Yamazaki, N.; Matsumoto, M.; Higashi, F. *J. Polym. Sci., Polym. Chem. Ed.* **1975,** *13,* 1343.
10. Venkatesan, D.; Srinivasan, M. *Polym. Int.* **1993,** *29,* 275.
11. Venkatesan, D.; Srinivasan, M. *J. Macromol. Sci., Pure and Appl. Chem.* **1993,** *30,* 801.

Poly(thioether ketone)s

See: Poly(aryl thioether ketone)s

Poly(thioether sulfone)s

See: Poly(aryl thioether sulfone)s

Polythioethers

See: *Supported Polythioethers and Polythiacrownethers*

Polythioglycidols

See: *Polyglycidols, Polythioglycidols*

Polythionylphosphazene

See: *Phosphorus-Containing Polymers (Overview)*

Polythiooxamides

See: *Coordination Polymers, Dithiooxamides*

Polythiophenes

See: *Conducting Polymers (For Langmuir-Blodgett Film Fabrication)*
Conducting Polymers (Self-Acid-Doped, Conjugated)
Conductive Elastomeric Blends
Electrically Conducting Composites
Poly(bithiophenes) (Advanced Derivatives)
Polythiophenes (Conducting Polymers)
Polythiophenes (Organometallic Synthesis)
Polythiophenes, Substituted (Regioselectivity, Electrical and Optical Properties)
Signal Transduction Composites (Biomaterials with Electroactive Polymers)
Thiophene Copolymers (Electrical and Third-Order NLO Properties

POLYTHIOPHENES (Conducting Polymers)

Kurt E. Gackeler* and N. Arsalani
Institute of Organic Chemistry
Faculty of Chemistry and Pharmacy
University of Tuebingen

Among the electrically conducting polymers, such as polyacetylene (PA), poly(p-phenylene) (PPP), polyaniline (PAN), polypyrrole (PPy), and their substituted derivatives, polythiophene (PTh) has rapidly become the subject of considerable interest.[1,2]

In this article the different methods of synthesis of electrically conducting polythiophenes and their properties are highlighted.

PREPARATION

Chemical Methods

Polythiophene (PTh)

The chemical synthesis of PTh has been performed mostly by coupling reactions of Grignard compounds.

*Author to whom correspondence should be addressed.

Substituted Polythiophenes (s-PThs)

Most s-PThs have been prepared by electrochemical polymerization. Only a few, especially poly(alkylthiophene)s (PAThs), have been prepared by chemical routes.

Copolymers and Composites of PThs

In order to prepare from electrically conducting polymers (ECPs) useful materials with good mechanical properties, environmental stability, and simple processability as well as conductivity, the preparation of composites, blends, and block or graft copolymers from ECPs and insulating polymers has been envisaged.

s-PThs with suitable processability represent a large group of ECPs that have been used for the preparation of a wide variety of polymer blends or composites with conventional polymers such as polyethylene (PE), polystyrene (PS), poly(vinyl chloride) (PVC), poly(ethylene-*co*-vinyl acetate) (EVA), poly(methyl methacrylate) (PMMA), polyacrylonitrile, and poly(phenylene oxide).[4-10]

Thiophene has been copolymerized with various aromatic compounds such as benzene, biphenyl, substituted benzene, or pyridine compounds by means of different chemical syntheses, including Grignard coupling, reductive coupling, Stetter reaction followed by ring closure, and oxidative dehydrogenation.[3,11-13]

Graft and block copolymers of thiophene with polystyrene have been prepared; they show good solubility and high conductivity.[14,15]

Electrochemical Methods

Chemical syntheses are the most adequate methods for the preparation of electrically conducting PThs and other ECPs with defined structures, but so far, most work in the preparation of ECPs and especially PThs has used electrochemical polymerization. Here, the polymer is directly obtained in its oxidized or reduced state, and its thickness as well as its doping level can be well controlled by the electrolysis time and potential, respectively.

Copolymers and Composites of PThs

Direct electropolymerization of thiophene from an electrolytic medium containing the dissolved host polymer–for example, PMMA, PVC, or polystyrene–is one method for preparing polythiophene composites.[19]

PROPERTIES

The conductivity, electroactivity, and energy gap (Eg) of ECPs depend on the extent of the conjugated π-systems. In the case of exclusive 2,5-linkage of the thiophenes and no defects in the conjugated backbone of the polymer, the effective mean conjugation length is increased; as a result, the conductivity and electroactivity are also augmented but the electron gap is decreased.

By chemical or electrochemical doping processes–the removal of electrons from the π-valence band (p-doping) or the addition of electrons to the π^*-conduction band (n-doping)–the Eg of PThs is decreased and therefore polymers with higher conductivity are obtained.

The introduction of an alkyl chain to the ring, especially at the 3-position, has the effect that the solubility of the polymer is increased. It has been found that PAThs with alkyl chains of hexyl or larger are readily processed from solution, and that those with larger alkyl chains can be processed from the melt.[20]

Solutions and films of PAThs show very interesting and potentially useful thermochromic and solvatochromic effects when they contain alkyl groups larger than methyl.

APPLICATIONS

The synthesis of PThs with new properties and the possibility of simple processability lead to wide applications in science and technology. PThs show different electrical and electronic properties at doped and undoped states with different types of applications. Existence of the electrochemical reversibility of the transition between the doped and the undoped states and the resulting change of color have also led to other fields of applications.

PTh and s-PThs have been used in rechargeable batteries and have shown high voltage, high energy, and power densities, although self-discharge exists also in these batteries.[21-23] The color changes of PThs and the variety of reduced, neutral, and oxidated forms have opened the potential of utilizing PThs for the production of electrochromic devices. Studies have confirmed that PThs offer good optical contrast, fast response times, and sufficient cycle life.[16,17,24]

The application of PThs as sensors is one of the most promising fields in chemistry and biology. The electrodes from PThs can be used as selective modified electrodes for electrocatalytic or electroanalytical purposes.[18,25] Biosensors that are based on PThs and that include the use of enzymes with specific responses have also been prepared.[26,27] PThs have also been utilized as active layers in semiconducting devices such as field-effect devices (diodes and transistors) and light-emitting diodes.[28-32] Other uses of PThs include antistatic and electromagnetic shield materials.[33,34]

REFERENCES

1. Billingham, N. C.; Calvert, P. D. *Adv. Polym. Sci.* **1989**, *90*, 1.
2. Arsalani, N.; Geckeler, K. E. *J. Prakt. Chem.* **1995**, *337*, 1.
3. Yamamoto, T.; Miyazaki, Y.; Fukuda, T.; Zhou, Z.-H.; Maruyama, T.; Kanbara, T.; Osakada, K. *Synth. Met.* **1993**, *55*, 1214.
4. Molton, J.; Smith, P. *J. Polym. Sci., Polym. Phys. Ed.* **1992**, *30(8)*, 871.
5. Hotta, S.; Rughooputh, S. D. D. V.; Heeger, A. J. *Synth. Met.* **1987**, *22*, 79.
6. Wang, H. J.; Toppare, L.; Fernandez, J. E. *Macromolecules* **1990**, *23(4)*, 1053.
7. Österholm, J.-E.; Laakso, J.; Nyholm, P.; Isotalo, H.; Stubb, H.; Inganäs, O.; Salaneck, W. R. *Synth. Met.* **1989**, *28*, C435.
8. Laakso, J.; Österholm, J. E.; Nyholm, P. *Synth. Met.* **1989**, *28*, C467.
9. Park, Y. H.; Kim, Y. K.; Lee, D. S. *J. Appl. Polym. Sci.* **1990**, *40*, 1487.
10. Schantz, S.; Ljungqvist, N. *Synth. Met.* **1993**, 55-57, 3483.
11. Montheard, J. P.; Pascal, T.; Seytre, G.; Boiteux, G. *Makromol. Chem., Rapid Commun.* **1985**, *6*, 679.
12. Pouwer, K. L.; Vries, T. R.; Havinga, E. E.; Meijer, E. W.; Wynberg, H. *J. Chem. Soc., Chem. Commun.* **1988**, 1432.
13. Ruiz, J. P.; Child, A. D.; Nayak, K.; Marynick, D. S.; Reynolds, J. R. *Synth. Met.* **1991**, 41-43, 783.
14. Francois, B.; Olinga, T. *Synth. Met.* **1993**, 55-57, 3489.
15. Olinga, T.; Francois, B. *J. Chim. Phys.* **1992**, *89*, 1072.
16. Roncali, J.; Garnier, F.; Garreau, R.; Lemaire, M. *J. Chem. Soc., Chem. Commun.* **1987**, 1500.
17. Roncali, J.; Garreau, R.; Yassar, A.; Marque, P.; Garnier, F.; Lemaire, M. *J. Phys. Chem.* **1987**, *91*, 6706.
18. Roncali, J.; Garreau, R.; Delabouglise, D.; Garnier, F.; Lemaire, M. *J. Chem. Soc., Chem. Commun.* **1989**, 679.
19. Roncali, J.; Garnier, F. *J. Phys. Chem.* **1988**, *92*, 833.
20. Gustafsson, G.; Inganäs, O.; Salaneck, W. R.; Laakso, J.; Loponen, M.; Taka, T.; Österholm, J.-E.; Stubb, H.; Hjertberg, T. In *Conjugated Polymers*; Bredas, J. L.; Silbey, R., Eds.; Kluwer Academic: The Netherlands, 1991; p 315.
21. Arabizzani, C.; Mastragostino, M.; Panero, S.; Prosperi, P.; Scrosati, B. *Synth. Met.* **1989**, *28*, C663.
22. Kawai, T.; Kuwabara, T.; Wang, S.; Yoshino, K. *J. Electrochem. Soc.* **1990**, *137*, 3793.
23. Li, F.; Albery, W. J. *J. Electroanal. Chem., Interfacial Electrochem.* **1991**, *302*, 279
24. Yashima, H.; Kobayashi, M.; Lee, K. B.; Chung, D.; Heeger, A. J.; Wudl, F. *J. Electrochem. Soc.* **1987**, *134*, 46.
25. Lemaire, M.; Delabouglise, D.; Garreau, R.; Guy, A.; Roncali, J. *J. Chem. Soc., Chem. Commun.* **1988**, 658.
26. Geckeler, K. E.; Müller, B. *Naturwissenschaften* **1993**, *80*, 18.
27. Genies, E. M.; Marchesiello, M. *Synth. Met.* **1993**, 55-57, 3677.
28. Tomozawa, H.; Braun, D.; Phillips, S.; Heeger, A. J.; Kroemer, H. *Synth. Met.* **1987**, *22*, 63.
29. Tsumura, A.; Koezuka, H.; Ando, T. *Synth. Met.* **1988**, *25*, 11.
30. Assadi, A.; Svensson, C.; Willander, M.; Inganäs, O. *Synth. Met.* **1989**, *28*, C863.
31. Uchida, M.; Ohmori, Y.; Morishima, C.; Yoshino, K. *Synth. Met.* **1993**, 55-57, 4180.
32. Ohmori, Y.; Morishima, C.; Uchida, M.; Yoshino, K. *Synth. Met.* **1993**, 55-57, 4180.
33. Kochem, K.-H.; ter Meer, H.-U.; Millauer, H. *Kunststoffe* **1992**, *82*, 575.
34. Colameri, N. F.; Schacklette, L. W. *IEEE Trans. Instrum. Meas.* **1992**, *41*, 9.

POLYTHIOPHENES (Organometallic Synthesis)

M. David Curtis and Mark D. McClain*
Macromolecular Science and Engineering Center
The University of Michigan

Conjugated polymers have been intensively studied since Shirakawa's discovery that the resistivity of *trans*-polyacetylene is drastically decreased upon partial chemical oxidation.[1] Since this effect was thought to originate from electron mobility in the partially filled π-bonding band, other polyconjugated systems have also been investigated, including polyphenylene, polypyrrole, and polythiophene (PT). All these polymers have a rigid rod structure and are insoluble and infusible, making it essentially impossible to process these polymers into flexible films or shaped objects. Alkylated PT derivatives, i.e., polyalkylthiophenes or PATs, show good environmental stability and are soluble in common organic solvents when the alkyl group has at least 4 carbon atoms. Hence, these compounds have been extensively studied.

The most widely studied system is the 3-alkylthiophenes, which are soluble in common organic solvents (THF, $CHCl_3$, CH_2Cl_2, toluene) when R ≥ Bu. Monosubstitution at the 3-position of the ring introduces asymmetry to the monomer, allowing regioisomerism in the polymer.

*Author to whom correspondence should be addressed.

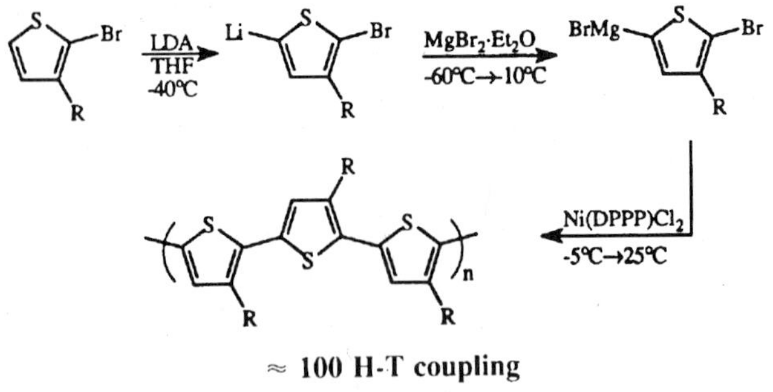

FIGURE 1. McCullough synthesis of regioregular PAT.

In general, syntheses of polyalkylthiophenes (PATs) can be divided into two methods: oxidative coupling (either electrolytic or chemical) and organometallic coupling.

PREPARATION

Most organometallic syntheses have their origins in the 2,5-dihalothiophenes from which thienyllithium, magnesium, zinc, and nickel intermediates can be generated. Using low temperatures and selective metallation, McCullough et al. recently prepared regioregular PAT (**Figure 1**)[2]

Another completely regioregular synthesis was discovered independently by Chen and Rieke by coupling zinciobromothiophenes formed from highly activated zinc (Zn*) and 2,5-dibromo-3-alkylthiophenes, with $Ni(DPPE)Cl_2$.[3,4] The regioselectivity of this coupling reaction was critically dependent on catalyst. $Pd(PPh_3)_4$ gave completely random coupling whereas $Ni(DPPE)Cl_2$ gave essentially 100% regioselectivity for HT coupling (**Figure 2**).

Stille-type cross-coupling of $-SnR_3$, $-B(OR)_2$, -OTf (triflate) or sulfinates with halides is also applicable to the synthesis of polythiophenes. Recently, a coupling scheme based on organomercurials was reported.[5]

PROPERTIES

Polythiophenes are very poor conductors of electricity unless they are partially oxidized (p-doped) or reduced (n-doped). The n-doped material is very air sensitive, whereas the p-doped material is stable in the absence of atmospheric moisture.

Unsubstituted polythiophene, regardless of the method used to synthesize it, is a red, insoluble, infusible solid. The maximum conductivity of pressed pellets of p-doped polythiophene

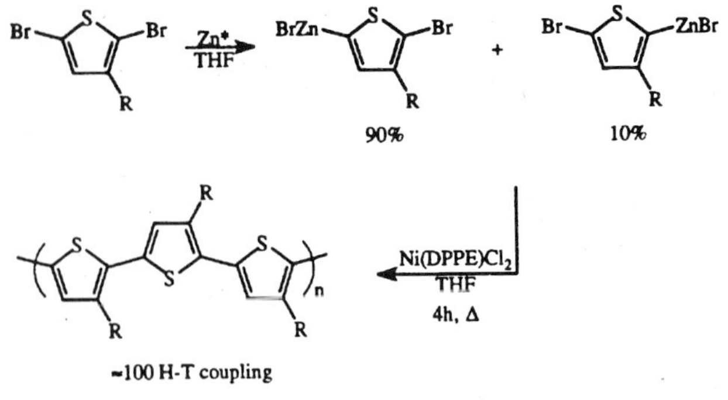

FIGURE 2. Rieke synthesis of regioregular PAT.

is ca. 10–100 S/cm.[6] Poly(3-alkylthiophene)s are soluble if the alkyl group is butyl or larger, and flexible, free-standing films may be cast from solution by a variety of methods, e.g., spin coating, Langmuir–Blodgett film technique, or simple evaporation of the polymer solution. P-doped films of these regiorandom poly(3-alkylthiophene)s typically have conductivities in the range of 10^{-2} S/cm. McCullough showed that thin films (10–25 μ thickness) of regioregular, HT-coupled PATs could have conductivities as high as 5×10^3 S/cm when doped with I_2.[7]

PATs are normally orange to red in color, but their electronic properties are influenced by the M_w, regioregularity, planarity, and degree of interchain interactions in the solid state.

Increasing the degree of HT links increases the planarity and, hence, the conjugation length, resulting in a shift of λ_{max} to longer wavelengths in the solution state where interchain interactions are absent. On going to the solid state, the regioregular polymer shows a shift of 72 nm as compared to no shift for the regiorandom polymer. These results are interpreted to mean that the side chains in the regioregular polymer are capable of packing together in a regular manner (side chain crystallization) and driving π–π stacking of the chain backbones in the solid state. This stacking causes the conjugated chain backbone to assume a more planar conformation, thus increasing the effective conjugation length.

APPLICATIONS

In the laboratory, conducting polymers, including polythiophenes, have been used to construct field effect transistors, Schottky diodes, electrochromic devices, LEDs, high efficiency solar cells, push–pull amplifiers, batteries, and NLO devices.[8-16] To our knowledge, none of these applications with polythiophenes or polythiophene copolymers as the electroactive phase have been brought to the market stage of development. However, as structure–property relationships continue to be developed, as new polymers bearing functional groups designed to enhance a particular desired property are synthesized, and as our theoretical understanding of the photo- and electrophysical processes increases, it is surely just a matter of time until devices based on highly conjugated polymers find their way into commercial applications.

REFERENCES

1. Shirakawa, H.; Louis, E. J.; MacDiarmid, A. G.; Chiang, C. K.; Heeger, A. J. *J. Chem. Soc., Chem. Commun.* **1977,** 578.
2. McCullough, R. D.; Lowe, R. D. *J. Chem. Soc., Chem. Commun.* **1992,** 70-72.
3. Chen, T. A.; Rieke, R. D. *J. Am. Chem. Soc.* **1992,** *114,* 10087-10088.
4. Chen, T. A.; Rieke, R. D. *Synth. Met.* **1993,** *60,* 175-177.
5. McClain, M. D.; Curtis, M. D. *Polym. Prepr. (Am. Chem. Soc., Div. Polym. Chem.)* **1994,** *35,* 307-308.
6. Bredas, J. L.; Street, G. B. *Acc. Chem. Res.* **1985,** *18,* 309-315.
7. McCullough, R. D.; Williams, S. P. *J. Am. Chem. Soc.* **1993,** *115,* 11608.
8. Tsumura, A.; Koezuka, H.; Ando, T. *Synth. Metals* **1988,** *25,* 11-23.
9. Garnier, F.; Hajlaoui, R.; Yassar, A.; Srivastava, P. *Science* **1994,** *265,* 1684-1686.
10. Tomozawa, H.; Braun, D.; Phillips, S.; Heeger, A. J.; Kroemer, H. *Synth. Metals* **1987,** *22,* 63-69.
11. Gustafsson, J. C.; Inganas, O.; Andersson, A. M. *Synth. Metals* **1994,** *62,* 17-21.

12. Burn, P. L.; Holmes, A. B.; Kraft, A.; Bradley, D. D. C.; Brown, A. R.; Friend, R. H.; Gymer, R. W. *Nature* **1992**, *356*, 47.
13. Sailor, M. J.; Ginsburg, E. J.; Gorman, C. B.; Kumar, A.; Grubbs, R. H.; Lewis, N. S. *Science* **1990**, *249*, 1146.
14. McCoy, C. H.; Wrighton, M. S. *Chem. Mater.* **1993**, *5*, 914-916.
15. Matsunaga, T. *Frontiers of Polymer Research*; Prasad, P. N.; Nigam, J. K., Eds.; Plenum: New York, 1991; 245-257.
16. Heeger, A. J. *Science and Applications of Conducting Polymers;* Salaneck, W. R.; Clark, D. T.; Samuelsen, E. J., Eds.; IOP: Bristol, UK, 1990; p 1.

POLYTHIOPHENES, SUBSTITUTED (Regioselectivity, Electrical, and Optical Properties)

Maria C. Gallazzi and G. Zerbi
Dipartimento di Chimica Industriale e Ingegneria Chimica
Politecnico di Milano

Polythienylenes (or polythiophenes) are an intensively studied class of polymers.[1] They belong to the class of "conductive polymers" whose common feature is a π-conjugated electronic structure. Chain molecules with delocalized electronic structure tend to be stiff with restricted flexibility and relatively strong interchain attractive interactions thus yielding rather insoluble and intractable materials.

Soluble polymers can be obtained by substituted thiophenes bearing long flexible chains of at least four carbon atoms in β position. Contrary to what happens in other families of conducting polymers (like polyacetylene), in substituted polythiophenes solubility and excellent electrical and optical properties are not at all mutually exclusive. In fact, one of the main drawbacks of unsubstituted polythiophene prepared either by electrochemical or oxidative chemical methods ($FeCl_3$ or $Cu(ClO_4)_2$) was the occurrence of a certain amount of wrong couplings: α–β and β–β′ together with the regular α–α′ linkages (**Figure 1a**).

The presence of a substituent in the β position greatly reduces the amount of the previously reported wrong couplings giving rise to more regular polymers with improved crystallinity with respect to the parent polymer. The substituted polymers have good electrical and optical properties and are also solution and melt processable.

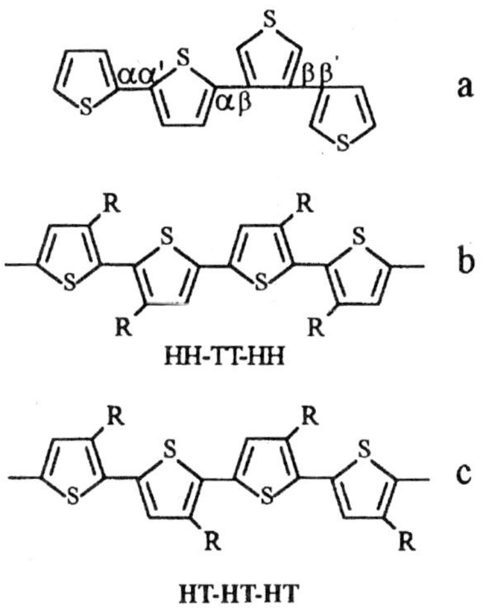

FIGURE 1. Type of linkages in polythiophene and polyalkylthiophene. (a) α and β linkages, (b) head-to-head, tail-to-tail linkages and (c) head-to-tail linkages.

The introduction of long alkyl chains not only makes the polymers soluble, but also favors such collective phenomena as thermochromism and solvatochromism.[2,3] Recently it has also been suggested that longer chains help in inducing a self-orientation of the regioregular polymers thus increasing electrical conductivity.

More recently it has been realized that the properties of the polymers are affected by the regiochemistry of the coupling, which can occur head-to-head (2-2′) or head-to-tail (2-5′) (**Figures 1b** and **1c**). The electrical and optical properties are indeed related to the "effective conjugation length," that is mainly affected by conformational distortions of the backbone induced by steric interactions of the side chains. These distortions become dominant in head-to-head coupling as shown by the fact that polymers prepared from 3,3′-dialkyl- or 4,4′-dialkyl-2-2′,-bithiophenes (in which the occurrence of head-to-head linkages is maximized (Figure 1b)) show reduced conjugation length (large blue shift in the UV-vis absorption maximum) and reduced conductivity (**Table 1**).[4-7]

TABLE 1. Relevance of the Head-to-Head Defects on Electrical and Optical Properties in Poly(alkylthiophene)s

Monomer	Polym. method	UV-vis λ_{max} (nm)		σ S/cm	Dopant	Dyads HH%[a]	Triads HH-TT%[b]	Reference
		$CHcl_3$	solid					
3,4-DHT	$FeCl_3$	315	315	—	—	—	—	16
3,3′-DHBT	$FeCl_3$	389	389	3.7	$NOPF_6$	100	—	5
3-HT	Zn/Pd	428	432	5	I_2	—	25	17
3-HT	Electroc.	440	480	1–10	ClO_4-	30	—	15
3-HT	$FeCl_3$	439	508	15	$NOPF_6$	—	10	5
3-HT	Li/Mg/Ni	442	504	150	I_2	—	1	18
3-HT	Mg/Ni	440	514	100	$FeCl_3$	<10	—	8
3-HT	Mg/Ni	450	527	—	—	<10	—	14
3-HT	Zn/Ni	456	560	1000	—	—	1.5	13

[a]Head-to-head dyads (HH) % calculated on the total α CH_2.
[b]Head-to-head, tail-to-tail triads % calculated on the total aromatic protons; DHT = dihexylthiophene; DHBT = dihexylbithiophene; HT = hexylthiophene.

Substituted thiophenes were mainly polymerized by three methods: electrochemical, chemical oxidation with $FeCl_3$ or $Cu(ClO_4)_2$, and reductive coupling of a Grignard reagent or of an organozinc intermediate catalyzed by nickel or palladium phosphine complexes.[1,9-13]

The relevance of the amount of head-to-head defects on the electrical and optical properties are apparent from the data reported in Table 1. As shown in Table 1 regioregular head-to-tail polyalkylthiophenes exhibit rather large electrical conductivity values, as large as in the case of some polyacetylenes.

REFERENCES

1. Roncali, J. *Chem. Rev.* **1992,** *92,* 711.
2. Inganäs, O.; Gustafsson, G.; Salaneck, W. R. *Synth. Met.* **1989,** *28,* C377.
3. Inganäs, O.; Gustafsson, G. *Synth. Met.* **1990,** *37,* 195.
4. Kriesche, B.; Hellberg, J.; Lilja, C. *J. Chem. Soc., Chem. Commun.* **1987,** 1476.
5. Souto Maior, R. M.; Hinkelman, K.; Eckert, H.; Wudl, F. *Macromolecules* **1990,** *23,* 1268.
6. Zagorska, M; Krische, B. *Polymer* **1990,** *31,* 1379.
7. Zagorska, M.; Kulszewicz-Bajer, I.; Pron, A.; Firlej, L.; Bernierand, P.; Galtier, M. *Synth. Met.* **1991,** *45,* 385.
8. Inganäs, O.; Salaneck, J. E.: Osterholm, J. E.; Laakso, J. *Synth. Met.* **1988,** *22,* 395.
9. Hotta, S. *Synth. Met.* **1987,** *22,* 103.
10. Sugimoto, R.; Takeda, S.; Gu, H. B.; Yoshino, K. *Chem. Express* **1986,** *1,* 635.
11. Inoue, M. B.; Velazquez, E. F.; Inoue, M. *Synth. Met.* **1988,** *24,* 223.
12. Jen, K.; Miller, G. G.; Elsenbaumer, R. L. *J. Chem. Soc., Chem. Commun.* **1986,** 1346.
13. Chin, T.; Rieke, R. D. *J. Am. Chem. Soc.* **1992,** *114,* 10087.
14. Gallazzi, M. C.; Castellani, L.; Zerbi, G.; Sozzani, P. *Synth. Met.* **1991,** 41-43, 495.
15. Leclerc, M.; Diaz, F. M.; Wegner, G. *Makromol. Chem.* **1989,** *190,* 3105.
16. Leclerc, M.; Daoust, G. *J. Chem. Soc., Chem. Commun.* **1990,** 273.
17. Chen, T.; Rieke, R. D. *Synth. Met.* **1993,** *60,* 175.
18. McCullough, R. D.; Lowe, R. D.; Jayaraman, M.; Anderson, D. L. *J. Org. Chem.* **1993,** *58,* 904.

Polythiophenylene

See: Poly(phenylene sulfide)

Polytriazoles

See: Polyoxadiazoles and Polytriazoles
Poly(1,2,4-triazole)s

POLY(1,2,4-TRIAZOLE)S

Jin Chul Jung
Department of Materials Science and Engineering
Pohang University of Science and Technology

Although no polytriazoles have been commercialized yet, aromatic polytriazoles still draw attention because the heterocyclic aromaticity of the triazole ring can provide excellent thermal stability to polymers.

A wide variety of new synthetic methods has followed to prepare various aromatic polytriazoles. Aromatic poly(1,2,4-triazole)s are characterized by high crystallinity, excellent thermal stability, and tautomerism of the mobile hydrogen atom attached to the heterocyclic ring.

SYNTHESIS

Poly(4-amino-1,2,4-triazole)s are commonly prepared by dehydration of dicarboxylic acid dihydrazides in the presence of a small excess of hydrazine. This reaction is conducted in an autoclave under high pressures and temperatures and is used to obtain an AB type poly(4-amino-1,2,4-triazole).

Polyaminotriazoles of the AB type can also be prepared from diamidrazones by elimination of ammonia.[1]

Another synthetic route to polyaminotriazoles is cyclodehydration of polyhydrazides in the presence of excess hydrazine.[2,3]

Poly(1,2,4-triazole)s without amino group substitution at the 4-position can be prepared by various methods. The first one was introduced by Abshire and Marvel.[4] They condensated 1,3-benzeneditetrazole with *N,N*-diphenyl isophthalimidoyl chloride to obtain an *N*-phenyl-substituted tetrazole ring-containing polymer, which was converted to poly(4-phenyl-1,2,4-triazole) by thermal elimination of nitrogen.

Another more general synthesis of 4-alkyl- or aryl-substituted poly(1,2,4-triazole)s can be conducted by condensation of *N,N′*-disubstituted bisimidoyl halides with dicarboxylic acid dihydrazides followed by thermal cyclodehydration of the polyacylamidrazones formed by polymerization.

When polyhydrazides are directly cyclodehydrated in the presence of a primary amine dissolved in PPA, it also gives 4-substituted poly(1,2,4-triazole)s.[5] A specific example is illustrated in **Equation 1**.

Another convenient synthetic approach to polytriazoles without 4-substitution is the condensation of diamidrazones with dicarboxylic acid chlorides.

Recently, Bae, Zin, and Jung have prepared a series of poly(1,2,4-triazole)s having linear (n-alkyloxy)methyl side chains whose chain length varied from (n-butyloxy)methyl to (n-dodecyloxy)methyl.[6] This synthetic path is illustrated in **Equation 2**.

PROPERTIES

Physical and solution properties of poly(4-amino-1,2,4-triazole)s are well described in review articles.[8,9] When the polyaminotriazoles are treated with nitrous acid, the amino group is eliminated to form deaminated poly(1,2,4-triazole)s, as shown in **Equation 3**.[7]

The aromaticity of the 1,2,4-triazole ring affords excellent thermal stability, high crystallinity, and rigidity to the 1,2,4-triazole-containing polymers, and para-linked wholly aromatic poly(1,2,4-triazole-3,5-diyl-1,4-phenylene) is completely insoluble in any organic solvents if it has high molecular weights. To obtain soluble polytriazole, it is desirable to incorporate some m-phenylene structure units into the main chain.

Poly(1,2,4-triazole-3,5-diyl-1,3-phenylene-1,2,4-triazole-3,5-diyl-1,4-phenylene) **24** is highly soluble in concentrated sulfuric or formic acid, and its molecular weight and viscosity values have been measured in 90% formic acid.[5]

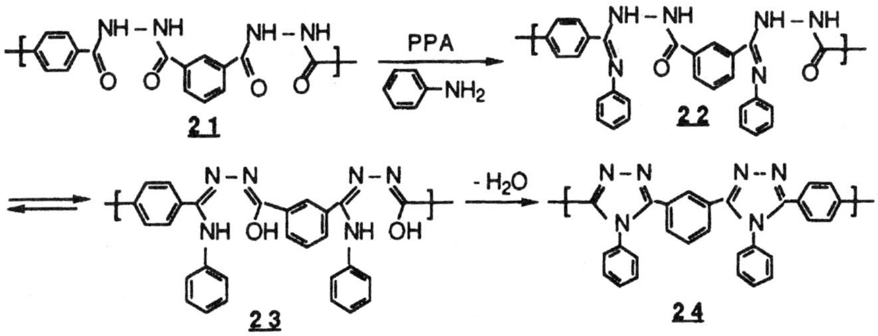

1

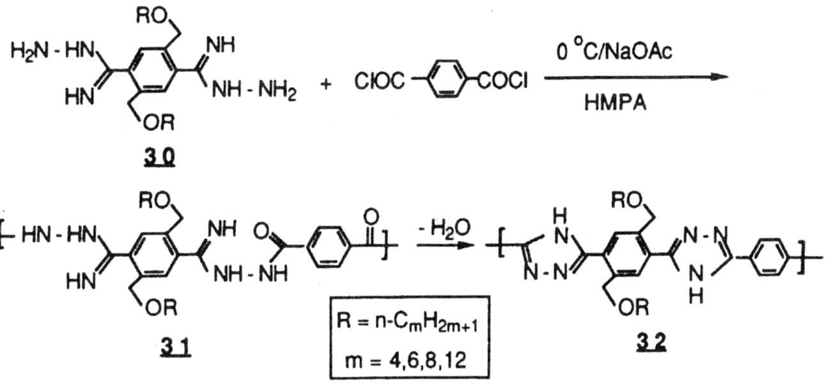

2

Thermal resistance measured by TGA under nitrogen revealed an initial decomposition temperature of 512°C.[5]

The polymer **24** with a molecular weight of 29,700 has been spun into fiber from concentrated formic acid.[5] Tenacity of the fiber was 2.52 g/denier and zero-strength temperature was 465°C.

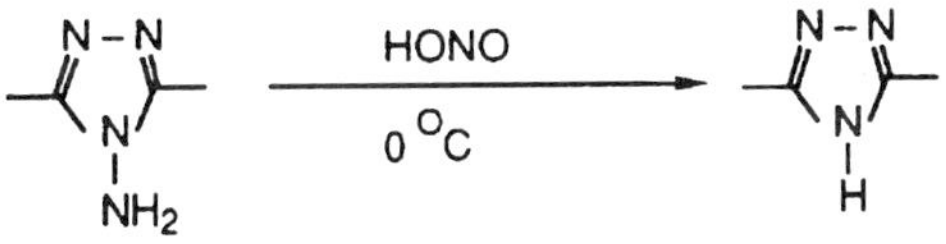

3

Though the polymers are wholly aromatic with all para-linked phenylenes, they were highly soluble in many common organic solvents such as DMF, DMSO, or HMPA.[6] This high solubility results from the pendant long side chains. With increasing side-chain length, T_g and thermal resistance decreased and the solubility increased.

In spite of the presence of the long side chains the polymers were slightly crystalline and the polymer with m = 12 was found to form a liquid crystalline structure in melt.

REFERENCES

1. Macura, K.; Lieser, T. *Ann.* **1969,** *564,* 64.
2. Frazer, A. H.; Sweeny, W.; Wallenberger, F. T. *J. Polym. Sci., Part A* **1964,** *2,* 1157.
3. Stille, J. K.; Arnold, F. E. *J. Polym. Sci., Part A* **1965,** *3,* 4284.
4. Abshire, C. J.; Marvel, C. S. *Makromol. Chem.* **1961,** *44/46,* 388.
5. Holsten, J. R.; Lilyquist, M. R. *J. Polym. Sci., Part A* **1965,** *3,* 3905.
6. Bae, C.; Zin, W.-C.; Jung, J. C. *IUPAC International Symposium on Functional and High Performance Polymers* Nov. 14-16, 1994, Taiwan preprint, pp 571.
7. Iwakura, Y.; Nakajima, M.; Kitani, E. *Makromol. Chem.* **1961,** *44/46,* 408.
8. Nakajima, M.; Hasegawa, M. "Polyaminotriazoles" *in Encycl. Polym. Sci. Technol.,* 1st ed.; Vol. 10, p 623.
9. Nishikubo, T.; Hasegawa, M. "Polyaminotriazoles" *in Encycl. Polym. Sci. Eng.,* 2nd ed.; Vol. 11, p 507.

Polytrimethylenimine

See: Polyalkylenimines

Poly(1-trimethylsiloxy-1,3-butadiene)

See: Alternating Ethylene–Vinyl Alcohol Copolymer

Polytrimethylsilylpropyne

See: Pervaporation Membranes (for Separating Organic Solvents)
Polysilylpropynes (Steric Effects on Material Properties)

Polytrimethylvinylsilane

See: Vinylsilanes (Isomerization Polymerization)

Polytrithiocarbonate

See: Phase-Transfer Catalyzed Polymer Synthesis

Polyureas

See: *Condensation Polymers (Synthesis from Silylated Diamines)*
Double Isomerization Polymerization
High Pressure Polymerization Processing (High Temperature Thermosetting Polymers)
Polybiurets
Polyureas (Second-Order Nonlinear Optical Properties
Reaction Injection Molding (RIM) Materials

POLYUREAS (Second-Order Nonlinear Optical Properties)

Hari Singh Nalwa
Hitachi Research Laboratory
Hitachi Ltd.

Second-order nonlinear optical materials have been considered the key elements to future photonics devices. A wide variety of organic materials have been investigated for second-order nonlinear optics (NLO), which include single crystals, Langmuir–Blodgett films, self-assembled systems, composites, organometallics, liquid crystals, and polymers.[1-4] Organic polymers having NLO chromophores either in the side chain or in the main chain have emerged as one of the most interesting materials for second-order nonlinear optics because they offer intrinsic tailorability, integration with other substrates, ease of processing and fabrication, high mechanical strength, and inexpensiveness. In chromophoric main-chain polymers, crosslinking assists in stabilization of orientation of dipoles induced during poling, which exhibits longer lifetime of persistent NLO activity than that of side-chain chromophoric polymers owing to the reduced relaxation process.[5-7] Polyureas in which urea molecules are covalently bonded together into a single polymer strand are a class of main-chain polymers. Urea moieties have a large dipole moment of 4.7 debye and can be aligned by electrical poling to produce a noncentrosymmetric structure desired for second-order nonlinear optical effects. Because of the ease of tailorability of urea moieties, polyureas with chromophoric main-chain as well as with chromophoric side chains have been prepared.

SYNTHESIS AND SPECTROSCOPY

A general procedure of preparing polyureas is outlined in **Scheme I**. Polyureas were characterized by UV/Visible and NMR spectroscopy, gel permeation chromatography (GPC), and thermal techniques. Polyureas are highly soluble in many organic solvents and could be cast into transparent flexible thin films. X-ray diffraction studies evidenced the amorphous nature of polyureas.

SECOND-HARMONIC GENERATION

Polyureas prepared from aliphatic and aromatic diamines showed the shortest cut-off wavelength of about 300 nm. Polyureas constitute a new class of chromophoric main-chain polymers and seem attractive NLO materials, usable as frequency doublers and for integrated electro-optical devices.

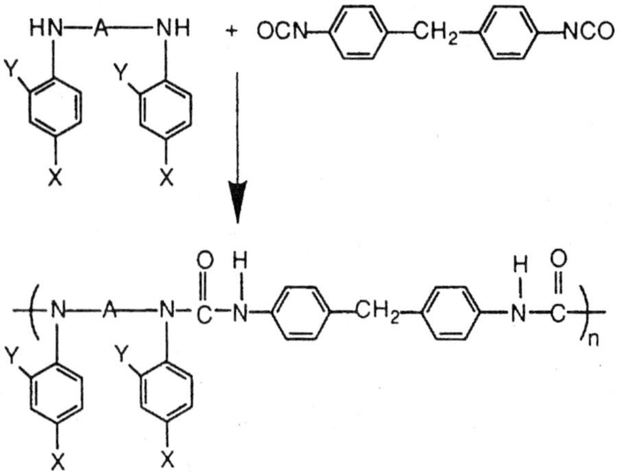

I

REFERENCES

1. Chemla, D. S.; Zyss, J., Eds. *Nonlinear Optical Properties of Organic Molecules and Crystals*; Academic: New York, 1987.
2. Nalwa, H. S. *J. Macromol. Sci. Rev. Chem. Phys.* **1991,** *31,* 341.
3. Nalwa, H. S.; Watanbe, T.; Kakuta, A.; Mukoh, A.; Miyata, S. *Electron. Lett.* **1992,** *28,* 1409.
4. Burland, D. M.; Miller, R. D.; Walsh, C. A. *Chem. Rev.* **1994,** *94,* 31.
5. Stenger-Smith, J. D.; Fischer, J. W.; Henry, R. A.; Hoover, J. M.; Lindsay, G. A.; Hayden, L. M. *Makromol. Chem., Rapid Commun.* **1990,** *11,* 141.
6. Kohler, W.; Robello, D. R.; Dao, P. T.; Willand, C. S.; Williams, D. J. *J. Chem. Phys.* **1990,** *93,* 9157.
7. Xu, C.; Wu, B.; Dalton, L. R.; Ranon, P. M.; Shi, Y.; Steier, W. H. *Macromolecules* **1992,** *25,* 6716.

POLYURETHANE CATALYSIS

Robert Becker*
Institut für Angewandte Chemie (ACA)

Lothar Thiele
Henkel KGaA

ISOCYANATE REACTIONS

Isocyanates are carbonyl compounds with double bonds. Their reaction mechanisms closely correspond to verified reaction mechanisms of common carbonyl compounds.[1] In a similar way one can transfer the well-known catalytic effects in the reactions of carbonyl compounds into the complicated isocyanate reactions.

It is possible to accelerate reactions of carbonyl compounds by Lewis bases and Lewis acids.

Lewis-Base Catalysis

To begin with, let us consider the isocyanate–water reaction accelerated by tertiary amines, the most important Lewis-base catalysts. The investigations in reference 4 show a nearly linear dependence of the logarithm of the rate constants on the logarithm of the basicity constants. These reactions follow *Broenstedt's* law of catalysis. This linearity rule, however, is valid only if the association (basicity) constants are considered, which are determined in the solvents used. Obviously the formation of

*Author to whom correspondence should be addressed.

associates between water or alcohol and the catalyst affects the reaction rate. This association increases with increasing acidity of the alcohols almost without exception. On the other hand, association increases with growing basicity of the amine. At equal basicity of the amine catalysts, the association constant drops due to the steric hindrance of the amine nitrogen.[3]

1,4-Diazabicyclo[2,2,2]octane (DABCO) and dimethyl piperazine (DMP) have nearly the same basicity. Nevertheless, there is a difference of the rate constants in favor of DABCO due to the greater steric hindrance of DMP. The difference is greater in isocyanate–alcohol reactions catalyzed by these amines than in isocyanate–water reactions.[2] This different influence of steric effects suggests a mechanism with alcohol activation rather than isocyanate activation by the amine.[4]

Lewis-Acid Catalysis

It is more difficult to find appropriate arguments in substantiation of the mechanism of Lewis-acid catalysts. A definite interpretation of the observed phenomena frequently seems to be problematic. However, because a synergism is observed with the simultaneous use of Lewis-base and Lewis-acid catalysts,[5-8] only activation of the isocyanates can be assumed for Lewis acids.

The isocyanate reactivity grows with increasing electrophilicity in the following order: aliphatic < aliphatic-aromatic < aromatic isocyanates.[9-12] Moreover, the Hammet constants of the substituted phenyl isocyanates show relatively high positive values (1.7–3.3). This indicates that an attack on the carbonyl carbon of the isocyanate becomes easier with increasing electrophilicity.[11-14]

In Lewis-acid catalysis one should expect that increasing acidity of the catalysts raises their activity. Hence, the acceleration effect should increase in the following order: tetraalkyltin compounds < dialkyltin compounds < tin tetrahalides. To the contrary, the maximum catalytic activity was observed with dialkyltin compounds.[15-17]

Insertion Catalysis

Organotin compounds differ in their mode of action in isocyanate-containing reaction mixtures. They can act in a further, third, mechanism of catalysis: the insertion mechanism. This is an insertion of isocyanate into the Me–X bond of certain metal compounds.

Insertion catalysts can catalyze side reactions of polyurethane formation to a much higher degree than Lewis-acid catalysts do. When the mixing of an organotin compound with isocyanate leads to increased temperatures, and a separation of isocyanurate crystals from the mixture is observed, an insertion reaction is indicated. The metal–nitrogen bond in the carbamate formed in the first reaction step is split by the alcohol, forming metal alcoholate and urethane. The alcoholate catalyzes all further reactions. Accordingly, organotin compounds accelerate urethane formation both as Lewis-acid and as insertion catalysts.

REFERENCES

1. Thiele, L.; Becker, R.; In Frisch, K. C.; Klempner, D., Eds.; *Advances in Urethane Science and Technology*; Technomic: Lancaster, PA, 1993; Vol. 12, pp 59-85.
2. Aleksandrova, J. V.; Krotova, L. D.; Raner, D.; Emelin, E. A.; Tarakanov, O. G. *Zh. Priklad. Khim. USSR* **1975,** *10,* 2245.
3. Kolbe, A.; Pracejus, H. *Ber. Bunsenges. Phys. Chem.* **1966,** *70,* 883.
4. Thiele, L. *Plaste Kautschuk* **1983,** *30,* 668.
5. Reegen, S. C.; Frisch, K. C. *J. Polym. Sci., A-1 Part* **1970,** 2883.
6. Cirkov, N. Ju.; Nesterov, O. V.; Entelis, S. G. *Kinetika Kataliz* USSR **1973,** *14,* 916.
7. Bechara, I. S. *ACS, Org. Coat., Plast. Chem. Pap.* **1980,** *43,* 914.
8. Thiele, L.; Becker, R.; Frommelt, H. *Faserforsch. Textiltechnik/Z. Polymerforsch.* **1977,** *28,* 405.
9. Timm, T. *Kautsch, Gummi, Kunstst.* **1983,** *36,* 257.
10. Bailey, M. E.; Kirss, V.; Spaunburgh, R. G. *Ind. Eng. Chem.* **1956,** *48,* 794.
11. Brock, F. H. *J. Org. Chem.* **1959,** *24,* 1802.
12. Kaplan, M. *J. Chem. Eng. Data* **1961,** *6,* 272.
13. Wongkamolsesh, K. Ph.D. Thesis, 1986, University of Detroit.
14. Entelis, S. G.; Nesterov, O. V.; Tiger, R. P. *Dokl. Akad. Nauk.* USSR **1968,** *178,* 661.
15. Thiele, L.; Becker, R.; Frommelt, H. *Faserforsch. u. Textiltechnik/Z. Polymerforsch.* **1977,** *28,* 343.
16. Nesterov, O. V.; Zabrodin, B.; Cirkov, J. N.; Entelis, S. G. *Kinetica Kataliz* USSR **1966,** *7,* 805.
17. Varentcova, V. N.; Goldstein, I. P.; Paleeva, I. E.; Tarakanov, O. G.; Gurjanova, E. N. *Zh. Obsc. Khim.* USSR **1982,** *52,* 1612.

POLYURETHANE CATALYSTS

Menas S. Vratsanos
Air Products and Chemicals, Incorporated

The term polyurethane is broadly used to describe materials that contain urethane, urea, or isocyanurate linkages as part of the polymer backbone. Polyurethanes are one of the most diverse classes of materials known. By controlling the components in the formulation, one can produce polyurethanes in the form of flexible foams, rigid foams, and elastomers. These various formulations are used in many areas, including coatings, adhesives, and structural components for the housing, transportation, and sporting goods industries.

For a given polyurethane formulation, the catalyst is most responsible for controlling the reaction time and for defining polymer architecture that influences the ultimate mechanical properties. Specifically, it is the catalyst's activity and selectivity towards each of the many reactions occurring in the formation of a polyurethane that determine the structure of the resulting material.

CHEMISTRY

Two basic components that typically make up a polyurethane formulation are a diisocyanate and a polyol. Additives such as catalysts, chain extenders, blowing agents, and surfactants are often used to modify the processing and properties of a polyurethane.

The reactions can be broken into three categories: blowing, gelling, and crosslinking. The blowing reaction, i.e., the reaction of isocyanate and water, produces carbamic acid (R-NH-COOH) as an intermediate. As the carbamic acid readily decomposes into an amine and carbon dioxide gas (CO_2), the polyurethane expands into a foam. Physical blowing agents such as chlorofluorocarbons (CFCs) and methylene chloride can also be used to produce foams, but these foams are generated by the vaporization of these liquids and not by a chemical reaction. The gelling reactions refer to those reactions that do not produce a volatile product, such as the formation of urethane or urea, and that yield linear polymers. The crosslinking reactions also do not generate a volatile product, and they make the polyurethane insoluble.

CLASSES OF CATALYSTS

The myriad of catalyst choices available to a polyurethane formulator presents a significant challenge. There are three major classes of catalysts: tertiary amines, nonprotonic salts, and organometallics.

Tertiary Amine Catalysts

Amine catalysts, especially tertiary amine catalysts, form the largest class of catalysts used in the manufacture of polyurethanes. Major suppliers of amine catalysts include Air Products and Chemicals, Inc.; The Huntsman Corporation; and OSi Specialties, Inc. Tertiary amines catalyze both the gelling and blowing reactions to different extents, but they are generally not very active isocyanurate catalysts. Increased catalyst basicity generally increases activity. In addition to steric hindrance, other geometric factors have been shown to play an important role in determining catalyst activity and selectivity.

Triethylene diamine (TEDA) or 1,4-diazabicyclo[2.2.2]-octane is the most widely employed tertiary amine used in the production of polyurethanes. Its unusually high activity in spite of its moderate basicity is due to a lack of steric hindrance. While TEDA is generally regarded as an excellent gelling catalyst, it is also a powerful catalyst for the blowing reaction. Bis(2-dimethylaminoethyl)ether (BDMAEE), another commonly used tertiary amine catalyst, is especially important in the production of high-resiliency foam. Derivatives of morpholine are also effective as catalysts.[1,2,4]

Acid-blocked amines are another important category of catalysts. These compounds, which are formed by the adduction of a tertiary amine with an acid, typically serve to reduce catalyst activity early in the polymerization process and thus allow more time for flow in the mold.

A growing trend that is driven by environmental concerns is the reduction of emissions and odors from volatile organic compounds from industrial processes. In polyurethanes, the catalyst–typically a tertiary amine and/or an organotin compound–is part of the formulation that does not become covalently bound to the polymer. An unincorporated catalyst may have sufficient vapor pressure to volatilize from the polyurethane. The incorporation of active hydrogens (such as primary and secondary hydroxyls and amines) into the catalyst structure has been extensively used to reduce odors, emissions, and vinyl staining.[6-22] By containing an isocyanate-reactive group, these non-fugitive catalysts can become part of the polyurethane network.

Nonprotonic Salt Catalysts

Nonprotonic salt compounds, such as quaternary ammonium and alkali metal carboxylates, comprise an important group of catalysts for polyurethanes. The active catalytic species for nonprotonic salts is the anion. In contrast, the active catalytic species in acid-blocked amine catalysts is the free amine (which is in equilibrium with the salt).

Quaternary ammonium carboxylates of the general structure $[RR'R''R'''N^+][R''''COO^-]$, as well as amines such as substituted hexahydro-*s*-triazines and 2,4,6-tris(dimethylaminomethyl)phenol, are useful in the formation of isocyanurate.[3,5,23,24]

Organometallic Catalysts

The organometallic compounds are another large class of catalysts used in the manufacture of polyurethanes. The most common catalysts in this class are the organotin compounds, which are salts of organic acids, especially dialkyltin dicarboxylates. Among the more common organotin catalysts are dibutyltin dilaurate $\{[CH_3(CH_2)_3]_2Sn[OCO(CH_2)_{10}CH_3]^2$, DBTDL$\}$, stannous octoate $\{Sn[OCOC_7H_{15}]^2\}$, dibutyltin diacetate $\{[CH_3(CH_2)_3]_2Sn[OCOCH_3]_2\}$, and dibutyltin dimercaptide $\{[CH_3(CH_2)_3]_2Sn[SC_{12}H_{25}]_2\}$. The organometallic catalysts tend to be more selective than the tertiary amines for the gelling reactions.

A synergy in catalyst activity has been observed when tertiary amines are used in combination with organotin compounds.[25]

Latent organotin compounds have been described by B. Jousseaume et al.,[26,27] who demonstrate in their work that bis(2-acyloxy alkyl)diorganotins, which show greatly reduced activity as catalysts at room temperature, readily decompose at higher temperatures and yield diorganotin dicarboxylates (active species) and olefins.

Some recent significant advances in tin catalysis technology have been developed at Air Products and Chemicals by Nichols and Dickenson,[28-31] who describe a family of tin-based catalysts that contain hydroxyl functionality and are more hydrolytically stable than those traditionally used.[28-31]

SUMMARY

Catalysts play an important role in reducing reaction time and controlling polymer properties in the manufacture of polyurethanes. Each catalyst has a unique activity and selectivity for the reactions which occur during the formation of a polyurethane. Of the many types of polyurethane catalysts, the two most commercially significant classes are tertiary amines and organotin compounds. The combinations of catalysts are used primarily to provide the proper balance between processing and physical property requirements. A variety of approaches have been developed to modify catalysts to meet changing environmental and manufacturing concerns, such as reduced odor and delayed activity.

ACKNOWLEDGMENT

The author wishes to thank Dr. J. E. Dewhurst for his help in preparing this article.

REFERENCES

1. Hostettler, F.; Cox, E. F. *Ind. Eng. Chem.* **1960,** *52,* 609.
2. Britain, J. W.; Gemeinhardt, P. G. *J. Appl. Polym. Sci.* **1960,** *4,* 207.
3. Bechara, I. *J. Cell. Plast.* **1979,** *15,* 102.
4. Fondots, D. C. *J. Cell. Plast.* **1975,** *11,* 250.
5. Nicholas, L.; Gmitter, G. T. *J. Cell. Plast.* **1965,** *1,* 85.
6. Zimmerman, R. L.; Austin, T. H. *J. Cell. Plast.* **1988,** *24,* 256.
7. Kopp, R.; Freitag, H. A. U.S. Patent 4 510 269, 1985.

8. Arai, S.; Hasegawa, M. U.S. Patent 4 590 223, 1986.
9. Yamasaki, H.; Morii, M.; Yokota, Y.; Mamada, A.; Nabata, Y. U.S. Patent 4 714 719, 1987.
10. Schiffauer, R.; Buchmuller, W. U.S. Patent 4 957 944, 1990.
11. Weider, R.; Scholz, U. U.S. Patent 5 134 217, 1992.
12. Bailey, F. E. Jr.; Hinney, H. R.; Matlock, P. L.; Schiffauer, R. U.S. Patent 5 200 434, 1993.
13. Savoca, A. C. L.; Listemann, M. L. U.S. Patent 5 143 944, 1992.
14. Savoca, A. C. L.; Listemann, M. L. U.S. Patent 5 194 609, 1993.
15. Listemann, M. L.; Savoca, A. C. L.; Minnich, K. E.; Lassila, K. R. U.S. Patent 5 233 039, 1993.
16. Savoca, A. C. L.; Louie, M.; Listemann, M. L. U.S. Patent 5 166 223, 1992.
17. Savoca, A. C. L.; Louie, M. U.S. Patent 5 322 940, 1994.
18. Abe, H.; Fukushima, T.; Sotoya, K.; Harada, S.; Kitagawa, H.; Morii, M.; Isayama, Y. U.S. Patent 5 315 041, 1994.
19. Casey, J. P.; Clift, S. M.; Kem, K. M. U.S. Patent 5 034 426, 1991.
20. Haas, P.; Freitag, H. A.; Avar, G.; Sommerfeld, C. D.; Illger, H. W. U.S. Patent 4 644 017, 1987.
21. Casey, J. P.; Carr, R. V. C.; Wasilczyk, G. J.; Petrella, R. G. U.S. Patent 5 071 809, 1991.
22. Tamano, Y.; Ishida, M.; Okuzono, S. U.S. Patent 5 229 430, 1993.
23. Ashida, K. In *Handbook of Polymer Foams and Foam Technology*; Klempner, D.; Frisch, K. C., Eds.; Hanser: Munich, 1991; Chapter 6.
24. DeLeon, A. *Adv. Urethane Sci. Technol.* **1981,** *8,* 1.
25. Wolfe, H. W. Jr. *Foam Bulletin* **1960,** 1.
26. Jousseaume, B.; Gouron, V.; Pereyre, M.; Frances, J. M. *Appl. Organomet. Chem.* **1991,** *5,* 135.
27. Frances, J. M.; Gouron, V.; Jousseaume, B.; Pereyre, M. U.S. Patent 5 075 468, 1991.
28. Nichols, J. D.; Dickenson, J. B. U.S. Patent 4 978 728, 1990.
29. Nichols, J. D.; Dickenson, J. B. U.S. Patent 4 981 924, 1991.
30. Nichols, J. D.; Dickenson, J. B. U.S. Patent 4 987 244, 1991.
31. Nichols, J. D.; Dickenson, J. B. U.S. Patent 5 089 583, 1992.

POLYURETHANE ELASTOMERS, SEGMENTED (Nonhydrogen Bonding Systems)

Claus D. Eisenbach,* Karl Fischer, Heidi Hayen, Hartmut Nefzger, Alexander Ribbe, and Edmund Stadler
Makromolekulare Chemie II
Bayreuth Institut für Makromolekülforschung

A series of segmented poly(ether urethane) elastomers with highly flexible polyoxytetramethylene soft segments and molecularly uniform, hydrogen-bond-free hard segments, based on piperazine, butanediol-bischloroformate and a variable constitutive unit in the center acting as a joint, have been synthesized. Nonhydrogen bond-forming systems are advantageous because of their chemical and thermal stability, which prevents the destruction of the precisely defined chain architecture during sample treatment at elevated temperature. Thermal, dynamic, mechanical, and X-ray scattering studies have shown that the size, shape, and packing order of the hard domains correlate with the primary structure of the hard segment and can be varied at will by the sample history. Depending on the conformation or the stereogeometry of the built-in central constitutive hard segment unit, chain-extended or chain-folded hard-segment crystallization can be achieved. The resulting cylindrical or lamellar hard domain shape has a distinct impact on the material properties, which can be reversibly controlled by the choice of the hard segment center unit and the thermal treatment of the sample.

Segmented poly(ether urethane) (PEU) elastomers are multiblock copolymers $(A\text{–}B)_n$ with an alternating sequence of polyurethane (PU) hard segments and, for example, polyether soft segments. Due to the limited miscibility of the hard and soft segments, these multiblock copolymers form microphase-separated multiphase systems.[1-7] At working temperature, the hard-segment segregation leads to domains that are glassy or crystalline and thus act as multifunctional, thermoreversible crosslinks and as fillers in a more or less continuous liquid soft segment phase.

Because this unique structure gives segmented PEUs very attractive properties, there has been considerable scientific and industrial interest in the structure–properties relationships. The overall properties of the sample material, for example, elasticity and mechanical behavior, are closely related to the extent of phase separation between hard and soft segments as reflected in the size, shape, and packing order of the hard domains.[4-20]

Nonhydrogen-forming segmented polyurethanes were first synthesized by Harrell by the condensation of a diol with bis(carbamic acid chloride).[22] A more advantageous method is the reaction of a secondary diamine with a diol-bis(chloroformate).[7,21,23,24] These PEUs possess a better thermal stability (up to 300°C) than do polyurethanes with a hydrogen bonding that is derived from the classical reaction based on the addition of isocyanates and alcohol groups.[25-27]

We report on the design, synthesis, and properties of the polyurethane elastomer PEU-Na-e without hydrogen bonding and having a specially designed architecture of molecularly uniform hard segments that permits control of the type of self-organization of structural elements as well as of the supramolecular structure and sample morphology (**Strucutre 1**).

CONCLUSIONS

Segmented poly(ether urethane) elastomers with hard segments based on piperazine and butanediol-bischloroformate exhibited a high chemical and thermal stability because of the lack of hydrogen bonding. Consequently, experimental data could be correlated more easily with specific molecular structures and morphologies because various thermal sample histories could be applied without altering the primary structure.

The study of a series of segmented poly(ether urethane)s with nonhydrogen bonding had shown that the morphology and the size and shape of the tailor-made hard domains could be controlled or triggered by both the polymer constitution and sample annealing. Only the building-in of a hard-segment center unit capable of forming a sharp and adjacent folded structure in an otherwise chain-extended crystallized structure was needed to change the superstructure of a macromolecular system. Extended or chain-folded hard domains as well as the resulting cylindrical or lamellar molecular packing had a great

*Author to whom correspondence should be addressed.

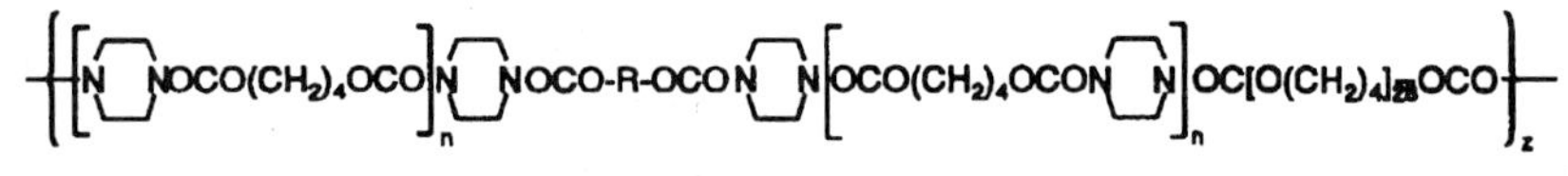

PEU-N

PEU-N with N=2n+1					
	a	b	c	d	e
R	$-(CH_2)_4-$	$-(CH_2)_6-$	$-(CH_2)_8-$	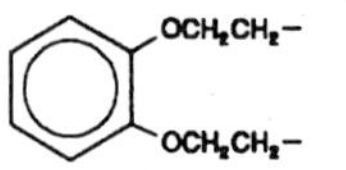	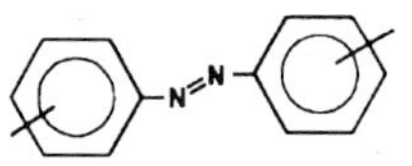

1

influence on the material properties. Consequently, the possibility of creating special supramolecular structures by small changes in the architecture of the macromolecules has opened interesting perspectives for the design and synthesis of novel polymer materials with new processing and material properties.

ACKNOWLEDGMENT

We gratefully acknowledge financial support of this work by the German Ministry of Research and Technology (Grant No. 03M40436), Bayer AG (Leverkusen), and the German Science Foundation (SFB 213, University of Bayreuth).

REFERENCES

1. Oertel, H. *Bayer Farben Revue* **1965,** *11,* 1.
2. Oertel, H. *Chemikar. Ztg.* **1974,** *98,* 344.
3. Estes, G. M.; Seymour, R. W.; Cooper, S. L. *Macromolecules* **1971,** *4,* 452.
4. Noshay, A.; McGrath, J. E. *Block Copolymers* Academic: New York, 1977; p 365.
5. Cooper, S. L.; Estes, G. *Multiphase Polymers*; Eds.; Advances in Chemistry, Series 176, American Chemical Society: Washington, DC, 1979.
6. Eisenbach, C. D.; Baumgartner, M.; Günter, In *Advances in Elastomers and Rubber Elasticity*; Lal, J.; Mark, J. E., Eds.; Plenum: New York, 1987; p 51.
7. Eisenbach, C. D.; Nefzger, H. In *Contemporary Topics in Polymer Science*; Culbertson, W. M. Ed.; Plenum: New York, 1989; Vol. 6, p 339.
8. Wilkes, G. L.; Emerson, J. A. *J. Appl. Phys.* **1976,** *47,* 4261.
9. Kwei, T. K. *J. Appl. Polym. Sci.* **1982,** *27,* 2891.
10. Camberlin, Y.; Pascault, J. P. *J. Polym. Sci., Polym. Phys. Ed.* **1984,** *22,* 1835.
11. Lee, H. S.; Wang, Y. K.; MacKnight, W. J.; Hsu, S. L. *Macromolecules* **1988,** *21,* 270.
12. Lee, H. S.; Hsu, S. L. *Macromolecules* **1989,** *22,* 1100.
13. Li, Y.; Gao, T.; Chu, B. *Macromolecules* **1992,** *25,* 1737.
14. Li, Y.; Gao, T.; Liu, J.; Linliu, K.; Desper, C. R.; Chu, B. *Macromolecules* **1992,** *25,* 7365.
15. Li, Y.; Reu, Z.; Zhao, M.; Yang, H.; Chu, B. *Macromolecules* **1993,** *26,* 612.
16. Li, Y.; Kang, W.; Stoffer, J. O.; Chu, B. *Macromolecules* **1994,** *27,* 612.
17. Meckel, W.; Goyert, W.; Wieder, W. *Thermoplastic Elastomers*; Legge, N. R. et al., Eds.; Munich: 1987; p 13.
18. Gronski, W.; Stoeppelmann, G. *Polymer Preprints* **1989,** *29,* 46.
19. Camargo, R. E.; Macosko, C. W.; Tirrell, M.; Wellinghoff, S. T. *Polymer* **1985,** *26,* 1145.
20. Serrano, M.; MacKnight, W. J.; Thomas, E. L.; Ottino, J. M. *Polymer* **1987,** *28,* 1667.
21. Eisenbach, C. D.; Nefzger, H. In *Handbook of Polymer Synthesis*; Kricheldorf, H. R. Ed.; Marcel Dekker: New York, 1992; p 685.
22. Harrell, Jr., L. L. *Macromolecules* **1969,** *2,* 607.
23. Meltzer, A. D.; Spiess, H. W.; Eisenbach, C. D.; Hayen, H. *Makromo. Chem., Rapid Commun.* **1991,** *12,* 261.
24. Eisenbach, C. D.; Hayen, H.; Popp, G. *Polymer Preprints* **1994,** 35/1, 583.
25. Kricheldorf, H. R.; Awe, J. *Makromol. Chem., Rapid Commun.* **1988,** *9,* 681.
26. Nefzger, H. Ph.D. Dissertation, University of Karlsruhe, 1987.
27. Bayer, O. *Das Diisocyanatpolyadditionsverfahren*; Carl Hanser Verlag: Munich, 1963; p 14.

POLYURETHANE IONOMERS

M. H. George,* R. Arshady, D. K. Kakati,
W. Al-Shahib, and R. Gosain
Department of Chemistry
Imperial College of Science, Technology and Medicine

Polyurethane ionomers are a class of polyurethanes containing a small mole% of ionic groups, usually less than 10 mol%, either near the backbone of the polymer or as distant pendant groups. Polyurethanes with a higher mole% of ionic groups should be referred to as ionic polymers, whereas those with a very high ionic content should be termed polyelectrolytes. However, in the literature, the name "polyurethane ionomers" is often applied loosely to all polyurethanes containing ions of any ionic mole% content. Polyurethane ionomers can be cationic, anionic, or zwitterionic, depending on the nature of the ionic groups.

PREPARATION

The various routes that introduce ionic groups into polyurethanes can be classified in two main methods:

- the modification of reactive functional groups present in polyurethanes, including modification of the –NH– groups in the polyurethane backbone; and

*Author to whom correspondence should be addressed.

- the introduction of ionic centers into the corresponding monomers (chain extenders) preceding step growth polymerization.

Polyurethane Ionomers via Functional Group Modification

The first reported synthesis of a polyurethane cationomer by functional group modification by Dieterich[1] used *N*-methylamino-2,2′-diethanol as a chain extender in the synthesis of segmented polyurethanes. The polyurethanes were then quaternized by using both bifunctional and monofunctional quaternizing agents.

Several workers have reported the use of the tertiary amine hard-segment chain extender, *N*-methyl diethanol amine (MDEA) to synthesize polyurethane ionomers.[2-7] The common two-step solution polymerization route was used for these segmented polyurethanes containing tertiary amine groups. Quaternization of the amine groups by alkyl chlorides yielded cationomers. When the MDEA chain-extended polyurethanes reacted with γ-propane sultone, polyurethane zwitterions were produced.

Carboxylated and sulfonated polyurethane ionomers have also been prepared by a nucleophilic reaction involving the urethane hydrogen.[8-10] Mixed anionomers containing both sulfonate and carboxylate have also been made in this way.[11] The base polyurethane was treated with NaH at –5 to 0°C to remove the urethane hydrogen. Subsequent nucleophilic reaction with the stoichiometric amount of γ-propane sultone or β-propiolactone at room temperature yielded the sulfonated and carboxylated ionomers, respectively.

A plasma deposition technique has also been developed to introduce sulfonate groups onto the surface of polyurethane films. Giroux et al. used a sulfur dioxide plasma to introduce sulfonate groups directly onto the film surface, thus avoiding the use of the carcinogenic propane sultone.[13]

Polyurethane Ionomers via Ionic Chain Extenders

The use of ionic chain extenders for the synthesis of polyurethane ionomers provides a wider choice of reactants, faster rates of reaction, as well as good control of the ionic content of the resultant polymer. The major problems are the limited solubility of some ionic diols in the solvents usually used in polyurethane synthesis, such as *N,N′*-dimethylformamide, dimethyl sulfoxide, and tetrahydrofuran.

Polyurethane anionomers have been synthesized by the chain extension of NCO-terminated prepolymers with the sodium salt of *N*-(2-amino ethyl)-2-amino ethane sulfonic acid.[14] The synthesis of ionic diols containing thiosulfate and sulfonate groups and their subsequent use for synthesizing anionic polyurethanes has been reported.[15,16]

Ionic diols, such as potassium and sodium 4-(1,2-dihydroxyethyl) benzene sulfonates, have been used in the preparation of polyurethane anionomers.[17] Recently, polyurethanes containing phosphate groups have been synthesized.[17,18]

PROPERTIES

It is commonly accepted that the presence of ionic groups in a solid ion-containing polyurethane (or other polymer) results in some aggregation of the ionic groups leading to ionic crosslinks in the solid state. For polyurethane ionomers, there can be hard domain/soft segment phase separation occurring as well, depending on their precise chemical structure.

It is often observed, for example, that the glass temperature, T_g, of the soft segment of an ionomeric polyurethane (or the T_g of other polymers) is greater than that of the corresponding non-ionic polyurethane (or other non-ionic polymer). In general, there are dramatic changes in the other solid-state properties of macromolecules containing ionic groups. It is usually found that hard/soft block phase separation of polyurethane ionomers is greater than for their non-ionic polymers analogues. However, some contrary evidence has been found.[24] There is usually a gradual increase in Young' modulus and ultimate stength when the percent of ionic groups in a low-ionic-content polyurethane ionomer increases.[11,18] As expected, there is also an increase in the equilibrium uptake of water in films as the %ionic content of the hard block of a polyurethane increases.[25]

The exact structure and morphology of polyurethane ionomers in the solid state remains controversial. Not only can there be the separation of hard-segment domains and soft segments, but there is also a necessary rearrangement and relocation of ions in the solid polyurethane ionomer. The ions are usually attached chemically to the hard segment of an ionomer molecule.

SEM studies of polyurethane ionomers have been carried out by several workers. In some cases, such as the surface morphologies of polyurethanes containing phosphoric acid and monosodium phosphate groups, coil-like structures of ~1–3 μm were noted.[19] Other workers had previously observed similar rod-like structures for different polyurethane ionomers by other workers.[26,27] Other researchers using SEM observed "granules" in the polyurethane samples with a size of 0.5–3 μm.[19] The size of both coil-like structures and granules could not be determined precisely. However, the interpretation of SEM for all of the polymers is strongly dependent on the choice of the casting solvent chosen as well as many other factors.[28,29]

In a study of polyurethane ionomers containing varying concentrations of mixed carboxylate and sulfonate anions, DSC, DMTA, and SAXS analysis indicated that the modulus enhancement of the mixed anionic pairs arose from a wide variety of factors, including aggregate packing, ionic anchoring, and degrees of phase separation.[10]

X-ray-absorption fine-structure spectroscopy (EXAFS) can be used to investigate specific atom environments in ionomers and has been applied to polyurethane ionomers.[12]

A small-angle neutron-scattering study (SANS) of sulfonated polyurethane ionomers has been reported.[30,31]

APPLICATIONS

The field of applications of polyurethane ionomers is growing rapidly, and many more applications will be developed in the future.

J. H. Braybrook and L. D. Hall recently surveyed the application of all polymer surfaces in medicine, including drug release systems, and H. J. Griesser reviewed the degradation of polyurethanes in biomedical applications.[32,33] Relatively nonbiodegradable polyurethanes have been used in the production of wound dressings and of artificial artery and heart materials.[34-36] The extension to the applications of polyurethane ionomers seems logical. Already, for example, sutures have been coated with an ionically-bonded block elastomeric copolymer of a quaternary polyurethane and a polyanionic polymer.[37] W. J. Passl has reviewed all of the polymers used in modern pharmaceutics, including the use of polyurethanes as semipermeable membranes.[38] The expected antistatic and improved adhesion of polyurethane ionomers could find application in the tissue adhesive area. The possibility of incorporating phospholipid analogues into polyurethanes (sometimes in the zwitterion form) is an exciting development since phospholipids are important building blocks of biological membranes.[22,23]

In recent years, the development of the artificial kidney has been of great interest. The development of machines containing polymer hollow fibers for the removal of toxic substances has been an exciting development, because it permits a reduction of the separator size. Polyurethane ionomer hollow fibers are possible candidates for future development.[39]

The separation of liquids using membranes is one of the most promising means of energy-saving separation technology. Based on the earlier synthesis of polyurethane cationomers, crosslinked cationomeric polyurethane membranes that were mechanically strong enough for pervaporation use have been synthesized recently.[21] These were applied to the separation of ethanol–water mixtures.[25]

Polyurethane ionomers are often hydrophilic enough to be readily dispersed in water, and several of their properties have been evaluated.

Some other applications of polyurethane ionomers have been reviewed.[40] In many applications, organic based systems are being replaced by aqueous systems because of the reduced cost and for environmental reasons. Polyurethane ionomer dispersions are particularly attractive because their chemistry is versatile and easily controlled. Typical real or possible applications include coatings/treatments for textiles, paper, leather, various plastics, and glass fibers and bottles. Polyurethane ionomers can also be used as adhesives involving textiles and other materials and as additives to various inks, sealants, and concrete.

The incorporation of ionic phosphorus components into polyurethane foams has already been mentioned. For low concentrations of a phosphorus-containing compound, often ionic in nature, fire resistance of the rigid foams increased, and physical properties were retained.[20]

REFERENCES

1. Dieterich, D.; Keberle, W.; Witt, H. *Angew. Chem. Internat. Ed.* **1970**, 9. 40.
2. Hwang, K. K. S.; Yang, C. Z.; Cooper, S. L. *Polym. Eng. Sci.* **1981,** *21,* 1027.
3. Yang, C. Z.; Hwang, K. K. S.; Cooper, S. L. *Makromol. Chem.* **1983,** *184,* 651.
4. Miller, J. A.; Yang, K. K. S.; Cooper, S. L. *J. Makromol. Sci. Phys.* **1983,** B22, 321.
5. Speckhard, T. A.; Hwang, K. K. S.; Yang, C. Z.; Laupan, W.-R.; Cooper, S. L. *J. Makromol. Sci. Phys.* **1984,** B21, 175.
6. Hsu, S. L.; Xia, H. X.; Szmant, H. H.; Frisch, K. C. *J. Appl. Polym. Sci.* **1984,** *29,* 2467.
7. Lelah, M. D.; Pierce, J. A.; Lambrecht, L. K.; Cooper, S. L. *J. Colloid & Inter. Sci.* **1985,** *104,* 422.
8. Hwang, K. K. S.; Speckhard, T. A.; Cooper, S. L. *J. Makromol. Sci. Phys.* **1984,** B23, 153.
9. Lee, D. C.; Register, R. A.; Yang, C. Z.; Cooper, S. L. *Macromolecules* **1988,** *21,* 998.
10. Visser, S. A.; Cooper, S. L. *Macromolecules* **1991,** *24,* 2576.
11. Visser, S. A.; Cooper, S. L. *Polymer* **1992,** *33,* 3790.
12. Ding, Y. S.; Register, R. A.; Yang, C. Z.; Cooper, S. L. *Polymer* **1989,** *30,* 1213.
13. Giroux, T. A.; Cooper, S. L. *J. Appl. Polym. Sci.* **1991,** *43,* 145.
14. Lorenz, O.; Rose, G. *Angew. Macromol. Chem.* **1983,** *118,* 91.
15. Chui, T. Y. T.; Lam, P. K. H.; George, M. H.; Barrie, J. A. *Polym. Commun.* **1988,** *29,* 317.
16. Chui, T. Y. T.; Coote, A. S.; George, M. H.; Barrie, J. A. *Polym. Commun.* **1988,** *29,* 40.
17. Lam, P. K. H.; George, M. H.; Barrie, J. A. *Polymer* **1989,** *30,* 2321.
18. Lam, P. K. H.; George, M. H.; Barrie, J. A. *Polymer Commun.* **1991,** *32,* 80.
19. Kakati, D. K.; George, M. H. *Polymer* **1993,** *34,* 4319.
20. Tashev, E.; Shenkov, S.; Troev, K.; Borissov, G.; Zabski, L.; Edlinski, Z. *Eur. Polym. J.* **1988,** *11,* 1101.
21. Varma, S. C.; Ahsan, M. A.; George, M. H.; Barrie, J. A. *Polym. Commun.* **1990,** *31,* 11.
22. Yamada, M.; Li, Y.; Nakaya, T. *Makromol. Chem., Rapid Commun.* **1995,** *16,* 25.
23. Yamada, M.; Li, Y.; Nakaya, T. *J. Macromol. Sci., Pure Applied Chem.* **1995,** A32, 1235.
24. Vilenski, V. A.; Kercha, Y. Y.; Shtompel, V. I.; Sukhorukova, S. A.; Levchenko, N. I.; Grekov, A. P. *Vysoko. Soedin.* **1990,** A32, 1511.
25. Ahsan, M. A.; Varma, S. C.; George, M. H.; Barrie, J. A. *Polym. Commun.* **1991,** *32,* 509.
26. Kakati, D. K. Ph.D. Thesis, University of London, 1993.
27. Chui, Y. T. Ph.D. Thesis, University of London, 1991.
28. Lam, P. K. H. Ph.D. Thesis, University of London, 1989.
29. *Developments in Block Copolymers*; Goodman, I., Ed.; Applied Science: London, 1982; Vol. 1, Chapter 1.
30. Visser, S. A.; Pruckmayr, G.; Cooper, S. L. *Macromolecules* **1990,** *24,* 6769.
31. 88 Register, R. A.; Pruckmayr, G.; Cooper, S. L. *Macromolecules* **1990,** *23,* 3023.
32. Braybrook, J. H.; Hall, L. D. *Progr. Polym. Sci.* **1990,** *15,* 715.
33. Griesser, H. J. *Polym. Degrad. Stab.* **1991,** *33,* 329.
34. Rosevear, A.; Kennedy, J. F.; Cabal, J. M. S. *Immobilized Enzymes and Cells* Adam Hilger, 1987.
35. Dombrow, B. A.; *Polyurethanes*; Reinhold: New York, 1965.
36. Grace, W. R. Br. Patent 1 541 100, 1979.
37. U.S. Patent 3 987 797 (1976) by Stephenson, M.
38. Passl, W. J. *Progr. Polym. Sci.* **1989,** *14,* 629.
39. Paul, D. *Progr. Polym. Sci.* **1989,** *14,* 597.
40. Davies, W. D. In *Adhesives for Water Based Coatings*; Karsa, D. R., Ed.; Royal Society of London: London, 1990; p 181.

Polyurethanes

See: Diisocyanates (In Situ Generation)
Foamed Plastics
High Pressure Polymerization Processing (High Temperature Thermosetting Polymers)
High Solids Polyurethane Coatings
Liquid Crystalline Polyurethanes

Phosphorus-Containing Polymers (From Dicarboxylic Acids with C-P Bonds)
Photocrosslinking (Overview)
Polyurethane Catalysis
Polyurethane Elastomers, Segmented (Non-Hydrogen Bonding Systems)
Polyurethane Ionomers
Polyurethanes (Overview)
Polyurethanes, Blocked Copolymer (Reactive Modifiers for Epoxy Resins)
Powder Coatings (Overview)
Quinone-Amine Polyurethanes
Reaction Injection Molding (RIM) Materials
Resins (Applications in Coatings Industry)
Rigid Polyurethane Foams
Step Polymerization Catalysts
Thermodegradable Polymers (Azo Groups in Main Chain)
Vegetal Biomass (1. Monomers and their Polymerization)
Water-Borne Coatings (Urethane/Acrylic Hybrid Polymers)
Wound Dressing Materials

POLYURETHANES (Overview)

M. S. Bhatnager
Professor of Chemistry (Retired)
Indian Institute of Technology

RAW MATERIALS

The raw materials for preparing polyurethanes are polyisocyanates, polyols, diamines, catalysts, additives, and blocking agents.[2-5] The polyisocyanates are either aliphatic like hexamethylene diisocyanates (HDI) (CAS Registry No. 822-06-0), isophorene diisocyanate (IPDI) (CAS Registry No. 4098-71-9), dimethyl diisocyanate (DMDI) (CAS Registry No. 68239-06-5), 4,4′-diisocyanato dicyclo hexylmethane (Desmodur), or aromatic like 2,4-toluene diisocyanate (TDI) (CAS Registry No. 584-84-9), 1,5-napthalene diisocyanate (NDI) (CAS Registry No. 3173-72-6), 4,4′-methylene diphenyl diisocyanate MDI (CAS Registry No. 101-68-8).

Polyols are either polyethers, such as propylene glycol (PG) and trimethylolpropane (TMP) combined with sucrose or polyesters, such as ethylene glycol, 1,2-propanediol, 1,4-butenediol, and diethylene glycol combined with glycerol. Polyethers are used to produce flexible and rigid foams and polyesters are used to produce elastomers, flexible foams, and coatings. Lewis acid and Lewis bases are used as catalysts. The additives are polysiloxane-polyether, carbodiamide piperazine, chloro-fluoro-hydrocarbons, and phosphorous and nitrogen compounds.

STRUCTURE AND MECHANISM

The structure of polyurethanes is highly influenced by intermolecular forces, such as hydrogen bonding, polarizability, van der Waals forces, stiffness of the chain, and crosslinking.[1,6-8] There may be crystalline regions between the flexible chains.

PREPARATION

Generally, polyurethane polymers are formed either by the reaction of bischloroformate with a diamine or by the reaction of a diisocyanate with a diol.[9-12]

Polyurethanes with a high molecular weight can be produced by one of two low-temperature-condensation methods: interfacial polycondensation or solution polycondensation.

TYPES OF SOLID POLYURETHANES

The different types of PU can be classified as linear, castable, millable, thermoplastic, cellular, sprayable, poromeric, and fibrous (DuPont's spandex).

Linear PUs are prepared by the action of aliphatic glycol and aliphatic diisocyanate. As shown in **Equation 1**, no crosslinking or branching takes place.

$$\text{-[- }(CH_2)_x\text{ OOCNH }(CH_2)_y\text{ NHCOO }(CH_2)_x\text{ OOCNH }(CH_2)_y\text{ NCOO -]-n} \qquad (1)$$

The reaction product of butanediol and hexamethylene diisocyanate yields a value of 4 and 6 respectively for x and y. The properties of such materials are similar to those of polyamides, which are of limited interest.

Castable PUs are obtained by extending polyol with diisocyanate to yield an isocyanate-terminated prepolymer. Glycol or diamine chain extenders then bring about further chain extension. A slight excess of diisocyanate is normally employed, and this enables the crosslinking to occur at the urethane or urea group.

In the one-shot material, the long-chain polyol and the chain extender are mixed without a chemical reaction taking place. The diisocyanate is then added and chain crosslinking takes place simultaneously.

The millable PUs are formed when a deficiency of diisocyanate results in a hydroxyl-terminated polymer that is relatively stable and noncrosslinked. After reaching the correct molecular weight, this product appears as a plastic gum that can be hardened on a rubber mill. Crosslinking can be effected by the addition to the mill of more diisocyanate, peroxide, or sulfur. Carbon black may also be added to the mill. The properties of the final product are similar to those of castable PUs with a restricted range.

The thermoplastic PUs are elastomeric and can be processed on modern plastic equipment. Chemically, they are similar to castable PUs, and an excess of diisocyanate is used to bring about crosslinking. The diisocyanate used results in thermostable crosslinks that break in an injection cylinder at 160°C and crosslink on cooling. These PUs can be injection molded, extruded, blow molded, and calendered.

The sprayable PUs are based on the one-shot castable elastomers and are sprayed at elevated temperatures. Unlike other coating materials, sprayable PUs do not require a solvent.

The thermoplastic or millable PUs can be modified to be more linear and then dissolved in methyl ethyl ketone (MEK), acetone, ethyl acetate (EtAc), etc. The solution can be brushed or sprayed. In this case, only thin coats are feasible.

The poromeric PUs have a porous structure. DuPont's foam, for example, is a leather-like material based on a polyurethane that contains randomly dispersed fibers and possesses a porous structure enabling it to breathe like leather.

Spandex consists of long-chain polymers containing 85% segmented polyurethane. These fibers are strong and resistant to weather.

PROPERTIES

Aliphatic polyurethanes melt at lower temperatures due to the greater flexibility imparted to the polymer chain by oxygen links. Because they degrade with the evolution of carbon dioxide above 200°C, they are difficult to produce by the melt method. A high-temperature-solution method can be employed. By and large, they are thermally unstable.

Aromatic or piperazine-ring polymers can be prepared by low-temperature methods, although their melt temperatures are quite high. They can be formed into shapes only by dissolving them in an organic solvent; for example, polyurethane made from piperazine and ethylene chloroformate decomposes at 245°C, but the crystalline polymer is soluble in chloroform, hot dimethylformamide, and cyclohexanone. Clear amorphous film can be hot pressed at 260°C for a short time, and fibers can be produced by using a chloroform methanol solution.

APPLICATIONS

Polyurethanes are used widely in products such as foams, elastomers, adhesives, coatings, and textiles.

REFERENCES

1. Bhatnagar, M. S. *"Textbook of High Polymers, Vol. 2,"* to be published, Navneet: Bombay.
2. Thaspliyal, B. P. et al. *Progr. Polym. Sci.* **1990,** *15,* 735.
3. Bialas, N. et al. *Makromol. Chem.* **1990,** *191,* 1843.
4. Senekar, S. D.; Dunlop, K. L.; Lowery, M. K.; Potter, T. A. Mobay Corps, EP Patent 453 914, 1990.
5. Zhou, G. et al. *J. Polym. Sci. A, Polym. Chem.* **1991,** *29,* 1097, 1105.
6. Ribbe, A. et al. *Makromol. Chem., Rapid Commun.* **1992,** 13.
7. Hespe, H. F. et al. *J. Appl. Polym. Sci.* **1992,** *44,* 2029.
8. Engle, L. P. et al. *J. Macromol. Sci. Rev. Macromol.-Chem. Phys.* **1993,** C33, 239.
9. Bose, D. et al. *Makromol. Chemie. Makromol. Simp.* **1991,** *50,* 191.
10. Papadimitrakopoulos, F. et al. *Polym. Prepr. (Am. Chem. Soc. Div. Polym. Chem.)* **1990,** *31,* 486.
11. Petrovic, Z. S. et al. *J. Appl. Polym. Sci.* **1991,** *42,* 79.
12. Behrand, D. et al. *Plaste Kautsch.* **1990,** *37,* 194.

POLYURETHANES, BLOCKED COPOLYMER (Reactive Modifiers for Epoxy Resins)

Karl-Heinz Hentschel
LS Marketing
Bayer AG

BLOCKED POLYURETHANE COPOLYMERS

This paper deals mainly with special blocked polyurethane copolymers (abbreviated subsequently as "BPUCs") that can be cured at ambient temperatures. I will also discuss blocked and unblocked polyurethane (PU) resins and copolymers that are especially well-suited for modifying epoxy resins in terms of flexibility, crack-bridging ability, etc.

Basic Principles of Blocked Polyurethane Copolymers

BPUCs have been used most frequently in applications in which they are mixed with compounds having acidic hydrogens and are stable at room temperature. A temperature increase in this case initiates chain elongation or a crosslinking polyisocyanate polyaddition reaction.[1]

Typically, BPUCs are produced from PU prepolymers that have pendant isocyanate groups based on aromatic or aliphatic diisocyanate monomers and a polyether or polyester polyol. In the final step, a blocking agent having acidic hydrogens such as phenol, caprolactam, malonate esters, methyl ethyl ketoxime, and isononyl phenol is added to the free isocyanate groups of the PU copolymers.[2]

Curing reactions of blocked PU copolymers often proceed via an elimination-addition mechanism.[3] Other mechanisms, including displacement reactions, prevail in some cases.

Blocked Polyurethane Copolymer Systems Curing at Ambient Temperature

Generally, a BPUC will react with a compound that bears reactive functional groups to form new chemical bonds, thus liberating the blocking agent. Urea linkages will form when polyamines are used as the compounds bearing the reactive functional groups (**Figure 1**). If the crosslinking reaction proceeds sufficiently fast at the ambient temperature, an epoxy resin can be incorporated into the system. The polyamine hardener will bring about both the cure of the BPUC and the epoxy resin more or less simultaneously and may even link both polymer types in a common polymer network if all the ingredients are chosen properly.

MODIFICATION OF EPOXIES USING BLOCKED POLYURETHANE COPOLYMERS

As modifying resins for making epoxies, flexible alkyl phenol blocked PU copolymers (Desmocap™, Bayer AG) have gained importance in the market, as have polyether urethane urea amines (Euredur™ 70 and 75, Witco) and polyether urethanes bearing pendant (meth)acrylate ester end groups and cured with polyamines via a Michael addition reaction (Ancarez™ 300, Anchor).

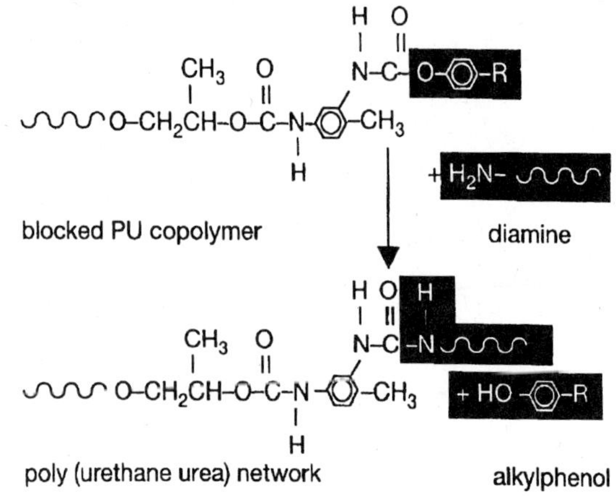

FIGURE 1. Cure mode (schematical) of a BPUC with a polyamine.

PRACTICAL APPLICATION

A particular advantage of BPUC-epoxy combinations cured with (cyclo)aliphatic diamines and/or polyamines is their suitability for crack-bridging and, depending on the formulation used, the wide scope for varying the flexibility. Hence these systems are widely applied for the protection of concrete substrates prone to be affected by cracks such as industrial floorings, car-park decking, sealers between concrete or brick supports, and ceramic paving tiles, surface sealing of bridges, and bridging for spectator areas of sports stadiums (with an additional top coat). In addition, they are used to provide flexibility to adhesives and cements.

TRENDS IN RESEARCH AND DEVELOPMENT

BPUC resins used as modifiers for epoxies are being offered more in special delivery forms in order to comply with the specific requirements of the customers. Ongoing and future developments have to focus on a significant decrease in the product viscosity, the need for lower amounts of material, and a total avoidance of the blocking agent.

CONCLUSIONS

Blocked PU copolymers, especially solvent-free copolymers containing alkylphenols as blocking agents (Desmocap, Bayer), in the past fifteen years have shown to help protect concrete surfaces. This is brought about by the high versatility of these copolymers in the flexibilizing of epoxy resins.

REFERENCES

1. Oertel, G. Ed. *Polyurethane Handbook. Chemistry-Raw-Materials-Processing-Application-Properties*, 2nd ed.; Hanser: New York, 1994.
2. Jacobs, R. L.; Long, J. W. *J. Elast. Plast.* **1979**, *11*, 15.
3. Mirgel, V.; Nachtkamp, K. *Farbe + Lack* **1983**, *12*, 928.

Poly(vinyl acetate)

See: *Photocrosslinking (Overview)*
Poly(vinyl alcohol) (High Molecular Weight)

Poly(vinyl alcohol)

See: *Bioartificial Materials*
Photocrosslinking (Overview)
Poly(vinyl alcohol) (High Molecular Weight)
Poly(vinyl alcohol) (High Strength and High Modulus)
Poly(vinyl alcohol) Fiber (High Strength and High Modulus)
Poly(vinyl alcohol) Fibers (High Modulus; by Vibrating Zone-Drawing)
Poly(vinyl alcohol)-Iodine Interactions
Poly(vinyl alcohol)-Ion Complex Gels
Poly(vinyl ketal)s
Superconducting Filaments, High Tc (Suspension and Solution Spinning from PVA Solution)
Suture Materials (Current Research and Development)

POLY(VINYL ALCOHOL) (High Molecular Weight)

Toheí Yamamoto* and Osamu Sangen
Faculty of Engineering
Himeji Insitute of Technology

Mikiharu Kamachi
Faculty of Science
Osaka University

Poly(vinyl alcohol) (PVA) is useful as an industrial, medical, or biomimetic material. A PVA of high molecular weight is important to the production of a spinning gel that has both a high degree of strength and a high modulus.[1] However, it is very difficult to prepare such a gel because chain transfer reactions control the molecular weight of precursors generally obtained by radical polymerization of vinyl acetate (VAc). Moreover, degradation occurs by hydrolysis of branched poly(vinyl acetate) (PVAc) formed by a chain transfer reaction.[2-4] Therefore, the highest degree of polymerization (P_n) of PVA ever found was 9440 as reported by Imai et al.[5] The highest P_n of PVA obtained commercially was only 2000, although in one special case a P_n of 6000 was obtained. In the 1980s, demand for a high molecular weight PVA increased. We have tried to synthesize high molecular weight PVAs with a P_n greater than 1×10^4. This study presents some good methods of obtaining high molecular weight PVAs.

We discovered an initiator-free photo-emulsion polymerization method that we applied to VAc at a low temperature (0°C) and obtained a high molecular weight PVA after the hydrolysis of PVAc.[6,7] When we applied this method to vinyl pivalate (VP), we obtained high molecular weight poly(VP)(PVP) which yielded a higher molecular weight PVA ($P_n = 2 \times 10^4$) after hydrolysis by a new saponification method.[8,9] Other methods employed to produce high molecular weight PVAs use a solution process. The first is the photopolymerization of VAc using trimethylbenzoyl diphenylphosphine oxide (TMDPO) or pivaloyl dimethylphosphate (PDM) which produces active radicals even at a very low temperature (–30°C).[10,11] A second method is polymerization of VP in dimethyl sulfoxide at a higher temperature using AIBN (α,α′-azobisisobutyronitrile) as an initiator; this yields a high molecular weight PVA after hydrolysis.[12]

PROPERTIES

Niklajev et al. found that the crystal melting point, strength, strength elongation, and softening point of PVA film hardly changed with the P_n. The T_g and the abrasion resistance of PVA film, however, did increase as the P_n rose. Fukae et al. also found that the crystal melting point of PVA film scarcely changed with P_n but changed with tacticity.[13] Matsuoi et al. studied the structure of high molecular weight PVA film.[14] The solution properties of high molecular weight PVAs were investigated by Sanefuji et al.,[15] Lechner et al.,[16] and Yamaura et al.[15-18]

APPLICATIONS

Because high molecular weight PVA has the properties described above, it is a good material for the production of

*Author to whom correspondence should be addressed.

fibers, films (e.g., for pervaporation), gels (e.g., for artificial crystalline lens and muscle).

REFERENCES

1. Sakurada, I.; Ito, T.; Nakamae, K. *J. Polym. Sci.* **1966,** C15, 75.
2. Sakurada, I.; Yoshizaki, O. *Kohbunshi Kagaku* **1975,** *14,* 339.
3. Nozakura, S.; Morishima, Y.; Iimura, H.; Irie, Y. *J. Polym. Sci., Polym. Chem. Ed.* **1976,** *14,* 759.
4. Morishima, Y.; Kim, W.; Nozakura, S. *Polym. J.* **1976,** *8,* 196.
5. Imai, K.; Shiomi, T.; Tezuka, Y.; Kawashima, T.; Jin, T. *J. Polym. Sci.* **1988,** *26,* 1961.
6. Yamamoto, T.; Seki, S.; Hirota, M.; Kamachi, M. *Polym. J.* **1987,** *19,* 1417.
7. Yamamoto, T.; Seki, S.; Fukae, R.; Sangen, O.; Kamachi, M. *Polym. J.* **1990,** *22,* 567.
8. Yamamoto, T.; Yoda, S.; Sangen, O.; Fukae, R.; Kamachi, M. *Polym. J.* **1989,** *21,* 1053.
9. Yamamoto, T.; Yoda, S.; Takase, H.; Saso, T.; Sangen, O.; Fukae, R.; Kamachi, M.; Sato, T. *Polym. J.* **1991,** *23,* 185.
10. Kamachi, M.; Kajiwara, A.; Yamamoto, T. *The 34th IUPAC Int. Symp. on Macromolecules, Prague*: Prague **1992,** 1-P82.
11. Kajiwara, A.; Morishima, Y.; Kamachi, M. *Polym. Preprints, Japan* **1991,** *40,* 1554.
12. Fukae, R.; Kako, T.; Yamamoto, T.; Satake, S.; Sangen, O; Kamachi, M.; Kawakami, K. *Polym. Preprints, Japan* **1993,** *42,* 197.
13. Fukae, R.; Yamamoto, T.; Snagen, O.; Saso, T.; Kako, T.; Kamachi, M. *Polym. J.* **1990,** *22,* 636.
14. Matsuoi, M.; Harashina, Y. *Polymer Preprints, Japan* **1992,** *41,* 1223.
15. Sanefuji, T.; Oikawa, H.; Kajiya, K.; Shimamura, K.; Yokoyama, F. *Nippon Rheology Gakkaishi* **1989,** *17,* 94.
16. Lechner, M. D.; Mattern, R. *Angew. Macromol. Chem.* **1984,** 123/124, 45.
17. Yamaura, K.; Shindo, N.; Matsuzawa, S. *Colloid Polym. Sci.* **1981,** *259,* 1143.
18. Yamaura, K.; Daimoh, M.; Tanigami, T.; Matsuzawa, S. *J. Appl. Polym. Sci.* **1988,** *36,* 1707.

POLY(VINYL ALCOHOL) (High Strength and High Modulus)

Kazuo Yamura
Faculty of Textile Science and Technology
Shinshu University

Poly(vinyl alcohol) (PVA) with a planar zigzag structure, such as polyethylene (PE), has the potential to induce high modulus and high strength. The crystal modulus of PVA and the strength of perfectly oriented PVA fiber are so high, they are estimated to be 250–300 GPa and ~30 GPa, respectively.[1,2] To this day, despite various attempts the limited values have not been obtained experimentally for PVA. A Young's modulus of 120 GPa is the highest value reached so far. Matsuo and Sawatari produced ultradrawn, ultrahigh molecular weight PE films with a Young's modulus of 216 GPa at 20°C, nearly equal to the crystal lattice modulus of PE.[3] If PVA fibers and films with high modulus and high strength nearly equal to the limited values are produced, they are useful as high-performance materials in comparison with PE, as PVA has a higher melting point (220–250°C) than PE, which melts at 125–140°C.

In this article various attempts to investigate the high mechanical characteristics, the Young's modulus, and strength values obtained for PVA are shown.

RELATIONSHIP BETWEEN MECHANICAL PROPERTIES AND DRAW RATIO

The maximum Young's modulus, strength at break, and draw ratio are 115 GPa, 3.8 GPa, and 120, respectively.[4,5] They do not reach the maximum values of ultrahigh molecular weight PE. This is considered to be caused by intermolecular hydrogen bonding, which interferes with drawing.

REFERENCES

1. Sakurada, T.; Itoh, T.; Nakamae, K. *J. Polym. Sci.* **1966,** C15, 75.
2. Tashiro, K.; Kobayashi, M.; Tadokoro, H. *J. Macromol. Sci.* **1977,** *10,* 731.
3. Matsuo, M.; Sawatari, C. *Macromolecules* **1986,** *19,* 2036.
4. Hyon, S-H.; Ikada, Y. U.S. Patent 4 765 937, 1988.
5. Kunugi, T.; Kawasumi, T.; Itoh, T. *J. Appl. Polym. Sci.* **1990,** *40,* 2101.

POLY(VINYL ALCOHOL) FIBER (High Strength and High Modulus)

Suong-Hyu Hyon, Won-Ill Cha, and Yoshito Ikada
Research Center for Biomedical Engineering
Kyoto University

Totally aromatic polyamide (Aramid) fibers have been the most industrially produced of the high-strength, high-modulus fibers. As the Aramid fibers are too expensive to be widely applied, however, there is a large demand for the development of other less costly high-strength and high-modulus fibers.

Many attempts have been made to develop high-strength, high-modulus fibers from conventional polymers, such as polyethylene (PE), polypropylene (PP), polyoxymethylene (POM), and poly(vinyl alcohol) (PVA).[1-21] Among these nonrigid polymers, PP and POM have been found to produce fibers with relatively low moduli because the theoretically attainable modulus is low due to their inherent spiral chain structure. PE and PVA have the potential to yield high-strength and high-modulus fiber, however, as they have high theoretically attainable moduli because of their planar zigzag structure. As PVA has a melting temperature as high as 230°C in contrast to PE, which has a low melting temperature around 130°C, it seems possible that high-strength and high-modulus fibers comparable to Aramid can be fabricated from PVA.

PREPARATION

PVA powders were kindly supplied by Unitika Co., LTD. Unless otherwise specified, PVA with a DP of 5000 was used.

PROPERTIES

Gelation of PVA Solutions

Recently we discovered that a high water content, and high-strength, transparent PVA hydrogel with approximately double the strength and 1.5 times higher elongation than that of conventional PVA hydrogels could be obtained from a PVA solution prepared from mixed solvents consisting of water and a water-miscible organic solvent, such as DMSO.[22,23]

Gel Spinning of PVA

This article describes gel spinning of PVA from spinning dope consisting of an H_2O/DMSO mixture to produce highly drawable PVA fibers. PVA fibers spun from 80 wt% DMSO solution were found to have the highest draw ratio (40 times) when drawn at 160°C and 200°C.

The yield stress of the gel fiber obtained from 80 wt% DMSO solution is much higher than that for the gel fiber prepared from 100 wt% DMSO. Moreover, the tensile strength to break is 1.5 times greater in the 80 wt% DMSO solution than in the 100 wt% DMSO solution. The crystallinity of undrawn PVA fibers prepared from 80 wt% DMSO solution is higher than that of undrawn PVA fiber prepared from 100 wt% DMSO. This result indicates that mixtures of H_2O/DMSO are necessary to fabricate undrawn PVA fibers of high strength. The undrawn fibers obtained from the mixed solvents were quite transparent. The maximum strength and modulus of 2.8 and 64 GPa, respectively, were achieved after hot drawing the gel fibers spun at a coagulating temperature of –20°C. With a rise in coagulating temperature, the ultimate tensile strength decreased drastically. The strongest PVA fibers were obtained by hot drawing the dry-gel fibers coagulated at the lowest temperature.

A presumable mechanism for the formation of high-strength and high-modulus fibers can be proposed as follows. When a homogeneous solution is obtained by complete dissolution of PVA in a mixed solvent of water and DMSO at a high temperature between 120°C and 140°C, is cooled below room temperature in methanol, the PVA chains may instantly undergo reduction in mobility, resulting in the formation of small nuclei due to local chain aggregation through hydrogen bonding between the PVA segments. Consequently, the solution sets to a gel. Formation of this network gel structure may allow spinning to high drawing and hence high chain orientation along the fiber axis to yield high-strength and high-modulus fibers. On the other hand, conventional gel spinning, which uses dopes prepared from a single organic solvent, such as DMSO, will not make such high drawing possible, probably because of insufficient formation of the three-dimensional gel structure.

In conclusion, the PVA chains in H_2O/DMSO mixtures may expand to a high degree and can produce a gel structure with many homogeneously distributed crosslinks when the PVA solubility is reduced by lowering the solution temperature. The exceedingly high drawing, realized by the favorable gel structure, may also lead to the formation of a PVA crystalline structure with high crystallinity and large lamella size.

APPLICATIONS

The high-strength and high-modulus PVA fibers obtained in this study are applicable for tire cords of radial tires, bullet-proof jackets, motor belts, ropes used in ship mooring, tension members for optical fibers, asbestos-substitute fibers, reinforcing fibers for fiber reinforced plastics (FRP), and textiles for furniture.

REFERENCES

1. Smith, P.; Lemstra, P. J. *J. Mater. Sci.* **1980,** *15,* 505.
2. Smith, P.; Lemstra, P. J. *Colloid Polym. Sci.* **1980,** *258,* 891.
3. Zwijnenburg, A.; Pennings, A. J. *Colloid Polym. Sci.* **1976,** *254,* 818.
4. Furuhata, K.; Yokokawa, T.; Miyasaka, K. *J. Polym. Sci., Polym. Phys. Ed.* **1983,** *22,* 133.
5. Kanamoto, T.; Ohama, T.; Tanaka, K.; Takeda, M.; Porter, R. S. *Polymer* **1987,** *28,* 1517.
6. Wang, L. H.; Ottani, S.; Porter, R. S. *Polymer* **1991,** *32,* 1776.
7. Matsuo, M.; Sawatari, C.; Nakano, T. *Polym. J.* **1986,** *18,* 759.
8. Peguy, A.; St. John Manley, R. *Polym. Commun.* **1984,** *25,* 39.
9. Baranov, A. O.; Prut, E. V. *J. Appl. Polym. Sci.* **1992,** *44,* 1557.
10. Nakagawa, K.; Konaka, T.; Yamakawa, S. *Polymer* **1985,** *26,* 84.
11. Takeuchi, Y.; Yamamoto, T.; Nakagawa, K.; Yamakawa, S. *J. Polym. Sci., Polym. Phys. Ed.* **1985,** *23,* 1193.
12. Cebe, P.; Grubb, D. T. *J. Mater. Sci.* **1985,** *20,* 4465.
13. Garrett, P. D.; Grubb, D. T. *Polym. Commun.* **1988,** *29,* 60.
14. Kunugi, T.; Kawasumi, T.; Ito, T. *J. Appl. Polym. Sci.* **1990,** *40,* 2101.
15. Masuo, F.; Yamaoka, K.; Kamikawa, H.; Sato, E. Jpn. Patent 9 768, 1962.
16. Matsubayashi, K.; Segawa, H. Jpn. Patent 535 769, 1964.
17. Tanaka, H.; Suzuki, M.; Ueda, F. U.S. Patent 4 603 083, 1986.
18. Kwon, Y. D.; Kavesh, S.; Prevorsek, D. C. U.S. Patent 4 440 711, 1982.
19. Wu, T. C.; West, J. C. U.S. Patent 4 463 138, 1982.
20. Yamaura, K.; Tanigami, T.; Hayashi, N.; Kosuda, K.; Okuda, S.; Takemura, Y.; Itoh, M.; Matsuzawa, S. *J. Appl. Polym. Sci.* **1990,** *40,* 905.
21. Fujiwara, H.; Shibayama, M.; Chen, J. H.; Nomura, S. *J. Appl. Polym. Sci.* **1989,** *37,* 1403.
22. Hyon, S.-H.; Cha, W.-I.; Ikada, Y. *Polym. Bull.* **1989,** *22,* 119.
23. Cha, W.-I.; Hyon, S.-H.; Ikada, Y. *Makromol. Chem.* **1992,** *193,* 1913.

POLY(VINYL ALCOHOL) FIBERS (High Modulus; by Vibrating Zone-Drawing)

Toshio Kunugi
Department of Applied Chemistry and Biotechnology
Yamanashi University

The first zone-drawing/zone-annealing method was invented in 1979 as a method for the preparation of high-modulus and high-strength fibers or films.[1] Since then the method has been applied to a variety of crystalline polymers and improved stepwise. The improved methods resulted in the multistep zone-drawing/zone-annealing method proposed in 1982, the vertical two-step zone drawing method of 1985, the high-temperature zone-drawing method suggested in 1987, and the vibrating zone-drawing method developed in 1992.[2-5] So far these methods, as well as the extremely high-tension annealing method proposed by Kunugi et al. in 1991, have been applied to nylons 6 (25 GPa) and 66 (13 GPa), poly(ethylene terephthalate) (41 GPa), polyethylene (232 GPa), polypropylene (27 GPa), polyoxymethylene (75 GPa), poly(vinyl alcohol) (115 GPa), poly(ether ether ketone) (20 GPa), polyimide (24 GPa), and polyarylate (83 GPa) fibers or films.[6-33] The values indicated in parentheses are the maximum moduli attained by the application of these methods at room temperature.

APPLICATIONS

As poly(vinyl alcohol) has a zigzag planar molecular conformation, the crystal modulus of 250 GPa is very high, making it comparable to that of polyethylene. Further, because of its inter- and intramolecular hydrogen bonds it has a remarkably high melting point of 245°C, as compared with polyethylene. The attained maximum modulus of PVA fibers could not exceed 60 GPa for a long time, however although the 200–300-fold superdrawing of polyethylene could not be attained, the PVA fiber obtained in the present study indicated a high modulus close to 100 GPa, despite a low draw ratio of 20-fold. Also, properties such as hot-water resistance were improved. PVA has additional characteristics, such as low density, low cost, alkali resistance, and good adhesiveness. The obtained PVA fibers may have many applications similar to high-modulus polyethylene fiber, aramid, and polyarylate fibers. It could be used for ropes, cloths, nets, belts, tirecord, and fibers for reinforcement of composite materials such as thermoplastics or concrete.

REFERENCES

1. Kunugi, T.; Suzuki, A.; Akiyama, I.; Hashimoto, M. *Polym. Prepr. Am. Chem. Soc. Div. Polym. Chem.* **1979,** *20,* 778.
2. Kunugi, T.; Ikuta, T.; Hasimoto, M. *Polymer* **1982,** *23,* 1983.
3. Kunugi, T. *Polym. Prepr. Jpn.* **1985,** *34,* 1093.
4. Kunugi, T.; Suzuki, A. *Polym. Prepr. Jpn.* **1987,** *36,* 1115.
5. Kunugi, T. *Abstracts of Papers* 34th IUPAC Intern. Symp. Macrom., Prague, 1992; Abstract, 2-P89.
6. Kunugi, T. *Polymer* **1982,** *(Commun.) 23,* 176.
7. Kunugi, T.; Akiyama, I.; Hashimoto, M. *Polymer* **1982,** *23,* 1193.
8. Kunugi, T.; Akiyama, I.; Hashimoto, M. *Polymer* **1982,** *23,* 1199.
9. Kunugi, T.; Suzuki, A.; Kubota, M. *Kobun-shi Ronbunshu* **1992,** *49,* 161.
10. Kunugi, T.; Suzuki, A.; Chida, K. *J. Polym. Sci., Part B, Polym. Phys. Ed.* **1992,** *31,* 803.
11. Kunugi, T.; Suzuki, A. *Abstracts of Papers* 2nd Pacific Polymer Conference Otsu 1991; Preprints 339.
12. Suzuki, A.; Maruyama, S.; Kunugi, T. *Kobun-shi Ronbunshu* **1992,** *49,* 741.
13. Kunugi, T.; Suzuki, A.; Hashimoto, M. *J. Appl. Polym. Sci.* **1981,** *26,* 213.
14. Kunugi, T.; Suzuki, A.; Hasimoto, M. *J. Appl. Polym. Sci.* **1981,** *26,* 1951.
15. Suzuki, A.; Kobayashi, K.; Kunugi, T. *Kobun-shi Ronbunshu* **1994,** *51,* 5.
16. Kunugi, T.; Suzuki, A. *Abstracts of Papers* 1st International Conference on the Frontier of Polymer Research, New Delhi, 1991; Abstract, HP VIIL4.
17. Kunugi, T.; Suzuki, A.; Tsuiki, T. *Abstracts of Papers* 4th the Society of Polymer Science, Japan International Polymer Conference, Yokohama, 1992; Prepr. 1C02, 142.
18. Kunugi, T.; Aoki, I.; Hashimoto, M. *Kobun-shi Ronbunshu* **1981,** *38,* 301.
19. Kunugi, T.; Oomori, S.; Mikami, S. *Polymer* **1988,** *29,* 814.
20. Kunugi, T. *J. Polym. Sci., Polym. Lett.* **1982,** *20,* 329.
21. Kunugi, T.; Ito, T.; Hasimoto, M. *J. Appl. Polym. Sci.* **1983,** *28,* 179.
22. Kunugi, T.; Suzuki, T. *Polym. Prepr. Jpn.* **1987,** *36,* 1115.
23. Kunugi, T.; Suzuki, A.; Sone, Y. *Abstracts of Papers* Polymer Processing, Japan, 3rd Annual Meeting, Tokyo, 1991; D209, p 353.
24. Kunugi, T.; Kawasumi, T.; Ito, T. *J. Appl. Polym. Sci.* **1990,** *40,* 2101.
25. Kunugi, T.; Kawasumi, T. *Abstracts of Papers* 32nd IUPAC International Symposium Macrom., Preprints, Kyoto, 1988; *2.7.23,* p 431.
26. Kunugi, T.; Suzuki, A. *Abstracts of Papers* 3rd Pacific Polymer Conference Preprints, Gold Coast, 1993; 817.
27. Kunugi, T.; Mizushima, A.; Hayakawa, T. *Polymer (Commun.)* **1986,** *27,* 176.
28. Kunugi, T.; Hayakawa, T.; Mizushima, A. *Polymer* **1991,** *32,* 808.
29. Kunugi, T.; Suzuki, A.; Itoda, J. *Kobun-shi Ronbunshu* **1990,** *47,* 961.
30. Kunugi, T. *Abstracts of Papers* 1st Pacific Polymer Conference, Preprints, Maui, 1989; 95.
31. Kunugi, T. *Abstracts of Papers* 2nd the Society of Polymer Science, Japan International Polymer Conference Prepr., Tokyo, 1986; 11.
32. Kunugi, T.; Kawasumi, K. *Abstracts of Papers* Sen-i Gakkai, Showa 62 Nen Annual Meeting, Preprints, Tokyo, 1987; 3C02, 22.
33. Suzuki, A.; Kirihara, K.; Kunugi, T. *Kobun-shi Ronbunshu* **1993,** *50,* 583.

POLY(VINLY ALCOHOL)–IODINE INTERACTIONS

Toshio Yokota* and Yoshiharu Kimura
Department of Polymer Science and Engineering
Kyoto Institute of Technology

Iodine–polymer interactions have been studied extensively. Of these studies, the interaction with starch (amylose) and poly(vinyl alcohol) (PVA) to form blue-colored complexes are the most well known.[1] With regard to amylose–iodine, the formation of a polyiodide complex inside the helical chain of amylose was first proposed by Freudenberg in 1939.[2] This inclusion complex was substantiated by Rundle and Baldwin in 1944.[3] A similar coloration mechanism was also proposed for the blue-colored aqueous solution of PVA–iodine by such pioneer researchers as Zwick and Tebelev.[4,5] Zwick thought that the polyiodide ions may be formed in the host of helical PVA molecules, whereas Tebelev insisted that it is formed in the voids of the molecular aggregates of PVA.

These chromophoric polyiodide ions are generally represented by the formula of I^-_{2n+1}- where *n* denotes the number of iodine molecules in bonding. Regarding the PVA–iodine complex, Yokota and Kimura proposed a discrete structure of the chromophoric polyiodine species and the mechanism of its formation.[6-10]

FORMATION

The formation of the blue complex between PVA and I_2 is generally much slower than that between starch and iodine. For inducing the former, solid iodine is to be used together with salts such as KI, because iodine alone is less soluble and less reactive in aqueous PVA solution. Yokota found that the similar blue-colored complex can be formed at 5°C after mixing an aqueous ethanolic solution of iodine with an aqueous soltuion of PVA containing boric acid.[6] Generally, the formation of the chromophore is greatly activated in the presence of boric acid in the PVA–I_2 system.[11]

*Author to whom correspondence should be addressed.

STOICHIOMETRY OF IODINE/IODIDE IN THE "BOUND" SPECIES

The major species bound in the PVA cage are pentaiodide ions, that is, I_5^- with linear configuration and/or $I_2{\cdot}I_3^-$ with distorted chain structure. Concerning the resonance Raman spectrum of the PVA–iodine solution, it was concluded that the former species is the major species for the chromophoric "bound" species.[12]

FORMATION AND DECOMPOSITION MECHANISM OF CHROMOPHORE

A plausible mechanism for the decomposition of the polyiodide ions can be deduced. The polyiodide ion formed in the PVA matrix is a linear pentaiodide ion (I_5^-) that is stabilized by delocalization of the negative charge in a hydrophobic cavity, which is formed by aggregation of the PVA molecules. Its decomposition to I^- and two I_2 molecules is considered to proceed in the hydrophobic–hydrophilic boundary layer of the cavity by localization of the negative charge on one iodine atom pushed out in the hydrophilic environment. After the decomposition, I^- is released into the aqueous phase while the other fragments, I_2, settle in the original cavity. So, I_2 molecules do not scramble among the PVA cavities. In the equilibrium state, free I^- attacks the I_2 bound in the PVA matrix, and the formation and decomposition reactions of I_5^- compete with each other.

REFERENCES

1. Staudinger, H.; Frey, K.; Starck, W. *Hochmolekulare Verbb.* **1927,** *69,* 1782.
2. Freudenberg, K.; Schaaf, E.; Dumpert, G.; Ploetz, T. *Naturwissenschaften* **1939,** *27,* 850.
3. Rundle, R. E.; Baldwin, R. R. *J. Am. Chem. Soc.* **1944,** *66,* 2116.
4. Zwick, M. M. *J. Polym. Sci.* **1965,** *9,* 2392.
5. Tebelev, L. G.; Mikulskii, G. F.; Korchagina, E. P.; Glikman, S. G. *Vysokomol. Soedin.* **1965,** *7,* 132.
6. Yokota, T.; Kimura, Y. *Makromol. Chem.* **1984,** *185,* 749.
7. Yokota, T.; Kimura, Y. *Makromol. Chem.* **1985,** *186,* 549.
8. Yokota, T.; Kimura, Y. *Makromol. Chem., Rapid Commun.* **1986,** *7,* 249.
9. Yokota, T.; Kimura, Y. *Makromol. Chem.* **1989,** *190,* 939.
10. Yokota, T.; Kimura, Y. *Makromol. Chem.* **1993,** *194,* 295.
11. West, C. D.; *J. Chem. Phys.* **1951,** *19,* 1432.
12. Inagaki, F.; Harada, I.; Shimanouchi, T.; Tasumi, M. *Bull. Chem. Soc. Jpn.* **1972,** *45,* 3384.

POLY(VINYL ALCOHOL)–ION COMPLEX GELS

Mitsuhiro Shibayama
Department of Polymer Science and Engineering
Kyoto Institute of Technology

Some water-soluble polymers undergo ion-complexation with inorganic ions. Typical examples are poly(vinyl alcohol) (PVA) with borate ions ($B(OH)_4^-$), polysaccharide with borate ions, polyvinylpyridine with Cu^{2+}, Fe^{3+}, or Co^{2+}, and PVA with Cu^{2+}.[1-11] The complexation depends on pH, ion concentration, temperature, and other environmental variables. Because the complexation plays a dramatic role in its solution properties, like gelation and viscosity thickening, numerous industrial applications have been developed.[12]

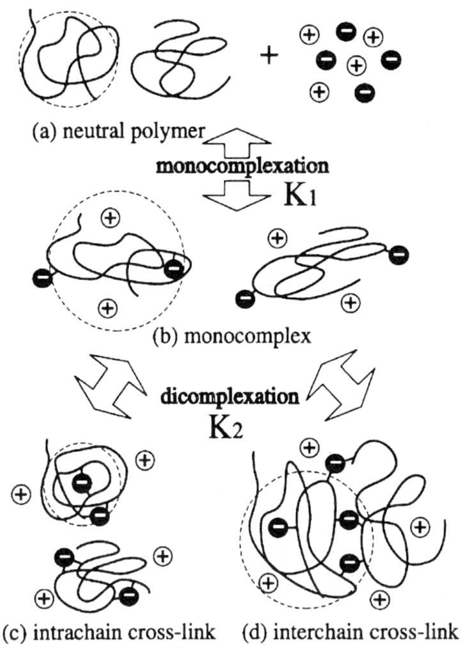

FIGURE 1. Scheme of polymer–ion complexation. Polymer chains (drawn with solid curves) are complexed with negatively charged ions ⊖, solid circles with minus sign; and monocomplexes (b), or dicomplexes (c) and (d), are formed. K_1 and K_2 are the complexation constants.

Figure 1 illustrates a scheme of polymer–ion complexation consisting of polymer chains and negatively charged ions. Without ions, polymer chains behave as neutral polymer chains and have their proper chain dimension, as illustrated by a dashed circle in Figure 1a. By adding ions, a complexation occurs, as shown in Figure 1b. This is a one-to-one type complexation, that is, monocomplexation, which occurs between a monomer unit of polymer chain, and one of the added ions, and the complexed polymer chain, which behaves as a polyelectrolyte. As a result, the polymer chain expands with respect to the neutral state. If the monocomplex site is capable of another complexation with another monomer unit of a polymer chain, however, a dicomplex is formed. This leads to an intrachain crosslinking as seen in Figure 1c and/or interchain crosslinking as in Figure 1d. The intrachain crosslink results in a chain contraction or precipitation. On the other hand, the interchain crosslinking causes a clustering or gelation of the polymer solution. Thus, these complexations provide various aspects of the physical properties of the complexed polymer solutions.

PVA–BORATE COMPLEXES

Sol–Gel Transitions

Borax ($Na_2B_4O_7{\cdot}10H_2O$) has long been used as a gelling agent of PVA aqueous solutions. Its application to fiber manufacturing has a particularly long history.[13] PVA fibers with a uniform cross-section were successfully produced and then the technique was modified to produce high-strength PVA fibers.[14] The structure of the PVA–borate ion complex has been investigated by several groups of researchers with ^{11}B-NMR.[5,6,15,16]

Viscosity

Viscosity behaviors of PVA–borate ion aqueous solutions were studied by Ochiai et al.[17] They found that the intrinsic viscosity, [η], of PVA solutions in the presence of borate ions oscillated when a small amount of passive salt was added. Leibler et al. successfully interpreted this intriguing viscosity behavior with a Flory type mean-field theory.[18]

Small-Angle Neutron Scattering

The microscopic structure of the PVA–borate ion complexes changes on the sol–gel transition. Small-angle neutron scattering (SANS) is a sensitive tool to detect the transition because a high contrast between the solute (PVA chains) and the solvent (deuterated water) can be obtained.

PVA–VANADATE GELS

PVA–borate ion complex gels are exclusively formed in a basic condition. In an acidic condition, vanadate ions are known to be a crosslinking agent of PVA chains.[19,20] In the case of PVA–vanadate ion complexes, the gelation mechanism is rather complicated because the orthovanadate ion in aqueous media undergoes complex hydrolysis–polymerization reactions on neutralization with acids and produces ions such as $V_2O_7^{4-}$, $HV_2O_7^{3-}$, $V_{10}O_{28}^{6-}$, $HV_{10}O_{28}^{5-}$, $H_2V_{10}O_{28}^{4-}$, etc.[21,22] These ions are exclusively formed in an aqueous solution by varying the vanadate ion concentration and the pH. However, it was found that only $H_2V_{10}O_{28}^{4-}$ ions, present around pH 2–3, are responsible for the gelation of PVA aqueous solutions.[23]

PVA–CONGO RED COMPLEXES

Different from the PVA–ion complexes discussed previously, Congo red is a crosslink agent that can be used in neutral or mild basic conditions.[24,25]

Sol–Gel Transition

PVA–CR complex solutions exhibit a unique sol–gel transition curve. The unique feature of the sol–gel transition behavior in the PVA–CR complex is caused by strong electrostatic interaction, where bound ions interact effectively with each other and deform the PVA chains. This aspect is more clearly depicted by viscometry of dilute solutions of the complex.

CONCLUDING REMARKS

We have shown several aspects of PVA–ion complexes from rheological, spectroscopic, thermodynamic, and structural points of view. The individual size of a polymer chain is strongly influenced by complexation. Monocomplexation expands the polymer chain, whereas dicomplexation leads to a contraction or an expansion. When these changes occur on a polymer chain they can be manifested macroscopically as swelling/shrinking phenomena of the corresponding polymer gels. The complexation, however, is highly affected by environmental variables, such as temperature, ionic strength, pH, etc. A strong acidic (or basic) environment or an addition of salt shields the electrostatic interaction between charged ions on a chain (electrostatic screening effect). The experimental evidence shown here clearly indicates a close connection among the rheological, thermodynamic, and microscopic structures of polymer–ion complexes. The further understanding of these polymer–ion complexes creates a new field in the science of polymer–ion complexes, where the molecular interactions play significant roles in sophisticated functions, namely, molecular recognition, motion, replication, etc.

ACKNOWLEDGMENT

This work is supported in part by the Ministry of Education, Science and Culture, Japan (Grant-in-Aid, No. 04805092, 05805080, 06651053, 07241242, and 08231245).

REFERENCES

1. Deuel, H.; Neukom, A. *Makromol. Chem.* **1949,** *3,* 13.
2. Saito, S.; Okuyama, H.; Kishimoto, H.; Fujiyama, Y. *Kolloid-Z.* **1955,** *144,* 41.
3. Lorand, J. P.; Edwards, J. O. *J. Org. Chem.* **1959,** *24,* 769.
4. Pezron, E.; Ricard, A.; Lafuma, F.; Audebert, R. *Macromolecules* **1988,** *21,* 1126.
5. Pezron, E.; Leibler, L.; Ricard, A.; Lafuma, F.; Audebert, R. *Macromolecules* **1989,** *22,* 1169.
6. Pezron, R.; Leibler, L.; Lafuma, F. *Macromolecules* **1989,** *22,* 2656.
7. Pezron, E.; Ricard, A.; Leibler, L. *J. Polym. Sci., Polym. Phys. Ed.* **1990,** *28,* 2445.
8. Tsuchida, E.; Nishikawa, H. *Adv. Polym. Sci.* **1977,** *24,* 1.
9. Saito, S.; Okuyama, H. *Kolloid-Z.* **1974,** *139,* 150.
10. Yokoi, H.; Kawata, S.; Iwaizumi, M. *J. Am. Chem. Soc.* **1986,** *108,* 3358.
11. Narisawa, M.; Ono, K.; Murakami, K. *Polymer* **1989,** *30,* 1540.
12. Menjivar, J. A. *Water Soluble Polymers*, Glass, L., Ed.; American Chemical Society: Washington, DC, p 209.
13. Arakawa, M. *Sen-i Gakkaishi (Jpn.)* **1960,** *16,* 849.
14. Fujiwara, H.; Shibayama, M.; Chen, J.; Nomura, S. *J. Appl. Polym. Sci.* **1989,** *37,* 1403.
15. Sinton, S. W. *Macromolecules* **1987,** *20,* 2430.
16. Shibayama, M.; Sato, M.; Kimura, Y.; Fujiwara, K.; Nomura, S. *Polymer* **1988,** *29,* 336.
17. Ochiai, H.; Kurita, Y.; Murakami, I. *Makromol. Chem.* **1984,** *185,* 167.
18. Leibler, L.; Pezron, E.; Pincus, P. A. *Polymer* **1988,** *29,* 1105.
19. Signaigo, F. K. U.S. Patent 2 518 193, 1986.
20. Kawakami, H.; Fujiwara, H.; Kinoshita, Y.; Jpn. Patent S47-40894, 1972.
21. Shiller, K.; Thilo, E. *Anorg. Z. Anorg. Allg. Chem.* **1961,** *261,* 310.
22. Murmann, K. L.; Giese, K. C. *Inorg. Chem.* **1978,** *17,* 1160.
23. Shibayama, M.; Adachi, A.; Ikkai, F.; Kurokawa, H.; Sakurai, S.; Nomura, S. *Macromolecules* **1993,** *26,* 623.
24. Beltman, H.; Lyklema, J. *Farady Disc. Chem. Soc.* **1974,** *57,* 92.
25. Shibayama, M.; Ikkai, F.; Moriwaki, R.; Nomura, S. *Macromolecules* **1994,** *27,* 1738.

Poly(vinyl chloride)

See: Foamed Plastics
Head-to-Head Polymers
Photocrosslinking (Overview)

Poly(vinyl chloride) (Dielectric Properties)
Poly(vinyl chloride) (Low Molecular Weight)
Poly(vinyl chloride) (Mechanisms of Stabilization)
Poly(vinyl chloride) (Structural Order)
Poly(vinyl chloride), Rigid (Soft Fillers)
Poly(vinyl chloride), Soft (Calcium Carbonate Filled)
Powder Coatings (Overview)
Precipitation Polymerization (Overview)
Vinyl Chloride (Minisuspension Polymerization)
Vinyl Chloride Copolymers

POLY(VINYL CHLORIDE) (Dielectric Properties)

P. A. M. Steeman*
DSM Research BV

J. Van Turnhout
TNO Plastics and Rubber Institute

Similar to the well-known mechanical relaxation processes, most (polar) polymers also exhibit relaxations in their electrical polarization when subjected to an electrical field. Dielectric relaxation phenomena are studied with dielectric spectroscopy. Poly(vinyl chloride) (PVC) is a polar polymer because of the presence of the polar carbon–chloride bond. Therefore, dielectric spectroscopy has been widely used to complement the mechanical relaxation data of PVC by investigating the pronounced dipolar polarization processes.[1-15]

PROPERTIES

The dipolar polarization processes in rigid poly(vinyl chloride) have been studied by many authors.[1-15] Detailed reviews are given by McCrum[1] et al. and by Hedvig.[2] Additionally, studies have been performed on substituted (hydrogenated or chlorinated) PVC.[5,6,10,15] Often the material is plasticized thus changing it from a brittle to a tough and flexible material. Plasticizing has a big effect on the dielectric properties because it shifts the glass transition to lower temperatures, sometimes even leading to peak splitting as a consequence of the demixing of polymer and plasticizer.[1,2,6,10] Moreover, the dielectric properties of PVC in solution, of copolymers, and of blends with other polymers, have been studied in detail.[16-22] Less attention has been paid to the dielectric properties in the rubbery state, where the dielectric losses from polarization are often overshadowed by strong conduction losses.[7,8,23]

Dipolar Relaxation Mechanisms in the Glassy State

In the dielectric loss curves, two relaxation processes are clearly observed. At –70°C(0.1 Hz) to 25°C(10 kHz) the weak and broad β-process, a secondary transition in the glassy state, is detected. At 75°C(0.1 Hz) to 100°C(10 kHz) the strong and sharp α-relaxation, which is related to the glass–rubber transition of the material, is found. Finally, at temperatures above the glass–rubber transition the material becomes electrically conductive, resulting in a sharp increase of the dielectric loss, especially at the lowest frequencies.

*Author to whom correspondence should be addressed.

Relaxation Mechanisms in the Rubbery State

In studying the effect of molar mass on the dielectric properties of PVC, a marked effect was observed in the rubbery state of the high molar mass samples: the occurrence of a *space–charge* relaxation.[23]

APPLICATIONS

PVC is widely used in a variety of electrical low-voltage applications, particularly for consumer-grade products. The best known are wire covering, insulation sleeving (including shrinking sleeves), cable sheathing, and insulating tapes. In most applications plasticized PVC or copolymers of vinyl chloride and vinyl acetate are used in order to obtain flexibility. Often (inorganic) fillers and pigments are added. PVC-based electrical insulation has the advantage of good chemical resistance; furthermore, it is self-extinguishing. The temperature limit of operation is about 60°C. It had a large market in gramophone records.

REFERENCES

1. McCrum, N. G.; Read, B. E.; Williams, G. *Anelastic and Dielectric Effects in Polymeric Solids*; Wiley: London, 1967.
2. Hedvig, P. *Dielectric Spectroscopy of Polymers*; Adam Hilger: Bristol, 1977.
3. Ishida, Y. *Kolloid-Z.* **1960,** *168,* 29.
4. Ishida, Y. *J. Polym. Sci.* **1969,** *1,* 1835.
5. Reddish, W. *J. Polym. Sci.* **1966,** C14, 123-137.
6. Kakutani, H.; Asahina, M.; *J. Polym. Sci.* **1969,** Part A2, *7,* 1473-1481.
7. Dufour, J.; Asch, G.; Pillonnet, A. *C. R. Acad. Sci. B* **1973,** *276,* 111-113.
8. Kopac, Z.; Schworer, U.; Stoll, B. *Colloid Polym. Sci.* **1976,** *254,* 957-963.
9. Kopac, Z.; Schworer, U.; Stoll, B. *Colloid Polym. Sci.* **1976,** *254,* 1053-1058.
10. Utracki, L. W.; Jukes, J. A. *J. Vinyl Technol.* **1984,** *6,* 85-93.
11. Naoki, M. *J. Chem. Phys.* **1989,** *91,* 5030-5036.
12. del Val, J. J.; Alegria, A.; Colmenero, J. *Makromol. Chem.* **1989,** *190,* 3257-3267.
13. del Val, J. J.; Colmenero, J.; Mijangos, C.; Millan, J. L. *Makromol. Chem.* **1989,** *190,* 893-905.
14. Havriliak, S. Jr.; Shortridge, T. J. *Macromolecules* **1990,** *23,* 648-653.
15. del Val, J. J.; Colmenero, J.; Mijangos, C.; Millan, J. L. *J. Polym. Sci., Polym. Phys. Ed.* **1994,** *32,* 871-880.
16. Salem, M. B.; Oun, A. M.; Omar, M. M. *Polym. Int.* **1991,** *26,* 209-218.
17. Singh, R.; Panwar, V. S.; Tandon, R. P.; Gupta, N. P.; Chandra, S. *J. Appl. Phys.* **1992,** *72,* 3410-3416.
18. Saito, S.; Nakjima, T. *J. Polym. Sci.* **1959,** *37,* 229.
19. Malik, T. M.; Prud'Homme, R. E. *Polym. Eng. Sci.* **1984,** *24,* 144-152.
20. Rellick, G. S.; Runt, J. *J. Polym. Sci., Polym. Phys. Ed.* **1986,** *24,* 279-302.
21. Rellick, G. S.; Runt, J. *J. Polym. Sci., Polym. Phys. Ed.* **1986,** *24,* 313-324.
22. Rellick, G. S.; Runt, J. *J. Polym. Sci., Polym. Phys. Ed.* **1988,** *26,* 1425-1438.
23. Steeman, P. A. M.; Gondard, C.; Scherrenberg, R. L. *J. Polym. Sci., Polym. Phys. Ed.* **1994,** *32,* 119-130.

POLY(VINYL CHLORIDE) (Low Molecular Weight)

Maria Obloj-Muzaj
Industrial Chemistry Research Institute

Roman Poturalski and Jan Zielinski
Nitrogen (Petrochemical) Works "Wloclawek"

There are various ways of improving the flow properties of PVC that vary in outcome and economic consequences, such as using flow modifiers, easy-flow polymer components, lubricants, and other agents in PVC processing; copolymerizing vinyl chloride with comonomers like vinyl acetate or propylene; processing at higher temperatures as long as PVC thermal stability is sufficiently high; and reducing the polymer molecular weight by high-temperature polymerization or by using chain length regulators.[1]

Many factors impede these ways such as the regulations limiting the amounts of some additives to be used for diverse applications, especially for medical purposes or for food contact, high cost of some additives, usually low thermal stability of copolymers, for example, with vinyl acetate, and higher pressures occurring during the polymerizations run at higher temperatures.[2]

Reduction of the molecular weight of PVC by the use of chain regulators seeemed to be of particular interest.

Few companies produce low molecular weight PVC, namely BASF–Germany (Vinoflex S 5015), Solvay–Belgium (Solvic 250 SA), Lucky–South Korea (LS-050), LVM–The Netherlands (Marvylan S-5002), and Nitrogen (Petrochemical) Works "Wloclawek" in Poland (Polanvil S-52 HF).

Their viscosity index is usually about 61 mg/g, what corresponds with melt flow index till about 200 g/10 min (at 463 K = 190°C, 21.6 kG); K-value about 50; bulk density in the range 0.52 (Solvic 250 SA) till 0.60 g/ccm (Polanvil S-52 HF); sieve analysis: minimum 85% above 0.063 (Sovic 250 SA), maximum 2% above 0.250 mm (Marvylan S-5002).

PREPARATION AND PROPERTIES

A high-flowability poly(vinyl chloride) also specially designed for injection molding for blow molding, extrusion, and calendering of rigid goods was developed at the Industrial Chemistry Research Institute (ICRI). It was prepared in the suspension polymerization process, using typical suspension stabilizers and radical initiators in the presence of chain length regulator. But polymerization has been specially controlled for building a very regular chain "head-to-tail." It was important to avoid the deterioration of PVC thermal stability.

APPLICATIONS

Processing of high-flow PVC is far easier than processing PVC grades with higher molecular weight. The surface quality of goods, fittings, tubes, bottles, and syringes was excellent, better than from other PVC grades.

On the basis of presentations and other investigations we can state that application of poly(vinyl chloride) with high flowability and high thermal stability for rigid goods has many advantages from the economical, ecological, and processing points of view: better yield of extrusion and injection molding; lower load on processing equipment–torques, injection pressures; very high glossy surface of goods; savings on additives like flow modifiers, lubricants, and thermal stabilizers (very important from the economic point of view); processing at lower temperatures (by 10–15°C); new applications; and the possibility of using environmentally more friendly stabilizers, free of heavy metals.

ACKNOWLEDGMENTS

The authors wish to thank their colleagues and coworkers, especially Krzysztof Kaczorowski, M. Sc. and Krystyna Lozowska, M. Sc. of the Nitrogen Works "Wloclawek," and Maria B. Wojtowicz M. Sc., Janusz Kloze, and Wanda Luniewska of the Industrial Chemistry Research Institute for valued cooperation and scientific or technical assistance.

REFERENCES

1. Ehrhardt, H.; Weber, Z. *Polimery* **1984,** 29(10-12), 480.
2. Obloj-Muzaj, M. *Polimery* **1991,** 36(7-9), 297.

POLY(VINYL CHLORIDE) (Mechanisms of Stabilization)

W. H. Starnes, Jr.
Applied Science Ph.D. Program
Department of Chemistry
College of William and Mary

The intrinsic molecular stability of PVC toward heat and light is much less than that of most polymers that are of commercial interest.

The initial instability of PVC is manifested primarily by a sequential dehydrochlorination process that forms conjugated polyene sequences (**Equation 1**). All of the evidence now available points to an ionic or quasi-ionic (polar concerted) mechanism for thermal polyene growth.[10]

Polyene sequences impart undesirable color to the polymer, even when the level of HCl loss is very low. They also undergo various thermal and photochemical transformations themselves, and both they and the undegraded parts of the polymer can experience air oxidation. Moreover, during mechanical processing that is performed without good resin lubrication, PVC undergoes C–C scission to form carbon-centered radicals that can initiate the oxidation chain reaction.[3]

$$-\!(\mathrm{CHClCH_2})_n\!- \;\rightarrow\; -\!(\mathrm{CH{=}CH})_m\!-\!(\mathrm{CHClCH_2})_{n-m}\!- + m\mathrm{HCl} \qquad 1$$

As a result of all of these destructive processes, which may occur separately or in concert, depending upon conditions, the polymer experiences crosslinking and/or chain scission and eventually loses the properties that make it a useful material. Hence it requires effective protection, either by the inclusion of additives that impart stability, or in other ways.

PREVENTION OF AIR OXIDATION AND PHOTODEGRADATION

Various common types of antioxidants and photostabilizers are able to shield PVC from the action of molecular oxygen and solar radiation. Small amounts of phenolic antioxidants can be incorporated into commercial stabilizer packages, but antioxidant activity also may be provided by additives that are introduced primarily in order to prevent reaction 1. The latter materials include organic phosphites,[1,5-7,11] which may scavenge oxygen-containing radicals or decompose peroxides that are autoxidation initiators. Organotin thiolates (mercaptides), which are very effective as heat stabilizers, also seem to have more than one function, in that they may be either peroxide decomposers themselves or serve as sources of thiols and thioethers that destroy peroxides or scavenge radicals that are autoxidation intermediates.[2,3,5,7,11]

Photostabilizers for polymers generally can be subsumed under one of four major categories: UV absorbers such as organic compounds, carbon black, titanium dioxide, and other pigments that can dissipate the absorbed radiant energy in innocuous ways; quenchers such as Ni(II) chelates, which deactivate the excited states of chromophores that can serve as photoinitiating species; radical scavengers such as the hindered amine light stabilizers (HALS), which trap the chain-carrying intermediates that are involved in photooxidation; and peroxide decomposers, which remove potential photooxidation initiators.

THERMAL STABILIZERS AND THEIR POSSIBLE MECHANISMS OF ACTION

Most of the thermal stabilizers for PVC are bases that, though weak, are able to scavenge the degradation catalyst, HCl. This might be the only significant function of some commonly used stabilizers such as certain complex salts of Pb(II), for example.[11]

A very important property of most, if not all, of the best stabilizers and stabilizer mixtures is their ability to replace labile chloride with groups that are less susceptible to removal by thermolysis. Such stabilizers frequently are organic metal salts that can be represented by the formula MY_2, where M is a metallic species such as R_2Sn^{+2}(R = alkyl), Ba^{+2}, Cd^{+2}, Ca^{+2}, Zn^{+2}, or Pb^{+2}, and Y is an organic anion such as thiolate or carboxylate.[1,2,5-7,11,12]

STABILIZATION BY ORGANIC ADDITIVES

Owing to their relatively low activities, nonmetallic organic compounds ordinarily cannot be used as primary heat stabilizers. They can, however, contribute significantly to the effectiveness of many stabilizer packages containing metal constituents. Several organic additives either have been used commercially or have been suggested for use in stabilizer blends. They include, inter alia, β-amino-crotonates, phosphites, epoxides, *N,N'*-diphenylthiourea, 2-phenylindole, β-diketones, β-ketocarboxylic esters, polyols, and dihydropyridines.[1,2,4-7,11]

STABILIZATION BY CHEMICAL MODIFICATION

If all labile structures were removed before processing and fabrication, articles made from PVC should have much greater intrinsic stability, and the need to stabilize them with additives thus should be less compelling. Hence many attempts to realize this elusive goal have been made. Early work on this problem was surveyed some time ago, and the improved stability that results from organotin pretreatment has been noted.[8] Thiols also are effective as chemical pretreatment agents, though they are not as helpful in this regard as the best organotins.[9]

REFERENCES

1. Andreas, H. *Plastics Additives Handbook*; 3rd ed.; Gächter, R.; Müller, H., Eds.; Hanser: New York, NY, 1990; Chapter 4.
2. Braun, D.; Bezdadea, E. *Encyclopedia of PVC*, 2nd ed.; Nass, L. I.; Heiberger, C. A., Eds.; Marcel Dekker: New York, NY, 1986; Vol. 1, Chapter 8.
3. Cooray, B. B.; Scott, G. *Dev. Polym. Stab.* **1980,** *2,* 53.
4. Guyot, A.; Michel, A. *Dev. Polym. Stab.* **1980,** *2,* 89.
5. Iván, B.; Kelen, T.; Tüdós, F. *Degradation and Stabilisation of Polymers;* Jellinek, H. H. G.; Kachi, H., Eds.; Elsevier: New York, NY, 1989a, Vol. 2, Chapter 8.
6. Minsker, K. S.; Kolesov, S. V.; Zaikov, G. E. *Degradation and Stabilization of Vinyl Chloride-Based Polymers*; Pergamon: New York, NY, 1988.
7. Owen, E. D. *Degradation and Stabilisation of PVC*; Owen, E. D., Ed.; Elsevier Applied Science: New York, NY, 1984; Chapter 5.
8. Starnes, Jr., W. H.; Plitz, I. M. *Macromolecules* **1976,** *9, 633,* 878.
9. Starnes, Jr., W. H.; Plitz, I. M.; Hische, D. C.; Freed, D. J.; Schilling, F. C.; Schilling, M. L. *Macromolecules* **1978,** *11,* 373.
10. Starnes, Jr., W. H.; Girois, S. *Polymer Yearbook* **1995,** *12,* 105.
11. Wypych, J. *Polyvinyl Chloride Stabilization*; Elsevier: New York, NY, 1986.
12. Yassin, A. A.; Sabaa, M. W. *J. Macromol. Sci., Rev. Macromol. Chem. Phys.* **1990,** *30,* 491.

POLY(VINYL CHLORIDE) (Structural Order)

Christian Gondard,* Rolf Scherrenberg, and Harry Reynaers
PVC Basic Research
Limburgse Vinyl Maatschappij N.V. H Hartlaan

The structural order in PVC, as manifested by a low degree of crystallinity, is of great importance for its mechanical as well as for its rheological properties. PVC exhibits relatively good mechanical properties regarding its molecular weight in the absence of a rigid skeleton or any steric hindrance to prevent molecular motions and flexibility. Therefore, one is forced to admit that special interactions and/or structural order are responsible for its mechanical and rheological behavior.

The long debate about the definition of the ordered structure in PVC is due to apparent deviations between the theoretical approach of crystallinity and results obtained by physical-analytical methods and the real mechanical and rheological properties.

Commercial PVC exhibits a low level of defects in chemical structure, such as short-chain branching and unsaturation. It has been suggested that the level of short chain-branching is in the range of 1 to 4 per thousand carbon atoms.[1-6] The overall level of unsaturation is assumed to be in the range of one double bond per thousand monomers, predominantly present at the chain end.[7-11] The content of internal head-to-head structure is very low as well and does not exceed 0.2 per 1000 monomer

*Author to whom correspondence should be addressed.

units.[12,13] Free radical polymerization of commercial PVC leads to a polymer with a quite perfect chain, which should favor the development of structural order.

The main debate about the microstructure of PVC focuses on the determination of the degree of crystallinity and the type of crystalline morphology.[14-16]

CRYSTALLINITY AND POLYMERIZATION CONDITIONS

The syndiotacticity (as well as the molecular weight) of PVC resin increases on lowering the polymerization temperature. Nevertheless for commercial PVC this range of variation is quite low. The difference in percentage of syndiotactic triads between the lowest and the highest industrial polymerization temperatures does not exceed 3%. But surprisingly, the WAXS crystallinity index for virgin powder and processed PVC resins as a function of their polymerization temperature, is almost 2.5 times higher for the highest K-value (low polymerization temperature) than for the lowest K-value (high polymerization temperature).[17] This observation can be understood by an increase of perfection and size of ordered structure, as detected by WAXS, rather than by the creation of ordered structure, owing to a 3% increase of syndiotactic triads.

According to our evidence, the build up of crystallinity is a continuous phenomenon occurring during the whole polymerization process and not, as often suggested, mainly at the earliest stages of polymerization when the so-called microdomains are formed.[17-19] The crystallinity index is almost constant for conversion levels between 9 and 90%.

PVC STRUCTURAL ORDER AND THERMAL TREATMENT

PVC powders exhibit no reproducible DSC thermogram on the first heating run. The thermograms of the compression-molded PVC-materials are more reproducible. Generally, thermograms of PVC processed in this way exhibit two different endotherms.[20-22] The first endotherm has been attributed to the melting of the crystalline fraction that has initially been melted during the processing operation, whereas the second one, situated at about 20°C above the processing temperature, has been associated with the original crystalline fraction that did not undergo melting during processing. From this point of view, the ratio between endotherms has been suggested to be a good estimate for the level of gelation of the processed sample. Hence, this method is often used to measure the degree of gelation as a function of processing conditions.[23-32] Gelation has to be understood as the disappearance of the original grain structure.

In spite of the lack of direct information about the amount of ordered structure at the processing temperature, rheology seems to prove the existence of remaining superstructure, which prevents PVC to exhibit newtonian flow at the processing temperature (190–200°C).

CRYSTALLINITY AND PLASTICIZED PVC

Because of the porous structure of its grain and its microstructure, PVC is used in a very broad application domain ranging from rubberlike material when blended with plasticizer, to rigid material. Non polymeric plasticizers of PVC are esters such as phthalates, (bis(2-ethylhexyl)phthalate (DOP) being the most extensively used), mellitate, adipate, and phosphite. Non polymeric plasticizers of PVC do not belong to solvents or even the best swelling agents of PVC.[17] The presence of unplasticized regions in plasticized PVC has been demonstrated by means of electron microscopy.[33,34] These unplasticized regions were attributed to ordered structure.[33,35,36]

REFERENCES

1. Bovey, F. A.; Schilling, F. C.; Starnes, Jr., W. H. *Macromolecules* **1975,** *8,* 437.
2. Bowmer, T. N.; Ho, S. Y.; O'Donnell, J. H.; Park, G. S.; Saleem, M. *Eur. Polym. J.* **1982,** *18,* 61.
3. Liebman, S. A.; Ahlstrom, D. H.; Starnes, Jr., W. H.; Schilling, F. C. *J. of Macromol. Sci.–Chem.* **1982,** A17(6), 935.
4. Hjertberg, T.; Sörvik, E. M. *Polymer* **1983,** *24,* 673.
5. Starnes, Jr., W. H.; Schilling, F. C.; Plitz, I. M.; Cais, R. E.; Freed, D. J.; Hartless, R. L.; Bovey, F. A. *Macromolecules* **1983,** *16,* 790.
6. Park, G. S. *J. of Vinyl Technology* **1985,** 7(2), 60.
7. Michel, A.; Schmidt, G.; Castaneda, E.; Guyot, A. *Die Angewandte Makromol. Chem.* **1975,** *47,* 61.
8. Hildenbrand, P.; Ahrens, W.; Brandstetter, F.; Simck, S. *J. of Macromol. Sci.–Chem.* **1982,** A17, 1093.
9. Hjertberg, T.; Sörvik, E. M. *J. of Macromol. Sci.–Chem.* **1982,** A17(6), 983.
10. Petiaud, R.; Pham, Q. T. *Makromol. Chem.* **1989,** *190,* 893.
11. Hjertberg, T.; Sörvik, E. M. *Polymer* **1983,** *24,* 685.
12. Starnes, Jr., W. H. *Pure & Appl. Chem.* **1985,** *57(7),* 1001.
13. Törnell, B. *Handbook of Polymer Sci. and Tech.;* Marcel Dekker, 1989; Vol. 1, p 347.
14. Hobson, R. J.; Windle, A. H. *Makromol. Chem. Theory Simul.* **1993,** *2,* 257.
15. Ballard, D. G. H.; Burgess, A. N.; Dekoninck, J. M.; Roberts, E. A. *Polymer* **1987,** *28,* 3.
16. Gilbert, M. *J. of Macromol. Sci.–Rev. in Macromol. Chem.* **1994,** C34(1), 77.
17. Scherrenberg, R. L.; Reynaers, H.; Gondard, C.; Booij, M. *Polym. Sci.; Part B.: Polym. Phys.* **1994,** *32,* 99.
18. Robinson, M. E. R.; Bower, D. I.; Allsopp, M. W.; Willis, H. A.; Zichy, V. *Polymer* **1978,** *19,* 1225.
19. Biais, R.; Geny, C.; Mordini, C.; Carrega, M. *The British Polym. J.* **1980,** December, 179.
20. Gray, A.; Gilbert, M. *Polymer* **1976,** *17,* 44.
21. Gilbert, M.; Vyvoda, J. C. *Polymer* **1981,** *22,* 1135.
22. Dawson, P. C.; Gilbert, M.; Maddams, W. F. *J. of Polym. Sci.: Part B: Polym. Phys.* **1991,** *29,* 1407.
23. Gilbert, M.; Ansari, K. E. *J. Appl. Polym. Sci.* **1982,** *27,* 2553.
24. Gilbert, M.; Hemsley, D. A.; Miadonye, A. *Plastics and Rubber Process. and Appl.* **1983,** *3,* 343.
25. Marshall, D. E.; Higgs, R. P.; Obande, O. P. *Plastics and Rubber Process. and Appl.* **1983,** *3,* 353.
26. Potente, H.; Schultheis, S. M. *Kunststoffe* **1987,** *77(4),* 401.
27. Covas, J. A.; Gilbert, M.; Marshall, D. E. *Plastics and Rubber Process. and Appl.* **1988,** *9,* 107.
28. Obande, O. P.; Gilbert, M. *Plastics and Rubber Process. and Appl.* **1988,** *10,* 231.
29. Teh, J. W.; Cooper, A. A.; Rudin, A.; Batiste, J. L. H. *J. of Vinyl Technology* **1989,** *11(1),* 33.

30. Teh, J. W.; Cooper, A. A.; Rudin, A.; Batiste, J. L. H. *Makromol. Chem., Macromol. Symp.* **1989,** *29,* 123.
31. Obande, O. P. *J. Appl. Polym. Sci.* **1991,** *42,* 1433.
32. Covas, J. A.; Gilbert, M. *Polym. Eng. Sci.* **1992,** 32(11), 743.
33. Gezovich, D. M.; Beil, P. H. *Inter. J. of Polym. Mater.* **1971,** *1,* 3.
34. Soni, P. L.; Geil, P. H.; Collins, E. A. *J. Macromol. Sci.–Phys.* **1981,** B20, 479.
35. Lebedev, V. P.; Derlyukova, L. Y.; Razinskaya, I. N.; Okladnov, N. A.; Shtarkman, B. P. *Polym. Sci. USSR* **1965,** *7,* 366.
36. Tabb, D. L.; Koenig, J. L. *Macromolecules* **1975,** 8(6), 929.

POLY(VINYL CHLORIDE), RIGID (Soft Fillers)

Frederick H. Axtell and Pranee Phinyocheep
Department of Chemistry
Faculty of Science
Mahidol University

The recycling of vulcanized rubber has historically required high cost processes such as devulcanization reactions, crumbing, cryogenic grinding and metal/textile/polymer separation.[2] The reuse of scrap tires has been for relatively low value applications such as asphalt toughening, carpet backing, and underlay extenders.[3,11] Reject rubber gloves have been used as a source for crumb rubber.[1] A recent patent reported a new recycling process in which entire reject sport shoes are shredded, then fine particles are air-separated and the scrap recycled as a filler in the outsole compound of a lower quality shoe.[8] Recently it has been reported that scrap rubber has been used as a toughening additive for plastics, which would be a higher value application for the scrap.[4,5,9,10,13] Other possible recycling strategies include the burning of scrap rubber for electric generation.[6]

This study used crosslinked elastomer dust generated during the manufacture of sports shoes. The surfaces of the sole components (midsole and outsole) are buffed prior to adhesion. The laminated sole edges are also buffed to the desired shape prior to shoe assembly. The use of this dust eliminates the need for the size reduction processes usually employed in rubber recycling. Thus a substantial cost savings can be made in the recycling process. The aim of this research is to study the use of this waste elastomer dust as an impact modifier in plastics, rather than the less demanding use as a low cost extender.

The use of the untreated scrap dust and the influence of scrap dust particle size on the performance of the poly(vinyl chloride) (PVC) compounds was studied. The impact properties of PVC are known to be strongly dependent on the fusion level and so this effect was investigated.

Further improvement in toughness may be obtainable by the selected use of interfacial agents to improve the dispersion of, and interaction between, the soft filler particles and the PVC matrix.[14] The use of coupling agents and compatibilizers to improve the toughness of the PVC composites containing soft fillers was studied.[7,12]

REFERENCES

1. Anon., *Polym. Rubb. Asia* June **1991,** 6(34), 4.
2. Axtell, F. H.; Sophanowong, W.; Rujiranontapong, K. *Plast. Rubb. Compos. Process. Applic.* **1994,** *21,* 53-61.
3. Blumenthal, M. H. *Biocycle* **1991,** 32(10), 47.
4. Deanin, R. D.; Hashemiolya, S. M. *Polym. Mater. Sci. Engng.* **1987,** *57,* 212-216.
5. Duhaime, J. R. M.; Baker, W. E. *Plast. Rubb. Compos. Process. Applic.* **1991,** *15(2),* 87-93.
6. Fisher, P. M. *Materials World* **1994,** 2(6), 310-312.
7. Jarukumjorn, K. A study of the effect of compatibilizers on the toughening of PVC with vulcanized rubber scrap. M.Sc. thesis, Mahidol University, 1993.
8. Nike, Inc., Atlanta (Reuter) in *Bangkok Post, Business Post, XLVII* (35) (4/2/1992) 18.
9. Pittolo, M.; Burford, R. P. *Rubb. Chem. Technol.* **1985,** *58,* 97-106.
10. Pramanik, P. Recycling Ground Rubber Tires into Plastic Products, Workshop Papers, Mississaugo, Ontario, 26th October 1993, Paper 4, PP12.8 (13) Corp. Ed., Ontario, Center for Materials Research; Ontario, Queens University; Ontario, Ministry of Environment and Energy.
11. Shaw, D. *Eur. Rubb. J.* **1991,** 173(9), 21.
12. Sophanowong, W. A study of the toughness of rigid PVC filled with recycled vulcanized elastomer dust. M.Sc. thesis, Mahidol University, 1993.
13. Vibien, P. Workshop papers Mississauga, Ontario. 26th October 1993, Paper 6, pp 12.8 (13), Corp Ed. Ontario, Center for Materials Research; Ontario, Queens University; Ontario, Ministry of Environment and Energy.
14. Xanthos, M. *Polym. Engng. Sci. and Science* **1988,** *28,* 1392-1400.

POLY(VINYL CHLORIDE), SOFT (Calcium Carbonate Filled)

Frederick H. Axtell and Jatuporn Ratanapaka
Department of Chemistry
Faculty of Science
Mahidol University

Many commercial poly(vinyl chloride) (PVC) compounds contain calcium carbonate filler to reduce raw materials costs. The technological challenge is to achieve the cost reduction without sacrificing the compound performance with regard to processability, electrical, mechanical, weathering, and appearance properties. The combined performance and cost criteria limit the amount and type of calcium carbonate filler that can be used in a given application. The addition of mineral fillers such as calcium carbonate affects the thermal properties of plastics, and also leads to a deterioration in mechanical properties owing to stress concentration effects.[1]

The wide natural abundance of marble, limestone, and chalk worldwide suitable for use in plastics has resulted in calcium carbonate being the largest volume filler used in plastics.[2] The plentiful sources also keep the costs down for the end user. In poly(vinyl chloride) the ground natural calcium carbonate grade is the most commonly used filler; for certain higher value applications coated ground or coated precipitate grades of calcium carbonate are used. However, the higher costs usually inhibit their commercial application as fillers in soft poly(vinyl chloride). Precipitated calcium carbonate is used in poly(vinyl chloride) plastisols as a viscosity modifier.

The advantages of calcium carbonate over other mineral fillers for use in plasticized poly(vinyl chloride) include: lower cost, non-toxicity, transparency, low moisture content, easy coating ability, tribological softness, and the acid-acceptance capability providing a degree of secondary stabilization for PVC during processing.

The quality of calcium carbonate varies depending on its mineral source and manufacturing process, so that a number of different grades are available commercially. These differ in their particle size distribution, surface treatment, and cost.

Poly(vinyl chloride) compounds containing different grades of calcium carbonate will show a range of performances.

REFERENCES

1. Birley, A. W.; Haworth, B.; Batchelor, J. *Physics of Plastics*; Hanser: Munich, Germany, 1992; p 50.
2. Ferringo, T. H. *Handbook of Fillers and Reinforcements for Plastics*; Katz, H. S.; Milewski, J. V., Eds.; Von Nostrand Reinhold: New York, NY, 1978; pp 31-35.

Poly(vinyl ether)s

See: Living Polymerization, Fast (Vinyl Ethers)
Personal Care Application Polymers (Acetylene-Derived)
Vinyl Ether Polymers

Poly(vinyl fluoride)

See: Engineering Thermoplastics (Survey of Industrial Polymers)

POLY(VINYL KETAL)S

Naofumi Nakamura
Tezukayama College

Acetalization reaction of poly(vinyl alcohol) (PVA) by aldehydes was carried out easily, but ketalization reaction was very difficult in aqueous media.[1] Other than an old patent by I. G. Farben AG.,[2] which reports successful results with cyclohexane, reports on ketalization reaction between PVA and ketones are not known. The author has succeeded in synthesizing poly(vinyl ketal) (PVKL) of any ketalization degree through ketalizing PVA with various ketones in nonaqueous media, mainly dimethyl sulfoxide (DMSO), under the presence of acidic catalyst.

PREPARATION

Reaction in Homogeneous System

The reaction between PVA and ketones proceeds as shown in **Equation 1**.

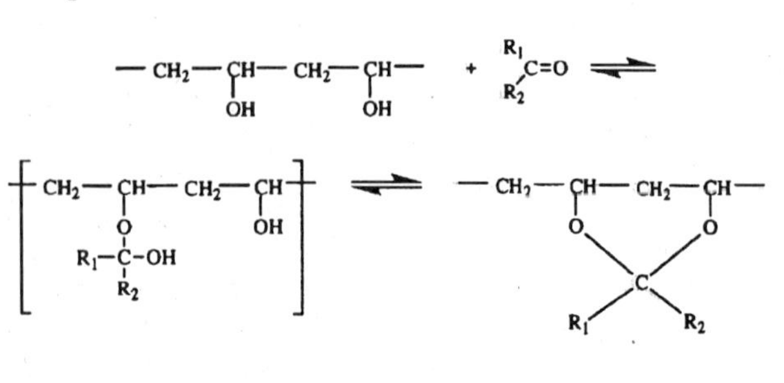

1

Ketalization reaction of PVA by 9 ketones [acetone, methyl ethyl ketone (MEK), methyl *n*-propyl ketone (nPK), methyl *i*-propyl ketone (iPK), methyl *n*-butyl ketone (nBK), methyl *i*-butyl ketone (iBK), methyl *t*-butyl ketone (tBK), cyclohexanone (Hex) and cyclopentanone (Pent)] were carried out.

Maximum Ketalization Degree[4,5]

The ketalization reaction with benzophenone does not proceed at all. Steric hindrance of bulky side chain of ketones seems to affect the ketalization. In cyclic ketones, maximum ketalization degree is *ca* 65 mol% for Pent and is lower than *ca* 80 mol% for Hex. This result shows the effect of ring flexibility.

PROPERTIES

Solubility in Organic Solvents[3-6]

In cases of PVKL (aliphatic ketones), it becomes soluble in alcohol while OH groups in PVA decrease, but in cases of highly ketalized PVKL (cyclic ketones), it becomes insoluble in alcohol. All PVKL dissolves in DMSO, except highly ketalized PVKL (Hex). In nonpolar solvents, such as acetone, ether, ethylacetate, etc., PVKL is insoluble.

Hydrophobicity Factor[4,5]

The hydrolysis of PVKL film proceeds in the following order: acetone > MEK > nPK > Pent > iPK ≒ nBK > iBK > Hex > tBK.

Moisture Regain[5,7]

The moisture regain is in the following order: acetone > MEK > nPK ≒ Pent > iPK ≒ nBK > Hex > iBK > tBK.

O_2 Permeability

The O_2 permeabilities of PVKL are decreased proportional to the ketalization degree. O_2 permeability ($cm^3 \cdot cm/cm^2$ sec cm Hg $\times$ 10^{-13}) of PVKL (acetone) are 4.4 (0 mol%), 12.2 (17.3 mol%), 18.5 (65.6 mol%), at 25°C.

APPLICATIONS

Applications of PVKL have not been studied, but the following applications seem to be useful: (1) adhesive agent–(a) use for glass binding; (b) use for the adhesive agent after dissolving of PVKL in alcohol; and (2) water soluble film–medical use (in order to prevent infection, textiles that patients have already used are put into a package made by water-soluble film, then washed thoroughly with water).

REFERENCES

1. Sakurada, I. *Polyvinyl Alcohol Fibers* Marcel Dekker: New York and Bassel, 1985.
2. Känzlein, G. *DRP* **1932,** *551,* 968 (to I.G. Farben AG).
3. Nakamura, N. *J. Appl. Polym. Sci.* **1987,** *33,* 2065.
4. Nakamura, N. *J. Appl. Polym. Sci.* **1991,** *42,* 1341.
5. Nakamura, N. *J. Appl. Polym. Sci.* **1993,** *47,* 1653.
6. Nakamura, N.; Aoshima, S.; Higashimura, T. *J. Appl. Polym. Sci.* **1988,** *36,* 1877.
7. Nakamura, N.; Suzuki, K. *J. Appl. Polym. Sci.* **1992,** *44,* 1135.

Polyvinylamine

See: *Flocculants (Organic, Overview)*
Polyvinylamine (Overview)
Polyvinylamine (Preparation, Properties, Flocculants (Organic, Overview) and Applications)

POLYVINYLAMINE (Overview)

R. K. Pinschmidt, Jr.* and D. J. Sagl
Air Products and Chemicals, Incorporated

Polyvinylamine (PVAm, CAS No. 26336-38-9; HCI salt, CAS No. 29792-49-2) (**Scheme I**) is the simplest practical polymeric analog of a primary amine. A predominately linear polymer with a repeating unit molecular weight of 43, PVAm is isomeric with polyethyleneimine (PEI, CAS No. 9002-98-6). The amine groups in PVAm are all primary, and they are directly pendant to an all carbon main chain. PEI, however, is a near random mix of primary, secondary, and tertiary amines in its more common highly branched form and all secondary in its linear form, with the amines as part of the main chain. PVAm is also related to the next higher homolog, polyallylamine (PAAm, CAS No. 30551-89-4), but differs from both PEI and PAAm substantially in properties and synthesis.

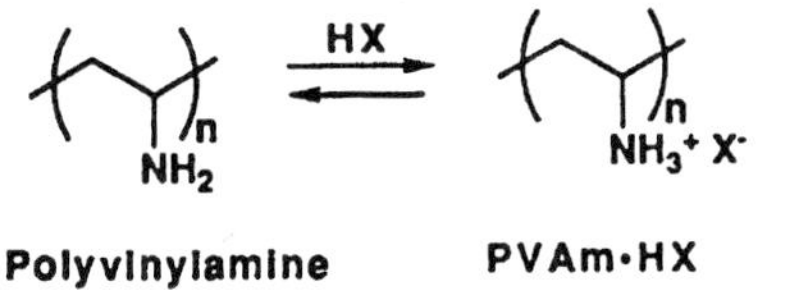

I

SYNTHESIS

Although PVAm is frequently reported in the general[2,3] and patent literature, a practical, flexible synthesis has proved surprisingly elusive.[1-3] The simplest potential precursor, vinylamine, is unstable relative to tautomeric acetaldehyde-ammonia imine and its condensation products.[4,5]

Polyvinylamine, like poly(vinyl alcohol) (CAS No. 9002-89-5), has thus been prepared via indirect means.[1] The two most common approaches include Hoffmann rearrangements from polyacrylamide (CAS No. 9003-05-8) and polymerization and hydrolysis of protected vinylamines.

N-Vinylamides have been reported frequently as precursor monomers for polyvinylamines, but hydrolysis of the derived polymers is again difficult with any but the simplest and most easily removed blocking groups.[7] *N*-vinylacetamide (NVA) was available as a developmental material in the late 1970s and has been reported as a precursor to PVAm by several groups before and since.[9-11] NVA, when pure, polymerizes and copolymerizes well. The resulting polymers can be hydrolyzed completely, albeit under vigorous conditions (e.g., excess 12N NCl for 48 hr.)[12] Strong base is also effective, but requires elevated temperatures in a pressure reactor.[13]

Polymerization and hydrolysis of the lowest amide homolog, *N*-vinylformamide (NVF), appear at present to offer the most practical route to preparing PVAm. NVF, although difficult to prepare in high purity owing to its high reactivity, polymerizes and copolymerizes readily to high or low molecular weight poly(vinylformamide)s that hydrolyze efficiently under mild stoichiometric acid or base conditions.[14,15]

PROPERTIES

Solution properties of water soluble unhydrolyzed PNVF or PNVA are similar to polyacrylamide (PAM), of similar molecular weight.

PVAm·HCl is soluble in water, formamide, ethylene glycol and some alcohol/water mixtures. It is insoluble in common solvents such as acetone, alcohols, dimethylformamide, and methylene chloride. A T_g for PVAm·HCl is not observed below its decomposition temperature (220–250°C). Poly(vinylamine) free base is soluble additionally in lower alcohols, has a T_g (dry) of 45°C and is amorphous. Poly(vinylamine·HCl) exhibits good stability in high salt environments, in strongly acidic environments and at elevated temperatures.

PVAm, with over 30% amine content, shows an extremely wide range of basicity as a function of extent of protonation.[6] Plots of $pK_{app(HA)n}$ vs. pH are particularly linear for PVAm and imply a high level of rigidity and salt impermeability.[19] The solution viscosity of fully protonated PVAm·HCl is high. This is attributed to the high cationic charge density on the polymer backbone, which should cause the polymer to adopt a relatively uncoiled conformation. As the pH increases, the cationic character decreases, resulting in decreased viscosity.

DERIVATIVES

PVAm has been demonstrated to react with acids, metal cations, epoxies, aminoplasts, isocyanates, aldehydes, anhydrides, carbon dioxide, lactones, acid chlorides, and Michael acceptors.[3,16,17] PVAm alkylation has been reported frequently.[3,20-22,33]

APPLICATIONS

PVAm and its derivatives have been found to be useful in a number of applications. The utility of PVAm is derived primarily from the cationic nature of the primary amine groups under neutral or acidic conditions and from the high reactivity of the amine in its free base form, for example, in applications requiring polymers with good crosslinkability. In its cationic form, PVAm adsorbs onto a wide variety of anionically charged natural and synthetic surfaces, such as cellulose fibers, organic waste, natural clays and minerals, as well as plasma treated polymer surfaces. The cationicity of PVAm and its derivatives makes them useful in such areas as papermaking and waste water treatment.[23]

Polyvinylaminals prepared by reaction of PVAm with aldehydes are reported to function as associative thickeners and for drag reduction.[24,25] A great deal of work was reported on binding colorants, dyes, and drugs to PVAm and its copolymers.[8,18,26] PVAm in crosslinked form has been used as a metal chelating resin both before and after functionalization.[27,28] In addition, PVAm crosslinked with divinylbenzene, and optionally converted to the tertiary amine form, has been employed as an anion exchange resin.[22,29]

*Author to whom correspondence should be addressed.

Some work has been done to explore the utility of high molecular weight PVAm in oil field applications, especially enhanced oil recovery.[16,31] PVAm has been used as an additive for bonding agents used to lay building materials,[32] and as a stationary phase for liquid chromatography suitable for resolving D,L mixtures of aminoacids.[5a,30] PVAm was also cited as a dye fixative for ink jet printing.[34,35]

REFERENCES

1. Augurt, T. A. In *Encyclopedia of Polymer Science and Engineering;* Kroschwitz, J. I.; Ed-in-Chief; John Wiley & Sons: New York, 1971; Vol. 14, p 251.
2. Ergozhin, E. E.; Mukhitdinova, E. A.; Khalikova, V. K. *Tr. Inst. Kim. Nauk, Akad. Nauk. Kaz. SSR* **1978,** *46,* 47.
3. Fischer, T.; Heitz, W. *Macromol. Chem. Phys.* **1994,** *195,* 679.
4. Nielsen, A. T.; Atkins, R. L.; Moore, D. W.; Scott, R.; Mallory, D.; LaBerge, J. M. *J. Org. Chem.* **1973,** *38,* 3288.
5. Gabriel, S. *Chem. Ber.* 1888, *21,* 1049; this early report of the synthesis of vinylamine was later shown to be incorrect [Howard, C. C.; Marchkwald, W. *Chem. Ber.* 1899, *32,* 2036.]

5a. Hanford, W. E.; Stevenson, H. B. U.S. Patent 2 365 340, 1944, to E. I. DuPont.

6. Katchalsky, A.; Mazur, J.; Spitnik, P. *J. Polym. Sci.* **1957,** *23,* 513.
7. Sawayama, S.; Sato, K.; Yamada, Y.; Ueno, N. *Mitsubishi Kasei R & D Rev.* **1993,** *7,* 55.
8. Wang, P. C.; Wingard, R. E. U.S. Patent 4 051 138, 1977, to Dynapol.
9. Dawson, D. J.; Gless, R. D.; Wingard, Jr., R. E. *J. Am. Chem. Soc.* **1976,** *98,* 5996.
10. Stackman, R. W.; Summerville, R. H. *Ind. Eng. Chem. Prod. Res. Dev.* **1985,** *24,* 242.
11. Marumo, K. *Kobunshi, Special Ed.–New Developments in Reactive Macromolecules* **1991,** *40,* (May).
12. Overberger, C. G.; Kikyotani, S. *J. Polym. Sci., Polym. Chem. Ed.* **1983,** *21,* 525.
13. Dawson, D. J.; Brock, P. J. U.S. Patent 4 393 174, 1983 to Dynapol.
14. Kurtz, P.; Disselnkoetter, H. *Ger Offenl.* 1 228 246, 1 224 304, 1965, to Bayer; see also Kurtz, P.; Disselnkoetter, H. *Liebigs Ann. Chem.* **1972,** *764,* 69.
15. Jenson, H.; Schmidt, E.; Mitzlaff, M.; Cramer, J.; Pistovius, R.; Pietsch, R.; Dehmer, K. U.S. Patent 4 322 271, 1982, to Hoechst.
16. Badesso, R. J.; Lai, T.-W.; Pinschmidt, Jr., R. K.; Sagl, D. J.; Vijayendran, B. R. *Amer. Chem. Soc. Polymer Preprints* **1991,** *32,* 110.
17. Badesso, R. J.; Pinschmidt, Jr., R. K.; Sagl, D. J. *Proc. Amer. Chem. Soc., Div. Polym. Mat. Sci. Engr.* **1993,** *69,* 251. Badesso, R. J.; Pinschmidt, Jr., R. K.; Sagl, D. J. In *Hydrophilic Polymers Performance with Environmental Acceptance* Glass, E. Ed.; American Chemical Society: Washington, DC, 1995, 489.
18. Dawson, D. J. *Polym. Prepr.* **1976,** *17,* 779.
19. Marinsky, J. A.; Reddy, M. M.; Ephraim, J.; Mathuthu, A. *SKB Tech. Rep.* **1988,** SKB 88–04.
20. Chang, C.; Muccio, D. D.; St. Pierre, T.; Chen, C. C.; Overberger, C. G. *Macromolecules* **1986,** *19,* 913.
21. Chen, J.-C.; Vasconcellos, S. R.; Walterich, Jr., G. C.; Chen, F. U.S. Patent 5 292 441, 1994, to Betz PaperChem.
22. Ito, T.; Kubota, H.; Kasigawi, T. Eur. Patent Appl. 9 311 397, 1993.
23. Linhart, F.; Auhorn, W. *Das Papier* **1992,** 10A, 000.
24. McAndrew, T. P.; Nordquist, A. F.; Pinschmidt, Jr., R. K.; Eichelberger, D. P. U.S. Patent 5 270 379, 1993, to Air Products and Chemicals.
25. Nordquist, A. F.; Pickering, T. L.; Pinschmidt, R. K.; Smigo, J. G. U.S. Patent 5 232 553, 1993, to Air Products and Chemicals.
26. Dawson, D. J. *Aldrichimica Acta* **1981,** *14,* 23.
27. Tbal, H.; Delporte, M.; Morcellet, J.; Morcellet, M. *Eur. Polym. J.* **1992,** *28,* 671.
28. Rbal, H.; Le Maguer, D.; Morcellet, J.; Delporte, M.; Morcellet, M. *Reactive Polymers* **1992,** *17,* 207.
29. Kubota, H.; Ito, T.; Fukuda, J.; Ungar, J. *Off. Proc. 54th Int. Water Conf.* **1993,** and references therein.
30. Yuan, Z.; He, B.-L. *Chinese Sci. Bull.* **1991,** *36,* 903.
31. Lai, T.-W.; Vijayendran, B. R. U.S. Patent 4 843 118, 1989, to Air Products and Chemicals.
32. Park, J.-H. U. K. Patent Appl. 2 275 476, 1993.
33. Harrington, IV, J. C.; Chen, J.-C.; Chen, F. U.S. Patent 5 269 942, 1993 to Betz PaperChem.
34. Pawolwski, N. E. U.S. Patent 5 230 733, 1993 to Hewlett Packard.
35. Takimoto, H.; Yoneyama, T.; Sano, H. Jpn. Patent 63280781, 1988, to Mitsubishi Kasei (CA 110; 214915).

POLYVINYLAMINE
(Preparation, Properties, and Applications)

Hiroshi Uyama* and Shiro Kobayashi
Department of Molecular Chemistry and Engineering
Faculty of Engineering
Tohoku University

Polyvinylamine (PVAm) is a polyamine having primary amino groups linked directly to the main chain. Polyamines and polyquaternary ammonium salts are important practically because of their use for various purposes.[1] Recently, PVAm has been manufactured industrially by some companies and used as a flocculating agent for the removal of particulate matter in turbid natural water. In the laboratory, PVAm is also an attractive starting polymer for the preparation of functional polymers.

PREPARATION

Recently, preparation of PVAm on an industrial scale has been achieved successfully with the development of a precursor synthesis. *N*-vinylformamide (NVF) and *N*-vinylacetamide (NVA) were polymerized radically, followed by acid hydrolysis to produce PVAm hydrochloride (**Scheme I**).[3-5] polyNVF was converted into PVAm under milder conditions than was polyNVA.

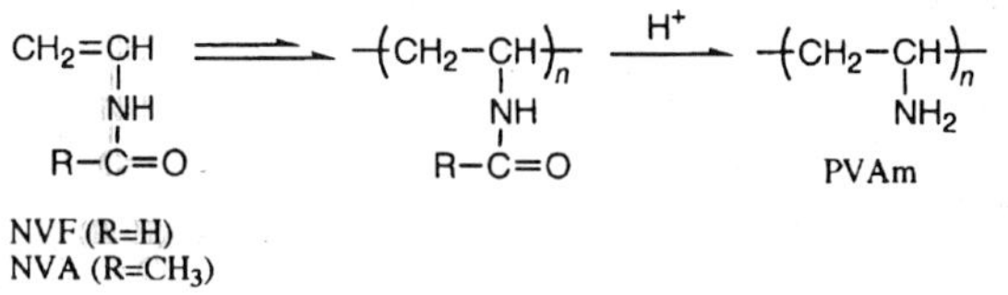

I

PROPERTIES

PVAm is not stable under air because it reacts with carbon dioxide in the atmosphere. Its acid salts are very stable, however. Solution properties of PVAm were investigated using a highly pure poly(vinylamine hydrochloride) sample (PVAm-1·HCl, molecular weight = 1×10^4). The pK_a value of this sample was 8.49 in a potassium chloride aqueous solution (ionic

*Author to whom correspondence should be addressed.

strength μ = 1.0 mol/L) at 25°C.[6] PVAm showed quantitative ionization at about pH 3, that is, all the amino groups attached to the polymer chain are protonated.

APPLICATIONS

PVAm bearing reactive primary amino groups is modified readily by electrophilic reagents. Methylated PVAm was synthesized by reacting with methyl iodide in methanol.[2] PVAm was also reacted with isocyanate and acyl chloride in high conversions of the amino group to give poly(*N*-vinylurea) and poly(*N*-vinylalkanamide), respectively.

PVAm gel was prepared, and its viscosity behaviors were applied to chemomechanical materials. PVAm was reacted with bis-epoxide as a crosslinking agent to give PVAm gel.[7]

The preparation of a polymeric catalyst bearing cobalt(II) phthalocyanine, coordinatively bound to PVAm, was reported.[8] PVAm acted as a polymeric ligand for the metal ion.

PVAm is expected to be used in industrial applications such as pigment retention aids in the manufacturing of fine paper, filtration aids, sludge dewatering agents, and emulsion breakers.

REFERENCES

1. Vorchheimer, N. *Encyclopedia of Polymer Science and Engineering*, 2nd ed.; John Wiley & Sons: New York, 1986; Vol. 11, pp 489-507.
2. Fischer, T.; Heitz, W. *Macromol. Chem. Phys.* **1994,** *195,* 679.
3. Itagaki, K.; Ito, T.; Ando, K.; Watanabe, J. *Jpn. Kokai* **1986,** *61,* 51007.
4. Dawson, D. J.; Gless, R. D.; Wingard, Jr., R. E. *J. Am. Chem. Soc.* **1976,** *98,* 5996.
5. Akashi, M.; Yashima, E.; Yamashita, T.; Miyauchi, N.; Sugita, S.; Marumo, K. *J. Polym. Sci., Polym. Chem. Ed.* **1990,** *28,* 3487.
6. Kobayashi, S.; Suh, K-D.; Shirokura, Y. *Macromolecules* **1989,** *22,* 2363.
7. Kobayashi, S.; Suh, K-D.; Shirokura, Y.; Fujioka, T. *Polym. J.* **1989,** *21,* 971.
8. Brouwer, W. M.; Piet, P.; German, A. L. *Makromol. Chem.* **1984,** *185,* 363.

Polyvinylcarbazole

See: *Charge Transport Polymers (for Organic Electroluminescent Devices)*
Poly(N-vinylcarbazole)

POLY(*N*-VINYLCARBAZOLE)

Maria Obloj-Muzaj,* Helena Kurowska, Henryk Dudziak, and Andrzej Molek
Industrial Chemistry Research Institute
Warsaw University of Technology

N-vinylcarbazole (NVK) is a very reactive monomer. It polymerizes radically, ionically, and thermally. Review of various NVK polymerizations can be found in the monograph *Poly*(*N-vinylcarbazole*)[1] and other early works;[2,3] papers on this subject, however, are not numerous.[1-3]

While the data concerning PVK synthesis stopped in the 1980s, many patent data touching PVK applications in photoconductive areas appeared. It was expected because this polymer is very interesting and useful for information carriers.

PREPARATION

The basic step in technology is radical homopolymerization of *N*-vinylcarbazole in water suspension. Solid monomer melts in polymerization conditions. The reaction is carried out periodically. The raw polymer, including some amount of unreacted monomer, is then separated by filtration. Residual monomer is removed from the polymer and regenerated. Polymer–after removing the residual monomer–is washed and dried under vacuum. The product is in the form of white pearls.

The polymer can be prepared by the suspension polymerization of *N*-vinylcarbazole in the presence of conventional free radical initiators and suspension stabilizer in mass process.

PROPERTIES AND APPLICATIONS

Poly(*N*-vinylcarbazole) is suitable for application in various photoconductive technics–mainly in xerographic layers and holographic layers. Application of PVK to layers operating in the infrared and X-ray ranges also appears possible. Our preliminary experiments have shown the PVK-based layer to be sensitive to X-rays.

Mechanical properties and stability of PVK layers depend on the molecular weight of the polymer. Higher molecular weight means better mechanical properties, but worse solubility. On the basis of our investigations on PVK with $\overline{M}_w$ in the range 0.5–3.0 million, we can state that PVK with $\overline{M}_w$ about 1.5 million has enough stability and mechanical properties in the xerographic layer and solubility sufficient for preparing photoconductive complexes.

Photoconductivity of the layer depends on its composition, the solvent used, and preparation conditions. Usually the skeleton of photoconductive complex contains helix-shaped PVK chains with a low molecular weight sensitizer inside the helix. It is possible to enhance the photoconductivity by doping this system with special additives. Preparing and conditioning of the layer is very important and has great influence on its final properties and stability. The electrophotographic properties of layers obtained from PVK–TNF (trinitrofluorenone) complexes of various molar ratio have been investigated.

The PVK–TNF complex may also be used as an imaging material in medical equipment utilizing X-rays and in X-ray flaw detectors.

Poly(*N*-vinylcarbazole) and other photoconductive polymers should be used in their specific area. Their mechanical or thermal properties should not be compared with properties of typical mass polymers as polyolefins, PVC and others, because the fields of application are quite different. Properties of PVK are sufficiently good for various photoconductive applications.

REFERENCES

1. Person, J. M.; Stolka, M. *Polymer Monographs,* Vol. 6: *Poly(N-vinylcarbazole*), Gordon and Breach: New York, 1981.
2. Penwell, R. C.; Ganguly, B. N.; Smith, T. W. *J. Polym. Sci., Macromol. Rev.* **1978,** *13,* 63.
3. Al-Abidin, Kusai, M. Z.; Jones, R. G. *J. Chem. Soc. Faraday Trans.* **1982,** *78(2),* 513.

*Author to whom correspondence should be addressed.

Poly(vinylene sulfide)

See: *Polysulfides (Prepared from Sulfur Dioxide)*

Poly(vinylidene fluoride)

See: *Engineering Thermoplastics (Survey of Industrial Polymers)*
Ferroelectric Polymers (Structural Phase Transitions)
Fluoropolymer Coatings (New Developments)
Magnetic Field Processing
Poly(vinylidene fluoride) (Overview)
Poly(vinylidene fluoride) (Synthesis, Microstructure, and Chain Conformations)
Poly(vinylidene fluoride) Phases and Morphology (Effect of Crystallization Conditions)
Powder Coatings (Overview)
Pyroelectricity
Vinylidene Fluoride-Based Thermoplastics (Overview and Commercial Aspects)
Vinylidene Fluoride-Based Thermoplastics (Applications)
Vinylidene Fluoride-Based Thermoplastics (Blends with Other Polymers)
Vinylidene Fluoride-Based Thermoplastics (Homopolymerization and Copolymerization)
Vinylidene Fluoride-Trifluoroethylene Copolymers (Ferroelectric-to-Paraelectric Phase Transition)

POLY(VINYLIDENE FLUORIDE) (Overview)

B.-J. Jungnickel
Deutsches Kunststoff-Institut (German Plastics Institute)

Poly(vinylidene fluoride), commonly abbrebiated as PVDF or PVF_2, is a polymer with increasing scientific attention and industrial importance because of its outstanding electrical properties, its chemical and weather resistance, its durability, and its biocompatibility. The polymer attracted particular attention in the late 1970s after the detection of its extraordinary electrical properties.[1-3] The material can carry a thermodynamically stable electric polarization (*ferroelectricity*), the strength of which changes with mechanical load (*piezoelectricity*), and temperature change (*pyroelectricity*).

The synthesis, the molecular, supermolecular, and crystalline structures, respectively, and the properties of PVDF already have been reviewed several times in detail.[4,5]

CHEMISTRY

PVDF, **Structure 1**, is made by radical polymerization in emulsion or suspension of the (gaseous) monomer 1,1-difluoroethylene ($CH_2 = CF_2$). The reaction is preferentially performed in water at elevated temperature (up to 150°C) and pressure [(1...100)MPa] with organic peroxides (e.g., benzoyl peroxide or diisopropyl peroxy-dicarbonate) as catalysts.

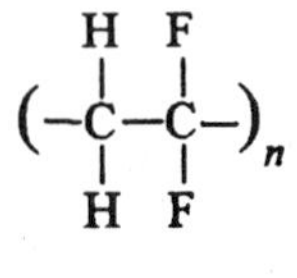

1

PHYSICAL STRUCTURE

Crystal Structures

PVDF exhibits a very complicated crystallization behavior. Three regular chain conformations have been detected so far, two of them being existent in the crystals in both parallel and antiparallel alignment of the corresponding electric moments of the conformation repeat units.[6]

Supermolecular Structure

PVDF crystallizes upon cooling of the melt predominantly in spherulitic supermolecular morphology.

PROPERTIES

General

PVDF exhibits a number of high-performance properties. It is a tough, hard, aging, heat resistant, and thermally stable engineering resin. It is, consequently, highly weather resistant, particularly against extreme temperature changes, and it is resistant against most chemicals. Under the influence of severe (more than 1MGy) uv, β, and γ irradiation, it crosslinks predominantly.[7] It has a high abrasion resistance and low friction coefficient to most other materials. PVDF is–as most halogenated polymers are–fire resistant; it is not flammable and, therefore, will not propagate flames, and it is nondipping and self-extinguishing at fire. It is tasteless, odorless, and nontoxic. PVDF is chemically stable up to 380°C where thermal decomposition sets on, and hydrogen fluoride is eliminated.

Thermal and Mechanical

A number of the thermal properties of PVDF are linked with the crystallization and the thermally induced morphological changes. The dynamic-mechanical behavior of α crystalline PVDF exhibits two distinct relaxations, namely the α relaxation around 80°C and the β relaxation at about –30°C. Sometimes, a much smaller relaxation (γ) can be detected at still lower temperature of –60°C which, however, in most instances merges into the β relaxation. A fourth weak relaxation (β′) is located at 40°C. The β relaxation is ascribed to the glass transition particularly in the crystal amorphous interface and the α relaxation shall be due to chain librations in the crystalline phase.

Electrical

The strong electric moment of the chain repeat unit, at the one hand, and the easy crystallizability of the PVDF, at the other hand, are responsible for a number of outstanding electric properties of this polymer.

The crystalline β, γ, and δ phases of PVDF, which all carry a permanent polarization, are ferroelectric. This is the basis of a number of technical applications of this material.

Beside being ferroelectric, PVDF is also piezoelectric and pyroelectric. The permanent polarization changes after mechanical distortions like tension, compression or action of hydrostatic pressure, and after a sudden temperature change.[1] It is important to note that these properties exist only in polarized materials, and that their strengths are roughly proportional to the permanent polarization. Nevertheless, there are a number of indications that the effects are linked with the semicrystallinity of the PVDF.

APPLICATIONS

General

PVDF is readily melt processable in injection molding and in extrusion. It can be spun into fibers or drawn into films. Since it is tasteless, odorless, and untoxic, PVDF can be used in the food and pharmaceutical industry or for packaging purposes. Its heat, chemical, and weather resistance, its low water absorbance, and its long-term aging stability makes it highly suited for protective and corrosion resistant coatings of metallic parts or other articles (pipes, fittings, pumps, valves) and of decorating finishes which are exposed to severe weather or harsh environmental conditions. Among these are chemical and petrochemical processing equipments, and equipments of the nuclear or electric industries.

Electric and Electronic

For all its mentioned properties and for its high dielectric constant and its low dielectric loss, PVDF can be used advantageously for covering and insulating of electric cables and wires, for electric and electronic jacketing, and for other molded parts in electric devices or electronic circuits. It is used as an insulating layer in condensators.

An increasing amount of PVDF is used as electret film in electronics and in transducer technology.[4] The ferroelectric and piezoelectric properties of PVDF enable its application in audiotransducers such as loudspeakers or headphones and microphones. It is particularly advantageous that the necessary sheets are mechanically flexible films which can easily be produced with thicknesses raning between some μm and 1mm, and that they can have arbitrary shape and bend. The low acoustic impedance against water and biological tissues enables under water applications like hydrophones, including ultrasonic transducers, and *in vivo* medical applications, for example, control of respiration or cardiac sounds.[9-11] Implanted PVDF films are reported to promote bone growth after fractures.[12] Other fields of application are robotics (tactile sensors, proximity transducers, and actuators) and electromechanical devices.[13] The pyroelectricity of PVDF is used in sensitive pyrodetectors as thermal and ir sensors, for example, intruder detection, pyroelectric vidicons, or simply ir and heat intensity measurement.

REFERENCES

1. Kawai, H. *Jap. J. Appl Phys.* **1968,** *8,* 975.
2. Bergman, J. G.; McFee, J. H.; Crane, G. R. *Appl. Phys. Lett.* **1971,** *18,* 203.
3. Lovinger, A. J.; Davis, G. T.; Furukawa, T.; Broadhurst, M. G. *Macromolecules* **1982,** *15,* 323.
4. Wang, T. T.; Herbert, J. M.; Glass, A. M., Eds.; *The Application of Ferroelectric Polymers*; Blackie: Glasgow and London, 1987.
5. Lovinger, A. J. *PVDF.* In *Developments in Crystalline Polymers–1,* Bassett, C. D., Ed.; Appl. Science: London and New Jersey, 1982.
6. Allegra, G.; Meille, S. V. *Eur. Polym. J.* **1983,** *19,* 1017.
7. Timmerman, R.; Greyson, W. *J. Appl. Polym. Sci.* **1962,** *6,* 456.
8. Lovinger, A. J.; Wang, T. T. *Polymer* **1979,** *20,* 725.
9. Powers, J. *Long Range Hydrophones* **1987,** Chapter 6, Reference 8.
10. Kraft, G.; Francesconi, R.; DeRossi, D.; Dario, P.; Ruschi, S.; Contini, C. *Proc. Int. Symp. Ambulatory Monitoring,* Padua/Italy, 1985.
11. Shuford, R. J.; Wide, A. F.; Ricca, J. J.; Thomas, G. R. *Polym. Eng. Sci.* **1976,** *16,* 25.
12. Ficat, J. J.; Escourrou, G.; Fauran, M. J.; Durrox, R.; Ficat, P.; Lacabanne, C.; Micheron, F. *Ferroelectrics* **1983,** *51,* 121.
13. Dario, P. *Transducers for Advanced Robotics* **1987,** Chapter 10, Reference 8.

POLY(VINYLIDENE FLUORIDE) (Synthesis, Microstructure and Chain Conformations)

Saverio Russo,* Maurizio Pianca, and Giovanni Moggi

Dipaffimento di Chimica e Industriale, Universitá, Genova

Ausimont CRS, Bollate

IMAG CNR, Genova

Poly(vinylidene fluoride) (PVDF) is one of the first discovered fluoropolymers and one of the most attractive in this class of materials for its very relevant industrial applications.[1]

Excellent reviews on PVDF preparation and properties have already been published and contain useful information on the known state of the art.[2-4]

POLYMERIZATION

Monomer

PVDF is synthesized by free radical polymerization of 1,1-difluoroethane, usually called vinylidene fluoride (VDF), a colorless, almost odorless gas, boiling at –82°C.

The First Reports on VDF Free Radical Polymerization

The first successful polymerization of VDF was described in the already mentioned patent, using VDF, benzoyl peroxide and water, at 89–91°C and pressures of 860–955 atm.[1] A subsequent patent reported the polymerization conditions in more detail.[5]

Emulsion Polymerization

The use of chemically stable surfactants not containing C–H bonds, which otherwise can act as transfer agents or inhibitors of VDF free radical polymerization, allowed to carry out the first successful VDF emulsion polymerization.[6]

High molecular weight (MW) (50,000–300,000) PVDF has been obtained operating in the presence of *t*-butyl peroxide or similar initiators.[8] The polymer showed high melting point (168–173°C) and high crystallinity values; indeed its heat of fusion, (58.5 J/g), corresponds to a crystallinity content of about 63% on the basis of the heat of fusion value of 93.2 J/g for 100% crystallinity.[9]

Suspension Polymerization

Suspension polymerizations are generally carried out in an aqueous system containing peroxide-based organic initiators with or without a colloidal dispersant.

Generally, the products obtained by suspension polymerization are used in fluid handling systems employed in chemical processes and related fields, for example, in solid and lined pipes, valves, pumps, tower packings, woven fiber cloths, and tank lines, whereas emulsion-born PVDF products and their organosol dispersions are extensively applied in coating applications.

*Author to whom correspondence should be addressed.

Laboratory Scale Polymerizations

The γ-ray initiated polymerization of VDF has been carried out in acetone, giving a polymer with MW higher than by conventional methods.[10] The γ-ray initiated polymerization of VDF has also been reported to be successful operating in the gas phase.[11]

MICROSTRUCTURE AND CONFORMATIONS

NMR Spectroscopy

^{19}F NMR spectroscopy is the most powerful tool for studying PVDF microstructure. A study of the distribution of head-to-head, head-to-tail and tail-to-tail enchainments of PVDF has been reported in detail.[12] By analyzing the ^{7}F NMR spectra at 188 MHz of two commercial products, the above reference clearly showed that those polymers had regiosequence distributions which were described by first order Markov chain statistics, thus implicating that the structural disorder in PVDF could not be characterized accurately by a single parameter such as the percent of head-to-head, tail-to-tail defect content.

Vibrational Spectroscopy

IR spectroscopy is a very important technique for the investigation of some microstructural characteristics of PVDF, giving information about the chain conformation of the polymer, from which it is possible to deduce the kind of crystalline form present in the sample. PVDF can exist mainly in three crystalline modifications, denoted as I (β) (orthorombic), II (α) (pseudo orthorombic), III (γ) (monoclinic), in which the chains assume the conformations TTTT, TGTG′ and TTTGTTTG′, respectively.[13,14] Two additional forms, namely IV (∈) and V (αρ) have recently been evidenced by means of X-ray diffraction.[15,16] The features of the vibrational spectra of PVDF have attracted a great deal of interest[13,17-20] and specific infrared absorptions, characteristic of the different polymorphous forms, have been employed extensively in the study of the crystallization process.[13,17-20]

ACKNOWLEDGMENTS

Financial support by MURST and CNR is gratefully acknowledged.

REFERENCES

1. Du Pont de Nemours, Brit. Patent 590 817, 1947. *Chem. Abstract* 42, 794a.
2. Lovinger, A. J. *Developments in Crystalline Polymers*, Vol. 1, Bassett, G. C. Ed., Elsevier Barking, UK, 1982, pp 195-273.
3. Dohany, J. E.; Humphrey, J. S. *Encyclopedia of Polymer Science and Engineering* Mark, H. F.; Bikales, N. M.; Overberger, C. G.; Menges, G., Eds.; John Wiley & Sons: New York, 1987; Vol. 17, pp 532-548.
4. Dohany, J. E.; Kirk-Othmer, *Encyclopedia of Chemical Technology* John Wiley & Sons: New York, 1994; Vol. 11, pp 694-712.
5. Du Pont de Nemours, U.S. Patent 2 435 537, 1948; Ford, T. A.; Hanford, W. E. *Chem. Abstract 42*, 3215a.
6. Pennsalt, U.S. Patent 3 031 437, 1962; Iserson, H. *Chem. Abstract 63*, 3639f.
7. Pennsalt, U.S. Patent 3 245 971, 1966; Iserson, H. *Chem. Abstract 64*, 19822a.
8. Diamond Shamrock, Ger. Offen. 2 063 248, 1971; Stallings, J. P. *Chem. Abstract 75*, 110713.
9. Welch, G. J.; Miller, R. L. *J. Polym. Sci., Polym. Phys. Ed.* **1976,** *14,* 1683.
10. Doll, W. W.; Lando, J. B. *J. Appl. Polym. Sci.* **1970,** *14,* 1767.
11. Madorskaya, L. Y.; Agapitov, A. P.; Samoilov, A. M.; Makeenko, T. G.; Otradine, G. A.; Loginova, N. N.; Ivanchev, S. S. *Vysokomol. Soedin.* **1989,** Ser. B, *31,* 737; *Chem Abstract 112*, 159033r.
12. Cais, K. E.; Sloane, N. J. A. *Polymer* **1983,** *24,* 179.
13. Hasegawa, R.; Kobayashi, M.; Tadokoro, H. *Polym. J.* **1972,** *3,* 591.
14. Bachmann, M. A.; Gordon, W. L.; Koenig, J. L.; Lando, J. B. *J. Appl. Phys.* **1979,** *50,* 6106.
15. Lovinger, A. J. *Macromolecules* **1981,** *14,* 322.
16. Lovinger, A. J. *Macromolecules* **1982,** *15,* 40.
17. Wentink Jr., T.; Willworth, L. J.; Phaneuf, J. P. *J. Polym. Sci.* **1961,** *55,* 551.
18. Cortili, G.; Zerbi, G. *Spectrochim. Acta* **1967,** 23A, 285.
19. Zerbi, G.; Cortili, G. *Spectrochim. Acta* **1970,** 26A, 733.
20. Kobayashi, M.; Tashiro, K.; Tadokoro, H. *Macromolecules* **1982,** *8,* 158.

POLY(VINYLIDENE FLUORIDE) PHASES AND MORPHOLOGY (Effect of Crystallization Conditions)

Rinaldo Gregório Jr.,* Marcelo Cestari, Nadia Chaves, P. S. Nociti, Jose Alexandre de Mendonça, and Alessandra de Almeida Lucas
Materials Engineering
Federal University of São Carlos

Poly(vinylidene fluoride) (PVDF or PVF_2) is a polycrystalline polymer which started drawing scientific interest in the seventies, because of its extraordinary pyro- and piezo-electric properties. These properties combined with both high elasticity and processing ability lend this material numerous technological applications. Another feature that distinguishes PVDF from other polymers is its polymorphism, that is, it may present at least four crystalline phases, namely α, β, γ and δ (or II, I, III, and IV, respectively). The α phase is the most common, being normally obtained by melt or solution crystallization at elevated or moderate supercooling (T < 160°C). From a technological viewpoint the β phase is the most desirable, being widely used in pyro and piezoelectrical applications. This phase is usually obtained by drawing of originally α phase films.

The effect of the crystallization conditions of PVDF on the crystalline phase and resulting morphologies has been widely investigated. Hasegawa et al. studied the formation of the α, β, and γ phases and the transitions among them under different conditions.[1] The unoriented α phase was obtained by melt crystallization at atmospheric pressure and from solution with a mixture of monochlorobenzene–dimethylformamide or acetone. Drawing of α phase films at ratios of 400%, realized at 50 and 150°C resulted in oriented β and α phase, respectively. Unoriented β phase was obtained by thermal treatment at high pressures of α phase (292°C and 4500 atm) and γ phase specimens (285°C and 4000 atm). The γ phase was obtained by crystallization from the melt at high pressure (4000 atm), by thermal treatment (285°C

*Author to whom correspondence should be addressed.

and 4000 atm) of α specimens and from solution with dimethyl sulfoxide. Osaki and Ishida demonstrated that annealing of α phase films at 185.8°C and atmospheric pressure leads to transformation into the γ phase.[2] They also showed that melt crystallization of PVDF yields the γ phase at temperatures between 160 and 180°C, a maximum fraction of this phase being obtained at 170°C. Prest and Luca in a morphological investigation using optical polarized light microscopy, observed that isothermal crystallization of PVDF for 15 hours and above 160°C resulted in two distinct spherulitic forms: small spherulites exhibiting a coarser ringed structure and large ones also ringed, but with a periodicity approximately double that of the smaller ones.[3] They also observed that part of the larger spherulites lost their birefringence near 172°C and part near 185°C. With the smaller ones this occurred at 178°C. They attributed this result to the different constituent crystalline forms of the spherulites: the large ones consisting of the α and γ phase, with melting temperature (T_m) 172 and 185°C, respectively, and the smaller ones of the β phase (T_m = 178°C). Later, Lovinger observed that the minimum crystallization temperature necessary for the two spherulitie forms to appear depends on the PVDF resin used and the thermal history of the sample.[4] Hsu and Geil demonstrated that thin films (<1000 Å) can be quenched from the melt in a complete amorphous state, by ultraquenching (–160°C).[5] From the amorphous state, crystallization occurs in the β phase at approximately –30°C. Quenching of thin films at 0°C or even at room temperature also yields the β phase, confirmed later for films thinner than 30 μm.[6]

In a recent study we carried out a systematic investigation of the effect of the crystallization temperature from solution with dimethylacetamide (DMA) of PVDF on its morphology and on the crystalline phase type and content.[7] It was observed that formation of the α, β, or γ phases depends exclusively on the time and temperature at which the process occurs.

ACKNOWLEDGMENT

The authors thank FAPESP, CNPg and Volkswagen Foundation for their financial support.

REFERENCES

1. Hasegawa, R.; Kobayashi, M.; Tadokoro, H. Molecular Conformation and Packing of Poly(vinylidene Fluoride). Stability of Three Crystalline Forms and the Effect of High Pressure. *Polym. J.* **1972,** 3(5), 591.
2. Osaki, S.; Ishida, Y. Effects of Annealing and Isothermal Crystallization upon Crystalline Forms of Poly(vinylidene Fluoride). *J. Polym. Phys. Ed.* **1975,** *13,* 1071.
3. Prest, W. M.; Luca, D. J. The Morphology and Thermal Response of High-temperature-crystallized Poly(vinylidene Fluoride). *J. Appl. Phys.* **1976,** 49(10), 5042.
4. Lovinger, A. J. Crystallization and Morphology of the Melt-solidified Poly(vinylidene Fluoride). *J. Polym. Sci., Polym. Phys. Ed.* **1980,** *18,* 793.
5. Hsu, C. C.; Geil, P. H. Morphology Structure-property Relationship in Ultraquenched Poly(vinylidene Fluoride). *J. Appl. Phys.* **1984,** *56(9),* 2404.
6. Yang, D.; Chen, Y. β-phase Formation of Poly(vinylidene Fluoride) from the Melt Induced by Quenching. *J. Mater. Sci. Lett.* **1978,** *6,* 599.
7. Gregorio, Jr., R.; Cestari, M. Effect of Crystallization Temperature on the Crystalline Phase Content and Morphology of Poly(vinylidene Fluoride). *J. Polym. Sci., Part B. Polym. Phys.* **1994,** *32,* 859.

Poly(vinylidene fluoride-trifluoroethylene)

See: *Ferroelectric Polymers (Structural Phase Transitions)*
Vinylidene Fluoride-Trifluoroethylene Copolymers (Ferroelectric-to-Paraelectric Phase Transition)

POLY(VINYLPYRIDINIUM SALT)S

Wilmer K. Fife
Department of Chemistry
Indiana University–Purdue University at Indianapolis

Poly(vinylpyridinium salt)s have progressed from laboratory curiosities with unexpected and interesting solution properties to materials with importance to a number of application areas. Several reviews are available that collectively provide a comprehensive description of these materials and their properties.[1-9] This chapter will highlight the synthesis and selected properties and applications of polymers derived from 1-alkyl-2-,3- and 4-vinylpyridiniums and their derivatives, compounds as shown in **Structures 1-3**, respectively. Poly(4-vinylpyridinium salt)s, **Structure 4**, were first synthesized by Fuoss and Strauss and reported in 1948.[10] These interesting polycations have been investigated for usefulness as polyelectrolytes and polysoaps, complexing agents for synthetic and natural polyanions, electrically conducting matrices for enzymes such as glucose oxidase, membranes for removal of biological and chemical contaminants from water, supports for anion exchange and a variety of chemical reagents, models for hydrolytic enzymes, and selectively permeable membranes.[1,3,8,9,11-21]

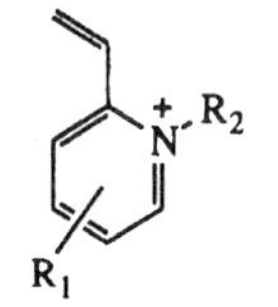

1

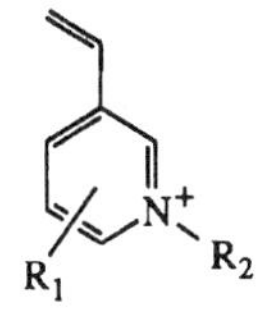

2

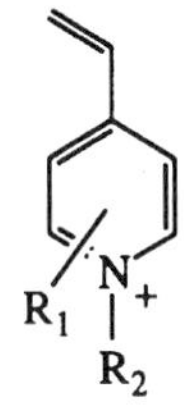

3

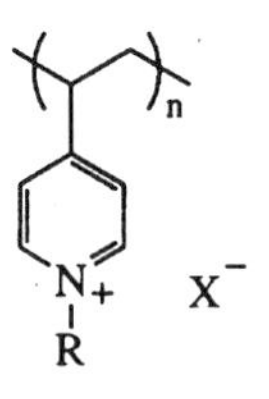

4

SYNTHESIS

Two general synthetic routes to the poly(vinylpyridinium salt)s have been utilized. The approach first described by Fuoss and Strauss involves radical initiated polymerization of a vinylpyridine monomer, for example 4-vinylpridine, followed by *N*-alkylation of the polymer with an alkyl halide or dialkyl sulfate.[10] The second procedure requires synthesis of an *N*-protium or *N*-alkyl-substituted vinylpyridinium monomer, **1**, **2**, or **3**, where R_2 = H, or alkyl. Polymerization of such a monomer leads directly to the corresponding poly(1-substituted-vinylpyridinium salt). Following the discovery by Shyluk in 1964 of spontaneous polymerization of 1,2-dimethyl-5-vinylpyridinium ions,[26] this second route to the polyvinylpyridiniums became well established in the 1970s from research led by Salamone in the United States, Ringsdorf in Germany, and Kabanov in Russia.[22,23,26-29]

Investigation of *N*-substituted-4-vinylpyridinium salts was restricted primarily to the *N*-protium and *N*-methyl monomers until 1993 when Fife and co-workers reported a general synthesis of 1-alkyl-4-vinylpyridiniums via treatment of 4-vinylpyridine with alkyl triflates ($ROSO_2CF_3$).[30] Homo- and copolymerization of these salts is readily accomplished by initiation with light, heat, free radicals, or bases–nucleophiles.[24,25]

PROPERTIES

Solubility

A dominant feature of polycations such as **Structure 4** is the influence of electrical charge on their physical and chemical behavior. However, the nature of the carbon skeleton can be decisive. Such attributes as solubility in water or organic solvents can be controlled by varying the structure of *N*-alkyl groups.[12-16] Polymers that carry small alkyl groups such as *N*-methyl or *N*-ethyl are charge-dominated, polyelectrolytes, while those with *N*-alkyl groups of six or more carbons are nonpolar, hydrocarbon-like materials.

Color

Another prominent and obvious property of polyvinylpyridiniums is their variable color. Fuoss and co-workers first noted that both monomeric and polymeric *N*-alkylpyridinium salts can be colorless, green, orange, or brown.[32,33] Boucher and Mollett have since established that the colors are due to formation of charge-transfer complexes and depend on the four-position substituent, counter-ion, temperature, and solvent used in preparation and purification.[34]

Chemical Properties

The polyvinylpyridiniums much like the monomolecular 1-alkylpyridiniums are stable to strong acids–electrophiles and oxidants, but readily undergo reactions with bases–nucleophiles and reducing agents.[35]The chemical properties of counter-ions have been widely utilized in both linear and crosslinked polyvinylpyridiniums. Several reagents that carry reactive counter ions are commercially available for use in industrial or laboratory scale oxidation, reduction, fluoride-ion exchange, and other applications which are considered in the applications section that follows.

APPLICATIONS

Poly(vinylpyridinium)s and Chemical Reagents

Polymer-supported reagents represent an important growth area in synthetic chemistry. **Table 1** contains a listing of polyvinylpyridinium-supported reagents that are currently available commercially.

TABLE 1. Polyvinylpyridinium—Supported Reagents

Support	Chemically Active Agent	Application	Reference
Homopolymer Supports			
Poly(2-vinylpyridinium) (Atactic, isotactic, syndiotactic)	$Cr_2O_7^{2-}$	Oxidation of alcohols to aldehydes or ketones	36
Poly(4-vinylpyridinium)	$Cr_2O_7^{2-}$		36
Poly(2-vinylpyridinium)	BH_3	Hydroboration/Reduction of aldehydes and ketones	37
Poly(4-vinylpyridinium)	SO_3	Sulfation, Sulfonation, Redox reactions	38,39
Solid-Phase Polymer Supports			
Poly(4-vinylpyridine-co-divinylbenzene-2 or 25%	Pyridine	Acid Scavenger, base/nucleophile	40
Poly(4-vinylpyridinium-co-divinylbenzene-2%)	$Cr_2O_7^{2-}$	Oxidant	41
Poly(4-vinylpyridinium-co-divinylbenzene-2%)	$ClCrO_3^-$	Oxidant	42,43
Poly(4-vinylpyridinium-co-divinylbenzene-2 or 25%)	$F(HF)_x^-$	Fluorinating Agent	44
Poly(4-vinylpyridinium-co-divinylbenzene-25%)	F	Anion Exchanger, F^-/X^-	45
Poly(4-vinylpyridinium-co-divinylbenzene-2%)	Br_3^-	Brominating Agent	46

Polyvinylpyridiniums as Catalysts

Polyvinylpyridiniums have been evaluated as catalysts for solvolysis of neutral and anionic substrates.[31]

Interpolyelectrolyte Complexes

The polyvinylpyridiniums form interpolyelectrolyte complexes (IPECs) with natural polyanions such as DNA and RNA, and with synthetic polyanions such as polyacrylate, polymethacrylate and poly(4-vinylbenzenesulfonate).[9]

Polyvinylpyridiniums as Selective Adsorbent Phases and Membranes

The polycationic nature of the poly(vinylpyridiniums) makes them likely candidates for use as stationary phases with ion-exchange properties and as films or membranes that exhibit discrimination toward potential transportable substrates.

Polyvinylpyridiniums as Electrically Conducting Media

Poly(vinylpyridiniums) can be used to "wire" enzymes and simpler chemical systems to electrodes. Heller and co-workers have developed an electrode that responds to glucose in aqueous solution via the enzyme, glucose oxidase, and poly(*N*-alkyl-4-vinylpyridinium) wiring.[17] Leidner and co-workers have demonstrated the ability of an electrostatically assembled interface to connect solution redox chemistry with a glassy carbon electrode.[47]

REFERENCES

1. Kunitaki, T.; Sinkai, S. *Adv. Phys. Org. Chem.* **1980,** *17,* 435.
2. Tsuchida, E.; Abe, K. In: Canton, H.-J., et al., Eds. *Advances in Polymer Science*; Springer-Verlag: New York; **1982,** *45,* 1.
3. Menger, F. M. In *Topics in Current Chemistry* (Biomimetic Bioorg. Chem. 3); Springer-Verlag: Berlin, 1986; Vol. 136.
4. Egorov, V. V.; Zubov, V. P. *Russ. Chem. Rev.* **1987,** *56,* 1153.
5. Philipp, B.; Dautzenberg, H.; Linow, K.-J.; Kötz, J.; Dawydoff, W. *Progr. Polym. Sci.* **1989,** *14,* 91.
6. Paleos, C. M. "Polymerization of Micelle-Forming Monomers." In: Paleos, C. M., Ed. *Polymerization in Organized Media* Gordon and Breach: Philadelphia, PA, 1992; Chapter 3, pp 183-214.
7. Kabanov, V. A. "Template Polymerization." In: Paleos, C. M., Ed. *Polymerization in Organized Media* Gordon and Breach: Philadelphia, PA, 1992; Chapter 7, pp 369-454.
8. Fife, W. K. "Molecular Association and Polymerization of 1-Alkyl-4-vinylpyridinium Ions." In: Dubin, P. et al., Eds.; *Macromolecular Complexes in Chemistry and Biology*; Springer-Verlag: Berlin, Heidelberg, 1994; Chapter 6, pp 71-93.
9. Kabanov, V. A. "Basic Properties of Soluble Interpolyelectrolyte Complexes Applied to Bioengineering and Cell Transformation" In: Dubin, P. et al., Eds.; *Macromolecular Complexes in Chemistry and Biology*; Springer-Verlag: Berlin, Heidelberg, 1994; Chapter 10, pp 151-174.
10. Fuoss, R. M.; Strauss, U. P. *J. Polym. Sci.* **1948,** *3,* 246.
11. Fuoss, R. M. *Science* **1948,** 545.
12. Strauss, U. P.; Gershfeld, N. L. *J. Phys. Chem.* **1954,** *58,* 747.
13. Strauss, U. P.; Gershfeld, N. L.; Crook, E. H. *J. Phys. Chem.* **1956,** *60,* 577.
14. Medalia, A. I.; Freedman, H. H.; Sinha, S. *J. Polym. Sci.* **1959,** *40,* 15.
15. Strauss, U. P.; Williams, B. L. *J. Phys. Chem.* **1961,** *65,* 1390.
16. Watermann, D.; Wall, F.T. *J. Phys. Chem.* **1960,** *64,* 581.
17. Heller, A. *Acc. Chem. Res.* **1990,** *23,* 128.
18. Kawabata, N.; Hayashi, T.; Matsumoto, T. *Appl. Environ. Microbiol.* **1983,** *46,* 203.
19. Akelah, A.; Sherrington, D. C. *Chem. Rev.* **1981,** *81,* 557.
20. Hodge, P.; Sherrington, D. C., Eds.; "Polymer-Supported Reactions in Organic Synthesis" Wiley: London, 1980.
21. Yoshikawa, M.; Suzuki, M.; Sanui, K.; Ogata, N. *J. Membr. Sci.* **1987,** *32,* 235.
22. Kabanov, V. A. *Pure Appl. Chem.* **1971,** *4,* 425.
23. Salamone, J. C.; Snider, B.; Fitch, W. L. *J. Polym. Sci., Part A-1* **1971,** *9,* 1493.
24. Ranganathan, P.; Fife, W. K.; Zeldin, M. *J. Polym. Sci., Part A, Polym. Chem.* **1990,** *28,* 2711.
25. Xin, Y.; Hu, Y.; Zeldin, M.; Fife, W. K. *Macromolecules* **1993,** *26,* 4670.
26. Shyluk, W. P. *J. Polym. Sci., Part A* **1964,** *2,* 2291.
27. Salamone, J. C.; Mahmud, M. U.; Watterson, A. C.; Olson, A. P. *J. Polym. Sci., Polym. Chem. Ed.* **1982,** *20,* 1153.
28. Martin, V.; Sutter, H.; Ringsdorf, H. *Makromol. Chem.* **1976,** *89,* 1976.
29. Kabanov, V. A.; Aliev, K. V.; Kargina, O. V.; Patrikeeva, T. I.; Kargin, V. A. *J. Polym. Sci., Part C* **1967,** 1079.
30. Fife, W. K.; Ranganathan, P.; Zeldin, M. *J. Org. Chem.* **1990,** *55,* 5610.
31. Letsinger, R. L.; Savereide, T. J. *J. Am. Chem. Soc.* **1962,** *84,* 3122.
32. Coleman, B. D.; Fuoss, R. M. *J. Am. Chem. Soc.* **1955,** *77,* 5472.
33. Kronick, P. L.; Fuoss, R. M. *J. Am. Chem. Soc.* **1955,** *77,* 6114.
34. Boucher, E. A.; Mollett, C. C. *J. Chem. Soc., Faraday Trans. I* **1982,** *78,* 1401.
35. Boulton, A. J.; McKillop, A., Eds. "Six-membered Rings with One Nitrogen Atom"; Part 2A. In *Comprehensive Heterocyclic Chemistry*; Katritzky, A. R.; Rees, C. W., Eds.; Pergamon: Oxford-New York, 1984; pp 1-98.
36. Tamani, B.; Goudarzian, N. *Polym. Bull.* (Berlin) **1990,** *23,* 295.
37. Menger, F. M.; Shinozaki, H.; Lee, H.-C. *J. Org. Chem.* **1980,** *45,* 2724.
38. Olah, G. A.; Vankar, Y. D.; Massoud, A. *Synthesis* **1979,** 984.
39. Hamada, Y.; Shiori, T. *Chem. Pharm. Bull.* **1982,** *30,* 1921.
40. Reilly Industries, Indianapolis, IN.
41. Frécht, J. M. J.; Darling, P.; Farall, M. J. *J. Org. Chem.* **1981,** *46,* 1728.
42. Fréchet, J. M. J.; Warnock, J.; Farall, M. J. *J. Org. Chem.* **1978,** *43,* 2618.
43. Bergbreiter, D. E.; Chendran, R. *J. Am. Chem. Soc.* **1985,** *107,* 4792.
44. Olah, G. A.; Li, X.-Y.; Wang, Q.; Prakash, S. *Synthesis* **1993,** 693.
45. Rubinsztajn, S.; Fife, W. K., unpublished results.
46. Fréchet, J. M. J.; Farall, M. J.; Nuyen, L. J. *J. Macromol. Sci. Chem.* **1977,** A11, 507.
47. Kasem, K. K.; Fife, W. K.; Zeldin, M.; Leidner, C. R. *J. Electroanal. Chem.* **1990,** *296,* 221.

Poly(N-vinylpyrrolidone)

See: Personal Care Application Polymers (Acetylene-Derived)
Poly(N-vinyl-2-pyrrolidone) Interactions

POLY(N-VINYL-2-PYRROLIDONE INTERACTIONS

Ligia Gargallo and Deodato Radić
Depto. Química Física
Facultad de Química
Pontificia Universidad Católica de Chile

POLY(N-VINYL-2-PYRROLIDONE)

General Physico-Chemical Considerations

The interaction of poly(*N*-vinyl-2-pyrrolidone) (PVP) with small consolutes in aqueous and non-aqueous media has been thoroughly studied.[1-50] This polymer shows a number of interesting properties. One of them is the high capability to interact with different kinds of molecules, such as iodine, detergents, drugs, dyes, aromatic compounds, carboxylic acids, and other compounds.[3-50] The biological behavior of poly(*N*-vinyl-2-pyrrolidone) and more specifically physiological and pharmacological aspects, are very important, because of the use of the polymer pharmaceutically and clinically in a variety of applications including that as a blood-plasma substitute, and because of its use in hair sprays and other aerosol products, droplets of which may be inhaled into the lungs.[51-56] This polymer has some similarities with proteins, especially serum albumin. In particular, PVP also shows strong interactions with anionic organic consolutes.

Poly(*N*-vinyl-2-pyrrolidone) (PVP) is one of the products developed by Reppe in his pioneering work in acetylene chemistry.[57-60] It is a water-soluble high polymer of many interesting properties.[60] Some of these, such as solubility, low toxicity, and physiological compatibility made Hecht and Weese utilize this compound as a plasma extender.[51]

BINDING PROPERTIES

Binding Properties of PVP with Small Cosolutes

Interaction of PVP with Methyl Orange and Its Homologs

The interaction of PVP with methyl orange, ethyl orange, propyl orange, and butyl orange has been studied by an equilibrium dialysis method at different temperatures (from 5°–35°C).[36] The binding constants and the thermodynamic parameters of the binding of methyl orange and its homologs by poly(*N*-vinyl-2-pyrrolidone) have been calculated.[36]

In the series of methyl orange and its homologs, increasing alkyl chain length of the dyes, in going from methyl orange to butyl orange, causes an increase in the first binding constant, that is, an increase in the free energy change and a decrease in the absolute magnitude in ΔH.

Interaction of PVP with Bromocresol Green

Studies involving the binding of methyl orange and its homologs onto poly(*N*-vinyl-2-pyrrolidone) show that substrates of greater hydrophobicity exhibited stronger binding.[36,63] It also demonstrated the significance of hydrophobic interactions.[61,62] Maruthamuthu and Dhandavel have reported the binding of Bromocresol Green (BCG) onto poly(*N*-vinyl-2-pyrrolidone) (PVP).[63]

Interaction of PVP with Aromatic Compounds

The binding equilibria for PVP in aqueous solutions with nitrobenzene, phenol, benzoic acid, sodium benzoate, and sodium naphthoate, have been studied over the temperature range 3–60° by the equilibrium dialysis method.[38] The binding constants at 30°, and, hence, the free energies of binding, increase as the size of the aromatic compound is increased, an increase in the number of polar groups, as in the series: phenol, resorcinol, phcloroglucinol, which also leads to an increase in the binding constant.

Interaction of PVP with Detergents

The interaction of sodium dodecyl sulfate (SDS) and aqueous poly(*N*-vinyl-2-pyrrolidone) was studied over a large range of SDS concentrations and several polymer concentrations by conductivity and viscosity measurements at 25°C.[7-11]

According to Fishman et al., the detergent would form submicellar clusters stabilized with the aid of PVP.[9] The adsorbed surfactant aggregates have aggregation numbers reported to be 30–50% lower than pure surfactant micelles.[64] Beyond a certain concentration of the surfactant that is higher than the critical micelle concentration (cmc), the system favors aggregation of any additional surfactant molecules into free micelles. This transition would occur because Coulombic repulsion among the adsorbed surfactant aggregates precludes any additional adsorption on the polymer.[24]

Interaction of PVP with Monoalkylxanthates

The interaction of poly(*N*-vinyl-2-pyrrolidone) with potassium monoalkyl xanthates was studied by UV-vis spectrophotometry and ultrafiltration measurements.[65]

PVP-Polymer Cooperative Interaction

Interpolymer Complex Formation

The cooperative interaction of two complementary polymers gives place to the formation of polycomplexes (PCs).[66,67] They have recently found several important applications: they are used for casting ultrafiltration membranes, in template polymerization, or in the design of polymeric materials to be used in interfaces.[68-70] PCs are stabilized through different types of interactions, but in this review we will be concerned only with hydrogen-bonded PCs because the system poly(*N*-vinyl-2-pyrrolidone) (PVP), a polybase, can interact with polyacids.[67,71-74]

The interaction of poly(monobenzyl itaconate) (PMBzI) and poly(*N*-vinyl-2-pyrrolidone) (PVP) in methanol solution was studied by Piérola et al.[74] A soluble interpolymer complex is formed whose properties depend on the method employed to prepare the sample.

The solution behavior of PVP is affected by specific interactions between the amide functionality and polar solvents (water, chloroform, alcohols, and amines), causing expansion of the polymer chain. Non-polar solvents such as benzene, toluene, and carbon tetrachloride also dissolve PVP if water is rigorously removed. The polymer readily forms complexes with a number of organic molecules and with molecular iodine.

PVP-INTERACTION IN BINARY SOLVENTS

When a polymer is dissolved in a binary solvent mixture, one of the solvents preferentially solvates the polymer, and brings about a change in the composition of solvent in the vicinity of the chain relative to the rest of the solution.[42,76-83]

The preferential adsorption behavior was studied of poly(*N*-vinyl-2-pyrrolidone) in mixtures containing aromatic isomeric components: 2-propanol–cumene, 2-propanol–mesitylene, 2-propanol–*p*-xylene, 2-propanol–ethylbenzene, and 2-propanol–toluene. These mixtures were chosen in order to investigate both the role played by the interaction involved in binary solvents and the steric effect of the solutes.[42]

CONFORMATIONAL BEHAVIOR OF PVP DEPENDENCE ON SOLVENT-INTERACTION

The properties of dilute solutions of poly(*N*-vinyl-2-pyrrolidone) and their unperturbed dimensions were determined several years ago by Meza and Gargallo.[75]

ACKNOWLEDGMENTS

The authors wish to express their appreciation to Mrs. Cristina Arriagada for her technical assistance in this work, and to the publishers and authors for permission to reproduce figures and tables from our publications as indicated specifically in the legends of them. We express our thanks to DIUC, Universidad Católica de Chile, and to FONDECYT for partial financial support.

REFERENCES

1. Molyneux, P. J. *Water Soluble Synthetic Polymers: Properties and Behavior*; 5th ed.; CRC. 1980; Vol. I (Chapters 1 and 4) and Vol. II (Chapter 2).
2. Flory, P. J. *Principles of Polymer Chemistry*; Cornell University; **1953,** p 523.
3. Siggia, S. *J. Am. Pharm. Assoc., Sci. Ed.* **1957,** *44,* 201.
4. Barkin, S.; Frank, H. P.; Eirich, F. R. In *Proceedings IUPAC Symposium on Macromolecular Chemistry*, Turin, 1954; pp 844-853.
5. PVP-Iodine, GAF Bulletin GAF: New York, 1962.
6. Schenck, H. V.; Sinak, P.; Haedicke, E. *J. Pharm. Sci.* **1979,** *68,* 1505.
7. Turro, N.; Baretz, B. H.; Kuo, P. L. *Macromolecules* **1984,** *17,* 1321.
8. Fischman, M. L.; Eirich, F. R. *J. Phys. Chem.* **1971,** *75,* 3135.
9. Fischman, M. L.; Eirich, F. R. *J. Phys. Chem.* **1975,** *79,* 2740.
10. Arai, H.; Murata, M.; Shinoda, K. *J. Colloid Interface Sci.* **1971,** *37,* 223.
11. Murata, M.; Arai, H. *J. Colloid Interface Sci.* **1973,** *44,* 475.
12. Cournoyer, R. F.; Sigia, S. *J. Polym. Sci., Polym. Chem. Ed.* **1974,** *12,* 603.
13. Kirsh, Y. E.; Soos, T. A.; Karaputadze, T. M. *Eur. Polym. J.* **1979,** *15,* 223.
14. Kirsh, Y. E.; Soos, T. A.; Karaputadze, T. M. *Eur. Polym. J.* **1983,** *19(7),* 639.
15. Molyneux, P. J.; Ahmed, G. S. *Kolloid-Z Polymere* **1973,** *251,* 310.
16. Saito, S. *J. Colloid Sci.* **1960,** *15,* 283.
17. Saito, S. *J. Colloid Interf. Sci.* **1967,** *24,* 227.
18. Saito, S. *Kolloid-Zu. Z. Polymere* **1967,** *215,* 16.
19. Saito, S. *Kolloid Zeitschrift* **1955,** 143(2), 66.
20. Bloor, D. M.; Wyn-Jones, E. *J. Chem. Soc., Faraday Trans 2* **1982,** *78,* 657.
21. Leung, R.; Shah, D. O. *J. Colloid Interf. Sci.* **1986,** 113(2), 484.
22. Aizawa, M.; Komatzu, T.; Nakagawa, T. *Bull. Chem. Soc. Jpn.* **1988,** *20,* 39.
23. García-López de Sa, T.; Garrido, L. M.; Allende, J. L. *Bull. Chem. Soc. Jpn.* **1988,** *20,* 39.
24. Cabane, B.; Duplessix, R. *Colloid and Surfaces* **1985,** *13,* 19.
25. Chibowski, S.; Szczypa, J. *Polish J. Chem. (Formerly Raczmiki Chemii)* **1982,** *56,* 359.
26. Lissi, E. A.; Abuin, E. *J. Colloid Interface Sci.* **1985,** 105(1), 1.
27. Higuchi, T.; Kuramoto, R. *J. Amer. Pharmaceut. Assoc. Sci. Edn.* **1954,** *43,* 393; 3, 98.
28. Guttmann, D.; Higuchi, T. *J. Amer. Pharmaceut. Assoc. Sci. Edn.* **1956,** *45,* 659.
29. Miyawaki, G. M.; Patel, N. K.; Kostenbuder, J. *J. Amer. Pharmaceut. Assoc. Sci. Edn.* **1959,** *48,* 315.
30. Bahal, C. K.; Kostenbauder, H. B. *J. Pharmaceut. Sci.* **1964,** *53,* 1027.
31. Scholtan, W. *Makromol. Chemie* **1953,** *11,* 131.
32. Oster, G. *J. Polym. Sci.* **1955,** *16,* 235.
33. Frank, H. P.; Barkin, S.; Eirich, F. R. *J. Phys. Chem.* **1957,** *61,* 1375.
34. Reeves, R. L.; Harkaway, S. A.; Sochor, A. R. *J. Polym. Sci., Polym. Chem. Ed.* **1981,** *19,* 2427.
35. Takagishi, T.; Imajo, K.; Nakagami, K.; Kuroki, N. *J. Polym. Sci., Polym. Chem. Ed.* **1977,** *15,* 31.
36. Takagishi, T.; Kuroki, N. *J. Polym. Sci., Polym. Chem. Ed.* **1973,** *11,* 1889.
37. Sardharwalla, I.; Lawton, J. B. *Polymer* **1985,** *26,* 751.
38. Molyneux, P.; Frank, H. J. *J. Am. Chem. Soc.* **1961,** *83,* 3169.
39. Molyneux, P.; Frank, H. J. *J. Am. Chem. Soc.* **1961,** *83,* 3175.
40. Bandyopadhyay, P.; Rodríguez, F. *Polymer* **1972,** *13,* 119.
41. Gargallo, L.; Radić, D. *Polymer* **1983,** *24,* 91.
42. Gargallo, L.; Radić, D. *Polym. Commun.* **1985,** *26,* 149.
43. Plaizier-Vercammen, J. A.; Néve, R. E. *J. Pharm. Sci.* **1982,** *71(5),* 552.
44. Plaizier-Vercammen, J. A. *J. Pharm. Sci.* **1983,** *72(9),* 1042.
45. Molyneux, P. J.; Vekavakayan Ondha, *J. Chem. Soc., Faraday Trans.* **1979,** *257,* 855.
46. Sheth, G. N. *J. Appl. Polym. Sci.* **1986,** *31,* 1227.
47. Sheth, G. N. *J. Appl. Polym. Sci.* **1986,** *32,* 4333.
48. Molyneux, P. J.; Carnarakis-Lentzos, M. *Colloid Polym. Sci.* **1979,** *257,* 855.
49. Molyneux, P.; Frank, H. P. *J. Am. Chem. Soc.* **1964,** *86,* 4753.
50. Léty-Sistel, C.; Sébille, B.; Quivoron, C. *J. Polym. Sci. Symposium* **1975,** *52,* 311.
51. Hecht, G.; Weese, H. *Münch. Med. Wochenschr.* **1943,** *90,* 11.
52. Thrower, W. R.; Campbell, H. *Lancet* **1951,** *260,* 1096.
53. Ravin, H. A.; Seligman, A. M.; Fine, J. *N. Engl. J. Med.* **1952,** *247,* 921.
54. Reynolds, J. E. F.; Prased, A. B. Eds. *Martindale: The Extra Pharmacopeia*; 28th Ed.; The Pharmaceutical: London, 1982.
55. Levy, G. B.; Caldas, I.; Fergus, D. *Anal. Chem.* **1952,** *24,* 1799.
56. Tsunemitsu, K.; Murakami, Y.; Toyoshima, K. *Polyvinyl Alcohol: Properties and Applications*; Finch, C. A., Ed.; John Wiley & Sons: London, 1973.
57. Coponhaver, J. W.; Bigelow, M. H. *Acetylene and Carbon Monoxide Chemistry*; Reinhold: New York, 1949.

58. Reppe, W. *Chemie und Technik der Acetylen-Druck-Reaktionen*, Weinheim: Verlag Chemie, 1951.
59. Reppe, W. *Neue Entwicklungen auf dem Gebiete der Chemie des Acetylens und Kohlenoxyde*; Springer Verlag: Berlin, 1949.
60. General Aniline and Film, *PVP, An Annotated Bibliography*, New York, 1951.
61. Nishide, H.; Yuasa, M.; Hasegawa, E.; Tsuchida, E. *Macromolecules* **1987**, *20*, 1913.
62. Maruthamuthu, M.; Sobhana, M. *J. Polym. Sci., Polym. Chem. Ed.* **1979**, *17*, 3159.
63. Maruthamuthu, M.; Dhandavel, R. *Makromol. Chem., Rapid Commun.* **1980**, *1*, 633.
64. Zana, R.; Lianos, P.; Lang, J. *J. Phys. Chem.* **1985**, *89*, 41.
65. Gargallo, L.; Pérez-Cotapos, J.; Santos, J.; Radic, D. *Langmuir* **1993**, *9*, 681.
66. Bekturov, E. A.; Bimendine, L. A. *Adv. Polym. Sci.* **1981**, *45*, 99.
67. Piérola, I. F.; Cáceres, M.; Cáceres, P.; Castellanos, M. A.; Núñez, J. *Eur. Polym. J.* **1988**, *24*, 895.
68. Desbrieres, J.; Rianudo, M. *Eur. Polym. J.* **1981**, *17*, 1265.
69. Smid, J.; Tan, Y. Y.; Challa, G. *Eur. Polym. J.* **1983**, *19*, 853.
70. Izumrudov, V. A.; Savitskii, A. P.; Bakeev, K. N.; Zezin, A. B.; Kabanov, V. A. *Makromol. Chem., Rapid Commun.* **1984**, *5*, 709.
71. Ohno, H.; Abe, K.; Tsuchida, E. *Makromol. Chem.* **1978**, *179*, 755.
72. Michaels, A. S.; Miekka, R. G. *J. Phys. Chem.* **1961**, *1*, 765.
73. Leiva, A.; Gargallo, L.; Radić, D. *Polym. Intern.* **1994**, *24*, 393.
74. Pérez-Dorado, A.; Piérola, I. F.; Baselga, J.; Gargallo, L.; Radić, D. *Makromol. Chem.* **1990**, *191*, 2905.
75. Meza, R.; Gargallo, L. *Eur. Polym. J.* **1977**, *13*, 235.
76. Dondos, A.; Benoit, H. *J. Polym. Sci., Polym. Phys. Ed.* **1977**, *15*, 137.
77. Hert, M.; Strazielle, C.; Benoit, H. *Makromol. Chem.* **1973**, *172*, 169.
78. Zivny, A.; Pouchly, J.; Solc, K. *Collect. Czech. Chem.* **1967**, *32*, 2753.
79. Gargallo, L.; Radić, D.; Katime, I. *Eur. Polym. J.* **1981**, *17*, 439.
80. Gargallo, L.; Radić, D.; Fernádez-Piérola, I. *Makromol. Chem., Rapid Commun.* **1982**, *3*, 409.
81. Radić, D.; Gargallo, L. *Polymer* **1981**, *22*, 1045.
82. Gargallo, L. *Makromol. Chem.* **1976**, *177*, 233.
83. Dondos, A. *Makromol. Chem.* **1977**, *178*, 7421.

POLYVINYLTETRAZOLES

Vadim V. Annenkov and Viktoriya A. Kruglova
Irkutsk State University

The first information about polyvinyltetrazoles (PVT) dates back more than 30 years. Originally interest in them depended on their high energy content that made it possible to use them as components of solid rocket propellants and explosives. At the same time, tetrazole derivatives have a number of peculiarities. They combine electron-donating pyridinic nitrogen atoms and an acceptor π-system within the same heterocycle. Tetrazoles with unsubstituted N–H protons are acids, so vinyltetrazoles are suitable subjects for the investigations of various factors' influences on the reactivity of monomers and for the synthesis of polymers with diverse properties.

Most of the known PVT are derivatives of *N*-vinyltetrazoles: 1-vinyltetrazole (1VT), 1-vinyl-5-methyltetrazole (1V5MT), 1-vinyl-5-aminotetrazole (VAT), and *C*-vinyltetrazoles: 5-vinyltetrazole (5VT), 1-methyl-5-vinyltetrazole (1M5VT), 2-methyl-5-vinyltetrazole (2M5VT), 5-isopropenyltetrazole (IPT).

PREPARATION OF POLYVINYLTETRAZOLES

Synthesis of Monomers

Synthesis of N-vinyltetrazoles

Early syntheses of *N*-vinyltetrazoles were carried out by the scheme including alkylation of tetrazoles by 2-chloroethanol, chlorination by thionyl chloride and dehydrohalogenation (**Scheme 1**).[1]

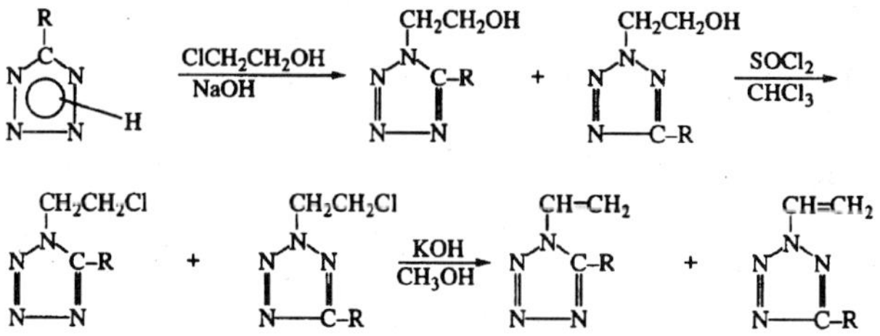

1

Polymerization of Vinyltetrazoles

Vinyltetrazoles readily polymerize by the radical mechanism in bulk, solution, and emulsion.[2] Usual initiators are applied, including redox initiators. The polymerization is also possible by the action of γ-rays, X-rays, electrons, or ultraviolet rays.

Copolymerization of Vinyltetrazoles

Vinyltetrazoles enter into the radical copolymerization with most vinyl monomers. Vinyltetrazoles' activities in the radical copolymerization with styrene are in a good agreement with their electronic structure.[3]

PROPERTIES OF POLYVINYLTETRAZOLES

Molecular weights of PVT depend on the synthesis conditions and vary from 30,000 to several millions. Density of the polymers is rather high (1.19–1.36 g/cm^3) and introduction of alkyl substituents into the ring decreases density. The glass-transition temperatures of poly-*C*-vinyltetrazoles are 373 K for p-5VT, 378 K for p-1M5VT, 328 K for p-2M5VT, 278-283 K for poly(2-propyl-5-vinyltetrazole), and 261-263 K for poly(2-butyl-5-vinyltetrazole).[4]

Chemical Properties of Polyvinyltetrazoles

Chemical properties of p-5VT and p-IPT are connected with the presence of acidic N–H proton and basic pyridinic nitrogen atoms. Alkylation of p-5VT by dimethylsulfate give copolymers of 1M5VT and 2M5VT with the isomer ratio 2:3. The reaction with methyl iodide provides only 70% degree of alkylation.[5]

Salts of p-5VT and p-IPT may be obtained readily by reaction of the polymers and bases. Interaction of p-5VT in water solutions with metal salts (Cu^{2+}, Ni^{2+}, Co^{2+}, Mn^{2+}) takes place by substitution of the acidic protons, such as in poly(acrylic acid).

Units of p-VAT contain different donor centers–amino group and pyridinic nitrogens. Its alkylation runs by nitrogen in the position 4 of the ring with obtaining of quaternized water-soluble polymer.

Thermal Properties of Polyvinyltetrazoles

In spite of the high energy content of PVT, thermal decomposition starts only beyond 470 K. Decomposition of p-5VT and its methylated derivatives is the most investigated. Destruction of p-1M5VT with release of nitrogen begins at 530 K through azidoazomethine rearrangement and penetration of the nitrene ~N: into C–H bond.

N_2 is the single gaseous product of the thermal decomposition of p-1VT. Cyanamide and carbodiimide fragments fixed on the polymer chain which are able to cyclotrimerize into triazines and isotriazines are identified in the condensed phase.[6]

Properties of Polyvinyltetrazoles Solutions

Solubility of poly(*N*-vinyltetrazoles) is restricted to the polar solvents such as DMFA and DMSO in contrast to polymers of *N*-vinylimidazoles and *N*-vinyltriazoles. This is because of the high polarity of tetrazole cycle leading to increased self-association of the polymers.[7] p-5VT is also insoluble in the most of organic solvents owing to the system of hydrogen and ionic bonds between tetrazole residues.[8]

Polyelectrolyte Properties of Poly(*C*-vinyltetrazoles)

As known, tetrazole and its *C*-derivatives possess significant N–H acidity comparable with that of carboxylic acids. So, polymers of 5VT and IPT show typical polyacid properties.[8a]

The tetrazole-containing polyacids are capable to form interpolymer complexes through hydrogen bonds with weak polybases in solution. The reaction of p-IPT with polyacrylamide (PAAm) and poly(ethylene oxide) (PEO) is displayed by increase of the solutions' pH and decrease of the viscosity. p-5VT reacts only with PAAm. Its lesser activity may be due to the involvement of p-5VT chains in the above-mentioned self association.

The reaction of p-5VT and p-IPT with a stronger polybase, poly(1-vinylimidazole), leads to precipitation of insoluble complex, stabilized by hydrogen bonds.[9]

AREAS OF POSSIBLE APPLICATION OF POLYVINYLTETRAZOLES

The initial practical interest in PVT was due to their high density, coupled with ability to destruct and to burn with evolution of large amounts of energy and gaseous products. So, they were recommended as components of solid fuel and pyrotechnical mixtures, igniters for rocket propellants, binders of plastic explosives and solid gas generating compositions.[2,10-12] For this purpose it was proposed to use the most of polyvinyltetrazoles, but p-2M5VT is the most promising in this field for its best solubility and compatibility with other components of burning compositions.[13,14]

Water-soluble tetrazole-containing polymers show a variable biological activity. Their acute toxicity varies from LD_{50} = 166 mg/kg for p-5VT to 1340 mg/kg for IPT-VPD copolymer. Similar to most of the polyacids, polymers of 5VT and IPT possess anticoagulant activity.[15] p-IPT is a direct blood anticoagulant with activity approaching to heparin but there are some differences in mechanisms of their action.

On the whole, works in the field of a practical application of the tetrazole-containing polyacids are in their infancy and will be progressing in the future.

REFERENCES

1. Finnegan, W. G.; Henry, R. A.; Skolnik, S. U.S. Patent 3 004 959, October 17, 1961.
2. Adicoff, A. U.S. Patent 3 036 086, May 22, 1962.
3. Kruglova, V. A.; Annenkov, V. V.; Ratovskii, G. V.; Shivernovskaya, O. A. *Vysokomol. Soedin, Ser. B* **1988,** *30,* 233.
4. Roshchupkin, V. P.; Nedel'ko, V. V.; Larikova,T. S.; Kurmas, S. V.; Afanas'ev, N. A.; Fronchek, E. V.; Korolev, G. V. *Vysokomol. Soedin, Ser. A* **1989,** *31,* 1726.
5. Kruglova, V. A.; Kizhnyaev, V. N.; Shafeev, M. A.; Busilova, S. R.; Gareev, G. A. *Fiz.-khim. Osnovy Sinteza i Pererab. Polimerov, Gor'kii* **1987,** 22.
6. Lesnikovich, A. I.; Levchik, S. V.; Ivashkevich, O. A.; Bolvanovich, E. E.; Gaponik, P. N.; Korsunskii, B. L.; Nedelko, V. V. *Thermochim. Acta* **1993,** *215,* 303.
7. Kizhnyaev, V. N.; Astakhov, M. B.; Smirnov, A. I. *Vysokomol. Soedin. Ser. A* **1994,** *36,* 104.
8. Kizhnyaev, V. N.; Kruglova, V. A.; Annenkov, V. V.; Vereshchagin, L. I. *Vysokomol. Soedin., Ser. B* **1989,** *31,* 420.

8a. Annenkov, V. V.; Kruglova, V. A. *J. Polym. Sci., Polym. Chem. Ed.* **1993,** *31,* 1903.

9. Annenkov, V. V.; Kruglova, V. A.; Shevtchuk, O. A. *Vysokomol. Soedin., Ser. B* **1990,** *32,* 723.
10. Finnegan, W. G.; Henry, R. A. U.S. Patent 3 055 911, September 25, 1962.
11. Burkardt, L. A.; Finnegan, W. G.; Smith, R. L. U.S. Patent 3 332 353, July 25, 1967.
12. Henry, R. A. U.S. Patent 3 383 389, May 14, 1968.
13. Chang, M. S.; Highby, J. H.; Mackenzie, G. L. U.S. Patent 3 954 528, May 4, 1976.
14. Mishra, I. B.; Kieft, L. J. V. U.S. Patent 4 875 949, Oct. 24, 1989.
15. Kruglova, V. A.; Annenkov, V. V.; Moskvitina, L. T.; Boyko, N. M.; Busilova, S. R.; Kazimirovskaya, V. B.; Kizhnyaev, V. N.; Levina, M. N. *Khim.-Farm. Zh.* **1989,** *23,* 195.

Polyxylylenes

See: *Aromatic Hydrocarbon-Based Polymers*
Poly(p-xylylene)s (Structure, Properties, and Applications)
Poly(p-xylylene)s, Coatings and Films
Vapor Deposition Polymerization ([2.2]Paracyclophane and Analogues)

POLY(*p*-XYLYLENE)S (Structure, Properties, and Applications)

Andreas Greiner
Philipps-Universität Marburg
Phyikalische Chemie/Polymere
Hans-Meerwein-Strasse

The synthesis as well as the chemical and physical properties of poly(*p*-xylylene) (PPX) have drawn considerable fundamental and economical interest. Intensive research on PPX with regard to structure–property relationships and applications was mostly driven by progress in the synthesis of PPX. The vapor phase pyrolysis of *p*-xylene yielded PPX as yellowish insoluble film-forming material.[1] Soluble linear PPX was obtained by the vapor phase pyrolysis of paracyclophane (vapor coating process) initiating intensive research on structure–property relationships and potential applications.[2] The synthesis and analysis of PPX other than by vapor phase pyrolysis are limited by the poor solubility of PPX.

A variety of excellent reviews have summarized the up-to-date data concerning the synthesis and structure–property relationships of PPX.[3]

GLASS TRANSITION TEMPERATURES (T_g), MELTING POINTS (T_M), AND SOLUBILITIES OF PPXS, AND THE CRYSTALLIZATION OF PPX

PPX prepared by vapor phase pyrolysis of paracyclophane is obtained as a hardly soluble, partially crystalline, film-forming material.[2] It is soluble above 250°C in exotic solvents as chlorinated biphenyls or benzylbenzoate. The glass transition of PPX is at 80°C[2] although this seems to be controversial.[5] The crystalline melting point of PPX is around 420°C where it starts to decompose. It has never been obtained completely amorphous. It exhibits polymorphism namely an α- and a β-modification with major peaks at $2\theta = 16.79°$ and 22.52° for the α-modification and $2\theta = 20.00°$ for the β-modification.[6] The transition from the α- to the β-modification is observed at 220°C[5] and is irreversible, unless it is annealed for a long term period at 352°C.[7]

An additional transition is observed at 270°C assigned as $\beta_1 \rightarrow \beta_2$ which is reversible.[4]

Based on solubility tests linear chains are claimed for PPX prepared by the pyrolysis of paracyclophane[2] but crosslinked PPX is expected on pyrolysis of xylenes. Linear chains of PPXs with substituents on the phenylene segments prepared by vapor phase pyrolysis of paracyclophanes, respectively, dehydrohalogenation of α-halo-xylenes were confirmed by gelpermeation chromatography. T_g of the phenyl-, the benzyl-, and the phenylethyl-substituted PPX shift from 107°C to 53°C, to 33°C. The completely amorphous benzoyl-substituted PPX shows the highest T_g reported for PPXs (T_g = 146°C). The phenyl-ethyl-substituted PPX is crystalline in contrast to the phenyl-, benzyl-, and benzoyl-substituted PPXs, which are completely amorphous.

MECHANICAL PROPERTIES

The Young's moduli of PPXs with substituents on the phenylene segments are significantly higher than the tensile modulus of PPX at room temperature. The Young's moduli of PPXs drop significantly at 200°C. The elongation at break is rather low for PPX, but significantly higher for the alkyl-substituted PPXs and the chloro-substituted PPX, respectively, which was also confirmed by the investigations of Spivack.[8]

A theoretical tensile strength at break of 23 GPa was calculated for PPX.[9] This value was verified by the investigation of PPX fibers drawn at 420°C and extrapolation of the data. The fibers are characterized by a tensile strength at break of 3.0 GPa, a Young's modulus of 102 GPa, and a strain at break of 3%. An ultimate tensile strength at break of 23 GPa and an ultimate Young's modulus of 208 GPa was extrapolated by plotting the reciprocal tensile strength at break and the reciprocal Young's modulus, respectively, versus the reciprocal draw ration.

ELECTRICAL PROPERTIES

The electrical properties of PPX are of interest for its high dielectric constant of 2.65 and for its low dissipation factor of 0.0002. The electrical properties of PPXs with substituents on the phenylene segments do not reach these values probably due to their more polar character.

OPTICAL PROPERTIES

No significant absorption is detected in the UV/Vis-spectra of PPX and chloro-substituted PPX above 300 nm which are, therefore, transparent.[3c]

THERMAL STABILITY OF PPXs

The thermal stability of PPXs depends on the synthetic route, on the substituents, on the thermal history of the sample, and on additives.

PPX prepared by Hofmann degradation is significantly more stable than PPX prepared by pyrolysis of xylene.[11] The thermal stability in air of PPXs with chloro-substituents on the phenylene segments is significantly higher than PPX.[3c,10,13,14] Degradation of PPX can be explained in terms of cleavage of carbon–carbon bonds and by reaction of terminal benzyl radicals with oxygen to peroxides resulting in deterioration of polymer properties.[3c,12]

APPLICATIONS OF PPXs

Most applications of PPXs are based on their coating properties: PPX coatings are transparent, pinhole free, insulating with low dissipation factor, noncorrosive, flexible, stress free, and biocompatible. Especially delicate articles like thermistors, circuit boards, or capacitors can be coated accurately by PPXs directly on polymerization. Even very delicate articles like butterfly wings can be coated by PPX preserving original structures. From here applications in a wide variety of conservation applications in museums may result.[14]

The promising electrical properties of PPX for applications as coating materials in microelectronics are shown by a comparative study of different materials such as PPX, acrylics, epoxides, silicones, polyurethanes, and glass.[15] PPX is useful as anticorrosion coating for high frequency resonators which was also demonstrated by a comparative study.[16] The electrical conductivity of uncoated polyacetylene doped with AsF_5 decreased by a factor of more than 20 in comparison to PPX coated polyacetylene doped with AsF_5.[17] Free standing, pinhole free films of PPX thinner than 1000 Å have been prepared and used for optical devices such as beam splitters in optical instruments.

REFERENCES

1. Szwarc, M. *Nature* **1947,** *160,* 403.
2. Gorham, W. F. *J. Polym. Sci., Part A-1* **1966,** *4,* 3027.

3. a) Errede, L. A.; Szwarc, M. *Quart. Rev.* **1958,** *12,* 301. b) Frazer, A. H.; In *High Temperature Resistant Polymers*; Wiley-Interscience: New York, 1968; c) Gorham, W. F.; In *Encyclopedia of Polymer Science and Technology*; Wiley-Interscience: New York, 1974; d) Niegisch, W. D. In *Encyclopedia of Polymer Science and Technology*; Wiley-Interscience: New York, 1974; e) Szwarc, M. *Polym. Eng. Sci.* **1976,** *16,* 473; f) Beach, W. F.; Lee, C.; Bassett, D. R.; Austin, T. M.; Olson, R. In *Encyclopedia of Polymer Science and Technology*; Wiley & Sons: New York, 1989; g) Iwatsuki, S. *Adv. Polym. Sci.* **1984,** *58,* 93.
4. Niegisch, W. D. *J. Appl. Phys.* **1966,** *37,* 4041.
5. Kirckpatrick, D. E.; Wunderlich, B. *Makromol. Chem.* **1985,** *186,* 2595.
6. Iwamoto, R. I.; Bopp, R. C.; Wunderlich, B. *J. Polym. Sci., Polym. Phys. Ed.* **1975,** *13,* 1925.
7. Kirckpatrick, D. E.; Wunderlich, B. *J. Polym. Sci.,Part B Polym. Phys. Ed.* **1986,** *24,* 931.
8. Spivack, M. A. *Rev. Sci. Instr.* **1972,** *43,* 985.
9. van der Werff, H.; Pennings, A. J. *Polym. Bull.* **1988,** *19,* 587.
10. Nowlin, T. E.; Foss Smith, D., Jr.; Cieloszyk, G. S. *J. Polym. Sci., Polym. Chem. Ed.* **1980,** *18,* 2103.
11. Schaefgen, J. R. *J. Polym. Sci.* **1959,** *51,* 133.
12. Shaw, R. G.; Yeh, Y. L.; Lewis, J. W. U.S. Patent 3 503 903, 1970.
13. Baker, T. E.; Fix, G. L.; Judge, J. S. *J. Electrochem. Soc.* **1980,** *127,* 1851.
14. Grattan, D. W. *Can. Chem. News* **1989,** 25.
15. Smetana, W.; Wiedermann, W.; Fasching, G. M.; *Kunststoffe* **1985,** *75,* 174.
16. Ostwald, R.; Bogenschütz, A. F. *Metalloberfläche* **1988,** *42,* 427.
17. Osterholm, J. E.; Yasuda, H. K.; Levenson, L. L. *J. Appl. Polym. Sci.* **1982,** *27,* 931.

POLY(*p*-XYLYLENE)S, COATINGS AND FILMS

Larissa Alexandrova* and Ricardo Vera-Graziano
Instituto de Investigaciones en Materiales, UNAM

Poly(*p*-xylylene)s, PPXs, are polymers of commercial importance with their main applications as engineering coatings in the electronics industry, not only because of their excellent properties, but rather because of the process by which they are formed. This is the so-called vapor deposition polymerization process, by which the monomer polymerizes spontaneously with 100% conversion during deposition from the gas phase on a substrate, and no solvent or catalyst is required. The solid polymer is formed directly from the gaseous monomer during deposition without any discernable intervening liquid phase. It is possible, therefore, to produce almost defect-free coatings of uniform thickness on the substrate–even on those showing complicated surface profiles.

This process was first developed by Gorham, using the cyclic dimer of *p*-xylylene, [2.2]paracyclophane, as the starting material. By means of its pyrolysis in vacuum, an extremely reactive monomer, *p*-xylylene, was generated and during vapor deposition of the latter on a substrate at room temperature high molecular weight, linear poly(*p*-xylylene) (PPX) was produced (see **Figure 1**).

PRODUCTION

At present commercial vapor deposition equipment is supplied by Union Carbide Co., and the term "Parylenes" is used to refer to the PPXs formed by the Gorham process.

*Author to whom correspondence should be addressed.

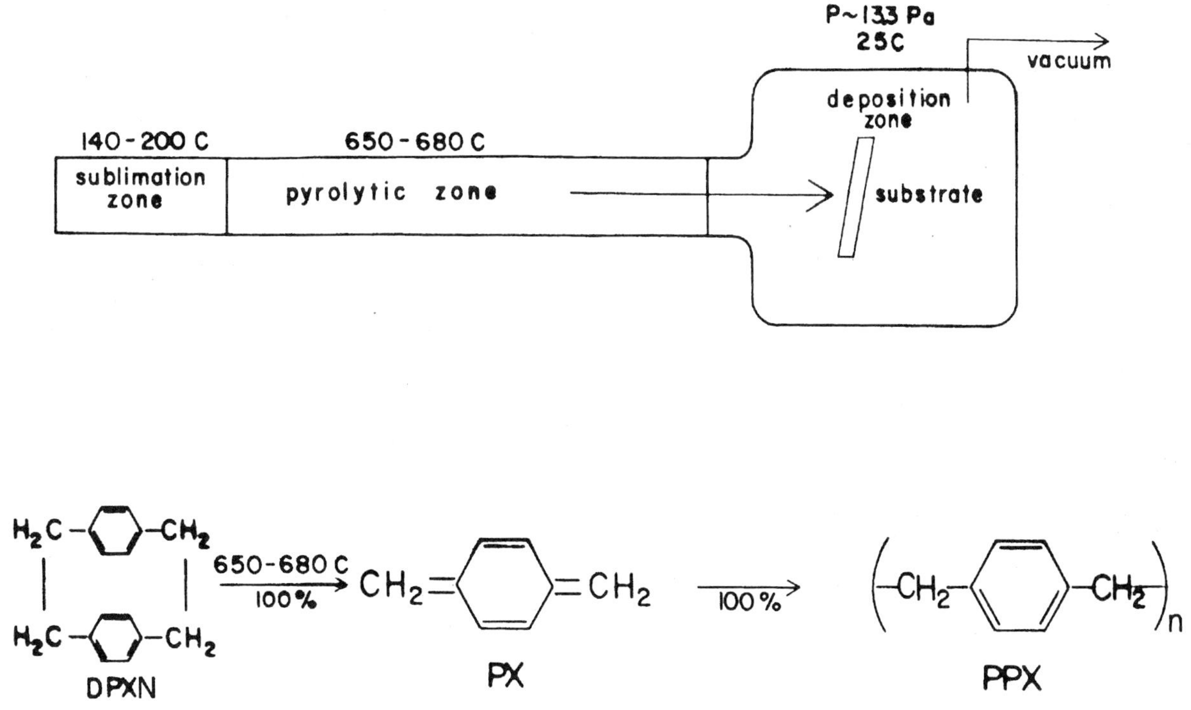

FIGURE 1. Parylene deposition apparatus and chemical scheme of the process.

The very mild pyrolytic conditions make the production of different substituted PPXs possible through the use of the proper substituted cyclophanes as starting materials. There are data available in the literature reporting the production of PPXs based on dibromo-, dicyano-, dimethyl-, and the other substituted cyclodi-*p*-xylylene (DPXN), dichloro-*p*-xylylene (DPXC), and tetrachlorodi-*p*-xylylene (DPXD).[1,2] All of these are supplied by Union Carbide Co., and the polymers based on them are known as Parylene N [CAS 25722-33-2], Parylene C [CAS 9052-19-1], and Parylene D [CAS 52261-45-7], respectively.[3]

FEATURES OF THE MONOMERS

The structure and reactivity of the PXs have been studied.[5,6] Theoretical calculations show that in the ground state PX has a planar tetraolefinic (or quinonoid) structure and is diamagnetic (singlet), while in the excited state it has a benzenoid structure and is biradical (triplet).

STRUCTURE AND MORPHOLOGY

Parylenes formed by the Gorham process, under typical conditions, are highly crystalline polymers. The crystal structure and the morphology of Parylene N are well known; not so its glass transition temperature (T_g).[7-10] According to recent data the T_g of Parylene N is 13°C, and the range over which T_g affects the heat capacity measurements extends from −33°C to +57°C.[8] The other Parylenes are supposed to have T_gs very close to the Parylene N one.

PPX exists in two crystalline forms, α and β. Coating, formed by the usual Gorham process, shows the α-crystalline structure, but it may change to β-form either by heating or stretching.[9,11] It has been found that the β-form is also polymorphous: two of its forms β_1 and β_2 exist.[9]

The structure and morphology of PPXs strongly depend on the substrate temperature. So, PPX produced in the temperature range −17° to +26°C is composed of α crystals with a 60% crystallinity. Outside this range, the crystals usually show the β-phase.

COPOLYMERS

Random copolymers are produced when using as starting materials asymmetric cyclodi-*p*-xylylenes (an odd number of substitutes in benzoic rings) or by simultaneous deposition of the jointly or separately pyrolyzed cyclodi-*p*-xylylenes with different substitutes. In this way, PX can be copolymerized with chloro-, bromo-, *n*-butyl, ethyl-, carbomethoxy-, and other substituted *p*-xylylenes.[4,12] Varying the substrate temperature results in copolymers of different structures.

The copolymers obtained in this way are linear soluble products. They form very transparent amorphous films with fusion points lower than the melting point of PPX. Copolymer coatings of PX with vinyl monomers, such as maleic anhydride, 4-vinylpyridine, *N*-vinylpyrrolidine, and styrene can also be obtained by the vapor deposition technique.[13]

PROPERTIES

Most mechanical properties of the PPXs are in the same range as those of conventional plastics. They vary considerably with aging and annealing.[14] However PPXs keep their elasticity even at very low temperatures. The bulk barrier properties of Parylenes are outstanding, among the best of organic polymeric coatings.

At room or near room temperature most PPXs are inert to organic solvents.[2-4] Even the best solvents (e.g., xylene) cause a swelling of the commercial PPXs of no more than 3%; concentrated sulfuric acid at 150°C does not produce any noticeable alteration of PPXN.

PPXs exhibit stable dielectric properties over a wide range of temperatures. The dielectric constant and dielectric loss are low and, most important, unaffected by absorption of atmospheric water vapor. The dielectric constant and dissipation factor of Parylenes do not change significantly.

Light absorption of PPXs in the visible range is minimal. They normally form transparent films, particularly PPXC.[3,4] The surface energies of the three commercial PPXs, estimated by contact angle measurements, were about the same, with a value of approximately 45 mJ/m². Gas-plasma treatments of the Parylene surface raise the surface energy and improve the adhesion properties.[15]

Investigations on the thermal and thermo-oxidative stability of commercial Parylenes have shown that the initial degradation temperatures were 270°C for PPXN and PPXC, but 320°C for PPXD at a heating rate of 10°/min in air. Under nitrogen atmosphere all polymers started to be destroyed at 450°C.

APPLICATIONS

Parylenes are in general used in the electronic and microelectronic industry as protective and insulation coatings for circuit boards, wiring assemblies, and hybrid circuits. The very low dielectric loss of Parylenes allows their application as dielectric materials in high quality capacitors. Because of the possibility of formation of uniform coatings of any desirable thickness, Parylenes have also been used as electroinsulation coatings in the manufacture of miniature transformers and motor armatures. Additionally they have been applied for better preservation of old books and documents, since relatively thin and very transparent coatings of Parylene reduce their brittleness and make them much stronger. This concept has been extended to other fragile artifacts. Since PPX coatings can be deposited onto very small particles of any shape, they are used for the encapsulation of strong reducers, oxidants, alkaline materials, their hydroxides, and other moisture sensitive compounds. Coated with only a small amount of Parylenes these substances become less moisture sensitive and remain free-flowing after exposures to ambient conditions.[3,4] Drug particles microencapsulated with PPXs provide effective control-release-activity.[16] PPXs have been proposed as protective coatings for magnetic and electroconductive polymer films and to provide corrosion protection for a number of thin, porous inorganic materials.[17,18] PPXs have very low friction coefficients, allowing the use of Parylene coatings as a dry lubricant on the bearing surfaces of miniature stepping motors.[3] Parylenes may be used in medicine for coating several medical instruments and accessories because of their good biocompatibility.[3,4]

The vacuum deposition technique for the formation of PPXs gives the possibility of their codeposition with any material which can be deposited by thermal evaporation. In this way new composite materials based on PPXs can be obtained. Thus, metal loaded poly(*p*-xylylene) films have been produced.[19] Since the process of producing PPXs is conducted within the walls of a vacuum chamber this does not pollute the environment and reduces to a minimum human contact with hazardous chemicals.

REFERENCES

1. Gorham, W. F. (to Union Carbide Cor.), U.S. Patent 3 342 754, September 19, 1967.
2. Gorham, W. F. *J. Polym. Sci.* **1966,** A-1, 4(2), 3027, and references therein.
3. Beach, W. F.; Lee, C.; Basset, D. R.; Austin, T. M.; Olson, R. In *Encyclopedia of Polym. Sci. and Eng., 17,* 2nd ed.; Wiley: New York, 1989; p 990.
4. Kardash, I. E.; Pebalk, A. V.; Pravednikov, A. N. Chemistry and Application of Poly(*p*-xylylene)s, Rev., *Uspehi Khimii,* Ser.Khim.i Tekhnol. Vysokomol.Soed., Itogi Nauki i Tekhniki, VINITI, Akad. Nauk. SSSR, 1984, *119,* 66.
5. Coulson, C. A. *Disc. Far. Soc.* **1947,** *2,* 36 and 2, 9; Namiott, A.; Daitkina, M.; Syrkin, Ya.K. *Dokl. Akad. Nauk SSSR* **1945,** *48(4),* 285.
6. Montgomery, L. K.; Huffman, J. C.; Jurczack, E. A.; Grendze, M. P. *J. Am. Chem. Soc.* **1986,** *108,* 6004, and references therein.
7. Kubo, S.; Wunderlich, B. *J. Polym. Sci., Poly. Phys. Ed.* **1972,** *10,* 1949.
8. Kirkpatrick, D. E.; Wunderlich, B. *Makromol. Chem.* **1985,** *186,* 2595.
9. Miller, K. J.; Hollinger, H. B.; Grebowic, J.; Wunderlich, B. *Macromolecules* **1990,** *23,* 3885, and references therein.
10. Katayama, K. I. et al. *Polymer* **1983,** *24,* 1155; *J. Polym. Sci., Polym. Phys. Ed.* **1984,** *22,* 669.
11. Murthy, N. S.; Hyo-gun-Kim *Polymer* **1984,** *25,* 1093.
12. Gorham, W. F. (to Union Carbide Corp.) U.S. Patent 3 288 728, November 29, 1966.
13. Kardash, I. E. et al. *Dokl. Akad. Nauk SSSR* **1991,** 319(1), 173.
14. Nowlin, T.; Smith, D. F.; Cieloszyk, G. S. *J. Polym. Sci., Polym. Chem. Ed.* **1980,** *18,* 2103.
15. Nowlin, T. E.; Smith, D. F. *J. Appl. Polym. Sci.* **1980,** *25(8),* 1691.
16. Versic, R. J.; Selenke, W. (to Dodge, Ronald T., Co.) *PCT Int. Appl.* WO 90 02 546, March 22, 1990.
17. Koichi, M.; Susumu, K.; Shibu, O. *Fujitsu Sci. Techn. J.* **1973,** 9(4), 153; Osterholm, J. E.; Yasuda, H. K.; Levenson, L. L. *J. Appl. Polym. Sci.* **1982,** *27(3),* 931.
18. Olson, R. A.; McGuire, P. F.; Frank, W. A. (to Union Carbide Chem. & Plast. Co., Inc.) Eur. Patent Appl. EP 448 016, September 25, 1991; Nemoto, M.; Nakayama, M.; Yajima, K.; Ueda, K.; Shibahra, M. (to TDK Corp.) *Jpn. Kokai Tokkyo Koho* JP 03 41 703 [91 41,703], February 22, 1991.
19. Bieg, K. W. *J. Vac. Sci. Technol.* **1981,** *18(3),* 1231.

Polyylides

See: Polyzwitterions (Overview)

Polyzwitterions

See: Polyzwitterions (Overview)
Zwitterionic Polymers (Dilute Solution and Rheological Properties)
Zwitterionic Polymers (Ionic Conductivity)

POLYZWITTERIONS (Overview)

Jean-Claude Galin
Institut Charles Sadron (CNRS-ULP)

Zwitterions may be defined as electrically neutral molecules which carry an equal number (most often one for low molecular weight compounds) of positively and negatively charged sites linked though covalent bonds and typically not electronically conjugated with each other. They are dipolar species (and not "dipolar ions," an incorrect although still surviving term) quite different from salts (or electrolytes) where the cation and anion are separate particles which can be completely dissociated in a medium of high enough dielectric permittivity. This zwitterion definition actually includes a very broad variety of structures, but the polyzwitterions discussed in this review will be restricted to those with net formal positive and negative charges stable under a wide range of physico-chemical conditions (pH, polarity...) and occurring in strictly equal concentration on each individual macromolecular chain.

SYNTHESIS OF POLYZWITTERIONS

Polyylides

Polyylides may be considered as the first and well-identified sub-class among the polyzwitterions. Many studies have been devoted to crosslinked resins functionalized with transient and reactive phosphonium ylides as immobilized Wittig reagents for alkene synthesis.[2] Strong delocalization of the negative charge is a requirement for resonance stabilization of the ylide structure, and it implies highly electrowithdrawing groups as substituents on the anionic site. Polymeric ylides were thus obtained by free-radical homopolymerization and copolymerization of the corresponding monomers or by functionalization of a polymeric precursor.

Polyzwitterions with Variable Intercharge Distances

Free Radical Polymerization and Copolymerization of Zwitterionic Monomers

The zwitterion synthesis previously discussed may be transposed to tertiary amine bearing monomers in the vinyl or (meth)acrylic series and some representative crystalline monomers selected from dozens of polymerizable species described in the literature are given in **Table 1**. The ammoniopropanesulfonate monomers 4, 8, and 18 are already industrial products [Rashigh A. G., Ludwigshafen, Germany].

In most cases, the dipolar moiety is disconnected from the polymerizable double bond by a long enough spacer and the zwitterionic monomer roughly shows fairly similar trends towards polymerization as its amino precursor. In the other cases, such as for 4-vinylpyridinioethanecarboxylate or 4-vinylpyridinioethane or butanesulfonates [monomers 17 and 20] the electronic distribution is significantly shifted with respect to that of 4-vinylpyridine, but without perturbing the polymerizability.[3,4] For hydrophilic monomers ($A^- = CO_2^-$, SO_3^-), polymerization is generally initiated in homogeneous aqueous solution by classical red-ox initiators ($K_2S_2O_8$, $K_2S_2O_8/NaS_2O_5$) or soluble azo derivatives such as 4,4′-azobiscyanovaleric acid.

SOLUTION PROPERTIES OF THE POLYZWITTERIONS

Hydrophilic Polyzwitterions in Aqueous Solution

Polyammonioalkanesulfonates are the most representative because of the stability of their dipolar moiety over a broad pH range. In most cases they are water insoluble at room temperature,

TABLE 1. (Meth)acrylic and Vinylheterocyclic Zwitterionic Monomers and Glass Transition Temperatures of their Corresponding Homopolymers

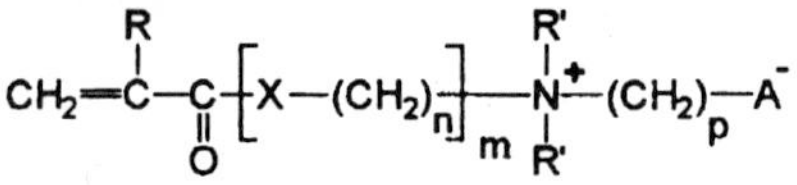

N°	R	X	n	m	R'	p	A^-	Tg °C	References
1	CH_3	O	2	1	CH_3	2	CO_2^-		29
2	H	O	2	1	CH_3	2	SO_3^-		19
3	H	O	2	1	CH_3	3	SO_3^-		6
4	CH_3	O	2	1	CH_3	3	SO_3^-	250[a]	47
5	CH_3	O	2	2	C_2H_5	3	SO_3^-	133[b]	5
6	CH_3	O	2	3	CH_3	3	SO_3^-		
7	H	NH	3	1	CH_3	3	SO_3^-		48
8	CH_3	NH	3	1	CH_3	3	SO_3^-	260[a]	47
9	CH_3	NH	0	1	CH_3	4	SO_3^-		47
10	CH_3	O	2	1	CH_3	2	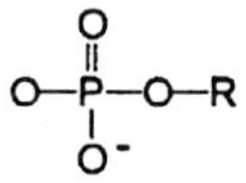		49
11	H	O	2	1	CH_3	2		130	
12	H	O	2	1	CH_3	3		115	
13	CH_3	O	2	1	CH_3	2	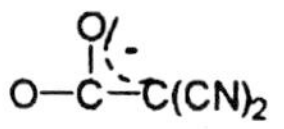	183	1,28
14	CH_3	O	2	1	CH_3	3		149	
15	CH_3	O	2	2	C_2H_5	3		89	

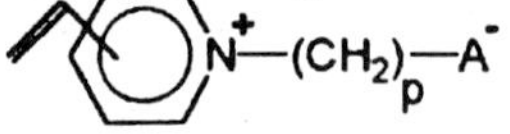

N°	Isomer	p	A^-	Tg °C	References
16	2	2	SO_3^-		19
17	4	3	SO_3^-		19
18	2	3	SO_3^-		5
19	4	3	SO_3^-		5
20	4	4	SO_3^-		11
21	2	2			1,28
22	2	3			
23	4	2	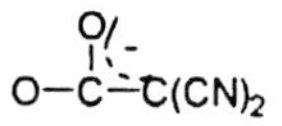	206	
24	4	3		187[c]	

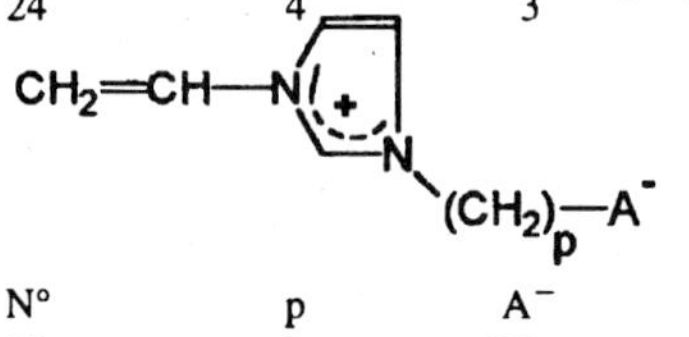

N°	p	A^-	Tg °C	References
25	2	SO_3^-		19
26	3	SO_3^-		50
27	4	SO_3^-		50
28	2	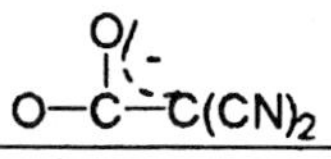	216	1,28
29	3		196[c]	

[a]*Source*: Reference 31.
[b]*Source*: Reference 30.
[c]*Source*: Reference 33a.

but a temperature increase or, better, the addition of an electrolyte generally results in solubilization.[9,11,15-19] Comparison of the minimum critical salt concentrations (CSC) necessary to dissolve the polyzwitterions allows estimation of the respective influences of the cation and of the anion.

Very similar "salting-in" effects in aqueous solution are also observed in the three following cases: (a) zwitterionic copolymers with neutral comonomers such as *N*-vinylpyrrolidone or 2-vinylpyridine;[20,21] (b) crosslinked zwitterionic hydrogels, where swelling is an increasing function of the salt concentrations;[8,9]

(c) formation of water-soluble polyzwitterion-polyelectrolyte "complexes" of no definite stoichiometry and of high viscosity.[22]

Hydrophobic Polyzwitterions in Organic Solvents

Copolymers

For copolymers associating a majority of weakly polar or apolar A units and a minority of zwitterionic B units in their chain, the so-called "zwitterionomers," the solution properties in nonpolar organic solvents are dominated by association through dipolar interactions between the unsolvated zwitterions, as clearly shown on three typical systems: a) semi-telechelic, telechelic or multifunctional star polyisoprenes bearing ammoniopropanesulfonate endgroups;[23,24] b) statistical styrene copolymers bearing ammonioalkanecarboxylate side groups N^+-$(CH_2)_p$-CO_2^- ($p = 1, 4$);[14] c) telechelic polystyrenes and polyisoprenes functionalized as in b).[25]

Polymer–Solvent Interactions and Unperturbed Dimensions

The morphological and hydrodynamical properties of polyzwitterions in dilute solution have been analyzed only in a few cases.[15,16,26-28]

BULK PROPERTIES OF POLYZWITTERIONS

Most studies are related to atactic polyammoniopropanesulfonates which are generally amorphous polymers.

Glass Transition Temperatures T_g's and Plasticization

As a general rule, strong intra and intermolecular dipolar interactions between the zwitterionic groups result in an efficient reduction of chain segment mobility which leads to very high T_g.[28,30,31] The ΔT_g increase with respect to the amino precursor polymer is in most cases higher than 100°C, and higher for ammoniopropanesulfonate than for ammonioalkoxydicyanoethenolate zwitterions. In the former series, T_g is generally not reached before decomposition ($T_g > 250$°C), except for a number of polysoaps.

Efficient plasticization of polyzwitterions requires highly polar diluents such as water or glycerol for ammonioalkylsulfonates and the liquid salt ethylammonium nitrate ($T_m \sim 14$°C) appears as a "universal" plasticizer.[28,30-33]

Miscible Blends with Inorganic Salts

Because of strong ion–dipole interactions, polyzwitterions show the unique property of solubilization of low molecular weight salts in unusually high amounts to yield completely amorphous blends. In many cases, solubilization is observed up to stoichiometric ratios. Incompatibility with NaCl and miscibility with NaI are generally observed for all the stoichiometric salt-polymer blends.

MICROPHASE SEPARATION IN ZWITTERIONOMERS

Since the pioneering work of Graiver et al., a rather limited number of studies have definitely demonstrated that zwitterionomers show a complex heterogeneous structure in bulk.[12,34-37]

APPLICATIONS

Because of their very specific properties either in bulk or in solution, polyzwitterions appear of definite technological interest in a number of fields: salt resistant water soluble (*co*)polymers for tertiary oil recovery ("antipolyelectrolyte" behavior); water soluble (*co*)polymers for drag reduction,[38] hydrogels for agricultural (water reservoirs) or biomedical (drug release) purposes;[8] chelating resins for metallic cations like Co^{++}, Cu^{++}, and Fe^{++},[13] dipolar copolymers of potential nonlinear optical properties.[38a] The zwitterionomers afford in many cases an alternative to the well-known ionomers considered for their specific viscous and viscoelastic properties in bulk and solution (thermoplastic elastomers, shear-thickening additive in nonpolar solvents, etc.).[39] Moreover, they lead to functionalized latexes when obtained by emulsion copolymerization.[7] *N*-butylacrylate or styrene-2-methacryloyloxyethylphosphorylcholine copolymers were recently developed as biocompatible materials showing enhanced properties for the design of permeable hydrogels;[40] hybridized and stabilized liposomes;[10,41] glucose biosensors; and protein adsorption resistant membranes for blood compatibility.[10,42-44]

Finally, polysquaraines and polycrococaines were synthesized by polycondensation: they are of much interest in the field of electro-active conjugated polymers with alternate donor and acceptor moieties in the main chain and characterized by a small band-gap.[45,46]

REFERENCES

1. Pujol-Fortin, M. L.; Galin, J. C. *Macromolecules* **1991,** *24,* 4523.
2. Hodge, P. In *Syntheses and Separations using Functional Polymers*; Sherrington, D. C.; Hodge, P., Eds.; Wiley: New York, 1988; Chapter 2.
3. Salcedo, R.; Cardoso, J.; Manero, O.; Monroy, V. M.; Escobar, J. L. V.; Rubio-Arroyo, M. F. *Polymer* **1989,** *30,* 1747.
4. Salcedo, R.; Alcala, T.; Cardoso, J.; Manero, O. *Polymer* **1991,** *32,* 2300.
5. Monroy-Soto, V. M.; Galin, J. C. *Polymer* **1984,** *25,* 121.
6. Liaw, D. J.; Lin, J. R.; Chung, K. C. *J. Macromol. Sci. Pure Appl. Chem.* **1993,** A30, 51.
7. Gingreau, C. Thesis; Strasbourg; December, 1993.
8. Huglin, M. B.; Rego, J. M. *Macromolecules* **1991,** *24,* 2556.
9. Itoh, Y.; Abe, K.; Senoh, S. *Makromol. Chem.* **1986,** *187,* 1691.
10. Kojima, M.; Ishihara, K.; Watanabe, A.; Nakabayashi, N. *Biomaterials* **1991,** *12,* 121.
11. Hart, R.; Timmerman, D. *J. Polym. Sci.* **1958,** *28,* 638.
12. Graiver, D.; Baer, E.; Litt, M.; Baney, R. H. *J. Polym. Sci., Polym. Chem. Ed.* **1979,** *17,* 3559.
13. Hamaide, T.; Germanaud, L.; Le Perchec, P. *Makromol. Chem.* **1986,** *187,* 1097.
14. Hamaide, T.; Gnambodoe, M.; Guyot, A. *Polymer* **1990,** *31,* 286.
15. Salamone, J. C.; Volksen, W.; Olson, A. P.; Israel, S. C. *Polymer* **1978,** *19,* 1157.
16. Monroy-Soto, V. M.; Galin, J. C. *Polymer* **1984,** *25,* 254.
17. Schulz, D. N.; Peiffer, D. G.; Agarwal, P. K.; Larabee, L.; Kaladas, J. J.; Soni, L.; Handwerker, B.; Garner, R. T. *Polymer* **1986,** *27,* 1734.
18. Liaw, D. J.; Lee, W. F.; Whung, Y. C.; Lin, M. C. *J. Appl. Polym. Sci.* **1987,** *34,* 999.
19. Wielema, T. A.; Engberts, J. B. F. N. *Eur. Polym. J.* **1987,** *23,* 947.
20. Schulz, D. N.; Kitano, K.; Danik, J. A.; Kaladas, J. J. In *Adv. Chem. Ser. 223,* 165; Glass, J. E., Ed.; American Chemical Society: Washington, 1989.

21. Cardoso, J.; Manero, O. *J. Polym. Sci., Polym. Phys. Ed.* **1991,** *29,* 639.
22. Knoesel, R.; Ehrmann, M.; Galin, J. C. *Polymer* **1993,** *34,* 1925.
23. Davidson, N. S.; Fetters, L. J.; Funk, W. G.; Graessley, W. W.; Hadjichristidis, N. *Macromolecules* **1988,** *21,* 112.
24. Wang, Z. G. *Langmuir* **1990,** *6,* 928.
25. Hamaide, T.; Guyot, A.; Charlier, P.; Jerome, R. *Polymer* **1991,** *32,* 1089.
26. Huglin, M. B.; Radwan, M. A. *Polymer International* **1991,** *26,* 97.
27. Huglin, M. B.; Radwan, M. A. *Makromol. Chem.* **1991,** *192,* 243.
28. Pujol-Fortin, M. L. Thesis, Strasbourg, January, 1991.
29. Asonova, T. A.; Zezin, A. B.; Razvodovskii, Y. F. *Polym. Sci. USSR* **1974,** *16,* 896.
30. Galin, M.; Marchal, E.; Mathis, A.; Meurer, B.; Monroy, Y. M.; Galin, J. C. *Polymer* **1987,** *28,* 2297.
31. Galin, J. C.; Galin, M. *J. Polym. Sci., Polym. Phys. Ed.* **1992,** *30,* 1103.
32. Galin, M.; Mathis, A.; Galin, J. C. *Macromolecules* **1993,** *26,* 4919.
33. Galin, M.; Mathis, A.; Galin, J. C. Polymex **1993,** International Symposium on Polymers Cancun; Preprint **1993,** p 261.
33a. Galin, M.; Marchal, E. *Polym. Adv. Techn.* **1996,** *7,* 50.
34. Graiver, D.; Litt, M.; Baer, E. *J. Polym. Sci., Polym. Chem. Ed.* **1979,** *17,* 3573.
35. Graiver, D.; Litt, M.; Baer, E. *J. Polym. Sci., Polym. Chem. Ed.* **1979,** *17,* 3589.
36. Graiver, D.; Litt, M.; Baer, E. *J. Polym. Sci., Polym. Chem. Ed.* **1979,** *17,* 3607.
37. Graiver, D.; Litt, M.; Baer, E. *J. Polym. Sci., Polym. Chem. Ed.* **1979,** *17,* 3625.
38. Mumick, P. S.; Welch, P. M.; Salazar, L. C.; McCormick, C. L. *Macromolecules* **1994,** *27,* 323.
38a. Combellas, C.; Petit, M. A.; Thiebault, A.; Froyer, G.; Bosc, D. *Makromol. Chem.* **1992,** *193,* 2445.
39. Tant, M. R.; Wilkes, G. L. *J. Macromol. Sci. Rev. Macromol. Chem. Phys.* **1988,** C28, 1.
40. Ishihara, K.; Ueda, T.; Nakabayoshi, N. *Polymer J.* **1990,** *22,* 355.
41. Ishihara, K.; Nakabayashi, N. *J. Polym. Sci., Polym. Chem. Ed.* **1991,** *29,* 831.
42. Ishihara, K.; Ohta, S.; Yoshikawa, T.; Nakabayashi, N. *J. Polym. Sci., Polym. Chem. Ed.* **1992,** *30,* 929.
43. Ueda, T.; Oshida, H.; Kurita, K.; Ishihara, K.; Nakabayashi, N. *Polymer J.* **1992,** *24,* 1259.
44. Ishihara, K.; Tsuji, T.; Kurosaki, T.; Nakabayoshi, N. *J. Biomed. Mat. Res.* **1994,** *28,* 225.
45. Havinga, E. E.; ter Hoeve, W.; Wynberg, H. *Polym. Bull.* **1992,** *29,* 119.
46. Havinga, E. E.; ter Hoeve, W.; Wynberg, H. *Synth. Met.* **1993,** 55-57, 299.
47. Bahr, U.; Wieden, H.; Rinkler, H. A.; Nischk, G. *Makromol. Chem.* **1972,** *161,* 1.
48. Lee, W. F.; Tsai, C. C. *Polymer* **1994,** *35,* 2210.
49. Yasuzawa, M.; Nakaya, T.; Imoto, M. *Makromol. Chem., Rapid Commun.* **1985,** *6,* 721; **1985,** *6,* 727.
50. Salamone, J. C.; Volksen, W.; Israel, S. C.; Olson, A. P.; Raia, D. C. *Polymer* **1977,** *18,* 1058.

Porogens, Polymeric

See: Macroporous Polymers (Polymeric Porogens to Control Pore Size)

Porphinatoiron Complexes

See: Blood Substitutes

Porphyrin Binding

See: Antibody-Polymer(Porphyrin) Complexes

POWDER COATINGS (Overview)

Shelby F. Thames, James W. Rawlins, and Samuel D. Pace II
Department of Polymer Science
The University of Southern Mississippi

The development of powder coatings (PC) is driven by the desire to reduce energy consumption and limit volatile organic-compound (VOC) emissions. PC were first introduced in the early 1950s and consisted of fluidized bed-applied, finely ground thermoplastic (TP) resins. The emergence of electrostatic spray-application processing in the early 1960s allowed PC to become commercially competitive with liquid coatings. It was during this period that a new dimension in the form of thermoset (TS) epoxy PC began development.

Energy conservation and VOCs continued to be a driving force behind the development of PC technology, and thus they gained widespread acceptance in the late 1970s. Their growth has outpaced all other coating types worldwide.

PC are applied as thin films (1–3 mils) for decorative finishes and thick films (5–40 mils) for protective finishes. During the early years of PC, the market was in primarily protective finishes. Currently, decorative finishes provide the largest powder market.

POWDER COATING RAW MATERIALS

PC components are classified as polymers (resins), curing agents, pigments, extenders, and additives. The polymer(s) form the continuous phase and dominate coating performance. Curing agents are reactive intermediates with functionality of two or greater and are designed to increase molecular weight during the curing step. Hiding pigments add opacity or color to the coating(s), and extenders function to modify cost, affect rheology, affect mechanical properties, and alter gloss. Additives alter storage stability, processing, curing rates, and often final film properties, and are usually added in amounts less than 5% by weight.

RESINS

Polymers are classified into two broad categories, thermoplastics and thermosets. TP are linear high molecular weight, nonreactive molecules that flow and level on application. Typically, high molecular weight polymers are difficult to grind because of their inherent toughness, resulting in larger particle sizes and thus, thicker films (5–40 mils). TS are low molecular weight, reactive molecules that will cure and increase in molecular weight, becoming insoluble during application. TS grind easily, producing very fine powders that allow thin film applications (0.5–4 mils).

THERMOPLASTIC POLYMERS

Many TP polymers have been evaluated in powder-coating processes, but few have the proper combination of physical properties, melt viscosity, thermal stability, and other necessary characteristics to hold large niches of the market. Most TP PC give poor adhesion. This is a result of the nature of the repeat units, low polarity, and crystallinity. Surface preparation and priming are therefore recommended for good performance.

Ethylene Acrylic Acid Copolymers

The copolymer synthesized from ethylene and acrylic acid (EAA) exhibits a synergistic combination of polymer properties.

With good adhesion and a low melting point (93°C), EAA powder facilitates flame-spray application and shop-applied fusion-bond processes.

Polyamides (Nylons)

Nylons 11 and 12, the most common polyamide PC resins in North America, confer similar properties. The more common, less expensive polyamides, nylon 6, nylon 6,6, and nylon 6,10, melt at temperatures higher than those normally required for film formation. It is for this reason that nylon 11 and 12 are preferred over other nylons as powder-coating resins.

Nylons are most frequently applied by a fluidized bed in thicknesses of 200–700 μm. Thinner films of 100–150 μm can be achieved via electrostatic spraying. Film formation by fusion occurs best at 220°C for nylons because higher temperatures tend to reduce molecular weight. High-gloss finishes are produced by water quenching, whereas matte finishes are obtained by a slower cooling rate.

Polyesters

Although TP polyesters typically exhibit poor outdoor weathering, chemical resistance, and mechanical performance, they offer an excellent porcelainlike finish. Properly formulated polyesters provide good adhesion to most metal substrates without primers; they offer scratch and abrasion resistance, good hardness, outstanding flexibility, and impact resistance.

Polyethylene

Low-density polyethylenes are inexpensive, easily applied by a fluidized bed or hot-fusion method, and provide chemical resistance, toughness, and outstanding electrical-insulating properties. The coatings are typically smooth, warm to the touch, show excellent low-temperature flexibility, and display medium gloss. Polyethylene coatings have good release properties, which allow viscous, sticky materials to be easily cleaned from their surfaces, facilitating their use in laboratory equipment.

Polyethylene films are limited by poor abrasion and minimum cut resistance, poor adhesion, and outdoor durability. They are useful for dish drainer racks, municipal pipe linings, display racks, plating racks, shower caddies, and refrigerator shelving.

Polypropylene

PC use isotactic polypropylene (PP) produced via Ziegler–Natta catalyst (100% isotactic) or anionic polymerization (>75% isotactic). As a surface coating, polypropylene offers many useful properties. It is inexpensive and provides medium-high gloss, outstanding chemical and solvent resistance, good surface hardness, and flexibility. When pigmented and stabilized for UV light exposure, PP also has excellent weatherability.

Poly(vinyl chloride)

Poly(vinyl chloride) (PVC) offers low cost and combined with other attributes plasticized PVC earns the distinction of being the largest-selling TP powder coating in North America. PVC PC have good exterior durability and a medium-soft, glossy finish. They bond well to most metal substrates when applied over a suitable primer and withstand the stress of metal fabrication operations, such as bending, embossing, and drawing.

Poly(vinylidene fluoride)

Poly(vinylidene fluoride) (PVDF) is noted for outstanding exterior durability. PVDF-based PC have exceptional electrical insulation properties and chemical resistance, with the exception of hydrocarbon solvents. In addition to weatherability, PVDF coatings are flexible, abrasion and corrosion resistant, and possess low surface energy, which makes them resistant to foreign-material retention.

THERMOSET POLYMERS

TS PC molecular weights range from 600–5000 g/mol. These materials grind easily when the T_g is 40°C or above. The three main types of TS polymers sold commercially are acrylics, epoxies, and polyesters.

Acrylics

Early acrylic powders established a good reputation for exterior durability combined with trouble-free application characteristics. Today, the bulk of acrylic PC sold in the United States are hydroxyl functional and are cured by urethane crosslinkers.

Acrylics are commonly used in appliance markets (ovens, washing machines, etc.) where alkali resistance is critical. Also, excellent electrostatic properties facilitate realistic and controllable thin-film applications.

Epoxies

Epoxy resins are very versatile and are used in a variety of applications. The commercially important epoxies are derived from bisphenol-A. The bisphenol-A-based resins must have molecular weights above 900 g/mol to be stable powders (melting point approximately 70°C). The major uses of epoxy PC are in domestic equipment, metal furniture, building components and fittings, automotive components, agricultural equipment, radiators, boilers, pipes, containers, concrete, reinforcing bars, and electrical equipment.

Polyesters

Because of their high versatility, polyesters are the most commonly used resins in the TS PC industry. A variety of monomers and monomer functionalities are available in both aromatic and aliphatic types. Stoichiometry controls the resin functionality. Polyesters are either acid functional or hydroxyl functional. Therefore, resin synthesis can be tailored to a specific end use.

Epoxy/Polyesters

Epoxy/polyester PC, also known as hybrids, incorporate epoxy resins and acid-functional polyester resins. Epoxy/polyesters were developed primarily to compensate for the poor weathering and brittle nature of epoxies.

Epoxy/polyesters are similar to the epoxy PC in terms of impact, chemical and corrosion resistance, and flexibility, but give marginally softer films and diminished resistance to bases and organic solvents.

Polyester/Urethanes

The polyester urethane powders of recent years are chemically comparable to solvent-based exterior-grade urethanes for aircraft, buses, railroad cars, and trucks; they combine outstanding thin-film appearance, toughness, and excellent weathering. Polyester/urethanes perform better as thin films (1–2 mil), whereas thicker films are less flexible and lower in impact resistance.

Typical applications include fluorescent light fixtures, wheels (steel and aluminum), patio furniture, playground equipment, chrome wheel fittings and trim, garden tractors, appliance panels, ornamental iron, air conditioner cabinets, restaurant furniture supports, and transformer cases.

Polyester/TGIC

TGIC is a low molecular weight trifunctional curing agent. This aliphatic epoxy imparts excellent outdoor stability, mechanical properties, and edge coverage at high film thicknesses.

Curing Agents (Hardeners)

Curing agents are generally low molecular weight multifunctional molecules that increase the molecular weight and form network structures. Functional groups can originate from epoxides, isocyanates, amines, phenols, acids, hydroxyls, anhydrides, and silanol, to name a few.

Triglycidyl Isocyanurate

Triglicidyl isocyanurate offers low molecular weight with triepoxy functionality. TGIC is by far the most important epoxy functional crosslinker. TGIC coatings exhibit outstanding outdoor durability, and with proper selection of binder and pigments, provide up to 10 years of performance.

Polyisocyanates

Polyisocyanate crosslinkers may be aromatic or aliphatic. Each has its own advantages; aliphatics are outdoor stable, and aromatics are more cost effective. Isophorone diisocyanate and hexamethylene diisocyanate are the more commercially important aliphatic isocyanates.

Amines

Dicyandiamide and Derivatives

Dicyandiamide or 1-cyanoquanidine is widely used for crosslinking epoxy resins.

Aliphatic and Aromatic Amines

Primary and secondary amines react rapidly with epoxy resins at room temperature and provide chemically resistant films. Poor storage stability and high vapor pressures have limited the use of amines in PC.

Polyphenols

Phenolic resins are produced by the reaction of a phenol and an aldehyde (with or without modification). Both resole (base catalyzed) and novolac (acid catalyzed) phenolics are used as hardeners for PC. Low cost is the main advantage of phenolics. However, these aromatic curing agents yellow easily, degrade rapidly outdoors, cure slowly with epoxies, and produce brittle films.

Acid Anhydrides

Acid anhydrides are used as curing agents for epoxy or hydroxy functional resins. The anhydride must be opened by an active hydrogen donor or a Lewis base in order to react with epoxy groups.

Amino Resins (Glycoluril, Melamine, Benzoguanimine)

Common in solvent-borne coatings, modified melamine resins are now finding utility in powder systems. Condensation occurs with hydroxyl groups at high temperatures (>175°C), evolving methanol. Melamine-crosslinked PC's performance is reduced at high film thicknesses because of blistering. Outgassing and the fact that most melamines are liquid have limited their utility in powder systems.

β-Hydroxyalkyl Amides

β-Hydroxyalkyl amides are a new alternative for crosslinking acid-functional resins. Aliphatic polyesters cured using β-hydroxyalkyl amides are termed "superdurable" PC. These systems promise weatherability equal to or superior to PVDF-based liquid and TP powder coatings.

ADDITIVES

Additives effect or enhance a variety of properties and produce innovative finishes in relatively low concentrations. Common PC additives include catalysts, degassing agents, flow additives, matting agents, waxes, and special-effect additives.

PIGMENTS, FILLERS, AND EXTENDERS

Most pigments used in liquid coatings can be used in powder systems. Care should be exercised over particle-size control and overall moisture content.[1]

Fillers not only lower the cost of PC but improve certain properties of the system, such as abrasion resistance, rheology, corrosion resistance, and weatherability. The most commonly used fillers are barytes (barium sulfate), blanc fixe (barium sulphate), calcite (calcium carbonate), dolomite (calcium manganese carbonate), and mica.

POWDER COATING MANUFACTURE

Early PC manufacture used ball mills, sigma blade mixers, or two-roll mills to homogenize the formulation. During the early 1960s, successful attempts to melt mix PC materials were achieved using single- and twin-screw extruders.

Current manufacturing processes consist of four distinct stages: premixing, melt mixing, grinding, and sizing.[2] Powder coating quality is also improved when the manufacturing area utilizes some type of humidity control.

POWDER COATING APPLICATION

PC are applied by several methods. The fluidized-bed process was the first commercially successful PC application method. Essentially all PC were applied by the fluidized-bed process from 1958–1965. This method was superseded by what is now the most common mode of powder coating application–the electrostatic powder-spray (EPS) process. EPS offers two advantages over the fluidized-bed process: substrates may be coated without preheating and thin films may be accurately produced (2 mils or less). EPS is available in two types: corona and triboelectric.

CURRENT AND FUTURE TRENDS

The PC industrial market share could increase from 1993's 6% to 50% by the year 2000, with the necessary technological improvements; for example: lower curing schedules; ability to coat and cure on nonmetallic substrates; thinner, smoother films; coil and blank coatings; can coatings; TGIC-free polyester coatings; increased outdoor durability; and improved mechanical properties.[1]

Processing

A new method of powder preparation has been developed by Mandral et al. to process powders without traditional extrusion or other remaining steps that include premixing, grinding, and sizing. This method utilizes carbon dioxide in its supercritical state. A claimed feature of this process is more uniform particle composition and narrower particle-size distributions.[3]

ACKNOWLEDGMENTS

We want to express our gratitude and appreciation to all of our co-workers who assisted in the development of this article. Special thanks to Debbie Ballard, Michael Blanton, and Sharathkumar Mendon.

REFERENCES

1. SCM Chemical *Powder Coat* **1995**, *6*, 10.
2. Misev, T. A. *Powder Coatings Chemistry and Technology*; John Wiley & Sons: London, UK, 1991; Chapter 5.
3. Mandel, F. S.; Green, C. D.; Scheibelhoffer, A. S. U.S. Patent 5 399 597, 1995.

Prebiotic Polymers

See: Paracyanogen

PRECERAMIC POLYMERS

Yitbarek H. Mariam and Ke Feng
Department of Chemistry and Center for Theoretical Studies of Physical Systems
Clark Atlanta University

The preceramic polymer pyrolysis technology that stemmed from Yajima's work on the polysilane→polycarbosilane (PCS) transformation is viewed as an important and unconventional processing technology for the fabrication of high-technology and high-temperature engineering and structural materials. The Yajima transformation is not unlike that for the preparation of carbon fibers and carbon/carbon composites, and further developments in the field have shown that the Yajima process possesses general applicability to the preparation of a variety of ceramic materials from polymeric and oligomeric precursors via pyrolysis. A variety of ceramics are accessible via the polymer route, and precursors that lead to SiC, Si_3N_4, Si_3N_4/SiC, BN, AlN, B_4C, SiO_2, Al_2O_3, TiN, and others have been developed.

Since ceramics have high melting temperatures (for example, > 2500°C for SiC) and are insoluble and very hard, machining of such materials is difficult and expensive. One primary advantage of the preceramic polymer route over the conventional powder processing technology is the ease of fabrication of useful forms such as coatings, fibers, and complex shapes.

LATENT REACTIVITY, CERAMIC YIELD, AND DENSITY CHANGES[1]

The pyrolytic conversion of precursors to ceramic materials involves volume changes as extraneous organic ligands are removed as gaseous products. This process may, and often does, create porosity/voids and densification-induced stress. If problems associated with porosity/voids and densification are to be minimized, the ceramic yield (= weight of ceramic residue × 100/weight of pyrolysis charge) should be in an acceptable range of 60–75% or greater. A precursor should have at least two inherent characteristics: latent reactivity and branched structures. The latent reactivity can provide the opportunity for crosslinking and thereby provide for both maintaining appropriate shape during processing and high ceramic yield. Linear polymers generally give low ceramic yield owing to backbone reactions that lead to volatilities. Branched structures can, however, slow backbone reactions by sterically hindered structures that require multiple bond ruptures.

SYNTHESIS AND PROPERTIES

Types of Polymers and Typical Reactions

Most polymers synthesized contain Si-C, Si-N, Si-O, B-N and Al-N linkages. The Si, C, N, B, and Al atoms may have various substituents including H, alkyl, vinyl, aryl, alkenyl, halide, etc.

Modification of SiC Precursors

Polysilane$[(MeSiH)_x(CHSi)_y]_n$ (where x+y = 1) gave a substantial amount of elemental Si or C when pyrolyzed (ceramic yield of 12–27%, 950°C/Ar with a composition of SiC (77%) and C (23%); and, when crosslinked a ceramic yield of 60% with a composition of SiC and Si (25.6%)). Use of catalytic quantities of group-IV-metal complexes resulted in crosslinking such that pyrolysis of these polysilanes gave close to stoichiometric SiC (≥ 95% wt) and only very little elemental Si (5%) and ZrC/TiC (< 2%).

Modification of Si_3N_4 Precursors

Optimal candidate precursors for Si_3N_4 can be -(H_2Si-NH)-, -(H_2Si-NHNH)-, -(MeSiH-NH)- and -(SiH_2-NMe)- because they can be converted into Si_3N_4 upon pyrolysis by losing H_2 and/or CH_4. The precursors can be prepared from ammonolysis of H_2SiCl_2 and $MeHSiCl_2$. But such systems are too unstable and/or of low molecular weight to be directly useful. $-(H_2Si\text{-}NMe)_n$- is more stable in the absence of air and moisture but gives only 38–40% yield because of its low molecular weight. Two approaches that have been undertaken to address such problems were developed in Laine's[2] and Seyferth's[3] groups. Use of transition metal catalysts, for example, $Ru_3(CO)_{12}$, increased both the molecular weight and ceramic yield as illustrated by the pyrolysis studies on MeNH-[-H_2Si-NMe-]$_x$-H oligomers and polymers.[4]

Pyrolysis Data

Most polysilanes, polycarbosilanes, and polysilazanes will need some pre-pyrolysis treatments such as UV radiation, electron or γ-ray irradiation, thermal crosslinking, oxidative or catalytic curing to give ceramic yields better than 80%. The presence of deleterious impurity elements, often in the form of SiO_2, excess Si and C, is very apparent in most cases.

Other Precursors

In addition to the representative classes of precursors discussed, other types of systems have also been investigated. These include: boron-containing systems such as borosiloxane polymers, carborane-siloxane polymers, $Si_xC_{x+y}B_z$ precursors such as those that can be prepared by hydroboration of -(RSiCH = CH_2)$_n$- to yield -(RSiCH$_2$CH$_2$-B-)$_n$- and precursors for Si-C-B-N ceramics; aluminum-containing systems such as siloxy-substituted alumoxanes as precursors for aluminosilicates [$(Al_2O_3)_x(SiO_2)_y$], products of diols and organoaluminum compounds as precursors for Al_2O_3, melt-processable poly[(acyloxy)aloxane] as alumina precursors (this latter system can be melt-spun into thin fibers); and transition-metal-containing systems. Some reports have also been made on preparation of P_3N_5, $SiPN_3$, and Sn_3N_4 from polymeric precursors; attempts have also been made to prepare C_3N_4 from pyrolysis of high nitrogen content organic materials.

Transition-Metal-Containing Systems

Transition metal ceramics have potential for applications as electronic, structural, or refractory materials. Such materials can be prepared by pyrolysis of organometallic precursors. Some of these include: TiN from poly(titanium imide), [=Ti-$(NR_2)_2$]$_n$; intermediates from ammonolysis of metal (M) halides that give M-nitrides (M=Al, Ti, Nb, etc.); W_2C from $Cp_2(CO_2Me)_2W_2(CO)_4$; titanium and niobium carbonitrides from titanium and niobium acetylides, respectively.

APPLICATIONS

The availability of preceramic polymers with different compositions and properties can lead to various uses and applications such as ceramic fibers, protecting coatings, ceramic powders, matrix of carbon and ceramic fibers, infiltration of porous materials, and monoliths. Their high tensile strength, relatively low densities, and high temperature capability provide ceramic materials with potentials for use in the fabrication of rocket nozzles, leading edges (such as nose cones of hypersonic air craft), various types of internal engine components (both jet and conventional), and for uses with nuclear reactors and similar extreme environments. Two specific areas of research on applications for preceramic polymers are ceramic fibers and composite materials (CM).

Ceramic Fibers

The production process of cermic fibers involves melt or solution spinning to prepare polymer fibers, curing of the polymer fibers, and pyrolysis of the cured fiber to obtain the desired ceramic fibers. An important application of ceramic fibers is the reinforcement of plastics, metals, glasses, and ceramic materials for applications that include structural engineering parts, space and aeronautics components, and sports articles. Some ceramic fibers that are commercially available include those based on Si-C-O, Si-Ti-C-O, Si-N-O, Si-N-C-O, and Si-C-N compositions.

Composite Materials[5]

The preparation of homogeneously mixed crystalline ceramic solid solutions and composites is an important area of application for preceramic polymers. The preparative route followed can be categorized into three general approaches. Mixtures of stable nitrides, carbides, silicides, and metal borides have been prepared by combining a preceramic polymer with an inorganic reactive filler such as pure, fine-grained metals. Composites prepared in this way include Ti/SiC, ZrN/SiC, MC/SiC (M = Al, Ti, V, Zr, Nb, Ta, W), WC/W_5Si_3, and TiB_2/TiN. In another approach, blending of two precursors has been used. Composites and/or solid solutions prepared in this way include 2H AlN/β-SiC or 2H SiC/AlN, Si_3N_4AlN, Si_3N_4BN, and β-Si_3N_4/β-SiC. The third approach involves use of single-source molecular compounds that contain all the elements of the desired composite.

SUMMARY

The preparation of homogeneous complex, multi-functional composite materials and solid solutions of two or more pure phases has not quite matured yet. Continued efforts in solution-based synthesis and processing techniques as well as in solid state synthesis are still necessary to meet future technological needs. Detailed characterization of the physical and chemical states of the materials should also continue in parallel with the synthesizing efforts. A multi-disciplinary approach to these efforts is certainly most beneficial to adequately address the challenges, and collaboration among materials chemists and engineers holds great promise for progress.

ACKNOWLEDGMENT

The authors thank the financial support of this work by DOE/BES (Grant No. DE-FG05-87ER45306) and NSF (Grant No. HRD-915407).

REFERENCES

1. a) Laine, R. M.; Babonneau, F. *Chem. Mater.* **1993,** 5(3), 260; b) Seyferth, D.; Strohmann, C.; Tracy, H. J.; Robinson, J. L. *Mater. Res. Soc. Symp. Proc.* **1992,** *249,* 3.
2. Blum, Y. D.; Laine, R. M. *Organometallics* **1986,** *6,* 2081.
3. Seyferth, D.; Wiseman, G. H. *J. Am. Ceram. Soc.* **1984,** *67,* C-132.
4. Laine, R. M.; Blum, Y. D.; Hamlin, R. D.; Chow, A. In *Ultrastructure Processing of Advanced Ceramics*; Mackenzie, J. D., Urlich, D. R. Eds.; Wiley: New York, 1988; p 761.
5. a) Paine, R. T.; Janik, J. F.; Fan, M. *Polyhedron* **1994,** *13(8),* 1225; b) Gogotsi, Y. G. *J. Mater. Sci.* **1994,** *29,* 2541; c) Mazdiyasni, K. S. *Mater. Sci. Eng.* **1991,** A144, 83; d) Mah, T.; Yu, Y. F.; Hermes, E. E.; Mazdiyasni, K. S. *Fiber Reinforced Ceramic Composites*; Noyes: Park Ridge, NJ, 1990; Chapter 10.

PRECIPITATION POLYMERIZATION (Overview)

Alain Guyot
CNRS Laboratoire de Chimie et Procédés de Polymerisation

The definition of precipitation polymerization is already a matter of discussion. A recent review by Arshady on heterogeneous polymerization processes indicates that polymerization processes leading to the formation of macroscopically apparent polymer precipitates are referred to as precipitation polymerization.[1] They further indicate that the distinction between various terminology such as employed–emulsion, suspension, dispersion, or precipitation should be made on the basis of four different criteria: the initial state of the polymerization mixture, the kinetics of the polymerization, the mechanism of particle formation, and the shape and size of the final polymer particles. The suspension polymerization can be easily distinguished because it is the direct conversion of monomer droplets, while all other processes involve the two steps of nucleation of particles and then their growth. Heterogeneous medium is used in emulsion (less than 1 micron instead of more than 10 microns for the suspension), and emulsion leads to mono-dispersed particles (which is obviously not always the case). The distinction between dispersion and precipitation polymerization, which both start with a homogeneous medium, may be less clear. It seems to be performed on the basis of the particle size and shape; the precipitation process gives larger and less regular particles.

DISPERSION POLYMERIZATION

Dispersion polymerization is actually a precipitation polymerization carried out in the presence of a suitable steric stabilizer in order to control the particle size. Although the process has been mainly studied in radical polymerization, a few experiments have been done in polycondensation for the production of polyesters or polyurethanes in the presence of specially designed stabilizers carrying either cationic groups or epoxy for ionic or covalent grafting onto the particles. References are given in the review of Arshady and Georges.[2] More recently, dispersion polymerization has been applied to anionic polymerization in the condition of living polymerization.

Excellent reviews of the dispersion polymerization with radical mechanism are available. The early work in the laboratories of ICI were concerned with the production of sub-micron particles from poly(methacrylic ester)s precipitating in hydrocarbons.[3,4] More recently, it has been shown that monodisperse particles of a few microns (up to 20) can be obtained when the polymerization is carried out in water-ethanol mixtures in the presence of high polymers soluble in the medium, typically styrene polymerization in the presence of polyvinylpyrolidone (PVP).[5]

PRECIPITATION POLYMERIZATION OF ACRYLONITRILE

Polyacrylonitrile is fully insoluble in its monomer. Then if the polymerization is carried out in bulk, precipitation of solid polymer takes place as soon as the polymerization is started; autoacceleration of the polymerization is observed as soon as the process is started up to a rather high degree of conversion. Solution polymerization of acrylonitrile in *N,N*-dimethylformamide (DMF) follows the expected kinetics of an homogeneous radical polymerization, while in bulk the initiator exponent is 0.9 higher than the normal value of 0.5[6] The more striking feature of bulk acrylonitrile polymerization is the accumulation of occluded radicals inside the polymer particles; the phenomenon is so important that the concentration of radicals may reach value high enough for their detection by Electron Spin Resonance (ESR) and their quantitative evaluation becomes possible.[7] In addition, when the temperature is raised to 60°C, even in the dark, a very fast polymerization takes place but stops soon after, while the ESR signal disappears.

These features led Bamford and Jenkins to formulate the so-called occlusion theory: the growing radicals, initiated in the liquid medium (where the radical generators are soluble), soon precipitate as high coiled macromolecules, which flocculate to produce polymer particles.[8]

The occluded radicals may be considered as living radicals and this opportunity has been used to synthesize block copolymers.

PRECIPITATION POLYMERIZATION OF VINYL CHLORIDE

The polymer of vinyl chloride (PVC) is mostly insoluble in its monomer. It is admitted that spontaneous precipitation takes place when the polymerization degree (DP) rises about 25 while for most of the polymer, the DP is around 1000.[9] The concentration of polymer dissolved in its monomer remains quite limited (less than 10^{-3}%). However, a rather large amount of monomer can be dissolved in the polymer, which is then swollen with its monomer. This amount is dependent on the temperature, and through it, on the pressure (the liquid monomer is boiling at −12°C, and develops a pressure of about 10 atm at 60°C; at 50°C, the polymer is swollen with about 32% of its weight of monomer, so that at that temperature, when conversion reaches 77%, there is no more liquid monomer phase around the polymer).[10] It is well established that the polymerization takes place mainly in the swollen polymer phase, where it is much more rapid than in the liquid monomer phase. The events that take place in the monomer phase are important only at the very beginning of polymerization where the first particles are nucleated, their number being kept constant up to the end of the process.

MORPHOLOGY

The morphology of poly(vinyl chloride) has been more extensively studied than that of polyacrylonitrile. The main reason is that PVC can be used directly as produced in the polymerization reactor, and in some of its applciations both the size of the grains and their porosity are important parameters that must be kept under control. Another reason is that most of the huge production of PVC involves either a true precipitation process (actually the bulk process) or a suspension process, but in the second case (actually the more important industrially) an actual precipitation mechanism is working inside the monomer droplets suspended in water. The various stages of the formation of the morphology have been described in detail by Allsopp using a nomenclature adopted during a special symposium devoted to PVC and described by Geil.[11,12]

CONCLUSION

It is still difficult to give a clear-cut definition of precipitation polymerization. The best illustrations of this kind of polymerization are given by the polymerization of acrylonitrile and the bulk polymerization of vinyl chloride. Both systems differ mainly by the degree of swelling of the polymer by its monomer; however, even in poly(vinyl chloride), the swelling is not at a high enough level to eliminate the heterogeneous character of the polymerization at the micronic scale. The main common feature is to give rise to aggregates at different levels with morphologies more or less far from spherical ones. Of course, there are some other polymers that can be described as obtained by precipitation polymerization, for instance vinylidene chloride polymers of fluorinated polymers.[13,14] However, it was not our purpose to give an exhaustive review, and in addition, the corresponding polymers are often referred to as produced by emulsion polymerization.

REFERENCES

1. Arshady, R. *Colloid and Polym. Sci.* **1992,** *270,* 717.
2. Arshady, R.; George, M. *Polym. Engin. Sci.* **1993,** *33,* 865.
3. Barett, K. E, J. *Dispersion Polymerization in Organic Media*; Wiley: London, 1975.
4. Walbridge, D. J. *Comprehensive Polym. Sci.*; Pergamon: Oxford, 1989; Vol. 4, Chapter 15.
5. Lok, P. K.; Ober, K. C. *Can. J. Chem.* **1985,** *63,* 209.
6. Vidotto, G.; Crosato-Arnaldi, A.; Talamini, G. *Makromol. Chem.* **1969,** *122,* 91.
7. Bamford, C. H.; Ingram, J. E.; Jenkins, A. D.; Symons, M. R. C. *Nature* **1955,** *175,* 894.
8. Bamford, C. H.; Jenkins, A. D. *J. Chim. Phys. Chim. Biol.* **1959,** *56,* 798.
9. Mickley, H. S.; Michaels, A. S.; Moore, A. L. *J. Polym. Sci.* **1962,** *60,* 121.
10. Berens, A. R. *Angew. Macromol. Chem.* **1975,** *47,* 97.
11. Allsopp, M. W. *Pure Applied Chem.* **1981,** *53,* 449.
12. Geil, P. H. *J. Macromol. Sci. Chem.* **1977,** B14, 171.
13. Wessling, R. A.; Harrison, I. R. *J. Polym. Sci., Polym. Phys. Ed.* **1971,** *9,* 3471 and **1973,** *11,* 875.
14. Storti, G.; Carra, S.; Morbidelli, M.; Aralla, V.; Broco, G.; Gianetti, E. *J. Fluorine Chem.* **1987,** *35,* 237.

PRECIPITATION POLYMERIZATION (Process Description and Preparation of Particles)

Harald D. H. Stöver
Department of Chemistry
McMaster University

Kai Li
Corporate Research and Development
3M Canada Incorporated

Precipitation polymerization may be defined as a polymerization that starts as a homogeneous solution of monomer and initiator in a solvent that is a precipitant for the polymer formed. In certain cases, such as vinyl chloride or acrylonitrile, the monomer itself is a precipitant for its own polymer so that precipitation polymerization can take place in bulk. Precipitation polymerization is distinct from the closely related dispersion polymerization process in that no steric or electrostatic stabilizers are present, and there is therefore the potential to produce polymers uncontaminated by any residual stabilizer or surfactant.

Precipitation polymerizations have been studied for many years. Some important commercial polymers such as poly(vinyl chloride), polyacrylonitrile, polytetrafluorethylene, polychlorotrifluoroethylene and poly(vinylidene dichloride) are manufactured by this process.[1,2] Owing to the absence of stabilizers, precipitation polymerization usually yields polymers as a coagulum or as aggregates,[3] rather than as fine microspheres. Recent developments in this field have shown, however, that at least for some systems this process can be refined to prepare narrow or even monodisperse polymer microspheres.

PREPARATION OF POLYMER PARTICLES

Kawaguchi et al. obtained monodisperse 1 μm size particles in the precipitation copolymerization of acrylamide with other comonomers.[4] They found that methacrylic acid as a comonomer was very effective in producing small, monodisperse microspheres. Naka and Yamamoto et al. reported the synthesis of monodisperse poly(diethylene glycol dimethacrylate) microspheres.[5,6] Highly crosslinked poly(divinylbenzene) microspheres were recently prepared by precipitation polymerization in acetonitrile by the authors.[7]

REFERENCES

1. Xie, T. Y.; Hamielec, A. E.; Wood, P. E.; Woods, D. R. *J. Vinyl Technol.* **1991,** *13(1),* 2-25.
2. Murray, D. L.; Piirma, I. *Macromolecules* **1993,** *26*(21), 5577-86.
3. Bamford, C. H.; Ledwith, A.; Sen Gupta, P. K. *J. Appl. Polym. Sci.* **1980,** *25,* 2559.
4. Kawaguchi, H.; Kawahare, M.; Yaguchi, N.; Hoshino, F.; Ohtsuka, Y. *Polym. J.* **1988,** *20,* 903.
5. Naka, Y.; Kaetsu, I.; Yamamoto, Y.; Hayashi, K. *J. Polym. Sci., Polym. Chem. Ed.* **1991,** *29,* 1197.
6. Naka, Y.; Nakahama, Y. *J. Polym. Sci., Part A: Polym. Chem.* **1992,** *30,* 1287-1298.
7. Li, K.; Stöver, H. D. H. *J. Polym. Sci., Polym. Chem.* **1993,** *31,* 3257.

Precipitation Resins

See: Ion-Selective Reagents (Dual Mechanism Bifunctional Polymers)

Prepregs

See: Epoxy Resins (High Performance Composite Applications)
Graphite Fiber Composites (Electrochemical Processing)
Magnetic Field Processing

Pressure Effects

See: Catalysts, Polymeric (Pressure Effects)
High-Pressure Polycondensation (For High-Temperature Polymers)

High-Pressure Polymerization Processing (High-Temperature Thermosetting Polymers)
High-Pressure Radical Polymerization
Nylons (by High-Pressure Solid-State Polycondensation)
Polymer-Polymer Mixtures (Under High Pressure; Physical Properties)

Pressure Sensitive Adhesives

See: Fluoro-Acrylate Adhesive Blends (Surface Segregation Behavior)
Polyterpene Resins
Silicone Release Coatings
Tack, Elastomers
Vinyl Ether Polymers

PRINTING INKS

Alessandro Gandini and Anne Blayo
Matériaux Polymères
Ecole Francaise de Papeterie et des Industries Graphiques

Printing inks are liquid suspensions of very variable viscosity, depending on the printing process in which they intervene.[1] They have a homogeneous appearance although, almost without exception, they are constituted of at least two phases. Whatever the printing process, the components of printing inks can be divided into three categories: the pigment (or, less frequently, soluble dyes), the vehicle, and the additives.

Pigments are used in the form of finely ground powders with an average size of about one micrometer that are obtained from natural sources (mineral or vegetal) or, more frequently, by synthetic methods. They are dispersed in a fluid phase, the vehicle, to form a colloidal suspension. Soluble dyes are also used, but much more rarely, for example in the case of some low-viscosity inks. Pigments and dyes provide the printing inks with their main visual characteristics and are therefore components of prime importance. The introduction of these particles into the vehicle is usually carried out either by dry grinding and dispersion or by a pretreatment involving their wetting in order to facilitate the mixing.[1,2] In both instances a surface treatment may be required to improve the quality of the resulting interface.

The fluid phase of an ink, generally designated by the terms **vehicle** or **varnish**, is constituted by a blend of oligomers and polymers in one or several solvents and diluents. The vehicle has two major roles: on the one hand, it must ensure the transport and the binding of the ink onto the substrate; on the other hand, after the ink has dried, the whole or part of the vehicle must form a continuous film at the surface of the substrate. In this way the pigment is first carried onto the printed area and thereafter permanently protected from the external agents. The choice of a vehicle determines the ink-drying mechanism and most of its rheological properties.

The additives are specific products, such as waxes, drying catalysts, anti-oxidants, defoamers, and rheological additives, always used in modest proportions (less than 10% by weight) in inks in order to enhance some of their properties, depending on the printing process, the substrate characteristics, the final use of the printed object, and so on. Compositions without added pigments are used for transparent coating and take the name of **varnishes**.

There are essentially two classes of printing inks: (i) Inks for the letterpress and the litho-offset processes, usually called **paste inks** because of their high viscosity (2 to 40 Pa.s at room temperature) and sometimes "buttery" consistency. They are also designated by the adjective "oleoresinous." (ii) Inks for the gravure and the flexographic processes, commonly called **liquid inks**, since they are characterized by a very fluid nature (0.05 to 1 Pa.s for gravure inks, 0.1 to 10 Pa.s for flexo inks).

A third type of ink falls outside these two categories, namely, screen-printing inks. The viscosity of these inks is intermediate between those of liquid and paste inks, namely 1.5 to 2 Pa.s at room temperature.

Today, the main printing processes are offset, flexography, and gravure, which covered, respectively, 46, 18, and 18% of the printing activities in North America in 1994, with an upward trend of both flexography and digital printing.

One specificity of printing inks is that they must remain liquid during printing and during the diverse operations that take place before it, but they must dry as quickly as possible after they have been spread as a thin film onto the substrate.

COMPOSITION OF INKS UNDERGOING CHEMICAL DRYING

Two main chemical drying processes are used for printing inks, namely, oxidation drying and radiation-induced curing. These mechanisms principally concern paste inks, although recent developments involve the use of radiation for the curing of some liquid inks.

Components

Vegetable Oils and "Standolies"

The vehicle of oxidation drying inks contains a high proportion of unsaturated vegetable or "drying" oils, unaltered or modified, in combination with other components. The drying aptitude of an oil is connected to the number and the relative position (presence or lack of conjugation) of the unsaturations along the fatty-acid chain. For ink purposes, it is convenient to classify oils in three categories, namely, the drying oils characterized by iodine values between 150 and 180; the semi-drying oils, which have iodine values ranging between 110 and 150; and the non-drying oils, which do not show marked increase in viscosity on exposure to air.

Linseed and tung oil are the best examples of the first category. The drying properties of linseed oil are due to its high content of the tri-unsaturated linolenic acid chains.

Castor oil is a non-drying oil extracted from the seeds of the shrub *Ricinus communis* (iodine value: 82–90).[1,3] It is slightly more viscous than other drying oils and is used for preparing alkyds and other resins capable of giving glossy films of excellent flexibility with a good adhesion on metals.

The second category of oils includes the use of semi-drying vegetable oils in the field of printing inks and has spurred renewed interest in the last decade.

Soybean oil is the vegetable oil most frequently used in printing inks after linseed oil; **sunflower and rapeseed oil** are more recent additives to printing inks.

Prepolymerization of Drying Oils ("Standolization")

Standolies are obtained by a controlled treatment of drying oils at high temperature (about 290°C) under an inert atmosphere or vacuum. This process has kept the old name of "standolization" from the time when oils were left for long periods at room temperature to thicken in contact with air. The reaction is an oligomerization among triglyceride macromonomers promoted by the free radicals arising from homolytic scissions at high temperature, and by Diels-Alder cycloadditions between unsaturated moieties and conjugated systems.

Resins

Resins are non-crystalline solid or liquid polymers of relatively high molecular weight. In printing inks, the resins have the fundamental role of binding the pigments on the substrate and moreover dictate the rheological properties of the ink before, during, and after printing (viscosity, viscoelasticity, tack). Indeed, the nature of the resin determines the properties of the dry film in terms of hardness, gloss, flexibility, and adhesion.

Natural resins used in paste inks are mainly rosin derivatives.

The most important synthetic resins for offset printing inks are rosin-modified phenol-formaldehyde resins, alkyd resins, and hydrocarbon resins.

Alkyd resins play a primary role in the composition of paste inks and other surface coating materials.

Hydrocarbon resins remain the most versatile complementary formulating tool, particularly in litho inks.

Radiation-Induced Curing and Associated Compositions

Printing inks and varnishes requiring radiation curing processes have undergone a major development in recent years, especially in the field of paste inks. This general technology is based on the quasi-instantaneous polymerization of the components of the system, induced by UV light or electrons. A highly crosslinked dry film is produced without any solvent release.

The field of radiation curing has been entirely covered by a six-volume set of books where the interested reader will find a comprehensive treatment of all the relevant aspects.[4]

RHEOLOGICAL PROPERTIES

In the printing processes, the rheological properties of the inks are of prime importance as they partly govern the transfer mechanisms and thus the quality of the printed product. Obviously, these rheological properties depend essentially on the composition of the ink vehicle. Two types of materials must be considered: liquid inks (for flexography and gravure) and paste inks (for offset). Although the former compositions are pigment dispersions, they may be regarded as Newtonian fluids, with viscosities lower than 10 Pa.s. On the contrary, the latter inks are non-Newtonian dispersions of high viscosity and their complex rheological behavior is a topic of great relevance.[5]

CONCLUSION

From the viewpoint of a polymer scientist, the major issue in printing inks relates to the need to adapt the structure, composition, and viscosity of polymers/oligomers mixtures to the specificities of each printing process and the corresponding drying mechanisms.

Today, the global trend concerning environmental constraints and enhanced safety consciousness imposes the following major research topics for the forthcoming years: the increasing use of non-volatile, biodegradable, and renewable raw materials, namely vegetable products and derivatives; the replacement of volatile solvents in liquid inks by water; and the development of UV/EB curing systems, which does not involve solvent evaporation, even in the case of liquid inks.

To these aspects one must add the requirements imposed by the increasing productivity of the presses, namely vehicle formulations with rheological properties capable of adapting the corresponding inks to these very high speeds.

REFERENCES

1. Leach, R. H.; Pierce, R. J.; Hickman, E. P.; Mackenzie, M. J.; Smith, H. G. *The Printing Ink Manual*; 5th ed., Blueprint: London, 1993.
2. Mercado, A. *Am. Ink. Maker* **1991,** *69(6),* 19; *69*(7), 36; *69*(8), 54.
3. Wicks, Z. W.; Jones, F. N.; Pappas, S. P. *Organic Coatings: Science and Technology*; John Wiley & Sons: New York, 1992; Vol. I.
4. Oldring, P. K. T. chief Ed., *Chemistry and Technology of UV and EB Formulation for Coatings, Inks and Paints* (Vol. 1 *Markets and Curing Equipment*; Vol. 2 *Prepolymers and Reactive Diluents for UV and EB Curable Formulations*; Vol. 3 *Photoinitiators for Free Radical and Cationic Polymerization*; Vol. 4 *Photophysics and Photochemistry*; Vol. 6 *Specialised Applications for UV and EB)*, SITA Technology: London, 1991.
5. Hayashi, T.; Morita, K.; Amari, T. *J. Jpn. Soc. Col. Mater.* **1993,** *66*(11), 655.

Processing Aids

See: Rubber Additives, Multifunctional

Prodrugs, Polymeric

See: Anticancer Polymeric Prodrugs, Targetable Polymeric Drugs

[1.1.1]PROPELLANES (New Vistas in their Polymer Chemistry)

Arnulf-Dieter Schlüter, R. Freudenberger, and R. Klopsch

Institut für Organische Chemie

Freie Universität Berlin

The discovery of a large-scale synthesis of the parent [1.1.1]propellane **1a** about ten years ago and the subsequent synthesis of substituted derivatives like **1b–d** laid the basis for what has become a fruitful area of polymer chemistry: the synthesis of polymers with bicyclo[1.1.1]pentane fragments in the backbone.[1-3] It was proven that [1.1.1]propellanes have potential as new and useful monomers. A number of polymers were prepared ranging from homopolymers like **2a–d**[2,4,5] to strictly alternating copolymers **3** and **4**[2,3,6,7] and polyamides like **5**.[3] The structural characteristic of all these polymers, the bicyclopentane fragment, is of interest as a constituent part of polymers, whenever the effect of linearly connecting and rigid units is of importance. In these respects it may be compared with the 1,4-phenylene unit. In the homopolymers these fragments are directly attached to each other, which results in hydrocarbon polymers whose structural simplicity resembles polyethylene (PE). On the materials properties side, polymers **2**, however, are quite different from PE in that they are not flexible but rather prototype rigid-rod polymers, like poly(*para*-phenylene). In copolymers, the bicyclopentane also plays the expected stiffening role, leading to a significant increase of glass transition temperatures as compared with the corresponding olefin homopolymers. Last but not least, polyamide **5** is expected to show comparable material properties to Kevlar® and Duron® owing to the similar rigidity of the backbones. The exploratory phase of the research on [1.1.1]propellanes and their use in polymer chemistry has already been reviewed.[8]

1

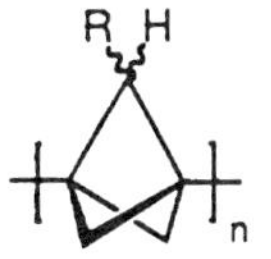

2

a: R = H
b: R = $(CH_2)_4$–CH_3
c: R = $(CH_2)_3$O–CH_3
d: R = CH_2–$C(CH_3)_3$

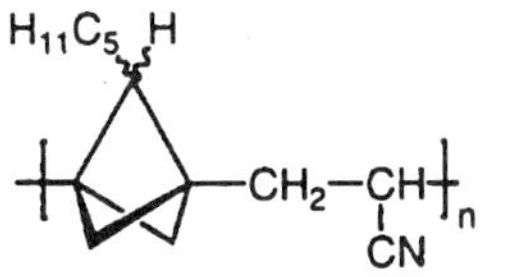

3

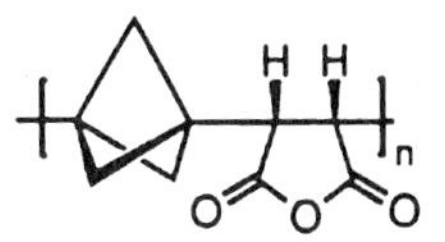

4

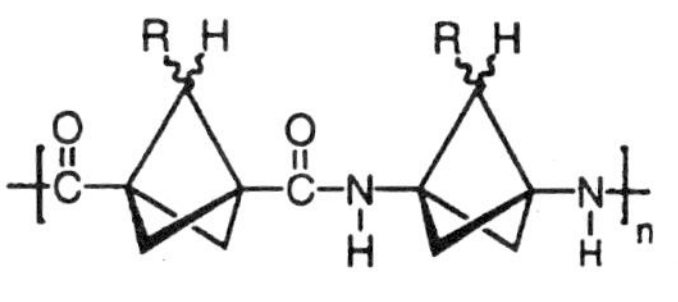

5

R = H, C_5H_{11}

PROPELLANE COPOLYMERS: THERMOPLASTICS FOR CONTACT LENSES?

Propellanes can be copolymerized with a broad variety of olefins in which the bicyclopentane unit causes a significant increase of T_g. Researchers from industry and ourselves have asked the question whether these interesting features of propellanes can be utilized to make thermoplastics with high oxygen permeability. All that would be required would be to copolymerize the propellane with a silicon-containing comonomer.[9]

Research in this direction is being continued. The future will show if mankind will get a chance to look at the world through strained hydrocarbon cages in polymer backbones.

FIRST STEPS TOWARD CYLINDRICALLY SHAPED DENDRITIC MACROMOLECULES

All dendrimers known today ideally have a spherical shape. This is where the polypropellanes come into play. If it were possible to decorate these rod-like polymers with chains carrying functional groups, they could then serve as a one-dimensional matrix for a variety of fragments. The attachment of dendritic fragments, for example, would result in a new molecular architecture characterized by a rigid backbone wrapped up by wedges that become increasingly more branched upon going from the inner to the outer regions. The envelope of these polymers would not be a sphere anymore but a cylinder. Such macrocylinders, in addition to their beauty, may find application in the areas of vesicles, membranes, and molecular reinforcement.

OUTLOOK

During the nine years since the first practicable synthesis of the parent [1.1.1]propellane was published, the polymer chemistry of this intriguing monomer has been developed in considerable depth. Polypropellanes, because of their unusual structure and despite the fact that they are expensive to make, are even now under consideration for certain applications and on the best way to become essential parts of strategies to new and exciting macromolecules. These directions will be heavily pursued. Besides this, future research on [1.1.1]propellanes will also be directed toward the accomplishment of a truly living anionic polymerization. This would allow for the synthesis not only of one-dimensional matrices in defined and predictable lengths but also of new block copolymers that consist of, for example, rod-like and flexible parts. Now that functionalized rods are available, the properties of these rods and of flexible polymers with the same substitution pattern ought to be compared. This would help to further integrate polypropellanes into the colorful picture of polymer science much like adding a missing piece to a mosaic.

ACKNOWLEDGMENTS

Financial support of this work by the Deutsche Forschungsgemeinschaft and the Fonds der Chemischen Industrie is gratefully acknowledged.

REFERENCES

1. Semmler, K.; Szeimies, G.; Belzner, J. *J. Am. Chem. Soc.* **1985,** *104,* 5239.
2. Optiz, K.; Schlüter, A.-D. *Angew. Chem. Int. Ed. Engl.* **1989,** *28,* 456.

3. Bothe, H.; Schlüter, A.-D. *Makromol. Chem., Rapid Commun.* **1988,** *9,* 529; *Chem. Ber.* **1991,** *124,* 587.
4. Kaszynski, P.; Michl, J. *J. Am. Chem. Soc.* **1988,** *110,* 5225.
5. Schlüter, A.-D. *Polym. Commun.* **1989,** *30,* 34.
6. Gosau, J.-M.; Schlüter, A.-D. *Chem. Ber.* **1990,** *123,* 2449.
7. Gosau, J.-M.; Bothe, H.; Schlüter, A.-D. *Polym. Bull. (Berlin)* **1991,** *25,* 293.
8. Bothe, H.; Gosau, M.; Schlüter, A.-D. *Makromol. Chem.* **1991,** *192,* 2497.
9. Herbrechtsmeier, P.; Schäfer, H.; Seiferling, B.; Schlüter, A.-D.; Bothe, H.; Freudenberger, R. Eur. Patent. Appl. E.P. 538188 Al 930421, 1992.

Propellants

See: *Cellulose Acetate Nitrate*
Telechelic Polyoxetanes
Thermodegradable Polymers (Azo Groups in Main Chain)

PROPYLENE POLYMERIZATION (Stochastic Analysis of the Propagation Mechanism)

Riichirô Chûjô
Department of Materials Engineering
Teikyo University of Science and Technology

Every process in natural science is stochastic. Free energy in an equilibrium state is composed of enthalpy and entropy terms. The efforts of polymer scientists to obtain any regular structures are equivalent to those to realize a much larger term than entropy one. The situation is quite analogous to the kinetic process, if terms having the dimension of energy are modified by a term such as activation enthalpy instead of enthalpy.

Propagation reaction in polymerization process is a typical stochastic reaction. Characterization of the reaction is, therefore, invaluable not only for obtaining the information on irregular structures distributed in regular ones, but also for performing propagation mechanisms on both regular and irregular structures.

Stereoregularities occurring in propagation reactions are the most important structural factors in vinyl and vinyl-type polymers. Among these polymers, polypropylene is an exclusive commodity. The stochastic analysis is, therefore, important in industry as well as in science. This article deals with the stochastic analysis of the propagation mechanism in propylene polymerization based on the numerical data on microtacticity of polymers obtained from NMR.

ANALYSIS IN DIAD AND TRIAD LEVELS

Stochastic analysis of the propagation mechansim from ^{13}C NMR data was done with the aids of three formulae (**Equation 1**).[7] Symbols m and r denote mesic and racemic diads, respectively. These are equivalent to the two-site model in which one site produces complete isotactic polymers, while the other produces polymers controlled by symmetric Bernoulian process. The symbol ω denotes the fraction of the complete isotactic parts and σ denotes the probability of isotactic addition in the symmetric Bernoullian parts.

$$(mm) = \omega + (1-\omega)\,\sigma^2$$
$$(mr) = 2(1-\omega)\,\sigma\,(1-\sigma)$$
$$(rr) = (1-\omega)(1-\sigma)^2 \qquad 1$$

EXTENSION OF THE THREE-SITE MODEL

The two-site model satisfactorily explains the major part of the catalyst-active sites and is a suitable tool for analyzing the effects of the donors on catalyst-active sites. Nevertheless, it does not completely describe the behavior of Mg-supported Ziegler–Natta catalyst. Probably there are two kinds of asymmetric site: one is originally asymmetric and the other is asymmetric produced by the addition of external donors. This idea was originated by Cheng.[2] According to his terminology, this model is E/E/B model (E and B are initials of enantiomorphic and Bernoullian, respectively). This terminology is, however, not recommended, because of the coexistence of the technical terms of catalytic chemistry and stochastic theory. If any terminology is required, α/α/σ should be much better. Analysis is done by Härkönen and others.[3] Similar analysis is done for samples without external donors.[4]

Owing to the development of catalysis chemistry, *r* diad in isotactic polymers appears only in the form of *rr* consecutive triad, because it appears as isolated *d* monoad in ll...l sequence or isolated *l* monoad in *dd...d* sequence. As a result four kinds among ten of pentad have sufficient intensities of ^{13}C NMR spectra. They are mmmm, mmmr, mmrr, and mrrm. Considering normalization condition, we can use three independent variables. The two-site model is, therefore, the most reliable. In order to avoid ambiguity of the three-site model due to the many (five) number of parameters, the two-site model must be applied to check the existence of asymmetric center before the application of the three-site model.

REFERENCES

1. Inoue, Y.; Nishioka, A.; Chûjô, R. *Makromol. Chem.* **1973,** *168,* 163.
2. Cheng, H. N. *J. Appl. Polym. Sci.* **1988,** *35,* 1639.
3. Härkönen, M.; Seppälä, L. V.; Salminen, H. *Polym. J.* **1995,** *27,* 256.
4. Paukkeri, R.; Iiskola, E.; Lehtinen, A.; Salminen, H. *Polymer* **1994,** *35,* 2636.

Prostheses

See: *Ligament Replacement, Artificial*
Ligament Replacement Polymers (Commercial Products)
Ligament Replacement Polymers (Technological and Design Aspects)

Protective Clothing

See: *Butyl Rubber (for Chemical Protective Clothing)*

Protective Colloids

See: *Colloids, Protective*

PROTEIN-BASED POLYMERIC MATERIALS (Synthesis and Properties)

Dan W. Urry,* D. T. McPherson, J. Xu, D. C. Gowda, and N. Jing
Laboratory of Molecular Biophysics
School of Medicine
University of Alabama at Birmingham

T. M. Parker
Bioelastics Research, Ltd.

H. Daniell and C. Guda
Department of Botany–Microbiology
Molecular Genetics Program
Auburn University

Protein-based polymers are high polymers of repeating peptide sequences. Their potential as materials derives from the capacity to control composition and chain length, the possibility for low cost production, and an extraordinary diversity of function. They have the potential to do all that proteins do in living organisms (this is a realizable potential as the principles of folding, assembly, and function become adequately understood), and they can be designed to do more than evolution has yet called upon proteins to do.

Through evolution proteins were never called upon to prevent adhesions after surgical procedures; to prevent adhesion to catheters and drainage tubes used in medical procedures; to function in soft tissue augmentation for cosmetic, reconstructive, and other reasons; to provide functional scaffoldings that could be remodeled into natural tissues; to provide matrices for controlled drug delivery; to function as biosensors in various assays; to function as transducers capable of both sensing and actuating, that is, interconverting free energies involving the intensive variables of temperature, pressure, mechanical force, chemical potential, electro-chemical potential and electromagnetic radiation; to be biodegradable plastics with controllable half-lives; to bc controllable super absorbents; to function in controlled release of agricultural enhancement factors, and so on. These examples and many more are possible uses for protein-based polymeric materials.[1,2]

PROTEIN-BASED POLYMER SYNTHESIS

Protein-based polymers may be synthesized using classical solution and solid phase chemical methods and using gene construction and expression in the cells of animals and plants.

Chemical Syntheses

The significant care required for correct and reproducible chemical syntheses are discussed followed by the detailed solution syntheses of composite pentamers, namely, their incorporation into 30mers of exact sequence that are then polymerized to form polytricosapeptides, or poly(30mers) of fixed sequence.

Microbial Biosyntheses

Biotechnology provides an alternate route to the synthesis of protein-based polymers, one by which the living organism, be it uni- or multi-cellular, becomes the synthesizer.[3-7]

*Author to whom correspondence should be addressed.

EXAMPLE OF A SEQUENCE DEPENDENT PROPERTY, THE PK$_A$

The greatest advantage of protein-based polymers over all other polymers is the capacity to control sequence. This is combined with the diversity and functional richness of the fundamental repeating unit and is further enhanced by the capacity to introduce enzyme specific sites.

By means of genetic engineering and expression in *E. coli*, the preparation of the poly(30mer) is not significantly more difficult than that of the poly(5mer), and, once the genetic engineering is achieved, more product is obtained simply by means of an additional fermentation. For chemical synthesis, on the other hand, each synthesis is as taxing as the previous one.

Properties of particular interest, such as an induced pK_a shift, which is a measure of an output of chemical energy, require control of sequence.

PRINCIPLES OF ENERGY CONVERSION BY PROTEIN-BASED POLYMERS

These are three underlying principals in protein-based polymer design and function. First, hydrophobic folding and assembly can be controlled by changing the temperature, T_T at which the hydrophic folding and assembly transition occurs. This results in related pairs of corollaries. Increasing functional hydrophobicity lowers T_T and decreasing functional hydrophobicity raises T_T. Consequently, lowering T_t from above to below an operating temperature drives hydrophobic folding and assembly, whereas raising T_t from below to above the operating temperature drives hydrophobic unfolding and disassembly. Second, increasing hydrophobicity, that is, lowering T_T raises the pK_a of a chemical couple when the charged species occurs at higher pH (for example, COOH/COO–) and lowers the pK_a of a chemical couple when the charged species occurs at lower pH (for example, NH_3^+/NH_2). Third, there are many ways that the value of T_T that is the functional hydrophobicity, can be changed for a given polymer composition, chain length, and concentration changing salt concentration, changing concentration of an organic solute, changing polymer side chain ionization, changing polymer phosphorylation, changing pressure, changing the redox state of an attached prosthetic group, changing absorption of light by an attached prosthetic group that is caused by the absorption to undergo a change in hydrophobicity, and changing neutralization of a charged side chain by a counter ion.[1,8,9]

MEANS OF PERFORMING CHEMICAL WORK

With the proper polymer design, each of the above energy inputs that resulted in a change in T_t with the performance of mechanical work should be able to perform chemical work. The protein-based polymer design required to convert various energy inputs to chemical energy can utilize a charged (polar)/uncharged (apolar) chemical couple as the actuator and, as the sensor, can use a chemical entity that changes its polarity (hydrophobicity) in response to a particular energy input. Thus a pair of functional groups are used that are coupled by virtue of being part of the same hydrophobic folding domain.

In principle, each of the following can perform chemical work when part of a properly designed protein-based polymer: changing the oxidative state of an attached redox couple (electro-chemical transduction), absorption of light by an attached

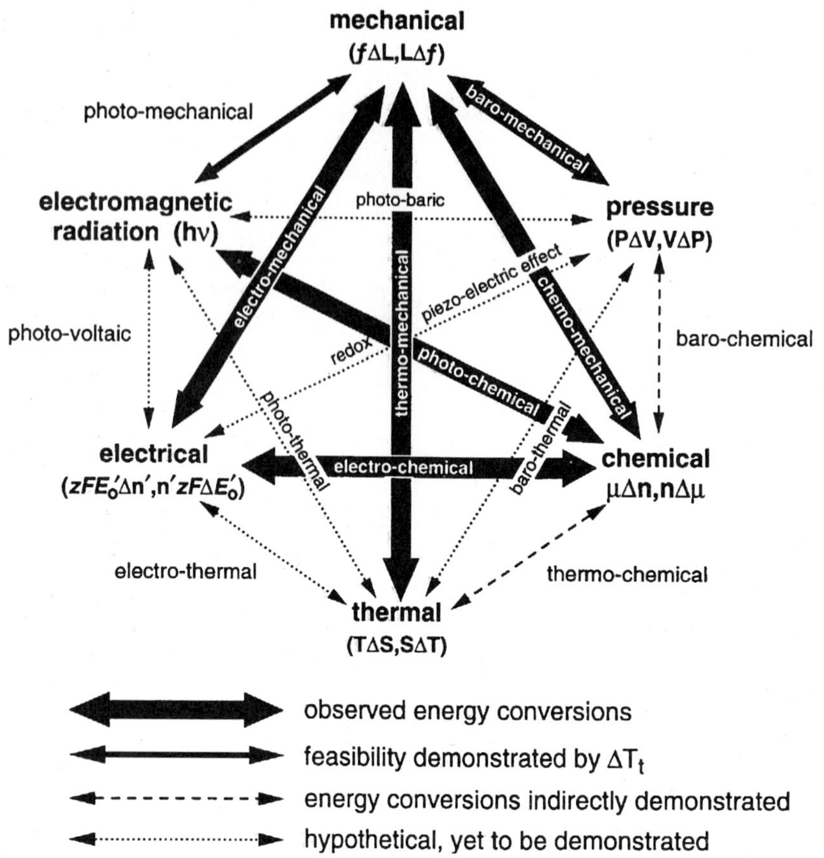

FIGURE 1. Demonstrated and putative pairwise energy conversions using the ΔT_t mechanism. *Source:* Reference 1. Adapted with permission.

chromophore that changes its hydrophobicity as when going from *trans* to *cis* (photo-chemical transduction), and the decrease in volume as waters of hydration form around aromatic side chains in response to an applied pressure (baro-chemical transduction). It is also possible to design a polymer capable of various forms of chemochemical transduction. The input of mechanical energy results in the performance of chemical work of proton uptake (that is, mechano-chemical transduction).[10,11]

All of the demonstrated and putative energy conversions of which protein-based polymers are considered capable of performing are shown in **Figure 1**.[1] It appears possible to design a protein-based polymer to sense any of six particular energy changes and to actuate, that is to output, in any of the six forms of free energy, the intensive variables of which are mechanical force, temperature, pressure, chemical potential, electrochemical potential, and light (that is, electromagnetic radiation). The potential and versatility of protein-based polymers can be seen by applications as far-ranging as the prevention of adhesions, tissue reconstruction, controlled delivery of therapeutic substances, and agricultural enhancement factors and biodegradable plastics. The future of protein-based polymers looks promising and exciting.

ACKNOWLEDGMENTS

This work was supported in part by Contract Nos. N00014-89-J-1970 from the Department of the Navy, Office of Naval Research; DAAK60-93-C-0094 from the U.S. Army Natick RD&E Center; DAAL03-93-C-0005 from the Department of the Army, Army Research Office; and P30-AI-277767 from the National Institutes of Health.

REFERENCES

1. Urry, D. W. *Angew. Chem. Int. Ed. Engl.* **1993,** *32,* 819-841.
2. Urry, D. W.; Nicol, A.; McPherson, D. T.; Xu, J.; Shewry, P. R.; Harris, C. M.; Parker, T. M.; Gowda, D. C. In *Handbook of Biomaterials and Bioengineering Part A: Materials*; Marcel Dekker: NY, 1995; Vol. 2, pp 1619-1673.
3. Deguchi, Y.; Krejchi, M. T.; Borbely, J.; Fournier, M. J.; Mason, T. L.; Tirrell, D. A. *Mat. Res. Soc. Symp. Proc.* **1993,** *292,* 205-210.
4. Capello, J. *Curr. Opin. Struc. Biol.* **1992,** *2,* 582-586.
5. McPherson, D. T.; Morrow, C.; Minehan, D. S.; Wu, J.; Hunter, E.; Urry, D. W. *Escherichia coli, Biotechnol, Prog.* **1992,** *8,* 347-352.
6. Murata, T.; Horinouchi, S.; Beppu, T.; *J. of Biotech.* **1993,** *28,* 301-312.
7. McPherson, D. T.; Xu, J.; Urry, D. W. *Protein Expression and Purification 7,* 51-57, February 1996.
8. Urry, D. W. *Porg. Biophys. Molec. Biol.* **1992,** *57,* 23-57.
9. Urry, D. W. *Int. J. Quant. Chem.: Quant. Biol. Symp.* **1994,** 3-15.
10. Urry, D. W.; Peng, S. Q.; Hayes, L.; Jaggard, J.; Harris, R. D. *Biopolymers* **1990,** *30,* 215-218.
11. Urry, D. W.; Peng. S. Q. *J. Am. Chem. Soc.* **1995,** 8478-8479.

Proteins

See: *Biopolymers (Overview)*
Biopolymers (Electrochemical Behavior)
Cellulose Derivative-Protein Complex (in Ice Cream)
Cellulose Derivative-Protein Complexes (Whey and Corn Steep Liquor, Enrichment of Macaroni)
Cellulose Derivatives (Reclaimation of Proteins from Cheese Whey)
Collagen
Connectin (Titin) (Large Filamentous Protein)
Gelatin (Molecular Weight Distributions)
Globular Proteins (Structures and Functions)
Gluten Plastic, Biodegradable
Hemoproteins
Keratins
Marine Adhesive Proteins, Synthetic
Monolayer and Langmuir-Blodgett Films (of Protein)
Nonthrombogenic Polymers (Multiblock Copolymers of Polyether and Polyamide)
Nucleohistone Complexes (Protein-Nucleic Acid Interactions)
Poly(α-amino acid) Spherical Particles
Protein-Based Polymeric Materials (Synthesis and Properties)
Radiation Induced Polymerization (Biomedical Applications)
Reverse Micelles (Microcontainers for Functional Polymers)
Self-Assembled Polymers (at Interfaces)
Silk (Its Formation, Structure, Character, and Utilization)
Silk (Physico-Chemical Properties)
Silk Fibers (Chemical Modification)
Silk Fibers (Grafting)
Silk Fibroin (Soft Tissue Compatible Polymer)
Spider Silk (Production of Polypeptide Polymers)
Textile Fibers (Structure and Properties)
Wool (Overview)
Wool Grafting
Wool Keratin

Pseudo-Poly(amino acid)s

See: *Controlled Drug Delivery Systems*
Pseudo-Poly(Amino Acid)s (Overview)

PSEUDO-POLY(AMINO ACID)S (OVERVIEW)

Joachim Kohn and Stephen Brocchini*
Department of Chemistry
Rutgers–The State University of New Jersey

Pseudo-poly(amino acid)s are derived from naturally occurring amino acids. Unlike conventional poly(amino acid)s where neighboring amino acids are linked by amide bonds to yield a peptide-like backbone, pseudo-poly(amino acid)s consist of building blocks of α-L-amino acids linked by non-amide bonds such as ester, iminocarbonate, and carbonate bonds. The term "pseudo-poly(amino acid)" is analogous to the term "pseudo-peptide" which was introduced in 1983 to designate a peptide containing non-amide backbone linkages.[1]

Although a virtually unlimited number of poly(amino acid)s can be derived from different amino acid sequences, conventional poly(amino acid)s have found very little practical use. Most conventional poly(amino acid)s decompose in the molten state, swell in moist environments, and degrade (albeit slowly) by hydrolytic and/or enzymatic mechanisms. These properties and the high cost of conventional poly(amino acid)s are the main obstacles for their commercial development.

The development of pseudo-poly(amino acid)s represents an attempt (i) to circumvent some of the unfavorable material properties of conventional poly(amino acid)s and (ii) to increase the range of amino-acid-derived polymers that can be considered for industrial and/or medical applications.

Pseudo-poly(amino acid)s were first described in 1984 and have since been evaluated for use in several medical applications.[2,3] Although a range of different pseudo-poly(amino acid)s has been prepared (**Table 1**), detailed studies of the physical properties, biological properties, and possible applications of these polymers have so far been conducted only for a select group of new tyrosine-derived polymers (polycarbonates, polyiminocarbonates, and polyarylates).

SYNTHESIS AND PROPERTIES OF TYROSINE-DERIVED POLYCARBONATES

One of the main differences between conventional poly(amino acid)s and pseudo-poly(amino acid)s is in their respective degree of order. While conventional poly(amino acid)s exhibit the α-helical and β-pleated sheet structures of peptides and proteins, the disruption of the regular peptide backbone by non-amide backbone linkages prevents the formation of crystalline or ordered domains. Consequently, all currently known pseudo-poly(amino acid)s are amorphous materials that are readily soluble in a variety of organic solvents. This general feature is clearly demonstrated by tyrosine-derived polycarbonates.

SYNTHESIS AND PROPERTIES OF TYROSINE-DERIVED POLYARYLATES

In this family of tyrosine-derived polyarylates, the glass transition temperatures decrease as a function of increasing length of either the pendant chain or the diacid.[11] The polyarylates are thermally stable polymers with thermal decomposition temperatures in the range of 300°C. Evidence from X-ray scattering and differential scanning calorimetry (DSC) indicates that tyrosine-derived polyarylates are amorphous. This combination of properties (T_g, T_d, and non-crystallinity) allows for the successful application of a wide range of processing techniques.

Tyrosine-derived polyarylates offer the ability to widely alter the polymeric properties by changes in either the backbone or the pendant chain structure. These polymers appear most adept at addressing medical implant needs where a slowly degrading, relatively flexible, and soft polymer is required.

*Author to whom correspondence should be addressed.

TABLE 1. Currently Available Pseudo-Poly(Amino Acid)s

Polymer	Structure	References
Poly(N-acyl trans-4-hydroxy-L-proline ester)	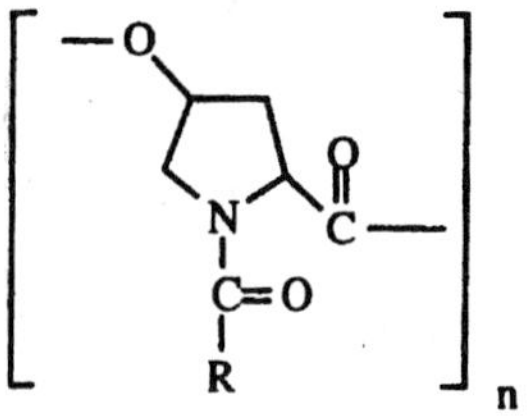	4,6,15
Poly(N-acyl L-serine ester)	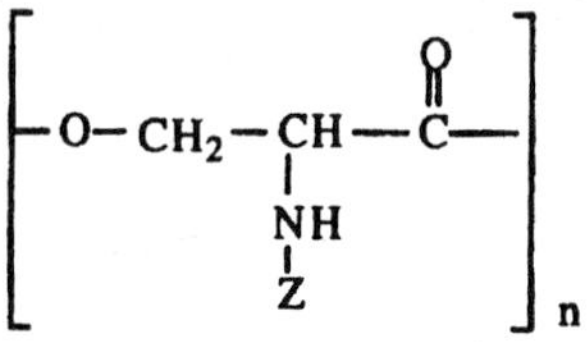	5,16-18
Tyrosine-derived polycarbonates	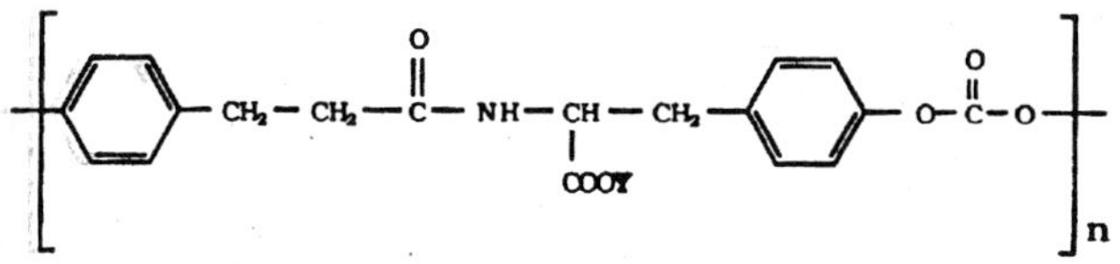 Y = ethyl → poly(DTE carbonate) Y = butyl → poly(DTB carbonate) Y = hexyl → poly(DTH carbonate) Y = octyl → poly(DTO carbonate)	8,9,14,19,20
Tyrosine-derived polyarylates	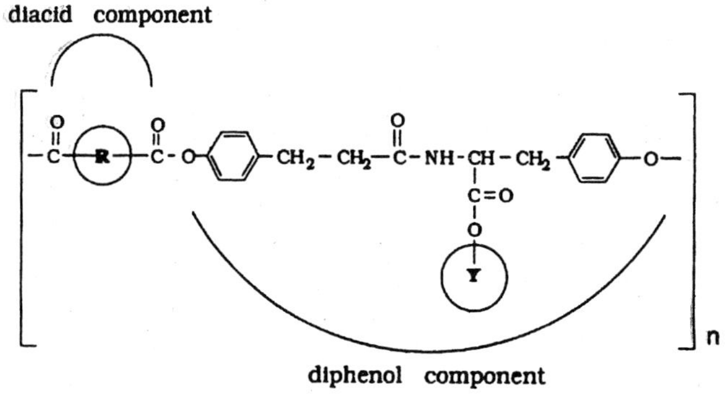	10,11,21
Tyrosine-derived polyiminocarbonates	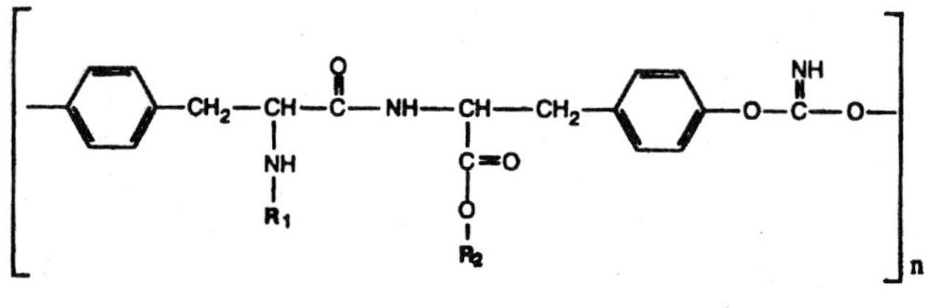	7,14,22,23

SYNTHESIS AND PROPERTIES OF TYROSINE-DERIVED POLYIMINOCARBONATES

Polyiminocarbonates can be regarded as backbone-modified analogues of polycarbonates in which the carbonyl oxygen (= O) has been replaced by an imino group (= NH). This chemical change has pronounced effects on the polymer properties. Most important is the significantly reduced hydrolytic and thermal stability of polyiminocarbonates as compared to polycarbonates.

An important property of polyiminocarbonates is their low thermal decomposition temperature (approximately 150°C).[7,12]

Tyrosine-derived polyiminocarbonates are as stiff and as strong as the corresponding tyrosine-derived polycarbonates.[7,13] Polyiminocarbonates are brittle and break without plastic deformation after less than 5% of elongation, while polycarbonates can, under appropriate conditions, be elongated by 400%.

APPLICATIONS

General Considerations

All currently available pseudo-poly(amino acid)s have been intentionally designed to degrade under physiological conditions. Although pseudo-poly(amino acid)s are expected to be environmentally friendly, the relatively high cost of pseudo-poly(amino acid)s makes their use as degradable packaging and consumer plastics unlikely. However, because of their apparent non-toxicity and high degree of tissue compatibility, these materials are promising candidates for use in medical applications, in particular, degradable medical implants and degradable drug delivery systems.

CONCLUSION

The concept of polymerizing derivatives of α-L-amino acids and dipeptides by bonds other than conventional amide linkages found in peptides in proteins is relatively new. The resulting polymers have been designated as "pseudo"-poly(amino acid)s since their backbones contain non-amide bonds. At this point, the limited experience gained from the detailed exploration of a small number of pseudo-poly(amino acid)s clearly indicates that pseudo-poly(amino acid)s combine the non-toxicity of the natural amino acids and dipeptides from which they are derived with favorable material properties not usually found in conventional poly(amino acid)s. Of particular importance is the improved processability exhibited by all pseudo-poly(amino acid)s.

Currently, the most intensively investigated pseudo-poly(amino acid)s are the tyrosine-derived polyiminocarbonates, polycarbonates, and polyarylates. Because of the relatively high strength and stiffness, tyrosine-derived polycarbonates are being investigated for use in orthopedic implants, while the softer, more flexible polyarylates seem to be particularly suited in controlled drug-release applications. Tyrosine-derived polyiminocarbonates may degrade too fast to find practical applications in medicine.

ACKNOWLEDGMENTS

A portion of the scientific work reviewed here was supported by NIH grants GM39455 and GM49894. Joachim Kohn also acknowledges support from NIH Research Career Development Award GM00550.

REFERENCES

1. Spatola, A. F. Peptide backbone modifications: A structure-activity analysis of peptides containing amide bond surrogates, conformational constraints and related backbone replacements. In *Chemistry and Biochemistry of Amino Acids, Peptides, and Proteins*; Weinstein, B. Ed.; Marcel Dekker: New York, NY, 1983; pp 267.
2. Kohn, J.; Langer, R. A new approach to the development of bioerodible polymers for controlled release applications employing naturally occurring amino acids, in *Polymeric Materials, Science and Engineering*; American Chemical Society: Washington, DC, 1984; pp 119.
3. Kohn, J. *Trends Polym. Sci.* **1993,** 1(7), 206.
4. Kohn, J.; Langer, R. *J. Am. Chem. Soc.* **1987,** *109,* 817.
5. Zhou, Q. X.; Kohn, J. *Macromolecules* **1990,** *23,* 3399.
6. Yu-Kwon, H.; Langer, R. *Macromolecules* **1989,** *22,* 3250.
7. Pulapura, S.; Li, C.; Kohn, J. *Biomaterials* 1990b, *11,* 666.
8. Ertel, S. I.; Kohn, J. *J. Biomed. Mater. Res.* **1994,** *28,* 919.
9. Pulapura, S.; Kohn, J. *Biopolymers* **1992,** *32,* 411.
10. Fiordeliso, J.; Bron, S.; Kohn, J. *J. Biomater. Sci. (Polymer. Ed.)* **1994,** 5(6), 497.
11. Kohn, J. *Tyrosine-based polyarylates: Polymers designed for the systematic study of structure-property correlations*, 20th Annual Meeting of the Society for Biomaterials; Boston, MA; Society for Biomaterials, Minneapolis, MN, 1994; pp 67.
12. Li, C.; Kohn, J. *Macromolecules* **1989,** *22(5),* 2029.
13. Engelberg, I.; Kohn, J. *Biomaterials* **1991,** *12*(3), 292.
14. Silver, F. H.; Marks, M.; Kato, Y. P.; Li, C.; Pulapura, S.; Kohn, J. *J. Long-Term Effects Med. Implants* **1992,** 1(4), 329.
15. Yu, H.; Lin, J.; Langer, R. *Preparation of hydroxyproline polyesters*; 14th International Symposium on Controlled Release of Bioactive Materials: Toronto, Canada, Controlled Release Society, Lincolnshire, IL 1987; pp 109.
16. Gelbin, M. E.; Kohn, J. *Polym. Preprints* **1991,** *32(2),* 241.
17. Gelbin, M. E.; Kohn, J. *J. Am. Chem. Soc.* **1992,** *114,* 3962.
18. Fietier, I.; Le Borgne, A.; Spassky, N. *Polym. Bull.* **1990,** *24,* 349.
19. Ertel, S. I.; Kohn, J. *A homologous series of tyrosine-derived polycarbonates: Surface characterization and degradation*, The 19th Annual Meeting of the Society of Biomaterials; Birmingham, AL, Society for Biomaterials, Minneapolis, MN 1993a; pp 133.
20. Ertel, S. I.; Parsons, R.; Kohn, J. *Investigations of a tyrosine-derived polycarbonate as a potential orthopaedic implant*, The 19th Annual Meeting of the Society of Biomaterials; Birmingham, AL, Society for Biomaterials, Minneapolis, MN 1993b; pp 17.
21. Bron, S.; Kohn, J. The effect of small changes in the chemical structure on the chain mobility in the glassy state of a new series of polyarylates, in *Proceed. Am. Chem. Soc., Div. Polym. Mat. Sci. Eng.*; American Chemical Society: Washington, DC, 1993; pp 37.
22. Haque, F.; Li, C.; Kohn, J. *Polyiminocarbonates: New biomaterials for the design of implantable release devices*; 16th International Symposium on Controlled Release of Bioactive Materials, Chicago, IL, Controlled Release Society, Lincolnshire, IL (USA), 1989; pp 115.
23. Kohn, J.; Li, C. U.S. Patent 5 140 094, issued 1992.

Pullulan

See: Hydrophobized Polysaccharides (Versatile Functions)

Pultruded Materials

See: Liquid Crystalline Polymers (Heat-Strengthened, Thermoplastic)

Pyroelectric Materials

See: Phthalocyanines, Polymeric
Polyamides, Alicyclic and Aromatic (Ferroelectric Hysteresis and Pyroelectricity)
Poly(vinylidene fluoride) (Overview)
Pyroelectricity

PYROELECTRICITY

Geoffrey R. Davies
Interdisciplinary Research Centre in Polymer Science and Technology
University of Leeds

There have been many reviews of pyroelectricity in polymers.[1-7] Most of these have concentrated upon poly(vinylidene fluoride) (PVDF) and its copolymers since these are the best commercially available pyroelectric polymers.

The phenomenon of pyroelectricity is the reversible change in polarization of a material caused by a change in temperature (where polarization is electric dipole moment per unit volume).[8] All pyroelectric materials are also piezoelectric to some degree though the converse is not true. Pyroelectricity is not a common phenomenon since pyroelectric materials lack a macroscopic center of symmetry and exhibit a permanent polarization. A single crystal must therefore belong to a restricted set of crystalline classes known as the polar classes and, in an aggregate, such crystals must be preferentially aligned to prevent their pyroelectric response averaging to zero. This preferential alignment is produced by the application of large electric field in an operation known as poling, which results in the aggregate possessing a permanent polarization. Pyroelectric polymers, whether semi-crystalline or amorphous, must be capable of exhibiting a significant permanent polarization since, as will be seen, the observed pyroelectric response is proportional to the polarization achieved.

The first requirement for a pyroelectric polymer is, therefore, that the chain must be polar. This rules out polymers with highly symmetric chains such as polyethylene or polytetrafluoroethylene and also asymmetric polymers that are only weakly polar such as polypropylene or polystyrene.

The second requirement is that the polymer must be capable of retaining a permanent polarization at the required operating temperature. A normal amorphous polymer must therefore have a glass transition temperature, T_g, well above its operating temperature. T_g is not a limitation for semi-crystalline polymers with polar crystalline phases, however. The glass transition temperature of PVDF is –50°C yet it can exhibit strong pyroelectricity well above room temperature owing to the polarization locked in the crystals. In such a ferroelectric semicrystalline polymer then it is the Curie temperature, T_c, that must be above the operating temperature in order for the crystalline phase to retain its polarization.

POSSIBLE MATERIALS

Pyroelectricity is closely related to piezoelectricity or ferroelectricity. In many cases, materials are reported as being ferroelectric or piezoelectric without their pyroelectric activity being reported, though they may reasonably be expected to display pyroelectric activity.

Semi-Crystalline, Ferroelectric Polymers

Poly(vinylidene fluoride) and Its Copolymers

The best known pyroelectric polymer is the semi-crystalline polymer poly(vinylidene fluoride) (PVDF), which is also piezoelectric and ferroelectric.

PVDF can exhibit at least four different crystalline phases, three of which are polar, but the highest piezoelectric and pyroelectric activity is seen in samples containing the β phase (or form I in an alternative notation), which has the highest polarization. In the β phase the polymer chains adopt an essentially all trans conformation with all the fluorine atoms on the same side of the chain creating a large dipole moment of about 2 Debye per repeat unit perpendicular to the chain. The crystallinity of PVDF is only about 50%, however, and perfect crystalline orientation is not achieved by poling, hence bulk polarizations around 50 mC/m² are obtained and such films exhibit room-temperature, pyroelectric coefficients of about 30μC/m²/K.

The Odd Nylons

A recent review summarizes the piezoelectric and pyroelectric studies on odd nylons.[9] To date, the pyroelectric coefficient of well-poled samples have not been reported but they may be expected to be comparable with PVDF since a sample polarization in excess of 100 mC/m² is obtainable for nylon 5. The materials may be of particular interest in high temperature applications since little decay of activity is observed after annealing at temperatures as high as 185°C (nylon 11) or 200°C (nylon 7).[10]

Copolymer of Acrylonitrile and Allylcyanide

The high dipole moment of the cyanide group might suggest that polyacrylonitrile would be a possible candidate for a piezoelectric and pyroelectric polymer. In fact, it displays a relatively weak piezoelectric activity, probably due to its tendency to form helical chains in a paracrystalline array.[11] It has recently been reported, however, that a copolymer of acrylonitrile and allylcyanide shows ferroelectric behavior with a high remnant polarization.[12] This is attributed to a ferroelectric paracrystalline phase that could lead to piezoelectric and pyroelectric activity.

Amorphous Polymers

Poly(vinyl chloride)

Poly(vinyl chloride) (PVC) is a simple example of an amorphous, polar polymer with a T_g above room temperature. Its pyroelectric response is not adequate for technological applications, but it has served as a simple model system for the study of the mechanisms of pyroelectricity.[13]

Alternating Copolymers of Vinylidene Cyanide

Alternating copolymers of vinylidene cyanide have received much attention since the first report of piezoelectric activity in a vinylidene cyanide/vinyl acetate copolymer.[14] This copolymer, which was subsequently shown to yield a pyroelectric coefficient of 3.1 μC/m²/K and, more recently, 6μC/m²/K, is a further example of an amorphous polar polymer.[15,16]

Poly(phenylether nitrile)

Poly(phenylether nitrile) is actually a semicrystalline polymer that, when quenched and poled near its T_g of 143°C, has been shown to exhibit a pyroelectric coefficient of 18μC/m²/K at 100°C.[17] This promising material is deserving of further study.

Langmuir–Blodgett Films

There have been many attempts to fabricate thin, pyroelectric polymeric films by the Langmuir–Blodget technique. To date, rather low pyroelectric coefficients of the order of 1 μC/m²/K are typically achieved. The technique is of interest, however, since asymmetric films can be produced without the need for poling.

Chiral Liquid Crystalline Polymers

Chiral liquid crystalline polymers are another interesting example of a polymeric system that can exhibit pyroelectric activity without the need for poling. Such systems have attracted a great deal of interest recently, mainly in view of their possible applications in electro-optic devices.[18,19] Shibaev and colleagues developed the first ferroelectric chiral liquid crystalline polymer and a large number of such systems are now known.[20-22,28]

Polymer/Ceramic Composites

There are now a number of ferroelectric ceramics such as lead-zirconate titanate (PZT) or barium titanate with attractive pyroelectric properties. The materials tend to be brittle, however, and difficult to form into thin films. A number of polymer/ceramic composites have therefore been developed that combine the desirable mechanical properties of polymers with the electrical properties of the ceramic. Such composites typically exhibit higher pyroelectric coefficients than conventional polymers and much lower permittivities than ceramics.[23] The dielectric noise figure of merit is improved and the materials are candidates for use in thermal imaging devices for example.[24]

MECHANISMS

The mechanisms by which pyroelectricity occurs in polymers are now broadly understood. In the case of amorphous polymers, there are two main contributions: the change in the libration angle of dipoles as the temperature increases and macroscopic thermal espansion. In the case of semi-crystalline polymers two further contributions arise: the change in permittivity of the two phases with temperature and the change in crystallinity with temperature. In ferroelectric systems co-operative interactions cause the spontaneous polarization to vary with temperature in a characteristic fashion, the polarization reducing at an increasing rate as the Curie temperature is approached. This leads to a peak in the pyroelectric response a little below the Curie temperature.

APPLICATIONS

Infrared Detection

The main device application for pyroelectric polymers is the detection of infrared radiation.[25,26]

Polarization Characterization

Investigation of periodically poled NLO polymers has also been performed by the new technique of scanning pyroelectric microscopy.[27]

Novel Thermal Sensor Applications

The ready availability of thin, pyroelectric films has prompted the development of a number of novel techniques that utilize these films as thermal sensors.

CONCLUSION

It is noticeable that virtually all applications of pyroelectric polymer films use a PVDF homopolymer or copolymer as the active element owing to their high response and commercial availability. Despite many years of research, no polymer has yet been found to compete commercially with PVDF-based materials.

REFERENCES

1. Das Gupta, D. K. *Ferroelectrics* **1981,** *33,* 75.
2. Joshi, J. C.; Dawar, A. L. *Phys. Stat. Sol. A–Appl. Research* **1982,** *70,* 353.
3. Broadhurst, M. G.; Davis, G. T. *Topics in Appl Phys.* **1987,** *33,* 285.
4. Danz, R.; Geiss, D. *Acta Polymerica* **1987,** *38,* 399.
5. Furukawa, T. *IEEE Transactions on Electrical Insulation* **1989,** *24, 3,* 375.
6. Mansingh, A.; Arora, A. K. *Indian J. Pure & Appl. Phys.* **1991,** *29,* 657.
7. Das Gupta, D. K. *Ferroelectrics* **1991,** *118,* 165.
8. Lang, S. B. *Sourcebook of Pyroelectricity*; Gordon & Breach: New York, 1974.
9. Scheinbeim, J. I.; Newman, B. A. *Trends in Polymer Science* **1993,** *1,* 394.
10. Takase, Y.; Lee, J. W.; Scheinbeim, J. I.; Newman, B. A. *Macromolecules* **1991,** *24,* 6644.
11. Ueda, H.; Carr, S. H. *Polymer Journal* **1984,** *16,* 661.
12. Tasaka, S.; Nakamura, T.; Inagaki, N. *Jpn. J. Appl. Phys.* **1992,** *31,* 2492.
13. Mopsik, F. I.; Broadhurst, M. G. *J. Appl. Phys.* **1975,** *46,* 4204.
14. Miyata, S.; Yoshikawa, M.; Tasako, S.; Ko, M. *Polymer* Journal **1980,** *12,* 857.
15. Tasaka, S.; Asato, K. M.; Yoshikawa, M.; Miyata, S.; Ko, M. *Ferroelectrics* **1984,** *57,* 267.
16. Wright, D. J.; *Ph.D. Thesis; University of Leeds*, England 1993.
17. Tasaka, S.; Toyama, T.; Inagaki, N. *Jpn. J. Appl. Phys.* **1994,** *33,* 5838.
18. Kozlovsky, M. V.; Beresnev, L. A. *Phase Transitions* **1992,** *40,* Pt A, 129.
19. Scherowsky, G. *Makromol. Chem., Macromol. Symp.* **1993,** *69,* 87.
20. Shibaev, V.; Koslovsky, M.; Beresnev, L.; Pozhidayev, E.; Yablonsky, S. *Polym. Bull.* **1984,** *12,* 299.
21. Shibaev, V.; Plate, N. *Pure Appl. Chem.* **1985,** *57,* 1589.
22. Le Barny, P.; Dubois, J. C. *Side Chain Liquid Crystalline Polymers*; McArdle, C. B. Ed.; Blackie: London, 1989.
23. Wang, Y. G.; Zhong, W. L.; Zhang, P. L. *J. Appl. Phys.* **1993,** *74,* 521.
24. Murphy, C. E.; Richardson, T.; Roberts, G. G. *Ferroelectrics* **1992,** *134,* 189.
25. Whatmore, R. W. *Rep. Prog. Phys.* **1986,** *49,* 1335.
26. Neumann, N.; Kohler, R.; Hofmann, G. *Ferroelectrics* **1991,** *118,* 319.
27. Yilmaz, S.; Bauer, S.; Wirges, W.; Gerhardmulthaupt, R. *Appl Phys. Letters* **1993,** *63,* 1724.
28. Kocot, A.; Wrzalik, R.; Vij, J. K.; Zentel, R. *J. Appl. Phys.* **1994,** *75,* 728.

Pyrolysis

See: Cellulose (Pyrolysis)

Q

QUINONE-AMINE POLYURETHANES

David E. Nikles* Jason L. Cain, Antony P. Chacko, Jeng-Li Liang, and Russell I. Webb
Department of Chemistry and Center for Materials for Information Technology
The University of Alabama

Amine-quinone polymers containing the 2,5-diamino-1,4-benzoquinone function group in the polymer chain have been the subject of research owing to their ability to inhibit the corrosion of iron. The synthesis and properties of this new class of polymers, amine–quinone polymers, has been reported by Erhan and co-workers.[1–6] These polymers were prepared by a multiple-step condensation polymerization of a diamine with benzoquinone. A number of polymers were prepared using either aliphatic or aromatic diamines. Depending on the choice of diamine, polymers could be prepared that were soluble in organic solvents and films could be cast by solution-coating techniques. It was reported that the amine–quinone polymers would displace moisture from the surface of iron, rendering it hydrophobic. This was significant because the rate of iron corrosion greatly increases when surface moisture is present. Erhan recognized the coatings containing amine–quinone polymers may inhibit surface corrosion of iron and, accordingly, has obtained patents on these new materials.[7,8]

Our interest in the reliability of metal particle tape (MP tape) for archival data storage has led to the exploration of means to enhance the stability of the iron particles used in MP tape. Erhan's reports have inspired us to devise methods to prepare monomers containing the 2,5-diamine-1,4-benzoquinone functional group and to use these monomers to prepare amine–quinone polymers. The purpose was to prepare polymers with physical properties that can be systematically controlled.

AMINE–QUINONE POLYURETHANES

The amine–quinone monomers were used as chain extenders to prepare a series of different urethane block copolymers by condensation with different oligomeric diols and different diisocyanates. The soft segment diol was either poly(caprolactone) diol with M_n = 1250, poly(1,2-propylene glycol) (Voranol 220-056, from Dow Chemical) with M_n = 2000, poly(1,2-butylene glycol) (B-2000, from Dow Chemical) break with M_n = 2000 or polytetrahydrofuran (Terathane 650, from DuPont) with M_n = 650. The diisocyanate monomer was either diphenylmethane-4,4′-diisocyanate (MDI), isophorone diisocyanate (IPDI), or tolylene-2,4-diisocyanate (TDI). The TDI was a mixture of isomers, 80% tolylene-2,4-diisocyanate and 20%

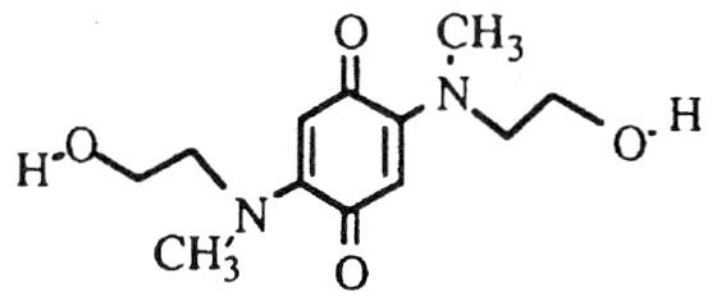

I

AQM-1

tolylene,2,6-diisocyanate. The chain extender was either 1,4-butane diol, AQM-1, or a mixture of 1,4-butane diol and AQM-1.

The polyurethanes were prepared by one of three different processes: a two-stage melt polymerization where the soft segment diol was first condensed with the diisocyanate, followed by chain extension with the hard segment diol; a two-stage solution polymerization where the polymerization scheme in process 1 (**Scheme I**) was done in a solvent, either tetrahydrofuran or *N,N*-dimethylforamide; or a modified melt polymerization where all the monomers were combined.

CONCLUSION

Amine–quinone polyurethanes can be prepared by the condensation of diol monomers containing the 2,5-diamino-1,4-benzoquinone functional group, and with diisocyanates in combination with other diol monomers. The 2,5-diamino-1,4-benzoquinone functional group survives the polymerization intact, giving red-brown colored polyurethanes. The amine–quinone polyurethane chemistry is no different from that for other polyurethanes, and polymers having physical properties ranging from elastomeric to thermoplastic can be prepared. The amine–quinone polymers can be used as binders in metal particle tape and serve to protect the iron pigments from corrosion.

REFERENCES

1. Kaleem, K.; Chertok, F.; Erhan, S. *Prog. Organ. Coatings* **1987**, *15*, 63.
2. Kaleem, K.; Chertok, F.; Erhan, S. *J. Polym. Sci., Pt. A* **1989**, *27*, 865.
3. Nithianandam, V. S.; Erhan, S. *J. Appl. Polym. Sci.* **1991**, *42*, 2385.
4. Nithianandam, V. S.; Kaleem, K.; Chertok, F.; Erhan, S. *J. Appl. Polym. Sci.* **1991**, *42*, 2893.
5. Nithianandam, V. S.; Chertok, F.; Erhan, S. *J. Appl. Polym. Sci.* **1991**, *42*, 2899.
6. Nithianandam, V. S.; Erhan, S. *Polymer* **1991**, *32*(6), 1146.
7. Erhan, S. Novel Polymers, U.S. Patent 4 882 413, 1989.
8. Erhan, S. Marine Paint Composition Comprising Quinone/Polyamine Polymers, U.S. Patent 4 981 946, 1990.

QUINONE-DIAMINE POLYMERS

Semih Erhan
Albert Einstein Medical Center

Polymers produced by reacting diamines and quinones have some rather unusual properties, such as extraordinary affinity toward metals, excellent anticorrosive action, high temperature performance, impressive abrasion, and impact resistance. Strangely, however, none of these characteristics have become known until recently, although the first reference that the reaction between amines and quinones could produce polymeric materials was made in 1918 by Suida.[1] There were several other

*Author to whom correspondence should be addressed.

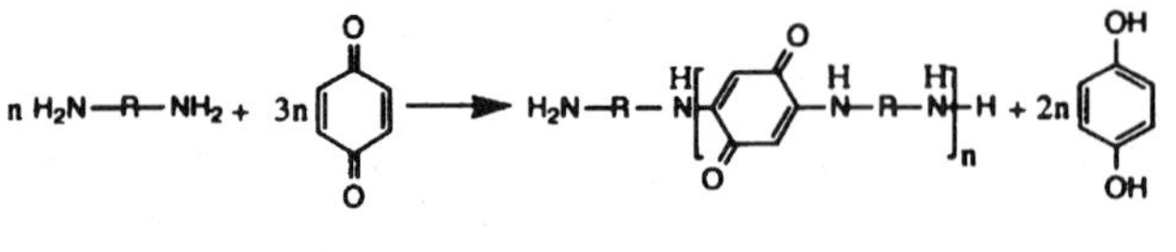

I

studies on these polymers, including a more recent study reporting that polymers prepared from p-benzoquinone and aliphatic diamines could transfer heat.[2–4] During the early 1980s, after a chance observation of an unexpectedly strong adhesion of these polymers toward glass, studies were initiated to explore their characteristics fully. These studies led to the realization of their unusually strong affinity toward metals, their being strong enough to displace water from wet, rusty metal surfaces as well as other properties.[5–20]

CHEMISTRY OF THE POLYMERS

The reaction between amines and quinones has been studied exhaustively since the late 19th century through early 20th century, because quinone tanning was the method of choice for the preparation of specialty leathers, such as chamois.[21]

If one uses a diamine in the reaction a linear polymer is obtained that ends with free amino groups at each terminus (**Scheme I**).

This reaction occurs very easily, sometimes even at 0°C and is over within 3–4 hours. The reaction can easily be followed by UV spectroscopy because the absorption maxima of hydroquinone, p-benzoquinone and diamine-substituted benzoquinones are 295, 245, and 345 nm, respectively. So far the amines used in the preparation of polymers are: ethylenediamine; 1,6-diaminohexane; 1,8-diaminooctane; 1,10-diaminodecane; 1,12-diaminododecane; lysine; 1,4-diaminocyclohexane; m- and p-phenylenediamine; tri-methylhexamethylenediamine; Jeffamine® D-400; Jeffamine® D-2000; 1,3-bis(3-aminophenoxy)benzene (APB); 1,3-bis(4-aminophenoxy)benzene (TPE-R); 1,4-bis(4-aminophenoxy)benzene (TPE-Q); 9-9-bis(4-aminophenyl)fluorene (FDA); 2,2-bis(4-aminophenyl)diisopropylbenzene (bisaniline); 4,4-bis(4-aminophenoxy)bisphenol A (BAPP); and 4,4-bis(aminophenoxy)diphenyl sulfone (BAPS).

The quinones used in these preparations are *o*- and *p*-benzoquinone, 2-methylbenzoquinone and 2-phenylbenzoquinone. Polymers prepared with poly(ether amine)s, such as the Jeffamines, are represented as poly(ether amine quinone)s (PEQs) and those prepared with other amines such as poly(amine quinone)s (PAQs), in our publications.

PROPERTIES OF THE POLYMERS

A wide variety of properties are exhibited by PAQs, depending solely on the chemical nature of their precursors and to a lesser extent on the solvents used. These include solubility, curing, solvent resistance, high and low temperature performance, and anticorrosive properties.

SOLUBILITY

All polymers are insoluble in water, but many are soluble in organic solvents, while quite a few are insoluble even in sulfuric acid.

The mechanism of binding is due to "chemisorption," whereby the free electrons of electronegative elements are contributed to election-poor, outer orbitals of metals, creating a very strong bond.

CURING

Although affinity of these polymers toward metals that are strong enough to display water is quite impressive, they require curing to exhibit the optimal characteristics.

PAQs can be cured either chemically, at ambient or higher temperatures or thermally, with or without a curing agent. Since each repeat unit carries two imino nitrogens, any multifunctional chemical that can react with active protons can cure them: diisocyanates, diepoxides, polyaziridines, and titanates as well as epoxy resins, to name a few.

Thermal cure, however, takes place by the substitution of the imino nitrogens of other chains onto the available sites on the quinone molecules for those polymers that do not require addition of a curing agent. Naturally one can also add diamines or other curing agents. Thermal curing can be achieved in 2–3 h at 110–150°C.

PROPERTIES OF THE CURED POLYMERS

Anticorrosive Characteristics

The most thoroughly studied characteristics of these polymers are their anticorrosive properties. All cured films exhibit a very strong anticorrosive activity when the coating is undamaged. Polymers can be applied using all standard methods, such as spraying, brushing, dipping, etc. The films formed are pinhole free. When the polymers are cast on a teflon-covered surface one can obtain free-standing polymer films. The polymers coat all organic coatings.

Abrasion and Impact Resistance

Jeffamine D-400 based polymer passes the ASTM D 968-81 test for abrasion not only with the standard sand but also when silicon carbide sand is used, after the standard sand, on the same panel carrying a film of 1 mil thickness.

High Temperature Performance

Some polymers prepared with aromatic diamines exhibit excellent high temperature performance. The polymers were prepared starting with a reactant ratio of quinone:diamine = 3:1, in THF.

REFERENCES

1. Suida, H. *Ann* **1918**, *416*, 164.
2. Szita, J.; Marvel, C. S. *J. Polym. Sci. Part 1* **1969**, *7*, 3203.
3. Ueda, M.; Sakai, N.; Imai, Y. *Macromol. Chem.* **1979**, *180*, 2813.
4. Jenkins, R. K.; Byrd, N. R.; Lister, J. L. *J. Appl. Polym. Sci.* **1969**, *13*, 1265.
5. Kaleem, K.; Chertok, F.; Erhan, S. *Prog. Org. Coating* **1987**, *15*, 63.
6. Kaleem, K.; Chertok, F.; Erhan, S. *J. Polym. Science Part A* **1989**, *27*, 865.

7. Kaleem, K.; Chertok, F.; Erhan, S. *New Polym., Materials* **1990**, *1*, 265.
8. Nithianandam, V. S.; Kaleem, K.; Chertok, F.; Erhan, S. *J. Appl. Polym. Science* **1991**, *42*, 2893.
9. Nithianandam, V. S.; Chertok, F.; Erhan, S. *J. Appl. Polym. Science* **1991**, *42*, 2899.
10. Nithianandam, V. S.; Chertok, F.; Erhan, S. *J. Coatings Technl.* **1991**, *63*, 47.
11. Nithianandam, V. S.; Chertok, F.; Erhan, S. *J. Coating Technol.* **1991**, *63*, 51.
12. Nithianandam, V. S.; Erhan, S. *J. Appl. Polym. Science* **1991**, *42*, 2385.
13. Nithianandam, V. S.; Erhan, S. *Polymer* **1991**, *32*, 1145.
14. Reddy, T. A.; Erhan, S. *Intl. J. Polym. Materials* **1993**, *19*, 109.
15. Reddy, T. A.; Erhan, S. *J. Polym. Science* **1994**, *32*, 557.
16. Reddy, T. A.; Erhan, S. *Intl. J. Polymeric Mater.* **1994**, *23*, 167.
17. Reddy, T. A.; Erhan, S. *J. Appl. Polym. Science* **1994**, *51*, 1591.
18. Reddy, T. A.; Macaione, D.; Erhan, S. *J. Polym. Science* **1994**, *32*, 1977.
19. Reddy, T. A.; Nithianandam, V. S.; Erhan, S. Proc. 3rd Intl. Symp. on Metallized Plastics; Fundamental and applied aspects, Mittal, K. S., Ed.; Plenum: NY, 1993; p 43.
20. U.S. Patents 4 831 107; 4 882 413; 4 981 946.
21. Hilpert, S.; Brauns, F. *Collegium* **1925**, 64.

R

RADIATION CURING

R. S. Davidson
Chemical Laboratory
The University of Kent at Canterbury

Radiation curing is a term used to describe processes involving polymerization or crosslinking reactions that have been initiated or brought about by electromagnetic radiation. The most commonly used types of radiation include UV and visible light and electron beams. It is possible to classify light induced reactions as photoinitiated polymerization reactions and photopolymerization reactions.[1] Electron beam radiation can be used to initiate polymerization and to induce crosslinking.[1a] Often the two processes co-occur.

RADIATION CURING PROCESSES

Free Radical

Free radical mediated curing processes still dominates the industry. This is in part owing to the ready availability of a wide range of materials thereby making it possible to produce a range of coatings having widely different properties such as hard, tough coatings to soft, highly flexible coatings.

The most ubiquitous of radical cured materials are acrylates and they cure via a chain process.[2]

Initiation is commonly brought about by benzoyl, substituted benzoyl and phosphinoyl radicals, with phenyl and alkyl radicals playing a less important role.[2a] Propagation is rapid with acrylates and hence rapid cure is commonly encountered. When di and higher functionalized acrylates are used cure occurs even more rapidly owing to the occurrence of crosslinking.

Methacrylates can also be polymerized in a similar manner but are slower to react.

Cationic Processes

The cationic polymerization of epoxides (oxiranes) and vinyl ethers is a well-known reaction.[3] For such reactions to be employed in radiation curing there is the necessity to employ either photoinitiators that liberate acid, or if eb radiation is employed, additives that will ultimately release acid.[1a]

Epoxides are polymerized under the influence of Bronsted and Lewis acids.[4]

The range of epoxides available for use in radiation curing is rather limited. There are some cycloaliphatic epoxides available and these are more reactive than glycidyl ethers.

The range of epoxide prepolymers is even more limited. Epoxidized polybutadiene and glycidyl ethers of novalak resins are available. Glycidyl ethers are less reactive than the cycloaliphatic epoxides. The ring opening of both species has a finite energy of activation. As a consequence, when epoxides are cured by UV radiation, the extent of cure is affected by the heat that is available from the lamps. Frequently it is found that cure is not complete when the coating leaves the irradiation unit. Since acid is present in the coating, polymerization often continues after irradiation and this phenomenon is known as "post cure."

Vinyl ethers cure rapidly in the presence of Lewis and Bronsted acids.[5] Use of polyfunctional vinyl ethers leads to crosslinking. A good range of vinyl ether reactive diluents is now available and some oligomeric (prepolymer) materials are also available.

DISTINCTION BETWEEN UV/VISIBLE AND ELECTRON BEAM CURING

UV Visible Curing

For an effective and acceptable rate of cure to be attained it is necessary to incorporate a photoinitiator. The function of the photoinitiator is to generate radicals that will initiate polymerization.

eb Curing

When high energy electrons enter a coating they lead to ionization of the bulk material, that is, the ionization process is not specific to a particular group as is the case with photoinitiation. Ionization of the bulk material can lead to the generation of radical sites in the backbone of a reactive diluent or prepolymer, rather than at the reactive group, for example, acrylate.[6] As a consequence eb-radiation induces polymerization not only via the reactive groups but by crosslinking involving carbon atoms not contained in the reactive groups.[7]

Distinction

eb Curing leads to crosslinking via the backbone of the diluent and prepolymer at an early stage in the curing process whereas with UV and visible light-initiated cure, crosslinking via such groups occurs at a much later period.

CHEMISTRY OF PHOTOINITIATORS

Introduction

Initiators have been classified as belonging to either the Type I or Type II class.[2a,8] Type I photoinitiators are ones that following excitation undergo a bond cleavage.

Type II photoinitiators are of themselves relatively inefficient but when used in the presence of a synergist (typically a tertiary amine), reaction rapidly ensues. If a tertiary amine is used as synergist, the initiating radical (an α-aminoalkyl radical) is derived from the amine.

Free Radical Initiators

Type I Photoinitiators

Excellent reviews are available detailing the initiators that are commercially available and describe their chemistry.[8]

There is a drive to produce Type I photoinitiators that absorb at > 400 nm because such compounds could be used in pigmented (titanium dioxide) systems.[9] For a Type I initiator to function it must absorb the light efficiently, have a quantum yield for α-scission as near as possible to one, and produce radicals that initiate polymerization efficiently.

Recently bisacylphosphine oxides have been introduced.[10,11] Although the light absorption properties of this class of compound are not that different from the monoacylphosphine oxides, advantages accrue from the fact that there is the potential to produce four initiating radicals from each mole of initiator.

Another type of initiator that has the potential to liberate two initiating radicals per mole of initiator is that based on dibenzoylmethane.[12]

During the last decade much interest has been shown in polymeric and polymerizable photoinitiators.[13]

Type II Photoinitiators

Many aromatic ketones (benzophenone, substituted benzophenones, benzils, fluorenone, xanthone, thioxanthones) will act as Type II initiators with their performance being dramatically enhanced by the use of an amine (tertiary)synergists. Of the amine synergists commercially available, alkanolamines (triethanolamine, *N,N*-dimethylethanolamine and *N*-methyldiethanolamine), and derivatives of p-N,N-dimethylaminobenzoic acid are in common use. Clearly one class of amine is aliphatic and the other aromatic and this has consequences with their use. Aliphatic amines are transparent down to ~260 nm and consequently a coating containing such an amine can in principle use light from the UV lamp down to ~260 nm. The aromatic amines display strong absorption around 300 nm and consequently screen much of the UV light.[14]

Cationic Photoinitiators

Cationic initiators are compounds that under the influence of UV or visible or eb radiation lead to the release of acid, which catalyzes the desired polymerization process.[3]

Diazonium salts were some of the earliest materials to be used but these have been replaced by the more thermally stable iodonium and sulfonium salts.[15]

To cure an epoxide or vinyl ether using eb it is necessary to use a cationic initiator.

The mechanism of photodecomposition of onium salts is complex, with in-cage and out-of-cage processes contributing. In the in-cage mechanism it was shown that the proton source was the initiator itself and not the solvent.[16]

A totally different type of acid-generating initiator is based on a derivative of benzoin.[17] Central to the mechanism is the rearrangement reaction of the intermediate radical that leads to release of the sulfonic acid.

Anionic Photoinitiators

There are very few initiators of this type. Tertiary amine salts of ketocarboxylic acids[18] were the earliest to be described. Newer systems based on peptide chemistry have been described and used in microlithography.[19]

Oxygen Inhibition of Cure

Oxygen decreases the efficiency of radical curing processes by quenching the triplet states of photoinitiators, scavenging radicals generated by the initiator system, and scavenging the growing macroradical. When thin films are being cured, oxygen inhibition manifests itself by the cured film having a surface that is tacky to the touch.

The effectiveness of amines in retarding oxygen inhibition is likely to be linked to their ionization potential (the lower the better) and steric effects (bulky groups reducing efficiency). An unfortunate drawback of incorporating tertiary amines into formulations is that they enhance the rate of photoyellowing of cured films.

Industrial Uses

The uses of radiation curing are extremely varied and include wood and cork coating, metal coating (manufacture of car-headlamp reflectors) plastic coating (PVC floor tiles, compact discs, coating glass [optical fibers]), production of printed circuit boards, and in the graphic arts.[20] More esoteric but valuable uses include use in the dental industry and the production of 3D models and prototypes.[21,22]

The industry is expected to grow over the next decade because it will, in many cases, replace conventional solvent-based formulations, play an important part in new technologies, (in the growth of optical fiber communication systems), and lead to new manufacturing processes.

REFERENCES

1. De Schryver, F. C. *Pure and Applied Chem.* **1973**, *34*, 213.
1a. Davidson, R. S. *Radiation Curing in Polymer Science and Technology*; Fouassier, J. P.; Rabek, J. F., Eds.; Elsevier Science: Essex, England, 1993, Vol. III, Chapter 9.
2. Jenkins, A. D.; Ledwith, A. *Structure Reactivity and Mechanism in Polymer Chemistry*; Wiley: London, UK, 1974.
2a. Hageman, H. J. In *Photopolymerization and Photoimaging Science and Technology*; Allen, N. S., Ed.; Elsevier: Essex, England, 1989, Chapter 1.
3. Pappas, S. P. *Photopolymerization and Photoimaging Science and Technology*; Allen, N. S., Ed.; Elsevier: Essex, England, 1989, Chapter 2.
4. Dreyfus, P.; Dreyfus, M. P. *Comprehensive Chemical Kinetics*; Bamford, C. H.; Tipper, C. F. H., Eds.; Elsevier: Amsterdam, 1976, Vol. 15, Chapter 4.
5. Sawamoto, M. *Progress in Polymer Science* **1991**, *16*, 111.
6. Batten, R. J.; Davidson, R. S.; Wilkinson, S. A. *Polymer Paints Colours J.* **1989**, *179*, 176.
7. Davidson, R. S.; Ellis, R.; Wilkinson, S.; Summersgill, C. *Eur. Polym. J.* **1987**, *23*, 105.
8. Dietliker, K. In *Chemistry and Technology of uv and eb Formulations for Coatings*; Inks and Paint, Vol. 3, ed. PKT Oldring SITA Technology Ltd.: London 1992.
9. Davidson, R. S. *J. Photochem. Photobiol. A: Chem.* **1993**, *73*, 81.
10. Rutsch, W.; Angerer, H.; Desobry, V.; Dietliker, K.; Husler, R. Proc. XVI *Conference on Organic Coatings: Science and Technology*, Athens 1990; 423.
11. Ellrich, K.; Herzig, C. Eur. Patent 184095.
12. Bosch, P.; del Monte, F.; Mateo, J. L.; Davidson, R. S. *J. Photochem., Photobiol. A. Chem.* **1994**, *78*, 79.
13. Davidson, R. S. *J. Photochem. Photobiol. A. Chem.* **1993**, *69*, 263.
14. Herlihy, S. L.; Battersby, G. C. *Proceedings Radtech '94*; North America 1994; Radtech International North America, Ill, USA, 1994; 156.
15. Sahyun, M. R. V.; DeVoe, R. J.; Olofson, P. M.; *Radiation Curing in Polymer Science and Technology* Vol. II, Fouassier, J. P.; Rabek, J. F., Eds.; Elsevier: Essex, England, 1993; Chapter 10.
16. Hacker, N. P. *Radiation Curing in Polymer Science and Technology* Vol. II, Fouassier, J. P.; Rabek, J. F., Eds.; Elsevier: Essex, England, 1993; Chapter 9.
17. Hageman, H. J.; Jansen, L. G. H. *Makromol. Chem.* **1988**, *189*, 2781.

18. Mayer, W.; Rudolf, H.; De Cleur, E. *Angew. Makromol. Chem.* **1981**, *93*, 83.
19. Cameron, J. F.; Frechet, J. M. *J. Org. Chem.* **1990**, *55*, 5919; *Pure and Appl. Chem.* **1991**, *61*, 1239.
20. Salim, M. S. In *Radiation Curing of Polymers*, II Randell, D. R., Ed. The Royal Society of Chemistry, Cambridge, England, 1991; p 3.
21. Linden, L. A. *Radiation Curing in Polymer Science and Technology* Vol. IV, Fouassier, J. P.; Rabek, J. F., Eds.; Elsevier: Essex, England, 1993; Chapter 13.
22. Bernhard, P.; Hofmann, M.; Hunziker, M.; Klingert, B.; Schultkers, A.; Steinmann, B. In *Radiation Curing in Polymer Science and Technology* Vol. IV, Fouassier, J. P.; Rabek, J. F., Eds.; Elsevier: Essex, England, 1993; Chapter 10.

Radiation-Induced Degradation

See: *DNA Degradation (Atomic Target Method using Synchrotron Radiation)*
Polytetrafluoroethylene (Effects of γ-Irradiation)

Radiation-Induced Polymerization

See: *Cyclic Siloxanes (Ring-Opening Polymerization)*
Inclusion Complexes (Overview)
Inclusion Polymerization
Polyethylene (Surface Functionalization)
Radiation Curing
Radiation-Induced Polymerization (Biomedical Applications)
Surface Modification (by Low Energy Electron Beam)
Wool Grafting

RADIATION-INDUCED POLYMERIZATION (Biomedical Applications)

Isao Kaetsu
Faculty of Science and Technology
Kinki University

The irradiation of ionizing radiation such as gamma-rays and electron beams causes the formation of chemically active species such as ions and radicals and initiates radiation chemical reactions such as polymerization, graft copolymerization, crosslinking, and degradation of polymer.[1] The characteristics of radiation-induced reactions can be summarized into the following two points: formation of active species and an initiation reaction can occur under a very broad range of conditions such as solid and viscous states, complexed multi-phases, low and high temperatures, low and high pressures, and so on, owing to the penetrating power of radiation through a reaction system; and concentration of active species can be changed and controlled widely by the energy and intensity of radiation. Therefore, two kinds of polymerization have been studied with great interest since the early stages of research in this field—the radiation-induced polymerizations in a solid state and at low temperature. Many workers have studied the radiation polymerization of crystalline monomers such as acrylamide, acrylic acid, vinyl carbazole, and a series of cyclic monomers such as trioxane, tetraoxane, diketene, and β-propiolactone.[2] However, the author found that certain monomers can be supercooled easily and stably and has studied the radiation polymerization in an amorphous or glassy solid state of a supercooled monomer in detail and systematically.[3,4] The main basic research in this area was almost finished by the early half of the 1970s.

However, since 1973 various kinds of applications have been developed using this polymerization technique of the author's group.[5,6] First it was applied to a successful cast polymerization process of organic glass and optical plastic lenses in the 1970s.[7] The two characteristics of this polymerization, less temperature rise during polymerization—and smaller amount of volume shrinkage with the polymerization—were advantageous for the control of heat and dimensional exactness required in the casting process. However, another kind of application to immobilization of biofunctional materials has been studied and developed extensively for biochemical and biomedical uses by means of polymerization in a supercooled state.[8] In this case, the characteristic use of the viscous or solid phase and of low temperatures in this polymerization was advantageous for safe and effective immobilization of biological components into a polymer. Thus, various components from molecules to microorganisms such as enzymes, antibodies, organellas, yeast, microbial cells and tissue cells were immobilized and investigated for the practical uses as bioreactors, diagnostic systems, materials for artificial organs, and so on.[9–14]

In the past twenty years, the most concentrated subject of application for the author's group has been drug delivery systems, a technique for immobilizing drugs and hormones into a polymer-controlled, sustained and targeted release.[15] The polymerization of supercooled monomer also showed the advantage of the preparation of drug-delivery systems. The drug-delivery systems have been effectively used in many clinical tests for local chemotherapy and hormone therapy.[16]

PREPARATION AND PROPERTIES

Supercooling Monomers

The monomers having a tendency to be supercooled can be summarized as follows: monomers having a bulky side chain, monomers having a side chain of strong hydrogen bonding, and multi-component monomeric systems or a mixture of monomer with supercooling monomer.

Many acrylates and methacrylates having a relatively large ester chain group can be supercooled. Therefore, a supercooling monomer is not a specific one, but rather common among the ester and ether monomers.

APPLICATIONS

The advantages of supercooled state polymerization can be summarized as follows: (1) large polymerization rate into 100% conversion owing to high viscosity; (2) high molecular weight of polymers owing to less termination; (3) less effect of polymerization retarder and inhibitor such as oxygen and impurity, owing to high viscosity; (4) easy control of temperature (heat) and dimension, resulting in fewer defects and strains in the products; (5) easy retention of original form, complex structure, and multi-phases through polymerization; and (6) easy retention of original activity of biological species. The application include both biochemical and biomedical applications.

Immobilization of Proteins for Diagnostic and Therapeutic Systems

The immobilization of several enzymes for medical uses was studied.[17]

Immobilization of Cells for Cell Culture

There are three kinds of immobilization techniques for different types of cells. The first is the adsorption of yeast on a radiation-polymerized hydrophilic gel. The second method is the adsorption of fungi into a radiation-reinforced fibrous network support such as gauze and non-bonded cloth. The third type of immobilization is the adsorption of tissue cells such as liver cells and glial cells on a heterogeneous, porous surface.

Immobilization of Drugs and Hormones for Drug Delivery Systems

Drug delivery systems have been studied and developed most actively by the author's group since 1974.[18,19] The drug delivery systems have two main functions, controlled release and targeting. For implantable systems, targeting can be achieved by a suitable implantation close to the target site so the main subject is the controlled release technique.

Very recently the author's group started the study and development of a signal-responsive delivery system that releases a chemical compound with chemical, biological, physiological, and psychological effects in response to a signal from the environment. The fundamental technique of this system is a combination of sensor and delivery systems into a micro-machine form. This promising new research field will be studied in detail in the near future.

REFERENCES

1. Chapiro, A. *Radiation Chemistry in Polymeric Systems* **1962**, Interscience, New York.
2. Okamura, S.; Hayashi, K.; Kitanishi, Y. *J. Polymer Sci.* **1962**, *58*, 925.
3. Kaetsu, I.; Tsuji, K.; Hayashi, K.; Okamura, S. *J. Polymer Sci.* **1967**, *5*, 1899.
4. Kaetsu, I.; Okubo, H.; Ito, A.; Hayashi, K. *J. Polymer Sci.* **1972**, *10*, 2203.
5. Kaetsu, I. *Radiation Processing of Polymers*; Singh, A.; Silverman, J., Eds.; Hanser: NY, 1992; p 150–185.
6. Kaetsu, I. *Advances in Polymer Sciences*; Okamura, S., Ed. Springer-Verlag: Berlin-Heidelberg, 1993; p 81–97.
7. Kaetsu, I.; Yoshii, F.; Okubo, H.; Ito, A. *Polymer Preprint* **1975**, *16*, 465.
8. Kaetsu, I. *Radiat. Phys. Chem.* **1981**, *18*, 343.
9. Kaetsu, I.; Kumakura, M.; Yoshida, M. *Biotechnol. Bioeng.* **1979**, *21*, 847.
10. Kumakura, M.; Kaetsu, I. *Appl. Biochem. Biotechnol.* **1983**, *8*, 87.
11. Fujimura, T.; Yoshii, F.; Kaetsu, I.; Inoue, Y.; Shibata, K. Z. *Naturforsch.* **1980**, *35c*, 477.
12. Fujimura, T.; Kaetsu, I. *J. Appl. Biochem. Biotechnol.* **1983**, *8*, 145.
13. Kumakura, M.; Kaetsu, I. *Biotechnol. Bioeng.* **1979**, *21*, 679.
14. Yoshii, F.; Kaetsu, I. *J. Appl. Biochem. Biotechnol.* **1983**, *8*, 115.
15. Yoshida, M.; Kumakura, M.; Kaetau, I. *J. Pharmaceu. Sci.* **1979**, *68*, 628.
16. Hanyu, F.; Nakamura, M.; Yamada, A.; Sakurai, Y.; Yoshida, M.; Kaetsu, I. *Cancer and Chemotherapy* **1980**, *7*, 1824.
17. Kaetsu, I.; Yoshida, M.; Yamada, A.; Sakurai, Y. *J. Biomed. Mater. Res.* **1980**, *14*, 199.
18. Kaetsu, I. Radiation Processing of Polymers (Ed. by Singh, A. and Silverman, J.) Chapter 8, p 149–185 1992, Hanser Publishers, New York.
19. Kaetsu, I. Advances in Polymer Science (Ed. by Okamura, S.) 1993, Vol. 105, p 81–97. Springer-Verlag, Berlin.

Radical Polymerization

See: Free Radical Polymerization

Radical Ring-Opening Polymerization

See: Ring-Opening Polymerization, Free Radical Ring-Opening Polymerization, Radical (with No Shrinkage in Volume)

RADIOPAQUE POLYMERS

Johannes Smid, Israel Cabasso, and Gautam Chatterjee
Polymer Research Institute
Faculty of Chemistry
College of Environmental Science and Forestry
State University of New York at Syracuse

Synthetic polymers are used in ever larger quantities in a variety of applications including medical devices and implants. In dentistry, polymeric resins find extensive application in removable dentures, temporary crown and bridge materials, restorative fillings, and impression compounds. Applications of this nature require non-destructive diagnostic evaluation procedures, for example, to detect cracks or leakages, or to retrieve plastic objects that have accidentally become dislodged in the body. For this reason, the need for radiopaque polymers has been argued on several occasions, among others by the American Dental Association, which has stressed its use in dental resins.[1]

Most conventional polymers are composed of the elements C, H, O, and N. This makes them essentially X-ray transparent. To convert them into X-ray opaque materials, elements of high atomic mass must be incorporated. This can be accomplished by either dispersing the radiopacifying agent physically into the polymer resin, or by permanently anchoring the additive to the polymer matrix, for example, by copolymerization with monomers containing elements such as bismuth. Physical mixing, with or without the aid of a solvent, can lead to heterogeneous or homogeneous blends. Complete solubilization can sometimes be achieved when the polymer contains moieties that chelate with the metal ion. For hydrophobic radiopacifiers such as triphenylbismuth, only weak interactions with the polymer matrix are needed in order to achieve compatibility.

SYNTHESIS AND PROPERTIES

Physical Blends

Hetergeneous Systems

A physical, heterogeneous dispersion of a heavy metal, or of a heavy atom organic or inorganic compound with a polymer, typifies this class of radiopaque materials. Gold gauze, lead foil, and fine wire have been inserted in poly(methyl methacrylate)

(PMMA) [9011-14-7] for dental applications.[9] Powdered glasses with barium, lead, or bismuth content have also been embedded frequently in polymer composites, sometimes by making a slurry with the monomer prior to polymerization.[3–7]

Homogeneous Systems

High boiling aliphatic and aromatic halides have been dissolved in PMMA to impart radiopacity.[2,7]

Cation chelating moieties on polymers generally promote the solubilization of metal salts. The chelation also reduces the problem of leaching.

Carbonyl and phosphonate-containing monomers and polymers solubilize a variety of heavy metal salts such as cupric nitrate, manganese dichloride, praseodymium chloride, bismuth trichloride, bismuth tribromide and uranyl nitrate hexahydrate.[8–16]

One of the most versatile radiopacifier agents is triphenylbismuth.[17–19] It forms miscible and often optically transparent blends of high radiopacity with a wide range of polymeric materials including polystyrenes, polyalkenes, polyacrylates, poly(vinyl chloride), and epoxy resins.

Polymers with Chemically Bound Radiopacifier

The first approach in this field was the synthesis of halogenated monomers. A copolymer of MMA and 35 wt% of 2,3-dibromopropyl methacrylate yielded a material with a radiopacity equivalent to that of aluminum.[2]

Spherical radiopaque hydrogel particles designed for endovascular occlusion have been reported.[21]

Another approach is the synthesis and subsequent polymerization of heavy metal-containing monomers. Copolymerization of MMA with zinc and barium acrylates produced dental resins of acceptable radiopacity.[2]

The monomer styryldiphenyl bismuth yields polymers with covalently-bound triphenylbismuth. In addition to organobismuth monomers, styrylaryl tin and lead monomers have also been reported but their higher toxicity in polymers makes them less attractive as radiopacifiers.[22]

APPLICATIONS

X-ray contrast polymers provide a quick, reliable and nondestructive alternative for the analysis of polymeric materials. This is especially helpful where plastics are employed for biomedical purposes such as body implants, catheters, sutures, and medical adhesives. Radiopaque materials have been used in the monitoring of fallopian tube blockage and in detecting changes in kidneys and other body functions.[20,23] The largest number of applications are in dentistry where radiopaque resins have been incorporated in dentures, and in filling and restorative materials, among others.[5–7] Other potential applications are in plastic wrappings or toys since their detection after accidental swallowing by infants would be greatly facilitated.[24] Radiopaque polymers for coding purposes, detection of mechanical deficiencies such as cracks and crazes, and in explosives may be other potential uses.

ACKNOWLEDGMENT

For those studies that were carried out in the laboratory of the authors, the financial support of the Department of Health and Human Services through the National Institute of Dental Research and of the Polymers Program of the National Science Foundation is gratefully acknowledged.

REFERENCES

1. Council on Dental Materials, Instruments and Equipment *J. Am. Dent. Assoc.* **1981**, *102*, 347.
2. Davy, K. W. M.; Causton, B. E. *J. Dent.* **1982**, *10*, 254.
3. Primack, J. E. *J. Prosthet. Dent.* **1972**, *28*, 363.
4. Chandler, H. H.; Bowen, R. L.; Paffenbarger, G. C. *J. Biomed. Mater. Res.* **1971**, *5*, 253, 335, 359.
5. Bowen, R. L.; Cleek, G. W. *J. Dent. Res.* **1972**, *51*, 177.
6. Elzay, R. P.; Pearson, G. O.; Irish, E. F. *J. Prosthet. Dent.* **1971**, *25*, 251.
7. Combe, E. C. *J. Dent. Res.* **1971**, *50*, 668.
8. Combe, E. C. *J. Dent.* **1972**, *1*, 93.
9. Smid, J.; Cabasso, I.; Rawls, H. R.; Obligin, A.; Delaviz, Y.; Sahni, S. K.; Zhang, Z. X. *Makromol. Chem. Rapid Commun.* **1987**, *8*, 543.
10. Cabasso, I.; Sahni, S.; Rawls, H. R.; Smid, J. U.S. Patent 4 882 392, issued November 21, 1989; *Chem. Abstract* **1990**, *112*, 165051m.
11. Delaviz, Y.; Cabasso, I.; Smid, J. *Recl. Trav. Chim. Pays-Bas* **1990**, *109*, 176.
12. Delaviz, Y.; Cabasso, I.; Smid, J. *Polymer* **1990**, *31*, 2083.
13. Delaviz, Y.; Zhang, Z. X.; Cabasso, I.; Smid, J. In Kleintjes, L. A.; Lemstra, P. J., Eds. *Integration of Polymer Science & Technology*; Elsevier: 1989; Vol. 3, p 279.
14. Rawls, H. R.; Starr, J.; Kasten, F. H.; Murray, M.; Smid, J.; Cabasso, I. *Dent. Mater.* **1990**, *6*, 1990.
15. Cabasso, I.; Smid, J. *The Bulletin of the Bismuth Institute* **1990**, *60*, 1.
16. Cabasso, I.; Smid, J.; Sahni, S. K. *J. Appl. Polym. Sci.* **1989**, *38*, 1653.
17. Delaviz, Y.; Zhang, Z. X.; Cabasso, I.; Smid, J. *J. Appl. Polym. Sci.* **1990**, *40*, 835.
18. Smid, J.; Delaviz, Y.; Cabasso, I. U.S. Patent 5 256 334, issued October 26, 1993. *Chem. Abstract* **1990**, *113*, 213266e.
19. Chatterjee, G.; Cabasso, I.; Smid, J. *J. Appl. Polym. Sci.* **1995**, *55*, 851.
20. Brit. Patent 1 204 862, issued September 9, 1970 to Pharmacia Aktienbolag; *Chem. Abstract* **1970**, *73*, 112951z.
21. Horak, D.; Metalov, M.; Svec, F.; Drobnik, J.; Kalal, J.; Borovicka, M.; Adamyan, A. A.; Voronkova, O. S.; Gumargalieva, K. Z. *Biomaterials* **1987**, *8*, 142.
22. Ignatious, F.; Delaviz, Y.; Cabasso, I.; Smid, J. *Makromol. Chem. Macromol. Symp.* **1992**, *60*, 247.
23. Hoffman, A. S. U.S. Patent 4 359 454, issued November 16, 1982 *Chem. Abstract* **1983**, *98*, 28669x.
24. Silberman-Hazony, R. In Mark, H. F.; Bikales, N. M.; Overberger, C. G.; Menges, G., Eds., *Encyclopedia of Polymer Science and Engineering*; 2nd ed., John Wiley & Sons: New York, NY, 1988; Vol. 14, p 1.

Ramie

See: Cellulosic Materials (Moisture Sorption Properties)

Rare Earth Catalysts

See: Butadiene Polymerization (Supported Lanthanide Catalysts)
Butadiene Polymerization, Neodymium-Catalyzed
Rare Earth-Initiated Polymerization (of Dienes)
Rare Earth Polymerization Initiators

RARE-EARTH-INITIATED POLYMERIZATION (of Dienes)

David J. Wilson
Research and Development Laboratory
EniChem Elastomers Ltd.

Initiating systems based on rare earth metals otherwise known as lanthanide or seltenen erden (SE) metals are capable of polymerizing and copolymerizing a number of monomers including acrylates, oxiranes and ethylenes and acetylenes.[1,2] It is in the preparation of high polymers from conjugated dienes, however, that such systems have received the most attention in academia and industry, both historically and at present. These systems are gaining preference over traditional Ziegler-Natta systems for the production of polydienes (Ti, Co, Ni) owing to the properties of the final polymer and the flexibility of their manufacture.[3] Besides systems based on uranium, which are commercially unsuitable, rare earth systems provide the only means whereby polybutadiene and butadiene-isoprene copolymers of very high *cis* stereoregularity (> 97% *cis*) may be prepared, and one of two means of preparing high *cis* polyisoprene, the other being lithium alkyls, which can give up to 97% *cis* content.[4] The very high *cis* stereoregularity is crucial to the improved physical and technological properties displayed by these polymers and in the provision of superior materials for tire and other applications.

The recognition that the highest activity was obtained from systems based on the rare earth metal neodymium (Nd) has led to an almost exclusive use of this rare earth in both academic and commercial spheres.[2,7] A number of useful reviews and discussions on rare earth initiating systems for polydienes already exist.[2,3,8–14]

SYNTHESIS AND PROPERTIES OF THE POLYMERS

Standard Rare Earth Initiating Systems

Nd-based initiating systems have the highest activity. The activity of rare earth systems for the polymerization of conjugated dienes follows the order of the rare earth metal Nd > Pr > Gd (or Ce) irrespective of the system used.[2,7,10] In addition the *cis* content of the polymer varies little with the type of rare earth.[2,7,10] The initiator activity follows closely the change in rare earth ion size.[13]

Although several examples of alcoholate, alkylphosphate, and chelate systems can be found in the literature, initiating systems based on $Nd(carboxylate)_3$ have received the most attention.[2,6,7,18]

Polydienes Prepared by Rare Earth Systems

Nd-based initiating systems are capable of polymerizing a number of conjugated dienes: 1,3-butadiene; 2-methylbutadiene(isoprene); 2-ethylbutadiene and 2-isopropylbutadiene, 2-cyclopropylbutadiene, 1,3-pentadiene(piperylene), 2-methyl-1,3-pentadiene, 2,3-dimethyl-1,3-pentadiene, hexadiene, and deuterated dienes.[22–29] They have also been used to prepare copolymers: poly(butadiene-*co*-isoprene); poly(butadiene-*co*-2-ethylbutadiene) and poly(butadiene-*co*-isopropylbutadiene), poly(butadiene-*co*-2-cyclopropylbutadiene), poly(butadiene-*co*-1,3-pentadiene), poly(butadiene-*co*-oxygen-containing aluminium compounds), poly(butadiene-*co*-styrene), poly(butadiene-*co*-2,4-hexadiene), poly(butadiene-*co*-ethylene, poly(butadiene-*co*-propylene, and poly(2-cyclopropylbutadiene-*co*-*trans*-5-methylheptatriene-1,2,6.[20–22,30–35] Activity decreases in the order butadiene > isoprene > piperylene, with polymer *cis* content decreasing in the same order, albeit only slightly from butadiene to isoprene.[2,7,16,17,19,26]

cis-Polybutadiene and -Polyisoprene

Preparation

The majority of work on the rare earth-initiated polymerization of dienes has concerned the stereospecific polymerization of butadiene or isoprene.[3] The homopolymerization of both of these dienes using binary systems and ternary systems based on $Nd(carboxylate)_3$ is useful as a direct comparison of these two dienes.[2,7,16,17,26,30,36,37] The polymerization of either butadiene or isoprene using $Nd(alcoholate)_3$ has been documented.[6,18,38] All these systems give polybutadiene and polyisoprene with a *cis* content of >98% and >94%, respectively.[7,17]

Polymerization Kinetics and Mechanism

For rare earth systems the polymerization mechanism and the mechanism for the formation of *cis/trans/vinyl* units in the polymer is established.[6,8,12,26]

cis-Poly(butadiene-co-isoprene)

An appreciable amount of work has been done on the copolymerization of butadiene Nd isoprene using rare earth initiating systems.[5,7,9,11,15,16,39] Although block copolymers have been prepared using Nd systems, the majority of interest has been in random copolymers.[5,15,16,39]

SYNTHESIS AND CHARACTERISTICS OF COMMERCIAL HIGH *cis* Nd-POLYBUTADIENE

Despite the interest in different dienes, copolymers and systems producing polymers of varied microstructure as covered in the preceding pages, only two types of material are of interest commercially: high *cis*-polybutadiene (> 92% *cis*) and polybutadiene-*ran*-polyisoprene, and only the former is in commercial production. Commercial Nd-based initiating systems are the hydrocarbon-soluble ternary systems based on $Nd(carboxylate)_3$.

The characteristics of broad molecular weight distribution, low branching index and very high *cis* content of polybutadiene prepared using Nd-based initiating systems, lead to improved physical and technological properties compared to polydienes prepared by conventional systems. Characterization of Nd-polybutadiene by gel permeation chromatography reveals a significant high molecular weight "tail," in some cases approaching biomodality.[7,8,40,41] This undoubtedly contributes to the improved physical properties of Nd-polybutadiene.

PROPERTIES AND APPLICATIONS OF COMMERCIAL HIGH *cis* Nd-POLYBUTADIENE

The most important physical properties of high *cis* polybutadiene with respect to their applications are strain crystallization, green strength, tack, banding behavior, abrasion resistance, "fatigue-to-failure" life, tear strength, and hysteresis.

The main use for Nd-based polybutadiene is in tire formulations where it is used with styrene-butadiene rubber (SBR) or natural rubber (NR) for tread and sidewall applications.[42,43]

Non-Tire Applications

There are a number of non-tire applications for commercial Nd-polybutadiene. The present non-documented uses include hoses, conveyor belts, and shoe soles. Nd-polybutadiene can also be used in place of Ni or Co-polybutadiene in gold ball formulations.[44,45]

High *cis* Polybutadiene-Improvements

In an attempt to further improve Nd-polybutadiene processibility and properties such as abrasion, a number of authors have used "post-reaction" materials to link the polymer chains. PCl_3 has been used to improve the processibility and improve the carbon black incorporation time of Nd-polybutadiene.[46] Phenyl tin chlorides have been used as coupling agents, and the resulting materials show enhanced abrasion resistance, in tire tread and sidewall applications, compared to the parent polymer.[47,48] In a series of patents Japanese workers have used butyl trimellitate as a coupling agent and the resulting materials have found use in high-impact polystyrene (HIPS) applications.[49] Trimesic acid chloride has also been used as a coupling agent by the same workers.[50]

REFERENCES

1. Yasuda, H.; Tamai, H. *Prog. Polym. Sci.* **1993**, *18*, 1097.
2. Shen, Z. *Inorg. Chim. Acta* **1987**, *140*, 7.
3. Porri, L.; Giarrusso, A. *Comprehensive Polymer Science*; Pergamon: Oxford, 1989; 4, Part II, Chapter 5.
4. Hsieh, H. L.; Tobolsky, A. V. *J. Polym. Sci.* **1957**, *25*, 245.
5. Bruzzone, M. *Elastomers and Rubber Elasticity*, ACS Symp. Ser. No. 193, 1982, Chapter 2.
6. Mazzei, A. *Makromol. Chem. Suppl.* **1981**, *4*, 61.
7. Shen, Z.; Ouyang, J.; Wang, F.; Hu, Z.; Yu, F.; Oian, B. *J. Polym. Sci. Polym. Chem.* **1980**, *18*, 3345.
8. Witte, J. *Angew. Makromol. Chem.* **1981**, *94*(1443), 119.
9. Marina, N. G.; Monakov, Y. B.; Rafikov, S. R.; Gadeleva, K. K. *Polym. Sci. U.S.S.R.* **1984b**, 26(6), 1251.
10. Marina, N. G.; Monakov, Y. B.; Sabirov, Z. M.; Tolstikov, G. A. *Polym. Sci.* **1991**, *33*(3), 387.
11. Duvakina, N. V.; Marina, N. G.; Monakov, Y. B.; Rafikov, S. R. *Polym. Sci. U.S.S.R.* **1989**, *31*(2), 243.
12. Porri, L.; Giarrusso, A.; Ricci, G. *Prog. Polym. Sci.* **1991**, *16*, 405.
13. Bruzzone, M.; Gordini, S. 140th Meet. ACS, Rubb. Div., Detroit, Michigan, October 8–11, 1991; paper B.
14. Wilson, D. J. *Makromol. Chem. Macromol. Symp.* **1993a**, *66*, 273.
15. Shen, Z.; Song, X.; Xiao, S.; Yang, J.; Kan, X. *J. Appl. Polym. Sci.* **1983**, *28*, 1585.
16. Hsieh, H. L.; Yeh, H. C. *Rubb. Chem. Technol.* **1985**, *58*, 117.
17. Yang, J. H.; Tsutsui, M.; Shen, Z.; Bergbreiter, D. E. *Macromols* **1982**, *15*, 230.
18. Pedretti, U.; Lugli, G.; Poggio, S.; Mazzei, A. German Patent 2 833 721, February 8, 1979.
19. Sylvester, G.; Stollfuss, B. 133rd Meet. Rubb. Div. ACS; Dallas, Texas, April 19–22, 1988; paper No. 32.
20. Bolognesi, A.; Destri, S.; Zi-nan, Z.; Porri, L. *Makromol. Chem., Rapid Commun.* **1984**, *5*, 679.
21. Skuratov, K. D.; Lobach, M. I.; Khachaturov, A. S.; Kormer, V. A. *Polymer Sci. U.S.S.R.* **1987**, *29*(7), 1544.
22. Marina, N. G.; Duvakina, N. V.; Monakov, Y. B.; Dzhemilev, U. M.; Rafikov, S. R. *Polym. Sci. U.S.S.R.* **1985**, *27*(6), 1348.
23. Monakov, Y. B.; Marina, N. G.; Khairullina, R. M.; Kozlova, O. I.; Tolstikov, G. A. *Inorg. Chim. Acta* **1988**, *142*, 161.
24. Bolognesi, A.; Destri, S.; Porri, L.; Wang, F. *Makromol. Chem., Rapid Commun.* **1982**, *3*, 187.
25. Cabassi, F.; Italia, S.; Ricci, G.; Porri, L. *Transition Metal Catalyzed Polymerizations Ziegler-Natta and Metathesis Polymerizations*; Cambridge University: New York, 1988; p 655.
26. Hsieh, H. L.; Yeh, H. C. *Ind. Eng. Chem. Prod. Res. Dev.* **1988**, *25*, 456.
27. Wang, F.; Bolognesi, A.; Immirzi, A.; Porri, L. *Makromol. Chem.* **1981**, *182*, 3617.
28. Destri, S.; Gatti, G.; Porri, L. *Makromol. Chem. Rap. Comm.* **1981**, *2*, 605.
29. Destri, S.; Bolognesi, A.; Porri, L.; Wang, F. *Makromol. Chem. Rapid Comm.* **1982**, *3*, 187.
30. Murinov, Y. B.; Monakov, Y. B. *Inorg. Chim. Acta* **1987**, *140*, 25.
31. Marina, N. G.; Duvakina, N. V.; Monakov, Y. B.; Rafikov, S. R. *Dokl. Akad. Nauk. SSSR* **1984a**, *274*(6), 1414.
32. Marina, N. G.; Duvakina, N. V.; Monakov, Y. B.; Kuchin, A. V.; Tolstikov, G. A. *Dokl. Akad. Nauk SSSR* **1984c**, *276*, 635.
33. Wu, Y.; Wu, G.; Qi, Y.; Wei, Y. *Hecheng Xiangjiao Gongye* **1992**, *15*(3), 154.
34. Wu, Y.; Wu, G.; Qi, Y.; Wei, Y. *Hecheng Xiangjiao Gongye* **1993**, *16*(3), 149.
35. Kudashev, R. K.; Glukhova, E. A.; Muslukhov, R. R.; Monakov, Y. B.; Rafikov, S. R. *Dokl. Akad. Nauk SSSR* **1989**, *308*(6), 1398.
36. Sanyagin, A. A.; Kormer, V. A. *Dokl. Akad. Nauk. SSSR* **1985**, *283*(5), 1209.
37. Ricci, G.; Boffa, G.; Porri, L. *Makromol. Chem. Rapid Commun.* **1986**, *7*, 355.
38. Huang, B.; Shi, Y.; Fang, W.; Zhong, C.; Tang, X. *Hecheng Xiangjiao Gongye* **1992**, *15*(1), 23.
39. Lobach, M. I.; Poletoyeva, I. A.; Khachaturov, A. S.; Druz, N. N.; Kormer, V. A. *Polymer* **1977**, *18*, 1196.
40. Wilson, D. J.; Jenkins, D. K. *Polym. Bull.* **1995**, *35*, 257.
41. Marquardt, P.; Reichert, K.-H. STEPOL '94, p 47 (see Biagini, et al., Reference 4.
42. Lauretti, E.; Miani, B.; Mistrali, F. *Tire Technology International '93*; U.K. and International: Dorking, 1993; p 72.
43. Colombo, L.; Busetti, S.; DiPasquale, A.; Miani, B. *Kaut. u. Gummi Kunst.* **1993**, *46*, 458.
44. Kakiuchi, S.; Saito, T.; Tomita, S. U.S. Patent 4 683 257, July 28, 1987.
45. Gendrau, P. M.; Cadoriga, L. C. Eur. Patent Appl., EP0386915A1, September 12, 1990.
46. Andreussi, P.; Bianchi, R.; Bruzzone, M. Eur. Patent App. EP 386 808A1, September 12, 1990.
47. Hattori, L.; Tsutsumi, F.; Sakasibara, M.; Makino, K. *J. Elast. & Plas.* **1991**, *23*, 135.
48. Ohaski, M.; Mori, H. Japanese Patent JP 0162 341, March 8, 1989.
49. Ikematsu, T.; Iwase, K. Jap. Patent JP 0 5279 515, October 26, 1993.
50. Ikematsu, T.; Myamoto, K. Jap. Patent JP 0559 103, March 9, 1993.

RARE EARTH POLYMERIZATION INITIATORS

Hajime Yasuda and Eiji Ihara
Department of Applied Chemistry
Faculty of Engineering
Hiroshima University

Rare earth metals including lanthanide elements have attracted much attention in various scientific fields in recent years. In particular, application of lanthanide complexes in synthetic reaction has been actively investigated.[1]

We have investigated application of lanthanide complexes for polymer syntheses and found that organolanthanide complex can polymerize some polar monomers in a living manner. High molecular weight poly(alkyl methacrylate), poly(alkyl acrylate), and polylactone with narrow molecular weight distribution were obtained.[2]

On the other hand, it has been widely known that lanthanide complexes can catalyze polymerization of olefins without any cocatalyst. Being such a single component system is the greatest advantage of lanthanide complex compared to Ziegler type catalyst based on group IV metals. It has been reported that lanthanide complexes based on pentamethylcyclopentadienyl (Cp*) ligand show high activity for ethylene polymerization comparable to that of Kaminsky type catalyst.[3]

PREPARATION AND PROPERTIES

General Considerations

All the rare earth metal initiators used below are highly air and moisture sensitive complexes.

Polymerization of Polar Monomers[2]

Alkyl Methacrylate

We have found that the organolanthanide complexes polymerize alkyl methacrylate monomers in a living manner. As initiators, samarium methyl and hydride complexes with pentamethylcyclopentadienyl ligand (Cp*) are most effective.

Characteristic points of this polymerization are: PMMAs with extremely narrow molecular weight distribution ($M_w/M_n < 1.05$) were always obtained by the reaction at −95°C to 40°C; by changing the feed ratio of monomer to initiator, PMMA with various chain length (up to $M_n > 500 \times 10^3$) can be obtained; and by lowering the reaction temperature highly syndiotactic PMMA (rr = 95% at −95°C) can be obtained.

Alkyl Acrylate Monomers

Alkyl acrylates can also be polymerized by lanthanide initiators in a living manner. The polymerization proceeds very rapidly and affords monodisperse poly(alkyl acrylate) in 10 min at 0°C.

Lanthanide initiated polymerization is the only way for synthesis of high molecular weight poly(alkyl acrylate) ($M_n > 100 \times 10^3$) with narrow molecular weight distribution.

Lactone

Lanthanide complexes can initiate ring opening polymerization of lactones. Numbers 6 and 7 membered cyclic monomers can be used. Lanthanide methyl and alkoxy complexes are the effective catalysts for this polymerization. This polymerization proceeds in a living manner and polylactones with narrow molecular weight distribution were obtained in high yield.

Polymerization of Non-polar Monomers

Ethylene

It is widely known that lanthanide complexes have polymerization activity for ethylene. Ethylene polymerization by di- and trivalent lanthanide complex with pentamethylcyclopentadienyl (Cp*) ligand has been reported.[3] In order to investigate the effect of ligand modification on the reactivity of lanthanide complexes to olefins, we synthesized lanthanide complexes with new ligand systems of bridged Cp rings with bulky substituents.

Copolymerization of Polar Monomers with Ethylene[4]

Lanthanide complexes have initiating ability of both polar and non-polar monomers. By using this unique character of lanthanide initiators, we attempted to prepare block copolymer of ethylene and polar monomers.

After the polymerization of ethylene, polar monomer was added. Then, the polymerization of the polar monomer occurred giving AB type block copolymer of ethylene-polar monomer.

By using this copolymerization method, we can easily add chemical reactivity to polyethylene. Such a modification of chemical reactivity of polyethylene has been desirable from an industrial point of view.

REFERENCES

1. Molander, G. A. *Chem. Rev.* **1992**, *92*, 29.
2. Yasuda, H.; Yamamoto, H.; Yamashita, M.; Yokota, K.; Nakamura, A.; Miyake, S.; Kai, Y.; Kanehisa, N. *Macromolecules* **1993**, *26*, 7134; Yasuda, H.; Yamamoto, H.; Takemoto, Y.; Yamashita, M.; Yokota, K.; Miyake, S.; Nakamura, A. *Makromol. Chem., Macromol. Symp.* **1993**, *67*, 187; Yasuda, H.; Ihara, E. *Yuki Gosei Kagaku Kyokaishi* **1993**, *51*, 931; *Chem. Abstr.* **1993**, *119*, 250529; Yasuda, H.; Ihara, E. *Nippon Gomu Kyokaishi* **1993**, *66*, 612; *Chem. Abstr.* **1994**, *121*, 10006.
3. Watson, P. L.; Herskovitz, T. *ACS Symp. Ser.* **1983**, *212*, 459; Jeske, G.; Lauke, H.; Mauermann, H.; Swepston, P. N.; Schumann, H.; Marks, T. J. *J. Am. Chem. Soc.* **1985**, *107*, 8091; Evans, W. J.; Ulibarri, T. A.; Ziller, J. W. *J. Am. Chem. Soc.* **1990**, *112*, 2314.
4. Yasuda, H.; Furo, M.; Yamamoto, H.; Nakamura, A.; Miyake, S.; Kibino, N. *Macromolecules* **1992**, *25*, 115.

Rayon

See: Cellulosic Materials (Moisture Sorption Properties)

REACTION INJECTION MOLDING (RIM) MATERIALS

Andrew N. Hrymak
Department of Chemical Engineering
McMaster University

Reaction injection molding (RIM) is a family of liquid reactive molding processes where two reactive liquid monomers or prepolymers are mixed and then injected into a closed mold.

RIM is characterized by very fast solidification due to crosslinking, microphase separation, or crystallization. Typically, reaction times for the material systems are between 1.5 and 20 seconds, with most commercial material systems at the low end of that range.

The nature of closed mold filling requires the material systems used in RIM to have fast reaction times, produce no by-products, completely react, and ideally allow a wide degree of property variation within a particular polymer chemistry. Polyurethanes (PU) are the major chemical systems used in RIM; other important RIM materials are polyureas, nylon, epoxy, and polyester. RIM is mixing-activated, as distinguished from thermoset molding material systems, which are thermally activated. This article introduces the RIM material systems. More information is available in Macosko and Sweeney.[1,2]

POLYURETHANES AND POLYUREAS

Variations on isocyanate chemistry are the most widely used RIM material systems in commercial practice. The key reaction for polyurethanes reacts a di- or multifunctional polyol with a di- or multifunctional isocyanate. Multifunctional polyols or isocyanate compounds lead to branched and crosslinked polymers.

Polyurea systems are available where polyether amines [e.g., amine terminated polyoxypropylene] and short diamines [e.g., 3,5-diethyltoluene (DETDA) are used.[4,5]

The most common isocyanates used in RIM processing are the isomers of 4,4′-diphenylmethane diisocyanate (MDI). Since nearly pure MDI (impurities include 2,4′ isomers and insoluble dimers) melts at 42°C, modified forms of MDI are used.

Urethane elastomers typically use polyether diols (or triols) ranging in molecular weight from 1000 to 10,000, such as poly(propylene oxide) (PPO) polyol capped with poly(ethylene oxide). In general, polyols can be polyesters with hydroxyl groups or polyethers.[3]

Chain extenders are low molecular weight species such as 1,4-butanediol and 1,2-ethanediol.

Tertiary amines and metal catalysts are used to accelerate the reaction times for polyol systems. Typical catalysts include 1,4-diazobicyclo-[2.2.2]octane (DABCO) and dibutyl tin dilaurate (DBTDL).

Other constituents in an RIM formulation include fillers, blowing agents, surfactants, mold release agents, and pigments. The usual blowing agents are dry air or nitrogen.

The addition of the low molecular weight diol or diamine (chain extender) is key to the structure development in an RIM system. Segmented block copolymers are formed from the urethanes, with the hard segment derived from the diisocyanate–chain extender reaction and the soft segment due to the reaction of diisocyanate and the oligomeric diol.

In automotive applications, polyureas provide physical properties superior to polyurethanes. In polyurethane–polyurea systems, the low molecular weight chain extender is a diamine, which is much more reactive than a diol. Although there is considerable work in all polyurea systems, they are rarely used for large mold applications because of their fast reaction times.[6]

The major advantage of polyurethane–urea systems over PU is that the modulus of the hard phase does not change as much with temperature. Another advantage of amine-extended systems is that they allow more effective use of internal mold release compounds that prevent the polymer from sticking to the mold surface.

POLYURETHANE–ISOCYANURATE

The trimerization of isocyanates to form isocyanurates is catalyzed by potassium salts or quaternary ammonium carboxylates.[7,8] Crosslinking of the isocyanurates leads to a polyisocyanurate.

NYLON 6

The reaction for nylon 6 is a base-catalyzed reaction of ε-caprolactam under anhydrous conditions. Acyl lactam and bisimide are used as initiators, and alkali and alkaline earth metals are used as catalysts. In the RIM system, both reactant streams contain molten caprolactam. Nylon systems build solid modulus through the crystallization, rather than through microphase separation and vitrification as polyurethane and polyurea systems do. Disadvantages of nylon RIM systems include water absorption and difficult processing (due to the need for high temperatures and humidity control). Further information is available in Hedrick et al., van Geenan et al., and Sibal et al.[9–11] Mixed nylon 6 and nylon 6,12 systems are described by Iobst and Garner.[12]

DICYCLOPENTADIENE

Dicyclopentadiene (DCP) was the first material system developed specifically for the RIM process.[13] It is formed by crosslinking through a two-part formulation with tungsten chloride and DCP in one side and diethylaluminum chloride and DCP in the other.[14] Molding studies with cyclopentadiene systems have been reported by Ng et al.[15,16]

EPOXIES

Epoxy resins have better mechanical, thermal, and electrical properties than other RIM materials. However, the reaction rate is slower and the heat release due to reaction is much greater than polyurethane, polyurea, or nylon 6 systems. The typical epoxy RIM system uses a diglycidyl ether of bisphenol A (DGEBA) epoxy with an amine curing agent. Epoxy RIM systems are used in moldings with long fibers, including an RIM–pultrusion process.[17]

STYRENICS

The reaction chemistry for styrenics is a chain-growth copolymerization between unsaturated polyester prepolymers and liquid vinyl monomers, such as styrene. Sytrenic RIM systems are characterized by their relatively low reactant viscosity and instability. They are used for electronic encapsulation.[18–20]

INTERPENETRATING POLYMER NETWORKS

Polyurethanes exhibit a high thermal expansion coefficient and low rigidity. Thus, reinforcing agents, or alternative polymers such as epoxies and styrenics, have been used for structural applications. The interpenetrating polymer network (IPN) approach includes a second reaction within the system to enhance the final properties and make up for the deficiencies in the polyurethane system, thus forming a polymer alloy.[21,22]

THE FUTURE

There are no obvious new major material systems on the horizon for commercial RIM processing. Isocyanate-based RIM material can be recycled to regrind filler or processed to a prepolymeric form amenable to reaction in an existing chemical system.[23,24] They are under increasing scrutiny in a number of jurisdictions because of adverse health effects from frequent exposure in the workplace.[25,26]

REFERENCES

1. Macosko, C. W. *RIM Fundamentals of Reaction Injection Molding*; Hanser: New York, NY, 1989.
2. Sweeney, F. M. *Reaction Injection Molding Machinery and Processes*; Marcel Dekker: New York, 1987.
3. Frisch, K. C.; Klempner, D. *Comprehensive Polymer Science*; Allen, G. et al., Eds.; Pergamon: Oxford, U.K. 1989; Vol. 5.
4. Grigsby, R. A.; Rice, D. M. *J. Cellular Plast.* **1986**, *22*, 484.
5. Ryan, A. J.; Stanford, J. L.; Birch, A. J. *Polymer* **1993**, *34*, 4874.
6. Pannone, M. C.; Macosko, C. W. *Polym. Eng. Sci.* **1988**, *28*, 660.
7. Vespoli, N. P.; Alberino, L. M. *Polym. Proc. Eng.* **1985**, *3*, 127.
8. Ryan, A. J.; Stanford, J. L.; Tao, X. Q. *Polymer* **1993**, *34*, 4020.
9. Hedrick, R. M.; Gabbert, J. D.; Wohl, M. H. *Reaction Injection Molding*; Kresta, J. E., Ed.; ACS Symposium Series 270; American Chemical Society: Washington, DC, 1985.
10. van Geenan, A. A.; Bongers, J. J. M.; van der Loos, L. M.; Vrinssen, C. H. (DSM) U.S. Patent 4 540 515, 1984.
11. Sibal, P. W.; Camargo, R. E.; Macosko, C. W. *Polym. Proc. Eng.* **1983**, *1*, 147.
12. Iobst, S. A.; Garner, D. P. *Abstract ACS* **1988**, *195*, JUN, 51-MACR.
13. Klosiewicz, D. W. (Hercules) U.S. Patent 4 400 340, 1983.
14. Matejka, L.; Houtman, C.; Macosko, C. W. *J. Appl. Polym. Sci.* **1985**, *30*, 2787.
15. Ng, H.; Manas-Zloczower, I.; Shmorhun, M. *Poly. Eng. Sci.* **1994**, *34*, 929.
16. Ng, H.; Manas-Zloczower, I.; Shmorhun, M. *Poly. Eng. Sci.* **1994**, *34*, 921.
17. Ishida, H.; Zimmerman, D. A. *Polym. Comp.* **1994**, *15*, 93.
18. Your, J. J. A.; Karles, G. D.; Ekerdt, J. G.; Trachtenberg, I.; Barlow, J. W. *Ind. Eng. Chem.* **1989**, *28*, 1456.
19. Christiansen, W. H.; Ekerdt, J. G.; Trachtenberg, I.; Barlow, J. W. *Ind. Eng. Chem. Res.* **1990**, *29*, 463.
20. Karbles, G. D.; Christiansen, W. H.; Ekerdt, J. G.; Trachtenberg, I.; Barlow, J. W. *Ind. Eng. Chem. Res.* **1991**, *30*, 646.
21. Hsu, T. J.; Lee, L. J. *J. Appl. Polym. Sci.* **1988**, *36*, 1157.
22. Nguyen, L. T.; Suh, N. P. *Adv. Chem. Ser.* **1984**, *208*, 311.
23. Farrissey, W. J.; Morgan, R. E.; Tabor, R. L.; Zawisza, M. *ACS Symp. Ser.* **1992**, *513*, 272.
24. Vanderwal, H. R. *J. Reinf. Plastics Comp.* **1994**, *13*, 87.
25. Goebell, J. *Kunststoffe - German Plast.* **1992**, *82*, 641.
26. Baur, X. et al. *Int. Arch. Occup. Environ. Health* **1994**, *66*, 141.

REACTIVE LIQUID RUBBERS (Epoxy Toughening Agents)

Rolf Mülhaupt
Freiburger Materialforschungszentrum und Institute für Makromolekulare Chemie der Albert-Ludwigs Universität

Crosslink density is an important parameter governing properties of cured epoxy resins. As a rule, stiffness, strength, glass transition temperature, creep resistance, thermal, thermo-oxidative, and environmental and chemical stability improve with increasing crosslink densities. However, this property improvement is paralleled by undesirable embrittlement evidenced by premature mechanical failure when epoxy materials are exposed to mechanical stresses, especially during impact. Therefore it is an important objective in epoxy resin development to overcome such brittleness limitations without impairing attractive stiffness/strength/stability/processability property balance. In order to improve toughness, that is, resistance to crack propagation, energy at the crack tip must be dissipated effectively.

Since crosslinked or high molecular weight, noncrosslinked rubbers are difficult to disperse in epoxy resins and would impair high filler levels typical for numerous applications, phase separation of telechelic elastomers, referred to as reactive liquid rubbers or elastomer prepolymers, during cure represents a widely applied process of commercial significance. In an alternative process, liquid rubbers can be polymerized prior to cure in the epoxy/hardener solution in order to preform colloidal rubber particle dispersions. In order to qualify as toughening agents, reactive liquid rubbers must meet the following criteria: miscibility with epoxy resin/hardener solution, low viscosity, complete phase separation during cure to prevent matrix flexibilization, formation of uniformly dispersed rubber microphases with average diameters < 1–5 μm, and good interfacial adhesion to the rigid epoxy matrix, and glass transition temperature of the dispersed rubber must be <0°C, preferably <40°C. Overviews on liquid-rubber toughened epoxy resins were presented by Riew and Gillham and Riew.[1,2] The role of compatibility in designing compatibilized liquid rubbers was reviewed by Mülhaupt and Shaw.[3–7]

In addition to volume fraction, connectivity, and cure kinetics, liquid rubber molecular architecture plays an important role in morphology control, which is the key to improved property balance of the liquid–rubber-toughened multiphase epoxy material. Compatibility can be improved by incorporating reactive groups. These groups react with epoxy resins during cure to produce epoxy-terminated liquid rubbers that are readily incorporated into the epoxy network. Depending on stoichiometry, this addition-type reaction can be used to achieve advancement reactions that enhance molecular weight and promote formation of ABA-type segmented liquid rubbers.

PROPERTIES AND APPLICATIONS

Nitrile and Acrylic Liquid Rubbers

Design of compatibilized liquid nitrile rubbers was pioneered by researchers at B.F. Goodrich during the 1970s.[8–10] In free radical processes using difunctional initiators, butadiene is copolymerized with acrylonitrile to produce di*c*arboxylic-acid-*t*erminated oligo(*b*utadiene-co-acrylo*n*itrile), abbreviated CTBN, with acrylonitrile content ranging between 0 and 30% and molecular weights 2000 < M_n < 5000 g/mol. Also amino- (ATBN) and vinyl-terminated nitrile liquid rubbers (VTBN) are available commercially.

Segmented nitrile rubbers were derived from CTBN via esterification with various linear or branched poly(alkylene oxides) or by copolymerizing ε-caprolactone onto carboxy end groups of nitrile and butadiene liquid rubbers.[11]

Similar concepts for liquid rubber compatibilization were applied to acrylic liquid rubbers based upon poly(butyl acrylate). For example, Kirshenbaum, Gazit, and Bell improved toughening agent performance of poly(butyl acrylate) liquid rubber by enhancing compatibility via incorporation of acrylonitrile and acrylic acid comonomers.[21]

POLYESTERS AND POLYETHERS

Hydroxy- or carboxy-terminated polyester and polyether as well as segmented liquid rubbers containing polyester and polyether segments are potential low-cost liquid rubbers and have found a broad range of applications as flexibilizers and toughening agents. Typically polyester liquid rubbers with molecular weights ranging between 1000 and 10,000 g/mole are produced in polycondensation reactions from flexible dicarboxylic acid or dicarboxlate esters respectively and diol building blocks, for example, adipic acid, sebacic acid, dimer and trimer fatty acids, ethane-1,2-diol, propane-1,3-diol, butane, 1,4-diol, hexane-1,6-diol, 2,2-dimethyl-propane-1,3-diol (neopentylglycol, decane-1,12-diol, and 2,2,4-trimethylhexane-1,6-diol). Similar to CTBN, carboxy-terminated polyesters are reacted with epoxy resins to prepare epoxy-terminated polyester with improved compatibility via advancement reaction.

Dihydroxy-terminated oligo(alkylene oxide) liquid rubbers, for example, oligo(propylene oxide) and oligotetrahydrofuran, are versatile intermediates for preparation of polyether liquid rubbers with various other end groups and for manufacturing segmented oligomers and polymers such as poly(ether ester)s and poly(ether urethane)s. Liquid oligoether diols are commercially available and are produced by anionic ring-opening polymerization of propylene oxide with bisalkoxide initiators or by cationic ring-opening polymerization of tetrahydrofuran, for example, by using trifluoromethanesulfonicacid anhydride as initiator.[13,14]

SILICONES AND HYBRID COMPOSITES

Silicone rubber offers attractive potential for epoxy toughening because of low glass transition temperatures <–100°C, extremely low water uptake, low dielectric constant, and excellent thermo-oxidative stability in comparison to nitrile rubbers, which contain olefinic unsaturation. In comparison to other rubbers, however, most silicones and also silicone liquid rubbers are immiscible with common epoxy resins and hardeners. Therefore, silicone resins must be compatibilized either via advancement reaction, involving silicone end groups, or via incorporation of more polar substituents, respectively. Silicone liquid rubber chemistry and application as epoxy toughening agents were reviewed by McGrath.[15–18]

LIQUID RUBBER BLENDS AND STRUCTURAL LAMINATES

Another recent development of compatibilized liquid rubbers for application in hybrid composites and structural laminates involves liquid rubber blend technology.

CONCLUSION

Molecular design of liquid rubbers aiming at improved compatibility with epoxy matrix is the key to a family of new structural materials that combine stiffness, strength, toughness, and thermal, oxidative, and chemical stability with high creep-resistance and fatigue life. Breakthroughs have led to novel generations of high–speed-impact resistant, liquid–rubber-toughened epoxy materials that compete successfully with established materials, especially metal alloys. In the future, controlled synthesis of segmented telechelics continues to play a key role in understanding of basic structure/property relationships and micromechanic mechanisms accounting for improved service life of advanced polymeric structural materials.

REFERENCES

1. Riew, C. K.; Gillham, J. K. *Rubber-Modified Thermoset Resins: Adv. Chem. Ser.*, American Chemical Society: Washington, 1984; 208.
2. Riew, C. K. *Rubber-Toughened Plastics*; **1989**, *Adv. Chem. Ser.* **1989**, American Chemical Society: Washington, 1989; 222.
3. Mülhaupt. R.; Powell, J. H. Adhes. 90, Int. Conf. 1990; 10/1-1-0/6; Plast. Rubber Institute: London, 1990; *Chem. Abstract* **1990**, *118*, 235, 366.
4. Mülhaupt, R.; Rüfenacht, W. Eur. Patent Appl. EP 353 190 1990, assigned to Ciba-Geigy AG *Chem. Abstract* **1990**, *113*, 25,098.
5. Mülhaupt, R.; Rüfenacht, W. Eur. Patent Appl. EP 384 896 1990; assigned to Ciba-Geigy AG *Chem. Abstract* **1990**, *114*, 280,551.
6. Mülhaupt, R.; Rüfenacht, W. Eur. Patent Appl. EP 381 625 1990; assigned to Ciba-Geigy AG *Chem. Abstract* **1990**, *114*, 63,513.
7. Shaw, S. J. In ollyer 1994 165.
8. Rowe, E. H.; Siebert, A. R.; Drake, R. S. *Mod. Plast.* **1970**, *47*, 110.
9. Drake, R. *Polym. Mater. Sci. Eng.* **1990**, *63*, 802.
10. Siebert, A. R. *Rubber-Toughened Plastics; Adv. Chem. Ser.* **1989**, *222*, American Chemical Society: Washington, DC, 1989; p 179.
11. Mülhaut, R.; Rüfenacht, W.; Powell, J. H.; Mechera, K. Eur. Patent Appl. EP 358 603 1989, assigned to Ciba-Geigy AG; *Chem. Abstract* **1989**, *113*, 98,997.
12. Kirshenbaum, S.; L.; Gazit, S.; Bell, J. P. Rubber-Modified Thermoset Resins; *Adv. Chem. Ser. 208*, American Chemical Society: Washington, DC, 1984; p 163.
13. Inoue, S.; Aida, T. *Ring-Opening Polymerization*; Ivin, K. J.; Saegusa, T., Eds.; Elsevier Applied Science: New York, 1984; Chapter 4.
14. *Ring-Opening Polymerization*; Brunelle, D. J., Ed.; Hanser: Munich, 1993.
15. Riffle, J. S.; Yilgor, I.; Tran, C.; Wilkes, G. L.; McGrath, J. E.; Banthia, A. K. *Am. Chem. Soc., Symp. Ser.* **1983**, *221*, *(Epoxy Resin Chemistry II)*; American Chemical Society: Washington, DC, 1983; 21.
16. Yilgor, I.; Riffle, J. S.; McGrath, J. E. *ACS Symp. Ser.* **1985**, *282*, *(Reactive Oligomers)*; American Chemical Society; Washington, DC, 1985; p 161.
17. Yilgor, I.; McGrath, J. E. *Adv. Polym. Sci.* **1986**, *86*, 1.
18. Yorkgitis, E. M.; Eiss, N. S., Jr.; Tran, C.; Wilkes, G. L.; McGrath, J. E. *Adv. Polym. Sci.* **1985**, *72*, 80.

Reactive Oligomers

See: Benzocyclobutene Homopolymerization (Chemistry and Applications)
Reactive Oligomers (Overview)

REACTIVE OLIGOMERS (Overview)

Yasuo Shimano*
Department of Chemical and Biological Engineering
Hachinohe National College of Technology

Shiro Kobayashi
Department of Molecular Chemistry and Engineering
Faculty of Engineering
Tohoku University

Reactive oligomers can be regarded as the oligomers having reactive functional groups at the oligomer ends, in the side chains, and/or in the main chain. The first attempt to prepare such oligomers may be credited to Staudinger and Luthy, who obtained α-ω-dimethyl ether and α,ω-dimethyl ester of oligooxymethylenes.[1] Among the reactive oligomers, end-functionalized oligomers are called telechelic oligomers, telechelic polymers, or telechelics, which can be classified as mono, di, tri, and polytelechelics depending on the functionality, that is, number of functional groups per number of oligomer or polymer chains.[2,3]

The reactive oligomers and polymers bearing polymerizable groups such as vinyl groups are called macromolecular monomers, macromonomers, or macromers.[4]

PREPARATION OF REACTIVE OLIGOMERS

Via Radical Polymerization

Radical polymerization can be utilized for the preparation of the telechelic oligomers in two methods, namely, initiation technique and chain transfer technique.[4–7]

Initiation Technique

By use of large amounts of initiators, the propagating radicals in radical polymerization can be terminated only by combination with polymer radicals or primary radicals (dead-end polymerization).[8]

Azo-Compounds. By use of AIBN as initiator, the telechelic oligomers of styrene, butadiene, and methyl methacrylate with two nitrile end-groups have been prepared.[9–11] The nitrile end-groups can be modified to provide several functional groups, for example, carboxyl groups and amine groups.[9,10]

Disulfides. In radical polymerizations, disulfides can be used as transfer agents or as both initiators and transfer agents giving rise to the telechelics.

Transfer Technique

By employing transfer agents with high transfer constants, telechelic oligomers and telomers have been prepared. Aliphatic halogen compounds and disulfides are the most usual transfer agents.

Via Anionic Polymerization

Utilizing living anionic polymerization, various end-reactive oligomers with well-defined structure, narrow molecular weight distribution, and controlled molecular weight can be produced by terminating the propagating species with suitable capping agents, and by employing the functionalized anionic initiators.

Hydroxyl End-Groups

Termination of the living anions with ethylene oxide can produce corresponding alcoholate, which, however, are subject to propagating ring-opening polymerization of the ethylene oxide.[12,13] Therefore, the termination should be followed by quenching immediately.[14] Using this method, hydroxyl terminated polystyrenes, polybutadienes, and polyisoprenes have been prepared.[12,15,16]

Carboxyl End-Groups

Direct termination of the propagating species with CO_2 can introduce carboxyl groups, but also produce considerable amounts of dimer ketones and trimer carbinols.[17] When the living species were transformed into Grignard salts prior to termination, carboxylates were formed quantitatively using gaseous CO_2.[18,19]

Amine End-Groups

Amine terminated mono-telechelics of polystyrene have initially been prepared employing potassium amide as an initiator.[20] As a blocked amino group initiator, p-lithio-*N,N*-bis(trimethylsilyl)aniline has been used to introduce primary amino group.[21]

Group Transfer Polymerization

Group transfer polymerizations or methyl methacrylate (MMA) initiated by silyl ketene acetals can produce living polyMMA having ω-functional groups arising from the initiators used.

Via Carbocationic Polymerization

Recently, a variety of telechelic oligomers with well-defined structure have been prepared via carbocationic polymerization of unsaturated monomers, especially by utilizing living polymerization technique or "inifer" technique.

Telechelic polyisobutylenes have been prepared utilizing the "inifer" method, in which termination and transfer reactions can be employed to introduce the functional end-groups.[22–25]

Via Ring-Opening Polymerization

Functionalized oligomers prepared via ring-opening polymerization have great importance for commercial use.

Cyclic Ethers

Ring-opening cationic and anionic polymerization of oxiranes has been utilized for the preparation of telechelic polyethers. α,ω-Dihydroxy poly(ethylene oxide)s have been prepared using hydroxide ion, or bifunctional alkoxide ion obtained from naphthyl sodium and ethylene oxide.[26,27]

In the cationic ring-opening living polymerization of tetrahydrofuran (THF), functional groups can be introduced to the polymer ends utilizing termination of the propagating species with nucleophiles.

Cyclic Siloxanes

Polysiloxanes possess unique properties such as low temperature flexibility, high gas permeability, excellent electrical

*Author to whom correspondence should be addressed.

properties, good biocompatibility, and interesting surface properties. A variety of bifunctional telechelic polysiloxanes have been prepared because of the importance of the polysiloxane segments in block copolymers.[28]

2-Oxazolines

Cationic ring-opening polymerization of 2-oxazolines (ROZO) proceeds via oxazolinium species to give rise to living polyROZO (PROZO). Polymerization of ROZO initiated by bifunctional initiator, such as bisoxazolinium salts, allylic dihalides, or xylylene dihalides (o-, m-, p-) yields living PROZO having an oxazolinium salt on both ends.[29,30]

Via Step-Growth Polymerization

Step-growth polymerization can provide a variety of reactive oligomers with characteristic features such as high glass-transition temperatures. The most widespread method for preparation of telechelics by step-growth polymerization is the one that is carried out in nonstoichiometric conditions.

Biphenylene end-capped polyquinolines, α,ω-diphenol oligoarylsulfones, and dihydroxy oligocarbonates have been prepared in these conditions.[31–36] Phase transfer catalyzed polycondensation of bisphenol A and 1,4-dichloro-2-butene (excess) has been carried out using nonstoichiometric conditions to yield chloro-terminated oligomers.[37]

APPLICATIONS OF REACTIVE OLIGOMERS

Reactive oligomers can become valuable components of a variety of polymer materials, such as block copolymers, networks, chain-extended polymers, and graft copolymers.

Properties

Thermoplastic elastomers. Thermoplastic elastomers are the most notable application of block copolymers. They have several primary advantages to conventional rubbers, such as easy processing, low processing cost, and possibility of recycling. Commercially available thermoplastic elastomers derived from functionalized oligomers contain mainly polyurethanes and polyesters.

Membranes. Various membranes are made from block copolymers for the purpose of gas or liquid separations, ultrafiltration, and desalination.[38] For example, polyester-*block*-polyurethanes have been utilized for aqueous dialysis and polyamide-*block*-poly(alkylene oxide)s are used for ultrafiltration.[39–41]

Biomedical Applications. A variety of biomedical potentials of the block copolymers derived from functionalized oligomers have been reported. Silicone-urethane block copolymers can be used in blood pumps, intra-aortic balloons, and heart replacements because of their unique properties such as strength, toughness, flexibility, and low interaction with plasma proteins.[38] Polyester-polyether copolymers are useful for surgical sutures, blood-transfusion tubes, and blood-storage containers.[42,43] Ethylene oxide-ethylene terephthalate block copolymers are used in surgical dressings.[44]

Chain Extension

Chain extension reaction of telechelic oligomers can provide valuable polymer materials such as polyurethanes and polyesters.

Crosslinking

Networks formed from telechelic oligomers by crosslinking are of great technological interest because of advantages such as low energy processability and better controlled properties. Linear or three-arm star hydroxyl telechelic polyisobutylenes are useful for preparation of polyurethane rubbery networks without free chain ends, differing from liquid polydienes.[45,46]

Graft Copolymers

When end-reactive oligomers are reacted with reactive functional groups combined with a second, preformed polymer, graft copolymers can be formed.

REFERENCES

1. Staudinger, H.; Luthy, M. *Helv. Chim. Acta* **1925**, *8*, 41.
2. Uraneck, C. A.; Hsieh, H. L.; Buck, O. G. *J. Polym. Sci.* **1960**, *46*, 535.
3. Goethals, E. J., Ed., *Telechelic Polymers*; CRC: Boca Raton, FL, 1989.
4. Nuyken, O. *Encyclopedia of Polymer Science and Engineering*; 2nd ed., Wiley-Interscience: New York, 1985; Vol. 16, 494.
5. Athey, Jr.; R. D. *Prog. Org. Coat.* **1979**, *7*, 289.
6. Gobran, R. H. In *High Polymers: Chemical Reactions of Polymers*; Fettes, E. M., Ed.; Wiley-Interscience: New York, 1964; Vol. XIX, Chapter 4.
7. Starks, C. M. *Free Radical Telomerization*; Academic: New York, 1974.
8. Bamford, C. H.; Tipper, C. F. *Comprehensive Chemical Kinetics*; Elsevier: Amsterdam, 1976; Vol. 14a.
9. Konter, W.; Bohmer, B.; Kohler, K. H.; Heitz, W. *Makromol. Chem.* **1981**, *182*, 2619.
10. Heitz, W.; Ball, P.; Lattekamp, M. Z. *Kautschuk Gummi Asbest Kunstst.* **1981**, *34*, 459.
11. Heitz, W.; Konter, W.; Guth, W.; Bomer, B. *Pure Appl. Chem. Macromol. Chem.* **1973**, *8*, 65.
12. Brody, H.; Richards, D. H.; Szwarc, M. *Chem. Ind.* **1958**, 1473.
13. Richards, D. H.; Szwarc, M. *Trans. Farady Soc.* **1959**, *55*, 1644.
14. Steiner, E. C.; Pelletier, R. R.; Trucks, R. O. *J. Am. Chem. Soc.* **1964**, *86*, 4678.
15. Hayashi, K.; Marvel, C. S. *J. Polym. Sci., Part A1* **1964**, *2*, 2571.
16. Morton, M.; Fetters, L. J.; Inomator, J.; Rubio, D. C.; Young, R. N. *Rubber Chem. Technol.* **1976**, *49*, 303.
17. Morton, M.; Mikesell, S. L. *J. Macromol. Sci. Chem.* **1973**, *A7*, 1391.
18. Burgess, F. J.; Richards, D. H. *Polymer* **1976**, *17*, 1020.
19. Mansson, P. *J. Polym. Sci., Polym. Chem. Ed.* **1980**, *18*, 1945.
20. Higginson, W. C. E.; Wooding, N. S. *J. Chem. Soc.* **1952**, 760.
21. Schulz, D. N.; Halasa, P. *J. Polym. Sci., Polym. Chem. Ed.* **1977**, *15*, 2401.
22. VerStrate, G.; Baldwin, F. P. *Polym. Prepr.* **1976**, *17*, 808.
23. U.S. Patent 4 278 822, 1981.
24. Kennedy, J. P.; Smith, R. A. *Polym. Prepr.* **1979**, *20*, 316.
25. Kennedy, J. P.; Smith, R. A. *J. Polym. Sci., Polym. Chem. Ed.* **1980**, *18*, 1523.
26. Perry, S.; Hibbert, H. *Can. J. Res.* **1933**, *8*, 102; *J. Am. Chem. Soc.* **1940**, *62*, 2599.
27. Cabasso, I.; Zilkha, A. *J. Macromol. Sci. Chem.* **1974**, *A8*, 1313.
28. *Silicon Compounds, Register Rev.*; Petrarch Systems Inc.: Bristol, PA 1982.
29. Kobayashi, S.; Igarashi, T.; Moriuchi, Y.; Saegusa, T. *Macromolecules* **1986**, *19*, 535.

30. Kobayashi, S.; Uyama, H.; Narita, Y. *Macromolecules* **1992**, *25*, 3232.
31. Stille, J. K. *Pure Appl. Chem.* **1978**, *50*, 273.
32. Garapon, J.; Stille, J. K. *Macromolecules* **1980**, *13*, 1361.
33. Droske, J. P.; Gaik, U. M.; Stille, J. K. *Macromolecules* **1984**, *17*, 10.
34. Pourdjavadi, A.; Madec, P. J.; Maréchal, E. *Eur. Polym. J.* **1984**, *20*, 305.
35. Gagnebien, D.; Madec, P. J.; Maréchal, E. *Eur. Polym. J.* **1985**, *21*, 273.
36. French Patent 7 336 841, 1973.
37. Percec, V., Rinaldi, P. L.; Auman, B. C. *Makromol. Chem.* **1984**, *185*, 617.
38. Riess, G.; Hurtrez, G.; Bahadur, P. *Encyclopedia of Polymer Science & Engineering*; 2nd ed., Wiley-Interscience: New York, 1985; Vol. 2, 324.
39. Thakore, Y. B. Thesis, University of Utah, Salt Lake City, Utah, 1981.
40. Japan Kokai Tokkyo Koho JP 81 37 007, 1981.
41. Japan Kokai Tokkyo Koho JP 76 079 684, 1976.
42. Japan Kokai Tokkyo Koho, JP 80 101 268, 1980.
43. U.S. Patent 4 314 561, 1982.
44. Seung, S. L. N.; Young, R. N. *J. Polym. Sci., Polym. Lett. Ed.* **1979**, *17*, 233.
45. Kennedy, J. P.; Ross, L. R.; Lackey, J. E.; Nuyken, O. *Polym. Bull.* **1981**, *4*, 67.
46. Miyabayashi, T.; Kennedy, J. P. *J. Appl. Polym. Sci.* **1986**, *31*, 2523.

Reactive Processing

See: *Methacrylic Resins (Improved Heat Resistance)*
Peroxide Initiators (Overview)
Polyamides, Lactam-Based
Poly(styrene-co-N-alkylmaleimide)s (by Reactive Extrusion; Miscibility in Blends)
Reaction Injection Molding (RIM) Materials
Reactive Processing, Thermoplastics

REACTIVE PROCESSING, THERMOPLASTICS

Robert J. Kumpf
Corporate Polymer Research
Bayer Corporation

E. H. Jonsson
Plastics Division Research and Development
Bayer AG

J. S. Wiggins
Plastics New Product Development
Bayer Corporation

In response to competitive pressures, resin producers have focused their R & D efforts toward improving the performance of existing resins by using computer-controlled injection molding, by improving manufacturing processes, by developing new polymer blends and additives and, increasingly, by moving chemistry from reaction vessels into processing equipment. This last strategy is often referred to as reactive processing.

The rationale for reactive processing is straightforward. Processing equipment has long been used to expand product portfolios through addition of additives, dyes, and pigments. This allows the chemical manufacturing plant to concentrate on base polymers, thereby minimizing the average cost of goods. Reactive processing simply takes the model a step further. The processing equipment is seen not only as a mixing device but also as a true reaction vessel where chemistry can be performed and controlled.

Reactive processing encompasses a rich collection of chemistries. The objectives of reactive processing are equally diverse, ranging from viscosity modification to the formation of microphase separated block copolymers. A number of excellent review articles have been published that collect and categorize the myriad chemical reactions that have been exploited in reactive processing. Particularly noteworthy are articles by Lui and Baker;[1] Brown;[2] Tzoganakis;[3] and Dagli.[4] Xanthos has edited a highly recommended and comprehensive book on reactive processing.[5]

REACTIVE PROCESSING—CHEMICAL STRATEGIES

In general, six chemical strategies have been employed in reactive processing: bulk polymerization; graft reactions; inter-chain copolymer formation; coupling reactions; and polymer functionalization.

Bulk Polymerization

A classic example of bulk polymerization via reactive processing is thermoplastic polyurethane (TPU). TPU elastomers are segmented block copolymers prepared from low molecular weight polyester or polyether polyols, a monomeric diol, and a diisocyanate. They are produced commercially by pumping monomers directly into extruders. High polymer forms in the extruder screws and is isolated as strands and finally as pellets.[6]

Graft Reactions

In graft reactions, a polymer is treated with a monomer, usually in the presence of an initiator.

One important example of this type of chemistry is the grafting of maleic anhydride (MAH) to polypropylene (PP).[7–9] A radical initiator such as dicumyl or benzoyl peroxide is utilized in the formation of the grafting site by hydrogen abstraction. The so-formed reactive species then reacts with the MAH to form pendant succinic or maleic anhydride groups.

Inter-Chain Copolymer Formation

A number of condensation polymers are "living" in that at higher temperatures they undergo inter-chain cross reactions. For example, polyesters and polycarbonates easily transesterify, polyamides can undergo catalyzed transamidation, and polyurethanes are subject to transurethanation reactions. These inter-chain reactions have been exploited to form copolymers via reactive processing. Researchers at Bayer Corp. have reported reactive processing routes to polycarbonate/poly(ether ketone) and polycarbonate/poly(ether sulfone) block copolymers in extruders and kneaders.[10–12]

In contrast to transesterification reactions, transetherification reactions between aromatic polyethers are not facile and occur mostly in solutions containing strong base. Mullins and coworkers found, however, that relatively large amounts of cesium fluoride (0.5–1.0 wt%) catalyze ether cleavage to give cesium phenolate endgroups that attack polycarbonate to give a copolymer.[22]

In some cases reactive processing is the only route to a particular material. The limited solubility of semicrystalline poly(ether sulfone)s hinders their use in classical solution reactions. The reactive processing route discovered by Mullins allow for the synthesis of block copolymers from semi-crystalline poly(ether sulfone)s.

Coupling Reactions

Condensation polymers are inherently end-functionalized. These endgroups can be used in reactive processing. For example, the amino and/or carboxylic acid functionality of nylons offer a "handle" for reactive processing. Triacca et al. have reported the reactive compatibilization of PA6/poly(styrene-co-acrylonitrile) (SAN) blends.[14]

REACTIVE PROCESSING APPLICATIONS—BLEND COMPATIBILIZATION

Clearly the objective of much reactive processing has been compatibilized blends. The success of super-tough nylons, produced via the reactive processing of PA 6,6 and EPDM, has validated this approach.

Miscible blends are thermodynamically stable, molecular level mixtures. Immiscible blends are separated into macroscopic phases with minimal interfacial adhesion and unstable morphologies. Compatibilized blends are also macro-phase separated. The presence of interfacial agents or chemical bonds, however, stabilizes the morphology and increases interfacial adhesion. Reactive processing is often employed to either couple the phases or form copolymers, which assemble at the interface and stabilize the blend morphology.

Nakayama et al. studied the compatibilization of polystyrene/poly(methyl methacrylate) blends.[15] They compared a preformed block copolymer added to the blend versus a block copolymer formed during processing. The reactive processing route was found to be more efficient; 1.4% wt *in situ* block copolymer gave smaller particle sizes and a narrower size distribution than 5% wt premade block copolymer.

ADVANCES IN PROCESS TECHNOLOGIES

Polymerization reactions and chemical modifications on polymers have been reported using virtually all conventional polymer processing equipment including batch-mixers, kneaders, single-screw extruders, and twin-screw extruders.[16–19] The key to success for performing chemical reactions and modifications in polymer processing equipment is a sound understanding of the chemical reactions being conducted, the kinetics associated with these reactions, and the proper choice and design of equipment to accommodate the chemical reaction requirements. It is important to carefully design reactive processing equipment as chemical reactors rather than simply as processing or compounding tools. Often, chemical reactions can be combined with compounding steps in efforts to "streamline" the overall process leading to economic advantages.

FUTURE TRENDS IN REACTIVE PROCESSING OF ENGINEERING POLYMERS

There are a number of incentives to consider reactive processing as a route to new materials: the cost of developing entirely new engineering resins using traditional polymerization chemistry, familiarity with the properties and processing attributes of the base resin, shortened product development cycles, and tailored physical properties.

The technical problems involved in reactive processing are not, however, trivial. Mixing materials with widely different viscosities is problematic. The final properties of the material are sensitive to myriad variables: processing temperature, residence times, shear rates, screw design, polymer molecular weight, and polymer miscibility.

The capital investment required to develop and introduce a new engineering thermoplastic has increased dramatically. This had led to an increased interest in reaction processing routes to new materials. Successful development of new engineering polymer grades by reactive processing will call for a multidisciplined approach with expertise required in chemistry, polymer science, polymer engineering, computer modeling, and statistical process control. With such an approach it may be economically possible to develop and produce specialty engineering thermoplastics for niche and emerging markets.

REFERENCES

1. Liu, N. C.; Baker, W. E. *Advances in Polymer Technology* **1992**, *11*(4), 249.
2. Brown, S. B. *Reactive Extrusion*; Xanthos, M., ed.; Hanser: New York, 1992; Chapter 4, pp 75–200.
3. Tzoganakis, C. *Advances in Polymer Technology* **1989**, *9*(4), 321.
4. Dagli, S. S.; Xanthos, M. *Polymer Engineering and Science* **1991**, *31*, 929.
5. Xanthos, M., Ed.; *Reactive Extrusion*; Hanser: New York, 1992.
6. Legge, N. R.; Holden, G.; Schroeder, H. E., Eds.; *Thermoplastic Elastomers*; Hanser: New York, 1987, p 23.
7. Gaylord, N.; Mehta, M. *J. Polym. Sci. Polym. Chem.* **1988**, *26*, 1189.
8. Kowalski, R. C. *Proceedings "Compalloy—89;"* New Orleans, LA, April 5–7, 1989.
9. Gaylord, N. G. U.S. Patent 4 506 056, 1985.
10. Kumpf, R. J.; Pielartzik, H.; Nerger, D.; Wehrman, R. U.S. Patent 5 169 907, 1992.
11. Kumpf, R. J.; Pielartzik, H.; Nerger, D.; Wehrman, R. U.S. Patent 5 221 727, 1993.
12. Kumpf, R. J.; Pielartzik, H.; Nerger, D.; Wherman, R. *Macromolecular Assembly in Polymeric Systems ACS Symposium Series*; Stroeve, P.; Balazs, A., Eds.; American Chemical Society: 1992; 493, pp 300–312.
13. Mullins, M. J.; Woo, E. European Patent 0 353 478 A1, 1990.
14. Triacca, V. J.; Ziaee, S.; Barlow, J. W.; Keskula, H.; Paul, D. R. *Polymer* **1991**, *32*(8), 1401.
15. Nakayama, A.; Inoue, T.; Guegan, P.; Macosko, C. W. *Polymer Preprints* **1993**, *34*, 840.
16. Sue, H. J.; Huang, J.; Yee, A. *Polymer* **1992**, *33*(22), 4868.
17. Rimer, S.; George, M. *Eur. Polym. J.* **1993**, *29*(2/3), 205.
18. Van Ballegooie, P.; Rudin, A. *Polym. Eng. and Sci.* **1988**, *28*(21), 1434.
19. Utracki, L.; Shi, Z. *Polym. Eng. and Sci.* **1992**, *32*(24), 1824.

Reagents

See: Azidation Polymer

Recycling

See: Compatibilizers, Polymeric (Recycling of Multilayer Structures)

Cyclic Imino Ethers (Ring-Opening Polymerization of 1,3-Oxazo Monomers)
Nylon-6 (Depolymerization)
Photodegradable Plastics
Polyaniline (Gold Recovery)
Polyethylene, Recycled
Poly(vinyl chloride), Rigid (Soft Fillers)
Recycling, Plastics
Recycling Techniques, Plastic Composites
Recycling, Thermodynamic (by Ring-Closing Depolymerization)

RECYCLING, PLASTICS

Constantine D. Papaspyrides* and John G. Poulakis
Laboratory of Polymer Technology (Special Chemical Technology)
Department of Chemical Engineering
National Technical University of Athens

There are many different types of plastic wastes, but for simplicity they may be divided broadly into the following four categories: single grades of plastic in the form of scrap that can be incorporated into the process from which it originated; Single grades of plastic that have been contaminated and cannot be incorporated into the process from which they originated and therefore have to be further processed for different end-use applications; mixed plastic waste streams with known composition and essentially free of non-plastic contaminants; and randomly collected or municipal refuse that is contaminated with non-plastic materials.[2]

There are four main approaches to plastics recycling (excluding as non-acceptable dumping on land or at sea with or without prior treatment) primary recycling, secondary recycling, tertiary or "chemical" recycling, and quaternary recycling.[3]

PRIMARY RECYCLING

Primary recycling is the recycling of clean, uncontaminated single-type waste. This recycling level remains the most popular as it ensures simplicity and low cost, especially when done "in-plant" and feeding with scrap of controlled history.[4] The recycled scrap or waste is either mixed with virgin material to assure product quality or used as second grade material.[5]

Primary Recycling and Chemistry

It should be emphasized that during processing or inservice use a plastic may undergo chemical reactions that can affect its physical properties. If these properties are affected to any marked extent the plastic may become unsuitable for its original use.

For example, in the case of PVC, during reprocessing in the melt phase, the loss of hydrogen chloride yields to unsaturation. This unsaturation, if not inhibited, leads to intense discoloration of the polymer by the unzipping process, which eliminates a molecule of hydrogen chloride in each step. On the other hand, polyolefins undergo mechanochemical chain scission in the presence of oxygen both at elevated temperatures (thermal oxidation) and during exposure to ultra-violet light (photo-oxidation). Hydroperoxides are formed in both cases. The further photolysis of in-chain hydroperoxides leads to rapid reduction in the molecular weight and a subsequent loss of tensile and impact strength in solid materials.[6,7]

GE Specialty Chemicals showed that multiple extrusion passes of restabilized HDPE did not conserve consistency of melt flow rate. To maintain melt flow consistency, a mixture of Ultranox 626 phosphite and Ultranox 276 hindered phenolic accomplished the target of reusing 100% post-consumer HDPE.

Work by Ciba Geigy's Additives for Plastics Recycling Group, Switzerland, shows similar improvements in recycled HDPE through the action of the company's Irganox phenolics and Irgafos phosphites.

Quantum Chemical's USA Division has included butylated hydroxy-toluene (BHT) antioxidants formulated specifically for recycling both LDPE and HDPE.[8]

PET can likewise gain from increased stabilization, though its individual requirements differ from polyolefins.

SECONDARY RECYCLING

There are two main approaches to secondary recycling. One approach is to separate the plastics from their contaminants and then segregate the plastics into generic types, one or more of which is then recycled into products produced from virgin or primary recycled material. The other approach is to separate the plastics from their associated contaminants and remelt them as a mixture without segregation.

Secondary Recycling by Type

The separation of plastics into types is based on differences in density, shape, color, physicochemical properties, and solubility.

Density

Air classification can be used to separate plastics on the basis of their bulk densities, thus film and foamed plastics may be separated from heavier forms of plastic material or paper.[9–12]

The densities of the major thermoplastics give the potential to separate them into types by a series of float-sink operations.[13,14] Water may be used to separate PP, LDPE, and HDPE from PS, PVC, and PET.

A different approach to separating mixed plastics wastes by density has been reported.[15–17] The process uses the properties of a fluid near its critical point to allow fine separations at mild temperatures and pressures. Carbon dioxide is the most commonly used supercritical fluid and can be compressed to densities in the range of 1000 k/cm^3. Since the operation of non-olefin thermoplastics will require fluid densities up to approximately 1400 k/cm^3, mixtures of carbon dioxide and sulfur hexafluoride, a very dense supercritical fluid, will be required.

The centrifugal field produced in a hydrocyclone has been extensively used for the separation of plastics.

The Mitsui Mining and Smelting Company has developed a process that depends on the wetting characteristics of the various plastics.[18,19]

*Author to whom correspondence should be addressed.

Shape

An interesting variation of the sink-float or hydrocyclone process is the addition of a step that granulates or grinds the PET bottles cryogenically. Because adhesive contaminants are freezed at cryogenic temperatures, whereas PET is not, adhesive contaminants in a cryogenic process become a fine powder. The fines are then easily removed from the coarser PET flake by screening. This process can produce a very clean recycled PET, free of adhesives and is used commercially by Western Environmental Plastics.[1] This technique is also widely applied to recycle scrap tires.

Color

Photoelectric sensors are used for the separation of mixed, whole, or baled plastic containers. After debaling and screening, the containers are manipulated into a single line presentation to an optical sensor that performs a three-class identification: Class 1: dairy HDPE and PP; Class 2; PET and PVC; and Class 3: mixed color HDPE containers. Another optical sensor can be used to further identify green and amber PET from clear PET containers, PP from dairy HDPE containers and mixed color HDPE according to seven color classifications. The Society of the Plastics Industry has devised a voluntary coding system (a triangle with a number corresponding to each polymer in the middle) to be imprinted on plastic containers, which will help to identify the resin used.[20]

Physicochemical Properties

A new dry method uses high-voltage drums for the separation of mixed plastic wastes, taking advantage of their different relative positions in the charging sequence.

Solubility

Solubility-based processes include stages of treating plastic waste with solvents so that the polymeric materials are dissolved and then recovered by several ways. This technology deals either with single-type plastic scrap or with mixtures.

Secondary Recycling of Plastics Mixtures by Remelting

Recycling of heterogeneous mixed plastics represents a very important goal. The possibility of obtaining secondary materials with good mechanical properties in a single recycling operation without separation steps is very appealing, but many obstacles must be overcome. In particular, the thermodynamic incompatibility of chemically different polymers leads to poor mechanical performance of the resultant mixtures, usually inferior to that of the original components. Approaches to secondary recycling of mixtures of plastics mainly include modification of the plastic waste to improve mechanical performance, and design and use of special equipment for melt compounding of mixed plastics.

TERTIARY OR "CHEMICAL" RECYCLING

Tertiary or "chemical" recycling has been defined as the breakdown of polymeric waste into reusable fractions for reincarnation as polymers, monomers, fuels, or chemicals. Chemical recycling mainly includes pyrolysis, gasification, hydrogenation, hydrolysis, glycolysis, and depolymerization:

Pyrolysis

Pyrolysis refers to the thermal decomposition of plastics at elevated temperatures with the exclusion of oxygen. This option yields a wide variety of gaseous, liquid, and solid products, depending on the raw materials, the process, and the reaction conditions.

Deutsche Reifen und Kunststoffepyrolyse GmbH employs a fluidized bed technique.[22,23] Using a feedstock of waste plastics the basic products formed are 44% gases, 26% oils, and 30% solid residue.[24,25] When the plant runs on old tires, the yield is 40% carbon black, 25% pyrolysis oil, 25% high BTU gases, and 10% steel.[26,27]

Procedyne Co. has used a fluidized-bed cracking furnace at 800°F and 250 psi to recover liquid-hydrocarbon fuel or chemicals from atactic polypropylene.[28]

Unlike the fluidized bed process, the rotary tubular process (demonstrated by KWU-Umwelttechnik) is geared exclusively to the recovery of fuel gas.

Batelle Columbus is working on a high temperature pyrolysis process, at 1100 to 1200°C. It is referred that at high temperatures, primarily ethylene and methane will be obtained when the feed is made up of polyolefins.[29]

Gasification

This technique coverts plastics waste to synthesis gas. Part of the feedstock is burned at pressures between 50 and 70 bars and the remaining is subjected to gasification temperatures of 1300 to 1500°C. The process is also referred to as partial oxidation owing to the addition of pure oxygen in the feed. The synthesis gas is made up of H_2, CO_2, CO, and H_2O.

Hydrogenation

As liquid fuels are produced by carbon hydrogenation and plastics are also carbon compounds, it could be feasible to hydrogenate them. The specific advantage of this technique is that the macromolecules are mainly converted to an oil and that very little gas, and also hardly any coke, is obtained.[30]

Hydrolysis

Unmixed (or sorted into individual grades) plastics produced by condensation reactions [polyurethanes, polyamides, polyesters, polycarbonates] can be split in their starting materials (monomers) or intermediates when exposed to the action of vapor water, high pressure, and elevated temperature.

Extensive work has been carried out on the hydrolysis of PUR foams.[31,32]

For polyesters, such as PET, hydrolysis merely reverses the polycondensation reaction. By treating PET with water in excess at 150–250°C in the presence of sodium acetate as catalyst ethylene glycol (EG) and terephthalic acid are yielded after 4 hr. Catalysts for hydrolysis are either acids (such as sulfuric) or bases (such as ammonium hydroxide).[34] An acid catalyst will promote the hydrolysis in 10–30 min at 60–95°C.[35]

Glycolysis/Alcoholysis—Methanolysis

Glycols instead of water can be used to convert polyurethanes at about 200°C into a polyhydroxy compounds liquid regenerate. This regenerate can replace part of the virgin polyol for the synthesis of a second generation product.

When recycled PET is heated with an excess of a glycol, such as propylene glycol (PG), in the presence of a catalyst, is converted through a transesterification reaction to short-chain fragments. Typical catalysts are amines, alkoxides, or metal salts of acetic acid.

Depolymerization

Molten metals or molten metal salts can be used for breaking down the macromolecules and conversion into monomers. For example, heating of PMMA up to 400–500°C yields in methyl methacrylate of purity as high as 98%. Molten Metal Tech., USA, is operating a steel bath at 1650°C to convert plastics waste or other contaminated materials with highly toxic substances from the chemical industry into gases (carbon monoxide and hydrogen), chemicals and alloys.[21,29,36,37]

QUATERNARY RECYCLING

The fourth option in plastic waste management is the recovery of its energy content. Incineration (combustion) aiming at the recovery of energy is currently the most effective way to reduce the volume of organic material that owing to a lack of other recycling possibilities, may then be ended in landfilling disposal.

Plastics (either thermoplastics or thermosettings) are actually high-yielding energy sources. There is an estimation that by burning 1 ton of waste approximately 250 liters of heating oil could be saved.[38]

At this point it should be emphasized that both tertiary and quaternary recycling to not preserve the value added during polymerization of plastics. At current polymer prices this poses a major disadvantage.

CONCLUSION

Yesterday's consumption should not, in any case, be tomorrow's waste. The world's fossil and material energy resources are finite and are being reduced by threatening rates. The need to obtain maximum benefits with minimum waste is an essential requirement for resource management. An important part of this management effort focuses on recovery and reuse of the materials produced, by giving them, where possible, a "second life." Plastics are regarded, sometimes without objective justification, as the sinner of the environment. However, there is no doubt that technology exists to utilize the material or energy contents of plastics ware.

Primary recycling is ideal for clean, uncontaminated single-type scrap but degradation during service life or reprocessing should be taken during service life or reprocessing should be taken under serious consideration. Secondary recycling by type can be accomplished by various methods but the cost associated with the separation and decontamination of the wastes poses undoubtedly an inherent obstacle. Dissolution-based techniques seem worth developing but cannot be considered yet as panacea. Secondary recycling of plastics mixtures by remelting excludes the throw-away connotation but is condemned to produce downgraded products owing to incompatibility problems. Compatibilization is effective only in specific cases of plastics mixtures. Tertiary or "chemical" recycling, using high investment processes, succeeds in recovering chemical products but crosses out the value added during polymerization. The latter is valid also for the last resort, quaternary recycling (energy recovery of plastics waste), that can substitute other energy sources and solve disposal problems. However, it is strongly accused of undesired emissions.

Biopolymers and biodegradable plastics (plastics containing additives which induce bio- or photo-degradation) may not replace easily the synthetic plastics establishment and may cause new problems without solving the old ones. They will not extend the life of a landfill, do not solve the esthetic problem posed by litter, and do not reduce (and may increase) the use of plastics.[39,40]

The aforementioned aspects strengthen the belief that no single option can provide a total solution to the problem and therefore an integrated approach, which takes into account a number of options, is essential. It is imperative for academia and for industry to continue developing recycling technologies and for governments to redefine the framework of plastics management for the protection or rehabilitation of the environment. Finally the society, beyond its ecological conscience, must be prepared to undertake some of the financial responsibility associated with the cost of applying the new recycling technologies on a large scale.

REFERENCES

1. Ehrig, R. J. "*Plastics Recycling: Products and Processes*" Hanser: Munich, Vienna, New York, Barcelona, 1992.
2. Drain, K. F.; Murphy, W. R.; Otterburn, M. S. *Conservation and Recycling* **1981**, *4*(4), 201.
3. Standard Guide—The Developments of Standards Relating to the Proper use of Recycled Plastics. Designation D 5033-90. American Society for Testing and Materials, Philadelphia, PA.
4. Warren, L. M.; Burns, R. *Plastics Technology* **1988**, *6*, 41.
5. Neale, C. W.; Hilyard, N. C.; Barber, P. *Conservation and Recycling* **1983**, *6*(3), 91.
6. Sadrmohaghegh, C.; Scott, G. *Eur. Polymer J.* **1980**, *16*, 1037.
7. Boldizar, A.; Gevert, T.; Jakubowicz, I.; Borgen, M.; Linde, M.; Hjertberg, T. *Advances in Recovery and Recycling*; Henstock, M. E.; Skov, H. R., Eds.; Vol. I, pp 331–335, Collected Papers of the ReC'93 International Recycling Congress, Geneva Switzerland, Hexagon Ltd, 1993.
8. Ogando, J. *Plastics Technology* **1993**, *July* 56.
9. Jensen, J. W.; Holman, J. L.; Stephenson, J. B. *Recycling and Disposal of Waste Plastics*; Chapter 7, Recycling and Disposal of Solid Wastes. Yen, T. F., Ed., Ann Arbor Science, 1974.
10. Emminger, H. *Kunststoffe* **1978**, *68*(5), 270.
11. Ansems, A. M. M. *Recycle '93 6th International Forum and Exposition* pp 5/3.1–5/3.12, March 1993.
12. Humber, N. *Recycle '93 6th International Forum and Exposition*; pp 17/2.1–17/2.9, March 1993.
13. De Gaspari, J. *Plastics Technology* **1990**, *May*, p 59.
14. Schroder, R. *IUPAC International Symposium—Recycling of Polymers*, Invited Lecture No. 7, Marbella–Spain, September 18–20, 1991.
15. Cobarr Spa, "Recycling of PET Using Supercritical Carbon Dioxide, e.g. for Use in Recycling of Drink Bottles to Give PET Reusable to in Food Packaging" U.S. Patent 5 049 647, 1991.
16. Super, M. S.; Enick, R. M.; Beckman, E. *Recycle '92 5th Annual International Forum and Exposition* pp 5/7.1–5/7.19, April 1992.
17. Beckman, E. J. *Recycle '93 6th Annual International Forum and Exposition* pp 14/4.1–14/4.16, March 1993.

18. Saitoh, K.; Takada, H.; Nagahama, T.; Izumi, S. "Separation of Large Extanses of Plastic Film" *Japan Kokai, 73*:81880 (1973).
19. Saitoh, K.; Nagano, I.; Izumi, S. "New Separation Technique for Waste Plastics" *Resour. Recov. Conserv.* **1976**, *2*, 127.
20. Wilkinson, S. L.; Dunphy, J. F. *Chemical Week* **1988**, *April 20*, 14.
21. Leaversuch, R. D. *Modern Plastics International* **1991**, *July* pp 26.
22. Steiner, K. W. *Kunststoffe* **1984**, *74*(4), 186.
23. Weber, A. *Kunststoffe* **1990**, *80*(4), 478.
24. Menges, G.; Eysmondt, B. V.; Feldhaus, A.; Offergeld, H. *Kunststoffe* **1988**, *78*(7), 573.
25. Kaminsky, W. *Recycle '93 6th Annual International Forum and Exposition* pp 7/4.1–7/4.20, March 1993.
26. *Chemical Engineering* **1982** *November 1*, 20E.
27. Buekens, A.; Tsyen, M. "Chemical Conversion of Plastics" *Plastics Recycling TVV Course Programme 1991–1992*, 16–18 September, 1992, Atomium Brussels, Belgium.
28. "Plastic Waste:" A Source for Chemicals and Fuels" *Chemical Week* **1982**, *August 11*, 34.
29. Menges, G.; Brandrup, J. *Recycle '93 6th Annual International Forum and Exposition* pp 10/4.1–10/4.20, March 1993.
30. "Hydrogenation, Pyrolysis, Alcoholysis/Glykolysis, Hydrolysis" *Bayer Research*, 4th ed. December 1990; pp 22.
31. Ionescu, M.; Zugravu, V.; Mihalache, I.; Mihis, B.; Mihai, S. *IUPAC International Symposium—Recycling of Polymers* Poster Contribution No. 4, Marbella—Spain, September 18–20, 1991.
32. Weigand, E. *Recycling and Reuse of Polyurethanes—Possibilities and Limitations* Technical Information, Bayer AG, Leverkusen, 1992.
33. Dutt, V.; Bastian, C. *Recycle '93 6th Annual International Forum and Exposition* pp 13/1.1–13/1.16 March 1993.
34. Lamparter, R. A.; Barna, B. A.; Johnsrud, D. R., U.S. Patent 4 542 239, 1985.
35. Mandoki, J. W. U.S. Patent 4 605 762, 1986.
36. Segui, E. D.; Alarcon, B. C. U.S. Patent 2 858 255, 1958.
37. Tatsumi, T.; Yoshihara, H.; Uesaka, G., U.S. Patent 3 886 202, 1975.
38. *The Future of Recycling*, Bayer Environment 2/91, pp 92.
39. Selke, S. E. M. *Packaging and the Environment* Technomic: Lancaster, PA, 1990, pp 119–139.
40. "Industry Opinions Harden in Degradable Plastics Issue" *Modern Plastics International* **1990**, *April* p 10.

RECYCLING TECHNIQUES, PLASTIC COMPOSITES

Michele Melchiorre and Karsten Löhr
Daimler-Benz AG
Research Center Ulm

The process of recycling includes a processing stage and subsequent utilization. Processing procedures are presented here that enable the recycling of used products and production waste.

The processing of secondary raw materials is referred to as waste processing or recycling. A characteristic of waste processing is the disintegration of the product form, that is the product is destroyed during processing.

Technical processes and methods are generally applied for processing. Thus for recycling, it is also possible to distinguish among mechanical, chemical, and thermal processes. For utilization, on the other hand, a distinction is made among the levels of working material, raw material, and energy levels. Accordingly, recycling can also comply with these designations.

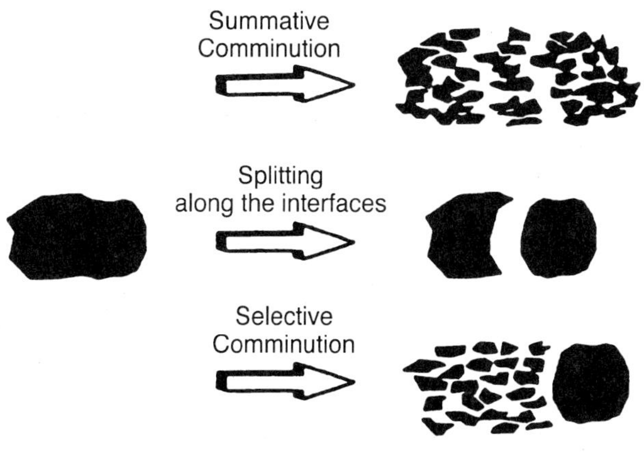

FIGURE 1. Principles of liberation.

LIBERATION

Comminution for Processing

Comminution has advanced to the most important procedure for the treatment of solid waste. The reasons for this correspond to those for raw materials processing: production of materials of desired particle size for handling, application, storage, and use; acceleration of material reactions by generation of high specific surfaces for chemical and thermal treatment; and liberation of marketable components of a composite material and obtaining valuable materials.

For comminution, solid materials are subjected to mechanical stresses to overcome the cohesive forces of the material. Basically, mechanical stress can be generated in the material in form of tension, pressure, and shear forces as well as bending and torsion moments.

With regard to the liberation of components from composite material, one can distinguish among three options (**Figure 1**).

Summative Comminution

Summative comminution achieves a reduction of particle size. Two different objectives can be followed for this, granulation, or changing to lumps or pieces, and liberation, or comminution to obtain particles of specific types.

In the case of granulation, specific particles of relatively narrow particle size ranges are generated from the material to be comminuted, which can generally consist of several materials. Normally this serves as the preliminary comminution of the material to prepare it for a subsequent process.

Alternatively, summative comminution can also take place for a period sufficient to split up the working material composite, and the individual components are liberated as specific material types. Two kinds of particles are always formed here: composite and liberated particles. Summative comminution takes place by cutting or shearing.

Splitting Along Interfaces

The splitting of a composite material along the interface is probably the best way of achieving liberation—provided that this option is given.

Selective Comminution

The initiation of fracture by defined mechanical stress enables a preferential comminution of only one component of composite materials while the other materials remain more or less undamaged.

Dissolution for Processing

When processing raw mineral materials, one distinguishes between pyrometallurgical and hydrometallurgical processes.

Corresponding to these mechanisms, it is also possible to distinguish between pyrolytic and hydrolytic procedures for plastics. The degradation of macromolecules to monomers can result fundamentally through a complete depolymerization for which pyrolytic procedures are mostly used. In addition, depolymerization to oligomers can also take place by a specific fragmentation of the chains at different positions. This takes place mostly by hydrolytic processes as a reversal of plastics synthesis. Finally, for plastics there is also a possibility of dissolution. Here the secondary crosslinking of the polymers is removed by solvents without a polymeric degradation and there is a transfer of the individual macromolecules into the solution.

Dissolution of Polymers

Similar to comminution, it is possible to distinguish between summative and selective dissolution. For summative dissolution, the composite bonding is removed by a dissolution of the entire material into a solution. With a selective dissolution, only one or several specific components of a composite material dissolve, while other components remain undissolved.

Hydrolysis and Glycolysis

In the case of plastics, aqueous media or organic liquids can be used for a wet chemical liberation of composite materials. The latter generally concerns high-grade glycols and generalizing; one can also speak of glycolysis. For hydrolytic or glycolytic degradation, certain functional groups of the polymeric macromolecules are broken. By this means, the polymeric chain length is reduced. The material becomes increasingly plastic and less viscous. The consistency can be compared with the states of wax, fats, and eventually oils.

Pyrolysis and Hydrogenation

For the pyrolytic liberation of plastics, thermal degradation reactions of the polymeric chains are used. This takes place differently to pure combustion under controlled atmospheric conditions. Pyrolysis generally takes place under exclusion of oxygen of with stoichiometric addition of oxygen to achieve a specified smouldering. The constant reaction procedure is enhanced by operation in a fluidized bed layer of quartz sand. The reaction in a hydrogen atmosphere is referred to as *hydrogenation.*

Similar to hydrolytic liberation, these reactions lead to a reduction of the polymer chain length. This ultimately results in formation of oils and gases that are discharged from the process.

SEPARATION

Sortation in Processing

For the mechanical separation into different kinds of material, it must be warranted that the particle can move within the mixture as unhindered as possible. The particles of a specific type must be able to move in a direction without dragging particles of foreign substances along with them.

Sortation According to Particle Type (Classifying or Sizing)

For these processes, different particles are separated on the basis of a specific particle characteristic (diameter, weight, cross-section, shape, etc.)

Processes related to particle characteristics can be distinguished according to whether separation takes place according to the geometric dimensions of the particle (sizing) or according to the physical flow behavior of the particle in a fluid in motion (current classifying).

Sortation According to Material Type

With separation processes related to the type of material, different material types are separated on the basis of a specific material characteristic (optical properties, electric conductivity, density, elasticity, etc.)

Rectifying in Processing

Physical phase transitions are utilized for rectifying dissolved materials. In general, these are concerned with phase separations of a first order, or among solid, liquid, and gaseous phases. Besides that, there are also techniques of utilizing phase transitions of a higher order, or recrystallization and remagnetizing that however, have not achieved any industrial significance.

Solid-Liquid Separation

The separation of particles from a liquid takes place as a result of steric or gravimetric effects. The relevant techniques are referred to as filtration and centrifugation or, respectively, sedimentation.

Liquid-Liquid Separation

A separation of different liquids on the basis of differences in their molecular size is referred to as dialysis.

Distillation

A liquid can be evaporated by heating to its boiling point. Different liquids generally possess a different boiling point. Consequently, a separation can be achieved by means of an exact temperature regulation.

TECHNICAL APPLICATIONS

Cable Processing

Cable scrap is processed by a combination of comminution and separation procedures. Already at an early stage, the exposed metal and plastic is separated from the shredded material. The residual material is then subjected to a further size reduction and in this way the consumption of cutting blades and energy is reduced.

Reprocessing Painted Plastic Parts

An application for dividing is found in automotive bumper recycling. The paint on bumpers has to be removed before recycling because it is incompatible with the thermoplastic substrate. Even an admixture of the mere 2% of paint leads to unsatisfactory results. Examinations showed that the mechanical strains acting on the material are focused on the small paint particles. This gives rise to micro-fractures and thus weakens the material considerably. As a consequence the paint has to be removed completely before high-technical reuse.

Composite Plastic Waste from Dashboards

Owing to conventional dashboard manufacturings, about 1 to 2 kg production residues per dashboard are produced. This waste is often a three-layer composite consisting of a thick-walled, glass fiber-reinforced thermoplastic (GRP) substrate, an intermediate polyurethane foam and a decorative ABS/PVC foil. As up to now no suitable method has been available to recycle this waste material, the whole amount of waste has to be disposed. The recovery of the three composite materials can be done by three dry mechanical processing steps. Owing to this recycling method an almost complete separation of the mixture into foil, foam, and supporting base can be achieved.

REFERENCES

1. *SME Mineral Processing Handbook*; Weiss, N. L., Ed.; Kingsport, TE, 1985.
2. Vesiling, A.; Rimer, A. E. Unit Operations in Resource Recovery Engineering; Prentice-Hall: NJ, 1981.
3. Schubert, H. Aufbereitung fester mineralischer Rohstoffe Bd. 1–3, VEB: Leibzig, 1989.
4. Gründer, W. Aufbereitungskunde Bd. 1–2, Hermann Hübner Verlag: Goslar, 1965.
5. Melchiorre, M.; Kettemann, B.-U.; Löhr, K. Material Recovery of Composite Products, Davos Recycle '94; Davos, Switzerland, May 1994.
6. Löhr, K. F.; Melchiorre, M. *Proc. 8th Europ. Symp. Comminution* 1994; 152–163.
7. Löhr, K. F.; Kettemann, B.-U.; Melchiorre, M. *Proc. R'95* Geneve, 1995.
8. Basta, N.; Fouhy, K.; Gilges, K.; Shanley, A.; Ushio, S. *Chem. Eng.* **1990**, *11*, 37–43.
9. Menges, G.; Emminger, H.; Lackner, G. *Int. J. Mat. Prod. Techn.* **1991**, *6*, *4*, 307–330.

RECYCLING, THERMODYNAMIC (by Ring-Closing Depolymerization)

Hartwig Höcker and Helmet Keul
Lehrstuhl für Textilchemie und Makromolekulare Chemie der Rheinisch-Westfälischen Technischen Hochschule Aachen

The chemical recycling with the aim of recovering valuable monomers or their precursors has been applied especially for polymers with heteroatoms in the main chain obtained by polycondensation and polyaddition polymerization.

The present contribution will focus on ring-closing depolymerization of polymers that are prepared by ring-opening polymerization. We shall call this method thermodynamic recycling, since thermodynamic parameters define whether the polymer or the cyclic monomer is the stable product under given conditions. The reverse of the ring-opening polymerization, the ring-closing depolymerization, can be realized in many cases in the presence of suitable catalysts and thus opens a gate to recycle polymers "thermodynamically," or through adjustment of a thermodynamic equilibrium.[1]

RING-CLOSING DEPOLYMERIZATION OF POLYESTERS, POLYCARBONATES, AND POLYSILOXANES

From the time of their discovery polymers containing heteroatoms in the main chain were found to be less stable than polymers containing only carbon atoms in the main chain except for very special chemical structures (high temperature stable polymers). Generally the thermal instability of the polymers was considered to be an undesired phenomenon, and studies have been aimed at the prevention rather than the promotion of depolymerization.[2] In the following several examples of the ring-closing depolymerization will be discussed.

Poly(ϵ-caprolactone) [poly(ECL)] is prepared by ring-opening polymerization of ϵ-caprolactone and is used as polymeric plastisizer and as additive to other polymers for the improvement of their properties.[3] Upon treatment of poly(ECL) in toluene solution with catalytic amounts of a suitable catalyst, such as 0.5 mol% Bu_2-$Sn(OMe)_2$ (with respect to ϵ-caprolactone repeating units), cyclic oligomers $(ECL)_x$ are formed; ϵ-caprolactone, however, is absent from this mixture.[4]

Catalyst ring closing depolymerization of poly(ECL) with $Bu_2Sn(OMe)_2$ as catalyst in the melt at 260°C with removal of the volatile materials by distillation results in a quantitative conversion of the polymer into volatile products. Gas-chromatographic analysis of the distillate obtained revealed 95.4wt% ECL beside 3.7wt% of its dimer and 2.9wt% of higher oligomers.

Poly(β-propiolactone) [poly(β-PL)] yields neither β-propiolactone nor the cyclic oligomers on thermolysis, but the typical ester pyrolysis product via intramolecular *cis* elimination, for example, acrylic acid or acyclic oligomers with acrylate and carboxylic acid end groups.[5]

Poly[(R)-3-hydroxybutyrate] {poly[(R)β-BL]}, the polymer produced by microorganisms via an enzymatic reaction or synthesized by ring-opening polymerization of (R)- or (S)-β-butyrolactone or by ring-opening polymerization of (R,R,R)–4,8,12-trimethyl-1,5,9-trioxacyclododeca-2,6,10-trione (the optically active cyclic trimer of (R)-β-butyrolactone) upon depolymerization in solution or in the melt, results in different products.[6–8] From the melt *cis* elimination is observed which leads mainly to crotonic acid or acyclic oligomers with a crotonate and a carboxylic acid end group. However, upon changing the reaction conditions, ring closing depolymerization leads to a homologous series of cyclic oligomers $[(R)\beta\text{-BL}]_x$.

Poly(2,2-dimethyl-3-propiolactone) [polypivalolactone, poly (PVL)] is by far more stable than the unsubstituted or β-substituted poly(β-propiolactone) owing to the substitution of the labile α-hydrogen atoms by methyl groups.[5]

For the back-biting depolymerization mechanism it was proven by TGA combined with the electron impact mass spectral analysis (EI-MS) on the evolving gases that under certain conditions all chains depolymerize simultaneously from the chain ends; hence the reaction called "living depolymerization" (cf. ring-closing depolymerization of copolymers).

Thermal depolymerization of aliphatic and aromatic polycarbonates has received continued attention in the literature.[10–12] It is believed that polycarbonates undergo thermal depolymerization by a number of ionic mechanisms, but also radical processes were taken into consideration.[13–15] Most important are the intramolecular transesterification reactions.

We will focus on the intramolecular process that leads to cyclic carbonates and was found to operate both in aliphatic and in totally aromatic polycarbonates.

McNeill et al. reported that the EI–mass spectra of the products of pyrolysis of poly(trimethylene carbonate) [poly(TMC)] and of poly(2,2-dimethyltrimetylene carbonate) [poly(DTC)] show the presence of molecular ion peaks.[14,16]

Our group has investigated the catalyzed depolymerization process of poly(DTC) with $Bu_2Sn(OMe)_2$ as a catalyst in vacuo.[17] The optimum temperature for the ring-closing depolymerization of poly(DTC) is between 240° and 260°C. Above this temperature degradation of the polymer takes place to form carbon dioxide and 2,2-dimethyloxetane.

RING-CLOSING DEPOLYMERIZATION OF COPOLYMERS

Manring et al. studied the thermal depolymerization of random and block copolymers of 2-methyl-2-propyl-3-propiolactone (MPPL) with 2-methyl-2-ethyl-3-propiolactone (MEPL) for mechanistic reasons.[9] They found support for a living depolymerization from the chain end. Poly(MPPL-stat-MEPL) with $COO^{-+}PBu_4$ endgroups (M_n = 68.400, M_w/M_n = 1.14) results in the simultaneous formation of MEPL, and MPPL, in the contrast, poly(MPPL)-block-poly(MEPL) results in the appearance of MEPL prior to MPPL. The opposite result is obtained when poly(MEPL)-block-poly(MPPL) with $COO^{-+}PBu_4$ end groups are depolymerized, that is, MPPL appears prior to MEPL.

The triblock copolymer poly(DTC)-b-poly(THF)-b-poly(DTC) was successfully depolymerized in a step-by-step fashion; first the polytetrahydrofuran [poly(THF)] block was depolymerized with $CF_3SO_3CH_3$ as a catalyst at 150°C and consecutively poly(DTC) was depolymerized with $Bu_2Sn(OMe)_2$. During the first step THF is removed and the residual material was identified as poly(DTC) with a monomodal distribution. This result proves that under the conditions of the first depolymerization step the DTC–DTC diads are stable.[18]

CONCLUSION

Ring-closing depolymerization of homo- and copolymers often leads to cyclic products used for the preparation of these polymers by ring-opening polymerization, for example, poly(ϵ-caprolactone) yields ϵ-caprolactone, poly(2,2-dimethyltrimethylene carbonate) yields 2,2-dimethyltrimethylene carbonate, and polytetrahydrofuran yields tetrahydrofuran. In other cases a series of cyclic oligomers is produced, such as, cyclic dimethyl siloxane oligomers from polydimethylsiloxane, or cyclic pivalolactone oligomers from polypivalolactone.

REFERENCES

1. Höcker, H. *KGK Kautschuk Gummi Kunststoffe* **1993**, *46*, 556.
2. Grassie, N. *Development in Polymer Degradation*; Elsevier: London, 1977; Vols. 1–6; Jellinek, H. H. G. *Degradation and Stabilization of Polymers*; Elsevier: New York, 1983; Howkins, W. L. *Polymer Degradation and Stabilization*; Springer: Berlin, 1984.
3. Jerome, R.; Teyssié, P. *Anionic ring-opening polymerization: Lactones*; In Allen, G.; Bevington, J. C., Eds.; *Comprehensive Polymer Science*; Pergamon: Oxford; Vol. 3, pp 501.
4. Nelissen, M.; Keul, H.; Höcker, H. *Macromol. Chem. Phys.* **1995**, *196*, 194.
5. Iwabuchi, S.; Jaacks, V.; Gahl, F. H.; Kern, W. *Makromol. Chem.* **1973**, *165*, 59.
6. Doi, Y. *Microbiol Polyesters*; VCH: Weinheim, 1990.
7. Zhang, Y.; Gross, R. A.; Lenz, R. W. *Macromolecules* **1990**, *23*, 3206; Hori, Y.; Suzuki, M.; Yamaguchi, A.; Nishishita, T. *Macromolecules* **1993**, *26*, 5533.
8. Melchiors, M.; Keul, H.; Höcker, H. *Macromol. Rapid Commun.* **1994**, *15*, 497.
9. Manring, L. E.; Blume, R. C.; Dee, G. T. *Macromolecules* **1990**, *23*, 1902.
10. Foti, S.; Giuffrida, M.; Maravigna, P.; Montaudo, P. *J. Polym. Sci., Polym. Chem. Ed.* **1983**, *21*, 1567.
11. Ballistreri, A.; Montaudo, G.; Puglisi, C.; Scamporrino, E.; Vitalini, D.; Cucinella, S. *J. Polym. Sci., Polym. Chem. Ed.* **1988**, *26*, 2113.
12. Montaudo, G.; Puglisi, C.; Samperi, F. *Polym. Deg. & Stab.* **1989**, *26*, 285.
13. McNeill, I. C. In *Comprehensive Polymer Science*; Eastmond, G. C.; Ledwith, A.; Russo, S.; Sigwalt, P., eds.; Pergamon: Oxford, 1989; Vol. 6, p 492.
14. McNeill, I. C.; Rincon, A. *Polym. Deg. & Stab.* **1989**, *24*, 171.
15. McNeill, I. C.; Rincon, A. *Polym. Deg. & Stab.* **1990**, *27*, 35.
16. McNeill, I. C.; Rincon, A. *Polym. Deg. and Stab.* **1989**, *24*, 59.
17. Keul, H.; Müller, A. J.; Höcker, H.; Sylvester, G.; Schön, N. *Makromol. Chem. Macromol. Symp.* **1993**, *67*, 289.
18. Müller, A. J.; Keul, H.; Höcker, H. *Eur. Polym. J.* **1993**, *29*, 1171.

Red Blood Cell Substitutes

See: *Blood Compatible Polymers*
Blood Substitutes
Microencapsulation (Artificial Cells)
Red Blood Cell Substitutes, Artificial (for Transfusion)

RED BLOOD CELL SUBSTITUTES, ARTIFICIAL (for Transfusion)

Thomas Ming Swi Chang
Artificial Cells and Organs Research Center
Faculty of Medicine
McGill University

The two major approaches for artificial blood are modified hemoglobin and perfluorochemicals. Most centers and industries are now carrying out research and development on the use of modified hemoglobin as red blood cell substitutes.

MODIFIED HEMOGLOBIN

Modified hemoglobin is especially useful in these emergency situations: it does not contain blood group antigens, therefore, it can be used without the need for crossmatching or typing. This would save much time and would permit on-the-spot transfusion as required, similar to giving intravenous salt solution. Furthermore, modified hemoglobin can be lyophilized and stored as a stable dried powder which can be reconstituted with the appropriate salt solution just before use.

What is modified hemoglobin? Hemoglobin molecules extracted from red blood cells are modified by microencapsulation or crosslinkage. This stabilizes the hemoglobin molecules and also allows the sterilization of the products to remove H.I.V. and other microorganisms. Rapid progress is being made towards clinical use.

Encapsulated Hemoglobin: Artificial Red Blood Cells[8–16]

This author was first to report the preparation of artificial red blood cells.[1] Microencapsulated hemoglobin or artificial red blood cells are now being extensively explored by many researchers around the world. Recent advances make possible the following results: The average half-time in the circulation is now up to 20 hours. The uptake is mainly by the reticuloendothelial system. It is possible to replace 90% of the red blood cells in rats with these artificial red blood cells; the animals with this percentage of exchange transfusion still remain viable. Studies also reported effectiveness in hemorrhagic shock. Extensive study by a number of groups shows that the artificial red blood cells are not toxic. To further improve stability and biodegradability, we are now using biodegradable polymer membrane to prepare artificial red blood cells of less than 200 nanometers (less than 0.2 micron) diameters.[13] Microencapsulation of hemoglobin to prepare artificial red blood cells is a rather ambitious approach.[1]

Crosslinked Hemoglobin

Hemoglobin contains many amino groups and most are on the surface of the hemoglobin molecule. I first reported the use of a bifunctional agent (diacid) to crosslink hemoglobin.[2,3] This was used first to form crosslinked hemoglobin membranes for artificial red blood cells, but I found that with decreasing size of artificial cells, all the hemoglobin molecules are crosslinked into polyhemoglobin.[2,3] Crosslinking prevents the breakdown of hemoglobin tetramers into dimers.[2,3]

Other crosslinkers are also being developed. Some of these are based on bifunctional dialdehydes derived from oxidizing the ring structures of sugar or nucleotides.

Conjugated hemoglobin is the crosslinking of hemoglobin to polymers.[2–4] The use of soluble polymers resulted in soluble conjugated hemoglobin with good circulation time.[14–16] These are now being used in clinical trials.

Present Status and Future Research and Development

Crosslinked hemoglobin including polyhemoglobin, intramolecularly crosslinked hemoglobin, and conjugated hemoglobin is likely to be the first modified hemoglobin ready for routine clinical use. Six North American companies are now carrying out clinical trials in human.[17–20] Some are at the Phase II clinical trial stage. The emphasis of clinical trial is now on studying the efficacy of modified hemoglobin in humans.[19] Animal studies have already demonstrated efficacy.[5–8]

PERFLUOROCHEMICALS

Of the synthetic organic materials, silicone and fluorocarbon are known for their ability to carry oxygen. In 1968, Geyer, Monroe and Taylor demonstrated that finely emulsified fluorocarbon could replace essentially all the blood of rats with the rats surviving and recovering.[21] This existing demonstration did not immediately lead to clinical application because F-tributylamine available at that time had a long retention time ($T_{1/2}$ of more than 800 days) in the reticuloendothelial system (RES) and, therefore, it could not be used clinically. Extensive work was carried out by Naito and Yokoyama, resulting in the development in 1976 of fluosol-DA 20 suitable for clinical testing.[22]

New Perfluorochemicals

Two new types of preparation of perfluorochemical (PFC) have been developed.[23–28] One is based on perfluoroctyl bromide ($C_8F_{17}Br$) and perfluorodichoroctane ($C_8F_{16}Cl_2$). Both types allow the use of higher concentrations of PFC.

Present Status and Future Perspectives of Perfluorochemicals

The biggest advantage of perfluorochemicals is that they are synthetic material that can be chemically produced in large amounts without having to depend on donor blood or other biological sources. The earlier problem of complement activation has been solved by changing the surfactant. Higher concentrations of the new perfluorochemicals can now be used to increase oxygen carrying capacity, although this is currently limited to the rather low dosage of 0.9 g/kg for human use, with the patients still breathing high oxygen. Further research and development are likely to resolve the problem related to side effects at higher dosage. Then the highest dosage will only be limited by that which would not cause significant RES suppression. In this regard, ever-improving perfluorochemicals with decreasing residual time in the reticuloendothelial system are being made available. This should result in lower oxygen levels required for breathing.

REFERENCES

1. Chang, T. M. S. Honours Report, McGill University, 1957; *J. Biomater. Artif. Cells Artif. Org.* **1988**, *16*, 1.
2. Chang, T. M. S. *Science* **1964**, *146*(3643), 524.
3. Chang, T. M. S. Ph.D. Thesis, McGill University, 1965.
4. Chang, T. M. S. *Artificial Cells*; Charles C. Thomas: Springfield, IL, 1972.
5. *Blood Substitutes and Oxygen Carriers*; Chang, T. M. S., Ed.; Marcel Dekker: New York, NY, 1992, p 784.
6. Chang, T. M. S.; Reiss, J. G.; Winslow, R., Eds.; *Artificial Cells, Blood Substitutes and Immobilization Biotechnology*, Int. J. **1994**, *22*, 123–360.
7. Winslow, R., Ed.; *Artificial Cells, Blood Substitutes and Immobilization Biotechnology*, Int. J. **1994**, *22*, 360–944.
8. Chang, T. M. S. *Biochem. Biophys. Res. Com.* **1971**, *44*, 1531.
9. Djordjevich, L.; Miller, I. F. *Exp. Hematol.* **1980**, *8*, 584.
10. Farmer, M. C.; Rudolph, A. S.; Vandegriff, K. D.; Havre, M. D.; Bayne, S. A.; Johnson, S. A. *J. Biomater. Artif. Cells Artif. Org.* **1988**, *16*, 289–299.

11. Rudolph, A. S. *Artificial Cells, Blood Substitutes and Immobilization Biotechnology*, Int. J. **1994**, *22*, 347–360.
12. Tsuchida, E. *Artificial Cells, Blood Substitutes and Immobilization Biotechnology*, Int. J. **1994**, *22*, 467–479.
13. Yu, W. P.; Chang, T. M. S. *Artificial Cells, Blood Substitutes and Immobilization Biotechnology*, Int. J. **1994**, *22*, 889–894.
14. Wong, J. T. *Biomaterials, Artificial Cells and Artificial Organs* **1988**, *16*, 237–245.
15. Nho, K.; Glower, D.; Bredehoeft, S.; Shankar, H.; Shorr, R.; Abuchowski, A. *J. Biomaterials, Artificial Cells and Immobilization Biotechnology* **1992**, *20*, 511–524.
16. Iwashita, Y. *J. Biomaterials, Artificial Cells and Immobilization Biotechnology* **1992**, *20*, 299–308.
17. Hughes, G. S.; Jacobs, E. *Abstracts of Papers*, 5th Int. Symposium on Blood Substitutes; San Diego, CA, 1993; #H15, p 134.
18. Shoemaker, S.; Gerber, M.; Evans, G.; Paik, L.; Scoggin, C. *Artificial Cells, Blood Substitutes and Immobilization Biotechnology*, Int. J. **1994**, *22*, 457–465.
19. Fratantoni, J. C. *Transfusion* **1994**, *34*, 712–713.
20. O'Donnell, J. K.; Swanson, M.; Pilder, S.; Martin, M.; Hoover, K.; Huntress, V.; Karet, C.; Pinkert, C.; Lago, W.; Logan, J. *J. Biomaterials, Artificial Cells and Immobilization Biotechnology* **1992**, *20*(1), 149.
21. Geyer, R. P.; Monroe, R. G.; Taylor, K. In *Organ Perfusion and Preservation*; Norman, J. C.; Forkman, J.; Hardison, W. G.; Rudolf, L. E.; Veith, F. J., Eds.; Appleton Century Crofts: New York, NY, 1968; pp 85–96.
22. Naito, R.; Yokoyama, K. In *Blood Substitutes and Plasma Expanders*; Jamieson, G. A.; Greenwalt, T. J., Eds.; Alan R Liss: New York, 1978; 81.
23. Faithfull, N. S. *J. Biomaterials, Artificial Cells and Artificial Organs* **1992**, *20*, 797–804.
24. Goodin, T. 2nd Ann. IBC Conf. Blood Substitutes and Related Products, Washington, DC, 1994.
25. Goodin, T. H.; Grossbard, E. B.; Kaufman, R. J.; Richard, T. J.; Kolata, R. J.; Allen, J. S.; Layton, T. E. *Crit. Care Med.* **1994**, *22*, 680–689.
26. Reiss, J. *Artificial Cells, Blood Substitutes and Immobilization Biotechnology*, Int. J. **1994**, *22*, 945–1511.
27. Reiss, J. G. *Vox Sang.* **1991**, *61*, 225–239.
28. Wahr, J. A.; Trouwborst, C. W.; Spence, R. K., et al. *Anesthesiology* **1994**, *80*, A397.

Redox Initiators

See: *Block Copolymers (from Macroinitiators)*
End Group Analysis
Redox Polymerization

REDOX POLYMERIZATION

A. Sezai Saraç
Istanbul Technical University

Generation of free radicals under mild conditions by one electron transfer reaction with redox initiation has found wide application for initiating polymerization reactions.[1,2] This particular technique has industrial importance, for example, in low-temperature emulsion polymerization.[3]

Redox polymerization elucidate redox mechanisms, by determination of polymer end groups generated during redox reactions by transient radical intermediates. They have a very short induction period and relatively low energy of activation (40–80 kJ/mol^{-1}).

REDOX INITIATION IN AQUEOUS MEDIA

Jellinek applied Fenton's reagent, the combination of H_2O_2 and ferrous salt, to the oxidation of many organic compounds including alcohols, glycols, aldehydes, ethers, esters, and amines.[4] **Equation 1** shows a mechanism for the reaction between hydrogen peroxide and ferrous ion which consists of a one-electron transfer from the ferrous ion to the peroxide with dissociation of the oxygen–oxygen bond and the creation of one hydroxyl radical and one hydroxyl ion.[5]

$$Fe^{2+} + H_2O_2 \rightarrow [HO \dot{-} OH]Fe^{3+} \rightarrow Fe^{3+} + OH^- + \cdot OH \qquad 1$$

In the presence of sufficient monomer, all the ·OH radicals generated can initiate polymerization.

Ferrous ions will promote the decomposition of a number of compounds, including alkyl and acyl peroxides.

With persulfate initiator several monomers (e.g., acrylonitrile, methacrylic acid, methacrylamide, methyl methacrylate, ethyl acrylate) can be grafted on wood fibers by initiation of free radicals on cysteine present in wool.[6,7]

The oxyacids of sulfur such as sulfite, bisulfite, bisulfate, thiosulfate, metabisulfite, dithionate, and sulfide form efficient redox systems in conjunction with persulfates.[8,9]

Polymerizations of acrylamide, acrylonitrile, methacrylamide, and methyl methacrylate with persulfate (peroxydisulfate) and several different reducing agents have been reported; reducing agents may be isoamyl alcohol, polyethylene glycol, ascorbic acid, or malonic acid.[10–16]

METAL ION OXIDANTS IN REDOX INITIATION

Numerous reducing agents like alcohols, thiols, ketones, aldehydes, acids, amines, and amides in combination with oxidizing metal ions participate in general single-electron transfer reactions for free-radical polymerization. Metal ions used mainly for this purpose are Mn(III) (and Mn(VII)), Ce(IV), V(V), Co(III), Cr(VI), and Fe(III).

REDOX INITIATION IN NONAQUEOUS MEDIA

Intermolecular Processes

Organic peroxides such as diacyl peroxides and tertiary amines form a complex in nonaqueous solution which cleaves into radicals and radical cations. Radicals are the main initiating species (**Schemes I** and **II**).[17,18]

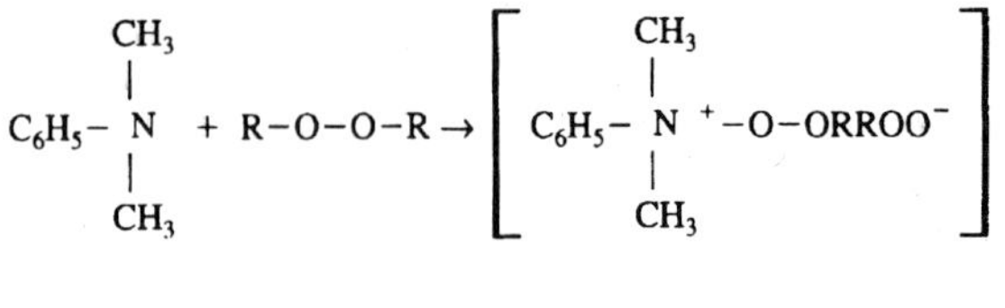

I

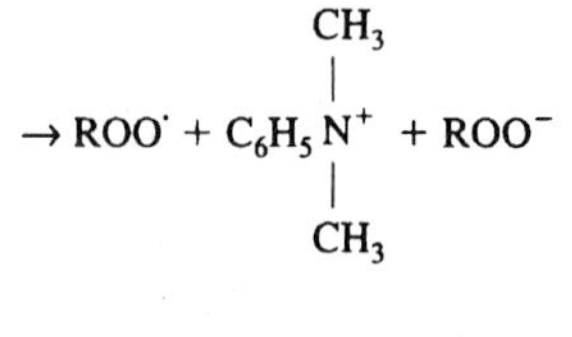

II

Other reducing agents, such as formic acids, sulfonic acids, and hydrazines instead of tertiary amines, have been reported.[19–21]

Intramolecular Processes

Metal chelates have been used for initiation of different types of polymerizations. Metal acetylacetonates such as $MoO_2(acac)_2$, $Mn(II)(acac)_2$, and $Co(II)(acac)_2$ have been reported as initiator for polymerization of chloral by ionic polymerization mechanism.

A wide range of organic compounds increase the rate of initiation by metal chelates.[22] Free-radical polymerization of acrylonitrile initiated by $Mn(acac)_3$ (Mn acetylacetonate) occurs readily at 25°C in Me_2SO solution, although there is little reaction in the bulk monomer. Piperidine and some pyridine derivatives increase the rates of polymerization of several vinyl monomers initiated by $Mn(acac)_3$ at 70°C. Kaeriyama reported analogous results for $Mn(acac)_3$ and $Fe(acac)_3$ and some carbonyl compounds. The effect of ketones depends on their nucleophilicity.[23]

REFERENCES

1. Misra, G. S.; Bajpai, V. D. N. *Prog. Polym. Sci.* **1982**, *8*, 61–131.
2. Misra, G. S. In *Encyclopedia of Polymer Science and Technology*; Wiley: New York, NY, 1988; Vol. 8, pp 61–131.
3. Vanderhoft, J. W. In *Vinyl Polymerization*; Ham, G. E., Ed.; Marcel Dekker: New York, NY, 1969; Chapter 1.
4. Jellinek, H. H. G. *Degradation of Vinyl Polymers*; Academic: New York, NY, 1955.
5. Jellinek, H. H. G. *Pure Appl. Chem.* **1962**, *4*, 419.
6. Wolfram, L. J.; Menkart, J. *Am. Dystuff Reptr.* **1967**, *56*, 110.
7. Arai, K.; Komine, S.; Negishi, M. *J. Polym. Sci. Part A-1* **1970**, *8*, 917.
8. Prochess, H.; Patat, F. *Makromol. Chem.* **1968**, *114*, 11.
9. Mark, P. C.; Ugelstad, J. *Makromol. Chem.* **1969**, *128*, 83.
10. Ariff, M.; Jainuddin, M. D.; Gopalan, V. *J. Polym. Sci.* **1985**, *23*, 2063.
11. Tajuddin, I.; Kader, K. S. Abdul *J. Macromol. Sci., Chem.* **1986**, *A23*(1), 149.
12. Narain, H.; Jagadale, S. M.; Ghatge, N. D. *J. Polym. Sci.* **1981**, *19*, 1225.
13. Sarpudeenm, A.; Tajuddin, I.; Anwarudding, Q. *Eur. Polym. J.* **1984**, *20*(10), 1019.
14. Lenka, S.; Dhal, A. K. *J. Polym. Sci.* **1981**, *19*, 2115.
15. Basha, P. G.; Ariff, M.; Jainudeen, M. D.; Gopalan, V. *J. Macromol. Sci., Chem.* **1986**, *A23*(4), 473.
16. Pantar, A. V.; Atreyi, M.; Rav, M. V. R. *Angew. Makromol. Chem.* **1985**, *129*, 163.
17. Horner, L. *J. Polym. Sci.* **1955**, *18*, 438.
18. Horner, L.; Kirmse, W. *Annalen* **1955**, *567*, 48.
19. Horner, L.; Schwerk, E. *Angew. Chem.* **1949**, *61*, 411.
20. Horner, L.; Schwerk, E. *Kunststoffe* **1949**, *39*, 292.
21. Orr, R. J.; Williams, H. L. *Discuss. Faraday Soc.* **1953**, *14*, 170, 237.
22. Kastning, E. G.; Nearmann, H.; Reid, H.; Berding, C. *Angew. Chem. Int. Ed. Engl.* **1965**, *4*, 322.
23. Kaeriyama, K. *Bull. Chem. Soc. Appl.* **1975**, *43*, 1511.

Redox Polymers

See: Dendrimers, Luminescent and Redox-Active Ion-Selective Reagents (Dual Mechanism Bifunctional Polymers)

REGENERATION TEMPLATES, ARTIFICIAL SKIN AND NERVES

Diane L. Ellis and I. V. Yannas
Masschusetts Institute of Technology
Department of Mechanical Engineering

The development of regeneration templates is an important area of polymer materials science which has had many recent advances. Regeneration templates contain polymers of natural origin which are developed into specific devices with optimized parameters for regeneration. They induce the synthesis of physiological tissue (tissue regeneration) when placed in contact with a well-defined lesion (wound) in adult mammalian tissue in which it has been previously demonstrated that spontaneous regeneration does not occur.[1–3] The biological activity of these polymers in regeneration templates has greatly contributed to the fields of wound healing and developmental biology where they are used in devices for tissue and organ repair.

The field of wound healing is an area under constant study incorporating significant contributions from both medicine and engineering. Most tissues consist of cells and a primarily insoluble extracellular matrix (ECM). The ECM provides stiffness, strength, stability of shape, and scaffolding to cells. Exact composition and structure of the ECM vary among tissues, but it typically contains hydrated macromolecular networks with glycoproteins such as collagen, fibronectin, elastin, and lamanin; and glycosaminoglycans (GAG) such as chondroitin 6-sulfate, dermatan sulfate, heparan sulfate, and hyaluronic acid.

The regeneration templates that we describe here were developed as chemical analogs of the ECM. Besides contributing to would healing, these analogs have excited interest in the process of wound repair and development. We will discuss those analogs that have been synthesized as regeneration templates which are graft copolymers of collagen and one of several possible glycosaminoglycans.

PREPARATION AND PROPERTIES

Our regeneration templates are based on collagen-glycosaminoglycan (CG) copolymers. The types of glycosaminoglycans grafted onto collagen include chondroitin 6-sulfate, chondroitin 4-sulfate, heparan sulfate, heparin, dermatan sulfate, and keratan sulfate.[6] Subsequent work has focused on a Type I collagen-*graft*-chondroitin 6-sulfate copolymer.

Collagen is a crystalline fibrous protein which accounts for approximately one-third of the total protein in vertebrates. Its main function is to mechanically reinforce connective tissues.

Type I collagen is used to prepare the CG copolymer. This is a fibrillar, insoluble collagen extracted from cowhide by lime treatment and mechanically fragmented.[7]

The glycosaminoglycan we used to prepare the templates is chondroitin 6-sulfate derived from shark cartilage (SIGMA). Once the collagen is dispersed in acetic acid, 0.13 g chondroitin 6-sulfate in 0.05 M acetic acid is slowly added to the blending solution with a peristaltic pump over 15 minutes. After an additional 15 minutes of blending, a white collagen–chondroitin 6-sulfate coprecipitate is removed from the blender.

The collagen–GAG coprecipitate is freeze-dried (lyophilized) into a porous foam.

After the freeze-drying process, the GAG must be grafted onto the collagen chains to prevent elution in *in vivo* studies because the ionic bond between the GAG and collagen is only stable at acidic pH. An effective procedure for grafting GAG onto collagen chains is exposure of the dry foams to 105°C under vacuum for 24 hours. At least 10 wt % (dry foam basis) of the GAG can be grafted to the collagen by vacuum dehydration The mechanism of crosslinking the GAG chains to the collagen can be postulated to be the product of the condensation of carboxyl groups of glucuronic acid residues on the GAG chain with ϵ-amino groups of lysyl residues on collagen, with water as a condensation product.

This crosslinking of the matrix foam is one method of controlling the *in vivo* degradation rate of the collagen–GAG matrix.[6] Further crosslinking is obtained by exposure to aqueous glutaraldehyde, which is a known crosslinking agent for proteins and has been used as a fixative for tissue samples for histological studies.

APPLICATIONS

Device Fabrication

Devices have been developed for skin regeneration and peripheral nerve regeneration. The skin regeneration template consists of a two-layer device for implantation. This device is grafted into the full-thickness skin wounds, and is referred to as Stage 1 when not seeded with cells prior to implantation, and as Stage 2 when seeded with autologous cells.[9–11] Stage 1 devices inhibit contraction in the wound site, but are eventually degraded and a scar results. They are useful for contraction delay and regeneration of the dermis at the wound site or both dermis and epidermis. Stage 2 devices are necessary for regeneration to take place.

Stage 1 devices are made by coating a porous collagen-*graft*-glycosaminoglycan 3 mm-thick sheet with a thin layer of a moisture-curing polydimethoxysiloxane prepolymer (Silastic Medical A, Dow Corning) under sterile conditions. The silicone layer is cured by immersing the device in 0.05 M acetic acid over 24 hours. The silicone is polymerized into a 0.025 mm-thick elastomeric membrane.

Stage 2 devices are Stage 1 devices that have been inoculated with autologous, uncultured epidermal cell suspensions.[9,10]

Biological Activity and Matrix Structure

Studies with the collagen–GAG matrices have focused on the regeneration of specific tissues known not to regenerate spontaneously. The activity of regeneration templates has been studied in two well-established models: the full thickness excised skin would in the guinea pig and in humans;[2,3] and the 10 and 15 mm transected gaps in the sciatic nerve of the rat.[4,5]

Certain collagen–GAG matrices in the Stage 2 device can induce synthesis of dermis in full thickness skin wounds of the guinea pig. These particular copolymers are capable of significantly delaying or arresting wound contraction.

The biological activity of the matrices described above depends critically on the average pore diameter of the porous matrix, the degradation rate of the matrix in the wound site, and the density of cells seeded into the matrix.

CONCLUSIONS

The necessary attributes of a regeneration template can be understood in terms of a highly-specific cell-matrix interaction which diverts the mechanism of wound healing away from contraction and scar synthesis toward the inhibition of contraction and tissue regeneration in the woundbed. The collage–GAG copolymer matrices are capable of significantly delaying the kinetics of wound contraction. The biological activity of these polymer matrices resides in their porous, cell-free, macromolecular structure. To be biologically active, the polymeric matrix must be resistant to rapid solubilization, sufficiently nondiffusible over a period of approximately 10 days (upper limit of degradation rate requirement). A minimum average pore diameter still allows for entry and migration of cells from the woundbed (lower limit of pore diameter requirement). The matrix must also possess a specific surface that is sufficiently extensive to allow a critical number of cell–matrix interactions (upper limit of pore diameter requirement).[1,8]

We conclude that a set of physicochemical parameters exist which serve to yield a degradable, porous, cell-free macromolecular matrix which modifies the mechanism of wound healing in skin and nerve wounds. These well-defined matrices are very simple chemical analogs of extracellular matrix which can be used to answer important questions about the aspects of tissue remodeling which distinguish conventional wound healing from normal development or from regeneration.

REFERENCES

1. Yannas, I. V. *Angew. Chem.* **1990**, *29*, 20.
2. Billingham, R. E.; Medawar, P. B. *J. Exp Biol.* **1951**, *28*, 385.
3. Billingham, R. E.; Medawar, P. B. *J. Anat.* **1955**, *89*, 114.
4. Lundborg, G.; Dahlin, L. B.; Danielson, N.; Gelberman, R. H.; Longo, F. M.; Powell, H. L.; Varon, S. *Exp. Neurol.* **1982**, *76*, 361.
5. Lundborg, G. *Acta Orthop. Scand.* **1987**, *58*, 145.
6. Yannas, I. V.; Burke, J. F.; Huang, C.; Gordon, P. L. *Polym. Prepr., Am. Chem. Soc. Div. Polym. Chem.* **1975**, *16*(2), 209.
7. Yannas, I. V.; Burke, J. F.; Gordon, P. L.; Huang, C.; Rubinstein, R. H. *J. Biomed. Mater. Res.* **1980**, *14*, 107.
8. Yannas, I. V.; Lee, E.; Orgill, D. P.; Skrabut, E. M.; Murphy, G. F. *Proc. Natl. Acad. Sci. U.S.A.* **1989**, *86*, 933.
9. Yannas, I. V.; Burke, J. F. *J. Biomed. Mater. Res.* **1980**, *14*, 65.
10. Yannas, I. V.; Burke, J. F.; Orgill, D. P.; Skrabut, E. M. *Science* **1982**, *215*, 174.
11. Yannas, I. V.; Lee, E.; Skrabut, E. M.; Orgill, D. P.; Murphy, G. F. *J. Cell Biol.* **1987**, *105*, 223a.

Reinforcing Agents

See: Additives (Property and Processing Modifiers)
Composites (Structure, Properties, and Manufacturing)
Fillers and Reinforcing Agents
Thermoplastic Composites

REISSERT POLYMERS

Harry W. Gibson
Department of Chemistry
Virginia Polytechnic Institute and State University

Jean-Pierre Leblanc
National Starch and Chemical Company

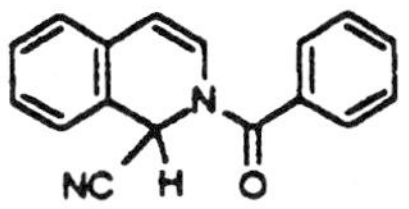

1

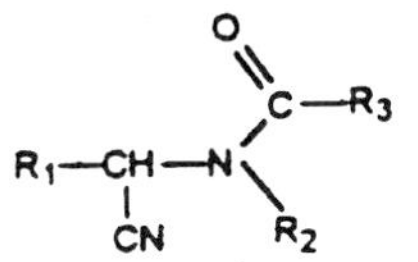

2

The title of this section was chosen to represent not only polymeric Reissert compounds—polymers containing the Reissert moiety—but also those generated using the Reissert chemistry. Derivatives of these species are also included since they justify the peculiar use of this chemistry. Both approaches have been investigated predominantly by Prof. Gibson and his group. Our format thus follows that of a published overview.[1]

PREPARATION AND PROPERTIES

Reissert Compounds

The first α-acyl aminonitrile (Reissert) compound, 2-benzoyl-1,2-dihydroisoquinaldonitrile (**Structure 1**), appeared in 1905.[2] Its discoverer, Reissert, obtained it by mixing isoquinoline with benzoyl chloride and potassium cyanide using pseudo-Schotten–Baumann conditions.

Open-chain or acyclic analogs (**Structure 2**) of traditional Reissert compounds were introduced half-a-century later.[3,4]

The attractiveness of using Reissert compound chemistry in polymers arises from their ability to donate a proton, and the nucelophilicity of the conjugate base.[5–8]

Polymeric Reissert Compounds

Jois and Gibson demonstrated the high conversion achieved with the Reissert chemistry. They quantitatively condensed adipoyl chloride and twice the molar amount of TMSCN (trimethylsilyl cyanide) and isoquinoline to prepare the dimer.[9] It was therefore natural to extend this reaction to bis(isoquinoline) monomers. Polymers with M_n up to 16×10^3 were obtained.

Surprisingly, in the polymer field, acyclic polymeric Reissert compounds appeared before the heterocycle-based polymers. Voznesenskaya et al. condensed bis(α-aminonitrile)s with diacid chlorides, and thermally cyclized the resulting polymers to poly(phenylene imidazolone)s, thereby imparting aromaticity to the final product.[10]

An alternate route to acyclic polymeric (Reissert compound)s consists of the condensation of dialdehydes with diamines in the presence of a stoichiometric amount of TMSCN to generate a poly(α-aminonitrile), which was acylated *in situ* to a polymeric Reissert compound.

One of the main goals of studying these polymers was to alter nylons, unsubstituted generic analogs of polymeric Reissert compounds. Thus, we copolymerized a bis(α-aminonitrile) with a diamine and a diacid chloride to generate nylon copolymers. Differences in nucleophilicity of α-aminonitriles and amines are sufficient to allow either the formation of block copolymers or, by adding the diamine after the bis(α-amininitrile), random copolymers. These random copolymers are amorphous as are, of course, the *N*-substituted polymeric Reissert compounds, whereas the block copolymers exhibit thermal changes of both components.[11]

Note that one can directly prepare substituted polymeric Reissert compounds by use of the Reissert anion chemistry. Acylation of a bis(α-aminonitrile) affords a bis(Reissert compound), which upon treatment with a base gives a dianionic-species, a suitable nucleophilic which can react with a dihaloalkane.

REFERENCES

1. Gibson, H. W.; Pandya, A. et al. *Makromol. Chem. Macromol. Symp.* **1992**, *54–55*, 413.
2. Reissert, A. *Chem. Ber.* **1905**, *38*, 1063.
3. McEwen, W. E.; Grossi, A. V. et al. *J. Org. Chem.* **1980**, *45*, 1301.
4. Cooney, J. V. *J. Heterocyclic Chem.* **1983**, *20*, 823.
5. McEwen, W. E.; Cobb, R. L. *Chem. Rev.* **1955**, *55*, 511.
6. Popp, F. D. *Heterocycles* **1968**, *9*, 1.
7. Popp, F. D. *Heterocycles* **1980**, *24*, 187.
8. Popp, F. D. *Quinolines*; Jones, G., Ed.; John Wiley & Sons: New York, NY, 1982; Part II, pp 353–375.
9. Jois, Y. H. R.; Gibson, H. W. *Macromolecules* **1993**, *26*, 6151.
10. Voznesenskaya, N. E.; Teleshov, E. N.; Pravednikov, A. P. *Vysokomol. Soedin. Ser. B* **1977**, *19*, 634; *Chem. Abstract* **1977**, *87*, 202161k.
11. Gibson, H. W.; Jois, Y. H. R. U.S. Patent 5 194 575, 1993.

Release Agents

See: *Additives (Property and Processing Modifiers)*
Polytetrafluoroethylene
Silicone Release Coatings

Renewable Resource Polymers

See: *Polyesters (Derived from Renewable Resources)*
Starch-Based Plastics (Measurement of Biodegradability)
Vegetal Biomass (1. Monomers and their Polymerization)
Vegetal Biomass (2. Oligomers and their Polymerization)
Vegetal Biomass (3. Polymers, Derivatives and Composites)

Resin Transfer Molding

See: *Epoxy Resins (High Performance Composite Applications)*

Resins

See: *Acetal Resins*
Acetal Resins (Homopolymers Copolymers)
Acetylene-Terminated Monomers (Catalyzed Cure)
Antibacterial Resins
Benzoxazine Monomers and Polymers (New Phenolic Resins by Ring-Opening Polymerization)
Bismaleimide Resins (Modification with Engineering Plastics)
Melamine Resins (Overview)

RESINS (Applications in Coatings Industry)

Amit Mukherjee and Subrata Ghosh
Berger Paints India Ltd.

Contrary to the general belief that coatings are used for decorative purposes, the major role of surface coating is to protect the surface from environment abuses. As early as 6000 years ago, surface coatings were used to protect and preserve substrates from corrosion.

The importance of resins in coatings was known even during the prehistoric age. Gum arabic, egg whites, yolks, gelatin, beeswax, and so forth have been well documented as binders.[1]

The resin is mainly responsible for the film properties of surface coatings. With increasing demand for specific surface coatings, growth in science and technology for manufacturing resin and resin systems has been exponential.

However, in the mid-1970s, stringent regulations on the levels of volatile organic compounds (VOC) of coating systems laid the foundation for extensive research in "greener" or "environmentally friendly" coatings, for example, high solid, waterborne, powder, and radiation-curable coatings.[2–6]

PREPARATION AND PROPERTIES

Alkyds

Alkyds are the most utilized resins in the coatings industry because they offer great versatility in formulations as well as desirable properties.[7] Alkyds are by definition "oil-modified polyesters" an additional advantage since one of their major starting materials is vegetable oil, an economical natural product.

In general, alkyds are known for their excellent wetting, adhesion, flow leveling, good gloss, and compatibility with a variety of resins. However, the presence of ester linkages in their structure makes them vulnerable to humidity.

Water-borne alkyd systems have also been widely investigated.[8] Resins with molecular weight 1000–1300 are ideal for high-solid coatings.[9,10]

Polyesters

Polyesters are similar in general chemical constitution to alkyds but they do not contain vegetable oil fatty acids in their structure. The resin is termed unsaturated or saturated depending upon the presence or absence of unsaturation in the monomers used.

In general, the saturated polyesters are fast curing, have excellent stain and detergent resistance, and are usually hard and flexible. A comprehensive list of the curing agents vis-à-vis the application of the cured polyesters is given by Huber et al.[11]

The unsaturated polyester gives hard films with good chemical and solvent resistance. The unsaturation present in these systems also makes them suitable for radiation curing when used together with vinyl or acrylic diluents.[12,13]

High-solid, water-dispersible, and water-soluble polyesters are being actively researched.[14,15]

Polyester used alone finds extensive application in powder coatings; other uses follow after the hybrids (epoxy–polyesters).[24] These coatings' films have outstanding adhesion, excellent chemical resistance, and unlimited shelf life.

Acrylic Resins

Depending on molecular weight and flow characteristics, the acrylic resins are divided into two groups, the thermoplasts (mol wt 50,000 to 150,000; reflow upon heating) and the thermosets (mol wt 20,000 to 30,000).

The thermoplasts have excellent resistance to ultraviolet light and are durable, transparent, and white in color. They are applied extensively in automotive topcoats and for damage repair finishes, often in conjunction with an external plasticizer to decrease their rigidity (e.g., butyl benzyl phthalate for polymethyl methacrylate).

The low molecular weight thermosetting acrylics (mol wt < 25,000) besides having low VOCs are chemical and alkali resistant, soften less at elevated temperature, and are more compatible with other resins than their thermoplastic counterparts. However, they have comparatively inferior resistance and mechanical properties.[16]

Various water-dispersible acrylates and a new class of emulsions involving the copolymerization of methacrylics with a heavily branched monocarboxylic containing nine carbon atoms provide an answer to the VOC regulations.[17,18]

The application of acrylates in powder coating is limited because of their incompatibility with the existing epoxy and polyester systems. However, several new thermosetting acrylics should expand the horizons of their practical applicability.[19]

Amino Resins

Amino resins are the products of condensation of formaldehyde with compounds containing two or more amine groups (e.g., urea, melamine, benzoguanamine, thiourea, and glycoluril) under basic conditions, although acidic conditions are also effective for melamine.[20] These condensation products are subsequently alkylated with butanol, propanol, or methanol to impart compatibility with other resins and solubility in organic solvents.

Pure amino resins films are brittle in nature and, therefore, seldom used alone. Instead, they are used as hardening or crosslinking agents for a variety of resins such as oil-modified alkyds, epoxies, acrylics, and saturated polyesters. The excellent color and color retention, hardness, and chemical resistance of the amino resins are thereby conferred to the coating system.

Polyurethanes (PUs)

Versatile and high performance materials, the PUs are a recognized industrial standard because of their excellent durability; light stability; abrasion, corrosion and chemical resistance; low stoving temperature; and good color and gloss retention.

Recent advances in PU dispersion have made them commercially viable for wood, plastics, masonry, and metals coatings.[22,23]

Recent technology taking advantage of the useful properties of both solvent- and water-borne systems has emerged: the emulsifier (acrylic polyols) form a skin over the polyisocyanates thereby protecting them from reacting with water. These acrylic polyols subsequently act as curing agents leading to PUs with properties comparable to those of solvent-borne PUs.[21]

Phenol-Formaldehyde (PF) Resins

The resoles are self-crosslinked by organic or inorganic acids or by the application of heat, whereas the novolacs are cured in the presence of an additional crosslinking agent such as hexamethylene tetraamine. The cured PF resins give brittle films and so are made flexible by reactive plasticizers such as alkyds, epoxies, or polyamides.[25]

PF base coatings are an important constituent in corrosion preventing formulations by virtue of their low vapor and oxygen diffusion, low conductance, and low moisture absorption.

Silicones

Silicone coatings are known for their excellent thermal stability, and their abrasion, chemical, and water resistance as well as their unique flexibility and low toxicity.

Pure silicon resins are rarely used alone because of their high cost. Instead, they are either reacted (through the hydroxy or methoxy functionalities present in their backbone), or blended with polyesters (to enhance weatherability), alkyds (for drying speed, to improve adhesion, and in some cases to make the alkyd water soluble), polyimides (to increase solubility in inexpensive solvents), epoxies, vinyl ethers, or acrylics (for radiation curable formulations).

Epoxy Resins

Epoxy resins have grown from being minor in the resin family in terms of usage, to wide use as high performance materials because of their good mechanical properties, excellent resistance to corrosion and chemicals, and good insulating properties, combined with excellent adhesion to different substrates.

The low molecular weight epoxies (mol wt 340–6000) are the most widely used resin systems for powder coating because of their ideal melting point, ability to crosslink by polyaddition mechanism (often without liberation of volatiles), good electrostatic spraying characteristics, and an almost similar performance profile to that of solvent-borne epoxies.[25a] Besides these powder coating formulations, high-solid epoxies and water-borne epoxy dispersions are investigated as VOC-compliant coatings.[26–29]

APPLICATIONS

Coatings applications are numerous. From marine vessels to high-speed aviation vehicles, from domestic house walls to nuclear power plants, suitable coatings with tailor-made properties are applied.

The future of resin research in the coatings industry is well-defined but nevertheless challenging. It will mainly revolve around ecological concerns; low toxicity, recyclability, low temperature cure, microbiodegradability, and so forth. Moreover, increased attention will be given to some promising generic systems such as fluoropolymers, silicones, interpenetrating networks, and liquid crystal polymers.

REFERENCES

1. Gooch, J. W. *Lead Based Paint Handbook*; Plenum: New York, NY, 1993; p 13.
2. Pilcher, G. R. *Eur. Coat. J.* **1993**, *1–2*, 52.
3. Hill, W.; Wicks, Z. W. *Prog. Org. Coat.* **1982**, *10*, 55.
4. Demmer, C. J. *JOCCA* **1982**, *65*, 249.
5. Bodner, E. D. *Eur. Coat. J.* **1993**, *9*, 642.
6. Weisman, J. *Proc. Conf. Radiation Curing Asia* **1986**, 11.
7. Tynsall, L. A. *Calculation Techniques in the Formulation of Alkyds and Related Resins*, 2nd ed.; Middlesex Paint Research Association, 1986.
8. Naser, A. M.; Gomma, A. Z. *JOCCA* **1978**, *61*, 23.
9. Takahasi, M. *Polym. Plast. Tech. Eng.* **1980**, *15*(1), 1.
10. Holmberg, K.; Johansson, J. A. *Proc. 8th Inf. Conf. Org. Coat. Sci. Tech.* **1987**, 255.
11. Huber, W. F.; Stoye, D. *Coatings Technology Handbook*; Satas, D., Ed.; Marcel Dekker: New York, NY, 1991; p 361.
12. Raffey, C. G. *Photopolymerization of Surface Coatings*; Wiley: London, U.K., 1982.
13. Dowbenko, R. *Prog. Org. Coat.* **1983**, *11*, 71.
14. Olsen, M. R.; Laxon, J. M.; Jones, F. N. *J. Coat. Tech.* **1983**, *55*, 45.
15. Payne, K. S.; Jones, F. N.; Brandenberger, L. W. *JCT* **1985**, *57*, 35.
16. Hare, C. H. *JPCL* **1993**, *10*(8), 69.
17. Bax, J. *J. Water-borne Coat.* **1981**, *4*(2), 14.
18. Scholten, H. *Eur. Coat. J.* **1991**, *12*, 865.
19. Yousuf, N. K. *Mod. Paint Coat.* **1989**, *79*(6), 48.
20. Berge, A. *3rd Int. Conf. Org. Coat. Sci. and Tech.*, Athens, Greece, 1977.
21. Hergenrother, R.; Ruttmann, G. *JPCL* **1990**, *7*(7), 40.
22. Lomex, R. *Eur. Coat. J.* **1991**, *10*, 630.
23. Mannock, H. *Surf. Coat. Int.* **1993**, *76*, 348.
24. Browde, G. L.; Chow, S. W. *Ind. Eng. Chem. Prod. Res. Div.* **1982**, *21*, 142.
25. Marlens, C. R. *Technology of Paints, Varnishes and Lacquers*; Reinhold: NY, 1968; Chapter VI.

25a. Bate, D. *The Science of Powder Coating: Chemistry, Formulation and Application*. London, Sita, 1990, Vol. 1.
26. Chatta, M. S.; Oene, H. Vane *Ind. Eng. Chem. Prod. Res. Div.* **1982**, *21*(3), 437.
27. Roy, A. G. *Am. Paint & Coat. J.* **1993**, *77*(36), 53.
28. Ramasri, M.; Shirsalkar, M. M. *JOCCA* **1986**, *69*, 248.
29. Stewart, R. M. *Mod. Paint & Coat.* **1992**, *82*, 39.

RESINS AND PAINTS (Analytical Pyrolysis)

Hans-Rolf Schulten and Hauke Wilcken
Department of Trace Analysis
Fachhochschule Fresenius

Nonvolatility of macromolecules is one of the principal problems of polymer analysis. Therefore, analytical pyrolysis, the controlled thermal degradation in vacuum and investigation of pyrolysis products by a combination of analytical techniques, is an important method for direct characterization of technical polymers and biopolymers.[1,2] Whereas nondestructive methods like UV-, IR- and NMR-spectroscopy offer information about functional groups and structural elements, pyrolysis, in combination with mass spectrometry (Py-MS), enables us to record large sequences of the polymeric chain.[3] In addition to intact monomers and oligomers which are thermally cleaved during high temperature pyrolysis, Py-MS detects additives and non-bonded, trapped polymer subunits at low temperatures.

Our article describes the application of pyrolysis-field ionization mass spectrometry with widely used technical materials such as resins and paints. Similar products have been analyzed in the same way by Py-FIMS (pyrolysis–field ionization mass spectrometry), like acrylic and methacrylic resins, polyamides, foils, and packing materials.[4–7]

Pyrolysis-field ionization mass spectrometry is well-suited for the investigation of resins and paints. This method allows the identification of different polymer subunits, like monomers, dimers, backbone fragments, and so on. Moreover, it enables the characterization, and thereby the differentiation, of the examined compounds. According to the described results, this technique is applicable to both research problems like structure determination and industrial problems like quality control. The combination with chemometrical evaluation methods is a powerful tool for visualization and simplification of information.

REFERENCES

1. Schulten, H.-R.; Lattimer, R. P. *Mass Spectrom. Rev.* **1984**, *3*, 231.
2. Schulten, H.-R. In *Analytical Pyrolysis* Roland Jones, C. E.; Cramers, C. A., Eds.; Elsevier: Amsterdam, The Netherlands, 1977; p 17.
3. Schulten, H.-R.; Plage, B. In *Analytiker-Taschenbuch* Borsdorf, R.; Fresenius, W.; Günzler, H.; Huber, W.; Kelker, H.; Lüderwald, I.; Tölg, G.; Wisser, H., Eds.; Springer-Verlag: Berlin, 1990; Vol. 9, p 225.
4. Plage, B.; Schulten, H.-R.; Schneider, J.; Ringsdorf, H. *Macromolecules* **1990**, *23*, 3417.
5. Bahr, Y.; Lüderwald, I.; Müller, R.; Schulten, H.-R. *Angew. Makromol. Chem.* **1984**, *120*, 163.
6. Schulten, H.-R.; Gundermann, H. *Neue Verpackung* **1986**, *39*, 46.
7. Schulten, H.-R.; Plage, B.; Gundermann, H. *Neue Verpackung* **1988**, *41*, 114.

Resists

See: *Chemical Amplification Resists*
Chemically Amplified Resists (for Deep-UV Lithography)
Chemically Amplified Resists (New Generations)
Cyanoacrylates
Functional Polymaleimides, N-Protected
Novolak Resins (for Microlithography)
Polyimides (Negative-type Photosensitive Precursors)
Polysilanes (Overview)
Polysilanes (Bearing Hydroxy Groups and Derivatives)
Polystyrene and Derivatives, Photolysis
Silacyclobutanes and Related Compounds (Ring-Opening Polymerization)
Sulfur Dioxide Copolymerization

Resorbable Materials

See: *Bioadsorable Polymers (Tissue Engineering)*
Bone Cements, Bioresorbable
Hybrid Artificial Organs

Retarders

See: *Rubber Additives, Multifunctional*

Reverse Micelles

See: *Enzyme-Catalyzed Polymerization*
Reverse Micelles (Microcontainers for Functional Polymers)

REVERSE MICELLES (Microcontainers for Functional Polymers)

Alexander V. Kabanov
Department of Pharmaceutical Sciences
College of Pharmacy
University of Nebraska Medical Center

The phenomenon of solubilization and functioning of biopolymers in surfactant reverse micelles was discovered in 1952 by Hanahan who reported a retention of lipolytic enzyme function in phospholipid solutions in an organic solvent.[1] In the 80s this field attracted the attention of numerous laboratories and consequently has developed very rapidly.[10–22] Such impressive progress results from two fundamental advantages of reversed micelle systems. First, these systems provide means for easy dissolution of hydrophobic and hydrophilic substances, therefore enabling a variety of substances to be combined within the same system. Second, reverse micelles ensure a compartmentalization of solubilized macromolecules, thus permitting the construction of supramolecular structures in a controlled manner. This new development opens fundamental possibilities in engineering of supramolecular polymer systems and the design of novel polymeric materials, as demonstrated in a number of recent publications.[25–27]

POLYMER-CONTAINING REVERSE MICELLES: PREPARATION AND BASIC PROPERTIES

Polymer-Free Systems

Reverse micelles represent supramolecular aggregates spontaneously formed in organic solutions in which the polar groups of the surfactant form a hydrophilic core, while the hydrophobic tails of the surfactant are exposed to the bulk solvent. These aggregates usually can solubilize substantial amounts of water (up to several percent), which results in the formation of an aqueous pool inside reverse micelles that is covered by a monolayer of the surfactant molecules (term "water-in-oil microemulsions" is also applicable to these systems). The molecular characteristics of reverse micelles strongly depend on the chemical structure of the surfactant, the solvent, and the concentration of the components of the system. Specifically, the reverse micelles of an anionic surfactant Aerosol OT (sodium bis-(2-ethylhexyl)sulfosuccinate) have a virtually spherical shape and are characterized by a very narrow size distribution.[28–30]

Preparation of Reverse Micelle Systems

There are several methods to entrap the polymer in the reverse micelle. The first method is to inject a stock polymer solution into a three-component system containing organic solvent plus surfactant.[3] A second approach is to dissolve a dry polymer in the organic solution containing reverse micelles and required amount of water.[9] A third approach is to extract polymer in the organic phase in a two-phase system containing approximately equal volumes of the aqueous polymer solutions and the organic solvent containing the surfactant.[5] Finally, in the case of the polymers capable of complexation with the surfactant, the polymer-surfactant complexes (e.g., "Folch-Pi" proteolipids, or complexes of polyelectrolytes with oppositely charged surfactants) can be produced in water (or in another common solvent) and then dissolved in the organic solvent that may or may not contain additional amounts of surfactant, or water.[27,31] All these methods result in spontaneous entrapment of the polymer in the surfactant shell; the obtained systems being thermodynamically stable and optically transparent.

Reverse Micelles Containing Synthetic Polymers

The complexes of synthetic polycations with Aerosol OT formed in organic solvents have recently been studied.[23–27] Depending on the amount of water added to the system, various polycation-surfactant structures are formed.

Enzymatic Catalysis in Reverse Micelles

The intensive development of micellar enzymology had its true origin in 1977 with the breakthrough discovery by Martinek, Levashov, Klyachko, and Berezin that water soluble enzymes (such as α-chymotrypsin) preserve their catalytic function in the Aerosol OT reverse micelles in aliphatic hydrocarbons.[3] This was shortly followed by similar publications by Luisi, Douzou, and Menger.[6–9] Numerous enzymes of different classes have been studied in various reverse micelle systems that time.[20]

When the geometric complementarity between the enzyme and reverse micelle is achieved, the surfactant matrix tightly compresses the enzyme molecule, thereby "freezing" its rotational and vibration motions. This can be directly demonstrated using spin labels covalently attached to the active sites of the enzymes.[35–36a] Basically, this phenomenon is similar to the formation of the globules from synthetic polyions, induced by the micellar matrix.[23] However, the most unexpected and nontrivial observation is that the micellar matrix stabilizes the most active conformation of the enzyme. In many cases the superactivity of the enzyme is observed under the optimal conditions, its catalytic activity in the reverse micelles being multiple or even hundreds of times higher than that in the homogeneous aqueous solution.[18,22]

APPLICATION OF REVERSED MICELLES AS MICROCONTAINERS FOR FUNCTIONAL POLYMERS

A New Strategy for the Study of Oligomeric Enzymes

The possibility of changing in a controlled manner the oligomeric composition of the multisubunit enzymes by varying the size of the micellar matrix opens new prospects for studying oligomeric enzymes and polyenzyme complexes.[32] Kabanov et al. have demonstrated the functional activity of individual subunits of several oligomeric enzymes, which can be produced in homogeneous aqueous media only under the denaturing conditions.[37,38]

A New Way in Homogeneous Immunoassay

Reverse micelles were used as modulators of the activity of the enzyme tag in a new homogeneous immunoassay.[39] This assay is based on a principle of geometric complementarity of solubilized enzyme and micellar matrix.

Cell-free Translation In Reverse Micelles

Although the major efforts in the field under discussion have been devoted to the studies of proteins solubilized in the reverse micelle, some important results with other biopolymer molecules such as DNA, RNA, and their complexes were also obtained.

Protein Separation

Reverse micelles have been utilized for the separation and extraction of proteins. This technique, developed primarily by the group of Hatton and some other authors, is based on the phase transfer of a protein from aqueous phase to organic phase containing the reverse micelles.[40–45]

Chemical Modification of Biopolymers with Water-Insoluble Reagents

Modification of a water-soluble polymer (e.g., protein) with a water-soluble reagent is needed in a number of applications. Specifically, artificial introduction of a lipid moiety (e.g., fatty acid) into a protein molecule permits, in some cases, a significant enhancement of its biological activity.[47] However attempts to carry out such reactions in aqueous media often meet with serious obstacles.[48] Under these conditions the reaction is uncontrollable. The reagent forms an emulsion, the reaction proceeds on the reagent drop surface and cannot be stopped when the protein is modified to a low degree. Utilization of reverse micelle systems as a medium for these reactions overcomes these difficulties.[48]

Tailoring of Polymer Conjugates

Synthesis of the conjugates of natural and synthetic macromolecules is very important for the purposes of medicine and biotechnology. To obtain such conjugates a chemical reaction is usually carried out in an aqueous solution. However, these methods are connected with inherent difficulties in controlling the direction of the reaction and the degree of conversion. Specifically, at low concentrations such reactions proceed with a low yield, while in the concentrated solutions, large insoluble aggregates are usually formed because of intermolecular crosslinking. These problems have been solved by using reverse micelles as matrix microreactors for conjugation reactions.[49] This method has been applied by Kabanov and co-workers for synthesis of the protein-polyelectrolyte, as well as protein-protein conjugates.[23,25]

Polymerization in Reverse Micelles

Another application of reverse micelle systems as matrix microreactors, developed by Levashov et al., consists in utilizing them for production of nanoparticles containing immobilized enzyme molecules.[50,51]

Enzymatic Transformation of Water-Insoluble Compounds

An obvious application of enzymes solubilized in reverse micelles is to utilize them in organic synthesis. The possibility of using reversed micelles in fine organic synthesis is discussed in the first publications by Martinek et al.[3,4]

CONCLUSION

This article briefly reviews the most important concepts and results related to reverse micelles as microcontainers for functional biopolymers. For advanced reading we recommend the following reviews: general concepts of micellar enzymology, enzyme kinetics in reverse micelles, relation to membranology, biotechnological applications, and bioseparation.[2,10–20,22,33,34,46]

REFERENCES

1. Hanahan, D. J. *J. Biol. Chem.* **1952**, *195*, 199.
2. Martinek, K.; Levashov, A. V.; Khmelnitski, Yu. L.; Klyachko N. L.; Berezin, I. V. *Eur. J. Biochem.* **1986**, *155*, 453.
3. Martinek, K.; Levashov, A. V.; Klyachko, N. L.; Berezin, I. V. *Dokl. Akad. Nauk. SSSR* **1977**, *236*, 920.
4. Martinek, K.; Levashov, A. V.; Khmelnitski, Yu. L.; Klyachko, N. L.; Berezin, I. V. *Science*; Washington, DC, 1982, *218*, 889.
5. Luisi, P. L.; Henninger, F.; Joppich, M. *Biochim. Biophys. Res. Commun.* **1977**, *74*, 1384.
6. Luisi, P. L.; Bonner, F. J.; Pellegrini, A.; Wiget, P.; Wolf, R. *Helv. Chim. Acta.* **1979**, *62*, 740.
7. Balny, C.; Keh, E.; Douzou, P. *Biochem. Soc. Trans.* **1978**, *6*, 1277.
8. Balny, C.; Douzou, P. *Biochimie* **1979**, *61*, 445.
9. Menger, F. M.; Yamada, K. *J. Amer. Chem. Soc.* **1979**, *101*, 6731.
10. Khmelnitsky, Yu. L.; Levashov, A. V.; Klyachko, N. L.; Martinek, K. *Russ. Chem. Rev.* **1984**, *53*, 319.
11. Levashov, A. V.; Klyachko, N. L.; Khmelnitski, Y. L.; Martinek, K. In *Surfactants in solutions* Mittal, K. L., Ed.; Plenum: New York, 1984; Vol. 2, p 1069.
12. Luisi, P. L. *Angew. Chem. Int. Engl. Edn.* **1985**, *24*, 439.
13. Luisi, P. L.; Magid, L. *CRC Crit. Rev. Biochem.* **1986**, *20*, 409.
14. Waks, M. *Proteins* **1986**, *1*, 14.
15. Luisi, P. L.; Laane, C. *Trends Biotechnol.* **1986**, *4*, 153.
16. Shield, J. W.; Ferguson, H. D.; Bommarius, A. S.; Hatton, T. A. *Ind. Eng. Chem. Fundam.* **1986**, *25*, 603.
17. Martinek, K.; Berezin, I. V.; Khmelnitski, Yu. L.; Klyachko, N. L.; Levashov, A. V. *Coll. Czech. Chem. Commun.* **1987**, *52*, 2589.
18. Martinek, K.; Klyachko, N. L.; Kabanov, A. V.; Khmelnitski, Yu. L.; Levashov, A. V. *Biochim. Biophys. Acta.* **1989**, *981*, 161.
19. Luisi, P. L.; Giomini, M.; Pileni, M. P.; Robinsin, B. H. *Biochim. Biophys. Acta.* **1989**, *981*, 161.
20. Pileny, M. P., Ed.; *Structure and Reactivity in Reverse Micelles*; Elsevier: Amsterdam, 1989.
21. Luisi, P. L. In *Kinetics and catalysts in Microheterogeneous Systems*; Gratzel, M.; Kalyanasundaram, K., Eds.; Marcel Dekker: New York, 1991; p 115.
22. Klyachko, N. L.; Levashov, A. V.; Kabanov, A. V.; Khmelnitsky, Yu. L.; Martinek, K. In *Kinetics and catalysts in Microheterogeneous Systems*; Gratzel, M.; Kalyanasundaram, K., Eds., Marcel Dekker: New York, 1991; p 135.
23. Kabanov, A. V.; Levashov, A. V.; Khrutskaya, M. M.; Kabanov, V. A. *Macromol. Chem.* **1990**, *191*, 2801.
24. Kabanov, A. V. *Micellar Systems for Supramolecular Engineering of Biologically Active Macromolecules*; Dr. Sc. Thesis, Moscow State University: Moscow 1990.
25. Kabanov, A. V. *Makromol. Chem., Macromol. Symp.* **1991**, *44*, 253.
26. Bakeev, K. N.; Chugunov, S. A.; Larina, T. A.; MacKnight, W. J.; Zezin, A. B.; Kabanov, V. A. *Polymer Sci.* **1994**, *36*, 200.
27. Kabanov, A. V.; Sergeev, V. G.; Foster, M. S.; Kasaikin, A. V.; Levashov, A. V.; Kabanov, V. A. *Macromolecules* **1995**, *28*, 3657.
28. Fletcher, P. D. I.; Howe, A. M.; Perrins, N. M.; Robinson, B. H.; Toprakcioglu, C.; Dore, J. C. In *Surfactants in Solution*; Mittal, K. L.; Lindmann, D., Eds.; Plenum: New York, 1984; p 1745.
29. Robinson, B. H.; Toprakchioglu, C.; Dore, J. C.; Cheiux, P. *J. Chem. Soc., Lond. Faraday Trans. (1)* **1984**, *80*, 13.
30. Clarke, J. H. R.; Nicholson, J. D.; Regan, K. N. *J. Chem. Soc., Faraday Trans. (1)* **1985**, *81*, 1173.
31. Binks, B. P.; Chatenay, D.; Nicot, C.; Urbach, W.; Waks, M. *Biophys. J.* **1989**, *55*, 949.
32. Kabanov, A. V.; Nametkin, S. N.; Evtushenko, G. N.; Chernov, N. N.; Klyachko, N. L.; Levashov, A. V.; Martinek, K. *Biochim. Biophys. Acta* **1989**, *996*, 147.
33. Kabanov, A. V.; Levashov, A. V.; Klyachko, N. L.; Nametkin, S. N.; Pshezhetskii, A. V.; Martinek, K. *J. Theor. Biol.* **1988**, *133*, 327.
34. Khmelnitsky, Y. L.; Kabanov, A. V.; Klyachko, N. L.; Levashov, A. V.; Martinek, K. In *Structure and Reactivity in Reverse Micelles*; Pileny, M.p. Ed., Elsevier: Amsterdam 1989; p 230.
35. Likhtenstein, G. I.; Belonogova, O. V.; Levashov, A. V.; Klyachko, N. L.; Khmelnitsky, Yu. L.; Martinek, K. *Biokhimiya* **1983**, *48*, 379.
36. Klyachko, N. L.; Bogdanova, N. G.; Koltover, V. K.; Martinek, K.; Levashov, A. V. *Biokhimiya* **1989**, *54*, 1224.
36a. Levashov, A. V.; Klyachko, N. L.; Bogdanova, N. G.; Martinek, K. *FEBS Lett.* **1990**, *268*, 238.
37. Kabanov, A. V.; Nametkin, S. N.; Klyachko, N. L.; Levashov, A. V. *FEBS Lett.* **1991**, *278*, 143.
38. Kabanov, A. V.; Nametkin, S. N.; Chernov, N. N.; Klyachko, N. L.; Levashov, A. V. *FEBS Lett.* **1991**, *295*, 73.
39. Kabanov, A. V.; Khrutskaya, M. M.; Eremin, S. A.; Klyachko, N. L.; Levashov, A. V. *Anal. Biochem.* **1989**, *181*, 145.
40. Goklen, K. E.; Hatton, T. A. *Biotechnol. Progr.* **1985**, *1*, 69.
41. Goklen, K. E.; Hatton, T. A. *Sep. Sci. Tech.* **1987**, *22*, 831.

42. Woll, J. M.; Hatton, T. A.; Yarmush, M. L. *Biotechnol. Progr.* **1989**, *5*, 57.
43. Dekker, M.; Van Riet, K.; Weijers, S. R.; Baltussen, J. W. A.; Laane, C.; Bijsterbosch, B. H. *Chem. Eng. J.* **1986**, *33*, B27.
44. Leser, M. E.; Wei, G.; Luisi, P. L.; Maestro, M. *Biochem. Biophys. Res. Commun.* **1986**, *135*, 629.
45. Shi, H. Q.; Xia, L.; Guo, R.; Shen, Z. Y. *Chin. J. Biotechnol.* **1989**, *5*, 183.
46. Hatton, T. A. In *Surfactant-Based Separations*; Scamchorn, J. F.; Harwell, J., Eds.; Marcel Dekker; New York, 1989; p 55.
47. Kabanov, A. V.; Alakhov, V. Yu. *J. Contr. Release* **1994**, *28*, 15.
48. Kabanov, A. V.; Levashov, A. V.; Martinek, K. *Ann. N.Y. Acad. Sci.* **1987**, *501*, 63.
49. Kabanov, A. v.; Alakhov, V. Yu.; Klinskii, E. Yu.; Khrutskaya, M. M.; Rakhnyanskaya, A. A.; Polinskii, A. S.; Yaroslavov, A. A.; Severin, E. S.; Levashov, A. V.; Kabanov, V. A. *Dokl. Acad. Nauk SSSR* **1988**, *302*, 735.
50. Abakumova, E. G.; Levashov, A. V.; Berezin, I. V.; Martinek, K. *Dokl. Acad. Nauk SSSR* **1985**, *283*, 136.
51. Khmelnitsky, Yu. L.; Neverova, I. N.; Gedrovich, A. V.; Polyakov, V. A.; Levashov, A. V.; Martinek, K. *Eur. J. Biochem.* **1992**, *210*, 751.

Reverse Osmosis Membranes

See: Polybenzimidazoles (Overview)

REVERSIBLY CROSSLINKED GELS

Stephen G. Weber* and Elizabeth T. Wise
Department of Chemistry
Chevron Science Center
University of Pittsburgh

The use of gels at this time is limited to specialized applications, including foods; controlled drug release or soft contact lenses; the production of various polymers, plastics, and membranes; and chromatographic and electrophoretic methods.[1–5]

Yet there is a trend toward the study of soft, wet substances similar to those found in many biological systems. For example, a gel comprises the vitreous humor that fills the interior of the eye and the material of the cornea, as well as the fluid that lubricates the skeletal joints.[4] Furthermore, gels are flexible and can be said to move in a fashion analogous to muscle.

In order to exploit the benefits of gels, a complete understanding of their nature and behavior is necessary. Ferry defined a gel as a "substantially diluted system which exhibits no steady state flow".[6] Because a gel is a physical state intermediate between a solid and a liquid, it has unique characteristics that are difficult to define. The two components of a gel are a liquid and a network of polymer molecules that prevents the liquid from flowing away and thus gives the gel its solidity. The gel, therefore, has viscous *and* elastic properties.

Although similarities between covalent and reversibly crosslinked gels exist, covalent crosslinks can be regarded as permanent; in reversible gels, the crosslinking interactions can be defined in terms on an equilibrium between the bonded and nonbonded states. Reversible gels can be formed and destroyed because a change in conditions (temperature, pH, solvent, ionic strength) can shift the equilibrium responsible for the crosslinking.

*Author to whom correspondence should be addressed.

TYPES OF CROSSLINKS

Separated Domains

Gelation accompanying crystallization from dilute or moderately dilute solutions was recognized some time ago.[7,8] Here, crystallites bind molecules together to form a three-dimensional network. The aggregated crystallites that act as tie points represent one region having particular properties, and the unassociated portion of the polymer chains represent another. Actually, gelation is observed for both lamellar (thin plate) and micellar (aggregation of molecules) junctions, and several systems exhibit polymorphic crystalline structures.[10] The crystals, formed on cooling, dissolve on heating. This destruction of the network is reversible; thus, the gels are referred to as thermoreversible gels.

Many studies have dealt with the gelation of linear weakly polar or nonpolar polymers in organic solvents resulting from crystallization.[7–16] Such polymers as polystyrene, poly(vinyl chloride), poly(vinyl alcohol), and poly(4-methylpentene) produce gels.[17–21] This phenomenon is also observed in the natural polymer, seaweed polysaccharide alginate.[22] Homopolymers of ethylene, ethylene-hydrogenated polybutadiene copolymers, and ethylene-vinyl acetate copolymers have also been investigated.[10]

Gelation in many polymers, including agarose, carrageenan, and gelatin, occurs after a transition from a disordered state (random coil) to an ordered helix conformation upon cooling.[23–29] Experimental conditions, such as temperature, solvent, or the presence of a salt, dictate the formation of a gel.

Complexation

Crosslinking may take place by specific interactions such as hydrogen bonds or hydrophobic interactions.

An example of work done on systems stabilized through hydrogen bonds involves polyacid-polybase couples (no proton transfer).[30] The polybase is one of the homopolymers, polyoxyethylene, poly(vinyl methyl ether), or polyvinylpyrrolidone, and the polyacid is poly(acrylic acid).

Aqueous solutions containing surfactants and water-soluble polymers have been the subject of a considerable number of studies.[31–35] When the surfactant concentration is high enough, two distinct macroscopic phases—a water-rich solution and a polymer-rich hydrogel—separate. Both regions contain hydrophobic microphases made up of surfactant molecules and polymer side chains.

Complexation may occur between metallic cations, such as Cu^{2+} or Cr^{3+}, or basic anions, such as $B(OH)_4^-$ or $Sb(OH)_6^-$, and synthetic polymers, including poly(vinyl alcohol) or polyacrylamide, or natural polymers, such as polysaccharides or proteins.[36–41] The formation of complexes that act as tie points governs the physical properties of the system, such as gelation, demixing, or rheological behavior. The strength and lifetime of these crosslinks can be controlled by the complexing ion, as well as by external conditions such as temperature, pH, and ionic strength.

Ion Pairs

Several reviews focus on a different type of nonbonding interaction, which is represented by ionomers.[42–47] It is suggested that ion pairs form and act as strong electric dipoles that interact to form multiples.[48] The multiplets then form aggregates that include both ionic and nonionic material.

EXPERIMENTAL TECHNIQUES

As is true for other colloidal/polymer solutions, direct visualization of gel structure is difficult. Indirect methods that include spectroscopy, scattering, and rheology are the predominant techniques employed in the study of gels. Yet the potential to "see" these microstructures through the application of newly developed methods such as scanning tunnelling microscopy may help elucidate the complex nature of reversibly crosslinked gels.

THEORETICAL MODELS

Because gels are so elusive, it is particularly difficult to develop theoretical models that explain and predict their behavior. In addition, the large variety of mechanisms that can result in gelation inhibits the goal of obtaining a general understanding of reversible gels. Some models originate from polymer theory, while others come from the theory of chemical equilibrium.

APPLICATIONS

Thermoreversible and pH-Sensitive Gels

Water-containing gels, or hydrogels, can be made sensitive to temperature and pH.[4,9,39,49–61] Such gels can be useful in the delivery of drugs because a change in the pore size of the gel network is induced by environmental changes.[2,56,62] Separation techniques such as chromatography or electrophoresis can benefit from the use of such polymers (e.g., guar, cellulose, and polyacrylamides) crosslinked by inorganic ions are numerous in the oil and gas industries.[26,40,58] Moreover, gelatin foods are made of hydrogels that "set" when chilled.[1]

Photosensitive and Electroactive Gels

The size of pores in gel membranes, and the size of particles of solvent-swollen gels can be controlled electrically.[56] Theoretically, a group of gel particles of 1-μm diameter could shrink to 4% of its original volume within 1 msec under an electric field of 5 V/mm. The corresponding change in pore size can be manipulated for use as a chemical valve or drug delivery system, and the rapid response could be suitable for use in an artificial muscle or robotic device.

Chemically Sensitive Gels

Relatively fewer systems are responsive to chemical species, rather than physical phenomena. Tanaka recently found that a lectin-loaded polymer gel reversibly swells and shrinks in the presence of different saccharides.[63] Another swollen gel, observed by Cussler, preferentially absorbs water while excluding substances dissolved in it.[64] At this time, applications for these systems are only envisioned.

FUTURE TRENDS

There appear to be three areas where further research will be directed: improvements in experimental techniques, refinement of quantitative models, and the development of chemically sensitive smart materials. There is a desire to "see" gel networks in real time, and there is a need to predict the behavior of these systems through reliable models.

Smart materials are systems that both sense *and* actuate and, therefore, mimic characteristics of living systems. There are numerous gels that respond to physical and chemical stimuli, often based on macroscopic, and relatively nonspecific, electrical interactions. Materials that operate on the basis of *molecular recognition* and competitive binding and thus are *specific* for a particular compound or class of compounds are much less studied.[63,66] A quantitative understanding of these systems will make it possible to predict material properties based on the concentration of the material's components and the concentration of the agonist and, thus, to apply these systems to particular needs.

REFERENCES

1. Rutenberg, M. W. In *Handbook of Water Soluble Gums and Resins*; Davidson, R. L., Ed.; McGraw-Hill: New York, 1980; Chapter 22.
2. Peppas, N. A. In *Medical Applications of Controlled Release*; Langer, R. S.; Wise, D. L., Eds.; CRC: Boca Raton, FL, 1984; Vol. 2, Chapter 10.
3. Kydonieus, A. F.; Smith, I. K.; Beroza, M. In *Controlled Release Polymeric Formulations*; Paul, D. R.; Harris, F. W., Eds.; ACS Symposium Series 33; American Chemical Society: Washington, DC, 1976; Chapter 24.
4. Tanaka, T. *Scientific American* **1981**, *244*(1), 124.
5. Cottrell, I. W.; Kovacs, P. In *Handbook of Water Soluble Gums and Resins*; Davidson, R. L., Ed.; McGraw-Hill: New York, 1980; Chapter 2.
6. Ferry, J. D. *Viscoelastic Properties of Polymers*: John Wiley & Sons: New York, 1980; p 529.
7. Flory, P. J. *Discuss. Faraday Soc.* **1974**, *57*, 7.
8. Mandelkern, L. *Crystallization of Polymers*; McGraw-Hill: New York, 1964; pp 113 and 308ff.
9. Keller, A. In *Structure-Property Relationships of Polymeric Solids* Hiltner, A., Ed.; Plenum: New York, 1983; p 29.
10. Domszy, R. C.; Alamo, R.; Edwards, C. O.; Mandelkern, L. *Macromolecules* **1986**, *19*, 310.
11. Takahashi, A.; Sakai, M.; Kato, T. *Polym. J.* **1980**, *12*, 335.
12. Mandelkern, L. *Discuss. Faraday Soc.* **1979**, *68*, 685.
13. Smith, P.; Lemstra, P. J.; Booig, H. C. *J. Polym. Sci., Polym. Phys. Ed.* **1981**, *19*, 877.
14. Lemstra, P. J.; Smith, P. *Br. Polym. J.* **1980**, 212.
15. Edwards, C. O.; Mandelkern, L. *J. Polym. Sci., Polym. Lett. Ed.* **1982**, *20*, 355.
16. Mandelkern, L.; Edwards, C. O.; Domszy, R. C.; Davidson, M. W. In *Microdomains in Polymer Solution*; Dubin, P., Ed.; Plenum: New York, 1985.
17. Lemstra, P. J.; Challa, G. *J. Polym. Sci., Polym. Phys. Ed.* **1975**, *13*, 1809.
18. Guenet, J. M.; Lotz, B.; Wittman, J. C. *Macromolecules* **1985**, *18*, 420.
19. Guerrero, S. J.; Keller, A.; Soni, P. L.; Geil, P. H. *J. Macromol. Sci., Phys.* **1981**, *B20*, 161.
20. Ohkura, M.; Kanaya, T.; Kaji, K. *Polymer* **1992**, *33*(17), 3686.
21. Charlet, G.; Phuong-Nguyen, H.; Delmas, G. *Macromolecules* **1984**, *17*, 1200.
22. Atkins, E. D. T.; Nieduszynski, I. A.; Mackie, W.; Parker, K. D.; Smolko, E. E. *Biopolymers* **1973**, *12*, 1865.

23. Clark, A. H.; Ross-Murphy, S. B. *Adv. Polym. Sci.* **1987**, *83*, 57.
24. Nishinari, K.; Watase, M.; Kohyama, K.; Nishinari, N.; Oakenfull, D.; Koide, S.; Ogino, K.; Williams, P. A.; Phillips, G. O. *Polym. J.* **1992**, *24*(9), 871.
25. Viebke, C.; Piculell, L.; Nilsson, S. *Macromolecules* **1994**, *27*, 4160.
26. Djabourov, M.; Leblond, J.; Papon, P. *J. Phys. Fr.* **1988**, *49*, 319.
27. Shukla, P. *Polymer* **1992**, *33*, 365.
28. Morris, E. R.; Rees, D. A.; Robinson, G. *J. Mol. Biol.* **1980**, *138*, 349.
29. Higgs, P. G.; Ball, R. C. *J. Phys. Fr.* **1989**, *50*, 3285.
30. Iliopoulos, I.; Audebert, R.; Quivoron, C. In *Reversible Polymeric Gels and Related Systems*; Russo, P. S., Ed.; ACS Symposium Series 350; American Chemical Society: Washington, DC, 1987; Chapter 6.
31. Dualeh, A. J.; Steiner, C. A. *Macromolecules* **1990**, *23*, 251.
32. Cabane, B.; Duplessix, R. *Colloids Surf.* **1985**, *13*, 19.
33. Turro, N.; Baretz, B. H.; Kuo, P.-L. *Macromolecules* **1984**, *17*, 1321.
34. Shih, L. B.; Sheu, E. Y.; Chen, S. H. *Macromolecules* **1988**, *21*, 1387.
35. Landoll, L. M. *J. Polym. Sci., Polym. Chem. Ed.* **1982**, *20*, 443.
36. Tsuchida, E.; Nishide, H. *Adv. Polym. Sci.* **1977**, *24*, 1.
37. Menjivar, J. A. In *Water Soluble Polymers*; Glass, L., Ed.; American Chemical Society: Washington, DC, 1986; p 209.
38. Pezron, E.; Ricard, A.; Lafuma, F.; Audebert, R. *Macromolecules* **1988**, *21*, 1121.
39. Pezron, E.; Leibler, L.; Ricard, A.; Audebert, R. *Macromolecules* **1988**, *21*, 1126.
40. Prud'homme, R. K.; Uhl, J. T.; Poinsalte, J. P.; Halverson, F. *Soc. Petrol. Eng. J.* **1983**, *23*, 804.
41. Ochiai, H.; Kurita, Y.; Murakami, I. *Makromol. Chem.* **1984**, *185*, 167.
42. Eisenberg, A.; King, M. *Ion Containing Polymers*; Academic: New York, 1978.
43. MacKnight, W. J.; Earnest, T. R. *Polym. Sci. Macromol. Rev.* **1981**, *16*, 41.
44. *Ionic Polymers*; Holiday, C., Ed.; Applied Science: London, 1975.
45. Eisenberg, A. *Ions in Polymers* Advances in Chemistry Series 187; American Chemical Society: Washington, DC, 1980.
46. *Developments in Ion Containing Polymers* Wilson, A. D.; Prosser, H. J., Eds.; Applied Science: London, 1983.
47. *Coulombic Interactions in Macromolecular Systems* Eisenberg, A.; Bailey, F. E., Eds.; ACS Symposium Series 302; American Chemical Society: Washington, DC, 1986.
48. Dreyfuss, B. *Macromolecules* **1985**, *18*, 284.
49. Djabourov, M.; Leblond, J. In *Reversible Polymeric Gels and Related Systems*; Russo, P. S., Ed.; ACS Symposium Series 350; American Chemical Society: Washington, DC, 1987; Chapter 14.
50. Chujo, Y.; Sada, K.; Saegusa, T. *Macromolecules* **1993**, *26*(24), 6315.
51. Chattopadhyay, A.; Bhattacharya, D.; Singh, R. P. *Materials Lett.* **1993**, *17*, 179.
52. Sato, T.; Tsujii, Y.; Fukuda, T.; Miyamoto, T. *Macromolecules* **1992**, *25*, 5970.
53. Kesavan, S.; Prud'homme, R. K. *Macromolecules* **1992**, *25*, 2026.
54. Coviello, T.; Burchard, W. *Macromolecules* **1992**, *25*, 1011.
55. Leibler, L.; Pezron, E.; Pincus, P. A. *Polymer* **1988**, *29*, 1105.
56. Osada, Y.; Ross-Murphy, S. B. *Scientific American* 268, **1993**, 82.
57. Sheppard, N. F. Jr.; Tucker, R. C.; Salehi-Had, S. *Sensors and Actuators B.* **1993**, *10*, 73.
58. Chatterji, J.; Borchardt, J. K. *J. Petrol. Tech.* **1981**, *33*, 2042.
59. Nakamura, K.; Itoh, T.; Sakurai, M.; Nakagawa, T. *Polym. J.* **1992**, *24*(12), 1419.
60. Kato, T.; Yokoyama, M.; Takahashi, A. *Colloid & Polym. Sci.* **1978**, *256*, 15.
61. Haas, H. C.; MacDonald, R. L.; Schuler, A. N. *J. Polym. Sci., Part A-1* **1971**, *9*, 959.
62. Hoffman, A. S.; Afrassiabi, A.; Dong, L. C. *J. Controlled Release* **1986**, *4*, 213.
63. Kokufata, E.; Zhang, Y.; Tanaka, T. *Nature (London)* **1991**, *351*, 302.
64. Cussler, E. L.; Stokar, M. R.; Vaarberg, J. E. *AIChE J.* **1984**, *30*, 578.
65. Ricka, J.; Tanaka, T. *Macromolecules* **1985**, *18*, 83.
66. Hilger, C.; Stadler, R. *Macromolecules* **1992**, *25*, 6670.

Rhamnogalacturonans

See: Hemicelluloses

Rhamsan Gum

See: Welan Microbial Polysaccharide

Ribonucleic Acid

See: RNA

RIGID POLYURETHANE FOAMS

Henri J. M. Grünbauer
Urethane Polymers Research and Development
Dow Benelux NV

Urethane foams represent a highly versatile class of expanded thermoset materials that find application in a wide range of industrial products such as carpet backings, mattresses, insulation panels, and refrigerators. Within this spectrum, closed-cell rigid foams are an important sub-class owing to their outstanding thermal insulation properties when expanded with suitable blowing agents of low thermal conductivity.[1–3] It is precisely this property that is presently endangered by the worldwide decision to abolish the use of ozone-depleting "hard" CFC-11 as the blowing agent of choice for thermal insulation foam.[4,5] At this very moment, polyurethane industries are going through a considerable effort to replace CFC-11 by non-ozone depleting (ODP) blowing agents that, preferably, exhibit low global warming potentials (GWP) as well. Against this background, an overview of the properties of rigid polyurethane foams focuses almost inevitably on what is known as "the CFC issue."

RAW MATERIALS

Isocyanates

Essential to the formation of urethane polymers is the presence of molecules bearing at least two highly reactive isocyanate groups. For rigid foams, mixtures of monomeric 4,4′-diisocyanato diphenyl methane or MDI and higher oligomers or PMDI are commonly used.

Polyols

Isocyanates react with any molecular entity possessing active hydrogens, such as hydroxylic, carboxylic, and amine groups. For rigid foams, primary and secondary hydroxyl groups containing polyether polyols are most important, followed by polyesters, which have the oldest application.

Additives

A variety of different additives is usually present in small quantities on the polyol side. Among these, catalysts and surfactants deserve special attention because they are indispensible to a proper control of the foaming process itself. The role of catalysts is to balance expansion and gelation of the foam in such a way that foam collapse—that is, excessive blowing as well as vitrification and too fast polymerization—can be avoided. Catalyst packages for rigid foams therefore usually contain more than one catalyst. A very large number of catalysts for rigid foams belong to the chemical class of tertiary amines.[6]

Polydimethylsiloxane derivatives are the surfactants of choice to control cell formation in rigids.[7] The presence of these so-called silicone surfactants serves a twofold purpose. First of all, they stabilize the foam after mixing polyol and isocyanate by lowering the surface tension of the gas liquid interface. Mechanistically, this effect arises from a preferred *accumulation* of surfactant molecules at the interface. A second, equally important role of surfactants is to stabilize the polymerizing liquid–gas interface during the roughly 30-fold volume increase of a rising foam.[8] Here, the mechanism is rather dynamic: the expanding foam continuously creates new surface area of high tension that needs to be stabilized by fast *migration* of surfactant toward the interface (Marangoni effect).

Blowing Agents

A significant portion of the excellent thermal insulation properties of traditional rigid PU foam originates from the very low thermal insulation value of CFC-11 vapor. Among the alternatives with similar boiling points, HCFC-141b has since long been recognized as a viable substitute for CFC-11 owing to its low ODP and GWP values in spite of a slightly higher thermal conductivity value.[9,10] The presence of chlorine in this type of molecular as well as the residual ODP value, however, limits their expected lifetime as CFC alternatives. Volatile hydrocarbons such as pentane are attractive alternatives too, provided that safe processing and handling conditions for this type of flammable blowing agent are implemented.[11] Significant processing modifications are also required for the use of low boiling alternatives, such as HFC-134A.[12] Perfluorinated hydrocarbons are a special class of molecule that is exemplified by perfluorohexane. Foams with much smaller cell sizes have been produced by using this type of compound in combination with specifically designed polyol formulations.[13]

Carbon dioxide represents a special case because of its "reactive blowing" capabilities. CO_2 can be generated by reacting isocyanates with, for example, carboxilic acids or water present as polyol constituents.[3,14]

CONCLUSION

Inherent to the very nature of rigid PU foam formation is the requirement of a proper balance between properties and actions of all ingredients present in the formulation. By replacing only one of these ingredients—the blowing agent—it is soon realized that all ingredients need to be adapted to the new situation. In other words, simple drop-in substitutes for CFC-11 do not exist. Instead, the CFC issue has created a need for fundamental research on all aspects of urethane-foaming in relation to the ultimate foam properties. It is through this type of research that development of innovative zero ODP/GWP formulations beyond present technology standards can be expected.

REFERENCES

1. Oertel, G., Ed., *Polyurethane Handbook*; 1st ed., Carl Hanser Verlag: Munich, 1985.
2. Woods, G. *The ICI Polyurethanes Book*; John Wiley & Sons: New York, 1987.
3. Gum, W. F.; Riese, W.; Ulrich, H. Eds.; *Reaction Polymers, Chemistry, Technology, Applications, Markets*; 1st ed., Carl Hanser Verlag: Munich, 1992.
4. Rowland, F. S. *Annu. Rev. Phys. Chem.* **1991**, *42*, 731.
5. Tominaga, T. *Pure & Appl. Chem.* **1992**, *64*, 529.
6. Malwitz, N.; Wong, S.-W.; Frisch, K. C.; Manis, P. A. *J. Cellular Plastics* **1987**, *23*, 461.
7. Grüning, B.; Koerner, G. *Tenside Surf. Det.* **1989**, *26*, 313.
8. Vincent, B. *In Surfactants*; Tadros, T. F., Ed.; Academic: London, 1984.
9. Jeffs, G. M. F.; Sparrow, D. J. *Proc. Utech '90 Conf. 31* 1990.
10. Dwyer, F. J.; Parker, R. C.; Thrun, K. M.; Zwolinski, L. M. *Proc. SPI/ISOPA Polyurethanes World Congress* 1991, 752.
11. Lerch, B. P.; Taubenmann, P. *Proc. SPI/ISOPA Polyurethanes World Congress* 1991, 160.
12. Grünbauer, H. J. M.; Broos, J. A. F.; Thoen, J. A.; Smits, G. F.; Lehnert, A. B.; Hoenke, M. S. *J. Reinforced Plastics and Composites* **1994**, *13*, 361.
13. Volkert, O. *Proc. SPI/ISOPA Polyurethanes World Congress* 1991, 740.
14. Grünbauer, H. J. M.; Thoen, J. A.; Folmer, J. C. W.; van Lieshout, H. C. *J. Cellular Plastics* **1992**, *28*, 37.

Rigid Poly(vinyl chloride)

See: Poly(vinyl chloride), Rigid (Soft Fillers)

RIGID-ROD POLYBENZOXALES AND POLYBENZOTHIAZOLES (PBZT and PBO)

Satish Kumar
School of Textile and Fiber Engineering
Georgia Institute of Technology

Rigid-rod polymers that have received the most attention are poly(p-phenylene benzobisthiazole) (PBZT) and poly(p-phenylene benzobisoxazole) (PBO). An economic synthesis route for PBO monomer may make it possible for the PBO fiber to be commercially viable.[2,3] Poly(2,5-benzoxazole) (ABPBO) and poly(2,6-benzothiazole) (ABPBZT) are semirigid ordered polymers. They also have high thermal stability, but their chain stiffness is much lower compared with rigid rod PBZT and PBO.

The synthesis routes of trans-PBZT and cis-PBO are shown in **Figures 1** and **2**. In both cases, condensation polymerization is carried out in polyphosphoric acid. These systems exhibit lyotropic liquid crystalline behavior. The monomer synthesis of DABDO and DABDT was developed by Wolfe.[1] A relatively more economic synthesis for PBO monomer, 4,6-diaminoresorcinol, has

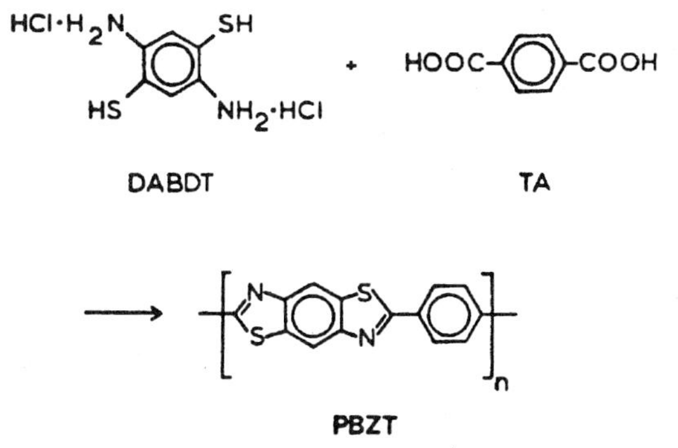

FIGURE 1. Synthesis of PBZT.

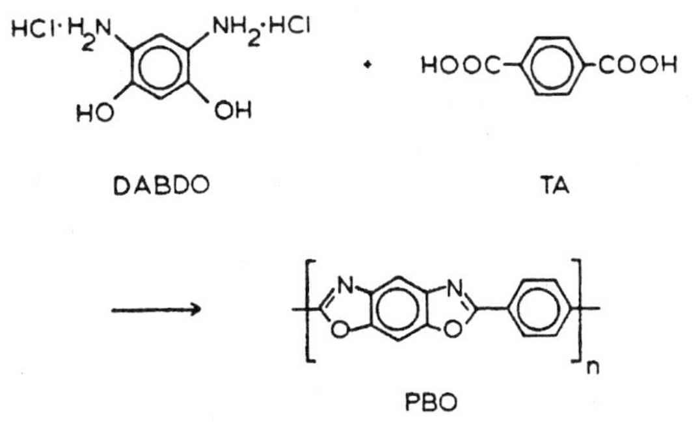

FIGURE 2. Synthesis of PBO.

been reported by Lysenko.[2] Ion–containing water-soluble rigid-rod polymers also are being synthesized.[4] Synthesis of modified PBZT exhibiting thermotropic liquid crystalline behavior recently has been reported by Kimura et al.[5]

PBZT and PBO are soluble in strong protic acids such as poly(phosphoric acid) (PPA), methanesulfonic acid (MSA) and chlorosulfonic acid (CSA).[6–8] These polymers now also have been found to dissolve in nitroalkanes (e.g., nitromethane) where solubilization is mediated via a Lewis acid (e.g., $AlCl_3$).[9–11] Both PBZT and PBO are characterized with persistent lengths of greater than 50 nm.[7,12–14] The high axial ratio of individual rigid-rod molecules of PBO and PBZT results in low critical concentration (about 5 wt%) for forming the nematic liquid crystalline phase.

Fiber spinning of rigid-rod polymers from solutions is done by the dry-wet spinning technique.

Structure formation during fiber spinning has been studied.[15–17] A network of oriented microfibrils is formed during coagulation of PBZT and PBO fibers.

Thermal stability of the PBZT, ABPBZT, and ABPBO is comparable to that of the thermal stability of the PBO, both in inert atmosphere and in air. PBZT and other ordered polymers are among the most radiation resistant.[18] Critical oxygen concentration (COC) is the minimum oxygen concentration for onset of combustion. A material with COC greater than 32.5 in the top and 21 in the bottom ignition mode is intrinsically noncombustible. The COC values for PBZT and PBO indicate that both these materials are intrinsically noncombustible.[19]

Both PBO and PBZT fibers have negative coefficients of thermal expansion along the fiber axis. The theoretical tensile modulus for cis-PBO fiber of ideal orientation is 690 GPa and that for trans-PBZT it is 620 GPa.[20]

Ordered polymers also are being investigated for nonlinear optical properties.[21,22] A χ^3 value of 4.0×10^{-10} esu (wavelength 602 nm) is reported for PBZT. The χ^3 value for PBO is comparable to the value of PBZT.[22] Other modified ordered polymers also have been made for improved nonlinear optical properties.[21] One restriction on the use of rigid-rod polymers is the ability to produce optical quality film. Progress made in this area has been reported recently by Lee and Wang.[23]

REFERENCES

1. Wolfe, J. F.; Sybert, P. D.; Sybert, J. R. U.S. Patent 4 533 692, U.S. Patent 4 533 693, and U.S. Patent 4 533 724, 1985 to SRI International.
2. Lysenko, Z. U.S. Patent 4 766 244, 1988, to Dow Chemical Co.
3. Ledbetter, H. D.; Rosenberg, S.; Hurtig, C. W. in Reference 7, p 253.
4. Dang, T. D.; Bai, S. J. et al. *J. Polymer Sci.* (Phys.) **1993**, *31*, 1941.
5. Kimura, K.; Meurer, D. L. et al. *Macromolecules* **1994**, *27*, 1303.
6. Berry, G. C.; Adams, W. W.; Eby, R. K.; McLemore, D. E., Eds.; *The Materials Science and Engineering of Rigid-rod Polymers* Materials Research Society Symposium Proc., Pittsburgh, PA, 1989; Vol. 134, p 181.
7. Wong, C.-P.; Ohnuma, H.; Berry, G. C. *J. Polym. Sci., Polym. Symposium* **1978**, *65*, 173.
8. Cotts, D. B.; Berry, G. C. *Macromolecules* **1981**, *14*, 930.
9. Venkatasubramanian, N.; Polk, M. B. et al. *J. Polymer Sci., Phys ed.* **1993**, *31*, 1965.
10. Connolly, J. W.; Dudis, D. S. et al., submitted to *Chemistry of Materials.*
11. Jenekhe, S. A.; Johnson, P. O. *Macromolecules* **1990**, *23*, 4419.
12. Aharoni, S. M. *Macromolecules* **1985**, *18*, 2624.
13. Crosby III, C. R.; Ford, N. C. Jr. et al. *J. Chem. Phys.* **1981**, *75*, 4298.
14. Arpin, M.; Strazielle, C. *Polymer* **1977**, *18*, 591.
15. Cohen, Y.; Thomas, E. L. *Polym. Engr. and Sci.* **1985**, *25*, 1093.
16. Cohen, Y.; Thomas, E. L. *Macromolecules* **1988**, *21*, 433.
17. Radler, M. J.; Landes, B. G. et al. *J. Polym. Sci* (Phys.) **1994**, *32*, 2567.
18. Kumar, S.; Adams, W. W. *Polymer* **1990**, *31*, 15.
19. Uy, W. C.; Mammone, J. F. *Canadian Textile Journal* **1988**, 54.
20. Wierschke, S. G.; Shoemaker, J. R. et al. *Polymer* **1992**, *33*, 3357.
21. Reinhardt, B. A. *Trends in Polymer Sci.* **1993**, *1*, 4.
22. Prasad, P. N. in Adams, W. W.; Eby, R. K.; McLemore, D. E., Eds.; *The Materials Science and Engineering of Rigid-rod Polymers* Materials Research Society Symposium Proc., Pittsburgh, PA, 1989; Vol. 134, p 635.
23. Lee, J. W.; Wang, C. S. *Polymer* **1994**, *35*, 3673.

Rigid-Rod Polymers

See: Molecular Composites (High Modulus Polyimide/Polyimide Films)
Molecular Composites (Liquid Crystalline Polyamide and Amorphous Polyimide)
Molecular Composites (Polyimide/Polyimide, Laminate Processability)
Molecular Composites (Third Generation Polymers)
Nonlinear Optical Polymers [Poly(arylene ethynylene) Derivatives]
Organometallic Polymers, Fluoroarylene Bridged (Rigid Rods to Segmented Chains)

Polyelectrolytes, Rigid-Rod [Poly(p-phenylene)-Based]
Polyimides, Soluble Rigid-Rods (via Pd-Catalyzed Polycondensation)
Polyisocyanides
Poly(phenylene ethynylene) (Synthesis and Optical Properties)
Poly[N,N′-(sulfo-p-phenylene)terephthalamide]
[1.1.1]Propellanes (New Vistas in their Polymer Chemistry)
Rigid-Rod Polybenzoxazoles and Polybenzothiazoles (PBZT and PBO)
Rigid-Rod Poly(p-phenylene benzobisthiazole)
Whiskers of Rigid-Rod Polymers

RIGID-ROD POLY(p-PHENYLENE BENZOBISTHIAZOLE)

Hyun Hoon Song
Department of Macromolecular Science
Han Nam University

My Dotrong
University of Dayton Research Institute

R. C. Evers
Wright Laboratory
Wright-Patterson Air Force Base

In the mid-1970s, the U.S. Air Force Materials Laboratory initiated the Ordered Polymer Program to develop high performance structural polymers. Following the initial synthesis of high molecular weight rigid-rod poly(p-phenylene benzobisthiazole) (PBT), subsequent research efforts established the principles for the synthesis, processing, and properties of PBT.[1] This polymer with a rigid molecular structure and an ordered ultrastructure represented a new class of materials with environmental stability, high strength, and modulus, as well as the potential for optoelectronic applications.

PREPARATION AND PROPERTIES

Synthesis

High molecular weight PBT was synthesized through polycondensation in poly(phosphoric acid) of terephthalic acid with highly purified 2,5-diamino-1,4-benzenedithiol dihydrochloride.

Solution Properties and Processing

PBT is intractable and is soluble only in strong protic acids such as PPA or MSA. Lewis acids in aprotic organic solvents also solubilize the polymer by forming complexes of the rigid chain backbone and the Lewis acid in the organic solvent.[2–4] The intrinsic viscosity of PBT in MSA depends on molecular weight to the power of 1.8, thus indicating that the molecules behave very much like rigid-rods in dilute solution.[5] Dilute solutions of PBT are isotropic, but as the polymer concentration increases to greater than 5 weight percent in MSA, the solutions become anisotropic and exhibit stir opalescence.[6] As the polymer concentration is increased beyond this critical value, the solution viscosity passes through a maximum and reaches a limiting value. This behavior is indicative of the lyotropic liquid crystalline nematic nature of the PBT solutions, which leads to the outstanding tensile properties of fibers and films spun from these solutions.

Fibers are usually prepared from the anisotropic solution by dry-jet wet spinning techniques.[7]

Thermal Properties

Owing to the aromatic-heterocyclic nature of PBT, it exhibits excellent thermal and thermooxidative stability, which is among the highest of any organic polymers. No melting or glass transition is observed up to the point of thermal degradation. The onset of weight loss in air under thermogravimetric analysis occurred in the range of 600°C with an extrapolated temperature of degradation at 620°C.

Mechanical Properties

Fibers spun from anisotropic PBT solutions have a microfibrillar structure with the chains highly oriented to the fiber axis, which results in excellent tensile strengths and moduli.[8] The specific tensile modulus and strength of PBT are among the highest recorded for organic fibers.

Even though PBT fibers have outstanding tensile properties, they suffer from relatively poor axial compressive strength, which severely limits their range of application.

Optical and Electrical Properties

The PBT molecular structure with extended π conjugation suggests an extensive electron delocalization. Efforts have recently focused in the elucidation of the electrical and optical behaviors of PBT. The band gap calculated for PBT in a coplanar conformation is 1.73 eV, which is very close to the corresponding values reported for polyacetylene.[9,10] The theoretical calculation also predicts that the band gaps increase markedly as the nonplanarity increases, which may well explain the low conductivity of $\sim 10^{12}$ S cm^{-1} in an undoped PBT.[11] The conductivity, however, increases up to 20 S cm^{-1} after electrochemical doping.[12]

PBT shows strong nonlinear optical effects. Since the polymer is centrosymmetric, the nonlinear optical response is third order. Rao et al. have reported $\chi^{(3)} = \sim 10^{-11}$ esu at both 585 and 605 nm with subpicosecond response time.[13] The PBT film exhibits optical anisotropy with birefringence as large as 0.5.[14,15] The anisotropy of $\chi^{(3)}$ was also observed in a biaxially oriented PBT film that yielded values from 5.4×10^{-12} to 7.2×10^{-12} esu depending on the orientation of the film.[13]

REFERENCES

1. Wolfe, J. F.; Loo, B. H.; Arnold, F. E. *Macromolecules* **1981**, *14*, 915.
2. Jenekhe, S. A.; Johnson, P. O. *Polym. Mater. Sci. Eng.* **1989**, *60*, 502.
3. Jenekhe, S. A.; Johnson, P. O.; Agrawal, A. K. *Macromolecules* **1989**, *22*, 3216.
4. Jenekhe, S. A.; Johnson, P. O. *Macromolecules* **1990**, *23*, 4419.
5. Lee, C. C.; Chu, S. G.; Berry, G. C. *J. Polym. Sci., Polym. Phys. Ed.* **1983**, *21*, 1573.
6. Wolfe, J. F.; Loo, B. H.; Arnold, F. E. *Macromolecules* **1981**, *14*, 915.
7. Choe, E. N.; Kim, S. N. *Macromolecules* **1981**, *14*, 920.
8. Allen, S. R.; Fillipov, A. G.; Farris, R. J.; Thomas, E. L.; Wong, C.; Berry, G. C.; Chenevey, E. C. *Macromolecules* **1981**, *14*, 1135.

9. Bhaumik, D.; Mark, J. E. *J. Polym. Sci., Polym. Phys. Ed.* **1983**, *21*, 1111, 2543.
10. "Symposium on Conducting Polymers," *ACS Polym. Preprints 23*, **1982.**
11. Baker, R. E.; Chen, D. Y. *Annu. Rep., Conf. Electr. Insul. Dielectr. Phenom.* **1981**, 351.
12. DePra, A.; Gaudiello, J. G.; Marks, T. J. *Macromolecules* **1988**, *21*, 2295.
13. Rao, D. N.; Swiatkiecz, J.; Chopra, P.; Ghoshal, S. K.; Prasad, P. N. *Appl. Phys. Lett.* **1986**, *48*, 1187.
14. Bai, S. J.; Spry, R. J.; Zelmon, D. E.; Ramabadran, U.; Jackson, J. *J. Polym. Sci., Poly. Phys. Ed.* **1992**, *30*, 1507.
15. Swiatkiewicz, J.; Prasad, P. N. Karasz, F. E.; Druy, M.; Glatoski, P. *Appl. Phys. Lett.* **1990**, *56*, 892.

RIM Materials

See: Polyamides, Lactam-Based
Reaction Injection Molding (RIM) Materials

Ring-Closing Depolymerization

See: Recycling, Thermodynamic (by Ring-Closing Depolymerization)

RING-OPENING-CLOSING POLYMERIZATION

Shiro Kobayashi*
Department of Molecular Chemistry and Engineering
Faculty of Engineering
Tohoku University

Jun-ichi Kadokawa
Department of Materials Science and Engineering
Faculty of Engineering
Yamagata University

*Author to whom correspondence should be addressed.

This article describes a new type of polymerization based on the novel concept that was born by the combination of ring-opening polymerization and cyclopolymerization (ring-opening-closing polymerization).

RING-OPENING-CLOSING ALTERNATING COPOLYMERIZATION (ROCAC) VIA ZWITTERION INTERMEDIATES

The general scheme of the "ring-opening-closing alternating copolymerization (ROCAC)" involving zwitterion intermediates is given as follows. The combination of a cyclic monomer A and a non-cyclic monomer B produces a copolymer that has a structure with a ring-opened unit from monomer A and a ring-closed unit from monomer B in alternating arrangement (**Structure 1**).[1–3]

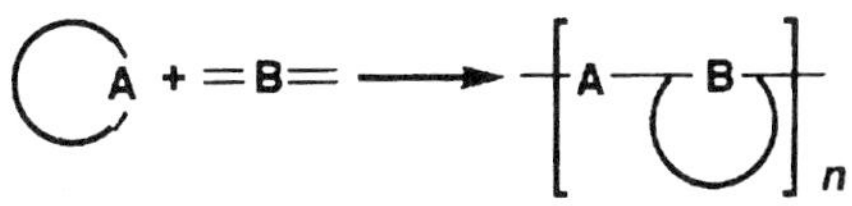

1

The first example of this novel type of copolymerization was the reaction of cyclic phosphonites (**1a–c**) with cis,trans- and cis,cis-muconic acid (**2a** and **2b**), which gave alternating copolymers **3a–c** having a repeating unit of ring-opened structure of **1** and ring-closed (lactone) structure of **2** (**Structure 2**).[4,5]

A second example of ROCAC is shown by the reaction of cyclic phosphonites (**1**) with aliphatic dialdehydes (**8**).[6,7] The reaction of 1:1 monomer feed ratio proceeded without any added catalyst to give an alternating copolymer **9**, whose repeating unit consists of one part formed by the ring-opening of monomer **1** and the other part (cyclic ether) by the ring-closing of monomer **8** (**Structure 3**).

Copolymerization of **1** with methacrylic and acrylic anhydrides (**16a** and **16b**, respectively) has been investigated.[8] A 1:1 mixture of monomer produced an alternating copolymer **17** with one part of the repeating unit coming from the ring opening of **1** and the other part (cyclic anhydride) from the ring-closing of **16** (**Structure 4**).

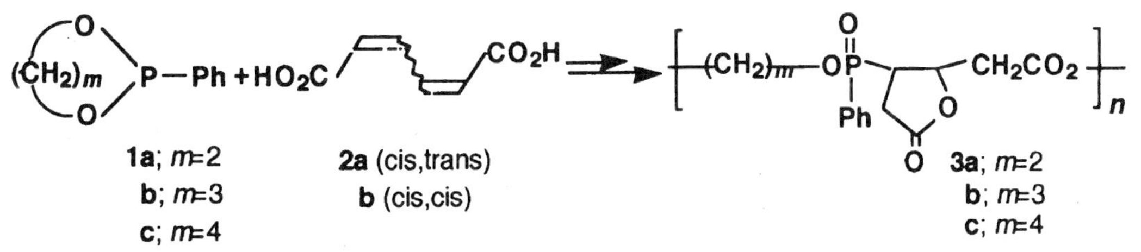

2

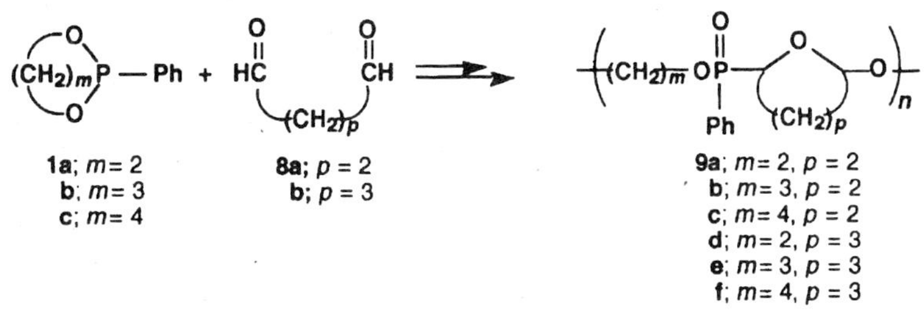

3

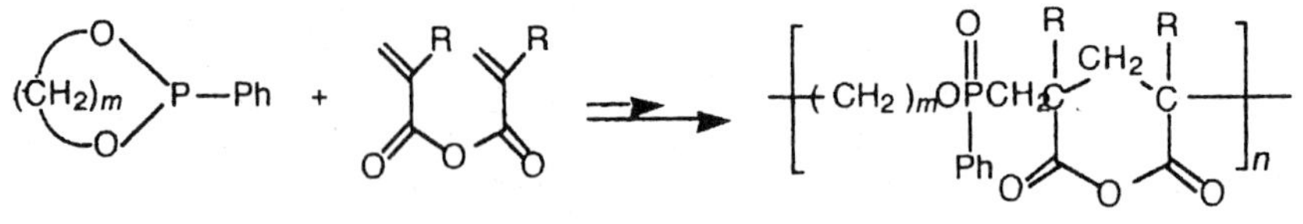

4

1a; m=2
b; m=3
c; m=4

16a; R=CH_3
b; R=H

17a; m=2, R=CH_3
b; m=3, R=CH_3
c; m=4, R=CH_3
d; m=2, R=H
e; m=3, R=H
f; m=4, R=H

The combination of monomers in ROCAC has been extended to the copolymerization of 2-methyl-2-oxazoline as monomer A with N-methyldiacrylamide as monomer B.[9]

RING-OPENING-CLOSING COPOLYMERIZATION (ROCC) VIA RADICAL PROCESS

Takahashi reported that radical copolymerization of 1,1-dichloro-2-vinylcyclopropane with maleic anhydride methyl acrylate, methyl methacrylate, and styrene affords copolymers containing a "ring-opened cyclized unit" through a radical 1,5-ring-opening reaction followed by addition of the propagating radical to MMA and an intramolecular ring-closing (cyclization).

Therefore, this radical copolymerization can be considered to belong to "ring-opening-closing copolymerization (ROCC)."

RING-OPENING-CLOSING POLYMERIZATION (ROCP)

The polymerization involving ring-opening and ring-closing in one monomer is possible according to the following scheme (**Structure 5**).

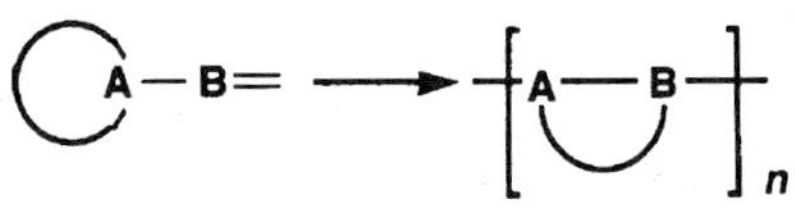

5

In the cationic cyclopolymerization of diepoxy monomer, 1,2-bis[2-(2,3-epoxypropoxy)ethoxy]benzene, the reaction occurred involving ring-opening of two epoxy groups and cyclization (ring-closing) to give polymer containing a crown ether unit.[10]

REFERENCES

1. Kobayashi, S. *Makromol. Chem., Macromol. Symp.* **1991**, *42/43*, 93.
2. Kobayashi, S.; Lundmark, S.; Kadokawa, J.; Uyama, H.; Shoda, S. *Makromol. Chem., Macromol. Symp.* **1993**, *73*, 137.
3. Lundmark, S.; Kobayashi, S. *Makromol. Chem., Macromol. Symp.* **1992**, *54/55*, 107.
4. Kobayashi, S.; Kadokawa, J.; Uyama, H.; Shoda, S.; Lundmark, S. *Macromolecules* **1990**, *23*, 3541.
5. Kobayashi, S.; Kadokawa, J.; Uyama, H.; Shoda, S.; Lundmark, S. *Macromolecules* **1992**, *25*, 5861.
6. Kobayashi, S.; Lundmark, S.; Kadokawa, J.; Albertsson, A. C. *Macromolecules* **1991**, *24*, 2129.
7. Kobayashi, S.; Lundmark, S.; Kadokawa, J.; Albertsson, A. C. *Macromolecules* **1992**, *25*, 5867.
8. Lundmark, S.; Kadokawa, J.; Kobayashi, S. *Macromolecules* **1992**, *25*, 5873.
9. Kadokawa, J.; Matsumura, Y.; Kobayashi, S. *Macromol. Chem. Phys.* **1994**, *195*, 3689.
10. Yokota, K.; Hashimoto, H.; Kakuchi, T.; Takada, Y. *Makromol. Chem., Rapid Commun.* **1984**, *5*, 115.

RING-OPENING COORDINATION POLYMERIZATION (by Soluble Multinuclear Alkoxides)

Philippe Dubois
Center for Education and Research on Macromolecules (CERM)
University of Liège - Institute of Chemistry

Although the ring-opening polymerization of oxirane into a polyether was discovered by Wurtz as early as 1859,[1] the polymerization of heterocyclic monomers, including lactones, lactams, oxiranes, oxetanes, tetrahydrofurans, dioxolanes, thiiranes, thietanes, aziridines, Leuch's anhydrides, cyclosiloxanes, and others, has been mainly achieved only in the last decades.

All these polymerizations have been promoted by using different types of initiators, belonging respectively to acid-base, ionic, and coordination catalysts.[2] In particular, coordination-type polynuclear compounds, containing several metal atoms linked together through oxygen bridges, have been very effective in controlling the chain-growth processes.[3,4] Unfortunately, the ill-defined structure of the active catalytic species has often prevented a deep understanding of their structural, mechanistic, and kinetic behavior in the mastering of the polymerization processes. In this frame, the design of coordination ring-opening catalysts, having a well-characterized composition that can be systematically modified in an easy reproducible way, should, and sometimes does, result in some interesting advances in the field.

Soluble μ-oxo-bimetallic trinuclear alkoxides having the general formula $[(RO)_X\text{-}M_1\text{-}O\text{-}M_2\text{-}O\text{-}M_1\text{-}(OR)_X]_n$ have been purposely synthesized. These compounds range among the best known initiators for the ring-opening polymerization of lactones, oxiranes, and thiiranes.[5–7] The corresponding polyesters, polyethers, and polythioethers may be obtained under very mild conditions, with high degrees of conversion, high polymerization rates, good stereoregularity and in the case of lactones, with an excellent control of the molecular weight and polydispersity of the polymeric chains.

BIMETALLIC μ-OXOALKOXIDES: SYNTHESIS AND PROPERTIES

The bimetallic μ-oxoalkoxides have a composition that closely corresponds to **Scheme I**,

$$(RO)_{p-1}M_2^{(P)} - O - M_1^{(II)} - O - M_2^{(p)}(OR)_{p-1} \qquad \text{I}$$

where $M_1^{(II)}$ may be Cr, Mn, Fe, Co, Zn, Mo; $M_2^{(P)}$ being either $Al^{(III)}$ or $Ti^{(IV)}$, and R, practically any n-, sec-, or ter-alkoxy group (most often propyl or butyl).

General Properties

The bimetallic μ-oxoalkoxides **I** display an amazingly high solubility in most organic solvents and particularly in hydrocarbons.[8]

This property can be explained by an intermolecular association giving an aggregate in which a core of μ-oxo bridged metals is surrounded by a lipophylic layer of alkoxy groups.

RING-OPENING POLYMERIZATION OF LACTONES

Kinetics and "Living" Behavior of the Homopolymerization

The bimetallic μ-oxoalkoxides prepared up to now are highly active in lactone ring-opening polymerization.

Excellent activity is observed under very mild experimental conditions and, as could be expected, the activity level of a given catalyst depends very much on its composition, probably to the extent where it controls the characteristics of the catalytic coordination aggregate. Another significant feature is the dependence of the order in catalyst on its structure and principally its mean degree of association.[10]

Mechanism of the Polymerization Reaction

The ring-opening polymerization is characterized by a "coordination-insertion" mechanism.[7,10]

Controlled Synthesis of Block Copolymers

The fast and living ring-opening polymerization of lactones by bimetallic μ-oxoalkoxides has led to successful block copolymerizations.

Macromolecular Engineering of Polylactones

Similar to the bimetallic μ-oxoalkoxides, it is well established that some aluminum alkoxides, including the commercially available $Al(O^iPr)_3$ are also very efficient initiators for the ring-opening polymerization of lactones such as ε-CL, δ-VL, β-PL, and β-BL.[11–13] Fundamental kinetic and structural investigations have confirmed the livingness of the ring-opening polymerization that fits in the "coordination-insertion" mechanism. This mechanism results in linear chains end-capped with an ester carrying the R "radical" of the initiator (R = iso-Pr for $(Al(Oiso\text{-}Pr)_3)$ and a hydroxyl end-group that results from the hydrolysis of the polyester growing sites.

The great versatility of aluminum alkoxides, functionalized or not, allows for the macromolecular engineering of the aliphatic polyesters, as illustrated by the perfectly controlled synthesis of telechelic polymers, block copolymers, and macromonomers, i.e., promoters of graft copolymers.[14–17]

RING-OPENING POLYMERIZATION OF OXIRANES

Catalytic Activity in Homopolymerization

In contrast to the inefficiency of simpler alkoxides, bimetallic μ-oxoalkoxides, in perfect agreement with Vandenberg's proposal, rank also among the most active catalysts for conversion of oxiranes and thiiranes into high molecular weight, often stereoregular, polyethers, and polysulfides.[5,18]

Specific Controls by Addition of Ligands

The proportion of isotactic polyether chains varies from 5 to 80% depending on the nature of M_2 and of the OR groups. This suggests that the more or less enantiomorphic character of the catalytic sites (symmetrically distributed) depends quite closely on the degree of association. This hypothesis could be confirmed by using alcohol ligands to dissociate the aggregates as shown by the sharp decrease of the stereospecificity degree of the isotactic fraction.

By the addition of an oligomeric diol or triol, e.g., poly(propylene glycol) 400 and polyglycerol 600, as a co-initiator, polyether chains with a molecular weight in the range of 6,000–13,000 have been synthesized. Such polyether-polyols constitute actually very interesting precursors in the manufacture of polyurethane foams with improved mechanical properties (higher load bearing properties, higher resilience, etc.).

Control of the Relative Reactivity in Copolymerization

Incorporation of a given monomer in a growing copolyether chain sharply depends on parameters other than its intrinsic reactivity, in relation to the structural features of the bimetallic μ-oxoalkoxides. A very interesting illustration of that dependence comes from solvent effects in the copolymerization of epichlorhydrin (ECH) and methyloxirane (MO).[9] Despite its slow rate of homopolymerization compared with MO, ECH is incorporated preferentially in these copolymers when working in an inert solvent such as n-heptane. However, in a more polar solvent (e.g., nitrobenzene), the situation is completely reversed and MO is preferred, with a corresponding increase of the overall polymerization rate.

POLYMERIZATION OF ISOCYANATES

Although many bimetallic μ-oxoalkoxides have been shown to be very active, the aluminum-zinc derivatives range between the most interesting ones. In addition to their high catalytic activity in ring-opening polymerization, they also promote the polymerization of heterocumulenes such as isocyanates.[19] Depending on the reaction temperature, isocyanates can be polymerized into polyamide-1, $\text{-(N(R)-C(O))}_n\text{-}$, of very high molecular weight.

It is interesting to mention also that another cumulene, i.e., carbon dioxide, can be copolymerized with methyloxirane to form a copolymer containing approximately 25% carbonate units. Once again, carbon dioxide is inserted into the Al-OR bond of the catalyst to produce the corresponding mixed carbonate.

CONCLUSIONS

Soluble bimetallic μ-oxoalkoxides rank among the best initiators known for lactone and oxirane ring-opening polymerization into high molecular weight polyesters and polyethers. In addition and owing to their high activity, their well-defined and characterized composition, and the possibility to modify systematically their structure, they offer an attractive and precious tool to study the mechanism, the kinetics, and the stereoselectivity of these ring-opening polymerization processes. Owing to their remarkable solubility and the topochemical influence of the aggregates on the active site behavior, these complex trinuclear alkoxides can be viewed as borderline catalysts between their homogeneous and heterogeneous counterparts. Further evidence can be found in the polymerization of ethylene and propylene catalyzed by similar μ-oxoalkoxide "supporting" transition metals in the presence of alkylaluminums.[20] Once again, the stereoregularity of polypropylene can so be controlled.

REFERENCES

1. Wurtz, A. *Annalen* **1859**, *10*, 110, Ber 1877.
2. Lundberg, R. D.; Cox, E. F. In *Ring-Opening Polymerization*; Frisch, K. C., Ed.; Marcel Dekker: New York, 1969; Vol. 2, 247.
3. Gurgliolo, A. E. *Rev. Macromol. Chem.* **1966**, *1*, 39.
4. Osgan, M.; Teyssié, Ph. *J. Polym. Sci.* **1970**, *8*, 319.
5. Kohler, N.; Osgan, M.; Teyssié, Ph. *Polym. Lett.* **1968**, *6*, 559.
6. Osgan, M.; Pasero, J. J.; Teyssié, Ph. *Polym. Lett.* **1970**, *88*, 319.
7. Hamitou, A.; Ouhadi, T.; Jérôme, R.; Teyssié, Ph. *J. Polym. Sci., Polym. Chem. Ed.* **1977**, *15*, 865.
8. Ouhadi, T.; Bioul, J. P.; Stevens, C.; Warin, R.; Hocks, L.; Teyssié, Ph. *Inorg. Chim. Acta* **1976**, *19*, 203.
9. Teyssié, Ph.; Ouhadi, T.; Bioul, J. P. *Int. Rev. Sci., Phys. Chem. Ser.* **1975**, *118*, 191.
10. Ouhadi, T.; Hamitou, A.; Jérôme, R.; Teyssié, Ph. *Macromolecules* **1976**, *9*, 927.
11. Ouhadi, T.; Stevens, C.; Teyssié, Ph. *Makromol. Chem. Suppl.* **1975**, *1*, 191.
12. Dubois, Ph.; Jérôme, R.; Teyssié, Ph. *Polym. Bull.* **1989**, *22*, 475.
13. Dubois, Ph., Ph.D. Thesis, University of Liège, Liège, Belgium, 1991.
14. Dubois, Ph.; Jérôme, R.; Teyssié, Ph. *Macromolecules* **1991**, *24*, 2266.
15. Dubois, Ph.; Jérôme, R.; Teyssié, Ph. *Makromol. Chem., Macromol. Symp.* **1990**, *42/43*, 103.
16. Dubois, Ph.; Degée, Ph.; Ropson, N.; Jérôme, R. "*Molecular Engineering of Polylactones and Polylactides by Ring-Opening Polymerization*"; In Macromolecular Design of Polymeric Materials, Hatada, K., Ed.; Marcel Dekker: 1996, Chapter 14, 247.
17. Löfgren, A.; Albertsson, A.-C.; Dubois, Ph.; Jérôme, R.; Teyssié, Ph. *Macromolecules* **1994**, *27*, 5556.
18. Vandenberg, E. J. *J. Polym. Sci.* **1969**, *A1*(7), 525.
19. Druet, J. P. Ph.D. Thesis, University of Liège, Liège, Belgium, 1976.
20. Herrmann, C.; Streck, R. Bad Nauheim Meeting, Fachgruppe Makromolekulare Chemie, Gesellschaft Deutscher Chemiker, April 1980.

Ring-Opening Metathesis Polymerization

See: *Metathesis Polymerization, Cycloolefins*
p-Phenylenevinylene Oligomers, Homo- and Copolymers (Metathesis Preparation)
Poly(cyclopentadienylene vinylene)
Polymerization Separation (Ring-Opening Polymerization)
Ring-Opening Metathesis Polymerization (Formation of Cyclic Butadiene Oligomers)
Silacyclobutanes and Related Compounds (Ring-Opening Polymerization)
Uniform Polymers

RING-OPENING METATHESIS POLYMERIZATION (Formation of Cyclic Butadiene Oligomers)

Emma Thorn-Csányi,* J. Hammer, and J. U. Zilles
Institut für Technische und Makromolekulare Chemie der Universität Hamburg

Cyclic olefins can be polymerized via ring-opening metathesis polymerization (ROMP) without any loss of double bonds.[1–5] The catalytic active species is a transition metal carbene complex that can be generated *in situ* from catalyst and cocatalyst (classical systems) or in the case when stable carbene complexes are added in substance.

The reaction product of ROMP generally consists of two parts: a high molecular weight fraction of linear polymer and a low molecular weight fraction containing a series of cyclic oligomers. The presence of this bimodal molecular weight distribution is due to the concurrent formation of polymer via chain growth and oligomers via chain degradation. If secondary metathesis processes (such as "back biting") are hindered compared to chain growth, living systems are observed.[7]

Polymer generation occurs only if the initial monomer amount exceeds the critical monomer concentration $[M]_c$. Below this value only cyclic oligomers are present in the reaction mixture at equilibrium. The nature of cyclic oligomers is determined by the double-bond distribution in the polymer chain. If a ring is thermodynamically favored it will be the main product regarding the oligomer fraction. In the case of polypentenamer, for example, the appreciably stable cyclopentene will be readily formed by degradation of the polymer chain; its equilibrium concentration depends upon the temperature.[8–10]

Contrary to the cyclopentene system, ROMP of 1Z,5Z-cyclooctadiene (COD) or degradation of 1,4-polybutadiene (BR) yields a broad variety of cycles with butadiene tetramers and pentamers as main products. No monomer and dimer were observed at equilibrium; the amount of trimer was insignificant.[6,11–13]

Investigations concerning the synthesis of p-phenylene vinylene/butadiene copolymers[14] using a stable tungsten carbene complex[15] taking into account our earlier results on ROMP of COD employing $WCl_6/EtOEt/SnBu_4$[16] gave us a valuable hint that the literature data of cyclic oligomers in the cyclobutene system had to be checked. This article deals with new knowledge concerning the distribution of cyclic oligomers in the cyclobutene system.[17]

CONCLUSION

ROMP of cyclic butadiene oligomers, ADMET of 1,5-hexadiene, and intramolecular metathesis degradation of 1,4-polybutadiene, led to preferred formation of cyclic butadiene trimers (especially of the all-E one) within the oligomeric product spectrum. The highest achievable concentration of all-E

*Author to whom correspondence should be addressed.

cyclododecatriene at thermodynamically controlled ring-chain equilibrium is ~0.29 mol(C_4H_6)/L at room temperature (in cyclohexane, methylcyclohexane).

The sterical arrangement of double bonds of cyclic oligomers is predominantly E. The highest amount of E content was found within the trimers (97%), followed by tetramers (84%) at room temperature. There exists a temperature dependence of E/Z double bond ratios (it is more distinctive than for BR), as our latest results show. This is further proof that the product distribution obtained is controlled thermodynamically.

The application of our knowledge concerning the composition of the oligomer fraction in the cyclobutene system to the 1-methylcyclobutene system opens a synthetic route to cyclic isoprene trimers via intramolecular metathesis degradation of 1,4-polyisoprene. These trimethylcyclododecatrienes are valuable intermediate products for the synthesis of fragrances and pheromones.[18,19] Our results show that the formation of cyclic trimers is also strongly preferred, especially that of the all-E isomer.

Additionally to already published syntheses of cyclic isoprene trimer,[20] further experiments concerning the exact oligomer distribution, including their sterical arrangement especially of trimers at thermodynamically controlled equilibrium, are in process.

Finally it should be mentioned that results concerning the oligomer distribution at equilibrium and its temperature dependence open a way to determine thermodynamic data concerning cyclic oligomers.

REFERENCES

1. Ivin, K. J. *Olefin Metathesis*; Academic: London, 1983.
2. Dragutan, V.; Balaban, A. T.; Dimonie, M. *Olefin Metathesis and Ring-Opening Polymerization of Cycloolefins*; John Wiley & Sons: Chichester, Editura Academiei, Bucarest, 1985.
3. Ivin, K. J. in *Encyclopedia of Polymer Science and Engineering*, Sec. Ed.; John Wiley & Sons: New York, 1988; Vol. 9, 634.
4. Ofstead, E. A.; Wagener, K. B. in *New Methods for Polymer Synthesis*; Plenum: New York, 1992.
5. Schrock, R. R. *Pure & Appl. Chem.* **1994**, *66*, 1447.
6. Calderon, N.; Scott, K. W.; Ofstead, E. A.; Judy, W. A.; Ward, J. P. *Am. Chem. Soc., Adv. Chem. Ser.* **1969**, *91*, 399.
7. Grubbs, R. H. in *Encyclop. Polym. Sci. Eng.*; Suppl. Vol. J. Wiley & Sons: New York, **1989**, 380.
8. Ofstead, E. A.; Calderon, N. *Makromol. Chem.* **1972**, *154*, 21.
9. Pampus, G.; Lehnert, G. *Makromol. Chem.* **1974**, *175*, 2605.
10. Makovetskii, K. L.; Red'kina, L. T. *Dokl. Akad. Nauk SSSR* **1976**, *231*, 143.
11. Dolgoplosk, B. A.; Tlenkopachev, M. A.; Kop'eva, I. A.; Bychkova, N. A.; Korslak, Y. V.; Timofeeva, G. I.; Tinyakova, E. I. *Dokl. Chem.* **1976**, *227*, 279.
12. Chauvin, Y.; Comereuc, D.; Zaborowski, G. *Makromol. Chem.* **1978**, *179*, 1285.
13. Höcker, H.; Suter, U. W. *Makromol. Chem.* **1988**, *189*, 1603.
14. Thorn-Csányi, E.; Pflug, K. P. *Makromol. Chem. Rapid Commun.* **1993**, *14*, 619.
15. Grubbs, R. H.; Johnson, K.; Virgil, S. C.; Ziller, J. W. *J. Am. Chem. Soc.* **1990**, *112*, 5384.
16. Schwartau, M. Thesis, University of Hamburg, Germany, 1986.
17. Thorn-Csányi, E.; Hammer, J.; Pflug, K. P.; Zilles, J. U. *Macromol. Chem. Phys.* **1995**, *196*, 1043.
18. Odinokov, V. N.; Kukovinets, O. S.; Sakharova, N. I.; Tolstikov, G. A. *Zh. Org. Khim* **1984**, *20*, 1866.
19. Odinokov, V. N.; Kukovinets, O. S.; Sakharova, N. I.; Tolstikov, G. A. *Zh. Org. Khim* **1989**, *25*, 29.
20. Thorn-Csányi, E.; Hammer, J.; Zilles, J. U. *Makromol. Rapid Commun.* **1994**, *15*, 797.

Ring-Opening Polymerization

See: *Benzocyclobutene-Based Monomers (Diels-Alder Ring-Opening Polymerization)*
Benzoxazine Monomers and Polymers (New Phenolic Resins by Ring-Opening Polymerization)
Bicyclic Acetals, Ring-Opening Polymerization
Bicyclic and Spirocyclic Monomers, Oxygen-Containing (Ring-Opening Polymerization)
Bicyclic Ethers, Ring-Opening Polymerization
Bicyclic Lactams, Ring-Opening Polymerization
Bicyclic Lactones, Ring-Opening Polymerization
Bicyclobutanes, Ring-Opening Polymerization
Cyclic Carbonates (Ring-Opening Polymerization)
Cyclic Imino Ethers (Ring-Opening Polymerization)
Cyclic Imino Ethers (Ring-Opening Polymerization of 1,3-Oxazo Monomers)
Cyclic Ketene Dithioacetals (Polymerization)
Cyclic Oligomers of Engineering Thermoplastics
Cyclic Siloxane (Ring-Opening Polymerization)
Diphenimides (Ring-Opening Polymerization)
Double Isomerization Polymerization
Electrochemical Polymerization
Glycoproteins (Synthesis by NCA Method)
Macrocyclic Aramids
Metal-Containing Silicon-Based Polymers
Metallocenophanes, Strained (Thermal Ring-Opening Polymerization)
Metathesis Polymerization, Cycloolefins
2-Oxazolines, Polymerization
p-Phenylenevinylene Oligomers, Homo- and Copolymers (Metathesis Preparation)
Poly(N-acylethylenimine)s
Poly(alkylene phosphate)s
Polyalkylenimines
Polyamides, Lactam-Based
Poly(cyclopentadienylene vinylene)
Polyesters (Synthesis from Lactones)
Poly(β-ketone)
Polymerization Separation (Ring-Opening Metathesis Polymerization)
Polysaccharides, Synthetic (Preparation and Biological Activities)
Polysilanes (Ring-Opening Polymerization of Cyclotetrasilanes)
Polysilazanes (Through Ring-Opening Polymerization)
Reactive Oligomers (Overview)
Recycling, Thermodynamic (by Ring-Closing Depolymerization)
Ring-Opening-Closing Polymerization
Ring-Opening Coordination Polymerization (by Soluble Multinuclear Aldoxides)
Ring-Opening Metathesis Polymerization (Formation of Cyclic Butadiene Oligomers)

Ring-Opening Polymerization, Anionic (With Expansion in Volume)
Ring-Opening Polymerization, Cationic (with Expansion in Volume)
Ring-Opening Polymerization, Free Radical
Ring-Opening Polymerization, Radical (with no Shrinkage in Volume)
Silacyclobutanes and Related Compounds (Ring-Opening Polymerization)
Siloxane-Containing Polymers
Sulfur Dioxide Copolymerization
Template Polymerization (Overview)
Transannular Polymerization
Ultrasonically Assisted Polymer Synthesis
Vegetal Biomass (1. Monomers and their Polymerization)
Zwitterionic Polymerization (Overview)
Zwitterionic Polymerization (Aziridine, Oxazoline Derivatives with Anhydrides)

RING-OPENING POLYMERIZATION, ANIONIC (with Expansion in Volume)

Takeshi Endo* and Fumio Sanda
Research Laboratory of Resources Utilization
Tokyo Institute of Technology

It is well known that vinyl and cyclic monomers are accompanied by volume shrinkage on polymerization. Either vinyl or cyclic monomers show a linear relationship between volume shrinkage on polymerization versus the reciprocal of the molecular weight.[1] As the molecular weight of monomer increases, the shrinkage decreases. Shrinkage of vinyl monomers is about two times larger than that of cyclic monomers. Therefore, monomers that undergo multiple ring-opening polymerization are expected to show slight shrinkage or expansion in volume. In the course of this concept, we have developed some monomers that undergo double ring-opening polymerization—namely, bicyclo orthoesters (BOEs), spiro orthoesters (SOEs), and spiro orthocarbonates (SOCs).[2] These monomers usually undergo cationic double ring-opening polymerization by Lewis acids such as boron trifluoride ether complex ($BF_3{\cdot}OEt_2$). Some monomers show volume expansion as large as 10% during polymerization.

Monomers that Undergo Anionic Polymerization with Expansion in Volume

Monomers that show volume expansion in cationic and radical polymerizations have been intensively investigated in the last 20 years; however, fewer studies have been done on monomers that show volume expansion on anionic polymerization.

γ-Butyrolactone Derivatives

We have found that bicyclobis(γ-butyrolactone) (**1**)[3,4] and spirobis(γ-butyrolactone) (**4**)[5] cannot undergo homopolymerizations but undergo alternating ring-opening copolymerizations with epoxide (**2**) to afford copolymers (**3,5**) via double ring-opening isomerization of **1** and **4** (**Structure 1**).

*Author to whom correspondence should be addressed.

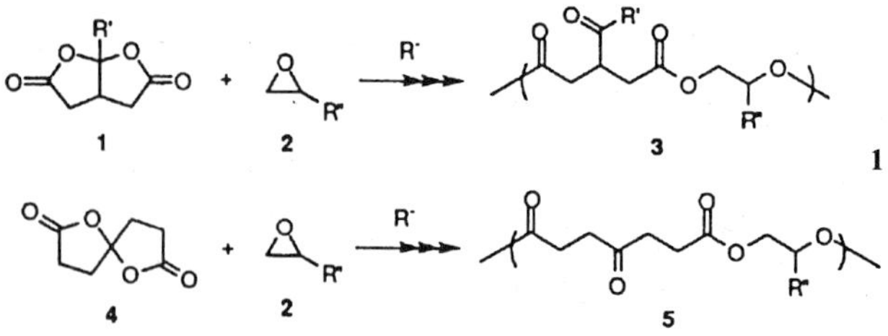

1

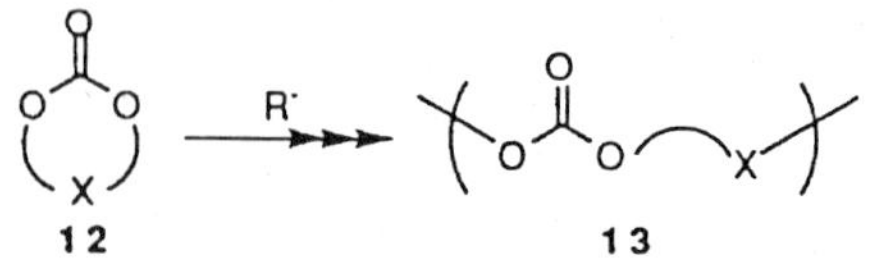

2

The volume change on copolymerization of **1** with glycidyl phenyl ether and bifunctional epoxides (**6**) was very small or slightly expandable.[6]

Cyclic Carbonate

Recently, a new type of monomer that undergoes ring-opening polymerization to show volume expansion on polymerization has been developed. Although cyclic carbonate (**12**) can undertake single ring-opening polymerization, which is different from BOEs, SOEs, and SOCs, **12** shows as large as 10% volume expansion on polymerization to afford poly(carbonate) (**13**) (**Structure 2**).[7]

Large-Membered Lactones

Few studies on anionic polymerization of large-membered lactones have been reported compared to studies on lactones bearing 4-, 6-, and 7-membered rings.[8] Recently, anionic polymerization of 12-, 13-, and 16-membered lactones initiated by metal alkoxides has been developed.[9] The larger the number of ring members becomes, the more the structure is like that of the corresponding polymer. Therefore, a large-membered lactone may be expected as a candidate monomer that shows no volume shrinkage on polymerization. Volume shrinkage on polymerization of 16-membered lactone was 6.4%, which was nearly the same as that of the 7-membered lactone. Synthesis of larger-membered lactones is expected in the future.

Summary

Here we have summarized the molecular design, synthesis, and polymerization behavior of monomers that undergo anionic ring-opening polymerization to show slight shrinkage or expansion in volume on polymerization. Anionic ring-opening polymerization is one of the most minute polymerization methods; therefore, molecular design for the introduction of function based on the controlled higher-order structure as well as expansion in volume into monomers and polymers may be strongly expected in the future.

REFERENCES

1. Endo, T.; Ogasawara, T. *Netsukoukaseijushi (Japanese)* **1984**, *5*, 30.
2. Takata, T.; Endo, T. In *Expanding Monomers: Synthesis, Characterization, and Application*; Sadhir, R. K.; Luck, R. M., Eds.; CRC: Boca Raton, FL, 1992.
3. Takata, T.; Tadokoro, A.; Endo, T. *Macromolecules* **1992**, *25*, 2782.

4. Tadokoro, A.; Takata, T.; Endo, T. *Macromolecules* **1993**, *26*, 4400.
5. Takata, T.; Tadokoro, A.; Chung, K.; Endo, T. *Macromolecules* **1995**, *28*, 1340.
6. Takata, T.; Chung, K.; Tadokoro, A.; Endo. T. *Macromolecules* **1993**, *26*, 6686.
7. Nemoto, H.; Takata, T.; Endo, T. *Polym. Prepr. Jpn.* **1990**, *39*, 284.
8. Koleske, J. V.; Lundberg, R. D. *J. Polym. Sci., Part A-2* **1972**, *10*, 323.
9. Nomura, R.; Ueno, A.; Endo, T. *Macromolecules* **1994**, *27*, 620.

RING-OPENING POLYMERIZATION, CATIONIC (with Expansion in Volume)

Takeshi Endo* and Fumiko Sanda
Research Laboratory of Resources Utilization
Tokyo Institute of Technology

In this article, recent works on monomers that undergo cationic ring-opening polymerization expected to show volume expansion are described. The studies since 1990 are summarized in this article because the studies previous to 1990 have been already reviewed elsewhere.[1,2]

MONOMERS THAT UNDERGO CATIONIC POLYMERIZATION WITH EXPANSION IN VOLUME

Spiro Orthoester (SOE)

SOEs have been developed as expandable monomers on cationic double ring-opening polymerization.[3] Recently, a new type of cationic polymerization of SOE, which proceeds via a single ring-opening process to give poly(cyclic orthoester), has been achieved. Namely, SOE bearing seven-membered cyclic ether (**1**) afforded single ring-opening polymer (**2**) and double ring-opening polymer (**3**). The obtained polymers easily depolymerized by treatment with hydrogen chloride at room temperature to recover the starting monomers in good yields.[3] The polymer obtained at 0°C was entirely **2**, whereas the polymer obtained at 120°C consisted of 25% **2** and 75% **3**. The unit ratio of **3** increased and the volume shrinkage on polymerization decreased as the polymerization temperature rose[4] (**Scheme I**).

Spiro Orthocarbonate (SOC)

SOEs generally show low shrinkage or slightly small expansion on polymerization. On the contrary, SOCs show larger volume expansion than SOEs. Therefore, SOCs are more effective than SOEs for lowering volume shrinkage of curing materials. Cationic polymerization behaviors of SOCs were very different from each other by the ring numbers and substituents.

Cyclic Carbonate

Recently, a new type of monomer that undergoes ring-opening polymerization to show volume expansion upon polymerization has been developed. Although the polymerization of cyclic carbonate (**20**) was single ring-opening polymerization that was different from BOEs, SOEs, and SOCs, **20** showed as large as 10% volume expansion in polymerization to afford polycarbonate (**21**).[5] The evaluation of dipole moments of the

*Author to whom correspondence should be addressed.

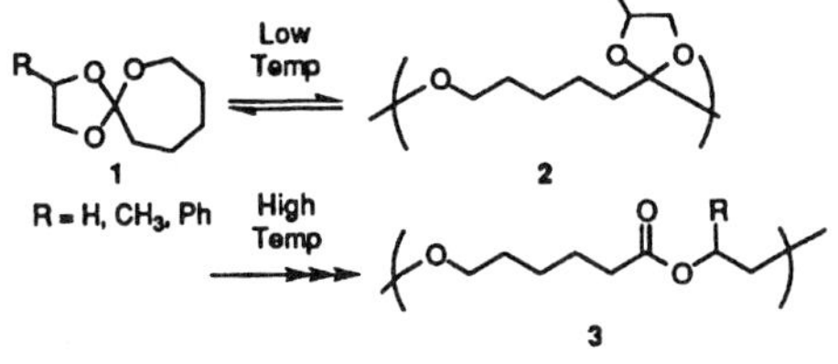

I

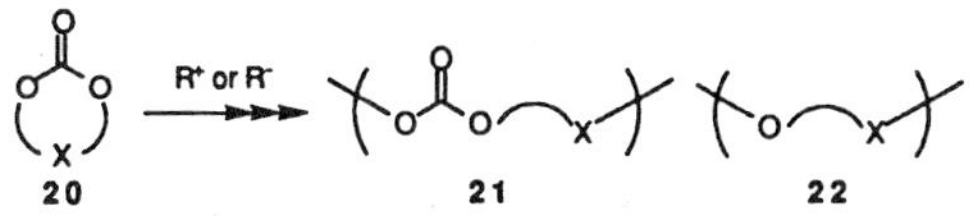

II

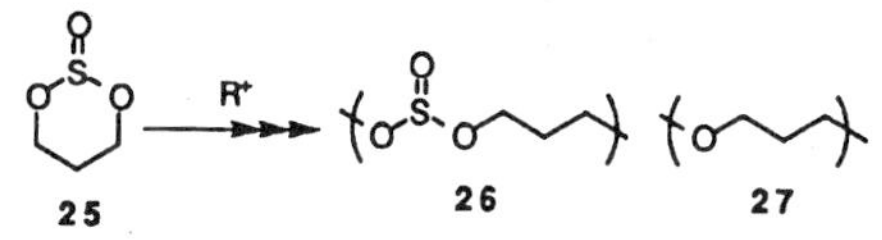

III

monomers and polymers calculated by the molecular orbital method suggested the mechanism of volume expansion of **20**. Namely, this volume expansion should be caused by the difference of large intermolecular interaction of the monomer and small intermolecular interaction of the polymer (**Scheme II**).[6]

Poly(ether) unit (**22**) was produced through decarboxylation in addition to **21** in the cationic ring-opening polymerization of **20** with usual Lewis acid catalyst such as $BF_3{\cdot}OEt_2$. We have found that cyclic carbonates undergo polymerization with alkyl halides as initiators without decarboxylation.[7]

Cyclic Sulfite

From the evaluation of dipole moments of several cyclic monomers and polymers, we have found that cyclic sulfite (**25**) with S=O moiety in equatorial position will be expected to show volume expansion similar to cyclic carbonates. Although anionic polymerization of **25** did not proceed, cationic polymerization of **25** afforded polymers (**26,27**) accompanying desufoxylation.[8] The studies on various kinds of cyclic sulfites to show volume expansion are now under investigation (**Scheme III**).

Alkenyl Methylene Dioxolane

Polymerization of monomers bearing two polymerizable carbon–carbon double bonds usually affords crosslinked polymers. However, we have succeeded in cationic polymerization of 1,3-dioxolanes having two polymerizable double bonds.[9] Cationic polymerization of 4-methylene-1,3-dioxolane bearing 2-ethenyl and 2-isopropenyl groups by CH_3SO_3H afforded a soluble poly(keto ether) that consisted of a carbon–carbon double bond in the side chain and a unit containing a vinyl ether group in the main chain.

SUMMARY

In this article, cationic ring-opening polymerization behaviors of SOE, SOC, cyclic carbonate, cyclic sulfite, and alkenyl methylene dioxolane have been summarized. SOE and SOC have been developed as expandable monomers during the last 20 years. Recent studies gradually clarified the detailed polymerization behaviors and mechanisms such as the equilibrium of single ring-opening process of SOE. The new type of expandable monomers, cyclic carbonate and sulfite, may also be candidate monomers for applications to non-shrinkage materials in the near future. Further research on molecular design, synthesis, and polymerization behaviors should enable them to be used for practical materials in the future.

REFERENCES

1. Takata, T.; Endo, T. Ionic Polymerization of Oxygen-containing Bicyclic, Spirocyclic, and Related Expandable Monomers; Sadhir, R. K.; Luck, R. M., Eds.; *Expanding Monomers: Synthesis, Characterization, and Application* CRC: Boca Raton, FL, 1992.
2. Brady, R. F. J. *J. Macromol. Sci., Rev. Macromol. Chem. Phys.* **1992**, *C32*, 135.
3. (a) Chikaoka, S.; Takata, T.; Endo, T. *Macromolecules* **1991**, *24*, 331. (b) Chikaoka, S.; Takata, T.; Endo, T. *Macromolecules* **1991**, *24*, 6557.
4. Chikaoka, S.; Takata, T.; Endo, T. *Macromolecules* **1992**, *25*, 625.
5. Nemoto, H.; Takata, T.; Endo, T. *Polym. Prepr. Jpn.* **1990**, *39*, 284.
6. (a) Ishii, T.; Amiya, S.; Nagata, S.; Sanda, F.; Takata, T.; Endo, T. *Polym. Prepr. Jpn.* **1992**, *41*, 1851. (b) Ariga, T.; Takata, T.; Endo, T. *Polym. Prepr. Jpn.* **1994**, *43*, 294.
7. Ariga, T.; Takata, T.; Endo, T. *J. Polym. Sci., Part A: Polym. Chem.* **1993**, *31*, 581.
8. Azuma, N.; Takata, T.; Sanda, F.; Endo, T. *Macromolecules* **1995**, *28*, 7331.
9. (a) Park, J.; Yokozawa, T.; Endo, T. *Makromol. Chem.* **1993**, *194*, 2017. (b) Park, J.; Yokozawa, T.; Endo, T. *J. Polym. Sci., Part A: Polym. Chem.* **1993**, *31*, 1141.

RING-OPENING POLYMERIZATION, FREE RADICAL

Bogdan C. Simionescu
Department of Macromolecules
"Gh. Asachi" Technical University

Simona Morariu
"P. Poni" Institute of Macromolecular Chemistry

The ionic polymerization of cyclic monomers is a well-known and very common process. Surprisingly, free radical ring-opening polymerization is quite rare. However, different researchers concentrated on the synthesis of cyclic monomers able to undergo free radical homo- and copolymerization, the extent of ring-opening influencing the structure of the obtained macromolecular compounds. The present work deals with the radical polymerization of different classes of cyclic compounds.

VINYLCYCLOPROPANE DERIVATIVES

The first report on the radical ring-opening polymerization of vinylcyclopropane appeared in 1965.[1] The resulting polymer was shown to contain about 80% 1,5-opened units and about 20% undetermined structural units, but no cyclopropane rings in the macromolecular structure.

Studying the behavior of 1,1-disubstituted 2-vinylcyclopropanes, Endo has shown that the rate-determining step of the polymerization is not the ring-opening step, but the radical addition step.[2–4] 2-Vinylcyclopropanes substituted at the 1-position by electron-withdrawing groups (ester, cyano, phenyl, etc.) undergo radical ring-opening polymerization yielding polymers containing only 1,5-type units.[2–8]

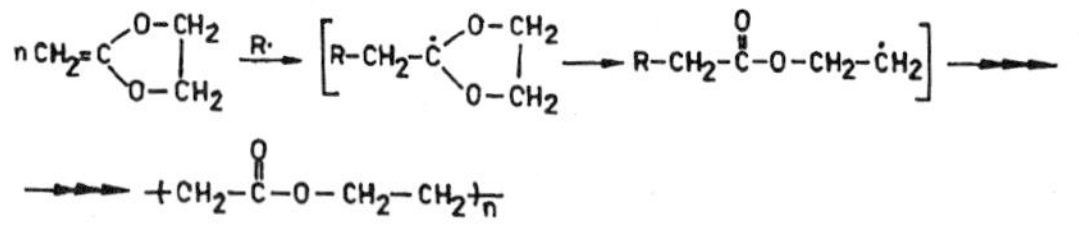

I

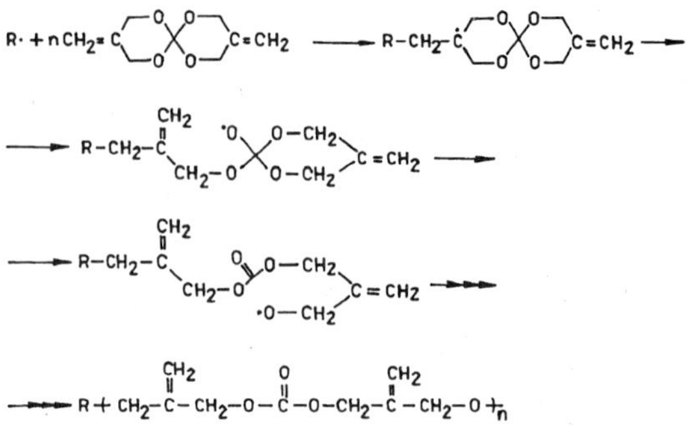

II

CYCLIC KETENE ACETALS

The carbon–oxygen double bond is about 50 kcal/mol more stable than a carbon–carbon double bond. As a consequence, it was estimated that unsaturated cyclic compounds possessing one oxygen atom into the ring could undergo free radical ring-opening polymerization. Thus, Bailey extended the number of monomers able to polymerize through free-radical initiation by synthesis of various methylene derivatives of 1,3-dioxolane.[9] In the presence of radical initiators and at high temperatures, 2-methylene-1,3-dioxolane yields a high molecular weight polyester according to **Scheme I.**[10]

UNSATURATED SPIRO ORTHO COMPOUNDS

Spiro ortho esters and spiro ortho carbonates are among cyclic monomers polymerizing by radical mechanism without shrinking. Their double ring-opening polymerization proceeds according to **Scheme II.**

The polymerization of 3,9-dimethylene-1,5,7,11-tetraoxospiro[5,5]undecane (Scheme VI) initiated by di-tert-butyl-peroxide at 130°C and stopped below 30% conversion yields a soluble polycarbonate possessing exo-methylene groups. At high conversion, this monomer produces a highly crosslinked resin, very similar to the material issued from the polymerization of diallyl carbonate.[11]

IMPORTANCE AND POTENTIAL APPLICATIONS OF UNSATURATED CYCLIC MONOMERS

A wide variety of biodegradable polymers can be prepared through the radical copolymerization of unsaturated cyclic monomers with classical vinyl monomers. Cyclic monomers, introduced in a limited amount in the chain, in order to maintain the main properties of the polymer issued from the vinyl monomer, provide biodegradable chemical links and thus transform many large scale industrial polymers in biodegradable materials. For example, Bailey reported the copolymerization of cyclic ketene acetals with conventional vinyl monomers; the incorporation of ester groups in the main chain transformed the high molecular weight compounds into biodegradable materials.[12] Similar results are obtained by the radical copolymerization of 2-(o-chlorophenyl)-4-methylene-1,3-dioxolane with styrene and/or methyl methacrylate, that is, the copolymers undergo bacterial degradation.[13]

The copolymerization of vinyl monomers with methylene derivatives of 1,3-dioxolane introduces photodegradable carbonyl groups in the polymer chain.[14] The hydrolysis of the copolymers containing ester units—obtained by the same approach—yields oligomers of the classical vinyl monomers ended by reactive terminal groups and able to be used in condensation reactions.[15] The strong nucleophilic character of cyclic ketene acetals allows their copolymerization with electrophilic vinyl monomers through a zwitterionic mechanism.[16,17]

The monomers presented here can be considered as unconventional monomers, at least from the point of view of the "classical" radical polymerization. Their synthesis and homo- and copolymerization represent a new research area in polymer chemistry with interesting potential applications. These monomers make possible the obtainment of bio- and photodegradable polymers, of oligomers and polymers with terminal functional groups, as well as of monomer mixtures able to expand during polymerization.

REFERENCES

1. Takahashi, T.; Yamashita, I. *J. Polym. Sci., Part B* **1965**, *3*, 251.
2. Endo, T.; Watanabe, M.; Suga, K.; Yokozawa, T. *J. Polym. Sci., Part A: Polym. Chem.* **1987**, *25*, 3039.
3. Endo, T.; Watanabe, M.; Suga, K.; Yokozawa, T. *Makromol. Chem.* **1989a**, *190*, 691.
4. Endo, T.; Watanabe, M.; Suga, K.; Yokozawa, T. *J. Polym. Sci., Part A: Polym. Chem.* **1989b**, *27*, 1435.
5. Cho, I.; Lee, J. Y. *Makromol. Chem., Rapid Commun.* **1984**, *5*, 263.
6. Takahashi, T. *J. Polym. Sci., Part A1* **1968**, *6*, 403.
7. Endo, T.; Suga, K. *J. Polym. Sci., Part A: Polym. Chem.* **1989c**, *27*, 1831.
8. Zhulin, V. M.; Gonkberg, M. G.; Volchek, A. R.; Nefedox, O. M.; Shashkov, A. S. *Vysokomol. Soedin., Ser. A* **1971**, *13*, 2153.
9. Bailey, W. J.; Chen, P. Y.; Chen, S. C.; Chiao, W. B.; Endo, T.; Gapud, B.; Kuruganti, V.; Lin, Y. N.; Ni, Z.; Pan, C. Y.; Shaffer, S. E.; Sidney, L.; Wu, S. R.; Yamamoto, N.; Yamazaki, N.; Yonezawa, K.; Zhou, L. L. *Makromol. Chem. Macromol. Symp.* **1986**, *6*, 81.
10. Bailey, W. J.; Ni, Z.; Wu, S. R. *J. Polym. Sci., Polym. Chem. Ed.* **1982a**, *20*, 3021.
11. Endo, T.; Bailey, W. J. *J. Polym. Sci., Polym. Lett. Ed.* **1975a**, *13*, 193.
12. Bailey, W. J.; Gapud, B. *Polym. Prepr.* **1984c**, *25*(1), 58.
13. Morariu, S.; Simionescu, B. C. in preparation.
14. Hiraguri, Y.; Endo, T. *J. Polym. Sci., Part C: Polym. Lett.* **1989d**, *27*, 1.
15. Bailey, W. J.; Gapud, B.; Lin, Y. N.; Ni, Z.; Wu, S. R. *Polym. Prepr.* **1984d**, *25*(1), 142.
16. Yokozawa, T.; Takagi, J.; Endo, T. *J. Polym. Sci., Part A: Polym. Chem.* **1989**, *27*, 291.
17. Yokozawa, T.; Takagi, J.; Endo, T. *J. Polym. Sci., Part C: Polym. Lett.* **1990b**, *28*, 279.

RING-OPENING POLYMERIZATION, RADICAL (with No Shrinkage in Volume)

Takeshi Endo* and Fumio Sanda
Research Laboratory of Resources Utilization
Tokyo Institute of Technology

*Author to whom correspondence should be addressed.

There are monomers that undergo radical ring-opening polymerization, although the number of such monomers is less than those used for cationic, anionic, and coordination polymerizations. Cyclic disulfide, tetrafluoroethylene sulfide, vinylcyclopropanes, vinyloxiranes, vinylthiirane, vinylcyclobutane, vinyltetrahydrofuran, methylene oxetane, methylene tetrahydrofuran, methylene tetrahydropyran, cyclic vinylsulfones, methylene spiro orthoesters, methylene spiro orthocarbonates, methylene dioxolanes, cyclic ketene acetals, and *o*-xylylene dimers are the monomers that have been reported to undergo radical ring-opening polymerization.[1–15]

We have developed monomers that not only undergo radical ring-opening polymerization but also show low shrinkage or expansion in volume of polymerization. The studies since 1990 are summarized in this article because the studies previous to 1990 have already been reviewed elsewhere.[16]

MONOMERS THAT UNDERGO RADICAL POLYMERIZATION WITH LOW SHRINKAGE IN VOLUME

Vinylcyclopropane

Radical polymerization of 1,1-disubstituted 2-vinylcyclopropanes (**1**) has been reported to afford polymers (**2**) selectively obtained by 1,5-ring-opening polymerizations.[3]

Radical copolymerization of 1,1-bis(ethoxycarbonyl)-2-vinylcyclopropane (ECVCP) and methyl methacrylate (MMA) afforded copolymer obtained by the copolymerization proceeded through ring-opening followed by intramolecular cyclization.[18] The degree of volume shrinkage on the copolymerization decreased as the content of ECVCP increased.

Methylene Dioxolane

Radical polymerization of 4-methylene-1,3-dioxolane (R′ = H) afforded poly(ketone) with complete elimination of benzophenone.[13] The copolymer obtained showed photodegradability.[19]

Spiro Orthocarbonate (SOC)

SOCs show volume expansion during cationic double ring-opening polymerization. Therefore, SOCs bearing exomethylene groups can be candidate monomers that show volume expansion on radical polymerization.

Unsymmetrical SOCs bearing exomethylene groups at α-position of ether oxygen (**15**) afforded polymers with ring-opened (**17**) and vinyl polymerized units (**18**). Similarly unsymmetrical SOCs (**16**) bearing exomethylene groups at β-position of ether oxygen also afforded polymers with two unit (**19, 20**).[20] The degree of ring-opening depended upon the ring number and steric hindrance of exomethylene group. Volume shrinkage of SOCs lowered as the degree of ring-opening increased (**Structure 1**).

Vinylcyclopropanone Cyclic Acetal

1,1-Disubstituted 2-vinylcyclopropanes (**1**) undergo radical ring-opening polymerization to give polymers bearing mainly a 1,5-ring-opening unit (**2**).[3,17] Furthermore, cyclic ketene acetals (**21**) undergo radical polymerization to give mainly polyesters (**22**) via ring-opening isomerization process (**Structure 2**).[14]

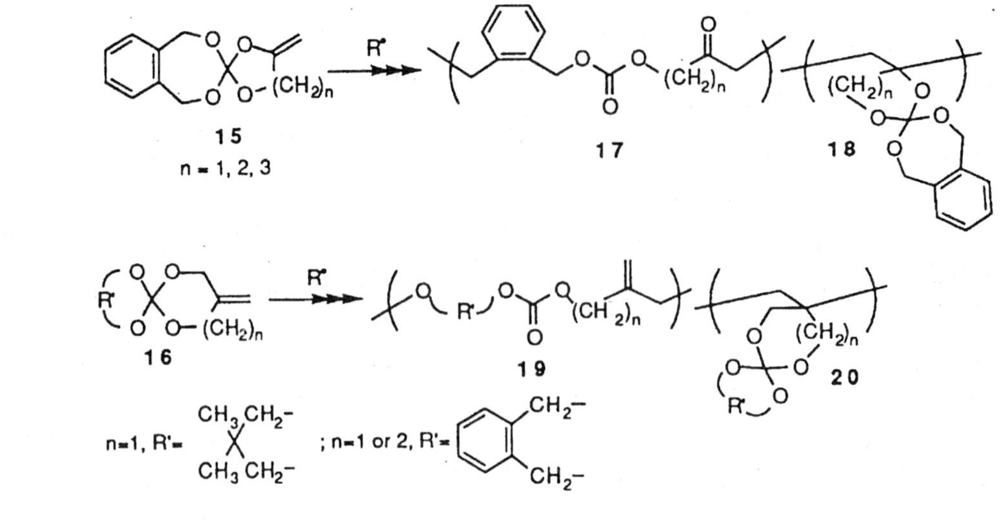

1

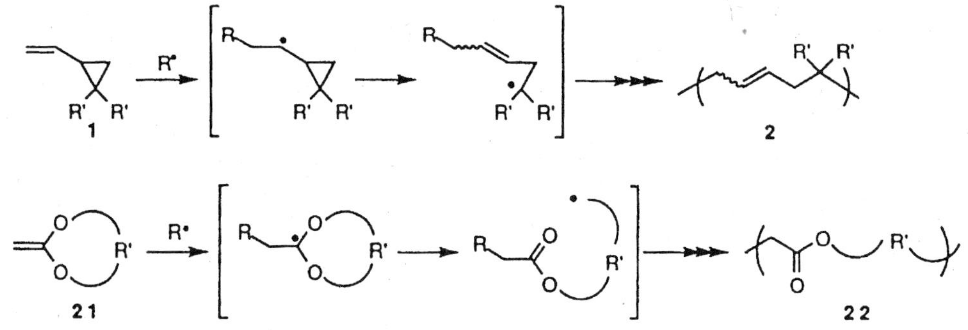

2

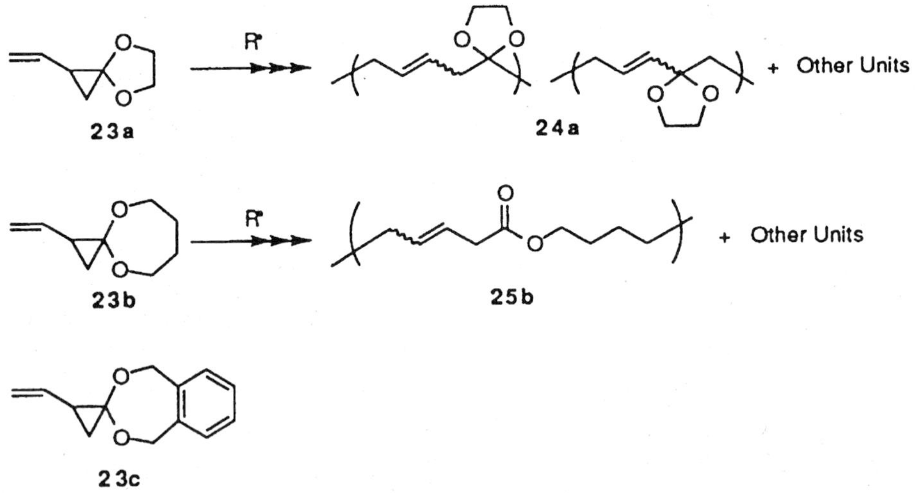

3

The radical polymerization mode of vinylcyclopropanone cyclic acetals (**23**), hybrid monomers containing vinylcyclopropane, and cyclic ketene acetal structures, depended on the number of ring members.[21] **23a**-bearing dioxolane ring afforded polymer (**24a**) formed by single ring-opening of vinylcyclopropane moicty. On the contrary, **23b**-bearing dioxepane ring mainly afforded polyester (**25b**) formed by double ring-opening, and **23c** also mainly afforded double ring-opening polymer. Volume shrinkage of **23c** was as small as 3.2% at 120°C (**Structure 3**).

SUMMARY

In this article, molecular design, synthesis, and polymerization behavior of monomers that undergo radical ring-opening polymerization to show how volume shrinkage or expansion on polymerization have been summarized. We wish to accomplish syntheses of monomers that show expansion in volume on polymerization according to the molecular design, which involves concepts such as large intermolecular interaction to develop a universal concept on volume expansion. For practical uses such as precision materials, adhesives, and so on, properties of polymers obtained are also important in addition to volume change on polymerization. We hope our work will become the basis for the development of functional materials that show expansion in volume during radical polymerization and curing processes.

REFERENCES

1. (a) Tobolsky, A. V.; Baysal, B. *J. Am. Chem. Soc.* **1953**, *75*, 1757. (b) Whiney, R. B.; Calvin, M. *J. Chem. Phys.* **1955**, 1750. (c) Barltrop, J. A.; Hayes, P. M.; Calvin, M. *J. Am. Chem. Soc.* **1954**, *76*, 4348.
2. Brasen, W. R.; Cripps, H. N.; Bottomley, C. G.; Farlow, M. W.; Krespan, C. G. *J. Org. Chem.* **1965**, *30*, 4188.
3. (a) Takahashi, T.; Yamashita, I. *Bull. Chem. Soc. Jpn.* **1964**, *37*, 131. (b) Takahashi, T.; Yamashita, I. *J. Polym. Sci.* **1965**, *B-3*, 251, (c) Takahashi, T. *J. Polym. Sci. A-1* **1968**, *6*, 403. (d) Lishanskii, I. S.; Zak, A. G.; Fedorova, E. F.; Khachaturov, A. S. *Vysokomolekul. Soedin. Ser. A* **1965**, *7*, 966, (e) Lishanskii, I. S.; Khachaturov, A. S. *Vysokomolekul. Soedin. Ser. A* **1967**, *9*, 1895. (f) Lishanskii, I. S.; Forma, O. S. *Vysokomolekul. Soedin Ser. A* **1969**, *11*, 1398. (g) Cho, I.; Ahn, K.-D. *J. Polym. Sci., Polym. Lett.* **1977**, *15*, 751. (h) Cho, I.; Ahn, K.-D. *J. Polym. Sci., Polym. Chem.* **1979**, *17*, 3169. (i) Cho, I.; Lee, J.-Y. *Makromol. Chem., Rapid Commun.* **1984**, *5*, 263. (j) Cho, I.; Song, S. S. *Makromol Chem., Rapid Commun.* **1989**, *10*, 85. (k) Endo, T.; Watanabe, M.; Suga, K.; Yokozawa, T. *J. Polym. Sci., Part A: Polym. Chem.* **1987**, *25*, 3039. (l) Endo, T.; Watanabe, M.; Suga, K.; Yokozawa, T. *Makromol. Chem.* **1989**, *190*, 691. (m) Endo, T.; Watanabe, M.; Suga, K.; Yokozawa, T. *J. Polym. Sci., Part A: Polym. Chem.* **1989**, *27*, 1435. (n) Endo, T.; Suga, K. *J. Polym. Sci., Part A: Polym. Chem.* **1989**, *27*, 1831.
4. (a) Cho, I.; Kim, J.-B. *J. Polym. Sci., Polym. Lett.* **1983**, *21*, 433. (b) Endo, T.; Kanda, N. *J. Polym. Sci., Polym. Chem.* **1985**, *23*, 1931. (c) Endo, T.; Suga, K.; Orikasa, Y.; Kojima, S. *J. Applied Polym. Sci.* **1989**, *37*, 1815. (d) Ito, O.; Ishizuka, T.; Iino, M.; Matsuda, M.; Endo, T.; Yokozawa, T. *Int. J. Chem. Kinet.* **1991**, *23*, 853. (e) Koizumi, T.; Nojima, Y.; Endo, T. *J. Polym. Sci., Part A: Polym. Chem.* **1993**, *31*, 3489.
5. (a) Lautenschlaeger, F.; Schnecko, H. *J. Polym. Sci., Part A-1* **1970**, *8*, 2579. (b) Suga, K.; Endo, T. *J. Polym. Sci., Part C: Polym. Lett.* **1989**, *27*, 381.
6. Hiraguri, Y.; Endo, T. *J. Polym. Sci., Polym. Lett.* **1989**, *27*, 333.
7. Salon, M. C.; Gandini, A.; Gey, C. *Makromol. Chem., Rapid Commun.* **1988**, *9*, 539.
8. Sidney, L. N.; Shaffer, S. E.; Bailey, W. J. *ACS Polym. Prepr.* **1981**, *22*(2), 373.
9. Bailey, W. J.; Chen, P. Y.; Chen, S.-C.; Chiao, W.; Endo, T.; Gapud, B.; Lin, Y.-N.; Ni, Z.; Pan, C.-Y.; Shaffer, S. E.; Sidney, L.; Wu, S.-R.; Yamamoto, N.; Yamazaki, N.; Yonezawa, K. *J. Mocromol. Sci. Chem.* **1984**, *A21*, 1611.
10. (a) Cho, I.; Kim, S.-K.; Lee, M.-H.; *J. Polym. Sci., Polym. Symp.* **1986**, *74*, 219. (b) Cho, I.; Lee, M.-H. *J. Polym. Sci., Polym. Lett.* **1987**, *25*, 309. (c) Cho, I.; Choi, S. Y. *Makromol. Chem., Rapid. Commun.* **1991**, *12*, 399.
11. (a) Endo, T.; Bailey, W. J. *J. Polym. Sci., Polym. Lett.* **1980**, *18*, 25. (b) Endo, T.; Okawara, M.; Yamazaki, N.; Bailey, W. J. *J. Polym. Sci., Polym. Chem.* **1981**, *19*, 1283. (b) Han, Y. K.; Choi, S. K. *J. Polym. Sci., Polym. Chem.* **1983**, *21*, 353. (c) Pan, C.-Y.; Wang, Y.; Huang, W.-Y. *Makromol. Chem., Rapid Commun.* **1986**, *7*, 627. (d) Pan, C.-Y.; Lu, S.-X.; Bailey, W. J. *Makromol. Chem.* **1987**, *188*, 1651. (e) Pan, C.-Y.; Wang, Y.; Bailey, W. J. *J. Polym. Sci., Part A: Polym. Chem.* **1988**, *26*, 2737.
12. (a) Endo, T.; Bailey, W. J. *J. Polym. Sci., Polym. Lett.* **1975**, *13*, 193. (b) Endo, T.; Bailey, W. J. *J. Polym. Sci., Polym. Chem.* **1975**, *13*, 2525. (c) Bailey, W. J.; Endo, T. *J. Polym. Sci., Polym. Symp.* **1978**, *64*, 17. (d) Thompson, V. P.; Williams, E. F.; Bailey, W. J. *J. Dent. Res.* **1979**, *58*, 1522. (e) Tagoshi, H.; Endo, T. *J. Polym. Sci., Part A: Polym. Chem.* **1989**, *27*, 1415. (f) Sugiyama, J.; Yokozawa, T.; Endo, T. *J. Polym. Sci., Part A: Polym. Chem.* **1990**, *28*, 3529. (g) Stansbury, J. W. *J. Dent. Res.* **1992**, *71*, 1408.
13. (a) Pan, C.-Y.; Wu, Z.; Bailey, W. J. *J. Polym. Sci., Part C: Polym. Lett.* **1987**, *25*, 243. (b) Hiraguri, Y.; Endo, T. *J. Am. Chem. Soc.* **1987**, *109*, 3779. (c) Pan, C.-Y.; Wu, Z.; Zhu, Q.-R.; Bailey, W. J. *J. Makromol. Sci., Chem.* **1987**, *A25*, 27. (d) Gong, M. S.; Chang, S.-I.; Cho, I. *Makromol. Chem., Rapid Commun.* **1989**, *10*, 201. (e) Cho, I.; Lee, T.-W. *Makromol. Chem., Rapid Commun.* **1989**, *10*, 453. (f) Hiraguri, Y.; Endo, T.; *J. Polym. Sci. Polym., Part A: Polym. Chem.* **1989**, *27*, 4403. (g) Hiraguri, Y.; Endo, T. *J. Polym. Sci., Part A: Polym. Chem.* **1989**, *27*, 2135. (h) Hiraguri, Y.; Endo, T. *J. Polym. Sci., Part A: Polym. Chem.* **1990**, *28*, 2881. (i) Cho, I.; Kim, B.-G.; Park, Y.-C.; Kim, C.-B.; Gong, M.-S. *Makromol. Chem., Rapid Commun.* **1991**, *12*, 141. (j) Hiraguri, Y.; Endo, T. *J. Polym. Sci., Part A: Polym. Chem.* **1992**, *30*, 689.
14. (a) Bailey, W. J.; Ni, Z.; Wu, S.-R. *J. Polym. Sci., Polym. Chem.* **1982**, *20*, 3021. (b) Bailey, W. J.; Wu, S.-R.; Ni, Z. *Makromol. Chem.* **1982**, *183*, 1913. (c) Cho, I.; Gong, M. S. *J. Polym. Sci., Polym. Lett.* **1982**, *20*, 361. (d) Bailey, W. J.; Ni, Z.; Wu, S.-R. *Macromolecules* **1982**, *15*, 711. (e) Bailey, W. J.; Wu, S.-R.; Ni, Z. *J. Macromol. Sci., Chem.* **1982**, *A18*, 973. (f) Endo, T.; Okawara, M.; Bailey, W. J.; Azuma, K.; Nate, K.; Yokono, H. *J. Polym. Sci., Polym. Lett.* **1983**, *21*, 373. (g) Endo, T.; Yako, N.; Azuma, K.; Nate, K. *Makromol. Chem.* **1985**, *186*, 1543. (h) Acar, M. H.; Nambu, Y.; Yamamoto, K.; Endo, T. *J. Polym. Sci., Part A: Polym. Chem.* **1989**, *27*, 4441. (i) Yokozawa, T.; Hayashi, R.; Endo, T. *J. Polym. Sci., Part A: Polym. Chem.* **1990**, *28*, 3739. (j) Hiraguri, Y.; Tokiwa, Y. *J. Polym. Sci., Part A: Polym. Chem.* **1993**, *31*, 3159.
15. Errede, L. A. *J. Polym. Sci.* **1961**, *49*, 253.
16. (a) Bailey, W. J.; Chou, J. L.; Feng, P.-Z.; Kuruganti, V.; Zhou, L.-L. *Acta Polym.* **1988**, *39*, 335. (b) Bailey, W. J.; Chou, J. L.; Feng, P.-Z.; Issari, B.; Kuruganti, V.; Zhou, L.-L. *J. Macromol. Sci., Chem.* **1988**, *A25*, 781. (c) Endo, T.; Yokozawa, T. in *New Method for Polymer Synthesis*; Mijs, W. J., Eds.; Plenum: New York, 1992, p 155.
17. (a) Sanda, F.; Takata, T.; Endo, T. *Macromolecules* **1993**, *26*, 1818. (b) Sanda, F.; Takata, T.; Endo, T. *Macromolecules* **1994**, *27*, 3986.
18. Sanda, F.; Takata, T.; Endo, T. *Macromolecules* **1994**, *27*, 3982.
19. Koizumi, T.; Hasegawa, Y.; Takata, T.; Endo, T. *J. Polym. Sci., Part A: Polym. Chem.* **1994**, *32*, 3193.
20. Sanda, F.; Takata, T.; Endo, T. *Macromolecules* **1993**, *26*, 737.
21. (a) Sanda, F.; Takata, T.; Endo, T. *J. Polym. Sci., Part A: Polym. Chem.* **1993**, *31*, 2659. (b) Sanda, F.; Takata, T.; Endo, T. *Macromolecules* **1994**, *27*, 1099.

RNA

See: Biopolymers (Overview)
Nucleic Acid Analogs

Rod-Like Polymers

See: Hairy Rod-Like Polymers
Polyimides, Precursors (Dye-Containing Side-Groups)

Rosins

See: Vegetal Biomass (2. Oligomers and their Polymerization)

Rotaxanes

See: Functionalized Polymers (Comb, Rotaxanic and Dendrimeric Structures)
Polyrotaxanes (Overview)

Rotomolding Polymers

See: Polyethylene (Commercial)

RUBBER ADDITIVES, MULTIFUNCTIONAL

Basudam Adhikari, Sukumar N. Maiti, and Tapan Kumar Khanra
Materials Science Centre
Indian Institute of Technology

Ingredients in a rubber compound control the processing and fabrication characteristics as well as the service performance. The nature and number of the additives to be used depend on the chemical nature of the base polymers and the properties and performance demands of the finished rubber products. Often, as many as thirty different additives are incorporated in a single rubber compound to obtain a broad spectrum of properties. Critical analysis of the state-of-the-art in relation to rubber compounding technology may generate the following conclusions:

1. Formulation of a rubber compound for a specific end use needs awareness of the properties, amounts, and sequence of incorporation of a large number of additives.
2. Incorporation and uniform dispersion of the components must be done by careful control of the mixing conditions to avoid premature interactions between the components.
3. Increase of the number of components enhances mixing period, affecting the production economy.
4. Reproducibility of the properties from batch to batch may be affected.
5. Complexity of the vulcanizate due to the presence of many ingredients makes the analysis of a rubber product very difficult for quality control.

These points arise owing to the curative, stabilization, and processing aid systems, although the interaction between the fillers and other components cannot be ruled out.

This situation demands minimization of the number of ingredients while deriving a broad spectrum of properties. Reports during the last 8–10 years show the initiation of research work in this direction to get multifunctional activities from a single ingredient. The results have shown some positive achievements for multifunctional rubber additives in commercial production.

Based on the function during processing, fabrication, and service performance, multifunctional additives reported so far may be grouped into the following categories:

1. Antidegradants cum cure activators and processing aids,
2. Accelerator cum retarder,
3. Antioxidant cum retarder,
4. Curing agents cum processing aids,
5. Curing agents cum antioxidants,
6. Accelerator cum antioxidants, and
7. Curing agent, processing aid, dispersant, and mold release agents.

CONCLUSION

The concept of multifunctional additives for processing and vulcanization of elastomers is novel. Some of these additives seem to have promising and interesting application potential; but they must be made commercially more attractive. Many developmental and associated problems still await solutions before they can be considered seriously for commercialization. One of the limitations is the problem of dispersion of such additives in the rubber compound for uniform performance of the vulcanizate. The cost of production of the multifunctional additives may also pose a problem for their competitiveness with the conventional rubber additives. Some of the multifunctional additives do not appear to be encouraging because they create more problems in compounding and vulcanization of rubbers rather than solving them. Moreover, the claims of improved performance for such additives are not always based on careful studies. Unless all these and other associated challenges are adequately met, the multifunctional rubber additives will exist as only laboratory curiosities.

Rubber-Modified Polymers

See: Blends (Amorphous Rubbery Mixtures)
Deformation and Fracture (Micromechanical Mechanisms)
Epoxies, Rubber-Modified
Epoxy Acrylate-Based Resins
Glassy Polymers (Toughening Via Dilatational Plasticity)
Poly(butylene terephthalate)/Polycarbonate Blends
Polystyrene, High Impact
Polystyrene, Rubber Toughened
Reactive Liquid Rubbers (Epoxy Toughening Agents)
Unsaturated Polyester Resins (Toughening with Elastomers)
Unsaturated Polyester Resins (Toughened with Liquid Rubber)

Rubber Reinforcing Agents

See: Carbon Black
Carbon Black (Graft Copolymers)

Rubber-Toughened Polymers

See: Rubber-Modified Polymers

Rubbers

See: Acrylic Elastomer Curing
Acrylic Rubber
Chloroprene Rubber

Diene Rubbers, Conventional (Ozone Degradation and Stabilization)
Epoxidized Natural Rubber
Interpenetrating Polymer Networks, Rubber-Based
Natural Rubber (Chemical Modification)
Natural Rubber (Hevea brasiliensis)
Natural Rubber Latex (Polymeric Flocculant for Tin Tailings Slurries)
Polybutadienes
Reactive Liquid Rubbers (Epoxy Toughening Agents)
Rubber Additives, Multifunctional
Tack, Elastomers
Vulcanization (Overview)

Ruthenium-Containing Polymers

See: *Ferrocene-Backbone Polymers*

S

SAN

See: Styrene-Acrylonitrile Copolymers

SBS

See: Styrene-Butadiene Copolymers

SCALE INHIBITORS, POLYMERIC

Ernest Senogles* and W. O. S. Doherty
Department of Molecular Sciences
James Cook University

O. L. Crees
Sugar Research Institute

Many industries experience scaling of metal heat exchanger surfaces. These encrustations are formed from the combined effects of several processes involving inorganic and/or organic molecules or ions. The rate and extent of these processes are influenced by physical, chemical, and sometimes, biological factors. The characteristics of the scale formed depend on a number of parameters, including the concentrations of lattice forming ions, the amounts of dissolved and suspended solids, the pH and flow properties of the solution, the rate of evaporation, and the operating temperature (and pressure) of the system.[1] Scale facilitates the corrosion of surfaces, restricts fluid flow, and—because it has low thermal conductivity—reduces the quantity of heat that is transferred across the metal fluid boundary.[2,3] As a consequence, scaling is a major operational problem, and the heat exchanger units have to be cleaned from time to time by chemical and/or mechanical means to maintain optimum output. This invariably adds to operation costs.

Naturally occurring polymers such as starch, gum tragacanth, carboxymethyl cellulose, algins, tannins, lignosulfonates, and gelatin were first used for scale control. A later approach was the use of chelating agents that form soluble complexes with lattice forming ions. Because stoichiometric amounts and a precise range of pH conditions are necessary, their applications are both costly and limited. More recently, substoichiometric amounts of synthetic water-soluble low molar mass polymers or copolymers, which are thermally stable and provide corrosion protection for the heat exchanger surfaces, have been employed.[1,4–12]

PREPARATION

Polymeric scale inhibitors may contain cationic, anionic, nonionic, or ampholytic groups. The majority of polymers are based on acrylic or maleic acid. Low molar mass samples are found to be the most effective; thus, it is usually necessary to use high initiator concentrations and/or radical transfer agents in the preparation of these polymers.

PROPERTIES

Mechanism and Kinetics of Crystallization

Scale inhibitors function in substoichiometric amounts by interfering in the nucleation and/or the crystal growth processes. The inhibitor is adsorbed at active sites on the nuclei or microcrystalline surfaces, blocking or restricting further growth. As a result, the onset of crystallization can be delayed and crystallization rates reduced. Crystal habit and size can also be affected. These physical changes may result in the formation of a scale which adheres less easily to a surface than that formed in the absence of the inhibitor. Low molar mass polymers are particularly effective inhibitors. They also function as dispersing and to a lesser extent chelating agents.

Chemical Composition and Structure

Polymers that possess functional groups capable of hydrogen bonding often are effective scale inhibitors.[13] Thus, polymers containing carboxylic, sulfonic, phosphoric, phosphinic, and phosphonic groups show high activity. Others containing hydroxyl, amino or amide groups are less effective.[14–16] Specific factors are also important.

Molar Mass

Several ingredients have shown that the lower the molar mass of a polymer, the more effective it is as a scale inhibitor.[4,16,18,19,23,24] Nevertheless, there is an optimum molar mass *below* which inhibitor efficiency is reduced. *Above* this molar mass, polymers become progressively worse inhibitors as a result of the increased bridging ability and decreased solution mobility of the larger molecules. Furthermore, as the molar mass increases, the sequestering ability of the polymer for ions such as calcium also rises because of the higher charge density per molecule. At the same time, the solubility of the metal-polymer complex formed declines, so that a smaller proportion of the polymer is available for crystal adsorption.[22] On the other hand, if the molar mass of the polymer is too low, adsorption onto active growth centers can be inadequate and lead to a reduced inhibitory capacity.

Concentration and pH

As polymer concentration increases, there is a rapid increase in inhibition efficiency, after which only a slight increase is observed. In most applications, the polymer is used at the parts per million (ppm) level and is introduced as an aqueous solution.

Changes in pH can alter the nature of the precipitate phase and the surface charge of the crystallites formed; both of these factors influence crystallization. It also affects the degree of ionization of polyelectrolytes.

Temperature and Thermal Stability

Temperature may directly affect the phase that predominates in the scale and the crystal morphology.[27] The ability of the

*Author to whom correspondence should be addressed.

polymer to inhibit scale formation decreases as the operating temperature increases.[21,22,24]

APPLICATIONS

Polymeric scale inhibitors find use in a wide range of applications; some important ones are summarized below.

Water Systems

The heating of water as in desalination of sea water by distillation, boilers, and process cooling systems, results in the precipitation of sparingly soluble salts and the agglomeration of suspended solids. Reviews and articles on cooling and boiling water systems are available.[1,20,28,29]

In some systems the use of a single inhibitor is not effective because of the multicomponent nature of the scale and in some cases its high calcium content. At high calcium ion concentrations polycarboxylates can be partially complexed thereby reducing their efficiency.[22] The ideal formulation is one that contains a threshold inhibitor, a corrosion inhibitor, a crystal modifier, a chelating agent, and a dispersant. A threshold inhibitor is one that can prevent precipitation for a long period of time; thus, it delays the nucleation process or its rate.

Geothermal and Oil Wells

For the majority of geothermal and oil wells, the commonly encountered deposits include calcium carbonate, calcium sulfate, strontium sulfate, barium sulfate (barite), calcium fluoride (fluorite), iron carbonate, iron oxide, iron sulfide, iron disulfide, silica, and silicates. Some naturally occurring radioactive compounds that include radium-226, radium-228, and potassium-40 have also been detected.[30–34] Radium sulfate usually co-precipitates with barite.

Many studies have been carried out on the inhibition of barium sulfate formation, since it is one of the most intractable scaling components because of its very low solubility.[25,26,33–35]

Sugar Evaporators

The scaling of sugar mill evaporators is a consequence of the inorganic and organic non-sugar impurities present in sugar juice.

The calcium content of cane juice is higher than that of beet juice and its pH over the evaporation process is usually between 6.8 and 7.4, whereas in the case of beet juice this range is 8–9. Because of these factors, scale inhibitors perform differently in these industries. Polymers based on acrylic or maleic acid are currently used.

In a comprehensive program designed to provide information on the performance of these polymers in the Australian cane sugar industry, our laboratory has carried out detailed analyses of the chemical composition of evaporator scales using various state-of-the-art techniques. A large number of components have been identified and the scale compositions varies with both mill and evaporator unit.[36] Major components often comprise one or more of the following: calcium oxalate, dihydrate, calcium oxalate monohydrate, hydroxyapatite, calcium sulfate dihydrate, silica, and organic matter. Studies on the formation of the first four of these from pure solutions of their lattice ions suggest that for the oxalates, sodium polymetaphosphate (CAS Registry No. 50813-16-6), sodium polyaspartate (CAS Registry No. 94525-01-6), and poly(maleic acid) interfere with the nucleation and/or crystal growth, so that slower growth occurs and the morphology is affected.[17] On the other hand, acrylic acid based polymers act as additional nucleation sites, promoting growth with the formation of small crystals. All the above-mentioned polymers prevent the growth of hydroxyapatite and calcium sulfate dihydrate nuclei for prolonged periods, the extent of which depends on the concentration employed.

We conclude from the research described above that polymers will be more effective in mills if they are added to the sugar juice over all the evaporation units rather than before the first. Furthermore, for optimum results, it is likely that different polymers will need to be added to the different units and that these will vary from mill to mill. A combination of polymers, as well as perhaps of other antiscalants, may provide the greatest degree of control of some scales.

REFERENCES

1. Hays, G. F.; Thomas, P. A.; Libutti, B. L. *Proc. Engineering Found, Conf.* Pennsylvania; Engineering Foundation: New York, 1982; 489.
2. Schesinger, H. Z. *Tech. Phys.* **1931**, *12*, 33.
3. Staub, S.; Paturau, M. *Principle of Sugar Technology* Elsevier: New York, 1963; Vol. 111, Chapter 11.
4. Smith, B. R. *Desalination* **1967**, *3*, 263.
5. Sexsmith, D. R.; Phelan, J. V.; Savinelli, E. A. *Int. Water Conf.* Pittsburgh, Pennsylvania, 1971.
6. Vetter, O. J. *J. Pet. Techn.* 997, 1972.
7. Logan, D. P.; Walker, J. L. *The Int. Congress on Techn. and Techn. Exchange*; International Technology Institute: Pennsylvania, PA, 1982.
8. Ralston, H.; Whitney, L.; Walker, J. U.S. Patent 4 452 703, 1984.
9. Gill, J. S.; Rey, S. P.; Wiernik, J. H. U.S. Patent 4 933 090, 1990.
10. Emmons, D. H.; Dodd, F. W.; Kinsella, M. A. U.S. Patent 5 167 828, 1992.
11. Perez, L. A. Canadian Patent 2 083 453, 1993.
12. Wood, L. L. U.S. Patent 5 286 810, 1994a.
13. Breen, P. J.; Downs, H. H.; Diel, B. N. *Spec. Publ.-R. Soc. Chem. (Chem. Oil Ind.: Dev. Appl.)* **1991**, *97*, 186.
14. McCartney, E. R.; Alexander, A. E. *J. Colloid Sci.* **1958**, *13*, 383.
15. Kuntze, R. A. *Nature* **1966**, *211*, 406.
16. Crawford, J. E.; Crematy, E. P.; Alexander, A. E. *Aust. J. Chem.* **1968**, *21*, 1067.
17. Crees, O. L.; Doherty, W. O. S.; Senogles, E. *Preprint 3rd Pacific Polymer Conf.* 801; Polymer Division Royal Australian Chemical Institute: Melborne, 1993.
18. Sexsmith, D. R.; Petrey, E. Q. *Desalination* **1973**, *13*, 87.
19. Solomon, D. H.; Rolfe, P. F. *Desalination* **1966**, *1*, 260.
20. Fivizzani, K. P.; Hoots, J. E.; Cloud, R. W. *Surface Reactive Peptides and Polymers; Discovery and Commercialization*; ACS Symposium Series 444; Sikes, C. S.; Wheeler, A. P., Eds.; The American Chemical Society: Washington, DC, 1991; Chapter 22.
21. Pierce, C. C.; Hoots, J. E. *Chem. Aspects Regul. Miner., Sym. Div. Ind. Eng. Chem.* American Chemical Society: 1988; 53.
22. Pierce, C. C.; Grattan, D. A. *Corrosion, Nat. Assoc. Corrosion Eng.*; St. Louis, MO; 1988; paper no. 205. National Association of Corrosion Engineering: Houston, TX, 1988.
23. Williams, F. V.; Ruehrwein, R. A. *J. Am. Chem. Soc.* **1957**, *79*, 4898.

24. Flesher, P.; Streatfield, E. L.; Pearce, A. S.; Hydes, O. D. *3rd. Int. Sym. on Fresh Water from the Sea* **1970**, *1*, 493.
25. Leung, W. H.; Nancollas, G. H. *J. Crystal Growth* **1978**, *44*, 163.
26. Leung, W. H.; Nancollas, G. H. *J. Inorg. Nucl. Chem.* **1978**, *40*, 1871.
27. Doherty, W. O. S.; Crees, O. L.; Senogles, E. *Cryst. Res. Technol.* **1994**, *29*(4), 517.
28. McCoy, J. W. *The Chemical Treatment of Boiler Water* Chemical: 1981.
29. McDonald, A. C.; Soos, J. L. U.S. Patent 4 931 206, 1990.
30. Smith, A. L. *J. Pet. Techn.* 697, 1987.
31. Gray, P. R. *J. Pet. Techn.* 12, 1993.
32. Gallup, D. L.; Featherstone, J. L. *Geothermal Resource Council Transaction* **1993**, *17*, 379.
33. Oddo, J. E. *Oil and Gas J.* 33, 1994.
34. Nancollas, G. H.; Purdie, N. *Trans. Faraday Soc.* **1963**, *59*, 735.
35. Liu, S. T.; Nancollas, G. H. *J. Colloid Interface Sci.* **1975**, *52*, 593.
36. Doherty, W. O. S.; Crees, O. L.; Senogles, E. *Cryst. Res. Technol.* **1993**, *28*(5), 603.

Scleroglucan

See: Water-Soluble Polymers (Oil Recovery Applications)

Sealants

See: Dental Sealants
Polysulfides (Prepared from Sulfur Dioxide)
Polysulfides (Use in Sealants)
Silicone Rubber Latex
Silicone Sealants

Seeded Polymerization

See: Monodisperse Polymer Particles (Methods of Preparation)
Monodisperse Polymer Particles (Preparation and Applications)

SELF-ASSEMBLED POLYMERS (at Interfaces)

Takahiro Seki
Research Laboratory of Resources Utilization
Tokyo Institute of Technology

This chapter will concentrate on the self-assembled polymeric films formed at interfaces, which is now starting to form a new and important field of ultrathin polymeric film technology.

We start with the molecular film preparation of low molecular mass compounds. In order to obtain molecularly controlled-ordered films, we may first consider the LB technique developed by Langmuir and Blodgett.[1] The alternative is the self-assembled (SA) chemosorption method proposed by J. Sagiv.[2] Both methods are based on the monomolecular adsorption and arrangement in the layer state at the interface between air and liquid phases (LB method) and between liquid and solid phases (SA method). Sagiv's method, which was reported in 1980, utilized the chemisorption of silane compounds on a glass or metal oxide surface. In 1983, R. G. Nuzzo and D. L. Allara introduced the procedure of the monolayer adsorption of sulfur-containing compounds on a gold surface.[3]

The same methodology seems to be, in principle, applicable for macromolecules. Indeed, many examples are reported with regard to polymer assembly at the interfaces. We have now a substantial accumulation of knowledge concerning polymeric LB films.[4–7] Nevertheless, the procedure for self-assembled polymers based on the direct chemisorption onto solid surfaces has been quite recently proposed, and no suitable introductory articles seem to have appeared in this new region. In this context, this chapter deals with the recent trends of studies on self-assembled polymer films formed mostly at the solid-liquid interface. For a detailed background of these areas, readers may refer to the excellent book written by A. Ulman.[8]

SELF-ASSEMBLY OF POLYMERS ON SOLID SURFACES

Monolayer Formation

To prepare a fixed self-assembled organic molecular film on a solid substrate, we have to consider molecules that contain a strongly binding moiety (frequently termed as a tethering or anchoring group) onto the solid surface. Similarly to the low molecular mass derivatives, the following two methods are feasible: the use of trichlorosilane derivatives for metal oxide surfaces and of thiol or disulfide derivatives for metal (mostly gold) surfaces. The concept of a spontaneous assembly of polymers onto a solid surface is shown in **Figure 1**. The sulfur/gold systems have attracted an increasing interest in recent years because more ordered and sophisticated monolayers are formed in these systems. Trichlorosilanes, in contrast, have a polymeric

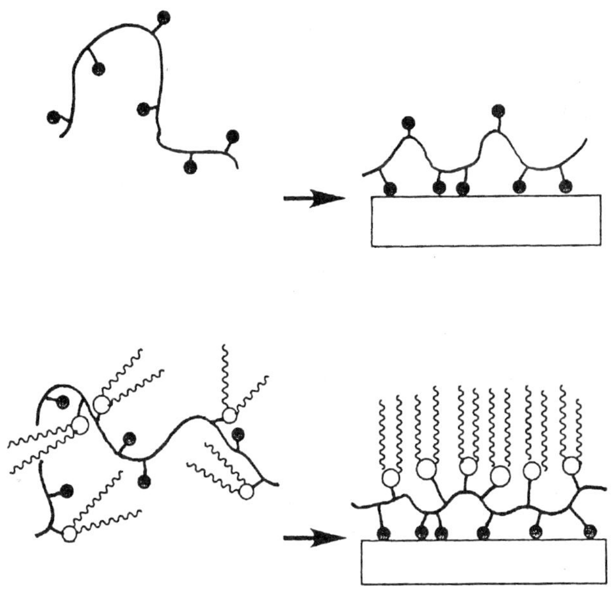

FIGURE 1. The preparative concept of a spontaneously self-assembled polymer film having an anchoring (tethering) unit on a solid substrate. The polymer bears the anchoring unit only (upper) or both the anchoring and the self-organizing units (lower). The shadowed circle represents the anchoring unit.

nature and often lead to disordered layers. Therefore, the fabrication of self-assembled ultrathin polymer films is mostly carried out in the sulfur/gold systems.

J. M. Stouffer and T. J. McCarthy first reported the spontaneous adsorption of sulfur-functionalized polymer onto gold surfaces using polystyrene derivatives.[9] Some years later T. J. Lenk et al. prepared self-assembled films of sulfur-containing poly(methyl methacrylate)s on gold.[10] The same strategy was reported concurrently by F. Sun et al.[11] employing poly(hydroxyethyl acrylate)s bearing a disulfide anchoring unit.

The use of rigid rod-like macromolecules provides a new type of self-assembled polymer film. The synthesis of solvent-soluble thiophenol-terminated oligoimides, the monolayer formation of gold, and the structural characterization were undertaken by Kwan et al.[12] They reported the formation of densely packed monolayers having the tilt angle about 55–60°C from the surface normal. This self-assembled film was found to have a very different main chain orientation from that in LB films, which lay down in the substrate plane.

Multilayer Formation

To our knowledge, the ordered multi-layer formation of sulfur-tethered polymers on gold surfaces has not been reported for synthetic polymers.

G. Decher et al. demonstrated that it is possible to build up multilayered thin films of polymers on charged surfaces via the alternate deposition of polycations and polyanions.[13–16]

M. F. Rubner and co-workers proposed the fabrication of heterostructured multilayers of conjugated polyions such as poly(thiophen acetic acid) and sulfonated polyaniline by applying Decher's assembling technique.[17] They also showed that the same method is applicable for the p-type-doped state of conjugated polymers having no fixed ionic groups.[18] Such molecularly-controlled conjugated-molecular films may be of use in sensors and tailor-designed electrical and optical devices.

SELF-ASSEMBLY OF POLYMERS BENEATH FLOATING MONOLAYERS

Self-Assembly via Electrostatic Interaction

Monolayers of a charged amphiphile are known to form a polyion-complex at the air-water interface with an oppositely charged polyelectrolyte dissolved in the aqueous subphase; these monolayers are well suited for LB deposition.[19–26] On the stoichiometry of the complex, the X-ray photoelectron spectroscopy (XPS) analysis of the transferred film revealed formation of almost exact 1:1 ionic complexation.[19,22,23]

Self-Assembly via Hydrogen Bonding

In comparison with the electrostatic interaction stated above, there are to date few examples of hydrogen bonding at the monolayer interface. M. Niwa and co-workers synthesized amphiphilic block copolymer composed of poly(acrylic acid) or polyoxyethylene and polystyrene.[27] Interaction between the block copolymer and a hydrogen-bond-forming hydrophilic polymer in the subphase was reversibly controlled by pH, and this brought about the changes in the monolayer areas of the block copolymer.

Self-Assembly via Specific Site Recognition of Protein

More recently, K. Fujita et al. constructed a bilayer of streptavidin bridged by a bis(biotinyl)-terminated polypeptide at the air-water interface.[28] The rigidity of this helical polypeptide was to play an important role in the formation of a well-defined bilayer structure of biological macromolecules.

H. Morgan et al. demonstrated the formation of monolayers and multilayers of alternate streptavidin and bifunctional biotinyl ligand by self-assembly on gold.[29]

REFERENCES

1. Blodgett, K. *J. Am. Chem. Soc.* **1937**.
2. Sagiv, J. *J. Am. Chem. Soc.* **1980**, *102*, 92.
3. Nuzzo, R. G.; Allara, D. L. *J. Am. Chem. Soc.* **1983**, *105*, 4481.
4. Barraud, A.; Palacin, S., Eds.; *Thin Solid Films* (Proc. 5th Int. Conf. on Langmuir-Blodgett Films, Paris) 1991; 210/211.
5. Leblanc, R. M.; Salesse, C., Eds.; *Thin Solid Films* (Proc. 6th Int. Conf. on Organized Molecular Films) Trois-Rivieres, 1994; 242–244.
6. *Adv. Mater.* **1991**, *3* (Special Issue on Organic Thin Films).
7. Ringsdorf, H.; Schlarb, B.; Venzmer, J. *Angew. Chem. Int. Ed. Engl.* **1988**, *27*, 113.
8. Ulman, A. *An Introduction to Ultrathin Organic Films: From Langmuir-Blodgett to Self-Assembly*; Academic: San Diego, CA, 1991.
9. Stouffer, J. M.; McCarthy, T. J. *Macromolecules* **1988**, *21*, 1204.
10. Lenk, T. J.; Hallmark, V. M.; Rabolt, J. F.; Haussling, L.; Ringsdorf, H. *Macromolecules* **1993**, *26*, 1230.
11. Sun, F.; Grainger, D. W. *J. Polym. Sci. A, Polym. Chem.* **1993**, *31*, 1729.
12. Kwan, W. S. V.; Atanasoska, L.; Miller, L. L. *Langmuir* **1991**, *7*, 1419.
13. Decher, G.; Hong, J. D. *Ber. Bunsenges. Phys. Chem.* **1991**, *95*, 1430.
14. Decher, G.; Hong, J.-D. *Macromol. Chem., Macromol. Symposium* **1991**, *46*, 321.
15. Decher, G.; Hong, J.-D.; Schmitt, J. *Thin Solid Films* **1992**, *210/211*, 831.
16. Decher, G.; Lvov, Y.; Schmitt, J. *Thin Solid Films* **1994**, *244*, 772.
17. Ferreira, M.; Cheung, J. H.; Rubner, M. F. *Thin Solid Films* **1994**, *244*, 806.
18. Cheung, J. H.; Fou, A. F.; Rubner, M. F. *Thin Solid Films* **1994**, *244*, 985.
19. Shimomura, M.; Kunitake, T. *Thin Solid Films* **1985**, *132*, 243.
20. Nishiyama, K.; Kurihara, M.; Fujihira, M. *Thin Solid Films* **1989**, *179*, 477.
21. Erdelen, C.; Laschewsky, A.; Ringsdorf, H.; Schneider, J.; Schuster, A. *Thin Solid Films* **1989**, *180*, 153.
22. Takahara, A.; Morotomi, N.; Hirooka, S.; Higashi, N.; Kunitake, T.; Kajiyama, T. *Macromolecules* **1989**, *22*, 617.
23. Higashi, N.; Kunitake, K. *Chem. Lett.* **1986**, 105.
24. Kunitake, T.; Higashi, N.; Kunitake, M.; Fukushige, Y. *Macromolecules* **1989**, *22*, 485.
25. Shimomura, M.; Aiba, S. *Nippon Kagaku Kaishi* **1993**, 905.
26. Seki, T.; Tohnai, A.; Tamaki, T.; Ueno, K. *J. Chem. Soc., Chem. Commun.* **1993**, 1876.
27. Niwa, M.; Hayashi, T.; Higashi, N. *Langmuir* **1990**, *6*, 263.
28. Fujita, K.; Kimura, S.; Imanishi, Y.; Rump, E.; van Esch, J.; Ringsdorf, H. *J. Am. Chem. Soc.* **1994**, *116*, 5479.
29. Morgan, H.; Tailor, D. M.; D'Silva, C.; Fukushima, H. *Thin Solid Films* **1992**, *210/211*, 773.

SELF-ASSEMBLED POLYMERS (Photopolymerizable Bolaform Amphiphiles)

Matthew Tirrell, H. Ted Davis, and Guangzhao Mao*
Department of Chemical Engineering and Materials Science
University of Minnesota

Volker Hessel and Helmut Ringsdorf
Institut für Organische Chemie
Universität Mainz

A bola is a long string with balls at the end used for entangling cattle. Its molecular counterparts, bolaform amphiphiles, are α,ω-dipolar molecules each containing two hydrophilic moieties connected by a hydrophobic chain. They can be viewed as two conventional amphiphiles covalently linked through their hydrophobic chain ends. Natural bolaform amphiphile archaebacteria survive at high temperature and in acidic condition.[1] Lipid membranes consisting of bolaform amphiphiles show unusual resistance against fusion.[2,3] Bolaform amphiphiles have attracted interest in fundamental research because of their unusual aggregation behavior at an air-water interface and in a liquid medium.[4–10] Cationic bolaform amphiphiles with strongly associating hydrophobic parts adsorb on negatively charged mica surfaces and result in surface charge reversal.[11,12]

Self-assembled monolayers are molecular assemblies that are formed spontaneously by the immersion of an appropriate substrate into a solution of an active amphiphile.[13] The strong amphiphile-substrate interactions—chemical, hydrogen-bonding, or electrostatic in origin—push molecules to self-assemble at a solid-liquid interface.

It was clear at the beginning of thin film research that the low stability of organic films against chemical, mechanical, and thermal degradation is a major obstacle for their applications. Polymeric films usually show higher integrity than monomeric films.

Molecules undergoing topochemical reactions are attractive candidates because this type of polymerization is a lattice controlled reaction that requires minimum translational movement of the reactive species. Topochemical photopolymerization of phenylene diacrylic acid and cinnamic acid derivatives has already been reported but was only successful for highly ordered molecules, for example, in the crystalline state and in monolayers.[14–17] We have studied bolaform amphiphiles containing phenylene diacrylic acid derivative groups, self-assembled by physisorption into mono- and multilayers, and subsequently UV irradiated. The photoreaction was followed by UV spectroscopy and the structure and stability of the self-assembled films before and after photopolymerization were studied using the atomic force microscope (AFM) and the surface forces apparatus (SFA).

PREPARATION

Surface Preparation

A freshly cleaved muscovite-mica piece was placed in the amphiphile aqueous solution. Bolaform amphiphile multilayers were deposited by alternating adsorption of the cationic bolaform amphiphile and an anionic polymer. The anionic polymer, poly(2-acrylamide-2-methyl-1-propanesulfonic acid), was purchased from Aldrich.

PROPERTIES

Influence of Alkyl Chain Length and Counterion on Monolayer Structure

AFM and SFA were used to study the effect of the alkyl chain length and the degree of counterion association on molecular packing in self-assembled monolayers.

We have shown, with the help of SFA and AFM, that bolaform amphiphile self-assembled monolayer interfacial properties can be controlled to a certain degree by varying, the amphiphile chemical structure, in particular, the hydrophobic chain length and the type of counterions.

Reversal of Surface Charge by Bolaform Amphiphile Adsorption

We have established the method of alternating the self-assembly of a cationic bolaform amphiphile and an anionic polymer multilayer.[10] The sequential adsorption requires the recovery of the surface charge with the adsorption of each layer. An end-on adsorption of bolaform amphiphiles maintains a finite charge separation within the layer and thus reverses the sign of the surface charge.

Improvement of Film Stability by Photopolymerization

Photopolymerization is expected to reduce the rate of film desorption in an organic solvent. We were able to compare the chemical stability of monomeric and photopolymerized amphiphile films.

The much slower dissolution process of the photoirradiated film corroborates with the notion that photopolymerization in the monolayer film increases the film stability and integrity.

CONCLUSIONS

Ultrathin organic multilayer films are appealing to scientific researchers not only because they serve as molecular level models but also because they can be transformed into superlattices through the successive deposition of alternating layers of different materials.[18] Films containing both polymers and surfactants offer many interesting possibilities for manipulating their structure and properties. Self-assembly via physical adsorption is a particularly simple means of creating such films. It needs neither the formation of oriented films nor special equipment. G. Decher et al. introduced the concept of alternate adsorption of dipolar cationic/anionic amphiphiles, cationic/anionic polymers, and mixed systems of both dipolar amphiphiles and polymers.[19–21] Self-assembly of multilayers using sequential adsorption requires a complete recovery of the surface charge or of the surface functional groups. A total charge reversal is achieved by using ionic amphiphiles with more than one charge group or ionic polymer. Photopolymerizable bolaform amphiphiles provides a means not only to reverse the surface charge but also to stabilize the molecular assembly by photopolymerization. The photopolymerization process might be optimized by controlling the mono- and multilayer structures through selective variation of the amphiphile chemical structure.

*Author to whom correspondence should be addressed.

REFERENCES

1. Gliozzi, A.; Rolandi, R.; de Rosa, M.; Gambacorta, A. *J. Membr. Biol.* **1983**, *75*, 45.
2. Fuhrhop, J.-H.; David, H.-H.; Mathieu, J.; Liman, U.; Winter, H.-J.; Boekema, E. *J. Am. Chem. Soc.* **1986**, *108*, 1785.
3. Ringsdorf, H.; Schlarb, B.; Venzmer, J. *Angew. Chem. Int. Ed. Engl.* **1988**, *27*, 113.
4. Hessel, V.; Ringsdorf, H.; Laversanne, R.; Nallet, F. *Recl. Trav. Chim. Pays-Bas. 112*, 339.
5. Hessel, V.; Ringsdorf, H.; Festag, R.; Wendorff, J. H. *Makromol. Chem. Rapid Commun.* **1993**, *14*, 707.
6. Zana, R.; Benrraou, M.; Rueff, R. *Langmuir* **1991**, *7*, 1072.
7. Alami, E.; Levy, H.; Zana, R.; Skoulios, A. *Langmuir* **1993**, *9*, 940.
8. Alami, E.; Beinert, G.; Marie, P.; Zana, R. *Langmuir* **1993**, *9*, 1465.
9. Devinsky, F.; Lacko, I.; Imam, T. *J. Colloid Interface Sci.* **1990**, *143*, 336.
10. Kim, J.-M.; Thompson, D. A. *Langmuir* **1992**, *8*, 637.
11. Mao, G.; Tsao, Y.; Tirrell, M.; Davis, H. T.; Hessel, V.; Ringsdorf, H. *Langmuir* **1993**, *9*, 3461.
12. Mao, G.; Tsao, Y.-H.; Tirrell, M.; Davis, H. T.; Hessel, V.; van Esch, J.; Ringsdorf, H. *Langmuir* **1994**, *10*, 4174.
13. An Introduction to Ultrathin Organic Films: From Langmuir-Blodgett to Self-Assembly; Ullman, A., Ed.; Academic: New York, 1991.
14. Hasagawa, M. *Pure Appl. Chem.* **1968**, *58*, 1179.
15. Schmidt, G. M. *J. Pure Appl. Chem.* **1971**, *27*, 647.
16. Reiser, A.; Egerton, P. L. *Photogr. Sci. Eng.* **1979**, *23*, 144.
17. Hakanishi, F.; Fang, P.; Xu, Y. *J. Photopolym. Sci. Technol. 1* **1991**, *4*, 65.
18. Swalen, J. D.; Allara, D. L.; Andrade, J. D.; Chandross, E. A.; Garoff, S.; Israelachvili, J.; McCarthy, T. J.; Murray, R.; Pease, R. F.; Rabolt, J. F.; Wynne, K. J.; Yu, H. *Langmuir* **1987**, *3*, 932.
19. Decher, G.; Hong, J. D. *Makromol. Chem., Makromol. Symp.* **1991**, *46*, 321.
20. Decher, G.; Hong, J. D.; Schmidt, J. *Thin Solid Films* **1992**, *210/211*, 831.
21. Decher, G.; Hong, J. D. *Ber. Bunsen-Ges. Phys. Chem.* **1991**, *95*, 1430.

Self-Assembly

See: *Calixarenes (as Supramolecular Cyclic Oligomers)*
Hydrophobized Polysaccharides (Versatile Functions)
Langmuir-Blodgett Films (Organic Polymers for Photonics Applications)
Lipid Membranes (High Stability from Archaebacterial Extermophiles)
Macrocyclic Aramids
Metallosupramolecular Helicates
Molecular Assemblies (Polymerization in Aqueous Solution)
Multilayer Films (Polyelectrolytes)
Nonlinear Optical Materials
Peptide-Based Nanotubes (New Class of Functional Biomaterials)
Phenolics (Linear and Cyclic Oligomers)
Polyrotaxanes (Overview)
Self-Assembled Polymers (at Interfaces)
Self-Assembled Polymers (Photopolymerizable Bolaform Amphiphiles)
Signal Transduction Composites (Biomaterials with Electroactive Polymers)
Supramolecular Mesomorphic Polymers (Assembling in Ultrathin Films)
Supramolecular Self-Assembly (Liquid Crystalline Polymers, Hydrogen Bonding)
Supramolecular Structures (Bipyridine Block Copolymers with Cu(I) Ions)
Ultrathin Films (Self-Assembled)

Semiconducting Polymers

See: *Conjugated Polymers (Insulating and Conducting Forms)*

Sensors

See: *Amphiphilic Gas Sensors (Based on Polymer-Coated Piezoelectric Crystals)*
Chemical Sensors
Fluorescent Polymers (2-Vinylquinoline Hydrogels and Soluble Polymers)
Ionene Polymers
Ion-Selective Electrodes and Biosensors
Ion-Selective Microchemical Sensors (Cardiology)
Polyphenylacetylene
Pyroelectricity

Sequential Polymerization

See: *Gradient Polymers*

Shish-Kebab Polymers

See: *Stacked Transition Metal Macrocycles (Semiconductive Properties)*

Shock Wave Polymerization

See: *High Pressure Radical Polymerization*

SIGNAL TRANSDUCTION COMPOSITES (Biomaterials with Electroactive Polymers)

Lynne A. Samuelson,* M. Ayyagari, and D. L. Kaplan
Biotechnology Division
U.S. Army Natick RD&E Center

R. Pande, S. Kamtekar, K. G. Chittibabu, H. Gao, K. A. Marx, J. Kumar, and S. K. Tripathy
Center for Advanced Materials
Department of Chemistry and Physics
University of Massachusetts, Lowell

This research focuses on the coupling of biological (receptor) macromolecules to a novel conducting-electroactive-polymer matrix. Any biotinylated macromolecule with a signal

*Author to whom correspondence should be addressed.

transduction function may be incorporated into a device design. These features include a polythiophene backbone that provides enhanced film ruggedness and stability, derivatization of the backbone with a surfactant group for facile monolayer assembly or hydrophobic surface binding, and a covalent attachment of a flexible biotin group for the specific, irreversible binding of the macromolecule. Our rationale in utilizing this coupling is to optimize the signal transduction by using the unusual optical and electronic properties of conducting electroactive polymers to serve as integrated "intelligent" pathways between the biomaterial and the desired stimulus for the potential biosensor and optical display. Currently, researchers are investigating a variety of approaches to address this need, including thin layers (Langmuir-Blodgett (LB) sandwiches) of preformed conjugated polymers, embedding sensing elements (enzymes and DNA) in conductive, electrochemically grown polymers, and the incorporation of electron mediators such as ferrocene into an LB assembly to provide intimate contact between the sensing and the transduction components.[1–6]

Our approach is to create organized monolayer assemblies that have a direct covalent linkage between the sensing or recognition element (biomaterial) and the transduction matrix (conducting electroactive polymer) using both LB and self-assembly fabrication methodologies.

The first system involves the specific cassette attachment of a light-harvesting protein (phycoerythrin) to the conducting electroactive polymer matrix using the LB technique for electro-optic signal transduction applications. The second involves immobilization of the same polymer and the enzyme, alkaline phosphatase, on the surface of glass capillaries using self-assembly techniques for the chemiluminescence-based detection of environmental pollutants such as organophosphorus-based insecticides (paraoxon and methyl parathion).

In both cases, a series of novel copolymers of polythiophene have been synthesized that are derivatized with long alkyl chains to optimize film formation and adhesion to hydrophobic surfaces and with biotin moieties to bind the biological materials of interest through biotin-streptavidin complexation. Streptavidin is a well-known tetramer protein which contains four identical and highly specific biotin binding sites.[7] This complex, once formed, is essentially irreversible with a stability comparable with that of a covalent bond.[8,9] Streptavidin-biotin complexation was used as our bridging system to "anchor" the biomolecules of interest to the polymer matrix.

Poly(3-alkylthiophene)s were chosen as the polymer system since they were known to be processible, functional polymers with reasonably good thermal and temporal stability. Our first studies with these materials involved the synthesis of a biotinylated copolymer of 3-alkylthiophene and 3-methanolthiophene to form poly(3-undecylthiophene-co-3-thiophenemethanol). This material was found to form fairly oriented and spatially organized protein molecular assemblies using the LB technique but was somewhat limited in its protein binding capacity.[10,11] We have since optimized this system for protein immobilization by building in flexibility in the copolymer by adding a longer spacer arm between the conjugated polymer backbone and the biotin group to maximize protein integration and binding into the relatively deep streptavidin pocket. To achieve this flexibility, we copolymerized 3-undecylthiophene with functionalizable 3-methanolthiophene using ferric chloride as a catalyst to form the biotinylated copolymer poly(3-undecylthiophene-co-3-thiophenecarboxaldehyde) biotin-LC-hydrazone.[12] This material, in conjunction with streptavidin-biotin complexation, was used for our generic-cassette-attachment methodology.

CONCLUSIONS

With the aim of ultimately creating assemblies with intelligent properties for signal transduction applications, we have described a "generic" methodology that involves immobilizing organized monolayers of biotinylated biological macromolecules to solid supports using both Langmuir–Blodgett and hydrophobic self-assembly techniques. In this study, we designed and synthesized a novel class of conducting electroactive copolymers of polythiophene, which have been derivatized with biotin for protein attachment through biotin-streptavidin complexation and with long alkyl tails for facile monolayer formation and subsequent binding to hydrophobic substrates for device fabrication. We have explored the potential versatility of this approach both with photoactive, algal, light-harvesting proteins for electro-optic signal transduction applications and with the enzyme, alkaline phosphatase (conjugated with streptavidin) as a sensing element in the detection of organophosphorus-based insecticides. This cassette-polythiophene copolymer-immobilization strategy has demonstrated that the incorporation of virtually any macromolecule that can be derivatized with streptavidin or biotin is feasible and thus is very promising as a vehicle for biosensor development.

ACKNOWLEDGMENTS

The authors acknowledge support from ARO Grant DAAL03-91-G-0064 and Contract DAAK60-93-K-0012. We also wish to thank Dr. Manjunath Kamath for providing the polymer and Ms. Kristie Ann VanderWiede for technical assistance.

REFERENCES

1. Decher, G.; Hong, J. D.; Schmidt, J. *Thin Solid Films* **1992**, *210–211*, 831.
2. Royappa, A. T.; Rubner, M. F. *Langmuir* **1992**, *8*, 3168.
3. Aizawa, M. *Proc. International Workshop on Intelligent Materials, Tsukuba, March 1989*; Technomic: Lancaster, PA, 1989; p 209.
4. Minehan, D. S.; Marx, K. A.; Lim, J. O.; Kamath, M.; Tripathy, S. K. *Proc. 1st Int. Conf. on Intelligent Materials, Kanagawa, March 1992*; Technomic: Lancaster-Basel, 1992; p 50.
5. Minehan, D. S.; Marx, K. A.; Tripathy, S. K. *Macromolecules* **1994**, *27*, 777.
6. Skotheim, T. A.; Lee, H. S.; Hale, P. D.; Karan, H. I.; Okamoto, Y.; Samuelson, L. A.; Tripathy, S. K. *Synthetic Metals* **1990**, *42*(1-2), 1433.
7. Weber, P. C.; Ohlendorf, D. H.; Wendoloski, J. J.; Salemme, F. R. *Science* **1989**, *243*, 85.
8. Green, N. M. *Advances in Protein Chemistry* **1975**, *29*, 85.
9. Gimlick, R. K.; Giese, R. W. *J. Biol. Chem.* **1988**, *263*, 210.
10. Kamath, M.; Chittibabu, K. G.; Kim, J. O.; Marx, K. A.; Kumar, J.; Tripathy, S. K.; Samuelson, L. A.; Kaplan, D. L. *J. Macromol. Sci. A-Pure Appl. Chem.* **1993**, *30*(8), 493.
11. Samuelson, L. A.; Wiley, B.; Kaplan, D. L.; Sengupta, S.; Kamath, M.; Lim, J. O.; Cazeca, M.; Kumar, J.; Marx, K. A.; Tripathy, S. K. *J. Intelligent Material Systems and Structures* **May 1994**, *5*(3), 305.
12. Chittibabu, K. G.; Kamath, M.; Lim, J. O.; Marx, K. A.; Kumar, J.; Tripathy, S. K. *Mat. Res. Soc. Symp. Proc.* **1994**, *330*, 185.

SILACYCLOBUTANES AND RELATED COMPOUNDS (Ring-Opening Polymerization)

Edward D. Babich
T. J. Watson Research Center
IBM Corporation

A special place in organosilicon chemistry belongs to the four-membered heterocycles with silicon atoms—silacyclobutanes—which should formally be considered as saturated compounds but whose reactivity reveals a very high degree of "unsaturation" comparable with that of carbon-carbon double bonds.

One of the most unusual properties of such molecules is their ability to form high molecular weight linear polymers. This transformation occurs irreversibly either thermally or under the influence of different catalysts with nearly quantitative yields. The unique character of this reaction lies in the fact that carbon analogs-cyclobutanes as well as silacyclopentanes and other silacycloalkanes do not polymerize thermally.

POLYMERIZATION METHODS

Thermal Process

The thermally initiated polymerization of the heterocycle **1a** (R = R' = Me; R" = H) which proceeds via ring-opening with the rupture of endocyclic Si-C bonds, shown in **Equation 1**, was first observed when it was heated in a sealed evacuated ampoule at 125–130°C.[1]

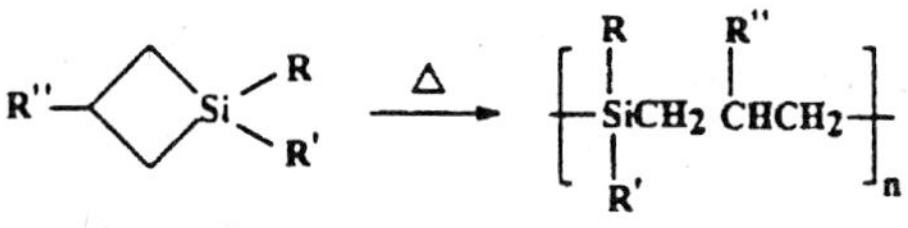

I III 1

Poly(1,1-dimethyl-1-silacyclobutane) (**IIIa**) (called also 1,1-dimethyl-1-siltrimethylene) was obtained as an elastic material soluble in a monomer as well as in benzene and chloroform.

Thermal ring-opening polymerization (TROP) of some other types of four-membered silicon heterocycles takes place in the same manner. Thus, compound **IIa** (R = R' = R" = R" = CH_3) forms soluble (poly-1,1,3,3-tetramethyl-1,3-disilacyclobutane) (**IVa**) (known also as dimethylsilmethylene) (**Equation 2**):[3]

II IV 2

Compound **V** polymerizes very easily at 60°C via scission of the Si-C(benzyl)-bonds forming poly(1,1-dimethyl-1-sila-2,3-benzocyclobutene (**VI**) (known also as poly{methylene(dimethylsilylene)}-1,2-phenylene) (**Equation 3**).[4,5]

A wide variety of polymers based on 1-sila- and 1,3-disilacyclobutanes have been prepared by TROP. Reactions shown in Equations 1–3 are typical for the majority of currently known

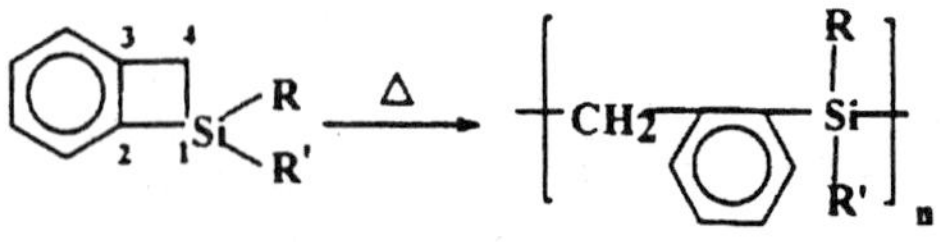

V VI 3

1-sila- and 1,3-disilacyclobutanes with different Si-alkyl and aryl substituents, as well as for 1-germa- and 1,3-digermacyclobutanes.[6–8] Polymerization occurs in a strictly selective manner by combining monomer units in a "head-to-tail" fashion. All classes of amines inhibit the reactions in Equations 1–3.[2,5]

Although all the details of the mechanism of the TROP process for silacyclobutanes are not known, there are some well-established experimental data that shed some light on it. Radical initiators and UV light do not cause polymerization and radical scavengers do not inhibit TROP.[2–4] Molecular weights are independent of the reaction time (conversion) and starting concentrations of monomers and increase as the temperature rises. The reactions in Equations 1 and 3 are distinguished by a narrow molecular weight distribution (less than 1.1).

Silacyclobutanes are strained-heterocyclic systems in which internal valence angles formed by silicon are much more deformed (up to 80° and more in 1-silacyclobutanes) compared to CCC- angles.[2]

Anionic Polymerization

Polymerization of 1-Silacyclobutanes

Anionic ring-opening polymerization (AROP) of four-membered rings with one silicon atom under the influence of an organolithium compound or of metallic lithium, hydroxides, or silanolates of the alkali metals also occurs via the rupture of endocyclic Si-C bonds. This leads to the formation of soluble linear heterochain polymers with structures that are identical to those of products of thermal polymerization, although with much lower molecular weights.[2,5,11–13]

Cationic Polymerization

Polymerization of Four-Membered Heterocycles with Silicon

This type of polymerization is not typical for silacyclobutanes. One of the main requirements for polymerization by a cationic mechanism is that positively charged species must survive long enough to ensure the growth of a polymer chain instead of rapidly collapsing through the reaction with an anion into the inactive products.[14] Numerous studies have shown that typical cationic initiators, such as Brønsted acids, iodine, and inorganic Lewis acids, react with silacyclobutanes to form stable addition products.[2] There are only two publications reporting the formation of polymers from 2,3-benzo-1-sila[5] and 1,2-disilacyclobutanes under the influence of $AlCl_3$. The molecular weights of these polymers are much lower than those of polymers produced by thermal processes.[9]

Coordination Polymerization

Transition Metal Catalysis

After thermal initiation, the use of transition metal catalysts can be considered as another universal approach to the synthesis of linear high molecular weight polymers from various silacyclobutanes and butenes.

In 1966, W. A. Kriner proposed a mechanism for ring-opening polymerization of silacyclobutanes catalyzed by transition metals that includes a zwitterion formation as an initiation step and an electrophylic propagation by the siliconium ion (a positive side of the zwitterion).[15] In 1968, G. Lewin and J. B. Carmichael obtained some experimental data that was in support of a zwitterionic mechanism.[16]

PROPERTIES

A classic investigation of the behavior of linear heterochain polymers with silicon was carried out by Finkel'shtein and coworkers in 1973, who demonstrated that such polymers without functional groups undergo depolymerization with quantitative yields.[17] Thus, poly(1,1-(dimethyl-1-sila-2:3-benzocyclobutene) **(VI)** forms the dimermeso-1,1,5,5-tetramethyl-1,5-disila-2:3,6:7-dibenzocycloocta-2,6-diene **(XXX)** upon pyrolysis at 390°C (**Equation 4**).

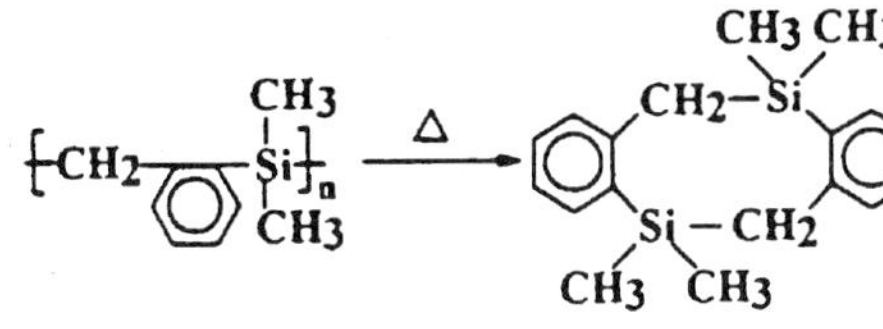

4

VI **XXX**

Poly(1,1-dimethyl-1-sila-3,4-benzocyclopentene) showed higher thermostability. Its pyrolysis at 450°C occurs with the formation of the monomer.

E-Beam Sensitivities and Oxygen Plasma Stabilities of Heterochain Polymers with Silicon

Similar to the behavior of PDMS heterochain polymers with $SiCH_2CH_2CH_2$, $SiCH_2CH(CH_3)CH_2$, and $SiCH_2SiCH_2$ backbones and pendant methyl groups are very sensitive to electron beam irradiation because it results in crosslinking even at low doses (0.5–3.0 $\mu C/cm^2$ at 25 KeV) as measured by the loss of solubility in organic solvents at room temperature.[18,19]

APPLICATIONS

Crosslinking of Polymers via Silacyclobutane Functional Groups

The use of silacyclobutane groups has been suggested as the basis of a new method of thermal hardening of different polymers. This was one of the first applications of the TROP process and was patented in 1964.[10,20] Linear polysiloxane oligomers containing 1-silacyclobutane end or pendant groups crosslink at 180–200°C.

When asymmetrically substituted 1,3-disilacyclobutane groups are used instead of 1-silacyclobutane groups, crosslinking occurs about ten times faster.[21]

The same approach to thermally crosslinkable polysiloxanes was the subject of several patents in the early 1970s.[22–24]

Ceramics

An interest in finding new sources of silicon fibers led to the investigation of linear poly(1,3-disilacyclobutane)s as SiC precursors. It was claimed in the patent that 85% ceramic yield is achievable from a prepolymer with $SiCH_2Si$ structure containing Si-H pendant groups.[25]

Resist Materials and Oxygen Plasma Barriers

High sensitivity to e-beam irradiation and much higher stability in oxygen plasmas make it possible to use poly(1-sila- and -1,3-disilacyclobutane)s in microlithography as negative tone resist materials for a double layer patterning technique.

CONCLUSIONS

Heterochain polymers on the basis of four-membered rings with silicon atoms have a variety of properties usable in different fields of modern technology. The main limitation to their practical use arises from the complicated and expensive methods used for the preparation of monomers that are based on Grignard reactions. The development of new approaches based on high temperature catalytic reactions similar to the "direct" synthesis of linear chlorosilanes will clear the way for their industrial use.

ACKNOWLEDGMENT

I wish to extend my sincere appreciation to Dr. Alessandro Callegary and Miss Katherina Babich for their help, encouragement, useful discussions, and valuable assistance.

REFERENCES

1. Vdovin, V. M.; Pushchevaya, K. S.; Petrov, A. D. *Doklady Akad. Nauk SSSR* **1961**, *141*, 843.
2. Nametkin, N. S.; Vdovin, V. M. *Izvestiya Akad. Nauk SSSR, Ser. Khim.*; Bulletin of the Academy of Sciences of the USSR, Division of Chemical Science **1974**, *23*(5), 1153.
3. Nametkin, N. S.; Vdovin, V. M. *J. Polym. Sci., C.* No. 4, 1043, 1963.
4. Nametkin, N. S.; Vdovin, V. M.; Finkel'shtein, E. Sh.; Yatsenko, M. S.; Ushakov, N. V. *Vysokomol. Soedin.*, Ser. B **1969**, *11*, 207.
5. Salamone, J. C.; Fitch, W. L. *J. Polym. Sci. A-1* **1971**, *9*, 1741.
6. Ushakov, N. V.; Finkel'shtein, E. Sh.; Babich, E. D. *Visokomol. Soedin* in press 1994.
7. Nametkin, N. S.; Kuz'min, O. V.; Zav'yalov, V. I.; Zueva, G. Y.; Babich, E. D.; Vdovin, V. M.; Chernysheva, T. I. *Izv. Akad. Nauk SSSR, Ser. Khim.* 1969, 976.
8. Mironov, V. F.; Mikhailynts, S. A.; Gar, T. K. *Zhurnal Obshchei Khimii* **1969**, *39*, No. 11, p 2601 (2542 in Engl. Transl.).
9. Shina, K. *J. Organometallic Chem.* **1986**, *310*, C57.
10. Babich, E. D. Ph.D. Thesis, Academy of Sciences of the USSR, A. V. Topchiev Institute of Petrochemical Synthesis, Moscow, 1966.
11. Gilman, H.; Atwell, W. H. *J. Am. Chem. Soc.* **1964**, *86*, 2687.
12. Ushakov, N. V.; Vdovin, V. M.; Pozdnyakova, M. V.; Pritula, N. A. *Izv. Akad. Nauk SSSR, Ser. Khim.* No. 9; Bulletin of the Academy of Sciences of the USSR, Division of Chemical Science, p 2125.
13. Ushakov, N. V.; Pritula, N. A.; Rebrov, A. I. *Russian Chemical Bulletin Izv. Akad. Nauk*, **1993**, *42*(8), 1372.
14. *Encyclopedia of Polymer Science and Technology*, 2nd ed.; John Wiley & Sons: New York: 1989; Vol. 2, 729.
15. Kriner, W. A. *J. Polym. Sci. Part A-1* **1966**, *14*, 444.
16. Lewin, G.; Carmichael, J. B. *J. Polym. Sci. Part A-1* **1968**, *6*, 1.
17. Finkel'shtein, E. Sh.; Sedov, V. V.; Nametkin, N. S.; Kusminskii, A. S.; Vdovin, V. M.; Yatsenko, M. S.; Oppengeim, V. D.; Gusel'nikov, L. E. *Vysokomol. Soedin. Ser. A* **1973**, *15*(1) 186.

18. Babich, E.; Paraszczak, J.; Hatzakis, M.; Rishton, S.; Grenon, B.; Linde, H. *Microelectr. Eng.* **1989**, *9*, 537.
19. Babich, E.; Paraszczak, J.; Hatzakis, M.; Shaw, J.; Grenon, B. *Microelectr. Eng.* **1985**, *3*, 279.
20. Nametkin, N. S.; Vdovin, V. M.; Babich, E. D.; Zav'ylov, V. I. USSR Patent 161 911; *Chem. Abstract* **1964**, *161*, 8480d.
21. Karel'skii, V. N.; Thesis. Ph.D. Academy of Sciences of the USSR, Topchiev Institute of Petrochemical Synthesis, Moscow, 1970.
22. Owen, W. J.; Jonas, D. A. (Midland Silicones Ltd.), Ger. Patent 2 110 878, 1971.
23. Jonas, D. A.; Owen, W. J. (Dow Corning Ltd.), U.S. Patent 3 719 696, 1973.
24. Jonas, D. A.; Owen, W. S. (Dow Corning Ltd.), U.S. Patent 3 694 427, 1972.
25. Smith, T. L.; Rouge, B. U.S. Patent 4 631 179, 1986.

Silane Coupling Agent

See: Coupling Agents

SILANE COUPLING AGENTS (Adhesion Promoters)

Peter G. Pape
Dow Corning Corporation

Silane coupling agents will change the interface between an organic polymer and an inorganic substate.[1] They are used in a variety of applications, including reinforced plastics, coatings, paints, inks, sealants, adhesives, and elastomers. Their use results in improved bonding and upgraded mechanical and electrical properties. Other benefits are also obtained, such as, improved wetting, surface smoothness, and viscosity reduction.

CHEMICAL DESCRIPTION

Typical commercial silane coupling agents are listed in **Table 1**.

TABLE 1. Representative Commercial Silane Coupling Agents

Organofunctional groups	Chemical structure
A. Vinyl	$CH_2{=}CHSi(OCH_3)_3$
B. Chloropropyl	$ClCH_2CH_2CH_2Si(OCH_3)_3$
C. Epoxy	0 /\ $CH_2CHCH_2OCH_2CH_2CH_2Si(OCH_3)_3$
D. Methacrylate	CH_3 0 | ‖ $CH_2{=}C-C-OCH_2CH_2CH_2Si(OCH_3)_3$
E. Primary amine	$H_2NCH_2CH_2CH_2Si(OC_2H_5)_3$
F. Diamine	$H_2NCH_2CH_2NHCH_2CH_2CH_2Si(OCH_3)_3$
G. Methyl	$CH_3Si(OCH_3)_3$
H. Cationic styryl	$CH_2{=}CHC_6H_4CH_2NHCH_2CH_2NH(CH_2)_3Si(OCH_3)_3$ | HCl
I. Phenyl	$C_6H_5Si(OCH_3)_3$

BONDING TO THE INORGANIC SUBSTRATE

Silane coupling agents that contain three inorganic reactive groups on silicon will bond well to most inorganic substrates, especially if the substrate contains silicon, aluminum, or most heavy metals in its structure. During application of a silane coupling agent, the alkoxy groups on silicon hydrolyze to silanols, either with added water or from residual water on the inorganic surface, and then the silanols coordinate with metal hydroxyl groups on the inorganic surface to form an oxane bond with the elimination of water.

BONDING TO THE ORGANIC POLYMER

The bond to the organic polymer is complex. In the case of a thermoset polymer, the silane bonds to the polymer through the chemical reactivities of the silane and polymer; that is, an epoxysilane or aminosilane will bond with an epoxy resin; an aminosilane will bond with a phenolic resin; a methacrylate silane will bond through styrene crosslinking with an unsaturated polyester resin. The bonding of thermoplastic polymers through a silane coupling agent is best explained by the interdiffusion and interpolymer network (IPN) formation in the interphase regions.[2]

OPTIMIZING PERFORMANCE

The chemical structure of the organic group in a silane coupling agent has a large effect on its effectiveness in a composite or as an adhesion promoter in coatings.

APPLICATIONS

All glass fiber that is used in reinforced plastics must have a coupling agent on the surface in order to prevent debonding of the polymer from the glass by the attack of moisture. Without a coupling agent, a glass reinforced composite becomes very weak during aging in ambient conditions.

SUMMARY

Silane coupling agents are critical chemicals in many applications in the plastics industry. Without them, many applications which require adhesion of an inorganic substrate to an organic resin would fail to meet required strength or durability requirements. Other chemical classes of interfacial modifiers often do not provide this type of durable chemical bonding.

REFERENCES

1. Plueddemann, E. P. *Silane Coupling Agents* 2nd ed., Plenum: New York, 1991.
2. Chaudhury, M. K.; Gentle, T. M.; Plueddemann, E. P. *J. Adhesion Sci. Technol.* **1987**, *1*, 23.

Silanol-Containing Polymers

See: Blends (Silanol Functional Groups)

Silicon Carbide

See: Inorganic Fibers (Surface Modification by Grafting)
Polysilanes (Overview)
Preceramic Polymers

Silicon-Containing Polymers

See: *Aminosiloxanes*
Anoxic Polymer Materials
Blends (Silanol Functional Groups)
Bridged Polysilsesquioxanes (Highly Porous Hybrid Organic-Inorganic Materials)
Contact Lenses, Gas Permeable
Ferrocene-Backbone Polymers
Fluorosilicon Oligomers
Fluorosilicones (Monomer Synthesis and Applications)
Functional Polysiloxanes
Inorganic Fibers (Surface Modification by Grafting)
Inorganic Nanostructured Materials
Inorganic/Organic Hybrid Polymers (High Temperature, Oxidatively Stable)
Ladder Polyphenylsilsesquioxane
Metal Complex Catalysts, Polymeric
Metal-Containing Silicon-Based Polymers
Model Silicone Networks
Organic/Inorganic Composite Materials
Organic/Inorganic Composites
Organic/Inorganic Polymer Hybrids
Polyborosiloxanes
Polydimethylsiloxane Networks, Monodisperse
Poly [(disilanylene) Oligophenylene]s
Poly(imide siloxane)s
Polymethylphenylsiloxane (Photophysics, Conformation and Dynamics)
Polyoxyarylenesiloxanes
Polysilanes (Overview)
Polysilanes (Bearing Hydroxy Groups and Derivatives)
Polysilanes (Ring-Opening Polymerization of Cyclotetrasilanes)
Polysilazanes (Through Ring-Opening Polymerization)
Polysilylpropynes (Steric Effects on Material Properties)
Preceramic Polymers
Reactive Liquid Rubbers (Epoxy Toughening Agents)
Resins (Applications in Coatings Industry)
Silacyclobutanes and Related Compounds (Ring-Opening Polymerization)
Silicon Nitride
Silicone Elastomers
Silicone Polymers, Organo-Modified (Application in Personal Care Products)
Silicone Rubber Latex
Silicone Sealants
Silicones
Siloxane-Containing Polymers
Siloxane Elastomers
Silsesquioxane-Based Polymers
Vinylsilanes (Isomerization Polymerization)

Silcone Networks

See: *Polydimethylsiloxane Networks, Monodisperse*
Silicone Elastomers

SILICON NITRIDE

Horst Lange*
PK-F. Bayer AG

Gerhard Wötting
Cremer Forschungsinstitut

Gerhard Winter
Hermann C. Starck Berlin GmbH + Co KG

SILICON NITRIDE AS CERAMIC MATERIAL

Silicon nitride is of great interest as a heavy-duty material for applications in chemical engineering, wear technology, metal working, energy technology, and, mainly, in engine and turbine construction.[1–13] This broad interest arises from a favorable combination of great hardness, wear resistance, chemical stability, low density, and high mechanical strength at temperatures up to 1300°C. The application of Si_3N_4 ceramic parts in combustion engines and gas turbines promises advantages like higher efficiency by better fuel exploitation through higher combustion temperatures, faster response of oscillating or rotating engine components (valves, pistons, turbochargers, and turbine rotors), through low density, and longer service intervals, and through a generally improved wear behavior.

Silicon nitride ceramics are produced either by nitriding silicon powder compacts or by sintering Si_3N_4-powder compacts.[14,15]

THE STRUCTURE OF SILICON NITRIDE—ORIGIN OF HARDNESS, DURABILITY, AND STRENGTH

Crystalline silicon nitride exists in two hexagonal modifications—α and β—which are built up from a three-dimensional network of SiN_4-tetrahedrons.[16,17] Rigid structure and strong covalent bonds are the cause of the extraordinary hardness, durability, and mechanical strength of this material.

SILICON NITRIDE WHISKERS

After the discovery that crystal whiskers show unusual mechanical strength, interest in properties, production, and applications of ceramic whiskers also grew soon.[23–25]

Chemically, the same methods as in the synthesis of spheroidal Si_3N_4-powder particles are used for the production of Si_3N_4-whiskers. However, additional measures are taken to favor general principles like vapor phase transport or crystallization by vapor-liquid-solid-mechanisms for whisker formation.

Whiskers from the Reaction of Si and N_2

In direct nitridation of elemental silicon, Si_3N_4-whiskers are formed by reaction of silicon vapor and N_2.

Whiskers from Carbothermal Reduction of SiO_2

This process, which may be run continuously in conveyor-type pusher-furnaces, yields whiskers from the gas-phase reaction of SiO and N_2 or NH_3 in CO-containing atmospheres.[26–28]

Whiskers from the Reaction of $SiCl_4$ with NH_3

The reaction of $SiCl_4$ with NH_3 leads to $Si(NH)_2$, which yields amorphous Si_3N_4 on calcination. If small amounts of Fe

*Author to whom correspondence should be addressed.

or SiO_2 are added to this material and the resulting mixture is heated to 1200–1500°C in the absence of oxygen, Si_3N_4-whiskers are formed by a vapor-liquid-solid-mechanism from Fe- or silicate-containing eutectical melt-droplets.[18,28]

Other Methods

Whiskers from the Reaction of SiS_2 with NH_3

Originally intended to produce high-purity Si_3N_4-powder the reaction of SiS_2 with NH_3 offers a good possibility to produce Si_3N_4-whiskers.[19–21]

Whiskers from the Reaction of SiC and NH_3

This reaction is remarkable since it leads to Si_3N_4-particles of unusual morphology. By heating a mixture of SiC- and Fe-powder in NH_3 to 1440°C followed by a short-term, temperature rise to 1480°C, Glemser and Horn obtained spiral-shaped α-Si_3N_4-whiskers.[22] Some of these spirals grew to a size noticeable by the naked eye.

THE MANUFACTURE OF SINTERED Si_3N_4 CERAMICS

Physical and Chemical Aspects of Si_3N_4 Sintering

"Dry" sintering techniques used in densification of oxide ceramics cannot be applied to the sintering of Si_3N_4 powder compacts. Strong covalent Si-*N*-bonds hinder practical useful mass transport by grain boundary or lattice diffusion. The activation of diffusion by simply rising the sintering temperature is limited by the appearance of an ever increasing thermal decomposition according to (**Equation 1**):

$$Si_3N_4(s) \rightarrow 3\ Si(l) + 2\ N_2(g);\ Si(l) \rightarrow Si(g) \qquad 1$$

Since a N_2-particle pressure of 1 bar is reached at about 1900°C several precautions have to be taken in order to prevent excessive decomposition during sintering of Si_3N_4 at high temperatures. Most important is the addition of compounds which form molten silicate phases and allow densification by liquid phase sintering mechanisms. Well-suited additives are, for example, earth alkaline oxides and rare earth oxides either alone or in combination with Al_2O_3 or AlN.[32–36]

A prolonged thermal treatment far beyond the moment, when densification is complete, favors grain growth by minimization of boundary energy ("Ostwald ripening") and also in needle-grained structures and leads to grain coarsening with growth of spheroidal grains. In this case structural development leads to the formation of globular grains, no matter whether α- or β-Si_3N_4 has been used as the starting material.

After sintering, in all these materials an amorphous or semicrystalline secondary phase between grain boundaries or in grain triple points remains. Because material properties at high temperatures are especially influenced by the secondary phase, the reduction, modification, or complete elimination of this phenomenon is of major interest to the optimization of Si_3N_4 ceramic properties.[37]

PHYSICAL AND CHEMICAL PROPERTIES

Physical Properties

Table 1 shows a compilation of important properties of completely densified Si_3N_4 ceramics. The scattering range of data is mainly due to structural differences, which are influenced by starting powder characteristics, amount and kind of sintering additives, way of powder shaping, sintering techniques as well as temperature, pressure, and soaking time during sintering. Fracture strength is surely the characteristic property, which is determined most often; usually it is considered as a placative criterion of quality.

TABLE 1. Physical Properties of Dense Si_3N_4 Material

Decomposition temperature	2173 K
Theoretical density (th. D.)	α-Phase: 3,168....3,188 gcm^{-3} β-Phase: 3,19....3,202 gcm^{-3}
Material density	95–100% th. D.
Thermal expansion coefficient (293–1473 K)	2,9....3,6 $10^{-6}\ K^{-1}$
Specific heat	700 $Jkg^{-1}K^{-1}$
Electrical resistance (RT)	$10^{13}\ \Omega$ cm
Microhardness (Vickers)	1400–1700 MNm^{-2}
Bending strength (4-Point, Rt)	600–1000 MPa
Fracture toughness (RT)	5–8 MPa.$m^{1/2}$
Elasticity modulus (RT)	280–320 GPA
Thermal conductivity (RT)	15–30 $Wm^{-1}K^{-1}$
Critical temperature difference in thermal shock (RT)	600–800 K

RT = room temperature.

A general comparison of fracture strength is difficult, since materials for specialized applications have been developed. Here, the main criterion of quality is not only fracture strength, but, for example, the strength level at temperature > 1000°C, fracture toughness, hardness, or wear resistance. Today, state of the art is a strength level of 800–1000 MPa at room temperature; however, materials with fracture strengths exceeding 1000 MPa have already been described.[38,39]

Chemical Properties

Due to possible broad applications of sintered Si_3N_4, its chemical properties have been investigated with special regard to corrosive behavior.[40–43]

Silicon nitride is inert to numerous molten metals (e.g., aluminum[Al], zinc[Zn], tin[Sn], lead[Pb], copper[Cu], silver[Ag], and cadmium[Cd]); however, in contact with transition molten metals like iron (Fe), cobalt (Co), nickel (Ni), vanadium (V), and chromium (Cr), it reacts readily to form metal silicides and N_2. In contact with molten salts, oxidic slags, and glasses, only slow corrosion occurs.

With the exception of hydrofluoric acid, Si_3N_4 ceramics are resistant to mineral acids. Mainly porous Si_3N_4 materials are attacked by strong caustic solutions or melts with formation of NH_3.

In the use of engine components (e.g., valves, turbo charger, gas turbines) the oxidation behavior of sintered Si_3N_4 is especially important. Since Si_3N_4 is coated with a thin, protective layer of SiO_2 on heating in air, a good oxidation resistance is shown up to 1400°C.

SUMMARY AND PREVIEW

Sintered silicon nitride is a promising material for applications in engine construction, mechanical engineering, wear and chemical technology, where high mechanical strength at high temperatures, corrosion- and wear-resistance, hardness, and low density are desired. Cutting tools, bearings, mill-liners, and engine components like valves, valve guides, turbocharger rotors, or gas turbines are prominent examples of applications where this material is already used routinely or where it is being tested as a basic material in development projects.

To date, the required high-quality Si_3N_4-powders are commercially produced mainly by direct nitridation, and to a lesser extent also by $SiCl_4/NH_3$-liquid-phase reaction ("diimide-process"). Among processes under development, the $SiCl_4/NH_3$-gas phase reaction seems to offer the best potential to furnish economically high-purity, very sintering-active Si_3N_4-powders on a large scale.

Chemical and physical characteristics of the starting material kind and amount of sintering additives, and sintering technique determine material properties. Their interdependence influences structural development as well as type and composition of silicate phases formed at grain boundaries.

Crystallization and reduction of the amount of these silicate phases are important to further improve material properties like fracture strength, creep, and corrosion resistance especially at high temperatures.

A manufacturing process free of faults from powder synthesis to finished component is required to produce high-quality, reliable ceramic components. Based on cheap, high-quality Si_3N_4 powders, easy to process, mainly injection molding as powder shaping technology has a great potential for the large-scale manufacturing of very complex components. For economical consolidation near net-shape pressureless sintering is of great importance. The development of economical working methods for finishing sintered components of this very hard material is a critical point for decreasing the still very high finishing costs.

The work on economical methods to produce high-quality Si_3N_4-powders, to steadily improve powder and material properties, and to manufacture reliable working ceramic components of superior quality can not be looked at separately; it is, in its entirety, a necessary requirement for a broad application of silicon nitride advanced ceramics in the future.

REFERENCES

1. Lenoe, E. M.; Katz, R. N.; Burke, J. J. *Ceramics for High Performance Applications, III*; Plenum: New York, 1979.
2. Riley, F. L. *Sprechsaal* **1985**, *118*, 225.
3. Bunk, W.; Böhmer, M.; Kibler, H. *Keramische Komponenten für Fahrzeuggasturbinen II*; Springer-Verlag: Berlin, 1984.
4. Walzer, P.; Langer, M. *Motortech. Z.* **1983**, *44*, 1.
5. Taguchi, M. *Adv. Ceram. Mater.* **1987**, *2*, 754.
6. Gugel, E. *Keram. Z.* **1987**, *39*, 698.
7. Jack, K. H. *Endeavour (New Series)* **1987**, *11*, 80.
8. Thümmler, F.; Grathwohl, G. *Keram. Z.* **1988**, *40*, 157.
9. Fingerle, D.; Fripan, M.; Dworak, U. *Z. wirtschaftl. Fert.* **1987**, *82*, 392.
10. Heinrich, J.; Huber, J.; Stingl, P. *Z. wirtschaftl. Fert.* **1987**, *82*, 487.
11. Knoch, H.; Hunold, K. *Techn. Mitt. Haus d. Technik, Essen* **1987**, *80*, 31.
12. Steinmann, D. *cav. 1988 (Nr. 2)* S. 11.
13. Uppenbrock, K. *Sprechsaal* **1988**, *121*, 135; *ibid.* **1988**, *121* 494; *ibid.* **1989**, *122*, 232.
14. Ziegler, G.; Heinrich, J.; Wötting, G. *J. Mater. Sci.* **1987**, *22*, 3041.
15. Boberski, C.; Hamminger, R.; Peuckert, M.; Aldinger, F.; Dillinger, R.; Heinrich, J.; Huber, J. *Angew. Chem. Adv. Mater.* **1989**, *101*, 1592.
16. Ruddlesden, S. N.; Popper, P. *Acta Cryst.* **1958**, *11*, 465.
17. Wild, S.; Grieveson, P.; Jack, K. H. *Special Ceramics* **1972**, *5*, 385.
18. Kohtoku, Y. In Somiya, S.; Nitomo, M.; Yoshimura M. (Hrsg.): *Silicon Nitride-1*; Elsevier Applied Science: London, 1989; S. 71 ff.
19. Morgan, P. E. D. *DOE-Report 83-004283*, **1983**.
20. Morgan, P. E. D.; Pugar, E. A. *J. Amer. Ceram. Soc.* **1985**, *68*, 699.
21. Morgan, P. E. D. U.S. Patent 4 552 740, 1985.
22. Glemser, O.; Horn, H.-G. *Naturwissenschaften* **1962**, *49*, 538.
23. Herring, C.; Galt, J. K. *Phys. Rev.* **1952**, *85*, 1060.
24. Hanna, S. B.; Abdel-Mohsen, F. F. *Angew. Makromolekul. Chem.* **1985**, *133*, 25.
25. Grewe, H.; Dreyer, K.; Kolaska, J. *cfi/DKG Ber.* **1987**, *303*.
26. Johnson, R. C.; Alley, J. K.; Warwick, W. H.; Shell, R. H. U.S. Patent 3 244 480, 1966.
27. Tanaka, M.; Kawabe, T. J. Patent 1 324 479, 1986.
28. Niwano, K. In: Somiya, S.; Mitomo, M.; Yoshimura, M. (Hrsg): *Silicon Nitride - 1*; Elsevier Applied Science: London, 1989; S. 117 ff.
29. Motojima, S.; Iwanaga, H. *Kinzoku* **1989**, *59*, 47.
30. Ziegler, G. *cfi/Ber. DKG* **1991**, *68*, 72.
31. Greil, P.; Hoffmann, M. J.; Weisskopf, K. L.; Petzow, G. *Fortschrittsber. DKG* **1986/87**, *2*, 27.
32. Terwilliger, G. R.; Langer, F. F. *J. Mater. Sci.* **1975**, *10*, 1169.
33. Prochazka, S.; Greskovich, C. D. *AMMRC Technical Report* **1978**, *78-32*, SRD-77-178.
34. Smith, J. T.; Quackenbush, C. L. In Somiya, S.; Saito, S. (Hrsg.): *Proc. Int. Symp. Factors in Densification and Sintering of Oxide and Nonoxide Ceramics*; Hakone: 1978; S. 426 ff.
35. Mitomo, M. *J. Mater. Sci.* **1976**, *11*, 1103.
36. Mitomo, M. In Somiya, S.; Mitomo, M.; Yoshimura, M. (Hrsg.) *Silicon Nitride-1*; Elsevier Applied Science: London, 1990; S. 1 ff.
37. Greil, P. *Science of Ceramics* **1988**, *14*, 645.
38. Ueno, K. *J. Ceram. Soc. Japan Intern. Ed.* **1989**, *97*, 80.
39. Komeya, K. wird veröffentlicht in: *Proc. 7th CIMTEC*, Montecatini **1990**.
40. Rabenau, A. *Ber. Dtsch. Keram. Ges.* **1963**, *40*, 6.
41. Wötting, G.; Ziegler, G. *Sprechsaal* **1990**, *123*.
42. Feld, H.; Gugel, E.; Nitzsche, H. G. *Werkstoffe und Korrosion* **1969**, 671.
43. Tressler, R. E.; McNallun, M. (Hrsg.): *Proc. Symp. Corrosion and Corrosive Degradation of Ceramics*; American Chemical Society: Westerville, 1989.

SILICONE ELASTOMERS

Stephen J. Clarson
Department of Materials Science and Engineering
Polymer Research Center
University of Cincinnati

James E. Mark
Department of Chemistry
Polymer Research Center
University of Cincinnati

Along with the many synthetic polymers that were developed around the time of the second World War, polydimethylsiloxane

(PDMS) -$[(CH_3)_2SiO]_y$- has recently celebrated its fiftieth year as a commercial material. The polysiloxane or silicone industry may be conveniently grouped into products based upon fluids, elastomers, and resins—with the elastomeric materials accounting for approximately 50% of the applications of siloxane polymers.[1] Details of the formulation, and fabrication and applications of silicone elastomers up to 1978 have been documented by Lynch and developments into the 1980s have been covered in a series of reviews and monographs from the Dow Corning Corporation's elastomers group.[2–5] It is of particular importance for the work described in this overview to note that during the last 20 years significant theoretical developments have led toward a more rigorous molecular understanding of rubber-like elastomeric behavior.[6–8] Such theories are increasingly realistic in the treatment of topological features such as junction fluctuations and entangling and also the detailed structural characteristics that distinguish an elastomeric chain of one type from another. On the experimental side, new methods have been developed for the synthesis of elastomeric networks having well defined structures and new techniques established for network characterization.[9,10] Although experimental studies of the mechanical properties and swelling characteristics remain the most widely used methods for studying polymer networks, information from strain birefringence, strain dichroism, segmental orientation, and small-angle neutron scattering investigations is now being utilized to provide a more complete understanding of the properties of elastomeric materials.[8,11,12]

PREPARATION OF SILICONE NETWORKS

A wide variety of methods are available for the crosslinking of siloxane polymer chains to give elastomeric materials. Many of these approaches give poorly controlled crosslink densities and the placements and distributions of the crosslink sites are ill-defined. Examples of such random crosslinking methods include peroxide thermolysis or high energy irradiation.[13,14]

There have been considerable efforts over the last fifteen years or so to prepare siloxane chains containing functional groups either at the chain ends or as substituents.[15] In the case where the subsequent reactions of these functional groups are stoichiometric and proceed to completion, such reactions have the capability of producing well defined elastomeric materials.[9,10,16–24] These elastomers may be classified as "model" networks in the sense that independent information obtained on the network precursor chains can give valuable information on the resulting network structures. One example of such a reaction is the end-linking of hydroxyl-terminated PDMS with tetraethoxysilane (TEOS) where the formation of a tetrafunctional crosslink and subsequent condensation reactions lead to gelation.

STEREOREGULAR SILOXANE ELASTOMERS

In a wide variety of their commercial applications silicone elastomers are filled with silica (SiO_2) or titania (TiO_2) in order to improve their mechanical properties.[28–35] Among the various reasons for such filler incorporation is the fact that the low melting temperature of PDMS makes it difficult to utilize self-reinforcing effects such as strain-induced crystallization, which is seen for many of the commercial organic elastomers having a regular structure.[25,36]

An exciting development in this area is the stereochemical control of the reaction used in the synthesis of the asymmetrically substituted polysiloxanes -$[(CH_3)RSiO]_y$-. In particular, the isolation of the cis-stereoisomeric trimer and subsequent polymerization under kinetic conditions leads to polymers that crystallize above room temperature for the R = 3,3,3-trifluoropropyl- and phenyl-substituted systems.[37–43] Subsequent crosslinking of these stereoregular polymers then leads to unfilled siloxane elastomers with properties far superior to their atactic analogs.[38,43,44] In particular such materials do exhibit strain-induced crystallization. This methodology, which has been pioneered by Saam and Kuo, could revolutionize the siloxane based elastomers business in a way similar to that seen for the α-olefin based polymers when stereochemical control was introduced.

SOME FUTURE DIRECTIONS

In the area of preparing silicone elastomers there continues to be a need for new cure chemistries and a better mechanistic understanding of the reactions currently employed. A number of very interesting computer simulations of the gelation process for siloxane network formation have been reported and, with increasingly powerful computers, more detailed studies are expected.[45–47] Well defined siloxane elastomers have been used as model systems for correlations with the predictions of the modern theories of rubber-like elasticity. There are a number of modern theories, such as the recent model proposed by Erman and Monnerie, which are in need of a full comparison with experiments.[48–50] Other experimental techniques such as strain birefringence, strain dichroism, segmental orientation, nuclear magnetic resonance, and small-angle neutron scattering investigations will also be further utilized to provide a more complete understanding of the properties of elastomeric materials. New theory is needed for non-Gaussian networks, and computational methods using rotational isomeric state theories have given valuable insights into these effects.[26,27,51,52] In the area of filled siloxane elastomers, there is still no rigorous molecular theory for polymer-filler interactions. Finally, as there are a number of common features between siloxane cure systems and sol-gel methods, their is still much work to be done on composite materials based on ceramics/glasses and siloxane polymers.[53]

REFERENCES

1. Rochow, E. G. In *"Siloxane Polymers"*, Clarson, S. J.; Semlyen, J. A., Eds.; Prentice Hall: Englewood Cliffs, New Jersey, 1993.
2. Lynch, W. *"Handbook of Silicone Rubber Fabrication"*; Van Nostrand Reinhold: New York, 1978.
3. Warrick, E. L.; Pierce, O. R.; Polmanteer, K. E.; Saam, J. C. *Rubber Chem. Tech.* **1979**, *52*, 437.
4. Polmanteer, K. E. *Rubber Chem. Tech.* **1981**, *54*, 1051.
5. Polmanteer, K. E. *Rubber Chem. Tech.* **1988**, *61*, 470.
6. Flory, P. J. *Proc. Royal. Soc. London Ser. A.* **1976**, *351*, 351.
7. Eichinger, B. E. *Ann. Rev. Phy. Chem.* **1983**, *34*, 359.
8. Mark, J. E.; Erman, B. *"Rubberlike Elasticity: A Molecular Primer"*; Wiley-Interscience: New York, 1988.
9. Mark, J. E. *Adv. Polym. Sci.* **1982**, *44*, 1.
10. Mark, J. E. *Acc. Chem. Res.* **1985**, *18*, 202.

11. Flory, P. J. *"Principles of Polymer Chemistry"*; Cornell University: Ithaca, New York, 1953.
12. Treloar, L. R. G. *"The Physics of Rubber Elasticity"*; 3rd. ed., Clarendon: Oxford, 1975.
13. Bueche, A. M. *J. Polym. Sci.* **1955**, *15*, 105.
14. Baker, D.; Charlesby, A.; Morris, J. *Polymer* **1968**, *9*, 437.
15. Lebrun, J. J.; Porte, H. In *"Comprehensive Polymer Science"*; Eastmond, G. C.; Ledwith, A.; Russo, S.; Sigwalt, P., Eds.; Pergamon: Oxford, 1989; Vol. 5, 593.
16. Mark, J. E.; Sullivan, J. L. *J. Chem. Phys.* **1977**, *66*, 1006.
17. Valles, E. M.; Macosko, C. W. *Macromolecules* **1979**, *12*, 673.
18. Meyers, K. O.; Bye, M. L.; Merrill, E. W. *Macromolecules* **1980**, *13*, 1045.
19. Mark, J. E.; Rahalkar, R. R.; Sullivan, J. L. *J. Chem. Phys.* **1979**, *70*, 1794.
20. Llorente, M. A.; Mark, J. E. *Rubber Chem. Tech.* **1980**, *53*, 988.
21. Mark, J. E. *Pure Appl. Chem.* **1981**, *53*, 1495.
22. Mark, J. E. *Polymer J.* **1985**, *17*, 265.
23. Mark, J. E. *Br. Polymer J.* **1985**, *17*, 144.
24. Xu, P.; Mark, J. E. *Rubber Chem. Tech.* **1991**, *63*, 276.
25. Clarson, S. J.; Mark, J. E.; Dodgson, K. *Polymer Communications* **1988**, *29*, 208.
26. Mark, J. E.; Curro, J. G. *J. Chem. Phys.* **1983**, *79*, 5705.
27. Curro, J. G.; Mark, J. E. *J. Chem. Phys.* **1984**, *80*, 4521.
28. Noll, W. *"Chemistry and Technology of Silicones"*; Academic: New York, 1968.
29. Mark, J. E.; Allcock, H. R.; West, R. *"Inorganic Polymers"*; Prentice Hall: Englewood Cliffs, NJ, 1992.
30. Warrick, E. L.; Pierce, O. R.; Polmanteer, K. E.; Saam, J. C. *Rubber Chem. Technol.* **1978**, *52*, 437.
31. Polmanteer, K. E. *Rubber Chem. Tech.* **1981**, *54*, 1051.
32. Polmanteer, K. E. *Rubber Chem. Tech.* **1988**, *61*, 470.
33. Lutz, M. A.; Polmanteer, K. E.; Chapman, H. L. *Rubber Chem. Tech.* **1985**, *58*, 939.
34. Chapman, H. L.; Lutz, M. A.; Polmanteer, K. E. *Rubber Chem. Tech.* **1985**, *58*, 953.
35. Polmanteer, K. E.; Chapman, H. L.; Lutz, M. A. *Rubber Chem. Tech.* **1985**, *58*, 965.
36. Clarson, S. J.; Dodgson, K.; Semlyen, J. A. *Polymer* **1985**, *26*, 930.
37. Kuo, C. M.; Saam, J. C.; Taylor, R. B. *Polym. International* **1994**, *33*, 187.
38. Battjes, K. P.; Kuo, C. M.; Miller, R. L.; Saam, J. C. *Macromolecules* **1995**, *28*, 790.
39. Curtis, M. D.; Scrinivas, T.; Elsheik, M. *Polym. Prepr.* **1984**, *25*(1), 224.
40. Momper, B.; Wagner, T.; Maschke, U.; Ballauff, M.; Fischer, E. W. *Polymer Communications* **1990**, *31*, 186.
41. Clarson, S. J. Results presented at the American Chemical Society Award Symposium in Applied Polymer Science in honor of James E. Mark, 207th ACS National Meeting, San Diego, CA, March 1994.
42. Zhang, X. K. Ph.D. Thesis University of Cincinnati, 1995.
43. Clarson, S. J.; Mark, J. E.; *Polymer Communications* **1989**, *30*, 275.
44. Zhang, X. K.; Clarson, S. J.; Bunning, T. J. Ms. in preparation, 1996.
45. Leung, Y. K.; Eichinger, B. E. *J. Chem. Phys.* **1984**, *80*, 3877.
46. Leung, Y. K.; Eichinger, B. E. *J. Chem. Phys.* **1984**, *80*, 3885.
47. Shy, L. Y.; Eichinger, B. E. *Macromolecules* **1986**, *19*, 2787.
48. Erman, B.; Monnerie, L. *Macromolecules* **1989**, *22*, 3342.
49. Fontaine, F.; Morland, C.; Noel, C.; Monnerie, L.; Erman, B. *Macromolecules* **1989**, *22*, 3348.
50. Fontaine, F.; Noel, C.; Monnerie, L.; Erman, B. *Macromolecules* **1989**, *22*, 3352.
51. Clarson, S. J.; Mark, J. E.; Sun, C. C.; Dodgson, K. *European Polymer Journal* **1992**, *28*, 823.
52. Mark, J. E. *Acc. Chem. Res.* **1994**, *27*, 271.
53. Mark, J. E.; *J. Inorganic and Organometallic Polym.* **1991**, *1*, 431.

SILICONE POLYMERS, ORGANO-MODIFIED (Application in Personal Care Products)

David T. Floyd and Klaus R. Jenni
Goldschmidt Chemical Corporation

In the general cosmetic area, silicones and especially organo-modified silicones have had a long history of use.[1–4] These silicone polymers are essentially non-toxic, nonirritating, and safe for use in all personal care products. In fact, the dimethicone copolyols are one of the few glycol polymers reviewed and approved by the Cosmetic Ingredient Review Panel as safe as presently used.[5]

These compounds are generally described as polyether-polysiloxane copolymers.[6,7] Their INCI (CTFA) name is dimethicone copolyol (**Figure 1**).

The polyether group in dimethicone copolyols can consist of homopolymeric chains of either ethylene or propylene oxide or can contain varying proportions of both ethylene and propylene oxide, as a copolymer of ethylene/propylene oxide. Through the chemical variations that can be realized in the composition of dimethicone copolyols, these products can provide unique surfactant properties. They can offer significant advantages as surface tension depressants, wetting agents, emulsifiers, and foam builders in hair treatment preparations.[8–13]

By understanding their surface activities and physical-chemical properties, the formulator can make choices to optimize the desired effect in a shampoo system. For example, combining a high molecular weight silicone polymer with a mid-range polymer containing a high proportion of propylene oxide will provide optimization of gloss, conditioning, and wet and dry combing.[14]

The majority of silicone polymers in use in industry today are the nonionic type. It is also possible to bond ionic side chains to polydimethylsiloxane either additionally or exclusively. This opens up a completely new range of property combinations that is still untapped for the most part.

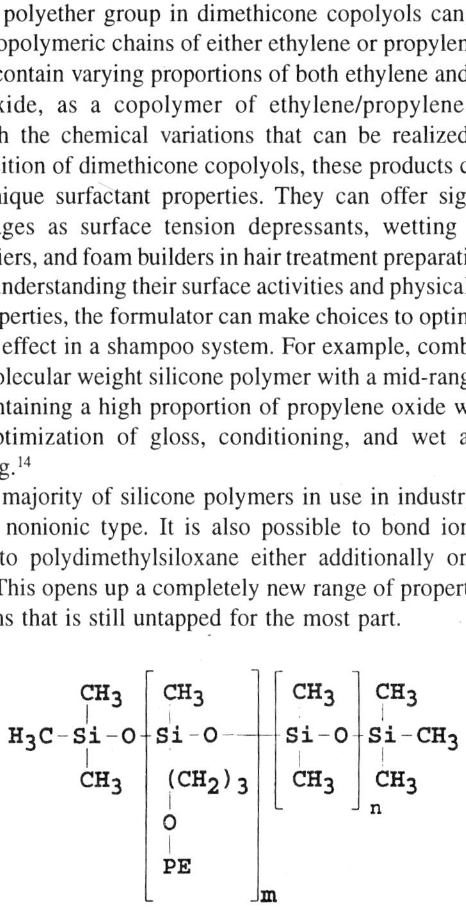

$$PE = (-C_2H_4O)_x(-C_3H_6O)_y\text{-}H$$

FIGURE 1. Polyether polysiloxane copolymer CTFA: dimethicone copolyol CAS # 64365-23-7; 68937-54-2; 68938-54-5.

Silicone betaines and silicone quats are excellent additives in hair and skin care preparations.[16] The silicone quats are more highly substantive than the silicone betaines. These polymers show good compatibility with anionic surfactants and, therefore, are suitable as additives in shampoos. Silicone betaines and silicone quats are good antistatics. In comparison to pure organic polymeric quats, silicone quats and betaines give hair a silky, silicone gloss and feel.

The silicone quaternary compounds are also good conditioning agents. They are exceptionally mild to skin and eyes, unlike their organic counterparts. Previous studies have discussed the substantivity of these compounds to hair substrates.[2,3,12,17] The high cost of silicone quats as sole conditioning agents often precludes their use. However, as small adjuvants in traditional conditioning bases at levels of 0.2–0.4%, they give dramatic improvements in conditioning and improved tactile properties.

Another class of conditioning agents are the thiosulfate modified siloxanes, often referred to as Bunte salts named after the German chemist who first synthesized the thiosulfate esters.[15]

The routes of synthesis for these polymers have been previously described.[2,3,12,17] A recent study has shown that these can be powerful conditioning agents on hair.[17]

Foaming emulsions requires the combination of an oily hydrocarbon typically dispersed in water with a hydrocarbon based surfactant at the interface. Organosilicone water-in-oil (W/O) emulsifiers are versatile substances with enhanced emulsion stabilizing activity. Because of this specific property they facilitate the manufacturing of mineral oil-free water-in-oil emulsions and light water-in-oil emulsions with low fatting properties that can be adjusted to cream or lotion consistencies.[18,19]

The synthesis of organosilicone water-in-oil emulsifiers is based on the linkage of polymethylsiloxane chains with alkyl side chains and polyglycol groups. The polymethylsiloxane chain possesses both hydrophobic and lipophobic properties. The alkyl side chains induce the necessary lipophilic, and the polyglycol groups the necessary hydrophilic characteristics onto the emulsifier molecule.

The high stabilizing potential of these silicone emulsifiers is most likely due to the strong absorption of the phase boundaries that results from the polymeric character and the presence of hydrophilic/lipophilic and hydrophobic/lipophobic regions of the molecule.

The water-in-oil formulations produced by the polyalkyl polyether polysiloxane copolymers offer new avenues for the product development chemist to explore.

Silicone polymers have been shown by studies to have good emollient properties.[20] Silicone polymers are also good protective emollients, providing skin protection, waterproofing, and resistance while reducing the overall irritation profiles of compounded formulations.[21,22]

One class of silicone polymers that is of interest is the polyalkyl polysiloxane copolymers, which are lipophilic, emollient polymers.

It is known that polyalkyl modified polysiloxanes are compatible with emollients, waxes, and other silicone polymers.[23] They are excellent additives for improving the gliding properties of emulsions.

CONCLUSION

Recent developments in silicone technology have introduced silicone polymers with specialized physical properties that can enhance personal care products. Poly(dimethyl organosiloxane)s can be specifically modified toward a desired application by the choice of a suitable organic moiety. Hydrophobic derivatives find application in emulsions or nonaqueous preparations. Hydrophilic nonionic and ionic products find uses in aqueous as well as anhydrous systems. In skin-care products, improved silky feel is obtained by the addition of a suitable silicone compound. In hair treatments, improved wet and dry combability, and improved gloss, combined with a silky feel, completes the desired properties. Silicone polymers in pharmaceutical and cosmetic bases can produce superior emulsions, contribute to barrier protection of the skin, and maximize the delivery of actives to the skin.

Silicone polymers have low toxicity profiles and their inclusion in many personal care products has shown that the resulting formulation itself has a better overall toxicity profile.

There is a large number of chemical methods available to modify silicones which, even in this article, cannot be totally covered. It is obvious to see that further product optimization will take place with the ever-increasing requirement for silicone compounds in personal care products.

REFERENCES

1. DiSapio, A.; Fridd, P. *Preprints 15th IFSCC Congress 1,* 89, London, 1988.
2. Schaefer, D. *Preprints 15th IFSCC Congress 1*, 103, London, 1988.
3. Floyd, D. *Cosmetic and Pharmaceutical Applications of Polymers*, Gebelein, C., Ed.; 49 Plenum: New York, 1991.
4. Starch, M.; Krosic, C. *Cosmetic Technology* 20, November, 1982.
5. Final report of the safety assessment for dimethicone copolyol, prepared by The Expert Panel of the Cosmetic Ingredient Review, April 1981.
6. Noll, W. *Chemistry and Technology of Silicones*; Academic: New York, 1968, 404.
7. Meals, R. *Encyclopedia of Chemical Technology* 2nd ed. **1969**, *18*, 221.
8. DP 1595730 Th. Goldschmidt AG.
9. DP 3133869 Th. Goldschmidt AG.
10. DP 2533074 Th. Goldschmidt AG.
11. David, W.; Jones, D. *Polymer Preprints* **1970**, *11*, 447.
12. Wolfes, W. *Preprints 14th IFSCC Congress* **1986**, *1*, 473.
13. Wendell, S.; DiSapio, A. *Cosmetics and Toiletries* **1983**, *98*, 103.
14. Floyd, D.; Leidreiter, H.; Sarnecki, B.; Maczkiewitz, U. *Preprints 17th IFSCC Congress, P*, 297, Yokahama, 1992.
15. U.S. Patent 4891166 Th. Goldschmidt AG.
16. Meyer, H. *Cosmetics and Toiletries Manufacturers & Suppliers XV*, 5, January 1990.
17. Grüning, B.; Leidreiter, H. *SÖFW* **1992**, *3*, 117.
18. Hameyer, P.; Gould, C. *Manufacturing Chemist* 20, January, 1990.
19. Hameyer, P. *HAPPI* **1991**, *28*(10), 88.
20. Brand, H.; Brand-Garnys, E. *Cosmetic and Toiletries* **1992**, *107*, 93.
21. DiSapio, A. *Drug & Cosmetic Industry 29*, May, 1994.
22. Goldember, R. *J. Society Cosmet. Chem.* **1979**, *30*, 415.
23. U.S. Patent 4275101 RCA Corp.

SILICONE RELEASE COATINGS

Michael J. Owen and J. Darrell Jones
Dow Corning Corporation

Release coatings are generally used to prevent things from sticking together! This simplistic statement embraces a wealth of technology and a global industry involving silicone and other polymers.

Silicones in many forms have excellent release properties but to present a non-stick surface and at the same time offer little or no transfer to the released surface requires a cured film surface free from migratory species.

In this application silicone generally means polydimethylsiloxane (PDMS) although fluorosilicone release coatings are now available for silicone-based adhesives. PDMS offers a dimethyl surface to the air of very low surface energy. PDMS release coatings have solid surface tensions in the 20 +/– 2 mN/m range.[1] Low surface energy alone, however, does not guarantee a good release surface. The silicone backbone is an extremely flexible polymer chain with virtually unhindered rotation about the siloxane bonds.[2] This flexible backbone together with low surface energy are the essential elements of the unique release capability of PDMS, but to obtain a non-migratory release surface the polymer chains must also be tied together in a coherent film. This involves the use of crosslinking, or curing, chemistry; the crux of silicone release-coating development.

RELEASE COATING MARKETS

By far the largest single use is to make release liners to carry and protect pressure-sensitive adhesives (PSAs) until the moment of application. The role of the release liner is to protect the adhesive until it is at the point of application when it must peel away cleanly with little application of force.

Other markets for release liners that are not PSA-based include carriers for oily and sticky masses such as sealants, gaskets and mastics, interleaving layers for rubber processing, and support and release base for casting of plastic films such as polyethylene, polyurethane, acetate, and vinyl films. Further examples of release liners include food processing aids such as baking or pan liners, food packaging materials such as gum wrappers, asphalt packaging in the form of coated drum interiors and release-coated inner layers of paper sacks, caul or separating sheets for the production of high pressure plastic laminates, and release liners for tiles and shingles.

CURING CHEMISTRY

In principle any of the multitude of crosslinking chemistries used to cure silicone elastomers could be used for release coatings, but in practice two types have become predominant: the condensation or tin-catalyzed systems shown in **Equation 1** for example, and the addition cure systems, shown in **Equation 2**.[3]

Historically, the condensation type systems came first and they are still widely used today.

Such coatings were the basis of the silicone release-liner business from the late 1950s until the mid-1970s. Environmental pressure on solvent emission then made the development of 100% solids coatings a paramount need that was answered by

$$-\overset{|}{\underset{|}{Si}}-OH + -\overset{|}{\underset{|}{Si}}-OR \longrightarrow -\overset{|}{\underset{|}{Si}}O\overset{|}{\underset{|}{Si}}- + ROH \tag{1}$$

$$-\overset{|}{\underset{|}{Si}}-CH_2=CH + -\overset{|}{\underset{|}{Si}}H \longrightarrow -\overset{|}{\underset{|}{Si}}CH_2CH_2\overset{|}{\underset{|}{Si}}- \tag{2}$$

addition-curing chemistry (Equation 2). The addition of a silicone hydride group across a vinyl group is catalyzed by a few ppm of platinum in the form of an organo or organosilicon complex.[4] This contrasts sharply with the 2–15% of organotin salt used in the condensation systems. Being an addition reaction, no byproducts are evolved during curing.

THE CURING STAGE

Neither electron-beam radiation curing nor UV curing have yet lived up to their promise, and thermal cure remains predominant. Although microwave radiation will evaporate water and IR will dry solvent, the most prevalent method by far for curing is hot, forced air.

FACTORS AFFECTING RELEASE PERFORMANCE

Release performance is normally measured by the force necessary to remove an adherend from the release surface. It is quoted as a force per lineal measurement, the latter being the width of the adherend at right angles to the direction of removal. This measurement can be affected by numerous fundamental and practical variables. The fundamental variables include control of the elastic nature of the silicone, surface nature of the silicone, chemical type of the adhesive, elastic nature of the adhesive, and other factors inherent in the materials. Practical considerations include delaminating speed and angle, the effect of imperfect coverage of the substrate by the silicone, state of cure, thickness of both the silicone and adhesive layers and temperature and humidity of the test environment.

Environmental pressures to eliminate solvents affect adhesives as much as coatings. Until the mid-1970s solvent-based rubber adhesives were at least 80% of the label and tape industrial sectors. It proved impossible to either emulsify these materials or make a solventless version. Major changes in chemistry and materials were required. Two competitive technologies emerged: water-based acrylic emulsions and rubber-based, hot-melt adhesives. These have very different peel rate profiles from solvent-based rubber adhesives. It has been possible to link the emergence of solventless silicone with these new adhesives and make products that satisfy current needs.

REFERENCES

1. Owen, M. J. *J. Coatings Technol.* **1981**, *53*, 49.
2. Tobolsky, A. V. In "Properties and Structures of Polymers," John Wiley & Sons: New York, 1960; p 67.
3. Thomas, D. R. In "Siloxane Polymers," Clarson, S. J.; Semlyen, J. A., Eds.; PTR Prentice Hall: Englewood Cliffs, NJ, 1993; p 567.
4. Speier, J. L. *Adv. Organomet. Chem.* **1979**, *17*, 407.

SILICONE RUBBER LATEX

Donald T. Liles
Designed Materials Development
Dow Corning Corporation

Silicone rubber latex is an aqueous dispersion of silicone rubber particles that forms elastomers upon removal of water. This process occurs at room temperature to form RTV (room temperature vulcanizing) silicone rubber that is useful as coatings, sealants, and coating additives. Film formation proceeds in the absence of coalescing solvents or plasticizers and, as a result, products based on silicone rubber latex can be formulated to have low VOC (volatile organic compound) levels.

DISCUSSION

Silicone emulsions were first reported in 1948 by Barker and Jerabek and they were prepared by mechanically dispersing the desired polymer in water in the presence of a surfactant such as a sodium alkyl sulfonate ester.[1]

In 1959, Hyde and Wehrly reported that high molecular weight PDMS emulsions could be prepared by emulsion polymerization of cyclic oligomers using surfactants and strong mineral acids.[2]

It is also possible to carry out emulsion copolymerizations of PDMS with other functional siloxanes such as methylhydrogensiloxane or methylvinylsiloxane or siloxane precursors like $(EtO)_2SiMeH$ to produce emulsion copolymers that can crosslink or react via these functionalities.[3,4]

Artificial emulsions of PDMS can be prepared by emulsifying silicone polymers in a water/surfactant mixture using a high-shear device. This process was first accomplished by using a volatile solvent to reduce viscosity of the polymer followed by removing the solvent after emulsification. More recently, however, means have been found to emulsify siloxane polymers without the use of solvents.[5] Using this procedure, it is possible to prepare polymer emulsions and silicone rubber latex-using nonionic surfactants.[6]

Crosslinking

Silicone emulsion polymers must be crosslinked in order for the latex to produce an elastomer.

Removal of water from precured silicone emulsions produces elastomeric films at room temperature without the use of coalescing solvents. Although the reasons for this phenomenon are not fully understood, an explanation is that silicone rubber particles are very soft (T_g of PDMS = –120°C) and high osmotic pressures that develop during the removal of water force the soft rubber particles together so that they become efficiently packed. Since the precured silicone latex particles are crosslinked, interparticle migration of polymer chains cannot occur. As a result, the original integrity of the particles is maintained throughout the life of the cured (dry) latex.

Reinforcing Filler and Other Additives

Silicone rubber requires reinforcement in order for it to have any appreciable mechanical properties and silica is the reinforcement of choice. The same is true for silicone rubber latex. Aqueous dispersions of silica are the preferred forms of silica for reinforcing silicone rubber latex because the dispersions are simply stirred into the latex.

Properties of Silicone Rubber Latex

Adhesion of silicone rubber latex varies from excellent to poor, depending upon the composition of the latex and the substrate. Straight, unmodified silicone rubber latex has rather poor to marginal adhesion to most substrates. However, adhesion is substantially improved by the addition of colloidal silica to various silane adhesion promoters, or both.

Generally, silicone rubber latex has poor dielectric properties, moderate high-temperature stability, good low-temperature properties and excellent resistance to weathering.

The surfactants remain in the cured latex indefinitely and lead to poor dielectric properties. Thus, cured silicone rubber latex behaves more like a semiconductor than an insulator.[7] These surfactants can be removed to some extent by water washing and also upon weathering, which results in improved dielectric properties.[7]

Exterior architectural coatings based on silicone rubber latex that have been in service for about 15 years have shown little signs of degradation.

Commercial Applications of Silicone Rubber Latex

Commercial applications of silicone rubber latex center around coatings, sealants or caulks, and additives. Coatings were the first products available and they were introduced in the early 1980s. Sealants were the next type of product based on silicone rubber latex and they were found commercially in the mid 1980s. In recent years silicone rubber latex has found use as an additive in organic latex systems as modifiers to achieve specific properties, such as higher moisture vapor permeability, improved durability, increased mar resistance, and increased hydrophobicity.

The largest volume use of silicone rubber latex coatings is in architectural elastomeric coatings. Such coatings are designed to bridge cracks in building facades and offer protection against water penetration, even against wind-driven rain.[8] These coatings have excellent durability since the coating binder is essentially silicone rubber.

Fabric coatings based on silicone rubber latex are also commercially important, as solvent-based coatings are becoming less desirable. Although solventless silicone rubber fabric coatings are commercially available, silicone rubber latex offers several advantages over these solventless fabric coatings such as lower viscosity, lower cost, and more flexibility in the choice of coating equipment. An automotive airbag coating based on silicone rubber latex has been described.[9]

Silicone rubber latex sealants are essentially silicone rubber latex compositions that have rheological properties such that they are non-slump, they can be delivered from conventional sealant cartridges or other sealant handling equipment, and they can fill cracks. In order to minimize shrinkage, it is desirable to formulate the sealant with solids as high as practical, which is on the order of 70–85 volume percent. This is usually accomplished by loading the silicone latex with fillers such as calcium carbonate or silica. The chief attribute of silicone latex sealants is their ease of use, which is a result of their low odor and water clean-up properties.[10]

SUMMARY

Silicone rubber latex is best prepared by crosslinking silicone emulsion polymer using a tin catalyst. Cured (dry) silicone rubber latex has poor dielectric properties, excellent low temperature flexibility, moderate thermal stability, and excellent weatherability properties. Commercial applications of silicone rubber latex include coatings, sealants, and coating additives.

REFERENCES

1. Barker, H.; Jerabek, R. British Patent 0 596 833, 1948.
2. Hyde, F.; Wehrly J. U.S. Patent 2 891 920, 1959.
3. Liles, D. U.S. Patent 4 962 153, 1990.
4. Saam, J.; Wegener, R. U.S. Patent 4 273 634, 1981.
5. Derian, P.; Feder, M.; Paillet, J.; Peignier, M.; Senechal, A.; Ulrich, J. French Patent 9 212 519, 1992.
6. Feder, M.; Ulrich, J. French Patent 9 212 518, 1992.
7. Smith, S. B. *Proc. Water-Borne and Higher-Solids Symp.* Story, R.; Thames, S., Eds., 81 1983.
8. Lefler, H.; III, *Science and Technology of Building Seals, Sealants, Glazing and Waterproofing*, 2nd Vol. ASTM STP 1200, Klosowski, J., Ed., 260–267 1992.
9. Inoue, Y.; Momii, K. U.S. Patent 5 254 621, 1993.
10. Liles, D.; Shephard, N. *Science and Technology of Building Seals, Sealants, Glazing and Waterproofing*; 2nd Vol., ASTM STP 1200; Klosowski, J., Ed.; 1992; 280–291.

SILICONE SEALANTS

Richard D. Grant
Dow Corning Australia Pry Ltd.

The earliest silicone-based sealant materials, developed in the 1940s, were simply blends of polydimethylsiloxane (PDMS, CAS No. 70131-67-8) and a filler.[1] In the early 1950s curable silicones began to appear, based on crosslinking hydroxyl terminated PDMS with alkyl silicates or polysilicates.[2] These materials were two-pack systems, as base and curing agent needed to be stored separately and properties of the cured product varied with the mix ratio. The term "room-temperature vulcanized" (RTV) was coined to describe those products. In the late 1950s and early 1960s one part, moisture-cured silicone sealants began to appear.

Silicone sealants have found utility in a wide range of applications, from structural glazing to perimeter sealing and conformal coatings on electronic components. This is due partly to their stability to degradative agents, both chemical and physical, and partly to the ability of the silicone polymer to be formulated into products with a diversity of physical properties. This enables the sealants to be individually tailored to specific end uses.

PREPARATION

The fundamental components of any curable silicone composition are a silicone polymer and a crosslinker. Depending on the nature of the crosslinker, a catalyst may or may not be essential in order for the crosslinking reaction to occur at practical rates; however, in practice almost all silicone sealants will contain some level of catalyst. In order to achieve the required strength, reinforcing fillers, primarily silica, are used, as well as various bulking fillers such as calcium carbonate.

Silicone Polymers

By far the most common polymer used in silicone sealants is polydimethylsiloxane (PDMS).

The molecular weight of the PDMS matrix influences the properties of the sealant in which it is used. In general, an increase in molecular weight causes a decrease in the modulus of the cured sealant, and a corresponding increase in the ultimate elongation. In addition, the increase in viscosity associated with an increase in molecular weight causes the uncured sealant to be thicker. This influences its ease of use, because in many applications the uncured sealant needs to be extruded into position through a narrow nozzle.

Crosslinkers

Crosslinkers for silicone sealants are generally of the form $R_nSi(OL)_{4-n}$, where R is a short chain alkyl group (usually methyl or ethyl), –OL is a leaving group, and n = 0 or 1.

The two-part systems were first developed in the 1950s and used alkyl silicate crosslinkers.[2]

More recently developed, the one-part, moisture-cured systems have great advantages in terms of ease of use, since mixing prior to use is not necessary.

Fillers

Many different fillers have been used in silicone sealants in order primarily to improve the physical properties of the cured product, and also to modify the rheology of the uncured materials. They fall broadly into two categories: reinforcing and nonreinforcing fillers, although there is some overlap between the two.

Other Components

Plasticizers

Although organic plasticizers have been used, the most common plasticizers for silicone sealants are endblocked PDMS polymers (CAS No. 63148-62-9).

Adhesion Promoters

These are typically alkoxy silanes or siloxanes. Often they will have a functionally terminated alkyl chain that can interact with a specific substrate in order to improve interfacial interactions with the silicone.

Catalysts

These are in general not true catalysts because they are consumed or transformed during cure, and are better designated accelerators. They are typically based on tin (either Sn II or Sn IV) or titanium, but the carboxylates of lead, zinc, zirconium, iron, cadmium, barium, calcium, and manganese have also been used.[3]

PROPERTIES

The most important properties demanded of a sealant are long term resistance to weathering and low property drift, high movement capability over the life of the seal, and good adhesion to the substrate. Other important properties include ease of handling and application, convenient rate of cure, resistance to staining, and, of course, price. The emphasis on various properties will vary according to the individual application. For example for structural glazing a higher modulus, high strength material is essential, whereas for weatherproofing applications, surface appearance and ease of application will take a higher priority.

APPLICATIONS

Silicone sealants have found utility wherever a high performance sealing product is required. A few of the multitude of such applications are outlined below.

Structural glazing involves the attachment of panes of glass to the facade of a building.[4] Frequently the sealant/adhesive will be the only form of attachment of the glass to the frame, although many designs involve support from metal channels in the frame. Clearly this application requires a product with the strength and flexibility to withstand movements in the building structure and the variable wind loads to which it is subjected, as well as supporting the dead load of the glass, and with very long service life. Experience has shown silicone sealants to perform extremely well in this application.

Perimeter seals do not have the strength requirements of a structural application but nevertheless still require a high dynamic movement capability, and excellent adhesion to avoid water seepage.

Paving applications, for example in road sealing, require very high movement capability owing to the large thermal joint-movement experienced, and the excellent environmental stability characteristic of silicones.

Other applications, such as domestic sealing (e.g., around shower enclosures) have generally less stringent performance requirements, but benefit from the ease of use, longevity, permanent flexibility, and inertness afforded by silicones.

REFERENCES

1. Klosowski, J. J.; Gant, G. A. L. In *Plastics, Mortars, Sealants and Caulking Compounds*; Seymour, R. B., Ed.; ACS Symposium Series 113; American Chemical Society: Washington, DC, 1979; pp 113–127.
2. Hyde, J. F. U.S. Patent 2 571 039, 1951.
3. Noll, W. *Chemistry and Technology of Silicones*; Academic: San Diego, 1968; pp 397–399.
4. Klosowski, J. M. *Sealants in Construction*; Marcel Dekker: New York, 1989; pp 215–261.

Silicones

See: *Model Silicone Networks*
Reactive Liquid Rubbers (Epoxy Toughening Agents)
Resins (Applications in Coatings Industry)
Silicone Elastomers
Silicone Polymers, Organo-Modified (Application in Personal Care Products)
Silicone Release Coatings
Silicone Rubber Latex
Silicone Sealants
Silicones (Overview)

SILICONES (Overview)

A. Tomanek
Wacker-Chemie GmbH

SILICONES—SYNTHESIS, PROPERTIES, AND APPLICATIONS

The synthetic route to silicones proceeds via methylchlorosilanes. In industry almost all methylchlorosilanes are made from silicon and methylchloride by the Müller-Rochow direct synthesis.

Syntheses other than the direct synthesis are used to obtain special silanes in smaller quantities. An extraordinary broad range of silicone products is opened by the hydrosilylation reaction. It consists of the addition of hydrogen containing silanes to products with double or triple bonds.

By this reaction practically every important organo function can be introduced into silicone compounds. Therefore, the hydrosilylation reaction is especially responsible for the high diversity of commercial silicones. The hydrosilylation is extensively used to synthesize organofunctional silicones bearing vinyl groups, amino groups, and others.

Organopolysiloxanes are obtained by hydrolysis of chlorosilanes. The resultant silanols immediately undergo polycondensation.

The mechanism of hydrolysis and the composition of the silane mixture govern the size of the polysiloxane molecules and the number of hydroxylic groups it contains. Both intramolecular and intermolecular condensation can occur with the formation of cyclic siloxanes and linear siloxanes.

Silicone rubber consists essentially of silicone polymers and fillers. The many silicone rubber grades differ mainly by the type of polymer employed. Silicone polymers for solid silicone rubber (HTV-silicone rubber) show pasty consistency corresponding to molecular weights of 500,000 to 1 million, whereas pourable silicone rubber (RTV-silicone rubber) is of liquid consistency corresponding to molecular weights of 10–20,000.

Crosslinking can be achieved by different reactions at high temperatures (HTV-rubber) and room temperature (RTV-rubber). The liquid RTV-silicone rubber crosslinks by both condensation and addition mechanisms.

In order to obtain acceptable mechanical values, addition of fillers is of highest importance. Among the different fillers, fumed silica exhibits the strongest reinforcing effect, which permits us to compound silicone rubber grades with tear strength of more than 12 N/mm^2. However, many other fillers are used, such as quartz flour for higher resistance to mineral fluids and chemicals, iron oxide to enhance thermal stability, and carbon black to obtain electrical conductivity.

Silicones have an excellent resistance to heat and cold, ranging from –60 to +250°C. The resistance to low temperatures manifests itself in low pouring points of silicone fluids, elastic behavior of silicone rubber at temperatures as low as –60°C, and a glass transition temperature of –120°C.

The flash point of silicone fluids showing viscosities higher than 100 m Pa sec lies above 300°C. Silicone rubber has a flash point of 750°C and excellent flame retardancy. When combustion takes place, toxic or aggressive gases are evolved; the left ash is an excellent insulator.

Silicone fluids as well as silicone rubber and resins are very efficiency insulating materials. Values for dielectric strength, resistivity, and dielectric constant do not change much over a wide temperature range. In particular, silicones are highly resistant to tracking. Silicones show hydrophobic properties and impart spontaneously water repellency to materials such as glass, ceramics, and building materials where they easily spread out to form thin, nearly molecular films. A closely related effect is the ability of silicones to act as release agents for tacky substances.

The great ease with which silicone fluids spread out over surfaces is due to the low intermolecular forces of attraction that finds its expression in surface tension values as low as 30 mN/m. This surface activity is a chief factor in the use of silicones as antifoam agents, foam stabilizers, and free-flowing agents in paints. Silicone rubber is highly resistant to ozone and radiation. This property is responsible for its exceptional weathering properties. Low modulus of elasticity and a high mechanical dissipation factor render silicone rubber an excellent medium for shock and sound absorption.

Because of its good flow characteristics, silicone rubber is easily processed by casting or extrusion techniques. Parts processed from silicone rubber by extrusion are transparent tubes for the food industry and medical applications, cable insulation for application at high temperature, and sealing materials for flat irons or oven doors.

In the rubber industry silicone products additionally function as mold-release agents to produce gloves and other rubber articles.

Even more widespread is the use of silicone release agents in the plastics industry. In particular, injection molding processes for the production of foamed articles depend highly on very sophisticated silicone release agents.

Many chemical consumer products have been extensively improved through the addition of silicones. Cleaners and polishes contain silicone fluids that result in higher gloss and make polishing easier.

Cosmetics as well largely profit from silicones because of their high compatibility with skin, their water-repellent effect, and the soft feeling they leave. The pleasant touch is also the reason that silicone fluids are widely used in hair-care cosmetics.

Of special interest are siloxane polyimide copolymers, which are more soluble, easier to process, and more resistant to water absorption thereby maintaining the heat resistance at levels as high as 500°C. In epoxy molding, compound for the encapsulation of electronics with a softer consistency and less susceptible to cracks is obtained by incorporation of aminosiloxanes in the form of interpenetrating networks.

Silk

See: Biospinning (Silk Formation, Multiple Spinning Mechanisms)
Silk (Its Formation, Structure, Character and Utilization)
Silk (Physico-Chemical Properties)
Silk Fibers (Chemical Modification)
Silk Fibers (Grafting)
Silk Fibroin (Soft Tissue Compatible Polymer)
Spider Silk (Production of Polypeptide Polymers)

SILK (Its Formation, Structure, Character, and Utilization)

Keiichi Komatsu
The Silk Science Research Institute

Silk is widely used as an excellent fiber for textiles. As a typical protein, silk is a good research material. Numerous studies on silk have been performed continuously for more than 100 years. These studies have contributed not only to the improvement of production processes and the quality of raw silk products, which has supported the development of sericulture and the silk industry, but also to the progress of research on protein and fiber.

MECHANISM OF COCOON FIBER FORMATION

It has long been known that the cocoon fiber is formed by the insolubilization caused only by a mechanical action, the silkworm's spinning, of the liquid silk stored in the silk gland of a mature silkworm. The insolubilization via the mechanical action which can be referred to as mechanical denaturation, is a characteristic of silk fibroin, and the process is not yet discovered in natural polymers including proteins and synthetic high polymers, including proteins.

Molecular theories have shown that the mechanism of fiber formation of liquid silk fibroin can be regarded as the unfolding of molecular chains due to the shearing stress caused by the passing of liquid silk through the spinneret. Further studies have been carried out on the molecular conformations of fibroin and sericin, and their changes at each of the silk glands have been studied in detail.

DISSOLUTION BEHAVIOR AND STRUCTURE OF SERICIN

Although there are many arguments as to whether sericin is composed of a single protein or not, it seems to be composed of more than one component, judged at least from its dissolution behavior in hot water. As the dissolution in hot water is an important practical problem that is closely related to the reliability of fiber from cocoons and the production process of raw silk, studies concerning its relation to higher-order structure have been performed.[1]

The dissolution curve determined by the UV absorption method showed three bends, indicating that the cocoon fiber, sericin, is composed of four different fractions that differ in dissolution velocity.

There are differences in amino acid composition among the four fractions of sericin, but they are not large enough to explain the difference of dissolution in hot water.[1]

AMINO ACID COMPOSITION AND CHEMICAL STRUCTURE OF FIBROIN

The characteristics of amino acid composition of fibroin indicate that amino acids with short side chains like glycine, alanine, and serine are abundant, reaching more than 80% when tyrosine is included, and that amino acids with polar side chains occur in small amounts. It is interesting to note that the amino acid composition of fibroin is apparently specific to silkworm species, so that fibroin can be classified accordingly.

FINE STRUCTURE OF SILK

The cocoon fiber that was spun by a mature silkworm is composed of two filaments. The cross section of filament approximates a triangle, but the shape varies. Assuming that the cross section is circular, its diameter and cross-sectional area are about 10 μm and 78 μm^2, respectively.

CRYSTAL STRUCTURE AND TRANSITION OF MOLECULAR CONFORMATION OF FIBROIN

There are two kinds of fibroin crystals: silk I and silk II. Silk II has a β structure as its molecular conformation and the silk fiber that it constitutes has been studied in detail. As for the silk I obtained by air-drying liquid silk in the silk gland below 50°C, it is difficult to orient crystallites without causing the β-transition, so that analysis of its crystal structure has not progressed.

CHARACTERISTICS AND USES OF SILK FIBROIN

Silk is an excellent textile fiber. The load-extension curve that characterizes the mechanical properties of fibroin filament differs according to the species of silkworm that has spun the fibroin filament.

The tensile strength and elongation of silk are sufficient for textile use, moreover, Young's modulus is well suited as a textile fiber. Silk's energy-of-impact strength can be 10 times larger than other textile fibers. In contrast, silk's durability is inferior to its mechanical strength because the abrasion strength, bending strength, and fatigue caused by repetitive stretching and relaxation are lower than those of other fibers.

It seems that these defects are derived from its fibril structure. But, the fibril structure contributes to silk's graceful luster, and its elegant aesthetic properties. Silk's ability to regain moisture is larger than that of other fibers except wool, and its thermal conductivity is smaller than that of other fibers.

These advantages make silk comfortable to wear. One can say that silk is the softest, most beautiful, and most comfortable fiber to wear.

The amino acid composition of fibroin is comparatively simple, it is rich in glycine, serine, alanine, and tyrosine. Considering the physiological function of these four amino acids, glycine and serine have been recently shown to reduce the amount of cholesterol in blood, alanine promotes the metabolism of alcohol by the liver, and tyrosine is effective in preventing dementia.

On the basis of this research, the soluble fibroin powder that is produced by hydrolysis with hydrochloric acid or protease is used as a functional food additive in the refrigeration of drinks, cakes, noodles, and other foods.[2] As previously mentioned, the randomly coiled fibroin molecule changes easily to the extended β structure under mild conditions, for example, the mechanical action of agitation of solution or stretching of film, the addition of hydrophobic solvents or salts, acidification, and heating above 50°C. These characteristics of fibroin have been applied to nontextile use as in the immobilization of enzymes; with hormones, antibodies, and other biological catalyzers; in cosmetics and hair-care products.[3–6]

REFERENCES

1. Komatsu, K. *Bull. Sericult. Exp. Sta.* **1975**, *26*, 135.
2. Hirabayashi, K.; Chen, K.; Sei, K. *New Food Industry* **1991**, *33*(11), 1.
3. Kitaguchi, M.; Asakura, T.; Sakai, H.; Komtsu, K. *Rept. Prog. Polymer Phys. Japan* **1989**, *32*, 617.
4. Asakura, T.; Kitaguchi, M.; Demura, M.; Sakai, H.; Komatsu, K. *J. Appl. Polym. Sci.* **1992**, *49*, 49.
5. Komatsu, K. *Dying Industry* **1989**, *37*, 1.
6. Nakayama, H.; Kanno, H. Proteins for cosmetic use; In *Bio New Materials*; Ogawara, M.; Matsunaga, T., Ed.; CMC: Tokyo, 1991; pp 302–306.

SILK (Physico-Chemical Properties)

Eisaku Iizuka
Department of Functional Polymer Science
Faculty of Textile Science and Technology
Shinshu University

Currently, studies on the structure and physicochemical properties of silk are unpopular. The silk that had been studied was primarily restricted to that of domestic silkworms (*Bombyx mori*). However, nontextile uses for silk are becoming popular, especially in Japan. Sericulture of nonmulberry silkworms is currently of interest.

CHARACTERIZATION

Amino Acid Composition and Sequence of Amino Acid Residues

The amino acid composition of silk is specific to the species, and the sequence of amino acid residue is also species specific.

Molecular Weight

Recent reports claim that the native *Bombyx mori* fibroin molecules consists of one heavy chain (H-chain) and one or two light chains (L-chains).[1] In sum, these reports state that the molecular weight of the H-chain is about 300,000 and that of the L-chain is about 30,000; their molecular weight as a whole is 370,000.[2] It has been suggested that the L-chain(s) is bridged to the H-chain by S–S bonds through 1/2 cystine residue (a residue of 3 to 8 per molecule).

STRUCTURES

Structure in Solution

The cotton effect characteristic of the β-form of proteins was first measured with *Bombyx mori* fibroin.[10,11] This enabled optical rotatory dispersion (ORD) and circular dichroism (CD) techniques to be used to analyze the β-form, as well as the α-helix of polypeptides, including proteins. Using these techniques the conformation of various silk proteins was analyzed.[3] *Bombyx mori* fibroin exhibits a larger negative CD band, near 199 nm, and a very small positive CD band at 230 nm. The former band is typical of the coiled (or unordered) conformation; however, its intensity is far smaller than the typical coiled conformation, even this protein has more than 40 mol % optically inactive glycine residue. This suggests the presence of some compact conformation in solution.

Structure in Solid State

The α-helix, β-form, polyglycine II, and collagen-like coiled coil have been distinguished in the structure of silk thread.[4]

Bombyx mori fibroin assumes a quasistable crystalline structure, known as the α-form (not the α-helix) or as Silk I.[5,6] This quasistable crystalline structure is considered to be closely related to the structure of silk fibroin in solution.

SPINNING SILK

Concentrated solutions of regenerated *Bombyx mori* fibroin (LiBr silk) apparently show a Newtonian behavior; a tremendous increase in shear stress occurs suddenly at a certain shear rate, at and above which the shear stress no longer vanishes when rotation of the plate of a cone-and-plate viscometer is stopped. This reveals the formation of some three-dimensional structure in solution.

One thing that should be mentioned is that the viscosity of silk fibroin in concentrated solutions is extremely low; it is only 0.9 poise at the polymer concentration of 10 g/dl. In addition, the load of drawing silk thread by the movement of a silkworm is very small (<0.01 g/denier). This leads to the idea that the concentrated solution of silk fibroin is liquid crystalline.[7] The mechanism of silk spinning silk has been investigated by other researchers.[8,9] It is safe to say that silk thread is produced through a joint process of liquid crystal and gel spinning.

PROPERTIES

Domestic Silk

It has been mentioned that silk fibroin forms fibers more easily with decreasing polymer concentrations. This suggests that silk thread becomes more ordered and more crystalline when it becomes thinner, this is elucidated in the experimental evidence. This has been the case for all varieties of silk thread tested.[10–12]

APPLICATIONS

Fibroin films can be used to immobilize enzymes.[13] It may be possible for cells to multiply on fibroin films, to filter gasses selectively, and to be used as artificial skin. Silk fibroin has an antibacterial effect, the mechanism of which is unknown, and its film is said to be useful in preventing food from rotting.

Silk powder can be obtained by freeze-drying or spray drying the fibroin solutions, or by crushing the degummed silk. Silk powder is used to moisturize cosmetics. Silk powder prepared from hydrolyzed fibroin is now popular as a food additive in Japan in anticipation of the cholesterol-controlling power of glycine and serine, and the regulatory power of alanine in alcohol metabolism.

CONCLUDING REMARKS

Since the first man-made fiber was called an "artificial silk," silk thread has taken the lead as a model for textile fibers. There are many examples showing that the fibrous structure of silk thread, including its macroscopic texture, is imitated by finished man-made fibers. The liquid–crystal spinning that is now popular in the textile industry dates back, so to speak, to an era before mankind appeared in the world. Silk proteins have now been spotlighted in the field of nontextile usage. An old Chinese adage says that to have doubt about time-honored things is to hit on an epoch-making idea.

REFERENCES

1. Sasaki, T.; Tashiro, Y. *J. Cell Biol.* **1976**, *70*, 648.
2. Shimura, K. *Experientia* **1983**, *39*, 455.
3. Iizuka, E. *Biochim. Biophys. Acta* **1968**, *160*, 454.
4. Rudall, K. M.; Kenchington, W. A. *Rev. Ent.* **1971**, *16*, 73.
5. Shimizu, M. *Bull. Seric. Exp. Stn. Japan* (In Japanese) **1941**, *10*, 475.
6. Kratky, O.; Schauenstein, E. *Discuss. Faraday Soc.* **1951**, *11*, 171.
7. Iizuka, E. *J. Appl. Polymer Sci. Appl. Polymer Symp.* **1985**, *41*, 173.
8. Magoshi, J. *Polymer* **1977**, *18*, 643.
9. Kataoka, K.; Uematsu, I. *Kobunshi Ronbunshu* (In Japanese) **1977**, *34*, 37.
10. Iizuka, E.; Li, Q. *J. Seric. Sci. Jpn.* (In Japanese) **1991a**, *60*, 425.
11. Iizuka, E.; Yang, L. X. *J. Seric. Sci. Jpn.* (In Japanese) **1993**, *62*, 111.
12. Iizuka, E.; Uegaki, K.; Takamatsu, H.; Okachi, Y.; Kawai, E. *J. Seric. Sci. Jpn.* **1994**, *63*, 64.
13. Demura, M.; Asakura, T. *Biotech. Bioeng.* **1989**, *33*, 598.

SILK FIBERS (Chemical Modification)

Masuhiro Tsukada*
National Institute of Sericultural and Entomological Science

Giuliano Freddi
Stazione sperimentale per la Seta

The reaction of selected chemical agents with silk fibers is a particularly attractive system that is used to obtain effective and specific modification of the fibrous substrate. Noticeable changes in the physical and chemical properties of the silk can be obtained, avoiding some of the drawbacks that arise from graft-copolymerization of vinyl monomers and from the loading of silk with large amounts of polymer, which is often needed to obtain the desired effects.

Various kinds of epoxides and acid anhydrides can be used as modifying agents. The first attempts to apply these reagents for the chemical modification of silk date back to the beginning of the 1970s, thanks to the scientific contribution of Shiozaki and Tanaka.[1–3] Their studies led to the industrial exploitation of the epoxide treatment, which became quite popular in Japan for its effectiveness in improving some of the inferior properties of silk, such as crease recovery, wash-and-wear behavior, light resistance, and so on.

PREPARATION

Degummed silk fibers can be treated in yarn or fabric form. Depending on the kind of epoxide used, a weight gain of up to 20% can be obtained with both domestic (*Bombyx mori*) and wild (*Antheraea pernyi, Philosamia cynthia ricini*) silk fibers.[2–5] Although a large number of epoxy compounds are available, ethylene glycol diglycidyl ether (EDGE) is the most popular for silk finishing on an industrial scale.

*Author to whom correspondence should be addressed.

REACTIVITY OF EPOXIDES AND ANHYDRIDES TOWARD SILK

As a protein, silk fibroin contains various nucleophiles, such as carboxyl, aliphatic hydroxyl, phenol, amino, imidazole, indole, amide groups belonging to the side chains of the constituent amino acids. All these groups represent potential reactive sites for the addition of epoxides and for the acylation with anhydrides.

Following the reaction with epoxides, the amount of tyrosine (Tyr), histidine (His), lysine (Lys), and arginine (Arg) recovered after acid hydrolysis significantly decreases.[1-7,9-11]

The reactivity of various acid anhydrides toward silk depends on both steric and chemical factors related to the characteristics of the anhydride substituent.[12,13]

An interesting aspect of the reaction of epoxides and anhydrides with silk fibers is the possibility that crosslinks are formed between adjacent fibroin chains. Bifunctional epoxides, such as EDGE, and some dibasic acid anhydrides, such as glutaric and phthalic anhydride, seem more effective in inducing crosslinking.[3,14,16,17] The physical and mechanical properties of silk fibers are significantly affected by the presence of crosslinks within the fiber matrix.

PROPERTIES

Crystalline Structure and Molecular Orientation

The amino acid side groups that are potentially reactive toward epoxides and acid anhydrides are mainly located in the amorphous regions of silk fibers, where the reaction is likely to occur. The β-sheet crystalline regions are not affected as demonstrated by the X-ray diffraction measurements.[6,7,15]

Thermal Behavior

The results of differential scanning calorimetry (DSC) and thermomechanical (TMA) analyses are consistent in detecting a slightly higher thermal stability of *B. mori* silk fibers modified with epoxides.[7,18] The formation of crosslinks between adjacent fibroin chains might be responsible for this effect.

The DSC and TMA pattern of silk fibers modified with acid anhydrides is substantially unaffected.[14,16]

Tensile Properties

Tensile strength and elongation at break of silk fibers modified with acid anhydrides do not show significant changes compared to the untreated sample.[15,16]

Silk fibers treated with bifunctional epoxides, exhibit a slight increase in tensile strength at low weight gains (5–10%).[7,18,19]

Chemical and Photochemical Resistance

The chemical resistance of silk fibers to hydrolytic attack is considerably improved by reaction with epoxides.[2,3,8,19]

Silk is sensitive to photodegradation induced by exposure to sunlight. The treatment with various epoxides can result in a significant decrease in the rate of photoyellowing, the protective effect is proportional to the amount of weight gain.[7,19] The reaction with glutaric anhydride is also effective in protecting silk against photoyellowing.[16]

APPLICATIONS

Similar to grafting, the application of chemical modification with epoxides and acid anhydrides is currently confined to the textile field. Using these techniques has the advantage of causing specific modifications to the fiber properties. A low weight gain is usually necessary (≤10%) to attain the required effect. Epoxides are effective in improving the crease-recovery behavior of silk fibers in both the dry and wet state, whereas acid anhydrides cause significant changes in fiber dyeability.[2,3,7,8,14,16,17,19,20] Epoxides have already been introduced into industrial-scale silk processing, acid anhydrides are still in the experimental stage.

Shiozaki and Tanaka first reported the considerable improvement in crease recovery of epoxide-treated silk fibers.[2,3]

Aside from changes in dyeability, the reaction of silk with acid anhydrides (glutaric, phthalic, o-sulfobenzoic) is effective in improving crease recovery, although the typical silk handle is partly impaired by a certain degree of fiber stiffness.[14,16] However, the application of these modifying agents under controlled treatment conditions may confer new characteristics on silk, leading to new opportunities for using modified silk fibers in several textile applications.

REFERENCES

1. Shiozaki, H.; Tanaka, Y. *Polym. Lett.* **1969**, *7*, 325.
2. Shiozaki, H.; Tanaka, Y. *Die Makromol. Chem.* **1971**, *143*, 25.
3. Shiozaki, H.; Tanaka, Y. *Die Makromol. Chem.* **1972**, *152*, 217.
4. Tanaka, Y.; Shiozaki, H. *J. Polym. Sci., Polym. Chem. Ed.* **1974**, *12*, 2741.
5. Shiozaki, H.; Tanaka, Y. *J. Seric. Sci. Jpn.* **1992**, *61*(3), 247.
6. Tsukada, M.; Gotoh, Y.; Freddi, G.; Matsumura, M.; Shiozaki, H.; Ishikawa, H. *J. Appl. Polym. Sci.* **1992**, *44*, 2203.
7. Tsukada, M.; Shiozaki, H.; Gotoh, Y.; Freddi, G. *J. Appl. Polym. Sci.* **1993**, *50*, 1841.
8. Xiaojun, P.; Jitao, W.; Jie, S. *J. Soc. Dyers and Colorists* **1993**, *109*, 19.
9. Tanaka, Y.; Shiozaki, H. *Proceedings of the 7th International Wool Textile Research Conference* Tokyo, **1985**; Vol. IV, p 441.
10. Tanaka, Y.; Shiozaki, H. *Sen-i Gakkaishi* **1989**, *45*(4), 147.
11. Shiozaki, H.; Tsukada, M.; Gotoh, Y.; Kasai, N.; Freddi, G. *J. Appl. Polym. Sci.* **1994**, *52*, 1037.
12. Shiozaki, H.; Tanaka, Y. *J. Seric. Sci. Jpn.* **1980**, *49*(4), 307.
13. Shiozaki, H.; Tsukada, M.; Matsumura, M. *J. Seric. Sci. Jpn.* **1988**, *57*(2), 165.
14. Tsukada, M.; Gotoh, Y.; Freddi, G.; Shiozaki, H. *J. Appl. Polym. Sci.* **1992**, *45*, 1189.
15. Tsukada, M.; Gotoh, Y.; Shiozaki, H.; Freddi, G.; Crighton, J. S. *J. Appl. Polym. Sci.* **1994**, *51*, 345.
16. Tsukada, M.; Shiozaki, H. *J. Appl. Polym. Sci.* **1989**, *37*, 2637.
17. Kamiishi, Y.; Tanaka, Y.; Arai, K. *Proceedings of the 8th International Wood Textile Research Conference*; Cranshaw, G. H., Ed.; Christchurch, 1990; Vol. I, p 346.
18. Tsukada, M.; Nagura, M.; Ishikawa, H.; Shiozaki, H. *J. Appl. Polym. Sci.* **1991**, *43*, 643.
19. Tanaka, Y.; Suda, N.; Nagasaka, T.; Onooka, R.; Tejima, T.; Nakasato, A.; Kamiishi, Y.; Shimomura, S.; Hatta, S. *Proceedings of the 7th International Wool Textile Research Conference* Tokyo, 1985; Vol. IV, p 461.
20. Freddi, G.; Tsukada, M.; Kato, H.; Shiozaki, H. *J. Appl. Polym. Sci.* **1994**, *52*, 769.

SILK FIBERS (Grafting)

Giuliano Freddi*
Stazione sperimentale per la Seta

Masuhiro Tsukada
National Institute of Sericultural and Entomological Science

The graft-copolymerization of vinyl monomers onto silk fibers (*B. mori*) started in the early 1960s in Japan. During the past three decades there has been a considerable academic and technological interest in this field. In recent years grafting has been regarded not only as an effective treatment to increase silk weight but also as a powerful method to improve the textile performances of silk, such as crease recovery, dimensional stability, rub resistance, photoyellowing, oil and water repellency, color fastness, etc.

PREPARATION

The graft-copolymerization of vinyl monomers onto silk fibers is carried out by the free radical initiation method. The primary step involves the formation of free radical sites on the backbone of the fibroin molecule. This can be achieved by both chemical and physical methods. The former are by far the most popular and imply the use of redox systems, while the latter include various methods such as irradiation (UV, γ-rays) and photo-initiation (charge transfer complexes).

REACTIVITY OF VINYL MONOMERS TOWARD SILK

The graft-copolymerization of MMA onto silk fibers was extensively investigated by Nayak and coworkers.[7–11,14,20] The accumulated kinetic data indicate that the radical mechanism is responsible for the initiation and propagation of the graft-copolymerization reaction.

The type of vinyl monomer and initiator, temperature, time, pH, reagent concentration, and polymerization medium can exert a significant influence on weight gain and grafting efficiency.[1,2,5–23,30,31]

PROPERTIES

Crystallinity and Molecular Orientation

Grafting takes place in the amorphous domains of the fibers, either for steric or chemical reasons, such as accessibility and availability of reactive sites. The crystalline domains are not directly affected by grafting, as demonstrated by wide-angle X-ray diffraction measurements.

Grafting with vinyl monomers may exert a noticeable effect on the molecular orientation of silk fibers. The degree of orientation of the crystalline regions, calculated on the basis of the X-ray measurements, remains unchanged for weight gain values not exceeding 60%.[24,28]

Thermal Behavior

The differential scanning calorimetry (DSC) profile of untreated silk fibers is characterized by a broad and large endothermic transition starting beyond 250°C, with a peak temperature at around 315°C, attributed to the thermal decomposition of silk fibroin with oriented β-sheet structure.[33] Silk fibers grafted with different vinyl monomers usually display a slight upward shift of the decomposition temperature.[3,24,25–27,29,31,32]

The increased thermal stability of grafted silk fibers is confirmed by thermogravimetric analysis (TGA).[24,26,32] Following grafting, the fibers exhibit higher values of weight retention beyond 300°C, in good agreement with DSC measurements.

The dynamic viscoelastic properties of silk fibers are significantly affected by grafting.[4,24,26,28,29,32]

The graft-copolymerization of vinyl monomers onto silk fibers generally results in an enhanced, thermally induced molecular motion of the fibroin chains. Interactions between the fibroin molecules and the grafted polymer chains occur in the amorphous regions, where grafting takes place.

APPLICATIONS

The application of the graft-copolymerization of vinyl monomers onto silk fibers is currently confined to the textile field. Since silk still plays a major role in fashion collections, in spite of its low tonnage share in the total world fiber production, grafting techniques should conform to fashion trends, and market requirements. For example, soft handle, elegant luster, beautiful drape, and brilliant color shades are highly desired in Western-style silk articles. Therefore, grafting becomes less important, except for certain articles such as necktie yarn and fabric, borders, and embroidery. However, Japanese traditional costumes (kimonos) often require heavy fabrics with a certain degree of fiber stiffness. Grafting with vinyl monomers may be effective in conferring the desired bulkiness and stiffness to silk.

MAA and HEMA are currently applied for silk grafting. Grafting on an industrial scale is generally carried out on degummed silk yarn in the bank or package form by means of spray-dyeing machines. To avoid changes in the affinity for most acid, metal complex, and reactive dyes, grafting is often carried out on dyed silk yarn. The graft-copolymerization of MAA onto silk fabrics has been recently reported.[22] Moreover, a new method for simultaneous dyeing and grafting of both silk yarns and fabrics has been proposed.[23a]

MAA-grafted silk fibers maintain the typical silk-like handle and drape.

The selection and use of grafting agents with special substituents that, once present in the fiber matrix, confer improved and/or new properties on silk, may give a strong impulse to the future development of grafting techniques.

REFERENCES

1. Lee, Y. W.; Song, K. E. *Res. Rep. ORD.* **1984**, *26*, 53.
2. Lee, Y. W.; Song, K. E. *Korean J. Seric. Sci.* **1985**, *27*, 47.
3. Tsukada, M. *J. Appl. Polym. Sci.* **1988**, *35*, 965.
4. Tsukada, M.; Kasai, N.; Freddi, G. *J. Appl. Polym. Sci.* **1993**, *50*, 885.
5. Kojima, K.; Iwabuchi, S.; Kojima, K.; Tarumi, N. *J. Polym. Sci., Polym. Letters* **1971**, *9*, 25.
6. Tsukada, M.; Yamamoto, T.; Nakabayashi, N.; Ishikawa, H.; Freddi, G. *J. Appl. Polym. Sci.* **1991**, *43*, 2115.
7. Nayak, P. L.; Lenka, S.; Pati, N. C. *Die Angew. Makromol. Chem.* **1978**, *68*, 117.

*Author to whom correspondence should be addressed.

8. Nayak, P. L.; Lenka, S.; Pati, N. C. *Die Angew. Makromol. Chem.* **1979**, *75*, 29.
9. Nayak, P. L.; Lenka, S.; Pati, N. C. *J. Appl. Polym. Sci.* **1979**, *23*, 1345.
10. Misra, S.; Nayak, P. L.; Sahu, G. *J. Appl. Polym. Sci.* **1982**, *27*, 1903.
11. Panda, G.; Pati, N. C.; Nayak, P. L. *J. Appl. Polym. Sci.* **1980**, *25*, 1479.
12. Padhi, N. P.; Tripathy, S. S.; Jena, S.; Singh, B. C. *J. Appl. Polym. Sci.* **1983**, *28*, 1811.
13. Samal, R. K.; Suryanarayan, G. V.; Das, P. C.; Panda, G.; Das, D. P.; Nayak, N. C. *J. Appl. Polym. Sci.* **1981**, *26*, 2221.
14. Samal, S.; Sahu, G.; Nayak, P. L. *J. Appl. Polym. Sci.* **1984**, *29*, 3283.
15. Samal, S.; Sahu, G.; Nayak, P. L. *J. Macromol. Sci. Chem.* **1984**, *A21*(5), 539.
16. Samal, S.; Sahu, G.; Nayak, P. L. *J. Macromol. Sci. Chem.* **1984**, *A21*(6–7), 725.
17. Panda, G.; Pradhan, A. K.; Pati, N. C.; Nayak, P. L. *J. Polym. Sci., Polym. Chem. Ed.* **1980**, *18*, 3311.
18. Nayak, P. L.; Lenka, S.; Mishra, M. K. *J. Polym. Sci., Polym. Chem. Ed.* **1980**, *18*, 2247.
19. Lenka, S.; Nayak, P. L.; Mishra, M. K. *Die Angew. Makromol. Chem.* **1980**, *84*, 183.
20. Nayak, P. L.; Lenka, S.; Pati, N. C. *Die Angew. Makromol. Chem.* **1980**, *85*, 29.
21. Mishra, M. K. *Colloid and Polymer Sci.* **1982**, *260*, 508.
22. Czerny, A. R.; Ubler, A. M.; Schindler, W. *Melliand Textilberichte* **1990**, *71*, 211.
23. Marzinkowski, J. M.; Trappe, J.; Hambsch, B. *Melliand Textilberichte* **1991**, *72*, 538.
23a. Schindlert, W.; Dietel, A. *Melliand Textilberichte* **1993**, *74*, 81.
24. Tsukada, M.; Freddi, G.; Ishiguro, Y.; Shiozaki, H. *J. Appl. Polym. Sci.* **1993**, *50*, 1519.
25. Tsukada, M.; Ishiguro, Y. *J. Seric. Sci. Jpn.* **1984**, *53*, 310.
26. Tsukada, M.; Freddi, G.; Monti, P.; Bertoluzza, A.; Shiozaki, H. *J. Appl. Polym. Sci.* **1993**, *49*, 1835.
27. Tsukada, M.; Shiozaki, H. *J. Appl. Polym. Sci.* **1990**, *39*, 1289.
28. Tsukada, M.; Freddi, G.; Shiozaki, H.; Pusch, N. *J. Appl. Polym. Sci.* **1993**, *49*, 593.
29. Tsukada, M.; Shiozaki, H.; Crighton, J. S.; Kasai, N. *J. Appl. Polym. Sci.* **1993**, *48*, 113.
30. Tsukada, M.; Shiozaki, H.; Crighton, J. S. *J. Appl. Polym. Sci.* **1993**, *48*, 1409.
31. Samal, R. K.; Nanda, C. N.; Satrusallya, S. C.; Nayak, P. L. *J. Appl. Polym. Sci.* **1983**, *28*, 1311.
32. Tsukada, M.; Freddi, G.; Monti, P.; Bertoluzza, A. *J. Appl. Polym. Sci.* **1993**, *49*, 1565.
33. Ishikawa, H.; Tsukada, M.; Doizume, T.; Hirabayashi, K.; Gakkaishi, Sen-i **1972**, *28*, 91.

SILK FIBROIN (Soft Tissue Compatible Polymer)

Norihiko Minoura
National Institute of Materials and Chemical Research

Soft-tissue compatible polymers are the materials of choice for a variety of biomedical needs ranging from artificial skin, corneas, and tendons to prostheses and implanted devices for artificial organs and drug delivery. Soft-tissue compatible polymers have to be nontoxic to be able to be used in the human body. For a number of biomedical applications it is desirable to have a soft-tissue compatible material that is noninert and also biodegradable. The material would be expected to adhere to tissue cells and to be assimilated by the tissue. Of late, significant research efforts are being made toward the identification of this type of biomaterial. Collagen is one of the most promising materials in this regard, and is currently used as artificial skin. Silk fibroin is a fibrous protein that is used for surgical sutures. In order to expand the application of silk fibroin, we have recently initiated studies on the preparation and properties of films of silk fibroin.

PROPERTIES

Mechanical Properties of Silk Fibroin Films

The silk fibroin films in the dry state were extremely brittle and unsuitable for practical uses.[1] There are several significant differences in the mechanical behavior between the dry and the wet states of the silk fibroin films. The elongation at break of the silk fibroin films treated with 50% methanol for 30 and 60 min increased up to 100 and 70%, respectively, whereas the breakdown tensile strength was reduced. Judging from the shape of the stress–strain curves, the silk fibroin films became soft in the presence of water. These results suggest that water in the films serves as a plasticizer.

Enzymatic Degradation of Silk Fibroin Films

The morphology of the films treated with aqueous methanol solutions consist of a domain structure composed of crystalline and amorphous regions: during the process of the enzyme degradation, silk fibroin molecules in the amorphous region of the film surface are digested by the enzyme and those in the crystalline region (or highly dense region) remain unaffected. In comparison with the silk fibroin films, surgical sutures, which are made from highly oriented molecules of silk fibroin protein, were not digested even after immersion in the enzyme solution, and the surface morphology of the suture did not change.[1,2] These results suggest that by properly controlling the degree of the molecular orientation of the silk fibroin, it is possible to develop silk fibroin materials of varying degrees of enzyme degradability.

Cell Attachment and Growth on Silk Fibroin Films

Fibroblast cells (L-929) were incubated on the film of the native *Bombyx mori* silk fibroin. The silk fibroin film shows a remarkable affinity for attachment of L-929 cells. Further, the attached cells were not only able to remain viable, they also proliferated on the surface of this film.

APPLICATIONS

Silk fibroin films have excellent properties as has been previously described. Furthermore, we have found that silk fibroin has high dissolved-oxygen permeability and high water-vapor permeability.[1,3] Therefore, silk fibroin has the potential to find application in the development of soft contact lenses, artificial corneas, artificial skin substitutes, and so on. Recently, we reported that silk fibroin film has drug permeability that is controlled by the surrounding pH.[4] This result may lead to the application of silk fibroin film in the development of controlled drug-delivery devices that respond to the conditions found in the body.

The properties and the functions of silk fibroin films can be improved and augmented by various kinds of chemical and physical modification. Of late, technology to introduce new genes into the silkworm has also been established. Because one silkworm can produce about 0.5 g silk fibroin in dry weight, a silkworm can be regarded as a bioreactor. Thus, using the genetic-engineering approach just mentioned, it is hoped that silk proteins with completely novel properties and functions can be obtained. Recently, it has been reported that a silklike polymer with "fibronectin cell attachment functionality" was developed via a genetic-engineering route.[5] All these recent developments indicate that silk fibroins form a class of soft-tissue compatible materials that are very promising.

REFERENCES

1. Minoura, N.; Tsukada, M.; Nagura, M. *Biomaterials* **1990**, *11*, 430.
2. Minoura, N. In *Biomedical Applications of Polymeric Materials*; Tsuruta, T. et al., Eds.; CRC: Boca Raton, FL, 1993, p 117.
3. Minoura, N.; Tsukada, M.; Nagura, M. *Polymer* **1990b**, *31*, 265.
4. Chen, J.; Minoura, N.; Tanioka, A. *Polymer* **1994**, *35*, 2853.
5. Anderson, J. P.; Cappello, J.; Martin, D. C. *Biopolymers* **1994**, *34*, 1049.

SILOXANE-CONTAINING POLYMERS

Bogdan C. Simionescu
Department of Macromolecules
"Gh. Asachi" Technical University

Valeria Harabagiu and Cristofor I. Simionescu
"P. Poni" Institute of Macromolecular Chemistry

The technological importance of block and graft copolymers has recently grown. Designed to combine the properties of two or more polymers, block and graft copolymers have extended the frontier of polymeric materials. Among this group siloxane-containing copolymers represent a unique class.

SYNTHESIS OF SILOXANE-CONTAINING BLOCK AND GRAFT COPOLYMERS

Siloxane-containing copolymers are hybrid materials whose components are obtained by quite different synthetic approaches.

Living Polymerization

Until recently, living anionic polymerization of cyclosiloxanes was hardly investigated. Nowadays, it was reconsidered as a powerful technique in the preparation of very well defined polysiloxanes.[6,7] Living anionic polymerization also allows the synthesis of siloxane–vinyl random and block copolymers. The first report dealt with living anionic copolymerization of octamethylcyclotetrasiloxane (D_4) and isoprene.[8] Over the next 10 years, organolithium, potassium, sodium compounds, lithium silanolates, and siloxanolates were used to initiate the anionic copolymerization of D_4 or hexamethylcyclotrisiloxane (D_3) with styrene or methyl methacrylate.

Successive anionic polymerization of differently substituted siloxane monomers was also used to obtain AB- or ABA-type block copolymers.

More complex synthetic approaches have been also proposed. The preparation of star polymers via living anionic polymerization followed by multifunctional termination using chlorosilanes was demonstrated by Hadjichristidis, whereas Dickstein obtained well-defined telechelic star polydimethylsiloxanes.[5,6] Similar pathways toward siloxane star polymers have been described in different reports.[7–10]

Ring-Opening Polymerization

The ring-opening polymerization of cyclic organic monomers, like lactones, lactams, or oxazolines, by using functional siloxane oligomers as initiators and comonomers was investigated quite recently. AB- or ABA-type copolymers with thermoplastic elastomer characteristics or with end uses as modifying agents in polymer blends were obtained.[11–14]

Radical Polymerization

Two directions can be distinguished when considering the preparation of siloxane copolymers by radical copolymerization of organic monomers: radical copolymerization of siloxane macromers with vinyl monomers, which results in graft copolymers, and radical polymerization of vinyl monomers in the presence of siloxane macroinitiators, which yields block copolymers.

Coupling by Hydrosilation

The hydrosilation reaction between Si-H functionalized polysiloxanes and organic polymers with terminal unsaturated groups is one of the earliest methods used in the synthesis of siloxane-containing copolymers. Pioneering developments in the field belong to Dow Corning Corporation. Starting in 1956, they prepared, characterized, and used polysiloxane (PDMS)–poly(ethylene oxide) (PEO) block and graft copolymers. Controlled synthesis of AB-, ABA-, $(AB)_n$-type PDMS–PEO block copolymers, or of PDMS–PEO graft copolymers, and of PDMS–PEO–PPO (polypropylene oxide) graft copolymers was later reported.[15–19] These copolymers were assessed as emulsifiers and stabilizers in many systems, they were used in cryobiology applications, or as a polymer matrix in conducting materials.[17–21]

The hydrosilation process remains a very useful route for linking together organic and siloxane sequences. It has been widely used over the last 10 years in the preparation of more sophisticated polymeric structures (liquid crystalline materials, unusual polyether–polysiloxane copolymers,[22–28] or in the modification of polymer shapes and properties.[29,30]

Step-Growth Polymerization

Step-growth polymerization is the most widely used method for the synthesis of different types of siloxane–organic copolymers. McGrath and co-workers have done prolific work in the field; they have prepared a large variety of segmented siloxane copolymers starting from silicofunctional and organofunctional oligosiloxanes.[1]

PROPERTIES AND APPLICATIONS OF SILOXANE CONTAINING BLOCK AND GRAFT COPOLYMERS

An excellent review[2] as well as recent studies[40–48] found in the literature are available concerning the properties and applications of siloxane-containing copolymers.[2] These polymeric materials generally represent microphase-separated systems because of the high incompatibility of PDMS (almost exclusively used as soft segment) with organic macromolecular compounds.[31–39] The most important driving force in phase separation of PDMS-containing copolymers is the very low solubility parameter of PDMS. This unusual characteristic, together with the very high flexibility of PDMS, produce hard and soft copolymers derived from properties ranging from thermoplastic elastomers to rubber-modified thermoplastics. However, the association of PDMS with hydrophilic-type organic segments gives rise to materials with amphiphilic properties.

Other interesting features of PDMS are very low surface tension and surface energy. As a consequence, the air-polymer surface of siloxane copolymers is practically fully composed of PDMS. Siloxane-containing block and graft copolymers are used in surface modification (they impart biocompatibility, hydrophobicity, release properties, oxygen and radiation resistance, surface finish and gloss, etc.) of conventional polymers.

REFERENCES

1. Yilgor, I.; McGrath, J. E. *Adv. Polym. Sci.* **1988**, *86*, 1.
2. Boileau, S. *ACS Symp. Ser.* **1985**, *286*, 23.
3. Maschke, U.; Wagner, T.; Coqueret, X. *Makromol. Chem.* **1992**, *193*, 2453.
4. Morton, M.; Rembaum, A.; Bostick, E. E. *J. Appl. Polym. Sci.* **1964**, *8*, 2707.
5. Hadjichristidis, N.; Guyot, A.; Fetters, L. J. *Macromolecules* **1978**, *11*, 889.
6. Dickstein, W. H.; Lillya, C. P. *Macromolecules* **1989**, *22*, 3886.
7. Wilczek, L.; Rubinsztajn, S.; Fortuniak, W. W.; Chojnowski, J.; Tverdokhlebova, I. I.; Volokova, R. V. *Bull. Pol. Acad. Sci.: Chem.* **1989**, *37*, 91.
8. Tverdokhlebova, I. I.; Larina, T. A.; Mamaeva, I. I.; Pertsova, N. V.; Wilczek, L.; Rubinsztajn, S.; Chojnowski, J. *Vysokomol. Soedin., Ser. B.* **1990**, *32*, 292.
9. Bhattacharya, S. K.; Smith, C. A.; Dickstein, W. H. *Macromolecules* **1992**, *25*, 1373.
10. Ogawa, T.; Suzuki, T.; Mita, I. *Macromol. Chem. Phys.* **1994**, *195*, 1973.
11. Yilgor, I.; Steckle, Jr., W. P.; Yilgor, E.; Freelin, R. G.; Riffle, J. S. *J. Polym. Sci.: Part A: Polym. Chem.* **1989**, *27*, 3673.
12. Lefebvre, P. M.; Jerome, R.; Teyssie, Ph. *J. Polym. Sci., Polym. Chem. Ed.* **1983**, *21*, 789.
13. Policastro, P. P.; Hernandez, P. K. *Polym. Bull.* **1986**, *16*, 43.
14. Way, T. F.; Padias, A. B.; Hall, Jr., H. K. *Polym. Bull.* **1990**, *24*, 157.
15. Galin, M.; Mathis, A. *Macromolecules* **1981**, *14*, 677.
16. Haesslin, H. W. *Makromol. Chem.* **1985**, *186*, 357.
17. MacFarlane, D. R. Bannister, D. J.; Barry, W. R.; Forsyth, M.; Jeffrey, R. L. *Polym. Prepr.* **1987**, *28*(1), 405.
18. Zhou, G.; Khan, I. M.; Smid, J. *Polym. Prepr.* **1989**, *30*(1), 416.
19. Giurgiu, D.; Hamciuc, V.; Harabagiu, V.; Ionescu, M.; Mihalache, I.; Ionescu, C. *Mem. Sect. St. Acad. Rom.* **1992**, *IV, XV*, 277.
20. Haesslin, H. W.; Eicke, H. F.; Riess, G. *Makromol. Chem.* **1984**, *185*, 2625.
21. Guegan, P.; Lestel, L.; Teyssie, D.; Boileau, S.; Cheradame, H. *Polym. Prepr.* **1993**, *34*(1), 209.
22. Percec, V.; Hsu, C. S. *Polym. Bull.* **1990**, *23*, 463.
23. Percec, V.; Tomazos, D. *Contemporary Topics in Polymer Science* Vol. 7; Plenum: New York, London, 1992, p 247.
24. Sellinger, A.; Laine, R. M. *Polym. Prepr.* **1993**, *34*(1), 288.
25. Poths, H.; Zentel, R. *Macromol. Rapid Commun.* **1994**, *15*, 433.
26. Wischerhoff, E.; Zentel, R.; Redmond, M.; Mondain-Monval, O.; Coles, H. *Macromol. Chem. Phys.* **1994**, *195*, 1593.
27. Crivello, J. V.; Fan, M. *Polym. Prepr.* **1993**, *34*(1), 213.
28. Crivello, J. V.; Sasaki, H. *J. Macromol. Sci.—Pure Appl. Chem.* **1993**, *A30*, 173.
29. Guo, X.; Rempel, G. L. *Macromolecules* **1992**, *25*, 883.
30. Gabor, A. H.; Lehner, E. A.; Long, T. E.; Mao, G.; Rauch, E. C.; Schell, B. A.; Ober, C. K. *Polym. Prepr.* **1993**, *34*(1), 284.
31. Smid, J.; Fish, D.; Khan, I. M.; Wu, E.; Zhou, G. *Adv. Chem. Ser.* **1990**, *224*, 113.
32. Brook, M. A.; Kremers, C. H.; Sebastian, T.; Yu, W. *J. Polym. Sci.: Part C: Polym. Lett.* **1989**, *27*, 229.
33. Kiefer, L. A.; Yoon, T. H.; Glass, T. E.; McGrath, J. E. *Polym. Prepr.* **1992**, *33*(2), 227.
34. Harabagiu, V.; Pinteala, M.; Cotzur, C.; Simionescu, B. C.; Simionescu, C. I. *Synth. Polym. J.* **1994**, *1*, 259.
35. Andolino Brandt, P. J.; Senger Elsbernd, C. L.; Patel, N.; York, G.; McGrath, J. E. *Polymer* **1990**, *31*, 180.
36. Matsukawa, K.; Inoue, H. *J. Polym. Sci.: Part C: Polym. Lett.* **1990**, *28*, 13.
37. Benrashid, R.; Nelson, G. L. *J. Polym. Sci.: Part A: Polym. Chem.* **1994**, *32*, 1847.
38. Kajiyama, M.; Kakimoto, M.; Imai, Y. *Macromolecules* **1990**, *23*, 1244.
39. Yang, C. Z.; Li, C.; Cooper, S. L. *J. Polym. Sci.: Part B: Polym. Phys.* **1991**, *29*, 75.
40. Hoover, J. M.; Smith, S. D.; DeSimone, J. M.; Ward, T. C.; McGrath, J. E. *Polym. Prepr.* **1988**, *29*(1), 166.
41. Feng, D.; Wilkes, G. L.; Crivello, J. V. *Polym. Prepr.* **1989**, *30*(2), 229.
42. Kukukyavuz, S.; Kukukyavuz, Z.; Erdogan, G. *Polymer* **1990**, *31*, 379.
43. Nagase, Y.; Naruse, A.; Matsui, K. *Polymer* **1990**, *31*, 121.
44. Ohyanagi, M.; Nishide, H.; Suenaga, K.; Tsuchida, E. *Polym. Bull.* **1990**, *23*, 637.
45. Shibayama, M.; Inoue, M.; Yamamoto, T.; Nomura, S. *Polymer* **1990**, *31*, 749.
46. Gorelova, M. M.; Pertsin, A. J.; Levin, V. Yu; Makarova, L. I.; Filimonova, L. V. *J. Appl. Polym. Sci.* **1992**, *45*, 2075.
47. Li, W.; Huang, B. *J. Polym. Sci.: Part B: Polym. Phys.* **1992**, *30*, 727.
48. Samseth, J.; Mortensen, K.; Burns, J. L.; Spontak, R. J. *J. Appl. Polym. Sci.* **1992**, *44*, 1245.

SILOXANE ELASTOMERS

Peter D. Folland and Janis G. Matisons*
Polymer Science Group
School of Chemical Technology
University of South Australia

In recent years, there has been increasing research into "high performance elastomers" based on inorganic polymers in order to bridge the gaps that organic elastomers have left and to penetrate into new fields.[1]

*Author to whom correspondence should be addressed.

Dvornic and Lenz[2] enumerate the properties that set siloxane elastomers apart from the organic elastomers as:

- high temperature resistance (constant usage at >250°C is possible);
- thermo-oxidative stability;
- hydrolytic stability;
- flexibility below minimum ambient temperatures [low glass transition temperature (T_g)];
- useful mechanical properties over a wide temperature range;
- constant electrical properties;
- chemical inertness;
- resistance to solvents (softening, swelling, and dissolving);
- relatively economic in production; and
- usable under normal processing procedures.

To these can be added the properties of weathering resistance and low surface tension for both siloxane polymers and elastomers. Other unusual properties of polydimethylsiloxanes (PDMS) are unexpectedly low values of vapor pressure, viscosity, temperature coefficient of viscosity, entropy of dilution, and excess volume of mixing; these types of siloxane also possess large permeabilities.[3] Such properties carry over into the elastomers based on PDMS.

PREPARATION

To convert liquid or waxy siloxane polymers into an integrated elastomeric matrix, the polymeric chains must be crosslinked.

The curing of a siloxane elastomer may be accomplished by heating in the presence of various catalysts or by high-energy curing. Low-temperature curing [also referred to as "low-temperature vulcanization" (LTV) or more commonly as "room-temperature vulcanization" (RTV)] can be further divided by the reaction type used to crosslink the siloxanes; these are an addition reaction (commonly called "hydrosilation" or "hydrosilylation") and a condensation reaction.

High-Temperature Cure

Siloxane elastomers, produced by high-temperature curing, are used in the general rubber industry, where the specialist equipment needed for the mixing, transfer, and heating of highly viscous polymer blends is common. Consequently, much of the information that is available for rubber processing also applies to the high-temperature processing of siloxanes using peroxide initiators.

Noll[4] details the vulcanization of silicone elastomers in his classic text on the chemistry and technology of silicones. Crosslinking is generally agreed to occur through the abstraction of hydrogen from a backbone methyl group by a peroxy radical, followed by the combination of two of these $SiC \cdot H_2$ radicals to form an ethyl bridge between two polymer chains.

Marsden[5] was the first to show that the incorporation of a small proportion of vinyl groups (typically 0.1–0.15%w/w) into the polymer allows curing to proceed with lower levels of (and a greater variety of) initiators.

Room-Temperature Cure

Condensation Curing

Siloxane RTV elastomers are derived as either one or two component packs. The basis of most cartridge packs of silicone elastomers on the commercial and industrial market are condensation cured or (crosslinked) siloxane elastomers.

Acetone-type, one-component RTV elastomers, for example, use propenoxysilane as a crosslinking agent and generate acetone in the curing process. This particular type of RTV elastomer is well suited for use in the electronics industry, especially because of its reversion resistance.

One-component packs exclude moisture or are formulated with moisture scavengers. On opening to the atmosphere, the prevailing humidity initiates the condensation reaction.

A typical example of the condensation route to siloxane elastomers uses an α,ω-silanol-terminated polydimethylsiloxane with methyltriacetoxysilane as the crosslinker. On exposure to humid air, curing occurs, either by direct reaction between the acetoxy and the silanol, or by the conversion of the acetoxy group to a silanol group prior to the condensation of both silanol groups.

Addition Curing

Siloxane elastomers produced from a multipack system are likely to be based on the hydrosilylation-addition cure. Unlike condensation crosslinking, where Si–O–Si bonds were being formed, the addition curing by hydrosilylation forms two Si-C linkages between the two siloxane chains.

Generally, the base polymer is an α,ω-divinyl-terminated polydimethylsiloxane with a molecular weight of between 50,000 and 300,000. Polymers with a few reactive vinyl groups on the siloxane backbone increase the tensile strength of the elastomer.

Platinum is still the transition metal commonly used to catalyze hydrosilylation reactions, and incorporation of various additives is still necessary to produce a commercial elastomer.

Other Cure Methods

Two other important methods used to crosslink siloxane elastomers are high-energy radiation curing and photocuring. High-energy radiation curing generates free radicals within the organic portion of the siloxane to facilitate crosslinking.

Photocuring uses UV radiation to initiate the formation of reactive intermediates from specific organofunctional groups attached to the siloxane backbone. The free radicals or reactive cations that are formed interact with each other and the other pendant organic groups to produce a crosslinked elastomer. The UV-visible light needed to initiate photocuring is ~250–550 nm. Thomas[6] lists the four main technologies used in photocuring as:

- thiol/thiolene chemistry,
- hydrosilane/nydrasilene addition,
- acrylate polymerization, and
- cationic polymerization.

Fillers

Fillers are an essential component in most siloxane elastomers because of the improved mechanical properties that are needed. Although defined as inert materials that are employed to extend the product, reduce cost, and provide increased hardness,[7] for siloxane elastomers the term "fillers" has also come to include reinforcing agents. Such fillers are fine powders used to increase the strength, hardness, and abrasion resistance of rubber.

The two most commonly encountered reinforcing fillers for siloxane elastomers are carbon black and amorphous silica (fumed and precipitated silica). Fumed silica (produced by the pyrolysis of silicon tetrachloride in an atmosphere of hydrogen and oxygen) is the reinforcing agent of choice because of its chemical affinity to and dispersion in siloxane elastomers, its large surface area (up to 400 m^2/g), color (white allows for final-product pigmentation), and effective reinforcement and property enhancement.[8]

PROPERTIES

A wide range of elastomers can be manufactured from polysiloxanes to meet a variety of specialized requirements.

Most commercial manufacturers determine which property (physical, chemical, electrical, etc.) is important in a particular situation and exploit formulating techniques to enhance that particular property. Tensile strength and elongation are the two key properties considered in siloxane elastomer formulations. In specific applications, subsidiary properties such as tear can also become very important. Another property of siloxane elastomers that is important to manufacturers and end users alike is its resistance to aging in adverse conditions.

APPLICATIONS

Peroxide-Cured Products

Siloxane elastomers cured using peroxide initiators have found markets in many household, commercial, and industrial products. Such products include nipples for infant bottles; insulation for electrical wires and cables; temperature- and oil-resistant components in automobile and aircraft engines; and tubing for industrial and medical requirements including catheters, vascular prostheses, and implants.[9] Even products that can withstand temperatures of up to 600°C for 3 h are capable of being produced in this way.[10]

Condensation-Cured Products

Condensation-cured products are generally one-pack cartridge systems, such as those used as sealants in the building industry. They are also used as adhesives in aerospace, marine, and building applications; for *in situ* gasket formation; and for the sealing of electrical components and in numerous high-temperature applications (e.g., microwave ovens). Protective coatings for electrical circuitry have become a growing application for condensation-cured siloxane elastomers.

Addition-Cured Products

The major uses of addition-cured siloxane elastomers are in the electronics industry in the potting and encapsulation of circuitry; in the automotive industry as flexible bonding agents, with minor use in the formation of intricate molds for plaster casting, etc. or in relieving stress between the inner and outer layers of windshields.

Other Cure Methods

Radiation and photocuring of siloxane elastomers are limited to areas such as surface coatings and adhesives. Radiation curing is also used in dental applications, such as in fillings, molds, and denture liners.[11]

SUMMARY

The term "siloxane elastomers" covers a wide range of products with a vast number of uses and is made more complex because several cure methods are available for their manufacture. Siloxane elastomers have unique physical and chemical properties that allow them to fill application areas that organic elastomers cannot. As more information is correlated on cure, fillers, crosslinkers, and other structure–property relationships, the role of siloxane elastomers in tomorrow's technology will become more significant. Still more work is needed in modeling/correlating filler–matrix interactions to mechanical performance. As such models become better able to predict product properties, new applications will emerge.

REFERENCES

1. Mark, J. E. *Die Angew. Makromol. Chem.* **1992**, *202/203*, 1.
2. Dvornic, P. R.; Lenz, R. W. *High Temperature Siloxane Elastomers*; Hüthig &Wepf Verlag: Basel, 1990.
3. *Silicon-Based Polymer Science. A Comprehensive Resource*; Zeigler, J. M.; Fearon, F. W. G., Eds.; Advances in Chemistry, 224; American Chemical Society: Washington, DC, 1990.
4. Noll, W. *Chemistry and Technology of Silicones*; Academic: Orlando, FL, 1968.
5. Marsden, J. U.S. Patent 2 448 565, 1948.
6. Thomas, D. R. *Siloxane Polymers*; Clarson, S. J.; Semlyen, J. A., Eds.; Prentice Hall: Englewood Cliffs, NJ, 1993.
7. *Hawley's Condensed Chemical Dictionary*; Sax, N. I.; Lewis, R. J., Eds.; Van Nostrand Reinhold: New York, 1987; 11th ed.
8. Pouchelon, A.; Vondracek, P. *Rubber Chem. Technol.* **1989**, *62*, 788.
9. Friedman, D. W.; Orland, P. J.; Greco, R. S. *Implantation Biology: The Host Response and Biomedical Devices*; Greco, R. S., Eds.; CRC: Boca Raton, FL, 1994.
10. Takahashi, M.; Yoshino, M.; Sato, T. Jpn. Kokai Tokkyo Koho JP 05 287 086, 1993, (CA 120: 166574).
11. Lai, J. H. U.S. Patent 5 268 396, 1992.

SILSESQUIOXANE-BASED POLYMERS

Joseph D. Lichtenhan
Phillips Laboratory
Edwards Air Force Base

Silsesquioxanes are a continually evolving class of compounds closely related to silicones through both composition and a shared system of nomenclature. Represented by the generic formula $(RSiO_{1.5})_n$ and as denoted by the name, each silicon atom is bound to an average of one and a half oxygens (sesqui) and to one hydrocarbon group (ane). The oxygen to silicon ratio of 1.5 in these materials is intermediate between that found in silicones and silica, thus making them interesting hybrid compositions with unique physical properties. They are

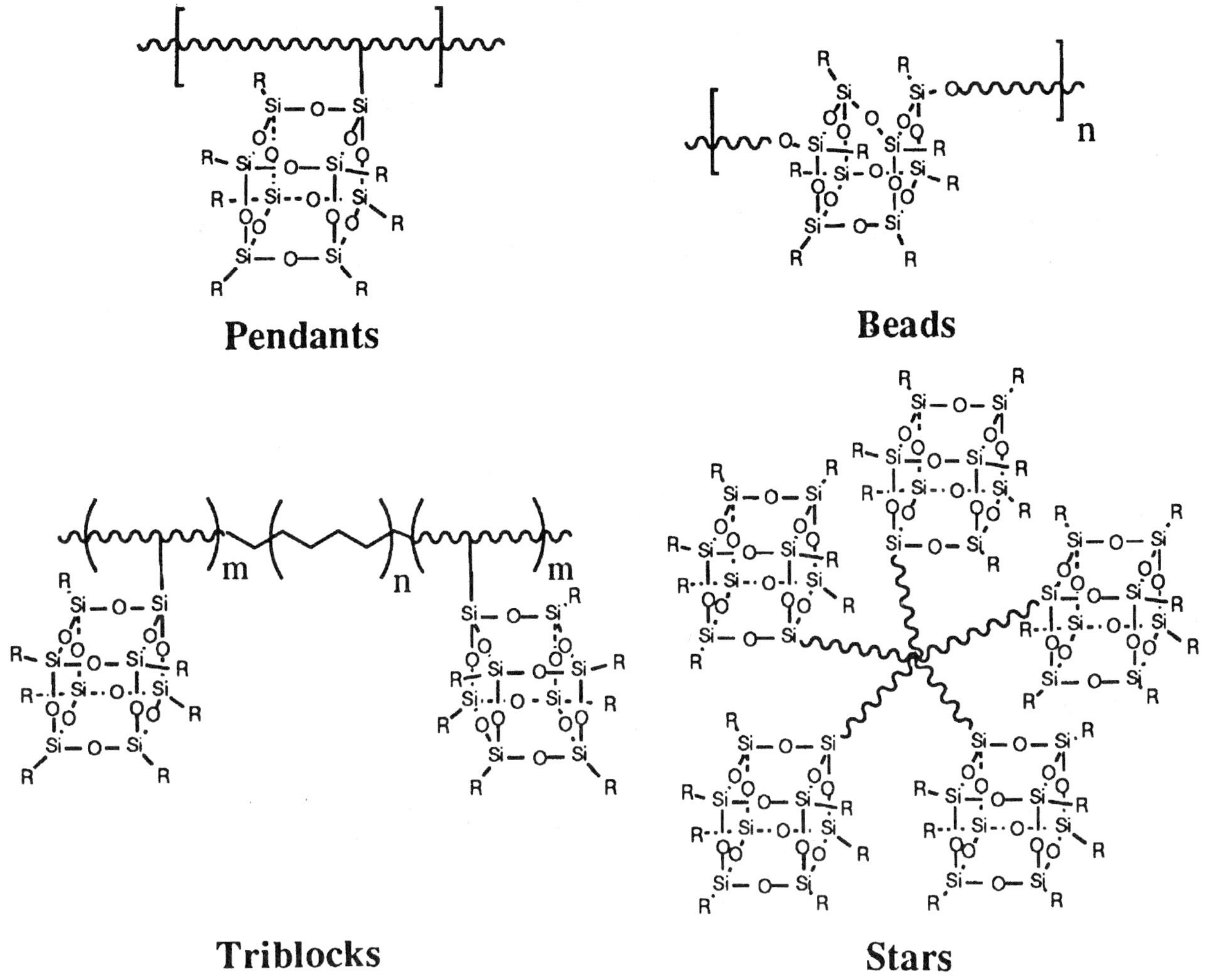

FIGURE 1. Architectures of silsesquioxane-based polymers derived from POSS macromers.

known to exist in the form of polycyclic oligomers and as polymers.[1]

Despite more than five decades of academic and industrial interest in silsesquioxanes, there continues to be a surge of activity in this field directed primarily toward the development of new synthetic methods, compounds, properties, and applications for this class of materials.[2] The most active areas involve the synthesis and functionalization of polyhedral oligomeric silsesquioxane (POSS) compounds and on exploring their use in catalysis, as nonchelating ligands, as precursors to silica surfaces and interfaces, as precursors to silicates, and as polymerizable reagents.[3–10]

PREPARATION AND PROPERTIES

Polyhedral Oligomeric Silsesquioxanes

The majority of POSS compounds have highly symmetrical, fully condensed silicon-oxygen frameworks with an equivalent organic functionality on each silicon atom. POSS molecules are normally crystalline solids, sublimable under vacuum, with a wide range of melting points and decomposition temperatures near to or exceeding 400°C.[1]

Chemistry of Fully Condensed POSS Systems

Several transformations of the organic groups attached to the silicon atoms within functionalized, fully condensed POSS frameworks have been reported. These have primarily involved standard organic transformations such as hydrogenation, chlorination, hydrosilation, esterification, acylation, and expoxidation and occur in high yield with retention of the polyhedral structure.[6,9–11]

Chemistry of Incompletely Condensed POSS Systems

Investigations into the chemistry of POSS compounds containing incompletely condensed frameworks have focused on reactions of the silanol groups. POSS molecules with one or two reactive functionalities are desirable as graftable or polymerizable macromonomers (macromers) for the preparation of linear POSS-based polymers and hybrid materials (**Figure 1**).[7,12]

Polyhedral Oligomeric Silsesquioxane-Based Polymers

Chemically and structurally well-defined POSS molecules have been used to prepare linear silsesquioxane-based polymers.[7,13] Four polymeric architectures can be envisioned that incorporate POSS segments into linear systems. These are referred to as pendant, bead, triblock, and star architectures (Figure 1).

The use of POSS macromers with discrete structures enables the possibility of establishing property relationships based on monomer and chain structure in silsesquioxane-based materials. While detailed structure–property relationships have only begun to be determined for polymeric POSS-based systems, it is anticipated that the behavior of POSS segments in linear polymers will be similar to that observed for conventional organic (rigid) hard blocks in other polymeric materials.[7,14] Incorporation of the POSS group should serve to reduce inter- and intramolecular chain mobility.

POSS-based polymers are unlike polymeric hydrocarbons in that they typically show very high decomposition temperatures and are resistant to combustion. This property is thought to be associated with the preceramic nature of the silsesquioxane component, which forms a glassy layer of SiO_xC_y during pyrolysis that may prevent or retard the diffusion of oxygen through the surface char.

Using both solution and melt techniques, POSS-based polymers can be processed into fibers, foams, films, and monolithic pieces.[7] When in the form of films or disks they have an appearance similar to that of clear, transparent plastics. POSS-based polymers also exhibit a range of miscibilities with hydrocarbon-based polymers.[15]

APPLICATIONS

Applications of silsesquioxane-based materials vary considerably throughout industry and academia.

REFERENCES

1. (a) Korchkov, V. P.; Martynova, T. N. *Russ. J. Appl. Chem.* **1985**, *58*, 1923. Translated from *Zhurnal Prikladnoi Khimii.* **1985**, *58*, 2089. (b) Ton'shin, A. M.; Kamaritskii, B. A.; Spektor, V. N. *Russian Chem. Rev.* **1983**, *52*, 775. Translated from *Uspekhi Khimii.* **1983**, *52*, 1365. (c) Voronkov, M. G.; Lavrent'yev, V. I. *Top. Curr. Chem.* **1982**, *102*, 199.
2. Earliest reports of chemistry relating to silsesquioxanes occurred in the 1850s by Buff, Wholer, and Landenburg. (Buff, H.; Wohler, F. *Ann.* **1857**, *104*, 94. Ladenburg, A. *Ber.* **1873**, *6*, 379. Reports on the polymeric nature of silsesquioxanes appeared in 1914. Meads, J. A.; Kipping, F. S. *J. Am. Chem. Soc.* **1914**, *105*, 679. Palmer, K. W.; Kipping, F. S. *J. Am. Chem. Soc.* **1930**, 1020. The first POSS was isolated in 1946. Scott, D. W. *J. Am. Chem. Soc.* **1946**, *68*, 356.
3. (a) Feher, F. J.; Tajima, T. L. *J. Am. Chem. Soc.* **1994**, *116*, 2145 and references therein. (b) Field, L. D.; Lindall, C. M.; Maschmeyer, T.; Masters, A. F. *Aust. J. Chem.* **1994**, *47*, 1127. (c) Herrmann, W. A.; Anwander, R.; Dufaud, V.; Scherer, W. *Angew. Chem. Int. Ed. Engl.* **1994**, *33*, 1285. (d) Winkhofer, N.; Voigt, A.; Dorn, H.; Roesky, H.; Steiner, A.; Stalke, D.; Reller, A. *Angew. Chem. Int. Ed. Engl.* **1994**, *33*, 1352. (e) Isobe, K.; Yagasaki, A. *Acc. Chem. Res.* **1993**, *26*, 524. (f) Edelmann, F. T. *Angew. Chem. Int. Ed. Engl.* **1992**, *31*, 586. (g) Igonin, V. A.; Shchegolikhina, O. I.; Lindeman, S. V.; Levitsky, M. M.; Struchkov, Yu. T.; Zhdanov, A. A. *J. Organomet. Chem.* **1992**, *423*, 351.
4. Budzichowski, T. A.; Feher, F. J. *Organometallics* **1991**, *10*, 812.
5. Banaszak Holl, M. M.; Lee, S.; McFeely, F. R. *Appl. Phys. Lett.* **1994**, *65*, 1097. Lee, S.; Makan, S.; Banaszak Holl, M. M.; McFeely, F. R. *J. Am. Chem. Soc.* Submitted 1994. Banaszak Holl, M. M.; McFeely, F. R. *Phys. Rev. Let.* **1993**, *71*, 2441. Bürgy, H.; Calzaferri, G.; Kamber, I. *Mikrochim. Acta.* **1988**, *1*, 401. Beer, R.; Bürgy, H.; Calzaferri, G.; Kamber, I. *J. Electron Spect. Rel. Phenom.* **1987**, *44*, 121.
6. Brevett, C. S.; Cagle, P. C.; Klemper, W. G.; Millar, D. M.; Ruben, G. C. *J. Inorg. Organomet. Polym.* **1992**, *1*, 335. Feher, F. J.; Weller, K. J. *Organometallics* **1990**, *9*, 2638. Feher, F. J.; Weller, K. J. *Inorg. Chem.* **1991**, *30*, 880. Agaskar, P. A.; Day, V. W.; Klemperer, W. G. *J. Am. Chem. Soc.* **1987**, *109*, 5554.
7. Haddad, T. S.; Lichtenhan, J. D. *J. Inorg. Organomet. Polym.* **1995**, *5*, 237. Lichtenhan, J. D.; Vu, N. Q.; Garter, J. A.; Gilman, J. W.; Feher, F. J. *Macromolecules* **1993**, *26*, 2141.
8. Weidner, R.; Zeller, N.; Deubzer, B.; Frey, V. U.S. Patent 5 047 492, 1991.
9. Sellinger, A.; Laine, R. M.; Chu, V.; Viney, C. *Polymer Sci. A* **1994**, *32*, 3069.
10. Bassindale, A. R.; Gentle, T. E. *J. Mater. Chem.* **1993**, *3*, 1319.
11. Calzaferri, G.; Imhof, R. *J. Chem. Soc. Dalton Trans.* **1992**, 3391, and references therein. Herren, D.; Bürgy, H.; Calzaferri, G. *Helv. Chim. Acta.* **1991**, *74*, 24 and references therein. Kreuzer, F.-H.; Maurer, R.; Spes, P. *Makromol. Chem., Macromol. Symp.* **1991**, *50*, 215. Feher, F. J.; Budzichowski, T. A. *J. Organomet. Chem.* **1989**, *373*, 153. Feher, F. J.; Budzichowski, T. A. *J. Organomet. Chem.* **1989**, *379*, 33.
12. (a) Feher, F. J.; Schwab, J. J.; Ziller, J. W. *J. Organomet. Chem.* submitted 1994. (b) Lichtenhan, J. D.; Feher, F. J.; Gilman, J. W., unpublished work, 1991.
13. (a) Lichtenhan, J. D.; Otonari, Y. A.; Carr, M. J. **1995**, *28*, 8435, 1992–4. (b) Lichtenhan, J. D.; Shockey, E. G., unpublished work, 1993. (c) Lichtenhan, J. D.; Bolf, A. B., unpublished work, 1993.
14. For example, see: Mark, J. E.; Allcock, H. R.; West, R. *Inorganic Polymers* Prentice Hall, New Jersey, 1992; pp 155–8. Sperling, L. H. *Introduction to Physical Polymer Science*; Wiley: 1986; pp 346–348.
15. Noel, C. J.; Lichtenhan, J. D., unpublished work, 1994.

Size Exclusion Chromatography Packings

See: Poly(α-amino acid) Spherical Particles

Skin Care Materials

See: Personal Care Polymers

Skin Regeneration

See: Regeneration Templates, Artificial Skin and Nerves

Skin Substitutes

See: Wound Dressing Materials

SMALL-BANDGAP POLYMERS (Special Class of Organic Conducting Compounds)

Jean-Marc Toussaint and Jean-Luc Brédas*
Chimie des Matériaux Nouveaux
Centre de Recherche en Electronique et Photonique Moléculaires
Université de Mons-Hainaut

In this brief review we focus on a special class of compounds, conjugated polymers, which exhibit small intrinsic

*Author to whom correspondence should be addressed.

bandgap energies. The second section of this article is devoted to the illustration of some of the properties and advantages that small-bandgap polymers offer. An overview of the different ways to design and synthesize small-bandgap compounds is presented in the third section. The main perspectives of the field are given in the final section.

PROPERTIES AND ADVANTAGES OF SMALL-BANDGAP POLYMERS

The bandgap of a material can be defined as the amount of energy required for the promotion of an electron from the highest occupied electronic band (called the valence band) to the first empty band above it (i.e., the conduction band). Thus, by "small-bandgap polymer", one usually means an energetic separation between the edges of these two bands (which also corresponds to the energy difference between the ionization potential and electron affinity values) that is on the order of 1 eV or smaller.

Most of the conjugated polymers exhibit bandgap energies lying between 1.5 and 4 eV, which qualifies them as insulators or wide-bandgap semiconductors; in order to become good electrical conductors they have to be treated via a redox (or doping) reaction. Thus, considering that all other conditions or external parameters are equivalent, small-bandgap polymers present *a priori* higher intrinsic electrical conductivities than traditional conjugated polymers because the smaller the bandgap, the higher the number of electrons that can jump from the valence band to the conduction band at room temperature and participate in the current.

Another attractive characteristic of small-bandgap polymers is related to their optical properties. Most of the large-bandgap conjugated polymers are colored because their lowest electronic transitions (i.e., the ones involving the smallest amount of energy) correspond to the visible region of the electromagnetic spectrum, somewhere in the ~1.5 to 3.0 eV range. However, if the bandgap energy is low, the electronic transitions of smallest energies can appear dominantly in the IR region and the corresponding material could be reasonably transparent in either its neutral or doped state. The availability of transparent conducting polymers would be a major step forward for the application of conjugated polymers, particularly as antistatic coatings, electromagnetic shielding elements, or IR sensors.[1,2]

According to a theoretical study by Agrawal et al., small-bandgap polymers could also represent an interesting class of compounds because of their nonlinear optical properties.[2a]

DESIGN AND STUDY OF SMALL-BANDGAP POLYMERS

Ladder-Type Polymers

The basic premise of this route is to try and construct fully fused-ring hydrocarbon structures in order to obtain polymer systems that more or less correspond to a one-dimensional graphite. Several low-bandgap polymers have been designed and studied according to this route, such as polypyrazinopyrazine (or paracyanogen), polypyridinopyridine, polyacene-like systems, or polyperinaphthalene.[3–9]

However, more recently exploration of new synthetic routes has allowed for the solubilization of substituted oligomers of polyperinaphthalene in common organic solvents.[9b] This progress in the design of ladder-type polymers not only gives rise to well-characterized materials, but it also makes possible the study of the physical properties as a function of molecular size. For example, based on measured π–π^* transition energies, a bandgap of 0.7 eV is predicted for the infinite polymer, a value that is in good agreement with the 0.5 eV calculated bandgap.[10,11]

Introduction of Quinoid Contributions

Wudl et al. have synthesized polyisothianaphthene, a polymer in which the repeating unit can be described as a thiophene ring onto which a benzene ring is fused along the β-β' bond.[12–14] The basic concept behind the synthesis is that the aromatic nature of the benzene ring could modify the geometric and/or electronic structure of the polythiophene backbone, namely, by enhancing the contributions of resonance forms that involve a quinoid structure along the conjugated backbone. From optical absorption measurements, it appears that polyisothianaphthene possesses a bandgap of ~1.0 eV, i.e., half that of its parent polymer, polythiophene. Moreover, on doping, polyisothianaphthene films become transparent because of their very small optical density in the visible region; polyisothianaphthene is the first example of a transparent conducting polymer.

Alternating Electron-Donor and Electron-Acceptor Moieties

A third route to low-bandgap polymers has been recently initiated by Havinga et al., who proposed building a conjugated backbone along which strong electron-donor and electron-acceptor groups regularly alternate.[15,16] The main idea is to couple units along a conjugated path with a low ionization potential to units with a large electron affinity so that the resulting conjugated polymer exhibits small-bandgap energy. This has led to the synthesis of polysquaraines and polycroconaines, which are stable polymers with bandgaps that can be as small as 0.5 eV.

Along the same line, Ferraris and Lambert have synthesized polydicyanomethylene-cyclopenta-dithiophene, a polymer made by the succession of bithiophene units in which the two rings are linked by a dicyanomethylene moiety.[17,18] Very interestingly, the experimental bandgap of this polymer, as deduced from absorption spectra, is on the order of 0.8 eV.

CONCLUSIONS

In this article we have focused on a special class of conjugated polymers, those with a small energy difference between the occupied and unoccupied electronic levels. We have shown that because of their interesting conducting and optical properties, small-bandgap polymers can lead to the more conventional applications of conjugated polymers, but can also trigger a whole series of new exploitations. We have discussed the state of the art in research for small-bandgap polymers by illustrating the three routes that are usually followed: the ladder-type polymer route, the introduction of quinoid contributions to the ground state of the polymer, and recently discovered mixing of strong electron-donor and electron-acceptor units along a conjugated backbone. It should be clear from our analysis that important advances have already been made in the design, synthesis, and characterization of small-bandgap polymers; however, it is obvious that this field is still in its infancy. New discoveries have still to appear, in particular if one is to achieve the development of truly metallic organic polymers.

ACKNOWLEDGMENTS

This work was supported in part by the European Commission through the ESPRIT Basic Action Research TOPFIT Project 7282; the Belgian Prime Minister Office of Science Policy "Programme d'Impulsion en Technologie de l'Information" and "Pôle d'Attraction Interuniversitaire en Chimie Supramoléculaire et Catalyse," FNRS/FRFC; and an IBM Academic Joint Study. Stimulating discussions with Drs. H. Braünling, D. de Leeuw, E. E. Havinga, M. Kertesz, K. Müllen, A. Tol, and R. J. Visser acknowledged.

REFERENCES

1. Jonas, F.; Schrader, L. D. *Synth. Met.* **1991**, *41–43*, 831.
2. Defieuw, G.; Sanijn, R.; Hoogmartens, I.; Vanderzande, D.; Gelan, J. *Synth. Met.* **1993**, *55–57*, 3702.
2a. Agrawal, G. P.; Cojan, C.; Flytzanis, C. *Phys. Rev. B* **1978**, *17*, 776.
3. Brédas, J. L.; Thémans, B.; André, J. M. *J. Chem. Phys.* **1983**, *78*, 6137.
4. Wuckel, L.; Lehmann, G. *Makromol. Chem. Macromol. Symp.* **1990**, *37*, 195.
5. Kao, J.; Lilly, A. C. *J. Am. Chem. Soc.* **1987**, *109*, 4149.
6. Yamabe, T.; Tanaka, K.; Ohzeki, K.; Yata, S. *J. Phys. Paris* **1983**, *44*, C3–645.
7. Kertesz, M.; Lee, Y. S.; Stewart, J. J. P. *Int. J. Quantum Chem.* **1989**, *35*, 305.
8. Murakami, M.; Yoshimura, S. *J. Chem. Soc. Chem. Commun.* **1984**, 1649.
9. Iqbal, Z.; Ivory, D. M.; Marti, J.; Brédas, J. L.; Baughman, R. H. *Mol. Cryst. Liq. Cryst.* **1985**, *118*, 103.
9b. Koch, K. H.; Fahnenstich, U.; Baumgarten, M.; Müllen, K. *Synth. Met.* **1991**, *41–43*, 1619.
10. Müllen, K. *Pure Appl. Chem.* **1993**, *65*, 89.
11. Baumgartner, M.; Koch, K. H.; Müllen, K. *J. Am. Chem. Soc.* **1994**, *116*, 7341.
12. Wudl, F.; Kobayashi, M.; Heeger, A. J. *J. Org. Chem.* **1984**, *49*, 3381.
13. Wudl, F.; Kobayashi, M.; Colaneri, N.; Boysel, M.; Heeger, A. J. *Mol. Cryst. Liq. Cryst.* **1985**, *118*, 195.
14. Kobayashi, M.; Colaneri, N.; Boysel, M.; Wudl, F.; Heeger, A. J. *J. Chem. Phys.* **1985**, *82*, 5717.
15. Havinga, E. E.; ten Hoeve, W.; Wynberg, H. *Polym. Bull.* **1992**, *29*, 119.
16. Havinga, E. E.; ten Hoeve, W.; Wynberg, H. *Synth. Met.* **1993**, *55–57*, 299.
17. Ferraris, J. P.; Lambert, T. L. *J. Chem. Soc. Chem. Commun.* **1991**, 1268.
18. Lambert, T. L.; Ferraris, J. P. *J. Chem. Soc. Chem. Commun.* **1991**, 752.

SMART HYDROGELS

Kinam Park* and Haesun Park
School of Pharmacy
Purdue University

Smart materials are synthetic materials that have multiple functions in addition to inherent structural properties.[1] The ultimate smart materials are those that function as living organisms do. Several smart materials have been prepared to date. Examples are shape memory alloys and polymers, conductive polymers, piezoelectric ceramics and polymers, muscle-like polymeric fibers, and environment-sensitive hydrogels.[3] We focus on smart hydrogels.

HYDROGELS AND SMART HYDROGELS

Hydrogels are three-dimensional polymer networks that swell in aqueous solutions while maintaining their structural integrity. A hydrogel's water content is usually more than 20% of its total weight.

Smart (or intelligent) hydrogels are those that respond to changes in environmental conditions by either expanding, shrinking, bending, or degrading their three-dimensional structures. The response of smart hydrogels to environmental changes is usually measured by changes in the hydrogel volume. Numerous studies have shown that the volume change of smart hydrogels in response to the alteration in environmental conditions is often discontinuous and abrupt. This phenomenon of drastic volume change is called volume collapse or volume phase transition.[2–4]

ENVIRONMENTAL FACTORS CONTROLLING THE SWELLING PROPERTIES OF HYDROGELS

Environmental stimuli that trigger the response of smart hydrogels include temperature, pH, ionic strength, salt type, electric field, external stress, solvent, light, pressure, specific ligand molecule, enzyme, and a combination of these. Of these parameters, temperature, pH, and specific ligand have been the variables of intensive research interest, primarily because they are the ones that are important in biological systems.

pH

The pH sensitivity of hydrogels results from the presence of weak acid or base functionality on the polymer backbone.[5–8]

The window of pH at which hydrogels show substantial volume change depends on the type of weak acid or base used. If hydrogels contain weak acid functionality, they will swell more as the pH of the medium increases. The reverse is true with hydrogels containing weak bases.

Temperature

The temperature-sensitive (also called thermosensitive or thermoresponsive) hydrogels undergo volume collapse (or shrinkage) upon heating above a certain temperature.[9] The thermosensitive property comes from the monomers that have the lower critical solution temperature (LCST) phenomenon in the uncrosslinked homopolymer form in aqueous solution.[10] The most commonly used LCST monomers are N-substituted acrylamide derivatives, such as *N*-isopropylacrylamide, *N*-*n*-propylacrylamide, and *N*-cyclopropylmethacrylamide, because they are easy to obtain or prepare.[10–14] Cellulose ethers, such as hydroxypropylcellulose, also possess a LCST in water.[14a] The hydrogels made of LCST monomers collapse at temperatures higher than the LCST. These hydrogels swell again as the temperature falls again below LCST.

*Author to whom correspondence should be addressed.

pH and Temperature

Hydrogels sensitive to both pH and temperature can be prepared using two different types of monomers.[15]

Electric Field

The electrosensitive hydrogels, which are capable of reversible contraction and expansion under an infinitesimal change in electric potential, are usually made of charged polymers, such as poly(acrylic acid) (PAA) and poly(2-acrylamide-2-methylpropanesulfonic acid) (PAMPS).[3,16–18]

The electric sensitivity of the polyelectrolyte gels occurs in the presence of ions in solution. In the presence of an applied electric field, the ions (both coions and counterions) move to the positive or negative electrode, while the polyions of the hydrogel cannot move. This results in a change in the ion concentration-dependent osmotic pressure, and the gel either swells or shrinks to reach its new equilibrium. Because the osmotic pressure of the positive side is different from that of the negative side, the net result of swelling or shrinking is bending of the gel.

Light

The light-sensitive hydrogels can be made by incorporating reversible photochromic moieties into the hydrogel structure as a polymer backbone, as a pendant group, or as a crosslinker.

Solvent

Hu et al. prepared hydrogels by using a monomer mixture of acrylamide and sodium methacrylate.[19] When the content of sodium methacrylate was ~5%, the gel showed volume phase transition when the acetone content was ~60%. As the volume content of acetone increases, the gel becomes collapsed because of the increased hydrophobic interaction between the polymer main chains and the decreased interaction between polymer chains and solvents.

Pressure

The pressure-sensitive hydrogels undergo swelling upon an increase in pressure. The swelling ratio of poly(*N*-isopropylacrylamide) gels increased from 6 to 14 when the pressure increased from 20 to 26 atm.

Biomolecules

Hydrogels that respond to specific molecules found in the body are of great interest. One such hydrogel is glucose-sensitive hydrogel, which has potential applications in the development of self-regulating insulin delivery systems. A ternary copolymer containing a phenylboronic acid group was used to prepare a glucose-responsive polymer complex.[20]

Recently, another type of glucose-sensitive sol–gel phase-reversible hydrogel was synthesized.[21–23]

APPLICATIONS

Drug Delivery

Most glucose-sensitive hydrogels utilize pH-sensitive polymers such as poly(diethylaminoethyl methacrylate) (PDEAEMA).[24–26] Glucose oxidase, which is often immobilized in a pH-responsive hydrogel membrane, transforms glucose into gluconic acid, which lowers the pH of the hydrogel membrane and protonates the amino moieties in the PDEAEMA hydrogel. The charged amino moieties repel each other and result in swelling of the hydrogel membrane, and this in turn results in more release of insulin. In addition to the pH-sensitive hydrogel membranes, the glucose-sensitive sol–gel phase-reversible hydrogels have the ability to control the pulsatile release of insulin.[20,23,27]

Thermosensitive hydrogels are another type of smart hydrogel frequently used in controlled drug delivery. By altering the temperature around the thermosensitive hydrogels, the release of drug from the gel can be turned on and off at will. Pioneering research in this approach has been done by Hoffman and Kim.[28,29]

Prosthetic Biomaterials

The best example for transformation of electrochemical stimuli into mechanical work (i.e., contraction) is probably human muscle tissue.[1] The demand for electrically driven muscle-like actuators made of polymer gels in advanced robotics has stimulated extensive research on polymeric gels capable of reversible contraction and expansion under physicochemical stimuli.[30] Smart materials that emulate the contractions and secretions of human organs in response to environmental changes in temperature, pH, or electric field may soon be used in medical implants, prosthetic muscles or organs, and robotic grippers.[31]

Extraction and Separation

The ability of smart hydrogels in solutions to reversibly swell and shrink with small changes in environmental conditions can be used to prepare purification devices.[32]

CONSIDERATIONS IN THE APPLICATION OF SMART HYDROGELS

Response Time

Although smart hydrogels with exciting properties exist and more of these hydrogels will be developed, the responses of most hydrogels to changes in environmental factors are too slow.

The fast response time is the key for the successful application of smart hydrogels in drug delivery, artificial muscles, or microswitches.

Structural Rigidity

For long-term applications smart hydrogels must have structural resilience to maintain their property of repeated expansion and collapse.

Biocompatibility

The selection of smart hydrogels to be implanted into the body is limited by the availability of biocompatible hydrogels. The ability of the implanted hydrogels may be far different from that of hydrogels tested in the laboratory.

Cost

Although research on smart hydrogels should not be hindered by the cost of the potential future products, the issue of cost needs to be kept in mind, because smart material systems that are not cost-effective will not survive no matter how good they are.[1]

REFERENCES

1. Studt, T. *Research and Development Magazine* **1992**, *April*, 55.
2. Tanaka, T. *Sci. Am.* **1981**, *244*, 124.
3. Tanaka, T.; Nishio, I.; Sun, S.-T.; Ueno-Nishio, S. *Science* **1982**, *218*, 467.
4. Tanaka, T. *ACS Symp. Ser.* **1992**, *480*, 1.
5. Beltran, S.; Baker, J. P.; Hooper, H. H.; Blanch, H. W.; Prausnitz, J. M. *Macromolecules* **1991**, *24*, 549.
6. Brannon-Peppas, L.; Peppas, N. A. *J. Controlled Rel.* **1991**, *16*, 319.
7. Pradny, M.; Kopecek, J. *Makromol. Chem.* **1990**, *191*, 1887.
8. Siegel, R. A.; Falamarzian, M.; Firestone, B. A.; Moxley, B. C. *J. Controlled Rel.* **1988**, *8*, 179.
9. Katono, H.; Sanui, K.; Ogata, N.; Okano, T.; Sakurai, Y. *Polym. J.* **1991**, *23*, 1179.
10. Badiger, M. V.; Rajamohanan, P. R.; Kulkarni, M. G.; Ganapathy, S.; Mashelkar, R. A. *Macromolecules* **1991**, *24*, 106.
11. Kawaguchi, H.; Fujimoto, K.; Mizuhara, Y. *Coll. Polym. Sci.* **1992**, *270*, 53.
12. Tsuji, T.; Konno, M.; Sato, S. *J. Chem. Eng. Jpn.* **1990**, *23*, 447.
13. Anon. *Macromolecules* **1990**, *23*, 4887.
14. Stevenson, W. T. K.; Sefton, M. V. *Trends Polym. Sci.* **1994**, *2*, 98.

14a. Harsh, D. C.; Gehrke, S. H. *J. Controlled Rel.* **1991**, *17*, 175.

15. Yoshida, M.; Yang, J-S.; Kumakura, M.; Hagiwara, M.; Katakai, R. *Eur. Polym. J.* **1991**, *27*, 997.
16. Kaetsu, I.; Uchida, K.; Morita, Y.; Okubo, M. *Radiat. Phys. Chem.* **1992**, *40*, 157.
17. Kurauchi, T.; Shiga, T.; Hirose, Y.; Okada, A. *Polymer Gels: Fundamentals and Biomedical Applications*; Plenum: New York, 1991; pp 237–246.
18. Osada, Y.; Okuzaki, H.; Hori, H. *Nature* **1992**, *355*, 242.
19. Hu, Y.; Horie, K.; Ushiki, H.; Yamashita, T.; Tsunomori, F. *Macromolecules* **1993**, *26*, 1761.
20. Kitano, S.; Koyama, Y.; Kataoka, K.; Okano, T.; Sakurai, Y. *J. Controlled Rel.* **1992**, *19*, 162.
21. Lee, S. J.; Park, K. *Polym. Prep.* **1994**, *35*, 391.
22. Lee, S. J.; Park, K. *Proc. Intern. Symp. Control. Rel. Bioact. Mater.* **1994**, *21*, 93.
23. Nakamae, K.; Miyata, T.; Jikihara, A.; Hoffman, A. S. *J. Biomater. Sci. Polym. Edn.* **1994**, *6*, 79.
24. Albin, G.; Horbett, T. A.; Ratner, B. D. *Pulsed and Self-Regulated Drug Delivery*; CRC: Boca Raton, FL, 1990; Chapter 9, pp 159–185.
25. Ishihara, K.; Kobayashi, M.; Shionohara, I. *Makromol. Chem. Rapid Commun.* **1983**, *4*, 327.
26. Klumb, L. A.; Horbett, T. A. *J. Controlled Rel.* **1993**, *27*, 95.
27. Shino, D.; Kataoka, K.; Koyama, Y.; Yokoyama, M.; Okano, T.; Sakurai, Y. In *Proceedings of the First International Conference on Intelligent Materials*; Technomic: Lancaster, PA, 1993; pp 301–304.
28. Dong, L. C.; Hoffman, A. S. *J. Controlled Rel.* **1990**, *13*, 21.
29. Bae, Y. H.; Okano, T.; Kim, S. W. *Pharm. Res.* **1991**, *8*, 624.
30. De Rossi, D.; Suzuki, M.; Osada, Y.; Morasso, P. *J. Intell. Mater. Syst. Struct.* **1992**, *3*, 75.
31. Constance, J. *Mech. Eng.* **1991**, *113*, 51.
32. Marchetti, M.; Cussler, E. L. *Sep. Purif. Methods* **1989**, *18*, 177.

Smart Materials

See: Intelligent Materials

Soft Tissue Compatible Polymers

See: Silk Fibroin (Soft Tissue Compatible Polymer)

Sol-Gel Polymerization

See: Organic/Inorganic Hybrid Polymers

Solid Polymer Electrolytes

See: Solid Polymer Electrolytes (Overview)
Solid Polymer Electrolytes (Preparation, Characterization, and Properties)
Solid Polymer Electrolytes (Polyether Blends and Composites)

SOLID POLYMER ELECTROLYTES (Overview)

Denis G. Fauteux* and M. F. van Buren
Arthur D. Little Incorporated

Solid polymer electrolytes (SPEs) are ionically conducting materials comprising a macromolecular system and a salt. SPE materials frequently also contain organic solvent plasticizers that may be the major component of the SPE. For the purposes of this review, plasticizer-containing SPEs will be termed gel SPEs as opposed to pure SPEs for those that do not contain any plasticizer. The macromolecular component of the SPE may be either a linear or a crosslinked polymer, a copolymer, or an interpenetrating polymer network. Desirable characteristics for an SPE macromolecule include good electrochemical stability in reduction and oxidation, an ability to coordinate and enable the transport of ionic species, a T_g well below the projected use temperature of the electrolyte, and chemical stability in contact with active electrode materials.

There has been almost exponential growth in research and development activity focused on the development of high ionic conductivity SPEs for their use in high energy density solid-state batteries.

This activity has been driven by two factors; first, both the domestic and foreign electronics industries have recognized portable energy supplies as a limiting factor in the design of new consumer electronic devices; and second, the development of a battery-powered automobile (EV) for the general public has become a major investment area for federal and state government research funds.

IONIC CONDUCTIVITY σ IN SPEs

In a true, unplasticized, SPE the electrolyte's salt is mobilized by dissolution in an ion-coordinating macromolecule. The archetypical SPE polymer and the one most frequently studied is poly(ethylene oxide) (PEO). A large variety of alkali, alkaline earth, transition, and rare earth metal salts are soluble in PEO, and salt solubilities in PEO may be as high as 2 M. In standard models of PEO-alkali metal salt solutions, the metal cation is surrounded by a PEO helix in which 3 to 6 oxygen atoms in the

*Author to whom correspondence should be addressed.

PEO chain are coordinated with the cation. Associated anions external to the PEO chain may also be coordinated with the cation. Ion mobility in an SPE is, therefore, strongly associated with polymer chain segment motions and local structural relaxations.

The reported σs of Li^+X^--PEO electrolytes range from about 5×10^{-5} S-cm^{-1} (X^- = triflate) to 7×10^{-3} S-cm^{-1} (X^- = bistrifluoromethane sulfonimide) at temperatures of 80 to 100°C, well above the melting point of PEO. At room temperature the reported σs for the same SPEs drop to 7×10^{-8} S-cm^{-1} and 7×10^{-5} S-cm^{-1}, respectively. $\sigma \approx 10^{-3}$ S-cm^{-1} is generally regarded as the level of SPE ionic conductivity needed by many common electronic devices to derive useful amounts of electronic current from electrochemical cells. Lower SPE ionic conductivities require excessively thin layers of SPE in the cell to attain similar electronic currents.

The reported cationic transference number (t_+) for many Li^+X^--PEO electrolytes is about 0.2 to 0.3; only 30% of the measured conductivity is attributable to Li^+ mobility.

SPE MATERIALS

Poly(alkaline oxide) Systems

PEO is unique among the poly(alkaline oxide) polymers in its ability to loosely complex Li^+ and has a measured ionic conductivity of ca. 10^{-3} S-cm^{-1} at 100°C. The conductivity maximum as a function of [Li^+X^-] for the PEO-LiX SPE system occurs at about 6–8 mole% lithium salt. PEO-LiX SPE systems have been well characterized by numerous workers including Robitaille and Fauteux, and PEO has remained the benchmark for SPE performance in lithium secondary batteries.[1] Poly(propylene oxide) (PPO) is decidedly inferior to PEO with respect to its properties as an SPE macromolecule. Other polyether systems with 1, 3, 4, or more methylene groups separating the ether oxygens are also substantially inferior to PEO as an SPE macromolecule owing to low salt solubility and low ionic conductivity.[2] However, PEO has two significant limitations as an SPE:

- Below PEO's T_m of 66°C the conductivity drops off sharply to about 10^{-6} S-cm^{-1}. This level of conductivity is too low for cells used at ambient temperature.
- The Li^+-PEO complex in time tends to form a crystalline phase domain in an amorphous PEO matrix. These crystalline areas are essentially non-conducting.

The current strategy, of most groups active in SPE development is to maintain use of PEO as the solid state solvent for Li^+ ions but to modify the polymer's structure so that the polymer retains a completely amorphous morphology at or preferably below normal room temperature. The approaches used to achieve this are to "freeze" the PEO matrix in an amorphous structure by crosslinking relatively low molecular weight PEO while in a molten state, to copolymerize PEO with comonomers such as propylene oxide, thereby suppressing formation of ordered PEO domains, to copolymerize PEO in block or graft copolymer systems, or to form an SPE gel based on PEO.

DIRECTIONS FOR FUTURE SPE MATERIALS DEVELOPMENT

The polymer electrolyte technology is still in its infancy. Until 1989–1990, only two major research and development activities dedicated to the development of a SPE-based lithium rechargeable battery technology have aggressively pursued its development. However, in North America the situation has changed significantly since the formation of the United States Advanced Battery Consortium (USABC) and its endorsement of the polymer electrolyte technology for the production of high-energy density battery systems for EV applications. We are currently witnessing major efforts to develop commercial products within a three-year time frame.

Thus far, lithium polymer electrolyte rechargeable batteries have exhibited several important performance limitations. They have not successfully met all of the power density, energy density, specific energy, low temperature performance, and cycle life expectations.

One must recognize that at this time ethylene oxide based polymers have not provided a polymer host structure having sufficient conductivity and cationic transference number to exceed competitive performance requirements. It is believed that more complex polymer host structures will be required in order to achieve improved performance.

It is believed that an approach based on the concept of polymer-alloys may provide valuable insight. Polymer alloys are loosely defined as polymeric-based multicomponent systems in which all of the components chemically interact with each other. These interactions should be strong enough to prevent phase separation and crystallization over a wide range of temperature and compositions. In addition, the polymer alloys will enable better control of the mobility of the ionic species present in the electrolyte while maintaining high ionic conductivity.

REFERENCES

1. Robitaille, C. D.; Fauteux, D. G. *J. Electrochem. Soc.* **1986**, *133*(2), 315.
2. Armand, M. *Solid State Ionics* **1983**, *9 and 10*, 745.

SOLID POLYMER ELECTROLYTES (Polyether Blends and Composites)

Zbigniew Florjańczyk* and Władysław Wieczorek
Faculty of Chemistry
Warsaw University of Technology

Solid polymer electrolytes are ionically conducting materials formed by the dissolution of salt by ion-coordinating macromolecules. The conductivity mechanism in these systems is not exactly known, but it is assumed that the charge transfer can proceed as a result of both ion motion in a polymer matrix and segmental motion of polymer chains with which the ions are bonded.

Since the early investigations by Fenton et al. and Armand et al. these materials have received considerable attention because of their wide applicability in electrochemical devices.[1–3] The main goal has been the development of solid-state primary and secondary batteries with a lithium or lithium alloy anode and a composite cathode based on intercalated materials. Polymer electrolytes are also sought for a variety of

*Author to whom correspondence should be addressed.

other applications such as capacitors, electrochemical sensors, photoelectrochemical sensors, photoelectrochromic devices (windows or displays), and fuel cells with proton polymeric electrolytes. From among the commercially available polymers, poly(ethylene oxide) (PEO) (CAS Registry No. 25322-68-3) exhibits the best properties as a solid solvent.

This work is devoted to a particular type of material composed of polyether (mainly PEO) complexes and solid additives—inorganic fillers or high molecular weight organic polymers.

GENERAL CHARACTERISTICS OF SYSTEMS

Blend-Based Polymeric Electrolytes

The phase structure of electrolytes obtained from PEO blends is complicated. At ambient temperature the electrolytes can contain crystalline PEO complexes with a salt, a PEO crystalline phase, and one or two amorphous phases of different glass transition temperatures.

The addition of certain polymers to PEO complexes can cause the appearance of an amorphous phase of a clearly higher elasticity than that in PEO-based electrolytes. The glass transition temperatures for these phases are close to that of PEO before the salt addition (~213 K). Thus, it can be assumed that the polymer added acts in these systems as a crosslinking inhibitor.

Isotactic PMMA (*i*-PMMA) (CAS Registry No. 25188-98-1) contrary to products of an atactic and syndiotactic structure, does not form miscible blends with PEO.[5] Therefore, it can be expected that in electrolytes containing this polymer the conducting amorphous phase does not contain the component responsible for hindering the ion mobility. The conductive flexible regions are probably separated by a nonconductive polymer, which lowers the probability of coordination of neighboring PEO chains around the cations. This results in an increase of mobility of the solid polymeric solvent and, hence, in higher conductivity values. The highest ambient temperature conductivities of electrolytes containing lithium salts are $\sim 9 \times 10^{-5}$ S cm^{-1}, which is higher than those estimated for lithium electrolytes containing a pure amorphous phase of PEO ($\sim 5 \times 10^{-5}$ S cm^{-1}).[4]

Still higher conductivity values ($\sim 10^{-4}$ S cm^{-1} at ambient temperature and in the 10^{-2}–10^{-3} S cm^{-1} range at 100°C can be achieved when using copolymers with a comb-like structure as the blend components, obtained from the polymerization of macro monomeric units containing poly(propylene glycol) (CAS Registry No. 25322-69-4) side chains in acrylic esters.

The highest conductivity values were obtained for systems containing 6–14 oxirane monomeric units in a side chain.

The addition of nonpolar polymers does not improve the conducting properties of PEO complexes. Gray et al. have shown that for electrolytes based on PEO–polystyrene blends a significant improvement of mechanical properties is accompanied by a relatively small decrease in ionic conductivity, up to a certain percolation limit.[6]

Composite Polyether–Polyacrylamide–Salt Electrolytes

High conductivity values can be achieved in systems in which the nonconducting component is not miscible with the amorphous PEO phase. A similar effect can be obtained when dispersing some solid polar polymers in a PEO–inorganic salt matrix.

DSC studies show that the degree of crystallinity of PEO is strongly reduced after the addition of polyacrylamide (PAAM), up to ~4% in samples of the highest conductivity. However, the favorable influence of PAAM on the electrolyte conductivity is observed also at 100°C. This permits us to assume that this compound can also accelerate the ion transport in completely amorphous systems. This hypothesis was confirmed by studies on the influence of PAAM on the conductivity of amorphous electrolytes containing $LiClO_4$ and poly(oxymethylene-oligooxyethylene) (OMPEO)—$[CH_2O(CH_2CH_2O)_n]$— in which $n \approx 8$ of molecular weight $\sim 10^5$ (a detailed account of this class of electrolytes is given in Reference 7). In these systems a few fold increase in conductivity can be achieved at an optimal composite composition, especially at ambient and subambient temperatures.

Composites with Inorganic Fillers

The degree of crystallinity of PEO in electrolytes may be decreased under the influence of various inorganic additives, such as α-Al_2O_3, SiO_2, γ-$LiAlO_2$, or zeolites.[7–11] However, contrary to systems with PAAM, a decrease in crystallinity is usually accompanied by a rise in the electrolyte glass transition temperature. The ambient temperature conductivity values of these systems generally do not exceed 10^{-5} S cm^{-1}, that is, lower than that reported for alkali salt complexes with OMPEO which is believed to be very close to the conductivity of an ideally amorphous PEO phase.[4]

APPLICATION OF BLEND-BASED AND COMPOSITE POLYMERIC ELECTROLYTES

In spite of numerous studies on blend-based and composite polymeric electrolytes there are only a few reports concerning the application of these systems in lithium polymer cells, electrochromic windows, or fuel cells.

The poly(propylene glycol)–PMMA blend-based electrolyte doped with a lithium salt is of particular interest for electrochromic devices. Low molecular weight amorphous poly(propylene glycol) has reasonably high ionic conductivity when complexed with lithium salts. Unfortunately, the poly(propylene glycol)–salt system is hydrophilic and has poor mechanical properties and adhesiveness. These problems have been overcome by blending with PMMA. Methyl methacrylate was polymerized in a poly(propylene glycol)–salt solution by using azobisisobutyronitrile as an initiator. The transparency of this electrolyte is excellent, and ionic conductivity is $\sim 7 \times 10^{-5}$ S cm^{-1} at room temperature.

Proton-conducting blend-based and composite polymeric electrolytes have also been studied by several research groups.[12–15] Various proton-conducting electrolytes based on blends of PEO with various acrylic and methacrylic polymers have been studied.[15,16] The role of methacrylic and acrylic additives is not only to modify the structure of the host PEO but also to act as a hydrophobic center, thus protecting the electrolytes against humidity. This is especially important for samples studied in a fuel cell where the use of moist gases and water cannot be avoided. Some of the electrolytes studied exhibit conductivities higher than 10^{-3} S cm^{-1} at ambient temperatures.

These electrolytes are thermally stable up to 100°C. This behavior seems to be sufficient for the application of these systems in ambient temperature fuel cells. The cells studied are capable of producing power densities exceeding 40 mW cm^{-2} and current densities higher than 100 mA cm^{-2}, but the voltage efficiency calculated as $V_{cell}/1.23$ V is low (0.35–0.4).[12] Other features that should be improved are the retention of the acid in the polymer matrix in the case of long-term applications, and the high electrode–electrolyte interfacial resistance.

REFERENCES

1. Fenton, P. F.; Parker, J. M.; Wright, P. V. *Polymer* **1992**, *14*, 1063.
2. Armand, M. B.; Chabagno, J. M.; Duclot, M. Presented at the 2nd International Conference on Solid Electrolytes. St. Andrews, 1978.
3. Armand, M. B.; Chabagno, J. M.; Duclot, M. In *Fast Ion Transport in Solids*; Vashita, P., et al., Eds.; Elsevier: New York, 1979; p 131.
4. *Polymer Electrolytes Reviews I*; MacCallumn, J. R.; Vincent, C. A., Eds.; Elsevier: London, England, 1987.
5. Silvestre, C.; Cimmino, S.; Matruscelli, E.; Karash, F.; MacKnight, W. J. *Polymer* **1987**, *28*, 1190.
6. Gray, F. M.; MacCallum, J. R.; Vincent, C. A. *Solid State Ionics* **1986**, *18/19*, 252.
7. Płocharski, J.; Wieczorek, W. *Solid State Ionics* **1988**, *28/30*, 979.
8. Wieczorek, W.; Such, K.; Wyciślik, H.; Płocharski, J. *Solid State Ionics* **1989**, *36*, 255.
9. Croce, F.; Scrosati, B.; Mariotto, G. *Chemistry of Materials* **1992**, *4*, 1134.
10. Wieczorek, W.; Such, K.; Przyłuski, J.; Florjańczyk, Z. *Synth. Metals* **1991**, *45*, 373.
11. Wieczorek, W. *Mater. Sci. Eng.* **1992**, *B15*, 108.
12. Przyłuski, J.; Dabrowska, A.; Styś, S.; Wieczorek, W. *Solid State Ionics* **1993**, *60*, 141.
13. Chandra, S.; Muraya, K. K.; Hashmi, S. A. In *Recent Advances in Fast Ion Conducting Materials and Devices*; Chowdari, B. V. R., et al., Eds.; World Scientific: Singapore, 1990; p 549.
14. Chandra, A.; Shrivastava, P. C.; Chandra, S. In *Solid State Ionics—Materials and Applications*; Chowdari, B. V. R., et al., Eds.; World Scientific: Singapore, 1992; p 397.
15. Dabrowska, A.; Wieczorek, W. *Mat. Sci. Eng.* **1994**, *B22*, 107.
16. Dabrowska, A.; Wieczorek, W. *Mat. Sci. Eng.* **1994**, *B22*, 117.

SOLID POLYMER ELECTROLYTES (Preparation, Characterization, and Properties)

Alessandro Gandini* and Jean-François Le Nest
Ecole Française de Papeterie et des Industries Graphiques (INPG)

Polymer electrolytes are macromolecular materials capable of sustaining the transport of (at least part of) the ionic species they contain. They differ from *polyelectrolytes*, because they are aimed essentially at applications like batteries, sensors, electrochromic devices, and supercapacitors and fulfill, therefore, the same basic roles as other liquid (e.g., aqueous solutions) or solid electrolytes (e.g., inorganic glasses).

*Author to whom correspondence should be addressed.

Although the field related to polymer electrolytes is young, because the first materials tested in that context were elaborated in the 1970s, a great deal of research has been devoted to its different facets since the early 1980s, as shown by the very large number of publications, including numerous monographs, and by the establishment of a biennial international symposium on these topics.[1–8]

The following treatment deals with materials which are to be inserted in electrochemical devices based mostly on lithium as the species to be oxidized and reduced at the corresponding electrodes because this element insures the highest energy density.

THE ELABORATION OF POLYMER ELECTROLYTES

Linear Polymers

The first polymer to be successfully tested as electrolyte was linear poly(ethylene oxide) (PEO). It provides an excellent solvation of alkali metal and other cations and can therefore dissolve high concentrations of the corresponding salts of strong acids. The pioneering studies on polymer electrolytes were conducted with this material and gave satisfactory results in terms of conductivity and film-forming ability, but showed problems related to crystallization and creep.

One way to avoid crystallization is to upset the order in the enchainment of monomer units within a linear structure, namely to prepare random or short-block copolymers. The reactivity of oxirane is much more pronounced than that of its homologues and therefore truly random copolymers are difficult to obtain and tend to have relatively low molar masses and thus poor mechanical properties. As for block copolymers, the PEO-PPO structures are well known, but not in sequences of short blocks so that crystallization remains a problem.

Another approach consists in using a high molar mass polymer bearing solvating chains as side groups.[9,10]

The use of some of the latter topologies has provided good results in terms of avoiding the appearance of crystal domains, but the mechanical behavior was not optimized, and of course, the problems of creep and high-temperature flow remained unsolved.

Crosslinked Polymers

A more comprehensive strategy to eliminate at the same time crystallization and flow consists in elaborating crosslinked structures with relatively short polyether chains among branching points. A whole family of polymer electrolyte networks has been prepared and characterized following different chemical and topological criteria.[11,12]

OPTIMIZATION AND FUTURE TRENDS

The best system should give the highest conductivity coupled with good mechanical properties, ease of processing into thin films, long-lived storage and electrochemical stability, and, in certain instances, unipolar mobility. Recent work has provided considerable improvements in some of these properties in the search for the optimal performance, as briefly summarized below.

The polymeric structures that seem to give the best behavior in this context are built following the criterion of maximum disorder. An example of such architecture is shown in **Scheme I**,

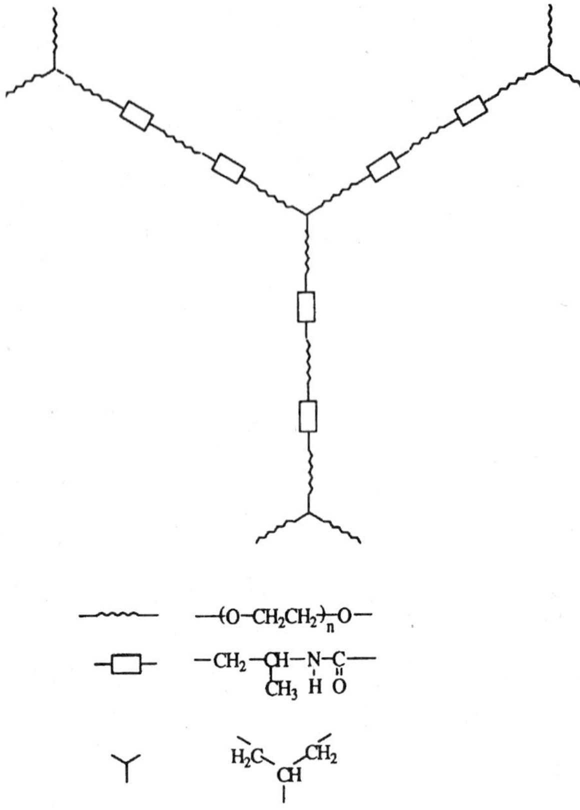

I

which depicts a portion of a network prepared from a oligoethylene oxide triol and an oligoethylene oxide diisocyanate synthesized from the corresponding diamine.[13] In this crosslinked topology the urethane moieties are placed along the PEO chains instead of being the crosslink points, which here are the flexible glyceridic groups. The overall structure is therefore characterized by a very high degree of disorder with PEO domains randomly interrupted by urethane sites. The addition of plasticizing salts like $LiN(CF_3SO_2)_2$ to this network gives the highest conductivity recorded to date with a polyether-based polymer electrolyte.[13]

The film-forming aptitudes of polyether structures can be improved considerably by introducing polysaccharidic chains into their networks. It is well known that certain cellulose and chitin derivatives give thin films with excellent mechanical properties and their incorporation into polymer electrolytes has shown that this tendency is preserved.

A different strategy to minimize the possibility of crystallization consists in adding fillers or plasticizers to the polymer electrolyte.[5–8] Thus, for example, propylene carbonate added to PEO-based materials gives rise to an increase in ionic conductivity, not only because it reduces the glass transition temperature and inhibits crystallization, but also because it increases the dielectric constant of the electrolyte.[14]

CONCLUSION

In fewer than 20 years this discipline has given rise to a new family of useful materials and to a body of fundamental knowledge relevant not only to polymer chemistry and physics, but also to electrochemistry. The growing interest in developing nonpolluting vehicles actioned by electrical power will find viable solutions if, among other problems, adequate polymer electrolytes will be found for the solid state batteries, which will provide the driving energy. In the same vein, considerable energy saving can be achieved by the use of electrochromic windows, in which again, polymer electrolytes can help providing optimum operating conditions.

REFERENCES

1. Wright, P. V. *Br. Polym. J.* **1975**, *7*, 319.
2. Armand, M.; Chabagno, J. M.; Duclot, M. *Fast Ion Transport in Solids*; Vashishta, P.; Mundy, J. N.; Shenoy, G. K., Eds.; North Holland: New York, 1979; p 131.
3. *Polymer Electrolyte Reviews I*; MacCallum, J. R.; Vincent, C. A., Eds.; Elsevier: London, 1987.
4. *Polymer Electrolyte Reviews I*; MacCallum, J. R.; Vincent, C. A., Eds.; Elsevier: London, 1989.
5. *Applications of Electroactive Polymers*; Scrosati, B., Ed.; Chapman and Hall: London, 1993.
6. *British Polymer Journal Special Issue*; Cameron, G. G., Ed.; 1988; 20, N°3.
7. *Second International Symposium on Polymer Electrolytes*; Scrosati, B., Ed.; Elsevier: London, 1990.
8. *Electrochimica Acta Special Issue* **1992**, *37*, No. 9; **1995**, *40*, No. 13–14.
9. Cowie, J. M. G.; Anderson, A. T.; Andrei, M.; Martin, A. C. S.; Roberts, C. ref. 8, p 1539.
10. Motogami, K.; Kono, M.; Mori, S.; Watanabe, M.; Ogata, N. ref. 8, p 1725.
11. Le Nest, J. F.; Gandini, A.; Cheradame, H. ref. 6, p 253.
12. Le Nest, J. F.; Gandini, A.; Schoenenberger, C. *Trends Polym. Sci.* **1994**, *2*, 432.
13. Le Nest, J. F.; Callens, S.; Gandini, A.; Armand, M. ref. 8, p 1585.
14. Huq, R.; Koksbang, R.; Tonder, P. E.; Farrington, G. C. ref. 8, p 1681.

SOLID-STATE POLYAMIDATION

Constantine D. Papaspyrides
Laboratory of Polymer Technology
(Special Chemical Technology)
Department of Chemical Engineering
National Technical University of Athens

Linear polyamides of aliphatic diamines and aliphatic diacids (referred to by the commercial name nylon) are produced by polycondensation. For the polyhexamethylene adipamide (nylon 6,6), the well-known reaction is shown in **Equation 1**:

$$nH_2N(CH_2)_6NH_2 + nHOOC(CH_2)_4COOH \leftrightarrow$$
$$-[-NH(CH_2)_6NHCO(CH_2)_4CO-]_n - +2nH_2O \qquad 1$$

This is a typical equilibrium reaction, and the water formed has to be removed from the system to facilitate a shift toward high molecular weight.

To ensure an equivalence between the diamine and the diacid end groups, an aqueous solution of the salt of these ingredients is formed. The salt formation reaction is highly exothermic, and it can be written as shown in **Equation 2**:[1–3]

$$H_2N(CH_2)_6NH_2 + HOOC(CH_2)_4COOH \rightarrow$$

$$\ldots H_3N^+(CH_2)_6NH_3{}^{+-}OOC(CH_2)_4COO^- \ldots \qquad 2$$

The salt can be isolated by striking its aqueous solution by typical nonsolvents (e.g., methanol). It comprises a distinct chemical compound, it is hygroscopic, and it is readily soluble in water. After drying, it has a melting point in the range of 191–193°C. The industrial production of nylon 6,6 involves solution polycondensation on an aqueous solution of the salt. Subsequently, the water is removed from the reactor and the polycondensation is continued in the melt state.[4]

WHAT IS A SOLID-STATE POLYAMIDATION?

During the past 50 years, extensive research has been carried out on solid-state polyamidation (SSP) of nylon salts. Heating the solid monomer at temperatures below its melting point has been proposed.[5] Routes to polymerization include heating in sealed vessels under inert atmosphere, in open vessels while inert gas passes, and in inert liquid media.[1,6–15]

In the first case, several milligrams of nylon salt are reacted in a tiny tube filled with inert gas, for example, a thermogravimetric analysis (TGA) chamber. In the second case, the salt is heated in an open system, such as a fluidized bed, and the passing inert gas serves also to effect removal of the byproduct of the reaction. An alternative way is to use inert liquid media to heat the salt. The term "inert" means that the liquid does not dissolve or interact with the salt or the polymer formed. The salt crystals are dispersed in the liquid that is heated at a temperature below the melting point of the salt.

ADVANTAGES OF SSP

The advantages of this alternative technology, which has not yet been applied on an industrial scale, could be summarized as follows:

- It uses simpler technology. As mentioned already, with current technology, water is added during the salt preparation stage and water is removed during the course of polymerization. This expensive cycle would be avoided.
- Reaction temperatures at much lower levels, which give better quality, are expected, compared with current technology with reaction temperatures above the melting point of the polymer formed. This leads to undesirable side reactions. Generally, cyclization reactions may occur while severe decomposition either of the reactants or the products cannot be excluded.
- In principle, lowering the reaction temperature leads to significant energy savings.

REACTION MECHANISM

It is very interesting to note that there are still severe discrepancies about the nature of the mechanism of SSP. Even for the same starting material, at the same temperature, different researchers provide completely different kinetic data that support different reaction models.

CONCLUSIONS

From a technological point of view the lesson of this review can be summarized in the following points:

- SSP of nylon salts may be accompanied, depending on the reaction conditions, by a typical transition of the process from the solid to the melt state. This behavior can be correlated with the role of the polycondensation water formed, and the model of water accumulation–hydration–transition to the melt state seems satisfactory.
- It is possible to prepare nylons, starting from the monomer salt, in real solid or quasisolid reaction conditions, by using catalysts. Intermediate stages of the melt reaction can be avoided, but the reaction temperature remains a critical parameter. The literature suggests that boric acid is the most effective catalyst. Again, the reaction behavior encountered seems compatible with the important role of the polycondensation water in SSP processes.

REFERENCES

1. Oya, S.; Tomioka, M.; Asaki, T. *Kobunshi Kagaku (Chemistry of High Polymers)* **1966**, *23*(254), 415.
2. Khripkov, E. G.; Baranova, S. A.; Kharitonov, V. M.; Kudryavtsev, G. I. *Vysokomol. Soedin.* **1972**, *14B*, 172.
3. Papaspyrides, C. D.; Kampouris, E. M. *Polymer* **1984**, *25*, 791.
4. Billmeyer, F. W.; Jr. *Textbook of Polymer Science* Interscience: New York, 1966.
5. Flory, P. J. U.S. Patent 2 172 374, 1939; *Chem. Abstract* **1940**, *34*, 198.
6. Volokhina, A. V.; Kudryavtsev, G. I.; Skuratov, S. M.; Bonetskaya, A. K. *J. Polym. Sci.* **1961**, *53*, 289.
7. Bruck, S. D. *Ind. Eng. Chem. Prod. Res. Dev.* **1963**, *2*(2), 119.
8. Volokhina, A. V.; Kudryavtsev, G. I.; Raeva, M. V.; Bogdanov, M. N.; Kalmykova, V. D.; Mandrosova, F. M.; Okrowchedlidze, N. P. *Khim. Volokna* **1964**, *6*, 30.
9. Macchi, E. M.; Morosoff, N.; Morawetz, H. *J. Polym. Sci., Part A-1* **1968**, *6*, 2033.
10. Frayer, P. D.; Lando, J. B. *Mol. Cryst. Liquid Cryst.* **1969**, *9*, 465.
11. Khripkov, E. G.; Kharitonov, V. M.; Kudryavtsev, G. I. *Khim. Volokna* **1970**, *6*, 615.
12. Dasgupta, P. Ph.D. Thesis, University of Akron, 1978.
13. Gaymans, R. J.; Schuijer, J. In *Polymerization Reactors and Processes*; ACS Symposium Series 104; American Chemical Society: Washington, DC, 1979; p 137.
14. Gaymans, R. J.; Amirtharaj, J.; Kamp, H. *J. Appl. Polym. Sci.* **1982**, *27*, 2513.
15. Kampouris, E. M. *Polymer* **1976**, *17*, 409.

Solid-State Polymerization

See: *Comb-Like Polymers*
Cyclopolymerization (Overview)
High-Pressure Polymerization Processing (High-Temperature Thermosetting Polymers)
Inclusion Complexes (Overview)

Nylons (by High-Pressure Solid-State Polycondensation)
Polyaddition, Radical
Polydiacetylenes, Unconventional
Poly(dichlorophenylene oxide)s (by Pyridine/Copper Complex)
Poly(ethylene terephthalate) (Chemistry and Preparation)
Solid-State Polyamidation
Solid-State Polymerization (Crosslinked Di-p-Ethynyl Substituted Polymers)
Topochemical Photopolymerization (Types of Topochemical Behavior)
Topochemical Polymerization (Diacetylenes)

SOLID STATE POLYMERIZATION (Crosslinked Di-p-Ethynyl Substituted Polymers)

Anastasios P. Melissaris
Materials Engineering Division
Research Institute
University of Dayton

Morton H. Litt
Department of Macromolecular Science
Case Western Reserve University

The problem in using high T_g polymers in composites is that the usual monomers are amorphous and must be polymerized at temperatures above the final T_g in order to get high conversion and good mechanical properties in the matrix. If the polymer is crosslinked, two further problems arise. First, as the polymerization continues the matrix shrinks, stressing the matrix/fiber interfacial bond. Second, as the composite cools from the high curing temperature, the matrix shrinks further. Because its coefficient of expansion is higher than that of the fiber, the shrinkage puts further stress on the interfacial bond. The stress can sometimes become so great that the composite loses its integrity. These problems can be minimized or avoided by polymerizing self-organizing systems. Monomers with terminal functional groups that form smectic liquid crystal layers are needed. Because all the terminal functional groups are packed into a very small fraction of the total volume, they can polymerize well to high conversion without needing to diffuse to find other reactive ends. Thus high T_g does not prevent high conversion at low temperatures. Similarly, polymerization in an ordered phase means that the molecules are well packed even before polymerization, so there should be little or no shrinkage.

We have found that the acetylene terminal group as para substituent in aromatic rigid rod monomers is a good candidate as a polymerizable functionality.[1] Rigid linear aromatic structures with terminal p-ethynyl groups could pack smectically with the acetylene groups interdigitated.

The objective of this research was to develop new monomers that could polymerize in the solid or liquid crystalline state at moderately elevated temperatures (130–250°C) and yield highly crosslinked, thermally stable resins suitable as matrices for high-temperature composites.

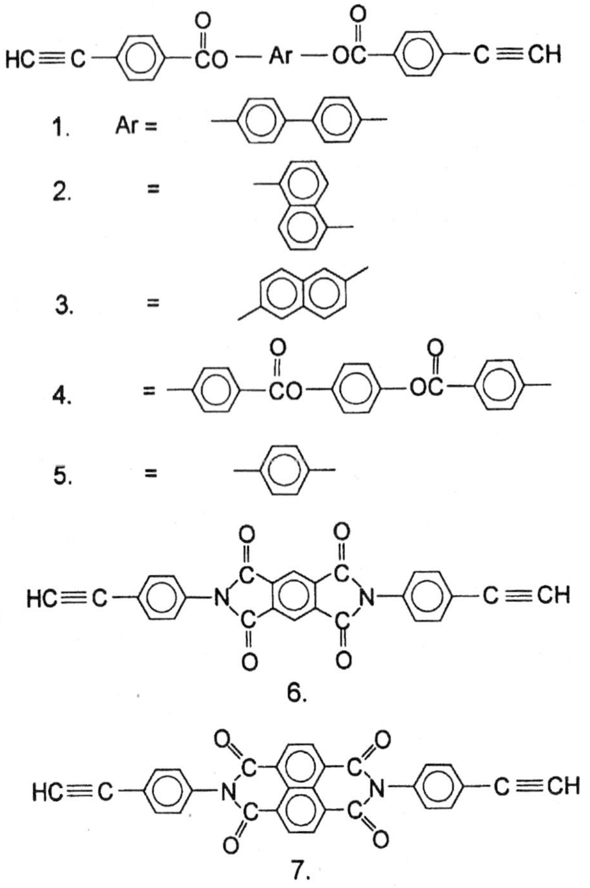

SCHEME I. Chemical structures of di-p-ethynyl substituted rigid rod monomers.

PREPARATION

The chemical structures of the monomers synthesized, **1–7**, are shown in **Scheme I**. Monomers **1–7** were prepared in high yield and high purity and their detailed syntheses were reported from our laboratory. The starting materials, p-ethynyl benzoyl chloride and p-ethynyl aniline were synthesized in large quantities by new, high yield methods developed in our laboratory.[2,3]

THERMAL POLYMERIZATION

The polymers are denoted by appending **p** to the monomer number, **1p–7p**. Highly crosslinked resins, **1p–7p**, were obtained by heating the corresponding monomers **1–5** at 220°C for 30 min and subsequently at 250°C for 1.5 h (postcuring) in N_2 or by heating **6** and **7** at 220°C for 30 min and subsequently at 280°C for 1.5 h.[1–3]

PROCESSING

We were interested in developing a processing technique for di-p-acetylene terminated rigid-rod monomers because we speculated that their polymers should have high moduli, little or no polymerization shrinkage, high T_g, and very low thermal expansion. Despite the reported difficulties we developed a processing technique that involves solid state polymerization under moderate pressure. Our goal was not to find the optimum processing conditions for each of the monomers, but to show that this class of monomers can be processed and yield polymers with unique properties. Monomers **1–7** are highly crystalline materials that polymerize in the solid state to yield polymers that keep a certain amount of crystallinity. Monomers **1–3** were processed under well defined conditions.[1] Because of manpower limitations we did not optimize the processing conditions for all monomers.

CONCLUSION AND APPLICATIONS

New di-p-ethynyl benzoyl ester and imide monomers were synthesized in both high yield and high purity. They polymerized in the crystalline state to 100% conversion, yielding highly crosslinked resins retaining gross order. The difference between this and previously reported work on ethynyl terminated monomers or oligomers is that these monomers are liquid crystalline or crystalline during polymerization and yield polymers that retain their overall orientation and low-order crystallinity. Because they polymerize in the solid state, they exhibit little or no polymerization shrinkage (0–2.3%). The modulus drops only slightly at the T_g because of the high crosslink density. EBEs polymerize to high conversion in spite of their rigid matrix because their lamellar organization brings all the ethynyl groups into a small volume. There may be further acceleration of the polymerization because of the interdigitation of their acetylene groups. The polymers had T_gs much higher than the polymerization temperature because EBEs are properly organized to polymerize in the solid state.

Solid-state processing of these monomers that neither melt nor have a softening point generated cohesive, high modulus, high-T_g polymers. Because these polymer plates have high T_gs, high moduli, low thermal expansion, very low polymerization shrinkage, and good thermal and thermo-oxidative stability, they may be good candidates for high-temperature composite matrices. Such composites should have good mechanical properties because the total matrix shrinkage for both polymerization and cooling from the reaction temperature is very low, thus minimizing the stress on the interfacial matrix-fiber bond.

REFERENCES

1. Melissaris, A. P.; Litt, M. H. *Macromolecules* **1994a**, *27*, 2675.
2. Melissaris, A. P.; Litt, M. H. *Macromolecules* **1993**, *26*, 6734.
3. Melissaris, A. P.; Litt, M. H. *Polymer* **1994b**, *35*, 3305.

SOLUBILITY PARAMETERS (A Synthetic Approach)

R. D. Sanderson and A. J. van Reenen
Institute for Polymer Science
University of Stellenbosch

T. R. du Toit
Plastics Technology Department
Cape Technikon

The solubility parameter theory is one of the best known and most used theories in polymer science by which solvent–polymer and polymer–polymer interaction can be predicted. This work gives a working knowledge of how the theory can be used to synthesize compatible polymers.[1]

THEORETICAL BACKGROUND

"Similia similibus solvuntur," like dissolves like, does not only apply to liquid mixtures but also to polymer solutions.

The solubility parameter (δ) is defined as the square root of the cohesive energy density (CED), which is a measurement of the intensity of intermolecular interactions in a pure liquid or solid. The solubility parameter relates to the heat of mixing through **Equation 1**:

$$\Delta H_{mix} = n_s V_s \Phi_p \left(\delta_s - \delta_p\right)^2 \qquad 1$$

where n_s, V_s, and δ_s are the number of molar volume, and total solubility parameter of the solvent, respectively. The polymer volume fraction and total solubility parameter are represented by Φ_p and δ_p. Because ΔG_{mix} decreases with a decrease in ΔH_{mix}, it is expected that a system where $(\delta_s - \delta_p)^2$ is small should have high or complete miscibility.[2]

The solubility parameter theory was developed by Scatchard and further expanded by Hildebrand.[3,4]

For many liquids and amorphous polymers, the cohesive energy is also dependent on the interaction between polar groups and hydrogen. Hansen divided the CED and subsequently the solubility parameter corresponding to three types of interactive forces (**Equation 2**):

$$\delta_t^2 = \delta_d^2 + \delta_p^2 + \delta_h^2 \qquad 2$$

where δ_t, is called the total solubility parameter and δ_d, δ_p, and δ_h are due to dispersion, polar, and hydrogen bonding forces, respectively.[5] The solubility region of a polymer and/or solvent can be depicted in three-dimensional vector space by a sphere with (δ_p, δ_p, δ_h) as the center.

The solubility parameters of most solvents can be determined from their molar energies or enthalpies of evaporation. Hoy published a comprehensive set of solvents and their solubility parameters.[6] Polymers on the other hand cannot be vaporized (because of their size and enormous cohesive energies) without decomposing. Therefore, their solubility parameters must be determined indirectly, either by theoretical predictions or by the solvent spectrum method. The latter is used in the paint industry and is a measurement of the dissolution behavior or degree of swelling in a range of solvents with known solubility parameters.[2] Hansen combined a solvent spectrum method with other methods of evaluation in deriving three-component solubility parameters for polymers.[5,7]

SYNTHESIZING A POLYMER TO MATCH THE SOLUBILITY PARAMETERS OF ANOTHER POLYMER

In the use of the Hildebrand or total solubility, parameter polymers are usually regarded as miscible when the δ_t values of the polymers are within 2 or 3 $MPa^{1/2}$ of one another.[8] When using three-dimensional solubility parameters, the ${}_{ij}R$ method, as used for solvents, can also be used to predict compatibility of polymers (**Equation 3**):

$$ {}_{ij}R = \left[4\left({}_c\delta_d - {}_j\delta_d\right)^2 + \left({}_i\delta_p - {}_j\delta_p\right)^2 + \left({}_i\delta_h - {}_j\delta_h\right)^2\right]^{1/2} \quad 3 $$

where ${}_{ij}R$ represents the distance between the solubility parameters of polymer i, (${}_i\delta_d$, ${}_i\delta_p$, ${}_i\delta_h$), and polymer j, (${}_j\delta_d$, ${}_j\delta_p$, ${}_j\delta_h$).[2] For a polymer to be compatible with another, its three-dimensional solubility parameters must be as close as possible to that of the matrix polymer. As an example, how to determine the composition of a polymer that would be compatible with polysulfone according to the solubility parameter theory is shown.

Several authors developed methods to determine the optimum multicomponent solvent composition for polymers.[8–10]

The terpolymerization of styrene, methyl methacrylate, and acrylonitrile was done in solution at 60°C by using benzoylperoxide as the initiator. The reaction was stopped before 10% conversion of the monomers took place to ensure random terpolymers.

The compatibility of the terpolymer and polysulfone was tested by mutually dissolving different ratios of terpolymer and polysulfone in chloroform. The solvent was evaporated off at room temperature under vacuum. Differential scanning calorimetry termograms of the blends features a broad T_g between 135 and 165°C for all ratios of terpolymer and polysulfone. This indicates that polysulfone and the prepared terpolymer are compatible in all ratios from solution.[12]

REFERENCES

1. Du Toit, T. R.; van Reenen, A. J.; Sanderson, R. D. *The Solubility Parameters of Polysulphone*, in press.
2. Barton, A. F. M. *Handbook of Solubility Parameters and Other Cohesion Parameters*; CRC: Boca Raton, FL, 1983.
3. Scatchard, G. *Chem. Rev.* **1931**, 321.
4. Hildebrand, J. H.; Scott, R. L. *The Solubility of Nonelectrolites*; Reinhold: New York, 1950.
5. Hansen, C. M. *J. Paint Technol.* **1967**, *39*, 104.
6. Hoy, K. L. *J. Paint Technol.* **1970**, *42*, 76.
7. Hansen, C. M. *J. Paint Technol.* **1967**, *39*, 505.
8. Schneier, B. *Polym. Lett.* **1972**, *10*, 245.
9. Behnken, D. W. *Polymer* **1983**, *24*, 193.
10. Rigby, Z. *Polymer* **1978**, *19*, 1229.
11. Froeling, P. E.; Hillegers, T. *Polymer* **1981**, *22*, 261.
12. Du Toit, T. R.; Van Reenen, A. J.; Sanderson, R. D., in press.

Sonochemical Polymerization

See: *Ultrasonic Degradation (Preparation of Block Copolymers)*
Ultrasonically Assisted Polymer Synthesis

Spacer Groups

See: *Liquid Crystalline Aromatic Polyesters*
Liquid Crystalline Polyacetylene Derivatives
Liquid Crystalline Polymers (Polyesters of Bibenzoic Acid)
Liquid Crystalline Polymers (Thermotropic Polyesters
Liquid Crystalline Polymers, Side Chain
Liquid Crystalline Polyurethanes
Polyoxyarylenesiloxanes

Spheres

See: *Colloidal Crystals, Polymeric*
Poly(α-amino acid) Spherical Particles
Spheres, Poly(butylene terephthalate)

SPHERES, POLY(BUTYLENE TEREPHTHALATE)

Mark E. Nichols
Ford Research Laboratory

Richard E. Robertson
Macromolecular Science and Engineering Center and Engineering Department of Materials Sciences
University of Michigan

The production of small polymer spheres of a narrow, well-controlled size distribution has been possible for some time. Spheres of both crystalline and amorphous polymers have been produced. The spheres of amorphous polymers usually are produced through the suspension polymerization of vinyl polymers, the most common of which is polystyrene.[1–3]

The other route to small polymeric spheres, which is not currently used commercially, is through the controlled crystallization of polymer solutions. For example, spheres have been formed by the crystallization of polyethylene from nitrobenzene and o-chlorophenol of poly(3,3-bis(chloromethyl)-oxacyclobutane) from xylene, and of poly(vinylidine fluoride) from a 50/50 mixture of acetone/xylene.[4–6]

While many systems may indeed conform to the behavior outlined for the polyethylene system, the more common case is likely to be where: 1) the crystallization/melting curve lies above the liquid-liquid demixing boundary (for UCST systems), and 2) crystallization can be suppressed or delayed long enough to first allow liquid-liquid demixing. If this is true, then a simple method for producing small spheres of a narrow size distribution may exist for many more polymers. This is the case for poly(butylene terephthalate) (CAS 26062-94-2), where we have produced microspheres by crystallizing from a moderately concentrated solution of the polymer in a liquid diglycidyl ether of bisphenol-A (DGEBA) epoxy prepolymer.[8,9] The spheres have a high surface and are of a uniform small size.

PREPARATION

One phase solution of PBT in liquid epoxy can be prepared by heating PBT beads in liquid epoxy resin at 230°C until the beads dissolve, typically 1h. The particular epoxy prepolymer used is of some importance. The viscosity at room temperature should not be too high, or separation of the particles from the solution could be difficult. Epon 828 (Shell Chemical Co.) has been found to work well. After heating, the solution is cooled to a specific temperature and held isothermally to allow the spheres to form. Typical temperatures and times of crystallization for 5% PBT solutions are 155°C for 20 min.[7]

PROPERTIES

Size and Morphology

The process detailed above could have widespread utility in producing spheres of polymers that are not easily produced by suspension polymerization methods. This is because the circumstances under which the spheres are produced could likely be reproduced for many polymers by the appropriate choice of solvent. The particles are, in general, a single spherulite.

Formation Mechanism

The above process produces particles with a diameter of approximately 35 μm "wet" and 23 μm dry.

The crystalline PBT microspheres have been suggested to form due to liquid-liquid phase separation of the polymer-solvent system, followed by rapid crystallization of the polymer-rich phase.[7]

APPLICATIONS

Currently small polymeric spheres are used as size markers, media in chromatography columns, substrates in medical diagnostic tests, and controlled release drug delivery systems.[9,10] For many of these applications smooth microspheres are best suited. However, for some of these applications the higher surface areas available with crystalline particles could prove useful. For spheres used in controlled release drug delivery or in medical diagnostic tests, the higher surface area could provide a more active surface for reaction or diffusion. It is also possible to functionalize the surface of microspheres for use as catalysts, in which case the higher surface area of the crystalline globules could provide a more reactive catalyst substrate.[11] In the field of polymer composites it is possible to electrostatically coat fibers with small thermoplastic particles. On heating this may provide better fiber wet-out than other methods of producing thermoplastic composites.

REFERENCES

1. Chiang, J.; Prud'homme, R. K. *J. Colloid Interface Sci.* **1988**, *122*, 284.
2. Tamai, H.; Nishida, Y.; Suzawa, T. *J. Colloid Interface Sci.* **1991**, *146*, 288.
3. Ferrick, M. R.; Murtagh, J.; Thomas, J. K. *Macromolecules* **1989**, *22*, 1515.
4. Hay, I. L.; Keller, A. *Kolloid Z. Z. Polym.* **1965**, *204*, 43.
5. Garber, C. A.; Greil, P. H. *J. Appl. Phys.* **1966**, *36*, 4034.
6. Sakaoku, K.; Peterlin, A. *J. Macromol. Sci. - Phys.* **1967**, *B1*, 401.
7. Nicholas, M. E.; Robertson, R. E. *J. Polym. Sci., Polym. Phys. Ed.* **1994**, *32*, 573.
8. Nicholas, M. E.; Robertson, R. E. *J. Polym. Sci., Polym. Phys. Ed.* **1994**, *32*, 1607.
9. Kamshead, J. T.; Treleaven, J. G.; Gibson, F. M.; Ugelstad, J.; Rembaum, A.; Phillip, T.; Caine, G. D. *Lancet* **1984**, *January 14*, 70.
10. Fitch, R. M.; Gajria, C.; Tarcha, P. J. *J. Coll. Interface. Sci.* **1979**, *71*, 107.
11. Fitch, R. M. *Macromolecules*; Benoit, H.; Rempp, P., Eds.; Pergamon: Oxford, 1982.

SPIDER SILK (Production of Polypeptide Polymers)

Charlene M. Mello and David L. Kaplan
Biotechnology Division
U.S. Army Natick Research Development and Engineering Center

Our group has focused on the production of a family silk recombinant proteins to improve our understanding of the structure–property relationships of these protein polymers.[1–3] We review the design, construction, and synthesis of a subset of this spider silk family.

BACKGROUND

The mechanical properties of dragline silk from the spider, *Nephila clavipes*, have been extensively characterized under quasi-static and high strain rates.[4,5] Some of the native dragline silk fibers were found to exhibit mechanical properties comparable to synthetic high-performance fibers and displayed an 11% elongation to break (three- to four-fold higher than synthetic high-performance fibers). In addition, single dragline silk fibers exhibit no evidence of tensile breaks on their outer surfaces or failure by kinking on the inner surface when tightly curved.[5] These observations are indicative of useful compressive properties, a common limitation with synthetic high-performance fibers.

These proteins perform a key natural function as structural fibers, to absorb impact energy from flying insects without breaking. They dissipate energy over a broad area and balance stiffness, strength, and extensibility.[6,7] In addition to their unusual mechanical properties and visual luster, silks also exhibit interesting interference patterns within the electromagnetic spectrum, unusual viscometric patterns related to processing, and piezoelectric properties.[8–11]

Silk proteins contain repetitive crystalline domains that influence higher order conformations. The characteristic antiparallel β-sheet structure of many silks, including spider and silkworm silks, assembles through hydrogen bonding and van der Waals interactions due to the close packing density of the short side chain amino acids in the polymer chains.[12,13] These fibers are insoluble in dilute acids and dilute alkali, resistant to most proteolytic enzymes, and hydrolyzed by concentrated sulfuric acid.[14–16] Lithium bromide has been found to solubilize spider dragline silk without hydrolysis.[1]

Candelas *et al.* investigated *in vivo* silk protein expression from excised major ampullate glands (source of the dragline fiber) of *N. clavipes* with radiolabelled amino acids and reported a single protein product.[17] More recent studies of the dragline fiber and pre-spun protein also identified a single protein with a molecular weight of 275 kDa determined by polyacrylamide gel electrophoresis.[1] Peptide sequences generated from selective chemical cleavage of the dragline fiber and the purified glandular protein revealed key sequences encompassing the repetitive domains of the dragline protein, AAGGAGQGGY and YLGSQGAGQ.[1] We have designed synthetic genes based on the repetitive regions of this dragline protein and protein sequence translated from a major ampullate gland cDNA sequence.[3,18]

SUMMARY

Polymers will continue to be major components for future materials, with increasing demands on performance. The ability to design and tailor structural proteins at the genetic level provides new opportunities to meet some of these stringent performance requirements. Increasing environmental awareness also demands structural polymers that efficiently and economically meet application needs in an environmentally sound fashion, in terms of both production and disposal. The production process described here operates in an environmentally compatible manner, and economics are expected to improve with scale and further technical advances. Biological systems have also evolved to control molecular architecture and provide efficient structural materials. However, because of other life-sustaining energy demands on the organisms, these materials may not represent their full performance potential. Therefore, through the use of genetic engineering in combination with mutagenesis, chemical modifications, the incorporation of unnatural amino acids, and the use of new backbone structures, a new family of structural protein polymers may be developed that are precisely tailored for specific applications.

ACKNOWLEDGMENTS

Special thanks are extended to Kevin McGrath, John Prince, Carla DiGirolamo, Steven Arcidiacono, Fred Allen, Michelle West, Jamie Parsons, and Paul Gauger for their technical contributions to our program, which have made this chapter possible. We also thank Gerard Barry and Timothy K. Ball (Monsanto) for the antispider dragline antibody.

REFERENCES

1. Mello, C. M.; Yeung, B.; Senecal, K.; Vouros, P.; Kaplan, D. L. In *Silk Polymers: Materials Science and Biotechnology*; American Chemical Society Symposium Series 544; Kaplan, D. L. et al., Eds.; 1994, 67–79.
2. Mello, C. M.; Arcidiacono, S.; Beckwitt, R.; Prince, J.; Senecal, K.; Kaplan, D. *Mat. Res. Soc. Symp.* **1994**, *330*, 37.
3. Prince, J. T.; McGrath, K. P.; DiGirolamo, C. M.; Kaplan, D. L. *Biochemistry* **1995**, *34*, 10879.
4. Cunniff, P. M.; Fossey, S. A.; Auerbach, M. A.; Song, J. W.; Kaplan, D. L.; Adams, W. W.; Eby, R. K.; Mahoney, D.; Vezie, D. L. *Polym. Adv. Technol.* **1994**, *5*, 401.
5. Cunniff, P. M.; Fossey, S. A.; Auerbach, M. A.; Song, J. W. In *Silk Polymers: Materials Science and Biotechnology*; American Chemical Society Symposium Series 544; Kaplan, D. L., et al., Eds.; American Chemical Society: Washington, DC, 1994; pp 234–251.
6. Gosline, J. M.; Denny, M. W.; DeMont, M. E. *Nature* **1984**, *309*, 551.
7. Gosline, J. M.; DeMont, M. E.; Denny, M. W. *Endeavour* **1986**, *10*(1), 37.
8. Craig, C. L.; Bernard, G. D. *Ecology* **1989**, *71*(2), 616.
9. Magoshi, J.; Magoshi, Y.; Nakamura, S. *J. Appl. Polym. Sci.* **1985**, *41*, 187.
10. Fukada, E. *J. Phys. Soc. Japan* **1956**, *12*, 1301.
11. Ando, Y.; Okano, R.; Nishida, K.; Miyata, S.; Fukada, E. *Reports on Prog. In Polymer Physics in Japan* **1980**, *23*, 775.
12. Lucas, F.; Shaw, J. B. T.; Smith, S. G. *J. Mol. Biol.* **1960**, *2*, 339.
13. Fossey, S. A.; Nemethy, G.; Gibson, K. D.; Scherga, H. A. *Biopolymers* **1991**, *31*, 1529.
14. Tillinghast, E. K.; Kavanagh, E. J. *J. Exp. Zool.* **1977**, *202*, 213.
15. Tillinghast, E. K. *Insect Biochem.* **1984**, *14*(1), 115.
16. Townley, M. A.; Tillinghast, E. K. *J. Arachnol.* **1988**, *16*, 303.
17. Xu, M.; Lewis, R. V. *Proc. Natl. Acad. Sci. USA* **1990**, *87*, 7120.
18. Sambrook, J.; Fritsch, E. F.; Maniatis, T. *Molecular Cloning, A Laboratory Manual* Cold Spring Harbor: 1989.

SPIROACETAL BACKBONE POLYMERS

Shigeo Nakamura
Department of Applied Chemistry
Faculty of Engineering
Kanagawa University

Spiro compounds are organic and have two adjacent rings in a molecule, an atom is shared by two rings. The shared atom is referred to as a tetravalent spiro atom such as carbon, silicon, and nitrogen in quaternary ammonium salts.

Polymers with spiroacetal moiety in their backbones are prepared either by reaction between cyclic diketone and tetrol or self-condensation of dihydroxyacetone, or by polycondensation of a monomer containing a spiroacetal moiety as a reactant.

In the former method, completely double-stranded spiroacetal backbone polymers are obtained, whereas partly double-stranded polymers are prepared in the latter process. They have extremely high melting temperature compared to corresponding linear polyethers due to restricted rotation, and their thermal and chemical stabilities are enhanced since two bonds must be cleaved within a ring to induce a decrease in molecular weight that deteriorates various properties.

PREPARATION

Completely Double-Stranded Spiroacetal Polymers

The first completely double-stranded spiroacetal backbone polymer **III** was prepared in a yield of 90% by refluxing pentaerythritol and 1,4-cyclohexanedione in benzene using p-toluenesulfonic acid as a catalyst (**Equation 1**).[1]

The self-condensation of dihydroxyacetone in the presence of acid catalyst in benzene gives highly crystalline polymer **VIII**, which is insoluble in any solvent[2] (**Equation 2**).

PROPERTIES

The completely double-stranded polymer in Equation 1 is highly crystalline, probably higher than 95%, and no amorphous halo occurs in the X-ray diffraction pattern.[1] Any evidence of crosslinking or branching is not observed. It is infusible and insoluble in conventional organic solvents. It dissolves only in hexafluoro-2-propanol at room temperature.

The polymer **VIII** does not melt and dissolve even in hexafluoro-2-propanol. This polymer is crystalline; therefore, insolubility is induced not by crosslinking but by high polarity.

As a method of modification of condensation polymers, diamines and diacids are partly substituted by those with spiroacetal moieties.

By partial incorporation of spiroacetal structures in condensation polymers, dyeability is increased as crystallinity is somewhat decreased without lowering melting temperature. Solubility and adhesive property are improved without deteriorating thermal stability.

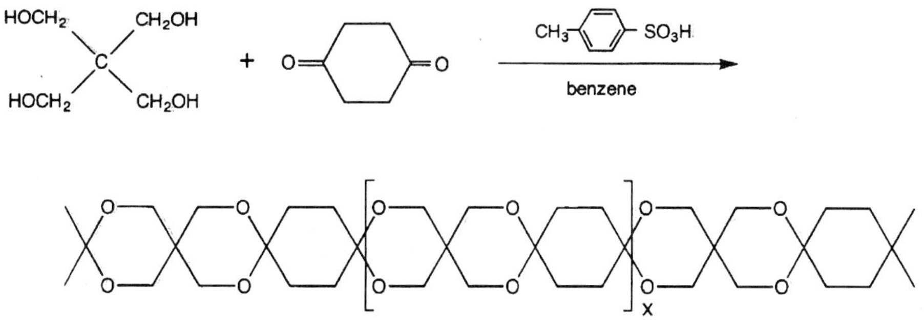

1

[III]

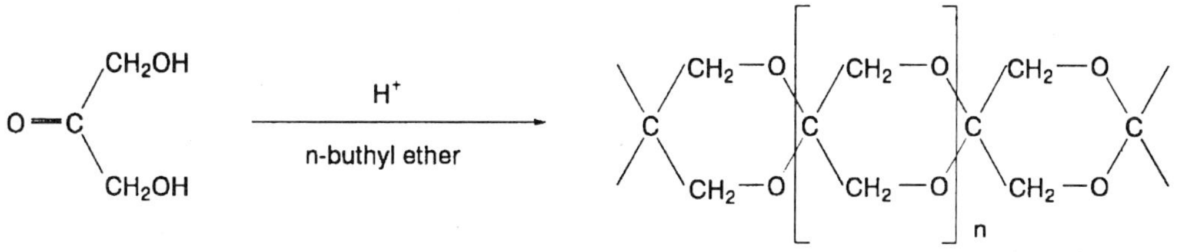

2

[VIII]

REFERENCES

1. Bailey, W. J.; Volpe, A. A. *J. Polym. Sci., Part A-1* **1970**, *8*, 2109.
2. Bailey, W. J.; Beam, C. F. Jr.; Haddad, I. *Polym. Prepr. Am. Chem. Soc. Div. Polym. Chem.* **1971**, *12*, 169.

Spirocyclic Monomers

See: *Bicyclic and Spirocyclic Monomers, Oxygen-Containing (Ring-Opening Polymerization)*
Ring-Opening Polymerization, Anionic (with Expansion in Volume)
Ring-Opening Polymerization, Cationic (with Expansion in Volume)
Ring-Opening Polymerization, Free Radical
Ring-Opening Polymerization, Radical (with no Shrinkage in Volume)

Spontaneous Copolymerization

See: *2-Oxazolines, Polymerization*
Periodic Polymers
Zwitterionic Polymerization (Overview)
Zwitterionic Polymerization (Aziridine, Oxazoline Derivatives with Anhydrides)

Stabilizers

See: *Additives (Types and Applications)*
Antioxidants and Stabilizers
Degradation (Weatherability)
Hindered Amine Light Stabilizers, Monomeric
Polyethylene (Stabilization and Compounding)
Polyolefins (Overview)
Poly(vinyl chloride) (Mechanisms of Stabilization)

STACKED TRANSITION METAL MACROCYCLES (Semiconductive Properties)

Michael Hanack,* Manuela Lang, and Rainer Polley
Institut für Organische Chemie
University of Tübingen

Recently, interest in phthalocyanines and their derivatives has increased because of their properties, which are useful for applications in material science.[1–7] Phthalocyanines and structurally related compounds (e.g., naphthalocyanines and porphyrins) are of interest, for example, in nonlinear optics, as liquid crystals, as Langmuir-Blodgett (LB) films, as photosensitizers, and as carrier generation materials in near-IR.[2,8–17] The substituted derivatives of phthalocyanines function as active components in various processes driven by visible light: photoredox reactions and photooxidation in solution, activity in cancer therapy, photoelectrochemical cells, photovoltaic cells, and electrophotographic applications.[1,18–27]

SYNTHESIS OF PHTHALOCYANINES

Hitherto, about 70 different elements could be used as central atoms in phthalocyanines, thereby, for example, controlling the oxidation potential of these compounds and leading to very different electrical properties. Most metallophthalocyanines can be prepared from phthalodinitrile derivatives and the corresponding metals or metal salts in high boiling solvents such as 1-chloronaphthalene or quinoline. They are also obtained by subsequent metal insertion into the phthalocyanine. The synthesis of naphthalocyaninato and phenanthrenocyaninato systems follows similar routes from the appropriate starting materials (e.g., 1,2- and 2,3-dicyanonaphthalene for the 1,2- and 2,3-NcM

*Author to whom correspondence should be addressed.

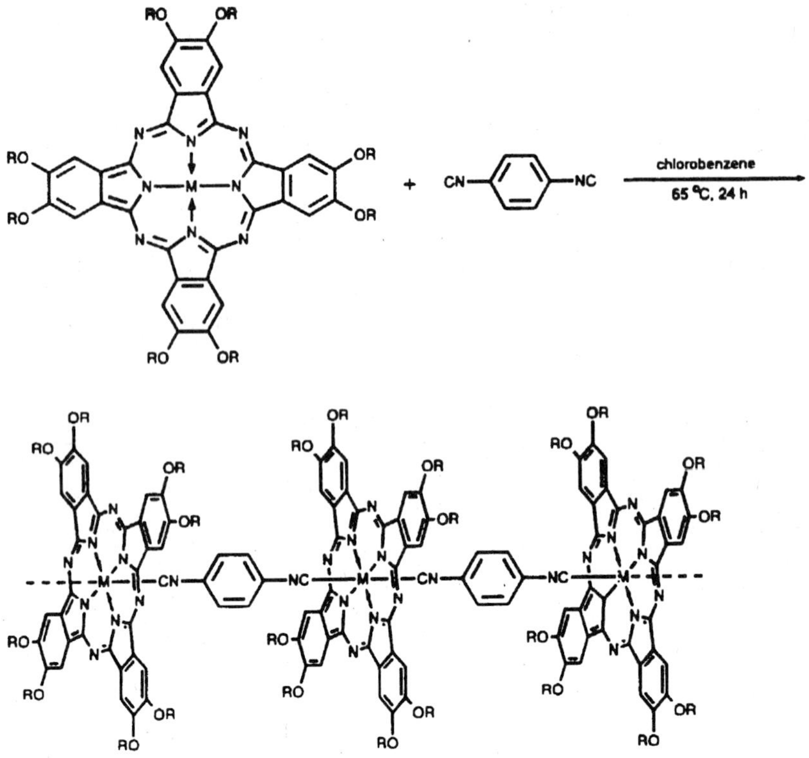

SCHEME I.

system, respectively).[28–33] A mild method to obtain phthalocyaninatometal complexes in high yields is by heating phthalodinitrile with metal salts in an alcohol (e.g., 1-pentanol) in the presence of 1,8-diazabicyclo[5.4.0]undec-1-ene.[34–36]

STACKED TRANSITION-METAL COMPLEXES WITH PHTHALO- AND NAPHTHALOCYANINES

A few years ago we developed a route to obtain a stacked arrangement of phthalocyaninato and naphthalocyaninato transition-metal compounds, which lead to coordination polymers where the macrocycle, the central metal atom, and the bridging ligand can be varied systematically. The stacking is achieved by bisaxially connecting the central transition-metal atoms of the macrocycles with bidentate bridging ligands (L). Such bridged macrocyclic metal compounds $[MacM(L)]_n$ have been synthesized and investigated by us in detail with respect to their physical properties.[2,3]

The bridging ligands (L) are linear π-electron containing organic molecules (e.g., pyrazine (pyz), *p*-diisocyanobenzene (dib), substituted *p*-diisocyanobenzenes, tetrazine (tz), substituted tetrazines, and others). If the oxidation state of the central metal atom is +3 (e.g., Co^{3+} and Fe^{3+}) charged bridging ligands such as cyanide (CN^-), thiocyanate (SCN^-), and others can also be used. As macrocycles, phthalocyanines, substituted phthalocyanines, 1,2- and 2,3-naphthalocyanines, phenanthrenocyanines, tetrabenzoporphyrins, and tetranaphthoporphyrins have been used.[2,3]

Many of the bridged macrocyclic metal complexes $[MacM(L)]_n$ ("shish-kebab polymers") can be prepared in high yields and with high purity by treating the metallomacrocycle, MacM, with the pure ligand or with the ligand in an appropriate solvent (e.g., acetone and chlorobenzene).

Powder Conductivities and Doping of Stacked Polymers

The powder conductivities of most of the bridged phthalocyaninato transition-metal complexes, $[PcM(L)]_n$, for M is Fe, Ru, Os, Co, Rh, and L is pyz, dib, etc. are low, in the range of 10^{-6}–10^{-7} S/cm. However, many of these compounds can be doped either chemically or electrochemically.

The conductivities can be increased to 10^{-3}–10^{-1} S/cm via doping, depending on the iodine content. In general, the complexes, $[MacM(L)]_n$, are practically insoluble in organic solvents. However, soluble oligomers, $[R_4PcM(L)]_n$ and $[R_8PcM(L)]_n$, have been prepared using substituted metallomacrocycles, R_4PcM, in which R = *t*-Bu, Et, OR′ (R′= C_5H_{11} – $C_{12}H_{25}$) and R_8PcM (R= C_5H_{11} – $C_{12}H_{25}$, OC_5H_{11} – $OC_{12}H_{25}$) and M = Fe, Ru.[37–47] **Scheme I** shows one of the routes to synthesize soluble bridged coordination polymers (i.e., $[(RO)_8PcFe(dib)]_n$).[48,49]

BRIDGED MACROCYCLIC COMPLEXES WITH TRIVALENT TRANSITION-METAL IONS

The central transition metals within the macrocycles may also have the oxidation number +3. An octahedral configuration of the metal in the macrocycles is also possible, for instance, for Fe^{3+}, Co^{3+}, and Rh^{3+}. For the formation of the corresponding polymers, CN^-, SCN^-, or N_3^- are suitable bridging ligands.[50–53]

Figure 1 schematically shows a cyano-bridged intrinsic semiconductive phthalocyanine complex that can be synthesized by the displacement of the axial anion X^- by CN^- in a

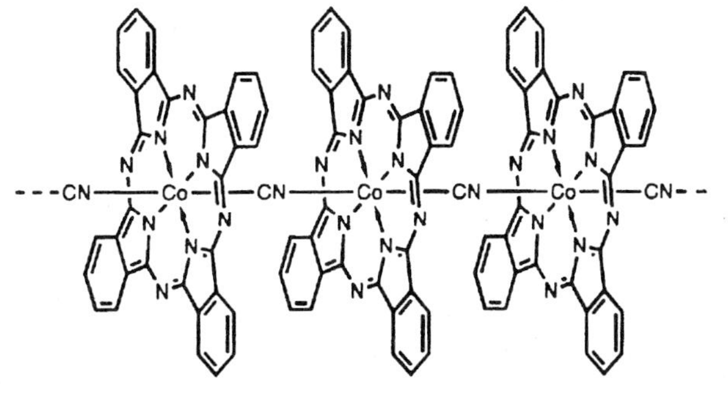

FIGURE 1. Schematic drawing of $[PcCo(CN)]_n$.

coordinatively unsaturated compound PcMX. This synthesis has been used for the preparation of $[PcMn(CN)]_n$, $[PcFe(CN)]_n$, and $[2,3\text{-}NcFe(CN)]_n$.

Although the main emphasis of spectroscopic and chemical characterization of $[PcM(CN)_n$ polymers has been directed toward the cobalt derivative, the results are generalizable. The polymers with other transition metals are found by X-ray diffraction studies to be isostructural to $[PcCo(CN)]_n$.

REFERENCES

1. *Phthalocyanines, Properties, and Applications*; Leznoff, C. C.; Lever, A. B. P., Eds.; VCH: New York, 1989–1993; Vol. 1–3.
2. Schultz, H.; Lehmann, H.; Rein, M.; Hanack, M. *Struct. Bonding* **1991**, *74*, 41.
3. Hanack, M.; Datz, A.; Fay, R.; Fischer, K.; Keppeler, U.; Koch, J.; Metz, J.; Mezger, M.; Schneider, O.; Schulze, H. J. In *Handbook of Conducting Polymers*; Skotheim, T. A., Ed.; Marcel Dekker: New York, 1986.
4. Hanack, M.; Lang, M. *Adv. Mater.* **1994**, *6*, 819.
5. Marks, T. J. *Science* **1985**, *227*, 881.
6. Marks, T. J. *Angew. Chem. Int. Ed. Engl.* **1990**, *29*, 857.
7. Hoffman, B. M.; Ibers, J. A. *Acc. Chem. Res.* **1983**, *16*, 15.
8. Casstevens, M.; Samok, M.; Pfleger, J.; Prasad, P. N. *J. Chem. Phys.* **1990**, *92*, 2019.
9. Simon, J.; Bassoul, P.; Norvez, S. *New J. Chem.* **1989**, *13*, 13.
10. Van der Pol, J. F.; Neeleman, E.; Zwikker, J. W.; Nolte, R. J. M.; Drenth, W.; Aerts, J.; Visser, R.; Picken, S. J. *Liq. Cryst.* **1989**, *6*, 577.
11. Simon, J. Sirlin, C. *Pure Appl. Chem.* **1989**, *61*, 1625.
12. Roberts, G. G.; Petty, M. C.; Baker, S.; Fowler, M. T.; Thomas, N. J. *Thin Solid Films* **1985**, *132*, 113.
13. Cook, M. J.; Dunn, A. J.; Daniel, F. M.; Hart, R. C. O.; Richardson, R. M.; Roser, S. J. *Thin Solid Films* **1988**, *159*, 395.
14. Palacin, S.; Lesieur, P.; Stefanelli, I.; Barraud, A. *Thin Solid Films* **1988**, *159*, 83.
15. Mohammad, M. A.; Ottenbreit, P.; Prass, W.; Schnurpfeil, G.; Wöhrle, D. *Thin Solid Films* **1992**, *213*, 285.
16. Kato, M.; Nishioka, Y.; Kaifu, K.; Kawamura, K.; Ohno, S. *Appl. Phys. Lett.* **1985**, *86*, 196.
17. Law, K. Y. *Chem. Rev.* **1993**, *93*, 449.
18. Darwent, J. R.; Douglas, P.; Harriman, A.; Porter, G.; Richoux, M. C. *Coord. Chem. Rev.* **1982**, *44*, 83.
19. Wöhrle, D.; Gitzel, J.; Krawczyk, G.; Tsuchida, E.; Ohno, H.; Nishisaka, T. *J. Macromol. Chem. A* **1988**, *25*, 1227.
20. Henderson, A.; Dougherty, T. J. *Photochem. Photobiol.* **1992**, *55*, 145.
21. Wöhrle, D.; Ardeschirpur, A.; Heuermann, A.; Müller, S.; Graschew, G.; Rinneberg, H.; Kohl, M.; Neukammer, J. *Makromol. Chem., Makromol. Symp.* **1992**, *59*, 17.
22. Klofta, T. J.; Danzinger, J.; Lee, P.; Pankow, J.; Nebesny, K. W.; Armstrong, N. R. *J. Phys. Chem.* **1987**, *91*, 5646.
23. Schlettwein, D.; Kaneko, M.; Yamada, A.; Wöhrle, D.; Jaeger, N. I. *J. Phys. Chem.* **1991**, *95*, 1748.
24. Simon, J. J.; André, H. J. *Molecular Semiconductors*; Springer Verlag: Berlin, Germany, 1985.
25. Wöhrle, D.; Meissner, D. *Adv. Mater.* **1991**, *3*, 129.
26. Loutfy, R. O.; Hsiao, C. K.; Hor, A. M.; DiPaola-Baranyl, G. J. *J. Imaging Sci.* **1985**, *29*, 148.
27. Takano, S.; Enokida, T.; Kabata, A. *Chem. Lett.* **1984**, 2037.
28. Deger, S.; Hanack, M. *Synth. Met.* **1986**, *13*, 319.
29. Hanack, M.; Deger, S.; Keppeler, U.; Lange, A.; Leverenz, A.; Rein, M. *Synth. Met.* **1987**, *19*, 739.
30. Deger, S.; Hanack, M. *Isr. J. Chem.* **1986**, *27*, 347.
31. Hanack, M.; Renz, G.; Strähle, J.; Schmid, S. *Chem. Ber.* **1988**, *121*, 1479.
32. Hanack, M.; Hirsch, A.; Lange, A.; Rein, M.; Renz, G.; Vermehren, P. *J. Mater. Res.* **1991**, *6*, 385.
33. Hanack, M.; Renz, G.; Strähle, J.; Schmid, S. *J. Org. Chem.* **1991**, *56*, 3501.
34. Tomoda, H.; Saito, S.; Shiraishi, S. *Chem. Lett.* **1983**, 313.
35. Tomoda, H.; Saito, S.; Ogawa, S.; Shiraishi, S. *Chem. Lett.* **1980**, 1277.
36. Wöhrle, D.; Schnurpfeil, G.; Knothe, G. *Dyes and Pigments* **1992**, *18*, 91.
37. Simon, J.; Bassoul, P. In *Phthalocyanines, Properties and Applications*; Leznoff, C. C.; Lever, A. B. P., Eds.; VCH: New York, 1993; Vol. 2, p 223.
38. Hanack, M.; Osio-Barcina, J.; Witke, E.; Pohmer, J. *Synthesis* **1992**, 211.
39. Hanack, M.; Hirsch, A.; Lehmann, H. *Angew. Chem.* **1990**, *102*, 1499; *Angew. Chem. Int. Ed. Engl.* **1990**, *29*, 1467.
40. Hanack, M.; Beck, A.; Lehmann, H. *Synthesis* **1987**, 703.
41. Kumada, M.; Tamao, K.; Sunitani, K. *Org. Synth.* **1978**, *58*, 127.
42. Cuellar, E. A.; Marks, T. J. *Inorg. Chem.* **1981**, *20*, 3766.
43. Otha, K.; Jaquemin, L.; Sirlin, C.; Bosio, L.; Simon, J. *New J. Chem.* **1988**, *12*, 751.
44. Cuellar, E. A.; Marks, T. J. *Inorg. Chem.* **1981**, *20*, 3766.
45. Masurel, D.; Sirlin, C.; Simon, J. *New J. Chem.* **1987**, *11*, 455.
46. Sauer, T.; Wegner, G. *Mol. Cryst. Liq. Cryst.* **1988**, *162B*, 97.
47. Pawlowski, G.; Hanack, M. *Synthesis* **1980**, 287.
48. Witke, E., Ph.D. Dissertation, Universität Tübingen, 1993.
49. Schmid, G.; Witke, E.; Schlick, U.; Knecht, S.; Hanack, M. *J. Mater. Chem.* **1995**, *34*, 3621.
50. Metz, J.; Hanack, M. *J. Am. Chem. Soc.* **1983**, *105*, 828.
51. Schwartz, M.; Hatfield, W. E.; Joesten, M. D.; Hanack, M.; Datz, A. *Inorg. Chem.* **1985**, *24*, 4198.
52. Hanack, M.; Hedtmann-Rein, C.; Datz, A.; Keppeler, U.; Münz, X. *Synth. Met.* **1987**, *19*, 787.
53. Hedtmann-Rein, C.; Hanack, M.; Peters, K.; Peters, E.-M.; Schnering, H. G. V. *Inorg. Chem.* **1987**, *26*, 2649.

Stain Blocking Polymers

See: Fibers (Stainproofing)

STAR-BLOCK COPOLYMERS

Koji Ishizu
Department of Polymer Science
Tokyo Institute of Technology

Theoretical treatments of the microphase separation of block copolymers are classified into types, namely, strong segregation theories and weak segregation theories.[1–8] de la Cruz and Sanchez have calculated the phase stability criteria and static structure factors in the weak segregation regime for n-arm star diblock copolymers [$(AB)_n$ star].[9] According to their results, as the arm number (n) increases, the core of the $(AB)_n$ type star will naturally become richer in A monomers and the monomers deep in the core will be screened effectively from interacting with B monomers. While the order-disorder transitions (ODT) for $(AB)_n$ star copolymers have been studied in the weak segregation regime, the quantitative nature of predictions has not yet been examined in detail for neat copolymers.[10,11]

We studied the self-micellization of $(AB)_n$ star copolymers as a parameter of n from the transmission electron microscopy (TEM) observation of microphase-separated structure in the strong segregation regime. Moreover, we made clear the packing structure of $(AB)_n$ star copolymers in solution and in the bulk film from the small-angle X-ray scattering (SAXS) measurements.[12]

PREPARATION

Poly[styrene(S)-*b*-isoprene(I)] diblock anions were prepared by the usual sequential anionic addition using *n*-butyllithium (n-BuLi) as an initiator in benzene in a sealed glass apparatus under a pressure of 10^{-6} mmHg. Next, isoprene monomer was added into this solution with vapor.

$(AB)_n$ star copolymers were prepared by crosslinking poly(S-*b*-I) diblock anions with an addition of a small amount of DVB (ca. 55%, *m*-/*p*-isomer = 2) in benzene.

In general, the spherical microdomains of diblock copolymers can be packed into one of the three cubic forms: simple cubic, FCC, and body-centered cubic (BCC). According to Ohta and Kawasaki, a BCC arrangement is only slightly more favored than the FCC arrangement.[3] So, the packing structure of star copolymers is the most efficient way of packing spheres.

Such chain conformation of $(AB)_n$ star copolymers is very similar to that of star-shaped polymers. Daoud and Cotton were the first to study the conformation and dimensions of star polymers using "thermal blob" model.[13] According to theoretical results, the central cores of star polymers to not interpenetrate with each other even beyond the overlap concentration (C*).[14] So, stars with many arms (the critical number of arms is estimated to be of order 10^2) are expected to form a crystalline array near the C*. Then, it can be expected that $(AB)_n$ star copolymers also form the structural ordering near the C*.

It is concluded that highly armed $(AB)_n$ star copolymers are packed in the lattice of a BCC structure near the C* concentration. So, the transformation of a C* solution into a continuous film for $(AB)_n$ star copolymers is very interesting from the point of view of hierarchical structure transformation of cubic lattice.

REFERENCES

1. Helfand, E. *Macromolecules* **1975**, *8*, 552.
2. Helfand, E.; Wassermann, Z. R. *Macromolecules* **1976**, *9*, 879.
3. Ohta, T.; Kawasaki, K. *Macromolecules* **1986**, *19*, 2621.
4. Leibler, L. *Macromolecules* **1980**, *13*, 1602.
5. Fredrickson, G. H.; Helfand, E. *J. Chem. Phys.* **1987**, *87*, 697.
6. Fredrickson, G. H. *Macromolecules* **1987**, *20*, 2535.
7. Green, P. F.; Christensen, T. H.; Russell, T. P.; Jerome, R. *J. Chem. Phys.* **1990**, *92*, 1478.
8. Green, P. F.; Christensen, T. M.; Russell, T. P. *Macromolecules* **1991**, *24*, 252.
9. de la Cruz, M. O.; Sanchez, I. C. *Macromolecules* **1986**, *19*, 2501.
10. Hashimoto, T.; Ijichi, Y.; Fetters, L. J. *J. Chem. Phys.* **1988**, *89*, 2463.
11. Ijichi, Y.; Hashimoto, T.; Fetters, L. J. *Macromolecules* **1989**, *22*, 2817.
12. Ishizu, K.; Uchida, S. *Polymer* **1994**, *35*, 4712.
13. Daoud, M.; Cotton, J. P. *J. Phys. (Les Ulis, Fr.)* **1982**, *43*, 531.
14. Witten, T. A.; Pincus, P. A.; Cates, M. E. *Europhys. Lett.* **1986**, *2*, 137.

STAR-BRANCHED AND CROSSLINKED POLYMERS (Cyclotriphosphazene Cores)

Ji Young Chang* and Man Jung Han
Department of Applied Chemistry
Ajou University

Star-branched polymers[1] and crosslinked polymers[2–5] with cyclotriphosphazene cores have also been reported. These polymers are of interest due to their regular structures and properties conferred from the cyclotriphosphazene ring.

STAR-BRANCHED POLYMERS

Preparation of Functional Cyclotriphosphazenes

A multifunctional initiator, hexakis[*p*-(bromomethyl)-phenoxy]cyclotriphosphazene for star-branched polymers and a monofunctional initiator, $N_3P_3(OC_6H_5)_5(OC_6H_4\text{-}p\text{-}CH_2Br)$ for linear polymers were prepared.[2]

Polymerization

Polymerization of 2-methyl-2-oxazoline initiated by these compounds was carried out in the bulk state (**Scheme I**).

Characterization and Properties of the Polymers

Three polymers with DPs of 5.3, 13.4, and 23 were obtained by varying the feed ratios of monomer to initiator—i.e., 30, 80, and 130, respectively. On the basis of six-armed star-branched structures of the polymers, DPs of the polymers were found to be close to the feed ratios. The calculated molecular weight of the linear polymer from DP was almost the same as that of the star-branched polymer with a DP of 5.3. However, the number-average molecular weight measured by GPC was much higher than that of the star-branched polymer (8200 *vs.* 4100), indicating that the linear polymer had a larger radius of gyration in a solution than the star-branched polymer.

The polymers were soluble in water and other polar solvents such as acetone, chloroform, methanol, and *N,N*-dimethylformamide. The linear polymer had a bulky hydrophobic head and

*Author to whom correspondence should be addressed.

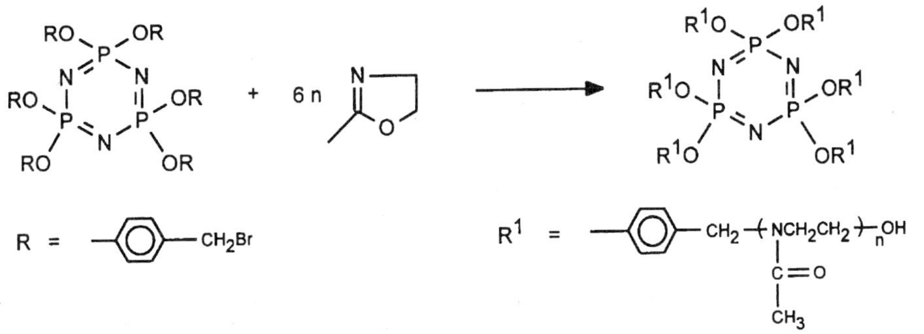

8

I

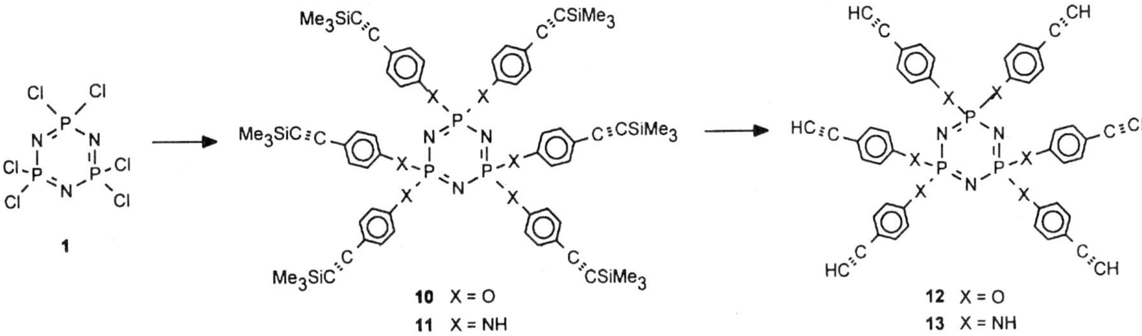

II

a hydrophilic chain. It exhibited nonionic surfactant behavior, forming micelles above a polymer concentration of about 0.1% by weight.

CROSSLINKED POLYMERS

Preparation of Thermally Curable Cyclotriphosphazenes

Hexakis(4-ethynylphenoxy)cyclotriphosphazene (**12**) (Chang, J. Y.; Ji, H. J.; Han, M. J., unpublished data) and hexakis(4-ethynylanilino)cyclotriphosphazene (**13**)[3] were prepared as described in **Scheme II**.

Polymerization and Properties

Thermal polymerization of compounds **12** and **13** was examined by differential scanning calorimetry (DSC). In the thermogram of compound **12**, a sharp endotherm for melting transition at 132°C and a strong exotherm between 180°C and 240°C due to inter- and intramolecular reaction of ethynyl groups were observed in the first scan. No significant endotherm or exotherm appeared in the second scan, which indicated that most ethynyl groups had reacted during the first scan.

Thermal polymerization of the ethynylanilino group-substituted cyclotriphosphazene (**13**) also took place in a similar way. However, no melting transition was observed before crosslinking. In the first scan, a strong exotherm was observed between 180°C and 280°C and no exotherm appeared in the second scan.

APPLICATION

Compounds **12** and **13** have potential uses as precursors of inflammable heat-resistant resins for composite matrices. Compound **12** especially is of interest since it has a wide "processing window" between a melting point and a cure onset temperature.

ACKNOWLEDGMENT

This work was supported by Korea Science and Engineering Foundation.

REFERENCES

1. Chang, J. Y.; Ji, H. J.; Han, M. J.; Rhee, S. B.; Cheong, S.; Yoon, M. *Macromolecules* **1994**, *27*, 1376.
2. Chang, J. Y.; Rhee, S. B.; Cheong, S.; Yoon, M. *Macromolecules* **1992**, *25*, 2666.
3. Kumar, D.; Fohlen, G. M.; Parker, J. A. *Macromolecules* **1983**, *16*, 1250.
4. Kumar, D.; Fohlen, G. M.; Parker, J. A. *J. Polym. Sci., Polym. Chem. Ed.* **1983**, *21*, 3155.
5. Kumar, D.; Gupta, A. D.; Khullar, M. *J. Poly. Sci. Polym. Chem. Ed.* **1993**, *31*, 707.

Star Polymers

See: *Branched Polymers*
Hyperbranched Polyester
Ionomers (Overview)
Living Anionic Polymerization
Macrocyclic Block Copolymers
Miktoarm Star Polymers
2-Oxazolines, Polymerization
Poly(ether amide) Triblock & Star Block Copolymers
Star-Block Copolymers
Star-Branched and Crosslinked Polymers (Cyclotriphosphazene Cores)
Star-Shaped Polymers
Stereoregular Polymethacrylates
Tailor-Made Polymers

Telechelic Oligomers (with Cyclic Onium Salt Groups)
Viscosity-Index Improves

STAR-SHAPED POLYMERS

Paul F. Rempp and Pierre J. Lutz
Institut Charles Sadron (CNRS/ULP)

Chain polymerization of vinylic monomers usually lead to linear macromolecules. However, if chain transfer to polymer occurs, as in free radical polymerization of ethylene or of vinyl acetate, branches are likely to be formed. The occurrence of branches has major consequences on the behavior of such polymers. It tends to impede or even prevent crystallization of polymers that can fit into a crystal lattice. Moreover, the mechanical behavior of amorphous polymers is affected by branching. Branched macromolecules are more compact than their linear homologues because their segment density is higher within the polymer coil. Consequently, the tendency of macromolecules to interpenetrate each other, in bulk as well as in solution, is greatly reduced.

It was, therefore, important to investigate systematically the effect of branching on the properties of polymeric species in solution as well as in bulk. Tailor-made branched polymers were needed for that purpose. Among them, star-shaped macromolecules[1–4] have been thoroughly investigated in a number of research groups. Star molecules are constituted of linear polymeric chains of similar size attached to one single multifunctional branch point, which can be polymeric itself. It is referred to as the core (or as the central nodulus) connecting the arms or branches.

In order to establish the influence of branching on polymer behavior, properties of a star polymer are to be compared with those of its linear homologue (of the same nature and molar mass), taking into account structural parameters such as number length of individual arms.

Because of their compactness, star-shaped molecules show a typical solution behavior. Owing to their high segment density, which implies a small hydrodynamic volume, they exhibit high elution volumes in size exclusion chromatography (GPC), low radii of gyration, low limiting viscosity numbers, and high translational diffusion coefficients as compared to corresponding figures relative to linear homologues.[5] Provided the number of branches is sufficiently high, the solution properties of star-shaped polymers are determined chiefly by the average length of their arms.[6]

Techniques best suited to synthesize star polymers are those involving active sites with long life-times, "living" polymerizations.[7,8] Most star polymers have been synthesized anionically, although recent attempts by cationic polymerization or by group transfer polymerization (GTP) have been successful.

Basically, several routes can be used to synthesize star polymers:[3,4] either living arms are made first, and used to generate the core; or the core is made first and subsequently serves as a plurifunctional initiator to grow the arms *from* the core; a three-step method, developed recently, combines the advantages of both procedures.

Beyond regular star polymers, various other kinds of star-shaped macromolecules have attracted interest and have lead to specific applications:

- Functional star polymers in which well-defined functions are located at the outer end of each branch;
- Star-block copolymers in which each branch is constituted of two homopolymeric blocks of different chemical nature, usually exhibiting different properties; and
- Heterostar copolymers in which the cores carry equal numbers of branches of two kinds, differing by chemical nature as well as average length.

REFERENCES

1. Bywater, S. *Adv. Polym. Sci.* **1979**, *30*, 89.
2. Martin, K. M.; Ward, T. C.; McGrath, J. E. *Anionic Polymerization* ACS Symposium Series, **1972**, *156*, 121.
3. Rein, D. H.; Rempp, P.; Lutz, P. *Makromol. Chem., Macromol. Symp.* **1993**, *67*, 237.
4. Rempp, P.; Lutz, P.; Franta, E. *J.M.S. Pure & Applied Chem.* **1994**, *A31*, 891.
5. Huber, K.; Bantle, S. et al. *Macromolecules* **1986**, *19*, 1404.
6. Orofino, T. A. *Polymer* **1961**, *2*, 295; **1962**, *2*, 305.
7. Szwarc, M. *Makromol. Chem. Rapid Commun.* **1992**, *13*, 141.
8. Rempp, P.; Lutz, P. *Makromol. Chem. Macromol. Symp.* **1993**, *67*, 1.

Starch

See: *Environmentally Degradable Polymers*
Polyesters (Derived from Renewable Resources)
Starch-Based Plastics (Measurement of Biodegradability)
Starch Biodegradation (in Starch-Plastic Blends)
Starch Graft Copolymers
Starch-Polymer Composites
Starch Polymer, Natural and Synthetic
Vegetal Biomass (3. Polymers, Derivatives and Composites)

STARCH-BASED PLASTICS (Measurement of Biodegradability)

Sherald H. Gordon,* S. H. Imam, and R. V. Greene
Biopolymer Research Unit
National Center for Agricultural Utilization Research
Agricultural Research Service
U.S. Department of Agriculture

Concern over the persistence and ecological effects of petrochemical-based plastic materials in the environment have led to research efforts by private, academic, and government laboratories to develop plastic-like polymers that degrade more rapidly. These biodegradable plastic formulations are intended for consumer products that require a finite lifetime. Examples include agricultural films used for weed control and water retention in crops, as well as packaging materials and

*Author to whom correspondence should be addressed.

other rigid containers that have limited durability and can degrade after use. Some types of plastics have been developed that decompose chemically, for example, after exposure to sunlight.[1]

An alternative approach taken recently at the U.S. Department of Agriculture (USDA) and other laboratories is development of plastics from polymeric materials occurring naturally or produced from renewable biomass resources, particularly agricultural feedstocks.[2,3] Starch, because of its abundance and low cost, is one of the most promising biopolymers for use in biodegradable plastics. Many plastic formulations containing corn starch and synthetic polymers have been prepared and studied extensively at this laboratory. Doane and co-workers[4] recognized the potential and advantages of using starch as a raw material in the 1970s when the need for replacing nonrenewable petroleum feedstocks was most urgent. Two comprehensive reviews of USDA research on starch-plastic materials are provided by Doane[5] and Swanson et al.[6]

TYPES OF STARCH-BASED PLASTIC MATERIALS

Research at this laboratory has been conducted under three basic approaches. In the first, starch was blended with synthetic polymers or with other biodegradable polymers and formed into films or injection-molded articles. Examples of these composite materials are blends of starch with polyethylene (PE) and poly(ethylene-co-acrylic acid) (EAA) containing up to 60% starch.[7,8] Another recent example is the combination of starch with other biodegradable polymers such as polyhydroxybutyrate-polyhydroxyvalerate (PHBV) to develop 100% biodegradable plastics.[9,10] PHBV, a polyester, is naturally produced by bacteria from agricultural raw materials.[11]

In the second approach, thermoplastic polymer chains were grafted onto starch, or starch was otherwise chemically modified, and the products were formed into films or injection-molded items. Examples of starch graft copolymers are starch-poly(methyl acrylate) or starch-polystyrene having a 50–60% grafted synthetic chains of about 500,000 molecular weight.[12] Another example is polyurethane plastic prepared by reacting polymeric isocyanates with starch-derived polyols.[13]

More recently, in an attractive third approach, starch itself has been proposed as a solitary thermoplastic material that can be cast into films or extruded with low molecular weight plasticizers. This results in low-cost articles with high-starch content (70–90%) and inherent biodegradability,[14] as well as good physical strength.[15]

QUANTITATIVE MEASUREMENT OF BIODEGRADATION

Although starch-based plastics are reportedly biodegradable, accurate quantitation of the extent and rate of their biodegradation in the environment is difficult.

A number of quantitative and semiquantitative techniques have been used successfully in USDA studies to evaluate two-component, starch-synthetic plastic blends. Among them are biochemical and chemical analyses[16–20] and fluorescence microscopy.[6,19] Cross-polarization/magic angle spinning ^{13}C solid-state NMR spectroscopy (CP/MAS ^{13}C-NMR) was used to estimate the extent of hydrolysis of starch in enzymatically degraded starch-PE-EAA injection-molded plastics.[20]

CONCLUSION

Despite great progress made recently, the field of biodegradable plastics is still in its infancy and, therefore, our present analytical tools must be considered rudimentary. Yet, some unexpected findings have already resulted, for example, discovery of V-type helical structures in starch that resist enzymatic degradation and the discovery that PHBV degradation is greatly accelerated in the presence of starch. It is expected that advance will be rapid in this area because new tools can be readily adapted and applied from other fields of research.

For researchers in the field, the future is exciting. Not only will they partake in the development of green technology to benefit humankind, but they also will enjoy the intellectual challenge of creating sophisticated methods and instrumentation to evaluate designer bioplastics for a cleaner, healthier, and improved environment.

REFERENCES

1. Johnson, R. *Proc. Symp. on Degradable Plastics*; Soc. Plastics Industry: Washington, DC, 1987.
2. Narayan, R.; ACS Symposium Series No. 575, American Chemical Society: Washington, 1994; Chapter 1.
3. Jane, J.; Lim, S. et al. ACS Symposium Series No. 575, American Chemical Society: Washington, 1994, Chapter 6.
4. Doane, W. M.; Swanson, C. L.; Fanta, G. F. *Emerging Technologies for Materials and Chemicals from Biomass*; Powell, R. M. et al., Eds.; American Chemical Society: 1992.
5. Doane, W. M. *Starch/Stärke, Washington, DC* **1992**, *44*, 293.
6. Swanson, C. L.; Shogren, R. L. *J. Environ. Polym. Degrad.* **1993**, *1*, 155.
7. Otey, F. H.; Westhoff, R. P.; Russell, C. R. *Ind. Eng. Chem. Prod. Res. Dev.* **1977**, *16*, 305.
8. Otey, F. H.; Westhoff, R. P.; Doane, W. M. *Ind. Eng. Chem. Prod. Res. Dev.* **1980**, *19*, 592.
9. Shogren, R. L. *J. Environ. Polym. Degrad.*, submitted, 1995.
10. Imam, S. H.; Gordon, S. H. et al. *J. Environ. Polym. Degrad.* **1995**, *3*, in press.
11. Luzier, W. D. *Proc. Natl. Acad. Sci.* **1992**, *89*, 839.
12. Bagley, E. B.; Fanta, G. F. et al. *Poly. Eng. Sci.* **1977**, *17*, 311.
13. Otey, F. H.; Westhoff, R. P. et al. *Ind. Eng. Chem. Prod. Res. Dev.* **1969**, *8*, 267.
14. Willett, J. L.; Jasberg, B. K.; Swanson, C. L. *ACS Symposium Series No. 575*; American Chemical Society: Washington, 1994; Chapter 3.
15. *Biodegradable Polymers and Plastics*; Van der Zee, M.; Vert, M. et al., Eds.; Royal Society of Chemistry: Cambridge, UK, 1992.
16. Imam, S. H.; Gould, J. M. et al. *Curr. Microbiol.* **1992**, *25*, 1.
17. Imam, S. H.; Burgess-Cassler, A. et al. *Curr. Microbiol.* **1991**, *22*, 365.
18. Shogren, R. L.; Thompson, A. R. et al. *J. Appl. Polym. Sci.* **1991**, *42*, 2279.
19. Shogren, R. L.; Thompson, A. R. et al. *J. Appl. Polym. Sci.* **1992**, *44*, 1971.
20. Imam, S. H.; Gordon, S. H. et al. *Biotechnol. Techniq.* **1993**, *7*, 791.

STARCH BIODEGRADATION (In Starch-Plastic Blends)

S. H. Imam,* Sherald H. Gordon, and R. V. Greene
Biopolymer Research Unit
National Center for Agricultural Utilization Research
Agricultural Research Service AFA
*U.S. Department of Agriculture***

K. A. Niño
Departamento de Microbiologia e Inmunologia
Facultad de Ciencias Biologicas
Universidad Autónoma de Neuvo León

Plastic formulations have been developed that incorporate starch and other biopolymers into environmentally degradable plastic products as a replacement for nonrenewable, petroleum-derived synthetic materials.[1–5] Efforts are ever intensifying to develop and commercialize environmentally responsible packaging and delivery materials from starch, as well as articles for single-use consumer applications. Research and development activities in this field are now global and have been reviewed recently.[1,6]

Currently, a number of methods are used to incorporate starch into plastics. Particularly, extrusion and injection-molding methods have been utilized to blend starch with a variety of polymers to produce thermoplastic products.[4,7] Blending of starch with other materials and its further processing exert considerable influence on the physical and mechanical properties of the starch-plastic product, thus, affecting both the rate and extent of starch degradation in such blends.

EXTRUDED PLASTIC FILMS

Starch degradation in extruded plastic-films containing up to 40% cornstarch (dry weight basis) in combination with polyethylene (PE) and poly(ethylene-co-acrylic acid) (EAA) has been extensively evaluated. These films prepared by the semidry method of Otey represent a plastic matrix composed of a hydrophobic synthetic polymer (PE), highly resistant to microbial degradation, and naturally biodegradable hydrophobic starch.[4] EAA, which is incorporated at relatively low levels (ca. 15–25%), has both hydrophobic domains and hydrophilic and makes the other two polymers compatible; hence, the term "compatabilization agent."

INJECTION-MOLDED PLASTIC COMPOSITES

Recent studies were conducted pertaining to the biodegradation of starch-PE-EAA composites prepared by injection molding methods.[8]

CONCLUSION

In practical terms, biodegradation can be defined as the process in which macromolecular material is either disintegrated or reduced to smaller molecular size components by living organisms or their hydrolytic enzymes, such that the principal end products are mainly water and carbon dioxide. In the case of the starch-PE-EAA plastic material, starch has been shown to be a biodegradable component of the plastic matrix that is readily susceptible to microbial breakdown under a variety of conditions. The synthetic polymers, PE and EAA in the matrix, however, remain nonbiodegradable. Despite a report describing the production of a polyethylene-degrading enzyme by a *Streptomyces* species, most of the literature indicates that synthetic polymers do not undergo biodegradation.[9] Depletion of starch may produce a considerably loose and disintegrated matrix and the presence of certain catalysts can initiate chemical oxidation, rendering the synthetic polymers in the matrix sufficiently unstable to undergo further degradation.[10] Peanasky et al. studied biodegradation of granual starch-polyethylene composites and found little degradation below the percolation threshold of about 40% starch.[11] A more recent study reports that in starch-polyethylene blends degradation varies considerably depending on environmental factors.[12] Particularly, in the landfill environment, degradation of such blends containing less than 30% starch is quite slow. It has also been suggested that microbial degradation of starch-polymers embedded in a plastic matrix is limited by the low rates of diffusion of oxygen, nutrients, and hydrolytic enzymes into the blends, and of breakdown products released from them.[13]

Data indicate that the degradation of starch-PE-EAA plastics in a given environment can occur by one or a combination of mechanisms such as biodegradation resulting directly from ubiquitous enzymatic and microbial activities, enzymatic or microbial activities associated with biofilms formed on the exposed surfaces, bioinjestion or bioconsumption of the material by small organisms such as insects, rodents, fish, etc., and possibly by the mechanical disintegration resulting with disassociation of starch, PE, and EAA components of the plastic matrix.

*Author to whom correspondence should be addressed.
**The mention of firm names or trade products does not imply that they are endorsed or recommended by the U.S. Department of Agriculture over other firms or similar products not mentioned. All programs and services of the U.S. Department of Agriculture are offered on a nondiscriminatory basis without regard to race, color, national origin, religion, sex, age, marital status, or handicap.

REFERENCES

1. Vert, M.; Feijen, J.; Albertsson, A.; Scott, G.; Chiellini, E. *Biodegradable Polymers and Plastics*; Royal Society of Chemistry: Cambridge, England, 1992.
2. Griffin, H. L. U.S. Patent 4 016 117, 1977.
3. Otey, F. H.; Westhoff, R. P.; Russell, C. R. *Ind. Eng. Chem. Dev. Prod. Res.* **1977**, *16*, 305.
4. Otey, F. H.; Westhoff, R. P. U.S. Patent 4 337 181, 1982.
5. Jane, J.; Lim, S.; Paetau, I.; Spence, K.; Wang, S. *Polymers from Agricultural Coproducts*; ACS Symposium Series No. 575, American Chemical Society: Washington, DC, 1994; Chapter VI.
6. Swanson, C. L.; Shogren, R. L.; Fanta, G. F.; Imam, S. H. *J. Environ. Polym. Degrad.* **1993**, *1*, 155.
7. Jasberg, B. K.; Swanson, C. L.; Shogren, R. L.; Doane, W. M. *J. Polym. Mat.* **1992**, *9*, 613.
8. Imam, S. H.; Gordon, S. H.; Burgess-Cassler, A.; Greene, R. V. *J. Environ. Polym. Degrad.* 1995, (accepted).
9. Pometto, A. L.; Lee, B.; Johnson, K. E. *Appl. Environ. Microbiol.* **1992**, *58*, 731.
10. Sung, W.; Nikolov, Z. L. *Ind. Eng. Chem. Res.* **1992**, *31*, 2332.
11. Peanasky, J. S.; Long, J. M.; Wool, R. P. *J. Polym. Sci. Part B. Polym. Phys. Ed.* **1991**, *29*, 565.

12. U.S. Congress, Office of Technology Assessment, *Biopolymer: Making Materials Nature's Way*; U.S. Government Printing Office: Washington, DC, 1993.
13. Cole, M. A. *Agricultural and Synthetic Polymers: Biodegradability and Utilization*; ACS Symposium Series No. 433, Chapter VIII, American Chemical Society: Washington, DC, 1990.

STARCH GRAFT COPOLYMERS

George F. Fanta
Plant Polymer Research
*National Center for Agricultural Utilization Research USDA**
Agricultural Research Service

Starch is a polysaccharide composed of repeating 1,4-α-D-glucopyranosyl units (often called anhydroglucose units, or AGU) and is one of the cheapest and most readily available of all natural polymers.

Use of starch as an extender and replacement for synthetic polymers is currently an active research area. Incorporation of starch into other polymers not only reduces our dependence on petrochemical-derived monomers but also provides materials in which the starch portion can biodegrade rapidly in the environment. Free radical graft polymerization is an excellent method for preparing these starch-synthetic polymer composites. Earlier reviews on starch graft copolymers have been published.[1–5]

PREPARATION

Graft copolymers are prepared by first generating free radicals on the polysaccharide backbone and then allowing these free radicals to serve as macroinitiators for the vinyl or acrylic monomer. Perhaps the most commonly used method of chemical initiation is the reaction of starch with ceric salts, such as ceric ammonium nitrate. This method, first published by Mino and Kaizerman, has been used with many different monomer systems.[6]

Other chemical initiating systems have also been used, although with less frequency. Some examples of these are trivalent manganese, cupric ion, and the formation of thiocarbonate substituents on the starch backbone followed by reaction with hydrogen peroxide.[9–12]

Acrylonitrile

Ceric-initiated graft polymerization of acrylonitrile onto starch occurs readily in water at room temperature with high grafting efficiency, i.e., with little formation of homopolymer. The conversion of monomer to polymer is nearly quantitative, and graft copolymers typically contain 50–60% grafted polyacrylonitrile (PAN).

Hydrolysis of starch-*g*-PAN with aqueous sodium (or potassium) hydroxide converts the nitrile substituents of PAN to mixtures of carboxamide and alkali metal carboxylate and yields a highly viscous graft copolymer dispersion.[13–14] Carboxamide to carboxylate ratios vary with saponification conditions, but are typically on the order of 1:2. Complete saponification to carboxylate does not occur.

Monomers Yielding Water Soluble Polymer Grafts

Graft polymerizations of ionic and/or water soluble monomers, such as acrylamide or acrylic acid, are efficiently initiated with electron beam irradiation or with gamma irradiation from cobalt-60.

Reyes et al. produced starch-*g*-polyacrylamide and starch-*g*-poly(acrylic acid) on a large scale by addition of electron beam preirradiated starch to solutions of monomer in either water or aqueous organic solvent sytems.[15]

A number of amine containing monomers have been graft polymerized onto starch, generally in the form of their quaternary ammonium or mineral acid salts.

Nitric acid salts of N,N-dimethylaminoethyl methacrylate [$CH_2{=}C(CH_3)CO_2CH_2CH_2N(CH_3)_2.HNO_3$; DMAEM.AHNO$_3$] and N-t-butylaminoethyl methacrylate [$CH_2{=}C(CH_3)CO_2CH_2CH_2NC(CH_3)_3.HNO_3$; TBAEMA.HNO$_3$] were graft polymerized onto starch with Fe^{+2}/hydrogen peroxide initiation to give products containing about 20% grafted polymer.[16]

Monomers Yielding Thermoplastic Polymers Grafts

Graft polymerizations of methyl methacrylate have been efficiently carried out with a number of chemical initiators such as ceric salts and the Fe^{+2}/hydrogen peroxide redox system. Early studies with this monomer were carried out by Brockway.[7,8]

Methyl acrylate also graft polymerizes readily to starch with ceric ammonium nitrate initiation, and graft copolymers with poly(methyl acrylate) (PMA) contents ranging from about 40 to 70% have been prepared.[17–19]

Unlike the acrylate and methacrylate esters, styrene graft polymerizes poorly in the presence of ceric salts. Graft copolymers have been prepared, however, when an active monomer such as acrylonitrile was included with styrene in the monomer system.[20]

Simultaneous cobalt-60 irradiation is a convenient technique for preparing starch-*g*-polystyrene. Polymerizations have been carried out by irradiating semisolid mixtures prepared from starch, styrene, water, and an organic solvent such as ethylene glycol, acetonitrile, ethanol, methanol, acetone, or dimethyl formamide.[21]

PROPERTIES

Copolymers with Hydrophilic Grafts

Graft copolymers with hydrophilic grafts, such as polyacrylamide, poly(acrylic acid), and the various cationic polymers derived from DMAEMA, swell in water at room temperature. When aqueous suspensions are heated, graft copolymers either dissolve or disperse to give smooth pastes.

Graft copolymers resulting from alkaline hydrolysis of starch-*g*-PAN have properties that are quite different from those of copolymers prepared by direct grafting of water soluble monomers. Taylor and Bagley observed that hydrolyzed starch-*g*-PAN (HSPAN) is in the form of a highly swollen but insoluble gel after reaction with alkali; and these authors have concluded that the thickening action of HSPAN is due to nearly complete absorption of interstitial water to give a system consisting of

*Names are necessary to report factually on available data; however, the USDA neither guarantees nor warrants the standard of the product, and the use of the name by USDA implies no approval of the product to the exclusion of others that may also be suitable.

highly swollen, deformable gel particles that are closely packed and in intimate contact.[22]

When isolated from water and dried, micron-sized particles of HSPAN gel agglomerate into macroparticles that do not revert back to the original smooth gel when placed back into water, but instead swell and absorb many times their weight of aqueous fluid.

If water dispersions of HSPAN are spread onto a tray and allowed to air-dry, continuous films are obtained. When placed in water, these films swell as a single entity and produce a continuous sheet of highly swollen gel having sufficient integrity to allow it to be manipulated without breaking.[24]

Modifications of the basic HSPAN technology have been published. In one such method, flour was simply substituted for starch in the graft polymerization reaction.[25] In another method, up to 10 mole percent of the acrylonitrile was replaced with a comonomer such as 2-acrylamido-2-methylpropanesulfonic acid.[23,25,26] These modifications not only yielded products with higher water absorbencies, but the use of comonomers reduced the reaction time needed for nitrile hydrolysis. In addition to the HSPAN family of products, starch graft copolymer absorbents have also been prepared by grafting water soluble monomers such as acrylic acid and sodium acrylate onto starch in the presence of a crosslinking agent, such as N,N′-methylenebisacrylamide.[27]

COPOLYMERS WITH HYDROPHOBIC GRAFTS

Graft copolymers with hydrophobic grafts do not swell appreciably in water, but remain as grainy solids, even after prolonged heating. Despite the insolubility of these polymers, Gugliemelli and co-workers were able to prepare graft copolymers in latex form by using a gelatinized cationic starch as the substance for grafting and then treating the resulting polymer dispersions with ultrasound.[28,29,30] Continuous films were obtained from these latex-like materials by air drying.

If starch-thermoplastic graft copolymers containing about 50% grafted polymer, by weight, are processed by extrusion, continuous plastics are formed, even though the starch portion of the polymer has not been melted.[17]

Although extruded plastics prepared from starch-*g*-polystyrene are extremely brittle, extrudates obtained from starch-*g*-PMA are tough and flexible owing to the low glass transition temperature (8°C) of PMA grafts.[17] Because of their excellent physical properties, the processing of these graft copolymers has been extensively studied.

APPLICATIONS

HSPAN

The absorbent properties of HSPAN have suggested numerous applications for these graft copolymers. In the area of personal care, HSPAN polymers have been tested in disposable soft goods such as diapers, adult incontinent pads, hospital underpads, and feminine napkins to increase absorbence of body fluids and to decrease the bulk of these items. HSPAN has also been sold as an absorbent for addition to ostomy and urinal bags to prevent leaks and odors. Use of HSPAN as a wound dressing has been patented,[31] and results of successful animal trials have been published.[32,33] The polymer has also been used as a perspiration absorbent in body powder formulations.[34]

Agricultural applications for HSPAN have also been successful. As a seed coating, the polymer absorbs water and holds it at the seed surface, thus increasing both rate of germination and the percentage of planted seeds that germinate.[35–37] HSPAN has also been applied as a gel slurry to the root zone of plants before transplanting to reduce transplant shock and to improve plant survival.[38] HSPAN has been added to soils to increase their water holding capacity[39–40] and has been used as a tackifier component in hydroseeding formulations.[41]

Other applications for HSPAN have also been found, one of which is the removal of water from organic solvents and fuels.[42] Disposable fuel filters containing HSPAN are being manufactured, and another company is marketing porous, HSPAN-containing bags that can be lowered to the bottom of fuel storage tanks to absorb condensed water. Finally, HSPAN has been used as a thickener for fire-fighting fluids and as an agar substitute.[43,44]

Copolymers with Hydrophilic Grafts

Hydrophilic graft copolymers prepared from either cationic or anionic monomer systems are easily dispersed in water and function as flocculants for suspended solids.

Graft copolymers prepared from cationic amine-containing monomers also function as retention aids in the production of mineral-filled paper, and their performance has been evaluated on a 32 inch pilot Fourdrinier machine.

Hydrophilic starch graft copolymers also have potential applications as thickeners.

Copolymers with Hydrophobic Grafts

Extrusion processing of starch-*g*-PMA yields tough, leathery plastics with tensile properties suitable for many commercial applications.

REFERENCES

1. Fanta, G. F. Block and Graft Copolymerization; Ceresa, R. J., Ed.; John Wiley & Sons: London, 1973; Vol. 1, Chapter 1 and 2.
2. Fanta, G. F.; Bagley, E. B. *Encycl. Polym. Sci. Technol., Starch, graft copolymers* **1977**, Suppl. *2*, 665.
3. Stannett, V. T.; Doane, W. M.; Fanta, G. F. *Absorbency*; Chatterjee, P. K., Ed.; Elsevier: Amsterdam, 1985, Chapter VIII.
4. Fanta, G. F.; Doane, W. M. *Modified Starches; Properties and Uses*; Wurzburg, O. B., Ed.; CRC: Boca Raton, FL, 1986; Chapter 10.
5. Doane, W. M.; Swanson, C. L.; Fanta, G. F. *Emerging Technologies for Materials and Chemical from Biomass*; Rowell, R. M.; Schultz, T. P.; Narayan, R., Eds.; ACS Symposium Series No. 476, 1992; Chapter 13.
6. Mino, G.; Kaizerman, S. *J. Polym. Sci., A new method for the preparation of graft copolymers. Polymerization initiated by ceric ion redox systems* **1958**, *12*, 242.
7. Brockway, C. E.; Moser, K. B. *J. Polym. Sci. Part A, Grafting of poly(methyl methacrylate) to granular corn starch* **1963**, *1*, 1025.
8. Brockway, C. E. *J. Polym. Sci. Part A. efficiency and frequency of grafting of methyl methacrylate to granular corn starch* **1964**, 2, 3721.
9. Mehrotra, R.; Råby, B. *J. Appl. Polym. Sci., Graft polymerization onto starch. I. Complexes of Mn^{3+} as initiators* **1977**, *21*, 1647.

10. Doba, T.; Rodehed, C.; Rånby, B. *Macromolecules, Mechanism of graft copolymerization onto polysaccharides initiated by metal ion oxidation reactions of model compounds for starch and cellulose* **1984**, *17*, 2512.
11. Imoto, M.; Morita, E.; Ouchi, T. *J. Polym. Sci. Polym. Symp., Vinyl polymerization CCCLXXVIII. Radical polymerization of methyl methacrylate with starch in aqueous solution of Cu (II) ion* **1980**, *68*, 1.
12. Faessinger, R. W.; Conte, J. S. U.S. Patent 3 359 224, 1967.
13. Weaver, M. O.; Gugliemelli, L. A.; Doane, W. M.; Russell, C. R. *J. Appl. Polym. Sci., Hydrolyzed starch-polyacrylonitrile graft copolymers: Effect of structure on properties* **1971**, *15*, 3015.
14. Gugliemelli, L. A.; Weaver, M. O.; Russell, C. R.; Rist, C. E. *J. Appl. Polym. Sci., Base-hydrolyzed starch-polyacrylonitrile (S-PAN) graft copolymer. S-PAN-1:1, PAN M.W. 794,000* **1969**, *13*, 2007.
15. Reyes, Z.; Clark, C. F.; Comas, M.; Russell, C. R.; Rist, C. E. *Nucl. Appl., Continuous production of graft copolymers of starch with acrylamide and acrylic acid by electron preirradiation* **1969**, *6*, 509.
16. Fanta, G. F.; Burr, R. C.; Doane, W. M.; Russell, C. R. *J. Appl. Polym. Sci., Influence of starch granule swelling on graft copolymer composition. A comparison of monomers* **1971**, *15*, 2651.
17. Bagley, E. B.; Fanta, G. F.; Burr, R. C.; Doane, W. M.; Russell, C. R. *Polym. Eng. Sci., Graft copolymers of polysaccharides with thermoplastic polymers. A new type of filled plastic* **1977**, *17*, 311.
18. Dennenberg, R. J.; Bothast, R. J.; Abbott, T. P. *J. Appl. Polym. Sci., A new biodegradable plastic made from starch graft poly(methyl acrylate) copolymer* **1978**, *22*, 459.
19. Swanson, C. L.; Fanta, G. F.; Fecht, R. G.; Burr, R. C. *Polymer Applications of Renewable Resource Materials*; Carraher, C. E., Jr.; Sperling, L. H., Eds.; Plenum: New York, 1983, p 59.
20. Gugliemelli, L. A.; Swanson, C. L.; Doane, W. M.; Russell, C. R. *J. Polym. Sci. Polym. Lett. Ed., Preparation of starch poly(styrene-co-acrylonitrile) graft polymers by cerium (IV) initiation* **1977**, *15*, 739.
21. Fanta, G. F.; Burr, R. C.; Doane, W. M.; Russell, C. R. *J. Polym. Sci. Appl., Graft polymerization of styrene onto starch by simultaneous cobalt-60 irradiation* **1977**, *21*, 425.
22. Taylor, N. W.; Bagley, E. B. *J. Appl. Polym. Sci., Dispersions or solutions? A mechanism for certain thickening agents* **1974**, *18*, 2747.
23. Fanta, G. F.; Doane W. M. *Agricultural and Synthetic Polymers, Biodegradability and Utilization*; Glass, J. E.; Swift, G., Eds.; ACS Symposium Series 433, ACS: Washington, DC, 1990; Chapter 24.
24. Weaver, M. O.; Bagley, E. B.; Fanta, G. F.; Doane, W. M. *Appl. Polym. Symp., Gel sheets produced by hydration of films from the potassium salt of hydrolyzed starch-polyacrylonitrile graft copolymer* **1974**, *25*, 97.
25. Fanta, G. F.; Burr, R. C.; Doane, W. M.; Russell, C. R. *Die Stärke, Absorbent polymers from starch and flour through graft polymerization of acrylonitrile and comonomer mixtures* **1978**, *30*, 237.
26. Fanta, G. F.; Burr, R. C.; Doane, W. M. *Starch/Stärke, Absorbents from saponified starch-g-poly(acrylonitrile-co-2-acrylamide-2-methylpropanesulfonic acid). Some practical considerations* **1987**, *39*, 322.
27. Masuda, F.; Nishida, K.; Nakamura, A. U.S. Patent 4 076 663, 1978.
28. Gugliemelli, L. A.; Swanson, C. L.; Baker, F. L.; Doane, W. M.; Russell, C. R. *J. Polym. Sci. Polym. Chem. Ed., Cationic starch-polyacrylonitrile graft copolymer latexes* **1974**, *12*, 2683.
29. Gugliemelli, L. A.; Swanson, C. L.; Doane, W. M.; Russell, C. R. *J. Appl. Polym. Sci., Latexes of starch-based graft polymers containing polymerized acrylonitrile* **1976**, *20*, 3175.
30. Gugliemelli, L. A.; Swanson, C. L.; Doane, W. M.; Russell, C. R. *J. Polym. Sci. Polym. Lett. Ed., Cationic starch graft-polychloroprene latexes* **1976**, *14*, 215.
31. Spence, W. R. U.S. Patent 4 226 232, 1980.
32. Veldez, H. *Equine Pract., A hydrogel preparation for cleansing and protecting equine wounds* **1980**, *2*, 33.
33. Geronemus, R. G.; Robins, P. *J. Dermatol. Surg. Oncol., The effect of two new dressings on epidermal wound healings* **1982**, *8*, 850.
34. Spence, W. R. U.S. Patent 4 272 514, 1981.
35. Deterling, D. *Progressive Former, "Super Slurper" gets your crop moving earlier* **1981**, *96*(2), 56K.
36. Tanzy, K. *Soybean Dig., Is there a Super Slurper in your future* **1981**, *41*(1), 306.
37. Kirkland, E. R.; Palanuik, E.; Ingman, H. V. U.S. Patent 4 495 724, 1985.
38. Hamilton, J. L.; Lowe, R. H. *Tobacco Sci., Use of a water absorbent polymer in tobacco seedling production and transplanting* **1982**, *26*, 17.
39. Shrader, W. D.; Mustejeran, A. *Coatings Plastics Prepr. ACS Div. Org. Coatings and Plastics Chem., Potentials of a polyacrylonitrile-starch polymer, Super Slurper, to modify water holding properties of soils* **1977**, *37*(1), 683.
40. Miller, D. E. *Soil Sci. Soc. Am. J., Effect of H-SPAN on water retained by soils after irrigation* **1979**, *43*, 628.
41. Kay, B. L. *Weeds Trees Turf, Mulch choices for erosion control and plant establishment* **1980**, *19*(8), 16.
42. Fanta, G. F.; Burr, R. C.; Orton, W. M.; Doane, W. L. *Science, Liquid phase dehydration of aqueous ethanol-gasoline mixtures* **1980**, *210*, 646.
43. Cooke, R. C. *Hortic. Sci., The use of an agar substitute in the initial growth of Boston ferns in vitro* **1977**, *12*, 339.
44. Leppla, N. C. *J. Ga. Entomol. Soc., Starch compounds as adjunct gelling agents in larval diet of the cabbage looper* **1976**, *11*, 251.

STARCH-POLYMER COMPOSITES

G. J. L. Griffin
Ecological Materials Research Institute
EPRON Industries Ltd.

In the past 20 years, commencing with the work at Brunel University in the UK[1] and at Peoria in the United States,[2] the growing demand from environmental campaign groups and green political activists has led to an evolution of a range of degradable plastics, several of which are based on starch composites. Degradable composites derived from poly(vinyl alcohol) (CA 9002-89-5) also have benefited from inclusion of starch as a filler to reduce the price and create a strong film suitable for use as water soluble laundry bags for hospitals.[2]

The U.S. Department of Agriculture (USDA) at Peoria biodegradable plastic project approached its target by adopting starch as the matrix polymer, gelatinizing it by the action of heat and water and then seeking synthetic polymeric additives which would make the mix processable on familiar plastics machinery.

The work at Brunel University in the UK was based on the complementary technique of using polyolefin polymers as the continuous phase with particulate starch additive as a filler. Extra additives ensured that the synthetic polymer phase oxidized on a controlled timetable and as a result the product was almost indistinguishable from the familiar polymer materials and could be processed on existing machinery. The patented[3] process reached the European market in 1974 in the form of LDPE shopping bags which are still in production.

The commercial history of both the Peoria and the Brunel work over the following years has been complex but these two types of products with their different properties and costs have found markets.

Growth of a field of scientific activity around degradable plastics has encouraged development of a wide range of new activities having no connection with the original target of dealing with waste plastics.

STARCH PROPERTIES

Size and Shape

Starch grains, as separated from the fiber and protein associated with them in living plant material, show size and shape characteristics that identify them with particular vegetable sources and major compilations of such data have been carried out.[5–6]

Thermal Properties

The early conviction that starch could not withstand high processing temperatures was challenged when it was shown at Brunel University that composites of starch and polypropylene containing as much as 30% ww of maize starch could be extruded with melt temperatures as high as 230°C. Success was attributable to the use of intensely dried starch (< 1% H_2O) and the fact that exposure time was limited to a few seconds transit time in screw extruders.

Optical Properties

Good quality commercial starches give the visual impression of pure white powders. Displacing air from starch powders by oils having matching refractive indices reveals that the grains are in fact transparent and the possibility of preparing transparent composites becomes interesting. Pilot plant trials at Brunel University did achieve clear films using plasticized PVC because the refractive index of the polymer matrix could be adjusted as desired by careful selection and blending of plasticizers.

Matrix/Particle Bonding

Examination of the earliest starch/polymer mixes made it clear that the degree of adhesion between polysaccharide surfaces and nonpolar olefin polymer matrix materials was modest. Using a combination of silicone treatment and small diameter starch particles such as rice or taro made clear the possibility of achieving starch-polymer composites in which starches could legitimately be described as "reinforcing fillers." Another observation that could prove important was the effect of starch particles on the stiffness of high density polyethylene.[7] This effect first showed up in a major study of the possible application of starch as a filler in compounds used for blowing bottles. The top-loading stress required to achieve a given vertical deformation is a standard test procedure for such bottles and the presence of starch loading up to about 15% showed a marked increase in the stiffness of the bottles, which implied that a decrease of wall thickness, and therefore weight, could be achieved.

Degradability

The general possibility of using composites based on mixtures of biodegradable filler phases in nondegradable matrix phases was put on record in early publications but without specific suggestions.[8] The fact that composites based on phenolic resins with 50% vv of woodflour filler (Bakelite) were very stable in outdoor environments was attributed to the nature of the resin, but there is evidence that filler particles capable of expanding when they absorb moisture can disrupt polymer matrices provided that the polymer used has a fairly low tensile strength and would be classified as a brittle material. This behavior has been exploited in development of shooting targets (clay pigeons or skeets) which will disintegrate into powder after exposure to weathering for a few weeks.[10]

Now that it is accepted that the addition of starch as a filler to plastics will enhance latent biodegradability in the matrix polymer, and given that the starch is generally less than half of the cost of the polymer, it is not surprising that many workers in the field of biodegradable plastics have sought to evaluate such blends. From the early work at Brunel University with polyester polyurethanes in the early 1970s, we now see polylactic acid, polyhydroxybutyrate and butyrate valerate materials, polycaprolactone, and poly(vinyl alcohol) all receiving the starch treatment.[4,9]

REFERENCES

1. Griffin, G. J. L. *Amer. Chem. Soc. Advances in Chem. Series* **1974**, *134*, 159–160.
2. Otey, F. H.; Mark, A. M. et al. *Ind. Eng. Chem. Prod. Res. & Dev.* **1974**, *13*, 90.
3. Griffin, G. J. L. U.K. Patent Applic'ns 23469 & 26259; U.K. Patents 1 485 033 & 1 487 050; U.S. Patents 4 021 388 & 4 016 117 etc.
4. Koenig, M. F.; Huang, S. J. *Proc. of Amer. Chem. Soc. Pol. Materials Div'n* **1992**, *67*, 290–2981.
5. Nägeli, W. *Die Starkekorner*; Schultess: Zürich, 1858.
6. Reichert, E. T. Carnegie Institute: Washington, **1913**, *173*.
7. Griffin, G. J. L.; Hashemi, S. A. *Iranian J. Pod. Sci.* 1992, I. 45–54.
8. Hueck, H. J. *Internat'l Biodteriorat'n Bulletin* **1974**, *10*(3), 87–90.
9. Griffin, G. J. L. U.S. Pat. 5, 460, 383.
10. Otey, F. H.; Westhoff, R. P.; Doane, W. M. *Ind. Eng. Chem. Prod. Res. and Dev.* **1980**, *19*, 592.

STARCH POLYMERS, NATURAL AND SYNTHETIC

Shinichi Kitamura
Department of Agricultural Chemistry
Kyoto Prefectural University

Starch consists of mainly two chemically distinguishable polysaccharides: amylose, an essentially linear (1→4)-α-D-glucan, and amylopectin, a branched glucan that contains largely α-(1→4) linkages, but with branches attached by α-(1→6) linkages.

When an aqueous suspension of starch granules is heated above a certain temperature, the granules swell and undergo gelatinization to form a viscous paste. Storing such a starch paste at low temperature results in precipitation at low concentration and gel formation at high concentration. These phenomena are often referred to as retrogradation. Gelatinization and retrogradation of starch are of importance in many industrial applications, especially in the food industry. Several monographs have appeared that review historical and current aspects of starch science and technology.[1–5]

Amylose fractions have been widely studied, largely because of their interesting solution properties, which include complex formation with iodine, butanol, and other organic reagents, as well as a tendency for molecular association (retrogradation).[6] Compared with amylose, solution properties of amylopectin fractions have been investigated to a considerably lesser extent, because of its ambiguous primary structure and very large molecular weight polydispersity.[7,8]

This article focuses on the solution properties of amylose and amylopectin and their characteristic properties, especially aggregation and gelling properties.

DILUTE SOLUTION PROPERTIES

Amylose

Many hydrodynamic studies of amylose solutions have been reported, but the results are not completely consistent.

Everett and Foster, and Banks and Greenwood proposed that amylose adopts a random coil conformation, while Cowie et al. have proposed a predominantly helical, semirigid chain.[10,12–14] It is difficult to present a rational explanation for these divergent conclusions. However, data on the molecular weight dependence, based on the z-average *mean-square* radius of gyration, $<S^2>_z$, for amyloses in DMSO suggest the presence of aggregation and polydispersity in the system for natural amyloses. It was concluded that without excluded-volume effects, the overall conformation of high molecular weight amylose in water and DMSO are similar and that it is a random coil.

Most recently we have shown that the experimental $<S^2>_z$ for synthetic amylose in DMSO can be stimulated by amylose model chains with excluded-volume chain sequence which have been generated by a Monte Carlo method using a conformational energy map of the dimer segment, that is, α-maltose.[11,15]

Amylopectin

Although amylopectin is the major component of starch, far fewer studies of solution properties have been published for this molecule.[7,8] Amylopectin molecules vary with respect to molecular weight, the number of branches per molecule, the length of branches, and the pattern of branches. These wide varieties make it difficult to prepare well-characterized samples.

In contrast to natural amylopectin samples, synthetic amylopectins enable us to systematically investigate the effects of molecular weight and the number of branches per molecules on solution properties.

CHARACTERISTIC PROPERTIES

Aggregation (Precipitation)

Aqueous amylose solutions of low concentration become increasingly cloudy owing to the progressive association of the polymer molecules into larger particles. The rate of aggregation is affected by many factors, including temperature, pH, additives, polymer concentration, and the molecular weight.[16]

The association rates of amyloses show a sharp maximum at a DP of 80, with shorter and longer molecules being much more soluble. Actually, synthetic amyloses with $M_w > 5 \times 10^5$ can be soluble in water and fairly stable for a long time.

Amylose retrogradation at low concentration appears to proceed through a two-step process; the first step consisting of formation of stable nuclei in the solution, and the second constituting further growth of the nuclei by the addition of amylose molecules to their surfaces.

Gelation

At high concentrations amylose solutions form a gel and the aging (retrogradation) of the gel then proceeds. Recently Gidlay and Bulpin reported a phase diagram for a water-amylose system as a function of concentration and DP.[9] For chains of DP > 250 gelation occurs at > 1.0% concentration. The gelling mechanism of amylose may be more complicated than precipitation from aqueous dilute solution because of extensive nucleation and the formation of very small particles. Because they are in close proximity, the amylose particles tends to aggregate and the dispersion forms a three-dimensional network throughout the solution.

Recently the mechanism of gelling and retrogradation of amylose and starch have been extensively studied on a molecular basis using both small and wide angle X-ray diffraction, NMR, differential scanning calorimetry, Raman spectroscopy, fluorescence probe techniques, and electron microscope observations.[9,17–24]

It has been suggested that starch gels are composites, containing gelatinized granules embedded in an amylose matrix.[18] In the initial stage, irreversible gelation and crystallization occurs within the amylose matrix. This is followed by a reversible crystallization involving amylopectin within a granule on storage. Long-term increases in the modulus of starch gels parallel the change in crystallinity. This is consistent with partial crystallization within the granules that would result in an increase in rigidity of the granules, thus enhancing their reinforcement of the amylose matrix.

REFERENCES

1. Banks, W.; Greenwood, C. T. *Starch and its Components*; Edinburgh University: Edinburgh, 1975.
2. Nakamura, M. *Denpun Kagaku (Starch science) Handbook*: Suzuki, S., Ed.; Azakura: Tokyo, 1977.
3. Whistler, R. L.; BeMiller, J. N. *Starch*; Paschall, E. F., Ed.; Academic: London, 1984.
4. *Starch: Properties and Potential*; Galliard, T., Ed.; John Wiley & Sons: Chichester, 1987.
5. *Developments in Carbohydrate Chemistry*; Alexander, J.; Zobel, H. F., Eds.; American Association of Cereal Chemists: Minnesota, 1992.
6. Kitamura, S.; Kuge, T. *Food Hydrocolloids* **1989**, *3*, 313.
7. Banks, W.; Gedgges, R.; Greenwood, C. T.; Jones, I. G. *Stärke/Starch* **1972**, *24*, 245.
8. Lelievre, J.; Lewis, J. A.; Marsden, K. *Carbohydr. Res.* **1986**, *153*, 195.
9. Gidley, M. J.; Bulpin, P. V. *Macromolecules* **1989**, *22*, 341.
10. Everett, W. W.; Foster, J. F. *J. Am. Chem. Soc.* **1959**, *81*, 3459.
11. Kitamura, S.; Okamoto, T.; Nakata, Y.; Hayashi, T.; Kuge, T. *Biopolymers* **1987**, *26*, 537.
12. Banks, W.; Greenwood, C. T. *Carbohydr. Res.* **1968**, *7*, 349, 414.
13. Cowie, J. M. G. *Makromol. Chem.* **1961**, *42*, 230.
14. Fujii, M.; Honda, K.; Fujita, H. *Biopolymers* **1973**, *12*, 1177.
15. Nakata, Y.; Kitamura, S.; Takeo, K.; Norisuye, T. *Polym. J.* **1994**, *26*, 1085.
16. Collison, R. In *Starch and its Derivatives*; Radley, J. A., Eds.; Chapman and Hall: London, 1968; Chapter 6.
17. Miles, M. J.; Morris, V. J.; Ring, S. G. *Carbohydr. Res.* **1985a**, *135*, 257.

18. Miles, M. J.; Morris, V. J.; Orford, P. D.; Ring, S. G. *Carbohydr. Res.* **1985b**, *135*, 271.
19. I'Anson, K. J.; Miles, M. J.; Morris, V. J.; Ring, S. G. *Carbohydr. Res.* **1988**, *8*, 45.
20. Müller, J. J.; Gernat, C.; Schulz, W.; Müller, E. Vorwerg, W.; Damaschun, G. *Biopolymers* **1995**, *35*, 271.
21. Nakazawa, F.; Noguchi, S.; Takahashi, J.; Takada, M. *Agric. Biol. Chem.* **1984**, *48*, 201.
22. Bulkin, B. J.; Kwak, Y.; Dea, I. C. M. *Carbohydr. Res.* **1987**, *160*, 95.
23. Hayashi, A.; Kotani, Y.; Cho, C.-H. *Agric. Biol. Chem.* **1984**, *48*, 949.
24. Leloup, V. M.; Colonna, P.; Ring, S. G.; Roberts, K.; Wells, B. *Carbohydr. Polym.* **1992**, *18*, 189.

Step-Growth Polymerization, Addition, and Condensation

See: Bisallenes, Polyadditions
Block Copolymers (by Step-Growth Polymerization)
Carbenoid Polycondensation
Carbonylation Polymerizations (Palladium-Catalyzed)
Condensation Polymers (Synthesis from Silylated Diamines)
Condensation Polymers (Using Phosphorus Pentoxide/Methanesulfonic Acid)
Conjugated Polymers (Prepared by Organometallic Processes)
Diketene Polymerization
Enol Ester Polymers
Functional Polymers (Telechelics and Macromonomers)
High Pressure Polycondensation (for High Temperature Polymers)
High Solids Polyurethane Coatings
Ladder Polymers (Cycloaddition Copolymerization of Diynes)
Ladder Polymers (Methods of Preparation)
Liquid Crystalline Polymers (Conjugated Chain)
Oxidative Coupling
Palladium-Catalyzed Synthesis (Monomers and Polymers)
Polyamines (Ru-Catalyzed Polycondensation of Diamines and Diols)
Polybenzimidazoles (Overview)
Polyimides, Soluble Rigid-Rods (via Pd-Catalyzed Polycondensation)
Step Polymerization Catalysts
Ultrasonically Assisted Polymer Synthesis

STEP POLYMERIZATION CATALYSTS

Fabrizio Parodi*
Saiag S.p.A.

Saverio Russo
Dipartimento diChimica e Chimica Industriale
Università di Genova

As summarized in recent overviews,[1,2] step polymerization comprises almost innumerable condensation and addition reaction derived from the immense patrimony of organic chemistry (including those pertaining to heterocyclic compounds).

Following a well accepted definition, step polymerization includes all chemical processes affording linear, branched, or crosslinked macromolecular compounds via a step-by-step succession of elementary condensation and/or addition reactions, independent of each other, among reactive centers brought by different monomeric, oligomeric, and polymeric molecules.

CATALYSIS IN STEP POLYMERIZATION

With the exception of a certain number of processes based on very fast and even violently exothermic reactions, use of catalysts or reaction media exerting catalytic effects is a general practice in step polymerizations.[3,4] Catalytic systems are used much more extensively than is required by the intrinsic needs of reaction feasibility, in order to lead the reactions themselves to completion in shorter times, as demanded by the industrial polymerization or resin-hardening processes.

An example of fast uncatalyzed polymerization is the synthesis of poly(quaternary ammonium halide)s from nonaromatic tertiary diamines and α,ω-dihaloalkanes or bis(benzyl halide)s.

Step polymerization is carried out by different experimental procedures of two fundamental types: homogeneous-phase polymerizations, i.e., solution and bulk (melt and solid-state) processes; and polymerizations in heterophase systems, implying the presence of liquid (aqueous or organic) solvents or suspending media.

Homogeneous polymerizations are the most important processes, particularly from an industrial standpoint, especially if one, besides synthesis of the large variety of thermoplastic condensation polymers, considers the curing processes of thermosetting resins, such as epoxies, phenol-, urea-, and melamine-formaldehyde resins, polyurethane, polyisocyanurate, and silicone resins.

As a consequence, homogeneous catalysis plays the most relevant role in the area of step polymerizations. Moreover, according to some review articles on the subject,[5,6] the industrial world production of condensation polymers, resins, and related monomers involves consumption of the highest amounts of catalysts in the whole area of industrial chemical processes based on homogeneous catalysis.

HOMOGENEOUS CATALYSTS

A systematic description of catalysts for step polymerizations can be given by grouping them into the three families of acids, bases, and metal compounds (the latter including promoters of the broadest spectrum of reactions, ranging from addition and substitution to oxidation processes).

Protic and Lewis Acid Catalysts

Protic or Brönsted acids were the first catalysts to find applications in step polymerizations, for example in direct polyesterification and in synthesis/hardening of phenol- and urea-formaldehyde resins.

Phenolic and Amino Resins

Brönsted acid catalysis is extensively used for synthesis of novolac-type and curing of resol-type phenol-formaldehyde resins.[9,10]

*Author to whom correspondence should be addressed.

Acids such as H_2SO_4, *p*-toluenesulfonic acid, HCl, and H_3PO_4 are used as catalysts. Oxalic acid also is used. For reaction injection processing, latent acid catalysts have been devised to be active at temperatures ≥ 150°C: monophenylmaleate and phenyl trifluoromethanesulfonate both of which liberate the corresponding free acid by hydrolysis.

Acid-catalyzed synthesis and hardening of urea-melamine- and benzoguanamine (2,6-diamino-1-phenyl-1,3,5-triazine)-formaldehyde resins[11] are analogous to those seen above for phenolics.

Furfuryl Alcohol Polymers

Another remarkable example of catalysis by protic acids is formation of thermosets from furfuryl alcohol (or various furfuryl alcohol-containing systems).[12]

Polyaroylation and Polysulfonylation

Applications of protic and Lewis acid catalysts can be found in syntheses of aromatic polyketones and polysulfones in particular, (poly(arylene ether ketone)s and poly(arylene ether sulfone)s) via electrophilic aromatic substitution routes: condensations of aroyl or arenesulfonyl chlorides with arylethers under Friedel-Crafts conditions; condensations of arenecarboxylic or arenesulfonic acids with arylethers; and direct Friedel-Crafts polycarbonylation or polysulfonylation of diarylethers with $COCl_2$ (or its substitutes, such as alkyl thiochloroformates and dialkyl dithiocarbonates), or SO_2Cl_2, respectively.

As described in several reviews, development of aromatic polyketones[13–15] and polysulfones[16] substantially followed parallel routes, with similar synthetic approaches and catalysis criteria being applied.

For Friedel-Crafts poly-aroylations and -sulfonylations, a variety of Lewis acids, related adducts and strong protic acids can be selected from the broad range of catalysts used in Friedel-Crafts chemistry.[17–19]

More recently, excellent polyaroylation catalysts (used in large excess) have proved to be 1:1 Lewis acid-Lewis base adducts, namely those of $AlCl_3$ with LiCl, N,N-dimethylformamide or quaternary ammonium halides. Polyaroylations and polysulfonylations also can be accomplished at moderate temperatures in solutions of BF_3 (catalyst) in anhydrous HF (solvent).

A breakthrough for these polycondensations has been the adoption of superacids such as CF_3SO_3H,[20] simultaneously working as catalysts and good polymer solvents, thus allowing smooth syntheses of high molecular weight polymers at, or even below, room temperature.

Siloxane Polymers

Brönsted acids are useful catalysts of the silanol-silanol condensation reaction involved in synthesis and crosslinking of polysiloxanes.[21]

Strong mineral acids and sulfonic acids are suitable.

Base Catalysts

The large amounts of tertiary amine catalysts consumed worldwide in polyurethane and epoxy resin technologies give an indication of the great industrial importance of bases in the catalysis of step polymerizations.

Polyesterification processes represent a wide field for base catalysis.[8] Strong rate enhancements in the Schotten-Baumann polyesterification and in the similar synthetic route to polycarbonates are induced by the tertiary amines (triethylamine and trialkylamines in general, pyridine, quinoline, etc.) normally added as HCl acceptors.

Strong nucleophiles are necessary as catalysts of both alcoholysis and phenolysis methods of polytransesterification and synthesis of polycarbonates.

Phenolysis methods (often referred to as diphenate processes) are used for melt synthesis of wholly-aromatic polyesters (or polyarylates)[22] and polycarbonates. Examples of catalysts are alkali metal hydroxides (generating phenoxides by salification of bisphenols), alkali metals (Li, Na), alkali metal phenoxides (Li and Na phenoxides), and potassium borophenoxide ($KBH_3OC_6H_5$, prepared from KBH_4 and phenol).

Acidolysis methods of polytransesterification, typically using acetic esters, are of known value for production of polyarylates and, particularly, of high-melting liquid-crystalline polyesters and copolyesters.[8,22]

Polyesters also are attainable from anhydrides and epoxides, a reaction very slow per se, but strongly accelerated by mild nucleophiles. This reaction, of limited interest for synthesis of linear polymers, is widely exploited for curing epoxy resins.

Phenolic and Amino Resins

Synthesis of such resins is another important area of application of nucleophiles as catalysts. Bases are optimum catalysts for preparation of resol-type phenolics under mild conditions.[9,10]

Alkali metal hydroxides and carbonates, NH_4OH, alkaline-earth metal hydroxides (easily removable at the end of the process by precipitation, e.g., as carbonates), as well as tertiary amines and alkanolamines are used as catalysts. Mg, Ca, and Ba hydroxides are especially valuable due to their attitude to favor *ortho*-substitutions by complexation mechanisms (preparation of "high-*ortho* resins").

Polymers from Isocyanates

Several reviews have been devoted to catalysis of isocyanate reactions in polymer syntheses.[23–25] Base catalysts play a primary role, though their selectivity is generally modest.

Lewis bases, and particularly tertiary amines, are efficient catalysts of all such reactions.

Solvents greatly influence both partial reaction orders and values of rate constants, in the sense of a general and progressive rate decrease as solvents of increasingly high dielectric constant are used.

Typical tertiary amines extensively used in polyurethane technology are N-methyl- and N-ethyl-morpholine, 1,4-dimethylpiperazine, and 1,4-diamino[2.2.2]bicyclooctane (DABCO). DABCO is a catalyst of choice for its efficiency, outstandingly higher (i.e., at least one order of magnitude) as compared to other amines.

Polyisocyanurates

Strong base catalyze trimerization of isocyanates where an anion is involved (Lewis bases act in a similar manner).

A plethora of compounds have been found suitable,[24,25] and some are commonly used as catalysts of choice: strong tertiary polyamines (e.g., N,N′,N″-tris(dimethylaminopropyl)*sym*-hexahydrotriazine), alkali metal carbonates and carboxylates (K acetate), alkali metal oxides and alkoxides (Na methoxide), and quaternary ammonium and phosphonium hydroxides (e.g., benzyltrimethylammonium hydroxide). The efficiency of such catalysts increases as their basicity becomes higher, and solvents of high dielectric constant further accelerate the reaction (N,N-dimethylformamide, N-methylpyrrolidone, dimethyl sulfoxide, etc.). Alcohols and phenols display remarkable cocatalytic effects, so that tertiary alkanolamines (such as isopropanolamines) and dialkylaminoalkyl phenols (e.g., 2, 4,6-tris(dimethylaminomethyl)phenol) are excellent commercial trimerization catalysts.

Metal Catalysts

Direct Polyesterification

As catalysts of direct polyesterification, a great number of metal compounds have been examined and their effectiveness evaluated with the related reaction mechanisms and kinetics.[7] Compounds of Group IV elements are known as very efficient catalysts, and among them, Ti^{IV}, Zi^{IV}, and Sn^{IV} derivatives are most commonly used.

Typical catalysts are dibutyltin oxide, dibutyltin dicarboxylates (diacetate), and, far more efficient, Ti and Zr tetraalkoxides. Among Ti and Zr tetraalkoxides,[26] those with not too bulky alkyls are largely preferred and extensively used: Ti and Zr tetraethoxides, tetraisopropoxides, and tetra-*n*-butoxides.

Polytransesterification Processes

Synthesis of polyesters by the alcoholysis method of transesterification can be performed by two different procedures: with equimolar, or nearly equimolar, amounts of reactants by a single-stage process or with an excess of the most volatile reactant as in syntheses of poly(ethylene terephthalate) (PET) and poly(butylene terephthalate) (PBT) from dimethylterephthalate and ethylene glycol or 1,4-butanediol, respectively.

A huge number of transition and nontransition metal compounds have been devised as catalysts. Typical catalysts for the first (transesterification) stage are acetates of Ca, Ba, Zn, Cd, Mn^{II}, Co^{II}, Pb^{II}, etc., whereas Sb_2O_3, GeO_2, and PbO are effective to promote the second (polycondensation) stage.

Phenic Resins

Carboxylates (usually acetates) of Mg, Ca, Zn, Cd, Mn^{II}, Cu^{II}, Co^{II}, Ni^{II}, Pb^{II}, etc., are useful catalysts (the so-called neutral catalysts) for manufacture of high-*ortho* resins (in particular novolacs) analogously to bases, but with better selectivity. Zinc acetate is particularly advantageous due to its capacity to promote nearly exclusively *ortho*-substitution.

Polymers from Isocyanates

Besides those catalyzed by bases, the reactions of isocyanates with protic nucleophiles are strongly promoted by metal compounds.

Acetylacetonates, 2-ethylhexanoates (*octoates*) and carboxylates in general of a great variety of both transition (Zn, Cu^{II}, Mn^{II}, Fe, Co^{II}, Ni^{II}, etc.) and non-transition metals (such as Bi^{III}, Pb^{II}, Sn), as well as the corresponding alkylmetal (e.g., dialkyltin) carboxylates, display strong catalytic effects. Sn^{II} 2-ethylhexanoate (*tin octoate*), dibutyltin diacetate and dilaurate are the most extensively used metal catalysts in polyurethane syntheses by virtue of their high efficiency (at least two orders of magnitude greater with respect to the best DABCO catalyst) and modest allophanate promotion.

Poly(phenylene ether)s

The most important method for preparation of these polymers is oxidative-coupling polymerization of 2,6-disubstituted (namely, 2,6-dialkyl) phenols, in particular 2,6-dimethylphenol, a subject reviewed by several authors.[27,28]

Although this polymerization may be accomplished by means of several oxidizing agents (e.g., MnO_2, PbO_2, AgO, and $NaBiO_3$), the preferred industrial way is catalytic oxidation with molecular oxygen (pure oxygen and/or air bubbled in an organic solution of reactant and catalyst).

Typical catalysts are combinations of a copper halide (generally Cu_2Cl_2 or $Cu_2\ Br_2$, though Cu^{II} and mixtures of Cu^{I} and Cu^{II} halides are also usable), a complexing agent such as a primary or secondary amine or aminoalcohol (e.g., di-*n*-butylamine, morpholine, ethanolamine), and optionally, a base (e.g., pyridine, *p*-dimethylaminopyridine, 1,2-dimethylaminoethane, triethylamine, NaOH, KOH, etc.).

PHASE TRANSFER CATALYSTS

Phase transfer catalysis (PTC) is the method to accelerate organic syntheses in two-phase systems (and to perform them under particularly mild conditions) by means of chemical agents (PT catalysts) which rapidly carry reactive species constituting, or contained in, one phase to react into the other phase through the interface between them.[29–31]

PTC can be applied successfully to a variety of polymerizations,[32] some of which are of wide industrial use:

- syntheses of polyesters (polyarylates) and polycarbonates from aqueous solutions of alkali metal bisphenoxides and diacyl chlorides or bis(chloroformate)s, respectively, dissolved in water-immiscible solvents; and
- the polyetherification process for production of epoxy resins from aqueous solutions of bisphenols + alkali metal hydroxides, and epichlorohydrin.

Other remarkable examples of PT-catalyzed processes are polyetherification via nucleophilic aromatic or aliphatic substitutions, as those involving condensation of alkali metal bisphenoxides with electron-withdrawn dihaloaromatics (syntheses of polyethersulfones) or α,ω-dihaloalkanes, respectively.

MICROWAVE IRRADIATION

In the field of step polymerization, microwave technology has been successfully applied to promote curing of various thermosetting resins and related composites: epoxyamine systems, polyimides, isocyanate-epoxy resins, and polyurethanes, etc.

Although experiments have been performed in different conditions and conflicting experimental evidences have been reported for some systems, enhancements (from 10- to 40-fold) of polymerization rates seem to be generally attainable.

Despite some controversial opinions on the reasons of such effects (e.g., strong and selective heating of reacting centers), two general conclusions can be drawn: slow reactions exhibit greater rate enhancements than inherently fast processes; and rate increases (and thus advantages over conventional heat treatments) become more and more remarkable as the reaction temperatures are decreased.

ACKNOWLEDGMENTS

The kind assistance of Dr. Stefania Bisbano in editing the numerous chemical formulae and reaction schemes is greatly appreciated. Partial support from M.U.R.S.T. (40% funds) and C.N.R. is gratefully acknowledged.

REFERENCES

1. Manaresi, P.; Munari, A. In *Comprehensive Polymer Science*; Eastmond, G. C.; Ledwith, A. et al. Eds.; Pergamon: Oxford, 1989.
2. Parodi, F. In *The Encyclopedia of Advanced Materials*; Bloor, D.; Brook, R. J. et al. Eds.; Pergamon: Oxford, 1994.
3. Jones, D. C.; White, T. R., Eds. *Kinetics and Mechanisms of Polymerization* Vol. 3, Marcel Dekker: New York, 1972.
4. Schildknecht, C. E.; Skeist, I., Eds., *Polymerization Processes*; Wiley-Interscience: New York, 1977.
5. Bikales, N. M. *Adv. Chem. Ser.* **1968**, *70*, 233.
6. Parshall, G. W. *Homogeneous Catalysis*; John Wiley & Sons: New York, 1980.
7. Fradet, A.; Maréchal, E. *Adv. Polym. Sci.* **1982**, *43*, 51.
8. Pilati, F. In *Comprehensive Polymer Science*; Eastmond, G. C.; Ledwith, A. et al. Eds.; Pergamon: Oxford, 1989.
9. Knop, A.; Pilato, L. A. *Phenolic Resins*; Springer-Verlag: Berlin, 1985.
10. Kopf, P. W. In *Encyclopedia of Polymer Science and Engineering* 2nd ed., Mark, H. F.; Bikales, N. M. et al. Eds.; John Wiley & Sons: New York, 1988.
11. Widmer, G. In *Encyclopedia of Polymer Science and Technology*; Bikales, N. M.; Mark, H. F. et al. Eds.; John Wiley & Sons: New York, 1965. Petersen, H. In *Houben-Weil-Methoden der Organischen Chemie*; Bartl, H.; Falbe, J., Eds.; Georg Thieme Verlag: Stuttgart, 1987.
12. Gandini, A. *Adv. Polym. Sci.* **1977**, *25*, 47. Gandini, A. In *Encyclopedia of Polymer Science and Engineering* 2nd ed.; Mark, H. F.; Bikales, N. M. et al. Eds.; John Wiley & Sons: New York, 1987.
13. Rose, J. B. In *Recent Advances in Mechanistic and Synthetic Aspects of Polymerization*; Fontanille, M.; Guyot, A., Eds.; D. Reidel: Dordrecht, 1987.
14. Mullins, M. J.; Woo, E. P. *J. Macromol. Sci., Rev. Macromol., Chem. Phys.* **1987**, *C27*, 313.
15. Staniland, P. A. In *Comprehensive Polymer Science*; Eastmond, G. C.; Ledwith, A. et al. Eds., Pergamon: Oxford, 1989.
16. Parodi, F. In *Comprehensive Polymer Science*; Eastmond, G. C.; Ledwith, A. et al. Eds., Pergamon: Oxford, 1989.
17. Olah, G. A., Ed. *Friedel-Crafts and Related Reactions*; Interscience: New York: Vol. 1, 1963 and Vol. 3, 1964.
18. Olah, G. A., Ed. *Friedel-Crafts Chemistry*; John Wiley & Sons: New York, 1973.
19. Olah, G. A.; Meidar, D. In *Kirk-Othmer Encyclopedia of Chemical Technology* 3rd ed., Mark, H. F.; Othmer, D. F. et al. Eds.; John Wiley & Sons: New York, 1980.
20. Olah, G. A.; Prakash, G. K. S.; Sommer, J. *Superacids* John Wiley & Sons: New York, 1985.
21. Lebrun, J. J.; Porte, H. In *Comprehensive Polymer Science*; Eastmond, G. C.; Ledwith, A. et al. Eds.; Pergamon: Oxford, 1989.
22. Dean, B. D.; Matzner, M.; Tibbitt, J. M. In *Comprehensive Polymer Science*; Eastmond, G. C.; Ledwith, A. et al. Eds.; Pergamon: Oxford, 1989.
23. Reegen, S. L.; Frisch, K. C. *Adv. Urethane Sci. Technol.* **1971**, *1*, 1.
24. Sayigh, A. A. R. *Adv. Urethane Sci. Technol.* **1974**, *3*, 141.
25. Entelis, S. G.; Tiger, R. P. *Adv. Urethane Sci. Technol.* **1981**, *8*, 19.
26. Bradley, D. C.; Mehrotra, R. C.; Gaur, D. P. *Metal Alkoxides*; Academic: London, 1978.
27. Chandra, R. *Progr. Polym. Sci.* **1982**, *8*, 469.
28. White, D. M. In *Comprehensive Polymer Science*; Eastmond, G. C.; Ledwith, A. et al. Eds.; Pergamon, Oxford, 1989.
29. Weber, W. P.; Gokel, G. W. *Phase Transfer Catalysis in Organic Synthesis*; Springer-Verlag: Berlin, 1977.
30. Starks, C. M.; Liotta, C. *Phase-Transfer Catalysis—Principles and Techniques*; Academic: New York, 1978.
31. Dehmlow, E. V.; Dehmlow, S. S. *Phase Transfer Catalysis*; Verlag Chemie: Weinheim, 1980.
32. Mahamat, O.; Majdoub, M. et al. In *Recent Advances in Mechanistic and Synthetic Aspects of Polymerization*; Fontanille, M.; Guyot, A. Eds.; D. Reidel: Dordrecht, 1987.

Stereo-Block Copolymers

See: Stereo-Block Copolymers (of L- and D-Lactides)
Uniform Polymers

STEREO-BLOCK COPOLYMERS (Of L- and D-Lactides)

Nobuhiko Yui
Japan Advanced Institute of Science and Technology

Biodegradable polymers have much potential for applications such as implantable carriers for drug delivery systems as well as for surgical repair materials.[1–5] Especially, polylactides have been applied for these purposes based on their good biodegradability, biocompatibility, high mechanical strength, and excellent shaping and molding properties.[6,7] Furthermore, a variation of degradation characteristics and mechanical properties of polylactides has been achieved by blending with more amorphous materials or making a random-copolymer with other lactones such as glycolide or ε-caprolactone.[8–11]

Poly(lactide)s exist in several enantiomeric forms, L, D, DL, and meso. For instance, the polymers derived from the L and D monomers are semi-crystalline, while those from the DL (racemic mixture of L and D) are amorphous. The crystalline structure of poly(L-lactide) (PLLA) and poly(D-lactide) (PDLA) has been reported to consist of left- and right-handed helical chain conformations, respectively.[12] Recently, the blending of PLLA and PDLA has been reported to give rise to the formation of a stereocomplex that shows a different crystalline structure and morphology from the pure polymers.[13,14] This complex shows a melting temperature that is approximately 50°C higher than that of PLLA and PDLA.

A stereocomplex formation is expected to occur in the case of block copolymers of L-lactide (LLA) and D-lactide (DLA). Furthermore, the degree of stereocomplex formation may be controlled by the composition of the block copolymers.

This article describes the synthesis, characterization, and properties of sequential di-block copolymers of PLLA and PDLA obtained by a living ring-opening polymerization of LLA and DLA initiated by aluminum isopropoxide, as schematically

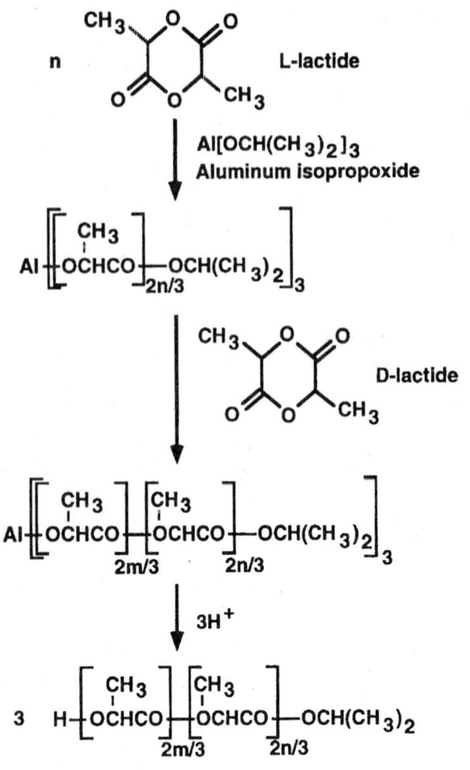

FIGURE 1. Synthetic route of di-block copolymer of L- and D-lactides by aluminum isopropoxide.

illustrated in **Figure 1**. The stereocomplex between two enantiomeric segments was formed in these block copolymers.

PROPERTIES

Thermal Properties of Block Copolymers of L- and D-Lactides

The thermal analysis of homopolymers and block copolymers was carried out with a Perkin-Elmer DSC4 differential scanning calorimeter (DSC) calibrated with pure indium. The block copolymer showed a T_m around 205°C, which is approximately 40°C higher than that of PLLA. This result provides strong evidence for the stereocomplex formation between PLLA and PDLA chains in the copolymer.

Furthermore, the results of DSC measurements for block copolymers were compared with those for blendmers. The block copolymers showed two endothermal peaks assigned to the melting of a stereocomplex in the first scan, and one endothermal peak for the melting of a stereocomplex in the third scan. Two endothermal peaks in the first scan are attributed not to different crystalline structure but to the variation of crystallite size in the stereocomplex, because the endothermal peak for the stereocomplex in the third scan was situated between two endothermal peaks in the first scan. In every scan, no crystallization of an enantiomeric excess of one of the segments was observed. However, the blendmers showed two distinct endothermal peaks assigned to the melting of a stereocomplex and crystallized enantiomeric excess in the first scan, and one endothermal peak for the fusion of a stereocomplex in the third scan. By further thermal treatments, the T_m of the stereocomplex decreased in blendmers, especially in the case of the higher molecular weights. The decrease in the T_m in the block copolymers was within a degree, although that in the blendmers was ~ 22°.

Solubility of Block Copolymers of L- and D-Lactides

The solubility of block copolymers was tested using toluene, tetrahydrofuran, toluene/chloroform (1/1), chloroform, dichloromethane, and tetrachloromethane, all of which are well known as solvents for poly(lactide)s. Chloroform was the only solvent that dissolves the block copolymers. The block copolymers generally showed a lower solubility than the blendmer, and much lower than the corresponding homopolymers. The low solubility of the block copolymers and the blendmers indicates the formation of a stereocomplex between PLLA and PDLA segments in these polymers.

It should be acceptable that the limited solubility of block copolymers is due to some difference in the stereocomplex formation between block copolymers and blendmers. The interpolymer complexation in the block copolymers is expected to cause network formation if a certain copolymer chain can associate with the PLLA and PDLA segments in different chains to form a stereocomplex. Because three-dimensional network formation is generally known to lead to more insoluble compounds, it may be reasonable that the stereocomplex network in block copolymers shows lower solubility than the complex in the blendmers.

In conclusion, the block copolymerization of LLA and DLA favors the stereocomplex formation between PLLA and PDLA segments, which is different from that in blendmers of PLLA and PDLA, and prevents the crystallization of the enantiomeric excess. These characteristics are considered to be strongly related to the definite mode of complexation, which may be different from that in the blendmers.[15]

REFERENCES

1. Kim, S. W.; Petersen, R. V.; Feijen, J. In: *Drug Design*; Ariens, E. J., Ed.; Academic: New York, 1980; Vol. 10, p 193.
2. Wood, D. A. *Int. J. Pharm.* **1980**, *7*, 1.
3. Heller, J. *CRC Crit. Rev. Ther. Drug Carrier Syst.* **1985**, *1*, 39.
4. Jeoung, S. Y.; Kim, S. W. *Arch. Pharm. Res.* **1986**, *9*, 63.
5. Holland, S. J.; Tighe, B. J.; Gould, P. L. *J. Controlled Release* **1986**, *4*, 155.
6. Kulkarni, R. K.; Moore, E. G.; Hegyeli, A. F.; Leonard, F. *J. Biomed. Mater. Res.* **1971**, *5*, 169.
7. Chabot, F.; Vert, M.; Chapelle, S.; Granger, P. *Polymer* **1983**, *24*, 53.
8. Feng, X. D.; Song, C. X.; Chen, W. Y. *J. Polym. Sci., Polym. Lett. Ed.* **1981**, *21*, 593.
9. Song, C. X.; Feng, X. D. *Macromolecules* **1984**, *17*, 2764.
10. Kricheldorf, H. R.; Mang, T.; Jonté, J. M. *Macromolecules* **1984**, *17*, 2173.
11. Vion, J.-M.; Jérôme, R.; Teyssié, P.; Aubin, M.; Prud'homme, R. E. *Macromolecules* **1986**, *19*, 1828.
12. De Santis, P.; Kovacs, A. J. *Biopolymers* **1968**, *6*, 299.
13. Ikada, Y.; Jamshidi, K.; Tsuji, H.; Hyon, S.-H. *Macromolecules* **1987**, *20*, 904.
14. US. 4719246 (1988), E.I. DuPont de Nemours and Co., Inv.: J. R. Murdock and G. L. Loomis; *Chem. Abstract* **1988**, *108*, 132811y.

15. Yui, N. *Yakugaku-Kenkyuu no Shinpo (Adv. Pharm. Sci.* in Japanese) **1993**, *9*, 171.

Stereocomplexes

See: *Stereo-Block Copolymers (of* L- *and* D-*Lactides)*
Stereocomplexes (Between Enantiomeric or Diastereomeric Polymers)
Stereoregular Polymethacrylates

STEREOCOMPLEXES (Between Enantiomeric or Diastereomeric Macromolecules)

Koichi Hatada, Tatsuki Kitayama, and Osamu Nakagawa
Department of Chemistry
Faculty of Engineering Science
Osaka University

It has been known that a pair of polymers that display the ability of polymer-polymer interaction stoichiometrically at the level of monomeric units can form polymer–polymer complexes. Their formation is often associated with highly selective interaction between macromolecules.[20]

Stereocomplex is one kind of polymer-polymer complex, in which stereochemical configuration of polymer chains is of prime importance for the complex formation, and can be divided into two categories; one is stereocomplex formed between enantiomeric macromolecules and the other is that formed between diastereomeric macromolecules. Typical examples reported in the literature are summarized in **Table 1**.

TABLE 1. Stereocomplexes From Stereoregular Polymers[a]

Component polymers	Reference
Stereocomplexes formed between enantiomeric macromolecules	
it-polymers of (R)- and (S)-α-ethyl-α-methyl-β-propiolactones	10 17
it-polymers of (R)- and (S)-*t*-butylthiiranes	8
it-polymers of (R)- and (S)-α-methylbenzyl methacrylates	11
Polymers of D-n and L-lactides	12
Polymers of γ-benzyl D- and L-glutamates	9
Stereocomplexes formed between diastereomeric macromolecules	23
it- and *st*-PMMAs	18
it-PMMA and *st*-poly(methacrylic acid)	19
it-PMMA and *st*-poly(isobutyl methacrylate)	2
it- and *st*-poly(methyl α-chloroacrylate)	7

[a]*it*-: isotactic, *st*-: syndiotactic.

STEREOCOMPLEX FORMED BETWEEN ENANTIOMERIC MACROMOLECULES

When highly optically pure *it*-poly(α-ethyl-α-methyl-β-propiolactone)s of opposite absolute configurations were mixed in equimolar amounts in solution, a new crystal was formed. Since the crystal structure and morphology are not similar to those of the individual *it*-polymers, the new crystal should be due to stereocomplex formation. The melting point of the complex is 40°C higher than those of the individual components and increased with an increase in the degree of isotacticity.[17]

D,L-lactide provided an amorphous and optically inactive polymer, while polymerizations of L- and D-lactide afforded crystalline and optically active polymers that gave a single endothermic peak around 180°C in the DSC thermogram. In the case of blend of poly(L-lactide) and poly(D-lactide) (50/50), the endothermic peak at 180°C in DSC thermogram disappeared and a new peak appeared around 230°C, suggesting a stereocomplex formation with a new crystalline structure.[12]

STEREOCOMPLEX FORMED BETWEEN DIASTEREOMERIC MACROMOLECULES

Stereocomplex Formation Between *it*- and *st*-PMMAs

It has been known that *it*- and *st*-PMMA chains associate to form a crystalline stereocomplex in certain solvents such as toluene, dimethylformamide, and acetone. The term "stereocomplex" was first used for this polymer complex by Liquori et al.[18]

Though the structure of the stereocomplex is still not clear, several scientists have proposed models of the structure Kusanagi et al. expected the structure to be closely related to the double-stranded helix of *it*-PMMA.[16] Bosscher et al. proposed that the stereocomplex is a double-stranded helix in which a helix of *it*-PMMA chain with a small radius (a 30/4 helical conformation) is surrounded by an *st*-PMMA helical chain with a larger radius (a 60/4 helical conformation).[3] Recently, Schomaker and Challa proposed another double-stranded helix model for the stereocomplex with a 9/1 helical symmetry and 18.9 Å pitch.[21] Vorenkamp et al. reported that *it*- and *st*-PMMAs formed the stereocomplex in a ratio of *it*-/*st*- = 2 at the level of monomeric units.[22]

Several applications of PMMA stereocomplex were reported, for example the susceptibility of syndiotactic or atactic PMMA to solvent-induced cracking and crazing was reduced and the fracture toughness was enhanced by blending with *it*-PMMA. The crystalline stereocomplex regions formed would act as pseudocrosslinks that would stabilize the crazes.[1] Another example for applications of PMMA stereocomplex is ultrafine filters or semi-permeable membranes because the complex is very porous and has a narrow distribution of pore sizes. In fact, a hollow fiber semipermeable membrane has been made from the complex and used in an artificial kidney.[13]

We recently found that *st*-polymethacrylates of primary and secondary esters form stereocomplexes with *it*-PMMA.[14,15] This finding is apparently consistent with the double-stranded helix model proposed by Challa and his co-workers.[5]

Template Polymerization

Radical polymerization of MMA in the presence of *it*-PMMA in dimethylformamide, acetone, or acetonitrile is an example of template polymerization that is also called replica or matrix polymerization. This polymerization gave a PMMA richer in syndiotacticity than that polymerized in the absence of an *it*-PMMA matrix.[4,6]

REFERENCES

1. Allen, P. E. M.; Host, D. M.; Truong, V. T.; Williams, D. R. G. *Eur. Polym. J.* **1985**, *21*, 603.

2. Bosscher, F.; Keekstra, D.; Challa, G. *Polymer* **1981**, *22*, 124.
3. Bosscher, F.; Ten Brinke, G. T.; Challa, G. *Macromolecules* **1982**, *15*, 1442.
4. Buter, R.; Tan, Y. Y.; Challa, G. *J. Polym. Sci.* **1973**, *A-1, 11*, 1003.
5. Challa, G.; de Boer, A.; Tan, Y. Y. *Int. J. Polym. Meter.* **1976**, *4*, 239.
6. Challa, G.; Tan, Y. Y. *Pure Appl. Chem.* **1981**, *53*, 627.
7. Dever, G. R.; Karasz, F. E.; MacKnight, W. J.; Lenz, R. W. *Macromolecules* **1975**, *8*, 349.
8. Dumas, P.; Spassky, N.; Sigwalt, P. *Makromol. Chem.* **1972**, *156*, 55.
9. Fukuzawa, T.; Uematsu, I. *Polym. J.* **1974**, *6*, 537.
10. Grenier, d.; Prud'homme, R. E. *Macromolecules* **1983**, *16*, 302.
11. Hatada, K.; Shimizu, S.; Terawaki, Y.; Ohta, K.; Yuki, H. *Polym. J.* **1981**, *8*, 811.
12. Ikada, Y.; Jamshidi, K.; Tsuji, H.; Hyon, S.-H. *Macromolecules* **1987**, *20*, 906.
13. Kimura, M.; Kobayashi, T.; Fujii, Y.; Takeyama, T.; Nakasatomi, M. *Polym. Prepr. Jpn* **1981**, *30*, 2432.
14. Kitayama, T.; Fujimoto, N.; Terawaki, Y.; Hatada, K. *Polym. Bull.* **1990**, *23*, 279.
15. Kitayama, T.; Fujimoto, N.; Hatada, K. *Polym. Bull.* **1991**, *26*, 629.
16. Kusanagi, H.; Tadokoro, H.; Chatani, T. *Macromolecules* **1976**, *9*, 531.
17. Lavallee, C.; Prud'homme, R. E. *Macromolecules* **1989**, *22*, 2438.
18. Liquori, A. M.; Anzuino, G.; Corio, V. M.; D'Alagni, M.; de Santis, P.; Savino, A. *Nature* **1965**, *206*, 358.
19. Lohmeyer, J. H. G. M.; Tan, Y. Y.; Lako, P.; Challa, G. *Polymer* **1978**, *19*, 1171.
20. Papisov, I. M.; Litmanovich, A. A. *Adv. Polym. Sci.* **1988**, *90*, 139.
21. Schomaker, E.; Challa, G. *Macromolecules* **1989**, *22*, 3337.
22. Vorenkamp, E. J.; Bosscher, F.; Challa, G. *Polymer* **1979**, *20*, 59.
23. Watanabe, W. H.; Ryan, C. F.; Fleischer, Jr. P. C.; Garrett, B. C. *J. Phys. Chem.* **1961**, *65*, 896.

Stereoelective Polymerization

See: *Optically Active Polymers (Overview)*

Stereoregular Polymerization

See: *Inclusion Complexes (Overview)*
Inclusion Polymerization
Metallocene Catalysts
Phenylacetylene (Stereospecific Living Polymerization, Rh Complexes)
Polyolefins (Overview)
Polypropylene (Commercial)
Polystyrene, Stereoregular
Polystyrene, Syndiotactic
Polystyrene, Syndiotactic (Sulfonation and Crystallization)
Rare Earth-Initiated Polymerization (of Dienes)
Stereoregular Polymerization Macromonomer
Stereoregular Polymethacrylates
Styrene, Stereospecific Polymerization
Topochemical Photopolymerization (Types of Topochemical Behavior)
Uniform Polymers
Zinc Catalysts (Polymerization of Styrene and Methyl Methacrylate)

STEREOREGULAR POLYMETHACRYLATE MACROMONOMER

Koichi Hatada,* Tatsuki Kitayama, and Osamu Nakagawa
Department of Chemistry
Faculty of Engineering Science
Osaka University

Poly(methyl methacrylate) (PMMA) is one of the typical polymers whose tacticity can be widely controlled by changing the reaction conditions.[5,15] The properties in bulk and in solution depend on the tacticity.[2,5,14,15] Therefore, control of stereoregularity of PMMA macromonomer should provide a promising way for controlling the properties of graft polymers and comb-like polymers.

Recently we prepared highly stereoregular PMMA macromonomers by vinylbenzylmagnesium chloride (VBzMgCl) initiated polymerization and by stereospecific living polymerizations, and studied their radical and anionic polymerizations and copolymerizations.[3,6–9,13]

PREPARATION OF STEREOREGULAR PMMA MACROMONOMER

Stereoregular PMMA macromonomer can be prepared through anionic polymerization. In order to introduce the functional end-group to the PMMA chain, the following two methods are useful. One is the polymerization of methyl methacrylate (MMA) by initiator containing a vinyl group (*Initiator method*); and the other is coupling reaction between the PMMA living anion and a vinyl-type compound carrying an appropriate functional group (*Terminator method*).[3,6–8]

POLYMERIZATION OF STEREOREGULAR PMMA MACROMONOMER

Polymerization of Styrene-Type PMMA Macromonomer

Radical polymerizations of the stereoregular PMMA macromonomers with styrene-type end-group (**Equation 1**) were carried out with 2,2′-azobisisobutyronitrile (AIBN) in toluene-d_8 at different temperatures, and the reaction was followed by ^{1}H NMR.[7,13] The first order plots of monomer consumption of the *it*- and *st*-PMMA macromonomers at 60°C indicate that the rate of polymerization (Rp) of the *it*-macromonomer is larger than that of *st*-macromonomer. Average degree of polymerization (DP) of the polymacromonomer obtained from *it*-macromonomer is higher than that from *st*-macromonomer. The larger reactivity of *it*-PMMA macromonomer than *st*-PMMA macromonomer may be due to the higher segmental mobility of *it*-PMMA chain than that of *st*-macromonomer, which was evidenced from ^{13}C-T_1 measurement.[2,13]

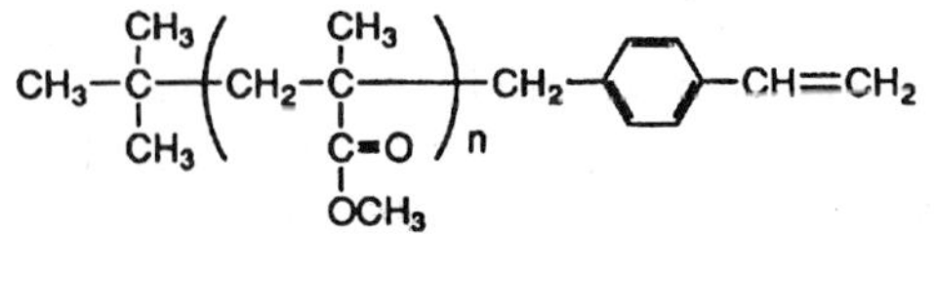

1

*Author to whom correspondence should be addressed.

Radical copolymerizations of *it*- (M_1) and *st*- (M_2) PMMA macromonomers (1/1 mol/mol) were carried out with AIBN in several solvents.[8] Number average molecular weights ($\overline{M}_n$'s) of *it*- and *st*-PMMA macromonomers used were 3690 and 4240, respectively. The ratios of M_1 to M_2 in the copolymers obtained are larger than units, indicating that the *it*-PMMA macromonomer is more reactive than *st*-PMMA macromonomer.

Polymerization of Methacrylate-Type PMMA Macromonomer

The polymacromonomers prepared from the stereoregular PMMA macromonomers with styrene-type end-group have stereoregular side chains. However, the stereoregularity in the main chain is not expected to be so high even in the case of Ziegler polymerization. In order to prepare stereoregular polymacromonomer with respect not only to the side chains but also to the main chain, stereospecific anionic polymerization and copolymerization of stereoregular PMMA macromonomers having methacryloyl function were studied.[9,10]

The direct determination of main-chain tacticity of the polymacromonomers from their ^{1}H NMR spectra is rather difficult owing to the overlap of the signals of the main-chain and side-chain methacrylate units. In order to avoid this difficulty, *it*- and *st*-macromonomers with MMA-d_8 units were polymerized under the same conditions as those for the non-deuterated macromonomer, and the resulting polymacromonomers were subjected to ^{1}H NMR measurements.[9,11] The main-chain tacticities of the polymacromonomers prepared with DPHLi in THF and with t-C_4H_9MgBr in toluene are predominantly syndiotactic and isotactic, respectively, though the stereoregularities are slightly lower than those for PMMAs obtained under the same conditions.[1,4] The knowledge on stereospecificity of polymerization of methacrylate is applicable to the polymerization of methacrylate-type macromonomers.

SOLUTION PROPERTIES OF STEREOREGULAR COMB-LIKE AND GRAFT POLYMERS

The comb-like polymers (polymacromonomers) and graft polymers showed peculiar behavior to their stereoregularities main chain and side chain.

Solution viscosities of the polymacromonomers were determined by GPC-differential viscometer (GPC-VIS) in THF at 40°C.[9,12]

Though log[η]-logMW plot for linear polymer is usually a straight line as expected from Mar-Houwink-Sakurada's equation, the plot for these polymacromonomers deviated from the linearity.

REFERENCES

1. Cao, Z.-K.; Okamoto, Y.; Hatada, K. *Kobunshi Ronbunshu* **1986**, *43*, 857.
2. Hatada, K.; Kitayama, T.; Okamoto, Y.; Ohta, K.; Umemura, Y.; Yuki, H. *Macromol. Chem.* **1978**, *179*, 485.
3. Hatada, K.; Nakanishi, H.; Ute, K.; Kitayama, T. *Polym. J.* **1986**, *18*, 581.
4. Hatada, K.; Ute, K.; Tanaka, K.; Okamoto, Y.; Kitayama, T. *Polym. J.* **1986**, *18*, 1037.
5. Hatada, K.; Kitayama, T.; Ute, K. *Prog. Polym. Sci.* **1988**, *13*, 189.
6. Hatada, K.; Shinozaki, T.; Ute, K.; Kitayama, T. *Polym. Bull.* **1988**, *19*, 231.
7. Hatada, K.; Kitayama, T.; Ute, K.; Masuda, E.; Shinozaki, T.; Yamamoto, M. *Polym. Bull.* **1989**, *21*, 165.
8. Hatada, K.; Kitayama, T. *Macromolecular Design: Concept and Practice (Macromonomers, Macroinitiators, Macroiniferters, Macroinimers, Macroiniters, and Macroinifers)*, Polymer Frontiers International: 1994, pp 85–127.
9. Hatada, K.; Kitayama, T.; Nakagawa, O.; Nishiura, T. *Macromolecular Engineering: Contemporary Themes*; Plenum: 1995, p 171–188.
10. Kitayama, T.; Nakagawa, O.; Kishiro, S.; Nishiura, T.; Hatada, K. *Polym. J.* **1993**, *25*, 707.
11. Kitayama, T.; Nakagawa, O.; Hatada, K., submitted to *Polym. J.*, 1995b.
12. Kitayama, T.; Nakagawa, O.; Hirotani, S.; Nishiura, T.; Hatada, K. submitted to *Polym. J.* 1995c.
13. Masuda, E.; Kishiro, S.; Kitayama, T.; Hatada, K. *Polym. J.* **1991**, *23*, 847.
14. Thompson, E. V. *J. Polym. Sci., A-2* **1966**, *4*, 199.
15. Yuki, H.; Hatada, K. *Adv. Polym. Sci.* **1979**, *31*, 1.

STEREOREGULAR POLYMETHACRYLATES

Koichi Hatada and Tatsuki Kitayama
Department of Chemistry
Faculty of Engineering
Osaka University

SYNTHESIS

Methacrylate is one of the most extensively studied classes of vinyl monomers with regard to stereospecificity of polymerization. Stereoregular poly(methyl methacrylate)s (PMMAs) were first prepared by Fox et al.[1,2] and Miller et al. Since then many papers have been published on stereospecific polymerization of methyl methacrylate (MMA) and other methacrylates. A variety of lithium, magnesium, and aluminum compounds have been used for preparation of highly stereoregular polymethacrylates. Data on the tacticity of poly(α-substituted acrylate)s prepared by radical and anionic initiators are thoroughly collected.[3,4] Isotactic PMMA is usually prepared by anionic initiators such as alkyllithium and Grignard reagents in nonpolar solvents such as toluene. For example, polymerization of MMA with C_6H_5MgBr at 30°C was claimed to give PMMAs with 100% isotacticity.[5]

t-C_4H_9MgBr gives highly isotactic PMMA with narrow molecular weight distribution. Polymerization is highly stereospecific and living.

Polymerizations of tertiary alkyl esters in toluene with alkyllithium give highly isotactic polymers. In particular, trimethylsilyl methacrylate gives isotactic poly(methacrylic acid) after terminating the reaction with methanol containing aqueous HCl.[6] A class of methacrylates that bear much bulkier ester group like triphenylmethyl (trityl) group form isotactic polymers irrespective of polymerization conditions. In particular, 1-phenylbenzosuberyl methacrylate gives an isotactic polymer with an isotactic triad content of 98% by radical polymerization.[7]

When chiral anionic initiators such as (–)-sparteine-fluorenyllithium complex are used, optically active polymers whose chirality arises from selectively built one-handed helical conformation are obtained.

Radical polymerization of MMA gives syndiotactic-rich PMMA (syndiotactic triad fraction ~62% at 60°C). Polymerizations of MMA in polar solvents with typical anionic initiators such as alkyllithium and Grignard reagent give syndiotactic PMMA. Anionic polymerization with bulky alkyllithiums such as 1,1-diphenylhexyllithium in THF at low temperature (≦–70°C) proceeds in a living manner to give syndiotactic-rich PMMA. Addition of several equivalents of LiCl per living chain end permits living polymerization at temperature up to –20°C.[8]

Organolanthanide complexes such as $[(C_5Me_5)_2SmH]_2$ give syndiotactic and high molecular weight PMMAs with narrow molecular weight distribution in toluene.[9]

The t-C_4H_9Li-R_3Al initiator systems are effective for syndiotactic living polymerizations of most of other methacrylates than MMA and give highly syndiotactic polymers with narrow molecular weight distribution.

STEREOREGULAR COPOLYMERS

Stereospecific living polymerizations with t-C_4H_9MgBr and t-C_4H_9Li-R_3Al have been successfully applied to copolymerization of MMA with other methacrylates, and block and random copolymers with high isotacticity or syndiotacticity can be prepared.

Sequential copolymerization of diphenylmethyl and triphenylmethyl methacrylates in THF with 1,1-diphenylhexyllithium at –78°C gives a block copolymer that comprises syndiotactic poly(diphenylmethyl methacrylate) block and isotactic poly(triphenylmethyl methacrylate) block. Hydrolysis of the block copolymer followed by methylation with diazomethane results in a stereoblock PMMA with narrow molecular weight distribution.[10]

Stereoblock PMMAs when annealed in the solid state form a stereocomplex easily as compared with the corresponding mixture of isotactic and syndiotactic PMMAs and the melting point of the complex is higher than that of the mixture.

TACTICITY DEPENDENT PROPERTIES

The glass transition temperature (T_g) of PMMA strongly depends on tacticity and is lower for isotactic PMMA than for syndiotactic PMMA.

The T_g of polymethacrylate also depends on the structure of ester group[11] and generally increases as the bulkiness of the side chain increases and its flexibility decreases.

Isotactic PMMA forms a double-strand helix in the crystal which consists of two chains with the same helix sense and direction (10/1 helix) but shifted 10.40 Å along the helix axis with respect to each other.[12] Syndiotactic PMMA forms isomorphous crystalline inclusion complexes with organic solvents such as acetone, chloroacetone, cyclohexanone, and benzene where the PMMA chains take a large helical structure. Crystallinity is lost on desorption of the solvent molecules.[13]

Marked differences in chemical reactivity are usually found among polymethacrylates of different tacticity. Isotactic PMMA is hydrolyzed more rapidly than syndiotactic PMMA.[14] Thermal degradation temperature (T_d) of the syndiotactic PMMA is higher than that of the isotactic PMMA with molecular weight less than 25,000, while the latter showed higher T_d than the former at higher molecular weight range.[11] The rate of anhydride formation was about 4 times higher for isotactic poly(methacrylic acid) than for atactic polymer.[15]

Syndiotactic sodium polymethacrylate is much more biodegradable than the isotactic polymer under aerobic conditions.

REFERENCES

1. Fox, T. G.; Garret, B. S. et al. *J. Am. Chem. Soc.* **1958**, *80*, 1768.
2. Miller, R. G. J.; Mills, B. et al. *Chem. Ind.* **1958**, 1323.
3. Yuki, H.; Hatada, K. *Adv. Polym. Sci.* **1979**, *31*, 1.
4. Hatada, K.; Kitayama, T.; Ute, K. *Prog. Polym. Sci.* **1988**, *13*, 189.
5. Nishioka, A.; Watanabe, H. et al. *J. Polym. Sci.* **1960**, *48*, 241.
6. Kitayama, T.; Zhang, Y.; Hatada, K. *Polym. J.* **1994**, *26*, 868.
7. Nakano, T.; Mori, M.; Okamoto, Y. *Macromolecules* **1993**, *26*, 867.
8. Varshney, S. K.; Hautekeer, J. P. et al. *Macromolecules* **1990**, *23*, 2618.
9. Yasuda, H.; Yamamoto, H. et al. *J. Am. Chem. Soc.* **1992**, *114*, 4908.
10. Doherty, M. A.; Hogan-Esch, T. E. *Makromol. Chem.* **1986**, *187*, 61.
11. Hatada, K.; Kitayama, T. et al. *J. Macromol. Sci., Pure Appl. Chem.* **1993**, *A30*, 645.
12. Kusanagi, H.; Tadokoro, H.; Chatani, Y. *Macromolecules* **1976**, *9*, 531.
13. Kusuyama, H.; Miyamoto, N. et al. *Polym. Commun.* **1983**, *24*, 119.
14. Semen, J.; Lando, J. B. *Macromolecules* **1969**, *2*, 570.
15. Geuskens, G.; Hellinckx, H.; David, C. *Eur. Polym. J.* **1971**, *7*, 561.
16. Abe, Y.; Matsumura, S. et al. *Yukagaku* **1984**, *33*, 228.

Stereoselective Polymerization

See: Rare Earth Polymerization Initiators

STIMULI-RESPONSIVE MICROCAPSULES

Kenji Kono and Toru Takagishi
Department of Applied Materials Science
College of Engineering
Osaka Prefecture University

Microcapsules can entrap a variety of chemicals in the inner spaces. Because encapsulated materials are separated by the capsule membrane from the environment, damage of the materials induced by a number of factors in the environment can be reduced.

Recently, as a new type of microcapsule, stimuli-responsive microcapsules, (**Figure 1**) have been proposed. Stimuli-responsive microcapsules possess a potential usefulness that conventional microcapsules never have.

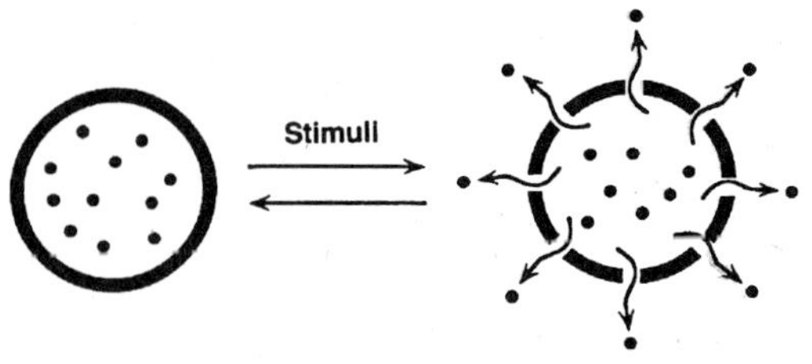

FIGURE 1. Schematic illustration of stimuli-responsive microcapsules.

PREPARATION AND PROPERTY

Surface-Modified Microcapsule

Among techniques for preparation of stimuli-responsive microcapsules, modification of microcapsule membranes is best studied. Rosenthal and Chang first reported incorporating lipids with a semipermeable microcapsule membrane and showed that lipid layers on the microcapsule act as an effective barrier to permeation of small molecules through the microcapsule membrane.[1] They prepared microcapsules with ultra-thin nylon membranes by interfacial polymerization and coated the microcapsule membranes with egg lecithin by incubating microcapsules in a solution of the lipid in tetradecane.

A number of synthetic lipids and biolipids form bilayer with gel-to-liquid crystalline phase transition at a certain temperature, the phase transition temperature (T_c). This phase transition involves conversion of a relatively ordered arrangement of hydrocarbon chains below the T_c to a relatively disordered arrangement above the T_c. Permeation of water soluble solutes such as metal ions are quite limited through the lipid bilayer membranes in the gel-state, whereas the permeation is enhanced above the T_c.

Okahata and his co-workers have used bilayers for temperature-sensitization of semipermeable nylon capsule membranes.[2] They prepared nylon capsules with 2 mm of diameter coated with the bilayer membranes of dialkyldimethylammonium bromide ($2C_nN^+2C_1$, n = 12–18).

Several attempts to render microcapsule membranes stimuli-responsive have also been made by modification of their surface with stimuli-sensitive polymers. Kokufuta et al. prepared polystyrene microcapsules with diameter of 10 μm and coated their surfaces with polyiminoethylene.[3] They also prepared pH-sensitive microcapsules using two kinds of copolymers of maleic acid with methyl vinyl ether and with styrene as the membrane surface-modifiers.[4] Based on a similar concept, Okahata et al. prepared thermo-responsive capsules and pH-responsive capsules by combining thermosensitive polymer grafts such as poly(N-isopropylacrylamide), and weak polyelectrolytes grafts, such as poly(methacrylic acid) and poly(4-vinylpyridine), respectively.[5,6]

Microcapsules Consisting of Stimuli-Sensitive Membranes

A number of stimuli-responsive microcapsules have been prepared by modification of the surface of stable semipermeable microcapsule membranes.

Miyauchi et al. prepared microcapsule with ~ 60 μm of diameter consisting of poly(L-lysine-*alt*-terephthalic acid), which has a large number of carboxyl groups.[7]

Microcapsules made of polyelectrolyte complex membrane was also shown to induce alteration of the membrane structure responding to external pH change.[9] When weak polyanions and/or weak polycations are used as microcapsule membrane components, structure of the membrane changes depending on pH. This type of microcapsule is used for encapsulation of living cells.[9]

Polyelectrolyte complex capsule is favorable for preparation of other stimuli-responsive capsules. Because introduction of molecules or groups with various functionalities into polyelectrolyte complex capsule membranes is possible by using functional molecule-attached polyelectrolytes as capsule membrane components, polyelectrolyte complex capsules with various functionalities can be made. According to the strategy, partly crosslinked poly(acrylic acid)polyethylenimine complex capsules containing a copolymer of acrylic acid and bis(4-(dimethylamino)phenyl)(4-vinylphenyl)methyl leucohydroxide with 6 mm of diameter were prepared.[10] Triphenylmethane leucohydroxide residues in the copolymer dissociate into ionic pairs under ultraviolet light irradiation and thermally recombine with the counter ion.

In a similar manner, various functional molecules, such as thermo-sensitive polymers and enzymes, can be incorporated into the capsule membrane.[11–14] Therefore, it is expected that various functional capsules can be made using polyelectrolyte complex capsules.

Microcapsules Containing Stimuli-Sensitive Molecules Inside

When functional molecules that can regulate adsorption and desorption of small molecules responding to stimuli are encapsulated in microcapsules with semipermeable membranes, release of the small molecules can be controlled by the stimuli. For example, Makino et al. prepared nylon microcapsules with 70 μm of diameter-containing concanavalin A and succinyl-amidophenyl-glucopyranoside insulin as a glucose-responsive insulin delivery system.[15]

Igari et al. have developed microencapsulated liposome as a novel drug-delivery system.[16]

A number of polymers with increased binding affinity for small molecules have been synthesized.[17–20] In general, binding affinity of these polymers vary depending on temperature.

APPLICATIONS

Microcapsules are being applied in a great many fields. Microcapsules with new functionalities should render their potential range of application more widely. One of the most suitable fields for application of stimuli-responsive microcapsules is drug-delivery system. Release of drugs from these microcapsules is controlled by external stimuli such as heat, electric field, and light, or responding to environmental changes, such as pH change and glucose concentration change. Therefore, they can supply a proper amount of drugs to a suitable site at a suitable time.

For this purpose, a number of responsive systems have been reported using polymers.[21,22] Because the release rate from microcapsules is generally controlled by diffusion rate of chemicals in their thin membranes, quick response of release to the stimuli can be expected compared with other systems such as polymer gels.

In addition, microcapsules coated with lipid bilayer membranes have a surface similar to biomembrane surfaces and hence, the surface of the microcapsule is considered to be highly biocompatible. In fact, when zwitterion-type lipid molecules such as phosphatidylcholine, which are major components of the external leaflet of the erythrocyte membrane, are used as a coating lipid, platelet adhesion onto the lipid membrane-coated capsule membrane was shown to be suppressed.[23] Moreover, when lipid membranes coated on the microcapsule membrane are combined with functional molecules such as membrane proteins, the resultant microcapsules can achieve biological functions.[24,25] The situations mentioned above might elevate the importance of the microcapsules for the therapeutic application.

REFERENCES

1. Rosenthal, A. M.; Chang, T. M. S. *J. Membr. Sci.* **1980**, *6*, 329.
2. Okahata, Y.; Lim, H. J.; Nakamura, G.; Hachiya, S. *J. Am. Chem. Soc.* **1983**, *105*, 4855.
3. Kokufuta, E.; Sodeyama, T.; Katano, T. *J. Chem. Soc., Chem. Commun.* **1986**, 641.
4. Kokufuta, E.; Shimizu, N.; Nakamura, I. *Biotech. Bioeng.* **1988**, *32*, 289.
5. Okahata, Y.; Noguchi, H.; Seki, T. *Macromolecules* **1986**, *19*, 498.
6. Okahata, Y.; Noguchi, H.; Seki, T. *Macromolecules* **1987**, *20*, 15.
7. Miyauchi, E.; Togawa, Y.; Makino, K.; Ohshima, H.; Kondo, T. *J. Microencaps.* **1992**, *9*, 329.
8. Kono, K.; Tabata, F.; Takagishi, T. *J. Membr. Sci.* **1993**, *76*, 233.
9. Goosen, M. F. A. *CRC Rev. Biocompatibility* **1987**, *3*, 1.
10. Kono, K.; Nishihara, Y.; Takagishi, T. *J. Appl. Polym. Sci.* **1995**, *56*, 707.
11. Heskins, M.; Guillet, J. E. *J. Macromol. Sci. Chem.* **1968**, *A2*, 1441.
12. Kono, K.; Hayashi, H.; Takagishi, T. *J. Controlled Release* **1994**, *30*, 69.
13. Albin, G.; Horbett, T. A.; Ratner, B. D. *J. Controlled Release* **1985**, *2*, 153.
14. Ito, Y.; Casolaro, M.; Kono, K.; Imanishi, Y. *J. Controlled Release* **1989**, *10*, 195.
15. Makino, K.; Mack, E. J.; Okano, T.; Kim, S. W. *J. Controlled Release* **1990**, *12*, 235.
16. Igari, Y.; Kibat, P. G.; Langer, R. *J. Controlled Release* **1990**, *14*, 263.
17. Takagishi, T.; Hayashi, A.; Kuroki, N. *J. Polym. Sci. Polym. Chem. Ed.* **1982**, *20*, 1533.
18. Takagishi, T.; Baba, T.; Hosokawa, H. *Macromol. Chem. Rapid Commun.* **1988**, *9*, 553.
19. Takagishi, T.; Kim, Y. J.; Hosokawa, T.; Morimoto, K.; Kono, K. *J. Polym. Sci. Polym. Chem. Ed.* **1993**, *31*, 365.
20. Takagishi, T.; Morimoto, K.; Okajima, S.; Chikusa, Y.; Kono, K. *Colloid Polym. Sci.* **1994**, *272*, 1627.
21. Kost, J.; Langer, R. *Trends Biotechnol.* **1992**, *10*, 127.
22. Stevenson, W. T. K.; Sefton, M. V. *Trends Polym. Sci.* **1994**, *2*, 98.
23. Kono, K.; Ito, Y.; Kimura, S.; Imanishi, Y. *Biomaterials* **1989**, *10*, 455.
24. Kono, K.; Kimura, S.; Imanishi, Y. *Bull. Chem. Soc. Jpn.* **1989**, *62*, 3587.
25. Kono, K.; Kimura, S.; Imanishi, Y. *J. Membr. Sci.* **1990**, *50*, 85.

Stimuli-Responsive Polymers

See: *Environmentally Responsive Gels*
Functional Immobilized Biocatalysts (Stimulus-Sensitive Gels)
Glucose Responsive Polymers (Medical Applications)
Intelligent Membranes
Intelligent Polymers (in Medicine and Biotechnology)
Smart Hydrogels
Stimuli-Responsive Microcapsules
Stimuli-Responsive Polymers (Thermodynamics in Drug Delivery Technology)

STIMULI-RESPONSIVE POLYMERS (Thermodynamics in Drug Delivery Technology)

Mario Casolaro
Department of Chemistry
University of Siena

In the last few years, interest in intelligent materials has grown. New polymers with both hydrophilic and hydrophobic characteristics, responsive to external physico-chemical (pH, redox, heat, etc.) and biochemical (glucose) signals, have been synthesized.[1–5] In the latter case, response is obtained with biological components such as enzymes (e.g., GOD glucose oxidase) that act as signal transducers in glucose-sensitive insulin releasing systems.[6–9] The presence of a large number of ionizable groups along the macromolecular chain significantly modifies conformational properties due to electrostatic forces. An important group of water-soluble, nonionic polymers, which form thermoreversible gels having a wide range of *LCST* (*Lower Critical Solution Temperature*), is based on N-alkylacrylamide homopolymers and copolymers with or without acidic/basic comonomers.[10–12]

Thermodynamic aspects of water soluble functional polymers incorporating carboxyl acid groups and hydrophobic moieties have recently been studied as stimulus-responsive polymers, contracting and lengthening reversibly in response to external conditions such as pH and temperature.[3,5,19] Macromolecular extension and contraction can be enlarged by constructing a three-dimensional polymer network in hydrogels[20] or membranes having chemical valve function.[5,8,19,21–27] Research has been performed on the feasibility of using such systems as switches for chemical separation and as drug delivery devices.[8,19,26,28]

PROPERTIES

Thermo-Sensitive Polymers

A polymer solution that shows phase separation on heating has a lower critical solution temperature (LCST).[10]

Among polymers showing LCST in aqueous solution, poly(N-isopropylacrylamide) (PNIPAAm) has been studied in crosslinked[13–18] and random copolymer networks such as thermosensitive hydrogels.[12] Changing the critical hydrophilic/hydrophobic balance causes a change in LCST, whereby thermosensitivity can be controlled by adding hydrophobic or hydrophilic comonomers such as butyl methacrylate or acrylamide. A useful feature of thermosensitive polymers is the possibility of controlling their LCST by varying the monomer composition. In general, incorporation of hydrophobic comonomers lowers the LCST, while it increases with hydrophilic or ionizable comonomers.[12,23] A major reason for the great interest in PNIPAAm is the fact that it has an LCST around 32°C, close to body temperature, thus making it an attractive system for pharmaceutical applications, such as controlled drug delivery and solute separation.[13,14] The behavior of PNIPAAm grafted on membranes was recently studied by Iwata et al. with a view to constructing temperature-sensitive devices to regulate filtration characteristics.[23]

Besides simple cloud-point measurements for the characterization of LCST phenomena, solution calorimetry seems to be a powerful technique devoid of complications due to variations in precipitated aggregate sizes.[32] This technique gives thermodynamic parameters that provide insight into the forces responsible for the transition. Use of the sensitive scanning microcalorimeter, developed to study structural transitions in proteins, lipids, and nucleic acids, allows LCST transition parameters to be determined with precision.[33–35]

We recently studied thermodynamic behavior of PNIPAAm during the protonation of carboxylate groups in copolymers with N-methacryloyl-L-valine (MAVA) or N-acryloyl-L-valine (AVA).[19,30,31]

Other Stimuli-Sensitive Polymers

Photoresponsive polymers are usually obtained by introducing photochromic molecules into macromolecules. Several synthetic polypeptides, like poly(L-glutamic acid) containing azobenzene or other molecules in a side group, have been prepared and studied for reversible variations in their conformation similar to photosensitive systems occurring in nature.[29] On irradiation, photochromic moieties undergo reversible stereochemical rearrangements which in turn induce structural changes in the macromolecular matrix. Photoresponse effects include light-induced variations in viscosity and solubility, photocontrol of membrane functions, photoregulation of binding and releasing of drugs, and photomechanical effects. Detailed reviews on this subject, including photosensitive gels, have been published.[36–38]

Other kinds of stimuli-responsive polymers are related to redox properties. They have been investigated for various purposes, such as polymeric reagents or electronic materials.[39] Two classes of such polymers can be distinguished, characterized by conjugated polymer backbones or separate redox-active moieties.[40] Two reagents based on lipoamide and nicotinamide derivatives have been considered with a view to application in the biomedical field and as components of drug delivery systems.[4,28]

APPLICATIONS

Chemical Valve Systems

A membrane with chemical valve function is a composite structure.[21] It consists of a porous substrate onto which a suitable stimuli-responsive polymer has been graft copolymerized. Expansion and contraction of the graft polymer chains is based on the specific interaction between micro- and macromolecular solutes contained in the permeant that gives rise to a significant conformational change in the grafted polymer.

The redox, temperature, and pH dependence of filtration characteristics through these composites was investigated, and it was found that water and solutes of different molecular weights permeate the membrane pores at rates dependent on the conformational state of the grafted polymers.[2,5,8,22,23,25–28] Several studies with membranes of 0.2 μm pore size and pH- and temperature-sensitive graft-polymers have been reported. Iwata et al. recently reported temperature-sensitive membranes of porous poly(vinylidene fluoride) grafted with PNIPAAm alone or copolymerized with acrylamide (hydrophilic) and n-butyl methacrylate (hydrophobic) and showed that the water flow rate may be switched on and off in a reversible manner as the temperature is cycled.[23] The temperature dependence of permeability changes according to the LCST change of the copolymer.

Permselectivity of pH-sensitive systems has been investigated for its ability to control solute separation of different molecular weight polyoxyethylene (POE) through porous cellulose membranes grafted with poly(acrylic acid), as a function of graft content.[8] The higher the graft content, the lower the permeation, and more solute permeated at low pH. Permeation also was controlled by pore size, and POE with a molecular weight from 6000 to 20,000 was the most suitable solute for permeation studies.

The permeability of insulin in relation to time showed a linear increase in solute concentration when diffusional experiments were performed through porous cellulose membrane grafted with polyelectrolytes.[8,26]

Self-Regulating Systems

Several polymer systems based on hydrogels in combination with the enzyme glucose oxidase (GOD) have been used to control insulin delivery at rates dependent on external glucose concentration.[6,7,16]

An alternative system, based on the chemical valve principle, is illustrated in **Figure 1**. A porous membrane with poly(carboxyl acid) grafted on the surface is prepared as a pH-responsive membrane.[8] The enzyme GOD is immobilized on the carboxyl group of the polymer to make a composite responsive to glucose concentrations. The GOD catalyzes the conversion of glucose to gluconic acid. The latter, being a relatively strong acid (logK 3.6),[41] protonates the basic carboxylate group of the polymer. In the absence of glucose and at physiological pH, carboxyl groups of the polymer are ionized and the repulsion between negative charges makes the polymer chains extend, obstructing the membrane pores. The presence of glucose

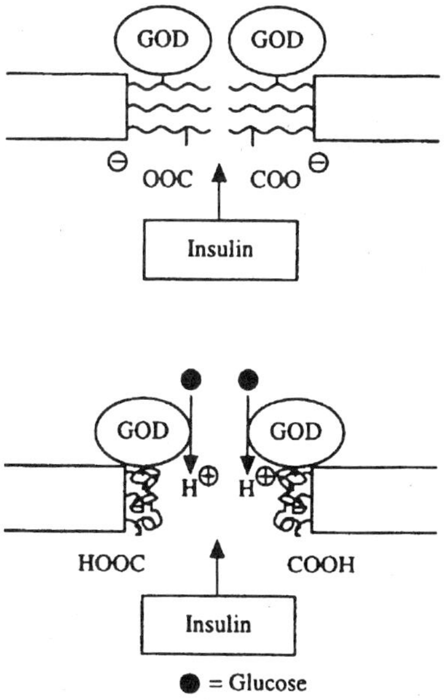

FIGURE 1. Principle of controlled release of insulin (GOD: glucose oxidase).

makes the polymer chains coil because transformation of glucose to gluconic acid by the enzyme GOD lowers pH of the microenvironment allowing protonation of the COO^- groups on the polymer.[26] The pores open for the prompt passage of insulin.

REFERENCES

1. Takagi, T.; Takahashi, K. et al. *Proceedings of the First International Conference On Intelligent Materials*; Technomic: Lancaster, PA, 1993.
2. Imanishi, Y. *Synthesis Of Biocomposite Materials* CRC: Boca Raton, FL, 1992.
3. Barbucci, R.; Casolaro, M.; Magnani, A. *Coord. Chem. Rev.* **1992**, *120*, 29.
4. Casolaro, M.; Busi, E. *Polymer* **1994**, *35*, 360.
5. Casolaro, M. *Reactive Polymers* **1994**, *23*, 71.
6. Ishihara, K.; Kobayashi, M. et al. *Polym. J.* **1984**, *16*, 625.
7. Kost, J.; Horbett, T. A. et al. *Biomed. Mat. Res.* **1985**, *19*, 1117.
8. Ito, Y.; Casolaro, M.; Kono, K.; Imanishi, Y. *J. Controlled Release* **1989**, *10*, 195.
9. Chung, D. J.; Ito, Y.; Imanishi, Y. *J. Controlled Release* **1992**, *18*, 45.
10. Heskins, M.; Guillet, J. E. *J. Macromol. Sci., Chem.* **1968**, *A2*(8), 1441.
11. Bae, Y. H.; Okano, T.; Kim, S. W. *J. Polym. Sci., Polym. Phys.* **1990**, *28*, 923.
12. Feil, H.; Bae, Y. H. et al. *Macromolecules* **1993**, *26*, 2496.
13. Hoffman, A. S. *J. Controlled Release* **1987**, *6*, 297.
14. Hoffman, A. S.; Afrassiabi, A.; Dong, L. C. *J. Controlled Release* **1986**, *4*, 213.
15. Bae, Y. H.; Okano, T. et al. *Makromol. Chem. Rapid Comm.* **1987**, *8*, 481.
16. Bae, Y. H.; Okano, T.; Kim, S. W. *J. Controlled Release* **1989**, *9*, 271.
17. Gutowska, A.; Bae, Y. H. et al. *J. Controlled Release* **1992**, *22*, 95.
18. Takeuchi, S.; Omodaka, I. et al. *Makromol. Chem.* **1993**, *194*, 1991.
19. Casolaro, M.; Barbucci, R. *Encyclopedia Handbook of Biomaterials and Bioengineering, Part A: Materials*, Vol. 2, Wise, D. L. et al., Eds.; Marcel Dekker, ch. 41, 1385–1412 (1995).
20. Harland, R. S.; Prud'homme, R. K. *Polyelectrolyte Gels*; ACS: Washington, 1992.
21. Osada, Y. *Adv. Polym. Sci.* **1987**, *82*, 1.
22. Iwata, H.; Matsuda, T. *J. Membr. Sci.* **1988**, *38*, 185.
23. Iwata, H.; Oodate, H. et al. *J. Membr. Sci.* **1991**, *55*, 119.
24. Maekawa, S.; Gong, J. P.; Osada, Y. *Makromol. Rapid Commun.* **1994**, *15*, 73.
25. Ito, Y.; Inaba, M.; Chung, D. J.; Imanishi, Y. *Macromolecules* **1992**, *25*, 7313.
26. Barbucci, R.; Casolaro, M.; Magnani, A. *J. Controlled Release* **1991**, *17*, 79.
27. Casolaro, M.; Barbucci, R. *Coll. Surf. A: Phys. Chem. Eng. Asp.* **1993**, *77*, 81.
28. Russo, N.; Anastassopoulou, J.; Barone, G. *Properties and Chemistry of Biomolecular Systems*; Kluwer: Dordrecht, 1994.
29. Pieroni, O.; Fissi, A. *J. Photochem. Photobiol. B: Biol.* **1992**, *12*, 125.
30. Casolaro, M.; Busi, E.; Landi, F. *The Development of Science For The Improvement Of Human Life: Proceedings of the 2nd Kyoto-Siena Symposium*, Kyoto, 1994.
31. Supramolecular Science, Casolaro, M. submitted.
32. Wolf, B. A. *Pure Appl. Chem.* **1985**, *57*, 323.
33. Hinz, H. J. *Methods Enzymol.* **1986**, *130*, 59.
34. Battistel, E.; Luisi, P. L.; Rialdi, G. *J. Phys. Chem.* **1988**, *92*, 6680.
35. Jackson, W.; Brandts, J. *Biochemistry* **1970**, *9*, 2294.
36. Smets, G. *Adv. Polym. Sci.* **1983**, *50*, 17.
37. Irie, M. *Adv. Polym. Sci.* **1990**, *94*, 27.
38. Pieroni, O.; Fissi, A.; Ciardelli, F. *Photochem. Photobiol.* **1986**, *44*, 789.
39. Laschewsky, A.; Ward, M. D. *Polymer* **1991**, *32*, 146.
40. Manecke, G.; Storck, W. *Encyclopedia of Polymer Science and Engineering*; Wiley: New York, **1986**, *5*, 725.
41. Martell, A. E.; Smith, R. M. *Critical Stability Constants*; Plenum: New York, 1974.

STRESS-INDUCED CHEMILUMINESCENCE IMAGING

Satoru Hosoda,* Yoshinori Seki, and Hayato Kihara
Petrochemicals Research Laboratory
Sumitomo Chemical Company Ltd.

Generally, degradation of polymer materials is caused by various factors such as heat, UV light, irradiation, ozone, mechanical stress, and electrical stress. The process of degradation is accompanied by very weak emission of chemiluminescence.[1,2] Since chemiluminescence has a high sensitivity to oxidative degradation,[3–5] its measurement has been used as a tool to evaluate stability of polymer materials against heat,[6,7] UV light,[8–10] electron beam,[11] and γ-ray irradiation.[12,13] Mechanical[14,15] and electrical stresses[16] also were reported to generate weak light emission, attributable to several processes including chemiluminescence from the bimolecular termination reaction of peroxy radicals. Mechanical stress-induced chemiluminescence (SICL) was measured by several authors[17–20] from the viewpoint of durability evaluation of polymer materials under stress. We have also reported some of the results on the CL imaging of polymers under heat and mechanical stress.[21]

EXPERIMENTAL

Chemilimunescence Measurement

Imaging experiments were carried out with an ARGUS 100 (Hamamatsu Photonics Co.) and a Quantum Microscope (Charles Evans and Associates), individually.

RESULTS AND DISCUSSION

Chemiluminescence Under the Mechanical Stress

Weak light emission has been reported for several kinds of polymers under mechanical stress. This luminescence is attributable to several processes including chemiluminescence caused by bimolecular termination of peroxy radicals. George et al.[19] reported results for SICL during deformation of nylon 6,6 fibers and concluded that the main part of chemiluminescence resulted from progressive scission of taut tie molecules in the amorphous phase.

Chemiluminescence Under the Electrical Stress

Many kinds of plastics and rubbers have been used to insulate electric cable. Low density polyethylene (PE) has been widely

*Author to whom correspondence should be addressed.

used as an insulator of power transmission cable through its high dielectricity and good processability. Many studies have focused on electrical breakdown under long-term high voltage application.

Bamji et al.[22,23] reported weak light emission when PE was charged by high voltage in nitrogen or *in vacuo*. They supposed the luminescence to be attributable to recombination of electron and hole injected into PE (electroluminescence). UV light thus emitted caused the chain scission of polyethylene molecules and finally generate the microcrack, which acts as a precursor of an electrical-tree breakdown. Since the luminescence appears in advance at the lower level of the charged voltage rather than that of an actual cable breakdown, electroluminescence is a premonitory phenomenon for power cable degradation. Kaminaga et al.[24] studied the degradation mechanism of PE power cable using luminescence measurement and spectrum analysis and found that the luminescence spectrum was different from that of thermal oxidation or of the void discharge.

We also measured electroluminescence from a PE sheet with needle-plane electrodes under charging an alternating current at room temperature and successfully obtained an image of electroluminescence. Luminescence was observed above some threshold value of voltage, and thereafter the total intensity of the luminescence increased with increasing the voltage.

Chemiluminescence imaging is a promising technique for visualizing and predicting sites of degradation induced by mechanical and electrical stresses in various polymer materials. This technique could be developed to study the effect of the causes that affect polymer degradation.

ACKNOWLEDGMENT

The authors express their appreciation to Sumitomo Chemical Co. Ltd. for permission to publish this paper.

REFERENCES

1. Ashby, G. E. *J. Polymer. Sci.* **1961**, *L*, 99.
2. Schard, M. P.; Russell, C. A. *J. Appl. Polym. Sci.* **1964**, *8*, 985.
3. Rychla, L. M.; Fodor, Zs.; Rychly, J. *Polym. Deg. Stab.* **1981**, *3*, 371.
4. George, G. A. *Development in Polymer Degradation* 3; Grassie, N., Ed.; Applied Science: New York, 1981.
5. Zlatkevich, L. *J. Polym. Sci., Polym. Lett. Ed.* **1983**, *21*, 571.
6. Hosoda, S.; Kihara, H. ANTEC'88, 941 **1988**.
7. Kihara, H.; Hosoda, S. *Polym. J.* **1990**, *22*, 763.
8. Osawa, Z.; Konoma, F. et al. *Polym. Photochemistry* **1986**, *7*, 337.
9. Chen, J.; Konoma, F.; Osawa, Z. *Polym. Photochemistry* **1986**, *7*, 469.
10. Kihara, H.; Yabe, T.; Hosoda, S. *Polymer Bull.* **1992**, *29*, 369.
11. Yoshii, F.; Sasaki, T. et al. *Ikigaku* **1986**, *56*, 252.
12. Yoshii, F.; Sasaki, T. et al. *J. Appl. Polym. Sci.* **1986**, *31*, 1343.
13. Itoh, M. *Nihon Gomukyoukaishi* **1985**, *58*, 461.
14. Streletskii, A.; Butyagin, P. *Polym. Sci., USSR* **1973**, *15*, 739.
15. Butyagin, P. U.; Yerofeyev, V. S. et al. *Polym. Sci. USSR* **1970**, *12*, 330.
16. Bamji, S.; Bulinki, A.; Densley, J. *Makromol. Chem. Macromol. Symp.* **1989**, *25*, 271.
17. Fanter, D.; Levy, R. *Chemtech.* **1979**, 682.
18. Monaco, S.; Richardson, J. et al. *Ind. Eng. Chem. Proc. Res. Dev.* **1982**, *21*, 546.
19. George, G. A.; Egglestone, G. T.; Riddell, S. Z. *J. Appl. Polym. Sci.* **1982**, *27*, 3999.
20. Kihara, H.; Seki, Y.; Hosoda, S. *Polymer '91, Prep.* **1991**, 329.
21. Hosoda, S.; Seki, Y.; Kihara, H. *Polymer* **1993**, *34*, 4602.
22. Bamji, S.; Bulinski, A.; Densly, J. *J. Appl. Phys.* **1987**, *61*, 694.
23. Bamji, S.; Bulinski, A.; Densly, J. *Makromol. Chem., Macromol. Symp.* **1990**, *25*, 271.
24. Kaminaga, K.; Shigetsugu, H. et al. *Conference Record of the 1994 IEEE International Symposium on Electrical Insulation* Pittsburgh, **1994**; 336.

Stresses

See: Injection Moldings (Residual Stresses)

Structural Adhesives

See: Reactive Liquid Rubbers (Epoxy Toughening Agents)

Styrene-Acrylic Emulsions

See: Floor Finishes

Styrene-Acrylonitrile Copolymers

See: Acrylonitrile-Butadiene-Styrene
Blends (Thermoplastic Polyurethane-SAN)
Bulk Polymerization

Styrene-Butadiene Copolymers

See: Styrene-Butadiene-Styrene Elastomer, Hydrogenated
Styrene-Butadiene-Styrene Triblock Copolymer
Styrenic Resins

STYRENE-BUTADIENE-STYRENE ELASTOMER, HYDROGENATED

Yasushi Kishimoto
Polyolefins Development Department
Asahi Chemical Industry Company, Ltd.

Hydrogenated styrenic block copolymer represented as SEBS is usually a 3-block copolymer with a rubbery block between two polystyrene blocks.

The mid-block has been selectively and quantitatively hydrogenated to remove the unsaturated double bonds from the polybutadiene to give an ethylene/butylene structure.

CHARACTERISTICS

Since SEBS has no unsaturated group in the polymer main chain, it has excellent weatherability and heat resistance and superior electrical and hygienic properties. It also has a good balance of other properties, such as tensile strength and elongation, usability at low temperatures, and good compatibility with various plastics, plasticizers, and fillers.

Compared with other thermoplastic elastomers (TPEs), SEBS is the most rubber-like, with a wide range of performance profiles through compounding.

Modified SEBS, developed recently,[1] has greater adhesive properties and is more compatible with plastics.

Modified SEBS, has the potential for use as a reactive modifier for polymers. For example, when modified SEBS is blended with polyamide (Nylon 6,66), the modified SEBS reacts with functional groups in polyamide to generate graft copolymer, which disperses SEBS in the Nylon matrix as small rubber particles to increase the impact resistance of the blend.[2]

APPLICATIONS

SEBS itself is usually used as a neat polymer, and its compounds are used for a wide range of applications.

In case of SEBS neat polymer, including modified SEBS, lower molecular weight SEBS is used in adhesives and sealants with tackifier or softener for diapers and sanitary goods. It is also used in forming multilayered films or sheets because it adheres well to general-purpose plastics, engineering plastics, and metals.

Another important use of SEBS polymer is as an impact modifier of plastics, (such as polystyrene and polypropyrene) rubber components, and binders for polymer alloys. In particular, modified SEBS is highly appropriate for use as an engineering plastics modifier, as mentioned above. Also, modified SEBS is superior for use in the modification of thermo-setting resins, such as FRP or epoxy.

However SEBS polymer with relatively higher molecular weight is used for thermoplastic elastomer compounds. SEBS compounds basically consist of SEBS polymer, polypropyrene, and a plasticizer, such as paraffinic oil and filler.

Because SEBS are excellent elastomeric materials with good processability and cost performance, they are used for molded parts or multilayer molded parts with injection molding, extrusion molding, or blow molding for automotive interiors or exteriors, such as air-bag housing; electrical appliances, such as gaskets; OA equipment, such as functional buttons and cushions; and for Food and Drug Administration-approved uses of medical equipment.[3]

Several new applications of SEBS as elastomeric materials are expected.

REFERENCES

1. Kishimoto, Y.; Hirata, A. Presented at the 28th Annual Meeting of the IISRP, Edinburgh 1987.
2. Kishimoto, Y.; Sakamoto, H.; Kato, K.; Morita, H. *Polymer Preprints (Japan)* **1991**, *40*(3), 779.
3. Asahi Chemical Industry Co. Ltd. Japanese Patent 94-200086A 1994.

STYRENE-BUTADIENE-STYRENE TRIBLOCK COPOLYMER

Ging-Ho Hsiue
Department of Chemical Engineering
National Tsing Hua University

The triblock copolymer, poly(styrene-*b*-butadiene-*b*-styrene) (SBS), has received great attention since its 1965 commercialization by Shell Chemical Company under the trademark Kraton®. SBS readily passed the test of elastomer. Moreover, SBS has the solubility and thermoplastic properties of polystyrene, while at ambient temperatures it has the toughness and resilience of vulcanized natural rubber or polybutadiene. These characteristics result from its unique molecular structure. At ambient temperature SBS is physically indistinguishable from conventional vulcanized elastomers, but chemically consists of individual chains with a complete absence of covalent crosslinks between the polymer chains. The thermoplastic rubber named as a thermoplastic elastomer really is a unique class of polymer.

PHYSICAL CHARACTERISTICS

SBS is commercially important because of its physical characteristics. A variety of adhesives, coatings, films, extrusions, and molded articles made from SBS can be obtained using many techniques, such as solution casting, extrusion, and injection molding.

MORPHOLOGY

Patent and scientific literature describes numerous molecular variations. The key requirement is that the molecules of the thermoplastic elastomer terminate with hard, glass-like endblocks, which are incompatible with the rubbery midblocks.[1–5] Consequently, all such polymers consist of two phases in the solid state—a continuous rubber phase and a basically discontinuous plastic phase that "locks" the rubber molecules in place.

MICROPHASE SEPARATION OF SBS

Many works have been published on the investigation of the physical structure of thermoplastic elastomer.[6–9] Two glass-transition temperature (T_g) peaks are found in dynamic mechanical tests with an SBS block copolymer. In contrast, only one peak is found in a random SBR copolymer with the same styrene/butadiene ratio.

PHYSICAL PROPERTIES

Stress-Strain Properties

The stress-strain relationships of SBS indicated that the triblock copolymer has high stress and strain values in comparison with the glassy and rubbery polymers. The stress-strain properties indicate that SBS has good hysteresis property.

Solubility and Solution Viscosity

The behavior of thermoplastic elastomer toward solvents is unique because of the two segments present in each molecule. Each segment retains it own solubility parameter. A given polymer will dissolve only in solvents whose solubility parameters are close to those of the polymer. In thermoplastic elastomer, two solubility parameters are involved, one for the endblock and one for the midblock. A "good" solvent for a thermoplastic elastomer must therefore be one that dissolves both endblocks and midblocks.

The best solvents for SBS have the solubility parameters between 7.4 (diethyl ether) and 10.3 (bromobenzene).

The solvents influence the viscosity of the solution. The SBS shows a maximum viscosity in solvents whose solubility parameters are either slightly higher than polystyrene or slightly lower than polybutadienes. Solutions in solvents with solubility parameters in between those of two homopolymers show a lower viscosity.

MODIFICATION OF SBS

SBS could be chemically and/or physically modified for various applications. The modification of SBS is feasible since it possesses an unsaturated polybutadiene structure.

GENERAL APPLICATIONS

The SBS with suitable physical and chemical properties can be utilized in many fields, such as (1) adhesives, caulks, sealants, and coatings, and (2) extrusion and moulding. These groups roughly coincide with the use of the polymers as solutions and as heated fluids, respectively. A notable exception to this correlation is the application of SBS in hot-melt adhesives.

REFERENCES

1. Ceresa, R. J. *Block and Graft Copolymerization*; John Wiley & Sons: New York, 1973; Vol. 1.
2. U.S. Patent 3 239 478, 1966.
3. Zelinski, R.; Childers, C. W. *Rubber Chem. and Tech.* **1968**, *41*, 161.
4. Marrs, O. L.; Edmonds, L. O. *Adhesives Age* **1971**, 14, 15.
5. Meier, D. J. *J. Polym. Sci. Part C* **1969**, *26*, 81.
6. Bradford, E. B.; Vanzo, E. *J. Polym. Sci. Part A-1* **1968**, *6*, 1661.
7. LeGrand, D. G. *Polymer Letters* **1970**, *8*, 195.
8. Folkes, M. J.; Keller, A. *Polymer* **1971**, *12*, 222.
9. Lewis, P. R.; Price, C. *Polymer* **1972**, *13*, 20.

Styrene-Ethylene-Butylene-Styrene Copolymer

See: Styrene-Butadiene-Styrene Elastomer, Hydrogenated

STYRENE-MALEIC ANHYDRIDE COPOLYMER

Kang I. Lee
Monsanto Company

Styrene-maleic anhydride copolymer has been a well-known polymeric system since the early 1930s.[1] Unlike most other polymers, styrene-maleic anhydride copolymeric materials are so chemically versatile that one can modify the material at one's will. This is because of the presence of anhydride functionalities in the polymer system.

PREPARATION

Polymerization

Styrene-maleic anhydride copolymers can be prepared by many different polymerization techniques. Since maleic anhydride can not be homopolymerized, no more than 50 mol% maleic anhydride can be incorporated into the copolymer. Many of the styrene-maleic anhydride copolymers with low maleic anhydride content are being made commercially by continuous mass polymerization techniques.

A batch mass polymerization technique is also being commercially employed for preparing high molecular weight styrene-maleic anhydride copolymers.[2]

Most commercial processes for preparing styrene-maleic anhydride copolymers with high maleic anhydride contents are conducted in aromatic hydrocarbon solvents by means of the precipitation polymerization technique.[3]

The rubber-modified styrene/maleic anhydride/various acrylate terpolymers are also extensively studied.[4] Recently a blend of the styrene-maleic anhydride copolymer with the poly(butylene terephthalate) was announced as a new product. It is claimed to have improved tensile strength and high-use-temperature over 100°C, while it is cheaper than the poly(butylene terephthalate). The application of this material was suggested for the automobile bumper area because of its excellent shock resistance.

Low molecular weight styrene-maleic anhydride copolymers are frequently added to numerous coatings formulations. The presence of styrene in the polymer provides the hardness of the coating, while the presence of the polar maleic group enhances the adhesion to the substrate. For example, styrene-maleic copolymers are being used as latex paint additives.

Low molecular weight styrene-maleic anhydride copolymeric materials have been used in floor polishes as leveling agents due to the low solution viscosity and the high functionality. The added advantage of using the styrene-maleic anhydride copolymeric material in polishes is its optical clearity.

Many of the 1:1 alternating styrene-maleic anhydride copolymers are being used as flattening agents in powder coatings. Styrene-maleic copolymers have been used for excellent pigment dispersant in many different paint applications. Also, the copolymer is being used in ink formulation as dispersing agents.

Esterified styrene-maleic copolymers are used for adhesive applications. An example is the beer bottle label adhesive formulation.

The water solubility of various styrene-maleic anhydride copolymeric derivatives provides very convenient means of applying the materials to various substrates.

The large application area of the styrene-maleic copolymer is in the fabric size area.

Modification

Much chemical modification work of the styrene-maleic anhydride copolymer has been done in order to change its degree of hydrophilicity and other properties. Examples include reactions with alcohols, ammonia, amines, and alkalies.

PROPERTIES

Styrene-maleic anhydride copolymers are known to have high heat distortion temperature. The heat stability of the polymer is much higher than homopolystyrene, but below a well-known engineering plastic, such as polycarbonate. The glass transition temperature of the 1:1 alternating styrene-maleic anhydride copolymer is as high as 170°C.

Most styrene-maleic copolymers have excellent adhesive properties. The polymer is very compatible with a number of different materials.

Moisture is an excellent plasticizer for the alkali salt of styrene-maleic anhydride copolymer.

APPLICATIONS

High molecular weight styrene-maleic anhydride copolymers with low anhydride contents are being used for the engineering plastics area where the high heat distortion temperature is required.

The amide acid ammonium salt of the styrene-maleic anhydride copolymer has received considerable attention in the area of paper surface sizing. In combination with starch, the material is applied on the surface of the paper to provide resistance to water and organic liquid. In addition to conventional sizing performance, the styrene-maleic-copolymer-treated paper improves the printability and porosity of the paper.

Many different ester forms of styrene-maleic anhydride copolymers are being used as the dry film photoresist binder applications.

The disodium salt and the half-amide, half-ammonium salts of styrene-maleic anhydride copolymer can be effective emulsifying agents in soap stabilized emulsion polymerizations. The synthetic emulsifying system based on the styrene-maleic copolymer can be precipitated in the acidification step and eliminate scum formation and pollution problems without harmful effects on the finished polymer.

REFERENCES

1. Wagner-Jauregg, T. *Berichte* **1930**, *6*, 3212.
2. Russell, G. T.; Napper, D. H.; Gilbert, R. G. *Macromolecules* **1988**, *21*, 2141.
3. Hobden, F. W. *J. Oil and Colour Chemist's Assoc.* **1958**, *41*, 24.
4. Hall, W. J.; Kruse, R. L.; Mendelson, R. A.; Trementozzi, Q. A. *The Effects of Hostile Environments on Coatings and Plastics, ACS Symposium Series* **1983**, *229*, 49.

Styrene-Maleic Anhydride Copolymers

See: *Anticancer Polymeric Prodrugs, Targetable*
Poly(styrene-co-N-alkylmaleimide)s (by Reactive Extrusion; Miscibility in Blends)
Styrene-Maleic Anhydride Copolymer

STYRENE, STEREOSPECIFIC POLYMERIZATION

Marcos Lopes Dias
Instituto de Macromoléculas
Universidade Federal do Rio de Janeiro

Stereospecific polymerization of styrene was first reported by Natta and co-workers.[1,2] They prepared isotactic polystyrene (i-PS) employing a coordination catalyst derived from titanium tetrachloride and alkyl-aluminum compounds. Since then several authors have reported on the synthesis of i-PS promoted by different types of polymerization systems. However, syndiotactic polystyrene (s-PS) was first prepared only recently by using transition metal compounds and aluminoxanes.[3]

ISOSPECIFIC POLYMERIZATION

Coordination Catalysis

Polymerization with Heterogeneous Catalysts

Styrene polymerization with the heterogeneous Ziegler-Natta type catalysts activated by alkyl-aluminum compounds generally produce a mixture of nonstereoregular and isotactic polymer.

The first system used successfully to obtain i-PS was $TiCl_4$/alkyl-aluminum (e.g., $Al(C_2H_5)_3$ or $Al(C_2H_5)_2Cl$).[1,2] It is well known that in the isospecific polymerization using heterogeneous catalysts the steric control is attributed to the presence of chiral centers on the catalyst particle surface.

The features of the styrene polymerization with solid $TiCl_3$-alkyl-aluminum compounds were also reported.[4–6] These types of catalysts are highly regioselective, and the monomer insertion is primary.

A lot of data has been reported on the stereospecific polymerization of propylene with $MgCl_2/TiCl_4$ supported catalysts. Soga has developed a series of heterogeneous catalysts modifying the Hercules catalyst formed by $TiCl_3$ and $(RCp)_2Ti(CH_3)_2$.[7,8] Using a Solvay type $TiCl_3$ and $Cp_2Ti(CH_3)_2$ (Cp = cyclopentadienyl) as a catalyst for styrene, an extremely high stereoregular i-PS (> 90% isotactic) was obtained.

Polymerization with Homogeneous Catalysts

Polystyrene with isotactic structure prepared by coordination homogeneous catalysts was first reported by Ascenso and coworkers.[9]

High-molecular-weight polystyrene can be obtained by using homogeneous catalysts from $Ni(acac)_2/N(C_2H_5)_3$ (acac = acetylacetonate) activated by methylaluminoxane (MAO).[10]

Anionic Inhibition

Isotactic polymerization of styrene is promoted by some anionic initiator systems. Early experiments showed that at very low temperatures butyllithium (LiC_4H_9) in hydrocarbons polymerizes styrene producing crystalline polystyrene in the absence of ethers and alkyl chlorides.[11] The presence of traces of water was supposed to be responsible for the formation of stereoregular structures.

Cazzaniga and Cohen have also investigated the anionic polymerization aiming to produce block copolymers with control of the i-PS sequences.[12,13]

SYNDIOSPECIFIC POLYMERIZATION

Only recently the polymerization of styrene to syndiotactic polymer was achieved by using coordination catalytic systems consisting of titanium compounds and MAO.[3,14–16] A variety of titanium compounds may be used. Examples are alkyl-titanium (tetrabenzyltitanium), monocyclopentadienyl derivatives (e.g., η^5-cyclopentadienyltrichloro titanium), alkoxy-titanium (e.g., tetrabutoxy titanium), and titanocenes (e.g., biscyclopentadienyl titanium dichloride). Independently of the ligand bonded to the Ti center, compounds of Ti(IV), Ti(III), and Ti(II) valence states give active catalysts. Only Ti(0) catalysts produces a-PS.

Zirconium compounds and MAO are also able to produce syndiotactic polystyrene.[17,18] Nevertheless, these Zr derivatives in general are less effective than Ti based catalysts. Usually, polymer yields are lower and higher polymerization temperatures are required. The s-PS are less stereoregular.[19,20]

PROPERTIES OF STEREOREGULAR POLYSTYRENE

In the case of polystyrene, the product of common use has atactic structure (a-PS) and is an important commodity plastic. a-PS is an amorphous material and its transparency, absence of color, and good processability permit us to obtain a range of plastic ware with gloss and different degree of opacity. However, it has low mechanical resistance. Due to high configuration regularity, s-PS obtained with titanium catalysts is a very interesting material since it has a high melting temperature (≈ 270°C) and high stiffness. This polymer also exhibits high impact strength and toughness and excellent chemical and water resistance.

From the melt state, both i-PS and s-PS are able to crystallize giving rise to spherulitic structures. The growth rate of these structures depends on the crystallization temperature.[21,22]

Melting temperature, T_m, of s-PS varies with the polymerization temperature, probably as a function of the changes in the molecular weight. T_m from nearly 250 to 271°C is reported. The equilibrium melting temperature $T_m°$ was found to be 275°C. Glass transition temperatures in the range of 90 to 100°C have been reported in literature.

The syndiotactic polymer shows a very interesting and complex polymorphic behavior. Evidence has been seen of various crystalline forms.[23–32]

APPLICATIONS

Due to excellent thermal and mechanical properties, s-PS is considered a new engineering polymer with a niche at the middle to top end of the performance polymer market. Potential applications include films and molded parts requiring stiffness, chemical and heat resistance and dimensional stability. Highly crystalline grades of this polymer have been developed by Idemitsu Petrochemical and Dow Chemical, its U.S. licensee and research partner. They have made claims of performances comparable in some respect to other engineering plastics like nylon 6,6 and poly(phenylene sulfide).

REFERENCES

1. Natta, G.; Corradini, P. *Rend. Accad. Naz. Lincei* **1955**, *18*(8), 19.
2. Natta, G.; Corradini, P. *Makromol. Chem.* **1955**, *16*, 77.
3. Ishihara, N.; Seimiya, T.; Kuramoto, M.; Uoi, M. *Macromolecules* **1986**, *19*, 2465.
4. Burnett, G. M.; Tait, P. J. T. *Polymer* **1960**, *1*, 151.
5. Ammendola, P.; Tancredi, T.; Zambelli, A. *Macromolecules* **1986**, *19*, 307.
6. Benaboura, A.; Deffieux, A.; Sigwalt, P. *Makromol. Chem.* **1987**, *188*, 21.
7. Soga, K.; Yanagihara, H. *Makromol. Chem., Rapid Commun.* **1988**, *9*, 23.
8. Soga, K.; Uozumi, T.; Yanagihara, H.; Shiono, T. *Makromol. Chem., Rapid Commun.* **1990**, *11*, 229.
9. Ascenso, J. R.; Dias, A. R.; Gomes, P. T.; Ramão, C. C.; Pham, Q.; Neibecker, D.; Tkatchenko, I. *Macromolecules* **1989**, *22*, 1000.
10. Longo, P.; Grassi, A.; Oliva, L.; Ammendola, P. *Makromol. Chem.* **1990**, *191*, 237.
11. Kern, R. J. *Nature* **1960**, *187*, 410.
12. Cazzaniga; Cohen, R. E. *Macromolecules* **1989**, *22*, 4125.
13. Cazzaniga; Cohen, R. E. *Macromolecules* **1991**, *24*, 5817.
14. Ammendola, P.; Pellecchia, C.; Longo, P.; Zambelli, A. *Gazz. Chim. Ital.* **1987**, *117*, 65.
15. Grassi, A.; Pellecchia, C.; Longo, P.; Zambelli, A. *Gazz. Chim. Ital.* **1987**, *117*, 249.
16. Ishihara, N.; Kuramoto, M.; Uoi, M. *Macromolecules* **1988**, *21*, 3356.
17. Zambelli, A.; Longo, P.; Pellecchia, C.; Grassi, A. *Macromolecules* **1987**, *20*, 2035.
18. Pellecchia, C.; Longo, P.; Grassi, A.; Ammendola, P.; Zambelli, A. *Makromol. Chem., Rapid Commun.* **1987**, *8*, 277.
19. Zambelli, A.; Oliva, L.; Pellecchia, C. *Macromolecules* **1989**, *22*, 2129.
20. Longo, P.; Proto, A.; Oliva, L. *Macromol. Chem. Rapid Commun.* **1994**, *15*, 151.
21. Boon, J.; Chalia, G.; van Krevelen, D. W. *J. Polym. Sci.* **1968**, *A-2*, 6, 1791.
22. Cimmino, S.; Di Pace, E.; Martuscelli, E.; Silvestre, C. *Polymer* **1991**, *32*(6), 1080.
23. Greis, O.; Xu, Y.; Asano, T.; Petermann, J. *Polymer* **1989**, *30*, 590.
24. Immirzi, A.; De Candia, F.; Iannelli, P.; Zambelli, A. *Makromol. Chem., Rapid Commun.* **1988**, *9*, 761.
25. Guerra, G.; Vitagliano, V. M.; De Rosa, C.; Petraccone, V.; Corradini, P. *Macromolecules* **1990**, *23*, 1539.
26. Capitani, D.; De Rosa, C.; Ferrando, A.; Grassi, A.; Segre, A. L. *Macromolecules* **1992**, *25*, 3874.
27. Wang, Y. K.; Savage, J. D.; Decai, Y.; Hsu, S. L. *Macromolecules* **1992**, *25*, 3659.
28. Vittoria, V.; Ruvolo Filho, A.; De Candia, F. *J. Macromol. Sci.* **1992**, *B.31*, 133.
29. Su, Z.; Morgan, R. J.; Lewis, D. N. *Polymer* **1992**, *33*, 660.
30. Chatani, Y.; Shimane, Y.; Inoue, Y.; Inagaki, T.; Ijitsu, T.; Yukinari, T.; Ishioka, T. *Polymer* **1992**, *33*, 488.
31. Corradini, P. *Makromol. Chem., Macrolom. Symp.* **1993**, *66*, 11.
32. Napolitano, R.; Pirozzi, B. *Macromolecules* **1993**, *26*, 7225.

STYRENESULFONIC ACID AND ITS SALTS (Polymerization and Applications)

Valery F. Kurenkov* and V. A. Myagchenkov
Kazan State Technological University

Poly(styrenesulfonic acid) (PSSA) and its salts are synthetic water-soluble polymers of great practical value, being highly effective flocculants, antistatics, emulsifiers, catalysts for various reactions, matrices for polymerization, ion-exchange resins (in a crosslinked form), and membranes, and are also convenient model compounds for the investigation of the properties of biopolymers. The useful properties of polymers and copolymers of styrenesulfonic acid (SSA) and its salts stimulate interest in the polymerization and copolymerization processes.

POLYMERIZATION

The homogeneous free radical polymerization of monovalent and divalent salts of SSA in aqueous, water-dioxane, water-dimethyl sulfoxide (DMSO), and DMSO-dioxane mixtures solutions have been investigated at 70°C under helium with ammonium persulfate (AP) or AIBN as initiators.[2–4]

COPOLYMERIZATION

The free radical copolymerization of Na-SSA with noniono-genic monomer has been investigated in various reaction media.[5,6]

*Author to whom correspondence should be addressed.

PROPERTIES

PSSA and its salts are strong polyelectrolytes. When an aqueous solution of polyelectrolytes is diluted, the reduced viscosity η_{sp}/C increases rapidly.

The behavior of polyelectrolytes based on PSSA in various ionizing media depends on the nature and the state of the ionogenic groups, and most important, on the character of interactions in the polyion-counterion system in polyelectrolyte solutions, which change the conformational state of the macromolecules, as well as the hydrodynamic and thermodynamic properties of polyelectrolyte solutions.[4,7–9]

APPLICATION

PSSA and its salts are used to improve the properties of commercial products and processes.[1,4] Ion exchange crosslinked PSSA resins are used with widely differing values of pH, because the ionic properties of PSSA are independent on pH. Latexes of the base of styrenesulfonated copolymers are stable to coagulation in acidic and basic pH. Copolymerization of SSA and its salts with vinyl monomers is used to improve wettability, dyeability, ion exchange, and polyelectrolyte and antistatic properties of copolymers.

PSSA and its salts are used as catalysts for Mayer reactions, hydrolysis, and acetalization. Polyelectrolyte complexes of PSSA and poly(vinylidene fluoride) or poly(vinylpyridinium bromide) are used as semipermeability and ion exchange membranes and films. Na-PSSA is an excellent flocculant for municipal and industrial waste treatment. (Co)polymers of SSA and its salts may be used to inhibit corrosion, to produce latexes and ion-exchange resins, as thickeners, emulsifiers, etc.

REFERENCES

1. Kangas, D. A. in *Functional Monomers*; Yacum, R. H. and Nyguist, E. B., Eds.; Marcel Dekker: New York, 1973; Chapter 4, 489.
2. Kurenkev, V. F.; Myagchenkov, V. A. *Eur. Polym. J.* **1979**, *15*, 849.
3. Kurenkov, V. F.; Myagchenkov, V. A. *Acta Polymerica* **1986**, *37*, 424.
4. Bekturov, E. A.; Myagchenkov, V. A. Kurenkov, V. F. *Polymers and Copolymers of Styrenesulfonic Acid* Nauka: Alma-Ata, 1989.
5. Mjagchenkov, V. A.; Kurenkov, V. F.; Frenkel, S. Ya. *Acta Polymerica* **1982**, *33*, 388.
6. Kurenkov, V. F.; Mjagchenkov, V. A. *Acta Polymerica* **1986**, *37*, 517.
7. Kurenkov, V. F.; Mjagchenkov, V. A. *Acta Polymerica* **1987**, *38*, 679.
8. Marina, N. G.; Monakov, Yu, B. in *Chemistry and Physico-Chemistry of High-Molecular Compounds* Ufa: AN SSSR, Bashkirian branch 1976; 116.
9. Bekturov, E. A.; Bakauova, Z. Kh. *Synthetic Water-Soluble Polymers in Solutions* Alma-Ata: Nauka, 1981; 36.

STYRENIC RESINS

José M. Sosa and Kenneth P. Blackmon
Fina Research and Technology Center

Worldwide capacity of polystyrene (PS) (CAS Registry No. 9003-53-6) in 1993 reached 12 billion Kg. In the United States, Huntsman, Dow, BASF, Fina, Novacor, and Chevron account for 85% of the nameplate capacity of 2.5 billion Kg of PS and impact modified PS (IPS) (CAS Registry No. 9003-55-8). Arco, BASF, Huntsman, and Scott Paper have a combined nameplate capacity of 400 million Kg of expandable polystyrene (EPS).

STYRENE-DERIVED POLYMERS

PS produced via a free-radical mechanism is an atactic polymer with a glass transition temperature (T_g) of 105°C. PS homopolymers, often referred to as crystal PS, are amorphous, relatively brittle materials that can be easily molded into complex shapes with excellent clarity.

IPS is a composite formed by polymerizing styrene in the presence of polybutadiene rubber or styrene-butadiene copolymers to yield rubber particles dispersed in a PS matrix. Copolymerization of styrene with acrylonitrile, α-methyl styrene, and maleic anhydride gives polymers with improved thermal and mechanical properties. Polymers with improved impact strength are made by copolymerizing styrene with butadiene to form random copolymers or by making IPS. Terpolymers of acrylonitrile, butadiene, and styrene give plastics (ABS) (CAS Registry No. 9003-56-6) with a good balance of thermal, mechanical, and impact properties that can compete in some applications with engineering resins.

PRODUCTION OF COMMERCIAL RESINS

Styrene-based resins are manufactured primarily by a bulk, or mass, process with diluents. Some resins are prepared by suspension (EPS) and emulsion (ABS); however, the trend has been to develop continuous mass processes where applicable, because of economics and other factors.[7,8]

CURRENT DEVELOPMENTS

Process

Initiator Technology

Peroxides and azo compounds have been used for many years as initiators for the free-radical polymerization of styrene.

Traditionally, monofunctional peroxides such as benzoyl peroxide, *t*-butyl perbenzoate, and *t*-butyl peroctoate have been used to reduce polymerization time or temperature, relative to thermally initiated systems. With these conventional inhibitors, however, it is not possible to increase the polymerization rate (e.g., by increasing reaction temperature) without sacrificing polymer molecular weight. Fortunately, the introduction of bifunctional initiators in recent years has allowed for optimization of the polymerization process in terms of rate and molecular weight.

Unsymmetrical bifunctional initiators contain labile peroxy groups having distinctly different thermal decomposition characteristics and, thus, may offer advantages in a particular polymerization temperature profile.[12]

Product

Extrusion Applications

The use of high-impact PS (HIPS) in extrusion/thermoforming applications (e.g., cups and containers) represents a major market segment for the resin.

Styrene-Butadiene Block Copolymers

Styrene-butadiene copolymers are produced by solution polymerization with anionic initiators. Depending on the properties that are desired, these elastomers can be produced with linear or star-shaped (radial) structures with 10–50% bound styrene.

High-Gloss Impact Polystyrene

In rubber-modified PS, physical properties such as impact strength, toughness, and gloss are largely influenced by rubber level, particle size, and morphology.

In general, impact PSs for use in high-gloss applications have small particles (< 1 μm) and, thus, moderate impact strengths (50–60 J/m).

Low Residual Volatiles

An unplasticized, high molecular weight (MW 310,000–350,000 g/mol) PS resin is typically used in the production of foam due to the requirements of high melt strength.

Environmental Stress Crack Resistance

In crystal PS, factors such as molecular weight and molecular weight distribution play a role in environmental stress crack resistance (ESCR) behavior, because craze fibril volume increases with an increase in molecular weight. In rubber-modified materials, ESCR can be optimized through control of rubber particle size and distribution, rubber content, plasticizer level, and matrix molecular weight.[4]

Lithographic Applications

The use of styrenic polymers in lithographic applications such as signs, back-lighting displays, floral tags, and promotional items is rapidly growing.

Outdoor Weathering

The outdoor weatherability of HIPS is marginal because of the susceptibility of polybutadiene to oxidation. The formation of free radicals can also lead to degradation in the styrenic phase.

The weatherability of styrenic polymers is significantly improved by the use of hindered amine light stabilizers with optionally, UV absorbers. HIPS materials containing appropriate stabilizer packages exhibit superior Yellowness indices in accelerated weathering tests (Xenon arc weatherometer with spray).[5]

In recent years, PS manufacturers have improved the commodity grades and have introduced grades that serve specialty markets. While PS manufacturers seek to replace ABS and poly(vinyl chloride) in some applications, polyolefin and poly(ethylene terephthalate) manufacturers have also penetrated markets that are traditionally PS markets. Other areas for development include the electronics, medical, and construction industries. The recycling of plastics, as well as health and environmental issues, will continue to challenge resin manufacturers to improve process technologies and to lower costs and improve quality.

REFERENCES

1. Simon, R. H. M., Chappelear, D. C. *Polymerization Reactors and Processes*; ACS Advances in Chemistry Series 91; American Chemical Society: Washington, DC, 1979.
2. Mallikarjun, R., Nauman, E. B. *Polym.-Plast. Tech. Eng.* **1989**, *28*(2), 137.
3. Kim, K. J.; Liang, W.; Choi, K. Y. *Ind. Eng. Chem. Res.* **1989**, *28*, 131.
4. Bubeck, R. A.; Arends, C. B.; Hall, E. L.; Vander Sande, J. B. *Polym. Eng. Sci.* **1981**, *21*(10), 624.
5. Fina Oil and Chemical Co., Deer Park, TX, unpublished results.

SULFUR DIOXIDE COPOLYMERIZATION

Zbigniew Florjańczyk
Faculty of Chemistry
Warsaw University of Technology

Sulfur dioxide is a monomer that does not undergo homopolymerization. However, it can form copolymers according to different polymerization mechanisms. In a majority of polymerizations, and also in many other organic reactions, SO_2 exhibits electrophilic properties that result from the presence of a partial positive charge on the sulfur atom (~1.47) and relatively low energy of the lowest unoccupied molecular orbital—LUMO (~–1.1 eV).[1]

The SO_2 copolymerization with various monomers containing unsaturated C—C bonds resulting in polysulfones is best known and of the greatest importance in the synthesis of polymers. References give a list of monographs related to this particular subject.[2–6]

FREE RADICAL COPOLYMERIZATION

General Reactivity and Mechanism of Propagation

In free radical copolymerization SO_2 behaves like a monomer of high general activity and strong electron-acceptor properties. Its Q and e parameters values are 0.93 and 3.3, respectively.[6] In reactions with monomers characterized by electron-donor properties and low resonance stabilization, such as alkenes, cycloalkenes, alkynes, allyl alcohol, allyl acetate, or vinyl acetate, copolymers of an alternating structure are formed (**Equation 1**).

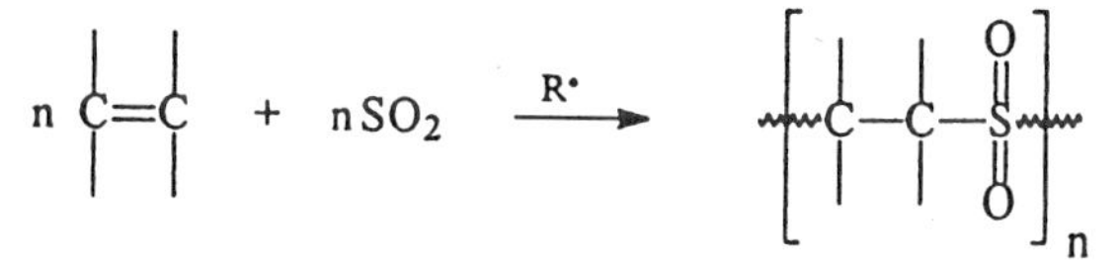

1

An essential limitation in carrying out the alternating copolymerization of SO_2 and electron donor monomers is the fact that the ceiling temperature (T_c) (that is the temperature at which the chain growth rate equalizes with that of depropagation) is relatively small for these systems.[7]

Several monomers form linear polysulfones containing less than 50 mol% of SO_2 monomeric units $[—(C—C)_m—S(O)_2—]_n$. The formation of such polymers was observed, among others, in the copolymerization of SO_2 with styrene and several of its derivatives, vinyl chloride, vinyl bromide, chloroprene, phenylacetylene, acrylamide, acrylic acid, alkyl acrylates, and aryl acrylates.[8–26]

TYPES OF INITIATORS AND POLYMERIZATION CONDITIONS

Owing to the low ceiling temperature values of many copolymerizations or terpolymerizations involving SO_2, various redox systems are usually used as initiators of these processes. Sulfur dioxide is most often the electron acceptor in these systems.

Hydrogen peroxide, alkyl hydroperoxides, alkyl peroxides, amine oxides, nitrates, sulfates, persulfates, chlorates and perchlorates of ammonium, silver, magnesium, and copper are used among others as electron donors.[27–33]

PROPERTIES AND APPLICATION OF ALIPHATIC POLYSULFONES

SO_2 copolymers with propylene and butadiene are thermoplastic materials of mechanical properties comparable with those of the olefin homopolymer. With an increase in the number of carbon atoms in the olefin, the copolymer density decreases (from about 1.5 in the case of the copolymer with propene to ca. 1.3 for poly(1-pentene sulfone) [*42255-65-2*] as well as the glass transition temperature). Thus, poly(1-decane sulfone) [*33990-98-6*] is an elastomer at room temperature. These copolymers did not find wide application, mainly owing to the too low thermal stability.

An interesting property of aliphatic sulfones is their high sensitivity to ionizing radiation (electron beam, γ or X-rays). Sulfur dioxide and olefins are the main products of the radiation-induced degradation of poly(alkene sulfone).

Poly(alkene sulfone)s, having a silicon bearing group, can be used as positive resists for direct-write lithography capable of resolving features below 0.5 μm.

Braun et al. showed that the alternating copolymerization of sulfur dioxide with many different mesogenic monomers is a very efficient method for the synthesis of liquid crystalline polysulfones.[34,35]

Some water soluble monomers of a zwitterionic structure, forming micelles and lyotropic liquid crystals and the concentrated regime, were found to undergo spontaneous alternating copolymerization with SO_2.[36]

RING-OPENING COPOLYMERIZATION

The copolymerization of SO_2 with oxiranes results in the formation of polymers containing sulfite monomeric units in the main chain (**Equation 1**).[6,36a–41]

1

Products of a composition close to equimolar are formed in anionic copolymerization, because the chain growth proceeds mainly due to the oxirane/SO_2 complex addition.[39,40]

STEP POLYMERIZATION

No systematic studies have been undertaken yet on the application of SO_2 in step polymerization. In the presence of alkali metals and crown ethers, poly(sulfonylmethylene-1,4-phenylenemethylene) can be obtained directly from SO_2 and 4-chloromethylbenzyl chloride.[42]

REFERENCES

1. Florjańczyk, Z.; Raducha, D. *Polish J. Chem.* **1995**, *69*, 48.
2. Fettes, E. M.; Davis, F. O. *Polyethers*; Interscience: New York, 1962; Vol. 3, pp 225–270.
3. Tokura, N. *Encyclopedia of Polymer Science and Technology*; Mark, H. F.; Gaylord, N. G., Eds.; Interscience: New York, 1968; Vol. 9, pp 460–485.
4. Fawcett, H. F. *Encyclopedia of Polymer Science and Engineering*; John Wiley & Sons: New York, 1987, Vol. 10, pp 408–432.
5. Florjańczyk, Z.; Floriańczyk, T.; Kłopotek, B. B. *Polimery-Tworzywa Wielkocząsteczkowe* **1987**, *32*, 213.
6. Florjańczyk, Z. *Prog. Polym. Sci.* **1991**, *16*, 509.
7. Dainton, F. S.; Ivin, K. J. *Nature* **1948**, *162*, 705.
8. Barb, W. G. *Proc. R. Soc. London, Ser A* **1952**, *212*, 177.
9. Herz, J.; Hummel, D.; Schneider, C. *Makromol. Chem.* **1963**, *63*, 12; **1963**, *64*, 95.
10. Cais, R. e.; O'Donnell, J. H.; Bovey, F. A. *Macromolecules* **1977**, *10*, 254.
11. Tokura, N.; Matsuda, M.; Shirai, I.; Shinina, K.; Ogawa, Y. *Bull. Chem. Soc. Jap.* **1962**, *35*, 1043.
12. Tokura, N.; Matsuda, M.; Ogawa, Y. *J. Polym. Sci Part A* **1963**, *1*, 2965.
13. Tokura, N.; Matsuda, M.; Arkawa, K. *J. Polym. Sci Part A* **1964**, *2*, 3355.
14. Ino, M.; Hara, A.; Tokura, T. *Makromol. Chem.* **1966**, *98*, 81.
15. Marvel, C. S.; Glavis, F. J. *J. Am. Chem. Soc.* **1938**, *60*, 2622.
16. Cais, R. E.; O'Donnell, J. H. *J. Macromol. Chem.-Chem. A* **1982**, *17*, 1407 and references cited therein.
17. Hrabak, F.; Błazek, J.; Weber, J. *Makromol. Chem.* **1966**, *97*, 9.
18. Matsuda, M.; Hara, Y. *J. Polym. Sci., Part A-1* **1972**, *10*, 837.
19. Kim, S. J.; Aiyama, F.; Matsuda, M. L. *Polym. Sci., Polym. Chem. Ed.* **1987**, *25*, 1641.
20. Firth, W. C.; Palmer, L. E. *Macromolecules* **1971**, *4*, 654.
21. Cais, R. E.; Stuk, J. *Polymer* **1978**, *19*, 179.
22. Tsonis, Ch. P.; Pasika, W. M. *Polymer* **1989**, *21*, 253.
23. Wazeer, M. I.; Ali, S. A.; Tsonis, Ch. *J. Polym. Sci., Polym. Phys. Ed.* **1988**, *26*, 1539.
24. Florjańczyk, Z.; Floriańczyk, T.; Klopotek, B. B. *Makromol. Chem.* **1988**, *188*, 2811.
25. Florjańczyk, Z.; Floriańczyk, T.; Zygado, E. *Makromol. Chem.* **1988**, *189*, 2719.
26. Florjańczyk, Z.; Zygadło, E. *Polymer* **1991**, *32*, 2853.
27. Fredenc, D. S.; Cogan, H. D.; Marvel, C. S. *J. Am. Chem. Soc.* **1934**, *56*, 1815.
28. Matsuda, M.; Thoi, H. *J. Macromol. Sci. A* **1977**, *11*, 1423.
29. Marvel, C. S.; Glavis, F. J. *J. Am. Chem. Soc.* **1938**, *60*, 2622.
30. Marvel, C. S.; Audrieth, L. F.; Sharkey, W. H. *J. Am. Chem. Soc.* **1942**, *64*, 1229.
31. Ivin, K. J. *Nature* **1957**, *180*, 90.
32. Sully, B. D. *J. Chem. Soc.* **1950**, 1948.
33. Frey, F. E.; Snow, R. D.; Schulze, W. A. U.S. Patent 2 280 818, 1942.
34. Braun, D.; Herr, R. P.; Arnold, N. *Makromol. Chem. Rapid Commun.* **1987**, *8*, 359.
35. Braun, D.; Arnold, N.; Liebmann, A.; Schmidtke, I. *Makromol. Chem.* **1993**, *194*, 2687.
36. Anton, P.; Laschewsky, A. *Makromol. Chem., Rapid Commun.* **1991**, *12*, 189.

36a. Razuwajew, G. A.; Etlis, W. S.; Grobow, N. *Zhur. Obsh. Khim.* **1961**, *31*, 1328.
37. Soga, K.; Hattori, I.; Kinoshita, J.; Ikeda, S. *J. Polym. Sci., Polym. Chem. Ed.* **1977**, *15*, 745.
38. Soga, K.; Kiyohara, K.; Hattori, I.; Ikeda, S. *Makromol. Chem.* **1981**, *181*, 2151.
39. Florjańczyk, Z.; Raducha, D. *Makromol. Chem., Rapid Commun.* **1991**, *12*, 393.
40. Florjańczyk, Z.; Raducha, D. *Makromol. Chem.* **1993**, *194*, 2605.
41. Florjańczyk, Z.; Raducha, D. *Makromol. Chem., Rapid Commun.* **1993**, *14*, 207.
42. Soga, K.; Nagata, N.; Hattori, I.; Ikeda, S. *Makromol. Chem.* **1980**, *181*, 2019.

Sulfur Dioxide Copolymers

See: *Polysulfides (Prepared from Sulfur Dioxide)*
Polysulfones
Sulfur Dioxide Copolymerization

Superabsortive Polymers

See: *Starch Graft Copolymers*
Superabsorptive Polymers (from Natural Polysaccharides and Polypeptides)

SUPERABSORPTIVE POLYMERS (from Natural Polysaccharides and Polypeptides)

Tomoko Ichikawa and Toshinari Nakajima
Faculty of Home Economics
Otsuma Women's University
School of Human Life and Environmental Science
Ochanomizu University

Superabsorptive polymers swell remarkably well to form hydrogel structures with absorbed water, which can amount to several hundred times the weight of the polymer itself.

Various natural polymer materials have been applied for water absorption such as cloth, pulp, or paper products made from plant and animal fibers of cotton, hemp, and silk. Water-absorptive polymer foods made from natural materials such as Konnyaku, agar, and tofu have also been used.

Synthetic polymers similar to these were first described by Dr. Fanta et al. at the Northern Laboratory of the U.S. Department of Agriculture. Such polymers were developed with a view to utilizing agricultural materials, and are typified by the hydrolyzed corn starch-acrylonitrile graft copolymer system.[1] Since then, super water-absorptive polymers have received considerable attention, and new such materials have been developed successively in the U.S.A. and in Japan.[2,3]

Today, superabsorptive polymers are composed of synthetic, semisynthetic, or natural polymers.

PREPARATION

In general, superabsorptive polymers possess a structure in which a water-soluble polymer have been made insoluble by some process, typically by means of crosslinking agents, resulting in polymers that have the capacity to absorb several hundred times their weight in pure water.

At present, more than 90% of areas in which superabsorptive polymers are used are related to sanitary products.

Water-absorptive polymers currently in use include hydrolysis products of starch-acrylonitrile graft polymers, carboxymethylcellulose, polycarboxylic acids, polyacrylamides, noncrosslinked polymer blends, crosslinked polyacrylate products, and other resins such as poly(vinyl alcohol)s and poly(ethylene oxide)s.

There are many foods with gel-like structures, and these structures are formed in general with thermally irreversible and ionic crosslinks. They are obtained by either physical (heating–cooling) or chemical (ion crosslinking with Ca^{2+}, Mg^{2+}) treatment of polymers with sol-solution structures.

PROPERTIES

Materials which are of increasing interest in superabsorptive polymers show high capacity for immediate swelling on pouring water onto them, and they have properties of gelling and holding the water absorbed. At the top-level, their water-absorbing power ranges from several hundred to 1,000 times their weight.

Water Absorption Properties of Polyionic Complexes from Polysaccharides and Polypeptides

Few studies have focused on the water-absorption behavior of polyionic polymer complexes.[4–17] Mixed polymers of natural origin with opposite ionic charge can behave as appropriate water-absorbing agents, and can retain "entrapped" water molecules.

Natural polysaccharides and polypeptides can be used for application as superabsorbing materials in the food industry. We performed a series of experiments on synthesis of polyionic complexes, and some physical properties characteristic of superabsorptive polymers were also determined.

For natural polyion complex formation, four kinds of combinations were selected: as the cationic component, chitosan, which is a chain of ring structures, or polylysine, which is a linear flexible chain, and as the anionic component, carboxymethyl cellulose (CMC), which is a chain of ring structures, or poly(glutamic acid), which is a linear flexible chain were used.[18–21] Several complex films were prepared from the above four components in different ratios by solvent-casting, and examined with regard to their water-absorption behavior.

The degree of film swelling in water was greater when the complex was made from chitosan or CMC of a higher ratio. The addition of electrolytes such as NaCl or $CaCl_2$ to the dipping solution markedly decreased the water absorptive ratio with the increase of the solute concentration.

The water absorption properties of polyionic complex films made from polysaccharides and polypeptides showed that to produce a polyion complex with a high water absorption capacity and high modulus upon swelling, one component should consist of polymerized rings such as CMC or chitosan to provide rigidity to the complex, and the other component should be a polymerized flexible linear molecule such as polylysine or poly(glutamic acid).

APPLICATION

Materials with high absorbing capacity for water or other aqueous fluids were used in a large number of sanitary products in the early years, and then uses as agricultural-gardening agents and in civil engineering followed. Further applications in cosmetic-toiletries, medical, and other fields also have been developed recently. These applications are based on the materials' ability to soak up liquid and so immobilize it.

Superabsorbent polymers possess specific functions that are not found in other polymer products, and so comprise a class of promising materials for a range of uses. Further interesting applications will be developed in the future. Water is indispensable to our daily life. Consequently, many superabsorptive polymers of natural origin will be developed for various applications concerned with water.

REFERENCES

1. Fanta, G. F.; Burr, R. C.; Russel, C. R.; Rist, C. E. *J. Appl. Polym. Sci.* **1967**, *11*, 457; U.S. Patent 3 981 100, 1976.
2. *Chemical Week* P24, July 21, 1974.
3. Masuda, F. *Chem. Econ. Engineer. Rev.* **1983**, *15*(11), 19.
4. Oyama, H. T.; Nakajika, T. *J. Polym. Sci.* **1983**, *21*, 2987.
5. Oyama, H.; Nakajika, T. *J. Appl. Polym. Sci.* **1984**, *29*, 2143.
6. Hirai, Y.; Nakajima, T. *J. Appl. Polym. Sci.* **1988**, *35*, 1325.
7. Hirai, Y.; Nakajima, T. *J. Macromol. Sci. Phys.* **1988**, *B27*, 359.
8. Hirai, Y.; Nakajima, T. *J. Appl. Polym. Sci.* **1989**, *37*, 2275.
9. Hirai, Y.; Nakajima, T. *J. Macromol. Sci.* **1989**, *A26*, 1465.
10. Hirai, Y.; Nakajima, T. *J. Macromol. Sci. Phys.* **1991**, *B30*, 141.
11. Hirai, Y.; Nakajima, T. *J. Macromol. Sci. Phys.* **1993**, *B32*, 337.
12. Kim, J. H.; Kim, J. Y.; Lee, Y. M. et al. *J. Appl. Polym. Sci.* **1992**, *45*, 1711.
13. Smid, J.; Fish, D. *Encycl. Polym. Sci. Tech.* 2nd ed., Vol. 11, 1988, P720.
14. Gelman, R. A.; Glaser, D. N.; Blackwell, J. *Bio-polymers* **1973**, 1223.
15. Nakajima, A.; Shinoda, K. *J. Colloid Interface Sci.* **1976**, *55*, 126.
16. Fukuda, H.; Kikuchi, Y. *Macromol. Chem.* **1979**, *180*, 1631.
17. Arguelles-Monal, W. *Macromol. Chem.* **1988**, *9*, 693.
18. Ichikawa, T.; Yoshikawa, E.; Kubota, H. et al. *Kobunshi Ronbunshu* **1990**, *47*, 709.
19. Ichikawa, T.; Araki, C.; Nakajima, T. *Kobunshi Ronbunshu* **1991**, *48*, 789.
20. Ichikawa, T.; Araki, C.; Nakajima, T. *Kobunshi Ronbunshu* **1992**, *49*, 617.
21. Ichikawa, T.; Mitsumura, Y.; Nakajima, T. *J. Appl. Polym. Sci.* **1994**, *54*, 105.

Supercoils

See: DNA (Bending and Twisting Elasticity)

SUPERCONDUCTING FILAMENTS, HIGH T_C (Suspension and Solution Spinning from PVA Solution)

Tomoko Goto
Nagoya Institute of Technology Japan

The discovery of the high temperature oxide superconductor in 1986 was one of the major discoveries of this century.[1] Following that, the discovery of the high critical temperature (T_c) oxide superconductor with T_c exceeding the liquid nitrogen temperatures, such as $Y_1Ba_2Cu_3O_x$ (123 superconductor), $Bi_2Sr_2Ca_2O_x$ (Bi2223), and $Tl_2Ba_2Ca_2Cu_3O_x$ (Tl2223) promised wide application in future technology involving superconducting magnets, power cables, and electric devices and sensors, because operating at temperature near 77 K (Kelvin) would greatly simplify and cheapen refrigeration.[2–4] Construction of powerful magnets or other practical devices necessitates the high values of superconducting limit of the critical temperature T_c, the critical magnetic field H_c, and the critical current density J_c.

Many efforts have been made to apply the new high T_c oxide superconductors for practical use; however, the development of conductors from high T_c superconductor has presented enormous challenges. Major application has been hindered by their relatively low J_c caused by grain boundary weak link, flux line movement, and extreme anisotropy.

We have studied the preparation of superconducting oxide of long filament using a textile fiber spinning technology from poly(vinyl alcohol) (PVA) solution for the precursor of the oxide, and developed two methods for the fabrication of the oxide superconducting long filaments. One is a suspension spinning using powder for a starting material, and the other is a homogeneous solution spinning.[5,6] A high J_c value of more than 10^4 A/cm^2 at 77 K and 0 T was attained for the melt-processed Y-system filaments after suspension spinning and solution spinning methods. Whereas suspension spinning can produce many kinds of high T_c superconducting filaments, solution spinning successfully fabricates only a Y-system superconducting filament. However, solution spinning can prepare more uniform and finer materials than suspension spinning does, and it introduces an effective flux pinning at 77 K.

REFERENCES

1. Bednorze, J. G.; Müller, K. *Z. Phys.* **1986**, *B 64*, 189.
2. Wu, M. K.; Ashburn, J. R.; Toung, C. J.; Hor, P. H.; Meng, R. L.; Gao, L.; Huang, Z. J.; Wang, Y. Q.; Chu, C. W. *Phys. Rev. Lett.* **1987**, *58*, 908.
3. Maeda, H.; Tanaka, Y.; Fukutomi, M.; Asano, T. *Jpn. J. Appl. Phys.* **1988**, *27*, L209.
4. Sheng, Z. Z.; Hermann, A. M. *Nature* **1988**, *332*, 138.
5. Goto, T.; Kada, M. *Jpn. J. Appl. Phys.* **1987**, *26*, L1527.
6. Goto, T.; Sugishita, T. *Physica C* **1990**, *171*, 441.

Supercooled State Polymerization

See: Radiation Induced Polymerization (Biomedical Applications)

Supercritical Fluids

See: Acrylamide, Inverse Emulsion Polymerization (in Supercritical Carbon Dioxide)
High Solids Coatings (Use of Supercritical Fluids)

Supermolecular Structures

See: Polypropylene, Isotactic (Supermolecular Structure)

Poly(vinylidene fluoride) (Overview)
Supramolecular Formation and Structure (PEO-PPO-PEO Block Copolymers)

Supported Catalysts

See: *Butadiene Polymerization (Supported Lanthanide Catalysts)*
Catalysts, Polymeric (Higher Activity than Monomeric Analogs)
Crown Ethers, Polymeric (Catalytic Activity)
Olefin Polymerization Catalysts (Experimental and Theoretical Characterization)
Olefin Polymerization Catalysts (Silica-Supported Titanium Tetrachloride)
Olefin Polymerization Catalysts, Soluble (Group 4, Cationic Active Centers)
Phase Transfer Catalysts, Polymeric (Overview)
Phase Transfer Catalysts, Polymeric (Structure and Activity)
Phosphinated Polystyrene-Bound Metal Complexes (Pd and Rh)
Polybenzimidazoles (Overview)
Propylene Polymerization (Stochastic Analysis of the Propagation Mechanism)
Supported Catalysts (Lewis Acid and Ziegler–Natta)
Supported Chiral Catalysts
Supported Chromium Polymerization Catalysts (on Silica, Phillips Catalysts)
Supported Lewis Acid Catalysts (on Polypropylene; Recoverable and Reusable)
Supported Thiazolium Salt Catalysts (for Acyloin Condensation)
Supported Ziegler–Natta Catalysts (on Silica, for Olefin Polymerization)
Supported Ziegler–Natta Catalysts (Titanium-Based Silica; Polymerization Properties)
Supported Ziegler–Natta Catalysts [Titanium (IV) Chloride on Reduced n-Type Oxides]
Tautomer Polymerization

SUPPORTED CATALYSTS (Lewis Acid and Ziegler–Natta)

Ruicheng Ran
Department of Chemistry
Mississippi State University

Since the 1970s, the functional polymeric materials including polymer–metal complexes, metal-containing polymeric materials, polymeric reagents, polymeric catalysts, and so forth, have been a very active area of study in polymer science. Many unique properties and applications of these materials have been given increased attention in recent years.[1–10]

It is well known that Lewis acids such as $AlCl_3$, $TiCl_4$, BF_3, BBr_3, $SnCl_4$, $GaCl_3$, $SbCl_5$, $FeCl_3$, $ZrCl_4$, and so on are very useful catalysts for many organic synthetic reactions and some polymerizations. All of these Lewis acids, however, are very easily hydrolyzed, evolving fumes even by moisture in air so that their use, storage, and separation from the reaction mixtures are inconvenient. To solve this problem, a series of complexes of Lewis acids with polymer carriers used as polymer-supported Lewis acid catalysts were developed.[11–49] These complexes showed not only very good catalytic activity in some organic synthesis reactions such as esterification, acetylation, etherification, and ketal formation, very easily separated from the reaction mixture by filtration, but also excellent stability, and they can be reused at least 6–10 times without losing their catalytic activities. They were also used as a catalyst for cationic polymerization of styrene, methyl vinyl ether, and α-methylstyrene.[22,35,38] A poly(α-methylstyrene) with high molecular weight was obtained by cationic polymerization catalyzed by polystyrene-supported gallium trichloride at room temperature. I prepared some polymer-supported Ziegler-Natta catalysts on that basis, by combining polymer-supported Lewis acid complexes with alkyl aluminum halides.[50,51] These catalysts showed very good stability and high catalytic yield for the polymerization of some α-olefins such as ethylene, propylene, isobutylene, 1,3-butadiene, isoprene, and their copolymerization. Compared with common Ziegler–Natta catalysts which have been developed by using inorganic carriers such as magnesium salts or silicates, the polymer-supported catalysts will not contaminate the polyolefins with metal ions.[52–69]

APPLICATION

In Organic Synthesis

The Lewis acids have been widely used as catalysts in organic reactions for a long time. All types of organic reactions which can be catalyzed by Lewis acids can be catalyzed by polymer-supported Lewis acid catalysts.

Catalytic Esterification

Esterification is an important organic reaction for organics synthesis both in the laboratory and in industry. Many esterification reactions of carboxylic acids with alcohols can be catalyzed by polymer-supported Lewis acid complexes in good yields.

Catalytic Ketal Formation

The results of the ketal formations of several cycloaliphatic ketones with an alcohol indicated that the polymer-supported Lewis acid catalysts showed very high catalytic activity with high yield in these reactions.

Catalytic Acetylation

Acetylation is widely used for both preparation of molecules and functional group protection in organic synthesis. It is clear that the polymer-supported Lewis acid catalysts showed very good catalytic activity in those acetylations.

Stability and Reuse Property

The polymer-supported Lewis acid catalysts are very stable for storage and applications. They can be stored for at least two years without losing their catalytic activities. They also can be reused many times with high catalytic yield and can be regenerated by treatment with Lewis acid.

Catalytic Etherification and Friedel–Crafts Alkylation

In etherification of aromatic alcohol, the polymer-supported Lewis acid complexes showed catalytic activity. In the alkylation of substituted benzenes, Lewis acids are typical catalysts. Polymer-supported Lewis acid complexes showed good catalytic activity in Friedel–Crafts alkylations.

Application of Polymer-Supported Lewis Acid Catalysts in Cationic Polymerization

The polymer-supported Lewis acid complexes are useful catalysts for cationic polymerization of many monomers such as α-methylstyrene, styrene, methylvinyl ether, propylene oxide, and so forth.

Application of Polymer-Supported Ziegler–Natta Catalysts in Polymerization of α-Olefin

Supported Ziegler–Natta catalysts for polymerization of α-olefin have been developed for the most part using inorganic carriers.[50,51] In most instances the carriers are magnesium salts or silicates. The major problem with these metal salt-supported catalysts, in addition to their high reactivity, is that they introduce metal ions as contamination into the polyolefin. This alters the physical properties of the polymer. A series of polymer-supported Ziegler–Natta catalysts based on the polymer-supported Lewis acid catalysts have been developed and used for polymerization of some olefins such as ethylene, propylene, isobutylene, isoprene, butadiene, and more.

Polymerization of Isobutylene

Polyisobutylene (PIB) [9003-27-4] or copolymers of isobutylene may be used as adhesives, sealants, insulating oils, etc., and are commercially prepared by cationic polymerization. Hence, we polymerized isobutylene catalyzed by the polymer-supported Ziegler–Natta catalysts under atmospheric pressure at 0°C. The polymer-supported catalysts showed good catalytic activity for the polymerization of isobutylene.

Copolymerization of Isobutylene with Isoprene

It is well known that the copolymer of isobutylene with isoprene (1.5–4.5 mole%) is called butyl rubber which is a commercially important polymer used for inner tubes, engine mounts and springs, chemical tank linings, protective clothing, hoses, gaskets, and electrical insulation. We carried out the copolymerization of isobutylene and isoprene in the presence of polymer-supported Ziegler–Natta catalysts under atmospheric pressure at 0°C.

In conclusion, the polymer-supported Lewis acid catalysts and the polymer-supported Ziegler–Natta catalysts are very easy to prepare, very stable for either storage or use, and convenient to separate from the reaction mixture. They can be reused and regenerated. The polymer-supported Lewis acid catalysts showed good catalytic activity for organic synthesis reactions such as esterification, acetylation, ketal formation and etherification, and for the cationic polymerization of α-methylstyrene. The polymer-supported Ziegler–Natta catalysts have the same or greater catalytic activity and catalytic yield in (co)polymerization of olefins such as butadiene, isobutylene, and isoprene.

REFERENCES

1. Fettes, E. M. *Chemical Reactions on Polymers*; Wiley-Interscience: New York, NY, 1964.
2. Pittman, C. U. Jr.; Smith, L. R. *J. Am. Chem. Soc.* **1975**, *97*(7), 1742.
3. Mathur, N. K.; Narang, C. K.; Williams, R. E. *Polymers as Aids in Organic Synthesis* Academic: New York, 1980.
4. Hodge, P.; Sherrington, D. C. *Polymer-Supported Reactions in Organic Chemistry*; Wiley-Interscience: New York, 1980.
5. Akelah, A. *Synthesis* **1981**, *6*, 413.
6. Pittman, C. U., Jr. In *Comprehensive Organometallic Chemistry*; Wilkinson, S. G.; Stone, F. G. A.; Abel, E. W.; Eds.; Pergamon: Oxford, U.K., 1982; Vol. 8, pp 553–611.
7. Hartly, F. R. *Supported Metal Complexes*; Reidel: Dordrecht, The Netherlands, 1985.
8. Smith, K. *Solid Supports and Catalysts in Organic Synthesis*; Ellis Horwood and Prentice Hall: New York, 1986.
9. Ford, W. T. *Polymeric Reagents and Catalysts* ACS Symposium Series 308; American Chemistry Society: Washington, DC, 1986.
10. Toshima, N.; Kanaka, K.; Komiyama, M.; Hirai, H. *J. Macromol. Sci., Chem.* **1988**, *A25*, 1349–1366.
11. Neckers, D. C.; Kooistra, D. A.; Green, G. W. *J. Am. Chem. Soc.* **1972**, *94*(26), 9284.
12. Blossey, E. C.; Turner, L. M.; Neckers, D. C. *Tetrahedron Lett.* **1973**, *21*, 1823.
13. Blossey, E. C.; Turner, L. M.; Neckers, D. C. *J. Org. Chem.* **1975**, *40*(7), 959.
14. Ran, Ruicheng; Shuojian, Jiang; Jia, Xinru *Chin. J. Appl. Chem.* **1985**, *2*(1), 29–33; *Chem. Abstract* **1985**, *103*(4), 27939v.
15. Ran, Ruicheng; Jiang, Shuojian; Shen, Ji *J. Mol. Sci.* **1985**, *2*, 221–4 *Chem. Abstract* **1988**, *109*, 21080q.
16. Ran, Ruicheng; Jiang, Shuojian, Shen, Ji *Chem. J. Chin. Univ.* **1986**, *7*(3), 281–6; *Chem. Abstract* **1987**, *106*(10), 69077e.
17. Ran, Ruicheng; Jiang, Shuojian *Polym. Commun.* **1985**, *5*, 377–9.
18. Ran, Ruicheng, Pei, Weiwei; Jia, Xinru; Shen, Ji; Jiang, S. *KEXUE TONGBAO* **1986**, *10*, 748–52.
19. Ran, Ruicheng; Li, Mingqian, Jia, Xinru *J. Chem. Educ.* **1985**, *1*, 27–8.
20. Ran, Ruicheng; Pei, Weiwei; Jia, Xinru; Shen, Ji; Jiang, S. *Polym. Commun.* **1986**, *5*, 379–83; *Chem. Abstract* **1988**, *108*(3), 21080q.
21. Ran, Ruicheng; Pei, Weiwei; Jia, Xinru; Shen, Ji; Jiang, S. *Polym. Commun.* **1986**, *6*, 453–7; *Chem. Abstract* **1988**, *108*(3), 21082s.
22. Ran, Ruicheng; Jia, Xinru; Li, Mingqian; Jiang, Shuojian *Petrochem. Technol.* **1986**, *15*(9), 532–9; *Chem. Abstract* **1987**, *106*(2), 5500e.
23. Ran, Ruicheng; Jia, Xinru; Pei, Weiwei; Jiang, Shuojian *Acta Sci. Nat. Univ. Beijing* **1986**, *6*, 29–35; *Chem. Abstract* **1988**, *108*(3), 21088y.
24. Ran, Ruicheng; Pei, Weiwei; Jia, Xinru; Shen, Ji; Jiang, S. *Chem. J. Chin. Univ.* **1986**, *7*(7), 645–50; *Chem. Abstract* **1987**, *106*(2), 6086m.
25. Ran, Ruicheng; Huang, Jin; Shen, Ji *Acta Polym. Sinica* (ch), **1987**, *3*, 312–6; *Chem. Abstract* **1989**, *110*(1), 7166g.
26. Ran, Ruicheng; Huang, Jin; Jia, Xinru; Shen, Ji *J. Cat. & Cat.* (ch), **1987**, *8*(4), 440–3; *Chem. Abstract* **1989**, *110*(1), 7168j.
27. Ran, Ruicheng; Jia, Xinru; Wu, Xianghong; Pei, Weiwei *Chem. J. Chin. Univ.* **1987**, *8*(12), 1141–5; *Chem. Abstract* **1988**, *109*, 21082s.
28. Ran, Ruicheng; Huang, Jin; Shen, Ji *Chin. J. Appl. Chem.* **1987**, *3*(5), 32–6; *Chem. Abstract* **1988**, *109*, 54437k.
29. Ran, Ruicheng; Huang, Jin; Shen, Ji *Acta Polym. Sinica* (ch), **1987**, *6*, 476–9; *Chem. Abstract* **1989**, *110*(11), 94630v.
30. Ran, Ruicheng; Jiang, Shuojian, Shen, Ji *J. Macromol. Sci., Chem.* **1987**, *A24*(6), 669–679; *Chem. Abstract* **1987**, *105*, 60145x.

31. Ran, Ruicheng; Pei, Weiwei; Jia, Xinru; Wu, Xianghong *J. Org. Chem.* (ch), **1987**, *4*, 286–72; *Chem. Abstract* **1988**, *108*(17), 150013u.
32. Ran, Ruicheng; Pei, Weiwei; Jia, Xinru; Shen, Ji; Jiang, S. *Sci. Bull.* (Eg), **1987**, *32*(6), 388–94; *Chem. Abstract* **1988**, *108*(5), 36846t.
33. Shen, Ji; Ran, Ruicheng *Chem. J. Chin. Univ.* **1988**, *9*(11), 1168–72; *Chem. Abstract* **1989**, *111*(2), 103229w.
34. Ran, Ruicheng; Wu, Xianghong; Jia, Xinru; Pei, Weiwei *Acta Polym. Sinica* (Ch), **1988**, *6*, 67–70; *Chem. Abstract* **1989**, *110*(1), 7170d.
35. Ran, Ruicheng; Jia, Xinru; Jiang, Shuojian *Petrochem. Technol.* (ch), **1988**, *17*(1), 15–20; *Chem. Abstract* **1988**, *108*(20), 168034m.
36. Pei, Weiwei; Lu, Xiaoho; Ran, Ruicheng *Acta Sci. Nat. Univ. Beijing* **1988**, *24*(2), 28–32; *Chem. Abstract* **1989**, *110*(7), 56683g.
37. Ran, Ruicheng; Pei, Weiwei; Jia, Xinru; Shen, Ji *J. Mol. Cat.* (ch), **1988**, *2*(2), 112–8; *Chem. Abstract* **1989**, *111*(1), 7011y.
38. Ran, Ruicheng; Jia, Xinru; Li, Mingqian; Jiang, S. *J. Macromol. Sci., Chem.* **1988**, *A25*(8), 907–22; *Chem. Abstract* **1988**, *109*, 150080j.
39. Ran, Ruicheng; Shen, Ji *J. Macromol. Sci., Chem.* **1988**, *A25*(8), 923–33; *Chem. Abstract* **1988**, *109*, 189478v.
40. Ran, Ruicheng; Mao, Guoping *J. Cat.* **1989**, *10*(1), 92–7; *Chem. Abstract* **1990**, *112*(7), 54947p.
41. Pei, Weiwei; Lu, Xiaoho; Ran, Ruicheng *HUAXUE XUEBAO* **1989**, *47*(1), 97–101; *Chem. Abstract* **1989**, *110*, 194186y.
42. Ran, Ruicheng; Mao, Guoping *Chem. J. Chin. Univ.* **1989**, *10*(7), 784–6; *Chem. Abstract* **1990**, *112*(19), 179267y.
43. Ran, Ruicheng; Mao, Guoping *Acta Sci. Nat. Univ. Beijing* **1989**, *24*(2), 28–32; *Chem. Abstract* **1991**, *114*(11), 101087m.
44. Ran, Ruicheng; Mao, Guoping *Acta Polym. Sinica* (Ch), **1989**, *6*, 64–69; *Chem. Abstract* **1990**, *113*(25), 230393m.
45. Ran, Ruicheng; Mao, Guoping *J. Chem. Reg.* (Ch), **1990**, *12*(2), 75–8; *Chem. Abstract* **1991**, *114*(3), 23504d.
46. Ran, Ruicheng; Fu, Dianking *J. Macromol. Sci., Chem.* **1990**, *A27*(5), 625–636.
47. Ran, Ruicheng; Mao, Goping *J. Macromol. Sci., Chem.* **1990**, *A27*(2), 125–36; *Chem. Abstract* **1990**, *113*(2), 6953m.
48. Ran, Ruicheng; Fu, Dianking *Chin. J. Polym. Sci.* **1991**, *9*(1), 79–85; *Chem. Abstract* **1993**, *116*, 173261c.
49. Ran, Ruicheng; Fu, Diankui; Ji, Shen; Wang, Qingyang *J. Polym. Sci., Polym. Chem. Ed.* **1993**, *31*, 2915–2921.
50. Ran, Ruicheng; *J. Polym. Sci., Polym. Chem. Ed.* **1993**, *31*, 1561–1569; *Chem. Abstract* **1993**, *118*, 255,404g.
51. Ran, Ruicheng; *Book of Abstracts* 208th National Meeting of the American Chemical Society: Washington, D.C., 1994, PWSE 279.
52. Fierri, R.; Chien, J. C. W. *J. Polym. Sci., Polym. Chem. Ed.* **1994**, *32*, 661–673.
53. Lee, S.; Brian, K. W. S.; Ripplinger, M. P.; Wooster, J. J. et al. U.S. Patent 5 231 151 A 930727.
54. Mitchell, K. E.; Miller, D. C.; Godbehere, D. W.; Hawley, G. R. U.S. Patent 5 235 011 A 930831.
55. Hara, D.; Sato, M.; Mori, M. Eur. Patent EP 530814 A1 930310.
56. Woo, S. I.; Kim, I. I. U.S. Patent 5192729 A 930309.
57. Sano, A.; Suzuki, K. et al. Eur. Patent EP 507504 A2 921007.
58. Bailly, J. C. A.; Behue, P. Eur. Patent EP 437080 A1 910717.
59. Gessell, D. E.; Hosman, D. P. U.S. Patent 4945142 A 900731.
60. Quijada, R.; Wanderley, A. M. R. *Stud. Surf. Sci. Catal.* **1986**, *25*, 419–29.
61. Li, Y.; Jun, D. *J. Macromol. Sci., Chem.* **1987**, *A24*, 227.
62. Chemenko, G. M.; Tinyakov, E. I.; Kakuliya, Ts. V.; Khananashvili, L. M.; Novikov, Yu. V.; Volipin, M. E. *Vysokomol. Soedin Ser.* **1983**, *B25*, 919.
63. Bedell, S. A.; Coleman, W. M.; Howell, W. R. Jr. U.S. Patent 4 623 707, 1980.
64. Collomb, J.; Duran, D. C.; Havas, L.; Morerol, F. R. M. Eur. Pat. Appl., EP 211 624, 1987.
65. Dow Chemical, Neth. Appl. NL 85 02 580, 1987.
66. Yano, T.; Inoue, T.; Ikai, S.; Shimizu, M.; Kai, Y.; Tamura, M. *J. Polym. Sci., Polym. Chem. Ed.* **1988**, *26*, 457–467, 477–490.
67. Spitz, R.; Duvanel, L.; Guyot, A. *Makromol. Chem.* **1988**, *189*, 549.
68. Chien, J. C. W.; Hu, Y. L. *J. Polym. Sci., Polym. Chem. Ed.* **1987**, *25*, 2847, 2881.
69. Sun, L.; Hsu, C. C.; Bacon, D. W. *J. Polym. Sci., Polym. Chem. Ed.* **1994**, *32*, 2127, 2135.

SUPPORTED CHIRAL CATALYSTS

Shinichi Itsuno
Department of Materials Science
Toyohashi University of Technology

Several asymmetric synthesis that use polymer-supported chiral catalysts have been reported. They are the most attractive type of polymer-supported species for applications in asymmetric synthesis, since a relatively small amount of polymer can be used to transform a large amount of reactants. Another advantage of performing asymmetric reactions with polymer-supported catalysts is that their use allows both the recycling of the catalysts and the easy separation of the low molecular weight chiral products. The polymer provides a unique microenvironment for the stereoselective reaction, which in favorable cases may result in enhanced stereoselectivity.

TRANSITION METAL CATALYSTS

Hydrogenation

Hydrogenation of N-acyl α-amino acids has been accomplished with rhodium bound to crosslinked polymers with chiral tertiary phosphine ligands. The polymeric rhodium catalysts gave results comparable to those obtained with a similar homogeneous catalyst system.[1]

Hydroformylation

Hydroformylation of styrene in the presence of ethyl orthoformate can be achieved using platinum catalysts containing polymer-supported chiral diphosphine ligand.[2–6]

Cross-Coupling

Asymmetric cross-coupling reaction of a secondary alkyl Grignard reagent with vinyl chloride can be achieved with a chiral nickel catalyst bound to crosslinked polystyrene beads.[7]

Dihydroxylation

Copolymerization of cinchona alkaloid olefin monomer with acrylonitrile followed by complexation with osmium tetroxide gave the polymeric catalysts in the asymmetric dihydroxylation of olefin.[8]

Epoxidation

Farrall et al. performed Sharpless epoxidation of geraniol with polymer-supported L-tartrate ester H to give epoxygeraniol with 49–65% ee, and the same product was obtained in 95% ee using the monomeric Sharpless catalyst.[9,10]

Reduction

Asymmetric catalytic reduction of aromatic ketones can be brought about through hydrogen transfer reaction. Optically active alkylaminomethylpyridines have been used as ligands in the Ir-catalyzed hydrogenation reaction. Butyrophenone can be reduced by the use of this type of chiral catalyst in a polymer-bound form in 86% ee.[18–21]

PHASE TRANSFER CATALYST

Chiral Quaternary Ammonium Salts

Quaternary ammonium salts are often used as phase transfer catalyst (PTC) of a two-phase system. Optically active quaternary ammonium salts have been applied for several types of asymmetric reactions that require PTC. Polymer-supported quaternary ammonium salts are usually prepared by the reaction of chiral amines or amino alcohols with chloromethylated polymers.[15,16]

Michael Reaction

Michael reaction of methyl 1-oxoindan-2-carboxylate with methyl vinyl ketone in the presence of polymer-supported chiral quaternary ammonium salt yielded the corresponding S-adduct in a 61% chemical yield with 27% ee.

Borohydride Reduction

The same polymeric chiral ammonium salts derived from quinine have been used as PTC of borohydride reduction of ketones.[17]

Hydroxylation

Oxidation of cyclic ketones with O_2 in the presence of polymeric quaternary ammonium salt gave α-hydroxyketone in 56% yield with 14% ee.

Darzens Reaction

Darzens reaction was catalyzed with chiral polymeric PTC to give a chiral epoxide in 23% ee.[18]

Epoxidation of Chalcone

Kobayashi performed epoxidation of chalcone with hydrogen peroxide in the presence of polymer-supported PTC. However, only low enantioselectivity was obtained.[19] Itsuno's reaction of amino acid *N*-carboxyanhydride with amino-methylated polystyrene afforded polymer-supported poly(amino acid).[20] This polymer also gave very high enantioselectivities (~99% ee) in the epoxidation of trans-chalcone. The separation of the catalyst from the reaction system was greatly simplified.[21]

BASE CATALYSTS

Michael Reaction

Hermann performed the first application of polystyrene-supported cinchona alkaloid in the Michael addition of methyl 1-oxoindan-2-carboxylate to methyl vinyl ketone.[22] The polymers were prepared by functionalization of crosslinked polystyrene followed by a reaction with quinine.

LEWIS ACID CATALYST

Reduction with Oxazaborolidine Catalyst

Reaction of an optically active α-amino alcohol with borane yields 1,3,2-oxazaborolidines that are a new generation of reduction catalysts.[23] Chiral oxazaborolidines catalyze borane reduction of ketones, oximes, and imines.[24,25] We believe that boron atoms in the borolidines act as chiral Lewis acids that activate these substrates.

Alkylation with Dialkylzinc

Like chiral oxazaborolidine catalyst in borane reductions, the chiral complex prepared from optically active α-amino alcohol with diethylzinc has proven to be an extremely efficient catalyst for alkylation of aldehyde.[26] Polymer-supported amino alcohol–$ZnEt_2$ afforded high enantioselectivities up to 99% ee.

Diels-Alder Reaction

Diels–Alder reaction is also activated by the addition of Lewis acid: chiral Lewis acids have been prepared and used for asymmetric Diels–Alder reactions. Extensive studies on designing chiral Lewis acid catalysts based on B, Al, and Ti elements have led to fruitful results in these reactions.[27–29] Polymer-supported Lewis acids were prepared from haloborane or borane with crosslinked polymers having a chiral moiety such as an amino alcohol, diol, or *N*-sulfonylamino acid.

REFERENCES

1. Baker, G. L.; Fritschel, S. J.; Stille, J. R. *J. Org. Chem.* **1981**, *46*, 2954.
2. Parrinello, G.; Deschenaux, R.; Stille, J. K. *J. Org. Chem.* **1986**, *51*, 4189.
3. Parrinello, G.; Stille, J. K. *J. Am. Chem. Soc.* **1987**, *109*, 7122.
4. Stille, J. K. In *Macromolecules*; Benoit, H.; Rempp, P., Eds.; Pergamon: Oxford, UK, 1982, p 99.
5. Stille, J. K.; Parrinello, G. J. *Mol. Cat.* **1983**, *21*, 203.
6. Stille, J. K. *J. Macromol. Sci., Chem.* **1984**, *A21*, 1689.
7. Hayashi, T.; Nagashima, N.; Kumada, M. *Tetrahedron Lett.* **1980**, *21*, 4623.
8. Kim, B. M.; Sharpless, K. B. *Tetrahedron Lett.* **1990**, *31*, 3003.
9. Katsuki, T.; Sharpless, K. B. *J. Am. Chem. Soc.* **1980**, *102*, 5974.
10. Farrall, M. J.; Alexis, M.; Trecarten, M. *Nouv. J. Chim.* **1983**, *7*, 449.
11. Kaschig, J. (CIBA-GEIGY) EP-B 0246 194 A2, 1985.
12. Kaschig, J. (CIBA-GEIGY) EP-B 0246 195 A1, 1985.
13. Kaschig, J. (CIBA-GEIGY) EP-B 0246 196 A1, 1985.
14. Kaschig, J. (CIBA-GEIGY) EP-B 0251 994 A1, 1986.
15. Tomoi, M.; Ford, W. T. *Synthesis and Separations using Functional Polymers*; Sherrington, D. C.; Hodge, P., Eds.; John Wiley & Sons, 1988, pp 181.
16. Itsuno, S.; Uchikoshi, K.; Ito, K. *J. Am. Chem. Soc.* **1990**, *112*, 8187.
17. Boyer, B.; Lamaty, G.; Roque, J-P; Solofo, J. *New J. Chem.* **1986**, *10*, 559.
18. Colonna, S.; Fornasier, R.; Pfeiffer, U. *J. Chem. Soc., Perkin Trans. 1* **1978**, *8*.
19. Kobayashi, N.; Iwai, K. *Makromol. Chem., Rapid Commun.* **1981**, *2*, 105.
20. Itsuno, S.; Koizumi, T.; Okumura, C.; Ito, K. *Synthesis* **1995**, 50.

21. Itsuno, S.; Sakakura, M.; Ito, K. *J. Org. Chem.* **1990**, *55*, 6047.
22. Hermann, K.; Wynberg, H. *Helv. Chim. Acta* **1977**, *60*, 2208.
23. Itsuno, S.; Hirao, A.; Nakahama, S.; Yamazaki, N. *J. Chem. Soc., Perkin Trans. 1* **1983**, 1674.
24. Itsuno, S.; Ito, K. In *Trends in Organic Chemistry*; Council of Scientific Integration, Research Trends: Trivandrum, India, 1991; Vol. 1, p 199.
25. Itsuno, S. *Kikan Kagaku Sosetsu* **1993**, *19*, 17.
26. Itsuno, S.; Fréchet, J. M. J. *J. Org. Chem.* **1987**, *52*, 4140.
27. Kagan, H. B.; Riant, O. *Chem. Rev.* **1992**, *92*, 1007.
28. Pindur, U.; Lutz, G.; Otto, C. *Chem. Rev.* **1993**, *93*, 741.
29. Deloux, L.; Srebnik, M. *Chem. Rev.* **1993**, *93*, 763.

SUPPORTED CHROMIUM POLYMERIZATION CATALYSTS (on Silica, Phillips Catalysts)

C. E. Marsden
Crosfield Ltd.

The Phillips catalyst originated in the 1950s when Hogan and Banks, working at Phillips Petroleum Company, discovered that ethylene could be converted to a solid polymer over a chromium oxide-silica-alumina catalyst.[1] They recognized two important features of this catalyst: chromium in the Cr(VI) state and a powdered substrate of silica or silica-alumina. Despite extensive study since then, precise details of the manner in which these catalysts control polymerization kinetics and polymer structure remain poorly understood. It has, however, been clearly demonstrated that the chemical composition of the catalyst and the porous structure of the support strongly influence the activity of the catalyst and, perhaps more importantly, the structural characteristics of the polymer produced.[2,3]

CATALYST PREPARATION

Phillips catalysts typically comprise 1 wt % chromium deposited on the surface of a silica or silica-containing support of appropriate porous structure and particle size distribution. The chromium, generally as Cr(III), is applied to the support using standard impregnation procedures.[4] Distribution of the chromium on the surface is not critical at this stage since during the high temperature activation procedure the chromium species are oxidized to Cr(VI) species which are mobile prior to reacting with surface hydroxyls.

Silica cogels such as silica-titanias, useful as supports for Phillips catalysts, are also produced via gel technology. Titanium is incorporated into the silica structure in the acid solution prior to hydrosol production.[5] A third type of Phillips catalysts is prepared by surface impregnation of the support or catalyst with modifiers such as titanium or aluminum.

THE ROLE OF THE SILICA SUPPORT

Active Site Formation

The silica support surface facilitates production of a highly active catalyst. Prior to polymerization, the catalyst must be activated at high temperature in an oxidizing atmosphere. During this treatment the chromium is chemically bonded to the surface by reaction with hydroxyl groups to form surface Cr(VI) species, which, upon exposure to ethylene in the reactor, are reduced to generate the active sites which catalyze the polymerization.

Nature of the Active Site

Polyethylenes produced using Phillips catalysts tend to have relatively broad molecular weight distribution (MWD). The ratio of the weight-average to the number-average molecular weight (M_w/M_n) for Cr/silica is typically 6–15.[3] The broad MWD has been attributed to the diversity of active sites created by the amorphous nature of the silica surface.

Fragmentation

The silica support fragments during polymerization.[6] This is an essential characteristic of a polymerization catalyst for which continued access to the internal active sites is maintained.

Influence of Support Structure on Activity and Polymer Properties

The porous structure of the support strongly influences the catalyst activity and controls the polymer structure providing opportunities for new and improved catalysts and polyethylene resins.

INFLUENCE OF CATALYST COMPOSITION ON PERFORMANCE

The chemical composition of the catalyst exerts a strong influence on the structure of the polyethylene produced. Numerous patents have claimed advantages in catalyst activity and polymer properties as a result of the addition of various modifiers to the basic Cr/silica catalyst. These include titanium, aluminum, zirconium, fluoride, boron, and phosphorus compounds, either added alone or in various combinations. Such modifiers may be incorporated into the catalyst during silica gel formation, as, for example, in the production of silica titania and silica zirconia cogels, or by surface impregnation.[5,7]

Titanium

Titanium has been incorporated into Phillips-type catalysts in a multitude of ways as documented in numerous patents. In the simplest method, a titanium ester is allowed to react with the hydroxyl groups of a predried silica in the absence or presence of chromium. Titanium contents of approximately 5% can be obtained by this procedure. Catalysts modified in this manner have higher activity and produce HDPE with higher MI and broader MWD than unmodified catalysts.

Titanium incorporated into the silica structure via cogelation also results in decreased induction time, increased catalyst activity, and increased MI. In contrast to titanium impregnation, however, catalysts of this type have been developed which impart a narrow MWD and extremely high environmental stress crack resistance (ESCR) to the polymer.

Aluminum

The addition of aluminum to Phillips catalysts via impregnation with aluminium alkoxide increases the polymer MI in proportion to the aluminum content, and improves the catalyst response to hydrogen.[8] Rekers et al. indicate that the molecular weight (MW) of polyethylene decreases and MWD broadens with the proportion of aluminum compound present.[9] They further claim that Al-modified catalysts activated under specific conditions produce high-density polyethylenes with improved ESCR.

Fluoride

The addition of inorganic fluorides, such as ammonium silicofluoride, to the Cr/silica catalyst prior to activation provides a means for production of polyethylenes with high MW and narrower MWD.

NONSILICA SUPPORTS

Several materials apart from silica have been investigated as potential supports for Phillips catalysts; alumina, aluminum phosphate, pillared silicate clays, and zeolites.

CATALYST ACTIVATION

Phillips catalyst only catalyze ethylene polymerization following calcination at high temperature. Conventionally this is achieved by treatment with dry air in a fluidized bed reactor using temperatures of 500–950°C and residence times of up to 12 h. As catalyst activation temperature is raised, catalyst activity and polymer MI increase and polymer MWD, as determined by HLMI/MI, narrows. This relationship is used commercially to control MW and other polymer properties.

POLYMERIZATION

Phillips catalysts are used commercially in both slurry and gas phase processes. Reaction temperatures used in the slurry process are typically in the range of 90–108°C.

Copolymerization of ethylene with various 1-olefins can be readily accomplished with Phillips catalysts. Comonomer incorporation is random, decreasing density, lowering crystallinity, and increasing flexibility and ESCR.

OUTLOOK FOR THE FUTURE

Phillips catalysts will continue to be commercially important for the production of polyethylene. The strong relationships between the chemical composition and porous structure of these catalysts and the polymer structure produced have facilitated the development and production of a wide range of polyethylene resins. The additional influences of activation procedures, whereby the surface chemistry of the active catalyst can be modified, and reactor conditions, such as the absence or presence of comonomer, cocatalyst, and so forth, extend the scope of the polyethylene structures substantially. As a consequence, some 35 years after the original patent, patents claiming improved catalyst systems and polymers are still being granted.[1] For example, a recent U.S. patent claims a linear low density polyethylene copolymer with a density in the range of ~0.89–0.915 g/cm^3, possessing improved toughness and impact resistance.[10] The patent holders used a high pore volume Cr/silica–titania, activated in dry air at 650°C for 5 hr, exposing it to CO at 350°C for 2 hr, followed by a N_2 purge and cooling. This catalyst was used to polymerize ethylene and 1 hexene in the presence of H_2 and triethyl boron cocatalyst. Future developments will probably be of this type, where existing or new support structures and compositions, combined with modified activation procedures and used under specific reaction conditions, will lead to new polyethylene structures with improved applications properties.

REFERENCES

1. Hogan, J. P.; Banks, R. L. U.S. Patent 2 825 721, 1958.
2. McDaniel, M. P. *J. Polym. Sci., Polym. Chem. Ed.* **1983**, *21*, 1217.
3. McDaniel, M. P. *Adv. in Catal.* **1985**, *33*, 47.
4. Marsden, C. E. *Preparation of Catalysts*; Elsevier: New York, NY, 1991, Vol. V, 215.
5. Dietz, R. E. U.S. Patent 3 887 494, 1975.
6. McDaniel, M. P. *J. Polym. Sci., Polym. Chem. Ed.* **1981**, *19*, 1967.
7. Dombro, R. A. U.S. Patent 4 476 243, 1986.
8. Frielingsdorf, H.; Gruber, W.; Mueller-Tamm, H. U.S. Patent 4 110 522, 1978.
9. Rekers, L. J.; French, R. J.; Mayhew, H. W. U.S. Patent 4 356 294, 1982.
10. McDaniel, M. P.; Benham, E. A. U.S. Patent 5 208 309, 1993.

SUPPORTED LEWIS ACID CATALYSTS (on Polypropylene; Recoverable and Reusable)

T. C. Chung
Department of Materials Science and Engineering
The Pennsylvania State University

The idea of immobilizing catalysts obviously can be applied to many reaction systems. Here, I will discuss a new class of immobilized Lewis acid catalysts.[1–3] My colleagues and I prepared them by chemical bonding between homogeneous Lewis acids such as $EtAlCl_2$, Et_2AlCl, and BF_3, and the hydroxylated polypropylene (PP–OH) substrate. The resulting immobilized Lewis acid catalysts are active in the carbocationic polymerization of isobutylene, as illustrated in **Equation 1**.

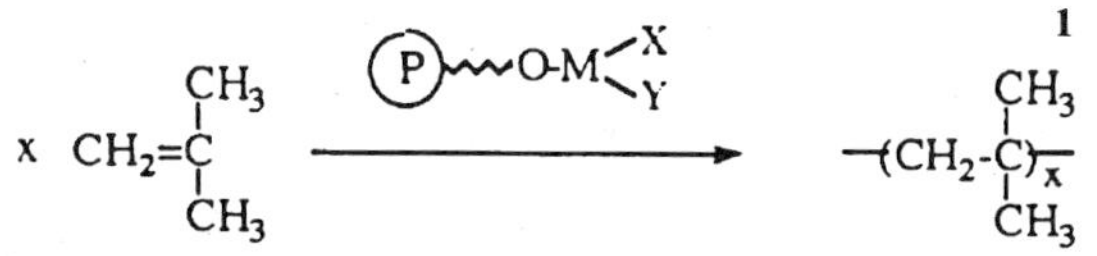

1

where P is the partially crystalline PP and M is a B or Al atom. The ligands (X and Y) can be either alkyl or halogen groups.

The catalysts can be recovered and reused for many reaction cycles without significantly losing their activity.

PREPARATION

Synthesis of Comb-like PP–OH

The preparation of PP–OH with comb-like structure involves a continuous polymerization process and a new functionalization chemistry, using borane comonomers and Ziegler–Natta catalysts.[4,5]

Immobilization Reactions

The hydroxylated polypropylene (PP–OH) was reacted with Lewis acids, such as $EtAlCl_2$, Et_2AlCl, and BF_3.

APPLICATION

The polypropylene immobilized catalysts were used as the Lewis acid catalysts in the carbocationic polymerization of isobutylene as shown in Equation 1.[1–3]

In the runs using powder form of PP–O–$AlCl_2$ catalyst, the quantitative conversion from monomer to polymer was completed within 15 minutes. The same catalyst activity was maintained for several reaction cycles. By contract, PP–O–$AlCl_2$ catalyst with chunk form (particle size > 5 mm) was sluggish. It required about 2–3 hr to obtain complete conversion; and the yields of PIB products are very dependent on the reaction time. This slow carbocationic polymerization of isobutylene is obviously related to the availability of catalysts. Despite the difference in reaction rate with particle size, the catalyst can be recovered and reused in subsequent reaction cycles. Elemental analysis and ^{6}Al NMR results show no significant changes in the aluminum species after more than 10 reaction cycles.

ACKNOWLEDGMENTS

The author would like to thank the Polymer Program of the National Science Foundation and Exxon Chemical Company for their financial support.

REFERENCES

1. Chung, T. C.; Kumar, A. *Polym. Bull.* **1992**, *28*, 123.
2. Chung, T. C.; Kumar, A.; Rhubright, D. *Polym. Bull.* **1993**, *30*, 385.
3. Chung, T. C. U.S. Patent 5 288 677, 1994.
4. Chung, T. C. *Macromolecules* **1988**, *21*, 865.
5. Ramakrishnan, S.; Berluche, E.; Chung, T. C. *Macromolecules* **1990**, *23*, 378.
6. Noth, H.; Wrackmeyer, B. *"Nuclear Magnetic Resonance Spectroscopy of Boron Compounds;"* Springer-Verlag: Heidelberg, Germany, 1978.

Supported Phthalocyaninates

See: Chemical Sensors

SUPPORTED POLYTHIOETHERS AND POLYTHIACROWN ETHERS

Kazuichi Tsuda and Keiji Yamashita
Nagoya Institute of Technology

Polymeric ligands with an acyclic polyether or a crown ether side chain have been intensively studied as ion-exchange resins, metal ion adsorbents, and polymeric phase-transfer catalysts.[1–6] The ether ligand has an affinity for hard cations such as alkali metal ions and protons. Compared with the oxygen atoms of the ether ligand, sulfur atoms in a thioether ligand prefer soft cations such as heavy metal ions. Murroy et al. reported that a sulfur atom has a strong affinity for Ag(I) and Hg(II) ions.[7] The results suggest that the polymer ligands having thioether units will be effective adsorbents for toxic heavy metal ions, Hg(II), and for a valuable noble metal ion, Ag(I) from industrial waste water. There are, however, few investigations about the thioether polymer and thiacrown ligands because of their poor solubility in ordinary solvents.[8–12] Since the insolubility obstructs the processing and designing of an effective polymer ligand, those with thioether units and thiacrown units were prepared by polymer modification reaction on an insoluble crosslinked resin.[8–10]

Soluble polymer ligands having tetrathioether side chains as a heavy metal ion adsorbent were synthesized by copolymerization of the corresponding thioether monomer with styrene (St).[11] We investigated the solubility and binding ability for heavy metal ions and antibacterial activity.

CONCLUSION

We have described the binding character and application of thioether and thiacrown side chain polymers. Thioether and thiacrown polymers resemble each other and have selective binding ability for soft acid, Hg(II) and Ag(I).

Thioether side-chain polymers are suitable as heavy metal ion adsorbents, because the syntheses of thioether polymer ligand are much easier than those of thiacrown polymers whose properties are no more advantageous. However, the thiacrown ether polymer is an efficient adsorbent for heavy metal ions under acidic conditions.

The thioether side-chain polymers and Hg(II) and Ag(I) ions form two kinds of complexes. These complexes are mono- or bidentate and that their formations do not involve an important change in the conformation of the ligand. At a low sulfur content in the polymer 1:1 complexes are formed, and at higher sulfur contents, 1:2 complexes. In the intermediate range, mixtures of both stoichiometries are encountered.

Insoluble thioether and thiacrown polymers are obtained by the reaction of chloromethylated polystyrene–DVB with thioether or thiacrown compounds. Soluble sulfide and thiacrown side-chain polymer ligands are readily prepared by radical polymerization of the corresponding sulfide or thiacrown monomers.

Additives do not obtain on the binding efficiency with increasing sulfurs in side chains, so, the simplest monosulfide side chain polymer which is the easiest to synthesize is the best way to achieve adsorbents, catalysts and so on.

The polymer-Ag(I) complexes are also soluble in ordinary organic solvents. The soluble thioether polymer–Ag(I) complexes can be impregnated into polyester cloth, which then exhibit strong antibacterial activity against *Staphylococcus aureus*.

REFERENCES

1. Sherrington, D. C. In *Polymer-Supported Reactions in Organic Synthesis*; Hodge, P.; Sherrington, D. C., Eds.; Wiley: Chichester, UK, 1980; p 157.
2. Hojo, N. *Chelate Resins-Ion Exchange Resins*; Nankodo: Tokyo, Japan, 1976.
3. Mckeazie, W. M.; Sherrington, D. C. *J. Chem. Soc., Chem. Commun.* **1978**, 541.
4. Yamada, S.; Takahashi, K.; Okahara, M. *J. Org. Chem.* **1979**, *44*, 1099.
5. Regen, S. L. *Angew. Chem., Int. Ed. English* **1979**, *18*, 421.
6. Tomoi, M.; Yanai, N.; Shiiki, S.; Kakiuchi, H. *J. Polym. Sci., Polym. Chem. Ed.* **1984**, *22*, 911.
7. Murroy, S. G.; Hartley, F. R. *Chem. Rev.* **1981**, *81*, 365.
8. Lauth, M.; Frere, Y.; Meurer, B.; Gramain, Ph. *React. Polym.* **1990**, *13*, 63.

9. Lauth, M.; Frere, Y.; Prevost, M.; Gramain, Ph. *React. Polym.* **1990**, *13*, 71.
10. Tomoi, M.; Abe, O.; Takazu, N.; Kakiuchi, H. *Makromol. Chem.* **1983**, *184*, 2431.
11. Yamashita, K.; Yamada, A.; Goto, T.; Iida, K.; Higashi, K.; Nango, M.; Tsuda, K. *React. Polym.* **1994**, *22*, 65.
12. Yamashita, K.; Tsuda, K., submitted for publication in *React. Polym.*

Supported Reagents

See: *Ion-Selective Reagents (Dual Mechanism Bifunctional Polymers)*
Poly(vinylpyridinium salts)

SUPPORTED THIAZOLIUM SALT CATALYSTS (for Acyloin Condensation)

Kazuichi Tsuda and Keiji Yamashita
Department of Applied Chemistry
Nagoya Institute of Technology

Breslow has reported that the thiazolium salt unit is an active site of a co-enzyme thiamine pyrophosphate, which controls much biochemical reaction such as decarboxylation and acyl transfer reaction.[1] On the basis of Breslow's study, many researchers have investigated the catalytic activity of thiazolium salts in not only biochemical reactions, but also organic reactions such as acyloin condensation and formose reactions.

From an industrial viewpoint, the acyloin condensation is an important reaction because we can get many chemicals by using this reaction. The thiazolium salt compounds have been considered an important catalyst for this reaction because they can be used easily and are less toxic than the cyanide ion, which is one of the catalysts for acyloin condensation. But the syntheses of thiazolium salt were not at all easy; so they were not favorable for industrial use. Therefore, it is reasonable to make a polymer-supported thiazolium salt catalyst because the polymer catalyst can be used more easily than the low molecular weight catalyst and it is recyclable after recovering. Furthermore, the polymer catalyst is expected to have some polymer effect that is commonly discussed in the study of the polymer catalyst.

PREPARATION

Vinylthiazole Monomer

Thiazolium salt polymers have been prepared by the following two methods: quarternization of a thiazole polymer prepared by the polymerization of the corresponding thiazole monomer with an alkyl halide, and quarternization of a commercializable thiazole such as 5-(2′-hydroxyethyl)-4-methylthiazole with a chloromethylated styryl resin.

Quarternization

The thiazolium salt polymers were prepared by quarternization of the corresponding thiazole polymers because the thiazolium salt monomer did not have polymerizability. Quarternization of 4,4′-vinylphenylthiazole and 5-[2′-(4″-vinylbenzyl)ethoxy]-4-methylthiazole polymers with methyl iodide was carried out in acetonitrile.[4,5] Their structures are shown in **Scheme I**. In the abbreviations PTS⁺ (x)-DVB(z) and BTS⁺ (x)-DVB(z) in this scheme, x means a mol % of the quarternized thiazole units and z means a mol % of the DVB units as a crosslinking point.

PROPERTIES

Acyloin Condensation in Organic Solvent

Schilling reported that their thiazolium salt polymers, had a sufficient activity for the benzoin condensation of furfural in MeOH.[3] He also reported that a thiazolium salt unit is necessary to the catalysis instead of the thiazole unit and the base as a co-catalyst.

Recycling of Polymer-Bound Thiazolium Salt Catalyst

Challa had detailed experiments about the recycling of the polymer bound thiazolium salt catalyst.[6] His thiazolium salt polymer was made from a chloromethylated Amberlite XAD-2 nonionic adsorbent. He investigated a batchwise reutilization and also their reactivity in their own steady-state flow stirred-tank reactor in which one can test the continuous production of benzoin from benzaldehyde.

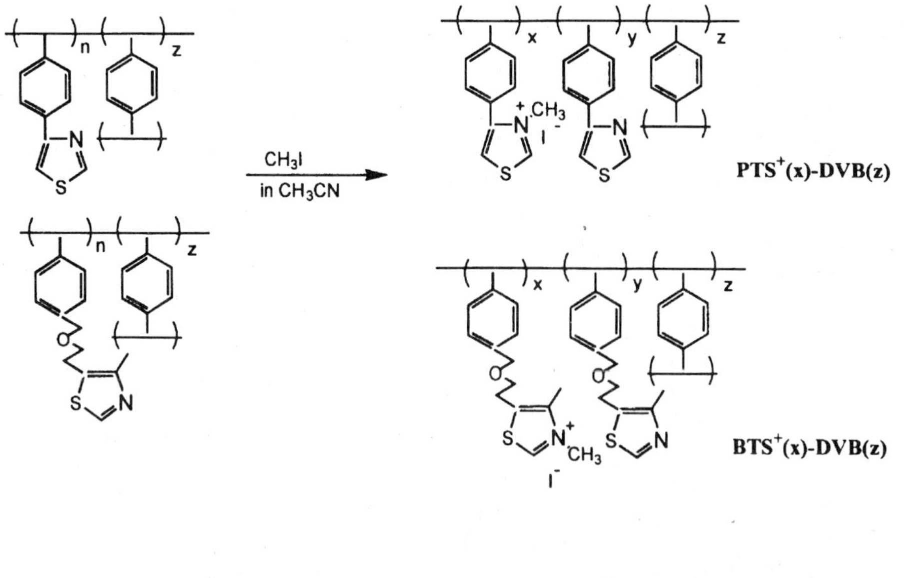

I

Acyloin Condensation in Buffer Solution

In the aqueous system, Tagaki, Shinkai, and Breslow have gotten highly active thiazolium salt catalysts in a molecular assembly system such as a cationic micellar and a cyclodextrin.[2,3,7] It was well known that a polymer main chain also supplies the hydrophobic domain in the aqueous system even in a hydrophilic polymer. The polymeric catalyst is regarded as one of the molecular assembly catalysts.

Yamashita also showed that PTS^+ (x)-DVB(z) (Scheme I) had a catalytic activity even in an aqueous system in which the low molecular weight analog, MPTI, no longer had their activity.[4,8]

APPLICATION

Catalysts of Formorse Reaction

Butlerow first focused on artificial synthesis of carbohydrates from simple compounds such as formaldehyde.[9] After his report many inorganic and organic bases have been used as catalysts.[10,11] Especially, thiazolium salts were reported as catalysts, which accelerated the selective syntheses of trioses such as glyceraldehyde (GOA) and dihydroxyacetone (DHA).[12]

Yamashita used PTS^+ polymer as a catalyst for the formose reaction.[13] PTS^+ (13) and crosslinked polymer catalyst, PTS^+ (15)-DVB(20), exhibited approximately the same catalytic activity as that of a low molecular weight analog. From the GC chromatograms of the products, the products were mainly DHA with a small amount of tetrose. The crosslinked polymer catalyst was easily recovered from the reaction mixture by filtration and the recycled polymer catalyst had similar activity to that of the organic polymer catalyst.

REFERENCES

1. Breslow, R. *J. Am. Chem. Soc.* **1957**, *79*, 1762. *ibid* **1958**, *80*, 3719. *ibid* **1959**, *81*, 3080.
2. Hilvert, D.; Breslow, R. *Bioorg. Chem.* **1984**, *12*, 206.
3. Schilling, C. L. Jr.; Mulvaney, J. E. *Macromolecules* **1968**, *1*, 445, 452.
4. Yamashita, K.; Osaki, T.; Sasaki, K.; Yokota, N.; Oshima, N.; Nango, M.; Tsuda, K. *J. Polym. Sci., Polym. Chem. Ed.* **1994**, *32*, 1711.
5. Yamashita, K.; Tokuda, H.; Tsuda, K. *J. Polym. Sci., Polym. Chem. Ed.* **1989**, *27*, 1333.
6. Berg, H. J.; Challa, G.; Pandit, U. K. *React. Polym.* **1989**, *11*, 127.
7. Shinkai, S.; Yamashita, T.; Kusano, Y.; Manabe, O. *J. Org. Chem.* **1980**, *45*, 4947.
8. Yamashita, K.; Watanabe, J.; Ikeda, R.; Abe, D.; Tsuda, K. *Macromolecules* **1989**, *22*, 4382.
9. Butlerow, A. *C. R. Acad. Sci.* **1861**, *53*, 145.
10. Shigemasa, Y.; Shimano, M.; Sakazawa, C.; Matsuura, T. *Bull. Chem. Soc. Jpn.* **1975**, *48*, 2099. Gabel, N. W.; Ponnanperuma, C. *Nature* **1967**, *216*, 463.
11. Castells, J.; Geijo, F.; Calaborra, F. L. *Tetrahedron Lett.* **1980**, *21*, 4517; Matsumoto, T.; Kamiyama, M.; Inoue, S. *Chem. Lett.* **1980**, 839.
12. Shigemasa, Y.; Sasaki, Y.; Ueda, N.; Nakashima, R. *Bull. Chem. Soc. Jpn.* **1984**, *57*, 2671; Matsumoto, T.; Yamamoto, H.; Inoue, S. *J. Am. Chem. Soc.* **1984**, *106*, 4829.
13. Yamashita, K.; Wakao, N.; Nango, M.; Tsuda, K. *J. Polym. Sci., Polym. Chem. Ed.* **1990**, *30*, 2247.

SUPPORTED ZIEGLER–NATTA CATALYSTS (on Silica, for Olefin Polymerization)

Meinolf Kersting
Polymer Research Division
BASF AG

Four decades after the discovery by Ziegler and Natta that transition metal compounds in combination with aluminum alkyls could polymerize ethylene and propylene, the polyolefin business has grown into a major industry. The development of catalyst systems with a high efficiency has been the key to rapid commercial development. These catalysts continue to attract attention from many laboratories around the world—both industrial and academic.[1–3]

It is the purpose of this review to discuss ZN catalysts supported on silica, and it will focus primarily on ethylene and propylene polymerization because the most attention has been devoted to these monomers.

THE SILICA SUPPORT

The use of amorphous silicas as supports for olefin polymerization catalysts has been widespread since the 1960s. They play an important role in the polymerization process and have found wide acceptance as supports for ZN catalysts and more recently, for metallocene catalyst systems.[4] Silica supports participate in the process of catalyst site formation and determine catalyst performance, catalyst flow properties, and the growth of the nascent polymer particle.

Pore Volume

Silica supports for ZN catalysts should have a high pore volume.[8,9]

Pore Size Distribution

The pore size distribution is another parameter that is important for olefin polymerization catalysts, in particular with respect to its fragmentation behavior.

Surface Modification

Surface modification of silica supports has been proven to be a versatile tool for optimizing the impregnation process of the catalyst components.

Catalyst Loading

The catalyst deposition on a silica support is the most important process step that determines the overall catalyst performance.

Morphology

From a technological viewpoint, the excellent ability of silica supports in controlling the morphology of the nascent polymer is one of the most important process parameters. This is due to the replication phenomenon in which each catalyst particle replicates itself into a nascent polymer particle.[11]

SILICA SUPPORTED Ti AND Zr CATALYST SYSTEMS

One of the main reasons to deposit the active components of a ZN catalyst on a silica support is to increase the activity of the catalyst. Ziegler–Natta catalysts based on $TiCl_4$ and supported on silica but lacking $MgCl_2$ were among the first catalytic systems investigated. Fixation of $TiCl_4$ to the silica gel skeleton was thought to give rather stable and well characterizable surface species. However, there is the deactivation of active species by bimolecular reduction mechanisms that might be slower owing to immobilization of the active sites on the silica surface.[12] Murray, as well as Chien were among the first who found that $TiCl_4$ can be chemically fixed on silica by reaction with hydroxyl groups present on the surface.[12–14]

SILICA SUPPORTED $MgCL_2/TiCl_4$ CATALYST SYSTEMS

A variety of compounds has been used to enhance the performance of $TiCl_4$-based ZN catalysts but the most effective support has resulted from the use of magnesium dichloride. The introduction of supported $MgCl_2/TiCl_4$ catalysts has been a milestone in industrial olefin polymerization. Most modern ZN catalysts today use $MgCl_2/TiCl_4$ catalyst systems. Despite their high activity, $MgCl_2$ supported $TiCl_4$ catalysts are not suitable for all polymerization processes. Especially for gas phase processes, $MgCl_2$ supported catalysts are often too friable unless they are prepolymerized prior to use are often too friable and yield polymers with a detrimental morphology, while silica has been proven a good support for gas phase polymerization catalysts.

SILICA-SUPPORTED METALLOCENE CATALYST SYSTEMS

Commercial interest in metallocene catalysts has been rapidly increasing in the last few years. The objective of companies involved in this area is to develop a catalyst that retrofits into existing polymerization technology.

There is an increasing recognition that silica-supported metallocene catalysts have some distinct advantages over homogeneous catalysts, in particular for those to be used in gas-phase processes.

CONTROL OF MORPHOLOGY

According to Galli there are five necessary prerequisites for a supported catalyst to produce a replica of itself.[6,7]

- a high surface area
- a high porosity of the catalyst particle
- free access of the monomer throughout the catalyst particle
- homogeneous distribution of the active centers throughout the catalyst particle
- mechanical strength high enough to withstand catalyst manipulations without fragmentation but with a certain fragility to allow fragmentation during polymerization

Silica-supported catalysts are almost ideally suited to fulfill these requirements, and a number of papers have appeared focusing on the morphological aspects of silica-supported ZN catalysts.[9–11,15–17]

The phenomenon of particle replication has been investigated in detail with silica supported chromium catalysts.[5,18–21] It is at present unclear whether analogous conclusions can be drawn from these results. Similar investigations on silica supported ZN catalysts are scarce.[17,22]

CONCLUSION

The variety of silica supports that are commercially available and the knowledge about each component used in supported catalyst preparation offer a tremendous resource to tailor these catalysts to any product or process requirements in the future.

Silica-supported Ziegler–Natta catalysts are expected to be in commercial use for the next one or two decades. Their dominance will undoubtedly be challenged by metallocene catalysts at some time in the future. Based on the broad experience with silica supports, their inherent advantages, and process requirements at least from gas-phase processes, it may be predicted that metallocene catalysts will also be supported on silica to retrofit into existing polymerization technology.

The results show that exciting opportunities exist where heterogeneous catalysis can teach new insights about homogeneous catalysis. The challenge lies in *building bridges* between these two research areas.

REFERENCES

1. Tait, P. J. T. *Comprehensive Polymer Science*; Pergamon: Oxford, 1989; Vol. 4, p 1.
2. van der Ven, S. *Polypropylene and other Polyolefins (Studies in Polymer Science 7)*; Elsevier; Amsterdam, 1990.
3. Gavens, P. D.; Bottrill, M.; Kelland, J. W.; McMeeking, J. *Comprehensive Organometallic Chemistry*; Wilkinson, G.; Stone, F. G. A.; Abel, E. W., Eds.; Pergamon: Oxford, 1982; Vol. 3, Chapter 22.5.
4. Young, G. W.; Ward, D. G.; Lu, J. P. Proceedings of the Third International Business Forum on Specialty Polyolefins, SPO '93, p 299.
5. McDaniel, M. P. *Adv. Catal.* **1985**, *33*, 47.
6. Galli, P.; Luciani, L.; Cecchin, E. *Angew. Makromol. Chem.* **1981**, *94*, 63.
7. Galli, P.; Barbè, P. C.; Noristi, L. *Angew. Makromol. Chem.* **1984**, *120*, 73.
8. Zakharov, V. A.; Yermakov, Y. I. *Catal. Rev.-Sci. Eng.* **1979**, *19*, 67.
9. Muñoz-Escalona, A.; Alarcón, C.; Albornoz, L. A.; Fuentes, A.; Sequera, J. A. *Transition Metals and Organometallics as Catalysts for Olefin Polymerization*; Kaminsky, W.; Sinn, H., Eds.; Springer: Berlin, 1988; p 417.
10. Muñoz-Escalona, A.; Gallardo, J. A.; Hernández, J. G.; Albornoz, L. A. *Transition Metal Catalyzed Polymerizations; Ziegler-Natta and Metathesis Polymerizations*; Quirk, R. P., Ed.; Cambridge University: 1988; p 512.
11. Muñoz-Escalona, A.; Hernández, J. G.; Gallardo, J. A.; Sustic, A. *Advances in Polyolefins*; Seymour, R. B.; Cheng, T., Eds.; Plenum: New York, 1987; p 179.
12. Chien, J. C. W. *J. Catal.* **1971**, *23*, 71.
13. Murray, J.; Sharp, M. J.; Hockey, J. A. *J. Catal.* **1970**, *18*, 52.
14. Maksimov, N. G.; Kushnareva, E. G.; Zakharov, V. A.; Anufrienko, V. F.; Zhdan, P. A.; Yermakov, Y. I. *Kinet. Katal.* **1974**, *15*, 738.
15. Muñoz-Escalona, A.; Hernandez, J. G.; Gallardo, J. A. *J. Appl. Polym. Sci.* **1984**, *29*, 1187.

15a. Wang, J.-G.; Pang, D.-R.; Cheng, X.-T.; Huang, B.-T. *Makromol. Chem.* **1983**, *194*, 211.

16. Carol, F. J.; Wagner, B. E.; Levine, I. J.; Goeke, G. L.; Noshay, A. *Advances in Polyolefins*; Seymour, R. B.; Cheng, T., Eds.; Plenum: New York, 1987; p 337.
17. Kersting, M.; Braun, H.-G. *J. Mater. Sci.*, to be published, 1996.
18. McDaniel, M. P. *J. Polym. Sci., Polym. Chem. Ed.* **1981**, *19*, 1967.

19. Niegisch, W. D.; Crisafulli, S. T.; Nagel, T. S.; Wagner, B. E. *Macromolecules* **1992**, *25*, 3910.
20. Weist, E. L.; Ali, A. H.; Naik, B. G.; Conner, W. C. *Macromolecules* **1989**, *22*, 3244.
21. Conner, W. C.; Webb, S. W.; Spanne, P.; Jones, K. W. *Macromolecules* **1990**, *23*, 4742.
22. Scheirs, J.; Bigger, S. W.; Delatycki, O. *J. Mater. Sci.* **1991**, *26*, 3171.

SUPPORTED ZIEGLER–NATTA CATATLYSTS (Titanium-Based Silica; Polymerization Properties)

Jonathan P. Blitz*
Department of Chemistry
Eastern Illinois University

Craig C. Meverden
Quantum Chemical Company

Ziegler–Natta catalyst systems have been tremendously important in the commercial production of polyolefins for four decades.

There are several practical reasons to support Ziegler–Natta catalyst species. First, active metal site dispersion can be increased by deposition on the surface of a carrier material. This can result in increased polymerization activity per metal atom and lead to polymers containing lower levels of residual catalyst. Another advantage of supported Ziegler–Natta catalysts is that the morphology of the polymer particles produced can be controlled by varying the morphology of the support.

Silica gel is a particularly useful material on which to support Ziegler–Natta catalysts. Silica particles can be porous with high surface areas and relatively uniform particle size and shape. In addition, silica surfaces consist of various silanol (Si-OH) and siloxane (Si-O-Si) groups that can chemically bond with compounds used in Ziegler–Natta formulations (e.g., transition metal salts and metal alkyls). Thus, the silica support can alter the molecular level structure of the catalyst, which may in turn affect catalytic behavior. In this case the silica gel serves not only as a support but as a reagent in the catalyst formulation.

Many industrially important Ziegler–Natta catalysts undoubtedly consist of many types of species, including those that are chemically bonded to silica surface groups and those that are deposited or precipitated onto the surface.[1–6] These catalysts are complex mixtures that are at best difficult to characterize. We have approached this very complex problem by systematically altering the chemistry of the silica surface, in an attempt to control the way in which the active species are bonded to the support.[7–9] If the mode of metal bonding to the surface can be varied in a controlled fashion, then the polymerization properties of the catalysts may be systematically compared.

CONCLUSIONS

An effort has been made to correlate molecular level structure of silica-supported Ziegler–Natta catalysts with polymerization behavior. We have shown that it is possible to synthesize silica-supported catalysts with known (and variable) structures. It is also possible to correlate catalytic structure of these catalysts with polymerization behavior. When $TiCl_4$ is bonded directly to the silica, these catalysts exhibit strikingly different behaviors depending on how the Ti is bonded to the surface. The Ti catalyst that is predominantly singly bonded to the surface exhibits much higher activity and greater comonomer incorporation under the conditions studied. Catalysts in which magnesium bonding to the surface is varied followed by $TiCl_4$ reaction with the bonded magnesium, also exhibit differences in catalytic behavior. These differences are much more subtle, however. Since the active species in all of these catalysts is titanium, changing the way Ti is chemically bonded to the surface exerts a large effect on catalytic behavior. By saturating the surface with magnesium prior to $TiCl_4$ addition, regardless of how the magnesium is bonded to the surface, one is exerting a secondary effect on the titanium species. Even so, significant differences in polymerization kinetics and polymer molecular weight are detected depending on how the magnesium is bonded to the silica gel.

*Author to whom correspondence should be addressed.

REFERENCES

1. U.S. Patent 4 263 171, 1981 to Chemplex Co.
2. U.S. Patent 4 530 913, 1985 to Chemplex Co.
3. U.S. Patent 4 481 301, 1984 to Mobil Corp.
4. U.S. Patent 4 562 169, 1985 to Mobil Corp.
5. U.S. Patent 4 434 242, 1984 to Cities Service Co.
6. U.S. Patent 4 558 024, 1985 to Exxon Research and Engineering Co.
7. Blitz, J. P. *Colloids and Surfaces* **1992**, *63*, 11.
8. Blitz, J. P. *Polym. Mater. Sci. Eng.* **1992**, *67*, 53.
9. Blitz, J. P. *New Advances in Polyolefins*; Plenum: New York, 1993; p 1.

SUPPORTED ZIEGLER–NATTA CATALYSTS (Titanium (IV) Chloride on Reduced N-Type Oxides)

Ulf Schuchardt,** Alexandre Espeleta, Soraya Jericó,
Inés Joekes, Marcos Antônio Nobre, and Annegret Noll
Instituto de Química
Universidade Estadual de Campinas

One of the principal interests in olefin polymerization is the improvement of the catalytic systems to yield higher polymerization activities and polymers with advanced properties. Today the majority of the commercial catalysts for polyolefin synthesis are supported on solids. In heterogeneous catalysis, the catalyst-support interaction plays an important role owing to the influence on the activity of the metal catalysts. Particular interest has been focused on the utilization of oxides, especially n-type oxides. N-type oxides contain oxygen on their surfaces that can be removed by heating under vacuum or with UV light.[2,3] The resulting oxygen deficiency provides better conditions for a strong catalyst-support interaction.[4]

**Author to whom correspondence should be addressed.

PREPARATION AND PROPERTIES

Catalyst Preparation and Characterization

The surfaces of n-type oxides, such as zinc(II), iron(III), titanium(IV), vanadium(V), chromium(VI), copper(II), or lanthanum(III) oxides, contain oxygen that can be easily removed by reduction. The reduction process should be carried out rapidly since the reduced oxides are unstable. Different kinds of reducing agents, such as $AlMe_3$, $AlEt_3$, and BuLi, were tested. BuLi was found to be the most suitable reducing agent as the reduction takes place immediately. The reduction is indicated by a color change of the oxide.

Polymerization and Polymer Properties

Polymerization of ethylene and propylene was carried out in an 1 L glass reactor with heptane as solvent. Using 300 to 400 mg of the pre-reduced catalyst and 1 mL of neat $AlEt_3$ as cocatalyst, the reactions were usually performed for 1 h at 50°C and 2 bar of ethylene or propylene.[5]

The polymerization activity of $TiCl_4$ supported on reduced TiO_2 is much higher than that of the unreduced system, and comparable to the activity of $TiCl_4$ supported on $MgCl_2$.

The low activity of $TiCl_4$ supported on reduced MgO is due to insufficient reduction of MgO, caused by its great chemical stability.[2] The activity of $TiCl_4$ supported on reduced ZnO is higher, comparable to that reported for $TiCl_4$ supported on other oxides, such as SiO_2, showing that the reduction of ZnO was not very effective. V_2O_5 does not behave like a n-type oxide, because reduction leads to an irreversible reduction of the vanadium and does not cause point defects. The resulting very weak catalyst-oxide interaction leads to the poor polymerization activity of this system. The other n-type oxides (MnO_2, La_2O_3, and TiO_2) represent, after reduction, excellent materials for the heterogenization of $TiCl_4$. Even considering the uncertain determination of the number of active sites, the polymerization activities of these catalysts are greater than 1000 kg/mol Ti-h-bar and, thus, in the same range of those reported for the commercially used $TiCl_4$ supported on $MgCl_2$.[1]

The polymerization of ethylene follows typical decay-type kinetics. The activity decreases continuously during the first hour followed by a constant but very low activity. Complexation of the active centers, rather than mass transport effects, is probably responsible for the deactivation.[7] In agreement with the activity decay, the molecular weight of polyethylene increases up to 60 min and then remains approximately constant. The linear polymer chain growth during the first hour of polymerization leads to polyethylene with a very high molecular weight (M_v of 2500 kg/mol).[8]

One of the most important characteristics of polypropylene is its stereoregularity. Isotactic polypropylene is highly crystalline and, therefore, displays very attractive mechanical properties. The isotacticity of the polypropylene obtained with $TiCl_4$ on reduced TiO_2 is in the range of 75–85%. The high isotacticity index is comparable to that obtained for polypropylene produced with commercial catalyst systems in the presence of Lewis bases.[9] The good stereospecific behavior of our catalyst without the use of Lewis bases implies the existence of asymmetric centers on the catalyst surface, caused by the previous reduction, together with improved electronic interactions between the catalyst and the support. The described properties, together with a relatively high molecular weight (M_v) of 300–400 kg/mol of the polypropylene, lead to materials that are suitable for new technical applications, such as sterilizable polymeric materials. Parts of these investigations led to a patent application.[6]

ACKNOWLEDGMENTS

This work was financed by Financiadora de Estudos e Projectos (FINEP) and by the Stifterverband für die Wissenschaft. Fellowships from the Coordenação de Aperfeiçoamento de Pessoal de Ensino Superior—Deutscher Akademischer Austauschdienst (Capes—DAAD), from the Conselho Nacional de Desenvolvimento Científico e Tecnológico (CNPq) and from the Deutsche Forschungsgemeinschaft (DFG) are gratefully acknowledged.

REFERENCES

1. van der Ven, S. *Polypropylene and Other Polyolefin: Polymerization and Characterization*; Elsevier: Amsterdam, 1990; Chapter 9.
2. Bond, G. C. *Heterogeneous Catalysis*: Clarendon: Oxford, 1974.
3. Howe, R. F.; Grätzel, M. *J. Phys. Chem.* **1985**, *89*, 4495.
4. Henrich, V. E. *Prog. Surf. Sci.* **1983**, *14*, 175.
5. Jericó, S.; Schuchardt, U.; Kaminsky, W.; Joekes, I. *J. Polym. Sci. Part A: Polym. Chem.* **1994**, *32*, 929.
6. Jericó de Carvalho, S. Ph.D. Thesis, Universidade Estadual de Campinas, 1992.
7. Schnauss, A.; Reichert, K.-H. *Makromol. Chem., Rapid Commun.* **1990**, *11*, 315.
8. Jericó de Carvalho, S.; Schuchardt, U. F.; Joekes, I.; Nobre, M. A., Braz. Patent 92/03055, 1992.
9. Pino, P.; Mühlhaupt, R. *Angew. Chem. Int. Ed. Engl.* **1980**, *19*, 857.

SUPRAMOLECULAR FORMATION AND STRUCTURE (PEO–PPO–PEO Block Copolymers)

Benjamin Chu* and Guangwei Wu
Department of Chemistry
State University of New York at Stony Brook

Poly(oxyethylene-oxypropylene-oxyethylene) (PEO–PPO–PEO) block copolymers are widely used nonionic surfactants.[1,2] The development of this type of copolymers can go back to the early 1950s.[3] By varying the total block length and the weight ratio of PEO to PPO, a variety of products with optimum properties has been designed and synthesized to meet the specific requirements in many different industrial applications.

PHYSICAL PROPERTIES AND APPLICATIONS

PEO–PPO–PEO block copolymers vary widely in their physical properties, so that it is usually not feasible to substitute one copolymer for another. The block copolymers vary from almost water-insoluble compounds to those that have no solution cloud point, even at the boiling point of water. The copolymer solubility in water increases with increasing weight

*Author to whom correspondence should be addressed.

percentage of the hydrophilic part (PEO) of the copolymer and with decreasing temperature. The cloud point increases with increasing weight percentage of the hydrophilic part (PEO) of the compound.

PEO–PPO–PEO block copolymers have found many applications in the chemical, pharmaceutical, cosmetics, and food industries. The widespread industrial use of PEO–PPO–PEO copolymers is due in part to their unusually low order of toxicity. The copolymers can be used as detergents or as emulsifying, wetting, thickening, coating, solubilizing, dispersing, and foaming agents.

SUPRAMOLECULAR FORMATION AND STRUCTURE

Aqueous Solution

The micellar structure formed by PEO–PPO–PEO copolymers is similar to that of other block copolymers that have much longer chain lengths.[25] Temperature plays an important role in the micellization of PEO–PPO–PEO copolymers in aqueous solution because of hydrogen bonding, which effectively changes the solvent quality of the components in the block copolymer. At high copolymer concentrations, "gelation" and quasi-crystallization of the micelles have been observed.

Critical Micelle Concentration (cmc) and Thermodynamics of Micelle Formation

The critical micelle concentration of PEO–PPO–PEO copolymers in aqueous solution depends on molecular composition and temperature. The cmc value increases with increasing PEO block length or decreasing PPO block length at fixed temperatures. For a copolymer of definite composition, the cmc value decreases with increasing temperature.

Micellar Formation and Micellar Structure

The micellar size is of the order of ~10 nm.[4,5] The aggregation number of the micelles increases with increasing temperature. When the length of the middle block (PPO) is fixed, the micellar size increases slightly with increasing PEO block length.

Crystallization and Gel Formation

At high micelle concentrations, ordered structures have been observed.[18–21,23,24] On the basis of the SANS pattern under a constant shear field, a body-center-cubic packing has been proposed to describe the ordered micellar structure.[18–21]

At high copolymer concentrations, transparent gels can be formed over an appropriate temperature range. The concentration for gel formation varies with copolymer composition, decreasing with increasing PPO block length.[26]

Copolymer in Xylene/–Water Mixtures

In comparison with aqueous solution, much less attention has been paid to the ternary system: copolymer/–organic solvent/–water, or copolymer in an organic solvent, even though some of the colloidal properties are expected to be similar to those in aqueous solution.[6–17]

Solubilization of Xylene by Copolymer in Aqueous Solution

Organic solvents that are compatible with the PPO core and be solubilized by PEO–PPO–PEO in aqueous solution.[22,27] The maximum solubilization amount is related to the copolymer concentration and temperature.

REFERENCES

1. Lundsted, L. G.; Schmolka, L. R. In *Block and Graft Copolymerization*; Ceresa, R. J., Ed.; John Wiley & Sons: New York, 1975; Vol. 2, Chapters 1 and 2.
2. Schmolka, I. R. In *Polymers for Controlled Drug Delivery*; Tarcha, P. I., Eds.; CRC: Boca Raton, FL, 1991; Chapter 10.
3. Vaughn, T. H.; Suter, H. R.; Lundsted, L. G.; Kramer, M. G. *J. Am. Oil Chem. Soc.* **1951**, *28*, 294.
4. Zhou, Z. K.; Chu, B. *Macromolecules* **1988**, *21*, 2548.
5. Zhou, Z. K.; Chu, B. *J. Colloid Interface Sci.* **1988**, *126*, 171.
6. Tonisakis, A.; Hilfiker, R.; Chu, B. *J. Colloid Interface Sci.* **1990**, *135*, 427.
7. Wu, G.; Zhou, Z. K.; Chu, B. *Macromolecules* **1993**, *26*, 2117.
8. Wu, G.; Zhou, Z. K.; Chu, B. *J. Polym. Sci., Polym. Phys. Ed.* **1993**, *31*, 2035.
9. Wu, G.; Chu, B. *Macromolecules* **1994**, *27*, 1766.
10. Wu, G.; Ying, Q.; Chu, B. *Macromolecules* **1994**, *27*, 5758.
11. Chu, B.; Wu, G.; Schneider, D. K. *J. Polym. Sci., Polym. Phys. Ed.* **1994**, *32*, 2605.
12. Wu, G.; Chu, B.; Schneider, D. K. *J. Phys. Chem.* **1995**, *98*, 12018.
13. Chu, B. *Langmuir* **1995**, *11*, 414.
14. Chu, B.; Wu, G. *Macromol. Symp.* **1994**, *87*, 55.
15. Chu, B.; Wu, G. *Macromol. Symp.* **1995**, *90*, 251.
16. Samii, A. A.; Lindman, B.; Karlström, G. *Progr. Colloid Polym. Sci.* **1990**, *82*, 280.
17. Tiberg, F.; Malmsten, M., Linse, P., Lindman, B. *Langmuir* **1991**, *7*, 2723.
18. Mortensen, K.; Brown, W.; Nordén, B. *Phys. Rev. Lett.* **1992**, *68*, 2340.
19. Mortensen, K. *Europhys. Lett.* **1992**, *19*, 599.
20. Mortensen, K. *Progr. Colloid Polym. Sci.* **1993**, *91*, 69.
21. Mortensen, K. *Progr. Colloid Polym. Sci.* **1993**, *93*, 72.
22. Hurter, P. N.; Hatton, T. A. *Langmuir* **1992**, *8*, 1291.
23. Wanka, G.; Hoffmann, H.; Ulbricht, W. *Colloid Polym. Sci.* **1990**, *268*, 101.
24. Wanka, G.; Hoffmann, H.; Ulbricht, W. *Macromolecules* **1994**, *27*, 4145.
25. Tuzar, Z.; Kratovchvíl, P. In *Surface and Colloid Science*; Matijevic, E., Ed.; Plenum: New York, 1993, Vol. 15, Chapter 1.
26. Schmolka, I. R. *Perfume and Cosmetic* **1967**, *82*, 25.
27. Nagarajan, R.; Barry, M.; Ruckenstein, E. *Langmuir* **1986**, *2*, 210.

SUPRAMOLECULAR MESOMORPHIC POLYMERS (Assembling in Ultrathin Films)

Vladimir V. Tsukruk
College of Engineering and Applied Science
Western Michigan University

"Mesomorphism" or liquid crystallinity, that is, the ability to form equilibrium phases intermediate between the crystalline state and isotropic fluid, offers a promising route toward supramolecular organization with a high level of complexity and flexibility. Due to a fine balance between the anisotropy of shape and intermolecular interactions, mesomorphic polymers can form a variety of supramolecular structures. The resulting supramolecular architecture includes frustrated smectics and bicontinuous phases, chiral smectics and biaxial nematics, and columnar ordering and hexatic structures.[1,2] These supramolecular structures have intriguing physical properties, such as ferroelectricity in chiral smectics and optical bistability in biaxial nematics.

For potential applications in the fields of nonlinear optics, optical storage, ferroelectric and piezoelectric sensors, conductive films, and "command" surfaces, functional polymeric materials must form supramolecular assemblies in ultrathin films with suitable symmetry and a suitable type of local molecular ordering. The use of mesomorphic ordering is considered a highly promising basis for the fabrication of suitable supramolecular architecture in ultrathin films of molecular and submicron thickness, which are the most appropriate form for many applications.

SUPRAMOLECULAR MESOMORPHIC POLYMERS

Among the most interesting and widely explored macromolecular architectures are the following: classic side-chain and main-chain polymeric liquid crystals (PLCs), side chains with lateral and paired attachments of mesogenic groups, combined polymers with mesogenic side groups and mesogenic groups in backbones, hybrid polymers with mesogenic groups of different shapes, main-chain polymers of "crank-shaft" shape, main-chain polymers with cross-shaped mesogenic groups, "hairy-rod" polymers, "sanidic" polymers with board-like groups, side-chain and main-chain polymers with discoid groups, and chains with pyramid-like segments.[2,3]

Three-dimensional polymers with controlled cascade architecture or dendrimers present a new class of nanoscopic supramolecular compounds or chemical clusters that may form entirely new supramolecular assemblies never before observed.[5,6]

FABRICATION OF ULTRATHIN FILMS

Ongoing research in the field of ultrathin films demonstrates that appropriate manipulations with macromolecules of various architecture could produce ultrathin films with a variety of ordered superstructures.

The LB technique combined with self-assembly is a promising way to build composite molecular films. Possible versions include the formation of stable monolayers at the air–water interface followed by their deposition and chemical binding to solid substrate or the physical adsorption of molecules from the water subphase on a Langmuir monolayer followed by the LB transfer of this composite bilayer film.[8,9]

ASSEMBLY IN ULTRATHIN FILMS: SOME RESULTS

To date, the formation of atomically smooth mono- and multilayers, continuous and discontinuous films, frustrated supramolecular structures, twisted multilayer ordering, segregated microphase structures, and hybrid layer-columnar order has been observed in various ultrathin films, including spin-coated, self-assembled, LB, single-component films; composite films built from different compounds; and composite films built by combining techniques.

Mesogenic groups with switchable conformations (e.g., azobenzenes) were incorporated into a polymer chain.[10] The possibility of controlled variations of molecular ordering in ultrathin films by light illumination was demonstrated; the technique is promising for digital and holographic optical storage. Controlled variation of molecular orientations at surfaces via photochemical changes of molecular organization was recently shown for ultrathin films.[11,12]

Much attention and effort have been focused on the fabrication of polymer films from chiral smectic polymers that have ferroelectric and piezoelectric properties useful for sensors.[13]

Only a few detailed studies of molecular films from discotic LCs are known, although LB films from discotic LCs were discussed several years ago.[14] Suresh and Karthaus speculated about the possibility of edge-on molecular packing of discotic molecules into layered structures.[15,16] A firm confirmation of this type of molecular packing was obtained only within the last three years.[17–21]

LB films from hairy-rod polymers based on polyglutamates (PGs) exhibits relatively low optical losses, a high level of birefringence, and high orientational order, and can form chiral phases. Photochromic LB films with internal molecular ordering controlled by light irradiation were recently fabricated from PGs with chromophore side groups (azobenzenes and spiropyranes) by Cooper and Menzel.[4,22]

Self-assembly through the tailoring of ionic interactions holds promise for future developments. In the past few years it has been demonstrated that an appropriate combination of polycation and polyanion polymers may lead to fabrication of highly organized and stable multilayered films.[7]

Asymmetric derivatives of fullerenes with distorted spherical symmetry might form molecular and multimolecular LB films with layered structures similar to classic smectics but with very different intralayer ordering. LB films from asymmetric derivatives of fullerenes and complexes of fullerenes with alkylamine chains ("hairy balls") were recently obtained.[23–25]

TRENDS AND PROBLEMS

A critical issue in film fabrication is the metastable nature of resulting assemblies at interfaces. The process of self-assembly itself involves the complex interaction of selective adsorption, chemical or physical binding to the surface, surface diffusion and aggregation, and in-plane and vertical microphase separations, which create the possibility for multiple local metastable configurations. In order to control the final surface morphology and supramolecular organization of composite polymer films, we must control the kinetics of the self-assembly and maintain conditions that provide the proper balance in the competition between these processes. To this end, we consider both "freeze and look" and *in situ* investigations of self-assembly kinetics in mesomorphic polymers very important.

ACKNOWLEDGMENTS

V. V. Tsukruk acknowledges the Faculty Research and Creative Activities Support Grant, Western Michigan University, and the Petroleum Research Fund, administered by the American Chemical Society, for support of the preparation of this article.

REFERENCES

1. *Side Chain Liquid Crystal Polymers*; McArdle, C. B., Ed.; Chapman and Hall: New York, 1989.

2. *Liquid Crystal Polymers: From Structures to Applications* Collyer, A. A., Ed.; Elsevier: Amsterdam, 1992.
3. Tsukruk, V. V.; Shilov, V. V. *Structure of Polymeric Liquid Crystals*; Naukova Dumka: Kiev, 1990.
4. Cooper, T.; Tondiglia, V.; Natarajan, L. V.; Shapiro, M.; Obermeier, K.; Crane, R. L. *Appl. Optics* **1993**, *32*, 674.
5. Frechet, J. M. *Science (Washington, DC)*, **1994**, *263*, 1711.
6. Xu, Z.; Moore, J. S. *Angew. Chem.* **1993**, *32*, 1354.
7. Decher, G.; Hong, J.-D. *Makromol. Chem., Macromol. Symp.* **1991**, *46*, 321.
8. Duschl, C.; Vogel, H.; Liley, M. *Angew. Chem. Int. Ed. Engl.* **1994**, *33*, 1274.
9. Tsukruk, V. V.; Bliznyuk, V. N.; Reneker, D. H.; Kirstein, S.; Möhwald, H. *Thin Solid Films* **1994**, *244*, 763.
10. Herman, O.; Wendorff, J. H.; Kreuder, W.; Ringsdorf, H. *Makromol. Chem. Rapid. Commun.* **1986**, *1*, 97.
11. Ishimura, K. et al. *Langmuir* **1993**, *9*, 3298.
12. Sasaki, T. et al. *J. Am. Chem. Soc.* **1994**, *116*, 625.
13. Fadel, H.; Percec, V.; Zheng, Q.; Advincula, R.; Duran, R. *Macromolecules* **1993**, *26*, 1650.
14. Laschewsky, A. *Adv. Mater.* **1989**, *28*, 1574.
15. Suresh, K. A.; Blumstein, A.; Rodhelez, F. *J. Phys. France* **1985**, *48*, 453.
16. Karthaus, O.; Ringsdorf, H.; Urban, C. *Makromol. Chem. Macromol. Symp.* **1991**, *46*, 347.
17. Karthaus, O.; Ringsdorf, H.; Tsukruk, V. V.; Wendorff, J. H. *Langmuir* **1992**, *8*, 2279.
18. Josefowicz, J. Y.; Maliszewskyj, N. C.; Idziak, S. H.; Heiney, P. A.; McCauley, J.; Smith, A. B. *Science (Washington, D.C.)* **1993**, *260*, 323.
19. Albouy, P. A.; Vandevyver, M.; Perez, X.; Ecoffet, C.; Markovitsi, D. et al. *Langmuir* **1992**, *8*, 2262.
20. Reiche, J.; Dietel, R.; Janietz, D.; Lemmetyinen, H.; Brehmer, L. *Thin Solid Films* **1992**, *226*, 265.
21. Nostrum, C. F.; Nolte, R.; Devillers, M. A.; Oostergetel, G. T. et al. *Macromolecules* **1993**, *26*, 3306.
22. Menzel, H.; Hallensleben, M. L.; Schmidt, A.; Knoll, W.; Fisher, T.; Stumpe, J. *Macromolecules* **1993**, *26*, 3644.
23. Maliszewskyj, N.; Heiney, P.; Jones, D. et al. *Langmuir* **1993**, *9*, 1439.
24. Williams, G.; Soi, A.; Hirsch, A. et al. *Thin Solid Films* **1993**, *230*, 73.
25. Wang, Y.; Vakhin, D.; Uphaus, R. et al. *Thin Solid Films* **1994**, *242*, 40.

SUPRAMOLECULAR SELF-ASSEMBLY (Liquid Crystalline Polymers, Hydrogen Bonding)

Takashi Kato
Institute of Industrial Science
The University of Tokyo

Jean M. J. Fréchet
Department of Chemistry
Baker Laboratory
Cornell University

In the design of liquid-crystalline polymers, only covalent bonds are generally used for the building of molecular structure.[1,2] However, noncovalent interactions also have great potential for the structure design of mesogenic functional polymers because the supramolecular self-assembly that directs highly functional materials can be achieved by molecular interactions.

H-Bond Donors

I

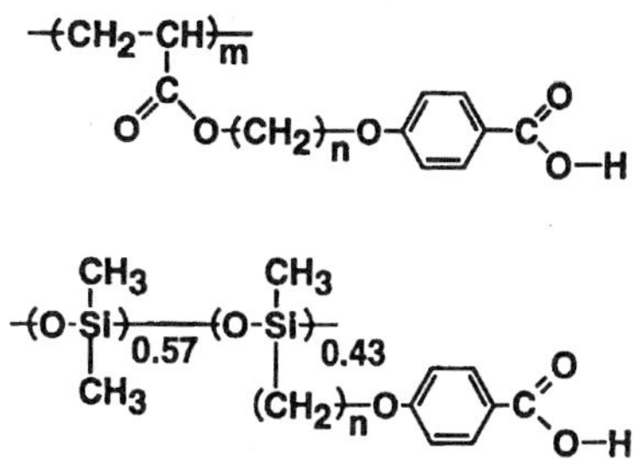

1a: n=6
1b: n=11

2a: n=5
2b: n=8

H-Bond Acceptors

II

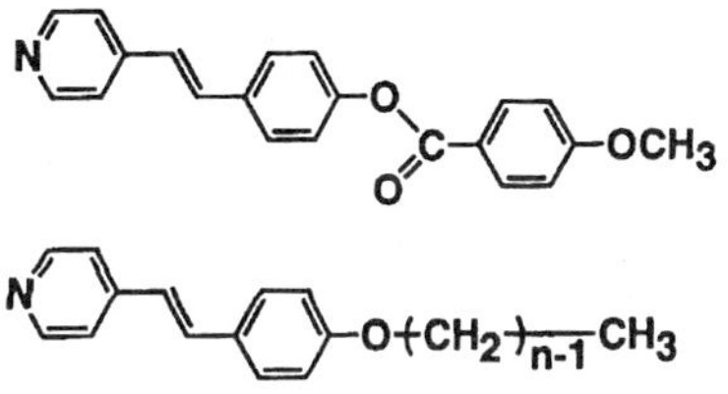

n=1
n=6

In this article, we report our approach to the design and the preparation of hydrogen-bonded liquid-crystalline side-chain polymers and networks that are obtained by supramolecular self-assembly of H-bonding donors and acceptors (**Schemes I** and **II**).

PROPERTIES

Supramolecular Self-Assembly of Side-Chain Liquid-Crystalline Polymers

Polyacrylates **1a** and **1b** and polysiloxanes **2a** and **2b** have benzoic acid moieties in the side chain. This group functions as H-bonding donor. Stilbazoles **3** and **4** possess a 4-pyridyl group that serves as H-bonding acceptor. The 1:1 self-assembly of these H-bonding donor and acceptor results in the formation of supramolecular liquid-crystalline side-chain polymers.[3-5,7,8] These polymer complexes behave as one single component. The extended mesogens that induce and stabilize mesophases are obtained in the side chain of the polymers.

Induction of a Ferroelectric Mesophase by Supramolecular Self-Assembly of Polymers and Chiral Compounds

Induction of the ferroelectric mesophases by supramolecular self-assembly through the hydrogen bonding has been observed for the complexes of functionalized polymers and chiral stilbazoles.

Supramolecular Liquid-Crystalline Polymer Network

A liquid-crystalline supramolecular network has been built by self-assembly of polymer **1a** and 4,4′-bipyridine through hydrogen bonding.[6] The complex clearly exhibits glassy-smectic A and smectic A–isotropic transitions at 95 and 205°C, respectively. The mesomorphic melt of the complex is viscous but shows shear flow between glass plates.

The dynamic nature of the hydrogen bonds of the intermolecular mesogen should contribute to the induction of a fluid mesophase and the reversibility of the network structure. Such behavior cannot be observed for the fully crosslinked polymers with covalent bonds. The use of the dynamic feature of hydrogen bonding may be a key to the design of highly functional materials by supramolecular self-assembly.

REFERENCES

1. *Side Chain Liquid Crystal Polymers*; McArdle, C. B., Ed.; Blackie: Glasgow, 1989.
2. Percec, V.; Tomazos, D. *Comprehensive Polymer Science, 1st Supplement* Allen, G., Ed.; Pergamon: Oxford, 1992; p 299.
3. Kato, T.; Fréchet, J. M. J. *Macromolecules* **1989**, *22*, 3818.
4. Kato, T.; Fréchet, J. M. J. *Macromol. Symp.* **1995**, *98*, 311.
5. Kato, T.; Kihara, H.; Uryu, T.; Fujishima, A.; Fréchet, J. M. J. *Macromolecules* **1992**, *25*, 6836.
6. Kato, T.; Kihara, H.; Uryu, T.; Kumar, U.; Fréchet, J. M. J. *Angew. Chem., Int. Ed. Engl.* **1994**, *33*, 1644.
7. Kumar, U.; Kato, T.; Fréchet, J. M. J. *J. Am. Chem. Soc.* **1992**, *114*, 6630.
8. Kumar, U.; Fréchet, J. M. J.; Kato, T.; Ujiie, S.; Iimura, K. *Angew. Chem., Int. Ed. Engl.* **1992**, *31*, 1531.

Supramolecular Structures

See: *Calixarenes (as Supramolecular Cyclic Oligomers)*
Comb-Like Amphiphilic Polymers (Supramolecular Structure)
Ionene Oligomers
Macromolecular Recognition (New Supramolecular Structures)
Metallosupramolecular Helicates
Polyrotaxanes (Overview)
Self-Assembled Polymers (at Interface)
Supramolecular Formation and Structure (PEO-PPO-PEO Block Copolymers)
Supramolecular Mesomorphic Polymers (Assembling in Ultrathin Films)
Supramolecular Self-Assembly (Liquid Crystalline Polymers, Hydrogen Bonding)
Supramolecular Structures (Bipyridine Block Copolymers with Cu(I) Ions)

SUPRAMOLECULAR STRUCTURES (Bipyridine Block Copolymers with Cu(I) Ions)

Claus D. Eisenbach* Andreas Göldel, Monika Terskan-Reinold, and Ulrich S. Schubert
Lehrstuhl für Makromolekulare Chemie II and Bayreuther Institut für Makromolekülforschung (BIMF)
Universität Bayreuth

It has been known for several years that oligo(bipyridines) spontaneously form helical metal complexes, whereby two oligo(bipyridine) strands wrap around Ag^I or Cu^I cations.[1] This self-organization process shows a highly cooperative effect and is very similar to the formation of the double-helix in nucleic acids. Oligo(bipyridines) containing two to five 2,2′-bipyridine (bpy) units[1] have been synthesized via chemistry based on well-described 6,6′-substituted-2,2′-(bpy) building blocks (e.g., 2,2′-(bpy) compounds: 6,6′ = CH_3, CH_2Cl, CH_2Br, or CH_2OH; 6 = CH_2Cl, CH_2Br, or CH_2OH and 6′ = CH_3. (bpy) compounds: 6,6′ = CH_3; 6 = CH_2Br, CH_2OH, 6′ = CH_3).[2–5] In this article we present a new way to create supramolecular structures in ABA triblock and $(AB)_n$ multiblock copolymer type macromolecules through bipyridine (bpy) moieties in the A block (segment) by using the specific self-assembly feature of bpy molecules in the presence of copper(I) (**Figure 1**). In addition to the spectroscopic proof of the formation of the mononuclear $[Cu(I)(bpy)_2]$ complex, the thermal and mechanical behavior revealed the novel superstructure of the polymer ion complex with elastomer properties.[6,7]

CONCLUSIONS

The copolymer systems with bipyridine building blocks showed features typical for the controlled self-organization of bpy units in the presence of Cu(I) ions. The formation of the $[Cu(I)(bpy)_2]$ complex was confirmed by UV-spectroscopy. The novel structures were reflected from the thermal and mechanical properties, and could be visualized by using transmission electron microscopy (TEM) technique. The architecture of the macromolecules and the bipyridine/Cu(I) ratio had a distinct influence on the resulting superstructure and morphology, in that microphase separated systems were obtained for the complexed multiblock copolymer with excess Cu(I) only, whereas the telechelic type triblock copolymer gave highly ordered superstructures.

Ongoing work in designing supramolecular macromolecular structures is focusing on the synthesis of block copolymer systems with oligo(bipyridine) building blocks and copolymers with sequential distribution of (oligo)bipyridine constitutional units along the chain. We recently developed a new synthetic procedure towards 6,6′–*bis*(functionalized)-oligo(bpy) units which facilitated their controlled incorporation along the macromolecule.[8] Such systems are expected to show novel material properties due to self-organization and cooperative features of the bpy units.

ACKNOWLEDGMENT

This study was supported by the German Ministry of Research and Technology (Grant No. 03C2013/4) and the Deutsche Forschungsgemeinschaft (SFB 213).

REFERENCES

1. Lehn, J.-M.; Rigault, A.; Siegel, J.; Harrowfield, J.; Chevrier, B.; Moras, D. *Proc. Natl. Acad. Sci. USA.* **1987**, *84*, 2565; Lehn, J.-M.; Rigault, A. *Angew. Chem.* **1988**, *100*, 1121; ibid *Int. Ed.* **1988**, *27*, 1095; Garrett, T. M.; Koert, U.; Lehn, J.-M.; Rigault, A.; Meyer, D.; Fischer, J. *J. Chem. Soc., Chem. Commun.* **1990**, 557; Koert, U.; Harding, M. M.; Lehn, J.-M. *Nature* **1990**, *346*, 339; Harding, M. M.; Koert, U.; Lehn, J.-M.; Marquis-Rigault, A.; Piquet, C.; Siegel, J. *Helv. Chim. Acta* **1991**, *74*, 594.
2. Newkome, G. R.; Pantaleo, D. C.; Puckett, W. E.; Ziefle, P. L.; Deutsch, W. A. *Inorg. Nucl. Chem.* **1981**, *43*, 1529.

*Author to whom correspondence should be addressed.

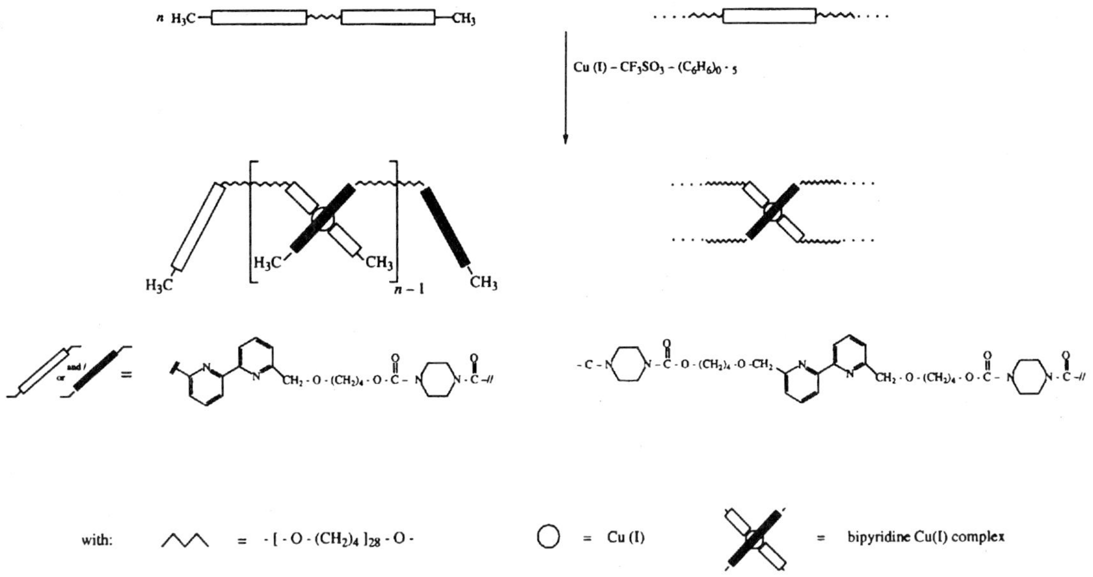

FIGURE 1. Schematic of the chain extension or the formation of tetrafunctional crosslinks of bipyridine Cu(I) complex by Cu(I) complexation of the bipyridine units in the ABA triblock or AB multiblock copolymer.

3. Newkome, G. R.; Puckett, W. E.; Kiefer, G. E.; Gupta, V. K.; Xia, Y.-J.; Coreil, M.; Hackney, M. A. *J. Org. Chem.* **1982**, *47*, 4116.
4. Newkome, G. R.; Gupta, V. G.; Fronczek, F. R. *Inorg. Chem.* **1983**, *22*, 171.
5. Rodriguez-Ubis, J.-C.; Alpha, B.; Plancherel, D.; Lehn, J.-M. *Helv. Chim. Acta* **1984**, *67*, 2264.
6. Eisenbach, C. D.; Schubert, U. S. *Macromolecules* **1993**, *26*, 7372.
7. Eisenbach, C. D.; Göldel, A.; Terskan-Reinold, M.; Schubert, U. S. *Makromol. Chem. Phys.* **1995**, *196*, 1077.
8. Eisenbach, C. D.; Schubert, U. S.; Baker, G. R.; Newkome, G. R. *J. Chem. Soc., Chem. Commun.* **1995**, 69.

SURFACE ACTIVE MONOMERS

Noritaka Ohtani
Department of Materials Engineering and Applied Chemistry
Akita University

Surface active monomers are polymerizable amphiphiles, composed of a hydrophilic portion and a hydrophobic portion in addition to a polymerizable group. In general, they show surface-active properties very similar to those of surfactants.

PREPARATION AND PROPERTIES

Structure and Synthesis

Surface active monomers are amphiphilic molecules, constituted of a hydrophilic head-group and one or two hydrophobic tails in addition to a polymerizable group. The hydrophobic tail is usually a hydrocarbon chain, which can be linear or branched. Like simple surfactants, surface active monomers so far synthesized are classified into four groups depending on the nature of the head-group, that is, anionic, cationic, nonionic, or zwitterionic.

Surface active monomers have a polymerizable group such as allyl, dienic, vinyl, etc. So far only addition polymerization has been employed for the micelle-forming polymerizable surfactants, while polycondensations of amino acid groups, etc. have been examined in other types of amphiphile assemblies such as monolayers or vesicles. The locus of the polymerizable group in an organized micelle is different according to the type of monomers.

Micellization of surface-active monomers also takes place in nonaqueous solvents. In this case, the micellar structures are the inverse of those formed in water; the head groups form the core while the hydrophobic tails are in contact with the apolar solvents. Schryver and co-workers have reported CMC and aggregation number of *N,N*-didodecyl-*N*-methyl-*N*-(2-(methacryloyloxy)ethyl)ammonium chloride, which is established to form reverse micelles in toluene in the presence of minimal water.[8]

Polymerization

Many authors have reported the influence of micellization on the radical polymerizations of surface active monomers. In water, normal micelles are formed above CMC. The rates of polymerization of surface active monomers are usually high above CMC.

All the homopolymers and copolymers derived from surface active monomers, if water-soluble, always exhibit characteristic features of polysoaps. Their solution viscosity is characterized by polyelectrolyte-like behavior in a similar manner to other types of ion-containing polymers.[1]

APPLICATIONS

Surface active monomers and their polymers are capable of modifying the nature of solid surface. The polymerization of surface active monomers, adsorbed on alumina, has been done in order to improve dispersion stability of the particles.[2] It is often attempted to polymerize the precursors of surface active monomers and then to introduce the amphiphilic nature to the preformed polymers. In this way, copolymers of cationic surface active monomers with some vinyl monomers have been successfully applied as a flocculant of clay dispersion.[3] Crosslinked polymers of cationic surface active monomers are good catalysts for phase-transfer reactions.[3–5] Some surface active monomers are useful as an emulsifier of emulsion polymerization.[7,8]

REFERENCES

1. Ohtani, N.; Inoue, Y.; Mizuoka, H.; Itoh, K. *J. Polym. Sci., Polym. Chem. Ed.* **1994**, *32*, 2589.
2. Esumi, K.; Nakao, T.; Ito, S. *J. Colloid Interface Sci.* **1993**, *156*, 256.
3. Serita, H.; Ohtani, N.; Kimura, C. *Yukagaku* **1981**, *30*, 96; *Kobunshi Ronbunshu* **1984**, *41*, 315.
4. Regen, S. L. *Angew. Chem. Int. Ed. Engl.* **1979**, *18*, 421.
5. Ohtani, N. *J. Synth. Org. Chem.* **1985**, *43*, 313.
6. Guillaume, J. L.; Pichot, C.; Guillot, J. *J. Polym. Sci., Polym. Chem. Ed.* **1990**, *28*, 137.
7. Nagai, K.; Satoh, H.; Kuramoto, N. *Polymer* **1992**, *33*, 5303; **1993**, *34*, 4969.
8. Voortmans, G.; Verbeeck, A.; Jackers, C.; De Schryver, F. C. *Macromolecules* **1988**, *21*, 1977.

Surface Modification

See: *Carbon/Carbon Composites (with Surface Treated Carbon Fibers)*
Carbon Fibers (Plasma Surface Treatments)
Fillers, Polypropylene
Fillers, Surface Modification
Fluoropolymers (Preparation by Fluorination of Polymers)
Fluoropolymers (Surface Modification by Eximer-Laser Irradiation)
Fluorosilicones (Monomer Synthesis and Applications)
Hydrophilic Polymers (for Friction Reduction)
Hydrophilic Surfaces (Plasma Treatment and Plasma Polymerization)
Inorganic Fibers (Surface Modification by Grafting)
Inorganic Surfaces, Grafting
Phospholipid-Immobilized Surfaces
Photomodification (Pendant Acyloxyimino Groups to Amino Groups)
Polyester Fabrics (Surface Modification by Low Temperature Plasma)
Polyethylene (Surface Functionalization)
Polypropylene (Nucleating Agents)
Poly(vinyl chloride), Soft (Calcium Carbonate Filled)
Silane Coupling Agents (Adhesion Promotors)
Styrene-Butadiene-Styrene Triblock Copolymer
Surface Modification (Overview)
Surface Modification (by Low Energy Electron Beam)
Surface-Modified Polymers
Ultrathin Films (by Plasma Deposition)
Wool Grafting

SURFACE MODIFICATION (Overview)

Fabio Garbassi and E. Occhiello
EniChem S.p.A.—Istituto Guido Donegani

Surface treatments are aimed at optimizing the interactions of plastics with the environment for improving processability (e.g., lowering friction during weaving of man-made fibers) or providing improved properties (e.g., lowering oxygen permeability of films), usually minimizing the effect on bulk properties, such as density and tensile strength.

In a recent book a first attempt at a general assessment of the field has been presented,[1] and in a number of other books accounts focussing more on specific subjects are available.[2–10]

CHEMICAL MODIFICATION OF SURFACES

The first group of surface modification techniques to be covered involves an actual change of the polymer chemistry in a surface layer whose thickness will depend on the function but will be anyway rather low owing to the low diffusivity of reactive species into the material. A list of typical examples of these techniques is reported in **Table 1**.

COATING OF SURFACES

The second group of surface modification techniques includes coating in its various forms. Its philosophy is based on the fact that the environment "feels" the coating rather than the substrate. The chemistry and the thickness of the coating, along with its affordability, will of course depend on the application. Some typical examples are reported in **Table 2**.

SURFACE MODIFICATION BY BULK FORMULATION

As opposed to coating processes, controlling surface-related properties by bulk formulation is very much in the hands of material manufacturers. Actually, compounding is one of the most important and most diffused methods for controlling surface behavior, yet typically it is not included in books dealing with surfaces. In **Table 3** some of the typical ways of controlling the surface acting on the bulk are reported and are neatly divided in two families: acting on the polymer itself and through additives.

TABLE 1. A Non-Comprehensive List of Chemical Modification Techniques

Technique	Property	Application
Hydrolysis	"Hand"	Polyester and polyamide fibers
CF_4/O_2 plasma	Access to underlying layers	Etching of photoresists
Corona treatment	Wettability/ adhesion of inks	Polyolefin films for packaging
Oxygen plasma	Paintability	Polyolefin bumpers
Na/NH_3 treatment	Adhesion	PTFE materials
Fluorination/ sulfonation	Barrier to fuels	HDPE tanks
Grafting	Biocompatibility	Soft-tissue prostheses
Derivatization	Recognition	Surface analysis

TABLE 2. A Non-Comprehensive List of Coating Techniques

Coating	Property	Application
Lubricants and antistats	Friction, antistaticity	Sizing of fibers
PVDC emulsion	Barrier	Packaging films
Crosslinked siloxanes	Wear	Plastic lenses
Chlorinated PE primers	Paintability	PP bumpers
Extrusion coating	Food contact, sealability	Packaging films
Evaporated Al layers	Barrier	Packaging films
Plasma-polymerized silica layers	Barrier	Packaging films
Magnetic coatings	Capability of magnetic storage	Tapes, floppy disks, etc.
Sputtered metal oxide layers	Optical filtering	Glazing
Pyrolitic carbon	Biocompatibility	Cardiac valves

TABLE 3. Selected Case of Surface Modification by Bulk Formulation

Material	Property	Application
Polyolefin formulations	Friction	Plastic films
Block copolymers	Biocompatibility	Artifical heart
Slip and anti-blocking agents	Friction	Plat films
Inorganic compounds (e.g., fibers)	Wear	Technical goods, e.g., gears
Ozone stabilizers	Oxidative stability	Tires

REFERENCES

1. Garbassi, F.; Morra, M.; Occhiello, E. *Polymer Surfaces—from Physics to Technology*; John Wiley & Sons: New York, 1994.
2. Wu, S. *Polymer Interface and Adhesion*; Marcel Dekker: New York, 1982.
3. Hollahan, J. R.; Bell, A. T., Eds.; *Techniques and Applications of Plasma Chemistry*; Wiley: New York, 1974.
4. Vossen, J. L.; Kern, W., Eds.; *Thin Film Processes*; Academic: New York, 1978.
5. Garbassi, F.; Occhiello, E., Eds.; *High Energy Density Technologies in Materials Science*; Kluwer: Dordrecht, 1990.
6. Mittal, K. L., Ed.; *Physicochemical Aspects of Polymer Surfaces*; Plenum: New York, 1983.
7. Andrade, J. D., Ed.; *Surface and Interfacial Aspects of Biomedical Polymers*; Plenum: New York, 1985.
8. Chan, C.-M. *Polymer Surface Modification and Characterization*; Hanser: Munich, 1994.
9. Feast, W. J.; Munro, H. S., Eds.; *Polymer Surfaces and Interfaces*; Wiley: Chichester, 1987.
10. Schlick, M. J., Ed.; *Surface Characteristics of Fibers and Textiles*; Marcel Dekker: New York, 1975.

SURFACE MODIFICATION, POLYOLEFIN FILMS (by Low Energy Electron Beam)

Masahiko Imai and Kazuo Ametani
Tokyo Metropolitan Isotope Research Center

Shin Hiwasa, Wataru Yamahara, Jun Tanaka, and Harumichi Shimizu
Saitama Institute of Technology

Polyethylene (PE) and polypropylene (PP) have been widely used for many kinds of products as polymer materials, but they have a weak point, which is, that it is difficult to bond other materials with them or to print on them. This weak point is somewhat improved by corona discharge treatment, but it is not sufficient. The modification of PE and PP films by radiation-induced grafting method using γ-ray or high or middle range energy electron beam has been reported by many researchers.[1–4] Mori reported about radiation-induced grafting using low energy electron beam (LEEB) by simultaneous irradiation method.[5,6] In this study, it was tried to improve surface properties of PE and PP films by some radiation techniques including grafting with LEEB.

PREPARATION

Oriented polypropylene film (OPP), cast film of polypropylene (CPP), including 0.1% butyl hydroxy toluene (BHT), and these films treated with corona discharge were used as PP films.

PROPERTIES

Table 1 shows the tensile strength and the peeling strength of modified LLDPE films. The peeling strength increased 2–3 times by the 4 methods. The tensile strength did not decrease by each method.

By pre-irradiation grafting-methods the tensile strength of PP films decreased, but those of PE films did not decrease. Therefore, it is desirable to modify PE by pre-irradiation grafting methods. One merit of pre-irradiation methods is that a company with no irradiation facility can use them. After irradiation, the sample is carried to another place for some period, then the grafting reaction can be initiated. Another merit of the pre-irradiation methods used in this experiment is that the sample is able to be irradiated in air. It is needless to use any bags or ampules to avoid air.

REFERENCES

1. Imai, M. *Polym. Sci.: Polym. Chem. Ed.* **1981**, *19*, 1689.
2. Imai, M. *Polym. Sci.: Polym. Chem. Ed.* **1982**, *20*, 15.
3. Kabanov, V. Y.; Aliv, R. E. *Radiat. Phys. Chem.* **1985**, *26*, 697.
4. Kaji, K. *J. Appl. Polym. Sci.* **1986**, *32*, 4405.
5. Mori, K.; Koshiishi, K.; Masuhara, K. *Kobunshi Ronbunshu* **1991**, *48*, 1.
6. Mori, K.; Koshiishi, K.; Masuhara, K. *Kobunshi Ronbunshu* **1991**, *48*, 547.

TABLE 1 Change of Properties by Modification (LLDPE)

Method	Tensile strength (kg/mm^2)	Peeling strength (kg/cm)
Raw material	3.21	0.07
γ- Ray irradiation	3.77	0.07
EB irradiation	3.76	0.07
γ-Ray pre-irradiation	3.01–3.95	0.09–0.28
EB pre-irradation	3.20–3.95	0.17–0.19
Spray	4.29–5.29	0.15–0.27
Coating	3.41–4.59	0.17–0.23

SURFACE-MODIFIED POLYMERS

Allan S. Hoffman
Center for Bioengineering, FL-20
University of Washington

Many diverse applications of polymers in "everyday" uses, in industrial uses, and in medical uses depends upon physical, chemical, and biological interactions of molecules of microorganisms at their surfaces.

If a polymer surface does not have the desired properties for a particular application, then undesirable surface events can lead to failure of the material and the device or system containing it.

It is also important for the surface chemist to realize that the surface of normal, "as fabricated" polymers usually do not have the same compositions as the bulk phase. This can be due to adsorption of contaminants from the surrounding environment (e.g., oils, soils, or bacteria), desorption of impurities and surface active compounds from the bulk phase (e.g., lubricants added as process aids), oxidation or hydrolysis of surface groups, and preferred molecular orientation of surface groups (e.g., to minimize surface free energy).

Many different methods for polymer surface modification have evolved over the past 50 years. They can be divided here into three categories: physicochemical, mechanical, and biological methods.

Because of their great versatility, RFGD treatments have been increasingly used to modify surfaces.[2–5] The plasma gas of RFGD contains vacuum UV radiation plus many reactive species, such as free radicals and energetic electrons and ions. The vacuum UV radiation plus free radicals can cause many different reactions on the polymer surface, depending on the gas or vapor used in the plasma, the reactor geometry, and operating conditions.[6,7] The major effects of RFGD include surface etching or ablation, surface crosslinking, and deposition of a new polymer coating on the substrate polymer.

METHODS FOR CONVERTING HYDROPHOBIC TO HYDROPHILIC SURFACES

One of the most active areas of surface modification has been to convert nonpolar, hydrophobic surfaces to polar, hydrophilic, and water-wettable surfaces.[1–3,5]

Attachment of PEO to surfaces has been one of the most extensively studied treatments.[8–13] If the molecular weight of the PEO is above ~500–1,000, the surface will resist protein adsorption and cell adhesion.[10] Such a surface has been called "nonfouling." A new method for immobilizing a PEO surfactant by crosslinking is hydrophobic "tail" to polymer surface groups in an argon RFGD has been applied to a number of different amphipathic compounds, in order to functionalize hydrophobic surfaces.[12,13]

BIOLOGICAL SURFACE MODIFICATION

Surfaces are modified biologically by immobilizing biomolecules and cells onto them. Many diverse biomolecules may be immobilized using a wide variety of techniques.[14–17]

Biomolecules are covalently linked to surfaces when it is desired to recover and recycle the immobilized biomolecule and/or when it is necessary to avoid desorption of the biomolecule.[14–17]

REFERENCES

1. Hoffman, A. S. *Ann. NY Acad. Sci.* **1987**, *516*, 96.
2. Gombotz, W. R.; Hoffman, A. S. In *Critical Reviews in Biocompatibility*; Williams, D., Ed.; CRC: Boca Raton, FL, 1987; Vol. 4.
3. Yasuda, H. *Plasma Polymerization*; Academic: Orlando, FL 1985.
4. Iwata, H. *J. Polym. Sci., Polym. Chem.* **1988**, *26*, 3309.
5. Terlingen, J. G. A. *J. Appl. Polym. Sci.* **1994**, *52*, 39.
6. Danilich, M. J. *J. Biomtls. Sci., Polym. Ed.* **1992**, *3*, 195.
7. Uchida, E.; Uyama, Y.; Ikada, Y. *J. Appl. Polym. Sci.* **1993**, *47*, 417.
8. *PEO Chemistry, Biotechnological and Biomedical Applications*; Harris, J. M., Ed.; Plenum: NY, 1992.
9. Merrill, E. W.; Salzman, E. W. *ASAIO J.* **1983**, *6*, 60.
10. Nagaoka, S. In *Polymers as Biomaterials*; Shalaby, S. W.; Hoffman, A. S.; Ratner, B. D.; Horbett, T. A., Eds.; Plenum: NY, 1985.
11. Desai, N. P.; Hubbell, J. A. *Biomaterials* **1991**, *12*, 144.
12. Terlingen, J. G. A. *J. Biomtls. Sci. Polym. Ed.* **1993**, *4*, 165.
13. Sheu, M. S.; Hoffman, A. S.; Feijen, J. *J. Adhes. Sci. Tech.* **1992**, *6*, 995.
14. Hoffman, A. S. *Clinical Mtls.* **1992**, *11*, 61.
15. Hoffman, A. S. *Artif. Organs* **1992**, *16*, 43.
16. Kim, S. W.; Feijen, J. In *Critical Reviews in Biocompatibility*; Williams, D., Ed.; CRC: Boca Raton, FL, 1985; Vol. 4, p 229.
17. Carr, P. W.; Bowers, L. D. *Immobilized Enzymes in Analytical and Clinical Chemistry*; John Wiley & Sons: NY, 1980.

Sufactants, Polymeric

See: Cyclic Imino Ethers (Ring-Opening Polymerization)
Cyclic Imino Ethers (Ring-Opening Polymerization of 1,3-Oxazo Monomers)
Micelle-Forming Monomers, Photopolymerization
Polysoaps
Surface Active Monomers
Surfactants, Polymeric (Overview)
Surfactants, Polymeric (Colloid Stabilization)
Surfactants, Polymeric (Compatibilizers for Blends)

SURFACTANTS, POLYMERIC (Overview)

Yuya Yamashita*
Professor Emeritus of Nagoya University

Koichi Ito
Department of Materials Science
Toyohashi University of Technology

Polymeric surfactants or polymeric surface-active agents may be most properly defined as the amphiphilic polymers that contain both hydrophobic and hydrophilic segments of some definite lengths in the form of block or graft copolymers.[1,2]

Typical industrial developments include poly(ethylene oxide)-poly(propylene oxide)-poly(ethylene oxide) block copolymer as a low foaming surfactant by Wyandotte, polydimethylsiloxane-poly(ethylene oxide) graft copolymer as a polyurethane-foam stabilizer by UCC, and lyophilic-lyophobic block or graft by ICI.[3–5] Current developments in polymer syntheses, particularly

*Author to whom correspondence should be addressed.

in living polymerizations, have made preparation of a variety of well-defined amphiphilic block and graft copolymers possible.

PREPARATION OF POLYMERIC SURFACTANTS

Living Polymerization

Living polymerization has been the best technique for the precise control of degree of polymerization with sharp distribution and for the active ("living") end-group for subsequent polymerization (block copolymerization).

Living polymerizations are very useful in that the living ends are readily transformed to a variety of functionalities.

Macromonomers are usually mono-end-functional polymers. They have been prepared most conveniently in living systems either by introducing a polymerizable group in the initiator fragment or by end-capping with appropriate reagents such as *p*-vinylbenzyl chloride or methacryloyl chloride for the anionic living ends and methacrylic acid or *p*-vinylbenzyl alcoholate for the cationic end.

Chain-Transfer Polymerization

Chain-transfer reaction can be a convenient process for preparing an end-functional polymer of a relatively low degree of polymerization. Thus a radical polymerization of a monomer, M, in the presence of an effective chain-transfer agent, XY, may produce an oligomer with a structure, $X\text{-}[M]_n\text{-}Y$, with a degree of polymerization approximated by $n = [M]/C_s[XY]$, where C_s is a chain-transfer constant. An example is polymerization of alkyl methacrylate in the presence of thioglycolic acid.

They have been found to be useful precursors either to block copolymers via end-group reactions or to graft copolymers via macromonomers.

Polymeric Initiators

Just like a living polymer that initiates a subsequent polymerization to give a block copolymer, many polymeric initiators containing azo or peroxy functions have been developed to prepare block or graft copolymers by a radical process.[7,8]

PROPERTIES AND APPLICATIONS

A huge number of amphiphilic block and graft copolymers have been prepared by a variety of methods including those described above and have been characterized in terms of their surface- or colloid-chemical properties.[1,2,6]

Surface properties, such as surface tension, contact angle (wettability), and more directly surface composition by means of ESCA have been studied with a number of polymeric surfactants to indicate the distinct surface accumulation of the segments of lower surface energy.

Macromonomers can be useful as reactive emulsifiers and dispersants when applied to emulsion and dispersion polymerizations. Thus a macromonomer, very soluble in the medium, is copolymerized in the presence of an excessive amount of a substrate monomer to afford graft copolymers that act as effective steric stabilizers.

A number of applications of block copolymers or macromonomers and graft copolymers therefrom to compatibilizers between different polymers to make polymer alloys have been reported and reviewed.[2,6,9]

REFERENCES

1. Yamashita, Y.; Ito, K. *Yukagaku* **1980**, *29*, 219.
2. Ito, K. *Hyomen* **1990**, *28*, 739.
3. Lundsted, L. G.; Schmolka, I. R. *Block and Graft Copolymerization*; Ceresa, R. J., Ed.; John Wiley & Sons: 1976; Vol. 2, Chapter 1, pp 1–111.
4. Plumb, J. B.; Atherton, J. H. *Block Copolymers*; Allport, J. C.; Janes, W. H., Eds.; Appl. Sci.: London, 1973; Chapter 6, pp 305–503.
5. Barrett, K. E. J. *Dispersion Polymerization in Organic Media*; John Wiley & Sons, 1975.
6. Yamashita, Y., Ed. *Chemistry and Industry of Macromonomers*; Hüthig & Wepf: Heidelberg, 1993.
7. Ueda, A.; Nagai, S. *Kobunshi* **1990**, *39*, 202.
8. Nuyken, O.; Weidner, R. *Adv. Polym. Sci.* **1986**, *73/74*, 145.
9. Tsuda, T.; Azuma, K. *Nippon Gomu Kyokaishi (J. Soc. Rubber Ind., Jpn.)* **1992**, *65*, 578.

SURFACTANTS, POLYMERIC (Colloid Stabilization)

Irja Piirma
The University of Akron

COLLOID STABILIZATION

A colloidal dispersion has to be stabilized to remain as a fine subdivision of particles. In nature, where the continuous phase is often water, stabilization is achieved by polymeric surface active materials. In biological systems, where ionic strength is high, polyelectrolytes are used to provide stability to the system.

Modes of Colloid Stabilization

Unprotected particles in a dispersion will rapidly aggregate if not provided with a potential energy barrier. The primary source of attraction between colloidal particles arises as a consequence of van der Waals forces between molecules.[1,2] These forces can have different sources of origin. The transitory dipole-transitory dipole forces are long-range attractive forces, called London dispersion forces, that are electrical in nature and originate from the electronic fluctuations of the atoms. It has been calculated that their magnitude of attraction is calculated to be ~5–10 nm.[3]

To overcome these attractive forces, stability can be provided to the colloidal particles by surrounding them with an electrical double layer (electrostatic stabilization), adsorbed or chemically attached oligomeric or polymeric materials (steric stabilization), or a free polymer in the dispersion medium (depletion stabilization). The last two are often combined under the heading of polymeric stabilization.[3]

POLYMERIC SURFACTANTS AND POLYMER COLLOID STABILIZATION

Nonaqueous Systems

While naturally occurring polymeric surface-active materials are operative almost exclusively in aqueous media, the development of synthetic polymeric surfactants arose first from the need for stabilizing materials which would be operative in nonaqueous systems.

Aqueous Systems

Environmental considerations have made water increasingly more important as the dispersion medium.

Polymeric Electrostatic Stabilization of Colloidal Particles

The oldest and, therefore, the most extensively studied polymerizable surfactants are the carboxylic acids, such as acrylic and methacrylic acids. Homopolymers of these carboxylic acids have been classified as polymeric surfactants since they are water soluble and surface active in the form of alkaline metal salts. They have found many applications as stabilizers in particulate dispersions and as flocculants in water and sewage purification treatment. The carboxylic acids are used in polymeric emulsions and dispersions as either chemically bound or physically adsorbed stabilizers.

Carboxylic acid homo- and copolymer are sufficiently surface active to be used as surfactants in emulsion polymerization. P. Roe studied the behavior of polymeric carboxylic surfactants that were copolymers of acrylonitrile and acrylic acid, with their molecular weight modified by *n*-octyl mercaptan.[4] The alkaline salts of these copolymers are water soluble and have shown considerable stabilizing capability with no foaming tendencies.

R. Evans et al. carried out a study of the incipient flocculation behavior of poly(vinyl acetate) latices with several carboxylic acid homopolymer and graft copolymer surfactants.[5]

R. Buscall studied the stabilization of polystyrene latices with a poly(acrylic acid) of molecular weight 16,000 as a function of the degree of acid neutralization.[6]

Carboxylic acids are not the only materials that have been used as homopolymers or have been incorporated into polymer chains as surface-active comonomers to provide self-stabilization of dispersion and emulsion particles. A system using the comonomer pair, styrene and sodium styrenesulfonate, has been extensively studied.[7–9]

Polymeric Nonionic Stabilization of Colloidal Particles

The utilization of polymeric nonionic surface active materials as stabilizers in aqueous emulsion polymerization has not been as widespread as their use in nonaqueous systems.

There is one polymeric surfactant, the poly(vinyl alcohol) (PVA), that has been an important dispersion stabilizer for decades. It is used either as the sole stabilizer or, more often, in conjunction with an ionic surfactant. Another group of amphipathic materials produced commercially (BASF Wyandotte Corp.) are the block polymer surfactants called PLURONIC®, TETRONIC®, and PLURADOT®.

In 1982, I. R. Piirma and M. C. Chang reported that a change in the hydrocarbon solvent–water ratio of the phase inversion had a drastic effect on the outcome of an emulsion polymerization.[11] The case in question is the polymerization recipe with a chemical structure and a combination of ingredients, monomer-to-water-to-nonionic surfactant such as to produce at the beginning of the polymerization reaction a water-in-oil rather than an oil-in-water type of emulsion. In this case, the monomer is the continuous phase at the beginning of the reaction. Phase inversion occurs when the monomer disappears as a result of polymerization, and the system becomes an oil-in-water type emulsion. Because of the change in the polymerization mechanism from dispersion to emulsion polymerization, the latex has a bimodal and broad particle size distribution.[12,13]

Piirma and co-workers have synthesized a variety of block and graft copolymers with polyoxyethylene as the hydrophile and have used these materials as the sole polymerization stabilizers.[10]

REFERENCES

1. Hamaker, H. C. *Physica* **1937**, *4*, 1058.
2. London, F. *Trans. Faraday Soc.* **1937**, *33*, 8.
3. Napper, D. H. *Polymeric Stabilization of Colloidal Dispersions*; Academic: London, 1983.
4. Roe, P. J. *Colloid Interface Sci.* **1971**, *37*, 93.
5. Evans, R.; Davison, J. B.; Napper, D. H. *J. Polym. Sci., Polym. Lett. Ed.* **1972**, *10*, 449.
6. Buscall, R. *J. Chem. Soc., Faraday Trans.* **1981**, *177*, 909.
7. Juang, J. S.; Krieger, I. M. *J. Polym. Sci.* **1976**, *14*, 2089.
8. Chonde, Y.; Krieger, I. M. *J. Appl. Polym. Sci.* **1981**, *26*, 1819.
9. Turner, S. R.; Weiss, R. A.; Lundberg, R. D. *J. Polym. Sci. Chem.* **1985**, *23*, 535.
10. Piirma, I. *Polymeric Surfactants, Surfactant Science Series # 42*; Marcel Dekker: New York, 1992; Chapter 7.
11. Piirma, I.; Chang, M. C. *J. Polym. Sci., Polym. Chem.* **1982**, *20*, 489.
12. Piirma, I.; Maw, T. S. *Polym. Bull.* **1984**, *11*, 497.
13. Piirma, I.; Lenzotti, J. R. *Br. Polym. J.* **1989**, *21*, 45.

SURFACTANTS, POLYMERIC (Compatibilizers for Blends)

Jaan Noolandi
Xerox Research Centre of Canada

The role of polymeric surfactants as compatibilizers for polymer blends must be considered in conjunction with the blending process. As an example, we will consider a dispersion obtained by mechanical mixing, where two incompatible homopolymers with an added diblock copolymer are mixed to disperse one phase into another. The result of the process can be characterized by such measurements as the particular size, shape, size and shape distribution, and homogeneity. These depend on material parameters—such as chain length, stiffness, polydispersity, chemical structure, and architecture (for the same chemical composition)—as well as on externally imposed parameters, such as the vessel size, shape, and mixing rate.[1]

Polymeric surfactants enter into the picture as surface-active components which are used to modify the nature of the dispersion.[2] Usually, a more homogeneous dispersion of smaller particles is desired, although the interfacial strength can also be an issue. The surfactants can be either formed by interfacial reactions during the mixing or added as a separate premade component.[3,4]

The most popular choices for polymeric surfactants are graft copolymers and multiblock copolymers. Of the multiblock copolymers, the Shell Kraton family, which has an *A-B-A* triblock structure, has been widely used as an added component. Diblock copolymers show better performance as compatibilizers, all other parameters being equal.[5]

MICELLE FORMATION OF DIBLOCK COPOLYMER AB IN HOMOPOLYMERS A AND B

The specifics of individual micelle structures depend very much on the chain lengths and on the interaction parameters of the components.[6] Considering only spherical micelles, the corona (shell of the outer blocks) can be compressed flat against the inner core or open and extended, depending on the chain lengths of the outer blocks and of the homopolymer matrix. There is a dynamic balance between the number of diblock copolymers in micelles and the number of diblock copolymers at the interfaces. This balance is kinetically affected by the mixing conditions, however, because the amount of new interfacial area that can be created depends not only on the molecular energetics but also on the difference of the viscosities of the bulk phases. This complicated problem is unsolved at present. However, a rough guide as to the optimum block lengths for an effective diblock copolymer surfactant for two incompatible homopolymers is to calculate the critical micelle concentration of the surfactant molecules in the bulk phases and to exceed this concentration by a few percent.

POLYMERIC SURFACTANTS THAT ARE INCOMPATIBLE WITH THE BULK PHASES

Binary blends of poly(ethylene terephthalate) (PET) and high density polyethylene (HDPE) normally exhibit poor mechanical properties. When a commercial block copolymer KRATON G1652® Thermoplastic Rubber, poly(styrene-*b*-ethylene-*co*-butylene-*b*-styrene), or S-EB-S is added to the blend, there is a remarkable improvement in ductility.[7] This is all the more remarkable because there is no miscibility between PET and the PS segments of the S-EB-S.

INTERFACIAL ADHESION

Interfacial strength or adhesion is related to the penetration or interdigitation of the polymer chains from the bulk phases across their mutual interface.[8] The presence of a polymeric surfactant at the interface may either increase or decrease the interfacial strength, depending on its own interpenetration into the two bulk phases. It is generally accepted that the polymeric surfactants are most effective as interfacial strengthening agents when the loops are big enough and the ends are long enough to form entanglements with the bulk phases.[9–12]

SPECIFIC INTERACTIONS

If a portion of a polymeric surfactant can be partly cocrystallized with one of the components of the polymer blend, the surfactant becomes effectively part of that component.[13] This amounts to a surface treatment of the dispersed phase.[14,15] The adhesion of the dispersed component in the host matrix again depends on the entanglement properties of the exposed part of the surfactant. Another study shows that A/B blends compatibilized with a C–D block copolymer, in which the copolymer components show an exothermic heat of mixing with the bulk phases A/B, are superior to systems A/A–B/B with respect to emulsifying efficiency.[16] Ionic interactions are also important for stabilizing an interface dictating the microphase behavior.[17,18] Finally, it should be mentioned that although most of our discussion has centered around dispersions, continuous morphologies can also be stabilized by block copolymer surfactants.[1]

REFERENCES

1. Blacher, S.; Brouers, F.; Fayt, R.; Teyssié, Ph. *J. Polym. Sci., Polym. Phys. Ed.* **1993**, *31*, 655.
2. Piirma, I. *Polymeric Surfactants* Surfactant Science Series; Marcel Dekker: New York, 1992; Vol. 42.
3. Paulus, G.; Jèrôme, R.; Teyssié, P. *British Polymer J.* **1987**, *19*, 361.
4. Guégan, P.; Macosko, C. W.; Ishizone, T.; Hirao, A.; Nakahama, S. *Macromolecules* **1994**, *27*, 4993.
5. Fayt, R.; Jèrôme, R.; Teyssié, Ph. *J. Polym. Sci., Polym. Phys. Ed.* **1989**, *27*, 775.
6. Whitmore, M. D.; Noolandi, J. *Macromolecules* **1985**, *18*, 657.
7. Paul, D. R. In *Thermoplastic Elastomers*; Hanser: New York, 1987; Chapter 12, sec. 6.
8. Brown, H. R. *Macromolecules* **1989**, *22*, 2859.
9. Noolandi, J.; Kavassalis, T. A. *Molecular Conformation and Dynamics of Macromolecules in Condensed Systems*; Studies in Polymer Science; Elsevier: New York, 1988; Vol. 2.
10. Fetters, L. J.; Lohse, D. J.; Richter, D.; Witten, T. A.; Zirkel, A. *Macromolecules* **1994**, *27*, 4639.
11. Gersappe, D.; Irvine, D.; Balazs, A. C.; Liu, Y.; Sokolov, J.; Rafailovich, M.; Schwarz, S.; Peiffer, D. G. *Science* **1994**, *265*, 1072.
12. Janarthanan, V.; Stein, R. S.; Garrett, P. D. *Macromolecules* **1994**, *27*, 4855.
13. Kim, K. U.; Meier, D. J. *Polymer* (Korea) **1989**, *13*, 40.
14. Fayt, R.; Hadjandreou, P.; Teyssié, Ph. *J. Polym. Sci., Polym. Phys. Ed.* **1985**, *23*, 337.
15. Tang, T.; Jing, X.; Huang, B. *J. Macromol. Sci., Phys.* **1994**, *B33*(3,4), 287–305.
16. Auschra, C.; Stadler, R.; Voigt-Martin, I. G. *Polymer* **1993**, *34*, 2081.
17. Russell, T. P.; Jèrôme, R.; Charlier, P.; Foucart, M. *Macromolecules* **1988**, *21*, 1709–1717.
18. Eisenbach, C. D.; Hofmann, J.; MacKnight, W. J. *Macromolecules* **1994**, *27*, 2162–3165.

Suspending Agents

See: Suspension Polymerization
Water-Swellable Polymers (Carbomer Resins)

SUSPENSION POLYMERIZATION

Jorge E. Puig and Eduardo Mendizábal
Departamento de Ingeniería Química
Universidad de Guadalajara

Suspension polymerization is also known as bead or pearl polymerization because smooth, spherical particles are produced suspended in an aqueous phase. Suspension polymerization has the advantage over emulsion polymerization of producing cleaner polymers that can be easily separated from the aqueous media by filtration or centrifugation. Because of its importance, suspension polymerization has been the subject of numerous reviews and papers.[1–6]

In this process, the monomer (or monomers) is (are) suspended as droplets in an aqueous phase with continuous agitation; in an inverse suspension polymerization, drops of a water-soluble monomer are dispersed in an organic media.[4] Usually, suspending agents are used to prevent coalescence of particles during the reaction.[4]

KINETICS ASPECTS

Polymerization takes place in the organic droplets by a free-radical mechanism.[5,7,8] The initiators are usually soluble in the monomer. Sometimes, two initiators that decompose at different temperatures are used to reduce the monomer content in the beads.

In suspension polymerization, each droplet acts as a small bulk polymerization reactor.[5,7] The kinetic mechanism is similar to that of bulk polymerization, inasmuch as the reaction rates are not influenced to a great extent by bead size under typical commercial polymerization conditions.[4,9]

BEAD SIZE AND PARTICLE SIZE DISTRIBUTION

Both modeling and experimental approaches for mass polymerization give good descriptions for the molecular weight distribution (MWD) in suspension polymerization.[10,11] However, there is little understanding of the evolution of PSD in this process.

Industrial suspension polymerization is carried out at conditions where turbulence (inertial subrange) is the main factor responsible for drop breakage and coalescence.[12–14]

SUSPENDING AGENTS

In order to prevent coagulation during suspension polymerization, especially during the second stage when particles become sticky, suspending agents or stabilizers are added to the reacting system. Suspending agents can be water-soluble polymers (also called "protective colloids") or finely divided inorganic salts.[4]

The early literature and patents describe the use of small amounts (0.1 to 1% with respect to water) of finely divided inorganic compounds such as talc, calcium and magnesium carbonates, silicates, and phosphates together with small amounts of surfactants.[4]

Water-soluble polymers, commonly used as protective colloids are biopolymers (gelatin, proteins, derivatives of cellulose, and polysaccharides) or synthetic polymers (poly(vinyl alcohol)s, polyvinylpyrrolidone, and other hydrophobic-hydrophilic block copolymers).[4] A good steric protective colloid is a diblock polymer (A_mB_n) with a segment compatible with the surface of the particle (A) and a segment compatible with the continuous phase (B).

Poly(vinyl alcohol)s are the most widely used protective colloids because they can be easily prepared by controlled partial hydrolysis of poly(vinyl acetate) (PVA).[1]

The selection of an appropriate suspension agent for a given reacting system is a time-consuming trial-and-error process.

For water-insoluble monomers (styrene, butyl acrylate, butyl methacrylate) the polymerization cannot be completed when the stability time of the nonreacting suspension was either too small or too large because of coalescence or agglomeration.[16] Mendizábal et al. found that high molecular weight (> 70,000) and partially hydrolyzed (88%) PVAs were the best protective colloids.[16] Highly hydrolyzed PVAs, regardless of their molecular weight, were poor suspension agents. Mendizábal et al. also found that increasing the PVA concentration yields a smaller bead size and a narrower PSD as a result of a decrease in the surface and interfacial tension.[16]

CONCLUDING REMARKS

The development of models for predicting PSD as well as the conditions for suspension stability at all reaction stages is needed because it offers the possibility of designing and optimizing suspension polymerization and for obtaining products with better characteristics. The model of Alvarez et al. appears to be a good starting point.[15]

In spite of the complexity of the stability phenomena of dispersed systems combined with chemical reactions, the stability time measurements of nonreacting suspensions seems to be a valuable tool for choosing the best protective colloid and its optimal concentration for a given suspension polymerization. More work is required to understand the relationship in nonreacting and reacting disperse systems.

REFERENCES

1. Dawkins, A. In *Comprehensive Polymer Science*; Eastwood, G. C. et al., Eds.; Pergamon: New York, 1989; Vol. 4, p 231.
2. Trommsdorff, E.; Schildknetch, C. E. In *High Polymers, Vol. X*; Schildknetch, C. E., Ed.; Interscience: New York, 1956.
3. Napper, D. H. *Polymeric Stabilization of Colloid Dispersions*; Academic: New York, 1983.
4. Munzer, M.; Trommsdorff, E. In *Polymerization Processes*; Schildknecht, C. E.; Skeist, I., Eds.; John Wiley & Sons: New York, 1977; p 106.
5. Grulke, E. A. In *Encyclopedia of Polymer Science & Engineering: Vol. 16*; Mark, H. F. et al., Eds.; John Wiley & Sons: New York, 1989; Vol. 16, p 443.
6. Yuan, H. G.; Kalfas, G.; Ray, W. H. *Rev. Macromol. Chem. Phys.* **1991**, *C31*, 215.
7. Odain, G. *Principles of Polymerization*; John Wiley & Sons: New York, 1981.
8. Mendizábal, E.; Jasso, C. F.; Oropeza, A. *Rev. Plasticos Moder.* **1989**, *399*, 391.
9. Hohenstein, W. P.; Marks, H. F. *J. Polym. Sci.* **1946**, *1*, 127.
10. Biesenberger, J. A.; Sebastian, D. H. *Principles of Polymerization Engineering*; John Wiley & Sons: New York, 1983.
11. Tirrell, M.; Galvan, R.; Laurence, R. L. In *Chemical Reaction and Reactor Engineering*; Carberry, J. J.; Varma, A. M., Eds.; Marcel Dekker: New York, 1987.
12. Shinnar, R. *J. Fluid Mech.* **1961**, *10*, 259.
13. Abrahamson, J. *Chem. Eng. Sci.* **1975**, *30*, 1371.
14. Wood, P. E. *Principles of Polymer Reactor Design and Operation: Mixing Effects in Suspension Polymerization*; Internal Report, Department of Chemical Engineering, McMaster University, Ontario, 1989.
15. Alvarez, J. J.; Hernández, M. *Chem. Eng. Sci.* **1994**, *49*, 99.
16. Mendizábal, E.; Castellanos-Ortega, J. R.; Puig, J. E. *Colloids and Surf.* **1992**, *63*, 209.

Suspension Polymerizations

See: Poly(methyl methacrylate)
Polystyrene (by Suspension Polymerization)
Poly(vinylidene fluoride) (Synthesis, Microstructure, and Chain Conformations)
Suspension Polymerization

Ultrasonically Assisted Polymer Synthesis
Vinyl Chloride Copolymers

SUTURE MATERIALS (Current Research and Development)

C. C. Chu
Fiber and Polymer Science Program
Department of Textiles and Apparel
Cornell University

One of the most important contributions of polymer science to medicine during the past several decades is the development of synthetic wound closure biomaterials, particularly absorbable sutures and clips. Although synthetic nonabsorbable suture materials are made from those well-known household names of polymers such as nylon, polyester, teflon, and polypropylene, the most important research in and development of polymeric suture materials are in the area of absorbable sutures, which degrade with time in both *in vitro* and *in vivo* environments. **Table 1** lists the types of commercial sutures, their trade names, and the polymers that the sutures are made from.

GENERAL BIODEGRADATION PROPERTIES

Because all existing synthetic absorbable sutures belong to mainly linear aliphatic polyesters, simple hydrolytic main-chain scissions have been suggested as the means of biodegradation.[1–4]

Molecular Weight

The molecular weight of these suture fibers decreases drastically with the duration of hydrolysis; increases in the rate of the weight loss depend on the chemical structure of the sutures.

Mechanical Properties and Mass Loss of Mass

Along with the loss of molecular weight, the tensile properties and the mass of absorbable sutures also decrease during hydrolysis.

The Level of Crystallinity and the Crystallite Structure

In all reported studies on the hydrolytic degradation of synthetic absorbable sutures, the level of crystallinity of the sutures as hydrolysis progresses always increases initially, reaches a maximum, and then decreases.

Orientation

One of the unique aspects of these absorbable sutures is the fiber orientation that does not exist in nonfibrous biodegradable materials. This anisotropy characteristic implies that the hydrolytic degradation process along the fiber axis (i.e., parallel to the fiber molecules) should be different from the direction perpendicular to the fiber axis because different ways of molecular packing and arrangement would result in a different rate of water diffusion through the materials, i.e., a different rate of hydrolysis. Up to the present time, there is only one reported study examining different rates of hydrolysis due to orientation of suture fibers.[5]

TABLE 1. Commercial Suture Materials and Their Constituent Polymers

Constituent polymers	Trade names
Absorbable:	
Collagen	Catgut
Regenerated collagen	Collagen
Poly(glycolic acid) or Polyglycolide	Dexon
Poly(glycolide-co-lactide) or Polyglactin 910	Vicryl
Poly(*p*-dioxanone)	PDSII
Poly(glycolide-co-trimethylene carbonate)	Maxon
Polyglycolide-co-ϵ-cparolactone or Polyglecaprone 25	Monocryl
Non-Absorbable:	
Polypeptides	Silk, Surgical Silk, Dermal, Virgin Silk
Celluosics	Cotton
Polyamides: nylon 66 and 6	Surgilon, Dermalon, Nurolon, Ethilon, Supramid
Polyesters: (butylene terephthalate)	Miralene
Polyesters: (ethylene terephthalate)	Dacron, TrCron, Ethibond, Mersilene, Ethiflex, Polydek, Tevdek, Mirafil
Polyesters: (tetramethylene ether) Terephthalate-co-Tetramethylene terephthalate	Novavil
Polypropylene (isotactic)	Prolene, Surgilene
Polytetrafluoroethylene (expanded)	Gore-Tex

NEW RESEARCH AND DEVELOPMENT ACTIVITIES

The current research and development in suture materials focus on: the chemical modification of existing absorbable suture materials either to improve their biodegradation and handling properties like antimicrobial or to render them biologically active acceleration of wound healing; the modification of existing nonabsorbable sutures for better handling and mechanical properties; and, to a lesser extent, the development of completely new nonabsorbable sutures for achieving better handling, biological, and/or mechanical properties.

High-Strength Poly(vinyl alcohol)

Poly(vinyl alcohol) in hydrogel form or surface modified with RGD (Arg-Gly-Asp) has been reported to have some biomedical use, such as synthetic vascular grafts and synthetic vitreous.[6–8] For suture use, PVA must be in fibrous form.

A Potentially Bioactive and More Pliable Polypropylene-Based Monofilament Suture

Among the non-absorbable monofilament sutures, polypropylene (PP) has been considered to be the most stable *in vivo*. However, existing PP sutures have one major drawback in their handling property, i.e., pliability, knot security. Very recently, Liu et al. reported the development of an innovative PP-based new monofilament suture that they claimed to be more flexible, have better handling property and bioactive potential through chemical bonding of the surface active suture with antibiotic, growth factor, and anti-clotting agents.[9,10]

This new PP-based suture is made by melt-blending of the nonionic isotactic PP with an ionic copolymer of either ethylene or propylene and ethyleneically unsaturated carboxylic acids (e.g., acrylic or methacrylic acid) during the fiber melt-spinning process.

Syndiotactic Polypropylene Sutures with Greater Flexibility

Liu claimed PP sutures from syndiotactic PP exhibited improved handling properties, such as greater flexibility, when compared with an isotactic PP suture.[12,13]

Elastomeric Poly(ether imide ester) Monofilament Suture

It is always desirable to have monofilament sutures that have the same handling properties as braided sutures of the same size. With this in mind, Liu et al. recently reported a new class of monofilament suture based on a [poly(ether imide ester).][11,14] They claimed that this new suture has excellent mechanical and handling properties, such as resistance to creep, good knot security, and physical stability.

Improved Biodegradation of Absorbable Sutures Through Copolymerization

The biodegradation of all absorbable sutures could be described in terms of both strength loss and weight loss profiles. Due to the inherent structural-property relationship in fibers, an improvement in the strength loss profile unfortunately also leads to a prolonged suture mass retention within biological systems. Such a prolonged suture mass retention is undesirable because it elicits a chronic inflammatory reaction and a predisposition to infection, granuloma formation, etc. Therefore, recent research efforts have been to develop new absorbable sutures from the random and segmented copolymerization of either existing absorbable suture components or new monomers chemically different from existing absorbable suture components.

An example of such efforts is the copolymerization of a PDS suture with glycolides, lactides, or morpholine-2,5-dione (MD).[16–19] Copolymer sutures made from PDS and morpholine-2,5-dione exhibit rather interesting biodegradation properties.[15]

REFERENCES

1. Chu, C. C. In *Biocompatible Polymer, Metals, and Composites*; Szycher, M., Ed.; Technomic: Lancaster, PA, 1983; Chapter 22.
2. Chu, C. C. In *CRC Critical Reviews in Biocompatibility*; Williams, D. F., Ed-in-Chief; CRC: Boca Raton, FL, 1985; Vol. 1, Issue 3, pp 261–322.
3. Chu, C. C. In *Concise Encyclopedia of Medical and Dental Materials*; Williams, D. F., Ed.; Pergamon: London, 1990; pp 346–353.
4. Gilding, D. K. In *CRC Biocompatibility of Clinical Implant Materials*; Vol. II, Williams, D. F., Ed.; CRC: Boca Raton, FL, 1981, Chapter 9.
5. Lin, H. L.; Chu, C. C.; Grubb, D. *J. Biomed. Mater. Res.* **1993**, *27*(2), 153.
6. Tamura, K.; Hitomi, S.; Natsume, T.; Kobayashi, T.; Kuwabara, O.; Ohonishi, T.; Nakamura, K. *Japan J. Artif. Organs* **1992**, *21*, 176.
7. Iio, K.; Minoura, N.; Aiba, S.; Nagura, M.; Kodama, M. *J. Biomed. Mater. Res.* **1994**, *28*(4), 459.
8. Tomita, N.; Nagata, N.; Ueda, Y.; Tamai, S.; Hyon, S. H.; Ikeuchi, K.; Ikada, Y. In *Biomaterial-Tissue Interface*; Doherty, P. J. et al., Eds.; *Advances in Biomaterials*; Elsevier Science: New York, 1992; Vol. 10.
9. Liu, C. K.; Brewer, J. U.S. Patent 5 284 489, 1994.
10. Liu, C. K.; Brewer, J.; Kokish, M. *20th Annual Meeting of the Society for Biomaterials*; Boston, MA, April 5–9, 1994; p 477.
11. Liu, C. K.; Brewer, J.; Kokish, M. In *20th Annual Meeting of the Society for Biomaterials*; Boston, MA, 1994; p 243.
12. Liu, C. K. U.S. Patent 5 269 807, 1993.
13. Liu, C. K. *13th Southern Biomedical Engineering Conference*; Washington, DC, 1994; pp 748–751.
14. Liu, C. K.; Brewer, J. U.S. Patent 5 225 485, 1993.
15. Chu, C. C. *Annals of Surgery* **1981**, *193*(3), 365.
16. Shalaby, S. W. In *Biomedical Polymers: Designed-to-Degrade Systems*; Shalaby, S. W., Ed.; Hanser: New York, 1994; Chapter 1.
17. Shalaby, S. W.; Koelmel, D. F. U.S. Patent 4 441 496, 1984.
18. Bezwada, R. S.; Shalaby, S. W.; Newman, H. D.; Kafrauy, A. U.S. Patent 4 653 497, 1987.
19. Bezwada, R. S.; Shalaby, S. W.; Newman, H. D.; Kafrauy, A. U.S. Patent 4 643 191, 1987.

Swollen-State Polymerization

See: Poly(ethylene terephthalate) (Ultra-High Molecular Weight)

Symplexes

See: Polyelectrolyte Complexes (Overview)

Syndiotactic Polypropylene

See: Metallocene Catalysts
Polypropylene (Commercial)

Syndiotactic Polystyrene

See: Metallocene Catalysts
Polystyrene, Stereoregular
Polystyrene, Syndiotactic
Polystyrene, Syndiotactic (Sulfonation and Crystallization)
Styrene, Stereospecific Polymerization
Thermoreversible Gels (Isotactic, Syndiotactic, and Atactic Polystyrene)

SYNZYMES

Junghun Suh
Department of Chemistry
Seoul National University

Since enzymes are the most effective homogeneous catalysts found in nature with respect to the degree of rate enhancement and accuracy, it has been a challenge for decades to prepare artificial enzymes, synthetic catalysts mimicking enzymes. Several types of artificial enzymes have been prepared by using various kinds of molecules. The word "synzymes" (*syn*thetic *enzymes*) is used to designated synthetic polymers with enzyme-like catalytic activities.[1]

A wide range of linear polymers and some branched polymers have been examined as synzymes. Examples of linear polymers used as backbones of synzymes are poly(4-vinylpyridine), poly(*N*-vinylimidazole), poly[4(5)-vinylimidazole], poly[5(6)-vinylbenzimidazole], and copolymers such as copoly[*N*-vinylpyrroline-4(5)-vinylimidazole].[2–7] Poly(ethylenimine) (PEI) is the most intensively investigated branched polymer.[8]

MICROENVIRONMENTS

Synzymes equipped with hydrophobic domains bind hydrophobic substrates to form synzyme-substrate complexes. When both hydrophobic groups and charged centers are present on polymers, the resulting hydrophobic polyelectrolytes may be considered as analogues of micelles, without the basic instability element of micelles. For example, the *N*-alkylated forms of poly(4-vinylpyridine), poly(*N*-vinylimidazole), or their copolymers possess local domains similar to micelles.[6,9] When apolar groups are attached to a part of the nitrogen atoms of PEI and some of the amines are protonated, micelle-like domains are present on the PEI surface.

Efforts have been made to design synzymes with enantioselectivity by introducing chiral microenvironments on polymers. For example, when PEI derivatives containing optically active L-histidine moieties were prepared, up to a 3.6-fold rate difference was observed for the hydrolysis of D- and L-*N*-carbobenzoxyamino acid *p*-nitrophenyl esters.[10]

BINDING SITES

The limited solubility of polymer derivatives in water and the lack of specific binding sites on the polymer skeletons were major obstacles faced in the early stage of synzyme studies. In order to achieve substantial improvements in catalytic capabilities beyond those of micelles, the construction of well-defined binding sites on the polymer backbones is necessary.

Macrocyclic metal centers constructed on PEI through metal-template condensation provide the polymer with binding sites capable of recognizing the anionic esters.

As for binding sites for certain hydrophobic moieties, β-cyclodextrin (CD) has been attached to PEI.[11] The PEI derivative containing CD manifests a high affinity towards *t*-butylphenyl compounds and accelerates deacylation of esters containing *t*-butylphenyl moieties.

CATALYTIC GROUPS

Various functional groups have been attached to synzymes as catalytic groups. Pyridyl, imidazolyl, and benzimidazolyl groups are related to the imidazole moiety of histidine, an amino acid residue of enzymes. PEI derivatives equipped with imidazolyl or *N,N*-dialkylaminopyridyl groups manifest high reactivity toward ester hydrolysis. These heterocyclic amines readily attack labile esters such as nitrophenyl carboxylates, producing *N*-acyl intermediates. Since the *N*-acyl intermediates are hydrolyzed faster than the ester substrates, effective catalysis is achieved for the ester hydrolysis.[9] As oxygen nucleophiles, oximes or hydroxamic acids have been attached to synzymes.

Metal ions play essential catalytic roles in metalloenzymes.[12] Macrocyclic metal centers have been built on PEI through the metal-template condensation of PEI with dicarbonyl compounds and exploited in biomimetic catalysis.

ADJUSTMENT OF CONFORMATION

Synthetic polymers have been successfully employed as macromolecular spacers in the design of effective hosts for organic compounds or metal ions.[13,14] The macromolecular host molecules are synthesized by using guest molecules or their analogues as templates. If the host selectively recognizes and stabilizes the transition state, effective catalysis is expected.

THE COMBINATION OF CATALYTIC ELEMENTS

Effective synzymes would be designed if two or more pieces of the catalytic elements discussed above are combined and incorporated into the polymer skeleton. The combination of electrostatic and hydrophobic properties of the microenvironment can lead to synergistic effects. The intrinsic reactivity of catalytic functional groups attached to synzymes may be improved by modifying the microenvironment. Two or more catalytic groups have been attached to some polymers.

Enzymes catalyze virtually every kind of chemical reaction. Although synzymes have been designed so far for a limited number of reactions, it will be a challenge to design synzymes mimicking synthetases, oxidoreductases, and other types of enzymes.

REFERENCES

1. Klotz, I. M.; Royer, G. P.; Scarpa, I. S. *Proc. Natl. Acad. Sci. U.S.A.* **1971**, *68*, 263.
2. Letsinger, R. L.; Savereide, T. J. *J. Am. Chem. Soc.* **1962**, *84*, 3122.
3. Overberger, C. G.; Pierre, T. St.; Yaroslavksy, C.; Yaroslavsky, S. *J. Am. Chem. Soc.* **1966**, *88*, 1184.
4. Kopple, K. D. *Biopolymers* **1968**, *6*, 1417.
5. Kunitake, T.; Shimada, F.; Aso, C. *J. Am. Chem. Soc.* **1969**, *91*, 2716.
6. Overberger, C. G.; Salamone, J. C. *Acc. Chem. Res.* **1969**, *2*, 217.
7. Kirsh, Y. E.; Pluzhnov, S. K.; Shomina, T. S.; Kabanov, V. A.; Kargin, V. A. *Vysokonol. Soedin., Ser. A.* **1970**, *12*, 186; *Chem. Abstract* **1970**, *72*, 86675.
8. Klotz, I. M.; Royer, G. P.; Sloniesky, A. R. *Biochemistry* **1969**, *8*, 4752.
9. Kunitake, T.; Shinkai, S. *Adv. Phys. Org. Chem.* **1980**, *17*, 435.
10. Kimura, Y.; Nango, M.; Ihara, Y.; Kuroki, N. *Chem. Lett.* **1984**, 429.
11. Suh, J.; Lee, S. H.; Zoh, K. D. *J. Am. Chem. Soc.* **1992**, *114*, 7917.
12. Suh, J. *Acc. Chem. Res.* **1992**, *25*, 273.
13. Shea, K. J.; Thompson, E. A.; Pandey, S. D.; Beauchamp, P. S. *J. Am. Chem. Soc.* **1980**, *102*, 3149.
14. Suh, J.; Lee, S. H.; Paik, H.-j. *Inorg. Chem.* **1994**, *33*, 3.

T

TACK, ELASTOMERS

Gary R. Hamed
College of Polymer Science and Polymer Engineering
The University of Akron

Tack is the joint strength that develops when two objects are *contacted* for a short time under little or nor pressure.[1–4] The phenomenon may be divided into two types: autohesive tack, or simply autohesion, when the two objects have the same composition, and adhesive tack, when they do not. Tack is important in the manufacture of many rubber goods, including tires, belts, and certain hoses that are made by plying together reinforced rubber layers. Additionally, tacky adhesives are commonly coated onto various substrates to make pressure sensitive tapes, labels, stickers, etc.

In order to be tacky, a material must simultaneously possess both a liquid-like and a solid-like character. The former imparts the ability for rapid bond formation, whereas the latter provides resistance to rupture upon loading. These requirements are met only for certain compositions that contain high molecular weight elastomers, that have mobile chains (i.e., above T_g) but are entangled with one another. Natural rubber, which is rather easily deformed at low strains and has a strain-induced strengthening mechanism (i.e., strain-induced crystallization) possesses high autohesion and is widely used to formulate pressure-sensitive adhesives.

AUTOHESION

For any elastomer type, there is a molecular weight range in which autohesion will be maximized. This can be understood by considering the extremes. At a sufficiently low molecular weight, all elastomers will rapidly form an autohesive bond, but because cohesive strength is low, autohesion also will be low and cohesion limited. On the other hand, when the molecular weight is too high, bond formation is slow and hence tack is low and bond formation is limited. Only at intermediate molecular weights is there the appropriate compromise between the ease of bond formation and retaining sufficient cohesion.

Low levels of tackifier (~ 5 phr), such as a *p*-tert-octyl phenolic resin, are often added to rubbery compositions to prevent the loss of autohesion with aging.

ADHESIVE TACK

A low resistance to deformation does not insure a high adhesive tack. The adhesive must also possess a high cohesive strength. This will prevent the adhesive from easily splitting apart during bond rupture. It is often required that the adhesive be strong enough so that a fracture will not occur within it when a joint is loaded, but rather fracture will proceed by interfacial separation between the adhesive and substrate. Although reduction of entanglements is needed to impart sufficient compliance to a hydrocarbon elastomer for good adhesive tack, the extent of dilution must not be so great as to eliminate entanglements, because the low cohesive strength of the adhesive would preclude high adhesive tack.

One limitation of pressure sensitive adhesives based on natural rubber is the relatively poor long-term shear resistance. This is due to an excessive creeping flow of the adhesive. Adhesives based on triblock copolymers with rubbery polyisoprene midblocks and glassy polystyrene and segments produce joints with improved shear resistance.[5] The end blocks phase-separate forming hard domains, which are covalently linked to the rubbery midblocks. This morphology prevents the rubber segments from easily disentangling and flowing apart.

Hydrocarbon elastomers required added tackifier in order to become useful pressure-sensitive adhesives. However, some polar elastomers, most notably polyacrylate types, are inherently tacky in their neat form.[6] Commercial materials often are complex systems containing multiple monomers, including, e.g., alkyl acrylates, vinyl acetate, and a carboxylic acid functionalized monomer.

REFERENCES

1. Hamed, G. R. *Rubber Chem. Technol.* **1981**, *54*, 576.
2. Barquins, M.; Maugis, D. *J. Adhesion* **1981**, *13*, 53.
3. Bothe, L.; Rehage, G. *Die Angewandte Makromolekulare Chemie* **1981**, *100*, 39.
4. Busse, W. F.; Lambert, J. M.; Verdery, R. B. *J. Appl. Physics* **1946**, *17*, 376.
5. Kraus, G.; Rollmann, K. W.; Gray, R. A. *J. Adhesion* **1979**, *10*, 221.
6. Fries, J. A. *Int. J. Adhesion* **1982**, *2*, 187.

TAILOR-MADE POLYMERS

Yves Gnanou
Laboratoire de Chimie des Polymères Organiques—ENSCPB
Université Bordeaux

Tailor-made polymers are designated by a molecular structure that is defined without any ambiguity and in which the fluctuation in size, composition, and functionality is very low.

Long restricted to linear macromolecules of controlled size and low dimensional fluctuation, the term tailor-made polymers now encompasses a variety of macromolecular structures and architectures. Chains organized in controlled topologies as in star-shaped, macrocyclic, comb-like polymers, or compounds with a different nature like block or graft copolymers, samples involving a precise terminal functionality, all structurally controlled polymers fall in the tailor-made category.

TAILOR-MADE LINEAR POLYMERS

The prerequisite for obtaining polymers that exhibit structures as close as possible to the expected structure is to work with long-lived species.

Even though living-chain polymerizations afford homogeneous populations of chains, a certain fluctuation in the size of the latter cannot be avoided. Tirrell was the first to show that strictly isomolecular synthetic macromolecules are accessible by genetic engineering. Artificial proteins, monodisperse derivatives of poly (α,L glutamic acid), and so on have been designed and elaborated by Tirrell's team.[1]

ω-FUNCTIONALIZED POLYMERS: TELECHELICS AND MACROMONOMERS

Low-molar-mass polymers that are characterized by a flawless functionality of their extremities are particularly valuable in macromolecular synthesis as intermediates for the elaboration of complex macromolecular assemblies. When these ω-functionalized oligomers carry functional groups at both chain ends that are designed to participate in step-growth processes, they are termed *telechelics* and are used in chain-extension reactions or for the synthesis of model networks.[2]

When the function fitting the chain end is a polymerizable group that is able to get involved in a chain-addition process, such ω-functionalized species are designated as *macromonomers.*[2] Macromonomers give rise to graft copolymers on copolymerization with a comonomer.

For these two types of reactive polymers, living polymerizations offer unmatched advantages as to the chain-end functionalization.

Telechelics by Living Processes

There are a number of ways to gain access to telechelic polymers, the recourse to living processes is imperative only when samples exhibiting both narrow molar-mass distribution and flawless functionality are searched for. Three strategies can be followed to generate well-defined telechelic polymers. The first pathway involves the use of an efficient difunctional initiator that is designed to give rise to difunctional living chains, which implies the deactivation of the difunctional living chain by an appropriate functional agent.[3] The access to telechelics can also be contemplated when combining an initiator fitted with the function that will end-cap one of the chain extremities and a difunctional coupling reagent. In the majority of cases, the function carried by the initiator has to be protected so it will not interfere with the polymerization.[10] If, instead of a coupling agent, a functional deactivator is used in association with the above-mentioned functional initiator, the possibility of synthesizing α,ω-heterodifunctional species exists.[5]

Macromonomers

The macromonomer method is currently viewed as one of the most powerful techniques used to elaborate two-phase systems.[2,6–8] Indeed, almost any kind of graft and backbone can be combined in a copolymer structure by mere copolymerization of a macromonomer with a comonomer. On incorporation into the copolymer, the macromonomer gives rise to a graft, whereas the comonomer contributes to build up the backbone on polymerization.

STAR-SHAPED POLYMERS

Star-shaped polymers consist of a central core to which a given number of chains or branches are attached. Star molecules containing a precise number of branches of controlled size are particularly valuable for the investigation of the effect because of branching on the properties in solution and in the bulk.

With regard to application, polymers with star architecture are generally used to reduce the intrinsic viscosity of polymer solutions. For instance, star molecules are purposely added to paints and coatings in order to increase their solid content without affecting viscosity or spraying properties.[9]

Two different strategies have been developed for the synthesis of tailor-made star polymers. The core-first method is based on the use of a plurifunctional initiator from which the star arms are grown. In the arm-first method, the central body is built in a second step either via copolymerization of living growing chains with a difunctional molecule or by deactivation of the living growing chains with a plurifunctional agent.[10,11]

DENDRIMERS AND POLYMERS WITH DENDRITIC STRUCTURES

Dendrimers are a new class of tailor-branched macromolecules whose main characteristics are the absence of any fluctuation in size and functionality.[12]

The elaboration of such perfect molecular assemblies demanded that specific methods of synthesis be purposely developed; no polymerization process—even the living ones—was appropriate for the objective sought. The principle of preparation of such dendritic molecules is based on the sequential iteration of two successive processes:

- from the reactive sites present on a substrate introduction of an equal number of branches; and
- modification of the termini of the branches newly introduced so as to create two or more terminal reactive functions on these loci.

CYCLIC POLYMERS

To access cyclic polymers of controlled size, a procedure based on end-to-end reactions has been developed. Reaction of a linear α,ω-difunctional chain with a small coupling agent fitted with antagonist functions is bound to give rise to the expected ring closure, in addition to linear polycondensation.[13]

Höcker and Rempp are the two pioneers of this method. They showed that reaction of living difunctional carbanionic polystyrene (PS) with 1,4-bis(bromomethyl)benzene yields the expected cyclic macromolecules along with polycondensates.[13–15]

Instead of resorting to a bimolecular process that involves two successive reactions, Deffieux and his co-workers have experimented with a new approach based on a unimolecular reaction.[16–18]

BLOCK COPOLYMERS

It is of paramount importance for the various applications contemplated that the block copolymers exhibit a minimal fluctuation in size and above all in composition. Methods based on living polymerizations are again the most appropriate for gaining access to copolymers that correspond to the above criteria. The sequential addition of monomers is the first of these methods of synthesis: it has been used extensively for the preparation of block copolymers. Polymerizations proceeding under living conditions through one of the different chain mechanisms have been applied. Methods that resort to the site-transformation techniques or to the use of a macroinitiator also use living polymerizations.

Other pathways have also been considered, for example, calling on condensation techniques among polymers carrying antagonist terminal functions.[6,19–21]

GRAFT COPOLYMERS

Graft copolymers whose structure can be well controlled experimentally also belong to the family of model polymers.

CONCLUSIONS

The emergence of living polymerization processes and the continuous spread of these techniques to many monomers has greatly helped to engineer molecularly a variety of tailor-made macromolecular architectures. Homopolymers with low fluctuation in size and functionally, polymers with often-complex topologies, and block and graft copolymers exhibiting low compositional heterogeneity are examples of tailor-made polymers that have been made available by virtue of living processes. When constructing macromolecules of such geometric beauty and perfection, it is essential to check at each step of the synthetic process that no side reaction has thwarted the pathway.

The examples of macromolecular architecture given above should not lead one to infer that living polymerizations offer a panacea for elaborating any type of tailor-made macromolecules. The polymers prepared by genetic engineering and the dendrimers are two unique examples of macromolecules that have been generated by methods other than living-chain polymerizations. Specific preparative schemes have been developed in order to synthesize these isomolecular macromolecules.

REFERENCES

1. Mc Grath, K. P.; Fournier, M. J.; Mason, T. L.; Tirrell, D. A. *J. Am. Chem. Soc.* **1992**, *114*, 727.
2. Percec, V.; Pugh, C.; Nuyken, O.; Pask, S. In *Comprehensive Polymer Science*; Allen, G.; Bevington, J., Eds.; Pergamon: New York, 1989; 6, 281.
3. Richards, D. *J. Polym. Sci. B* **1968**, *6*, 417.
4. Sogah, D.; Webster, O. *J. Polym. Sci., Polym. Lett. Ed.* **1983**, *21*, 927.
5. Degée, P.; Dubois, P.; Jérome, R.; Teyssié, Ph. *Macromolecules* **1992**, *25*, 4242.
6. Rempp, P.; Franta, E. *Adv. Polym. Sci.* **1984**, *1*, 58.
7. Gnanou, Y. *Ind. J. Technology* **1993**, *31*, 317.
8. Chujo, Y. *Chemistry and Industry of Macromonomers* Yamashita, Y., Ed.; Hüthig & Wepf: Basel, 1993.
9. Simms, J.; Spinelli, M. *J. Coating Technol.* **1987**, *59*, 752, 16.
10. Cloutet, E.; Fillaut, J.-L.; Gnanou, Y.; Astruc, D. *Chem. Com.* **1994**, 2433.
11. Bauer, B.; Fetters, L. *Rubber Chem. Tech.* **1978**, *51*, 406.
12. Tomalia, D.; Durst, H. *Topics in Current Chemistry* **1993**, *165*, 193.
13. Hild, G.; Kohler, A.; Rempp, P. *Eur. Polym. J.* **1980**, *16*, 525.
14. Geiser, D.; Höcker, H. *Macromolecules* **1980**, *13*, 653.
15. Hild, G.; Strazielle, C.; Rempp, P. *Eur. Polym. J.* **1983**, *19*, 721.
16. Schappacher, M.; Deffieux, A. *Makromol. Chem., Rapid Com.* **1991**, *12*, 447.
17. Schappacher, M.; Deffieux, A. *Macromolecules* **1992**, *25*, 6744.
18. Rique-Lurbet, L.; Schappacher, M.; Deffieux, A. *Macromolecules* **1994**, *27*, 6318.
19. Mougin, N.; Rempp, P.; Gnanou, Y. *J. Polym. Sci., Poly. Chem. Ed.* **1993**, *31*, 1253.
20. Mougin, N.; Rempp, P.; Gnanou, Y. *Makromol. Chem.* **1993**, *194*, 2553.
21. Chaumont, P.; Beinert, G.; Herz, J.; Rempp, P. *Polymer* **1981**, *22*, 663.

Tannins

See: Vegetal Biomass (2. Oligomers and their Polymerization)

Tar

See: Coal Tar Pitch

TAUTOMER POLYMERIZATION

Seizo Masuda* and Tahei Tomida
Department of Chemical Science and Technology
Faculty of Engineering
University of Tokushim

Isomerization polymerizations are defined as addition polymerizations that are accompanied by intramolecular isomerization during reaction.[1,2] These polymerizations involve rearrangement of propagating species or monomers.

Structural isomers which differ significantly in the relative positions of their atoms are called tautomers. Tautomerism is divided into two classes: cationotropy and anionotropy. When tautomerization involves only the shift of a proton, it is called prototropy or proton tautomerization. Prototropy involves interconversion, such as internal olefin, ketimine–enamine, keto–enol, nitro–acinitro (oxime–nitroso), and lactam–lactam.

2-Butene (internal olefin type tautomer) is polymerized by the Ziegler–Natta catalysis to give poly(1-butene).[3]

The Ziegler–Natta catalyst also serves as a tautomerization catalyst. The internal olefin-type tautomer leads to the polymer without possessing the structure of the original monomer. Endo et al. have published a series of studies on monomer-isomerization polymerization.[4]

PREPARATION OF UNSATURATED KETO–ENOL TAUTOMERS

Methacryloylacetone was prepared by condensation of ethyl methacrylate and acetone.[5,6] However, the yield was not very good and this reaction could not be applied to the preparation of acryloylacetone.

Reaction of (meth)acryloyl chloride with sodium acetylacetonate leads to (meth)acryloylacetylacetone.[7,8]

TAUTOMER PROPERTIES

Though acryloylacetone (AA) has three theoretically possible forms, one ketonic and two enolic, the selective decoupled ^{13}C-NMR spectrum revealed that it occurs only in the enolic form (2, **Scheme I**).[9] That is to say, the equilibrium of AA shifts to the enol.

The stronger the substituent, R, in the structure CH_2=CH–CO–CH_2–R pulls on the electron, the more the enolic form predominates.

*Author to whom correspondence should be addressed.

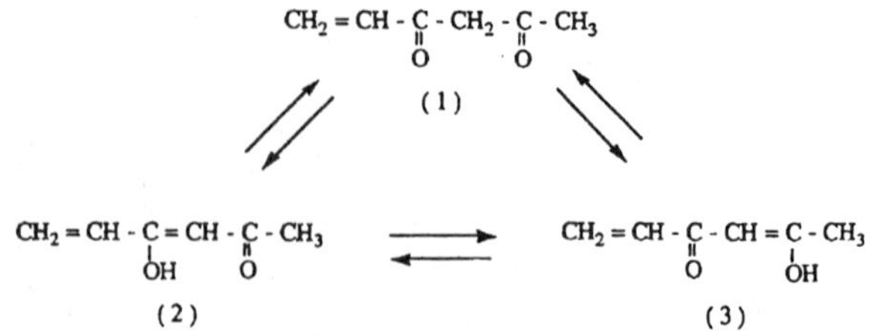

$$CH_2 = CH - \overset{O}{\underset{\parallel}{C}} - CH_2 - \overset{O}{\underset{\parallel}{C}} - CH_3$$

(1)

I

$$CH_2 = CH - \overset{OH}{\underset{\mid}{C}} = CH - \overset{O}{\underset{\parallel}{C}} - CH_3 \quad \longrightarrow \quad CH_2 = CH - \overset{O}{\underset{\parallel}{C}} - CH = \overset{OH}{\underset{\mid}{C}} - CH_3$$

(2)   (3)

APPLICATIONS OF THE POLYMERS

Adsorption and Separation of Metal Ions

It is well known that β-diketones such as acetylacetone possess a complexing affinity for a number of metal ions.

The acid-dissociation constant and binding constants of polyacryloylacetone (PAA), which were prepared by ionic polymerization of methyl vinyl ketone and its simultaneous condensation with acetic anhydride, were determined by use of a modified Bjerrum technique.[10] On the basis of the displacement of a proton from the chelate acid by the metal, it was found that the polymer is equivalent to its monomeric analog (acetylacetone) for Cu(II), and is bound more strongly with UO_2(II) by three orders of magnitude than its monomeric analog.

Tomida et al. also reported on a dissociation constant and overall binding constants of the polymer from AA.[11] The values obtained are different from those indicated by Hoeschele. The difference is probably a result of the solvent used. The adsorption equilibria of PAA with metal ions conform to the Langmuir isotherm. The maximum adsorption is shown in the following order (**Equation 1**):

$$Hg(II) \gg Cu(II) > Ni(II) > Co(II) > Cd(II) \qquad 1$$

The magnitude of the stability constants is also in agreement with this order. The separation of Cu(II) from Co(II) and Ni(II) was successfully achieved by a stepwise decrease in pH of the eluent. Pairs of light rare-earth metal ions were also separated by chromatography using glass beads coated with PAA as the stationary phase and HCl solutions as the eluent.[12]

Metal Complexes of Diketone Polymer as Catalysts

In the last two decades, much work on metal complexes of polymers as catalysts has been done in connection with hybridization of catalysts.[13,14] However, there are not many reports on diketone polymers as catalysts. Polymer-supported oxobis(pentane-2,4-dionato)vanadium(IV) could catalyze the oxidation of dimethyl sulfoxide, dibutyl sulfide, and cyclohexane in the presence of *t*-butyl hydroperoxide.[15]

Other Research

In connection with electrical energy and laser systems, the fluorescence properties of lanthanide–polymer complexes are of considerable interest.[16] Okamoto et al. reported that the electroconductivity of poly(β-diketone) silylated with dichlorodimethylsilane is ~10^{-9} Ω^{-1} cm^{-1}.[17]

REFERENCES

1. Kennedy, P. J.; Otsu, T. *Adv. Polym. Sci.* **1970**, *7*, 382.
2. Cesca, S. *Encycl. Polym. Sci. Eng.* **1989**, *8*, 463.
3. Otsu, T.; Shimizu, A.; Imoto, M. *J. Polym. Sci.* **1966**, *A1, 4*, 1579.
4. Endo, K.; Fujii, K.; Otsu, T. *Macromol. Chem. Phys.* **1994**, *195*, 1913.
5. Teyssie, Ph.; Smets, G. *Makromol. Chem.* **1958**, *26*, 245.
6. Despic, A. R.; Kosanovic, Dj. *Makromol. Chem.* **1959**, *29*, 151.
7. Degeiso, R. C.; Donaruma, L. G. *J. Appl. Polym. Sci.* **1963**, *7*, 1515.
8. Soutif, J. C.; Rao, C. J.; Casals, P.; Broose, J. C. *Makromol. Chem., Rapid Commun.* **1985**, *6*, 197.
9. Masuda, S.; Tanaka, M.; Ota, T. *Chem. Lett.* **1984**, 1327.
10. Hoeschelel, G. K.; Andelman, J. B.; Gregor, H. P. *J. Phy. Chem.* **1958**, *62*, 1239.
11. Tomida, T.; Tomida, M.; Nishihara, Y.; Nakabayashi, I.; Okazaki, T.; Masuda, S. *Polymer* **1990**, *31*, 102.
12. Tomida, T.; Inagawa, K.; Masuda, S. *J. Chem. Eng. Japan* **1993**, *26*, 79.
13. Ciardelli, F.; Carlini, C.; Pertici, P.; Valentini, G. *J. Macromo. Sci.-Chem.* **1989**, *A26*, 327.
14. Tsuchida, E. *Macromol. Rev.* **1982**, *16*, 397.
15. Bhaduri, S.; Ghosh, A.; Khwaja, H. *J. Chem. Soc., Dalton Trans.* **1981**, 447.
16. Okamoto, Y. *J. Macromo. Sci.-Chem.* **1987**, *A24*, 455.
17. Okamoto, Y.; Hwang, E. F.; Wang, M. C. *J. Poly. Sci., Poly. Lett. Ed.* **1985**, *23*, 285.

TELECHELIC OLIGOMERS (with Cyclic Onium Salt Groups)

Yasuyuki Tezuka
Department of Organic and Polymeric Materials
Tokyo Institute of Technology

The term "telechelics" was produced from the Greek words *tele* = end and *chelos* = claw, and referred initially to oligomers possessing two identical reactive groups at both chain ends of a linear macromolecule for the application of crosslinking and chain extension reactions.[1]

Telechelics prepared by means of a living polymerization technique, possessing the uniformity in the molecular weight distribution, are preferred in order to achieve strict structural control over the final product. In addition, the living polymerization technique can offer a means to introduce functional groups quantitatively at the chain ends by appropriate termination reaction.

TELECHELICS WITH CYCLIC ONIUM SALT GROUPS

Preparation

By making use of telechelics having cyclic onium salt groups, a novel reaction process has been developed to achieve an efficient synthesis of well-defined multidimensional and multicomponent macromolecular structures.[1]

Thus a series of moderately strained cyclic onium salt groups such as 4-membered (azetidinium), 5-membered cyclic (pyrrolidinium), and 6-membered bicyclic (quinuclidinium) ammonium salts as well as 5-membered cyclic sulfonium (tetrahydrothiophenium) salt groups were introduced either at a single chain end or both chain ends of polytetrahydrofuran, poly(THF), through the reaction of the oxonium salt end group of a living poly(THF) produced with triflic acid ester or a triflic anhydride as initiator, respectively, with the corresponding cyclic amines or a sulfide.

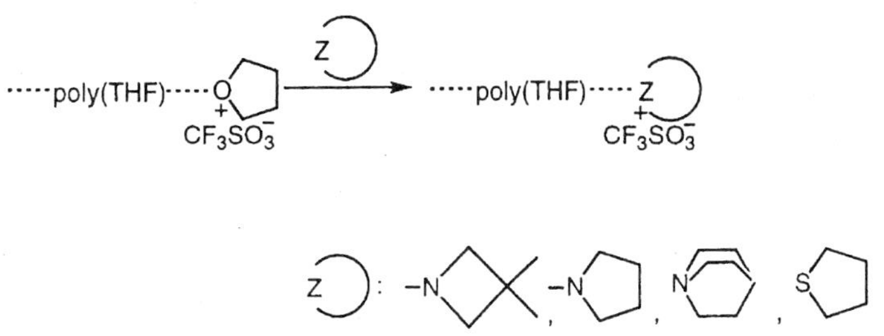

I

The selective and quantitative nucleophilic substitution of oxonium salt produced uniform size mono- and bifunctional telechelic poly(THF)s (**Scheme I**).[2,3]

Synthesis of Star Polymers, Polymacromonomers, and Model Networks

A simple precipitation of a THF solution of a new telechelic polymer, the poly(THF) having a 4-membered cyclic onium salt group, into an aqueous solution containing an excess amount of plurifunctional carboxylate salt, such as sodium 1,2,4,5-benzenetetracarboxylate, resulted in the product comprising a star polymer with a branch number of only four without the contamination of those with two or three branches. Thus the simple procedure provided a star polymer having a branch number exactly equal to the functionality of the carboxylate used in the reaction. A star polymer of a defined number of branches was obtained in an almost pure form after three-times repetition of the precipitation treatment.[4]

A model network, a crosslinked macromolecule possessing defined branch numbers at the crosslinking points and controlled segment length between the crosslinking points, is produced in principle by employing bifunctional telechelic polymers, including living polymers, by procedures applied for synthesis of a star polymer. Thus the reaction of tri- and tetracarboxylates with a bifunctional telechelic poly(THF) having cyclic onium salt groups was performed to result in gel products in almost quantitative yields.[4]

The swelling degree of the gel products obtained with tetrafunctional carboxylate was substantially different from that obtained with the trifunctional one. This result evidently indicates a significance of the control over the structural parameters of network polymers to modulate the fundamental gel property.

REFERENCES

1. Tezuka, Y. *Prog. Polym. Sci.* **1992**, *17*, 471.
2. Tezuka, Y.; Goethals, E. J. *Eur. Polym. J.* **1982**, *18*, 991.
3. Tezuka, Y.; Goethals, E. J. *Makromol. Chem.* **1987**, *188*, 783.
4. Tezuka, Y.; Goethals, E. J. *Makromol. Chem.* **1987**, *188*, 791.

TELECHELIC POLYOXETANES

Hemant Desai
Defence Research Agency

Early work on telechelic polymers centered around polybutadiene, with applications ranging from elastomeric networks in rocket-propellant binders to coatings; this work is well reviewed by Athey.[1,2] Research to prepare novel substituted telechelic oxetanes has been limited because of the relatively high cost of the monomers, compared to other cyclic ethers, such as oxiranes. Hence, only poly(3,3-bis(chloromethyl) oxetane) has been significantly used in industrial applications, largely as a chemically resistant thermoplastic for laboratory apparatus.[3]

Nevertheless, some research has been carried out to prepare oxetane-based materials for applications, such as rocket propellant and explosive compositions. Much of the syntheses of these types of substituted polyoxetane has been investigated by Manser.

A convenient, scaleable, and, above all, safe procedure of producing hydroxytelechelic poly(3,3-(nitratomethyl) methyl oxetane) (polyNMMO) is presented. This uses an incremental monomer-addition (IMA) technique, but proceeds predominantly under active chain end (ACE) conditions. A procedure for the production of a linear, hydroxytelechelic oligomer of NMMO by an activated monomer mechanism (AMM) is also given.

PREPARATION

Scheme I illustrates the overall mechanisms that are believed to operate in the polymerization of NMMO and strained cyclic ethers in general. Incremental monomer addition has not been widely used to control molecular weight, but this communication shows that telechelic oxetanes may be prepared by this method. Moreover, in the field of energetic monomers, such as NMMO, it is desirable to control the exotherm because of ring opening so that decomposition of the energetic moiety is not effected, hence the risk of explosion during polymerization is minimized. "All-monomer-in" (AMI) techniques are riskier than IMA methods that enable safer scale-up and production of these materials. A further advantage of IMA addition for the ACE polymerization using a $BF_3 \cdot OEt_2$/alcohol system is that the alcohol fragment is readily incorporated into the polymer backbone so that difunctional (butandiol as alcohol), trifunctional (glycerol or metriol), and tetrafunctional hydroxy-terminated polymers may be formed simply by varying the alcohol.

APPLICATIONS

The main applications of polyNMMO, both as hydroxytelechelic polymer and nitratotelechelic oligomer, are expected to be as ingredients in propellant, pyrotechnic, or explosive compositions. In propellant compositions, a T_g of –25°C is deemed to be high for a binder holding together an energetic filler because it may be used in extreme environmental conditions, either on Earth or in space. In these conditions it is imperative that the binder does not become glassy, as this tends to increase the sensitivity of the composition. However, a suitable plasticizer can reduce the T_g below a generally acceptable

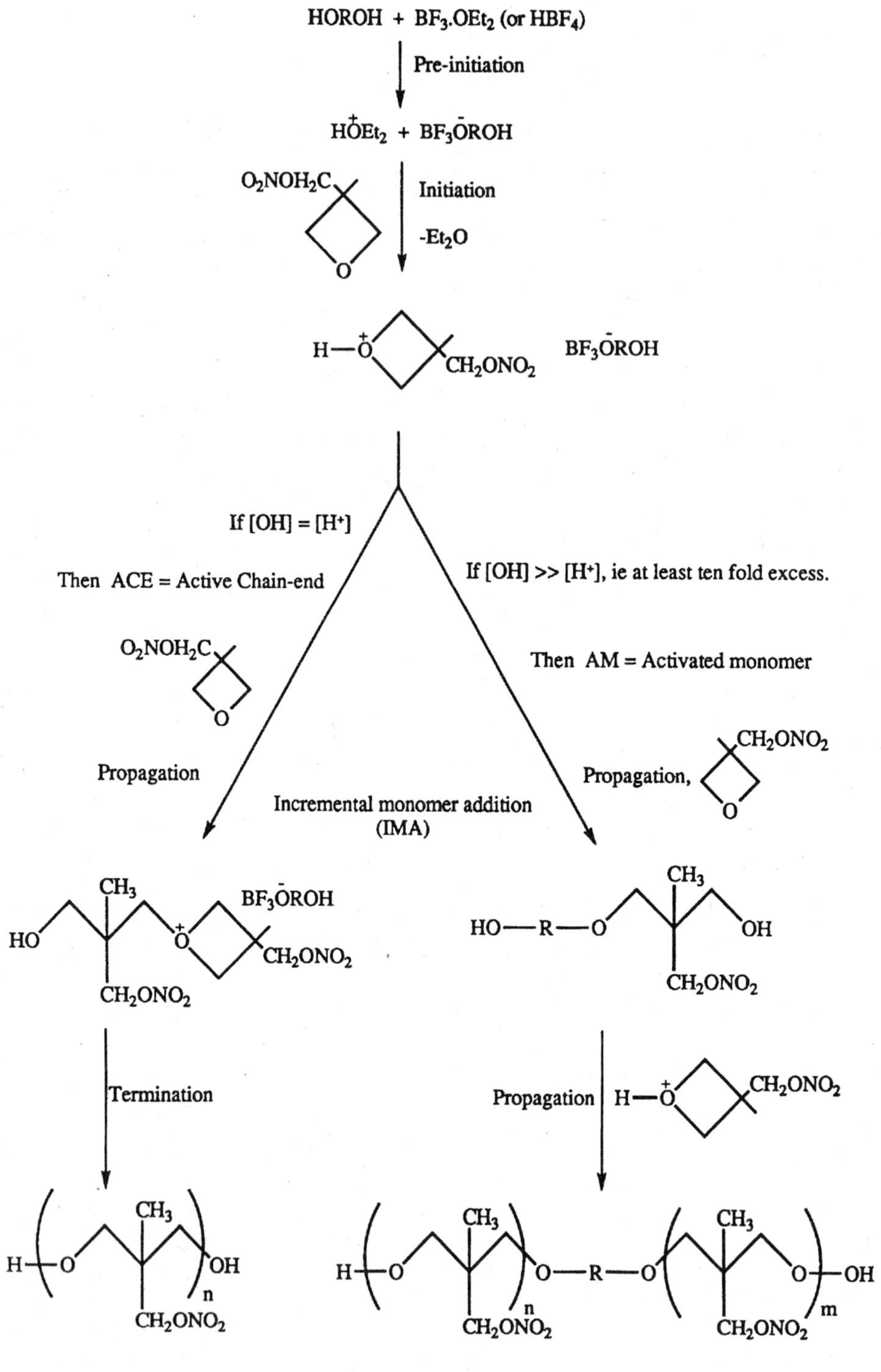

SCHEME I. Overall mechanisms for the polymerization of NMMO.

figure of –50°C and, of course, the use of an energetic binder can offer the advantage of better performance. Similar considerations apply to explosive compositions, although in some cases safety is a higher priority. The use of polyNMMO, cured with isocyanates, in forming polymer-bonded explosives (PBXs) is expected to make such materials significantly safer, with respect to impact and heat, compared to materials in which inert binders, such as hydroxy-terminated polybutadienes are used. This is a result of a lower requirement of the high-energy filler in order to achieve a given performance.

The nitratotelechelic oligomer may be used as an energetic and fully miscible plasticizer for the parent homopolymer, that is, PolyNMMO, other nitrate-ester based polymers, or energetic binders in general. Another potential application of this type of compound is as a fuel additive for automobile engines, where it may serve to oxidize carbonaceous compounds thereby cleaning the engine.

REFERENCES

1. Athey, Jr., R. D. *Prog. Organic Coatings* **1979**, 7, 289–329.
2. Athey, Jr., R. D. *J. Coatings Technol.* **1982**, Vol. 54, 690, p 47–50.
3. Boardman, H. In *Encyclopaedia of Polymer Science and Technology* Mark, M. F.; Gaylord, N. G., Eds.; Wiley-Interscience: New York, 1968, Vol. 9, p 668.

Telechelics

See: *Aminosiloxanes*
Block Copolymer-Like Structures (End-Group Ionic Interactions in Blends)
Carbenoid Polycondensation
End Functionalized Oligomers (by Chain-Transfer Technique)
Free Radical Iniferters
Functional Polymers (By Carbocationic Polymerization)
Functional Polymers (Telechelics and Macromonomers)
Functionalized Polymers (Comb, Rotaxanic, and Dendrimeric Structures)
Functionalized Polymers (Telechelics and Macromonomers, via Cationic Polymerization)
Graft and Network Copolymers (by Macromolecular Ion Coupling Reactions)
Ionomers (Overview)
Ionomers (Solution Behavior)
Reactive Oligomers (Overview)
Tailor-Made Polymers
Telechelic Oligomers (with Cyclic Onium Salt Groups)
Telechelic Polyoxetanes

Template Polymerization

See: *Ladder Polymers (Methods of Preparation)*
Matrix Polymerization (Overview)
Optically Active Polymers (Overview)
Poly(vinylpyridinium salts)
Stereocomplexes (Between Enantiomeric or Diastereomeric Polymers)
Template Polymerization (Overview)
Template Polymerization (Kinetics)

TEMPLATE POLYMERIZATION (Overview)

Stefan Polowiński
Department of Physical Chemistry of Polymers
Technical University of Łódź

Template or matrix polymerization belongs to the general field of polymerization of organized systems and can be defined as the synthesis of polymers in which specific interactions between performed macromolecules and a growing chain take place.

The influence of the template on the process and the product is usually called "template effect" or "chain effect." It can be revealed by comparing the template process and products of the reaction with conventional polymerization carried out under the same conditions, but in this system the template was replaced by a low molecular weight analog.

CHAIN-TEMPLATE POLYMERIZATION

In the case of chain-template polymerization, the reacting system includes: template, monomer, solvent, and initiator. If the template contains groups with a structure complementary to the groups existing in the monomer, a complex template-monomer could be formed. In such a complex, monomer units and template are connected by strong intermolecular forces (e.g., hydrogen bonds). The monomer can also be connected with the template by covalent bonds.

For both systems with covalent bonds or strong intermolecular forces ordering occurs before polymerization starts. Polymerization in this case is called template polymerization type I.[2–4] The first example of template polymerization was published by Kämmerer and colleagues.[5–7]

An example of type-I template polymerization is the synthesis suggested by Bamford that was examined by Jantas.[1,8] Poly(vinyl alcohol) esterified by methacrylic chloride was used as a template.

Another mechanism of template polymerization takes place if there is no preferential adsorption of monomer onto the template, but growing oligomer, which is created in the bulk, can complex with the template. This mechanism is called the type-II, or "pick-up" mechanism.[2,3]

Many radical polymerization systems were examined by Challa et al.[9–11] A number of articles have been devoted to polymerization of methyl methacrylate (MM) in the presence of poly(methyl methacrylate) (PMM). The role of tacticity of the template used was stressed.

Ferguson and Shah described polymerization of acrylic acid (AA) in the presence of poly(ethylene imine) (PEI) and poly(ethylene oxide) (PEO).[12]

Polymerization of methacrylic acid (MA) in aqueous solutions in the presence of poly(vinylpyrrolidone) (PVP) was investigated by Shavit and Cohen.[13]

Template polymerization of methacrylic acid in an aqueous system using PVP as a template has been described.[14–17]

TEMPLATE COPOLYMERIZATION

The most interesting problem concerning template copolymerization deals with the possibility of synthesizing copolymers with a defined sequence of units.

Synthesis of various multimonomers and their copolymerization with styrene, acrylonitrile, or acrylic acid has been described.[18–26] The problem of the relation between reaction conditions and structure of copolymers obtained from multimethacrylate and styrene, as well as from multimethacrylate and acrylic acid, has been discussed.[27]

As a result of copolymerization of methacrylic acid with styrene carried out in the presence of PEG as the template, it was determined that changing the concentration of the template changes the reactivity ratio, both r_1 and r_2. Additionally, regarding the copolymerization of methacrylic acid with methyl methacrylate, it was found that the presence of PEG significantly influences the reactivity ratios of both monomers.[28] Copolymerization of methacrylic acid with acrylic acid in the presence of the same template (PEG) was also investigated.[29]

REFERENCES

1. Bamford, C. H. In *Development in Polymerization-2*; Haward, R. N., Ed.; Applied Science: London, 1979.
2. Challa, G.; Tan, Y. Y. *Pure Appl. Chem.* **1981**, *53*, 627.
3. Tan, Y. Y.; Challa, G. In *Encyclopedia of Polymer Science and Engineering*; Mark, H. F., et al., Eds.; John Wiley & Sons: New York, 1989, Vol. 16, 554.
4. Tan, Y. Y. In *Comprehensive Polymer Science*; Allen, G.; Bevington, J. C., Eds.; Pergamon: New York, 1989; Vol. 3, p 245.
5. Kämmerer, H.; Jung, A. *Makromol. Chem.* **1966**, *101*, 284.
6. Kämmerer, H.; Ozaki, S. *Makromol. Chem.* **1966**, *91*, 1.
7. Kämmerer, H.; Shukla, I.; Onder, N.; Schurmann, G. *J. Polym. Sci., Polym. Symp.* **1967**, *22*, 213.
8. Jantas, R.; Połowiński, S. *J. Polym. Sci., Polym. Chem. Ed.* **1986**, *24*, 1819.
9. Buter, R.; Tan, Y. Y.; Challa, G. *J. Polym. Sci., Polym. Chem. Ed.* **1972**, *10*, 1031.
10. Buter, R.; Tan, Y. Y.; Challa, G. *J. Polym. Sci.* **1973**, *11*, 1003, 1013, 2975.
11. Gons, J.; Vorenkamp, E. J.; Challa, G. *J. Polym. Sci., Polym. Chem. Ed.* **1975**, *13*, 1699.
12. Ferguson, J.; Shah, S. A. O. *Eur. Polym. J.* **1968**, *4*, 611.
13. Shavit, N.; Cohen, J. In *Polymerization in Organized Systems*; Elias, H. G., Ed.; Gordon and Breach: London, 1977; p 213.
14. Matuszewska-Czerwik, J.; Połowiński, S. *Eur. Polym. J.* **1990**, *26*, 549.
15. Matuszewska-Czerwik, J.; Połowiński, S. *Eur. Polym. J.* **1991**, *27*, 743–746.
16. Matuszewska-Czerwik, J.; Połowiński, S. *Eur. Polym. J.* **1991**, *27*, 133.
17. Matuszewska-Czerwik, J.; Połowiński, S. *Eur. Polym. J.* **1992**, *28*, 1481.
18. Połowiński, S.; Janowska, G. *Eur. Polym. J.* **1975**, *11*, 183.
19. Połowiński, S. *Eur. Polym. J.* **1978**, *14*, 563.
20. Połowiński, S. *Polimery* **1972**, *17*, 409.
21. Połowiński, S.; Janowska, G. *Polimery* **1972**, *17*, 464.
22. Jantas, R.; Połowiński, S. *J. Pol. Sci., Polym. Chem. Ed.* **1986**, *24*, 1819.
23. Jantas, R. *Polymer International* **1993**, *32*, 357.
24. Jantas, R.; Połowiński, S. *Acta Polym.* **1989**, *40*, 225.
25. Jantas, R.; Połowiński, S.; Strobin, G. *Polymer International* **1995**, *37*, 315.
26. Jantas, R. *Acta Polym.* **1991**, *42*, 539.
27. Jantas, R. J. M. S. *Pure Appl. Chem.* **1992**, *A29*, 557.
28. Połowiński, S. *Eur. Polym. J.* **1983**, *19*, 679.
29. Połowiński, S. *Acta Polym.* **1992**, *43*, 99.

TEMPLATE POLYMERIZATION (Kinetics)

James Ferguson
University of Strathclyde

J. Smyllie
Glasgow Caledonian University

The term *template polymerization*, often called *replica* or *matrix polymerization*, will be applied to those systems in which polymer chains propagate along macromolecular templates for at least part of their growth. Such propagation may affect various features of the formed polymer as compared with the situation in the absence of template, that is, the blank polymerization.

There are three main characteristics associated with template polymerization:

1. Structural and conformational features in the template should be reflected in corresponding features in the polymer being formed, that is, the daughter polymer.[1–4]
2. Enhancement in the rate of polymerization in the presence of template polymer as compared to that of the blank polymerization.[5–8]
3. Formation of a polymer–polymer complex during polymerization.[2,5,7,9]

MECHANISM OF TEMPLATE POLYMERIZATION

In homopolymerization (or copolymerization), the mechanism of polymerization of monomer can be of various types (e.g., free radical, ionic, or step growth) depending on the reaction conditions and monomer.

In a review paper by Tsuchida and Osada, classification of template polymerization was made according to the mode of interaction between the component species in the system.[10] However in another review, Challa and Tan have suggested that the mechanism of template polymerization for the majority of template systems can be classified into two main idealized types.[11]

In Mechanism I, monomer is pre-absorbed by the template, which upon initiation results in propagation taking place along the template exclusively. In Mechanism II, active oligomers created in "free" solution complex with template macromolecules upon reaching a critical chain length, and continue to propagate along the template by adding monomer from the surrounding solution. This formation of a critical chain length of the polymeric radical is essential for a mechanism of type II.

TYPE I TEMPLATE POLYMERIZATION

An interesting example of a type I template reaction is the polymerization of 4-vinylpyridine (4-VPy) in the presence of polyacids. Investigations into this were a direct result of an extension on the findings of Kabanov et al. on the reaction between 4-VPy and alkyl halides in organic media.[7,12]

Salamone, Snider, and Fitch have questioned the mechanism of initiation proposed by Kabanov et al. for the spontaneous polymerization of 4-VPy in the presence of alkyl halides and polymeric acids, even though the propagation process suggested was similar.[7,13,14] It was also shown that in the presence of weak acids, including PAA in aqueous solution, the formation of an ionene polymer occurred.

TYPE II TEMPLATE POLYMERIZATION

There are numerous examples of type II reactions, a prime example being the stereospecific free-radical polymerization of methyl methacrylate (MMA) in the presence of tactic poly(methyl methacrylate) (PMMA) in dimethylformamide (DMF). This system has been extensively studied by Challa and his co-workers, who have shown that the presence of the preformed polymer not only influences the reaction kinetics but also the microstructure of the newly formed polymer.[3,4,6,11,15–18]

N-vinylpyrrolidone template polymerizations are also generally of type II. They have been examined using both poly(methacrylic acid) (PMMA) and poly(acrylic acid) (PAA) templates. Bartels and co-workers studied the influence of PMAA on the polymerization of *N*-vinylpyrrolidone (NVP) in dimethylformamide (DMF) at 60°C.[19] It was found that the presence of PMAA produced a rate enhancement that was more pronounced with increasing chain length and syndiotacticity of the template. Later work by Challa et al., involving interferometry in combination with equilibrium dialysis, showed that no preferential adsorption of NVP by PMAA occurred.[11] Indeed, the solvent DMF, rather than the monomer, was shown to be preferentially associated.

The polymerization of NVP in the presence of PAA has been studied by Ferguson and Rajan, also in DMF, because it had been previously shown that in aqueous solvents the monomer was hydrolyzed to pyrrolidone and acetaldehyde.[20]

KINETIC STUDIES ON TEMPLATE POLYMERIZATIONS

Ferguson and Połowiński have discussed the problem of the interpretation of rate constant data from template polymerization systems.[21]

They noted that the rate constant for termination, k_t, for template polymerization is often lower by a few orders of magnitude than k_t for the blank reaction. Also, the rate constant for propagation, k_p, apparently decreases in the presence of the template. However k_p^2/k_t is usually higher for the template process, and this is given as the reason that the overall rate of the template polymerization is higher.

A critical inspection of the kinetic results in template polymerization is very important. First of all the rate of template polymerization should be compared not only with the polymerization rate of the same monomer in the same solvent, but also with the system in which the low molecular weight analogue of the template is present. Change in chemical composition of the solvent used can also change the velocity of polymerization. Another very important parameter that should be taken into account is the solution viscosity. The possibility of the Tromsdorff effect creating an enhancement in rate of polymerization must always be considered. The template, as a macromolecular compound, changes the viscosity in comparison with the viscosity of a polymerizing system in a pure solvent. It is well known that increase of viscosity can change the rate constant of termination and hence the overall rate of polymerization. In addition, in many systems, as a product of template polymerization, insoluble complex is formed. In that case it is obvious that the character of the polymerization and its kinetics change. Description of polymerization kinetics in heterogeneous systems is complicated, more so because the structure of the complex formed is not always very well defined. In very dilute systems, fortunately, specific interactions do lead to a precise complex structure.

REFERENCES

1. Ferguson, J.; Al-Alawi, S.; Granmayeh, R. *Eur. Polym. J.* **1983**, *19*, 475.
2. Miura, Y.; Kakui, T.; Kinoshita, M. *Makromolek. Chem.* **1975**, *176*, 1567.
3. Buter, R.; Tan, Y. Y.; Challa, G. *J. Polym. Sci. A1* **1972**, *10*, 1031.
4. Buter, R.; Tan, Y. Y.; Challa, G. *J. Polym. Sci., Polym. Chem.* **1973**, *11*, 1003.
5. Ferguson, J.; Shah, S. A. O. *Eur. Polym. J.* **1968**, *4*, 343.
6. Gons, J.; Vorenkamp, E. J.; Challa, G. *J. Polym. Sci.; Polym. Chem.* **1975**, *13*, 1699.
7. Kabanov, V. A.; Aliev, K. V.; Kargina, O. V.; Patrikeeva, T. I.; Kargin, V. A. *J. Polym. Sci.* **1967**, *C16*, 1079.
8. Bamford, C. H.; Shiiki, Z. *Polymer* **1968**, *9*, 5961.
9. Shima, K.; Kakui, Y.; Kinoshita, M.; Imoto, M. *Makromolek. Chem.* **1972**, *154*, 247.
10. Tsuchida, E.; Osada, Y. *J. Polym. Sci.; Polym. Chem.* **1975**, *13*, 559.
11. Challa, G.; Tan, Y. Y. *Pure Appl. Chem.* **1981**, *53*, 627.
12. Kabanov, V. A. *Pure Appl. Chem.* **1967**, *15*, 391.
13. Salamone, J. C.; Snider, B.; Fitch, W. L. *J. Polym. Sci., Polym. Lett.* **1971**, *9*, 13.
14. Salamone, J. C.; Snider, B.; Fitch, W. L. *J. Polym. Sci., Part A-1.* **1971**, *9*, 1493.
15. Buter, R.; Tan, Y. Y.; Challa, G. *J. Polym. Sci.; Polym. Chem.* **1973**, *11*, 1013.
16. Buter, R.; Tan, Y. Y.; Challa, G. *J. Polym. Sci.; Polym. Chem.* **1973**, *11*, 2975.
17. Gons, J.; Vorenkamp, E. J.; Challa, G. *J. Polym. Sci.; Polym. Chem.* **1977**, *15*, 3031.
18. Gons, J.; Straatman, L. J. P.; Challa, G. *J. Polym. Sci.; Polym. Chem.* **1978**, *16*, 427.
19. Bartels, T.; Tan, Y. Y.; Challa, G. *J. Polym. Sci.; Polym. Chem.* **1977**, *15*, 341.
20. Ferguson, J.; Ventkatraman, S. R. *Europ. Polym. J.* **1979**, *15*, 627.
21. Ferguson, J.; Połowiński, S. *Polymer Yearbook*, 1994.

Terpene Resins

See: *Polyterpene Resins*
Vegetal Biomass (1. Monomers and their Polymerization)

Tetrafluoroethylene-Containing Polymers

See: *Engineering Thermoplastics (Survey of Industrial Polymers)*
Ferroelectric Polymers (Structural Phase Transitions)
Fluorinated Plastics, Amorphors
Fluoropolymers (Surface Modification by Eximer-Laser Irradiation)
Perfluorinated Ionomers (Overview)
Polytetrafluoroethylene
Tetrafluoroethylene Copolymers (Overview)

TETRAFLUOROETHYLENE COPOLYMERS (Overview)

Teruo Takakura
Functional Products Research and Development Center
Chemicals General Division
ASAHI Glass Company, Ltd.

Polytetrafluoroethylene (PTFE) [9002-84-0], the most widely used fluoropolymer, was discovered in 1938. It has excellent thermal, chemical, electrical, and surface properties and has found a variety of applications in industrial fields. Many types of fluoropolymers have since been developed to cover the disadvantages of PTFE or to meet market needs.

From the 1950s through the 1970s, major efforts were directed toward improving processability and mechanical properties of PTFE; valuable copolymers of tetrafluoroethylene (TFE) and other fluorinated and non-fluorinated ethylenic monomers were developed and commercialized. Copolymers of TFE/hexafluoropropylene (FEP) [25067-11-2], TFE/perfluoro(alkyl vinyl ether) (PFA) [26655-00-5] are melt-processable perfluoroplastics. Alternating copolymers of TFE/ethylene (ETFE) [25038-71-5] or TFE/propylene gave unique partially fluorinated plastics or elastomers.

Sophisticated fluoropolymers have been commercialized since the mid-1970s. Copolymers of TFE and perfluoro(alkyl vinyl ether) having sulfonic ester or carboxylic ester produced materials of ion-exchange membranes for electrolytic processes in the chlor-alkali industry.[1] Soluble fluoroplastics, copolymers of fluoroethylene (TFE or chlorotrifluoro ethylene) and alkyl vinyl ether (FEVE) were also commercialized during the same period. The introduction of the hydroxy group to FEVE by copolymerizing hydroxyalkyl vinyl ether led to crosslinking coating materials, such as weather-resistant paints.[2] Other fluoropolymers commercialized during this time were thermoplastic fluoroelastomers and transparent fluoroplastics. Thermoplastic elastomers are block, or graft, copolymers that consist of an elastomeric soft segment, such as a copolymer of vinyldene fluoride (VdF) and hexafluoropropylene (HFP), and a fluoroplastic hard segment, such as poly(vinylidene fluoride) (PVdF) [24937-79-9] or ETFE. Transparent fluoroplastics are novel fluoropolymers with a cyclic structure in the main chain; they opened a new category of amorphous perfluororesins.[3] They exhibit chemical and electrical properties similar to those of PTFE, high transparency to infrared to ultraviolet light, and a very low refractive index.

Copolymers of TFE (Commercial)

Detailed properties of thermoplastic fluororesins should be referred to in other books[4] or brochures, but some properties are summarized to clarify the difference of each copolymer, together with other fluorinated thermoplastics. PFA and FEP are melt-processable perfluorinated polymers. Maintaining almost the same chemical resistance and flammability as PTFE, they have improved mechanical properties such as flexural modulus and izod impact strength, ETFE, the alternating copolymer of TFE and ethylene, has less chemical resistance and flammability compared to PTFE, but has superior mechanical properties, even to PFA and FEP. PVdF, which is not a copolymer of TFE but has the same molar ratio of C, f, and H as ETFE, shows characteristic properties owing to its alternating structure of CF_2 and CH_2.

FEP, COPOLYMERS OF TFE AND HFP

The chief advantage of FEP over PTFE is its lower melt viscosity. FEP has similar mechanical properties to PTFE and PFA at room temperature. FEP melts at about 275°C and its maximum service temperature is 200°C. On the other hand PFA melts above 300°C and can be used over the same temperature range as PTFE.

FEP is used in valves and tanks in the chemical industry, heat-exchangers, wires and cables, and flexible printed circuits. Recently, the use of FEP as an insulating material for local area networks has been increasing swiftly.

ETFE, COPOLYMERS OF TFE AND ETHYLENE

Copolymers of TFE and ethylene have a nearly 1:1 alternating structure. Consequently, they exhibit a unique combination of mechanical, chemical, and electrical properties together with excellent weathering. To improve its poor thermal-stress crack resistance, a small amount of termonomer with a bulky side chain is introduced into the copolymer.

REFERENCES

1. *Modern Chlor-alkali Technology*; Jackson, C. Ed.; Ellis Horwood, Chishcster, 1983, Vol. 2.
2. Munekata, S. *Progress in Organic Coatings* **1988**, *16*, 113.
3. Nakamura, M.; Oharu, K.; Sugiyama, N. *Preprints (15th Symposium on Fluorine Chemistry)* **1990**, *76*; Resnic, P. R. *Polymer Preprints (ACS Div. Polym. Chem.)* **1990**, *31*, 312.
4. Kirk-Othmer, *Encyclopedia of Chemical Technology*, 3rd ed.; "Fluorine Compounds, Organic," Wiley-Interscience: Vol. 11, p 1.

TEXTILE FIBERS (Shingosen)

Miyoshi Okamoto
Toray Industries, Incorporated

Kanji Kajiwara*
Kyoto Institute of Technology

The market for polyester fiber first appeared in the early 1960s, and expanded from 20,000–30,000 tons per year in 1960 to 5 million tons in 1993. The general demand for good natural fibers has fueled this expansion. There were quite a few attempts in the 1960s to mimic natural silk with fibers having a non-circular cross-section that affords the luster of silk and good resilience. Toray developed a silk-like polyester fabric (Sillook™) by the alkali weight reduction of T-shaped cross-sectional fiber. Sillook™ was a commercial success from 1970 till 1985.

Because the physical characteristics of polyester best imitate silk (for example, the high Young's modulus), the majority of research in this area focused on polyester filament yarn.

Fashion trends evolved from causal to feminine, and thin fabrics woven from polyester appeared in the market in late 1988. These new fabrics are termed *shingosen* (literally, new synthetic fiber) and surpass silk to some extent with respect to hand feel, drape, and shape retention.

Shingosen development started with the pursuit of silk-like fiber from polyester.

PROPERTIES (CONCEPT AND DEFINITION OF SHINGOSEN)

The term *shingosen* appeared in the latter half of 1987, and it was publicly recognized in 1988. No explicit definition was

*Author to whom correspondence should be addressed.

given for shingosen, which refers to products of novel fabric characteristics developed after 1986. Shingosen is made primarily of filaments and can be classified according to four types. It possesses a characteristic texture and hand feel markedly different from conventional synthetic and even natural fibers. Shingosen is the product not of a new technology, but of the integration of existing technologies from spinning to finishing.

Commercially available shingosen products are classified into four categories: (1) peach skin, (2) new silky, (3) new spun, and (4) dry touch, where each class is visually presented by the scanning electron microscopic pictures.

Peach skin is considered a natural extension of the ultra-fine fiber application stimulated by the commercial success of Ecsaine™ (an artificial suede-like leather).

New silky is represented by Sillook Royal S™, which has a fine longitudinal slit on polyester filament with a tri-lobal cross-section produced by the combination of the conjugate spinning and partial dissolution technology. The slit has the effect of silk-scrooping and the tri-lobal cross-section affords crispness to the fabric.

The new spun type, represented by Malor™, is a worsted-like fabric where a dual yarn structure is constituted of thick filaments of uneven deniers in the core surrounded by ultra-fine filaments. The fabric of new spun type is voluminous and has an elasticity suitable to tailoring into a suit, as well as into a kimono.

The dry touch type feels dry and cool owing to microcraters and micro-voids produced by inorganic microparticles dispersed in polymer prior to spinning and to surface dissolution after spinning. The example (CEO α™) shows how the rough surface and non-circular cross-section afford cool and dry, hemp-like hand feel. Minute gaps between filaments of various cross-sections promote good moisture absorbency.

APPLICATIONS

Shingosen is applied to many men's and ladies' fashions, ranging from formal suits to casual shirts.

TEXTILE FIBERS (Structure and Properties)

M. A. Wilding and J. W. S. Hearle
Department of Textiles
UMIST

The overwhelming majority of the fiber market, which includes general textile products, consists of "partially ordered, partially oriented assemblies of linear polymer molecules." It divides into three categories—natural, regenerated, and synthetic polymer fibers—and consists of six chemical types—cellulosic, protein, polyamide, polyester, acrylic, and polypropylene. A new generation of high-performance fibers comprises highly oriented and highly crystalline linear polymer materials and the 2-D polymer system in carbon fibers.

POLYMER FIBER STRUCTURES

Protein Fibers

Wool and hair are composed of keratin, a mixture of proteins.

Silk is extruded by the silkworm as a natural block copolymer, fibroin, in which simple sequences form crystalline regions with the β–lattice, and other blocks form amorphous regions.

Cellulose Fibers

The cellulose molecule comprises glucose residues linked by oxygen, –O–, atoms.

Natural Plant Fibers

Cotton is one of the purest native cellulose fibers, although small amounts of other substances (hemicellulose, gums, etc.) are also present.

Flax, jute, sisal, hemp, and similar native cellulose fibers are less pure than cotton, and are multi-cellular, so that they can be easily fibrillated. The helical angle of fibrils in the cell walls is usually much smaller (ca 10°), which gives higher intrinsic stiffness.

Regenerated Fibers

Viscose rayon fibers are produced by converting cellulose into a derivative soluble in sodium hydroxide. Extrusion into an acid bath causes precipitation, which is followed by stretching.

The derivative fiber, cellulose acetate, is also dry spun from solution. The molecules are only partially acetylated, and their irregularity leads to very low crystallinity.

"First-Generation" Synthetic Polymer Fibers

Melt-Spur Fibers

The principal melt-spun fibers are the condensation polymers, nylon 6, nylon 66, and poly(ethylene terephthalate), and the addition polymer, poly(propylene).

There has been limited use of other polyamides, polyesters, and polyolefins, some copolymerization and blending, and the incorporation of other chemical groups to give special features such as particular dye receptivity.

Unlike natural fibers, where there is a clearly defined cellular architecture dictated by genetics, in synthetic fibers the structure is controlled by thermomechanical operations during the manufacturing process. This leads to an enormous diversity of possible fiber morphologies.

What is known with some certainty is that the melt-spun fibers have a density that is equivalent to a two-phase combination of ordered (crystalline) and disordered (amorphous) polymer molecules, with the ordered fraction occupying somewhere between 30% and 50% of the fiber mass.

Fibers from Solution

Acrylic fibers have acrylonitrile as their main monomer, copolymerized with up to 15% of various minor components.

Acrylic fibers are formed from solution, usually by wet-spinning into a coagulating bath, although dry-spinning (solvent evaporation) has been used.

There has been limited use of other vinyl fibers, such as poly(vinyl chloride) and poly(vinyl alcohol).

"Second-Generation" (High Performance) Synthetic Polymer Fibers

Fibers from Liquid-Crystals of Rigid-Rod Molecules

One group of the new, high-modulus, high-tenacity (HM-HT) fibers consists of the aramids, typified by *Kevlar* and *Twaron*:

poly(phenylene terephthalamide)$[Co.(C_6H_4).CO.NH.(C_6H_4).NH]_n$

Technora is a related copolymer. Some other polymers, such as polybenzoxazole (PBO), with chains containing multiple rings have also been made into fibers by similar routes, but are not commercially available.

The aramids decompose at about 600°C before they melt. They must be spun from solution in concentrated sulfuric acid by a dry-jet wet-spinning process. The molecules themselves are inherently extremely rigid (rod-like), and in solution they aggregate, in much the way that logs in a river tend to cluster together, to form liquid crystals, which line up on extrusion. The solid fiber is very highly-oriented and chain-extended; it contains virtually no amorphous polymer.

Fibers from Flexible Chain Polymers

The macromolecules of certain polymers—notably polyethylene—are sufficiently flexible to enable high levels of chain orientation and chain extension, combined with high crystallinity, to be achieved under carefully-controlled conditions of drawing. The first successful experiments were conducted by Ward and co-workers, who reported extension ratios in solid polymer in excess of 30X.[1] High-modulus polyethylene (HMPE) fibers, such as *Certran*, made by this route have mechanical properties approaching those of aramids, though with melting points below 150°C, which renders them unsuitable for some applications. Higher moduli and strengths are achieved with the gel-spun polyethylene fibers, *Spectra* and *Dyneema*, which can utilize ultra-high molecular weight grades of polyethylene.

Carbon Fiber

Carbon fiber is virtually in a class of its own, as it is based on two-dimensional molecular sheets. It is a modified fiber in the sense that it is generally produced from a precursor such as polyacrylonitrile, pitch, or other fibers.

PROPERTIES AND DEPENDENCE ON STRUCTURE

Moisture Absorption

Most textile fibers are affected in some way by the presence of moisture. An important technical characteristic of a fiber is its moisture regain, defined as the weight of water contained within a specimen expressed as a percentage of its oven-dry weight. An alternative measure is the moisture content, expressed as a percentage of the total weight.

Moisture absorption has important implications for the processing and use of textile materials. Among many effects, it makes dyeing easier, may increase comfort, reduces static electric charges, may lead to set of creases on drying, and causes changes of dimensions.

Tensile Stress–Strain Relations

Textile fibers are used in a great variety of ways in products of diverse construction and complexity, but their tensile properties almost always dictate the most appropriate choice.

The tensile behavior of fibers such as nylon and polyester is believed to be dominated by deformation in the amorphous regions between crystallites, but there has been little detailed analysis of the mechanics of the composite system. In ordinary rayon, with a micellar structure, the initial modulus is fairly high in the dry state because of hydrogen bonding between cellulose molecules in the amorphous regions, but becomes extremely low in wet state.

The HM-HT fibers generally show almost linear stress-strain curves up to low breaking extensions. To a first approximation, the modulus corresponds to the theoretical value for polymer chains, modified to some extent by defects and misorientation.

In cotton, where there is no significant amorphous phase, one might also expect a straight line approximating to the behavior of the pure cellulose crystal, but the most striking feature of the tensile curve is its concavity. Cotton is unusual in being stronger wet than dry.

In wool, the tensile stress-strain curve falls into three regions, based on the response of the crystalline fibrils and the crosslinked matrix.

Elastic Recovery

The ability of a fiber to recover from an imposed extension is of major technological importance in many applications.

Generally, elastic recovery is good up to the yield point but then becomes worse. Wool is exceptional in showing good recovery from high strains due to the reversal of the α–β crystal transformation.

Mechanical Anisotropy

Textile fibers invariably display some degree of orientation within their fine structure. As a result, the mechanical properties are anisotropic.

Time-Dependence of Mechanical Properties

Most fibers show considerable time-dependence in mechanical properties. This is conveniently shown by creep under constant load, but analogous effects are found in stress relaxation at constant deformation, in rate effects in tensile testing, and in the complicated time-dependent circumstances of use.

Thermo-Mechanical Responses

The interaction of temperature with the dimensions and mechanical properties of fibers, referred to as the thermomechanical behavior, is key to certain post-production treatments, notably heat-setting, and to their stability for various end-use applications, such as the ability to withstand washing and ironing.

Fracture and Fatigue

Simple fiber rupture is rarely the main cause of unserviceability in a textile product: the effects of wear, including flex fatigue and abrasion, may be of greater significance in bringing about a gradual loss of mechanical strength and other performance characteristics, as well as change of appearance. Nevertheless, life-expectancy is ultimately limited by how long the component fibers remain intact.

Optical Properties

The optical properties of fibers are a function of their detailed fine structure. The refractive index, for example, depends on molecular polarizability, but is also a measure of the regularity and packing density of the chain molecules. It can be used to assess such features as crystallinity. Birefringence, the difference between the refractive indices along and parallel to the fiber axis, is closely associated with the degree of chain orientation.

Refractive index and birefringence have a direct bearing on such practical appearance qualities as the luster of fabrics.

Friction

The importance of fiber friction is difficult to overstate: drafting of staple sliver requires frictional forces to induce fiber orientation, and it is virtually friction alone which is responsible for the integrity of a staple yarn. Friction also stabilizes the yarns within a woven structure.

The chemistry of a fiber determines, to some extent, its frictional behavior: polyolefin fibers usually display low friction owing to the inert, oily nature of their molecules. On the other hand, where the molecules contain strongly polar groups the potential for adhesion to other surfaces tends to result in high frictional coefficients.

Electrical Properties

For the most part, polymeric fibers have poor electrical conductivity, which means that electrostatic charges, once developed, are slow to leak away. This can give rise to numerous problems in the processing and use of textile fibers.

In contrast to the majority, certain modern fibers are significantly conducting. These include carbon fibers and high-performance polyethylene in which the main chain C–C bonds are virtually continuous and highly aligned, leading to a significantly increased dielectric constant in the axial direction. The electrical properties of fibers are especially important in that they are often exploited in the measurement of moisture content, yarn regularity, and various other technological characteristics. See, for example, Hearle[2] and Woods.[3]

REFERENCES

1. Capaccio, G.; Ward, I. M. *Properties of Ultra-high Modulus Linear Polyethylenes. Nat. Phys. Sci.* **1973**, *243*, p 143.
2. Hearle, J. W. S. *Structure, Properties, and Uses, in Fiber Structure* Hearle, J. W. S.; Peters, R. H., Eds.; Butterworth & Co.: London, 1963.
3. Woods, H. J. *Physics of Fibers*, The Institute of Physics and John Wright & Sons: Bristol, UK **1955**.

Textiles

See: *Cotton (Non-Formaldehyde Crosslinking Agents)*
Flame Retardant Finishing
Textile Fibers (Shingosen)
Textile Fibers (Structure and Properties)

Thermal Degradation

See: *Degradation (Thermal, Polystyrene and Related Vinyl Polymers)*
Thermodegradable Polymers (Azo Groups in Main Chain)

Thermal Paper

See: *Thermosensitive Paper (Thermosensitive Dyes, Thermochromic Compounds)*

Thermal Polymerization

See: *Liquid Crystalline State Polymerization*
Silacyclobutanes and Related Compounds (Ring-Opening Polymerization)

Thermochromic Polymers

See: *Conjugated Polymers (Insulating and Conducting Forms)*
Polysilanes (Overview)

THERMODEGRADABLE POLYMERS (Azo Groups in Main Chain)

Yukio Shimura
Kantogakuin University

Azo compounds as well as organic peroxides are widely used as initiators for radical polymerization. Among them, azobisisobutyronitrile is well known to initiate polymerization through butyronitrile radicals produced by the elimination of nitrogen molecules. The azo compounds, generally aliphatic, are thermally and photochemically decomposed to give organic radicals and accordingly have been studied as macroinitiators to give block copolymers and other polymer matrixes.[1–3] The simple utilization of the elimination process of nitrogen molecules results in the cleavage of the azo molecules. Accordingly, the incorporation of the azo compounds to polymer main chains can open the way to thermo- and photodegradable polymers.[4–8]

PROPERTIES

Thermal Decomposition Properties of 4,4′-Azobis-4-cyanopentanoic acid, Azobis(2-cyanopropanol), and Thermodegradable Polyamide and Polyurethanes[2,5]

Thermogravimetric curves of ACPA, ACP and their polyamide and polyurethane are shown in **Figure 1**. The initial weight loss of ACP starts at about 122°C and reaches 14.5% at about 155°C. The polyurethanes start decomposition nearly at the same temperature.

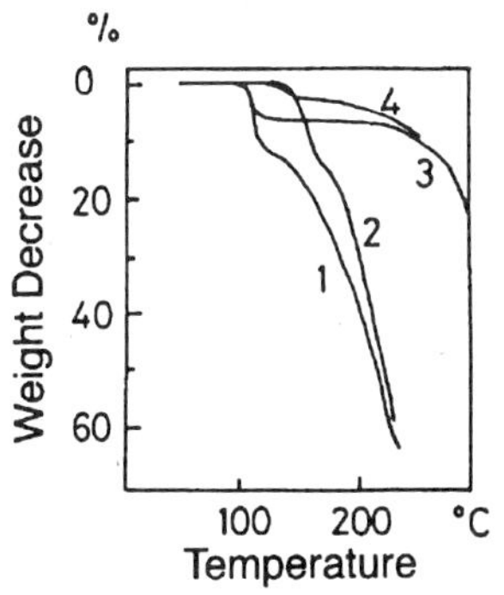

FIGURE 1. Thermo-gravimetry of azo compounds and thermodegradable polymers. 1. 4,4′-azobis-4-cyanaopentanoic acid (ACPA); 2. Azobis-2-cyanopentanol (ACP); 3. Polyamide from ACPA and hexamethylene diamine; 4. Polyurethane from polycaprolactonediol, MDI, and ACP.

APPLICATIONS

- Macroinitiators[1]
- Microcapsules for Printing Inks[7]
- Binders for Explosives and Propellant[8]
- Films[6]
- Adhesives[6]

REFERENCES

1. Ueda, A.; Nagai, S. *Nihon Secchaku Gakkaishi* **1990**, *26*, 112.
2. Ueda, A.; Shiotsu, Y.; Hidaka, Y.; Nagai, S. *Kobunshi Ronbunshu* **1976**, *33*, 131.
3. Ueda, A.; Nagai, S. *Kobunshi Ronbunshu* **1986**, *43*, 97.
4. Kenley, R. A.; Manser, G. E. *Macromolecules* **1985**, *18*, 127.
5. Shimura, Y.; Chen, D. *Macromolecules* **1993**, *26*, 5004.
6. Shimura, Y.; Chen, D., unpublished results.
7. Nuyken, O.; Dauth, J.; Pekruhn, W. *Angew. Makromol. Chem.* **1991**, *187*, 207.
8. Hendry, D. G.; Hill, M. E.; Peters, H. M. U.S. Patent 3 909 497, 1975.

THERMOPLASTIC COMPOSITES

Aurelio Savadori*
EniChem

J. Schuster
Institut für Verbundwerkstoffe GmbH
Universität Kaiserslautern

Mineral fibers and fibers are an approved way for reinforcing polymers improving their mechanical and thermal properties such as stiffness, strength, and thermal resistance. These materials, which consist of two or more physically distinct and separable components, usually with superior properties, are commonly named composites.[1] Today, it is possible to identify five families of products on the market:

- Mineral-filled polymers
- Short-fiber-reinforced polymers
- Long-fiber-reinforced polymers
- Continuous-fiber-reinforced (Advanced Thermoplastic Composites, TPCs) polymers
- Discontinuous aligned-fiber-reinforced polymers

There are many applications based on either semicrystalline or amorphous polymers containing mineral fillers, short fibers, or long fibers. In contrast, the market for continuous fiber products is still embryonic, focusing on applications in aerospace, defense, and sporting goods where performance is more important than price.

The most widely used reinforcing fiber is glass. Thanks to economics, carbon and aramid fibers possess high specific tensile strength and, in the case of carbon fibers, high specific modulus. The mechanical performance increases from mineral-filled to short, long and continuous fibers. Continuous fiber has superior elastic modulus, impact performance, and tensile strength due to the nature of the fiber network, and higher fiber volume. Therefore, continuous fibers are preordained candidates for structural applications in competition with continuous fiber reinforced thermosetting materials.

*Author to whom correspondence should be addressed.

PRODUCTION PROCESSES

Wetting mineral fillers or impregnating fibers with thermoplastic polymers is difficult due to higher polymer viscosities compared to thermosets.

Mineral-Filled and Short-Fibers Reinforced Systems

The most common way to impregnate these products is through melt-compounding.[2]

Long Fiber Reinforced Systems

There are two distinct product forms on the market: pellets and sheets (thermoplastic composite sheets). The pellets are obtained from processes similar to those for impregnating continuous fiber, whereas the sheets are obtained by two systems called:

- Wet slurry process
- Dry process

Continuous Fiber Reinforced Systems

The final products can be classified as follows:

- Preimpregnated Product Forms (PPF) obtained through:
  - melt impregnation
  - solution processing
  - slurry deposition
  - dry powder impregnation with on-line consolidation

The main impregnation technologies have been reviewed recently by Savadori, Cutolo, and Hepola.[3,4]

PROPERTIES

The most important polymer properties that can be modified by a reinforcement are:

- density
- elastic modulus
- mechanical resistance (tensile strength, impact strength, fatigue behavior)
- thermal expansion
- creep

PROCESSING

Mechanisms

The processing technologies for thermoplastic composites depend on the specific types of product forms of each material (pellets, sheets, tow). Mineral-filled and short-fiber, long-fiber (pellets) reinforced polymer are transformed into finished parts by injection molding extrusion. The most important problem of composite manufacturing is the viscosity increase due to the filler (particles or fibers). The influence of filler content and aspect ratio on rheology of those products has been described by Fisa.[5]

Technology of Stamping

Stamping is the most commonly used processing technology for producing parts either with long fiber or continuous fiber reinforced sheets.

APPLICATIONS

Among the inorganic fillers, calcium carbonate, kaolin, and talc have the greatest potential of application. Considering rigid PVC, calcium carbonate is used for extruded pipes and floor tiles. Regarding polypropylene, calcium and talc are the most important fillers. Major applications are in the automotive and appliance markets because of the increased stiffness and high-temperature creep resistance. All that is more evident with the talc than calcium carbonate which, on the other hand, gives the polymer better impact performance. In the automotive industry, applications of talc-filled PP include, for example, fan shrouds, heater housing and ducts, fluid pump parts, battery heat shields, and dash boards. In the appliance industry, pump housings, washing machines, and dishwasher components (inner tubs) are made of talc-filled PP. Short glass fibers are the most commonly used fiber material. Representative applications are electrical insulation connectors, automotive instrument panel (dashboards), radiator fans, business machines housing, dishwasher components, and the inner-tub of washing machines.

Long-fiber reinforced materials are currently used most often in the automotive market: bumper back-up beams, seat frames, floor panels, battery traps, and spare wheel wells.[7,8] The continuous-fiber composites, which represent today only a small portion of the total market of thermoplastic composites, are mainly focused on the military aircraft industry where the high performance properties and advantages of thermoplastics are appreciated. Due to the secrecy of military programs it is not easy to get accurate information but a few examples are reported in the literature.[6] In commercial aircraft the material most used seems to be glass/PEI for interiors followed by CF/PEEK.[8,9] Fischer reports interesting applications in tennis rackets (PA and carbon fibers). Due to the viscoelastic behavior of thermoplastic matrices, racket stiffness increases with the force of the ball's impact.[10]

REFERENCES

1. Hull, D. An Introduction to Composite Materials; Cambridge University, 1981.
2. Bigg, D. M. *Thermoplastic Matrix Composites in International Encyclopedia of Composites*; Lee, S. M., Ed.; VCH: NY, 1991; Vol. 6, 11.
3. Savadori, A.; Cutolo, D. *Makromol. Chem. Macromol. Symp.* **1993**, *68*, 109.
4. Hepola, P. J. CCM Report 93-40, University of Delaware, Center for Composite Materials, 1993.
5. Fisa, B. *Injection Molding of Thermoplastic Composite Materials Technology in Composite Materials Technology: Process and Properties*; Mollick, P. K.; Newman, S., Eds.; Carl Hanser Verlag: München, 1990.
6. Cogswell, F. N. *Thermoplastic Aromatic Polymer Composites*; Butterworth Heinemann Ltd: Oxford, 1992.
7. Döring, E. *Proceedings "Verbundwerk 1991"*; Wiesbaden, Germany, 1991.
8. Gysin, H. *Proceedings "Verbundwerk 1991"*; Wiesbaden, Germany, 1991.
9. Harper, R. C. *SAMPE Journal* **1992**, *28*, 3, 9.
10. Fischer, K.-J. Advanced Composites, 30, 1993.

Thermoplastic Elastomers

See: *Living Radical Polymerization, Iodine Transfer*
Macromonomer Thermoplastic Elastomers
Metallocene Catalysts (Group 4 Elements, New Polymeric Materials)
Poly(phenylene sulfide) (Elastomer Toughened)
Polypropylene Blends and Composites
Reactive Oligomers (Overview)
Reactive Processing, Thermoplastics
Styrene-Butadiene-Styrene Elastomer, Hydrogenated
Styrene-Butadiene-Styrene Triblock Copolymer
Thermoplastic Elastomers (Overview)

THERMOPLASTIC ELASTOMERS (Overview)

Geoffrey Holden
Holden Polymer Consulting, Incorporated

The use of thermoplastic elastomers has significantly increased since they were first produced about 35 years ago. Two books have covered this subject in detail. One concentrates on the scientific aspects of these polymers,[1] whereas the other concentrates on their end uses.[2]

Their outstanding advantage can be summarized in a single phrase: they allow rubber-like articles to be produced with the rapid processing techniques developed by the thermoplastic industry. They have many of the physical properties of rubbers, e.g., softness, flexibility, and resilience. However, they achieve their properties by solidification, as opposed to crosslinking in vulcanized rubbers.

At higher temperatures, the properties of thermoplastic elastomers are usually not as good as those of the conventional vulcanized rubbers. Thermoplastic elastomers are, therefore, applied in areas where these properties are less important, e.g., footwear, wire insulation, adhesives, and polymer blending.

CLASSIFICATION AND STRUCTURE

Thermoplastic elastomers can be divided into three basic types:

1. Styrenic Thermoplastic Elastomers
2. Hard Polymer/Elastomer Combinations
3. Multiblock Polymers with Crystalline Hard Segments.

At least one elastomeric phase and one hard phase must be present and the hard phase (or phases) must become soft and fluid at higher temperatures so that the material as a whole can flow as a thermoplastic.

Styrenic Thermoplastic Elastomers

These are based on simple molecules, such as an A–B–A block copolymer, where A is a polystyrene and B is an elastomer segment.

In commercial applications, three elastomeric midsegments have been used for many years—polybutadiene, polyisoprene, and poly(ethylene–butylene). The corresponding block copolymers will be referred to as S–B–S, S–I–S, and S–EB–S. Recently, polymers with poly(ethylene–propylene) mid-segments (S–EP–S) have been introduced.

Hard Polymer/Elastomer Combinations

Some thermoplastic elastomers are not block copolymers, but fine dispersions of a hard thermoplastic polymer and an elastomer. The two materials usually form interdispersed co-continuous phases. Polypropylene is often chosen as the hard thermoplastic because it is cheap, solvent-resistant, and has a high crystal melting point (165°C).

Combinations with ethylene–propylene–diene monomer (EPDM) or ethylene–propylene copolymer (EPR) are the most important commercial products based on polypropylene; other elastomers that can be used include nitrile, butyl, and natural rubbers.

Multiblock Polymers with Crystalline Hard Segments

These polymers have multiblock $(A–B)_n$ structures, in which the hard (A) segments are crystalline thermoplastics, whereas the softer, elastomeric (B) segments are amorphous. In the best-known types, the hard segments are thermoplastic polyurethanes, thermoplastic polyesters, or thermoplastic polyamides and the soft segments are either polyesters or polyethers. Recently, similar materials have been introduced in which the hard segments are polyethylene and the soft segments are either homopolymers of α-olefins, such as poly(l-octene) or copolymers of mixed α-olefins.

STRUCTURE/PROPERTY RELATIONSHIPS

With such a variety of materials as described above, the properties of thermoplastic elastomers can cover an exceptionally wide range. Some are very soft and rubbery where others are hard and tough, and in fact they approach the ill-defined interface between elastomers and flexible thermoplastics.

Since most thermoplastic elastomers are phase-separated systems, they show many of the characteristics of the individual polymers that constitute the phases. For example, each phase has its own glass transition temperature (T_g) (or crystal melting point (T_m), if it is crystalline), and these in turn determine the temperatures at which a particular thermoplastic elastomer goes through transitions in its physical properties.

Hard Phase

The choice of polymer in the hard phase strongly influences the oil and solvent resistance of the thermoplastic elastomers. Even if the elastomer phase resists a particular oil or solvent, if the oil or solvent swells the hard phase, all the useful physical properties of the thermoplastic elastomer will be lost. In most thermoplastic elastomers, this hard phase is crystalline and resistant to oils and solvents. Styrenic thermoplastic elastomers are an exception. As pure polymers, they have poor oil and solvent resistance (although this can be improved by compounding).

Soft Elastomer Phase

In the styrenic thermoplastic elastomers, analogous S–B–S, S–I–S, and S–EB–S polymers have somewhat different properties. S–B–S polymers are lowest in cost, S–I–S equivalents are the softest, and the S–EB–S polymers are the most stable but also the highest priced.

Hard/Soft Phase Ratio

The hardness of these materials depends on the ratio of the volume of the hard phase to that of the softer elastomer phase. In the styrenic thermoplastic elastomers, this ratio can vary throughout quite wide limits. Thus, in an S–B–S block copolymer, as the ratio of the S to B segments increases, the phase morphology changes from a dispersion of spheres of S in a continuous phase of B to a dispersion of rods of S in a continuous phase of B and then to a lamellar or "sandwich" structure in which both S and B are continuous. If the proportion of S is increased still further, the effect is reversed; S now becomes disperse and B continuous. As the polystyrene phase predominates, the block copolymer gets harder and stiffer until eventually it becomes a clear flexible thermoplastic (e.g., Phillips K–Resin).

APPLICATIONS

Styrenic Thermoplastic Elastomers

These materials differ from the other thermoplastic elastomers in at least two significant ways. First, both the hard and soft phases are amorphous, and thus the pure polymers are soluble in common solvents such as toluene. Second, in their various end uses, these polymers are always compounded with large amounts of ingredients such as other polymers, oils, resins, and fillers.

Typical applications include footwear, wire and cable insulation, and automotive and pharmaceutical items. Processing (e.g., molding or extrusion) these compounded products is simple. Usually, compounds based on S–B–S block copolymers are processed under conditions suitable for polystyrene whereas those based on S–EB–S block copolymers are processed under conditions suitable for polypropylene. Another major application of styrenic thermoplastic rubbers is in adhesives, sealants, and coatings.

Hard Polymer/Elastomer Combinations

Almost all applications for these hard polymer/elastomer combinations are as replacements for vulcanized rubber. Polypropylene/EPDM or EPR combinations have been used to make injection molded bumpers for automobiles, where a combination of toughness, low temperature flexibility, and low cost makes them very attractive. PVC/nitrile rubber blends are widely used in Japan and are becoming important in North America and Europe.

Multiblock Polymers with Crystalline Hard Segments

The very tough materials based on polyurethane, polyester, or polyamide hard segments are generally regarded as premium products. Because of their crystalline hard segments and polar elastomer segments, they have excellent oil resistance. Thus they are used in such demanding applications as blow-molded boots for automobile steering gear assemblies, grease seals, drive belts, and hydraulic hoses. They can also be blended with polar polymers such as PVC. The polymers with polyethylene hard segments cost less. Their suggested applications include wire and cable insulation and blends with polypropylene, either to improve impact resistance or as the soft phase in a hard polymer/elastomer combination.

REFERENCES AND FURTHER READING

1. *Thermoplastic Elastomers–A Comprehensive Review* (Legge, N. R.; Holden, G.; Schroeder, H. E., Eds.), Hanser & Oxford University: Munich/New York, 1987.
2. *Handbook of Thermoplastic Elastomers*, 2nd Ed.; Walker, B. M.; Rader, C. P., Eds.; Van Nostrand Reinhold: New York, 1988.
3. Blum, H. R. paper presented at the 4th International Conference on Thermoplastic Elastomer Markets and Products sponsored by Schotland Business Research, Orlando, FL., February 13–15, 1991.

Thermoplastic Modifiers

See: Epoxy Resins (Modification with Engineering Plastics)

Thermoplastics

See: Acrylonitrile-Acrylic Elastomer-Styrene Terpolymer
Blends (Thermoplastic Polyurethane-SAN)
Cellulose-Filled Composites
Engineering Thermoplastics (Survey of Industrial Polymers)
Lignin-Based Polymers
Melt Spinning (Polymer Formation Under High Stress Conditions)
Polyimide, NEW-TPI (Semicrystalline Thermoplastic)
Polyoxymethylene Fiber, Superdrawn
Powder Coatings (Overview)
Vinylidene Fluoride-Based Thermoplastics (Overview and Commercial Aspects)
Vinylidene Fluoride-Based Thermoplastics (Applications)
Vinylidene Fluoride-Based Thermoplastics (Blends with Other Polymers)
Vinylidene Fluoride-Based Thermoplastics (Homopolymerization and Copolymerization)

THERMOREVERSIBLE GELATION (Polyacrylonitrile Solutions)

Zahir Bashir
Courtaulds

Gelation caused by chemical crosslinking is *irreversible* because the junctions of the network are permanent chemical bonds. Such gelation can occur during three-dimensional polymerization of multi-functional organic or inorganic monomers in solution.

Physical gelation usually occurs when the solvency of the medium worsens, either through addition of a non-solvent, as in coagulation, or through a change in temperature. The latter case leads to *thermoreversible gelation*, whereby a polymer solution gels on changing the temperature; this gelation is reversible, as the junctions of the solvent-swollen network may be removed by altering the temperature, causing the gel to melt and reform the solution.[1–4]

PREPARATION AND PROPERTIES OF POLYACRYLONITRILE GELS

Irreversible Gelation of PAN Solutions

Adding bases, such as sodium hydroxide, to a PAN–dimethyl sulfoxide (DMSO) solution causes gelation. This gelation is *irreversible* and some have suggested that it is caused by chemical crosslinking through the nitrile group.[5]

Reversible-Gelation of PAN–DMF, PAN–DMSO, and PAN–DMA Solutions

Most detailed gelation studies on PAN have been on thermoreversible gels. The gelation of solutions involving the common organic solvents *N,N*-dimethylformamide (DMF), dimethyl sulfoxide (DMSO), and *N,N*-dimethylacetamide (DMA) will be considered.[6–12]

Bisschops[6,7] examined the changes in the rheological properties occurring when PAN–DMF solutions were cooled below 0°C.

Labudzinska et al.[8,9] investigated the gelation of PAN–DMF solutions with small amounts of water by X-ray and light-scattering experiments.

Paul[10,11] investigated the reversible gelation of dimethylacetamide (DMA) solutions of a PAN copolymer containing 7.7% vinyl acetate.

Thermoreversible Gelation of Polyacrylonitrile from Propylene Carbonate

Propylene carbonate (PC) is an interesting solvent for PAN because chain-folded single crystals similar to those of polyethylene have been grown from *dilute* solutions in this solvent by Holland et al.[13] Later work by Bashir showed that when concentrated solutions (about 5–20%) were made by dissolving the PAN powder in PC at 180°C and cooling, the solution gelled.[14] These gels were unlike those from DMF, being formed quite rapidly[14] the gelation time being on the order of seconds or minutes, rather than hours or weeks. The PAN–PC gels were always cream-colored and transparent gels could not be obtained.

The properties of the 20% PAN–PC gel were examined by DSC, X-ray[14] and nuclear magnetic resonance (NMR).[14a] DSC showed an unambiguous gel melting endotherm. Further, the gelation of PAN from PC solutions occurs rapidly so that even at a cooling rate of –10°C/min., and exothermic gel-crystallization transition was readily observable,[14] whereas the gelation of DMF solutions occurs so slowly that the associated exotherm is not measurable.[12]

Polarized infrared studies on orientated gel films showed dichroism in the solvent vibration bands, indicating solvent orientation.[15] The nature of the dichroism was such that it suggested that PC was complexed with the chain by dipole–dipole association between the carbonyl group of the solvent and the nitrile.[15] Hence, polymer–solvent co-crystallization was felt to be a plausible concept.[15]

Gelation of Solutions Containing Other Organic Solvents with Carbonyl Groups

Gelation from two other organic solvents, γ-butyrolactone (γ-BL) and ethylene carbonate (EC), has also been studied. These two solvents are similar to PC in that both are ring compounds with carbonyl groups. Bashir et al. have shown evidence suggestive of the formation of solvated crystallites from γ-BL and EC solutions and concluded that the phenomenon must be quite general and should occur with solvents containing carbonyl groups that can complex with the nitriles.[16]

Gelation from Aqueous Ionic Solvents and from Pure Water

While organic solvents with polar groups such as the carbonyl are one class of solvent for PAN, aqueous ionic solutions such as NaSCN (aq.) and $ZnCl_2$ (aq.) are another. Gel films made from NaSCN (aq.) gave diffraction patterns indicating that solvated crystallites can be formed with aqueous ionic solvents as well.[17]

Pure water does not dissolve or swell PAN to any significant degree at room temperature. However, Coxe, Porosoff, as well as others[18–21] have shown that pure water can dissolve PAN above about 185°C—that is, at a temperature far higher than the boiling point of water at atmospheric pressure. There are several patents on water-plasticized melt spinning of PAN.[18,19,21] Experiments in thick-walled tubes by Frushour[20] have shown that a transparent solution is indeed obtained when PAN is heated with water to temperatures above 185°C. On cooling this PAN-water solution *under pressure*, a solid-like material is formed. This material is in effect a PAN-water gel or a water-plasticized solid, though it has not been considered as such. Unlike the formation of thermoreversible gels from the solvents discussed previously, the reversible gelation of PAN in water can only be demonstrated under pressure because water would evaporate at the dissolution temperature unless pressure is applied.

APPLICATIONS OF THERMOREVERSIBLE GELATION

"Gel spinning" is a term that has been used in the synthetic organic-fiber industry, originally to describe the production of ultra high modulus fibers of polyethylene (PE) following a process first enunciated by Smith and Lemstra.[22–24] However, the term "gel spinning" is a misnomer when used in connection with the spinning of fibers from high molecular weight linear organic polymers. In reality what is extruded is not a *gel*, but a 10–15% *solution* of very high molecular weight PE in xylene, decalin, or paraffin wax at high temperature.

The question arises as to whether ultra high modulus fibers can be made from atactic PAN in a similar way. Apart from the intrinsic merit of a high modulus reinforcement fiber, there would also be interest in its potential for producing superior carbon fibers. The literature indicates several attempts at "gel spinning" of high molecular weight PAN using conventional coagulation or casting routes.[25–28] As the atactic PAN chain cannot adopt an *all-trans* conformation and has a greater cross-sectional area than the PE chain, one cannot expect the ultra high modulus of PE from this polymer.[29,30] This is confirmed by the fact that the best fibers produced from high molecular weight PAN have moduli no greater than 20–30 GPa, whether "gel spinning" or any other route is used.[25–28]

In the author's view, a *binary* system that is closely analogous to polyethylene-xylene is PAN–PC, owing to the facts that a solution can be obtained at moderately high temperatures (well below the boiling point of the solvent PC) and the polymer *crystallizes rapidly* from it. Thus, Atureliya and Bashir have shown that it is possible to plasticize PAN powder with PC and form a free flowing powder that can be fed to an extruder, where it is melted and forced through a fiber spinning die.[31] The filaments solidify on-line spontaneously due to the gelation induced by crystallization; as in the gel spinning of PE, these filaments can be reheated and drawn just above the gel melting point,[31] though complete solvent removal may be more difficult because of the high boiling point of PC.

Thermoreversible gels have also been considered for use as solid electrolytes, owing to potential applications in solid-state batteries, fuel cells, chemical sensors, and display materials. One can envisage extruding (spinning, molding, thermoforming) the appropriate ternary composition of PAN, EC, and $LiClO_4$ at an elevated temperature; on cooling, the system would gel, giving a shaped article with conducting properties. The challenge in this type of solid electrolyte is how to obtain the highest conductivity and sufficient mechanical rigidity in the gel, as well as have a material whose electrical and mechanical properties do not change rapidly due to plasticizer evaporation from the gel.

Thermally induced phase separation has also been studied in order to obtain membranes with controlled porosity after solvent extraction from the gel. Such membranes have potential uses in various kinds of filtration applications. In PAN membranes, there is the additional interest in carbonizing the material to obtain a carbon membrane or foam.

REFERENCES

1. Rose, P. I. *Encyclo. Of Polymer Science and Engineering*; Mark, H. F.; Bikales, N. M.; Overberger, C. G.; Menges, G.; Kroschwitz, J. I., Eds.; John Wiley & Sons: New York, 1987; Vol. 7, p 488.
2. Smith, P.; Lemstra, P. J.; Booij, H. C. *J. Polym. Sci. (Phys. Ed.)* **1981**, *19*, 877.
3. Daniel, C.; Dammer, C.; Guenet, J-M. *Polymer* **1994**, *35*, 4243.
4. Tager, A. *Physical Chemistry of Polymers* 2nd ed.; Mir: Moscow, 1978; Chapter 15, p 486.
5. Bashir, Z.; Manns, G.; Service, D. M.; Bott, D. C.; Herbert, I. R.; Ibbett, R. N.; Church, S. P. *Polymer* **1991**, *32*, 1826.
6. Bisschops, J. *J. Polym. Sci.* **1954**, *12*, 583.
7. Bisschops, J. *J. Polym. Sci.* **1955**, *17*, 89.
8. Labudzinska, A.; Ziabicki, A. *Kolloid Z.* **1971**, *243*, 21.
9. Labudzinska, A.; Wasiak, A.; Ziabicki, A. *J. Polym. Sci. (part C)* **1967**, *16*, 2835.
10. Paul, D. R. *J. Applied Polym. Sci.* **1967**, *11*, 439.
11. Paul, D. R. *J. Applied Polym. Sci.* **1967**, *11*, 1719.
12. Beckmann, J.; Zenke, D. *Colloid and Polymer Science* **1993**, *271*, 436.
13. Holland, V. F.; Mitchell, S. B.; Hunter, W. L.; Lindenmeyer, P. H. *J. Polym. Sci.* **1962**, *62*, 145.

14. Bashir, Z. *Journal of Polymer Sci. (Phys.)* **1992**, *30*, 1299.
14a. Herbert, I. R.; Tipping, A.; Bashir, Z. *J. Polym. Sci. Phys. Ed.* **1993**, *31*, 1459.
15. Bashir, Z.; Atureliya, S. K.; Church, S. P. *J. Materials Science* **1993**, *28*, 2721.
16. Bashir, Z.; Church, S. P.; Price, D. M. *Acta Polymerica* **1993**, *44*, 211.
17. Allen, R. A.; Ward, I. M.; Bashir, Z. unpublished work 1993.
18. Coxe, C. D. U.S. Patent 2 585 444, 1948.
19. Porosoff, H. U.S. Patent 4 163 770, 1979.
20. Frushour, B. G. *Polym. Bulletin* **1982**, *7*, 1.
21. Zwick, M. M. U.S. Patent 4 301 112, 1981.
22. Smith, P.; Lemstra, P. J. U.S. Patent 4 344 908, 1982.
23. Smith, P.; Lemstra, P. J. *J. Materials Sci.* **1980**, *15*, 505.
24. Zwijnenburg, A.; Pennings, A. J. *Colloid & Polymer Sci.* **1976**, *254*, 868.
25. Inoue, T.; Yamamoto, J. Kuraray, Co. Ltd., Japanese patent application Sho 60-165221, 1985.
26. Kameda, T.; Kanamoto, T. *Nihon Reoroji Gakkaishi (J. of the Soc. of Rheology, Japan)* **1993**, *21*, 156.
27. Dobretsov, S. L.; Lomonosova, N. V.; Stelmakh, V. P.; Frenkel, S. Ya. *Vysokomol. Soyed. SSSR* **1972**, *A14*(5), 1143.
28. Maslowski, E.; Urbanska, A. *America's Textile International, Fiber World, FW 2* September, 1989.
29. Allen, R.; Ward, I. M.; Bashir, Z. *Polymer* **1994**, *35*, 2063.
30. Allen, R.; Ward, I. M.; Bashir, Z. *Polymer* **1994**, *35*, 4035.
31. Atureliya, S. K.; Bashir, Z. *Polymer* **1993**, *34*, 5116.
32. Voice, A. M.; Southall, J. P.; Rogers, V.; Matthews, K. H.; Davies, G. R.; McIntyre, J. E.; Ward, I. M. *Polymer* **1994**, *35*, 3363.

Thermoreversible Gels

See: *Acryloyl-L-Proline Alkyl Ester Hydrogels*
Colloidal Dispersions, Nanogels
Enzyme-Poly(N-isopropylacrylamide) Conjugates (Thermoreversibly Soluble/Insoluble)
Poly(N-isopropylacrylamide) Hydrogels
Stimuli-Responsive Polymers (Thermodynamics in Drug Delivery Technology)
Thermoreversible Gelation (Polyacrylonitrile Solutions)
Thermoreversible Gels (Isotactic, Syndiotactic, and Atactic Polystyrene)

THERMOREVERSIBLE GELS (Isotactic, Syndiotactic, and Atactic Polystyrene)

Masamichi Kobayashi and Toshinori Yoshioka
Department of Macromolecular Science
Faculty of Science
Osaka University

There are many polystyrene samples with different stereoregularity covering the range from highly isotactic (IPS) to highly syndiotactic polystyrene (SPS), including various noncrystalline atactic polystyrenes (APS) of different tacticities between the two extremes. Therefore, we are able to elucidate the effects of tacticity on the gelation behavior and the resultant structure formed in gels.

In the present article, we review the recent studies of polystyrene gels made from the structured perspective.

ISOTACTIC POLYSTYRENE GELS

Structural studies of physical gels of polystyrene dispersed in organic solvents date back to the work done by Keller and co-workers on IPS gels dispersed in decalin.[1] They showed that transparent gels were obtained on cooling decalin solutions of IPS prepared at an elevated temperature under sufficiently high super-cooling.

Molecular Conformation in IPS/Carbon Disulfide (CS_2) Gels Revealed by Infrared Spectroscopy

In infrared (IR) spectra of highly crystalline IPS and SPS samples, there appear many absorption bands associated with the regular sequences of specific types of skeletal conformation [the TG form of IPS and the TT or TTGG form of SPS].

As an IPS/CS_2 solution cools, it turns into a transparent gel below −40°C. Some IR bands begin to appear at the gelation temperature and get stronger on further cooling. The frequencies, relative intensities, and shapes of the bands appearing in the cooled gel are close to those of the conformation-sensitive bands of crystalline IPS, indicating that the TG-type (not near-planar type) ordered sequences are formed and grow along with gelation.

IPS/Decalin Gels

Unlike the thermoreversible process of the IPS/CS_2 system, gelation of the IPS/decalin system is complicated. When an IPS/decalin solution prepared at an elevated temperature is allowed to stand at room temperature, it turns first into a transparent gel, and then gradually becomes turbid. The latter process includes deposition of the trigonal crystallites of IPS.[2]

SYNDIOTACTIC POLYSTYRENE GELS

Syndiotactic polystyrene is a typical stereoregular polymer situated at the counterpart of IPS. Like IPS, this polymer is highly crystallizable. Despite the similarity in chemical structure and crystallizability, these two stereoregular polymers differ from each other in gelation behavior.

Molecular Conformation Formed in Gels

In crystalline phases, SPS molecules assume two types of ordered skeletal conformation, the TT type constructing a planar chain (α-form) and the TTGG type constructing a (2/1) helix (β-form), although various crystal modifications belonging to each type have been found so far.[4–9] This polymer is soluble in various organic solvents at elevated temperatures, and the hot solutions turn into transparent gels by allowing them to stand at room temperature.

The kinetics of the conformational ordering in the SPS/$CHCl_3$ system were investigated by time-resolved IR spectroscopy.[10]

Formation of Gel-Networks

The states of polymer aggregates formed in SPS/$CDCl_3$ gels and their changes with gelation time t were investigated by time-resolved measurements of SANS.[11] The result tells us that the conformational ordering of SPS molecules proceeds in parallel to the growth of polymer aggregates.

Atactic Polystyrene Gels

Wellinghoff et al.[12] proposed that APS gels consisted of a continuous network of polymer-rich phase crosslinked by glassy microdomains formed through spinodal phase transition of concentrated solution. From the phase diagrams obtained for APS/solvent systems, Baer and coworkers concluded that gelation occurred through the segment-segment interactions between overlapped APS molecules.[13,14] On the contrary, Guenet and co-workers proposed a different mechanism based on the appearance of enhanced low-angle light scattering. They claimed that formation of microcrystals including both chain segments and solvent molecules (polymer-solvent complex crystals) is responsible for the gelation.[15,16] Temperature dependence of chain trajectory in APS/CS_2 gels was investigated by means of SANS.[17] The result suggested that APS molecules assumed a rather compact form in gel rather than an expanded form found in CS_2 solution (a good solvent for APS). Commercial APS is comparatively syndiotactic-rich with ca. 65% diad syndiotacticity. The conformational ordering on gelation of this sample dispersed in CS_2 (C = 20 g/dL) was investigated by IR spectroscopy. Unlike IPS and SPS gels, we observed no significant spectral change. Only the 572 cm^{-1} band due to the TTGG conformation exhibited a slight increase in intensity, suggesting that short TTGG sequences are formed in the syndiotactic portions of APS molecules.[3]

REFERENCES

1. Girolamo, M.; Keller, A.; Miyasaka, K.; Overbergh, N. *J. Polym. Sci. Polym. Phys. Ed.* **1976**, *14*, 39.
2. Sundararajan, P. R. *Macromolecules* **1979**, *12*, 575.
3. Nakaoki, T.; Kobayashi, M. *J. Mol. Struct.* **1991**, *242*, 315.
4. Immirzi, A.; de Candia, F.; Ianneli, P.; Zambelli, A.; Vittoria, V. *Makromol. Chem. Rapid Commun.* **1988**, *9*, 761; Vittoria, V.; de Candia, F.; Ianneli, P.; Immirzi, A. *ibid* **1988**, *9*, 765.
5. Kobayashi, M.; Nakaoki, T.; Ishihara, N. *Macromolecules* **1989**, *22*, 4377.
6. Guerra, G.; Vitagliano, V. M.; De Rosa, C.; Petraccone, V.; Corradini, P. *Macromolecules* **1990**, *23*, 1539.
7. Corradini, P.; Napolitano, R.; Pirozzi, B. *Eur. Polym. J.* **1990**, *26*, 157.
8. Greis, O.; Xu, Y.; Asano, T.; Petermann, K. *Polymer* **1989**, *30*, 590.
9. Chatani, Y.; Shimane, Y.; Inoue, Y.; Inagaki, T.; Ishioka, T.; Ijitsu, T.; Yukinari, T. *Polymer* **1992**, *33*, 488.
10. Kobayashi, M.; Kozasa, T. *Appl. Spectrosc.* **1993**, *47*, 1417.
11. Kobayashi, M.; Yoshioka, T.; Imai, M.; Itoh, Y. *Macromolecules* **1995**, *28*, 7376.
12. Wellinghoff, S.; Shaw, J.; Baer, E. *Macromolecules* **1979**, *12*, 932.
13. Tan, H.-M.; Moet, A.; Hiltner, A.; Baer, E. *Macromolecules* **1983**, *16*, 28.
14. Boyer, R.; Baer, E.; Hiltner, A. *Macromolecules* **1985**, *18*, 427.
15. Guenet, J.-M.; Willmott, N. F. F.; Ellsmore, P. A. *Polym. Commun.* **1983**, *24*, 230.
16. Gan, J. Y. S.; François, J.; Guenet, J.-M. *Macromolecules* **1986**, *19*, 173.
17. Izumi, Y.; Katano, S.; Funahshi, S.; Furusawa, M.; Arai, M. *Physica B.* **1992**, *180 & 181*, 539, 545.

THERMOSENSITIVE PAPER (Thermosensitive Dyes, Thermochromic Compounds)

Masato Nanasawa
Department of Applied Chemistry and Biotechnology
Yamanashi University

Changes in temperature may alter dyes' physical and chemical properties, which may lead to color fading, undesirable in paints and inks. Recently, these color changes have been used for optical recording devices. These thermoresponsive dyes are called thermosensitive dyes, including thermochromic (reversible color change) dyes.

Thermal recording is classified into physical and chemical processes. Dyes and waxes dispersed in a polymer melt transfer the picture to paper; this process is called thermal transfer recording. In the chemical process, pigment and developer coated on thermal (thermosensitive) paper melt and the picture is directly developed on paper by acid–base reaction. Thermochromic dyes are a promising application in new erasable direct-read-after-write (EDRAW) systems, but are not yet in practical use.

PREPARATION AND PROPERTIES

Thermal Paper (Thermosensitive Paper)

Although several processes have been proposed and used for thermal paper, the leuco dye and acid developer process discovered by the NCR Company is now the most common because it affords a snow-white paper and a vivid developed picture. The picture develops through the reaction of colorless leuco dye with acid component in polymer matrices. The reaction mechanism is shown in **Figure 1**.

Developer and Sensitizer

Phenol and benzoic acid derivatives are used as acid developers and they react with leuco dyes above a mixed melting point as shown in Figure 1. An acid developer for thermal paper is essential to increase the forward reaction above the mixed melting point and also to prevent the back reaction (bleaching reaction) at ambient temperature.

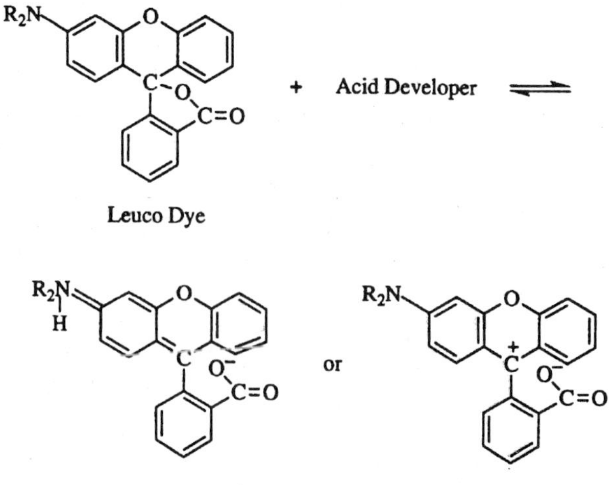

FIGURE 1. Color development of fluoran dye with acid developer.

Binding Polymers

Polymers for thermal paper are used as film-forming material to bind dyes on paper and also play important roles such as separation of dyes from developer in order to produce white paper, creating a reaction field between dyes and developer at elevated temperature, and preserving colored species.

Erasable Thermal Paper

Thermal paper used in word processors and facsimile machines is not stable enough for long storage. The forward and back reaction in Figure 1 gradually proceeds by itself, so the paper is normally discarded within a couple of years.

Thermochromic Dye

Metal complexes of triphenylmethane derivatives (metallophthalein) indicate thermochromism by protolytic dissociation in buffer solution,[1] and the developed color is relatively stable in organic media. The thermochromic behavior of metallophthalein is dependent on the kind of phtalein dyes, bivalent metal ions, and basic developer.

Glass transition temperature (T_g) of polymer also affects the sensitivity and stability of color species.[2] Half-recovery time of developed color using a flexible polymer (T_g: 90°) is less than several hours, whereas the developed color in rigid polymer is stable for a long time and can be bleached by storage in 84% RH desiccator.

Thermosensitive Dyes

The rate of ring closure reaction (bleaching) in Figure 1 is slower than that of ring-opening reaction (color development). When both acidic and basic groups are introduced into a single molecule, the reaction rate can be thermally controlled with the developer.[3]

REFERENCES

1. Nakada, S. et al. *Bull. Chem. Soc. Jpn.* **1981**, *54*, 2913.
2. Nanasawa, M. et al. *Polym. J.* **1988**, *20*, 715.
3. Watanabe, J. *Prc. Polym. Mater. Jpn.* **1993**, *2*, 181.

Thermosets

See: *Benzocyclobutene Homopolymerization (Chemistry and Applications)*
Cellulose-Filled Composites
Composites, Thermosetting Polymers
Cyanate-Based Resins (Polycyanurates)
Cyanate Ester Resins
Cyclomatrix Phosphazene Polymers
Dental Polymers (Overview)
Epoxies, Rubber-Modified
Epoxy Acrylate-Based Resins
Epoxy-Imide, Phosphorylated Epoxy-Imide Polymers
Epoxy Resins (Overview)
Epoxy Resins (Curing Reactions)
Epoxy Resins (High Performance Composite Applications)
High Pressure Polymerization Processing (High Temperature Thermosetting Polymers)
High Temperature Phthalonitrile Resins
Inorganic/Organic Hybrid Polymers (High Temperature, Oxidatively Stable)
Lignin-Based Polymers
Liquid Crystalline Elastomers
Nonlinear Optical Polymers, Thermoset
Ordered Epoxy Networks
Poly(aryl ether ketone)s, Amine-Terminated Polybutadienes
Powder Coatings (Overview)
Thermosets (Main Chain Liquid Crystalline Polymers)
Urea-Formaldehyde Adhesive Resins

THERMOSETS (Main Chain Liquid Crystalline Polymers)

Wei-Fang A. Su
Westinghouse Science and Technology Center

Although large numbers of thermoplastic liquid crystalline polymers have been synthesized, relatively few thermoset liquid crystalline polymers have been prepared. However, with mesogenic groups along the polymers main chain and crosslinking capability between chains, thermosetting liquid crystalline polymers should exhibit desirable properties in the uniaxial and transverse directions and be good candidates for self-reinforcing composite materials.

Thermoset main chain liquid crystalline polymers can be prepared from appropriate polymers or monomers containing a crosslinking functionality. Liquid crystalline monomers with acrylate, acetylenic, nadimide, maleimide, cyanate, and epoxy functionalities have been synthesized.[1–6] Liquid crystalline polyesters have been crosslinked through stilbene and tolane functionalities.[7]

Epoxy is the most important resin used in today's fiber-reinforced composites. It is thought that the development of highly ordered liquid crystallinity epoxies will result in high-performance composites, even when a resin is used without reinforcing material in molecularly self-reinforced composites.

PREPARATIONS AND PROPERTIES

Biphenol Epoxy

We prepared biphenol epoxy (CAS registry number 2461-46-3) with a yield of 75% according to the steps shown in **Scheme I**.[8]

Dihydroxy Phenyl Benzoate Epoxy

Muller et al. reported the synthesis of dihydroxy phenyl benzoate epoxy with a yield of 71% from 4,4′-dihydroxy phenyl benzoate and epichlorohydrin in alkali.

Methylstilbene-Containing Epoxy

Barclay et al. reported the synthesis and characterization of a liquid crystalline epoxy based on dihydroxymethylstilbene (CAS registry number 131090-75-0 or 130756-91-1).[9]

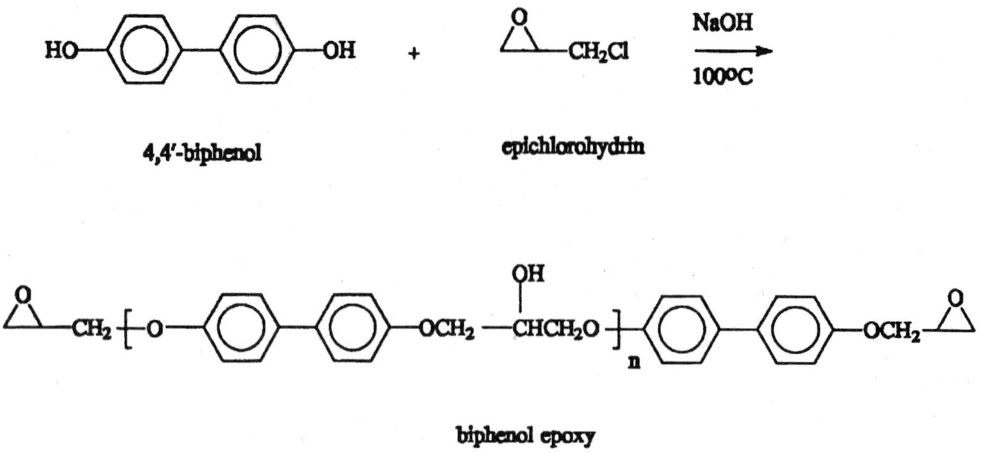

SCHEME I. Synthesis of biphenol epoxy.

APPLICATIONS

When liquid crystalline epoxies are cured with noncrystalline base or acid curing agents, they exhibit better properties, including dimensional stability, high modulus, a low thermal expansion coefficient, and a high dielectric strength compared with the conventional bisphenol A epoxy. The properties of liquid crystalline thermosets should be further enhanced if appropriate liquid crystalline curing agents are employed. The anisotropic properties associated with conventional thermoplastic liquid crystalline polymers are reduced or eliminated when the thermoset liquid crystalline monomer or prepolymer are crosslinked with curing agents. They have potential applications as matrices in high-performance, advanced composites and as molecularly self-reinforced composites.

The structure of methylstilbene mesogen is noncentrally symmetric and exhibits a second-order, nonlinear optical polarizability useful for nonlinear optics. The crosslinking epoxy end group of methylstilbene-containing epoxy locks the nonlinear functionality in place and solves the long-term stability problem observed in nonlinear optical, molecularly dispersed polymer systems or noncrosslinking nonlinear optical polymers.

REFERENCES

1. Broer, D. J.; Finkelmann, H.; Kondo, K. *Makromol. Chem.* **1988**, *189*, 185–194.
2. Ando, M.; Uryu, T. *J. Polym. Sci., Part A: Polym. Chem.* **1990**, *28*, 2575–2584.
3. Conciatori, A. B.; Choe, F. W.; Farrow, G. U.S. Patent 4 440 945, 1984.
4. Hoyt, A. E.; Benicewicz, B. C. *J. Polym. Sci., Part A: Polym. Chem.* **1990**, *28*, 3403–3415.
5. Hoyt, A. E.; Benicewicz, B. C. *J. Polym. Sci., Part A: Polym. Chem.* **1990**, *28*, 3417–3427.
6. Barclay, G. G.; Ober, C. K.; Papathomas, K. I. et al. *Macromolecules* **1992**, *25*, 2947–2954.
7. Calundann, G. W.; Rasoul, H. A.; Hall, H. K. U.S. Patent 4 654 412, 1987.
8. Su, W-F. A. *J. Polym. Sci., Part A: Polym. Chem.* **1993**, *31*, 3251–3256.
9. Barclay, G. G.; Ober, C. K.; Papathomas, K. I. et al. *J. Polym. Sci., Part A: Polym. Chem.* **1992**, *30*, 1831–1843.

Thermotropic Liquid Crystalline Polymers

See: *Liquid Crystalline Polymer-Based Blends (Thermotropic)*
Liquid Crystalline Polymers (Heat-Strengthened, Thermotropic)
Liquid Crystalline Polymers (Main-Chain, Thermotropic)
Liquid Crystalline Polymers (Polyesters)
Liquid Crystalline Polymers (Polyesters of Bibenzoic Acid)
Liquid Crystalline Polymers (Rheology)
Liquid Crystalline Polymers (Thermotropic Polyesters)
Liquid Crystalline Polymers (Viologen: Thermotropic and Lyotropic)
Liquid Crystalline Polymers (Wholly Aromatic Polyesters)
Liquid Crystalline Polymers, Side Chain
Liquid Crystalline Polyurethanes

Thiazole-Containing Polymers

See: *Supported Thiazolium Salt Catalysts (for Acyloin Condensation)*

Thickening Agents

See: *Additives (Property and Processing Modifiers)*
Water-Swellable Polymers (Carbomer Resins)

Thin Films

See: *Block Copolymer Thin Films*
Charge Transport Polymers (for Organic Electroluminescent Devices)
Enzyme-Catalyzed Polymerization
Photoreactive Langmuir-Blodgett-Kuhn Assemblies (Functionalized Liquid-Crystalline Side Chain Polymers)

THIOPHENE COPOLYMERS (Electrical and Third-Order Nonlinear Optical Properties)

Hari Singh Nalwa
Hitachi Research Laboratory

Conjugated polymers have been extensively studied for their potential applications in electronics and photonics.[1–3] Among them, polythiophenes are a special class of organic π-conjugated polymeric materials. In particular much attention has been focused on polythiophenes because of their excellent environmental stability, thermal stability, processability, mechanical strength, and ease of fabrication.[4] The copolymers of 3-methylthiophene and methyl methacrylate discussed here are an example of α-substitution to the polythiophene backbone. Methyl methacrylate (MMA) segments are nonconjugated polar species, which is introduced at the α-positions would interrupt the delocalization of π-electron along the conjugated backbone. The nonconjugated MMA segments play a very important role in introducing solution processability. It can be viewed positively because the increased components of methyl methacrylate segments would be useful in tailoring optical transparency.

The copolymers of 3-methylthiophene and methyl methacrylate can be obtained because an intermediate Grignard reagent of 3-methylthiophene provides the feasibility of the forming blocks of either poly(3-methylthiophene) or poly(methyl methacrylate) depending upon the synthetic strategies.[5,6] The degree of polymerization of each conjugated 3-methylthiophene and non-conjugated methyl methacrylate segment is controlled by the synthetic route. The 3-methylthiophene blocks were polymerized by the cross-coupling reaction using a nickel-phosphine complex catalysts 1,3-bis(diphenylphosphinopropane) nickel(II) bromide $NiBr_2$ (dppp) in dry 2-methyltetrahydrofuran.[8,9] Copolymerization occurs via a Grignard reagent catalyzed polymerization process of methyl methacrylate.[10] The copolymer obtained by this synthetic route was red powder and is referred to (3-MeTh/MMA)-red copolymer thereafter. In the second synthetic strategy, the polymerization of methyl methacrylate was carried out from the intermediate Grignard reagent, and then allowed to copolymerize itself using $NiBr_2$ (dppp) catalyst. The copolymer obtained by this route was a yellow powder and is referred to as (3-MeTH/MMA)-yellow copolymer.

PROPERTIES

Electrical Conductivity

The electrical conductivity of undoped copolymers was on the order of 10^{-11} to 10^{-13} S/cm at room temperature. When the red copolymer was exposed to iodine vapors, its electrical conductivity increased by more than eleven orders of magnitude. The iodine doped red copolymer showed an electrical conductivity of 6.5 s/cm. In the case of yellow copolymer, the conductivity after iodine doping reached up to 7.2×10^{-5} S/cm. The five orders of magnitude difference in electrical conductivity occurs from the presence of more conjugated 3-methylthiophene blocks in red copolymer.[6] In yellow copolymer, the lower conductivity occurs because the nonconjugated methyl methacrylate segments that prevent π-electron delocalization. The electrical conductivity of 750 S/cm has been reported for electrochemically prepared poly(3-methylthiophene). In another study, the 200 nm thick films of poly(3-methylthiophene) showed an electrical conductivity of 2000 S/cm.[11] The extended π-electron conjugation plays a key role in determining the magnitude of conductivity. Relatively lower conductivity of copolymers appears because of the interrupted π-conjugation by polar nonconjugated methyl methacrylate segments, which act as carrier traps and hence circumvent charge-carrier mobility.

Third-Order Optical Nonlinearity and Its Dynamics

Third harmonic generation (THG) measurements at 1.860 μm fundamental wavelength were performed using a difference-frequency generation of a Q-switched Nd:YAG laser and a tunable dye laser. The $\chi^{(3)}$ value was evaluated by comparing the THG signal with standard fused silica where the $\chi^{(3)}$ of 3.0×10^{-14} esu was used as a reference.[12] The THG experiments yielded the $\chi^{(3)}$ of 5.0×10^{-13} esu for the yellow copolymer and 1.3×10^{-11} esu for the red copolymer at 1.860 μm.[13] Surprisingly the magnitude of the $\chi^{(3)}$ differs by about 2 orders. The large $\chi^{(3)}$ of the red copolymer originates as a result of resonant enhancement, because the copolymer absorbs at harmonic frequency (3ω). The yellow copolymer has no absorption at the harmonic frequency, therefore its $\chi^{(3)}$ has no resonance effect.

These copolymers form an important class of electronic and photonic materials as they exhibit desired optical transparency, solubility, and good environmental stability, coupled with large conductivity and third-order optical nonlinearity. These copolymers may have potential for uses in electronic and photonic devices.

REFERENCES

1. Skotheim, T. A., Ed. *Handbook of Conductive Polymers* Marcel Dekker: New York, 1986.
2. Chemla, D. S.; Zyss, J., Eds. *Nonlinear Optical Properties of Organic Molecules and Crystals*; Academic: New York, 1987.
3. Nalwa, M. S.; Miyata, S., Eds. *Nonlinear Optics of Organic Molecules and Polymers*, CRC: Florida, 1996.
4. Roncali, J. *Chem. Rev.* **1992**, 92.
5. Huang, W.; Park, J. M. *J. Chem. Soc. Chem. Commun.* **1987**, 856.
6. Nalwa, H. S. *Polymer* **1991**, *32*, 745.
7. Barker, J. M.; Huddlestone, P. R.; Wood, M. I. *Synth. Commun.* **1975**, *5*, 59.
8. Kobayashi, M.; Chen, J.; Moraes, T. C.; Heeger, A. J.; Wudl, F. *Synth. Met.* **1984**, *1*, 77.
9. Jen, K. Y.; Miller, G. G.; Elsenbaumer, R. L. *J. Chem. Soc. Chem. Commun.* **1986**, 1346.
10. Goode, W. E.; Owen, F. H.; Fellman, R. P.; Snyder, W. H.; Moore, J. E. *J. Polym. Sci.* **1960**, *46*, 317.
11. Roncali, J.; Yasser, A.; Garnier, F. *Synth. Met.* **1989**, *28*, 275.
12. Hermann, J. P. *Opt. Commun.* **1973**, *9*, 74.
13. Nalwa, H. S. *Thin Solid Films* **1993**, *235*, 175.

Thiourea Inclusion Complexes

See: Inclusion Polymerization

Tin-Containing Polymers

See: Ferrocene-Backbone Polymers
Organo-Tin Polymers

Silacyclobutanes and Related Compounds (Ring-Opening Polymerization)

Tissue Adhesives

See: *Cyanoacrylates*
Poly(2-cyanoacrylate)s

Tissue Engineering

See: *Bioabsorbable Polymers (Tissue Engineering)*

Tissue Regeneration

See: *Hybrid Artificial Organs*

Titin

See: *Connectin (Titin) (Large Filamentous Protein)*

TOPOCHEMICAL PHOTOPOLYMERIZATION (Types of Topochemical Behavior)

Masaki Hasegawa
Department of Materials Science and Technology
Faculty of Engineering
Toin University of Yokohama

HISTORY OF TOPOCHEMICAL [2 + 2]PHOTOREACTIONS

Among these topochemical organic reactions, [2 + 2] photodimerization of olefin crystals is one of the most intensively investigated subjects. Formation of α-truxillic and β-truxinic acids from two types of cinnamic acid crystals has been known for a long time,[1,2] and is interpreted in terms of a crystal lattice controlled reaction.[3] In any polymerization systems investigated, however, no clear-cut example of a topochemical polymerization, in which all elementary steps proceed under a crystal lattice-controlled process, were found before the discovery of the photopolymerization of 2,5-distyrylpyrazine (DSP) crystal (**Scheme I**).[4]

Crystallographic study of DSP and poly-DSP demonstrated that the polymerization proceeded with retention of the space group of starting DSP crystal (Pbca).[5,6] As well as the first example of topochemical reaction in a pure sense, the four-center type photopolymerization of DSP crystal was the first example of photopolymerization via a step-growth mechanism.[7,8]

Of further interest was that the crystallographic result of DSP and poly-DSP had revealed a very important future possibility that an "absolute" asymmetric synthesis can be achieved if any prochiral molecule, unsymmetric diolefin derivative for example, crystallized into a chiral crystal and if the reaction of the chiral crystal proceeded in the same manner as the DSP crystal with the retention of crystal lattice.[9,10] Such types of absolute asymmetric syntheses with a high enantiomeric yield were performed by topochemical [2+2]photoreaction of unsymmetric diolefin crystals.[11–14]

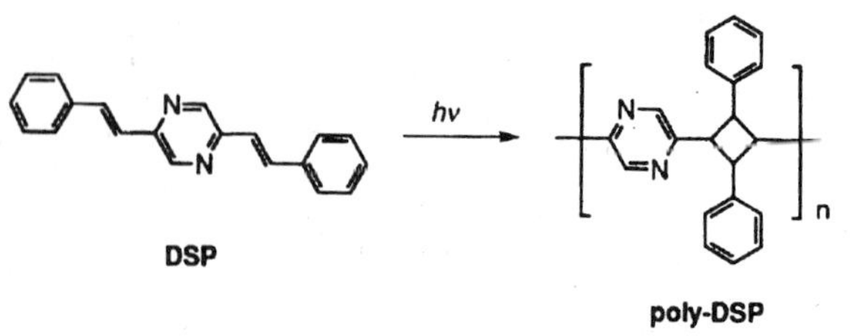

I

ABSOLUTE ASYMMETRIC SYNTHESIS AND AMPLIFICATION OF ASYMMETRY

Along with the guidepost[9,10] based on the crystal-to-crystal transition from DSP to poly-DSP crystal, the absolute asymmetric synthesis was achieved by the topochemical reaction of the chiral crystal of an achiral diolefin compound in the absence of any external chiral reagents.[11–13]

Such growth of chiral crystals without seeding is regarded as amplification of asymmetry. That is, the crystal seeding only causes the crystal formation with the same chirality, but the amplification of asymmetry without seeding always occurs with an equal probability. This phenomenon may be a key stage in the generation of molecules having a single chirality with a large excess if the assumption is made that the diffusion of the chiral molecules in nature is much faster than the next occurrence of amplification of asymmetry. Thus, the entire topochemical process, including the crystallization process, could be a model of the generation of chiral homogeneity in prebiotic era in nature.

CONCLUDING REMARKS

Although several types of topochemical organic reactions have been newly developed, based on empirical molecular design, achieving the desired type of crystal structure in any given case is very difficult because the factors that control the crystal packing are not yet satisfactorily understood; not only a slight change in the chemical structure of the starting compounds but, even for the same compound, a different crystallization procedure causes a drastic variation in the packing arrangement of the molecules and in the topochemical photoreaction mode.

REFERENCES

1. Liebermann, C. *Ber. Dtsch. Chem. Ges.* **1889**, *22*, 124.
2. Liebermann, C.; Bergami, O. *Ber. Dtsch. Chem. Ges.* **1889**, *22*, 782.
3. Bernstein, H. I.; Quimby, W. C. *J. Am. Chem. Soc.* **1943**, *65*, 1845.
4. Hasegawa, M.; Suzuki, Y. *J. Polym. Sci., Part B* **1967**, *5*, 813.
5. Sasada, Y.; Shimanouchi, H.; Nakanishi, H.; Hasegawa, M. *Bull. Chem. Soc. Jpn.* **1971**, *44*, 1262.
6. Nakanishi, H.; Hasegawa, M.; Sasada, Y. *J. Polym. Sci., Part A-2* **1972**, *10*, 1537.
7. Meyer, W.; Lieser, G.; Wegner, G. *J. Polym. Sci., Polym. Phys.* **1978**, *16*, 1365.
8. Nakanishi, H.; Hasegawa, M.; Sasada, Y. *J. Polym. Sci., Polym. Lett.* **1979**, *17*, 459.
9. Wegner, G. *Polymer Preprints, Div. of Polym. Chem., Am. Chem. Soc.* April, 1972, *13* No. 1, p 470.
10. Wegner, G. *Adv. Chem. Ser.* **1973**, *129*, 255.
11. Addadi, L.; van Mil, J.; Lahav, M. *J. Am. Chem. Soc.* **1982**, *104*, 3422.

12. Hasegawa, M.; Chung, C.-M.; Muro, N.; Maekawa, Y. *J. Am. Chem. Soc.* **1990**, *112*, 5676.
13. Chung, C.-M.; Hasegawa, M. *J. Am. Chem. Soc.* **1991**, *113*, 7311.
14. Chung, C.-M.; Kunita, A.; Hayashi, K.; Nakamura, F.; Saigo, K.; Hasegawa, M. *J. Am. Chem. Soc.* **1991**, *113*, 7316.

Topochemical Polymerization

See: *Solid-State Polymerization*
Topochemical Photopolymerization (Types of Photochemical Behavior)
Topochemical Polymerization (Diacetylenes)

TOPOCHEMICAL POLYMERIZATION (Diacetylenes)

Shuji Okada and Hachiro Nakanishi
Institute for Chemical Reaction Science
Tohoku University

Hiro Matsuda
National Institute of Materials and Chemical Research

Several diacetylenes, that is, butadiynes according to IUPAC nomenclature, have been known to become colored materials by exposing them to light.[1–3] This phenomenon was first explained as topochemical solid-state polymerization of butadiynes via 1,4-addition as shown in **Figure 1** by Wegner.[4]

PROPERTIES

Optical Properties

Polydiacetylenes display a characteristic blue or red, and metallic luster appears when high conversion is achieved. Typical polydiacetylenes show sharp excitonic absorption around 500–700 nm in the solid state. Third-order nonlinear optical properties of polydiacetylenes are one of the most interesting subjects. From the studies of the third-order hyperpolarizability (γ) of the series of linear π-conjugated molecules without large polarization, it was clarified that the longer π-electron system molecules have, the larger γ values they show. As the most extended π-conjugation system, the third-order nonlinear optical susceptibility ($\chi^{(3)}$) for third-harmonic generation of polydiacetylene single crystals was first evaluated among many conjugated polymers to be near 10^{-9} esu in three-photon resonant region.[6] Enhancement of $\chi^{(3)}$ values was realized by polydiacetylenes with π-conjugation between polymer backbone and pendant side chains, and about one order of enhancement was observed.[7]

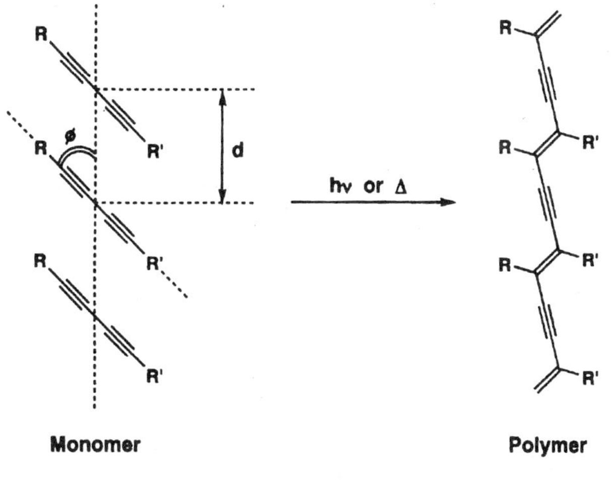

FIGURE 1. Solid-state polymerization scheme of butadiyne monomer to polymer (polydiacetylene). Appropriate alignment of butadiynes for solid-state polymerization in crystals is d = 5 Å and $\phi = 45°$.

Electrical Properties

Undoped polydiacetylenes are insulators. For single crystalline polydiacetylenes conventional chemical doping procedure is not applicable, though they are effective for polydiacetylenes in multilayers and soluble ones. This is due to nearly perfect crystal lattice of the polymers, which disturbs immersion of dopants into crystals. Then, doping during monomer crystallization or polymerization,[8] and electrochemical doping in solid state[9] was investigated and found to be effective in increasing conductivity to semiconducting level.

Magnetic Properties

N-(2,2,6,6-tetramethyl-4-piperidyl-1-oxyl)-10,12-pentacosadiynamide was synthesized as a solid-state polymerizable butadiyne monomer with stable radicals.[10]

Mechanical Properties

Maximum Young's modulus of polydiacetylenes was experimentally obtained to be 62 GPa.[11] Because polydiacetylenes are well aligned in one direction in crystals, mechanical properties along the π-conjugated main chain direction are essentially high, whereas those in perpendicular direction to main chain are extremely low. Aiming at increasing mechanical properties even in perpendicular direction to polydiacetylene backbones, two dimensionally crosslinked polymers were prepared using nylon salt from 10,12-docosadiynedioic acid and 1,5-diaminopentane.[5]

Applications

Polydiacetylenes have not been put to practical use so far, though there are many applications considered using properties mentioned: The application which comes close to real commercialization was the time-temperature indicators.[12]

One more interesting subject is polydiacetylene microcrystals in dimension less than micrometer prepared by the reprecipitation method.[13,14] When microcrystals can be obtained with controlled size and shape, especially in the main chain direction, polymerization is expected to proceed continuously until the polymerization ends reach both edges of the microcrystals. This is because there are small numbers of defects in small numbers of molecules in microcrystals. In such condition, synthesis of mono-dispersed polydiacetylenes might be realized. Qualitative molecular weight control of a polydiacetylene was demonstrated: Polymers from larger sized microcrystals have higher molecular weight than that from smaller microcrystals. Further investigation is required for the polydiacetylenes to appear on the market.

REFERENCES

1. Dunitz, J. D.; Robertson, J. M. *J. Chem. Soc.* **1947**, 1145.
2. Armitage, J. B.; Jones, E. R. H.; Whiting, M. C. *J. Chem. Soc.* **1993**, *1952*.
3. Black, H. K.; Weedon, B. C. L. *J. Chem. Soc.* **1953**, 1785.
4. Wegner, G. *Z. Naturforsch* **1969**, *24b*, 824.
5. Nakanishi, H.; Matsuda, H.; Kato, M. *Mat. Res. Soc. Int. Mtg. Adv. Mat.* **1989**, *1*, 291.
6. Sauteret, C.; Hermann, J. P.; Frey, R.; Pradère, F.; Ducuing, J.; Baughman, R. H.; Chance, R. R. *Phys. Rev. Lett.* **1976**, *36*, 956.
7. Matsuda, H.; Okada, S.; Nakanishi, H. *Mol. Cryst. Liq. Cryst.* **1992**, *217*, 43.
8. Nakanishi, H.; Mizutani, F.; Kato, M.; Hasumi, K. *J. Polym. Sci., Polym. Lett. Ed.* **1983**, *21*, 983.
9. Nakanishi, H.; Matsuda, H.; Tanaka, Y.; Kato, M. *Polym. Preprints* **1984**, *25*, 244.
10. Takaragi, S.; Matsuda, H.; Okada, S.; Nakanishi, H.; Kato, M. *Extended Abstracts of the 56th Spring Meeting*; The Chemical Society of Japan, Tokyo, 1988, p 991.
11. Galiotis, C.; Young, R. J. *Polymer* **1983**, *24*, 1023.
12. Patel, G. N.; Yee, C. K. U.S. Patent 4 228 126, 1980.
13. Kasai, K.; Nalwa, H. S.; Okada, S.; Matsuda, H.; Oikawa, H.; Minami, N.; Kakuta, A.; Ono, K.; Mukoh, A.; Nakanishi, H. *Jpn. J. Appl. Phys.* **1992**, *31*, L1131.
14. Nalwa, H. S.; Kasai, K.; Okada, S.; Matsuda, H.; Oikawa, H.; Minami, N.; Kakuta, A.; Mukoh, A.; Ono, K.; Nakanishi, H. *Polym. Adv. Technol.* **1995**, *6*, 69.

Toughened Polymers

See: *Deformation and Fracture (Micromechanical Mechanisms)*
Epoxies, Rubber-Modified
Epoxy Acrylate-Based Resins
Glassy Polymers (Toughening via Dilatational Plasticity)
Polyamides (Impact Toughness and Toughening)
Poly(aryl ether ketone)s, Amine-Terminated
Polystyrene, Rubber Toughened
Reactive Liquid Rubbers (Epoxy Toughening Agents)
Unsaturated Polyester Resins (Toughening with Elastomers)
Unsaturated Polyester Resins (Toughening with Liquid Rubber)

TRANSANNULAR POLYMERIZATION

Deyue Yan
Department of Applied Chemistry
Shanghai Jiao Tong University

In the early 1960s transannular polymerizations of several cyclodiolefins were reported by Kennedy and Hinlicky,[1] Sartori et al.,[2] Graham et al.,[3] and Reichel and Marvel.[4] Later Pinazzi and his colleagues[5] cxtended transannular polymerization to dicyclo (6-1.0) nonene-4, which consists of one carbon-carbon double bond and one cyclopropane. Recently we found that 1,5-cyclooctadiene[6,7] and 1,5,9-cyclododecatriene[8–10] can polymerize via a transannular mechanism in the presence of BF_3, $AlCl_3$, $AlEtCl_2$, or $AlEt_3Cl_3$. Almost all of the polymers produced in transannular polymerizations possess low molecular weights (several thousand) except that which is generated from 2-carbethoxybicyclo[2.2.1]-2,5-heptadiene. Typical examples of the transannular polymerizations of several monomers are demonstrated in **Table 1**.

REFERENCES

1. Kennedy, J. P.; Hinlicky, J. A. *Polymer* **1965**, *6*, 133.
2. Sartori, G.; Valvassori, A.; Turba, V.; Lachi, M. P. *Chim. Ind. (Milan)* **1963**, *45*, 1529.
3. Graham, P. J.; Buhle, E. L.; Pappas, N. *J. Org. Chem.* **1961**, *26*, 4658.
4. Reichel, B.; Marvel, C. S. *J. Polyml. Sci., Part. A* **1963**, *1*, 2935.
5. Pinazzi, C. P.; Brasse, J. C.; Pleurdeau, A.; Brossas, J. *Polym. Prepr. Am. Chem. Soc. Div. Polym. Chem.* **1972**, *13*, 445.
6. Yuan, C.; Yan, D.; Xu, X.; Wang, F. *Acta Chimica Sinica* **1988**, *46*, 919.
7. Yuan, C.; Yan, D.; Xu, X.; Wang, F.; Gong, Z. *Makromol. Chem.* **1988**, *189*, 733.
8. Yan, D.; Hu, X. *J. Polym. Sci., Part C* **1988**, *26*, 65.
9. Yan, D.; Yuan, C.; Wang, F. *Acta Chimica Sinica* **1990**, *48*, 830.
10. Yan, D.; Yuan, C. *Macromolecules* **1990**, *23*, 3554.

Transition Metal Catalysts

See: *Butadiene Polymerization (Supported Lanthanide Catalysts)*
Butadiene Polymerization, Neodymium-Catalyzed
Catalysts (Olefin Polymerization)
Catalysts, Polymeric (Higher Activity than Monomeric Analogs)
Dendrimers, Luminescent and Redox-Active
Diyne Cycloaddition Copolymerization (Transition Metal Catalyzed)
Electron Transfer Reactions (Novel Polymerization by Divalent Samarium)
Ladder Polymers (Cycloaddition Copolymerization of Cyclic Diynes)
Liquid Crystalline Polymers (Conjugated Chain)
Living Coordination Polymerization
Metallocene/Aluminoxane Catalysts (Olefin Polymerization)
Metallocene Catalysts
Metallocene Catalysts (Cationic Group 4 Metal Alkyl Complexes, Olefin Polymerization)
Metallocene Catalysts (Group 4 Elements, New Polymeric Materials)
Monomer-Isomerization Polymerization (2-Butene with Ziegler–Natta Catalysts)
Olefin-Carbon Monoxide, Alternating
Olefin-Carbon Monoxide Copolymers (Overview)
Olefin Polymerization by 2, ω-Linkage (Migratory Nickel-Phosphorane Catalysts)
Olefin Polymerization Catalysts (Experimental and Theoretical Characterization)
Olefin Polymerization Catalysts (Silica-Supported Titanium Tetrachloride)
Olefin Polymerization Catalysts, Soluble (Group 4, Cationic Active Centers)

TABLE 1. List of Transannular Polymerization

Monomer	Polymer	Catalyst	Solvent	T(°C)	Reference
		$AlCl_3$	CH_3CH_2Cl	−128	1
		$AlBr_3$, $EtAlCl_2$, etc.	n-heptane	−78	2
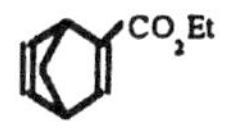	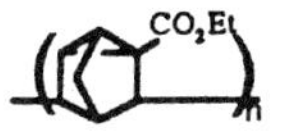 and 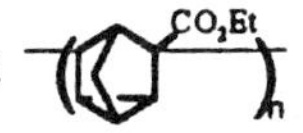	Azodiisobutyronitrile	Benzene	60	3
	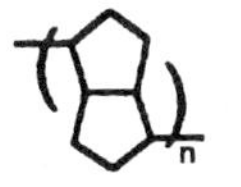	$Al(isobutyl)_3$ and $TiCl_4$ $BF_3{\cdot}OEt_2$, $AlCl_3$ $AlEtCl_2$, $Al_2Et_3Cl_3$	Benzene or heptane CH_2Cl_2 CH_2Cl_2	25, 45, or 0 Room temp. −18	4 6 7
	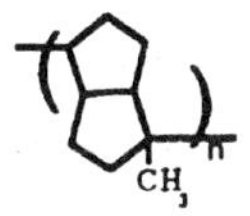	Cationic type			5
	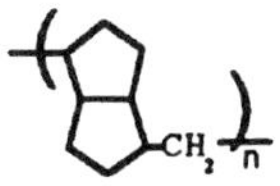	Ziegler-Natta type			5
		$AlEt_3$ and $TiCl_4$ $BF_3{\cdot}OEt_2$, $AlCl_3$ $Al_2Et_3Cl_3$, $AlEtCl_2$	CH_2Cl_2 CH_2Cl_2 CH_2Cl_2	Room temp. 24 −18	8 9 10

Phenylacetylene (Stereospecific Living Polymerization, Rh Complexes)
Phosphinated Polystyrene-Bound Metal Complexes (Pd and Rh)
Polyamines (Ru-Catalyzed Polycondensation of Diamines and Diols)
Polyelectrolytes, Rigid-Rod [Poly(p-Phenylene)-Based]
Polyethylene (Commercial)
Polyethylene (High Density Preparation)
Poly(1,1′-Ferrocene-Arylene)s (Pd-Catalyzed)
Polyimides, Soluble Rigid-Rods (via Pd-Catalyzed Polycondensation)
3,4-Polyisoprene, Crystalline
Polyphenylacetylene
Polyphenylacetylene-Based Permselective Membranes
Poly(phenylene ethynylene) (Synthesis and Optical Properties)
Poly(phenylene vinylene)s (by the Heck Coupling Reaction)
Silacyclobutanes and Related Compounds (Ring-Opening Polymerization)
Styrene, Stereospecific Polymerization
Supported Chiral Catalysts
Supported Chromium Polymerization Catalysts (on Silica, Phillips Catalysts)
Supported Ziegler–Natta Catalysts (on Silica, for Olefin Polymerization)
Supported Ziegler–Natta Catalysts (Titanium-Based Silica; Polymerization Properties)
Supported Ziegler–Natta Catalysts (Titanium (IV) Chloride on Reduced N-Type Oxides)
Zinc Catalysts (Polymerization of Styrene and Methyl Methacrylate)

Triboelectricity

See: Mechano Ions (Macro Ionic Products by Mechanical Fracture)

Trifluoroethylene Copolymers

See: Ferroelectric Polymers (Structural Phase Transitions)
Vinylidene Fluoride-Trifluoroethylene Copolymers (Ferroelectric-to-Paraelectric Phase Transition)

TRISUBSTITUTED ETHYLENE COPOLYMERS

Gregory B. Kharas
Chemistry Department
DePaul University

Trisubstituted ethylenes (TSE, $CHR^1{=}CR^2R^3$) attracted attention of polymer chemists in the forties when the versatility and commercial importance of radical copolymerization of olefins stimulated numerous studies concerned with the relation between monomer structure and reactivity.[1,2]

POLYMER SYNTHESIS

Electron-poor monomers copolymerize readily with monosubstituted electron-rich monomers such as styrene,[4] vinyl acetate,[3,4] vinyl ethers,[3,5] N-vinylcarbazole,[3] and N-vinylpyrrolidone.[6] Polar interactions between dimethyl cyanofumarate and α-methylstyrene, indene, or furan are regarded as essential for copolymerization of the TSE with 1,1- and 1,2-disubstituted monomers.[7] However, all attempts to copolymerize these same trisubstituted carboxylate monomers with mono-substituted comonomers of similar polarity, such as methyl acrylate or acrylonitrile, were unsuccessful[3] and monomers like ethyl 2-cyano-3-phenyl-2-propenoate (ECPP) or 2-phenyl-1,1-dicyanoethene (PDE) showed very low reactivity in copolymerization with acrylonitrile.[8]

Copolymerization of electron-deficient TSE with electron-rich monomers results in copolymers with enhanced alternation of the monomer units in the polymer chain. The tendency toward alternation increases as the difference in polarity between the two monomers increases. By varying the electron withdrawing substituents of a TSE monomer and choosing appropriate monosubstituted alkene, the copolymers with 1:1 composition can be prepared, as for example in the case of copolymers of MCPP or PDE with vinyl ethers,[5] where neither TSE monomers or vinyl ethers form homopolymers by free radical initiation.

PROPERTIES

All statistical and alternating copolymers of electrophilic TSE monomers with donor monomers such as styrene, vinyl acetate, and N-vinylpyrrolidone, are amorphous polymers with glass transition raised by the presence of the electron-withdrawing substituents that restrict conformational mobility.[6,9,10] Thermal stability of an alternating PDE-N-vinylpyrrolidone copolymer is the same when heated in air or nitrogen and the onset of decomposition of the copolymer is 60°C lower than that of polyvinylpyrrolidone (320°C). Physical and mechanical properties of styrene copolymers with PDE and ECPP were studied by Narkis.[11] Both copolymers showed increased softening points in comparison to polystyrene, but whereas copolymers styrene-PDE (34 mol% of PDE) were inferior to polystyrene in flexural strength and the impact resistance, the copolymer styrene-ECPP behaved similarly to polystyrene.

Ito et al. have reported that alternating copolymers of trisubstituted alkenes carrying cyano- and carbomethoxy groups can undergo chain scission reactions as a result of the electron-withdrawing groups and sever steric hindrance.[12] Such polymers could prove useful for the manufacture of microelectronic test structures and devices.

REFERENCES

1. Alfrey, T., Jr.; Greenberg, S. *J. Polym. Sci.* **1948**, *3*, 297.
2. Doak, K. K. *J. Am. Chem. Soc.* **1948**, *70*, 1525.
3. Hall, H. K., Jr.; Dally, S. R. *Macromolecules* **1975**, *8*, 22.
4. Kharas, G. B.; Kohn, D. H. *J. Polym. Sci., Polym. Chem. Ed.* **1983**, *21*, 1457.
5. Kharas, G. B.; Ajbani, H. *J. Polym. Sci.: Part A: Polym. Chem. Ed.* **1993**, *31*, 2295.
6. Kharas, G. B.; Kohn, D. H. *J. Polym. Sci., Polym. Chem. Ed.* **1984**, *22*, 583.
7. Hall, H. K., Jr.; Righettini, R. F. *Polym. Bull.* **1986**, *16*, 405.
8. Ronel, S. H.; Kohn, D. H. *J. Polym. Sci. A-1* **1969**, *7*, 2209.
9. Kharas, G. B. *J. Appl. Polym. Sci.* **1988**, *35*, 733.
10. Hall, H. K.; Ykman, P. *Macromolecules* **1977**, *10*, 464.
11. Narkis, M.; Kohn, D. *J. Polym. Sci. A-1* **1967**, *5*, 1049.
12. Ito, H.; Hrusa, C.; Hall, H. K.; Padias, A. B. *J. Polym. Sci., Polym. Chem. Ed.* **1986**, *24*, 955.

Tubular Polymers

See: *Macromolecular Recognition (New Supramolecular Structures)*
Peptide-Based Nanotubes (New Class of Functional Biomaterials)

U

ULTRA-STRONG POLYETHYLENE FIBERS

Dusan C. Prevorsek
AlliedSignal Senior Science Fellow, Emeritus

Ultra-strong polyethylene (USPE) fibers are the latest development in the field of high performance fibers. Their high strength, which greatly surpassed that of the existing higher-melting aramid (Kevlar), soon caught the attention of the scientific community as, through numerous laboratory experiments and the work of Pennings, it was revealed that flexible polymers such as PE can be transformed into very strong fibers that match the strength of aramid fibers without relying on the presence of a liquid crystalline state.[1] Until then, it had been believed that presence of a liquid crystalline state was essential to the productions of such fibers.

By way of background, Kevlar-type aramid fibers were discovered by a small group of DuPont scientists headed by Morgan and their manufacturing process was invented by Kwolek.[2] The USPE fibers, however, are the ultimate product of the ideas and work of many diverse researchers.[3]

The most significant technological breakthrough in the development of USPE fibers came through Pennings evolutionary research into other methods of producing strong fibers, such as the Free Growth Method and the Surface Growth Method. Pennings' group later recorded the first successful experiments with "gel spinning."[4,5]

SOLUTION PROCESSING OF USPE

General Background

Two large-scale process development efforts followed Pennings' group's demonstration that USPE fibers could be successfully produced using the solution processing method. One, headed by Smith and Lemstra at DSM, led to DYNEEMA (U.S. Patent dated August 17, 1982) and the other, developed by Kavesh and Prevorsek at AlliedSignal, led to a group of products sold under the trade name SPECTRA (U.S. Patent dated November 1, 1983).[6,7]

The principal difference between the DSM and the AlliedSignal process is the type of solvent and the way it is removed.

Another difference lies in the processes themselves. For example, the Stamicarbon (DSM) process involves spinning an approximately 5% solution of very high molecular weight PE (generally M_w if 2–3 × 10^6). Fibers with gradually decreasing solvent undergo multi-step drawing at gradually increasing temperatures. The AlliedSignal Spectra process involves spinning of very high molecular weight PE using non-volatile paraffin oil as a solvent. After taking up the cooled solution that is now transformed into a gel, the non-volatile solvent is removed by a volatile solvent. There are several choices of volatile solvents, including low-boiling fluorocarbons. Depending on the molecular weight, processing conditions, and especially the total draw ratio, fibers with tenacity of 2.7 to 3.5 N/tex are obtained.

APPLICATIONS

The most important advantage of Spectra-type USPE fiber over other high performance fibers is the combination of the low density (0.97 g/cm), very high modulus, and exceptionall high strength. Their specific strength is approximately 35 higher than that of high modulus aramid or S-glass fibers, abo two times that of high modulus carbon fibers, and about times that of steel. Its free-breaking length in air is greater th 300 km and, since its density is less than unity, its free breaki length in water is infinite. Because of their unmatched fatig and abrasion resistance, Spectra fibers are used in fish ne fishing lines, and teeters for satellites. Their moisture insen tivity makes these fibers particularly well suited for otl marine applications such as sail cloth, cordage, composites boat hulls, water skis, and sailboards. Spectra 900 halyards a Spectra 1000 sails have been used in competitive yacht rac including The America's Cup and Australian Admiral's C races.

The microwave transparency of USPE fibers is comparab to that of quartz and much lower than that of aramid, E-glas and carbon fibers. Spectra composites are also outstanding f use in radar antennae because of their unique mechanical pro erties and impact and moisture resistance.

In addition, Spectra composites' high energy absorptio potential at ballistic rate deformations, coupled with thei unmatched damage tolerance, make them the optimum materia for a variety of light-weight armor applications such as militar helmets and body armor. Spectra composites recently testec successfully for use as light-weight explosion blast-proof aircraft luggage containers.

However, in selecting USPE fibers for other specific applications and for purposes of design, one must be aware that the advantages of their mechanical properties are restricted, depending on their use, to relatively low temperatures. The ceiling temperature of Spectra fibers and composites varies a great deal. At constant loads, the ceiling temperature of USPE fibers could be as low as 90°C, but at ballistic rates of deformations, Spectra armor retains its energy absorption potential almost up to the melting point of the fiber.

REFERENCES

1. Pennings, A. J.; Menninger, K. E. In *Ultra High Modulus Polymers*; Cifferi, A.; Ward, I. M., Eds.; Applied Science: London, 1979; p 117.
2. Schaefgen, J. R.; Bair, T. J.; Ballau, J. W.; Kwolek, S. K.; Morgan, P. A.; Panar, M.; Zimmerman, J. In *Ultra High Modulus Polymers*; Ciferri, A.; Ward, I. M., Eds.; Applied Science: London, 1979; p 173.
3. Porter, R. S.; Kanamoto, T. In *History of Synthetic Fibers*; Seymour, R., Ed.
4. Zwijnenburg, A.; Pennings, A. *J. Coll. Polym. Sci.* **1975**, *253*, 452.
5. Smith, P.; Lemstra, P. J.; Kalb, B.; Pennings, A. J. *Poly. Bull.* **1979**, *1*, 733.
6. Smith, P.; Lemstra, P. J. U.S. Patents 4 344 908; 4 422 993; 4 430 383.
7. Kavesh, S.; Prevorsek, D. C. U.S. Patents 4 413 110; 4 536 536; 4 663 101.

Ultradrawing

See: Polyethylene, High Performance (Gel-Like Spherulite Press Method and Ultradrawing)

Ultrafiltration Membranes

See: Polyoxadiazoles and Polytriazoles

ULTRAHIGH MOLECULAR WEIGHT POLYMERS

Silvia Ioan, Maria Bercea, and Cristofor I. Simionescu
Department of Macromolecules
"P. Poni" Institute of Macromolecular Chemistry

Bogdan C. Simionescu
"Gh. Asachi" Technical University

Over the last decades, ultrahigh molecular weight (u.h.m.w.) polymers ("pleistomers," "pleistos"—"very many," so "very many mers") have penetrated in different domains including the advanced technologies.[1]

Many present and possible applications, on the industrial scale of u.h.m.w. polymers arise from the unusual properties they induce to their solutions.

SYNTHESIS

The synthesis of polymers with extremely high molecular weights has been of great interest to the scientific community. Systematic preparation of such polymers requires refined experimental techniques that must meet the most stringent specifications, irrespective of the mode of synthesis.

The preparative methods of u.h.m.w. vinyl polymers include anionic polymerizations, anionic polymerization with narrow molecular weight distributions, emulsion polymerization, suspension polymerization, thermal polymerization at low temperature, and plasma-induced polymerization.[2–10] The processing of u.h.m.w. polyesters, polyamides, polyethylene, and polyphosphazene has been reported and underlines their technological importance.[11–15]

SOLUTION PROPERTIES OF ULTRAHIGH MOLECULAR WEIGHT POLYMERS

Concentration Regimes

The thermodynamic properties of polymer solutions are usually discussed from the point of view of their dependence on two parameters, polymer concentration c and molecular weight M. Experimental and theoretical work has established the existence of three concentration regimes: dilute, semidilute, and concentrated, with the critical concentrations c* and c**, respectively. For u.h.m.w. polymers the selection of concentration regimes in studying the thermodynamic properties is more complicated as compared to usual length polymers. A series of papers suggested the existence of a new critical concentration, $c' < c^*$, which separates the dilute solutions in extremely dilute and dilute.[16–21] Both c* and c′ vary as $[\eta]^{-1}$ or M^{-a}, "a" being the Mark–Houwink exponent, which is constant only for certain molecular weights (M) domains, and $[\eta]$ being the limiting viscosity number.[22]

Dilute Solutions

Theta Conditions

Phase equilibria studies of binary systems evidenced the existence—for most polymers—of two critical temperatures at which the macromolecules are unperturbed, the endothermal theta temperature, Θ_+, and the exothermal theta temperature, Θ_-. In practice, frequently Θ_+ is determined, the other temperature being less accessible. A mathematical expression, allowing the prediction of the endo- and exothermal theta temperatures starting from a limited number of experiments on the variation of the second virial coefficient (A_2) with temperature (T), was proposed for high molecular weight polymers.[24,25]

Unperturbed Dimensions

For the determination of the unperturbed dimensions of u.h.m.w. polymers (i.e., the root mean square radius of gyration, $\langle S_0^2\rangle^{1/2}$, or the root mean square end-to-end distance, $\langle r_0^2\rangle^{1/2}$) the following methods are available:

1. *Light scattering method*, which requires measurements in theta conditions. The processing of experimental data for u.h.m.w. polymers is very difficult owing to the curvatures appearing in the well known Zimm representation. The methods of Berry and Fujita were used in the literature for linearizing the light scattering data.[1,4,26–32]
2. *Extrapolation methods* at zero molecular weight (applied to experimental data determined in good solvents), linear only on domains of molecular weights.[33–39]

Excluded Volume Effect

Theoretical investigations have established different equations in which the second virial coefficient A_2 and the expansion factor α_s are functions of short- and long-range interactions through the interpenetration function $\Psi(z)$. The equations belong to the so-called two-parameter theory.[40,41] Part of them became classical relations on excluded volume for linear, flexible uncharged polymers, and proved to be accurate in the small, excluded volume regime. Other theories and equations try to describe properties of long chain polymers in good solvents at infinite dilution. One can cite the lattice calculation of Domb and Barrett used over the whole range of the excluded volume interaction, and/or the renormalization group theory of Douglas and Freed, which represents an approximate method for converting two parameter calculations in three dimensions, to renormalization group expressions.[42–44] Each approach has a limited applicability and validity, so that it is necessary to test its reliability by use of experimental data on u.h.m.w. polymers in good solvents. Douglas and Freed made a critical presentation of the existing theories as compared to experimental data and their interpretation over a large domain of the excluded volume.[44] Later, a number of theoretical and experimental investigations of the problem were developed by several authors.[45–51]

Conformational Characteristics

The results reported for u.h.m.w. polymers were obtained by processing the experimental data yielded by static light scattering (according to the Berry, Fujita, and Suzuki methods recommended by different authors), dynamic light scattering or/and viscometric measurements.[4,6,8,23,26–28,38,52–60]

Shear Rheology of Entangled Ultrahigh Molecular Weight Polymer Solutions

For u.h.m.w. polymers the network state is realized at lower mass concentrations than for usual ones.[23,61] Viscoelastic behavior in network polymer solutions is a reflection of dynamic processes at the molecular level.[62]

APPLICATIONS

In fundamental research, u.h.m.w. polymers allow the verification of many generally accepted theories in polymer physics—theories issued based on experimental data recorded for normal length polymers (with the molecular weights much lower than 5.10^6), as well as the formulation of new or improved mathematical relations to be used in macromolecular science.

As for their practical significance, the addition of very small amounts of certain u.h.m.w. synthetic or natural (modified) polymers to a solvent was proved to give rise to a wide range of interesting and unusual flow phenomena, such as turbulent drag reduction, decreased heat transfer coefficients, and high resistance to elongational flow.[63–66] Also, they can be employed as thickening agents in food technology, the pharmaceutical industry, and enhanced oil recovery.[67] In the mechanical separation of solids and liquids, long chain polymeric flocculating agents prove to be more effective at low dosage than conventional agents.[68,69] Literature indicates the use of u.h.m.w. polymers at high-performance fibers and in composites.[70,71]

REFERENCES

1. Simionescu, C. I.; Simionescu, B. C.; Ioan, S. *J. Macromol. Sci.-Chem.* **1985**, *A22*, 765.
2. McIntyre, D.; Fetters, L. J.; Slagowski, E. *Science* **1971**, *176*, 1041.
3. Fukuda, M.; Fukutomi, M.; Kato, Y.; Hashimoto, T. *J. Polym. Sci., Polym. Phys. Ed.* **1974**, *12*, 871.
4. Miyaki, Y.; Einaga, Y.; Fujita, H. *Macromolecules* **1978**, *11*, 1180.
5. Wu, C.; He, J.; Shi, F. *Gaofenzi Xuebao* **1991**, *1*, 121.
6. Appelt, B.; Meyerhoff, G. *Macromolecules* **1980**, *13*, 657.
7. Stickler, M.; Meyerhoff, G. *Makromol. Chem.* **1978**, *179*, 2729.
8. Simionescu, B. C.; Ioan, S.; Flondor, A.; Simionescu, C. I. *Angew. Makromol. Chem.* **1987**, *152*, 121.
9. Simionescu, B. C.; Leanca, M.; Ananiescu, C.; Simionescu, C. I. *Polym. Bull.* **1980**, *3*, 437.
10. Osada, Y.; Shen, M. *Polym. Prepr.* **1979**, *20*, 36.
11. Tate, S.; Narisawa, H.; Watanabe, Y.; Chiba, A. Jpn. Pat. 88 108022, **1988**.
12. Watanabe, Y.; Narisawa, H.; Tate, S.; Chiba, A. Jpn. Pat. 89 85217, 1989.
13. Fujimoto, A.; Mori, T.; Hiruta, S. *Nippon Kagaku Kaishi* **1988**, *3*, 337.
14. Rotzinger, B. P.; Chanzy, H. D.; Smith, P. *Polymer* **1989**, *30*, 1814.
15. Klein, J. A.; Bell, A. T.; Soong, D. S. *Macromolecules* **1987**, *20*, 782.
16. Dondos, A.; Skondras, P.; Pierri, E.; Benoit, H. *Makromol. Chem.* **1983**, *184*, 2153.
17. Dondos, A.; Tsitsilianis, C.; Staikos, G. *Eur. Polym. J.* **1989**, *30*, 1690.
18. Dondos, A.; Tsitsilianis, C. *Polymer International* **1992**, *28*, 151.
19. Dondos, A.; Papanagopoulos, D. *Makromol. Chem., Rapid Commun.* **1993**, *14*, 7.
20. Pierri, E.; Dondos, A. *Eur. Polym. J.* **1987**, *23*, 347.
21. Papanagopoulos, D.; Dondos, A. *Macromol. Chem. Phys.* **1994**, *195*.
22. Wales, M. *J. Polym. Sci., Polym. Phys. Ed.* **1980**, *18*, 907.
23. Bercea, M. Ph.D. Thesis, "Gh. Asachi" Technical University, Jassy, Romania, 1994.
24. Simionescu, B. C.; Ioan, C.; Ioan, S.; Simionescu, C. I. *Macromol. Chem. Phys. Macromol. Symp.* 1994.
25. Simionescu, B. C.; Ioan, S.; Ioan, C.; Simionescu, C. I. *Synth. Polym. J.* 1994.
26. Berry, G. C. *J. Chem. Phys.* **1966**, *44*, 4550.
27. Fujita, H. *Polym. J.* **1970**, *1*, 337.
28. Miyaki, Y.; Einaga, Y.; Fujita, H.; Fukuda, M. *Macromolecules* **1980**, *13*, 588.
29. Simionescu, C. I.; Simionescu, B. C.; Ioan, S. *Makromol. Chem., Rapid Commun.* **1983**, *4*, 549.
30. Simionescu, C. I.; Simionescu, B. C.; Ioan, S. *Makromol. Chem.* **1983**, *184*, 829.
31. Simionescu, C. I.; Simionescu, B. C.; Ioan, S. *J. Polym. Sci., Polym. Lett. Ed.* **1985**, *23*, 121.
32. Simionescu, B. C.; Ioan, S.; Bercea, M.; Simionescu, C. I. *Eur. Polym. J.* **1991**, *27*, 589.
33. Cowie, J. M. G. *Polymer* **1966**, *7*, 487.
34. Lovell, P. A. *Comprehensive Polymer Science: Polymer Characterization*; Pergamon: Oxford, 1989; Vol. 1.
35. Gavara, R.; Campos, A.; Figueruelo, J. E. *Makromol. Chem.* **1990**, *191*, 1915.
36. Mays, J. W.; Hadjichristidis, N.; Lidner, J. S. *J. Polym. Sci., Part B: Polym. Phys.* **1990**, *28*, 1881.
37. Cotts, P. M.; Ferline, S.; Dagli, G.; Pearson, D. S. *Macromolecules* **1991**, *24*, 6730.
38. Simionescu, C. I.; Ioan, S.; Bercea, M.; Simionescu, B. C. *Mem. Sectiilor. St. Acad. Rom.* **1991**, *XIV*, 29; and references cited therein.
39. Zisenis, M.; Springer, J. *Polymer* **1993**, *34*, 2363.
40. Yamakawa, H. *Modern Theory of Polymer Solutions*; Harper and Row: New York, 1971.
41. Yamakawa, H. *Pure Appl. Chem.* **1972**, *31*, 179.
42. Domb, C.; Barrett, A. J. *Polymer* **1976**, *17*, 179.
43. Barrett, A. J.; Domb, C. *Proc. R. Soc. London* **1981**, *376*, 361.
44. Douglas, J. F.; Freed, K. F. *Macromolecules* **1984**, *17*, 1854, 2344.
45. Fujita, H.; Norisuye, T. *Macromolecules* **1985**, *18*, 1637.
46. Fujita, H. *Polymer Solutions*; Elsevier: Amsterdam, 1990.
47. Huber, K.; Stockmayer, W. H. *Macromolecules* **1987**, *20*, 1400.
48. Freed, K. F.; Wang, S. Q.; Roovers, J.; Douglas, J. F. *Macromolecules* **1988**, *21*, 2219.
49. Freed, K. F. *Renormalization Group Theory of Macromolecules*; Wiley-Interscience: New York, 1987.
50. Lindner, J. S.; Wilson, W. W.; Mays, J. W. *Macromolecules* **1988**, *21*, 3304.
51. des Cloizeaux, J.; Jannink, G. *Polymers in Solution*; Oxford University: Oxford, 1990.
52. Simionescu, C. I.; Simionescu, B. C.; Ioan, S.; Neamtu, I. *Polymer* **1987**, *28*, 165.
53. Suzuki, H. *British Polym. J.* **1979**, *11*, 35.
54. Ioan, S.; Simionescu, B. C.; Simionescu, C. I. *Polym. Bull.* **1982**, *6*, 421.
55. Ioan, S.; Sava, C.; Simionescu, B. C. *Bull. I. P. Iasi* **1985**, *XXXI*(XXXV), 117.
56. Tsunashima, Y.; Hirata, M.; Nemoto, N.; Kurata, M. *Macromolecules* **1988**, *21*, 1107.

57. Simionescu, C. I.; Simionescu, B. C.; Ioan, S. *Macromolecules* **1985**, *18*, 1995.
58. Nemoto, N.; Makita, Y.; Tsunashima, Y.; Kurata, M. *Macromolecules* **1984**, *17*, 425.
59. Tsunashima, Y.; Hirata, M.; Nemoto, N.; Kurata, M. *Macromolecules* **1987**, *20*, 1992.
60. Bhatt, M.; Jamieson, A. M. *Macromolecules* **1989**, *22*, 2724.
61. Schurz, J. *Prog. Polym. Sci.* **1991**, *16*, 1; and references cited therein.
62. Graessley, W. W. *Polymers in Solution*; Plenum: New York, London, 1986.
63. Kulicke, W. M.; Gräger, H.; Kötter, M. *Adv. Polym. Sci.* **1989**, *89*, 1.
64. Hayashi, H. *J. S. M. E. International J.* **1991**, *34*, 1.
65. Jones, D. M.; Walters, K.; Williams, P. R. *Rheol. Acta* **1987**, *26*, 20.
66. Ferguson, J.; Walters, K.; Wolff, C. *Rheol. Acta* **1990**, *29*, 571.
67. Kulicke, W. M.; Otto, M.; Baar, A. *Makromol. Chem.* **1993**, *194*, 751; and references cited therein.
68. Montani, R.; Nowakowski, T.; Knapick, E. *Pulp. Pap.* **1985**, *59*, 212.
69. Kulicke, W. M.; Budirahardjo, R.; Prescher, M. *Chem. Ing. Tech.* **1989**, *61*, 828.
70. Maslowski, E. *Hem. Vlakna* **1988**, *28*, 8.
71. Zachariades, A. E. U.S. Patent 4 944 974, 1990.

ULTRASONIC DEGRADATION (Preparation of Block Copolymers)

Gareth J. Price,* Peter J. West, and Paul F. Smith
School of Chemistry
University of Bath

Most people regard the application of high intensity ultrasound to enhance chemical processes—sometimes termed *sonochemistry*—as a new development. In fact, the application of ultrasound to polymers considerably predates other chemical applications. As long ago as the 1930s the reduction in the viscosity of solutions of natural polymers such as agar, starch, and gelatin on sonication was noted by Szalay, Gyorgi, and Flosdorf and Chambers.[1–3] Sporadic interest has been shown since then but, with the recent development of readily available equipment, particularly for operating on a large scale, there is currently considerable research in the area,[4] alongside that in other branches of chemistry.[5,6]

Ultrasound is most easily defined as sound with a frequency above the threshold of human hearing, and the threshold for ultrasound is often taken to be 20 kHz.

ULTRASONIC DEGRADATION OF POLYMERS IN SOLUTION

At the outset, it should be stressed that in this article the term *degradation* is taken simply to mean a lowering of the polymer molecular weight rather than any oxidative or other process.

The first effects of ultrasound on polymers in solution were noted in the 1920s. After considerable discussion as to whether changes in solution viscosity were permanent or due to thixotropic effects, it was realized that the process was one of chain breakage and consequent reduction in molecular weights. The early reports prompted a large body of work over the succeeding two decades, which has been reviewed by Basedow and Ebert and by Price, aimed at characterizing the process in terms of the rate of bond cleavage for a wide range of polymers and the effect of the solution and ultrasound parrameters.[7,8]

Mechanism of Degradation

While there is still some debate about the precise origins of the degradation, it has been shown to be a direct consequence of cavitation. Under conditions which suppress cavitation, no degradation has been found.[10] The mechanism can briefly be best described as the polymer chain being caught in the rapid flow of solvent molecules caused by the collapse of cavitation bubbles. A second cause of solvent movement is the shock waves generated after the implosion of the bubbles. The chains are thus subjected to extremely large shear forces resulting in the stretching of the chain and, if the force is sufficiently large, breakage of a bond in the chain.

Applications of Ultrasonic Degradation

At its most straightforward level, the degradation can be used as an additional processing parameter to control the molecular weight distribution. The faster degradation of higher molecular weight species narrows the distribution markedly with consequent modification of the physical properties of the polymer.

A second application of the degradation utilizes these macromolecular radicals as initiating species in the preparation of block and graft copolymers and end-capped materials. This approach was used in early work by Melville and Murray to show the presence of radical intermediates in the degradation.[11]

A large number of workers have sonicated mixtures of two polymers dissolved in a common solvent, cross-reaction between the two types of radicals forming a block copolymer. Henglein prepared poly(acrylonitrile-*b*-acrylamide) in aqueous solution and a large number of similar reactions have since been carried out.[7,8,12] Nakagawa and co-workers sonicated mixtures of PMMA and polystyrene in toluene and produced a block copolymer with a molecular weight lower than either of the homopolymers.[13] Malhotra and Gauthier investigated the degradation of polystyrene in the presence of several poly(alkyl methacrylate)s and isolated the copolymers by selective dissolution.[14] A number of more recent studies have been performed by Xu and co-workers largely involving water soluble materials.[9,15,16] For example, copolymers of poly(ethylene oxide), PEO, with polyacrylonitrile and polyacrylamide, PAM, the latter of which are difficult to prepare by conventional means, were prepared in reasonable yield.

A second approach favored in our work at Bath is to sonicate a polymer dissolved in a solution containing the second monomer. This allows better control over the block length and, more importantly, the block material to be recovered by simple precipitation rather than extraction with two separate solvent/non-solvent combinations.

SUMMARY AND CONCLUSIONS

The ultrasonic degradation will take place whenever polymers in solution are irradiated. Thus, care should be taken, for example, when sonication is used to promote solubility before analysis. The degradation affects all macromolecules in solution

*Author to whom correspondence should be addressed.

and provides an extra parameter in controlling the molecular weight and polydispersity during polymer synthesis or processing. Recent work has been aimed at scaling up sonochemical processes and equipment is now available to sonicate large volumes in flow systems at pilot-plant or low production levels.[17,18] Currently, the production of block copolymers has not reached commercially significant levels, but recent advances in the area may change this situation in the near future.

REFERENCES

1. Szalay, A. S. *Phys. Chem. A.* **1993**, *164*, 234.
2. Gyorgi, A. S. *Nature* **1933**, *131*, 278.
3. Flosdorf, E. W.; Chambers, L. A. *J. Amer. Chem. Soc.* **1933**, *55*, 3051.
4. Price, G. J. In *Current Trends in Sonochemistry*; Price, G. J., Ed.; R. S. C. Special Publication, *116*, R.S.C.: Cambridge, 1992; p 87.
5. Suslick, K. S. *Ultrasound: Its chemical, physical and biological effects* V.C.H.: New York, 1990.
6. Mason, T. J. *Practical Sonochemistry* Ellis Horwood: Chichester, 1991.
7. Basedow, A. M.; Ebert, K. *Adv. Polym. Sci.* **1977**, *22*, 83.
8. Price, G. J. *Advances in Sonochemistry* **1990**, *1*, 231.
9. Chen, K.; Shen, Y.; Li, H.; Xu, X. *Gaofenzi Tongxun* **1985**, *6*, 401.
10. Weissler, A. *J. Appl. Phys.* **1950**, *21*, 171.
11. Melville, H. W.; Murray, A. *Trans. Farad. Soc.* **1950**, *46*, 996.
12. Henglein, A. *Makromol. Chem.* **1955**, *15*, 188.
13. Nagakawa, K.; Kawase, S.; Kakurai, T. *Kobonshi Ronbunshu* **1976**, *33*, 1976.
14. Malhotra, S. L.; Gauthier, J. *J. Macromol. Sci. Chem.* **1982**, *A18*, 783.
15. Hu, X.; Xu, X. *Huagong Xuebao* **1982**, *4*, 319.
16. Chen, K.; Shen, Y.; Li, H.; Xu, X. *J. Macromol. Sci. Macromol. Chem.* **1985**, *A22*, 455.
17. Berlan, J.; Mason, T. J. *Ultrasonics* **1992**, *30*, 203.
18. Martin, P. D. Reference 4, Ch. 12, p. 158.

ULTRASONICALLY ASSISTED POLYMER SYNTHESIS

Gareth J. Price
School of Chemistry
University of Bath

In recent years, the wider availability of sonochemical apparatus and the advent of advanced equipment for macromolecular characterization have led to renewed interest, and ultrasound has been used in a number of areas of polymer chemistry.[1]

In the broadest sense, ultrasound can be defined as sound with a frequency above the threshold of human hearing, which is often taken to be ~18 kHz. Perhaps, the simplest working definition is that the lower frequency limit of ultrasound is 20 kHz. There is no formal upper limit, but in practice, generation in liquids of frequencies above 5–10 MHz is difficult. Clearly at these frequencies there is no direct coupling into chemical bonds, although attenuation of the sound wave can lead to heating. The promotion of reactions is due to an indirect effect known as *cavitation.*[2]

Whatever the precise origin of the chemical effects, it is certain that species exist in high energy, dissociated, vibrationally excited forms inside collapsing cavitation bubbles. Three "zones" in the reaction system can thus be identified: the center of the bubble containing the reactive species formed under the harsh conditions; the bulk liquid in which no *primary* sonochemical activity takes place although subsequent reaction with sonochemically generated species may occur; and the interfacial region where there are large temperature and pressure gradients. An additional factor of importance in polymer sonochemistry is the large shear gradients caused in this region by rapid movement of solvent molecules around the cavities.

RADICAL POLYMERIZATION INITIATED BY ULTRASOUND

Sonication can produce high concentrations of excited-state species during cavitation. Water is particularly susceptible to cavitation,[18] and the possibility of using the H· and OH· produced to initiate polymerization was suggested some years ago when Lindstrom and Lamm produced polyacrylonitrile in aqueous solution using this method.[3] It is now known that radicals can be produced as a result of cavitation in virtually all organic liquids under sonication at sufficiently high intensities.[4,5]

It has become clear over the past few years that vinyl monomers can be polymerized solely by irradiation with ultrasound and this has led to more detailed study of the various parts of the reaction. A series of papers detailing some mechanistic studies of the sonochemical polymerization has been published by Kruus and co-workers.[6–9] Work in the author's laboratory has also focused on similar monomers. High molecular weight polymer is formed at early stages of the reaction, but the average value at longer times falls exponentially.

There are two possible routes by which sonochemical initiation could take place. Sonication of pure liquids could produce initiating radicals or could accelerate the decomposition of added initiators such as peroxides or azocompounds dissolved in the liquid. Both of these mechanisms have been studied in detail recently.

Kruus et al., as well as our group at Bath, have studied the effect of experimental conditions on the rate of polymerization and on the properties of the final polymers produced. For example, the rate was found to be proportional to the monomer concentration, [M], and to depend on the square root of the ultrasound intensity, I_{us}. The final molecular weight varied inversely with [M] and scaled as $I_{us}^{-1/2}$. In contrast to the initiation, higher temperatures lead to the usual increase in the rate of polymerization. The main termination reactions are relatively insensitive to temperature so that the effect on the propagation reaction is dominant. Despite the effect on the rate, there was little variation in the final conversion achieved, again suggesting that the polymerization is limited by the suppression of cavitation as the solution viscosity rises. From our kinetic results, we concluded that the ultrasound has little, if any, significant effect on the propagation or termination reactions of growing radicals, but only on the rate of radical production (assuming a factor of two or three to be insignificant compared to the several orders of magnitude acceleration of the initiation reactions).

SUSPENSION AND EMULSION POLYMERIZATION

There are various aspects of this heterogeneous process where ultrasound could be beneficial. First, it could be used to form and maintain the emulsion since it is known that sonication can be an extremely efficient method of mixing and dispersing. Second, the production of radicals in the aqueous phase could be used as the method of initiation. Hence, there is scope for the reduction or elimination of the additives used in the polymerization such as emulsifiers and initiators.

RING OPENING POLYMERIZATIONS

Probably the most commercially significant reaction of this type in terms of the amount of polymer produced is the reaction of ϵ-caprolactam to nylon 6. This has been studied by Ragaini et al., who showed that ultrasound enhanced the initial ring opening phase allowing a single-step polymerization without the need to add water to start the reaction.[10] High molecular weight materials were formed in shorter reaction times with narrower distributions of molecular weight than when using the conventional process. More recently, Stoessel and co-workers have reported the use of ultrasound in the ring opening of polycarbonate oligomers.[11]

CONDENSATION POLYMERIZATIONS

There have been few reports of the application of ultrasound to condensation reactions. Watanabe et al. applied ultrasound from a cleaning bath to the preparation of aromatic polyformals from, for example, bisphenol-A and methylene bromide.[12]

The sonicated reactions gave considerably higher yields of polymers with much higher inherent viscosities, effects attributed to promotion of the transport of reactions to the surface of the powdered base, which is typical of sonochemical acceleration in heterogeneous processes.

ELECTROCHEMICALLY PROMOTED POLYMERIZATIONS

One area of chemistry to which ultrasound is beginning to be applied with some success is electrochemistry.[13,14] Topare and co-workers published a preliminary study of electropolymerization in which copolymers of isoprene and α-methylstyrene were prepared.[15,16]

Ito and co-workers extensively studied the electrochemical polymerization of thiophene under the influence of ultrasound from a cleaning bath operating at 45 kHz.[17]

POLYMERIZATION EMPLOYING ORGANOMETALLIC REAGENTS

The effect of ultrasound on several types of reaction including the preparation of polyphenylenes has been investigated but two other examples will serve here to illustrate potential uses.[18,19]

Ziegler–Natta Polymerizations

As part of our program, a preliminary study was made of the effect of ultrasound on the heterogeneous, Ziegler–Natta polymerization of styrene using a $TiCl_4/Al(C_2H_5)_3$ catalyst system.[20] Maintaining the sonication throughout the polymerization increased the rate of reaction while the NMR spectra of all the polymers showed them to be virtually identical and, as expected, to be almost exclusively isotactic suggesting that ultrasound has no significant effect on the catalyst structure or the coordination reactions.

Polyorganosilanes

A number of silanes have been studied but the preparation of polymethylphenylsilane will be used as an example here. Ultrasound was first applied to the synthesis of this polymer by Matyjaszewski and co-workers who produced materials with monomodal molecular weight distributions and polydispersities as low as 1.2, albeit in rather low yield (11–15%) using ultrasound at 60°C in toluene.[21,22] Conversely, Miller et al. reported somewhat conflicting results in that the sonication method did not yield polymers with a monomodal distribution unless diglyme or 15-crown-5 were added to the solvent.[23]

We carried out this polymerization under a wide range of conditions in an attempt to solve this apparent difference. The use of higher intensities resulted in narrower molecular weight distributions.

CONCLUSIONS AND FUTURE PROSPECTS

While there has been a relatively small amount of work related to sonochemical polymerization performed, the benefits offered in a range of reaction types warrant further studies in the area, in terms of acceleration of rate and increased yields, but particularly in the control of molecular weight during polymerization. The ability to control the molecular weight without adding chain transfer agents as well as switching off the initiation during the polymerization to prevent heat build-up in highly exothermic reactions may be advantages that can be exploited in particular cases.

It is difficult to envisage high tonnage production of polyolefins by Ziegler polymerization under ultrasound, but there may well be some cases where relatively small amounts of high value materials needing specific molecular weight distributions are required where sonochemistry can play a part. Perhaps the most promising (and so-far least studied) examples are those which can be classified as *organometallic* polymerization. These, and other heterogeneous reactions can be used in the production of a range of highly functionalized materials, such as conducting and/or NLO polymers, side-chain liquid crystals, or dendrimers. These are low-volume, high-value products and it is in this general area where ultrasound may have a major role in the near future.

REFERENCES

1. Price, G. J. In *Current Trends in Sonochemistry*; Price, G. J., Ed.; R.S.C.: Special Publication 116, R.S.C.: Cambridge, 1992a; p 87.
2. Leighton, T. G. *The Acoustic Bubble*; Academic: London, 1994.
3. Lindstrom, O.; Lamm, O. *J. Phys. Colloid Chem.* **1951**, *55*, 1139.
4. Riesz, P.; Misik, V. *J. Phys. Chem.* **1994**, *98*, 1634.
5. Hart, E.; Henglein, A. *J. Phys. Chem.* **1986**, *90*, 5889.
6. Kruus, P. *Ultrasonics* **1983**, *21*, 193.
7. Kruus, P.; Patraboy, T. J. *J. Phys. Chem.* **1985**, *89*, 3379.
8. Kruus, P.; O'Neill, M. L.; Robertson, D. *Ultrasonics* **1990**, *28*, 304.
9. Kruus, P. *Advances in Sonochemistry* **1991**, *2*, 1.
10. Ragaini, V. Italian Patent Appl. 20478-A/90.
11. Stoessel, S. J. *J. Appl. Poly. Sci.* **1993**, *48*, 505.
12. Watanabe, S.; Matsubara, I.; Kakimoto, M.; Imai, Y. *Polym. J.* **1993**, *25*, 989.
13. Mason, T. J.; Lorimer, J. P.; Walton, D. J. *Ultrasonics* **1990**, *28*, 251.

14. Walton, D. J.; Chyla, A.; Lorimer, J. P.; Mason, T. J. *Synthetic Commun.* **1990**, *20*, 1843.
15. Akbulut, U.; Topare, L.; Yurttas, B. *Polymer* **1986**, *27*, 803.
16. Topare, L.; Eren, S.; Akbulut, U. *Polymer Commun.* **1987**, *28*, 36.
17. Ito, M.; Osawa, S.; Tanake, K.; Kuwano, J. *J. Polym. Sci. Polym. Phys.* **1992**, *30*, 19.
18. Price, G. J. *J. Chem. Soc. Chem. Commun.* **1992**, 1209.
19. David, L. D. U.S. Patent 4 576 688, 1986.
20. Price, G. J.; Patel, A. M. *Polymer Commun.* **1992**, *33*, 4435.
21. Kim, H. K.; Matyjaszewski, K. *J. Amer. Chem. Soc.* **1989**, *110*, 3321.
22. Kim, H. K.; Uchida, H.; Matyjaszewski, K. *Polym. Prepr.* **1989**, *30*(2), 119.
23. Miller, R. D.; Thompson, D.; Sooriyakumaran, R.; Fickes, G. N. *J. Polym. Sci. Polym. Chem.* **1991**, *29*, 813.

Ultrathin Films

See: *Conducting Polymers (For Langmuir-Blodgett Film Fabrication)*
Langmuir-Blodgett Film Polymerization
Multilayer Films (Polyelectrolytes)
Self-Assembled Polymers (Photopolymerizable Bolaform Amphiphiles)
Supramolecular Mesomorphic Polymers (Assembling in Ultrathin Films)
Ultrathin Films (by Plasma Deposition)
Ultrathin Films (Self-Assembled)

ULTRATHIN FILMS (by Plasma Deposition)

Buddy D. Ratner
Center for Bioengineering and Department of Chemical Engineering
University of Washington

Many useful techniques have evolved for depositing thin polymeric films.[1–4] In this article, I propose that deposition of thin polymeric films from low temperature, nonequilibrium plasmas is among the most versatile of these methods.[5,6] Plasma deposition can be used to form overlayers 10–1000 Å thick are that smooth, pinhole free, and exhibit tenacious adhesion to their substrate.

TYPES OF PLASMA DEPOSITED THIN POLYMERIC LAYERS AND A FEW EXAMPLES

Fluoropolymer Depositions

A plasma fluoropolymer coating can be deposited by introducing tetrafluoroethylene (TFE) into the reactor. Where conventional polymers from TFE contain only CF_2 units, the plasma deposited fluoropolymer films will have a distribution of functional groups including CF_3, CF_2, CF and hydrocarbon species. Fluoropolymer plasma films from TFE are extremely nonwettable, even more so than Teflon. Films of this type have been described in many publications.[7–10] RF-plasma fluoropolymer deposits have demonstrated low reactivity with blood platelets suggesting medical applications.[11,12] They also exhibit a tight binding of proteins so they may have application for immunoassays and protein separations.[13] Under some conditions, plasma depositions from fluoro monomers can form films with a chain orientation normal to the substrate surface.[9] However, in most cases plasma deposits are amorphous and unoriented.

Methane

Methane and other saturated alkanes are readily deposited in a plasma to form hydrocarbon-rich films.[14,15] As the deposition power is increased or as hydrogen is blended in with the methane in the plasma, the organic nature of the film decreases, and the film structure approaches a graphite or diamond-like character.

Triethylene Glycol Dimethyl Ether

Poly(ethylene oxide)-like plasma deposited films can be prepared from low molecular weight precursors such as triethylene glycol dimethyl ether (triglyme) and tetraethylene glycol dimethyl ether (tetraglyme).[16] The plasma deposition reactions contribute to good thin film qualities, while the condensation brings in intact monomer contributing to structure retention. Glyme films, deposited under appropriate mild conditions, show extremely low protein adsorption and cell attachment.[16] The chemical structure of these films, as analyzed by both ESCA and SIMS, suggests poly(ethylene oxide)-like structures.[16,17]

N-Vinylpyrrolidone

N-vinylpyrrolidone (NVP) can be plasma deposited to produce films with some retention of NVP structure.[18] However, the fraction of nitrogen in the film decreases with increasing power density in the deposition plasma, and a substantial carbonyl group concentration is noted.

SURFACE STRUCTURES: RETENTION OR LOSS OF CHEMICAL STRUCTURE

Plasma-deposited thin films from organic monomer precursors can range in character from similar to conventionally polymerized material to almost inorganic in nature.

Gas-phase species in plasma have sufficient energy to break any organic bond. Thus, plasma-deposited films generally have a scrambled chemical structure that has little resemblance to the starting monomer. Films deposited from common monomers under typical plasma reaction conditions may have useful properties, but their chemistry is ill-defined and multifunctional. To develop plasma-deposited films with a surface structure resembling the starting monomer, monomer can be condensed onto a cold stage under conditions with low plasma energy density. By using this technology, poly(ethylene oxide)-like films,[16] poly(2-hydroxyethyl methacrylate) deposits,[19] novel fluoropolymer deposits,[19] and polystyrene coatings[20] can be synthesized. Monomers containing double bonds, in general, produce plasma-deposited films with considerable structure retention, particularly where plasma powers are kept low.

CONCLUSIONS AND PERSPECTIVES

The primary problems associated with plasma deposited thin films are reproducibility in manufacture, quality control, film stability, and durability. Abrasion and deformation of plasma deposits on plastic or elastomeric substrates can lead to film deadhesion or cracking.

An impediment to wider application of plasma depositions is problems associated with reproducibility and reactor specificity. If a reactor configuration is altered, reaction parameters have to be redeveloped, often a tedious process. By monitoring the gas phase species in the plasma, the optical emission from the plasma, and the surface chemistry during deposition, pattern recognition methods might be used to adjust plasma parameters to a state that will produce surfaces (as assessed by ESCA and SIMS) of desired characteristics.

The structure of plasma-deposited films is extremely complex and characterized by a broad distribution of chemical functional groups and chain crosslinks. Before we truly understand how plasma deposited films perform in a variety of applications, we must understand what their structure is. Although much progress has been made in the area, particularly using ESCA and SIMS,[21,22] much remains to be learned. Newer techniques such as those using synchrotron radiation (EXAFS, NEXAFS, and X-ray absorption), scanning probe microscopies (STM and AFM), advanced SIMS methods (time-of-flight, surface analysis by laser ionization, and tandem SIMS) and high resolution electron energy loss spectroscopy (HREELS) will advance understanding of plasma thin films. Because of the complexity of these coatings, the surface characterization of plasma deposited films will be a fertile subject for investigation for many years.

ACKNOWLEDGMENT

Support from NIH grant RR01296 was received for many of the studies presented in this article.

REFERENCES

1. Soutif, J. C.; Brosse, J. C. *Reactive Polymers* **1990**, *12*, 3–29.
2. Ratner, B. D. "Surface modification of polymers: chemical, biological and surface analytical challenges"; *Biosens. Bioelect.* 1995, in press.
3. Baglin, J. E. E. *NATO ASI Ser.-Appl. Sci.* **1988**, *144*, 369.
4. Garbassi, F.; Morra, M.; Occhiello, E. *Polymer Surfaces: From Physics to Technology*; John Wiley & Sons: Chichester, UK, 1994.
5. d'Agostino, R. *Plasma Deposition, Treatment and Etching of Polymers*; Academic: San Diego, 1990.
6. Yasuda, H. *Plasma Polymerization*; Academic: New York, 1985.
7. Masuoka, T.; Yasuda, H. *J. Polym. Sci., Polym. Chem. Ed.* **1982**, *20*, 2633.
8. Morosoff, N.; Yasuda, H. "Plasma polymerization of tetrafluoroethylene in a capacitively coupled discharge with internal electrodes;" in *ACS Symposium Series*; Shen, M.; Bell, A. T., Eds.; American Chemical Society: Washington, DC, 1979, Vol. 108, p 163.
9. Castner, D. G.; Lewis, K. B. *Langmuir* **1993**, *9*, 537.
10. d'Agostino, R.; Cramarossa, F.; Fracassi, F.; Illuzzi, F. "Plasma polymerization of fluorocarbons"; in *Plasma Deposition, Treatment and Etching of Polymers*; d'Agostino, R., Ed.; Academic: Boston, 1990, p 95.
11. Garfinkle, A. M.; Hoffman, A. S. et al. *Trans. Am. Soc. Artif. Int. Organs* **1984**, *30*, 432.
12. Hoffman, A. S.; Ratner, B. D. et al. "The small diameter vascular graft—a biomaterials challenge" in *Polymers in Medicine II—Biomedical and Pharmaceutical Applications*; Chiellini, E.; Giusti, P. et al., Eds.; Plenum: New York, 1985; p 157.
13. Hoffman, A. S.; Kiaei, D. et al. *Clinical Materials* **1991**, *8*, 3.
14. Kaplan, S.; Dilks, A. *Thin Solids Films* **1981**, *84*, 419.
15. Ho, C.-P.; Yasuda, H. *J. Appl. Polym. Sci.* **1990**, *39*, 1541.
16. Lopez, G. P.; Ratner, B. D. et al. *J. Biomed. Mater. Res.* **1992**, *26*, 415.
17. Lopez, G. P.; Ratner, B. D.; Briggs, D. *Trans. Soc. Biomat.* **1991**, *14*, 277.
18. Marchant, R. E.; Yu, D.; Khoo, C. *J. Polym. Sci.* **1989**, *27*, 881.
19. Lopez, G. P.; Ratner, B. D. *J. Polym. Sci., Polym. Chem. Ed.* **1992**, *30*, 2415.
20. Leggett, G. J.; Ratner, B. D.; Vickerman, J. C. *Surf. Interface Anal.* **1995**, *23*, 22.
21. Chilkoti, A.; Ratner, B. D.; Briggs, D. *Chem. Mater.* **1991**, *3*, 51.
22. Chilkoti, A.; Ratner, B. D.; Briggs, D. *Anal. Chem.* **1991**, *63*, 1612.

ULTRATHIN FILMS (Self-Assembled)

Toyoki Kunitake*
Faculty of Engineering
Kyushu University

Burm-Jong Lee
Department of Chemistry
Inje University

Ultrathin polymer films attract much attention from the standpoint of fundamental research and their application potentials. In the former aspect, the formation of molecularly thin films is necessarily related to self-assembly of component molecules. Biological membranes are pronounced examples of this kind in that they are composed of molecular bilayer organizations. In practical applications, ultrathin films would be novel materials suited for efficient materials' separation and for electronic and optoelectronic uses.[1]

Polymer synthesis on two-dimensional (2D) templates is probably the single most important method to obtain molecularly thin polymer films. Synthetic bilayer membranes possess 2D molecular organizations that are essentially identical to those of biolipid bilayers. Therefore, they constitute convenient matrices for preparation of 2D polymers, either in the form of aqueous (and non-aqueous) dispersions, or in the form of cast films.

SYNTHETIC BILAYER MEMBRANE

The bilayer assembly similar to that of biological membranes is formed from a variety of synthetic amphiphiles. The first example of these amphiphiles was provided by didodecyldimethylammonium bromide.[2]

In a recent review[3] the molecular design of bilayer-forming molecules, their aggregate morphologies, and some applications are described. Other excellent reviews have been published on various aspects of synthetic bilayer membranes: Fendler prepared an extensive list of amphiphiles that are capable of bilayer formation (up to 1982)[4]; routes to functional vesicle membranes have been discussed by Fuhrhop and Mathieu;[5] an extensive

*Author to whom correspondence should be addressed.

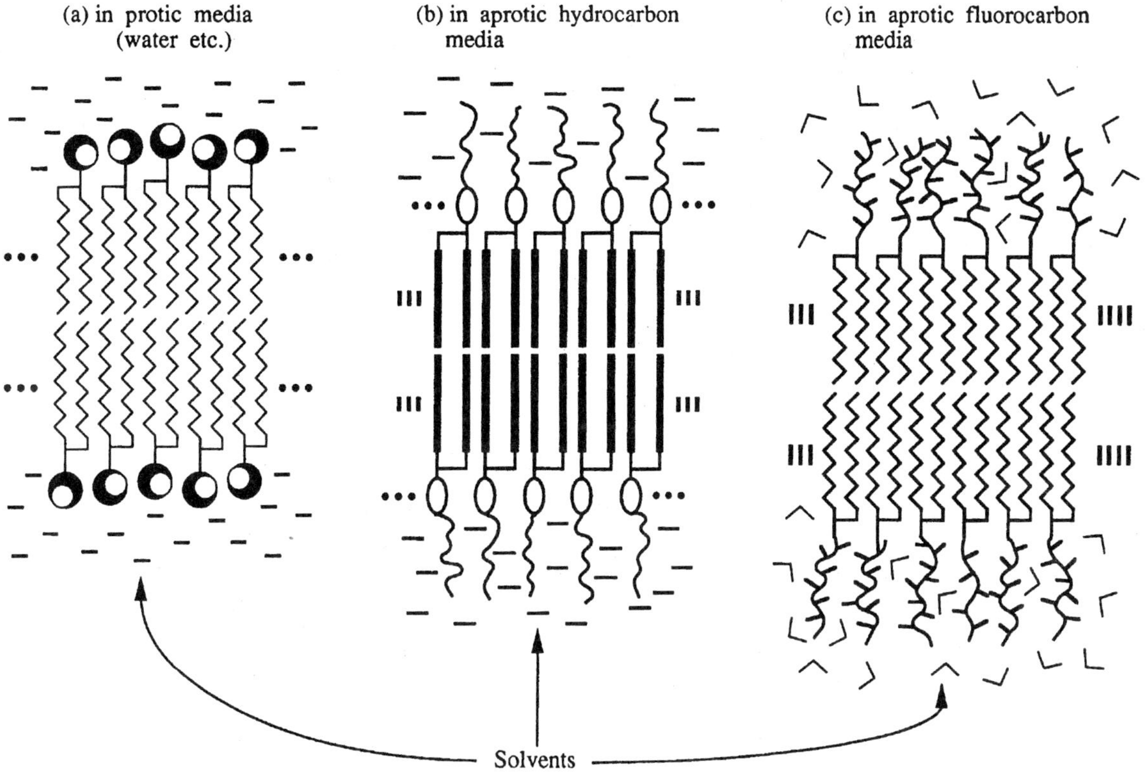

FIGURE 1. Generalized bilayer concept.

review on polymeric bilayer membranes was published by Ringsdorf, Schlarb, and Venzmer.[6]

Figure 1 represents the generalized bilayer concept. Self-assembling bilayers can be formed in protic media, in aprotic hydrocarbon media, or in aprotic fluorocarbon media. The common driving force of the molecular assembly is the solute/solvent immiscibilities (enthalpic force) that arise from differences in cohesive energy between solute (amphiphile) and solvent. The cohesive energy of the solvent is greater than that of the solute in the case of a and b, but this relation is reversed in c. Therefore, the magnitude of the cohesive energy of the solute *per se* is not relevant for promoting effective molecular assembly.

MULTILAYERED 2D POLYMER NETWORKS

The outstanding self-assembling property of bilayer membranes is fully demonstrated by the formation of free-standing cast films. Transparent films with regular, multilayer structures are readily obtained when aqueous dispersions of synthetic bilayer membranes are spread on solid supports, as is the case with some phospholipids. It is remarkable that flexible, self-supporting films are obtainable from non-covalent molecular aggregates. These cast films are stable enough to accommodate a variety of hydrophilic guest molecules (small molecules, proteins, nanoparticles) at the interbilayer space without significant disruption of the regular structure.

Synthetic bilayer membranes are self-assembling molecular systems. Therefore, incorporation of guest molecules into the interbilayer space of cast films does not destroy the original bilayer assemblage. This unique property allows the use of widely variable guest molecules for the subsequent immobilization. The large structural variety of synthetic bilayer membranes further enhances the flexibility of the system.

MULTILAYERED 2D INORGANIC POLYMERS

When mixed dispersions of alkoxysilanes (e.g., $(MeO)_3SiMe$) and bilayer components are cast and subjected to ammonia treatment, extensive hydrolysis and condensation of the alkoxysilane unit proceed, and stable films of polysiloxane 2D-network are formed upon extraction of matrix bilayers. The SEM observation of the cross section of the film demonstrates that ultrathin layers with 20-Å or 2-nm thickness are produced parallel to the film plane under proper conditions.[7] The morphology of the cross-section is highly variable (thin layer, plate, pores, etc.) depending on the casting conditions.[8]

BUILD-UP OF POLYION-COMPLEXED MULTILAYERS

The formation of polyion-complexed monolayers at the air-water interface has been used as a distinctive process in the fabrication of LB films.[9] By employing these polyion complexes, the preparation of LB films with special functionalities became feasible. For example, Higashi and Kunitake succeeded in forming the LB film of an otherwise intractable fluorocarbon monolayer by complexing with potassium polystyrene-sulfonate.[10] The LB films of azobenzene amphiphiles as polyion complexes were prepared by spreading monolayers on subphases containing anionic[9] or cationic[11] polymers. Ringsdorf et al. reported the domain formation of fatty acid monolayers on a polyethylenimine-containing subphase.[12] Thin films of polyimide[13,14] and poly(arylene vinylene)[15] were prepared by the LB method by electrostatically attaching hydrophobic chains to the precursor polymers which, by themselves, are incapable of monolayer formation. Recently, Nakashima et al. reported that cationic siloxane copolymers could form stable monolayers at the air-water interface in the presence of potassium poly(vinyl sulfate) in the subphase.[16]

These results are basis for build-up of alternate films of polycations and polyanions as proposed by Decher and co-workers.[17]

2D POLYMER NETWORKS PREPARED BY THE LANGMUIR-BLODGETT TECHNIQUE

2D polyion complexation between oppositely-charged polymers at the air-water interface and subsequent covalent crosslinking of the resulting LB films give rise to molecularly thin molecular networks. For example, copolymers of maleic acid and octadecyl vinyl ether form stable monolayers on aqueous polyallylamine.[18]

EPITAXIAL ADSORPTION OF POLYMER CHAINS ON GRAPHITE

Polymerization-induced epitaxy (PIE), as we define it,[19] is the synthesis of polymeric molecules in the presence of a solid substrate, where the resulting polymer on the solid surface has a structure that is strongly influenced by the atomic arrangement of the substrate surface.

In the case of a pyrolytic graphite (HOPG) substrate, a resulting film is adsorbed on the surface by van der Waal's interaction only. Solution polymerization of various monomers is conducted with graphite plates immersed in the solution, and then the graphite is thoroughly washed by good solvents of polymers. In all cases, polymer chains are two-dimensionally close-packed and commensurate with the graphite lattice.

REFERENCES

1. Ulman, A. *Ultrathin Organic Films*; Academic: Boston, 1991, Chapter 5.
2. Kunitake, T.; Okahata, Y. *J. Am. Chem. Soc.* **1977**, *99*, 3860.
3. Kunitake, T. *Angew. Chem. Int. (Engl. ed.)* **1992**, *31*, 709.
4. Fendler, J. H. *Membrane Mimetic Chemistry*; Wiley-Interscience: New York, 1982, Chapter 6.
5. Fuhrhop, J.-H.; Mathieu, J. *Angew. Chem. Int. (Engl. ed.)* **1984**, *23*, 100.
6. Ringsdorf, H.; Schlarb, B.; Venzmer, J. *Angew. Chem. Int. (Engl. ed.)* **1988**, *27*, 113.
7. Sakata, K.; Kunitake, T. *J. Chem. Soc., Chem. Commun.* **1990**, 504.
8. Sakata, K.; Kunitake, T. *Chem. Lett.* **1989**, 2159.
9. Shimomura, M.; Kunitake, T. *Thin Solid Films* **1985**, *132*, 243.
10. Higashi, N.; Kunitake, T. *Chem. Lett.* **1986**, 105.
11. Umemura, J.; Hishiro, Y.; Kawai, T.; Takenaka, T.; Gotoh, Y.; Fujimura, M. *Thin Solid Films* **1989**, *178*, 281.
12. Chi, L. F.; Johnston, R. R.; Ringsdorf, H. *Langmuir* **1991**, *7*, 2323.
13. Suzuki, M.; Kakimoto, M.; Konishi, T.; Imai, Y.; Iwamoto, M.; Hino, T. *Chem. Lett.* **1986**, 395.
14. Baker, S.; Seki, A.; Seto, J. *Thin Solid Films* **1989**, *180*, 263.
15. Era, M.; Kamiyama, K.; Yoshiura, K.; Momii, T.; Murata, H.; Tokoto, S.; Tsutsui, T.; Saito, S. *Thin Solid Films* **1989**, *179*, 1.
16. Kunitake, M.; Nishi, T.; Yamamoto, H.; Nasu, K.; Manabe, O.; Nakashima, N. *Langmuir* **1994**, *10*, 3207.
17. Decher, G.; Hong, J. D.; Schmitt, J. *Thin Solid Films* **1992**, *210/211*, 831; Schmitt, J.; Grünewald, T.; Kajer, K.; Pershan, P.; Decker, G.; Lösche, M. *Macromolecules* **1993**, *26*, 7058.
18. Ueno, T.; Kunitake, T. *Chem. Lett.* **1990**, 1927.
19. Sano, M.; Sasaki, D. Y.; Kunitake, T. *J. Chem. Soc., Chem. Commun.* **1992**, 1326.

Ultraviolet Stabilizers

See: Photostabilizers

Uniaxial Sheets

See: Liquid Crystalline Polymers (Heat-Strengthened, Thermotropic)

UNIFORM POLYMERS

Koichi Hatada* and Tatsuki Kitayama
Department of Chemistry
Faculty of Engineering Science
Osaka University

A uniform polymer has been defined in an International Union of Pure and Applied Chemistry (IUPAC) document as "a polymer composed of molecules uniform with respect to relative molecular mass and constitution."[1]

PREPARATION AND PROPERTIES OF UNIFORM POLYMERS

An early attempt to obtain a uniform polymer involved the isolation of uniform polyoxymethylene diacetates with degrees of polymerization (DP) ranging from 1 to 26 by Staudinger and his co-workers.[2–4]

Very pure uniform polyoxymethylene diacetates ($8 \leq DP \leq 16$) were isolated recently by means of supercritical fluid chromatography (SFC), which is one of the promising techniques for the separation and isolation of uniform oligomers or polymers. These uniform polyoxymethylenes were evaluated for their physical properties and single crystal structures.[5]

Uniform polymethylenes were prepared by a stepwise reaction of oligomers up to a carbon number of 390.[6,7]

Uniform cycloalkanes were prepared to a molecular weight (MW) of 2805 using metathesis polymerization of cyclooctene with $WCl_6/EtAlCl_2$ followed by GPC fractionation and subsequent hydrogenation of the uniform cycloalkene.[8–10]

Oxidative polymerization of tetracosa-1,23-diyne followed by chromatographic separation and hydrogenation over a Pd/C catalyst gave a series of uniform linear alkanes up to $C_{384}H_{770}$.[11] The same reaction under high dilution conditions provided a mixture of cyclic oligomers which were fractionated chromatographically and then hydrogenated to give uniform cycloalkanes up to $C_{288}H_{576}$ (MW = 4040).[11]

*Author to whom correspondence should be addressed.

Uniform polyoxyethylenes (DP = 54) was prepared by the reaction of the α,ω-ditosylate and the monosodium salt of an 18mer at a 1:2 ratio.[12]

The preparation by a convergent growth method of dendritic macromolecules that are nearly uniform with respect to molecular weight was reported by several authors.[13,14]

SFC is one of the most promising techniques for the separation of polymers or high DP oligomers since supercritical mobile phase offers the advantage of high diffusibilities along with widely controllable power of solvation.[15–17]

Highly isotactic and highly syndiotactic PMMAs with narrow molecular weight distribution can be prepared by stereospecific living polymerizations with t-C_4H_9MgBr and t-C_4H_9Li–(n-$C_4H_9)_3Al$ (Li/Al = 1/3 mol/mol), respectively.[18–21] By SFC fractionation of these PMMA, uniform PMMAs with high stereoregularities were isolated up to a 100mer and their properties were investigated.[22–26]

Glass transition temperatures (T_gs) of the uniform PMMAs are higher than those for the nonuniform PMMAs with similar $\overline{DP}$s. The linear relation between T_g and reciprocal molecular weight indicates that the T_gs of isotactic and syndiotactic PMMAs with infinite molecular weight are 51.9 and 125.7°C, respectively. The reciprocal melting point of the uniform isotactic PMMA increased linearly with an increase in the reciprocal DP. Extrapolation of the linear relation to 1/DP = 0 gave a melting point for isotactic PMMA with infinite DP as 171.1°C.[24,26]

Application of Uniform Polymers

Synthetic polymers usually have distributions of molecular weight, stereoregularity, and copolymer composition, and their properties are always observed as averaged values. This is one of the characteristic features of polymer science. However, our knowledge of polymer science can be enhanced greatly with the study of uniform polymers. One of the synthetic applications of uniform polymer is the use of end-functionalized uniform polymer as building blocks to construct uniform polymer architectures with higher molecular weight.

The thermal degradation behavior of the uniform PMMAs was studied by thermogravimetry and the SFC analysis of the degradation products to estimate zip length. A determination of zip length on polymers with a broad molecular weight distribution is rather difficult, requiring massive amounts of calculations to fit the calculated molecular weight distribution to the observed distribution.[23] Early results for the uniform polymer have confirmed that it is extremely useful for the elucidation of zip length in thermal degradation.[21]

Uniform polymethylene was used to study the crystal thickness and folding behavior of lamellar polymethylene crystals. It was concluded that: (1) chain folding is found in all uniform polymethylenes starting with $C_{150}H_{302}$; (2) the fold length is always an integral reciprocal of the extended chain length; and (3) the end groups are located at the lamellar surface and the fold itself must be sharp and adjacently re-entrant.[7] It should be noted that this conclusion could not have been obtained without the uniform polymethylenes.

Uniform polymers are very useful as standard materials for polymer characterization. For example, uniform PMMAs are useful as GPC standards for the calibration of molecular weight and for the direct evaluation of instrumental spreading.[25]

Developments in synthetic methods and in separation science now allows us to obtain synthetic uniform polymers with repeat units up to 100 for PMMA (MW = 10070) and 388 for polymethylene (MW = 5473). Our knowledge of polymer science can be enhanced greatly with the study of uniform polymers although almost all of the experiments in polymer science have been carried out using nonuniform polymers. It is a challenging subject to prepare large quantities of uniform polymers by synthetic methods.

REFERENCES

1. International Union of Pure and Applied Chemistry, Macromolecular Division: Commission to Macromolecular Nomenclature; *Compendium of Macromolecular Nomenclature*; Blackwell Sci.: Oxford, U.K. 1991; p 521.
2. Staudinger, H.; Luthy, M. *Helv. Chim. Acta* **1925**, *8*, 41.
3. Staudinger, H.; Johner, H.; Signer, R.; Mie, G.; Hengstenberg, J. *Z. Physik. Chem.* **1927**, *126*, 425.
4. Staudinger, H.; Signer, R.; Johner, H.; Luthy, M.; Kern, W.; Russidis, D.; Schweitzer, O. *Ann. Chem.* **1929**, *474*, 145.
5. Ute, K.; Takahashi, T.; Matsui, K.; Hatada, K. *Polym. J.* **1993**, *25*, 1275.
6. Bidd, I.; Whitting, M. C. *J. Chem. Soc., Chem. Commun.* **1985**, 543.
7. Ungar, G.; Steiny, J.; Keller, A.; Bidd, I.; Whitting, M. C. *Science* **1985**, *229*, 386.
8. Sato, H.; Okimoto, K.; Tanaka, Y. *J. Macromol. Sci., Chem.* **1977**, *A11*, 767.
9. Ando, I.; Yamanobe, T.; Asakura, T. *Prog. NMR Spectrosc.* **1990**, *22*, 349.
10. Takenaka, T.; Yamanobe, T.; Komoto, T.; Ando, I.; Sato, H. *Solid State Commun.* **1987**, *61*, 563.
11. Lee, K. S.; Wegner, G. *Makromol. Chem., Rapid Commun.* **1985**, *6*, 203.
12. Kinugasa, S.; Takatsu, A.; Nakanishi, H.; Nakahara, H.; Hattori, S. *Macromolecules* **1992**, *25*, 4848.
13. Hawker, C. J.; Frechet, J. M. J. *J. Am. Chem. Soc.* **1990**, *112*, 7638.
14. Miller, T. M.; Neeman, T. X.; Zayas, R.; Bair, H. E. *J. Am. Chem. Soc.* **1992**, *114*, 1018.
15. Jentoft, R. E.; Gouw, T. H. *J. Polym. Sci.* **1969**, *B7*, 811.
16. Klesper, E.; Hartmann, W. *Eur. Polym. J.* **1978**, *14*, 77.
17. Hirata, Y.; Nakata, F. *J. Chromatogr.* **1984**, *295*, 315.
18. Hatada, K.; Ute, K.; Tanaka, K.; Kitayama, T.; Okamoto, Y. *Polym. J.* **1985**, *17*, 977.
19. Hatada, K.; Ute, K.; Tanaka, K.; Okamoto, Y.; Kitayama, T. *Polym. J.* **1986**, *18*, 1037.
20. Kitayama, T.; Shinozaki, T.; Masuda, E.; Yamamoto, M.; Hatada, K. *Polym. Bull.* **1988**, *20*, 505.
21. Kitayama, T.; Shinozaki, T.; Sakamoto, T.; Yamamoto, M.; Hatada, K. *Makromol. Chem. Suppl.* **1989**, *15*, 167.
22. Hatada, K.; Ute, K.; Nishimura, T.; Kashiyama, M.; Saito, T.; Takeuchi, M. *Polym. Bull.* **1990**, *23*, 157.
23. Ute, K.; Miyatake, N.; Asada, T.; Hatada, K. *Polym. Bull.* **1992**, *28*, 561.
24. Miyatake, N.; Osugi, Y.; Ute, K.; Hatada, K. *Polym. Prepr. Jpn.* **1993**, *42*, 342.
25. Ute, K.; Miyatake, N.; Osugi, Y.; Hatada, K. *Polym. J.* **1993**, *25*, 1153.

26. Hatada, K.; Ute, K.; Kitayama, T.; Nishiura, T.; Miyatake, N. *Makromol. Chem., Macromol. Symp.* **1994**, *85*, 325.
27. Hatada, K.; Nishiura, T.; Kitayama, T.; Tsubota, M. *Polym. Bull.* **1996**, *36*, 399.

UNSATURATED POLYESTER RESINS (Overview)

Güngör Gündüz
Kimya Mühendisliği Bölümü
Orta Doğu Teknik Üniversitesi

Unsaturated polyesters are mixtures of polyester prepolymers with some aliphatic unsaturation on the backbone and a vinyl monomer. The backbone resin is made of glycols, saturated acids, and unsaturated acids that provide available sites for reaction with vinyl monomers so that a three-dimensional network is formed.

The unsaturated polyesters have tremendous versatility with very low costs. They are easily handled, can be used alone, mixed or processed with other polymers, and compounded with a very large variety of fillers from powders to fibers because they have excellent wetting properties. The ease of moldability, availability, and low cost make them one of the most important polymers used industrially, and they are the most widely used thermosets in polymeric composites.

RAW MATERIALS

There are a number of potential raw materials for making unsaturated polyesters. One or more of a variety can be used:

- Glycols:propylene, ethylene, diethylene, glycerol, etc.
- Saturated acids: phthalic anhydride (or acid), adipic acid, sebacic acid, etc.
- Unsaturated acids: maleic anhydride (or acid), fumaric acid, itaconic acid, etc.
- Monomers: styrene, α-methylstyrene, methyl acrylate, methyl methacrylate, diallyl phthalate, triallyl cyanurate, etc.

The most common industrial unsaturated polyester prepolymer is made from 1,3-propylene glycol (PG), phthalic anhydride (PA), and maleic anhydride (MA), and the most common vinyl monomer used is styrene to make the final viscous solution.

CURING

Initiators and Accelerators

Curing of unsaturated polyester resin is achieved by using an initiator (or catalyst) and an accelerator. A wide range of peroxides and azo and azine compounds can be used as initiators depending on the curing temperature. For room-temperature curing, methyl ethyl ketone peroxide is used, and for moderate temperature (~60–90°C) curing benzoyl peroxide is used. For hot press or oven curing (~130–150°C) di-t-butyl peroxide or t-butyl perbenzoate is used.

To accelerate the decomposition of peroxides, some metal compounds, tertiary amines, and mercaptans can be used. Cobalt naphthenate (CoNp) and cobalt octanoate (CoOc) are the most widely used metal compounds as accelerators.

Crosslinking

Styrene polymerizes and connects the MA moieties on the prepolymer chains. Hence, a three-dimensional network is formed owing to crosslinking. The MA introduced as the raw material transforms into fumarate in the course of polyesterification reaction. Fumarate and maleate are *trans-* and *cis-* forms, respectively, and the fumarate has a high affinity to react with styrene. MA usually gives large fractions of fumarate when reacted with PG. The copolymerization of styrene with fumarate forms microgels even at conversion levels as low as 3–4%. The crosslinking reactions then proceed as intramicrogel and intermicrogel reactions.[1–5]

Thickening

Sheet and bulk molding compounds are chemically thickened compositions. This is achieved with group IIA metal oxides and hydroxides, especially with MgO, which is added at about 5%.

STRUCTURE–PROPERTY RELATIONS

In industrial prepolymer synthesis the MW is kept constant, around 1100, otherwise gelation can take place. The styrene/prepolymer ratio can be changed as wished and products of different properties can be produced. High mechanical properties are achieved at a styrene/MA ratio of around 2.

Effect of Alcohols

The use of linear and relatively long alcohols such as diethylene (DEG) or 1,4-butylene glycol usually gives great flexibility to the chain, so they can be used together with PG to make flexible chains.

Effect of Acids

The ortho- and iso-phthalic acids are the most commonly used saturated acids while MA is the most commonly used unsaturated acid in the industry. The effects of acids are more or less similar to the effects of glycols such that aliphatic acids such as adipic acid or sebacic acid impart flexibility, and an increase in the MW of aliphatic acid further increases flexibility of the chain formed.

Chain Length

The increase in the chain length of prepolymer greatly enhances the chemical resistance. This is due to a decrease in the amount of hydrophilic –OH and –COOH end groups.

Isomerization

The mechanical properties of the final cured product highly depend on the interaction of the double bonds on the backbone with vinyl monomer. It is known that fumarates copolymerize much faster than maleates. Therefore maleate-to-fumarate transition must be achieved as much as possible during resin synthesis.

Vinyl Monomers

The vinyl monomer acts as solvent for the prepolymer and copolymerizes with the unsaturated bond on the prepolymer backbone, yielding a thermoset product. The monomer to be added must copolymerize, have relatively low vapor pressure, low heat of reaction, and also low cost. Styrene fulfills these requirements and it is the most widely used monomer in the industry.

Glass Transition Temperature

Glass transition temperature (T_g) depends on the chemical compounds used in the synthesis of prepolymer and the vinyl monomer in making the resin. Besides the chemical effects, T_g increases with an increase in the MW of the prepolymer. T_g depends also on styrene/prepolymer ratio, and the maximum T_g is achieved as 135°C when this ratio is between 2 and 3.[6]

FILLERS, ADDITIVES, MODIFIERS

Fillers are used to lower the cost and modify the properties of the polymer. They can be used for toughening, reinforcement, plasticization, wear resistance, environmental stability, and flame retardancy. They permit high curing temperatures and also reduce cracking and crazing.

FLAME RETARDANT POLYESTERS

Flame retardant compounds either cool the pyrolysis zone of the burning polymer, such as alumna hydrate, which evolves water above 250°C, or inhibit free radical combustion, such as halogenated organic compounds, or else form a protective layer on the polymer surface, such as phosphorous compounds. The use of halogenated compounds finds more applications than the other two methods.

APPLICATIONS

Unsaturated polyesters have a very wide area of application either in the form of pure resin or in the compounded form with fillers. Some special chemicals such as plasticizers, antioxidants, heat and ultraviolet stabilizers, and flame retarders can be added to the resin depending on the area of application. One of the early uses of unsaturated polyesters was to produce cast items such as knife and umbrella handles, encapsulation of electronic assemblies, and embedding decorative, zoological, geological, and anatomical specimens. The most important casting application is the manufacture of pearl buttons. Another important application is to manufacture floor-tiles by mixing unsaturated polyester resin with fillers such as limestone, baryte, silica, china clay, and pigments.[7] Polyester compounds can be formulated for the manufacture of bathroom fixtures such as tubs and lavatories, and also vanity tops, bar tops, kitchen tops, and panels. Coating of concrete, pipe, or cast-iron for corrosion control can be successfully accomplished by using unsaturated polyester resins. However polymer impregnation of concrete can give both corrosion resistance and outstanding mechanical properties.[8]

The development of bulk and sheet molding compounds by using glass fiber in the composition extended the application of unsaturated polyesters to a variety of areas such as transportation, electrical appliances, building, and construction. Unsaturated polyesters have a leading role in the development of glass fiber reinforced products.

Sheet molding compounds or prepregs can be made by mixing unsaturated polyester resin, filler, chopped strand mat, pigment, and catalyst and then shaping them in the form of a pliable sheet. Prepregs then can be pressed to give the shape of the mold.

Unsaturated polyester-based composites have extensive applications in manufacture of tanks, containers, automobile bumpers, aircraft components, office, household and outdoor furniture, caskets, and sporting equipment. Unsaturated polyesters find also large applications in the optical and electrical industries. However new polymers developed have partly substituted unsaturated polyesters in these areas.

REFERENCES

1. Yang, Y. S.; Suspene, L. *Polym. Eng. Sci.* **1991**, *31*, 321.
2. Huang, Y. J.; Chen, C. J. *J. App. Poly. Sci.* **1993**, *47*, 1533.
3. Huang, Y. J.; Lu, T. J.; Hwu, W. *Polym. Eng. Sci.* **1993**, *33*, 1.
4. Hsu, C. P.; Lee, L. J. *Polymer* **1993**, *34*, 4497.
5. Yang, Y. S.; Lee, L. J. *Polymer* **1988**, *29*, 1793.
6. Lucas, J. C.; Borrajo, J.; Williams, J. J. *Polymer* **1993**, *34*, 3216.
7. Doyle, E. N. *The Development and Use of Polyester Products*; McGraw-Hill: 1967.
8. Gündüz, G. *Steel-Fiber Reinforced Polymer Impregnated Concrete*; in *Handbook of Ceramics and Composites*; Cheremisinoff, N. P., Ed.; Marcel Dekker: 1992; Chapter 6.

UNSATURATED POLYESTER RESINS (Polyester Composites)

Carlos F. Jasso-Gastinel
Department of Chemical Engineering
University of Guadalajara

Ovidio Laguna
Polymer Science and Technology Institute (CSIC)

Since their development in the 1940s, unsaturated polyester resins have been widely accepted as matrices for composite materials. Their importance within thermoset polymers is mainly due to low cost, along with high strength/weight ratio. In addition, effortless variations in formulation to look for extra specific properties as well as easiness and versatility of processing, make them very attractive for various industrial applications replacing wood and metal articles.

PREPARATION

Blending of the polyester resin with other polymers, creating a new type of compounds, has been developing a new family of materials. Such hybrid products can be of the interpenetrating polymer network (IPN) type, or simply physical blends with high molecular weight polymers like PVC.[1,2] Those technologies opened the field for numerous investigations in recent years using isocyanates to form a binder material useful for making reinforced composites for example.[3]

Flame Resistance

To reduce high flammability of polyester resins, halogenated components, such as chlorendic anhydride, have been used. Dense smoke generation by combustion decreases using methyl methacrylate monomer and/or increasing crosslinking density with the presence of divinyl benzene.[4]

Chemical Resistance

The type and concentration of ester groups, crosslinking density, and average molecular weight influence the chemical resistance of polyester composites for industrial applications. Chemical vulnerability exists mainly on account of the ester group; for that reason it is common to accomplish protection using bulky aromatic glycols or diacids. On that line, bisphenol-A systems and vinyl ester resins give better protection than isophthalic systems.

Crosslinking Reaction

Because crosslinking kinetics are very important to control thermosetting processing time, conversion studies have multiplied using various approaches. However, because resin, monomer, initiator, and activator concentrations may affect reaction rate, it is advisable to choose particular formulations through experimental design. Jasso et al. used a factorial composite design to study gel and cure times including interaction effects of the variables involved in commercial formulations.[5]

REINFORCEMENT

Properties of unsaturated polyester resins are generally modified by the inclusion of continuous (filaments, fibers, wires, tapes, etc.) or discontinuous (flakes, particulates, whiskers, etc.) geometries of organic and/or inorganic materials. The resultant composite material will improve one or more characteristics in properties such as mechanical (strength, stiffness, creep, etc.), chemical (corrosion, flammability, etc.), or thermal (expansion coefficient, heat capacity, etc.). In addition, reinforcements simplify formability and may help in cost reduction.

Glass fiber in different formulations sizes and designs is by far the most common reinforcement material.

For both continuous and discontinuous reinforcements, the higher the ability for interaction (in physicochemical meaning) with the polymer, the more cooperative will be the composite performance.

FABRICATION PROCESSES

A wide variety of processing techniques to accomplish different purposes are currently in use for unsaturated polyester resins. Fabrication can be achieved using room temperature or heating even above 100°C. Pressure can be 0–2.76 MPa (400 psi) at the "low level" or up to 13.78 MPa (2000 psi) at the "high level." Technology may be complicated to meet specific requirements or simple for huge articles (yacht hulls, roofs, etc.).

Polyester molding compounds are available using short fibers (3–6 mm long) to make "bulk molding compounds" (BMC) and long fibers (> 25 mm long) to make "sheet molding compounds" (SMC). The former are widely used because of easy mixing. SMC exhibit good strength properties, thanks to fiber length. Using thickener and reinforcement, the properties of the "thick molding compounds" (TMC) are expected to fall between those of the BMC and SMC. A review looking at techniques and trends was presented by Kamal and Ryan.[6]

CHARACTERIZATION AND PROPERTIES

By blending rubber or mixing reactive liquid rubber with unsaturated polyester resins, toughened SMCs can be prepared and be considered for engineering applications.[7] Fatigue mechanisms of fiber filled composites are generally associated with initiation of microcracks in the matrix, debonding of fiber matrix interface, and fiber ends cracking. Such fatigue behavior will be reflected on changes in size and shape of hysteresis loops of SMCs.[8] Energy dissipation during dynamic deformation is higher in discontinuous than in continuous fiber composites.[9,10] That means that fiber ends and matrix fiber interfaces contribute greatly to energy dissipation. On such a property the aspect ratio arises again as a variable. Nevertheless, heat generation at high frequencies and loads can not be dissipated easily, raising the temperature, which in turn will further increase damping, decreasing strength. That sequence leads to macroscopic failure.

Studies on unsaturated polyester resins reinforced with cellulosic materials have shown the feasibility of making BMC type materials.[11] Of course these composites show lower strength properties than glass reinforced polyesters but they possess comparable specific stiffness[15] and impact resistance.[11,12]

Although Satyanarayana et al. have reported the preparation of cellulosic reinforced unsaturated polyester resins consumer articles, demonstrating the potential of cellulosic fibers for nondemanding applications, more research has to be done on fiber chemical treatment and coupling agents-diluted resin interaction.[11]

On thermal properties, an important general feature of reinforcements is their influence in the increase of heat distortion temperature as well as glass transition temperature and decrease of thermal expansion coefficient. The same kind of modifications are expected to increase crosslink density. The inclusion of tough (rubbery) components in the composite enhancing mechanical properties can also decrease brittleness at low temperatures.

Dielectric character of unsaturated polyester resins becomes benefited by low dipolar crosslinking monomers. Reinforcements increase resistance to failure especially including inorganic fillers. In general, isophthalic resins can improve performance in electrical applications, and vinyl toluene monomer can contribute for better thermal resistance. For specific properties' data, the reader can look at the *Modern Plastics Encyclopedia*.[4]

APPLICATIONS

The broad variety of end-use applications for reinforced unsaturated polyester resins in conjunction is much larger in volume than non-reinforced resins, in spite of impressive growth in artificial marble production.

Some important fields of application of unsaturated polyester resins have well established markets and are still expected to grow along with markets where specific properties are important (like corrosion, oxidation, or electrical resistance). Major fields include marine industry, transportation, and construction (residential and industrial). Electrical applications are expected to grow more rapidly with automation of SMC/BMC products because these composites provide better electrical properties than epoxy and phenolic resins at high temperatures. Good surface appearance (Class A finish) for decorative parts and protective layers for the finished product are sometimes needed. A protective layer (gel coat) is mineral-filled resin applied as a first layer to the mold. It can be clear or pigmented, hiding defects especially for marine and sanitary ware.

In recent times, interest in thermoset recycling has appeared for contributions to tertiary recycling, which is "the recovery of chemical products from waste plastics." That is a much better option than purely grinding reinforced thermosets to use the grind as reinforcing material (for secondary recycling). Pyrolisis of SMC can be used along with filler recovery.[13] Tesoro and Wu reported characterization of soluble products obtained from a neutral hydrolysis of a cured unreinforced polyester resin.[14] The oligomer obtained was explored as an acid curing agent for epoxy resin with promising results. Recycling technologies for thermosets including unsaturated polyester resins are emerging and will refrain in environmental negative impact. Traditional reinforced unsaturated polyester resins are expected to increase in use in conjunction with hybrid formulations that extend their possible applications by expanding their properties.

REFERENCES

1. Klempner, D.; Frisch, K. C.; Frisch, H. L. *Polym. Eng. Sci.* **1974**, *17*(9), 646.
2. Hosti, G. *Rev. Gen. Caout. Plast.* Ed. Plastiques, **1968**, *5*, 207.
3. Rust, D.; Williams, P. *SPI/RPC* Session 10F 1988.
4. Miller, D.; Petrella, R.; Manca, A. *Mod. Plast.* September 1976, 96; *Modern Plastics Encyclopedia*; McGraw-Hill: (revised annually).
5. Jasso, C. F.; Velazquez, S.; Mendizabal, E., unpublished results.
6. Kamal, M. R.; Ryan, E. R. *Adv. Polym. Technol.* **1984**, *4*, 323.
7. Tong, S. N.; Wu, P. T. K. *Polym-Plast Technol. Eng.* **1988**, *27*(4), 519.
8. Wang, S. S.; Chim, E. S. M. *J. Compos. Mater.* **1983**, *17*, 114.
9. Gibson, R. F.; Yau, A. *J. Compos. Mater.* **1980**, *14*, 155.
10. Gibson, R. F. *J. Eng. Mater. and Perf.* **1992**, *1*(1), 11.
11. Owolabi, O.; Czvikovszky, T.; Kovacs, I. *J. Appl. Polym. Sci.* **1985**, *30*, 1827.
12. Jasso, C. F.; Hernández, H. R.; Mendizabal, E. *Polimex 93 prepr., Int. Symp. on Polym.*, 1993; 287.
13. Vernyi, B. *Plast. News* 26 February 1990, p 1,7.
14. Tesoro, G.; Wu, Y. *Adv. in Polym. Technol.* **1993**, *12*, 185.
15. Satyanarayana, K. G.; Sukuraman, K.; Kulkarni, A. G.; Pillai, S. G. K.; Rohatgi, P. K. *Compos.* **1986**, *17*, 4, 329.

UNSATURATED POLYESTER RESINS (Toughening with Elastomers)

Douglas S. McBain
Specialty Polymers Division
GenCorp

Unsaturated polyester resins are used widely for glass fiber-reinforced composites due to low cost, performance properties, and the ease with which they are processed with fillers and reinforcements. Improvements in the physical properties and impact resistance of cured unsaturated polyester matrix composites are highly desirable, as increased toughness would be of great commercial value.

The chemical constituents of the unsaturated polyester backbone can be manipulated effectively to modify some end-use properties.[1–3]

Other physical and chemical approaches to the modification of unsaturated polyester resin properties have been reported. These approaches include the use of dispersed elastomers, both liquid and solid, to enhance the toughness of the cured resin matrix.

LIQUID RUBBER MODIFICATION

The predominant materials that have characterized the liquid rubber additive approach have been polybutadiene and butadiene-acrylonitrile copolymers and the more polar polyepichlorohydrin.[4–24] Terminal and pendant functionality of these liquid rubbers has included hydroxyl, carboxyl, amine, and vinyl groups.

SOLID RUBBER MODIFICATION

The use of elastomers to modify physically and enhance the toughness and crack-resistance of unsaturated polyester resin molding compositions has been reported frequently.

Generally speaking, addition of an incompatible elastomer results in a decrease in modulus and loss of strength. Those materials with functional groups (terminal or pendant) such as carboxyl, amino, or vinyl, which can react with the polyester matrix, are preferred to their nonfunctionalized counterparts and display better properties in the cured mixture.

URETHANE MODIFICATION

The use of soluble, low molecular weight polyether urethanes as toughening agents for unsaturated polyester systems also has been reported.[25–27]

SUMMARY

The use of elastomers, predominantly the low molecular weight reactive liquid rubbers, for improving the toughness and impact resistance of unsaturated polyester resins has been persistently reported. The improvements in properties and toughness found with added elastomers are relatively less than those obtained in other thermosetting resins, such as epoxies.

Many of the elastomeric additives that seem promising in laboratory studies may not perform similarly or adequately when the system is cured under actual plant processing conditions, leading to their limited viability for actual commercial use.

REFERENCES

1. Boenig, H. V. *Unsaturated Polyesters: Structure and Properties*; Elsevier: Amsterdam, 1964.
2. Burns, R. *Polyester Molding Compounds*; Marcel Dekker: New York, 1982.
3. Pritchard, G.; Rhoads, G. V. *Poly. Composites* **1981**, *2*, 179.
4. Crosbie, G. A.; Phillips, M. G. *37th Ann. Conf. Comp. Inst. SPI*, 1982.
5. Crosbie, G. A.; Phillips, M. G. *J. Mat. Sci.* **1985**, *20*, 563.
6. Drake, R. S.; Siebert, A. R. *42nd Ann. Conf. Comp. Inst. SPI*, 1987.
7. Lee, B. L.; Howard, F. H. *36th Ann. Conf. Comp. Inst. SPI*, 1981.
8. Lee, B. L.; Riew, C. K.; Siebert, A. R. *Poly. Mat. Sci. Eng.* **1990**, *63*, 686.
9. McCarthy, W. J.; Bowerman, H. H. *29th Ann. Conf. Comp. Inst. SPI*, 1974.
10. McGarry, F. J.; Sultan, J. N. *42nd Ann. Conf. Comp. Inst. SPI*, 1969.

11. McGarry, F. J.; Rowe, E. H.; Riew, C. K. *32nd Ann. Conf. Comp. Inst. SPI*, 1977.
12. McGarry, F. J.; Rowe, E. H.; Riew, C. K. *Poly. Eng. Sci.* **1978**, *18*, 78.
13. Nichols, C. S.; Horning, G. J. *38th Ann. Conf. Comp. Inst. SPI*, 1983.
14. Riew, C. K.; Rowe, E. H.; Backderf, R. H.; Guiley, C. D. *30th Ann. Conf. Comp. Inst. SPI*, 1975.
15. Riew, C. K.; Goodrich, B. F. U.S. Patent 4 274 994, Re. 31,577, 1984.
16. Rowe, E. H.; Howard, F. H. *33rd Ann. Conf. Comp. Inst. SPI*, 1978.
17. Rowe, E. H. *34th Ann. Conf. Comp. Inst. SPI*, 1979.
18. Rowe, E. H.; McGarry, F. J. *35th Ann. Conf. Comp. Inst. SPI*, 1980.
19. Siebert, A. R.; Rowe, E. H.; Riew, C. K. *27th Ann. Conf. Comp. Inst. SPI*, 1972.
20. Siebert, A. R.; B. F. Goodrich Co. U.S. Patent 5 268 452, 1985.
21. Suspene, L.; Yang, Y. S.; Pascault, J. P. *Additive Effects on the Toughening of Unsaturated Polyester Resins*; In *Rubber-Toughened Plastics: Adv. Chem. Ser.*; Riew, C. K., Ed.; American Chemical Society: Washington, DC, 1989; Vol. 222.
22. Tetlow, P. D.; Mandell, J. F.; McGarry, F. J. *34th Ann. Conf. Comp. Inst. SPI*, 1979.
23. Ullett, J. S.; Chartoff, R. P. *Poly. Mat. Sci. Eng.* **1994**, *70*, 100.
24. Ullett, J. S.; Chartoff, R. P. *Poly. Mat. Sci. Eng.* **1994**, *70*, 291.
25. Kim, D. S.; Park, C. E. *50th Ann. Conf. SPE*, 2606, 1992.
26. Kim, D. S.; Cho, K.; An, J. H.; Park, C. E. *J. Mat. Sci. Letters* **1992**, *11*, 1197.
27. Kim, D. S.; Cho, K.; An, J. H.; Park, C. E. *J. Mat. Sci. Letters* **1994**, *29*, 1854.

UNSATURATED POLYESTER RESINS (Toughening with Liquid Rubber)

Jill S. Ullett and R. P. Chartoff
The Center for Basic and Applied Polymer Research
The University of Dayton

Unsaturated polyester (UP) resins are one of the most important commercial thermosetting resins and are used in a variety of applications including architectural panels, boats, auto body panels, and appliance housings. Vinyl esters (VE) are used often when chemical resistance is required. Like other thermosets, unsaturated polyester and VE resins crosslink to form a brittle matrix with low fracture toughness and impact strength values.

The objective for liquid rubber additions is to generate a dispersed rubbery phase which will increase energy dissipation during fracture without adversely affecting the stiffness and heat deflection properties of the matrix.

MORPHOLOGICAL FEATURES REQUIRED FOR TOUGHNESS

In general, a rubber-modified thermoset should have: a significant number (or volume fraction) of second-phase particles, good dispersion of the particles throughout the matrix, particle sizes that are not too big nor too small, and a particle matrix interface that is strong enough. Ullett and Chartoff[5] observed a significant dependence of fracture toughness measured in compact tension on the volume fraction of rubbery second-phase particles in the unsaturated polyester and VE resins they studied.

DISPERSION OF RUBBER PHASE

A well-distributed second phase is important for maximizing improvements in fracture toughness while minimizing the deleterious effects of a particulate phase.

RESIN RUBBER COMPATIBILITY AND PHASE SEPARATION

Ideally, one would like a level of compatibility between the liquid rubber and unreacted resin such that they form one phase when mixed. In order to obtain compatible resin–rubber blends, researchers have tried varying resin chemistry including end and pendent groups, and have synthesized block copolymers to be used as toughening additives or as compatibilizing agents for commercially available liquid rubbers.

LIQUID RUBBER ADDITIVES

As described in the literature, many types of liquid elastomers have been used to modify various thermosetting resins. Poly(butadiene-acrylonitrile) (NBR) is one of the most often used rubbers and has been modified in a variety of ways to improve its compatibility with various resins. Focusing on polyester resins, blends containing NBR with carboxyl end groups,[2–6] with vinyl end groups,[1–4,6–9] and with amine end groups[2,5,10] have been reported.

AFFECTS OF LIQUID RUBBER ADDITION ON BLEND MECHANICAL PROPERTIES

Addition of liquid rubbers in quantities typically used for toughening unsaturated polyester resins generally result in slight to moderate reductions (<5% to 25%) in tensile and flexural stiffness.[1,3,8,10,11] Tensile strength may increase[8,10,12] or decrease[1,10–12] depending on many factors such as rubber concentration and second-phase particle size.

AFFECTS OF LIQUID RUBBER ADDITION ON BLEND GLASS TRANSITION TEMPERATURES

Small to moderate additions of liquid rubbers to unsaturated polyester resins may result in an increase, a decrease, or in no change of the cured unsaturated polyester resin glass transition temperature.

FUTURE STUDIES

Although some progress has been made in understanding the effects of morphology on fracture toughness, more work is needed to better define the relative importance of particle size, interparticle spacing, particle structure, and particle-matrix adhesion. Old and new techniques should be explored to improve resin–rubber compatibility so that particle size can be better controlled and manipulated. Further research in the area of rubber-modified fiber-reinforced molding compounds, such as SMC (sheet molding compound), is indeed. The toughening mechanisms active in liquid rubber-modified SMC materials are not well understood. Interactions between the rubbery additive and each of the typical SMC components (e.g., glass, filler, and low profile additive) have not been adequately researched. Interfacial effects between the rubber and glass fibers and between the rubber and filler particles need to be investigated.

APPLICATIONS

Applications for liquid rubber modified vinyl esters exist in the corrosion prevention market.[13] For example, toughened vinyl esters are used as primer coatings for storage tanks. The principal application for liquid rubber modified unsaturated polyester resins is in short fiber-reinforced molding compounds used to make automotive body parts.[13]

REFERENCES

1. Ullett, J. S.; Chartoff, R. P. *Poly. Engr. and Sci.* accepted for publication, 1994.
2. Tetlow, P. D.; Mandell, J. F.; McGarry, F. J. *Prod. 34th Annu. Conf. RP/CI, SPI* session 23F, 1979.
3. Crosbie, G. A.; Phillips, M. G. *J. of Mat. Sci.* **1985b**, *20*, 563.
4. Kostanski, L. K.; Krolikowski, W. *Int. Poly. Sci. and Tech.* (translated by Niesiolowska D.) **1985**, *12*, T/131.
5. Grossman, R. F. *Rubber Toughened Plastics*; Riew, C. K.; Ed.; ACS Series 222, American Chemical Society: New York, 1989; Chapter 21.
6. Ullett, J. S. Ph.D. Dissertation, University of Dayton, 1992.
7. McGarry, F. J.; Rowe, E. H.; Riew, C. K. *Poly. Engr. and Sci.* **1978**, *18*, 78.
8. Nichols, C. S.; Horning, G. T. *Proceedings 38th Annu. Conf. RP/CI, SPI* session 19-B, 1983.
9. Drake, R. S.; Siebert, A. R. *42nd Annu. Conf. Comp. Inst., SPI* session 11d., 1987.
10. McGarry, F. J.; Subramanium, R. *Proceedings of the American Chemical Society, Division of Polymeric Materials: Science and Engineering*; American Chemical Society: Washington, 1993; Vol. 70, 95.
11. Gatward, C. H.; Hogg, P. J.; Hull, D. *6th Int. Conf. on Deformation Yield and Fracture of Polymers*; The Plastics and Rubber Institute: Cambridge, 1985; paper 32.1.
12. Crosbie & Phillips, 1983.
13. Egan, D.; Goodrich, B. F. Co., personal communication.

Unsaturated Polyesters

See: Carbenoid Polycondensation
Cardo Polyesters (Heat Resistant and Photosensitive)
Dimer Acid-Based Polyesters
Peroxide Initiators (Overview)
Resins (Applications in Coatings Industry)
Unsaturated Polyester Resins (Overview)
Unsaturated Polyester Resins (Polyester Composites)
Unsaturated Polyester Resins (Toughening with Elastomers)
Unsaturated Polyester Resins (Toughening with Liquid Rubber)

UREA–FORMALDEHYDE ADHESIVE RESINS*

Anthony H. Conner
Forest Products Laboratory
USDA Forest Service

Approximately 1 million metric tons of urea–formaldehyde resin are produced annually. More than 70% of this urea–formaldehyde resin is used by the forest products industry for a variety of purposes.[1] The resin is used in the production of an adhesive for bonding particleboard (61% of the urea–formaldehyde used by the industry), medium-density fiberboard (27%), hardwood plywood (5%), and a laminating adhesive for bonding (7%), for example, furniture case goods, overlays to panels, and interior flush doors.

Urea–formaldehyde resins are the most prominent examples of the class of thermosetting resins usually referred to as amino resins.[2,3]

CHEMISTRY OF UREA–FORMALDEHYDE RESIN FORMATION

Urea–formaldehyde resins are formed by the reaction of urea and formaldehyde. The overall reaction of urea with formaldehyde is quite complex and, although initially studied early in this century, is not completely understood at the present time.[4]

The synthesis of a urea–formaldehyde resin takes place in two stages. In the first stage, urea is hydroxymethylolated by the addition of formaldehyde to the amino groups.

The second stage of urea–formaldehyde resin synthesis consists of the condensation of the methylolureas to low molecular weight polymers.

FORMALDEHYDE EMISSION FROM UREA–FORMALDEHYDE RESINS

Few issues in the forest products industry rival the debate and concern over the emission of formaldehyde from products bonded with urea–formaldehyde adhesive resins.

The evolution of formaldehyde from urea–formaldehyde materials is incontrovertible. We can conclude that the reactions leading to the formation of the urea–formaldehyde products formed during urea–formaldehyde resin synthesis and cure are reversible. In the forward direction, water is eliminated; therefore, the reverse reactions can be viewed as hydrolysis, which leads to the release of formaldehyde.[5] Because most, if not all, of these reactions are catalyzed by acid, the use of an acid catalyst to hasten bond cure unfortunately also increases the rate of hydrolysis and formaldehyde liberation.

The most widely used approach for reducing formaldehyde emission levels has emphasized decreasing the mole ratio of F/U.

Recent research has suggested possible new methods to lower formaldehyde emission levels. This research involves two strategies: the modification of the chemistry of urea–formaldehyde resins and the replacement of the formaldehyde component in urea–formaldehyde resins with a less volatile aldehyde or its chemical equivalent.

CONCLUSION

Urea–formaldehyde resin is a major commercial adhesive, especially within the forest products industry. It offers a number of advantages when compared with other adhesive systems. However, despite the fact that great strides have been made to offset its major disadvantage by lowering the formaldehyde emission levels of products bonded with urea–formaldehyde adhesive resin, the industry still faces the possibility of more restrictive regulations on formaldehyde in dwellings. Moreover, tighter restrictions on formaldehyde levels in the workplace are also likely. New research efforts are needed to address this concern if urea–formaldehyde adhesive resins are to maintain their prominent position as a versatile adhesive system.

REFERENCES

1. White, J. T. *For. Prod. J.* **1995**, *45*(3), 21.
2. Williams, L. L. In *Kirk-Othmer Encyclopedia of Chemical Technology*, 4th ed.; Kroschwitz, J. I.; Howe-Grant, M., Eds.; John Wiley & Sons: New York, NY; 1991; Vol. 2, pp 604–637.
3. Updegraff, I. H. In *Handbook of Adhesives*, 3rd ed.; Skeist, I., Ed.; van Nostrand Reinhold: New York, 1990; 341–346.
4. Pizzi, A. In *Wood Adhesives: Chemistry and Technology*; Pizzi, A., Ed.; Marcel Dekker: New York, 1983.
5. Myers, G. E. In *Wood Adhesives in 1985: Status and Needs*; Christiansen, A. W. et al., Eds.; Forest Prod. Res. Soc.: Madison, WI, 1986.

UREA–FORMALDEHYDE GLUE RESINS

Manfred Dunky
Krems Chemie AG

Urea–formaldehyde glue resins are the most important type of urea–formaldehyde-resins, and are thermosetting duromers consisting of oligomer and linear or branched polymer molecules, which after the hardening process, are an insoluble three-dimensional network and not further able to be thermoformed by melting. In the stage of application, urea–formaldehyde-glue resins are still, more or less, soluble or dispersed in water. The expression "glue resin" is valid especially for use in the wood-working industry.

CHEMISTRY OF UREA–FORMALDEHYDE-GLUE RESINS

Methylolation of urea and condensation are the two main reactions during synthesis of a urea–formaldehyde-glue resin. Both types of reactions depend on molar ratio F/U, on pH, on temperature, on concentration of the reaction mixture, and on reaction time. The reaction takes place in aqueous medium because of good solubility of both urea and formaldehyde.

Co-Condensation with Melamine

The low resistance of pure urea–formaldehyde-glue resins against attach of humidity and water, especially at higher temperatures, can be improved by the addition of melamine in different forms. Because of the high costs, pure melamine-formaldehyde-glue resins are not in use.

PRODUCTION OF UREA–FORMALDEHYDE-RESINS

Production of urea–formaldehyde-resins is influenced by several parameters and needs precise control of course, speed, and extent of reaction: molar ratio F/U, concentration of reaction mix, temperature, pH, type and amount of catalysts (acids, bases, ammonium salts), reaction time, and impurities of raw materials (e.g., content of methanol and formic acid in formaldehyde). Usually a three-stage process (at least) as described by Horn, is used for the production of glue resins.[3]

PARAMETERS FOR UREA–FORMALDEHYDE-GLUE RESINS FOR THE WOOD-WORKING INDUSTRY

Influence of Molar Ratio F/U on Properties of Glue Resins

Since formaldehyde is the real reactive component in urea–formaldehyde-resins, the molar ratio F/U has critical influence on hardening reaction with steeply increasing content of free formaldehyde with higher F/U. The hardening reaction is induced by acid generating reaction of free formaldehyde of resin with the hardener (ammonium salt); therefore, a higher reactivity and a tighter network formation can be examined at higher F/U.[1]

Influence of Degree of Condensation

At a given solid content and molar ratio viscosity of glue resins increases with degree of condensation, that means with lower compatibility with water or higher turbidity point. The degree of condensation equals with higher molar masses.[2]

RESINS FOR PARTICLEBOARD AND MEDIUM-DENSITY FIBERBOARD

Particleboards and medium-density fiberboards with low subsequent formaldehyde emission need urea–formaldehyde-glue resins with molar ratios F/U near equimolar composition. In fact, F/U of modern urea–formaldehyde-glue resins is 1.02 to 1.08. At the higher end of this range, and especially for resins with still higher F/U, addition of formaldehyde catchers, special pretreatment of fresh or glued particles, or post-treatment of boards is necessary.

GLUE RESINS FOR PLYWOOD, BLOCKBOARD AND FURNITURE MANUREA–FORMALDEHYDEACTURE

Use of urea–formaldehyde-resins gives glue lines without special resistance to humidity, water and weather. To get better resistance use of melamine together with the urea–formaldehyde-resin (urea–formaldehyde + MF, Murea–formaldehyde, MUPF, urea–formaldehyde + MF – powder, urea–formaldehyde + powdered melamine) or use of phenolic/resorcinolic resins is necessary. For industrial use liquid urea–formaldehyde-resins are mixed with fillers/extenders (e.g., rye flour, wheat flour, beechwood flour, and others), hardeners (mostly special commercial ones), and formaldehyde scavengers (often commercial hardener/catcher combinations).

SUBSEQUENT FORMALDEHYDE-EMISSION FROM WOOD-BASED PANELS

The main task for the urea–formaldehyde-resin industry for the last 15 years, starting at the end of the 1970s, was to decrease subsequent formaldehyde emission from wood-based panels, especially particleboards. Today chemical industry can be proud to claim that this problem has been solved in conjunction with the wood-working industry. This was done by consequent decreasing molar ratio F/U without dramatic loss of glue line bond strength, and even at much lower specific pressing times due to improved presses (general use of continuous press technology since mid-1980s), and higher press temperatures.

SUMMARY AND PROSPECT

Urea–formaldehyde-glue resins are one of the oldest synthetic adhesives for wood, but there is still great potential for development. Future urea–formaldehyde-glue resins will be of decisive importance for the wood-based panels industry. The consequent and permanent improvement of urea–formaldehyde-glue resins, and the possibility of mixed application with other compounds, guarantee better and new forms of application. The low price of urea–formaldehyde-resins, as compared with other adhesives, emphasizes the great importance of urea–formaldehyde-glue resins.

REFERENCES

1. Dunky, M. *Holzforsch. Holzverwert.* **1985**, *37*, 75.
2. Dunky, M. *Holz Roh Werkst.* **1990**, *48*, 371.
3. Horn, V.; Benndorf, G.; Rädler, K. P. *Plaste Kautsch.* **1978**, *25*, 570.

Urea-Formaldehyde Resins

See: *Urea-Formaldehyde Adhesive Resins*
Urea-Formaldehyde Glue Resins

Urea Inclusion Complexes

See: *Inclusion Complexes (Overview)*
Inclusion Polymerization
Molecular Complexes [Poly(ethylene oxide) and Urea]

Urethane Dispersions

See: *Water-Borne Coatings (Urethane/Acrylic Hybrid Polymers)*

Urushi

See: *Japanese Lacquer: Japan: Urushi (Properties of Urushi Liquid and Urushi Film)*

UV-Curing

See: *Photoinduced Polymerization*
Photoinitiators (for Free Radical Polymerization)
Photoinitiators (for Photocuring)
Photomodification (Pendant Acyloxyimino Groups to Amino Groups)

V

Vanadate Ion Complex Gels

See: Poly(vinyl alcohol)-Ion Complex Gels

Vapor Deposition Polymerization

See: Poly(p-xylylene)s, Coatings and Films

VAPOR DEPOSITION POLYMERIZATION ([2.2]Paracyclophane and Analogues)

Masataka Kubo* and Shouji Iwatsuki
Instrumental Analysis Center
Mie University

VAPOR DEPOSITION POLYMERIZATION OF [2.2]PARACYCLOPHANE

In 1966, Gorham developed a vapor deposition polymerization technique in which [2.2]paracyclophane (**1**) was pyrolyzed in reduced pressure at 600°C, and the pyrolyzed gas was condensed on a solid surface below 30°C to give poly(*p*-xylylene) (**7**) as a colorless transparent film (**Scheme I**).[1] The vapor deposition method is advantageous for exclusively giving *p*-xylylene (**6**) in a quantitative yield due to a mild pyrolysis temperature of 600°C and degradation ability of [2.2]paracyclophane. The polymer film obtained was soluble in hot chlorinated biphenyls and benzyl benzoate, indicating freedom from crosslinkage.

PROPERTIES OF POLY(*p*-XYLYLENE)

The vapor deposition coating process was developed by Union Carbide Corporation which commercially manufactures poly(*p*-xylylene)s under the trade name Parylene. The film obtained by this process is free of pinholes and can be adjusted to a thickness of several submicrons to several millimeters. Poly(*p*-xylylene) (Parylene N) is a dielectric exhibiting a very low dissipation factor, a high dielectric strength, and a dielectric constant invariable with frequency. It is used as a dielectric of a plastic-film capacitor. Poly(2-chloro-*p*-xylylene) (Parylene C) exhibits a very low permeability to moisture and other corrosive gases and is useful for coating critical electric assemblies. Chow et al. prepared poly($\alpha,\alpha,\alpha',\alpha'$-tetrafluoro-*p*-xylylene) by vacuum pyrolysis of 1,1,2,2,9,9,10,10-octafluoro[2.2]paracyclophane at 720°C.[2] This polymer also exhibits physical and electrical properties similar to those of Parylene polymers. It is extremely resistant to sunlight even after exposure for 3600 h whereas Parylene N changes to a brittle material after exposure for 535 h.

VAPOR DEPOSITION POLYMERIZATION OF HETEROPHANE[7]

Heterophanes such as [2.2](2,5)thiophenophane,[2] [2.2](2,5)-furanophane,[2] and [2.2](2,5)furano(2,5)-thiophenophane[3] were expected to be applicable to vapor deposition polymerization similarly to [2.2]paracyclophane.

VAPOR DEPOSITION POLYMERIZATION OF 1,9-DISUBSTITUTED [2.2]PARACYCLOPHANE

Some 1,9-disubstituted [2.2]paracyclophanes were found to be subject to vapor deposition polymerization. 1,9-Dichloro[2.2]paracyclophane was sublimed under pressure of 0.2 mmHg at 120°C followed by vacuum pyrolysis at 580°C. The pyrolyzed gas condensed on a glass surface at 20°C to give a colorless transparent film of poly(7-chloro-*p*-xylylene) in quantitative yield.[5] Subsequently, the film was converted into π-conjugated poly(1,4-phenylene vinylene) by heating it under nitrogen at 300°C.[5] This method is considered to be a novel approach for preparation of poly(1,4-phenylene vinylene) film.

1,9-Bis(trimethylsilyloxy)[2.2]paracyclophane undergoes vapor deposition to give a film of poly(7-trimethylsilyloxy-*p*-xylylene) in quantitative yield.[6] The film was able to be readily converted to poly(7-hydroxy-*p*-xylylene) by hydrolysis.[6]

REFERENCES

1. Gorham, W. F. *J. Polym. Sci.* **1966**, *4*, 3027.
2. Chow, S. W.; Leob, W. E.; White, C. E. *J. Polym. Sci.* **1969**, *13*, 2325.
3. Winberg, N. E.; Fawcett, F. S. et al. *J. Am. Chem. Soc.* **1960**, *82*, 1428.
4. Fletcher, J. R.; Sutherland, I. O. *Chem. Commun.* **1969**, 1504.
5. Iwatsuki, S.; Kubo, M.; Kumeuchi, K. *Chem. Lett.* **1991**, 1071.
6. Iwatsuki, S.; Kubo, M.; Toyota, K. *Polymer Bull* **1995**, *35*, 615.
7. Kubo, M.; Yamashita, H.; Iwatsuki, S. *Kobunshi Ronbunshu* **1989**, *46*, 241.

Varnishes

See: Melamine Resins (Overview)
Printing Inks

Vascular Grafts

See: Blood Compatible Polymers

VEGETAL BIOMASS (1. Monomers and Their Polymerization)

Alessandro Gandini** and Mohamed Naceur Belgacem
Matériaux Polymères
Ecole Française de Papeterie et des Industries Graphiques

This vegetal biomass constitutes an indispensible source of food, feed, and fuel, but of course also of polymeric materials.

This article is devoted to a brief survey of the exploitation of monomeric, oligomeric, and polymeric substances obtained from the biomass in order to prepare novel polymeric materials and is subdivided into three corresponding chapters. Papermaking and the use of wood as structural material will not be covered here because the aim of this article is to emphasize polymer synthesis, that is, the intervention of macromolecular chemistry in the elaboration of original structures and consequently

*Author to whom correspondence should be addressed.

**Author to whom correspondence should be addressed.

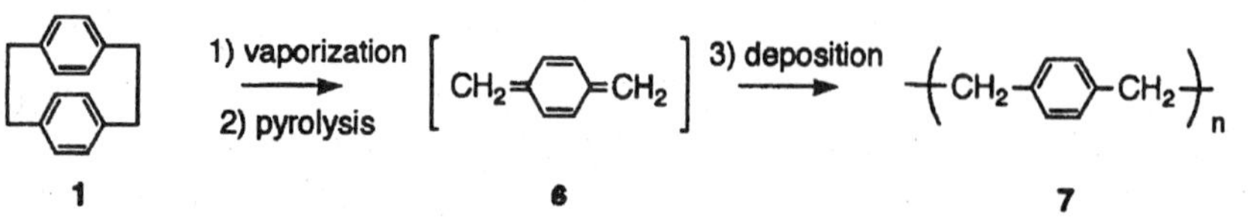
I

materials with interesting properties and promising applications. This topic was dealt with in more detail in a recent review on polymers from renewable resources to which the interested reader is referred.[1]

MONOSACCHARIDES AND DISACCHARIDES

Condensation Reactions

Polysaccharides can be obtained by polycondensation of monosaccharides, such as the pentoses and hexoses.[2]

Because sugars are molecules possessing five hydroxy functions, the synthesis of linear polymers derived from them could not be possible without masking three of these functions. Nature gives much better results and indeed these studies have only been aimed at mechanistic aspects and the synthesis of model compounds rather than materials.

Ring Opening Reactions

This type of polymerization is a cationic process promoted by acidic initiators which requires the use of two types of monomeric structures, namely either internal orthoesters or anhydrosugars.[5,6]

DIOLS AND TRIOLS

Glycerol is a trifunctional monomer that is obtained by the hydrolysis of natural fats and oils. It is used essentially in the synthesis of polyesters (alkyd resins) and crosslinked polyurethanes.

Ethylene glycol and propylene glycol, which are conventional diols in petrochemistry, can also be obtained in an alternative process from polysaccharides and sugars by a catalyzed splitting reaction conducted in a reducing hydrogen atmosphere.[7] These diols are, of course, used as monomers in polycondensation reactions to give polyesters and polyurethanes.

TERPENES

Biomass is also a source of alkenyl monomers in the form of terpenes, which are unsaturated hydrocarbons present in many essential oils. They are an important source of resinous oligomers. Representative terpenes are (i) α-pinene, (ii) β-pinene, (iii) limonene, (iv) myrcene, (v) 3-carene, and (vi) camphene, the most abundant being α- and β-pinene, which are also the most frequently used monomers.

Terpene resins typically are synthesized from a mixture of 75% β-pinene, and 25% α-pinene and limonene, with or without comonomers, like styrene.[9] These materials are mostly used as adhesives, mastics, sealants, printing ink vehicles, paints, and varnishes.[8]

FURANIC MONOMERS FROM PENTOSES AND HEXOSES

Certain sugars, oligosaccharides (hemicelluloses) and polysaccharides (cellulose, starch, etc.) can be easily converted into furanic compounds by acidic hydrolysis: pentoses lead to the formation of 2-furancarbaldehyde or furfural and hexoses give 5-hydroxymethyl-2-furancarbaldehyde. These two compounds can be chemically modified through simple reactions to give a large variety of monomers polymerizable by chain or stepwise reactions.[1] The most important sources of hemicelluloses are agricultural wastes such as corn cobs, rice hulls, sugarcane bagasse, olive residues, and cotton seeds, and, also, wood and by-products of paper mills.[10]

Furfural and Derivatives

Furfural is prepared by acid-catalyzed dehydration-cyclization of pentoses.[11]

A large variety of furfural-derived monomers can be obtained: vinylic, 2-furfurylidene ketone and homologues, acrylic, and oxiranes. The individual syntheses have been reported in detail and are summarized and referenced in a recent review.[1]

Another family of furanic difunctional monomers can be obtained from monofunctional compounds derived from furfural, namely furoic acid and derivatives, furyl isocyanate, furfuryl amine, and furfuryl isocyanate, by acid-catalysed condensation with ketones or aldehydes.[3,4,13]

The difuranic difunctional compounds thus obtained are diacids, diacid chlorides, diisocyanates, and diamines (these latter compounds in turn can be a source of diisocyanates by phosgenation), which therefore are suitable for polycondensation polymerizations.

Hydroxymethyl Furfural and Its Derivatives

Hydroxymethyl furfural obtained preferably from fructose, is another source of difunctional monomers for stepwise polymerization.[11]

POLYMERIZATION AND COPOLYMERIZATION OF FURANIC MONOMERS

Chain and step reactions involving furanic monomers by themselves or in conjunction with other monomers are reviewed here in a succinct form.[1,14–16]

Polyaddition Systems

Cationic Polymerization

The cationic polymerization of furan and 5-methyl furan gives highly conjugated polymers with complicated structures. Contrary to the well-characterized polypyrrol and polythiophene which possess regular linear conjugated structures, polyfuran is, in fact, a crosslinked material which has been the source of contradictory reports.[17,18]

Anionic Polymerization

2-Furyloxiranes are particularly sensitive to anionic activation, more so than its aliphatic and aromatic homologues and the corresponding polyether (DP up to *ca* 100) can be readily obtained with conventional initiators.[19] 2-Furyl and 2-furfuryl isocyanate give polyisocyanates, through the activation of the C=N bond by anionic polymerization.[20]

Free Radical Polymerization

This mode of polymerization is not well suited to furanic vinylic monomers because the primary and the polymeric radicals tend to add onto the C5 position of the furan ring more readily than to the correct site, namely its vinylic unsaturation.[1,14–16,21] 2-Furfuryl acrylate and methacrylate are much more adapted to free radical polymerizations precisely because with these monomers the initiation and propagation steps occur on the expected acrylic moiety giving a stabilized radical, without detectable side reactions on the heterocycle. Copolymers of these monomers with conventional counterparts like styrene and methyl methacrylate can, therefore, also be prepared with no difficulty.

Polycondensation Systems

Furfuryl alcohol is the most important furanic monomer from the point of view of industrial production because materials derived from it have found a range of applications, mostly for foundry cores and molds, corrosion resistant materials, and precursors to graphitic composition and adhesives. The mechanism of polymerization of furfuryl alcohol is complicated and important investigations in the last 30 years have dealt with this problem. In fact the products of these acid-catalyzed polycondensations are black crosslinked materials with complex intermediate and ultimate structures.[1,12,14–16] The basic condensation reaction occurs between OH groups and mobile hydrogen atom at the C5 position of the furan ring to give Structure **46** or by sporadic OH-OH condensation to give occasional CH_2OCH_2 linkages, as shown in **Scheme I**.

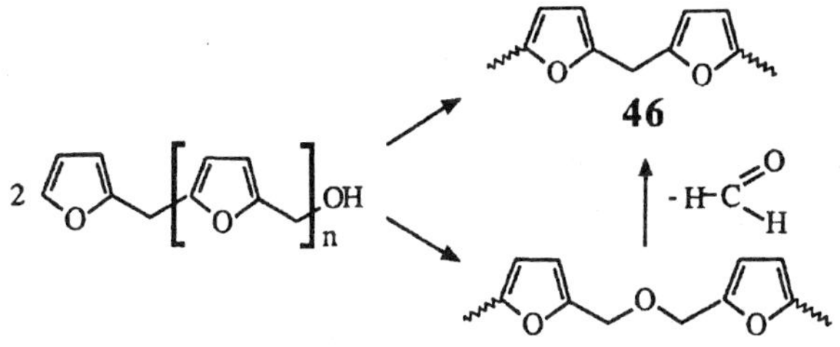

Rigid-chain polyesters from 2,5-furandicarboxylic acid chloride and hydroquinone can be prepared, but are intractable because of their insolubility and their decomposition before melting. Indeed, model trimeric structures were investigated to obtain thermotropic compounds.[22] Aliphatic comonomers, however, can be introduced to reduce chain stiffness, but the possibility of liquid crystalline behavior is thereby lost.

2,5-Furandicarboxylic acid was also used for the preparation of polyamides in conjunction with aromatic or furanic diamines. The polymers obtained were crystalline materials with good thermal properties. The polyamide prepared from 2,5-furandicarboxylic acid and *p*-phenylenediamine displayed lyotropic liquid crystal properties provided its molecular mass was above 50,000. The behavior of this polymer was close to that of fully aromatic polyamides in terms of thermal and mechanical features.[23]

In recent years, the synthesis and characterization of furnic polyurethanes has only been tackled systematically.[24–26]

Three families of polyurethanes were prepared, respectively, from: (i) furanic diisocyanates and aliphatic or benzylic diols; (ii) aliphatic or aromatic diisocyanates and furanic diols; and (iii) entirely furanic monomers. All these materials possessed high molecular weights and manifested high structural regularity. It was also shown that the presence of furanic moiety improved the thermal properties of these polymers. The glass-transition temperatures of furanic polyurethanes are somewhat lower than those of equivalent aromatic structures, but higher than those of aliphatic counterparts.

The Chemical Modification of Furanic Polymers

Thanks to its more pronounced dienic character compared with its homologues thiophene and pyrrole, the furan ring presents two interesting reactions: (i) the Diels–Alder cycloaddition with dienophiles such as maleic anhydride and maleimides and (ii) the electrophilic substitution reaction at the C2 or C5 positions already evoked. The application of these reactions to furan moieties present in the polymers and copolymers described above, whether they are present as side groups or as an integral part of the chain, can give rise to interesting structures and therefore to novel materials.

Examples of these chemical modifications are the introduction of polar sites into a nonpolar macromolecule, the synthesis of grafted structures from copolymers bearing lateral furan rings using cationic polymerization and the crosslinking of furanic polymers and copolymers with bis-dienophiles. Since the Diels–Alder reaction is thermoreversible, the latter modification can be conceived as a way of preparing a network, which would revert to a thermoplastic material upon heating.

REFERENCES

1. Gandini, A. *Comprehensive Polymer Science* Supp. 1; Aggarwal, S. L.; Russo, S. L., Eds.; Pergamon: Oxford, UK, 1992; p 527.
2. Yalpani, M. *Polysaccharides* Elsevier: New York, 1988.
3. Cawse, J. L.; Stanford, J. L.; Still, R. H. *Macromol. Chem.* **1984**, *185*, 697.
4. Cawse, J. L.; Stanford, J. L.; Still, R. H. *Macromol. Chem.* **1984**, *185*, 709.
5. Shuerch, C. *Adv. Carbohyd. Chem. Biochem.* **1981**, *39*, 157.
6. Schuerch, C. *Encyclopedia of Polymer Science and Engineering* Mark, H. F.; Bikales, N. M.; Overberger, C. G.; Menges, G., Eds.; Wiley, New York, 1988; Vol. 3.
7. Marini, L.; Casale, B. *Biomass for Energy, Industry and Environment* Grassi, G.; Collina, A.; Zibetta, H.; Eds.; Elsevier, London, 1992; p 1269.
8. Keszler, B.; Kennedy, J. P. *Adv. Polym. Sci.* **1990**, *100*, 1.
9. Vredenburgh, W.; Foley, K. F.; Scarlatti, A. N. *Encyclopedia of Polymer Science and Engineering* Mark, H. F.; Bikales, N. M.; Overberger, C. G.; Menges, G.; Eds.; Wiley: New York, 1988; Vol. 7, p 758.
10. *Making and Marketing of Furfural: Added Value for Agro-Industrial Wastes*; UNCTAD/GATT, Genova, 1979.
11. Theander, O.; Nelson, D. A. *Adv. Carbohydr. Chem. Biochem.* **1988**, *46*, 273.
12. McKillip, W. J. *ACS Symp. Ser.* **1989**, *385*, 408.
13. Pennanen, S.; Nyman, G. *Acta Chem. Scand.* **1972**, *26*, 1018.
14. Gandini, A. *Adv. Polym. Sci.* **1977**, *25*, 47.

15. Gandini, A. *Encyclopedia of Polymer Science and Engineering* Mark, H. F.; Bikales, N. M.; Overberger, C. G.; Menges, G., Eds.; Wiley, New York, 1988; Vol. 7; p 454.
16. Gandini, A. *ACS Symp. Ser.* **1989**, *433*, 195.
17. Tourillon, G.; Garnier, F. *J. Electroanal. Chem. Interfacial Electrochem.* **1982**, *135*, 173.
18. Tediar, F. *Eur. Polym. J.* **1985**, *21*, 317.
19. Salon, M. C.; Amri, H.; Gandini, A. *Polym. Commun.* **1990**, *31*, 210.
20. Hui, Z.; Gandini, A. *Polym. Bull.* **1989**, *21*, 255.
21. Rieumont, J.; Vega, R.; Davidenko, N.; Paz, J. A. *Eur. Polym. J.* **1988**, *24*, 909.
22. Maccio, D.; Costa, G.; Valenti, B.; Gandini, A. *Heterocycles* **1993**, *36*, 1225.
23. Mitiakoudis, A.; Gandini, A. *Macromolecules* **1991**, *24*, 830.
24. Belgacem, M. N.; Quillerou, J.; Gandini, A.; Roux, G.; Rivero, J. *Eur. Polym. J.* **1989**, *25*, 1125.
25. Belgacem, M. N.; Quillerou, J.; Gandini, A. *Eur. Polym. J.* **1993**, *29*, 1217.
26. Boufi, S.; Belgacem, M. N.; Quillerou, J.; Gandini, A. *Macromolecules* **1993**, *26*, 6706.

VEGETAL BIOMASS (2. Oligomers and Their Polymerization)

Alessandro Gandini* and Mohamed Naceur Belgacem
Matériaux Polymères
Ecole Française de Papeterie et des Industries Graphiques

The term "oligomers" used here refers to natural structures which are either oligomeric as such, like triglycerides or rosins, or the products of polymer fragmentations usually at the stage of isolation of refining, as with lignins. In both instances one deals typically with molecular masses ranging from a few hundred to a few thousand.[1]

DRYING OILS

Among the vast array of vegetal oils, which are glyceride esters of fatty acids, one finds, broadly speaking, three types of structures with respect to the sensitivity to atmospheric thickening (drying) provoked by oxidative polymerization: (i) saturated aliphatic chains as with esters of palmitic and stearic acid are nondrying and, therefore, inert to free-radical attack; (ii) modestly unsaturated chains, as with esters of oleic acid, which are too sluggish in their oxidative polymerization to be useful as drying agents, and (iii) multiply unsaturated chains as with esters of linoleic, linolenic, and eleostearic acid, which respond adequately to the basic requirements of drying at a correspondingly increasing rate.[2–4]

The interest of drying oils as materials stems from their aptitude to give elastic films or skins under the effect of atmospheric oxygen. This polymerization occurs thanks to the high reactivity of C-H bonds situated next to a C=C unsaturation.[5]

NON-DRYING OILS

Although castor oil is not in itself a drying material, it can be used as macromonomer in polycondensation reactions thanks to its three hydroxy groups per ricinoleic triglyceride molecule. Elastomeric polyurethanes are one of the major applications of this natural product, which also can be used in the preparation of polyesters and polyethers.

TANNINS

Chemically, tannins, or condensed tannins, are phenolic-based natural products.[6] As most natural substances, they are a mixture of oligo- and poly-flavonoids with average molecular masses between 1000 and 4000 depending on the source from which they are extracted.

The curing of formaldehyde-based prepolymers containing tannins at room temperature and in a basic medium gives materials with good adhesive properties, especially for plywood.[7,8]

ROSINS

Rosins are the most important resins extracted from plants, with an annual production of about 10^5 tons, essentially from pine trees.[3,9] They have been used for a long time in various traditional lubricating, coating, and sealing applications. Their major component is abietic acid, which possesses two-conjugated unsaturations and is therefore sensitive to oxidative polymerization following the reaction mechanism for drying oils.

LIGNINS

Lignins are the second most important renewable resource after cellulose. Lignins are amorphous crosslinked polymeric materials, which play a paramount role in insuring the cohesion of cellulosic fibers in actic as a matrix in natural woody composites. The term, lignins, used in the plural, indicates that there is no unique structure for these polymers which bear, however, common features based on phenolic and aliphatic moieties.

Lignin is generally obtained by specific degradative dissolution from woods and plants during the pulping processes mostly associated with papermaking.[10] As traditional pulping processes (soda, kraft, and sulfite cooking) are carried out at relative high pressures and temperatures under acidic or basic conditions and for relatively long reaction times (4–6 hours), native lignins are subjected to drastic chemical modifications from the point of view of the decrease of molecular mass and of the nature of the ultimate functions they will possess. Recently, some novel pulping technologies and biomass refineries like the organosolv and steam explosion processes have provided original sources of fragmented lignins.[11–18]

Lignins as Macromonomers

Lignins can be used as obtained from the pulping processes or chemically modified depending on the strategy adopted and the applications sought for the corresponding materials.

Lignins rich in guaiacyl moieties (i.e., containing substantial proportions of labile aromatic C-H bonds) have been used in the preparation of phenolic-type resins by self-condensation reactions. Both kraft and sulfite lignins were tested in this context.[19]

More frequently, however, lignins have been incorporated as coreactants in phenol-formaldehyde resins.

The partial replacement of phenols by lignins in phenol–formaldehyde resins seems to give materials with good adhesive properties if the lignins are functionalized with a preliminary,

*Author to whom correspondence should be addressed.

simple, and low-cost procedure to render them more soluble and more reactive.

Another area in which lignins were investigated is grafting reactions in polymerization processes initiated by free radicals using chemical catalysis or gamma rays. Recently, kraft lignins were successfully grafted with styrene in solution of dimethyl sulfoxide and using calcium chloride and hydrogen peroxide as an initiator system.[20] Acrylamide seems to give much better results particularly when cerium ions are used as initiators.[21–23]

A second way of exploiting lignins as core oligomers is by anionic grafting with oxiranes.[24,25]

Various lignins have been successfully epoxidized and used as co-macromonomers with diamines or dianhydrides in the presence of elastomeric macromonomers bearing the NH_2 function as end groups to obtain rubber-modified tough epoxy networks.[26]

Recent work describes the preparation of lignin derivatives such as sulfonate esters, alkanoates, and functional ethers and the use of lignins as comonomers in the preparation of star-shaped copolymers with poly(ethylene glycol) methyl ether, and styrene.[27–31]

Lignins as Additives for Polymers

Lignins can be incorporated as fillers or additives to polymer formulations. The incorporation of lignins in phenol-formaldehyde resins has been the most important way of valorizing lignins as such, particularly organosolv structures.

Polymeric blends involving lignins have been the object of some studies including those using lignins as such and those using chemically modified lignins. The former systems concern mostly kraft lignin mixed with polyurethanes and epoxide resins.[32–35]

Recently, other trends in the valorization of lignins have been published. Thus, it was reported that the controlled pyrolysis of lignins can yield activated carbons, carbon fibers, and carbonaceous adsorbents.[36–39] Lignin-based composites with poly(vinyl chloride) as a matrix were also reported.[40,41]

Finally, traditional kraft and lignosulfonate lignins have been used as stabilizers against various forms of degradation of polymers. Thanks to their phenolic character and ability to trap free radicals, lignins can substitute other more costly additives to stabilize polyolefins against thermal, oxidative, and photolytic degradation.[42,43]

REFERENCES

1. Gandini, A. *Comprehensive Polymer Science* Supp. 1; Aggarwal, S. L.; Russo, S., Eds.; Pergamon: Oxford, UK; 1992; p 527.
2. Gunstone, F. D.; Harwood, J. L.; Padley, F. B. *The Lipid Handbook*; Chapman and Hall: 1986.
3. Solomon, D. H. *The Chemistry of Organic Film Formers*; Krieger: Huntington, New York, 1977.
4. Gunstone, F. D. *Chemistry and Biochemistry of Fatty acids and their Glycerides*; Chapman and Hall: London, 1967.
5. Porter, N. A.; Lehman, L. S.; Weber, B. A.; Smith, K. J. *J. Am. Chem. Soc.* **1981**, *103*, 6447.
6. Pizzi, A. *Wood Adhesives*; Marcel Dekker: New York, 1983; Chapter IV.
7. Pizzi, A. *Renewable Resource Materials*; Carraher, C. E., Jr.; Sperling, L. H., Eds.; Plenum: New York, 1986; p 323.
8. Pizzi, A. *Advanced Wood Adhesive Technology*; Pizzi, A., Ed.; Marcel Dekker: 1994.
9. Rowe, J. W. *Natural Products of Woody Plants*; Springer Verlag: Berlin, Heidelberg, 1989; Chapter X and XI.
10. Casey, J. P. *Pulp and Paper Chemistry and Chemical Technology*; 3rd ed.; Wiley: New York, 1980; Vol. 1.
11. Vasquez, G.; Antorrena, G.; Gonzalez, J. *Biomass for Energy and Industry, 7th E.C. Conference*; Hall, D. O.; Grassi, G.; Scheer, H., Eds.; James & James Science: London, 1994; p 1189.
12. Papatheofanous, M. G.; Koukios, E. G. *Biomass for Energy and Industry, 7th E.C. Conference*; Hall, D. O.; Grassi, G.; Scheer, H., Eds.; James & James Science: London, 1994; p 1123.
13. Sarkanen, K. V. *Progress in Biomass Conversion*; Sarkanen, K. V.; Tillman, D. A., Eds.; Academic: New York, 1980.
14. Lora, J. H.; Creamaer, A. W.; Wu, L. C. F.; Goyal, G. C. *Cellulosics: Chemical, Biochemical and Material Aspects*; Kennedy, J. F.; Phillips, G. O.; Williams, P. A., Eds.; Ellis Horwood: New York, 1993; p 251.
15. Hua, X.; Kokta, B. V.; Kaliaguine, S. *Cellulosics: Pulp, Fibre and Environmental Aspects*; Kennedy, J. F.; Phillips, G. O.; Williams, P. A., Eds.; Ellis Horwood: New York, 1993; p 37.
16. Tomasec, M.; Kokta, B. V. *Cellulosics: Pulp, Fibre and Environmental Aspects*; Kennedy, J. F.; Phillips, G. O.; Williams, P. A., Eds.; Ellis Horwood: New York, 1993; p 46.
17. Carrasco, F.; Garceau, J. J.; Ahmed, A.; Kokta, B. V. *Cellulosics: Pulp, Fibre and Environmental Aspects*; Kennedy, J. F.; Phillips, G. O.; Williams, P. A., Eds.; Ellis Horwood: New York, 1993; p 51.
18. Ben, Y.; Kokta, B. V.; Doucet, J. *Cellulosics: Pulp, Fibre and Environmental Aspects*; Kennedy, J. F.; Phillips, G. O.; Williams, P. A., Eds.; Ellis Horwood: New York, 1993; p 57.
19. Nimz, H. H. *Wood Adhesives*; Pizzi, A., Ed.; Marcel Dekker: New York, 1983; Chapter V.
20. Meister, J. J.; Chen, M.-J. *Macromolecules* **1991**, *14*, 6843.
21. Meister, J. J.; Lathia, A.; Chang, F. *J. Polym. Sci.: Polym. Chem. Ed.* **1991**, *29*, 1465.
22. Meister, J. J.; Patil, D. R. *Macromolecules* **1985**, *18*, 1559.
23. Meister, J. J.; Patil, D. R.; Augustin, C.; Lai, J. Z. *ACS Symp. Ser.* **1989**, *397*, 294.
24. Wu, L. C. F.; Glasser, W. G. *J. Appl. Polym. Sci.* **1984**, *29*, 1111.
25. de Oliveria, W.; Glasser, W. G. *ACS Symp. Ser.* **1989**, *397*, 414.
26. Glasser, W. G. *Wood Processing and Utilisation*; Kennedy, J. F.; Phillips, G. O.; Williams, A. P. A., Eds.; Horwood: Chichester, 1989; p 163.
27. Dhara, K.; Jain, R. K.; Glasser, W. G. *Holzforschung* **1993**, *47*, 402.
28. Jain, R. K.; Glasser, W. G. *Holzforschung* **1993**, *47*, 225.
29. Jain, R. K.; Glasser, W. G. *Holzforschung* **1993**, *47*, 325.
30. Huth, S. P.; Cole, W. J. W. *Holzforschung* **1994**, *48*, 23.
31. Oliveira, W.; Glasser, W. G. *J. Wood Chem. Technol.* **1994**, *14*, 119.
32. Feldman, D.; Banu, D.; Natahsohn, A.; Wang, J. *J. Appl. Polym. Sci.* **1991**, *42*, 1537.
33. Natahsohn, A.; Lacasse, M.; Banu, D.; Feldman, D. *J. Appl. Polym. Sci.* **1990**, *40*, 899.
34. Feldman, D.; Lacasse, M. A. *J. Appl. Polym. Sci.* **1994**, *51*, 701.
35. Wang, J.; Banu, D.; Feldman, D. *J. Adhes. Sci. Technol.* **1992**, *6*, 587.
36. Rodriguez-Mirasol, J.; Cordero, T.; Rodriguez, J. J. *Carbon* **1993**, *31*, 87.
37. Yamaguchi, T.; Sato, Y. *Nippon Kagaku Kaishi* **1993**, *3*, 133.
38. Sudo, K.; Shimizu, K.; Nakashima, N.; Yokoyama, A. *J. Appl. Polym. Sci.* **1993**, *48*, 1485.
39. Simitzis, J.; Sfyrakis, J. *J. Analyt. Appl. Pyrlysis* **1993**, *26*, 37.
40. Feldman, D.; Banu, D.; el-Raghi, S. *J. Macromolec. Sci.* **1994**, *31*, 55.

41. Oliveira, W.; Glasser, W. G. *J. Appl. Polym. Sci.* **1994**, *51*, 563.
42. De Paoli, M. A.; Furlan, L. T. *Polym. Degradation Stab.* **1985**, *13*, 129.
43. Kosikova, B.; Miklosevy, K.; Demianova, V. *Europ. Polym. J.* **1993**, *29*, 1995.

VEGETAL BIOMASS (3. Polymers, Derivatives, and Composites)

Alessandro Gandini* and Mohamed Naceur Belgacem
Matériaux Polymères
Ecole Française de Papeterie et des Industries Graphiques

Nature provides remarkable ways of making high polymers through photosynthesis. Most of them are polysaccharides, polyterpenes, or polyphenolic structures. This article deals with the exploitation of these macromolecules through chemical modification or through their inclusion in various matrices.

CELLULOSE

Polysaccharides are the most important natural polymers.[1] They consist of monosaccharides' residues joined by glycosidic bonds. Cellulose, the most abundant renewable polymer and also one of the most interesting in terms of its applications in materials, is a polysaccharide based on glucosic units.[2] The repeat unit of cellulose is cellobiose.

Cellulose can be found in a pure, isolated form, as in cotton, in which case it does not need chemical refining.[3]

In addition to the very good mechanical properties of cellulose, which are mainly due to its high crystallinity, its relatively high thermal stability, water affinity, and slow biodegradability provide important materials such as cotton textiles and paper-based[4] products, which owe their remarkable character to cellulosic fibers.[3,4]

Chemical Modification

The partial or total substitution of the OH groups of cellulose with small moieties can induce important changes in its specific properties. This operation leads to the formation of a large variety of materials usually called cellulose derivatives, which owe their interest to such features as improved solubility, easier processing, or greater flexibility.

Inorganic Esters

Nitrocellulose (also called cellulose nitrate) is the oldest cellulose derivative. It is obtained by the nitration of cellulose with a mixture of concentrated sulfuric and nitric acid. Depending on the degree of substitution of hydroxy groups of cellulose, cellulose nitrates have found many areas of applications, such as propellants, explosives, lacquers, and adhesives. However, because of their flammability, the explosive tendencies of the dry materials, and the discovery of similar or polymers that perform better, their use today is relegated mostly to explosives and propellants.[6]

Other, less important inorganic esters of cellulose are cellulose phosphates and sulfates.[1,6] Cellulose phosphates are manufactured in the form of sodium or ammonium salts and used as medical and pharmaceutical products, flame-retardants, and food additives.

Organic Esters

The most important representatives of cellulose carboxylates are the triacetate, the triacetate-propionate, and the acetate-butyrate.[1,6,7] Unlike cellulose nitrates, the production of cellulose carboxylates is growing.

Cellulose triacetate is used for photographic films, transparent or pigmented films, and separation membranes. Its ability to crystallize also allows its conversion to fibers. Interest in cellulose diacetate is also growing because of its use in cigarette filters.

Ethers

The etherification of the hydroxy functions of cellulose gives technically useful products.[1,6,7–9] The most important families of cellulose ethers are the alkyl and carboxyalkyl derivatives that are obtained by etherification of cellulose with alkyl halides or sulfates and ω-halocarboxylic acids or salts, respectively. Sodium carboxymethyl cellulose is obtained by the treatment of cellulose with sodium chloroacetate in an alkaline medium. These products are water-soluble anionic polyelectrolytes used as thickeners and food additives that can also be readily cast into films.

Hydroxyethyl and hydroxypropyl cellulose are synthesized from cellulose and ethylene and propylene oxide, respectively. They are nonionic water-soluble products with good film-forming properties. They have found applications as adhesives, paper, food additives, thickeners, and encapsulation materials. Methyl and ethyl cellulose are industrial commodities that possess similar properties and that find similar applications. Cellulose derivatives bearing both ether and hydroxyalkyl groups, such as ethylhydroxyethylcellulose, are also commercially available and play useful specific roles in similar areas.

Cationic cellulose ethers in which the positive charge is typically held by ammonium moieties are also commercial products used in the paper and pharmaceutical industries.

Xanthates

The conventional procedure for the synthesis of cellulose xanthates involves two steps: (i) the production of alkali cellulose by treating cellulose with a concentrated sodium hydroxide solution, and (ii) the reaction of the former with carbon disulfide to produce xanthated cellulose, usually called viscose.[6,7] Viscose is essentially used in the production of textile fibers called rayon and as a precursor to regenerated cellulose, called cellophane.

Cellulosic Liquid Crystals

Cellulose and many of its derivatives show lyotropic cholesteric liquid crystal behavior.[10–13] The preparation and characterization of high-modulus fibers and films from these mesophases have aroused considerable interest because of the obvious gain in mechanical, optical, and other properties linked to high supramolecular ordering.

*Author to whom correspondence should be addressed.

Grafting and Crosslinking

Grafting by Radical Polymerization

Persulfates and the redox systems Fe^{++}/H_2O_2 are typical initiators used for the activation of cellulose. The highest grafting efficiencies have been reported with cerium(IV) as initiator because free radicals are formed directly only on the cellulose macromolecules.

Gamma-ray irradiation was also applied to free-radical grafting of cellulose and cellulose derivatives.

Grafting by Ionic Polymerizations

Acrylonitrile was grafted anionically from the sodium derivative Cell-O^- Na^+ in aprotic media. Chain extention of the hydroxy groups of cellulose with ethylene oxide occurs in the presence of basic catalysts, but only short grafts are obtained.[14–17]

The natural OH groups of cellulose or COCl groups introduced by functionalization have been used to initiate the ring-opening polymerization of lactones and lactams and to obtain polyester- and polyamide-grafted materials.[14]

Grafted by Polycondensation Reactions

Hydroxy functions along the chains of cellulose and celluloses derivatives can also be involved in grafting by polycondensation processes. The most efficient method consists in preparing polyesters, polyamides, and polyurethanes that are then grafted onto cellulose by the condensation of COCl or NCO groups with its primary OH groups.

The most promising areas of application for cellulosic grafted materials are connected with papermaking (specific modification of the properties of papers and cardboards) and polyelectrolyte activity and water superabsorbancy for sanitary uses, soil conditioning, medical applications, and the drying of organic liquids.[18] Cellulose and cellulose ethers grafted with polyacrylamide and partly converted into anionic acrylate moieties can replace similar nonbiodegradable crosslinked polyacrylates in superabsorbancy.

Cellulose-Based Polymeric Blends

Within the important and growing field of polymer blends, the use of natural polymers is still modest, but cellulose has been used in numerous combinations with synthetics.

Cellulosic Fibers in Composites

Wood is a remarkable example of composite materials based on polymers. The reinforcing elements in that assembly are, of course, the cellulosic fibers. The use of these same fibers in a coarse form in synthetic composite materials such as particle board is practically as old as polymer technology and remains a very important use of wood chips and wastes. More recently, attention has been focused on the elaboration of composites that contain cellulosic fibers added as pure individual elements in different types of thermoplastic and thermoset matrices in order to obtain substantial improvements in the mechanical properties of the resulting materials. The number of conferences and related papers devoted to this subject in the past few years testify to its vitality.[5,19,20]

OTHER VEGETAL POLYSACCHARIDES

Starch is the main food-reserve polysaccharide contained in plants and is therefore a very widely distributed natural organic polymer. Although starch can be quantitatively converted to a single sugar, D-glucose, it is in fact a heterogeneous material composed of two different polysaccharides, namely, linear amylose (15–25%) and highly branched amylopectin. Starch and cellulose are very similar polysaccharides in terms of chemical structure, but the latter has been studied much more extensively because of the enormous relevance of its applications in materials.

Starch has also been the subject of numerous investigations about its chemical modification and grafting in order to obtain useful derivatives.[21] Thus, for example, carboxymethyl starch, starch xanthates, and starch grafted with polyelectrolytic strands for superabsorbing materials are commercially available products.

The biodegradability of starch has sparked a considerable amount of research aimed at developing new polymeric materials that are degradable in nature.

ANIMAL POLYSACCHARIDES: CHITIN AND CHITOSAN

Chitin is a natural polysaccharide that has N-acetyl sidegroups. It is the most abundant polysaccharide from animal sources and is found mostly in the shells of marine invertebrates. Deacetylation of chitin leads to the formation of chitosan, which is therefore its primary amino derivative.

Partial deacetylation of chitin gives amide-amine copolymers of variable composition, which today are available commercially. The inclusion of these polymers derived from animal sources in the present context is justified by the interest in using them in materials, as such or in a chemically modified form.[22,23]

Chitin, chitosan, and their numerous derivatives have a strong chelating character that makes them particularly interesting in applications such as ion collectors (recovery of traces of metals from solutions) and chromatographic substrates. They are also used as textile additives, dialysis membranes, and in medical, surgical, and pharmaceutical applications.

CHEMICAL MODIFICATION OF WOOD AND OTHER LIGNOCELLULOSICS

The main reasons for chemically modifying wood, apart from specific preservation treatments, are aimed at its plasticization and at the improvement of its dimensional stability and weathering, namely, resistance to moisture, light, and oxygen. In essence, these modifications partially or totally destroy the crystalline character of cellulose, that is, they destroy intermacromolecular hydrogen bonds by establishing new moieties with a nonpolar or less pronounced polar character.[24] These processes will affect the hydroxy groups and will therefore not spare lignins and hemicelluloses.

The reagents used for softening the wood structure include organic halides, anhydrides, oxiranes, and isocyanates. The replacement of the OH groups gives rise to various new sites (esters, ethers, urethanes), but more importantly to the inclusion in the macrostructure of aliphatic or aromatic moieties of different sizes. It follows that, depending on the nature, specific structure, (that is, aliphatic chain length attached to the reactive group) and amount added relative to the wood substrate, a whole variety of materials can be obtained; they will range in mechanical and thermal properties from semicrystalline to amorphous, from rigid to elastomeric, from water-swelled to hydrophobic, from thermoset to thermoplastic. This last change implies that lignin has been partly depolymerized and its network structure destroyed.[25–27]

Two limiting examples of these treatments show the range of applications of the resulting materials. The antishrink properties linked to the decrease in wood's water affinity can be greatly improved by partial acetylation, which, however, does not destroy the main supramolecular features of the substrate.[25] Conversely, the esterification of wood with important proportions of long-chain aliphatic anhydrides gets rid of most hydrogen bonding, cellulose crystallinity, and the crosslinked character of lignin. These drastic effects, coupled with the plasticizing role of the aliphatic chains, produce a softened thermoplastic material that can be cast into films, moulded into objects, and expanded to yield foams.[27] The properties of these chemically modified woods can be further modulated by grafting with appropriate alkenyl monomers that give vitrous or elastomeric branches.

Lignocellulosic fibers derived from wood or agricultural wastes, or, indeed, from plants yielding special morphologies (e.g., coconut, agave, flax, kenaf) are used in polymeric composites following the same criteria discussed above for pure cellulosic fibers.

REFERENCES

1. Yalpani, M. *Polysaccharides*; Elsevier: New York, 1988.
2. Brown, R. M. In *Cellulose and Other Polymer Systems: Biogenesis, Structure and Degradation*; Brown, R. M., Ed.; Plenum: New York, 1982.
3. Brown, H. B.; Ware, J. O. *Cotton*; McGraw-Hill: New York, 1958.
4. Casey, J. P. *Pulp and Paper Chemistry and Chemical Technology*, 3rd ed.; Wiley: New York, 1980; Vol. 1.
5. Rowell, R. M.; Laufenberg, T. L.; Rowell, J. K. *Materials Research Society Symp. Proc.* **1992**, *266*, 47–194.
6. Bikales, N. M.; Segal, L. *Cellulose and Cellulose Derivatives*; Wiley: New York, 1971.
7. Nevell, T. P.; Zeronian, S. H. *Cellulose Chemistry and its Applications*; Horwood: Chichester, UK, 1985.
8. Davidson, R. L. *Handbook of Water-Soluble Polymers*; McGraw-Hill: New York, 1980.
9. Dönges, R. *Br. Polym. J.* **1990**, *23*, 315.
10. Gray, D. G. In *Cellulosics Utilization*; Inagaki, H.; Phillips, G. O., Eds.; Elsevier: London, 1989; 123.
11. Gilbert, R. D. *ACS Symp. Ser.* **1990**, *433*, 259.
12. Zugenmaier, P. In *Cellulosics: Chemical, Biochemical and Material Aspects*; Kennedy, J. F.; Phillips, G. O.; Williams, P. A.; Ellis Horwood: New York, 1993; 105.
13. Revol, J.-F.; Giasson, J.; Guo, J.-X.; Hanley, S. J.; Harkness, B.; Marchessault, R. H.; Gray, D. G. In *Cellulosics: Chemical, Biochemical and Material Aspects*; Kennedy, J. F.; Phillips, G. O.; Williams, P. A., Eds.; Ellis Horwood: New York, 1993; 115.
14. Hebeish, A.; Guthrie, J. T. *The Chemistry and Technologie of Cellulosic Copolymers*; Springer Verlag: Berlin, 1981.
15. *ACS Symp. Ser.* **1982**, *187*.
16. Narayan, R. *Cellulosics Utilization*; Inagaki, H.; Phillips, G. O., Eds.; Elsevier: London, 1989; 110.
17. Stannett, V. T. In *Cellulose: Structural and Functional Aspects*; Kennedy, J. F.; Phillips, G. O.; Williams, P. A., Eds.; Horwood: Chichester, UK, 1989.
18. Chatterjee, P. K. *Absorbency*; Elsevier: Amsterdam, 1985.
19. *Pacific Rim Bio-Based Composites Symposium*; Plakett, D. V.; Dunningham, E. A., Eds.; Rotura: New Zealand, 1992.
20. Felix, J. M.; Gatenholm, P. *J. Appl. Polym. Sci.* **1991**, *42*, 609.
21. Fanta, G. F.; Doane, W. M. *ACS Symp. Ser.* **1990**, *433*, 288.
22. Muzzarelli, R. A. A. *Chitin*; Pergamon: Oxford, 1977.
23. Skjak-Bæk, G.; Anthonsen, T.; Sandford, P. *Chitin and Chitosan*; Elsevier Applied Science: London, 1988.
24. *Chemical Modification of Lignocellulosics*; Plakett, D. V.; Dunningham, E. A., Eds.; Rotura: New Zealand, 1992.
25. Rowell, R. M. In *Wood and Cellulosic Chemistry*; Hon, D. N. S.; Shiraishi, N., Eds.; Marcel Dekker: New York, 1991; Chapter 15.
26. Hon, D. N. S.; Xing, L. M. *ACS Symp. Ser.* **1990**, *489*, 113.
27. Shiraishi, N. In *Wood and Cellulosic Chemistry*; Hon, D. N. S.; Shiraishi, N., Eds.; Marcel Dekker: New York, 1991; Chapter 18.

Vegetal Oils

See: Vegetal Biomass (2. Oligomers and their Polymerization)

VESICLES, POLYMERIZATION IN

Jenci Kurja,* Hennie F. Zirkzee, and Anton L. German
Department of Polymer Chemistry and Technology
Eindhoven University of Technology

Roeland J. M. Nolte
Department of Organic Chemistry
University of Nijmegen

Ian A. Maxwell
Memtec Ltd.

It has been known for some time that one can perform a free radical type of polymerization in micelles—i.e., an emulsion polymerization[1,2]—and in inverse micelles—i.e., an inverse emulsion polymerization.[2,3] Concomitantly, the question arises whether one can polymerize in other surfactant structures such as vesicles—i.e., performing polymerization in vesicles.

POLYMERIZABLE VESICLES VS. POLYMERIZATION IN VESICLES

Polymerizable vesicles are defined here as vesicles in which the surfactant molecules are polymerized. An excellent overview of possible polymerizable vesicles is presented by Ringsdorf.[5]

Polymerization in vesicles is defined as the polymerization of a monomer in the hydrophobic vesicle bilayer, with the restriction that the monomer is not the original surfactant molecule.[5,6]

The primary goal of performing polymerization in vesicles is to use the vesicle structure in order to obtain a morphology that is similar to that of hollow latex particles.[6]

Monomer Concentrations in the System

The total monomer content in the vesicle system prior to polymerization is a very crucial parameter.

*Author to whom correspondence should be addressed.

The Initiator System

The initiators used in a conventional emulsion polymerization can be divided into three groups: dissociation, redox, and photoinitiators. The dissociative initiators can be divided into water-soluble initiators, such as peroxysulfates and ammonium peroxysulfate, and oil-soluble dissociative initiators, such as benzoyl peroxide and 2,2′-azobisisobutyronitrile (AIBN). However, a fraction of the above-mentioned initiators may dissolve in the aqueous phase and cause secondary nucleation—i.e., the formation of latex particles—during polymerization in vesicles.[6]

The choice of the initiator system is restricted because of the requirement that during the polymerization process the monomer swollen vesicle structure is retained and the formation of new particles is prevented—i.e., only vesicular nucleation occurs.[6] This means that all monomer swollen vesicles should be polymerized "simultaneously", resulting in polymer-containing vesicles. If this is not the case, monomer and surfactant from nonreacting monomer swollen vesicles will diffuse to polymerizing vesicles.[6]

Summarizing

To perform real polymerization in vesicles, a swelling technique for the incorporation of monomer in the vesicle bilayer combined with characterization procedures for vesicle stability are required. The type and amount of initiator are critical in ensuring vesicular nucleation and that polymerization takes place in the vesicle bilayer.

KINETIC PROCESSES DURING POLYMERIZATION IN VESICLES

Radical polymerizations of unsaturated monomers in host systems such as vesicles have some similarities with a conventional emulsion polymerization, where traditional surfactants are used. These surfactants can form micelles, which will act as loci (host system) for polymerization. The procedure for polymerization in vesicles is that initially monomer is incorporated into the vesicle bilayer by bilayer swelling. After swelling, polymerization is performed by adding a free radical initiator or a photoinitiating system. The procedure for polymerization in vesicles is presented schematically in **Figure 1**.

Another item that should be considered when one wants to perform polymerization in vesicles is the total amount of vesicle surface offered. If the vesicle surface is not sufficiently large, radicals formed in the aqueous phase can cause secondary nucleation leading to the formation of unwanted latex particles.[6] The formation of vesicles is governed by the self-organization of the amphiphilic molecules in water.[7]

CONCLUSIONS

Reliable swelling experiments are required to adjust the monomer concentration in the vesicle bilayer before polymerization in vesicles can be performed. It is recognized that the choice of the initiator system is a very crucial parameter to ensure exclusively polymerization in vesicles—i.e., avoiding secondary nucleation at all times. The characterization of the vesicle system is also very important. At each point during the swelling and polymerization process, proper control of the vesicle structure is crucial to ensure polymerization in vesicles.

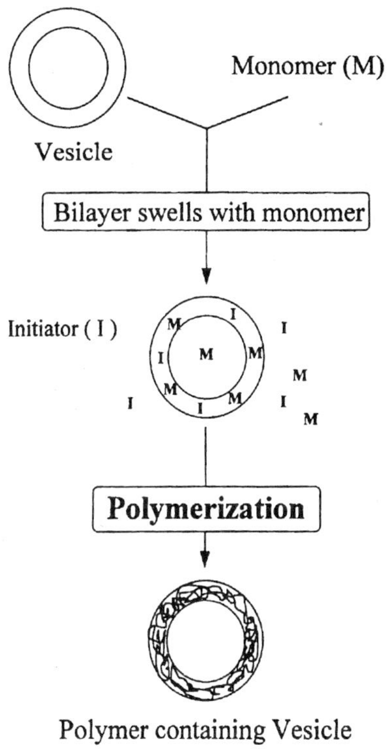

FIGURE 1. Schematic route for polymerization in vesicles.

It has become obvious that to perform successful polymerization in vesicles the chain of knowledge between thermodynamics—i.e., vesicle bilayer swelling, polymerization kinetics, characterization of the vesicle system, and the process control of the polymerization process—has to be closed.

The concepts and experiments presented show good perspectives for polymerization in vesicles that may lead to new polymeric materials based on polymerization in vesicles.

REFERENCES

1. Blackley, D. C. *Emulsion Polymerization*; Applied Science: London, 1975.
2. Barton, J.; Capek, I. *Radical Polymerization in Disperse Systems*; Ellis Horwood: New York, 1994.
3. Candau, F. *Scientific Methods for the Study of Polymer Colloids and Their Applications*; Candau, F.; Ottewill, R. H., Eds.; Kluwer Academic: Dordrecht, 1988.
4. Ringsdorf, H.; Schlarb, B.; Venzmer, J. *Angew. Chem.* **1988**, *100*, 117.
5. Murtagh, J.; Thomas, J. K. *Faraday Discuss. Chem. Soc.* **1986**, *81*, 127.
6. Kurja, J.; Nolte, R. J. M.; Maxwell, I. A.; German, A. L. *Polymer* **1993**, *34*, 2045.
7. Israelachvili, J. N.; Mitchell, D. J.; Ninham, B. W. *J. Chem. Soc. Faraday Trans.* **1976**, *72*, 1525.

Vinyl Acetate Copolymers

See: Alternating Ethylene-Vinyl Alcohol Copolymer
Controlled Drug Delivery Systems
Ethylene-Vinyl Acetate Copolymer (Chemical Modification, Compatibilizer in Blends)
Vinyl Chloride Copolymers

VINYL CHLORIDE (Minisuspension Polymerization)

Preben C. Mørk,* B. Sæthre, and J. Ugelstad
Department of Industrial Chemistry
University of Trondheim

Low-viscosity plastisols for production of paste poly(vinyl chloride) (PVC) require broader particle size distributions (0.2–3 μm) than those obtainable by traditional emulsion polymerization techniques. This was improved upon by the introduction of mixed surfactant/fatty alcohol emulsifier systems, which resulted in small, stable monomer droplets where initiation of polymerization took place.[1]

A miniemulsion is a relatively stable oil-in-water emulsion with droplet sizes in the range of about 50–500 nm, intermediate in size between conventional emulsions and microemulsions.

When the polymer is insoluble in its own monomer, as is the case for PVC, polymerization takes place mainly in small polymer particles suspended in the dispersed vinyl chloride (VCM) droplets. These particles are agglomerates of so-called microdomains, formed by precipitation and coagulation of single polymer radicals.[2] Thus, the microcosmos in which polymerization takes place is in effect a suspension, the volume of which corresponds to that of the miniemulsion droplet. For this reason, it is felt that the term "minisuspension" polymerization more adequately describes the system.

Procedures for production of PVC latexes by initiation in stable, submicron monomer droplets prepared by various swelling methods have been described by Ugelstad.[3–5]

The kinetics and mechanisms of VCM polymerization have been reviewed by Ugelstad et al.[7] and was recently discussed by Xie et al.[2]

MINIEMULSION PREPARATION

Stable miniemulsions of monomers may be prepared by adding monomer to a mixed emulsifier system consisting of an anionic emulsifier and a long-chain, highly water-insoluble alkanol under ordinary stirring.

The importance and versatility of the miniemulsion polymerization technique were markedly enhanced as Ugelstad introduced methods for preparation of stable miniemulsions based on diffusion.[6]

THERMODYNAMICS OF SWELLING

The thermodynamics and kinetic aspects of the preparation of miniemulsions based on activated swelling have been thoroughly discussed by Ugelstad et al.[8–10]

VCM MINIEMULSIONS: PREPARATION AND STABILITY

The long-time stability of both monodisperse and polydisperse miniemulsions of VCM has been studied by Sæthre.[6]

MINISUSPENSION POLYMERIZATION

Sæthre carried out a comprehensive investigation of how the size characteristics of a PVC latex obtained by minisuspension polymerization depend on those of the miniemulsion employed.[6] To our knowledge, this is the only published work using stable, preformed, and well-defined miniemulsions of VCM.

In conclusion, it may be stated that the size characteristics of the PVC latexes produced by minisuspension polymerization are closely related to those of the monomer miniemulsion employed. The extent of secondary particle formation during polymerization depends on emulsifier concentration as well as on type and amount of initiator. It increases markedly when the concentration of free emulsifier exceeds the CMC value. Below the CMC, the number of new particles formed may be kept below 1% by weight. Both monodisperse and polydisperse latexes with particles in the size range of 0.2–1.5 μm may easily be prepared.

REFERENCES

1. Ugelstad, J.; Ellingsen, T.; Kaggerud, K. H. *Adv. Org. Coatings Sci. Techn.* **1980**, *2*, 1.
2. Xie, T. Y.; Hamielec, A. E. Wood, P. E.; Woods, D. R. *Polymer* **1991**, *32*, 537.
3. Ugelstad, J. U.S. Patent 4 113 687, 1978.
4. Ugelstad, J. U.S. Patent 4 186 120, 1980.
5. Ugelstad, J. U.S. Patent 4 336 173, 1982.
6. Sæthre, B. Thesis, University of Trondheim, Trondheim, Norway, 1994.
7. Ugelstad, J.; Mørk, P. C.; Hansen, F. K. *Pure Appl. Chem.* **1981**, *53*, 323.
8. Ugelstad, J.; Mørk, P. C.; Kaggerud, K. H.; Ellingsen, T.; Berge, A. *Adv. Colloid Interface Sci.* **1980**, *13*, 101.
9. Ugelstad, J.; Mørk, P. C.; Berge, A.; Ellingsen, T.; Khan, A. A. In *Emulsion Polymerization*; Piirma, I., Ed.; Academic: New York, 1982; p 383.
10. Ugelstad, J.; Mørk, P. C.; Mfutakamba, H. R.; Soleimany, E.; Nordhuus, I.; Schmid, R.; Berge, A.; Ellingsen, T.; Aune, O.; Nustad, K. In *Science and Technology of Polymer Colloids*; Poehlein, G. W.; Ottewill, R. H.; Goodwin, J. M., Eds.; Nato ASI Series 67; M. Nijhoff: Boston, 1983; p 51.

VINYL CHLORIDE COPOLYMERS

Peter V. Smallwood
Zeneca Resins

Poly(vinyl chloride) (PVC) is one of the world's major bulk commodity polymers. Yet, because it degrades below its melting point, it cannot be processed unless it contains processing agents or plasticizers and thermal stabilizers. An alternative method of increasing its processability is to copolymerize it with other monomers to reduce its melt viscosity, thereby "internally plasticizing" the polymer.

COPOLYMERIZATION OF VC

Most monomers react well with VC but some (e.g., styrene, α-methyl styrene, and butadiene) react rapidly but form stable radicals that do not propagate.

VC copolymers are manufactured commercially by free radical polymerization using suspension, emulsion, and solution polymerization. Polymerization and copolymerization of VC by

*Author to whom correspondence should be addressed.

anionic polymerization have been extensively investigated but have not proved commercially successful. The transition metal-based coordination catalysts invariably abstract -Cl from the polymerization chain, terminating propagation.

VINYL ACETATE

Most VA copolymers are made by the batch suspension process, which is cheaper and gives a cleaner product than emulsion polymerization.

Copolymers containing different amounts of VA and of varying molecular weights are produced. As with PVC homopolymers, molecular weights are measured as "K values," which are calculated from the viscosity of a 0.5% solution of the copolymer in cyclohexanone. The higher the K value, the higher the molecular weight.[1] Increasing the VA content of the polymer reduces melt viscosity for easier processing but also reduces the melting point and thermal stability. Copolymers containing about 15% VA with a low molecular weight (K value ~60) have the best processing properties and are used in vinyl flooring and vinyl records. VC is a chain transfer agent and the molecular weight of VC polymers decreases as the polymerization temperature increases.

Some VA copolymers are made by the emulsion route (e.g., Vinnolit) to give paste resins for very low temperature applications (~130°C). Thus, they are used to make foam backing for cushioned vinyl flooring, automotive sealants, and underbody protective coatings.

OLEFIN COPOLYMERS

Olefin copolymers are also made to reduce the melt viscosity of PVC, without the loss of mechanical properties or thermal stability of VA copolymers. However, they are more expensive to produce. VC-ethylene copolymers tend to be made by emulsion polymerization (e.g., Air Products, Tosoh, and Vinamul) and are used in surface coating applications.

GRAFT COPOLYMERS

When the two polymers are compatible (e.g., PVC and (PMMA) poly(methyl methacrylate) or PVC and styrene/butadiene copolymers), physical blends give the same performance as graft copolymers. VC-graft copolymers give advantages when the two polymers are partially compatible. Thus, VC is grafted onto ethylene-VA copolymers (EVA) and acrylates [e.g., poly(butyl acrylate)s or PBAs]. PVC-EVA or acrylic copolymers, containing up to 10% of the grafted polymer, are used to increase the impact strength of rigid PVC, particularly where good weatherability is required (e.g., in window frames).

VINYLIDENE CHLORIDE (VdC) COPOLYMERS

Small amounts of VC (10–20%) are used (e.g., Dow, Kureha, and Solvay) to modify the properties of VdC made by suspension and emulsion polymerization.[2,3] These copolymers were originally used in molding applications in which the products require good chemical resistance, such as gasoline filters, valves, and chemical process equipment. Extruded films, tubing, and multilayer films that utilize the excellent barrier properties of the crystallized VdC copolymer are increasing in importance.

CHLOROTRIFLUOROETHYLENE (CTFE) COPOLYMERS

Very small quantities of a CTFE-VC copolymers (e.g., Oxy 461) are manufactured by Occidental. The copolymer has excellent heat, light, and chemical stability and can be dissolved in solvents and used in coating applications or processed as a powder.

SURFACE COATING AND ADHESIVE APPLICATIONS

VC copolymers are used in solvent and, more recently, waterborne applications. VC-VA copolymers for use in solvent-borne applications must be homogeneous and are made using monomer feed processes by emulsion or solution polymerization (Union Carbide's UCAR range and Wacker's Vinnol grades). Copolymerization of small amounts of a carboxylic acid-containing monomer improves adhesion, particularly to metals.

For environmental reasons, it is becoming increasingly more important to reduce the VOC, or amount of volatile organic solvent emitted during the handling and drying of the surface coating. This is achieved by maximizing the solids content of the solution without increasing the solution viscosity unacceptably. The biggest reduction in VOC can be achieved by dispersing the copolymer in water.

VC-ethylene copolymers (e.g., Air Products Airflex and Vinamul) are used in interior and exterior trade paints, particularly in the United States. VC-VA-ethylene terpolymers have a greater resistance to alkalis than straight VC-VA copolymers and are used in house and masonry paints. Ethylene copolymers are also used in textile coatings, where the fire-retardant properties of PVC are important, and in adhesives.[4]

REFERENCES

1. Smallwood, P. V. In *Encyclopedia of Polymer Science and Engineering*; John Wiley & Sons: 1989; Vol. 17 and Supplement.
2. Meeks, M. R. *Macromol. Synth.* **1972**, *4*, 137.
3. Wessling, R. A. *Polyvinylidene Chloride*; Gordon and Breach Science: 1977.
4. Farmer, D. B.; Edser, M. H. *Chem. Ind. (Lond)* 1983.

Vinyl Chloride Polymers

See: *Peroxide Initiators (Overview)*
Powder Coatings (Overview)
Precipitation Polymerization (Overview)
Vinyl Chloride (Minisuspension Polymerization)
Vinyl Chloride Copolymers

Vinyl Esters

See: *Epoxy Acrylate-Based Resins*
Unsaturated Polyester Resins (Toughened with Liquid Rubber)

Vinyl Ether Polymers

See: *Living Polymerization, Fast (Vinyl Ethers)*
Personal Care Application Polymers (Acetylene-Derived)
Vinyl Ether Polymers (Overview)

VINYL ETHER POLYMERS (Overview)

Jürgen Barwich*
EDM/KE

Helmut W. J. Müller
EDM/KH
BASF Aktiengesellschaft

Polymers and copolymers of vinyl ethers are used for various applications. Homopolymers and copolymers with acrylates are used as raw materials for producing pressure-sensitive adhesive compounds.

PRODUCTION OF POLYMERS

Homopolymers

Vinyl ethers belong to the group of readily polymerizable compounds. Cationic initiators such as BF_3 and $AlCl_3$ allow rapid polymerization.

Anionic and free-radical initiators have no significance for polymerizing vinyl ethers on an industrial scale.

Copolymers

Copolymers of vinyl ethers and acrylates are obtained by free-radical emulsion polymerization. Polymerization takes place after the introduction of the monomer mixture into the slightly alkaline water, 70–90°C, containing a dispersing agent. Potassium peroxodisulfate is used as initiator.

PROPERTIES

Depending on the degree of polymerization, poly(vinyl ether)s are viscous oils, tacky soft resins, or rubber-like substances. Vinyl ether polymers are very resistant to hydrolysis and are tasteless and odorless.

The solubility is influenced by the alkyl group. The solubility of poly(vinyl methyl ether) in water is due to hydrogen bonding between the water molecules and ether oxygen.

STABILIZATION

The influence of oxygen, heat, and light causes chain cleavage, crosslinking, and reactions with oxygen.

These aging processes affect vinyl ether polymers in an undesirable way as well. Antioxidants must therefore be used to obtain adhesive compounds with constant properties.

APPLICATION

General

In addition to rubber and polyisobutylene, vinyl ether polymers belong to the classical intermediates for pressure-sensitive adhesive compounds. They are suitable for producing adhesive compounds for pressure-sensitive medical articles, pressure-sensitive labels, and film.

Vinyl Methyl Ether Polymers

Vinyl methyl ether polymers are applied as blending components for pressure-sensitive adhesive compounds based on other raw materials, usually aqueous dispersion-type adhesives. They are used primarily for achieving defined hydrophilic properties and increasing the adhesiveness.

One of the methods of increasing the water-vapor permeability of adhesive compounds is to add vinyl methyl ether polymers. This possibility of modification can be exploited, for instance, in the production of pressure-sensitive medical articles.[1]

Vinyl Ethyl Ether Polymers

Vinyl ethyl ether polymers are very similar to the substantially more widely used vinyl isobutyl ether polymers with regard to properties and processability. One difference is that the water-vapor permeability of these polymers is about equivalent to the average loss of water through the human skin.[1] Because of this property, vinyl ethyl ether polymers are particularly suitable for use in the medical field.

Vinyl Isobutyl Ether Polymers

Vinyl isobutyl ether polymers belong to the oldest and most widely used group of polyvinyl ethers. They are employed as adhesives compounds applied from a solution or melt and as blending components for increasing the adhesion.

OUTLOOK

The significance of vinyl ether polymers has changed over the course of time. However, these polymers still have a firm foothold as products for special applications in the adhesives field. The fact that poly(vinyl ether)s can also be used for adhesive compounds suggested for new technologies indicates that they will maintain this position in the future as well. They are recommended, for instance, for use as an adhesion-increasing component in solventfree, heat-crosslinking pressure-sensitive polyurethane systems[2] and solvent-free, radiation-curing pressure-sensitive polyacrylate adhesives.[3]

The properties of acrylic copolymers can be varied to a great extent by judicious selection of the monomers. In view of this possibility and the growing demand for new products, it is to be expected that the use of acrylic/vinyl ether copolymers will be applied on an increasing scale.

REFERENCES

1. Seymour, D. E.; Da Costa, N. M.; Hodgson, M. E.; Dow, J. Br. Patent 1 280 631, 1968.
2. Hagenweiler, K.; Scholz, K. Deutsche Offenlegungsschrift 2 328 430, 1973.
3. Steuben, K. C. *Adhesives Age* **1977**, *20*(6), 16.

Vinyl Polybutadienes

See: *Polybutadienes*

VINYL POLYPEROXIDES (Oxygen Copolymerization)

Kaushal Kishore** and K. Shanmugananda Murthy
Inorganic and Physical Chemistry
IDL-Nitro Nobel Basic Research Institute
Indian Institute of Science

*Author to whom correspondence should be addressed.

**Author to whom correspondence should be addressed.

Polyperoxides form a narrow but important class of polymers that are generally the polymeric counterparts of simple organic peroxides. Polyperoxides can have peroxy linkages either in the main chain or the side chain. The main-chain polyperoxides are obtained either by free-radical polymerization or by the condensation method.[1–4] The latter method is generally employed to prepare acid polyperoxides. Vinyl polyperoxides, in which the vinyl monomer and oxygen alternate, are synthesized by oxygen copolymerization. Acid polyperoxides, in contrast to vinyl polyperoxides, are more hazardous and are nearly explosive. However, poly(butadiene peroxide)[5] and poly(1,1,4,4-tetrafluoro-1,3-butadiene peroxide)[6] are exceptions. Among the polyperoxides reported thus far, poly(styrene peroxide) (PSP) has been most widely studied.

PREPARATION OF POLYPEROXIDES

Kinetics of Oxygen Copolymerization

The synthesis of polyperoxides falls into the broad area of autooxidation, which usually refers to slow oxidative processes brought about by oxygen at moderate temperatures. Besides forming polyperoxides, which are alternating copolymers of oxygen and vinyl monomers, many side reactions also occur, leading to various oxidation products, depending upon the partial pressure of oxygen and the temperature of the reaction.

>150 mm of Oxygen

In this pressure range, polyperoxide is the main product.

Synthetic Methods

The oxygen copolymerization of vinyl monomers can be carried out either by photopolymerization in the absence or presence of photoinitiators or by thermal polymerization with or without initiators.[9,15]

Characterization

For formation of the 1:1 alternating polyperoxide of vinyl monomer and oxygen has been confirmed using elemental analyses and spectroscopic techniques. The structure of the polyperoxide is given as **Structure 1**:[12] In the case of dienes such as 1,3-butadiene, chloroprene, isoprene, etc., both 1,2 and 1,4 structures are possible.[3]

PROPERTIES OF POLYPEROXIDES

Stability and Physical Properties

Polymeric peroxides are either colorless or yellow viscous oils or solids that are undistillable.[8] They have a number-average molecular weight in the range 300–5000.[3] These peroxides are stable in the cold and dark. All the polyperoxides degrade highly exothermically, the rate of decomposition increases with temperature, and they are nearly explosive above 100°C.[12] Controlled heating and dilution with solvents reduce their explosive character. Among the polyperoxides reported, indene and 1,1-diphenylethylene are the most stable and those of acrylonitrile, methyl acrylate, methyl vinyl ketone, and chloroprene are the least stable, the latter decomposing even at 0°C.[11]

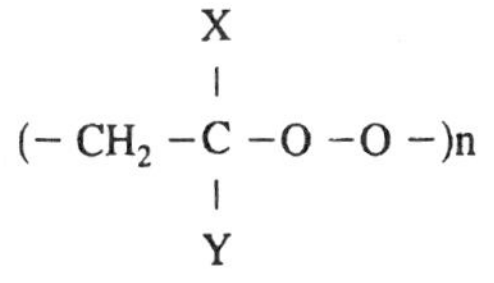

X = H, alkyl, phenyl, etc.,
Y = phenyl, Cl, –COOR, CN, vinyl, etc.
where R = alkyl

1

Thermal Decomposition

Polyperoxides generally degrade through breakage of the weak peroxy links, forming macroalkoxy radicals that can undergo two types of reactions: (i) unimolecular decomposition and (ii) disproportionation. In PSP, there are competing processes. In PSP, the main disproportionation products are α-hydroxy acetophenone, phenylglycol, and small quantities of phenylglyoxal. In unimolecular decomposition, which is a chain unzipping process, the macroalkoxy radicals undergo scission at a bond β to the radical center, forming carbonyl compounds: benzaldehyde and formaldehyde.

Photodecomposition

Photodecomposition studies have been carried out for poly(α-methylstyrene peroxide) (PMSP), PSP and polystyrene containing random peroxy linkages.[7,9–11,16]

APPLICATIONS

Initiators

Unlike acid polyperoxides, the initiating capabilities of the vinyl polyperoxides have been studied less.[17] PSP and PMSP have been examined as initiators for radical polymerization.[13,18] They are less efficient than AIBN and BPO.

Block Copolymers

The use of vinyl polyperoxides as initiators for radical polymerization results in a polymer that has polyperoxide segments in the backbone.[13] Such polymers can be used as initiators to make block copolymers with another vinyl monomer.[14]

Fuels

Polyperoxides degrade quite exothermically and exhibit the unique phenomenon of autopyrolysis. They also support a flame in air (autocombustion) with a burning rate comparable to that of solid propellants.[12,18] Such fast-burning characteristics make polyperoxides potential special fuel materials.

PERSPECTIVES AND CONCLUSIONS

Polyperoxides are potentially cheap materials, as peroxy groups come from abundantly available oxygen. They are, therefore, attractive candidates for commercial exploitation. However, they are generally formed in very low yields and have very low molecular weight.

The occurrence of competing reactions of unimolecular decomposition and disproportionation, as observed in PSP, has not been established in other similar polyperoxides due to the lack of data. In PMSP, unlike PSP, the photodecomposition at 30°C yields exclusively unimolecular decomposition products: acetophenone and formaldehyde.

Chemical reactions of polyperoxides have not been studied, with the exception of decomposition and reduction reactions. Reactive O–O bonds in these polymers are quite attractive for use as reagents for functional group modifications. In copolyperoxides, knowledge of the reactivity of -MO_2· radicals, which is more important than the reactivity of monomers in deciding the structural features, can be helpful in synthesizing multicomponent peroxide polymers.

ACKNOWLEDGMENT

K. S. M. thanks the management of IDL Chemicals Ltd. and Dr. G. D. Prasad, Chief Executive, INBRI division for sponsoring him for the Ph.D. program.

REFERENCES

1. Kishore, K.; Gayathri, V.; Ravindran, K. *J. Macromol. Sci. Chem.* **1981**, *A16*, 1359.
2. Mukundan, T.; Kishore, K. *Prog. Polym. Sci.* **1990**, *15*, 475.
3. Mogilevich, M. M. *Russ. Chem. Rev.* **1979**, *48*, 199.
4. Simonescu, C. I.; Comanita, E.; Pastravanu, M.; Dimitriu, S. *Prog. Polym. Sci.* **1986**, *12*, 1.
5. Alexander, D. S. *Ind. Eng. Chem.* **1959**, *51*, 733.
6. Putnam, R. E.; Sharkey, W. H. *J. Polym. Sci.* **1966**, *A1*(4), 2289.
7. George, G. A.; Hodgeman, D. K. *Eur. Polym. J.* **1977**, *13*, 63.
8. Hendry, D. G. *Encycl. Polym. Sci. Technol.* **1968**, *9*, 807.
9. Miller, A. A.; Mayo, F. R. *J. Am. Chem. Soc.* **1956**, *78*, 1017–1034.
10. Russell, G. A. *J. Am. Chem. Soc.* **1956**, *78*, 1035.
11. Mayo, F. R.; Miller, A. A.; Russell, G. A. *J. Am. Chem. Soc.* **1958**, *80*, 2265–2507.
12. Kishore, K.; Mukundan, T. *Nature* **1986**, *324*, 130.
13. Murthy, K. S.; Kishore, K.; Mohan, V. K. *Macromolecules* **1994**, *27*, 7109.
14. Subramaniam, K.; Kishore, K., Indian Institute of Science, unpublished data.
15. Kodaira, T.; Hashimoto, K.; Sakanaka, Y.; Tanihata, S.; Ikeda, K. *Bull. Chem. Soc. Jpn.* **1978**, *51*, 1487.
16. Weir, N. A.; Milkie, T. H. *Macromol. Chem.* **1978**, *179*, 1989.
17. Kuchanov, S. I. *Russ. Chem. Rev.* (Engl. Trans.) **1991**, *60*, 689; *Uspekhi Khimil* **1991**, *60*, 1346.
18. Mukundan, T.; Kutty, K. A. S.; Kishore, K. *Fuel* **1993**, *72*, 688.

Vinylcyclopropane Polymerization

See: *Ring-Opening Polymerization, Free Radical*
Ring-Opening Polymerization, Radical (with No Shrinkage in Volume)

Vinylidene Chloride Copolymers

See: *Vinyl Chloride Copolymers*

Vinylidene Cyanide Copolymers

See: *Ferroelectric Polymers (Structural Phase Transitions)*
Pyroelectricity

VINYLIDENE FLUORIDE-BASED THERMOPLASTICS (Overview and Commercial Aspects)

J. Steve Humphrey, Jr.
Elf Atochem North America, Incorporated

Poly(vinylidene fluoride) (PVDF) is the addition polymer of vinylidene fluoride (VDF).

The gem-difluoro substitution on alternating carbons in PVDF confers unique characteristics within the homologous series of fluorinated PE, which includes polymers derived from vinyl fluoride (VF), trifluoroethylene (VF_3) and tetrafluoroethylene (TFE).

In PVDF structure, however, significantly improved oxidation resistance, weathering resistance, and high-temperature chemical resistance and mechanical properties are manifested compared to PE, together with other unique characteristics related to the high dipole moment of the gem-difluoro group.

PVDF can be processed to provide products that respond to pressure (piezoelectric) or temperature (pyroelectric) fluctuations by voltage changes, again because of unique polar characteristics. The polar characteristics and triboelectric properties are the basis for the use of PVDF in toner formulations for photocopiers.

Precautions are in order when working with molten PVDF because toxic hydrogen fluoride is the byproduct of any thermal decomposition.

Addition of comonomers such as hexafluoropropylene (HFP) in the addition polymerization of fluorinated olefins creates linear copolymers that usually have significantly different properties from the parent PVDF. Many different fluorinated comonomers have been used to tailor the properties of processability of resinous PVDF, but the most important commercial fluorinated copolymers are those based upon HFP, TFE, and chlorotrifluoroethylene (CTFE).

Flexibilized PVDF resins based upon copolymer technology have been commercialized, the HFP copolymers being particularly important for wire and cable applications where millions of pounds of the relatively expensive PVDF are used annually because of their performance in fire environments. The stress-cracking resistance of PVDF in chemical environments is improved by low levels of HFP incorporation in the polymer.

Blends of PVDF with numerous other polymers have been studied and several compounds commercialized; generally the compatibility is based upon the blended polymer having carbonyl functionality that interacts via hydrogen bonding to the PVDF. Acrylic-type polymers are particularly important in formulating weatherable coatings for metals based upon PVDF as the nearly indestructible binder; this application annually involves millions of pounds of PVDF in premium, high-quality paints for coil steel and aluminum.

PVDF has been developed commercially since the mid-1960s. It is a relatively high-priced resin with a high specific gravity (1.76). The commercial success has been achieved by selling its performance enhancement against less costly polymers and its adequate performance against higher priced fluoropolymers in many applications.

A critical assessment of the scientific information available up to 1981 is contained in an excellent review article by Lovinger.[1] Tashiro provides a more recent scientific review of the literature, analyzing and resolving different interpretations of certain PVDF properties and behavior, especially the relationship between the polymer structure and the complex morphology.[2]

REFERENCES

1. Lovinger, A. J. In *Developments in Crystalline Polymers*; Bassett, D. C., Ed.; Applied Science: NJ, 1982; Vol. 1; p 195.
2. Tashiro, K. In *Ferroelectric Polymers*; Nalwa, H. S., Ed.; Marcel Dekker: New York; p 63.

VINYLIDENE FLUORIDE-BASED THERMOPLASTICS (Applications)

J. Steve Humphrey, Jr. and Edward J. Bartoszek
Elf Atochem North America, Incorporated AFA

The family of PVDF polymers, homopolymers, and copolymers, is broad and can be tailored in properties to fit most applications. These partially crystalline polymers are commercially produced by emulsion and suspension polymerization techniques.

One of the major applications for PVDF is in exterior architectural paints. PVDF does not absorb in the ultraviolet range above 200 nm, thus it is essentially transparent to UV. Because it does not absorb UV, PVDF does not degrade in sunlight. In addition, it has excellent moisture, abrasion, and fungus resistance, good mechanicals, and the ability to be formulated and applied from liquid dispersion systems. This combination makes PVDF an excellent choice for exterior paints.

PVDF polymers possess good resistance to most chemicals. Ketones, esters, and some highly polar solvents can swell or dissolve PVDF, especially at elevated temperatures. This may be a limitation for some CPI (chemical process industry) applications but is the key factor for coatings applications.

PVDF versatility in coatings extends beyond liquid systems. In addition to capability for powder coatings, PVDF formulations lend themselves well to application by standard melt processable techniques.

PVDF and PVDF compositions can be extruded into film or sheet and subsequently applied to another substrate by lamination, thermoforming, or molding approaches.

The CPI market segment is another major area utilizing PVDF homopolymers and copolymers.

The electrical–electronic (E/E) market is another major user of PVDF resins. The scope of applications covers plenum cables (telephone, fire alarm, signal, co-axial, fiber optic, power-limited, etc.), aircraft wiring, solder sleeves, computer back-panel wiring, tye-wraps, automotive, and others.

PVDF resins have a high dielectric constant and relatively high dissipation factor. This factor makes the resins unattractive as the *primary* layer of plenum cable constructions because the absorption of electromagnetic energy is very high, especially in high-speed data transmission cables.

PVDF parts have found significant use in nuclear radiation environments. PVDF resins are also used in piezoelectric and pyroactive components. It is beyond the scope of this article to delve into the mechanism of piezoelectric behavior. However, Kepler[1] and Kepler and Anderson[2] are good review papers covering this subject.

Some emerging new applications involving PVDF entail the use of PVDF to enhance the performance of other materials. One example is the incorporation of a low level of PVDF into a base material such as PVC for siding, and applying a capstock of PVDF compounded composition. Another specialty application has developed recently: PVDF is being used as a binder for electroactive materials in lithium ion batteries.[3,4]

REFERENCES

1. Kepler, R. G. *Piezoelectricity, pyroelectricity, and ferroelectricity in organic materials. Ann. Rev. Phys. Chem.* **1978**, *29*, 497–518.
2. Kepler, R. G.; Anderson, R. A. *Piezoelectricity and pyroelectricity in PVDF. J. Appl. Phys.* **1978**, *49*(8), 4490–4494.
3. Gozdz, A. S.; Schmutz, C. N.; Tarascon, J-M. "Rechargeable Lithium Intercalation Battery with Hybrid Polymeric Electrolyte." U.S. Patent 5 296 318, March 22, 1994.
4. Gozdz, A. S.; Schmutz, C. N.; Tarascon, J.-M.; Warren, P. *"Method of Making an Electrolyte Activatable Lithium-Ion Rechargeable Cell."* U.S. Patent 5 456 000, October, 10, 1995.

VINYLIDENE FLUORIDE-BASED THERMOPLASTICS (Blends with Other Polymers)

J. Steve Humphrey, Jr. and X. Drujon
Elf Atochem North America, Incorporated

In comparison with other fluoropolymers, PVDF associates with a large number of other polymers to form miscible blends. This distinctive behavior is due to the polarity of the PVDF repeat unit. As a matter of fact, all polymers reported to be miscible with PVDF have carbonyl functionalities in their side groups.

BLENDS OF PVDF WITH POLY(METHYL METHACRYLATE)

Due to cost and performance advantages, blends of PVDF with poly(methyl methacrylate) (PMMA) are used in a variety of applications, the major one being KYNAR 500®, PVDF-based architectural coatings (Elf Atochem registered trademark) which are applied as paints to aluminum or steel used for the facade of high-rise buildings, sports stadiums, and so on. These blends are also the most extensively studied system. Hydrogen bonding between the carbonyl oxygen of PMMA and the protons of PVDF was demonstrated using FTIR.[1]

The existence of an upper critical solution temperature, located below the melting point of PVDF, has been suggested by several authors.[2] For coating formulations, a composition containing 70 wt % of PVDF and 30 wt % of acrylic resin is recommended.

BLENDS OF PVDF WITH OTHER POLYMERS

A list of homopolymers miscible with PVDF is given in **Table 1**. Miscibility is demonstrated by the presence of an intermediate transition between the glass transitions of the two polymers (DSC, DMA), by a depression of the melting point of PVDF (DSC), or more dramatically by a lower critical solution temperature (LCST). All of these polymers have in common a carbonyl group in their structure, which gives rise to specific interactions. Many of the blends containing PVDF show a cloud point on heating.

TABLE 1. Homopolymers Compatible with PVDF

Homopolymer	B (Cal/cm^3)
Poly(methyl methacrylate)	−3.85
Poly(ethyl methacrylate)	−2.85
Poly(methoxymethyl methacrylate)	
Poly(acetonyl methacrylate)	
Poly(methyl acrylate)	−3.10
Poly(ethyl acrylate)	−1.71
Poly(vinyl methyl ketone)	−2.02
Poly(*N*-vinyl-2-pyrrolidone)	−5.78
Poly(*N,N*-dimethylacrylamide)	−5.78
Poly(ε-caprolactone)	
Poly(neopentyl glycol adipate)	−2.09
Poly(vinyl acetate)	−3.88

PRACTICAL ASPECTS OF PVDF BLENDS WITH OTHER POLYMERS

In addition to the coating applications for long-term protection of metal surfaces in the building and construction industries, there is a body of practical applications where blending with other polymers is practiced. PVDF is used as a processing aid in polyolefins, as a component in adhesive compounds, in ink and toner formulation, and many other applications where the polar PVDF homo- or copolymers resins have brought a special property related to the unique functional feature of the highly polar, alternating gemdifluoro substitution along the chain.

REFERENCES

1. Leonard, C.; Halary, J. L.; Mennerie, L. *Macromolecules* **1985**, *18*, 718.
2. Tomura, H.; Saito, H.; Inerie, T. *Macromolecules* **1992**, *25*(5), 1611.

VINYLIDENE FLUORIDE-BASED THERMOPLASTICS (Homopolymerization and Copolymerization)

J. Steve Humphrey, Jr. and X. Drujon
Elf Atochem North America, Incorporated

POLYMERIZATION

The free radical polymerization of vinylidene fluoride (VDF) can be initiated in a variety of conditions. Bulk and solution polymerization have been reported with numerous initiators, or under the influence of radiation. However, owing to the large heat of polymerization of VDF, only emulsion and suspension polymerizations are of commercial significance.

The fluorine atom is relatively small and provides little stabilization to the free radical of a propagating chain. Hence, the polymerization of VDF is not completely regiospecific, and the polymer contains occasional reversed monomer units.

COPOLYMERIZATION

VDF is a strongly polar molecule with a small tendency for conjugation and resonance stabilization. In consequence, most nonfluorinated monomers are poor partners for VDF in free radical copolymerization.

From a commercial point of view, prime interest has been concentrated on other fluorinated monomers such as TFE, HFP, and CTFE that help retain or enhance the desirable thermal, chemical, and mechanical properties of PVDF.

PVDF GRAFT AND BLOCK COPOLYMERS

Given the difficulty of combining VDF with functionalized monomers by simple copolymerization, much work has been directed toward the synthesis of block or graft copolymers. Two strategies have been used with success. Hydroxy-ended oligomers of PVDF can be prepared by polymerization of VDF in methanol.[1] Similarly, sulfate terminated PVDF can be prepared by persulfate initiated emulsion polymerization of VDF in the presence of alcohols; these end groups can be displaced with acrylic or methacrylic acid derivatives.[2] The resulting macromonomers are copolymerizable.

Another approach is to introduce "reactive handles" on the PVDF backbone. Grafting of a variety of monomers has been induced following irradiation, ozonolysis, or base induced dehydrofluorination of PVDF.[3–5] See Nalwa's book for extensive reviews of crystalline structures and properties as well as more details on the various polymerization approaches for vinylidene fluoride-based thermoplastic resins.[6]

REFERENCES

1. Oku, J.; Chan, R. J. H.; Hall, H. K. Jr.; Hughes, O. R. *Polymer Bull.* **1986**, *16*, 481–485.
2. Nicholas, P. O. U.S. Patent 5 194 508, 1993; 5 244 993, 1993; and 5 244 298, 1993, to B. F. Goodrich Company.
3. Svarfvar, B. L.; Ekman, K. B.; Sundell, M. J.; Nasman, J. H. *Polym. Preprints (Am. Chem. Sci., Div. Polymer Chem.)* **1993**, *34*(2), 640–641.
4. Boutevin, B.; Robin, J. J.; Serdani, A. *Eur. Polym. J.* **1992**, *28*(12), 1507–1511.
5. Brennan, J. V.; McCarthy, T. J. *Polymer Preprints (Am. Chem. Sci., Div. Polymer Chem.)* **1989**, *30*(2), 152.
6. *Ferroelectric Polymers*; Nalwa, H. S., Ed.; Marcel Dekker: New York, 1995.

VINYLIDENE FLUORIDE-TRIFLUOROETHYLENE COPOLYMERS (Ferroelectric-to-Paraelectric Phase Transition)

Roberto Luiz Moreira
Departamento de Física
Instituto de Ciências Exatas
Universidade Federal de Minas Gerais

Ferroelectricity was first observed by Valasek in 1920 in Rochelle salt crystals.[1] This phenomenon has since been observed in an increasing number of crystals (single crystals, mixed crystals, and solid solutions) and ceramics.[2] Ferroelectric materials necessarily possess a polar axis and thus belong to a polar symmetry class, i.e., a pyroelectric class. Thus, they show very interesting correlated properties, such as pyroelectricity, piezoelectricity, and second harmonic generation (shg), that give them many technological applications.[3] Only in 1977–78 was ferroelectricity observed in a polymeric material, poly(vinylidene fluoride) [PVDF],[4,5] whose strong piezoelectric, pyroelectric, and shg properties had been known since 1969–71.[6–8]

PVDF homopolymer is a semicrystalline material (the degree of crystallinity is ~50–60%) that can present four different well-studied crystalline phases at room temperature, α, β, γ, and δ.[9–10] The first of these phases is an antipolar phase whereas the others show polar axes. The β phase is ferroelectric, since the high electrical fields needed to give a sample macroscopic polarization (and reversal polarizations) transform phases γ and δ into β.

Although some polymeric materials have been classified as piezoelectric polymers, true piezoelectricity and ferroelectricity (those that originate from a polar crystalline phase) undoubtedly occur only in PVDF and some of its copolymers and in blends with compatible polymers.[11]

Ferroelectric materials represent currently, but not necessarily, a phase transition to a higher-temperature paraelectric phase (the prototype phase) at a critical temperature.[3] This is not the case for the PVDF homopolymer, where the ferroelectric phase is stable until it reaches its melting point. The occurrence of a ferroelectric-to-paraelectric phase transition in polymeric materials has been observed in two PVDF copolymers: poly(vinylidene fluoride-trifluoroethylene)[12–14] and poly(vinylidene fluoride-tetrafluoroethylene).[15,16] In the second of these copolymers, the transition temperatures are very close to the melting point (for a copolymer with 25 mol % of tetrafluoroethylene, the paraelectric phase is observed only when cooling the material from the melt, in a short temperature range below the recrystallization temperature). This would explain why the only reported studies on this kind of phase transition in this copolymer are, to our knowledge, the above-mentioned references.

Ferroelectric transitions, however, have been investigated extensively in poly(vinylidene fluoride-trifluoroethylene) copolymers [P(VDF-TrFE)], for materials ranging from 18–63 mol% of trifluoroethylene (TrFE).[17–27] It is now well established that the phase transition occurs because the gauche sequences appear in the all-*trans* low-temperature phase.

No significant problems arise in measuring the phase transition of these semicrystalline materials. However, interpreting the data can be delicate, due to the peculiarities of these systems, such as the influence of thermal history, crystallinity, microstructure, and the presence of defects in the bulk of the material.[28–31] Moreover, the ferroelectric and paraelectric phases coexist in very large temperature intervals (up to 40°C), giving rise to a blurring of the observed physical anomalies around the phase transition region.

ACKNOWLEDGMENTS

The studies by the author and his collaborators referred to in this article were supported by funds from the Brazilian government agencies CNPq, FINEP, and FAPEMIG. The author is grateful to Dr. J. Florêncio, Jr. for a critical reading of the manuscript.

REFERENCES

1. Valasek, J. *Phys. Rev.* **1920**, *15*, 537.
2. *Landolt-Börnstein New Series Group III*; Springer-Verlag: Berlin-Heidelberg, 1982; Vol. 16.
3. Lines, M. E.; Glass, A. M. *Principles and Applications of Ferroelectric and Related Materials*; Oxford University: Oxford, 1979.
4. Latour, M. *Polymer* **1977**, *18*, 278.
5. Naegele, D.; Yoon, D. Y. *Appl. Phys. Lett.* **1978**, *33*, 132.
6. Kaway, H. *Jpn. J. Appl. Phys.* **1969**, *8*, 975.
7. Bergman, J. G. Jr.; McFee, J. H.; Crane, G. R. *Appl. Phys. Lett.* **1971**, *18*, 203.
8. Glass, A. M.; McFee, J. H.; Bergman, J. G. Jr. *J. Appl. Phys.* **1971**, *42*, 5219.
9. Broadhurst, M. G.; Davis, G. T. In *Topics in Applied Physics—Electrets*; Sessler, G. M., Ed.; Springer-Verlag: Berlin, 1980; Vol. 33, Chapter 5.
10. Lovinger, A. J. In *Developments in crystalline polymers—I*; Basset, D. C., Ed.; Applied Science: London, 1982, Chapter 5.
11. Lovinger, A. J. *Science* **1983**, *220*, 1115.
12. Furukawa, T.; Date, M.; Fukada, E.; Tajitsu, Y.; Chiba, A. *Jpn. J. Appl. Phys.* **1980**, *19*, L109.
13. Furukawa, T.; Johnson, G. E.; Bair, H. E.; Tajitsu, Y.; Chiba, A.; Fukada, E. *Ferroelectrics* **1981**, *32*, 61.
14. Yamada, T.; Ueda, T.; Kitayama, T. *J. Appl. Phys.* **1981**, *52*, 948.
15. Lovinger, A. J. *Macromolecules* **1983**, *16*, 1529.
16. Lovinger, A. J.; Johnson, G. E.; Bair, H. E.; Anderson, E. W. *J. Appl. Phys.* **1984**, *56*, 2412.
17. Lovinger, A. J.; Furukawa, T.; Davis, G. T.; Broadhurst, M. G. *Polymer* **1983**, *24*, 1225.
18. Lovinger, A. J.; Furukawa, T.; Davis, G. T.; Broadhurst, M. G. *Polymer* **1983**, *24*, 1233.
19. Tashiro, K.; Takano, K.; Kobayashi, M.; Chatani, Y.; Tadokoro, H. *Ferroelectrics* **1984**, *57*, 297.
20. Krüger, J. K.; Petzelt, J.; Legrand, J. F. *Colloid and Polymer Sci.* **1986**, *264*, 791.
21. Ikeda, S.; Koninami, H.; Koyama, K.; Wada, Y. *Ferroelectrics* **1987**, *76*, 421.
22. Tashiro, K.; Kobayashi, M. *Polymer* **1988**, *29*, 426.
23. Tashiro, K. *Phase Transitions* **1989**, *18*, 213.
24. Moreira, R. L.; Saint-Grégoire, P.; Latour, M. *Phase Transitions* **1989**, *14*, 243.
25. Legrand, J. F. *Ferroelectrics* **1989**, *91*, 303.
26. Furukawa, T. *Phase Transitions* **1989**, *18*, 143.
27. Furukawa, T. *Ferroelectrics* **1990**, *104*, 229.
28. Moreira, R. L.; Saint-Grégoire, P.; Lopez, M.; Latour, M. *J. Polym. Sci., Polym. Phys. Ed.* **1989**, *27*, 709.
29. Moreira, R. L.; Almairac, R.; Latour, M. *J. Phys.: Condens. Matter* **1989**, *1*, 4273.
30. Green, J. S.; Farmer, B. L.; Rabolt, J. F. *J. Appl. Phys.* **1986**, *60*, 2690.
31. Stack, G. M.; Ting, R. Y. *J. Polym. Sci., Polym. Phys. Ed.* **1988**, *26*, 55.

VINYLSILANES (Isomerization Polymerization)

Jun-ichi Oku
Department of Applied Chemistry
Nagoya Institute of Technology

Anionic polymerization of vinylsilanes (CAS Registry No. 119045-92-0) was first reported by Nametkine et al.[1] They

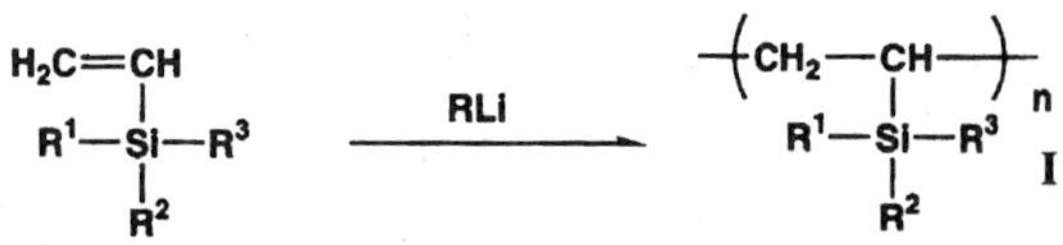

I

revealed that the polymerization of trimethylvinylsilane (TMVS) (CAS Registry No. 25036-32-3) and some other vinylsilanes with alkyllithium in heptane at temperatures between –60 and 75°C results in good yields of white powdery polymers. They described the polymer structure produced by vinyl polymerization, as shown in **Scheme I**.

However, we found that the addition of *N,N,N',N'*-tetramethylethylenediamine (TMEDA) to the polymerization system of TMVS with butyllithium (BuLi) changes the appearance of the resulting polymer from white powder to viscous liquid. The change is due to the isomerization of the propagating end.[2]

PREPARATION AND PROPERTIES

Trimethylvinylsilane

The anionic polymerization of TMVS with *n*- or *s*-BuLi in hexane in the presence of the amount of TMEDA equimolar to the initiator produced a viscous liquid polymer, although its M_n was high.[2] Studies revealed that the proton abstraction occurs intramolecularly and that the abstractions within the end unit and from the penultimate unit are both possible (**Scheme II**).[3]

The content of isomerized units (CIU), the fraction of the monomer units that have the isomerized structure in all the monomer units, can be calculated from the peak areas of silylmethylene and silylmethine protons in the ^1H NMR spectrum of the polymer. The CIU varied with polymerization conditions, such as polymerization temperature, monomer concentration, and the proportion of added TMEDA to the initiator.[4] At a high temperature (60°C), the CIU rose to 100%. PTMVS obtained without TMEDA contains some isomerized units (CIU: 6–14%).

Dimethylphenylvinylsilane

The polymerization of dimethylphenylvinylsilane (DMPVS) (CAS Registry No. 26744-16-1), which has a phenyl group, with BuLi/TMEDA in hexane resulted in a polymer with isomerized monomer units similar to those of TMVS.[5]

Vinylsilane

Recently, isomerization polymerization of vinylsilane, $CH_2 = CHSiH_3$, was reported.[6] Polyvinylsilane obtained with $LiAlH_4$ or *n*-BuLi consists of isomerized units, $CH(CH_3)SiH_2$.

APPLICATIONS

Polyvinylsilane is expected to serve as a starting material for β-silicon carbide.[6] Because polyvinylsilanes produced by the isomerization polymerization contain silicon atoms in the backbones, their physical properties would be very different from carbon-chain polymers. Many new applications of these polymers will appear in the near future.

REFERENCES

1. Nametkine, N. S.; Topchiev, A. V.; Dourgarian, S. G. *J. Polym. Sci., Part C* **1963**, *4*, 1053.
2. Asami, R.; Oku, J.; Takeuchi, M.; Nakamura, K.; Takaki, M. *Polym. J.* **1988**, *20*, 699.
3. Oku, J.; Hasegawa, T.; Nakamura, K.; Takeuchi, M.; Takaki, M.; Asami, R. *Polym. J.* **1991**, *23*, 195.
4. Oku, J.; Hasegawa, T.; Takeuchi, T.; Takaki, M. *Polym. J.* **1991**, *23*, 1377.
5. Oku, J.; Hasegawa, T.; Kubota, Y.; Takaki, M.; Asami, R. *Polym. Bull.* **1992**, *28*, 505.
6. Ito, M. *Muki Kobunshi 2* Sangyo Tosho: Tokyo, Japan, 1993; Chapter 3, section 2 (in Japanese).

Vinyltetrazole Copolymers

See: Polyvinyltetrazoles

Viologen Polymers

See: Ionene Oligomers
Liquid Crystalline Polymers (Viologen: Thermotropic and Lyotropic)
Viologen Polymers (Electron-Transfer Catalysts)

VIOLOGEN POLYMERS (Electron-Transfer Catalysts)

Yu Xian Da and Li Shuben
Lanzhou Institute of Chemical Physics
Academia Sinica

Liu Futian
The Research Institute of Polymer Chemistry
Nankai University

The discovery of viologens about 60 years ago[1] has led to enormous advances, due to their reversible electrochemical behavior and potential applications. Michaelis first reported on

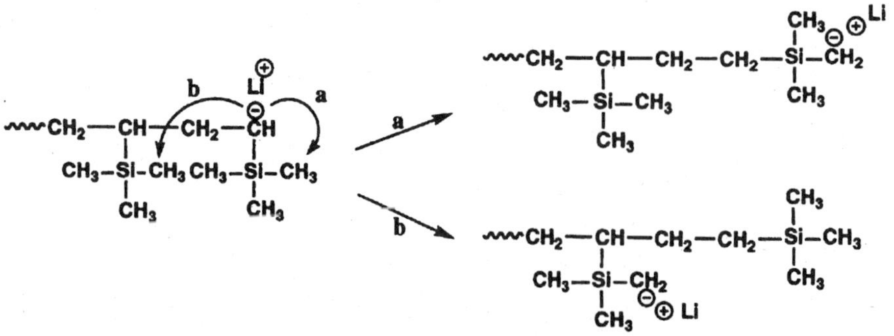

II

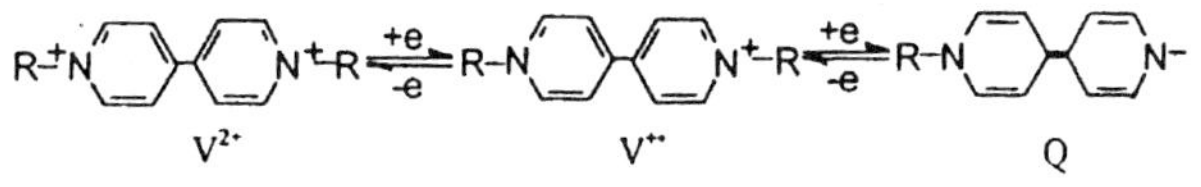

FIGURE 1. Reversible electrochemical behaviors of viologens.

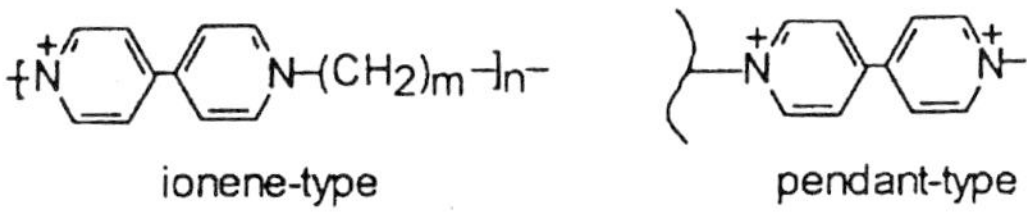

FIGURE 2. Classification of viologen polymers.

the electrochemical behavior of a class of compounds formally known as 1,1′-disubstituted 4,4′-bipyridinium ions (**Figure 1**), which he christened the "viologens." An important application of viologens is as electron-transfer catalyst (ETC) in hydrogen production by photoreduction of protons.[2–5]

PREPARATION OF VIOLOGEN POLYMERS

Classification of Viologen Polymers

Viologen polymers are usually classified into two main types, the ionene-type and the pendant-type, according to the positions of viologen groups in the polymeric backbone (**Figure 2**).

Preparation of the Ionene-Type Viologen Polymers

The ionene-type viologen polymers were generally prepared by the condensation of simple dipyridine and dihalides, the "Menschutkin Reaction.[9–15]

Preparation of Pendant-Type Viologen Polymers

The pendant-type viologen polymers are usually synthesized in two different ways: polymerization of monomers with viologen moiety or copolymerization of the monomers with viologen structure with other monomers and by the polymeric functional reaction to incorporate the viologen structure onto the polymer chain.

Polymers with pendant viologen moiety can also be prepared by polymer functionalization using the polymer material available.[6–8,17–19]

PROPERTIES OF VIOLOGEN POLYMERS

Polyalkylviologens as ionene-type viologen polymers showed the dimer-monomer equilibrium of viologen radicals when reduced in aqueous solution of sodium dithionate ($Na_2S_2O_4$).

The authors[20] have applied the ionic surfactant sodium lauryl sulfate to retard the formation of the dimer form of polyalkylviologen radicals in H_2O~CH_2Cl_2 two-phase systems. The CH_2Cl_2 insoluble polyalkylviologen radicals could be introduced into the CH_2Cl_2 phase, but the existing forms of viologen radicals are not altered significantly. Further study[21] based on the formation of polyalkylviologen-polyanion complexes showed that the complexes had good redox properties.

Viologen Polymers as Electron-Transfer Catalysts

Viologens and viologen polymers have received much attention as ETC in redox systems, particularly in hydrogen production by photoreduction. The development of new conversion systems composed of polymers has become an active research field in the area of solar energy conversion. Also under study are reactions using viologens and their polymers (PV^{2+}) as the electron mediators for the reduction of organic compounds to make good use of inexpensive and readily available reducing agents such as zinc and sodium dithionite ($Na_2S_2O_2$).

Reduction of Azobenzene in the Presence of PV^{2+}

Azobenzene that is not reduced by Zn or $Na_2S_2O_4$ at room temperature could be reduced in high yields[7,22] to hydrazobenzene under mild conditions in the presence of viologen polymers.

Debromination of vic-Dibromides and Diphenyl Bromomethane Using PV^{2+} as ETC

The debrominations of diphenylbromomethane (Ph_2CHBr) using PV^{2+} as ETC were carried out in our investigations.[21,23–25]

Reduction of Activated Carbonyl Compounds

Viologens and viologen polymers also mediate the electron transfer from solid reductants (zinc powder) to activated carbonyl compounds such as α-ketoesters,[16] quinones,[26] α-diketones and diphenyl ketones.[27]

OUTLOOK

Viologen polymers are attracting much attention as functional redox materials to construct the electron-transfer systems for the reduction of organic compounds. Combining the photoreduction of viologen polymers with their electron-transfer behavior might help to utilize the solar energy for photochemical redox reactions. The single-electron transfer from an electron donor to an electron acceptor through the reversible redox behavior of viologen polymers is a useful model in biological studies. The viologen polymers with good membrane-forming capacity are attractive for use in constructing electron-transport membranes to separate the oxidation-reduction reactions, which would result in a product with no easily separated by-product and continuous reactions. The redox reversibility of viologen polymers has been considered for use as one component of secondary batteries. The research on the viologen polymers is progressing well.

REFERENCES

1. Michaelis, L.; Hill, E. S. *J. Gen. Physiol.* **1953**, *16*, 859.
2. Fox, M. A.; Dulay, A. T. *Chem. Rev.* **1993**, *93*, 341.
3. Kaneko, M.; Yamada, A. *Adv. Polym. Sci.* **1984**, *55*, 2.
4. Richoux, M. C. *Photogeneration of Hydrogen*; Academic: 1982.
5. Frank, A. J.; Willner, I.; Goren, Z.; Degani, Y. *J. Am. Chem. Soc.* **1987**, *109*, 3568.
6. Ageishi, A.; Endo, T.; Okawara, M. *J. Polym. Sci., Polym. Chem. Ed.* **1983**, *21*, 175.
7. Saotome, Y.; Endo, T.; Okawara, M. *Macromolecules* **1983**, *16*, 881.
8. Endo, T.; Ageishi, K.; Okawara, M. *J. Org. Chem.* **1986**, *51*, 4309.
9. Renbaum, A.; Baumgartner, W.; Eisenberg, A. *J. Polym. Sci. Part B.* **1968**, *6*, 159.
10. Noguchi, H.; Renbaum, A. *J. Polym. Sci. Part B.* **1969**, *7*, 383.
11. Renbaum, A.; Hermann, A. M.; Steward, F. E.; Gutmann, F. *J. Phys. Chem.* **1969**, *73*, 513.

12. Renbaum, A.; Singer, S.; Keyer, H. *J. Polym. Sci. Part B.* **1969**, *7*, 395.
13. Factor, A.; Heinsohn, G. E. *J. Polym. Sci. Part B.* **1971**, *9*, 289.
14. Simon, M. S.; Moore, D. T. *J. Polym. Sci., Polym. Chem. Ed.* **1975**, *13*, 1.
15. Sato, H.; Tamamura, T. *J. Appl. Polym. Sci.* **1979**, *24*, 2075.
16. Okawara, M.; Hirose, T.; Kamiya, N. *J. Polym. Sci., Polym. Chem. Ed.* **1979**, *17*, 927.
17. Liu, F. T.; Yu, X. D.; Li, S. B. *Eur. Polym. J.* **1994**, *30*, 289.
18. Liu, F. T.; Yu, X. D.; Li, S. B. *J. Polym. Sci., Polym. Chem. Ed.* **1993**, *31*, 3245.
19. Sato, T.; Nambu, Y.; Endo, T. *J. Polym. Sci., Polym. Lett.* **1989**, *27*, 289.
20. Liu, F. T.; Yu, X. D.; Li, S. B., unpublished data.
21. Liu, F. T.; Yu, X. D.; Li, S. B. *J. Polym. Sci., Polym. Chem. Ed.* **1994**, *32*, 1043.
22. Liu, F. T.; Song, N.; Yu, X. D.; Li, S. B. *Makromol. Chem. Phys.* **1994**, *195*, 2169.
23. Liu, F. T.; Yu, X. D.; Li, S. B. *Chinese Chem. Lett.* **1993**, *4*, 577.
24. Liu, F. T.; Feng, L. B.; Yu, X. D.; Li, S. B. *Chinese Chem. Lett.* **1993**, *4*, 959.
25. Liu, F. T.; Yu, X. D.; Li, S. B.; Feng, L. B. *Eur. Polym. J.*
26. Takeishi, M.; Yabe, Y.; Hayama, S. *Makromol. Chem., Rapid Commun.* **1981**, *2*, 374.
27. Endo, T.; Saotome, Y.; Okawara, M. *Tetrahedron Lett.* **1985**, *26*, 4525.

Viscose Rayon

See: *Textile Fibers (Structure and Properties)*

VISCOSITY-INDEX IMPROVERS

Abhimanyu O. Patil
Corporate Research Laboratory
Exxon Research and Engineering Company

Viscosity index improvers (VIIs) or viscosity modifiers (VMs) are lubricating-oil additives, usually high-molecular-weight oil soluble polymers, that permit one to formulate oils with desirable viscosity-temperature, shear-rate characteristics.

VISCOSITY IMPROVERS

The oldest viscosity modifiers were based on polyisobutylene polymers. They have been withdrawn from commercial use due to oxidative and mechanical instability. Currently, the principal monofunctional viscosity modifiers are ethylene-propylene copolymers (EP) also known as olefin copolymers (OCP), styrene-hydrogenated diene copolymers, and polymethacrylates (PMA).

POLYMETHACRYLATES

Typically, these polymers are made by free-radical polymerization of alkyl acrylates, where the alkyl group could be methyl, butyl, dodecyl, or octadecyl.

The concept of combining two polymers by making comb-shaped polyalkylmethacrylates with olefin copolymer side-chain was discussed recently.[1] These comb polymers enable one to take advantage of (PMAs) for low-temperature properties as well as hydrocarbon polymers for their good thickening efficiency.

ETHYLENE-PROPYLENE COPOLYMERS

Because of their relatively low cost and good thickening efficiencies, ethylene-propylene copolymers [9010-79-1] have achieved a major position in the VII market.

Manipulation of composition, structure, and morphology of ethylene-propylene copolymers results in interesting solution properties of the polymers. At room temperature and above, the molecular weight of the copolymer is an important factor affecting the viscosity of the oil. With decreasing temperature, the composition and structure have the most pronounced effect on the viscosity of the oil.[2]

HYDROGENATED DIENE-STYRENE COPOLYMERS

The hydrogenated isoprene-styrene viscosity modifiers are diblock copolymers with one continuous block of oil-insoluble polystyrene followed by one continuous block of oil-soluble hydrogenated isoprene. Depending on the average molecular weight, the styrene-isoprene ratio, the temperature, and the paraffin content of the oil, these polymers form micelles that impart unique properties to the oil.[3]

Multiarm star-branched hydrogenated polyisoprenes are also well known.[4,5] These structures are very shear-stable for their thickening power as measured by the SSI test.

MULTIFUNCTIONAL VISCOSITY IMPROVERS

The idea of extending the use of VII polymers to control sludge, varnish, and oxidation in engines has stimulated the design of multifunctional polymers. Polymers that contain polar functional groups such as amines act as a dispersant for sludge suspension. Polymers with this characteristic that also modify viscosity are termed multifunctional viscosity modifiers or dispersant viscosity modifiers. Further modification of the polymer to introduce antioxidant characteristic is also desirable, and yields antioxidant dispersant viscosity modifiers.

We recently reported the direct polymerization of amine-functionalized ethylene-propylene copolymers and the use of these polymers as multifunctional viscosity modifiers.[6–8]

The most commonly used procedure to prepare multifunctional viscosity improver (MFVI) is the chemical modification of preformed polymers. This approach has much more synthetic flexibility, and the patent literature covers many reactions.[9–13]

The MFVIs are frequently used in place of simple VIIs in motor oil because they decrease the amount of dispersant required to maintain a desired performance level.

Lube-oil technology offers fascinating research opportunities, from molecule synthesis to polymer science. Many opportunities remain to produce a structure that is optimum in terms of mechanical stability, low-temperature viscometric properties, fuel economy, and multifunctionalities, including dispersancy and antioxidancy.

ACKNOWLEDGMENTS

I would like to thank G. Ver Strate, K. Lewtas, G. W. Schriver, and D. N. Schulz for their comments and Nancy Jagoe for help with literature searches.

REFERENCES

1. Omeis, J.; Pennewiss, H. *Polym. Prepr., Am. Chem. Soc., Div. Polym. Chem.* **1994**, *35*, 714.
2. Rubin, I. D. *Polym. Prepr., Am. Chem. Soc., Div. Polym. Chem.* **1991**, *32*, 84.
3. Watanabe, H.; Kotaka, T. *Macromolecules* **1983**, *16*, 1783.
4. Eckert, R. U.S. Patent 4 116 917, 1978.
5. Eckert, R. J.; Covey, D. F. *Lub. Sci.* **1988**, *65*, 1.
6. Patil, A. O.; Datta, S.; Lundberg, R. D. *Polym. Prepr., Am. Chem. Soc., Div. Polym. Chem.* **1990**, *31*, 456.
7. Patil, A. O.; Datta, S.; Gardiner, J. B.; Lundberg, R. D. U.S. Patent 5 030 370, 1991.
8. Patil, A. O.; Datta, S. *Polym. Prepr., Am. Chem. Soc., Div. Polym. Chem.* **1994**, *35*, 716.
9. Bloch, R.; McCrary, T. J.; Brownawell, D. W. U.S. Patent 4 517 104, 1985.
10. Gardiner, J. B.; Dick, M. N. U.S. Patent 4 780 228, 1988.
11. Kapuscinski, M. M.; Jones, R. E. U.S. Patent 4 699 723, 1987.
12. Kapuscinski, M. M.; Nalesnik, T. E.; Biggs, R. T.; Chafetz, H.; Liu, C. S. U.S. Patent 4 948 524, 1990.
13. Nalesnik, T. E.; Brock, P. W. U.S. Patent 4 863 623, 1989.

VITREOUS BODY, ARTIFICIAL

Traian V. Chirila,* Ye Hong, and Paul D. Dalton
Department of Biomaterials and Polymer Research
Lions Eye Institute

The vitreous body (*corpus vitreus*), also termed vitreous humor, vitreus, or vitreous, is a clear and transparent mass (gel, liquid, or a mixture of both) that fills the posterior cavity of the eye in vertebrates, and is located between the image-forming lens and the light-absorbing retina. The human vitreous body is a hydrogel with a very high water content (>99%) that provides an adequate support for the retina and allows the diffusion of metabolic solutes. It is accepted that the vitreous body possesses a unique macromolecular organization, very likely a double-network system consisting of a scaffold of randomly spaced rod-like collagen fibers filled and entangled with a network of very large, coiled-up macromolecules of hyaluronic acid (hyaluronan).

POLYMERS AS VITREOUS SUBSTITUTES

The use of polymers as vitreous substitutes began later than their use for other intraocular prostheses, as previously described in several preliminary reviews and recently in an in-depth review.[2–7]

Poly(N-vinyl-2-pyrrolidinone) (PVP) (*CAS Reg. No. 9003-39-8*) was the first synthetic polymer proposed and tested as a vitreous substitute. Following the injection of PVP aqueous solutions into the rabbit vitreous cavity, the vitreous body became opaque.[8] In human patients, it was found that PVP induced less inflammation than air or saline.[9]

Polyacrylamide (PAAm) (*CAS Reg. No. 9003-05-8*) has evolved in an interesting way as a vitreous substitute. Müller-Jensen and Köhler polymerized the monomer (acrylamide, inside the vitreous cavity of rabbit eyes, apparently ignoring the well-known toxicity of this compound.[10] However, the ensuing pathologic complications convinced the investigators that the polymer, rather than the monomer, should be implanted into the eye.[11] The dialyzed gels led to many fewer complications, which emphasized the harmful effects of residual monomer. PAAm has not been tried since as a vitreous substitute.

Hydrogels of poly(glyceryl monomethacrylate) (PGMA) (*CAS Reg. No. 28474-30-8*) were described by Refojo.[12,13] Despite initial optimism, it was later acknowledged that the procedure of implanting dried gels had two main drawbacks: first, the surgery is too traumatic; second, the swelling rate of the implant is too low.[4] An injectable PGMA gel was later developed that was injected into the rabbit vitreous cavity with promising results.[14]

Poly(vinyl alcohol) (PVA) (*CAS Reg. No. 9002-89-5*) after crosslinking by γ-irradiation, has been proposed as a vitreous substitute and investigated for many years by Yamauchi and coworkers.[15–18] Crosslinked PVA hydrogels remain valid contenders for vitreous substitution.

Poly(2-hydroxyethyl acrylate) (PHEA) (*CAS Reg. No. 26022-14-0*) was specifically developed as a material for vitreous substitution.[6] The crosslinked gel displayed remarkable physical characteristics, including injectability. Unfortunately, extensive *in vivo* experiments have shown severe complications, including vitritis, membrane formation, and chorioretinal atrophy. These findings preclude the further use of PHEA as a vitreous substitute.

Aqueous solutions of modified natural polymers, such as commercial polysaccharides (gums, cellulose, derivatives, dextrans) were tested as vitreous substitutes with generally poor results.[3,19–21]

RECENT WORK ON ARTIFICIAL VITREOUS BODY

The lesson that can be learned from the previous trials is that any polymer obtainable as a water-insoluble (but water-swellable) transparent gel can presumably function as a vitreous substitute provided at least that (i) its fragmentation during surgical manipulation is minor, (ii) it does not induce toxic reaction, and (iii) it has viscoelastic properties.

Certain hydrogels with high and very high equilibrium water content have been investigated recently. Polymerization and copolymerization in aqueous solutions of methyl 2-methyoxy-2-(1-oxopropenylamino)acetate (*CAS Reg. No. 77402-03-0*), better known as methyl acrylamidoglycolate methyl ether (MAGME), resulted in hydrogels with a very high water content and good optical properties.[22]

PMAGME gel was injected in the vitreous cavity of rabbits, and the clinical observation was continued for 12 months.[23] Despite acceptable clarity, no rise in intraocular pressure, and no significant inflammation, the pallor of the optic nerve dish was seen after six months in all eyes. Histological examination of the central retina by light microscopy indicated dramatic pathological changes, including extensive disorganization of the inner layers of the retina, loss of nerve fibers, proliferation of glial cells, severe damage of the optic nerve, and retinal detachment. Analysis by immunohistochemistry of labeled retinal sections revealed loss of ganglion cells and extensive pathological changes of the Müller cells and astrocytes.

*Author to whom correspondence should be addressed.

Another approach investigated was to assess as vitreous substitutes the homopolymer of NVP or its copolymers with 2-hydroxyethyl methacrylate (HEMA), all crosslinked.[24–26]

This gel was injected in the vitreous cavity of rabbits and the eyes were followed up for four weeks. Clinically, there were no complications over this period. Histologic examination of the enucleated eyes indicated that the neural elements of the retina were not affected. However, macrophages, multinucleate giant cells, and plasma cells were present in the vitreous cavity. Round particles were seen in the cavity and even within the layers of the retina. These particles could be a result of gel fragmentation as an incipient stage of its degradation, but further investigations are necessary to assess this possibility.

ACKNOWLEDGMENTS

The financial support of the National Health and Medical Research Council of Australia through grant 940707 made this article possible and it is gratefully acknowledged. The authors are also grateful to Professor Ian Constable for his support and encouragement. The collaboration of Sarojini Vijayasekaran and Sjakon Tahija and the editorial assistance of Ruth Gutteridge are kindly acknowledged.

REFERENCES

1. Smith, R. *Br. J. Ophthalmol.* **1952**, *36*, 385.
2. Peyman, G. A.; Ericson, E. S.; May, D. R. *Surv. Ophthalmol.* **1972**, *88*, 540.
3. Kishimoto, M.; Yamanouchi, U.; Mori, S.; and Nakamori, F. *Nippon Ganka Gakkai Zasshi* **1964**, *68*, 1145.
4. Refojo, M. F. *Polym. Sci. Technol.* **1975**, *8*, 313.
5. Balazs, E. A.; Hulsch, E. In *Advances in Vitreous Surgery*; Thomas, C. C., ed., Springfield, IL, 1976; Chapter 57.
6. Chan, I. M.; Tolentino, F. I.; Refojo, M. F.; Fournier, G.; Albert, D. M. *Retina* **1984**, *4*, 51.
7. Chirila, T. V.; Tahija, S.; Hong, Y.; Vijayasekaran, S.; Constable, I. J. *J. Biomater. Applications* **1994**, *9*, 121.
8. Scuderi, G. *Ann. Ottalmol. Clin, Oculistica* **1954**, *80*, 213.
9. Hayano, S.; Yoshino, T. *Rinsho Ganka* **1959**, *13*, 449.
10. Muller-Jensen, K.; Koehler, H. *Ber. Deut. Ophthalmol. Ges.* **1968**, *68*, 181.
11. Muller-Jensen, K.; Klein, J.; Bauer-Stab, G. *Ber. Deut. Ophthalmol. Ges.* **1969**, *69*, 530.
12. Refojo, M. F. *J. Appl. Polym. Sci.* **1965**, *9*, 3161.
13. Refojo, M. F. *Polym. Sci. Technol.* **1975**, *8*, 313.
14. Hogen-Esch, T. E.; Shah, K. R.; Fitzgerald, C. R. *J. Biomed. Mater. Res.* **1976**, *10*, 975.
15. Yamauchi, A.; Matsuzawa, Y.; Nishioka, K.; Hara, Y.; Kamiya, S. *Kobunshi Ronbunshu* **1977**, *34*, 261.
16. Hara, Y.; Nishioka, K.; Kamiya, S.; Yamauchi, A.; Matsuzwa, Y. *Nippon Ganka Kiyou* **1977**, *28*, 576.
17. Hara, Y.; Hara, T.; Hatanaka, O. L.; Hirai, H.; Ichiba, S.; Kamiya, S.; Nako, S. *Nippon Ganka Kiyou* **1984**, *35*, 1340.
18. Yamauchi, A. In *Polymer Gels*; Plenum: New York, 1991; pp 127–134.
19. Mori, S. *Nippon Ganka Gakkai Zasshi* **1965**, *69*, 1441.
20. Mori, S. *Nippon Ganka Gakkai Zasshi* **1967**, *71*, 22.
21. Gombos, G. M.; Berman, E. R. *Acta Ophthalmol.* **1967**, *45*, 794.
22. Chirila et al., unpublished.
23. Chirila, T. V.; Constable, I. J.; Hong, Y.; Vijayasekaran, S.; Humphrey, M. F.; Dalton, P. D.; Tahija, S. G.; Maley, M. A. L.; Cuypers, M. J. H.; Sharp, C.; Moore, S. R.; Davies, M. J. *Cells Mater.* **1995**, *5*, 83.
24. Dalton, P. D.; Chirila, T. V.; Hong, Y. Jefferson, A. *Polym. Gels Netw.*, in press.
25. Hong, Y.; Chirila, T. V.; Vijayasekaran, S.; Dalton, P. D.; Tahija, S. G.; Cuypers, M. J. H.; Constable, I. J. *J. Biomed. Mater. Res.*, in press.
26. Vijayasekaran, P. D.; Chirila, T. V.; Hong, Y.; Tajiha, S. G.; Dalton, P. D.; Constable, Il J.; McAllister, I. L. *J. Biomater. Sci. Polym. Edn.*; in press.

Vitreous Substitutes

See: Vitreous Body, Artificial

Volume Expansion

See: Ring-Opening Polymerization, Anionic (with Expansion in Volume)
Ring-Opening Polymerization, Cationic (with Expansion in Volume)
Ring-Opening Polymerization, Radical (with No Shrinkage in Volume)

Vulcanization

See: Chlorosulfonated Polyethylene
Natural Rubber (Chemical Modification)
Rubber Additives, Multifunctional
Vulcanization (Overview)

VULCANIZATION (Overview)

A. P. Kuriakose
Department of Polymer Science and Rubber Technology
Cochin University of Science and Technology

Vulcanized rubber is an ideal construction material because of its high elasticity, strength, resistance to abrasion, impermeability to gases, and resistance to chemicals and solvents. Introduced by Charles Goodyear in 1944, the process of sulfur vulcanization (U.S. Patent No. 3633) has changed very little in spite of rapid progress in rubber technology during this period. The first discovery that led to the modern process of vulcanization was that adding sulfur to rubber then heating the mixture, greatly improved the properties of the rubber. Then, around 1910, Oenslanger found that adding aniline to a rubber-sulfur formulation greatly increased the rate of vulcanization and improved the final vulcanizate's properties.[1] This was quickly followed by the discovery of many organic accelerators that could be used in the process of rubber vulcanization.

One hundred and fifty years after the discovery of sulfur vulcanization sulfur remains unchallenged as the principal vulcanizing or curing ingredient for natural rubber and diene synthetic rubbers. Structural features of a sulfur-vulcanized network can be represented as follows (**Structure 1**).

The search for vulcanizing agent to use as an alternative to sulfur was carried out mainly as an effort to improve the heat

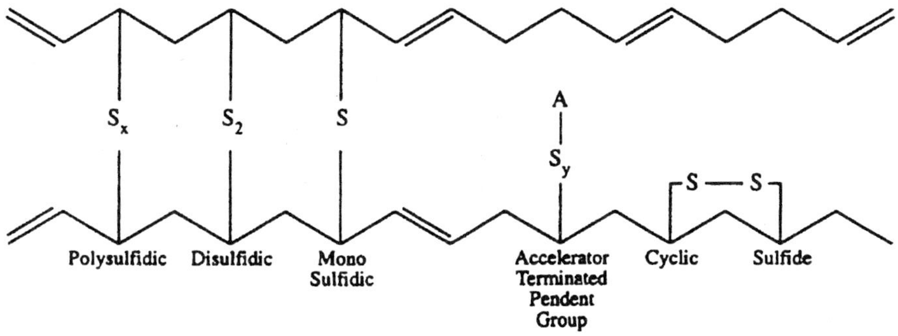

1

resistance of the vulcanizates. Among the different nonsulfur vulcanization systems, the best known is still the free-radical method that employs organic peroxides. Di-*tert*-butyl peroxide and dicumyl peroxide are commercially important for peroxide vulcanization systems. The mechanism of peroxide vulcanization has been described in detail by Loan and Bateman.[2,3]

The sulfur-donor system is another method of vulcanization that does not use elemental sulfur. Here, also, the effective crosslinks are formed through sulfur atoms. In principle, the process involves the splitting of S atoms from organic accelerators, which in turn crosslinks the rubber to form vulcanizates that are basically similar to other sulfur vulcanizates. Accelerators of the thiuram type and thiourea derivatives are examples of this class. The technology and mechanism of sulfur-donor vulcanization systems have been discussed in detail in the literature.[4–6]

Other nonsulfur vulcanization systems include crosslinking with reactive olefin groups or with a reactive nitrogen function, resin vulcanization, and vulcanization with compounds containing reactive chlorine. Vulcanization can also be achieved by irradiation, exposure to ultraviolet light, and metal-oxide-initiated processes.

Much controversy exists about the fundamental chemistry of accelerated sulfur vulcanization systems, which are important to industry. Accelerated systems include those with a single accelerator, binary accelerator systems, and sulfur-donor systems.

In the area of sulfur-donor systems, too, investigators continue to argue about polar and ionic mechanisms. Duchacek and co-workers[8,9] proposed a purely polar mechanism, while Kruger et al.[10] suggested a purely radical mechanism.[8–10] Coleman et al. concluded that the initial formation of sulfurating species is ionic in nature, while the actual crosslinking species are radicals.[7]

Various investigators have arrived at similar conclusions about binary-accelerator systems, although here, too, arguments for polar and radical mechanisms continue.

Thiourea and its derivatives have known advantageous properties, especially when used in binary systems along with a primary accelerator.

The author and co-workers investigated different binary systems containing thiourea derivatives as secondary accelerators.[14–19] One aim of the study was to conclusively prove the existence of the ionic mechanism described in such systems. Also, although some thiourea derivatives have been tried in rubber vulcanization, the acceleration activity of the dithiobiuret derivatives of thiourea has not been investigated. The results of these studies indicates a nucleophilic reaction mechanism in the different systems under review, as suggested earlier by Philpot and Kemperman.[11–13]

The effect of amidinothiourea/thiourea as a secondary accelerator in the sulfur vulcanization of a synthetic rubber, styrene butadiene rubber, was also investigated. As with dithiobiuret derivatives, natural rubber vulcanization using amidinothiourea follows a polar mechanism whereas a synthetic rubber such as SBR does not follow this pattern. A mixed polar/free-radical mechanism is most likely in the latter. The study on filled vulcanizates of SBR also yields generally similar conclusions.

REFERENCES

1. Hofmann, W. *Vulcanization and Vulcanizing Agents*; MacLaren: London, 1967.
2. Loan, L. D. *Rubber Chem. Technol.* **1967**, *40*, 149.
3. Bateman, L.; Moore, C. G.; Porter, M.; Saville, B. *The Chemistry and Physics of Rubber-Like Substances*; Bateman, L., Ed.; MacLaren: London, 1963.
4. Scheele, W.; Lorenz, O. *Kautsch Gummi.* **1955**, *8*, WT 85.
5. Kemperman, T. *Tech. Notes Rubber Ind.* **1965**, *37*, 5.
6. McCall, E. *J. Rubber Res. Inst. Malysia* **1969**, *22*, 354.
7. Coleman, M. M.; Shelton, J. R.; Koenig, J. L. *Rubber Chem. Technol.* **1973**, *46*, 957.
8. Duchacek, V. *J. Appl. Polym. Sci.* **1972**, *16*, 3245.
9. Duchacek, V.; Bhattacharya, T. B.; Kutta, A. *J. Polym. Sci., Polym. Chem. Ed.* **1989**, *27*, 1089.
10. Kruger, F. W. H.; McGill, W. J. *J. Appl. Polym. Sci.* **1992**, *45*, 563.
11. Philpot, M. W. *Proc. IV IRI Rubb. Tech. Conference*; Publisher: London, 1962; Preprint 39.
12. Kemperman, T. *Rubber Chem. Technol.* **1988**, *61*, 422.
13. Kemperman, T. *Kautsch Gummi Kunstat.* **1967**, *20*, 126.
14. Mathew, G.; Viswanatha Pillai, P.; Kuriakose, A. P. *Rubber Chem. Technol.* **1992**, *65*, 277–292.
15. Kuriakose, A. P.; Mathew, G. *Indian J. of Technol.* **1986**, *26*, 344–347.
16. Mathew, G.; Kuriakose, B.; Kuriakose, A. P. *Kautsch Gummi Kunst* **1992**, *45*, 490–493.
17. Mathew, G.; Kuriakose, A. P. *J. Appl. Polym. Sci.* **1993**, *49*, 2009–2017.
18. Mathew, G.; Mathew, N. M.; Kuriakose, A. P. *Polym. Plast. Technol. and Eng.* **1993**, *32*, 5439.
19. Mathew, C.; Mini, V. T. E.; Kuriakose, A. P.; Francis, J. *J. Appl. Polym. Sci.* **1994**, *54*, 1033–41.

W

Water-Absorptive Polymers

See: *Starch Graft Copolymers*
Superabsorptive Polymers (from Natural Polysaccharides and Polypeptides)
Water-Swellable Polymers (Carbomer Resins)

WATER-BORNE COATINGS (Urethane/Acrylic Hybrid Polymers)

Bruce A. Gruber
Air Products and Chemicals, Incorporated

New water-borne technologies for coatings are being introduced at a rapid rate. One example of such technology is a novel family of urethane plus acrylic polymers that exhibit performance characteristics that are superior to polyurethane/polyacrylic blends. These water-borne systems are synthesized in innovative ways aimed at increased polymer–polymer compatibility between the urethane and acrylic portions and are based on graft copolymerization, block copolymerization, or interpenetration of polymer networks (IPNs). The materials are water-borne colloids and can be differentiated from blends, in that at least one polymer is synthesized in the presence of the other and there is some degree of interaction between the two polymers either through covalent bonds, as in grafts, or physical entanglements, as in IPNs. These polymer systems are referred to as hybrids, alloys, or copolymers.

WATER-BORNE URETHANES AND ACRYLICS

Water-borne coating systems are gaining increased acceptance within the pain and coatings industry as environmental and safety, worker right-to-know, and hazardous waste regulations drive efforts to reduce emissions.

Two polymer systems that have found applications as water-borne coatings are acrylic emulsions and urethane dispersions.

Because of these differentiated properties, urethanes and acrylics have established their own identify in specific coating applications. Acrylic emulsions are used in exterior applications where their nonyellowing characteristics, as well as excellent weatherability, are outstanding. Water-borne urethanes are suited to uses where good flexibility and toughness are important, such as leather and plastic coatings. A major drawback to wider use of water-borne urethanes, despite their excellent properties, has been their high cost.

POLYMER BLENDS

Blending acrylic emulsions and urethane dispersions is widely practiced in the coatings industry. Water-borne urethanes are frequently blended with acrylic emulsions to add toughness to the latter as well as facilitate coalescence of hard acrylic resins with reduced amounts of cosolvent. Likewise, acrylic emulsions are added to urethane dispersions primarily to lower the cost of urethane.

WATER-BORNE URETHANE/ACRYLIC HYBRID POLYMERS

Workers began synthesizing water-borne urethane/acrylics in new and innovative ways aimed at increased polymer–polymer compatibility based on graft copolymerization, block copolymerization, or interpenetration of two networks.

In 1969, researchers reported a vinyl-containing copolymer latex aimed at combining nonblocking of vinyl acetate, without its brittleness and poor toughness and the toughness of urethanes without the high cost.[2]

A vinyl-terminated urethane oligomer was used to prepare self-dispersing water-borne urethanes from either aliphatic or aromatic isocyanates and containing small amounts of process solvent.[3] The self-dispersing character eliminates the need for surfactant emulsification.

A water-reducible fully reacted polyurethane has been used for the *in situ* polymerization of one or more ethylenically unsaturated monomers.[4]

By using a combination of several techniques, a vinyl-terminated, self-dispersing urethane prepolymer was added to water and chain extended; then monomers were added in a delay process to form an acrylic grafted urethane.[6]

These examples produce, in principle, water-borne materials in which both urethane and acrylic polymers are present in the same dispersion particle.

To achieve a more homogeneous mixture, isocyanate-terminated urethane prepolymers were prepared in the presence of liquid polymerizable monomers as solvent for the prepolymer reaction.[7] After chain extension, the acrylate monomers were polymerized. The resulting aqueous dispersions yielded films with unexpectedly improved mechanical properties.

In other processes, hybrid-type materials have been prepared when a water-soluble acrylic is modified with a grafted urethane-containing side chain.[8] Coatings prepared from these polymers are cured with melamines and exhibit high gloss and combine high hardness with good flexibility when used as top coats over hydro base coats.

An interesting technique to achieve interaction between a urethane dispersion and an acrylic emulsion is to create a covalent bond during cure.

Acrylic emulsions containing trimethoxysilyl[9] or *N*-methylolacrylamide[10,11] functionality have been blended with urethane dispersions to develop a combination of properties.

Finally, an example of a block copolymer of urethane and acrylic is given in a patent by Tobolsky.[12]

APPLICATIONS

Blending water-borne urethane dispersions with acrylic emulsions results in coating systems with lower VOCs and reduced overall cost. It also may bring a broader range of properties to the blended system; however, in many cases properties lie on the "rule of blends" average axis.

Some of the approaches cited have now begun to find use in commercial products for flooring, furniture, plastic substrates, and metal coatings. The coatings literature refers to these interacting systems as hybrids, alloys, or copolymers.[1,5,8,13–18]

MARKET OPPORTUNITIES FOR URETHANE-ACRYLIC COPOLYMERS

Copolymers have been found to exhibit more urethane-like properties at equivalent acrylic loading than equivalent blends, lowering the overall cost since urethane properties can be obtained with higher acrylic content. The materials exhibit high gloss, good mar and stain resistance, and excellent adhesion to wood and plastics. Compared with physical blends of acrylics and urethanes, the copolymers are more stable and have higher solids and better gloss.[5] Other advantages of hybrid-type materials include reduced water sensitivity, improved colloidal stability, and reduced cosolvent requirements.[14] Alloys can be made that are totally solvent free and certain resistance properties are better than the values provided by each polymer system separately.[15]

Coatings from unpigmented hybrid dispersions form films with good adhesion, flexibility, and hardness. Pigmented versions showed increased hardness with good flexibility and toughness, providing coatings formulators with an attractive alternative for the development of industrial coating applications.[16] These materials are also being used for aqueous wood lacquers, where key properties include rapid dry, rapid property development, good ultimate hardness, excellent stain and chemical resistance, and low VOC formulations.[17,18]

REFERENCES

1. Arnoldus, R. *Polym. Paint Colour J.* **1988**, *178*, 861.
2. Isaacs, P. K. U.S. Patent 3 472 807, 1969.
3. Trovati, A. U.S. Patent 4 497 932, 1985.
4. Guagliardo, M. U.S. Patent 4 318 833, 1982.
5. Coogan, R.; Pollano, G. *Am. Paint Coatings J.* **1990**, *74*(57), 38.
6. Zom, W. M.; Goos, H. C.; Stijntjes, J. U.S. Patent 4 730 021, 1988.
7. Loewigkeit, P.; Van Dyk, K. A. U.S. Patent 4 644 030, 1987.
8. Metzger, C. W.; Hauefler, H.; Munch, J.; Freese, K-H. Orth U. Eur. Patent 0 365 098, 1989.
9. Werner, E. R.; Konsza, E. E. U.S. Patent 5 204 404, 1993.
10. Den Hartog, H. C.; Konsza, E. E.; Matthews, J. F.; Werner, E. R. Jr. U.S. Patent 4 954 559, 1990.
11. Den Hartog, H. C.; Konsza, E. E.; Matthews, J. F.; Werner, E. R., Jr. U.S. Patent 5 006 413, 1991.
12. Tobolsky, A. V. U.S. Patent 3 865 898, 1975.
13. Jansse, P. L. *J. Oil Colour Chem. Assoc.* **1989**, *12*, 478.
14. Jansse, P. L. *Polym. Paint Colour J.* **1991**, *181*, 398.
15. Loewrigkeit, P. *Am. Paint Coatings J. Conv. Daily* **1990**, *75*(17), 42.
16. Hegedus, C. R.; Kloiber, K. A. Presented at the Water-Borne, Higher-Solids, and Powder Coatings Symposium, New Orleans, February 9–11, 1994.
17. Derby, R. *Modern Woodworking* **1994**, *6*(3).
18. Derby, R.; Gruber, B. A.; Klein, R. J. *Am. Paint Coatings J.* **1995**, January 30; p 57.

Water-Repellent Agents

See: *Cotton Fabric (Water-Repellent Finishes)*

Water-Soluble Polymers

See: *Biologically Active Agricultural Polymers (Mechanism of Action on Plants)*
Cellulosic Ethers, Cationic
Cellulosic Liquid Crystals
Colloidal Metal
Colloids, Protective
Cyclic Imino Ethers (Ring-Opening Polymerization)
Drag-Reducing Polymers
Flocculants (Organic, Overview)
Food Polymers
Globular Polybases, Microemulsion-like
Guar (Graft Copolymerization)
Gums (Overview)
Hair and Skin Care Biomaterials (Cosmetology and Dermatology)
Hair and Skin Cosmetics
Hydrophilic Polymers (for Friction Reduction)
Hydroxyethyl Cellulose (Preparation and Use in Protein Precipitation)
Hydroxypropyl Cellulose (Crosslinked, Cholesteric Liquid Crystal)
Immobilized Microbial Cells
Inverse-Emulsion/Suspension Polymerization
Living Radical Polymerization, Iodine Transfer
Micelle-Polymer Complexes
Nucleic Acid Analogs
Personal Care Applications Polymers (Acetylene-Derived)
Polyampholytes (Overview)
Polyampholytes (Properties in Aqueous Solution)
Poly(1,3-dioxolane)
Polyelectrolyte Complexes (Overview)
Polyelectrolyte Complexes (In Immunology)
Polyelectrolyte Complexes (Targeting of Nucleic Acids)
Polyelectrolytes (Association of Hydrophobic Counterions)
Polyelectrolytes, Cationic
Polyelectrolytes, Rigid-Rod [Poly(p-phenylene)-Based]
Poly(ethylene oxide) (Overview)
Poly(ethylene oxide) (Applications in Drug Delivery)
Polyethylenimine (Overview)
Polyglycidols, Polythioglycidols
Poly[N,N′-(Sulfo-p-phenylene)terephthalamide]
Poly(vinyl alcohol)-Ion Complex Gels
Poly(N-vinyl-2-pyrrolidone) Interactions
Polyvinylamine (Overview)
Polyvinylamine
Poly(vinylpyridinium salt)s
Polyvinyltetrazoles
Polyzwitterions (Overview)
Scale Inhibitors, Polymeric
Starch Graft Copolymers
Starch Polymers, Natural and Synthetic
Styrenesulfonic Acid and its Salts (Polymerization and Applications)
Superabsorptive Polymers (from Natural Polysaccharides and Polypeptides)
Supramolecular Formation and Structure (PEO-PPO-PEO Block Copolymers)

Surface Active Monomers
Water-Soluble Polymers (Oil Recovery Applications)
Water-Soluble Polymers (Rheological Properties)
Welan Microbial Polysaccharide
Xanthan, Depolymerization
Xanthan Gum (Overview)
Xanthan Gum (Conformation and Solution Properties)
Zwitterionic Polymers (Dilute Solution and Rheological Properties)

WATER-SOLUBLE POLYMERS (Oil Recovery Applications)

Ahmad Moradi-Araghi* and Iqbal Ahmed
Phillips Research Center
Phillips Petroleum Company

The use of water to increase oil production is the least expensive oil recovery technique and is referred to as waterflooding. The possibility of using water-soluble polymers is conceived from the fact that the water-to-oil mobility ratio of a conventional waterflooding project could be reduced if the viscosity of the injection water could be raised and thereby sweep efficiency of the injection fluid could be improved.

The concept of using high viscosity water to increase the efficiency of water flooding has existed since early to mid-1940s. Detling of Shell Development obtained a patent covering the use of several additives, including water-soluble polymers for viscous water flooding.[1] Other patents have been granted covering specific water-soluble polymers or specific process for viscous water flooding.[2–27] This type of oil-field application of water-soluble polymers is generally referred to as mobility control.

Polymer-augmented viscous water flooding requires the injection of a large volume of aqueous polymer solution. This is capital intensive and often needs a long time to produce results due to large distance between wells. An alternative popular option is the diversion of flooding fluids by correcting the permeability of the reservoir and modifying the injection profile in the injection wells using hydrogels, which are generally prepared by the crosslinking of water-soluble polymers, or using hydrogels to deal with the water-coning problem in producing wells.[27–58] Water-coning is a term given to the mechanism of the entry of bottom water into producing wells.[59] In this report, water-soluble polymers used for mobility control, reservoir permeability corrections, injection profile modification in injection wells, and water shut-off in producing wells will be considered. These polymers may also be used in other oil-field applications such as filtration control, flocculation control, shale stabilization, and drilling mud additives for fluid loss control.[27,48]

POLYMERS

Acrylamide polymers and acrylamide-based copolymers have been the focus of the research of synthetic water-soluble polymers for oil recovery applications. In fact, most of the worldwide polymer-aided oil recovery projects to date have been carried out with polyacrylamides or partially hydrolyzed polyacrylamides.[47,55,64–72]

In recent years, the water-soluble polymer research efforts for oil recovery applications have been aimed at developing acrylamide-based polymers that are resistant to thermal hydrolysis as well as possessing similar viscosity building properties in oilfield brines.[25–27,32–34,39,42,46,48,53,61,62] These research efforts have produced commercial products based on acrylamide, which can tolerate the hostile environmental conditions of high hardness and high temperatures without precipitation and viscosity loss.[73]

Although the modified or unmodified natural polymers such as carboxymethyl cellulose and hydroxyethyl cellulose have received attention in earlier research efforts.[2,25] there has been a considerable amount of interest in recent years in developing a natural or biopolymer-based product for high salinity and high temperature reservoir applications.[40,41,45,51,52,61,74,75] Xanthan and scleroglucan are favorable candidates.

POLYMER GELS

In recent years, gels produced by an *in situ* crosslinking reaction of water-soluble polymers in oil-bearing formations have become popular in the enhanced oil recovery processes to overcome such problems as "water coning" and to reduce permeability in high permeable zones. The excess of water production can significantly shorten the economic life of a well. The polymeric gels can be used to block such water intrusion into the producing wells.

Gels can be effective in redirecting the injection water flow to less permeable zones containing oil by placement in high permeable streaks or fractures near the injection wells. This process is known as injection profile modification or injection profile control. This type of gel treatment increases the oil production and decreases the water production.

Gels used in oil recovery applications are hydrogels—i.e., the polymer networks that possess the ability to swell in water and retain a significant fraction of water within their structures—but these will not dissolve in water. These gels typically consist of about 0.5–3% of crosslinked water-soluble polymers that hold 99.5–97% water in an equilibrium state.

Although the gelation of water-soluble polymers can be achieved by a number of multivalent metal ions such as chromium, aluminum, titanium, and zirconium,[63] the gel formation by aluminum(III) and chromium(III) has been studied in detail and also used in field treatments. Metallic crosslinking occurs through ionic bonding between positive multivalent metal ions and the negative sites of a polymer such as the carboxylate groups of partially hydrolyzed polyacrylamide or biopolymers.[39,43,51,81] The gelation of polyacrylamides using aluminum citrate as a crosslinking agent was first reported in 1974.[28] Since then, this gel system has been applied in a number of fields with impressive economic success.[47,76–78]

The most commonly applied polymer/organic crosslinking system for oil recovery is based on the well-known condensation reaction of phenols with aldehydes. The polyacrylamide/phenol/formaldehyde gel system has been studied in detail.[34,39,56,79–81] Moradi-Araghi studied a number of relatively

*Author to whom correspondence should be addressed.

low toxicity organic compounds to produce gels for oilfield applications.[56]

ECONOMIC ASPECTS

The economic predictions of the polymer-aided oil recovery are somewhat complex.[82] The cost estimation not only is sensitive to the polymer price but also is sensitive to the crude oil price as well as to the amount of incremental oil recovered. Lewin and Associates estimated the cost of oil produced from polymer floods in the range of $30.00–$46.00 per barrel when used in a true tertiary stage (i.e., after waterflood) and $22.00–$28.00 per barrel when used in a secondary stage at a polymer price of $2.83/lb.[82] The U.S. National Petroleum Council has also studied economic feasibility of chemical enhanced oil recovery processes and concluded that the chemical EOR processes have the potential to recover a significant amount of original oil in place (0.4 billion barrels of 341 billion barrels of unrecovered United States proven reserve) at a crude oil price of $20 per barrel.[83] Thus, full-scale polymer-augmented waterflooding will not continue in the future, until a large and sustained increase in crude oil price occurs. However, the application of gels is usually a one-time treatment and requires a much smaller volume. Hence, the pay-off period is much shorter. Also, there is another economic incentive to use gels in oil recovery applications. It is estimated that more than 20 billion barrels of water are produced each year in the United States by oilfield operations.[84] This is an average of 7 barrels of water for each barrel of oil produced. The cost of water disposal is typically in the range of $0.25–$1.50 per barrel, depending on the methods of transportation. Therefore, there is an increasing need to reduce water production, and so the area of near-wellbore gel treatments or profile control is fairly active today. The use of gel technology in conjunction with new oil-well technology, such as the horizontal well, may prove beneficial even in this time of low crude oil price.

ACKNOWLEDGMENT

We thank Phillips Petroleum Company for permission to publish this review paper.

REFERENCES

1. Detling, K. D. U.S. Patent 2 341 500, 1944.
2. Binder, G. G. Jr.; West, R. C.; Andresen, K. H. U.S. Patent 2 731 414, 1956.
3. Sandiford, B. B.; Keller, H. F. Jr. U.S. Patent 2 827 964, 1958.
4. von Engelhardt, W.; Trommsdorff, E.; Tunn, W. U.S. Patent 2 842 492, 1958.
5. Kolodny, E. R. U.S. Patent 3 002 960, 1961.
6. Patton, J. T. U.S. Patent 3 020 207, 1962.
7. Zerweck, W.; Bulian, W.; Kunze, W.; Kiesewetter, A. U.S. Patent 3 020 953, 1962.
8. Roper, R. U.S. Patent 3 025 237, 1962.
9. Mckennon, K. R. U.S. Patent 3 039 529, 1962.
10. Patton, J. T. U.S. Patent 3 042 611, 1962.
11. Roper, R. U.S. Patent 3 070 158, 1962.
12. Meadors, G. V.; Russell, C. D. U.S. Patent 3 076 504, 1963.
13. Stright, P. L.; Turbak, A. F. U.S. Patent 3 079 336, 1963.
14. Turbak, A. F. U.S. Patent 3 079 337, 1963.
15. Turbak, A. F. U.S. Patent 3 085 063, 1963.
16. Maurer, J. J. Jr. U.S. Patent 3 087 539, 1963.
17. Sandiford, B. B.; Keller, H. F. Jr. U.S. Patent 3 116 791, 1964.
18. Leavitt, R. L. U.S. Patent 3 989 592, 1976.
19. Schawab, F. C.; Sheppard, E. W.; Chen, C. S. H. U.S. Patent 4 110 232, 1978.
20. Hunter, W. D. U.S. Patent 4 217 230, 1980.
21. Landoll, L. M. U.S. Patent 4 529 523, 1985.
22. Wu, M. M.; Ball, L. E. U.S. Patent 4 540 498, 1985.
23. Chen, C. S. H.; Williams, A. L. U.S. Patent 4 577 000, 1986.
24. Bock, J.; Valint, P. I.; Pace, S. J. U.S. Patent 4 702 319, 1987.
25. Bock, J.; Pace, S. J.; Schulz, D. N. U.S. Patent 4 702 759, 1987.
26. Evani, S. U.S. Patent 4 814 096, 1989.
27. Stahl, G. A.; Westerman, I. J.; Hsieh, H. L.; Moradi-Araghi, A. U.S. Patent 4 951 921, 1990.
28. Needham, R. B.; Threlkeld, C. B.; Gall, J. W. *SPE of AIME Improved Oil Recovery Symp. Preprint* Paper No. SPE-4747; 1974; pp 139–148.
29. Knight, B. L.; Gogarty, W. B. Fr. Patent 2 399 536, 1979.
30. Sandiford, B. B. U.S. Patent 4 332 297, 1982.
31. Falk, D. O. U.S. Patent 4 435 528, 1984.
32. Moradi-Araghi, A.; Cleveland, D. H.; Westerman, I. J. *SPE Paper 16273*; SPE International Symposium on Oilfield Chemistry; San Antonio, TX, 1987; 319.
33. Smith, R. E. U.S. Patent 4 706 754, 1987.
34. Moradi-Araghi, A.; Beardmore, D. H.; Stahl, G. A. In *Water-Soluble Polymers for Petroleum Recovery*; Stahl, G. A.; Schulz, D. N., Eds.; Plenum, New York, 1988; pp 299–312.
35. Summers, L. E.; Purkaple, J. D.; Allison, J. D. In *Water-Soluble Polymers for Petroleum Recovery*; Stahl, G. A.; Schulz, D. N., Eds.; Plenum: New York, 1988; pp 313–327.
36. Allison, J. D.; Purkaple, J. D. U.S. Patent 4 773 481, 1988.
37. Allison, J. D. U.S. Patent 4 773 482, 1988.
38. Paul, J. M. U.S. Patent 4 773 483, 1988.
39. Moradi-Araghi, A.; Bjornson, G.; Doe, P. H. *SPE Paper 18500*; SPE International Symposium on Oilfield Chemistry; Houston, TX, 1989; 367.
40. Strom, E. T.; Paul, J. M.; Phelps, C. H.; Sampath, K. *SPE Paper 19633*; 64th Annual SPE Technical Conference and Exhibition; San Antonio, TX, 1989; 69.
41. Sampath, K.; Jones, L. G.; Strom, E. T.; Phelps, C. H.; Chiou, C. S. *SPE Paper 19867*; 64th Annual SPE Technical Conference and Exhibition; San Antonio, TX, 1989; 765.
42. Mumallah, N. A. U.S. Patent 4 917 186, 1990.
43. Sydansk, R. D. *SPE Paper 20214*; SPE/DOE 7th Symposium on Enhanced Oil Recovery; Tulsa, OK, 1990; 397.
44. Lockhart, T. P.; Albonico, P.; Burrafato, G. *J. Appl. Polym. Sci.* **1991**, *43*, 1527.
45. Kohler, N.; Zaitoun, A. *SPE Paper 21000*; SPE International Symposium on Oilfield Chemistry; Anaheim, CA, 1991; 37.
46. Hsieh, H. L.; Moradi-Araghi, A.; Stahl, G. A.; Westerman, I. J. *Makromol. Chem., Macromol. Symp.* **1992**, *64*, 121.
47. Moffit, P. D.; Zornes, D. R. *SPE Paper 24933*; 67th Annual SPE Technical Conference and Exhibition; Washington, DC, 1992; 813.
48. Stahl, G. A.; Westerman, I. J.; Hsieh, H. L.; Moradi-Araghi, A.; Hedges, J. H.; Bjornson, G. U.S. Patent 5 080 809, 1992.
49. Hunter, B. L.; Buell, R. S.; Abate, T. A. *SPE Paper 24031*; SPE Western Regional Meeting; Bakersfield, CA, 1992; 41.

50. Lockhart, T. P.; Albonico, P. *SPE/DOE Paper 24194*; SPE/DOE 8th Symposium on Enhanced Oil Recovery; Tulsa, OK, 1992; 397.
51. Fielde, I.; Stavland, A. *SPE Paper 25223*; SPE International Symposium on Oilfield Chemistry; New Orleans, LA, 1993; 695.
52. Kohler, N.; Rahbari, R.; Han, M.; Zaitoun, A. *SPE Paper 25225*; SPE International Symposium on Oilfield Chemistry; New Orleans, LA, 1993; 709.
53. Ahmed, I.; Hsieh, H. L.; Moradi-Araghi, A.; Patel, B. U.S. Patent 5 270 382, 1993.
54. Seright, R. S. *SPE/DOE Paper 27740*; SPE/DOE 9th Symposium on Improved Oil Recovery; Tulsa, OK, 1994; 19.
55. Gruenenfelder, M. A.; Zaitaoun, A.; Kohler, N.; Ali, A. S.; Linser, T. M. *SPE/DOE Paper 27770*; SPE/DOE 9th Symposium on Improved Oil Recovery; Tulsa, 1994; 387.
56. Moradi-Araghi, A. *SPE/DOE Paper 27826*; SPE/DOE 9th Symposium on Improved Oil Recovery, Tulsa, OK; 1994; 483.
57. Bartosek, M.; Mennella, A.; Lockhart, T. P. *SPE/DOE Paper 27828*; SPE/DOE 9th Symposium on Improved Oil Recovery; Tulsa, OK, 1994; 505.
58. Whittington, L. E.; Naae, D. G.; Braun, R. W. *SPE/DOE Paper 27830*; SPE/DOE 9th Symposium on Improved Oil Recovery; Tulsa, OK, 1994; 529.
59. Muskat, M.; Wyckoff, R. D. *Trans. AIME* **1935**, *114*, 144.
60. Seright, R. S.; Henrici, B. J. *SPE Reservoir Eng.* **1990**, *5*, 52.
61. McCormick, C. L.; Hester, R. D. *DOE Report DOE/BC/10844-20*; Bartlesville, OK, 1990; (available from National Technical Information Services, U.S. Department of Commerce).
62. McCormick, C. L.; Johnson, C. B. In *Water-Soluble Polymers for Petroleum Recovery*; Stahl, G. A.; Schulz, D. N., Eds.; Plenum: New York, 1988; pp 161–180.
63. Ahmed, I.; Moradi-Araghi, A. *Trends in Polym. Sci.* **1994**, *2*(3), 92.
64. "Polymers Raise Waterflood Efficiency" *Oil & Gas J.* **1964**, *62*(18), 46.
65. Sandiford, B. B. *J. Pet. Tech.* **1964**, *16*, 917.
66. Trantham, J. C., Ed. *Doe Report DOE/BETC/TPR-78/8*; Bartlesville, OK, 1978; (available from National Technical Information Service, U.S. Department of Commerce).
67. Manning, R. K.; Pope, G. A.; Lake L. W.; Paul, G. W. *DOE Report DOE/BC/10327-19*; Bartlesville, OK, 1983; (available from National Technical Information Service, U.S. Department of Commerce).
68. Sydansk, R. D.; Moore, P. E. *Oil & Gas J.* **1992**, *90*(3), 40.
69. Moffitt, P. D. *J. Pet. Tech.* **1993**, *45*, 356.
70. Southwell, G. P.; Psey, S. M. *SPE/DOE Paper 27779*; SPE/DOE 9th Symposium on Improved Oil Recovery; Tulsa, OK, 1994; 513.
71. Mack, J. C.; Smith, J. E. *SPE/DOE Paper 27780*; SPE/DOE 9th Symposium on Improved Oil Recovery; Tulsa, OK, 1994; 527.
72. Delamaide, E.; Corlay, P.; Demin, W. *SPE/DOE Paper 27819*; SPE/DOE 9th Symposium on Improved Oil Recovery; Tulsa, OK, 1994; 401.
73. Anon. *HE Polymer* [*Brochure*]; Drilling Specialities: Bartlesville, OK, USA, 1992.
74. Ryles, R. G. *SPE Reservoir Eng.* **1988**, *3*, 23.
75. Abdu, M. K.; Chung, H. S.; Phelps, C. H.; Klarick, T. M. *SPE/DOE Paper 12642*; SPE/DOE 4th Symposium on Improved Oil Recovery; Tulsa, OK, 1984; 137.
76. Zornes, D. R.; Long, H. Q.; Cornelius, A. J. *SPE Paper 14113*; International Meeting on Petroleum Engineering; Beijing, China, 1986; 311.
77. Doll, T. E.; Hanson, M. T. *SPE Paper 15162*; Rocky Mountain Regional Meeting of the SPE; Billings, MT, 1986; 281.
78. Mack, J. C.; Warren, J. *J. Pet. Tech.* **1984**, *36*, 1145.
79. Falk, D. O. U.S. Patent 4 485 875, 1984.
80. Swanson, B. L. U.S. Patent 4 440 228, 1984.
81. Moradi-Araghi, A.; Bjornson, G.; Doe, P. H. *SPE Adv. Tech. Series* **1993**, *1*(1), 140.
82. Lewin and Associates, *DOE report DOE/ET/12072-2*; Washington, DC, 1981; (available from National Technical Information Service, U.S. Department of Commerce).
83. *"Enhanced Oil Recovery"*; National Petroleum Council: (USA), 1984.
84. *Report to Congress: Management of Wastes from the Exploration, Development and Production of Crude Oil, Natural Gas, and Geothermal Energy*; U.S. Environmental Protection Agency; U.S. Government Printing Office: Washington, DC, 1987; **1**, Oil and Gas: Executive Summaries (PB 88-146220 and PB 88-146253).

WATER-SOLUBLE POLYMERS (Rheological Properties)

Kam Chiu Tam*
Advanced Materials Research Center
School of Mechanical and Production Engineering
Nanyang Technological University

C. Tiu
Department of Chemical Engineering
Monash University

Water-soluble polymers find wide application in the industry today because of their abilities to modify the flow properties of many water-based products and are gaining greater prominence because of very stringent environmental regulations by governmental monitoring agencies and pressure groups. One such example is the replacement of solvent-based with water-borne paint products. The rheological properties of these materials have become a subject of much attention by both industrial and academic research laboratories.

Water-soluble polymers are derived from either biological or synthetic sources. The former forms a class of polymers called biopolymers, which includes polysaccharides, polypeptides, and polynucleotides. Such polymers are commonly found in food products and their presence serves many purposes, one of which is modification of the rheological properties. Synthetic water-soluble polymers have also been developed to meet the growing demand of many different kinds of industrial applications. One key benefit of such systems is their resistance to attack by living organisms. Generally, such polymers are either ionic or non-ionic in nature. The solubilities of such polymers are the result of a large number of polar or hydrogen-bonding functional groups present on the polymer chains. Molyneux[1] provides an excellent review of a wide range of water-soluble synthetic polymers.

The rheological properties of water-soluble polymers have been extensively studied by many research groups. **Table 1** provides a brief summary of rheological studies on water-soluble polymers.

*Author to whom correspondence should be addressed.

TABLE 1. Representative Studies on Rheology of Water-Soluble Polymer Solutions

Investigators	Polymer types[a]	Molecular weight (in millions)	Polymer conc. (wt%)	Apparatus	Remarks
Bruce and Schwarz[2]	PAAm	1.0–7.0	0.005–1.19	Weissenberg rheogoniometer	One of the earliest works on rheological study of polyacrylamide solutions.
Darby[3]	PAAm	4.0	0.01–0.05	Weissenberg rheogoniometer	Attempted to measure zero shear viscosity of dilute polymer solutions. Some extrapolation necessary to obtain η_0.
Tsai and Darby[33]	PAAm	4.0	0.01–0.05	Weissenberg rheogoniometer	Apparent viscosity and first normal stress measured for dilute solutions over a wide shear rate range. Model developed to predict transient response.
Whitcomb and Macosko[4]	XG	7.6	0.01–1.0	Weissenberg rheogoniometer, capillary tube	Rheological study performed over wide shear rate range. Yield stress observed for concentrated solutions. Rodlike conformation is proposed.
Argumedo et al.[5]	PAAm	7.0	0.012–2.0	Weissenberg rheogoniometer	Carreau model used for determining the relaxation time constants. Zero shear viscosity determined by extrapolation.
Chiou and Gordon[6]	PAAm	1.0–7.0	3×10^{-4}–4×10^{-3}	Modified Zimm–Crothers viscometer	One of earliest reported attempts to measure zero shear viscosity. Low shear rate viscosity down to 0.1 s^{-1} measured. Relaxation time determined from Zimm model.
Cho and Harnett[7]	PAAm, PEO	4.0–7.0	0.1–1.0	Weissenberg, rheogoniometer, capillary tube	Effects of shear and solvents on the flow properties of polymer solutions studied.
Kulicke et al.[8]	PAAm	0.51–5.3	0.10–5.0	Wiessenberg rheogoniometer, Contraves LS 100, Rheometric RMS	Comprehensive review of polyacrylamide behavior ranging from molecular properties to rheology.
Chauveteau et al.[9]	PAAm	7.0	0.018–0.1	Capillary tube	Flow of water-soluble polymer solutions in packed bed. Apparent shear thickening behavior observed.
Rochefort and Middleman[10]	XG	3.5	0.05–0.50	Rheometric RMS	Studies on the helix transition of xanthan gum. Steady shear and oscillatory tests performed at various temperatures.
Ait-Kadi et al.[11]	PAAm	7.0	0.017–0.085	Weissenberg rheogoniometer	Steady shear and stress growth behavior examined. PAAm prepared in water/glycerol salt solutions.

[a]PAAm, polyacrylamide; XG, xanthan gum; PEO, poly(ethylene oxide); CMC, carboxymethyl cellulose; EHEC, ethylhydroxyethyl cellulose; HPC, hydroxpropyl cellulose; HMHEC, hydrophobically modified hydroxypropyl cellulose; MC, methyl cellulose; HEUR, hydrophobic ethoxylated urethane associative polymer.

STEADY SHEAR PROPERTIES

Polymer solutions can be broadly classified into dilute, semi-dilute, and concentrated solutions. They generally exhibit non-Newtonian behavior, where the shear thinning characteristic is among the commonly observed features of polymer solutions. The nature and shape of the viscosity curves depend on a number of factors such as polymer concentration, solvent power, molecular weight and distribution, and molecular structure of polymer molecules. A typical flow curve exhibits two Newtonian plateaus—one at the low shear (commonly referred to the "zero shear condition") and the other at the higher shear (infinite shear condition). The shear thinning region lies between the two limiting conditions and may extend over a few decades of shear rates.

Effects of Polymer Concentrations

Separan AP30, a liner macromolecule, is generally flexible and exhibits marked shear thinning behavior. Similarly, the viscosity curve for xanthan gum (Keltrol) also demonstrates shear thinning behavior even at very low polymer concentrations. The

TABLE 1. Representative Studies on Rheology of Water-Soluble Polymer Solutions (Continued)

Investigators	Polymer types[a]	Molecular weight (in millions)	Polymer conc. (wt%)	Apparatus	Remarks
Suto et al.[12]	HPC	0.093	35–55	Capillary rheometer	HPC at these concentrations exhibits lyotropic liquid crystalline behavior.
Georgelos and Torkelson[13]	PEO	4.0	0.025–0.50	Capillary tube	Shear thickening behavior reported for PEO in water and methanol.
Tam and Tiu[14]	PAAm, PEO, CMC, XG	0.70–4.0	5×10^{-3}–1.0	Contraves LS30 sinus	Steady shear and oscillatory properties measured in dilute to semidilute regime. Effects of salts on rheological properties reported.
Carlsson et al.[15]	EHEC, HPC, MC	0.15–0.66	1.0	Bohlin VOR	Interactions with different types of surfactants studied. Phase behavior and rheological properties examined.
Milas et al.[16]	XG	5.25	0.0125–0.50	Rheometric RFS	Steady shear and oscillatory data presented. Detailed characterization of the critical concentration regime.
Vissmann and Bewersdorff[17]	PAAm, PEO	1.0–6.0	2×10^{-3}–2×10^{-1}	Contraves LS2, Haake RV3	Studied effects of preshearing and elongational properties of drag-reducing polymer solutions.
Jenkins et al.[18]	HEUR	0.017–0.10	0.50–5.0	Bohlin VOR	Studied effects of hydrophobic end caps on the rheological properties of associative polymers. Steady shear and oscillatory properties measured.
Tam and Tiu[19,20]	PAAm, PEO, CMC, XG	0.7–8.0	1×10^{-3}–1.0	Contraves LS30 sinus	Steady shear data of various water-soluble polymers examined. Viscosity master curves based on modified Graessley's model.
Liang and Mackay[21]	XG	4.0	0.01–0.05	Rheometric RFS	Xanthan gum prepared in frutose solution. Shear, oscillation, and stress relaxation tests carried out to determine viscous and elastic contributions of polymer molecules.
Annable et al.[22]	HEUR	0.011–0.033	1.0–5.0	Bohlin VOR, Bohlin CS	Hydrophobic end caps' effects on rheological properties. Rheology depends on strength of association.
Ortiz et al.[23]	PEO	0.3–5.0	0.5–7.0	Weissenberg rheogoniometer, Bohlin VOr, rheometric SR	Polymer solutions prepared in water and glycerol. Modified rheological model developed for shear and oscillatory data.
Vlassopoulos and Schowalter[24]	PAAm, PEO	7–17	5×10^{-5}–1×10^{-1}	Rheometric SR, Contraves LS30	Characterization of dilute drag reducing polymer solutions. Shear thinning observed at very low concentrations.

[a]PAAm, polyacrylamide; XG, xanthan gum; PEO, poly(ethylene oxide); CMC, carboxymethyl cellulose; EHEC, ethylhydroxyethyl cellulose; HPC, hydroxpropyl cellulose; HMHEC, hydrophobically modified hydroxypropyl cellulose; MC, methyl cellulose; HEUR, hydrophobic ethoxylated urethane associative polymer.

viscosity curves of Separan AP30 and Keltrol are identical at low polymer concentrations despite their structural differences. However, in the semidilute or concentrated region ($c > 2000$ ppm). Keltrol exhibits larger viscosity levels and shear thins at much lower shear rates compared with Separan AP30.

Temperature Effects

The rheological properties of polymer solutions depend strongly on temperature. Hence, an ability to predict the variation of the steady shear and dynamic properties with respect to temperature is critical in many applications. The viscosity of many Newtonian liquids such as benzene exhibits an Arrhenius-type exponential dependence on the temperature.[25]

Effects of Electrolytes

A small alteration in the strength of the macroion/counterion interaction could have a significant effect on the properties of the polyelectrolytes. The steady and dynamic shear viscosities of polyelectrolyte solutions are generally many orders of magnitude larger than the non-ionic polymers of similar structure and molecular weight. These large differences in properties are largely due to the change in polymer coil conformation as a result of the electrostatic repulsion between fixed charges along the polymer chains. Various polyion/counterion systems have been studied such as sodium polyacrylate/NaCl,[26] polystyrenesulfonate/KCl,[27] poly(acrylic acid)/NaBr and poly(methacrylic acid)/NaCl.[28] Recent studies by Tam and Tiu[14,29] on partially hydrolyzed polyacrylamide in mono- and divalent salt solutions indicated that the ionic charges of the counterions significantly alter the rheological properties of the polymer solutions. The effects of counterion charges on water-soluble polymers are dependent on the presence of charges and flexibility of the polymer chains.[19] Good coverage on this subject can be found in the works of Rice and Nagasawa[30] and Morawetz.[31]

OSCILLATORY PROPERTIES

The elastic properties can be determined by different types of unidirectional shear flow experiments such as normal stress measurements, small-amplitude oscillation, stress growth, and relaxation.[32] In the case of dilute polymer solutions, the storage and loss moduli are determined from small-amplitude oscillatory measurements as they normally cannot be measured by the normal force method.

Effects of Polymer Concentrations

The effect of polymer concentrations on the elastic properties can be studied by examining the relaxation time. The incentives for measuring the elastic properties for dilute polymer solutions stem from the belief that much of the anomalous behavior, such as drag reduction in turbulent flow or pressure drop enhancement in flows through packed bed, may be due to these elastic properties.[34]

Temperature Effects

The effects of temperature on dynamic properties are similar to those of steady shear behavior.

Effects of Electrolytes

Similar to the steady shear properties, the presence of electrolytes will significantly alter the elastic properties of water-soluble polymer solutions. Some studies have been carried out for concentrated systems such as poly(methacrylic acid),[35] sodium polyacrylate,[36,37] polystyrenesulfone,[38] and poly(*m*-methyl-2-vinylpyridinium chloride).[39] The change in the polymer coil conformation and the excluded volume due to charge screening by electrolytes will have a large effect on the elastic properties of polyelectrolytes. Divalent salts have a much greater effect on the relaxation time as the shielding effect is much stronger, resulting in a more compact polymer coil at a given salt concentration. Such behavior will result in a reduction in the effectiveness of the polymer in applications such as drag reduction and enhanced oil recovery where high salt contents are encountered.

APPLICATIONS

Water-soluble polymers are used in various applications ranging from protein separation in biotechnology to thickening in coatings and pharmaceutical applications. However, only applications directly related to rheological properties will be discussed here.

One of the prime uses of water-soluble polymers is in modifying the flow properties of cosmetic, pharmaceutical, and coating products. For example, thickening agents such as hydroxyethyl cellulose are widely used in the formulation of water-borne coatings. The rheology of such products can be adjusted by using the appropriate amount and types of thickeners. New and innovative thickeners such as the hydrophobically modified ethylene oxide urethane (HEUR)[40,41] and hydrophobically modified alkali-swellable emulsions (HASE)[42–44] have recently been developed. An excellent source of information on this topic can be found in the *Handbook of Coatings Additives* by Calbo.[45]

High molecular weight polymers have been found to exhibit drag reducing properties under turbulent flow conditions even in very small quantities.[46–48] The potential use in long distant transport of crude oil and minerals have been the subject of intensive research in the 1960s and 1970s.[49–51] However, harnessing such technology has been restricted by economic constraints and also the polymer's inability to resist shear degradation.

Water-soluble polymers are an ideal choice in controlling the stability of the water/oil interface in order to reduce the occurrence of viscous fingering. A great deal of attention have been focused on identifying potential water-soluble polymers that can efficiently control the flow properties of displacement fluids for enhanced oil recovery.[52,53] The candidates commonly chosen are relatively rigid xanthan gum and the more flexible polyacrylamide. These polymers exhibit better stability to shear and temperature effects, which can be rather severe in an oil well. However, one common problem with water-soluble polymers is their sensitivity to electrolytes commonly present in an oil well. The flow properties will change with the concentrations and types of ionic species present and this will affect the efficiency of such polymers in controlling the mobility of the oil/water interface in an oil well.

REFERENCES

1. Molyneux, P. *Water Soluble Synthetic Polymers: Properties and Behavior*; CRC: Boca Raton, FL, 1985.
2. Bruce, C.; Schwarz, W. H. *J. Polym. Sci., Part A2* **1969**, *7*, 909.
3. Darby, R. *Trans. Soc. Rheol.* **1970**, *14*, 185.
4. Whitcomb, P. J.; Macosko, C. W. *J. Rheol.* **1978**, *22*, 493.
5. Argumedo, A.; Tung, T. T.; Chang, K. I. *J. Rheol.* **1978**, *22*, 449.
6. Chiou, C. S.; Gordon, R. J. *AIChE J.* **1980**, *26*, 852.
7. Cho, Y. I.; Hartnett, J. P. *Adv. Heat Transfer* **1982**, *15*, 59.
8. Kulicke, W. M.; Kniewske, R.; Klein, J. *Prog. Polym. Sci.* **1982**, *8*, 373.
9. Chauveteau, G.; Moan, M. J.; Mageur, A. *J. Non-Newt. Fluid Mech.* **1984**, *16*, 315.
10. Rochefort, W. E.; Middleman, S. *J. Rheol.* **1987**, *31*, 337.
11. Ait-Kadi, A.; Carreau, P. J.; Chauveteau, G. *J. Rheol.* **1987**, *31*, 537.
12. Suto, S.; Ohshiro, M.; Ito, R.; Karasawa, M. *Polymer* **1987**, *28*, 236.
13. Georgelos, P. N.; Torkelson, J. M. *J. Non-Newt. Fluid Mech.* **1988**, *27*, 191.
14. Tam, K. C.; Tiu, C. *J. Rheol.* **1989**, *33*, 257.
15. Carlsson, A.; Karlstrom, G.; Lindman, B. *Colloids and Surfaces* **1990**, *47*, 147.
16. Milas, M.; Rinaudo, M.; Knipper, M.; Schuppiser, J. L. *Macromolecules* **1990**, *23*, 2506.

17. Vissmann, K.; Bewersdorff, H.-W. *J. Non-Newt. Fluid Mech.* **1990**, *34*, 289.
18. Jenkins, R. D.; Silebi, C. A.; El-Asser, M. S. *Polymers as Rheology Modifiers, ACS Symposium Series*; Schultz, D. N.; Glass, J. E., Ed. **1991**, *462*, 222.
19. Tam, K. C.; Tiu, C. *Polym. Plast. Tech. Eng.* **1993**, *32*, 123.
20. Tam, K. C.; Tiu, C. *J. Non-Newt. Fluid Mech.* **1993**, *46*, 275.
21. Liang, C. H.; Mackay, M. E. *J. Rheol.* **1993**, *37*, 149.
22. Annable, T.; Buscall, R.; Ettelaie, R.; Whittlestone, D. *J. Rheol.* **1993**, *37*, 695.
23. Ortiz, M.; De Kee, D.; Carreau, P. J. *J. Rheol.* **1994**, *38*, 519.
24. Vlassopoulos, D.; Schowalter, W. R. *J. Rheol.* **1994**, *38*, 1427.
25. Bird, R. B.; Stewart, W. E.; Lightfoot, E. N. *Transport Phenomena*; Wiley: New York, 1960.
26. Takahashi, A.; Nagasawa, M. *J. Am. Chem. Soc.* **1964**, *86*, 543.
27. Nagasawa, M.; Eguchi, Y. *J. Phys. Chem.* **1967**, *71*, 880.
28. Noda, I.; Suge, T.; Nagasawa, M. *J. Phys. Chem.* **1970**, *74*, 710.
29. Tam, K. C.; Tiu, C. *Colloid Polym. Sci.* **1990**, *268*, 911.
30. Rice, S. A.; Nagasawa, M. *Polyelectrolyte Solutions*; Academic: London, 1961.
31. Morawetz, H. *Macromolecules in Solution*; Wiley: New York, 1975.
32. Bird, R. B.; Armstrong, R. C.; Hassager, O. *Dynamics of Polymeric Liquid: Fluid Mechanics* 2nd ed.; Wiley: New York, 1987; Vol. 1.
33. Tsai; Darby, *J. Rheol.* **1978**, *22*, 219.
34. Lumley, J. L. *J. Polym. Sci. Macromol. Rev.* **1973**, *7*, 283.
35. Silberberg, A.; Mijnlieff, P. F. *J. Polym. Sci., Part A2* **1970**, *8*, 1089.
36. Nishida, N. *J. Polym. Sci., Part A2* **1966**, *4*, 8459.
37. Sakai, M.; Noda, I.; Nagasawa, M. *J. Polym. Sci., Part A2* **1972**, *10*, 1047.
38. Rosser, R. B.; Nemoto, N.; Schrag, J. L.; Ferry, J. D. *J. Polym. Sci., Polym. Phys. Ed.* **1978**, *16*, 1031.
39. Yamaguchi, M.; Wakutsu, M.; Takahashi, Y.; Noda, I. *Macromolecules* **1992**, *25*, 470.
40. Glass, J. E. *Water Soluble Polymer: Beauty with Performance*; ACS Symposium Series 213; American Chemical Society: Washington, DC, 1986.
41. Schulz, D. N.; Glass, J. E. *Polymer as Rheology Modifiers*; ACS Symposium Series 462: American Chemical Society: Washington, DC, 1991.
42. Shay, G. D. *Polymers in Aqueous Media: Performance through Association*; Advances in Chemistry 223; Glass, J. E., Ed.; American Chemical Society: Washington, DC, 1989; p 457.
43. Schaller, E. J. *Surface Coatings Aust.* **1985**, *22*, 6.
44. Lesota, S.; Lewandowski, E. W.; Schaller, E. J. *J. Coatings Technol.* **1989**, *61*, 135.
45. Calbo, L. J. *Handbook of Coatings Additives: Vol. 2*; Marcel Dekker: New York, 1992.
46. Toms, B. A. *Proc. First Int. Congr. On Rheology: Vol. II.* **1948**, *61*, 135.
47. Mysels, K. J. U.S. Patent 2 492 173, 1949.
48. Virk, P. S. *AIChE J.* **1975**, *21*, 625.
49. Sellin, R. J. H.; Hoyt, J.; Pollert, J.; Scrivener, O. *J. Hydraulic Res.* **1982**, *20*, 235.
50. Kulicke, W. M.; Kotter, M.; Grager, H. *Adv. Polym. Sci.* **1989**, *89*, 1.
51. Morgan, S. E.; McCormick, C. L. *Prog. Polym. Sci.* **1990**, *15*, 103.
52. Gupta, S. P.; Trushenki, S. P. *Soc. Pet. Eng. J.* **1978**, *55*, 345.
53. McCormick, C. L.; Hester, R. D.; Neidlinger, H. H.; Wildman, G. C. In *Surface Phenomena in Enhanced Oil Recovery*; Plenum: New York, 1981; p 741.

WATER-SWELLABLE POLYMERS (Carbomer Resins)

Manzer J. Durrani* and P. A. Manji
The B.F. Goodrich Company

Carbomers are synthetic high molecular weight resins. Carbopol® resins belong to the carbomer family, whose unique chemistry allows them to be used as highly efficient thickeners, suspending aids for insoluble ingredients, stabilizers for emulsions, and controlled release agents (that form hydrogels) for pharmaceutical products.

CARBOPOL RESIN CHEMISTRY

Carbopol resins are crosslinked, water-swellable polymers of acrylic acid with molecular weights greater than 1 billion daltons. The carboxyl groups with a pK_a of 6.5 are responsible for much of the product's activity.

The Carbopol resins are crosslinked with polyalkenyl polyethers. They are very mild acids, and the carboxylic groups within the resin provide the acidic qualities and the principal chemical sites that affect their physical thickening characteristics. The thickening properties come about from two mechanisms depending on the solvent being thickened. The most common is an aqueous dispersion of Carbopol resin, in which acid groups are neutralized into their appropriate salts; the polymer expands to give an extensive gel-like structure. A second mechanism for thickening the Carbopol resins operates when they are dispersed in a nonaqueous, polar solvent. In this case, a hydroxyl donor is added to expand the resin through hydrogen bonding.

Pharmaceuticals

Water-swellable polymers are used in a diverse range of pharmaceutical applications. Carbopol resins provide controlled release in tablets and bioadhesion in buccal, ophthalmic, intestinal, nasal, vaginal, and rectal applications. These polymers thicken at very low concentrations (<1%) to produce a wide range of viscosities and flow properties. This makes them valuable in topical lotions, in creams and gels, in oral suspensions, and in transdermal gel reservoirs. In addition, Pemulen® polymeric emulsifiers function as primary emulsifiers in topical oil-in-water systems with essentially no need for irritating surfactants. Noveon® AA-1 USP (polycarbophil) functions as a bioadhesive agent and Noveon CA-1 USP and CA-2 USP (calcium polycarbophil) provide bulking action in chewable laxative tablets.

Due to their crosslinked nature, these polymers swell but do not dissolve in water. In a pharmaceutical tablet, they tend to form microgels composed of clusters of the ultimate swollen 0.2-μm particles (which swell about 1000 times). Due to the convoluted and dense nature of the crosslinking, these ultimate particles may be technically single molecules. On swelling, they form the brick and mortar for the highly viscous but stable gel. After a time (typically between 10 and 20 h), osmotic forces cause fissures and cracks within the "microgel diffusion barrier," which results in "attrition" of gel clusters (similar to "popping

*Author to whom correspondence should be addressed.

of popcorn"). The gel clusters then disperse in the dissolution fluid.

Recently, there has been much interest in the development of novel polymers exhibiting bioadhesive activity, as a basis for novel controlled-release systems for drugs. These devices are developed for topical, buccal or nasal, ocular, and intestinal or rectal application.[1–5] Carbopol 934P NF resin and Noveon AA-1 (polycarbophil) are being extensively formulated in bioadhesive or mucoadhesive drug-delivery systems.

CARBOPOL RESINS FOR STABILITY

In addition to thickening, Carbopol resins provide an unusual flow property to liquids by providing a yield value. Many water-soluble resins or dispersions behave as pseudoplastic liquids. Carbopol resin solutions are different in that they exhibit plastic flow. A plastic liquid does not flow until a minimum applied stress is reached.

A neutralized Carbopol resin gel demonstrates this yield value phenomenon by behaving as a stiff gel under little or no mechanical force but readily flowing upon applying mechanical work such as stirring, pumping, or spreading the gel. The yield value provides stability to emulsions and suspensions.

SUMMARY

Carbopol resins are safe and effective ingredients for pharmaceutical and personal care products. They are highly efficient thickeners in water and compound solvents. Carbopol resin polymers can function as secondary emulsifiers, whereas Pemulen polymeric emulsifiers function as primary emulsifiers. The yield value of swollen Carbopol resins is the key to their performance in stabilizing emulsions and suspensions. Carbopol resins are effective sustained-release matrices for various drugs. They control the rate of release via swelling of the polymer and diffusion of the drug.

REFERENCES

1. Ishida, M.; Nambu, N.; Nagai, T. *Chem. Pharm. Bull.* **1983**, *31*, 1010.
2. Nagai, T.; Nishimoto, Y.; Nambu, N.; Suzuki, Y.; Sekine, K. *J. Contr. Rel.* **1984**, *1*, 15.
3. Hui, H. W.; Robinson, J. R. *Int. J. Pharm.* **1985**, *26*, 203.
4. Park, K.; Ch'ng, H. S.; Robinson, J. R. In *Recent Advances in Drug Delivery System*; Anderson, J. M.; Kim, S. W., Eds.; Plenum: New York, 1984; Vol. 163.
5. Robinson, J. R.; Langer, M.; Veillard, M. *Ann N. Y. Acad. Sci.* **1987**, *507*, 307.

Waxes

See: *Floor Finishes*

Weatherability

See: *Degradation (Weatherability)*
Injection Moldings (Residual Stresses)
Weathering of Polymers (Methodology and Limitations of Accelerated Testing)

WEATHERING OF POLYMERS (Methodology and Limitations of Accelerated Testing)

John Scheirs*
Polymer Analysis Laboratory
ExcelPlas Australia

Warren D. Ketola
3M Traffic Control Materials Division

Richard M. Fischer
3M Weathering Resource Center

For years the accepted method for establishing the exterior durability performance for polymeric systems has been through extended exposure at outdoor test facilities, usually located in warm sunny climates. This approach was acceptable when material performance could often be established in a few months of exposure. However, with the ever-improving durability of modern plastics and coatings systems, exterior exposures often take years to establish meaningful performance trends. To hasten results for critical product development programs, researchers are forced to use more severe environments than the real world to accelerate material failures. Historically, material scientists have been reluctant to use "accelerated lab testing" to make durability judgments. This hesitancy has been justified since accelerated testing has been fraught with results that has often had little in common with outdoor, real-time exposure data for identical materials.

The linking of increased customer expectations along with more durable polymer systems has placed considerable pressure on those working in the field of weathering testing to improve the results obtained from accelerated tests. Significant improvements have been made on several fronts to improve accelerated testing, in general. These include improved testing equipment, better experimental test design and data analysis, new analytical tools and instrumentation for monitoring degradation processes, and improved weathering test procedures.

POLYMER PHOTODEGRADATION

The Grotthus–Draper law (the first law of photochemistry states that "only light absorbed by a material can have any effect on it." Therefore, the absorption properties of the polymer are critical to the effect of light on it. The more light is absorbed, the higher the potential for degradation if the energy of the light absorbed is sufficient to break some of the chemical bonds.[1] Short wavelengths usually cause more damage than long wavelengths. Because of the higher energy of these photons, they are capable of breaking more types of bonds. Once photoinitiation has occurred, photo and thermal oxidation often is the primary process leading to loss of physical properties for a polymer system.

PRIMARY WEATHERING STRESSES

As stated previously, UV radiation is of sufficient energy to break chemical bonds in polymers. This is especially true with short-wavelength UV radiation. However, many pure polymers

*Author to whom correspondence should be addressed.

(e.g., polyethylene) do not absorb solar UV radiation but contain residual initiators or chromophores produced by processing and formulation conditions. These chromophores can absorb solar UV radiation and form free radicals, which begin the degradation chain reaction.[1]

Heat can cause degradation of polymers such as PVC that are thermally unstable. However, temperatures reached by polymers in exterior exposure are more likely to accelerate the rate of photoinitiated degradation reactions. For example, the rate of photodegradation of polyethylene and polyamides doubles for each 10°C increase in surface temperature.[2]

The most common indicators of moisture in any location are average rainfall and relative humidity. Although this information may be useful for climate comparisons, it does not give any indication of the time surfaces are actually wet. Results from a multiyear study of time of wetness showed that, in many locations, surfaces are wet for 6–8 h per day and that this surface wetness is caused by dew, not rain.[3] Prolonged wetting by dew can influence polymer degradation in several ways: water can extract UV absorbers and stabilizers added to protect the polymer, water can be a direct participant in degradation reactions such as hydrolysis, and water absorbed by a polymer can act as a plasticizer, increasing the mobility of polymer chains and enhancing the probability of reaction with reactive species. However, the most important effect of water on polymers may result from the generation of mechanical stresses caused as water-swollen polymers begin to dry.[2] Tensile stresses are created in the outer layer as it dries and is restricted from shrinking because the inner layers are still swollen. These tensile stresses can result in the formation of microcracks. In addition, the repeated swelling and shrinking caused by cyclic wetting and drying can also induce cracking.

LABORATORY-ACCELERATED WEATHERING

Although outdoor exposures are relatively inexpensive, considerable time is often needed to detect differences between materials, especially with development of more stable polymers and advances in the development of UV absorbers and stabilizers. Researchers often need to obtain results rapidly and must rely on laboratory-accelerated weathering tests that attempt to simulate the relative performance rankings obtained in longer-term outdoor exposures. There are five light sources that are used for laboratory-accelerated weathering of polymers: carbon arc, fluorescent UV lamps, xenon arc, medium-pressure mercury lamps, and metal halide lamps. All of these light sources produce UV radiation of intensities and wavelengths that can greatly accelerate polymer degradation. However, the usefulness of the results from laboratory-accelerated weathering tests depends on how well the light source simulates sunlight in the wavelengths known to produce the degradation of polymer or product features of interest and how well the temperature and moisture conditions used simulate the effects of heat and moisture found in outdoor exposures.

ANALYTICAL METHODS

Over the past 10 years several analytical techniques have been successfully employed in following the chemical changes in polymeric materials during the weathering process. The general approach is to use a "sensitive" analytical technique that can detect weathering-induced changes before typical physical changes become evident. These "typical" changes might be gloss loss, color shift, or changes in modulus. The critical assumption is that these early changes, often some chemical change or rate of degradation product formation, will directly relate to the actual weathering performance of the material. This could, hopefully, shorten exposure times from years to weeks or even days.

Croll reports on a survey of four techniques [scanning electron microscopy (SEM), electron spin resonance spectroscopy (ESR), X-ray photoelectron spectroscopy (XPS or ESCA), Fourier transform IR spectroscopy (FTIR)] for applicability in determining the weatherability of organic coatings.[4] Promising results were reported from all three spectroscopy methods. SEM was found to be quite sensitive in detecting early weathering changes but was difficult to quantify. ESR has been evaluated in several studies by other researchers,[5–9] as have XPS[10–12] and FTIR.[13–17]

As with any highly accelerated test, the analytical approach also assumes greater risk for predicting natural weathering results. For example, autoacceleration of the degradation processes may not be evident during these "early" stages of weathering, or early indications of durability may later disappear because of a transient stabilizer. All accelerated test procedures must finally be judged by comparison to "real-world" weathering results.

SERVICE LIFE PREDICTIONS

A new approach to developing a service life prediction for materials in a weathering environment using a reliability-based methodology has been proposed and active research efforts are underway. First pioneered by the aerospace and electronics industries, a reliability-based methodology may be the only technically reasonable approach to deal with the variability in weathering testing.[18,19] This approach requires precise measurement of the primary critical stresses (light, specimen temperature, and wet time) to develop realistic models to convert accelerated test results to an actual service life prediction. Currently, the instrumentation and systems are not in place to reproducibly collect these data in either accelerated or actual outdoor weathering. Reducing a reliability-based program to practice for a broad range of materials will require substantial time and effort.

REFERENCES

1. Searle, N. D. In *Encyclopedia of Polymer Science and Engineering*; 2nd Ed.; Mark, H., Ed.; 1989; Vol. 17; p 796.
2. Kockott, D. *Polym. Degrad. Stabil.* **1989**, *25*, 181.
3. Gutmann, H.; Sereda, P. *Metals Corrosion in the Atmosphere*; ASTM STP 435; American Society for Testing and Materials; Philadelphia, PA, 1968; pp 326–359.
4. Croll, S. *Prog. Org. Coat.* **1987**, *15*, 223.
5. Gerlock, J.; Bauer, D.; Briggs, L. *Polymer. Preprints* **1984**, *25*, 30.
6. Bauer, D.; Gerlock, J.; Dickie, R. *Prog. Org. Coat.* **1987**, *15*, 209.
7. Okamoto, S.; Ohya-Nishiguchi, H. *J. Japan Soc. Col. Mat.* **1990**, *63*, 392.
8. Sagawa, C.; Suzuki, T.; Tsujita, T.; Maeda, K. *Proceedings of the 2nd International Conference on Fluorine in Coatings*; Salford; Paint Research Association: Middlesex, United Kingdom, 1994; paper 29.

9. Sommer, A.; Zirngiebl, E.; Kahl, L.; Schonfelder, M. *Prog. Org. Ctng.* **1991**, *19*, 79.
10. Jernberg, P.; Lala, D.; Sjostrom, C. In *Durability of Building Materials and Components*; Baker, J. M. et al., Eds.; Chapman & Hall: 1990; Vol. 5, pp 515–526.
11. Wilson, G.; Skerry, B. B. *Poly. Mat. Sci. Eng.* **1993**, *68*, 72.
12. Sjostrom, C. In *Durability of Building Materials and Components*; Nagataki, S. et al., Eds.; Chapman & Hall: 1987; Vol. 4, pp 567–574.
13. Webb, J. *Effect of Ultraviolet Radiation on Transparent Polymer Films as Determined Using In-Situ FTIR-RA Spectroscopy*; SERI/TR—225-2177, DE84013039; SERI: Golden, CO; August 1984.
14. McEwen, D.; Verma, M.; Turner, R. *J. Coat. Tech.* **1987**, *59*, No. 755, 123.
15. Hartshorn, J. In *Proceedings of ACS Division of Polymeric Materials*; New Orleans; American Chemical Society: Washington, DC, 1987; Vol. 57, p 880.
16. Bauer, D.; Paputa Peck, M.; Carter III, R. *J. Coat. Tech.* **1987**, *59*, No. 755, 103.
17. Carter, R.; McCallum, J. *Poly. Deg. Stabil.* **1984**, *4*, 145.
18. McKnight, M.; Martin, J.; Masters, L. *J. Prot. Coat. Lngs.* **1985**, *2*, 18.
19. Fischer, R.; Ketola, W.; Martin, J.; Jorgensen, G.; Mertzel, E.; Pernisz, U.; Zerlaut, G. *Crit. Rev. Surf. Chem.* **1993**, *2*(4), 311.

Welan Gum

See: Welan Microbial Polysaccharide

WELAN MICROBIAL POLYSACCHARIDE

Christina T. Andrade* and Léa Lopes
Instituto de Macromoléculas (IMA/UFRJ)

Michel Milas and Marguerite Rinaudo
Centre de Recherches sur les Macromolécules Végétales (CERMAV/CNRS)

The majority of polysaccharides currently used for food, textile, pharmaceutic, and other industrial applications is of plant origin;[1–3] however, over the last 20 years, polysaccharides from microbial source are gaining increasing importance in a wide range of industrial processes.[4–9]

The first bacterial polysaccharide produced on a commercial scale was dextran [9004-54-0], in the early 1940s. Dextran, a polymer of glucose units joined predominantly through α-1,6-glycosidic linkages, is extracellularly synthesized by *Leuconostoc mesenteroides*.[16,17] The polysaccharide and its derivatives have wide applications in the food, drug, cosmetic, and photographic industries. Dextran is used as an ultrapure inert ingredients in pharmaceuticals and blood plasma products. The second fermentation glycopolymer to be commercialized was xanthan gum [11138-66-2], which is produced by the bacterium *Xanthomonas campestris*. Xanthan is a nongelling heteropolysaccharide whose primary structure[19,20] consists of a cellulosic backbone with a trisaccharide side-chain on alternate glucose residues. The existence of two different ordered conformations in solution (native and renatured) has been reported;[21,22] however, the secondary structure of xanthan is still a matter of debate.[23–25] Xanthan is employed commercially in food and non-food products, mainly as thickening, stabilizing, and suspending agent.[13,15] This polyelectrolyte is also used in textile printing and dying as a print paste thickener,[26,27] as a carrier in agricultural sprays,[28] in gelled detergents and explosives, in water-based and emulsion paints,[29] as a drag-reducing polymer fluid in oil recovery,[30,31] and in many other applications.

Some other bacterial and fungal polysaccharides have been the object of much research. Succinoglycan [73667- 50-2] is a branched exopolysaccharide synthesized by several strains of *Alcaligenes, Pseudomonas, Agrobacterium,* and *Rhizobium*.[32] Succinoglycan has many analogies with xanthan gum.[14,33] Another bacterial polymer that has some similarities with xanthan is the polysaccharide secreted by *Acetobacter xylinium*, acetan [110865-71-9]. The structures of xanthan gum and acetan are identical with respect to the backbone, backbone-lateral chain linkage, and the first two sugar residues in the side chain.[34] In the helical confirmation, acetan forms thixotropic fluids characteristic of xanthan samples.[8,35] Curdlan [54724-00-4] is a β-D-glucan polymer. Curdlan synthesized by *Alcaligenes faecalis* var. *myxogenes* and also by strains of *Agrobacterium*, forms a firm, resilient and thermally irreversible gel when heated in aqueous suspension. Pullulan is the generic name given to the exocellular α-D-glucan elaborated by many strains of the dimorphic fungus *Aureobasidium pullulans*.[38] Pullulan is a linear, high molecular weight polymer made of maltotriose repeating units that readily dissolves in water to give solutions, whose rheological properties depend on the microbial strain and on the duration of the fermentation process.[5] Commercial interest in this polysaccharide is related to its ability to produce strong, nontoxic, and nondigestible films, appropriate for food packaging.[4,5,10,11]

At the beginning of the 1980s, the search for useful microbial polymers yielded a new group of bacterial polysaccharides developed by Kelco, division of Merck, U.S. The gellan family consists of gellan [71010-52-1], welan [96949-22-3], rhamsan [96949-21-2], S-198, and S-657 gums.[39–43] All members of the gellan gum family, in the native form, contain O-acetyl groups as substituents.[39–47] In the native state, gellan forms a soft and elastic gel, but after deacetylation this polysaccharide forms hard and brittle gel due to enhancement of the intermolecular interactions.[48–51]

Welan gum, commercialized by Kelco under the name BIOZAN® (formerly known as S-130), has a monosaccharide side chain on the O-3 position of the 4-linked glucose;[46] this can be either α-L-rhamnose or α-L-mannose. Preliminary X-ray results from native and deacetylated welan led to a two-fold helical conformation, in aqueous solution as in the solid state. Analysis of these X-ray patterns also indicated that the crystallinity and chain conformation of the molecules are not seriously influenced by acetyl group.[48] However, more recent X-ray studies[54a,55] and computer modeling analysis[56] of welan demonstrate that this polysaccharide may adopt the same three-fold double helical structure as that characterized for gellan.[51a]

PREPARATION AND PROPERTIES OF WELAN GUM

Welan gum is industrially produced by a strain of *Alcaligenes* (ATCC31555) in aerobic submerged fermentation.[52,56a,57]

*Author to whom correspondence should be addressed.

The stiffness parameter of welan chain is in the same range as those obtained for xanthan[54,59] and DNA[60] and welan and xanthan gums are among the stiffest polysaccharides known.[61]

Welan gum was primarily promoted[52,57] in the market for its excellent stability towards pH changes and shear and for retention of viscosity, even at a temperature of 140°C. Unlike gellan and xanthan, welan shows no evidence[53,54,62,63] of conformational transition on heating and cooling between 0°C and 100°C. The optical activity of this polysaccharide is independent of both salt concentration and temperature.[57a,58,62,64] Moreover, rheological data reported for welan[53,57a] showed almost no influence of ionic strength on the viscosities of their solutions. One interpretation of this behavior was that welan would exist as disordered coils in solution.[53] Nevertheless, evidence from NMR linewidths[63,65] and from the near salt independence of macromolecular dimension (as characterized by intrinsic viscosity)[63,64] strongly indicates that welan adopts an ordered conformation in aqueous solution, similar to that found for xanthan in its ordered conformations.[60] The existence of a highly stiff and stable conformation for welan is also supported by the large mean chain extension per backbone sugar residue found by the radius of gyration and by the value of 1.41 calculated for the Mark–Houwink exponent.[54] This value is very close to those estimated for other very stiff and multistrand microbial polymers.[64] The ordered conformation of welan is much more stable than that of gellan; however, it was recently demonstrated that the native double-helix structure of welan can be disturbed by using dimethyl sulfoxide as solvent. The welan molecules undergo a rapid conformational transition to disordered coils in dimethyl sulfoxide, and the renaturation from the disordered state forms shorter helices connected in a stable crosslinked network.[66]

Investigations of the dependence of the intrinsic viscosity on ionic strength, light-scattering, and potentiometric proton titration results also suggested a very weak polyelectrolyte character for welan polysaccharide.[53,54,56,64] However, recent rheological studies showed a typical polyelectrolyte behavior for welan, analogous to that characterized for xanthan.[60] The presence of acetyl groups in the NaOH-treated welan, established by infrared spectrometry, shows the strong resistance of these substituents to hydrolysis.[65,67] The acetyl groups are probably stabilized by the secondary structure of the polymer. The decrease of the viscosity as a function of Na^+ concentration confirms the polyelectrolyte behavior of this bacterial polymer[60] and raises the question of whether the low electrolyte character[53,54,56,64] may have resulted from the presence of aggregates (or microgels).

The molecular weight[59,67] of welan is approximately $\langle M_w \rangle \approx 1 \times 10^6$. The solution properties of this gum are governed basically by intramolecular interactions.[53,54,62] Although detailed rheological investigations have not been carried out yet for welan and rhamsan, the viscoelastic behavior of their semidilute solutions is characteristic of a typical *weak-gel* system.[63,64,66]

APPLICATIONS

Welan gum was introduced to the oil field, in 1985, as a drilling fluid.[11,55] The chain stiffness of welan leads to highly viscous and salt-tolerant solutions that are important in oil field destiny.[61] This polysaccharide has a very good compatibility with calcium and could have potential use in cementing applications. Welan also has utility as an anti-washout additive for underwater concrete placement.[55] The high viscosity at low shear rates and low viscosity at high shear rates, plus the high viscosity at low polymer concentrations, contribute to the excellent suspending capacity of welan. Actually, welan gum is a powerful rheological/suspending agent.[11,52]

REFERENCES

1. Mantell, C. L. *The Water-Soluble Gums*; Hafner: New York, 1965.
2. Whistler, R. L. In *Industrial Gums: Polysaccharides and Their Derivatives*; Whistler, R. L.; BeMiller, L. N., Eds.; 2nd ed.; Academic: New York, 1973; Chapter I.
3. Lopes, L.; Andrade, C. T.; Mano, E. B. *Ciência Hoje* **1991**, *12*, 65.
4. Wells, J. *Extracellular Microbial Polysaccharides*; ACS Symposium Series 45; American Chemical Society: Washington, DC, 1977; p 299.
5. Sandford, P. A. In *Polysaccharides in Food*; Blanshard, J. M. V.; Mitchell, J. R., Eds.; Butterworths: London, 1979; Chapter 16.
6. O'Neill, M. A.; Morris, V. J.; Selvendran, R. R. In *Gums and Stabilizers for the Food Industry 3*; Phillips, G. O. et al., Eds.; Elsevier Applied Science: London, 1986; p 29.
7. Sutherland, I. W. In *Biomedical and Biotechnological Advance in Industrial Polysaccharides*; Crescenzi, V. et al., Eds.; Gordon and Breach Science: New York, 1989; p 123.
8. Morris, V. J. In *Biotechnology and Polymers*; Gebelein, C. G., Ed.; Plenum: New York, 1991; p 135.
9. Parolis, H.; Wilkinson, S. G. *Carbohydr. Res.* **1992**, *231*, vii.
10. Kennedy, J. F.; Griffiths, A. J.; Atkins, D. P. In *Gums and Stabilizers for the Food Industry 2*; Phillips, G. O. et al., Eds.; Pergamon: Oxford, 1984; p 417.
11. Crescenzi, V. *TRIP* **1994**, *2*, 104.
12. Shamel, R. E. In *Biotechnological Polymers*; Gebelein, C. G., Ed.; Technomic: Pennsylvania, 1993; p 17.
13. McNeely, W. H.; Kang, K. S. In *Industrial Gums: Polysaccharides and Their Derivatives* 2nd ed.; Whistler, R. L.; BeMiller, J. N., Eds.; Academic: New York, 1973; Chapter XXI.
14. Crescenzi, V.; Dentini, M.; Coviello, T. In *Industrial Polysaccharides: The Impact of Biotechnology and Advanced Methodologies*; Stivala, S. S. et al., Eds.; Gordon and Breach Science: New York, 1987; p 69.
15. Symes, K. C. *Fd. Chem.* **1980**, *6*, 63.
16. Jeanes, A. R.; Haynes, W. C.; Wilham, C. A.; Rankin, J. C.; Melvin, E. H.; Austin, M. J.; Cluskey, J. E.; Fisher, B. E.; Tsuchiya, H. M.; Rist, C. E. *J. Am. Chem. Soc.* **1954**, *76*, 5041.
17. Murphy, P. T.; Whistler, R. L. In *Industrial Gums: Polysaccharides and Their Derivatives* 2nd ed.; Whistler, R. L.; BeMiller, J. N., Eds.; Academic: New York, 1973; Chapter XXIII.
18. Slodki, M. E. *Encyclopedia of Chemical Technology*; Standen, A., Ed.; Interscience: John Wiley & Sons: New York, 1981; p 439.
19. Jansson, P. E.; Kenne, L.; Lindberg, B. *Carbohydr. Res.* **1976**, *45*, 275.
20. Melton, L. D.; Mindt, L.; Rees, D.; Sanderson, G. R. *Carbohydr. Res.* **1976**, *46*, 245.
21. Milas, M.; Rinaudo, M. *Carbohydr. Res.* **1986**, *158*, 191.
22. Milas, M.; Viehweg, H.; Weiss, A. *Carbohydr. Polymers* **1990**, *13*, 119.
23. Sato, T.; Kojima, S.; Norisuye, T.; Fujita, M. *Polymer J.* **1984**, *16*, 423.

24. Goodall, D. M. In *Frontiers in Carbohydrate Research 2*; Chandrasekaran, R., Ed.; Elsevier Science: Essex, 1992; p 141.
25. Bezemer, L.; Ubbink, J. B.; Kooker, J. A.; Kuil, M. E.; Leyte, J. C. *Macromolecules* **1993**, *26*, 6436.
26. Fitzgerald, E. E. *Book Pap., Natl. Tech. Conf.; AATCC* **1983**, 306.
27. Bajaj, P.; Chavan, R. B.; Manjeet, B. *JMS: Rev. Macromol. Chem. Phys.* **1984**, *C24*, 387.
28. Colegrove, G. T. *Ind. Enq. Chem. Prod. Res. Dev.* **1983**, *22*, 456.
29. Sutherland, I. W. In *Biotechnology*; Rehm, H. J.; Reed, G., Eds.; Verlag Chemie: Weinheim, 1983; Vol. 3, Chapter 3.
30. Morgan, S. E.; McCormick, C. L. *Prog. Polym. Sci.* **1990**, *15*, 103.
31. Ahmed, I.; Moradi-Araghi, A. *TRIP* **1994**, *2*, 92.
32. Glazebrook, J.; Reed, J. W.; Reuber, T. L.; Walker, G. C. *Int. J. Biol. Macromol.* **1990**, *12*, 67.
33. Rinaudo, M.; Milas, M.; Tinland, B. In *Novel Biodegradable Microbial Polymers*; (NATO ASI Series); Dawes, E. A., Ed.; Kluwer Academic: Dordrecht, 1990; p 349.
34. Couso, R. O.; Ielpi, L.; Dankert, M. *J. Gen. Microbiol.* **1987**, *133*, 2123.
35. Morris, V. J. *Agro-Food-Industry Hi-Tech* 1992, May/June, 3.
36. Harada, T.; Misaki, A.; Saito, M. *Archs. Biochem. Biophys.* **1968**, *124*, 292.
37. Harada, T. In *Polysaccharides in Food*; Blanshard, J. M. V.; Mitchell, J. R., Eds.; Butterworths: London, 1979; Chapter 18.
38. Wallenfels, K.; Bender, H.; Keilich, G.; Freudenberger, D. *Biochem. Z.* **1965**, *341*, 433.
39. Jansson, P. E.; Lindberg, B.; Widmalm, G.; Sandford, P. A. *Carbohydr. Res.* **1985**, *139*, 217.
40. Jansson, P. E.; Lindberg, B.; Lindberg, J.; Maeka-Wa, E.; Sandford, P. A. *Carbohydr. Res.* **1986**, *156*, 157.
41. Jansson, P. E.; Kumar, N. S.; Lindberg, B. *Carbohydr. Res.* **1986**, *156*, 165.
42. Chowdhury, T. A.; Lindberg, B.; Lindquist, U.; Baird, J. *Carbohydr. Res.* **1987**, *161*, 127.
43. Chowdhury, T. A.; Lindberg, B.; Lindquist, U.; Baird, J. *Carbohydr. Res.* **1987**, *164*, 117.
44. Jansson, P. E.; Lindberg, B.; Sandford, P. A. *Carbohydr. Res.* **1983**, *124*, 135.
45. O'Neill, M. A.; Selvendran, R. R.; Morris, V. J. *Carbohydr. Res.* **1983**, *124*, 123.
46. O'Neill, M. A.; Selvendran, R. R.; Morris, V. J.; Eagles, J. *Carbohydr. Res.* **1986**, *147*, 295.
47. Kuo, M. S.; Mort, A. J. *Carbohydr. Res.* **1986**, *156*, 173.
48. Attwool, P. T.; Atkins, E.; Miles, M. J.; Morris, V. J. *Carbohydr. Res.* **1986**, *148*, C1.
49. Atkins, E. In *Industrial Polysaccharides: Progress in Biotechnology 3*; Yalpani, M., Ed.; Elsevier Science: Amsterdam, 1987; p 177.
50. Tako, M.; Sakae, A.; Nakamura, S. *Agric. Biol. Chem.* **1989**, *53*, 771.
51. Quinn, F. X.; Hatakeyama, T.; Yoshida, H.; Takahashi, M.; Hatakeyama, H. *Polymer Gels and Networks* **1993**, *1*, 93.
51a. Chandrasekaran, R.; Radha, A.; Thailambal, V. G. *Carbohydr. Res.* **1992**, *224*, 1.
52. Kang, K. S.; Veeder, G. T.; Cottrell, I. W. In *Microbial Polysaccharides: Progress in Industrial Microbiology*; Bushell, M. E., Ed.; Elsevier Science: Amsterdam, 1983; p. 231.
53. Crescenzi, V.; Dentini, M.; Dea, I. C. M. *Carbohydr. Res.* **1987**, *160*, 283.
54. Urbani, R.; Brant, D. A. *Carbohydr. Polymers* **1989**, *11*, 169.
54a. Cairns, P.; Miles, M. J.; Morris, V. J. *Carbohydr. Polymers* **1991**, *14*, 367.
55. Chandrasekaran, R.; Radha, A.; Lee, E. J. *Carbohydr. Res.* **1994**, *252*, 183.
56. Lee, E. J.; Chandrasekaran, R. *Carbohydr. Res.* **1991**, *214*, 11.
56a. Kang, K. S.; Veeder, G. T. U.S. Patent 4 342 866, 1982.
57. Sandford, P. A.; Cottrell, I. W.; Pettitt, D. J. *Pure and Appl. Chem.* **1984**, *56*, 879.
57a. Tako, M.; Kiriaki, M. *Agric. Biol. Chem.* **1990**, *54*, 3079.
58. Tako, M. In *Carbohydrates and Carbohydrates Polymers: Analysis, Biotechnology, Modification, Antiviral, Biomedical and Other Applications*; Yalpani, M., Ed.; ATL: New York, 1993; Chapter 19.
59. Kwon, B.; Foss, P. A.; Rha, C. In *Industrial Polysaccharides: Progress in Biotechnology 3*; Yalpani, M., Ed.; Elsevier Science: Amsterdam, 1987; p 253.
60. Campana, S.; Andrade, C. T.; Milas, M.; Rinaudo, M. *Int. J. Biol. Macromol.* **1990**, *12*, 379.
61. Stokke, B. T.; Smidsrod, O.; Marthinsen, A. B. L.; Algsaeter, A. In *Water Soluble Polymers for Petroleum Recovery*; Stahl, G. A.; Schulz, D. N., Eds.; Plenum: New York, 1988; p 243.
62. Crescenzi, V.; Dentini, M.; Coviello, T.; Rizzo, R. *Carbohydr. Res.* **1986**, *149*, 425.
63. Robinson, G.; Manninng, C. E.; Morris, E. R. In *Food Polymers, Gels and Colloids*; Dickinson, E., Ed.; Royal Society of Chemistry: Cambridge, 1991; p 22.
64. Cesàro, A.; Gamini, A.; Navarini, L. *Polymer* **1992**, *33*, 4001.
65. Lopes, L.; Andrade, C. T.; Milas, M.; Rinaudo, M. *International Symposium on Polymers: POLIMEX-93* (Preprints), Cancún, 1993; p 279.
66. Hember, M. W. N.; Richardson, R. K.; Morris, E. R. *Carbohydr. Res.* **1994**, *252*, 209.
67. Lopes, L.; Milas, M.; Rinaudo, M. *Int. J. Biol. Macromol.* **1994**, *16*, 253.

Whey

See: *Cellulose Derivative-Protein Complex (in Ice Cream)*
Cellulose Derivative-Protein Complexes (Whey and Corn Steep Liquor, Enrichment of Macaroni)
Cellulose Derivatives (Reclamation of Proteins from Cheese Whey)

Whiskers

See: *Carbon Whisker (Surface Modification by Grafting)*
Magnetic Field Processing
Silicon Nitride
Whiskers of Rigid-Rod Polymers

WHISKERS OF RIGID-ROD POLYMERS

Yuhiko Yamashita* and Kunio Kimura
Faculty of Environmental Science and Technology
Okayama University

The extended chain crystals of both flexible and rigid-rod polymers, the so-called polymer whiskers, have been obtained

*Author to whom correspondence should be addressed.

by solid-state polymerization of the monomers and polymerization in solution. Examples of the former have included (1) oxacyclo-alkanes such as 1,3,5-trioxane, (2) diolefins undergoing "four-center type" photopolymerization, and (3) monomers with conjugated triple bonds such as substituted diacetylenes.[1–6] The first whisker obtained by the polymerization in solution system was polyoxymethylene.[7–9]

The preparation of polymer whiskers by the polymerization in solution system is of great interest for the processing of infusible aromatic polymers into extended chain crystals. That is, even infusible polymers can be processed by building monomers into some special crystal form one-by-one while polymerizing.

PREPARATION

Historical Progress on Preparation of Aromatic Polyester Whiskers

In 1976, Economy et al. developed a new synthetic method of poly(p-oxybenzoyl) (POB) that was a high-temperature polymerization of p-acetoxybenzoic acid (p-ABA) in an aromatic heat exchange medium.[10] Kricheldorf et al. prepared POB under a variety of reaction conditions. They found that the polycondensations in solution led to polymers with high molecular weights in the range of 20,000–50,000 though, at the beginning, oligomer crystals are precipitated with a degree of polymerization in the range of 6–8. From these findings, a new reaction mechanism was discussed.[11,12] We also obtained fibrillar crystals of POB applying a low concentration of monomer and vigorous stirring.[13] We found that whiskers of POB can be formed by using liquid paraffin and Therm S 800 (a mixture of isomers of diethylbiphenyl) as solvents.[14,15]

The Morphological and Structural Features of POB Whiskers[15–18]

The scanning electron micrograph of the whiskers is shown in **Figure 1**. The whiskers are 30–70 μm in length and 1–2 μm in diameter.

Control of the Whisker Length[19]

The changes of the length and the width of the whiskers prepared at 330°C in liquid paraffin are followed as a function of polymerization time. The crystals grow laterally, as well as longitudinally, at the beginning of the whisker growth; after a certain time, they grow only longitudinally.

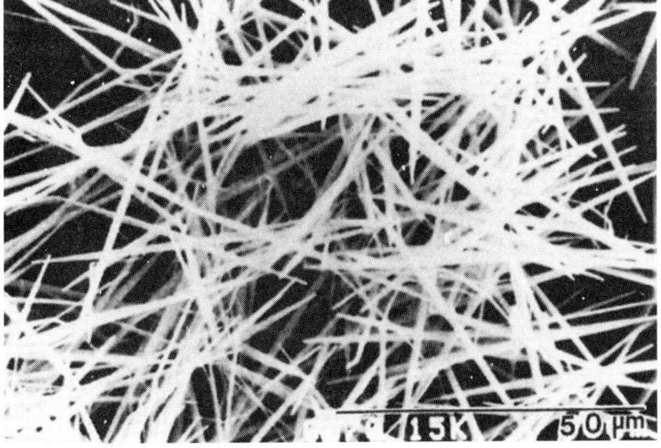

FIGURE 1. Scanning electron micrograph of POB whiskers.

The very narrow distribution of both length and width indicates that the lamellar crystals with screw distributions serve as primary nuclei for the growth of whiskers and the oligomers produced after the formation of primary nuclei are exhausted only in the growth of whiskers.

Whiskers of Other Aromatic Polyesters

From the success on preparation of POB whiskers, preparation of whiskers for other aromatic polyesters including poly(2-oxy-6-naphthoyl) (PON), poly(4′-oxy-4-biphenylcarbonyl (POBP), and poly(4-mercaptobenzoyl) (PMB)[20–22] was carried out by high temperature solution polymerization.

PROPERTIES

Thermal Properties of POB Whiskers

It is known that POB crystals show a reversible first-order solid-solid transition at around 350°C, and it is regarded as a transition to a pseudohexagonal packing of polymer molecules by a rotation of 1,4-phenylene ring.[23] The transition temperatures of the whiskers are higher than those of the crystals with other morphology. The whiskers display significantly higher thermal stability than the other crystals on the basis of TGA.

APPLICATION

Whiskers of rigid-rod polyesters are new materials and their suitable application fields are now developing. Most applications will be based on excellent mechanical properties such as strength and modulus, outstanding thermal stabilities, and good chemical resistances. Furthermore, these whiskers have lower densities than conventional ceramic whiskers. Therefore, the candidates for applications are reinforcements of organic composites used in items such as tennis rackets, shafts of golf clubs, and fishing rods. They will also be applied as reinforcements of dental materials such as dentures because of their white color. They will be used, it is hoped, in substitutions for asbestos.

REFERENCES

1. Andrew, E. H.; Martin, G. E. *J. Mater. Sci.* **1973**, *8*, 1315.
2. Patell, Y. R.; Schultz, J. M. *J. Macromol. Sci. Phys.* **1973**, *B7*, 445.
3. Iguchi, M.; Nakanishi, H.; Hasegawa, M. *J. Polym. Chem., A1* **1968**, *6*, 1055.
4. Nakanishi, H.; Suzuki, V.; Suzuki, F.; Hasegawa, M. *J. Polym. Sci. A1* **1969**, *7*, 753.
5. Wegner, G. *Makromol. Chem.* **1972**, *154*, 35.
6. Bloor, D.; Koski, L.; Stevens, G. C.; Preston, F. H.; Ando, D. J. *J. Mater. Sci.* **1975**, *10*, 1678.
7. Iguchi, M. *Br. Polym. J.* **1973**, *5*, 195.
8. Iguchi, M.; Murase, I.; Watanabe, K. *Br. Polym. J.* **1974**, *6*, 61.
9. Iguchi, M.; Murase, I. *Makromol. Chem.* **1975**, *176*, 2113.
10. Economy, J.; Storm, R. S.; Markovich, V. I.; Cottis, G. G.; Nowak, B. E. *J. Polym. Sci., Polym. Chem. Ed.* **1976**, *14*, 2207.
11. Kricheldorf, H. R.; Schwarz, G. *Makromol. Chem.* **1983**, *184*, 475.
12. Hieser, G.; Schwarz, G.; Kricheldorf, H. R. *J. Polymer Sci. Polym. Phys. Ed.* **1983**, *21*, 1599.

13. Yamashita, Y.; Akagi, K.; Monobe, K.; Shimamura, K. *Proceedings of 1983 Korea Japan Joint Meeting on Textile Science and Technology* **1983**, 130.
14. Kato, Y.; Endo, S.; Kimura, K.; Yamashita, Y.; Tsugita, H.; Monobe, K. *Kobunshi Ronbunshu* **1987**, *44*, 35.
15. Yamashita, Y.; Kato, Y.; Endo, S.; Kimura, K. *Makromol. Chem. Rapid Commun.* **1988**, *9*, 687.
16. Yamashita, Y.; Kato, Y.; Endo, S.; Kimura, K.; Tsugita, H.; Monobe, K. *Koubunshi Ronbunshu* **1987**, *44*, 41.
17. Kato, Y.; Yamashita, Y.; Kimura, K.; Endo, S.; Kajisaki, K. *Koubunshi Ronbunshu* **1988**, *45*, 973.
18. Kato, Y.; Yamashita, Y.; Kimura, K.; Endo, S.; Ohta, T. *Koubunshi Ronbunshu* **1990**, *47*, 583.
19. Kimura, K.; Endo, S.; Kato, Y.; Yamashita, Y. *Polymer* **1994**, *35*, 123.
20. Kimura, K.; Endo, S.; Kato, Y.; Yamashita, Y. *Polymer* **1993**, *34*, 1054.
21. Kimura, K.; Endo, S.; Kato, Y.; Inaba, T.; Yamashita, Y. *Macromolecules* **1995**, *28*, 255.
22. Kimura, K.; Yamashita, Y. *Polymer* **1994**, *35*, 3311.
23. Kricheldorf, H. R.; Conradi, A. *Macromolecules* **1989**, *22*, 14.

Wire and Cable Coatings

See: *Polyethylene (Commercial)*

Wood

See: *Carbon Composites*
Vegetal Biomass (3. Polymers, Derivatives and Composites)
Wood (Chemical Modification)
Wood, Chemically Modified (Thermoplasticization and Solubilization)
Wood Composites (High Performance)
Wood Graft Copolymers
Wood-Plastic Composites
Wood-Polymer Composites

WOOD (Chemical Modification)

Makoto Ohkoshi
Forestry and Forest Products Research Institute

Wood is used in various areas because of many desirable properties such as high specific strength, low heat conductivity, electric insulation, ability of controlling atmospheric humidity, beautiful appearance, and so on. However, wood also has undesirable properties. For example, wood swells and shrinks with changing moisture content; wood is biologically degraded by microorganisms; wood undergoes flaming combustion with increasing temperature; wood is degraded by action of acids and bases; and wood is photochemically degraded by ultraviolet light. The chemical modification of wood can improve these undesirable properties.

WOOD-POLYMER COMPOSITES (WPC)

Wood-polymer composites (WPC) are produced by impregnating the vinyl monomers into wood and polymerizing them. The monomers used are methyl methacrylate, styrene, unsaturated polyester, a mixture of those, etc. The initiation of polymerization is done by generation of free radicals with irradiation of radiant ray or heating with catalyst (initiator). Benzoyl peroxide and azobisisobutyronitrile are frequently used as catalysts in the concentration range of 0.2 to 1%.

In the wood-polymer composites the mechanical properties are increased because the cell cavities in wood are filled with a hard polymer. The increases are especially large in the compressive strength perpendicular to grain and the hardness.

The resistance to moisture is also increased in the wood-polymer composites because the resin filled in cell cavities retards or prevents the movement of moisture in wood.

REACTIONS WITH CHEMICALS

The cell wall polymers (cellulose, hemicellulose, and lignin) have many hydroxyl groups that work as the reaction sites with chemicals. The chemical reagents can chemically bond with the cell wall polymers, bringing about the substitution of hydroxyl groups. The bonded chemicals decrease the hygroscopicity of wood and prevents the biological enzymatic reaction and makes the wood resistant to the attacks of microorganisms and termites.

THERMOPLASTICIZATION OF WOOD

Wood, lacks the thermoplasticity and solubility in organic solvents unlike plastics. The reasons are that cellulose in wood is a crystalline polymer, lignin is a three-dimensional macromolecule, and chemical bonds are formed between the main components of wood.

Wood can also be converted into a thermoplastic material if the reaction with chemicals is carried out under the same conditions as the crystalline cellulose in wood is derivatized. Thus, wood is made thermoplastic by the esterification reaction with a fatty acid in a nonaqueous cellulose solvent medium (N_2O_4-dimethylformamide), a trifluoroacetic anhydride-fatty acid mixture, or a fatty acid chloride-pyridine-dimethylformamide system. The degree of thermoplasticity of wood depends upon the reaction method, the molecular size of substituent groups introduced, and the degree of substitution.

The thermoplasticized wood can be dissolved in neutral aqueous solvents, organic solvents, or organic solutions. For example, the esterified wood with a series of aliphatic acids can be dissolved by the treatment of 200–270°C for 20–150 minutes in benzyl ether, styrene oxide, resorcinol, benzaldehyde, aqueous phenol solution, chloroform-dioxane, mixture, benzene-acetone mixture, etc. The potential applications of the dissolved wood are preparations of wood-based adhesives, moldings, fibers, among others.

WOOD, CHEMICALLY MODIFIED (Thermoplasticization and Solubilization)

Mitsuhiro Morita* and Isao Sakata
Department of Forest Products
Faculty of Agriculture
Kyushu University

In this article, the thermoplasticization of wood by cyanoethylation and the improvement of the plasticity and the solubility

*Author to whom correspondence should be addressed.

of this chemically modified wood is studied with a halogen treatment, an ozone treatment, and other treatments with some metal salts.[1–5]

THERMAL FLOW OF CYANOETHYLATED WOOD (CE-WOOD) AND CHLORINATED-CYANOETHYLATED WOOD (CL-CE-WOOD)

Modification of cellulose by cyanoethylation (CEC) has been carried out to improve resistance to rotting, heat, and abrasion. Further, CEC is useful as an insulating material or dielectric. It is expected, therefore, that cyanoethylation of wood (CE-wood) will result in improvements in characteristics similar to those obtained by cyanoethylation of cellulose.

The temperature (T_f) at which the wood could flow was measured by a simple mechanical flow tester. The values of CE-wood dropped from 270°C to 240°C with an increase in the nitrogen content from 4% to 9%, respectively. At 4% nitrogen content, chlorination had no effect on T_f. However, CE-wood with a nitrogen content of crashing approximately 5% developed thermal flow after treatment with the chlorine solution. The T_f value for Cl-CE-wood decreased remarkably with an increase in the amount of combined chlorine up to about 2–3%, and the value for such a slightly chlorinated CE-wood was about 100°C lower than that of the original CE-wood.

EFFECT OF CHLORINATION OF CHEMICALLY MODIFIED WOODS

The thermoplasticity and solubility of chemically modified wood were improved considerably by chlorination. Chlorine is utilized commercially as a bleaching and/or delignifying agent in multistage pulp-bleaching sequences. The reactions occurring when a chlorine agent is applied to wood are grouped into substitution, oxidation, and hydrolysis reaction.

The effect due to the structural modification of the lignin moiety in the CE-wood caused by the chlorine treatment in addition to an insufficient internal plasticizing effect previously provided by the introduction of cyanoethyl group to wood polymer molecule may play important roles in the lowering of T_f and in the high solubility in solvents.

EFFECT OF TREATMENTS WITH OXIDIZING AGENTS ON THE SOLUBILITY OF CHEMICALLY MODIFIED WOODS

It is found that the solubility of chemically modified wood can be improved by the treatment that splits the crosslinkage of lignin.

The solubility of the ozone-treated CE-wood increased with extension of the treatment time, but leveled off at around 20 minutes treatment time. After 30 minutes of ozone treatment a solubility of 88% in m-cresol was obtained.

It was found that the T_f of chemically modified wood was lowered by adding a small amount of a transition-metal halide, such as ferric chloride, cupric chloride, or zinc chloride.

APPLICATION OF CHEMICALLY MODIFIED WOOD

Based on the thermoplasticization of wood, several attempts have been made for its application. Examples of this are preparation of films, moldings, moldable boards, and others.

There are many potential applications in dissolution or liquefaction of chemically modified wood. For example, the Cl-CE-wood becomes a transparent uniform-film by casting it from an organic solution. The film thus obtained had rigid characteristics, but it was found that the Cl-CE-wood was plasticized effectively by blending cyanoethylated poly(vinyl alcohol) (CE-PVA), and that the Cl-CE-wood/CE-PVA blended-film had great mechanical strength.[6]

CHEMICAL CONVERSION OF CE-WOOD TO FUNCTIONAL MATERIALS

Since the nitrile group in CEC or CE-wood is high in reactivity, these cyanoethylated substances are expected to behave as reaction intermediates. For example, CE-wood reacts with hydroxylamine to give wood bearing the amidoxime group that tends to adsorb heavy metal ions such as uranium, or mercury ion.[7]

Conversion of CE-Wood to CB-Wood

CE-wood treated with chlorine is soluble only in cresol, DMF, and a few other organic solvents. However, Cl-CE-wood is readily converted to a water soluble material, CB-wood, by treatment with alkaline hydrogen peroxide. Over 90% of the nitrile groups in the sample are converted to amide group, which corresponds to a carbomoylethyl content of 6 mmol/g.

Viscosity Behavior of CE-Wood and Its Derivatives

Brestkin indicated that the CEC/DMF system formed a liquid crystalline state at high concentrations.[8] It seemed likely that the transition point in the concentration dependence indicated the start of formation of a liquid crystalline state. Accordingly, the structure of the solution was studied by a polarization microscopic method. Solutions for Cl-CE-wood and CEC revealed formation of a colored band structure above the concentrations to the transition points. It was also found that concentrated aqueous solutions of CB-wood formed liquid crystals. It is interesting to note that liquid crystalline behavior was found not only with the cellulose derivatives but also with the chemically modified wood, which contained a substantial proportion of lignin.

Gelation of CE-Wood

It is also well known that the nitrile group chelates with Cu(I). The viscosity for CE-wood and CEC in DMF increased with increasing addition of Cu(I).

In conclusion, we think that the studies of thermoplasticization and dissolution of wood are new fields for the chemical modification of wood with large future potential. To achieve the progress in these fields, more fundamental and critical studies should be made. Application studies for these woody materials to industrial development should also be made.

REFERENCES

1. Morita, M.; Sakata, I. *J. Appl. Polym. Sci.* **1986**, *31*, 831.
2. Morita, M.; Shigematsu, M.; Sakata, I. *Cellulose Chem. Technol.* **1987**, *21*, 255.
3. Morita, M.; Sakata, I. *Mokuzai Gakkaishi* **1988**, *34*, 917.
4. Morita, M.; Sakata, I. *Mokuzai Gakkaishi* **1988**, *34*, 910.

5. Yamawaki, T.; Morita, M.; Sakata, I. *Mokuzai Gakkaishi* **1991**, *37*, 449.
6. Yamawaki, Y.; Morita, M.; Sakata, I. *J. Appl. Polymer Sci.* **1990**, *40*, 1757.
7. Morita, M.; Higuchi, M.; Sakata, I. *J. Appl. Polym. Sci.* **1987**, *34*, 1013.
8. Brestkin, Y. V.; Volkova, L. A.; Kutsenko, L. I. Me'tser, Y. A.; Shepelevskii, A. A.; Frenkel, Y. *Polym. U.S.S.R.* **1986**, *28*, 35.

WOOD COMPOSITES (High Performance)

Shigehisa Ishihara
Wood Research Institute
Kyoto Institute

Wood is the most ancient but still the most widely used structural material in the world because it is comfortable and suitable for human life. However, because it is a natural cellular composite material of botanical origin, wood not only has many defects, such as knots, check, and decay, it is restricted in heavy construction because of the dimension and quality of the trunk. In order to avoid such defects and to enhance the quality and enlarge the dimension of wood as a raw material, wood composites or wood-based materials are developed.

GLUED-LAMINATED TIMBER (GLULAM)

Glued-laminated timber or glulam refers to two or more layers of wood glued together with the grain of all layers or laminations approximately parallel longitudinally. Glued-laminated timber for structural purposes is an engineered, stress-rated product comprised of assemblies of suitably selected and prepared wood laminations securely bonded together with approved adhesives. Durable synthetic resin adhesives such as phenolresorcinol, resorcinol, and melamine resin types are used for structural laminating.

The more laminae glued together, the glued-laminated timber improves structural performance or strength properties in laminating materials of laminae thickness, its structural and seasoning grades, and positioning or placement of the laminations in the member.

Advantages of glued-laminated timbers in heavy building constructions or bridges are endurance under outdoor exposure and design or stress for fastenings.

LAMINATED VENEER LUMBER (LVL)

Laminated veneer lumber (LVL) is manufactured by lamination of veneers into 2–10 cm in thicknesses of common solid sawn lumber. LVL, also known as parallel laminated veneer (PLV), has anisotropically superior physical properties such as the highest strength-to-weight ratio in tension among common structural materials and the highest dimensional stability among common wood composites.

LVLs have many applications. Because strength-reducing defects exist, LVLs increase the feasibility of light trusses and I-sections. Flanges of I-section beams, combined with steel-pipe lattice webs and Z-section beams, have a web composed of plywood or hardboard glued into a machine-grooved in the LVL flanges. I-section beams are widely used as joists and rafters in light-frame constructions. Open-web trusses of LVL steel pipe combinations are used as second floor joists, roof trusses in large buildings, and arch roof trusses of exhibition halls. LVLs also are used in the manufacture of housing components, furniture, and fitting industries, because of their ease of processing, uniform properties, strength, fire retardancy, extended service life, good adaptability to curved parts, small radius, stability, and better edge appearance. Frames of doors, and windows, arch door frames, box beams, door rails, staircases, chairs, beds, and counter tables are popular applications and uses of LVLs.

WOOD-BASED PARTICLE PANEL MATERIALS

Wood-based particle panel materials are classified under the generic term composite board, which includes particle-board, flakeboard, waferboard, oriented-strand board (OSB), and mineral-bonded board. In general, particleboard is used as a generic term for all particle panel products.

Particle panel products are defined as any wood-based panel product made of pieces of wood smaller than veneer sheets but larger than wood fiber. Sources of raw material include planer shaving, sawdust, plywood mill waste, and round logs or wood waste, such as branches, broken logs, and tops. Thermosetting urea-formaldehyde, urea-melamine-formaldehyde, and phenol-formaldehyde adhesives are the major types of binders used.

PLYWOOD

Plywood is a glued wood panel made of thin layers of wood veneer with the grain of adjacent layers at an angle, usually 90°. Each layer consists of a single thin veneer, or ply, or of two or more plies laminated together parallel with grain direction. The usual construction has an odd number of layers. This cross-lamination gives plywood its characteristics and makes it a versatile building material. Plywood is classified as construction, industrial, and decorative plywood.

Exterior plywood is bonded with waterproof (exterior) glues such as phenol-formaldehyde resin (PF) and melamine-based phenol formaldehyde resin (MPF). Interior plywood is bonded with interior glues such as urea-formaldehyde resin (UF) and urea-based melamine formaldehyde resin (UMF), intermediate glues such as UF and melamine-based urea formaldehyde resin (MUF), and exterior (waterproof) such as PF and MUF resins.

Plywood offers several advantages compared with solid wood of the same species. The superiority of plywood to solid wood of the same species or another wood composite is in its higher dimensional stability in varying climatic changes of moisture and temperature.

FIBERBOARD

Fiberboard is a category of the large class of reconstitution boards made from wood fiber or fiber bundles. The term fibers or fiber bundles refers to any wood or other lignocellulosic material of the same size as a wood cell, the biological building block of all cellulosic material. These boards in final form retain some of the properties of the original wood but due to manufacturing methods, gain new and different properties.

Most strength properties of fiberboard are strongly affected by raw material, board density, type of binder, binder content, and heat treatment, and therefore are controllable within the manufacturing process of all fiberboards related to hygroscopicity of the fibers in the wood cell wall which are much more difficult to control. Prominent consequences of moisture content fluctuations are dimensional changes and alterations of moist strength properties. Mechanical and physical properties of fiberboard are viscoelastic due to wood cell wall components and construction.

WOOD GRAFT COPOLYMERS

John J. Meister* and Meng-Jiu Chen
Department of Chemistry
Institute of New Mexico Mining and Technology

In our research, based on the successful graft copolymerization of lignin and ethene monomers, wood materials that contained lignin including wood pulp, wood veneer, and wood filament were grafted with styrene and methyl methacrylate.[2–9] The reaction was a free radical polymerization coinitiated by halide ions and a hydroperoxide in a polar aprotic solvent at 30°C. The results show that after the reaction, virtually all the wood materials have significant weight increases, the surface properties of the wood have been completely changed from hydrophilic to hydrophobic, and the Fourier transform infrared (FTIR) spectra of the unextractable products exhibit absorbance peaks characteristic of both wood and polymer. Grafting permanently attaches the polymerized monomer to wood, promotes the dispersity of wood fibers in a polymeric matrix, improves the adhesion of wood to thermoplastics, plasticizes the wood, and makes properties of the wood a combination proportioned according to the weight fractions of wood and sidechain in the grafted product. Grafting provides a promising method to chemically change the surface properties of wood and internally plasticize wood. Grafted wood can be used to make composites of wood and many hydrophobic polymeric matrixes without the use of coupling, adhesion promoting, or dispersing agents.

PREPARATION

Materials

Materials that can be grafted with this technology are portions of vascular plants. The lignin content of the wood pulps was 25–29%, and the lignin content of the wood filament was 16%. Ethene monomers used were styrene and methyl methacrylate.

Basic Reaction Rules

Yield, grafting, and properties of the product depend on both the reactivity of the wood material and the monomer. Different wood pulps had varied reaction activity in the copolymerization because they were made from different species of trees by different processes. In the copolymerization with wood, styrene is more reactive than methyl methacrylate.

PROPERTIES

Thermal Properties

The results of Thermogravimetric Analysis (TGA, Du Pont Instrument Model 951) show that below 200°C, wood is less thermally stable than polystyrene and poly(methyl methacrylate) thermal stability of the grafted wood pulps is always better than that of ungrafted wood pulps, and the more grafted polymer that exists, the better the thermal stability of the grafted wood pulps.

Mechanical Properties

The tensile strength, elongation, and modulus of the copolymerization products are higher than those of the polystyrene recovered from the copolymerization products by benzene extraction, and that as the content of wood pulp in the composites increases, the tensile strength, elongation, and modulus also increase.

APPLICATIONS

The major problem in preparing composites of wood fibers and thermoplastic polymers has been the poor interfacial adhesion between the hydrophilic wood fillers and the hydrophobic polymeric matrix.[10–12] A related problem in making wood-thermoplastic composites is the poor dispersion of wood in the polymeric matrix.[11,14]

An alternative method to overcome these problems is to graft short polymer segments onto the fiber surface.[1,10,11,13,15] In our research, grafted wood materials have been obtained. The copolymerization product of wood fibers is a thermoplastic composite material that contains homopolymer as a continuous matrix phase and uniformly dispersed grafted wood fibers as a reinforcing material. The mechanical properties of the composite materials are better than those of the homopolymeric matrix. The grafted wood fibers can be recovered by solvent extraction of the copolymerization product and the surface properties of the grafted wood fibers can be completely changed from hydrophilic to hydrophobic. Therefore, it is predictable that wood-thermoplastic composites with good mechanical properties can also be made by mixing the grafted wood fibers with commercial polymers, and the two major problems in making wood-thermoplastic composites, poor adhesion and poor dispersity, can be solved. This wood grafting technique provides a promising method to make wood-thermoplastic composites of good quality without the use of coupling, adhesion promoting, or dispersing agents.

REFERENCES

1. Abou-Zeid, N. Y.; Higazy, A.; Hebeish, A. *Die Angewandte Makromolekulare Chemie* **1984**, *121*, 69.
2. Meister, J. J.; Patil, D. R.; Channell, H. *J. Appl. Polym. Sci.* **1984**, *29*, 3457.
3. Meister, J. J.; Patil, D. R. *Macromolecules* **1985**, *18*, 1559.
4. *Renewable-Resource Materials: New Polymer Sources*; Meister, J. J.; Carraher, C. E.; Sperling, L. H., Eds.; Plenum: New York, 1986; pp 305–322.
5. *Lignin, Properties and Uses*; Meister, J. J.; Richards, E. G.; Sarkanen, S.; Glasser, W., Eds.; Symposium Series #397; American Chemical Society: Washington, DC, 1989; pp 58–81.
6. Meister, J. J.; Chen, M.-J. *Macromolecules* **1991**, *24*(26), 6843.
7. Meister, J. J.; Chen, M.-J. *J. of Applied Polym. Sci.* **1993**, *49*, 935.
8. Meister, J. J.; Chen, M.-J. *J. Environ. Polym. Degrad.* **1994**, *2*(2), 137–152.
9. Meister, J. J.; Chen, M.-J. *J. Wood Chem. Tech.* **1994**, *14*(3), 403–428.

*Author to whom correspondence should be addressed.

10. Raj, R. G.; Kokta, B. V.; Daneault, C. *Makromol. Chem., Macromol. Symp.* **1989**, *28*, 187.
11. Maldas, D.; Kokta, B. V. *Reinforced Plastics and Composites* **1991**, *10*, 42.
12. Woodhams, R. T.; Thomas, G.; Rodgers, D. K. *Polym. Eng. Sci.* **1984**, *24*, 1166.
13. Dalvag, H.; Klason, C.; Stromvall, H.-E. *Intern J. Polym. Maters.* **1985**, *11*, 9.
14. Selke, S. E.; Nieman, K.; Childress, J.; Keal, M.; Bimpson, R. *Advanced Technology Applications to Eastern Hardwood Utilization*; M.S.U.: Lansing, MI, May 30–31, 1990.
15. Kokta, B. V.; Dembele, F.; Daneault, T. C. *Polym. Inter. Sci. Eng.* **1985**, *52*, 99.

WOOD-PLASTIC COMPOSITES

K. M. Idriss Ali* and Mubarak A. Khan
Radiation and Polymer Chemistry Laboratory
Institute of Nuclear Science and Technology
Bangladesh Atomic Energy Commission

In a real sense, modification of various physical and mechanical properties of wood and its products by chemical treatment started in the 1930s.[1] But the momentum gained after World War II when there was a real break-through of scientific activities in almost all fields. There are many countries in the world where there is an acute dearth of high quality timber and low grade wood is being used as a substitute. As a result, quality products with the desired properties are seldom available. All these accumulated factors led to the development of wood-plastic composites.

ANATOMY OF WOOD AND ITS DRAWBACKS

Wood is a natural product composed of a highly polymerized carbohydrate (cellulose, 50–60%), lignin (25%), solid material (24%), gums, resins, and so on, with 1% ash. Though the versatility of wood is well recognized, there are some inherent characteristics that tend to restrict its versatile and potential applications. The most important drawback is the dimensional instability at variations of moisture contents, resulting in contractions and expansions in some anatomical directions of wood, which ultimately lead to warping and asymmetrical distortions. Wood has also relatively low resistance to abrasion and weather conditions. It tends to decay in the presence of fungi, bacteria, microbes, and insects and it is easily combustible.

PLASTIC AND ITS RELATION TO WOOD

The combination of wood and plastic makes a very promising product for industrial application. By combining plastic with wood through chemical treatment, radial and tangential strengths can be enhanced to the same degree as that of its longitudinal strength, which is the real unilateral strength of the wood. The combination also makes the product more resistant to abrasion, biological attack, and weather conditions. It also improves its electrical and thermal insulating capabilities for probable use as insulators. The best way to make the combination is to alloy or impregnate the building blocks of plastic, that is, monomer molecules, into the void spaces of wood and then to polymerize them among the fibers of wood and even in the saturated cell walls.

REVIEW OF THE INITIAL WORKS ON WPC

Pioneering work on wood-plastic composite (WPC) was done by the Brookhaven National Laboratory. The first report appeared in 1956 on this topic gamma radiation on the polymer formation in the wood matrix.[2–4] Different types of wood substrates and monomers were used to find the best types for suitable production of WPC for use in some specific applications. Research and development activity on wood-plastic composite not only flourished in the U.S., but also took place in other parts of the world.

MECHANISM OF WPC

In general, wood-plastic composite formation is thought to occur through free radical reactions leading to polymerization. The free radical can be generated through the reaction of a heat-sensitive catalyst, as well as by the Co-60 gamma radiation.

RECENT TRENDS IN WPC

Having established the preparation mode of the wood-plastic composite using vinyl monomers, particularly methyl methacrylate (MMA), styrene (ST), vinyl acetate (VAc), etc., under ionizing radiation at competitively reasonable production costs, attention was obviously drawn towards improvement of the characteristic properties of the composites. As such, various impregnating solutions were evolved, i.e., vinyl monomer to which other monomers, oligomers, organic and inorganic salts, and acids were added as additives and co-additives in order to develop specific properties. These included fire retardancy, insect repellency, microbe resistance, chemical corrosion resistance, abrasion resistance, high compact, tensile and bending strengths, enhanced hardness, and toughness so that these composites can be used for desired applications.

OTHER METHODS OF POLYMERIZATION

In place of gamma irradiation, many used a high-energy electron beam as a source of polymerization of the impregnating solution into the wood substrates.[6–10] Not only the electron beam was used as a means of radiation for polymerization; sometimes plasma ray treatment was also adopted.[9]

GRAFTING

During wood-plastic composite formation, polymerization occurs mainly in the void spaces of the wood substrates. It was observed that copolymerization also occurred between the monomer unit and the cell-wall structure under the electron beam irradiation.[6,7]

THERMAL POLYMERIZATION

Wood-plastic composite preparation by the application of high-energy radiation (gamma ray, electron beam, plasma treatment, etc.) involves high initial cost. The thermal polymerization process of impregnating the solution into the wood matrix is an alternative method.

*Author to whom correspondence should be addressed.

Among many different vinyl monomers used during the 1970s and 1980s, MMA appears to be the preferred monomer for both the radiation and thermal (catalyst-heat) processes.[11] Sometimes the MMA and ST mixture is used for the composite preparation by the heat-catalyst method.

CROSSLINKING EFFECT

The crosslinking agent has an advantage in that this forms a gel at the beginning of the polymerization so that the monomer cannot expand out of the wood matrix, owing to the rise of temperature caused by exothermic reaction. The crosslinking agents induce a high temperature gradient so that the polymer does not melt during the sanding process and, thus, it prevents loading of sand papers.[13]

ROLE OF ADDITIVES AND CO-ADDITIVES

In the pursuit of versatile and diversified applications of wood-plastic composites prepared with soft- and hardwoods, low grade and high quality timbers, in the presence of MMA, ST, or any other suitable monomers, various additives and co-additives, such as monomers, oligomers, organic and inorganic salts, dyes, and acids, are being used in minute amounts. This is done to induce specific characteristics in the composites to enhance different tensile strengths and hardness, fire retardancy, insect and microbe repellancy, water and moisture absorption resistance, and coloring of the end products.

Many workers used several different monomers to make wood-plastic composites.[11,12]

Sometimes special chemicals are incorporated in the impregnating solution to develop certain characteristics for the composites. Recently, a minute amount (1% v/v) of mono-, di-, and tri-functional monomers, acid, copper sulfate, lithium nitrate, urea, epoxy, polyester, and urethane oligomers were used with a series of bulk monomers mixed with the swelling agent, methanol, for the preparation of a wood-plastic composite with simul wood (soft wood).[14,15]

WATER-SOLUBLE MONOMER

Conventional monomers for wood-plastic composite are MMA, ST, and other vinyl monomers that are generally insoluble in water. Because the wood usually contains 6–10% moisture, and swelling of the cell wall structure with a swelling solvent reduces the heat of polymerization, resulting in better composite formation, it is desirable to search for a water-soluble monomer that can be used for some specific purposes. Khan et al. prepared wood-plastic composites with simul wood and acrylamide (AM), a water-soluble monomer at all proportions, in the presence of a series of additives and co-additives.[16,17] These experiments opened a new era of wood-plastic composites for future generations.

ELECTRICAL AND THERMAL PROPERTIES

Electrical and thermal behaviors of a material are very much dependent on the nature of conduction ability. Plastic and dry wood are good insulators. Wood-plastic composite prepared with wood and MMA, butyl methacrylate, MMA-ST, and styrene-acrylonitrile showed that conductivity (reciprocal of resistivity) of the composite was much lower than that of the untreated parent wood. However, the conductivities of both treated and untreated wood samples increase with moisture content and, at the fiber saturation point (30%), the increase is very minimum and attains a plateau.[10,18]

The composites are normally burned with a greater generation of smoke or with a candle-like localized flame, depending on the types of wood and the impregnating solutions.

CHEMICAL CORROSION

Generally, wood-plastic composites are more resistant to chemical corrosion, compared to the parent wood.

APPLICATIONS

The cost of the wood-plastic composite is, undoubtedly, more than that of the parent wood. But the composite is essential in some specific applications where the untreated wood is not very successful. This is because of the enhanced and improved physical, chemical, and mechanical properties attained by the composite over the normal wood. Moreover, since the composite has high-impact resistance, compression strength, hardness, and abrasion resistance, it can be very successfully used as flooring material in public places like churches, shopping malls, schools, pavilions, and fairs with minimum cost of maintenance. This material can have potential applications in such broad areas as construction materials, (flooring on basketball courts, public places, parquet, stair treads, railway sleepers, handrails, kitchen cabinets, pillars and beams of rural huts, boats, and window sashes and frames), as furniture articles, (tables, chairs, bedsteads, benches, indoor panels and decorative, and tools) as industrial materials (wooden molds, textile shuttles, pegs, electrical motor parts, tool handles, mountings, knobs, brushes); as sporting goods, (handles of rackets, skis, baseball bats, billiard sticks, and hockey sticks) and as household items (kitchen articles, like salad bowls, spoon handles, knife handles, cutting boards), and rulers, musical instruments, and toys.

REFERENCES

1. Meyer, J. A. *Chemistry of the Solid Wood*; Rowell, R., Ed.; Advances in Chemistry Series 207; American Chemical Society: Washington, DC, 1984; Chapter 6.
2. Ballantine, S.; Manowitz, B. *USAEC Report BNL 389*; P. 19, Brookhaven National Laboratory: May 1956.
3. Karpov, V. L.; Malinsky, Y. M.; Serenkov, V. I.; Klimanova, R. S.; Freidin, S. A. *Nucleonics* **1960**, *18*(3), 88.
4. Kenaga, D. L.; Fennessey, J. P.; Sannett, V. T. *Forest Prod. J.* **1962**, *12*(4), 161.
5. Iannazzi, F. D.; Levins, P. L.; Perry, Jr., F. G.; Lindstrom, R. S. *US-AEC Report TID 21434*; Arthur D. Little: September 1, 1964; 108.
6. Yoshizawa, S.; Hemda, T.; Fukuoka, M.; Hashizamic, Y.; Takamuram, T. *Radiat. Phys. Chem.* **1981**, *18*(5–6), 1185.
7. Takashi, H.; Was, S. I.; Yoshinori, H. *Org. Coat. Plast. Chem.* **1981**, *45*, 375.
8. Handa, T.; Sea, I.; Ishii, T.; Hashizume, Y. *Polym. Sci. Technol.* **1983**, *20*, 67.
9. Cho, D. C.; Sjooblom, E. *Polym. Material Sci. & Eng. Proceeding of ACS Division of Polymeric Materials Sci. & Eng.*; 62 Spring Meeting: Boston, Massachusetts, 1990.
10. Khan, M. A.; Ali, K. M. I.; Wang, W. *Radiat. Phys. Chem.* **1991**, *38*(3), 303.

11. Langwing, J. E.; Meyer, J. A.; Dawidso, R. W. *For. Prod. J.* **1969**, *19*(11), 57.
12. Khan, M. A.; Ali, K. M. I.; Ahmad, M. U. *J. Appl. Polym. Sci.* **1992**, *45*, 2113.
13. Meyer, J. A. *For. Prod. J.* **1968**, *18*(5), 89.
14. Ali, K. M. I.; Khan, M. A.; Husain, M. *Polym. Plast. Technol. Eng.* **1994**, *33*(4), 477.
15. Khan, M. A.; Ali, K. M. I. *J. Appl. Polym. Sci.* **1993**, *49*, 1989.
16. Khan, M. A.; Ali, K. M. I. *Radiat. Phys. Chem.* **1992**, *40*(6), 433.
17. Ali, K. M. I.; Khan, M. A.; Husain, M. *Radiat. Phys. Chem.* **1994**, *44*(4), 421.
18. Yap, M. G. S.; Que, Y. T.; Chia, L. H. C.; Chem, H. S. O. *J. Appl. Polym. Sci.* **1991**, *43*, 2057.

WOOD-POLYMER COMPOSITES

Takashi Taniguchi
Faculty of Education
Niigata University

Keizo Okamura
Faculty of Agriculture
Kyoto University

DEFINITION

Wood-polymer composite (WPC) is a wood impregnated with polymers in order to strengthen the properties of the natural wood. In the narrow sense, WPC is a composite made by impregnating wood with a polymerizable monomer (mainly vinyl-monomer) or prepolymer, and polymerizing them to a solid.

During the past three decades, researchers have impregnated wood with a variety of chemicals to produce WPC and investigated their properties. There have been many interesting review articles about WPC materials, processes, properties, and products.[1–10]

MONOMERS

The most widely used industrial monomers, such as methyl methacrylate (MMA) and styrene (St) are the cheapest WPC base chemicals and are easy for engineering. The lower the viscosity of monomer is the easier the impregnation becomes. To a monomer solution, soluble, light-stable dyes can be added to color the final WPC. The color emphasizes the grain structure of the particular species. The third component, chemicals such as a crosslinker, an absorbent of ultraviolet rays, a flame retarder, and preservatives can be added to improve the properties of WPC and the organic solvent for swelling the cell wall to control the impregnation area of monomers.

PROPERTIES OF WPC

Mechanical Properties

The most pronounced mechanical properties improved by WPC are compression strength and hardness. These properties are improved 4 to 5 times that of untreated samples, if attentive selection is made to the wood species and monomers. This is the reason that the first commercial use of WPC was flooring. The properties of WPC are improved 1.5 to 2 times for tension, bending, and impact strengths, compared to the untreated wood.

Dimensional Stability

The introduction of polymers generally reduces swelling and shrinkage from water, and promotes dimensional stability. WPC especially, shows more resistance to water-absorption.

Manufactured Properties

WPC has a fine finishing surface from sanding or buffing without surface coating. Woodworking machines cannot be used well; metal- or plastic-working machines are better suited for WPC.

Decay Resistance

WPC shows an increase in decay resistance. The rate of weight loss of white rot and brown rot decreases accompanied by the increase of polymer loading.

APPLICATIONS

Commercial production of WPC began in the mid-1960s using the radiation process in the U.S. and in the early 1970s using the catalyst-heat process in Japan. Now there are many companies, more than five in the U.S., more than six in Japan, that produce WPC products, mainly flooring. The former is the flooring for heavy duty walking resistant to footwear, and the latter is mainly for light duty walking. If the cost of goods installation, easy maintenance, and long-life of WPC flooring is considered, the total cost of WPC flooring becomes one of the least expensive.

WPC IN THE FUTURE

In Japan and the U.S., many companies have produced the WPC laminated flooring by the catalyst-heat process. The lamination of the thin WPC veneer and wooden-board will grow to a larger market. It will be possible to get new materials that have the same properties of WPC with low polymer loading, changing the kind of polymer and increasing the polymer content of present wooden boards, such as particleboard and waferboard. Wood, as a renewable resource, has environmental advantages and better balanced properties, compared with other synthetic materials. Further, WPC has joined its properties with the merit of long-life usage. Future use of WPC will depend upon the new creation by the producers and the needs of the market.

REFERENCES

1. Meyer, J. A. *Polym. Plast. Technol. Eng.* **1977**, *9*(2), 181.
2. Meyer, J. A. *Wood Sci.* **1981**, *14*(2), 49.
3. Meyer, J. A. *Forest Prod. J.* **1982**, *32*(1), 24.
4. Meyer, J. A. *The Chemistry of Solid Wood*; In Rowell, R. M., Ed.; **1984**, 257.
5. Meyer, J. A. *Technologist* **1987**, *1*(3), 4.
6. Schneider, M. H. *Wood and Fiber Sci.* **1994**, *26*(1), 142.
7. Murayama, T. *Plastic Age (Japan)* **1978**, *24*(10), 95.
8. Gotoh, T. *Industrial Materials (Japan)* **1978**, *26*, 101.
9. Taniguchi, T.; Nakato, K. *Polymer Digest (Japan)* **1982**, *34*(10), 2.
10. Taniguchi, T. *Wood Industry (Japan)* **1993**, *48*(7), 304.

Wool

See: Wool (Overview)
Wool Grafting
Wool Keratin

WOOL (Overview)

Anthony P. Pierlot
CSIRO Division of Wool Technology

STRUCTURE

Wool, the fiber obtained from the fleece of sheep, has a large variation in length and diameter depending on breed, health, and diet. With annual shearing, the fiber length varies from about 300 mm for coarse wool (40 μm diameter) to 70 mm for the finest merino wool (16 μm diameter).

Wool belongs to a group of proteins known as hard keratin. Unstretched wool gives an α-diffraction pattern, while stretched fibers show a β-pattern. At the microscopic level, wool exhibits a well-defined hierarchical structure with many discrete components ranging from the molecular level to the macroscopic fiber. A more detailed description of the structure of the wool fiber and of the protein chains is discussed.[1–3]

FIBER MECHANICAL PROPERTIES

Wool fibers absorb moisture from the surrounding air to reach an equilibrium that depends on the relative humidity of the environment. This property, which is highly desirable in textiles, has been attributed to the amorphous regions of the fiber. Fibers absorb about 15% moisture at 65% relative humidity and have a saturated mass increase of about 35%.[4]

Although wool has a complex structure, a two phase-composite model can largely account for the mechanical properties of the fiber.[5,6]

Summary

A model with two phases, consisting of a crystalline phase with low water penetrability oriented parallel to the fiber axis embedded in an amorphous, water-sensitive matrix, can generally describe the mechanical properties of the wool fiber. A glass transition exists in the amorphous matrix that results in a large change in the properties of this phase with either temperature or water content. Unfortunately, as knowledge of the bulk mechanical properties is generally insensitive to events occurring at the molecular level, within and between the two phases, no one particular molecular model has gained wide acceptance.

TEXTILE APPLICATIONS OF FIBER PROPERTIES

The water-sensitive nature of the T_g of wool and the low wet-transition temperature allows easy transition of the fiber matrix regions between the two sites during processing, wear, laundering, and pressing. This has important consequences, not only in fabric performance, but also in fiber processing.

Pleats in wool that are durable to laundering and pressing can be obtained by heating wool to a temperature about 100°C above the T_g.

Wool's fiber properties, for example, moisture absorption, are often utilized in apparel fabric design. Sometimes, however, when worn in contact with the skin, wool results in the unpleasant sensation of prickle and itch. This response to wool has sometimes wrongly been assumed by the consumer to be an allergic reaction. True allergies to wool are extremely rare and this sensation is due to mechanical irritation by coarse fiber ends and not from immunological factors.

NON-APPAREL APPLICATIONS

Apart from its traditional use for clothing and floor coverings, the wool fiber itself can now be found in applications as diverse as thermal insulation, weed mats, and as a means to soak up oil spills. Wool has been used for thermal insulation in houses in Switzerland for many decades and is being increasingly used in Australia for both homes and commercial buildings. Wool is naturally fire-resistant and the increased use is being driven by consumer demand for natural products in their homes.

Wool has a hydrophilic interior, but hydrophobic surface, and this natural affinity for oil makes it an ideal material for cleaning up oil spills because it can adsorb up to 40 times its own weight in oil.[7]

ACKNOWLEDGMENT

The author wishes to thank colleagues within the Division for many helpful discussions and acknowledges the financial assistance of Australian Wool Growers and the Australian Government through the International Wool Secretariat.

REFERENCES

1. Fraser, R. D. B.; MacRae, T. P.; Rogers, G. E. *Keratins—Their Composition, Structure and Biosynthesis*; C. C. Thomas: Springfield, 1972.
2. Rogers, G. E.; Reis, P. J.; Ward, K. A.; Marshall, R. C., Eds. *The Biology of Wool and Hair*; Chapman and Hall: London, 1989.
3. *Cellular and Molecular Biology of Intermediate Filaments*; Goldman, R. D.; Steinert, P. M., Eds.; Plenum: New York, 1990.
4. Watt, I. C.; D'Arcy, R. L. *J. Text. Inst.* **1979**, *70*, 298.
5. Feughelman, M. *Encycl. Polym. Sci. Eng.* **1987**, *8*, 566.
6. Hearle, J. W. S.; Susutoglu, M. *Proc. 7th Int. Wool Text. Res. Conf.* Tokyo **1985**, *1*, 214.
7. Fisher, G. *Tech. Text. Int.* February 8, 1993.

WOOL GRAFTING

Ian M. Russell* and Julie M. Kure
CSIRO Division of Wool Technology
Department of Applied Chemistry, RMIT

Anthony P. Pierlot
CSIRO Division of Wool Technology

Robert A. Shanks
Department of Applied Chemistry, RMIT

The blending of synthetic fibers with wool has been widely adopted by the textile industry as a means by which the desirable characteristics of both fibers can be incorporated into a fabric or, at times, as a cost reduction technique. An alternative approach that has been considered, at least in research laboratories, has been to modify the properties of the wool fiber by grafting the synthetic polymer into the wool peptide chains.

*Author to whom correspondence should be addressed.

Experience in grafting of polymers into wool has generally shown that large additions are necessary to achieve any appreciable change in physical property. These large additions can be readily achieved, but in most cases, the resultant fiber has shown properties inferior to the original wool fiber.

METHODS OF GRAFTING

Polymer Addition

Initiation of polymerization in wool has usually been achieved by irradiation or by allowing the fibers to absorb a chemical initiator before or during polymerization.[2] Thermally induced (spontaneous) grafting has also been achieved. Grafting both onto the fiber surface and into the fiber are possible. Watt and Arai give excellent reviews on the chemicals used for both internal and external applications to wool, while Nayak gives a detailed investigation of methods and mechanisms of grafting.[3–5]

PROPERTIES

Fiber modulus and some performance properties, such as wrinkling, deteriorate when wool is wet. Part of the benefit from blending of wool with hydrophobic fibers such as polyester relies on improving wear characteristics at high moisture levels. Grafting of polymer into wool under dry conditions has been attempted to create a new network that has reduced swelling when wet.[8]

Stress–strain curves have been examined for wool-polyacrylonitrile copolymers. Wolfram and Menkart found that grafting increased the breaking strength and elongation at break of dry fibers.[6]

Cross-section analysis of fibers grafted with 100% polystyrene showed irregular peripheral structures with cracks indicating damage to the structure of wool.[1] The heat settability of the copolymer improved, but poor crease recovery was obtained.

Internal grafting with linear polymers decreased the wrinkle recovery of wool most likely as a result of the incorporation of soft, low T_g polymers or damage to the crystalline regions.[7,9] Arai and Wasley have given excellent reviews on the properties of grafted polymers.[4,10]

ACKNOWLEDGMENTS

The authors wish to thank colleagues within the Division for many helpful discussions and acknowledge the financial assistance of the Australian Wool Growers and the Australian Government through the International Wool Secretariat. JMK acknowledges financial assistance through a scholarship from the International Wool Secretariat.

REFERENCES

1. Ishibashi, H.; Oku, M. *3rd Int. Wool Text. Res. Conf.* Paris, **1965**, *III*, 385.
2. Watt, I. C. *J. Macromol. Sci.—Chem.* **1970a**, *A4*, 1079.
3. Watt, I. C. *J. Macromol. Sci.—Revs. Macromol. Chem.* **1970b**, *C5*, 175.
4. Arai, K. In *Block and Graft Copolymerization*; Vol. I, Ceresa R. J.; Ed., John Wiley & Sons: London, 1973.
5. Nayak, P. L. *J. Macromol. Sci.—Rev. Macromol. Chem.* **1976**, *C14*, 193.
6. Wolfram, L. J.; Menkart, J. *Amer. Dyestuff Reptr.* **1967**, *56*, 110.
7. Russell, I. M.; Evans, D. J. *J. Appl. Poly. Sci.* **1990**, *40*, 1951.
8. Watt, I. C.; Leeder, J. D. *3rd Int. Wool Text. Res. Conf.*; Paris, **1965**, *III*, 39.
9. Jones, F. W.; Leeder, J. D. *J. Text. Inst.* **1974**, *65*, 627.
10. Wasley, W. L. In *Block and Graft Copolymerization*; Vol. I; Ceresa, R. J., Ed.; John Wiley & Sons: London, 1973.

WOOL KERATIN

Yoshiharu Nakamura
Faculty of Engineering
Fukui University

Keratins are a generic term referring to insoluble proteins that construct the bulk of the horny layer of the epidermis, covering the body surface of higher vertebrates and of epidermal appendages, such as hair, nails, claws, and horns of mammals, scales of reptiles, and feathers of birds. The chemical composition of keratins is specialized by containing the diamino acid residue cystine, in a considerable amount, in their polypeptide chains. As a result, polypeptide chains can be crosslinked. This distinguishes keratins from other fibrous, structural proteins and endows them with durability with respect to physical and chemical attack, except for encounters with reducing or oxidizing agents. Details are referred to in the literature.[1–3]

MORPHOLOGICAL STRUCTURE OF KERATIN FIBERS

The tertiary structure of polypeptides composing keratins and the morphological structure of keratin fibers have been studied by a number of researchers by means of X-ray diffraction, infrared spectroscopy, and electron microscopy. At high angle, X-ray diffraction measurements unstretched wool fibers give a characteristic pattern, called the α-pattern, whereas stretched fibers give a different pattern, the β-pattern. Some keratins, particularly feathers, project the β-pattern, even in the unstretched condition.

The Bendit and Feughelman's review on morphological structures together with mechanical properties can be, even now, helpful.[5] On the morphological structure of the wool surface, Makinson's book and Leeder's review, should be consulted.[4,6]

MECHANICAL PROPERTIES OF KERATIN FIBERS

The mechanical properties of keratin fibers are remarkably affected by the relative humidity, temperature, and duration at which the fibers were tested. Keratin fibers are characterized by being able to recover their mechanical properties if the extension is no farther than the yield region.

WOOL FIBERS AS AN ECOMATERIAL

Wool, a natural fiber, is a renewable resource, and can be recycled. Because wool keratin is biodegradable in spite of being insoluble in water, it is also regarded as an ecomaterial. It is, in addition, possible to increase its ecological value by carrying out chemical modification to the wool fiber that contains functional side groups attributing to acidic, basic, and polar amino acid residues.

In everyday living conditions, it is recognized that ammonia, amine, and mercaptan gases are representative among various odors. Particularly, the odor of the mercaptan is the strongest among them. For the decomposition of mercaptan

X

X-ray Contrast Polymers

See: Radiopaque Polymers

XANTHAN, DEPOLYMERIZATION

Bjørn E. Christensen* and Olav Smidsrød
Norwegian Biopolymer Laboratory
Department of Biotechnology
University of Trondheim, NTH

Bjørn T. Stokke
Norwegian Biopolymer Laboratory
Department of Physics and Mathematics
University of Trondheim, NTH

Xanthan is a high molecular-weight, water-soluble, bacterial polysaccharide. It is built up of pentasaccharide repeating units, and the backbone, containing β-1,4-linked glucose residues, corresponds to a cellulosic chain. However, the presence of trisaccharide side-chains on every second backbone residues strongly modulates the polymer properties compared to cellulose, by ensuring solubility in aqueous solutions. The side-chains may contain substituents like *O*-acetate and pyruvate diketal. The latter contributes to the polyelectrolyte character in addition to the carboxylate group of the glucuronic acid.

Compared with a wide range of water-soluble polymers, xanthan is particularly stable toward degradation.[1] This is the basis for selecting xanthan as viscosifyer when long term stability is required, particularly in oil field applications. The enhanced stability of xanthan as compared with other water-soluble polymers is clearly associated with the ordered, double-stranded form.[2–5] For this reason a high ionic strength (e.g., sea water) can be used to maintain T_m at values well above 100°C, and thus keep xanthan in its ordered state at such temperatures.

The enhanced stability of xanthan can qualitatively be attributed to the double-stranded structure. The main feature is that a double-stranded structure may tolerate several breaks in the polymer backbone without a correspondingly large decrease in molecular weight due to interchain, non-covalent bonds, which to a certain extent may prevent strand separation and a break of the duplex.

In recent papers the depolymerization of xanthan by acid hydrolysis has been investigated.[4,6–8] In addition, xanthan can be effectively depolymerized by free radical mechanisms (also called oxidative-reductive depolymerization), for example, in the presence of H_2O_2 and Fe^{2+}, as well as by other methods.[5] As a general rule, the response of xanthan toward depolymerizing agents is largely governed by two interdepending factors, namely the conformational state and the structure of the side chains. The latter becomes important because changes in the chemical structure may in principle influence the conformational state. In the case of xanthan it was found, however, that the conformational properties were basically unmodified during acid hydrolysis even though quite severe changes occurred in the side chains (total removal of the terminal β-1,4-linked mannose residues and up to 46% removal of the entire side-chains).[4,6–8] Hence, the depolymerization behavior is largely determined by the extent of chain cleavage of the cellulosic backbone, in addition to external conditions influencing the order–disorder characteristics.

*Author to whom correspondence should be addressed.

DEGRADATION MECHANISM

Acid Hydrolysis

Depolymerization of polysaccharides proceeds through cleavage of glycosidic linkages. These are, in most cases—including xanthan—formed by a reaction between a hemiacetal (the sugar ring) and a primary or secondary hydroxyl group. The corresponding acetals (glycosides) can be hydrolyzed by an acid-catalyzed mechanism[9] Different glycosides vary considerably in their susceptibility toward acid hydrolysis. This is also seen in xanthan. When degraded in moderately strong acids (pH 1–4) the most labile glycosidic linkage is that between the terminal β-D-mannose and the glucuronic acid.[4]

Organic substituents found in xanthan, i.e., the *O*-acetate and pyruvate diketal, particularly the latter, are rapidly hydrolyzed in acid. This permits the preparation of acetate- and pyruvate-free xanthan with high molecular weight.

Oxidative-Reductive Depolymerization (ORD)

It is well known that polysaccharides can be depolymerized in the presence of reducing agents and molecular oxygen.

Xanthan is readily depolymerized by ORD.[5] Compared with acid hydrolysis, the extent of reaction in the side-chains is smaller at the same degree of depolymerization (unpublished results). ORD can more easily be performed in different conformational states by varying the temperature and/or the ionic strength.[5]

Enzymic Depolymerization

The only enzymes that seem to depolymerize xanthan to a reasonable extent are cellulases. They degrade xanthan only in its disordered conformation, typically in salt free solutions at 50°C.[13–15]

Mechanically Induced Depolymerization

Polymer depolymerization can also be induced by mechanical means. Both ultrasonic treatment or high shear rates can be utilized to obtain depolymerization. Depolymerization of xanthan by ultrasound has been used to produce molecules of lowered M_w for the purpose of performing physical studies in aqueous solution.[16–18] The shear rates used for mechanical depolymerization are typical in the order of 10^6 s^{-1} which is higher than commonly encountered shear rates during an injection process.[19] One of the reasons for using xanthan in connection with production of oil is its stability against mechanical degradation.[20] Xanthan also shows better performance than synthetic polymers in drag-reduction because of its resistance against mechanical degradation.[21]

ACKNOWLEDGMENT

This work was supported by VISTA (grant V6312), Statoil and by the Research Council of Norway (grant BT27472).

REFERENCES

1. Davison, P.; Mentzer, E. *J. Soc. Pet. Eng.* **1982**, *22*, 353.
2. Ash, S. G.; Clarke-Sturman, A. J.; Calvert, R.; Nisbet, T. M. *J. Soc. Pet. Eng.*, 1983; paper 12085.
3. Seright, R. S.; Henrici, B. J. *J. Soc. Pet. Eng.*, 1986; paper 14946.
4. Christensen, B. E.; Smidsrød, O. *Carbohydr. Res.* **1991**, *214*, 55.
5. Hjerde, T.; Kristiansen, T. S.; Stokke, B. T.; Smidsrød, O.; Christensen, B. E. *Carbohydr. Polym.* **1994**, *24*, 265.
6. Christensen, B. E.; Knudsen, K. D.; Smidsrød, O.; Kitamura, S.; Takeo, K. *Biopolymers* **1993a**, *33*, 151.
7. Christensen, B. E.; Smidsrød, O.; Elgsaeter, A.; Stokke, B. T. *Macromolecules* **1993b**, *26*, 6111.
8. Christensen, B. E.; Smidsrød, O.; Stokke, B. T. *Front. Biomed. Biotechnol.* **1993c**, *1*, 166.
9. BeMiller, J. N. *Adv. Carbohydr. Chem.* **1976**, *22*, 25.
10. Smidsrød, O.; Haug, A.; Larsen, B. *Acta Chem. Scand.* **1963**, *17*, 2628.
11. Smidsrød, O.; Haug, A.; Larsen, B. *Acta Chem. Scand.* **1965**, *19*, 10.
12. Herp, A. *The Carbohydrates*; Academic: New York, 1980; Vol. Ib, pp 1276–1287.
13. Rinaudo, M.; Milas, M. *Int. J. Biol. Macromol.* **1980**, *2*, 45.
14. Sutherland, I. W. *Carbohydr. Res.* **1984**, *131*, 93.
15. Cheetham, N. W.; Mashimba, E. N. M. *Carbohydr. Polym.* **1991**, *15*, 195.
16. Paradossi, G.; Brant, D. A. *Macromolecules* **1982**, *15*, 874.
17. Sato, T.; Norisuye, T.; Fujita, H. *Macromolecules* **1984**, *17*, 2696.
18. Milas, M.; Rinaudo, M.; Tinland, B. *Carbohydr. Polym.* **1986**, *6*, 95.
19. Kojima, T.; Tabata, K.; Ikumoto, T.; Yanaki, T. *Agric. Biol. Chem.* **1984**, *48*, 915.
20. Sorbie, K. S. *Polymer-Improved Oil Recovery*; Blackie: Glasgow, 1991.
21. Rochefort, S.; Middleman, S. *Polymer flow interaction. AIP Conference Proceedings*; Rabin, Ed., **1985**, *137*, 117.

Xanthan Gum

See: *Gums (Overview)*
Water-Soluble Polymers (Oil Recovery Applications)
Water-Soluble Polymers (Rheological Properties)
Welan Microbial Polysaccharide
Xanthan, Depolymerization
Xanthan Gum (Overview)
Xanthan Gum (Conformational and Solution Properties)

XANTHAN GUM (Overview)

Francis X. Quinn
Institut Français du Pétrole
Rueil-Malmaison

Xanthan gum is the generic name given to the anionic water-soluble extra-cellular hetero-polysaccharides produced by bacteria of the genus *Xanthomonas*. Synthesized in bulk by culture fermentation, xanthan gum is most commonly derived from *Xanthomonas campestris*, where the repeat unit consisting of a linear backbone of (1 → 4)-β-D-glucose residues has a trisaccharide side group attached at C3 of every alternate glucose residue.

The weight-average molecular weight (M_w) of xanthan is generally in the range of 2×10^6 to 6×10^6, approximately 2000 to 6000 repeating units. Native xanthan adopts a double-helical conformation. Solutions above a threshold concentration from hydrogels that exhibit a gel-sol transition. Meso-phase and liquid crystalline structures can be observed in aqueous xanthan systems over a wide range of water contents.

Xanthan increases the viscosity of aqueous solutions even at low polysaccharide concentrations. Although the viscosity of these solutions is almost independent of temperature over a broad temperature range, pseudoplastic flow (or shear thinning) is observed at high shear rates. Owing to these unique physical properties, xanthan gum is widely used in food and industrial applications as a thickening, suspending, and gelling agent, as well as a stabilizer for emulsions and dispersions.

DEGRADATION

Xanthan degradation can be divided into two categories: enzymatic and chemical. Degradation is generally monitored by measuring the reduction in solution viscosity and by analyzing the liberated products using techniques such as HPLC, UV absorbance, and mass spectroscopy.

Enzymatic Degradation

Extra-cellular polysaccharides are substrates for degradative enzymes where the enzymes exhibit a high degree of substrate specificity.[1] A number of enzymes known to degrade xanthan gum (xanthanases) have been isolated from soil, water, and sewage enrichment cultures.

Chemical Degradation

The thermal stability of xanthan is largely determined by the conformation of the polymer, and, thus, by the temperature, pH, and salinity of the solution. At 90°C in 1 g/l aqueous NaCl xanthan degrades rapidly (<2 days). Degradation is progressively inhibited by increasing the salt concentration up to 50 g/l where degradation at that temperature is no longer observed. The presence of divalent cations reinforces high-order structures further improving resistance to thermal degradation. Xanthan degradation can occur through acid or base catalyzed hydrolysis, resulting in the loss of acetyl and pyruvate substituents, and cleavage of glycosidic linkages.

STRUCTURE

Following biosynthesis and excretion from *Xanthomonas campestris*, xanthan adopts a double-helical conformation. X-ray fiber diffraction studies reveal that the polysaccharide forms a 5_1 double-helix of diameter 2.5 nm and pitch 4.7 nm, where the anti-parallel configuration is favored over a parallel arrangement of the strands.[2] The trisaccharide side-chain is folded along the backbone, this configuration being stabilized by a number of intramolecular hydrogen bonds. This double-helical organization is very rigid, and measured values of the xanthan persistence length (the tendency to maintain linearity) in this conformation vary from 120–150 nm, compared to 60 nm for DNA.

The so-called order-disorder transformation of xanthan observed as a function of temperature or salinity has been the focal point of much debate, having been studied using a broad selection of experimental techniques.

Cations, especially Ca^{2+}, promote xanthan aggregation and gelation even at low cationic concentrations. The integrity of xanthan hydrogels is maintained by junction zones composed of oriented bundles of double-helices where a combination of electrostatic, Van der Waals, and hydrogen bond interactions stabilize the high order structure. Xanthan hydrogels exhibit a gel–sol transition, with the transition temperature depending on the degrees of acetylation and pyruvylation, the concentration and the molecular weight of xanthan, and the nature of the cations present.

PROPERTIES

Xanthan forms high viscosity solutions at low shear rates, even at low polysaccharide concentrations. The solution viscosity increases in the presence of salt, above a threshold salt concentration. The degrees of pyruvate and acetyl substitution respectively increase and decrease the viscosity of xanthan solutions. In addition, elimination of the terminal mannose residue reduces the thickening ability compared to standard xanthan, whereas elimination of both the terminal mannose and the adjoining glucuronic acid residues has the opposite effect.[3] In the presence of a small applied force, xanthan solutions resist flow up to a critical value (yield point) owing to aggregates and chain entanglements. Once the yield point has been passed the solutions become pseudoplastic, the viscosity progressively decreasing as the shear stress is increased. Shear thinning results from the gradual disruption of the solution structure and alignment of the polymer chains in the direction of the shear force. Provided that the applied stress is insufficient to degrade the solution, xanthan solutions regain their original state once the shear stress is released.

STORAGE

Anti-microbial agents can be added to xanthan solutions during storage, with sodium azide, glutaraldehyde, and formaldehyde being commonly used products.

APPLICATIONS

Xanthan gum is widely used in food, personal care products, and pharmaceutical formulations where, owing to its effect on the rheological properties of aqueous solutions, it imparts desired properties such as stability, improved texture, and controlled release of active ingredients. Other advantages of xanthan incorporation in food products include reduced ice formation on freezing and elimination of syneresis for annealed formulations. Recently developed, tough, flexible, and biodegradable, xanthan films may be used in food wrapping applications or as a substrate for artificial skin.[4] Xanthan is approved by the FDA for use as a stabilizer, emulsifier, and foam enhancer, as well as a thickening, suspending, and bodifying agent.

Xanthan has a number of important oil field applications. It is ideal for formulating drilling fluids with or without added solids. The same xanthan characteristics are used to formulate completion and workover fluids.

Enhanced oil recovery (EOR) requires maintaining the reservoir pressure through re-injection of a suitable liquid. By increasing the viscosity of brines, xanthan increases sweep efficiency thereby aiding uniform dispersal of the flooding solution and optimizing recovery. Xanthan gels containing metal ions, particularly chromium, are used for controlling flow diversion of polymer flooding in profile modification applications.[5,6]

REFERENCES

1. Sutherland, I. W. *Biotechnology of Microbial Exopolysaccharides*; Cambridge: Cambridge, 1990; Chapter 4.
2. Okuyama, K.; Arnott, S.; Moorhouse, R.; Walkinshaw, M. D.; Atkins, E. D. T.; Wolf-Ummish, C. In *Fiber Diffraction Methods*; French, A. D.; Gardner, K. H., Eds.; American Chemical Society: Washington, DC, 1980; Chapter 26.
3. Hassler, R. A.; Doherty, D. H. *Biotech. Prog.* **1990**, *6*, 182.
4. Quinn, F. X.; Hatakeyama, T.; Nakamura, K. Japan Patent appl. 1 18C1, 381, 1994(c).
5. Hughes, D. S.; Teeuw, D.; Cottrell, I. W.; Tollas, J. M. *Soc. Petrol. Eng.* 1988; paper 17400.
6. Lund, T.; Smidsrod, O.; Stokke, B. J.; Elgsaeter, A. *Carbohydr. Polym.* **1988**, *8*, 245.

XANTHAN GUM (Conformation and Solution Properties)

Takashi Norisuye* and Akio Teramoto
Department of Macromolecular Science
Osaka University

Early viscometric and spectroscopic studies showed that xanthan in aqueous salt of low ionic strength undergoes an order–disorder conformation change with rising temperature.[1–3] The nature of the ordered conformation, however, has long been the subject of considerable debate, in that two different helical conformations, single-stranded and double-stranded, were proposed and each was supported by different experiments.[3–6] It is now widely accepted that the ordered conformation is the double-stranded helix and that its change to the disordered conformation is associated with breaking of the helix.

Double-helical xanthan in aqueous NaCl has a high rigidity[7,8] and its concentrated solution forms a cholesteric mesophase.[9,10] When its semidilute solution, once heated to a predominantly disordered state, is cooled to room temperature, the polymer forms a weak gel-like network structure and exhibits enhanced viscoelasticity and shear thinning (or non-Newtonian behavior).[11–13]

PROPERTIES

Double-Helical Conformation

Moorhouse et al. found that xanthan in the crystalline state assumes a helical conformation having five-fold symmetry and a pitch of 4.7 nm, that is, 0.47 nm per main-chain glucose residue.[15] However, they were unable to determine whether the 5_1 helix consists of a single chain or paired chains. Okuyama

*Author to whom correspondence should be addressed.

et al. reinvestigated the fiber density and possible packing arrangements of crystalline xanthan, and concluded the helix to be double-stranded.[16]

Before the X-ray work of Okuyama et al., Holzwarth and Prestridge had proposed a double-helix for the ordered conformation in solution on the basis of electron microscopy.[4]

Conformation Changes

A recent analysis of molecular weight data shows that the complete separation of dimers is a very slow process in support of the work of Liu and Norisuye and that, upon heating of pure water solutions at 95°C for 9 h, about 70% of dimers dissociate into single chains.[19]

Polyelectrolyte Behavior

The experimental [η] was virtually independent of C_s in the range between 0.005 and 0.1 M for M_w below 10^6.

In contrast to this behavior of [η], the second virial coefficient A_2 and the third virial coefficient (for an Na sample with $M_w = 2.3 \times 10^5$), both associated with inter-dimer interactions, increase markedly as C_s decreases from 1 to 0.005 M.[56] This contrast is one of the typical features of rod-like polyelectrolytes. The observed C_s dependence of A_2 was found to be described rather accurately by the theory for line-charge rods.[20]

Liquid Crystal Formation

Xanthan, a rigid polyelectrolyte of double-helical conformation, is naturally expected to form a cholesteric mesophase at higher concentrations because of a definite sense of the helix. In fact, it was found that aqueous solutions of the polysaccharide show typical features of cholesterics and the coexistence of isotropic and anisotropic phases, namely biphasic at some concentrations.[9,18,21] The remarkable C_s dependence indicates that electrostatic interactions in addition to the asymmetric shape play a major role in the liquid crystallinity of xanthan as in the second and third virial coefficients discussed above.

By taking into account such interactions in detail, in addition to the molecular asymmetry, Sato and Teramoto developed a statistical mechanical theory for a uniformly charged, wormlike cylinder model to explain the observed phase behavior.[23]

The cholesteric structure of Na xanthan was also studied in some detail with respect to the cholesteric pitch of xanthan liquid crystals, which varied with polymer concentration, the sample's molecular weight, temperature, and solvent.[22] The inversion of helix sense with changing temperature, a phenomenon that had been noted so far only in neutral polymers, was found to occur and was subjected to a thermodynamic analysis.

Dynamic Properties of Moderately Concentrated Solutions

Solutions with polymer mass concentrations c high enough but lower than c_i are called concentrated isotropic solutions or, conventionally, moderately concentrated or semidilute solutions. While static data on such xanthan solutions are so far very limited,[24,25] a considerable amount of a dynamic and rheological data have been accumulated since the early work of Whitcomb and Macosko.[26]

Autoclaved samples may be more promising for use as viscosity control agents and enhanced oil-recovery agents because their semidilute solutions can have larger η values at low shear rates and stronger shear rate dependence than do those of double-stranded samples.[12,13] This enhancement in shear thinning or pseudoplasticity was interpreted as due to the disruption of junction zones by shear. Data of the dynamic viscosity and the dynamic storage and loss moduli for thermally treated solutions are also consistent with weakly gel-like solutions and show enhanced viscoelasticity.[11–14,25,27]

When a xanthan solution is mixed with a galactomannan such as locust bean gum at a high T and cooled to a T below which the ordered xanthan conformation was predominant, the mixture forms a firm, rubbery gel despite the fact that neither polymer gels by itself.[3,28] Similar gelation is known to occur for mixtures of xanthan and a glucomannan (konjac mannan); konjac mannan also does not gel in water.[29,30] These observations indicate that xanthan dimers (or chains) at a certain conformational stage prefer to associate with galactomannan or glucomannan chains rather than other xanthan dimers (or chains). Annable et al. concluded from mechanical spectroscopy, electron spin resonance, and differential scanning calorimetry that the gelation of xanthan with konjac mannan occurs only after the alignment of side chains, or after the local intrachain ordering in the early process of the disorder–order conformation change.[30] Thus, both the xanthan gelation and the above-mentioned network formation of semidilute solutions appear to arise from the capacity of interchain association that such locally ordered xanthan chains possess. It is probably this capacity that gives rise to the interesting rheological behavior of xanthan gum, resulting in its widespread use as a food additive.

REFERENCES

1. Rees, D. A. *Biochem. J.* **1972**, *126*, 257.
2. Holzwarth, G. *Biochemistry* **1976**, *15*, 4333.
3. Morris, E. R.; Rees, D. A.; Young, G.; Walkinshaw, M. D.; Darke, D. *J. Mol. Biol.* **1977**, *110*, 1.
4. Holzwarth, G.; Prestridge, E. B. *Science* **1977**, *197*, 757.
5. Holzwarth, G. *Carbohydr. Res.* **1978**, *66*, 173.
6. Sato, T.; Norisuye, T.; Fujita, H. *Polym. J.* **1984**, *16*, 341.
7. Sato, T.; Norisuye, T.; Fujita, H. *Macromolecules* **1984**, *17*, 2696.
8. Sho, T.; Sato, T.; Norisuye, T. *Biophys. Chem.* **1986**, *25*, 307.
9. Maret, G.; Milas, M.; Rinaudo, M. *Polym. Bull.* **1981**, *4*, 291.
10. Sato, T.; Kakihara, T.; Teramoto, A. *Polymer* **1990**, *31*, 824.
11. Ross-Murphy, S. B.; Morris, V. J.; Morris, E. R. *Faraday Symp. Chem. Soc.* **1983**, *18*, 115.
12. Oviatt, Jr., H. W.; Brant, D. A. *Int. J. Biol. Macromol.* **1993**, *15*, 3.
13. Oviatt, Jr., H. W.; Brant, D. A. *Macromolecules* **1994**, *27*, 2402.
14. Kratky, O.; Porod, G. *Rec. Trav. Chim. Pays-Bas* **1949**, *68*, 1106.
15. Moorhouse, R.; Walkinshaw, M. D.; Arnott, S. *Extracellular Microbial Polysaccharides*; Sandford, P. A.; Laskin, A. I., Eds.; ACS Symp. Ser. No. 45, 1977; p 90.
16. Okuyama, K.; Arnott, S.; Moorhouse, R.; Walkinshaw, M. D.; Atkins, E. D. T.; Wolf-Ullish, Ch. *Fiber Diffraction Methods* French, A. D.; Gardner, K. H., Eds.; ACS Symp. Ser. No. 141, 1980; p 411.
17. Paoletti, S.; Cesàro, A.; Delben, F. *Carbohydr. Res.* **1983**, *123*, 173.
18. Milas, M.; Rinaudo, M. *Solution Properties of Polysaccharides*; Brant, D. A., Ed.; ACS Symp. Ser. No. 150, 1981; p 25.

19. Kawakami, K.; Okabe, Y.; Norisuye, T. *Carbohydr. Res.* **1991**, *14*, 189.
20. Fixman, M.; Skolnick, J. *Macromolecules* **1978**, *11*, 863.
21. Milas, M.; Rinaudo, M. *Polym. Bull.* **1983**, *10*, 271.
22. Sato, Y.; Sato, T.; Teramoto, A. *Polymer Preprints, Japan* **1992**, *41*, 1411.
23. Sato, T.; Teramoto, A. *Physica A* **1991**, *176*, 72.
24. Coviello, T.; Burchard, W.; Dentini, M.; Crescenzi, V. *Macromolecules* **1987**, *20*, 1102.
25. Kojima, T.; Berry, G. C. *Polymer* **1988**, *29*, 2249.
26. Whitcomb, P. J.; Macosko, C. W. *J. Rheol.* **1978**, *22*, 493.
27. Frangou, S. A.; Morris, E. R.; Rees, D. A.; Richardson, R. K.; Ross-Murphy, S. B. *J. Polym. Sci., Polym. Lett. Ed.* **1982**, *20*, 531.
28. Dea, I. C. M.; Morris, E. R.; Rees, D. A.; Welsh, E. J.; Barnes, H. A.; Price, J. *Carbohydr. Res.* **1977**, *57*, 249.
29. Williams, P. A.; Day, D. H.; Langdon, M. J.; Phillips, G. O.; Nishinari, K. *Food Hydrocoll.* **1991**, *4*, 489.
30. Annable, P.; Williams, P. A.; Nishinari, K. *Macromolecules* **1994**, *27*, 4204.

Xerogels

See: *Bridged Polysilsesquioxanes (Highly Porous Hybrid Organic-Inorganic Materials)*
Electrically Conducting Composites
Poly(ethylene oxide) (Applications in Drug Delivery)

Xylan

See: *Hemicelluloses*

Xyloglucans

See: *Hemicelluloses*

Ylid Monomers and Polymers

See: *Aminimides, Monomers and Polymers*
Polyzwitterions (Overview)

Z

Ziegler–Natta Catalysts

See: *Block Copolymers (by Changing Polymerization Mechanism)*
Functionalized Polyolefins (Using Borane Monomers and Transition Metal Catalysts)
Gas Phase Olefin Polymerization
Liquid Crystalline Polyacetylene Derivatives
Monomer-Isomerization Polymerization (2-Butene with Ziegler–Natta Catalysts)
Olefin Polymerization Catalysts (Experimental and Theoretical Characterization)
Olefin Polymerization Catalysts (Silica-Supported Titanium Tetrachloride)
Polyethylene (High Density Preparation)
Polyolefins (Overview)
Polypropylene (Commercial)
Polypropylene (Controlled Degradation)
Polystyrene, Stereoregular
Styrene, Stereospecific Polymerization
Supported Catalysts (Lewis Acid and Ziegler–Natta)
Supported Ziegler–Natta Catalysts (on Silica, for Olefin Polymerization)
Supported Ziegler–Natta Catalysts (Titanium-Based Silica; Polymerization Properties)
Supported Ziegler–Natta Catalysts [Titanium (IV) Chloride on Reduced n-Type Oxides]
Ultrasonically Assisted Polymer Synthesis

ZINC CATALYSTS (Polymerization of Styrene and Methyl Methacrylate)

Franco M. Rabagliati
Departamento de Ciencias Químicas
Facultad de Química y Biología
Universidad de Santiago de Chile

Organozinc compounds have been used as catalysts for several monomer polymerizations. Combinations of dialkylzinc, or alkyl halide, with $TiCl_4$ and other compounds of transition metals were employed in stereospecific polymerization of olefins.[1–4]

Dialkylzinc compounds induce anionic-coordination polymerization of acrylic esters, and polymerization of vinyl ethers goes through cationic coordination when it is used for organozinc compounds in presence of a co-catalyst.[5–9]

Complexes of calcium with tetraethylzinc and of butyllithium with diethylzinc, Et_2Zn-BuLi, are effective catalysts in stereospecific polymerization of methyl methacrylate, styrene, and other conjugated vinyl monomers.[10,11]

The most important application of organozinc compounds as polymerization catalysts is for epoxide monomers, such as ethylene oxide and propylene oxide.

Diphenylzinc was also used in combination with ketones, Ph_2Zn-ketone systems, as initiators of ethylene oxide and of propylene oxide monomers.[12]

There are only a few examples of styrene and of methyl methacrylate polymerization initiated by zinc catalysts, and in all cases the zinc compound is combined with a second compound.

The polymerization of styrene by Ph_2Zn-H_2O systems has shown to be greatly influenced by molar ratio H_2O–Ph_2Zn, solvent employed, and the temperature at which polymerization was performed.[13] The polystyrenes, PSt, obtained with these systems are of high molecular weight with broad molecular weight distributions, MWD. With progressing polymerization, the abundance of PSt portions, at low and higher molecular weights, move to the higher regions, centering at around 3.5×10^5 for larger conversions.

Diphenylzinc was used in combination with some alcohols for styrene polymerization.[14] There is was established that conversion to polystyrene increases when moving from *n*- to *tert*-butanol in agreement with the reactivity of these alcohols with diphenylzinc.

Diphenylzinc has been used by itself and in a molar ratio of 1:1 with water, in combination with zirconocenes, and also in a three-component mixture. MAO-zirconocene-Ph_2Zn, as initiator of styrene polymerization in toluene at 60°C.[15] In all cases, high molecular weight polystyrenes were obtained.

We studied the polymerization of MMA using both Ph_2Zn by itself and also combined with water.[16] Our conclusions indicate that these systems behave differently. Ph_2Zn-H_2O systems polymerize MMA in a way similar to the one postulated for oxirane and for styrene with the molar ratio H_2O/Ph_2Zn, being the determinant for conversion to polymer. Furthermore, obtained PMMA was found to be syndiotatic-rich for all the molar ratios H_2O/Ph_2Zn used.

For polymerization of MMA using only Ph_2Zn, without any co-catalyst, at 120°C in benzene, there was proof that indicates some sort of association between monomer and Ph_2Zn. We concluded that the monomer forms a complex with Ph_2Zn, Zn...O=C, which weakens the vinylic bond of monomer and facilitates its free radical polymerization. A similar situation was postulated by Imoto et al. when polymerizing MMA and acrylonitrile by AIBN in the presence of zinc chloride.[17,18]

ACKNOWLEDGMENTS

Financial support from Dirección de Investigación Científica y Tecnológica, Universidad de Santiago de Chile (DICYT-USACH) and from Fondo Nacional de Desarrollo Científico y Tecnológico (FONDECYT) are gratefully acknowledged.

REFERENCES

1. Boor, I. *J. Polym. Sci.* **1965**, *A3*, 995.
2. Boor, I.; Short, G. *J. Polym. Sci.* **1971**, *A1*, 249.
3. Cesca, S.; Bertolini, G.; Santi, G.; Duranti, P. *J. Polym. Sci.* **1971**, *A1*, 1575.
4. Gray, A.; Callaer, A.; Edgecombe, F. *Can. J. Chem.* **1963**, *41*, 1502.
5. Ikeda, M.; Hirano, T.; Tsuruta, T. *Makromol. Chem.* **1971**, *150*, 127.
6. Tsuruta, T.; Kawakami, Y.; Tsushima, R. *Makromol. Chem.* **1971**, *149*, 135.
7. Imai, H.; Saegusa, T.; Furukawa, J. *Makromol. Chem.* **1965**, *81*, 92.

8. Nakano, S.; Iwasaki, K.; Fukutani, H. *J. Polym. Sci.* **1968,** *A1,* 1209.
9. Spassky, N.; Sigwalt, P. *Bull. Soc. Chim. Fran.* **1967,** 4617.
10. Furukawa, J.; Tsuruta, T.; Inoue, S.; Kawasaki, A.; Kawabata, N. *J. Polym. Sci.* **1959,** *35,* 268.
11. Labbe, G.; Smets, G. *J. Polym. Sci.* **1967,** *A1,* 1359.
12. a) López, F.; Calcagno, M. P.; Contreras, J. M.; Torrellas, Z.; Rabagliati, F. M. *Polymer Bull.* **1989,** *21,* 287. b) López, F.; Calcagno, M. P.; Contreras, J. M.; Ramírez, M.; Felisola, K.; Rabagliati, F. M. *Polymer Bull.* **1989,** *21,* 449. c) López, F.; Calcagno, M. P.; Contreras, J. M.; Torrellas, Z.; Felisola, K.; Rabagliati, F. M.; Katime, I. A. *Polym. Int.* **1991,** *24,* 105. d) Calcagno, M. P.; López, F.; Contreras, J. M.; Ramírez, M.; Rabagliati, F. M. *Eur. Polym. J.* **1991,** *27,* 751.
13. Rabagliati, F. M.; Terraza, C. A. *Polym. Bull.* **1992,** *28,* 511.
14. Rabagliati, F. M.; Terraza, C. A.; Ramírez, L. A.; Contreras, J. M. *Macromol. Symp.* **1994,** *84,* 103.
15. Rabagliati, F. M.; Terraza, C. A.; Cuevas, M. V.; Quijada, J. R. *Polym. Bull.* In press, 1996, *36.*
16. a) Rabagliati, F. M.; Ramírez, L. A.; Terraza, C. A. *Bol. Soc. Chil. Quím.* **1994,** *39,* 57. b) Rabagliati, F. M.; Terraza, C. A.; Ramírez, L. A.; Contreras, J. M. *Makromol. Symp.* **1994,** *84,* 103.
17. Imoto, M.; Otsu, T.; Harada, Y. *Makromol. Chem.* **1963,** *65,* 180.
18. Imoto, M.; Otsu, T.; Nakabayashi, M. *Makromol. Chem.* **1963,** *65,* 194.

Zwitterionic Polymerization

See: *Cyclic Imino Ethers (Ring-Opening Polymerization)*
Group-Transfer Alternating Copolymerization
2-Oxazolines, Polymerization
Periodic Polymers
Poly(N-acylethylenimine)s
Ring-Opening-Closing Polymerization
Zwitterionic Polymerization (Overview)
Zwitterionic Polymerization (Aziridine, Oxazoline Derivatives with Anhydrides)

ZWITTERIONIC POLYMERIZATION (Overview)

Masato Suzuki
Research Laboratory of Resources Utilization
Tokyo Institute of Technology

Takeo Saegusa
Kansai Research Institute
Kyoto Research Park

The no-catalyst copolymerization is conducted by a combination of a monomer, having nucleophilic reactivity (M_N), with that having electrophilic reactivity (M_E). **Equations 1–4** show the conceptional reaction scheme of it. When an M_N monomer is mixed with an M_E monomer, a spontaneous reaction (sometimes heating is necessary) occurs to generate a zwitterion **1**, namely "genetic zwitterion" (Equation 1). Two molecules of **1** react with each other to produce the first "macrozwitterion" **2** (Equation 2). Successive reactions of **1** with **2** (Equation 3) and reactions among macrozwitterions of various sizes (Equation 4) give rise to production of an alternating copolymer.

$$M_N + M_E \longrightarrow {}^{+}M_N \text{——} M_E^{-} \qquad (1)$$

$$1 + 1 \longrightarrow {}^{+}M_N \text{——} M_EM_N \text{——} M_E^{-} \qquad (2)$$

$$2 + nx1 \longrightarrow {}^{+}M_N \text{——}(M_EM_N)_{n+1}\text{——} M_E^{-} \qquad (3)$$

$${}^{+}M_N \text{——}(M_EM_N)_m\text{——} M_E^{-} + {}^{+}M_N \text{——}(M_EM_N)_n\text{——} M_E^{-} \longrightarrow {}^{+}M_N \text{——}(M_EM_N)_{m+n+1}\text{——} M_E^{-} \qquad (4)$$

Many combinations of M_N and M_E monomers have been explored for the no-catalyst copolymerization.

Scheme I shows a representative example; 2-oxazoline **3** spontaneously reacts with β-propiolactone **4** at room temperature to produce the alternating copolymer **7** in a quantitative yield.[1,2]

Acrylic acid **8** acts as an interesting M_E monomer (**Scheme II**).[3] The Michael-type addition of **3** to **8** generates a zwitterionic intermediate **9**, however, which instantly isomerizes to the more stable zwitterion **5** through proton transfer. It should be noticed that this genetic zwitterion is common with that in Scheme I. Thus, the copolymer produced from **3** with **8** has the same structure as that from **3** with **4**.

Not only compounds having a nucleophilic N atom, but also some phosphorus(III) compounds, have been successfully employed to cause no-catalyst copolymerizations as M_N monomers.

Benzoquinone derivatives copolymerize with some cyclic and acyclic phosphorus compounds.[7,9,10]

Copolymerizations of cyclic disulfide with phosphorus compounds proceed by way of nucleophilic cleavage of S-S bonds.[4,5,8]

A periodic terpolymer **26**, which has 1:1:1 regulated sequences from three monomers, is successfully produced when **17** and methyl acrylate (or acrylonitrile) are copolymerized under CO_2 atmosphere (**Scheme III**).[6]

Similarly, combinations of aziridine with carbonyl sulfide and ethylene oxide with sulfur dioxide produce alternating copolymers via zwitterionic intermediates.[11,12]

REFERENCES

1. Saegusa, T.; Ikeda, H.; Fujii, H. *Macromolecules* **1972,** *5,* 354.
2. Saegusa, T.; Kobayashi, S.; Kimura, Y. *Macromolecules* **1974,** *7,* 1.
3. Saegusa, T.; Kobayashi, S.; Kimura, Y. *Macromolecules* **1974,** *7,* 139.
4. Kobayashi, S.; Chow, T. Y.; Saegusa, T. *Polym. Bull.* **1983,** *9,* 588.
5. Kobayashi, S.; Chow, T. Y.; Takehara, K.; Okawa, M.; Saegusa, T. *Polym. Prep. Jpn.* **1982,** *31,* 318.
6. Saegusa, T.; Kobayashi, S.; Kimura, Y. *Macromolecules* **1977,** *10,* 68.
7. Saegusa, T.; Kobayashi, T.; Kobayashi, S. *Macromolecules* **1981,** *14,* 463.
8. Kobayashi, S.; Chow, T. Y.; Takehara, K.; Okawa, M.; Saegusa, T. *Polym. Prep. Jpn.* **1982,** *31,* 1217.
9. Saegusa, T.; Kobayashi, T.; Chow, T. Y.; Kobayashi, S. *Macromolecules* **1979,** *12,* 533.
10. Kobayashi, S.; Okawa, M.; Niwano, M.; Saegusa, T. *Polym. Bull.* **1981,** *5,* 331.
11. Yokota, H.; Kondo, M. *J. Polym. Sci. Part A-1* **1971,** *9,* 13.
12. Soga, K.; Kiyohara, K.; Hattori, I.; Ikeda, S. *Makromol. Chem.* **1980,** *181,* 2151.

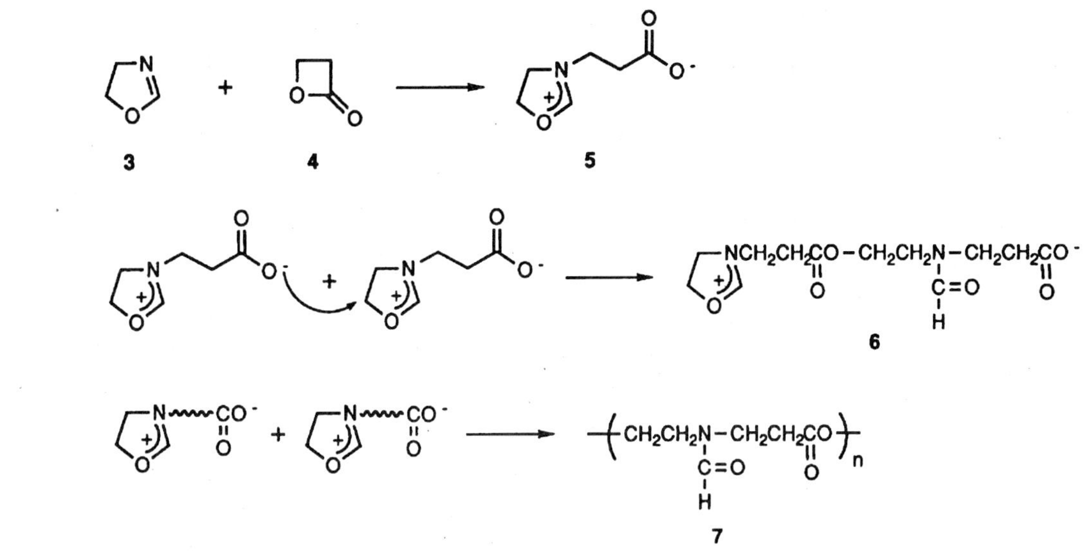

I

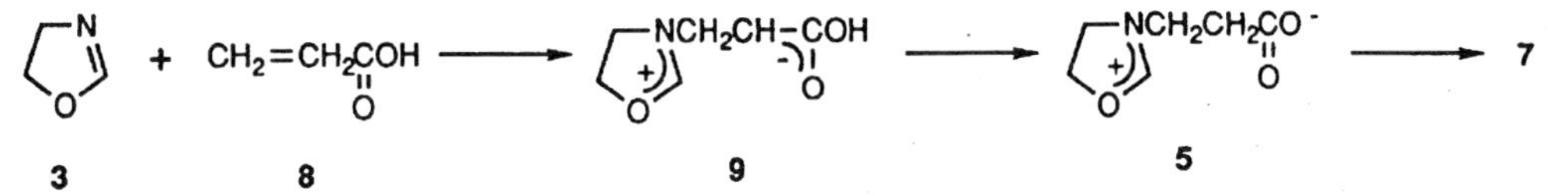

II

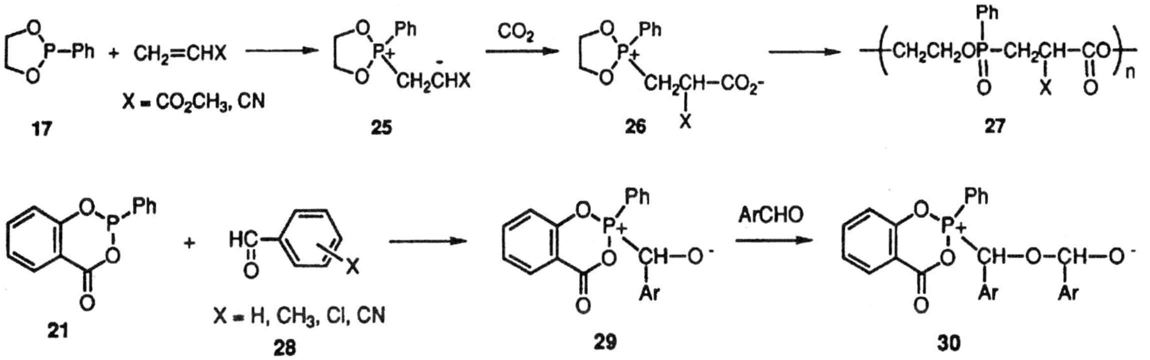

III

ZWITTERIONIC POLYMERIZATION (Aziridine, Oxazoline Derivatives with Anhydrides)

S. Amalia Pooley, B. L. Rivas, and G. S. Canessa
Departamento de Polímeros
Facultad de Ciencias Químicas
Universidad de Concepción

Most polymerization reactions require initiator, catalyst, or high energy radiation, but there have been reports on copolymerizations that occur in the absence of any added initiator. Copolymerization of this type is based on a specific combination of two monomers: one monomer has nucleophilic reactivity (MN) and the other has electrophilic reactivity (ME). The interaction of the monomers produces a zwitterion ($^{+}MNME^{-}$) which is the so-called "genetic zwitterion," responsible for both initiation and propagation.

If the growth reaction involves only "polyaddition" and "polycondensation" reactions, an alternating copolymer is achieved.

This type of copolymerization is sometimes called "no-catalyst alternating copolymerization via zwitterion intermediates" or "spontaneous alternating copolymerization via zwitterion intermediates." These no-catalyst copolymerizations have been developed mainly by Saegusa et al., Odian et al., and the present authors.[1–26]

ANALYSIS OF COPOLYMERIZATION SYSTEMS

Following the structure of the obtained copolymers when some nucleophilic monomers such as 2-methyl-2-oxazoline (MOX), 2-ethyl-2-oxazoline (EtOX), aziridine (Az), acetyl-aziridine (AAz), 2-methyl-aziridine (MAz) and N-(2-hydroxyethyl)aziridine (HEAz) react with aliphatic and aromatic anhydrides, will be analyzed (see structures) (**Schemes I–III**).

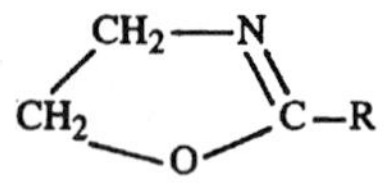

I

$R = CH_3$ MOX
$R = CH_2CH_3$ EtOX

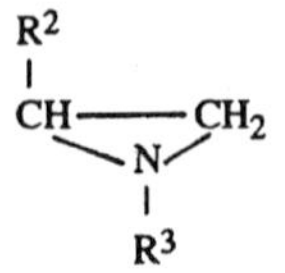

II

$R^2 = H$	$R^3 = H$	Az
$R^2 = CH_3$	$R^3 = H$	MAz
$R^2 = H$	$R^3 = CO\text{-}CH_3$	AAz
$R^2 = H$	$R^3 = CH_2CH_2OH$	HEAz

CONCLUSIONS

In general, the obtained copolymers are not alternating, because homopropagation reactions involving reaction ion–dipole reaction were favored.

The propagation reaction between anhydrides and oxazoline or aziridine derivatives proceeded by ring-opening reactions. For the EtOX/GAn copolymers, cyclic monomeric units of EtOX were also determined.

The reactivity of MOX and EtOX with SAn or PhAn are very similar and these copolymers show, in general, a partial hydrolysis of the acyl group from imide linkages.

By ^{1}H-NMR spectroscopy it was possible to determine the Az-homodyad and the Az-heterodyad concentration in copolymers Az–SAn.

By the crosslinking reaction of Man with AAz (MOX or Az), insoluble copolymers were produced. Copolymers Az–MAn showed properties as resins with sorption ability to bind copper(II), and particularly uranium(VI).

The copolymerization of AAz and MOX (structural isomers) with MAn yield copolymers of similar structure composition, showing both systems' similar chemical reactivity.

The copolymerization behavior of PhAn with Az and MAz is essentially the same; nevertheless, owing to the MAz molecule asymmetry, two different monomeric units of MAz were found.

In the copolymerization of HEAz with PhAn, a partial esterification of the 2-hydroxyethyl lateral group was observed.

ACKNOWLEDGMENTS

The authors thank "Dirección de Investigación" of the Universidad de Concepción and FONDECYT (Grant 1940981) for the financial support.

REFERENCES

1. Saegusa, T. *Chemtech* **1975**, *5*, 295.
2. Saegusa, T.; Kobayashi, S.; Kimura, Y. *Pure Appl. Chem.* **1976**, *48*, 307.
3. Kobayashi, S.; Isobe, M.; Saegusa, T. *Macromolecules* **1982**, *15*, 703.
4. Odian, G.; Gunatillake, P. A. *Macromolecules* **1984**, *17*, 1297.
5. Odian, G.; Gunatillake, P. A. *Macromolecules* **1985**, *18*, 605.
6. Rivas, B. L.; Pooley, S. A. *An. Quim. Sec. C* **1983**, *79*, 62.
7. Rivas, B. L.; Canessa, G. S.; Pooley, S. A. *Polym. Bull. (Berlin)* **1983**, *9*, 417.
8. Canessa, G. S.; Pooley, S. A.; Parra, M.; Rivas, B. L. *Polym. Bull. (Berlin)* **1984**, *11*, 465.
9. Rivas, B. L.; Canessa, G. S.; Pooley, S. A. *Polym. Bull. (Berlin)* **1985**, *13*, 65.
10. Rivas, B. L.; Canessa, G. S.; Pooley, S. A.; Maturana, H. A.; Angne, U. *Eur. Polym. J.* **1985**, *21*, 939.
11. Rivas, B. L.; Canessa, G. S.; Pooley, S. A. *Polym. Bull. (Berlin)* **1985**, *13*, 103.
12. Rivas, B. L.; Canessa, G. S.; Pooley, S. A. *Polym. Bull. (Berlin)* **1985**, *13*, 519.
13. Rivas, B. L.; Canessa, G. S.; Pooley, S. A. *Makromol. Chem.* **1986**, *187*, 71.
14. Rivas, B. L.; Canessa, G. S.; Pooley, S. A. *Makromol. Chem.* **1987**, *188*, 149.
15. Rivas, B. L.; Canessa, G. S.; Pooley, S. A. *Makromol. Chem. Rapid Commun.* **1987**, *8*, 365.
16. Rivas, B. L.; del C. Pizarro, G.; Canessa, G. S. *Polym. Bull. (Berlin)* **1988**, *19*, 123.
17. Rivas, B. L.; Canessa, G. S.; Pooley, S. A. *Eur. Polym. J.* **1989**, *25*, 225.
18. Rivas, B. L.; Canessa, G. S.; Pooley, S. A. *Makromol. Chem.* **1986**, *190*, 2665.
19. Rivas, B. L.; Canessa, G. S.; Pooley, S. A.; Maturana, H. A. *Polym. Bull. (Berlin)* **1989**, *22*, 173.
20. Rivas, B. L.; Canessa, G. S.; Pooley, S. A. *Polym. Bull. (Berlin)* **1990**, *23*, 171.
21. Rivas, B. L.; Canessa, G. S.; Pooley, S. A. *Bol. Soc. Chil. Quím.* **1991**, *36*, 23.
22. Rivas, B. L.; Canessa, G. S.; Pooley, S. A. *Bol. Soc. Chil. Quím.* **1992**, *37*, 225.
23. Rivas, B. L.; Canessa, G. S.; Pooley, S. A. *Eur. Polym. J.* **1992**, *28*, 43.
24. Rivas, B. L.; Canessa, G. S.; Pooley, S. A. *Eur. Polym. J.* **1993**, *29*, 1089.
25. Rivas, B. L.; Canessa, G. S.; Pooley, S. A. *Eur. Polym. J.* **1993**, *29*, 1239.
26. Pooley, S. A.; Canessa, G. S.; Rivas, B. L.; Espejo, E. *Bol. Soc. Chil. Quím.* **1994**, *39*.

Zwitterionic Polymers

See: *Polyzwitterions (Overview)*
Zwitterionic Polymers (Dilute Solution and Rheological Properties)
Zwitterionic Polymers (Ionic Conductivity)

$G = (CH_2)_2$; SAn $G = (CH_2)_3$; GAn G = (benzene ring) PhAn

III

ZWITTERIONIC POLYMERS (Dilute Solution and Rheological Properties)

Judith Cardoso*
Universidad Autónoma Metropolitana-Iztapalapa

O. Manero and M. Valdés
Instituto de Investigaciones en Materiales, UNAM

T. Orta
Instituto de Ingeniería, UNAM

Zwitterionic polymers contain charged pendant groups along the macromolecular chain. These groups possess two charges of opposing signs joined by covalent bonds, which induce a strong dipolar structure that gives rise to interesting properties.

The synthesis, characterization, and dilute solution properties of five atactic, aliphatic, or aromatic polysulfopropylbetaines have been described.[1,2] These studies have illustrated specific bulk properties such as high-chain rigidity and very unusual and unique antipolyelectrolyte behavior characterized by chain expansion in dilute solution. These properties also have been observed by other authors.[1–5]

In this article attention will be given to systems with intermediate ionic content (around 60%) or intermediate charge density. In this case, the zwitterionic structures are indeed copolymers with the following structures (**Structure 1** and **2**). In both cases, x is the degree of quaternization or percentage of zwitterionic groups.

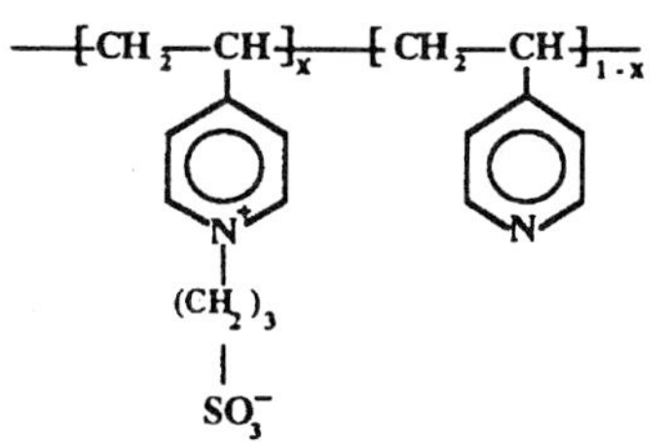

1

P4VP—SB

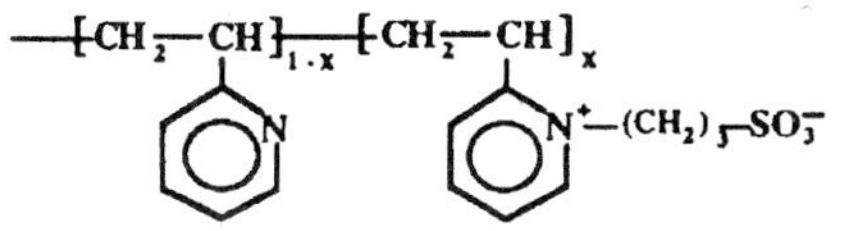

2

P2VP—SB

PREPARATION

Polymers were prepared by free-radical polymerization according to the technique illustrated in Reference 6.

Dilute Solution Properties

The hydrodynamic behavior of the P2VP-SB copolymer (molecular weights of 100,000 and 200,000) with monovalent (NaCl, NaI) and divalent salts ($CaCl_2$) was studied.[6]

In apparent difference with results from polymers with 100% ionic groups, an initial increase in the intrinsic viscosity was observed up to concentrations of about 0.5 M, with saturation at higher molarity, for the copolymer with the monovalent salts. Qualitative differences were observed in the copolymer with the divalent salt ($CaCl_2$), where an increasing intrinsic viscosity up to 2 M with no signs of saturation was shown.

Rheological Properties

Rheological measurements performed on aqueous solutions of P2VP-SB at several concentrations and temperatures showed the variation of modulus and viscosity with frequency upon making time-temperature data reduction at 293 K.

Strong viscoelastic effects were observed with behavior similar to that exhibited by non-ionic, flexible polymers.

Shear-thickening behavior has also been observed in low charge density polyampholytes in nonaqueous solvents.[7] In this case, the shear behavior of polystyrene with sulfonated groups was profoundly affected by ionic associations. Shear-thickening phenomena have been explained to occur when the shear-rate modifies and subsequently increases the ratio of inter- to intramolecular ionic associations. Consequently, these increased associations induce the formation of structures with higher resistance to flow.

Dilatant shear-viscosity behavior has been also observed in concentrated suspensions.[8,9] One of the explanations given for the increase in viscosity at moderately high shear-rates involved the ordering of particles into layers coincident with surfaces of constant shear.

STUDIES ON P4VP-SB-SALT SOLUTIONS

Aqueous solutions of P4VP-SB-salt systems with three inorganic salts ($LiClO_4$, $CaCl_2$, and NaCl) at two polymer concentrations (18% and 25%) were prepared.

This polymer did not dissolve in pure water, and addition of salt was necessary. In contrast, P2VP-SB dissolved quite easily up to concentrations of 30%. The rheological studies were undertaken in the linear viscoelastic regime.

At low salt concentrations, Newtonian behavior was observed for both polymer concentrations. With increasing salt concentration, viscoelastic behavior was apparent at a critical salt content.

Several studies have dealt with polymers containing both anionic and cationic groups.[10–20] These materials exhibit a complex solution behavior in which, in striking contrast to conventional polyelectrolytes, they become more soluble as the ionic strength of the solution is increased. This manifestation reflects the strong interaction of the polyampholytes with the dissolved inorganic salts.

APPLICATIONS

The observed increase of the intrinsic viscosity with salt concentration pertains to several applications. For instance, in enhanced oil recovery operations, surfactants and polymer solutions are injected in the oil wells to facilitate the extraction of the remaining oil which, in several cases, may be quite abundant. Solutions of polyelectrolytes have been used; however, in media of high salinity, as those occurring in most wells,

*Author to whom correspondence should be addressed.

polyelectrolytes behave as Newtonian liquids. In contrast, zwitterionic polymers preserve their viscoelastic properties in these media. Their consistency and chemical stability are also properties that may be useful in drilling fluids. Moreover, surfactants with zwitterionic structures form a variety of fluid microstructures that are responsible for ultralow interfacial tensions in the presence of oil.[21–23]

REFERENCES

1. Monroy Soto, V. M.; Galin, J. C. *Polymer* **1984**, *25*, 121.
2. Monroy Soto, V. M.; Galin, J. C. *Polymer* **1984**, *25*, 254.
3. Schulz, D. N.; Pfeiffer, D. G.; Agarwal, P. K.; Larabee, J.; Kaladas, J. J.; Soni, L.; Handweker, B.; Garner, R. T. *Polymer* **1986**, *27*, 1734.
4. Salamone, J. C.; Volksen, W.; Israel, S. C.; Olson, A. P.; Raia, D. C. *Polymer* **1977**, *18*, 1058.
5. Salamone, J. C.; Rice, W. C. In *Polyampholytes, Encyclopedia of Polymers Science and Engineering*; 2nd ed., Mark, H. F.; Bikales, N. M.; Overbergerz, C. G.; Menges, G., Eds.; John Wiley & Sons: New York, 1988; Vol. 11.
6. Cardoso, J.; Manero, O. *J. Polym. Sci: Part B: Polym. Phys.* **1991**, *29*, 639.
7. Peiffer, D. G.; Landberg, R. D.; Duvdevani, I. *Polymer* **1986**, *27*, 1453.
8. Hoffman, R. L. *Trans. Soc. Rheol.* **1972**, *16*, 155.
9. Hoffman, R. L. *Adv. Coll. and Interf. Sci.* **1982**, *17*, 161.
10. Rice, S. A.; Harris, F. E. *J. Polym. Sci.* **1956**, *24*, 326.
11. Alfrey Jr., T.; Pinner, S. H. *J. Polym. Sci.* **1957**, *23*, 533.
12. Hart, R.; Timmerman, D. *J. Polym. Sci.* **1958**, *28*, 638.
13. Salamone, J. C.; Watterson, A. C.; Hsu, T. D.; Tsai, C. C.; Mahmud, M. U.; Wisniewski, A. W.; Israel, S. C. *J. Polym. Sci.* **1978**, *C64*, 299.
14. Salamone, J. C.; Watterson, A. C.; Hsu, T. D.; Tsai, C. C.; Mahmud, M. U. *J. Polym. Sci., Polym. Lett.; Edn.* **1977**, *15*, 487.
15. Salamone, J. C.; Watterson, A. C.; Olson, A. P. *Polymeric Amines and Ammonium Salts*; Goethals, E. J., Ed.; Pergamon: Oxford, 1980.
16. Salamone, J. C.; Tsai, C. C.; Olson, A. P.; Watterson, A. C. *Adv. Chem. Ser.* **1980**, *187*, 337.
17. Salamone, J. C.; Tsai, C. C.; Olson, A. P.; Watterson, A. C. *J. Polym. Chem. Sci. Polym. Chem. Edn.* **1980**, *18*, 2983.
18. Watterson, A. C.; Moniz, W. B.; Stone, T.; Nagabhushanam, T.; Salamone, J. C. IUPAC 28th Macromolecular Symposium; Univ. of Massachusetts, Amherst, 1982; p 22.
19. Salamone, J. C.; Tsai, C. C.; Anwaruddin, Q.; Olson, A. P.; Arnold, M.; Nagabhushanam, T.; Sawan, S.; Watterson, A. C. *J. Polym. Sci., Polym. Edn.* **1984**, *22*, 2005.
20. Peiffer, D. G.; Lundberg, R. D. *Polymer* **1985**, *26*, 1058.
21. Faingold, S. I.; Kuusk, A. *Eesti NSV. Read Akad. Toim, Reem* **1985**, *34*, 210.
22. Hein, H. *Chim Oggi* **1985**, 33.
23. Moriya, M.; Nishimura, A. *J. Am. Chem. Soc.* **1986**, *63*, 263.

ZWITTERIONIC POLYMERS (Ionic Conductivity)

Judith Cardoso* and O. Manero
Instituto de Investigaciones en Materiales, UNAM
Universidad Autónoma Metropolitana-Iztapalapa

*Author to whom correspondence should be addressed.

Recently, there has been an increasing interest in the study of compounds that possess zwitterionic structures. There are several reasons to undertake research on such structures. Among them we may mention their unique chemical structure, a straightforward synthetic route, high ionic contents, interesting dilute solution, and solid state properties, the presence of highly polar groups, and the possibility to obtain copolymers of varying proportions.[1–5]

With regard to solid-state properties, one extremely interesting fact is the ability of the sulfobetaine derivatives to form "homogeneous solid solutions" between mineral salts and the polymeric matrix. Dielectric properties show high permitivities and strong dipolar moments, which suggest applications in dry batteries and piezoelectricity.

ZWITTERIONIC DERIVATIVES OF THE POLYVINYLPYRIDINES

Evidence that polymers with zwitterionic structures may form clusters has been proposed in studies on solid samples by transmission electron microscopy.[6] In this case, clusters are formed from attractive interactions among the ionic groups located in the same or different polymer chains. The size of clusters and agglomerates are a little bigger compared to typical values reported for the size of the clusters in the ionomer case.[7]

The relationship between the ionic content of random zwitterionic copolymers with T_g in a range of low content of ionic groups display similar behavior exhibited by ionomers, the variation of T_g with ionic content is linear up to 12–15%. P4VP-SB copolymers show a higher T_g, but the increase in T_g with ionic content is higher for the P2VP-SB copolymers.

Results illustrate that copolymers with low zwitterionic content display similarities with ionomers, such as the association of dipole segments in multiplets.

ZWITTERIONIC DERIVATIVES OF THE POLY(*N,N*-DIMETHYLAMINOETHYL METHACRYLATE)

It is interesting that the sulfobetaine dipole moment has been estimated to be between 6 and > 20 D, but upon quaternization, these values are increased by a factor of nearly 2 with respect to those of their tertiary amino precursors.[8]

Up-to-date literature searches reveal that poly(ethylene oxide) [25322-68-3] is by far the best-known, ion-supporting polymer that may be combined with a large amount of inorganic and organic salts.[9–12] In these systems, conductivity increases with salt content within a short range of salt proportions, but it shows a pronounced decrease for higher salt contents.

However, polymers of the sulfobetaine type possess the ability to dissolve salts with various network energies at high proportions. For instance, the system polysulfobetaine/lithium perchlorate showed no substantial segregation up to equimolar ratios (1:1 M salt per mol of monomer).

Among the sulfobetaine derivatives those obtained from the poly(*N,N*-dimethylaminoethyl methacrylate) are the compounds that present the more flexible structures.

Further observations of the phase behavior of zwitterionic-salt systems show the melting and absorption of the salt by the polymer, similar to those made in other systems, such as in the PEO–$LiCF_3SO_3$ mix.[13] These observations indicate that as soon as the crystalline phase melts, it is subsequently dissolved into the polymer phase. At this point, conductivity increases steeply, manifesting an enhanced ionic mobility previously limited by the potential barriers in the crystalline phase.

The mobility and flexibility of the amorphous structure of the polymeric matrix enhance ionic conductivity, and this is reflected in the strong correlation between T_g of the polymer–salt system with conductivity. This may also be shown by considering the variation of conductivity with temperature and salt concentration, and how these relations are affected by the variation of T_g with salt concentration.

The ionic conductivity variation with temperature may be analyzed in the framework of the configuration entropy model or the free volume theory.[14,15]

In the case of the zwitterionic–salt system, addition of salts results in a decrease of the potential barrier with the corresponding increase in conductivity. In this regard, the maximum in the energy barrier is closely associated with the region of minimum in conductivity (salt–monomer ratio of 0.2).

As observed, the thermal expansion coefficient presents a small increment as salt concentration is increased. Most polymers give values around 0.0005 K^{-1}. With respect to f_{Tg} (fractional free volume at $T = T_g$) there is an increase of 4 times between $r = 0.2$, and $r = 1$, which contrasts with the small increase in T_g about 14°C. Similarly, a rise of 3 times in the fractional free volume at T = 240°C is observed along the salt concentration range. These estimations explain why conductivity increases with salt content. The fractional free volume actually increases with salt concentration, and this increment agrees with the proposed mechanism of solvation and shielding of the polymer dipoles by the salt ions. The action of these ions prevents the polymer dipoles from forming close-packing structures and clustering, allowing the presence of vacancies resulting from the separation of the polymer ionic groups. There is additional evidence in dielectric relaxation analysis performed on these polymers that show a Debye regime at intermediate salt concentrations, which means that the dielectric response at these concentrations is produced by non-interacting dipoles.[16]

The increased vacancies concentration gives a mechanism for a continuous and random distribution of the free volume and, therefore, allows long-range diffusion of ion carriers through the structure.

REFERENCES

1. Salamone, J. C.; Volken, W.; Olson, A. P.; Israel, S. C. *Polymer* **1978**, *19*, 1157.
2. Faingold'd, S. I.; Kuusk, A. *Eesti NSV. Read. Akad. Toim. Keem.* **1985**, *34*, 210.
3. Kalpakei, B.; Chan, K. *Standard Oil Co. (Ohio) US PUS* 4 554, 1974.
4. Monroy, V. M.; Galin, J. C. *Polymer* **1984**, *25*, 254.
5. Monroy, V. M.; Marshal, E.; Mathis, A.; Maurer, B.; Galin, J. C. *Annual Meeting of the Society of Plastics Engineers*; New Orleans, 1984.
6. Castaño, V. M.; González, A. E.; Cardoso, J.; Monroy, V. M. *J. Mat. Res.* **1990**, *51*, 654.
7. MacKnight, W. J.; Earnest, T. R. *J. Polym. Sci. Macromol. Rev.* **1981**, *16*, 41.
8. Galin, M.; Marchal, E.; Mathis, A.; Maurer, B.; Monroy, V.; Galin, J. *Polymer* **1987**, *28*, 1937.
9. Bannister, D. J.; Davies, G. R.; Ward, I. M.; McIntire, J. E. *Polymer* **1984**, *25*, 1291.
10. *Polymer Electrolyte Review 1*; Mac Callum, J. R.; Vincent, C. A., Eds.; Elsevier Applied Science: London, 1987.
11. Fenton, D. E.; Parker, J. M.; Wright, P. V. *Polymer* **1973**, *14*, 589.
12. Wright, P. V. *Br. Polym. J.* **1975**, *7*, 319.
13. Sorensen, P. R.; Jacobsen, T. *Polym. Bull.* **1983**, *9*, 47.
14. Adam, G.; Gibbs, J. H. *J. Chem. Phys.* **1965**, *43*, 139.
15. Cohen, M. H.; Thurnbull, D. *J. Chem. Phys.* **1959**, *31*, 1164.
16. Cardoso, J.; Huanosta, A.; Manero, O. *J. Appl. Polym. Sci.: Polym. Symp.* **1991**, *49*, 155.

INDEX

Materials are indexed in both monomeric and polymeric forms, depending on the reference in the text. Readers are encouraged to make use of the cross-references that appear in the encyclopedia between articles. These references are to titles of related interest that can be found within the encyclopedia.

A

B

C

E

F

H

I

J

K

L

M

P

Q

R

S

T

U

V

W

X

Y

Z